Peterson's Graduate Programs in the Physical Sciences, Mathematics, Agricultural Sciences, the Environment & Natural Resources

2007

Book 4

PETERSON'S

A nelnet. COMPANY

PETERSON'S

A nelnet. COMPANY

About Peterson's, a Nelnet company

Peterson's (www.petersons.com) is a leading provider of education information and advice, with books and online resources focusing on education search, test preparation, and financial aid. Its Web site offers searchable databases and interactive tools for contacting educational institutions, online practice tests and instruction, and planning tools for securing financial aid. Peterson's serves 110 million education consumers annually.

For more information, contact Peterson's, 2000 Lenox Drive, Lawrenceville, NJ 08648; 800-338-3282; or find us on the World Wide Web at www.petersons.com/about.

Editor: Fern A. Oram; Production Editor: Susan W. Dilts; Copy Editors: Bret Bollmann, Michael Haines, Sally Ross, Jill C. Schwartz, Mark D. Snider, Pam Sullivan, Valerie Bolus Vaughan; Research Associate: Amy Weber; Programmer: Phyllis Johnson; Manufacturing Manager: Ray Golaszewski; Composition Manager: Linda M. Williams; Client Relations Representatives: Danielle Groncki, Mimi Kaufman, Karen Mount, Eric Wallace.

ISSN 1093-8443
ISBN-13: 978-0-7689-2155-7
ISBN-10: 0-7689-2155-4

Printed in the United States of America

10 9 8 7 6 5 4 3 2 1 09 08 07

Forty-first Edition

CONTENTS

A Note from the Peterson's Editors

The six volumes of *Peterson's Graduate and Professional Programs*, the only annually updated reference work of its kind, provide wide-ranging information on the graduate and professional programs offered by accredited colleges and universities in the United States, U.S. territories, and Canada and by those institutions outside the United States that are accredited by U.S. accrediting bodies. More than 44,000 individual academic and professional programs at more than 2,000 institutions are listed. *Peterson's Graduate and Professional Programs* have been used for more than forty years by prospective graduate and professional students, placement counselors, faculty advisers, and all others interested in postbaccalaureate education.

Book 1: *Graduate & Professional Programs: An Overview*, contains information on institutions as a whole, while Books 2 through 6 are devoted to specific academic and professional fields.

Book 2: *Graduate Programs in the Humanities, Arts & Social Sciences*

Book 3: *Graduate Programs in the Biological Sciences*

Book 4: *Graduate Programs in the Physical Sciences, Mathematics, Agricultural Sciences, the Environment & Natural Resources*

Book 5: *Graduate Programs in Engineering & Applied Sciences*

Book 6: *Graduate Programs in Business, Education, Health, Information Studies, Law & Social Work*

The books may be used individually or as a set. For example, if you have chosen a field of study but do not know what institution you want to attend or if you have a college or university in mind but have not chosen an academic field of study, it is best to begin with Book 1.

Book 1 presents several directories to help you identify programs of study that might interest you; you can then research those programs further in Books 2 through 6. The *Directory of Graduate and Professional Programs by Field* lists the 476 fields for which there are program directories in Books 2 through 6 and gives the names of those institutions that offer graduate degree programs in each.

For geographical or financial reasons, you may be interested in attending a particular institution and will want to know what it has to offer. You should turn to the *Directory of Institutions and Their Offerings*, which lists the degree programs available at each institution, again, in the 476 academic and professional fields for which Books 2 through 6 have program directories. As in the *Directory of Graduate and Professional Programs by Field*, the level of degrees offered is also indicated.

All books in the series include advice on graduate education, including topics such as admissions tests, financial aid, and accreditation. **The Graduate Adviser** includes two essays and information about accreditation. The first essay, "The Admissions Process," discusses general admission requirements, admission tests, factors to consider when selecting a graduate school or program, when and how to apply, and how admission decisions are made. Special information for international students and tips for minority students are also included. The second essay, "Financial Support," is an overview of the broad range of support available at the graduate level. Fellowships, scholarships, and grants; assistantships and internships; federal and private loan programs, as well as Federal Work-Study; and the GI bill are detailed. This essay concludes with advice on applying for need-based financial aid. "Accreditation and Accrediting Agencies" gives information on accreditation and its purpose and lists first institutional accrediting agencies and then specialized accrediting agencies relevant to each volume's specific fields of study.

With information on more than 44,000 graduate programs in 476 disciplines, *Peterson's Graduate and Professional Programs* give you all the information you need about the programs that are of interest to you in three formats: **Profiles** (capsule summaries of basic information), **Announcements** (information that an institution or program wants to emphasize, written by administrators), and **Close-Ups** (also written by administrators, with more expansive information than the **Profiles**, emphasizing different aspects of their programs). By using these various formats of program information, coupled with **Appendixes** and **Indexes** covering directories and subject areas for all six books, you will find that these guides provide the most comprehensive, accurate, and up-to-date graduate study information available.

Peterson's publishes a full line of resources with information you need to guide you through the graduate admissions process. Peterson's publications can be found at your local bookstore or library—or visit us on the Web at www.petersons.com.

Colleges and universities will be pleased to know that Peterson's helped you in your selection. Admissions staff members are more than happy to answer questions, address specific problems, and help in any way they can. The editors at Peterson's wish you great success in your graduate program search!

THE GRADUATE ADVISER

The Admissions Process

Generalizations about graduate admissions practices are not always helpful because each institution has its own set of guidelines and procedures. Nevertheless, some broad statements can be made about the admissions process that may help you plan your strategy.

General Requirements

Graduate schools and departments have requirements that applicants for admission must meet. Typically, these requirements include undergraduate transcripts (which provide information about undergraduate grade point average and course work applied toward a major), admission test scores, and letters of recommendation. Most graduate programs also ask for an essay or personal statement that describes your personal reasons for seeking graduate study. In some fields, such as art and music, portfolios or auditions may be required in addition to other evidence of talent. Some institutions require that the applicant have an undergraduate degree in the same subject as the intended graduate major.

Most institutions evaluate each applicant on the basis of the applicant's total record, and the weight accorded any given factor varies widely from institution to institution and from program to program.

Admission Tests

The major testing program used in graduate admissions is the Graduate Record Examinations (GRE)® testing program, sponsored by the GRE Board and administered by Educational Testing Service, Princeton, New Jersey.

The Graduate Record Examinations testing program consists of a General Test and eight Subject Tests. The General Test measures verbal reasoning, quantitative reasoning, and analytical writing skills. It is offered as a computer-adaptive test (CAT) in the United States, Canada, and many other countries. In the CAT, the computer determines which question to present next by adjusting to your previous responses. Paper-based General Test administrations are offered in some parts of the world.

The computer-adaptive General Test consists of a 30-minute verbal section, a 45-minute quantitative section, and a 75-minute analytical writing section. In addition, an unidentified verbal or quantitative section that doesn't count toward a score may be included and an identified research section that is not scored may also be included.

The paper-based General Test consists of two 30-minute verbal sections, two 30-minute quantitative sections, and a 75-minute analytical writing section. In addition, an unidentified verbal or quantitative section that doesn't count toward a score may be included.

The Subject Tests measure achievement and assume undergraduate majors or extensive background in the following eight disciplines:

- Biochemistry, Cell and Molecular Biology
- Biology
- Chemistry
- Computer Science
- Literature in English
- Mathematics
- Physics
- Psychology

The Subject Tests are available three times per year as paper-based administrations around the world. Testing time is 2 hours and 50 minutes. You can obtain more information about the GRE tests by visiting the ETS Web site at www.ets.org or consulting the *GRE Information and Registration Bulletin*. The *Bulletin* can be obtained at many undergraduate colleges. You can also download it from the ETS Web site or obtain it by contacting Graduate Record Examinations, Educational Testing Service, PO Box 6000, Princeton, NJ 08541-6000, telephone 1-609-771-7670.

A revised GRE General Test will be introduced in fall 2007 and is designed to increase test validity, provide faculty with better information regarding applicants' performance, address security concerns, increase worldwide access to the test, and make better use of advances in technology and psychometric design. Changes planned to the verbal reasoning section include greater emphasis on higher cognitive skills and less dependence on vocabulary; more text-based materials, such as reading passages; a broader selection of reading passages; emphasis on skills related to graduate work, such as complex reasoning; and expansion of computer-enabled tasks (e.g., clicking on a sentence in a passage to highlight it). In addition, there will be two 40-minute sections rather than one 30-minute section. Changes to the quantitative reasoning section include quantitative reasoning skills that are closer to skills generally used in graduate school, an increase in the proportion of questions involving real-life scenarios and data interpretation, a decrease in the proportion of geometry questions, and better use of technology (e.g., on-screen calculator). In addition, there will be two 40-minute sections rather than one 45-minute section. Changes to the analytical writing section include new, more focused prompts that reduce the possibility of reliance on memorized materials. The issue and argument tasks will each be 30 minutes in length and essay responses will be made available to designated score recipients. It is anticipated that the new score scale range will be 110 to 150, in 1-point increments. Final specification of the score scale will be determined based on data from the initial revised General Test administrations in fall 2007. A concordance table will be available in early January 2008 to assist score users in determining the relationship between old and new verbal and quantitative scores. The score scale for the analytical writing section will continue to be 0 to 6, in half-point increments. An expanded Internet-based testing network will also become available worldwide. Be sure to check the GRE Web site (www.gre.org) for updated information.

If you expect to apply for admission to a program that requires any of the GRE tests, you should select a test date well in advance of the application deadline. Scores on the computer-adaptive General Test are reported within ten to fifteen days; scores on the paper-based General Test and the Subject Tests are reported within six weeks.

Another testing program, the Miller Analogies Test (MAT), is administered at more than 400 Controlled Testing Centers in the United States, Canada, and other countries. The MAT computer-based test is now available. Testing time is 60 minutes. The test consists of 120 partial analogies. You can obtain the *Miller Analogies Test Candidate Information Booklet,* which contains a list of test centers and instructions for taking the test from http://harcourtassessment.com/HAIWEB/Cultures/en-US/dotCom/milleranalogies.com.htm or by calling Harcourt Assessment, Inc., at 1-800-622-3231.

Check the specific requirements of the programs to which you are applying.

Factors Involved in Selecting a Graduate School or Program

Selecting a graduate school and a specific program of study is a complex matter. Quality of the faculty; program and course offerings; the nature, size, and location of the institution; admission requirements; cost; and the availability of financial assistance are among the many factors that affect one's choice of institution. Other considerations are job placement and achievements of the program's graduates and the institution's resources, such as libraries, laboratories, and computer facilities. If you are to make the best possible choice, you need to learn as much as you can about the schools and programs you are considering before you apply. The following steps may help you narrow your choices.

- Talk to alumni of the programs or institutions you are considering to get their impressions of how well they were prepared for work in their fields of study.

- Remember that graduate school requirements change, so be sure to get the most up-to-date information possible.
- Talk to department faculty and the graduate adviser at your undergraduate institution. They often have information about programs of study at other institutions.
- Visit the Web sites of the graduate schools in which you are interested to request a graduate catalog. Contact the department chair in your chosen field of study for additional information about the department and the field.
- Visit as many campuses as possible. Call ahead for an appointment with the graduate adviser in your field of interest and be sure to check out the facilities and talk to students.

When and How to Apply

You should begin the application process at least one year before you expect to begin your graduate study. Find out the application deadline for each institution (many are provided in the **Profile** section of this volume). Go to the institution Web site and find out if you can apply online. If not, request a paper application form. Fill out this form thoroughly and neatly. Assume that the school needs all the information it is requesting and that the admissions officer will be sensitive to the neatness and overall quality of what you submit. Do not supply more information than the school requires.

The institution may ask at least one question that will require a three- or four-paragraph answer. Compose your response on the assumption that the admissions officer is interested in both what you think and how you express yourself. Keep your statement brief and to the point, but, at the same time, include all pertinent information about your past experiences and your educational goals. Individual statements vary greatly in style and content, which helps admissions officers to differentiate among applicants. Many graduate departments give considerable weight to the statement in making their admissions decisions, so be sure to take the time to prepare a thoughtful and concise statement.

If recommendations are a part of the admissions requirements, carefully choose the individuals you ask to write them. It is generally best to ask current or former professors to write the recommendations, provided they are able to attest to your intellectual ability and motivation for doing the work required of a graduate student. It is advisable to provide stamped, preaddressed envelopes to people being asked to submit recommendations on your behalf.

Completed applications, including references and transcripts and admission test scores, should be received at the institution by the specified date.

Be advised that institutions do not usually make admissions decisions until all materials have been received. Enclose a self-addressed postcard with your application, requesting confirmation of receipt. Allow at least 10 days for the return of the postcard before making further inquiries.

If you plan to apply for financial support, it is imperative that you file your application early.

How Admission Decisions Are Made

The program you apply to is directly involved in the admissions process. Although the final decision is usually made by the graduate dean (or an associate) or by the faculty admissions committee, recommendations from faculty members in your intended field are important. At some institutions, an interview is incorporated into the decision process.

A Special Note for International Students

In addition to the steps already described, there are some special considerations for international students who intend to apply for graduate study in the United States. All graduate schools require an indication of competence in English. The purpose of the Test of English as a Foreign Language (TOEFL) is to evaluate the English proficiency of people who are nonnative speakers of English and want to study at colleges and universities where English is the language of instruction. The TOEFL is administered by Educational Testing Service (ETS) under the general direction of a policy board established by the College Board and the Graduate Record Examinations Board.

The TOEFL is administered as a computer-based test and the TOEFL iBT is administered as an Internet-based test throughout most of the world and these are available year-round by appointment only. It is not necessary to have previous computer experience to take the tests. The computer-based test consists of four sections—listening, reading, structure, and writing. Total testing time is approximately 4 hours. The TOEFL iBT consists of four sections—reading, listening, speaking, and writing. Total testing time is approximately 4 hours.

The TOEFL is offered in the paper-based format in areas of the world where computer-based testing is not available. The paper-based TOEFL consists of three sections—listening comprehension, structure and written expression, and reading comprehension. Testing time is approximately 3 hours.

The Test of Written English (TWE) is also given. TWE is a 30-minute essay that measures the examinee's ability to compose in English. Examinees receive a TWE score separate from their TOEFL score. The *Information Bulletin* contains information on local fees and registration procedures.

Additional information and registration materials are available from TOEFL Services, Educational Testing Service, P.O. Box 6151, Princeton, New Jersey 08541-6151. Telephone: 1-609-771-7100. E-mail: toefl@ets.org. World Wide Web: http://www.toefl.org.

International students should apply especially early because of the number of steps required to complete the admissions process. Furthermore, many United States graduate schools have a limited number of spaces for international students, and many more students apply than the schools can accommodate.

International students may find financial assistance from institutions very limited. The U.S. government requires international applicants to submit a certification of support, which is a statement attesting to the applicant's financial resources. In addition, international students *must* have health insurance coverage.

Tips for Minority Students

Indicators of a university's values in terms of diversity are found both in its recruitment programs and its resources directed to student success. Important questions: Does the institution vigorously recruit minorities for its graduate programs? Is there funding available to help with the costs associated with visiting the school? Are minorities represented in the institution's brochures or Web site or on their faculty rolls? What campus-based resources or services (including assistance in locating housing or career counseling and placement) are available? Is funding available to members of underrepresented groups?

At the program level, it is particularly important for minority students to investigate the "climate" of a program under consideration. How many minority students are enrolled and how many have graduated? What opportunities are there to work with diverse faculty and mentors whose research interests match yours? How are conflicts resolved or concerns addressed? How interested are faculty in building strong and supportive relations with students? "Climate" concerns should be addressed by posing questions to various individuals, including faculty members, current students, and alumni.

Information is also available through various organizations, such as the Hispanic Association of Colleges and Universities (HACU), and publications, such as *DIVERSE: Issues in Higher Education* and *Hispanic Outlook* magazine. There are also books devoted to this topic, such as *The Multicultural Student's Guide to Colleges* by Robert Mitchell.

Financial Support

The range of financial support at the graduate level is very broad. The following descriptions will give you a general idea of what you might expect and what will be expected of you as a financial support recipient.

Fellowships, Scholarships, and Grants

These are usually outright awards of a few hundred to many thousands of dollars with no service to the institution required in return. Fellowships and scholarships are usually awarded on the basis of merit and are highly competitive. Grants are made on the basis of financial need or special talent in a field of study. Many fellowships, scholarships, and grants not only cover tuition, fees, and supplies but also include stipends for living expenses with allowances for dependents. However, the terms of each should be examined because some do not permit recipients to supplement their income with outside work. Fellowships, scholarships, and grants may vary in the number of years for which they are awarded.

In addition to the availability of these funds at the university or program level, many excellent fellowship programs are available at the national level and may be applied for before and during enrollment in a graduate program. A listing of many of these programs can be found at the Council of Graduate Schools' Web site: http://www.cgsnet.org/ResourcesForStudents/fellowships.htm.

Assistantships and Internships

Many graduate students receive financial support through assistantships, particularly involving teaching or research duties. It is important to recognize that such appointments should not be simply employment relationships but rather should constitute an integral and important part of a student's graduate education. As such, the appointments should be accompanied by strong faculty mentoring and increasingly responsible apprenticeship experiences. The specific nature of these appointments in a given program should be considered in selecting that graduate program.

TEACHING ASSISTANTSHIPS

These usually provide a salary and full or partial tuition remission and may also provide health benefits. Unlike fellowships, scholarships, and grants, which require no service to the institution, teaching assistantships require recipients to provide the institution with a specific amount of undergraduate teaching, ideally related to the student's field of study. Some teaching assistants are limited to grading papers, compiling bibliographies, taking notes, or monitoring laboratories. At some graduate schools, teaching assistants must carry lighter course loads than regular full-time students.

RESEARCH ASSISTANTSHIPS

These are very similar to teaching assistantships in the manner in which financial assistance is provided. The difference is that recipients are given basic research assignments in their disciplines rather than teaching responsibilities. The work required is normally related to the student's field of study; in most instances, the assistantship supports the student's thesis or dissertation research.

ADMINISTRATIVE INTERNSHIPS

These are similar to assistantships in application of financial assistance funds, but the student is given an assignment on a part-time basis, usually as a special assistant with one of the university's administrative offices. The assignment may not necessarily be directly related to the recipient's discipline.

RESIDENCE HALL AND COUNSELING ASSISTANTSHIPS

These assistantships are frequently assigned to graduate students in psychology, counseling, and social work. Duties can vary from being available in a dean's office for a specific number of hours for consultation with undergraduates to living in campus residences and being responsible for both counseling and administrative tasks or advising student activity groups. Residence hall assistantships often include a room and board allowance and, in some cases, tuition assistance and stipends.

Health Insurance

The availability and affordability of health insurance is an important issue and one that should be considered in an applicant's choice of institution and program. While often included with assistantships and fellowships, this is not always the case and, even if provided, the benefits may be limited. It is important to note that the U.S. government requires international students to have health insurance.

The GI Bill

This provides financial assistance for students who are veterans of the United States armed forces. If you are a veteran, contact your local Veterans Administration office to determine your eligibility and to get full details about benefits. There are a number of programs that offer educational benefits to current military enlistees. Some states have tuition assistance programs for members of the National Guard. Contact the VA office at the college for more information.

Federal Work-Study Program (FWS)

Employment is another way some students finance their graduate studies. The federally funded Federal Work-Study Program provides eligible students with employment opportunities, usually in public and private nonprofit organizations. Federal funds pay up to 75 percent of the wages, with the remainder paid by the employing agency. FWS is available to graduate students who demonstrate financial need. Not all schools have these funds, and some only award them to undergraduates. Each school sets its application deadline and work-study earnings limits. Wages vary and are related to the type of work done. You must file the Free Application for Federal Student Aid (FAFSA) to be eligible for this program.

Loans

Many graduate students borrow to finance their graduate programs when other sources of assistance (which do not have to be repaid) prove insufficient. You should always read and understand the terms of any loan program before submitting your application.

FEDERAL LOANS

Federal Stafford Loans. The Federal Stafford Loan Program offers government-sponsored, low-interest loans to students through a private lender such as a bank, credit union, or savings and loan association.

There are two components of the Federal Stafford Loan program. Under the *subsidized* component of the program, the federal government pays the interest on the loan while you are enrolled in graduate school on at least a half-time basis. Under the *unsubsidized* component of the program, you pay the interest on the loan from the day proceeds are issued. Eligibility for the federal subsidy is based on demonstrated financial need as determined by the financial aid office from the information you provide on the FAFSA. A cosigner is not required, since the loan is not based on creditworthiness.

Although *unsubsidized* Federal Stafford Loans may not be as desirable as *subsidized* Federal Stafford Loans from the student's perspective, they are a useful source of support for those who may not qualify for the subsidized loans or who need additional financial assistance.

Graduate students may borrow up to $18,500 per year through the Stafford Loan Program, up to a cumulative maximum of $138,500, including undergraduate borrowing. This may include up to $8500 in Subsidized Stafford Loans annually, depending on eligibility, up to a cumulative maximum of $65,500, including undergraduate borrowing. The amount of the loan borrowed through the *unsubsidized* Stafford Program equals the total amount of the loan (as much $18,500) minus your eligibility for a Subsidized Stafford Loan (as much as $8500). You may borrow up to the cost of attendance at the school in which you are enrolled or will attend, minus estimated financial assistance from other federal, state, and private sources, up to a maximum of $18,500. The interest rate for the Federal Stafford Loans is fixed at 6.8%.

Two fees may be deducted from the loan proceeds upon disbursement: a guarantee fee of up to 1 percent, which is deposited in an insurance pool to ensure repayment to the lender if the borrower defaults, and a federally mandated 3 percent origination fee, which is used to offset the administrative cost of the Federal Stafford Loan Program. Many lenders do offer reduced-fee or "zero fee" loans.

Under the *subsidized* Federal Stafford Loan Program, repayment begins six months after your last date of enrollment on at least a half-time basis. Under the *unsubsidized* program, repayment of interest begins within thirty days from disbursement of the loan proceeds, and repayment of the principal begins six months after your last enrollment on at least a half-time basis. Some borrowers may choose to defer interest payments while they are in school. The accrued interest is added to the loan balance when the borrower begins repayment. There are several repayment options.

Federal Direct Loans. Some schools participate in the Department of Education's William D. Ford Direct Lending Program instead of the Federal Stafford Loan Program. The two programs are essentially the same except that with the Direct Loans, schools themselves provide the loans with funds from the federal government. Terms and interest rates are virtually the same except that there are a few additional repayment options with Federal Direct Loans.

Federal Perkins Loans. The Federal Perkins Loan is available to students demonstrating financial need and is administered directly by the school. Not all schools have these funds, and some may award them to undergraduates only. Eligibility is determined from the information you provide on the FAFSA. The school will notify you of your eligibility.

Eligible graduate students may borrow up to $6000 per year, up to a maximum of $40,000, including undergraduate borrowing (even if your previous Perkins Loans have been repaid). The interest rate for Federal Perkins Loans is 5 percent, and no interest accrues while you remain in school at least half-time. There are no guarantee, loan, or disbursement fees. Repayment begins nine months after your last date of enrollment on at least a half-time basis and may extend over a maximum of ten years with no prepayment penalty.

Deferring Your Federal Loan Repayments. If you borrowed under the Federal Stafford Loan Program or the Federal Perkins Loan Program for previous undergraduate or graduate study, your repayments may be deferred when you return to graduate school, depending on when you borrowed and under which program.

There are other deferment options available if you are temporarily unable to repay your loan. Information about these deferments is provided at your entrance and exit interviews. If you believe you are eligible for a deferment of your loan repayments, you must contact your lender to complete a deferment form. The deferment must be filed prior to the time your repayment is due, and it must be refiled when it expires if you remain eligible for deferment at that time.

SUPPLEMENTAL (PRIVATE) LOANS

Many lending institutions offer supplemental loan programs and other financing plans, such as the ones described here, to students seeking additional assistance in meeting their educational expenses. Some loan programs target all types of graduate students; others are designed specifically for business, law, or medical students. In addition, you can use private loans not specifically designed for education to help finance your graduate degree.

If you are considering borrowing through a supplemental or private loan program, you should carefully consider the terms and be sure to "read the fine print." Check with the program sponsor for the most current terms that will be applicable to the amounts you intend to borrow for graduate study. Most supplemental loan programs for graduate study offer unsubsidized, credit-based loans. In general, a credit-ready borrower is one who has a satisfactory credit history or no credit history at all. A creditworthy borrower generally must pass a credit test to be eligible to borrow or act as a cosigner for the loan funds.

Many supplemental loan programs have minimum and maximum annual loan limits. Some offer amounts equal to the cost of attendance minus any other aid you will receive for graduate study. If you are planning to borrow for several years of graduate study, consider whether there is a cumulative or aggregate limit on the amount you may borrow. Often this cumulative or aggregate limit will include any amounts you borrowed and have not repaid for undergraduate or previous graduate study.

The combination of the annual interest rate, loan fees, and the repayment terms you choose will determine how much you will repay over time. Compare these features in combination before you decide which loan program to use. Some loans offer interest rates that are adjusted monthly, some quarterly, some annually. Some offer interest rates that are lower during the in-school, grace, and deferment periods, and then increase when you begin repayment. Some programs include a loan "origination" fee, which is usually deducted from the principal amount you receive when the loan is disbursed, and must be repaid along with the interest and other principal when you graduate, withdraw from school, or drop below half-time study. Sometimes the loan fees are reduced if you borrow with a qualified cosigner. Some programs allow you to defer interest and/or principal payments while you are enrolled in graduate school. Many programs allow you to capitalize your interest payments; the interest due on your loan is added to the outstanding balance of your loan, so you don't have to repay immediately, but this increases the amount you owe. Other programs allow you to pay the interest as you go, which reduces the amount you later have to repay.

Some examples of supplemental programs follow. The private loan market is very competitive and your financial aid office can help you evaluate these and other programs.

CitiAssist Loans. Offered by Citibank, these no-fee loans help graduate students fill the gap between the financial aid they receive and the money they need for school. Visit www.studentloan.com for more loan information from Citibank.

EXCEL Loan. This program, sponsored by Nellie Mae, is designed for students who are not ready to borrow on their own and wish to borrow with a creditworthy cosigner. Visit www.nelliemae.com for more information.

Key Alternative Loan. This loan can bridge the gap between education costs and traditional funding. Visit www.key.com/html/H-1.3html for more information.

Graduate Access Loan. Sponsored by the Access Group, this is for graduate students enrolled at least half-time. The Web site is www.accessgroup.com.

Signature Student Loan. A loan program for students who are enrolled at least half-time, this is sponsored by Sallie Mae. Visit www.salliemae.com for more information.

Applying for Need-Based Financial Aid

Schools that award federal and institutional financial assistance based on need will require you to complete the FAFSA and, in some cases, an institutional financial aid application.

If you are applying for federal student assistance, you **must** complete the FAFSA. A service of the U.S. Department of Education, it is free to all applicants. Most applicants apply online at www.fafsa.ed.gov. Paper applications are available at the financial aid office of your local college.

After your FAFSA information has been processed, you will receive a Student Aid Report (SAR). If you provided an e-mail address on the FAFSA, this will be sent to you electronically; otherwise, it will be mailed to your home address.

Follow the instructions on the SAR if you need to correct information reported on your original application. If your situation changes after you file your FAFSA, contact your financial aid officer to discuss amending your information. You can also appeal your financial aid award if you have extenuating circumstances.

If you would like more information on federal student financial aid, visit the FAFSA Web site or download the most recent version of *The Student Guide* at http://studentaid.ed.gov/students/publications/student_guide/index.html. This guide is also available in Spanish.

The U.S. Department of Education also has a toll-free number for questions concerning federal student aid programs. The number is 1-800-4-FED AID (1-800-433-3243). If you are hearing impaired, call toll-free, 1-800-730-8913.

Summary

Remember that these are generalized statements about financial assistance at the graduate level. Because each institution allots its aid differently, you should communicate directly with the school and the specific department of interest to you. It is not unusual, for example, to find that an endowment vested within a specific department supports one or more fellowships. You may fit its requirements and specifications precisely.

Accreditation and Accrediting Agencies

Colleges and universities in the United States, and their individual academic and professional programs, are accredited by nongovernmental agencies concerned with monitoring the quality of education in this country. Agencies with both regional and national jurisdictions grant accreditation to institutions as a whole, while specialized bodies acting on a nationwide basis—often national professional associations—grant accreditation to departments and programs in specific fields.

Institutional and specialized accrediting agencies share the same basic concerns: the purpose an academic unit—whether university or program—has set for itself and how well it fulfills that purpose, the adequacy of its financial and other resources, the quality of its academic offerings, and the level of services it provides. Agencies that grant institutional accreditation take a broader view, of course, and examine university-wide or college-wide services with which a specialized agency may not concern itself.

Both types of agencies follow the same general procedures when considering an application for accreditation. The academic unit prepares a self-evaluation, focusing on the concerns mentioned above and usually including an assessment of both its strengths and weaknesses; a team of representatives of the accrediting body reviews this evaluation, visits the campus, and makes its own report; and finally, the accrediting body makes a decision on the application. Often, even when accreditation is granted, the agency makes a recommendation regarding how the institution or program can improve. All institutions and programs are also reviewed every few years to determine whether they continue to meet established standards; if they do not, they may lose their accreditation.

Accrediting agencies themselves are reviewed and evaluated periodically by the U.S. Department of Education and the Council for Higher Education Accreditation (CHEA). Recognized agencies adhere to certain standards and practices, and their authority in matters of accreditation is widely accepted in the educational community.

This does not mean, however, that accreditation is a simple matter, either for schools wishing to become accredited or for students deciding where to apply. Indeed, in certain fields the very meaning and methods of accreditation are the subject of a good deal of debate. For their part, those applying to graduate school should be aware of the safeguards provided by regional accreditation, especially in terms of degree acceptance and institutional longevity. Beyond this, applicants should understand the role that specialized accreditation plays in their field, as this varies considerably from one discipline to another. In certain professional fields, it is necessary to have graduated from a program that is accredited in order to be eligible for a license to practice, and in some fields the federal government also makes this a hiring requirement. In other disciplines, however, accreditation is not as essential, and there can be excellent programs that are not accredited. In fact, some programs choose not to seek accreditation, although most do.

Institutions and programs that present themselves for accreditation are sometimes granted the status of candidate for accreditation, or what is known as "preaccreditation." This may happen, for example, when an academic unit is too new to have met all the requirements for accreditation. Such status signifies initial recognition and indicates that the school or program in question is working to fulfill all requirements; it does not, however, guarantee that accreditation will be granted.

Institutional Accrediting Agencies—Regional

MIDDLE STATES ASSOCIATION OF COLLEGES AND SCHOOLS
Accredits institutions in Delaware, District of Columbia, Maryland, New Jersey, New York, Pennsylvania, Puerto Rico, and the Virgin Islands.
Jean Avnet Morse, Executive Director
Middle States Commission on Higher Education
3624 Market Street
Philadelphia, Pennsylvania 19104
Telephone: 267-284-5025
Fax: 215-662-5950
E-mail: jmorse@msche.org
World Wide Web: www.msche.org

NEW ENGLAND ASSOCIATION OF SCHOOLS AND COLLEGES
Accredits institutions in Connecticut, Maine, Massachusetts, New Hampshire, Rhode Island, and Vermont.
Barbara E. Brittingham, Interim Director
Commission on Institutions of Higher Education
209 Burlington Road
Bedford, Massachusetts 01730
Telephone: 781-541-5447
Fax: 781-271-0950
E-mail: bbrittingham@neasc.org
World Wide Web: www.neasc.org

NORTH CENTRAL ASSOCIATION OF COLLEGES AND SCHOOLS
Accredits institutions in Arizona, Arkansas, Colorado, Illinois, Indiana, Iowa, Kansas, Michigan, Minnesota, Missouri, Nebraska, New Mexico, North Dakota, Ohio, Oklahoma, South Dakota, West Virginia, Wisconsin, and Wyoming.
Steven D. Crow, Executive Director
The Higher Learning Commission
30 North LaSalle Street, Suite 2400
Chicago, Illinois 60602
Telephone: 312-263-0456
Fax: 312-263-7462
E-mail: scrow@hlcommission.org
World Wide Web: www.ncahigherlearningcommission.org

NORTHWEST COMMISSION ON COLLEGES AND UNIVERSITIES
Accredits institutions in Alaska, Idaho, Montana, Nevada, Oregon, Utah, and Washington.
Sandra E. Elman, President
8060 165th Avenue, NE, Suite 100
Redmond, Washington 98052
Telephone: 425-558-4224
Fax: 425-376-0596
E-mail: selman@nwccu.org
World Wide Web: www.nwccu.org

SOUTHERN ASSOCIATION OF COLLEGES AND SCHOOLS
Accredits institutions in Alabama, Florida, Georgia, Kentucky, Louisiana, Mississippi, North Carolina, South Carolina, Tennessee, Texas, and Virginia.
Belle S. Wheelan, President
Commission on Colleges
1866 Southern Lane
Decatur, Georgia 30033
Telephone: 404-679-4512
Fax: 404-679-4558
E-mail: bwheelan@sacscoc.org
World Wide Web: www.sacscoc.org

WESTERN ASSOCIATION OF SCHOOLS AND COLLEGES
Accredits institutions in California, Guam, and Hawaii.
Ralph A. Wolff, President and Executive Director
The Senior College Commission
985 Atlantic Avenue, Suite 100
Alameda, California 94501
Telephone: 510-748-9001
Fax: 510-748-9797
E-mail: rwolff@wascsenior.org
World Wide Web: www.wascsenior.org/wasc/

Institutional Accrediting Agencies—Other

ACCREDITING COUNCIL FOR INDEPENDENT COLLEGES AND SCHOOLS
Sheryl L. Moody, Executive Director
750 First Street, NE, Suite 980
Washington, DC 20002
Telephone: 202-336-6780
Fax: 202-842-2593
E-mail: smoody@acics.org
World Wide Web: www.acics.org

DISTANCE EDUCATION AND TRAINING COUNCIL
Accrediting Commission
Michael P. Lambert, Executive Director
1601 18th Street, NW
Washington, DC 20009
Telephone: 202-234-5100 Ext. 101
Fax: 202-332-1386
E-mail: detc@detc.org
World Wide Web: www.detc.org

Specialized Accrediting Agencies

[Only Book 1 of *Peterson's Graduate and Professional Programs* Series includes the complete list of specialized accrediting groups recognized by the U.S. Department of Education and the Council on Higher Education Accreditation (CHEA). The lists in Books 2, 3, 4, 5, and 6 are abridged.]

DIETETICS
Beverly E. Mitchell, Director
American Dietetic Association
Commission on Accreditation for Dietetics Education (CADE-ADA)
120 South Riverside Plaza, Suite 2000
Chicago, Illinois 60606
Phone: 312-899-4872
Fax: 312-899-4817
E-mail: bmitchell@eatright.org
Web: www.eatright.org/cade

FORESTRY
Michael T. Goergen Jr.
Executive Vice President and CEO
Society of American Foresters
5400 Grosvenor Lane
Bethesda, Maryland 20814
Telephone: 301-897-8720
Fax: 301-897-3690
E-mail: goergenm@safnet.org
World Wide Web: www.safnet.org

How to Use These Guides

As you identify the particular programs and institutions that interest you, you can use both Book 1 and the specialized volumes (Books 2–6) to obtain detailed information—Book 1 for information on the institutions overall and Books 2 through 6 for details about the individual graduate units and their degree programs.

Books 2 through 6 are divided into sections that contain one or more directories devoted to programs in a particular field. If you do not find a directory devoted to your field of interest in a specific book, consult *Directories and Subject Areas in Books 2–6* (located at the end of each volume). After you have identified the correct book, consult the *Directories and Subject Areas in This Book* index, which shows (as does the more general directory) what directories cover subjects not specifically named in a directory or section title. This index in Book 2, for example, will tell you that if you are interested in sculpture, you should see the directory entitled Art/Fine Arts. The Art/Fine Arts entry will direct you to the proper page.

Books 2 through 6 have a number of general directories. These directories have entries for the largest unit at an institution granting graduate degrees in that field. For example, the general Engineering and Applied Sciences directory in Book 5 consists of **Profiles** for colleges, schools, and departments of engineering and applied sciences.

General directories are followed by other directories, or sections, that give more detailed information about programs in particular areas of the general field that has been covered. The general Engineering and Applied Sciences directory, in the previous example, is followed by nineteen sections with directories in specific areas of engineering, such as Chemical Engineering, Industrial/Management Engineering, and Mechanical Engineering.

Because of the broad nature of many fields, any system of organization is bound to involve a certain amount of overlap. Environmental studies, for example, is a field whose various aspects are studied in several types of departments and schools. Readers interested in such studies will find information on relevant programs in Book 3 under Ecology and Environmental Biology; in Book 4 under Environmental Management and Policy and Natural Resources; in Book 5 under Energy Management and Policy and Environmental Engineering; and in Book 6 under Environmental and Occupational Health. To help you find all of the programs of interest to you, the introduction to each section of Books 2 through 6 includes, if applicable, a paragraph suggesting other sections and directories with information on related areas of study.

Directory of Institutions with Programs in the Physical Sciences, Mathematics, Agricultural Sciences, the Environment & Natural Resources

This directory lists institutions in alphabetical order and includes beneath each name the academic fields in the physical sciences, mathematics, agricultural sciences, the environment, and natural resources in which each institution offers graduate programs. The degree level in each field is also indicated, provided that the institution has supplied that information in response to *Peterson's Annual Survey of Graduate and Professional Institutions.* An *M* indicates that a master's degree program is offered; a *D* indicates that a doctoral degree program is offered; a *P* indicates that the first professional degree is offered; an *O* signifies that other advanced degrees (e.g., certificates or specialist degrees) are offered; and an * (asterisk) indicates that a **Close-Up** and/or **Announcement** is located in this volume. See the index, *Close-Ups and Announcements*, for the specific page number.

Profiles of Academic and Professional Programs in Books 2–6

Each section of **Profiles** has a table of contents that lists the Program Directories, **Announcements**, and **Close-Ups.** Program Directories consist of the **Profiles** of programs in the relevant fields, with **Announcements** following if programs have chosen to include them. **Cross-Discipline Announcements**, if any programs have chosen to submit such entries, and **Close-Ups,** which are more individualized statements, again if programs have chosen to submit them, are also listed.

The **Profiles** found in the 476 directories in Books 2 through 6 provide basic data about the graduate units in capsule form for quick reference. To make these directories as useful as possible, **Profiles** are generally listed for an institution's smallest academic unit within a subject area. In other words, if an institution has a College of Liberal Arts that administers many related programs, the **Profile** for the individual program (e.g., Program in History), not the entire College, appears in the directory.

There are some programs that do not fit into any current directory and are not given individual **Profiles**. The directory structure is reviewed annually in order to keep this number to a minimum and to accommodate major trends in graduate education.

The following outline describes the **Profile** information found in the guides and explains how best to use that information. Any item that does not apply to or was not provided by a graduate unit is omitted from its listing. The format of the **Profiles** is constant, making it easy to compare one institution with another and one program with another. A description of the information in the **Profiles** in Books 2 through 6 follows; the Book 1 **Profile** description is found in that Guide's "How to Use This Guide" article.

Identifying Information. The institution's name, in boldface type, is followed by a complete listing of the administrative structure for that field of study. (For example, University of Akron, Buchtel College of Arts and Sciences, Department of Theoretical and Applied Mathematics, Program in Mathematics.) The last unit listed is the one to which all information in the **Profile** pertains. The institution's city, state, and zip code follow.

Offerings. Each field of study offered by the unit is listed with all postbaccalaureate degrees awarded. Degrees that are not preceded by a specific concentration are awarded in the general field listed in the unit name. Frequently, fields of study are broken down into subspecializations, and those appear following the degrees awarded; for example, "Offerings in secondary education (M.Ed.), including English education, mathematics education, science education." Students enrolled in the M.Ed. program would be able to specialize in any of the three fields mentioned.

Professional Accreditation. Some **Profiles** indicate whether a program is professionally accredited. Because it is possible for a program to receive or lose professional accreditation at any time, students entering fields in which accreditation is important to a career should verify the status of programs by contacting either the chairperson or the appropriate accrediting association.

Jointly Offered Degrees. Explanatory statements concerning programs that are offered in cooperation with other institutions are included in the list of degrees offered. This occurs most commonly on a regional basis (for example, two state universities offering a cooperative Ph.D. in special education) or where the specialized nature of the institutions encourages joint efforts (a J.D./M.B.A. offered by a law school at an institution with no formal business programs and an institution with a business school but lacking a law school). Only programs that are truly cooperative are listed; those involving only limited course work at another institution are not. Interested students should contact the heads of such units for further information.

Part-Time and Evening/Weekend Programs. When information regarding the availability of part-time or evening/weekend study appears in the **Profile**, it means that students are able to earn a degree exclusively through such study.

Postbaccalaureate Distance Learning Degrees. A postbaccalaureate distance learning degree program signifies that course requirements can be fulfilled with minimal or no on-campus study.

Faculty. Figures on the number of faculty members actively involved with graduate students through teaching or research are separated into full- and part-time as well as men and women whenever the information has been supplied.

Students. Figures for the number of students enrolled in graduate and professional programs pertain to the semester of highest enrollment from the 2005–06 academic year. These figures are broken down into full- and part-time and men and women whenever the data have been supplied. Information on the number of matriculated students enrolled in the unit who are members of a minority group or are international students appears here. The average age of the matriculated students is followed by the number of applicants, the percentage accepted, and the number enrolled for fall 2005.

Degrees Awarded. The number of degrees awarded in the calendar year is listed, as is the percentage of students in those degree programs who entered university research/teaching, business/industry, or government service or continued full-time study. Many doctoral programs offer a terminal master's degree if students leave the program after completing only part of the requirements for a doctoral degree; that is indicated here. All degrees are classified into one of four types: master's, doctoral, first professional, and other advanced degrees. A unit may award one or several degrees at a given level; however, the data are only collected by type and may therefore represent several different degree programs.

Median Time to Degree. If provided, information on the median amount of time required to earn the degree for full-time and part-time students is listed here. Also provided is the percentage of students who began their doctoral program in 1997 and received their degree in eight years or less.

Degree Requirements. The information in this section is also broken down by type of degree, and all information for a degree level pertains to all degrees of that type unless otherwise specified. Degree requirements are collected in a simplified form to provide some very basic information on the nature of the program and on foreign language, thesis or dissertation, comprehensive exam, and registration requirements. Many units also provide a short list of additional requirements, such as fieldwork or an internship. No information is listed on the number of courses or credits required for completion or whether a minimum or maximum number of years or semesters is needed. For complete information on graduation requirements, contact the graduate school or program directly.

Entrance Requirements. Entrance requirements are broken down into the four degree levels of master's, doctoral, first professional, and other advanced degrees. Within each level, information may be provided in two basic categories: entrance exams and other requirements. The entrance exams are identified by the standard acronyms used by the testing agencies, unless they are not well known. Other entrance requirements are quite varied, but they often contain an undergraduate or graduate grade point average (GPA). Unless otherwise stated, the GPA is calculated on a 4.0 scale and is listed as a minimum required for admission. Additional exam requirements/recommendations for international students may be listed here. Application deadlines for domestic and international students, the application fee, and whether electronic applications are accepted may be listed here. Note that the deadline should be used for reference only; these dates are subject to change, and students interested in applying should contact the graduate unit directly about application procedures and deadlines.

Expenses. The typical cost of study for the 2005–06 academic year is given in two basic categories: tuition and fees. Cost of study may be quite complex at a graduate institution. There are often sliding scales for part-time study, a different cost for first-year students, and other variables that make it impossible to completely cover the cost of study for each graduate program. To provide the most usable information, figures are given for full-time study for a full year where available and for part-time study in terms of a per-unit rate (per credit, per semester hour, etc.). Occasionally, variances may be noted in tuition and fees for reasons such as the type of program, whether courses are taken during the day or evening, whether courses are at the master's or doctoral level, or other institution-specific reasons. Expenses are usually subject to change; for exact costs at any given time, contact your chosen schools and programs directly. Keep in mind that the tuition of Canadian institutions is usually given in Canadian dollars.

Financial Support. This section contains data on the number of awards administered by the institution and given to graduate students during the 2005–06 academic year. The first figure given represents the total number of students receiving financial support enrolled in that unit. If the unit has provided information on graduate appointments, these are broken down into three major categories: *fellowships* give money to graduate students to cover the cost of study and living expenses and are not based on a work obligation or research commitment, *research assistantships* provide stipends to graduate students for assistance in a formal research project with a faculty member, and *teaching assistantships* provide stipends to graduate students for teaching or for assisting faculty members in teaching undergraduate classes. Within each category, figures are given for the total number of awards, the average yearly amount per award, and whether full or partial tuition reimbursements are awarded. In addition to graduate appointments, the availability of several other financial aid sources is covered in this section. *Tuition waivers* are routinely part of a graduate appointment, but units sometimes waive part or all of a student's tuition even if a graduate appointment is not available. *Federal Work-Study* is made available to students who demonstrate need and meet the federal guidelines; this form of aid normally includes 10 or more hours of work per week in an office of the institution. *Institutionally sponsored loans* are low-interest loans available to graduate students to cover both educational and living expenses. *Career-related internships or fieldwork* offer money to students who are participating in a formal off-campus research project or practicum. Grants, scholarships, traineeships, unspecified assistantships, and other awards may also be noted. The availability of financial support to part-time students is also indicated here.

Some programs list the financial aid application deadline and the forms that need to be completed for students to be eligible for financial awards. There are two forms: FAFSA, the Free Application for Federal Student Aid, which is required for federal aid, and the PROFILE®.

Faculty Research. Each unit has the opportunity to list several keyword phrases describing the current research involving faculty members and graduate students. Space limitations prevent the unit from listing complete information on all research programs. The total expenditure for funded research from the previous academic year may also be included.

Unit Head and Application Contact. The head of the graduate program for each unit is listed with academic title and telephone and fax numbers and e-mail address if available. In addition to the unit head, many graduate programs list a separate contact for application and admission information, which follows the listing for the unit head. If no unit head or application contact is given, you should contact the overall institution for information on graduate admissions.

Announcements and Close-Ups

The **Announcements** and **Close-Ups** are supplementary insertions submitted by deans, chairs, and other administrators who wish to offer an additional, more individualized statement to readers. A number of graduate school and program administrators have attached **Announcements** to the end of their **Profile** listings. In them you will find information that an institution or program wants to emphasize. The **Close-Ups** are by their very nature more expansive and flexible than the **Profiles**, and the administrators who have written them may emphasize different aspects of their programs. All of these **Close-Ups** are organized in the same way (with the exception of a few that describe research and training opportunities instead of degree programs), and in each one you will find information on the same basic topics, such as programs of study, research facilities, tuition and fees, financial aid, and application procedures. If an institution or program has submitted a **Close-Up**, a boldface cross-reference appears below its **Profile**. As with the **Announcements**, all of the **Close-Ups** in the guides have been submitted by choice; the absence of an **Announcement** or **Close-Up** does not reflect any type of editorial judgment on the part of Peterson's and their presence in the guides should not be taken as an indication of status, quality, or approval. Statements regarding a university's objectives and accomplishments are a reflection of its own beliefs and are not the opinions of the Peterson's editors.

Cross-Discipline Announcements

In addition to the regular directories that present **Profiles** of programs in each field of study, many sections in Books 2 through 6 contain special notices under the heading **Cross-Discipline Announcements**. Appearing at the end of many **Profile** sections, these **Cross-Discipline Announcements** inform you about programs that you may find of interest described in a different section. A biochemistry department, for example, may place a notice under **Cross-Discipline Announcements** in the Chemistry section (Book 4) to alert chemistry students to that course of study. **Cross-Discipline Announcements**, also written by administrators to highlight their programs, will be helpful to you not only in finding out about programs in fields related to your own but also in locating departments that are actively recruiting students with a specific undergraduate major.

Appendixes

This section contains two appendixes. The first, *Institutional Changes Since the 2006 Edition*, lists institutions that have closed, moved, merged, or changed their name or status since the last edition of the guides. The second, *Abbreviations Used in the Guides*, gives abbreviations of degree names, along with what those abbreviations stand for. These appendixes are identical in all six volumes of *Peterson's Graduate and Professional Programs.*

Indexes

There are three indexes presented here. The first index, *Close-Ups and Announcements*, gives page references for all programs that have chosen to place **Close-Ups** and **Announcements** in this volume. It is arranged alphabetically by institution; within institutions, the arrangement is alphabetical by subject area. It is not an index to all programs in the book's directories of **Profiles**; readers must refer to the directories themselves for **Profile** information on programs that have not submitted the additional, more individualized statements. The second index, *Directories and Subject Areas in Books 2–6*, gives book references for the directories in Books 2-6, for example, "Industrial Design—Book 2," and also includes cross-references for subject area names not used in the directory structure, for example, "Computing Technology (see Computer Science)." The third index, *Directories and Subject Areas in This Book*, gives page references for the directories in this volume and cross-references for subject area names not used in this volume's directory structure.

Data Collection Procedures

The information published in the directories and **Profiles** of all the books is collected through *Peterson's Annual Survey of Graduate and Professional Institutions.* The survey is sent each spring to more than 2,000 institutions offering postbaccalaureate degree programs, including accredited institutions in the United States, U.S. territories, and Canada and those institutions outside the United States that are accredited by U.S. accrediting bodies. Deans and other administrators complete these surveys, providing information on programs in the 476 academic and professional fields covered in the guides as well as overall institutional information. While every effort has been made to ensure the accuracy and completeness of the data, information is sometimes unavailable or changes occur after publication deadlines. All usable information received in time for publication has been included. The omission of any particular item from a directory or **Profile** signifies either that the item is not applicable to the institution or program or that information was not available. **Profiles** of programs scheduled to begin during the 2006–07 academic year cannot, obviously, include statistics on enrollment or, in many cases, the number of faculty members. If no usable data were submitted by an institution, its name, address, and program name appear in order to indicate the availability of graduate work.

Criteria for Inclusion in This Guide

To be included in this guide, an institution must have full accreditation or be a candidate for accreditation (preaccreditation) status by an institutional or specialized accrediting body recognized by the U.S. Department of Education or the Council for Higher Education Accreditation (CHEA). Institutional accrediting bodies, which review each institution as a whole, include the six regional associations of schools and colleges (Middle States, New England, North Central, Northwest, Southern, and Western), each of which is responsible for a specified portion of the United States and its territories. Other institutional accrediting bodies are national in scope and accredit specific kinds of institutions (e.g., Bible colleges, independent colleges, and rabbinical and Talmudic schools). Program registration by the New York State Board of Regents is considered to be the equivalent of institutional accreditation, since the board requires that all programs offered by an institution meet its standards before recognition is granted. A Canadian institution must be chartered and authorized to grant degrees by the provincial government, affiliated with a chartered institution, or accredited by a recognized U.S. accrediting body. This guide also includes institutions outside the United States that are accredited by these U.S. accrediting bodies. There are recognized specialized or professional accrediting bodies in more than fifty different fields, each of which is authorized to accredit institutions or specific programs in its particular field. For specialized institutions that offer programs in one field only, we designate this to be the equivalent of institutional accreditation. A full explanation of the accrediting process and complete information on recognized institutional (regional and national) and specialized accrediting bodies can be found online at www.chea.org or at www.ed.gov/admins/finaid/accred/index.html.

DIRECTORY OF INSTITUTIONS WITH PROGRAMS IN THE PHYSICAL SCIENCES, MATHEMATICS, AGRICULTURAL SCIENCES, THE ENVIRONMENT & NATURAL RESOURCES

ACADIA UNIVERSITY

Applied Mathematics	M
Chemistry	M
Geology	M
Statistics	M

ADELPHI UNIVERSITY

Environmental Management and Policy	M

AIR FORCE INSTITUTE OF TECHNOLOGY

Applied Mathematics	M,D
Applied Physics	M,D
Astrophysics	M,D
Environmental Management and Policy	M
Optical Sciences	M,D
Planetary and Space Sciences	M,D

ALABAMA AGRICULTURAL AND MECHANICAL UNIVERSITY

Agricultural Sciences—General	M,D
Agronomy and Soil Sciences	M,D
Animal Sciences	M,D
Applied Physics	M,D
Environmental Sciences	M,D
Food Science and Technology	M,D
Optical Sciences	M,D
Physics	M,D
Plant Sciences	M,D

ALABAMA STATE UNIVERSITY

Mathematics	M,O

ALASKA PACIFIC UNIVERSITY

Environmental Sciences	M

ALBANY STATE UNIVERSITY

Water Resources	M

ALCORN STATE UNIVERSITY

Agricultural Sciences—General	M
Agronomy and Soil Sciences	M
Animal Sciences	M

AMERICAN UNIVERSITY

Chemistry	M
Environmental Management and Policy	M,D,O
Environmental Sciences	M
Marine Sciences	M
Mathematics	M
Physics	M
Statistics	M,O

AMERICAN UNIVERSITY OF BEIRUT

Agronomy and Soil Sciences	M
Animal Sciences	M
Aquaculture	M
Chemistry	M
Environmental Management and Policy	M
Environmental Sciences	M
Food Science and Technology	M
Geology	M
Mathematics	M
Physics	M
Plant Sciences	M

ANDREWS UNIVERSITY

Mathematics	M

ANGELO STATE UNIVERSITY

Agricultural Sciences—General	M
Animal Sciences	M

ANTIOCH NEW ENGLAND GRADUATE SCHOOL

Environmental Management and Policy	M,D
Environmental Sciences	M,D*

ANTIOCH UNIVERSITY SEATTLE

Environmental Management and Policy	M

APPALACHIAN STATE UNIVERSITY

Applied Physics	M
Mathematics	M

ARIZONA STATE UNIVERSITY

Applied Mathematics	M,D
Astronomy	M,D
Biostatistics	M,D
Chemistry	M,D
Computational Sciences	M,D
Geosciences	M,D
Mathematics	M,D
Physics	M,D
Statistics	M,D

ARIZONA STATE UNIVERSITY AT THE POLYTECHNIC CAMPUS

Environmental Management and Policy	M

ARKANSAS STATE UNIVERSITY

Agricultural Sciences—General	M,O
Chemistry	M,O
Environmental Sciences	D
Mathematics	M

ARKANSAS TECH UNIVERSITY

Fish, Game, and Wildlife Management	M

AUBURN UNIVERSITY

Agricultural Sciences—General	M,D
Agronomy and Soil Sciences	M,D
Animal Sciences	M,D
Aquaculture	M,D
Chemistry	M,D
Fish, Game, and Wildlife Management	M,D
Food Science and Technology	M,D
Forestry	M,D
Geology	M*
Horticulture	M,D
Hydrology	M,D
Mathematics	M,D
Physics	M,D*

BALL STATE UNIVERSITY

Chemistry	M
Geology	M
Geosciences	M
Mathematics	M
Natural Resources	M
Physics	M
Statistics	M

BARD COLLEGE

Environmental Management and Policy	M,O*

BAYLOR UNIVERSITY

Chemistry	M,D
Environmental Management and Policy	M
Geology	M,D
Geosciences	M,D
Limnology	M,D
Mathematics	M,D
Physics	M,D
Statistics	M,D

BEMIDJI STATE UNIVERSITY

Environmental Management and Policy	M

BERNARD M. BARUCH COLLEGE OF THE CITY UNIVERSITY OF NEW YORK

Mathematical and Computational Finance	M*
Statistics	M

BOISE STATE UNIVERSITY

Environmental Management and Policy	M
Geology	M
Geophysics	M,D
Geosciences	M

BOSTON COLLEGE

Chemistry	M,D
Geology	M
Geophysics	M
Inorganic Chemistry	M,D
Mathematics	M
Organic Chemistry	M,D
Physical Chemistry	M,D
Physics	M,D

BOSTON UNIVERSITY

Astronomy	M,D
Biostatistics	M,D
Chemistry	M,D
Environmental Management and Policy	M,D,O
Geosciences	M,D
Mathematical and Computational Finance	M,D
Mathematics	M,D
Photonics	M,D
Physics	M,D*

BOWLING GREEN STATE UNIVERSITY

Astronomy	M
Chemistry	M,D
Geology	M
Mathematics	M,D,O*
Physics	M
Statistics	M,D,O

BRADLEY UNIVERSITY

Chemistry	M

BRANDEIS UNIVERSITY

Chemistry	M,D
Inorganic Chemistry	M,D
Mathematics	M,D*
Organic Chemistry	M,D
Physical Chemistry	M,D
Physics	M,D

BRIGHAM YOUNG UNIVERSITY

Agricultural Sciences—General	M,D
Agronomy and Soil Sciences	M
Analytical Chemistry	M,D
Animal Sciences	M
Astronomy	M,D
Chemistry	M,D*
Fish, Game, and Wildlife Management	M,D
Food Science and Technology	M
Geology	M
Inorganic Chemistry	M,D
Mathematics	M,D
Organic Chemistry	M,D
Physical Chemistry	M,D
Physics	M,D
Plant Sciences	M
Statistics	M

BROCK UNIVERSITY

Chemistry	M
Geosciences	M
Mathematics	M,D
Physics	M
Statistics	M,D

BROOKLYN COLLEGE OF THE CITY UNIVERSITY OF NEW YORK

Applied Physics	M,D
Chemistry	M,D
Geology	M,D
Mathematics	M,D
Physics	M,D

BROWN UNIVERSITY

Applied Mathematics	M,D*
Biostatistics	M,D
Chemistry	M,D
Environmental Management and Policy	M
Geosciences	M,D
Mathematics	M,D
Physics	M,D

BRYN MAWR COLLEGE

Chemistry	M,D
Mathematics	M,D
Physics	M,D

BUCKNELL UNIVERSITY

Chemistry	M
Mathematics	M

BUFFALO STATE COLLEGE, STATE UNIVERSITY OF NEW YORK

Chemistry	M

CALIFORNIA INSTITUTE OF TECHNOLOGY

Applied Mathematics	M,D
Applied Physics	M,D
Astronomy	D
Chemistry	M,D
Computational Sciences	M,D
Geochemistry	M,D
Geology	M,D
Geophysics	M,D
Mathematics	D
Physics	D

Planetary and Space Sciences M,D

CALIFORNIA POLYTECHNIC STATE UNIVERSITY, SAN LUIS OBISPO
Agricultural Sciences—General M
Chemistry M
Environmental Management and Policy M
Forestry M
Mathematics M

CALIFORNIA STATE POLYTECHNIC UNIVERSITY, POMONA
Agricultural Sciences—General M
Animal Sciences M
Applied Mathematics M
Chemistry M
Environmental Sciences M
Food Science and Technology M
Mathematics M

CALIFORNIA STATE UNIVERSITY, BAKERSFIELD
Geology M
Hydrology M

CALIFORNIA STATE UNIVERSITY CHANNEL ISLANDS
Mathematics M

CALIFORNIA STATE UNIVERSITY, CHICO
Environmental Sciences M
Geology M
Hydrogeology M
Hydrology M

CALIFORNIA STATE UNIVERSITY, EAST BAY
Biostatistics M
Chemistry M
Geology M
Marine Sciences M
Mathematics M
Statistics M

CALIFORNIA STATE UNIVERSITY, FRESNO
Agricultural Sciences—General M
Animal Sciences M
Chemistry M
Food Science and Technology M
Geology M
Marine Sciences M
Mathematics M
Physics M
Plant Sciences M

CALIFORNIA STATE UNIVERSITY, FULLERTON
Analytical Chemistry M
Applied Mathematics M
Chemistry M
Environmental Management and Policy M
Environmental Sciences M
Geochemistry M
Geology M
Inorganic Chemistry M
Mathematics M
Organic Chemistry M
Physical Chemistry M
Physics M
Statistics M

CALIFORNIA STATE UNIVERSITY, LONG BEACH
Applied Mathematics M,D,O
Chemistry M
Geology M
Geosciences M
Mathematics M
Physics M

CALIFORNIA STATE UNIVERSITY, LOS ANGELES
Analytical Chemistry M
Applied Mathematics M
Chemistry M
Geology M
Inorganic Chemistry M
Mathematics M
Organic Chemistry M
Physical Chemistry M
Physics M

CALIFORNIA STATE UNIVERSITY, MONTEREY BAY
Marine Sciences M

CALIFORNIA STATE UNIVERSITY, NORTHRIDGE
Chemistry M
Geology M
Mathematics M
Physics M

CALIFORNIA STATE UNIVERSITY, SACRAMENTO
Chemistry M
Marine Sciences M
Mathematics M
Statistics M

CALIFORNIA STATE UNIVERSITY, SAN BERNARDINO
Mathematics M

CALIFORNIA STATE UNIVERSITY, SAN MARCOS
Mathematics M

CALIFORNIA UNIVERSITY OF PENNSYLVANIA
Geosciences M

CARLETON UNIVERSITY
Chemistry M,D
Geosciences M,D
Mathematics M,D
Physics M,D

CARNEGIE MELLON UNIVERSITY
Chemistry M,D*
Computational Sciences M,D
Mathematical and Computational Finance M,D
Mathematics M,D
Physics D
Statistics M,D*

CASE WESTERN RESERVE UNIVERSITY
Analytical Chemistry M,D
Applied Mathematics M,D
Astronomy M,D
Biostatistics M,D
Chemistry M,D
Geology M,D
Geosciences M,D
Inorganic Chemistry M,D
Mathematics M,D
Organic Chemistry M,D
Physical Chemistry M,D
Physics M,D
Statistics M,D

THE CATHOLIC UNIVERSITY OF AMERICA
Acoustics M,D
Chemistry M
Physics M,D

CENTRAL CONNECTICUT STATE UNIVERSITY
Chemistry M
Geosciences M
Mathematics M
Physics M
Statistics M

CENTRAL EUROPEAN UNIVERSITY
Environmental Management and Policy M,D

CENTRAL MICHIGAN UNIVERSITY
Chemistry M
Mathematics M,D
Physics M

CENTRAL MISSOURI STATE UNIVERSITY
Applied Mathematics M
Mathematics M

CENTRAL WASHINGTON UNIVERSITY
Chemistry M
Environmental Management and Policy M
Geology M
Mathematics M

CHAPMAN UNIVERSITY
Food Science and Technology M

CHICAGO STATE UNIVERSITY
Mathematics M

CHRISTOPHER NEWPORT UNIVERSITY
Applied Physics M
Environmental Sciences M
Physics M

CITY COLLEGE OF THE CITY UNIVERSITY OF NEW YORK
Atmospheric Sciences M,D
Chemistry M,D
Environmental Sciences M,D
Geosciences M,D
Mathematics M
Physics M,D*

CLAREMONT GRADUATE UNIVERSITY
Applied Mathematics M,D
Computational Sciences M,D
Mathematics M,D
Statistics M,D

CLARK ATLANTA UNIVERSITY
Applied Mathematics M
Chemistry M,D
Inorganic Chemistry M,D
Organic Chemistry M,D
Physical Chemistry M,D
Physics M

CLARKSON UNIVERSITY
Analytical Chemistry M,D
Chemistry M,D
Environmental Sciences M,D
Inorganic Chemistry M,D
Mathematics M,D
Organic Chemistry M,D
Physical Chemistry M,D
Physics M,D

CLARK UNIVERSITY
Chemistry M,D
Environmental Management and Policy M
Physics M,D

CLEMSON UNIVERSITY
Agricultural Sciences—General M,D
Animal Sciences M,D*
Applied Mathematics M,D
Aquaculture M,D
Astronomy M,D
Astrophysics M,D
Atmospheric Sciences M,D
Chemistry M,D*
Computational Sciences M,D
Environmental Management and Policy M,D
Environmental Sciences M,D*
Fish, Game, and Wildlife Management M,D*
Food Science and Technology M,D*
Forestry M,D*
Hydrogeology M*
Mathematics M,D*
Physics M,D*
Plant Sciences M,D*
Statistics M,D

CLEVELAND STATE UNIVERSITY
Analytical Chemistry M,D
Chemistry M,D
Condensed Matter Physics M
Environmental Management and Policy M
Environmental Sciences M,D
Inorganic Chemistry M,D
Mathematics M
Optical Sciences M
Organic Chemistry M,D
Physical Chemistry M,D
Physics M

COASTAL CAROLINA UNIVERSITY
Marine Sciences M

COLLEGE OF CHARLESTON
Environmental Sciences M
Mathematics M,O

COLLEGE OF STATEN ISLAND OF THE CITY UNIVERSITY OF NEW YORK

Chemistry D
Environmental Sciences M
Physics D

COLLEGE OF THE ATLANTIC

Environmental Management and Policy M

THE COLLEGE OF WILLIAM AND MARY

Chemistry M
Computational Sciences M
Marine Sciences M,D
Physics M,D

COLORADO SCHOOL OF MINES

Applied Physics M,D
Chemistry M,D
Environmental Sciences M,D
Geochemistry M,D,O
Geology M,D,O
Geophysics M,D,O
Geosciences M,D,O
Hydrogeology M,D,O
Mathematics M,D
Physics M,D

COLORADO STATE UNIVERSITY

Agricultural Sciences—General M,D
Agronomy and Soil Sciences M,D
Animal Sciences M,D
Atmospheric Sciences M,D
Chemistry M,D
Environmental Management and Policy M,D
Fish, Game, and Wildlife Management M,D
Food Science and Technology M,D
Forestry M,D
Geology M,D
Geophysics M,D
Geosciences M,D
Horticulture M,D
Hydrogeology M,D
Hydrology M,D
Mathematics M,D
Physics M,D
Plant Sciences M,D
Range Science M,D
Statistics M,D
Water Resources M,D

COLORADO STATE UNIVERSITY-PUEBLO

Chemistry M

COLUMBIA UNIVERSITY

Applied Mathematics M,D,O
Applied Physics M,D,O*
Astronomy M,D
Atmospheric Sciences M,D*
Biostatistics M,D
Chemical Physics M,D
Chemistry M,D
Environmental Management and Policy M*
Geochemistry M,D
Geodetic Sciences M,D
Geophysics M,D
Geosciences M,D
Inorganic Chemistry M,D
Mathematics M,D*
Meteorology M*
Oceanography M,D
Optical Sciences M,D,O
Organic Chemistry M,D
Physics M,D
Planetary and Space Sciences M,D
Plasma Physics M,D,O
Statistics M,D

COLUMBUS STATE UNIVERSITY

Environmental Sciences M

CONCORDIA UNIVERSITY (CANADA)

Chemistry M,D
Environmental Management and Policy O
Mathematics M,D

CORNELL UNIVERSITY

Agronomy and Soil Sciences M,D
Analytical Chemistry D
Animal Sciences M,D
Applied Mathematics M,D*
Applied Physics M,D
Astronomy D
Astrophysics D
Atmospheric Sciences M,D
Biometrics M,D
Chemical Physics D
Chemistry D*
Computational Sciences M,D
Environmental Management and Policy M,D
Environmental Sciences M,D
Fish, Game, and Wildlife Management M,D
Food Science and Technology M,D
Forestry M,D
Geochemistry M,D
Geology M,D
Geophysics M,D
Geosciences M,D
Horticulture M,D
Hydrology M,D
Inorganic Chemistry D
Limnology D
Marine Geology M,D
Marine Sciences M,D
Mathematics D
Mineralogy M,D
Natural Resources M,D
Oceanography D
Organic Chemistry D
Paleontology M,D
Physical Chemistry D
Physics M,D
Planetary and Space Sciences D
Plant Sciences M,D
Statistics M,D
Theoretical Chemistry D
Theoretical Physics M,D

CREIGHTON UNIVERSITY

Atmospheric Sciences M
Physics M

DALHOUSIE UNIVERSITY

Agricultural Sciences—General M
Applied Mathematics M,D
Chemistry M,D
Environmental Management and Policy M
Food Science and Technology M,D
Geosciences M,D
Marine Affairs M
Mathematics M,D
Oceanography M,D
Physics M,D
Statistics M,D

DARTMOUTH COLLEGE

Astronomy M,D
Chemistry D
Geosciences M,D
Mathematics D*
Physics M,D

DELAWARE STATE UNIVERSITY

Chemistry M
Mathematics M
Physics M

DEPAUL UNIVERSITY

Applied Physics M
Chemistry M
Mathematical and Computational Finance M,D
Mathematics M
Physics M
Statistics M

DOWLING COLLEGE

Mathematics M

DREXEL UNIVERSITY

Biostatistics M,D
Chemistry M,D
Environmental Management and Policy M
Environmental Sciences M,D
Food Science and Technology M,D
Mathematics M,D
Physics M,D

DUKE UNIVERSITY

Chemistry D
Environmental Management and Policy M,D*
Environmental Sciences M,D
Forestry M
Geology M,D
Marine Affairs M
Marine Sciences M*
Mathematics D
Natural Resources M,D*
Paleontology D
Physics M,D
Statistics D
Water Resources M

DUQUESNE UNIVERSITY

Chemistry M,D*
Environmental Management and Policy M,O
Environmental Sciences M,O
Mathematics M

EAST CAROLINA UNIVERSITY

Applied Mathematics M
Chemistry M
Environmental Management and Policy D
Geology M
Marine Affairs D
Mathematics M
Physics M,D*

EASTERN ILLINOIS UNIVERSITY

Chemistry M
Mathematics M

EASTERN KENTUCKY UNIVERSITY

Chemistry M
Geology M,D
Mathematics M

EASTERN MICHIGAN UNIVERSITY

Chemistry M
Mathematics M
Physics M
Statistics M

EASTERN NEW MEXICO UNIVERSITY

Chemistry M
Mathematics M

EASTERN WASHINGTON UNIVERSITY

Mathematics M

EAST TENNESSEE STATE UNIVERSITY

Chemistry M
Mathematics M

ÉCOLE POLYTECHNIQUE DE MONTRÉAL

Applied Mathematics M,D
Mathematics M,D
Optical Sciences M,D,O

EMBRY-RIDDLE AERONAUTICAL UNIVERSITY (FL)

Planetary and Space Sciences M

EMORY UNIVERSITY

Biostatistics M,D*
Chemistry D
Condensed Matter Physics D
Mathematics M,D*
Physics D

EMPORIA STATE UNIVERSITY

Chemistry M
Geosciences M
Mathematics M
Physics M

THE EVERGREEN STATE COLLEGE

Environmental Management and Policy M*

FAIRFIELD UNIVERSITY

Mathematics M

FAIRLEIGH DICKINSON UNIVERSITY, COLLEGE AT FLORHAM

Chemistry M

FAYETTEVILLE STATE UNIVERSITY

Mathematics M

FISK UNIVERSITY

Chemistry M
Physics M

FLORIDA AGRICULTURAL AND MECHANICAL UNIVERSITY

Animal Sciences	M
Chemistry	M
Environmental Sciences	M,D
Food Science and Technology	M
Physics	M,D
Plant Sciences	M

FLORIDA ATLANTIC UNIVERSITY

Applied Mathematics	M,D
Chemistry	M,D
Environmental Sciences	M
Geology	M
Mathematics	M,D
Physics	M,D

FLORIDA GULF COAST UNIVERSITY

Environmental Management and Policy	M
Environmental Sciences	M

FLORIDA INSTITUTE OF TECHNOLOGY

Applied Mathematics	M,D
Chemistry	M,D
Environmental Management and Policy	M,D
Environmental Sciences	M,D*
Marine Affairs	M,D
Marine Sciences	M,D
Meteorology	M,D
Oceanography	M,D*
Physics	M,D
Planetary and Space Sciences	M,D

FLORIDA INTERNATIONAL UNIVERSITY

Chemistry	M,D
Environmental Management and Policy	M
Environmental Sciences	M
Geosciences	M,D
Mathematics	M
Physics	M,D
Statistics	M

FLORIDA STATE UNIVERSITY

Analytical Chemistry	M,D
Applied Mathematics	M,D
Biostatistics	M,D
Chemical Physics	M,D
Chemistry	M,D*
Food Science and Technology	M,D
Geology	M,D
Geophysics	D
Inorganic Chemistry	M,D
Mathematical and Computational Finance	M,D
Mathematics	M,D
Meteorology	M,D
Oceanography	M,D*
Organic Chemistry	M,D
Physical Chemistry	M,D
Physics	M,D*
Statistics	M,D

FORT HAYS STATE UNIVERSITY

Geology	M

FORT VALLEY STATE UNIVERSITY

Animal Sciences	M

FRAMINGHAM STATE COLLEGE

Food Science and Technology	M*

FRIENDS UNIVERSITY

Environmental Management and Policy	M

FROSTBURG STATE UNIVERSITY

Fish, Game, and Wildlife Management	M

FURMAN UNIVERSITY

Chemistry	M

GANNON UNIVERSITY

Environmental Management and Policy	M
Environmental Sciences	O

GEORGE MASON UNIVERSITY

Applied Physics	M
Atmospheric Sciences	M,D,O
Chemistry	M
Computational Sciences	M,D,O*
Environmental Sciences	M,D
Geodetic Sciences	M,D,O
Mathematics	M,D
Physics	M,D
Statistics	M

GEORGETOWN UNIVERSITY

Analytical Chemistry	M,D
Biostatistics	M
Chemical Physics	M,D
Chemistry	M,D
Inorganic Chemistry	M,D
Organic Chemistry	M,D
Physical Chemistry	M,D
Theoretical Chemistry	M,D

THE GEORGE WASHINGTON UNIVERSITY

Analytical Chemistry	M,D
Applied Mathematics	M,D
Biostatistics	M,D
Chemistry	M,D
Environmental Management and Policy	M,D
Geology	M,D
Geosciences	M,D
Inorganic Chemistry	M,D
Mathematics	M,D
Organic Chemistry	M,D
Physical Chemistry	M,D
Physics	M,D
Statistics	M,D,O

GEORGIA INSTITUTE OF TECHNOLOGY

Applied Mathematics	M,D
Atmospheric Sciences	M,D
Chemistry	M,D
Environmental Management and Policy	M,D
Environmental Sciences	M,D
Geochemistry	M,D
Geophysics	M,D
Geosciences	M,D*
Hydrology	M,D
Mathematical and Computational Finance	M,D
Mathematics	M,D
Natural Resources	M,D
Physics	M,D
Statistics	M,D

GEORGIAN COURT UNIVERSITY

Mathematics	M,O

GEORGIA SOUTHERN UNIVERSITY

Mathematics	M

GEORGIA STATE UNIVERSITY

Astronomy	D
Chemistry	M,D
Environmental Management and Policy	M,O
Geology	M,O
Geosciences	M,O
Hydrology	M,O
Mathematics	M
Physics	M,D

GODDARD COLLEGE

Environmental Management and Policy	M

GOVERNORS STATE UNIVERSITY

Analytical Chemistry	M

GRADUATE SCHOOL AND UNIVERSITY CENTER OF THE CITY UNIVERSITY OF NEW YORK

Chemistry	D
Environmental Sciences	D
Geosciences	D
Mathematics	D
Physics	D

GRAND VALLEY STATE UNIVERSITY

Biostatistics	M

HAMPTON UNIVERSITY

Applied Mathematics	M
Chemistry	M
Physics	M,D

HARDIN-SIMMONS UNIVERSITY

Environmental Management and Policy	M
Mathematics	M

HARVARD UNIVERSITY

Applied Mathematics	M,D
Applied Physics	M,D
Astronomy	D
Astrophysics	D
Biostatistics	M,D
Chemical Physics	D
Chemistry	D*
Environmental Management and Policy	M,O
Environmental Sciences	M,D
Forestry	M
Geosciences	M,D
Inorganic Chemistry	D
Mathematics	D
Organic Chemistry	D
Physical Chemistry	D
Physics	D*
Planetary and Space Sciences	M,D
Statistics	M,D
Theoretical Physics	D

HOFSTRA UNIVERSITY

Applied Mathematics	M
Mathematics	M

HOWARD UNIVERSITY

Analytical Chemistry	M,D
Applied Mathematics	M,D
Atmospheric Sciences	M,D
Chemistry	M,D
Environmental Sciences	M,D
Inorganic Chemistry	M,D
Mathematics	M,D
Organic Chemistry	M,D
Physical Chemistry	M,D
Physics	M,D

HUMBOLDT STATE UNIVERSITY

Environmental Sciences	M
Natural Resources	M

HUNTER COLLEGE OF THE CITY UNIVERSITY OF NEW YORK

Applied Mathematics	M
Environmental Sciences	M,O
Geosciences	M,O
Mathematics	M
Physics	M,D

ICR GRADUATE SCHOOL

Astrophysics	M
Geology	M
Geophysics	M

IDAHO STATE UNIVERSITY

Chemistry	M
Environmental Sciences	M
Geology	M,O
Geophysics	M,O
Geosciences	M,O
Hydrology	M,O
Mathematics	M,D*
Physics	M*

ILLINOIS INSTITUTE OF TECHNOLOGY

Analytical Chemistry	M,D
Applied Mathematics	M,D
Chemistry	M,D
Environmental Management and Policy	M,D
Food Science and Technology	M
Inorganic Chemistry	M,D
Organic Chemistry	M,D
Physical Chemistry	M,D
Physics	M,D

ILLINOIS STATE UNIVERSITY

Agricultural Sciences—General	M
Chemistry	M
Hydrology	M
Mathematics	M

INDIANA STATE UNIVERSITY

Geosciences	M,D
Mathematics	M

INDIANA UNIVERSITY BLOOMINGTON

Analytical Chemistry	M,D
Applied Mathematics	M,D
Astronomy	M,D
Astrophysics	D
Chemistry	M,D
Environmental Sciences	M,D,O*
Geochemistry	M,D
Geology	M,D
Geophysics	M,D
Geosciences	M,D

Hydrogeology M,D
Inorganic Chemistry M,D
Mathematics M,D
Mineralogy M,D
Physical Chemistry M,D
Physics M,D*
Statistics M,D

INDIANA UNIVERSITY NORTHWEST
Environmental Sciences M,O

INDIANA UNIVERSITY OF PENNSYLVANIA
Applied Mathematics M
Chemistry M
Mathematics M
Physics M

INDIANA UNIVERSITY–PURDUE UNIVERSITY FORT WAYNE
Applied Mathematics M,O
Mathematics M,O
Statistics M,O

INDIANA UNIVERSITY–PURDUE UNIVERSITY INDIANAPOLIS
Applied Mathematics M,D
Chemistry M,D
Environmental Management and Policy M
Geology M
Mathematics M,D
Physics M,D
Statistics M,D

INDIANA UNIVERSITY SOUTH BEND
Applied Mathematics M

INSTITUTO TECNOLOGICO DE SANTO DOMINGO
Environmental Sciences M

INSTITUTO TECNOLÓGICO Y DE ESTUDIOS SUPERIORES DE MONTERREY, CAMPUS CIUDAD DE MÉXICO
Environmental Sciences M,D

INSTITUTO TECNOLÓGICO Y DE ESTUDIOS SUPERIORES DE MONTERREY, CAMPUS ESTADO DE MÉXICO
Environmental Management and Policy M,D

INSTITUTO TECNOLÓGICO Y DE ESTUDIOS SUPERIORES DE MONTERREY, CAMPUS IRAPUATO
Environmental Management and Policy M,D

INSTITUTO TECNOLÓGICO Y DE ESTUDIOS SUPERIORES DE MONTERREY, CAMPUS MONTERREY
Agricultural Sciences—General M,D
Chemistry M,D
Organic Chemistry M,D
Statistics M,D

INTER AMERICAN UNIVERSITY OF PUERTO RICO, SAN GERMÁN CAMPUS
Applied Mathematics M
Environmental Sciences M

IOWA STATE UNIVERSITY OF SCIENCE AND TECHNOLOGY
Agricultural Sciences—General M,D
Agronomy and Soil Sciences M,D
Animal Sciences M,D
Applied Mathematics M,D
Applied Physics M,D
Astronomy M,D
Astrophysics M,D
Biostatistics M,D
Chemistry M,D
Condensed Matter Physics M,D
Environmental Management and Policy M,D
Environmental Sciences M,D
Fish, Game, and Wildlife Management M,D
Food Science and Technology M,D*
Forestry M,D
Geology M,D
Geosciences M,D
Horticulture M,D
Mathematics M,D
Meteorology M,D
Natural Resources M,D
Physics M,D
Statistics M,D
Water Resources M,D

JACKSON STATE UNIVERSITY
Chemistry M,D
Environmental Sciences M,D
Mathematics M

JACKSONVILLE STATE UNIVERSITY
Mathematics M

JAMES MADISON UNIVERSITY
Mathematics M
Statistics M

JOHN CARROLL UNIVERSITY
Chemistry M
Mathematics M
Physics M

THE JOHNS HOPKINS UNIVERSITY
Applied Mathematics M,D
Applied Physics M
Astronomy D
Biostatistics M,D
Chemistry D
Environmental Management and Policy M*
Environmental Sciences M
Mathematics D
Physics D*
Statistics M,D

KANSAS STATE UNIVERSITY
Agricultural Sciences—General M,D
Agronomy and Soil Sciences M,D
Analytical Chemistry M,D
Animal Sciences M,D

Chemistry M,D
Environmental Management and Policy M
Food Science and Technology M,D
Geology M
Horticulture M,D
Inorganic Chemistry M,D
Mathematics M,D
Organic Chemistry M,D
Physical Chemistry M,D
Physics M,D
Range Science M,D
Statistics M,D

KEAN UNIVERSITY
Computational Sciences M
Environmental Management and Policy M
Mathematics M
Statistics M

KENT STATE UNIVERSITY
Analytical Chemistry M,D
Applied Mathematics M,D
Chemical Physics M,D
Chemistry M,D*
Geology M,D
Inorganic Chemistry M,D
Mathematics M,D
Organic Chemistry M,D
Physical Chemistry M,D
Physics M,D*

KENTUCKY STATE UNIVERSITY
Aquaculture M

KUTZTOWN UNIVERSITY OF PENNSYLVANIA
Mathematics M

LAKEHEAD UNIVERSITY
Chemistry M
Forestry M
Geology M
Mathematics M
Physics M
Statistics M

LAMAR UNIVERSITY
Chemistry M
Environmental Management and Policy M,D
Mathematics M

LAURENTIAN UNIVERSITY
Applied Physics M
Chemistry M
Geology M

LEHIGH UNIVERSITY
Applied Mathematics M,D
Chemistry M,D*
Computational Sciences M,D
Environmental Sciences M,D*
Geology M,D
Geosciences M,D
Mathematics M,D*
Photonics M,D
Physics M,D*
Statistics M,D

LEHMAN COLLEGE OF THE CITY UNIVERSITY OF NEW YORK
Mathematics M
Plant Sciences D

LOMA LINDA UNIVERSITY
Biostatistics M
Geosciences M,D

LONG ISLAND UNIVERSITY, BROOKLYN CAMPUS
Chemistry M

LONG ISLAND UNIVERSITY, C.W. POST CAMPUS
Applied Mathematics M
Environmental Management and Policy M
Environmental Sciences M
Mathematics M

LOUISIANA STATE UNIVERSITY AND AGRICULTURAL AND MECHANICAL COLLEGE
Agricultural Sciences—General M,D
Agronomy and Soil Sciences M,D
Animal Sciences M,D
Astronomy M,D
Astrophysics M,D
Chemistry M,D
Environmental Management and Policy M
Environmental Sciences M,D
Fish, Game, and Wildlife Management M,D
Food Science and Technology M,D
Forestry M,D
Geology M,D
Geophysics M,D
Horticulture M,D
Marine Affairs M,D
Mathematics M,D
Natural Resources M,D*
Oceanography M,D
Physics M,D
Statistics M

LOUISIANA STATE UNIVERSITY HEALTH SCIENCES CENTER
Biometrics M

LOUISIANA TECH UNIVERSITY
Chemistry M
Computational Sciences M,D
Mathematics M
Physics M,D
Statistics M

LOYOLA MARYMOUNT UNIVERSITY
Environmental Sciences M

LOYOLA UNIVERSITY CHICAGO
Chemistry M,D
Mathematics M
Statistics M

MARQUETTE UNIVERSITY
Analytical Chemistry M,D
Chemical Physics M,D
Chemistry M,D
Inorganic Chemistry M,D
Mathematics M,D
Organic Chemistry M,D
Physical Chemistry M,D
Statistics M,D

MARSHALL UNIVERSITY

Chemistry	M
Environmental Sciences	M
Mathematics	M
Physics	M

MARYWOOD UNIVERSITY

Food Science and Technology	M,O

MASSACHUSETTS COLLEGE OF PHARMACY AND HEALTH SCIENCES

Chemistry	M,D

MASSACHUSETTS INSTITUTE OF TECHNOLOGY

Atmospheric Sciences	M,D
Chemistry	D
Computational Sciences	M
Environmental Sciences	M,D,O
Geochemistry	M,D
Geology	M,D
Geophysics	M,D
Geosciences	M,D
Hydrology	M,D,O
Inorganic Chemistry	D
Marine Geology	M,D
Mathematics	D
Oceanography	M,D,O
Organic Chemistry	M,D,O
Physical Chemistry	D
Physics	D
Planetary and Space Sciences	M,D
Plasma Physics	M,D,O

MCGILL UNIVERSITY

Agricultural Sciences—General	M,D,O
Agronomy and Soil Sciences	M,D
Animal Sciences	M,D
Applied Mathematics	M,D
Atmospheric Sciences	M,D
Biostatistics	M,D,O
Chemistry	M,D
Computational Sciences	M,D
Fish, Game, and Wildlife Management	M,D
Food Science and Technology	M,D
Forestry	M,D
Geosciences	M,D
Mathematics	M,D
Meteorology	M,D
Natural Resources	M,D
Oceanography	M,D
Physics	M,D
Planetary and Space Sciences	M,D
Plant Sciences	M,D,O
Statistics	M,D

MCMASTER UNIVERSITY

Analytical Chemistry	M,D
Astrophysics	D
Chemical Physics	M,D
Chemistry	M,D
Geochemistry	M,D
Geology	M,D
Geosciences	M,D
Inorganic Chemistry	M,D
Mathematics	M,D
Organic Chemistry	M,D
Physical Chemistry	M,D
Physics	D
Statistics	M

MCNEESE STATE UNIVERSITY

Chemistry	M
Environmental Sciences	M
Mathematics	M
Statistics	M

MEDICAL COLLEGE OF WISCONSIN

Biostatistics	D*

MEDICAL UNIVERSITY OF SOUTH CAROLINA

Biometrics	P,M,D
Biostatistics	P,M,D

MEMORIAL UNIVERSITY OF NEWFOUNDLAND

Aquaculture	M
Chemistry	M,D
Computational Sciences	M
Condensed Matter Physics	M,D
Environmental Sciences	M
Fish, Game, and Wildlife Management	M,O
Food Science and Technology	M,D
Geology	M,D
Geophysics	M,D
Geosciences	M,D
Marine Affairs	M,D,O
Marine Sciences	M,O
Mathematics	M,D
Natural Resources	M
Oceanography	M,D
Physics	M,D
Statistics	M,D

MIAMI UNIVERSITY

Analytical Chemistry	M,D
Chemistry	M,D
Environmental Sciences	M
Geology	M,D
Inorganic Chemistry	M,D
Mathematics	M*
Organic Chemistry	M,D
Physical Chemistry	M,D
Physics	M
Statistics	M

MICHIGAN STATE UNIVERSITY

Agricultural Sciences—General	M,D
Agronomy and Soil Sciences	M,D
Animal Sciences	M,D
Applied Mathematics	M,D
Astronomy	M,D
Astrophysics	M,D
Chemical Physics	M,D
Chemistry	M,D
Environmental Management and Policy	M,D
Environmental Sciences	M,D
Fish, Game, and Wildlife Management	M,D
Food Science and Technology	M,D
Forestry	M,D
Geology	M,D
Geosciences	M,D
Horticulture	M,D
Mathematics	M,D
Physics	M,D
Plant Sciences	M,D
Statistics	M,D

MICHIGAN TECHNOLOGICAL UNIVERSITY

Chemistry	M,D
Computational Sciences	D
Environmental Management and Policy	M
Forestry	M,D
Geology	M,D
Geophysics	M
Mathematics	M,D
Physics	M,D

MIDDLE TENNESSEE STATE UNIVERSITY

Chemistry	M,D
Geosciences	O
Mathematics	M

MILLERSVILLE UNIVERSITY OF PENNSYLVANIA

Geosciences	M

MINNESOTA STATE UNIVERSITY MANKATO

Astronomy	M
Environmental Sciences	M
Mathematics	M
Physics	M
Statistics	M

MISSISSIPPI COLLEGE

Chemistry	M
Mathematics	M

MISSISSIPPI STATE UNIVERSITY

Agricultural Sciences—General	M,D
Agronomy and Soil Sciences	M,D
Animal Sciences	M
Chemistry	M,D
Fish, Game, and Wildlife Management	M
Food Science and Technology	M,D
Forestry	M,D
Geosciences	M
Mathematics	M,D
Physics	M,D
Plant Sciences	M,D
Statistics	M,D

MISSOURI STATE UNIVERSITY

Agricultural Sciences—General	M
Chemistry	M
Environmental Management and Policy	M
Geology	M
Geosciences	M
Mathematics	M
Plant Sciences	M

MONTANA STATE UNIVERSITY

Agricultural Sciences—General	M,D
Animal Sciences	M,D
Chemistry	M,D
Environmental Management and Policy	M,D
Environmental Sciences	M,D
Fish, Game, and Wildlife Management	M,D
Geosciences	M,D
Mathematics	M,D
Natural Resources	M,D
Physics	M,D
Plant Sciences	M,D
Range Science	M,D
Statistics	M,D

MONTANA TECH OF THE UNIVERSITY OF MONTANA

Geochemistry	M
Geology	M
Geosciences	M
Hydrogeology	M

MONTCLAIR STATE UNIVERSITY

Applied Mathematics	M,O
Chemistry	M
Environmental Management and Policy	M,D*
Environmental Sciences	M,D,O
Food Science and Technology	M,O
Geosciences	M,D,O
Mathematics	M,O
Statistics	M,O
Water Resources	M,D,O

MONTEREY INSTITUTE OF INTERNATIONAL STUDIES

Environmental Management and Policy	M

MOREHEAD STATE UNIVERSITY

Environmental Management and Policy	M

MORGAN STATE UNIVERSITY

Chemistry	M
Mathematics	M
Physics	M

MOUNT ALLISON UNIVERSITY

Chemistry	M

MURRAY STATE UNIVERSITY

Agricultural Sciences—General	M
Chemistry	M
Geosciences	M
Marine Sciences	M
Mathematics	M

NAROPA UNIVERSITY

Environmental Management and Policy	M

NAVAL POSTGRADUATE SCHOOL

Applied Mathematics	M,D
Applied Physics	M,D
Mathematics	M,D
Meteorology	M,D
Oceanography	M,D
Physics	M,D

NEW JERSEY INSTITUTE OF TECHNOLOGY

Applied Mathematics	M
Applied Physics	M,D
Chemistry	M,D

- Environmental Management and Policy M,D
- Environmental Sciences M,D
- Mathematics D
- Statistics M

NEW MEXICO HIGHLANDS UNIVERSITY

- Chemistry M
- Environmental Management and Policy M

NEW MEXICO INSTITUTE OF MINING AND TECHNOLOGY

- Applied Mathematics M,D
- Astrophysics M,D
- Atmospheric Sciences M,D
- Chemistry M,D
- Environmental Sciences M,D
- Geochemistry M,D
- Geology M,D
- Geophysics M,D
- Geosciences M,D
- Hydrology M,D
- Mathematical Physics M,D
- Mathematics M,D
- Physics M,D

NEW MEXICO STATE UNIVERSITY

- Agricultural Sciences—General M
- Agronomy and Soil Sciences M,D
- Animal Sciences M,D
- Astronomy M,D
- Chemistry M,D
- Fish, Game, and Wildlife Management M
- Geology M
- Horticulture M,D
- Mathematics M,D
- Physics M,D
- Plant Sciences M
- Range Science M,D
- Statistics M

NEW YORK INSTITUTE OF TECHNOLOGY

- Environmental Management and Policy M,O

NEW YORK MEDICAL COLLEGE

- Biostatistics M

NEW YORK UNIVERSITY

- Chemistry M,D*
- Food Science and Technology M,D
- Mathematical and Computational Finance M,D
- Mathematics M,D*
- Physics M,D
- Statistics M,D

NICHOLLS STATE UNIVERSITY

- Mathematics M

NORFOLK STATE UNIVERSITY

- Optical Sciences M

NORTH CAROLINA AGRICULTURAL AND TECHNICAL STATE UNIVERSITY

- Agricultural Sciences—General M
- Chemistry M
- Environmental Sciences M
- Plant Sciences M

NORTH CAROLINA CENTRAL UNIVERSITY

- Chemistry M
- Geosciences M
- Mathematics M

NORTH CAROLINA STATE UNIVERSITY

- Agricultural Sciences—General M,D
- Agronomy and Soil Sciences M,D
- Animal Sciences M,D
- Applied Mathematics M,D
- Atmospheric Sciences M,D
- Biometrics M,D
- Chemistry M,D
- Fish, Game, and Wildlife Management M
- Food Science and Technology M,D
- Forestry M,D
- Geosciences M,D*
- Horticulture M,D
- Marine Sciences M,D
- Mathematical and Computational Finance M*
- Mathematics M,D
- Meteorology M,D
- Natural Resources M
- Oceanography M,D
- Physics M,D
- Statistics M,D

NORTH DAKOTA STATE UNIVERSITY

- Agricultural Sciences—General M,D
- Agronomy and Soil Sciences M,D
- Animal Sciences M,D
- Applied Mathematics M,D
- Chemistry M,D
- Environmental Management and Policy M,D
- Environmental Sciences M,D
- Food Science and Technology M,D
- Mathematics M,D
- Physics M,D
- Plant Sciences M,D
- Range Science M,D
- Statistics M,D,O

NORTHEASTERN ILLINOIS UNIVERSITY

- Chemistry M
- Environmental Management and Policy M
- Geosciences M
- Mathematics M

NORTHEASTERN UNIVERSITY

- Analytical Chemistry M,D
- Applied Mathematics M,D
- Chemistry M,D*
- Inorganic Chemistry M,D
- Mathematics M,D*
- Organic Chemistry M,D
- Physical Chemistry M,D
- Physics M,D*

NORTHERN ARIZONA UNIVERSITY

- Applied Physics M
- Chemistry M
- Environmental Management and Policy M,O
- Environmental Sciences M,O
- Forestry M,D
- Geology M
- Geosciences M
- Mathematics M
- Statistics M

NORTHERN ILLINOIS UNIVERSITY

- Chemistry M,D
- Geology M,D
- Mathematics M,D
- Physics M,D
- Statistics M

NORTHERN MICHIGAN UNIVERSITY

- Chemistry M

NORTHWESTERN UNIVERSITY

- Applied Mathematics M,D
- Astronomy M,D
- Astrophysics M,D
- Chemistry D
- Computational Sciences M
- Geology M,D
- Geosciences M,D
- Mathematics D
- Physics M,D
- Statistics M,D

NORTHWEST MISSOURI STATE UNIVERSITY

- Agricultural Sciences—General M

NOVA SCOTIA AGRICULTURAL COLLEGE

- Agricultural Sciences—General M
- Agronomy and Soil Sciences M
- Animal Sciences M
- Aquaculture M
- Environmental Management and Policy M
- Environmental Sciences M
- Food Science and Technology M
- Horticulture M
- Water Resources M

NOVA SOUTHEASTERN UNIVERSITY

- Environmental Sciences M
- Marine Affairs M
- Marine Sciences M
- Oceanography M,D*

OAKLAND UNIVERSITY

- Applied Mathematics M,D
- Chemistry M,D
- Environmental Sciences M,D
- Mathematics M
- Physics M,D
- Statistics M,O

OGI SCHOOL OF SCIENCE & ENGINEERING AT OREGON HEALTH & SCIENCE UNIVERSITY

- Environmental Sciences M,D*

THE OHIO STATE UNIVERSITY

- Agricultural Sciences—General M,D
- Agronomy and Soil Sciences M,D
- Animal Sciences M,D
- Astronomy M,D
- Atmospheric Sciences M,D
- Biostatistics D
- Chemical Physics M,D
- Chemistry M,D
- Environmental Sciences M,D
- Food Science and Technology M,D
- Geodetic Sciences M,D
- Geology M,D
- Horticulture M,D
- Mathematics M,D*
- Natural Resources M,D
- Optical Sciences M,D
- Physics M,D
- Statistics M,D

OHIO UNIVERSITY

- Astronomy M,D
- Environmental Management and Policy M
- Geochemistry M
- Geology M
- Geophysics M
- Hydrogeology M
- Mathematics M,D*
- Physics M,D*

OKLAHOMA STATE UNIVERSITY

- Agricultural Sciences—General M,D
- Agronomy and Soil Sciences M,D
- Animal Sciences M,D
- Applied Mathematics M,D
- Chemistry M,D
- Environmental Sciences M,D
- Food Science and Technology M,D
- Forestry M
- Geology M
- Horticulture M,D
- Mathematics M,D*
- Natural Resources M,D
- Photonics M,D
- Physics M,D
- Plant Sciences M,D
- Statistics M,D

OLD DOMINION UNIVERSITY

- Analytical Chemistry M
- Chemistry M
- Mathematics M,D
- Oceanography M,D
- Organic Chemistry M
- Physical Chemistry M
- Physics M,D

OREGON HEALTH & SCIENCE UNIVERSITY

- Biostatistics M

OREGON STATE UNIVERSITY

- Agricultural Sciences—General M,D
- Agronomy and Soil Sciences M,D
- Analytical Chemistry M,D
- Animal Sciences M,D
- Atmospheric Sciences M,D
- Biometrics M,D
- Chemistry M,D
- Environmental Management and Policy M,D
- Environmental Sciences M,D
- Fish, Game, and Wildlife Management M,D

Food Science and Technology	M,D
Forestry	M,D
Geology	M,D
Geophysics	M,D
Geosciences	M,D
Horticulture	M,D
Inorganic Chemistry	M,D
Marine Affairs	M
Marine Sciences	M
Mathematics	M,D
Oceanography	M,D
Organic Chemistry	M,D
Physical Chemistry	M,D
Physics	M,D*
Range Science	M,D
Statistics	M,D

PACE UNIVERSITY

Environmental Sciences	M*

THE PENNSYLVANIA STATE UNIVERSITY HARRISBURG CAMPUS

Environmental Sciences	M

THE PENNSYLVANIA STATE UNIVERSITY UNIVERSITY PARK CAMPUS

Acoustics	M,D
Agricultural Sciences—General	M,D
Agronomy and Soil Sciences	M,D
Animal Sciences	M,D
Applied Mathematics	M,D
Astronomy	M,D
Astrophysics	M,D
Chemistry	M,D*
Environmental Management and Policy	M
Environmental Sciences	M
Fish, Game, and Wildlife Management	M,D
Food Science and Technology	M,D
Forestry	M,D
Geosciences	M,D*
Horticulture	M,D
Mathematics	M,D
Meteorology	M,D
Physics	M,D*
Statistics	M,D

PITTSBURG STATE UNIVERSITY

Applied Physics	M
Chemistry	M
Mathematics	M
Physics	M

PLYMOUTH STATE UNIVERSITY

Environmental Management and Policy	M
Meteorology	M

POLYTECHNIC UNIVERSITY, BROOKLYN CAMPUS

Chemistry	M,D
Environmental Sciences	M
Mathematics	M,D*
Physics	M,D

POLYTECHNIC UNIVERSITY, LONG ISLAND GRADUATE CENTER

Chemistry	M,D

POLYTECHNIC UNIVERSITY OF PUERTO RICO

Environmental Management and Policy	M

POLYTECHNIC UNIVERSITY, WESTCHESTER GRADUATE CENTER

Chemistry	M
Mathematical and Computational Finance	M,O

PONTIFICAL CATHOLIC UNIVERSITY OF PUERTO RICO

Chemistry	M

PORTLAND STATE UNIVERSITY

Chemistry	M,D
Environmental Management and Policy	M,D
Environmental Sciences	M,D
Geology	M,D
Mathematics	M,D,O
Physics	M,D
Statistics	M,D

PRAIRIE VIEW A&M UNIVERSITY

Agricultural Sciences—General	M
Agronomy and Soil Sciences	M
Animal Sciences	M
Chemistry	M
Mathematics	M

PRESCOTT COLLEGE

Environmental Management and Policy	M

PRINCETON UNIVERSITY

Applied Mathematics	M,D*
Applied Physics	M,D
Astrophysics	D
Atmospheric Sciences	D
Chemical Physics	D
Chemistry	M,D*
Computational Sciences	D
Environmental Management and Policy	M,D
Geology	D
Geophysics	D
Geosciences	D
Mathematical Physics	D
Mathematics	D
Oceanography	D
Photonics	D
Physics	D
Plasma Physics	D
Statistics	M,D

PURDUE UNIVERSITY

Agricultural Sciences—General	M,D
Agronomy and Soil Sciences	M,D
Analytical Chemistry	M,D
Animal Sciences	M,D
Aquaculture	M,D
Atmospheric Sciences	M,D
Chemistry	M,D
Environmental Management and Policy	M,D
Fish, Game, and Wildlife Management	M,D
Food Science and Technology	M,D
Forestry	M,D*
Geosciences	M,D
Horticulture	M,D
Inorganic Chemistry	M,D
Mathematics	M,D*
Natural Resources	M,D
Organic Chemistry	M,D
Physical Chemistry	M,D
Physics	M,D
Statistics	M,D,O*

PURDUE UNIVERSITY CALUMET

Mathematics	M

QUEENS COLLEGE OF THE CITY UNIVERSITY OF NEW YORK

Chemistry	M
Environmental Sciences	M
Geology	M
Mathematics	M
Physics	M,D*

QUEEN'S UNIVERSITY AT KINGSTON

Chemistry	M,D
Geology	M,D
Mathematics	M,D
Physics	M,D
Statistics	M,D

RENSSELAER POLYTECHNIC INSTITUTE

Analytical Chemistry	M,D
Applied Mathematics	M
Applied Physics	M,D
Astrophysics	M,D
Chemistry	M,D
Environmental Management and Policy	M,D
Environmental Sciences	M,D
Geochemistry	M,D
Geology	M,D
Geophysics	M,D
Geosciences	M,D
Inorganic Chemistry	M,D
Mathematics	M,D
Organic Chemistry	M,D
Physical Chemistry	M,D
Physics	M,D
Statistics	M

RHODE ISLAND COLLEGE

Mathematics	M,O

RICE UNIVERSITY

Applied Mathematics	M,D
Applied Physics	M,D
Astronomy	M,D
Biostatistics	M,D
Chemistry	M,D
Computational Sciences	M,D
Environmental Management and Policy	M
Environmental Sciences	M,D
Geophysics	M
Geosciences	M,D
Inorganic Chemistry	M,D
Mathematical and Computational Finance	M,D
Mathematics	M,D
Organic Chemistry	M,D
Physical Chemistry	M,D
Physics	M,D
Statistics	M,D

RIVIER COLLEGE

Mathematics	M

ROCHESTER INSTITUTE OF TECHNOLOGY

Applied Mathematics	M
Chemistry	M
Environmental Management and Policy	M
Environmental Sciences	M
Optical Sciences	M,D
Statistics	M,O

ROOSEVELT UNIVERSITY

Chemistry	M
Mathematics	M

ROSE-HULMAN INSTITUTE OF TECHNOLOGY

Optical Sciences	M

ROWAN UNIVERSITY

Mathematics	M

ROYAL MILITARY COLLEGE OF CANADA

Chemistry	M,D
Environmental Sciences	M,D
Mathematics	M
Physics	M

ROYAL ROADS UNIVERSITY

Environmental Management and Policy	M

RUSH UNIVERSITY

Geology	M,D,O

RUTGERS, THE STATE UNIVERSITY OF NEW JERSEY, CAMDEN

Chemistry	M
Mathematics	M

RUTGERS, THE STATE UNIVERSITY OF NEW JERSEY, NEWARK

Analytical Chemistry	M,D
Applied Physics	M,D
Chemistry	M,D
Environmental Sciences	M,D
Geology	M
Inorganic Chemistry	M,D
Mathematics	D
Organic Chemistry	M,D
Physical Chemistry	M,D

RUTGERS, THE STATE UNIVERSITY OF NEW JERSEY, NEW BRUNSWICK/PISCATAWAY

Analytical Chemistry	M,D
Animal Sciences	M,D
Applied Mathematics	M,D
Atmospheric Sciences	M,D
Biostatistics	M,D
Chemistry	M,D*
Condensed Matter Physics	M,D
Environmental Sciences	M,D
Food Science and Technology	M,D
Geology	M,D
Horticulture	M,D
Inorganic Chemistry	M,D
Mathematics	M,D
Oceanography	M,D*
Organic Chemistry	M,D
Physical Chemistry	M,D

Physics M,D
Plant Sciences M,D
Statistics M,D
Theoretical Physics M,D
Water Resources M,D

SACRED HEART UNIVERSITY
Chemistry M

ST. CLOUD STATE UNIVERSITY
Environmental Management and Policy M
Mathematics M
Statistics M

ST. FRANCIS XAVIER UNIVERSITY
Chemistry M
Geology M
Geosciences M
Physics M

ST. JOHN'S UNIVERSITY (NY)
Applied Mathematics M
Chemistry M
Mathematics M
Statistics M
Theoretical Physics M,O

SAINT JOSEPH COLLEGE
Chemistry M

SAINT JOSEPH'S UNIVERSITY
Environmental Management and Policy M,O

SAINT LOUIS UNIVERSITY
Chemistry M
Geophysics M,D
Geosciences M,D
Mathematics M,D
Meteorology M,D

SAINT MARY-OF-THE-WOODS COLLEGE
Environmental Management and Policy M

SAINT MARY'S UNIVERSITY
Astronomy M

SAINT MARY'S UNIVERSITY OF MINNESOTA
Environmental Management and Policy M,O

SAINT XAVIER UNIVERSITY
Mathematics M

SALEM STATE COLLEGE
Mathematics M

SAM HOUSTON STATE UNIVERSITY
Agricultural Sciences—General M
Chemistry M
Computational Sciences M
Mathematics M
Statistics M

SAN DIEGO STATE UNIVERSITY
Applied Mathematics M
Astronomy M
Biometrics D
Biostatistics M,D
Chemistry M,D
Computational Sciences M,D
Geology M
Mathematics M,D
Physics M
Statistics M

SAN FRANCISCO STATE UNIVERSITY
Astrophysics M
Chemistry M
Environmental Management and Policy M
Geosciences M
Marine Sciences M*
Mathematics M
Physics M

SAN JOSE STATE UNIVERSITY
Chemistry M
Environmental Management and Policy M
Geology M
Marine Sciences M
Mathematics M
Meteorology M
Physics M

SANTA CLARA UNIVERSITY
Applied Mathematics M

SAVANNAH STATE UNIVERSITY
Marine Sciences M

THE SCRIPPS RESEARCH INSTITUTE
Chemistry D

SETON HALL UNIVERSITY
Analytical Chemistry M,D
Chemistry M,D
Inorganic Chemistry M,D
Organic Chemistry M,D
Physical Chemistry M,D

SHIPPENSBURG UNIVERSITY OF PENNSYLVANIA
Environmental Management and Policy M

SIMON FRASER UNIVERSITY
Applied Mathematics M,D
Chemical Physics M,D
Chemistry M,D
Environmental Management and Policy M,D
Geosciences M
Mathematics M,D
Physics M,D
Statistics M,D

SLIPPERY ROCK UNIVERSITY OF PENNSYLVANIA
Environmental Management and Policy M

SMITH COLLEGE
Chemistry M

SOUTH DAKOTA SCHOOL OF MINES AND TECHNOLOGY
Atmospheric Sciences M,D*
Chemistry M,D
Environmental Sciences D
Geology M,D
Paleontology M
Physics M,D
Water Resources D

SOUTH DAKOTA STATE UNIVERSITY
Agricultural Sciences—General M,D
Agronomy and Soil Sciences M,D
Analytical Chemistry M,D
Animal Sciences M,D
Atmospheric Sciences D*
Chemistry M,D
Environmental Sciences D
Fish, Game, and Wildlife Management M,D
Inorganic Chemistry M,D
Mathematics M
Organic Chemistry M,D
Physical Chemistry M,D
Physics M
Plant Sciences M,D
Water Resources D

SOUTHEAST MISSOURI STATE UNIVERSITY
Chemistry M
Environmental Management and Policy M
Mathematics M

SOUTHERN CONNECTICUT STATE UNIVERSITY
Chemistry M
Mathematics M

SOUTHERN ILLINOIS UNIVERSITY CARBONDALE
Agricultural Sciences—General M
Agronomy and Soil Sciences M
Animal Sciences M
Applied Physics M,D*
Chemistry M,D*
Environmental Sciences D*
Forestry M
Geology M,D
Horticulture M
Mathematics M,D*
Physics M,D
Plant Sciences M,D
Statistics M,D

SOUTHERN ILLINOIS UNIVERSITY EDWARDSVILLE
Chemistry M
Environmental Management and Policy M
Environmental Sciences M
Mathematics M
Physics M

SOUTHERN METHODIST UNIVERSITY
Applied Mathematics M,D
Chemistry M,D
Computational Sciences M,D
Environmental Sciences M,D
Geology M,D
Geophysics M,D
Mathematics M,D
Physics M,D
Statistics M,D

SOUTHERN OREGON UNIVERSITY
Mathematics M

SOUTHERN UNIVERSITY AND AGRICULTURAL AND MECHANICAL COLLEGE
Agricultural Sciences—General M
Analytical Chemistry M
Chemistry M
Environmental Sciences M
Forestry M
Inorganic Chemistry M
Mathematics M
Organic Chemistry M
Physical Chemistry M
Physics M

STANFORD UNIVERSITY
Applied Physics M,D
Chemistry D*
Computational Sciences M,D
Environmental Management and Policy M
Environmental Sciences M,D,O
Geophysics M,D,O
Geosciences M,D,O
Mathematical and Computational Finance M,D
Mathematics M,D
Physics D
Statistics M,D

STATE UNIVERSITY OF NEW YORK AT BINGHAMTON
Analytical Chemistry M,D
Applied Physics M
Chemistry M,D
Geology M,D
Inorganic Chemistry M,D
Mathematics M,D
Organic Chemistry M,D
Physical Chemistry M,D
Physics M
Statistics M,D

STATE UNIVERSITY OF NEW YORK AT BUFFALO
Biostatistics M,D
Chemistry M,D
Geology M,D
Mathematics M,D
Physics M,D

STATE UNIVERSITY OF NEW YORK AT NEW PALTZ
Chemistry M
Geology M
Mathematics M

STATE UNIVERSITY OF NEW YORK AT OSWEGO
Chemistry M

STATE UNIVERSITY OF NEW YORK COLLEGE AT BROCKPORT
Computational Sciences M
Mathematics M*

STATE UNIVERSITY OF NEW YORK COLLEGE AT CORTLAND
Mathematics M

STATE UNIVERSITY OF NEW YORK COLLEGE AT ONEONTA
Geosciences M

STATE UNIVERSITY OF NEW YORK COLLEGE AT POTSDAM

Mathematics M

STATE UNIVERSITY OF NEW YORK COLLEGE OF ENVIRONMENTAL SCIENCE AND FORESTRY

Chemistry M,D
Environmental Management and Policy M,D
Environmental Sciences M,D
Fish, Game, and Wildlife Management M,D
Forestry M,D
Hydrology M,D
Natural Resources M,D
Organic Chemistry M,D
Plant Sciences M,D
Water Resources M,D

STATE UNIVERSITY OF NEW YORK, FREDONIA

Chemistry M
Mathematics M

STEPHEN F. AUSTIN STATE UNIVERSITY

Chemistry M
Environmental Sciences M
Forestry M,D
Geology M
Mathematics M
Physics M
Statistics M

STEVENS INSTITUTE OF TECHNOLOGY

Analytical Chemistry M,D,O
Applied Mathematics M,D
Chemistry M,D,O
Marine Affairs M
Mathematics M,D
Organic Chemistry M,D,O
Photonics O
Physical Chemistry M,D,O
Physics M,D,O
Statistics M,O

STONY BROOK UNIVERSITY, STATE UNIVERSITY OF NEW YORK

Applied Mathematics M,D*
Atmospheric Sciences M,D*
Chemistry M,D*
Environmental Management and Policy M,O*
Geosciences M,D*
Marine Sciences M,D*
Mathematics M,D*
Physics M,D*
Planetary and Space Sciences M,D
Statistics M,D

SUL ROSS STATE UNIVERSITY

Animal Sciences M
Chemistry M
Fish, Game, and Wildlife Management M
Geology M*
Range Science M

SYRACUSE UNIVERSITY

Chemistry M,D
Geology M,D
Mathematics M,D
Physics M,D
Statistics M

TARLETON STATE UNIVERSITY

Agricultural Sciences—General M
Environmental Sciences M
Mathematics M

TAYLOR UNIVERSITY

Environmental Sciences M

TEMPLE UNIVERSITY

Applied Mathematics M,D
Chemistry M,D*
Computational Sciences M,D
Geology M
Mathematics M,D
Physics M,D*
Statistics M,D

TENNESSEE STATE UNIVERSITY

Agricultural Sciences—General M,D
Chemistry M
Mathematics M

TENNESSEE TECHNOLOGICAL UNIVERSITY

Chemistry M
Environmental Sciences D
Fish, Game, and Wildlife Management M
Mathematics M

TEXAS A&M INTERNATIONAL UNIVERSITY

Mathematics M
Physics M

TEXAS A&M UNIVERSITY

Agricultural Sciences—General M,D
Agronomy and Soil Sciences M,D
Animal Sciences M,D
Applied Physics M,D
Chemistry M,D*
Fish, Game, and Wildlife Management M,D
Food Science and Technology M,D
Forestry M,D
Geology M,D
Geophysics M,D
Horticulture M,D
Hydrology M,D
Mathematics M,D
Meteorology M,D
Natural Resources M,D
Oceanography M,D
Physics M,D
Plant Sciences M,D
Range Science M,D
Statistics M,D
Water Resources M,D

TEXAS A&M UNIVERSITY AT GALVESTON

Marine Sciences M

TEXAS A&M UNIVERSITY–COMMERCE

Agricultural Sciences—General M
Chemistry M
Geosciences M
Mathematics M
Physics M

TEXAS A&M UNIVERSITY–CORPUS CHRISTI

Environmental Sciences M

TEXAS A&M UNIVERSITY–KINGSVILLE

Agricultural Sciences—General M,D
Agronomy and Soil Sciences M,D
Animal Sciences M
Chemistry M
Fish, Game, and Wildlife Management M,D
Geology M
Mathematics M
Plant Sciences M,D
Range Science M

TEXAS CHRISTIAN UNIVERSITY

Astronomy M,D
Astrophysics M,D
Chemistry M,D
Environmental Sciences M
Geology M
Geosciences M
Mathematics M
Physics M,D

TEXAS SOUTHERN UNIVERSITY

Chemistry M
Mathematics M

TEXAS STATE UNIVERSITY-SAN MARCOS

Applied Mathematics M
Chemistry M
Environmental Management and Policy M
Fish, Game, and Wildlife Management M
Mathematics M
Physics M

TEXAS TECH UNIVERSITY

Agricultural Sciences—General M,D
Agronomy and Soil Sciences M,D
Animal Sciences M,D
Applied Physics M,D
Atmospheric Sciences M,D
Chemistry M,D
Environmental Management and Policy D
Environmental Sciences M,D
Fish, Game, and Wildlife Management M,D
Food Science and Technology M,D
Geosciences M,D
Horticulture M,D
Mathematics M,D
Physics M,D
Plant Sciences M,D
Range Science M,D

TEXAS WOMAN'S UNIVERSITY

Chemistry M
Food Science and Technology M,D
Mathematics M

TOWSON UNIVERSITY

Applied Mathematics M
Environmental Management and Policy M
Environmental Sciences M,O

TRENT UNIVERSITY

Chemistry M
Environmental Management and Policy M,D
Physics M

TROPICAL AGRICULTURE RESEARCH AND HIGHER EDUCATION CENTER

Agricultural Sciences—General M,D
Environmental Management and Policy M,D
Forestry M,D
Water Resources M,D

TROY UNIVERSITY

Environmental Management and Policy M

TUFTS UNIVERSITY

Analytical Chemistry M,D
Biostatistics M,D
Chemistry M,D*
Environmental Management and Policy M,D,O
Environmental Sciences M,D
Inorganic Chemistry M,D
Mathematics M,D
Organic Chemistry M,D
Physical Chemistry M,D
Physics M,D

TULANE UNIVERSITY

Applied Mathematics M,D
Biostatistics M,D
Chemistry M,D
Geology M,D*
Mathematics M,D
Paleontology M,D
Physics M,D
Statistics M,D

TUSKEGEE UNIVERSITY

Agricultural Sciences—General M
Agronomy and Soil Sciences M
Animal Sciences M
Chemistry M
Environmental Sciences M
Food Science and Technology M
Plant Sciences M

UNIVERSIDAD DE LAS AMÉRICAS-PUEBLA

Food Science and Technology M

UNIVERSIDAD DEL TURABO

Environmental Management and Policy M

UNIVERSIDAD METROPOLITANA

Environmental Management and Policy M

UNIVERSIDAD NACIONAL PEDRO HENRIQUEZ URENA

Animal Sciences	P,M,D
Environmental Management and Policy	P,M,D
Horticulture	P,M,D

UNIVERSITÉ DE MONCTON

Astronomy	M
Chemistry	M
Food Science and Technology	M
Mathematics	M
Physics	M

UNIVERSITÉ DE MONTRÉAL

Chemistry	M,D
Environmental Management and Policy	O
Mathematics	M,D
Physics	M,D
Statistics	M,D

UNIVERSITÉ DE SHERBROOKE

Chemistry	M,D,O
Environmental Sciences	M,O
Mathematics	M,D
Physics	M,D

UNIVERSITÉ DU QUÉBEC À CHICOUTIMI

Environmental Management and Policy	M
Geosciences	M
Mineralogy	D

UNIVERSITÉ DU QUÉBEC À MONTRÉAL

Atmospheric Sciences	M,D,O
Chemistry	M
Environmental Sciences	M,D
Geology	M,D,O
Geosciences	M,D,O
Mathematics	M,D
Meteorology	M,D,O
Mineralogy	M,D,O
Natural Resources	M,D,O

UNIVERSITÉ DU QUÉBEC À RIMOUSKI

Fish, Game, and Wildlife Management	M,D,O
Marine Affairs	M,O
Oceanography	M,D

UNIVERSITÉ DU QUÉBEC À TROIS-RIVIÈRES

Chemistry	M
Environmental Sciences	M,D
Mathematics	M

UNIVERSITÉ DU QUÉBEC, INSTITUT NATIONAL DE LA RECHERCHE SCIENTIFIQUE

Environmental Management and Policy	M,D
Geosciences	M,D
Hydrology	M,D

UNIVERSITÉ LAVAL

Agricultural Sciences—General	M,D,O
Agronomy and Soil Sciences	M,D
Animal Sciences	M,D
Chemistry	M,D
Environmental Management and Policy	M,D
Environmental Sciences	M,D
Food Science and Technology	M,D
Forestry	M,D
Geodetic Sciences	M,D
Geology	M,D
Geosciences	M,D
Mathematics	M,D
Oceanography	D
Physics	M,D
Statistics	M

UNIVERSITY AT ALBANY, STATE UNIVERSITY OF NEW YORK

Atmospheric Sciences	M,D
Biostatistics	M,D
Chemistry	M,D
Environmental Management and Policy	M
Environmental Sciences	M
Geology	M,D
Geosciences	M,D
Mathematics	M,D
Physics	M,D
Statistics	M,D,O

THE UNIVERSITY OF AKRON

Applied Mathematics	M,D
Chemistry	M,D
Geology	M
Geophysics	M
Geosciences	M
Mathematics	M
Physics	M
Statistics	M

THE UNIVERSITY OF ALABAMA

Applied Mathematics	M,D
Chemistry	M,D
Geology	M,D
Mathematics	M,D
Physics	M,D
Statistics	M,D

THE UNIVERSITY OF ALABAMA AT BIRMINGHAM

Applied Mathematics	M,D
Biometrics	M,D
Biostatistics	M,D
Chemistry	M,D
Mathematics	M,D
Physics	M,D

THE UNIVERSITY OF ALABAMA IN HUNTSVILLE

Applied Mathematics	M,D
Atmospheric Sciences	M,D
Chemistry	M
Environmental Sciences	M,D
Mathematics	M,D
Optical Sciences	D
Physics	M,D

UNIVERSITY OF ALASKA ANCHORAGE

Environmental Sciences	M

UNIVERSITY OF ALASKA FAIRBANKS

Astrophysics	M,D
Atmospheric Sciences	M,D
Chemistry	M,D
Environmental Management and Policy	M
Environmental Sciences	M,D
Fish, Game, and Wildlife Management	M,D
Geology	M,D
Geophysics	M,D
Limnology	M,D
Marine Sciences	M,D*
Mathematics	M,D
Oceanography	M,D
Physics	M,D
Statistics	M,D

UNIVERSITY OF ALBERTA

Agricultural Sciences—General	M,D
Agronomy and Soil Sciences	M,D
Applied Mathematics	M,D,O
Astrophysics	M,D
Biostatistics	M,D,O
Chemistry	M,D
Condensed Matter Physics	M,D
Environmental Management and Policy	M,D
Environmental Sciences	M,D
Forestry	M,D
Geophysics	M,D
Geosciences	M,D
Mathematical and Computational Finance	M,D,O
Mathematical Physics	M,D,O
Mathematics	M,D,O
Natural Resources	M,D
Physics	M,D
Statistics	M,D,O

THE UNIVERSITY OF ARIZONA

Agricultural Sciences—General	M,D
Agronomy and Soil Sciences	M,D
Animal Sciences	M,D
Applied Mathematics	M,D*
Applied Physics	M
Astronomy	M,D
Atmospheric Sciences	M,D
Chemistry	M,D
Environmental Management and Policy	M
Environmental Sciences	M,D
Fish, Game, and Wildlife Management	M,D
Forestry	M,D
Geosciences	M,D
Hydrology	M,D
Mathematics	M,D
Natural Resources	M,D
Optical Sciences	M,D
Physics	M,D
Planetary and Space Sciences	M,D*
Plant Sciences	M,D
Range Science	M,D
Water Resources	M,D

UNIVERSITY OF ARKANSAS

Agricultural Sciences—General	M,D
Agronomy and Soil Sciences	M,D
Animal Sciences	M,D
Applied Physics	M
Chemistry	M,D
Food Science and Technology	M,D
Geology	M
Horticulture	M
Mathematics	M,D
Photonics	M,D*
Physics	M,D
Plant Sciences	D
Statistics	M

UNIVERSITY OF ARKANSAS AT LITTLE ROCK

Applied Mathematics	M
Chemistry	M
Mathematics	M
Statistics	M

UNIVERSITY OF ARKANSAS AT MONTICELLO

Forestry	M
Natural Resources	M

THE UNIVERSITY OF BRITISH COLUMBIA

Agricultural Sciences—General	M,D
Agronomy and Soil Sciences	M,D
Animal Sciences	M,D
Applied Mathematics	M,D
Astronomy	M,D
Atmospheric Sciences	M,D
Chemistry	M,D
Environmental Management and Policy	M,D
Food Science and Technology	M,D
Forestry	M,D
Geology	M,D
Geophysics	M,D
Marine Sciences	M,D
Mathematics	M,D
Oceanography	M,D
Physics	M,D
Plant Sciences	M,D
Statistics	M,D

UNIVERSITY OF CALGARY

Analytical Chemistry	M,D
Astronomy	M,D
Chemistry	M,D
Environmental Management and Policy	M,D
Geology	M,D
Geophysics	M,D
Inorganic Chemistry	M,D
Mathematics	M,D
Organic Chemistry	M,D
Physical Chemistry	M,D
Physics	M,D
Statistics	M,D
Theoretical Chemistry	M,D

UNIVERSITY OF CALIFORNIA, BERKELEY

Applied Mathematics	D
Astrophysics	D
Biostatistics	M,D
Chemistry	M,D
Environmental Management and Policy	M,D
Environmental Sciences	M,D
Forestry	M,D
Geology	M,D
Geophysics	M,D
Mathematics	M,D
Physics	D
Range Science	M
Statistics	M,D

UNIVERSITY OF CALIFORNIA, DAVIS

Agricultural Sciences—General	M
Agronomy and Soil Sciences	M,D
Animal Sciences	M,D
Applied Mathematics	M,D
Atmospheric Sciences	M,D
Biostatistics	M,D

Chemistry	M,D
Environmental Sciences	M,D
Food Science and Technology	M,D
Geology	M,D
Horticulture	M
Hydrology	M,D
Mathematics	M,D
Physics	M,D
Statistics	M,D

UNIVERSITY OF CALIFORNIA, IRVINE

Chemistry	M,D
Environmental Management and Policy	M,D
Geosciences	M,D
Mathematics	M,D
Physics	M,D

UNIVERSITY OF CALIFORNIA, LOS ANGELES

Astronomy	M,D
Astrophysics	M,D
Atmospheric Sciences	M,D
Biometrics	M,D*
Biostatistics	M,D
Chemistry	M,D
Environmental Sciences	D*
Geochemistry	M,D
Geology	M,D
Geophysics	M,D
Geosciences	M,D
Hydrology	M,D
Mathematics	M,D
Physics	M,D*
Planetary and Space Sciences	M,D
Statistics	M,D

UNIVERSITY OF CALIFORNIA, RIVERSIDE

Agronomy and Soil Sciences	M,D
Chemistry	M,D
Environmental Sciences	M,D
Geology	M,D*
Mathematics	M,D
Physics	M,D
Plant Sciences	M,D
Statistics	M,D
Water Resources	M,D

UNIVERSITY OF CALIFORNIA, SAN DIEGO

Applied Mathematics	M,D
Applied Physics	M,D
Chemistry	M,D*
Geosciences	M,D
Marine Sciences	M
Mathematics	M,D
Oceanography	M,D
Photonics	M,D
Physics	M,D
Statistics	M,D

UNIVERSITY OF CALIFORNIA, SAN FRANCISCO

Chemistry	D*

UNIVERSITY OF CALIFORNIA, SANTA BARBARA

Applied Mathematics	M,D
Chemistry	M,D
Environmental Management and Policy	M,D
Environmental Sciences	M,D*
Geology	M,D
Geophysics	M,D
Geosciences	M,D
Marine Sciences	M,D
Mathematical and Computational Finance	M,D
Mathematics	M,D
Physics	D
Statistics	M,D

UNIVERSITY OF CALIFORNIA, SANTA CRUZ

Astronomy	D
Astrophysics	D
Chemistry	M,D
Environmental Management and Policy	D
Geosciences	M,D
Marine Sciences	M,D
Mathematics	M,D
Physics	M,D

UNIVERSITY OF CENTRAL ARKANSAS

Mathematics	M

UNIVERSITY OF CENTRAL FLORIDA

Applied Mathematics	M,D,O
Chemistry	M,D
Computational Sciences	M,D
Mathematics	M,D,O
Optical Sciences	M,D
Photonics	M,D
Physics	M,D
Statistics	M,O

UNIVERSITY OF CENTRAL OKLAHOMA

Applied Mathematics	M
Chemistry	M
Mathematics	M
Physics	M
Statistics	M

UNIVERSITY OF CHICAGO

Applied Mathematics	M,D
Astronomy	M,D
Astrophysics	M,D
Atmospheric Sciences	M,D
Chemistry	D
Environmental Management and Policy	M,D
Environmental Sciences	M,D
Geophysics	M,D
Geosciences	M,D
Mathematical and Computational Finance	M
Mathematics	M,D
Paleontology	M,D
Physics	M,D
Planetary and Space Sciences	M,D
Statistics	M,D

UNIVERSITY OF CINCINNATI

Analytical Chemistry	M,D
Applied Mathematics	M,D
Biostatistics	M,D
Chemistry	M,D
Environmental Sciences	M,D
Geology	M,D
Inorganic Chemistry	M,D
Mathematics	M,D
Organic Chemistry	M,D
Physical Chemistry	M,D
Physics	M,D
Statistics	M,D

UNIVERSITY OF COLORADO AT BOULDER

Applied Mathematics	M,D
Astrophysics	M,D
Atmospheric Sciences	M,D
Chemical Physics	M,D
Chemistry	M,D
Environmental Management and Policy	M,D
Geology	M,D
Geophysics	M,D
Mathematical Physics	M,D
Mathematics	M,D
Oceanography	M,D
Optical Sciences	M,D
Physics	M,D
Plasma Physics	M,D

UNIVERSITY OF COLORADO AT COLORADO SPRINGS

Applied Mathematics	M
Environmental Sciences	M

UNIVERSITY OF COLORADO AT DENVER AND HEALTH SCIENCES CENTER

Biostatistics	M,D

UNIVERSITY OF COLORADO AT DENVER AND HEALTH SCIENCES CENTER—DOWNTOWN DENVER CAMPUS

Applied Mathematics	M,D
Chemistry	M
Environmental Sciences	M,O
Mathematics	M

UNIVERSITY OF CONNECTICUT

Agricultural Sciences—General	M,D
Agronomy and Soil Sciences	M,D
Animal Sciences	M,D
Applied Mathematics	M
Chemistry	M,D
Environmental Management and Policy	M,D
Geology	M,D
Marine Sciences	M,D
Mathematical and Computational Finance	M
Mathematics	M,D
Natural Resources	M,D
Oceanography	M,D
Physics	M,D
Plant Sciences	M,D
Statistics	M,D

UNIVERSITY OF DAYTON

Applied Mathematics	M
Chemistry	M
Optical Sciences	M,D

UNIVERSITY OF DELAWARE

Agricultural Sciences—General	M,D
Agronomy and Soil Sciences	M,D
Animal Sciences	M,D
Applied Mathematics	M,D
Astronomy	M,D
Atmospheric Sciences	D
Chemistry	M,D
Environmental Management and Policy	M,D*
Food Science and Technology	M,D
Geology	M,D*
Horticulture	M
Marine Affairs	M,D
Marine Geology	M,D
Marine Sciences	M,D
Mathematics	M,D*
Oceanography	M,D
Physics	M,D
Plant Sciences	M,D
Statistics	M

UNIVERSITY OF DENVER

Applied Mathematics	M,D
Chemistry	M,D
Environmental Management and Policy	M
Mathematics	M,D
Physics	M,D
Statistics	M

UNIVERSITY OF DETROIT MERCY

Chemistry	M
Mathematics	M

THE UNIVERSITY OF FINDLAY

Environmental Management and Policy	M

UNIVERSITY OF FLORIDA

Agricultural Sciences—General	M,D
Agronomy and Soil Sciences	M,D
Animal Sciences	M,D
Aquaculture	M,D
Astronomy	M,D
Biostatistics	M
Chemistry	M,D
Fish, Game, and Wildlife Management	M,D
Food Science and Technology	M,D
Forestry	M,D
Geology	M,D
Geosciences	M,D
Horticulture	M,D
Limnology	M,D
Marine Sciences	M,D
Mathematics	M,D
Natural Resources	M,D
Physics	M,D
Plant Sciences	D
Statistics	M,D
Water Resources	M,D

UNIVERSITY OF GEORGIA

Agricultural Sciences—General	M,D
Agronomy and Soil Sciences	M,D
Analytical Chemistry	M,D
Animal Sciences	M,D
Applied Mathematics	M,D
Astronomy	M,D
Chemistry	M,D
Food Science and Technology	M,D
Forestry	M,D
Geology	M,D
Horticulture	M,D
Inorganic Chemistry	M,D
Marine Sciences	M,D
Mathematics	M,D
Natural Resources	M,D
Oceanography	M,D
Organic Chemistry	M,D
Physical Chemistry	M,D
Physics	M,D
Statistics	M,D

UNIVERSITY OF GUAM

Environmental Sciences	M

UNIVERSITY OF GUELPH

Agricultural Sciences—General	M,D,O
Agronomy and Soil Sciences	M,D
Animal Sciences	M,D
Applied Mathematics	M,D
Aquaculture	M
Atmospheric Sciences	M,D
Chemistry	M,D
Environmental Management and Policy	M,D
Environmental Sciences	M,D
Food Science and Technology	M,D
Horticulture	M,D
Mathematics	M,D
Natural Resources	M,D
Physics	M,D
Statistics	M,D

UNIVERSITY OF HAWAII AT MANOA

Agricultural Sciences—General	M,D
Animal Sciences	M
Astronomy	M,D
Chemistry	M,D
Environmental Management and Policy	M,D,O
Food Science and Technology	M
Geochemistry	M,D
Geology	M,D
Geophysics	M,D
Horticulture	M,D
Hydrogeology	M,D
Marine Geology	M,D
Mathematics	M,D
Meteorology	M,D
Natural Resources	M,D
Oceanography	M,D
Physics	M,D
Planetary and Space Sciences	M,D
Plant Sciences	M,D

UNIVERSITY OF HOUSTON

Chemistry	M,D
Geology	M,D
Geophysics	M,D
Mathematics	M,D
Physics	M,D

UNIVERSITY OF HOUSTON–CLEAR LAKE

Chemistry	M
Environmental Management and Policy	M
Environmental Sciences	M
Mathematics	M
Statistics	M

UNIVERSITY OF IDAHO

Agronomy and Soil Sciences	M,D
Animal Sciences	M,D
Chemistry	M,D
Environmental Management and Policy	M,D
Environmental Sciences	M,D
Fish, Game, and Wildlife Management	M,D
Food Science and Technology	M,D
Forestry	M,D
Geology	M,D
Hydrology	M
Mathematics	M,D
Natural Resources	M,D
Physics	M,D
Plant Sciences	M,D
Range Science	M,D
Statistics	M

UNIVERSITY OF ILLINOIS AT CHICAGO

Applied Mathematics	M,D
Biostatistics	M,D
Chemistry	M,D
Geochemistry	M,D
Geology	M,D
Geophysics	M,D
Geosciences	M,D
Hydrogeology	M,D
Mathematics	M,D*
Mineralogy	M,D
Paleontology	M,D
Physics	M,D
Statistics	M,D
Water Resources	M,D

UNIVERSITY OF ILLINOIS AT SPRINGFIELD

Environmental Management and Policy	M
Environmental Sciences	M

UNIVERSITY OF ILLINOIS AT URBANA–CHAMPAIGN

Agricultural Sciences—General	M,D
Agronomy and Soil Sciences	M,D
Animal Sciences	M,D
Applied Mathematics	M,D
Astronomy	M,D
Atmospheric Sciences	M,D
Chemistry	M,D
Environmental Sciences	M,D
Food Science and Technology	M,D
Geochemistry	M,D
Geology	M,D
Geophysics	M,D
Geosciences	M,D
Mathematics	M,D
Natural Resources	M,D
Physics	M,D
Statistics	M,D

THE UNIVERSITY OF IOWA

Applied Mathematics	D
Astronomy	M
Biostatistics	M,D
Chemistry	M,D
Computational Sciences	D
Geosciences	M,D
Mathematics	M,D
Physics	M,D
Statistics	M,D,O

UNIVERSITY OF KANSAS

Applied Mathematics	M,D
Astronomy	M,D
Chemistry	M,D
Environmental Sciences	M,D
Geology	M,D
Mathematics	M,D
Physics	M,D*
Statistics	M,D
Water Resources	M

UNIVERSITY OF KENTUCKY

Agricultural Sciences—General	M,D
Agronomy and Soil Sciences	M,D
Animal Sciences	M,D
Applied Mathematics	M,D
Astronomy	M,D
Chemistry	M,D
Forestry	M
Geology	M,D
Mathematics	M,D
Physics	M,D
Plant Sciences	M
Statistics	M,D

UNIVERSITY OF LETHBRIDGE

Agricultural Sciences—General	M,D
Chemistry	M,D
Computational Sciences	M,D
Environmental Sciences	M,D
Mathematics	M,D
Physics	M,D

UNIVERSITY OF LOUISIANA AT LAFAYETTE

Geology	M
Mathematics	M,D
Physics	M

UNIVERSITY OF LOUISIANA AT MONROE

Geosciences	M

UNIVERSITY OF LOUISVILLE

Analytical Chemistry	M,D
Applied Mathematics	M,D
Biostatistics	M,D
Chemical Physics	M,D
Chemistry	M,D
Inorganic Chemistry	M,D
Mathematics	M,D
Organic Chemistry	M,D
Physical Chemistry	M,D
Physics	M

UNIVERSITY OF MAINE

Agricultural Sciences—General	M,D
Agronomy and Soil Sciences	M,D
Animal Sciences	M
Chemistry	M,D
Environmental Management and Policy	M,D
Environmental Sciences	M,D
Fish, Game, and Wildlife Management	M,D
Food Science and Technology	M,D
Forestry	M,D
Geology	M,D
Geosciences	M,D
Horticulture	M
Marine Affairs	M
Marine Sciences	M,D
Mathematics	M
Natural Resources	M,D
Oceanography	M,D
Physics	M,D
Plant Sciences	M,D

UNIVERSITY OF MANITOBA

Agricultural Sciences—General	M,D
Agronomy and Soil Sciences	M,D
Animal Sciences	M,D
Chemistry	M,D
Computational Sciences	M
Environmental Management and Policy	M,D
Food Science and Technology	M
Geology	M,D
Geophysics	M,D
Horticulture	M,D
Mathematics	M,D
Physics	M,D
Statistics	M,D

UNIVERSITY OF MARYLAND

Environmental Sciences	M,D
Marine Sciences	M,D*

UNIVERSITY OF MARYLAND, BALTIMORE COUNTY

Applied Mathematics	M,D
Applied Physics	M,D
Astrophysics	M,D
Atmospheric Sciences	M,D
Chemistry	M,D*
Environmental Sciences	M,D
Marine Sciences	M,D
Optical Sciences	M,D
Physics	M,D
Statistics	M,D

UNIVERSITY OF MARYLAND, COLLEGE PARK

Agricultural Sciences—General	P,M,D
Agronomy and Soil Sciences	M,D
Analytical Chemistry	M,D
Animal Sciences	M,D
Applied Mathematics	M,D
Astronomy	M,D
Chemical Physics	M,D
Chemistry	M,D
Environmental Sciences	M,D
Food Science and Technology	M,D
Geology	M,D*
Horticulture	D
Inorganic Chemistry	M,D
Marine Sciences	M,D
Mathematics	M,D
Meteorology	M,D
Natural Resources	M,D
Oceanography	M,D
Organic Chemistry	M,D
Physical Chemistry	M,D
Physics	M,D*
Statistics	M,D

UNIVERSITY OF MARYLAND EASTERN SHORE

Agricultural Sciences—General	M,D
Environmental Sciences	M,D
Food Science and Technology	M,D
Marine Sciences	M,D

UNIVERSITY OF MARYLAND UNIVERSITY COLLEGE

Environmental Management and Policy	M,O

UNIVERSITY OF MASSACHUSETTS AMHERST

Agronomy and Soil Sciences	M,D
Animal Sciences	M,D
Applied Mathematics	M
Astronomy	M,D
Chemistry	M,D

Fish, Game, and Wildlife Management	M,D
Food Science and Technology	M,D
Forestry	M,D
Geosciences	M,D
Marine Sciences	M
Mathematics	M,D
Physics	M,D
Plant Sciences	M,D
Statistics	M,D

UNIVERSITY OF MASSACHUSETTS BOSTON

Applied Physics	M
Chemistry	M
Environmental Sciences	M,D
Marine Sciences	D

UNIVERSITY OF MASSACHUSETTS DARTMOUTH

Chemistry	M
Marine Sciences	M,D
Physics	M

UNIVERSITY OF MASSACHUSETTS LOWELL

Applied Mathematics	M,D
Applied Physics	M,D
Chemistry	M,D
Computational Sciences	M,D
Environmental Management and Policy	M,D,O
Environmental Sciences	M,D,O
Mathematics	M,D
Optical Sciences	M,D
Physics	M,D

UNIVERSITY OF MEDICINE AND DENTISTRY OF NEW JERSEY

Biostatistics	M,D,O
Environmental Sciences	D

UNIVERSITY OF MEMPHIS

Applied Mathematics	M,D
Chemistry	M,D
Geology	M,D
Geophysics	M,D
Geosciences	M,D
Mathematics	M,D
Physics	M
Statistics	M,D

UNIVERSITY OF MIAMI

Chemistry	M,D
Environmental Management and Policy	M,D
Fish, Game, and Wildlife Management	M,D
Geophysics	M,D
Inorganic Chemistry	M,D
Marine Affairs	M
Marine Geology	M,D
Marine Sciences	M,D
Mathematics	M,D
Meteorology	M,D*
Oceanography	M,D
Organic Chemistry	M,D
Physical Chemistry	M,D
Physics	M,D*
Statistics	M

UNIVERSITY OF MICHIGAN

Analytical Chemistry	D
Applied Physics	D
Astronomy	M,D
Atmospheric Sciences	M,D
Biostatistics	M,D
Chemistry	D
Environmental Management and Policy	M,D
Forestry	M,D,O
Geochemistry	M,D
Geology	M,D
Inorganic Chemistry	D
Marine Geology	M,D
Mathematics	M,D
Mineralogy	M,D
Natural Resources	M,D,O
Oceanography	M,D
Organic Chemistry	D
Physical Chemistry	D
Physics	M,D*
Planetary and Space Sciences	M,D
Statistics	M,D

UNIVERSITY OF MICHIGAN–DEARBORN

Applied Mathematics	M
Computational Sciences	M
Environmental Sciences	M

UNIVERSITY OF MINNESOTA, DULUTH

Applied Mathematics	M*
Chemistry	M
Computational Sciences	M
Geology	M,D
Physics	M*

UNIVERSITY OF MINNESOTA, TWIN CITIES CAMPUS

Agricultural Sciences—General	M,D
Agronomy and Soil Sciences	M,D
Animal Sciences	M,D
Astronomy	M,D
Astrophysics	M,D
Biostatistics	M,D
Chemistry	M,D
Computational Sciences	M,D
Environmental Management and Policy	M,D
Fish, Game, and Wildlife Management	M,D
Food Science and Technology	M,D
Forestry	M,D
Geology	M,D
Geophysics	M,D
Mathematics	M,D
Natural Resources	M,D
Physics	M,D
Plant Sciences	M,D
Statistics	M,D
Water Resources	M,D

UNIVERSITY OF MISSISSIPPI

Chemistry	M,D
Computational Sciences	M,D
Mathematics	M,D
Physics	M,D

UNIVERSITY OF MISSOURI–COLUMBIA

Agricultural Sciences—General	M,D
Agronomy and Soil Sciences	M,D
Analytical Chemistry	M,D
Animal Sciences	M,D
Applied Mathematics	M
Astronomy	M,D
Atmospheric Sciences	M,D
Chemistry	M,D
Fish, Game, and Wildlife Management	M,D
Food Science and Technology	M,D
Forestry	M,D
Geology	M,D
Horticulture	M,D
Inorganic Chemistry	M,D
Mathematics	M,D
Organic Chemistry	M,D
Physical Chemistry	M,D
Physics	M,D
Plant Sciences	M,D
Statistics	M,D

UNIVERSITY OF MISSOURI–KANSAS CITY

Analytical Chemistry	M,D
Chemistry	M,D*
Geology	M,D
Geosciences	M,D
Inorganic Chemistry	M,D
Mathematics	M,D
Organic Chemistry	M,D
Physical Chemistry	M,D
Physics	M,D
Statistics	M,D

UNIVERSITY OF MISSOURI–ROLLA

Applied Mathematics	M
Chemistry	M,D
Geochemistry	M,D
Geology	M,D
Geophysics	M,D
Hydrology	M,D
Mathematics	M,D
Physics	M,D
Water Resources	M,D

UNIVERSITY OF MISSOURI–ST. LOUIS

Applied Mathematics	M,D,O
Applied Physics	M,D
Astrophysics	M,D
Chemistry	M,D
Environmental Management and Policy	M,D,O
Inorganic Chemistry	M,D
Mathematics	M,D,O
Organic Chemistry	M,D
Physical Chemistry	M,D
Physics	M,D

THE UNIVERSITY OF MONTANA

Analytical Chemistry	M,D
Chemistry	M,D
Environmental Management and Policy	M,D
Environmental Sciences	M*
Fish, Game, and Wildlife Management	M,D
Forestry	M,D
Geology	M,D
Geosciences	M,D
Inorganic Chemistry	M,D
Mathematics	M,D
Natural Resources	M,D
Organic Chemistry	M,D
Physical Chemistry	M,D

UNIVERSITY OF NEBRASKA AT OMAHA

Mathematics	M

UNIVERSITY OF NEBRASKA–LINCOLN

Agricultural Sciences—General	M,D
Agronomy and Soil Sciences	M,D
Analytical Chemistry	M,D
Animal Sciences	M,D
Astronomy	M,D
Biometrics	M
Chemistry	M,D
Food Science and Technology	M,D
Geosciences	M,D
Horticulture	M,D
Inorganic Chemistry	M,D
Mathematics	M,D
Natural Resources	M,D
Organic Chemistry	M,D
Physical Chemistry	M,D
Physics	M,D
Statistics	M,D

UNIVERSITY OF NEVADA, LAS VEGAS

Applied Mathematics	M,D
Chemistry	M,D
Computational Sciences	M,D
Environmental Sciences	M,D
Geosciences	M,D
Mathematics	M,D
Physics	M,D
Statistics	M,D
Water Resources	M

UNIVERSITY OF NEVADA, RENO

Agricultural Sciences—General	M,D
Animal Sciences	M
Atmospheric Sciences	M,D
Chemical Physics	D
Chemistry	M,D
Environmental Management and Policy	M
Environmental Sciences	M,D
Geochemistry	M,D,O
Geology	M,D,O
Geophysics	M,D,O
Hydrogeology	M,D
Hydrology	M,D
Mathematics	M
Physics	M,D

UNIVERSITY OF NEW BRUNSWICK FREDERICTON

Chemistry	M,D
Forestry	M,D
Geodetic Sciences	M,D,O
Geology	M,D
Hydrology	M,D
Mathematics	M,D
Physics	M,D
Statistics	M,D
Water Resources	M,D

UNIVERSITY OF NEW BRUNSWICK SAINT JOHN

Environmental Management and Policy	M

UNIVERSITY OF NEW ENGLAND

Marine Sciences	M

UNIVERSITY OF NEW HAMPSHIRE

Agronomy and Soil Sciences	M

Animal Sciences M,D
Applied Mathematics M,D
Atmospheric Sciences
Chemistry M,D
Environmental Management and Policy M
Fish, Game, and Wildlife Management M
Forestry M
Geochemistry M
Geology M
Geosciences M
Hydrology M
Mathematics M,D
Natural Resources D
Oceanography M,D
Physics M,D
Statistics M,D
Water Resources M

UNIVERSITY OF NEW HAVEN

Environmental Sciences M*

UNIVERSITY OF NEW MEXICO

Chemistry M,D
Geosciences M,D
Mathematics M,D
Optical Sciences M,D
Physics M,D
Planetary and Space Sciences M,D
Statistics M,D
Water Resources M

UNIVERSITY OF NEW ORLEANS

Chemistry M,D
Geology M
Geophysics M
Mathematics M
Physics M,D

THE UNIVERSITY OF NORTH CAROLINA AT CHAPEL HILL

Astronomy M,D
Astrophysics M,D
Atmospheric Sciences M,D
Biostatistics M,D
Chemistry M,D
Environmental Management and Policy M,D
Environmental Sciences M,D
Geology M,D
Marine Sciences M,D
Mathematics M,D*
Physics M,D
Statistics M,D

THE UNIVERSITY OF NORTH CAROLINA AT CHARLOTTE

Applied Mathematics D
Applied Physics M,D
Chemistry M
Geosciences M
Mathematical and Computational Finance M
Mathematics M,D*
Optical Sciences M,D

THE UNIVERSITY OF NORTH CAROLINA AT GREENSBORO

Chemistry M
Mathematics M

THE UNIVERSITY OF NORTH CAROLINA WILMINGTON

Chemistry M
Geology M*
Geosciences M

Marine Sciences M
Mathematics M

UNIVERSITY OF NORTH DAKOTA

Atmospheric Sciences M
Chemistry M,D
Fish, Game, and Wildlife Management M,D
Geology M,D
Geosciences M,D
Mathematics M
Physics M,D
Planetary and Space Sciences M

UNIVERSITY OF NORTHERN BRITISH COLUMBIA

Environmental Management and Policy M,D,O
Mathematics M,D,O
Natural Resources M,D,O

UNIVERSITY OF NORTHERN COLORADO

Chemistry M,D
Geosciences M
Mathematics M,D

UNIVERSITY OF NORTHERN IOWA

Chemistry M
Environmental Sciences M
Mathematics M

UNIVERSITY OF NORTH FLORIDA

Mathematics M
Statistics M

UNIVERSITY OF NORTH TEXAS

Chemistry M,D
Environmental Sciences M,D
Mathematics M,D
Physics M,D

UNIVERSITY OF NORTH TEXAS HEALTH SCIENCE CENTER AT FORT WORTH

Biostatistics M,D

UNIVERSITY OF NOTRE DAME

Applied Mathematics M,D
Chemistry M,D
Geosciences M,D
Inorganic Chemistry M,D
Mathematics M,D*
Organic Chemistry M,D
Physical Chemistry M,D
Physics D*

UNIVERSITY OF OKLAHOMA

Astrophysics M,D
Chemistry M,D
Environmental Sciences M,D
Geology M,D
Geophysics M
Mathematics M,D*
Meteorology M,D
Natural Resources M,D
Physics M,D
Water Resources M,D

UNIVERSITY OF OKLAHOMA HEALTH SCIENCES CENTER

Biostatistics M,D

UNIVERSITY OF OREGON

Chemistry M,D
Environmental Management and Policy M,D*
Geology M,D
Mathematics M,D
Physics M,D

UNIVERSITY OF OTTAWA

Chemistry M,D
Geosciences M,D
Mathematics M,D
Physics M,D
Statistics M,D

UNIVERSITY OF PENNSYLVANIA

Astrophysics M,D
Biostatistics M,D
Chemistry M,D
Environmental Management and Policy M
Geology M,D
Mathematics M,D
Physics M,D
Statistics M,D

UNIVERSITY OF PITTSBURGH

Applied Mathematics M,D
Biostatistics M,D
Chemistry M,D*
Environmental Management and Policy M
Geology M,D
Mathematical and Computational Finance M,D
Mathematics M,D
Physics M,D*
Planetary and Space Sciences M,D
Statistics M,D

UNIVERSITY OF PRINCE EDWARD ISLAND

Chemistry M

UNIVERSITY OF PUERTO RICO, MAYAGÜEZ CAMPUS

Agricultural Sciences—General M
Agronomy and Soil Sciences M
Animal Sciences M
Applied Mathematics M
Chemistry M,D
Computational Sciences M
Food Science and Technology M
Geology M
Horticulture M
Marine Sciences M,D*
Mathematics M
Oceanography M,D
Physics M
Statistics M

UNIVERSITY OF PUERTO RICO, MEDICAL SCIENCES CAMPUS

Biostatistics M

UNIVERSITY OF PUERTO RICO, RÍO PIEDRAS

Chemistry M,D*
Mathematics M,D
Physics M,D

UNIVERSITY OF REGINA

Analytical Chemistry M,D

Chemistry M,D
Geology M,D
Inorganic Chemistry M,D
Mathematics M,D
Organic Chemistry M,D
Physical Chemistry M,D
Physics M,D
Statistics M,D

UNIVERSITY OF RHODE ISLAND

Animal Sciences M,D
Applied Mathematics M,D,O
Aquaculture M,D
Chemistry M,D
Environmental Management and Policy M,D
Environmental Sciences M,D
Fish, Game, and Wildlife Management M,D
Food Science and Technology M,D
Geosciences M
Marine Affairs M,D
Marine Sciences M,D
Mathematics M,D
Natural Resources M,D
Oceanography M,D
Physics M,D
Plant Sciences M,D
Statistics M,D,O

UNIVERSITY OF ROCHESTER

Astronomy M,D
Biostatistics M,D
Chemistry M,D
Geology M,D
Geosciences M,D*
Mathematics M,D
Optical Sciences M,D
Physics M,D*
Statistics M,D

UNIVERSITY OF ST. THOMAS (MN)

Environmental Management and Policy M

UNIVERSITY OF SAN DIEGO

Geosciences M,D,O
Marine Affairs M
Marine Sciences M

UNIVERSITY OF SAN FRANCISCO

Chemistry M
Environmental Management and Policy M

UNIVERSITY OF SASKATCHEWAN

Agricultural Sciences—General M,D
Agronomy and Soil Sciences M,D
Animal Sciences M,D
Chemistry M,D
Food Science and Technology M,D
Geology M,D,O
Mathematics M,D
Physics M,D
Plant Sciences M,D
Statistics M,D

THE UNIVERSITY OF SCRANTON

Chemistry M

UNIVERSITY OF SOUTH ALABAMA

Marine Sciences M,D
Mathematics M

UNIVERSITY OF SOUTH CAROLINA

Astronomy	M,D
Biostatistics	M,D
Chemistry	M,D
Environmental Management and Policy	M
Environmental Sciences	M,D
Geology	M,D
Geosciences	M,D
Marine Sciences	M,D
Mathematics	M,D
Physics	M,D
Statistics	M,D,O

THE UNIVERSITY OF SOUTH DAKOTA

Chemistry	M
Computational Sciences	D
Mathematics	M
Statistics	D

UNIVERSITY OF SOUTHERN CALIFORNIA

Applied Mathematics	M,D
Biometrics	M
Biostatistics	M,D*
Chemical Physics	D
Chemistry	M,D
Geosciences	M,D
Marine Sciences	D
Mathematics	M,D
Oceanography	D
Physics	M,D
Statistics	M

UNIVERSITY OF SOUTHERN MAINE

Statistics	M

UNIVERSITY OF SOUTHERN MISSISSIPPI

Analytical Chemistry	M,D
Chemistry	M,D
Computational Sciences	D
Food Science and Technology	M,D
Geology	M
Hydrology	M,D
Inorganic Chemistry	M,D
Marine Sciences	M,D
Mathematics	M
Organic Chemistry	M,D
Physical Chemistry	M,D
Physics	M

UNIVERSITY OF SOUTH FLORIDA

Analytical Chemistry	M,D
Applied Mathematics	M,D
Applied Physics	M,D
Biostatistics	M,D
Chemistry	M,D
Computational Sciences	M,D
Environmental Management and Policy	M
Environmental Sciences	M
Geology	M,D
Inorganic Chemistry	M,D
Marine Sciences	M,D
Mathematics	M,D
Oceanography	M,D
Organic Chemistry	M,D
Physical Chemistry	M,D
Physics	M,D

THE UNIVERSITY OF TENNESSEE

Agricultural Sciences—General	M,D
Analytical Chemistry	M,D
Animal Sciences	M,D
Applied Mathematics	M,D
Chemical Physics	M,D
Chemistry	M,D
Environmental Management and Policy	M,D
Fish, Game, and Wildlife Management	M
Food Science and Technology	M,D
Forestry	M
Geology	M,D
Inorganic Chemistry	M,D
Mathematics	M,D
Organic Chemistry	M,D
Physical Chemistry	M,D
Physics	M,D
Plant Sciences	M
Statistics	M,D
Theoretical Chemistry	M,D

THE UNIVERSITY OF TENNESSEE AT CHATTANOOGA

Environmental Sciences	M

THE UNIVERSITY OF TENNESSEE AT MARTIN

Agricultural Sciences—General	M
Food Science and Technology	M

THE UNIVERSITY OF TENNESSEE SPACE INSTITUTE

Applied Mathematics	M
Physics	M,D

THE UNIVERSITY OF TEXAS AT ARLINGTON

Chemistry	M,D*
Environmental Sciences	M,D
Geology	M,D
Geosciences	M,D
Mathematics	M,D
Physics	M,D

THE UNIVERSITY OF TEXAS AT AUSTIN

Analytical Chemistry	M,D
Applied Mathematics	M,D
Astronomy	M,D
Chemistry	M,D
Computational Sciences	M,D
Environmental Management and Policy	M
Geology	M,D
Geosciences	M,D
Inorganic Chemistry	M,D
Marine Sciences	M,D*
Mathematics	M,D
Organic Chemistry	M,D
Physical Chemistry	M,D
Physics	M,D*
Statistics	M

THE UNIVERSITY OF TEXAS AT BROWNSVILLE

Mathematics	M
Physics	M

THE UNIVERSITY OF TEXAS AT DALLAS

Applied Mathematics	M,D
Chemistry	M,D
Geosciences	M,D
Mathematics	M,D*
Physics	M,D
Statistics	M,D

THE UNIVERSITY OF TEXAS AT EL PASO

Chemistry	M
Environmental Sciences	M,D
Geology	M,D
Geophysics	M
Mathematics	M
Physics	M
Statistics	M

THE UNIVERSITY OF TEXAS AT SAN ANTONIO

Chemistry	M
Environmental Sciences	M,D
Geology	M,D
Statistics	M

THE UNIVERSITY OF TEXAS AT TYLER

Mathematics	M

THE UNIVERSITY OF TEXAS HEALTH SCIENCE CENTER AT HOUSTON

Biometrics	M,D
Biostatistics	M,D

THE UNIVERSITY OF TEXAS OF THE PERMIAN BASIN

Geology	M

THE UNIVERSITY OF TEXAS–PAN AMERICAN

Mathematics	M

UNIVERSITY OF THE INCARNATE WORD

Mathematics	M

UNIVERSITY OF THE SCIENCES IN PHILADELPHIA

Chemistry	M,D

THE UNIVERSITY OF TOLEDO

Analytical Chemistry	M,D
Applied Mathematics	M,D
Chemistry	M,D*
Geology	M,D
Geosciences	M,D
Inorganic Chemistry	M,D
Mathematics	M,D*
Organic Chemistry	M,D
Physical Chemistry	M,D
Physics	M,D*
Statistics	M,D

UNIVERSITY OF TORONTO

Astronomy	M,D
Chemistry	M,D
Forestry	M,D
Geology	M,D
Mathematics	M,D
Physics	M,D
Statistics	M,D

UNIVERSITY OF TULSA

Chemistry	M
Geology	M
Geosciences	M,D
Mathematics	M

UNIVERSITY OF UTAH

Biostatistics	M,D
Chemical Physics	M,D
Chemistry	M,D
Computational Sciences	M
Environmental Sciences	M
Geology	M,D
Geophysics	M,D
Mathematics	M,D*
Meteorology	M,D
Physics	M,D
Statistics	M,D

UNIVERSITY OF VERMONT

Agricultural Sciences—General	M,D
Agronomy and Soil Sciences	M,D
Animal Sciences	M,D
Biostatistics	M
Chemistry	M,D
Environmental Management and Policy	M,D
Forestry	M,D
Geology	M
Horticulture	M,D
Mathematics	M,D
Natural Resources	M,D
Physics	M
Plant Sciences	M,D
Statistics	M

UNIVERSITY OF VICTORIA

Astronomy	M,D
Astrophysics	M,D
Chemistry	M,D
Condensed Matter Physics	M,D
Geophysics	M,D
Geosciences	M,D
Mathematics	M,D
Oceanography	M,D
Physics	M,D
Statistics	M,D
Theoretical Physics	M,D

UNIVERSITY OF VIRGINIA

Astronomy	M,D
Chemistry	M,D
Environmental Sciences	M,D
Mathematics	M,D
Physics	M,D
Statistics	M,D

UNIVERSITY OF WASHINGTON

Applied Mathematics	M,D
Applied Physics	M,D
Astronomy	M,D
Atmospheric Sciences	M,D
Biostatistics	M,D
Chemistry	M,D
Environmental Management and Policy	M,D*
Fish, Game, and Wildlife Management	M,D
Forestry	M,D
Geology	M,D
Geophysics	M,D
Horticulture	M,D
Hydrology	M,D
Marine Affairs	M
Marine Geology	M,D
Mathematics	M,D
Oceanography	M,D
Physics	M,D
Statistics	M,D

UNIVERSITY OF WATERLOO

Applied Mathematics	M,D
Biostatistics	M,D
Chemistry	M,D
Environmental Management and Policy	M
Geosciences	M,D
Mathematics	M,D
Physics	M,D
Statistics	M,D

THE UNIVERSITY OF WESTERN ONTARIO

Applied Mathematics	M,D
Astronomy	M,D
Biostatistics	M,D
Chemistry	M,D
Environmental Sciences	M,D
Geology	M,D
Geophysics	M,D
Geosciences	M,D
Mathematics	M,D
Physics	M,D
Plant Sciences	M,D
Statistics	M,D

UNIVERSITY OF WEST FLORIDA

Environmental Sciences	M
Marine Affairs	M
Mathematics	M

UNIVERSITY OF WINDSOR

Chemistry	M,D
Environmental Sciences	M,D
Geosciences	M,D
Mathematics	M,D
Physics	M,D
Statistics	M,D

UNIVERSITY OF WISCONSIN–GREEN BAY

Environmental Management and Policy	M
Environmental Sciences	M

UNIVERSITY OF WISCONSIN–LA CROSSE

Marine Sciences	M

UNIVERSITY OF WISCONSIN–MADISON

Agricultural Sciences—General	M,D
Agronomy and Soil Sciences	M,D
Animal Sciences	M,D
Astronomy	D
Atmospheric Sciences	M,D
Biometrics	M
Chemistry	M,D
Environmental Management and Policy	M,D
Environmental Sciences	M,D
Food Science and Technology	M,D
Forestry	M,D
Geology	M,D
Geophysics	M,D
Horticulture	M,D
Limnology	M,D
Marine Sciences	M,D
Mathematics	M,D
Oceanography	M,D
Physics	M,D
Plant Sciences	M,D
Statistics	M,D
Water Resources	M

UNIVERSITY OF WISCONSIN–MILWAUKEE

Chemistry	M,D
Geology	M,D
Mathematics	M,D
Physics	M,D

UNIVERSITY OF WISCONSIN–RIVER FALLS

Agricultural Sciences—General	M

UNIVERSITY OF WISCONSIN–STEVENS POINT

Natural Resources	M

UNIVERSITY OF WISCONSIN–STOUT

Food Science and Technology	M

UNIVERSITY OF WYOMING

Agricultural Sciences—General	M,D
Agronomy and Soil Sciences	M,D
Animal Sciences	M,D
Atmospheric Sciences	M,D
Chemistry	M,D
Food Science and Technology	M
Geology	M,D
Geophysics	M,D
Mathematics	M,D
Natural Resources	M,D
Range Science	M,D
Statistics	M,D
Water Resources	M,D

UTAH STATE UNIVERSITY

Agricultural Sciences—General	M,D
Agronomy and Soil Sciences	M,D
Animal Sciences	M,D
Applied Mathematics	M,D
Chemistry	M,D*
Environmental Management and Policy	M,D
Fish, Game, and Wildlife Management	M,D
Food Science and Technology	M,D
Forestry	M,D
Geology	M
Mathematics	M,D
Meteorology	M,D
Natural Resources	M
Physics	M,D
Plant Sciences	M,D
Range Science	M,D
Statistics	M,D
Water Resources	M,D

VANDERBILT UNIVERSITY

Analytical Chemistry	M,D
Astronomy	M,D
Chemistry	M,D*
Environmental Management and Policy	M,D
Environmental Sciences	M
Inorganic Chemistry	M,D
Mathematics	M,D
Organic Chemistry	M,D
Physical Chemistry	M,D
Physics	M,D
Theoretical Chemistry	M,D

VASSAR COLLEGE

Chemistry	M

VERMONT LAW SCHOOL

Environmental Management and Policy	M

VILLANOVA UNIVERSITY

Chemistry	M*
Mathematics	M
Statistics	M

VIRGINIA COMMONWEALTH UNIVERSITY

Analytical Chemistry	M,D
Applied Mathematics	M
Applied Physics	M
Biostatistics	M,D
Chemical Physics	M,D
Chemistry	M,D
Environmental Management and Policy	M
Environmental Sciences	M
Inorganic Chemistry	M,D
Mathematics	M,O
Organic Chemistry	M,D
Physical Chemistry	M,D
Physics	M
Statistics	M,O

VIRGINIA POLYTECHNIC INSTITUTE AND STATE UNIVERSITY

Agricultural Sciences—General	M,D
Agronomy and Soil Sciences	M,D
Animal Sciences	M,D
Applied Mathematics	M,D
Applied Physics	M,D
Chemistry	M,D
Environmental Sciences	M,D
Fish, Game, and Wildlife Management	M,D
Food Science and Technology	M,D
Forestry	M,D
Geology	M,D
Geophysics	M,D
Geosciences	M,D
Horticulture	M,D
Mathematical Physics	M,D
Mathematics	M,D
Natural Resources	M,D
Physics	M,D
Statistics	M,D

VIRGINIA STATE UNIVERSITY

Mathematics	M
Physics	M

WAKE FOREST UNIVERSITY

Analytical Chemistry	M,D
Chemistry	M,D
Inorganic Chemistry	M,D
Mathematics	M
Organic Chemistry	M,D
Physical Chemistry	M,D
Physics	M,D*

WASHINGTON STATE UNIVERSITY

Agricultural Sciences—General	M
Agronomy and Soil Sciences	M,D
Analytical Chemistry	M,D
Animal Sciences	M,D
Applied Mathematics	M,D
Chemistry	M,D
Environmental Sciences	M,D*
Food Science and Technology	M,D
Geochemistry	M,D
Geology	M,D
Horticulture	M,D
Hydrology	M,D
Inorganic Chemistry	M,D
Mathematics	M,D*
Natural Resources	M,D
Organic Chemistry	M,D
Physical Chemistry	M,D
Physics	M,D
Statistics	M

WASHINGTON STATE UNIVERSITY TRI-CITIES

Atmospheric Sciences	M,D
Chemistry	M
Environmental Sciences	M,D
Geosciences	M,D
Water Resources	M,D

WASHINGTON STATE UNIVERSITY VANCOUVER

Environmental Sciences	M

WASHINGTON UNIVERSITY IN ST. LOUIS

Chemistry	M,D
Geochemistry	M,D
Geology	M,D
Geophysics	M,D
Geosciences	M,D
Mathematics	M,D
Physics	M,D*
Planetary and Space Sciences	M,D
Statistics	M,D

WAYNE STATE UNIVERSITY

Applied Mathematics	M,D
Chemistry	M,D
Food Science and Technology	M,D
Geology	M
Mathematics	M,D
Physics	M,D
Statistics	M,D

WEBSTER UNIVERSITY

Environmental Management and Policy	M,D

WESLEYAN UNIVERSITY

Astronomy	M
Chemical Physics	M,D
Chemistry	M,D*
Geosciences	M
Inorganic Chemistry	M,D
Mathematics	M,D*
Organic Chemistry	M,D
Physics	M,D
Theoretical Chemistry	M,D

WESLEY COLLEGE

Environmental Management and Policy	M

WEST CHESTER UNIVERSITY OF PENNSYLVANIA

Astronomy	M
Chemistry	M
Geology	M
Mathematics	M

WESTERN CAROLINA UNIVERSITY

Chemistry	M
Mathematics	M

WESTERN CONNECTICUT STATE UNIVERSITY

Environmental Sciences	M
Geosciences	M
Mathematics	M
Planetary and Space Sciences	M

WESTERN ILLINOIS UNIVERSITY

Chemistry	M
Mathematics	M
Physics	M

WESTERN KENTUCKY UNIVERSITY

Agricultural Sciences—General	M
Chemistry	M
Geology	M
Mathematics	M

WESTERN MICHIGAN UNIVERSITY

Applied Mathematics	M
Biostatistics	M
Chemistry	M,D
Computational Sciences	M
Geology	M,D
Geosciences	M
Mathematics	M,D
Physics	M,D
Statistics	M,D

WESTERN WASHINGTON UNIVERSITY

Chemistry	M
Environmental Sciences	M
Geology	M
Mathematics	M

WEST TEXAS A&M UNIVERSITY

Agricultural Sciences—General	M,D
Animal Sciences	M
Chemistry	M
Environmental Sciences	M
Mathematics	M
Plant Sciences	M

WEST VIRGINIA UNIVERSITY

Agricultural Sciences—General	M,D
Agronomy and Soil Sciences	M,D
Analytical Chemistry	M,D
Animal Sciences	M,D
Applied Mathematics	M,D
Applied Physics	M,D
Chemical Physics	M,D
Chemistry	M,D
Condensed Matter Physics	M,D
Environmental Management and Policy	M,D
Fish, Game, and Wildlife Management	M
Food Science and Technology	M,D
Forestry	M,D
Geology	M,D
Geophysics	M,D
Horticulture	M
Hydrogeology	M,D
Hydrology	M,D
Inorganic Chemistry	M,D
Mathematics	M,D
Natural Resources	D
Organic Chemistry	M,D
Paleontology	M,D
Physical Chemistry	M,D
Physics	M,D
Plant Sciences	M,D
Plasma Physics	M,D
Statistics	M,D
Theoretical Chemistry	M,D
Theoretical Physics	M,D

WICHITA STATE UNIVERSITY

Applied Mathematics	M,D
Chemistry	M,D
Environmental Sciences	M
Geology	M
Mathematics	M,D
Physics	M
Statistics	M,D

WILFRID LAURIER UNIVERSITY

Mathematics	M

WILKES UNIVERSITY

Mathematics	M

WILLIAM PATERSON UNIVERSITY OF NEW JERSEY

Limnology	M

WOODS HOLE OCEANOGRAPHIC INSTITUTION

Geochemistry	M,D,O
Geophysics	M,D,O
Marine Geology	M,D,O
Oceanography	M,D,O*

WORCESTER POLYTECHNIC INSTITUTE

Applied Mathematics	M,D
Chemistry	M,D
Mathematics	M,D*
Physics	M,D
Statistics	M,D

WRIGHT STATE UNIVERSITY

Applied Mathematics	M
Chemistry	M
Environmental Management and Policy	M
Environmental Sciences	M,D
Geochemistry	M
Geology	M
Geophysics	M
Hydrogeology	M
Mathematics	M
Physics	M
Statistics	M

YALE UNIVERSITY

Applied Mathematics	M,D
Applied Physics	M,D
Astronomy	M,D
Biostatistics	M,D
Chemistry	D
Environmental Management and Policy	M,D
Environmental Sciences	M,D
Forestry	M,D*
Geochemistry	D
Geology	D
Geophysics	D
Geosciences	D
Inorganic Chemistry	D
Mathematics	M,D
Meteorology	D
Mineralogy	D
Oceanography	D
Organic Chemistry	D
Paleontology	D
Physical Chemistry	D
Physics	D*
Statistics	M,D

YORK UNIVERSITY

Applied Mathematics	M,D
Astronomy	M,D
Chemistry	M,D
Environmental Management and Policy	M,D
Geosciences	M,D
Mathematics	M,D
Physics	M,D
Planetary and Space Sciences	M,D
Statistics	M,D

YOUNGSTOWN STATE UNIVERSITY

Chemistry	M
Environmental Management and Policy	M,O
Mathematics	M

ACADEMIC AND PROFESSIONAL PROGRAMS IN THE PHYSICAL SCIENCES

Section 1
Astronomy and Astrophysics

This section contains a directory of institutions offering graduate work in astronomy and astrophysics, followed by in-depth entries submitted by institutions that chose to prepare detailed program descriptions. Additional information about programs listed in the directory but not augmented by an in-depth entry may be obtained by writing directly to the dean of a graduate school or chair of a department at the address given in the directory.

For programs offering related work, see also in this book Geosciences, Meteorology and Atmospheric Sciences, and Physics. In Book 3, see Biological and Biomedical Sciences and Biophysics; and in Book 5, see Aerospace/Aeronautical Engineering, Energy and Power Engineering (Nuclear Engineering), Engineering and Applied Sciences, and Mechanical Engineering and Mechanics.

CONTENTS

Program Directories

Close-Ups

See:

Astronomy

Arizona State University, Division of Graduate Studies, College of Liberal Arts and Sciences, Division of Natural Sciences and Mathematics, Department of Physics and Astronomy, Tempe, AZ 85287. Offers MNS, MS, PhD. *Degree requirements:* For master's, thesis, oral and written exams; for doctorate, thesis/dissertation. *Entrance requirements:* For master's and doctorate, GRE.

Boston University, Graduate School of Arts and Sciences, Department of Astronomy, Boston, MA 02215. Offers MA, PhD. *Students:* 38 full-time (20 women); includes 3 minority (1 American Indian/Alaska Native, 1 Asian American or Pacific Islander, 1 Hispanic American), 8 international. Average age 27. 76 applicants, 39% accepted, 13 enrolled. In 2005, 5 master's, 1 doctorate awarded. Terminal master's awarded for partial completion of doctoral program. *Degree requirements:* For master's, one foreign language, thesis or alternative, comprehensive exam, registration; for doctorate, one foreign language, thesis/dissertation, comprehensive exam, registration. *Entrance requirements:* For master's and doctorate, GRE General Test, GRE Subject Test (physics), 3 letters of recommendation. Additional exam requirements/recommendations for international students: Required—TOEFL (minimum score 550 paper-based; 213 computer-based). *Application deadline:* For fall admission, 1/15 for domestic students, 1/15 for international students. Application fee: $60. *Expenses:* Tuition: Full-time $31,530; part-time $985 per credit. Required fees: $316; $40 per semester. Tuition and fees vary according to course level and program. *Financial support:* In 2005–06, 37 students received support, including 1 fellowship with full tuition reimbursement available (averaging $16,500 per year), 27 research assistantships with full tuition reimbursements available (averaging $16,000 per year), 8 teaching assistantships with full tuition reimbursements available (averaging $16,000 per year); Federal Work-Study and unspecified assistantships also available. Support available to part-time students. Financial award application deadline: 1/15; financial award applicants required to submit FAFSA. *Unit head:* James Jackson, Chairman, 617-353-6499, Fax: 617-353-6463, E-mail: jackson@bu.edu. *Application contact:* Susanna Lamey, Department Administrator, 617-363-2625, Fax: 617-353-5704, E-mail: slamey@bu.edu.

Bowling Green State University, Graduate College, College of Arts and Sciences, Department of Physics and Astronomy, Bowling Green, OH 43403. Offers physics (MAT, MS); physics and astronomy (MAT). *Faculty:* 7 full-time (0 women), 1 part-time/adjunct (0 women). *Students:* 8 full-time (1 woman), 10 part-time (7 women); includes 1 Asian American or Pacific Islander, 4 international. Average age 34. 25 applicants, 36% accepted, 3 enrolled. In 2005, 10 degrees awarded. *Degree requirements:* For master's, thesis or alternative. *Entrance requirements:* For master's, GRE General Test. Additional exam requirements/recommendations for international students: Required—TOEFL. Application fee: $30. Electronic applications accepted. *Financial support:* In 2005–06, 10 teaching assistantships with full tuition reimbursements (averaging $13,214 per year) were awarded; research assistantships with full tuition reimbursements, career-related internships or fieldwork, institutionally sponsored loans, and unspecified assistantships also available. Financial award applicants required to submit FAFSA. *Faculty research:* Computational physics, solid-state physics, materials science, theoretical physics. *Unit head:* Dr. John Laird, Chair, 419-372-7244. *Application contact:* Dr. Lewis Fulcher, Graduate Coordinator, 419-372-2635.

Brigham Young University, Graduate Studies, College of Physical and Mathematical Sciences, Department of Physics and Astronomy, Provo, UT 84602-1001. Offers physics (MS, PhD); physics and astronomy (PhD). Part-time programs available. *Faculty:* 31 full-time (0 women). *Students:* 38 full-time (7 women), 3 part-time (1 woman); includes 3 minority (1 American Indian/Alaska Native, 2 Hispanic Americans), 9 international. Average age 28. 22 applicants, 68% accepted, 11 enrolled. In 2005, 14 master's, 1 doctorate awarded. Terminal master's awarded for partial completion of doctoral program. *Median time to degree:* Of those who began their doctoral program in fall 1997, 100% received their degree in 8 years or less. *Degree requirements:* For master's, thesis/dissertation, registration; for doctorate, thesis/dissertation, comprehensive exam, registration. *Entrance requirements:* For master's and doctorate, GRE Subject Test in physics, minimum GPA of 3.0 in last 60 hours. Additional exam requirements/recommendations for international students: Required—TOEFL (minimum score 550 paper-based; 213 computer-based). *Application deadline:* For fall admission, 1/15 priority date for domestic students, 1/15 priority date for international students. Application fee: $50. Electronic applications accepted. *Financial support:* In 2005–06, 2 fellowships with full tuition reimbursements (averaging $18,000 per year), 10 research assistantships with full tuition reimbursements (averaging $18,000 per year), 25 teaching assistantships with full tuition reimbursements (averaging $16,000 per year) were awarded; career-related internships or fieldwork, institutionally sponsored loans, and tuition waivers (partial) also available. Support available to part-time students. Financial award application deadline: 1/15. *Faculty research:* Acoustics; astrophysics; atomic, molecular, and optical physics; plasma; theoretical and mathematical physics. Total annual research expenditures:$994,000. *Unit head:* Dr. Scott D. Sommerfeldt, Chair, 801-422-2205, Fax: 801-422-0553, E-mail: scott_sommerfeldt@byu.edu. *Application contact:* Dr. Ross L. Spencer, Graduate Coordinator, 801-422-2341, Fax: 801-422-0553, E-mail: ross_spencer@byu.edu.

California Institute of Technology, Division of Physics, Mathematics and Astronomy, Department of Astronomy, Pasadena, CA 91125-0001. Offers PhD. *Degree requirements:* For doctorate, one foreign language, thesis/dissertation, candidacy and final exams. *Entrance requirements:* For doctorate, GRE General Test, GRE Subject Test. Additional exam requirements/recommendations for international students: Required—TOEFL. *Faculty research:* Observational and theoretical astrophysics, cosmology, radio astronomy, solar physics.

Case Western Reserve University, School of Graduate Studies, Department of Astronomy, Cleveland, OH 44106. Offers MS, PhD. Part-time programs available. *Degree requirements:* For doctorate, thesis/dissertation. *Entrance requirements:* For master's and doctorate, GRE General Test, GRE Subject Test (physics). Additional exam requirements/recommendations for international students: Required—TOEFL. *Faculty research:* Ground-based optical astronomy, high- and low-dispersion spectroscopy, theoretical astrophysics, galactic structure.

Clemson University, Graduate School, College of Engineering and Science, Department of Physics and Astronomy, Program in Physics, Clemson, SC 29634. Offers astronomy and astrophysics (MS, PhD); atmospheric physics (MS, PhD); biophysics (MS, PhD). Part-time programs available. *Students:* 53 full-time (15 women), 2 part-time; includes 2 minority (1 Asian American or Pacific Islander, 1 Hispanic American), 19 international. 46 applicants, 41% accepted, 10 enrolled. In 2005, 5 master's, 4 doctorates awarded. Terminal master's awarded for partial completion of doctoral program. *Degree requirements:* For master's, thesis or alternative; for doctorate, thesis/dissertation. *Entrance requirements:* For master's and doctorate, GRE General Test. Additional exam requirements/recommendations for international students: Required—TOEFL. *Application deadline:* For fall admission, 2/15 for domestic students. Applications are processed on a rolling basis. Application fee: $50. *Financial support:* Fellowships, research assistantships, teaching assistantships available. Financial award application deadline: 6/1; financial award applicants required to submit FAFSA. *Faculty research:* Radiation physics, solid-state physics, nuclear physics, radar and lidar studies of atmosphere. *Unit head:* Dr. Brad Myer, Head, 864-656-5320. *Application contact:* Dr. Miguel Larsen, Coordinator, 864-656-5309, Fax: 864-656-0805, E-mail: mlarsen@clemson.edu.

See Close-Ups on pages 331 and 333.

Columbia University, Graduate School of Arts and Sciences, Division of Natural Sciences, Department of Astronomy, New York, NY 10027. Offers M Phil, MA, PhD. Part-time programs available. *Faculty:* 9 full-time, 2 part-time/adjunct. *Students:* 14 full-time (5 women), 1 part-time; includes 1 minority (Asian American or Pacific Islander), 5 international. Average age 30. 47 applicants, 19% accepted. In 2005, 1 degree awarded. *Degree requirements:* For doctorate, thesis/dissertation. *Entrance requirements:* For master's and doctorate, GRE General Test, major in astronomy or physics. Additional exam requirements/recommendations for international students: Required—TOEFL. Application fee: $75. *Expenses:* Tuition: Full-time $31,448. Tuition and fees vary according to course level, course load, campus/location and program. *Financial support:* Fellowships, teaching assistantships, Federal Work-Study and institutionally sponsored loans available. Support available to part-time students. Financial award application deadline: 1/5; financial award applicants required to submit FAFSA. *Faculty research:* Theoretical astrophysics, x-ray astronomy, radio astronomy. *Unit head:* David Helfand, Chair, 212-854-2150, Fax: 212-854-8121, E-mail: dv@astro.columbia.edu. *Application contact:* Information Contact, 212-854-6850, Fax: 212-854-8121.

Cornell University, Graduate School, Graduate Fields of Arts and Sciences, Field of Astronomy and Space Sciences, Ithaca, NY 14853-0001. Offers astronomy (PhD); astrophysics (PhD); general space sciences (PhD); infrared astronomy (PhD); planetary studies (PhD); radio astronomy (PhD); radiophysics (PhD); theoretical astrophysics (PhD). *Faculty:* 38 full-time (3 women). *Students:* 31 full-time (10 women); includes 1 minority (Asian American or Pacific Islander), 11 international. 110 applicants, 21% accepted, 7 enrolled. In 2005, 6 doctorates awarded. *Degree requirements:* For doctorate, thesis/dissertation, comprehensive exam. *Entrance requirements:* For doctorate, GRE General Test, GRE Subject Test (physics), 3 letters of recommendation. Additional exam requirements/recommendations for international students: Required—TOEFL (minimum score 600 paper-based; 250 computer-based). *Application deadline:* For fall admission, 1/15 for domestic students. Application fee: $60. Electronic applications accepted. *Financial support:* In 2005–06, 31 students received support, including 3 fellowships with full tuition reimbursements available, 19 research assistantships with full tuition reimbursements available, 9 teaching assistantships with full tuition reimbursements available; institutionally sponsored loans, scholarships/grants, health care benefits, tuition waivers (full and partial), and unspecified assistantships also available. Financial award applicants required to submit FAFSA. *Faculty research:* Observational astrophysics, planetary sciences, cosmology, instrumentation, gravitational astrophysics. *Unit head:* Director of Graduate Studies, 607-255-4341. *Application contact:* Graduate Field Assistant, 607-255-4341, E-mail: oconnor@astro.cornell.edu.

Dartmouth College, School of Arts and Sciences, Department of Physics and Astronomy, Hanover, NH 03755. Offers MS, PhD. *Faculty:* 23 full-time (6 women), 3 part-time/adjunct (1 woman). *Students:* 42 full-time (13 women); includes 1 minority (African American), 18 international. Average age 26. 117 applicants, 23% accepted, 10 enrolled. In 2005, 3 master's, 5 doctorates awarded. Terminal master's awarded for partial completion of doctoral program. *Degree requirements:* For master's and doctorate, thesis/dissertation. *Entrance requirements:* For master's and doctorate, GRE General Test, GRE Subject Test. Additional exam requirements/recommendations for international students: Required—TOEFL. *Application deadline:* For fall admission, 2/1 for domestic students. Application fee: $15. *Expenses:* Tuition: Full-time $31,770. *Financial support:* In 2005–06, 43 students received support, including fellowships with full tuition reimbursements available (averaging $21,000 per year), research assistantships with full tuition reimbursements available (averaging $21,000 per year); Federal Work-Study, institutionally sponsored loans, scholarships/grants, and tuition waivers (full) also available. *Faculty research:* Matter physics, plasma and beam physics, space physics, astronomy, cosmology. Total annual research expenditures: $3.8 million. *Unit head:* Robert Caldwell, Chair, Graduate Admissions, 603-646-2742, Fax: 603-646-1446, E-mail: robert.caldwell@dartmouth.edu. *Application contact:* Jean Blandin, Administrative Assistant, 603-646-2854, Fax: 603-646-1446, E-mail: jean.blandin@dartmouth.edu.

Georgia State University, College of Arts and Sciences, Department of Physics and Astronomy, Program in Astronomy, Atlanta, GA 30303-3083. Offers PhD. *Faculty:* 17 full-time (0 women). *Students:* 25; includes 5 Asian Americans or Pacific Islanders, 2 Hispanic Americans. Average age 25. 13 applicants, 54% accepted. In 2005, 1 degree awarded. *Degree requirements:* For doctorate, 2 foreign languages, thesis/dissertation, exam. *Entrance requirements:* For doctorate, GRE General Test, GRE Subject Test. Additional exam requirements/recommendations for international students: Required—TOEFL. *Application deadline:* For fall admission, 7/1 for domestic students, 7/1 for international students; for spring admission, 11/15 for domestic students, 11/15 for international students. Applications are processed on a rolling basis. Application fee: $50. Electronic applications accepted. *Expenses:* Tuition, state resident: full-time $4,368; part-time $182 per term. Tuition, nonresident: full-time $8,732; part-time $728 per term. Required fees: $46 per hour. *Financial support:* In 2005–06, fellowships with tuition reimbursements (averaging $22,000 per year), research assistantships with tuition reimbursements (averaging $17,500 per year), teaching assistantships with tuition reimbursements (averaging $16,500 per year) were awarded; Federal Work-Study, institutionally sponsored loans, tuition waivers (full), and unspecified assistantships also available. Financial award application deadline: 5/1; financial award applicants required to submit FAFSA. *Faculty research:* Extragalactic photometry, theoretical astrophysics, young stellar objects. Total annual research expenditures: $1.3 million. *Unit head:* Dr. Douglas Gies, Director of Graduate Studies, 404-651-2279, Fax: 404-651-1366, E-mail: gies@chara.gsu.edu.

Harvard University, Graduate School of Arts and Sciences, Department of Astronomy, Cambridge, MA 02138. Offers astronomy (PhD); astrophysics (PhD). *Faculty:* 17 full-time (1 woman). *Students:* 41 full-time (10 women); includes 8 minority (all Asian Americans or Pacific Islanders), 8 international. Average age 25. 85 applicants, 18% accepted, 10 enrolled. In 2005, 6 doctorates awarded. *Degree requirements:* For doctorate, thesis/dissertation, paper, research project, teaching 2 semesters. *Entrance requirements:* For doctorate, GRE General Test, GRE Subject Test (physics). Additional exam requirements/recommendations for international students: Required—TOEFL. *Application deadline:* For fall admission, 12/15 for domestic students. Application fee: $80. Electronic applications accepted. *Expenses:* Tuition: Full-time $28,752. Full-time tuition and fees vary according to program and student level. *Financial support:* In 2005–06, 20 research assistantships (averaging $21,840 per year), 8 teaching assistantships (averaging $4,475 per year); fellowships, career-related internships or fieldwork, Federal Work-Study, institutionally sponsored loans, scholarships/grants, and health care benefits also available. Financial award application deadline: 12/30. *Faculty research:* Atomic and molecular physics, electromagnetism, solar physics, nuclear physics, fluid dynamics. *Unit head:* Peg Herlihy, Administrator, 617-495-3752. *Application contact:* 617-495-5315, E-mail: admiss@fas.harvard.edu.

Indiana University Bloomington, Graduate School, College of Arts and Sciences, Department of Astronomy, Bloomington, IN 47405-7000. Offers astronomy (MA, PhD); astrophysics (PhD). PhD offered through the University Graduate School. Part-time programs available. *Faculty:* 8 full-time (3 women). *Students:* 7 full-time (1 woman), 6 part-time (2 women), 2 international. Average age 26. In 2005, 1 master's, 3 doctorates awarded. Terminal master's awarded for partial completion of doctoral program. *Degree requirements:* For master's, thesis, oral exam; for doctorate, thesis/dissertation, written qualifying exam. *Entrance requirements:* For master's and doctorate, GRE General Test, GRE Subject Test (physics), BA or BS in science. Additional exam requirements/recommendations for international students: Required—TOEFL. *Application deadline:* For fall admission, 1/15 priority date for domestic students, 12/15 priority date for international students; for spring admission, 9/1 priority date for domestic students, 9/1 priority date for international students. Applications are processed on a rolling basis. Application fee: $50 ($60 for international students). Electronic applications accepted. *Expenses:* Tuition, state resident: full-time $5,437; part-time $227 per credit hour. Tuition, nonresident: full-time $15,836; part-time $660 per credit hour. Required fees: $821. Tuition and fees vary according to campus/location and program. *Financial support:* In 2005–06, 2 fellowships, 2 research assistantships, 7 teaching assistantships were awarded; Federal Work-Study and tuition waivers (full and partial) also available. Support available to part-time students. Financial award application deadline: 5/2. *Faculty research:* Galaxies and cosmology, stellar astronomy, cataclysmic variables, high-energy astrophysics, globular clusters. *Unit head:* Dr. Catherine Pilachowski, Chair, 812-855-6911. *Application contact:* Brenda Records, Secretary, 812-855-6912, Fax: 812-855-8725, E-mail: brecords@indiana.edu.

Iowa State University of Science and Technology, Graduate College, College of Liberal Arts and Sciences, Department of Physics and Astronomy, Ames, IA 50011. Offers applied physics (MS, PhD); astrophysics (MS, PhD); condensed matter physics (MS, PhD); high energy physics (MS, PhD); nuclear physics (MS, PhD); physics (MS, PhD). Part-time programs available. *Faculty:* 44 full-time, 3 part-time/adjunct. *Students:* 79 full-time (14 women), 6 part-time; includes 1 minority (Asian American or Pacific Islander), 56 international. 161 applicants, 34% accepted, 23 enrolled. In 2005, 5 master's, 7 doctorates awarded. Terminal master's awarded for partial completion of doctoral program. *Degree requirements:* For master's, thesis (for some programs); for doctorate, thesis/dissertation. *Entrance requirements:* For master's and doctorate, GRE General Test, GRE Subject Test (physics). Additional exam requirements/recommendations for international students: Required—TOEFL (paper score 550; computer score 213) or IELTS (score 6.5). *Application deadline:* For fall admission, 2/15 priority date for domestic students, 2/15 priority date for international students; for spring admission, 10/15 for domestic students, 10/15 for international students. Applications are processed on a rolling basis. Application fee: $30 ($70 for international students). Electronic applications accepted. *Expenses:* Tuition, state resident: full-time $6,410. Tuition, nonresident: full-time $16,422. Tuition and fees vary according to program. *Financial support:* In 2005–06, 48 research assistantships with full tuition reimbursements (averaging $14,626 per year), 30 teaching assistantships with full tuition reimbursements (averaging $14,533 per year) were awarded; fellowships, Federal Work-Study, institutionally sponsored loans, scholarships/grants, health care benefits, and unspecified assistantships also available. Support available to part-time students. Financial award application deadline: 2/15. *Faculty research:* Condensed-matter physics, including superconductivity and new materials; high-energy and nuclear physics; astronomy and astrophysics; atmospheric and environmental physics. Total annual research expenditures: $8.8 million. *Unit head:* Dr. Eli Rosenberg, Chair, 515-294-5441, Fax: 515-294-6027, E-mail: phys_astro@iastate.edu. *Application contact:* Dr. Steven Kawaler, Director of Graduate Education, 515-294-9728, E-mail: phys_astro@iastate.edu.

The Johns Hopkins University, Zanvyl Krieger School of Arts and Sciences, Henry A. Rowland Department of Physics and Astronomy, Baltimore, MD 21218-2699. Offers astronomy (PhD); physics (PhD). *Faculty:* 34 full-time (3 women), 17 part-time/adjunct (2 women). *Students:* 98 full-time (23 women), 1 part-time; includes 9 minority (1 African American, 1 American Indian/Alaska Native, 5 Asian Americans or Pacific Islanders, 2 Hispanic Americans), 49 international. Average age 28. 267 applicants, 22% accepted, 21 enrolled. *Degree requirements:* For doctorate, thesis/dissertation, comprehensive exam, registration. *Entrance requirements:* For doctorate, GRE General Test, GRE Subject Test. Additional exam requirements/recommendations for international students: Required—TOEFL (minimum score 600 paper-based; 250 computer-based). *Application deadline:* For fall admission, 1/15 for domestic students, 1/15 for international students. Application fee: $60. Electronic applications accepted. *Expenses:* Tuition: Full-time $30,960. Tuition and fees vary according to degree level and program. *Financial support:* In 2005–06, 9 fellowships with tuition reimbursements (averaging $2,500 per year), 53 research assistantships with full tuition reimbursements (averaging $20,000 per year), 43 teaching assistantships with full tuition reimbursements (averaging $15,000 per year) were awarded; career-related internships or fieldwork, Federal Work-Study, and institutionally sponsored loans also available. Financial award application deadline: 4/15; financial award applicants required to submit FAFSA. *Faculty research:* High-energy physics, condensed-matter astrophysics, particle and experimental physics, plasma physics. Total annual research expenditures: $26.2 million. *Unit head:* Dr. Jonathan A. Bagger, Chair, 410-516-7346, Fax: 410-516-7239, E-mail: bagger@jhu.edu. *Application contact:* Carmelita D. King, Academic Affairs Administrator, 410-516-7344, Fax: 410-516-7239, E-mail: jazzy@pha.jhu.edu.

See Close-Up on page 347.

Louisiana State University and Agricultural and Mechanical College, Graduate School, College of Basic Sciences, Department of Physics and Astronomy, Baton Rouge, LA 70803. Offers astronomy (PhD); astrophysics (PhD); physics (MS, PhD). *Faculty:* 45 full-time (1 woman). *Students:* 72 full-time (8 women), 4 part-time (1 woman); includes 2 African Americans, 2 Asian Americans or Pacific Islanders, 1 Hispanic American, 33 international. Average age 28. 99 applicants, 23% accepted, 76 enrolled. In 2005, 6 master's, 4 doctorates awarded. Terminal master's awarded for partial completion of doctoral program. *Degree requirements:* For master's, thesis or alternative; for doctorate, thesis/dissertation. *Entrance requirements:* For master's and doctorate, GRE General Test, minimum GPA of 3.0. Additional exam requirements/recommendations for international students: Required—TOEFL (minimum score 550 paper-based; 213 computer-based). *Application deadline:* For fall admission, 1/25 priority date for domestic students, 5/15 priority date for international students. Applications are processed on a rolling basis. Application fee: $25. Electronic applications accepted. *Financial support:* In 2005–06, 72 students received support, including 3 fellowships with full tuition reimbursements available (averaging $15,000 per year), 40 research assistantships with full and partial tuition reimbursements available (averaging $16,240 per year), 27 teaching assistantships with full and partial tuition reimbursements available (averaging $17,652 per year); institutionally sponsored loans, tuition waivers (full and partial), and unspecified assistantships also available. Financial award application deadline: 3/15; financial award applicants required to submit FAFSA. *Faculty research:* Experimental and theoretical atomic, nuclear, particle, cosmic-ray, low-temperature, and condensed-matter physics. Total annual research expenditures: $5.8 million. *Unit head:* Dr. Roger McNeil, Chair, 225-578-2261, Fax: 225-578-5855, E-mail: mcneil@phys.lsu.edu. *Application contact:* Dr. James Matthews, Graduate Adviser, 225-578-8598, Fax: 225-578-5855, E-mail: jmatth5@lsu.edu.

Michigan State University, The Graduate School, College of Natural Science, Department of Physics and Astronomy, East Lansing, MI 48824. Offers astrophysics and astronomy (MS, PhD); physics (MS, PhD). *Faculty:* 52 full-time (3 women). *Students:* 135 full-time (23 women), 6 part-time (3 women); includes 6 minority (2 American Indian/Alaska Native, 3 Asian Americans or Pacific Islanders, 1 Hispanic American), 70 international. Average age 28. 252 applicants, 8% accepted. In 2005, 18 master's, 19 doctorates awarded. *Degree requirements:* For master's, qualifying exam, thesis optional; for doctorate, thesis/dissertation, qualifying exam, comprehensive exam. *Entrance requirements:* For master's, GRE General Test (recommended), minimum GPA of 3.0 in science/math courses, coursework equivalent to a major in physics or astronomy, 3 letters of recommendation; for doctorate, GRE General Test (recommended), minimum GPA of 3.0 in science/math courses, coursework equivalent to a major in physics or astronomy, 3 letters of recommendation, research experience (recommended). Additional exam requirements/recommendations for international students: Required—TOEFL (minimum score 550 paper-based; 213 computer-based), Michigan State University ELT (85), Michigan ELAB (83). *Application deadline:* For fall admission, 12/27 for domestic students; for spring admission, 9/30 for domestic students. Application fee: $50. Electronic applications accepted. *Expenses:* Tuition, state resident: part-time $330 per credit hour. Tuition, nonresident: part-time $685 per credit hour. Tuition and fees vary according to program. *Financial support:* In 2005–06, 20 fellowships with tuition reimbursements (averaging $8,526 per year), 88 research assistantships with tuition reimbursements (averaging $14,281 per year), 40 teaching assistantships with tuition reimbursements (averaging $12,680 per year) were awarded; scholarships/grants and unspecified assistantships also available. *Faculty research:* Nuclear and accelerator physics, high energy physics, condensed matter physics, biophysics, astrophysics and astronomy. Total annual research expenditures: $6.4 million. *Unit head:* Dr. Wolfgang W. Bauer, Chairperson, 517-355-9200 Ext. 2015, Fax: 517-355-4500, E-mail: bauer@pa.msu.edu. *Application contact:* Debbie Simmons, Graduate Secretary, 517-355-9200 Ext. 2032, Fax: 517-353-4500, E-mail: grd_chair@pa.msu.edu.

Minnesota State University Mankato, College of Graduate Studies, College of Science, Engineering and Technology, Department of Physics and Astronomy, Mankato, MN 56001. Offers physics (MS); physics and astronomy (MT). *Students:* 3 full-time (1 woman), 2 part-time. Average age 34. In 2005, 2 degrees awarded. *Degree requirements:* For master's, one foreign language, thesis or alternative, comprehensive exam. *Entrance requirements:* For master's, minimum GPA of 3.0 during previous 2 years. Additional exam requirements/recommendations for international students: Required—TOEFL. *Application deadline:* For fall admission, 7/1 for domestic students; for spring admission, 11/1 for domestic students. Applications are processed on a rolling basis. Application fee: $40. Electronic applications accepted. *Expenses:* Tuition, state resident: part-time $243 per credit. Tuition, nonresident: part-time $400 per credit. Required fees: $30 per credit. *Financial support:* Research assistantships, teaching assistantships with full tuition reimbursements, Federal Work-Study and unspecified assistantships available. Support available to part-time students. Financial award application deadline: 3/15; financial award applicants required to submit FAFSA. *Unit head:* Dr. Mark Pickar, Chairperson, 507-389-5743. *Application contact:* 507-389-2321, E-mail: grad@mnsu.edu.

New Mexico State University, Graduate School, College of Arts and Sciences, Department of Astronomy, Las Cruces, NM 88003-8001. Offers MS, PhD. Part-time programs available. *Faculty:* 9 full-time (2 women), 4 part-time/adjunct (1 woman). *Students:* 24 full-time (10 women), 2 part-time (1 woman); includes 2 minority (1 Asian American or Pacific Islander, 1 Hispanic American), 5 international. Average age 27. 35 applicants, 20% accepted, 7 enrolled. In 2005, 3 master's, 2 doctorates awarded. Terminal master's awarded for partial completion of doctoral program. *Degree requirements:* For master's and doctorate, thesis/dissertation. *Entrance requirements:* For master's and doctorate, GRE General Test, GRE Subject Test (advanced physics). Additional exam requirements/recommendations for international students: Required—TOEFL. *Application deadline:* For fall admission, 2/15 priority date for domestic students, 2/15 priority date for international students. Applications are processed on a rolling basis. Application fee: $30 ($50 for international students). Electronic applications accepted. *Expenses:* Tuition, state resident: full-time $3,156; part-time $175 per credit. Tuition, nonresident: full-time $12,510; part-time $565 per credit. Required fees: $1,050. *Financial support:* In 2005–06, 4 fellowships with partial tuition reimbursements, 10 research assistantships with tuition reimbursements, 11 teaching assistantships with partial tuition reimbursements were awarded; scholarships/grants, health care benefits, and unspecified assistantships also available. Financial award application deadline: 3/1. *Faculty research:* Planetary systems, accreting binary stars, stellar populations, galaxies, interstellar medium. *Unit head:* Dr. James Murphy, Head, 505-646-5333, Fax: 505-646-1602, E-mail: murphy@nmsu.edu. *Application contact:* Dr. Chris Churchill, Assistant Professor, 505-646-1913, Fax: 505-646-1602, E-mail: cwc@nmsu.edu.

Northwestern University, The Graduate School, Judd A. and Marjorie Weinberg College of Arts and Sciences, Department of Physics and Astronomy, Evanston, IL 60208. Offers astrophysics (PhD); physics (MS, PhD). Admissions and degrees offered through The Graduate School. *Faculty:* 30 full-time (4 women), 2 part-time/adjunct (0 women). *Students:* 113 (26 women). Average age 24. 140 applicants, 21% accepted, 12 enrolled. In 2005, 4 master's, 10 doctorates awarded. *Median time to degree:* Of those who began their doctoral program in fall 1997, 50% received their degree in 8 years or less. *Degree requirements:* For doctorate, thesis/dissertation, qualifying exam. *Entrance requirements:* For doctorate, GRE General Test, GRE Subject Test. Additional exam requirements/recommendations for international students: Required—TOEFL. Application fee: $60 ($75 for international students). *Financial support:* In 2005–06, 57 students received support, including 9 fellowships with full tuition reimbursements available (averaging $12,906 per year), 28 research assistantships with partial tuition reimbursements available (averaging $18,732 per year), 20 teaching assistantships with full tuition reimbursements available (averaging $13,329 per year); career-related internships or fieldwork, Federal Work-Study, and institutionally sponsored loans also available. Financial award application deadline: 1/15; financial award applicants required to submit FAFSA. *Faculty research:* Nuclear and particle physics, condensed-matter physics, nonlinear physics, astrophysics. Total annual research expenditures: $6.3 million. *Unit head:* Melvin Ulmer, Chair, 847-491-5633, Fax: 847-491-9982, E-mail: physics-astronomy@northwestern.edu. *Application contact:* Mayda Velasco, Admission Officer, 847-467-7099, Fax: 847-491-9982, E-mail: physics-astronomy@northwestern.edu.

The Ohio State University, Graduate School, College of Mathematical and Physical Sciences, Department of Astronomy, Columbus, OH 43210. Offers MS, PhD. *Degree requirements:* For master's and doctorate, thesis/dissertation, comprehensive exam. *Entrance requirements:* For master's and doctorate, GRE General Test, GRE Subject Test (physics). Additional exam requirements/recommendations for international students: Required—TOEFL (minimum score 600 paper-based; 250 computer-based). Electronic applications accepted.

Ohio University, Graduate Studies, College of Arts and Sciences, Department of Physics and Astronomy, Athens, OH 45701-2979. Offers astronomy (MS, PhD); physics (MS, PhD). Part-time programs available. *Faculty:* 29 full-time (3 women), 1 (woman) part-time/adjunct. *Students:* 67 full-time (21 women), 57 international. Average age 24. 95 applicants, 42% accepted, 19 enrolled. In 2005, 12 master's, 12 doctorates awarded. Terminal master's awarded for partial completion of doctoral program. *Median time to degree:* Of those who began their doctoral program in fall 1997, 100% received their degree in 8 years or less. *Degree requirements:* For master's, thesis or alternative; for doctorate, thesis/dissertation, comprehensive exam. *Entrance requirements:* For master's and doctorate, minimum GPA of 3.0. Additional exam requirements/recommendations for international students: Required—TOEFL (minimum score 600 paper-based; 250 computer-based), IELT (minimum score 7), TWE (minimum score 4). *Application deadline:* For fall admission, 4/1 priority date for domestic students, 4/1 priority date for international students. Applications are processed on a rolling basis. Application fee: $45. Electronic applications accepted. *Financial support:* In 2005–06, 2 fellowships with full tuition reimbursements (averaging $9,400 per year), 31 research assistantships with full tuition reimbursements (averaging $19,000 per year), 33 teaching assistantships with full tuition reimbursements (averaging $15,600 per year) were awarded; scholarships/grants and unspecified assistantships also available. Financial award application deadline: 4/1. *Faculty research:* Nuclear physics, condensed-matter physics, nonlinear systems, acoustics, astrophysics. Total annual research expenditures: $3.2 million. *Unit head:* Dr. Joseph Shields, Chair, 740-593-0336, Fax: 740-593-6433, E-mail: shields@helios.phy.ohiou.edu. *Application contact:* Dr. Daniel S. Carman, Graduate Admissions Chair, 740-593-3964, Fax: 740-593-0433, E-mail: gradapp@phy.ohiou.edu.

See Close-Up on page 355.

The Pennsylvania State University University Park Campus, Graduate School, Eberly College of Science, Department of Astronomy and Astrophysics, State College, University Park, PA 16802-1503. Offers MS, PhD. *Students:* 25 full-time (5 women), 2 part-time (1 woman); includes 2 minority (both Hispanic Americans), 11 international. *Entrance requirements:* For master's and doctorate, GRE General Test. Application fee: $45. *Expenses:* Tuition, state resident: full-time $12,518; part-time $522 per credit. Tuition, nonresident: full-time $23,004; part-time $959 per credit. Required fees: $484. Tuition and fees vary according to course load, campus/location and program. *Unit head:* Dr. Lawrence W. Ramsey, Head, 814-865-0418, Fax: 814-863-3399, E-mail: lwr@astro.psu.edu.

Rice University, Graduate Programs, Wiess School of Natural Sciences, Department of Physics and Astronomy, Houston, TX 77251-1892. Offers physics (MA); physics and astronomy (MS, PhD). *Degree requirements:* For master's and doctorate, thesis/dissertation. *Entrance requirements:* For master's and doctorate, GRE General Test, GRE Subject Test (physics), minimum GPA of 3.0. Additional exam requirements/recommendations for international students: Required—TOEFL (minimum score 600 paper-based; 250 computer-based). Electronic applications accepted. *Faculty research:* Atomic, solid-state, and molecular physics; biophysics; medium- and high-energy physics, magnetospheric physics, planetary atmospheres, astrophysics.

Saint Mary's University, Faculty of Science, Department of Astronomy and Physics, Halifax, NS B3H 3C3, Canada. Offers M Sc. Part-time programs available. *Degree requirements:* For master's, thesis. *Entrance requirements:* For master's, honors degree. Additional exam requirements/recommendations for international students: Required—TOEFL. *Faculty research:* Young stellar objects, interstellar medium, star clusters, galactic structure, early-type galaxies.

Astronomy

San Diego State University, Graduate and Research Affairs, College of Sciences, Department of Astronomy, San Diego, CA 92182. Offers MS. *Students:* 5 full-time (0 women), 13 part-time (5 women); includes 4 minority (2 Asian Americans or Pacific Islanders, 2 Hispanic Americans), 2 international. Average age 30. 19 applicants, 58% accepted, 8 enrolled. In 2005, 6 degrees awarded. *Degree requirements:* For master's, thesis. *Entrance requirements:* For master's, GRE General Test, letters of reference, TA/GA application. Additional exam requirements/recommendations for international students: Required—TOEFL. *Application deadline:* For fall admission, 5/1 for domestic students, 5/1 for international students; for spring admission, 11/1 for domestic students, 10/1 for international students. Applications are processed on a rolling basis. Application fee: $55. Electronic applications accepted. *Financial support:* In 2005–06, 2 research assistantships were awarded; teaching assistantships, unspecified assistantships also available. Financial award applicants required to submit FAFSA. *Faculty research:* CCD, classical and dwarf novae, photometry, interactive binaries. Total annual research expenditures:$183,088. *Unit head:* Allen W. Shafter, Chair, 619-594-6182, Fax: 619-594-1413, E-mail: shafter@sciences.sdsu.edu. *Application contact:* William Welsh, Graduate Coordinator, 619-594-2288, Fax: 619-594-1413, E-mail: wwelsh@mail.sdsu.edu.

Texas Christian University, College of Science and Engineering, Department of Physics and Astronomy, Fort Worth, TX 76129-0002. Offers physics (MA, MS, PhD), including astrophysics (PhD), business (PhD), physics (PhD). Part-time and evening/weekend programs available. *Degree requirements:* For doctorate, thesis/dissertation, qualifying exams. *Entrance requirements:* For doctorate, GRE General Test. Additional exam requirements/recommendations for international students: Required—TOEFL. *Application deadline:* For fall admission, 3/1 for domestic students; for spring admission, 12/1 for domestic students. Applications are processed on a rolling basis. Application fee: $0. *Expenses:* Tuition: Part-time $740 per credit hour. *Financial support:* Fellowships, teaching assistantships available. Financial award application deadline: 3/1. *Unit head:* Dr. T W Zerda, Chairperson, 817-257-7375. *Application contact:* Dr. Bonnie Melhart, Associate Dean, College of Science and Engineering, E-mail: b.melhart@tcu.edu.

Université de Moncton, Faculty of Science, Department of Physics and Astronomy, Moncton, NB E1A 3E9, Canada. Offers M Sc. Part-time programs available. *Degree requirements:* For master's, thesis. *Entrance requirements:* For master's, proficiency in French. Electronic applications accepted. *Faculty research:* Thin films, optical properties, solar selective surfaces, microgravity and photonic materials.

The University of Arizona, Graduate College, College of Science, Department of Astronomy, Tucson, AZ 85721. Offers MS, PhD. *Degree requirements:* For doctorate, thesis/dissertation. *Entrance requirements:* For master's and doctorate, GRE General Test, GRE Subject Test. Additional exam requirements/recommendations for international students: Required—TOEFL. *Faculty research:* Astrophysics, submillimeter astronomy, infrared astronomy, NICMOS, SIRTF.

The University of British Columbia, Faculty of Graduate Studies, Faculty of Science, Program in Astronomy, Vancouver, BC V6T 1Z1, Canada. Offers M Sc, PhD.

University of Calgary, Faculty of Graduate Studies, Faculty of Science, Department of Physics and Astronomy, Calgary, AB T2N 1N4, Canada. Offers M Sc, PhD. Part-time programs available. *Faculty:* 22 full-time (1 woman), 12 part-time/adjunct (0 women). *Students:* 56 full-time (15 women), 4 part-time (2 women). Average age 28. 75 applicants, 100 enrolled. In 2005, 7 master's, 2 doctorates awarded. *Degree requirements:* For master's, thesis; for doctorate, thesis/dissertation, oral candidacy exam, written qualifying exam. *Entrance requirements:* For master's and doctorate, GRE General Test, GRE Subject Test. Additional exam requirements/recommendations for international students: Required—TOEFL (minimum score 550 paper-based; 213 computer-based). *Application deadline:* For fall admission, 3/1 for domestic students, 3/1 for international students. For winter admission, 7/1 for domestic students. Applications are processed on a rolling basis. Application fee: $130. Electronic applications accepted. *Financial support:* Fellowships with full and partial tuition reimbursements, research assistantships, teaching assistantships, institutionally sponsored loans available. Financial award application deadline:2/1. *Faculty research:* Astronomy and astrophysics, mass spectrometry, atmospheric physics, space physics, medical physics. Total annual research expenditures: $4.6 million. *Unit head:* Dr. A. R. Taylor, Head, 403-220-5385, Fax: 403-289-3331, E-mail: russ@ras.ucalgary.ca. *Application contact:* Dr. R. I. Thompson, Chairman, Graduate Affairs, 403-220-5407, Fax: 403-289-3331, E-mail: gradinfo@ucalgary.ca.

University of California, Los Angeles, Graduate Division, College of Letters and Science, Department of Physics and Astronomy, Program in Astronomy, Los Angeles, CA 90095. Offers MAT, MS, PhD. *Degree requirements:* For doctorate, thesis/dissertation, oral and written qualifying exams. *Entrance requirements:* For master's, GRE General Test, GRE Subject Test (physics), minimum GPA of 3.0; for doctorate, GRE General Test, GRE Subject Test (physics), minimum undergraduate GPA of 3.0. Electronic applications accepted.

University of California, Santa Cruz, Division of Graduate Studies, Division of Physical and Biological Sciences, Program in Astronomy and Astrophysics, Santa Cruz, CA 95064. Offers PhD. *Faculty:* 23 full-time (3 women). *Students:* 36 full-time (18 women); includes 4 minority (1 African American, 1 Asian American or Pacific Islander, 2 Hispanic Americans), 2 international. 137 applicants, 23% accepted, 7 enrolled. In 2005, 2 doctorates awarded. *Degree requirements:* For doctorate, one foreign language, thesis/dissertation, qualifying exam. *Entrance requirements:* For doctorate, GRE General Test, GRE Subject Test. *Application deadline:* For fall admission, 1/15 for domestic students. Application fee: $60. *Expenses:* Tuition, nonresident: full-time $14,694. *Financial support:* Fellowships, research assistantships, teaching assistantships, Federal Work-Study and institutionally sponsored loans available. Financial award application deadline: 1/15. *Faculty research:* Stellar structure and evolution, stellar spectroscopy, the interstellar medium, galactic structure, external galaxies and quasars. *Unit head:* Dr. Stephen Thorsett, Chairperson, 831-459-2976, E-mail: thorsett@ucolick.ucsc.edu. *Application contact:* Judy L. Glass, Reporting Analyst for Graduate Admissions, 831-459-5906, Fax: 831-459-4843, E-mail: jlglass@ucsc.edu.

University of Chicago, Division of the Physical Sciences, Department of Astronomy and Astrophysics, Chicago, IL 60637-1513. Offers SM, PhD. *Faculty:* 33 full-time (2 women), 4 part-time/adjunct (0 women). *Students:* 31 full-time (8 women); includes 10 minority (1 African American, 6 Asian Americans or Pacific Islanders, 3 Hispanic Americans). Average age 25. 110 applicants, 14% accepted, 5 enrolled. In 2005, 3 master's, 1 doctorate awarded. Terminal master's awarded for partial completion of doctoral program. *Degree requirements:* For master's, candidacy exam; for doctorate, thesis/dissertation, dissertation for publication. *Entrance requirements:* For master's, department candidacy examination, minimum GPA of 3.0; for doctorate, GRE General Test, GRE Subject Test, minimum GPA of 3.0. Additional exam requirements/recommendations for international students: Required—TOEFL (minimum score 600 paper-based; 250 computer-based); Recommended—IELT. *Application deadline:* For fall admission, 1/12 priority date for domestic students, 1/12 priority date for international students. Application fee: $55. Electronic applications accepted. *Financial support:* In 2005–06, 4 fellowships (averaging $4,250 per year), 71 research assistantships with full tuition reimbursements (averaging $21,996 per year), 19 teaching assistantships with full tuition reimbursements (averaging $18,594 per year) were awarded; career-related internships or fieldwork, Federal Work-Study, institutionally sponsored loans, and tuition waivers (partial) also available. Financial award application deadline: 12/28. *Faculty research:* Quasi-stellar object absorption lines, fluid dynamics, interstellar matter, particle physics, cosmology. *Unit head:* Angela Olinto, Chairman, 773-702-8203, Fax: 773-702-8212. *Application contact:* Laticia Rebeles, Graduate Student Affairs Administrator, 773-702-9808, Fax: 773-702-8212, E-mail: lrebeles@oddjob.uchicago.edu.

University of Delaware, College of Arts and Sciences, Department of Physics and Astronomy, Newark, DE 19716. Offers MS, PhD. Part-time programs available. *Faculty:* 33 full-time (2 women), 6 part-time/adjunct (1 woman). *Students:* 90 full-time (21 women); includes 4 minority (2 African Americans, 1 Asian American or Pacific Islander, 1 Hispanic American), 68 international. Average age 26. 123 applicants, 33% accepted, 20 enrolled. In 2005, 3 master's, 1 doctorate awarded. Terminal master's awarded for partial completion of doctoral program. *Degree requirements:* For master's and doctorate, thesis/dissertation. *Entrance requirements:* For master's and doctorate, GRE General Test, GRE Subject Test. Additional exam requirements/recommendations for international students: Required—TOEFL (minimum score 600 paper-based; 250 computer-based). *Application deadline:* For fall admission, 2/1 for domestic students. Application fee: $60. Electronic applications accepted. *Financial support:* In 2005–06, 82 students received support, including 2 fellowships with full tuition reimbursements available (averaging $19,000 per year), 27 research assistantships with full tuition reimbursements available (averaging $19,000 per year), 25 teaching assistantships with full tuition reimbursements available (averaging $19,000 per year); career-related internships or fieldwork, Federal Work-Study, institutionally sponsored loans, and corporate sponsorships also available. Financial award application deadline: 3/1. *Faculty research:* Magnetoresistance and magnetic materials, ultrafast optical phenomena, superfluidity, elementary particle physics, stellar atmospheres and interiors. Total annual research expenditures: $6.9 million. *Unit head:* Dr. George Hadjipanayis, Chair, 302-831-3361. *Application contact:* Dr. Norbert Mulders, Information Contact, 302-831-1995, E-mail: grad.physics@udel.edu.

University of Florida, Graduate School, College of Liberal Arts and Sciences, Department of Astronomy, Gainesville, FL 32611. Offers MS, PhD. *Faculty:* 20 full-time (2 women), 1 part-time/adjunct (0 women). *Students:* 33 (15 women); includes 4 minority (1 Asian American or Pacific Islander, 3 Hispanic Americans) 9 international. In 2005, 5 master's, 2 doctorates awarded. Terminal master's awarded for partial completion of doctoral program. *Degree requirements:* For master's, thesis (terminal MS); for doctorate, one foreign language, thesis/dissertation. *Entrance requirements:* For master's and doctorate, GRE General Test, minimum GPA of 3.0. Additional exam requirements/recommendations for international students: Required—TOEFL (minimum score 550 paper-based; 213 computer-based). *Application deadline:* For fall admission, 1/31 for domestic students. Applications are processed on a rolling basis. Application fee: $30. Electronic applications accepted. *Expenses:* Tuition, state resident: full-time $6,234. Tuition, nonresident: full-time $21,359. Tuition and fees vary according to program. *Financial support:* In 2005–06, fellowships (averaging $16,000 per year), 13 research assistantships (averaging $20,369 per year), 9 teaching assistantships (averaging $19,494 per year) were awarded; tuition waivers (full) and unspecified assistantships also available. Financial award application deadline: 1/31. *Faculty research:* Cosmology, photometry, variable and binary stars, dynamical solar system astronomy, infrared. Total annual research expenditures: $575,000. *Unit head:* Dr. Stanley F. Dermott, Chairman, 352-392-2052 Ext. 203, Fax: 352-392-5089, E-mail: dermott@astro.ufl.edu. *Application contact:* Dr. John P. Oliver, Coordinator, 352-392-2052 Ext. 206, Fax: 352-392-5089, E-mail: oliver@astro.ufl.edu.

University of Georgia, Graduate School, College of Arts and Sciences, Department of Physics and Astronomy, Athens, GA 30602. Offers physics (MS, PhD). *Faculty:* 24 full-time (2 women). *Students:* 54 full-time, 3 part-time; includes 6 minority (4 African Americans, 1 American Indian/Alaska Native, 1 Asian American or Pacific Islander), 27 international. 120 applicants, 23% accepted, 17 enrolled. In 2005, 4 master's, 2 doctorates awarded. *Degree requirements:* For master's, thesis; for doctorate, one foreign language, thesis/dissertation. *Entrance requirements:* For master's and doctorate, GRE General Test. *Application deadline:* For fall admission, 7/1 for domestic students; for spring admission, 11/15 for domestic students. Application fee: $50. Electronic applications accepted. *Financial support:* Fellowships, research assistantships, teaching assistantships, unspecified assistantships available. *Unit head:* Dr. Heinz-Bernd Schüttler, Head, 706-542-2485, Fax: 706-542-2492, E-mail: hbs@physast.uga.edu. *Application contact:* Dr. F. Todd Baker, Graduate Coordinator, 706-542-0979, Fax: 706-542-2492, E-mail: tbaker@physast.uga.edu.

University of Hawaii at Manoa, Graduate Division, Colleges of Arts and Sciences, College of Natural Sciences, Department of Physics and Astronomy, Honolulu, HI 96822. Offers astronomy (MS, PhD); physics (MS, PhD). *Faculty:* 66 full-time (5 women), 4 part-time/adjunct (1 woman). *Students:* 64 full-time (17 women), 3 part-time; includes 9 minority (1 African American, 6 Asian Americans or Pacific Islanders, 2 Hispanic Americans), 21 international. 176 applicants, 23% accepted, 12 enrolled. In 2005, 11 master's, 5 doctorates awarded. *Median time to degree:* Of those who began their doctoral program in fall 1997, 100% received their degree in 8 years or less. *Degree requirements:* For master's, qualifying exam or thesis; for doctorate, thesis/dissertation, oral comprehensive and qualifying exams. *Entrance requirements:* For master's and doctorate, GRE General Test, GRE Subject Test. Application fee: $50. *Expenses:* Tuition, state resident: full-time $8,400; part-time $200 per credit hour. Tuition, nonresident: full-time $11,088; part-time $462 per credit hour. Tuition and fees vary according to program. *Financial support:* In 2005–06, 43 research assistantships, 19 teaching assistantships were awarded. *Unit head:* Dr. Michael Peters, Chairperson, 808-956-7087, Fax: 808-956-7107, E-mail: mwp@phys.hawaii.edu. *Application contact:* Dr. Joshua Barnes, Graduate Chair, Astronomy, 808-956-8138, Fax: 808-956-4604, E-mail: barnes@ifa.hawaii.edu.

University of Illinois at Urbana–Champaign, Graduate College, College of Liberal Arts and Sciences, Department of Astronomy, Champaign, IL 61820. Offers MS, PhD. *Faculty:* 12 full-time (1 woman). *Students:* 17 full-time (8 women), 14 part-time (6 women), 21 international. 65 applicants, 14% accepted, 6 enrolled. In 2005, 1 master's awarded. *Degree requirements:* For doctorate, thesis/dissertation. *Entrance requirements:* For master's and doctorate, GRE General Test, GRE Subject Test, minimum GPA of 3.0. *Application deadline:* Applications are processed on a rolling basis. Application fee: $50 ($60 for international students). Electronic applications accepted. *Financial support:* In 2005–06, 22 research assistantships, 15 teaching assistantships were awarded; fellowships Financial award application deadline: 2/15. *Unit head:* You-Hua Chu, Chair, 217-333-5535, Fax: 217-244-7638, E-mail: yhchu@uiuc.edu. *Application contact:* Jeri Cochran, Administrative Assistant, 217-333-9784, Fax: 217-244-7638, E-mail: jcochran@uiuc.edu.

The University of Iowa, Graduate College, College of Liberal Arts and Sciences, Department of Physics and Astronomy, Program in Astronomy, Iowa City, IA 52242-1316. Offers MS. *Students:* 1 full-time (0 women). 3 applicants, 0% accepted, 0 enrolled. In 2005, 2 degrees awarded. *Degree requirements:* For master's, exam, thesis optional. *Entrance requirements:* For master's, GRE General Test, GRE Subject Test, minimum GPA of 3.0. Additional exam requirements/recommendations for international students: Required—TOEFL (minimum score 550 paper-based; 213 computer-based). *Application deadline:* For fall admission, 2/1 priority date for domestic students, 2/1 priority date for international students. Application fee: $60 ($85 for international students). Electronic applications accepted. *Expenses:* Tuition, state resident: part-time $1,882 per term. Tuition, nonresident: full-time $17,338; part-time $4,907 per term. Tuition and fees vary according to course load and program. *Financial support:* In 2005–06, 1 teaching assistantship with partial tuition reimbursement was awarded; fellowships, research assistantships with partial tuition reimbursements Financial award applicants required to submit FAFSA. *Unit head:* Thomas Boggess, Chair, Department of Physics and Astronomy, 319-335-1688, Fax: 319-335-1753.

University of Kansas, Graduate School, College of Liberal Arts and Sciences, Department of Physics and Astronomy, Lawrence, KS 66045. Offers computational physics and astronomy (MS); physics (MS, PhD). *Faculty:* 30. *Students:* 43 full-time (12 women), 9 part-time; includes 2 minority (both Hispanic Americans), 29 international. Average age 29. 62 applicants, 31% accepted. In 2005, 4 master's, 3 doctorates awarded. *Degree requirements:* For master's, thesis (for some programs); for doctorate, thesis/dissertation, comprehensive exam. *Entrance requirements:* Additional exam requirements/recommendations for international students: Required—TOEFL, TWE; Recommended—TSE. *Application deadline:* For fall admission, 3/1 priority date for domestic students, 3/1 priority date for international students; for spring admission, 10/1 priority date for domestic students, 10/1 priority date for international students. Applications are processed on a rolling basis. Application fee: $55 ($60 for international students). Electronic applications accepted. *Expenses:* Tuition, state resident: full-time $4,859. Tuition, nonresident: full-time $1,200. Required fees: $589. Tuition and fees vary according to program. *Financial support:* Fellowships with tuition reimbursements, research assistantships with full and partial tuition reimbursements, teaching assistantships with full and partial tuition reimbursements available. Financial award application deadline: 3/1. *Faculty research:* Condensed-matter, cosmology, elementary particles, nuclear physics, space physics. *Unit head:* Dr. Stephen J. Sanders, Chair, 785-864-4626, Fax: 785-864-5262. *Application contact:* Patricia Marvin, Graduate Admission Specialist, 785-864-4626, Fax: 785-864-5262, E-mail: physics@ku.edu.

See Close-Up on page 371.

University of Kentucky, Graduate School, Graduate School Programs from the College of Arts and Sciences, Program in Physics and Astronomy, Lexington, KY 40506-0032. Offers MS, PhD. *Faculty:* 26 full-time (2 women). *Students:* 53 full-time (16 women), 4 part-time (1 woman); includes 1 minority (Asian American or Pacific Islander), 35 international. 120 applicants, 27% accepted, 25 enrolled. In 2005, 5 master's, 2 doctorates awarded. *Median time to degree:* Of those who began their doctoral program in fall 1997, 54% received their degree in 8 years or less. *Degree requirements:* For master's, thesis optional; for doctorate, thesis/dissertation, comprehensive exam. *Entrance requirements:* For master's, GRE General Test, minimum undergraduate GPA of 2.5; for doctorate, GRE General Test, minimum graduate GPA of 3.0. Additional exam requirements/recommendations for international students: Required—TOEFL (minimum score 550 paper-based; 213 computer-based). *Application deadline:* For fall admission, 7/17 priority date for domestic students, 2/1 priority date for international students; for spring admission, 12/13 priority date for domestic students, 6/15 priority date for international students. Applications are processed on a rolling basis. Application fee: $40 ($55 for international students). Electronic applications accepted. *Expenses:* Tuition, state resident: full-time $3,159; part-time $331 per credit hour. Tuition, nonresident: full-time $6,984; part-time $756 per credit hour. Tuition and fees vary according to course load, degree level and program. *Financial support:* In 2005–06, 3 fellowships with full tuition reimbursements, 21 research assistantships with full tuition reimbursements (averaging $13,674 per year), 32 teaching assistantships with full tuition reimbursements (averaging $13,674 per year) were awarded; Federal Work-Study, institutionally sponsored loans, scholarships/grants, traineeships, health care benefits, tuition waivers (partial), and unspecified assistantships also available. Support available to part-time students. Financial award application deadline:3/15. *Faculty research:* Astrophysics, active galactic nuclei, and radio astronomy; Rydbert atoms, and electron scattering; TOF spectroscopy, hyperon interactions and muons; particle theory, lattice gauge theory, quark, and skyrmion models. Total annual research expenditures: $3.1 million. *Unit head:* Dr. Thomas Troland, Director of Graduate Studies, 859-257-8620, Fax: 859-323-2846, E-mail: troland@asta.pa.uky.edu. *Application contact:* Dr. Brian Jackson, Senior Associate Dean, 859-257-8176, Fax: 859-323-1928, E-mail: lance.brunner@uky.edu.

See Close-Up on page 373.

University of Maryland, College Park, Graduate Studies, College of Computer, Mathematical and Physical Sciences, Department of Astronomy, College Park, MD 20742. Offers MS, PhD. Part-time and evening/weekend programs available. *Faculty:* 69 full-time (12 women), 13 part-time/adjunct (4 women). *Students:* 38 full-time (21 women), 3 part-time; includes 3 minority (1 African American, 2 Asian Americans or Pacific Islanders), 8 international. 94 applicants, 18% accepted, 7 enrolled. In 2005, 8 master's, 5 doctorates awarded. Terminal master's awarded for partial completion of doctoral program. *Median time to degree:* Of those who began their doctoral program in fall 1997, 50% received their degree in 8 years or less. *Degree requirements:* For master's, thesis or alternative, written exam; for doctorate, thesis/dissertation, research project. *Entrance requirements:* For master's, GRE General Test, GRE Subject Test in physics, minimum GPA of 3.0, 3 letters of recommendation; for doctorate, GRE General Test, GRE Subject Test in physics, 3 letters of recommendation. *Application deadline:* For fall admission, 1/15 for domestic students, 1/15 for international students. Applications are processed on a rolling basis. Application fee: $60. Electronic applications accepted. *Financial support:* In 2005–06, 12 fellowships with full tuition reimbursements (averaging $4,116 per year), 30 research assistantships with tuition reimbursements (averaging $20,689 per year), 17 teaching assistantships with tuition reimbursements (averaging $14,798 per year) were awarded; career-related internships or fieldwork, Federal Work-Study, and scholarships/grants also available. Support available to part-time students. Financial award applicants required to submit FAFSA. *Faculty research:* Solar radio astronomy, plasma and high-energy astrophysics, galactic and extragalactic astronomy. Total annual research expenditures:$9 million. *Unit head:* Dr. Lee Mundy, Chair, 301-405-1508, Fax: 301-314-9067. *Application contact:* Dean of Graduate School, 301-405-4190, Fax: 301-314-9305.

University of Massachusetts Amherst, Graduate School, College of Natural Sciences and Mathematics, Department of Astronomy, Amherst, MA 01003. Offers MS, PhD. Part-time programs available. *Faculty:* 14 full-time (0 women). *Students:* 24 full-time (6 women), 19 international. Average age 29. 60 applicants, 0% accepted. In 2005, 4 master's, 2 doctorates awarded. Terminal master's awarded for partial completion of doctoral program. *Degree requirements:* For doctorate, thesis/dissertation. *Entrance requirements:* For master's and doctorate, GRE General Test, GRE Subject Test (physics). Additional exam requirements/recommendations for international students: Required—TOEFL (minimum score 530 paper-based; 197 computer-based). *Application deadline:* For fall admission, 2/1 priority date for domestic students, 2/1 priority date for international students; for spring admission, 10/1 for domestic students, 10/1 for international students. Applications are processed on a rolling basis. Application fee: $40 ($65 for international students). Electronic applications accepted. *Expenses:* Tuition, state resident: part-time $110 per credit. Tuition, nonresident: part-time $414 per credit. Required fees: $2,824 per term. One-time fee: $250 part-time. Full-time tuition and fees vary according to course load, campus/location, program and reciprocity agreements. *Financial support:* In 2005–06, research assistantships with full tuition reimbursements (averaging $16,375 per year), teaching assistantships with full tuition reimbursements (averaging $8,282 per year) were awarded; fellowships with full tuition reimbursements, career-related internships or fieldwork, Federal Work-Study, scholarships/grants, traineeships, and unspecified assistantships also available. Support available to part-time students. Financial award application deadline:2/1. *Unit head:* Dr. Ronald Snell, Director, 413-545-2194, Fax: 413-545-4223, E-mail: snell@astro.umass.edu. *Application contact:* Dr. Donald Candela, Chair, Admissions Committee, 413-545-2407, E-mail: candela@phast.umass.edu.

University of Michigan, Horace H. Rackham School of Graduate Studies, College of Literature, Science, and the Arts, Department of Astronomy, Ann Arbor, MI 48109. Offers MS, PhD. *Faculty:* 13 full-time (4 women). *Students:* 20 full-time (9 women); includes 5 minority (3 Asian Americans or Pacific Islanders, 2 Hispanic Americans). 65 applicants, 23% accepted, 3 enrolled. In 2005, 4 master's, 2 doctorates awarded. Terminal master's awarded for partial completion of doctoral program. *Degree requirements:* For master's, comprehensive exam, registration; for doctorate, thesis/dissertation, oral defense of dissertation, preliminary exam. *Entrance requirements:* For master's and doctorate, GRE General Test, GRE Subject Test. *Application deadline:* For fall admission, 1/5 for domestic students, 1/5 for international students. Application fee: $55. Electronic applications accepted. *Expenses:* Tuition, state resident: full-time $14,082; part-time $894 per credit hour. Tuition, nonresident: full-time $28,500; part-time $1,675 per credit hour. Required fees: $189; $189 per unit. *Financial support:* In 2005–06, 4 fellowships with full tuition reimbursements (averaging $18,656 per year), 8 research assistantships with full tuition reimbursements (averaging $18,656 per year), 5 teaching assistantships with full tuition reimbursements (averaging $18,656 per year) were awarded; institutionally sponsored loans, scholarships/grants, health care benefits, and unspecified assistantships also available. Financial award application deadline: 1/5; financial award applicants required to submit FAFSA. *Faculty research:* Extragalactic and galactic astronomy, cosmology, star and planet formation, high energy astrophysics. Total annual research expenditures: $2.1 million. *Unit head:* Dr. Douglas O. Richstone, Chair, 734-764-3440, Fax: 734-763-6317, E-mail: dor@umich.edu. *Application contact:* Sarah J. Lloyd, Administrative Associate, 734-764-3440, Fax: 734-763-6317, E-mail: sarlloyd@umich.edu.

University of Minnesota, Twin Cities Campus, Graduate School, Institute of Technology, School of Physics and Astronomy, Department of Astronomy, Minneapolis, MN 55455-0213. Offers astrophysics (MS, PhD). Terminal master's awarded for partial completion of doctoral program. *Degree requirements:* For master's, thesis optional; for doctorate, thesis/dissertation. *Entrance requirements:* For master's and doctorate, GRE General Test, GRE Subject Test. *Expenses:* Tuition, state resident: full-time $8,748; part-time $729 per credit. Tuition, nonresident: full-time $15,848; part-time $1,321 per credit. Full-time tuition and fees vary according to class time, course load, program and reciprocity agreements. *Faculty research:* Evolution of stars and galaxies; the interstellar medium; cosmology; observational, optical, infrared, and radio astronomy; computational astrophysics.

University of Missouri–Columbia, Graduate School, College of Arts and Sciences, Department of Physics and Astronomy, Columbia, MO 65211. Offers MS, PhD. *Faculty:* 23 full-time (5 women). *Students:* 33 full-time (6 women), 10 part-time (2 women), 26 international. In 2005, 4 master's, 4 doctorates awarded. Terminal master's awarded for partial completion of doctoral program. *Degree requirements:* For doctorate, one foreign language, thesis/dissertation. *Entrance requirements:* For master's and doctorate, GRE General Test, minimum GPA of 3.0. *Application deadline:* For fall admission, 4/15 for domestic students. Applications are processed on a rolling basis. Application fee: $45 ($60 for international students). *Financial support:* Research assistantships, teaching assistantships, institutionally sponsored loans available. *Unit head:* Dr. H. R. Chandrasekhar, Director of Graduate Studies, 573-882-6086, E-mail: chandra@missouri.edu.

University of Nebraska–Lincoln, Graduate College, College of Arts and Sciences, Department of Physics and Astronomy, Lincoln, NE 68588. Offers astronomy (MS, PhD); physics (MS, PhD). *Degree requirements:* For master's, thesis optional; for doctorate, thesis/dissertation, comprehensive exam. *Entrance requirements:* For master's and doctorate, GRE General Test. Additional exam requirements/recommendations for international students: Required—TOEFL (minimum score 550 paper-based; 213 computer-based). Electronic applications accepted. *Faculty research:* Electromagnetics of solids and thin films, photoionization, ion collisions with atoms, molecules and surfaces, nanostructures.

The University of North Carolina at Chapel Hill, Graduate School, College of Arts and Sciences, Department of Physics and Astronomy, Chapel Hill, NC 27599. Offers physics (MS, PhD). Terminal master's awarded for partial completion of doctoral program. *Degree requirements:* For master's, comprehensive exam, registration; for doctorate, thesis/dissertation, comprehensive exam, registration. *Entrance requirements:* For master's and doctorate, GRE General Test, minimum GPA of 3.0. Electronic applications accepted. *Faculty research:* Observational astronomy, fullerenes, polarized beams, nanotubes, nucleosynthesis in stars and supernovae, superstring theory, ballistic transport in semiconductors, gravitation.

University of Rochester, The College, Arts and Sciences, Department of Physics and Astronomy, Rochester, NY 14627-0250. Offers physics (MA, MS, PhD); physics and astronomy (PhD). Part-time programs available. Terminal master's awarded for partial completion of doctoral program. *Degree requirements:* For master's, thesis (for some programs), comprehensive exam; for doctorate, thesis/dissertation, qualifying exam, comprehensive exam. *Entrance requirements:* For master's and doctorate, GRE General Test. Additional exam requirements/recommendations for international students: Required—TOEFL.

See Close-Up on page 391.

University of South Carolina, The Graduate School, College of Arts and Sciences, Department of Physics and Astronomy, Columbia, SC 29208. Offers IMA, MAT, MS, PSM, PhD. IMA and MAT offered in cooperation with the College of Education. Part-time programs available. Terminal master's awarded for partial completion of doctoral program. *Degree requirements:* For master's, thesis, comprehensive exam, registration; for doctorate, one foreign language, thesis/dissertation, comprehensive exam, registration. *Entrance requirements:* For master's and doctorate, GRE General Test, GRE Subject Test. Additional exam requirements/recommendations for international students: Required—TOEFL (minimum score 570 paper-based; 230 computer-based). Electronic applications accepted. *Faculty research:* Condensed matter, intermediate-energy nuclear physics, foundations of quantum mechanics, astronomy/astrophysics.

The University of Texas at Austin, Graduate School, College of Natural Sciences, Department of Astronomy, Austin, TX 78712-1111. Offers MA, PhD. *Entrance requirements:* For master's and doctorate, GRE General Test, GRE Subject Test (physics). Additional exam requirements/recommendations for international students: Required—TOEFL. Electronic applications accepted. *Faculty research:* Stars, interstellar medium, galaxies, planetary astronomy, cosmology.

University of Toronto, School of Graduate Studies, Physical Sciences Division, Department of Astronomy, Toronto, ON M5S 1A1, Canada. Offers M Sc, PhD. Part-time programs available. *Degree requirements:* For doctorate, thesis/dissertation, qualifying exam, thesis defense. *Entrance requirements:* For master's, minimum B average, bachelor's degree in astronomy or equivalent, 3 letters of reference; for doctorate, GRE General Test and Subject Test in physics (strongly recommended), minimum B+ average, master's degree in astronomy or equivalent, or demonstrated research competance, 3 letters of reference.

University of Victoria, Faculty of Graduate Studies, Faculty of Science, Department of Physics and Astronomy, Victoria, BC V8W 2Y2, Canada. Offers astronomy and astrophysics (M Sc, PhD); condensed matter physics (M Sc, PhD); experimental particle physics (M Sc, PhD); medical physics (M Sc, PhD); ocean physics (PhD); ocean physics and geophysics (M Sc); theoretical physics (M Sc, PhD). *Faculty:* 16 full-time (0 women), 13 part-time/adjunct (1 woman). *Students:* 45 full-time, 8 international. Average age 25. 66 applicants, 45% accepted, 10 enrolled. In 2005, 4 master's, 1 doctorate awarded. *Median time to degree:* Of those who began their doctoral program in fall 1997, 100% received their degree in 8 years or less. *Degree requirements:* For master's, thesis, registration; for doctorate, thesis/dissertation, Candidacy exam, comprehensive exam, registration. *Entrance requirements:* For master's and doctorate, GRE. Additional exam requirements/recommendations for international students: Required—TOEFL (minimum score 575 paper-based; 233 computer-based), IELT (minimum score 7). *Application deadline:* For fall admission, 5/31 priority date for domestic students, 12/15 priority date for international students. Applications are processed on a rolling basis. Application fee: $75 ($125 for international students). Electronic applications accepted. Tuition and fees charges are reported in Canadian dollars. *Expenses:* Tuition, area resident: Full-time $4,492 Canadian dollars; part-time $749 Canadian dollars per term. International tuition: $5,346 Canadian dollars full-time. Required fees: $4,492 Canadian dollars; $749 Canadian dollars per term. Tuition and fees vary according to course load, campus/location and program. *Financial support:* In 2005–06, 3 students received support; fellowships, research assistantships, teaching assistantships, career-related internships or fieldwork, institutionally sponsored loans, and awards available. Financial award application deadline: 2/15. *Faculty research:* Old stellar populations; observational cosmology and large scale structure; cp violation; atlas. *Unit head:* Dr. J. Michael Ronev, Chair, 250-721-7698, Fax: 250-721-7715, E-mail: chair@phys.uvic.ca. *Application contact:* Dr. Chris J. Pritchet, Graduate Adviser, 250-721-7744, Fax: 250-721-7715, E-mail: pritchet@uvic.ca.

University of Virginia, College and Graduate School of Arts and Sciences, Department of Astronomy, Charlottesville, VA 22903. Offers MS, PhD. *Faculty:* 14 full-time (2 women). *Students:* 30 full-time (9 women); includes 1 minority (Asian American or Pacific Islander), 10 international. Average age 27. 56 applicants, 20% accepted, 5 enrolled. In 2005, 6 degrees awarded. *Degree requirements:* For master's, one foreign language, thesis; for doctorate, variable foreign language requirement, thesis/dissertation. *Entrance requirements:* For master's and doctorate, GRE General Test, GRE Subject Test. Additional exam requirements/recommendations for international students: Recommended—TOEFL (minimum score 630 paper-based). *Application deadline:* Applications are processed on a rolling basis. Application fee: $40. Electronic applications accepted. *Expenses:* Tuition, state resident: full-time $7,731. Tuition, nonresident: full-time $18,672. Required fees: $1,479. Full-time tuition and fees vary according to degree level and program. *Financial support:* Applicants required to submit FAFSA. *Unit head:* Robert T. Rood, Chairman, 434-924-7494, Fax: 434-924-3104.

Astronomy

University of Virginia (continued)

Application contact: Peter C. Brunjes, Associate Dean for Graduate Programs and Research, 434-924-7184, Fax: 434-924-6737, E-mail: grad-a-s@virginia.edu.

University of Washington, Graduate School, College of Arts and Sciences, Department of Astronomy, Seattle, WA 98195. Offers MS, PhD. Terminal master's awarded for partial completion of doctoral program. *Degree requirements:* For doctorate, thesis/dissertation. *Entrance requirements:* For master's and doctorate, GRE General Test, GRE Subject Test, minimum GPA of 3.0. Additional exam requirements/recommendations for international students: Required—TOEFL. *Faculty research:* Solar system dust, space astronomy, high-energy astrophysics, galactic and extragalactic astronomy, stellar astrophysics.

The University of Western Ontario, Faculty of Graduate Studies, Physical Sciences Division, Department of Physics and Astronomy, Program in Astronomy, London, ON N6A 5B8, Canada. Offers M Sc, PhD. Terminal master's awarded for partial completion of doctoral program. *Degree requirements:* For master's, thesis optional; for doctorate, thesis/dissertation, comprehensive exam. *Entrance requirements:* For master's, GRE Physics Test, honors B Sc degree, minimum B average (Canadian), A—(international); for doctorate, M Sc degree, minimum B average (Canadian), A—(international). Additional exam requirements/recommendations for international students: Required—TOEFL (minimum score 580 paper-based; 237 computer-based). *Faculty research:* Observational and theoretical astrophysics spectroscopy, photometry, spectro-polarimetry, variable stars, cosmology.

University of Wisconsin–Madison, Graduate School, College of Letters and Science, Department of Astronomy, Madison, WI 53706-1380. Offers PhD. *Degree requirements:* For doctorate, thesis/dissertation, comprehensive exam. *Entrance requirements:* For doctorate, GRE General Test, GRE Subject Test (physics), bachelor's degree in related field. Additional exam requirements/recommendations for international students: Required—TOEFL. Electronic applications accepted. *Faculty research:* Kinematics, evolution of galaxies, cosmic distance, scale and large-scale structures, interstellar intergalactic medium, star formation and evolution, solar system chemistry and dynamics.

Vanderbilt University, Graduate School, Department of Physics and Astronomy, Nashville, TN 37240-1001. Offers astronomy (MS); physics (MA, MAT, MS, PhD). *Faculty:* 65 full-time (3 women). *Students:* 61 full-time (14 women), 2 part-time; includes 3 minority (2 African Americans, 1 Hispanic American), 31 international. 153 applicants, 20% accepted, 22 enrolled. In 2005, 3 master's, 5 doctorates awarded. *Degree requirements:* For master's, thesis; for doctorate, thesis/dissertation, final and qualifying exams. *Entrance requirements:* For master's, GRE General Test; for doctorate, GRE General Test, GRE Subject Test. *Application deadline:* For fall admission, 1/15 for domestic students, 1/15 for international students. Application fee: $0. Electronic applications accepted. *Expenses:* Tuition: Full-time $15,396; part-time $1,283 per semester hour. Required fees: $2,202; $1,101 per semester. One-time fee: $30. Tuition and fees vary according to course load, program and student level. *Financial support:* Fellowships with full and partial tuition reimbursements, research assistantships with full tuition reimbursements, teaching assistantships with full tuition reimbursements, career-related internships or fieldwork, Federal Work-Study, and institutionally sponsored loans available. Financial award application deadline: 1/15. *Faculty research:* Experimental and theoretical physics, free electron laser, living-state physics, heavy-ion physics, nuclear structure. *Unit head:* Robert J. Scherrer, Chair, 615-322-2828, Fax: 615-343-7263. *Application contact:* Charles F. Maguire, Director of Graduate Studies, 615-322-2828, Fax: 615-343-7263, E-mail: charles.f.maguire@vanderbilt.edu.

Wesleyan University, Graduate Programs, Department of Astronomy, Middletown, CT 06459-0260. Offers MA. *Faculty:* 2 full-time (0 women). *Students:* 6 full-time (5 women); includes 2 minority (both African Americans) Average age 22. In 2005, 4 degrees awarded. *Degree requirements:* For master's, thesis. *Entrance requirements:* For master's, GRE General Test, GRE Subject Test. *Application deadline:* For fall admission, 1/5 for domestic students. Applications are processed on a rolling basis. Application fee: $0. *Expenses:* Tuition: Full-time $24,732. One-time fee: $20 full-time. *Financial support:* In 2005–06, 3 teaching assistantships were awarded Financial award application deadline: 4/15; financial award applicants required to submit FAFSA. *Faculty research:* Observational-theoretical astronomy and astrophysics. *Unit head:* William Herbst, Chairman, 860-685-3672, E-mail: wherbst@wesleyan.edu. *Application contact:* Linda Shettleworth, Information Contact, 860-685-2130, E-mail: shettleworth@wesleyan.edu.

West Chester University of Pennsylvania, Graduate Studies, College of Arts and Sciences, Department of Geology and Astronomy, West Chester, PA 19383. Offers physical science (MA). Part-time and evening/weekend programs available. *Students:* 3 full-time (0 women), 10 part-time (3 women); includes 1 African American. Average age 30. 7 applicants, 86% accepted, 3 enrolled. In 2005, 6 degrees awarded. *Degree requirements:* For master's, thesis optional. *Entrance requirements:* For master's, GRE General Test, interview. *Application deadline:* For fall admission, 4/15 for domestic students; for spring admission, 10/15 for domestic students. Applications are processed on a rolling basis. Application fee: $35. *Expenses:* Tuition, state resident: full-time $2,944; part-time $327 per credit. Tuition, nonresident: full-time $4,711; part-time $523 per credit. Required fees: $54 per semester. *Financial support:* In 2005–06, 1 research assistantship with full tuition reimbursement (averaging $5,000 per year) was awarded; unspecified assistantships also available. Support available to part-time students. Financial award application deadline: 2/15; financial award applicants required to submit FAFSA. *Faculty research:* Developing and using a meteorological data station. *Unit head:* Dr. Gil Wiswall, Chair, 610-436-2727, E-mail: gwiswall@wcupa.edu. *Application contact:* Dr. Steven Good, Information Contact, 610-436-2203, E-mail: sgood@wcupa.edu.

Yale University, Graduate School of Arts and Sciences, Department of Astronomy, New Haven, CT 06520. Offers MS, PhD. *Degree requirements:* For doctorate, thesis/dissertation. *Entrance requirements:* For doctorate, GRE General Test, GRE Subject Test (physics).

York University, Faculty of Graduate Studies, Faculty of Pure and Applied Science, Program in Physics and Astronomy, Toronto, ON M3J 1P3, Canada. Offers M Sc, PhD. Part-time and evening/weekend programs available. *Faculty:* 51 full-time (5 women), 2 part-time/adjunct (0 women). *Students:* 42 full-time (8 women), 4 part-time (1 woman). 54 applicants, 20% accepted, 10 enrolled. In 2005, 4 master's, 2 doctorates awarded. *Degree requirements:* For master's, thesis or alternative, registration; for doctorate, thesis/dissertation, comprehensive exam, registration. *Application deadline:* Applications are processed on a rolling basis. Application fee: $80. Electronic applications accepted. *Expenses:* Tuition, state resident: full-time $3,190; part-time $798 per term. International tuition: $7,515 full-time. Required fees: $217. Tuition and fees vary according to program. *Financial support:* In 2005–06, fellowships (averaging $11,388 per year), research assistantships (averaging $11,216 per year), teaching assistantships (averaging $8,084 per year) were awarded; fee bursaries also available. *Unit head:* Helen Freedhoff, Director, 416-736-5249.

Astrophysics

Air Force Institute of Technology, Graduate School of Engineering and Management, Department of Engineering Physics, Dayton, OH 45433-7765. Offers applied physics (MS, PhD); electro-optics (MS, PhD); materials science (PhD); nuclear engineering (MS, PhD); space physics (MS). Part-time programs available. *Faculty:* 20 full-time (1 woman). *Students:* 86 full-time (9 women), 4 part-time (all women). Average age 31. In 2005, 36 master's, 4 doctorates awarded. *Degree requirements:* For master's and doctorate, thesis/dissertation. *Entrance requirements:* For master's and doctorate, GRE General Test, minimum GPA of 3.0, U.S. citizenship. *Application deadline:* ; for spring admission, 3/1 for domestic students. Applications are processed on a rolling basis. Application fee: $0. *Financial support:* Fellowships with full tuition reimbursements, research assistantships with full and partial tuition reimbursements, scholarships/grants and unspecified assistantships available. Support available to part-time students. Financial award application deadline:3/15. *Faculty research:* High-energy lasers, space physics, nuclear weapon effects, semiconductor physics. Total annual research expenditures: $2.1 million. *Unit head:* Dr. Robert L. Hengehold, Head, 937-255-3636 Ext. 4502, Fax: 937-255-2921, E-mail: robert.hengehold@afit.edu. *Application contact:* Dr. David E. Weeks, Associate Professor of Physics, 937-255-3636 Ext. 4561, Fax: 937-255-2921, E-mail: david.weeks@afit.edu.

Clemson University, Graduate School, College of Engineering and Science, Department of Physics and Astronomy, Program in Physics, Clemson, SC 29634. Offers astronomy and astrophysics (MS, PhD); atmospheric physics (MS, PhD); biophysics (MS, PhD). Part-time programs available. *Students:* 53 full-time (15 women), 2 part-time; includes 2 minority (1 Asian American or Pacific Islander, 1 Hispanic American), 19 international. 46 applicants, 41% accepted, 10 enrolled. In 2005, 5 master's, 4 doctorates awarded. Terminal master's awarded for partial completion of doctoral program. *Degree requirements:* For master's, thesis or alternative; for doctorate, thesis/dissertation. *Entrance requirements:* For master's and doctorate, GRE General Test. Additional exam requirements/recommendations for international students: Required—TOEFL. *Application deadline:* For fall admission, 2/15 for domestic students. Applications are processed on a rolling basis. Application fee: $50. *Financial support:* Fellowships, research assistantships, teaching assistantships available. Financial award application deadline: 6/1; financial award applicants required to submit FAFSA. *Faculty research:* Radiation physics, solid-state physics, nuclear physics, radar and lidar studies of atmosphere. *Unit head:* Dr. Brad Myer, Head, 864-656-5320. *Application contact:* Dr. Miguel Larsen, Coordinator, 864-656-5309, Fax: 864-656-0805, E-mail: mlarsen@clemson.edu.

See Close-Ups on pages 331 and 333.

Cornell University, Graduate School, Graduate Fields of Arts and Sciences, Field of Astronomy and Space Sciences, Ithaca, NY 14853-0001. Offers astronomy (PhD); astrophysics (PhD); general space sciences (PhD); infrared astronomy (PhD); planetary studies (PhD); radio astronomy (PhD); radiophysics (PhD); theoretical astrophysics (PhD). *Faculty:* 38 full-time (3 women). *Students:* 31 full-time (10 women); includes 1 minority (Asian American or Pacific Islander), 11 international. 110 applicants, 21% accepted, 7 enrolled. In 2005, 6 doctorates awarded. *Degree requirements:* For doctorate, thesis/dissertation, comprehensive exam. *Entrance requirements:* For doctorate, GRE General Test, GRE Subject Test (physics), 3 letters of recommendation. Additional exam requirements/recommendations for international students: Required—TOEFL (minimum score 600 paper-based; 250 computer-based). *Application deadline:* For fall admission, 1/15 for domestic students. Application fee: $60. Electronic applications accepted. *Financial support:* In 2005–06, 31 students received support, including 3 fellowships with full tuition reimbursements available, 19 research assistantships with full tuition reimbursements available, 9 teaching assistantships with full tuition reimbursements available; institutionally sponsored loans, scholarships/grants, health care benefits, tuition waivers (full and partial), and unspecified assistantships also available. Financial award applicants required to submit FAFSA. *Faculty research:* Observational astrophysics, planetary sciences, cosmology, instrumentation, gravitational astrophysics. *Unit head:* Director of Graduate Studies, 607-255-4341. *Application contact:* Graduate Field Assistant, 607-255-4341, E-mail: oconnor@astro.cornell.edu.

Harvard University, Graduate School of Arts and Sciences, Department of Astronomy, Cambridge, MA 02138. Offers astronomy (PhD); astrophysics (PhD). *Faculty:* 17 full-time (1 woman). *Students:* 41 full-time (10 women); includes 8 minority (all Asian Americans or Pacific Islanders), 8 international. Average age 25. 85 applicants, 18% accepted, 10 enrolled. In 2005, 6 doctorates awarded. *Degree requirements:* For doctorate, thesis/dissertation, paper, research project, teaching 2 semesters. *Entrance requirements:* For doctorate, GRE General Test, GRE Subject Test (physics). Additional exam requirements/recommendations for international students: Required—TOEFL. *Application deadline:* For fall admission, 12/15 for domestic students. Application fee: $80. Electronic applications accepted. *Expenses:* Tuition: Full-time $28,752. Full-time tuition and fees vary according to program and student level. *Financial support:* In 2005–06, 20 research assistantships (averaging $21,840 per year), 8 teaching assistantships (averaging $4,475 per year); fellowships, career-related internships or fieldwork, Federal Work-Study, institutionally sponsored loans, scholarships/grants, and health care benefits also available. Financial award application deadline: 12/30. *Faculty research:* Atomic and molecular physics, electromagnetism, solar physics, nuclear physics, fluid dynamics. *Unit head:* Peg Herlihy, Administrator, 617-495-3752. *Application contact:* 617-495-5315, E-mail: admiss@fas.harvard.edu.

ICR Graduate School, Graduate Programs, Santee, CA 92071. Offers astro/geophysics (MS); biology (MS); geology (MS); science education (MS). Part-time programs available. *Faculty:* 5 full-time (1 woman), 1 part-time/adjunct (0 women). *Students:* 6 full-time (2 women), 6 part-time (3 women). Average age 45. In 2005, 4 degrees awarded. *Degree requirements:* For master's, thesis (for some programs), comprehensive exam (for some programs). *Entrance requirements:* For master's, minimum undergraduate GPA of 3.0, bachelor's degree in science or science education. *Application deadline:* Applications are processed on a rolling basis. Application fee: $30. *Expenses:* Tuition: Part-time $150 per semester hour. *Faculty research:* Age of the earth, limits of variation, catastrophe, optimum methods for teaching. Total annual research expenditures: $200,000. *Unit head:* Dr. Kenneth B. Cumming, Dean, 619-448-0900, Fax: 619-448-3469. *Application contact:* Dr. Jack Kriege, Registrar, 619-448-0900 Ext. 6016, Fax: 619-448-3469, E-mail: jkriege@icr.org.

Indiana University Bloomington, Graduate School, College of Arts and Sciences, Department of Astronomy, Program in Astrophysics, Bloomington, IN 47405-7000. Offers PhD. *Students:* 3 full-time (1 woman), 1 part-time, 1 international. Average age 25. *Degree requirements:* For doctorate, thesis/dissertation, written qualifying exam. *Entrance requirements:* For doctorate, GRE General Test, GRE Subject Test (physics), BA or BS in science. Additional exam requirements/recommendations for international students: Required—TOEFL. *Application deadline:* For fall admission, 1/15 priority date for domestic students, 12/15 priority date for international students; for spring admission, 9/1 priority date for domestic students, 9/1 priority date for international students. Applications are processed on a rolling basis. Application fee: $50 ($60 for international students). Electronic applications accepted. *Expenses:* Tuition, state resident: full-time $5,437; part-time $227 per credit hour. Tuition, nonresident: full-time $15,836; part-time $660 per credit hour. Required fees: $821. Tuition and fees vary according to campus/location and program. *Financial support:* Research assistantships, teaching assistantships available. Financial award application deadline: 5/2. *Faculty research:* Nuclear astrophysics, cosmic-ray physics, astrophysical fluid dynamics, active galactic nuclei, high-energy astrophysics. *Application contact:* Brenda Records, Secretary, 812-855-6912, Fax: 812-855-8725, E-mail: brecords@indiana.edu.

Iowa State University of Science and Technology, Graduate College, College of Liberal Arts and Sciences, Department of Physics and Astronomy, Ames, IA 50011. Offers applied physics (MS, PhD); astrophysics (MS, PhD); condensed matter physics (MS, PhD); high energy physics (MS, PhD); nuclear physics (MS, PhD); physics (MS, PhD). Part-time programs available. *Faculty:* 44 full-time, 3 part-time/adjunct. *Students:* 79 full-time (14 women), 6 part-time; includes 1 minority (Asian American or Pacific Islander), 56 international. 161 applicants, 34% accepted, 23 enrolled. In 2005, 5 master's, 7 doctorates awarded. Terminal master's awarded for partial completion of doctoral program. *Degree requirements:* For master's, thesis (for some programs); for doctorate, thesis/dissertation. *Entrance requirements:* For master's and doctorate, GRE General Test, GRE Subject Test (physics). Additional exam requirements/recommendations for international students: Required—TOEFL (paper score 550; computer score 213) or IELTS (score 6.5). *Application deadline:* For fall admission, 2/15 priority date for domestic students, 2/15 priority date for international students; for spring admission, 10/15 for domestic students, 10/15 for international students. Applications are processed on a rolling basis. Application fee: $30 ($70 for international students). Electronic applications accepted. *Expenses:* Tuition, state resident: full-time $6,410. Tuition, nonresident: full-time $16,422. Tuition and fees vary according to program. *Financial support:* In 2005–06, 48 research assistantships with full tuition reimbursements (averaging $14,626 per year), 30 teaching assistantships with full tuition reimbursements (averaging $14,533 per year) were awarded; fellowships, Federal Work-Study, institutionally sponsored loans, scholarships/grants, health care benefits, and unspecified assistantships also available. Support available to part-time students. Financial award application deadline: 2/15. *Faculty research:* Condensed-matter physics, including superconductivity and new materials; high-energy and nuclear physics; astronomy and astrophysics; atmospheric and environmental physics. Total annual research expenditures: $8.8 million. *Unit head:* Dr. Eli Rosenberg, Chair, 515-294-5441, Fax: 515-294-6027, E-mail: phys_astro@iastate.edu. *Application contact:* Dr. Steven Kawaler, Director of Graduate Education, 515-294-9728, E-mail: phys_astro@iastate.edu.

Louisiana State University and Agricultural and Mechanical College, Graduate School, College of Basic Sciences, Department of Physics and Astronomy, Baton Rouge, LA 70803. Offers astronomy (PhD); astrophysics (PhD); physics (MS, PhD). *Faculty:* 45 full-time (1 woman). *Students:* 72 full-time (8 women), 4 part-time (1 woman); includes 2 African Americans, 2 Asian Americans or Pacific Islanders, 1 Hispanic American, 33 international. Average age 28. 99 applicants, 23% accepted, 76 enrolled. In 2005, 6 master's, 4 doctorates awarded. Terminal master's awarded for partial completion of doctoral program. *Degree requirements:* For master's, thesis or alternative; for doctorate, thesis/dissertation. *Entrance requirements:* For master's and doctorate, GRE General Test, minimum GPA of 3.0. Additional exam requirements/recommendations for international students: Required—TOEFL (minimum score 550 paper-based; 213 computer-based). *Application deadline:* For fall admission, 1/25 priority date for domestic students, 5/15 priority date for international students. Applications are processed on a rolling basis. Application fee: $25. Electronic applications accepted. *Financial support:* In 2005–06, 72 students received support, including 3 fellowships with full tuition reimbursements available (averaging $15,000 per year), 40 research assistantships with full and partial tuition reimbursements available (averaging $16,240 per year), 27 teaching assistantships with full and partial tuition reimbursements available (averaging $17,652 per year); institutionally sponsored loans, tuition waivers (full and partial), and unspecified assistantships also available. Financial award application deadline: 3/15; financial award applicants required to submit FAFSA. *Faculty research:* Experimental and theoretical atomic, nuclear, particle, cosmic-ray, low-temperature, and condensed-matter physics. Total annual research expenditures: $5.8 million. *Unit head:* Dr. Roger McNeil, Chair, 225-578-2261, Fax: 225-578-5855, E-mail: mcneil@phys.lsu.edu. *Application contact:* Dr. James Matthews, Graduate Adviser, 225-578-8598, Fax: 225-578-5855, E-mail: jmatth5@lsu.edu.

McMaster University, School of Graduate Studies, Faculty of Science, Department of Physics and Astronomy, Hamilton, ON L8S 4M2, Canada. Offers astrophysics (PhD); physics (PhD). Part-time programs available. *Degree requirements:* For doctorate, thesis/dissertation, comprehensive exam. *Entrance requirements:* For doctorate, minimum B+ average. Additional exam requirements/recommendations for international students: Required—TOEFL (minimum score 550 paper-based; 213 computer-based). *Faculty research:* Condensed matter, astrophysics, nuclear, medical, nonlinear dynamics.

Michigan State University, The Graduate School, College of Natural Science, Department of Physics and Astronomy, East Lansing, MI 48824. Offers astrophysics and astronomy (MS, PhD); physics (MS, PhD). *Faculty:* 52 full-time (3 women). *Students:* 135 full-time (23 women), 6 part-time (3 women); includes 6 minority (2 American Indian/Alaska Native, 3 Asian Americans or Pacific Islanders, 1 Hispanic American), 70 international. Average age 28. 252 applicants, 8% accepted. In 2005, 18 master's, 19 doctorates awarded. *Degree requirements:* For master's, qualifying exam, thesis optional; for doctorate, thesis/dissertation, qualifying exam, comprehensive exam. *Entrance requirements:* For master's, GRE General Test (recommended), minimum GPA of 3.0 in science/math courses, coursework equivalent to a major in physics or astronomy, 3 letters of recommendation; for doctorate, GRE General Test (recommended), minimum GPA of 3.0 in science/math courses, coursework equivalent to a major in physics or astronomy, 3 letters of recommendation, research experience (recommended). Additional exam requirements/recommendations for international students: Required—TOEFL (minimum score 550 paper-based; 213 computer-based), Michigan State University ELT (85), Michigan ELAB (83). *Application deadline:* For fall admission, 12/27 for domestic students; for spring admission, 9/30 for domestic students. Application fee: $50. Electronic applications accepted. *Expenses:* Tuition, state resident: part-time $330 per credit hour. Tuition, nonresident: part-time $685 per credit hour. Tuition and fees vary according to program. *Financial support:* In 2005–06, 20 fellowships with tuition reimbursements (averaging $8,526 per year), 88 research assistantships with tuition reimbursements (averaging $14,281 per year), 40 teaching assistantships with tuition reimbursements (averaging $12,680 per year) were awarded; scholarships/grants and unspecified assistantships also available. *Faculty research:* Nuclear and accelerator physics, high energy physics, condensed matter physics, biophysics, astrophysics and astronomy. Total annual research expenditures: $6.4 million. *Unit head:* Dr. Wolfgang W. Bauer, Chairperson, 517-355-9200 Ext. 2015, Fax: 517-355-4500, E-mail: bauer@pa.msu.edu. *Application contact:* Debbie Simmons, Graduate Secretary, 517-355-9200 Ext. 2032, Fax: 517-353-4500, E-mail: grd_chair@pa.msu.edu.

New Mexico Institute of Mining and Technology, Graduate Studies, Department of Physics, Socorro, NM 87801. Offers astrophysics (MS, PhD); atmospheric physics (MS, PhD); instrumentation (MS); mathematical physics (PhD). *Degree requirements:* For master's, thesis optional; for doctorate, thesis/dissertation. *Entrance requirements:* For master's, GRE General Test; for doctorate, GRE General Test, GRE Subject Test. Additional exam requirements/recommendations for international students: Required—TOEFL (minimum score 540 paper-based; 207 computer-based). *Faculty research:* Cloud physics, stellar and extra-galactic processes.

Northwestern University, The Graduate School, Judd A. and Marjorie Weinberg College of Arts and Sciences, Department of Physics and Astronomy, Evanston, IL 60208. Offers astrophysics (PhD); physics (MS, PhD). Admissions and degrees offered through The Graduate School. *Faculty:* 30 full-time (4 women), 2 part-time/adjunct (0 women). *Students:* 113 (26 women). Average age 24. 140 applicants, 21% accepted, 12 enrolled. In 2005, 4 master's, 10 doctorates awarded. *Median time to degree:* Of those who began their doctoral program in fall 1997, 50% received their degree in 8 years or less. *Degree requirements:* For doctorate, thesis/dissertation, qualifying exam. *Entrance requirements:* For doctorate, GRE General Test, GRE Subject Test. Additional exam requirements/recommendations for international students: Required—TOEFL. Application fee: $60 ($75 for international students). *Financial support:* In 2005–06, 57 students received support, including 9 fellowships with full tuition reimbursements available (averaging $12,906 per year), 28 research assistantships with partial tuition reimbursements available (averaging $18,732 per year), 20 teaching assistantships with full tuition reimbursements available (averaging $13,329 per year); career-related internships or fieldwork, Federal Work-Study, and institutionally sponsored loans also available. Financial award application deadline: 1/15; financial award applicants required to submit FAFSA. *Faculty research:* Nuclear and particle physics, condensed-matter physics, nonlinear physics, astrophysics. Total annual research expenditures: $6.3 million. *Unit head:* Melvin Ulmer, Chair, 847-491-5633, Fax: 847-491-9982, E-mail: physics-astronomy@northwestern.edu. *Application contact:* Mayda Velasco, Admission Officer, 847-467-7099, Fax: 847-491-9982, E-mail: physics-astronomy@northwestern.edu.

The Pennsylvania State University University Park Campus, Graduate School, Eberly College of Science, Department of Astronomy and Astrophysics, State College, University Park, PA 16802-1503. Offers MS, PhD. *Students:* 25 full-time (5 women), 2 part-time (1 woman); includes 2 minority (both Hispanic Americans), 11 international. *Entrance requirements:* For master's and doctorate, GRE General Test. Application fee: $45. *Expenses:* Tuition, state resident: full-time $12,518; part-time $522 per credit. Tuition, nonresident: full-time $23,004; part-time $959 per credit. Required fees: $484. Tuition and fees vary according to course load, campus/location and program. *Unit head:* Dr. Lawrence W. Ramsey, Head, 814-865-0418, Fax: 814-863-3399, E-mail: lwr@astro.psu.edu.

Princeton University, Graduate School, Department of Astrophysical Sciences, Princeton, NJ 08544-1019. Offers astrophysical sciences (PhD); plasma physics (PhD). *Degree requirements:* For doctorate, thesis/dissertation. *Entrance requirements:* For doctorate, GRE General Test, GRE Subject Test (physics). Additional exam requirements/recommendations for international students: Required—TOEFL (minimum score 600 paper-based; 250 computer-based). Electronic applications accepted. *Faculty research:* Theoretical astrophysics, cosmology, galaxy formation, galactic dynamics, interstellar and intergalactic matter.

Rensselaer Polytechnic Institute, Graduate School, School of Science, Department of Physics, Applied Physics and Astronomy, Troy, NY 12180-3590. Offers physics (MS, PhD). *Faculty:* 24 full-time (3 women), 3 part-time/adjunct (0 women). *Students:* 59 full-time (11 women); includes 36 minority (all Asian Americans or Pacific Islanders) Average age 28. 90 applicants, 28% accepted. In 2005, 11 master's, 4 doctorates awarded. *Degree requirements:* For doctorate, thesis/dissertation. *Entrance requirements:* For master's and doctorate, GRE General Test, GRE Subject Test. Additional exam requirements/recommendations for international students: Required—TOEFL (minimum score 600 paper-based; 250 computer-based). *Application deadline:* For fall admission, 1/15 priority date for domestic students, 1/15 priority date for international students; for spring admission, 8/15 priority date for domestic students, 8/15 priority date for international students. Applications are processed on a rolling basis. Application fee: $75. Electronic applications accepted. *Expenses:* Tuition: Full-time $31,000; part-time $1,320 per credit. Required fees: $1,623. *Financial support:* In 2005–06, 10 fellowships with tuition reimbursements (averaging $25,000 per year), 27 research assistantships with tuition reimbursements (averaging $18,700 per year), 16 teaching assistantships with tuition reimbursements (averaging $19,000 per year) were awarded; career-related internships or fieldwork and institutionally sponsored loans also available. Financial award application deadline: 2/1. *Faculty research:* Astrophysics, condensed matter, nuclear physics, optics, physics education. Total annual research expenditures: $4.5 million. *Unit head:* Dr. G. C. Wang, Chair, 518-276-8387, Fax: 518-276-6680, E-mail: wangg@rpi.edu. *Application contact:* Dr. Toh-Ming Lu, Chair, Graduate Recruitment Committee, 518-276-8391, Fax: 518-276-6680, E-mail: mcquade@rpi.edu.

See Close-Up on page 363.

San Francisco State University, Division of Graduate Studies, College of Science and Engineering, Department of Physics and Astronomy, San Francisco, CA 94132-1722. Offers physics (MS). Part-time programs available. *Degree requirements:* For master's, thesis, registration. *Entrance requirements:* For master's, minimum GPA of 2.5 in last 60 units. Additional exam requirements/recommendations for international students: Required—TOEFL (minimum score 550 paper-based; 213 computer-based). Electronic applications accepted. *Faculty research:* Quark search, thin-films, dark matter detection, search for planetary systems, low temperature.

Texas Christian University, College of Science and Engineering, Department of Physics and Astronomy, Fort Worth, TX 76129-0002. Offers physics (MA, MS, PhD), including astrophysics (PhD), business (PhD), physics (PhD). Part-time and evening/weekend programs available. *Degree requirements:* For doctorate, thesis/dissertation, qualifying exams. *Entrance requirements:* For doctorate, GRE General Test. Additional exam requirements/recommendations for international students: Required—TOEFL. *Application deadline:* For fall admission, 3/1 for domestic students; for spring admission, 12/1 for domestic students. Applications are processed on a rolling basis. Application fee: $0. *Expenses:* Tuition: Part-time $740 per credit hour. *Financial support:* Fellowships, teaching assistantships available. Financial award application deadline: 3/1. *Unit head:* Dr. T W Zerda, Chairperson, 817-257-7375. *Application contact:* Dr. Bonnie Melhart, Associate Dean, College of Science and Engineering, E-mail: b.melhart@tcu.edu.

University of Alaska Fairbanks, College of Natural Sciences and Mathematics, Department of Physics, Fairbanks, AK 99775-7520. Offers atmospheric science (MS, PhD); computational physics (MS); general physics (MS); physics (MAT, PhD); space physics (PhD). Part-time programs available. *Faculty:* 14 full-time (1 woman). *Students:* 24 full-time (5 women), 2 part-time; includes 2 minority (1 African American, 1 Hispanic American), 4 international. Average age 29. 16 applicants, 56% accepted, 3 enrolled. In 2005, 4 master's, 1 doctorate awarded. Terminal master's awarded for partial completion of doctoral program. *Degree requirements:* For master's, thesis or alternative, comprehensive exam, registration; for doctorate, thesis/dissertation, comprehensive exam, registration. *Entrance requirements:* For master's, GRE General Test, BS in physics; for doctorate, GRE General Test. Additional exam requirements/recommendations for international students: Required—TOEFL (minimum score 550 paper-based; 213 computer-based). *Application deadline:* For fall admission, 3/1 for domestic students, 3/1 for international students. Applications are processed on a rolling basis. Application fee: $50. Electronic applications accepted. *Expenses:* Tuition, state resident: full-time $4,392; part-time $244 per credit. Tuition, nonresident: full-time $8,964; part-time $498 per credit. Required fees: $800; $5 per credit. $48 per contact hour. Tuition and fees vary according to course level, course load, campus/location and reciprocity agreements. *Financial support:* In 2005–06, 14 research assistantships with tuition reimbursements (averaging $10,507 per year), 4 teaching assistantships with tuition reimbursements (averaging $11,027 per year) were awarded; fellowships with tuition reimbursements, Federal Work-Study, scholarships/grants, and unspecified assistantships also available. Financial award applicants required to submit FAFSA. *Faculty research:* Atmospheric and ionospheric radar studies, space plasma theory, magnetospheric dynamics, space weather and auroral studies, turbulence and complex systems. *Unit head:* John D. Craven, Chair, 907-474-7339, Fax: 907-474-6130, E-mail: physics@uaf.edu.

University of Alberta, Faculty of Graduate Studies and Research, Department of Physics, Edmonton, AB T6G 2E1, Canada. Offers astrophysics (M Sc, PhD); condensed matter (M Sc, PhD); geophysics (M Sc, PhD); medical physics (M Sc, PhD); subatomic physics (M Sc, PhD). *Faculty:* 36 full-time (3 women), 7 part-time/adjunct (0 women). *Students:* 56 full-time (6 women), 16 part-time (2 women), 25 international. 85 applicants, 35% accepted. In 2005, 7 master's, 10 doctorates awarded. *Degree requirements:* For master's and doctorate, thesis/dissertation. *Entrance requirements:* For master's and doctorate, minimum GPA of 7.0 on a 9.0 scale. Additional exam requirements/recommendations for international students: Required—TOEFL. *Application deadline:* For fall admission, 2/15 for domestic students. Applications are processed on a rolling basis. Tuition and fees charges are reported in Canadian dollars. *Expenses:* Tuition, state resident: part-time $562 Canadian dollars per term. Tuition, nonresident: full-time $3,375 Canadian dollars. Required fees: $573 Canadian dollars; $84 Canadian dollars per term. *Financial support:* In 2005–06, 45 students received support, including 6 fellowships with partial tuition reimbursements available, 40 teaching assistantships; research assistantships, career-related internships or fieldwork, institutionally sponsored loans, and scholarships/grants also available. Financial award application deadline: 2/15. *Faculty research:* Cosmology, astroparticle physics, high-intermediate energy, magnetism, superconductivity.

Astrophysics

University of Alberta (continued)

Total annual research expenditures: $3.1 million. *Unit head:* Dr. R. Marchand, Associate Chair, 780-492-1072, E-mail: assoc-chair@phys.ualberta.ca. *Application contact:* Lynn Chandler, Program Advisor, 780-492-1072, Fax: 780-492-0714, E-mail: lynn@phys.ualberta.ca.

University of California, Berkeley, Graduate Division, College of Letters and Science, Department of Astronomy, Berkeley, CA 94720-1500. Offers astrophysics (PhD). *Degree requirements:* For doctorate, thesis/dissertation, qualifying exam. *Entrance requirements:* For doctorate, GRE General Test, GRE Subject Test, minimum GPA of 3.0. *Faculty research:* Theory, cosmology, radio astronomy, extra solar planets, infrared instrumentation.

University of California, Los Angeles, Graduate Division, College of Letters and Science, Department of Earth and Space Sciences, Program in Geophysics and Space Physics, Los Angeles, CA 90095. Offers MS, PhD. *Degree requirements:* For master's, comprehensive exams or thesis; for doctorate, thesis/dissertation, oral and written qualifying exams. *Entrance requirements:* For master's, GRE General Test, minimum GPA of 3.0; for doctorate, GRE General Test, minimum undergraduate GPA of 3.0. Electronic applications accepted.

University of California, Santa Cruz, Division of Graduate Studies, Division of Physical and Biological Sciences, Program in Astronomy and Astrophysics, Santa Cruz, CA 95064. Offers PhD. *Faculty:* 23 full-time (3 women). *Students:* 36 full-time (18 women); includes 4 minority (1 African American, 1 Asian American or Pacific Islander, 2 Hispanic Americans), 2 international. 137 applicants, 23% accepted, 7 enrolled. In 2005, 2 doctorates awarded. *Degree requirements:* For doctorate, one foreign language, thesis/dissertation, qualifying exam. *Entrance requirements:* For doctorate, GRE General Test, GRE Subject Test. *Application deadline:* For fall admission, 1/15 for domestic students. Application fee: $60. *Expenses:* Tuition, nonresident: full-time $14,694. *Financial support:* Fellowships, research assistantships, teaching assistantships, Federal Work-Study and institutionally sponsored loans available. Financial award application deadline: 1/15. *Faculty research:* Stellar structure and evolution, stellar spectroscopy, the interstellar medium, galactic structure, external galaxies and quasars. *Unit head:* Dr. Stephen Thorsett, Chairperson, 831-459-2976, E-mail: thorsett@ucolick.ucsc.edu. *Application contact:* Judy L. Glass, Reporting Analyst for Graduate Admissions, 831-459-5906, Fax: 831-459-4843, E-mail: jlglass@ucsc.edu.

University of Chicago, Division of the Physical Sciences, Department of Astronomy and Astrophysics, Chicago, IL 60637-1513. Offers SM, PhD. *Faculty:* 33 full-time (2 women), 4 part-time/adjunct (0 women). *Students:* 31 full-time (8 women); includes 10 minority (1 African American, 6 Asian Americans or Pacific Islanders, 3 Hispanic Americans). Average age 25. 110 applicants, 14% accepted, 5 enrolled. In 2005, 3 master's, 1 doctorate awarded. Terminal master's awarded for partial completion of doctoral program. *Degree requirements:* For master's, candidacy exam; for doctorate, thesis/dissertation, dissertation for publication. *Entrance requirements:* For master's, department candidacy examination, minimum GPA of 3.0; for doctorate, GRE General Test, GRE Subject Test, minimum GPA of 3.0. Additional exam requirements/recommendations for international students: Required—TOEFL (minimum score 600 paper-based; 250 computer-based); Recommended—IELT. *Application deadline:* For fall admission, 1/12 priority date for domestic students, 1/12 priority date for international students. Application fee: $55. Electronic applications accepted. *Financial support:* In 2005–06, 4 fellowships (averaging $4,250 per year), 71 research assistantships with full tuition reimbursements (averaging $21,996 per year), 19 teaching assistantships with full tuition reimbursements (averaging $18,594 per year) were awarded; career-related internships or fieldwork, Federal Work-Study, institutionally sponsored loans, and tuition waivers (partial) also available. Financial award application deadline: 12/28. *Faculty research:* Quasi-stellar object absorption lines, fluid dynamics, interstellar matter, particle physics, cosmology. *Unit head:* Angela Olinto, Chairman, 773-702-8203, Fax: 773-702-8212. *Application contact:* Laticia Rebeles, Graduate Student Affairs Administrator, 773-702-9808, Fax: 773-702-8212, E-mail: lrebeles@oddjob.uchicago.edu.

University of Colorado at Boulder, Graduate School, College of Arts and Sciences, Department of Astrophysical and Planetary Sciences, Boulder, CO 80309. Offers astrophysics (MS, PhD); planetary science (MS, PhD). *Faculty:* 19 full-time (3 women). *Students:* 70 full-time (34 women), 29 part-time (6 women); includes 6 minority (4 Asian Americans or Pacific Islanders, 2 Hispanic Americans), 9 international. Average age 29. 59 applicants, 100% accepted. In 2005, 16 master's, 14 doctorates awarded. Terminal master's awarded for partial completion of doctoral program. *Degree requirements:* For master's, thesis or alternative, comprehensive exam; for doctorate, one foreign language, thesis/dissertation. *Entrance requirements:* For master's, GRE General Test, GRE Subject Test, minimum undergraduate GPA of 3.0; for doctorate, GRE General Test, GRE Subject Test. *Application deadline:* For fall admission, 1/15 priority date for domestic students, 12/1 priority date for international students. Applications are processed on a rolling basis. Application fee: $50 ($60 for international students). *Financial support:* In 2005–06, fellowships (averaging $5,307 per year), research assistantships (averaging $15,770 per year), teaching assistantships (averaging $14,900 per year) were awarded; tuition waivers (full) also available. Support available to part-time students. Financial award application deadline: 2/1. *Faculty research:* Stellar and extragalactic astrophysics cosmology, space astronomy, planetary science. Total annual research expenditures: $8.1 million. *Unit head:* James Green, Chair, 303-492-8915, Fax: 303-492-3822, E-mail: jgreen@casa.colorado.edu. *Application contact:* Graduate Program Assistant, 303-492-8914, Fax: 303-492-3822, E-mail: apsgradsec@colorado.edu.

University of Maryland, Baltimore County, Graduate School, College of Natural Sciences and Mathematics, Department of Physics, Program in Applied Physics, Baltimore, MD 21250. Offers astrophysics (PhD); optics (MS, PhD); quantum optics (PhD); solid state physics (MS, PhD). *Expenses:* Tuition, state resident: part-time $395 per credit. Tuition, nonresident: part-time $652 per credit. Required fees: $82 per credit. Tuition and fees vary according to course load, program and reciprocity agreements. *Application contact:* Dr. Terrance L Worchesky, Director, 410-455-6779, Fax: 410-455-1072, E-mail: workchesk@umbc.edu.

University of Minnesota, Twin Cities Campus, Graduate School, Institute of Technology, School of Physics and Astronomy, Department of Astronomy, Minneapolis, MN 55455-0213. Offers astrophysics (MS, PhD). Terminal master's awarded for partial completion of doctoral program. *Degree requirements:* For master's, thesis optional; for doctorate, thesis/dissertation. *Entrance requirements:* For master's and doctorate, GRE General Test, GRE Subject Test. *Expenses:* Tuition, state resident: full-time $8,748; part-time $729 per credit. Tuition, nonresident: full-time $15,848; part-time $1,321 per credit. Full-time tuition and fees vary according to class time, course load, program and reciprocity agreements. *Faculty research:* Evolution of stars and galaxies; the interstellar medium; cosmology; observational, optical, infrared, and radio astronomy; computational astrophysics.

University of Missouri–St. Louis, College of Arts and Sciences, Department of Physics and Astronomy, St. Louis, MO 63121. Offers applied physics (MS); astrophysics (MS); physics (PhD). Part-time and evening/weekend programs available. *Faculty:* 13. *Students:* 10 full-time (2 women), 15 part-time (1 woman); includes 1 minority (Asian American or Pacific Islander), 6 international. Average age 30. In 2005, 3 master's awarded. Terminal master's awarded for partial completion of doctoral program. *Degree requirements:* For master's, thesis optional; for doctorate, thesis/dissertation. *Entrance requirements:* For master's, 2 letters of recommendation; for doctorate, GRE General Test, 2 letters of recommendation. Additional exam requirements/recommendations for international students: Required—TOEFL (minimum score 550 paper-based; 213 computer-based). *Application deadline:* For fall admission, 7/1 for domestic students; for spring admission, 12/1 priority date for domestic students. Applications are processed on a rolling basis. Application fee: $35 ($40 for international students). Electronic applications accepted. *Expenses:* Tuition, state resident: part-time $263 per credit hour. Tuition, nonresident: part-time $680 per credit hour. Required fees: $53 per credit hour. Tuition and fees vary according to program. *Financial support:* In 2005–06, 3 research assistantships with full and partial tuition reimbursements (averaging $13,333 per year), 11 teaching assistantships with full and partial tuition reimbursements (averaging $11,575 per year) were awarded; fellowships with full tuition reimbursements, career-related internships or fieldwork also available. *Faculty research:* Biophysics, atomic physics, nonlinear dynamics, materials science. *Unit head:* Dr. Ricardo Flores, Director of Graduate Studies, 314-516-5931, Fax: 314-516-6152, E-mail: flores@jinx.umsl.edu. *Application contact:* 314-516-5458, Fax: 314-516-5310, E-mail: gradadm@umsl.edu.

The University of North Carolina at Chapel Hill, Graduate School, College of Arts and Sciences, Department of Physics and Astronomy, Chapel Hill, NC 27599. Offers physics (MS, PhD). Terminal master's awarded for partial completion of doctoral program. *Degree requirements:* For master's, comprehensive exam, registration; for doctorate, thesis/dissertation, comprehensive exam, registration. *Entrance requirements:* For master's and doctorate, GRE General Test, minimum GPA of 3.0. Electronic applications accepted. *Faculty research:* Observational astronomy, fullerenes, polarized beams, nanotubes, nucleosynthesis in stars and supernovae, superstring theory, ballistic transport in semiconductors, gravitation.

University of Oklahoma, Graduate College, College of Arts and Sciences, Department of Physics and Astronomy, Norman, OK 73019-0390. Offers astrophysics (MS, PhD); physics (MS, PhD). Part-time programs available. *Faculty:* 29 full-time (4 women), 1 part-time/adjunct (0 women). *Students:* 51 full-time (15 women), 4 part-time (1 woman); includes 3 minority (2 African Americans, 1 Asian American or Pacific Islander), 27 international. 7 applicants, 86% accepted, 5 enrolled. In 2005, 4 master's, 8 doctorates awarded. Terminal master's awarded for partial completion of doctoral program. *Degree requirements:* For master's, thesis or alternative, departmental qualifying exam; for doctorate, thesis/dissertation, comprehensive, departmental qualifying, oral, and written exams. *Entrance requirements:* For master's and doctorate, GRE General Test, GRE Subject Test, 3 letters of recommendation. Additional exam requirements/recommendations for international students: Required—TOEFL (minimum score 600 paper-based; 250 computer-based). *Application deadline:* For fall admission, 3/1 for domestic students. Application fee: $40 ($90 for international students). *Expenses:* Tuition, state resident: full-time $3,029; part-time $126 per credit hour. Tuition, nonresident: full-time $10,807; part-time $450 per credit hour. Required fees: $1,231; $44 per credit hour. Tuition and fees vary according to course load and program. *Financial support:* In 2005–06, 9 students received support, including 2 fellowships with full tuition reimbursements available (averaging $4,000 per year), 23 research assistantships with partial tuition reimbursements available (averaging $13,909 per year), 34 teaching assistantships with partial tuition reimbursements available (averaging $13,748 per year); Federal Work-Study, scholarships/grants, health care benefits, tuition waivers (full), and unspecified assistantships also available. Financial award application deadline: 3/1; financial award applicants required to submit FAFSA. *Faculty research:* Atomic, molecular, and chemical physics; high energy physics; solid state and applied physics; astrophysics. Total annual research expenditures: $4.6 million. *Unit head:* Dr. Ryan Doezema, Chair, 405-325-3961, Fax: 405-325-7557, E-mail: rdoezema@ou.edu. *Application contact:* Debbie Barnhill, Graduate Studies Coordinator, 405-325-3961, Fax: 405-325-7557, E-mail: dbarnhill@ou.edu.

University of Pennsylvania, School of Arts and Sciences, Graduate Group in Physics and Astronomy, Philadelphia, PA 19104. Offers medical physics (MS); physics (PhD). Part-time programs available. *Degree requirements:* For doctorate, thesis/dissertation, oral, preliminary, and final exams. *Entrance requirements:* For doctorate, GRE General Test, GRE Subject Test (recommended). Additional exam requirements/recommendations for international students: Required—TOEFL; Recommended—TSE. Electronic applications accepted. *Faculty research:* Astrophysics, condensed matter experiment, condensed matter theory, particle experiment, particle theory.

University of Victoria, Faculty of Graduate Studies, Faculty of Science, Department of Physics and Astronomy, Victoria, BC V8W 2Y2, Canada. Offers astronomy and astrophysics (M Sc, PhD); condensed matter physics (M Sc, PhD); experimental particle physics (M Sc, PhD); medical physics (M Sc, PhD); ocean physics (PhD); ocean physics and geophysics (M Sc); theoretical physics (M Sc, PhD). *Faculty:* 16 full-time (0 women), 13 part-time/adjunct (1 woman). *Students:* 45 full-time, 8 international. Average age 25. 66 applicants, 45% accepted, 10 enrolled. In 2005, 4 master's, 1 doctorate awarded. *Median time to degree:* Of those who began their doctoral program in fall 1997, 100% received their degree in 8 years or less. *Degree requirements:* For master's, thesis, registration; for doctorate, thesis/dissertation, Candidacy exam, comprehensive exam, registration. *Entrance requirements:* For master's and doctorate, GRE. Additional exam requirements/recommendations for international students: Required—TOEFL (minimum score 575 paper-based; 233 computer-based), IELT (minimum score 7). *Application deadline:* For fall admission, 5/31 priority date for domestic students, 12/15 priority date for international students. Applications are processed on a rolling basis. Application fee: $75 ($125 for international students). Electronic applications accepted. Tuition and fees charges are reported in Canadian dollars. *Expenses:* Tuition, area resident: Full-time $4,492 Canadian dollars; part-time $749 Canadian dollars per term. International tuition: $5,346 Canadian dollars full-time. Required fees: $4,492 Canadian dollars; $749 Canadian dollars per term. Tuition and fees vary according to course load, campus/location and program. *Financial support:* In 2005–06, 3 students received support; fellowships, research assistantships, teaching assistantships, career-related internships or fieldwork, institutionally sponsored loans, and awards available. Financial award application deadline: 2/15. *Faculty research:* Old stellar populations; observational cosmology and large scale structure; cp violation; atlas. *Unit head:* Dr. J. Michael Ronev, Chair, 250-721-7698, Fax: 250-721-7715, E-mail: chair@phys.uvic.ca. *Application contact:* Dr. Chris J. Pritchet, Graduate Adviser, 250-721-7744, Fax: 250-721-7715, E-mail: pritchet@uvic.ca.

Section 2
Chemistry

This section contains a directory of institutions offering graduate work in chemistry, followed by in-depth entries submitted by institutions that chose to prepare detailed program descriptions. Additional information about programs listed in the directory but not augmented by an in-depth entry may be obtained by writing directly to the dean of a graduate school or chair of a department at the address given in the directory.

For programs offering related work, see also in this book Geosciences and Physics. In Book 3, see Biological and Biomedical Sciences, Biochemistry, Biophysics, Nutrition, and Pharmacology and Toxicology; in Book 5, see Engineering and Applied Sciences; Agricultural Engineering; Chemical Engineering; Geological, Mineral/Mining, and Petroleum Engineering; Materials Sciences and Engineering; and Pharmaceutical Engineering; and in Book 6, see Pharmacy and Pharmaceutical Sciences.

CONTENTS

Analytical Chemistry

Brigham Young University, Graduate Studies, College of Physical and Mathematical Sciences, Department of Chemistry and Biochemistry, Provo, UT 84602-1001. Offers analytical chemistry (MS, PhD); biochemistry (MS, PhD); inorganic chemistry (MS, PhD); organic chemistry (MS, PhD); physical chemistry (MS, PhD). *Faculty:* 35 full-time (2 women). *Students:* 100 full-time (33 women); includes 2 minority (both Asian Americans or Pacific Islanders), 68 international. Average age 29. 85 applicants, 44% accepted, 21 enrolled. In 2005, 9 master's, 9 doctorates awarded. *Median time to degree:* Of those who began their doctoral program in fall 1997, 87% received their degree in 8 years or less. *Degree requirements:* For master's, thesis, registration; for doctorate, thesis/dissertation, degree qualifying exam. *Entrance requirements:* For master's, pass 1 (biochemistry), 4 (chemistry) of 5 area exams, GRE General Test, minimum GPA of 3.0 in last 60 hours; for doctorate, pass 1 (biochemistry) or 4 (chemistry) of 5 area exams, GRE General Test, minimum GPA of 3.0 in last 60 hours. Additional exam requirements/recommendations for international students: Required—TOEFL, TWE. *Application deadline:* For fall admission, 2/1 priority date for domestic students, 2/1 priority date for international students. Applications are processed on a rolling basis. Application fee: $50. Electronic applications accepted. *Financial support:* In 2005–06, 100 students received support, including 12 fellowships with full tuition reimbursements available (averaging $20,500 per year), 47 research assistantships with full tuition reimbursements available (averaging $20,500 per year), 39 teaching assistantships with full tuition reimbursements available (averaging $20,400 per year); institutionally sponsored loans, scholarships/grants, health care benefits, tuition waivers (full), and unspecified assistantships also available. Financial award application deadline: 2/1. *Faculty research:* Separation science, molecular recognition, organic synthesis and biomedical application, biochemistry and molecular biology, molecular spectroscopy. Total annual research expenditures: $3.9 million. *Unit head:* Dr. Paul B. Farnsworth, Chair, 801-422-6502, Fax: 801-422-0153, E-mail: paul_farnsworth@byu.edu. *Application contact:* Dr. David V. Dearden, Graduate Coordinator, 801-422-2355, Fax: 801-422-0153, E-mail: david_dearden@byu.edu.

See Close-Up on page 97.

California State University, Fullerton, Graduate Studies, College of Natural Science and Mathematics, Department of Chemistry and Biochemistry, Fullerton, CA 92834-9480. Offers analytical chemistry (MS); biochemistry (MS); geochemistry (MS); inorganic chemistry (MS); organic chemistry (MS); physical chemistry (MS). Part-time programs available. *Students:* 19 full-time (6 women), 20 part-time (9 women); includes 20 minority (1 African American, 13 Asian Americans or Pacific Islanders, 6 Hispanic Americans), 8 international. Average age 28. 36 applicants, 47% accepted, 4 enrolled. In 2005, 11 degrees awarded. *Degree requirements:* For master's, thesis, departmental qualifying exam. *Entrance requirements:* For master's, minimum GPA of 2.5 in last 60 units of course work, major in chemistry or related field. Application fee: $55. *Expenses:* Tuition, area resident: Part-time $2,270 per year. Tuition, state resident: full-time $2,572; part-time $339 per unit. Tuition, nonresident: full-time $339; part-time $339 per unit. International tuition: $339 full-time. *Financial support:* Teaching assistantships, career-related internships or fieldwork, Federal Work-Study, institutionally sponsored loans, and scholarships/grants available. Support available to part-time students. Financial award application deadline: 3/1. *Unit head:* Dr. Maria Linder, Chair, 714-278-3621. *Application contact:* Dr. Gregory Williams, Adviser, 714-278-2170.

California State University, Los Angeles, Graduate Studies, College of Natural and Social Sciences, Department of Chemistry and Biochemistry, Los Angeles, CA 90032-8530. Offers analytical chemistry (MS); biochemistry (MS); chemistry (MS); inorganic chemistry (MS); organic chemistry (MS); physical chemistry (MS). Part-time and evening/weekend programs available. *Faculty:* 2 full-time (0 women). *Students:* 33 full-time (21 women), 21 part-time (10 women); includes 38 minority (3 African Americans, 21 Asian Americans or Pacific Islanders, 14 Hispanic Americans). In 2005, 5 degrees awarded. *Degree requirements:* For master's, one foreign language. *Entrance requirements:* Additional exam requirements/recommendations for international students: Required—TOEFL. *Application deadline:* For fall admission, 6/30 for domestic students; for spring admission, 2/1 for domestic students. Applications are processed on a rolling basis. Application fee: $55. *Expenses:* Tuition, area resident: Full-time $3,617. Tuition, state resident: full-time $3,617. Tuition, nonresident: full-time $9,719. International tuition: $9,719 full-time. *Financial support:* Federal Work-Study available. Support available to part-time students. Financial award application deadline: 3/1. *Faculty research:* Intercalation of heavy metal, carborane chemistry, conductive polymers and fabrics, titanium reagents, computer modeling and synthesis. *Unit head:* Dr. Wayne Tikkanen, Chair, 323-343-2300, Fax: 323-343-6490.

Case Western Reserve University, School of Graduate Studies, Department of Chemistry, Cleveland, OH 44106. Offers analytical chemistry (MS, PhD); inorganic chemistry (MS, PhD); organic chemistry (MS, PhD); physical chemistry (MS, PhD). Part-time programs available. Terminal master's awarded for partial completion of doctoral program. *Degree requirements:* For doctorate, thesis/dissertation. *Entrance requirements:* For master's and doctorate, GRE General Test, GRE Subject Test. Additional exam requirements/recommendations for international students: Required—TOEFL. *Faculty research:* Electrochemistry, synthetic chemistry, chemistry of life process, spectroscopy, kinetics.

Clarkson University, Graduate School, School of Arts and Sciences, Department of Chemistry, Potsdam, NY 13699. Offers analytical chemistry (MS, PhD); inorganic chemistry (MS, PhD); organic chemistry (MS, PhD); physical chemistry (MS, PhD). *Faculty:* 9 full-time (2 women). *Students:* 34 full-time (9 women), 1 (woman) part-time; includes 2 Asian Americans or Pacific Islanders, 22 international. Average age 27. 52 applicants, 62% accepted. In 2005, 3 master's, 2 doctorates awarded. *Median time to degree:* Of those who began their doctoral program in fall 1997, 100% received their degree in 8 years or less. *Degree requirements:* For doctorate, thesis/dissertation, departmental qualifying exam. *Entrance requirements:* For master's, GRE. Additional exam requirements/recommendations for international students: Required—TOEFL. *Application deadline:* For fall admission, 5/15 for domestic students; for spring admission, 10/15 priority date for domestic students. Applications are processed on a rolling basis. Application fee: $25 ($35 for international students). Electronic applications accepted. *Expenses:* Tuition: Full-time $20,160; part-time $840 per hour. Required fees: $215. *Financial support:* In 2005–06, 25 students received support, including 2 fellowships (averaging $25,000 per year), 11 research assistantships (averaging $19,032 per year), 12 teaching assistantships (averaging $19,032 per year); scholarships/grants and tuition waivers (partial) also available. *Faculty research:* Nanomaterial, surface science, polymers chemical biosensing, protein biochemistry drug design/delivery, molecular neuroscience and immunobiology. Total annual research expenditures: $1.6 million. *Unit head:* Dr. Phillip A. Christiansen, Division Head, 315-268-6669, Fax: 315-268-6610, E-mail: pac@clarkson.edu. *Application contact:* Donna Brockway, Graduate Admissions International Advisor/Assistant to the Provost, 315-268-6447, Fax: 315-268-7994, E-mail: brockway@clarkson.edu.

Cleveland State University, College of Graduate Studies, College of Science, Department of Chemistry, Cleveland, OH 44115. Offers analytical chemistry (MS); clinical chemistry (MS, PhD); clinical/bioanalytical chemistry (PhD); environmental chemistry (MS); inorganic chemistry (MS); organic chemistry (MS); physical chemistry (MS). Part-time and evening/weekend programs available. *Faculty:* 13 full-time (1 woman), 59 part-time/adjunct (12 women). *Students:* 46 full-time (22 women), 18 part-time (6 women); includes 4 minority (2 African Americans, 2 Asian Americans or Pacific Islanders), 35 international. Average age 31. 47 applicants, 62% accepted, 7 enrolled. In 2005, 2 master's, 5 doctorates awarded. *Median time to degree:* Of those who began their doctoral program in fall 1997, 67% received their degree in 8 years or less. *Degree requirements:* For master's, thesis (for some programs); for doctorate, thesis/dissertation. *Entrance requirements:* For master's and doctorate, GRE General Test, GRE Subject Test. Additional exam requirements/recommendations for international students: Required—TOEFL (minimum score 525 paper-based; 197 computer-based). *Application deadline:* For fall admission, 1/15 priority date for domestic students, 1/15 priority date for international students. Applications are processed on a rolling basis. Application fee: $30. Electronic applications accepted. *Expenses:* Tuition, state resident: full-time $10,700. Tuition, nonresident: full-time $14,628. Tuition and fees vary according to program. *Financial support:* In 2005–06, 44 students received support, including fellowships with full tuition reimbursements available (averaging $18,000 per year), research assistantships with full tuition reimbursements available (averaging $16,000 per year), teaching assistantships with full tuition reimbursements available (averaging $14,000 per year) Financial award application deadline: 1/15. *Faculty research:* Metalloenzyme mechanisms, dependent RNAse L and interferons, application of HPLC/LPCC to clinical systems, structure-function relationships of factor Va, MALDI-TOF based DNA sequencing. Total annual research expenditures: $3 million. *Unit head:* Dr. Lily M. Ng, Chair, 216-687-2467, E-mail: l.ng@csuohio.edu. *Application contact:* Richelle P. Emery, Administrative Coordinator, 216-687-2457, Fax: 216-687-9298, E-mail: r.emery@csuohio.edu.

Cornell University, Graduate School, Graduate Fields of Arts and Sciences, Field of Chemistry and Chemical Biology, Ithaca, NY 14853-0001. Offers analytical chemistry (PhD); bio-organic chemistry (PhD); biophysical chemistry (PhD); chemical biology (PhD); chemical physics (PhD); inorganic chemistry (PhD); materials chemistry (PhD); organic chemistry (PhD); organometallic chemistry (PhD); physical chemistry (PhD); polymer chemistry (PhD); theoretical chemistry (PhD). *Faculty:* 44 full-time (2 women). *Students:* 190 full-time (73 women); includes 23 minority (4 African Americans, 8 Asian Americans or Pacific Islanders, 11 Hispanic Americans), 65 international. 339 applicants, 35% accepted, 49 enrolled. In 2005, 23 doctorates awarded. *Degree requirements:* For doctorate, thesis/dissertation, comprehensive exam. *Entrance requirements:* For doctorate, GRE General Test, GRE Subject Test (chemistry), 3 letters of recommendation. Additional exam requirements/recommendations for international students: Required—TOEFL (minimum score 600 paper-based; 250 computer-based). *Application deadline:* For fall admission, 1/10 for domestic students. Application fee: $60. Electronic applications accepted. *Financial support:* In 2005–06, 185 students received support, including 33 fellowships with full tuition reimbursements available, 85 research assistantships with full tuition reimbursements available, 67 teaching assistantships with full tuition reimbursements available; institutionally sponsored loans, scholarships/grants, health care benefits, tuition waivers (full and partial), and unspecified assistantships also available. Financial award applicants required to submit FAFSA. *Faculty research:* Analytical, organic, inorganic, physical, materials, chemical biology. *Unit head:* Director of Graduate Studies, 607-255-4139, Fax: 607-255-4137. *Application contact:* Graduate Field Assistant, 607-255-4139, Fax: 607-255-4137, E-mail: chemgrad@cornell.edu.

See Close-Up on page 109.

Florida State University, Graduate Studies, College of Arts and Sciences, Department of Chemistry and Biochemistry, Tallahassee, FL 32306. Offers analytical chemistry (MS, PhD); biochemistry (MS, PhD); chemical physics (MS, PhD); inorganic chemistry (MS, PhD); organic chemistry (MS, PhD); physical chemistry (MS, PhD). Part-time programs available. *Faculty:* 36 full-time (6 women), 2 part-time/adjunct (0 women). *Students:* 144 full-time (50 women); includes 17 minority (7 African Americans, 1 American Indian/Alaska Native, 6 Asian Americans or Pacific Islanders, 3 Hispanic Americans), 58 international. Average age 25. 288 applicants, 27% accepted, 26 enrolled. In 2005, 14 master's, 15 doctorates awarded. Terminal master's awarded for partial completion of doctoral program. *Median time to degree:* Of those who began their doctoral program in fall 1997, 88% received their degree in 8 years or less. *Degree requirements:* For master's, thesis (for some programs), cumulative and diagnostic exams, comprehensive exam (for some programs), registration; for doctorate, thesis/dissertation, cumulative and diagnostic exams, comprehensive exam (for some programs), registration. *Entrance requirements:* For master's and doctorate, GRE General Test, minimum B average in undergraduate course work. Additional exam requirements/recommendations for international students: Required—TOEFL (minimum score 515 paper-based; 213 computer-based). *Application deadline:* For fall admission, 4/15 priority date for domestic students, 4/15 priority date for international students. Applications are processed on a rolling basis. Application fee: $30. Electronic applications accepted. *Financial support:* In 2005–06, 1 fellowship with tuition reimbursement (averaging $18,000 per year), 56 research assistantships with tuition reimbursements (averaging $19,000 per year), 83 teaching assistantships with tuition reimbursements (averaging $19,000 per year) were awarded; career-related internships or fieldwork, Federal Work-Study, institutionally sponsored loans, and traineeships also available. Financial award application deadline: 2/15; financial award applicants required to submit FAFSA. *Faculty research:* Spectroscopy, computational chemistry, nuclear chemistry, separations, synthesis. Total annual research expenditures: $6.5 million. *Unit head:* Dr. Naresh Dalal, Chairman, 850-644-3398, Fax: 850-644-8281. *Application contact:* Dr. Oliver Steinbock, Chair, Graduate Admissions Committee, 888-525-9286, Fax: 850-644-8281, E-mail: gradinfo@chem.fsu.edu.

See Close-Up on page 117.

Georgetown University, Graduate School of Arts and Sciences, Department of Chemistry, Washington, DC 20057. Offers analytical chemistry (MS, PhD); biochemistry (MS, PhD); chemical physics (MS, PhD); inorganic chemistry (MS, PhD); organic chemistry (MS, PhD); physical chemistry (MS, PhD); theoretical chemistry (MS, PhD). Terminal master's awarded for partial completion of doctoral program. *Degree requirements:* For master's, thesis (for some programs), qualifying exam; for doctorate, thesis/dissertation, comprehensive exam. *Entrance requirements:* For master's and doctorate, GRE General Test. Additional exam requirements/recommendations for international students: Required—TOEFL.

The George Washington University, Columbian College of Arts and Sciences, Department of Chemistry, Washington, DC 20052. Offers analytical chemistry (MS, PhD); inorganic chemistry (MS, PhD); materials science (MS, PhD); organic chemistry (MS, PhD); physical chemistry (MS, PhD). Part-time and evening/weekend programs available. Terminal master's awarded for partial completion of doctoral program. *Degree requirements:* For master's, thesis or alternative, comprehensive exam; for doctorate, thesis/dissertation, general exam. *Entrance requirements:* For master's and doctorate, GRE General Test, interview, minimum GPA of 3.0. Additional exam requirements/recommendations for international students: Required—TOEFL (minimum score 550 paper-based; 213 computer-based). Electronic applications accepted.

Governors State University, College of Arts and Sciences, Program in Analytical Chemistry, University Park, IL 60466-0975. Offers MS. Part-time and evening/weekend programs available. *Students:* 9 full-time, 17 part-time. Average age 29. *Degree requirements:* For master's, thesis or alternative. *Application deadline:* For fall admission, 7/15 for domestic students; for spring admission, 11/10 for domestic students. Applications are processed on a rolling basis. Application fee: $25. *Expenses:* Tuition: Part-time $157 per semester hour. Required fees: $480; $240 per semester. *Financial support:* Research assistantships, career-related internships or fieldwork, Federal Work-Study, institutionally sponsored loans, and scholarships/grants available. Support available to part-time students. Financial award application deadline: 5/1. *Faculty research:* Electrochemistry, photochemistry, spectrochemistry, biochemistry. *Unit head:* Dr. Sandra A Mayfield, Interim Dean, College of Arts and Sciences, 708-534-4101.

Howard University, Graduate School of Arts and Sciences, Department of Chemistry, Washington, DC 20059-0002. Offers analytical chemistry (MS, PhD); atmospheric (MS, PhD); biochemistry (MS, PhD); environmental (MS, PhD); inorganic chemistry (MS, PhD); organic chemistry (MS, PhD); physical chemistry (MS, PhD); polymer chemistry (MS, PhD). Part-time programs available. *Degree requirements:* For master's, one foreign language, thesis, teaching experience, comprehensive exam, registration; for doctorate, 2 foreign languages, thesis/dissertation, teaching experience, comprehensive exam, registration. *Entrance requirements:* For master's, GRE General Test, minimum GPA of 2.7; for doctorate, GRE General Test, minimum GPA of 3.0. *Faculty research:* Stratospheric aerosols, liquid crystals, polymer coatings, terrestrial and extraterrestrial atmospheres, amidogen reaction.

Illinois Institute of Technology, Graduate College, College of Science and Letters, Department of Biological, Chemical and Physical Sciences, Chemistry Division, Chicago, IL 60616-3793. Offers analytical chemistry (M Ch, MS, PhD); chemistry (M Chem); inorganic chemistry (MS, PhD); materials and chemical synthesis (M Ch); organic chemistry (MS, PhD); physical chemistry (MS, PhD); polymer chemistry (MS, PhD). Part-time and evening/weekend programs available. Postbaccalaureate distance learning degree programs offered (no on-campus study). Terminal master's awarded for partial completion of doctoral program. *Degree requirements:* For master's, thesis (for some programs), comprehensive exam; for doctorate, thesis/dissertation, comprehensive exam. *Entrance requirements:* For master's and doctorate, GRE General Test, minimum undergraduate GPA of 3.0. Additional exam requirements/recommendations for international students: Required—TOEFL (minimum score 550 paper-based; 213 computer-based). Electronic applications accepted. *Faculty research:* Organic and inorganic chemistry, polymers research, physical chemistry, analytical chemistry.

Indiana University Bloomington, Graduate School, College of Arts and Sciences, Department of Chemistry, Bloomington, IN 47405-7000. Offers analytical chemistry (PhD); biological chemistry (PhD); chemistry (MAT); inorganic chemistry (PhD); physical chemistry (PhD). PhD offered through the University Graduate School. *Faculty:* 29 full-time (2 women). *Students:* 110 full-time (38 women), 35 part-time (11 women); includes 8 minority (3 African Americans, 3 Asian Americans or Pacific Islanders, 2 Hispanic Americans), 57 international. Average age 26. In 2005, 7 master's, 20 doctorates awarded. Terminal master's awarded for partial completion of doctoral program. *Degree requirements:* For master's and doctorate, thesis/dissertation. *Entrance requirements:* For master's and doctorate, GRE General Test, GRE Subject Test. Additional exam requirements/recommendations for international students: Required—TOEFL. *Application deadline:* For fall admission, 1/15 priority date for domestic students, 12/15 priority date for international students; for spring admission, 9/1 priority date for domestic students, 9/1 priority date for international students. Applications are processed on a rolling basis. Application fee: $50 ($60 for international students). *Expenses:* Tuition, state resident: full-time $5,437; part-time $227 per credit hour. Tuition, nonresident: full-time $15,836; part-time $660 per credit hour. Required fees: $821. Tuition and fees vary according to campus/location and program. *Financial support:* In 2005–06, 23 fellowships with full tuition reimbursements, 57 research assistantships with full tuition reimbursements, 78 teaching assistantships with full tuition reimbursements were awarded; Federal Work-Study and institutionally sponsored loans also available. *Faculty research:* Synthesis of complex natural products, organic reaction mechanisms, organic electrochemistry, transitive-metal chemistry, solid-state and surface chemistry. Total annual research expenditures: $7.7 million. *Unit head:* Dr. David Clemmer, Chairperson, 812-855-2268. *Application contact:* Dr. Jack K. Crandall, Chairperson of Admissions, 812-855-2068, Fax: 812-855-8300, E-mail: chemgrad@indiana.edu.

Kansas State University, Graduate School, College of Arts and Sciences, Department of Chemistry, Manhattan, KS 66506. Offers analytical chemistry (MS); biological chemistry (MS); chemistry (PhD); inorganic chemistry (MS); materials chemistry (MS); organic chemistry (MS); physical chemistry (MS). *Faculty:* 14 full-time (1 woman). *Students:* 65 full-time (21 women), 5 part-time (2 women); includes 1 minority (African American), 48 international. 63 applicants, 67% accepted, 22 enrolled. In 2005, 3 master's, 5 doctorates awarded. Terminal master's awarded for partial completion of doctoral program. *Degree requirements:* For master's and doctorate, thesis/dissertation. *Entrance requirements:* For master's and doctorate, GRE, minimum GPA of 3.0. Additional exam requirements/recommendations for international students: Required—TOEFL (minimum score 550 paper-based; 213 computer-based). *Application deadline:* For fall admission, 2/1 priority date for domestic students, 2/1 priority date for international students; for spring admission, 10/1 for domestic students, 8/1 for international students. Applications are processed on a rolling basis. Application fee: $30 ($55 for international students). *Expenses:* Tuition, state resident: full-time $5,160; part-time $215. Tuition, nonresident: full-time $12,816; part-time $534. International tuition: $12,816 full-time. Required fees: $564. *Financial support:* In 2005–06, 30 research assistantships (averaging $11,956 per year), 26 teaching assistantships with full tuition reimbursements (averaging $11,945 per year) were awarded; fellowships, institutionally sponsored loans and scholarships/grants also available. Support available to part-time students. Financial award application deadline: 3/1; financial award applicants required to submit FAFSA. *Faculty research:* Nanotechnologies, functional materials, bio-organic and bio-physical processes, sensors and separations, synthesis and synthetic methods. Total annual research expenditures: $2.2 million. *Unit head:* Eric Maatta, Head, 785-532-6665, Fax: 785-532-6666, E-mail: eam@ksu.edu. *Application contact:* Christer Aakeröy, Director, 785-532-6096, Fax: 785-532-6666, E-mail: aakeroy@ksu.edu.

Kent State University, College of Arts and Sciences, Department of Chemistry, Kent, OH 44242-0001. Offers analytical chemistry (MS, PhD); biochemistry (MS, PhD); chemistry (MA, MS, PhD); inorganic chemistry (MS, PhD); organic chemistry (MS, PhD); physical chemistry (MS, PhD). Terminal master's awarded for partial completion of doctoral program. *Degree requirements:* For master's and doctorate, thesis/dissertation, comprehensive exam, registration. *Entrance requirements:* For master's and doctorate, placement exam, GRE General Test, GRE Subject Test (recommended), minimum GPA of 2.75. Additional exam requirements/recommendations for international students: Required—TOEFL (minimum score 575 paper-based; 230 computer-based). Electronic applications accepted. *Faculty research:* Biological chemistry, materials chemistry, molecular spectroscopy.

See Close-Up on page 121.

Marquette University, Graduate School, College of Arts and Sciences, Department of Chemistry, Milwaukee, WI 53201-1881. Offers analytical chemistry (MS, PhD); bioanalytical chemistry (MS, PhD); biophysical chemistry (MS, PhD); chemical physics (MS, PhD); inorganic chemistry (MS, PhD); organic chemistry (MS, PhD); physical chemistry (MS, PhD). Part-time programs available. Terminal master's awarded for partial completion of doctoral program. *Degree requirements:* For master's, comprehensive exam; for doctorate, thesis/dissertation, cumulative exams. *Entrance requirements:* For master's and doctorate, GRE Subject Test. Additional exam requirements/recommendations for international students: Required—TOEFL. *Faculty research:* Inorganic complexes, laser Raman spectroscopy, organic synthesis, chemical dynamics, biophysiology.

McMaster University, School of Graduate Studies, Faculty of Science, Department of Chemistry, Hamilton, ON L8S 4M2, Canada. Offers analytical chemistry (M Sc, PhD); chemical physics (M Sc, PhD); chemistry (M Sc, PhD); inorganic chemistry (M Sc, PhD); organic chemistry (M Sc, PhD); physical chemistry (M Sc, PhD); polymer chemistry (M Sc, PhD). Part-time programs available. Terminal master's awarded for partial completion of doctoral program. *Degree requirements:* For master's, thesis/dissertation; for doctorate, thesis/dissertation, comprehensive exam. *Entrance requirements:* For master's, minimum B+ average. Additional exam requirements/recommendations for international students: Required—TOEFL (minimum score 550 paper-based; 213 computer-based).

Miami University, Graduate School, College of Arts and Sciences, Department of Chemistry and Biochemistry, Oxford, OH 45056. Offers analytical chemistry (MS, PhD); biochemistry (MS, PhD); chemical education (MS, PhD); chemistry (MS, PhD); inorganic chemistry (MS, PhD); organic chemistry (MS, PhD); physical chemistry (MS, PhD). Part-time programs available. *Degree requirements:* For master's, thesis, final exam; for doctorate, thesis/dissertation, final exams, comprehensive exam. *Entrance requirements:* For master's, minimum undergraduate GPA of 3.0 during previous 2 years or 2.75 overall; for doctorate, minimum undergraduate GPA of 2.75, 3.0 graduate. Additional exam requirements/recommendations for international students: Required—TOEFL (minimum score 550 paper-based; 213 computer-based), TWE(minimum score 4). Electronic applications accepted.

Northeastern University, College of Arts and Sciences, Department of Chemistry and Chemical Biology, Boston, MA 02115-5096. Offers analytical chemistry (PhD); chemistry (MS, PhD); inorganic chemistry (PhD); organic chemistry (PhD); physical chemistry (PhD). Part-time and evening/weekend programs available. Terminal master's awarded for partial completion of doctoral program. *Degree requirements:* For master's, thesis (for some programs); for doctorate, thesis/dissertation, qualifying exam in specialty area. *Entrance requirements:* Additional exam requirements/recommendations for international students: Required—TOEFL. Electronic applications accepted. *Faculty research:* Electron transfer, theoretical chemical physics, analytical biotechnology, mass spectrometry, materials chemistry.

See Close-Up on page 129.

Old Dominion University, College of Sciences, Program in Chemistry, Norfolk, VA 23529. Offers analytical chemistry (MS); biochemistry (MS); clinical chemistry (MS); environmental chemistry (MS); organic chemistry (MS); physical chemistry (MS). Part-time and evening/weekend programs available. *Faculty:* 15 full-time (4 women). *Students:* 5 full-time (3 women), 8 part-time (3 women); includes 8 minority (4 African Americans, 1 American Indian/Alaska Native, 1 Asian American or Pacific Islander, 2 Hispanic Americans). Average age 27. 19 applicants, 74% accepted. In 2005, 1 degree awarded. *Degree requirements:* For master's, thesis, comprehensive exam. *Entrance requirements:* For master's, GRE General Test, minimum GPA of 3.0 in major, 2.5 overall. Additional exam requirements/recommendations for international students: Required—TOEFL. *Application deadline:* For fall admission, 7/1 for domestic students; for spring admission, 11/1 for domestic students. Applications are processed on a rolling basis. Application fee: $30. *Expenses:* Tuition, state resident: part-time $263 per credit hour. Tuition, nonresident: part-time $661 per credit hour. Required fees: $39 per semester. Part-time tuition and fees vary according to campus/location. *Financial support:* In 2005–06, research assistantships with tuition reimbursements (averaging $15,000 per year), teaching assistantships with tuition reimbursements (averaging $15,000 per year) were awarded; fellowships, career-related internships or fieldwork and scholarships/grants also available. Financial award application deadline: 2/15; financial award applicants required to submit FAFSA. *Faculty research:* Organic and trace metal biogeochemistry, materials chemistry, bioanalytical chemistry, computational chemistry. Total annual research expenditures: $854,968. *Unit head:* Dr. John R. Donat, Graduate Program Director, 757-683-4098, Fax: 757-683-4628, E-mail: chemgpd@odu.edu.

Oregon State University, Graduate School, College of Science, Department of Chemistry, Corvallis, OR 97331. Offers analytical chemistry (MS, PhD); chemistry (MA, MAIS); inorganic chemistry (MS, PhD); nuclear and radiation chemistry (MS, PhD); organic chemistry (MS, PhD); physical chemistry (MS, PhD). Part-time programs available. *Faculty:* 19 full-time (3 women), 6 part-time/adjunct (1 woman). *Students:* 74 full-time (30 women), 3 part-time; includes 6 minority (1 African American, 1 Asian American or Pacific Islander, 4 Hispanic Americans), 35 international. Average age 28. In 2005, 4 master's, 10 doctorates awarded. Terminal master's awarded for partial completion of doctoral program. *Degree requirements:* For master's and doctorate, one foreign language, thesis/dissertation. *Entrance requirements:* For master's and doctorate, minimum GPA of 3.0 in last 90 hours of course work. Additional exam requirements/recommendations for international students: Required—TOEFL. *Application deadline:* For fall admission, 3/1 for domestic students. Applications are processed on a rolling basis. Application fee: $50. *Expenses:* Tuition, area resident: Full-time $8,139; part-time $301 per credit. Tuition, state resident: full-time $8,139; part-time $501 per credit. Tuition, nonresident: full-time $14,376; part-time $532 per credit. International tuition: $14,376 full-time. Required fees: $1,266. *Financial support:* Fellowships, research assistantships, teaching assistantships, institutionally sponsored loans available. Support available to part-time students. Financial award application deadline: 2/1. *Faculty research:* Solid state chemistry, enzyme reaction mechanisms, structure and dynamics of gas molecules, chemiluminescence, nonlinear optical spectroscopy. *Unit head:* Dr. Douglas Keszler, Chair, 541-737-2081, Fax: 541-737-2062. *Application contact:* Carolyn Brumley, Graduate Secretary, 541-737-6707, Fax: 541-737-2062, E-mail: carolyn.brumley@orst.edu.

Purdue University, Graduate School, School of Science, Department of Chemistry, West Lafayette, IN 47907. Offers analytical chemistry (MS, PhD); biochemistry (MS, PhD); chemical education (MS, PhD); inorganic chemistry (MS, PhD); organic chemistry (MS, PhD); physical chemistry (MS, PhD). *Accreditation:* NCATE (one or more programs are accredited). *Faculty:* 43 full-time (9 women), 8 part-time/adjunct (2 women). *Students:* 293 full-time (114 women), 39 part-time (14 women); includes 53 minority (25 African Americans, 6 Asian Americans or Pacific Islanders, 22 Hispanic Americans), 138 international. Average age 28. 458 applicants, 36% accepted, 57 enrolled. In 2005, 9 master's, 36 doctorates awarded. Terminal master's awarded for partial completion of doctoral program. *Degree requirements:* For master's and doctorate, thesis/dissertation. *Entrance requirements:* Additional exam requirements/recommendations for international students: Required—TOEFL. *Application deadline:* For fall admission, 4/1 priority date for domestic students, 3/1 priority date for international students; for spring admission, 10/1 priority date for domestic students, 9/1 priority date for international students. Applications are processed on a rolling basis. Application fee: $55. Electronic applications accepted. *Financial support:* In 2005–06, 2 fellowships with partial tuition reimbursements (averaging $18,000 per year), 55 teaching assistantships with partial tuition reimbursements (averaging $18,000 per year) were awarded; research assistantships with partial tuition reimbursements, tuition waivers (partial) also available. Support available to part-time students. Financial award applicants required to submit FAFSA. *Unit head:* Dr. Timothy S Zwier, Head, 765-494-5203. *Application contact:* R. E. Wild, Chairman, Graduate Admissions, 765-494-5200, E-mail: wild@purdue.edu.

Rensselaer Polytechnic Institute, Graduate School, School of Science, Department of Chemistry and Chemical Biology, Troy, NY 12180-3590. Offers analytical chemistry (MS, PhD); biochemistry (MS, PhD); inorganic chemistry (MS, PhD); organic chemistry (MS, PhD); physical chemistry (MS, PhD); polymer chemistry (MS, PhD). Part-time and evening/weekend programs available. *Faculty:* 19 full-time (3 women). *Students:* 49 full-time (28 women), 2 part-time (1 woman), 33 international. Average age 24. 80 applicants, 23% accepted, 7 enrolled. In 2005, 1 master's, 14 doctorates awarded. Terminal master's awarded for partial completion of doctoral program. *Median time to degree:* Of those who began their doctoral program in fall 1997, 100% received their degree in 8 years or less. *Degree requirements:* For master's, thesis (for some programs), registration; for doctorate, thesis/dissertation, comprehensive exam, registration. *Entrance requirements:* For master's, GRE General Test, GRE Subject Test (strongly recommended); for doctorate, GRE General Test, GRE Subject Test (chemistry or biochemistry strongly recommended). Additional exam requirements/recommendations for international students: Required—TOEFL (minimum score 600 paper-based). *Application deadline:* For fall admission, 2/1 for domestic students; for spring admission, 11/15 for domestic students. Applications are processed on a rolling basis. Application fee: $75. Electronic applications accepted. *Expenses:* Tuition: Full-time $31,000; part-time $1,320 per credit. Required fees: $1,623. *Financial support:* In 2005–06, 49 students received support, including 1 fellowship with full tuition reimbursement available (averaging $30,000 per year), 25 research assistantships with full tuition reimbursements available (averaging $21,500 per year), 30 teaching assistantships with full tuition reimbursements available (averaging $21,500 per year); institutionally sponsored loans and tuition waivers (full and partial) also available. Financial award application deadline: 2/1. *Faculty research:* Synthetic polymer and biopolymer chemistry, physical chemistry of polymeric systems, bioanalytical chemistry, synthetic and computational drug design, protein folding and protein design. Total annual research expenditures: $1.9 million. *Unit head:* Dr. Linda B. McGown, Chair, 518-276-4856, Fax: 518-276-4887, E-mail: mcgowl@rpi.edu. *Application contact:* Beth E. McGraw, Department Admissions Assistant, 518-276-6456, Fax: 518-276-4887, E-mail: mcgrae@rpi.edu.

See Close-Up on page 137.

Rutgers, The State University of New Jersey, Newark, Graduate School, Program in Chemistry, Newark, NJ 07102. Offers analytical chemistry (MS, PhD); biochemistry (MS, PhD); inorganic chemistry (MS, PhD); organic chemistry (MS, PhD); physical chemistry (MS, PhD).

Analytical Chemistry

Rutgers, The State University of New Jersey, Newark (continued)

Part-time and evening/weekend programs available. *Faculty:* 22 full-time (5 women). *Students:* 24 full-time (12 women), 28 part-time (10 women); includes 31 minority (2 African Americans, 29 Asian Americans or Pacific Islanders). 69 applicants, 43% accepted, 13 enrolled. In 2005, 7 master's, 9 doctorates awarded. Terminal master's awarded for partial completion of doctoral program. *Degree requirements:* For master's, cumulative exams, thesis optional; for doctorate, thesis/dissertation, exams, research proposal. *Entrance requirements:* For master's and doctorate, GRE General Test, minimum undergraduate B average. Additional exam requirements/recommendations for international students: Required—TOEFL. *Application deadline:* For fall admission, 7/1 for domestic students; for spring admission, 12/1 for domestic students. Applications are processed on a rolling basis. Application fee: $50. Electronic applications accepted. *Expenses:* Tuition, state resident: full-time $10,440; part-time $435 per credit. Tuition, nonresident: full-time $15,520; part-time $637 per credit. *Financial support:* In 2005–06, 35 students received support, including 5 fellowships with partial tuition reimbursements available (averaging $18,000 per year), 6 research assistantships (averaging $16,988 per year), 20 teaching assistantships with partial tuition reimbursements available (averaging $16,988 per year); Federal Work-Study and institutionally sponsored loans also available. Financial award application deadline: 3/1. *Faculty research:* Medicinal chemistry, natural products, isotope effects, biophysics and biorganic approaches to enzyme mechanisms, organic and organometallic synthesis. *Unit head:* Prof. W. Philip Huskey, Chairman and Program Director, 973-353-5741, Fax: 973-353-1264, E-mail: huskey@andromeda.rutgers.edu.

Rutgers, The State University of New Jersey, New Brunswick/Piscataway, Graduate School, Program in Chemistry and Chemical Biology, New Brunswick, NJ 08901-1281. Offers analytical chemistry (MS, PhD); biological chemistry (PhD); chemistry education (MST); inorganic chemistry (MS, PhD); organic chemistry (MS, PhD); physical chemistry (MS, PhD). Part-time and evening/weekend programs available. *Faculty:* 48 full-time. *Students:* 104 full-time (41 women), 22 part-time (4 women); includes 11 minority (2 African Americans, 6 Asian Americans or Pacific Islanders, 3 Hispanic Americans), 51 international. Average age 29. 108 applicants, 51% accepted, 22 enrolled. In 2005, 13 master's, 10 doctorates awarded. Terminal master's awarded for partial completion of doctoral program. *Degree requirements:* For master's, thesis or alternative, exam, comprehensive exam, registration; for doctorate, thesis/dissertation, cumulative exams, 1 year residency, comprehensive exam, registration. *Entrance requirements:* For master's and doctorate, GRE General Test, GRE Subject Test. Additional exam requirements/recommendations for international students: Required—TOEFL. *Application deadline:* For fall admission, 4/15 priority date for domestic students, 4/1 priority date for international students; for spring admission, 12/1 priority date for domestic students, 12/1 priority date for international students. Applications are processed on a rolling basis. Application fee: $50. Electronic applications accepted. *Expenses:* Tuition, state resident: full-time $10,440; part-time $435 per credit. Tuition, nonresident: full-time $15,520; part-time $647 per credit. Required fees: $129 per credit. Tuition and fees vary according to program. *Financial support:* In 2005–06, 104 students received support, including 8 fellowships with full tuition reimbursements available (averaging $22,000 per year), 23 research assistantships with full tuition reimbursements available (averaging $18,347 per year), 59 teaching assistantships with full tuition reimbursements available (averaging $18,347 per year); career-related internships or fieldwork, Federal Work-Study, traineeships, and health care benefits also available. Financial award application deadline: 3/1; financial award applicants required to submit FAFSA. *Faculty research:* Biophysical organic/bioorganic, inorganic/bioinorganic, theoretical, and solid-state/surface chemistry. Total annual research expenditures: $14.5 million. *Unit head:* Dr. Roger Jones, Director and Chair, 732-445-4900, Fax: 732-445-5312, E-mail: jones@rutchem.rutgers.edu. *Application contact:* Dr. Martha A. Cotter, Vice Chair, 732-445-2259, Fax: 732-445-5312, E-mail: cotter@rutchem.rutgers.edu.

See Close-Up on page 139.

Seton Hall University, College of Arts and Sciences, Department of Chemistry and Biochemistry, South Orange, NJ 07079-2694. Offers analytical chemistry (MS, PhD); biochemistry (MS, PhD); chemistry (MS, PhD); inorganic chemistry (MS, PhD); organic chemistry (MS, PhD); physical chemistry (MS, PhD). Part-time and evening/weekend programs available. *Students:* 19 full-time (6 women), 46 part-time (17 women). Average age 34. 33 applicants, 79% accepted, 14 enrolled. In 2005, 11 master's, 7 doctorates awarded. Terminal master's awarded for partial completion of doctoral program. *Degree requirements:* For master's, formal seminar, thesis optional; for doctorate, thesis/dissertation, annual seminars, comprehensive exam. *Entrance requirements:* Additional exam requirements/recommendations for international students: Required—TOEFL. *Application deadline:* For fall admission, 7/1 priority date for domestic students, 7/1 priority date for international students; for spring admission, 11/1 priority date for domestic students, 11/1 priority date for international students. Applications are processed on a rolling basis. Application fee: $50. Electronic applications accepted. *Financial support:* In 2005–06, 1 research assistantship, 19 teaching assistantships were awarded; Federal Work-Study also available. *Faculty research:* DNA metal reactions; chromatography; bioinorganic, biophysical, organometallic, polymer chemistry; heterogeneous catalyst. *Unit head:* Dr. Nicholas Snow, Chair, 973-761-9414, Fax: 973-761-9772, E-mail: snownich@shu.edu. *Application contact:* Dr. Stephen Kelty, Director of Graduate Studies, 973-761-9129, Fax: 973-761-9772, E-mail: keltyste@shu.edu.

See Close-Up on page 141.

South Dakota State University, Graduate School, College of Arts and Science and College of Agriculture and Biological Sciences, Department of Chemistry, Brookings, SD 57007. Offers analytical chemistry (MS, PhD); biochemistry (MS, PhD); chemistry (MS, PhD); inorganic chemistry (MS, PhD); organic chemistry (MS, PhD); physical chemistry (MS, PhD). *Degree requirements:* For master's, thesis, oral exam; for doctorate, thesis/dissertation, preliminary oral and written exams, research tool. *Entrance requirements:* For master's, bachelor's degree in chemistry or equivalent. Additional exam requirements/recommendations for international students: Required—TOEFL. *Faculty research:* Environmental chemistry, computational chemistry, organic synthesis and photochemistry, novel material development and characterization.

Southern University and Agricultural and Mechanical College, Graduate School, College of Sciences, Department of Chemistry, Baton Rouge, LA 70813. Offers analytical chemistry (MS); biochemistry (MS); environmental sciences (MS); inorganic chemistry (MS); organic chemistry (MS); physical chemistry (MS). *Faculty:* 9 full-time (2 women), 3 part-time/adjunct (2 women). *Students:* 20 full-time (13 women), 8 part-time (5 women); all minorities (20 African Americans, 8 Asian Americans or Pacific Islanders). Average age 23. 30 applicants, 70% accepted, 14 enrolled. In 2005, 3 master's awarded. *Degree requirements:* For master's, thesis. *Entrance requirements:* For master's, GMAT or GRE General Test. Additional exam requirements/recommendations for international students: Required—TOEFL (minimum score 525 paper-based; 193 computer-based). *Application deadline:* For fall admission, 4/15 for domestic students; for spring admission, 11/1 priority date for domestic students. Applications are processed on a rolling basis. Application fee: $5. *Financial support:* In 2005–06, 31 research assistantships (averaging $7,000 per year), 10 teaching assistantships (averaging $7,000 per year) were awarded; scholarships/grants also available. Financial award application deadline: 4/15. *Faculty research:* Synthesis of macrocyclic ligands, latex accelerators, anticancer drugs, biosensors, absorption isotheums, isolation of specific enzymes from plants. Total annual research expenditures:$400,000. *Unit head:* Dr. Ella Kelley, Chair, 225-771-3990, Fax: 225-771-3992.

State University of New York at Binghamton, Graduate School, School of Arts and Sciences, Department of Chemistry, Binghamton, NY 13902-6000. Offers analytical chemistry (PhD); chemistry (MA, MS); inorganic chemistry (PhD); organic chemistry (PhD); physical chemistry (PhD). Part-time programs available. Terminal master's awarded for partial completion of doctoral program. *Degree requirements:* For master's, thesis or alternative, oral exam, seminar presentation; for doctorate, thesis/dissertation, cumulative exams. *Entrance requirements:* For master's and doctorate, GRE General Test, GRE Subject Test. Additional exam requirements/recommendations for international students: Required—TOEFL. Electronic applications accepted.

Stevens Institute of Technology, Graduate School, Arthur E. Imperatore School of Sciences and Arts, Department of Chemistry and Chemical Biology, Hoboken, NJ 07030. Offers chemical biology (MS, PhD); chemistry (MS, PhD), including analytical chemistry (MS, PhD, Certificate), chemical biology (MS, PhD, Certificate), organic chemistry, physical chemistry, polymer chemistry (MS, PhD, Certificate); chemistry and chemical biology (Certificate), including analytical chemistry (MS, PhD, Certificate), bioinformatics, biomedical chemistry, chemical biology (MS, PhD, Certificate), chemical physiology, polymer chemistry (MS, PhD, Certificate). Part-time and evening/weekend programs available. *Students:* 17 full-time (8 women), 28 part-time (16 women); includes 12 minority (9 Asian Americans or Pacific Islanders, 3 Hispanic Americans), 14 international. 51 applicants, 75% accepted.Terminal master's awarded for partial completion of doctoral program. *Degree requirements:* For master's, thesis or alternative; for doctorate, one foreign language, thesis/dissertation; for Certificate, project or thesis. *Entrance requirements:* Additional exam requirements/recommendations for international students: Required—TOEFL. *Application deadline:* Applications are processed on a rolling basis. Application fee: $50. Electronic applications accepted. *Expenses:* Tuition: Part-time $920 per credit hour. Tuition and fees vary according to program. *Financial support:* Fellowships, research assistantships, teaching assistantships, Federal Work-Study and institutionally sponsored loans available. Financial award application deadline: 4/1. *Unit head:* Dr. Francis T. Jones, Director, 201-216-5518, Fax: 201-216-8240.

Tufts University, Graduate School of Arts and Sciences, Department of Chemistry, Medford, MA 02155. Offers analytical chemistry (MS, PhD); bioorganic chemistry (MS, PhD); environmental chemistry (MS, PhD); inorganic chemistry (MS, PhD); organic chemistry (MS, PhD); physical chemistry (MS, PhD). *Faculty:* 17 full-time. *Students:* 55 (27 women); includes 5 minority (1 African American, 4 Asian Americans or Pacific Islanders) 27 international. 45 applicants, 87% accepted, 17 enrolled. In 2005, 4 master's, 4 doctorates awarded. Terminal master's awarded for partial completion of doctoral program. *Degree requirements:* For master's and doctorate, thesis/dissertation. *Entrance requirements:* For master's and doctorate, GRE General Test, GRE Subject Test. Additional exam requirements/recommendations for international students: Required—TOEFL (minimum score 600 paper-based; 250 computer-based), TSE. *Application deadline:* For fall admission, 1/15 for domestic students, 12/30 for international students; for spring admission, 10/15 for domestic students, 9/15 for international students. Applications are processed on a rolling basis. Application fee: $65. Electronic applications accepted. *Expenses:* Tuition: Full-time $32,360. Tuition and fees vary according to program. *Financial support:* Research assistantships with full and partial tuition reimbursements, teaching assistantships with full and partial tuition reimbursements, Federal Work-Study, scholarships/grants, and tuition waivers (partial) available. Financial award application deadline: 1/15; financial award applicants required to submit FAFSA. *Unit head:* Mary Jane Shultz, Chair, 617-627-3477, Fax: 617-627-3443. *Application contact:* Samuel Kounaves, Information Contact, 617-627-3441, Fax: 617-627-3443.

See Close-Up on page 151.

University of Calgary, Faculty of Graduate Studies, Faculty of Science, Department of Chemistry, Calgary, AB T2N 1N4, Canada. Offers analytical chemistry (M Sc, PhD); applied chemistry (M Sc, PhD); inorganic chemistry (M Sc, PhD); organic chemistry (M Sc, PhD); physical chemistry (M Sc, PhD); polymer chemistry (M Sc, PhD); theoretical chemistry (M Sc, PhD). *Faculty:* 31 full-time (2 women), 2 part-time/adjunct (0 women). *Students:* 80 full-time (38 women). Average age 25. 250 applicants, 6% accepted, 15 enrolled. In 2005, 7 master's, 7 doctorates awarded. *Degree requirements:* For master's, thesis; for doctorate, thesis/dissertation, candidacy exam. *Entrance requirements:* For master's, minimum GPA of 3.0; for doctorate, honors B Sc degree with minimum GPA of 3.7 or M Sc with minimum GPA of 3.3. Additional exam requirements/recommendations for international students: Required—TOEFL (minimum score 580 paper-based; 237 computer-based). *Application deadline:* For fall admission, 12/1 priority date for domestic students, 12/1 priority date for international students. Applications are processed on a rolling basis. Application fee: $100 ($130 for international students). Electronic applications accepted. *Financial support:* In 2005–06, 25 students received support, including research assistantships (averaging $11,000 per year), teaching assistantships (averaging $6,530 per year); fellowships, scholarships/grants also available. Financial award application deadline:12/1. *Faculty research:* Chemical analysis, chemical dynamics, synthesis theory. *Unit head:* Dr. Brian Keay, Head, 403-220-5340, E-mail: info@chem.ucalgary.ca. *Application contact:* Bonnie E. King, Graduate Program Administrator, 403-220-6252, E-mail: bking@ucalgary.ca.

University of Cincinnati, Division of Research and Advanced Studies, McMicken College of Arts and Sciences, Department of Chemistry, Cincinnati, OH 45221. Offers analytical chemistry (MS, PhD); biochemistry (MS, PhD); inorganic chemistry (MS, PhD); organic chemistry (MS, PhD); physical chemistry (MS, PhD); polymer chemistry (MS, PhD); sensors (PhD). Part-time and evening/weekend programs available. Terminal master's awarded for partial completion of doctoral program. *Degree requirements:* For master's, thesis optional; for doctorate, thesis/dissertation, comprehensive exam, registration. *Entrance requirements:* For master's and doctorate, GRE General Test. Additional exam requirements/recommendations for international students: Required—TOEFL (minimum score 580 paper-based; 237 computer-based); Recommended—TSE(minimum score 50). Electronic applications accepted. *Faculty research:* Biomedical chemistry, laser chemistry, surface science, chemical sensors, synthesis.

University of Georgia, Graduate School, College of Arts and Sciences, Department of Chemistry, Athens, GA 30602. Offers analytical chemistry (MS, PhD); inorganic chemistry (MS, PhD); organic chemistry (MS, PhD); physical chemistry (MS, PhD). *Faculty:* 22 full-time (2 women). *Students:* 155 full-time, 3 part-time; includes 12 minority (4 African Americans, 7 Asian Americans or Pacific Islanders, 1 Hispanic American), 87 international. 147 applicants, 56% accepted, 35 enrolled. In 2005, 2 master's, 19 doctorates awarded. Terminal master's awarded for partial completion of doctoral program. *Median time to degree:* Of those who began their doctoral program in fall 1997, 100% received their degree in 8 years or less. *Degree requirements:* For master's, thesis; for doctorate, one foreign language, thesis/dissertation. *Entrance requirements:* For master's and doctorate, GRE General Test. Additional exam requirements/recommendations for international students: Required—TOEFL (minimum score 213 computer-based), TSE(minimum score 50). *Application deadline:* For fall admission, 7/1 for domestic students; for spring admission, 11/15 for domestic students. Application fee: $50. Electronic applications accepted. *Financial support:* Fellowships, research assistantships, teaching assistantships, unspecified assistantships available. *Unit head:* Dr. John L. Stickney, Head, 706-542-2726, Fax: 706-542-9454, E-mail: stickney@chem.uga.edu. *Application contact:* Dr. Donald Kurtz, Information Contact, 706-542-2010, Fax: 706-542-9454, E-mail: kurtz@chem.uga.edu.

University of Louisville, Graduate School, College of Arts and Sciences, Department of Chemistry, Louisville, KY 40292-0001. Offers analytical chemistry (MS, PhD); biochemistry (MS, PhD); chemical physics (PhD); inorganic chemistry (MS, PhD); organic chemistry (MS, PhD); physical chemistry (MS, PhD). *Students:* 37 full-time (16 women), 16 part-time (7 women); includes 3 minority (2 African Americans, 1 Hispanic American), 24 international. Average age 28. In 2005, 1 master's, 8 doctorates awarded. *Degree requirements:* For master's, thesis/dissertation; for doctorate, thesis/dissertation, comprehensive exam. *Entrance requirements:* For master's and doctorate, GRE General Test. Additional exam requirements/recommendations for international students: Required—TOEFL. *Application deadline:* Applications are processed on a rolling basis. Application fee: $50. *Expenses:* Tuition, state resident: full-time $3,003; part-time $334 per credit hour. Tuition, nonresident: full-time $8,277; part-time $920 per credit hour. Tuition and fees vary according to course load, degree level and program. *Financial support:* In 2005–06, 33 teaching assistantships with tuition reimbursements were awarded; fellowships, research assistantships *Unit head:* Dr. George R. Pack, Chair, 502-852-6798, Fax: 502-852-8149, E-mail: george.pack@louisville.edu.

University of Maryland, College Park, Graduate Studies, College of Chemical and Life Sciences, Department of Chemistry and Biochemistry, Chemistry Program, College Park, MD 20742. Offers analytical chemistry (MS, PhD); inorganic chemistry (MS, PhD); organic chemistry (MS, PhD); physical chemistry (MS, PhD). Part-time and evening/weekend programs available. *Students:* 102 full-time (43 women), 4 part-time (2 women); includes 7 minority (2 African Americans, 3 Asian Americans or Pacific Islanders, 2 Hispanic Americans), 46 international. 128 applicants, 24% accepted, 30 enrolled. In 2005, 6 master's, 12 doctorates awarded. Terminal master's awarded for partial completion of doctoral program. *Median time to degree:* Of those who began their doctoral program in fall 1997, 33% received their degree in 8 years or less. *Degree requirements:* For master's, thesis optional; for doctorate, thesis/dissertation, 2 seminar presentations, oral exam. *Entrance requirements:* For master's and doctorate, GRE General Test, GRE Subject Test (recommended), minimum GPA of 3.0, 3 letters of recommendation. Additional exam requirements/recommendations for international students: Required—TOEFL, TSE. *Application deadline:* For fall admission, 4/1 for domestic students, 2/1 for international students; for spring admission, 10/21 for domestic students, 6/1 for international students. Applications are processed on a rolling basis. Application fee: $60. Electronic applications accepted. *Financial support:* In 2005–06, 10 fellowships (averaging $4,900 per year) were awarded; research assistantships, teaching assistantships with partial tuition reimbursements Financial award applicants required to submit FAFSA. *Faculty research:* Environmental chemistry, nuclear chemistry, lunar and environmental analysis, x-ray crystallography. *Application contact:* Dean of Graduate School, 301-405-4190, Fax: 301-314-9305.

University of Michigan, Horace H. Rackham School of Graduate Studies, College of Literature, Science, and the Arts, Department of Chemistry, Ann Arbor, MI 48109. Offers analytical chemistry (PhD); inorganic chemistry (PhD); material chemistry (PhD); organic chemistry (PhD); physical chemistry (PhD). *Faculty:* 48 full-time (8 women), 3 part-time/adjunct (all women). *Students:* ; includes 39 minority (4 African Americans, 1 American Indian/Alaska Native, 29 Asian Americans or Pacific Islanders, 5 Hispanic Americans), 66 international. Average age 26. 456 applicants, 29% accepted, 44 enrolled. In 2005, 37 degrees awarded. *Degree requirements:* For doctorate, thesis/dissertation, oral defense of dissertation, organic cumulative proficiency exams. *Entrance requirements:* For doctorate, GRE General Test, GRE Subject Test (recommended), 3 letters of recommendation. Additional exam requirements/recommendations for international students: Required—TOEFL. *Application deadline:* For fall admission, 1/31 for domestic students, 1/1 for international students. Applications are processed on a rolling basis. Application fee: $60 ($75 for international students). Electronic applications accepted. *Expenses:* Tuition, area resident: Full-time $14,082; part-time $894 per credit hour. Tuition, state resident: full-time $14,082; part-time $896 per credit hour. Tuition, nonresident: full-time $28,500; part-time $1,675 per credit hour. Required fees: $189; $189. *Financial support:* In 2005–06, 10 fellowships with full tuition reimbursements (averaging $20,000 per year), 70 research assistantships with full tuition reimbursements (averaging $19,000 per year), 120 teaching assistantships with full tuition reimbursements (averaging $20,000 per year) were awarded. Financial award applicants required to submit FAFSA. *Faculty research:* Biological catalysis, protein engineering, chemical sensors, de novo metalloprotein design, supramolecular architecture. Total annual research expenditures: $8 million. *Unit head:* Dr. Carol A. Fierke, Chair, 734-763-9681, Fax: 734-647-4847. *Application contact:* Linda Deitert, Assistant Director Graduate Studies, 734-764-7278, Fax: 734-647-4865, E-mail: chemadmissions@umich.edu.

University of Missouri–Columbia, Graduate School, College of Arts and Sciences, Department of Chemistry, Columbia, MO 65211. Offers analytical chemistry (MS, PhD); inorganic chemistry (MS, PhD); organic chemistry (MS, PhD); physical chemistry (MS, PhD). *Faculty:* 21 full-time (5 women). *Students:* 58 full-time (17 women), 44 part-time (16 women); includes 5 minority (3 African Americans, 2 Asian Americans or Pacific Islanders), 59 international. In 2005, 11 master's, 9 doctorates awarded. *Degree requirements:* For master's, thesis; for doctorate, one foreign language, thesis/dissertation. *Entrance requirements:* For master's and doctorate, GRE General Test, minimum GPA of 3.0. *Application deadline:* For fall admission, 4/1 for domestic studentsFor winter admission, 11/1 for domestic students. Applications are processed on a rolling basis. Application fee: $45 ($60 for international students). *Financial support:* Fellowships, research assistantships, teaching assistantships, institutionally sponsored loans available. *Unit head:* Dr. Sheryl A. Tucker, Director for Graduate Studies, 573-882-1729.

University of Missouri–Kansas City, College of Arts and Sciences, Department of Chemistry, Kansas City, MO 64110-2499. Offers analytical chemistry (MS, PhD); inorganic chemistry (MS, PhD); organic chemistry (MS, PhD); physical chemistry (MS, PhD); polymer chemistry (MS, PhD). PhD offered through the School of Graduate Studies. Part-time and evening/weekend programs available. *Faculty:* 14 full-time (2 women), 2 part-time/adjunct (0 women). *Students:* Average age 38. 13 applicants, 8% accepted, 1 enrolled. In 2005, 2 degrees awarded. *Degree requirements:* For master's, thesis (for some programs); for doctorate, thesis/dissertation. *Entrance requirements:* For master's, equivalent of American Chemical Society approved bachelor's degree in chemistry; for doctorate, GRE General Test, equivalent of American Chemical Society approved bachelor's degree in chemistry. Additional exam requirements/recommendations for international students: Required—TOEFL (minimum score 580 paper-based; 237 computer-based), TWE. *Application deadline:* For fall and spring admission, 4/15For winter admission, 10/15 for domestic students. Applications are processed on a rolling basis. Application fee: $25. Electronic applications accepted. *Expenses:* Tuition, state resident: full-time $4,738; part-time $263 per credit hour. Tuition, nonresident: full-time $12,235; part-time $679 per credit hour. Required fees: $582. Tuition and fees vary according to course load, program and student level. *Financial support:* In 2005–06, fellowships with partial tuition reimbursements (averaging $18,156 per year), research assistantships with partial tuition reimbursements (averaging $18,515 per year), teaching assistantships with partial tuition reimbursements (averaging $16,434 per year) were awarded; Federal Work-Study, institutionally sponsored loans, and scholarships/grants also available. Support available to part-time students. Financial award application deadline: 2/15. *Faculty research:* Molecular spectroscopy, characterization and synthesis of materials and compounds, computational chemistry, natural products, drug delivery systems and anti-tumor agents. Total annual research expenditures: $1.1 million. *Unit head:* Dr. Jerry Jean, Chairperson, 816-235-2273, Fax: 816-235-5502, E-mail: jeany@umkc.edu. *Application contact:* Graduate Recruiting Committee, 816-235-2272, Fax: 816-235-5502, E-mail: umkc-chemdept@umkc.edu.

The University of Montana, Graduate School, College of Arts and Sciences, Department of Chemistry, Missoula, MT 59812-0002. Offers chemistry (MS, PhD), including environmental/analytical chemistry, inorganic chemistry, organic chemistry, physical chemistry. *Faculty:* 16 full-time (2 women). *Students:* 36 full-time (9 women), 7 part-time (2 women), 9 international. 25 applicants, 44% accepted, 6 enrolled. In 2005, 1 master's, 2 doctorates awarded. Terminal master's awarded for partial completion of doctoral program. *Degree requirements:* For master's, thesis (for some programs); for doctorate, thesis/dissertation. *Entrance requirements:* For master's and doctorate, GRE General Test. Additional exam requirements/recommendations for international students: Required—TOEFL (minimum score 575 paper-based; 230 computer-based). *Application deadline:* For fall admission, 2/15 for domestic students. Applications are processed on a rolling basis. Application fee: $45. *Expenses:* Tuition, state resident: part-time $267 per credit. Tuition, nonresident: part-time $665 per credit. Part-time tuition and fees vary according to course load and degree level. *Financial support:* In 2005–06, 13 research assistantships with tuition reimbursements (averaging $14,000 per year), 12 teaching assistantships with full tuition reimbursements (averaging $14,000 per year) were awarded; Federal Work-Study, scholarships/grants, and unspecified assistantships also available. Financial award application deadline: 3/1; financial award applicants required to submit FAFSA. *Faculty research:* Reaction mechanisms and kinetics, inorganic and organic synthesis, analytical chemistry, natural products. Total annual research expenditures: $789,952. *Unit head:* Dr. Edward Rosenberg, Chair, 406-243-2592, Fax: 406-243-4227.

University of Nebraska–Lincoln, Graduate College, College of Arts and Sciences, Department of Chemistry, Lincoln, NE 68588. Offers analytical chemistry (PhD); chemistry (MS); inorganic chemistry (PhD); organic chemistry (PhD); physical chemistry (PhD). *Degree requirements:* For master's, one foreign language, departmental qualifying exam, thesis optional; for doctorate, one foreign language, thesis/dissertation, departmental qualifying exams, comprehensive exam. *Entrance requirements:* For master's and doctorate, GRE. Additional exam requirements/recommendations for international students: Required—TOEFL (minimum score 550 paper-based; 213 computer-based). Electronic applications accepted. *Faculty research:* Bioorganic and bioinorganic chemistry, biophysical and bioanalytical chemistry, structure-function of DNA and proteins, organometallics, mass spectrometry.

University of Regina, Faculty of Graduate Studies and Research, Faculty of Science, Department of Chemistry and Biochemistry, Regina, SK S4S 0A2, Canada. Offers analytical chemistry (M Sc, PhD); biochemistry (M Sc, PhD); inorganic chemistry (M Sc, PhD); organic chemistry (M Sc, PhD); physical chemistry (M Sc, PhD). Part-time programs available. *Faculty:* 10 full-time (2 women), 4 part-time/adjunct (1 woman). *Students:* 14 full-time (6 women), 3 part-time (2 women). 21 applicants, 33% accepted. *Degree requirements:* For master's and doctorate, thesis/dissertation, departmental qualifying exam. *Entrance requirements:* For master's and doctorate, GRE. Additional exam requirements/recommendations for international students: Required—TOEFL (minimum score 580 paper-based; 237 computer-based). *Application deadline:* For fall admission, 1/1 for domestic studentsFor winter admission, 7/1 for domestic students. Applications are processed on a rolling basis. Application fee: $60 ($100 for international students). *Financial support:* In 2005–06, 4 fellowships (averaging $14,886 per year), 1 research assistantship (averaging $12,750 per year), 5 teaching assistantships (averaging $13,501 per year) were awarded; scholarships/grants also available. Financial award application deadline: 6/15. *Faculty research:* Organic synthesis, organic oxidations, ionic liquids theoretical/computational chemistry, protein biochemistry/biophysics, environmental analytical, photophysical/photochemistry. *Unit head:* Dr. Andrew G. Wee, Head, 306-585-4767, Fax: 306-585-4894, E-mail: chem.chair@uregina.ca. *Application contact:* Dr. Allan East, Associate Professor, 306-585-4003, Fax: 306-585-4894, E-mail: allan.east@uregina.ca.

University of Southern Mississippi, Graduate School, College of Science and Technology, Department of Chemistry and Biochemistry, Hattiesburg, MS 39406-0001. Offers analytical chemistry (MS, PhD); biochemistry (MS, PhD); inorganic chemistry (MS, PhD); organic chemistry (MS, PhD); physical chemistry (MS, PhD). *Degree requirements:* For master's and doctorate, thesis/dissertation, comprehensive exam. *Entrance requirements:* For master's, GRE General Test, minimum GPA of 2.75 in last 60 hours; for doctorate, GRE General Test, minimum GPA of 3.5. Additional exam requirements/recommendations for international students: Required—TOEFL. *Faculty research:* Plant biochemistry, photo chemistry, polymer chemistry, x-ray analysis, enzyme chemistry.

University of South Florida, College of Graduate Studies, College of Arts and Sciences, Department of Chemistry, Tampa, FL 33620-9951. Offers analytical chemistry (MS, PhD); biochemistry (MS, PhD); inorganic chemistry (MS, PhD); organic chemistry (MS, PhD); physical chemistry (MS, PhD); polymer chemistry (PhD). Part-time programs available. *Faculty:* 22. *Students:* 102 full-time (45 women), 15 part-time (6 women); includes 13 minority (5 Asian Americans or Pacific Islanders, 8 Hispanic Americans), 60 international. 80 applicants, 71% accepted, 23 enrolled. In 2005, 2 master's awarded. Terminal master's awarded for partial completion of doctoral program. *Degree requirements:* For master's, thesis; for doctorate, 2 foreign languages, thesis/dissertation, colloquium. *Entrance requirements:* For master's and doctorate, GRE General Test, minimum GPA of 3.0 in last 30 hours of chemistry course work, 3 letters of recommendation. Additional exam requirements/recommendations for international students: Required—TOEFL (minimum score 550 paper-based; 213 computer-based), TSE(minimum score 50). *Application deadline:* For fall admission, 5/1 priority date for domestic students, 3/1 priority date for international students; for spring admission, 10/1 priority date for domestic students, 8/1 priority date for international students. Applications are processed on a rolling basis. Application fee: $30. Electronic applications accepted. *Financial support:* In 2005–06, 91 students received support; fellowships with partial tuition reimbursements available, research assistantships with partial tuition reimbursements available, teaching assistantships with partial tuition reimbursements available, career-related internships or fieldwork, institutionally sponsored loans, scholarships/grants, health care benefits, and unspecified assistantships available. Financial award application deadline: 6/30. *Faculty research:* Synthesis, bio-organic chemistry, bioinorganic chemistry, environmental chemistry, NMR. *Unit head:* Dr. Michael Zaworotko, Chairperson, 813-974-4129, Fax: 813-974-3203, E-mail: xtal@usf.edu. *Application contact:* Dr. Brian Space, Graduate Coordinator, 813-974-3397, Fax: 813-974-3203.

The University of Tennessee, Graduate School, College of Arts and Sciences, Department of Chemistry, Knoxville, TN 37996. Offers analytical chemistry (PhD); chemical physics (PhD); environmental chemistry (MS, PhD); inorganic chemistry (MS, PhD); organic chemistry (MS, PhD); physical chemistry (MS, PhD); polymer chemistry (MS, PhD); theoretical chemistry (PhD). Part-time programs available. Terminal master's awarded for partial completion of doctoral program. *Degree requirements:* For master's and doctorate, thesis/dissertation. *Entrance requirements:* For master's and doctorate, GRE General Test, minimum GPA of 2.7. Additional exam requirements/recommendations for international students: Required—TOEFL. Electronic applications accepted.

The University of Texas at Austin, Graduate School, College of Natural Sciences, Department of Chemistry and Biochemistry, Austin, TX 78712-1111. Offers analytical chemistry (MA, PhD); biochemistry (MA, PhD); inorganic chemistry (MA, PhD); organic chemistry (MA, PhD); physical chemistry (MA, PhD). *Entrance requirements:* For master's and doctorate, GRE General Test.

The University of Toledo, Graduate School, College of Arts and Sciences, Department of Chemistry, Toledo, OH 43606-3390. Offers analytical chemistry (MS, PhD); biological chemistry (MS, PhD); inorganic chemistry (MS, PhD); organic chemistry (MS, PhD); physical chemistry (MS, PhD). Part-time programs available. *Faculty:* 18. *Students:* 80 full-time (35 women), 8 part-time (5 women); includes 3 minority (2 African Americans, 1 Asian American or Pacific Islander), 43 international. Average age 27. 32 applicants, 75% accepted, 19 enrolled. In 2005, 7 master's awarded. *Degree requirements:* For master's and doctorate, thesis/dissertation. *Entrance requirements:* For master's and doctorate, GRE General Test, GRE Subject Test. Additional exam requirements/recommendations for international students: Required—TOEFL. *Application deadline:* For fall admission, 8/1 for domestic students. Applications are processed on a rolling basis. Application fee: $45. Electronic applications accepted. *Expenses:* Tuition, area resident: Full-time $3,312; part-time $308 per credit hour. Tuition, state resident: full-time $3,312. Tuition, nonresident: full-time $6,616; part-time $735 per credit hour. International tuition: $6,616 full-time. *Financial support:* In 2005–06, 1 research assistantship with full tuition reimbursement (averaging $4,000 per year), 55 teaching assistantships with full tuition reimbursements (averaging $13,336 per year) were awarded; fellowships, Federal Work-Study and institutionally sponsored loans also available. Support available to part-time students. Financial award application deadline: 4/1; financial award applicants required to submit FAFSA. *Faculty research:* Enzymology, materials chemistry, crystallography, theoretical chemistry. *Unit head:* Dr. Alan Pinkerton, Chair, 419-530-7902, Fax: 419-530-4033, E-mail: apinker@uoft02.utoledo.edu. *Application contact:* Charlene Morlock-Hansen, 419-530-2100, E-mail: charlene.hanson@utoledo.edu.

See Close-Up on page 167.

Vanderbilt University, Graduate School, Department of Chemistry, Nashville, TN 37240-1001. Offers analytical chemistry (MAT, MS, PhD); inorganic chemistry (MAT, MS, PhD); organic chemistry (MAT, MS, PhD); physical chemistry (MAT, MS, PhD); theoretical chemistry (MAT, MS, PhD). *Faculty:* 32 full-time (6 women), 3 part-time/adjunct (2 women). *Students:* 96 full-time (40 women); includes 4 minority (2 African Americans, 1 Asian American or Pacific Islander, 1 Hispanic American), 20 international. 223 applicants, 21% accepted, 22 enrolled. In 2005, 4 master's, 12 doctorates awarded. *Degree requirements:* For master's, thesis or alternative; for doctorate, thesis/dissertation, area, qualifying, and final exams. *Entrance requirements:* For master's and doctorate, GRE General Test, GRE Subject Test (recommended).

Analytical Chemistry

Vanderbilt University (continued)
Additional exam requirements/recommendations for international students: Required—TOEFL. *Application deadline:* For fall admission, 1/15 for domestic students, 1/15 for international students. Application fee: $0. Electronic applications accepted. *Expenses:* Tuition: Full-time $15,396; part-time $1,283 per semester hour. Required fees: $2,202; $1,101 per semester. One-time fee: $30. Tuition and fees vary according to course load, program and student level. *Financial support:* Fellowships with full and partial tuition reimbursements, research assistantships with full tuition reimbursements, teaching assistantships with full tuition reimbursements, Federal Work-Study, institutionally sponsored loans, and traineeships available. Financial award application deadline:1/15. *Faculty research:* Chemical synthesis; mechanistic, theoretical, bioorganic, analytical, and spectroscopic chemistry. *Unit head:* Ned A. Porter, Chair, 615-322-2861, Fax: 615-343-1234. *Application contact:* Charles M. Lukehart, Director of Graduate Studies, 615-322-2861, Fax: 615-343-1234, E-mail: charles.m.lukehart@vanderbilt.edu.

Virginia Commonwealth University, Graduate School, College of Humanities and Sciences, Department of Chemistry, Richmond, VA 23284-9005. Offers analytical (MS, PhD); chemical physics (PhD); inorganic (MS, PhD); organic (MS, PhD); physical (MS, PhD). Part-time programs available. *Faculty:* 18 full-time (6 women). *Students:* 36 full-time (13 women), 11 part-time (1 woman); includes 13 minority (6 African Americans, 1 American Indian/Alaska Native, 5 Asian Americans or Pacific Islanders, 1 Hispanic American), 13 international. 47 applicants, 74% accepted. In 2005, 2 master's, 8 doctorates awarded. Terminal master's awarded for partial completion of doctoral program. *Degree requirements:* For master's, thesis; for doctorate, thesis/dissertation, comprehensive cumulative exams, research proposal. *Entrance requirements:* For master's, GRE General Test, 30 undergraduate credits in chemistry; for doctorate, GRE General Test. *Application deadline:* For fall admission, 3/15 for domestic students; for spring admission, 11/15 for domestic students. Applications are processed on a rolling basis. Application fee: $50. *Expenses:* Tuition, state resident: full-time $3,185; part-time $405 per credit. Tuition, nonresident: full-time $7,952; part-time $940 per credit. Required fees: $751 per semester hour. Tuition and fees vary according to course load and program. *Financial support:* Fellowships, research assistantships, teaching assistantships, career-related internships or fieldwork and institutionally sponsored loans available. Support available to part-time students. Financial award application deadline: 7/1. *Faculty research:* Physical, organic, inorganic, analytical, and polymer chemistry; chemical physics. *Unit head:* Dr. Fred M. Hawkridge, Chair, 804-828-1298, Fax: 804-828-8599, E-mail: fmhawkri@vcu.edu. *Application contact:* Dr. M. Samy El-Shall, Chair, Graduate Recruiting and Admissions Committee, 804-828-3518, E-mail: mselshal@vcu.edu.

Wake Forest University, Graduate School, Department of Chemistry, Winston-Salem, NC 27109. Offers analytical chemistry (MS, PhD); inorganic chemistry (MS, PhD); organic chemistry (MS, PhD); physical chemistry (MS, PhD). Part-time programs available. *Faculty:* 17 full-time (4 women). *Students:* 32 full-time (17 women), 2 part-time; includes 1 minority (African American), 14 international. Average age 28. 60 applicants, 42% accepted, 12 enrolled. In 2005, 1 master's, 6 doctorates awarded. *Median time to degree:* Of those who began their doctoral program in fall 1997, 0% received their degree in 8 years or less. *Degree requirements:* For master's, one foreign language, thesis, comprehensive exam, registration; for doctorate, 2 foreign languages, thesis/dissertation, comprehensive exam, registration. *Entrance requirements:* For master's and doctorate, GRE General Test. Additional exam requirements/recommendations for international students: Required—TOEFL (minimum score 213 computer-based). *Application deadline:* For fall admission, 1/15 for domestic students, 1/15 for international students. Application fee: $45. Electronic applications accepted. *Financial support:* In 2005–06, 32 students received support, including 1 fellowship with full tuition reimbursement available (averaging $21,500 per year), 12 research assistantships with full tuition reimbursements available (averaging $19,500 per year), 17 teaching assistantships with full tuition reimbursements available (averaging $19,500 per year); scholarships/grants and tuition waivers (full and partial) also available. Support available to part-time students. Financial award application deadline: 1/15; financial award applicants required to submit FAFSA. *Unit head:* Dr. Bruce King, Director, 336-758-5774, Fax: 335-758-4656, E-mail: kingsb@wfu.edu.

Washington State University, Graduate School, College of Sciences, Department of Chemistry, Pullman, WA 99164. Offers analytical chemistry (MS, PhD); biological systems (MS, PhD); inorganic chemistry (MS, PhD); organic chemistry (MS, PhD); physical chemistry (MS, PhD). *Faculty:* 33. *Students:* 38 full-time (19 women), 2 part-time (1 woman); includes 3 minority (2 Asian Americans or Pacific Islanders, 1 Hispanic American), 11 international. Average age 29. 55 applicants, 33% accepted, 9 enrolled. In 2005, 5 master's, 7 doctorates awarded. Terminal master's awarded for partial completion of doctoral program. *Degree requirements:* For master's, oral exam, teaching experience, thesis optional; for doctorate, thesis/dissertation, oral exam, written exam, comprehensive exam. *Entrance requirements:* For master's and doctorate, GRE General Test, minimum GPA of 3.0, 3 letters of recommendation. *Application deadline:* For fall admission, 3/1 priority date for domestic students, 3/1 priority date for international students; for spring admission, 10/1 priority date for domestic students, 7/1 priority date for international students. Applications are processed on a rolling basis. Application fee: $35. *Expenses:* Tuition, state resident: full-time $6,295; part-time $336 per credit. Tuition, nonresident: full-time $15,949; part-time $819 per credit. Required fees: $933. Part-time tuition and fees vary according to campus/location and program. *Financial support:* In 2005–06, 38 students received support, including 3 fellowships (averaging $8,977 per year), 12 research assistantships with full and partial tuition reimbursements available (averaging $15,999 per year), 23 teaching assistantships with full and partial tuition reimbursements available (averaging $14,747 per year); career-related internships or fieldwork, Federal Work-Study, institutionally sponsored loans, scholarships/grants, health care benefits, unspecified assistantships, and summer support also available. Financial award application deadline: 4/1; financial award applicants required to submit FAFSA. *Faculty research:* Environmental chemistry, materials chemistry, radio chemistry, bio-organic, computational chemistry. Total annual research expenditures: $3.8 million. *Unit head:* Sue B. Clark, Chair, Admissions Committee, 509-335-8866, Fax: 509-335-8867, E-mail: sclark@mail.wsu.edu. *Application contact:* Admissions Committee.

West Virginia University, Eberly College of Arts and Sciences, Department of Chemistry, Morgantown, WV 26506. Offers analytical chemistry (MS, PhD); inorganic chemistry (MS, PhD); organic chemistry (MS, PhD); physical chemistry (MS, PhD); theoretical chemistry (MS, PhD). Part-time programs available. Postbaccalaureate distance learning degree programs offered (no on-campus study). *Faculty:* 14 full-time (1 woman), 5 part-time/adjunct (3 women). *Students:* 45 full-time (15 women), 3 part-time (2 women); includes 1 minority (Hispanic American), 26 international. Average age 28. 84 applicants, 58% accepted, 11 enrolled. In 2005, 5 master's, 3 doctorates awarded. Terminal master's awarded for partial completion of doctoral program. *Degree requirements:* For master's and doctorate, thesis/dissertation, registration. *Entrance requirements:* For master's, GRE General Test, GRE Subject Test (recommended), minimum GPA of 2.5; for doctorate, GRE General Test, GRE Subject Test (recommended), minimum GPA of 2.75. Additional exam requirements/recommendations for international students: Required—TOEFL. *Application deadline:* For fall admission, 3/1 for domestic students. Applications are processed on a rolling basis. Application fee: $50. Electronic applications accepted. *Expenses:* Tuition, area resident: Full-time $4,582; part-time $258 per credit hour. Tuition, state resident: full-time $4,582; part-time $258 per credit hour. Tuition, nonresident: full-time $1,382; part-time $741 per credit hour. International tuition: $1,382 full-time. *Financial support:* In 2005–06, fellowships (averaging $2,000 per year), research assistantships with full tuition reimbursements (averaging $13,000 per year), teaching assistantships with full tuition reimbursements (averaging $12,000 per year) were awarded; tuition waivers (full and partial) also available. Financial award application deadline: 2/1; financial award applicants required to submit FAFSA. *Faculty research:* Analysis of proteins, drug interactions, solids and effluents by advanced separation methods; new synthetic strategies for complex organic molecules; synthesis and structural characterization of metal complexes for polymerization catalysis. *Unit head:* Dr. Harry O. Finklea, Chair, 304-293-3435 Ext. 4453, Fax: 304-293-4904, E-mail: harry.finklea@mail.wvu.edu.

Chemistry

Acadia University, Faculty of Pure and Applied Science, Department of Chemistry, Wolfville, NS B4P 2R6, Canada. Offers M Sc. *Faculty:* 1 full-time (0 women). *Students:* 1 full-time (0 women). Average age 26. 5 applicants, 40% accepted, 1 enrolled. *Degree requirements:* For master's, thesis. *Entrance requirements:* For master's, GRE. Additional exam requirements/recommendations for international students: Required—TOEFL (minimum score 580 paper-based; 237 computer-based). *Application deadline:* For fall admission, 2/1 for domestic students. Application fee: $50. Electronic applications accepted. *Financial support:* In 2005–06, 1 teaching assistantship (averaging $8,000 per year) was awarded Financial award application deadline: 2/1. *Faculty research:* Atmospheric chemistry, chemical kinetics, bioelectrochemistry of proteins, organosilicon dewdrimers, self assembling monolayers. *Unit head:* Dr. John M. Roscoe, Head, 902-585-1353, Fax: 902-585-1114, E-mail: john.roscoe@acadiau.ca. *Application contact:* Avril Bird, Secretary, 902-585-1242, Fax: 902-585-1114, E-mail: avril.bird@acadiau.ca.

American University, College of Arts and Sciences, Department of Chemistry, Program in Chemistry, Washington, DC 20016-8001. Offers MS. *Students:* 14 full-time (12 women), 33 part-time (27 women); includes 24 minority (13 African Americans, 7 Asian Americans or Pacific Islanders, 4 Hispanic Americans), 3 international. Average age 26. In 2005, 9 degrees awarded. *Degree requirements:* For master's, thesis, comprehensive exam. *Entrance requirements:* For master's, GRE, minimum GPA of 3.0. *Application deadline:* For fall admission, 2/1 for domestic students; for spring admission, 10/1 priority date for domestic students. Applications are processed on a rolling basis. Application fee: $50. *Expenses:* Tuition: Full-time $17,802; part-time $989 per credit. Required fees: $380. *Financial support:* In 2005–06, 28 students received support, including fellowships with full tuition reimbursements available (averaging $14,500 per year), research assistantships with full tuition reimbursements available (averaging $8,500 per year), teaching assistantships with full tuition reimbursements available (averaging $8,500 per year); career-related internships or fieldwork, Federal Work-Study, scholarships/grants, and traineeships also available. Financial award application deadline: 2/1.

American University of Beirut, Graduate Programs, Faculty of Arts and Sciences, Beirut, Lebanon. Offers anthropology (MA); Arabic language and literature (MA); archaeology (MA); biology (MS); business administration (MBA); chemistry (MS); computer science (MS); economics (MA); education (MA); English language (MA); English literature (MA); environmental policy planning (MSES); finance and banking (MFB); financial economics (MFE); geology (MS); history (MA); mathematics (MS); Middle Eastern studies (MA); philosophy (MA); physics (MS); political studies (MA); psychology (MA); public administration (MA); sociology (MA). *Degree requirements:* For master's, one foreign language, thesis (for some programs), comprehensive exam, registration. *Entrance requirements:* For master's, GRE, letter of recommendation.

Arizona State University, Division of Graduate Studies, College of Liberal Arts and Sciences, Division of Natural Sciences and Mathematics, Department of Chemistry and Biochemistry, Tempe, AZ 85287. Offers MNS, MS, PhD. *Degree requirements:* For master's, thesis; for doctorate, one foreign language, thesis/dissertation. *Entrance requirements:* For master's and doctorate, GRE. Additional exam requirements/recommendations for international students: Required—TOEFL, TSE.

Arkansas State University, Graduate School, College of Sciences and Mathematics, Department of Chemistry and Physics, Jonesboro, State University, AR 72467. Offers chemistry (MS); chemistry education (MSE, SCCT). Part-time programs available. *Faculty:* 13 full-time (2 women). *Students:* Average age 26. 3 applicants, 100% accepted, 3 enrolled. In 2005, 2 degrees awarded. *Degree requirements:* For master's, thesis or alternative, comprehensive exam. *Entrance requirements:* For master's, GRE General Test or MAT, appropriate bachelor's degree; for SCCT, GRE General Test or MAT, interview, master's degree. Additional exam requirements/recommendations for international students: Required—TOEFL (minimum score 213 computer-based). *Application deadline:* For fall admission, 7/1 for domestic students; for spring admission, 11/15 priority date for domestic students. Applications are processed on a rolling basis. Application fee: $15 ($25 for international students). Electronic applications accepted. *Expenses:* Tuition, state resident: full-time $3,232; part-time $180 per hour. Tuition, nonresident: full-time $8,164; part-time $454 per hour. Required fees: $716; $37 per hour. $25 per semester. Tuition and fees vary according to course load and program. *Financial support:* Teaching assistantships, career-related internships or fieldwork, scholarships/grants, and unspecified assistantships available. Financial award application deadline: 7/1; financial award applicants required to submit FAFSA. *Unit head:* Dr. Michael Panigot, Interim Chair, 870-972-3086, Fax: 870-972-3089, E-mail: mpanigot@astate.edu.

Auburn University, Graduate School, College of Sciences and Mathematics, Department of Chemistry, Auburn University, AL 36849. Offers MS, PhD. Part-time programs available. *Faculty:* 21 full-time (1 woman). *Students:* 48 full-time (19 women), 21 part-time (9 women); includes 5 minority (4 African Americans, 1 American Indian/Alaska Native), 53 international. 32 applicants, 69% accepted, 13 enrolled. In 2005, 1 master's, 9 doctorates awarded. *Degree requirements:* For master's, thesis (for some programs); for doctorate, thesis/dissertation, oral and written exams. *Entrance requirements:* For master's and doctorate, GRE General Test. *Application deadline:* For fall admission, 7/7 for domestic students; for spring admission, 11/24 for domestic students. Applications are processed on a rolling basis. Application fee: $25 ($50 for international students). Electronic applications accepted. *Financial support:* Fellowships, research assistantships, teaching assistantships available. Financial award application deadline: 3/15. *Unit head:* Dr. S. D. Worley, Interim Chair, 334-844-4043, Fax: 334-844-4043. *Application contact:* Dr. Stephen L. McFarland, Acting Dean of the Graduate School, 334-844-4700.

Ball State University, Graduate School, College of Sciences and Humanities, Department of Chemistry, Muncie, IN 47306-1099. Offers MA, MS. *Faculty:* 10. *Students:* 2 full-time (1 woman), 9 part-time (4 women), 4 international. Average age 22. 19 applicants, 32% accepted, 3 enrolled. In 2005, 5 degrees awarded. *Entrance requirements:* For master's, GRE General Test. Application fee: $25 ($35 for international students). *Expenses:* Tuition, state resident: full-time $6,246. Tuition, nonresident: full-time $16,006. *Financial support:* In 2005–06, research assistantships with full tuition reimbursements (averaging $9,460 per year), 12 teaching assistantships with full tuition reimbursements (averaging $9,460 per year) were awarded. Financial award application deadline: 3/1. *Faculty research:* Synthetic and analytical chemistry, biochemistry, theoretical chemistry. *Unit head:* Dr. Robert Morris, Chair, 765-285-8060, Fax: 765-285-2351.

Baylor University, Graduate School, College of Arts and Sciences, Department of Chemistry and Biochemistry, Waco, TX 76798. Offers chemistry (MS, PhD). Part-time programs available. *Faculty:* 14 full-time (2 women). *Students:* 48 full-time (19 women), 2 part-time (1 woman); includes 2 minority (1 Asian American or Pacific Islander, 1 Hispanic American), 31 international. In 2005, 2 master's, 8 doctorates awarded. Terminal master's awarded for partial completion of doctoral program. *Degree requirements:* For master's, thesis/dissertation; for doctorate, thesis/dissertation, comprehensive exam. *Entrance requirements:* For master's and doctorate, GRE General Test, GRE Subject Test. Additional exam requirements/recommendations for international students: Required—TOEFL. *Application deadline:* For fall admission, 8/1 for domestic students. Applications are processed on a rolling basis. Application fee: $25. *Financial support:* In 2005–06, 20 students received support; fellowships, research assistantships, teaching assistantships, Federal Work-Study, institutionally sponsored loans, and tuition waivers (full) available. Support available to part-time students. *Unit head:* Dr. Marianna Busch, Chair, 254-710-3311, Fax: 254-710-2403, E-mail: marianna_busch@baylor.edu. *Application contact:* Dr. Carlos Manzanares, Director of Graduate Studies, 254-710-4247, Fax: 254-710-2403, E-mail: carlos_manzanares@baylor.edu.

Boston College, Graduate School of Arts and Sciences, Department of Chemistry, Chestnut Hill, MA 02467-3800. Offers biochemistry (PhD); inorganic chemistry (PhD); organic chemistry (PhD); physical chemistry (PhD); science education (MST). MST is offered through the School of Education for secondary school science teaching. Part-time programs available. *Students:* 118 full-time (45 women); includes 12 minority (2 African Americans, 8 Asian Americans or Pacific Islanders, 2 Hispanic Americans), 35 international. 212 applicants, 32% accepted, 17 enrolled. In 2005, 6 master's, 10 doctorates awarded. *Degree requirements:* For doctorate, thesis/dissertation, qualifying exam. *Entrance requirements:* For doctorate, GRE General Test, GRE Subject Test. Additional exam requirements/recommendations for international students: Required—TOEFL (minimum score 550 paper-based; 213 computer-based). *Application deadline:* For fall admission, 1/2 for domestic students. Application fee: $70. Electronic applications accepted. *Financial support:* Fellowships with full tuition reimbursements, research assistantships with full tuition reimbursements, teaching assistantships with full tuition reimbursements, Federal Work-Study available. Support available to part-time students. Financial award application deadline: 3/1; financial award applicants required to submit FAFSA. *Unit head:* Dr. David McFadden, Chairperson, 617-552-3605, E-mail: david.mcfadden@bc.edu. *Application contact:* Dr. Marc Snapper, Graduate Program Director, 617-552-8096, Fax: 617-552-0833, E-mail: marc.snapper@bc.edu.

Boston University, Graduate School of Arts and Sciences, Department of Chemistry, Boston, MA 02215. Offers MA, PhD. *Students:* 102 full-time (44 women), 3 part-time; includes 8 minority (1 American Indian/Alaska Native, 3 Asian Americans or Pacific Islanders, 4 Hispanic Americans), 42 international. Average age 27. 265 applicants, 31% accepted, 25 enrolled. In 2005, 11 master's, 17 doctorates awarded. Terminal master's awarded for partial completion of doctoral program. *Degree requirements:* For master's, one foreign language, registration; for doctorate, one foreign language, thesis/dissertation, comprehensive exam, registration. *Entrance requirements:* For master's and doctorate, GRE General Test, GRE Subject Test (recommended), 3 letters of recommendation. Additional exam requirements/recommendations for international students: Required—TOEFL (minimum score 550 paper-based; 213 computer-based). *Application deadline:* For fall admission, 7/1 for domestic students, 7/1 for international students; for spring admission, 10/15 for domestic students, 10/15 for international students. Application fee: $60. *Expenses:* Tuition: Full-time $31,530; part-time $985 per credit. Required fees: $316; $40 per semester. Tuition and fees vary according to course level and program. *Financial support:* In 2005–06, 115 students received support, including 6 fellowships with full tuition reimbursements available (averaging $16,500 per year), 57 research assistantships with full tuition reimbursements available (averaging $16,000 per year), 48 teaching assistantships with full tuition reimbursements available (averaging $16,000 per year); Federal Work-Study, scholarships/grants, and tuition waivers (full) also available. Support available to part-time students. Financial award application deadline: 1/15; financial award applicants required to submit FAFSA. *Unit head:* Guilford Jones, Chairman, 617-353-2498, Fax: 617-353-6466, E-mail: jones@bu.edu. *Application contact:* Matt Vigneau, Academic Administrator, 617-353-6548, Fax: 617-353-6466, E-mail: mvigneau@bu.edu.

Bowling Green State University, Graduate College, College of Arts and Sciences, Center for Photochemical Sciences, Bowling Green, OH 43403. Offers PhD. *Faculty:* 10 full-time (1 woman), 2 part-time/adjunct (both women). *Students:* 51 full-time (19 women), 1 part-time, 50 international. Average age 27. 55 applicants, 38% accepted, 7 enrolled. In 2005, 12 degrees awarded. *Degree requirements:* For doctorate, thesis/dissertation, comprehensive exam. *Entrance requirements:* For doctorate, GRE General Test. Additional exam requirements/recommendations for international students: Required—TOEFL. *Application deadline:* For fall admission, 1/1 for domestic students. Application fee: $30. Electronic applications accepted. *Financial support:* In 2005–06, 17 research assistantships with full tuition reimbursements (averaging $15,503 per year), 43 teaching assistantships with full tuition reimbursements (averaging $13,696 per year) were awarded; Federal Work-Study, tuition waivers (full), and unspecified assistantships also available. Financial award applicants required to submit FAFSA. *Faculty research:* Laser-initiated photopolymerization, spectroscopic and kinetic studies, optoelectronics of semiconductor multiple quantum wells, electron transfer processes, carotenoid pigments. *Unit head:* Dr. Douglas C. Neckers, Executive Director, 419-372-2033. *Application contact:* Dr. Phil Castellano, Graduate Program Coordinator, 419-372-7513, Fax: 419-372-0366.

Bowling Green State University, Graduate College, College of Arts and Sciences, Department of Chemistry, Bowling Green, OH 43403. Offers MAT, MS. Part-time programs available. *Faculty:* 16 full-time (1 woman), 3 part-time/adjunct (0 women). *Students:* 7 full-time (2 women), 1 part-time, 6 international. Average age 27. 27 applicants, 19% accepted, 0 enrolled. In 2005, 11 degrees awarded. *Degree requirements:* For master's, thesis or alternative. *Entrance requirements:* For master's, GRE General Test. Additional exam requirements/recommendations for international students: Required—TOEFL. *Application deadline:* Applications are processed on a rolling basis. Application fee: $30. Electronic applications accepted. *Financial support:* In 2005–06, 1 research assistantship with full tuition reimbursement (averaging $9,018 per year), 7 teaching assistantships with full tuition reimbursements (averaging $9,018 per year) were awarded; Federal Work-Study, tuition waivers (full), and unspecified assistantships also available. Financial award applicants required to submit FAFSA. *Faculty research:* Organic, inorganic, physical, and analytical chemistry; biochemistry; surface science. *Unit head:* Dr. Michael Ogawa, Chair, 419-372-2033. *Application contact:* Tom Kinstle, Graduate Program Coordinator, 419-372-2658, Fax: 419-372-0366.

Bradley University, Graduate School, College of Liberal Arts and Sciences, Department of Chemistry and Biochemistry, Peoria, IL 61625-0002. Offers chemistry (MS). Part-time and evening/weekend programs available. *Students:* 3 applicants, 67% accepted, 1 enrolled. *Degree requirements:* For master's, thesis, comprehensive exam. *Entrance requirements:* For master's, 2 letters of recommendation. Additional exam requirements/recommendations for international students: Required—TOEFL (minimum score 550 paper-based; 213 computer-based). *Application deadline:* For fall admission, 5/15 for domestic students; for spring admission, 10/15 priority date for domestic students. Applications are processed on a rolling basis. Application fee: $40. *Financial support:* Scholarships/grants, tuition waivers (partial), and unspecified assistantships available. Financial award application deadline: 4/1. *Unit head:* Dr. Kurt Field, Chairperson, 309-677-3024. *Application contact:* Dr. Kristi McQuade, Graduate Coordinator, 309-677-3022, E-mail: mcquade@bradley.edu.

Brandeis University, Graduate School of Arts and Sciences, Department of Chemistry, Waltham, MA 02454-9110. Offers inorganic chemistry (MS, PhD); organic chemistry (MS, PhD); physical chemistry (MS, PhD). *Faculty:* 20 full-time (3 women). *Students:* 42 full-time (20 women), 1 part-time; includes 2 minority (1 African American, 1 Hispanic American), 34 international. Average age 25. 87 applicants, 10% accepted. In 2005, 5 master's, 3 doctorates awarded. Terminal master's awarded for partial completion of doctoral program. *Degree requirements:* For master's, 1 year of residency; for doctorate, one foreign language, thesis/dissertation, 3 years of residency, 2 seminars, qualifying exams. *Entrance requirements:* For master's and doctorate, GRE General Test, resumé, letters of recommendation. Additional exam requirements/recommendations for international students: Required—TOEFL (minimum score 600 paper-based; 250 computer-based). *Application deadline:* For fall admission, 1/15 for domestic students. Applications are processed on a rolling basis. Application fee: $55. Electronic applications accepted. *Financial support:* In 2005–06, 25 fellowships (averaging $23,000 per year), 21 research assistantships (averaging $23,000 per year), 19 teaching assistantships (averaging $23,000 per year) were awarded; institutionally sponsored loans and scholarships/grants also available. Financial award application deadline: 4/15; financial award applicants required to submit CSS PROFILE or FAFSA. *Faculty research:* Oscillating chemical reactions, molecular recognition systems, protein crystallography, synthesis of natural product spectroscopy and magnetic resonance. Total annual research expenditures: $1,965. *Unit head:* Dr. Peter Jordan, Chair, 781-736-2540, Fax: 781-736-2516, E-mail: jordan@brandeis.edu. *Application contact:* Charlotte Haygazian, Graduate Admissions Secretary, 781-736-2500, Fax: 781-736-2516, E-mail: chemadm@brandeis.edu.

Brigham Young University, Graduate Studies, College of Physical and Mathematical Sciences, Department of Chemistry and Biochemistry, Provo, UT 84602-1001. Offers analytical chemistry (MS, PhD); biochemistry (MS, PhD); inorganic chemistry (MS, PhD); organic chemistry (MS, PhD); physical chemistry (MS, PhD). *Faculty:* 35 full-time (2 women). *Students:* 100 full-time (33 women); includes 2 minority (both Asian Americans or Pacific Islanders), 68 international. Average age 29. 85 applicants, 44% accepted, 21 enrolled. In 2005, 9 master's, 9 doctorates awarded. *Median time to degree:* Of those who began their doctoral program in fall 1997, 87% received their degree in 8 years or less. *Degree requirements:* For master's, thesis, registration; for doctorate, thesis/dissertation, degree qualifying exam. *Entrance requirements:* For master's, pass 1 (biochemistry), 4 (chemistry) of 5 area exams, GRE General Test, minimum GPA of 3.0 in last 60 hours; for doctorate, pass 1 (biochemistry) or 4 (chemistry) of 5 area exams, GRE General Test, minimum GPA of 3.0 in last 60 hours. Additional exam requirements/recommendations for international students: Required—TOEFL, TWE. *Application deadline:* For fall admission, 2/1 priority date for domestic students, 2/1 priority date for international students. Applications are processed on a rolling basis. Application fee: $50. Electronic applications accepted. *Financial support:* In 2005–06, 100 students received support, including 12 fellowships with full tuition reimbursements available (averaging $20,500 per year), 47 research assistantships with full tuition reimbursements available (averaging $20,500 per year), 39 teaching assistantships with full tuition reimbursements available (averaging $20,400 per year); institutionally sponsored loans, scholarships/grants, health care benefits, tuition waivers (full), and unspecified assistantships also available. Financial award application deadline: 2/1. *Faculty research:* Separation science, molecular recognition, organic synthesis and biomedical application, biochemistry and molecular biology, molecular spectroscopy. Total annual research expenditures: $3.9 million. *Unit head:* Dr. Paul B. Farnsworth, Chair, 801-422-6502, Fax: 801-422-0153, E-mail: paul_farnsworth@byu.edu. *Application contact:* Dr. David V. Dearden, Graduate Coordinator, 801-422-2355, Fax: 801-422-0153, E-mail: david_dearden@byu.edu.

See Close-Up on page 97.

Brock University, Graduate Studies, Faculty of Mathematics and Science, Program in Chemistry, St. Catharines, ON L2S 3A1, Canada. Offers M Sc. Part-time programs available. *Faculty:* 14 full-time (2 women), 1 part-time/adjunct (0 women). *Students:* 9 full-time (5 women), 5 international. 42 applicants, 12% accepted. In 2005, 2 degrees awarded. *Degree requirements:* For master's, thesis. *Entrance requirements:* For master's, honors B Sc in chemistry. Additional exam requirements/recommendations for international students: Required—TOEFL. *Application deadline:* Applications are processed on a rolling basis. Application fee: $75. Electronic applications accepted. *Financial support:* Fellowships, research assistantships, teaching assistantships, career-related internships or fieldwork, scholarships/grants, and unspecified assistantships available. Support available to part-time students. *Faculty research:* Bioorganic chemistry, trace element analysis, organic synthesis, electrochemistry, structural inorganic chemistry.

Brooklyn College of the City University of New York, Division of Graduate Studies, Department of Chemistry, Brooklyn, NY 11210-2889. Offers applied chemistry (MA); chemistry (MA, PhD). The department offers courses at Brooklyn College that are creditable toward the CUNY doctoral degree. Part-time programs available. *Degree requirements:* For master's, one foreign language, thesis or alternative. *Entrance requirements:* For master's, 2 letters of recommendation. Additional exam requirements/recommendations for international students: Required—TOEFL.

Brown University, Graduate School, Department of Chemistry, Providence, RI 02912. Offers biochemistry (PhD); chemistry (Sc M, PhD). *Degree requirements:* For master's, thesis; for doctorate, one foreign language, thesis/dissertation, cumulative exam.

Bryn Mawr College, Graduate School of Arts and Sciences, Department of Chemistry, Bryn Mawr, PA 19010-2899. Offers MA, PhD. *Faculty:* 8. *Students:* 1 full-time (0 women), 6 part-time (4 women); includes 2 minority (1 African American, 1 Asian American or Pacific Islander). 3 applicants, 33% accepted, 1 enrolled. In 2005, 3 master's, 1 doctorate awarded. *Degree requirements:* For master's, one foreign language, thesis, registration; for doctorate, 2 foreign languages, thesis/dissertation, comprehensive exam, registration. *Entrance requirements:* For master's and doctorate, GRE General Test, GRE Subject Test. Additional exam requirements/recommendations for international students: Required—TOEFL (minimum score 600 paper-based; 250 computer-based). *Application deadline:* For fall admission, 1/13 for domestic students, 1/13 for international students. Application fee: $30. *Financial support:* Research assistantships with full tuition reimbursements, teaching assistantships with partial tuition reimbursements, Federal Work-Study, scholarships/grants, and tuition waivers (partial) available. Support available to part-time students. Financial award application deadline: 1/13. *Unit head:* Dr. William Malachowski, Chair, 610-526-5104. *Application contact:* Lea R. Miller, Secretary, 610-526-5072, Fax: 610-526-5076, E-mail: lrmiller@brynmawr.edu.

Bucknell University, Graduate Studies, College of Arts and Sciences, Department of Chemistry, Lewisburg, PA 17837. Offers MA, MS. Part-time programs available. *Degree requirements:* For master's, thesis. *Entrance requirements:* For master's, GRE General Test, GRE Subject Test, minimum GPA of 2.8. Additional exam requirements/recommendations for international students: Required—TOEFL.

Buffalo State College, State University of New York, Graduate Studies and Research, Faculty of Natural and Social Sciences, Department of Chemistry, Buffalo, NY 14222-1095. Offers chemistry (MA); secondary education (MS Ed), including chemistry. Part-time and evening/weekend programs available. *Degree requirements:* For master's, thesis (for some programs), project. *Entrance requirements:* For master's, minimum GPA of 2.6, New York teaching certificate (MS Ed). Additional exam requirements/recommendations for international students: Required—TOEFL (minimum score 550 paper-based; 213 computer-based).

California Institute of Technology, Division of Chemistry and Chemical Engineering, Program in Chemistry, Pasadena, CA 91125-0001. Offers MS, PhD. *Faculty:* 22 full-time (2 women). *Students:* 221 full-time (83 women). Average age 24. 395 applicants, 31% accepted, 37 enrolled. In 2005, 3 master's, 38 doctorates awarded. Terminal master's awarded for partial completion of doctoral program. *Degree requirements:* For master's and doctorate, thesis/dissertation. *Entrance requirements:* Additional exam requirements/recommendations for international students: Required—TOEFL; Recommended—IELT, TWE, TSE. *Application deadline:* For fall admission, 1/1 for domestic students, 1/1 for international students. Application fee: $50. Electronic applications accepted. *Financial support:* In 2005–06, 226 students received support; fellowships, research assistantships, teaching assistantships, Federal Work-Study, institutionally sponsored loans, scholarships/grants, traineeships, health care benefits, and unspecified assistantships available. Financial award application deadline: 1/1. *Faculty research:* Genetic structure and gene expression, organic synthesis, reagents and mechanisms,

Chemistry

California Institute of Technology (continued)
homogeneous and electrochemical catalysis. *Unit head:* Dr. Douglas C. Rees, Executive Officer, 626-395-8393, Fax: 626-744-9524, E-mail: dcrees@caltech.edu. *Application contact:* Dian Buchness, Graduate Secretary, 626-395-6110, Fax: 626-568-8824, E-mail: dianb@its.caltech.edu.

California Polytechnic State University, San Luis Obispo, College of Science and Mathematics, Department of Chemistry and Biochemistry, San Luis Obispo, CA 93407. Offers polymers and coating science (MS). Part-time programs available. *Faculty:* 1 full-time (0 women), 2 part-time/adjunct (1 woman). *Students:* 3 full-time (1 woman), 3 part-time (1 woman). 5 applicants, 80% accepted, 4 enrolled. *Degree requirements:* For master's, comprehensive oral exam. *Entrance requirements:* For master's, minimum GPA of 2.5 in last 90 quarter units of course work. Additional exam requirements/recommendations for international students: Required—TOEFL, TWE. *Application deadline:* For fall admission, 6/1 for domestic students, 11/30 for international studentsFor winter admission, 8/1 for domestic students; for spring admission, 12/1 for domestic students. Applications are processed on a rolling basis. Application fee: $55. Electronic applications accepted. *Expenses:* Tuition, nonresident: full-time $226; part-time $226 per unit. Required fees: $4,827; $1,063 per unit. *Financial support:* Career-related internships or fieldwork and Federal Work-Study available. Support available to part-time students. Financial award application deadline: 3/2; financial award applicants required to submit FAFSA. *Unit head:* Dr. Ray Fernando, Graduate Coordinator, 805-756-2395, E-mail: rhfernan@calpoly.edu.

California State Polytechnic University, Pomona, Academic Affairs, College of Science, Program in Chemistry, Pomona, CA 91768-2557. Offers MS. Part-time programs available. *Students:* 14 full-time (8 women), 11 part-time (4 women); includes 15 minority (1 African American, 7 Asian Americans or Pacific Islanders, 7 Hispanic Americans), 2 international. Average age 27. 18 applicants, 67% accepted, 8 enrolled. In 2005, 1 degree awarded. *Degree requirements:* For master's, thesis. *Entrance requirements:* For master's, GRE General Test. *Application deadline:* For fall admission, 5/1 for domestic studentsFor winter admission, 10/15 for domestic students; for spring admission, 1/20 for domestic students. Applications are processed on a rolling basis. Application fee: $55. Electronic applications accepted. *Expenses:* Tuition, nonresident: full-time $9,021. International tuition: $9,021 full-time. Required fees: $3,597. *Financial support:* In 2005–06, 2 students received support. Career-related internships or fieldwork, Federal Work-Study, and institutionally sponsored loans available. Support available to part-time students. Financial award application deadline: 3/2; financial award applicants required to submit FAFSA. *Unit head:* Dr. Dennis R. Livesay, Graduate Coordinator, 909-869-4409.

California State University, East Bay, Academic Programs and Graduate Studies, College of Science, Department of Chemistry, Hayward, CA 94542-3000. Offers biochemistry (MS), including biochemistry, chemistry; chemistry (MS). *Students:* 25. 8 applicants, 50% accepted. In 2005, 1 degree awarded. *Degree requirements:* For master's, comprehensive exam or thesis. *Entrance requirements:* For master's, minimum GPA of 2.5 in field during previous 2 years of course work. Additional exam requirements/recommendations for international students: Required—TOEFL (minimum score 550 paper-based; 213 computer-based). *Application deadline:* For fall admission, 5/31 for domestic students, 4/30 for international studentsFor winter admission, 9/30 for domestic students. Applications are processed on a rolling basis. Application fee: $55. Electronic applications accepted. *Financial support:* Career-related internships or fieldwork, Federal Work-Study, and institutionally sponsored loans available. Support available to part-time students. Financial award application deadline: 3/2. *Unit head:* Dr. Richard Luibrand, Chair, 510-885-3452, E-mail: rich.luibrand@csueastbay.edu. *Application contact:* Deborah Baker, Associate Director, 510-885-3286, Fax: 510-885-4777, E-mail: deborah.baker@csueastbay.edu.

California State University, Fresno, Division of Graduate Studies, College of Science and Mathematics, Department of Chemistry, Fresno, CA 93740-8027. Offers MS. Part-time programs available. *Degree requirements:* For master's, thesis or alternative. *Entrance requirements:* For master's, GRE General Test, minimum GPA of 2.5. Additional exam requirements/recommendations for international students: Required—TOEFL. Electronic applications accepted. *Faculty research:* Genetics, viticulture, DNA, soils, molecular modeling, analysis of quinone.

California State University, Fullerton, Graduate Studies, College of Natural Science and Mathematics, Department of Chemistry and Biochemistry, Fullerton, CA 92834-9480. Offers analytical chemistry (MS); biochemistry (MS); geochemistry (MS); inorganic chemistry (MS); organic chemistry (MS); physical chemistry (MS). Part-time programs available. *Students:* 19 full-time (6 women), 20 part-time (9 women); includes 20 minority (1 African American, 13 Asian Americans or Pacific Islanders, 6 Hispanic Americans), 8 international. Average age 28. 36 applicants, 47% accepted, 4 enrolled. In 2005, 11 degrees awarded. *Degree requirements:* For master's, thesis, departmental qualifying exam. *Entrance requirements:* For master's, minimum GPA of 2.5 in last 60 units of course work, major in chemistry or related field. Application fee: $55. *Expenses:* Tuition, area resident: Part-time $2,270 per year. Tuition, state resident: full-time $2,572; part-time $339 per unit. Tuition, nonresident: full-time $339; part-time $339 per unit. International tuition: $339 full-time. *Financial support:* Teaching assistantships, career-related internships or fieldwork, Federal Work-Study, institutionally sponsored loans, and scholarships/grants available. Support available to part-time students. Financial award application deadline: 3/1. *Unit head:* Dr. Maria Linder, Chair, 714-278-3621. *Application contact:* Dr. Gregory Williams, Adviser, 714-278-2170.

California State University, Long Beach, Graduate Studies, College of Natural Sciences and Mathematics, Department of Chemistry and Biochemistry, Long Beach, CA 90840. Offers biochemistry (MS); chemistry (MS). Part-time programs available. *Students:* 23 full-time (15 women), 16 part-time (6 women); includes 23 minority (1 African American, 17 Asian Americans or Pacific Islanders, 5 Hispanic Americans). Average age 28. 37 applicants, 49% accepted, 12 enrolled. In 2005, 3 degrees awarded. *Degree requirements:* For master's, thesis, departmental qualifying exam. *Application deadline:* For fall admission, 7/1 for domestic students; for spring admission, 12/1 for domestic students. Applications are processed on a rolling basis. Application fee: $55. Electronic applications accepted. *Expenses:* Tuition, area resident: Full-time $3,102; part-time $1,800 per semester hour. Tuition, state resident: part-time $339 per semester hour. Tuition, nonresident: full-time $339. Required fees: $779. *Financial support:* Research assistantships, teaching assistantships, Federal Work-Study, institutionally sponsored loans, scholarships/grants, and unspecified assistantships available. Financial award application deadline: 3/2. *Faculty research:* Enzymology, organic synthesis, molecular modeling, environmental chemistry, reaction kinetics. *Unit head:* Dr. Douglas D. McAbee, Chair, 562-985-4941, Fax: 562-985-8557, E-mail: dmcabee@csulb.edu. *Application contact:* Dr. Lijuan Li, Graduate Coordinator, 562-985-5068, Fax: 562-985-2315, E-mail: lli@csulb.edu.

California State University, Los Angeles, Graduate Studies, College of Natural and Social Sciences, Department of Chemistry and Biochemistry, Major in Chemistry, Los Angeles, CA 90032-8530. Offers MS. *Students:* 29 full-time (20 women), 19 part-time (8 women); includes 34 minority (2 African Americans, 20 Asian Americans or Pacific Islanders, 12 Hispanic Americans). In 2005, 5 degrees awarded. *Degree requirements:* For master's, one foreign language. *Entrance requirements:* Additional exam requirements/recommendations for international students: Required—TOEFL. *Application deadline:* For fall admission, 6/30 for domestic students; for spring admission, 2/1 for domestic students. Applications are processed on a rolling basis. Application fee: $55. *Expenses:* Tuition, area resident: Full-time $3,617. Tuition, state resident: full-time $3,617. Tuition, nonresident: full-time $9,719. International tuition: $9,719 full-time. *Financial support:* Application deadline: 3/1. *Faculty research:* Transition-metal chemistry, electrochemistry, chromatography, computer modeling of reactions. *Unit head:* Dr. Jamil Momand, Head, 323-343-2361.

California State University, Northridge, Graduate Studies, College of Science and Mathematics, Department of Chemistry, Northridge, CA 91330. Offers MS. *Degree requirements:* For master's, thesis. *Entrance requirements:* For master's, GRE General Test or minimum GPA of 2.5. Additional exam requirements/recommendations for international students: Required—TOEFL.

California State University, Sacramento, Graduate Studies, College of Natural Sciences and Mathematics, Department of Chemistry, Sacramento, CA 95819-6048. Offers MS. Part-time programs available. *Students:* 8 full-time (6 women), 19 part-time (11 women); includes 6 minority (1 African American, 5 Asian Americans or Pacific Islanders), 1 international. Average age 31. 16 applicants, 75% accepted, 9 enrolled. *Degree requirements:* For master's, thesis or alternative, departmental qualifying exam, writing proficiency exam. *Entrance requirements:* For master's, minimum GPA of 2.5 during previous 2 years of course work, BA in chemistry or equivalent. Additional exam requirements/recommendations for international students: Required—TOEFL. *Application deadline:* Applications are processed on a rolling basis. Application fee: $55. Electronic applications accepted. *Expenses:* Tuition, area resident: Full-time $3,624; part-time $339 per unit. Tuition, state resident: full-time $3,624; part-time $339 per unit. Tuition, nonresident: full-time $13,824; part-time $339 per unit. International tuition: $13,824 full-time. Required fees: $276 per semester hour. *Financial support:* Career-related internships or fieldwork and Federal Work-Study available. Support available to part-time students. Financial award application deadline: 3/1. *Unit head:* Susan Crawford, Chair, 916-278-6684, Fax: 916-278-4986.

Carleton University, Faculty of Graduate Studies, Faculty of Science, Department of Chemistry, Ottawa, ON K1S 5B6, Canada. Offers M Sc, PhD. *Degree requirements:* For master's, thesis/dissertation; for doctorate, thesis/dissertation, comprehensive exam. *Entrance requirements:* For master's, honors degree; for doctorate, M Sc. Additional exam requirements/recommendations for international students: Required—TOEFL. *Application deadline:* Applications are processed on a rolling basis. Application fee: $75 Canadian dollars. *Financial support:* Fellowships, research assistantships, teaching assistantships, institutionally sponsored loans, scholarships/grants, and unspecified assistantships available. *Faculty research:* Bio-organic chemistry, analytical toxicology, theoretical and physical chemistry, inorganic chemistry. *Unit head:* G. W. Buchanan, Chair, 613-520-2600 Ext. 3840, Fax: 613-520-3749, E-mail: cns@carleton.ca. *Application contact:* P.R. Sundararajan, Associate Chair, Graduate Studies, 613-520-2600 Ext. 3605, Fax: 613-520-3749, E-mail: cns@carleton.ca.

Carnegie Mellon University, Mellon College of Science, Department of Chemistry, Pittsburgh, PA 15213-3891. Offers chemical instrumentation (MS); chemistry (MS, PhD); colloids, polymers and surfaces (MS); polymer science (MS). Part-time programs available. Terminal master's awarded for partial completion of doctoral program. *Degree requirements:* For doctorate, thesis/dissertation, departmental qualifying and oral exams, teaching experience. *Entrance requirements:* For master's, GRE General Test; for doctorate, GRE General Test, GRE Subject Test. Additional exam requirements/recommendations for international students: Required—TOEFL. Electronic applications accepted. *Faculty research:* Physical and theoretical chemistry, chemical synthesis, biophysical/bioinorganic chemistry.

See Close-Up on page 101.

Case Western Reserve University, School of Graduate Studies, Department of Chemistry, Cleveland, OH 44106. Offers analytical chemistry (MS, PhD); inorganic chemistry (MS, PhD); organic chemistry (MS, PhD); physical chemistry (MS, PhD). Part-time programs available. Terminal master's awarded for partial completion of doctoral program. *Degree requirements:* For doctorate, thesis/dissertation. *Entrance requirements:* For master's and doctorate, GRE General Test, GRE Subject Test. Additional exam requirements/recommendations for international students: Required—TOEFL. *Faculty research:* Electrochemistry, synthetic chemistry, chemistry of life process, spectroscopy, kinetics.

The Catholic University of America, School of Arts and Sciences, Department of Chemistry, Washington, DC 20064. Offers MS. Part-time programs available. *Faculty:* 5 full-time (3 women), 2 part-time/adjunct (0 women). *Students:* 1 applicant, 0% accepted, 0 enrolled. *Degree requirements:* For master's, one foreign language, thesis or alternative, comprehensive exam. *Entrance requirements:* For master's, GRE General Test, GRE Subject Test, 3 letters of recommendation. Additional exam requirements/recommendations for international students: Required—TOEFL (minimum score 580 paper-based; 237 computer-based). *Application deadline:* For fall admission, 2/1 for domestic students; for spring admission, 11/15 priority date for domestic students. Applications are processed on a rolling basis. Application fee: $55. Electronic applications accepted. *Expenses:* Tuition: Full-time $24,800; part-time $940 per credit. Required fees: $1,090; $285 per term. Part-time tuition and fees vary according to course load and program. *Financial support:* Fellowships, research assistantships, teaching assistantships, career-related internships or fieldwork, Federal Work-Study, scholarships/grants, tuition waivers (full and partial), and unspecified assistantships available. Support available to part-time students. Financial award application deadline: 2/1; financial award applicants required to submit FAFSA. *Faculty research:* Theoretical chemistry; bioinorganic chemistry; chemical kinetics; synthetic organic chemistry; inorganic, bio-organic, and physical organic chemistry. *Unit head:* Dr. Greg Brewer, Chair, 202-319-5385, Fax: 202-319-5381, E-mail: brewer@cua.edu.

Central Connecticut State University, School of Graduate Studies, School of Arts and Sciences, Department of Chemistry, New Britain, CT 06050-4010. Offers MS. Part-time and evening/weekend programs available. *Faculty:* 8 full-time (2 women), 2 part-time/adjunct (0 women). *Students:* 1 full-time (0 women), 3 part-time (2 women); includes 1 minority (Hispanic American) Average age 36. 7 applicants, 86% accepted, 4 enrolled. *Degree requirements:* For master's, thesis or alternative, comprehensive exam. *Entrance requirements:* For master's, minimum GPA of 2.7. Additional exam requirements/recommendations for international students: Required—TOEFL. *Application deadline:* For fall admission, 7/1 for domestic students; for spring admission, 12/1 for domestic students. Applications are processed on a rolling basis. Application fee: $50. Electronic applications accepted. *Expenses:* Tuition, area resident: Full-time $3,780; part-time $362 per credit. Tuition, state resident: full-time $5,670; part-time $362 per credit. Tuition, nonresident: full-time $10,530; part-time $362 per credit. International tuition: $10,530 full-time. Required fees: $3,064. One-time fee: $62 part-time. Tuition and fees vary according to degree level and program. *Financial support:* Career-related internships or fieldwork, Federal Work-Study, scholarships/grants, and unspecified assistantships available. Support available to part-time students. Financial award applicants required to submit FAFSA. *Unit head:* Dr. Guy Crundwell, Chair, 860-832-2675.

Central Michigan University, College of Graduate Studies, College of Science and Technology, Department of Chemistry, Mount Pleasant, MI 48859. Offers chemistry (MS); teaching chemistry (MA). *Accreditation:* NCATE (one or more programs are accredited). *Faculty:* 17 full-time (2 women). *Students:* 4 full-time (2 women), 13 part-time (7 women). Average age 28. In 2005, 9 degrees awarded. *Degree requirements:* For master's, thesis or alternative. *Application deadline:* Applications are processed on a rolling basis. Application fee: $35 ($45 for international students). *Expenses:* Tuition, area resident: Part-time $325 per credit hour. Tuition, state resident: part-time $603 per credit hour. Tuition and fees vary according to degree level and reciprocity agreements. *Financial support:* In 2005–06, 7 research assistantships with tuition reimbursements, 6 teaching assistantships with tuition reimbursements were awarded; fellowships with tuition reimbursements, career-related internships or fieldwork and Federal Work-Study also available. Financial award application deadline: 3/7. *Faculty research:* Biochemistry, analytical and organic-inorganic chemistry, polymer chemistry. *Unit head:* Dr. David Ash, Chairperson, 989-774-3981, Fax: 989-774-3883. *Application contact:* Dr. Philip J. Squattrito, Graduate Program Coordinator, 989-774-4407, E-mail: squat1pj@cmich.edu.

Central Washington University, Graduate Studies, Research and Continuing Education, College of the Sciences, Department of Chemistry, Ellensburg, WA 98926. Offers MS. Part-time programs available. *Faculty:* 9 full-time (4 women). *Students:* 5 full-time (3 women), 1 part-time, 1 international. 3 applicants, 100% accepted, 3 enrolled. In 2005, 1 degree awarded. *Degree requirements:* For master's, thesis. *Entrance requirements:* For master's, GRE General Test, minimum GPA of 3.0. Additional exam requirements/recommendations for international students: Required—TOEFL (minimum score 550 paper-based; 213 computer-based). *Application deadline:* For fall admission, 4/1 for domestic studentsFor winter admission, 10/1 for domestic students; for spring admission, 1/1 for domestic students. Applications are processed on a rolling basis. Application fee: $50. *Expenses:* Tuition, state resident: full-time $1,968; part-time $197 per credit. Tuition, nonresident: full-time $4,320; part-time $432 per credit. Required fees: $623. Tuition and fees vary according to degree level. *Financial support:* In 2005–06, 2 research assistantships with partial tuition reimbursements (averaging $8,100 per year), 5 teaching assistantships with partial tuition reimbursements (averaging $8,100 per year) were awarded; career-related internships or fieldwork and Federal Work-Study also available. Financial award application deadline: 3/1; financial award applicants required to submit FAFSA. *Unit head:* Dr. Martha Kurtz, Chair, 509-963-2811, Fax: 509-963-1050. *Application contact:* Justine Eason, Admissions Program Coordinator, 509-963-3103, Fax: 509-963-1799, E-mail: masters@cwu.edu.

City College of the City University of New York, Graduate School, College of Liberal Arts and Science, Division of Science, Department of Chemistry, Program in Chemistry, New York, NY 10031-9198. Offers MA, PhD. *Students:* 14 applicants, 43% accepted, 3 enrolled. In 2005, 2 degrees awarded. Terminal master's awarded for partial completion of doctoral program. *Degree requirements:* For doctorate, one foreign language, thesis/dissertation. *Entrance requirements:* For master's and doctorate, GRE. Additional exam requirements/recommendations for international students: Required—TOEFL (minimum score 500 paper-based; 173 computer-based). *Application deadline:* For fall admission, 5/1 for domestic students; for spring admission, 11/1 for domestic students. Application fee: $125. *Financial support:* Federal Work-Study available. Financial award application deadline: 6/1. *Faculty research:* Laser spectroscopy, bioorganic chemistry, polymer chemistry and crystallography, electroanalytical chemistry, ESR of metal clusters. *Application contact:* Teresa Bandosz, MA Advisor, 212-650-6017, Fax: 212-650-6107, E-mail: tbandosz@ccny.cuny.edu.

Clark Atlanta University, School of Arts and Sciences, Department of Chemistry, Atlanta, GA 30314. Offers inorganic chemistry (MS, PhD); organic chemistry (MS, PhD); physical chemistry (MS, PhD); science education (DA). Part-time programs available. *Degree requirements:* For master's, one foreign language, thesis, comprehensive exam; for doctorate, 2 foreign languages, thesis/dissertation, cumulative exam. *Entrance requirements:* For master's, GRE General Test, minimum GPA of 2.5; for doctorate, GRE General Test, GRE Subject Test, minimum graduate GPA of 3.0.

Clarkson University, Graduate School, School of Arts and Sciences, Department of Chemistry, Potsdam, NY 13699. Offers analytical chemistry (MS, PhD); inorganic chemistry (MS, PhD); organic chemistry (MS, PhD); physical chemistry (MS, PhD). *Faculty:* 9 full-time (2 women). *Students:* 34 full-time (9 women), 1 (woman) part-time; includes 2 Asian Americans or Pacific Islanders, 22 international. Average age 27. 52 applicants, 62% accepted. In 2005, 3 master's, 2 doctorates awarded. *Median time to degree:* Of those who began their doctoral program in fall 1997, 100% received their degree in 8 years or less. *Degree requirements:* For doctorate, thesis/dissertation, departmental qualifying exam. *Entrance requirements:* For master's, GRE. Additional exam requirements/recommendations for international students: Required—TOEFL. *Application deadline:* For fall admission, 5/15 for domestic students; for spring admission, 10/15 priority date for domestic students. Applications are processed on a rolling basis. Application fee: $25 ($35 for international students). Electronic applications accepted. *Expenses:* Tuition: Full-time $20,160; part-time $840 per hour. Required fees: $215. *Financial support:* In 2005–06, 25 students received support, including 2 fellowships (averaging $25,000 per year), 11 research assistantships (averaging $19,032 per year), 12 teaching assistantships (averaging $19,032 per year); scholarships/grants and tuition waivers (partial) also available. *Faculty research:* Nanomaterial, surface science, polymers chemical biosensing, protein biochemistry drug design/delivery, molecular neuroscience and immunobiology. Total annual research expenditures: $1.6 million. *Unit head:* Dr. Phillip A. Christiansen, Division Head, 315-268-6669, Fax: 315-268-6610, E-mail: pac@clarkson.edu. *Application contact:* Donna Brockway, Graduate Admissions International Advisor/Assistant to the Provost, 315-268-6447, Fax: 315-268-7994, E-mail: brockway@clarkson.edu.

Clark University, Graduate School, Department of Chemistry, Worcester, MA 01610-1477. Offers MA, PhD. *Faculty:* 8 full-time (2 women), 1 part-time/adjunct (0 women). *Students:* 16 full-time (4 women); includes 1 minority (Hispanic American), 14 international. Average age 29. 11 applicants, 100% accepted, 2 enrolled. In 2005, 2 master's awarded. Terminal master's awarded for partial completion of doctoral program. *Degree requirements:* For master's, thesis or alternative; for doctorate, one foreign language, thesis/dissertation. *Entrance requirements:* For master's and doctorate, GRE General Test. Additional exam requirements/recommendations for international students: Required—TOEFL. *Application deadline:* For fall admission, 2/15 for domestic students. Applications are processed on a rolling basis. Application fee: $50. *Expenses:* Tuition: Full-time $29,300. Required fees: $30. *Financial support:* In 2005–06, fellowships with tuition reimbursements (averaging $18,750 per year), 4 research assistantships with full tuition reimbursements (averaging $18,750 per year), 12 teaching assistantships with full tuition reimbursements (averaging $18,750 per year) were awarded; tuition waivers (full) also available. *Faculty research:* Nuclear chemistry, molecular biology simulation, NMR studies, anthrax edema, biochemistry. Total annual research expenditures: $530,000. *Unit head:* Dr. Mark Turnbull, Chair, 508-793-7116. *Application contact:* Rene Baril, Department Secretary, 528-793-7173, Fax: 528-793-8861, E-mail: chemistry@clarku.edu.

Clemson University, Graduate School, College of Engineering and Science, Department of Chemistry, Clemson, SC 29634. Offers MS, PhD. *Students:* 99 full-time (32 women), 3 part-time (1 woman); includes 10 minority (4 African Americans, 2 Asian Americans or Pacific Islanders, 4 Hispanic Americans), 41 international. Average age 25. 43 applicants, 67% accepted, 13 enrolled. In 2005, 5 master's, 6 doctorates awarded. *Degree requirements:* For master's and doctorate, one foreign language, thesis/dissertation. *Entrance requirements:* For master's and doctorate, GRE General Test. Additional exam requirements/recommendations for international students: Required—TOEFL. *Application deadline:* For fall admission, 6/1 for domestic students, 4/15 for international students. Application fee: $50. *Financial support:* Fellowships, research assistantships, teaching assistantships, unspecified assistantships available. Financial award applicants required to submit FAFSA. *Faculty research:* Fluorine chemistry, organic synthetic methods and natural products, metal and non-metal clusters, analytical spectroscopies, polymers. Total annual research expenditures: $1 million. *Unit head:* Dr. Luis Echegoyen, Chair, 864-656-5017, Fax: 864-656-6613, E-mail: luis@clemson.edu. *Application contact:* Dr. Steve Stuart, Coordinator, 864-656-5013, Fax: 864-656-6613, E-mail: ss@clemson.edu.

See Close-Up on pages 105 and 107.

Cleveland State University, College of Graduate Studies, College of Science, Department of Chemistry, Cleveland, OH 44115. Offers analytical chemistry (MS); clinical chemistry (MS, PhD); clinical/bioanalytical chemistry (PhD); environmental chemistry (MS); inorganic chemistry (MS); organic chemistry (MS); physical chemistry (MS). Part-time and evening/weekend programs available. *Faculty:* 13 full-time (1 woman), 59 part-time/adjunct (12 women). *Students:* 46 full-time (22 women), 18 part-time (6 women); includes 4 minority (2 African Americans, 2 Asian Americans or Pacific Islanders), 35 international. Average age 31. 47 applicants, 62% accepted, 7 enrolled. In 2005, 2 master's, 5 doctorates awarded. *Median time to degree:* Of those who began their doctoral program in fall 1997, 67% received their degree in 8 years or less. *Degree requirements:* For master's, thesis (for some programs); for doctorate, thesis/dissertation. *Entrance requirements:* For master's and doctorate, GRE General Test, GRE Subject Test. Additional exam requirements/recommendations for international students: Required—TOEFL (minimum score 525 paper-based; 197 computer-based). *Application deadline:* For fall admission, 1/15 priority date for domestic students, 1/15 priority date for international students. Applications are processed on a rolling basis. Application fee: $30. Electronic applications accepted. *Expenses:* Tuition, state resident: full-time $10,700. Tuition, nonresident: full-time $14,628. Tuition and fees vary according to program. *Financial support:* In 2005–06, 44 students received support, including fellowships with full tuition reimbursements available (averaging $18,000 per year), research assistantships with full tuition reimbursements available (averaging $16,000 per year), teaching assistantships with full tuition reimbursements available (averaging $14,000 per year) Financial award application deadline: 1/15. *Faculty research:* Metalloenzyme mechanisms, dependent RNAse L and interferons, application of HPLC/LPCC to clinical systems, structure-function relationships of factor Va, MALDI-TOF based DNA sequencing. Total annual research expenditures: $3 million. *Unit head:* Dr. Lily M. Ng, Chair, 216-687-2467, E-mail: l.ng@csuohio.edu. *Application contact:* Richelle P. Emery, Administrative Coordinator, 216-687-2457, Fax: 216-687-9298, E-mail: r.emery@csuohio.edu.

College of Staten Island of the City University of New York, Graduate Programs, Program in Polymer Chemistry, Staten Island, NY 10314-6600. Offers PhD. *Expenses:* Tuition, area resident: Full-time $3,200; part-time $270 per credit. Tuition, nonresident: full-time $500; part-time $500 per credit. International tuition: $500 full-time. Required fees: $328; $101 per semester. *Faculty research:* Computational and experimental biophysical studies of immunologically active truncated human thioredoxin, applicability of unique capping agents in producing novel metallic nanoparticles for links and coatings, peptide-cell interactions in saccharomyces cerevisiae, molecular structure and function of protective plant polymers. Total annual research expenditures:$677,916. *Unit head:* Dr. Nan–Loh Yang, Coordinator, 718-982-3899, Fax: 718-982-3910, E-mail: yang@mail.csi.cuny.edu. *Application contact:* Emmanuel Esperance, Deputy Director of Office of Recruitment and Admissions, 718-982-2259, Fax: 718-982-2500, E-mail: admissions@mail.csi.cuny.edu.

The College of William and Mary, Faculty of Arts and Sciences, Department of Chemistry, Williamsburg, VA 23187-8795. Offers MA, MS. Part-time programs available. *Faculty:* 16 full-time (3 women), 3 part-time/adjunct (1 woman). *Students:* 9 full-time (4 women), 2 part-time (1 woman); includes 1 minority (African American), 2 international. Average age 25. 4 applicants, 100% accepted, 3 enrolled. In 2005, 3 degrees awarded. *Degree requirements:* For master's, thesis (for some programs), comprehensive exam. *Entrance requirements:* For master's, GRE General Test, minimum GPA of 2.5. *Application deadline:* For fall admission, 1/15 for domestic students. Applications are processed on a rolling basis. Application fee: $30. *Expenses:* Tuition, area resident: Full-time $5,828; part-time $245 per credit. Tuition, state resident: full-time $5,828; part-time $245 per credit. Tuition, nonresident: full-time $17,980; part-time $685 per credit. International tuition: $17,980 full-time. Required fees:$3,051. Tuition and fees vary according to program. *Financial support:* In 2005–06, 1 research assistantship with full tuition reimbursement (averaging $12,700 per year), 3 teaching assistantships with full tuition reimbursements (averaging $12,700 per year) were awarded. Financial award application deadline: 5/1; financial award applicants required to submit FAFSA. *Faculty research:* Organic, inorganic, physical, polymer and analytic chemistry; biochemistry. Total annual research expenditures:$673,263.*Unit head:* Dr. Gary Rice, Chair, 757-221-2540, Fax: 757-221-2715. *Application contact:* Dr. Christopher J. Abelt, Graduate Director, 757-221-2551, Fax: 757-221-2715, E-mail: cjabel@wm.edu.

Colorado School of Mines, Graduate School, Department of Chemistry and Geochemistry, Program in Chemistry, Golden, CO 80401-1887. Offers applied chemistry (PhD); chemistry (MS). Part-time programs available. *Students:* 30 full-time (13 women), 7 part-time (2 women); includes 3 minority (2 African Americans, 1 Hispanic American), 2 international. 25 applicants, 84% accepted, 10 enrolled. In 2005, 5 master's, 2 doctorates awarded. *Degree requirements:* For master's, thesis/dissertation; for doctorate, thesis/dissertation, comprehensive exam. *Entrance requirements:* For master's and doctorate, GRE General Test. Additional exam requirements/recommendations for international students: Required—TOEFL (minimum score 550 paper-based; 213 computer-based). *Application deadline:* For fall admission, 1/1 priority date for domestic students, 1/1 priority date for international students; for spring admission, 9/1 priority date for domestic students, 9/1 priority date for international students. Application fee: $50. *Expenses:* Tuition, state resident: full-time $7,240; part-time $362 per credit hour. Tuition, nonresident: full-time $19,840; part-time $992 per credit hour. Required fees: $895. *Financial support:* In 2005–06, 7 students received support, including 1 fellowship with full tuition reimbursement available (averaging $9,600 per year), 11 research assistantships with full tuition reimbursements available (averaging $9,600 per year), 13 teaching assistantships with full tuition reimbursements available (averaging $9,600 per year); scholarships/grants, health care benefits, and unspecified assistantships also available. Financial award applicants required to submit FAFSA. *Application contact:* Pat MacCarthy, Professor, 303-273-3626, Fax: 303-273-3629, E-mail: pmaccart@mines.edu.

Colorado State University, Graduate School, College of Natural Sciences, Department of Chemistry, Fort Collins, CO 80523-0015. Offers MS, PhD. Part-time programs available. *Faculty:* 28 full-time (6 women). *Students:* 61 full-time (20 women), 76 part-time (24 women); includes 14 minority (2 American Indian/Alaska Native, 10 Asian Americans or Pacific Islanders, 2 Hispanic Americans), 27 international. Average age 27. 117 applicants, 64% accepted, 39 enrolled. In 2005, 2 master's, 8 doctorates awarded. Terminal master's awarded for partial completion of doctoral program. *Median time to degree:* Of those who began their doctoral program in fall 1997, 100% received their degree in 8 years or less. *Degree requirements:* For master's and doctorate, thesis/dissertation. *Entrance requirements:* For master's and doctorate, GRE General Test, minimum GPA of 3.2. Additional exam requirements/recommendations for international students: Required—TOEFL. *Application deadline:* For fall admission, 2/1 for domestic students, 2/1 for international students; for spring admission, 11/15 priority date for domestic students, 11/15 priority date for international students. Applications are processed on a rolling basis. Application fee: $50. Electronic applications accepted. *Expenses:* Tuition, state resident: full-time $3,690; part-time $205 per credit. Tuition, nonresident: full-time $14,958; part-time $831 per credit. Required fees: $1,061. *Financial support:* In 2005–06, 18 fellowships with full tuition reimbursements (averaging $3,000 per year), 92 research assistantships with full tuition reimbursements (averaging $20,000 per year), 39 teaching assistantships with full tuition reimbursements (averaging $20,000 per year) were awarded; unspecified assistantships also available. *Faculty research:* Analytical chemistry, inorganic chemistry, organic chemistry, physical chemistry, materials and biological chemistry. Total annual research expenditures: $7.5 million. *Unit head:* Dr. Anthony K. Rappé, Chairman, 970-491-6292, Fax: 970-491-1801, E-mail: rappe@lamar.colostate.edu. *Application contact:* Dr. Ellen R. Fisher, Chair, Graduate Recruiting, 970-491-5250, E-mail: erfisher@lamar.colostate.edu.

Colorado State University-Pueblo, College of Science and Mathematics, Pueblo, CO 81001-4901. Offers applied natural science (MS), including biochemistry, biology, chemistry. Part-time and evening/weekend programs available. *Faculty:* 15 full-time (6 women). *Students:* 14 full-time (9 women), 5 part-time (1 woman); includes 2 minority (both Hispanic Americans) 8 applicants, 100% accepted, 8 enrolled. In 2005, 9 degrees awarded. *Degree requirements:* For master's, thesis (for some programs), internship report (if non-thesis), comprehensive exam (for some programs), registration. *Entrance requirements:* For master's, GRE General Test, minimum GPA of 3.0. Additional exam requirements/recommendations for international students: Required—TOEFL (minimum score 500 paper-based; 173 computer-based). *Application deadline:* For fall admission, 6/15 priority date for domestic students, 6/15 priority date for international students; for spring admission, 10/15 priority date for domestic students, 10/15 priority date for international students. Applications are processed on a rolling basis. Application fee: $35. *Expenses:* Tuition, state resident: full-time $2,177; part-time $121 per credit hour. Tuition, nonresident: full-time $10,157; part-time $564 per credit hour. Required fees: $490; $41 per credit hour. *Financial support:* In 2005–06, 9 students received support, including 1 fellowship (averaging $1,000 per year), 3 teaching assistantships with partial tuition reimbursements available (averaging $9,000 per year); research assistantships, career-related internships or fieldwork, scholarships/grants, and unspecified assistantships also available. Financial award application deadline: 6/1; financial award applicants required to submit FAFSA. *Faculty research:* Fungal cell walls, molecular biology, bioactive materials synthesis, forensic chemistry,

Chemistry

Colorado State University-Pueblo (continued)
atomic force microscopy-surface chemistry. Total annual research expenditures: $400,000. *Unit head:* Dr. Kristina Proctor, Dean, 719-549-2340, Fax: 719-549-2732, E-mail: kristina.proctor@colostate-pueblo.edu. *Application contact:* Dr. Melvin Druelinger, Director, MSANS Program, 719-549-2325, Fax: 719-549-2071, E-mail: mel.druelinger@colostate-pueblo.edu.

Columbia University, Graduate School of Arts and Sciences, Division of Natural Sciences, Department of Chemistry, New York, NY 10027. Offers chemical physics (M Phil, PhD); inorganic chemistry (M Phil, MA, PhD); organic chemistry (M Phil, MA, PhD). *Faculty:* 17 full-time. *Students:* 115 full-time (35 women). Average age 27. 452 applicants, 20% accepted. In 2005, 13 master's, 20 doctorates awarded. *Degree requirements:* For master's, comp exams (MS); foreign language, teaching experience, oral/written exams (M Phil); for doctorate, one foreign language, thesis/dissertation. *Entrance requirements:* For master's and doctorate, GRE General Test, GRE Subject Test. Additional exam requirements/recommendations for international students: Required—TOEFL. Application fee: $75. *Expenses:* Tuition: Full-time $31,448. Tuition and fees vary according to course level, course load, campus/location and program. *Financial support:* Fellowships, teaching assistantships, Federal Work-Study and institutionally sponsored loans available. Support available to part-time students. Financial award application deadline: 1/5; financial award applicants required to submit FAFSA. *Faculty research:* Biophysics.*Unit head:* James Valentini, Chair, 212-854-7590, Fax: 212-932-1289, E-mail: jjvi@columbia.edu.

Concordia University, School of Graduate Studies, Faculty of Arts and Science, Department of Chemistry and Biochemistry, Montréal, QC H3G 1M8, Canada. Offers chemistry (M Sc, PhD). *Students:* 52 full-time (24 women), 4 part-time (3 women). In 2005, 7 master's, 5 doctorates awarded. *Degree requirements:* For master's and doctorate, thesis/dissertation. *Entrance requirements:* For master's, honors degree in chemistry; for doctorate, M Sc in biochemistry, biology, or chemistry. *Application deadline:* For fall admission, 6/1 for domestic studentsFor winter admission, 10/1 for domestic students; for spring admission, 4/1 for domestic students. Application fee: $50. *Expenses:* Tuition, state resident: full-time $834; part-time $334 per term. Tuition, nonresident: full-time $2,200; part-time $880 per term. Required fees: $680 per term. Tuition and fees vary according to degree level and program. *Financial support:* Teaching assistantships available. *Faculty research:* Bioanalytical, bio-organic, and inorganic chemistry; materials and solid-state chemistry. *Unit head:* Dr. Marcus Lawrence, Chair, 514-848-2424 Ext. 3355, Fax: 514-848-2868. *Application contact:* Dr. Cameron Skinner, Director, 514-848-2424 Ext. 3341, Fax: 514-848-2868.

Cornell University, Graduate School, Graduate Fields of Arts and Sciences, Field of Chemistry and Chemical Biology, Ithaca, NY 14853-0001. Offers analytical chemistry (PhD); bio-organic chemistry (PhD); biophysical chemistry (PhD); chemical biology (PhD); chemical physics (PhD); inorganic chemistry (PhD); materials chemistry (PhD); organic chemistry (PhD); organometallic chemistry (PhD); physical chemistry (PhD); polymer chemistry (PhD); theoretical chemistry (PhD). *Faculty:* 44 full-time (2 women). *Students:* 190 full-time (73 women); includes 23 minority (4 African Americans, 8 Asian Americans or Pacific Islanders, 11 Hispanic Americans), 65 international. 339 applicants, 35% accepted, 49 enrolled. In 2005, 23 doctorates awarded. *Degree requirements:* For doctorate, thesis/dissertation, comprehensive exam. *Entrance requirements:* For doctorate, GRE General Test, GRE Subject Test (chemistry), 3 letters of recommendation. Additional exam requirements/recommendations for international students: Required—TOEFL (minimum score 600 paper-based; 250 computer-based). *Application deadline:* For fall admission, 1/10 for domestic students. Application fee: $60. Electronic applications accepted. *Financial support:* In 2005–06, 185 students received support, including 33 fellowships with full tuition reimbursements available, 85 research assistantships with full tuition reimbursements available, 67 teaching assistantships with full tuition reimbursements available; institutionally sponsored loans, scholarships/grants, health care benefits, tuition waivers (full and partial), and unspecified assistantships also available. Financial award applicants required to submit FAFSA. *Faculty research:* Analytical, organic, inorganic, physical, materials, chemical biology. *Unit head:* Director of Graduate Studies, 607-255-4139, Fax: 607-255-4137. *Application contact:* Graduate Field Assistant, 607-255-4139, Fax: 607-255-4137, E-mail: chemgrad@cornell.edu.

See Close-Up on page 109.

Dalhousie University, Faculty of Graduate Studies, College of Arts and Science, Faculty of Science, Department of Chemistry, Halifax, NS B3H 4R2, Canada. Offers M Sc, PhD. Part-time programs available. *Faculty:* 26 full-time (4 women), 10 part-time/adjunct (2 women). *Students:* 70 full-time (29 women); includes 19 minority (2 African Americans, 7 Asian Americans or Pacific Islanders, 10 Hispanic Americans), 6 international. Average age 25. 120 applicants, 25% accepted, 16 enrolled. In 2005, 6 master's, 13 doctorates awarded. Terminal master's awarded for partial completion of doctoral program. *Median time to degree:* Of those who began their doctoral program in fall 1997, 100% received their degree in 8 years or less. *Degree requirements:* For master's and doctorate, thesis/dissertation. *Entrance requirements:* Additional exam requirements/recommendations for international students: Required—TOEFL (minimum score 580 paper-based; 237 computer-based), IELT(minimum score 7). *Application deadline:* For fall admission, 6/1 priority date for domestic students, 4/1 priority date for international studentsFor winter admission, 10/31 for domestic students; for spring admission, 2/28 for domestic students. Applications are processed on a rolling basis. Application fee: $70. *Financial support:* In 2005–06, fellowships with full tuition reimbursements (averaging $10,980 per year), teaching assistantships (averaging $3,247 per year) were awarded; scholarships/grants also available. Financial award application deadline: 4/15. *Faculty research:* Analytical, inorganic, organic, physical, and theoretical chemistry. Total annual research expenditures: $2.5 million. *Unit head:* Dr. J. A. Pincock, Chair, 902-494-3707, Fax: 902-494-1310, E-mail: james.pincock@dal.ca. *Application contact:* Dr. N. Burford, Graduate Coordinator, 902-494-3306, Fax: 92-494-1310, E-mail: neil.burford@dal.ca.

Dartmouth College, School of Arts and Sciences, Department of Chemistry, Hanover, NH 03755. Offers PhD. *Faculty:* 18 full-time (3 women), 1 part-time/adjunct (0 women). *Students:* 44 full-time (18 women); includes 2 minority (1 American Indian/Alaska Native, 1 Hispanic American), 16 international. Average age 27. 117 applicants, 13% accepted, 7 enrolled. In 2005, 3 doctorates awarded. *Degree requirements:* For doctorate, thesis/dissertation, departmental qualifying exams. *Entrance requirements:* For doctorate, GRE General Test, GRE Subject Test. Additional exam requirements/recommendations for international students: Required—TOEFL. *Application deadline:* For fall admission, 1/15 for domestic students. Application fee: $25. Electronic applications accepted. *Expenses:* Tuition: Full-time $31,770. *Financial support:* In 2005–06, 43 students received support, including fellowships with full tuition reimbursements available (averaging $21,000 per year), research assistantships with full tuition reimbursements available (averaging $21,000 per year); Federal Work-Study, institutionally sponsored loans, scholarships/grants, traineeships, tuition waivers (full), and unspecified assistantships also available. *Faculty research:* Organic and polymer synthesis, bioinorganic chemistry, magnetic resonance parameters. Total annual research expenditures: $2.8 million. *Unit head:* Dr. Joseph J. BelBruno, Chair, 603-646-2501. *Application contact:* Deborah Carr, Administrative Assistant, 603-646-2501, Fax: 603-646-3946, E-mail: deborah.a.carr@dartmouth.edu.

Delaware State University, Graduate Programs, Department of Chemistry, Dover, DE 19901-2277. Offers applied chemistry (MS); chemistry (MS). Part-time and evening/weekend programs available. *Entrance requirements:* For master's, GRE, minimum GPA of 3.0 in major, 2.75 overall. Electronic applications accepted. *Faculty research:* Chemiluminescence, environmental chemistry, forensic chemistry, heteropoly anions anti-cancer and antiviral agents, low temperature infrared studies of lithium salts.

DePaul University, College of Liberal Arts and Sciences, Department of Chemistry, Chicago, IL 60604-2287. Offers biochemistry (MS); chemistry (MS); polymer chemistry and coatings technology (MS). Part-time and evening/weekend programs available. *Faculty:* 11 full-time (4 women), 4 part-time/adjunct (1 woman). *Students:* 9 full-time (5 women), 9 part-time (3 women); includes 3 minority (1 American Indian/Alaska Native, 1 Asian American or Pacific Islander, 1 Hispanic American), 3 international. Average age 27. 6 applicants, 100% accepted, 4 enrolled. In 2005, 2 master's awarded. *Degree requirements:* For master's, thesis (for some programs), oral exam for selected programs. *Entrance requirements:* For master's, BS in chemistry or equivalent. Additional exam requirements/recommendations for international students: Required—TOEFL (minimum score 590 paper-based; 243 computer-based). *Application deadline:* For fall admission, 7/15 for domestic students, 5/1 for international studentsFor winter admission, 11/15 for domestic students; for spring admission, 2/15 for domestic students. Applications are processed on a rolling basis. Application fee: $35. Electronic applications accepted. *Financial support:* In 2005–06, 4 students received support, including 6 teaching assistantships with tuition reimbursements available (averaging $8,000 per year) Financial award application deadline: 4/1. *Faculty research:* Polymers, DNA sequencing, computational chemistry, water pollution, diffusion kinetics. Total annual research expenditures: $30,000. *Unit head:* Dr. Wendy S. Wolbach, Chair, 773-325-7420, Fax: 773-325-7421, E-mail: wwolbach@condor.depaul.edu. *Application contact:* Kavitha Chinthada, Director of Graduate Admissions, 773-325-7885, Fax: 773-325-7311, E-mail: kchintha@depaul.edu.

Drexel University, College of Arts and Sciences, Department of Chemistry, Philadelphia, PA 19104-2875. Offers MS, PhD. Part-time programs available. Terminal master's awarded for partial completion of doctoral program. *Degree requirements:* For master's, thesis optional; for doctorate, one foreign language, thesis/dissertation. *Entrance requirements:* For master's and doctorate, GRE. Additional exam requirements/recommendations for international students: Required—TOEFL, TSE (financial award applicants for teaching assistantships). Electronic applications accepted. *Faculty research:* Inorganic, analytical, organic, physical, and atmospheric polymer chemistry.

Duke University, Graduate School, Department of Chemistry, Durham, NC 27708. Offers PhD. *Faculty:* 23 full-time. *Students:* 108 full-time (46 women); includes 7 minority (3 African Americans, 1 American Indian/Alaska Native, 2 Asian Americans or Pacific Islanders, 1 Hispanic American), 51 international. 254 applicants, 31% accepted, 27 enrolled. In 2005, 18 doctorates awarded. *Degree requirements:* For doctorate, one foreign language, thesis/dissertation. *Entrance requirements:* For doctorate, GRE General Test, GRE Subject Test (recommended). Additional exam requirements/recommendations for international students: Required—IELT (preferred) or TOEFL. *Application deadline:* For fall admission, 12/31 for domestic students, 12/31 for international students. Application fee: $75. Electronic applications accepted. *Financial support:* Fellowships, research assistantships, teaching assistantships available. Financial award application deadline: 12/31. *Unit head:* Jie Liu, Director of Graduate Studies, 919-660-1549, Fax: 919-660-1607, E-mail: dgs@chem.duke.edu.

Duquesne University, Bayer School of Natural and Environmental Sciences, Department of Chemistry and Biochemistry, Pittsburgh, PA 15282-0001. Offers biochemistry (MS, PhD); chemistry (MS, PhD). Part-time programs available. *Faculty:* 17 full-time (5 women). *Students:* 36 full-time (19 women), 5 part-time (2 women); includes 1 minority (Hispanic American), 11 international. Average age 27. 34 applicants, 53% accepted, 8 enrolled. In 2005, 5 master's, 9 doctorates awarded. Terminal master's awarded for partial completion of doctoral program. *Degree requirements:* For master's, thesis (for some programs), comprehensive exam (for some programs), registration; for doctorate, thesis/dissertation, registration. *Entrance requirements:* For master's and doctorate, GRE General Test. Additional exam requirements/recommendations for international students: Required—TOEFL, TSE. *Application deadline:* For fall admission, 2/15 priority date for domestic students, 2/15 priority date for international students; for spring admission, 10/1 priority date for domestic students, 10/1 priority date for international students. Applications are processed on a rolling basis. Application fee: $0. *Expenses: Contact institution.* Tuition and fees vary according to degree level and program. *Financial support:* In 2005–06, 1 fellowship with full and partial tuition reimbursement (averaging $19,100 per year), 8 research assistantships with full tuition reimbursements (averaging $18,850 per year), 26 teaching assistantships with full tuition reimbursements (averaging $18,850 per year) were awarded; institutionally sponsored loans, scholarships/grants, and unspecified assistantships also available. Financial award application deadline: 5/1; financial award applicants required to submit FAFSA. *Faculty research:* Computational physical chemistry, bioinorganic chemistry, analytical chemistry, biophysics, synthetic organic chemistry. Total annual research expenditures: $1.5 million. *Unit head:* Dr. Jeffry Madura, Chair, 412-396-6341, Fax: 412-396-5683, E-mail: madura@duq.edu. *Application contact:* Mary Ann Quinn, Assistant to the Dean Graduate Affairs, 412-396-6339, Fax: 412-396-4881, E-mail: gradinfo@duq.edu.

See Close-Up on page 113.

East Carolina University, Graduate School, Thomas Harriot College of Arts and Sciences, Department of Chemistry, Greenville, NC 27858-4353. Offers MS. Part-time programs available. *Faculty:* 16 full-time (2 women). *Students:* 13 full-time (6 women), 6 part-time (4 women); includes 3 minority (1 African American, 2 Asian Americans or Pacific Islanders), 4 international. Average age 25. 8 applicants, 38% accepted, 3 enrolled. In 2005, 4 degrees awarded. *Degree requirements:* For master's, one foreign language, thesis, comprehensive exam. *Entrance requirements:* For master's, GRE General Test. Additional exam requirements/recommendations for international students: Required—TOEFL. *Application deadline:* For fall admission, 6/1 for domestic students; for spring admission, 10/15 for domestic students. Applications are processed on a rolling basis. Application fee: $50. *Expenses:* Tuition, area resident: Full-time $2,516. Tuition, state resident: full-time $2,516. Tuition, nonresident: full-time $12,832. International tuition: $12,832 full-time. *Financial support:* Teaching assistantships, Federal Work-Study available. Financial award application deadline:6/1. *Faculty research:* Organometallic, natural-product syntheses; chemometrics; electroanalytical method development; microcomputer adaptations for handicapped students. *Unit head:* Dr. Art L Rodriquez, Interim Chair, 252-328-9700, Fax: 252-328-6210, E-mail: rodrigueza@ecu.edu. *Application contact:* Dr. Paul D. Tschetter, Interim Dean of Graduate School, 252-328-6012, Fax: 252-328-6071, E-mail: gradschool@mail.ecu.edu.

Eastern Illinois University, Graduate School, College of Sciences, Department of Chemistry, Charleston, IL 61920-3099. Offers MS. *Faculty:* 13 full-time (2 women). In 2005, 5 degrees awarded. *Degree requirements:* For master's, thesis. *Entrance requirements:* For master's, GRE General Test. *Application deadline:* For fall admission, 7/31 for domestic students. Applications are processed on a rolling basis. Application fee: $30. *Expenses:* Tuition, state resident: part-time $150 per credit hour. Tuition, nonresident: part-time $452 per credit hour. Required fees: $738. *Financial support:* In 2005–06, 2 research assistantships with tuition reimbursements (averaging $7,200 per year), 2 teaching assistantships with tuition reimbursements (averaging $7,200 per year) were awarded. *Unit head:* Dr. Doug Klarup, Chairperson, 217-581-6227, E-mail: dgklarup@eiu.edu. *Application contact:* Dr. Barbara Lawrence, Coordinator, 217-581-2720, E-mail: balawrence@eiu.edu.

Eastern Kentucky University, The Graduate School, College of Arts and Sciences, Department of Chemistry, Richmond, KY 40475-3102. Offers MS. Part-time and evening/weekend programs available. *Entrance requirements:* For master's, GRE General Test, minimum GPA of 2.5. *Faculty research:* Organic synthesis, surface chemistry, inorganic chemistry, analytical chemistry.

Eastern Michigan University, Graduate School, College of Arts and Sciences, Department of Chemistry, Ypsilanti, MI 48197. Offers MS, MS/PhD. Evening/weekend programs available. *Faculty:* 17 full-time (5 women). *Students:* 2 full-time (both women), 28 part-time (12 women); includes 3 minority (1 American Indian/Alaska Native, 2 Asian Americans or Pacific Islanders), 15 international. Average age 28. In 2005, 6 degrees awarded. *Degree requirements:* For master's, thesis. *Entrance requirements:* For master's, GRE General Test. Additional exam requirements/recommendations for international students: Required—TOEFL. *Application deadline:* For fall admission, 5/15 priority date for domestic students, 5/1 priority date for international studentsFor winter admission, 10/15 for domestic students; for spring admission, 3/15 for domestic students. Applications are processed on a rolling basis. Application fee: $35.

Expenses: Tuition, state resident: full-time $7,838; part-time $327 per credit hour. Tuition, nonresident: full-time $15,770; part-time $657 per credit hour. Required fees: $33 per credit hour. $40 per term. Tuition and fees vary according to course level, course load and degree level. *Financial support:* In 2005–06, fellowships (averaging $4,000 per year), research assistantships with full tuition reimbursements (averaging $8,950 per year), teaching assistantships with full tuition reimbursements (averaging $8,950 per year) were awarded; career-related internships or fieldwork, Federal Work-Study, and institutionally sponsored loans also available. Support available to part-time students. Financial award applicants required to submit FAFSA. *Unit head:* Dr. Maria Milletti, Interim Head, 734-487-0106. *Application contact:* Dr. Krish Rengan, Coordinator, 734-487-0106.

Eastern New Mexico University, Graduate School, College of Liberal Arts and Sciences, Department of Physical Sciences, Portales, NM 88130. Offers chemistry (MS). Part-time programs available. *Faculty:* 5 full-time (0 women). *Students:* 2 full-time (0 women), 9 part-time (4 women), 7 international. Average age 28. 14 applicants, 7% accepted. In 2005, 1 degree awarded. *Degree requirements:* For master's, field exam, thesis optional. *Entrance requirements:* For master's, minimum GPA of 2.5. *Application deadline:* For fall admission, 8/20 for domestic students. Applications are processed on a rolling basis. Application fee: $10. Electronic applications accepted. *Expenses:* Tuition, area resident: Full-time $2,316; part-time $97 per credit hour. Tuition, state resident: full-time $2,316; part-time $97 per credit hour. Tuition, nonresident: full-time $7,872; part-time $328 per credit hour. Required fees: $33 per credit hour. *Financial support:* In 2005–06, 2 research assistantships (averaging $7,700 per year), 9 teaching assistantships (averaging $7,700 per year) were awarded; fellowships, career-related internships or fieldwork and Federal Work-Study also available. Support available to part-time students. Financial award application deadline: 3/1. *Faculty research:* Synfuel, electrochemistry, protein chemistry. *Unit head:* Dr. Newton Hilliard, Graduate Coordinator, 505-562-2463, E-mail: newton.hilliard@enmu.edu.

East Tennessee State University, School of Graduate Studies, College of Arts and Sciences, Department of Chemistry, Johnson City, TN 37614. Offers MS. Part-time and evening/weekend programs available. *Faculty:* 7 full-time (1 woman). *Students:* 10 full-time (3 women), 6 part-time (2 women); includes 1 minority (Asian American or Pacific Islander), 10 international. Average age 31. 23 applicants, 43% accepted, 6 enrolled. In 2005, 3 degrees awarded. *Degree requirements:* For master's, thesis, comprehensive exam. *Entrance requirements:* For master's, GRE. Additional exam requirements/recommendations for international students: Required—TOEFL (minimum score 550 paper-based; 213 computer-based). *Application deadline:* For fall admission, 7/15 for domestic students; for spring admission, 11/1 for domestic students. Applications are processed on a rolling basis. Application fee: $25 ($35 for international students). *Expenses:* Tuition, nonresident: full-time $9,312; part-time $404 per hour. Required fees: $261 per hour. *Financial support:* In 2005–06, 9 teaching assistantships with full tuition reimbursements (averaging $8,000 per year) were awarded; research assistantships with full tuition reimbursements, Federal Work-Study and institutionally sponsored loans also available. Financial award application deadline: 7/1; financial award applicants required to submit FAFSA. *Faculty research:* Development of luminescence techniques for chemical analysis, new functional materials and biosensor technology, synthesis of theoretically significant organic molecules and synthetic metals, synthesis and study of phosphatase enzyme models. Total annual research expenditures:$12,500. *Unit head:* Dr. Jeffrey G. Wardeska, Interim Chair, 423-439-4367, Fax: 423-439-5835, E-mail: rd1jeff@etsu.edu.

Emory University, Graduate School of Arts and Sciences, Department of Chemistry, Atlanta, GA 30322-1100. Offers PhD. *Faculty:* 19 full-time (1 woman), 4 part-time/adjunct (0 women). *Students:* 157 full-time (66 women); includes 74 minority (9 African Americans, 1 American Indian/Alaska Native, 59 Asian Americans or Pacific Islanders, 5 Hispanic Americans). 368 applicants, 18% accepted. In 2005, 13 doctorates awarded. *Median time to degree:* Of those who began their doctoral program in fall 1997, 99% received their degree in 8 years or less. *Degree requirements:* For doctorate, thesis/dissertation, comprehensive exam. *Entrance requirements:* For doctorate, GRE General Test, 3 letters of recommendation, curriculum vitae. Additional exam requirements/recommendations for international students: Required—TOEFL. *Application deadline:* For fall admission, 1/3 priority date for domestic students, 1/3 priority date for international students. Application fee: $50 ($0 for international students). Electronic applications accepted. *Expenses:* Tuition: Full-time $14,400. Required fees: $217. *Financial support:* In 2005–06, 15 fellowships with full tuition reimbursements (averaging $26,000 per year), 75 research assistantships with full tuition reimbursements (averaging $21,000 per year), 60 teaching assistantships with full tuition reimbursements (averaging $21,000 per year) were awarded; Federal Work-Study, institutionally sponsored loans, scholarships/grants, and health care benefits also available. Financial award application deadline: 1/3; financial award applicants required to submit FAFSA. *Faculty research:* Organometallic synthesis and catalysis, synthesis of natural products, x-ray crystallography, mass spectrometry, analytical neurochemistry. Total annual research expenditures: $5.1 million. *Unit head:* Dr. Joel Bowman, Chairman, 404-727-6585. *Application contact:* Dr. Tim Lian, Director of Graduate Studies, 404-727-6649, Fax: 404-727-6586, E-mail: gradchem@emory.edu.

Emporia State University, School of Graduate Studies, College of Liberal Arts and Sciences, Department of Physical Sciences, Emporia, KS 66801-5087. Offers chemistry (MS); earth science (MS); physical science (MS); physics (MS). *Faculty:* 15 full-time (2 women), 1 (woman) part-time/adjunct. *Students:* 5 full-time (1 woman), 19 part-time (5 women); includes 1 minority (African American) 8 applicants, 88% accepted, 5 enrolled. In 2005, 6 degrees awarded. *Degree requirements:* For master's, comprehensive exam or thesis. *Entrance requirements:* For master's, physical science qualifying exam, appropriate undergraduate degree. Additional exam requirements/recommendations for international students: Required—TOEFL. *Application deadline:* For fall admission, 8/15 for domestic students. Applications are processed on a rolling basis. Application fee: $30 ($75 for international students). Electronic applications accepted. *Expenses:* Tuition, state resident: full-time $2,890; part-time $132 per credit. Tuition, nonresident: full-time $9,258; part-time $422 per credit. Required fees: $626; $41 per credit. Tuition and fees vary according to degree level. *Financial support:* In 2005–06, research assistantships with full tuition reimbursements (averaging $6,492 per year), 8 teaching assistantships with full tuition reimbursements (averaging $6,492 per year) were awarded; Federal Work-Study, institutionally sponsored loans, health care benefits, and unspecified assistantships also available. Financial award application deadline: 3/15; financial award applicants required to submit FAFSA. *Faculty research:* Bredigite, larnite, and dicalcium silicates–Marble Canyon. *Unit head:* Dr. DeWayne Backhus, Chair, 620-341-5330, Fax: 620-341-6055, E-mail: dbackhus@emporia.edu.

Fairleigh Dickinson University, College at Florham, Maxwell Becton College of Arts and Sciences, Department of Chemistry and Geological Sciences, Program in Chemistry, Madison, NJ 07940-1099. Offers MS. *Students:* 24 full-time (10 women), 16 part-time (7 women), 33 international. Average age 27. 68 applicants, 59% accepted, 12 enrolled. In 2005, 5 degrees awarded. *Application deadline:* Applications are processed on a rolling basis. Application fee: $40. *Unit head:* Dr. William Fordham, Acting Chair, Department of Chemistry and Geological Sciences, 973-443-8779, Fax: 973-443-8766, E-mail: fordham@fdu.edu.

Fisk University, Graduate Programs, Department of Chemistry, Nashville, TN 37208-3051. Offers MA. *Degree requirements:* For master's, thesis, comprehensive exam. *Entrance requirements:* For master's, GRE General Test, minimum GPA of 3.0. *Faculty research:* Environmental studies, lithium compound synthesis, HIU compound synthesis.

Florida Agricultural and Mechanical University, Division of Graduate Studies, Research, and Continuing Education, College of Arts and Sciences, Department of Chemistry, Tallahassee, FL 32307-3200. Offers MS. *Degree requirements:* For master's, thesis optional. *Entrance requirements:* For master's, GRE General Test, minimum GPA of 3.0.

Florida Atlantic University, Charles E. Schmidt College of Science, Department of Chemistry and Biochemistry, Boca Raton, FL 33431-0991. Offers MS, MST, PhD. Part-time programs available. *Faculty:* 13 full-time (2 women). *Students:* 48 full-time (26 women), 6 part-time (4 women); includes 11 minority (2 African Americans, 2 Asian Americans or Pacific Islanders, 7 Hispanic Americans), 18 international. Average age 31. 24 applicants, 25% accepted, 5 enrolled. In 2005, 8 master's, 2 doctorates awarded. Terminal master's awarded for partial completion of doctoral program. *Degree requirements:* For master's, thesis/dissertation; for doctorate, thesis/dissertation, comprehensive exam. *Entrance requirements:* For master's, GRE General Test, minimum GPA of 3.0; for doctorate, GRE, minimum GPA of 3.0. Additional exam requirements/recommendations for international students: Required—TOEFL. *Application deadline:* For fall admission, 7/1 priority date for domestic students, 2/15 priority date for international students; for spring admission, 4/1 priority date for domestic students, 1/15 priority date for international students. Applications are processed on a rolling basis. Application fee: $30. *Expenses:* Tuition, area resident: Full-time $4,394; part-time $244 per credit. Tuition, state resident: full-time $4,394; part-time $244 per credit. Tuition, nonresident: full-time $16,441; part-time $912 per credit. International tuition: $16,441 full-time. *Financial support:* In 2005–06, 2 research assistantships with full tuition reimbursements (averaging $14,000 per year), 24 teaching assistantships with full tuition reimbursements (averaging $14,000 per year) were awarded; fellowships, Federal Work-Study also available. *Faculty research:* Polymer synthesis and characterization, spectroscopy, geochemistry, environmental chemistry, biomedical chemistry. Total annual research expenditures: $1.2 million. *Unit head:* Dr. Gregg B. Fields, Chair, 561-297-2093, Fax: 561-297-2759, E-mail: fieldsg@fau.edu. *Application contact:* Dr. Salvatore D. Lepore, Professor, 561-297-0330, Fax: 561-297-2759, E-mail: slepore@fau.edu.

Florida Institute of Technology, Graduate Programs, College of Science, Department of Chemistry, Melbourne, FL 32901-6975. Offers MS, PhD. Part-time programs available. *Faculty:* 4 full-time (0 women). *Students:* 12 full-time (7 women), 2 part-time (1 woman); includes 1 minority (African American), 13 international. Average age 30. 26 applicants, 62% accepted, 2 enrolled. In 2005, 3 master's, 1 doctorate awarded. Terminal master's awarded for partial completion of doctoral program. *Degree requirements:* For master's, thesis, oral defense of thesis; for doctorate, one foreign language, thesis/dissertation, oral defense of dissertation, dissertation research publishable to standards, comprehensive exam, registration. *Entrance requirements:* For master's, minimum GPA of 3.0; for doctorate, minimum GPA of 3.2, resumé, 3 letters of recommendation. Additional exam requirements/recommendations for international students: Required—TOEFL (minimum score 550 paper-based; 213 computer-based). *Application deadline:* Applications are processed on a rolling basis. Application fee: $50. Electronic applications accepted. *Expenses:* Tuition: Part-time $825 per credit. *Financial support:* In 2005–06, 19 teaching assistantships with full and partial tuition reimbursements (averaging $8,179 per year) were awarded; fellowships with full and partial tuition reimbursements, research assistantships with full and partial tuition reimbursements, career-related internships or fieldwork and tuition remissions also available. Financial award application deadline: 3/1; financial award applicants required to submit FAFSA. *Faculty research:* Energy storage applications, marine and organic chemistry, stereochemistry, medicinal chemistry, environmental chemistry. Total annual research expenditures:$847,085. *Unit head:* Dr. Michael W. Babich, Department Head, 321-674-8046, Fax: 321-674-8951, E-mail: babich@fit.edu. *Application contact:* Carolyn P. Farrior, Director of Graduate Admissions, 321-674-7118, Fax: 321-723-9468, E-mail: cfarrior@fit.edu.

Florida International University, College of Arts and Sciences, Department of Chemistry, Miami, FL 33199. Offers chemistry (MS, PhD); forensic science (MS). Part-time and evening/weekend programs available. *Faculty:* 22 full-time (2 women). *Students:* 46 full-time (21 women), 6 part-time (4 women); includes 13 minority (1 African American, 2 Asian Americans or Pacific Islanders, 10 Hispanic Americans), 33 international. Average age 31. 52 applicants, 33% accepted, 9 enrolled. In 2005, 12 master's, 1 doctorate awarded. *Degree requirements:* For master's and doctorate, thesis/dissertation. *Entrance requirements:* For master's and doctorate, GRE General Test. Additional exam requirements/recommendations for international students: Required—TOEFL. *Application deadline:* For fall admission, 4/1 for domestic students; for spring admission, 10/1 for domestic students. Applications are processed on a rolling basis. Application fee: $25. *Expenses:* Tuition, area resident: Part-time $239 per credit. Tuition, state resident: full-time $4,294; part-time $869 per credit. Tuition, nonresident: full-time $15,641; part-time $869 per credit. International tuition: $15,641 full-time. Required fees: $252; $126 per term. Tuition and fees vary according to program. *Financial support:* Research assistantships, teaching assistantships, Federal Work-Study, institutionally sponsored loans, and tuition waivers (full and partial) available. Support available to part-time students. Financial award application deadline:4/1. *Faculty research:* Organic synthesis and reaction catalysis, environmental chemistry, molecular beam studies, organic geochemistry, bioinorganic and organometallic chemistry. *Unit head:* Dr. Stanislaw Wnuk, Chairperson, 305-348-2606, Fax: 305-348-3772, E-mail: stanislaw.wnuk@fiu.edu.

Florida State University, Graduate Studies, College of Arts and Sciences, Department of Chemistry and Biochemistry, Tallahassee, FL 32306. Offers analytical chemistry (MS, PhD); biochemistry (MS, PhD); chemical physics (MS, PhD); inorganic chemistry (MS, PhD); organic chemistry (MS, PhD); physical chemistry (MS, PhD). Part-time programs available. *Faculty:* 36 full-time (6 women), 2 part-time/adjunct (0 women). *Students:* 144 full-time (50 women); includes 17 minority (7 African Americans, 1 American Indian/Alaska Native, 6 Asian Americans or Pacific Islanders, 3 Hispanic Americans), 58 international. Average age 25. 288 applicants, 27% accepted, 26 enrolled. In 2005, 14 master's, 15 doctorates awarded. Terminal master's awarded for partial completion of doctoral program. *Median time to degree:* Of those who began their doctoral program in fall 1997, 88% received their degree in 8 years or less. *Degree requirements:* For master's, thesis (for some programs), cumulative and diagnostic exams, comprehensive exam (for some programs), registration; for doctorate, thesis/dissertation, cumulative and diagnostic exams, comprehensive exam (for some programs), registration. *Entrance requirements:* For master's and doctorate, GRE General Test, minimum B average in undergraduate course work. Additional exam requirements/recommendations for international students: Required—TOEFL (minimum score 515 paper-based; 213 computer-based). *Application deadline:* For fall admission, 4/15 priority date for domestic students, 4/15 priority date for international students. Applications are processed on a rolling basis. Application fee: $30. Electronic applications accepted. *Financial support:* In 2005–06, 1 fellowship with tuition reimbursement (averaging $18,000 per year), 56 research assistantships with tuition reimbursements (averaging $19,000 per year), 83 teaching assistantships with tuition reimbursements (averaging $19,000 per year) were awarded; career-related internships or fieldwork, Federal Work-Study, institutionally sponsored loans, and traineeships also available. Financial award application deadline: 2/15; financial award applicants required to submit FAFSA. *Faculty research:* Spectroscopy, computational chemistry, nuclear chemistry, separations, synthesis. Total annual research expenditures: $6.5 million. *Unit head:* Dr. Naresh Dalal, Chairman, 850-644-3398, Fax: 850-644-8281. *Application contact:* Dr. Oliver Steinbock, Chair, Graduate Admissions Committee, 888-525-9286, Fax: 850-644-8281, E-mail: gradinfo@chem.fsu.edu.

See Close-Up on page 117.

Furman University, Graduate Division, Department of Chemistry, Greenville, SC 29613. Offers MS. *Faculty:* 9 full-time (3 women). *Students:* 4 full-time (0 women), 2 part-time (1 woman). Average age 23. 6 applicants, 100% accepted, 6 enrolled. In 2005, 5 degrees awarded. *Degree requirements:* For master's, thesis, comprehensive exam. *Entrance requirements:* For master's, GRE General Test, GRE Subject Test. *Application deadline:* For fall admission, 8/1 for domestic students, 8/1 for international studentsFor winter admission, 12/1 for domestic students; for spring admission, 2/1 for domestic students. Applications are processed on a rolling basis. Application fee: $50. *Expenses:* Tuition: Full-time $315; part-time $315 per credit. *Financial support:* In 2005–06, 6 students received support, including 6 fellowships (averaging $5,557 per year); research assistantships, scholarships/grants and unspecified assistantships also available. Financial award application deadline: 8/1. *Faculty research:* Computer-assisted chemical analysis, DNA-metal interactions, laser-initiated reactions, nucleic acid chemistry and biochemistry. *Unit head:* Dr. Lon B. Knight, Professor, 864-294-3372, Fax: 864-294-3559, E-mail: lon.knight@furman.edu. *Application contact:* Myra Crumley, Information Contact, 864-294-2056, Fax: 864-294-3559, E-mail: myra.crumley@furman.edu.

Chemistry

George Mason University, College of Arts and Sciences, Department of Chemistry, Fairfax, VA 22030. Offers MS. *Faculty:* 17 full-time (3 women), 3 part-time/adjunct (1 woman). *Students:* 5 full-time (3 women), 17 part-time (8 women); includes 1 African American, 5 Asian Americans or Pacific Islanders, 3 international. Average age 24. 23 applicants, 87% accepted, 13 enrolled. In 2005, 3 degrees awarded. *Degree requirements:* For master's, thesis or alternative. *Entrance requirements:* For master's, GRE General Test, minimum GPA of 3.0 in last 60 hours of course work. *Application deadline:* For fall admission, 5/1 for domestic students; for spring admission, 11/1 for domestic students. Electronic applications accepted. *Expenses:* Tuition, area resident: Full-time $5,244; part-time $219 per credit. Tuition, state resident: part-time $651 per credit. Tuition, nonresident: full-time $15,636. International tuition: $15,636 full-time. Required fees: $1,524; $65 per credit. *Financial support:* Research assistantships available. Support available to part-time students. Financial award application deadline: 3/1; financial award applicants required to submit FAFSA. *Unit head:* Gregory Foster, Chairperson, 703-993-1081, Fax: 703-993-1070, E-mail: gfoster@gmu.edu.

Georgetown University, Graduate School of Arts and Sciences, Department of Chemistry, Washington, DC 20057. Offers analytical chemistry (MS, PhD); biochemistry (MS, PhD); chemical physics (MS, PhD); inorganic chemistry (MS, PhD); organic chemistry (MS, PhD); physical chemistry (MS, PhD); theoretical chemistry (MS, PhD). Terminal master's awarded for partial completion of doctoral program. *Degree requirements:* For master's, thesis (for some programs), qualifying exam; for doctorate, thesis/dissertation, comprehensive exam. *Entrance requirements:* For master's and doctorate, GRE General Test. Additional exam requirements/recommendations for international students: Required—TOEFL.

The George Washington University, Columbian College of Arts and Sciences, Department of Chemistry, Washington, DC 20052. Offers analytical chemistry (MS, PhD); inorganic chemistry (MS, PhD); materials science (MS, PhD); organic chemistry (MS, PhD); physical chemistry (MS, PhD). Part-time and evening/weekend programs available. Terminal master's awarded for partial completion of doctoral program. *Degree requirements:* For master's, thesis or alternative, comprehensive exam; for doctorate, thesis/dissertation, general exam. *Entrance requirements:* For master's and doctorate, GRE General Test, interview, minimum GPA of 3.0. Additional exam requirements/recommendations for international students: Required—TOEFL (minimum score 550 paper-based; 213 computer-based). Electronic applications accepted.

Georgia Institute of Technology, Graduate Studies and Research, College of Sciences, School of Chemistry and Biochemistry, Atlanta, GA 30332-0001. Offers MS, MS Chem, PhD. Terminal master's awarded for partial completion of doctoral program. *Degree requirements:* For master's, thesis (for some programs); for doctorate, thesis/dissertation. *Entrance requirements:* For master's and doctorate, GRE General Test, GRE Subject Test, minimum GPA of 2.7. Additional exam requirements/recommendations for international students: Required—TOEFL. Electronic applications accepted. *Faculty research:* Inorganic, organic, physical, and analytical chemistry.

Georgia State University, College of Arts and Sciences, Department of Chemistry, Atlanta, GA 30303-3083. Offers MS, PhD. Part-time programs available. Terminal master's awarded for partial completion of doctoral program. *Degree requirements:* For master's, one foreign language, thesis or alternative, exam; for doctorate, one foreign language, thesis/dissertation, exam. *Entrance requirements:* For master's, GRE General Test, departmental supplemental form; for doctorate, GRE General Test. Additional exam requirements/recommendations for international students: Required—TOEFL. Electronic applications accepted. *Expenses:* Tuition, area resident: Full-time $4,368; part-time $182 per term. Tuition, state resident: full-time $4,368; part-time $182 per term. Tuition, nonresident: full-time $8,732; part-time $728 per term. Required fees: $46 per hour. *Faculty research:* DNA, AIDS, drug design, biothermodynamics, biological electron transfer and NMR applied to biochemical systems.

Graduate School and University Center of the City University of New York, Graduate Studies, Program in Chemistry, New York, NY 10016-4039. Offers PhD. *Faculty:* 64 full-time (5 women). *Students:* 132 full-time (48 women); includes 21 minority (8 African Americans, 7 Asian Americans or Pacific Islanders, 6 Hispanic Americans), 96 international. Average age 34. 78 applicants, 36% accepted, 23 enrolled. In 2005, 15 degrees awarded. *Degree requirements:* For doctorate, one foreign language, thesis/dissertation. *Entrance requirements:* For doctorate, GRE General Test, GRE Subject Test. Additional exam requirements/recommendations for international students: Required—TOEFL. *Application deadline:* For fall admission, 4/15 for domestic students; for spring admission, 11/15 for domestic students. Application fee: $125. Electronic applications accepted. *Financial support:* In 2005–06, 79 students received support, including 78 fellowships, 1 teaching assistantship; research assistantships, career-related internships or fieldwork, Federal Work-Study, institutionally sponsored loans, and tuition waivers (full and partial) also available. Financial award application deadline: 2/1; financial award applicants required to submit FAFSA. *Unit head:* Dr. Gerald Koeppl, Executive Officer, 212-817-8136, Fax: 212-817-1507, E-mail: gkoeppl@gc.cuny.edu.

Hampton University, Graduate College, Department of Chemistry, Hampton, VA 23668. Offers MS. Part-time and evening/weekend programs available. *Degree requirements:* For master's, thesis. *Entrance requirements:* For master's, GRE General Test.

Harvard University, Graduate School of Arts and Sciences, Department of Chemistry and Chemical Biology, Cambridge, MA 02138. Offers biochemical chemistry (PhD); inorganic chemistry (PhD); organic chemistry (PhD); physical chemistry (PhD). *Students:* 188 full-time (36 women). 346 applicants, 24% accepted. In 2005, 33 doctorates awarded. *Degree requirements:* For doctorate, thesis/dissertation, cumulative exams. *Entrance requirements:* For doctorate, GRE General Test, GRE Subject Test. Additional exam requirements/recommendations for international students: Required—TOEFL. *Application deadline:* For fall admission, 12/31 for domestic students. Application fee: $60. *Expenses:* Tuition: Full-time $28,752. Full-time tuition and fees vary according to program and student level. *Financial support:* Fellowships, research assistantships, teaching assistantships, career-related internships or fieldwork, Federal Work-Study, and institutionally sponsored loans available. Financial award application deadline: 12/30. *Unit head:* Betsey Cogswell, Administrator, 617-495-5696, Fax: 617-495-5264. *Application contact:* Graduate Admissions Office, 617-496-3208.

See Close-Up on page 119.

Howard University, Graduate School of Arts and Sciences, Department of Chemistry, Washington, DC 20059-0002. Offers analytical chemistry (MS, PhD); atmospheric (MS, PhD); biochemistry (MS, PhD); environmental (MS, PhD); inorganic chemistry (MS, PhD); organic chemistry (MS, PhD); physical chemistry (MS, PhD); polymer chemistry (MS, PhD). Part-time programs available. *Degree requirements:* For master's, one foreign language, thesis, teaching experience, comprehensive exam, registration; for doctorate, 2 foreign languages, thesis/dissertation, teaching experience, comprehensive exam, registration. *Entrance requirements:* For master's, GRE General Test, minimum GPA of 2.7; for doctorate, GRE General Test, minimum GPA of 3.0. *Faculty research:* Stratospheric aerosols, liquid crystals, polymer coatings, terrestrial and extraterrestrial atmospheres, amidogen reaction.

Idaho State University, Office of Graduate Studies, College of Arts and Sciences, Department of Chemistry, Pocatello, ID 83209. Offers MNS, MS. MS students must enter as undergraduates. *Degree requirements:* For master's, one foreign language, thesis (for some programs), comprehensive exam, registration. *Entrance requirements:* For master's, GRE General Test, minimum GPA of 3.0 in all upper division classes. Additional exam requirements/recommendations for international students: Required—TOEFL (minimum score 550 paper-based; 213 computer-based). *Faculty research:* Low temperature plasma, organic chemistry, physical chemistry, inorganic chemistry, analytical chemistry.

Illinois Institute of Technology, Graduate College, College of Science and Letters, Department of Biological, Chemical and Physical Sciences, Chemistry Division, Chicago, IL 60616-3793. Offers analytical chemistry (M Ch, MS, PhD); chemistry (M Chem); inorganic chemistry (MS, PhD); materials and chemical synthesis (M Ch); organic chemistry (MS, PhD); physical chemistry (MS, PhD); polymer chemistry (MS, PhD). Part-time and evening/weekend programs available. Postbaccalaureate distance learning degree programs offered (no on-campus study). Terminal master's awarded for partial completion of doctoral program. *Degree requirements:* For master's, thesis (for some programs), comprehensive exam; for doctorate, thesis/dissertation, comprehensive exam. *Entrance requirements:* For master's and doctorate, GRE General Test, minimum undergraduate GPA of 3.0. Additional exam requirements/recommendations for international students: Required—TOEFL (minimum score 550 paper-based; 213 computer-based). Electronic applications accepted. *Faculty research:* Organic and inorganic chemistry, polymers research, physical chemistry, analytical chemistry.

Illinois State University, Graduate School, College of Arts and Sciences, Department of Chemistry, Normal, IL 61790-2200. Offers MS. *Faculty:* 21 full-time (5 women). *Students:* 38 full-time (11 women), 11 part-time (5 women); includes 2 minority (1 African American, 1 Hispanic American), 14 international. 24 applicants, 79% accepted. In 2005, 13 degrees awarded. *Degree requirements:* For master's, thesis. *Entrance requirements:* For master's, GRE General Test, minimum GPA of 2.6 in last 60 hours of course work. *Application deadline:* Applications are processed on a rolling basis. Application fee: $30. *Expenses:* Tuition, area resident: Full-time $3,060. Tuition, state resident: full-time $3,060; part-time $170 per credit hour. Tuition, nonresident: full-time $6,390; part-time $355 per credit hour. International tuition: $6,390 full-time. Required fees: $1,411; $47 per credit hour. *Financial support:* In 2005–06, 9 research assistantships (averaging $14,929 per year), 21 teaching assistantships (averaging $8,232 per year) were awarded; tuition waivers (full) and unspecified assistantships also available. Financial award application deadline: 4/1. *Faculty research:* Science Teaching Excellence Partnership (STEP), a new technique for chemical imaging of model neurons, CAREER: crystallography and rare-earth educational and research activities, octahedral hexanuclear clusters: fundamental studies and inquiry into potential applications. Total annual research expenditures: $376,614. *Unit head:* Dr. Neil Skaggs, Acting Chair, 309-438-7661.

Indiana University Bloomington, Graduate School, College of Arts and Sciences, Department of Chemistry, Bloomington, IN 47405-7000. Offers analytical chemistry (PhD); biological chemistry (PhD); chemistry (MAT); inorganic chemistry (PhD); physical chemistry (PhD). PhD offered through the University Graduate School. *Faculty:* 29 full-time (2 women). *Students:* 110 full-time (38 women), 35 part-time (11 women); includes 8 minority (3 African Americans, 3 Asian Americans or Pacific Islanders, 2 Hispanic Americans), 57 international. Average age 26. In 2005, 7 master's, 20 doctorates awarded. Terminal master's awarded for partial completion of doctoral program. *Degree requirements:* For master's and doctorate, thesis/dissertation. *Entrance requirements:* For master's and doctorate, GRE General Test, GRE Subject Test. Additional exam requirements/recommendations for international students: Required—TOEFL. *Application deadline:* For fall admission, 1/15 priority date for domestic students, 12/15 priority date for international students; for spring admission, 9/1 priority date for domestic students, 9/1 priority date for international students. Applications are processed on a rolling basis. Application fee: $50 ($60 for international students). *Expenses:* Tuition, state resident: full-time $5,437; part-time $227 per credit hour. Tuition, nonresident: full-time $15,836; part-time $660 per credit hour. Required fees: $821. Tuition and fees vary according to campus/location and program. *Financial support:* In 2005–06, 23 fellowships with full tuition reimbursements, 78 teaching assistantships with full tuition reimbursements, 57 research assistantships with full tuition reimbursements were awarded; Federal Work-Study and institutionally sponsored loans also available. *Faculty research:* Synthesis of complex natural products, organic reaction mechanisms, organic electrochemistry, transitive-metal chemistry, solid-state and surface chemistry. Total annual research expenditures: $7.7 million. *Unit head:* Dr. David Clemmer, Chairperson, 812-855-2268. *Application contact:* Dr. Jack K. Crandall, Chairperson of Admissions, 812-855-2068, Fax: 812-855-8300, E-mail: chemgrad@indiana.edu.

Indiana University of Pennsylvania, Graduate School and Research, College of Natural Sciences and Mathematics, Department of Chemistry, Program in Chemistry, Indiana, PA 15705-1087. Offers MA, MS. Part-time programs available. *Degree requirements:* For master's, thesis optional. *Entrance requirements:* For master's, 2 letters of recommendation. Additional exam requirements/recommendations for international students: Required—TOEFL.

Indiana University–Purdue University Indianapolis, School of Science, Department of Chemistry, Indianapolis, IN 46202-2896. Offers MS, PhD, MD/PhD. Part-time and evening/weekend programs available. *Faculty:* 10 full-time (2 women). *Students:* 13 full-time (6 women), 25 part-time (16 women); includes 6 minority (4 African Americans, 1 American Indian/Alaska Native, 1 Hispanic American), 6 international. Average age 28. In 2005, 9 degrees awarded. Terminal master's awarded for partial completion of doctoral program. *Degree requirements:* For master's, thesis (for some programs); for doctorate, thesis/dissertation. *Entrance requirements:* For master's and doctorate, minimum GPA of 3.0. Additional exam requirements/recommendations for international students: Required—TOEFL, GRE (international applicants). *Application deadline:* Applications are processed on a rolling basis. Application fee: $50 ($60 for international students). *Expenses:* Tuition, area resident: Full-time $5,159; part-time $215 per credit hour. Tuition, state resident: full-time $5,159; part-time $215 per credit hour. Tuition, nonresident: full-time $14,890; part-time $620 per credit hour. International tuition: $14,890 full-time. Required fees: $614. Tuition and fees vary according to campus/location and program. *Financial support:* Fellowships with partial tuition reimbursements, research assistantships with partial tuition reimbursements, teaching assistantships with partial tuition reimbursements, career-related internships or fieldwork, institutionally sponsored loans, tuition waivers (partial), and co-op positions available. Financial award application deadline:3/1. *Faculty research:* Analytical, biological, inorganic, organic, and physical chemistry. Total annual research expenditures: $1.6 million. *Unit head:* Dr. Frank Schultz, Chair, 317-274-6872, Fax: 317-274-4701. *Application contact:* Eric Long, Information Contact, 317-274-6888, Fax: 317-274-4701, E-mail: long@chem.iupui.edu.

Instituto Tecnológico y de Estudios Superiores de Monterrey, Campus Monterrey, Graduate and Research Division, Program in Natural and Social Sciences, Monterrey, Mexico. Offers biotechnology (MS); chemistry (MS, PhD); communications (MS); education (MA). Part-time programs available. *Degree requirements:* For master's and doctorate, one foreign language, thesis/dissertation. *Entrance requirements:* For master's, PAEG; for doctorate, PAEG, master's degree in related field. Additional exam requirements/recommendations for international students: Required—TOEFL. *Faculty research:* Cultural industries, mineral substances, bioremediation, food processing, CQ in industrial chemical processing.

Iowa State University of Science and Technology, Graduate College, College of Liberal Arts and Sciences, Department of Chemistry, Ames, IA 50011. Offers MS, PhD. *Faculty:* 29 full-time, 2 part-time/adjunct. *Students:* 166 full-time (61 women), 4 part-time (2 women); includes 4 minority (2 Asian Americans or Pacific Islanders, 2 Hispanic Americans), 99 international. 63 applicants, 62% accepted, 34 enrolled. In 2005, 5 master's, 25 doctorates awarded. *Degree requirements:* For master's and doctorate, thesis/dissertation. *Entrance requirements:* Additional exam requirements/recommendations for international students: Required—GRE General Test, TOEFL (paper score 570; computer score 230) or IELTS (score 6.5). *Application deadline:* For fall admission, 2/1 priority date for domestic students, 2/1 priority date for international students. Applications are processed on a rolling basis. Application fee: $30 ($70 for international students). Electronic applications accepted. *Expenses:* Tuition, state resident: full-time $6,410. Tuition, nonresident: full-time $16,422. Tuition and fees vary according to program. *Financial support:* In 2005–06, 100 research assistantships with full and partial tuition reimbursements (averaging $17,846 per year), 66 teaching assistantships with full and partial tuition reimbursements (averaging $17,869 per year) were awarded; fellowships, scholarships/grants, health care benefits, and unspecified assistantships also available. *Unit head:* Dr. Jacob Petrich, Chair, 515-294-7812, Fax: 515-294-0105, E-mail: chemgrad@iastate.edu. *Application contact:* Lynette Edsall, Information Contact, 800-521-2436, E-mail: chemgrad@iastate.edu.

Jackson State University, Graduate School, School of Science and Technology, Department of Chemistry, Jackson, MS 39217. Offers MS, PhD. Part-time and evening/weekend programs available. *Degree requirements:* For master's and doctorate, thesis/dissertation, comprehensive exam. *Entrance requirements:* For master's, GRE General Test; for doctorate, MAT. Additional exam requirements/recommendations for international students: Required—TOEFL. *Faculty research:* Electrochemical and spectroscopic studies on charge transfer and energy transfer processes, spectroscopy of trapped molecular ions, respirable mine dust.

John Carroll University, Graduate School, Department of Chemistry, University Heights, OH 44118-4581. Offers MS. Part-time and evening/weekend programs available. *Degree requirements:* For master's, research essay or thesis. *Entrance requirements:* For master's, bachelor's degree in chemistry. Electronic applications accepted. *Faculty research:* Protein-nucleic acid interactions, protein-surface interactions, butyllithium compounds, copper proteins, magnetic materials.

The Johns Hopkins University, Zanvyl Krieger School of Arts and Sciences, Department of Chemistry, Baltimore, MD 21218-2699. Offers PhD. *Faculty:* 20 full-time (2 women). *Students:* 123 full-time (49 women); includes 1 minority (African American), 38 international. Average age 23. 69 applicants, 35% accepted, 24 enrolled. In 2005, 11 doctorates awarded. Terminal master's awarded for partial completion of doctoral program. *Degree requirements:* For doctorate, thesis/dissertation, oral exams, comprehensive exam, registration. *Entrance requirements:* For doctorate, GRE General Test, GRE Subject Test. Additional exam requirements/recommendations for international students: Required—TOEFL (minimum score 600 paper-based; 250 computer-based); Recommended—TSE. *Application deadline:* For fall admission, 1/5 for domestic students. Applications are processed on a rolling basis. Application fee: $60. Electronic applications accepted. *Expenses:* Tuition: Full-time $30,960. Tuition and fees vary according to degree level and program. *Financial support:* Fellowships, research assistantships, teaching assistantships, career-related internships or fieldwork, Federal Work-Study, institutionally sponsored loans, traineeships, and unspecified assistantships available. Financial award application deadline: 4/15; financial award applicants required to submit FAFSA. *Faculty research:* Experimental physical, biophysical, inorganic/materials, organic/bioorganic theoretical. *Application contact:* Jean Goodwin, Academic Program Coordinator, 410-516-5250, Fax: 410-516-8420, E-mail: jeang@jhu.edu.

Kansas State University, Graduate School, College of Arts and Sciences, Department of Chemistry, Manhattan, KS 66506. Offers analytical chemistry (MS); biological chemistry (MS); chemistry (PhD); inorganic chemistry (MS); materials chemistry (MS); organic chemistry (MS); physical chemistry (MS). *Faculty:* 14 full-time (1 woman). *Students:* 65 full-time (21 women), 5 part-time (2 women); includes 1 minority (African American), 48 international. 63 applicants, 67% accepted, 22 enrolled. In 2005, 3 master's, 5 doctorates awarded. Terminal master's awarded for partial completion of doctoral program. *Degree requirements:* For master's and doctorate, thesis/dissertation. *Entrance requirements:* For master's and doctorate, GRE, minimum GPA of 3.0. Additional exam requirements/recommendations for international students: Required—TOEFL (minimum score 550 paper-based; 213 computer-based). *Application deadline:* For fall admission, 2/1 priority date for domestic students, 2/1 priority date for international students; for spring admission, 10/1 for domestic students, 8/1 for international students. Applications are processed on a rolling basis. Application fee: $30 ($55 for international students). *Expenses:* Tuition, state resident: full-time $5,160; part-time $215. Tuition, nonresident: full-time $12,816; part-time $534. International tuition: $12,816 full-time. Required fees: $564. *Financial support:* In 2005–06, 30 research assistantships (averaging $11,956 per year), 26 teaching assistantships with full tuition reimbursements (averaging $11,945 per year) were awarded; fellowships, institutionally sponsored loans and scholarships/grants also available. Support available to part-time students. Financial award application deadline: 3/1; financial award applicants required to submit FAFSA. *Faculty research:* Nanotechnologies, functional materials, bio-organic and bio-physical processes, sensors and separations, synthesis and synthetic methods. Total annual research expenditures: $2.2 million. *Unit head:* Eric Maatta, Head, 785-532-6665, Fax: 785-532-6666, E-mail: eam@ksu.edu. *Application contact:* Christer Aakeröy, Director, 785-532-6096, Fax: 785-532-6666, E-mail: aakeroy@ksu.edu.

Kent State University, College of Arts and Sciences, Department of Chemistry, Kent, OH 44242-0001. Offers analytical chemistry (MS, PhD); biochemistry (MS, PhD); chemistry (MA, MS, PhD); inorganic chemistry (MS, PhD); organic chemistry (MS, PhD); physical chemistry (MS, PhD). Terminal master's awarded for partial completion of doctoral program. *Degree requirements:* For master's and doctorate, thesis/dissertation, comprehensive exam, registration. *Entrance requirements:* For master's and doctorate, placement exam, GRE General Test, GRE Subject Test (recommended), minimum GPA of 2.75. Additional exam requirements/recommendations for international students: Required—TOEFL (minimum score 575 paper-based; 230 computer-based). Electronic applications accepted. *Faculty research:* Biological chemistry, materials chemistry, molecular spectroscopy.

See Close-Up on page 121.

Lakehead University, Graduate Studies, Faculty of Social Sciences and Humanities, Department of Chemistry, Thunder Bay, ON P7B 5E1, Canada. Offers M Sc. Part-time and evening/weekend programs available. *Degree requirements:* For master's, thesis, oral examination. *Entrance requirements:* For master's, minimum B+ average. Additional exam requirements/recommendations for international students: Required—TOEFL. *Faculty research:* Physical inorganic chemistry, photochemistry, physical chemistry.

Lamar University, College of Graduate Studies, College of Arts and Sciences, Department of Chemistry and Physics, Beaumont, TX 77710. Offers chemistry (MS). Part-time programs available. *Faculty:* 12 full-time (1 woman). *Students:* 10 full-time (3 women), 11 part-time (4 women); includes 2 minority (both African Americans), 18 international. Average age 25. 42 applicants, 55% accepted, 5 enrolled. In 2005, 11 degrees awarded. *Degree requirements:* For master's, thesis, practicum. *Entrance requirements:* For master's, GRE General Test, minimum GPA of 2.5 in last 60 hours of course work. Additional exam requirements/recommendations for international students: Required—TOEFL, TWE, TSE. *Application deadline:* For fall admission, 8/1 for domestic students, 7/1 for international students; for spring admission, 12/1 for domestic students, 11/1 for international students. Applications are processed on a rolling basis. Application fee: $25 ($50 for international students). *Expenses:* Tuition, state resident: part-time $137 per semester hour. Tuition, nonresident: part-time $413 per semester hour. Required fees: $102 per semester hour. Tuition and fees vary according to course load. *Financial support:* In 2005–06, 6 students received support, including 5 teaching assistantships with partial tuition reimbursements available (averaging $9,000 per year); tuition waivers (partial) and unspecified assistantships also available. Financial award application deadline: 4/1. *Faculty research:* Environmental chemistry, surface chemistry, polymer chemistry, organic synthesis, computational chemistry. Total annual research expenditures: $750,000. *Unit head:* Dr. Richard S. Lumpkin, Chair, 409-880-8267, Fax: 409-880-8270, E-mail: lumpkines@hal.lamar.edu. *Application contact:* Dr. Paul Bernazzani, Graduate Advisor, 409-880-8272, Fax: 409-880-8270, E-mail: bernazzapx@hal.lamar.edu.

Laurentian University, School of Graduate Studies and Research, Programme in Chemistry and Biochemistry, Sudbury, ON P3E 2C6, Canada. Offers M Sc. Part-time programs available. *Degree requirements:* For master's, thesis or alternative. *Entrance requirements:* For master's, honors degree with minimum second class. *Faculty research:* Cell cycle checkpoints, kinetic modeling, toxicology to metal stress, quantum chemistry, biogeochemistry metal speciation.

Lehigh University, College of Arts and Sciences, Department of Chemistry, Bethlehem, PA 18015-3094. Offers chemistry (MS, PhD); clinical chemistry (MS); pharmaceutical chemistry (MS, PhD); polymer science and engineering (MS, PhD). Part-time programs available. Postbaccalaureate distance learning degree programs offered (no on-campus study). *Faculty:* 16 full-time (2 women), 1 part-time/adjunct. *Students:* 28 full-time (14 women), 98 part-time (55 women); includes 20 minority (12 African Americans, 5 Asian Americans or Pacific Islanders, 3 Hispanic Americans), 14 international. Average age 22. 73 applicants, 60% accepted, 45 enrolled. In 2005, 22 master's, 5 doctorates awarded. Terminal master's awarded for partial completion of doctoral program. *Degree requirements:* For master's, seminar, thesis optional; for doctorate, thesis/dissertation, 2 seminars, comprehensive exam, registration. *Entrance requirements:* For master's and doctorate, GRE General Test, bachelor's degree in chemistry or related field. Additional exam requirements/recommendations for international students: Required—TOEFL (minimum score 550 paper-based; 213 computer-based), TSE(minimum score 50); Recommended—TWE(minimum score 4). *Application deadline:* For fall admission, 7/15 priority date for domestic students, 7/15 priority date for international studentsFor winter admission, 12/1 for domestic students; for spring admission, 4/30 for domestic students. Applications are processed on a rolling basis. Application fee: $60. Electronic applications accepted. *Financial support:* In 2005–06, 25 students received support, including 4 fellowships (averaging $19,200 per year), 3 research assistantships with full tuition reimbursements available (averaging $18,840 per year), 17 teaching assistantships (averaging $18,840 per year) Financial award application deadline: 1/15. *Faculty research:* Surfaces and interfaces, polymers, drug conjugates, organo-metallics. Total annual research expenditures: $1.4 million. *Unit head:* Dr. Robert H. Flowers, Chairman, Fax: 610-758-6536. *Application contact:* Dr. Rebecca Miller, Graduate Coordinator, 610-758-3471, Fax: 610-758-6536, E-mail: inluchem@lehigh.edu.

Long Island University, Brooklyn Campus, Richard L. Conolly College of Liberal Arts and Sciences, Department of Chemistry, Brooklyn, NY 11201-8423. Offers MS. Part-time and evening/weekend programs available. *Degree requirements:* For master's, thesis or alternative. *Entrance requirements:* For master's, 2 letters of recommendation. Additional exam requirements/recommendations for international students: Required—TOEFL (minimum score 500 paper-based; 173 computer-based). Electronic applications accepted. *Faculty research:* Clinical chemistry, free radicals, heats of hydrogenation.

Louisiana State University and Agricultural and Mechanical College, Graduate School, College of Basic Sciences, Department of Chemistry, Baton Rouge, LA 70803. Offers MS, PhD. Part-time programs available. *Faculty:* 31 full-time (4 women). *Students:* 166 full-time (69 women), 3 part-time; includes 35 African Americans, 1 Asian American or Pacific Islander, 1 Hispanic American, 91 international. Average age 29. 146 applicants, 40% accepted, 87 enrolled. In 2005, 3 master's, 17 doctorates awarded. Terminal master's awarded for partial completion of doctoral program. *Degree requirements:* For master's, thesis (for some programs); for doctorate, thesis/dissertation, general exam. *Entrance requirements:* For master's and doctorate, GRE General Test, minimum GPA of 3.0. Additional exam requirements/recommendations for international students: Required—TOEFL (minimum score 550 paper-based; 213 computer-based). *Application deadline:* For fall admission, 3/1 priority date for domestic students, 5/15 priority date for international students; for spring admission, 8/1 for domestic students, 10/15 for international students. Applications are processed on a rolling basis. Application fee: $25. Electronic applications accepted. *Financial support:* In 2005–06, 161 students received support, including 26 fellowships with full tuition reimbursements available (averaging $15,144 per year), 57 research assistantships with full and partial tuition reimbursements available (averaging $22,404 per year), 77 teaching assistantships with full and partial tuition reimbursements available (averaging $20,935 per year); unspecified assistantships also available. Financial award application deadline: 7/1; financial award applicants required to submit FAFSA. *Faculty research:* Free radicals, bioinorganic chemistry, polymers, synthesis, spectroscopy. Total annual research expenditures: $7.7 million. *Unit head:* Dr. Andren Maverick, Chair, 225-578-7623, Fax: 225-578-3458. *Application contact:* Dr. Steven Watkins, Director of Graduate Studies, 225-578-3359, Fax: 225-578-3458, E-mail: swatkins@lsu.edu.

Louisiana Tech University, Graduate School, College of Engineering and Science, Department of Chemistry, Ruston, LA 71272. Offers MS. Part-time programs available. *Degree requirements:* For master's, thesis. *Entrance requirements:* For master's, GRE General Test, minimum GPA of 3.0 in last 60 hours. Additional exam requirements/recommendations for international students: Required—TOEFL. *Faculty research:* Vibrational spectroscopy, quantum studies of chemical reactions, enzyme kinetics, synthesis of transition metal compounds, NMR spectrometry.

Loyola University Chicago, Graduate School, Department of Chemistry, Chicago, IL 60611-2196. Offers MS, PhD. Part-time and evening/weekend programs available. *Faculty:* 12 full-time (2 women). *Students:* 26 full-time (12 women), 8 part-time (5 women); includes 5 minority (2 African Americans, 3 Asian Americans or Pacific Islanders), 9 international. Average age 33. 44 applicants, 75% accepted, 11 enrolled. In 2005, 4 master's, 4 doctorates awarded. Terminal master's awarded for partial completion of doctoral program. *Degree requirements:* For master's, thesis (for some programs); for doctorate, thesis/dissertation, comprehensive exam. *Entrance requirements:* For master's, GRE General Test, GRE Subject Test; for doctorate, GRE General Test, GRE Subject Test, entrance exams. Additional exam requirements/recommendations for international students: Required—TOEFL (minimum score 550 paper-based; 213 computer-based). *Application deadline:* For fall admission, 8/1 for domestic students; for spring admission, 12/1 for domestic students. Applications are processed on a rolling basis. Application fee: $40. Electronic applications accepted. *Expenses:* Tuition: Full-time $11,610; part-time $645 per credit. Required fees: $55 per semester. *Financial support:* In 2005–06, 20 students received support, including 10 fellowships with full tuition reimbursements available (averaging $24,000 per year), 10 teaching assistantships with full tuition reimbursements available (averaging $15,000 per year); research assistantships with full and partial tuition reimbursements available, Federal Work-Study, scholarships/grants, traineeships, and unspecified assistantships also available. Financial award application deadline: 2/1; financial award applicants required to submit FAFSA. *Faculty research:* Magnetic resonance of membrane/protein systems, organometallic catalysis, novel synthesis of natural products. Total annual research expenditures: $682,510. *Unit head:* Dr. Ken Olsen, Chair, 773-508-3121, Fax: 773-508-3086, E-mail: kolsen@luc.edu. *Application contact:* Julie D. Petry, Graduate Program Coordinator, 773-508-3104, Fax: 773-508-3086, E-mail: jpetry@luc.edu.

Marquette University, Graduate School, College of Arts and Sciences, Department of Chemistry, Milwaukee, WI 53201-1881. Offers analytical chemistry (MS, PhD); bioanalytical chemistry (MS, PhD); biophysical chemistry (MS, PhD); chemical physics (MS, PhD); inorganic chemistry (MS, PhD); organic chemistry (MS, PhD); physical chemistry (MS, PhD). Part-time programs available. Terminal master's awarded for partial completion of doctoral program. *Degree requirements:* For master's, comprehensive exam; for doctorate, thesis/dissertation, cumulative exams. *Entrance requirements:* For master's and doctorate, GRE Subject Test. Additional exam requirements/recommendations for international students: Required—TOEFL. *Faculty research:* Inorganic complexes, laser Raman spectroscopy, organic synthesis, chemical dynamics, biophysiology.

Marshall University, Academic Affairs Division, Graduate College, College of Science, Department of Chemistry, Huntington, WV 25755. Offers MS. *Faculty:* 5 full-time (0 women). *Students:* 7 full-time (1 woman); includes 1 minority (African American), 1 international. Average age 25. In 2005, 4 degrees awarded. *Degree requirements:* For master's, thesis. *Financial support:* Career-related internships or fieldwork available. *Unit head:* Dr. Michael Castellani, Chairperson, 304-696-6486, E-mail: castella@marshall.edu. *Application contact:* Information Contact, 304-746-1900, Fax: 304-746-1902, E-mail: services@marshall.edu.

Massachusetts College of Pharmacy and Health Sciences, Graduate Studies, Program in Chemistry, Boston, MA 02115-5896. Offers MS, PhD. Terminal master's awarded for partial completion of doctoral program. *Degree requirements:* For master's, thesis, oral defense of thesis; for doctorate, one foreign language, thesis/dissertation, oral defense of dissertation, qualifying exam, comprehensive exam, registration. *Entrance requirements:* For master's and doctorate, GRE General Test, minimum QPA of 3.0. Additional exam requirements/recommendations for international students: Required—TOEFL (minimum score 600 paper-based; 230 computer-based). *Faculty research:* Analytical chemistry, medicinal chemistry, organic chemistry, neurochemistry.

Chemistry

Massachusetts Institute of Technology, School of Science, Department of Chemistry, Cambridge, MA 02139-4307. Offers biological chemistry (PhD, Sc D); inorganic chemistry (PhD, Sc D); organic chemistry (PhD, Sc D); physical chemistry (PhD, Sc D). *Faculty:* 30 full-time (6 women). *Students:* 242 full-time (84 women), 3 part-time (2 women); includes 30 minority (5 African Americans, 1 American Indian/Alaska Native, 19 Asian Americans or Pacific Islanders, 5 Hispanic Americans), 77 international. Average age 26. 476 applicants, 27% accepted, 54 enrolled. In 2005, 48 doctorates awarded. *Degree requirements:* For doctorate, thesis/dissertation, comprehensive exam. *Entrance requirements:* For doctorate, GRE General Test. Additional exam requirements/recommendations for international students: Required—TOEFL (minimum score 577 paper-based; 233 computer-based). *Application deadline:* For fall admission, 12/15 for domestic students, 12/15 for international students. Application fee: $70. Electronic applications accepted. *Expenses:* Tuition: Full-time $32,100. Required fees: $200. Part-time tuition and fees vary according to course load. *Financial support:* In 2005–06, 210 students received support, including 36 fellowships with tuition reimbursements available (averaging $29,141 per year), 143 research assistantships with tuition reimbursements available (averaging $24,550 per year), 59 teaching assistantships with tuition reimbursements available (averaging $25,155 per year); career-related internships or fieldwork, Federal Work-Study, institutionally sponsored loans, scholarships/grants, health care benefits, and unspecified assistantships also available. *Faculty research:* Synthetic organic chemistry, enzymatic reaction mechanisms, inorganic and organometallic spectroscopy, high resolution NMR spectroscopy. Total annual research expenditures: $21.5 million. *Unit head:* Prof. Timothy Swager, Department Head, 617-253-1801, E-mail: tswager@mit.edu. *Application contact:* Susan Brighton, Graduate Administrator, 617-253-1845, Fax: 617-258-0241, E-mail: brighton@mit.edu.

McGill University, Faculty of Graduate and Postdoctoral Studies, Faculty of Science, Department of Chemistry, Montréal, QC H3A 2T5, Canada. Offers chemical biology (M Sc); chemistry (M Sc, PhD). *Degree requirements:* For master's, thesis, research project (chemistry biology); for doctorate, thesis/dissertation, oral examination. *Entrance requirements:* For master's, minimum GPA of 3.0, academic background in chemistry. Additional exam requirements/recommendations for international students: Required—TOEFL (minimum score 577 paper-based; 233 computer-based), IELT(minimum score 7). Electronic applications accepted.

McMaster University, School of Graduate Studies, Faculty of Science, Department of Chemistry, Hamilton, ON L8S 4M2, Canada. Offers analytical chemistry (M Sc, PhD); chemical physics (M Sc, PhD); chemistry (M Sc, PhD); inorganic chemistry (M Sc, PhD); organic chemistry (M Sc, PhD); physical chemistry (M Sc, PhD); polymer chemistry (M Sc, PhD). Part-time programs available. Terminal master's awarded for partial completion of doctoral program. *Degree requirements:* For master's, thesis/dissertation; for doctorate, thesis/dissertation, comprehensive exam. *Entrance requirements:* For master's, minimum B+ average. Additional exam requirements/recommendations for international students: Required—TOEFL (minimum score 550 paper-based; 213 computer-based).

McNeese State University, Graduate School, College of Science, Department of Biological and Environmental Sciences and Department of Chemistry, Program in Environmental and Chemical Sciences, Lake Charles, LA 70609. Offers MS. Evening/weekend programs available. *Faculty:* 10 full-time (1 woman). *Students:* 27 full-time (14 women), 8 part-time (5 women); includes 11 minority (9 African Americans, 2 Hispanic Americans), 9 international. In 2005, 7 degrees awarded. *Degree requirements:* For master's, thesis or alternative, comprehensive exam. *Entrance requirements:* For master's, GRE General Test. *Application deadline:* For fall admission, 7/15 for domestic students. Applications are processed on a rolling basis. Application fee: $20 ($30 for international students). *Expenses:* Tuition, area resident: Full-time $2,226; part-time $193 per hour. Tuition, state resident: full-time $2,226. Required fees: $862; $106 per hour. Tuition and fees vary according to course load. *Financial support:* Application deadline: 5/1. *Unit head:* Dr. Harold Stevenson, Coordinator, 337-475-5663, Fax: 337-475-5677, E-mail: hstevens@mcneese.edu.

McNeese State University, Graduate School, College of Science, Department of Chemistry, Lake Charles, LA 70609. Offers environmental and chemical sciences (MS). Evening/weekend programs available. *Degree requirements:* For master's, thesis or alternative, comprehensive exam. *Entrance requirements:* For master's, GRE General Test. *Application deadline:* For fall admission, 7/15 for domestic students. Applications are processed on a rolling basis. Application fee: $20 ($30 for international students). *Expenses:* Tuition, area resident: Full-time $2,226; part-time $193 per hour. Tuition, state resident: full-time $2,226. Required fees: $862; $106 per hour. Tuition and fees vary according to course load. *Financial support:* Teaching assistantships available. Financial award application deadline: 5/1. *Unit head:* Dr. Ron W. Darbeau, Head, 337-475-5776, Fax: 337-475-5950, E-mail: rdarbeau@mcneese.edu. *Application contact:* Dr. Harold Stevenson, Coordinator, 337-475-5663, Fax: 337-475-5677, E-mail: hstevens@mcneese.edu.

Memorial University of Newfoundland, School of Graduate Studies, Department of Chemistry, St. John's, NL A1C 5S7, Canada. Offers chemistry (M Sc, PhD); instrumental analysis (M Sc). Part-time programs available. *Students:* 50 full-time (14 women), 1 part-time, 30 international. 38 applicants, 18% accepted, 5 enrolled. In 2005, 5 degrees awarded. *Degree requirements:* For master's, thesis, research seminar, American Chemical Society Exam; for doctorate, thesis/dissertation, seminars, oral thesis defense, American Chemical Society Exam, comprehensive exam. *Entrance requirements:* For master's, B Sc or honors degree in chemistry (preferred); for doctorate, master's degree in chemistry or honors bachelor's degree. *Application deadline:* Applications are processed on a rolling basis. Application fee: $40 Canadian dollars. Electronic applications accepted. *Expenses:* Tuition: Part-time $733 per term. Tuition and fees vary according to degree level and program. *Financial support:* Fellowships, research assistantships, teaching assistantships available. *Faculty research:* Analytical/environmental chemistry; medicinal electrochemistry; inorganic, marine, organic, physical, and theoretical/computational chemistry, environmental science and instrumental analysis. *Unit head:* Dr. Robert Davis, Head, 709-737-8772, Fax: 709-737-3702, E-mail: rdavis@mun.ca. *Application contact:* Viola Martin, Secretary, 709-737-8773, Fax: 709-737-3702, E-mail: gradchem@mun.ca.

Miami University, Graduate School, College of Arts and Sciences, Department of Chemistry and Biochemistry, Oxford, OH 45056. Offers analytical chemistry (MS, PhD); biochemistry (MS, PhD); chemical education (MS, PhD); chemistry (MS, PhD); inorganic chemistry (MS, PhD); organic chemistry (MS, PhD); physical chemistry (MS, PhD). Part-time programs available. *Degree requirements:* For master's, thesis, final exam; for doctorate, thesis/dissertation, final exams, comprehensive exam. *Entrance requirements:* For master's, minimum undergraduate GPA of 3.0 during previous 2 years or 2.75 overall; for doctorate, minimum undergraduate GPA of 2.75, 3.0 graduate. Additional exam requirements/recommendations for international students: Required—TOEFL (minimum score 550 paper-based; 213 computer-based), TWE(minimum score 4). Electronic applications accepted.

Michigan State University, The Graduate School, College of Natural Science, Department of Chemistry, East Lansing, MI 48824. Offers chemical physics (PhD); chemistry (MS, PhD); chemistry-environmental toxicology (PhD); computational chemistry (MS). *Faculty:* 35 full-time (4 women). *Students:* 221 full-time (100 women), 4 part-time (2 women); includes 13 minority (4 African Americans, 1 American Indian/Alaska Native, 6 Asian Americans or Pacific Islanders, 2 Hispanic Americans), 129 international. Average age 27. 138 applicants, 66% accepted. In 2005, 9 master's, 19 doctorates awarded. *Degree requirements:* For master's, oral defense of thesis, thesis optional; for doctorate, thesis/dissertation, oral defense of dissertation, comprehensive exam. *Entrance requirements:* For master's, GRE General Test, bachelor's degree in chemistry; course work in chemistry, physics, and calculus; 3 letters of recommendation; for doctorate, GRE General Test, minimum GPA of 3.0; bachelor's or master's degree in chemistry; coursework in chemistry, physics, and calculus; 3 letters of recommendation. Additional exam requirements/recommendations for international students: Required—TOEFL (minimum score 550 paper-based; 213 computer-based), Michigan State University ELT (85), Michigan ELAB (83). *Application deadline:* For fall admission, 12/27 for domestic students. Application fee: $50. Electronic applications accepted. *Expenses:* Tuition, state resident: part-time $330 per credit hour. Tuition, nonresident: part-time $685 per credit hour. Tuition and fees vary according to program. *Financial support:* In 2005–06, 85 fellowships with tuition reimbursements (averaging $3,132 per year), 71 research assistantships with tuition reimbursements (averaging $15,305 per year), 138 teaching assistantships with tuition reimbursements (averaging $15,041 per year) were awarded; scholarships/grants and unspecified assistantships also available. *Faculty research:* Analytical chemistry, inorganic and organic chemistry, nuclear chemistry, physical chemistry, theoretical and computational chemistry. Total annual research expenditures: $8.7 million. *Unit head:* Dr. John L. McCracken, Chairperson, 517-355-9715 Ext. 346, Fax: 517-353-1793, E-mail: chair@chemistry.msu.edu. *Application contact:* Deborah Roper, Graduate Admissions Secretary, 517-355-9715 Ext. 362, Fax: 517-353-1793, E-mail: gradoff@crm.msu.edu.

Michigan Technological University, Graduate School, College of Sciences and Arts, Department of Chemistry, Houghton, MI 49931-1295. Offers MS, PhD. Part-time programs available. *Faculty:* 17 full-time (4 women), 3 part-time/adjunct (0 women). *Students:* 26 full-time (10 women), 9 part-time (6 women), 28 international. Average age 30. 46 applicants, 52% accepted, 10 enrolled. In 2005, 2 master's, 2 doctorates awarded. Terminal master's awarded for partial completion of doctoral program. *Median time to degree:* Of those who began their doctoral program in fall 1997, 38% received their degree in 8 years or less. *Degree requirements:* For master's, thesis/dissertation, registration; for doctorate, thesis/dissertation, comprehensive exam, registration. *Entrance requirements:* Additional exam requirements/recommendations for international students: Required—TOEFL (minimum score 550 paper-based; 213 computer-based). *Application deadline:* For fall admission, 2/1 for domestic students. Applications are processed on a rolling basis. Application fee: $40 ($45 for international students). Electronic applications accepted. *Expenses:* Tuition: Full-time $11,232; part-time $468 per credit. Required fees: $754; $377 per semester. Full-time tuition and fees vary according to course load, degree level and program. *Financial support:* In 2005–06, 26 students received support, including fellowships with full tuition reimbursements available (averaging $9,542 per year), 2 research assistantships with full tuition reimbursements available (averaging $9,542 per year), 23 teaching assistantships with full tuition reimbursements available (averaging $9,542 per year); career-related internships or fieldwork, Federal Work-Study, scholarships/grants, health care benefits, tuition waivers (partial), unspecified assistantships, and co-op also available. Financial award applicants required to submit FAFSA. *Faculty research:* Inorganic chemistry, physical/theoretical chemistry, bio/organic chemistry, polymer/materials chemistry, analytical/environmental chemistry. Total annual research expenditures:$169,063. *Unit head:* Dr. Sarah A. Green, Chair, 906-487-2048, Fax: 906-487-2061, E-mail: sgreen@mtu.edu. *Application contact:* Celine Grace, Office Assistant 5, 906-487-2048, Fax: 906-487-2061, E-mail: cegrace@mtu.edu.

Middle Tennessee State University, College of Graduate Studies, College of Basic and Applied Sciences, Department of Chemistry, Murfreesboro, TN 37132. Offers MS, DA. Part-time and evening/weekend programs available. Postbaccalaureate distance learning degree programs offered. *Degree requirements:* For master's, one foreign language, thesis, comprehensive exam; for doctorate, thesis/dissertation, comprehensive exam. *Entrance requirements:* For master's and doctorate, GRE General Test. Additional exam requirements/recommendations for international students: Required—TOEFL (minimum score 525 paper-based; 195 computer-based). Electronic applications accepted. *Faculty research:* Chemical education; computational chemistry and visualization; materials science and surface modifications; biochemistry, antibiotics and leukemia; environmental chemistry and toxicology.

Mississippi College, Graduate School, College of Arts and Sciences, Program in Combined Sciences, Department of Chemistry, Clinton, MS 39058. Offers MCS.

Mississippi State University, College of Arts and Sciences, Department of Chemistry, Mississippi State, MS 39762. Offers MS, PhD. *Faculty:* 16 full-time (3 women), 1 part-time/adjunct (0 women). *Students:* 37 full-time (15 women), 5 part-time (1 woman); includes 1 minority (Hispanic American), 32 international. Average age 30. 74 applicants, 14% accepted, 4 enrolled. In 2005, 3 master's, 3 doctorates awarded. Terminal master's awarded for partial completion of doctoral program. *Degree requirements:* For master's and doctorate, thesis/dissertation, comprehensive oral or written exam. *Entrance requirements:* For master's and doctorate, minimum GPA of 2.75. Additional exam requirements/recommendations for international students: Required—TOEFL. *Application deadline:* For fall admission, 7/1 for domestic students; for spring admission, 11/1 for domestic students. Applications are processed on a rolling basis. Application fee: $30. *Expenses:* Tuition, area resident: Full-time $4,312; part-time $240 per hour. Tuition, state resident: full-time $4,312; part-time $240 per hour. Tuition, nonresident: full-time $9,772; part-time $543 per hour. International tuition: $10,102 full-time. Tuition and fees vary according to course load. *Financial support:* In 2005–06, 25 teaching assistantships with full tuition reimbursements (averaging $11,805 per year) were awarded; research assistantships with full tuition reimbursements, Federal Work-Study, institutionally sponsored loans, scholarships/grants, and unspecified assistantships also available. Financial award applicants required to submit FAFSA. *Faculty research:* Spectroscopy, fluorometry, NMR, organic and inorganic synthesis, electrochemistry. Total annual research expenditures: $6.2 million. *Unit head:* Dr. Keith T. Mead, Head, 662-325-3584, Fax: 662-325-1618, E-mail: kmead@ra.msstate.edu. *Application contact:* Philip G. Bonfanti, Director of Admissions, 662-325-4104, Fax: 662-325-8872, E-mail: admit@msstate.edu.

Missouri State University, Graduate College, College of Natural and Applied Sciences, Department of Chemistry, Springfield, MO 65804-0094. Offers chemistry (MNAS, MS); secondary education (MS Ed), including chemistry. Part-time programs available. *Faculty:* 12 full-time (1 woman). *Students:* 8 full-time (4 women), 8 part-time (4 women), 2 international. Average age 30. 9 applicants, 44% accepted, 1 enrolled. In 2005, 8 degrees awarded. *Degree requirements:* For master's, thesis, comprehensive exam. *Entrance requirements:* For master's, GRE General Test (MS, MNAS), minimum undergraduate GPA of 3.0 (MS and MNAS), 9-12 teacher certification (MS Ed). Additional exam requirements/recommendations for international students: Required—TOEFL (minimum score 550 paper-based; 213 computer-based), IELT(minimum score 6). *Application deadline:* For fall admission, 7/20 for domestic students; for spring admission, 12/20 priority date for domestic students. Applications are processed on a rolling basis. Application fee: $30. Electronic applications accepted. *Expenses:* Tuition, state resident: full-time $3,402; part-time $189 per credit. Tuition, nonresident: full-time $6,804; part-time $378 per credit. International tuition: $6,804 full-time. Required fees: $207 per semester. Part-time tuition and fees vary according to course level, course load and program. *Financial support:* In 2005–06, research assistantships with full tuition reimbursements (averaging $8,750 per year), 7 teaching assistantships with full tuition reimbursements (averaging $8,750 per year) were awarded; Federal Work-Study, scholarships/grants, and unspecified assistantships also available. Financial award application deadline: 3/31; financial award applicants required to submit FAFSA. *Faculty research:* Chemistry of environmental systems, mechanisms of organic and organometallic reactions, enzymatic activity in lipid and protein reactions, computational chemistry, polymer properties. Total annual research expenditures: $80,000. *Unit head:* Dr. Paul Toom, Acting Head, 417-836-5506, Fax: 417-836-6934, E-mail: chemistry@missouristate.edu.

Montana State University, College of Graduate Studies, College of Letters and Science, Department of Chemistry and Biochemistry, Bozeman, MT 59717. Offers biochemistry (MS, PhD); chemistry (MS, PhD). Part-time programs available. *Faculty:* 16 full-time (4 women), 10 part-time/adjunct (6 women). *Students:* 1 full-time (0 women), 57 part-time (16 women); includes 1 minority (Hispanic American), 9 international. Average age 27. 33 applicants, 79% accepted, 18 enrolled. In 2005, 3 master's, 5 doctorates awarded. *Degree requirements:* For master's, thesis (for some programs), comprehensive exam, registration; for doctorate, thesis/dissertation, comprehensive exam, registration. *Entrance requirements:* For master's and doctorate, GRE General Test. Additional exam requirements/recommendations for international students: Required—TOEFL (minimum score 550 paper-based; 213 computer-based). *Application deadline:* For fall admission, 7/15 priority date for domestic students, 5/15 priority date for international students; for spring admission, 12/1 priority date for domestic students, 10/1 priority date for international students. Applications are processed on a rolling basis. Application fee: $30. Electronic applications accepted. *Expenses:* Tuition, state resident: full-time $4,132. Tuition, nonresident: full-time $1,132. *Financial support:* In 2005–06, 51 students received support, including 27 research assistantships with full tuition reimbursements available (averaging $22,955 per year), 24 teaching assistantships with full tuition reimbursements available (averaging $22,742 per year); unspecified assistantships also available. Financial award application deadline: 3/1; financial award applicants required to submit FAFSA. *Faculty research:* Structure, mechanism, nanotechnology, spectroscopy, synthesis. Total annual research expenditures: $5.4 million. *Unit head:* Dr. David Singel, Interim Department Head, 406-994-3960, Fax: 406-994-5407, E-mail: rchds@chemistry.montana.edu.

Montclair State University, The Graduate School, College of Science and Mathematics, Department of Chemistry and Biochemistry, Montclair, NJ 07043-1624. Offers chemistry (MS), including biochemistry. Part-time and evening/weekend programs available. *Faculty:* 11 full-time (2 women), 1 (woman) part-time/adjunct. *Students:* 8 full-time (5 women), 13 part-time (8 women); includes 8 minority (3 African Americans, 4 Asian Americans or Pacific Islanders, 1 Hispanic American), 4 international. 18 applicants, 72% accepted, 12 enrolled. In 2005, 5 degrees awarded. *Degree requirements:* For master's, comprehensive exam. *Entrance requirements:* For master's, GRE General Test, 24 credits of course work in undergraduate in chemistry, 2 letters of recommendation. Additional exam requirements/recommendations for international students: Required—TOEFL (minimum score 83 computer-based). *Application deadline:* Applications are processed on a rolling basis. Application fee: $60. Electronic applications accepted. *Expenses:* Tuition: Full-time $3,001; part-time $409 per credit. Required fees: $56 per credit. Tuition and fees vary according to course load, degree level and program. *Financial support:* In 2005–06, 3 research assistantships with full tuition reimbursements (averaging $7,000 per year) were awarded; Federal Work-Study, scholarships/grants, and unspecified assistantships also available. Support available to part-time students. Financial award application deadline: 3/1; financial award applicants required to submit FAFSA. *Faculty research:* Antimicrobial compounds; marine bacteria. Total annual research expenditures: $5,000. *Unit head:* Dr. Jeff Toney, Chair, 973-655-7121. *Application contact:* Dr. Mark Whitener, Adviser, 973-655-7166, E-mail: whitenerm@mail.montclair.edu.

Morgan State University, School of Graduate Studies, School of Computer, Mathematical, and Natural Sciences, Interdisciplinary Program in Science, Baltimore, MD 21251. Offers biology (MS); chemistry (MS); physics (MS). *Students:* 9 (2 women); includes 4 minority (all African Americans) 4 international. In 2005, 5 degrees awarded. *Degree requirements:* For master's, thesis, oral defense of thesis, comprehensive exam. *Entrance requirements:* For master's, GRE General Test, minimum GPA of 2.5. *Application deadline:* For fall admission, 2/1 for domestic students; for spring admission, 10/1 priority date for domestic students. Applications are processed on a rolling basis. Application fee: $0. *Expenses:* Tuition, state resident: part-time $272 per credit. Tuition, nonresident: part-time $478 per credit. Required fees: $58 per credit. *Financial support:* Fellowships, research assistantships, career-related internships or fieldwork, Federal Work-Study, institutionally sponsored loans, scholarships/grants, health care benefits, and unspecified assistantships available. Support available to part-time students. *Unit head:* Dr. Juarine Stewart, Dean, 443-885-4515, Fax: 443-885-8215. *Application contact:* Dr. James E. Waller, Admissions and Program Officer, 443-885-3185, Fax: 443-885-8226, E-mail: jwaller@moae.morgan.edu.

Mount Allison University, Faculty of Science, Department of Chemistry, Sackville, NB E4L 1E4, Canada. Offers M Sc. *Degree requirements:* For master's, thesis. *Entrance requirements:* For master's, honors degree in chemistry. *Faculty research:* Biophysical chemistry of model biomembranes, organic synthesis, fast-reaction kinetics, physical chemistry of micelles.

Murray State University, College of Science, Engineering and Technology, Department of Chemistry, Murray, KY 42071-0009. Offers MAT, MS. Part-time programs available. *Degree requirements:* For master's, thesis (for some programs). *Entrance requirements:* For master's, GRE General Test. Additional exam requirements/recommendations for international students: Required—TOEFL.

New Jersey Institute of Technology, Office of Graduate Studies, College of Science and Liberal Arts, Department of Chemistry and Environmental Science, Program in Chemistry, Newark, NJ 07102. Offers MS, PhD. Part-time and evening/weekend programs available. *Students:* 9 full-time (8 women), 8 part-time (4 women); includes 3 minority (2 Asian Americans or Pacific Islanders, 1 Hispanic American), 10 international. Average age 33. 37 applicants, 32% accepted, 5 enrolled. In 2005, 3 master's, 1 doctorate awarded. *Degree requirements:* For doctorate, thesis/dissertation. *Entrance requirements:* For master's, GRE General Test; for doctorate, GRE General Test, minimum graduate GPA of 3.5. Additional exam requirements/recommendations for international students: Required—TOEFL (minimum score 550 paper-based; 213 computer-based). *Application deadline:* For fall admission, 6/5 for domestic students; for spring admission, 10/15 for domestic students. Applications are processed on a rolling basis. Application fee: $60. Electronic applications accepted. *Expenses:* Tuition, state resident: full-time $9,620; part-time $520 per credit. Tuition, nonresident: full-time $13,542; part-time $715 per credit. Required fees: $78; $54 per credit. $78 per year. Tuition and fees vary according to course load. *Financial support:* Fellowships with full and partial tuition reimbursements, research assistantships with full and partial tuition reimbursements, teaching assistantships with full and partial tuition reimbursements, career-related internships or fieldwork, Federal Work-Study, institutionally sponsored loans, and unspecified assistantships available. Financial award application deadline: 3/15. *Faculty research:* Medical instrumentation, prosthesis design, biodegradation of hazardous waste, orthopedic biomechanics, image processing. *Application contact:* Kathryn Kelly, Director of Admissions, 973-596-3300, Fax: 973-596-3461, E-mail: admissions@njit.edu.

New Mexico Highlands University, Graduate Studies, College of Arts and Sciences, Department of Natural Sciences, Las Vegas, NM 87701. Offers applied chemistry (MS); biology (MS); environmental science and management (MS). Part-time programs available. *Faculty:* 11 full-time (4 women), 7 part-time/adjunct (2 women). *Students:* 15 full-time (6 women), 6 part-time (3 women); includes 4 minority (1 American Indian/Alaska Native, 3 Hispanic Americans), 13 international. Average age 29. 6 applicants, 100% accepted, 6 enrolled. In 2005, 8 degrees awarded. *Degree requirements:* For master's, thesis, comprehensive exam, registration. *Entrance requirements:* For master's, minimum undergraduate GPA of 3.0. Additional exam requirements/recommendations for international students: Required—TOEFL (minimum score 540 paper-based; 190 computer-based). *Application deadline:* For fall admission, 8/1 for domestic students. Applications are processed on a rolling basis. Application fee: $15. *Expenses:* Tuition, state resident: full-time $2,280; part-time $101 per credit. Tuition, nonresident: full-time $3,420; part-time $151 per credit. One-time fee: $20 full-time. *Financial support:* In 2005–06, 4 students received support, including 13 teaching assistantships (averaging $11,500 per year); research assistantships with full and partial tuition reimbursements available, Federal Work-Study, institutionally sponsored loans, scholarships/grants, and unspecified assistantships also available. Support available to part-time students. Financial award application deadline: 3/1. *Unit head:* Dr. Merritt Helvenston, Chair, 505-454-3263, Fax: 505-454-3103, E-mail: merritt@nmhu.edu. *Application contact:* Diane Trujillo, Administrative Assistant Graduate Studies, 505-454-3266, Fax: 505-454-3558, E-mail: dtrujillo@nmhu.edu.

New Mexico Institute of Mining and Technology, Graduate Studies, Department of Chemistry, Socorro, NM 87801. Offers biochemistry (MS); chemistry (MS); environmental chemistry (PhD); explosives technology and atmospheric chemistry (PhD). Part-time programs available. *Degree requirements:* For master's and doctorate, thesis/dissertation. *Entrance requirements:* For master's, GRE General Test; for doctorate, GRE General Test, GRE Subject Test. Additional exam requirements/recommendations for international students: Required—TOEFL (minimum score 540 paper-based; 207 computer-based). Electronic applications accepted. *Faculty research:* Organic, analytical, environmental, and explosives chemistry.

New Mexico State University, Graduate School, College of Arts and Sciences, Department of Chemistry and Biochemistry, Las Cruces, NM 88003-8001. Offers MS, PhD. Part-time programs available. *Faculty:* 19 full-time (1 woman), 2 part-time/adjunct (0 women). *Students:* 54 full-time (17 women), 8 part-time; includes 13 minority (1 American Indian/Alaska Native, 2 Asian Americans or Pacific Islanders, 10 Hispanic Americans), 29 international. Average age 30. 32 applicants, 34% accepted, 7 enrolled. In 2005, 7 master's, 3 doctorates awarded. *Degree requirements:* For master's and doctorate, thesis/dissertation. *Entrance requirements:* For master's and doctorate, GRE, BS in chemistry or biochemistry, minimum GPA of 3.0. Additional exam requirements/recommendations for international students: Required—TOEFL. *Application deadline:* For fall admission, 7/1 for domestic students; for spring admission, 11/1 for domestic students. Applications are processed on a rolling basis. Application fee: $30 ($50 for international students). *Expenses:* Tuition, state resident: full-time $3,156; part-time $175 per credit. Tuition, nonresident: full-time $12,510; part-time $565 per credit. Required fees: $1,050. *Financial support:* In 2005–06, 1 fellowship, 16 research assistantships, 20 teaching assistantships were awarded; career-related internships or fieldwork and Federal Work-Study also available. Support available to part-time students. Financial award application deadline: 3/1. *Faculty research:* Clays, surfaces, and water structure; electroanalytical and environmental chemistry; organometallic synthesis and organobiomimetics; molecular genetics and enzymology of stress; spectroscopy and reaction kinetics. *Unit head:* Dr. Aravamudan Gopalan, Head, 505-646-5877, Fax: 505-646-2649, E-mail: agopalan@nmsu.edu. *Application contact:* Dr. James Herndon, Associate Professor, Chemistry, 505-646-2738, Fax: 505-646-2649.

New York University, Graduate School of Arts and Science, Department of Chemistry, New York, NY 10012-1019. Offers MS, PhD. *Faculty:* 23 full-time (1 woman), 3 part-time/adjunct. *Students:* 91 full-time (37 women), 5 part-time; includes 10 minority (2 African Americans, 8 Asian Americans or Pacific Islanders), 68 international. Average age 29. 111 applicants, 28% accepted, 15 enrolled. In 2005, 8 master's, 8 doctorates awarded. *Degree requirements:* For master's, thesis or alternative; for doctorate, one foreign language, thesis/dissertation. *Entrance requirements:* For master's and doctorate, GRE General Test, GRE Subject Test. Additional exam requirements/recommendations for international students: Required—TOEFL, TSE. *Application deadline:* For fall admission, 1/4 for domestic students. Application fee: $80. *Financial support:* Fellowships with tuition reimbursements, research assistantships with tuition reimbursements, teaching assistantships with tuition reimbursements, career-related internships or fieldwork, Federal Work-Study, institutionally sponsored loans, scholarships/grants, health care benefits, unspecified assistantships, and teaching fellowships available. Financial award application deadline: 1/4; financial award applicants required to submit FAFSA. *Faculty research:* Biomolecular chemistry, theoretical and computational chemistry, physical chemistry, nanotechnology, bio-organic chemistry. *Unit head:* Nicholas Geacintov, Chairman, 212-998-8400, Fax: 212-260-7905, E-mail: grad.chem@nyu.edu. *Application contact:* Mark Tuckerman, Director of Graduate Studies, 212-998-8400, Fax: 212-260-7905, E-mail: grad.chem@nyu.edu.

See Close-Up on page 125.

North Carolina Agricultural and Technical State University, Graduate School, College of Arts and Sciences, Department of Chemistry, Greensboro, NC 27411. Offers MS. Part-time and evening/weekend programs available. *Degree requirements:* For master's, thesis or alternative, qualifying exam, comprehensive exam. *Entrance requirements:* For master's, GRE General Test, minimum GPA of 3.0. *Faculty research:* Tobacco pesticides.

North Carolina Central University, Division of Academic Affairs, College of Arts and Sciences, Department of Chemistry, Durham, NC 27707-3129. Offers MS. *Degree requirements:* For master's, one foreign language, thesis, comprehensive exam. *Entrance requirements:* For master's, GRE, minimum GPA of 3.0 in major, 2.5 overall. Additional exam requirements/recommendations for international students: Required—TOEFL.

North Carolina State University, Graduate School, College of Physical and Mathematical Sciences, Department of Chemistry, Raleigh, NC 27695. Offers MCH, MS, PhD. Part-time programs available. Terminal master's awarded for partial completion of doctoral program. *Degree requirements:* For master's, thesis (for some programs); for doctorate, thesis/dissertation. *Entrance requirements:* For master's and doctorate, GRE General Test (recommended). Electronic applications accepted. *Faculty research:* Biological chemistry, electrochemistry, organic/inorganic materials, natural products, organometallics.

North Dakota State University, The Graduate School, College of Science and Mathematics, Department of Chemistry, Fargo, ND 58105. Offers biochemistry (MS, PhD); chemistry (MS, PhD). *Faculty:* 13 full-time (1 woman), 1 part-time/adjunct (0 women). *Students:* 26 full-time (6 women), 12 international. Average age 24. 33 applicants, 58% accepted, 10 enrolled. In 2005, 3 master's, 3 doctorates awarded. Terminal master's awarded for partial completion of doctoral program. *Degree requirements:* For master's and doctorate, thesis/dissertation. *Entrance requirements:* For master's and doctorate, GRE General Test. Additional exam requirements/recommendations for international students: Required—TOEFL (minimum score 600 paper-based; 247 computer-based), GRE Subject. *Application deadline:* For fall admission, 6/1 for domestic students. Applications are processed on a rolling basis. Application fee: $45 ($60 for international students). Electronic applications accepted. *Financial support:* In 2005–06, 2 fellowships with tuition reimbursements (averaging $19,000 per year), 11 research assistantships with tuition reimbursements (averaging $19,000 per year), 13 teaching assistantships with tuition reimbursements (averaging $19,000 per year) were awarded; Federal Work-Study, institutionally sponsored loans, and scholarships/grants also available. Financial award application deadline: 4/15. *Faculty research:* Analytical, syntheticorganic, inorganic, physical, and theoretical chemistry. Total annual research expenditures: $1.6 million. *Unit head:* Dr. John Hershberger, Chair, 701-231-8225, Fax: 701-231-8831, E-mail: john.hershberger@ndsu.nodak.edu. *Application contact:* Dr. Seth Rasmussen, Chair, Graduate Admissions, 701-231-8747, Fax: 701-231-8831, E-mail: seth.rasmussen@ndsu.nodak.edu.

Northeastern Illinois University, Graduate College, College of Arts and Sciences, Department of Chemistry, Program in Chemistry, Chicago, IL 60625-4699. Offers MS. Part-time and evening/weekend programs available. *Degree requirements:* For master's, final exam or thesis. *Entrance requirements:* For master's, minimum B average 2 semesters chemistry; 2 semesters calculus, organic chemistry, physical chemistry, and physics; 1 semester analytic chemistry; minimum GPA 2.75. *Faculty research:* Liquid chromatographic separation of pharmaceuticals, Diels-Alder reaction products, organogermanium chemistry, mass spectroscopy.

Northeastern University, College of Arts and Sciences, Department of Chemistry and Chemical Biology, Boston, MA 02115-5096. Offers analytical chemistry (PhD); chemistry (MS, PhD); inorganic chemistry (PhD); organic chemistry (PhD); physical chemistry (PhD). Part-time and evening/weekend programs available. Terminal master's awarded for partial completion of doctoral program. *Degree requirements:* For master's, thesis (for some programs); for doctorate, thesis/dissertation, qualifying exam in specialty area. *Entrance requirements:* Additional exam requirements/recommendations for international students: Required—TOEFL. Electronic applications accepted. *Faculty research:* Electron transfer, theoretical chemical physics, analytical biotechnology, mass spectrometry, materials chemistry.

See Close-Up on page 129.

Northern Arizona University, Graduate College, College of Engineering and Natural Science, Department of Chemistry and Biochemistry, Flagstaff, AZ 86011. Offers chemistry (MS). Part-time programs available. *Degree requirements:* For master's, thesis. *Faculty research:* Biochemistry of exercise, organic and inorganic mechanism studies, inhibition of ice mutation, polymer separation.

Northern Illinois University, Graduate School, College of Liberal Arts and Sciences, Department of Chemistry and Biochemistry, De Kalb, IL 60115-2854. Offers chemistry (MS, PhD). *Faculty:* 16 full-time (1 woman), 3 part-time/adjunct (1 woman). *Students:* 46 full-time (19 women), 8 part-time (5 women); includes 5 minority (2 African Americans, 1 Asian American or

Chemistry

Northern Illinois University (continued)

Pacific Islander, 2 Hispanic Americans), 21 international. Average age 28. 63 applicants, 46% accepted, 9 enrolled. In 2005, 3 master's, 2 doctorates awarded. Terminal master's awarded for partial completion of doctoral program. *Degree requirements:* For master's, research seminar, thesis optional; for doctorate, one foreign language, thesis/dissertation, candidacy exam, dissertation defense, research seminar. *Entrance requirements:* For master's, GRE General Test, bachelor's degree in mathematics or science, minimum GPA of 2.75; for doctorate, GRE General Test, bachelor's degree in mathematics or science; minimum undergraduate GPA of 2.75, 3.2 graduate. Additional exam requirements/recommendations for international students: Required—TOEFL (minimum score 550 paper-based; 213 computer-based). *Application deadline:* For fall admission, 6/1 for domestic students, 5/1 for international students; for spring admission, 11/1 for domestic students, 10/1 for international students. Applications are processed on a rolling basis. Application fee: $30. Electronic applications accepted. *Expenses:* Tuition, area resident: Full-time $4,565; part-time $19 per credit hour. Tuition, state resident: full-time $4,565; part-time $191 per credit hour. Tuition, nonresident: full-time $9,129; part-time $382 per credit hour. *Financial support:* In 2005–06, 8 research assistantships with full tuition reimbursements, 37 teaching assistantships with full tuition reimbursements were awarded; fellowships with full tuition reimbursements, career-related internships or fieldwork, Federal Work-Study, scholarships/grants, tuition waivers (full), and unspecified assistantships also available. Support available to part-time students. Financial award applicants required to submit FAFSA. *Faculty research:* Viscoelastic properties of polymers, lig and buding tocytochrome coxidases, computational inorganic chemistry, chemistry of organosilanes. *Unit head:* Dr. James Erman, Chair, 815-753-1181, Fax: 815-753-4802, E-mail: jerman@niu.edu. *Application contact:* Dr. Jon Carnahan, Director, Graduate Studies, 815-753-6879, E-mail: carnahan@niu.edu.

Northern Michigan University, College of Graduate Studies, College of Arts and Sciences, Department of Chemistry, Marquette, MI 49855-5301. Offers biochemistry (MS); chemistry (MS). Part-time programs available. *Degree requirements:* For master's, thesis. *Entrance requirements:* For master's, GRE General Test, minimum GPA of 3.0.

Northwestern University, The Graduate School, Judd A. and Marjorie Weinberg College of Arts and Sciences, Department of Chemistry, Evanston, IL 60208. Offers PhD. Admissions and degrees offered through The Graduate School. *Faculty:* 27 full-time (3 women). *Students:* 231 full-time (87 women); includes 29 minority (3 African Americans, 21 Asian Americans or Pacific Islanders, 5 Hispanic Americans), 82 international. 348 applicants, 45% accepted, 36 enrolled. In 2005, 31 doctorates awarded. *Median time to degree:* Of those who began their doctoral program in fall 1997, 68% received their degree in 8 years or less. *Degree requirements:* For doctorate, thesis/dissertation. *Entrance requirements:* For doctorate, GRE General Test, GRE Subject Test (chemistry). Additional exam requirements/recommendations for international students: Required—TOEFL, TSE. *Application deadline:* Applications are processed on a rolling basis. Application fee: $60 ($75 for international students). Electronic applications accepted. *Financial support:* In 2005–06, 17 fellowships with full tuition reimbursements (averaging $11,673 per year), 68 research assistantships with partial tuition reimbursements (averaging $12,342 per year), 36 teaching assistantships with full tuition reimbursements (averaging $12,042 per year) were awarded; institutionally sponsored loans, scholarships/grants, and tuition waivers (full and partial) also available. Financial award application deadline: 12/31; financial award applicants required to submit FAFSA. *Faculty research:* Inorganic, organic, physical, environmental, materials, and chemistry of life processes. Total annual research expenditures: $14.9 million. *Unit head:* Prof. Hilary R. Godwin, Chair, 847-467-3543, Fax: 847-491-7713, E-mail: h-godwin@northwestern.edu. *Application contact:* Thomas Meade, Admissions Officer, 847-491-2481, Fax: 847-491-7713, E-mail: tmeade@northwestern.edu.

Oakland University, Graduate Study and Lifelong Learning, College of Arts and Sciences, Department of Chemistry, Rochester, MI 48309-4401. Offers chemistry (MS); health and environmental chemistry (PhD). *Faculty:* 4 full-time (2 women). *Students:* 16 full-time (8 women), 24 part-time (17 women); includes 2 minority (1 American Indian/Alaska Native, 1 Asian American or Pacific Islander), 13 international. Average age 31. 21 applicants, 90% accepted, 10 enrolled. In 2005, 12 master's awarded. *Degree requirements:* For master's and doctorate, thesis/dissertation. *Entrance requirements:* For master's, minimum GPA of 3.0 for unconditional admission; for doctorate, GRE Subject Test, minimum GPA of 3.0 for unconditional admission. *Application deadline:* For fall admission, 7/15 for domestic studentsFor winter admission, 12/1 for domestic students; for spring admission, 3/15 for domestic students. Applications are processed on a rolling basis. Application fee: $30. Electronic applications accepted. *Expenses:* Tuition, area resident: Full-time $9,192; part-time $383 per credit. Tuition, state resident: full-time $9,192; part-time $383 per credit. Tuition, nonresident: full-time $15,990; part-time $666 per credit. International tuition: $15,990 full-time. *Financial support:* Federal Work-Study, institutionally sponsored loans, and tuition waivers (full) available. Financial award application deadline: 3/1; financial award applicants required to submit FAFSA. *Faculty research:* Engineering self-assembling FVS for piezoimmunosensors, development of a novel GCxGC system, interactions in open-shell species, radiation damage to DNA-free radical mechanisms, Hydrophilic Xenoestrogens: Response and oxidation removal. Total annual research expenditures:$771,595. *Unit head:* Dr. Mark W. Severson, Chair, 248-370-2320, Fax: 248-370-2321, E-mail: severson@oakland.edu. *Application contact:* Dr. Kathleen W. Moore, Coordinator, 248-370-2338, Fax: 248-370-2321, E-mail: kmoore@oakland.edu.

The Ohio State University, Graduate School, College of Mathematical and Physical Sciences, Department of Chemistry, Columbus, OH 43210. Offers MS, PhD. *Degree requirements:* For master's, thesis optional; for doctorate, thesis/dissertation. *Entrance requirements:* For master's and doctorate, GRE General Test, GRE Subject Test (chemistry). Additional exam requirements/recommendations for international students: Required—TOEFL (minimum score 600 paper-based; 250 computer-based). Electronic applications accepted.

Oklahoma State University, College of Arts and Sciences, Department of Chemistry, Stillwater, OK 74078. Offers MS, PhD. *Faculty:* 20 full-time (1 woman), 1 (woman) part-time/adjunct. *Students:* 16 full-time (8 women), 34 part-time (15 women); includes 2 minority (1 American Indian/Alaska Native, 1 Asian American or Pacific Islander), 35 international. Average age 29. 44 applicants, 52% accepted, 9 enrolled. In 2005, 7 master's, 2 doctorates awarded. *Degree requirements:* For master's and doctorate, thesis/dissertation. *Entrance requirements:* For master's, GRE (recommended), placement exam; for doctorate, GRE (recommended), placement exams. Additional exam requirements/recommendations for international students: Required—TOEFL. *Application deadline:* For fall admission, 6/1 priority date for domestic students, 3/1 priority date for international students. Applications are processed on a rolling basis. Application fee: $40 ($75 for international students). Electronic applications accepted. *Expenses:* Tuition, state resident: full-time $4,253; part-time $139 per credit hour. Tuition, nonresident: full-time $12,569; part-time $485 per credit hour. Required fees: $43 per credit hour. One-time fee: $20 part-time. Tuition and fees vary according to course load and program. *Financial support:* In 2005–06, 3 research assistantships (averaging $16,048 per year), 42 teaching assistantships (averaging $17,786 per year) were awarded; fellowships, Federal Work-Study, scholarships/grants, health care benefits, tuition waivers (partial), and unspecified assistantships also available. Support available to part-time students. Financial award application deadline: 3/1. *Faculty research:* Materials science, surface chemistry, and nanoparticles; theoretical physical chemistry; synthetic and medicinal chemistry; bioanalytical chemistry; electromagnetic (UV, VIS, IR, Raman), mass, and x-ray spectroscopes. *Unit head:* Dr. Neil Purdie, Head, 405-744-5920, Fax: 405-744-6007, E-mail: npurdie@okstate.edu.

Old Dominion University, College of Sciences, Program in Chemistry, Norfolk, VA 23529. Offers analytical chemistry (MS); biochemistry (MS); clinical chemistry (MS); environmental chemistry (MS); organic chemistry (MS); physical chemistry (MS). Part-time and evening/weekend programs available. *Faculty:* 15 full-time (4 women). *Students:* 5 full-time (3 women), 8 part-time (3 women); includes 8 minority (4 African Americans, 1 American Indian/Alaska Native, 1 Asian American or Pacific Islander, 2 Hispanic Americans). Average age 27. 19 applicants, 74% accepted. In 2005, 1 degree awarded. *Degree requirements:* For master's, thesis, comprehensive exam. *Entrance requirements:* For master's, GRE General Test, minimum GPA of 3.0 in major, 2.5 overall. Additional exam requirements/recommendations for international students: Required—TOEFL. *Application deadline:* For fall admission, 7/1 for domestic students; for spring admission, 11/1 for domestic students. Applications are processed on a rolling basis. Application fee: $30. *Expenses:* Tuition, state resident: part-time $263 per credit hour. Tuition, nonresident: part-time $661 per credit hour. Required fees: $39 per semester. Part-time tuition and fees vary according to campus/location. *Financial support:* In 2005–06, research assistantships with tuition reimbursements (averaging $15,000 per year), teaching assistantships with tuition reimbursements (averaging $15,000 per year) were awarded; fellowships, career-related internships or fieldwork and scholarships/grants also available. Financial award application deadline: 2/15; financial award applicants required to submit FAFSA. *Faculty research:* Organic and trace metal biogeochemistry, materials chemistry, bioanalytical chemistry, computational chemistry. Total annual research expenditures: $854,968. *Unit head:* Dr. John R. Donat, Graduate Program Director, 757-683-4098, Fax: 757-683-4628, E-mail: chemgpd@odu.edu.

Old Dominion University, Darden College of Education, Programs in Secondary Education, Norfolk, VA 23529. Offers biology (MS Ed); chemistry (MS Ed); English (MS Ed); instructional technology (MS Ed); library science (MS Ed); secondary education (MS Ed). *Accreditation:* NCATE. Part-time and evening/weekend programs available. Postbaccalaureate distance learning degree programs offered (minimal on-campus study). *Faculty:* 28 full-time (11 women). *Students:* 50 full-time (37 women), 149 part-time (93 women); includes 24 minority (15 African Americans, 2 American Indian/Alaska Native, 2 Asian Americans or Pacific Islanders, 5 Hispanic Americans), 2 international. Average age 37. 44 applicants, 95% accepted. In 2005, 119 degrees awarded. *Degree requirements:* For master's, thesis optional. *Entrance requirements:* For master's, GRE General Test, or MAT, PRAXIS I for master's with licensure, minimum GPA of 2.8, teaching certificate. Additional exam requirements/recommendations for international students: Required—TOEFL. *Application deadline:* Applications are processed on a rolling basis. Application fee: $40. Electronic applications accepted. *Expenses:* Tuition, state resident: part-time $263 per credit hour. Tuition, nonresident: part-time $661 per credit hour. Required fees: $39 per semester. Part-time tuition and fees vary according to campus/location. *Financial support:* In 2005–06, 58 students received support, including 2 research assistantships with tuition reimbursements available (averaging $6,777 per year), 3 teaching assistantships with tuition reimbursements available (averaging $5,333 per year); fellowships, career-related internships or fieldwork, Federal Work-Study, institutionally sponsored loans, scholarships/grants, and tuition waivers (partial) also available. Support available to part-time students. Financial award application deadline: 2/15; financial award applicants required to submit FAFSA. *Faculty research:* Mathematics retraining, writing project for teachers, geography teaching, reading. *Unit head:* Dr. Robert Lucking, Graduate Program Director, 757-683-5545, Fax: 757-683-5862, E-mail: ecisgpd@odu.edu.

Oregon State University, Graduate School, College of Science, Department of Chemistry, Corvallis, OR 97331. Offers analytical chemistry (MS, PhD); chemistry (MA, MAIS); inorganic chemistry (MS, PhD); nuclear and radiation chemistry (MS, PhD); organic chemistry (MS, PhD); physical chemistry (MS, PhD). Part-time programs available. *Faculty:* 19 full-time (3 women), 6 part-time/adjunct (1 woman). *Students:* 74 full-time (30 women), 3 part-time; includes 6 minority (1 African American, 1 Asian American or Pacific Islander, 4 Hispanic Americans), 35 international. Average age 28. In 2005, 4 master's, 10 doctorates awarded. Terminal master's awarded for partial completion of doctoral program. *Degree requirements:* For master's and doctorate, one foreign language, thesis/dissertation. *Entrance requirements:* For master's and doctorate, minimum GPA of 3.0 in last 90 hours of course work. Additional exam requirements/recommendations for international students: Required—TOEFL. *Application deadline:* For fall admission, 3/1 for domestic students. Applications are processed on a rolling basis. Application fee: $50. *Expenses:* Tuition, area resident: Full-time $8,139; part-time $301 per credit. Tuition, state resident: full-time $8,139; part-time $501 per credit. Tuition, nonresident: full-time $14,376; part-time $532 per credit. International tuition: $14,376 full-time. Required fees: $1,266. *Financial support:* Fellowships, research assistantships, teaching assistantships, institutionally sponsored loans available. Support available to part-time students. Financial award application deadline: 2/1. *Faculty research:* Solid state chemistry, enzyme reaction mechanisms, structure and dynamics of gas molecules, chemiluminescence, nonlinear optical spectroscopy. *Unit head:* Dr. Douglas Keszler, Chair, 541-737-2081, Fax: 541-737-2062. *Application contact:* Carolyn Brumley, Graduate Secretary, 541-737-6707, Fax: 541-737-2062, E-mail: carolyn.brumley@orst.edu.

The Pennsylvania State University University Park Campus, Graduate School, Eberly College of Science, Department of Chemistry, State College, University Park, PA 16802-1503. Offers MS, PhD. *Students:* 235 full-time (93 women), 4 part-time (2 women); includes 11 minority (1 African American, 6 Asian Americans or Pacific Islanders, 4 Hispanic Americans), 57 international. *Entrance requirements:* For master's and doctorate, GRE General Test. Application fee: $45. *Expenses:* Tuition, state resident: full-time $12,518; part-time $522 per credit. Tuition, nonresident: full-time $23,004; part-time $959 per credit. Required fees: $484. Tuition and fees vary according to course load, campus/location and program. *Unit head:* Dr. Ayusman Sen, Head, 814-865-6553, Fax: 814-863-8403, E-mail: axs20@psu.edu. *Application contact:* Dana Coval-Dinant, Graduate Student Recruiting Manager, 814-865-1383, Fax: 814-865-3228, E-mail: dmc6@psu.edu.

See Close-Up on page 131.

Pittsburg State University, Graduate School, College of Arts and Sciences, Department of Chemistry, Pittsburg, KS 66762. Offers MS. *Students:* 4. *Degree requirements:* For master's, thesis or alternative. Application fee: $0 ($40 for international students). *Expenses:* Tuition, state resident: full-time $2,015; part-time $170 per credit hour. Tuition, nonresident: full-time $4,953; part-time $415 per credit hour. International tuition: $4,953 full-time. Tuition and fees vary according to course load, campus/location and program. *Financial support:* Research assistantships, teaching assistantships, career-related internships or fieldwork and Federal Work-Study available. *Unit head:* Dr. Charles Blatchley, Chairperson, 620-235-4398. *Application contact:* Marvene Darraugh, Administrative Officer, 620-235-4220, Fax: 620-235-4219, E-mail: mdarraug@pittstate.edu.

Polytechnic University, Brooklyn Campus, Department of Chemical and Biological Sciences and Engineering, Major in Chemistry, Brooklyn, NY 11201-2990. Offers MS. Part-time and evening/weekend programs available. *Students:* 3 full-time (all women), 7 part-time (3 women); includes 1 minority (Asian American or Pacific Islander), 4 international. Average age 32. 15 applicants, 47% accepted, 2 enrolled. In 2005, 1 master's awarded. *Degree requirements:* For master's, thesis (for some programs), comprehensive exam (for some programs), registration. *Entrance requirements:* For master's, GRE General Test, GRE Subject Test. Additional exam requirements/recommendations for international students: Required—TOEFL (minimum score 550 paper-based; 213 computer-based); Recommended—IELT(minimum score 7). *Application deadline:* For fall admission, 7/15 priority date for domestic students, 4/1 priority date for international students; for spring admission, 12/15 priority date for domestic students, 10/1 priority date for international students. Applications are processed on a rolling basis. Application fee: $55. Electronic applications accepted. *Expenses:* Tuition: Part-time $950 per unit. Required fees: $330 per semester. *Financial support:* Fellowships, research assistantships, teaching assistantships, institutionally sponsored loans available. Support available to part-time students. Financial award applicants required to submit FAFSA. *Faculty research:* Optical rotation of light by plastic films, supramolecular chemistry, unusual stereochemical opportunities, polyaniline copolymers.

Polytechnic University, Brooklyn Campus, Department of Chemical and Biological Sciences and Engineering, Major in Materials Chemistry, Brooklyn, NY 11201-2990. Offers PhD. Part-time and evening/weekend programs available. *Students:* 7 full-time (2 women), 4 part-time, 9 international. Average age 32. 6 applicants, 67% accepted, 0 enrolled. In 2005, 6 doctorates awarded. Terminal master's awarded for partial completion of doctoral program. *Degree requirements:* For doctorate, thesis/dissertation, comprehensive exam, registration. *Entrance requirements:* Additional exam requirements/recommendations for international students: Required—TOEFL (minimum score 550 paper-based; 213 computer-based); Recommended—IELT(minimum score 7). *Application deadline:* For fall admission, 7/15 priority date for domestic students, 4/1 priority date for international students; for spring admission, 12/15 priority date for domestic students, 10/1 priority date for international students. Applications are processed on a rolling basis. Application fee: $55. Electronic applications accepted. *Expenses:* Tuition: Part-time $950 per unit. Required fees: $330 per semester. *Financial support:* Fellowships, research assistantships, teaching assistantships, institutionally sponsored loans available. Support available to part-time students. Financial award applicants required to submit FAFSA.

Polytechnic University, Long Island Graduate Center, Graduate Programs, Department of Chemical Engineering, Chemistry and Material Science, Major in Chemistry, Melville, NY 11747. Offers MS, PhD. *Degree requirements:* For master's, thesis (for some programs), comprehensive exam (for some programs), registration; for doctorate, thesis/dissertation. *Entrance requirements:* Additional exam requirements/recommendations for international students: Required—TOEFL (minimum score 550 paper-based; 213 computer-based); Recommended—IELT(minimum score 7). *Application deadline:* For fall admission, 7/15 priority date for domestic students, 4/1 priority date for international students; for spring admission, 12/15 priority date for domestic students, 10/1 priority date for international students. Applications are processed on a rolling basis. Application fee: $55. Electronic applications accepted. *Expenses:* Tuition: Part-time $950 per unit. Required fees: $330 per semester.

Polytechnic University, Westchester Graduate Center, Graduate Programs, Department of Chemical Engineering, Chemistry, and Materials Science, Major in Chemistry, Hawthorne, NY 10532-1507. Offers MS. *Students:* 2 full-time (1 woman), 8 part-time (3 women); includes 2 minority (1 African American, 1 Asian American or Pacific Islander). Average age 33. 1 applicant, 100% accepted, 0 enrolled. In 2005, 2 degrees awarded. *Degree requirements:* For master's, thesis (for some programs), comprehensive exam (for some programs), registration. *Entrance requirements:* Additional exam requirements/recommendations for international students: Required—TOEFL (minimum score 550 paper-based; 213 computer-based); Recommended—IELT(minimum score 7). *Application deadline:* For fall admission, 7/15 priority date for domestic students, 4/1 priority date for international students; for spring admission, 12/15 priority date for domestic students, 10/1 priority date for international students. Applications are processed on a rolling basis. Application fee: $55. Electronic applications accepted. *Expenses:* Tuition: Part-time $950 per unit. Required fees: $330 per term.

Pontifical Catholic University of Puerto Rico, College of Sciences, Department of Chemistry, Ponce, PR 00717-0777. Offers MS. Part-time and evening/weekend programs available. *Degree requirements:* For master's, thesis. *Entrance requirements:* For master's, GRE General Test, minimum GPA of 3.0, minimum 37 credits in chemistry. Electronic applications accepted.

Portland State University, Graduate Studies, College of Liberal Arts and Sciences, Department of Chemistry, Portland, OR 97207-0751. Offers MA, MS, PhD. Part-time programs available. *Faculty:* 13 full-time (2 women), 4 part-time/adjunct (2 women). *Students:* 4 full-time (2 women), 3 part-time (2 women), 2 international. Average age 34. 6 applicants, 50% accepted, 1 enrolled. In 2005, 3 degrees awarded. *Degree requirements:* For master's, one foreign language, thesis; for doctorate, one foreign language, thesis/dissertation, cumulative exams, seminar presentations. *Entrance requirements:* For master's, GRE General Test, GRE Subject Test, minimum GPA of 3.0 in upper-division course work or 2.75 overall, 2 letters of recommendation. Additional exam requirements/recommendations for international students: Required—TOEFL (minimum score 550 paper-based; 213 computer-based). *Application deadline:* For fall admission, 2/15 priority date for domestic students, 2/15 priority date for international studentsFor winter admission, 9/1 for domestic students; for spring admission, 11/1 for domestic students. Applications are processed on a rolling basis. Application fee: $50. *Expenses:* Tuition, state resident: full-time $6,648; part-time $231 per credit. Tuition, nonresident: full-time $11,319; part-time $231 per credit. Required fees: $686; $67 per credit. *Financial support:* In 2005–06, 1 research assistantship with full tuition reimbursement (averaging $13,500 per year), 26 teaching assistantships with full tuition reimbursements (averaging $13,465 per year) were awarded; career-related internships or fieldwork, Federal Work-Study, scholarships/grants, tuition waivers (partial), and unspecified assistantships also available. Support available to part-time students. Financial award application deadline: 3/1; financial award applicants required to submit FAFSA. *Faculty research:* Synthetic inorganic chemistry, atmospheric chemistry, organic photochemistry, enzymology, analytical chemistry. Total annual research expenditures: $1 million. *Unit head:* Kevin A. Reynolds, Chair, 503-725-3811, Fax: 503-725-3888. *Application contact:* Abbey Lawrence, Department Secretary, 503-725-8756, Fax: 503-725-3888, E-mail: aslawren@pdx.edu.

Prairie View A&M University, Graduate School, College of Arts and Sciences, Department of Chemistry, Prairie View, TX 77446-0519. Offers MS. Part-time and evening/weekend programs available. *Faculty:* 8 part-time/adjunct (0 women). *Students:* 1 full-time (0 women), 2 part-time (1 woman); includes 2 minority (both African Americans) Average age 30. 3 applicants, 100% accepted, 3 enrolled. In 2005, 2 degrees awarded. *Degree requirements:* For master's, thesis. *Entrance requirements:* For master's, GRE General Test. *Application deadline:* For fall admission, 4/1 for domestic students; for spring admission, 10/1 for domestic students. Applications are processed on a rolling basis. Application fee: $50. Electronic applications accepted. *Expenses:* Tuition, state resident: full-time $1,440; part-time $80 per credit. Tuition, nonresident: full-time $6,444; part-time $358 per credit. *Financial support:* In 2005–06, 3 research assistantships (averaging $15,000 per year), 2 teaching assistantships with partial tuition reimbursements (averaging $12,000 per year) were awarded; fellowships, career-related internships or fieldwork, Federal Work-Study, institutionally sponsored loans, and tuition waivers (full and partial) also available. Support available to part-time students. Financial award application deadline: 4/1; financial award applicants required to submit FAFSA. *Faculty research:* Material science, environmental characterization (surface phenomena), activation of plasminogens, polymer modifications. Total annual research expenditures:$300,000. *Unit head:* Dr. Remi R. Oki, Head, 936-857-2616, Fax: 936-857-2095, E-mail: aroki@pvamu.edu.

Princeton University, Graduate School, Department of Chemistry, Princeton, NJ 08544-1019. Offers chemistry (PhD); industrial chemistry (MS). *Degree requirements:* For doctorate, thesis/dissertation, general exams. *Entrance requirements:* For master's, GRE General Test; for doctorate, GRE General Test, GRE Subject Test (recommended). Additional exam requirements/recommendations for international students: Required—TOEFL (minimum score 250 computer-based). Electronic applications accepted. *Faculty research:* Chemistry of interfaces, organic synthesis, organometallic chemistry, inorganic reactions, biostructural chemistry.

See Close-Up on page 133.

Purdue University, Graduate School, School of Science, Department of Chemistry, West Lafayette, IN 47907. Offers analytical chemistry (MS, PhD); biochemistry (MS, PhD); chemical education (MS, PhD); inorganic chemistry (MS, PhD); organic chemistry (MS, PhD); physical chemistry (MS, PhD). *Accreditation:* NCATE (one or more programs are accredited). *Faculty:* 43 full-time (9 women), 8 part-time/adjunct (2 women). *Students:* 293 full-time (114 women), 39 part-time (14 women); includes 53 minority (25 African Americans, 6 Asian Americans or Pacific Islanders, 22 Hispanic Americans), 138 international. Average age 28. 458 applicants, 36% accepted, 57 enrolled. In 2005, 9 master's, 36 doctorates awarded. Terminal master's awarded for partial completion of doctoral program. *Degree requirements:* For master's and doctorate, thesis/dissertation. *Entrance requirements:* Additional exam requirements/recommendations for international students: Required—TOEFL. *Application deadline:* For fall admission, 4/1 priority date for domestic students, 3/1 priority date for international students; for spring admission, 10/1 priority date for domestic students, 9/1 priority date for international students. Applications are processed on a rolling basis. Application fee: $55. Electronic applications accepted. *Financial support:* In 2005–06, 2 fellowships with partial tuition reimbursements (averaging $18,000 per year), 55 teaching assistantships with partial tuition reimbursements (averaging $18,000 per year) were awarded; research assistantships with partial tuition reimbursements, tuition waivers (partial) also available. Support available to part-time students. Financial award applicants required to submit FAFSA. *Unit head:* Dr. Timothy S Zwier, Head, 765-494-5203. *Application contact:* R. E. Wild, Chairman, Graduate Admissions, 765-494-5200, E-mail: wild@purdue.edu.

Queens College of the City University of New York, Division of Graduate Studies, Mathematics and Natural Sciences Division, Department of Chemistry and Biochemistry, Flushing, NY 11367-1597. Offers biochemistry (MA); chemistry (MA). Part-time and evening/weekend programs available. *Faculty:* 14 full-time (4 women). *Students:* 20 applicants, 65% accepted, 9 enrolled. In 2005, 3 degrees awarded. *Degree requirements:* For master's, comprehensive exam. *Entrance requirements:* For master's, GRE, previous course work in calculus and physics, minimum GPA of 3.0. Additional exam requirements/recommendations for international students: Required—TOEFL. *Application deadline:* For fall admission, 4/1 for domestic students; for spring admission, 11/1 for domestic students. Applications are processed on a rolling basis. Application fee: $125. *Expenses:* Tuition, state resident: part-time $270 per credit. Tuition, nonresident: part-time $500 per credit. Required fees: $112 per year. *Financial support:* Career-related internships or fieldwork, Federal Work-Study, institutionally sponsored loans, tuition waivers (partial), and adjunct lectureships available. Support available to part-time students. Financial award application deadline: 4/1; financial award applicants required to submit FAFSA. *Unit head:* Dr. William Hersh, Chairperson, 718-997-4144. *Application contact:* Graduate Adviser, 718-997-4100.

Queen's University at Kingston, School of Graduate Studies and Research, Faculty of Arts and Sciences, Department of Chemistry, Kingston, ON K7L 3N6, Canada. Offers M Sc, PhD. Part-time programs available. *Degree requirements:* For master's, thesis (for some programs); for doctorate, thesis/dissertation, comprehensive exam. *Entrance requirements:* Additional exam requirements/recommendations for international students: Required—TOEFL (minimum score 580 paper-based). *Faculty research:* Medicinal/biological chemistry, materials chemistry, environmental/analytical chemistry, theoretical/computational chemistry.

Rensselaer Polytechnic Institute, Graduate School, School of Science, Department of Chemistry and Chemical Biology, Troy, NY 12180-3590. Offers analytical chemistry (MS, PhD); biochemistry (MS, PhD); inorganic chemistry (MS, PhD); organic chemistry (MS, PhD); physical chemistry (MS, PhD); polymer chemistry (MS, PhD). Part-time and evening/weekend programs available. *Faculty:* 19 full-time (3 women). *Students:* 49 full-time (28 women), 2 part-time (1 woman), 33 international. Average age 24. 80 applicants, 23% accepted, 7 enrolled. In 2005, 1 master's, 14 doctorates awarded. Terminal master's awarded for partial completion of doctoral program. *Median time to degree:* Of those who began their doctoral program in fall 1997, 100% received their degree in 8 years or less. *Degree requirements:* For master's, thesis (for some programs), registration; for doctorate, thesis/dissertation, comprehensive exam, registration. *Entrance requirements:* For master's, GRE General Test, GRE Subject Test (strongly recommended); for doctorate, GRE General Test, GRE Subject Test (chemistry or biochemistry strongly recommended). Additional exam requirements/recommendations for international students: Required—TOEFL (minimum score 600 paper-based). *Application deadline:* For fall admission, 2/1 for domestic students; for spring admission, 11/15 for domestic students. Applications are processed on a rolling basis. Application fee: $75. Electronic applications accepted. *Expenses:* Tuition: Full-time $31,000; part-time $1,320 per credit. Required fees: $1,623. *Financial support:* In 2005–06, 49 students received support, including 1 fellowship with full tuition reimbursement available (averaging $30,000 per year), 25 research assistantships with full tuition reimbursements available (averaging $21,500 per year), 30 teaching assistantships with full tuition reimbursements available (averaging $21,500 per year); institutionally sponsored loans and tuition waivers (full and partial) also available. Financial award application deadline: 2/1. *Faculty research:* Synthetic polymer and biopolymer chemistry, physical chemistry of polymeric systems, bioanalytical chemistry, synthetic and computational drug design, protein folding and protein design. Total annual research expenditures: $1.9 million. *Unit head:* Dr. Linda B. McGown, Chair, 518-276-4856, Fax: 518-276-4887, E-mail: mcgowl@rpi.edu. *Application contact:* Beth E. McGraw, Department Admissions Assistant, 518-276-6456, Fax: 518-276-4887, E-mail: mcgrae@rpi.edu.

See Close-Up on page 137.

Rensselaer Polytechnic Institute, Graduate School, School of Science, Department of Earth and Environmental Sciences, Troy, NY 12180-3590. Offers environmental chemistry (MS, PhD); geochemistry (MS, PhD); geology (MS, PhD); geophysics (MS, PhD); petrology (MS, PhD). Part-time programs available. Terminal master's awarded for partial completion of doctoral program. *Degree requirements:* For master's, thesis (for some programs), comprehensive exam; for doctorate, thesis/dissertation, comprehensive exam. *Entrance requirements:* For master's and doctorate, GRE General Test. Additional exam requirements/recommendations for international students: Required—TOEFL. Electronic applications accepted. *Expenses:* Tuition: Full-time $31,000; part-time $1,320 per credit. Required fees: $1,623. *Faculty research:* Mantel geochemistry, contaminant geochemistry, seismology, GPS geodesy, remote sensing petrology.

See Close-Up on page 661.

Rice University, Graduate Programs, Wiess School of Natural Sciences, Department of Chemistry, Houston, TX 77251-1892. Offers chemistry (MA); inorganic chemistry (PhD); organic chemistry (PhD); physical chemistry (PhD). Terminal master's awarded for partial completion of doctoral program. *Degree requirements:* For master's and doctorate, thesis/dissertation. *Entrance requirements:* For master's and doctorate, GRE General Test, minimum GPA of 3.0. Additional exam requirements/recommendations for international students: Required—TOEFL. *Faculty research:* Nanoscience, biomaterials, nanobioinformatics, fullerene pharmaceuticals.

Rochester Institute of Technology, Graduate Enrollment Services, College of Science, Department of Chemistry, Rochester, NY 14623-5603. Offers MS. Part-time and evening/weekend programs available. *Students:* 11 full-time (7 women), 4 international. 7 applicants, 43% accepted, 1 enrolled. In 2005, 8 degrees awarded. *Entrance requirements:* For master's, American Chemical Society Exam, GRE, minimum GPA of 3.0. Additional exam requirements/recommendations for international students: Required—TOEFL. *Application deadline:* For fall admission, 3/1 for domestic students. Applications are processed on a rolling basis. Application fee: $50. *Expenses:* Tuition: Full-time $25,392; part-time $713 per credit. Required fees: $183; $61 per term. *Financial support:* Teaching assistantships, career-related internships or fieldwork, Federal Work-Study, institutionally sponsored loans, and tuition waivers (full and partial) available. Support available to part-time students. *Faculty research:* Organic polymer chemistry, magnetic resonance and imaging, inorganic coordination polymers, biophysical chemistry, physical polymer chemistry. *Unit head:* Dr. Terence Morrill, Head, 585-475-2497, E-mail: tcmsch@rit.edu.

Rochester Institute of Technology, Graduate Enrollment Services, College of Science, Department of Medical Sciences, Rochester, NY 14623-5603. Offers clinical chemistry (MS). *Students:* 3 full-time (1 woman), 6 part-time (3 women), 1 international. 8 applicants, 38% accepted, 2 enrolled. In 2005, 1 degree awarded. *Entrance requirements:* For master's, minimum GPA of 3.0. *Application deadline:* For fall admission, 3/1 for domestic students. Applications are processed on a rolling basis. Application fee: $50. *Expenses:* Tuition: Full-time $25,392; part-time $713 per credit. Required fees: $183; $61 per term. *Financial support:* Teaching assistantships available. *Unit head:* Dr. Richard Doolittle, Head, 585-475-2978, E-mail: rldsbi@rit.edu.

Roosevelt University, Graduate Division, College of Arts and Sciences, Department of Science and Mathematics, Program in Biotechnology and Chemical Science, Chicago, IL 60605-1394. Offers MS. Part-time and evening/weekend programs available. *Students:* 13 full-time (11 women), 22 part-time (13 women); includes 12 minority (9 African Americans, 3 Asian Americans or Pacific Islanders), 6 international. Average age 28. 28 applicants, 82% accepted, 12 enrolled. In 2005, 14 degrees awarded. *Degree requirements:* For master's, thesis optional.

Chemistry

Roosevelt University (continued)

Entrance requirements: For master's, minimum GPA of 2.7, undergraduate course work in science and mathematics. *Application deadline:* For fall admission, 6/1 for domestic students; for spring admission, 11/1 for domestic students. Applications are processed on a rolling basis. Application fee: $25 ($35 for international students). *Expenses:* Tuition: Full-time $12,384; part-time $688 per credit hour. *Financial support:* In 2005–06, 1 student received support, including 1 teaching assistantship; tuition waivers (partial) also available. Support available to part-time students. Financial award application deadline: 2/15. *Faculty research:* Phase-transfer catalysts, bioinorganic chemistry, long chain dicarboxylic acids, organosilicon compounds, spectroscopic studies. Total annual research expenditures:$1,000. *Unit head:* Cornelius Watson, Head, 312-341-3670. *Application contact:* Joanne Canyon-Heller, Coordinator of Graduate Admission, 877-APPLY RU, Fax: 312-281-3356, E-mail: applyru@roosevelt.edu.

Royal Military College of Canada, Division of Graduate Studies and Research, Science Division, Department of Chemistry and Chemical Engineering, Kingston, ON K7K 7B4, Canada. Offers chemical engineering (M Eng, MA Sc, PhD); chemistry (M Sc, PhD). *Degree requirements:* For master's, thesis/dissertation, registration; for doctorate, thesis/dissertation, comprehensive exam, registration. Electronic applications accepted.

Rutgers, The State University of New Jersey, Camden, Graduate School of Arts and Sciences, Program in Chemistry, Camden, NJ 08102-1401. Offers MS. Part-time and evening/weekend programs available. *Degree requirements:* For master's, thesis (for some programs). *Entrance requirements:* Additional exam requirements/recommendations for international students: Required—TOEFL; Recommended—TWE, TSE. Electronic applications accepted. *Faculty research:* Organic and inorganic synthesis, enzyme biochemistry, trace metal analysis, theoretical and molecular modeling.

Rutgers, The State University of New Jersey, Newark, Graduate School, Program in Chemistry, Newark, NJ 07102. Offers analytical chemistry (MS, PhD); biochemistry (MS, PhD); inorganic chemistry (MS, PhD); organic chemistry (MS, PhD); physical chemistry (MS, PhD). Part-time and evening/weekend programs available. *Faculty:* 22 full-time (5 women). *Students:* 24 full-time (12 women), 28 part-time (10 women); includes 31 minority (2 African Americans, 29 Asian Americans or Pacific Islanders). 69 applicants, 43% accepted, 13 enrolled. In 2005, 7 master's, 9 doctorates awarded. Terminal master's awarded for partial completion of doctoral program. *Degree requirements:* For master's, cumulative exams, thesis optional; for doctorate, thesis/dissertation, exams, research proposal. *Entrance requirements:* For master's and doctorate, GRE General Test, minimum undergraduate B average. Additional exam requirements/recommendations for international students: Required—TOEFL. *Application deadline:* For fall admission, 7/1 for domestic students; for spring admission, 12/1 for domestic students. Applications are processed on a rolling basis. Application fee: $50. Electronic applications accepted. *Expenses:* Tuition, state resident: full-time $10,440; part-time $435 per credit. Tuition, nonresident: full-time $15,520; part-time $637 per credit. *Financial support:* In 2005–06, 35 students received support, including 5 fellowships with partial tuition reimbursements available (averaging $18,000 per year), 6 research assistantships (averaging $16,988 per year), 20 teaching assistantships with partial tuition reimbursements available (averaging $16,988 per year); Federal Work-Study and institutionally sponsored loans also available. Financial award application deadline: 3/1. *Faculty research:* Medicinal chemistry, natural products, isotope effects, biophysics and biorganic approaches to enzyme mechanisms, organic and organometallic synthesis. *Unit head:* Prof. W. Philip Huskey, Chairman and Program Director, 973-353-5741, Fax: 973-353-1264, E-mail: huskey@andromeda.rutgers.edu.

Rutgers, The State University of New Jersey, New Brunswick/Piscataway, Graduate School, Program in Chemistry and Chemical Biology, New Brunswick, NJ 08901-1281. Offers analytical chemistry (MS, PhD); biological chemistry (PhD); chemistry education (MST); inorganic chemistry (MS, PhD); organic chemistry (MS, PhD); physical chemistry (MS, PhD). Part-time and evening/weekend programs available. *Faculty:* 48 full-time. *Students:* 104 full-time (41 women), 22 part-time (4 women); includes 11 minority (2 African Americans, 6 Asian Americans or Pacific Islanders, 3 Hispanic Americans), 51 international. Average age 29. 108 applicants, 51% accepted, 22 enrolled. In 2005, 13 master's, 10 doctorates awarded. Terminal master's awarded for partial completion of doctoral program. *Degree requirements:* For master's, thesis or alternative, exam, comprehensive exam, registration; for doctorate, thesis/dissertation, cumulative exams, 1 year residency, comprehensive exam, registration. *Entrance requirements:* For master's and doctorate, GRE General Test, GRE Subject Test. Additional exam requirements/recommendations for international students: Required—TOEFL. *Application deadline:* For fall admission, 4/15 priority date for domestic students, 4/1 priority date for international students; for spring admission, 12/1 priority date for domestic students, 12/1 priority date for international students. Applications are processed on a rolling basis. Application fee: $50. Electronic applications accepted. *Expenses:* Tuition, state resident: full-time $10,440; part-time $435 per credit. Tuition, nonresident: full-time $15,520; part-time $647 per credit. Required fees: $129 per credit. Tuition and fees vary according to program. *Financial support:* In 2005–06, 104 students received support, including 8 fellowships with full tuition reimbursements available (averaging $22,000 per year), 23 research assistantships with full tuition reimbursements available (averaging $18,347 per year), 59 teaching assistantships with full tuition reimbursements available (averaging $18,347 per year); career-related internships or fieldwork, Federal Work-Study, traineeships, and health care benefits also available. Financial award application deadline: 3/1; financial award applicants required to submit FAFSA. *Faculty research:* Biophysical organic/bioorganic, inorganic/bioinorganic, theoretical, and solid-state/surface chemistry. Total annual research expenditures: $14.5 million. *Unit head:* Dr. Roger Jones, Director and Chair, 732-445-4900, Fax: 732-445-5312, E-mail: jones@rutchem.rutgers.edu. *Application contact:* Dr. Martha A. Cotter, Vice Chair, 732-445-2259, Fax: 732-445-5312, E-mail: cotter@rutchem.rutgers.edu.

See Close-Up on page 139.

Rutgers, The State University of New Jersey, New Brunswick/Piscataway, Graduate School, Program in Environmental Sciences, New Brunswick, NJ 08901-1281. Offers air resources (MS, PhD); aquatic biology (MS, PhD); aquatic chemistry (MS, PhD); atmospheric science (MS, PhD); chemistry and physics of aerosol and hydrosol systems (MS, PhD); environmental chemistry (MS, PhD); environmental microbiology (MS, PhD); environmental toxicology (PhD); exposure assessment (PhD); fate and effects of pollutants (MS, PhD); pollution prevention and control (MS, PhD); water and wastewater treatment (MS, PhD); water resources (MS, PhD). *Faculty:* 81 full-time, 7 part-time/adjunct. *Students:* 49 full-time (27 women), 48 part-time (19 women); includes 10 minority (3 African Americans, 6 Asian Americans or Pacific Islanders, 1 Hispanic American), 24 international. Average age 32. 79 applicants, 41% accepted, 15 enrolled. In 2005, 8 master's, 10 doctorates awarded. Terminal master's awarded for partial completion of doctoral program. *Degree requirements:* For master's, thesis or alternative, oral final exam, comprehensive exam; for doctorate, thesis/dissertation, thesis defense, qualifying exam, comprehensive exam. *Entrance requirements:* For master's and doctorate, GRE General Test. Additional exam requirements/recommendations for international students: Required—TOEFL. *Application deadline:* For fall admission, 3/1 for domestic students; for spring admission, 11/1 for domestic students. Applications are processed on a rolling basis. Application fee: $50. Electronic applications accepted. *Expenses:* Tuition, state resident: full-time $10,440; part-time $435 per credit. Tuition, nonresident: full-time $15,520; part-time $647 per credit. Required fees: $129 per credit. Tuition and fees vary according to program. *Financial support:* In 2005–06, 10 fellowships with full tuition reimbursements (averaging $21,887 per year), 34 research assistantships with full tuition reimbursements (averaging $19,367 per year), 3 teaching assistantships with full tuition reimbursements (averaging $17,583 per year) were awarded; career-related internships or fieldwork and Federal Work-Study also available. Financial award application deadline: 1/15; financial award applicants required to submit FAFSA. *Faculty research:* Atmospheric sciences; biological waste treatment; contaminant fate and transport; exposure assessment; air, soil and water quality. Total annual research expenditures: $5.7 million. *Unit head:* John Reinfelder, Director, 732-932-8013, Fax: 732-932-8644, E-mail: reinfelder@envsci.rutgers.edu. *Application contact:* Dr. Paul J. Lioy, Graduate Admissions Committee, 732-932-0150, Fax: 732-445-0116, E-mail: plioy@eohsi.rutgers.edu.

Sacred Heart University, Graduate Studies, College of Arts and Sciences, Faculty of Chemistry, Fairfield, CT 06825-1000. Offers MS. Part-time and evening/weekend programs available. *Degree requirements:* For master's, thesis optional. *Entrance requirements:* For master's, bachelor's degree in related area, minimum GPA of 2.75. Additional exam requirements/recommendations for international students: Required—TOEFL (minimum score 550 paper-based).

St. Francis Xavier University, Graduate Studies, Department of Chemistry, Antigonish, NS B2G 2W5, Canada. Offers M Sc. *Students:* 5 applicants, 40% accepted. *Degree requirements:* For master's, thesis, registration. *Entrance requirements:* For master's, minimum undergraduate B average, undergraduate major in chemistry or related field. Additional exam requirements/recommendations for international students: Required—TOEFL (minimum score 580 paper-based; 236 computer-based). *Application deadline:* For fall admission, 9/1 for domestic students. Applications are processed on a rolling basis. Application fee: $40. *Faculty research:* Photoelectron spectroscopy, synthesis and properties of surfactants, nucleic acid synthesis, transition metal chemistry, colloids. Total annual research expenditures:$200,000. *Unit head:* Dr. Bernard V. Liengme, III, Professor, 902-867-2361, Fax: 902-867-2414, E-mail: bliengme@stfx.ca. *Application contact:* 902-867-2219, Fax: 902-867-2329, E-mail: admit@stfx.ca.

St. John's University, St. John's College of Liberal Arts and Sciences, Department of Chemistry, Queens, NY 11439. Offers MS. Part-time and evening/weekend programs available. *Faculty:* 10 full-time (4 women), 23 part-time/adjunct (5 women). *Students:* 6 full-time (3 women), 17 part-time (10 women); includes 10 minority (7 Asian Americans or Pacific Islanders, 3 Hispanic Americans), 3 international. Average age 30. 21 applicants, 67% accepted, 9 enrolled. In 2005, 6 degrees awarded. *Degree requirements:* For master's, thesis optional. *Entrance requirements:* For master's, minimum GPA of 3.0. Additional exam requirements/recommendations for international students: Required—TOEFL (minimum score 500 paper-based; 173 computer-based). *Application deadline:* For fall admission, 5/1 priority date for domestic students, 5/1 priority date for international students; for spring admission, 11/1 priority date for domestic students, 11/1 priority date for international students. Applications are processed on a rolling basis. Application fee: $40. Electronic applications accepted. *Expenses:* Tuition: Full-time $8,760; part-time $730 per credit. Required fees: $250; $125 per term. Tuition and fees vary according to program. *Financial support:* Research assistantships, teaching assistantships, scholarships/grants available. Support available to part-time students. Financial award application deadline: 3/1; financial award applicants required to submit FAFSA. *Faculty research:* Synthsis and reactions of a-lactams, NMR spectroscopy or nucleosides, analytical chemistry, environment chemistry and photochemistry of transition metal complexes. *Unit head:* Dr. Richard Rosso, Chair, 718-990-5216, E-mail: rossor@stjohns.edu. *Application contact:* Matthew Whelan, Director, Office of Admissions, 718-990-2000, Fax: 718-990-2096, E-mail: admissions@stjohns.edu.

Saint Joseph College, Graduate Division, Department of Biology, West Hartford, CT 06117-2700. Offers biology (MS), including general biology, molecular and cellular biology; biology/chemistry (MS). MS biology (including general biology; molecular and cellular biology) offered online only. Part-time and evening/weekend programs available. Postbaccalaureate distance learning degree programs offered (no on-campus study). *Faculty:* 4 full-time (2 women), 5 part-time/adjunct (3 women). *Students:* Average age 34. 34 applicants, 100% accepted, 11 enrolled. In 2005, 13 degrees awarded. *Degree requirements:* For master's, thesis or alternative, comprehensive exam. *Entrance requirements:* For master's, 2 letters of recommendation. *Application deadline:* Applications are processed on a rolling basis. Application fee: $50. Electronic applications accepted. *Expenses:* Tuition: Part-time $540 per credit. Required fees: $25 per credit. *Financial support:* Career-related internships or fieldwork, health care benefits, and unspecified assistantships available. Support available to part-time students. Financial award application deadline: 7/15; financial award applicants required to submit FAFSA. *Faculty research:* Neurology, cardiology, immunology, mircobiology. *Unit head:* Dr. Charles Morgan, Chair, 860-231-5335, E-mail: cmorgan@sjc.edu.

Saint Joseph College, Graduate Division, Department of Chemistry, West Hartford, CT 06117-2700. Offers biology/chemistry (MS); chemistry (MS). Part-time and evening/weekend programs available. *Faculty:* 5 full-time (1 woman). *Students:* 1 (woman) full-time, 4 part-time (3 women); includes 2 minority (1 Asian American or Pacific Islander, 1 Hispanic American). Average age 30. 4 applicants, 100% accepted, 2 enrolled. In 2005, 3 degrees awarded. *Degree requirements:* For master's, thesis optional. *Entrance requirements:* For master's, 2 letters of recommendation. *Application deadline:* Applications are processed on a rolling basis. Application fee: $50. Electronic applications accepted. *Expenses:* Tuition: Part-time $540 per credit. Required fees: $25 per credit. *Financial support:* Career-related internships or fieldwork, health care benefits, and unspecified assistantships available. Support available to part-time students. Financial award application deadline: 7/15; financial award applicants required to submit FAFSA. *Faculty research:* Regulation of cancer cells, selective oxidation of organic concept mapping in general chemistry. *Unit head:* Dr. Peter Markow, Chair, 860-231-5240, Fax: 860-233-5695, E-mail: pmarkow@sjc.edu.

Saint Louis University, Graduate School, College of Arts and Sciences and Graduate School, Department of Chemistry, St. Louis, MO 63103-2097. Offers MS, MS-R. Part-time programs available. *Faculty:* 16 full-time (3 women), 1 part-time/adjunct (0 women). *Students:* 12 full-time (9 women), 12 part-time (6 women); includes 6 minority (3 African Americans, 1 American Indian/Alaska Native, 1 Asian American or Pacific Islander, 1 Hispanic American), 3 international. Average age 27. 26 applicants, 92% accepted, 12 enrolled. In 2005, 9 degrees awarded. *Degree requirements:* For master's, thesis (for some programs), comprehensive exam. *Entrance requirements:* For master's, letters of recommendation, resumé. Additional exam requirements/recommendations for international students: Required—TOEFL (minimum score 550 paper-based; 213 computer-based). *Application deadline:* For fall admission, 7/1 for domestic students, 7/1 for international students; for spring admission, 11/1 for domestic students, 11/1 for international students. Applications are processed on a rolling basis. Application fee: $40. *Expenses:* Tuition: Part-time $760 per credit hour. Required fees: $55 per semester. *Financial support:* In 2005–06, 17 students received support, including 5 research assistantships with full tuition reimbursements available (averaging $11,300 per year), 5 teaching assistantships with full tuition reimbursements available (averaging $11,300 per year); health care benefits, tuition waivers (partial), and unspecified assistantships also available. Financial award application deadline: 6/1; financial award applicants required to submit FAFSA. *Faculty research:* Analytical chemistry, physical chemistry, inorganic chemistry, organic chemistry, biochemistry. Total annual research expenditures: $500,000. *Unit head:* Dr. Steven Buckner, Chairperson, 314-977-2850, Fax: 314-977-2521, E-mail: bucknes@slu.edu. *Application contact:* Gary Behrman, Associate Dean of the Graduate School, 314-977-3827, E-mail: behrmang@slu.edu.

Sam Houston State University, College of Arts and Sciences, Department of Chemistry, Huntsville, TX 77341. Offers MS. Part-time programs available. *Faculty:* 8 full-time (2 women). *Students:* 4 full-time (3 women), 3 part-time (1 woman), 2 international. Average age 26. In 2005, 1 degree awarded. *Degree requirements:* For master's, thesis (for some programs). *Entrance requirements:* For master's, GRE General Test. Additional exam requirements/recommendations for international students: Required—TOEFL (minimum score 550 paper-based; 213 computer-based). *Application deadline:* For fall admission, 8/1 for domestic students; for spring admission, 12/1 for domestic students. Applications are processed on a rolling basis. Application fee: $20. *Financial support:* Research assistantships, teaching assistantships, Federal Work-Study, institutionally sponsored loans, and tuition waivers (partial) available. Support available to part-time students. Financial award application deadline: 5/31; financial award applicants required to submit FAFSA. *Unit head:* Dr. Rick White, Chair, 936-294-1060, Fax: 936-294-4996, E-mail: chm_rcw@shsu.edu. *Application contact:* Dr. Tom Chasteen, Advisor, 936-294-1533, Fax: 936-299-1585.

San Diego State University, Graduate and Research Affairs, College of Sciences, Department of Chemistry, San Diego, CA 92182. Offers MA, MS, PhD. *Students:* 13 full-time (5 women), 53 part-time (23 women); includes 14 minority (1 African American, 9 Asian Americans or Pacific Islanders, 4 Hispanic Americans), 22 international. Average age 30. 66 applicants, 53% accepted, 18 enrolled. In 2005, 15 master's, 7 doctorates awarded. Terminal master's awarded for partial completion of doctoral program. *Degree requirements:* For doctorate, thesis/dissertation. *Entrance requirements:* For master's, GRE General Test, bachelor's degree in related field, 3 letters of reference; for doctorate, GRE General Test, GRE Subject Test. Additional exam requirements/recommendations for international students: Required—TOEFL. *Application deadline:* For fall admission, 5/1 for domestic students, 5/1 for international students; for spring admission, 11/1 for domestic students, 10/1 for international students. Applications are processed on a rolling basis. Application fee: $55. Electronic applications accepted. *Expenses:* Tuition: Full-time $3,704. Required fees: $1,201 per term. Tuition and fees vary according to course load and campus/location. *Financial support:* Fellowships, research assistantships, teaching assistantships, scholarships/grants and unspecified assistantships available. Financial award applicants required to submit FAFSA. *Faculty research:* Nonlinear, laser, and electrochemistry; surface reaction dynamics; catalysis, synthesis, and organometallics; proteins, enzymology, and gene expression regulation. Total annual research expenditures: $1.3 million. *Unit head:* Carl Carrano, Chair, 619-594-5595, Fax: 619-594-4634, E-mail: carrano@sciences.sdsu.edu. *Application contact:* Tom Cole, Graduate Adviser, 619-594-5579, E-mail: tcole@sciences.sdsu.edu.

San Francisco State University, Division of Graduate Studies, College of Science and Engineering, Department of Chemistry and Biochemistry, San Francisco, CA 94132-1722. Offers chemistry (MS), including biochemistry. Part-time programs available. *Degree requirements:* For master's, thesis, ACS exams in 3 chemical disciplines (including physical chemistry), essay test. *Entrance requirements:* For master's, minimum GPA of 2.5 in last 60 units. Additional exam requirements/recommendations for international students: Required—TOEFL (minimum score 550 paper-based; 213 computer-based). Electronic applications accepted. *Faculty research:* Physical chemistry of macromolecules, physical and synthetic organic chemistry, membrane and enzyme biochemistry, organometallic chemistry.

San Jose State University, Graduate Studies and Research, College of Science, Department of Chemistry, San Jose, CA 95192-0001. Offers MA, MS. Part-time and evening/weekend programs available. *Students:* 12 full-time (7 women), 17 part-time (13 women); includes 9 minority (1 African American, 7 Asian Americans or Pacific Islanders, 1 Hispanic American), 9 international. Average age 31. 24 applicants, 42% accepted, 6 enrolled. In 2005, 6 degrees awarded. *Degree requirements:* For master's, thesis or alternative. *Entrance requirements:* For master's, GRE, minimum B average. *Application deadline:* For fall admission, 6/29 for domestic students; for spring admission, 11/30 for domestic students. Applications are processed on a rolling basis. Application fee: $59. Electronic applications accepted. *Expenses:* Tuition, nonresident: part-time $339 per unit. Required fees: $1,286 per semester. Tuition and fees vary according to course load and degree level. *Financial support:* In 2005–06, 8 teaching assistantships were awarded; career-related internships or fieldwork, Federal Work-Study, and institutionally sponsored loans also available. Support available to part-time students. Financial award application deadline: 6/5; financial award applicants required to submit FAFSA. *Faculty research:* Intercalated compounds, organic/biochemical reaction mechanisms, complexing agents in biochemistry, DNA repair, metabolic inhibitors. *Unit head:* Brad Stone, Chair, 408-924-4500, Fax: 408-924-4945.

The Scripps Research Institute, Kellogg School of Science and Technology, Program in Chemistry and Chemical Biology, La Jolla, CA 92037. Offers chemical biology (PhD); chemistry (PhD). *Faculty:* 20 full-time (0 women). *Students:* 104 full-time (29 women). Average age 22. 291 applicants, 23% accepted, 24 enrolled. In 2005, 11 degrees awarded. *Degree requirements:* For doctorate, thesis/dissertation. *Entrance requirements:* For doctorate, GRE General Test, GRE Subject Test. Additional exam requirements/recommendations for international students: Required—TOEFL. *Application deadline:* For fall admission, 1/1 for domestic students, 1/1 for international students. Application fee: $0. *Expenses:* Tuition: Full-time $5,000. *Financial support:* Institutionally sponsored loans and stipends available. *Faculty research:* Synthetic organic chemistry and natural product synthesis, bio-organic chemistry and molecular design, biocatalysis and protein design, chemical biology. *Unit head:* Dr. James R. Williamson, Associate Dean, 858-784-8740. *Application contact:* Marylyn Rinaldi, Administrative Director, 858-784-8469, Fax: 858-784-2802, E-mail: mrinaldi@scripps.edu.

Seton Hall University, College of Arts and Sciences, Department of Chemistry and Biochemistry, South Orange, NJ 07079-2694. Offers analytical chemistry (MS, PhD); biochemistry (MS, PhD); chemistry (MS); inorganic chemistry (MS, PhD); organic chemistry (MS, PhD); physical chemistry (MS, PhD). Part-time and evening/weekend programs available. *Students:* 19 full-time (6 women), 46 part-time (17 women). Average age 34. 33 applicants, 79% accepted, 14 enrolled. In 2005, 11 master's, 7 doctorates awarded. Terminal master's awarded for partial completion of doctoral program. *Degree requirements:* For master's, formal seminar, thesis optional; for doctorate, thesis/dissertation, annual seminars, comprehensive exam. *Entrance requirements:* Additional exam requirements/recommendations for international students: Required—TOEFL. *Application deadline:* For fall admission, 7/1 priority date for domestic students, 7/1 priority date for international students; for spring admission, 11/1 priority date for domestic students, 11/1 priority date for international students. Applications are processed on a rolling basis. Application fee: $50. Electronic applications accepted. *Financial support:* In 2005–06, 1 research assistantship, 19 teaching assistantships were awarded; Federal Work-Study also available. *Faculty research:* DNA metal reactions; chromatography; bioinorganic, biophysical, organometallic, polymer chemistry; heterogeneous catalyst. *Unit head:* Dr. Nicholas Snow, Chair, 973-761-9414, Fax: 973-761-9772, E-mail: snownich@shu.edu. *Application contact:* Dr. Stephen Kelty, Director of Graduate Studies, 973-761-9129, Fax: 973-761-9772, E-mail: keltyste@shu.edu.

See Close-Up on page 141.

Simon Fraser University, Graduate Studies, Faculty of Science, Department of Chemistry, Burnaby, BC V5A 1S6, Canada. Offers chemical physics (M Sc, PhD); chemistry (M Sc, PhD). *Degree requirements:* For master's and doctorate, thesis/dissertation. *Entrance requirements:* For master's, minimum GPA of 3.0; for doctorate, minimum GPA of 3.5. Additional exam requirements/recommendations for international students: Required—TOEFL or IELTS. *Faculty research:* Organic chemistry, nuclear chemistry, physical chemistry, inorganic chemistry, theoretical chemistry.

Smith College, Graduate Programs, Department of Chemistry, Northampton, MA 01063. Offers MAT. Part-time programs available. *Faculty:* 9 full-time (6 women). *Entrance requirements:* For master's, GRE General Test, GRE Subject Test. *Application deadline:* For fall admission, 4/1 for domestic students, 1/15 for international students; for spring admission, 12/1 for domestic students. Application fee: $60. *Expenses:* Tuition: Full-time $30,520; part-time $955 per credit. *Financial support:* Career-related internships or fieldwork and institutionally sponsored loans available. Support available to part-time students. Financial award application deadline: 1/15; financial award applicants required to submit CSS PROFILE or FAFSA. *Unit head:* Cristina Suarez, Chair, 413-585-3838.

South Dakota School of Mines and Technology, Graduate Division, College of Engineering, Doctoral Program in Materials Engineering and Science, Rapid City, SD 57701-3995. Offers chemical engineering (PhD); chemistry (PhD); civil engineering (PhD); electrical engineering (PhD); mechanical engineering (PhD); metallurgical engineering (PhD); physics (PhD). Part-time programs available. *Faculty:* 6 full-time (0 women), 1 part-time/adjunct (0 women). *Students:* 3 full-time (0 women), 4 part-time, 2 international. In 2005, 5 degrees awarded. *Degree requirements:* For doctorate, thesis/dissertation. *Entrance requirements:* For doctorate, minimum graduate GPA of 3.0. Additional exam requirements/recommendations for international students: Required—TOEFL, TWE. *Application deadline:* For fall admission, 7/1 priority date for domestic students, 4/1 priority date for international students; for spring admission, 11/1 for domestic students, 9/1 for international students. Applications are processed on a rolling basis. Application fee: $35. Electronic applications accepted. *Expenses:* Tuition, area resident: Part-time $116 per credit hour. Tuition, state resident: full-time $2,084. Tuition, nonresident: full-time $6,146; part-time $341 per credit hour. Required fees: $1,805; $100 per credit hour. *Financial support:* In 2005–06, 1 fellowship (averaging $1,000 per year), 6 research assistantships with partial tuition reimbursements (averaging $15,390 per year), teaching assistantships with partial tuition reimbursements (averaging $4,482 per year) were awarded; Federal Work-Study and institutionally sponsored loans also available. Support available to part-time students. Financial award application deadline: 5/15. *Faculty research:* Thermophysical properties of solids, development of multiphase materials and composites, concrete technology, electronic polymer materials. *Unit head:* Dr. Duane C. Hrncir, Dean, 605-394-1237. *Application contact:* Jeannette R. Nilson, Program Assistant-Research and Graduate Education, 800-454-8162 Ext. 1206, Fax: 605-394-5360, E-mail: graduate_admissions@silver.sdsmt.edu.

South Dakota School of Mines and Technology, Graduate Division, College of Engineering, Master's Program in Materials Engineering and Science, Rapid City, SD 57701-3995. Offers chemistry (MS); metallurgical engineering (MS); physics (MS). *Faculty:* 6 full-time (0 women), 1 part-time/adjunct (0 women). *Students:* 9 full-time (1 woman), 3 part-time; includes 1 minority (Hispanic American), 6 international. Average age 26. In 2005, 4 degrees awarded. *Entrance requirements:* For master's, GRE General Test. Additional exam requirements/recommendations for international students: Required—TOEFL, TWE. *Application deadline:* For fall admission, 7/1 priority date for domestic students, 4/1 priority date for international students; for spring admission, 11/1 for domestic students, 9/1 for international students. Applications are processed on a rolling basis. Application fee: $35. Electronic applications accepted. *Expenses:* Tuition, area resident: Part-time $116 per credit hour. Tuition, state resident: full-time $2,084. Tuition, nonresident: full-time $6,146; part-time $341 per credit hour. Required fees: $1,805; $100 per credit hour. *Financial support:* In 2005–06, 15 research assistantships with partial tuition reimbursements (averaging $11,400 per year), 11 teaching assistantships with partial tuition reimbursements (averaging $4,063 per year) were awarded; fellowships Financial award application deadline: 5/15. *Unit head:* Dr. Daniel Heglund, 605-394-1241. *Application contact:* Jeannette R. Nilson, Program Assistant-Research and Graduate Education, 800-454-8162 Ext. 1206, Fax: 605-394-5360, E-mail: graduate_admissions@silver.sdsmt.edu.

South Dakota State University, Graduate School, College of Arts and Science and College of Agriculture and Biological Sciences, Department of Chemistry, Brookings, SD 57007. Offers analytical chemistry (MS, PhD); biochemistry (MS, PhD); chemistry (MS, PhD); inorganic chemistry (MS, PhD); organic chemistry (MS, PhD); physical chemistry (MS, PhD). *Degree requirements:* For master's, thesis, oral exam; for doctorate, thesis/dissertation, preliminary oral and written exams, research tool. *Entrance requirements:* For master's, bachelor's degree in chemistry or equivalent. Additional exam requirements/recommendations for international students: Required—TOEFL. *Faculty research:* Environmental chemistry, computational chemistry, organic synthesis and photochemistry, novel material development and characterization.

Southeast Missouri State University, School of Graduate Studies, Department of Chemistry, Cape Girardeau, MO 63701-4799. Offers MNS. Part-time and evening/weekend programs available. *Faculty:* 10 full-time (1 woman). *Students:* 2 full-time (both women), 11 part-time (8 women); includes 4 minority (1 African American, 2 American Indian/Alaska Native, 1 Hispanic American). Average age 25. 5 applicants, 80% accepted. In 2005, 1 degree awarded. *Degree requirements:* For master's, thesis or alternative. *Entrance requirements:* For master's, GRE General Test, minimum undergraduate GPA of 2.75, minimum of 30 hours of undergraduate course work in science or mathematics. Additional exam requirements/recommendations for international students: Required—TOEFL (minimum score 550 paper-based; 213 computer-based). *Application deadline:* For fall admission, 8/1 for domestic students, 4/1 for international students; for spring admission, 11/21 for domestic students, 9/1 for international students. Applications are processed on a rolling basis. Application fee: $20 ($100 for international students). Electronic applications accepted. *Expenses:* Tuition, state resident: full-time $1,676; part-time $186 per hour. Tuition, nonresident: full-time $3,052; part-time $339 per hour. Required fees: $114; $13 per hour. Tuition and fees vary according to course load, degree level and campus/location. *Financial support:* In 2005–06, 10 students received support, including 6 research assistantships with full tuition reimbursements available (averaging $6,600 per year), 3 teaching assistantships with full tuition reimbursements available (averaging $6,600 per year); unspecified assistantships also available. Financial award applicants required to submit FAFSA. *Faculty research:* Crystallography, heavy metal detection, asymmetric synthesis, organic reactions with supported reagents. Total annual research expenditures: $27,259. *Unit head:* Dr. Phillip Crawford, Chairperson, 573-651-2166, Fax: 573-651-2223, E-mail: pcrawford@semo.edu. *Application contact:* Marsha L. Arant, Senior Administrative Assistant, Office of Graduate Studies, 573-651-2192, Fax: 573-651-2001, E-mail: marant@semo.edu.

Southern Connecticut State University, School of Graduate Studies, School of Arts and Sciences, Department of Chemistry, New Haven, CT 06515-1355. Offers MS, MLS/MS. Part-time and evening/weekend programs available. *Degree requirements:* For master's, thesis or alternative. *Entrance requirements:* For master's, interview, undergraduate work in chemistry. Electronic applications accepted.

Southern Illinois University Carbondale, Graduate School, College of Science, Department of Chemistry and Biochemistry, Carbondale, IL 62901-4701. Offers MS, PhD. Part-time programs available. *Faculty:* 18 full-time (1 woman), 2 part-time/adjunct (0 women). *Students:* 36 full-time (18 women), 19 part-time (10 women); includes 4 minority (all African Americans), 39 international. Average age 25. 53 applicants, 21% accepted, 6 enrolled. In 2005, 2 master's, 3 doctorates awarded. Terminal master's awarded for partial completion of doctoral program. *Degree requirements:* For master's, one foreign language, thesis; for doctorate, variable foreign language requirement, thesis/dissertation. *Entrance requirements:* For master's, minimum GPA of 2.7; for doctorate, GRE General Test, minimum GPA of 3.25. Additional exam requirements/recommendations for international students: Required—TOEFL. *Application deadline:* Applications are processed on a rolling basis. Application fee: $0. *Financial support:* In 2005–06, 17 research assistantships with full tuition reimbursements, 23 teaching assistantships with full tuition reimbursements were awarded; fellowships with full tuition reimbursements, Federal Work-Study, institutionally sponsored loans, and tuition waivers (full) also available. Support available to part-time students. *Faculty research:* Materials, separations, computational chemistry, synthetics. Total annual research expenditures:$1 million. *Unit head:* Lori Vermeuler, Chair, 618-453-6482, Fax: 618-453-6408. *Application contact:* Steve Scheiner, Chair, Graduate Admissions Committee, 618-453-6476, Fax: 618-453-6408, E-mail: scheiner@chem.siu.edu.

Announcement: The SIUC Department of Chemistry and Biochemistry offers opportunities in the following: forensic chemistry; environmental chemistry; design and fabrication of new materials; nanomaterials; biomaterials; smart materials; biomedical chemistry; computational examinations of environmental, biological, and catalytic chemistries; high-tech instrumentation; new methods for molecular recognition and separations; trace analyses.

See Close-Up on page 143.

Southern Illinois University Edwardsville, Graduate Studies and Research, College of Arts and Sciences, Department of Chemistry, Edwardsville, IL 62026-0001. Offers MS. Part-time and evening/weekend programs available. *Students:* 9 full-time (5 women), 12 part-time (8 women); includes 3 minority (2 African Americans, 1 Asian American or Pacific Islander), 7 international. Average age 33. 20 applicants, 55% accepted. In 2005, 5 degrees awarded. *Degree requirements:* For master's, thesis or alternative, final paper. *Entrance requirements:* Additional exam requirements/recommendations for international students: Required—TOEFL. *Application deadline:* For fall admission, 7/21 for domestic students, 6/1 for international students; for spring admission, 12/8 for domestic students, 10/1 for international students. Application fee: $30. Electronic applications accepted. *Expenses:* Tuition, state resident:

Chemistry

Southern Illinois University Edwardsville (continued)

part-time $190 per semester hour. Tuition, nonresident: part-time $380 per semester hour. Tuition and fees vary according to course load, reciprocity agreements and student level. *Financial support:* In 2005–06, 1 research assistantship with full tuition reimbursement, 15 teaching assistantships with full tuition reimbursements were awarded; fellowships with full tuition reimbursements, Federal Work-Study, institutionally sponsored loans, and unspecified assistantships also available. Support available to part-time students. Financial award application deadline: 3/1. *Unit head:* Dr. Robert P. Dixon, Chair, 618-650-3576, E-mail: rdixon@siue.edu. *Application contact:* Dr. Michael Shaw, Director, 618-650-3579, E-mail: michsha@siue.edu.

Southern Methodist University, Dedman College, Department of Chemistry, Dallas, TX 75275. Offers MS, PhD. Part-time programs available. *Degree requirements:* For master's, thesis, comprehensive exam (for some programs), registration; for doctorate, thesis/dissertation, presentation, comprehensive exam, registration. *Entrance requirements:* For master's and doctorate, GRE General Test, bachelor's degree in chemistry, minimum GPA of 3.0. Additional exam requirements/recommendations for international students: Required—TOEFL. *Faculty research:* Organic and inorganic synthesis, theoretical chemistry, organometallic chemistry, inorganic polymer chemistry, fundamental quantum chemistry.

Southern University and Agricultural and Mechanical College, Graduate School, College of Sciences, Department of Chemistry, Baton Rouge, LA 70813. Offers analytical chemistry (MS); biochemistry (MS); environmental sciences (MS); inorganic chemistry (MS); organic chemistry (MS); physical chemistry (MS). *Faculty:* 9 full-time (2 women), 3 part-time/adjunct (2 women). *Students:* 20 full-time (13 women), 8 part-time (5 women); all minorities (20 African Americans, 8 Asian Americans or Pacific Islanders). Average age 23. 30 applicants, 70% accepted, 14 enrolled. In 2005, 3 master's awarded. *Degree requirements:* For master's, thesis. *Entrance requirements:* For master's, GMAT or GRE General Test. Additional exam requirements/recommendations for international students: Required—TOEFL (minimum score 525 paper-based; 193 computer-based). *Application deadline:* For fall admission, 4/15 for domestic students; for spring admission, 11/1 priority date for domestic students. Applications are processed on a rolling basis. Application fee: $5. *Financial support:* In 2005–06, 31 research assistantships (averaging $7,000 per year), 10 teaching assistantships (averaging $7,000 per year) were awarded; scholarships/grants also available. Financial award application deadline: 4/15. *Faculty research:* Synthesis of macrocyclic ligands, latex accelerators, anticancer drugs, biosensors, absorption isotheums, isolation of specific enzymes from plants. Total annual research expenditures:$400,000. *Unit head:* Dr. Ella Kelley, Chair, 225-771-3990, Fax: 225-771-3992.

Stanford University, School of Humanities and Sciences, Department of Chemistry, Stanford, CA 94305-9991. Offers PhD. *Degree requirements:* For doctorate, thesis/dissertation. *Entrance requirements:* For doctorate, GRE General Test, GRE Subject Test. Additional exam requirements/recommendations for international students: Required—TOEFL. Electronic applications accepted.

See Close-Up on page 145.

State University of New York at Binghamton, Graduate School, School of Arts and Sciences, Department of Chemistry, Binghamton, NY 13902-6000. Offers analytical chemistry (PhD); chemistry (MA, MS); inorganic chemistry (PhD); organic chemistry (PhD); physical chemistry (PhD). Part-time programs available. Terminal master's awarded for partial completion of doctoral program. *Degree requirements:* For master's, thesis or alternative, oral exam, seminar presentation; for doctorate, thesis/dissertation, cumulative exams. *Entrance requirements:* For master's and doctorate, GRE General Test, GRE Subject Test. Additional exam requirements/recommendations for international students: Required—TOEFL. Electronic applications accepted.

State University of New York at Buffalo, Graduate School, College of Arts and Sciences, Department of Chemistry, Buffalo, NY 14260. Offers chemistry (MA, PhD); medicinal chemistry (MS, PhD). Part-time programs available. *Faculty:* 33 full-time (3 women), 5 part-time/adjunct (2 women). *Students:* 129 full-time (46 women), 13 part-time (4 women); includes 15 minority (7 African Americans, 2 American Indian/Alaska Native, 1 Asian American or Pacific Islander, 5 Hispanic Americans), 43 international. Average age 24. 383 applicants, 10% accepted, 27 enrolled. In 2005, 7 master's, 20 doctorates awarded. Terminal master's awarded for partial completion of doctoral program. *Median time to degree:* Of those who began their doctoral program in fall 1997, 100% received their degree in 8 years or less. *Degree requirements:* For master's, thesis or alternative, project; for doctorate, thesis/dissertation. *Entrance requirements:* For master's and doctorate, GRE General Test, GRE Subject Test. Additional exam requirements/recommendations for international students: Required—TOEFL. *Application deadline:* For fall admission, 3/1 priority date for domestic students, 3/1 priority date for international students. Applications are processed on a rolling basis. Application fee: $0 ($35 for international students). Electronic applications accepted. *Financial support:* In 2005–06, 10 fellowships with full tuition reimbursements, 50 research assistantships with full tuition reimbursements, 75 teaching assistantships with full tuition reimbursements were awarded; Federal Work-Study, institutionally sponsored loans, and unspecified assistantships also available. Financial award application deadline: 6/15; financial award applicants required to submit FAFSA. *Faculty research:* Synthesis, materials, environmental, analytical bio-organic, protein structure. Total annual research expenditures: $9.3 million. *Unit head:* Dr. Jim D. Atwood, Chairman, 716-645-6800 Ext. 2015, Fax: 716-645-6963, E-mail: chechair@buffalo.edu. *Application contact:* Dr. Huw M. L. Davies, Director of Graduate Studies, 716-645-6800 Ext. 2030, Fax: 716-645-6963, E-mail: hdavies@buffalo.edu.

State University of New York at New Paltz, Graduate School, School of Science and Engineering, Department of Chemistry, New Paltz, NY 12561. Offers MA, MAT, MS Ed. Part-time and evening/weekend programs available. *Faculty:* 6 full-time (3 women), 2 part-time/adjunct (0 women). In 2005, 1 degree awarded. *Degree requirements:* For master's, thesis. *Entrance requirements:* For master's, GRE General Test, minimum GPA of 3.0. Additional exam requirements/recommendations for international students: Required—TOEFL (minimum score 550 paper-based; 213 computer-based). *Application deadline:* For fall admission, 5/15 priority date for domestic students, 5/15 priority date for international students; for spring admission, 11/15 for domestic students, 11/15 for international students. Applications are processed on a rolling basis. Application fee: $50. *Expenses:* Tuition, state resident: full-time $3,450; part-time $288 per credit hour. Tuition, nonresident: full-time $3,550; part-time $455 per credit hour. Required fees: $27 per credit. $130 per semester. *Financial support:* Federal Work-Study and institutionally sponsored loans available. *Unit head:* Dr. Daniel Freedman, Chair, 845-257-3790. *Application contact:* Dr. Preeti Dhar, Coordinator, 845-257-3797, E-mail: dharp@newpaltz.edu.

State University of New York at Oswego, Graduate Studies, College of Arts and Sciences, Department of Chemistry, Oswego, NY 13126. Offers MS. Part-time programs available. *Faculty:* 4 full-time. *Students:* 8 full-time (3 women), 2 international. Average age 25. 10 applicants, 100% accepted. *Degree requirements:* For master's, thesis, comprehensive exam. *Entrance requirements:* For master's, GRE General Test, GRE Subject Test, BA or BS in chemistry. Additional exam requirements/recommendations for international students: Required—TOEFL (minimum score 550 paper-based; 213 computer-based). *Application deadline:* For fall admission, 4/1 for domestic students; for spring admission, 10/1 for domestic students. Applications are processed on a rolling basis. Application fee: $50. *Expenses:* Tuition, state resident: full-time $6,900; part-time $288 per credit. Tuition, nonresident: full-time $10,920; part-time $455 per credit. Tuition and fees vary according to program. *Financial support:* In 2005–06, 7 students received support, including 7 teaching assistantships with full tuition reimbursements available (averaging $11,000 per year); career-related internships or fieldwork, Federal Work-Study, institutionally sponsored loans, scholarships/grants, health care benefits, and unspecified assistantships also available. Support available to part-time students. Financial award application deadline: 4/1; financial award applicants required to submit FAFSA. *Unit head:* Dr. Kenneth Hyde, Chair, 315-312-3048. *Application contact:* Dr. Joseph Leferre, Graduate Coordinator, 315-312-3048.

State University of New York College of Environmental Science and Forestry, Faculty of Chemistry, Syracuse, NY 13210-2779. Offers biochemistry (MS, PhD); environmental and forest chemistry (MS, PhD); organic chemistry of natural products (MS, PhD); polymer chemistry (MS, PhD). *Faculty:* 14 full-time (1 woman). *Students:* 28 full-time (15 women), 8 part-time (2 women); includes 1 minority (American Indian/Alaska Native), 11 international. Average age 28. 23 applicants, 74% accepted, 2 enrolled. In 2005, 2 master's, 5 doctorates awarded. *Degree requirements:* For master's, thesis/dissertation, registration; for doctorate, thesis/dissertation, comprehensive exam, registration. *Entrance requirements:* For master's and doctorate, GRE General Test, GRE Subject Test, minimum GPA of 3.0. Additional exam requirements/recommendations for international students: Required—TOEFL (minimum score 550 paper-based; 213 computer-based). *Application deadline:* For fall admission, 2/1 priority date for domestic students, 2/1 priority date for international students; for spring admission, 11/1 priority date for domestic students, 11/1 priority date for international students. Applications are processed on a rolling basis. Application fee: $60. Electronic applications accepted. *Expenses:* Tuition, area resident: Full-time $6,900; part-time $288 per credit. Tuition, state resident: full-time $10,920. Tuition, nonresident: full-time $10,920; part-time $455 per credit. International tuition: $10,920 full-time. Required fees: $395; $32 per credit. $20 per term. One-time fee: $145. *Financial support:* In 2005–06, 35 students received support, including 3 fellowships with full tuition reimbursements available (averaging $9,446 per year), 13 research assistantships with full tuition reimbursements available (averaging $12,500 per year), 10 teaching assistantships with full tuition reimbursements available (averaging $11,540 per year); Federal Work-Study, institutionally sponsored loans, scholarships/grants, health care benefits, and unspecified assistantships also available. Financial award application deadline: 6/30; financial award applicants required to submit FAFSA. *Faculty research:* Polymer chemistry, biochemistry. Total annual research expenditures: $1.8 million. *Unit head:* Dr. John P. Hassett, Chair, 315-470-6827, Fax: 315-470-6856, E-mail: jphasset@syr.edu. *Application contact:* Dr. Dudley J. Raynal, Dean, Instruction and Graduate Studies, 315-470-6599, Fax: 315-470-6978, E-mail: esfgrad@esf.edu.

State University of New York, Fredonia, Graduate Studies, Department of Chemistry and Biochemistry, Fredonia, NY 14063-1136. Offers chemistry (MS); curriculum and instruction science education (MS Ed). Part-time and evening/weekend programs available. *Faculty:* 1 full-time (0 women). *Students:* 1 (woman) full-time. Average age 24. In 2005, 1 degree awarded. *Degree requirements:* For master's, thesis optional. *Application deadline:* For fall admission, 8/5 for domestic students; for spring admission, 12/1 for domestic students. Application fee: $50. *Expenses:* Tuition, state resident: full-time $3,456; part-time $288 per credit hour. Tuition, nonresident: full-time $5,460; part-time $455 per credit hour. Required fees: $543; $45 per credit hour. *Financial support:* Research assistantships, teaching assistantships with partial tuition reimbursements, tuition waivers (full and partial) available. Support available to part-time students. Financial award application deadline: 3/15. *Unit head:* Dr. Thomas Janik, Chairman, 716-673-3281, E-mail: thomas.janik@fredonia.edu.

Stephen F. Austin State University, Graduate School, College of Sciences and Mathematics, Department of Chemistry, Nacogdoches, TX 75962. Offers MS. Part-time programs available. *Faculty:* 10 full-time (5 women), 37 part-time/adjunct (12 women). *Students:* 2 full-time (0 women), 1 international. Average age 25. 2 applicants, 100% accepted. *Degree requirements:* For master's, comprehensive exam. *Entrance requirements:* For master's, GRE General Test, minimum GPA of 2.8 in last 60 hours, 2.5 overall. Additional exam requirements/recommendations for international students: Required—TOEFL. *Application deadline:* For fall admission, 8/1 for domestic students; for spring admission, 12/15 for domestic students. Applications are processed on a rolling basis. Application fee: $0 ($50 for international students). *Expenses:* Tuition, state resident: full-time $2,628; part-time $146 per credit hour. Tuition, nonresident: full-time $7,596; part-time $422 per credit hour. International tuition: $7,596 full-time. Required fees: $900; $170. *Financial support:* In 2005–06, 2 teaching assistantships (averaging $8,100 per year) were awarded; fellowships, research assistantships, Federal Work-Study, institutionally sponsored loans, and unspecified assistantships also available. Support available to part-time students. Financial award application deadline: 3/1. *Faculty research:* Synthesis and chemistry of ferrate ion, properties of fluoroberyllates, polymer chemistry. *Unit head:* Dr. Michael Janusa, Chair, 936-468-3606, E-mail: janusama@sfasu.edu.

Stevens Institute of Technology, Graduate School, Arthur E. Imperatore School of Sciences and Arts, Department of Chemistry and Chemical Biology, Hoboken, NJ 07030. Offers chemical biology (MS, PhD); chemistry (MS, PhD), including analytical chemistry (MS, PhD, Certificate), chemical biology (MS, PhD, Certificate), organic chemistry, physical chemistry, polymer chemistry (MS, PhD, Certificate); chemistry and chemical biology (Certificate), including analytical chemistry (MS, PhD, Certificate), bioinformatics, biomedical chemistry, chemical biology (MS, PhD, Certificate), chemical physiology, polymer chemistry (MS, PhD, Certificate). Part-time and evening/weekend programs available. *Students:* 17 full-time (8 women), 28 part-time (16 women); includes 12 minority (9 Asian Americans or Pacific Islanders, 3 Hispanic Americans), 14 international. 51 applicants, 75% accepted.Terminal master's awarded for partial completion of doctoral program. *Degree requirements:* For master's, thesis or alternative; for doctorate, one foreign language, thesis/dissertation; for Certificate, project or thesis. *Entrance requirements:* Additional exam requirements/recommendations for international students: Required—TOEFL. *Application deadline:* Applications are processed on a rolling basis. Application fee: $50. Electronic applications accepted. *Expenses:* Tuition: Part-time $920 per credit hour. Tuition and fees vary according to program. *Financial support:* Fellowships, research assistantships, teaching assistantships, Federal Work-Study and institutionally sponsored loans available. Financial award application deadline: 4/1. *Unit head:* Dr. Francis T. Jones, Director, 201-216-5518, Fax: 201-216-8240.

Stony Brook University, State University of New York, Graduate School, College of Arts and Sciences, Department of Chemistry, Stony Brook, NY 11794. Offers MAT, MS, PhD. MAT offered through the School of Professional Development and Continuing Studies. *Faculty:* 26 full-time (5 women). *Students:* 150 full-time (67 women), 2 part-time (1 woman); includes 14 minority (2 African Americans, 11 Asian Americans or Pacific Islanders, 1 Hispanic American), 104 international. Average age 28. 312 applicants, 26% accepted. In 2005, 19 master's, 14 doctorates awarded. Terminal master's awarded for partial completion of doctoral program. *Degree requirements:* For master's, thesis; for doctorate, one foreign language, thesis/dissertation. *Entrance requirements:* For master's and doctorate, GRE General Test. Additional exam requirements/recommendations for international students: Required—TOEFL. *Application deadline:* For fall admission, 1/15 for domestic students. Application fee: $50. *Expenses:* Tuition, area resident: Full-time $6,900; part-time $288. Tuition, state resident: full-time $6,900. Tuition, nonresident: full-time $10,920; part-time $455. International tuition: $10,920 full-time. Required fees: $704. *Financial support:* In 2005–06, 8 fellowships, 68 research assistantships, 56 teaching assistantships were awarded. Total annual research expenditures: $7.7 million. *Unit head:* Dr. Michael White, Chairman, 631-632-7880, Fax: 631-632-7960. *Application contact:* Dr. Scott Sieburth, Director, 631-632-7851, Fax: 631-632-7960, E-mail: ssieburth@notes.cc.sunysb.edu.

Announcement: Announcing Stony Brook's ACES Project, funded by the Dreyfus Foundation. ACES provides workshops and seminars for graduate and postdoctoral students, focusing on development of the nonlaboratory skills that enable career success. Activities are coordinated with Project WISE, the GAANN Fellowship Program, and the Center for Excellence in Learning and Teaching.

See Close-Up on page 147.

Sul Ross State University, School of Arts and Sciences, Department of Geology and Chemistry, Alpine, TX 79832. Offers MS. Part-time programs available. *Degree requirements:* For master's, thesis optional. *Entrance requirements:* For master's, GRE General Test, minimum GPA of 2.5 in last 60 hours of undergraduate work.

Syracuse University, Graduate School, College of Arts and Sciences, Department of Chemistry, Syracuse, NY 13244. Offers MS, PhD. *Students:* 58 full-time (25 women), 3 part-time; includes 8 minority (2 African Americans, 3 Asian Americans or Pacific Islanders, 3 Hispanic Americans), 25 international. 93 applicants, 62% accepted, 14 enrolled. *Degree requirements:* For master's, one foreign language, thesis (for some programs), comprehensive exam, registration; for doctorate, one foreign language, thesis/dissertation, comprehensive exam, registration. *Entrance requirements:* For master's and doctorate, GRE General Test. Additional exam requirements/recommendations for international students: Required—TOEFL. *Application deadline:* For fall admission, 1/10 for domestic students. Applications are processed on a rolling basis. Application fee: $65. Electronic applications accepted. *Financial support:* Fellowships with full tuition reimbursements, research assistantships with full and partial tuition reimbursements, teaching assistantships with full tuition reimbursements, tuition waivers (full) available. *Faculty research:* Synthetic organic chemistry, biophysical spectroscopy, solid state in organic chemistry, biochemistry, organometallic chemistry. *Unit head:* Dr. Jon Zubieta, Director, 315-443-2547, Fax: 315-443-4070. *Application contact:* Joyce Lagoe, 315-443-4109, E-mail: jalagoe@syr.edu.

Temple University, Graduate School, College of Science and Technology, Department of Chemistry, Philadelphia, PA 19122-6096. Offers MA, PhD. Evening/weekend programs available. *Faculty:* 13 full-time (3 women). *Students:* 1 full-time (0 women), 56 part-time (17 women); includes 6 minority (2 African Americans, 3 Asian Americans or Pacific Islanders, 1 Hispanic American), 37 international. 100 applicants, 25% accepted, 8 enrolled. In 2005, 8 master's, 2 doctorates awarded. Terminal master's awarded for partial completion of doctoral program. *Degree requirements:* For master's, thesis (for some programs); for doctorate, thesis/dissertation, teaching experience. *Entrance requirements:* For master's and doctorate, GRE General Test, minimum GPA of 3.0. Additional exam requirements/recommendations for international students: Required—TOEFL (minimum score 575 paper-based; 230 computer-based). *Application deadline:* For fall admission, 2/15 for domestic students, 12/15 for international students; for spring admission, 9/15 for domestic students, 8/1 for international students. Applications are processed on a rolling basis. Application fee: $50. Electronic applications accepted. *Expenses:* Tuition, state resident: full-time $8,694; part-time $483 per credit. Tuition, nonresident: full-time $12,672; part-time $704 per credit. International tuition: $12,672 full-time. Required fees: $500; $122 per semester. Tuition and fees vary according to course level, campus/location and program. *Financial support:* Fellowships, research assistantships, teaching assistantships available. Financial award application deadline: 1/15; financial award applicants required to submit FAFSA. *Faculty research:* Polymers, nonlinear optics, natural products, materials science, enantioselective synthesis. Total annual research expenditures: $691,921. *Unit head:* Dr. Robert Levis, Chair, 215-204-5241, Fax: 215-204-1532, E-mail: rjlevis@temple.edu.

Tennessee State University, Graduate School, College of Arts and Sciences, Department of Chemistry, Nashville, TN 37209-1561. Offers MS. *Faculty:* 5 full-time (1 woman). *Students:* 6 full-time (2 women), 2 part-time (1 woman); all minorities (6 African Americans, 1 Asian American or Pacific Islander, 1 Hispanic American), 2 international. Average age 28. 13 applicants, 38% accepted. In 2005, 3 degrees awarded. *Degree requirements:* For master's, thesis. *Entrance requirements:* For master's, GRE General Test, GRE Subject Test, minimum GPA of 3.0, BS in engineering or science. *Application deadline:* Applications are processed on a rolling basis. Application fee: $15. *Financial support:* In 2005–06, 7 teaching assistantships (averaging $19,501 per year) were awarded; research assistantships, unspecified assistantships also available. Financial award application deadline: 5/1. *Faculty research:* Binding benzol pyrenemetabolites to DNA. *Unit head:* Dr. Carlos L. Lee, Head, 615-963-5004.

Tennessee Technological University, Graduate School, College of Arts and Sciences, Department of Chemistry, Cookeville, TN 38505. Offers MS. Part-time programs available. *Faculty:* 16 full-time (1 woman). *Students:* 7 full-time (6 women), 5 part-time (3 women); includes 5 minority (all Asian Americans or Pacific Islanders) Average age 28. 21 applicants, 57% accepted, 4 enrolled. In 2005, 5 degrees awarded. *Degree requirements:* For master's, thesis. *Entrance requirements:* For master's, GRE General Test. Additional exam requirements/recommendations for international students: Required—TOEFL. *Application deadline:* For fall admission, 3/1 for domestic students; for spring admission, 8/1 for domestic students. Application fee: $25 ($30 for international students). *Expenses:* Tuition, state resident: full-time $8,421; part-time $307 per hour. Tuition, nonresident: full-time $22,389; part-time $711 per hour. *Financial support:* In 2005–06, 1 research assistantship (averaging $10,000 per year), 6 teaching assistantships (averaging $7,500 per year) were awarded; career-related internships or fieldwork also available. Financial award application deadline: 4/1. *Unit head:* Dr. Jeffrey Boles, Interim Chairperson, 931-372-3421, Fax: 931-372-3434, E-mail: jboles@tntech.edu. *Application contact:* Dr. Francis O. Otuonye, Associate Vice President for Research and Graduate Studies, 931-372-3233, Fax: 931-372-3497, E-mail: fotuonye@tntech.edu.

Texas A&M University, College of Science, Department of Chemistry, College Station, TX 77843. Offers MS, PhD. *Faculty:* 40 full-time (4 women). *Students:* 278 full-time (101 women), 10 part-time (6 women); includes 46 minority (6 African Americans, 2 American Indian/Alaska Native, 14 Asian Americans or Pacific Islanders, 24 Hispanic Americans), 122 international. Average age 24. 309 applicants, 42% accepted, 61 enrolled. In 2005, 10 master's, 28 doctorates awarded. Terminal master's awarded for partial completion of doctoral program. *Degree requirements:* For master's and doctorate, thesis/dissertation. *Entrance requirements:* For master's and doctorate, GRE General Test. Additional exam requirements/recommendations for international students: Required—TOEFL; Recommended—TSE. *Application deadline:* For fall admission, 3/1 for domestic students. Applications are processed on a rolling basis. Application fee: $75 for international students. Electronic applications accepted. *Expenses:* Tuition, state resident: full-time $4,488; part-time $187 per credit hour. Tuition, nonresident: full-time $11,112; part-time $463 per credit hour. Required fees: $1,974. *Financial support:* In 2005–06, fellowships with full tuition reimbursements (averaging $21,600 per year), research assistantships with full tuition reimbursements (averaging $18,600 per year), teaching assistantships with full tuition reimbursements (averaging $18,600 per year) were awarded. Financial award application deadline: 3/1; financial award applicants required to submit FAFSA. *Faculty research:* Biological chemistry, spectroscopy, structure and bonding, reactions and mechanisms, theoretical chemistry. *Unit head:* Dr. Emile A. Schweikert, Head, 979-845-2011, Fax: 979-845-4719. *Application contact:* Dr. Michael P. Rosynek, Graduate Advisor, 979-845-5345, Fax: 979-854-5211, E-mail: gradmail@mail.chem.tamu.edu.

See Close-Up on page 149.

Texas A&M University–Commerce, Graduate School, College of Arts and Sciences, Department of Chemistry, Commerce, TX 75429-3011. Offers M Ed, MS. Part-time programs available. *Faculty:* 6 full-time (0 women). *Students:* 7 full-time (2 women), 2 part-time (1 woman); includes 4 minority (2 Asian Americans or Pacific Islanders, 2 Hispanic Americans), 2 international. Average age 36. In 2005, 1 degree awarded. *Degree requirements:* For master's, thesis (for some programs), comprehensive exam. *Entrance requirements:* For master's, GRE General Test. *Application deadline:* For fall admission, 6/1 for domestic students; for spring admission, 11/1 priority date for domestic students. Applications are processed on a rolling basis. Application fee: $0 ($25 for international students). Electronic applications accepted. *Financial support:* In 2005–06, research assistantships (averaging $7,875 per year), teaching assistantships (averaging $7,875 per year) were awarded; Federal Work-Study, institutionally sponsored loans, and scholarships/grants also available. Financial award application deadline: 5/1; financial award applicants required to submit FAFSA. *Faculty research:* Analytical organic. Total annual research expenditures: $65,000. *Unit head:* Dr. Don R. Lee, Interim Head, 903-886-5378, Fax: 903-886-5997. *Application contact:* Tammi Thompson, Graduate Admissions Adviser, 843-886-5167, Fax: 843-886-5165, E-mail: tammi_thompson@tamu-commerce.edu.

Texas A&M University–Kingsville, College of Graduate Studies, College of Arts and Sciences, Department of Chemistry, Kingsville, TX 78363. Offers MS. Part-time programs available. *Degree requirements:* For master's, thesis or alternative, comprehensive exam. *Entrance requirements:* For master's, GRE General Test, minimum GPA of 3.0. Additional exam requirements/recommendations for international students: Required—TOEFL. *Faculty research:* Organic heterocycles, amino alcohol complexes, rare earth arsine complexes.

Texas Christian University, College of Science and Engineering, Department of Chemistry, Fort Worth, TX 76129-0002. Offers MA, MS, PhD. Part-time and evening/weekend programs available. *Degree requirements:* For master's, one foreign language, thesis optional; for doctorate, one foreign language, thesis/dissertation, cumulative exams. *Entrance requirements:* For master's and doctorate, GRE General Test. Additional exam requirements/recommendations for international students: Required—TOEFL. *Application deadline:* For fall admission, 3/1 for domestic students; for spring admission, 12/1 for domestic students. Applications are processed on a rolling basis. Application fee: $0. *Expenses:* Tuition: Part-time $740 per credit hour. *Financial support:* Fellowships, teaching assistantships, unspecified assistantships available. Financial award application deadline:3/1. *Unit head:* Dr. Jeffrey Coffer, Chairperson, 817-257-7195. *Application contact:* Dr. Bonnie Melhart, Associate Dean, College of Science and Engineering, E-mail: b.melhart@tcu.edu.

Texas Southern University, Graduate School, School of Science and Technology, Department of Chemistry, Houston, TX 77004-4584. Offers MS. *Faculty:* 3 full-time (0 women), 1 part-time/adjunct (0 women). *Students:* 8 full-time (3 women), 4 part-time (2 women); includes 11 minority (8 African Americans, 3 Asian Americans or Pacific Islanders), 1 international. Average age 33. 2 applicants, 100% accepted, 2 enrolled. In 2005, 2 degrees awarded. *Degree requirements:* For master's, one foreign language, thesis, comprehensive exam. *Entrance requirements:* For master's, GRE General Test, minimum GPA of 2.5. Additional exam requirements/recommendations for international students: Required—TOEFL. *Application deadline:* For fall admission, 7/15 for domestic students. Applications are processed on a rolling basis. Application fee: $50 ($75 for international students). *Expenses:* Tuition, area resident: Full-time $1,728; part-time $1,152 per credit hour. Tuition, state resident: full-time $1,728; part-time $1,152 per credit hour. Tuition, nonresident: full-time $6,174; part-time $4,116 per credit hour. International tuition: $6,174 full-time. Required fees: $2,122. Tuition and fees vary according to course load and degree level. *Financial support:* In 2005–06, 2 research assistantships (averaging $9,000 per year) were awarded; fellowships, teaching assistantships, institutionally sponsored loans also available. Financial award application deadline: 5/1. *Faculty research:* Analytical and physical chemistry, geochemistry, inorganic chemistry, biochemistry, organic chemistry. *Unit head:* Dr. John Sapp, Head, 713-313-7003, Fax: 713-313-7824, E-mail: sapp_jb@tsu.edu.

Texas State University-San Marcos, Graduate School, College of Science, Department of Chemistry and Biochemistry, Program in Chemistry, San Marcos, TX 78666. Offers MA, MS. *Students:* 8 full-time (4 women), 6 part-time; includes 7 minority (1 African American, 2 Asian Americans or Pacific Islanders, 4 Hispanic Americans), 1 international. Average age 29. 9 applicants, 89% accepted, 6 enrolled. In 2005, 3 degrees awarded. *Degree requirements:* For master's, thesis (for some programs), comprehensive exam. *Entrance requirements:* For master's, GRE General Test, minimum GPA of 2.75 in last 60 hours of course work. Additional exam requirements/recommendations for international students: Required—TOEFL. *Application deadline:* For fall admission, 6/15 for domestic students, 6/1 for international students; for spring admission, 10/15 for domestic students, 10/1 for international students. Applications are processed on a rolling basis. Application fee: $40 ($90 for international students). *Expenses:* Tuition, area resident: Full-time $3,168; part-time $116 per credit. Tuition, state resident: full-time $3,168; part-time $176 per credit. Tuition, nonresident: full-time $8,136; part-time $452 per credit. International tuition: $8,136 full-time. Required fees: $1,112; $74 per credit. Full-time tuition and fees vary according to course load. *Financial support:* In 2005–06, 12 students received support; research assistantships, teaching assistantships, career-related internships or fieldwork, Federal Work-Study, and institutionally sponsored loans available. Support available to part-time students. Financial award application deadline: 4/1; financial award applicants required to submit FAFSA. *Faculty research:* Metal ions in biological systems, cancer chemotherapy, absorption of pesticides on solid surfaces, polymer chemistry, biochemistry of nucleic acids. *Application contact:* Dr. J. Michael Willoughby, Dean of Graduate School, 512-245-2581, Fax: 512-245-8365, E-mail: gradcollege@txstate.edu.

Texas Tech University, Graduate School, College of Arts and Sciences, Department of Chemistry and Biochemistry, Lubbock, TX 79409. Offers chemistry (MS, PhD). Part-time programs available. *Faculty:* 22 full-time (1 woman). *Students:* 80 full-time (29 women), 4 part-time (1 woman); includes 3 minority (1 Asian American or Pacific Islander, 2 Hispanic Americans), 63 international. Average age 28. 85 applicants, 55% accepted, 25 enrolled. In 2005, 7 master's, 4 doctorates awarded. *Degree requirements:* For master's and doctorate, thesis/dissertation. *Entrance requirements:* For master's and doctorate, GRE General Test. Additional exam requirements/recommendations for international students: Required—TOEFL (minimum score 550 paper-based; 213 computer-based). *Application deadline:* Applications are processed on a rolling basis. Application fee: $50 ($60 for international students). Electronic applications accepted. *Expenses:* Tuition, area resident: Full-time $4,296. Tuition, state resident: full-time $4,296. Tuition, nonresident: full-time $10,920. International tuition: $10,920 full-time. Required fees: $1,992. Tuition and fees vary according to program. *Financial support:* In 2005–06, 14 students received support, including 27 research assistantships with partial tuition reimbursements available (averaging $16,049 per year), 47 teaching assistantships with partial tuition reimbursements available (averaging $15,932 per year); career-related internships or fieldwork, Federal Work-Study, and institutionally sponsored loans also available. Support available to part-time students. Financial award application deadline: 4/15; financial award applicants required to submit FAFSA. *Faculty research:* Ionic and molecular recognition with synthetic host molecules, ultratrace atmospheric analysis, plant biochemistry and phytochemical signaling, theoretical and computational chemistry, synthesis and applications of ionic liquids. Total annual research expenditures: $3.3 million. *Unit head:* Dr. Dominick J. Casadone, Chair, 806-742-3067, Fax: 806-742-1289, E-mail: chemchair@ttu.edu. *Application contact:* Kathy Jones, Assistant Advisor, 806-742-1844, Fax: 806-742-1289, E-mail: kathy.jones@ttu.edu.

Texas Woman's University, Graduate School, College of Arts and Sciences, Department of Chemistry and Physics, Denton, TX 76201. Offers chemistry (MS); chemistry teaching (MS); science teaching (MS). Part-time programs available. *Students:* 1 (woman) full-time. Average age 24. In 2005, 2 degrees awarded. *Degree requirements:* For master's, thesis, comprehensive exam. *Entrance requirements:* For master's, GRE General Test, bachelor's degree in chemistry or equivalent, 2 reference contacts. Additional exam requirements/recommendations for international students: Required—TOEFL (minimum score 550 paper-based; 213 computer-based). *Application deadline:* Applications are processed on a rolling basis. Application fee: $30 ($50 for international students). Electronic applications accepted. *Expenses:* Tuition, state resident: full-time $2,934; part-time $163. Tuition, nonresident: full-time $7,974; part-time $152. International tuition: $7,974 full-time. *Financial support:* In 2005–06, research assistantships (averaging $9,378 per year), teaching assistantships (averaging $9,378 per year) were awarded; career-related internships or fieldwork, Federal Work-Study, institutionally sponsored loans, scholarships/grants, traineeships, health care benefits, and unspecified assistantships also available. Support available to part-time students. Financial award application deadline: 3/1; financial award applicants required to submit FAFSA. *Faculty research:* Mechanisms and kinetics of organic reactions, mechanisms of enzyme catalysis, chelation chemistry of macrocyclic ligands. *Unit head:* Dr. Jack Gill, Interim Chair, 940-898-2550, Fax: 940-898-2548, E-mail: jgill@mail.twu.edu. *Application contact:* Samuel Wheeler, Coordinator of Graduate Admissions, 940-898-3188, Fax: 940-898-3081, E-mail: wheelersr@twu.edu.

Trent University, Graduate Studies, Program in Applications of Modeling in the Natural and Social Sciences, Department of Chemistry, Peterborough, ON K9J 7B8, Canada. Offers M Sc.

Chemistry

Trent University (continued)
Part-time programs available. *Degree requirements:* For master's, thesis. *Entrance requirements:* For master's, honours degree. *Faculty research:* Synthetic-organic chemistry, mass spectrometry and ion storage.

Tufts University, Graduate School of Arts and Sciences, Department of Chemistry, Medford, MA 02155. Offers analytical chemistry (MS, PhD); bioorganic chemistry (MS, PhD); environmental chemistry (MS, PhD); inorganic chemistry (MS, PhD); organic chemistry (MS, PhD); physical chemistry (MS, PhD). *Faculty:* 17 full-time. *Students:* 55 (27 women); includes 5 minority (1 African American, 4 Asian Americans or Pacific Islanders) 27 international. 45 applicants, 87% accepted, 17 enrolled. In 2005, 4 master's, 4 doctorates awarded. Terminal master's awarded for partial completion of doctoral program. *Degree requirements:* For master's and doctorate, thesis/dissertation. *Entrance requirements:* For master's and doctorate, GRE General Test, GRE Subject Test. Additional exam requirements/recommendations for international students: Required—TOEFL (minimum score 600 paper-based; 250 computer-based), TSE. *Application deadline:* For fall admission, 1/15 for domestic students, 12/30 for international students; for spring admission, 10/15 for domestic students, 9/15 for international students. Applications are processed on a rolling basis. Application fee: $65. Electronic applications accepted. *Expenses:* Tuition: Full-time $32,360. Tuition and fees vary according to program. *Financial support:* Research assistantships with full and partial tuition reimbursements, teaching assistantships with full and partial tuition reimbursements, Federal Work-Study, scholarships/grants, and tuition waivers (partial) available. Financial award application deadline: 1/15; financial award applicants required to submit FAFSA. *Unit head:* Mary Jane Shultz, Chair, 617-627-3477, Fax: 617-627-3443. *Application contact:* Samuel Kounaves, Information Contact, 617-627-3441, Fax: 617-627-3443.

See Close-Up on page 151.

Tulane University, Graduate School, Department of Chemistry, New Orleans, LA 70118-5669. Offers MA, MS, PhD. Terminal master's awarded for partial completion of doctoral program. *Degree requirements:* For master's and doctorate, thesis/dissertation. *Entrance requirements:* For master's, GRE General Test, minimum B average in undergraduate course work; for doctorate, GRE General Test. Additional exam requirements/recommendations for international students: Required—TOEFL; Recommended—TSE. Electronic applications accepted. *Faculty research:* Enzyme mechanisms, organic synthesis, photochemistry, theory of polymer dynamics.

Tuskegee University, Graduate Programs, College of Agricultural, Environmental and Natural Sciences, Department of Chemistry, Tuskegee, AL 36088. Offers MS. *Faculty:* 6 full-time (1 woman). *Students:* 5 full-time (3 women), 2 part-time; includes 4 minority (all African Americans), 2 international. Average age 31. In 2005, 1 degree awarded. *Degree requirements:* For master's, thesis. *Entrance requirements:* For master's, GRE General Test. Additional exam requirements/recommendations for international students: Required—TOEFL (minimum score 500 paper-based; 173 computer-based). *Application deadline:* For fall admission, 7/15 for domestic students. Applications are processed on a rolling basis. Application fee: $25 ($35 for international students). *Expenses:* Tuition: Full-time $12,400. Required fees: $300; $490 per credit. *Financial support:* Fellowships, teaching assistantships, Federal Work-Study and institutionally sponsored loans available. Support available to part-time students. Financial award application deadline: 4/15. *Unit head:* Dr. Gregory Pritchett, Head, 334-727-8836.

Université de Moncton, Faculty of Science, Department of Chemistry and Biochemistry, Moncton, NB E1A 3E9, Canada. Offers biochemistry (M Sc); chemistry (M Sc). Part-time programs available. *Degree requirements:* For master's, one foreign language, thesis. *Entrance requirements:* For master's, minimum GPA of 3.0. Electronic applications accepted. *Faculty research:* Environmental contaminants, natural products synthesis, nutraceutical, organic catalysis, molecular biology of cancer.

Université de Montréal, Faculty of Graduate Studies, Faculty of Arts and Sciences, Department of Chemistry, Montréal, QC H3C 3J7, Canada. Offers M Sc, PhD. *Faculty:* 41 full-time (8 women). *Students:* 175 full-time (60 women). 97 applicants, 32% accepted, 30 enrolled. In 2005, 22 master's, 10 doctorates awarded. *Degree requirements:* For master's, thesis; for doctorate, thesis/dissertation, general exam. *Entrance requirements:* For master's, B Sc in chemistry or the equivalent; for doctorate, M Sc in chemistry or equivalent. *Application deadline:* For fall and spring admission, 2/1For winter admission, 11/1 for domestic students. Application fee: $30. Electronic applications accepted. *Faculty research:* Analytical, inorganic, physical, and organic chemistry. *Unit head:* Robert Prud'homme, Chairman, 514-343-6730, Fax: 514-343-7586. *Application contact:* Andre Beauchamp, Professor, 514-343-6446, Fax: 514-343-7586, E-mail: andre.beauchamp@umontreal.ca.

Université de Sherbrooke, Faculty of Sciences, Department of Chemistry, Sherbrooke, QC J1K 2R1, Canada. Offers M Sc, PhD, Diploma. *Faculty:* 16 full-time (1 woman), 1 part-time/adjunct (0 women). *Students:* 57 full-time (15 women). 42 applicants, 29% accepted. In 2005, 10 master's, 9 doctorates awarded. *Degree requirements:* For master's and doctorate, thesis/dissertation. *Entrance requirements:* For doctorate, master's degree. *Application deadline:* For fall admission, 6/30 for domestic students. Applications are processed on a rolling basis. Application fee: $50. Electronic applications accepted. *Financial support:* Fellowships, research assistantships, teaching assistantships available. *Faculty research:* Organic, electro-, theoretical, and physical chemistry. *Unit head:* Dr. Andre Bandrauk, Chairman, 819-821-7088, Fax: 819-821-8017, E-mail: andre.bandrauk@usherbrooke.ca.

Université du Québec à Montréal, Graduate Programs, Program in Chemistry, Montréal, QC H3C 3P8, Canada. Offers M Sc. Part-time programs available. *Degree requirements:* For master's, thesis. *Entrance requirements:* For master's, appropriate bachelor's degree or equivalent and proficiency in French.

Université du Québec à Trois-Rivières, Graduate Programs, Program in Chemistry, Trois-Rivières, QC G9A 5H7, Canada. Offers M Sc. Part-time programs available. *Degree requirements:* For master's, thesis. *Entrance requirements:* For master's, appropriate bachelor's degree, proficiency in French.

Université Laval, Faculty of Sciences and Engineering, Department of Chemistry, Programs in Chemistry, Québec, QC G1K 7P4, Canada. Offers M Sc, PhD. Part-time programs available. Terminal master's awarded for partial completion of doctoral program. *Degree requirements:* For master's, thesis/dissertation; for doctorate, thesis/dissertation, comprehensive exam. *Entrance requirements:* For master's and doctorate, knowledge of French, comprehension of written English. Electronic applications accepted.

University at Albany, State University of New York, College of Arts and Sciences, Department of Chemistry, Albany, NY 12222-0001. Offers MS, PhD. *Students:* 17 full-time (6 women), 15 part-time (6 women). Average age 28. In 2005, 2 master's, 3 doctorates awarded. *Degree requirements:* For master's, one foreign language, thesis, major field exam; for doctorate, 2 foreign languages, thesis/dissertation, cumulative exams, oral proposition. *Entrance requirements:* For doctorate, GRE. Additional exam requirements/recommendations for international students: Required—TOEFL (minimum score 550 paper-based; 213 computer-based). *Application deadline:* For fall admission, 6/1 for domestic students, 6/1 for international students. Applications are processed on a rolling basis. Application fee: $60. Electronic applications accepted. *Financial support:* Research assistantships, teaching assistantships, minority assistantships available. Financial award application deadline: 6/1. *Faculty research:* Synthetic, organic, and inorganic chemistry; polymer chemistry; ESR and NMR spectroscopy; theoretical chemistry; physical biochemistry. *Unit head:* Dr. John Welch, Chair, 518-442-4400.

University at Albany, State University of New York, School of Public Health, Department of Environmental Health and Toxicology, Albany, NY 12222-0001. Offers environmental and occupational health (MS, PhD); environmental chemistry (MS, PhD); toxicology (MS, PhD). *Students:* 24 full-time (15 women), 17 part-time (9 women). Average age 31. In 2005, 1 master's, 2 doctorates awarded. *Degree requirements:* For master's and doctorate, thesis/dissertation. *Entrance requirements:* For master's and doctorate, GRE General Test, GRE Subject Test. Additional exam requirements/recommendations for international students: Required—TOEFL (minimum score 550 paper-based; 213 computer-based). *Application deadline:* For fall admission, 1/1 for domestic students. Applications are processed on a rolling basis. Application fee: $60. Electronic applications accepted. *Financial support:* Fellowships, research assistantships available. Financial award application deadline: 2/1. *Unit head:* Dr. Laurence Kaminsky, Chair, 518-473-7553. *Application contact:* Caitlin Reid, Assistant to the Chair, E-mail: reid@wadsworth.org.

The University of Akron, Graduate School, Buchtel College of Arts and Sciences, Department of Chemistry, Akron, OH 44325. Offers MS, PhD. Part-time and evening/weekend programs available. *Faculty:* 18 full-time (2 women), 2 part-time/adjunct (0 women). *Students:* 69 full-time (36 women), 8 part-time (1 woman); includes 9 minority (1 African American, 3 Asian Americans or Pacific Islanders, 5 Hispanic Americans), 35 international. Average age 29. 37 applicants, 51% accepted, 8 enrolled. In 2005, 2 master's, 18 doctorates awarded. Terminal master's awarded for partial completion of doctoral program. *Degree requirements:* For master's, one foreign language, thesis, seminar presentation; for doctorate, 2 foreign languages, thesis/dissertation, cumulative exams, oral exam. *Entrance requirements:* For master's and doctorate, minimum GPA of 2.75. Additional exam requirements/recommendations for international students: Required—TOEFL (minimum score 550 paper-based; 213 computer-based), Michigan English Language Assessment Battery. *Application deadline:* For fall admission, 3/1 for domestic students. Applications are processed on a rolling basis. Application fee: $30 ($40 for international students). Electronic applications accepted. *Expenses:* Tuition, area resident: Full-time $5,816; part-time $323 per credit. Tuition, state resident: full-time $5,816; part-time $323 per credit. Tuition, nonresident: full-time $9,976; part-time $554 per credit. Required fees: $794; $43 per credit. $12 per term. Tuition and fees vary according to course load, degree level and program. *Financial support:* In 2005–06, 17 research assistantships with full tuition reimbursements, 43 teaching assistantships with full tuition reimbursements were awarded; fellowships with full tuition reimbursements, tuition waivers (full) also available. Support available to part-time students. Total annual research expenditures: $1.3 million. *Unit head:* Dr. Michael Taschner, Interim Chair, 330-972-6135, E-mail: mtaschner@uakron.edu.

The University of Alabama, Graduate School, College of Arts and Sciences, Department of Chemistry, Tuscaloosa, AL 35487. Offers MS, PhD. Postbaccalaureate distance learning degree programs offered (minimal on-campus study). *Faculty:* 19 full-time (1 woman). *Students:* 69 full-time (23 women), 13 part-time (5 women); includes 7 minority (5 African Americans, 1 Asian American or Pacific Islander, 1 Hispanic American), 43 international. Average age 26. 137 applicants, 18% accepted, 12 enrolled. In 2005, 4 master's, 12 doctorates awarded. *Median time to degree:* Of those who began their doctoral program in fall 1997, 80% received their degree in 8 years or less. *Degree requirements:* For master's, thesis (for some programs), registration; for doctorate, thesis/dissertation, exams, research proposal, oral defense, comprehensive exam, registration. *Entrance requirements:* For master's and doctorate, GRE General Test, MAT, minimum GPA of 3.0. *Application deadline:* For fall admission, 7/15 priority date for domestic students, 1/15 priority date for international students. Applications are processed on a rolling basis. Application fee: $25. Electronic applications accepted. *Expenses:* Tuition, area resident: Full-time $2,432. Tuition, nonresident: full-time $6,758. *Financial support:* In 2005–06, 5 fellowships with full tuition reimbursements (averaging $18,900 per year), 41 research assistantships with full tuition reimbursements (averaging $18,900 per year), 29 teaching assistantships with full tuition reimbursements (averaging $18,900 per year) were awarded; career-related internships or fieldwork, Federal Work-Study, and scholarships/grants also available. Financial award application deadline: 7/14. *Faculty research:* Synthetic chemistry, environmental chemistry and green manufacturing, materials science and information technology, biological chemistry and biomolecular products. Total annual research expenditures: $2.6 million. *Unit head:* Dr. Joseph S. Thrasher, Chair, 205-348-8436, Fax: 205-348-9104, E-mail: gradchem@bama.ua.edu. *Application contact:* Dr. Carolyn J. Cassady, Director, Graduate Recruiting, 205-348-8443, Fax: 205-348-9104, E-mail: ccassady@bama.ua.edu.

The University of Alabama at Birmingham, School of Natural Sciences and Mathematics, Department of Chemistry, Birmingham, AL 35294. Offers MS, PhD. *Faculty:* 15 full-time, 6 part-time/adjunct. *Students:* 20 full-time (6 women), 1 part-time; includes 2 minority (1 African American, 1 Asian American or Pacific Islander), 7 international. 36 applicants, 39% accepted. In 2005, 3 master's, 2 doctorates awarded. *Degree requirements:* For master's and doctorate, thesis/dissertation. *Entrance requirements:* For master's and doctorate, GRE General Test. Additional exam requirements/recommendations for international students: Required—TOEFL. *Application deadline:* Applications are processed on a rolling basis. Application fee: $35 ($60 for international students). *Expenses:* Tuition, state resident: part-time $170 per credit hour. Tuition, nonresident: full-time $4,612; part-time $425 per credit hour. International tuition: $10,732 full-time. Required fees: $11 per credit hour. $124 per term. Tuition and fees vary according to course load, degree level and program. *Financial support:* In 2005–06, 10 fellowships with full tuition reimbursements (averaging $13,500 per year), 6 research assistantships with full tuition reimbursements (averaging $13,500 per year), 4 teaching assistantships with full tuition reimbursements (averaging $13,500 per year) were awarded; career-related internships or fieldwork, Federal Work-Study, institutionally sponsored loans, tuition waivers (full and partial), and unspecified assistantships also available. Support available to part-time students. Financial award application deadline: 5/1; financial award applicants required to submit FAFSA. *Faculty research:* General and biochemical synthesis; spectroscopic studies of chemical systems; analysis using chromatography, GC/MS, and designed electrode system. *Unit head:* Dr. David E. Graves, Chair, 205-934-8276, Fax: 205-934-2543.

The University of Alabama in Huntsville, School of Graduate Studies, College of Science, Department of Chemistry, Huntsville, AL 35899. Offers MS. Part-time and evening/weekend programs available. *Faculty:* 15 full-time (2 women). *Students:* 8 full-time (2 women), 2 part-time (both women); includes 2 minority (1 African American, 1 American Indian/Alaska Native), 2 international. Average age 26. 8 applicants, 100% accepted, 6 enrolled. In 2005, 3 degrees awarded. *Degree requirements:* For master's, thesis or alternative, oral and written exams, comprehensive exam, registration. *Entrance requirements:* For master's, GRE General Test, minimum GPA of 3.0. Additional exam requirements/recommendations for international students: Required—TOEFL (minimum score 550 paper-based; 213 computer-based). *Application deadline:* For fall admission, 5/30 for domestic students; for spring admission, 10/10 priority date for domestic students, 7/10 priority date for international students. Applications are processed on a rolling basis. Application fee: $40. *Expenses:* Tuition, state resident: full-time $5,866; part-time $244 per credit hour. Tuition, nonresident: full-time $12,060; part-time $500 per credit hour. Tuition and fees vary according to course load. *Financial support:* In 2005–06, 8 students received support, including 1 research assistantship with full and partial tuition reimbursement available (averaging $22,500 per year), 7 teaching assistantships with full and partial tuition reimbursements available (averaging $10,093 per year); fellowships with full and partial tuition reimbursements available, career-related internships or fieldwork, Federal Work-Study, institutionally sponsored loans, scholarships/grants, health care benefits, tuition waivers (full and partial), and unspecified assistantships also available. Support available to part-time students. Financial award application deadline: 4/1; financial award applicants required to submit FAFSA. *Faculty research:* Kinetics and bonding, organic nonlinear optical materials, x-ray crystallography, crystal growth in space, polymers, Raman spectroscopy. Total annual research expenditures: $725,401. *Unit head:* Dr. William Setzer, Chair, 256-824-6153, Fax: 256-824-6349, E-mail: wsetzer@matsci.uah.edu.

University of Alaska Fairbanks, College of Natural Sciences and Mathematics, Department of Chemistry and Biochemistry, Fairbanks, AK 99775-7520. Offers biochemistry and molecular biology (MS, PhD); chemistry (MA, MS); environmental chemistry (MS, PhD). Part-time programs available. *Faculty:* 13 full-time (3 women). *Students:* 30 full-time (13 women), 8 part-time (4 women); includes 6 minority (1 African American, 1 American Indian/Alaska Native, 2 Asian Americans or Pacific Islanders, 2 Hispanic Americans), 9 international. Average age 29. 35 applicants, 63% accepted, 9 enrolled. In 2005, 6 master's, 4 doctorates awarded. Terminal master's awarded for partial completion of doctoral program. *Degree requirements:* For master's, thesis, seminar, comprehensive exam, registration; for doctorate, thesis/dissertation, comprehensive exam, registration. *Entrance requirements:* For master's, GRE General Test; for doctorate, GRE General Test, GRE Subject Test (biology or chemistry). Additional exam requirements/recommendations for international students: Required—TOEFL (minimum score 550 paper-based; 213 computer-based). *Application deadline:* For fall admission, 6/1 for domestic students, 3/1 for international students; for spring admission, 12/1 for domestic students, 9/1 for international students. Applications are processed on a rolling basis. Application fee: $50. Electronic applications accepted. *Expenses:* Tuition, state resident: full-time $4,392; part-time $244 per credit. Tuition, nonresident: full-time $8,964; part-time $498 per credit. International tuition: $8,964 full-time. Required fees: $800; $5 per credit. $48 per contact hour. Tuition and fees vary according to course level, course load, campus/location and reciprocity agreements. *Financial support:* In 2005–06, 6 research assistantships with tuition reimbursements (averaging $9,595 per year), 16 teaching assistantships with tuition reimbursements (averaging $9,100 per year) were awarded; fellowships with tuition reimbursements, Federal Work-Study and scholarships/grants also available. Financial award applicants required to submit FAFSA. *Faculty research:* Atmospheric aerosols; plant chemistry; hibernation and neuroprotection; transition metal based drugs for diabetes; liganogated ion channels. *Unit head:* Dr. Thomas Clausen, Chair, 907-474-5510, Fax: 907-474-5640, E-mail: fychem@uaf.edu.

University of Alberta, Faculty of Graduate Studies and Research, Department of Chemistry, Edmonton, AB T6G 2E1, Canada. Offers M Sc, PhD. Part-time programs available. *Faculty:* 29 full-time (1 woman), 1 part-time/adjunct (0 women). *Students:* 137 full-time (39 women), 18 part-time (5 women), 37 international. 121 applicants, 69% accepted, 29 enrolled. In 2005, 9 master's, 28 doctorates awarded. Terminal master's awarded for partial completion of doctoral program. *Degree requirements:* For master's and doctorate, thesis/dissertation. *Entrance requirements:* For master's and doctorate, minimum GPA of 6.5 on 9 point scale. *Application deadline:* Applications are processed on a rolling basis. *Expenses: Contact institution.* Tuition and fees charges are reported in Canadian dollars. *Financial support:* In 2005–06, 30 fellowships (averaging $21,175 per year), 19 research assistantships (averaging $20,500 per year), 96 teaching assistantships with partial tuition reimbursements (averaging $20,500 per year) were awarded; scholarships/grants also available. *Faculty research:* Synthetic inorganic and organic chemistry, chemical biology and biochemical analysis, materials and surface chemistry, spectroscopy and instrumentation, computational chemistry. Total annual research expenditures: $10 million Canadian dollars. *Unit head:* Dr. Martin Cowie, Chair, 780-492-3249. *Application contact:* Ilona Baker, Department Office, 780-492-4414, Fax: 780-492-8231, E-mail: grad@chem.ualberta.ca.

The University of Arizona, Graduate College, College of Science, Department of Chemistry, Tucson, AZ 85721. Offers MA, MS, PhD. Part-time programs available. Terminal master's awarded for partial completion of doctoral program. *Degree requirements:* For master's, thesis (for some programs), registration; for doctorate, thesis/dissertation, comprehensive exam, registration. *Entrance requirements:* For master's and doctorate, American Chemical Society Exam. Additional exam requirements/recommendations for international students: Required—TOEFL (minimum score 550 paper-based; 213 computer-based), SPEAK test or TSE. Electronic applications accepted. *Faculty research:* Analytical, inorganic, organic, physical chemistry, biological chemistry.

University of Arkansas, Graduate School, J. William Fulbright College of Arts and Sciences, Department of Chemistry and Biochemistry, Fayetteville, AR 72701-1201. Offers chemistry (MS, PhD). *Students:* 47 full-time (19 women), 11 part-time (3 women); includes 4 minority (2 African Americans, 1 American Indian/Alaska Native, 1 Asian American or Pacific Islander), 19 international. 56 applicants, 20% accepted. In 2005, 3 master's, 5 doctorates awarded. *Degree requirements:* For master's and doctorate, one foreign language, thesis/dissertation. Application fee: $40 ($50 for international students). *Financial support:* In 2005–06, 2 fellowships with tuition reimbursements, 28 research assistantships, 22 teaching assistantships were awarded; career-related internships or fieldwork and Federal Work-Study also available. Support available to part-time students. Financial award application deadline: 4/1; financial award applicants required to submit FAFSA. *Unit head:* Dr. Bill Durham, Departmental Chairperson, 479-575-4601, Fax: 479-575-4049, E-mail: cehminfo@uark.edu.

University of Arkansas at Little Rock, Graduate School, College of Science and Mathematics, Department of Chemistry, Little Rock, AR 72204-1099. Offers MA, MS. Part-time and evening/weekend programs available. *Degree requirements:* For master's, thesis (MS). *Entrance requirements:* For master's, minimum GPA of 2.7.

The University of British Columbia, Faculty of Graduate Studies, Faculty of Science, Department of Chemistry, Vancouver, BC V6T 1Z1, Canada. Offers M Sc, PhD. Terminal master's awarded for partial completion of doctoral program. *Degree requirements:* For master's, thesis/dissertation; for doctorate, thesis/dissertation, comprehensive exam. *Entrance requirements:* For master's and doctorate, GRE General Test, GRE Subject Test. Additional exam requirements/recommendations for international students: Required—TOEFL (minimum score 580 paper-based; 237 computer-based). Electronic applications accepted. *Faculty research:* Organic, physical, analytical, inorganic, and bio-chemical projects.

University of Calgary, Faculty of Graduate Studies, Faculty of Science, Department of Chemistry, Calgary, AB T2N 1N4, Canada. Offers analytical chemistry (M Sc, PhD); applied chemistry (M Sc, PhD); inorganic chemistry (M Sc, PhD); organic chemistry (M Sc, PhD); physical chemistry (M Sc, PhD); polymer chemistry (M Sc, PhD); theoretical chemistry (M Sc, PhD). *Faculty:* 31 full-time (2 women), 2 part-time/adjunct (0 women). *Students:* 80 full-time (38 women). Average age 25. 250 applicants, 6% accepted, 15 enrolled. In 2005, 7 master's, 7 doctorates awarded. *Degree requirements:* For master's, thesis; for doctorate, thesis/dissertation, candidacy exam. *Entrance requirements:* For master's, minimum GPA of 3.0; for doctorate, honors B Sc degree with minimum GPA of 3.7 or M Sc with minimum GPA of 3.3. Additional exam requirements/recommendations for international students: Required—TOEFL (minimum score 580 paper-based; 237 computer-based). *Application deadline:* For fall admission, 12/1 priority date for domestic students, 12/1 priority date for international students. Applications are processed on a rolling basis. Application fee: $100 ($130 for international students). Electronic applications accepted. *Financial support:* In 2005–06, 25 students received support, including research assistantships (averaging $11,000 per year), teaching assistantships (averaging $6,530 per year); fellowships, scholarships/grants also available. Financial award application deadline:12/1. *Faculty research:* Chemical analysis, chemical dynamics, synthesis theory. *Unit head:* Dr. Brian Keay, Head, 403-220-5340, E-mail: info@chem.ucalgary.ca. *Application contact:* Bonnie E. King, Graduate Program Administrator, 403-220-6252, E-mail: bking@ucalgary.ca.

University of California, Berkeley, Graduate Division, College of Chemistry, Department of Chemistry, Berkeley, CA 94720-1500. Offers PhD. *Degree requirements:* For doctorate, thesis/dissertation, qualifying exam. *Entrance requirements:* For doctorate, GRE General Test, GRE Subject Test, minimum GPA of 3.0. Additional exam requirements/recommendations for international students: Required—TOEFL, TSE. Electronic applications accepted. *Faculty research:* Analytical bioinorganic, bio-organic, biophysical environmental, inorganic and organometallic.

University of California, Berkeley, Graduate Division, College of Natural Resources, Group in Agricultural and Environmental Chemistry, Berkeley, CA 94720-1500. Offers MS, PhD. Terminal master's awarded for partial completion of doctoral program. *Degree requirements:* For master's, exam or thesis; for doctorate, thesis/dissertation, qualifying exam, seminar presentation. *Entrance requirements:* For master's and doctorate, GRE General Test, minimum GPA of 3.0.

University of California, Davis, Graduate Studies, Graduate Group in Agricultural and Environmental Chemistry, Davis, CA 95616. Offers MS, PhD. *Faculty:* 50 full-time. *Students:* 48 full-time (28 women); includes 6 minority (1 African American, 3 Asian Americans or Pacific Islanders, 2 Hispanic Americans), 19 international. Average age 30. 40 applicants, 60% accepted, 13 enrolled. In 2005, 3 master's, 5 doctorates awarded. *Median time to degree:* Of those who began their doctoral program in fall 1997, 66.7% received their degree in 8 years or less. *Degree requirements:* For master's and doctorate, thesis/dissertation. *Entrance requirements:* For master's and doctorate, GRE General Test, minimum GPA of 3.0. Additional exam requirements/recommendations for international students: Required—TOEFL (minimum score 550 paper-based; 213 computer-based). *Application deadline:* For fall admission, 1/15 for domestic students, 1/15 for international students. Application fee: $60. Electronic applications accepted. *Expenses:* Tuition, state resident: full-time $8,960. Tuition, nonresident: full-time $14,694. International tuition: $14,694 full-time. *Financial support:* In 2005–06, 42 students received support, including 11 fellowships with full and partial tuition reimbursements available (averaging $14,798 per year), 16 research assistantships with full and partial tuition reimbursements available (averaging $16,174 per year), 10 teaching assistantships with partial tuition reimbursements available (averaging $12,179 per year); Federal Work-Study, institutionally sponsored loans, scholarships/grants, tuition waivers (full and partial), and unspecified assistantships also available. Financial award application deadline: 1/15; financial award applicants required to submit FAFSA. *Unit head:* Andrew Clifford, Graduate Group Chair, 530-752-3376, E-mail: ajclifford@ucdavis.edu. *Application contact:* Peggy Royale, Graduate Administrative Assistant, 530-752-1415, Fax: 530-752-4759, E-mail: pbroyale@ucdavis.edu.

University of California, Davis, Graduate Studies, Program in Chemistry, Davis, CA 95616. Offers MS, PhD. *Faculty:* 59 full-time. *Students:* 166 full-time (64 women); includes 23 minority (3 African Americans, 11 Asian Americans or Pacific Islanders, 9 Hispanic Americans), 43 international. Average age 28. 220 applicants, 48% accepted, 34 enrolled. In 2005, 10 master's, 18 doctorates awarded. Terminal master's awarded for partial completion of doctoral program. *Median time to degree:* Of those who began their doctoral program in fall 1997, 50% received their degree in 8 years or less. *Degree requirements:* For master's and doctorate, thesis/dissertation. *Entrance requirements:* For master's, minimum GPA of 3.0; for doctorate, GRE, minimum GPA of 3.0. Additional exam requirements/recommendations for international students: Required—TOEFL (minimum score 550 paper-based; 213 computer-based). *Application deadline:* For fall admission, 1/15 for domestic students, 1/15 for international students. Applications are processed on a rolling basis. Application fee: $60. Electronic applications accepted. *Expenses:* Tuition, state resident: full-time $8,960. Tuition, nonresident: full-time $14,694. International tuition: $14,694 full-time. *Financial support:* In 2005–06, 161 students received support, including 17 fellowships with full and partial tuition reimbursements available (averaging $9,751 per year), 18 research assistantships with full and partial tuition reimbursements available (averaging $13,738 per year), 93 teaching assistantships with partial tuition reimbursements available (averaging $15,098 per year); Federal Work-Study, institutionally sponsored loans, scholarships/grants, tuition waivers (full and partial), and unspecified assistantships also available. Financial award application deadline: 1/15; financial award applicants required to submit FAFSA. *Faculty research:* Analytical, biological, organic, inorganic, and theoretical chemistry. *Unit head:* David R. Britt, Chair, 530-752-6377, Fax: 530-752-8995, E-mail: rdbritt@ucdavis.edu. *Application contact:* Carol Barnes, Graduate Program Staff, 530-752-0953, E-mail: cbarnes@ucdavis.edu.

University of California, Irvine, Office of Graduate Studies, School of Physical Sciences, Department of Chemistry, Irvine, CA 92697. Offers chemical and material physics (PhD); chemical and materials physics (MS); chemistry (MS, PhD). *Degree requirements:* For doctorate, thesis/dissertation. *Entrance requirements:* For master's and doctorate, GRE General Test, GRE Subject Test, minimum GPA of 3.0. Additional exam requirements/recommendations for international students: Required—TOEFL (minimum score 550 paper-based; 213 computer-based). Electronic applications accepted. *Faculty research:* Analytical, organic, inorganic, physical, and atmospheric chemistry; biogeochemistry and climate; synthetic chemistry.

University of California, Los Angeles, Graduate Division, College of Letters and Science, Department of Chemistry and Biochemistry, Program in Chemistry, Los Angeles, CA 90095. Offers MS, PhD. *Entrance requirements:* For master's, GRE General Test, GRE Subject Test, minimum GPA of 3.0; for doctorate, GRE General Test, GRE Subject Test, minimum undergraduate GPA of 3.0. Electronic applications accepted.

University of California, Riverside, Graduate Division, Department of Chemistry, Riverside, CA 92521-0102. Offers MS, PhD. *Faculty:* 25 full-time (2 women). *Students:* 101 full-time (36 women); includes 20 minority (1 American Indian/Alaska Native, 10 Asian Americans or Pacific Islanders, 9 Hispanic Americans), 45 international. Average age 29. In 2005, 13 master's, 13 doctorates awarded. Terminal master's awarded for partial completion of doctoral program. *Degree requirements:* For master's, comprehensive exams or thesis; for doctorate, thesis/dissertation, qualifying exams, 3 quarters of teaching experience, research proposition. *Entrance requirements:* For master's and doctorate, GRE General Test, minimum GPA of 3.0. Additional exam requirements/recommendations for international students: Required—TOEFL (paper score 550; computer score 213) or TSE (greatly preferred); Recommended—TSE(minimum score 50). *Application deadline:* For fall admission, 5/1 for domestic students, 2/1 for international studentsFor winter admission, 9/1 for domestic students; for spring admission, 12/1 for domestic students. Applications are processed on a rolling basis. Application fee: $60 ($75 for international students). Electronic applications accepted. *Expenses:* Tuition, nonresident: full-time $14,694. Required fees: $9,009. Full-time tuition and fees vary according to program. *Financial support:* In 2005–06, research assistantships with full tuition reimbursements (averaging $14,000 per year), teaching assistantships with full tuition reimbursements (averaging $15,000 per year) were awarded; fellowships with full tuition reimbursements, career-related internships or fieldwork, Federal Work-Study, institutionally sponsored loans, and tuition waivers (full and partial) also available. Financial award application deadline: 2/1; financial award applicants required to submit FAFSA. *Faculty research:* Analytical, inorganic, organic, and physical chemistry; chemical physics. Total annual research expenditures: $4 million. *Unit head:* Prof. Christopher Y. Switzer, Chair, 951-827-3488, Fax: 951-827-4713, E-mail: chemchr@ucr.edu. *Application contact:* Prof. Michael J. Marsella, Graduate Adviser, 800-445-3153, Fax: 951-827-4713, E-mail: gradchem@ucr.edu.

University of California, San Diego, Graduate Studies and Research, Department of Chemistry and Biochemistry, La Jolla, CA 92093. Offers chemistry (MS, PhD). *Degree requirements:* For doctorate, thesis/dissertation. *Entrance requirements:* For doctorate, GRE General Test, GRE Subject Test. Electronic applications accepted.

See Close-Up on page 153.

University of California, San Francisco, School of Pharmacy and Graduate Division, Chemistry and Chemical Biology Graduate Program, San Francisco, CA 94143. Offers PhD. *Faculty:* 48 full-time (9 women), 1 part-time/adjunct (0 women). *Students:* 51 full-time (18 women); includes 18 minority (3 African Americans, 11 Asian Americans or Pacific Islanders, 4 Hispanic Americans), 2 international. Average age 27. 136 applicants, 13% accepted. In 2005, 9 degrees awarded. *Median time to degree:* Of those who began their doctoral program in fall 1997, 100% received their degree in 8 years or less. *Degree requirements:* For doctorate, thesis/dissertation. *Entrance requirements:* For doctorate, GRE General Test, GRE Subject Test, minimum GPA of 3.0. Additional exam requirements/recommendations for international students: Required—TOEFL. *Application deadline:* For fall admission, 1/3 for domestic students. Applications are processed on a rolling basis. Application fee: $80. Electronic applications accepted. *Financial support:* In 2005–06, 21 fellowships with partial tuition reimbursements (averaging $16,880 per year), 13 research assistantships with full tuition reimbursements (averaging $25,000 per year), 3 teaching assistantships with partial tuition reimbursements (averaging $7,383 per year) were awarded; institutionally sponsored loans, scholarships/grants, traineeships, and tuition waivers (full) also available. Financial award application deadline: 1/10. *Faculty research:* Biochemistry; macromolecular structure; cellular and molecular pharmacology; physical chemistry and computational biology; synthetic chemistry. *Unit head:* Dr. Charles S. Craik, Director, 415-476-1913, Fax: 415-502-4690. *Application contact:* Christine Olson, Graduate Program Coordinator, 415-476-1914, Fax: 415-514-1546, E-mail: ccb@picasso.ucsf.edu.

See Close-Up on page 155.

Chemistry

University of California, Santa Barbara, Graduate Division, College of Letters and Sciences, Division of Mathematics, Life, and Physical Sciences, Department of Chemistry and Biochemistry, Santa Barbara, CA 93106. Offers MA, MS, PhD. *Students:* 222 applicants, 51% accepted, 35 enrolled.Terminal master's awarded for partial completion of doctoral program. *Degree requirements:* For master's, thesis or alternative; for doctorate, thesis/dissertation. *Entrance requirements:* For master's and doctorate, GRE. Additional exam requirements/recommendations for international students: Required—TOEFL (minimum score 550 paper-based; 213 computer-based). *Application deadline:* For fall admission, 5/1 for domestic students, 5/1 for international studentsFor winter admission, 11/1 for domestic students; for spring admission, 2/1 for domestic students. Applications are processed on a rolling basis. Application fee: $60. Electronic applications accepted. *Financial support:* In 2005–06, 10 fellowships with full tuition reimbursements (averaging $12,000 per year), 35 research assistantships with full tuition reimbursements (averaging $17,388 per year), 53 teaching assistantships with full tuition reimbursements (averaging $17,471 per year) were awarded; career-related internships or fieldwork, Federal Work-Study, institutionally sponsored loans, scholarships/grants, and traineeships also available. Financial award application deadline: 1/15; financial award applicants required to submit FAFSA. *Faculty research:* Organic, biological, inorganic, physical, and materials chemistry. Total annual research expenditures: $7 million. *Unit head:* Dr. Stanley M. Parsons, Chair, 805-893-2056, Fax: 805-893-4120, E-mail: chemchair@chem.ucsb.edu. *Application contact:* Deedrea Anne Edgar, Manager, Student Affairs and External Relations, 805-893-2638, Fax: 805-893-4120, E-mail: edgar@chem.ucsb.edu.

University of California, Santa Cruz, Division of Graduate Studies, Division of Physical and Biological Sciences, Department of Chemistry and Biochemistry, Santa Cruz, CA 95064. Offers MS, PhD. *Faculty:* 22 full-time (3 women). *Students:* 92 full-time (43 women); includes 27 minority (1 African American, 2 American Indian/Alaska Native, 14 Asian Americans or Pacific Islanders, 10 Hispanic Americans), 10 international. 142 applicants, 41% accepted, 21 enrolled. In 2005, 2 master's, 10 doctorates awarded. *Degree requirements:* For doctorate, one foreign language, thesis/dissertation, qualifying exam. *Entrance requirements:* For master's and doctorate, GRE General Test, GRE Subject Test. *Application deadline:* For fall admission, 1/15 for domestic students. Application fee: $60. *Expenses:* Tuition, nonresident: full-time $14,694. Required fees: $9,437. *Financial support:* Fellowships, research assistantships, teaching assistantships, Federal Work-Study and institutionally sponsored loans available. Financial award application deadline: 1/15. *Faculty research:* Marine chemistry; biochemistry; inorganic, organic, and physical chemistry. *Unit head:* Dr. Joseph Konopelski, Chair, 831-459-2067. *Application contact:* Evie Alloy, Department Assistant, 831-459-2023, E-mail: gradinfo@chemistry.ucsc.edu.

University of Central Florida, College of Sciences, Department of Chemistry, Orlando, FL 32816. Offers MS, PhD. Part-time and evening/weekend programs available. *Faculty:* 18 full-time (3 women), 3 part-time/adjunct (1 woman). *Students:* 30 full-time (14 women), 38 part-time (12 women); includes 10 minority (3 African Americans, 5 Asian Americans or Pacific Islanders, 2 Hispanic Americans), 19 international. Average age 33. 73 applicants, 75% accepted, 37 enrolled. In 2005, 10 degrees awarded. *Degree requirements:* For master's, thesis, final exam. *Entrance requirements:* For master's, GRE General Test, minimum GPA of 3.0 in last 60 hours. Additional exam requirements/recommendations for international students: Required—TOEFL. *Application deadline:* For fall admission, 7/15 for domestic students; for spring admission, 12/1 for domestic students. Application fee: $30. Electronic applications accepted. *Expenses:* Tuition, state resident: full-time $5,788. Tuition, nonresident: full-time $21,927. Required fees: $241 per credit hour. *Financial support:* In 2005–06, 8 fellowships with partial tuition reimbursements (averaging $2,140 per year), 17 research assistantships with partial tuition reimbursements (averaging $8,300 per year), 20 teaching assistantships with partial tuition reimbursements (averaging $8,400 per year) were awarded; career-related internships or fieldwork, Federal Work-Study, institutionally sponsored loans, tuition waivers (partial), and unspecified assistantships also available. Financial award application deadline: 3/1; financial award applicants required to submit FAFSA. *Faculty research:* Physical and synthetic organic chemistry, lasers, polymers, biochemical action of pesticides, environmental analysis. Total annual research expenditures: $645,000. *Unit head:* Dr. Kevin Belfield, Chair, 407-823-2246, Fax: 407-823-2252, E-mail: kbelfield@mail.ucf.edu. *Application contact:* Dr. Andres Campiglia, Coordinator, 407-823-3289, Fax: 407-823-2252, E-mail: acampigl@mail.ucf.edu.

University of Central Oklahoma, College of Graduate Studies and Research, College of Mathematics and Science, Department of Chemistry, Edmond, OK 73034-5209. Offers MS. Part-time programs available. *Faculty:* 6 full-time (1 woman), 1 part-time/adjunct (0 women). *Students:* 24 full-time (19 women), 23 part-time (17 women); includes 2 minority (1 African American, 1 Hispanic American), 1 international. Average age 28. In 2005, 26 degrees awarded. *Entrance requirements:* For master's, GRE General Test. *Application deadline:* Applications are processed on a rolling basis. Application fee: $25. Electronic applications accepted. *Expenses:* Tuition, area resident: Full-time $2,988; part-time $125 per credit hour. Tuition, state resident: full-time $2,988; part-time $125 per credit hour. Tuition, nonresident: full-time $4,728; part-time $197 per credit hour. International tuition: $4,728 full-time. Required fees: $716; $16 per credit hour. *Financial support:* Application deadline: 3/31; *Unit head:* Dr. Cheryl Frech, Chairperson, 405-974-5476.

University of Chicago, Division of the Physical Sciences, Department of Chemistry, Chicago, IL 60637-1513. Offers PhD. *Faculty:* 24 full-time (2 women). *Students:* 169 full-time (56 women); includes 21 minority (2 African Americans, 12 Asian Americans or Pacific Islanders, 7 Hispanic Americans), 81 international. Average age 24. 271 applicants, 40% accepted, 31 enrolled. In 2005, 34 degrees awarded. *Median time to degree:* Of those who began their doctoral program in fall 1997, 100% received their degree in 8 years or less. *Degree requirements:* For doctorate, thesis/dissertation, comprehensive exam, registration. *Entrance requirements:* For doctorate, GRE General Test, GRE Subject Test. Additional exam requirements/recommendations for international students: Required—TOEFL (minimum score 600 paper-based; 250 computer-based), IELT(minimum score 7). *Application deadline:* For fall admission, 12/31 for domestic students, 12/31 for international students. Applications are processed on a rolling basis. Application fee: $55. Electronic applications accepted. *Expenses:* Contact institution. *Financial support:* In 2005–06, 155 students received support, including 15 fellowships with full tuition reimbursements available (averaging $27,215 per year), 123 research assistantships with full tuition reimbursements available (averaging $24,029 per year), 31 teaching assistantships with full tuition reimbursements available (averaging $25,464 per year); institutionally sponsored loans, scholarships/grants, traineeships, health care benefits, and unspecified assistantships also available. Financial award application deadline: 1/15; financial award applicants required to submit FAFSA. *Faculty research:* Organic, inorganic, physical, biological chemistry. Total annual research expenditures: $8 million. *Unit head:* Dr. Michael D. Hopkins, Chairman, 773-702-8639, Fax: 773-702-6594, E-mail: chem-chair@uchicago.edu. *Application contact:* Dr. Vera Dragisich, Executive Officer, 773-702-7250, Fax: 773-702-6594, E-mail: v-dragisich@uchicago.edu.

University of Cincinnati, Division of Research and Advanced Studies, McMicken College of Arts and Sciences, Department of Chemistry, Cincinnati, OH 45221. Offers analytical chemistry (MS, PhD); biochemistry (MS, PhD); inorganic chemistry (MS, PhD); organic chemistry (MS, PhD); physical chemistry (MS, PhD); polymer chemistry (MS, PhD); sensors (PhD). Part-time and evening/weekend programs available. Terminal master's awarded for partial completion of doctoral program. *Degree requirements:* For master's, thesis optional; for doctorate, thesis/dissertation, comprehensive exam, registration. *Entrance requirements:* For master's and doctorate, GRE General Test. Additional exam requirements/recommendations for international students: Required—TOEFL (minimum score 580 paper-based; 237 computer-based); Recommended—TSE(minimum score 50). Electronic applications accepted. *Faculty research:* Biomedical chemistry, laser chemistry, surface science, chemical sensors, synthesis.

University of Colorado at Boulder, Graduate School, College of Arts and Sciences, Department of Chemistry and Biochemistry, Boulder, CO 80309. Offers biochemistry (PhD); chemistry (MS). *Faculty:* 40 full-time (8 women). *Students:* 125 full-time (54 women), 61 part-time (24 women); includes 13 minority (1 African American, 1 American Indian/Alaska Native, 7 Asian Americans or Pacific Islanders, 4 Hispanic Americans), 13 international. Average age 27. 31 applicants, 100% accepted. In 2005, 12 master's, 23 doctorates awarded. *Degree requirements:* For master's, comprehensive exam or thesis; for doctorate, thesis/dissertation, cumulative exam, comprehensive exam. *Entrance requirements:* For master's, GRE General Test, GRE Subject Test, minimum GPA of 2.75; for doctorate, GRE General Test, GRE Subject Test, minimum GPA of 3.0. *Application deadline:* For fall admission, 2/28 priority date for domestic students, 12/1 priority date for international students. Applications are processed on a rolling basis. Application fee: $50 ($60 for international students). *Financial support:* In 2005–06, fellowships with full tuition reimbursements (averaging $6,060 per year), research assistantships with full tuition reimbursements (averaging $14,958 per year), teaching assistantships with full tuition reimbursements (averaging $13,932 per year) were awarded; institutionally sponsored loans, traineeships, and tuition waivers (full) also available. Support available to part-time students. Financial award application deadline: 2/28. *Faculty research:* Biochemistry, atmospheric chemistry, analytical chemistry, biophysical chemistry, chemical physics. Total annual research expenditures: $15.9 million. *Unit head:* Veronica Vaida, Chair, 303-492-6533, Fax: 303-492-5894, E-mail: chemdir@colorado.edu. *Application contact:* Graduate Program Assistant, 303-492-8978, Fax: 303-492-5894, E-mail: chem_gradstudents@colorado.edu.

University of Colorado at Denver and Health Sciences Center—Downtown Denver Campus, College of Liberal Arts and Sciences, Department of Chemistry, Denver, CO 80217-3364. Offers MS. Part-time programs available. *Faculty:* 9 full-time (2 women). *Students:* 6 full-time (4 women), 13 part-time (4 women); includes 7 minority (2 African Americans, 1 American Indian/Alaska Native, 3 Asian Americans or Pacific Islanders, 1 Hispanic American), 3 international. Average age 31. 9 applicants, 67% accepted, 6 enrolled. In 2005, 3 degrees awarded. *Degree requirements:* For master's, thesis or alternative, comprehensive exam, registration. *Entrance requirements:* For master's, undergraduate degree in chemistry, minimum GPA of 2.75. Additional exam requirements/recommendations for international students: Required—TOEFL (minimum score 525 paper-based; 197 computer-based). *Application deadline:* For fall admission, 1/15 for domestic students; for spring admission, 12/23 priority date for domestic students. Applications are processed on a rolling basis. Application fee: $50 ($75 for international students). Electronic applications accepted. *Expenses:* Tuition, state resident: part-time $325 per credit hour. Tuition, nonresident: part-time $1,077 per credit hour. Required fees: $145 per credit hour. One-time fee: $115 part-time. Tuition and fees vary according to course level and program. *Financial support:* Research assistantships, teaching assistantships, Federal Work-Study available. Financial award application deadline: 4/1; financial award applicants required to submit FAFSA. *Faculty research:* Protein electrochemistry, indoor air quality, atmospheric carbonul analysis, chemical education. *Unit head:* Prof. Douglas F. Dyckes, Program Director, 303-556-3203, Fax: 303-556-4776.

University of Connecticut, Graduate School, College of Liberal Arts and Sciences, Department of Chemistry, Field of Chemistry, Storrs, CT 06269. Offers MS, PhD. *Faculty:* 24 full-time (4 women). *Students:* 103 full-time (40 women), 6 part-time (3 women); includes 5 minority (2 African Americans, 2 Asian Americans or Pacific Islanders, 1 Hispanic American), 71 international. Average age 28. 99 applicants, 62% accepted, 32 enrolled. In 2005, 6 master's, 17 doctorates awarded. Terminal master's awarded for partial completion of doctoral program. *Degree requirements:* For master's, comprehensive exam; for doctorate, thesis/dissertation. *Entrance requirements:* For master's and doctorate, GRE General Test, GRE Subject Test. Additional exam requirements/recommendations for international students: Required—TOEFL (minimum score 550 paper-based; 213 computer-based). *Application deadline:* For fall admission, 2/1 priority date for domestic students, 2/1 priority date for international students; for spring admission, 11/1 for domestic students, 10/1 for international students. Applications are processed on a rolling basis. Application fee: $55. Electronic applications accepted. *Expenses:* Tuition, state resident: part-time $444 per credit hour. Tuition, nonresident: part-time $1,154 per credit hour. Tuition and fees vary according to course load. *Financial support:* In 2005–06, 48 research assistantships with full tuition reimbursements, 48 teaching assistantships with full tuition reimbursements were awarded; fellowships, Federal Work-Study, scholarships/grants, health care benefits, and unspecified assistantships also available. Financial award application deadline: 2/1; financial award applicants required to submit FAFSA. *Application contact:* Emilie Hogrebe, Graduate Program Coordinator, 860-486-3219, Fax: 860-480-2981, E-mail: hogrebe@nucleus.chem.uconn.edu.

University of Connecticut, Graduate School, School of Engineering, Polymer Program, Storrs, CT 06269-3136. Offers chemical engineering (MS, PhD); chemistry (MS, PhD); polymer science (MS, PhD). Part-time programs available. *Faculty:* 13 full-time (1 woman), 3 part-time/adjunct (0 women). *Students:* 67 full-time (16 women), 2 part-time; includes 3 minority (1 African American, 2 Asian Americans or Pacific Islanders), 56 international. Average age 27. 79 applicants, 13% accepted, 5 enrolled. In 2005, 9 degrees awarded. Terminal master's awarded for partial completion of doctoral program. *Median time to degree:* Of those who began their doctoral program in fall 1997, 100% received their degree in 8 years or less. *Degree requirements:* For master's, one foreign language, thesis (for some programs), registration; for doctorate, one foreign language, thesis/dissertation, comprehensive exam, registration. *Entrance requirements:* For master's, GRE; for doctorate, GRE General Test. Additional exam requirements/recommendations for international students: Required—TOEFL (minimum score 550 paper-based; 213 computer-based). *Application deadline:* For fall admission, 4/1 priority date for domestic students, 1/31 priority date for international students; for spring admission, 11/1 priority date for domestic students. Applications are processed on a rolling basis. Application fee: $55. Electronic applications accepted. *Expenses:* Tuition, state resident: part-time $444 per credit hour. Tuition, nonresident: part-time $1,154 per credit hour. Tuition and fees vary according to course load. *Financial support:* In 2005–06, 69 students received support, including 57 research assistantships with tuition reimbursements available (averaging $23,650 per year), 12 teaching assistantships with tuition reimbursements available (averaging $23,650 per year); fellowships with tuition reimbursements available, career-related internships or fieldwork, scholarships/grants, health care benefits, and unspecified assistantships also available. Financial award application deadline: 4/1. *Faculty research:* Sensors, photonics, processing morphology, synthesis. Total annual research expenditures: $3.3 million. *Unit head:* Thomas Seery, Director, 860-486-1337, Fax: 860-486-4745, E-mail: seery@mail.ims.uconn.edu. *Application contact:* Michelle L. Cahill, Coordinator for Graduate Admissions, 860-486-4613, Fax: 860-486-4745, E-mail: grad@ims.uconn.edu.

University of Dayton, Graduate School, College of Arts and Sciences, Department of Chemistry, Dayton, OH 45469-1300. Offers MS. Part-time programs available. *Faculty:* 6 full-time (0 women). *Students:* 6 full-time (5 women), 3 part-time; includes 1 minority (African American), 6 international. Average age 22. 37 applicants, 22% accepted, 5 enrolled. *Degree requirements:* For master's, thesis, registration. *Entrance requirements:* For master's, GRE, ACS standardized exams. Additional exam requirements/recommendations for international students: Required—TOEFL (minimum score 550 paper-based; 213 computer-based), TWE. *Application deadline:* For fall admission, 3/1 for domestic students, 3/1 for international students. Applications are processed on a rolling basis. Application fee: $0. Electronic applications accepted. *Expenses:* Tuition: Part-time $567 per credit hour. Required fees: $25 per term. Tuition and fees vary according to degree level and program. *Financial support:* In 2005–06, 8 teaching assistantships with full tuition reimbursements (averaging $9,275 per year) were awarded; scholarships/grants, health care benefits, tuition waivers (partial), and unspecified assistantships also available. Financial award applicants required to submit FAFSA. *Faculty research:* Organic synthesis, medicinal chemistry, enzyme purification, physical organic, materials chemistry and nanotechnology. *Unit head:* Dr. David Johnson, Director, 937-229-2631. *Application contact:* E. Eavers.

University of Delaware, College of Arts and Sciences, Department of Chemistry and Biochemistry, Newark, DE 19716. Offers biochemistry (MA, MS, PhD); chemistry (MA, MS, PhD). Part-time programs available. *Faculty:* 32 full-time (5 women), 2 part-time/adjunct (0 women). *Students:* 143 full-time (61 women), 11 part-time (3 women); includes 15 minority (4 African Americans, 8 Asian Americans or Pacific Islanders, 3 Hispanic Americans), 56 international. Average age 27. 147 applicants, 44% accepted, 28 enrolled. In 2005, 7 master's, 14 doctorates awarded. Terminal master's awarded for partial completion of doctoral program. *Degree requirements:* For master's, one foreign language, thesis (for some programs); for doctorate, one foreign language, thesis/dissertation, cumulative exam. *Entrance requirements:* For master's and doctorate, GRE General Test. Additional exam requirements/recommendations for international students: Required—TOEFL (minimum score 600 paper-based; 260 computer-based), TSE(minimum score 50). *Application deadline:* For fall admission, 3/31 priority date for domestic students, 3/31 priority date for international students. Applications are processed on a rolling basis. Application fee: $60. Electronic applications accepted. *Financial support:* In 2005–06, 89 students received support, including 4 fellowships with full tuition reimbursements available (averaging $21,750 per year), 72 research assistantships with full tuition reimbursements available (averaging $21,750 per year), 50 teaching assistantships with full tuition reimbursements available (averaging $21,750 per year) Financial award application deadline: 3/31. *Faculty research:* Protein studies; mechanism of enzymes; synthesis, electronic structure, and bonding of organic, inorganic and organometallic compounds; spectroscopy including single molecule spectroscopy. Total annual research expenditures: $7.8 million. *Unit head:* Dr. Charles G. Riordan, Chairman, 302-831-1247, Fax: 302-831-6335, E-mail: riordan@udel.edu. *Application contact:* Dr. Andrew Teplyakov, Graduate Coordinator, 302-831-1969, Fax: 302-831-6335, E-mail: andrewt@udel.edu.

University of Denver, Faculty of Natural Sciences and Mathematics, Department of Chemistry, Denver, CO 80208. Offers MA, MS, PhD. Part-time programs available. *Faculty:* 11 full-time (1 woman). *Students:* 5 full-time (1 woman), 13 part-time (7 women); includes 1 minority (Asian American or Pacific Islander), 5 international. 25 applicants, 68% accepted. In 2005, 6 master's, 2 doctorates awarded. Terminal master's awarded for partial completion of doctoral program. *Degree requirements:* For master's and doctorate, thesis/dissertation. *Entrance requirements:* For master's and doctorate, GRE General Test. Additional exam requirements/recommendations for international students: Required—TOEFL, TSE. *Application deadline:* Applications are processed on a rolling basis. Application fee: $45. *Expenses:* Tuition: Full-time $27,756; part-time $771 per credit. Required fees: $174. *Financial support:* In 2005–06, 12 research assistantships with full and partial tuition reimbursements (averaging $15,750 per year), 12 teaching assistantships with full and partial tuition reimbursements (averaging $15,750 per year) were awarded; career-related internships or fieldwork, Federal Work-Study, institutionally sponsored loans, and scholarships/grants also available. Support available to part-time students. Financial award application deadline: 3/1; financial award applicants required to submit FAFSA. *Faculty research:* Atmospheric chemistry, magnetic resonance, molecular spectroscopy, laser photolysis, biophysical chemistry. Total annual research expenditures: $1.7 million. *Unit head:* Dr. Lawrence Berliner, Chairperson, 303-871-2436. *Application contact:* 303-871-2435, E-mail: chem-info@du.edu.

University of Detroit Mercy, College of Engineering and Science, Department of Chemistry and Biochemistry, Detroit, MI 48219-0900. Offers macromolecular chemistry (MS). Evening/weekend programs available. *Degree requirements:* For master's; thesis. *Entrance requirements:* For master's, GRE General Test, minimum GPA of 3.0. *Faculty research:* Polymer and physical chemistry, industrial aspects of chemistry.

University of Florida, Graduate School, College of Liberal Arts and Sciences, Department of Chemistry, Gainesville, FL 32611. Offers MS, MST, PhD. *Faculty:* 51 full-time (7 women), 1 part-time/adjunct (0 women). *Students:* 284 (126 women); includes 27 minority (10 African Americans, 10 Asian Americans or Pacific Islanders, 7 Hispanic Americans) 141 international. In 2005, 8 master's, 43 doctorates awarded. *Degree requirements:* For master's, thesis; for doctorate, one foreign language, thesis/dissertation. *Entrance requirements:* For master's and doctorate, GRE General Test, minimum GPA of 3.0. Additional exam requirements/recommendations for international students: Required—TOEFL (minimum score 550 paper-based; 213 computer-based). *Application deadline:* For fall admission, 6/1 for domestic students. Applications are processed on a rolling basis. Application fee: $30. Electronic applications accepted. *Expenses:* Tuition, state resident: full-time $6,234. Tuition, nonresident: full-time $21,359. Tuition and fees vary according to program. *Financial support:* In 2005–06, 130 research assistantships (averaging $20,641 per year), 114 teaching assistantships (averaging $21,096 per year) were awarded; fellowships, institutionally sponsored loans and unspecified assistantships also available. *Faculty research:* Organic, analytical, physical, inorganic, and biological chemistry. *Unit head:* Dr. David E. Richardson, Chair, 352-392-0541, Fax: 352-392-3255, E-mail: der@chem.ufl.edu. *Application contact:* Dr. Ben Smith, Coordinator, 352-392-8180, Fax: 352-392-8758, E-mail: bwsmith@chem.ufl.edu.

University of Georgia, Graduate School, College of Arts and Sciences, Department of Chemistry, Athens, GA 30602. Offers analytical chemistry (MS, PhD); inorganic chemistry (MS, PhD); organic chemistry (MS, PhD); physical chemistry (MS, PhD). *Faculty:* 22 full-time (2 women). *Students:* 155 full-time, 3 part-time; includes 12 minority (4 African Americans, 7 Asian Americans or Pacific Islanders, 1 Hispanic American), 87 international. 147 applicants, 56% accepted, 35 enrolled. In 2005, 2 master's, 19 doctorates awarded. Terminal master's awarded for partial completion of doctoral program. *Median time to degree:* Of those who began their doctoral program in fall 1997, 100% received their degree in 8 years or less. *Degree requirements:* For master's, thesis; for doctorate, one foreign language, thesis/dissertation. *Entrance requirements:* For master's and doctorate, GRE General Test. Additional exam requirements/recommendations for international students: Required—TOEFL (minimum score 213 computer-based), TSE(minimum score 50). *Application deadline:* For fall admission, 7/1 for domestic students; for spring admission, 11/15 for domestic students. Application fee: $50. Electronic applications accepted. *Financial support:* Fellowships, research assistantships, teaching assistantships, unspecified assistantships available. *Unit head:* Dr. John L. Stickney, Head, 706-542-2726, Fax: 706-542-9454, E-mail: stickney@chem.uga.edu. *Application contact:* Dr. Donald Kurtz, Information Contact, 706-542-2010, Fax: 706-542-9454, E-mail: kurtz@chem.uga.edu.

University of Guelph, Graduate Program Services, College of Physical and Engineering Science, Guelph-Waterloo Centre for Graduate Work in Chemistry and Biochemistry, Guelph, ON N1G 2W1, Canada. Offers biochemistry (M Sc, PhD); chemistry (M Sc, PhD). Part-time programs available. *Faculty:* 56 full-time (8 women). *Students:* 189 full-time (41 women), 7 part-time (3 women). In 2005, 28 master's, 17 doctorates awarded. *Degree requirements:* For master's and doctorate, thesis/dissertation. *Application deadline:* Applications are processed on a rolling basis. Application fee: $75. *Financial support:* Fellowships, research assistantships, teaching assistantships available. *Faculty research:* Inorganic, analytical, biological, physical/theoretical, polymer, and organic chemistry. *Unit head:* Dr. A. L. Schwan, Director, 519-824-4120 Ext. 53848, Fax: 519-823-8097, E-mail: gwc@uoguelph.ca. *Application contact:* A. Wetmore, Administrative Assistant, 519-824-4120 Ext. 53848, Fax: 519-823-8097, E-mail: gwc@uoguelph.ca.

University of Hawaii at Manoa, Graduate Division, Colleges of Arts and Sciences, College of Natural Sciences, Department of Chemistry, Honolulu, HI 96822. Offers MS, PhD. *Faculty:* 15 full-time (3 women). *Students:* 38 full-time (11 women), 2 part-time (1 woman); includes 10 minority (1 American Indian/Alaska Native, 9 Asian Americans or Pacific Islanders), 10 international. Average age 28. 47 applicants, 51% accepted, 6 enrolled. In 2005, 1 master's, 2 doctorates awarded. *Degree requirements:* For master's and doctorate, thesis/dissertation. *Entrance requirements:* For master's and doctorate, GRE General Test, GRE Subject Test. *Application deadline:* For fall admission, 5/1 for domestic students, 3/1 for international students; for spring admission, 9/1 for domestic students, 8/1 for international students. Applications are processed on a rolling basis. Application fee: $50. *Expenses:* Tuition, state resident: full-time $8,400; part-time $200 per credit hour. Tuition, nonresident: full-time $11,088; part-time $462 per credit hour. Tuition and fees vary according to program. *Financial support:* In 2005–06, 5 research assistantships (averaging $17,050 per year), 34 teaching assistantships (averaging $14,640 per year) were awarded. Support available to part-time students. *Faculty research:* Marine natural products, biophysical spectroscopy, zeolites, organometallic hydrides, new visual pigments, theory of surfaces. Total annual research expenditures: $2.1 million. *Unit head:* Thomas Bopp, Chair, 808-956-7480, Fax: 808-956-5908, E-mail: tbopp@hawaii.edu. *Application contact:* Thomas Bopp, Chair, 808-956-7480, Fax: 808-956-5908, E-mail: tbopp@hawaii.edu.

University of Houston, College of Natural Sciences and Mathematics, Department of Chemistry, Houston, TX 77204. Offers MA, MS, PhD. Part-time programs available. *Faculty:* 21 full-time (0 women), 1 part-time/adjunct (0 women). *Students:* 114 full-time (39 women), 9 part-time (6 women); includes 11 minority (3 African Americans, 3 Asian Americans or Pacific Islanders, 5 Hispanic Americans), 95 international. Average age 29. 31 applicants, 90% accepted, 23 enrolled. In 2005, 6 master's, 8 doctorates awarded. Terminal master's awarded for partial completion of doctoral program. *Degree requirements:* For master's, thesis; for doctorate, thesis/dissertation, oral presentation. *Entrance requirements:* For master's and doctorate, GRE General Test. Additional exam requirements/recommendations for international students: Required—TOEFL, TSE. *Application deadline:* For fall admission, 7/20 for domestic students; for spring admission, 11/20 for domestic students. Applications are processed on a rolling basis. Application fee: $0 ($75 for international students). Electronic applications accepted. *Financial support:* In 2005–06, 29 teaching assistantships with full tuition reimbursements (averaging $13,700 per year) were awarded; fellowships with full tuition reimbursements, research assistantships with full tuition reimbursements, career-related internships or fieldwork, Federal Work-Study, institutionally sponsored loans, scholarships/grants, health care benefits, and unspecified assistantships also available. Support available to part-time students. Financial award application deadline: 3/10. *Faculty research:* Materials, molecular design, surface science, structural chemistry, synthesis. *Unit head:* Dr. David Wayne Hoffman, Chairperson, 713-743-2701, Fax: 713-743-2709, E-mail: hoffman@uh.edu. *Application contact:* Dr. Thomas Albright, Chair, Graduate Committee, 713-743-3270, Fax: 713-743-2709, E-mail: albright@uh.edu.

University of Houston–Clear Lake, School of Science and Computer Engineering, Program in Chemistry, Houston, TX 77058-1098. Offers MS. Part-time and evening/weekend programs available. *Entrance requirements:* For master's, GRE General Test. Additional exam requirements/recommendations for international students: Required—TOEFL (minimum score 550 paper-based; 213 computer-based).

University of Idaho, College of Graduate Studies, College of Science, Department of Chemistry, Moscow, ID 83844-2282. Offers MAT, MS, PhD. *Students:* 31 full-time (11 women), 7 part-time (2 women); includes 2 minority (both Asian Americans or Pacific Islanders), 13 international. Average age 31. In 2005, 5 master's, 4 doctorates awarded. *Degree requirements:* For master's, thesis or alternative; for doctorate, one foreign language, thesis/dissertation. *Entrance requirements:* For master's, minimum GPA of 2.8; for doctorate, minimum undergraduate GPA of 2.8, 3.0 graduate. *Application deadline:* For fall admission, 8/1 for domestic students; for spring admission, 12/15 for domestic students. Application fee: $55 ($60 for international students). *Expenses:* Tuition, nonresident: full-time $8,770; part-time $130 per credit. Required fees: $4,508; $217 per credit. *Financial support:* Fellowships, research assistantships, teaching assistantships available. Financial award application deadline: 2/15. *Unit head:* Dr. Peter R. Griffiths, Chair, 208-885-6552.

University of Illinois at Chicago, College of Pharmacy and Graduate College, Research & Graduate Studies, College of Pharmacy, Chicago, IL 60607-7128. Offers forensic science (MS); medicinal chemistry (MS, PhD); pharmaceutics (MS, PhD); pharmacodynamics (MS, PhD); pharmacognosy (MS, PhD); pharmacy administration (MS, PhD). Terminal master's awarded for partial completion of doctoral program. *Degree requirements:* For master's and doctorate, variable foreign language requirement, thesis/dissertation. *Entrance requirements:* For master's and doctorate, GRE General Test. Additional exam requirements/recommendations for international students: Required—TOEFL. Electronic applications accepted. Expenses: Contact institution.

University of Illinois at Chicago, Graduate College, College of Liberal Arts and Sciences, Department of Chemistry, Chicago, IL 60607-7128. Offers MS, PhD. Part-time programs available. Terminal master's awarded for partial completion of doctoral program. *Degree requirements:* For master's, thesis or cumulative exam; for doctorate, one foreign language, thesis/dissertation, cumulative exams. *Entrance requirements:* For master's and doctorate, GRE Subject Test, minimum GPA of 3.0. Additional exam requirements/recommendations for international students: Required—TOEFL. Electronic applications accepted.

University of Illinois at Urbana–Champaign, Graduate College, College of Liberal Arts and Sciences, School of Chemical Sciences, Department of Chemistry, Champaign, IL 61820. Offers MS, PhD. *Faculty:* 40 full-time (5 women), 2 part-time/adjunct (1 woman). *Students:* 312 full-time (104 women), 3 part-time (1 woman); includes 23 minority (4 African Americans, 16 Asian Americans or Pacific Islanders, 3 Hispanic Americans), 71 international. 376 applicants, 24% accepted, 72 enrolled. In 2005, 15 master's, 36 doctorates awarded. *Degree requirements:* For doctorate, one foreign language, thesis/dissertation. *Entrance requirements:* For master's, GRE General Test, GRE Subject Test, minimum GPA of 3.0. *Application deadline:* ; for spring admission, 10/25 for domestic students. Applications are processed on a rolling basis. Application fee: $50 ($60 for international students). Electronic applications accepted. *Financial support:* In 2005–06, 95 fellowships, 170 research assistantships, 229 teaching assistantships were awarded. Financial award application deadline: 2/15. *Unit head:* Steven C. Zimmerman, Interim Head, 217-333-5071, Fax: 217-244-5943, E-mail: sczimmer@uiuc.edu. *Application contact:* Dot Houchens, Assistant to the Head, 217-244-0618, Fax: 217-244-5943, E-mail: dorothyh@uiuc.edu.

The University of Iowa, Graduate College, College of Liberal Arts and Sciences, Department of Chemistry, Iowa City, IA 52242-1316. Offers MS, PhD, JD/PhD. *Faculty:* 33 full-time, 4 part-time/adjunct. *Students:* 68 full-time (31 women), 60 part-time (23 women); includes 12 minority (5 African Americans, 5 Asian Americans or Pacific Islanders, 2 Hispanic Americans), 66 international. 153 applicants, 25% accepted, 33 enrolled. In 2005, 12 master's, 13 doctorates awarded. *Degree requirements:* For master's, exam, thesis optional; for doctorate, thesis/dissertation, comprehensive exam, registration. *Entrance requirements:* For master's and doctorate, GRE General Test, minimum GPA of 3.0. Additional exam requirements/recommendations for international students: Required—TOEFL (minimum score 550 paper-based; 213 computer-based). *Application deadline:* For fall admission, 1/15 priority date for domestic students, 1/15 priority date for international students; for spring admission, 10/15 priority date for domestic students, 10/15 priority date for international students. Applications are processed on a rolling basis. Application fee: $60 ($85 for international students). Electronic applications accepted. *Expenses:* Tuition, state resident: part-time $1,882 per term. Tuition, nonresident: full-time $17,338; part-time $4,907 per term. Tuition and fees vary according to course load and program. *Financial support:* In 2005–06, 7 fellowships, 58 research assistantships with partial tuition reimbursements, 52 teaching assistantships with partial tuition reimbursements were awarded. Financial award applicants required to submit FAFSA. *Unit head:* David Wiemer, Chair, 319-335-1350, Fax: 319-335-1270.

University of Kansas, Graduate School, College of Liberal Arts and Sciences, Department of Chemistry, Lawrence, KS 66045. Offers MA, MS, PhD. *Faculty:* 28. *Students:* 80 full-time (42 women), 23 part-time (9 women); includes 8 minority (2 American Indian/Alaska Native, 2 Asian Americans or Pacific Islanders, 4 Hispanic Americans), 29 international. Average age 28. 96 applicants, 39% accepted. In 2005, 1 master's, 16 doctorates awarded. *Degree requirements:* For master's, thesis/dissertation; for doctorate, thesis/dissertation, comprehensive exam. *Entrance requirements:* For master's and doctorate, GRE General Test. Additional exam requirements/recommendations for international students: Required—TOEFL, TSE. *Application deadline:* For fall admission, 5/31 priority date for domestic students, 5/31 priority date for international students; for spring admission, 11/15 priority date for domestic students, 10/15 priority date for international students. Applications are processed on a rolling basis. Application fee: $55 ($60 for international students). Electronic applica-

Chemistry

University of Kansas (continued)

tions accepted. *Expenses:* Tuition, state resident: full-time $4,859. Tuition, nonresident: full-time $1,200. Required fees: $589. Tuition and fees vary according to program. *Financial support:* Fellowships with full tuition reimbursements, research assistantships with full and partial tuition reimbursements, teaching assistantships with full and partial tuition reimbursements, scholarships/grants, traineeships, and tuition waivers (full) available. Financial award application deadline: 4/15. *Faculty research:* Organometallic and inorganic synthetic methodology, bioanalytical chemistry, computational materials science, proteomics, biophysical chemistry. *Unit head:* Prof. Joseph Heppert, Chair, 785-864-4670, Fax: 785-864-5396, E-mail: jheppert@ku.edu. *Application contact:* Prof. Brian B. Laird, Associate Chair for Graduate Studies, 785-864-4632, Fax: 785-864-5396, E-mail: blaird@ku.edu.

University of Kentucky, Graduate School, Graduate School Programs from the College of Arts and Sciences, Program in Chemistry, Lexington, KY 40506-0032. Offers MS, PhD. Part-time programs available. *Faculty:* 26 full-time (3 women), 2 part-time/adjunct (0 women). *Students:* 110 full-time (33 women), 2 part-time; includes 9 minority (4 African Americans, 3 Asian Americans or Pacific Islanders, 2 Hispanic Americans), 56 international. Average age 28. 142 applicants, 42% accepted, 38 enrolled. In 2005, 3 master's, 10 doctorates awarded. Terminal master's awarded for partial completion of doctoral program. *Median time to degree:* Of those who began their doctoral program in fall 1997, 73% received their degree in 8 years or less. *Degree requirements:* For master's, thesis optional; for doctorate, thesis/dissertation, comprehensive exam. *Entrance requirements:* For master's, GRE General Test, minimum undergraduate GPA of 2.5; for doctorate, GRE General Test, minimum graduate GPA of 3.0. Additional exam requirements/recommendations for international students: Required—TOEFL (minimum score 550 paper-based; 213 computer-based). *Application deadline:* For fall admission, 7/17 priority date for domestic students, 2/1 priority date for international students; for spring admission, 12/13 priority date for domestic students, 6/15 priority date for international students. Applications are processed on a rolling basis. Application fee: $40 ($55 for international students). Electronic applications accepted. *Expenses:* Tuition, state resident: full-time $3,159; part-time $331 per credit hour. Tuition, nonresident: full-time $6,984; part-time $756 per credit hour. Tuition and fees vary according to course load, degree level and program. *Financial support:* In 2005–06, 99 students received support, including 9 fellowships, 42 research assistantships with full tuition reimbursements available (averaging $7,100 per year), 48 teaching assistantships with full tuition reimbursements available (averaging $7,100 per year); career-related internships or fieldwork, Federal Work-Study, institutionally sponsored loans, and tuition waivers (partial) also available. Support available to part-time students. Financial award application deadline: 3/15. *Faculty research:* Analytical, inorganic, organic, and physical chemistry; biological chemistry; nuclear chemistry; radiochemistry; materials chemistry. *Unit head:* Dr. Mark Meier, Director of Graduate Studies, 859-257-3838, Fax: 859-323-1069, E-mail: meier@pop.uky.edu. *Application contact:* Dr. Brian Jackson, Senior Associate Dean, 859-257-8176, Fax: 859-323-1928, E-mail: lance.brunner@uky.edu.

University of Lethbridge, School of Graduate Studies, Lethbridge, AB T1K 3M4, Canada. Offers accounting (MScM); addictions counseling (M Sc); agricultural biotechnology (M Sc); agricultural studies (M Sc, MA); anthropology (MA); archaeology (MA); art (MA); biochemistry (M Sc); biological sciences (M Sc); biomolecular science (PhD); biosystems and biodiversity (PhD); Canadian studies (MA); chemistry (M Sc); computer science (M Sc); computer science and geographical information science (M Sc); counseling psychology (M Ed); dramatic arts (MA); earth, space, and physical science (PhD); economics (MA); educational leadership (M Ed); English (MA); environmental science (M Sc); evolution and behavior (PhD); exercise science (M Sc); finance (MScM); French (MA); French/German (MA); French/Spanish (MA); general education (M Ed); general management (MScM); geography (M Sc, MA); German (MA); health sciences (M Sc, MA); history (MA); human resource management and labour relations (MScM); individualized multidisciplinary (M Sc, MA); information systems (MScM); international management (MScM); kinesiology (M Sc, MA); management (M Sc, MA); marketing (MScM); mathematics (M Sc); music (MA); Native American studies (MA); neuroscience (M Sc, PhD); new media (MA); nursing (M Sc); philosophy (MA); physics (M Sc); policy and strategy (MScM); political science (MA); psychology (M Sc, MA); religious studies (MA); sociology (MA); theoretical and computational science (PhD); urban and regional studies (MA). Part-time and evening/weekend programs available. *Faculty:* 250. *Students:* 193 full-time, 145 part-time. 35 applicants, 100% accepted, 35 enrolled. In 2005, 40 degrees awarded. *Degree requirements:* For doctorate, thesis/dissertation, comprehensive exam. *Entrance requirements:* For master's, GMAT (M Sc management), bachelor's degree in related field, minimum GPA of 3.0 during previous 20 graded semester courses, 2 years teaching or related experience (M Ed); for doctorate, master's degree, minimum graduate GPA of 3.5. Additional exam requirements/recommendations for international students: Required—TOEFL. Application fee: $60 Canadian dollars. *Expenses:* Tuition, nonresident: part-time $531 per course. Required fees: $83 per year. Tuition and fees vary according to degree level and program. *Financial support:* Fellowships, research assistantships, teaching assistantships, scholarships/grants, health care benefits, and unspecified assistantships available. *Faculty research:* Movement and brain plasticity, gibberellin physiology, photosynthesis, carbon cycling, molecular properties of main-group ring components. *Unit head:* Dr. Shamsul Alam, Dean, 403-329-2121, Fax: 403-329-2097, E-mail: inquiries@uleth.ca. *Application contact:* Kathy Schrage, Administrative Assistant, Office of the Academic Vice President, 403-329-2121, Fax: 403-329-2097, E-mail: inquiries@uleth.ca.

University of Louisville, Graduate School, College of Arts and Sciences, Department of Chemistry, Louisville, KY 40292-0001. Offers analytical chemistry (MS, PhD); biochemistry (MS, PhD); chemical physics (PhD); inorganic chemistry (MS, PhD); organic chemistry (MS, PhD); physical chemistry (MS, PhD). *Students:* 37 full-time (16 women), 16 part-time (7 women); includes 3 minority (2 African Americans, 1 Hispanic American), 24 international. Average age 28. In 2005, 1 master's, 8 doctorates awarded. *Degree requirements:* For master's, thesis/dissertation; for doctorate, thesis/dissertation, comprehensive exam. *Entrance requirements:* For master's and doctorate, GRE General Test. Additional exam requirements/recommendations for international students: Required—TOEFL. *Application deadline:* Applications are processed on a rolling basis. Application fee: $50. *Expenses:* Tuition, state resident: full-time $3,003; part-time $334 per credit hour. Tuition, nonresident: full-time $8,277; part-time $920 per credit hour. Tuition and fees vary according to course load, degree level and program. *Financial support:* In 2005–06, 33 teaching assistantships with tuition reimbursements were awarded; fellowships, research assistantships *Unit head:* Dr. George R. Pack, Chair, 502-852-6798, Fax: 502-852-8149, E-mail: george.pack@louisville.edu.

University of Maine, Graduate School, College of Liberal Arts and Sciences, Department of Chemistry, Orono, ME 04469. Offers MS, PhD. *Faculty:* 12 full-time (2 women). *Students:* 26 full-time (5 women), 8 part-time (2 women); includes 3 minority (all Asian Americans or Pacific Islanders), 22 international. Average age 30. 24 applicants, 75% accepted, 7 enrolled. In 2005, 3 master's, 1 doctorate awarded. Terminal master's awarded for partial completion of doctoral program. *Degree requirements:* For master's, thesis; for doctorate, thesis/dissertation, oral exam. *Entrance requirements:* For master's and doctorate, GRE General Test. Additional exam requirements/recommendations for international students: Required—TOEFL. *Application deadline:* For fall admission, 2/1 for domestic students. Applications are processed on a rolling basis. Application fee: $50. Electronic applications accepted. *Financial support:* In 2005–06, 3 research assistantships with tuition reimbursements (averaging $21,700 per year), 15 teaching assistantships with tuition reimbursements (averaging $14,500 per year) were awarded; tuition waivers (full and partial) also available. Financial award application deadline: 3/1. *Faculty research:* Quantum mechanics, insect chemistry, organic synthesis. *Unit head:* Francois Amar, Chair, 207-581-1168. *Application contact:* Scott G. Delcourt, Associate Dean of the Graduate School, 207-581-3219, Fax: 207-581-3232, E-mail: graduate@maine.edu.

University of Manitoba, Faculty of Graduate Studies, Faculty of Science, Department of Chemistry, Winnipeg, MB R3T 2N2, Canada. Offers M Sc, PhD. *Degree requirements:* For master's, thesis; for doctorate, one foreign language, thesis/dissertation.

University of Maryland, Baltimore County, Graduate School, College of Natural Sciences and Mathematics, Department of Chemistry and Biochemistry, Program in Chemistry, Baltimore, MD 21250. Offers MS, PhD. Part-time and evening/weekend programs available. *Faculty:* 20 full-time (4 women), 2 part-time/adjunct (0 women). *Students:* 55 full-time (30 women), 2 part-time (1 woman); includes 19 minority (10 African Americans, 1 American Indian/Alaska Native, 8 Asian Americans or Pacific Islanders), 22 international. Average age 26. 51 applicants, 67% accepted, 13 enrolled. In 2005, 4 master's, 7 doctorates awarded. *Degree requirements:* For doctorate, thesis/dissertation, comprehensive exam. *Entrance requirements:* For master's, GRE General Test, minimum GPA of 3.0; for doctorate, GRE General Test, GRE Subject Test, minimum GPA of 3.0. Additional exam requirements/recommendations for international students: Required—TOEFL (minimum score 550 paper-based; 213 computer-based). *Application deadline:* For fall admission, 6/1 for domestic students, 1/1 for international students; for spring admission, 11/1 for domestic students, 5/1 for international students. Applications are processed on a rolling basis. Application fee: $50. Electronic applications accepted. *Expenses:* Tuition, state resident: part-time $395 per credit. Tuition, nonresident: part-time $652 per credit. Required fees: $82 per credit. Tuition and fees vary according to course load, program and reciprocity agreements. *Financial support:* In 2005–06, fellowships with full tuition reimbursements (averaging $24,000 per year), research assistantships with full tuition reimbursements (averaging $21,000 per year), teaching assistantships with full tuition reimbursements (averaging $21,000 per year) were awarded; tuition waivers (full) also available. *Faculty research:* Bio-organic chemistry and enzyme catalysis, protein-nucleic acid interactions. Total annual research expenditures: $4.9 million. *Unit head:* Dr. William R. LaCourse, Director, Graduate Program, 410-455-2491, Fax: 410-455-2608, E-mail: chemgrad@umbc.edu. *Application contact:* Patricia Gagne, Graduate Coordinator, 410-455-2491, Fax: 410-455-2608, E-mail: pgagne1@umbc.edu.

See Close-Up on page 157.

University of Maryland, College Park, Graduate Studies, College of Chemical and Life Sciences, Department of Chemistry and Biochemistry, Chemistry Program, College Park, MD 20742. Offers analytical chemistry (MS, PhD); inorganic chemistry (MS, PhD); organic chemistry (MS, PhD); physical chemistry (MS, PhD). Part-time and evening/weekend programs available. *Students:* 102 full-time (43 women), 4 part-time (2 women); includes 7 minority (2 African Americans, 3 Asian Americans or Pacific Islanders, 2 Hispanic Americans), 46 international. 128 applicants, 24% accepted, 30 enrolled. In 2005, 6 master's, 12 doctorates awarded. Terminal master's awarded for partial completion of doctoral program. *Median time to degree:* Of those who began their doctoral program in fall 1997, 33% received their degree in 8 years or less. *Degree requirements:* For master's, thesis optional; for doctorate, thesis/dissertation, 2 seminar presentations, oral exam. *Entrance requirements:* For master's and doctorate, GRE General Test, GRE Subject Test (recommended), minimum GPA of 3.0, 3 letters of recommendation. Additional exam requirements/recommendations for international students: Required—TOEFL, TSE. *Application deadline:* For fall admission, 4/1 for domestic students, 2/1 for international students; for spring admission, 10/21 for domestic students, 6/1 for international students. Applications are processed on a rolling basis. Application fee: $60. Electronic applications accepted. *Financial support:* In 2005–06, 10 fellowships (averaging $4,900 per year) were awarded; research assistantships, teaching assistantships with partial tuition reimbursements Financial award applicants required to submit FAFSA. *Faculty research:* Environmental chemistry, nuclear chemistry, lunar and environmental analysis, x-ray crystallography. *Application contact:* Dean of Graduate School, 301-405-4190, Fax: 301-314-9305.

University of Massachusetts Amherst, Graduate School, College of Natural Sciences and Mathematics, Department of Chemistry, Amherst, MA 01003. Offers MS, PhD. Part-time programs available. *Faculty:* 29 full-time (4 women). *Students:* 127 full-time (55 women), 7 part-time (3 women); includes 6 minority (2 African Americans, 4 Hispanic Americans), 90 international. Average age 29. 196 applicants, 36% accepted, 30 enrolled. In 2005, 8 master's, 14 doctorates awarded. Terminal master's awarded for partial completion of doctoral program. *Degree requirements:* For master's, thesis; for doctorate, one foreign language, thesis/dissertation. *Entrance requirements:* Additional exam requirements/recommendations for international students: Required—TOEFL (minimum score 530 paper-based; 197 computer-based). *Application deadline:* For fall admission, 2/1 priority date for domestic students, 2/1 priority date for international students; for spring admission, 10/1 for domestic students, 10/1 for international students. Applications are processed on a rolling basis. Application fee: $40 ($65 for international students). Electronic applications accepted. *Expenses:* Tuition, state resident: part-time $110 per credit. Tuition, nonresident: part-time $414 per credit. Required fees: $2,824 per term. One-time fee: $250 part-time. Full-time tuition and fees vary according to course load, campus/location, program and reciprocity agreements. *Financial support:* In 2005–06, fellowships with full tuition reimbursements (averaging $14,324 per year), research assistantships with full tuition reimbursements (averaging $13,167 per year), teaching assistantships with full tuition reimbursements (averaging $11,077 per year) were awarded; career-related internships or fieldwork, Federal Work-Study, scholarships/grants, traineeships, and unspecified assistantships also available. Support available to part-time students. Financial award application deadline: 2/1. *Unit head:* Dr. Bret Jackson, Director, 413-545-2318, Fax: 413-545-4490, E-mail: jackson@chem.umass.edu.

University of Massachusetts Boston, Office of Graduate Studies and Research, College of Science and Mathematics, Program in Chemistry, Boston, MA 02125-3393. Offers MS. Part-time and evening/weekend programs available. *Degree requirements:* For master's, thesis, oral exams, comprehensive exam. *Entrance requirements:* For master's, GRE General Test, GRE Subject Test, minimum GPA of 2.75. *Faculty research:* Synthesis and mechanisms of organic nitrogen compounds, application of spin resonance in the study of structure and dynamics, chemical education and teacher training, new synthetic reagents, structural study of inorganic solids by infrared and Raman spectroscopy.

University of Massachusetts Dartmouth, Graduate School, College of Arts and Sciences, Department of Chemistry, North Dartmouth, MA 02747-2300. Offers MS. Part-time programs available. *Faculty:* 14 full-time (2 women), 4 part-time/adjunct (3 women). *Students:* 7 full-time (3 women), 10 part-time (4 women); includes 4 minority (1 African American, 2 Asian Americans or Pacific Islanders, 1 Hispanic American), 10 international. Average age 29. 16 applicants, 81% accepted, 6 enrolled. In 2005, 2 degrees awarded. *Degree requirements:* For master's, thesis or alternative, comprehensive exam (for some programs). *Entrance requirements:* For master's, GRE. Additional exam requirements/recommendations for international students: Required—TOEFL (minimum score 550 paper-based). *Application deadline:* For fall admission, 5/1 for domestic students, 3/1 for international students; for spring admission, 11/1 for domestic students, 9/1 for international students. Application fee: $35 ($55 for international students). Electronic applications accepted. *Expenses:* Tuition, state resident: full-time $2,071; part-time $86 per credit. Tuition, nonresident: full-time $8,099; part-time $337 per credit. Required fees: $9,437; $393 per credit. *Financial support:* In 2005–06, 2 research assistantships with full tuition reimbursements (averaging $13,975 per year), 14 teaching assistantships with full tuition reimbursements (averaging $13,000 per year) were awarded; Federal Work-Study also available. Support available to part-time students. Financial award application deadline: 3/1; financial award applicants required to submit FAFSA. *Faculty research:* Spectrometric analysis, pesticides and DNA, wound-healing, Arabian sea dentrification, oceanic nitrogen fixation. Total annual research expenditures: $1.3 million. *Unit head:* Dr. Timothy Su, Director, 508-999-8238, Fax: 508-999-9167, E-mail: tsu@umassd.edu. *Application contact:* Carol Novo, Graduate Admissions Officer, 508-999-8604, Fax: 508-999-8183, E-mail: graduate@umassd.edu.

University of Massachusetts Lowell, Graduate School, College of Arts and Sciences, Department of Chemistry, Lowell, MA 01854-2881. Offers biochemistry (PhD); chemistry (MS, PhD); environmental studies (PhD); polymer sciences (MS, PhD). Terminal master's awarded for partial completion of doctoral program. *Degree requirements:* For master's, thesis; for doctorate, 2 foreign languages, thesis/dissertation. *Entrance requirements:* For master's and doctorate, GRE General Test. Electronic applications accepted.

University of Memphis, Graduate School, College of Arts and Sciences, Department of Chemistry, Memphis, TN 38152. Offers MS, PhD. Part-time programs available. *Faculty:* 11 full-time (2 women), 4 part-time/adjunct (1 woman). *Students:* 27 full-time (11 women), 4 part-time (3 women); includes 5 minority (2 African Americans, 2 Asian Americans or Pacific Islanders, 1 Hispanic American), 10 international. 44 applicants, 27% accepted. In 2005, 2 master's, 2 doctorates awarded. *Degree requirements:* For master's and doctorate, thesis/dissertation, comprehensive exam. *Entrance requirements:* For master's, GRE General Test, 32 undergraduate hours in chemistry; for doctorate, GRE General Test. *Application deadline:* For fall admission, 8/1 for domestic students; for spring admission, 12/1 for domestic students. Applications are processed on a rolling basis. Application fee: $25 ($50 for international students). *Financial support:* In 2005–06, 26 students received support, including 6 research assistantships with full tuition reimbursements available, 17 teaching assistantships with full tuition reimbursements available *Faculty research:* Computational chemistry, molecular spectroscopy, heterocyclic compounds, photochromic materials, nanoscale materials. Total annual research expenditures: $663,363. *Unit head:* Dr. Peter K. Bridson, Chairman, 901-678-4414, Fax: 901-678-3447, E-mail: pbridson@memphis.edu. *Application contact:* Dr. Roger V. Lloyd, Coordinator of Graduate Studies, 901-678-2632, Fax: 901-678-3447, E-mail: rlloyd@memphis.edu.

University of Miami, Graduate School, College of Arts and Sciences, Department of Chemistry, Coral Gables, FL 33124. Offers chemistry (MS); inorganic chemistry (PhD); organic chemistry (PhD); physical chemistry (PhD). *Faculty:* 11 full-time (1 woman). *Students:* 38 full-time (6 women), 28 international. Average age 27. 62 applicants, 27% accepted, 5 enrolled. In 2005, 1 master's, 5 doctorates awarded. Terminal master's awarded for partial completion of doctoral program. *Degree requirements:* For master's, comprehensive exam; for doctorate, thesis/dissertation, comprehensive exam. *Entrance requirements:* For master's and doctorate, GRE General Test. Additional exam requirements/recommendations for international students: Required—TOEFL (minimum score 550 paper-based; 213 computer-based). *Application deadline:* For fall admission, 1/15 for domestic students, 1/15 for international students. Applications are processed on a rolling basis. Application fee: $50. Electronic applications accepted. *Financial support:* In 2005–06, 38 students received support, including 1 fellowship with full tuition reimbursement available (averaging $20,000 per year), 10 research assistantships with full tuition reimbursements available (averaging $20,000 per year), 27 teaching assistantships with full tuition reimbursements available (averaging $20,000 per year); tuition waivers (full) also available. Financial award application deadline: 5/1; financial award applicants required to submit FAFSA. *Faculty research:* Supramolecular chemistry, electrochemistry, surface chemistry, catalysis, organometalic. *Unit head:* Dr. Vaidyanathan Ramamurthy, Chairman, 305-284-2282, Fax: 305-284-4571. *Application contact:* Eva Johnson, Graduate Secretary, 305-284-2094, Fax: 305-284-4571, E-mail: evaj@miami.edu.

University of Michigan, Horace H. Rackham School of Graduate Studies, College of Literature, Science, and the Arts, Department of Chemistry, Ann Arbor, MI 48109. Offers analytical chemistry (PhD); inorganic chemistry (PhD); material chemistry (PhD); organic chemistry (PhD); physical chemistry (PhD). *Faculty:* 48 full-time (8 women), 3 part-time/adjunct (all women). *Students:* ; includes 39 minority (4 African Americans, 1 American Indian/Alaska Native, 29 Asian Americans or Pacific Islanders, 5 Hispanic Americans), 66 international. Average age 26. 456 applicants, 29% accepted, 44 enrolled. In 2005, 37 degrees awarded. *Degree requirements:* For doctorate, thesis/dissertation, oral defense of dissertation, organic cumulative proficiency exams. *Entrance requirements:* For doctorate, GRE General Test, GRE Subject Test (recommended), 3 letters of recommendation. Additional exam requirements/recommendations for international students: Required—TOEFL. *Application deadline:* For fall admission, 1/31 for domestic students, 1/1 for international students. Applications are processed on a rolling basis. Application fee: $60 ($75 for international students). Electronic applications accepted. *Expenses:* Tuition, area resident: Full-time $14,082; part-time $894 per credit hour. Tuition, state resident: full-time $14,082; part-time $896 per credit hour. Tuition, nonresident: full-time $28,500; part-time $1,675 per credit hour. Required fees: $189; $189. *Financial support:* In 2005–06, 10 fellowships with full tuition reimbursements (averaging $20,000 per year), 70 research assistantships with full tuition reimbursements (averaging $19,000 per year), 120 teaching assistantships with full tuition reimbursements (averaging $20,000 per year) were awarded. Financial award applicants required to submit FAFSA. *Faculty research:* Biological catalysis, protein engineering, chemical sensors, de novo metalloprotein design, supramolecular architecture. Total annual research expenditures: $8 million. *Unit head:* Dr. Carol A. Fierke, Chair, 734-763-9681, Fax: 734-647-4847. *Application contact:* Linda Deitert, Assistant Director Graduate Studies, 734-764-7278, Fax: 734-647-4865, E-mail: chemadmissions@umich.edu.

University of Minnesota, Duluth, Graduate School, College of Science and Engineering, Department of Chemistry, Duluth, MN 55812-2496. Offers MS. Part-time programs available. *Faculty:* 30 full-time (3 women). *Students:* 24 full-time (6 women), 3 part-time (2 women); includes 2 minority (1 American Indian/Alaska Native, 1 Asian American or Pacific Islander), 13 international. Average age 24. 25 applicants, 56% accepted, 12 enrolled. In 2005, 13 degrees awarded. *Degree requirements:* For master's, thesis, registration. *Entrance requirements:* For master's, bachelor's degree in chemistry, minimum GPA of 3.0. Additional exam requirements/recommendations for international students: Required—TOEFL (minimum score 550 paper-based; 213 computer-based). *Application deadline:* For fall admission, 7/15 for domestic students; for spring admission, 11/15 for domestic students. Applications are processed on a rolling basis. Application fee: $55 ($75 for international students). *Financial support:* In 2005–06, 24 students received support, including 1 research assistantship with full tuition reimbursement available (averaging $13,000 per year), 23 teaching assistantships with full tuition reimbursements available (averaging $13,000 per year); Federal Work-Study, institutionally sponsored loans, scholarships/grants, health care benefits, and unspecified assistantships also available. Support available to part-time students. Financial award application deadline: 3/15. *Faculty research:* Physical, inorganic, organic, and analytical chemistry; biochemistry and molecular biology. Total annual research expenditures: $500,000. *Unit head:* Dr. Viktor V. Zhdankin, Director of Graduate Studies, 218-726-6902, Fax: 218-726-7394.

University of Minnesota, Twin Cities Campus, Graduate School, Institute of Technology, Department of Chemistry, Minneapolis, MN 55455-0213. Offers MS, PhD. Part-time programs available. Terminal master's awarded for partial completion of doctoral program. *Degree requirements:* For master's, thesis or alternative; for doctorate, thesis/dissertation, preliminary candidacy exams. *Entrance requirements:* For master's and doctorate, GRE General Test. Additional exam requirements/recommendations for international students: Required—TOEFL. *Expenses:* Tuition, state resident: full-time $8,748; part-time $729 per credit. Tuition, nonresident: full-time $15,848; part-time $1,321 per credit. Full-time tuition and fees vary according to class time, course load, program and reciprocity agreements. *Faculty research:* Analytical, biological, inorganic, organic, and physical chemistry.

University of Minnesota, Twin Cities Campus, School of Public Health, Division of Environmental Health Sciences, Area in Environmental Chemistry, Minneapolis, MN 55455-0213. Offers MS, PhD. *Degree requirements:* For doctorate, thesis/dissertation. *Entrance requirements:* For master's and doctorate, GRE General Test. *Application deadline:* For fall admission, 3/1 for domestic students. Applications are processed on a rolling basis. Application fee: $55 ($75 for international students). *Expenses:* Tuition, state resident: full-time $8,748; part-time $729 per credit. Tuition, nonresident: full-time $15,848; part-time $1,321 per credit. Full-time tuition and fees vary according to class time, course load, program and reciprocity agreements. *Financial support:* Fellowships, research assistantships available. Financial award application deadline: 3/1. *Application contact:* Kathy Soupir, Major Coordinator, 612-625-0622, Fax: 612-626-4837, E-mail: soupi001@umn.edu.

University of Mississippi, Graduate School, College of Liberal Arts, Department of Chemistry and Biochemistry, Oxford, University, MS 38677. Offers MS, DA, PhD. *Faculty:* 19 full-time (2 women). *Students:* 34 full-time (13 women), 4 part-time (2 women); includes 2 minority (both African Americans), 23 international. 35 applicants, 46% accepted, 10 enrolled. *Degree requirements:* For master's, thesis; for doctorate, one foreign language, thesis/dissertation. *Entrance requirements:* For master's, GRE General Test, minimum GPA of 3.0; for doctorate, GRE General Test. Additional exam requirements/recommendations for international students: Required—TOEFL. *Application deadline:* For fall admission, 4/1 for domestic students; for spring admission, 10/1 for domestic students. Applications are processed on a rolling basis. Application fee: $25. Electronic applications accepted. *Expenses:* Tuition, area resident: Full-time $4,320; part-time $240 per credit hour. Tuition, state resident: full-time $4,320; part-time $240 per credit hour. Tuition, nonresident: full-time $9,744; part-time $301 per credit hour. Tuition and fees vary according to program. *Financial support:* Scholarships/grants available. Financial award application deadline: 3/1; financial award applicants required to submit FAFSA. *Unit head:* Dr. Charles Hussey, Chairman, 662-915-7301, Fax: 662-915-7300, E-mail: chemistry@olemiss.edu.

University of Missouri–Columbia, Graduate School, College of Arts and Sciences, Department of Chemistry, Columbia, MO 65211. Offers analytical chemistry (MS, PhD); inorganic chemistry (MS, PhD); organic chemistry (MS, PhD); physical chemistry (MS, PhD). *Faculty:* 21 full-time (5 women). *Students:* 58 full-time (17 women), 44 part-time (16 women); includes 5 minority (3 African Americans, 2 Asian Americans or Pacific Islanders), 59 international. In 2005, 11 master's, 9 doctorates awarded. *Degree requirements:* For master's, thesis; for doctorate, one foreign language, thesis/dissertation. *Entrance requirements:* For master's and doctorate, GRE General Test, minimum GPA of 3.0. *Application deadline:* For fall admission, 4/1 for domestic studentsFor winter admission, 11/1 for domestic students. Applications are processed on a rolling basis. Application fee: $45 ($60 for international students). *Financial support:* Fellowships, research assistantships, teaching assistantships, institutionally sponsored loans available. *Unit head:* Dr. Sheryl A. Tucker, Director for Graduate Studies, 573-882-1729.

University of Missouri–Kansas City, College of Arts and Sciences, Department of Chemistry, Kansas City, MO 64110-2499. Offers analytical chemistry (MS, PhD); inorganic chemistry (MS, PhD); organic chemistry (MS, PhD); physical chemistry (MS, PhD); polymer chemistry (MS, PhD). PhD offered through the School of Graduate Studies. Part-time and evening/weekend programs available. *Faculty:* 14 full-time (2 women), 2 part-time/adjunct (0 women). *Students:* Average age 38. 13 applicants, 8% accepted, 1 enrolled. In 2005, 2 degrees awarded. *Degree requirements:* For master's, thesis (for some programs); for doctorate, thesis/dissertation. *Entrance requirements:* For master's, equivalent of American Chemical Society approved bachelor's degree in chemistry; for doctorate, GRE General Test, equivalent of American Chemical Society approved bachelor's degree in chemistry. Additional exam requirements/recommendations for international students: Required—TOEFL (minimum score 580 paper-based; 237 computer-based), TWE. *Application deadline:* For fall and spring admission, 4/15For winter admission, 10/15 for domestic students. Applications are processed on a rolling basis. Application fee: $25. Electronic applications accepted. *Expenses:* Tuition, state resident: full-time $4,738; part-time $263 per credit hour. Tuition, nonresident: full-time $12,235; part-time $679 per credit hour. Required fees: $582. Tuition and fees vary according to course load, program and student level. *Financial support:* In 2005–06, fellowships with partial tuition reimbursements (averaging $18,156 per year), research assistantships with partial tuition reimbursements (averaging $18,515 per year), teaching assistantships with partial tuition reimbursements (averaging $16,434 per year) were awarded; Federal Work-Study, institutionally sponsored loans, and scholarships/grants also available. Support available to part-time students. Financial award application deadline: 2/15. *Faculty research:* Molecular spectroscopy, characterization and synthesis of materials and compounds, computational chemistry, natural products, drug delivery systems and anti-tumor agents. Total annual research expenditures: $1.1 million. *Unit head:* Dr. Jerry Jean, Chairperson, 816-235-2273, Fax: 816-235-5502, E-mail: jeany@umkc.edu. *Application contact:* Graduate Recruiting Committee, 816-235-2272, Fax: 816-235-5502, E-mail: umkc-chemdept@umkc.edu.

University of Missouri–Rolla, Graduate School, College of Arts and Sciences, Department of Chemistry, Rolla, MO 65409-0910. Offers chemistry (MS, PhD); chemistry education (MST). Terminal master's awarded for partial completion of doctoral program. *Degree requirements:* For doctorate, one foreign language, thesis/dissertation. *Entrance requirements:* For master's and doctorate, minimum GPA of 3.0. Electronic applications accepted. *Faculty research:* Structure and properties of materials; bioanalytical, environmental, and polymer chemistry.

University of Missouri–St. Louis, College of Arts and Sciences, Department of Chemistry and Biochemistry, St. Louis, MO 63121. Offers chemistry (MS, PhD), including inorganic chemistry, organic chemistry, physical chemistry. Part-time and evening/weekend programs available. *Faculty:* 16. *Students:* 28 full-time (20 women), 16 part-time (7 women); includes 6 minority (5 African Americans, 1 Hispanic American), 23 international. Average age 32. 2,298 applicants, 74% accepted. In 2005, 9 master's, 4 doctorates awarded. Terminal master's awarded for partial completion of doctoral program. *Degree requirements:* For master's, thesis optional; for doctorate, thesis/dissertation. *Entrance requirements:* For doctorate, GRE General Test, 3 letters of recommendation. Additional exam requirements/recommendations for international students: Required—TOEFL (minimum score 550 paper-based; 213 computer-based). *Application deadline:* For fall admission, 7/1 for domestic students; for spring admission, 12/7 priority date for domestic students. Applications are processed on a rolling basis. Application fee: $35 ($40 for international students). Electronic applications accepted. *Expenses:* Tuition, state resident: part-time $263 per credit hour. Tuition, nonresident: part-time $680 per credit hour. Required fees: $53 per credit hour. Tuition and fees vary according to program. *Financial support:* In 2005–06, 9 research assistantships with full and partial tuition reimbursements (averaging $16,500 per year), 17 teaching assistantships with full and partial tuition reimbursements (averaging $16,500 per year) were awarded; fellowships with full and partial tuition reimbursements *Faculty research:* Metallaborane chemistry, serum transferrin chemistry, natural products chemistry, organic synthesis. *Unit head:* Dr. Cynthia Dupureur, Director of Graduate Studies, 314-516-5311, Fax: 314-516-5342, E-mail: gradchem@umsl.edu. *Application contact:* 314-516-5458, Fax: 314-516-5310, E-mail: gradadm@umsl.edu.

The University of Montana, Graduate School, College of Arts and Sciences, Department of Chemistry, Missoula, MT 59812-0002. Offers chemistry (MS, PhD), including environmental/analytical chemistry, inorganic chemistry, organic chemistry, physical chemistry. *Faculty:* 16 full-time (2 women). *Students:* 36 full-time (9 women), 7 part-time (2 women), 9 international. 25 applicants, 44% accepted, 6 enrolled. In 2005, 1 master's, 2 doctorates awarded. Terminal master's awarded for partial completion of doctoral program. *Degree requirements:* For master's, thesis (for some programs); for doctorate, thesis/dissertation. *Entrance requirements:* For master's and doctorate, GRE General Test. Additional exam requirements/recommendations for international students: Required—TOEFL (minimum score 575 paper-based; 230 computer-based). *Application deadline:* For fall admission, 2/15 for domestic students. Applications are processed on a rolling basis. Application fee: $45. *Expenses:* Tuition, state resident: part-time $267 per credit. Tuition, nonresident: part-time $665 per credit. Part-time tuition and fees vary according to course load and degree level. *Financial support:* In 2005–06, 13 research assistantships with tuition reimbursements (averaging $14,000 per year), 12 teaching assistantships with full tuition reimbursements (averaging $14,000 per year) were awarded; Federal Work-Study, scholarships/grants, and unspecified assistantships also available. Financial award application deadline: 3/1; financial award applicants required to submit FAFSA. *Faculty research:* Reaction mechanisms and kinetics, inorganic and organic synthesis, analytical chemistry, natural products. Total annual research expenditures: $789,952. *Unit head:* Dr. Edward Rosenberg, Chair, 406-243-2592, Fax: 406-243-4227.

University of Nebraska–Lincoln, Graduate College, College of Arts and Sciences, Department of Chemistry, Lincoln, NE 68588. Offers analytical chemistry (PhD); chemistry (MS); inorganic chemistry (PhD); organic chemistry (PhD); physical chemistry (PhD). *Degree requirements:* For master's, one foreign language, departmental qualifying exam, thesis optional; for doctorate, one foreign language, thesis/dissertation, departmental qualifying exams, comprehensive exam. *Entrance requirements:* For master's and doctorate, GRE. Additional exam requirements/recommendations for international students: Required—TOEFL (minimum score 550 paper-based; 213 computer-based). Electronic applications accepted. *Faculty*

Chemistry

University of Nebraska–Lincoln (continued)
research: Bioorganic and bioinorganic chemistry, biophysical and bioanalytical chemistry, structure-function of DNA and proteins, organometallics, mass spectrometry.

University of Nevada, Las Vegas, Graduate College, College of Science, Department of Chemistry, Las Vegas, NV 89154-9900. Offers biochemistry (MS); chemistry (MS); environmental science/chemistry (PhD); radiochemistry (PhD). Part-time programs available. *Faculty:* 19 full-time (3 women), 5 part-time/adjunct (1 woman). *Students:* 17 full-time (6 women), 18 part-time (13 women); includes 6 minority (3 Asian Americans or Pacific Islanders, 3 Hispanic Americans), 11 international. 30 applicants, 70% accepted, 17 enrolled. In 2005, 7 degrees awarded. *Degree requirements:* For master's, thesis. *Entrance requirements:* For master's, GRE General Test, minimum GPA of 3.0 in last 2 years or 2.75 cumulative. Additional exam requirements/recommendations for international students: Required—TOEFL (minimum score 550 paper-based; 213 computer-based). *Application deadline:* For fall admission, 6/15 for domestic students, 5/1 for international students; for spring admission, 11/15 for domestic students, 10/1 for international students. Application fee: $60 ($75 for international students). Electronic applications accepted. *Expenses:* Tuition, state resident: part-time $150 per credit. Tuition, nonresident: part-time $315 per credit. Tuition and fees vary according to course load, program and reciprocity agreements. *Financial support:* In 2005–06, 3 research assistantships with partial tuition reimbursements (averaging $1,000 per year), 9 teaching assistantships with full tuition reimbursements (averaging $10,000 per year) were awarded; career-related internships or fieldwork, Federal Work-Study, institutionally sponsored loans, scholarships/grants, health care benefits, and unspecified assistantships also available. Support available to part-time students. Financial award application deadline: 3/1. *Unit head:* Dr. Spencer Steinberg, Chair, 702-895-3510. *Application contact:* Graduate Coordinator, 702-895-3753, Fax: 702-895-4180, E-mail: gradcollege@unlv.edu.

University of Nevada, Reno, Graduate School, College of Science, Department of Chemistry, Reno, NV 89557. Offers MS, PhD. Terminal master's awarded for partial completion of doctoral program. *Degree requirements:* For master's, thesis; for doctorate, one foreign language, thesis/dissertation. *Entrance requirements:* For master's, GRE, minimum GPA of 2.75; for doctorate, GRE, minimum GPA of 3.0. Additional exam requirements/recommendations for international students: Required—TOEFL.

University of New Brunswick Fredericton, School of Graduate Studies, Faculty of Science, Department of Chemistry, Fredericton, NB E3B 5A3, Canada. Offers M Sc, PhD. Part-time programs available. *Degree requirements:* For master's and doctorate, thesis/dissertation. *Entrance requirements:* For master's and doctorate, minimum GPA of 3.0. Additional exam requirements/recommendations for international students: Required—TOEFL, TWE. Electronic applications accepted. *Faculty research:* X-ray crystallography, fluorine, electrochemistry, quantum chemical abinitio, spectroscopy, theoretical organic synthesis, NMR, pulp and paper.

University of New Hampshire, Graduate School, College of Engineering and Physical Sciences, Department of Chemistry, Durham, NH 03824. Offers MS, MST, PhD. *Faculty:* 14 full-time. *Students:* 33 full-time (11 women), 17 part-time (7 women); includes 2 minority (both African Americans), 21 international. Average age 26. 38 applicants, 63% accepted, 11 enrolled. In 2005, 3 master's, 4 doctorates awarded. Terminal master's awarded for partial completion of doctoral program. *Degree requirements:* For master's, thesis; for doctorate, one foreign language, thesis/dissertation. *Entrance requirements:* Additional exam requirements/recommendations for international students: Required—TOEFL (minimum score 550 paper-based; 213 computer-based). *Application deadline:* For fall admission, 4/1 priority date for domestic students, 4/1 priority date for international studentsFor winter admission, 12/1 for domestic students. Applications are processed on a rolling basis. Application fee: $60. *Expenses:* Tuition, state resident: full-time $8,010; part-time $445 per credit hour. Tuition, nonresident: full-time $19,730; part-time $810 per credit hour. Required fees: $322 per semester. Tuition and fees vary according to course load and program. *Financial support:* In 2005–06, 2 fellowships, 11 research assistantships, 34 teaching assistantships were awarded; Federal Work-Study, scholarships/grants, and tuition waivers (full and partial) also available. Support available to part-time students. Financial award application deadline: 2/15. *Faculty research:* Analytical, physical, organic, and inorganic chemistry. *Unit head:* Dr. Chris Bauer, Chairperson, 603-862-1550. *Application contact:* Cindi Rohwer, Coordinator, 603-862-1550, E-mail: chem.dept@unh.edu.

University of New Mexico, Graduate School, College of Arts and Sciences, Department of Chemistry, Albuquerque, NM 87131-2039. Offers MS, PhD. *Faculty:* 22 full-time (4 women), 6 part-time/adjunct (1 woman). *Students:* 78 full-time (25 women), 5 part-time (3 women); includes 8 minority (1 African American, 1 American Indian/Alaska Native, 3 Asian Americans or Pacific Islanders, 3 Hispanic Americans), 52 international. Average age 31. 62 applicants, 37% accepted, 18 enrolled. In 2005, 5 master's, 19 doctorates awarded. *Degree requirements:* For master's and doctorate, thesis/dissertation, comprehensive exam. *Entrance requirements:* For master's and doctorate, department entrance exams. *Application deadline:* For fall admission, 7/30 for domestic students; for spring admission, 11/30 for domestic students. *Expenses:* Tuition: Full-time $3,388; part-time $238 per credit hour. Required fees: $385 per term. Tuition and fees vary according to course load and program. *Financial support:* In 2005–06, 10 students received support, including fellowships (averaging $4,000 per year), 24 research assistantships (averaging $11,920 per year), 42 teaching assistantships (averaging $13,891 per year); Federal Work-Study and health care benefits also available. Financial award application deadline: 3/1; financial award applicants required to submit FAFSA. *Faculty research:* Analytical, inorganic, organic, and physical chemistry; biological chemistry. Total annual research expenditures: $2.7 million. *Unit head:* Dr. Cary Morrow, Chair, 505-277-5319, Fax: 505-277-2609, E-mail: cmorrow@unm.edu. *Application contact:* Karen McElveny, Program Coordinator, 505-277-1779, Fax: 505-277-2609, E-mail: kamc@unm.edu.

University of New Orleans, Graduate School, College of Sciences, Department of Chemistry, New Orleans, LA 70148. Offers MS, PhD. *Degree requirements:* For master's and doctorate, variable foreign language requirement, thesis/dissertation, departmental qualifying exam. *Entrance requirements:* For master's and doctorate, GRE General Test. Additional exam requirements/recommendations for international students: Required—TOEFL (minimum score 550 paper-based; 213 computer-based). Electronic applications accepted. *Faculty research:* Synthesis and reactions of novel compounds, high-temperature kinetics, calculations of molecular electrostatic potentials, structures and reactions of metal complexes.

The University of North Carolina at Chapel Hill, Graduate School, College of Arts and Sciences, Department of Chemistry, Chapel Hill, NC 27599. Offers MA, MS, PhD. *Degree requirements:* For master's, thesis (for some programs), comprehensive exam; for doctorate, thesis/dissertation, comprehensive exam. *Entrance requirements:* For master's and doctorate, GRE General Test, GRE Subject Test, minimum GPA of 3.0.

The University of North Carolina at Charlotte, Graduate School, College of Arts and Sciences, Department of Chemistry, Charlotte, NC 28223-0001. Offers MS. Part-time programs available. *Faculty:* 14 full-time (4 women), 1 part-time/adjunct (0 women). *Students:* 3 full-time (2 women), 23 part-time (13 women); includes 5 minority (3 African Americans, 1 Asian American or Pacific Islander, 1 Hispanic American), 3 international. Average age 27. 19 applicants, 63% accepted, 9 enrolled. In 2005, 7 degrees awarded. *Degree requirements:* For master's, thesis. *Entrance requirements:* For master's, GRE General Test or MAT, minimum GPA of 3.0 in undergraduate major, 2.75 overall. Additional exam requirements/recommendations for international students: Required—TOEFL (minimum score 557 paper-based; 220 computer-based). *Application deadline:* For fall admission, 7/1 for domestic students, 5/1 for international students; for spring admission, 11/1 for domestic students, 10/1 for international students. Applications are processed on a rolling basis. Application fee: $55. Electronic applications accepted. *Expenses:* Tuition, area resident: Full-time $2,504; part-time $157 per credit. Tuition, state resident: full-time $2,504; part-time $157 per credit. Tuition, nonresident: full-time $12,711; part-time $794 per credit. International tuition: $12,711 full-time. Required fees: $1,424; $89 per credit. Tuition and fees vary according to course load and program. *Financial support:* In 2005–06, 10 research assistantships (averaging $5,885 per year), 25 teaching assistantships (averaging $10,935 per year) were awarded; fellowships, career-related internships or fieldwork, Federal Work-Study, institutionally sponsored loans, scholarships/grants, and unspecified assistantships also available. Support available to part-time students. Financial award application deadline: 4/1; financial award applicants required to submit FAFSA. *Faculty research:* Biophysical organic chemistry and biochemistry; polymers, biomaterials and nanostructures; materials chemistry; synthetic organic and inorganic chemistry; bioanalytical chemistry and mass spectrometry. *Unit head:* Dr. Bernadette T. Donovan-Merkert, Chair, 704-687-4765, Fax: 704-687-3151, E-mail: bdonovan@email.uncc.edu. *Application contact:* Kathy B. Giddings, Director of Graduate Admissions, 704-687-3366, Fax: 704-687-3279, E-mail: gradadm@email.uncc.edu.

The University of North Carolina at Greensboro, Graduate School, College of Arts and Sciences, Department of Chemistry and Biochemistry, Greensboro, NC 27412-5001. Offers biochemistry (MS); chemistry (MS). *Students:* 6 full-time, 19 part-time. *Degree requirements:* For master's, one foreign language, thesis. *Entrance requirements:* For master's, GRE General Test. Additional exam requirements/recommendations for international students: Required—TOEFL. *Application deadline:* For fall admission, 6/15 for domestic students; for spring admission, 3/15 priority date for domestic students. Applications are processed on a rolling basis. *Expenses:* Tuition, state resident: part-time $302 per credit hour. Tuition, nonresident: part-time $1,683 per credit hour. Required fees: $51 per credit hour. Tuition and fees vary according to course load and program. *Financial support:* Fellowships, research assistantships with full tuition reimbursements, teaching assistantships with full tuition reimbursements, career-related internships or fieldwork, scholarships/grants, and traineeships available. *Faculty research:* Synthesis of novel cyclopentadienes, molybdenum hydroxylase-cata ladder polymers, vinyl silicones. *Unit head:* Dr. Jerry Walsh, Interim Head, 336-334-5714, Fax: 336-334-5402, E-mail: jlwalsh@uncg.edu. *Application contact:* Michelle Harkleroad, Director of Graduate Admissions, 336-334-4884, Fax: 336-334-4424, E-mail: mbharkle@uncg.edu.

The University of North Carolina Wilmington, College of Arts and Sciences, Department of Chemistry, Wilmington, NC 28403-3297. Offers MS. Part-time programs available. *Faculty:* 14 full-time (2 women), 2 part-time/adjunct (0 women). *Students:* 16 full-time (10 women), 21 part-time (11 women); includes 2 minority (1 African American, 1 Asian American or Pacific Islander). Average age 29. 23 applicants, 70% accepted, 13 enrolled. In 2005, 6 degrees awarded. *Degree requirements:* For master's, thesis, comprehensive exam. *Entrance requirements:* For master's, GRE General Test, minimum B average in undergraduate major. *Application deadline:* For fall admission, 6/1 for domestic students. Applications are processed on a rolling basis. Application fee: $45. *Financial support:* In 2005–06, 11 teaching assistantships were awarded; career-related internships or fieldwork and Federal Work-Study also available. Support available to part-time students. Financial award application deadline: 3/15. *Unit head:* Dr. James H. Reeves, Chairman, 910-962-3456, E-mail: reeves@uncw.edu. *Application contact:* Dr. Robert D. Roer, Dean, Graduate School, 910-962-4117, Fax: 910-962-3787, E-mail: roer@uncw.edu.

University of North Dakota, Graduate School, College of Arts and Sciences, Department of Chemistry, Grand Forks, ND 58202. Offers MS, PhD. *Faculty:* 14 full-time (5 women). *Students:* 5 full-time (4 women), 26 part-time (6 women). 26 applicants, 19% accepted, 4 enrolled. In 2005, 3 degrees awarded. Terminal master's awarded for partial completion of doctoral program. *Degree requirements:* For master's, thesis/dissertation, final exam; for doctorate, thesis/dissertation, final exam, comprehensive exam. *Entrance requirements:* For master's and doctorate, GRE General Test, GRE Subject Test, minimum GPA of 3.0. Additional exam requirements/recommendations for international students: Required—TOEFL (minimum score 550 paper-based; 213 computer-based). *Application deadline:* For fall admission, 2/15 priority date for domestic students, 2/15 priority date for international students; for spring admission, 10/15 priority date for domestic students, 10/15 priority date for international students. Applications are processed on a rolling basis. Application fee: $35. Electronic applications accepted. *Financial support:* In 2005–06, 11 research assistantships with full tuition reimbursements (averaging $11,822 per year), 17 teaching assistantships with full tuition reimbursements (averaging $9,800 per year) were awarded; fellowships, Federal Work-Study, institutionally sponsored loans, scholarships/grants, and tuition waivers (full and partial) also available. Support available to part-time students. Financial award application deadline: 3/15; financial award applicants required to submit FAFSA. *Faculty research:* Synthetic and structural organometallic chemistry, photochemistry, theoretical chemistry, chromatographic chemistry, x-ray crystallography. *Unit head:* Dr. Harmon B. Abrahamson, Graduate Director, 701-777-4427, Fax: 701-777-2331, E-mail: harmon_abrahamson@und.nodak.edu. *Application contact:* Brenda Halle, Admissions Specialist, 701-777-2947, Fax: 701-777-3619, E-mail: brendahalle@mail.und.edu.

University of Northern Colorado, Graduate School, College of Natural and Health Sciences, School of Chemistry, Earth Sciences and Physics, Program in Chemistry, Greeley, CO 80639. Offers chemistry education (PhD); chemistry: education (MS); chemistry: research (MS). *Accreditation:* NCATE (one or more programs are accredited). Part-time programs available. *Degree requirements:* For master's, thesis or alternative, comprehensive exam; for doctorate, thesis/dissertation, comprehensive exam. *Entrance requirements:* For master's, 3 letters of reference; for doctorate, GRE General Test, 3 letters of reference. Electronic applications accepted. *Expenses:* Tuition, state resident: full-time $4,968; part-time $207 per credit hour. Tuition, nonresident: full-time $14,688; part-time $612 per credit hour. Required fees: $645; $32 per credit hour.

University of Northern Iowa, Graduate College, College of Natural Sciences, Department of Chemistry, Cedar Falls, IA 50614. Offers MA, MS. Part-time programs available. *Faculty:* 3 full-time (0 women). *Students:* 2 full-time (1 woman), (both international). 5 applicants, 40% accepted, 1 enrolled. In 2005, 1 degree awarded. *Degree requirements:* For master's, thesis or alternative, comprehensive exam (for some programs). *Entrance requirements:* Additional exam requirements/recommendations for international students: Required—TOEFL (minimum score 500 paper-based; 180 computer-based). *Application deadline:* For fall admission, 8/1 for domestic students. Applications are processed on a rolling basis. Application fee: $30 ($50 for international students). Electronic applications accepted. *Expenses:* Tuition, state resident: full-time $5,708. Tuition, nonresident: full-time $13,532. Required fees: $712. *Financial support:* Career-related internships or fieldwork, Federal Work-Study, scholarships/grants, and tuition waivers (full and partial) available. Support available to part-time students. Financial award application deadline: 2/1. *Unit head:* Dr. Shoshanna Coon, Interim Head, 319-273-2437, Fax: 319-273-7127, E-mail: shoshanna.coon@uni.edu.

University of North Texas, Robert B. Toulouse School of Graduate Studies, College of Arts and Sciences, Department of Chemistry, Denton, TX 76203. Offers MS, PhD. Part-time and evening/weekend programs available. *Faculty:* 19 full-time (4 women). *Students:* 60 full-time (25 women), 24 part-time (13 women); includes 9 minority (3 African Americans, 1 American Indian/Alaska Native, 3 Asian Americans or Pacific Islanders, 2 Hispanic Americans), 38 international. Average age 31. 46 applicants, 70% accepted, 13 enrolled. In 2005, 5 master's, 7 doctorates awarded. Terminal master's awarded for partial completion of doctoral program. *Degree requirements:* For master's, comprehensive exam; for doctorate, one foreign language, thesis/dissertation, comprehensive exam. *Entrance requirements:* For master's and doctorate, GRE General Test. *Application deadline:* For fall admission, 7/15 for domestic students. Application fee: $50 ($75 for international students). *Expenses:* Tuition, area resident: Full-time $3,258; part-time $181 per semester hour. Tuition, state resident: full-time $3,258; part-time $181 per semester hour. Tuition, nonresident: full-time $8,226; part-time $451 per semester hour. International tuition: $8,226 full-time. Required fees: $1,219; $68 per semester hour. *Financial support:* Fellowships, research assistantships, teaching assistantships, career-related internships or fieldwork, Federal Work-Study, and institutionally sponsored loans available. Financial award application deadline: 4/1. *Faculty research:* Analytical, inorganic, physical, and organic chemistry and materials. Total annual research expenditures: $2 million. *Unit head:* Dr. Ruthanne D. Thomas, Chair, 940-565-3515, Fax: 940-565-4318, E-mail: rthomas@

unt.edu. *Application contact:* Dr. Thomas R. Cundari, Graduate Adviser, 940-565-3554, Fax: 940-565-4318, E-mail: chem@unt.edu.

University of Notre Dame, Graduate School, College of Science, Department of Chemistry and Biochemistry, Notre Dame, IN 46556. Offers biochemistry (MS, PhD); inorganic chemistry (MS, PhD); organic chemistry (MS, PhD); physical chemistry (MS, PhD). *Faculty:* 30 full-time (3 women). *Students:* 148 full-time (52 women); includes 5 minority (1 African American, 1 Asian American or Pacific Islander, 3 Hispanic Americans), 65 international. 144 applicants, 38% accepted, 21 enrolled. In 2005, 7 master's, 6 doctorates awarded. Terminal master's awarded for partial completion of doctoral program. *Median time to degree:* Of those who began their doctoral program in fall 1997, 54% received their degree in 8 years or less. *Degree requirements:* For master's, thesis, comprehensive exam; for doctorate, thesis/dissertation, qualifying exam. *Entrance requirements:* For master's and doctorate, GRE General Test, GRE Subject Test (strongly recommended). Additional exam requirements/recommendations for international students: Required—TOEFL. *Application deadline:* For fall admission, 2/1 for domestic students. Applications are processed on a rolling basis. Application fee: $50. Electronic applications accepted. *Financial support:* In 2005–06, 148 students received support, including 19 fellowships with full tuition reimbursements available (averaging $22,000 per year), 57 research assistantships with full tuition reimbursements available (averaging $15,250 per year), 53 teaching assistantships with full tuition reimbursements available (averaging $16,000 per year); tuition waivers (full) also available. Financial award application deadline: 2/1. *Faculty research:* Reaction design and mechaninistic studies; reactive intermediates; synthesis, structure and reactivity of organometallic. cluster complexes, and biologically active natural products; bioorganic chemistry; enzymology. Total annual research expenditures: $9.4 million. *Unit head:* Dr. Richard E. Taylor, Director of Graduate Studies, 574-631-7058, Fax: 574-631-6652, E-mail: taylor.61@nd.edu. *Application contact:* Dr. Terrence J. Akai, Director of Graduate Admissions, 574-631-7706, Fax: 574-631-4183, E-mail: gradad@nd.edu.

University of Oklahoma, Graduate College, College of Arts and Sciences, Department of Chemistry and Biochemistry, Norman, OK 73019-0390. Offers MS, PhD. Part-time programs available. *Faculty:* 27 full-time (5 women). *Students:* 114 full-time (45 women), 12 part-time (4 women); includes 4 minority (3 African Americans, 1 Asian American or Pacific Islander), 79 international. 55 applicants, 71% accepted, 23 enrolled. In 2005, 15 master's, 4 doctorates awarded. Terminal master's awarded for partial completion of doctoral program. *Degree requirements:* For master's, thesis optional; for doctorate, thesis/dissertation. *Entrance requirements:* For master's, GRE, BS in chemistry; for doctorate, GRE. Additional exam requirements/recommendations for international students: Required—TOEFL (minimum score 550 paper-based; 213 computer-based). *Application deadline:* For fall admission, 4/1 priority date for domestic students, 4/1 priority date for international students; for spring admission, 9/1 priority date for domestic students, 9/1 priority date for international students. Applications are processed on a rolling basis. Application fee: $40 ($90 for international students). *Expenses:* Tuition, area resident: Full-time $3,029; part-time $126 per credit hour. Tuition, state resident: full-time $3,029; part-time $126 per credit hour. Tuition, nonresident: full-time $10,807; part-time $450 per credit hour. International tuition: $10,807 full-time. Required fees: $1,231; $44 per credit hour. Tuition and fees vary according to course load and program. *Financial support:* In 2005–06, 22 students received support, including 6 fellowships with full tuition reimbursements available (averaging $5,000 per year), 36 research assistantships with partial tuition reimbursements available (averaging $13,115 per year), 74 teaching assistantships with partial tuition reimbursements available (averaging $13,936 per year); career-related internships or fieldwork, Federal Work-Study, scholarships/grants, traineeships, and tuition waivers (partial) also available. Support available to part-time students. Financial award application deadline: 4/1; financial award applicants required to submit FAFSA. *Faculty research:* Genomics, mechanisms of enzyme action bacterial transport processes, protein and nucleic acid structure and function. Total annual research expenditures: $3.1 million. *Unit head:* Dr. Glenn Dryhurst, Chair, 405-325-4811, Fax: 405-325-6111, E-mail: gdryhurst@ou.edu. *Application contact:* Ariene Crawford, Graduate Recruiting Secretary, 405-325-2946, Fax: 405-325-6111, E-mail: admission@chemdept.chem.ou.edu.

University of Oregon, Graduate School, College of Arts and Sciences, Department of Chemistry, Eugene, OR 97403. Offers biochemistry (MA, MS, PhD); chemistry (MA, MS, PhD). *Faculty:* 34 full-time (7 women), 7 part-time/adjunct (3 women). *Students:* 114 full-time (42 women), 7 part-time (3 women); includes 8 minority (2 African Americans, 6 Asian Americans or Pacific Islanders), 15 international. 18 applicants, 89% accepted. In 2005, 8 master's, 17 doctorates awarded. Terminal master's awarded for partial completion of doctoral program. *Degree requirements:* For doctorate, thesis/dissertation. *Entrance requirements:* For master's and doctorate, GRE General Test, minimum GPA of 3.0. Additional exam requirements/recommendations for international students: Required—TOEFL. *Application deadline:* For fall admission, 1/10 for domestic students. Applications are processed on a rolling basis. Application fee: $50. *Financial support:* In 2005–06, 54 teaching assistantships were awarded; Federal Work-Study and institutionally sponsored loans also available. Financial award application deadline: 4/15. *Faculty research:* Organic chemistry, organometallic chemistry, inorganic chemistry, physical chemistry, materials science, biochemistry, chemical physics, molecular or cell biology. *Unit head:* Tom Dyke, Head, 541-346-4603. *Application contact:* Lynde Ritzow, Graduate Admissions Coordinator, 541-346-4789, E-mail: lynde@oregon.uoregon.edu.

University of Ottawa, Faculty of Graduate and Postdoctoral Studies, Faculty of Science, Ottawa-Carleton Chemistry Institute, Ottawa, ON K1N 6N5, Canada. Offers M Sc, PhD. *Faculty:* 22 full-time (2 women). *Students:* 78 full-time, 9 part-time. 61 applicants, 31% accepted, 15 enrolled. In 2005, 8 master's, 13 doctorates awarded. *Degree requirements:* For master's, thesis, seminar; for doctorate, thesis/dissertation, 2 seminars, comprehensive exam. *Entrance requirements:* For master's, honors B Sc degree or equivalent, minimum B average; for doctorate, honors B Sc with minimum B average or M Sc in chemistry with minimum B+ average. *Application deadline:* For fall admission, 3/1 priority date for domestic students, 2/15 priority date for international studentsFor winter admission, 11/15 for domestic students; for spring admission, 4/1 for domestic students. Applications are processed on a rolling basis. Application fee: $75. Electronic applications accepted. *Expenses:* Tuition: Part-time $260 per credit. Tuition and fees vary according to course load and program. *Financial support:* Fellowships, research assistantships with full tuition reimbursements, teaching assistantships with full tuition reimbursements, career-related internships or fieldwork, Federal Work-Study, scholarships/grants, traineeships, tuition waivers (full and partial), and unspecified assistantships available. Financial award application deadline: 2/15. *Faculty research:* Organic chemistry, physical chemistry, inorganic chemistry. *Unit head:* Dr. Alain St. Amant, Chair, 613-562-5728 Ext. 5199, Fax: 613-562-5665, E-mail: lise@science.uottawa.ca. *Application contact:* Lise Maisonneuve, Graduate Studies Administrator, 613-562-5800 Ext. 6335, Fax: 613-562-5486, E-mail: lise@science.uottawa.ca.

University of Pennsylvania, School of Arts and Sciences, Graduate Group in Chemistry, Philadelphia, PA 19104. Offers MS, PhD. *Degree requirements:* For doctorate, thesis/dissertation. *Entrance requirements:* For doctorate, GRE General Test, GRE Subject Test, previous graduate course work in organic, inorganic, and physical chemistry each with a lab, differential and integral calculus, and general physics with a lab. Additional exam requirements/recommendations for international students: Required—TOEFL. Electronic applications accepted.

University of Pittsburgh, School of Arts and Sciences, Department of Chemistry, Pittsburgh, PA 15260. Offers MS, PhD. Part-time and evening/weekend programs available. *Faculty:* 27 full-time (3 women), 9 part-time/adjunct (2 women). *Students:* 190 full-time (63 women), 4 part-time (1 woman); includes 9 minority (2 African Americans, 4 Asian Americans or Pacific Islanders, 3 Hispanic Americans), 99 international. Average age 24. 345 applicants, 28% accepted, 29 enrolled. In 2005, 11 master's, 23 doctorates awarded. Terminal master's awarded for partial completion of doctoral program. *Degree requirements:* For master's and doctorate, thesis/dissertation, comprehensive exam. *Entrance requirements:* For master's and doctorate, GRE General Test, GRE Subject Test. Additional exam requirements/recommendations for international students: Required—TOEFL (minimum score 550 paper-based; 213 computer-based). *Application deadline:* For fall admission, 2/1 for domestic students. Applications are processed on a rolling basis. Application fee: $50. Electronic applications accepted. *Expenses:* Tuition, state resident: full-time $13,194; part-time $537 per credit. Tuition, nonresident: full-time $25,012; part-time $1,026 per credit. Required fees: $700; $164 per term. Tuition and fees vary according to campus/location and program. *Financial support:* In 2005–06, 194 students received support, including 4 fellowships with tuition reimbursements available (averaging $16,000 per year), 16 research assistantships with tuition reimbursements available (averaging $19,927 per year), 74 teaching assistantships with tuition reimbursements available (averaging $20,322 per year); Federal Work-Study, scholarships/grants, and health care benefits also available. Financial award application deadline: 2/1. *Faculty research:* Analytical, inorganic, organic, physical, and surface chemistry. Total annual research expenditures: $12.6 million. *Unit head:* Dr. David H Waldeck, Chairman, 412-624-0415, Fax: 412-624-1649, E-mail: chemchr@pitt.edu. *Application contact:* Fran Nagy, Graduate Program Administrator, 412-624-8501, Fax: 412-624-8611, E-mail: fnagy@pitt.edu.

See Close-Up on page 161.

University of Prince Edward Island, Faculty of Science, Charlottetown, PE C1A 4P3, Canada. Offers biology (M Sc); chemistry (M Sc). *Degree requirements:* For master's, thesis. *Entrance requirements:* Additional exam requirements/recommendations for international students: Required—TOEFL (minimum score 550 paper-based; 213 computer-based), Canadian Academic English Language Assessment, Michigan English Language Assessment Battery, Canadian Test of English for Scholars and Trainees. Tuition charges are reported in Canadian dollars. *Expenses:* Tuition, area resident: Full-time $3,816 Canadian dollars. International tuition: $6,356 Canadian dollars full-time. Part-time tuition and fees vary according to course level, degree level, campus/location, program and student level. *Faculty research:* Ecology and wildlife biology, molecular, genetics and biotechnology, organametallic, bio-organic, supramolecular and synthetic organic chemistry, neurobiology and stoke materials science.

University of Puerto Rico, Mayagüez Campus, Graduate Studies, College of Arts and Sciences, Department of Chemistry, Mayagüez, PR 00681-9000. Offers MS, PhD. *Faculty:* 30. *Students:* 18 full-time (13 women), 67 part-time (43 women); includes 50 minority (all Hispanic Americans), 35 international. 28 applicants, 50% accepted, 11 enrolled. In 2005, 9 degrees awarded. *Degree requirements:* For master's, one foreign language, thesis, comprehensive exam; for doctorate, thesis/dissertation, comprehensive exam. *Application deadline:* For fall admission, 2/15 for domestic students; for spring admission, 9/15 for domestic students. Applications are processed on a rolling basis. Application fee: $20. *Expenses:* Tuition, state resident: full-time $900; part-time $100 per credit. International tuition: $4,655 full-time. Part-time tuition and fees vary according to course level and course load. *Financial support:* In 2005–06, 59 students received support, including 2 fellowships (averaging $1,200 per year), 20 research assistantships (averaging $1,500 per year), 37 teaching assistantships (averaging $927 per year); Federal Work-Study and institutionally sponsored loans also available. *Faculty research:* Biochemistry, spectroscopy, food chemistry, physical chemistry, electrochemistry. Total annual research expenditures: $170,258. *Unit head:* Dr. María A. Aponte, Director, 787-265-3849.

University of Puerto Rico, Río Piedras, College of Natural Sciences, Department of Chemistry, San Juan, PR 00931-3300. Offers MS, PhD. Part-time and evening/weekend programs available. *Students:* 134 full-time (78 women), 6 part-time (4 women); includes 139 minority (1 Asian American or Pacific Islander, 138 Hispanic Americans). Average age 27. In 2005, 4 master's, 6 doctorates awarded. *Degree requirements:* For master's and doctorate, one foreign language, thesis/dissertation, comprehensive exam. *Entrance requirements:* For master's, GRE General Test, GRE Subject Test, EXADEP, interview, minimum GPA of 3.0, letter of recommendation; for doctorate, GRE General Test, GRE Subject Test, minimum GPA of 3.0, letter of recommendation. Additional exam requirements/recommendations for international students: Required—TOEFL. *Application deadline:* For fall admission, 2/1 for domestic students, 2/1 for international students. Application fee: $17. *Expenses:* Tuition, state resident: part-time $100 per credit. Tuition, nonresident: part-time $294 per credit. Required fees: $72 per term. *Financial support:* Fellowships, research assistantships, teaching assistantships, Federal Work-Study, institutionally sponsored loans, and tuition waivers (partial) available. Financial award application deadline:5/31. *Faculty research:* Organometallic synthesis, transition metal chemistry, organic air pollutants. *Unit head:* Dr. Nestor M. Carballeira, Coordinator, 787-764-0000 Ext. 4818, Fax: 787-756-8242.

See Close-Up on page 163.

University of Regina, Faculty of Graduate Studies and Research, Faculty of Science, Department of Chemistry and Biochemistry, Regina, SK S4S 0A2, Canada. Offers analytical chemistry (M Sc, PhD); biochemistry (M Sc, PhD); inorganic chemistry (M Sc, PhD); organic chemistry (M Sc, PhD); physical chemistry (M Sc, PhD). Part-time programs available. *Faculty:* 10 full-time (2 women), 4 part-time/adjunct (1 woman). *Students:* 14 full-time (6 women), 3 part-time (2 women). 21 applicants, 33% accepted. *Degree requirements:* For master's and doctorate, thesis/dissertation, departmental qualifying exam. *Entrance requirements:* For master's and doctorate, GRE. Additional exam requirements/recommendations for international students: Required—TOEFL (minimum score 580 paper-based; 237 computer-based). *Application deadline:* For fall admission, 1/1 for domestic studentsFor winter admission, 7/1 for domestic students. Applications are processed on a rolling basis. Application fee: $60 ($100 for international students). *Financial support:* In 2005–06, 4 fellowships (averaging $14,886 per year), 1 research assistantship (averaging $12,750 per year), 5 teaching assistantships (averaging $13,501 per year) were awarded; scholarships/grants also available. Financial award application deadline: 6/15. *Faculty research:* Organic synthesis, organic oxidations, ionic liquids theoretical/computational chemistry, protein biochemistry/biophysics, environmental analytical, photophysical/photochemistry. *Unit head:* Dr. Andrew G. Wee, Head, 306-585-4767, Fax: 306-585-4894, E-mail: chem.chair@uregina.ca. *Application contact:* Dr. Allan East, Associate Professor, 306-585-4003, Fax: 306-585-4894, E-mail: allan.east@uregina.ca.

University of Rhode Island, Graduate School, College of Arts and Sciences, Department of Chemistry, Kingston, RI 02881. Offers MS, PhD. In 2005, 3 master's, 7 doctorates awarded. *Application deadline:* For fall admission, 4/15 for domestic students. Applications are processed on a rolling basis. Application fee: $35. *Expenses:* Tuition, state resident: full-time $5,522; part-time $307 per credit. Tuition, nonresident: full-time $15,992; part-time $888 per credit. Required fees: $1,786; $73 per credit. One-time fee: $80 part-time. *Unit head:* Dr. William Euler, Chairperson, 401-874-5090.

University of Rochester, The College, Arts and Sciences, Department of Chemistry, Rochester, NY 14627-0250. Offers MS, PhD. Terminal master's awarded for partial completion of doctoral program. *Degree requirements:* For doctorate, thesis/dissertation, qualifying exam. *Entrance requirements:* For master's and doctorate, GRE General Test. Additional exam requirements/recommendations for international students: Required—TOEFL.

University of San Francisco, College of Arts and Sciences, Department of Chemistry, San Francisco, CA 94117-1080. Offers MS. Part-time and evening/weekend programs available. *Faculty:* 5 full-time (1 woman), 1 part-time/adjunct (0 women). *Students:* 11 full-time (5 women), 2 part-time (1 woman); includes 5 minority (4 Asian Americans or Pacific Islanders, 1 Hispanic American), 5 international. Average age 30. 19 applicants, 63% accepted, 2 enrolled. In 2005, 3 degrees awarded. *Degree requirements:* For master's, thesis. *Entrance requirements:* For master's, GRE General Test, GRE Subject Test, BS in chemistry or related field. *Application deadline:* Applications are processed on a rolling basis. Application fee: $55 ($65 for international students). *Expenses:* Tuition: Part-time $925 per unit. Tuition and fees vary according to degree level, campus/location and program. *Financial support:* In 2005–06, 12 students received support; fellowships, research assistantships, teaching assistantships, career-related internships or fieldwork, Federal Work-Study, institutionally sponsored loans, and tuition waivers (partial) available. Support available to part-time students. Financial award application deadline: 3/2; financial award applicants required to submit FAFSA. *Faculty research:* Organic

Chemistry

University of San Francisco (continued)
photochemistry, genetics of chromatic adaptation, electron transfer processes in solution, metabolism of protein hormones. Total annual research expenditures: $75,000. *Unit head:* Dr. Kim Summerhays, Chair, 415-422-6157, Fax: 415-422-5157.

University of Saskatchewan, College of Graduate Studies and Research, College of Arts and Sciences, Department of Chemistry, Saskatoon, SK S7N 5A2, Canada. Offers M Sc, PhD. *Degree requirements:* For master's and doctorate, thesis/dissertation, registration. *Entrance requirements:* Additional exam requirements/recommendations for international students: Required—TOEFL.

The University of Scranton, Graduate School, Department of Chemistry, Program in Chemistry, Scranton, PA 18510. Offers MA, MS. Part-time and evening/weekend programs available. *Faculty:* 10 full-time (3 women), 1 part-time/adjunct (0 women). *Students:* 4 full-time (1 woman), 1 part-time. Average age 24. 3 applicants, 100% accepted. In 2005, 1 degree awarded. *Degree requirements:* For master's, thesis (for some programs), capstone experience, comprehensive exam (for some programs), registration. *Entrance requirements:* For master's, minimum GPA of 2.75. Additional exam requirements/recommendations for international students: Required—TOEFL. *Application deadline:* Applications are processed on a rolling basis. Application fee: $50. *Expenses:* Tuition: Part-time $647 per credit. Required fees: $25 per term. *Financial support:* In 2005–06, 5 students received support, including 5 teaching assistantships with full tuition reimbursements available (averaging $8,600 per year); career-related internships or fieldwork, Federal Work-Study, unspecified assistantships, and teaching fellowships also available. Support available to part-time students. Financial award application deadline: 3/1. *Unit head:* Dr. Christopher A. Baumann, Director, 570-941-6389, Fax: 570-941-7510, E-mail: cab@scranton.edu.

The University of Scranton, Graduate School, Department of Chemistry, Program in Clinical Chemistry, Scranton, PA 18510. Offers MA, MS. Part-time and evening/weekend programs available. *Faculty:* 10 full-time (3 women), 1 part-time/adjunct (0 women). *Students:* 2 full-time (1 woman), 3 part-time (1 woman); includes 1 minority (Asian American or Pacific Islander), 1 international. Average age 26. 3 applicants, 100% accepted. In 2005, 5 degrees awarded. *Degree requirements:* For master's, thesis (for some programs), capstone experience, comprehensive exam (for some programs), registration. *Entrance requirements:* For master's, minimum GPA of 2.75. Additional exam requirements/recommendations for international students: Required—TOEFL. *Application deadline:* Applications are processed on a rolling basis. Application fee: $50. *Expenses:* Tuition: Part-time $647 per credit. Required fees: $25 per term. *Financial support:* In 2005–06, 4 students received support, including 4 teaching assistantships with full tuition reimbursements available (averaging $8,600 per year); career-related internships or fieldwork, Federal Work-Study, unspecified assistantships, and teaching fellowships also available. Support available to part-time students. Financial award application deadline: 3/1. *Unit head:* Dr. Christopher A. Baumann, Director, 570-941-6389, Fax: 570-941-7510, E-mail: cab@scranton.edu.

University of South Carolina, The Graduate School, College of Arts and Sciences, Department of Chemistry and Biochemistry, Columbia, SC 29208. Offers IMA, MAT, MS, PhD. IMA and MAT offered in cooperation with the College of Education. Part-time programs available. Terminal master's awarded for partial completion of doctoral program. *Degree requirements:* For master's and doctorate, thesis/dissertation, comprehensive exam, registration. *Entrance requirements:* For master's and doctorate, GRE General Test. Additional exam requirements/recommendations for international students: Required—TOEFL. Electronic applications accepted. *Faculty research:* Spectroscopy, crystallography, organic and organometallic synthesis, analytical chemistry, materials.

The University of South Dakota, Graduate School, College of Arts and Sciences, Department of Chemistry, Vermillion, SD 57069-2390. Offers MA, MNS. *Faculty:* 8 full-time (1 woman), 3 part-time/adjunct (1 woman). *Students:* 11 full-time (3 women). In 2005, 1 degree awarded. *Degree requirements:* For master's, thesis, comprehensive exam. *Entrance requirements:* For master's, GRE (international students), minimum GPA of 2.7. Additional exam requirements/recommendations for international students: Required—TOEFL (minimum score 550 paper-based; 213 computer-based), IBT 79. *Application deadline:* For fall admission, 3/15 for domestic students. Applications are processed on a rolling basis. Application fee: $35. Electronic applications accepted. *Expenses:* Tuition, state resident: part-time $116 per credit hour. Tuition, nonresident: part-time $341 per credit hour. Required fees: $85 per credit hour. Tuition and fees vary according to course load, program and reciprocity agreements. *Financial support:* In 2005–06, 4 research assistantships with partial tuition reimbursements (averaging $11,000 per year), 6 teaching assistantships with partial tuition reimbursements (averaging $11,000 per year) were awarded; Federal Work-Study and unspecified assistantships also available. Support available to part-time students. Financial award applicants required to submit FAFSA. *Faculty research:* Electrochemistry, photochemistry, inorganic synthesis, environmental and solid-state chemistry. *Unit head:* Dr. Mary Berry, Chair, 605-677-5487, Fax: 605-677-6397, E-mail: mberry@usd.edu. *Application contact:* Andrew Sykes, Information Contact, 605-677-5487, Fax: 605-677-6397, E-mail: asykes@usd.edu.

University of Southern California, Graduate School, College of Letters, Arts and Sciences, Department of Chemistry, Program in Chemistry, Los Angeles, CA 90089. Offers MA, MS, PhD. *Faculty:* 27 full-time (3 women). *Students:* 120 full-time (44 women); includes 8 minority (1 African American, 5 Asian Americans or Pacific Islanders, 2 Hispanic Americans), 80 international. Average age 27. 800 applicants, 11% accepted, 25 enrolled. In 2005, 3 master's, 16 doctorates awarded. Terminal master's awarded for partial completion of doctoral program. *Degree requirements:* For master's, qualifying exam, thesis optional; for doctorate, thesis/dissertation, qualifying exam. *Entrance requirements:* For master's and doctorate, GRE General Test. *Application deadline:* For fall admission, 3/1 for domestic students. Applications are processed on a rolling basis. Application fee: $0. *Expenses:* Tuition: Full-time $25,416; part-time $1,059 per unit. Required fees: $484; $484 per year. Tuition and fees vary according to course load and program. *Financial support:* In 2005–06, 120 students received support, including fellowships with full tuition reimbursements available (averaging $22,000 per year), research assistantships with tuition reimbursements available (averaging $22,000 per year), teaching assistantships with tuition reimbursements available (averaging $22,000 per year); Federal Work-Study, institutionally sponsored loans, scholarships/grants, and health care benefits also available. Financial award application deadline: 3/1. *Faculty research:* Organic chemistry. *Application contact:* Heather Connor, Graduate Advisor, 213-740-6855, Fax: 213-740-2701, E-mail: hconnor@usc.edu.

University of Southern Mississippi, Graduate School, College of Science and Technology, Department of Chemistry and Biochemistry, Hattiesburg, MS 39406-0001. Offers analytical chemistry (MS, PhD); biochemistry (MS, PhD); inorganic chemistry (MS, PhD); organic chemistry (MS, PhD); physical chemistry (MS, PhD). *Degree requirements:* For master's and doctorate, thesis/dissertation, comprehensive exam. *Entrance requirements:* For master's, GRE General Test, minimum GPA of 2.75 in last 60 hours; for doctorate, GRE General Test, minimum GPA of 3.5. Additional exam requirements/recommendations for international students: Required—TOEFL. *Faculty research:* Plant biochemistry, photo chemistry, polymer chemistry, x-ray analysis, enzyme chemistry.

University of South Florida, College of Graduate Studies, College of Arts and Sciences, Department of Chemistry, Tampa, FL 33620-9951. Offers analytical chemistry (MS, PhD); biochemistry (MS, PhD); inorganic chemistry (MS, PhD); organic chemistry (MS, PhD); physical chemistry (MS, PhD); polymer chemistry (PhD). Part-time programs available. *Faculty:* 22. *Students:* 102 full-time (45 women), 15 part-time (6 women); includes 13 minority (5 Asian Americans or Pacific Islanders, 8 Hispanic Americans), 60 international. 80 applicants, 71% accepted, 23 enrolled. In 2005, 2 master's awarded. Terminal master's awarded for partial completion of doctoral program. *Degree requirements:* For master's, thesis; for doctorate, 2 foreign languages, thesis/dissertation, colloquium. *Entrance requirements:* For master's and doctorate, GRE General Test, minimum GPA of 3.0 in last 30 hours of chemistry course work, 3 letters of recommendation. Additional exam requirements/recommendations for international students: Required—TOEFL (minimum score 550 paper-based; 213 computer-based), TSE(minimum score 50). *Application deadline:* For fall admission, 5/1 priority date for domestic students, 3/1 priority date for international students; for spring admission, 10/1 priority date for domestic students, 8/1 priority date for international students. Applications are processed on a rolling basis. Application fee: $30. Electronic applications accepted. *Financial support:* In 2005–06, 91 students received support; fellowships with partial tuition reimbursements available, research assistantships with partial tuition reimbursements available, teaching assistantships with partial tuition reimbursements available, career-related internships or fieldwork, institutionally sponsored loans, scholarships/grants, health care benefits, and unspecified assistantships available. Financial award application deadline: 6/30. *Faculty research:* Synthesis, bio-organic chemistry, bioinorganic chemistry, environmental chemistry, NMR. *Unit head:* Dr. Michael Zaworotko, Chairperson, 813-974-4129, Fax: 813-974-3203, E-mail: xtal@usf.edu. *Application contact:* Dr. Brian Space, Graduate Coordinator, 813-974-3397, Fax: 813-974-3203.

The University of Tennessee, Graduate School, College of Arts and Sciences, Department of Chemistry, Knoxville, TN 37996. Offers analytical chemistry (MS, PhD); chemical physics (PhD); environmental chemistry (MS, PhD); inorganic chemistry (MS, PhD); organic chemistry (MS, PhD); physical chemistry (MS, PhD); polymer chemistry (MS, PhD); theoretical chemistry (PhD). Part-time programs available. Terminal master's awarded for partial completion of doctoral program. *Degree requirements:* For master's and doctorate, thesis/dissertation. *Entrance requirements:* For master's and doctorate, GRE General Test, minimum GPA of 2.7. Additional exam requirements/recommendations for international students: Required—TOEFL. Electronic applications accepted.

The University of Texas at Arlington, Graduate School, College of Science, Department of Chemistry and Biochemistry, Arlington, TX 76019. Offers applied chemistry (PhD); chemistry (MS). Part-time programs available. *Faculty:* 8 full-time (0 women). *Students:* 56 full-time (21 women), 2 part-time; includes 6 minority (3 African Americans, 2 Asian Americans or Pacific Islanders, 1 Hispanic American), 44 international. 61 applicants, 43% accepted, 19 enrolled. In 2005, 2 master's, 4 doctorates awarded. Terminal master's awarded for partial completion of doctoral program. *Median time to degree:* Of those who began their doctoral program in fall 1997, 100% received their degree in 8 years or less. *Degree requirements:* For master's, thesis optional; for doctorate, thesis/dissertation, internship, oral defense of dissertation. *Entrance requirements:* For master's, GRE General Test, minimum GPA of 3.0 in last 60 hours of course work; for doctorate, GRE General Test. Additional exam requirements/recommendations for international students: Required—TOEFL (minimum score 550 paper-based; 213 computer-based). *Application deadline:* For fall admission, 6/16 for domestic students. Applications are processed on a rolling basis. Application fee: $35 ($50 for international students). *Expenses:* Tuition, area resident: Full-time $3,350. Tuition, state resident: full-time $3,350. Tuition, nonresident: full-time $8,318. International tuition: $8,448 full-time. Required fees: $1,277. Full-time tuition and fees vary according to course level and program. *Financial support:* In 2005–06, 40 students received support, including 4 fellowships (averaging $1,000 per year), 13 research assistantships (averaging $19,000 per year), 27 teaching assistantships (averaging $19,000 per year); career-related internships or fieldwork, Federal Work-Study, institutionally sponsored loans, scholarships/grants, health care benefits, tuition waivers (partial), and unspecified assistantships also available. Financial award application deadline: 6/1; financial award applicants required to submit FAFSA. *Unit head:* Dr. Edward Bellion, Chairman, 817-272-3171, Fax: 817-272-3808. *Application contact:* Dr. Rasika Dias, Graduate Adviser, 817-272-3813, Fax: 817-272-3808, E-mail: dias@uta.edu.

Announcement: The Department of Chemistry and Biochemistry offers a program leading to the PhD in chemistry. In addition to the traditional PhD curriculum and dissertation, this program offers a paid industrial internship at a major US corporation and a series of survey courses in various aspects of chemistry. Graduates of this program have been 100% successful in obtaining employment after completion of this degree. The program is ideally suited for students interested in a career in the chemical industry or in academics. The department is active in a wide range of modern chemical/biochemical/materials research areas. Visit the department's Web site at http://utachem.uta.edu.

The University of Texas at Austin, Graduate School, College of Natural Sciences, Department of Chemistry and Biochemistry, Austin, TX 78712-1111. Offers analytical chemistry (MA, PhD); biochemistry (MA, PhD); inorganic chemistry (MA, PhD); organic chemistry (MA, PhD); physical chemistry (MA, PhD). *Entrance requirements:* For master's and doctorate, GRE General Test.

The University of Texas at Dallas, School of Natural Sciences and Mathematics, Programs in Chemistry, Richardson, TX 75083-0688. Offers MS, PhD. Part-time and evening/weekend programs available. *Faculty:* 17 full-time (1 woman). *Students:* 51 full-time (19 women), 7 part-time (5 women); includes 17 minority (1 African American, 6 Asian Americans or Pacific Islanders, 10 Hispanic Americans), 26 international. Average age 30. 52 applicants, 48% accepted, 18 enrolled. In 2005, 14 master's, 10 doctorates awarded. *Degree requirements:* For master's, thesis or internship; for doctorate, research practica. *Entrance requirements:* For master's and doctorate, GRE General Test, minimum GPA of 3.0 in upper-level course work in field. Additional exam requirements/recommendations for international students: Required—TOEFL (minimum score 550 paper-based; 213 computer-based). *Application deadline:* For fall admission, 7/15 for domestic students; for spring admission, 11/15 for domestic students. Applications are processed on a rolling basis. Application fee: $50 ($100 for international students). Electronic applications accepted. *Expenses:* Tuition, area resident: Full-time $5,450; part-time $303 per credit. Tuition, state resident: full-time $5,450; part-time $303 per credit. Tuition, nonresident: full-time $12,648; part-time $703 per credit. International tuition: $12,648 full-time. Tuition and fees vary according to program. *Financial support:* In 2005–06, 24 research assistantships with tuition reimbursements (averaging $14,928 per year), 19 teaching assistantships with tuition reimbursements (averaging $13,419 per year) were awarded; fellowships, career-related internships or fieldwork, Federal Work-Study, institutionally sponsored loans, and scholarships/grants also available. Support available to part-time students. Financial award application deadline: 4/30; financial award applicants required to submit FAFSA. *Faculty research:* Organic photochemistry, bioinorganic chemistry, organic solid-state and polymer chemistry, environmental chemistry, scanning probe microscopy. Total annual research expenditures: $5.2 million. *Unit head:* Dr. John P. Ferraris, Department Head, 972-883-2516, Fax: 972-883-2925, E-mail: chemistry@utdallas.edu. *Application contact:* Dr. Warren J Goux, Graduate Advisor, 972-883-2660, Fax: 972-883-2925, E-mail: wgoux@utdallas.edu.

The University of Texas at El Paso, Graduate School, College of Science, Department of Chemistry, El Paso, TX 79968-0001. Offers MS. Part-time and evening/weekend programs available. *Degree requirements:* For master's, thesis. *Entrance requirements:* For master's, GRE General Test, minimum GPA of 3.0. Additional exam requirements/recommendations for international students: Required—TOEFL. Electronic applications accepted.

The University of Texas at San Antonio, College of Sciences, Department of Chemistry, San Antonio, TX 78249-0617. Offers MS. *Degree requirements:* For master's, thesis optional. *Entrance requirements:* For master's, GRE General Test, minimum GPA of 3.0 in all undergraduate chemistry courses, 2 letters of recommendation. Additional exam requirements/recommendations for international students: Required—TOEFL (minimum score 500 paper-based; 173 computer-based). Electronic applications accepted.

University of the Sciences in Philadelphia, College of Graduate Studies, Program in Chemistry, Biochemistry and Pharmacognosy, Philadelphia, PA 19104-4495. Offers biochemistry (MS, PhD); chemistry (MS, PhD); medicinal chemistry (MS, PhD); pharmacognosy (MS, PhD). Part-time programs available. *Faculty:* 13 full-time (3 women). *Students:* 8 full-time (4 women),

7 part-time (3 women); includes 2 minority (both Asian Americans or Pacific Islanders), 5 international. Average age 28. In 2005, 3 master's, 4 doctorates awarded. *Degree requirements:* For master's, thesis/dissertation, qualifying exams; for doctorate, thesis/dissertation, qualifying exams, comprehensive exam. *Entrance requirements:* For master's and doctorate, GRE General Test, GRE Subject Test. Additional exam requirements/recommendations for international students: Required—TOEFL, TWE, TSE. *Application deadline:* Applications are processed on a rolling basis. Application fee: $50. *Expenses:* Tuition: Part-time $999 per credit. Tuition and fees vary according to program. *Financial support:* In 2005–06, 19 students received support, including 1 fellowship with full tuition reimbursement available (averaging $19,000 per year), 1 research assistantship with full tuition reimbursement available (averaging $19,000 per year), 12 teaching assistantships with full tuition reimbursements available (averaging $18,500 per year); institutionally sponsored loans, scholarships/grants, and tuition waivers (full) also available. Financial award application deadline: 5/1. *Faculty research:* Organic and medicinal synthesis, mass spectroscopy use in protein analysis, study of analogues of taxol, cholesteryl esters. Total annual research expenditures: $341,700. *Unit head:* Dr. James McKee, Director, 215-596-8847, Fax: 215-596-8543, E-mail: jmckee@usip.edu. *Application contact:* Joyce D'Angelo, Administrative Assistant, 215-596-8937, E-mail: j.dangel@usip.edu.

The University of Toledo, Graduate School, College of Arts and Sciences, Department of Chemistry, Toledo, OH 43606-3390. Offers analytical chemistry (MS, PhD); biological chemistry (MS, PhD); inorganic chemistry (MS, PhD); organic chemistry (MS, PhD); physical chemistry (MS, PhD). Part-time programs available. *Faculty:* 18. *Students:* 80 full-time (35 women), 8 part-time (5 women); includes 3 minority (2 African Americans, 1 Asian American or Pacific Islander), 43 international. Average age 27. 32 applicants, 75% accepted, 19 enrolled. In 2005, 7 master's awarded. *Degree requirements:* For master's and doctorate, thesis/dissertation. *Entrance requirements:* For master's and doctorate, GRE General Test, GRE Subject Test. Additional exam requirements/recommendations for international students: Required—TOEFL. *Application deadline:* For fall admission, 8/1 for domestic students. Applications are processed on a rolling basis. Application fee: $45. Electronic applications accepted. *Expenses:* Tuition, area resident: Full-time $3,312; part-time $308 per credit hour. Tuition, state resident: full-time $3,312. Tuition, nonresident: full-time $6,616; part-time $735 per credit hour. International tuition: $6,616 full-time. *Financial support:* In 2005–06, 1 research assistantship with full tuition reimbursement (averaging $4,000 per year), 55 teaching assistantships with full tuition reimbursements (averaging $13,336 per year) were awarded; fellowships, Federal Work-Study and institutionally sponsored loans also available. Support available to part-time students. Financial award application deadline: 4/1; financial award applicants required to submit FAFSA. *Faculty research:* Enzymology, materials chemistry, crystallography, theoretical chemistry. *Unit head:* Dr. Alan Pinkerton, Chair, 419-530-7902, Fax: 419-530-4033, E-mail: apinker@uoft02.utoledo.edu. *Application contact:* Charlene Morlock-Hansen, 419-530-2100, E-mail: charlene.hanson@utoledo.edu.

See Close-Up on page 167.

University of Toronto, School of Graduate Studies, Physical Sciences Division, Department of Chemistry, Toronto, ON M5S 1A1, Canada. Offers M Sc, PhD. *Degree requirements:* For master's, thesis; for doctorate, thesis/dissertation, oral exam, thesis defense. *Entrance requirements:* For master's, bachelor's degree in chemistry or a related field; for doctorate, master's degree in chemistry or a related field.

University of Tulsa, Graduate School, College of Engineering and Natural Sciences, Department of Chemistry, Tulsa, OK 74104-3189. Offers MS. *Faculty:* 7 full-time (0 women). *Students:* 7 full-time (6 women), 4 part-time (3 women); includes 2 minority (both Asian Americans or Pacific Islanders), 3 international. Average age 35. 6 applicants, 33% accepted, 1 enrolled. In 2005, 2 degrees awarded. *Degree requirements:* For master's, thesis (for some programs). *Entrance requirements:* For master's, GRE General Test. Additional exam requirements/recommendations for international students: Required—TOEFL (minimum score 550 paper-based; 213 computer-based), IELT(minimum score 6). *Application deadline:* Applications are processed on a rolling basis. Application fee: $30. Electronic applications accepted. *Expenses:* Tuition: Full-time $12,132; part-time $674 per credit hour. Required fees: $60; $3 per credit hour. *Financial support:* In 2005–06, 6 students received support, including 4 research assistantships with full and partial tuition reimbursements available (averaging $12,950 per year), 2 teaching assistantships with full tuition reimbursements available (averaging $10,000 per year); fellowships, career-related internships or fieldwork, Federal Work-Study, scholarships/grants, tuition waivers (full and partial), and unspecified assistantships also available. Support available to part-time students. Financial award application deadline: 2/1; financial award applicants required to submit FAFSA. *Faculty research:* Nanotechnology, polymer sensors, natural products chemistry, quantum dots, automotive catalyst materials. Total annual research expenditures: $1 million. *Unit head:* Dr. Dale C. Teeters, Chairperson, 918-631-3147, Fax: 918-631-3404, E-mail: dale-teeters@utulsa.edu.

University of Utah, The Graduate School, College of Science, Department of Chemistry, Salt Lake City, UT 84112-1107. Offers chemical physics (PhD); chemistry (M Phil, MA, MS, PhD); science teacher education (MS). Part-time programs available. Postbaccalaureate distance learning degree programs offered. *Faculty:* 27 full-time (4 women), 2 part-time/adjunct (0 women). *Students:* 151 full-time (66 women), 21 part-time (6 women); includes 9 minority (1 African American, 4 Asian Americans or Pacific Islanders, 4 Hispanic Americans), 76 international. Average age 28. 42 applicants, 79% accepted, 30 enrolled. In 2005, 14 master's, 22 doctorates awarded. Terminal master's awarded for partial completion of doctoral program. *Median time to degree:* Of those who began their doctoral program in fall 1997, 100% received their degree in 8 years or less. *Degree requirements:* For master's, 26 hours coursework, 10 hours research, thesis optional; for doctorate, thesis/dissertation, 18 hours coursework, 14 hours research. *Entrance requirements:* For master's and doctorate, GRE General Test, minimum GPA of 3.0. Additional exam requirements/recommendations for international students: Required—TOEFL (minimum score 620 paper-based; 260 computer-based), TSE. *Application deadline:* For fall admission, 7/1 for domestic students, 4/1 for international studentsFor winter admission, 3/15 for domestic students; for spring admission, 11/1 for domestic students. Applications are processed on a rolling basis. Application fee: $45 ($65 for international students). Electronic applications accepted. *Expenses:* Tuition, area resident: Full-time $2,932; part-time $2,212 per term. Tuition, state resident: full-time $2,932; part-time $2,212 per term. Tuition, nonresident: full-time $10,350; part-time $7,812 per term. International tuition: $10,350 full-time. Required fees: $590; $516 per term. Tuition and fees vary according to course load and program. *Financial support:* In 2005–06, 172 students received support, including research assistantships with tuition reimbursements available (averaging $21,500 per year), teaching assistantships with tuition reimbursements available (averaging $20,500 per year); fellowships with tuition reimbursements available, scholarships/grants and tuition waivers (full) also available. Financial award application deadline: 7/1; financial award applicants required to submit FAFSA. *Faculty research:* Biological, theoretical, inorganic, organic, and physical-analytical chemistry. Total annual research expenditures: $11.1 million. *Unit head:* Peter B. Armentrout, Chair, 801-581-6681, Fax: 801-581-8433, E-mail: armentrout@chemistry.utah.edu. *Application contact:* Jo Hoovey, Graduate Coordinator, 801-581-4393, Fax: 801-581-5408, E-mail: jhoovey@chem.utah.edu.

University of Vermont, Graduate College, College of Arts and Sciences, Department of Chemistry, Burlington, VT 05405. Offers chemistry (MS, MST, PhD); chemistry education (MAT). *Accreditation:* NCATE (one or more programs are accredited). *Students:* 42 (13 women); includes 1 minority (African American) 14 international. 33 applicants, 70% accepted, 6 enrolled. In 2005, 3 master's, 3 doctorates awarded. *Degree requirements:* For master's, one foreign language, thesis; for doctorate, 2 foreign languages, thesis/dissertation. *Entrance requirements:* For master's and doctorate, GRE General Test. Additional exam requirements/recommendations for international students: Required—TOEFL (minimum score 550 paper-based; 213 computer-based). *Application deadline:* For fall admission, 4/1 for domestic students. Applications are processed on a rolling basis. Application fee: $40. Electronic applications accepted. *Expenses:* Tuition, area resident: Part-time $410 per credit hour. Tuition, nonresident: part-time $1,034 per credit hour. *Financial support:* Fellowships, research assistantships, teaching assistantships available. Financial award application deadline: 3/1. *Unit head:* Dr. D. Matthews, Chairperson, 802-656-2594. *Application contact:* Dr. G. Friestad, Coordinator, 802-656-2594.

University of Victoria, Faculty of Graduate Studies, Faculty of Science, Department of Chemistry, Victoria, BC V8W 2Y2, Canada. Offers M Sc, PhD. *Faculty:* 16 full-time (3 women), 6 part-time/adjunct (1 woman). *Students:* Average age 23. 46 applicants, 30% accepted, 10 enrolled. In 2005, 2 master's, 3 doctorates awarded. *Median time to degree:* Of those who began their doctoral program in fall 1997, 100% received their degree in 8 years or less. *Degree requirements:* For master's, thesis, registration; for doctorate, thesis/dissertation, Candidacy Exam. *Entrance requirements:* For master's and doctorate, GRE Subject Test (85th percentile). Additional exam requirements/recommendations for international students: Required—TOEFL (minimum score 575 paper-based; 233 computer-based), IELT(minimum score 7). *Application deadline:* For fall admission, 2/1 priority date for domestic students, 12/15 priority date for international students. Applications are processed on a rolling basis. Application fee: $75 ($125 for international students). Electronic applications accepted. Tuition and fees charges are reported in Canadian dollars. *Expenses:* Tuition, area resident: Full-time $4,492 Canadian dollars; part-time $749 Canadian dollars per term. International tuition: $5,346 Canadian dollars full-time. Required fees: $4,492 Canadian dollars; $749 Canadian dollars per term. Tuition and fees vary according to course load, campus/location and program. *Financial support:* In 2005–06, 8 fellowships were awarded; research assistantships, teaching assistantships, career-related internships or fieldwork, institutionally sponsored loans, and awards also available. Financial award application deadline: 2/15. *Faculty research:* Laser spectroscopy and dynamics; inorganic, organic, and organometallic synthesis; electro and surface chemistry. Total annual research expenditures: $900,000. *Unit head:* Dr. Thomas M. Fyles, Chair, 250-721-7150, Fax: 250-721-7147, E-mail: chemhead@uvic.ca. *Application contact:* Dr. David J. Berg, Graduate Adviser, 250-721-7161, Fax: 250-721-7147, E-mail: djberg@uvic.ca.

University of Virginia, College and Graduate School of Arts and Sciences, Department of Chemistry, Charlottesville, VA 22903. Offers MA, MS, PhD. *Faculty:* 31 full-time (4 women). *Students:* 120 full-time (51 women), 2 part-time (1 woman); includes 13 minority (7 African Americans, 4 Asian Americans or Pacific Islanders, 2 Hispanic Americans), 22 international. Average age 25. 97 applicants, 73% accepted, 34 enrolled. In 2005, 11 master's, 14 doctorates awarded. *Degree requirements:* For master's and doctorate, thesis/dissertation. *Entrance requirements:* For master's and doctorate, GRE General Test, GRE Subject Test. *Application deadline:* Applications are processed on a rolling basis. Application fee: $40. Electronic applications accepted. *Expenses:* Tuition, state resident: full-time $7,731. Tuition, nonresident: full-time $18,672. Required fees: $1,479. Full-time tuition and fees vary according to degree level and program. *Financial support:* Applicants required to submit FAFSA. *Unit head:* Ian Harrison, Chairman, 434-924-3344, Fax: 434-924-3710, E-mail: chem@virginia.edu. *Application contact:* Peter C. Brunjes, Associate Dean for Graduate Programs and Research, 434-924-7184, Fax: 434-924-6737, E-mail: grad-a-s@virginia.edu.

See Close-Up on page 171.

University of Washington, Graduate School, College of Arts and Sciences, Department of Chemistry, Seattle, WA 98195. Offers MS, PhD. Terminal master's awarded for partial completion of doctoral program. *Degree requirements:* For master's, thesis (for some programs); for doctorate, thesis/dissertation. *Entrance requirements:* For master's and doctorate, GRE Subject Test, minimum GPA of 3.0. Additional exam requirements/recommendations for international students: Required—TOEFL, TSE. *Faculty research:* Biopolymers, material science and nanotechnology, organometallic chemistry, analytical chemistry, bioorganic chemistry.

University of Waterloo, Graduate Studies, Faculty of Science, Guelph-Waterloo Centre for Graduate Work in Chemistry and Biochemistry, Waterloo, ON N2L 3G1, Canada. Offers chemistry (M Sc, PhD). Part-time programs available. *Faculty:* 35 full-time (7 women), 22 part-time/adjunct (1 woman). *Students:* 108 full-time (41 women), 6 part-time (5 women). 128 applicants, 20% accepted, 17 enrolled. In 2005, 17 master's, 4 doctorates awarded. *Degree requirements:* For master's, project or thesis; for doctorate, thesis/dissertation. *Entrance requirements:* For master's, GRE, honors degree, minimum B average; for doctorate, GRE, master's degree, minimum B average. Additional exam requirements/recommendations for international students: Required—TOEFL, TWE. *Application deadline:* Applications are processed on a rolling basis. Application fee: $75 Canadian dollars. Electronic applications accepted. *Financial support:* Research assistantships, teaching assistantships available. *Faculty research:* Polymer, physical, inorganic, organic, and theoretical chemistry. *Unit head:* Dr. J. F. Honek, Director, 519-888-4567 Ext. 3945, Fax: 519-746-4806, E-mail: gwc@uwaterloo.ca. *Application contact:* A. Wetmore, Administrative Assistant, 519-888-4567 Ext. 3945, Fax: 519-746-4806, E-mail: gwc@uoguelph.ca.

The University of Western Ontario, Faculty of Graduate Studies, Physical Sciences Division, Department of Chemistry, London, ON N6A 5B8, Canada. Offers M Sc, PhD. *Degree requirements:* For master's and doctorate, thesis/dissertation. *Entrance requirements:* For master's, minimum B+ average, honors B Sc in chemistry; for doctorate, M Sc or equivalent in chemistry. Additional exam requirements/recommendations for international students: Required—TOEFL (paper score 570; computer score 230) or IELTS (paper score 6). *Faculty research:* Materials, inorganic, organic, physical and theoretical chemistry.

University of Windsor, Faculty of Graduate Studies and Research, Faculty of Science, Department of Chemistry and Biochemistry, Windsor, ON N9B 3P4, Canada. Offers M Sc, PhD. Part-time programs available. *Faculty:* 21 full-time (1 woman), 5 part-time/adjunct (1 woman). *Students:* 59 full-time (26 women). 71 applicants, 39% accepted. In 2005, 8 master's, 7 doctorates awarded. *Degree requirements:* For master's, thesis/dissertation; for doctorate, thesis/dissertation, comprehensive exam. *Entrance requirements:* For master's and doctorate, minimum B average. Additional exam requirements/recommendations for international students: Required—TOEFL (minimum score 560 paper-based; 220 computer-based), GRE. *Application deadline:* For fall admission, 7/1 for domestic students. Applications are processed on a rolling basis. Application fee: $55. Electronic applications accepted. *Financial support:* In 2005–06, 54 teaching assistantships (averaging $8,956 per year) were awarded; research assistantships, Federal Work-Study, scholarships/grants, unspecified assistantships, and bursaries also available. Financial award application deadline: 2/15. *Faculty research:* Molecular biology/recombinant DNA techniques (PCR, cloning mutagenesis), No/02 detectors, western immunoblotting and detection, CD/NMR protein/peptide structure determination, confocal/electron microscopes. *Unit head:* Dr. Douglas Stephan, Head, 519-253-3000 Ext. 3537, Fax: 519-973-7098, E-mail: stephan@uwindsor.ca. *Application contact:* Marlene Bezaire, Graduate Secretary, 519-253-3000 Ext. 3520, Fax: 519-971-7098, E-mail: spsgrad@uwindsor.ca.

University of Wisconsin–Madison, Graduate School, College of Engineering, Program in Environmental Chemistry and Technology, Madison, WI 53706-1380. Offers MS, PhD. Part-time programs available. *Faculty:* 12 full-time (2 women). *Students:* 15 full-time (5 women), 2 part-time, 4 international. 66 applicants, 11% accepted, 4 enrolled. In 2005, 1 master's, 2 doctorates awarded. Terminal master's awarded for partial completion of doctoral program. *Median time to degree:* Of those who began their doctoral program in fall 1997, 100% received their degree in 8 years or less. *Degree requirements:* For master's, thesis or alternative; for doctorate, thesis/dissertation. *Entrance requirements:* For master's and doctorate, GRE General Test. Additional exam requirements/recommendations for international students: Required—TOEFL. *Application deadline:* For fall admission, 1/1 priority date for domestic students, 1/1 priority date for international students. Application fee: $50. Electronic applications accepted. *Financial support:* In 2005–06, 15 students received support, including 1 fellowship with tuition reimbursement available (averaging $18,000 per year), 14 research assistantships with tuition reimbursements available (averaging $18,000 per year); Federal Work-Study and institutionally sponsored loans also available. Financial award application deadline: 1/1. *Faculty research:* Chemical limnology, chemical remediation, geochemistry, photocatalysis, water quality. Total

Chemistry

University of Wisconsin–Madison (continued)
annual research expenditures: $1 million. *Unit head:* Dr. Marc A. Anderson, Chair, 608-263-3264, E-mail: nanopor@wisc.edu. *Application contact:* Mary Possin, Student Services Coordinator, 608-263-3264, Fax: 608-265-2340, E-mail: mcpossin@wisc.edu.

University of Wisconsin–Madison, Graduate School, College of Letters and Science, Department of Chemistry, Madison, WI 53706-1380. Offers MS, PhD. Part-time programs available. Terminal master's awarded for partial completion of doctoral program. *Degree requirements:* For master's, thesis (for some programs); for doctorate, thesis/dissertation, cumulative exams, research proposal, seminar. *Entrance requirements:* For master's and doctorate, GRE, minimum GPA of 3.0. Additional exam requirements/recommendations for international students: Required—TOEFL. Electronic applications accepted. *Faculty research:* Analytical, inorganic, organic, physical, and macromolecular chemistry.

University of Wisconsin–Milwaukee, Graduate School, College of Letters and Sciences, Department of Chemistry, Milwaukee, WI 53201-0413. Offers MS, PhD. *Faculty:* 17 full-time (1 woman). *Students:* 52 full-time (24 women), 20 part-time (8 women); includes 5 minority (2 African Americans, 3 Asian Americans or Pacific Islanders), 39 international. 33 applicants, 58% accepted, 11 enrolled. In 2005, 5 master's, 1 doctorate awarded. *Degree requirements:* For master's, thesis or alternative; for doctorate, thesis/dissertation. *Application deadline:* For fall admission, 1/1 for domestic students; for spring admission, 9/1 for domestic students. Applications are processed on a rolling basis. Application fee: $45 ($75 for international students). *Expenses:* Tuition, area resident: Part-time $716 per credit. Tuition, state resident: part-time $776 per credit. Tuition, nonresident: part-time $1,614. Required fees: $229 per term. Tuition and fees vary according to course load and program. *Financial support:* In 2005–06, 1 fellowship, 6 research assistantships, 55 teaching assistantships were awarded; career-related internships or fieldwork and unspecified assistantships also available. Support available to part-time students. Financial award application deadline: 4/15. *Unit head:* Peter Geissinger, Representative, 414-229-5565, Fax: 414-229-5530, E-mail: geissing@uwm.edu.

University of Wyoming, Graduate School, College of Arts and Sciences, Department of Chemistry, Laramie, WY 82070. Offers MS, PhD. *Faculty:* 15 full-time (2 women), 4 part-time/adjunct (0 women). *Students:* 29 full-time (5 women), 4 part-time (2 women); includes 2 minority (1 Asian American or Pacific Islander, 1 Hispanic American), 14 international. Average age 29. 16 applicants, 13% accepted. In 2005, 1 master's, 7 doctorates awarded. *Median time to degree:* Of those who began their doctoral program in fall 1997, 71% received their degree in 8 years or less. *Degree requirements:* For master's and doctorate, thesis/dissertation. *Entrance requirements:* For master's and doctorate, GRE General Test, minimum GPA of 3.0. Additional exam requirements/recommendations for international students: Required—TOEFL (minimum score 600 paper-based). *Application deadline:* For fall admission, 4/15 priority date for domestic students, 2/28 priority date for international students. Applications are processed on a rolling basis. Application fee: $50. Electronic applications accepted. *Expenses:* Tuition, state resident: full-time $3,720; part-time $155 per credit hour. Tuition, nonresident: full-time $10,704; part-time $446 per credit hour. Required fees: $666; $162 per semester. Tuition and fees vary according to course load and program. *Financial support:* In 2005–06, 27 research assistantships with full tuition reimbursements (averaging $19,500 per year), 15 teaching assistantships with full tuition reimbursements (averaging $19,500 per year) were awarded; fellowships, traineeships and tuition waivers (full and partial) also available. Financial award application deadline: 3/1. *Faculty research:* Organic chemistry, inorganic chemistry, analytical chemistry, physical chemistry. Total annual research expenditures: $2.2 million. *Unit head:* Dr. Edward Clennan, Head, 307-766-4363, Fax: 307-766-2807, E-mail: chemistry@uwyo.edu. *Application contact:* Dave Anderson, Graduate Admissions Coordinator, 307-766-4363, Fax: 307-766-2807, E-mail: chemistry@uwyo.edu.

Utah State University, School of Graduate Studies, College of Science, Department of Chemistry and Biochemistry, Logan, UT 84322. Offers biochemistry (MS, PhD); chemistry (MS, PhD). Part-time programs available. *Faculty:* 16 full-time (3 women). *Students:* 134 full-time (43 women), 3 part-time (2 women); includes 3 minority (all African Americans), 89 international. Average age 28. 40 applicants, 58% accepted, 14 enrolled. In 2005, 3 master's, 4 doctorates awarded. Terminal master's awarded for partial completion of doctoral program. *Degree requirements:* For master's and doctorate, thesis/dissertation, oral and written exams. *Entrance requirements:* For master's and doctorate, GRE General Test, minimum GPA of 3.0. Additional exam requirements/recommendations for international students: Required—TOEFL. *Application deadline:* For fall admission, 4/15 priority date for domestic students, 4/15 priority date for international students; for spring admission, 10/15 for domestic students, 10/15 for international students. Applications are processed on a rolling basis. Application fee: $50 ($60 for international students). *Financial support:* In 2005–06, 29 research assistantships with partial tuition reimbursements (averaging $17,000 per year), 16 teaching assistantships with partial tuition reimbursements (averaging $17,000 per year) were awarded; fellowships with tuition reimbursements, Federal Work-Study, institutionally sponsored loans, and tuition waivers (partial) also available. Support available to part-time students. Financial award application deadline: 4/15. *Faculty research:* Analytical, inorganic, organic, and physical chemistry; iron in asbestos chemistry and carcinogenicity; dicopper complexes; photothermal spectrometry; metal molecule clusters. Total annual research expenditures: $2.1 million. *Unit head:* Dr. Steve Scheiner, Head, 435-797-7419, Fax: 435-797-3390, E-mail: scheiner@cc.usu.edu. *Application contact:* Dr. Lisa M. Berreau, Admissions Chair, 435-797-1625, Fax: 435-797-3390, E-mail: berreau@cc.usu.edu.

See Close-Up on page 173.

Vanderbilt University, Graduate School, Department of Chemistry, Nashville, TN 37240-1001. Offers analytical chemistry (MAT, MS, PhD); inorganic chemistry (MAT, MS, PhD); organic chemistry (MAT, MS, PhD); physical chemistry (MAT, MS, PhD); theoretical chemistry (MAT, MS, PhD). *Faculty:* 32 full-time (6 women), 3 part-time/adjunct (2 women). *Students:* 96 full-time (40 women); includes 4 minority (2 African Americans, 1 Asian American or Pacific Islander, 1 Hispanic American), 20 international. 223 applicants, 21% accepted, 22 enrolled. In 2005, 4 master's, 12 doctorates awarded. *Degree requirements:* For master's, thesis or alternative; for doctorate, thesis/dissertation, area, qualifying, and final exams. *Entrance requirements:* For master's and doctorate, GRE General Test, GRE Subject Test (recommended). Additional exam requirements/recommendations for international students: Required—TOEFL. *Application deadline:* For fall admission, 1/15 for domestic students, 1/15 for international students. Application fee: $0. Electronic applications accepted. *Expenses:* Tuition: Full-time $15,396; part-time $1,283 per semester hour. Required fees: $2,202; $1,101 per semester. One-time fee: $30. Tuition and fees vary according to course load, program and student level. *Financial support:* Fellowships with full and partial tuition reimbursements, research assistantships with full tuition reimbursements, teaching assistantships with full tuition reimbursements, Federal Work-Study, institutionally sponsored loans, and traineeships available. Financial award application deadline:1/15. *Faculty research:* Chemical synthesis; mechanistic, theoretical, bioorganic, analytical, and spectroscopic chemistry. *Unit head:* Ned A. Porter, Chair, 615-322-2861, Fax: 615-343-1234. *Application contact:* Charles M. Lukehart, Director of Graduate Studies, 615-322-2861, Fax: 615-343-1234, E-mail: charles.m.lukehart@vanderbilt.edu.

Vassar College, Graduate Programs, Poughkeepsie, NY 12604. Offers chemistry (MA, MS). Applicants accepted only if enrolled in undergraduate programs at Vassar College. Part-time programs available. *Degree requirements:* For master's, thesis. *Entrance requirements:* For master's, GRE General Test, bachelor's degree in related field. Application fee: $60. *Financial support:* Career-related internships or fieldwork available. *Unit head:* Alexander M. Thompson, Dean of Studies, 914-437-5257, E-mail: thompson@vassar.edu.

Villanova University, Graduate School of Liberal Arts and Sciences, Department of Chemistry, Villanova, PA 19085-1699. Offers MS. Part-time and evening/weekend programs available. *Faculty:* 5 full-time (2 women), 1 (woman) part-time/adjunct. *Students:* 10 full-time (4 women), 26 part-time (8 women); includes 5 minority (2 African Americans, 2 Asian Americans or Pacific Islanders, 1 Hispanic American). Average age 26. 24 applicants, 83% accepted. In 2005, 12 degrees awarded. *Degree requirements:* For master's, thesis (for some programs), comprehensive exam. *Entrance requirements:* For master's, GRE General Test, GRE Subject Test, minimum GPA of 3.0. Additional exam requirements/recommendations for international students: Required—TOEFL. *Application deadline:* For fall admission, 8/1 priority date for domestic students, 8/1 priority date for international students; for spring admission, 12/1 for domestic students, 12/1 for international students. Application fee: $50. Electronic applications accepted. *Expenses: Contact institution. Financial support:* Research assistantships, Federal Work-Study available. Financial award applicants required to submit FAFSA. *Unit head:* Dr. Barry Selinsky, Chair, 610-519-4840.

See Close-Up on page 175.

Virginia Commonwealth University, Graduate School, College of Humanities and Sciences, Department of Chemistry, Richmond, VA 23284-9005. Offers analytical (MS, PhD); chemical physics (PhD); inorganic (MS, PhD); organic (MS, PhD); physical (MS, PhD). Part-time programs available. *Faculty:* 18 full-time (6 women). *Students:* 36 full-time (13 women), 11 part-time (1 woman); includes 13 minority (6 African Americans, 1 American Indian/Alaska Native, 5 Asian Americans or Pacific Islanders, 1 Hispanic American), 13 international. 47 applicants, 74% accepted. In 2005, 2 master's, 8 doctorates awarded. Terminal master's awarded for partial completion of doctoral program. *Degree requirements:* For master's, thesis; for doctorate, thesis/dissertation, comprehensive cumulative exams, research proposal. *Entrance requirements:* For master's, GRE General Test, 30 undergraduate credits in chemistry; for doctorate, GRE General Test. *Application deadline:* For fall admission, 3/15 for domestic students; for spring admission, 11/15 for domestic students. Applications are processed on a rolling basis. Application fee: $50. *Expenses:* Tuition, state resident: full-time $3,185; part-time $405 per credit. Tuition, nonresident: full-time $7,952; part-time $940 per credit. Required fees: $751 per semester hour. Tuition and fees vary according to course load and program. *Financial support:* Fellowships, research assistantships, teaching assistantships, career-related internships or fieldwork and institutionally sponsored loans available. Support available to part-time students. Financial award application deadline: 7/1. *Faculty research:* Physical, organic, inorganic, analytical, and polymer chemistry; chemical physics. *Unit head:* Dr. Fred M. Hawkridge, Chair, 804-828-1298, Fax: 804-828-8599, E-mail: fmhawkri@vcu.edu. *Application contact:* Dr. M. Samy El-Shall, Chair, Graduate Recruiting and Admissions Committee, 804-828-3518, E-mail: mselshal@vcu.edu.

Virginia Polytechnic Institute and State University, Graduate School, College of Science, Department of Chemistry, Blacksburg, VA 24061. Offers MS, PhD. *Faculty:* 34 full-time (6 women). *Students:* 112 full-time (34 women), 5 part-time (2 women); includes 12 minority (6 African Americans, 4 Asian Americans or Pacific Islanders, 2 Hispanic Americans), 55 international. Average age 27. 131 applicants, 37% accepted, 30 enrolled. In 2005, 9 master's, 14 doctorates awarded. *Entrance requirements:* Additional exam requirements/recommendations for international students: Required—TOEFL (minimum score 600 paper-based; 250 computer-based), GRE. *Application deadline:* Applications are processed on a rolling basis. Application fee: $45. Electronic applications accepted. *Expenses:* Tuition, state resident: full-time $6,558; part-time $364 per credit. Tuition, nonresident: full-time $11,296; part-time $628 per credit. Required fees: $1,419; $468 per credit. $234 per term. *Financial support:* In 2005–06, 2 fellowships with full tuition reimbursements (averaging $2,126 per year), 38 research assistantships with full tuition reimbursements (averaging $16,689 per year), 72 teaching assistantships with full tuition reimbursements (averaging $13,902 per year) were awarded; career-related internships or fieldwork, Federal Work-Study, scholarships/grants, tuition waivers (partial), and unspecified assistantships also available. Financial award application deadline: 4/1. *Faculty research:* Analytical, inorganic, organic, physical, and polymer chemistry. *Unit head:* Dr. Joe Merola, Head, 540-231-4510, Fax: 540-231-3255. *Application contact:* Mark Anderson, 540-231-3869, Fax: 540-231-3255, E-mail: maander6@vt.edu.

Wake Forest University, Graduate School, Department of Chemistry, Winston-Salem, NC 27109. Offers analytical chemistry (MS, PhD); inorganic chemistry (MS, PhD); organic chemistry (MS, PhD); physical chemistry (MS, PhD). Part-time programs available. *Faculty:* 17 full-time (4 women). *Students:* 32 full-time (17 women), 2 part-time; includes 1 minority (African American), 14 international. Average age 28. 60 applicants, 42% accepted, 12 enrolled. In 2005, 1 master's, 6 doctorates awarded. *Median time to degree:* Of those who began their doctoral program in fall 1997, 0% received their degree in 8 years or less. *Degree requirements:* For master's, one foreign language, thesis, comprehensive exam, registration; for doctorate, 2 foreign languages, thesis/dissertation, comprehensive exam, registration. *Entrance requirements:* For master's and doctorate, GRE General Test. Additional exam requirements/recommendations for international students: Required—TOEFL (minimum score 213 computer-based). *Application deadline:* For fall admission, 1/15 for domestic students, 1/15 for international students. Application fee: $45. Electronic applications accepted. *Financial support:* In 2005–06, 32 students received support, including 1 fellowship with full tuition reimbursement available (averaging $21,500 per year), 12 research assistantships with full tuition reimbursements available (averaging $19,500 per year), 17 teaching assistantships with full tuition reimbursements available (averaging $19,500 per year); scholarships/grants and tuition waivers (full and partial) also available. Support available to part-time students. Financial award application deadline: 1/15; financial award applicants required to submit FAFSA. *Unit head:* Dr. Bruce King, Director, 336-758-5774, Fax: 335-758-4656, E-mail: kingsb@wfu.edu.

Washington State University, Graduate School, College of Sciences, Department of Chemistry, Pullman, WA 99164. Offers analytical chemistry (MS, PhD); biological systems (MS, PhD); inorganic chemistry (MS, PhD); organic chemistry (MS, PhD); physical chemistry (MS, PhD). *Faculty:* 33. *Students:* 38 full-time (19 women), 2 part-time (1 woman); includes 3 minority (2 Asian Americans or Pacific Islanders, 1 Hispanic American), 11 international. Average age 29. 55 applicants, 33% accepted, 9 enrolled. In 2005, 5 master's, 7 doctorates awarded. Terminal master's awarded for partial completion of doctoral program. *Degree requirements:* For master's, oral exam, teaching experience, thesis optional; for doctorate, thesis/dissertation, oral exam, written exam, comprehensive exam. *Entrance requirements:* For master's and doctorate, GRE General Test, minimum GPA of 3.0, 3 letters of recommendation. *Application deadline:* For fall admission, 3/1 priority date for domestic students, 3/1 priority date for international students; for spring admission, 10/1 priority date for domestic students, 7/1 priority date for international students. Applications are processed on a rolling basis. Application fee: $35. *Expenses:* Tuition, state resident: full-time $6,295; part-time $336 per credit. Tuition, nonresident: full-time $15,949; part-time $819 per credit. Required fees: $933. Part-time tuition and fees vary according to campus/location and program. *Financial support:* In 2005–06, 38 students received support, including 3 fellowships (averaging $8,977 per year), 12 research assistantships with full and partial tuition reimbursements available (averaging $15,999 per year), 23 teaching assistantships with full and partial tuition reimbursements available (averaging $14,747 per year); career-related internships or fieldwork, Federal Work-Study, institutionally sponsored loans, scholarships/grants, health care benefits, unspecified assistantships, and summer support also available. Financial award application deadline: 4/1; financial award applicants required to submit FAFSA. *Faculty research:* Environmental chemistry, materials chemistry, radio chemistry, bio-organic, computational chemistry. Total annual research expenditures: $3.8 million. *Unit head:* Sue B. Clark, Chair, Admissions Committee, 509-335-8866, Fax: 509-335-8867, E-mail: sclark@mail.wsu.edu. *Application contact:* Admissions Committee.

Washington State University Tri-Cities, Graduate Programs, Program in Chemistry, Richland, WA 99352-1671. Offers MS. *Faculty:* 13. *Students:* Average age 35. 1 applicant, 100% accepted, 1 enrolled. In 2005, 2 degrees awarded. *Degree requirements:* For master's, thesis, registration. *Entrance requirements:* For master's, GRE, minimum GPA of 3.0, 3 letters of recommendation. Additional exam requirements/recommendations for international students: Required—TOEFL. *Application deadline:* For fall admission, 3/1 for domestic students, 3/1 for international students; for spring admission, 10/1 for domestic students, 10/1 for international students. Application fee: $35. *Expenses:* Tuition, state resident: full-time $6,295;

part-time $336 per credit. Tuition, nonresident: full-time $15,949; part-time $819 per credit. International tuition: $15,949 full-time. Required fees: $429. Full-time tuition and fees vary according to campus/location and program. Part-time tuition and fees vary according to course load and program. *Financial support:* In 2005–06, 1 student received support. Total annual research expenditures: $25,469. *Application contact:* 509-372-7250, Fax: 509-372-7100.

Washington University in St. Louis, Graduate School of Arts and Sciences, Department of Chemistry, St. Louis, MO 63130-4899. Offers MA, PhD. Terminal master's awarded for partial completion of doctoral program. *Degree requirements:* For master's, thesis optional; for doctorate, thesis/dissertation. *Entrance requirements:* For master's and doctorate, GRE General Test, GRE Subject Test. Electronic applications accepted.

Wayne State University, Graduate School, College of Liberal Arts and Sciences, Department of Chemistry, Detroit, MI 48202. Offers MA, MS, PhD. *Faculty:* 21 full-time (2 women). *Students:* 155 full-time (64 women), 6 part-time (4 women); includes 13 minority (6 African Americans, 7 Asian Americans or Pacific Islanders), 104 international. Average age 29. 297 applicants, 26% accepted, 46 enrolled. In 2005, 9 master's, 16 doctorates awarded. *Degree requirements:* For master's, thesis (for some programs); for doctorate, thesis/dissertation. *Entrance requirements:* Additional exam requirements/recommendations for international students: Required—TOEFL (minimum score 550 paper-based; 213 computer-based); Recommended—TWE(minimum score 6). *Application deadline:* For fall admission, 7/1 for domestic students, 6/1 for international students. Applications are processed on a rolling basis. Application fee: $30 ($50 for international students). Electronic applications accepted. *Expenses:* Tuition, state resident: part-time $338 per credit hour. Tuition, nonresident: part-time $746 per credit hour. Required fees: $24 per credit hour. Full-time tuition and fees vary according to program. *Financial support:* In 2005–06, 7 fellowships, 59 research assistantships (averaging $15,693 per year), 75 teaching assistantships (averaging $14,997 per year) were awarded. Financial award application deadline: 7/1. *Faculty research:* Natural products synthesis, molecular biology, molecular mechanics calculations, organometallic chemistry, experimental physical chemistry. Total annual research expenditures: $5.6 million. *Unit head:* James Rigby, Chair, 313-577-3472, Fax: 313-577-8822, E-mail: aa392@wayne.edu. *Application contact:* Charles Winter, Graduate Director, 313-577-5224, E-mail: chw@chem.wayne.edu.

Wesleyan University, Graduate Programs, Department of Chemistry, Middletown, CT 06459-0260. Offers biochemistry (MA, PhD); chemical physics (MA, PhD); inorganic chemistry (MA, PhD); organic chemistry (MA, PhD); physical chemistry (MA, PhD); theoretical chemistry (MA, PhD). *Faculty:* 13 full-time (2 women), 2 part-time/adjunct (1 woman). *Students:* 41 full-time (20 women); includes 1 minority (Asian American or Pacific Islander), 22 international. Average age 26. In 2005, 2 master's, 2 doctorates awarded. Terminal master's awarded for partial completion of doctoral program. *Degree requirements:* For master's and doctorate, one foreign language, thesis/dissertation. *Entrance requirements:* For master's, GRE General Test, GRE Subject Test; for doctorate, GRE Subject Test. *Application deadline:* For fall admission, 3/1 for domestic students. Applications are processed on a rolling basis. Application fee: $0. *Expenses:* Tuition: Full-time $24,732. One-time fee: $20 full-time. *Financial support:* Fellowships, research assistantships, teaching assistantships, institutionally sponsored loans available. *Unit head:* Dr. Phillip Bolton, Chair, 860-685-2668. *Application contact:* Karen Karpa, Information Contact, 860-685-2573, Fax: 860-685-2211, E-mail: kkarpa@wesleyan.edu.

See Close-Up on page 177.

West Chester University of Pennsylvania, Graduate Studies, College of Arts and Sciences, Department of Chemistry, West Chester, PA 19383. Offers chemistry (M Ed, MS); clinical chemistry (MS); physical science (MA). Part-time and evening/weekend programs available. *Students:* 4 full-time (2 women), 4 part-time (2 women); includes 2 minority (1 African American, 1 Asian American or Pacific Islander). In 2005, 5 degrees awarded. *Degree requirements:* For master's, one foreign language, comprehensive exam. *Entrance requirements:* For master's, GRE General Test (recommended). *Application deadline:* For fall admission, 4/15 for domestic students; for spring admission, 10/15 for domestic students. Applications are processed on a rolling basis. Application fee: $35. *Expenses:* Tuition, state resident: full-time $2,944; part-time $327 per credit. Tuition, nonresident: full-time $4,711; part-time $523 per credit. Required fees: $54 per semester. *Financial support:* In 2005–06, 3 research assistantships with full tuition reimbursements (averaging $5,000 per year) were awarded Support available to part-time students. Financial award application deadline: 2/15; financial award applicants required to submit FAFSA. *Faculty research:* Solid phase rates into monodisperse polymers and palladium-mediated rates into novel materials. *Unit head:* Dr. James Falcone, Chair, 610-436-2631, E-mail: jfalcone@weupa.edu. *Application contact:* Dr. Naseer Ahmad, Graduate Coordinator, 610-436-2476, E-mail: anaseer@wcupa.edu.

Western Carolina University, Graduate School, College of Arts and Sciences, Department of Chemistry and Physics, Cullowhee, NC 28723. Offers chemistry (MAT, MS); comprehensive education-chemistry (MA Ed). Part-time and evening/weekend programs available. *Degree requirements:* For master's, variable foreign language requirement, thesis, comprehensive exam. *Entrance requirements:* For master's, GRE General Test. Additional exam requirements/recommendations for international students: Required—TOEFL (minimum score 550 paper-based; 213 computer-based).

Western Illinois University, School of Graduate Studies, College of Arts and Sciences, Department of Chemistry, Macomb, IL 61455-1390. Offers MS. Part-time programs available. *Students:* 29 full-time (14 women), 1 part-time; includes 5 minority (3 African Americans, 2 Asian Americans or Pacific Islanders), 21 international. Average age 25. 47 applicants, 87% accepted. In 2005, 4 degrees awarded. *Degree requirements:* For master's, thesis or alternative. *Entrance requirements:* Additional exam requirements/recommendations for international students: Required—TOEFL (minimum score 530 paper-based; 197 computer-based). *Application deadline:* Applications are processed on a rolling basis. Application fee: $30. Electronic applications accepted. *Expenses:* Tuition, state resident: full-time $3,599; part-time $200 per semester hour. Tuition, nonresident: full-time $7,198; part-time $400 per semester hour. Required fees: $890; $49 per semester hour. Tuition and fees vary according to campus/location. *Financial support:* In 2005–06, 17 students received support, including 17 research assistantships with full tuition reimbursements available (averaging $6,288 per year) Financial award applicants required to submit FAFSA. *Unit head:* Dr. Vivian Incera, Chairperson, 309-298-1538. *Application contact:* Dr. Barbara Baily, Director of Graduate Studies/Associate Provost, 309-298-1806, Fax: 309-298-2345, E-mail: grad-office@wiu.edu.

Western Kentucky University, Graduate Studies, Ogden College of Science and Engineering, Department of Chemistry, Bowling Green, KY 42101-3576. Offers chemistry (MA Ed, MS). *Faculty:* 8 full-time (2 women). *Students:* 3 full-time (2 women), 5 part-time (2 women), 2 international. Average age 32. 11 applicants, 27% accepted, 2 enrolled. In 2005, 4 degrees awarded. *Degree requirements:* For master's, thesis, comprehensive exam. *Entrance requirements:* For master's, GRE General Test, minimum GPA of 2.75. Additional exam requirements/recommendations for international students: Required—TOEFL (minimum score 555 paper-based; 213 computer-based). *Application deadline:* For fall admission, 7/1 priority date for domestic students, 5/15 priority date for international students; for spring admission, 11/1 for domestic students, 9/15 for international students. Applications are processed on a rolling basis. Application fee: $35. *Expenses:* Tuition, area resident: Full-time $5,816; part-time $299 per credit hour. Tuition, state resident: full-time $5,816; part-time $299 per credit hour. Tuition, nonresident: full-time $6,356; part-time $326 per credit hour. *Financial support:* In 2005–06, 9 students received support, including 3 research assistantships with partial tuition reimbursements available (averaging $6,000 per year), 6 teaching assistantships with partial tuition reimbursements available (averaging $6,000 per year); Federal Work-Study, institutionally sponsored loans, tuition waivers (partial), and service awards also available. Support available to part-time students. Financial award application deadline: 4/1; financial award applicants required to submit FAFSA. *Faculty research:* Catatonic surfactants, directed orthometalation reactions, thermal stability and degradation mechanisms, co-firing refused derived fuels, laser fluorescence. Total annual research expenditures:$202,172. *Unit head:* Dr. Cathleen J Webb, Head, 270-745-3457, Fax: 270-745-5361, E-mail: cathleen.webb@wku.edu.

Western Michigan University, Graduate College, College of Arts and Sciences, Department of Chemistry, Kalamazoo, MI 49008-5202. Offers MA, PhD. *Degree requirements:* For master's, thesis, departmental qualifying and oral exams; for doctorate, thesis/dissertation.

Western Washington University, Graduate School, College of Sciences and Technology, Department of Chemistry, Bellingham, WA 98225-5996. Offers MS. Part-time programs available. *Faculty:* 16. *Students:* 11 full-time (6 women), 4 part-time (3 women); includes 2 minority (both Asian Americans or Pacific Islanders) 10 applicants, 80% accepted, 7 enrolled. In 2005, 2 degrees awarded. *Degree requirements:* For master's, thesis (for some programs). *Entrance requirements:* For master's, GRE General Test, minimum GPA of 3.0 in last 60 semester hours or last 90 quarter hours. Additional exam requirements/recommendations for international students: Required—TOEFL (minimum score 567 paper-based; 227 computer-based). *Application deadline:* For fall admission, 6/1 for domestic studentsFor winter admission, 10/1 for domestic students; for spring admission, 2/1 for domestic students. Applications are processed on a rolling basis. Application fee: $50. *Expenses:* Tuition, area resident: Part-time $188 per credit. Tuition, state resident: full-time $5,628; part-time $539 per credit. Tuition, nonresident: full-time $16,176. Required fees: $624. *Financial support:* In 2005–06, research assistantships with partial tuition reimbursements (averaging $11,415 per year), 7 teaching assistantships with partial tuition reimbursements (averaging $11,928 per year) were awarded; career-related internships or fieldwork, Federal Work-Study, institutionally sponsored loans, scholarships/grants, tuition waivers (partial), and unspecified assistantships also available. Support available to part-time students. Financial award application deadline: 2/15; financial award applicants required to submit FAFSA. *Faculty research:* Bio-, organic, inorganic, physical, analytical chemistry. *Unit head:* Dr. Mark Wicholas, Chair, 360-650-3071. *Application contact:* Dr. Mark Bussell, Graduate Program Adviser, 360-650-3145.

West Texas A&M University, College of Agriculture, Nursing, and Natural Sciences, Department of Mathematics, Physical Sciences and Engineering Technology, Program in Chemistry, Canyon, TX 79016-0001. Offers MS. Part-time programs available. *Degree requirements:* For master's, thesis optional. *Entrance requirements:* For master's, GRE General Test. Additional exam requirements/recommendations for international students: Required—TOEFL (minimum score 550 paper-based). Electronic applications accepted. *Faculty research:* Biochemistry; inorganic, organic, and physical chemistry; vibrational spectroscopy; magnetic susceptibilities; carbene chemistry.

West Virginia University, Eberly College of Arts and Sciences, Department of Chemistry, Morgantown, WV 26506. Offers analytical chemistry (MS, PhD); inorganic chemistry (MS, PhD); organic chemistry (MS, PhD); physical chemistry (MS, PhD); theoretical chemistry (MS, PhD). Part-time programs available. Postbaccalaureate distance learning degree programs offered (no on-campus study). *Faculty:* 14 full-time (1 woman), 5 part-time/adjunct (3 women). *Students:* 45 full-time (15 women), 3 part-time (2 women); includes 1 minority (Hispanic American), 26 international. Average age 28. 84 applicants, 58% accepted, 11 enrolled. In 2005, 5 master's, 3 doctorates awarded. Terminal master's awarded for partial completion of doctoral program. *Degree requirements:* For master's and doctorate, thesis/dissertation, registration. *Entrance requirements:* For master's, GRE General Test, GRE Subject Test (recommended), minimum GPA of 2.5; for doctorate, GRE General Test, GRE Subject Test (recommended), minimum GPA of 2.75. Additional exam requirements/recommendations for international students: Required—TOEFL. *Application deadline:* For fall admission, 3/1 for domestic students. Applications are processed on a rolling basis. Application fee: $50. Electronic applications accepted. *Expenses:* Tuition, area resident: Full-time $4,582; part-time $258 per credit hour. Tuition, state resident: full-time $4,582; part-time $258 per credit hour. Tuition, nonresident: full-time $1,382; part-time $741 per credit hour. International tuition: $1,382 full-time. *Financial support:* In 2005–06, fellowships (averaging $2,000 per year), research assistantships with full tuition reimbursements (averaging $13,000 per year), teaching assistantships with full tuition reimbursements (averaging $12,000 per year) were awarded; tuition waivers (full and partial) also available. Financial award application deadline: 2/1; financial award applicants required to submit FAFSA. *Faculty research:* Analysis of proteins, drug interactions, solids and effluents by advanced separation methods; new synthetic strategies for complex organic molecules; synthesis and structural characterization of metal complexes for polymerization catalysis. *Unit head:* Dr. Harry O. Finklea, Chair, 304-293-3435 Ext. 4453, Fax: 304-293-4904, E-mail: harry.finklea@mail.wvu.edu.

Wichita State University, Graduate School, Fairmount College of Liberal Arts and Sciences, Department of Chemistry, Wichita, KS 67260. Offers MS, PhD. *Degree requirements:* For master's, variable foreign language requirement, thesis; for doctorate, thesis/dissertation, comprehensive exam. *Entrance requirements:* For master's and doctorate, GRE. Additional exam requirements/recommendations for international students: Required—TOEFL. Electronic applications accepted. *Faculty research:* Biochemistry; analytic, inorganic, organic, and polymer chemistry.

Worcester Polytechnic Institute, Graduate Studies and Enrollment, Department of Chemistry and Biochemistry, Worcester, MA 01609-2280. Offers biochemistry (MS); bioscience administration (MS); chemistry (MS, PhD). *Faculty:* 11 full-time (2 women). *Students:* 18 full-time (8 women); includes 1 minority (Asian American or Pacific Islander), 9 international. 20 applicants, 70% accepted, 7 enrolled. In 2005, 2 master's, 2 doctorates awarded. *Degree requirements:* For master's, thesis/dissertation; for doctorate, thesis/dissertation, comprehensive exam. *Entrance requirements:* For master's and doctorate, GRE General Test, 3 letters of recommendation. Additional exam requirements/recommendations for international students: Required—TOEFL (minimum score 550 paper-based; 213 computer-based). *Application deadline:* For fall admission, 1/15 for domestic students; for spring admission, 10/15 priority date for domestic students. Applications are processed on a rolling basis. Application fee: $70. Electronic applications accepted. *Expenses:* Tuition: Part-time $997 per credit hour. *Financial support:* In 2005–06, 15 students received support, including 3 research assistantships with full tuition reimbursements available, 10 teaching assistantships with full and partial tuition reimbursements available; fellowships with full tuition reimbursements available, career-related internships or fieldwork, institutionally sponsored loans, scholarships/grants, health care benefits, and unspecified assistantships also available. Financial award application deadline: 1/15. *Faculty research:* Plant biochemistry, membrane biophysics, photochemistry, organic synthesis, materials synthesis. Total annual research expenditures:$176,018. *Unit head:* Dr. James W Pavlik, Interim Head, 508-831-5371, Fax: 508-831-5933, E-mail: jwpavlik@wpi.edu. *Application contact:* Dr. Jose Arguello, Graduate Coordinator, 508-831-5326, Fax: 508-831-5933, E-mail: arguello@wpi.edu.

See Close-Up on page 181.

Wright State University, School of Graduate Studies, College of Science and Mathematics, Department of Chemistry, Dayton, OH 45435. Offers chemistry (MS); environmental sciences (MS). Part-time and evening/weekend programs available. *Degree requirements:* For master's, oral defense of thesis, seminar. *Entrance requirements:* Additional exam requirements/recommendations for international students: Required—TOEFL. *Faculty research:* Polymer synthesis and characterization, laser kinetics, organic and inorganic synthesis, analytical and environmental chemistry.

Yale University, Graduate School of Arts and Sciences, Department of Chemistry, New Haven, CT 06520. Offers biophysical chemistry (PhD); inorganic chemistry (PhD); organic chemistry (PhD); physical chemistry (PhD). *Degree requirements:* For doctorate, thesis/dissertation. *Entrance requirements:* For doctorate, GRE General Test, GRE Subject Test. Additional exam requirements/recommendations for international students: Required—TOEFL.

See Close-Up on page 183.

Chemistry

York University, Faculty of Graduate Studies, Faculty of Pure and Applied Science, Program in Chemistry, Toronto, ON M3J 1P3, Canada. Offers M Sc, PhD. Part-time and evening/weekend programs available. *Faculty:* 28 full-time (5 women), 8 part-time/adjunct (0 women). *Students:* 49 full-time (21 women), 15 part-time (4 women). 85 applicants. In 2005, 6 master's, 6 doctorates awarded. *Degree requirements:* For master's, thesis or alternative, registration; for doctorate, thesis/dissertation, registration. *Application deadline:* Applications are processed on a rolling basis. Application fee: $80. Electronic applications accepted. *Expenses:* Tuition, state resident: full-time $3,190; part-time $798 per term. International tuition: $7,515 full-time. Required fees: $217. Tuition and fees vary according to program. *Financial support:* In 2005–06, fellowships (averaging $6,839 per year), research assistantships (averaging $8,342 per year), teaching assistantships (averaging $10,654 per year) were awarded; fee bursaries also available. *Unit head:* Rene Fournier, Director, 416-736-5246.

Youngstown State University, Graduate School, College of Arts and Sciences, Department of Chemistry, Youngstown, OH 44555-0001. Offers MS. Part-time programs available. *Degree requirements:* For master's, thesis. *Entrance requirements:* For master's, bachelor's degree in chemistry, minimum GPA of 2.7. Additional exam requirements/recommendations for international students: Required—TOEFL. *Faculty research:* Analysis of antioxidants, chromatography, defects and disorder in crystalline oxides, hydrogen bonding, novel organic and organometallic materials.

Inorganic Chemistry

Boston College, Graduate School of Arts and Sciences, Department of Chemistry, Chestnut Hill, MA 02467-3800. Offers biochemistry (PhD); inorganic chemistry (PhD); organic chemistry (PhD); physical chemistry (PhD); science education (MST). MST is offered through the School of Education for secondary school science teaching. Part-time programs available. *Students:* 118 full-time (45 women); includes 12 minority (2 African Americans, 8 Asian Americans or Pacific Islanders, 2 Hispanic Americans), 35 international. 212 applicants, 32% accepted, 17 enrolled. In 2005, 6 master's, 10 doctorates awarded. *Degree requirements:* For doctorate, thesis/dissertation, qualifying exam. *Entrance requirements:* For doctorate, GRE General Test, GRE Subject Test. Additional exam requirements/recommendations for international students: Required—TOEFL (minimum score 550 paper-based; 213 computer-based). *Application deadline:* For fall admission, 1/2 for domestic students. Application fee: $70. Electronic applications accepted. *Financial support:* Fellowships with full tuition reimbursements, research assistantships with full tuition reimbursements, teaching assistantships with full tuition reimbursements, Federal Work-Study available. Support available to part-time students. Financial award application deadline: 3/1; financial award applicants required to submit FAFSA. *Unit head:* Dr. David McFadden, Chairperson, 617-552-3605, E-mail: david.mcfadden@bc.edu. *Application contact:* Dr. Marc Snapper, Graduate Program Director, 617-552-8096, Fax: 617-552-0833, E-mail: marc.snapper@bc.edu.

Brandeis University, Graduate School of Arts and Sciences, Department of Chemistry, Waltham, MA 02454-9110. Offers inorganic chemistry (MS, PhD); organic chemistry (MS, PhD); physical chemistry (MS, PhD). *Faculty:* 20 full-time (3 women). *Students:* 42 full-time (20 women), 1 part-time; includes 2 minority (1 African American, 1 Hispanic American), 34 international. Average age 25. 87 applicants, 10% accepted. In 2005, 5 master's, 3 doctorates awarded. Terminal master's awarded for partial completion of doctoral program. *Degree requirements:* For master's, 1 year of residency; for doctorate, one foreign language, thesis/dissertation, 3 years of residency, 2 seminars, qualifying exams. *Entrance requirements:* For master's and doctorate, GRE General Test, resumé, letters of recommendation. Additional exam requirements/recommendations for international students: Required—TOEFL (minimum score 600 paper-based; 250 computer-based). *Application deadline:* For fall admission, 1/15 for domestic students. Applications are processed on a rolling basis. Application fee: $55. Electronic applications accepted. *Financial support:* In 2005–06, 25 fellowships (averaging $23,000 per year), 21 research assistantships (averaging $23,000 per year), 19 teaching assistantships (averaging $23,000 per year) were awarded; institutionally sponsored loans and scholarships/grants also available. Financial award application deadline: 4/15; financial award applicants required to submit CSS PROFILE or FAFSA. *Faculty research:* Oscillating chemical reactions, molecular recognition systems, protein crystallography, synthesis of natural product spectroscopy and magnetic resonance. Total annual research expenditures: $1,965. *Unit head:* Dr. Peter Jordan, Chair, 781-736-2540, Fax: 781-736-2516, E-mail: jordan@brandeis.edu. *Application contact:* Charlotte Haygazian, Graduate Admissions Secretary, 781-736-2500, Fax: 781-736-2516, E-mail: chemadm@brandeis.edu.

Brigham Young University, Graduate Studies, College of Physical and Mathematical Sciences, Department of Chemistry and Biochemistry, Provo, UT 84602-1001. Offers analytical chemistry (MS, PhD); biochemistry (MS, PhD); inorganic chemistry (MS, PhD); organic chemistry (MS, PhD); physical chemistry (MS, PhD). *Faculty:* 35 full-time (2 women). *Students:* 100 full-time (33 women); includes 2 minority (both Asian Americans or Pacific Islanders), 68 international. Average age 29. 85 applicants, 44% accepted, 21 enrolled. In 2005, 9 master's, 9 doctorates awarded. *Median time to degree:* Of those who began their doctoral program in fall 1997, 87% received their degree in 8 years or less. *Degree requirements:* For master's, thesis, registration; for doctorate, thesis/dissertation, degree qualifying exam. *Entrance requirements:* For master's, pass 1 (biochemistry), 4 (chemistry) of 5 area exams, GRE General Test, minimum GPA of 3.0 in last 60 hours; for doctorate, pass 1 (biochemistry) or 4 (chemistry) of 5 area exams, GRE General Test, minimum GPA of 3.0 in last 60 hours. Additional exam requirements/recommendations for international students: Required—TOEFL, TWE. *Application deadline:* For fall admission, 2/1 priority date for domestic students, 2/1 priority date for international students. Applications are processed on a rolling basis. Application fee: $50. Electronic applications accepted. *Financial support:* In 2005–06, 100 students received support, including 12 fellowships with full tuition reimbursements available (averaging $20,500 per year), 47 research assistantships with full tuition reimbursements available (averaging $20,500 per year), 39 teaching assistantships with full tuition reimbursements available (averaging $20,400 per year); institutionally sponsored loans, scholarships/grants, health care benefits, tuition waivers (full), and unspecified assistantships also available. Financial award application deadline: 2/1. *Faculty research:* Separation science, molecular recognition, organic synthesis and biomedical application, biochemistry and molecular biology, molecular spectroscopy. Total annual research expenditures: $3.9 million. *Unit head:* Dr. Paul B. Farnsworth, Chair, 801-422-6502, Fax: 801-422-0153, E-mail: paul_farnsworth@byu.edu. *Application contact:* Dr. David V. Dearden, Graduate Coordinator, 801-422-2355, Fax: 801-422-0153, E-mail: david_dearden@byu.edu.

See Close-Up on page 97.

California State University, Fullerton, Graduate Studies, College of Natural Science and Mathematics, Department of Chemistry and Biochemistry, Fullerton, CA 92834-9480. Offers analytical chemistry (MS); biochemistry (MS); geochemistry (MS); inorganic chemistry (MS); organic chemistry (MS); physical chemistry (MS). Part-time programs available. *Students:* 19 full-time (6 women), 20 part-time (9 women); includes 20 minority (1 African American, 13 Asian Americans or Pacific Islanders, 6 Hispanic Americans), 8 international. Average age 28. 36 applicants, 47% accepted, 4 enrolled. In 2005, 11 degrees awarded. *Degree requirements:* For master's, thesis, departmental qualifying exam. *Entrance requirements:* For master's, minimum GPA of 2.5 in last 60 units of course work, major in chemistry or related field. Application fee: $55. *Expenses:* Tuition, area resident: Part-time $2,270 per year. Tuition, state resident: full-time $2,572; part-time $339 per unit. Tuition, nonresident: full-time $339; part-time $339 per unit. International tuition: $339 full-time. *Financial support:* Teaching assistantships, career-related internships or fieldwork, Federal Work-Study, institutionally sponsored loans, and scholarships/grants available. Support available to part-time students. Financial award application deadline: 3/1. *Unit head:* Dr. Maria Linder, Chair, 714-278-3621. *Application contact:* Dr. Gregory Williams, Adviser, 714-278-2170.

California State University, Los Angeles, Graduate Studies, College of Natural and Social Sciences, Department of Chemistry and Biochemistry, Los Angeles, CA 90032-8530. Offers analytical chemistry (MS); biochemistry (MS); chemistry (MS); inorganic chemistry (MS); organic chemistry (MS); physical chemistry (MS). Part-time and evening/weekend programs available. *Faculty:* 2 full-time (0 women). *Students:* 33 full-time (21 women), 21 part-time (10 women); includes 38 minority (3 African Americans, 21 Asian Americans or Pacific Islanders, 14 Hispanic Americans). In 2005, 5 degrees awarded. *Degree requirements:* For master's, one foreign language. *Entrance requirements:* Additional exam requirements/recommendations for international students: Required—TOEFL. *Application deadline:* For fall admission, 6/30 for domestic students; for spring admission, 2/1 for domestic students. Applications are processed on a rolling basis. Application fee: $55. *Expenses:* Tuition, area resident: Full-time $3,617. Tuition, state resident: full-time $3,617. Tuition, nonresident: full-time $9,719. International tuition: $9,719 full-time. *Financial support:* Federal Work-Study available. Support available to part-time students. Financial award application deadline: 3/1. *Faculty research:* Intercalation of heavy metal, carborane chemistry, conductive polymers and fabrics, titanium reagents, computer modeling and synthesis. *Unit head:* Dr. Wayne Tikkanen, Chair, 323-343-2300, Fax: 323-343-6490.

Case Western Reserve University, School of Graduate Studies, Department of Chemistry, Cleveland, OH 44106. Offers analytical chemistry (MS, PhD); inorganic chemistry (MS, PhD); organic chemistry (MS, PhD); physical chemistry (MS, PhD). Part-time programs available. Terminal master's awarded for partial completion of doctoral program. *Degree requirements:* For doctorate, thesis/dissertation. *Entrance requirements:* For master's and doctorate, GRE General Test, GRE Subject Test. Additional exam requirements/recommendations for international students: Required—TOEFL. *Faculty research:* Electrochemistry, synthetic chemistry, chemistry of life process, spectroscopy, kinetics.

Clark Atlanta University, School of Arts and Sciences, Department of Chemistry, Atlanta, GA 30314. Offers inorganic chemistry (MS, PhD); organic chemistry (MS, PhD); physical chemistry (MS, PhD); science education (DA). Part-time programs available. *Degree requirements:* For master's, one foreign language, thesis, comprehensive exam; for doctorate, 2 foreign languages, thesis/dissertation, cumulative exam. *Entrance requirements:* For master's, GRE General Test, minimum GPA of 2.5; for doctorate, GRE General Test, GRE Subject Test, minimum graduate GPA of 3.0.

Clarkson University, Graduate School, School of Arts and Sciences, Department of Chemistry, Potsdam, NY 13699. Offers analytical chemistry (MS, PhD); inorganic chemistry (MS, PhD); organic chemistry (MS, PhD); physical chemistry (MS, PhD). *Faculty:* 9 full-time (2 women). *Students:* 34 full-time (9 women), 1 (woman) part-time; includes 2 Asian Americans or Pacific Islanders, 22 international. Average age 27. 52 applicants, 62% accepted. In 2005, 3 master's, 2 doctorates awarded. *Median time to degree:* Of those who began their doctoral program in fall 1997, 100% received their degree in 8 years or less. *Degree requirements:* For doctorate, thesis/dissertation, departmental qualifying exam. *Entrance requirements:* For master's, GRE. Additional exam requirements/recommendations for international students: Required—TOEFL. *Application deadline:* For fall admission, 5/15 for domestic students; for spring admission, 10/15 priority date for domestic students. Applications are processed on a rolling basis. Application fee: $25 ($35 for international students). Electronic applications accepted. *Expenses:* Tuition: Full-time $20,160; part-time $840 per hour. Required fees: $215. *Financial support:* In 2005–06, 25 students received support, including 2 fellowships (averaging $25,000 per year), 11 research assistantships (averaging $19,032 per year), 12 teaching assistantships (averaging $19,032 per year); scholarships/grants and tuition waivers (partial) also available. *Faculty research:* Nanomaterial, surface science, polymers chemical biosensing, protein biochemistry drug design/delivery, molecular neuroscience and immunobiology. Total annual research expenditures: $1.6 million. *Unit head:* Dr. Phillip A. Christiansen, Division Head, 315-268-6669, Fax: 315-268-6610, E-mail: pac@clarkson.edu. *Application contact:* Donna Brockway, Graduate Admissions International Advisor/Assistant to the Provost, 315-268-6447, Fax: 315-268-7994, E-mail: brockway@clarkson.edu.

Cleveland State University, College of Graduate Studies, College of Science, Department of Chemistry, Cleveland, OH 44115. Offers analytical chemistry (MS); clinical chemistry (MS, PhD); clinical/bioanalytical chemistry (PhD); environmental chemistry (MS); inorganic chemistry (MS); organic chemistry (MS); physical chemistry (MS). Part-time and evening/weekend programs available. *Faculty:* 13 full-time (1 woman), 59 part-time/adjunct (12 women). *Students:* 46 full-time (22 women), 18 part-time (6 women); includes 4 minority (2 African Americans, 2 Asian Americans or Pacific Islanders), 35 international. Average age 31. 47 applicants, 62% accepted, 7 enrolled. In 2005, 2 master's, 5 doctorates awarded. *Median time to degree:* Of those who began their doctoral program in fall 1997, 67% received their degree in 8 years or less. *Degree requirements:* For master's, thesis (for some programs); for doctorate, thesis/dissertation. *Entrance requirements:* For master's and doctorate, GRE General Test, GRE Subject Test. Additional exam requirements/recommendations for international students: Required—TOEFL (minimum score 525 paper-based; 197 computer-based). *Application deadline:* For fall admission, 1/15 priority date for domestic students, 1/15 priority date for international students. Applications are processed on a rolling basis. Application fee: $30. Electronic applications accepted. *Expenses:* Tuition, state resident: full-time $10,700. Tuition, nonresident: full-time $14,628. Tuition and fees vary according to program. *Financial support:* In 2005–06, 44 students received support, including fellowships with full tuition reimbursements available (averaging $18,000 per year), research assistantships with full tuition reimbursements available (averaging $16,000 per year), teaching assistantships with full tuition reimbursements available (averaging $14,000 per year) Financial award application deadline: 1/15. *Faculty research:* Metalloenzyme mechanisms, dependent RNAse L and interferons, application of HPLC/LPCC to clinical systems, structure-function relationships of factor Va, MALDI-TOF based DNA sequencing. Total annual research expenditures: $3 million. *Unit head:* Dr. Lily M. Ng, Chair, 216-687-2467, E-mail: l.ng@csuohio.edu. *Application contact:* Richelle P. Emery, Administrative Coordinator, 216-687-2457, Fax: 216-687-9298, E-mail: r.emery@csuohio.edu.

Columbia University, Graduate School of Arts and Sciences, Division of Natural Sciences, Department of Chemistry, New York, NY 10027. Offers chemical physics (M Phil, PhD); inorganic chemistry (M Phil, MA, PhD); organic chemistry (M Phil, MA, PhD). *Faculty:* 17 full-time. *Students:* 115 full-time (35 women). Average age 27. 452 applicants, 20% accepted. In 2005, 13 master's, 20 doctorates awarded. *Degree requirements:* For master's, comp exams (MS); foreign language, teaching experience, oral/written exams (M Phil); for doctorate, one foreign language, thesis/dissertation. *Entrance requirements:* For master's and doctorate, GRE General Test, GRE Subject Test. Additional exam requirements/recommendations for international students: Required—TOEFL. Application fee: $75. *Expenses:* Tuition: Full-time $31,448. Tuition and fees vary according to course level, course load, campus/location and program. *Financial support:* Fellowships, teaching assistantships, Federal Work-Study and institutionally sponsored loans available. Support available to part-time students. Financial award application deadline: 1/5; financial award applicants required to submit FAFSA. *Faculty research:* Biophysics. *Unit head:* James Valentini, Chair, 212-854-7590, Fax: 212-932-1289, E-mail: jjvi@columbia.edu.

Cornell University, Graduate School, Graduate Fields of Arts and Sciences, Field of Chemistry and Chemical Biology, Ithaca, NY 14853-0001. Offers analytical chemistry (PhD); bio-organic chemistry (PhD); biophysical chemistry (PhD); chemical biology (PhD); chemical physics (PhD); inorganic chemistry (PhD); materials chemistry (PhD); organic chemistry (PhD); organometallic chemistry (PhD); physical chemistry (PhD); polymer chemistry (PhD); theoretical chemistry (PhD). *Faculty:* 44 full-time (2 women). *Students:* 190 full-time (73 women); includes 23 minority (4 African Americans, 8 Asian Americans or Pacific Islanders, 11 Hispanic Americans), 65 international. 339 applicants, 35% accepted, 49 enrolled. In 2005, 23 doctorates awarded. *Degree requirements:* For doctorate, thesis/dissertation, comprehensive exam. *Entrance requirements:* For doctorate, GRE General Test, GRE Subject Test (chemistry), 3 letters of recommendation. Additional exam requirements/recommendations for international students: Required—TOEFL (minimum score 600 paper-based; 250 computer-based). *Application deadline:* For fall admission, 1/10 for domestic students. Application fee: $60. Electronic applications accepted. *Financial support:* In 2005–06, 185 students received support, including 33 fellowships with full tuition reimbursements available, 85 research assistantships with full tuition reimbursements available, 67 teaching assistantships with full tuition reimbursements available; institutionally sponsored loans, scholarships/grants, health care benefits, tuition waivers (full and partial), and unspecified assistantships also available. Financial award applicants required to submit FAFSA. *Faculty research:* Analytical, organic, inorganic, physical, materials, chemical biology. *Unit head:* Director of Graduate Studies, 607-255-4139, Fax: 607-255-4137. *Application contact:* Graduate Field Assistant, 607-255-4139, Fax: 607-255-4137, E-mail: chemgrad@cornell.edu.

See Close-Up on page 109.

Florida State University, Graduate Studies, College of Arts and Sciences, Department of Chemistry and Biochemistry, Tallahassee, FL 32306. Offers analytical chemistry (MS, PhD); biochemistry (MS, PhD); chemical physics (MS, PhD); inorganic chemistry (MS, PhD); organic chemistry (MS, PhD); physical chemistry (MS, PhD). Part-time programs available. *Faculty:* 36 full-time (6 women), 2 part-time/adjunct (0 women). *Students:* 144 full-time (50 women); includes 17 minority (7 African Americans, 1 American Indian/Alaska Native, 6 Asian Americans or Pacific Islanders, 3 Hispanic Americans), 58 international. Average age 25. 288 applicants, 27% accepted, 26 enrolled. In 2005, 14 master's, 15 doctorates awarded. Terminal master's awarded for partial completion of doctoral program. *Median time to degree:* Of those who began their doctoral program in fall 1997, 88% received their degree in 8 years or less. *Degree requirements:* For master's, thesis (for some programs), cumulative and diagnostic exams, comprehensive exam (for some programs), registration; for doctorate, thesis/dissertation, cumulative and diagnostic exams, comprehensive exam (for some programs), registration. *Entrance requirements:* For master's and doctorate, GRE General Test, minimum B average in undergraduate course work. Additional exam requirements/recommendations for international students: Required—TOEFL (minimum score 515 paper-based; 213 computer-based). *Application deadline:* For fall admission, 4/15 priority date for domestic students, 4/15 priority date for international students. Applications are processed on a rolling basis. Application fee: $30. Electronic applications accepted. *Financial support:* In 2005–06, 1 fellowship with tuition reimbursement (averaging $18,000 per year), 56 research assistantships with tuition reimbursements (averaging $19,000 per year), 83 teaching assistantships with tuition reimbursements (averaging $19,000 per year) were awarded; career-related internships or fieldwork, Federal Work-Study, institutionally sponsored loans, and traineeships also available. Financial award application deadline: 2/15; financial award applicants required to submit FAFSA. *Faculty research:* Spectroscopy, computational chemistry, nuclear chemistry, separations, synthesis. Total annual research expenditures: $6.5 million. *Unit head:* Dr. Naresh Dalal, Chairman, 850-644-3398, Fax: 850-644-8281. *Application contact:* Dr. Oliver Steinbock, Chair, Graduate Admissions Committee, 888-525-9286, Fax: 850-644-8281, E-mail: gradinfo@chem.fsu.edu.

See Close-Up on page 117.

Georgetown University, Graduate School of Arts and Sciences, Department of Chemistry, Washington, DC 20057. Offers analytical chemistry (MS, PhD); biochemistry (MS, PhD); chemical physics (MS, PhD); inorganic chemistry (MS, PhD); organic chemistry (MS, PhD); physical chemistry (MS, PhD); theoretical chemistry (MS, PhD). Terminal master's awarded for partial completion of doctoral program. *Degree requirements:* For master's, thesis (for some programs), qualifying exam; for doctorate, thesis/dissertation, comprehensive exam. *Entrance requirements:* For master's and doctorate, GRE General Test. Additional exam requirements/recommendations for international students: Required—TOEFL.

The George Washington University, Columbian College of Arts and Sciences, Department of Chemistry, Washington, DC 20052. Offers analytical chemistry (MS, PhD); inorganic chemistry (MS, PhD); materials science (MS, PhD); organic chemistry (MS, PhD); physical chemistry (MS, PhD). Part-time and evening/weekend programs available. Terminal master's awarded for partial completion of doctoral program. *Degree requirements:* For master's, thesis or alternative, comprehensive exam; for doctorate, thesis/dissertation, general exam. *Entrance requirements:* For master's and doctorate, GRE General Test, interview, minimum GPA of 3.0. Additional exam requirements/recommendations for international students: Required—TOEFL (minimum score 550 paper-based; 213 computer-based). Electronic applications accepted.

Harvard University, Graduate School of Arts and Sciences, Department of Chemistry and Chemical Biology, Cambridge, MA 02138. Offers biochemical chemistry (PhD); inorganic chemistry (PhD); organic chemistry (PhD); physical chemistry (PhD). *Students:* 188 full-time (36 women). 346 applicants, 24% accepted. In 2005, 33 doctorates awarded. *Degree requirements:* For doctorate, thesis/dissertation, cumulative exams. *Entrance requirements:* For doctorate, GRE General Test, GRE Subject Test. Additional exam requirements/recommendations for international students: Required—TOEFL. *Application deadline:* For fall admission, 12/31 for domestic students. Application fee: $60. *Expenses:* Tuition: Full-time $28,752. Full-time tuition and fees vary according to program and student level. *Financial support:* Fellowships, research assistantships, teaching assistantships, career-related internships or fieldwork, Federal Work-Study, and institutionally sponsored loans available. Financial award application deadline: 12/30. *Unit head:* Betsey Cogswell, Administrator, 617-495-5696, Fax: 617-495-5264. *Application contact:* Graduate Admissions Office, 617-496-3208.

See Close-Up on page 119.

Howard University, Graduate School of Arts and Sciences, Department of Chemistry, Washington, DC 20059-0002. Offers analytical chemistry (MS, PhD); atmospheric (MS, PhD); biochemistry (MS, PhD); environmental (MS, PhD); inorganic chemistry (MS, PhD); organic chemistry (MS, PhD); physical chemistry (MS, PhD); polymer chemistry (MS, PhD). Part-time programs available. *Degree requirements:* For master's, one foreign language, thesis, teaching experience, comprehensive exam, registration; for doctorate, 2 foreign languages, thesis/dissertation, teaching experience, comprehensive exam, registration. *Entrance requirements:* For master's, GRE General Test, minimum GPA of 2.7; for doctorate, GRE General Test, minimum GPA of 3.0. *Faculty research:* Stratospheric aerosols, liquid crystals, polymer coatings, terrestrial and extraterrestrial atmospheres, amidogen reaction.

Illinois Institute of Technology, Graduate College, College of Science and Letters, Department of Biological, Chemical and Physical Sciences, Chemistry Division, Chicago, IL 60616-3793. Offers analytical chemistry (M Ch, MS, PhD); chemistry (M Chem); inorganic chemistry (MS, PhD); materials and chemical synthesis (M Ch); organic chemistry (MS, PhD); physical chemistry (MS, PhD); polymer chemistry (MS, PhD). Part-time and evening/weekend programs available. Postbaccalaureate distance learning degree programs offered (no on-campus study). Terminal master's awarded for partial completion of doctoral program. *Degree requirements:* For master's, thesis (for some programs), comprehensive exam; for doctorate, thesis/dissertation, comprehensive exam. *Entrance requirements:* For master's and doctorate, GRE General Test, minimum undergraduate GPA of 3.0. Additional exam requirements/recommendations for international students: Required—TOEFL (minimum score 550 paper-based; 213 computer-based). Electronic applications accepted. *Faculty research:* Organic and inorganic chemistry, polymers research, physical chemistry, analytical chemistry.

Indiana University Bloomington, Graduate School, College of Arts and Sciences, Department of Chemistry, Bloomington, IN 47405-7000. Offers analytical chemistry (PhD); biological chemistry (PhD); chemistry (MAT); inorganic chemistry (PhD); physical chemistry (PhD). PhD offered through the University Graduate School. *Faculty:* 29 full-time (2 women). *Students:* 110 full-time (38 women), 35 part-time (11 women); includes 8 minority (3 African Americans, 3 Asian Americans or Pacific Islanders, 2 Hispanic Americans), 57 international. Average age 26. In 2005, 7 master's, 20 doctorates awarded. Terminal master's awarded for partial completion of doctoral program. *Degree requirements:* For master's and doctorate, thesis/dissertation. *Entrance requirements:* For master's and doctorate, GRE General Test, GRE Subject Test. Additional exam requirements/recommendations for international students: Required—TOEFL. *Application deadline:* For fall admission, 1/15 priority date for domestic students, 12/15 priority date for international students; for spring admission, 9/1 priority date for domestic students, 9/1 priority date for international students. Applications are processed on a rolling basis. Application fee: $50 ($60 for international students). *Expenses:* Tuition, state resident: full-time $5,437; part-time $227 per credit hour. Tuition, nonresident: full-time $15,836; part-time $660 per credit hour. Required fees: $821. Tuition and fees vary according to campus/location and program. *Financial support:* In 2005–06, 23 fellowships with full tuition reimbursements, 57 research assistantships with full tuition reimbursements, 78 teaching assistantships with full tuition reimbursements were awarded; Federal Work-Study and institutionally sponsored loans also available. *Faculty research:* Synthesis of complex natural products, organic reaction mechanisms, organic electrochemistry, transitive-metal chemistry, solid-state and surface chemistry. Total annual research expenditures: $7.7 million. *Unit head:* Dr. David Clemmer, Chairperson, 812-855-2268. *Application contact:* Dr. Jack K. Crandall, Chairperson of Admissions, 812-855-2068, Fax: 812-855-8300, E-mail: chemgrad@indiana.edu.

Kansas State University, Graduate School, College of Arts and Sciences, Department of Chemistry, Manhattan, KS 66506. Offers analytical chemistry (MS); biological chemistry (MS); chemistry (PhD); inorganic chemistry (MS); materials chemistry (MS); organic chemistry (MS); physical chemistry (MS). *Faculty:* 14 full-time (1 woman). *Students:* 65 full-time (21 women), 5 part-time (2 women); includes 1 minority (African American), 48 international. 63 applicants, 67% accepted, 22 enrolled. In 2005, 3 master's, 5 doctorates awarded. Terminal master's awarded for partial completion of doctoral program. *Degree requirements:* For master's and doctorate, thesis/dissertation. *Entrance requirements:* For master's and doctorate, GRE, minimum GPA of 3.0. Additional exam requirements/recommendations for international students: Required—TOEFL (minimum score 550 paper-based; 213 computer-based). *Application deadline:* For fall admission, 2/1 priority date for domestic students, 2/1 priority date for international students; for spring admission, 10/1 for domestic students, 8/1 for international students. Applications are processed on a rolling basis. Application fee: $30 ($55 for international students). *Expenses:* Tuition, state resident: full-time $5,160; part-time $215. Tuition, nonresident: full-time $12,816; part-time $534. International tuition: $12,816 full-time. Required fees: $564. *Financial support:* In 2005–06, 30 research assistantships (averaging $11,956 per year), 26 teaching assistantships with full tuition reimbursements (averaging $11,945 per year) were awarded; fellowships, institutionally sponsored loans and scholarships/grants also available. Support available to part-time students. Financial award application deadline: 3/1; financial award applicants required to submit FAFSA. *Faculty research:* Nanotechnologies, functional materials, bio-organic and bio-physical processes, sensors and separations, synthesis and synthetic methods. Total annual research expenditures: $2.2 million. *Unit head:* Eric Maatta, Head, 785-532-6665, Fax: 785-532-6666, E-mail: eam@ksu.edu. *Application contact:* Christer Aakeröy, Director, 785-532-6096, Fax: 785-532-6666, E-mail: aakeroy@ksu.edu.

Kent State University, College of Arts and Sciences, Department of Chemistry, Kent, OH 44242-0001. Offers analytical chemistry (MS, PhD); biochemistry (MS, PhD); chemistry (MA, MS, PhD); inorganic chemistry (MS, PhD); organic chemistry (MS, PhD); physical chemistry (MS, PhD). Terminal master's awarded for partial completion of doctoral program. *Degree requirements:* For master's and doctorate, thesis/dissertation, comprehensive exam, registration. *Entrance requirements:* For master's and doctorate, placement exam, GRE General Test, GRE Subject Test (recommended), minimum GPA of 2.75. Additional exam requirements/recommendations for international students: Required—TOEFL (minimum score 575 paper-based; 230 computer-based). Electronic applications accepted. *Faculty research:* Biological chemistry, materials chemistry, molecular spectroscopy.

See Close-Up on page 121.

Marquette University, Graduate School, College of Arts and Sciences, Department of Chemistry, Milwaukee, WI 53201-1881. Offers analytical chemistry (MS, PhD); bioanalytical chemistry (MS, PhD); biophysical chemistry (MS, PhD); chemical physics (MS, PhD); inorganic chemistry (MS, PhD); organic chemistry (MS, PhD); physical chemistry (MS, PhD). Part-time programs available. Terminal master's awarded for partial completion of doctoral program. *Degree requirements:* For master's, comprehensive exam; for doctorate, thesis/dissertation, cumulative exams. *Entrance requirements:* For master's and doctorate, GRE Subject Test. Additional exam requirements/recommendations for international students: Required—TOEFL. *Faculty research:* Inorganic complexes, laser Raman spectroscopy, organic synthesis, chemical dynamics, biophysiology.

Massachusetts Institute of Technology, School of Science, Department of Chemistry, Cambridge, MA 02139-4307. Offers biological chemistry (PhD, Sc D); inorganic chemistry (PhD, Sc D); organic chemistry (PhD, Sc D); physical chemistry (PhD, Sc D). *Faculty:* 30 full-time (6 women). *Students:* 242 full-time (84 women), 3 part-time (2 women); includes 30 minority (5 African Americans, 1 American Indian/Alaska Native, 19 Asian Americans or Pacific Islanders, 5 Hispanic Americans), 77 international. Average age 26. 476 applicants, 27% accepted, 54 enrolled. In 2005, 48 doctorates awarded. *Degree requirements:* For doctorate, thesis/dissertation, comprehensive exam. *Entrance requirements:* For doctorate, GRE General Test. Additional exam requirements/recommendations for international students: Required—TOEFL (minimum score 577 paper-based; 233 computer-based). *Application deadline:* For fall admission, 12/15 for domestic students, 12/15 for international students. Application fee: $70. Electronic applications accepted. *Expenses:* Tuition: Full-time $32,100. Required fees: $200. Part-time tuition and fees vary according to course load. *Financial support:* In 2005–06, 210 students received support, including 36 fellowships with tuition reimbursements available (averaging $29,141 per year), 143 research assistantships with tuition reimbursements available (averaging $24,550 per year), 59 teaching assistantships with tuition reimbursements available (averaging $25,155 per year); career-related internships or fieldwork, Federal Work-Study, institutionally sponsored loans, scholarships/grants, health care benefits, and unspecified assistantships also available. *Faculty research:* Synthetic organic chemistry, enzymatic reaction mechanisms, inorganic and organometallic spectroscopy, high resolution NMR spectroscopy. Total annual research expenditures: $21.5 million. *Unit head:* Prof. Timothy Swager, Department Head, 617-253-1801, E-mail: tswager@mit.edu. *Application contact:* Susan Brighton, Graduate Administrator, 617-253-1845, Fax: 617-258-0241, E-mail: brighton@mit.edu.

McMaster University, School of Graduate Studies, Faculty of Science, Department of Chemistry, Hamilton, ON L8S 4M2, Canada. Offers analytical chemistry (M Sc, PhD); chemical physics (M Sc, PhD); chemistry (M Sc, PhD); inorganic chemistry (M Sc, PhD); organic chemistry (M Sc, PhD); physical chemistry (M Sc, PhD); polymer chemistry (M Sc, PhD). Part-time programs available. Terminal master's awarded for partial completion of doctoral program.

Inorganic Chemistry

McMaster University (continued)
Degree requirements: For master's, thesis/dissertation; for doctorate, thesis/dissertation, comprehensive exam. *Entrance requirements:* For master's, minimum B+ average. Additional exam requirements/recommendations for international students: Required—TOEFL (minimum score 550 paper-based; 213 computer-based).

Miami University, Graduate School, College of Arts and Sciences, Department of Chemistry and Biochemistry, Oxford, OH 45056. Offers analytical chemistry (MS, PhD); biochemistry (MS, PhD); chemical education (MS, PhD); chemistry (MS, PhD); inorganic chemistry (MS, PhD); organic chemistry (MS, PhD); physical chemistry (MS, PhD). Part-time programs available. *Degree requirements:* For master's, thesis, final exam; for doctorate, thesis/dissertation, final exams, comprehensive exam. *Entrance requirements:* For master's, minimum undergraduate GPA of 3.0 during previous 2 years or 2.75 overall; for doctorate, minimum undergraduate GPA of 2.75, 3.0 graduate. Additional exam requirements/recommendations for international students: Required—TOEFL (minimum score 550 paper-based; 213 computer-based), TWE(minimum score 4). Electronic applications accepted.

Northeastern University, College of Arts and Sciences, Department of Chemistry and Chemical Biology, Boston, MA 02115-5096. Offers analytical chemistry (PhD); chemistry (MS, PhD); inorganic chemistry (PhD); organic chemistry (PhD); physical chemistry (PhD). Part-time and evening/weekend programs available. Terminal master's awarded for partial completion of doctoral program. *Degree requirements:* For master's, thesis (for some programs); for doctorate, thesis/dissertation, qualifying exam in specialty area. *Entrance requirements:* Additional exam requirements/recommendations for international students: Required—TOEFL. Electronic applications accepted. *Faculty research:* Electron transfer, theoretical chemical physics, analytical biotechnology, mass spectrometry, materials chemistry.

See Close-Up on page 129.

Oregon State University, Graduate School, College of Science, Department of Chemistry, Corvallis, OR 97331. Offers analytical chemistry (MS, PhD); chemistry (MA, MAIS); inorganic chemistry (MS, PhD); nuclear and radiation chemistry (MS, PhD); organic chemistry (MS, PhD); physical chemistry (MS, PhD). Part-time programs available. *Faculty:* 19 full-time (3 women), 6 part-time/adjunct (1 woman). *Students:* 74 full-time (30 women), 3 part-time; includes 6 minority (1 African American, 1 Asian American or Pacific Islander, 4 Hispanic Americans), 35 international. Average age 28. In 2005, 4 master's, 10 doctorates awarded. Terminal master's awarded for partial completion of doctoral program. *Degree requirements:* For master's and doctorate, one foreign language, thesis/dissertation. *Entrance requirements:* For master's and doctorate, minimum GPA of 3.0 in last 90 hours of course work. Additional exam requirements/recommendations for international students: Required—TOEFL. *Application deadline:* For fall admission, 3/1 for domestic students. Applications are processed on a rolling basis. Application fee: $50. *Expenses:* Tuition, area resident: Full-time $8,139; part-time $301 per credit. Tuition, state resident: full-time $8,139; part-time $501 per credit. Tuition, nonresident: full-time $14,376; part-time $532 per credit. International tuition: $14,376 full-time. Required fees: $1,266. *Financial support:* Fellowships, research assistantships, teaching assistantships, institutionally sponsored loans available. Support available to part-time students. Financial award application deadline: 2/1. *Faculty research:* Solid state chemistry, enzyme reaction mechanisms, structure and dynamics of gas molecules, chemiluminescence, nonlinear optical spectroscopy. *Unit head:* Dr. Douglas Keszler, Chair, 541-737-2081, Fax: 541-737-2062. *Application contact:* Carolyn Brumley, Graduate Secretary, 541-737-6707, Fax: 541-737-2062, E-mail: carolyn.brumley@orst.edu.

Purdue University, Graduate School, School of Science, Department of Chemistry, West Lafayette, IN 47907. Offers analytical chemistry (MS, PhD); biochemistry (MS, PhD); chemical education (MS, PhD); inorganic chemistry (MS, PhD); organic chemistry (MS, PhD); physical chemistry (MS, PhD). *Accreditation:* NCATE (one or more programs are accredited). *Faculty:* 43 full-time (9 women), 8 part-time/adjunct (2 women). *Students:* 293 full-time (114 women), 39 part-time (14 women); includes 53 minority (25 African Americans, 6 Asian Americans or Pacific Islanders, 22 Hispanic Americans), 138 international. Average age 28. 458 applicants, 36% accepted, 57 enrolled. In 2005, 9 master's, 36 doctorates awarded. Terminal master's awarded for partial completion of doctoral program. *Degree requirements:* For master's and doctorate, thesis/dissertation. *Entrance requirements:* Additional exam requirements/recommendations for international students: Required—TOEFL. *Application deadline:* For fall admission, 4/1 priority date for domestic students, 3/1 priority date for international students; for spring admission, 10/1 priority date for domestic students, 9/1 priority date for international students. Applications are processed on a rolling basis. Application fee: $55. Electronic applications accepted. *Financial support:* In 2005–06, 2 fellowships with partial tuition reimbursements (averaging $18,000 per year), 55 teaching assistantships with partial tuition reimbursements (averaging $18,000 per year) were awarded; research assistantships with partial tuition reimbursements, tuition waivers (partial) also available. Support available to part-time students. Financial award applicants required to submit FAFSA. *Unit head:* Dr. Timothy S Zwier, Head, 765-494-5203. *Application contact:* R. E. Wild, Chairman, Graduate Admissions, 765-494-5200, E-mail: wild@purdue.edu.

Rensselaer Polytechnic Institute, Graduate School, School of Science, Department of Chemistry and Chemical Biology, Troy, NY 12180-3590. Offers analytical chemistry (MS, PhD); biochemistry (MS, PhD); inorganic chemistry (MS, PhD); organic chemistry (MS, PhD); physical chemistry (MS, PhD); polymer chemistry (MS, PhD). Part-time and evening/weekend programs available. *Faculty:* 19 full-time (3 women). *Students:* 49 full-time (28 women), 2 part-time (1 woman), 33 international. Average age 24. 80 applicants, 23% accepted, 7 enrolled. In 2005, 1 master's, 14 doctorates awarded. Terminal master's awarded for partial completion of doctoral program. *Median time to degree:* Of those who began their doctoral program in fall 1997, 100% received their degree in 8 years or less. *Degree requirements:* For master's, thesis (for some programs), registration; for doctorate, thesis/dissertation, comprehensive exam, registration. *Entrance requirements:* For master's, GRE General Test, GRE Subject Test (strongly recommended); for doctorate, GRE General Test, GRE Subject Test (chemistry or biochemistry strongly recommended). Additional exam requirements/recommendations for international students: Required—TOEFL (minimum score 600 paper-based). *Application deadline:* For fall admission, 2/1 for domestic students; for spring admission, 11/15 for domestic students. Applications are processed on a rolling basis. Application fee: $75. Electronic applications accepted. *Expenses:* Tuition: Full-time $31,000; part-time $1,320 per credit. Required fees: $1,623. *Financial support:* In 2005–06, 49 students received support, including 1 fellowship with full tuition reimbursement available (averaging $30,000 per year), 25 research assistantships with full tuition reimbursements available (averaging $21,500 per year), 30 teaching assistantships with full tuition reimbursements available (averaging $21,500 per year); institutionally sponsored loans and tuition waivers (full and partial) also available. Financial award application deadline: 2/1. *Faculty research:* Synthetic polymer and biopolymer chemistry, physical chemistry of polymeric systems, bioanalytical chemistry, synthetic and computational drug design, protein folding and protein design. Total annual research expenditures: $1.9 million. *Unit head:* Dr. Linda B. McGown, Chair, 518-276-4856, Fax: 518-276-4887, E-mail: mcgowl@rpi.edu. *Application contact:* Beth E. McGraw, Department Admissions Assistant, 518-276-6456, Fax: 518-276-4887, E-mail: mcgrae@rpi.edu.

See Close-Up on page 137.

Rice University, Graduate Programs, Wiess School of Natural Sciences, Department of Chemistry, Houston, TX 77251-1892. Offers chemistry (MA); inorganic chemistry (PhD); organic chemistry (PhD); physical chemistry (PhD). Terminal master's awarded for partial completion of doctoral program. *Degree requirements:* For master's and doctorate, thesis/dissertation. *Entrance requirements:* For master's and doctorate, GRE General Test, minimum GPA of 3.0. Additional exam requirements/recommendations for international students: Required—TOEFL. *Faculty research:* Nanoscience, biomaterials, nanobioinformatics, fullerene pharmaceuticals.

Rutgers, The State University of New Jersey, Newark, Graduate School, Program in Chemistry, Newark, NJ 07102. Offers analytical chemistry (MS, PhD); biochemistry (MS, PhD); inorganic chemistry (MS, PhD); organic chemistry (MS, PhD); physical chemistry (MS, PhD). Part-time and evening/weekend programs available. *Faculty:* 22 full-time (5 women). *Students:* 24 full-time (12 women), 28 part-time (10 women); includes 31 minority (2 African Americans, 29 Asian Americans or Pacific Islanders). 69 applicants, 43% accepted, 13 enrolled. In 2005, 7 master's, 9 doctorates awarded. Terminal master's awarded for partial completion of doctoral program. *Degree requirements:* For master's, cumulative exams, thesis optional; for doctorate, thesis/dissertation, exams, research proposal. *Entrance requirements:* For master's and doctorate, GRE General Test, minimum undergraduate B average. Additional exam requirements/recommendations for international students: Required—TOEFL. *Application deadline:* For fall admission, 7/1 for domestic students; for spring admission, 12/1 for domestic students. Applications are processed on a rolling basis. Application fee: $50. Electronic applications accepted. *Expenses:* Tuition, state resident: full-time $10,440; part-time $435 per credit. Tuition, nonresident: full-time $15,520; part-time $637 per credit. *Financial support:* In 2005–06, 35 students received support, including 5 fellowships with partial tuition reimbursements available (averaging $18,000 per year), 6 research assistantships (averaging $16,988 per year), 20 teaching assistantships with partial tuition reimbursements available (averaging $16,988 per year); Federal Work-Study and institutionally sponsored loans also available. Financial award application deadline: 3/1. *Faculty research:* Medicinal chemistry, natural products, isotope effects, biophysics and biorganic approaches to enzyme mechanisms, organic and organometallic synthesis. *Unit head:* Prof. W. Philip Huskey, Chairman and Program Director, 973-353-5741, Fax: 973-353-1264, E-mail: huskey@andromeda.rutgers.edu.

Rutgers, The State University of New Jersey, New Brunswick/Piscataway, Graduate School, Program in Chemistry and Chemical Biology, New Brunswick, NJ 08901-1281. Offers analytical chemistry (MS, PhD); biological chemistry (PhD); chemistry education (MST); inorganic chemistry (MS, PhD); organic chemistry (MS, PhD); physical chemistry (MS, PhD). Part-time and evening/weekend programs available. *Faculty:* 48 full-time. *Students:* 104 full-time (41 women), 22 part-time (4 women); includes 11 minority (2 African Americans, 6 Asian Americans or Pacific Islanders, 3 Hispanic Americans), 51 international. Average age 29. 108 applicants, 51% accepted, 22 enrolled. In 2005, 13 master's, 10 doctorates awarded. Terminal master's awarded for partial completion of doctoral program. *Degree requirements:* For master's, thesis or alternative, exam, comprehensive exam, registration; for doctorate, thesis/dissertation, cumulative exams, 1 year residency, comprehensive exam, registration. *Entrance requirements:* For master's and doctorate, GRE General Test, GRE Subject Test. Additional exam requirements/recommendations for international students: Required—TOEFL. *Application deadline:* For fall admission, 4/15 priority date for domestic students, 4/1 priority date for international students; for spring admission, 12/1 priority date for domestic students, 12/1 priority date for international students. Applications are processed on a rolling basis. Application fee: $50. Electronic applications accepted. *Expenses:* Tuition, state resident: full-time $10,440; part-time $435 per credit. Tuition, nonresident: full-time $15,520; part-time $647 per credit. Required fees: $129 per credit. Tuition and fees vary according to program. *Financial support:* In 2005–06, 104 students received support, including 8 fellowships with full tuition reimbursements available (averaging $22,000 per year), 23 research assistantships with full tuition reimbursements available (averaging $18,347 per year), 59 teaching assistantships with full tuition reimbursements available (averaging $18,347 per year); career-related internships or fieldwork, Federal Work-Study, traineeships, and health care benefits also available. Financial award application deadline: 3/1; financial award applicants required to submit FAFSA. *Faculty research:* Biophysical organic/bioorganic, inorganic/bioinorganic, theoretical, and solid-state/surface chemistry. Total annual research expenditures: $14.5 million. *Unit head:* Dr. Roger Jones, Director and Chair, 732-445-4900, Fax: 732-445-5312, E-mail: jones@rutchem.rutgers.edu. *Application contact:* Dr. Martha A. Cotter, Vice Chair, 732-445-2259, Fax: 732-445-5312, E-mail: cotter@rutchem.rutgers.edu.

See Close-Up on page 139.

Seton Hall University, College of Arts and Sciences, Department of Chemistry and Biochemistry, South Orange, NJ 07079-2694. Offers analytical chemistry (MS, PhD); biochemistry (MS, PhD); chemistry (MS); inorganic chemistry (MS, PhD); organic chemistry (MS, PhD); physical chemistry (MS, PhD). Part-time and evening/weekend programs available. *Students:* 19 full-time (6 women), 46 part-time (17 women). Average age 34. 33 applicants, 79% accepted, 14 enrolled. In 2005, 11 master's, 7 doctorates awarded. Terminal master's awarded for partial completion of doctoral program. *Degree requirements:* For master's, formal seminar, thesis optional; for doctorate, thesis/dissertation, annual seminars, comprehensive exam. *Entrance requirements:* Additional exam requirements/recommendations for international students: Required—TOEFL. *Application deadline:* For fall admission, 7/1 priority date for domestic students, 7/1 priority date for international students; for spring admission, 11/1 priority date for domestic students, 11/1 priority date for international students. Applications are processed on a rolling basis. Application fee: $50. Electronic applications accepted. *Financial support:* In 2005–06, 1 research assistantship, 19 teaching assistantships were awarded; Federal Work-Study also available. *Faculty research:* DNA metal reactions; chromatography; bioinorganic, biophysical, organometallic, polymer chemistry; heterogeneous catalyst. *Unit head:* Dr. Nicholas Snow, Chair, 973-761-9414, Fax: 973-761-9772, E-mail: snownich@shu.edu. *Application contact:* Dr. Stephen Kelty, Director of Graduate Studies, 973-761-9129, Fax: 973-761-9772, E-mail: keltyste@shu.edu.

See Close-Up on page 141.

South Dakota State University, Graduate School, College of Arts and Science and College of Agriculture and Biological Sciences, Department of Chemistry, Brookings, SD 57007. Offers analytical chemistry (MS, PhD); biochemistry (MS, PhD); chemistry (MS, PhD); inorganic chemistry (MS, PhD); organic chemistry (MS, PhD); physical chemistry (MS, PhD). *Degree requirements:* For master's, thesis, oral exam; for doctorate, thesis/dissertation, preliminary oral and written exams, research tool. *Entrance requirements:* For master's, bachelor's degree in chemistry or equivalent. Additional exam requirements/recommendations for international students: Required—TOEFL. *Faculty research:* Environmental chemistry, computational chemistry, organic synthesis and photochemistry, novel material development and characterization.

Southern University and Agricultural and Mechanical College, Graduate School, College of Sciences, Department of Chemistry, Baton Rouge, LA 70813. Offers analytical chemistry (MS); biochemistry (MS); environmental sciences (MS); inorganic chemistry (MS); organic chemistry (MS); physical chemistry (MS). *Faculty:* 9 full-time (2 women), 3 part-time/adjunct (2 women). *Students:* 20 full-time (13 women), 8 part-time (5 women); all minorities (20 African Americans, 8 Asian Americans or Pacific Islanders). Average age 23. 30 applicants, 70% accepted, 14 enrolled. In 2005, 3 master's awarded. *Degree requirements:* For master's, thesis. *Entrance requirements:* For master's, GMAT or GRE General Test. Additional exam requirements/recommendations for international students: Required—TOEFL (minimum score 525 paper-based; 193 computer-based). *Application deadline:* For fall admission, 4/15 for domestic students; for spring admission, 11/1 priority date for domestic students. Applications are processed on a rolling basis. Application fee: $5. *Financial support:* In 2005–06, 31 research assistantships (averaging $7,000 per year), 10 teaching assistantships (averaging $7,000 per year) were awarded; scholarships/grants also available. Financial award application deadline: 4/15. *Faculty research:* Synthesis of macrocyclic ligands, latex accelerators, anticancer drugs, biosensors, absorption isotheums, isolation of specific enzymes from plants. Total annual research expenditures:$400,000. *Unit head:* Dr. Ella Kelley, Chair, 225-771-3990, Fax: 225-771-3992.

State University of New York at Binghamton, Graduate School, School of Arts and Sciences, Department of Chemistry, Binghamton, NY 13902-6000. Offers analytical chemistry (PhD); chemistry (MA, MS); inorganic chemistry (PhD); organic chemistry (PhD); physical chemistry (PhD). Part-time programs available. Terminal master's awarded for partial completion of doctoral program. *Degree requirements:* For master's, thesis or alternative, oral exam, seminar presentation; for doctorate, thesis/dissertation, cumulative exams. *Entrance requirements:*

For master's and doctorate, GRE General Test, GRE Subject Test. Additional exam requirements/recommendations for international students: Required—TOEFL. Electronic applications accepted.

Tufts University, Graduate School of Arts and Sciences, Department of Chemistry, Medford, MA 02155. Offers analytical chemistry (MS, PhD); bioorganic chemistry (MS, PhD); environmental chemistry (MS, PhD); inorganic chemistry (MS, PhD); organic chemistry (MS, PhD); physical chemistry (MS, PhD). *Faculty:* 17 full-time. *Students:* 55 (27 women); includes 5 minority (1 African American, 4 Asian Americans or Pacific Islanders) 27 international. 45 applicants, 87% accepted, 17 enrolled. In 2005, 4 master's, 4 doctorates awarded. Terminal master's awarded for partial completion of doctoral program. *Degree requirements:* For master's and doctorate, thesis/dissertation. *Entrance requirements:* For master's and doctorate, GRE General Test, GRE Subject Test. Additional exam requirements/recommendations for international students: Required—TOEFL (minimum score 600 paper-based; 250 computer-based), TSE. *Application deadline:* For fall admission, 1/15 for domestic students, 12/30 for international students; for spring admission, 10/15 for domestic students, 9/15 for international students. Applications are processed on a rolling basis. Application fee: $65. Electronic applications accepted. *Expenses:* Tuition: Full-time $32,360. Tuition and fees vary according to program. *Financial support:* Research assistantships with full and partial tuition reimbursements, teaching assistantships with full and partial tuition reimbursements, Federal Work-Study, scholarships/grants, and tuition waivers (partial) available. Financial award application deadline: 1/15; financial award applicants required to submit FAFSA. *Unit head:* Mary Jane Shultz, Chair, 617-627-3477, Fax: 617-627-3443. *Application contact:* Samuel Kounaves, Information Contact, 617-627-3441, Fax: 617-627-3443.

See Close-Up on page 151.

University of Calgary, Faculty of Graduate Studies, Faculty of Science, Department of Chemistry, Calgary, AB T2N 1N4, Canada. Offers analytical chemistry (M Sc, PhD); applied chemistry (M Sc, PhD); inorganic chemistry (M Sc, PhD); organic chemistry (M Sc, PhD); physical chemistry (M Sc, PhD); polymer chemistry (M Sc, PhD); theoretical chemistry (M Sc, PhD). *Faculty:* 31 full-time (2 women), 2 part-time/adjunct (0 women). *Students:* 80 full-time (38 women). Average age 25. 250 applicants, 6% accepted, 15 enrolled. In 2005, 7 master's, 7 doctorates awarded. *Degree requirements:* For master's, thesis; for doctorate, thesis/dissertation, candidacy exam. *Entrance requirements:* For master's, minimum GPA of 3.0; for doctorate, honors B Sc degree with minimum GPA of 3.7 or M Sc with minimum GPA of 3.3. Additional exam requirements/recommendations for international students: Required—TOEFL (minimum score 580 paper-based; 237 computer-based). *Application deadline:* For fall admission, 12/1 priority date for domestic students, 12/1 priority date for international students. Applications are processed on a rolling basis. Application fee: $100 ($130 for international students). Electronic applications accepted. *Financial support:* In 2005–06, 25 students received support, including research assistantships (averaging $11,000 per year), teaching assistantships (averaging $6,530 per year); fellowships, scholarships/grants also available. Financial award application deadline:12/1. *Faculty research:* Chemical analysis, chemical dynamics, synthesis theory. *Unit head:* Dr. Brian Keay, Head, 403-220-5340, E-mail: info@chem.ucalgary.ca. *Application contact:* Bonnie E. King, Graduate Program Administrator, 403-220-6252, E-mail: bking@ucalgary.ca.

University of Cincinnati, Division of Research and Advanced Studies, McMicken College of Arts and Sciences, Department of Chemistry, Cincinnati, OH 45221. Offers analytical chemistry (MS, PhD); biochemistry (MS, PhD); inorganic chemistry (MS, PhD); organic chemistry (MS, PhD); physical chemistry (MS, PhD); polymer chemistry (MS, PhD); sensors (PhD). Part-time and evening/weekend programs available. Terminal master's awarded for partial completion of doctoral program. *Degree requirements:* For master's, thesis optional; for doctorate, thesis/dissertation, comprehensive exam, registration. *Entrance requirements:* For master's and doctorate, GRE General Test. Additional exam requirements/recommendations for international students: Required—TOEFL (minimum score 580 paper-based; 237 computer-based); Recommended—TSE(minimum score 50). Electronic applications accepted. *Faculty research:* Biomedical chemistry, laser chemistry, surface science, chemical sensors, synthesis.

University of Georgia, Graduate School, College of Arts and Sciences, Department of Chemistry, Athens, GA 30602. Offers analytical chemistry (MS, PhD); inorganic chemistry (MS, PhD); organic chemistry (MS, PhD); physical chemistry (MS, PhD). *Faculty:* 22 full-time (2 women). *Students:* 155 full-time, 3 part-time; includes 12 minority (4 African Americans, 7 Asian Americans or Pacific Islanders, 1 Hispanic American), 87 international. 147 applicants, 56% accepted, 35 enrolled. In 2005, 2 master's, 19 doctorates awarded. Terminal master's awarded for partial completion of doctoral program. *Median time to degree:* Of those who began their doctoral program in fall 1997, 100% received their degree in 8 years or less. *Degree requirements:* For master's, thesis; for doctorate, one foreign language, thesis/dissertation. *Entrance requirements:* For master's and doctorate, GRE General Test. Additional exam requirements/recommendations for international students: Required—TOEFL (minimum score 213 computer-based), TSE(minimum score 50). *Application deadline:* For fall admission, 7/1 for domestic students; for spring admission, 11/15 for domestic students. Application fee: $50. Electronic applications accepted. *Financial support:* Fellowships, research assistantships, teaching assistantships, unspecified assistantships available. *Unit head:* Dr. John L. Stickney, Head, 706-542-2726, Fax: 706-542-9454, E-mail: stickney@chem.uga.edu. *Application contact:* Dr. Donald Kurtz, Information Contact, 706-542-2010, Fax: 706-542-9454, E-mail: kurtz@chem.uga.edu.

University of Louisville, Graduate School, College of Arts and Sciences, Department of Chemistry, Louisville, KY 40292-0001. Offers analytical chemistry (MS, PhD); biochemistry (MS, PhD); chemical physics (PhD); inorganic chemistry (MS, PhD); organic chemistry (MS, PhD); physical chemistry (MS, PhD). *Students:* 37 full-time (16 women), 16 part-time (7 women); includes 3 minority (2 African Americans, 1 Hispanic American), 24 international. Average age 28. In 2005, 1 master's, 8 doctorates awarded. *Degree requirements:* For master's, thesis/dissertation; for doctorate, thesis/dissertation, comprehensive exam. *Entrance requirements:* For master's and doctorate, GRE General Test. Additional exam requirements/recommendations for international students: Required—TOEFL. *Application deadline:* Applications are processed on a rolling basis. Application fee: $50. *Expenses:* Tuition, state resident: full-time $3,003; part-time $334 per credit hour. Tuition, nonresident: full-time $8,277; part-time $920 per credit hour. Tuition and fees vary according to course load, degree level and program. *Financial support:* In 2005–06, 33 teaching assistantships with tuition reimbursements were awarded; fellowships, research assistantships *Unit head:* Dr. George R. Pack, Chair, 502-852-6798, Fax: 502-852-8149, E-mail: george.pack@louisville.edu.

University of Maryland, College Park, Graduate Studies, College of Chemical and Life Sciences, Department of Chemistry and Biochemistry, Chemistry Program, College Park, MD 20742. Offers analytical chemistry (MS, PhD); inorganic chemistry (MS, PhD); organic chemistry (MS, PhD); physical chemistry (MS, PhD). Part-time and evening/weekend programs available. *Students:* 102 full-time (43 women), 4 part-time (2 women); includes 7 minority (2 African Americans, 3 Asian Americans or Pacific Islanders, 2 Hispanic Americans), 46 international. 128 applicants, 24% accepted, 30 enrolled. In 2005, 6 master's, 12 doctorates awarded. Terminal master's awarded for partial completion of doctoral program. *Median time to degree:* Of those who began their doctoral program in fall 1997, 33% received their degree in 8 years or less. *Degree requirements:* For master's, thesis optional; for doctorate, thesis/dissertation, 2 seminar presentations, oral exam. *Entrance requirements:* For master's and doctorate, GRE General Test, GRE Subject Test (recommended), minimum GPA of 3.0, 3 letters of recommendation. Additional exam requirements/recommendations for international students: Required—TOEFL, TSE. *Application deadline:* For fall admission, 4/1 for domestic students, 2/1 for international students; for spring admission, 10/21 for domestic students, 6/1 for international students. Applications are processed on a rolling basis. Application fee: $60. Electronic applications accepted. *Financial support:* In 2005–06, 10 fellowships (averaging $4,900 per year) were awarded; research assistantships, teaching assistantships with partial tuition reimbursements Financial award applicants required to submit FAFSA. *Faculty research:* Environmental chemistry, nuclear chemistry, lunar and environmental analysis, x-ray crystallography. *Application contact:* Dean of Graduate School, 301-405-4190, Fax: 301-314-9305.

University of Miami, Graduate School, College of Arts and Sciences, Department of Chemistry, Coral Gables, FL 33124. Offers chemistry (MS); inorganic chemistry (PhD); organic chemistry (PhD); physical chemistry (PhD). *Faculty:* 11 full-time (1 woman). *Students:* 38 full-time (6 women), 28 international. Average age 27. 62 applicants, 27% accepted, 5 enrolled. In 2005, 1 master's, 5 doctorates awarded. Terminal master's awarded for partial completion of doctoral program. *Degree requirements:* For master's, comprehensive exam; for doctorate, thesis/dissertation, comprehensive exam. *Entrance requirements:* For master's and doctorate, GRE General Test. Additional exam requirements/recommendations for international students: Required—TOEFL (minimum score 550 paper-based; 213 computer-based). *Application deadline:* For fall admission, 1/15 for domestic students, 1/15 for international students. Applications are processed on a rolling basis. Application fee: $50. Electronic applications accepted. *Financial support:* In 2005–06, 38 students received support, including 1 fellowship with full tuition reimbursement available (averaging $20,000 per year), 10 research assistantships with full tuition reimbursements available (averaging $20,000 per year), 27 teaching assistantships with full tuition reimbursements available (averaging $20,000 per year); tuition waivers (full) also available. Financial award application deadline: 5/1; financial award applicants required to submit FAFSA. *Faculty research:* Supramolecular chemistry, electrochemistry, surface chemistry, catalysis, organometalic. *Unit head:* Dr. Vaidyanathan Ramamurthy, Chairman, 305-284-2282, Fax: 305-284-4571. *Application contact:* Eva Johnson, Graduate Secretary, 305-284-2094, Fax: 305-284-4571, E-mail: evaj@miami.edu.

University of Michigan, Horace H. Rackham School of Graduate Studies, College of Literature, Science, and the Arts, Department of Chemistry, Ann Arbor, MI 48109. Offers analytical chemistry (PhD); inorganic chemistry (PhD); material chemistry (PhD); organic chemistry (PhD); physical chemistry (PhD). *Faculty:* 48 full-time (8 women), 3 part-time/adjunct (all women). *Students:* ; includes 39 minority (4 African Americans, 1 American Indian/Alaska Native, 29 Asian Americans or Pacific Islanders, 5 Hispanic Americans), 66 international. Average age 26. 456 applicants, 29% accepted, 44 enrolled. In 2005, 37 degrees awarded. *Degree requirements:* For doctorate, thesis/dissertation, oral defense of dissertation, organic cumulative proficiency exams. *Entrance requirements:* For doctorate, GRE General Test, GRE Subject Test (recommended), 3 letters of recommendation. Additional exam requirements/recommendations for international students: Required—TOEFL. *Application deadline:* For fall admission, 1/31 for domestic students, 1/1 for international students. Applications are processed on a rolling basis. Application fee: $60 ($75 for international students). Electronic applications accepted. *Expenses:* Tuition, area resident: Full-time $14,082; part-time $894 per credit hour. Tuition, state resident: full-time $14,082; part-time $896 per credit hour. Tuition, nonresident: full-time $28,500; part-time $1,675 per credit hour. Required fees: $189; $189. *Financial support:* In 2005–06, 10 fellowships with full tuition reimbursements (averaging $20,000 per year), 70 research assistantships with full tuition reimbursements (averaging $19,000 per year), 120 teaching assistantships with full tuition reimbursements (averaging $20,000 per year) were awarded. Financial award applicants required to submit FAFSA. *Faculty research:* Biological catalysis, protein engineering, chemical sensors, de novo metalloprotein design, supramolecular architecture. Total annual research expenditures: $8 million. *Unit head:* Dr. Carol A. Fierke, Chair, 734-763-9681, Fax: 734-647-4847. *Application contact:* Linda Deitert, Assistant Director Graduate Studies, 734-764-7278, Fax: 734-647-4865, E-mail: chemadmissions@umich.edu.

University of Missouri–Columbia, Graduate School, College of Arts and Sciences, Department of Chemistry, Columbia, MO 65211. Offers analytical chemistry (MS, PhD); inorganic chemistry (MS, PhD); organic chemistry (MS, PhD); physical chemistry (MS, PhD). *Faculty:* 21 full-time (5 women). *Students:* 58 full-time (17 women), 44 part-time (16 women); includes 5 minority (3 African Americans, 2 Asian Americans or Pacific Islanders), 59 international. In 2005, 11 master's, 9 doctorates awarded. *Degree requirements:* For master's, thesis; for doctorate, one foreign language, thesis/dissertation. *Entrance requirements:* For master's and doctorate, GRE General Test, minimum GPA of 3.0. *Application deadline:* For fall admission, 4/1 for domestic studentsFor winter admission, 11/1 for domestic students. Applications are processed on a rolling basis. Application fee: $45 ($60 for international students). *Financial support:* Fellowships, research assistantships, teaching assistantships, institutionally sponsored loans available. *Unit head:* Dr. Sheryl A. Tucker, Director for Graduate Studies, 573-882-1729.

University of Missouri–Kansas City, College of Arts and Sciences, Department of Chemistry, Kansas City, MO 64110-2499. Offers analytical chemistry (MS, PhD); inorganic chemistry (MS, PhD); organic chemistry (MS, PhD); physical chemistry (MS, PhD); polymer chemistry (MS, PhD). PhD offered through the School of Graduate Studies. Part-time and evening/weekend programs available. *Faculty:* 14 full-time (2 women), 2 part-time/adjunct (0 women). *Students:* Average age 38. 13 applicants, 8% accepted, 1 enrolled. In 2005, 2 degrees awarded. *Degree requirements:* For master's, thesis (for some programs); for doctorate, thesis/dissertation. *Entrance requirements:* For master's, equivalent of American Chemical Society approved bachelor's degree in chemistry; for doctorate, GRE General Test, equivalent of American Chemical Society approved bachelor's degree in chemistry. Additional exam requirements/recommendations for international students: Required—TOEFL (minimum score 580 paper-based; 237 computer-based), TWE. *Application deadline:* For fall and spring admission, 4/15For winter admission, 10/15 for domestic students. Applications are processed on a rolling basis. Application fee: $25. Electronic applications accepted. *Expenses:* Tuition, state resident: full-time $4,738; part-time $263 per credit hour. Tuition, nonresident: full-time $12,235; part-time $679 per credit hour. Required fees: $582. Tuition and fees vary according to course load, program and student level. *Financial support:* In 2005–06, fellowships with partial tuition reimbursements (averaging $18,156 per year), research assistantships with partial tuition reimbursements (averaging $18,515 per year), teaching assistantships with partial tuition reimbursements (averaging $16,434 per year) were awarded; Federal Work-Study, institutionally sponsored loans, and scholarships/grants also available. Support available to part-time students. Financial award application deadline: 2/15. *Faculty research:* Molecular spectroscopy, characterization and synthesis of materials and compounds, computational chemistry, natural products, drug delivery systems and anti-tumor agents. Total annual research expenditures: $1.1 million. *Unit head:* Dr. Jerry Jean, Chairperson, 816-235-2273, Fax: 816-235-5502, E-mail: jeany@umkc.edu. *Application contact:* Graduate Recruiting Committee, 816-235-2272, Fax: 816-235-5502, E-mail: umkc-chemdept@umkc.edu.

University of Missouri–St. Louis, College of Arts and Sciences, Department of Chemistry and Biochemistry, St. Louis, MO 63121. Offers chemistry (MS, PhD), including inorganic chemistry, organic chemistry, physical chemistry. Part-time and evening/weekend programs available. *Faculty:* 16. *Students:* 28 full-time (20 women), 16 part-time (7 women); includes 6 minority (5 African Americans, 1 Hispanic American), 23 international. Average age 32. 2,298 applicants, 74% accepted. In 2005, 9 master's, 4 doctorates awarded. Terminal master's awarded for partial completion of doctoral program. *Degree requirements:* For master's, thesis optional; for doctorate, thesis/dissertation. *Entrance requirements:* For doctorate, GRE General Test, 3 letters of recommendation. Additional exam requirements/recommendations for international students: Required—TOEFL (minimum score 550 paper-based; 213 computer-based). *Application deadline:* For fall admission, 7/1 for domestic students; for spring admission, 12/7 priority date for domestic students. Applications are processed on a rolling basis. Application fee: $35 ($40 for international students). Electronic applications accepted. *Expenses:* Tuition, state resident: part-time $263 per credit hour. Tuition, nonresident: part-time $680 per credit hour. Required fees: $53 per credit hour. Tuition and fees vary according to program. *Financial support:* In 2005–06, 9 research assistantships with full and partial tuition reimbursements (averaging $16,500 per year), 17 teaching assistantships with full and partial tuition reimbursements (averaging $16,500 per year) were awarded; fellowships with full and partial tuition reimbursements *Faculty research:* Metallaborane chemistry, serum transferrin chemistry, natural products chemistry, organic synthesis. *Unit head:* Dr. Cynthia Dupureur, Director of Graduate Studies, 314-516-5311, Fax: 314-516-5342, E-mail: gradchem@umsl.edu. *Application contact:* 314-516-5458, Fax: 314-516-5310, E-mail: gradadm@umsl.edu.

Inorganic Chemistry

The University of Montana, Graduate School, College of Arts and Sciences, Department of Chemistry, Missoula, MT 59812-0002. Offers chemistry (MS, PhD), including environmental/analytical chemistry, inorganic chemistry, organic chemistry, physical chemistry. *Faculty:* 16 full-time (2 women). *Students:* 36 full-time (9 women), 7 part-time (2 women), 9 international. 25 applicants, 44% accepted, 6 enrolled. In 2005, 1 master's, 2 doctorates awarded. Terminal master's awarded for partial completion of doctoral program. *Degree requirements:* For master's, thesis (for some programs); for doctorate, thesis/dissertation. *Entrance requirements:* For master's and doctorate, GRE General Test. Additional exam requirements/recommendations for international students: Required—TOEFL (minimum score 575 paper-based; 230 computer-based). *Application deadline:* For fall admission, 2/15 for domestic students. Applications are processed on a rolling basis. Application fee: $45. *Expenses:* Tuition, state resident: part-time $267 per credit. Tuition, nonresident: part-time $665 per credit. Part-time tuition and fees vary according to course load and degree level. *Financial support:* In 2005–06, 13 research assistantships with tuition reimbursements (averaging $14,000 per year), 12 teaching assistantships with full tuition reimbursements (averaging $14,000 per year) were awarded; Federal Work-Study, scholarships/grants, and unspecified assistantships also available. Financial award application deadline: 3/1; financial award applicants required to submit FAFSA. *Faculty research:* Reaction mechanisms and kinetics, inorganic and organic synthesis, analytical chemistry, natural products. Total annual research expenditures: $789,952. *Unit head:* Dr. Edward Rosenberg, Chair, 406-243-2592, Fax: 406-243-4227.

University of Nebraska–Lincoln, Graduate College, College of Arts and Sciences, Department of Chemistry, Lincoln, NE 68588. Offers analytical chemistry (PhD); chemistry (MS); inorganic chemistry (PhD); organic chemistry (PhD); physical chemistry (PhD). *Degree requirements:* For master's, one foreign language, departmental qualifying exam, thesis optional; for doctorate, one foreign language, thesis/dissertation, departmental qualifying exams, comprehensive exam. *Entrance requirements:* For master's and doctorate, GRE. Additional exam requirements/recommendations for international students: Required—TOEFL (minimum score 550 paper-based; 213 computer-based). Electronic applications accepted. *Faculty research:* Bioorganic and bioinorganic chemistry, biophysical and bioanalytical chemistry, structure-function of DNA and proteins, organometallics, mass spectrometry.

University of Notre Dame, Graduate School, College of Science, Department of Chemistry and Biochemistry, Notre Dame, IN 46556. Offers biochemistry (MS, PhD); inorganic chemistry (MS, PhD); organic chemistry (MS, PhD); physical chemistry (MS, PhD). *Faculty:* 30 full-time (3 women). *Students:* 148 full-time (52 women); includes 5 minority (1 African American, 1 Asian American or Pacific Islander, 3 Hispanic Americans), 65 international. 144 applicants, 38% accepted, 21 enrolled. In 2005, 7 master's, 6 doctorates awarded. Terminal master's awarded for partial completion of doctoral program. *Median time to degree:* Of those who began their doctoral program in fall 1997, 54% received their degree in 8 years or less. *Degree requirements:* For master's, thesis, comprehensive exam; for doctorate, thesis/dissertation, qualifying exam. *Entrance requirements:* For master's and doctorate, GRE General Test, GRE Subject Test (strongly recommended). Additional exam requirements/recommendations for international students: Required—TOEFL. *Application deadline:* For fall admission, 2/1 for domestic students. Applications are processed on a rolling basis. Application fee: $50. Electronic applications accepted. *Financial support:* In 2005–06, 148 students received support, including 19 fellowships with full tuition reimbursements available (averaging $22,000 per year), 57 research assistantships with full tuition reimbursements available (averaging $15,250 per year), 53 teaching assistantships with full tuition reimbursements available (averaging $16,000 per year); tuition waivers (full) also available. Financial award application deadline: 2/1. *Faculty research:* Reaction design and mechanistic studies; reactive intermediates; synthesis, structure and reactivity of organometallic. cluster complexes, and biologically active natural products; bioorganic chemistry; enzymology. Total annual research expenditures: $9.4 million. *Unit head:* Dr. Richard E. Taylor, Director of Graduate Studies, 574-631-7058, Fax: 574-631-6652, E-mail: taylor.61@nd.edu. *Application contact:* Dr. Terrence J. Akai, Director of Graduate Admissions, 574-631-7706, Fax: 574-631-4183, E-mail: gradad@nd.edu.

University of Regina, Faculty of Graduate Studies and Research, Faculty of Science, Department of Chemistry and Biochemistry, Regina, SK S4S 0A2, Canada. Offers analytical chemistry (M Sc, PhD); biochemistry (M Sc, PhD); inorganic chemistry (M Sc, PhD); organic chemistry (M Sc, PhD); physical chemistry (M Sc, PhD). Part-time programs available. *Faculty:* 10 full-time (2 women), 4 part-time/adjunct (1 woman). *Students:* 14 full-time (6 women), 3 part-time (2 women). 21 applicants, 33% accepted. *Degree requirements:* For master's and doctorate, thesis/dissertation, departmental qualifying exam. *Entrance requirements:* For master's and doctorate, GRE. Additional exam requirements/recommendations for international students: Required—TOEFL (minimum score 580 paper-based; 237 computer-based). *Application deadline:* For fall admission, 1/1 for domestic studentsFor winter admission, 7/1 for domestic students. Applications are processed on a rolling basis. Application fee: $60 ($100 for international students). *Financial support:* In 2005–06, 4 fellowships (averaging $14,886 per year), 1 research assistantship (averaging $12,750 per year), 5 teaching assistantships (averaging $13,501 per year) were awarded; scholarships/grants also available. Financial award application deadline: 6/15. *Faculty research:* Organic synthesis, organic oxidations, ionic liquids theoretical/computational chemistry, protein biochemistry/biophysics, environmental analytical, photophysical/photochemistry. *Unit head:* Dr. Andrew G. Wee, Head, 306-585-4767, Fax: 306-585-4894, E-mail: chem.chair@uregina.ca. *Application contact:* Dr. Allan East, Associate Professor, 306-585-4003, Fax: 306-585-4894, E-mail: allan.east@uregina.ca.

University of Southern Mississippi, Graduate School, College of Science and Technology, Department of Chemistry and Biochemistry, Hattiesburg, MS 39406-0001. Offers analytical chemistry (MS, PhD); biochemistry (MS, PhD); inorganic chemistry (MS, PhD); organic chemistry (MS, PhD); physical chemistry (MS, PhD). *Degree requirements:* For master's and doctorate, thesis/dissertation, comprehensive exam. *Entrance requirements:* For master's, GRE General Test, minimum GPA of 2.75 in last 60 hours; for doctorate, GRE General Test, minimum GPA of 3.5. Additional exam requirements/recommendations for international students: Required—TOEFL. *Faculty research:* Plant biochemistry, photo chemistry, polymer chemistry, x-ray analysis, enzyme chemistry.

University of South Florida, College of Graduate Studies, College of Arts and Sciences, Department of Chemistry, Tampa, FL 33620-9951. Offers analytical chemistry (MS, PhD); biochemistry (MS, PhD); inorganic chemistry (MS, PhD); organic chemistry (MS, PhD); physical chemistry (MS, PhD); polymer chemistry (PhD). Part-time programs available. *Faculty:* 22. *Students:* 102 full-time (45 women), 15 part-time (6 women); includes 13 minority (5 Asian Americans or Pacific Islanders, 8 Hispanic Americans), 60 international. 80 applicants, 71% accepted, 23 enrolled. In 2005, 2 master's awarded. Terminal master's awarded for partial completion of doctoral program. *Degree requirements:* For master's, thesis; for doctorate, 2 foreign languages, thesis/dissertation, colloquium. *Entrance requirements:* For master's and doctorate, GRE General Test, minimum GPA of 3.0 in last 30 hours of chemistry course work, 3 letters of recommendation. Additional exam requirements/recommendations for international students: Required—TOEFL (minimum score 550 paper-based; 213 computer-based), TSE(minimum score 50). *Application deadline:* For fall admission, 5/1 priority date for domestic students, 3/1 priority date for international students; for spring admission, 10/1 priority date for domestic students, 8/1 priority date for international students. Applications are processed on a rolling basis. Application fee: $30. Electronic applications accepted. *Financial support:* In 2005–06, 91 students received support; fellowships with partial tuition reimbursements available, research assistantships with partial tuition reimbursements available, teaching assistantships with partial tuition reimbursements available, career-related internships or fieldwork, institutionally sponsored loans, scholarships/grants, health care benefits, and unspecified assistantships available. Financial award application deadline: 6/30. *Faculty research:* Synthesis, bio-organic chemistry, bioinorganic chemistry, environmental chemistry, NMR. *Unit head:* Dr. Michael Zaworotko, Chairperson, 813-974-4129, Fax: 813-974-3203, E-mail: xtal@usf.edu. *Application contact:* Dr. Brian Space, Graduate Coordinator, 813-974-3397, Fax: 813-974-3203.

The University of Tennessee, Graduate School, College of Arts and Sciences, Department of Chemistry, Knoxville, TN 37996. Offers analytical chemistry (MS, PhD); chemical physics (PhD); environmental chemistry (MS, PhD); inorganic chemistry (MS, PhD); organic chemistry (MS, PhD); physical chemistry (MS, PhD); polymer chemistry (MS, PhD); theoretical chemistry (PhD). Part-time programs available. Terminal master's awarded for partial completion of doctoral program. *Degree requirements:* For master's and doctorate, thesis/dissertation. *Entrance requirements:* For master's and doctorate, GRE General Test, minimum GPA of 2.7. Additional exam requirements/recommendations for international students: Required—TOEFL. Electronic applications accepted.

The University of Texas at Austin, Graduate School, College of Natural Sciences, Department of Chemistry and Biochemistry, Austin, TX 78712-1111. Offers analytical chemistry (MA, PhD); biochemistry (MA, PhD); inorganic chemistry (MA, PhD); organic chemistry (MA, PhD); physical chemistry (MA, PhD). *Entrance requirements:* For master's and doctorate, GRE General Test.

The University of Toledo, Graduate School, College of Arts and Sciences, Department of Chemistry, Toledo, OH 43606-3390. Offers analytical chemistry (MS, PhD); biological chemistry (MS, PhD); inorganic chemistry (MS, PhD); organic chemistry (MS, PhD); physical chemistry (MS, PhD). Part-time programs available. *Faculty:* 18. *Students:* 80 full-time (35 women), 8 part-time (5 women); includes 3 minority (2 African Americans, 1 Asian American or Pacific Islander), 43 international. Average age 27. 32 applicants, 75% accepted, 19 enrolled. In 2005, 7 master's awarded. *Degree requirements:* For master's and doctorate, thesis/dissertation. *Entrance requirements:* For master's and doctorate, GRE General Test, GRE Subject Test. Additional exam requirements/recommendations for international students: Required—TOEFL. *Application deadline:* For fall admission, 8/1 for domestic students. Applications are processed on a rolling basis. Application fee: $45. Electronic applications accepted. *Expenses:* Tuition, area resident: Full-time $3,312; part-time $308 per credit hour. Tuition, state resident: full-time $3,312. Tuition, nonresident: full-time $6,616; part-time $735 per credit hour. International tuition: $6,616 full-time. *Financial support:* In 2005–06, 1 research assistantship with full tuition reimbursement (averaging $4,000 per year), 55 teaching assistantships with full tuition reimbursements (averaging $13,336 per year) were awarded; fellowships, Federal Work-Study and institutionally sponsored loans also available. Support available to part-time students. Financial award application deadline: 4/1; financial award applicants required to submit FAFSA. *Faculty research:* Enzymology, materials chemistry, crystallography, theoretical chemistry. *Unit head:* Dr. Alan Pinkerton, Chair, 419-530-7902, Fax: 419-530-4033, E-mail: apinker@uoft02.utoledo.edu. *Application contact:* Charlene Morlock-Hansen, 419-530-2100, E-mail: charlene.hanson@utoledo.edu.

See Close-Up on page 167.

Vanderbilt University, Graduate School, Department of Chemistry, Nashville, TN 37240-1001. Offers analytical chemistry (MAT, MS, PhD); inorganic chemistry (MAT, MS, PhD); organic chemistry (MAT, MS, PhD); physical chemistry (MAT, MS, PhD); theoretical chemistry (MAT, MS, PhD). *Faculty:* 32 full-time (6 women), 3 part-time/adjunct (2 women). *Students:* 96 full-time (40 women); includes 4 minority (2 African Americans, 1 Asian American or Pacific Islander, 1 Hispanic American), 20 international. 223 applicants, 21% accepted, 22 enrolled. In 2005, 4 master's, 12 doctorates awarded. *Degree requirements:* For master's, thesis or alternative; for doctorate, thesis/dissertation, area, qualifying, and final exams. *Entrance requirements:* For master's and doctorate, GRE General Test, GRE Subject Test (recommended). Additional exam requirements/recommendations for international students: Required—TOEFL. *Application deadline:* For fall admission, 1/15 for domestic students, 1/15 for international students. Application fee: $0. Electronic applications accepted. *Expenses:* Tuition: Full-time $15,396; part-time $1,283 per semester hour. Required fees: $2,202; $1,101 per semester. One-time fee: $30. Tuition and fees vary according to course load, program and student level. *Financial support:* Fellowships with full and partial tuition reimbursements, research assistantships with full tuition reimbursements, teaching assistantships with full tuition reimbursements, Federal Work-Study, institutionally sponsored loans, and traineeships available. Financial award application deadline:1/15. *Faculty research:* Chemical synthesis; mechanistic, theoretical, bioorganic, analytical, and spectroscopic chemistry. *Unit head:* Ned A. Porter, Chair, 615-322-2861, Fax: 615-343-1234. *Application contact:* Charles M. Lukehart, Director of Graduate Studies, 615-322-2861, Fax: 615-343-1234, E-mail: charles.m.lukehart@vanderbilt.edu.

Virginia Commonwealth University, Graduate School, College of Humanities and Sciences, Department of Chemistry, Richmond, VA 23284-9005. Offers analytical (MS, PhD); chemical physics (PhD); inorganic (MS, PhD); organic (MS, PhD); physical (MS, PhD). Part-time programs available. *Faculty:* 18 full-time (6 women). *Students:* 36 full-time (13 women), 11 part-time (1 woman); includes 13 minority (6 African Americans, 1 American Indian/Alaska Native, 5 Asian Americans or Pacific Islanders, 1 Hispanic American), 13 international. 47 applicants, 74% accepted. In 2005, 2 master's, 8 doctorates awarded. Terminal master's awarded for partial completion of doctoral program. *Degree requirements:* For master's, thesis; for doctorate, thesis/dissertation, comprehensive cumulative exams, research proposal. *Entrance requirements:* For master's, GRE General Test, 30 undergraduate credits in chemistry; for doctorate, GRE General Test. *Application deadline:* For fall admission, 3/15 for domestic students; for spring admission, 11/15 for domestic students. Applications are processed on a rolling basis. Application fee: $50. *Expenses:* Tuition, state resident: full-time $3,185; part-time $405 per credit. Tuition, nonresident: full-time $7,952; part-time $940 per credit. Required fees: $751 per semester hour. Tuition and fees vary according to course load and program. *Financial support:* Fellowships, research assistantships, teaching assistantships, career-related internships or fieldwork and institutionally sponsored loans available. Support available to part-time students. Financial award application deadline: 7/1. *Faculty research:* Physical, organic, inorganic, analytical, and polymer chemistry; chemical physics. *Unit head:* Dr. Fred M. Hawkridge, Chair, 804-828-1298, Fax: 804-828-8599, E-mail: fmhawkri@vcu.edu. *Application contact:* Dr. M. Samy El-Shall, Chair, Graduate Recruiting and Admissions Committee, 804-828-3518, E-mail: mselshal@vcu.edu.

Wake Forest University, Graduate School, Department of Chemistry, Winston-Salem, NC 27109. Offers analytical chemistry (MS, PhD); inorganic chemistry (MS, PhD); organic chemistry (MS, PhD); physical chemistry (MS, PhD). Part-time programs available. *Faculty:* 17 full-time (4 women). *Students:* 32 full-time (17 women), 2 part-time; includes 1 minority (African American), 14 international. Average age 28. 60 applicants, 42% accepted, 12 enrolled. In 2005, 1 master's, 6 doctorates awarded. *Median time to degree:* Of those who began their doctoral program in fall 1997, 0% received their degree in 8 years or less. *Degree requirements:* For master's, one foreign language, thesis, comprehensive exam, registration; for doctorate, 2 foreign languages, thesis/dissertation, comprehensive exam, registration. *Entrance requirements:* For master's and doctorate, GRE General Test. Additional exam requirements/recommendations for international students: Required—TOEFL (minimum score 213 computer-based). *Application deadline:* For fall admission, 1/15 for domestic students, 1/15 for international students. Application fee: $45. Electronic applications accepted. *Financial support:* In 2005–06, 32 students received support, including 1 fellowship with full tuition reimbursement available (averaging $21,500 per year), 12 research assistantships with full tuition reimbursements available (averaging $19,500 per year), 17 teaching assistantships with full tuition reimbursements available (averaging $19,500 per year); scholarships/grants and tuition waivers (full and partial) also available. Support available to part-time students. Financial award application deadline: 1/15; financial award applicants required to submit FAFSA. *Unit head:* Dr. Bruce King, Director, 336-758-5774, Fax: 335-758-4656, E-mail: kingsb@wfu.edu.

Washington State University, Graduate School, College of Sciences, Department of Chemistry, Pullman, WA 99164. Offers analytical chemistry (MS, PhD); biological systems (MS, PhD); inorganic chemistry (MS, PhD); organic chemistry (MS, PhD); physical chemistry (MS, PhD). *Faculty:* 33. *Students:* 38 full-time (19 women), 2 part-time (1 woman); includes 3 minority (2 Asian Americans or Pacific Islanders, 1 Hispanic American), 11 international. Average age 29.

55 applicants, 33% accepted, 9 enrolled. In 2005, 5 master's, 7 doctorates awarded. Terminal master's awarded for partial completion of doctoral program. *Degree requirements:* For master's, oral exam, teaching experience, thesis optional; for doctorate, thesis/dissertation, oral exam, written exam, comprehensive exam. *Entrance requirements:* For master's and doctorate, GRE General Test, minimum GPA of 3.0, 3 letters of recommendation. *Application deadline:* For fall admission, 3/1 priority date for domestic students, 3/1 priority date for international students; for spring admission, 10/1 priority date for domestic students, 7/1 priority date for international students. Applications are processed on a rolling basis. Application fee: $35. *Expenses:* Tuition, state resident: full-time $6,295; part-time $336 per credit. Tuition, nonresident: full-time $15,949; part-time $819 per credit. Required fees: $933. Part-time tuition and fees vary according to campus/location and program. *Financial support:* In 2005–06, 38 students received support, including 3 fellowships (averaging $8,977 per year), 12 research assistantships with full and partial tuition reimbursements available (averaging $15,999 per year), 23 teaching assistantships with full and partial tuition reimbursements available (averaging $14,747 per year); career-related internships or fieldwork, Federal Work-Study, institutionally sponsored loans, scholarships/grants, health care benefits, unspecified assistantships, and summer support also available. Financial award application deadline: 4/1; financial award applicants required to submit FAFSA. *Faculty research:* Environmental chemistry, materials chemistry, radio chemistry, bio-organic, computational chemistry. Total annual research expenditures: $3.8 million. *Unit head:* Sue B. Clark, Chair, Admissions Committee, 509-335-8866, Fax: 509-335-8867, E-mail: sclark@mail.wsu.edu. *Application contact:* Admissions Committee.

Wesleyan University, Graduate Programs, Department of Chemistry, Middletown, CT 06459-0260. Offers biochemistry (MA, PhD); chemical physics (MA, PhD); inorganic chemistry (MA, PhD); organic chemistry (MA, PhD); physical chemistry (MA, PhD); theoretical chemistry (MA, PhD). *Faculty:* 13 full-time (2 women), 2 part-time/adjunct (1 woman). *Students:* 41 full-time (20 women); includes 1 minority (Asian American or Pacific Islander), 22 international. Average age 26. In 2005, 2 master's, 2 doctorates awarded. Terminal master's awarded for partial completion of doctoral program. *Degree requirements:* For master's and doctorate, one foreign language, thesis/dissertation. *Entrance requirements:* For master's, GRE General Test, GRE Subject Test; for doctorate, GRE Subject Test. *Application deadline:* For fall admission, 3/1 for domestic students. Applications are processed on a rolling basis. Application fee: $0. *Expenses:* Tuition: Full-time $24,732. One-time fee: $20 full-time. *Financial support:* Fellowships, research assistantships, teaching assistantships, institutionally sponsored loans available. *Unit head:* Dr. Phillip Bolton, Chair, 860-685-2668. *Application contact:* Karen Karpa, Information Contact, 860-685-2573, Fax: 860-685-2211, E-mail: kkarpa@wesleyan.edu.

See Close-Up on page 177.

West Virginia University, Eberly College of Arts and Sciences, Department of Chemistry, Morgantown, WV 26506. Offers analytical chemistry (MS, PhD); inorganic chemistry (MS, PhD); organic chemistry (MS, PhD); physical chemistry (MS, PhD); theoretical chemistry (MS, PhD). Part-time programs available. Postbaccalaureate distance learning degree programs offered (no on-campus study). *Faculty:* 14 full-time (1 woman), 5 part-time/adjunct (3 women). *Students:* 45 full-time (15 women), 3 part-time (2 women); includes 1 minority (Hispanic American), 26 international. Average age 28. 84 applicants, 58% accepted, 11 enrolled. In 2005, 5 master's, 3 doctorates awarded. Terminal master's awarded for partial completion of doctoral program. *Degree requirements:* For master's and doctorate, thesis/dissertation, registration. *Entrance requirements:* For master's, GRE General Test, GRE Subject Test (recommended), minimum GPA of 2.5; for doctorate, GRE General Test, GRE Subject Test (recommended), minimum GPA of 2.75. Additional exam requirements/recommendations for international students: Required—TOEFL. *Application deadline:* For fall admission, 3/1 for domestic students. Applications are processed on a rolling basis. Application fee: $50. Electronic applications accepted. *Expenses:* Tuition, area resident: Full-time $4,582; part-time $258 per credit hour. Tuition, state resident: full-time $4,582; part-time $258 per credit hour. Tuition, nonresident: full-time $1,382; part-time $741 per credit hour. International tuition: $1,382 full-time. *Financial support:* In 2005–06, fellowships (averaging $2,000 per year), research assistantships with full tuition reimbursements (averaging $13,000 per year), teaching assistantships with full tuition reimbursements (averaging $12,000 per year) were awarded; tuition waivers (full and partial) also available. Financial award application deadline: 2/1; financial award applicants required to submit FAFSA. *Faculty research:* Analysis of proteins, drug interactions, solids and effluents by advanced separation methods; new synthetic strategies for complex organic molecules; synthesis and structural characterization of metal complexes for polymerization catalysis. *Unit head:* Dr. Harry O. Finklea, Chair, 304-293-3435 Ext. 4453, Fax: 304-293-4904, E-mail: harry.finklea@mail.wvu.edu.

Yale University, Graduate School of Arts and Sciences, Department of Chemistry, New Haven, CT 06520. Offers biophysical chemistry (PhD); inorganic chemistry (PhD); organic chemistry (PhD); physical chemistry (PhD). *Degree requirements:* For doctorate, thesis/dissertation. *Entrance requirements:* For doctorate, GRE General Test, GRE Subject Test. Additional exam requirements/recommendations for international students: Required—TOEFL.

See Close-Up on page 183.

Organic Chemistry

Boston College, Graduate School of Arts and Sciences, Department of Chemistry, Chestnut Hill, MA 02467-3800. Offers biochemistry (PhD); inorganic chemistry (PhD); organic chemistry (PhD); physical chemistry (PhD); science education (MST). MST is offered through the School of Education for secondary school science teaching. Part-time programs available. *Students:* 118 full-time (45 women); includes 12 minority (2 African Americans, 8 Asian Americans or Pacific Islanders, 2 Hispanic Americans), 35 international. 212 applicants, 32% accepted, 17 enrolled. In 2005, 6 master's, 10 doctorates awarded. *Degree requirements:* For doctorate, thesis/dissertation, qualifying exam. *Entrance requirements:* For doctorate, GRE General Test, GRE Subject Test. Additional exam requirements/recommendations for international students: Required—TOEFL (minimum score 550 paper-based; 213 computer-based). *Application deadline:* For fall admission, 1/2 for domestic students. Application fee: $70. Electronic applications accepted. *Financial support:* Fellowships with full tuition reimbursements, research assistantships with full tuition reimbursements, teaching assistantships with full tuition reimbursements, Federal Work-Study available. Support available to part-time students. Financial award application deadline: 3/1; financial award applicants required to submit FAFSA. *Unit head:* Dr. David McFadden, Chairperson, 617-552-3605, E-mail: david.mcfadden@bc.edu. *Application contact:* Dr. Marc Snapper, Graduate Program Director, 617-552-8096, Fax: 617-552-0833, E-mail: marc.snapper@bc.edu.

Brandeis University, Graduate School of Arts and Sciences, Department of Chemistry, Waltham, MA 02454-9110. Offers inorganic chemistry (MS, PhD); organic chemistry (MS, PhD); physical chemistry (MS, PhD). *Faculty:* 20 full-time (3 women). *Students:* 42 full-time (20 women), 1 part-time; includes 2 minority (1 African American, 1 Hispanic American), 34 international. Average age 25. 87 applicants, 10% accepted. In 2005, 5 master's, 3 doctorates awarded. Terminal master's awarded for partial completion of doctoral program. *Degree requirements:* For master's, 1 year of residency; for doctorate, one foreign language, thesis/dissertation, 3 years of residency, 2 seminars, qualifying exams. *Entrance requirements:* For master's and doctorate, GRE General Test, resumé, letters of recommendation. Additional exam requirements/recommendations for international students: Required—TOEFL (minimum score 600 paper-based; 250 computer-based). *Application deadline:* For fall admission, 1/15 for domestic students. Applications are processed on a rolling basis. Application fee: $55. Electronic applications accepted. *Financial support:* In 2005–06, 25 fellowships (averaging $23,000 per year), 21 research assistantships (averaging $23,000 per year), 19 teaching assistantships (averaging $23,000 per year) were awarded; institutionally sponsored loans and scholarships/grants also available. Financial award application deadline: 4/15; financial award applicants required to submit CSS PROFILE or FAFSA. *Faculty research:* Oscillating chemical reactions, molecular recognition systems, protein crystallography, synthesis of natural product spectroscopy and magnetic resonance. Total annual research expenditures: $1,965. *Unit head:* Dr. Peter Jordan, Chair, 781-736-2540, Fax: 781-736-2516, E-mail: jordan@brandeis.edu. *Application contact:* Charlotte Haygazian, Graduate Admissions Secretary, 781-736-2500, Fax: 781-736-2516, E-mail: chemadm@brandeis.edu.

Brigham Young University, Graduate Studies, College of Physical and Mathematical Sciences, Department of Chemistry and Biochemistry, Provo, UT 84602-1001. Offers analytical chemistry (MS, PhD); biochemistry (MS, PhD); inorganic chemistry (MS, PhD); organic chemistry (MS, PhD); physical chemistry (MS, PhD). *Faculty:* 35 full-time (2 women). *Students:* 100 full-time (33 women); includes 2 minority (both Asian Americans or Pacific Islanders), 68 international. Average age 29. 85 applicants, 44% accepted, 21 enrolled. In 2005, 9 master's, 9 doctorates awarded. *Median time to degree:* Of those who began their doctoral program in fall 1997, 87% received their degree in 8 years or less. *Degree requirements:* For master's, thesis, registration; for doctorate, thesis/dissertation, degree qualifying exam. *Entrance requirements:* For master's, pass 1 (biochemistry), 4 (chemistry) of 5 area exams, GRE General Test, minimum GPA of 3.0 in last 60 hours; for doctorate, pass 1 (biochemistry) or 4 (chemistry) of 5 area exams, GRE General Test, minimum GPA of 3.0 in last 60 hours. Additional exam requirements/recommendations for international students: Required—TOEFL, TWE. *Application deadline:* For fall admission, 2/1 priority date for domestic students, 2/1 priority date for international students. Applications are processed on a rolling basis. Application fee: $50. Electronic applications accepted. *Financial support:* In 2005–06, 100 students received support, including 12 fellowships with full tuition reimbursements available (averaging $20,500 per year), 47 research assistantships with full tuition reimbursements available (averaging $20,500 per year), 39 teaching assistantships with full tuition reimbursements available (averaging $20,400 per year); institutionally sponsored loans, scholarships/grants, health care benefits, tuition waivers (full), and unspecified assistantships also available. Financial award application deadline: 2/1. *Faculty research:* Separation science, molecular recognition, organic synthesis and biomedical application, biochemistry and molecular biology, molecular spectroscopy. Total annual research expenditures: $3.9 million. *Unit head:* Dr. Paul B. Farnsworth, Chair, 801-422-6502, Fax: 801-422-0153, E-mail: paul_farnsworth@byu.edu. *Application contact:* Dr. David V. Dearden, Graduate Coordinator, 801-422-2355, Fax: 801-422-0153, E-mail: david_dearden@byu.edu.

See Close-Up on page 97.

California State University, Fullerton, Graduate Studies, College of Natural Science and Mathematics, Department of Chemistry and Biochemistry, Fullerton, CA 92834-9480. Offers analytical chemistry (MS); biochemistry (MS); geochemistry (MS); inorganic chemistry (MS); organic chemistry (MS); physical chemistry (MS). Part-time programs available. *Students:* 19 full-time (6 women), 20 part-time (9 women); includes 20 minority (1 African American, 13 Asian Americans or Pacific Islanders, 6 Hispanic Americans), 8 international. Average age 28. 36 applicants, 47% accepted, 4 enrolled. In 2005, 11 degrees awarded. *Degree requirements:* For master's, thesis, departmental qualifying exam. *Entrance requirements:* For master's, minimum GPA of 2.5 in last 60 units of course work, major in chemistry or related field. Application fee: $55. *Expenses:* Tuition, area resident: Part-time $2,270 per year. Tuition, state resident: full-time $2,572; part-time $339 per unit. Tuition, nonresident: full-time $339; part-time $339 per unit. International tuition: $339 full-time. *Financial support:* Teaching assistantships, career-related internships or fieldwork, Federal Work-Study, institutionally sponsored loans, and scholarships/grants available. Support available to part-time students. Financial award application deadline: 3/1. *Unit head:* Dr. Maria Linder, Chair, 714-278-3621. *Application contact:* Dr. Gregory Williams, Adviser, 714-278-2170.

California State University, Los Angeles, Graduate Studies, College of Natural and Social Sciences, Department of Chemistry and Biochemistry, Los Angeles, CA 90032-8530. Offers analytical chemistry (MS); biochemistry (MS); chemistry (MS); inorganic chemistry (MS); organic chemistry (MS); physical chemistry (MS). Part-time and evening/weekend programs available. *Faculty:* 2 full-time (0 women). *Students:* 33 full-time (21 women), 21 part-time (10 women); includes 38 minority (3 African Americans, 21 Asian Americans or Pacific Islanders, 14 Hispanic Americans). In 2005, 5 degrees awarded. *Degree requirements:* For master's, one foreign language. *Entrance requirements:* Additional exam requirements/recommendations for international students: Required—TOEFL. *Application deadline:* For fall admission, 6/30 for domestic students; for spring admission, 2/1 for domestic students. Applications are processed on a rolling basis. Application fee: $55. *Expenses:* Tuition, area resident: Full-time $3,617. Tuition, state resident: full-time $3,617. Tuition, nonresident: full-time $9,719. International tuition: $9,719 full-time. *Financial support:* Federal Work-Study available. Support available to part-time students. Financial award application deadline: 3/1. *Faculty research:* Intercalation of heavy metal, carborane chemistry, conductive polymers and fabrics, titanium reagents, computer modeling and synthesis. *Unit head:* Dr. Wayne Tikkanen, Chair, 323-343-2300, Fax: 323-343-6490.

Case Western Reserve University, School of Graduate Studies, Department of Chemistry, Cleveland, OH 44106. Offers analytical chemistry (MS, PhD); inorganic chemistry (MS, PhD); organic chemistry (MS, PhD); physical chemistry (MS, PhD). Part-time programs available. Terminal master's awarded for partial completion of doctoral program. *Degree requirements:* For doctorate, thesis/dissertation. *Entrance requirements:* For master's and doctorate, GRE General Test, GRE Subject Test. Additional exam requirements/recommendations for international students: Required—TOEFL. *Faculty research:* Electrochemistry, synthetic chemistry, chemistry of life process, spectroscopy, kinetics.

Clark Atlanta University, School of Arts and Sciences, Department of Chemistry, Atlanta, GA 30314. Offers inorganic chemistry (MS, PhD); organic chemistry (MS, PhD); physical chemistry (MS, PhD); science education (DA). Part-time programs available. *Degree requirements:* For master's, one foreign language, thesis, comprehensive exam; for doctorate, 2 foreign languages, thesis/dissertation, cumulative exam. *Entrance requirements:* For master's, GRE General Test, minimum GPA of 2.5; for doctorate, GRE General Test, GRE Subject Test, minimum graduate GPA of 3.0.

Clarkson University, Graduate School, School of Arts and Sciences, Department of Chemistry, Potsdam, NY 13699. Offers analytical chemistry (MS, PhD); inorganic chemistry (MS, PhD); organic chemistry (MS, PhD); physical chemistry (MS, PhD). *Faculty:* 9 full-time (2 women). *Students:* 34 full-time (9 women), 1 (woman) part-time; includes 2 Asian Americans or Pacific Islanders, 22 international. Average age 27. 52 applicants, 62% accepted. In 2005, 3 master's,

Organic Chemistry

Clarkson University (continued)

2 doctorates awarded. *Median time to degree:* Of those who began their doctoral program in fall 1997, 100% received their degree in 8 years or less. *Degree requirements:* For doctorate, thesis/dissertation, departmental qualifying exam. *Entrance requirements:* For master's, GRE. Additional exam requirements/recommendations for international students: Required—TOEFL. *Application deadline:* For fall admission, 5/15 for domestic students; for spring admission, 10/15 priority date for domestic students. Applications are processed on a rolling basis. Application fee: $25 ($35 for international students). Electronic applications accepted. *Expenses:* Tuition: Full-time $20,160; part-time $840 per hour. Required fees: $215. *Financial support:* In 2005–06, 25 students received support, including 2 fellowships (averaging $25,000 per year), 11 research assistantships (averaging $19,032 per year), 12 teaching assistantships (averaging $19,032 per year); scholarships/grants and tuition waivers (partial) also available. *Faculty research:* Nanomaterial, surface science, polymers chemical biosensing, protein biochemistry drug design/delivery, molecular neuroscience and immunobiology. Total annual research expenditures: $1.6 million. *Unit head:* Dr. Phillip A. Christiansen, Division Head, 315-268-6669, Fax: 315-268-6610, E-mail: pac@clarkson.edu. *Application contact:* Donna Brockway, Graduate Admissions International Advisor/Assistant to the Provost, 315-268-6447, Fax: 315-268-7994, E-mail: brockway@clarkson.edu.

Cleveland State University, College of Graduate Studies, College of Science, Department of Chemistry, Cleveland, OH 44115. Offers analytical chemistry (MS); clinical chemistry (MS, PhD); clinical/bioanalytical chemistry (PhD); environmental chemistry (MS); inorganic chemistry (MS); organic chemistry (MS); physical chemistry (MS). Part-time and evening/weekend programs available. *Faculty:* 13 full-time (1 woman), 59 part-time/adjunct (12 women). *Students:* 46 full-time (22 women), 18 part-time (6 women); includes 4 minority (2 African Americans, 2 Asian Americans or Pacific Islanders), 35 international. Average age 31. 47 applicants, 62% accepted, 7 enrolled. In 2005, 2 master's, 5 doctorates awarded. *Median time to degree:* Of those who began their doctoral program in fall 1997, 67% received their degree in 8 years or less. *Degree requirements:* For master's, thesis (for some programs); for doctorate, thesis/dissertation. *Entrance requirements:* For master's and doctorate, GRE General Test, GRE Subject Test. Additional exam requirements/recommendations for international students: Required—TOEFL (minimum score 525 paper-based; 197 computer-based). *Application deadline:* For fall admission, 1/15 priority date for domestic students, 1/15 priority date for international students. Applications are processed on a rolling basis. Application fee: $30. Electronic applications accepted. *Expenses:* Tuition, state resident: full-time $10,700. Tuition, nonresident: full-time $14,628. Tuition and fees vary according to program. *Financial support:* In 2005–06, 44 students received support, including fellowships with full tuition reimbursements available (averaging $18,000 per year), research assistantships with full tuition reimbursements available (averaging $16,000 per year), teaching assistantships with full tuition reimbursements available (averaging $14,000 per year) Financial award application deadline: 1/15. *Faculty research:* Metalloenzyme mechanisms, dependent RNAse L and interferons, application of HPLC/LPCC to clinical systems, structure-function relationships of factor Va, MALDI-TOF based DNA sequencing. Total annual research expenditures: $3 million. *Unit head:* Dr. Lily M. Ng, Chair, 216-687-2467, E-mail: l.ng@csuohio.edu. *Application contact:* Richelle P. Emery, Administrative Coordinator, 216-687-2457, Fax: 216-687-9298, E-mail: r.emery@csuohio.edu.

Columbia University, Graduate School of Arts and Sciences, Division of Natural Sciences, Department of Chemistry, New York, NY 10027. Offers chemical physics (M Phil, PhD); inorganic chemistry (M Phil, MA, PhD); organic chemistry (M Phil, MA, PhD). *Faculty:* 17 full-time. *Students:* 115 full-time (35 women). Average age 27. 452 applicants, 20% accepted. In 2005, 13 master's, 20 doctorates awarded. *Degree requirements:* For master's, comp exams (MS); foreign language, teaching experience, oral/written exams (M Phil); for doctorate, one foreign language, thesis/dissertation. *Entrance requirements:* For master's and doctorate, GRE General Test, GRE Subject Test. Additional exam requirements/recommendations for international students: Required—TOEFL. Application fee: $75. *Expenses:* Tuition: Full-time $31,448. Tuition and fees vary according to course level, course load, campus/location and program. *Financial support:* Fellowships, teaching assistantships, Federal Work-Study and institutionally sponsored loans available. Support available to part-time students. Financial award application deadline: 1/5; financial award applicants required to submit FAFSA. *Faculty research:* Biophysics. *Unit head:* James Valentini, Chair, 212-854-7590, Fax: 212-932-1289, E-mail: jjvi@columbia.edu.

Cornell University, Graduate School, Graduate Fields of Arts and Sciences, Field of Chemistry and Chemical Biology, Ithaca, NY 14853-0001. Offers analytical chemistry (PhD); bio-organic chemistry (PhD); biophysical chemistry (PhD); chemical biology (PhD); chemical physics (PhD); inorganic chemistry (PhD); materials chemistry (PhD); organic chemistry (PhD); organometallic chemistry (PhD); physical chemistry (PhD); polymer chemistry (PhD); theoretical chemistry (PhD). *Faculty:* 44 full-time (2 women). *Students:* 190 full-time (73 women); includes 23 minority (4 African Americans, 8 Asian Americans or Pacific Islanders, 11 Hispanic Americans), 65 international. 339 applicants, 35% accepted, 49 enrolled. In 2005, 23 doctorates awarded. *Degree requirements:* For doctorate, thesis/dissertation, comprehensive exam. *Entrance requirements:* For doctorate, GRE General Test, GRE Subject Test (chemistry), 3 letters of recommendation. Additional exam requirements/recommendations for international students: Required—TOEFL (minimum score 600 paper-based; 250 computer-based). *Application deadline:* For fall admission, 1/10 for domestic students. Application fee: $60. Electronic applications accepted. *Financial support:* In 2005–06, 185 students received support, including 33 fellowships with full tuition reimbursements available, 85 research assistantships with full tuition reimbursements available, 67 teaching assistantships with full tuition reimbursements available; institutionally sponsored loans, scholarships/grants, health care benefits, tuition waivers (full and partial), and unspecified assistantships also available. Financial award applicants required to submit FAFSA. *Faculty research:* Analytical, organic, inorganic, physical, materials, chemical biology. *Unit head:* Director of Graduate Studies, 607-255-4139, Fax: 607-255-4137. *Application contact:* Graduate Field Assistant, 607-255-4139, Fax: 607-255-4137, E-mail: chemgrad@cornell.edu.

See Close-Up on page 109.

Florida State University, Graduate Studies, College of Arts and Sciences, Department of Chemistry and Biochemistry, Tallahassee, FL 32306. Offers analytical chemistry (MS, PhD); biochemistry (MS, PhD); chemical physics (MS, PhD); inorganic chemistry (MS, PhD); organic chemistry (MS, PhD); physical chemistry (MS, PhD). Part-time programs available. *Faculty:* 36 full-time (6 women), 2 part-time/adjunct (0 women). *Students:* 144 full-time (50 women); includes 17 minority (7 African Americans, 1 American Indian/Alaska Native, 6 Asian Americans or Pacific Islanders, 3 Hispanic Americans), 58 international. Average age 25. 288 applicants, 27% accepted, 26 enrolled. In 2005, 14 master's, 15 doctorates awarded. Terminal master's awarded for partial completion of doctoral program. *Median time to degree:* Of those who began their doctoral program in fall 1997, 88% received their degree in 8 years or less. *Degree requirements:* For master's, thesis (for some programs), cumulative and diagnostic exams, comprehensive exam (for some programs), registration; for doctorate, thesis/dissertation, cumulative and diagnostic exams, comprehensive exam (for some programs), registration. *Entrance requirements:* For master's and doctorate, GRE General Test, minimum B average in undergraduate course work. Additional exam requirements/recommendations for international students: Required—TOEFL (minimum score 515 paper-based; 213 computer-based). *Application deadline:* For fall admission, 4/15 priority date for domestic students, 4/15 priority date for international students. Applications are processed on a rolling basis. Application fee: $30. Electronic applications accepted. *Financial support:* In 2005–06, 1 fellowship with tuition reimbursement (averaging $18,000 per year), 56 research assistantships with tuition reimbursements (averaging $19,000 per year), 83 teaching assistantships with tuition reimbursements (averaging $19,000 per year) were awarded; career-related internships or fieldwork, Federal Work-Study, institutionally sponsored loans, and traineeships also available. Financial award application deadline: 2/15; financial award applicants required to submit FAFSA. *Faculty research:* Spectroscopy, computational chemistry, nuclear chemistry, separations, synthesis. Total annual research expenditures: $6.5 million. *Unit head:* Dr. Naresh Dalal, Chairman, 850-644-3398, Fax: 850-644-8281. *Application contact:* Dr. Oliver Steinbock, Chair, Graduate Admissions Committee, 888-525-9286, Fax: 850-644-8281, E-mail: gradinfo@chem.fsu.edu.

See Close-Up on page 117.

Georgetown University, Graduate School of Arts and Sciences, Department of Chemistry, Washington, DC 20057. Offers analytical chemistry (MS, PhD); biochemistry (MS, PhD); chemical physics (MS, PhD); inorganic chemistry (MS, PhD); organic chemistry (MS, PhD); physical chemistry (MS, PhD); theoretical chemistry (MS, PhD). Terminal master's awarded for partial completion of doctoral program. *Degree requirements:* For master's, thesis (for some programs), qualifying exam; for doctorate, thesis/dissertation, comprehensive exam. *Entrance requirements:* For master's and doctorate, GRE General Test. Additional exam requirements/recommendations for international students: Required—TOEFL.

The George Washington University, Columbian College of Arts and Sciences, Department of Chemistry, Washington, DC 20052. Offers analytical chemistry (MS, PhD); inorganic chemistry (MS, PhD); materials science (MS, PhD); organic chemistry (MS, PhD); physical chemistry (MS, PhD). Part-time and evening/weekend programs available. Terminal master's awarded for partial completion of doctoral program. *Degree requirements:* For master's, thesis or alternative, comprehensive exam; for doctorate, thesis/dissertation, general exam. *Entrance requirements:* For master's and doctorate, GRE General Test, interview, minimum GPA of 3.0. Additional exam requirements/recommendations for international students: Required—TOEFL (minimum score 550 paper-based; 213 computer-based). Electronic applications accepted.

Harvard University, Graduate School of Arts and Sciences, Department of Chemistry and Chemical Biology, Cambridge, MA 02138. Offers biochemical chemistry (PhD); inorganic chemistry (PhD); organic chemistry (PhD); physical chemistry (PhD). *Students:* 188 full-time (36 women). 346 applicants, 24% accepted. In 2005, 33 doctorates awarded. *Degree requirements:* For doctorate, thesis/dissertation, cumulative exams. *Entrance requirements:* For doctorate, GRE General Test, GRE Subject Test. Additional exam requirements/recommendations for international students: Required—TOEFL. *Application deadline:* For fall admission, 12/31 for domestic students. Application fee: $60. *Expenses:* Tuition: Full-time $28,752. Full-time tuition and fees vary according to program and student level. *Financial support:* Fellowships, research assistantships, teaching assistantships, career-related internships or fieldwork, Federal Work-Study, and institutionally sponsored loans available. Financial award application deadline: 12/30. *Unit head:* Betsey Cogswell, Administrator, 617-495-5696, Fax: 617-495-5264. *Application contact:* Graduate Admissions Office, 617-496-3208.

See Close-Up on page 119.

Howard University, Graduate School of Arts and Sciences, Department of Chemistry, Washington, DC 20059-0002. Offers analytical chemistry (MS, PhD); atmospheric (MS, PhD); biochemistry (MS, PhD); environmental (MS, PhD); inorganic chemistry (MS, PhD); organic chemistry (MS, PhD); physical chemistry (MS, PhD); polymer chemistry (MS, PhD). Part-time programs available. *Degree requirements:* For master's, one foreign language, thesis, teaching experience, comprehensive exam, registration; for doctorate, 2 foreign languages, thesis/dissertation, teaching experience, comprehensive exam, registration. *Entrance requirements:* For master's, GRE General Test, minimum GPA of 2.7; for doctorate, GRE General Test, minimum GPA of 3.0. *Faculty research:* Stratospheric aerosols, liquid crystals, polymer coatings, terrestrial and extraterrestrial atmospheres, amidogen reaction.

Illinois Institute of Technology, Graduate College, College of Science and Letters, Department of Biological, Chemical and Physical Sciences, Chemistry Division, Chicago, IL 60616-3793. Offers analytical chemistry (M Ch, MS, PhD); chemistry (M Chem); inorganic chemistry (MS, PhD); materials and chemical synthesis (M Ch); organic chemistry (MS, PhD); physical chemistry (MS, PhD); polymer chemistry (MS, PhD). Part-time and evening/weekend programs available. Postbaccalaureate distance learning degree programs offered (no on-campus study). Terminal master's awarded for partial completion of doctoral program. *Degree requirements:* For master's, thesis (for some programs), comprehensive exam; for doctorate, thesis/dissertation, comprehensive exam. *Entrance requirements:* For master's and doctorate, GRE General Test, minimum undergraduate GPA of 3.0. Additional exam requirements/recommendations for international students: Required—TOEFL (minimum score 550 paper-based; 213 computer-based). Electronic applications accepted. *Faculty research:* Organic and inorganic chemistry, polymers research, physical chemistry, analytical chemistry.

Instituto Tecnológico y de Estudios Superiores de Monterrey, Campus Monterrey, Graduate and Research Division, Program in Natural and Social Sciences, Monterrey, Mexico. Offers biotechnology (MS); chemistry (MS, PhD); communications (MS); education (MA). Part-time programs available. *Degree requirements:* For master's and doctorate, one foreign language, thesis/dissertation. *Entrance requirements:* For master's, PAEG; for doctorate, PAEG, master's degree in related field. Additional exam requirements/recommendations for international students: Required—TOEFL. *Faculty research:* Cultural industries, mineral substances, bioremediation, food processing, CQ in industrial chemical processing.

Kansas State University, Graduate School, College of Arts and Sciences, Department of Chemistry, Manhattan, KS 66506. Offers analytical chemistry (MS); biological chemistry (MS); chemistry (PhD); inorganic chemistry (MS); materials chemistry (MS); organic chemistry (MS); physical chemistry (MS). *Faculty:* 14 full-time (1 woman). *Students:* 65 full-time (21 women), 5 part-time (2 women); includes 1 minority (African American), 48 international. 63 applicants, 67% accepted, 22 enrolled. In 2005, 3 master's, 5 doctorates awarded. Terminal master's awarded for partial completion of doctoral program. *Degree requirements:* For master's and doctorate, thesis/dissertation. *Entrance requirements:* For master's and doctorate, GRE, minimum GPA of 3.0. Additional exam requirements/recommendations for international students: Required—TOEFL (minimum score 550 paper-based; 213 computer-based). *Application deadline:* For fall admission, 2/1 priority date for domestic students, 2/1 priority date for international students; for spring admission, 10/1 for domestic students, 8/1 for international students. Applications are processed on a rolling basis. Application fee: $30 ($55 for international students). *Expenses:* Tuition, state resident: full-time $5,160; part-time $215. Tuition, nonresident: full-time $12,816; part-time $534. International tuition: $12,816 full-time. Required fees: $564. *Financial support:* In 2005–06, 30 research assistantships (averaging $11,956 per year), 26 teaching assistantships with full tuition reimbursements (averaging $11,945 per year) were awarded; fellowships, institutionally sponsored loans and scholarships/grants also available. Support available to part-time students. Financial award application deadline: 3/1; financial award applicants required to submit FAFSA. *Faculty research:* Nanotechnologies, functional materials, bio-organic and bio-physical processes, sensors and separations, synthesis and synthetic methods. Total annual research expenditures: $2.2 million. *Unit head:* Eric Maatta, Head, 785-532-6665, Fax: 785-532-6666, E-mail: eam@ksu.edu. *Application contact:* Christer Aakeröy, Director, 785-532-6096, Fax: 785-532-6666, E-mail: aakeroy@ksu.edu.

Kent State University, College of Arts and Sciences, Department of Chemistry, Kent, OH 44242-0001. Offers analytical chemistry (MS, PhD); biochemistry (MS, PhD); chemistry (MA, MS, PhD); inorganic chemistry (MS, PhD); organic chemistry (MS, PhD); physical chemistry (MS, PhD). Terminal master's awarded for partial completion of doctoral program. *Degree requirements:* For master's and doctorate, thesis/dissertation, comprehensive exam, registration. *Entrance requirements:* For master's and doctorate, placement exam, GRE General Test, GRE Subject Test (recommended), minimum GPA of 2.75. Additional exam requirements/recommendations for international students: Required—TOEFL (minimum score 575 paper-based; 230 computer-based). Electronic applications accepted. *Faculty research:* Biological chemistry, materials chemistry, molecular spectroscopy.

See Close-Up on page 121.

Marquette University, Graduate School, College of Arts and Sciences, Department of Chemistry, Milwaukee, WI 53201-1881. Offers analytical chemistry (MS, PhD); bioanalytical chemistry (MS, PhD); biophysical chemistry (MS, PhD); chemical physics (MS, PhD); inorganic chemistry (MS, PhD); organic chemistry (MS, PhD); physical chemistry (MS, PhD). Part-time programs available. Terminal master's awarded for partial completion of doctoral program. *Degree requirements:* For master's, comprehensive exam; for doctorate, thesis/dissertation, cumulative exams. *Entrance requirements:* For master's and doctorate, GRE Subject Test. Additional exam requirements/recommendations for international students: Required—TOEFL. *Faculty research:* Inorganic complexes, laser Raman spectroscopy, organic synthesis, chemical dynamics, biophysiology.

Massachusetts Institute of Technology, School of Engineering, Department of Civil and Environmental Engineering, Cambridge, MA 02139-4307. Offers biological oceanography (PhD, Sc D); chemical oceanography (PhD, Sc D); civil and environmental engineering (M Eng, SM, PhD, Sc D, CE); civil and environmental systems (PhD, Sc D); civil engineering (PhD, Sc D); coastal engineering (PhD, Sc D); construction engineering and management (PhD, Sc D); environmental biology (PhD, Sc D); environmental chemistry (PhD, Sc D); environmental engineering (PhD, Sc D); environmental fluid mechanics (PhD, Sc D); geotechnical and geoenvironmental engineering (PhD, Sc D); hydrology (PhD, Sc D); information technology (PhD, Sc D); oceanographic engineering (PhD, Sc D); structures and materials (PhD, Sc D); transportation (PhD, Sc D). *Faculty:* 34 full-time (3 women). *Students:* 179 full-time (66 women), 1 part-time; includes 15 minority (1 African American, 9 Asian Americans or Pacific Islanders, 5 Hispanic Americans), 116 international. Average age 26. 374 applicants, 38% accepted, 71 enrolled. In 2005, 76 master's, 27 doctorates, 1 other advanced degree awarded. *Degree requirements:* For master's and CE, thesis/dissertation; for doctorate, thesis/dissertation, comprehensive exam. *Entrance requirements:* For master's and doctorate, GRE General Test. Additional exam requirements/recommendations for international students: Required—TOEFL (minimum score 577 paper-based; 233 computer-based). *Application deadline:* For fall admission, 1/2 for domestic students, 1/2 for international students. Application fee: $70. Electronic applications accepted. *Expenses:* Tuition: Full-time $32,100. Required fees: $200. Part-time tuition and fees vary according to course load. *Financial support:* In 2005–06, 134 students received support, including 38 fellowships with tuition reimbursements available (averaging $23,116 per year), 86 research assistantships with tuition reimbursements available (averaging $22,765 per year), 13 teaching assistantships with tuition reimbursements available (averaging $19,735 per year); career-related internships or fieldwork, Federal Work-Study, institutionally sponsored loans, scholarships/grants, health care benefits, and unspecified assistantships also available. Total annual research expenditures: $12.4 million. *Unit head:* Prof. Patrick Jaillet, Department Head, 617-452-3379, Fax: 617-452-3294, E-mail: jaillet@mit.edu. *Application contact:* Graduate Admissions, 617-253-7119, Fax: 617-258-6775, E-mail: cee-admissions@mit.edu.

Massachusetts Institute of Technology, School of Science, Department of Chemistry, Cambridge, MA 02139-4307. Offers biological chemistry (PhD, Sc D); inorganic chemistry (PhD, Sc D); organic chemistry (PhD, Sc D); physical chemistry (PhD, Sc D). *Faculty:* 30 full-time (6 women). *Students:* 242 full-time (84 women), 3 part-time (2 women); includes 30 minority (5 African Americans, 1 American Indian/Alaska Native, 19 Asian Americans or Pacific Islanders, 5 Hispanic Americans), 77 international. Average age 26. 476 applicants, 27% accepted, 54 enrolled. In 2005, 48 doctorates awarded. *Degree requirements:* For doctorate, thesis/dissertation, comprehensive exam. *Entrance requirements:* For doctorate, GRE General Test. Additional exam requirements/recommendations for international students: Required—TOEFL (minimum score 577 paper-based; 233 computer-based). *Application deadline:* For fall admission, 12/15 for domestic students, 12/15 for international students. Application fee: $70. Electronic applications accepted. *Expenses:* Tuition: Full-time $32,100. Required fees: $200. Part-time tuition and fees vary according to course load. *Financial support:* In 2005–06, 210 students received support, including 36 fellowships with tuition reimbursements available (averaging $29,141 per year), 143 research assistantships with tuition reimbursements available (averaging $24,550 per year), 59 teaching assistantships with tuition reimbursements available (averaging $25,155 per year); career-related internships or fieldwork, Federal Work-Study, institutionally sponsored loans, scholarships/grants, health care benefits, and unspecified assistantships also available. *Faculty research:* Synthetic organic chemistry, enzymatic reaction mechanisms, inorganic and organometallic spectroscopy, high resolution NMR spectroscopy. Total annual research expenditures: $21.5 million. *Unit head:* Prof. Timothy Swager, Department Head, 617-253-1801, E-mail: tswager@mit.edu. *Application contact:* Susan Brighton, Graduate Administrator, 617-253-1845, Fax: 617-258-0241, E-mail: brighton@mit.edu.

McMaster University, School of Graduate Studies, Faculty of Science, Department of Chemistry, Hamilton, ON L8S 4M2, Canada. Offers analytical chemistry (M Sc, PhD); chemical physics (M Sc, PhD); chemistry (M Sc, PhD); inorganic chemistry (M Sc, PhD); organic chemistry (M Sc, PhD); physical chemistry (M Sc, PhD); polymer chemistry (M Sc, PhD). Part-time programs available. Terminal master's awarded for partial completion of doctoral program. *Degree requirements:* For master's, thesis/dissertation; for doctorate, thesis/dissertation, comprehensive exam. *Entrance requirements:* For master's, minimum B+ average. Additional exam requirements/recommendations for international students: Required—TOEFL (minimum score 550 paper-based; 213 computer-based).

Miami University, Graduate School, College of Arts and Sciences, Department of Chemistry and Biochemistry, Oxford, OH 45056. Offers analytical chemistry (MS, PhD); biochemistry (MS, PhD); chemical education (MS, PhD); chemistry (MS, PhD); inorganic chemistry (MS, PhD); organic chemistry (MS, PhD); physical chemistry (MS, PhD). Part-time programs available. *Degree requirements:* For master's, thesis, final exam; for doctorate, thesis/dissertation, final exams, comprehensive exam. *Entrance requirements:* For master's, minimum undergraduate GPA of 3.0 during previous 2 years or 2.75 overall; for doctorate, minimum undergraduate GPA of 2.75, 3.0 graduate. Additional exam requirements/recommendations for international students: Required—TOEFL (minimum score 550 paper-based; 213 computer-based), TWE(minimum score 4). Electronic applications accepted.

Northeastern University, College of Arts and Sciences, Department of Chemistry and Chemical Biology, Boston, MA 02115-5096. Offers analytical chemistry (PhD); chemistry (MS, PhD); inorganic chemistry (PhD); organic chemistry (PhD); physical chemistry (PhD). Part-time and evening/weekend programs available. Terminal master's awarded for partial completion of doctoral program. *Degree requirements:* For master's, thesis (for some programs); for doctorate, thesis/dissertation, qualifying exam in specialty area. *Entrance requirements:* Additional exam requirements/recommendations for international students: Required—TOEFL. Electronic applications accepted. *Faculty research:* Electron transfer, theoretical chemical physics, analytical biotechnology, mass spectrometry, materials chemistry.

See Close-Up on page 129.

Old Dominion University, College of Sciences, Program in Chemistry, Norfolk, VA 23529. Offers analytical chemistry (MS); biochemistry (MS); clinical chemistry (MS); environmental chemistry (MS); organic chemistry (MS); physical chemistry (MS). Part-time and evening/weekend programs available. *Faculty:* 15 full-time (4 women). *Students:* 5 full-time (3 women), 8 part-time (3 women); includes 8 minority (4 African Americans, 1 American Indian/Alaska Native, 1 Asian American or Pacific Islander, 2 Hispanic Americans). Average age 27. 19 applicants, 74% accepted. In 2005, 1 degree awarded. *Degree requirements:* For master's, thesis, comprehensive exam. *Entrance requirements:* For master's, GRE General Test, minimum GPA of 3.0 in major, 2.5 overall. Additional exam requirements/recommendations for international students: Required—TOEFL. *Application deadline:* For fall admission, 7/1 for domestic students; for spring admission, 11/1 for domestic students. Applications are processed on a rolling basis. Application fee: $30. *Expenses:* Tuition, state resident: part-time $263 per credit hour. Tuition, nonresident: part-time $661 per credit hour. Required fees: $39 per semester. Part-time tuition and fees vary according to campus/location. *Financial support:* In 2005–06, research assistantships with tuition reimbursements (averaging $15,000 per year), teaching assistantships with tuition reimbursements (averaging $15,000 per year) were awarded; fellowships, career-related internships or fieldwork and scholarships/grants also available. Financial award application deadline: 2/15; financial award applicants required to submit FAFSA. *Faculty research:* Organic and trace metal biogeochemistry, materials chemistry, bioanalytical chemistry, computational chemistry. Total annual research expenditures: $854,968. *Unit head:* Dr. John R. Donat, Graduate Program Director, 757-683-4098, Fax: 757-683-4628, E-mail: chemgpd@odu.edu.

Oregon State University, Graduate School, College of Science, Department of Chemistry, Corvallis, OR 97331. Offers analytical chemistry (MS, PhD); chemistry (MA, MAIS); inorganic chemistry (MS, PhD); nuclear and radiation chemistry (MS, PhD); organic chemistry (MS, PhD); physical chemistry (MS, PhD). Part-time programs available. *Faculty:* 19 full-time (3 women), 6 part-time/adjunct (1 woman). *Students:* 74 full-time (30 women), 3 part-time; includes 6 minority (1 African American, 1 Asian American or Pacific Islander, 4 Hispanic Americans), 35 international. Average age 28. In 2005, 4 master's, 10 doctorates awarded. Terminal master's awarded for partial completion of doctoral program. *Degree requirements:* For master's and doctorate, one foreign language, thesis/dissertation. *Entrance requirements:* For master's and doctorate, minimum GPA of 3.0 in last 90 hours of course work. Additional exam requirements/recommendations for international students: Required—TOEFL. *Application deadline:* For fall admission, 3/1 for domestic students. Applications are processed on a rolling basis. Application fee: $50. *Expenses:* Tuition, area resident: Full-time $8,139; part-time $301 per credit. Tuition, state resident: full-time $8,139; part-time $501 per credit. Tuition, nonresident: full-time $14,376; part-time $532 per credit. International tuition: $14,376 full-time. Required fees: $1,266. *Financial support:* Fellowships, research assistantships, teaching assistantships, institutionally sponsored loans available. Support available to part-time students. Financial award application deadline: 2/1. *Faculty research:* Solid state chemistry, enzyme reaction mechanisms, structure and dynamics of gas molecules, chemiluminescence, nonlinear optical spectroscopy. *Unit head:* Dr. Douglas Keszler, Chair, 541-737-2081, Fax: 541-737-2062. *Application contact:* Carolyn Brumley, Graduate Secretary, 541-737-6707, Fax: 541-737-2062, E-mail: carolyn.brumley@orst.edu.

Purdue University, Graduate School, School of Science, Department of Chemistry, West Lafayette, IN 47907. Offers analytical chemistry (MS, PhD); biochemistry (MS, PhD); chemical education (MS, PhD); inorganic chemistry (MS, PhD); organic chemistry (MS, PhD); physical chemistry (MS, PhD). *Accreditation:* NCATE (one or more programs are accredited). *Faculty:* 43 full-time (9 women), 8 part-time/adjunct (2 women). *Students:* 293 full-time (114 women), 39 part-time (14 women); includes 53 minority (25 African Americans, 6 Asian Americans or Pacific Islanders, 22 Hispanic Americans), 138 international. Average age 28. 458 applicants, 36% accepted, 57 enrolled. In 2005, 9 master's, 36 doctorates awarded. Terminal master's awarded for partial completion of doctoral program. *Degree requirements:* For master's and doctorate, thesis/dissertation. *Entrance requirements:* Additional exam requirements/recommendations for international students: Required—TOEFL. *Application deadline:* For fall admission, 4/1 priority date for domestic students, 3/1 priority date for international students; for spring admission, 10/1 priority date for domestic students, 9/1 priority date for international students. Applications are processed on a rolling basis. Application fee: $55. Electronic applications accepted. *Financial support:* In 2005–06, 2 fellowships with partial tuition reimbursements (averaging $18,000 per year), 55 teaching assistantships with partial tuition reimbursements (averaging $18,000 per year) were awarded; research assistantships with partial tuition reimbursements, tuition waivers (partial) also available. Support available to part-time students. Financial award applicants required to submit FAFSA. *Unit head:* Dr. Timothy S Zwier, Head, 765-494-5203. *Application contact:* R. E. Wild, Chairman, Graduate Admissions, 765-494-5200, E-mail: wild@purdue.edu.

Rensselaer Polytechnic Institute, Graduate School, School of Science, Department of Chemistry and Chemical Biology, Troy, NY 12180-3590. Offers analytical chemistry (MS, PhD); biochemistry (MS, PhD); inorganic chemistry (MS, PhD); organic chemistry (MS, PhD); physical chemistry (MS, PhD); polymer chemistry (MS, PhD). Part-time and evening/weekend programs available. *Faculty:* 19 full-time (3 women). *Students:* 49 full-time (28 women), 2 part-time (1 woman), 33 international. Average age 24. 80 applicants, 23% accepted, 7 enrolled. In 2005, 1 master's, 14 doctorates awarded. Terminal master's awarded for partial completion of doctoral program. *Median time to degree:* Of those who began their doctoral program in fall 1997, 100% received their degree in 8 years or less. *Degree requirements:* For master's, thesis (for some programs), registration; for doctorate, thesis/dissertation, comprehensive exam, registration. *Entrance requirements:* For master's, GRE General Test, GRE Subject Test (strongly recommended); for doctorate, GRE General Test, GRE Subject Test (chemistry or biochemistry strongly recommended). Additional exam requirements/recommendations for international students: Required—TOEFL (minimum score 600 paper-based). *Application deadline:* For fall admission, 2/1 for domestic students; for spring admission, 11/15 for domestic students. Applications are processed on a rolling basis. Application fee: $75. Electronic applications accepted. *Expenses:* Tuition: Full-time $31,000; part-time $1,320 per credit. Required fees: $1,623. *Financial support:* In 2005–06, 49 students received support, including 1 fellowship with full tuition reimbursement available (averaging $30,000 per year), 25 research assistantships with full tuition reimbursements available (averaging $21,500 per year), 30 teaching assistantships with full tuition reimbursements available (averaging $21,500 per year); institutionally sponsored loans and tuition waivers (full and partial) also available. Financial award application deadline: 2/1. *Faculty research:* Synthetic polymer and biopolymer chemistry, physical chemistry of polymeric systems, bioanalytical chemistry, synthetic and computational drug design, protein folding and protein design. Total annual research expenditures: $1.9 million. *Unit head:* Dr. Linda B. McGown, Chair, 518-276-4856, Fax: 518-276-4887, E-mail: mcgowl@rpi.edu. *Application contact:* Beth E. McGraw, Department Admissions Assistant, 518-276-6456, Fax: 518-276-4887, E-mail: mcgrae@rpi.edu.

See Close-Up on page 137.

Rice University, Graduate Programs, Wiess School of Natural Sciences, Department of Chemistry, Houston, TX 77251-1892. Offers chemistry (MA); inorganic chemistry (PhD); organic chemistry (PhD); physical chemistry (PhD). Terminal master's awarded for partial completion of doctoral program. *Degree requirements:* For master's and doctorate, thesis/dissertation. *Entrance requirements:* For master's and doctorate, GRE General Test, minimum GPA of 3.0. Additional exam requirements/recommendations for international students: Required—TOEFL. *Faculty research:* Nanoscience, biomaterials, nanobioinformatics, fullerene pharmaceuticals.

Rutgers, The State University of New Jersey, Newark, Graduate School, Program in Chemistry, Newark, NJ 07102. Offers analytical chemistry (MS, PhD); biochemistry (MS, PhD); inorganic chemistry (MS, PhD); organic chemistry (MS, PhD); physical chemistry (MS, PhD). Part-time and evening/weekend programs available. *Faculty:* 22 full-time (5 women). *Students:* 24 full-time (12 women), 28 part-time (10 women); includes 31 minority (2 African Americans, 29 Asian Americans or Pacific Islanders). 69 applicants, 43% accepted, 13 enrolled. In 2005, 7 master's, 9 doctorates awarded. Terminal master's awarded for partial completion of doctoral program. *Degree requirements:* For master's, cumulative exams, thesis optional; for doctorate, thesis/dissertation, exams, research proposal. *Entrance requirements:* For master's and doctorate, GRE General Test, minimum undergraduate B average. Additional exam requirements/recommendations for international students: Required—TOEFL. *Application deadline:* For fall admission, 7/1 for domestic students; for spring admission, 12/1 for domestic students. Applications are processed on a rolling basis. Application fee: $50. Electronic applications accepted. *Expenses:* Tuition, state resident: full-time $10,440; part-time $435 per credit. Tuition, nonresident: full-time $15,520; part-time $637 per credit. *Financial support:* In 2005–06, 35 students received support, including 5 fellowships with partial tuition reimbursements available (averaging $18,000 per year), 6 research assistantships (averaging $16,988 per year), 20 teaching assistantships with partial tuition reimbursements available (averaging $16,988 per year); Federal Work-Study and institutionally sponsored loans also available. Financial award application deadline: 3/1. *Faculty research:* Medicinal chemistry, natural

Organic Chemistry

Rutgers, The State University of New Jersey, Newark (continued)
products, isotope effects, biophysics and biorganic approaches to enzyme mechanisms, organic and organometallic synthesis. *Unit head:* Prof. W. Philip Huskey, Chairman and Program Director, 973-353-5741, Fax: 973-353-1264, E-mail: huskey@andromeda.rutgers.edu.

Rutgers, The State University of New Jersey, New Brunswick/Piscataway, Graduate School, Program in Chemistry and Chemical Biology, New Brunswick, NJ 08901-1281. Offers analytical chemistry (MS, PhD); biological chemistry (PhD); chemistry education (MST); inorganic chemistry (MS, PhD); organic chemistry (MS, PhD); physical chemistry (MS, PhD). Part-time and evening/weekend programs available. *Faculty:* 48 full-time. *Students:* 104 full-time (41 women), 22 part-time (4 women); includes 11 minority (2 African Americans, 6 Asian Americans or Pacific Islanders, 3 Hispanic Americans), 51 international. Average age 29. 108 applicants, 51% accepted, 22 enrolled. In 2005, 13 master's, 10 doctorates awarded. Terminal master's awarded for partial completion of doctoral program. *Degree requirements:* For master's, thesis or alternative, exam, comprehensive exam, registration; for doctorate, thesis/dissertation, cumulative exams, 1 year residency, comprehensive exam, registration. *Entrance requirements:* For master's and doctorate, GRE General Test, GRE Subject Test. Additional exam requirements/recommendations for international students: Required—TOEFL. *Application deadline:* For fall admission, 4/15 priority date for domestic students, 4/1 priority date for international students; for spring admission, 12/1 priority date for domestic students, 12/1 priority date for international students. Applications are processed on a rolling basis. Application fee: $50. Electronic applications accepted. *Expenses:* Tuition, state resident: full-time $10,440; part-time $435 per credit. Tuition, nonresident: full-time $15,520; part-time $647 per credit. Required fees: $129 per credit. Tuition and fees vary according to program. *Financial support:* In 2005–06, 104 students received support, including 8 fellowships with full tuition reimbursements available (averaging $22,000 per year), 23 research assistantships with full tuition reimbursements available (averaging $18,347 per year), 59 teaching assistantships with full tuition reimbursements available (averaging $18,347 per year); career-related internships or fieldwork, Federal Work-Study, traineeships, and health care benefits also available. Financial award application deadline: 3/1; financial award applicants required to submit FAFSA. *Faculty research:* Biophysical organic/bioorganic, inorganic/bioinorganic, theoretical, and solid-state/surface chemistry. Total annual research expenditures: $14.5 million. *Unit head:* Dr. Roger Jones, Director and Chair, 732-445-4900, Fax: 732-445-5312, E-mail: jones@rutchem.rutgers.edu. *Application contact:* Dr. Martha A. Cotter, Vice Chair, 732-445-2259, Fax: 732-445-5312, E-mail: cotter@rutchem.rutgers.edu.

See Close-Up on page 139.

Seton Hall University, College of Arts and Sciences, Department of Chemistry and Biochemistry, South Orange, NJ 07079-2694. Offers analytical chemistry (MS, PhD); biochemistry (MS, PhD); chemistry (MS); inorganic chemistry (MS, PhD); organic chemistry (MS, PhD); physical chemistry (MS, PhD). Part-time and evening/weekend programs available. *Students:* 19 full-time (6 women), 46 part-time (17 women). Average age 34. 33 applicants, 79% accepted, 14 enrolled. In 2005, 11 master's, 7 doctorates awarded. Terminal master's awarded for partial completion of doctoral program. *Degree requirements:* For master's, formal seminar, thesis optional; for doctorate, thesis/dissertation, annual seminars, comprehensive exam. *Entrance requirements:* Additional exam requirements/recommendations for international students: Required—TOEFL. *Application deadline:* For fall admission, 7/1 priority date for domestic students, 7/1 priority date for international students; for spring admission, 11/1 priority date for domestic students, 11/1 priority date for international students. Applications are processed on a rolling basis. Application fee: $50. Electronic applications accepted. *Financial support:* In 2005–06, 1 research assistantship, 19 teaching assistantships were awarded; Federal Work-Study also available. *Faculty research:* DNA metal reactions; chromatography; bioinorganic, biophysical, organometallic, polymer chemistry; heterogeneous catalyst. *Unit head:* Dr. Nicholas Snow, Chair, 973-761-9414, Fax: 973-761-9772, E-mail: snownich@shu.edu. *Application contact:* Dr. Stephen Kelty, Director of Graduate Studies, 973-761-9129, Fax: 973-761-9772, E-mail: keltyste@shu.edu.

See Close-Up on page 141.

South Dakota State University, Graduate School, College of Arts and Science and College of Agriculture and Biological Sciences, Department of Chemistry, Brookings, SD 57007. Offers analytical chemistry (MS, PhD); biochemistry (MS, PhD); chemistry (MS, PhD); inorganic chemistry (MS, PhD); organic chemistry (MS, PhD); physical chemistry (MS, PhD). *Degree requirements:* For master's, thesis, oral exam; for doctorate, thesis/dissertation, preliminary oral and written exams, research tool. *Entrance requirements:* For master's, bachelor's degree in chemistry or equivalent. Additional exam requirements/recommendations for international students: Required—TOEFL. *Faculty research:* Environmental chemistry, computational chemistry, organic synthesis and photochemistry, novel material development and characterization.

Southern University and Agricultural and Mechanical College, Graduate School, College of Sciences, Department of Chemistry, Baton Rouge, LA 70813. Offers analytical chemistry (MS); biochemistry (MS); environmental sciences (MS); inorganic chemistry (MS); organic chemistry (MS); physical chemistry (MS). *Faculty:* 9 full-time (2 women), 3 part-time/adjunct (2 women). *Students:* 20 full-time (13 women), 8 part-time (5 women); all minorities (20 African Americans, 8 Asian Americans or Pacific Islanders). Average age 23. 30 applicants, 70% accepted, 14 enrolled. In 2005, 3 master's awarded. *Degree requirements:* For master's, thesis. *Entrance requirements:* For master's, GMAT or GRE General Test. Additional exam requirements/recommendations for international students: Required—TOEFL (minimum score 525 paper-based; 193 computer-based). *Application deadline:* For fall admission, 4/15 for domestic students; for spring admission, 11/1 priority date for domestic students. Applications are processed on a rolling basis. Application fee: $5. *Financial support:* In 2005–06, 31 research assistantships (averaging $7,000 per year), 10 teaching assistantships (averaging $7,000 per year) were awarded; scholarships/grants also available. Financial award application deadline: 4/15. *Faculty research:* Synthesis of macrocyclic ligands, latex accelerators, anticancer drugs, biosensors, absorption isotheums, isolation of specific enzymes from plants. Total annual research expenditures:$400,000. *Unit head:* Dr. Ella Kelley, Chair, 225-771-3990, Fax: 225-771-3992.

State University of New York at Binghamton, Graduate School, School of Arts and Sciences, Department of Chemistry, Binghamton, NY 13902-6000. Offers analytical chemistry (PhD); chemistry (MA, MS); inorganic chemistry (PhD); organic chemistry (PhD); physical chemistry (PhD). Part-time programs available. Terminal master's awarded for partial completion of doctoral program. *Degree requirements:* For master's, thesis or alternative, oral exam, seminar presentation; for doctorate, thesis/dissertation, cumulative exams. *Entrance requirements:* For master's and doctorate, GRE General Test, GRE Subject Test. Additional exam requirements/recommendations for international students: Required—TOEFL. Electronic applications accepted.

State University of New York College of Environmental Science and Forestry, Faculty of Chemistry, Syracuse, NY 13210-2779. Offers biochemistry (MS, PhD); environmental and forest chemistry (MS, PhD); organic chemistry of natural products (MS, PhD); polymer chemistry (MS, PhD). *Faculty:* 14 full-time (1 woman). *Students:* 28 full-time (15 women), 8 part-time (2 women); includes 1 minority (American Indian/Alaska Native), 11 international. Average age 28. 23 applicants, 74% accepted, 2 enrolled. In 2005, 2 master's, 5 doctorates awarded. *Degree requirements:* For master's, thesis/dissertation, registration; for doctorate, thesis/dissertation, comprehensive exam, registration. *Entrance requirements:* For master's and doctorate, GRE General Test, GRE Subject Test, minimum GPA of 3.0. Additional exam requirements/recommendations for international students: Required—TOEFL (minimum score 550 paper-based; 213 computer-based). *Application deadline:* For fall admission, 2/1 priority date for domestic students, 2/1 priority date for international students; for spring admission, 11/1 priority date for domestic students, 11/1 priority date for international students. Applications are processed on a rolling basis. Application fee: $60. Electronic applications accepted. *Expenses:* Tuition, area resident: Full-time $6,900; part-time $288 per credit. Tuition, state resident: full-time $10,920. Tuition, nonresident: full-time $10,920; part-time $455 per credit. International tuition: $10,920 full-time. Required fees: $395; $32 per credit. $20 per term. One-time fee: $145. *Financial support:* In 2005–06, 35 students received support, including 3 fellowships with full tuition reimbursements available (averaging $9,446 per year), 13 research assistantships with full tuition reimbursements available (averaging $12,500 per year), 10 teaching assistantships with full tuition reimbursements available (averaging $11,540 per year); Federal Work-Study, institutionally sponsored loans, scholarships/grants, health care benefits, and unspecified assistantships also available. Financial award application deadline: 6/30; financial award applicants required to submit FAFSA. *Faculty research:* Polymer chemistry, biochemistry. Total annual research expenditures: $1.8 million. *Unit head:* Dr. John P. Hassett, Chair, 315-470-6827, Fax: 315-470-6856, E-mail: jphasset@syr.edu. *Application contact:* Dr. Dudley J. Raynal, Dean, Instruction and Graduate Studies, 315-470-6599, Fax: 315-470-6978, E-mail: esfgrad@esf.edu.

Stevens Institute of Technology, Graduate School, Arthur E. Imperatore School of Sciences and Arts, Department of Chemistry and Chemical Biology, Hoboken, NJ 07030. Offers chemical biology (MS, PhD); chemistry (MS, PhD), including analytical chemistry (MS, PhD, Certificate), chemical biology (MS, PhD, Certificate), organic chemistry, physical chemistry, polymer chemistry (MS, PhD, Certificate); chemistry and chemical biology (Certificate), including analytical chemistry (MS, PhD, Certificate), bioinformatics, biomedical chemistry, chemical biology (MS, PhD, Certificate), chemical physiology, polymer chemistry (MS, PhD, Certificate). Part-time and evening/weekend programs available. *Students:* 17 full-time (8 women), 28 part-time (16 women); includes 12 minority (9 Asian Americans or Pacific Islanders, 3 Hispanic Americans), 14 international. 51 applicants, 75% accepted.Terminal master's awarded for partial completion of doctoral program. *Degree requirements:* For master's, thesis or alternative; for doctorate, one foreign language, thesis/dissertation; for Certificate, project or thesis. *Entrance requirements:* Additional exam requirements/recommendations for international students: Required—TOEFL. *Application deadline:* Applications are processed on a rolling basis. Application fee: $50. Electronic applications accepted. *Expenses:* Tuition: Part-time $920 per credit hour. Tuition and fees vary according to program. *Financial support:* Fellowships, research assistantships, teaching assistantships, Federal Work-Study and institutionally sponsored loans available. Financial award application deadline: 4/1. *Unit head:* Dr. Francis T. Jones, Director, 201-216-5518, Fax: 201-216-8240.

Tufts University, Graduate School of Arts and Sciences, Department of Chemistry, Medford, MA 02155. Offers analytical chemistry (MS, PhD); bioorganic chemistry (MS, PhD); environmental chemistry (MS, PhD); inorganic chemistry (MS, PhD); organic chemistry (MS, PhD); physical chemistry (MS, PhD). *Faculty:* 17 full-time. *Students:* 55 (27 women); includes 5 minority (1 African American, 4 Asian Americans or Pacific Islanders) 27 international. 45 applicants, 87% accepted, 17 enrolled. In 2005, 4 master's, 4 doctorates awarded. Terminal master's awarded for partial completion of doctoral program. *Degree requirements:* For master's and doctorate, thesis/dissertation. *Entrance requirements:* For master's and doctorate, GRE General Test, GRE Subject Test. Additional exam requirements/recommendations for international students: Required—TOEFL (minimum score 600 paper-based; 250 computer-based), TSE. *Application deadline:* For fall admission, 1/15 for domestic students, 12/30 for international students; for spring admission, 10/15 for domestic students, 9/15 for international students. Applications are processed on a rolling basis. Application fee: $65. Electronic applications accepted. *Expenses:* Tuition: Full-time $32,360. Tuition and fees vary according to program. *Financial support:* Research assistantships with full and partial tuition reimbursements, teaching assistantships with full and partial tuition reimbursements, Federal Work-Study, scholarships/grants, and tuition waivers (partial) available. Financial award application deadline: 1/15; financial award applicants required to submit FAFSA. *Unit head:* Mary Jane Shultz, Chair, 617-627-3477, Fax: 617-627-3443. *Application contact:* Samuel Kounaves, Information Contact, 617-627-3441, Fax: 617-627-3443.

See Close-Up on page 151.

University of Calgary, Faculty of Graduate Studies, Faculty of Science, Department of Chemistry, Calgary, AB T2N 1N4, Canada. Offers analytical chemistry (M Sc, PhD); applied chemistry (M Sc, PhD); inorganic chemistry (M Sc, PhD); organic chemistry (M Sc, PhD); physical chemistry (M Sc, PhD); polymer chemistry (M Sc, PhD); theoretical chemistry (M Sc, PhD). *Faculty:* 31 full-time (2 women), 2 part-time/adjunct (0 women). *Students:* 80 full-time (38 women). Average age 25. 250 applicants, 6% accepted, 15 enrolled. In 2005, 7 master's, 7 doctorates awarded. *Degree requirements:* For master's, thesis; for doctorate, thesis/dissertation, candidacy exam. *Entrance requirements:* For master's, minimum GPA of 3.0; for doctorate, honors B Sc degree with minimum GPA of 3.7 or M Sc with minimum GPA of 3.3. Additional exam requirements/recommendations for international students: Required—TOEFL (minimum score 580 paper-based; 237 computer-based). *Application deadline:* For fall admission, 12/1 priority date for domestic students, 12/1 priority date for international students. Applications are processed on a rolling basis. Application fee: $100 ($130 for international students). Electronic applications accepted. *Financial support:* In 2005–06, 25 students received support, including research assistantships (averaging $11,000 per year), teaching assistantships (averaging $6,530 per year); fellowships, scholarships/grants also available. Financial award application deadline:12/1. *Faculty research:* Chemical analysis, chemical dynamics, synthesis theory. *Unit head:* Dr. Brian Keay, Head, 403-220-5340, E-mail: info@chem.ucalgary.ca. *Application contact:* Bonnie E. King, Graduate Program Administrator, 403-220-6252, E-mail: bking@ucalgary.ca.

University of Cincinnati, Division of Research and Advanced Studies, McMicken College of Arts and Sciences, Department of Chemistry, Cincinnati, OH 45221. Offers analytical chemistry (MS, PhD); biochemistry (MS, PhD); inorganic chemistry (MS, PhD); organic chemistry (MS, PhD); physical chemistry (MS, PhD); polymer chemistry (MS, PhD); sensors (PhD). Part-time and evening/weekend programs available. Terminal master's awarded for partial completion of doctoral program. *Degree requirements:* For master's, thesis optional; for doctorate, thesis/dissertation, comprehensive exam, registration. *Entrance requirements:* For master's and doctorate, GRE General Test. Additional exam requirements/recommendations for international students: Required—TOEFL (minimum score 580 paper-based; 237 computer-based); Recommended—TSE(minimum score 50). Electronic applications accepted. *Faculty research:* Biomedical chemistry, laser chemistry, surface science, chemical sensors, synthesis.

University of Georgia, Graduate School, College of Arts and Sciences, Department of Chemistry, Athens, GA 30602. Offers analytical chemistry (MS, PhD); inorganic chemistry (MS, PhD); organic chemistry (MS, PhD); physical chemistry (MS, PhD). *Faculty:* 22 full-time (2 women). *Students:* 155 full-time, 3 part-time; includes 12 minority (4 African Americans, 7 Asian Americans or Pacific Islanders, 1 Hispanic American), 87 international. 147 applicants, 56% accepted, 35 enrolled. In 2005, 2 master's, 19 doctorates awarded. Terminal master's awarded for partial completion of doctoral program. *Median time to degree:* Of those who began their doctoral program in fall 1997, 100% received their degree in 8 years or less. *Degree requirements:* For master's, thesis; for doctorate, one foreign language, thesis/dissertation. *Entrance requirements:* For master's and doctorate, GRE General Test. Additional exam requirements/recommendations for international students: Required—TOEFL (minimum score 213 computer-based), TSE(minimum score 50). *Application deadline:* For fall admission, 7/1 for domestic students; for spring admission, 11/15 for domestic students. Application fee: $50. Electronic applications accepted. *Financial support:* Fellowships, research assistantships, teaching assistantships, unspecified assistantships available. *Unit head:* Dr. John L. Stickney, Head, 706-542-2726, Fax: 706-542-9454, E-mail: stickney@chem.uga.edu. *Application contact:* Dr. Donald Kurtz, Information Contact, 706-542-2010, Fax: 706-542-9454, E-mail: kurtz@chem.uga.edu.

University of Louisville, Graduate School, College of Arts and Sciences, Department of Chemistry, Louisville, KY 40292-0001. Offers analytical chemistry (MS, PhD); biochemistry (MS, PhD); chemical physics (PhD); inorganic chemistry (MS, PhD); organic chemistry (MS, PhD); physical chemistry (MS, PhD). *Students:* 37 full-time (16 women), 16 part-time (7 women); includes 3 minority (2 African Americans, 1 Hispanic American), 24 international. Average age 28. In 2005, 1 master's, 8 doctorates awarded. *Degree requirements:* For master's, thesis/dissertation; for doctorate, thesis/dissertation, comprehensive exam. *Entrance requirements:* For master's and doctorate, GRE General Test. Additional exam requirements/recommendations for international students: Required—TOEFL. *Application deadline:* Applications are processed on a rolling basis. Application fee: $50. *Expenses:* Tuition, state resident: full-time $3,003; part-time $334 per credit hour. Tuition, nonresident: full-time $8,277; part-time $920 per credit hour. Tuition and fees vary according to course load, degree level and program. *Financial support:* In 2005–06, 33 teaching assistantships with tuition reimbursements were awarded; fellowships, research assistantships *Unit head:* Dr. George R. Pack, Chair, 502-852-6798, Fax: 502-852-8149, E-mail: george.pack@louisville.edu.

University of Maryland, College Park, Graduate Studies, College of Chemical and Life Sciences, Department of Chemistry and Biochemistry, Chemistry Program, College Park, MD 20742. Offers analytical chemistry (MS, PhD); inorganic chemistry (MS, PhD); organic chemistry (MS, PhD); physical chemistry (MS, PhD). Part-time and evening/weekend programs available. *Students:* 102 full-time (43 women), 4 part-time (2 women); includes 7 minority (2 African Americans, 3 Asian Americans or Pacific Islanders, 2 Hispanic Americans), 46 international. 128 applicants, 24% accepted, 30 enrolled. In 2005, 6 master's, 12 doctorates awarded. Terminal master's awarded for partial completion of doctoral program. *Median time to degree:* Of those who began their doctoral program in fall 1997, 33% received their degree in 8 years or less. *Degree requirements:* For master's, thesis optional; for doctorate, thesis/dissertation, 2 seminar presentations, oral exam. *Entrance requirements:* For master's and doctorate, GRE General Test, GRE Subject Test (recommended), minimum GPA of 3.0, 3 letters of recommendation. Additional exam requirements/recommendations for international students: Required—TOEFL, TSE. *Application deadline:* For fall admission, 4/1 for domestic students, 2/1 for international students; for spring admission, 10/21 for domestic students, 6/1 for international students. Applications are processed on a rolling basis. Application fee: $60. Electronic applications accepted. *Financial support:* In 2005–06, 10 fellowships (averaging $4,900 per year) were awarded; research assistantships, teaching assistantships with partial tuition reimbursements Financial award applicants required to submit FAFSA. *Faculty research:* Environmental chemistry, nuclear chemistry, lunar and environmental analysis, x-ray crystallography. *Application contact:* Dean of Graduate School, 301-405-4190, Fax: 301-314-9305.

University of Miami, Graduate School, College of Arts and Sciences, Department of Chemistry, Coral Gables, FL 33124. Offers chemistry (MS); inorganic chemistry (PhD); organic chemistry (PhD); physical chemistry (PhD). *Faculty:* 11 full-time (1 woman). *Students:* 38 full-time (6 women), 28 international. Average age 27. 62 applicants, 27% accepted, 5 enrolled. In 2005, 1 master's, 5 doctorates awarded. Terminal master's awarded for partial completion of doctoral program. *Degree requirements:* For master's, comprehensive exam; for doctorate, thesis/dissertation, comprehensive exam. *Entrance requirements:* For master's and doctorate, GRE General Test. Additional exam requirements/recommendations for international students: Required—TOEFL (minimum score 550 paper-based; 213 computer-based). *Application deadline:* For fall admission, 1/15 for domestic students, 1/15 for international students. Applications are processed on a rolling basis. Application fee: $50. Electronic applications accepted. *Financial support:* In 2005–06, 38 students received support, including 1 fellowship with full tuition reimbursement available (averaging $20,000 per year), 10 research assistantships with full tuition reimbursements available (averaging $20,000 per year), 27 teaching assistantships with full tuition reimbursements available (averaging $20,000 per year); tuition waivers (full) also available. Financial award application deadline: 5/1; financial award applicants required to submit FAFSA. *Faculty research:* Supramolecular chemistry, electrochemistry, surface chemistry, catalysis, organometalic. *Unit head:* Dr. Vaidyanathan Ramamurthy, Chairman, 305-284-2282, Fax: 305-284-4571. *Application contact:* Eva Johnson, Graduate Secretary, 305-284-2094, Fax: 305-284-4571, E-mail: evaj@miami.edu.

University of Michigan, Horace H. Rackham School of Graduate Studies, College of Literature, Science, and the Arts, Department of Chemistry, Ann Arbor, MI 48109. Offers analytical chemistry (PhD); inorganic chemistry (PhD); material chemistry (PhD); organic chemistry (PhD); physical chemistry (PhD). *Faculty:* 48 full-time (8 women), 3 part-time/adjunct (all women). *Students:* ; includes 39 minority (4 African Americans, 1 American Indian/Alaska Native, 29 Asian Americans or Pacific Islanders, 5 Hispanic Americans), 66 international. Average age 26. 456 applicants, 29% accepted, 44 enrolled. In 2005, 37 degrees awarded. *Degree requirements:* For doctorate, thesis/dissertation, oral defense of dissertation, organic cumulative proficiency exams. *Entrance requirements:* For doctorate, GRE General Test, GRE Subject Test (recommended), 3 letters of recommendation. Additional exam requirements/recommendations for international students: Required—TOEFL. *Application deadline:* For fall admission, 1/31 for domestic students, 1/1 for international students. Applications are processed on a rolling basis. Application fee: $60 ($75 for international students). Electronic applications accepted. *Expenses:* Tuition, area resident: Full-time $14,082; part-time $894 per credit hour. Tuition, state resident: full-time $14,082; part-time $896 per credit hour. Tuition, nonresident: full-time $28,500; part-time $1,675 per credit hour. Required fees: $189; $189. *Financial support:* In 2005–06, 10 fellowships with full tuition reimbursements (averaging $20,000 per year), 70 research assistantships with full tuition reimbursements (averaging $19,000 per year), 120 teaching assistantships with full tuition reimbursements (averaging $20,000 per year) were awarded. Financial award applicants required to submit FAFSA. *Faculty research:* Biological catalysis, protein engineering, chemical sensors, de novo metalloprotein design, supramolecular architecture. Total annual research expenditures: $8 million. *Unit head:* Dr. Carol A. Fierke, Chair, 734-763-9681, Fax: 734-647-4847. *Application contact:* Linda Deitert, Assistant Director Graduate Studies, 734-764-7278, Fax: 734-647-4865, E-mail: chemadmissions@umich.edu.

University of Missouri–Columbia, Graduate School, College of Arts and Sciences, Department of Chemistry, Columbia, MO 65211. Offers analytical chemistry (MS, PhD); inorganic chemistry (MS, PhD); organic chemistry (MS, PhD); physical chemistry (MS, PhD). *Faculty:* 21 full-time (5 women). *Students:* 58 full-time (17 women), 44 part-time (16 women); includes 5 minority (3 African Americans, 2 Asian Americans or Pacific Islanders), 59 international. In 2005, 11 master's, 9 doctorates awarded. *Degree requirements:* For master's, thesis; for doctorate, one foreign language, thesis/dissertation. *Entrance requirements:* For master's and doctorate, GRE General Test, minimum GPA of 3.0. *Application deadline:* For fall admission, 4/1 for domestic studentsFor winter admission, 11/1 for domestic students. Applications are processed on a rolling basis. Application fee: $45 ($60 for international students). *Financial support:* Fellowships, research assistantships, teaching assistantships, institutionally sponsored loans available. *Unit head:* Dr. Sheryl A. Tucker, Director for Graduate Studies, 573-882-1729.

University of Missouri–Kansas City, College of Arts and Sciences, Department of Chemistry, Kansas City, MO 64110-2499. Offers analytical chemistry (MS, PhD); inorganic chemistry (MS, PhD); organic chemistry (MS, PhD); physical chemistry (MS, PhD); polymer chemistry (MS, PhD). PhD offered through the School of Graduate Studies. Part-time and evening/weekend programs available. *Faculty:* 14 full-time (2 women), 2 part-time/adjunct (0 women). *Students:* Average age 38. 13 applicants, 8% accepted, 1 enrolled. In 2005, 2 degrees awarded. *Degree requirements:* For master's, thesis (for some programs); for doctorate, thesis/dissertation. *Entrance requirements:* For master's, equivalent of American Chemical Society approved bachelor's degree in chemistry; for doctorate, GRE General Test, equivalent of American Chemical Society approved bachelor's degree in chemistry. Additional exam requirements/recommendations for international students: Required—TOEFL (minimum score 580 paper-based; 237 computer-based), TWE. *Application deadline:* For fall and spring admission, 4/15For winter admission, 10/15 for domestic students. Applications are processed on a rolling basis. Application fee: $25. Electronic applications accepted. *Expenses:* Tuition, state resident: full-time $4,738; part-time $263 per credit hour. Tuition, nonresident: full-time $12,235; part-time $679 per credit hour. Required fees: $582. Tuition and fees vary according to course load, program and student level. *Financial support:* In 2005–06, fellowships with partial tuition reimbursements (averaging $18,156 per year), research assistantships with partial tuition reimbursements (averaging $18,515 per year), teaching assistantships with partial tuition reimbursements (averaging $16,434 per year) were awarded; Federal Work-Study, institutionally sponsored loans, and scholarships/grants also available. Support available to part-time students. Financial award application deadline: 2/15. *Faculty research:* Molecular spectroscopy, characterization and synthesis of materials and compounds, computational chemistry, natural products, drug delivery systems and anti-tumor agents. Total annual research expenditures: $1.1 million. *Unit head:* Dr. Jerry Jean, Chairperson, 816-235-2273, Fax: 816-235-5502, E-mail: jeany@umkc.edu. *Application contact:* Graduate Recruiting Committee, 816-235-2272, Fax: 816-235-5502, E-mail: umkc-chemdept@umkc.edu.

University of Missouri–St. Louis, College of Arts and Sciences, Department of Chemistry and Biochemistry, St. Louis, MO 63121. Offers chemistry (MS, PhD), including inorganic chemistry, organic chemistry, physical chemistry. Part-time and evening/weekend programs available. *Faculty:* 16. *Students:* 28 full-time (20 women), 16 part-time (7 women); includes 6 minority (5 African Americans, 1 Hispanic American), 23 international. Average age 32. 2,298 applicants, 74% accepted. In 2005, 9 master's, 4 doctorates awarded. Terminal master's awarded for partial completion of doctoral program. *Degree requirements:* For master's, thesis optional; for doctorate, thesis/dissertation. *Entrance requirements:* For doctorate, GRE General Test, 3 letters of recommendation. Additional exam requirements/recommendations for international students: Required—TOEFL (minimum score 550 paper-based; 213 computer-based). *Application deadline:* For fall admission, 7/1 for domestic students; for spring admission, 12/7 priority date for domestic students. Applications are processed on a rolling basis. Application fee: $35 ($40 for international students). Electronic applications accepted. *Expenses:* Tuition, state resident: part-time $263 per credit hour. Tuition, nonresident: part-time $680 per credit hour. Required fees: $53 per credit hour. Tuition and fees vary according to program. *Financial support:* In 2005–06, 9 research assistantships with full and partial tuition reimbursements (averaging $16,500 per year), 17 teaching assistantships with full and partial tuition reimbursements (averaging $16,500 per year) were awarded; fellowships with full and partial tuition reimbursements *Faculty research:* Metallaborane chemistry, serum transferrin chemistry, natural products chemistry, organic synthesis. *Unit head:* Dr. Cynthia Dupureur, Director of Graduate Studies, 314-516-5311, Fax: 314-516-5342, E-mail: gradchem@umsl.edu. *Application contact:* 314-516-5458, Fax: 314-516-5310, E-mail: gradadm@umsl.edu.

The University of Montana, Graduate School, College of Arts and Sciences, Department of Chemistry, Missoula, MT 59812-0002. Offers chemistry (MS, PhD), including environmental/analytical chemistry, inorganic chemistry, organic chemistry, physical chemistry. *Faculty:* 16 full-time (2 women). *Students:* 36 full-time (9 women), 7 part-time (2 women), 9 international. 25 applicants, 44% accepted, 6 enrolled. In 2005, 1 master's, 2 doctorates awarded. Terminal master's awarded for partial completion of doctoral program. *Degree requirements:* For master's, thesis (for some programs); for doctorate, thesis/dissertation. *Entrance requirements:* For master's and doctorate, GRE General Test. Additional exam requirements/recommendations for international students: Required—TOEFL (minimum score 575 paper-based; 230 computer-based). *Application deadline:* For fall admission, 2/15 for domestic students. Applications are processed on a rolling basis. Application fee: $45. *Expenses:* Tuition, state resident: part-time $267 per credit. Tuition, nonresident: part-time $665 per credit. Part-time tuition and fees vary according to course load and degree level. *Financial support:* In 2005–06, 13 research assistantships with tuition reimbursements (averaging $14,000 per year), 12 teaching assistantships with full tuition reimbursements (averaging $14,000 per year) were awarded; Federal Work-Study, scholarships/grants, and unspecified assistantships also available. Financial award application deadline: 3/1; financial award applicants required to submit FAFSA. *Faculty research:* Reaction mechanisms and kinetics, inorganic and organic synthesis, analytical chemistry, natural products. Total annual research expenditures: $789,952. *Unit head:* Dr. Edward Rosenberg, Chair, 406-243-2592, Fax: 406-243-4227.

University of Nebraska–Lincoln, Graduate College, College of Arts and Sciences, Department of Chemistry, Lincoln, NE 68588. Offers analytical chemistry (PhD); chemistry (MS); inorganic chemistry (PhD); organic chemistry (PhD); physical chemistry (PhD). *Degree requirements:* For master's, one foreign language, departmental qualifying exam, thesis optional; for doctorate, one foreign language, thesis/dissertation, departmental qualifying exams, comprehensive exam. *Entrance requirements:* For master's and doctorate, GRE. Additional exam requirements/recommendations for international students: Required—TOEFL (minimum score 550 paper-based; 213 computer-based). Electronic applications accepted. *Faculty research:* Bioorganic and bioinorganic chemistry, biophysical and bioanalytical chemistry, structure-function of DNA and proteins, organometallics, mass spectrometry.

University of Notre Dame, Graduate School, College of Science, Department of Chemistry and Biochemistry, Notre Dame, IN 46556. Offers biochemistry (MS, PhD); inorganic chemistry (MS, PhD); organic chemistry (MS, PhD); physical chemistry (MS, PhD). *Faculty:* 30 full-time (3 women). *Students:* 148 full-time (52 women); includes 5 minority (1 African American, 1 Asian American or Pacific Islander, 3 Hispanic Americans), 65 international. 144 applicants, 38% accepted, 21 enrolled. In 2005, 7 master's, 6 doctorates awarded. Terminal master's awarded for partial completion of doctoral program. *Median time to degree:* Of those who began their doctoral program in fall 1997, 54% received their degree in 8 years or less. *Degree requirements:* For master's, thesis, comprehensive exam; for doctorate, thesis/dissertation, qualifying exam. *Entrance requirements:* For master's and doctorate, GRE General Test, GRE Subject Test (strongly recommended). Additional exam requirements/recommendations for international students: Required—TOEFL. *Application deadline:* For fall admission, 2/1 for domestic students. Applications are processed on a rolling basis. Application fee: $50. Electronic applications accepted. *Financial support:* In 2005–06, 148 students received support, including 19 fellowships with full tuition reimbursements available (averaging $22,000 per year), 57 research assistantships with full tuition reimbursements available (averaging $15,250 per year), 53 teaching assistantships with full tuition reimbursements available (averaging $16,000 per year); tuition waivers (full) also available. Financial award application deadline: 2/1. *Faculty research:* Reaction design and mechaninistic studies; reactive intermediates; synthesis, structure and reactivity of organometallic. cluster complexes, and biologically active natural products; bioorganic chemistry; enzymology. Total annual research expenditures: $9.4 million. *Unit head:* Dr. Richard E. Taylor, Director of Graduate Studies, 574-631-7058, Fax: 574-631-6652, E-mail: taylor.61@nd.edu. *Application contact:* Dr. Terrence J. Akai, Director of Graduate Admissions, 574-631-7706, Fax: 574-631-4183, E-mail: gradad@nd.edu.

University of Regina, Faculty of Graduate Studies and Research, Faculty of Science, Department of Chemistry and Biochemistry, Regina, SK S4S 0A2, Canada. Offers analytical chemistry (M Sc, PhD); biochemistry (M Sc, PhD); inorganic chemistry (M Sc, PhD); organic chemistry (M Sc, PhD); physical chemistry (M Sc, PhD). Part-time programs available. *Faculty:* 10 full-time (2 women), 4 part-time/adjunct (1 woman). *Students:* 14 full-time (6 women), 3 part-time (2 women). 21 applicants, 33% accepted. *Degree requirements:* For master's and doctorate, thesis/dissertation, departmental qualifying exam. *Entrance requirements:* For master's and doctorate, GRE. Additional exam requirements/recommendations for international students: Required—TOEFL (minimum score 580 paper-based; 237 computer-based). *Application deadline:* For fall admission, 1/1 for domestic studentsFor winter admission, 7/1 for domestic students. Applications are processed on a rolling basis. Application fee: $60 ($100 for international students). *Financial support:* In 2005–06, 4 fellowships (averaging $14,886 per year), 1 research assistantship (averaging $12,750 per year), 5 teaching assistantships (averaging $13,501 per year) were awarded; scholarships/grants also available. Financial award application deadline: 6/15. *Faculty research:* Organic synthesis, organic oxidations, ionic liquids theoretical/computational chemistry, protein biochemistry/biophysics, environmental analytical, photophysical/photochemistry. *Unit head:* Dr. Andrew G. Wee, Head, 306-585-4767, Fax: 306-585-4894, E-mail: chem.chair@uregina.ca. *Application contact:* Dr. Allan East, Associate Professor, 306-585-4003, Fax: 306-585-4894, E-mail: allan.east@uregina.ca.

University of Southern Mississippi, Graduate School, College of Science and Technology, Department of Chemistry and Biochemistry, Hattiesburg, MS 39406-0001. Offers analytical chemistry (MS, PhD); biochemistry (MS, PhD); inorganic chemistry (MS, PhD); organic chemistry (MS, PhD); physical chemistry (MS, PhD). *Degree requirements:* For master's and doctorate,

Organic Chemistry

University of Southern Mississippi (continued)
thesis/dissertation, comprehensive exam. *Entrance requirements:* For master's, GRE General Test, minimum GPA of 2.75 in last 60 hours; for doctorate, GRE General Test, minimum GPA of 3.5. Additional exam requirements/recommendations for international students: Required—TOEFL. *Faculty research:* Plant biochemistry, photo chemistry, polymer chemistry, x-ray analysis, enzyme chemistry.

University of South Florida, College of Graduate Studies, College of Arts and Sciences, Department of Chemistry, Tampa, FL 33620-9951. Offers analytical chemistry (MS, PhD); biochemistry (MS, PhD); inorganic chemistry (MS, PhD); organic chemistry (MS, PhD); physical chemistry (MS, PhD); polymer chemistry (PhD). Part-time programs available. *Faculty:* 22. *Students:* 102 full-time (45 women), 15 part-time (6 women); includes 13 minority (5 Asian Americans or Pacific Islanders, 8 Hispanic Americans), 60 international. 80 applicants, 71% accepted, 23 enrolled. In 2005, 2 master's awarded. Terminal master's awarded for partial completion of doctoral program. *Degree requirements:* For master's, thesis; for doctorate, 2 foreign languages, thesis/dissertation, colloquium. *Entrance requirements:* For master's and doctorate, GRE General Test, minimum GPA of 3.0 in last 30 hours of chemistry course work, 3 letters of recommendation. Additional exam requirements/recommendations for international students: Required—TOEFL (minimum score 550 paper-based; 213 computer-based), TSE(minimum score 50). *Application deadline:* For fall admission, 5/1 priority date for domestic students, 3/1 priority date for international students; for spring admission, 10/1 priority date for domestic students, 8/1 priority date for international students. Applications are processed on a rolling basis. Application fee: $30. Electronic applications accepted. *Financial support:* In 2005–06, 91 students received support; fellowships with partial tuition reimbursements available, research assistantships with partial tuition reimbursements available, teaching assistantships with partial tuition reimbursements available, career-related internships or fieldwork, institutionally sponsored loans, scholarships/grants, health care benefits, and unspecified assistantships available. Financial award application deadline: 6/30. *Faculty research:* Synthesis, bio-organic chemistry, bioinorganic chemistry, environmental chemistry, NMR. *Unit head:* Dr. Michael Zaworotko, Chairperson, 813-974-4129, Fax: 813-974-3203, E-mail: xtal@usf.edu. *Application contact:* Dr. Brian Space, Graduate Coordinator, 813-974-3397, Fax: 813-974-3203.

The University of Tennessee, Graduate School, College of Arts and Sciences, Department of Chemistry, Knoxville, TN 37996. Offers analytical chemistry (MS, PhD); chemical physics (PhD); environmental chemistry (MS, PhD); inorganic chemistry (MS, PhD); organic chemistry (MS, PhD); physical chemistry (MS, PhD); polymer chemistry (MS, PhD); theoretical chemistry (PhD). Part-time programs available. Terminal master's awarded for partial completion of doctoral program. *Degree requirements:* For master's and doctorate, thesis/dissertation. *Entrance requirements:* For master's and doctorate, GRE General Test, minimum GPA of 2.7. Additional exam requirements/recommendations for international students: Required—TOEFL. Electronic applications accepted.

The University of Texas at Austin, Graduate School, College of Natural Sciences, Department of Chemistry and Biochemistry, Austin, TX 78712-1111. Offers analytical chemistry (MA, PhD); biochemistry (MA, PhD); inorganic chemistry (MA, PhD); organic chemistry (MA, PhD); physical chemistry (MA, PhD). *Entrance requirements:* For master's and doctorate, GRE General Test.

The University of Toledo, Graduate School, College of Arts and Sciences, Department of Chemistry, Toledo, OH 43606-3390. Offers analytical chemistry (MS, PhD); biological chemistry (MS, PhD); inorganic chemistry (MS, PhD); organic chemistry (MS, PhD); physical chemistry (MS, PhD). Part-time programs available. *Faculty:* 18. *Students:* 80 full-time (35 women), 8 part-time (5 women); includes 3 minority (2 African Americans, 1 Asian American or Pacific Islander), 43 international. Average age 27. 32 applicants, 75% accepted, 19 enrolled. In 2005, 7 master's awarded. *Degree requirements:* For master's and doctorate, thesis/dissertation. *Entrance requirements:* For master's and doctorate, GRE General Test, GRE Subject Test. Additional exam requirements/recommendations for international students: Required—TOEFL. *Application deadline:* For fall admission, 8/1 for domestic students. Applications are processed on a rolling basis. Application fee: $45. Electronic applications accepted. *Expenses:* Tuition, area resident: Full-time $3,312; part-time $308 per credit hour. Tuition, state resident: full-time $3,312. Tuition, nonresident: full-time $6,616; part-time $735 per credit hour. International tuition: $6,616 full-time. *Financial support:* In 2005–06, 1 research assistantship with full tuition reimbursement (averaging $4,000 per year), 55 teaching assistantships with full tuition reimbursements (averaging $13,336 per year) were awarded; fellowships, Federal Work-Study and institutionally sponsored loans also available. Support available to part-time students. Financial award application deadline: 4/1; financial award applicants required to submit FAFSA. *Faculty research:* Enzymology, materials chemistry, crystallography, theoretical chemistry. *Unit head:* Dr. Alan Pinkerton, Chair, 419-530-7902, Fax: 419-530-4033, E-mail: apinker@uoft02.utoledo.edu. *Application contact:* Charlene Morlock-Hansen, 419-530-2100, E-mail: charlene.hanson@utoledo.edu.

See Close-Up on page 167.

Vanderbilt University, Graduate School, Department of Chemistry, Nashville, TN 37240-1001. Offers analytical chemistry (MAT, MS, PhD); inorganic chemistry (MAT, MS, PhD); organic chemistry (MAT, MS, PhD); physical chemistry (MAT, MS, PhD); theoretical chemistry (MAT, MS, PhD). *Faculty:* 32 full-time (6 women), 3 part-time/adjunct (2 women). *Students:* 96 full-time (40 women); includes 4 minority (2 African Americans, 1 Asian American or Pacific Islander, 1 Hispanic American), 20 international. 223 applicants, 21% accepted, 22 enrolled. In 2005, 4 master's, 12 doctorates awarded. *Degree requirements:* For master's, thesis or alternative; for doctorate, thesis/dissertation, area, qualifying, and final exams. *Entrance requirements:* For master's and doctorate, GRE General Test, GRE Subject Test (recommended). Additional exam requirements/recommendations for international students: Required—TOEFL. *Application deadline:* For fall admission, 1/15 for domestic students, 1/15 for international students. Application fee: $0. Electronic applications accepted. *Expenses:* Tuition: Full-time $15,396; part-time $1,283 per semester hour. Required fees: $2,202; $1,101 per semester. One-time fee: $30. Tuition and fees vary according to course load, program and student level. *Financial support:* Fellowships with full and partial tuition reimbursements, research assistantships with full tuition reimbursements, teaching assistantships with full tuition reimbursements, Federal Work-Study, institutionally sponsored loans, and traineeships available. Financial award application deadline:1/15. *Faculty research:* Chemical synthesis; mechanistic, theoretical, bioorganic, analytical, and spectroscopic chemistry. *Unit head:* Ned A. Porter, Chair, 615-322-2861, Fax: 615-343-1234. *Application contact:* Charles M. Lukehart, Director of Graduate Studies, 615-322-2861, Fax: 615-343-1234, E-mail: charles.m.lukehart@vanderbilt.edu.

Virginia Commonwealth University, Graduate School, College of Humanities and Sciences, Department of Chemistry, Richmond, VA 23284-9005. Offers analytical (MS, PhD); chemical physics (PhD); inorganic (MS, PhD); organic (MS, PhD); physical (MS, PhD). Part-time programs available. *Faculty:* 18 full-time (6 women). *Students:* 36 full-time (13 women), 11 part-time (1 woman); includes 13 minority (6 African Americans, 1 American Indian/Alaska Native, 5 Asian Americans or Pacific Islanders, 1 Hispanic American), 13 international. 47 applicants, 74% accepted. In 2005, 2 master's, 8 doctorates awarded. Terminal master's awarded for partial completion of doctoral program. *Degree requirements:* For master's, thesis; for doctorate, thesis/dissertation, comprehensive cumulative exams, research proposal. *Entrance requirements:* For master's, GRE General Test, 30 undergraduate credits in chemistry; for doctorate, GRE General Test. *Application deadline:* For fall admission, 3/15 for domestic students; for spring admission, 11/15 for domestic students. Applications are processed on a rolling basis. Application fee: $50. *Expenses:* Tuition, state resident: full-time $3,185; part-time $405 per credit. Tuition, nonresident: full-time $7,952; part-time $940 per credit. Required fees: $751 per semester hour. Tuition and fees vary according to course load and program. *Financial support:* Fellowships, research assistantships, teaching assistantships, career-related internships or fieldwork and institutionally sponsored loans available. Support available to part-time students. Financial award application deadline: 7/1. *Faculty research:* Physical, organic, inorganic, analytical, and polymer chemistry; chemical physics. *Unit head:* Dr. Fred M. Hawkridge, Chair, 804-828-1298, Fax: 804-828-8599, E-mail: fmhawkri@vcu.edu. *Application contact:* Dr. M. Samy El-Shall, Chair, Graduate Recruiting and Admissions Committee, 804-828-3518, E-mail: mselshal@vcu.edu.

Wake Forest University, Graduate School, Department of Chemistry, Winston-Salem, NC 27109. Offers analytical chemistry (MS, PhD); inorganic chemistry (MS, PhD); organic chemistry (MS, PhD); physical chemistry (MS, PhD). Part-time programs available. *Faculty:* 17 full-time (4 women). *Students:* 32 full-time (17 women), 2 part-time; includes 1 minority (African American), 14 international. Average age 28. 60 applicants, 42% accepted, 12 enrolled. In 2005, 1 master's, 6 doctorates awarded. *Median time to degree:* Of those who began their doctoral program in fall 1997, 0% received their degree in 8 years or less. *Degree requirements:* For master's, one foreign language, thesis, comprehensive exam, registration; for doctorate, 2 foreign languages, thesis/dissertation, comprehensive exam, registration. *Entrance requirements:* For master's and doctorate, GRE General Test. Additional exam requirements/recommendations for international students: Required—TOEFL (minimum score 213 computer-based). *Application deadline:* For fall admission, 1/15 for domestic students, 1/15 for international students. Application fee: $45. Electronic applications accepted. *Financial support:* In 2005–06, 32 students received support, including 1 fellowship with full tuition reimbursement available (averaging $21,500 per year), 12 research assistantships with full tuition reimbursements available (averaging $19,500 per year), 17 teaching assistantships with full tuition reimbursements available (averaging $19,500 per year); scholarships/grants and tuition waivers (full and partial) also available. Support available to part-time students. Financial award application deadline: 1/15; financial award applicants required to submit FAFSA. *Unit head:* Dr. Bruce King, Director, 336-758-5774, Fax: 335-758-4656, E-mail: kingsb@wfu.edu.

Washington State University, Graduate School, College of Sciences, Department of Chemistry, Pullman, WA 99164. Offers analytical chemistry (MS, PhD); biological systems (MS, PhD); inorganic chemistry (MS, PhD); organic chemistry (MS, PhD); physical chemistry (MS, PhD). *Faculty:* 33. *Students:* 38 full-time (19 women), 2 part-time (1 woman); includes 3 minority (2 Asian Americans or Pacific Islanders, 1 Hispanic American), 11 international. Average age 29. 55 applicants, 33% accepted, 9 enrolled. In 2005, 5 master's, 7 doctorates awarded. Terminal master's awarded for partial completion of doctoral program. *Degree requirements:* For master's, oral exam, teaching experience, thesis optional; for doctorate, thesis/dissertation, oral exam, written exam, comprehensive exam. *Entrance requirements:* For master's and doctorate, GRE General Test, minimum GPA of 3.0, 3 letters of recommendation. *Application deadline:* For fall admission, 3/1 priority date for domestic students, 3/1 priority date for international students; for spring admission, 10/1 priority date for domestic students, 7/1 priority date for international students. Applications are processed on a rolling basis. Application fee: $35. *Expenses:* Tuition, state resident: full-time $6,295; part-time $336 per credit. Tuition, nonresident: full-time $15,949; part-time $819 per credit. Required fees: $933. Part-time tuition and fees vary according to campus/location and program. *Financial support:* In 2005–06, 38 students received support, including 3 fellowships (averaging $8,977 per year), 12 research assistantships with full and partial tuition reimbursements available (averaging $15,999 per year), 23 teaching assistantships with full and partial tuition reimbursements available (averaging $14,747 per year); career-related internships or fieldwork, Federal Work-Study, institutionally sponsored loans, scholarships/grants, health care benefits, unspecified assistantships, and summer support also available. Financial award application deadline: 4/1; financial award applicants required to submit FAFSA. *Faculty research:* Environmental chemistry, materials chemistry, radio chemistry, bio-organic, computational chemistry. Total annual research expenditures: $3.8 million. *Unit head:* Sue B. Clark, Chair, Admissions Committee, 509-335-8866, Fax: 509-335-8867, E-mail: sclark@mail.wsu.edu. *Application contact:* Admissions Committee.

Wesleyan University, Graduate Programs, Department of Chemistry, Middletown, CT 06459-0260. Offers biochemistry (MA, PhD); chemical physics (MA, PhD); inorganic chemistry (MA, PhD); organic chemistry (MA, PhD); physical chemistry (MA, PhD); theoretical chemistry (MA, PhD). *Faculty:* 13 full-time (2 women), 2 part-time/adjunct (1 woman). *Students:* 41 full-time (20 women); includes 1 minority (Asian American or Pacific Islander), 22 international. Average age 26. In 2005, 2 master's, 2 doctorates awarded. Terminal master's awarded for partial completion of doctoral program. *Degree requirements:* For master's and doctorate, one foreign language, thesis/dissertation. *Entrance requirements:* For master's, GRE General Test, GRE Subject Test; for doctorate, GRE Subject Test. *Application deadline:* For fall admission, 3/1 for domestic students. Applications are processed on a rolling basis. Application fee: $0. *Expenses:* Tuition: Full-time $24,732. One-time fee: $20 full-time. *Financial support:* Fellowships, research assistantships, teaching assistantships, institutionally sponsored loans available. *Unit head:* Dr. Phillip Bolton, Chair, 860-685-2668. *Application contact:* Karen Karpa, Information Contact, 860-685-2573, Fax: 860-685-2211, E-mail: kkarpa@wesleyan.edu.

See Close-Up on page 177.

West Virginia University, Eberly College of Arts and Sciences, Department of Chemistry, Morgantown, WV 26506. Offers analytical chemistry (MS, PhD); inorganic chemistry (MS, PhD); organic chemistry (MS, PhD); physical chemistry (MS, PhD); theoretical chemistry (MS, PhD). Part-time programs available. Postbaccalaureate distance learning degree programs offered (no on-campus study). *Faculty:* 14 full-time (1 woman), 5 part-time/adjunct (3 women). *Students:* 45 full-time (15 women), 3 part-time (2 women); includes 1 minority (Hispanic American), 26 international. Average age 28. 84 applicants, 58% accepted, 11 enrolled. In 2005, 5 master's, 3 doctorates awarded. Terminal master's awarded for partial completion of doctoral program. *Degree requirements:* For master's and doctorate, thesis/dissertation, registration. *Entrance requirements:* For master's, GRE General Test, GRE Subject Test (recommended), minimum GPA of 2.5; for doctorate, GRE General Test, GRE Subject Test (recommended), minimum GPA of 2.75. Additional exam requirements/recommendations for international students: Required—TOEFL. *Application deadline:* For fall admission, 3/1 for domestic students. Applications are processed on a rolling basis. Application fee: $50. Electronic applications accepted. *Expenses:* Tuition, area resident: Full-time $4,582; part-time $258 per credit hour. Tuition, state resident: full-time $4,582; part-time $258 per credit hour. Tuition, nonresident: full-time $1,382; part-time $741 per credit hour. International tuition: $1,382 full-time. *Financial support:* In 2005–06, fellowships (averaging $2,000 per year), research assistantships with full tuition reimbursements (averaging $13,000 per year), teaching assistantships with full tuition reimbursements (averaging $12,000 per year) were awarded; tuition waivers (full and partial) also available. Financial award application deadline: 2/1; financial award applicants required to submit FAFSA. *Faculty research:* Analysis of proteins, drug interactions, solids and effluents by advanced separation methods; new synthetic strategies for complex organic molecules; synthesis and structural characterization of metal complexes for polymerization catalysis. *Unit head:* Dr. Harry O. Finklea, Chair, 304-293-3435 Ext. 4453, Fax: 304-293-4904, E-mail: harry.finklea@mail.wvu.edu.

Yale University, Graduate School of Arts and Sciences, Department of Chemistry, New Haven, CT 06520. Offers biophysical chemistry (PhD); inorganic chemistry (PhD); organic chemistry (PhD); physical chemistry (PhD). *Degree requirements:* For doctorate, thesis/dissertation. *Entrance requirements:* For doctorate, GRE General Test, GRE Subject Test. Additional exam requirements/recommendations for international students: Required—TOEFL.

See Close-Up on page 183.

Physical Chemistry

Boston College, Graduate School of Arts and Sciences, Department of Chemistry, Chestnut Hill, MA 02467-3800. Offers biochemistry (PhD); inorganic chemistry (PhD); organic chemistry (PhD); physical chemistry (PhD); science education (MST). MST is offered through the School of Education for secondary school science teaching. Part-time programs available. *Students:* 118 full-time (45 women); includes 12 minority (2 African Americans, 8 Asian Americans or Pacific Islanders, 2 Hispanic Americans), 35 international. 212 applicants, 32% accepted, 17 enrolled. In 2005, 6 master's, 10 doctorates awarded. *Degree requirements:* For doctorate, thesis/dissertation, qualifying exam. *Entrance requirements:* For doctorate, GRE General Test, GRE Subject Test. Additional exam requirements/recommendations for international students: Required—TOEFL (minimum score 550 paper-based; 213 computer-based). *Application deadline:* For fall admission, 1/2 for domestic students. Application fee: $70. Electronic applications accepted. *Financial support:* Fellowships with full tuition reimbursements, research assistantships with full tuition reimbursements, teaching assistantships with full tuition reimbursements, Federal Work-Study available. Support available to part-time students. Financial award application deadline: 3/1; financial award applicants required to submit FAFSA. *Unit head:* Dr. David McFadden, Chairperson, 617-552-3605, E-mail: david.mcfadden@bc.edu. *Application contact:* Dr. Marc Snapper, Graduate Program Director, 617-552-8096, Fax: 617-552-0833, E-mail: marc.snapper@bc.edu.

Brandeis University, Graduate School of Arts and Sciences, Department of Chemistry, Waltham, MA 02454-9110. Offers inorganic chemistry (MS, PhD); organic chemistry (MS, PhD); physical chemistry (MS, PhD). *Faculty:* 20 full-time (3 women). *Students:* 42 full-time (20 women), 1 part-time; includes 2 minority (1 African American, 1 Hispanic American), 34 international. Average age 25. 87 applicants, 10% accepted. In 2005, 5 master's, 3 doctorates awarded. Terminal master's awarded for partial completion of doctoral program. *Degree requirements:* For master's, 1 year of residency; for doctorate, one foreign language, thesis/dissertation, 3 years of residency, 2 seminars, qualifying exams. *Entrance requirements:* For master's and doctorate, GRE General Test, resumé, letters of recommendation. Additional exam requirements/recommendations for international students: Required—TOEFL (minimum score 600 paper-based; 250 computer-based). *Application deadline:* For fall admission, 1/15 for domestic students. Applications are processed on a rolling basis. Application fee: $55. Electronic applications accepted. *Financial support:* In 2005–06, 25 fellowships (averaging $23,000 per year), 21 research assistantships (averaging $23,000 per year), 19 teaching assistantships (averaging $23,000 per year) were awarded; institutionally sponsored loans and scholarships/grants also available. Financial award application deadline: 4/15; financial award applicants required to submit CSS PROFILE or FAFSA. *Faculty research:* Oscillating chemical reactions, molecular recognition systems, protein crystallography, synthesis of natural product spectroscopy and magnetic resonance. Total annual research expenditures: $1,965. *Unit head:* Dr. Peter Jordan, Chair, 781-736-2540, Fax: 781-736-2516, E-mail: jordan@brandeis.edu. *Application contact:* Charlotte Haygazian, Graduate Admissions Secretary, 781-736-2500, Fax: 781-736-2516, E-mail: chemadm@brandeis.edu.

Brigham Young University, Graduate Studies, College of Physical and Mathematical Sciences, Department of Chemistry and Biochemistry, Provo, UT 84602-1001. Offers analytical chemistry (MS, PhD); biochemistry (MS, PhD); inorganic chemistry (MS, PhD); organic chemistry (MS, PhD); physical chemistry (MS, PhD). *Faculty:* 35 full-time (2 women). *Students:* 100 full-time (33 women); includes 2 minority (both Asian Americans or Pacific Islanders), 68 international. Average age 29. 85 applicants, 44% accepted, 21 enrolled. In 2005, 9 master's, 9 doctorates awarded. *Median time to degree:* Of those who began their doctoral program in fall 1997, 87% received their degree in 8 years or less. *Degree requirements:* For master's, thesis, registration; for doctorate, thesis/dissertation, degree qualifying exam. *Entrance requirements:* For master's, pass 1 (biochemistry), 4 (chemistry) of 5 area exams, GRE General Test, minimum GPA of 3.0 in last 60 hours; for doctorate, pass 1 (biochemistry) or 4 (chemistry) of 5 area exams, GRE General Test, minimum GPA of 3.0 in last 60 hours. Additional exam requirements/recommendations for international students: Required—TOEFL, TWE. *Application deadline:* For fall admission, 2/1 priority date for domestic students, 2/1 priority date for international students. Applications are processed on a rolling basis. Application fee: $50. Electronic applications accepted. *Financial support:* In 2005–06, 100 students received support, including 12 fellowships with full tuition reimbursements available (averaging $20,500 per year), 47 research assistantships with full tuition reimbursements available (averaging $20,500 per year), 39 teaching assistantships with full tuition reimbursements available (averaging $20,400 per year); institutionally sponsored loans, scholarships/grants, health care benefits, tuition waivers (full), and unspecified assistantships also available. Financial award application deadline: 2/1. *Faculty research:* Separation science, molecular recognition, organic synthesis and biomedical application, biochemistry and molecular biology, molecular spectroscopy. Total annual research expenditures: $3.9 million. *Unit head:* Dr. Paul B. Farnsworth, Chair, 801-422-6502, Fax: 801-422-0153, E-mail: paul_farnsworth@byu.edu. *Application contact:* Dr. David V. Dearden, Graduate Coordinator, 801-422-2355, Fax: 801-422-0153, E-mail: david_dearden@byu.edu.

See Close-Up on page 97.

California State University, Fullerton, Graduate Studies, College of Natural Science and Mathematics, Department of Chemistry and Biochemistry, Fullerton, CA 92834-9480. Offers analytical chemistry (MS); biochemistry (MS); geochemistry (MS); inorganic chemistry (MS); organic chemistry (MS); physical chemistry (MS). Part-time programs available. *Students:* 19 full-time (6 women), 20 part-time (9 women); includes 20 minority (1 African American, 13 Asian Americans or Pacific Islanders, 6 Hispanic Americans), 8 international. Average age 28. 36 applicants, 47% accepted, 4 enrolled. In 2005, 11 degrees awarded. *Degree requirements:* For master's, thesis, departmental qualifying exam. *Entrance requirements:* For master's, minimum GPA of 2.5 in last 60 units of course work, major in chemistry or related field. Application fee: $55. *Expenses:* Tuition, area resident: Part-time $2,270 per year. Tuition, state resident: full-time $2,572; part-time $339 per unit. Tuition, nonresident: full-time $339; part-time $339 per unit. International tuition: $339 full-time. *Financial support:* Teaching assistantships, career-related internships or fieldwork, Federal Work-Study, institutionally sponsored loans, and scholarships/grants available. Support available to part-time students. Financial award application deadline: 3/1. *Unit head:* Dr. Maria Linder, Chair, 714-278-3621. *Application contact:* Dr. Gregory Williams, Adviser, 714-278-2170.

California State University, Los Angeles, Graduate Studies, College of Natural and Social Sciences, Department of Chemistry and Biochemistry, Los Angeles, CA 90032-8530. Offers analytical chemistry (MS); biochemistry (MS); chemistry (MS); inorganic chemistry (MS); organic chemistry (MS); physical chemistry (MS). Part-time and evening/weekend programs available. *Faculty:* 2 full-time (0 women). *Students:* 33 full-time (21 women), 21 part-time (10 women); includes 38 minority (3 African Americans, 21 Asian Americans or Pacific Islanders, 14 Hispanic Americans). In 2005, 5 degrees awarded. *Degree requirements:* For master's, one foreign language. *Entrance requirements:* Additional exam requirements/recommendations for international students: Required—TOEFL. *Application deadline:* For fall admission, 6/30 for domestic students; for spring admission, 2/1 for domestic students. Applications are processed on a rolling basis. Application fee: $55. *Expenses:* Tuition, area resident: Full-time $3,617. Tuition, state resident: full-time $3,617. Tuition, nonresident: full-time $9,719. International tuition: $9,719 full-time. *Financial support:* Federal Work-Study available. Support available to part-time students. Financial award application deadline: 3/1. *Faculty research:* Intercalation of heavy metal, carborane chemistry, conductive polymers and fabrics, titanium reagents, computer modeling and synthesis. *Unit head:* Dr. Wayne Tikkanen, Chair, 323-343-2300, Fax: 323-343-6490.

Case Western Reserve University, School of Graduate Studies, Department of Chemistry, Cleveland, OH 44106. Offers analytical chemistry (MS, PhD); inorganic chemistry (MS, PhD); organic chemistry (MS, PhD); physical chemistry (MS, PhD). Part-time programs available. Terminal master's awarded for partial completion of doctoral program. *Degree requirements:* For doctorate, thesis/dissertation. *Entrance requirements:* For master's and doctorate, GRE General Test, GRE Subject Test. Additional exam requirements/recommendations for international students: Required—TOEFL. *Faculty research:* Electrochemistry, synthetic chemistry, chemistry of life process, spectroscopy, kinetics.

Clark Atlanta University, School of Arts and Sciences, Department of Chemistry, Atlanta, GA 30314. Offers inorganic chemistry (MS, PhD); organic chemistry (MS, PhD); physical chemistry (MS, PhD); science education (DA). Part-time programs available. *Degree requirements:* For master's, one foreign language, thesis, comprehensive exam; for doctorate, 2 foreign languages, thesis/dissertation, cumulative exam. *Entrance requirements:* For master's, GRE General Test, minimum GPA of 2.5; for doctorate, GRE General Test, GRE Subject Test, minimum graduate GPA of 3.0.

Clarkson University, Graduate School, School of Arts and Sciences, Department of Chemistry, Potsdam, NY 13699. Offers analytical chemistry (MS, PhD); inorganic chemistry (MS, PhD); organic chemistry (MS, PhD); physical chemistry (MS, PhD). *Faculty:* 9 full-time (2 women). *Students:* 34 full-time (9 women), 1 (woman) part-time; includes 2 Asian Americans or Pacific Islanders, 22 international. Average age 27. 52 applicants, 62% accepted. In 2005, 3 master's, 2 doctorates awarded. *Median time to degree:* Of those who began their doctoral program in fall 1997, 100% received their degree in 8 years or less. *Degree requirements:* For doctorate, thesis/dissertation, departmental qualifying exam. *Entrance requirements:* For master's, GRE. Additional exam requirements/recommendations for international students: Required—TOEFL. *Application deadline:* For fall admission, 5/15 for domestic students; for spring admission, 10/15 priority date for domestic students. Applications are processed on a rolling basis. Application fee: $25 ($35 for international students). Electronic applications accepted. *Expenses:* Tuition: Full-time $20,160; part-time $840 per hour. Required fees: $215. *Financial support:* In 2005–06, 25 students received support, including 2 fellowships (averaging $25,000 per year), 11 research assistantships (averaging $19,032 per year), 12 teaching assistantships (averaging $19,032 per year); scholarships/grants and tuition waivers (partial) also available. *Faculty research:* Nanomaterial, surface science, polymers chemical biosensing, protein biochemistry drug design/delivery, molecular neuroscience and immunobiology. Total annual research expenditures: $1.6 million. *Unit head:* Dr. Phillip A. Christiansen, Division Head, 315-268-6669, Fax: 315-268-6610, E-mail: pac@clarkson.edu. *Application contact:* Donna Brockway, Graduate Admissions International Advisor/Assistant to the Provost, 315-268-6447, Fax: 315-268-7994, E-mail: brockway@clarkson.edu.

Cleveland State University, College of Graduate Studies, College of Science, Department of Chemistry, Cleveland, OH 44115. Offers analytical chemistry (MS); clinical chemistry (MS, PhD); clinical/bioanalytical chemistry (PhD); environmental chemistry (MS); inorganic chemistry (MS); organic chemistry (MS); physical chemistry (MS). Part-time and evening/weekend programs available. *Faculty:* 13 full-time (1 woman), 59 part-time/adjunct (12 women). *Students:* 46 full-time (22 women), 18 part-time (6 women); includes 4 minority (2 African Americans, 2 Asian Americans or Pacific Islanders), 35 international. Average age 31. 47 applicants, 62% accepted, 7 enrolled. In 2005, 2 master's, 5 doctorates awarded. *Median time to degree:* Of those who began their doctoral program in fall 1997, 67% received their degree in 8 years or less. *Degree requirements:* For master's, thesis (for some programs); for doctorate, thesis/dissertation. *Entrance requirements:* For master's and doctorate, GRE General Test, GRE Subject Test. Additional exam requirements/recommendations for international students: Required—TOEFL (minimum score 525 paper-based; 197 computer-based). *Application deadline:* For fall admission, 1/15 priority date for domestic students, 1/15 priority date for international students. Applications are processed on a rolling basis. Application fee: $30. Electronic applications accepted. *Expenses:* Tuition, state resident: full-time $10,700. Tuition, nonresident: full-time $14,628. Tuition and fees vary according to program. *Financial support:* In 2005–06, 44 students received support, including fellowships with full tuition reimbursements available (averaging $18,000 per year), research assistantships with full tuition reimbursements available (averaging $16,000 per year), teaching assistantships with full tuition reimbursements available (averaging $14,000 per year) Financial award application deadline: 1/15. *Faculty research:* Metalloenzyme mechanisms, dependent RNAse L and interferons, application of HPLC/LPCC to clinical systems, structure-function relationships of factor Va, MALDI-TOF based DNA sequencing. Total annual research expenditures: $3 million. *Unit head:* Dr. Lily M. Ng, Chair, 216-687-2467, E-mail: l.ng@csuohio.edu. *Application contact:* Richelle P. Emery, Administrative Coordinator, 216-687-2457, Fax: 216-687-9298, E-mail: r.emery@csuohio.edu.

Cornell University, Graduate School, Graduate Fields of Arts and Sciences, Field of Chemistry and Chemical Biology, Ithaca, NY 14853-0001. Offers analytical chemistry (PhD); bio-organic chemistry (PhD); biophysical chemistry (PhD); chemical biology (PhD); chemical physics (PhD); inorganic chemistry (PhD); materials chemistry (PhD); organic chemistry (PhD); organometallic chemistry (PhD); physical chemistry (PhD); polymer chemistry (PhD); theoretical chemistry (PhD). *Faculty:* 44 full-time (2 women). *Students:* 190 full-time (73 women); includes 23 minority (4 African Americans, 8 Asian Americans or Pacific Islanders, 11 Hispanic Americans), 65 international. 339 applicants, 35% accepted, 49 enrolled. In 2005, 23 doctorates awarded. *Degree requirements:* For doctorate, thesis/dissertation, comprehensive exam. *Entrance requirements:* For doctorate, GRE General Test, GRE Subject Test (chemistry), 3 letters of recommendation. Additional exam requirements/recommendations for international students: Required—TOEFL (minimum score 600 paper-based; 250 computer-based). *Application deadline:* For fall admission, 1/10 for domestic students. Application fee: $60. Electronic applications accepted. *Financial support:* In 2005–06, 185 students received support, including 33 fellowships with full tuition reimbursements available, 85 research assistantships with full tuition reimbursements available, 67 teaching assistantships with full tuition reimbursements available; institutionally sponsored loans, scholarships/grants, health care benefits, tuition waivers (full and partial), and unspecified assistantships also available. Financial award applicants required to submit FAFSA. *Faculty research:* Analytical, organic, inorganic, physical, materials, chemical biology. *Unit head:* Director of Graduate Studies, 607-255-4139, Fax: 607-255-4137. *Application contact:* Graduate Field Assistant, 607-255-4139, Fax: 607-255-4137, E-mail: chemgrad@cornell.edu.

See Close-Up on page 109.

Florida State University, Graduate Studies, College of Arts and Sciences, Department of Chemistry and Biochemistry, Tallahassee, FL 32306. Offers analytical chemistry (MS, PhD); biochemistry (MS, PhD); chemical physics (MS, PhD); inorganic chemistry (MS, PhD); organic chemistry (MS, PhD); physical chemistry (MS, PhD). Part-time programs available. *Faculty:* 36 full-time (6 women), 2 part-time/adjunct (0 women). *Students:* 144 full-time (50 women); includes 17 minority (7 African Americans, 1 American Indian/Alaska Native, 6 Asian Americans or Pacific Islanders, 3 Hispanic Americans), 58 international. Average age 25. 288 applicants, 27% accepted, 26 enrolled. In 2005, 14 master's, 15 doctorates awarded. Terminal master's awarded for partial completion of doctoral program. *Median time to degree:* Of those who began their doctoral program in fall 1997, 88% received their degree in 8 years or less. *Degree requirements:* For master's, thesis (for some programs), cumulative and diagnostic exams, comprehensive exam (for some programs), registration; for doctorate, thesis/dissertation, cumulative and diagnostic exams, comprehensive exam (for some programs), registration. *Entrance requirements:* For master's and doctorate, GRE General Test, minimum B average in undergraduate course work. Additional exam requirements/recommendations for international students: Required—TOEFL (minimum score 515 paper-based; 213 computer-based). *Application deadline:* For fall admission, 4/15 priority date for domestic students, 4/15 priority date for international students. Applications are processed on a rolling basis. Application fee: $30. Electronic applications accepted. *Financial support:* In 2005–06, 1 fellowship with tuition reimbursement (averaging $18,000 per year), 56 research assistantships with tuition reimburse-

Physical Chemistry

Florida State University (continued)

ments (averaging $19,000 per year), 83 teaching assistantships with tuition reimbursements (averaging $19,000 per year) were awarded; career-related internships or fieldwork, Federal Work-Study, institutionally sponsored loans, and traineeships also available. Financial award application deadline: 2/15; financial award applicants required to submit FAFSA. *Faculty research:* Spectroscopy, computational chemistry, nuclear chemistry, separations, synthesis. Total annual research expenditures: $6.5 million. *Unit head:* Dr. Naresh Dalal, Chairman, 850-644-3398, Fax: 850-644-8281. *Application contact:* Dr. Oliver Steinbock, Chair, Graduate Admissions Committee, 888-525-9286, Fax: 850-644-8281, E-mail: gradinfo@chem.fsu.edu.

See Close-Up on page 117.

Georgetown University, Graduate School of Arts and Sciences, Department of Chemistry, Washington, DC 20057. Offers analytical chemistry (MS, PhD); biochemistry (MS, PhD); chemical physics (MS, PhD); inorganic chemistry (MS, PhD); organic chemistry (MS, PhD); physical chemistry (MS, PhD); theoretical chemistry (MS, PhD). Terminal master's awarded for partial completion of doctoral program. *Degree requirements:* For master's, thesis (for some programs), qualifying exam; for doctorate, thesis/dissertation, comprehensive exam. *Entrance requirements:* For master's and doctorate, GRE General Test. Additional exam requirements/recommendations for international students: Required—TOEFL.

The George Washington University, Columbian College of Arts and Sciences, Department of Chemistry, Washington, DC 20052. Offers analytical chemistry (MS, PhD); inorganic chemistry (MS, PhD); materials science (MS, PhD); organic chemistry (MS, PhD); physical chemistry (MS, PhD). Part-time and evening/weekend programs available. Terminal master's awarded for partial completion of doctoral program. *Degree requirements:* For master's, thesis or alternative, comprehensive exam; for doctorate, thesis/dissertation, general exam. *Entrance requirements:* For master's and doctorate, GRE General Test, interview, minimum GPA of 3.0. Additional exam requirements/recommendations for international students: Required—TOEFL (minimum score 550 paper-based; 213 computer-based). Electronic applications accepted.

Harvard University, Graduate School of Arts and Sciences, Department of Chemistry and Chemical Biology, Cambridge, MA 02138. Offers biochemical chemistry (PhD); inorganic chemistry (PhD); organic chemistry (PhD); physical chemistry (PhD). *Students:* 188 full-time (36 women). 346 applicants, 24% accepted. In 2005, 33 doctorates awarded. *Degree requirements:* For doctorate, thesis/dissertation, cumulative exams. *Entrance requirements:* For doctorate, GRE General Test, GRE Subject Test. Additional exam requirements/recommendations for international students: Required—TOEFL. *Application deadline:* For fall admission, 12/31 for domestic students. Application fee: $60. *Expenses:* Tuition: Full-time $28,752. Full-time tuition and fees vary according to program and student level. *Financial support:* Fellowships, research assistantships, teaching assistantships, career-related internships or fieldwork, Federal Work-Study, and institutionally sponsored loans available. Financial award application deadline: 12/30. *Unit head:* Betsey Cogswell, Administrator, 617-495-5696, Fax: 617-495-5264. *Application contact:* Graduate Admissions Office, 617-496-3208.

See Close-Up on page 119.

Howard University, Graduate School of Arts and Sciences, Department of Chemistry, Washington, DC 20059-0002. Offers analytical chemistry (MS, PhD); atmospheric (MS, PhD); biochemistry (MS, PhD); environmental (MS, PhD); inorganic chemistry (MS, PhD); organic chemistry (MS, PhD); physical chemistry (MS, PhD); polymer chemistry (MS, PhD). Part-time programs available. *Degree requirements:* For master's, one foreign language, thesis, teaching experience, comprehensive exam, registration; for doctorate, 2 foreign languages, thesis/dissertation, teaching experience, comprehensive exam, registration. *Entrance requirements:* For master's, GRE General Test, minimum GPA of 2.7; for doctorate, GRE General Test, minimum GPA of 3.0. *Faculty research:* Stratospheric aerosols, liquid crystals, polymer coatings, terrestrial and extraterrestrial atmospheres, amidogen reaction.

Illinois Institute of Technology, Graduate College, College of Science and Letters, Department of Biological, Chemical and Physical Sciences, Chemistry Division, Chicago, IL 60616-3793. Offers analytical chemistry (M Ch, MS, PhD); chemistry (M Chem); inorganic chemistry (MS, PhD); materials and chemical synthesis (M Ch); organic chemistry (MS, PhD); physical chemistry (MS, PhD); polymer chemistry (MS, PhD). Part-time and evening/weekend programs available. Postbaccalaureate distance learning degree programs offered (no on-campus study). Terminal master's awarded for partial completion of doctoral program. *Degree requirements:* For master's, thesis (for some programs), comprehensive exam; for doctorate, thesis/dissertation, comprehensive exam. *Entrance requirements:* For master's and doctorate, GRE General Test, minimum undergraduate GPA of 3.0. Additional exam requirements/recommendations for international students: Required—TOEFL (minimum score 550 paper-based; 213 computer-based). Electronic applications accepted. *Faculty research:* Organic and inorganic chemistry, polymers research, physical chemistry, analytical chemistry.

Indiana University Bloomington, Graduate School, College of Arts and Sciences, Department of Chemistry, Bloomington, IN 47405-7000. Offers analytical chemistry (PhD); biological chemistry (PhD); chemistry (MAT); inorganic chemistry (PhD); physical chemistry (PhD). PhD offered through the University Graduate School. *Faculty:* 29 full-time (2 women). *Students:* 110 full-time (38 women), 35 part-time (11 women); includes 8 minority (3 African Americans, 3 Asian Americans or Pacific Islanders, 2 Hispanic Americans), 57 international. Average age 26. In 2005, 7 master's, 20 doctorates awarded. Terminal master's awarded for partial completion of doctoral program. *Degree requirements:* For master's and doctorate, thesis/dissertation. *Entrance requirements:* For master's and doctorate, GRE General Test, GRE Subject Test. Additional exam requirements/recommendations for international students: Required—TOEFL. *Application deadline:* For fall admission, 1/15 priority date for domestic students, 12/15 priority date for international students; for spring admission, 9/1 priority date for domestic students, 9/1 priority date for international students. Applications are processed on a rolling basis. Application fee: $50 ($60 for international students). *Expenses:* Tuition, state resident: full-time $5,437; part-time $227 per credit hour. Tuition, nonresident: full-time $15,836; part-time $660 per credit hour. Required fees: $821. Tuition and fees vary according to campus/location and program. *Financial support:* In 2005–06, 23 fellowships with full tuition reimbursements, 57 research assistantships with full tuition reimbursements, 78 teaching assistantships with full tuition reimbursements were awarded; Federal Work-Study and institutionally sponsored loans also available. *Faculty research:* Synthesis of complex natural products, organic reaction mechanisms, organic electrochemistry, transitive-metal chemistry, solid-state and surface chemistry. Total annual research expenditures: $7.7 million. *Unit head:* Dr. David Clemmer, Chairperson, 812-855-2268. *Application contact:* Dr. Jack K. Crandall, Chairperson of Admissions, 812-855-2068, Fax: 812-855-8300, E-mail: chemgrad@indiana.edu.

Kansas State University, Graduate School, College of Arts and Sciences, Department of Chemistry, Manhattan, KS 66506. Offers analytical chemistry (MS); biological chemistry (MS); chemistry (PhD); inorganic chemistry (MS); materials chemistry (MS); organic chemistry (MS); physical chemistry (MS). *Faculty:* 14 full-time (1 woman). *Students:* 65 full-time (21 women), 5 part-time (2 women); includes 1 minority (African American), 48 international. 63 applicants, 67% accepted, 22 enrolled. In 2005, 3 master's, 5 doctorates awarded. Terminal master's awarded for partial completion of doctoral program. *Degree requirements:* For master's and doctorate, thesis/dissertation. *Entrance requirements:* For master's and doctorate, GRE, minimum GPA of 3.0. Additional exam requirements/recommendations for international students: Required—TOEFL (minimum score 550 paper-based; 213 computer-based). *Application deadline:* For fall admission, 2/1 priority date for domestic students, 2/1 priority date for international students; for spring admission, 10/1 for domestic students, 8/1 for international students. Applications are processed on a rolling basis. Application fee: $30 ($55 for international students). *Expenses:* Tuition, state resident: full-time $5,160; part-time $215. Tuition, nonresident: full-time $12,816; part-time $534. International tuition: $12,816 full-time. Required fees: $564. *Financial support:* In 2005–06, 30 research assistantships (averaging $11,956 per year), 26 teaching assistantships with full tuition reimbursements (averaging $11,945 per year) were awarded; fellowships, institutionally sponsored loans and scholarships/grants also available. Support available to part-time students. Financial award application deadline: 3/1; financial award applicants required to submit FAFSA. *Faculty research:* Nanotechnologies, functional materials, bio-organic and bio-physical processes, sensors and separations, synthesis and synthetic methods. Total annual research expenditures: $2.2 million. *Unit head:* Eric Maatta, Head, 785-532-6665, Fax: 785-532-6666, E-mail: eam@ksu.edu. *Application contact:* Christer Aakeröy, Director, 785-532-6096, Fax: 785-532-6666, E-mail: aakeroy@ksu.edu.

Kent State University, College of Arts and Sciences, Department of Chemistry, Kent, OH 44242-0001. Offers analytical chemistry (MS, PhD); biochemistry (MS, PhD); chemistry (MA, MS, PhD); inorganic chemistry (MS, PhD); organic chemistry (MS, PhD); physical chemistry (MS, PhD). Terminal master's awarded for partial completion of doctoral program. *Degree requirements:* For master's and doctorate, thesis/dissertation, comprehensive exam, registration. *Entrance requirements:* For master's and doctorate, placement exam, GRE General Test, GRE Subject Test (recommended), minimum GPA of 2.75. Additional exam requirements/recommendations for international students: Required—TOEFL (minimum score 575 paper-based; 230 computer-based). Electronic applications accepted. *Faculty research:* Biological chemistry, materials chemistry, molecular spectroscopy.

See Close-Up on page 121.

Marquette University, Graduate School, College of Arts and Sciences, Department of Chemistry, Milwaukee, WI 53201-1881. Offers analytical chemistry (MS, PhD); bioanalytical chemistry (MS, PhD); biophysical chemistry (MS, PhD); chemical physics (MS, PhD); inorganic chemistry (MS, PhD); organic chemistry (MS, PhD); physical chemistry (MS, PhD). Part-time programs available. Terminal master's awarded for partial completion of doctoral program. *Degree requirements:* For master's, comprehensive exam; for doctorate, thesis/dissertation, cumulative exams. *Entrance requirements:* For master's and doctorate, GRE Subject Test. Additional exam requirements/recommendations for international students: Required—TOEFL. *Faculty research:* Inorganic complexes, laser Raman spectroscopy, organic synthesis, chemical dynamics, biophysiology.

Massachusetts Institute of Technology, School of Science, Department of Chemistry, Cambridge, MA 02139-4307. Offers biological chemistry (PhD, Sc D); inorganic chemistry (PhD, Sc D); organic chemistry (PhD, Sc D); physical chemistry (PhD, Sc D). *Faculty:* 30 full-time (6 women). *Students:* 242 full-time (84 women), 3 part-time (2 women); includes 30 minority (5 African Americans, 1 American Indian/Alaska Native, 19 Asian Americans or Pacific Islanders, 5 Hispanic Americans), 77 international. Average age 26. 476 applicants, 27% accepted, 54 enrolled. In 2005, 48 doctorates awarded. *Degree requirements:* For doctorate, thesis/dissertation, comprehensive exam. *Entrance requirements:* For doctorate, GRE General Test. Additional exam requirements/recommendations for international students: Required—TOEFL (minimum score 577 paper-based; 233 computer-based). *Application deadline:* For fall admission, 12/15 for domestic students, 12/15 for international students. Application fee: $70. Electronic applications accepted. *Expenses:* Tuition: Full-time $32,100. Required fees: $200. Part-time tuition and fees vary according to course load. *Financial support:* In 2005–06, 210 students received support, including 36 fellowships with tuition reimbursements available (averaging $29,141 per year), 143 research assistantships with tuition reimbursements available (averaging $24,550 per year), 59 teaching assistantships with tuition reimbursements available (averaging $25,155 per year); career-related internships or fieldwork, Federal Work-Study, institutionally sponsored loans, scholarships/grants, health care benefits, and unspecified assistantships also available. *Faculty research:* Synthetic organic chemistry, enzymatic reaction mechanisms, inorganic and organometallic spectroscopy, high resolution NMR spectroscopy. Total annual research expenditures: $21.5 million. *Unit head:* Prof. Timothy Swager, Department Head, 617-253-1801, E-mail: tswager@mit.edu. *Application contact:* Susan Brighton, Graduate Administrator, 617-253-1845, Fax: 617-258-0241, E-mail: brighton@mit.edu.

McMaster University, School of Graduate Studies, Faculty of Science, Department of Chemistry, Hamilton, ON L8S 4M2, Canada. Offers analytical chemistry (M Sc, PhD); chemical physics (M Sc, PhD); chemistry (M Sc, PhD); inorganic chemistry (M Sc, PhD); organic chemistry (M Sc, PhD); physical chemistry (M Sc, PhD); polymer chemistry (M Sc, PhD). Part-time programs available. Terminal master's awarded for partial completion of doctoral program. *Degree requirements:* For master's, thesis/dissertation; for doctorate, thesis/dissertation, comprehensive exam. *Entrance requirements:* For master's, minimum B+ average. Additional exam requirements/recommendations for international students: Required—TOEFL (minimum score 550 paper-based; 213 computer-based).

Miami University, Graduate School, College of Arts and Sciences, Department of Chemistry and Biochemistry, Oxford, OH 45056. Offers analytical chemistry (MS, PhD); biochemistry (MS, PhD); chemical education (MS, PhD); chemistry (MS, PhD); inorganic chemistry (MS, PhD); organic chemistry (MS, PhD); physical chemistry (MS, PhD). Part-time programs available. *Degree requirements:* For master's, thesis, final exam; for doctorate, thesis/dissertation, final exams, comprehensive exam. *Entrance requirements:* For master's, minimum undergraduate GPA of 3.0 during previous 2 years or 2.75 overall; for doctorate, minimum undergraduate GPA of 2.75, 3.0 graduate. Additional exam requirements/recommendations for international students: Required—TOEFL (minimum score 550 paper-based; 213 computer-based), TWE(minimum score 4). Electronic applications accepted.

Northeastern University, College of Arts and Sciences, Department of Chemistry and Chemical Biology, Boston, MA 02115-5096. Offers analytical chemistry (PhD); chemistry (MS, PhD); inorganic chemistry (PhD); organic chemistry (PhD); physical chemistry (PhD). Part-time and evening/weekend programs available. Terminal master's awarded for partial completion of doctoral program. *Degree requirements:* For master's, thesis (for some programs); for doctorate, thesis/dissertation, qualifying exam in specialty area. *Entrance requirements:* Additional exam requirements/recommendations for international students: Required—TOEFL. Electronic applications accepted. *Faculty research:* Electron transfer, theoretical chemical physics, analytical biotechnology, mass spectrometry, materials chemistry.

See Close-Up on page 129.

Old Dominion University, College of Sciences, Program in Chemistry, Norfolk, VA 23529. Offers analytical chemistry (MS); biochemistry (MS); clinical chemistry (MS); environmental chemistry (MS); organic chemistry (MS); physical chemistry (MS). Part-time and evening/weekend programs available. *Faculty:* 15 full-time (4 women). *Students:* 5 full-time (3 women), 8 part-time (3 women); includes 8 minority (4 African Americans, 1 American Indian/Alaska Native, 1 Asian American or Pacific Islander, 2 Hispanic Americans). Average age 27. 19 applicants, 74% accepted. In 2005, 1 degree awarded. *Degree requirements:* For master's, thesis, comprehensive exam. *Entrance requirements:* For master's, GRE General Test, minimum GPA of 3.0 in major, 2.5 overall. Additional exam requirements/recommendations for international students: Required—TOEFL. *Application deadline:* For fall admission, 7/1 for domestic students; for spring admission, 11/1 for domestic students. Applications are processed on a rolling basis. Application fee: $30. *Expenses:* Tuition, state resident: part-time $263 per credit hour. Tuition, nonresident: part-time $661 per credit hour. Required fees: $39 per semester. Part-time tuition and fees vary according to campus/location. *Financial support:* In 2005–06, research assistantships with tuition reimbursements (averaging $15,000 per year), teaching assistantships with tuition reimbursements (averaging $15,000 per year) were awarded; fellowships, career-related internships or fieldwork and scholarships/grants also available. Financial award application deadline: 2/15; financial award applicants required to submit FAFSA. *Faculty research:* Organic and trace metal biogeochemistry, materials chemistry, bioanalytical chemistry, computational chemistry. Total annual research expenditures: $854,968. *Unit head:* Dr. John R. Donat, Graduate Program Director, 757-683-4098, Fax: 757-683-4628, E-mail: chemgpd@odu.edu.

Oregon State University, Graduate School, College of Science, Department of Chemistry, Corvallis, OR 97331. Offers analytical chemistry (MS, PhD); chemistry (MA, MAIS); inorganic chemistry (MS, PhD); nuclear and radiation chemistry (MS, PhD); organic chemistry (MS, PhD); physical chemistry (MS, PhD). Part-time programs available. *Faculty:* 19 full-time (3 women), 6 part-time/adjunct (1 woman). *Students:* 74 full-time (30 women), 3 part-time; includes 6 minority (1 African American, 1 Asian American or Pacific Islander, 4 Hispanic Americans), 35 international. Average age 28. In 2005, 4 master's, 10 doctorates awarded. Terminal master's awarded for partial completion of doctoral program. *Degree requirements:* For master's and doctorate, one foreign language, thesis/dissertation. *Entrance requirements:* For master's and doctorate, minimum GPA of 3.0 in last 90 hours of course work. Additional exam requirements/recommendations for international students: Required—TOEFL. *Application deadline:* For fall admission, 3/1 for domestic students. Applications are processed on a rolling basis. Application fee: $50. *Expenses:* Tuition, area resident: Full-time $8,139; part-time $301 per credit. Tuition, state resident: full-time $8,139; part-time $501 per credit. Tuition, nonresident: full-time $14,376; part-time $532 per credit. International tuition: $14,376 full-time. Required fees: $1,266. *Financial support:* Fellowships, research assistantships, teaching assistantships, institutionally sponsored loans available. Support available to part-time students. Financial award application deadline: 2/1. *Faculty research:* Solid state chemistry, enzyme reaction mechanisms, structure and dynamics of gas molecules, chemiluminescence, nonlinear optical spectroscopy. *Unit head:* Dr. Douglas Keszler, Chair, 541-737-2081, Fax: 541-737-2062. *Application contact:* Carolyn Brumley, Graduate Secretary, 541-737-6707, Fax: 541-737-2062, E-mail: carolyn.brumley@orst.edu.

Purdue University, Graduate School, School of Science, Department of Chemistry, West Lafayette, IN 47907. Offers analytical chemistry (MS, PhD); biochemistry (MS, PhD); chemical education (MS, PhD); inorganic chemistry (MS, PhD); organic chemistry (MS, PhD); physical chemistry (MS, PhD). *Accreditation:* NCATE (one or more programs are accredited). *Faculty:* 43 full-time (9 women), 8 part-time/adjunct (2 women). *Students:* 293 full-time (114 women), 39 part-time (14 women); includes 53 minority (25 African Americans, 6 Asian Americans or Pacific Islanders, 22 Hispanic Americans), 138 international. Average age 28. 458 applicants, 36% accepted, 57 enrolled. In 2005, 9 master's, 36 doctorates awarded. Terminal master's awarded for partial completion of doctoral program. *Degree requirements:* For master's and doctorate, thesis/dissertation. *Entrance requirements:* Additional exam requirements/recommendations for international students: Required—TOEFL. *Application deadline:* For fall admission, 4/1 priority date for domestic students, 3/1 priority date for international students; for spring admission, 10/1 priority date for domestic students, 9/1 priority date for international students. Applications are processed on a rolling basis. Application fee: $55. Electronic applications accepted. *Financial support:* In 2005–06, 2 fellowships with partial tuition reimbursements (averaging $18,000 per year), 55 teaching assistantships with partial tuition reimbursements (averaging $18,000 per year) were awarded; research assistantships with partial tuition reimbursements, tuition waivers (partial) also available. Support available to part-time students. Financial award applicants required to submit FAFSA. *Unit head:* Dr. Timothy S Zwier, Head, 765-494-5203. *Application contact:* R. E. Wild, Chairman, Graduate Admissions, 765-494-5200, E-mail: wild@purdue.edu.

Rensselaer Polytechnic Institute, Graduate School, School of Science, Department of Chemistry and Chemical Biology, Troy, NY 12180-3590. Offers analytical chemistry (MS, PhD); biochemistry (MS, PhD); inorganic chemistry (MS, PhD); organic chemistry (MS, PhD); physical chemistry (MS, PhD); polymer chemistry (MS, PhD). Part-time and evening/weekend programs available. *Faculty:* 19 full-time (3 women). *Students:* 49 full-time (28 women), 2 part-time (1 woman), 33 international. Average age 24. 80 applicants, 23% accepted, 7 enrolled. In 2005, 1 master's, 14 doctorates awarded. Terminal master's awarded for partial completion of doctoral program. *Median time to degree:* Of those who began their doctoral program in fall 1997, 100% received their degree in 8 years or less. *Degree requirements:* For master's, thesis (for some programs), registration; for doctorate, thesis/dissertation, comprehensive exam, registration. *Entrance requirements:* For master's, GRE General Test, GRE Subject Test (strongly recommended); for doctorate, GRE General Test, GRE Subject Test (chemistry or biochemistry strongly recommended). Additional exam requirements/recommendations for international students: Required—TOEFL (minimum score 600 paper-based). *Application deadline:* For fall admission, 2/1 for domestic students; for spring admission, 11/15 for domestic students. Applications are processed on a rolling basis. Application fee: $75. Electronic applications accepted. *Expenses:* Tuition: Full-time $31,000; part-time $1,320 per credit. Required fees: $1,623. *Financial support:* In 2005–06, 49 students received support, including 1 fellowship with full tuition reimbursement available (averaging $30,000 per year), 25 research assistantships with full tuition reimbursements available (averaging $21,500 per year), 30 teaching assistantships with full tuition reimbursements available (averaging $21,500 per year); institutionally sponsored loans and tuition waivers (full and partial) also available. Financial award application deadline: 2/1. *Faculty research:* Synthetic polymer and biopolymer chemistry, physical chemistry of polymeric systems, bioanalytical chemistry, synthetic and computational drug design, protein folding and protein design. Total annual research expenditures: $1.9 million. *Unit head:* Dr. Linda B. McGown, Chair, 518-276-4856, Fax: 518-276-4887, E-mail: mcgowl@rpi.edu. *Application contact:* Beth E. McGraw, Department Admissions Assistant, 518-276-6456, Fax: 518-276-4887, E-mail: mcgrae@rpi.edu.

See Close-Up on page 137.

Rice University, Graduate Programs, Wiess School of Natural Sciences, Department of Chemistry, Houston, TX 77251-1892. Offers chemistry (MA); inorganic chemistry (PhD); organic chemistry (PhD); physical chemistry (PhD). Terminal master's awarded for partial completion of doctoral program. *Degree requirements:* For master's and doctorate, thesis/dissertation. *Entrance requirements:* For master's and doctorate, GRE General Test, minimum GPA of 3.0. Additional exam requirements/recommendations for international students: Required—TOEFL. *Faculty research:* Nanoscience, biomaterials, nanobioinformatics, fullerene pharmaceuticals.

Rutgers, The State University of New Jersey, Newark, Graduate School, Program in Chemistry, Newark, NJ 07102. Offers analytical chemistry (MS, PhD); biochemistry (MS, PhD); inorganic chemistry (MS, PhD); organic chemistry (MS, PhD); physical chemistry (MS, PhD). Part-time and evening/weekend programs available. *Faculty:* 22 full-time (5 women). *Students:* 24 full-time (12 women), 28 part-time (10 women); includes 31 minority (2 African Americans, 29 Asian Americans or Pacific Islanders). 69 applicants, 43% accepted, 13 enrolled. In 2005, 7 master's, 9 doctorates awarded. Terminal master's awarded for partial completion of doctoral program. *Degree requirements:* For master's, cumulative exams, thesis optional; for doctorate, thesis/dissertation, exams, research proposal. *Entrance requirements:* For master's and doctorate, GRE General Test, minimum undergraduate B average. Additional exam requirements/recommendations for international students: Required—TOEFL. *Application deadline:* For fall admission, 7/1 for domestic students; for spring admission, 12/1 for domestic students. Applications are processed on a rolling basis. Application fee: $50. Electronic applications accepted. *Expenses:* Tuition, state resident: full-time $10,440; part-time $435 per credit. Tuition, nonresident: full-time $15,520; part-time $637 per credit. *Financial support:* In 2005–06, 35 students received support, including 5 fellowships with partial tuition reimbursements available (averaging $18,000 per year), 6 research assistantships (averaging $16,988 per year), 20 teaching assistantships with partial tuition reimbursements available (averaging $16,988 per year); Federal Work-Study and institutionally sponsored loans also available. Financial award application deadline: 3/1. *Faculty research:* Medicinal chemistry, natural products, isotope effects, biophysics and biorganic approaches to enzyme mechanisms, organic and organometallic synthesis. *Unit head:* Prof. W. Philip Huskey, Chairman and Program Director, 973-353-5741, Fax: 973-353-1264, E-mail: huskey@andromeda.rutgers.edu.

Rutgers, The State University of New Jersey, New Brunswick/Piscataway, Graduate School, Program in Chemistry and Chemical Biology, New Brunswick, NJ 08901-1281. Offers analytical chemistry (MS, PhD); biological chemistry (PhD); chemistry education (MST); inorganic chemistry (MS, PhD); organic chemistry (MS, PhD); physical chemistry (MS, PhD). Part-time and evening/weekend programs available. *Faculty:* 48 full-time. *Students:* 104 full-time (41 women), 22 part-time (4 women); includes 11 minority (2 African Americans, 6 Asian Americans or Pacific Islanders, 3 Hispanic Americans), 51 international. Average age 29. 108 applicants, 51% accepted, 22 enrolled. In 2005, 13 master's, 10 doctorates awarded. Terminal master's awarded for partial completion of doctoral program. *Degree requirements:* For master's, thesis or alternative, exam, comprehensive exam, registration; for doctorate, thesis/dissertation, cumulative exams, 1 year residency, comprehensive exam, registration. *Entrance requirements:* For master's and doctorate, GRE General Test, GRE Subject Test. Additional exam requirements/recommendations for international students: Required—TOEFL. *Application deadline:* For fall admission, 4/15 priority date for domestic students, 4/1 priority date for international students; for spring admission, 12/1 priority date for domestic students, 12/1 priority date for international students. Applications are processed on a rolling basis. Application fee: $50. Electronic applications accepted. *Expenses:* Tuition, state resident: full-time $10,440; part-time $435 per credit. Tuition, nonresident: full-time $15,520; part-time $647 per credit. Required fees: $129 per credit. Tuition and fees vary according to program. *Financial support:* In 2005–06, 104 students received support, including 8 fellowships with full tuition reimbursements available (averaging $22,000 per year), 23 research assistantships with full tuition reimbursements available (averaging $18,347 per year), 59 teaching assistantships with full tuition reimbursements available (averaging $18,347 per year); career-related internships or fieldwork, Federal Work-Study, traineeships, and health care benefits also available. Financial award application deadline: 3/1; financial award applicants required to submit FAFSA. *Faculty research:* Biophysical organic/bioorganic, inorganic/bioinorganic, theoretical, and solid-state/surface chemistry. Total annual research expenditures: $14.5 million. *Unit head:* Dr. Roger Jones, Director and Chair, 732-445-4900, Fax: 732-445-5312, E-mail: jones@rutchem.rutgers.edu. *Application contact:* Dr. Martha A. Cotter, Vice Chair, 732-445-2259, Fax: 732-445-5312, E-mail: cotter@rutchem.rutgers.edu.

See Close-Up on page 139.

Seton Hall University, College of Arts and Sciences, Department of Chemistry and Biochemistry, South Orange, NJ 07079-2694. Offers analytical chemistry (MS, PhD); biochemistry (MS, PhD); chemistry (MS); inorganic chemistry (MS, PhD); organic chemistry (MS, PhD); physical chemistry (MS, PhD). Part-time and evening/weekend programs available. *Students:* 19 full-time (6 women), 46 part-time (17 women). Average age 34. 33 applicants, 79% accepted, 14 enrolled. In 2005, 11 master's, 7 doctorates awarded. Terminal master's awarded for partial completion of doctoral program. *Degree requirements:* For master's, formal seminar, thesis optional; for doctorate, thesis/dissertation, annual seminars, comprehensive exam. *Entrance requirements:* Additional exam requirements/recommendations for international students: Required—TOEFL. *Application deadline:* For fall admission, 7/1 priority date for domestic students, 7/1 priority date for international students; for spring admission, 11/1 priority date for domestic students, 11/1 priority date for international students. Applications are processed on a rolling basis. Application fee: $50. Electronic applications accepted. *Financial support:* In 2005–06, 1 research assistantship, 19 teaching assistantships were awarded; Federal Work-Study also available. *Faculty research:* DNA metal reactions; chromatography; bioinorganic, biophysical, organometallic, polymer chemistry; heterogeneous catalyst. *Unit head:* Dr. Nicholas Snow, Chair, 973-761-9414, Fax: 973-761-9772, E-mail: snownich@shu.edu. *Application contact:* Dr. Stephen Kelty, Director of Graduate Studies, 973-761-9129, Fax: 973-761-9772, E-mail: keltyste@shu.edu.

See Close-Up on page 141.

South Dakota State University, Graduate School, College of Arts and Science and College of Agriculture and Biological Sciences, Department of Chemistry, Brookings, SD 57007. Offers analytical chemistry (MS, PhD); biochemistry (MS, PhD); chemistry (MS, PhD); inorganic chemistry (MS, PhD); organic chemistry (MS, PhD); physical chemistry (MS, PhD). *Degree requirements:* For master's, thesis, oral exam; for doctorate, thesis/dissertation, preliminary oral and written exams, research tool. *Entrance requirements:* For master's, bachelor's degree in chemistry or equivalent. Additional exam requirements/recommendations for international students: Required—TOEFL. *Faculty research:* Environmental chemistry, computational chemistry, organic synthesis and photochemistry, novel material development and characterization.

Southern University and Agricultural and Mechanical College, Graduate School, College of Sciences, Department of Chemistry, Baton Rouge, LA 70813. Offers analytical chemistry (MS); biochemistry (MS); environmental sciences (MS); inorganic chemistry (MS); organic chemistry (MS); physical chemistry (MS). *Faculty:* 9 full-time (2 women), 3 part-time/adjunct (2 women). *Students:* 20 full-time (13 women), 8 part-time (5 women); all minorities (20 African Americans, 8 Asian Americans or Pacific Islanders). Average age 23. 30 applicants, 70% accepted, 14 enrolled. In 2005, 3 master's awarded. *Degree requirements:* For master's, thesis. *Entrance requirements:* For master's, GMAT or GRE General Test. Additional exam requirements/recommendations for international students: Required—TOEFL (minimum score 525 paper-based; 193 computer-based). *Application deadline:* For fall admission, 4/15 for domestic students; for spring admission, 11/1 priority date for domestic students. Applications are processed on a rolling basis. Application fee: $5. *Financial support:* In 2005–06, 31 research assistantships (averaging $7,000 per year), 10 teaching assistantships (averaging $7,000 per year) were awarded; scholarships/grants also available. Financial award application deadline: 4/15. *Faculty research:* Synthesis of macrocyclic ligands, latex accelerators, anticancer drugs, biosensors, absorption isotheums, isolation of specific enzymes from plants. Total annual research expenditures:$400,000. *Unit head:* Dr. Ella Kelley, Chair, 225-771-3990, Fax: 225-771-3992.

State University of New York at Binghamton, Graduate School, School of Arts and Sciences, Department of Chemistry, Binghamton, NY 13902-6000. Offers analytical chemistry (PhD); chemistry (MA, MS); inorganic chemistry (PhD); organic chemistry (PhD); physical chemistry (PhD). Part-time programs available. Terminal master's awarded for partial completion of doctoral program. *Degree requirements:* For master's, thesis or alternative, oral exam, seminar presentation; for doctorate, thesis/dissertation, cumulative exams. *Entrance requirements:* For master's and doctorate, GRE General Test, GRE Subject Test. Additional exam requirements/recommendations for international students: Required—TOEFL. Electronic applications accepted.

Stevens Institute of Technology, Graduate School, Arthur E. Imperatore School of Sciences and Arts, Department of Chemistry and Chemical Biology, Hoboken, NJ 07030. Offers chemical biology (MS, PhD); chemistry (MS, PhD), including analytical chemistry (MS, PhD, Certificate), chemical biology (MS, PhD, Certificate), organic chemistry, physical chemistry, polymer chemistry (MS, PhD, Certificate); chemistry and chemical biology (Certificate), including analytical chemistry (MS, PhD, Certificate), bioinformatics, biomedical chemistry, chemical biology (MS, PhD, Certificate), chemical physiology, polymer chemistry (MS, PhD, Certificate). Part-time and evening/weekend programs available. *Students:* 17 full-time (8 women), 28 part-time (16 women); includes 12 minority (9 Asian Americans or Pacific Islanders, 3 Hispanic Americans), 14 international. 51 applicants, 75% accepted.Terminal master's awarded for partial completion of doctoral program. *Degree requirements:* For master's, thesis or alternative; for doctorate, one foreign language, thesis/dissertation; for Certificate, project or thesis. *Entrance requirements:* Additional exam requirements/recommendations for international students: Required—TOEFL. *Application deadline:* Applications are processed on a rolling basis. Application fee: $50. Electronic applications accepted. *Expenses:* Tuition: Part-time $920 per credit hour. Tuition and fees vary according to program. *Financial support:* Fellowships, research assistantships, teaching assistantships, Federal Work-Study and institutionally sponsored loans available. Financial award application deadline: 4/1. *Unit head:* Dr. Francis T. Jones, Director, 201-216-5518, Fax: 201-216-8240.

Tufts University, Graduate School of Arts and Sciences, Department of Chemistry, Medford, MA 02155. Offers analytical chemistry (MS, PhD); bioorganic chemistry (MS, PhD); environmental chemistry (MS, PhD); inorganic chemistry (MS, PhD); organic chemistry (MS, PhD); physical chemistry (MS, PhD). *Faculty:* 17 full-time. *Students:* 55 (27 women); includes 5 minority (1 African American, 4 Asian Americans or Pacific Islanders) 27 international. 45 applicants, 87% accepted, 17 enrolled. In 2005, 4 master's, 4 doctorates awarded. Terminal master's awarded

Physical Chemistry

Tufts University (continued)
for partial completion of doctoral program. *Degree requirements:* For master's and doctorate, thesis/dissertation. *Entrance requirements:* For master's and doctorate, GRE General Test, GRE Subject Test. Additional exam requirements/recommendations for international students: Required—TOEFL (minimum score 600 paper-based; 250 computer-based), TSE. *Application deadline:* For fall admission, 1/15 for domestic students, 12/30 for international students; for spring admission, 10/15 for domestic students, 9/15 for international students. Applications are processed on a rolling basis. Application fee: $65. Electronic applications accepted. *Expenses:* Tuition: Full-time $32,360. Tuition and fees vary according to program. *Financial support:* Research assistantships with full and partial tuition reimbursements, teaching assistantships with full and partial tuition reimbursements, Federal Work-Study, scholarships/grants, and tuition waivers (partial) available. Financial award application deadline: 1/15; financial award applicants required to submit FAFSA. *Unit head:* Mary Jane Shultz, Chair, 617-627-3477, Fax: 617-627-3443. *Application contact:* Samuel Kounaves, Information Contact, 617-627-3441, Fax: 617-627-3443.

See Close-Up on page 151.

University of Calgary, Faculty of Graduate Studies, Faculty of Science, Department of Chemistry, Calgary, AB T2N 1N4, Canada. Offers analytical chemistry (M Sc, PhD); applied chemistry (M Sc, PhD); inorganic chemistry (M Sc, PhD); organic chemistry (M Sc, PhD); physical chemistry (M Sc, PhD); polymer chemistry (M Sc, PhD); theoretical chemistry (M Sc, PhD). *Faculty:* 31 full-time (2 women), 2 part-time/adjunct (0 women). *Students:* 80 full-time (38 women). Average age 25. 250 applicants, 6% accepted, 15 enrolled. In 2005, 7 master's, 7 doctorates awarded. *Degree requirements:* For master's, thesis; for doctorate, thesis/dissertation, candidacy exam. *Entrance requirements:* For master's, minimum GPA of 3.0; for doctorate, honors B Sc degree with minimum GPA of 3.7 or M Sc with minimum GPA of 3.3. Additional exam requirements/recommendations for international students: Required—TOEFL (minimum score 580 paper-based; 237 computer-based). *Application deadline:* For fall admission, 12/1 priority date for domestic students, 12/1 priority date for international students. Applications are processed on a rolling basis. Application fee: $100 ($130 for international students). Electronic applications accepted. *Financial support:* In 2005–06, 25 students received support, including research assistantships (averaging $11,000 per year), teaching assistantships (averaging $6,530 per year); fellowships, scholarships/grants also available. Financial award application deadline:12/1. *Faculty research:* Chemical analysis, chemical dynamics, synthesis theory. *Unit head:* Dr. Brian Keay, Head, 403-220-5340, E-mail: info@chem.ucalgary.ca. *Application contact:* Bonnie E. King, Graduate Program Administrator, 403-220-6252, E-mail: bking@ucalgary.ca.

University of Cincinnati, Division of Research and Advanced Studies, McMicken College of Arts and Sciences, Department of Chemistry, Cincinnati, OH 45221. Offers analytical chemistry (MS, PhD); biochemistry (MS, PhD); inorganic chemistry (MS, PhD); organic chemistry (MS, PhD); physical chemistry (MS, PhD); polymer chemistry (MS, PhD); sensors (PhD). Part-time and evening/weekend programs available. Terminal master's awarded for partial completion of doctoral program. *Degree requirements:* For master's, thesis optional; for doctorate, thesis/dissertation, comprehensive exam, registration. *Entrance requirements:* For master's and doctorate, GRE General Test. Additional exam requirements/recommendations for international students: Required—TOEFL (minimum score 580 paper-based; 237 computer-based); Recommended—TSE(minimum score 50). Electronic applications accepted. *Faculty research:* Biomedical chemistry, laser chemistry, surface science, chemical sensors, synthesis.

University of Georgia, Graduate School, College of Arts and Sciences, Department of Chemistry, Athens, GA 30602. Offers analytical chemistry (MS, PhD); inorganic chemistry (MS, PhD); organic chemistry (MS, PhD); physical chemistry (MS, PhD). *Faculty:* 22 full-time (2 women). *Students:* 155 full-time, 3 part-time; includes 12 minority (4 African Americans, 7 Asian Americans or Pacific Islanders, 1 Hispanic American), 87 international. 147 applicants, 56% accepted, 35 enrolled. In 2005, 2 master's, 19 doctorates awarded. Terminal master's awarded for partial completion of doctoral program. *Median time to degree:* Of those who began their doctoral program in fall 1997, 100% received their degree in 8 years or less. *Degree requirements:* For master's, thesis; for doctorate, one foreign language, thesis/dissertation. *Entrance requirements:* For master's and doctorate, GRE General Test. Additional exam requirements/recommendations for international students: Required—TOEFL (minimum score 213 computer-based), TSE(minimum score 50). *Application deadline:* For fall admission, 7/1 for domestic students; for spring admission, 11/15 for domestic students. Application fee: $50. Electronic applications accepted. *Financial support:* Fellowships, research assistantships, teaching assistantships, unspecified assistantships available. *Unit head:* Dr. John L. Stickney, Head, 706-542-2726, Fax: 706-542-9454, E-mail: stickney@chem.uga.edu. *Application contact:* Dr. Donald Kurtz, Information Contact, 706-542-2010, Fax: 706-542-9454, E-mail: kurtz@chem.uga.edu.

University of Louisville, Graduate School, College of Arts and Sciences, Department of Chemistry, Louisville, KY 40292-0001. Offers analytical chemistry (MS, PhD); biochemistry (MS, PhD); chemical physics (PhD); inorganic chemistry (MS, PhD); organic chemistry (MS, PhD); physical chemistry (MS, PhD). *Students:* 37 full-time (16 women), 16 part-time (7 women); includes 3 minority (2 African Americans, 1 Hispanic American), 24 international. Average age 28. In 2005, 1 master's, 8 doctorates awarded. *Degree requirements:* For master's, thesis/dissertation; for doctorate, thesis/dissertation, comprehensive exam. *Entrance requirements:* For master's and doctorate, GRE General Test. Additional exam requirements/recommendations for international students: Required—TOEFL. *Application deadline:* Applications are processed on a rolling basis. Application fee: $50. *Expenses:* Tuition, state resident: full-time $3,003; part-time $334 per credit hour. Tuition, nonresident: full-time $8,277; part-time $920 per credit hour. Tuition and fees vary according to course load, degree level and program. *Financial support:* In 2005–06, 33 teaching assistantships with tuition reimbursements were awarded; fellowships, research assistantships *Unit head:* Dr. George R. Pack, Chair, 502-852-6798, Fax: 502-852-8149, E-mail: george.pack@louisville.edu.

University of Maryland, College Park, Graduate Studies, College of Chemical and Life Sciences, Department of Chemistry and Biochemistry, Chemistry Program, College Park, MD 20742. Offers analytical chemistry (MS, PhD); inorganic chemistry (MS, PhD); organic chemistry (MS, PhD); physical chemistry (MS, PhD). Part-time and evening/weekend programs available. *Students:* 102 full-time (43 women), 4 part-time (2 women); includes 7 minority (2 African Americans, 3 Asian Americans or Pacific Islanders, 2 Hispanic Americans), 46 international. 128 applicants, 24% accepted, 30 enrolled. In 2005, 6 master's, 12 doctorates awarded. Terminal master's awarded for partial completion of doctoral program. *Median time to degree:* Of those who began their doctoral program in fall 1997, 33% received their degree in 8 years or less. *Degree requirements:* For master's, thesis optional; for doctorate, thesis/dissertation, 2 seminar presentations, oral exam. *Entrance requirements:* For master's and doctorate, GRE General Test, GRE Subject Test (recommended), minimum GPA of 3.0, 3 letters of recommendation. Additional exam requirements/recommendations for international students: Required—TOEFL, TSE. *Application deadline:* For fall admission, 4/1 for domestic students, 2/1 for international students; for spring admission, 10/21 for domestic students, 6/1 for international students. Applications are processed on a rolling basis. Application fee: $60. Electronic applications accepted. *Financial support:* In 2005–06, 10 fellowships (averaging $4,900 per year) were awarded; research assistantships, teaching assistantships with partial tuition reimbursements Financial award applicants required to submit FAFSA. *Faculty research:* Environmental chemistry, nuclear chemistry, lunar and environmental analysis, x-ray crystallography. *Application contact:* Dean of Graduate School, 301-405-4190, Fax: 301-314-9305.

University of Miami, Graduate School, College of Arts and Sciences, Department of Chemistry, Coral Gables, FL 33124. Offers chemistry (MS); inorganic chemistry (PhD); organic chemistry (PhD); physical chemistry (PhD). *Faculty:* 11 full-time (1 woman). *Students:* 38 full-time (6 women), 28 international. Average age 27. 62 applicants, 27% accepted, 5 enrolled. In 2005, 1 master's, 5 doctorates awarded. Terminal master's awarded for partial completion of doctoral program. *Degree requirements:* For master's, comprehensive exam; for doctorate, thesis/dissertation, comprehensive exam. *Entrance requirements:* For master's and doctorate, GRE General Test. Additional exam requirements/recommendations for international students: Required—TOEFL (minimum score 550 paper-based; 213 computer-based). *Application deadline:* For fall admission, 1/15 for domestic students, 1/15 for international students. Applications are processed on a rolling basis. Application fee: $50. Electronic applications accepted. *Financial support:* In 2005–06, 38 students received support, including 1 fellowship with full tuition reimbursement available (averaging $20,000 per year), 10 research assistantships with full tuition reimbursements available (averaging $20,000 per year), 27 teaching assistantships with full tuition reimbursements available (averaging $20,000 per year); tuition waivers (full) also available. Financial award application deadline: 5/1; financial award applicants required to submit FAFSA. *Faculty research:* Supramolecular chemistry, electrochemistry, surface chemistry, catalysis, organometalic. *Unit head:* Dr. Vaidyanathan Ramamurthy, Chairman, 305-284-2282, Fax: 305-284-4571. *Application contact:* Eva Johnson, Graduate Secretary, 305-284-2094, Fax: 305-284-4571, E-mail: evaj@miami.edu.

University of Michigan, Horace H. Rackham School of Graduate Studies, College of Literature, Science, and the Arts, Department of Chemistry, Ann Arbor, MI 48109. Offers analytical chemistry (PhD); inorganic chemistry (PhD); material chemistry (PhD); organic chemistry (PhD); physical chemistry (PhD). *Faculty:* 48 full-time (8 women), 3 part-time/adjunct (all women). *Students:* ; includes 39 minority (4 African Americans, 1 American Indian/Alaska Native, 29 Asian Americans or Pacific Islanders, 5 Hispanic Americans), 66 international. Average age 26. 456 applicants, 29% accepted, 44 enrolled. In 2005, 37 degrees awarded. *Degree requirements:* For doctorate, thesis/dissertation, oral defense of dissertation, organic cumulative proficiency exams. *Entrance requirements:* For doctorate, GRE General Test, GRE Subject Test (recommended), 3 letters of recommendation. Additional exam requirements/recommendations for international students: Required—TOEFL. *Application deadline:* For fall admission, 1/31 for domestic students, 1/1 for international students. Applications are processed on a rolling basis. Application fee: $60 ($75 for international students). Electronic applications accepted. *Expenses:* Tuition, area resident: Full-time $14,082; part-time $894 per credit hour. Tuition, state resident: full-time $14,082; part-time $896 per credit hour. Tuition, nonresident: full-time $28,500; part-time $1,675 per credit hour. Required fees: $189; $189. *Financial support:* In 2005–06, 10 fellowships with full tuition reimbursements (averaging $20,000 per year), 70 research assistantships with full tuition reimbursements (averaging $19,000 per year), 120 teaching assistantships with full tuition reimbursements (averaging $20,000 per year) were awarded. Financial award applicants required to submit FAFSA. *Faculty research:* Biological catalysis, protein engineering, chemical sensors, de novo metalloprotein design, supramolecular architecture. Total annual research expenditures: $8 million. *Unit head:* Dr. Carol A. Fierke, Chair, 734-763-9681, Fax: 734-647-4847. *Application contact:* Linda Deitert, Assistant Director Graduate Studies, 734-764-7278, Fax: 734-647-4865, E-mail: chemadmissions@umich.edu.

University of Missouri–Columbia, Graduate School, College of Arts and Sciences, Department of Chemistry, Columbia, MO 65211. Offers analytical chemistry (MS, PhD); inorganic chemistry (MS, PhD); organic chemistry (MS, PhD); physical chemistry (MS, PhD). *Faculty:* 21 full-time (5 women). *Students:* 58 full-time (17 women), 44 part-time (16 women); includes 5 minority (3 African Americans, 2 Asian Americans or Pacific Islanders), 59 international. In 2005, 11 master's, 9 doctorates awarded. *Degree requirements:* For master's, thesis; for doctorate, one foreign language, thesis/dissertation. *Entrance requirements:* For master's and doctorate, GRE General Test, minimum GPA of 3.0. *Application deadline:* For fall admission, 4/1 for domestic studentsFor winter admission, 11/1 for domestic students. Applications are processed on a rolling basis. Application fee: $45 ($60 for international students). *Financial support:* Fellowships, research assistantships, teaching assistantships, institutionally sponsored loans available. *Unit head:* Dr. Sheryl A. Tucker, Director for Graduate Studies, 573-882-1729.

University of Missouri–Kansas City, College of Arts and Sciences, Department of Chemistry, Kansas City, MO 64110-2499. Offers analytical chemistry (MS, PhD); inorganic chemistry (MS, PhD); organic chemistry (MS, PhD); physical chemistry (MS, PhD); polymer chemistry (MS, PhD). PhD offered through the School of Graduate Studies. Part-time and evening/weekend programs available. *Faculty:* 14 full-time (2 women), 2 part-time/adjunct (0 women). *Students:* Average age 38. 13 applicants, 8% accepted, 1 enrolled. In 2005, 2 degrees awarded. *Degree requirements:* For master's, thesis (for some programs); for doctorate, thesis/dissertation. *Entrance requirements:* For master's, equivalent of American Chemical Society approved bachelor's degree in chemistry; for doctorate, GRE General Test, equivalent of American Chemical Society approved bachelor's degree in chemistry. Additional exam requirements/recommendations for international students: Required—TOEFL (minimum score 580 paper-based; 237 computer-based), TWE. *Application deadline:* For fall and spring admission, 4/15For winter admission, 10/15 for domestic students. Applications are processed on a rolling basis. Application fee: $25. Electronic applications accepted. *Expenses:* Tuition, state resident: full-time $4,738; part-time $263 per credit hour. Tuition, nonresident: full-time $12,235; part-time $679 per credit hour. Required fees: $582. Tuition and fees vary according to course load, program and student level. *Financial support:* In 2005–06, fellowships with partial tuition reimbursements (averaging $18,156 per year), research assistantships with partial tuition reimbursements (averaging $18,515 per year), teaching assistantships with partial tuition reimbursements (averaging $16,434 per year) were awarded; Federal Work-Study, institutionally sponsored loans, and scholarships/grants also available. Support available to part-time students. Financial award application deadline: 2/15. *Faculty research:* Molecular spectroscopy, characterization and synthesis of materials and compounds, computational chemistry, natural products, drug delivery systems and anti-tumor agents. Total annual research expenditures: $1.1 million. *Unit head:* Dr. Jerry Jean, Chairperson, 816-235-2273, Fax: 816-235-5502, E-mail: jeany@umkc.edu. *Application contact:* Graduate Recruiting Committee, 816-235-2272, Fax: 816-235-5502, E-mail: umkc-chemdept@umkc.edu.

University of Missouri–St. Louis, College of Arts and Sciences, Department of Chemistry and Biochemistry, St. Louis, MO 63121. Offers chemistry (MS, PhD), including inorganic chemistry, organic chemistry, physical chemistry. Part-time and evening/weekend programs available. *Faculty:* 16. *Students:* 28 full-time (20 women), 16 part-time (7 women); includes 6 minority (5 African Americans, 1 Hispanic American), 23 international. Average age 32. 2,298 applicants, 74% accepted. In 2005, 9 master's, 4 doctorates awarded. Terminal master's awarded for partial completion of doctoral program. *Degree requirements:* For master's, thesis optional; for doctorate, thesis/dissertation. *Entrance requirements:* For doctorate, GRE General Test, 3 letters of recommendation. Additional exam requirements/recommendations for international students: Required—TOEFL (minimum score 550 paper-based; 213 computer-based). *Application deadline:* For fall admission, 7/1 for domestic students; for spring admission, 12/7 priority date for domestic students. Applications are processed on a rolling basis. Application fee: $35 ($40 for international students). Electronic applications accepted. *Expenses:* Tuition, state resident: part-time $263 per credit hour. Tuition, nonresident: part-time $680 per credit hour. Required fees: $53 per credit hour. Tuition and fees vary according to program. *Financial support:* In 2005–06, 9 research assistantships with full and partial tuition reimbursements (averaging $16,500 per year), 17 teaching assistantships with full and partial tuition reimbursements (averaging $16,500 per year) were awarded; fellowships with full and partial tuition reimbursements *Faculty research:* Metallaborane chemistry, serum transferrin chemistry, natural products chemistry, organic synthesis. *Unit head:* Dr. Cynthia Dupureur, Director of Graduate Studies, 314-516-5311, Fax: 314-516-5342, E-mail: gradchem@umsl.edu. *Application contact:* 314-516-5458, Fax: 314-516-5310, E-mail: gradadm@umsl.edu.

The University of Montana, Graduate School, College of Arts and Sciences, Department of Chemistry, Missoula, MT 59812-0002. Offers chemistry (MS, PhD), including environmental/analytical chemistry, inorganic chemistry, organic chemistry, physical chemistry. *Faculty:* 16 full-time (2 women). *Students:* 36 full-time (9 women), 7 part-time (2 women), 9 international. 25 applicants, 44% accepted, 6 enrolled. In 2005, 1 master's, 2 doctorates awarded. Terminal master's awarded for partial completion of doctoral program. *Degree requirements:* For master's, thesis (for some programs); for doctorate, thesis/dissertation. *Entrance requirements:* For master's and doctorate, GRE General Test. Additional exam requirements/recommendations for international students: Required—TOEFL (minimum score 575 paper-based; 230 computer-based). *Application deadline:* For fall admission, 2/15 for domestic students. Applications are processed on a rolling basis. Application fee: $45. *Expenses:* Tuition, state resident: part-time $267 per credit. Tuition, nonresident: part-time $665 per credit. Part-time tuition and fees vary according to course load and degree level. *Financial support:* In 2005–06, 13 research assistantships with tuition reimbursements (averaging $14,000 per year), 12 teaching assistantships with full tuition reimbursements (averaging $14,000 per year) were awarded; Federal Work-Study, scholarships/grants, and unspecified assistantships also available. Financial award application deadline: 3/1; financial award applicants required to submit FAFSA. *Faculty research:* Reaction mechanisms and kinetics, inorganic and organic synthesis, analytical chemistry, natural products. Total annual research expenditures: $789,952. *Unit head:* Dr. Edward Rosenberg, Chair, 406-243-2592, Fax: 406-243-4227.

University of Nebraska–Lincoln, Graduate College, College of Arts and Sciences, Department of Chemistry, Lincoln, NE 68588. Offers analytical chemistry (PhD); chemistry (MS); inorganic chemistry (PhD); organic chemistry (PhD); physical chemistry (PhD). *Degree requirements:* For master's, one foreign language, departmental qualifying exam, thesis optional; for doctorate, one foreign language, thesis/dissertation, departmental qualifying exams, comprehensive exam. *Entrance requirements:* For master's and doctorate, GRE. Additional exam requirements/recommendations for international students: Required—TOEFL (minimum score 550 paper-based; 213 computer-based). Electronic applications accepted. *Faculty research:* Bioorganic and bioinorganic chemistry, biophysical and bioanalytical chemistry, structure-function of DNA and proteins, organometallics, mass spectrometry.

University of Notre Dame, Graduate School, College of Science, Department of Chemistry and Biochemistry, Notre Dame, IN 46556. Offers biochemistry (MS, PhD); inorganic chemistry (MS, PhD); organic chemistry (MS, PhD); physical chemistry (MS, PhD). *Faculty:* 30 full-time (3 women). *Students:* 148 full-time (52 women); includes 5 minority (1 African American, 1 Asian American or Pacific Islander, 3 Hispanic Americans), 65 international. 144 applicants, 38% accepted, 21 enrolled. In 2005, 7 master's, 6 doctorates awarded. Terminal master's awarded for partial completion of doctoral program. *Median time to degree:* Of those who began their doctoral program in fall 1997, 54% received their degree in 8 years or less. *Degree requirements:* For master's, thesis, comprehensive exam; for doctorate, thesis/dissertation, qualifying exam. *Entrance requirements:* For master's and doctorate, GRE General Test, GRE Subject Test (strongly recommended). Additional exam requirements/recommendations for international students: Required—TOEFL. *Application deadline:* For fall admission, 2/1 for domestic students. Applications are processed on a rolling basis. Application fee: $50. Electronic applications accepted. *Financial support:* In 2005–06, 148 students received support, including 19 fellowships with full tuition reimbursements available (averaging $22,000 per year), 57 research assistantships with full tuition reimbursements available (averaging $15,250 per year), 53 teaching assistantships with full tuition reimbursements available (averaging $16,000 per year); tuition waivers (full) also available. Financial award application deadline: 2/1. *Faculty research:* Reaction design and mechanistic studies; reactive intermediates; synthesis, structure and reactivity of organometallic. cluster complexes, and biologically active natural products; bioorganic chemistry; enzymology. Total annual research expenditures: $9.4 million. *Unit head:* Dr. Richard E. Taylor, Director of Graduate Studies, 574-631-7058, Fax: 574-631-6652, E-mail: taylor.61@nd.edu. *Application contact:* Dr. Terrence J. Akai, Director of Graduate Admissions, 574-631-7706, Fax: 574-631-4183, E-mail: gradad@nd.edu.

University of Regina, Faculty of Graduate Studies and Research, Faculty of Science, Department of Chemistry and Biochemistry, Regina, SK S4S 0A2, Canada. Offers analytical chemistry (M Sc, PhD); biochemistry (M Sc, PhD); inorganic chemistry (M Sc, PhD); organic chemistry (M Sc, PhD); physical chemistry (M Sc, PhD). Part-time programs available. *Faculty:* 10 full-time (2 women), 4 part-time/adjunct (1 woman). *Students:* 14 full-time (6 women), 3 part-time (2 women). 21 applicants, 33% accepted. *Degree requirements:* For master's and doctorate, thesis/dissertation, departmental qualifying exam. *Entrance requirements:* For master's and doctorate, GRE. Additional exam requirements/recommendations for international students: Required—TOEFL (minimum score 580 paper-based; 237 computer-based). *Application deadline:* For fall admission, 1/1 for domestic studentsFor winter admission, 7/1 for domestic students. Applications are processed on a rolling basis. Application fee: $60 ($100 for international students). *Financial support:* In 2005–06, 4 fellowships (averaging $14,886 per year), 1 research assistantship (averaging $12,750 per year), 5 teaching assistantships (averaging $13,501 per year) were awarded; scholarships/grants also available. Financial award application deadline: 6/15. *Faculty research:* Organic synthesis, organic oxidations, ionic liquids theoretical/computational chemistry, protein biochemistry/biophysics, environmental analytical, photophysical/photochemistry. *Unit head:* Dr. Andrew G. Wee, Head, 306-585-4767, Fax: 306-585-4894, E-mail: chem.chair@uregina.ca. *Application contact:* Dr. Allan East, Associate Professor, 306-585-4003, Fax: 306-585-4894, E-mail: allan.east@uregina.ca.

University of Southern Mississippi, Graduate School, College of Science and Technology, Department of Chemistry and Biochemistry, Hattiesburg, MS 39406-0001. Offers analytical chemistry (MS, PhD); biochemistry (MS, PhD); inorganic chemistry (MS, PhD); organic chemistry (MS, PhD); physical chemistry (MS, PhD). *Degree requirements:* For master's and doctorate, thesis/dissertation, comprehensive exam. *Entrance requirements:* For master's, GRE General Test, minimum GPA of 2.75 in last 60 hours; for doctorate, GRE General Test, minimum GPA of 3.5. Additional exam requirements/recommendations for international students: Required—TOEFL. *Faculty research:* Plant biochemistry, photo chemistry, polymer chemistry, x-ray analysis, enzyme chemistry.

University of South Florida, College of Graduate Studies, College of Arts and Sciences, Department of Chemistry, Tampa, FL 33620-9951. Offers analytical chemistry (MS, PhD); biochemistry (MS, PhD); inorganic chemistry (MS, PhD); organic chemistry (MS, PhD); physical chemistry (MS, PhD); polymer chemistry (PhD). Part-time programs available. *Faculty:* 22. *Students:* 102 full-time (45 women), 15 part-time (6 women); includes 13 minority (5 Asian Americans or Pacific Islanders, 8 Hispanic Americans), 60 international. 80 applicants, 71% accepted, 23 enrolled. In 2005, 2 master's awarded. Terminal master's awarded for partial completion of doctoral program. *Degree requirements:* For master's, thesis; for doctorate, 2 foreign languages, thesis/dissertation, colloquium. *Entrance requirements:* For master's and doctorate, GRE General Test, minimum GPA of 3.0 in last 30 hours of chemistry course work, 3 letters of recommendation. Additional exam requirements/recommendations for international students: Required—TOEFL (minimum score 550 paper-based; 213 computer-based), TSE(minimum score 50). *Application deadline:* For fall admission, 5/1 priority date for domestic students, 3/1 priority date for international students; for spring admission, 10/1 priority date for domestic students, 8/1 priority date for international students. Applications are processed on a rolling basis. Application fee: $30. Electronic applications accepted. *Financial support:* In 2005–06, 91 students received support; fellowships with partial tuition reimbursements available, research assistantships with partial tuition reimbursements available, teaching assistantships with partial tuition reimbursements available, career-related internships or fieldwork, institutionally sponsored loans, scholarships/grants, health care benefits, and unspecified assistantships available. Financial award application deadline: 6/30. *Faculty research:* Synthesis, bio-organic chemistry, bioinorganic chemistry, environmental chemistry, NMR. *Unit head:* Dr. Michael Zaworotko, Chairperson, 813-974-4129, Fax: 813-974-3203, E-mail: xtal@usf.edu. *Application contact:* Dr. Brian Space, Graduate Coordinator, 813-974-3397, Fax: 813-974-3203.

The University of Tennessee, Graduate School, College of Arts and Sciences, Department of Chemistry, Knoxville, TN 37996. Offers analytical chemistry (MS, PhD); chemical physics (PhD); environmental chemistry (MS, PhD); inorganic chemistry (MS, PhD); organic chemistry (MS, PhD); physical chemistry (MS, PhD); polymer chemistry (MS, PhD); theoretical chemistry (PhD). Part-time programs available. Terminal master's awarded for partial completion of doctoral program. *Degree requirements:* For master's and doctorate, thesis/dissertation. *Entrance requirements:* For master's and doctorate, GRE General Test, minimum GPA of 2.7. Additional exam requirements/recommendations for international students: Required—TOEFL. Electronic applications accepted.

The University of Texas at Austin, Graduate School, College of Natural Sciences, Department of Chemistry and Biochemistry, Austin, TX 78712-1111. Offers analytical chemistry (MA, PhD); biochemistry (MA, PhD); inorganic chemistry (MA, PhD); organic chemistry (MA, PhD); physical chemistry (MA, PhD). *Entrance requirements:* For master's and doctorate, GRE General Test.

The University of Toledo, Graduate School, College of Arts and Sciences, Department of Chemistry, Toledo, OH 43606-3390. Offers analytical chemistry (MS, PhD); biological chemistry (MS, PhD); inorganic chemistry (MS, PhD); organic chemistry (MS, PhD); physical chemistry (MS, PhD). Part-time programs available. *Faculty:* 18. *Students:* 80 full-time (35 women), 8 part-time (5 women); includes 3 minority (2 African Americans, 1 Asian American or Pacific Islander), 43 international. Average age 27. 32 applicants, 75% accepted, 19 enrolled. In 2005, 7 master's awarded. *Degree requirements:* For master's and doctorate, thesis/dissertation. *Entrance requirements:* For master's and doctorate, GRE General Test, GRE Subject Test. Additional exam requirements/recommendations for international students: Required—TOEFL. *Application deadline:* For fall admission, 8/1 for domestic students. Applications are processed on a rolling basis. Application fee: $45. Electronic applications accepted. *Expenses:* Tuition, area resident: Full-time $3,312; part-time $308 per credit hour. Tuition, state resident: full-time $3,312. Tuition, nonresident: full-time $6,616; part-time $735 per credit hour. International tuition: $6,616 full-time. *Financial support:* In 2005–06, 1 research assistantship with full tuition reimbursement (averaging $4,000 per year), 55 teaching assistantships with full tuition reimbursements (averaging $13,336 per year) were awarded; fellowships, Federal Work-Study and institutionally sponsored loans also available. Support available to part-time students. Financial award application deadline: 4/1; financial award applicants required to submit FAFSA. *Faculty research:* Enzymology, materials chemistry, crystallography, theoretical chemistry. *Unit head:* Dr. Alan Pinkerton, Chair, 419-530-7902, Fax: 419-530-4033, E-mail: apinker@uoft02.utoledo.edu. *Application contact:* Charlene Morlock-Hansen, 419-530-2100, E-mail: charlene.hanson@utoledo.edu.

See Close-Up on page 167.

Vanderbilt University, Graduate School, Department of Chemistry, Nashville, TN 37240-1001. Offers analytical chemistry (MAT, MS, PhD); inorganic chemistry (MAT, MS, PhD); organic chemistry (MAT, MS, PhD); physical chemistry (MAT, MS, PhD); theoretical chemistry (MAT, MS, PhD). *Faculty:* 32 full-time (6 women), 3 part-time/adjunct (2 women). *Students:* 96 full-time (40 women); includes 4 minority (2 African Americans, 1 Asian American or Pacific Islander, 1 Hispanic American), 20 international. 223 applicants, 21% accepted, 22 enrolled. In 2005, 4 master's, 12 doctorates awarded. *Degree requirements:* For master's, thesis or alternative; for doctorate, thesis/dissertation, area, qualifying, and final exams. *Entrance requirements:* For master's and doctorate, GRE General Test, GRE Subject Test (recommended). Additional exam requirements/recommendations for international students: Required—TOEFL. *Application deadline:* For fall admission, 1/15 for domestic students, 1/15 for international students. Application fee: $0. Electronic applications accepted. *Expenses:* Tuition: Full-time $15,396; part-time $1,283 per semester hour. Required fees: $2,202; $1,101 per semester. One-time fee: $30. Tuition and fees vary according to course load, program and student level. *Financial support:* Fellowships with full and partial tuition reimbursements, research assistantships with full tuition reimbursements, teaching assistantships with full tuition reimbursements, Federal Work-Study, institutionally sponsored loans, and traineeships available. Financial award application deadline:1/15. *Faculty research:* Chemical synthesis; mechanistic, theoretical, bioorganic, analytical, and spectroscopic chemistry. *Unit head:* Ned A. Porter, Chair, 615-322-2861, Fax: 615-343-1234. *Application contact:* Charles M. Lukehart, Director of Graduate Studies, 615-322-2861, Fax: 615-343-1234, E-mail: charles.m.lukehart@vanderbilt.edu.

Virginia Commonwealth University, Graduate School, College of Humanities and Sciences, Department of Chemistry, Richmond, VA 23284-9005. Offers analytical (MS, PhD); chemical physics (PhD); inorganic (MS, PhD); organic (MS, PhD); physical (MS, PhD). Part-time programs available. *Faculty:* 18 full-time (6 women). *Students:* 36 full-time (13 women), 11 part-time (1 woman); includes 13 minority (6 African Americans, 1 American Indian/Alaska Native, 5 Asian Americans or Pacific Islanders, 1 Hispanic American), 13 international. 47 applicants, 74% accepted. In 2005, 2 master's, 8 doctorates awarded. Terminal master's awarded for partial completion of doctoral program. *Degree requirements:* For master's, thesis; for doctorate, thesis/dissertation, comprehensive cumulative exams, research proposal. *Entrance requirements:* For master's, GRE General Test, 30 undergraduate credits in chemistry; for doctorate, GRE General Test. *Application deadline:* For fall admission, 3/15 for domestic students; for spring admission, 11/15 for domestic students. Applications are processed on a rolling basis. Application fee: $50. *Expenses:* Tuition, state resident: full-time $3,185; part-time $405 per credit. Tuition, nonresident: full-time $7,952; part-time $940 per credit. Required fees: $751 per semester hour. Tuition and fees vary according to course load and program. *Financial support:* Fellowships, research assistantships, teaching assistantships, career-related internships or fieldwork and institutionally sponsored loans available. Support available to part-time students. Financial award application deadline: 7/1. *Faculty research:* Physical, organic, inorganic, analytical, and polymer chemistry; chemical physics. *Unit head:* Dr. Fred M. Hawkridge, Chair, 804-828-1298, Fax: 804-828-8599, E-mail: fmhawkri@vcu.edu. *Application contact:* Dr. M. Samy El-Shall, Chair, Graduate Recruiting and Admissions Committee, 804-828-3518, E-mail: mselshal@vcu.edu.

Wake Forest University, Graduate School, Department of Chemistry, Winston-Salem, NC 27109. Offers analytical chemistry (MS, PhD); inorganic chemistry (MS, PhD); organic chemistry (MS, PhD); physical chemistry (MS, PhD). Part-time programs available. *Faculty:* 17 full-time (4 women). *Students:* 32 full-time (17 women), 2 part-time; includes 1 minority (African American), 14 international. Average age 28. 60 applicants, 42% accepted, 12 enrolled. In 2005, 1 master's, 6 doctorates awarded. *Median time to degree:* Of those who began their doctoral program in fall 1997, 0% received their degree in 8 years or less. *Degree requirements:* For master's, one foreign language, thesis, comprehensive exam, registration; for doctorate, 2 foreign languages, thesis/dissertation, comprehensive exam, registration. *Entrance requirements:* For master's and doctorate, GRE General Test. Additional exam requirements/recommendations for international students: Required—TOEFL (minimum score 213 computer-based). *Application deadline:* For fall admission, 1/15 for domestic students, 1/15 for international students. Application fee: $45. Electronic applications accepted. *Financial support:* In 2005–06, 32 students received support, including 1 fellowship with full tuition reimbursement available (averaging $21,500 per year), 12 research assistantships with full tuition reimbursements available (averaging $19,500 per year), 17 teaching assistantships with full tuition reimbursements available (averaging $19,500 per year); scholarships/grants and tuition waivers (full and partial) also available. Support available to part-time students. Financial award application deadline: 1/15; financial award applicants required to submit FAFSA. *Unit head:* Dr. Bruce King, Director, 336-758-5774, Fax: 335-758-4656, E-mail: kingsb@wfu.edu.

Washington State University, Graduate School, College of Sciences, Department of Chemistry, Pullman, WA 99164. Offers analytical chemistry (MS, PhD); biological systems (MS, PhD); inorganic chemistry (MS, PhD); organic chemistry (MS, PhD); physical chemistry (MS, PhD). *Faculty:* 33. *Students:* 38 full-time (19 women), 2 part-time (1 woman); includes 3 minority (2 Asian Americans or Pacific Islanders, 1 Hispanic American), 11 international. Average age 29. 55 applicants, 33% accepted, 9 enrolled. In 2005, 5 master's, 7 doctorates awarded. Terminal master's awarded for partial completion of doctoral program. *Degree requirements:* For master's, oral exam, teaching experience, thesis optional; for doctorate, thesis/dissertation, oral exam, written exam, comprehensive exam. *Entrance requirements:* For master's and doctorate, GRE General Test, minimum GPA of 3.0, 3 letters of recommendation. *Application deadline:* For fall admission, 3/1 priority date for domestic students, 3/1 priority date for international students; for spring admission, 10/1 priority date for domestic students, 7/1 priority date for international

Physical Chemistry

Washington State University (continued)

students. Applications are processed on a rolling basis. Application fee: $35. *Expenses:* Tuition, state resident: full-time $6,295; part-time $336 per credit. Tuition, nonresident: full-time $15,949; part-time $819 per credit. Required fees: $933. Part-time tuition and fees vary according to campus/location and program. *Financial support:* In 2005–06, 38 students received support, including 3 fellowships (averaging $8,977 per year), 12 research assistantships with full and partial tuition reimbursements available (averaging $15,999 per year), 23 teaching assistantships with full and partial tuition reimbursements available (averaging $14,747 per year); career-related internships or fieldwork, Federal Work-Study, institutionally sponsored loans, scholarships/grants, health care benefits, unspecified assistantships, and summer support also available. Financial award application deadline: 4/1; financial award applicants required to submit FAFSA. *Faculty research:* Environmental chemistry, materials chemistry, radio chemistry, bio-organic, computational chemistry. Total annual research expenditures: $3.8 million. *Unit head:* Sue B. Clark, Chair, Admissions Committee, 509-335-8866, Fax: 509-335-8867, E-mail: sclark@mail.wsu.edu. *Application contact:* Admissions Committee.

West Virginia University, Eberly College of Arts and Sciences, Department of Chemistry, Morgantown, WV 26506. Offers analytical chemistry (MS, PhD); inorganic chemistry (MS, PhD); organic chemistry (MS, PhD); physical chemistry (MS, PhD); theoretical chemistry (MS, PhD). Part-time programs available. Postbaccalaureate distance learning degree programs offered (no on-campus study). *Faculty:* 14 full-time (1 woman), 5 part-time/adjunct (3 women). *Students:* 45 full-time (15 women), 3 part-time (2 women); includes 1 minority (Hispanic American), 26 international. Average age 28. 84 applicants, 58% accepted, 11 enrolled. In 2005, 5 master's, 3 doctorates awarded. Terminal master's awarded for partial completion of doctoral program. *Degree requirements:* For master's and doctorate, thesis/dissertation, registration. *Entrance requirements:* For master's, GRE General Test, GRE Subject Test (recommended), minimum GPA of 2.5; for doctorate, GRE General Test, GRE Subject Test (recommended), minimum GPA of 2.75. Additional exam requirements/recommendations for international students: Required—TOEFL. *Application deadline:* For fall admission, 3/1 for domestic students. Applications are processed on a rolling basis. Application fee: $50. Electronic applications accepted. *Expenses:* Tuition, area resident: Full-time $4,582; part-time $258 per credit hour. Tuition, state resident: full-time $4,582; part-time $258 per credit hour. Tuition, nonresident: full-time $1,382; part-time $741 per credit hour. International tuition: $1,382 full-time. *Financial support:* In 2005–06, fellowships (averaging $2,000 per year), research assistantships with full tuition reimbursements (averaging $13,000 per year), teaching assistantships with full tuition reimbursements (averaging $12,000 per year) were awarded; tuition waivers (full and partial) also available. Financial award application deadline: 2/1; financial award applicants required to submit FAFSA. *Faculty research:* Analysis of proteins, drug interactions, solids and effluents by advanced separation methods; new synthetic strategies for complex organic molecules; synthesis and structural characterization of metal complexes for polymerization catalysis. *Unit head:* Dr. Harry O. Finklea, Chair, 304-293-3435 Ext. 4453, Fax: 304-293-4904, E-mail: harry.finklea@mail.wvu.edu.

Yale University, Graduate School of Arts and Sciences, Department of Chemistry, New Haven, CT 06520. Offers biophysical chemistry (PhD); inorganic chemistry (PhD); organic chemistry (PhD); physical chemistry (PhD). *Degree requirements:* For doctorate, thesis/dissertation. *Entrance requirements:* For doctorate, GRE General Test, GRE Subject Test. Additional exam requirements/recommendations for international students: Required—TOEFL.

See Close-Up on page 183.

Theoretical Chemistry

Cornell University, Graduate School, Graduate Fields of Arts and Sciences, Field of Chemistry and Chemical Biology, Ithaca, NY 14853-0001. Offers analytical chemistry (PhD); bio-organic chemistry (PhD); biophysical chemistry (PhD); chemical biology (PhD); chemical physics (PhD); inorganic chemistry (PhD); materials chemistry (PhD); organic chemistry (PhD); organometallic chemistry (PhD); physical chemistry (PhD); polymer chemistry (PhD); theoretical chemistry (PhD). *Faculty:* 44 full-time (2 women). *Students:* 190 full-time (73 women); includes 23 minority (4 African Americans, 8 Asian Americans or Pacific Islanders, 11 Hispanic Americans), 65 international. 339 applicants, 35% accepted, 49 enrolled. In 2005, 23 doctorates awarded. *Degree requirements:* For doctorate, thesis/dissertation, comprehensive exam. *Entrance requirements:* For doctorate, GRE General Test, GRE Subject Test (chemistry), 3 letters of recommendation. Additional exam requirements/recommendations for international students: Required—TOEFL (minimum score 600 paper-based; 250 computer-based). *Application deadline:* For fall admission, 1/10 for domestic students. Application fee: $60. Electronic applications accepted. *Financial support:* In 2005–06, 185 students received support, including 33 fellowships with full tuition reimbursements available, 85 research assistantships with full tuition reimbursements available, 67 teaching assistantships with full tuition reimbursements available; institutionally sponsored loans, scholarships/grants, health care benefits, tuition waivers (full and partial), and unspecified assistantships also available. Financial award applicants required to submit FAFSA. *Faculty research:* Analytical, organic, inorganic, physical, materials, chemical biology. *Unit head:* Director of Graduate Studies, 607-255-4139, Fax: 607-255-4137. *Application contact:* Graduate Field Assistant, 607-255-4139, Fax: 607-255-4137, E-mail: chemgrad@cornell.edu.

See Close-Up on page 109.

Georgetown University, Graduate School of Arts and Sciences, Department of Chemistry, Washington, DC 20057. Offers analytical chemistry (MS, PhD); biochemistry (MS, PhD); chemical physics (MS, PhD); inorganic chemistry (MS, PhD); organic chemistry (MS, PhD); physical chemistry (MS, PhD); theoretical chemistry (MS, PhD). Terminal master's awarded for partial completion of doctoral program. *Degree requirements:* For master's, thesis (for some programs), qualifying exam; for doctorate, thesis/dissertation, comprehensive exam. *Entrance requirements:* For master's and doctorate, GRE General Test. Additional exam requirements/recommendations for international students: Required—TOEFL.

University of Calgary, Faculty of Graduate Studies, Faculty of Science, Department of Chemistry, Calgary, AB T2N 1N4, Canada. Offers analytical chemistry (M Sc, PhD); applied chemistry (M Sc, PhD); inorganic chemistry (M Sc, PhD); organic chemistry (M Sc, PhD); physical chemistry (M Sc, PhD); polymer chemistry (M Sc, PhD); theoretical chemistry (M Sc, PhD). *Faculty:* 31 full-time (2 women), 2 part-time/adjunct (0 women). *Students:* 80 full-time (38 women). Average age 25. 250 applicants, 6% accepted, 15 enrolled. In 2005, 7 master's, 7 doctorates awarded. *Degree requirements:* For master's, thesis; for doctorate, thesis/dissertation, candidacy exam. *Entrance requirements:* For master's, minimum GPA of 3.0; for doctorate, honors B Sc degree with minimum GPA of 3.7 or M Sc with minimum GPA of 3.3. Additional exam requirements/recommendations for international students: Required—TOEFL (minimum score 580 paper-based; 237 computer-based). *Application deadline:* For fall admission, 12/1 priority date for domestic students, 12/1 priority date for international students. Applications are processed on a rolling basis. Application fee: $100 ($130 for international students). Electronic applications accepted. *Financial support:* In 2005–06, 25 students received support, including research assistantships (averaging $11,000 per year), teaching assistantships (averaging $6,530 per year); fellowships, scholarships/grants also available. Financial award application deadline:12/1. *Faculty research:* Chemical analysis, chemical dynamics, synthesis theory. *Unit head:* Dr. Brian Keay, Head, 403-220-5340, E-mail: info@chem.ucalgary.ca. *Application contact:* Bonnie E. King, Graduate Program Administrator, 403-220-6252, E-mail: bking@ucalgary.ca.

The University of Tennessee, Graduate School, College of Arts and Sciences, Department of Chemistry, Knoxville, TN 37996. Offers analytical chemistry (MS, PhD); chemical physics (PhD); environmental chemistry (MS, PhD); inorganic chemistry (MS, PhD); organic chemistry (MS, PhD); physical chemistry (MS, PhD); polymer chemistry (MS, PhD); theoretical chemistry (PhD). Part-time programs available. Terminal master's awarded for partial completion of doctoral program. *Degree requirements:* For master's and doctorate, thesis/dissertation. *Entrance requirements:* For master's and doctorate, GRE General Test, minimum GPA of 2.7. Additional exam requirements/recommendations for international students: Required—TOEFL. Electronic applications accepted.

Vanderbilt University, Graduate School, Department of Chemistry, Nashville, TN 37240-1001. Offers analytical chemistry (MAT, MS, PhD); inorganic chemistry (MAT, MS, PhD); organic chemistry (MAT, MS, PhD); physical chemistry (MAT, MS, PhD); theoretical chemistry (MAT, MS, PhD). *Faculty:* 32 full-time (6 women), 3 part-time/adjunct (2 women). *Students:* 96 full-time (40 women); includes 4 minority (2 African Americans, 1 Asian American or Pacific Islander, 1 Hispanic American), 20 international. 223 applicants, 21% accepted, 22 enrolled. In 2005, 4 master's, 12 doctorates awarded. *Degree requirements:* For master's, thesis or alternative; for doctorate, thesis/dissertation, area, qualifying, and final exams. *Entrance requirements:* For master's and doctorate, GRE General Test, GRE Subject Test (recommended). Additional exam requirements/recommendations for international students: Required—TOEFL. *Application deadline:* For fall admission, 1/15 for domestic students, 1/15 for international students. Application fee: $0. Electronic applications accepted. *Expenses:* Tuition: Full-time $15,396; part-time $1,283 per semester hour. Required fees: $2,202; $1,101 per semester. One-time fee: $30. Tuition and fees vary according to course load, program and student level. *Financial support:* Fellowships with full and partial tuition reimbursements, research assistantships with full tuition reimbursements, teaching assistantships with full tuition reimbursements, Federal Work-Study, institutionally sponsored loans, and traineeships available. Financial award application deadline:1/15. *Faculty research:* Chemical synthesis; mechanistic, theoretical, bioorganic, analytical, and spectroscopic chemistry. *Unit head:* Ned A. Porter, Chair, 615-322-2861, Fax: 615-343-1234. *Application contact:* Charles M. Lukehart, Director of Graduate Studies, 615-322-2861, Fax: 615-343-1234, E-mail: charles.m.lukehart@vanderbilt.edu.

Wesleyan University, Graduate Programs, Department of Chemistry, Middletown, CT 06459-0260. Offers biochemistry (MA, PhD); chemical physics (MA, PhD); inorganic chemistry (MA, PhD); organic chemistry (MA, PhD); physical chemistry (MA, PhD); theoretical chemistry (MA, PhD). *Faculty:* 13 full-time (2 women), 2 part-time/adjunct (1 woman). *Students:* 41 full-time (20 women); includes 1 minority (Asian American or Pacific Islander), 22 international. Average age 26. In 2005, 2 master's, 2 doctorates awarded. Terminal master's awarded for partial completion of doctoral program. *Degree requirements:* For master's and doctorate, one foreign language, thesis/dissertation. *Entrance requirements:* For master's, GRE General Test, GRE Subject Test; for doctorate, GRE Subject Test. *Application deadline:* For fall admission, 3/1 for domestic students. Applications are processed on a rolling basis. Application fee: $0. *Expenses:* Tuition: Full-time $24,732. One-time fee: $20 full-time. *Financial support:* Fellowships, research assistantships, teaching assistantships, institutionally sponsored loans available. *Unit head:* Dr. Phillip Bolton, Chair, 860-685-2668. *Application contact:* Karen Karpa, Information Contact, 860-685-2573, Fax: 860-685-2211, E-mail: kkarpa@wesleyan.edu.

See Close-Up on page 177.

West Virginia University, Eberly College of Arts and Sciences, Department of Chemistry, Morgantown, WV 26506. Offers analytical chemistry (MS, PhD); inorganic chemistry (MS, PhD); organic chemistry (MS, PhD); physical chemistry (MS, PhD); theoretical chemistry (MS, PhD). Part-time programs available. Postbaccalaureate distance learning degree programs offered (no on-campus study). *Faculty:* 14 full-time (1 woman), 5 part-time/adjunct (3 women). *Students:* 45 full-time (15 women), 3 part-time (2 women); includes 1 minority (Hispanic American), 26 international. Average age 28. 84 applicants, 58% accepted, 11 enrolled. In 2005, 5 master's, 3 doctorates awarded. Terminal master's awarded for partial completion of doctoral program. *Degree requirements:* For master's and doctorate, thesis/dissertation, registration. *Entrance requirements:* For master's, GRE General Test, GRE Subject Test (recommended), minimum GPA of 2.5; for doctorate, GRE General Test, GRE Subject Test (recommended), minimum GPA of 2.75. Additional exam requirements/recommendations for international students: Required—TOEFL. *Application deadline:* For fall admission, 3/1 for domestic students. Applications are processed on a rolling basis. Application fee: $50. Electronic applications accepted. *Expenses:* Tuition, area resident: Full-time $4,582; part-time $258 per credit hour. Tuition, state resident: full-time $4,582; part-time $258 per credit hour. Tuition, nonresident: full-time $1,382; part-time $741 per credit hour. International tuition: $1,382 full-time. *Financial support:* In 2005–06, fellowships (averaging $2,000 per year), research assistantships with full tuition reimbursements (averaging $13,000 per year), teaching assistantships with full tuition reimbursements (averaging $12,000 per year) were awarded; tuition waivers (full and partial) also available. Financial award application deadline: 2/1; financial award applicants required to submit FAFSA. *Faculty research:* Analysis of proteins, drug interactions, solids and effluents by advanced separation methods; new synthetic strategies for complex organic molecules; synthesis and structural characterization of metal complexes for polymerization catalysis. *Unit head:* Dr. Harry O. Finklea, Chair, 304-293-3435 Ext. 4453, Fax: 304-293-4904, E-mail: harry.finklea@mail.wvu.edu.

Cross-Discipline Announcements

Iowa State University of Science and Technology, Graduate College, College of Agriculture and College of Liberal Arts and Sciences, Department of Biochemistry, Biophysics, and Molecular Biology, Ames, IA 50011.

Department offers graduate degree programs in biochemistry; biophysics; genetics; molecular, cellular, and developmental biology; plant physiology; immunobiology; and toxicology. The design of the Molecular Biology Building enhances the opportunities for collaboration among faculty members in the life sciences. Students with degrees in chemistry are strongly encouraged to apply.

Massachusetts Institute of Technology, School of Engineering, Biological Engineering Division, Cambridge, MA 02139-4307.

Program provides opportunities for study and research at the interface of biology and engineering leading to specialization in bioengineering and applied biosciences. The areas include understanding how biological systems operate, especially when perturbed by genetic, chemical, or materials interventions or subjected to pathogens or toxins, and designing innovative biology-based technologies in diagnostics, therapeutics, materials, and devices for application to human health and diseases, as well as other societal problems and opportunities.

University of Wisconsin–Madison, Graduate School, Training Program in Biotechnology, Madison, WI 53706-1380.

The University of Wisconsin–Madison offers a predoctoral training program in biotechnology. Trainees receive PhDs in their major field (for example, chemistry) while receiving extensive cross-disciplinary training through the minor degree. Trainees participate in industrial internships and a weekly student seminar series with other program participants. These experiences reinforce the cross-disciplinary nature of the program. Students choose a major and a minor professor from a list of more than 130 faculty members in 40 different departments who conduct research related to biotechnology.

BRIGHAM YOUNG UNIVERSITY

Department of Chemistry and Biochemistry

Programs of Study The Department of Chemistry and Biochemistry at Brigham Young University (BYU) offers courses of study leading to Ph.D. and M.S. degrees in the areas of analytical, inorganic, organic, and physical chemistry and biochemistry. The research experience is the major element of the graduate programs. Most students complete their Ph.D. research in four to five years. Research groups often include cross-disciplinary collaboration with faculty members and students in biology, engineering, and physics as well as with other areas within chemistry and biochemistry. Department faculty members are highly involved in each student's progress and foster a strong tradition of mentoring. All chemistry students must pass proficiency exams demonstrating competence at the undergraduate level in at least four subject areas by the end of their first year; biochemistry students must prove proficiency in biochemistry. An individualized schedule of graduate courses is established for each student based on his or her needs and interests. Most of this course work is taken during the first year, with the remainder completed in the second year. Depending on the area of study chosen by the student, either a form of comprehensive examination or several periodic cumulative examinations are required. Also, all students present annual reviews and a research proposal to their faculty advisory committee. An active seminar schedule provides exposure to recent developments worldwide. Successful defense of a dissertation or thesis completes a student's training.

Research Facilities Research activities occupy more than 50 percent of a 192,000-square-foot building. The University library, where the science collection includes more than 500,000 volumes and about 9,000 journal subscriptions, is located about 150 yards away. In addition to instruments and facilities used by individual research groups, special research facilities used by the entire Department include NMR (200, 300, and 500 MHz), mass spectrometry (high-resolution, quadrupole, ion cyclotron resonance, ToF-SIMS, and MALDI), X-ray diffraction (powder and single crystal), spectrophotometry (IR, visible, UV, X-ray, and γ-ray), lasers (YAG, excimer, and dye), environmental analysis (PIXE; PIGE; trace gases; X-ray fluorescence; XPS(ESCA); chromatography, including capillary column GC/MS, ion, and supercritical fluid; particle size analyzers; environmental chambers; ICP; and capillary electrophoresis), thermodynamics (calorimeters of all types, including temperature and pressure scanning, titration, flow, heat conduction, power compensation, combustion, and metabolic), and molecular biology (DNA synthesizer and sequencer, phosphorimager, tissue culture facility, recombinant DNA facility, and ultracentrifuges). All computing facilities are fully networked, including computational chemistry and laboratory workstations as well as office personal computers, with convenient connection to supercomputing facilities and the Internet. Fully staffed shops for glassblowing, machining, and electronics also serve research needs.

Financial Aid The Department provides full financial assistance to all students through teaching and research assistantships, fellowships, and tuition scholarships. The twelve-month stipend for beginning students for the 2006–07 year is $20,000 (taxable). The amount of the stipend is adjusted annually.

Cost of Study Full tuition scholarships are provided to all graduate students making satisfactory progress. Books average about $200 per semester.

Living and Housing Costs The University Housing Office assists students in locating accommodations for on- and off-campus housing. Monthly rent and utilities range from $220 to $300 for a single student and from $450 to $650 for families. Other expenses range from $150 to $300 for single students and from $250 to $500 for families.

Student Group BYU has about 33,000 full-time students, with about 3,000 full-time graduate students. Students come from various academic and ethnic backgrounds and many geographic areas. The Department averages 90 graduate students. Thirty-nine percent are women; currently, 67 percent are international students from eleven countries.

Student Outcomes BYU graduate degrees lead to a wide range of independent careers, with former students serving in academia, government, and industrial positions. About half of recent Ph.D.'s have continued their training in postdoctoral positions at leading research institutions, with the remainder finding employment directly with regional or national firms, in academia, or at government labs.

Location BYU's beautiful 560-acre campus, with all the cultural and sports programs of a major university, is located in Provo, Utah (population 110,500), a semiurban area at the foot of Utah Valley's Wasatch Mountains. Outdoor recreational areas for skiing (snow and water), hiking, and camping are nearby, including nine spectacular national parks, six beautiful national monuments, fourteen major ski resorts, and forty-five diverse state parks. The Utah Symphony, Ballet West, Pioneer Theater Company, and the Utah Jazz basketball team are located in Salt Lake City, 45 miles to the north.

The University Brigham Young University is one of the largest privately owned universities in the United States. Established in 1875 as Brigham Young Academy and sponsored by the Church of Jesus Christ of Latter-day Saints, BYU has a tradition of high standards in moral integrity and academic scholarship. Along with extensive undergraduate programs, BYU offers graduate degrees in a variety of disciplines through fifty-three graduate departments, including the Marriott School of Management and the J. Reuben Clark Law School. The Department of Chemistry and Biochemistry is one of the leading research departments at BYU.

Applying Applicants should apply online at http://www.byu.edu/gradstudies. They must submit transcripts, letters of recommendation, and a $50 application fee. International students must pass the Test of English as a Foreign Language (TOEFL) or the International English Language Testing System (IELTS). The TOEFL score requirement is 580 (237 computer-based test) or higher, although preference is given to students with scores of 600 or higher. The IELTS requirement is an overall band score of 7.0, with a minimum band score of 6.0 in each module. The GRE General Test is required; GRE Subject Tests in chemistry or biochemistry are highly recommended. All application materials should be received by the Office of Graduate Studies no later than February 1 to be considered for admission the following fall. Applicants are not discriminated against on the basis of race, color, national origin, religion, gender, or handicap.

Correspondence and Information

Dr. David V. Dearden
Coordinator, Graduate Admissions
C101 BNSN (Benson Science Building)
Brigham Young University
Provo, Utah 84602

Phone: 801-422-4845
Fax: 801-422-0153
E-mail: chemguide@byu.edu
Web site: http://grads.chem.byu.edu

Brigham Young University

THE FACULTY AND THEIR RESEARCH

Merritt B. Andrus, Associate Professor; Ph.D., Utah, 1991; postdoctoral study at Harvard. Organic chemistry: synthetic organic chemistry that includes asymmetric catalytic methods, natural product synthesis, and combinational libraries.
Matthew C. Asplund, Assistant Professor; Ph.D., Berkeley, 1998; postdoctoral study at Pennsylvania. Physical chemistry: time-resolved infrared spectroscopy, reaction dynamics, surface chemistry.
Daniel E. Austin, Assistant Professor; Ph.D., Caltech, 2002; Staff Scientist, Sandia National Labs, 2002–05. Analytical chemistry: mass spectrometry, ion mobility, microparticle acceleration and impact ionization, instrumentation for planetary science and astrobiology.
David M. Belnap, Assistant Professor; Ph.D. Purdue, 1995; postdoctoral study at National Institutes of Health. Biochemistry: structure, function, assembly, and disassembly of viruses and cellular macromolecules; three-dimensional electron microscopy (cryoelectron microscopy and 3-D image reconstruction).
Juliana Boerio-Goates, Professor; Ph.D., Michigan, 1979; postdoctoral study at Michigan. Physical chemistry: structural and thermodynamic studies of phase transitions in molecular and ionic crystals, thermodynamics of biological materials, energetics of nanomaterials.
Gregory F. Burton, Professor; Ph.D., Virginia Commonwealth, 1989; postdoctoral study at Virginia Commonwealth, 1989–91. Biochemistry: molecular mechanisms of HIV/AIDS pathogenesis focusing on contributions of the follicular dendritic cell reservoir of HIV.
Allen R. Buskirk, Assistant Professor; Ph.D., Harvard, 2004. Biochemistry: directed molecular evolution of proteins and nucleic acids, structure/function of small RNAs in bacteria, protein-protein interactions and allostery in signaling pathways.
Steven L. Castle, Assistant Professor; Ph.D., Scripps Research Institute, 2000. NIH Fellowship, 2000–02, California, Irvine. Organic chemistry: development of new synthetic methods, natural product synthesis.
David V. Dearden, Professor; Ph.D., Caltech, 1989. NRC Fellowship, 1989, U.S. National Institute of Standards and Technology. Analytical/physical chemistry: host-guest molecular recognition in ion-molecule reactions, Fourier transform ion cyclotron resonance mass spectrometry with electrospray ionization.
Delbert J. Eatough, Professor; Ph.D., Brigham Young, 1967. Physical chemistry: environmental atmospheric chemistry of SO_x, NO_x, and organics; environmental analytical techniques; tracers and source apportionment of atmospheric pollutants; indoor atmospheric chemistry.
Paul B. Farnsworth, Professor; Ph.D., Wisconsin, 1981; postdoctoral study at Indiana. Analytical chemistry: fundamental and applied measurements on inductively coupled plasmas, elemental mass spectrometry, fluorescence microscopy.
Steven A. Fleming, Professor; Ph.D., Wisconsin, 1984. NIH Fellowship, 1984–86, Colorado State. Organic chemistry: photochemistry of aromatic compounds, rearrangements of small ring heterocycles, synthesis of natural products and novel compounds, determination of mechanisms of thermal rearrangements and photorearrangements.
Steven R. Goates, Professor; Ph.D., Michigan, 1981; postdoctoral study at Columbia. Analytical chemistry: analysis of complex samples by optical and especially laser-based methods, supersonic jet spectroscopy, spectroscopy of solid-state phenomena, investigation of chromatographic processes.
Steven W. Graves, Associate Professor; Ph.D., Yale, 1978; postdoctoral fellow, Tufts University School of Medicine, 1978–81; clinical chemistry fellow, Washington University School of Medicine, 1981–83. Biochemistry and bioanalytical/clinical chemistry and physiology: mechanisms of hypertensive complications of pregnancy, Na^+ pump regulation in disease, clinical assay development, proteomics of preterm birth and proteomics of preeclampsia.
Jaron C. Hansen, Assistant Professor; Ph.D., Purdue, 2002; postdoctoral study at Caltech Jet Propulsion Laboratory. Analytical/physical chemistry: kinetics and spectroscopy of atmospherically important molecules, development of novel kinetic techniques and instrumentation.
Roger G. Harrison, Associate Professor; Ph.D., Utah, 1993; postdoctoral study at Minnesota. Inorganic chemistry: host-guest molecular recognition, catalytic enzyme model complexes, metal-assembled cages, CdSe nanocrystals.
Steven R. Herron, Assistant Research Professor; Ph.D., California, Irvine, 2001. Staff Scientist, 2001–04, California State, Fullerton. Biochemistry: X-ray crystallography and protein structure-function.
John D. Lamb, Eliot A. Butler Professor; Ph.D., Brigham Young, 1978. Program Manager, Separations and Analysis, Office of Basic Energy Sciences, U.S. D.O.E., 1982–84. Inorganic chemistry: macrocyclic ligand chemistry, liquid membrane separations, ion chromatography, calorimetry.
Milton L. Lee, H. Tracy Hall Professor; Ph.D., Indiana, 1975; postdoctoral study at MIT. Analytical chemistry: capillary separations, microfluidics, mass spectrometry instrumentation, ion mobility spectrometry.
Matthew R. Linford, Assistant Professor; Ph.D., Stanford, 1996; postdoctoral study at the Max Planck Institute for Surface and Colloid Science. Analytical chemistry: functionalizing and patterning silicon with alkyl monolayers, time-of-flight secondary ion mass spectrometry, chemometrics, ultrathin organic and polymer films.
Noel L. Owen, Professor; Ph.D., Cambridge, 1964; postdoctoral study at Harvard. Physical chemistry: FTIR and NMR spectroscopy, molecular structure of bio-active natural products extracted from medicinally important plants and plant endophytes, spectroscopic studies of solid wood and wood polymers.
Matt A. Peterson, Associate Professor; Ph.D., Arizona, 1992. NIH Fellowship, 1993–94, Colorado State. Organic chemistry: synthetic methods, nucleosides and nucleotides, enediynes, enzyme inhibitors, development of antiviral and anticancer agents.
Morris J. Robins, J. Rex Goates Professor; Ph.D., Arizona State, 1965. Cancer Research Fellowship, 1965–66, Roswell Park Memorial Institute. Organic chemistry: chemistry of nucleic acid components, nucleoside analogues, and related biomolecules; design of mechanism-based enzyme inhibitors; development of anticancer and antiviral agents.
Paul B. Savage, Professor; Ph.D., Wisconsin, 1993. NIH Fellowship, 1994–95, Ohio State. Organic and bioorganic chemistry: development of membrane active antibiotics, glycolipid immunology.
Eric T. Sevy, Assistant Professor; Ph.D., Columbia, 1999; postdoctoral study at MIT. Physical chemistry: chemical and molecular dynamics of collisional relaxation, energy transfer processes, and atmospheric and combustion chemistry; high-resolution transient spectroscopy; photolithography.
Randall B. Shirts, Associate Professor; Ph.D., Harvard, 1979; postdoctoral study at Joint Institute for Laboratory Astrophysics and University of Colorado. Physical chemistry: theoretical chemistry, statistical mechanics of nanoscale systems, nonlinear dynamics, semiclassical quantization, laser-molecule interaction, quantum-classical correspondence.
Daniel L. Simmons, Professor; Ph.D., Wisconsin, 1986. NIH and Leukemia Society Fellowships, 1986–89, Harvard. Biochemistry: molecular mechanisms of neoplastic transformation by Rous sarcoma virus, messenger RNA splicing and translation mechanisms, prostaglandins and signal transduction.
Craig D. Thulin, Assistant Professor; Ph.D., Washington (Seattle), 1995; postdoctoral study at Vollum Institute for Advanced Biomedical Research and at Brigham Young. Biochemistry: proteomics and protein chemistry of degenerative diseases and signal transduction pathways.
Heidi R. Vollmer-Snarr, Assistant Professor; D.Phil., Oxford, 2000; NIH Fellowships, 2001–02, Sloan-Kettering Cancer Center and Columbia. Organic chemistry: synthesis, bio-organic and natural products chemistry, applicable to cancer and macular degeneration.
Gerald D. Watt, Professor; Ph.D., Brigham Young, 1968. NIH Fellowship, 1968, Yale. Biochemistry: nitrogenase, ferritins, metalloproteins.
Barry M. Willardson, Associate Professor; Ph.D., Purdue, 1990; postdoctoral study at Los Alamos National Laboratory. Biochemistry: regulation of G-protein-mediated signal transduction by phosducin-like proteins.
Brian F. Woodfield, Associate Professor; Ph.D., Berkeley, 1994; postdoctoral study at Naval Research Laboratory, Material Physics Branch, Washington, D.C. Physical chemistry: solid-state physics and thermodynamic properties of materials at low temperatures; systems of interest include nanomaterials, novel magnetic systems, heavy metals, and other technologically important materials.
Adam T. Woolley, Assistant Professor; Ph.D., Berkeley, 1997; Runyon-Winchell Cancer Research Fund Fellowship, 1998–2000, Harvard. Analytical chemistry: scanning probe microscopy of biomolecules with carbon nanotube tips, self-assembly and nanofabrication from biomolecular templates, microfabrication of miniaturized biochemical analysis systems.
Earl M. Woolley, Professor; Ph.D., Brigham Young, 1969. NRC Canada Fellowship, 1969–70, Lethbridge. Physical chemistry: thermodynamics of reactions in mixed aqueous-organic solvents, intermolecular hydrogen bonding reactions, electrolyte solutions, and of ionic and nonionic reactions in solution including surfactants; calorimetric methods.
S. Scott Zimmerman, Professor; Ph.D., Florida State, 1973. NIH Fellowship, 1973–77, Cornell. Biochemistry: molecular modeling and computational chemistry of biologically important molecules.

SELECTED PUBLICATIONS

Ma, Y., C. Song, J. We, and **M. B. Andrus.** Asymmetric, arylboronic acid addition to enones using novel dicyclophane imidazolium carbene-rhodium catalysts. *Angew. Chem. Int. Ed.* 42:5871–4, 2003.

Andrus, M. B., et al. Total synthesis of (+)-geldanamycin and (-)-ortho-quinogeldanamycin, asymmetric glycolate aldol reactions and biological evaluation. *J. Org. Chem.* 68:8162–9, 2003.

Andrus, M. B., J. Liu, E. L. Meredith, and E. Nartey. Synthesis of resveratrol using a direct decarbonylative heck approach from resorcylic acid. *Tetrahedron Lett.* 44: 4819–22, 2003.

Pan, T., et al. **(M. C. Asplund).** Fabrication of calcium fluoride capillary electrophoresis microdevices for on-chip infrared detection. *J. Chromatogr. A* 1027 (1–2):231–5, 2004.

Husseini, G. A., et al. **(M. C. Asplund, E. T. Sevy,** and **M. R. Linford).** Photochemical lithography: Creation of patterned, acid chloride functionalized surfaces using UV light and gas-phase oxalyl chloride. *Langmuir* 19(11):4856–8, 2003.

Lua, Y.-Y., et al. **(M. C. Asplund, S. A. Fleming,** and **M. R. Linford).** Amine-reactive monolayers on scribe silicon with controlled levels of functionality: The reaction of scribed silicon with mono- and diepoxides. *Angew. Chem. Int. Ed.* 42: 4046–9, 2003.

Asplund, M. C., M. T. Zanni, and R. M. Hochstrasser. Two-dimensional infrared spectroscopy of peptides by phase-controlled femtosecond vibrational photon echoes. *Proc. Natl. Acad. Sci. U.S.A.* 97:8219–24, 2000.

Bubeck, D., et at. **(D. M. Belnap).** The structure of the poliovirus 135S cell entry intermediate at 10-Angstrom resolution reveals the location of an externalized polypeptide that binds to membranes. *J. Virol.* 79:7745–55, 2005.

Heymann, J. B., M. Chagoyen, and **D. M. Belnap.** Common conventions for interchange and archiving of three-dimensional electron microscopy information in structural biology. *J. Struct. Biol.* 151:196–207, 2005.

Belnap, D. M., et al. Diversity of core antigen epitopes of hepatitis B virus. *Proc. Natl. Acad. Sci. U.S.A.* 100:10884–9, 2003.

Belnap, D. M., et al. Three-dimensional structure of poliovirus receptor bound to poliovirus. *Proc. Natl. Acad. Sci. U.S.A.* 97:73–8, 2000.

Li, G., L. Li, **J. Boerio-Goates,** and **B. F Woodfield.** High purity anatase TiO2 nanocrystals: Near room-temperature synthesis, grain growth kinetics, and surface hydration chemistry. *J. Am. Chem. Soc.* 127:8659–66, 2005.

Li, G., **J. Boerio-Goates, B. F. Woodfield,** and L. Li. Evidence of linear lattice expansion and covalency enhancement in rutile TiO2 nanocrystals. *Appl. Phys. Lett.* 85:2059–61, 2004.

Boerio-Goates, J., et al. Thermochemistry of inosine. *J. Chem. Thermodyn.* 37: 1239–49, 2005.

E. Bakken, and **J. Boerio-Goates** et al. Entropy of oxidation and redox energetics of CaMnO3-δ. *Solid State Ionics* 176:2261–7, 2005.

Estes, J. D., et al. **(G. F. Burton).** Follicular dendritic cell regulation of CXCR4-mediated germinal center CD4 T cell migration. *J. Immunol.* 173:6169–78, 2004.

Burton, G. F., et al. Follicular dendritic cell (FDC) contributions to HIV pathogenesis. *Semin. Immunol.* 14:275–84, 2002.

Estes, J. D., et al. **(G. F. Burton).** Follicular dendritic cell-mediated up-regulation of CXCR4 expression on CD4+T cells and HIV pathogenesis. *J. Immunol.* 169: 2313–22, 2002.

Smith, B. A., et al. **(G. F. Burton).** Persistence of infectious human immunodeficiency virus on follicular dendritic cells. *J. Immunol.* 166:690–6, 2001.

Buskirk, A. R., and D. R. Liu. Creating small-molecule-dependent switches to modulate biological functions. *Chem. Biol.* 12(2):151–61, 2005. Review.

Buskirk, A. R., A. Landrigan, and D. R. Liu. Engineering a ligand-dependent RNA transcriptional activator. *Chem. Biol.* 11(8):1157–63, 2004.

Buskirk, A. R., Y. C. Ong, Z. J. Gartner, and D. R. Liu. Directed evolution of ligand dependence: small- molecule-activated protein splicing. *Proc. Natl. Acad. Sci. U.S. A.*101(29):10505–10, 2004.

Buskirk, A. R., P. D. Kehayova, A. Landrigan, and D. R. Liu. In vivo evolution of an RNA-based transcriptional activator. *Chem. Biol.* 10(6):533–40, 2003.

Srikanth, G. S. C., and **S. L. Castle.** Advances in radical conjugate additions. *Tetrahedron* 61:10377–441, 2005.

He, L., et al. **(S. L. Castle).** Synthesis of B-substituted x-amino acids via Lewis acid promoted enantioselective radical conjugate additions. *J. Org. Chem.* 70:8140–7, 2005.

Reeder, M. D., et al. **(S. L. Castle).** Synthesis of the core structure of acutumine. *Org. Lett.* 7:1089–92, 2005.

Zhang, H., et al. **(D. V. Dearden).** Cucurbit[6]uril pseudorotaxanes: Distinctive gas phase dissociation and reactivity. *J. Am. Chem. Soc.* 125:9284–5, 2003.

Anderson, J. D., E. S. Paulsen, and **D. V. Dearden.** Alkali metal binding energies of dibenzo-18-crown-6: Experimental and computational results. *Int. J. Mass Spectrom.* 227:63–76, 2003.

Zhou, L., et al. **(D. V. Dearden** and **M. L. Lee).** Incorporation of a venturi device in electrospray ionization. *Anal. Chem.* 75:5978, 2003.

Liang, Y., J. S. Bradshaw, and **D. V. Dearden.** The thermodynamic basis for enantiodiscrimination: Gas phase measurement of the enthalpy and entropy of chiral amine recognition by dimethyldiketopyridino-18-crown-6. *J. Phys. Chem. A* 106:9665–71, 2002.

Pope, C. A., et al. **(D. J. Eatough).** Ambient particulate air pollution, heart rate variability, and blood markers of inflammation in a panel of elderly subjects. *Environ. Health Perspect.* 112(3):339–45, 2004.

Eatough, D. J., et al. Semi-volatile particulate organic material in southern Africa during SAFARI-2000. *J. Geophys. Res.* 108(D13):8479–786, 2003.

Long, R. W., et al. **(D. J. Eatough).** The measurement of PM2.5, including semi-volatile components, in the EMPACT program: Results from the Salt Lake City study. *Atmos. Environ.* 37:4407–17, 2003.

Macedone, J. H., A. A. Mills, and **P. B. Farnsworth.** Optical measurements of ion trajectories through the vacuum interface of an inductively coupled plasma mass spectrometer. *Appl. Spectrosc.* 58:463–7, 2004.

Macedone, J. H., D. J. Gammon, and **P. B. Farnsworth.** Factors affecting analyte transport through the sampling orifice of an inductively coupled plasma mass spectrometer. *Spectrochimica Acta* 56B:1687–95, 2001.

Fleming, S. A., R. Liu, J. T. Redd. Asymmetric dihydroxylation of disubstituted allenes. *Tetrahedron Lett.* 46:8095–8, 2005.

Fleming, S. A. Photocycloaddition of alkenes to excited alkenes. *Synthetic Organic Photochemistry, Molecular and Supramolecular Photochemistry* 12:141–60, 2004.

Fleming, S. A., S. C. Ward, C. Mao, and E. E. Parent. Photocyclization. *Spectrum* 15(3):8–14, 2002.

Goates, S. R., D. A. Schofield, and C. D. Bain. A study of nonionic surfactants at the air-water interface by sum-frequency spectroscopy and ellipsometry. *Langmuir* 15:1400–9, 1999.

Ji, Q., et al. **(S. R. Goates).** New optical design for laser flash photolysis studies in supercritical fluids. *Rev. Sci. Instrum.* 66:222–6, 1995.

Bently-Lewis, R, **S. W. Graves,** and E. W. Seely. The renin-aldosterone response to stimulation and suppression during normal pregnancy. *Hypertens Pregnancy* 24:1–16, 2005.

Merrell, K., et al. **(S. W. Graves).** Analysis of low-abundance, low-molecular-weight serum proteins using mass spectrometry. *J. Biomolec. Techniques* 14:235–44, 2004.

Esplin, M. S., M. B. Fausett, D. S. Faux, and **S. W. Graves.** Changes in the isoforms of the sodium pump in the placenta and myometrium of women in labor. *Am. J. Obstet. Gynecol.* 188:759–64, 2003.

Harrison, R. G., J. L. Burrows, and L. D. Hansen. Selectric guest encapsulation by a cobalt-assembled cage molecule. *Chem. Eur. J.* 11:5881–8, 2005.

Gardner, J. S., et al. **(R. G. Harrison** and **J. D. Lamb).** Anion binding by a tetradipicolylamine-substituted resorcinarene cavitand. *Inorg. Chem.* 44:4295–300, 2005.

Harrison, R. G., et al. Cation control of pore and channel size in cage-based metal-organic porous materials: *Inorg. Chem.* 41:838–43, 2002.

Reinheimer, E. W., et al. **(S. R. Herron).** Crystal structures of diphosphinated group 6 Fischer alkoxy carbenes. *J. Chem. Crystallogr.* 33:503–14, 2003.

Herron, S. R., et al. Characterization and implications of Ca^{2+} binding to pectate lyase C. *J. Biol. Chem.* 278:12271–7, 2003.

Herron, S. R., et al. Structure and function of pectic enzymes: Virulence factors of plant pathogens. *Proc. Natl. Acad. Sci. U.S.A.* 97:8762–9, 2000.

Lamb, J. D., and J. S. Gardner. Application of Macrocyclic Ligands to Analytical Chromatography. In *Macrocyclic Chemistry: Current Trends and Future Perspectives,* pp. 349–63, ed. Karsten Gloe. Massachusetts: Kluwer Academic Publishers, 2005.

Gardner, J., J. Walker, and **J. D. Lamb.** Permeability and durability effects of cellulose polymer variation in polymer inclusion membranes. *J. Membrane Sci.* 229:87–93, 2004.

Richens, D. A., et al. **(J. D. Lamb).** Use of mobile phase 18-crown-6 to improve peak resolution between mono- and divalent metal and amine cations in ion chromatography. *J. Chromatogr. A* 1016:155–64, 2003.

Nazarenko, A. Y., et al. **(J. D. Lamb).** Crystal structure of 13,27-dichloro-29,30-dihydroxy-3,9,17,23-tetramethyl-6,20-dioxa-3,9,17,23-tetraazatri-cyclo $[23.3.1.1^{11.}{}_{15}]$triaconta-1(29),11,13,15(30),25,27-hexaene-(aqua)magnesium hydrate, $Mg[C_{28}H_{40}C_{12}N_4O_4(H_2O)\ H_2O$. *Zeitschrift fur Kristallographie NCS* 218:1–3, 2003.

Paul, P. U., L. Li, **M. L. Lee,** and **P. B. Farnsworth.** Compact detector for proteins based on two-photon excitation of native fluorescence. *Anal. Chem.* 77(11):3690–3, 2005.

Gu, B., J. M. Armenta, and **M. L. Lee.** Preparation and evaluation of poly (polyethylene glycol methyl ether acrylate-co-polyethylene glycol diacrylate) monolith for protein analysis. *J. Chromatogr. A* 1079:382, 2005.

Yue, B., E. D. Lee, A. L. Rockwood, and **M. L. Lee.** Electron ionization in superimposed magnetic and radio frequency quadrupolar electric fields. *Anal. Chem.* 77:4160, 2005.

Peterson, Z. D., **M. L. Lee,** and S. W. Graves. Determination of serotonin and its precursors in human plasma by capillary electrophoresis-electrospray time-of-flight mass spectrometry. *J Chromatog. B* 810:101–10, 2004.

Wacaser, B. A., et al. **(M. R. Linford).** Chemomechanical surface patterning and functionalization of silicon surfaces using an atomic force microscope. *Appl. Phys. Lett.* 82(5):808–10, 2003.

Niederhauser, T. L., et al. **(M. R. Linford).** Arrays of chemomechanically patterned patches of homogeneous and mixed monolayers of 1-alkenes and alcohols on single silicon surfaces. *Angew. Chem. Int.* 41(13):2353–6, 2002.

Brigham Young University

Owen, N. L., and N. J. Hundley. Endophytes–the chemical synthesizers inside plants. *Sci. Progress* 87:79–99, 2004.

Cannon, J. G., R. A. Burton, S. G. Wood, and **N. L. Owen.** Naturally occurring fish poisons from plants. *J. Chem. Educ.* 81:1457–61, 2004.

Moore, A. K., and **N. L. Owen.** Infrared spectroscopic studies of solid wood. *Appl. Spectrosc. Rev.* 36:65–86, 2001.

Zhao, Z., et al. **(M. A. Peterson).** Bergman cycloaromatization of imidazole-fused enediynes: The remarkable effect of *N*-aryl substitution. *Tetrahedron Lett.* 45:3621–4, 2004.

Peterson, M. A., A. Bowman, and S. Morgan. Efficient preparation of N-benzyl secondary amines via benzylamine-borane mediated reductive amination. *Synthetic Commun.* 32:443–8, 2002.

Kumarasinghe, E. S., **M. A. Peterson,** and **M. J. Robins.** Synthesis of 5,6-bis (alkyny-1-yl)pyrimidines and related nucleosides. *Tetrahedron Lett.* 41:8741–5, 2000.

Janeba, Z., P. Francom, and **M. J. Robins.** Efficient syntheses of 2-chloro-2'-deoxyadenosine (cladribine) from 2'-deoxyguanosine. *J. Org. Chem.* 68:989–92, 2003.

Nowak, I., and **M. J. Robins.** Protection of the amino group of adenosine and guanosine derivatives by elaboration into a 2,5-dimethylpyrrole moiety. *Org. Lett.* 5:3345–8, 2003.

Robins, M. J. Ribonucleotide reductases: Radical chemistry and inhibition at the active site. *Nucleosides Nucleotides Nucl. Acids* 22:519–34, 2003.

Mattner, J., et al. **(P. B. Savage).** Both exogenous and endogenous glycolipid antigens activate NKT cells during microbial infections. *Nature* 434:525–8, 2005.

Ding, B., et al. **(P. B. Savage).** Origins of cell selectivity of cationic steroid antibiotics. *J. Am. Chem. Soc.* 126:13642–8, 2004.

Goff, R. D., et al. **(P. B. Savage).** Effects of lipid chain lengths in a-galactosylceramides on cytokine release by natural killer T cells. *J. Am. Chem. Soc.* 126:13602–3, 2004.

Sevy, E. T., et al. Competition between photochemistry and energy transfer in ultraviolet-excited diazabenzenes: I. Photofragmentation studies of pyrazine at 248 nm and 266 nm. *J. Chem. Phys.* 112:5829–43, 2000.

Sevy, E. T., C. A. Michaels, H. C. Tapalian, and G. W. Flynn. Competition between photochemistry and energy transfer in ultraviolet-excited diazabenzenes: II. Identifying the dominant energy donor for "supercollisions." *J. Chem. Phys.* 112: 5844–51, 2000.

Sevy, E. T., S. M. Rubin, Z. Lin, and G. W. Flynn. Translational and rotational excitation of the CO_2 (00^00) vibrationless state in the collisional quenching of highly vibrationally excited 2-methylpyrazine: Kinetics and dynamics of large energy transfers. *J. Chem. Phys.* 113:4912–32, 2000.

Shirts, R. B., and M. R. Shirts. Deviations from the Boltzmann distribution in small microcanonical quantum systems: Two approximate one-particle energy distributions. *J. Chem. Phys.* 117:5564–75, 2002.

Sohlberg, K., and **R. B. Shirts.** The symmetry of approximate Hamiltonians generated in Birkhoff-Gustavson normal form. *Phys. Rev. A* 54:416–22, 1996.

Sohlberg, K., and **R. B. Shirts.** Semiclassical quantization of nonintegrable system: Pushing the Fourier method into the chaotic regime. *J. Chem. Phys.* 101:7763–78, 1994.

Ballif, B. A., et al. **(D. L. Simmons).** Interaction of cyclooxygenases with an apoptosis- and autoimmunity-associated protein. *Proc. Natl. Acad. Sci. U.S.A.* 93: 5544–9, 1996.

Lu, X., et al. **(D. L. Simmons).** Nonsteroidal anti-inflammatory drugs cause apoptosis and induce cyclooxygenases in chicken embryo fibroblasts. *Proc. Natl. Acad. Sci. U.S.A.* 92:7961–5, 1995.

Xie, W., et al. **(D. L. Simmons).** Expression of a mitogen-responsive gene encoding prostaglandin synthase is regulated by mRNA splicing. *Proc. Natl. Acad. Sci. U.S.A.* 88:2692–6, 1991.

McLaughlin, J. N., et al. **(C. D. Thulin).** Regulation of cytosolic chaperonin-mediated protein folding by phosducin-like protein. *Proc. Natl. Acad. Sci. U.S.A.* 99:7962–7, 2002.

McLaughlin, J. N., et al. **(C. D. Thulin).** Regulation of angiotensin II–induced G protein signaling by phosducin-like protein. *J. Biol. Chem.* 277:34885–95, 2002.

Thulin, C. D., et al. **(B. M. Willardson).** Modulation of the G-protein regulator phosducin by calcium-calmodulin dependent protein kinase II phosphorylation and 14-3-3 protein binding. *J. Biol. Chem.* 276:23805–15, 2001.

Vollmer-Snarr, H. R., et al. Amino-Retinoid Compounds in the Human Retinal Pigment Epithelium. In *Retinal Degenerations: Mechanisms and Experimental Therapy,* eds. M. M. LaVail, J. G. Hollyfield, and R. E. Anderson. New York: Kluwer Academic/ Plenum Publishers, in press.

G. Karan, et al. **(Vollmer-Snarr, H. R.).** Lipofuscin accumulation, abnormal electrophysiology and photoreceptor degeneration in mutant ELOVL4 transgenic mice: A model for Stargardt macular degeneration. *Proc. Natl. Acad. Sci. U.S.A.* 102:4164–9, 2005.

Jockusch, S., et al. **(H. R. Vollmer-Snarr).** Photochemistry of A1E, a retinoid with a conjugated pyridinium moiety: Competition between pericyclic photooxygenation and pericyclization. *J. Am. Chem. Soc.* 126:4646–52, 2004.

Lindsay, S., et al. **(G. D. Watt).** Kinetic studies of iron deposition in horse spleen ferritin using O2 as oxidant. *Biochem. Biophys. Acta* 1621:57–66, 2003.

Lindsay, S., D. Brosnahan, and **G. D. Watt.** Hydrogen peroxide formation during iron deposition in horse spleen ferritin using O2 as an oxidant. *Biochemistry* 40: 3340–7, 2001.

Nyborg, A. C., J. L. Johnson, A. Gunn, and **G. D. Watt.** Evidence for a two-electron transfer using the all-ferrous Fe protein during nitrogenase catalysis. *J. Biol. Chem.* 275:39307–12, 2000.

Lukov, G. L., et al. **(B. M. Willardson).** Role of the isoprenyl pocket of the G protein $\beta\gamma$ subunit complex in the binding of phosducin and phosducin-like protein. *Biochemistry* 43:5651–60, 2004.

McLaughlin, J. N., et al. **(B. M. Willardson).** Regulatory interaction of phosducin-like protein with the cytosolic chaperonin complex. *Proc. Natl. Acad. Sci. U.S.A.* 99:7962–7, 2002.

Crawford, M. K., et al. **(B. F. Woodfield** and **J. Boerio-Goates).** Structure and properties of the integer-spin frustrated antiferromagnet GeNi2O4. *Phys. Rev. B* 68(22):220408/1–4, 2003.

Piccione, P. M., **B. F. Woodfield,** and **J. Boerio-Goates** et al. Entropy of pure molecular sieves. *J. Phys. Chem. B* 105:6025–30, 2001.

Woodfield, B. F., et al. **(J. Boerio-Goates).** Critical phenomena at the antiferromagnetic transition in MnO. *Phys. Rev. B* 60:7335, 1999.

Woodfield, B. F., et al. Superconducting-normal phase transition in $(Ba_{1-x}K_x)BiO_3$, x=0.40, 0.47. *Phys. Rev. Lett.* 83:4622–5, 1999.

Brown, B. R., S. P. Ziemer, T. L. Niederhauser, and **E. M. Woolley.** Apparent molar volumes and apparent molar heat capacities of aqueous D(+)-cellobiose, D(+)-maltose, and sucrose from 278.15 K to 393.15 K and at the pressure 0.4 MPa. *J. Chem. Thermodyn.* 37(8):843–53, 2005.

Ziemer, S. P., T. L. Niederhauser, and **E. M. Woolley.** Thermodynamics of complexation of aqueous 18-crown-6 with barium ion: Apparent molar volumes and apparent molar heat capacities of the aqueous (18-crown-6 + barium nitrate) complex at temperatures (278.15 to 393.15) K, at molalities (0.02 to 0.33) mol·kg-1, and at the pressure 0.35 MPa.

Ziemer, S. P., T. L. Niederhauser, and **E. M. Woolley.** Thermodynamics of complexation of aqueous 18-crown-6 with sodium ion: Apparent molar volumes and apparent molar heat capacities of the aqueous (18-crown-6 + sodium chloride) complex at temperatures (278.15 to 393.15) K, at molalities (0.02 to 0.30) mol·kg-1, and at the pressure 0.35 MPa. *J. Chem. Thermodyn.* 37(10):1071–84 [YJCHT 1453], 2005.

Liu, J., T. Pan, **A. T. Woolley,** and **M. L. Lee.** Surface-modified poly(methyl methacrylate) capillary electrophoresis microchips for protein and peptide analysis. *Anal. Chem.* 76:6948, 2004.

Becerril, H. A., R. M. Stoltenberg, C. F. Monson, and **A. T. Woolley.** Ionic surface masking for low background in single- and double-stranded DNA-templated silver and copper nanorods. *J. Mater. Chem.* 14:611–6, 2004.

Kelly, R. T., and **A. T. Woolley.** Thermal bonding of polymeric capillary electrophoresis microdevices in water. *Anal. Chem.* 75:1941–5, 2003.

Xin, H., and **A. T. Woolley.** DNA-templated nanotube localization. *J. Am. Chem. Soc.* 125:8710–1, 2003.

Harris, D. G., J. Shao, B. D. Marrow, and **S. S. Zimmerman.** Molecular modeling of the binding of 5-substituted 2'-deoxyuridine substrates to thymidine kinase of herpes simplex virus type-1. *Nucleosides Nucleotides Nucl. Acids* 23:555–65, 2004.

Harris, D. G., et al. **(S. S. Zimmerman).** Procedure for selecting starting conformations for energy minimization of nucleosides and nucleotides. *Nucleosides Nucleotides Nucl. Acids* 21:803–12, 2002.

Harris, D. G., et al. **(S. S. Zimmerman).** Kinetic and molecular modeling of nucleoside and nucleotide inhibition of malate dehydrogenase. *Nucleosides Nucleotides Nucl. Acids* 21:813–23, 2002.

Carnegie Mellon

CARNEGIE MELLON UNIVERSITY

Mellon College of Science
Department of Chemistry

Programs of Study

The Department of Chemistry offers programs leading to the M.S. and Ph.D. degrees. The majority of students are admitted to the Ph.D. program. Terminal master's programs in polymer science and in colloids, polymers, and surfaces are offered. The graduate program is highly individualized to allow exploration of interdisciplinary interests. Research is carried out in bioorganic, bioinorganic, computational, and theoretical chemistry; materials chemistry; polymer science; magnetochemistry; molecular biophysics; inorganic chemistry; green chemistry; NMR spectroscopy; optical and laser spectroscopy; photochemistry; physical chemistry; organic chemistry; and nuclear chemistry.

The graduate program at Carnegie Mellon emphasizes close interaction with faculty members in small research groups. Students in the Ph.D. program choose a research adviser during the first year and typically complete the course requirements within the first three semesters so that the focus throughout the program is on their thesis research. As part of their development as scientists, students also deliver a formal seminar, defend a research progress report, and develop an original research proposal. To support their overall professional development, students are required to assist in undergraduate teaching for two semesters. Additional opportunities to develop teaching and mentoring skills are available for those interested in future faculty careers. The final step in the program is the Ph.D. dissertation, which includes results of original research and constitutes a scientific contribution that is worthy of publication.

There are excellent opportunities for interdisciplinary programs with the Departments of Biological Sciences, Chemical Engineering, Materials Science, and Physics, along with the Biotechnology Program and the Center for Fluorescence Research in Biomedical Sciences.

Research Facilities

The Department of Chemistry is located in the Mellon Institute Building, a spacious and dramatic eight-story structure located near the main campus of Carnegie Mellon and directly adjacent to the University of Pittsburgh campus. The Department has the most modern instrumentation, to which students have hands-on access. This includes the Center for Molecular Analysis, with LC-Q electrospray/ion-trap and MALDI/TOF mass spectrometry and access to high-field NMR (one 500-MHz and two 300-MHz), as well as major laser spectroscopy and chemical synthesis laboratories. Extensive computational facilities are readily available. These include state-of-the-art computers at the Pittsburgh Supercomputing Center, which is housed in the Mellon Institute Building along with the Department of Biological Sciences and the Center for Fluorescence Research in Biomedical Sciences. The library of the Mellon Institute is exceptional.

Financial Aid

All U.S. doctoral degree students are guaranteed financial aid for the first academic year, usually as teaching assistants, with a stipend of $1850 per month and a tuition scholarship. In addition, competitive fellowships are available, which pay an additional $2000–$4000 per year. Research assistantships are also available for succeeding years. International students may be admitted without being granted financial aid.

Cost of Study

Tuition is $31,800 for the 2006–07 academic year.

Living and Housing Costs

Pittsburgh provides an attractive and reasonably priced living environment. On-campus housing is limited, but the Off-Campus Housing Office assists students in finding suitable accommodations. Most graduate students prefer to live in nearby rooms and apartments, which are readily available.

Student Group

Graduate enrollment at Carnegie Mellon totals 4,394 and includes students from all parts of the United States and many other countries. All students in the Department of Chemistry are receiving financial aid. Upon completing the Ph.D., a few students (15 percent) go directly into academic positions, but most continue as postdoctoral fellows (40 percent) or take industrial jobs (45 percent).

Location

Pittsburgh is in a large metropolitan area of 1.26 million people. The city is the headquarters for twelve Fortune 500 corporations, and there is a large concentration of research laboratories in the area. Carnegie Mellon is located in the Oakland neighborhood, one of the cultural and civic centers of Pittsburgh. The campus covers 90 acres and adjoins Schenley Park, the largest city park. The city's cultural and recreational opportunities are truly outstanding.

The University

Carnegie Mellon was first established in 1900 as the Carnegie Technical School through a gift from Andrew Carnegie. In 1912, the name was changed to Carnegie Institute of Technology. The Mellon Institute was founded by A. W. and R. B. Mellon; it carried out both pure research and applied research in cooperation with local industry. In 1967, the two entities merged to form Carnegie-Mellon University. Four colleges—the Carnegie Institute of Technology, the Mellon College of Science, the College of Fine Arts, and the College of Humanities and Social Sciences—offer both undergraduate and graduate programs. The University has assets in excess of $1 billion, a total enrollment of 9,891, and 1,421 faculty members.

Applying

Completed applications and credentials for graduate study in chemistry should be submitted by January 15 for decision by mid-April. However, admission decisions are made on a continuous basis, and applications are considered at any time. In addition to the application form, transcripts from all college-level institutions attended, three letters of recommendation, and an official report of the applicant's scores on the General Test and the Subject Test in chemistry of the Graduate Record Examinations are required. An official report of the score from the Test of English as a Foreign Language (TOEFL) is required for international students. A more detailed description of programs is given online and in the Departmental brochure, which is sent on request.

Correspondence and Information

Committee for Graduate Admissions
Department of Chemistry
Carnegie Mellon University
4400 Fifth Avenue
Pittsburgh, Pennsylvania 15213

Phone: 412-268-3150
Fax: 412-268-1061
E-mail: vb0g@andrew.cmu.edu
Web site: http://www.chem.cmu.edu

Carnegie Mellon University

THE FACULTY AND THEIR RESEARCH

Catalina Achim, Assistant Professor; Ph.D., Carnegie Mellon, 1998. Synthesis and structural and spectroscopic characterization of coordination compounds; intramolecular and intermolecular electron-transfer properties of mixed-valence complexes; magnetochemistry of clusters; bioinorganic chemistry; investigation of stereochemistry, molecular recognition, and reactivity properties of polynuclear complexes.

Bruce A. Armitage, Associate Professor; Ph.D., Arizona, 1993. Bioorganic and supramolecular chemistry: design and synthesis of functional DNA/RNA analogs, nucleic acid chemistry, photochemistry in supramolecular assemblies of molecules, development of probes for RNA structure and function, sensors for hybridization of nucleic acid probes.

Mark E. Bier, Associate Research Professor and Director, Center for Molecular Analysis; Ph.D., Purdue, 1988. Bioanalytical mass spectrometry (MS), real-time monitoring of enzyme reactions by electrospray-MS, development of novel ion trap analyzer designs, femtomole level protein identification, educational MS software: http://mass-spec.chem.cmu.edu/VMSL/.

Emile L. Bominaar, Associate Research Professor; Ph.D., Amsterdam (Netherlands), 1986. Theoretical inorganic chemistry and spectroscopy: spin effects on electron transfer in transition-metal compounds and metalloproteins, magnetooptics.

Terrence J. Collins, Thomas Lord Professor; Ph.D., Auckland (New Zealand), 1978. Green chemistry, inorganic chemistry, design of catalysts for activation of hydrogen peroxide, chemistry to eliminate persistent pollutants, bioinorganic chemistry of high-oxidation-state transition metal species, magnetic properties of multinuclear transition metal ions.

Neil M. Donahue, Assistant Professor; Ph.D., MIT, 1991. Kinetics and mechanisms of atmospherically important reactions, especially hydrocarbon oxidation mechanisms; spectroscopy and in-situ measurement of gas-phase free radicals, including FTIR absorption, diode laser absorption, and cavity ringdown spectroscopy; quantum-mechanical and dynamical studies of reactivity; barrier height control in radical-molecule reactions, dynamics, and spectroscopy of highly vibrationally excited intermediates.

Rebecca J. Freeland, Associate Dean and Associate Head; Ph.D., Carnegie Mellon, 1986. Chemical education; graduate education in science; instructional design and assessment of educational software for introductory chemistry; support for faculty, future faculty, and teaching assistants in improving teaching.

Andrew J. Gellman, Thomas Lord Professor and Head, Chemical Engineering; Ph.D., Berkeley, 1985. Surface science and surface chemistry, with particular interest in the areas of heterogeneous catalysis and tribology.

Roberto Gil, Research Scientist and Director of the NMR Facility.

Susan T. Graul, Lecturer; Ph.D., Purdue, 1989. Physical organic chemistry. Gas-phase ion chemistry: reaction mechanisms and dynamics, cluster ions and highly charged ions, photoinduced and collisionally activated dissociation processes, statistical theory and molecular orbital calculations.

Michael P. Hendrich, Associate Professor; Ph.D., Illinois, 1988. Biophysical and bioinorganic chemistry, transition metals associated with fundamental processes of living systems, electronic structure, high-frequency electron paramagnetic resonance, magnetochemistry.

Colin Horwitz, Research Professor; Ph.D., Northwestern, 1986. Inorganic and bioinorganic chemistry: synthesis and characterization of coordination complexes, oxidation chemistry.

Morton Kaplan, Professor; Ph.D., MIT, 1960. Nuclear chemistry, nuclear physics, and chemical physics: nuclear reactions of heavy ions and high-energy projectiles, ultrarelativistic nuclear collisions, recoil properties of radioactive products, Mössbauer resonance, perturbed angular correlations of gamma rays, statistical theory of nuclear reactions and light-particle emission.

Paul J. Karol, Professor; Ph.D., Columbia, 1967. Nuclear chemistry and physical chemistry: high-energy nuclear reactions; chemical separations, especially column chromatography; positronium lifetime quenching.

Hyung J. Kim, Professor and Head; Ph.D., SUNY at Stony Brook, 1988. Theoretical chemistry, equilibrium and nonequilibrium statistical mechanics, chemical reaction dynamics and spectroscopy in condensed phases, molecular dynamics computer simulations and quantum chemistry in solution.

Tomasz Kowalewski, Assistant Professor; Ph.D., Polish Academy of Sciences, 1988. Physical and biophysical chemistry, physical chemistry of macromolecules, nanostructure in soft condensed matter, glass transition, self-assembly, hydrophobic interaction, nanoscale polymer assemblies, protein misfolding, physicochemical aspects of Alzheimer's disease, protein-DNA interactions, nanoscale manipulation of matter and structure-property studies of polymers, scanning probe techniques.

Maria Kurnikova, Assistant Professor; Ph.D., Pittsburgh, 1998. Theoretical chemistry.

Miguel Llinás, Professor; Ph.D., Berkeley, 1971. Molecular biophysics: structural dynamics and functional studies of proteins in solution by NMR spectroscopy, plasminogen and blood coagulation proteins.

Danith Ly, Assistant Professor; Ph.D., Georgia Tech, 1998. Chemical genetics; functional genomics; combinatorial chemistry; characterizing the molecular basis of human embryonic stem cells and other intermediate (adult) stem cells; elucidating the mechanisms of human physiologic processes, pathology, and aging; design and synthesis of artificial transcription factors; application of combinatorial approach to the discovery of drug and novel protein functions.

Krzysztof Matyjaszewski, J. C. Warner Professor; Ph.D., Polish Academy of Sciences, 1976. Polymer organic chemistry: kinetics and thermodynamics of ionic reactions, cationic polymerization, radical polymerization, ring-opening polymerization, living polymers, inorganic and organometallic polymers, electronic materials.

Richard D. McCullough, Professor and Dean; Ph.D., Johns Hopkins, 1988. Organic and materials chemistry: Self-assembly and synthesis of highly conductive organic polymers and oligomers; conjugated polymer sensors, nanoelectronic assembly, and nanowires; synthesis and development of organic-inorganic hybrid magnetic and electronic materials; crystal engineering and self-assembly.

Eckard Münck, Professor; Ph.D., Darmstadt Technical (Germany), 1967. Active sites of metalloproteins, in particular sites containing iron-sulfur clusters or oxo-bridged iron dimers; study of synthetic clusters which mimic structures in proteins; magnetochemistry: Heisenberg and double-exchange; Mössbauer spectroscopy and electron paramagnetic resonance.

Gary D. Patterson, Professor; Ph.D., Stanford, 1972. Chemical physics and polymer science: application of light-scattering spectroscopy to problems of structure and dynamics in amorphous materials, physics and chemistry of liquids and solutions, conformational statistics and molecular dynamics of polymers, nature and dynamics of the glass transition, structure and dynamics of biopolymers, colloid science, complex fluids.

Linda A. Peteanu, Associate Professor; Ph.D., Chicago, 1989. Physical chemistry; biophysical chemistry; laser spectroscopy; application of resonance Raman and Stark effect-based spectroscopies to the study of ultrafast photochemical and biological reactions: proton transfer, electron transfer, charge transfer, and *cis-trans* isomerizations; effect of solvent environment on reactive excited states.

Stuart W. Staley, Professor; Ph.D., Yale, 1964. Physical organic chemistry: synthesis, dynamic NMR studies of electron transfer systems and theoretically interesting molecules, X-ray diffraction and NMR studies of organic crystals, electronic structure calculations of molecular geometries and properties.

Charles H. Van Dyke, Associate Professor; Ph.D., Pennsylvania, 1964. Synthetic inorganic chemistry and chemical education.

Lynn M. Walker, Associate Professor; Ph.D., Delaware, 1995. Rheology of complex fluids, rheo-optics and rheo-SANS of assembled macromolecular solutions, rheo-optics of immiscible polymer blends in complex flows, viscoelasticity and effects on atomization and spraying.

Newell R. Washburn, Assistant Professor; Ph.D. Berkeley, 1998. Synthesis of hybrid materials for tissue engineering, spectroscopic characterization of cytokine dynamics in extracellular matrices, combinatorial screening of cell-material interactions.

David Yaron, Associate Professor; Ph.D., Harvard, 1990. Theoretical chemistry, electronic structure of conducting polymers and nonlinear optical materials, theoretical description of large-amplitude vibrational motions.

SELECTED PUBLICATIONS

Berlinguette, C. P., A. Dragulescu-Andrasi, **C. Achim,** and K. R. Dunbar. A charge-transfer-induced spin-transition in a discrete complex: The role of extrinsic factors in stabilizing three electronic isomeric forms of a cyanide-bridged Co/Fe cluster. *J. Am. Chem. Soc.*127(18)6766–79, 2005.

Berlinguette, C. P., A. Dragulescu-Andrasi, **C. Achim,** and K. R. Dunbar. The unprecedented charge-transfer based spin transition in the cyanide-bridged trigonal bipyramidal cluster $\{[Co(tmphen)_2]_3[Fe(CN)_6]_2]\}$. *J. Am. Chem. Soc.* 126:6222–3, 2004.

Popescu, D.-L., T. J. Parolin, and **C. Achim.** Metal ion incorporation in modified PNA duplexes. *J. Am. Chem. Soc.* 125:6534–5, 2003.

Zhou, H.-C., et al. **(C. Achim).** High-nuclearity, sulfide-rich, molybdenum-iron-sulfur clusters: Reevaluation and extension. *Inorg. Chem.* 41:3191–201, 2002.

Cadieux, E., et al. **(C. Achim** and **E. Münck).** Biochemical, Mössbauer, and EPR studies of the diiron cluster of phenol hydroxylase from *Pseudomonas sp.* strain CF 600. *Biochemistry* 41:10680–91, 2002.

Tomlinson, A., et al. **(B. Armitage** and **D. Yaron).** A structural model for cyanine dyes templated into the minor groove of DNA. *J. Phys. Chem.*, in press.

Wang, Y., B. A. Armitage, and G. C. Berry. Reversible association of PNA-terminated poly(2-hydroxyethyl acrylate) from ATRP. *Macromolecules* 38:5846–8, 2005.

Marin, V. L., and **B. A. Armitage.** RNA guanine quadruplex invasion by complementary and homologous PNA probes. *J. Am. Chem. Soc.* 127:8032–3, 2005.

Renikuntla, B., and **B. A. Armitage.** Role reversal: Transfer of chirality from an aggregated cyanine dye to a peptide nucleic acid double helix. *Langmuir* 21:5362–6, 2005.

Curten, B., et al. **(M. E. Bier).** Synthesis, photophysical, photochemical, and biological properties of a stable form of caged GABA, 4-[[(2H-1-Benzopyran-2-one-7-amino-4-methoxy)carbonyl]amino] butanoic acid (BC204). *Photochem. Photobiol.* 81(3):641–8, 2005.

Yang, G., and **M. E. Bier.** Investigation of a rapid scan on an electrospray ion trap mass spectrometer. *Anal. Chem.*, 77:1663–71, 2005.

Ramsey, B. G., and **M. E. Bier.** An LDI mass spectrometry investigation of triarylboranes and tri-9-anthrylborane photolysis products. *J. Organomet. Chem.* 609:962–71, 2005.

Datta, B., **M. E. Bier,** S. Roy, and **B. A. Armitage.** Quadruplex formation by a guanine-rich PNA oligomer. *J. Am. Chem. Soc.* 127:4199–207, 2004.

Stoian, S., et al. **(E. L. Bominaar** and **E. Münck).** Mössbauer, EPR, and crystallographic characterization of a high-spin Fe(I) diketiminate complex with orbital degeneracy. *Inorg. Chem.* 44:4915–22, 2005.

Eckert, N. A., et al. **(E. L. Bominaar** and **E. Münck).** Synthesis, structure, and spectroscopy of an oxodiiron(II) complex. *J. Am. Chem. Soc.* 127:9344–5, 2005.

Stubna, A., et al. **(E. L. Bominaar** and **E. M. Münck).** A structural and Mössbauer study of complexes with $Fe_2(\mu\text{-}O(H))_2$ cores: Stepwise oxidation from $Fe^{II}(\mu\text{-}OH)_2Fe^{II}$ through $Fe^{II}(\mu\text{-}OH)_2Fe^{III}$ to $Fe^{III}((\text{-}O)(\mu\text{-}OH)Fe^{III}$. *Inorg. Chem.* 43:3067–79, 2004.

Tiago de Oliveira, F., et al. **(E. L. Bominaar** and **E. Münck).** Antisymmetric exchange in $[2Fe\text{-}2S]^{1+}$ clusters: EPR of the Rieske protein from *Thermus thermophilus* at pH 14. *J. Am. Chem. Soc.* 126:5338–9, 2004.

Bominaar, E. L., C. Achim, and S. A. Borshch. Theory for electron transfer from a mixed-valence dimer with paramagnetic sites to a mononuclear acceptor. *J. Chem. Phys.* 110:11411, 1999.

Bominaar, E. L., C. Achim, and J. Peterson. Theory for magnetic linear dichroism of electronic transitions between twofold-degenerate molecular spin levels. *J. Chem. Phys.* 109:942, 1998.

Bominaar, E. L., et al. **(E. Münck).** Double exchange and vibronic coupling in mixed-valence systems. Electronic structure if exchange-coupled siroheme–$[Fe_4S_4]^{2+}$ chromophore in oxidized *E. coli* sulfite reductase. *J. Am. Chem. Soc.* 117:6976, 1995.

Collins, T. J., and C. Walter. Little green molecules. *Sci. Am.*, in press.

Collins, T. J. TAML oxidant activators: A new approach to the activation of hydrogen peroxide for environmentally significant problems. *Acc. Chem. Res.* 35:782–90, 2002.

Collins, T. J., and **C. P. Horwitz** et al. TAML® catalytic oxidant activators in the pulp and paper industry. *Adv. Sustainability Green Chem. Eng.* 823:47–60, 2002.

Donahue, N. M., A. L. Robinson, C. O. Stanier, and S. N. Pandis. The coupled partitioning, dilution, and chemical aging of semivolatile organics. *Environ. Sci. Technol.*, in press.

Robinson, A. L., **N. M. Donahue,** and W. F. Rogge. Photochemical oxidation and changes in molecular composition of organic aerosol in the regional context. *J. Geophys. Res.*, in press.

Hartz, K. E., et al. **(N. M. Donahue).** Cloud condensation nuclei activation of limited solubility organic aerosol. *Atmos. Environ.*, in press.

Sage, A. M., and **N. M. Donahue.** Deconstructing experimental rate constant measurements: Obtaining intrinsic reaction parameters, kinetic isotope effects, and tunneling coefficients from kinetic data for OH + methane, ethane, and cyclohexane. *J. Photochem. Photobiol.* 176:238–49, 2005.

R. Freeland. *Collected Wisdom: Strategies & Resources for TAs.* Carnegie Mellon: Eberly Center for Teaching Excellence, 1998.

Hannah, K. C., **R. R. Gil,** and **B. A. Armitage.** ^{1}H-NMR and optical spectroscopic investigation of the sequence-dependent dimerization of a symmetrical cyanine dye in the DNA minor groove. *Biochemistry* (44):15924–9, 2005.

Desmarchelier, C., et al. **(R. R. Gil).** Antioxidant and free-radical scavenging activities of *Misodendrum punctulatum,* myzodendrone, and structurally related phenols. *Phytother. Res.* 19:1043–7, 2005.

Huang, J., **R. R. Gil,** and **K. Matyjaszewski.** Synthesis and characterization of copolymers of 5,6-benzo-2-methylene-1,3-dioxepane and n-butyl acrylate. *Polymer* 46(25):11698–706, 2005.

Iovu, M.C., E. E. Sheina, **R. R. Gil,** and **R. D. McCullough.** Experimental evidence for the quasi-"living" nature of the Grignard metathesis method for the synthesis of regioregular poly(3-alkylthiophenes). *Macromolecules* 38(21):8649–56, 2005.

Pierce, B. S., and **M. P. Hendrich.** Local and global effects of metal binding within the small subunit of ribonucleotide reductase. *J. Am. Chem. Soc.* 127:3610–23, 2005.

Hooper, A. B., et al. **M. P. Hendrich.** The oxidation of ammonia as an energy source in bacteria. In *Respiration in Archaea and Bacteria, the Netherlands Series: Advances in Photosynthesis and Respiration,* vol. 2, pp. 16:121–47, ed. D. Zannoni. Kluwer Scientific, 2005.

Golombek, A. P., and **M. P. Hendrich.** Quantitative analysis of dinuclear manganese (II) EPR spectra. *J. Magn. Reson.* 165:33–48, 2003.

Pierce, B. S., T. E. Elgren, and **M. P. Hendrich.** Mechanistic implications for the formation of the diiron cluster in ribonucleotide reductase provided by quantitative EPR spectroscopy. *J. Am. Chem. Soc.* 125:8748–59, 2003.

Ghosh, A., et al. **(M. P. Hendrich, C. P. Horwitz,** and **T. J. Collins).** Understanding the mechanism of H^+-induced demetalation as a design strategy for robust iron(III) peroxide-activating catalysts. *J. Am. Chem. Soc.* 125:12378–9, 2003.

Upadhyay, A. K., et al. **(M. P. Hendrich).** Spectroscopic characterization and assignment of reduction potentials in the tetraheme cytochrome c554 from *Nitrosomonas europaea. J. Am. Chem. Soc.* 125:1738–47, 2003.

Mondal, S., et al. **(C. P. Horwitz** and **T. J. Collins).** Oxidation of sulfur components in diesel fuel using Fe-TAML® catalysts and hydrogen peroxide. *Catal. Today,* in press.

Horwitz, C. P., and **T. J. Collins** et al. Iron TAML® catalysts in the pulp and paper industry. In *Feedstocks for the Future: Renewables for the Production of Chemicals and Materials,* ACS Symposium Series 921, eds. J. J. Bozell and M. K. Patel. Oxford University Press, 2006.

Sen Gupta, S., et al. **(C. P. Horwitz** and **T. J. Collins).** Rapid total destruction of chlorophenols by activated hydrogen peroxide. *Science* 296:326–8, 2002.

Horwitz, C. P., and **T. J. Collins.** Modifying the chemistry of a macrocyclic cobalt complex by remote site manipulation. *J. Phys. Chem. B* 105:8821–8, 2001.

Kaplan, M. The STAR Collaboration: Experimental and theoretical challenges in the search for the quark gluon plasma: The STAR Collaboration's critical assessment of the evidence from RHIC collisions. *Nucl. Phys. A,* in press.

Kaplan, M. The STAR Collaboration: Phi meson production in Au + Au and p + p collisons at s_{NN}=200 GeV. *Phys. Lett. B* 612:181, 2005.

Kaplan, M. The STAR Collaboration: Open charm yields in d + Au collisions at s_{NN}=200 GeV. *Phys. Lett.* 94:062301, 2005.

Kaplan, M. The STAR Collaboration: Azimuthal anisotropy and correlations at large transverse momenta in p + p and Au + Au collisions at s_{NN}=200 GeV. *Phys. Rev. Lett.* 93:252301, 2004.

Kaplan, M. The STAR Collaboration: Pseudorapidity asymmetry and centrality dependence of charged hadron spectra in d + Au Collisions at s_{NN}=200 GeV. *Phys. Rev. C* 70:064907, 2004.

Karol, P. J. The heavy elements. In *The Periodic Table: Into the 21st Century,* chap. 9, pp. 235–61, 2005.

Karol, P. J. The kilogram and the mole redux. *J. Chem. Educ.* 81:212, 2005.

Karol, P. J. SI basic units: The kilogram and the mole. *J. Chem. Educ.* 80:800, 2004.

Karol, P. J. The periodic table. In *Chemistry: Foundations and Applications*, vol. 3, pp. 227–8. Macmillan Reference USA, 2004.

Karol, P. J., H. Nakahara, B. W. Retling, and E. Vogt. On the claims for discovery of elements 110, 111, 112, 114, 116, and 118. *Pure Appl. Chem.* 75:1601–11, 2003.

Manjari, S. R., and **H. J. Kim.** Temperature- and pressure-dependence of the outer-sphere reorganization free energy for electron transfer reactions: A continuum approach. *J. Phys. Chem. B* 110:494–500, 2006.

Manjari, S. R., and **H. J. Kim.** On the temperature and pressure dependences of cavities in the dielectric continuum picture. *J. Chem. Phys.* 123:014504, 2005.

Shim, Y., M. Y. Choi, and **H. J. Kim.** A molecular dynamics computer simulation study of room-temperature ionic liquids: I. Equilibrium and nonequilibrium solvation structure and free energetics. *J. Chem. Phys.* 122:044510, 2005.

Shim, Y., M. Y. Choi, and **H. J. Kim.** A molecular dynamics computer simulation study of room-temperature ionic liquids: II. Equilibrium and nonequilibrium solvation dynamics. *J. Chem. Phys.* 122:044511, 2005.

Carnegie Mellon University

Kruk, M., et al. **(T. Kowalewski** and **K. Matyjaszewski).** Synthesis of mesoporous carbons using ordered and disordered mesoporous silica templates and polyacrylonitrile as carbon precursor. *J. Phys. Chem. B* 109:9216–25, 2005.

Legleiter, J., and **T. Kowalewski.** Insights into fluid tapping-mode atomic force microscopy provided by numerical simulations. *Appl. Phys. Lett.* 87:163120–3, 2005.

Wu, W., et al. **(T. Kowalewski** and **K. Matyjaszewski).** Self-assembly of pODMA-b-ptBA-b-pODMA triblock copolymers in bulk and on surfaces: A quantitative SAXS/AFM comparison. *Langmuir* 21:9721–7, 2005.

Coalson, R. D., and **M. G. Kurnikova.** Poisson-Nernst-Planck theory of ion permeation through biological channels. In *Handbook of Ion Channels: Dynamics, Structure, and Applications*, eds. S.-H. Chung, O. Andersen, and V. Krishnamurthy. Springer Verlag, in press.

Coalson, R., and **M. Kurnikova.** Poisson-Nernst-Planck theory approach to the calculation of current through biological ion channels. *IEEE Trans. Nanobiosci.* 4:81–93, 2005.

Speranskiy, K., and **M. Kurnikova.** Investigation of the glutamate ligand binding to the glutamate receptor by theoretical methodologies. *Biochemistry* 44(34):11508–17, 2005.

Mamonova, T., et al. **(M. Kurnikova).** Protein flexibility using constraints from molecular dynamics simulations. *Phys. Biol.* 2:S137–47, 2005.

Vranken, W. F., et al. **(M. Llinás).** The CCPN data model for NMR spectroscopy: Development of a software pipeline for HTP projects. *Proteins: Struct., Funct., Bioinform.*, in press.

Grishaev, A., et al. **(M. Llinás).** ABACUS, a direct method for protein NMR structure computation via assembly of fragments. *Proteins: Struct., Funct., Bioinform.*, in press.

Grishaev, A., and **M. Llinás.** Protein structure elucidation from minimal NMR data: The CLOUDS approach. *Methods Enzymol.* 394:261–95, 2005.

Dragulescu-Andrasi, A., P. Zhou, G. He, and **D. H. Ly.** Cell-permeable GPNA containing appropriate backbone stereochemistry and spacing binds sequence-specifically to RNA. *Chem. Comm.* 244, 2005.

Lagerholm, C. B., et al. **(D. H. Ly).** Optical coding of cells with quantum dots. *Nano Lett.* 4:2019, 2004.

Zhou, P., et al. **(D. H. Ly).** Novel binding and efficient cellular uptake of guanidined-based peptide nucleic acids (G-PNA). *J. Am. Chem. Soc.* 125:6878, 2003.

Simbulan-Rosenthal, C. M., and **D. H. Ly** et al. Misregulation of gene expression in primary fibroblasts lacking poly (ADP-ribose) polymerase. *Proc. Natl. Acad. Sci. U.S.A.* 97:1274, 2000.

Ly, D. H., D. J. Lockhart, P. G. Schultz, and R. A. Lerner. Mitotic misregulation and human aging. *Science* 285:2486, 2000.

Tang, C., et al. **(K. Matyjaszewski** and **T. Kowalewski).** Long-range ordered thin films of block copolymers prepared by zone-casting and their thermal conversion into ordered nanostructured carbon. *J. Am. Chem. Soc.* 127:6918–9, 2005.

Wu, W., **K. Matyjaszewski,** and **T. Kowalewski.** Monitoring surface thermal transitions of ABA triblock copolymers with crystalline segments using phase contrast tapping mode atomic force microscopy. *Langmuir* 21:1143–8, 2005.

Min, K., H. Gao, and **K. Matyjaszewski.** Preparation of homopolymers and block copolymers in miniemulsion by ATRP using activators generated by electron transfer (AGET). *J. Am. Chem. Soc.* 127:3825, 2005.

Lutz, J.-F., and **K. Matyjaszewski.** NMR monitoring of chain-end functionality in atom transfer radical polymerization of styrene. *J. Polym. Sci. Polym. Chem. Ed.* 43:897, 2005.

Zhang, Y., et al. **K. Matyjaszewski.** Structure and properties of poly(*n*-butyl acrylate-*b*-sulfone-*b*-*n*-butyl acrylate) triblock copolymers prepared by ATRP. *Macromol. Chem. Phys.* 206:33–42, 2005.

Jeffries-El, M., G. Sauvé, and **R. D. McCullough.** In situ end group functionalization of regioregular poly(3-alkylthiophene) using the Grignard metathesis polymerization method. *Adv. Mater.* 16:1017–9, 2004.

Sheina, E. E., et al. **(R. D. McCullough).** Chain-growth mechanism for regioregular nickel-initiated cross-coupling polymerizations. *Macromolecules* 37:3526–8, 2004.

Zhai, L., and **R. D. McCullough.** Regioregular polythiophene/gold nanoparticle hybrid materials. *J. Mater. Chem.* 14:141–3, 2004.

Ewbank, P. C., et al. **(R. D. McCullough).** Regioregular poly(thiophene-3-alkanoic acid)s: Water soluble conducting polymers suitable for chromatic chemosensing in solution and solid state. *Tetrahedron* 60:11269–75, 2004.

Ghosh, A., et al. **(E. Münck, C. P. Horwitz,** and **T. J. Collins).** Catalytically active μ-oxodiiron(IV) oxidants from iron(III) and dioxygen. *J. Am. Chem. Soc.* 127:2505–13, 2005.

Jensen, M. P., et al. **(E. Münck).** High-valent nonheme iron: Two distinct iron(IV) intermediates derived from a common iron(II) procursor. *J. Am. Chem. Soc.* 127: 10512–25, 2005.

Savin, D. A., **G. D. Patterson,** and J. R. Stevens. Evidence for the g-relaxation in the light-scattering spectrum of poly(n-hexyl methacrylate) near the glass transition. *J. Polym. Sci. Part B: Polym. Phys.* 43(12):1504–19, 2005.

Cao, R., Z. Gu, **G. D. Patterson,** and **B. A. Armitage.** A recoverable enzymatic microgel based on biomolecular recognition. *J. Am. Chem. Soc.* 126:726–7, 2004.

Gu, Z., **G. D. Patterson,** R. Cao, and **B. A. Armitage.** Self-assembled supramolecular microgels: Fractal structure and aggregation mechanism. *J. Polym. Sci., Part B: Polym. Phys.* 41:3037–46, 2003.

Savin, D. A., et al. **(G. D. Patterson, T. Kowalewski,** and **K. Matyjaszewski).** Synthesis and characterization of silica-graft polystytrene hybrid nanoparticles: Effect of constraint on the glass-transition temperature of spherical polymer brushes. *J. Polym. Sci., Part B: Polym. Phys.* 40:2667–76, 2002.

Liu, L. A., **L. A. Peteanu,** and **D. J. Yaron.** Effects of dsorder-induced symmetry breaking on the electroabsorption properties of a model dendrimer. *J. Phys. Chem. B* 108:16841–9, 2004.

Bandal, P. R., D. M. K. Lam, **L. Peteanu,** and M. Van Der Auweraer. Excited state localization in a 3-fold symmetric molecule as probed by electroabsorption spectroscopy. *J. Phys. Chem. B* 108:16834–40, 2004.

Wachsmann-Hogiu, S., et al. **(L. A. Peteanu** and **D. J. Yaron).** The effects of structural and micro-environmental disorder on the electronic properties of MEH-PPV and related oligomers. *J. Phys. Chem. B* 107:5133–43, 2003.

Chowdhury, A., et al. **(L. Peteanu** and **D. J. Yaron).** Stark spectroscopy of size-selected helical H-aggregates of a cyanine dye templated by duplex DNA: Effect of exciton coupling on electronic polarizabilites. *J. Phys. Chem. B* 107:3351–62, 2003.

Premvardhan, L. L., and **L. A. Peteanu.** Electronic properties of small model compounds that undergo excited-state intramolecular proton transfer as measured by electroabsorption spectroscopy. *J. Photochem. Photobiol. A* 154:69–80, 2002.

Staley, S. W., S. A. Vignon, and B. Eliasson. Conformational analysis and kinetics of ring inversion for methylene- and dimethylsilyl-bridged dicyclooctatetraene. *J. Org. Chem.* 66:3871–7, 2001.

Boman, P., B. Eliasson, R. A. Grimm, and **S. W. Staley.** Bond shift and charge transfer dynamics in methylene- and dimethylsilyl-bridged dicyclooctatetraene dianions. *J. Chem. Soc.* 2:1130–8, 2001.

Staley, S. W., and J. D. Kehlbeck. Effect of para substituents on the rate of bond shift in arylcyclooctatetraene. *J. Org. Chem.* 66:5572–9, 2001.

Staley, S. W., and J. D. Kehlbeck. Identification of an unsymmetrical isomer from the attempted synthesis of 1,8-dicyclooctatetraenylnaphthalene. *J. Org. Chem.* 66: 7389, 2001.

Staley, S. W., M. L. Peterson, and L. M. Wingert. Relationship between molecular structure, polarization, and crystal packing in 6-arylfulvenes. In *Anisotropic Organic Materials: Approaches to Polar Order,* chap. 8, eds. R. Glaser and P. Kaszynski. Washington, D.C.: American Chemical Society, 2001.

Lin-Gibson, S., R. L. Jones, **N. R. Washburn,** and F. Horkay. Structure-property relationships of photopolymerizable poly(ethylene glycol) dimethacrylate hydrogels. *Macromolecules* 38:2897–902, 2005.

Lin-Gibson, S., et al. **(N. R. Washburn).** Synthesis and characterization of PEG dimethacrylates and their hydrogels. *Biomacromolecules* 5:1280–7, 2004.

Mei, Y., et al. **(N. R. Washburn).** Solid-phase ATRP synthesis of peptide-polymer hybrids. *J. Am. Chem. Soc.* 126:3472–6, 2004.

Washburn, N. R., et al. High-throughput investigation of osteoblast response to polymer crystallinity: influence of nanometer-scale roughness on proliferation. *Biomaterials* 25:215–24, 2004.

Liu, L., **D. Yaron,** M. I. Sluch, and M. A. Berg. Modeling the effects of torsional disorder on the spectra of poly- and oligo-(p-phenyleneethynylenes). *J. Phys. Chem.* , in press.

Evans, K. L., M. Karabinos, G. Leinhardt, and **D. Yaron.** Chemistry in the field and chemistry in the classroom: A cognitive disconnect? *J. Chem. Educ.*, in press.

Janesko, B. G., and **D. Yaron.** Functional group basis sets. *J. Chem. Theor. Comput.* 1:267, 2005.

CLEMSON UNIVERSITY

CLEMSON UNIVERSITY

Department of Chemistry
Master of Science in Chemistry

Programs of Study

The Department of Chemistry offers programs leading to the Master of Science (M.S.), with an emphasis in analytical, inorganic, organic, physical chemistry, or chemistry education. Individual programs of study involve an intensive concentration in one of the traditional areas of chemistry or a concentration in a combination of areas. The degree program is research based, which means that it involves the completion and defense of an original research project. In addition to the course work, the research project comprises the bulk of the effort involved in the pursuit of an advanced chemistry degree.

M.S. students must complete at least 30 semester hours, 24 of which are course work (typically eight courses) and 6 of which are research and thesis preparation. Placement examinations and consultations with a faculty advisory committee during the new-student orientation are used to select first-year study courses. Students normally choose a thesis adviser and committee during the first semester and formulate the remaining course program after consulting with them. All degree candidates must present at least one research-based seminar to the Department as part of their degree program. Most students also present their work outside of Clemson, for example, at a national or regional meeting of a scientific society, such as the American Chemical Society. The final stage of the graduate degree program involves the writing and defense before the degree committee of a thesis describing the student's original research project.

Research Facilities

The Department of Chemistry is housed in the Howard L. Hunter Chemistry Laboratory, which includes more than 50,000 square feet of laboratory space for research and teaching. One of the finest research facilities in the Southeast, this building was completed in 1987 and accommodates about 100 graduate students, postdoctoral scientists, and visiting scientists. It includes a satellite chemistry library that houses the field's most important journals and supplements extensive holdings in the University's central library. Several chemistry research groups also occupy space in other on- and off-campus buildings.

The Department maintains a broad range of multiple-user instruments for chemistry research. Major research instrumentation holdings include three Fourier-transform NMR spectrometers; X-ray powder, single-crystal, and thin-film diffractometers; an electron spin resonance (ESR) spectrometer; gas chromatography/mass spectrometer systems; a thermal analysis system; and other state-of-the-art equipment maintained by individual faculty members in support of their research programs or through the Department's research partners.

Clemson University provides a diverse and extensive computing infrastructure supported locally within the Department of Chemistry and by the Office of Computer and Network Services (CNS) and the Division of Computing and Information Technology (DCIT). Various laboratories in the Department have high-speed SGI, Sun, and Linux workstations as well as a twenty-eight-processor cluster for parallel computations. PC and Macintosh computers are available in all departmental research labs and in many computer labs around the campus. The College of Engineering and Science has recently installed a 512-processor distributed Beowulf cluster that makes Clemson one of the top supercomputing sites in the Southeast. Clemson also participates in the high-speed Internet2 and partners with the Center for Advanced Engineering Fibers and Films, which has a state-of-the-art virtual reality laboratory and recently received a $1.3-million grant from the Keck Foundation to create a virtual visualization and design lab.

The Laser Laboratory is managed by Ya-Ping Sun and his research group. The laboratory is equipped with a CW Mode locked Nd: Yag Laser, a 20-Hz Q-Switched Nd: Yag Laser, and two synchronous pumped Dye Lasers. The laser configuration is capable of conducting pump probe experiments in the nanosecond time scale region up to the subpicosecond time scale region.

The Nuclear Magnetic Resonance Resource Center affords easy access to modern high-resolution NMR instruments for students, postdoctoral scientists, and faculty members. The primary instrumentation includes three multinuclear high-field spectrometers that are used for routine measurements as well as for advanced one- and two-dimensional NMR experiments in molecular structure determination, molecular dynamics, and chemical kinetics and thermodynamics.

Clemson's Electronic Imaging and Analytical Services (EIAS) group is one of the Southeast's premier analytical imaging and surface analysis facilities. Area researchers both on and off campus can take advantage of a broad range of capabilities, including scanning electron microscopy, transmission electron microscopy, and high-vacuum surface analysis. The EIAS facility is widely used in a number of areas but particularly in nanomaterial and nanotechnology research, which depend critically on the availability of tools that can characterize materials with submicrometer to subnanometer spatial resolution.

The Molecular Structure Center, under the direction of Don Vanderveer, provides the Department of Chemistry with methods of X-ray diffraction analysis, the most reliable and unambiguous means for determining the structure of ordered materials. The center maintains four separate diffractometer systems for performing both powder and single-crystal diffraction experiments. These include two Rigaku diffractometers. One is a sealed tube system equipped with a CCD area detector; the other has a detector that uses a powerful 18-kw rotating anode source. The center also has a conventional four-circle diffractometer with a sealed tubesource. A Scintag 2000 system with a germanium detector and a seven-position automatic sample changer is used for powder diffraction. Data processing and analysis are done on numerous PCs running Microsoft Windows and Red Hat Linux. The center has access to many electronic databases, including Cambridge Structural Database, Inorganic Crystal Structure Database, NIST Crystal Data File, and Powder Diffraction File.

Financial Aid

All chemistry graduate students at Clemson are supported by either teaching or research assistantships during the full course of their studies. Students in the first year are normally supported as teaching assistants in undergraduate laboratory sections. Stipends for teaching assistantships are competitive and change frequently. In fall 2005, the stipend was $18,000 for twelve months. Research assistantships are often available to support students working on funded research projects. Department and University fellowships that can supplement the stipend for well-qualified applicants are also available.

Cost of Study

Tuition for 2006–07 is $4643 per semester for in-state students and $9255 per semester for nonresidents. Off-campus rates are $535 per hour for in-state students and $918 per hour for nonresidents. Graduate assistants pay a flat fee of $1079 per semester and $348 per summer session. Graduate fellows pay South Carolina resident fees.

Living and Housing Costs

On-campus housing is available. For information, students should visit http://www.housing.clemson.edu. The cost of living in Clemson is quite low compared to the national average. Students who choose to live off the campus typically spend $300–$400 per month for rent, depending on location, amenities, roommates, etc.

Student Group

There are 101 full-time and 3 part-time graduate students enrolled in the Department, of whom 33 are women and 39 are international students.

Student Outcomes

Graduates seek and find employment in a very broad range of areas where chemistry is important, including private industry, national laboratories, and education. Public and private-sector employers that have recently hired Clemson chemistry graduates include Advanced Photonic Crystals (Charlotte, North Carolina), IRIX Pharmaceuticals (Florence, South Carolina), Milliken Chemical Co. (Spartanburg, South Carolina), Savannah River Lab (U.S. Department of Engineering) (Aiken, South Carolina), Michelin Tire Co. (Greenville, South Carolina), Dupont Chemical Company (Richmond, Virginia), Merck Chemical (Rahway, New Jersey), Clariant Corp. (Aiken, South Carolina), Micromass Inc. (Beverly, Massachusetts), Chiron Corp. (San Francisco, California), Toyota Co. (Japan), U.S. Army Research Lab (Adelphi, Maryland), NASA Langley Research Center (Hampton, Virginia), Transtech Pharma (North Carolina), Idaho National Engineering and Environmental Lab (U.S. Department of Engineering), Beckman Coulter (Allendale, New Jersey), Cree Semiconductors (North Carolina), National Institutes of Standards and Technology (Gaithersburg, Maryland), and Shire Laboratories (Maryland). Graduates have found employment at universities such as Harvard, Kennesaw State, the University of Connecticut, and UCLA.

Location

Clemson is a small, beautiful college town near the Blue Ridge Mountains and Lake Hartwell in upstate South Carolina. The Upstate is one of the country's fastest-growing areas and is the midpoint of the Charlotte-to-Atlanta I-85 corridor, a multistate area along Interstate 85 that runs from metro Atlanta to Richmond, Virginia, and encompasses Charlotte, North Carolina, and North Carolina's Research Triangle. Atlanta and Charlotte are each a 2-hour's drive away. Many financial institutions and other industries have national headquarters for a major presence in the Upstate, including Wachovia, Bank of America, BMW, Bon Secours St. Francis Health System, Bosch North America, Bowater, Charter Communications, Ernst and Young, Fluor Corporation, IBM, Microsoft, Michelin of North America, and many others.

The University and The Department

Clemson is classified by the Carnegie Foundation as Doctoral/Research University–Extensive, a category comprising less than 4 percent of all universities in America. The University's mission is to fulfill the covenant between its founder and the people of South Carolina to establish a "high seminary of learning" through its responsibilities of teaching, research, and extended public service. The University has identified eight areas of academic emphasis that create collaborations that, in turn, help fulfill the University's mission.

Chemistry has been taught at Clemson since the institution was founded in 1889. Seventy years later, chemistry became one of the University's first Ph.D. programs. Today, the Department of Chemistry is one of Clemson's largest departments.

Applying

Most students who apply to the Clemson graduate chemistry program have completed a B.S. or B.A. undergraduate degree in chemistry or have acquired the equivalent background. In most cases, the undergraduate program included at least two formal courses and the associated labs each in general chemistry, analytical chemistry, and physical chemistry and at least one formal course in inorganic chemistry. Experience in research is also desirable.

An applicant must complete an application form and supply transcripts from the undergraduate program and any prior graduate programs, test scores from the GRE general exam (verbal, quantitative, and analytical), at least two letters of recommendation from people familiar with the student's background, a completed personal statement form, and a completed financial assistance form. International students must also submit a test score for the TOEFL exam. The TSE exam is not required for international students; however, a TSE score above 50 greatly improves the chances of admission. Applications are processed most quickly if all materials are sent directly to the Department of Chemistry by March 1.

The fee for making an official application through the Clemson graduate school is normally $50; however, for applicants who are U.S. citizens or permanent residents, or who have received a degree from a U.S. college or university, or who possess sufficient English language proficiency that they are qualified to teach laboratory sections, the fee is covered by the Department of Chemistry. For all other applicants, application materials submitted to the Department of Chemistry should be considered preliminary only. An official application, with a paid application fee, is required for the application to be given formal consideration.

Correspondence and Information

Steve Stuart, Graduate Director
369 Hunter Laboratories
P.O. Box 340973
Clemson University
Clemson, South Carolina 29634-0973

Phone: 864-656-3095
864-656-5013
888-539-8854 (toll-free)
Fax: 864-656-6613
E-mail: chemgradprogram@chemed.ces.clemson.edu
Web site: http://chemistry.clemson.edu

Clemson University

THE FACULTY AND THEIR RESEARCH

Rudolph Abramovitch, Professor of Organic Chemistry; Ph.D., King's College (London), 1953; D.Sc., London, 1964. The chemistry of electron-deficient reactive intermediates.
The 1,2,4-triazolyl cation: Thermolytic and photolytic studies. *J. Org. Chem.* 66(4):1242–51, 2001 (with Beckert et al.).

Jeffrey R. Appling, Associate Professor of Chemistry Education; Ph.D., Georgia Tech, 1985. The development of new materials and methods to improve both instruction in general chemistry and the study skills of general chemistry students.
Math Survival Guide: Tips for Science Students, 2nd ed. New York: John Wiley & Sons, 2003 (with Richardson).

Dev P. Arya, Associate Professor of Bio-Organic and Medicinal Chemistry; Ph.D., Northeastern, 1996. Nucleic acid therapeutics and synthesis of small-molecule carbohydrate mimetics.
From triplex to B-form duplex stabilization: Reversal of target selectivity by aminoglycoside dimers. *Bioorg. Med. Chem. Lett.* 14:4643–6, 2004 (with Coffee and Xue).

Julia Brumaghim, Assistant Professor of Bio-Inorganic Chemistry; Ph.D., Illinois at Urbana–Champaign, 1999. The biological chemistry of the Fenton reaction and the antioxidant properties of selenium.
Encapsulation and stabilization of reactive aromatic diazonium ions and the tropylium ion within a supramolecular host. *Eur. J. Org. Chem.* 2004(24):5115–8, 2004 (with Michels, Pagliero, and Raymond).

Kenneth A. Christensen, Assistant Professor of Analytical Chemistry; Ph.D., Michigan, 1997. The molecular recognition and host cellular responses to microbial toxin proteins, including potential biowarfare agents.
Fluorescence resonance energy transfer-based stoichiometry in living cells. *Biophys. J.* 83:3652–64, 2002 (with Hoppe and Swanson).

George Chumanov, Associate Professor of Analytical Chemistry; Ph.D., Moscow State, 1988. The preparation and modification of different nanoparticles, including one-, two-, and three-dimensional regular structures; the investigation of their properties using different spectrochemical techniques together with optical, electron, atomic force, and scanning tunneling microscopy; and the development of new materials and devices for environmental and biomedical diagnostic applications.
Vacuum deposition of silver island film on chemically modified surfaces. *J. Vac. Sci. Technol. A* 21:723–7, 2003 (with Malynych).

Melanie Cooper, Alumni Distinguished Professor of Chemistry Education; Ph.D., Manchester (England), 1978. Investigations into the factors that affect student problem-solving strategies, the development and assessment of laboratory curricula, teaching strategies for large enrollment courses, investigating how group work affects student achievement, and whether gender has an effect on student problem-solving strategies.
A Chemists' Guide to Effective Teaching. Upper Saddle River, N.J.: Pearson Education, 2005 (with Greenbowe and Pienta).

Karen Creager, Lecturer of Analytical Chemistry; Ph.D., North Carolina at Chapel Hill, 1989. Photoelectrochemistry of semiconductor surfaces.
Electrochemical olefin epoxidation with manganese meso-tetraphenyl porphyrin catalyst and hydrogen peroxide generation at polymer-coated electrodes. *Inorg. Chem.* 29:1000, 1990 (with Nishihara, Murray, and Collman).

Stephen Creager, Professor of Analytical Chemistry; Ph.D., North Carolina at Chapel Hill, 1987. Electrochemical science and technology.
Redox potentials and kinetics of the Ce^{3+}/Ce^{4+} redox reaction and solubility of cerium sulfates in sulfuric acid solutions. *J. Power Sources* 109(2):431–8, 2002 (with Paulenova, Navratil, and Wei).

Darryl DesMarteau, Tobey-Beaudrot Professor of Inorganic Chemistry; Ph.D., Washington (Seattle), 1966. The study of fluorinated molecules and macromolecules.
Synthesis and comparison of N-fluorobis (perfluoroalkyl) sulfonyl imides with different perfluoroalkyl groups. *J. Fluorine Chem.* 111(2):253–7, 2001 (with Zhang).

Karl Dieter, Professor of Organic Chemistry; Ph.D., Pennsylvania, 1978. Organocopper, organopalladium, and organosulfur chemistry and asymmetric synthesis.
A study of factors affecting alpha- (N-carbamoyl) alkylcuprate chemistry. J. Org. Chem. 66(7):2302–11, 2001 (with Topping and Nice).

Brian N. Dominy, Assistant Professor; Ph.D., Scripps Research Institute, 2001. Physical chemistry.

Lourdes Echegoyen, Lecturer of Analytical and Physical Chemistry; Ph.D., Miami (Florida), 1990. Electrochemically enhanced transport across artificial membranes and the preparation and characterization of novel supramolecular assemblies using crown-ether derivatives.
The electrochemistry of C60 and related compounds. In *Organic Electrochemistry,* eds. H. Lund and O. Hammerich, 4th ed. New York: Marcel Dekker, 2000, p. 323 (with Luis Echegoyen).

Luis Echegoyen, Department Chair and Professor of Physical Chemistry; Ph.D., Puerto Rico, 1974. Fullerene chemistry, electrochemistry, and supramolecular chemistry.
Carbon nanotube doped polyaniline. *Adv. Mat.* 14:1480–3, 2002 (with Zengin et al.).

Lucy Eubanks, Lecturer of Chemistry Education; M.S.N.S., Seattle, 1967.
Chemistry in Context, 5th ed. New York: McGraw-Hill, 2005 (with Middlecamp, Pienta, Heltzel, and Weaver).

Grant Goodyear, Assistant Professor of Chemistry. The study of supercritical fluids near the critical point, the electronic structure of liquid semiconductors.

John W. Huffman, Professor of Organic Chemistry; Ph.D., Harvard, 1957. The synthesis of analogues and metabolites of delta-9-tetrahydrocannabinol (THC), the principal active component of marijuana.
Inhibition of glioma growth in vivo by selective activation of the CB2 cannabinoid receptor. *Cancer Res.* 61(15):5784–9, 2001 (with Sanchez et al.).

Shiou-Jyh Hwu, Professor of Inorganic Chemistry; Ph.D., Iowa State, 1985. Synthesis of solids containing confined transition-metal-oxide frameworks, synthesis of conducting solids in nonaqueous media, synthesis of a new class of hybrid materials via salt inclusion.
Vacancy ordering transitions in one-dimensional lattice gas with Coulomb interactions. *Phys. Rev. B: Condens. Matter* 6304(4):045405, 2001 (with King, Kuo, and Skove).

John Kaup, Lecturer of Physical and Analytical Chemistry; Ph.D., Utah, 1997. Analytical and physical laboratories.

Arkady Kholodenko, Professor of Physical Chemistry; Ph.D., Chicago, 1982. Theory of liquid crystalline semiflexible polymer solutions; statistical mechanics of disordered systems, including glasses and random copolymers; theory of knots and links, with applications to condensed matter and biological systems; theory of quantum and classical chaos.
"New" Veneziano amplitudes from "old" Fermat (hyper)surfaces. *Int. J. Geometry Phys.* 19:1655–703, 2004.

Alex Kitaygorodskiy, Lecturer and NMR Facility Director; Ph.D., Russian Academy of Sciences, 1979. Outer-sphere complexation of metal complexes.
Soluble dendron-functionalized carbon nanotubes: Preparation, characterization, and properties. *Chem. Mater.* 13(9):2864–9, 2001 (with Sun et al.).

Joseph W. Kolis, Professor of Inorganic Chemistry; Ph.D., Northwestern, 1984. The synthesis and chemistry of novel inorganic compounds with unusual structures and properties.
Observation of the interplay of microstructure and thermopower in the Al71Pd21Mn8-xRex quasicrystalline. *Phys. Rev. B: Condens. Matter* 6305(5):052202, 2001 (with Pope et al.).

R. Kenneth Marcus, Professor of Analytical Chemistry; Ph.D., Virginia, 1986. The development and application of new plasma techniques for the atomic spectroscopic analysis of diverse materials.
Collisional dissociation in plasma source mass spectrometry: A potential alternative to chemical reactions for isobar removal. *J. Anal. At. Spectrom.* 19:591–9, 2004.

Jason McNeill, Assistant Professor of Physical Chemistry; Ph.D., Berkeley, 1999. Understanding charge-molecule interactions in pi-conjugated organic semiconductors and molecular electronic materials.
Field-induced photoluminescence modulation of MEH-PPV under near-field optical excitation. *J. Phys. Chem. B* 105(1):76–82, 2001 (with O'Connor, Adams, Barbara, and Kammer).

William T. Pennington, Professor of Inorganic Chemistry; Ph.D., Arkansas, 1983. The crystal chemistry of materials involving halogen bonding (Lewis acid–Lewis base interactions).
Solid polymer electrolytes from polyanionic lithium salts based on the LiTFSI anion structure. *J. Electrochem. Soc.* 151:A1363–8, 2004 (with Geiculescu et al.).

Dvora Perahia, Assistant Professor of Physical Chemistry; Ph.D., Weizmann (Israel), 1991. The study of polymers and complex fluids.
Bulk and interfacial studies of a new and versatile semifluorinated lyotropic liquid crystalline polymer. *Macromolecules* 34(12):3954–61, 2001 (with Traiphol, Shah, and Smith).

Dennis W. Smith Jr., Associate Professor of Organic Chemistry; Ph.D., Florida, 1992. Thermal polymerization of bis-(ortho-diynylarene) (BODA) monomers to polyarlyene thermosets, cyclopolymerization of trifluorovinyl ether monomers to polymers containing the perfluorocyclobutane (PFCB) linkage, and renewable resource/biodegradable polymers based on polylactic acid derivatives.
A novel network polymer for templated carbon photonic crystal structures. *Langmuir* 19:7153–6, 2003 (with Perpall et al.).

Steven J. Stuart, Associate Professor of Physical Chemistry; Ph.D., Columbia, 1995. Computer simulations of complex systems.
An iterative variable-timestep algorithm for molecular dynamics simulations. *Mol. Sim.* 29:177–86, 2003 (with Hicks and Mury).

Ya-Ping Sun, Professor of Materials and Organic Chemistry; Ph.D., Florida State, 1989. The development of nanostructures and nanomaterials for optical, electronic, and biomedically significant applications, with emphasis on carbon nanotubes, dendritic nanostructures, semiconductor and metal nanoparticles, and polymeric nanocomposites.
Soluble dendron-functionalized carbon nanotubes: Preparation, characterization, and properties. *Chem. Mater.* 13(9):2864–9, 2001 (with Huang et al.).

CLEMSON UNIVERSITY

CLEMSON UNIVERSITY

Department of Chemistry
Doctor of Philosophy in Chemistry

Programs of Study

The Department of Chemistry offers programs leading to the Doctor of Philosophy (Ph.D.), with an emphasis in analytical, inorganic, organic, physical chemistry, or chemistry education. Individual programs of study involve an intensive concentration in one of the traditional areas of chemistry or a concentration in a combination of areas. The degree program is research based, which means that it involves the completion and defense of an original research project. In addition to the course work, the research project comprises the bulk of the effort involved in the pursuit of an advanced chemistry degree.

Ph.D. students must complete a core sequence of four courses and a selection of other courses relevant to their degree program within the first two years of study. Students must also demonstrate a comprehensive knowledge of their major area by satisfactory performance on a series of written cumulative examinations. All degree candidates must present at least three research-based seminars to the Department as part of their degree program, one of which is usually on a literature topic. The final stage of the graduate degree program involves the writing and defense before the degree committee of a dissertation describing the student's original research project.

Research Facilities

The Department of Chemistry is housed in the Howard L. Hunter Chemistry Laboratory, which includes more than 50,000 square feet of laboratory space for research and teaching. One of the finest research facilities in the Southeast, this building was completed in 1987 and accommodates about 100 graduate students, postdoctoral scientists, and visiting scientists. It includes a satellite chemistry library that houses the field's most important journals and supplements extensive holdings in the University's central library. Several chemistry research groups also occupy space in other on- and off-campus buildings.

The Department maintains a broad range of multiple-user instruments for chemistry research. Major research instrumentation holdings include three Fourier-transform NMR spectrometers; X-ray powder, single-crystal, and thin-film diffractometers; an electron spin resonance (ESR) spectrometer; gas chromatography/mass spectrometer systems; a thermal analysis system; and other state-of-the-art equipment maintained by individual faculty members in support of their research programs or through the Department's research partners.

Clemson University provides a diverse and extensive computing infrastructure supported locally within the Department of Chemistry and by the Office of Computer and Network Services (CNS) and the Division of Computing and Information Technology (DCIT). Various laboratories in the Department have high-speed SGI, Sun, and Linux workstations as well as a twenty-eight-processor cluster for parallel computations. PC and Macintosh computers are available in all departmental research labs and in many computer labs around the campus. The College of Engineering and Science has recently installed a 512-processor distributed Beowulf cluster that makes Clemson one of the top supercomputing sites in the Southeast. Clemson also participates in the high-speed Internet2 and partners with the Center for Advanced Engineering Fibers and Films, which has a state-of-the-art virtual reality laboratory and recently received a $1.3-million grant from the Keck Foundation to create a virtual visualization and design lab.

The Laser Laboratory is managed by Ya-Ping Sun and his research group. The laboratory is equipped with a CW Mode locked Nd: Yag Laser, a 20-Hz Q-Switched Nd: Yag Laser, and two synchronous pumped Dye Lasers. The laser configuration is capable of conducting pump probe experiments in the nanosecond time scale region up to the subpicosecond time scale region.

The Nuclear Magnetic Resonance Resource Center affords easy access to modern high-resolution NMR instruments for students, postdoctoral scientists, and faculty members. The primary instrumentation includes three multinuclear high-field spectrometers that are used for routine measurements as well as for advanced one- and two-dimensional NMR experiments in molecular structure determination, molecular dynamics, and chemical kinetics and thermodynamics.

Clemson's Electronic Imaging and Analytical Services (EIAS) group is one of the Southeast's premier analytical imaging and surface analysis facilities. Area researchers both on and off campus can take advantage of a broad range of capabilities, including scanning electron microscopy, transmission electron microscopy, and high-vacuum surface analysis. The EIAS facility is widely used in a number of areas but particularly in nanomaterial and nanotechnology research, which depend critically on the availability of tools that can characterize materials with submicrometer to subnanometer spatial resolution.

The Molecular Structure Center, under the direction of Don Vanderveer, provides the chemistry department with methods of X-ray diffraction analysis, the most reliable and unambiguous means for determining the structure of ordered materials. The center maintains four separate diffractometer systems for performing both powder and single-crystal diffraction experiments. These include two Rigaku diffractometers. One is a sealed tube system equipped with a CCD area detector; the other has a detector that uses a powerful 18-kw rotating anode source. The center also has a conventional four-circle diffractometer with a sealed tubesource. A Scintag 2000 system with a germanium detector and a seven-position automatic sample changer is used for powder diffraction. Data processing and analysis are done on numerous PCs running Microsoft Windows and Red Hat Linux. The center has access to many electronic databases, including Cambridge Structural Database, Inorganic Crystal Structure Database, NIST Crystal Data File, and Powder Diffraction File.

Financial Aid

All chemistry graduate students at Clemson are supported by either teaching or research assistantships during the full course of their studies. Students in the first year are normally supported as teaching assistants in undergraduate laboratory sections. Stipends for teaching assistantships are competitive and change frequently. In fall 2005, the stipend was $18,000 for twelve months. Research assistantships are often available to support students working on funded research projects. Department and University fellowships that can supplement the stipend for well-qualified applicants are also available.

Cost of Study

Tuition for 2006–07 is $4643 per semester for in-state students and $9255 per semester for nonresidents. Off-campus rates are $535 per hour for in-state students and $918 per hour for nonresidents. Graduate assistants pay a flat fee of $1079 per semester and $348 per summer session. Graduate fellows pay South Carolina resident fees.

Living and Housing Costs

On-campus housing is available. For information, students should visit http://www.housing.clemson.edu. The cost of living in Clemson is quite low compared to the national average. Students who choose to live off the campus typically spend $300–$400 per month for rent, depending on location, amenities, roommates, etc.

Student Group

There are 101 full-time and 3 part-time graduate students enrolled in the Department, of whom 33 are women and 39 are international students.

Student Outcomes

Graduates seek and find employment in a very broad range of areas where chemistry is important, including private industry, national laboratories, and education. Public and private-sector employers that have recently hired Clemson chemistry graduates include Advanced Photonic Crystals (Charlotte, North Carolina), IRIX Pharmaceuticals (Florence, South Carolina), Milliken Chemical Co. (Spartanburg, South Carolina), Savannah River Lab (U.S. Department of Engineering) (Aiken, South Carolina), Michelin Tire Co. (Greenville, South Carolina), Dupont Chemical Company (Richmond, Virginia), Merck Chemical (Rahway, New Jersey), Clariant Corp. (Aiken, South Carolina), Micromass Inc. (Beverly, Massachusetts), Chiron Corp. (San Francisco, California), Toyota Co. (Japan), U.S. Army Research Lab (Adelphi, Maryland), NASA Langley Research Center (Hampton, Virginia), Transtech Pharma (North Carolina), Idaho National Engineering and Environmental Lab (U.S. Department of Engineering), Beckman Coulter (Allendale, New Jersey), Cree Semiconductors (North Carolina), National Institutes of Standards and Technology (Gaithersburg, Maryland), and Shire Laboratories (Maryland). Graduates have found employment at universities such as Harvard, Kennesaw State, the University of Connecticut, and UCLA.

Location

Clemson is a small, beautiful college town near the Blue Ridge Mountains and Lake Hartwell in upstate South Carolina. The Upstate is one of the country's fastest-growing areas and is the midpoint of the Charlotte-to-Atlanta I-85 corridor, a multistate area along Interstate 85 that runs from metro Atlanta to Richmond, Virginia, and encompasses Charlotte, North Carolina, and North Carolina's Research Triangle. Atlanta and Charlotte are each a 2-hour's drive away. Many financial institutions and other industries have national headquarters for a major presence in the Upstate, including Wachovia, Bank of America, BMW, Bon Secours St. Francis Health System, Bosch North America, Bowater, Charter Communications, Ernst and Young, Fluor Corporation, IBM, Microsoft, Michelin of North America, and many others.

The University and The Department

Clemson is classified by the Carnegie Foundation as Doctoral/Research University–Extensive, a category comprising less than 4 percent of all universities in America. The University's mission is to fulfill the covenant between its founder and the people of South Carolina to establish a "high seminary of learning" through its responsibilities of teaching, research, and extended public service. The University has identified eight areas of academic emphasis that create collaborations that, in turn, help fulfill the University's mission.

Chemistry has been taught at Clemson since the institution was founded in 1889. Seventy years later, chemistry became one of the University's first Ph.D. programs. Today, the Department of Chemistry is one of Clemson's largest departments.

Applying

Most students who apply to the Clemson graduate chemistry program have completed a B.S. or B.A. undergraduate degree in chemistry or have acquired the equivalent background. In most cases, the undergraduate program included at least two formal courses and the associated labs each in general chemistry, analytical chemistry, and physical chemistry and at least one formal course in inorganic chemistry. Experience in research is also desirable.

An applicant must complete an application form and supply transcripts from the undergraduate program and any prior graduate programs, test scores from the GRE general exam (verbal, quantitative, and analytical), at least two letters of recommendation from people familiar with the student's background, a completed personal statement form, and a completed financial assistance form. International students must also submit a test score for the TOEFL exam. The TSE exam is not required for international students; however, a TSE score above 50 greatly improves the chances of admission. Applications are processed most quickly if all materials are sent directly to the Department of Chemistry by March 1.

The fee for making an official application through the Clemson graduate school is normally $50; however, for applicants who are U.S. citizens or permanent residents, or who have received a degree from a U.S. college or university, or who possess sufficient English language proficiency that they are qualified to teach laboratory sections, the fee is covered by the Department of Chemistry. For all other applicants, application materials submitted to the Department of Chemistry should be considered preliminary only. An official application, with a paid application fee, is required for the application to be given formal consideration.

Correspondence and Information

Steve Stuart, Graduate Director
369 Hunter Laboratories
P.O. Box 340973
Clemson University
Clemson, South Carolina 29634-0973
Phone: 864-656-3095
864-656-5013
888-539-8854 (toll-free)
Fax: 864-656-6613
E-mail: chemgradprogram@chemed.ces.clemson.edu
Web site: http://chemistry.clemson.edu

Clemson University

THE FACULTY AND THEIR RESEARCH

Rudolph Abramovitch, Professor of Organic Chemistry; Ph.D., King's College (London), 1953; D.Sc., London, 1964. The chemistry of electron-deficient reactive intermediates.
The 1,2,4-triazolyl cation: Thermolytic and photolytic studies. *J. Org. Chem.* 66(4):1242–51, 2001 (with Beckert et al.).
Jeffrey R. Appling, Associate Professor of Chemistry Education; Ph.D., Georgia Tech, 1985. The development of new materials and methods to improve both instruction in general chemistry and the study skills of general chemistry students.
Math Survival Guide: Tips for Science Students, 2nd ed. New York: John Wiley & Sons, 2003 (with Richardson).
Dev P. Arya, Associate Professor of Bio-Organic and Medicinal Chemistry; Ph.D., Northeastern, 1996. Nucleic acid therapeutics and synthesis of small-molecule carbohydrate mimetics.
From triplex to B-form duplex stabilization: Reversal of target selectivity by aminoglycoside dimers. *Bioorg. Med. Chem. Lett.* 14:4643–6, 2004 (with Coffee and Xue).
Julia Brumaghim, Assistant Professor of Bio-Inorganic Chemistry; Ph.D., Illinois at Urbana–Champaign, 1999. The biological chemistry of the Fenton reaction and the antioxidant properties of selenium.
Encapsulation and stabilization of reactive aromatic diazonium ions and the tropylium ion within a supramolecular host. *Eur. J. Org. Chem.* 2004(24):5115–8, 2004 (with Michels, Pagliero, and Raymond).
Kenneth A. Christensen, Assistant Professor of Analytical Chemistry; Ph.D., Michigan, 1997. The molecular recognition and host cellular responses to microbial toxin proteins, including potential biowarfare agents.
Fluorescence resonance energy transfer-based stoichiometry in living cells. *Biophys. J.* 83:3652–64, 2002 (with Hoppe and Swanson).
George Chumanov, Associate Professor of Analytical Chemistry; Ph.D., Moscow State, 1988. The preparation and modification of different nanoparticles, including one-, two-, and three-dimensional regular structures; the investigation of their properties using different spectrochemical techniques together with optical, electron, atomic force, and scanning tunneling microscopy; and the development of new materials and devices for environmental and biomedical diagnostic applications.
Vacuum deposition of silver island film on chemically modified surfaces. *J. Vac. Sci. Technol. A* 21:723–7, 2003 (with Malynych).
Melanie Cooper, Alumni Distinguished Professor of Chemistry Education; Ph.D., Manchester (England), 1978. Investigations into the factors that affect student problem-solving strategies, the development and assessment of laboratory curricula, teaching strategies for large enrollment courses, investigating how group work affects student achievement, and whether gender has an effect on student problem-solving strategies.
A Chemists' Guide to Effective Teaching. Upper Saddle River, N.J.: Pearson Education, 2005 (with Greenbowe and Pienta).
Karen Creager, Lecturer of Analytical Chemistry; Ph.D., North Carolina at Chapel Hill, 1989. Photoelectrochemistry of semiconductor surfaces.
Electrochemical olefin epoxidation with manganese meso-tetraphenyl porphyrin catalyst and hydrogen peroxide generation at polymer-coated electrodes. *Inorg. Chem.* 29:1000, 1990 (with Nishihara, Murray, and Collman).
Stephen Creager, Professor of Analytical Chemistry; Ph.D., North Carolina at Chapel Hill, 1987. Electrochemical science and technology.
Redox potentials and kinetics of the Ce^{3+}/Ce^{4+} redox reaction and solubility of cerium sulfates in sulfuric acid solutions. *J. Power Sources* 109(2):431–8, 2002 (with Paulenova, Navratil, and Wei).
Darryl DesMarteau, Tobey-Beaudrot Professor of Inorganic Chemistry; Ph.D., Washington (Seattle), 1966. The study of fluorinated molecules and macromolecules.
Synthesis and comparison of N-fluorobis (perfluoroalkyl) sulfonyl imides with different perfluoroalkyl groups. *J. Fluorine Chem.* 111(2):253–7, 2001 (with Zhang).
Karl Dieter, Professor of Organic Chemistry; Ph.D., Pennsylvania, 1978. Organocopper, organopalladium, and organosulfur chemistry and asymmetric synthesis.
A study of factors affecting alpha- (N-carbamoyl) alkylcuprate chemistry. J. Org. Chem. 66(7):2302–11, 2001 (with Topping and Nice).
Brian N. Dominy, Assistant Professor; Ph.D., Scripps Research Institute, 2001. Physical chemistry.
Lourdes Echegoyen, Lecturer of Analytical and Physical Chemistry; Ph.D., Miami (Florida), 1990. Electrochemically enhanced transport across artificial membranes and the preparation and characterization of novel supramolecular assemblies using crown-ether derivatives.
The electrochemistry of C60 and related compounds. In *Organic Electrochemistry,* eds. H. Lund and O. Hammerich, 4th ed. New York: Marcel Dekker, 2000, p. 323 (with Luis Echegoyen).
Luis Echegoyen, Department Chair and Professor of Physical Chemistry; Ph.D., Puerto Rico, 1974. Fullerene chemistry, electrochemistry, and supramolecular chemistry.
Carbon nanotube doped polyaniline. *Adv. Mat.* 14:1480–3, 2002 (with Zengin et al.).
Lucy Eubanks, Lecturer of Chemistry Education; M.S.N.S., Seattle, 1967.
Chemistry in Context, 5th ed. New York: McGraw-Hill, 2005 (with Middlecamp, Pienta, Heltzel, and Weaver).
Grant Goodyear, Assistant Professor of Chemistry. The study of supercritical fluids near the critical point, the electronic structure of liquid semiconductors.
John W. Huffman, Professor of Organic Chemistry; Ph.D., Harvard, 1957. The synthesis of analogues and metabolites of delta-9-tetrahydrocannabinol (THC), the principal active component of marijuana.
Inhibition of glioma growth in vivo by selective activation of the CB2 cannabinoid receptor. *Cancer Res.* 61(15):5784–9, 2001 (with Sanchez et al.).
Shiou-Jyh Hwu, Professor of Inorganic Chemistry; Ph.D., Iowa State, 1985. Synthesis of solids containing confined transition-metal-oxide frameworks, synthesis of conducting solids in nonaqueous media, synthesis of a new class of hybrid materials via salt inclusion.
Vacancy ordering transitions in one-dimensional lattice gas with Coulomb interactions. *Phys. Rev. B: Condens. Matter* 6304(4):045405, 2001 (with King, Kuo, and Skove).
John Kaup, Lecturer of Physical and Analytical Chemistry; Ph.D., Utah, 1997. Analytical and physical laboratories.
Arkady Kholodenko, Professor of Physical Chemistry; Ph.D., Chicago, 1982. Theory of liquid crystalline semiflexible polymer solutions; statistical mechanics of disordered systems, including glasses and random copolymers; theory of knots and links, with applications to condensed matter and biological systems; theory of quantum and classical chaos.
"New" Veneziano amplitudes from "old" Fermat (hyper)surfaces. *Int. J. Geometry Phys.* 19:1655–703, 2004.
Alex Kitaygorodskiy, Lecturer and NMR Facility Director; Ph.D., Russian Academy of Sciences, 1979. Outer-sphere complexation of metal complexes.
Soluble dendron-functionalized carbon nanotubes: Preparation, characterization, and properties. *Chem. Mater.* 13(9):2864–9, 2001 (with Sun et al.).
Joseph W. Kolis, Professor of Inorganic Chemistry; Ph.D., Northwestern, 1984. The synthesis and chemistry of novel inorganic compounds with unusual structures and properties.
Observation of the interplay of microstructure and thermopower in the Al71Pd21Mn8-xRex quasicrystalline. *Phys. Rev. B: Condens. Matter* 6305(5):052202, 2001 (with Pope et al.).
R. Kenneth Marcus, Professor of Analytical Chemistry; Ph.D., Virginia, 1986. The development and application of new plasma techniques for the atomic spectroscopic analysis of diverse materials.
Collisional dissociation in plasma source mass spectrometry: A potential alternative to chemical reactions for isobar removal. *J. Anal. At. Spectrom.* 19:591–9, 2004.
Jason McNeill, Assistant Professor of Physical Chemistry; Ph.D., Berkeley, 1999. Understanding charge-molecule interactions in pi-conjugated organic semiconductors and molecular electronic materials.
Field-induced photoluminescence modulation of MEH-PPV under near-field optical excitation. *J. Phys. Chem. B* 105(1):76–82, 2001 (with O'Connor, Adams, Barbara, and Kammer).
William T. Pennington, Professor of Inorganic Chemistry; Ph.D., Arkansas, 1983. The crystal chemistry of materials involving halogen bonding (Lewis acid–Lewis base interactions).
Solid polymer electrolytes from polyanionic lithium salts based on the LiTFSI anion structure. *J. Electrochem. Soc.* 151:A1363–8, 2004 (with Geiculescu et al.).
Dvora Perahia, Assistant Professor of Physical Chemistry; Ph.D., Weizmann (Israel), 1991. The study of polymers and complex fluids.
Bulk and interfacial studies of a new and versatile semifluorinated lyotropic liquid crystalline polymer. *Macromolecules* 34(12):3954–61, 2001 (with Traiphol, Shah, and Smith).
Dennis W. Smith Jr., Associate Professor of Organic Chemistry; Ph.D., Florida, 1992. Thermal polymerization of bis-(ortho-diynylarene) (BODA) monomers to polyarlyene thermosets, cyclopolymerization of trifluorovinyl ether monomers to polymers containing the perfluorocyclobutane (PFCB) linkage, and renewable resource/biodegradable polymers based on polylactic acid derivatives.
A novel network polymer for templated carbon photonic crystal structures. *Langmuir* 19:7153–6, 2003 (with Perpall et al.).
Steven J. Stuart, Associate Professor of Physical Chemistry; Ph.D., Columbia, 1995. Computer simulations of complex systems.
An iterative variable-timestep algorithm for molecular dynamics simulations. *Mol. Sim.* 29:177–86, 2003 (with Hicks and Mury).
Ya-Ping Sun, Professor of Materials and Organic Chemistry; Ph.D., Florida State, 1989. The development of nanostructures and nanomaterials for optical, electronic, and biomedically significant applications, with emphasis on carbon nanotubes, dendritic nanostructures, semiconductor and metal nanoparticles, and polymeric nanocomposites.
Soluble dendron-functionalized carbon nanotubes: Preparation, characterization, and properties. *Chem. Mater.* 13(9):2864–9, 2001 (with Huang et al.).

CORNELL UNIVERSITY

Field of Chemistry and Chemical Biology

Program of Study

The diverse research specialties of the graduate faculty include both traditional areas and interdisciplinary expertise in biotechnology, chemical communication, polymer science, and molecular engineering. Cornell's graduate program is designed to provide broad training in the fundamentals of chemistry and research methods and to culminate in the award of the doctorate. Several nationally renowned research centers and facilities at Cornell foster inquiry in traditional areas of the discipline and at the interface between chemistry and other physical and biological sciences, engineering, materials science, and mathematics. Graduate students enrolled in the field of chemistry and chemical biology select a major research concentration in one of the following subfields: analytical, bioorganic, biophysical, chemical biology, inorganic, materials, organic, organometallic, physical/chemical physics, polymer, or theoretical. Students also choose a minor concentration from these subfields or a related graduate field, such as materials science or biochemistry. Once major and minor concentrations have been selected, students choose permanent special committees consisting of the research adviser and 2 additional faculty members. First-year graduate students take proficiency examinations in inorganic, organic, and physical chemistry; a level of instruction suited to the student's background and future objectives is then recommended. The number of formal courses required depends on a student's previous preparation, chosen concentration, and the advice of the special committee. Students concentrating in organic chemistry prepare and defend a research proposal. Every student takes an admission-to-candidacy examination within the first two years of study. The Ph.D. is awarded upon completion of an original research project (directed by the chair of the special committee) and successful defense of the thesis.

Research Facilities

Research in the Department of Chemistry and Chemical Biology is supported by several departmental facilities, including the Nuclear Magnetic Resonance Facility, the X-Ray Diffraction Facility, the Mass Spectrometry Facility, and the Glass Shop. Additional facilities at Cornell include the Center for High Energy Synchrotron Studies, the Cornell Nanofabrication Facility, the Center for Theory and Simulation, and the Biotechnology Program in the New York State Center for Advanced Technology as well as specialized facilities associated with the Cornell Center for Materials Research and the Cornell Nanobiotechnology Center.

Financial Aid

Fellowships, scholarships, loans, and teaching and research assistantships are available. Nearly all Ph.D. students have assistantship or fellowship support sufficient to cover tuition, fees, health insurance, and living expenses. Fellowships and scholarships are also offered by state and national government agencies, foundations, and private corporations.

Cost of Study

Tuition and fees for students attending Cornell during the 2006–07 academic year are $32,800. Some increase for 2007–08 is anticipated.

Living and Housing Costs

Living expenses for the calendar year are typically $17,732 to $18,000 for single students. Additional expenses may include travel and medical and dental costs. The University maintains married student housing and graduate dormitory accommodations on and near the campus. Privately owned accommodations are available nearby.

Student Group

There are 214 graduate students currently enrolled in the Ph.D. program in the Department of Chemistry and Chemical Biology.

Location

Ithaca, a city of about 30,000 people, is located on Cayuga Lake in the beautiful Finger Lakes region of central New York State. The home of both Ithaca College and Cornell, the city is one of the country's great educational communities, offering cultural advantages that rival those of many large cities. Recreational activities, including hiking, cycling, boating, and skiing, abound. Light industry and technical and consulting firms are active in the area.

The University and The Department

Cornell University, founded in 1865 by Ezra Cornell, is composed of fourteen colleges and schools. Several of these schools and colleges were established under the land-grant college system and are part of the State University of New York; others are privately endowed. The Department of Chemistry and Chemical Biology occupies more than 300,000 square feet of space in Baker Laboratory and the adjacent S. T. Olin Laboratory.

Applying

Only students who expect to complete a doctoral program (Ph.D.) should apply. Application to the graduate field of chemistry and chemical biology at Cornell is accepted for fall admission only. Early submission of applications is strongly encouraged, with applications being evaluated as early as December 1. Completed applications received before January 10 may also serve as applications for Cornell fellowships. Transcripts of all grades (whether or not a degree has been conferred) and three letters of recommendation are required. Applications must include GRE General Test scores and one GRE Subject Test score in chemistry. International applicants must demonstrate proficiency in English, usually by submitting scores on the TOEFL. Minimum scores of 600 on the paper-based test (250 computer-based test, 77 Internet-based test), with a score in each of the three categories of at least 60 on the paper-based test (25, computer-based test), are required for application consideration.

Correspondence and Information

Graduate Coordinator
Chemistry and Chemical Biology
Baker Laboratory
Cornell University
Ithaca, New York 14853-1301
Phone: 607-255-4139
Fax: 607-255-4139
E-mail: chemgrad@cornell.edu
Web site: http://www.chem.cornell.edu

Cornell University

THE FACULTY AND THEIR RESEARCH

H. D. Abruña, Ph.D., North Carolina. Electrochemistry, chemically modified electrodes, redoxactive dendrimers, biosensors, X-ray-based techniques, OLED's, fuel cells, molecular electronics.
B. A. Baird, Ph.D., Cornell. Biophysical approaches to elucidate signal transduction mechanisms of cell membrane receptors involved in immunological responses.
T. P. Begley, Ph.D., Caltech. Bioorganic chemistry, enzymatic reaction mechanisms, DNA photochemistry, prolyl-4-hydroxylase, thiamine biosynthesis.
J. T. Brenna, Ph.D., Cornell. High-precision isotope ratio mass spectrometry, molecular and elemental mass spectrometry, mammalian metabolomics and polyunsaturated fatty acid biosynthesis.
R. A. Cerione, Ph.D., Rutgers. Structure and function of small molecular G proteins, cellular signal transduction.
G. K.-L. Chan, Ph.D., Cambridge. Theoretical chemical physics, electronic structure and dynamics, numerical renormalization group theory, conjugated systems, spin systems.
P. Chen, Ph.D. Stanford. Single molecule spectroscopy, bioinorganic/biophysical chemistry, metallochaperones, metalloregulatory proteins, live cell imaging of metal trafficking, metalloprotein folding.
P. J. Chirik, Ph.D., Caltech. Synthetic inorganic and organotransition metal chemistry directed toward small molecule activation and catalysis.
G. W. Coates, Ph.D., Stanford. Stereoselective catalysis, organic and polymer synthesis, reaction mechanisms.
D. B. Collum, Ph.D., Columbia. Organotransition metal and organolithium reaction mechanism, development of organometallic chemistry for organic synthesis.
B. R. Crane, Ph.D., Scripps Research Institute. Biophysical and bioinorganic chemistry, structural principles of redox and photochemistry in biological catalysis and regulation.
H. F. Davis, Ph.D., Berkeley. Chemical reaction dynamics using laser and molecular beam techniques.
F. J. DiSalvo, Ph.D., Stanford. Solid-state chemistry, synthesis and structure, electrical and magnetic properties, materials for fuel cells.
S. E. Ealick, Ph.D., Oklahoma. X-ray crystallography of macromolecules, applications of synchrotron radiation, enzymes of nucleotide metabolism, enzyme mechanism, structure-based drug design.
J. R. Engstrom, Ph.D., Caltech. Surface science, semiconductor surface chemistry, molecular-beam scattering, thin-film deposition, microfluidics.
G. S. Ezra, Ph.D., Oxford. Theoretical chemical physics, intramolecular dynamics, semiclassical mechanics, electron correlation.
R. C. Fay, Ph.D., Illinois. Textbook author.
J. H. Freed, Ph.D., Columbia. Theoretical and experimental studies of molecular dynamics and structure by magnetic resonance spectroscopy.
B. Ganem, Ph.D., Columbia. Synthetic organic and bioorganic chemistry, methods and reactions for the synthesis of rare natural products and biologically active compounds.
E. P. Giannelis, Ph.D., Michigan State. Materials chemistry, polymer nanocomposites, self-assembling systems.
G. P. Hess, Ph.D., Berkeley. Reaction mechanisms of membrane-bound proteins on cell surfaces required for signal transmission in the nervous system, transient kinetics and laser pulse-photolysis measurements.
M. A. Hines, Ph.D., Stanford. Fundamental studies of chemical etching, new methods of nanofabrication, properties of nanomechanical systems, scanning tunneling microscopy.
R. Hoffmann, Ph.D., Harvard. Electronic structure of organic, organometallic, and inorganic molecules and extended structures, transition states and reaction intermediates.
P. L. Houston, Ph.D., MIT. Chemical kinetics and reaction dynamics.
S. Lee, Ph.D., Chicago. Materials chemistry, synthesis structure and electronic structure of entended solids.
H. Lin, Ph.D., Columbia. Chemical biology, biochemistry, enzymatic reaction mechanisms, protein posttranslational modification (ADP-ribosylation, diphthamide, C-mannosylation), proteomics.
R. F. Loring, Ph.D., Stanford. Theoretical chemical physics, dynamics in small molecule and polymeric liquids.
J. A. Marohn, Ph.D., Caltech. Scanned-probe microscopy investigations of novel magnetic and electronic materials.
F. W. McLafferty, Ph.D., Cornell. Mass spectrometry, characterization of large biomolecules.
D. T. McQuade, Ph.D., Wisconsin. Design and synthesis of self-assembling small molecules and polymers for application in catalysis, sensing, and biomaterials.
J. Meinwald, Ph.D., Harvard. Organic chemistry; insect chemical ecology; characterization, biosynthesis, and synthesis of natural products.
J. T. Njarðarson, Ph.D., Yale. Organic chemistry, development of new synthetic methods, natural product synthesis.
C. K. Ober, Ph.D., Massachusetts. Polymer synthesis, self-assembling materials, microlithography, liquid crystalline polymers.
J. Park, Ph.D., Berkeley. Physical chemistry, single-molecule electron transport spectroscopy, optical spectroscopy on single molecules/nanostructures, nanoscale biosensor.
H. A. Scheraga, Ph.D., Duke. Physical chemical studies of proteins and aqueous solutions.
D. Y. Sogah, Ph.D., UCLA. Supramolecular chemistry, polypeptide-based polymers, group transfer polymerization; living free-radical polymerization; hyperbranched and helical polymers; organic-inorganic nanocomposites, nanobiotechnology.
D. S. Tan, Ph.D., Harvard. Diversity-oriented organic synthesis of small molecule libraries related to natural products, high-throughput screening for biological probes.
D. A. Usher, Ph.D., Cambridge. Polynucleotide analogs, template reactions, chemical evolution.
B. Widom, Ph.D., Cornell. Statistical mechanics of phase transitions, critical phenomena, and interfaces.
U. Wiesner, Ph.D., Mainz. Topics at the interface between polymer science and solid-state chemistry, nanostructured hybrids, nanobiotechnology, nanoparticles, bioimaging, fuel cells.
C. F. Wilcox, Ph.D., UCLA. Elucidation of the relationship between molecular structure and energy by theoretical analysis and synthesis of selected novel molecules.
P. T. Wolczanski, Ph.D., Caltech. Synthesis and reactivity of transition metal complexes, materials preparation.
D. B. Zax, Ph.D., Berkeley. Studies of novel synthetic and naturally occurring polymeric materials, structure and dynamical studies in solids.

Analytical Chemistry
H. D. Abruña, J. T. Brenna, R. A. Cerione, J. R. Engstrom, M. A. Hines, J. A. Marohn, F. W. McLafferty, D. B. Zax.

Bioorganic Chemistry
T. P. Begley, G. W. Coates, B. Ganem, F. W. McLafferty, D. T. McQuade, J. Meinwald, D. Y. Sogah, D. S. Tan, D. A. Usher.

Biophysical Chemistry
B. A. Baird, R. A. Cerione, P. Chen, B. R. Crane, S. E. Ealick, J. H. Freed, G. P. Hess, H. A. Scheraga.

Chemical Biology
B. A. Baird, T. P. Begley, R. A. Cerione, P. Chen, B. R. Crane, S. E. Ealick, B. Ganem, G. P. Hess, H. Lin, D. T. McQuade, J. T. Njarðarson, H. A. Scheraga, D. S. Tan, D. A. Usher.

Inorganic Chemistry
H. D. Abruña, P. Chen, P. J. Chirik, G. W. Coates, D. B. Collum, B. R. Crane, F. J. DiSalvo, R. C. Fay, R. Hoffmann, S. Lee, P. T. Wolczanski, D. B. Zax.

Materials Chemistry
H. D. Abruña, G. W. Coates, F. J. DiSalvo, J. R. Engstrom, E. P. Giannelis, M. A. Hines, R. Hoffmann, P. L. Houston, S. Lee, R. F. Loring, J. A. Marohn, D. T. McQuade, C. K. Ober, D. Y. Sogah, U. Wiesner, P. T. Wolczanski, D. B. Zax.

Organic Chemistry
T. P. Begley, P. J. Chirik, G. W. Coates, D. B. Collum, B. Ganem, H. Lin, D. T. McQuade, J. Meinwald, J. T. Njarðarson, D. Y. Sogah, D. S. Tan, D. A. Usher, C. F. Wilcox, P. T. Wolczanski.

Organometallic Chemistry
P. J. Chirik, G. W. Coates, D. B. Collum, P. T. Wolczanski.

Physical Chemistry/Chemical Physics
H. D. Abruña, B. A. Baird, G. K.-L. Chan, P. Chen, H. F. Davis, F. J. DiSalvo, J. R. Engstrom, G. S. Ezra, J. H. Freed, M. A. Hines, R. Hoffmann, P. L. Houston, R. F. Loring, J. A. Marohn, J. Park, H. A. Scheraga, B. Widom, U. Wiesner, D. B. Zax.

Polymer Chemistry
P. J. Chirik, G. W. Coates, E. P. Giannelis, R. Hoffmann, R. F. Loring, J. A. Marohn, D. T. McQuade, C. K. Ober, H. A. Scheraga, D. Y. Sogah, U. Wiesner.

Theoretical Chemistry
G. K.-L. Chan, G. S. Ezra, J. H. Freed, R. Hoffmann, S. Lee, R. F. Loring, H. A. Scheraga, B. Widom, C. F. Wilcox.

SELECTED PUBLICATIONS

Orth, R., et al. **(B. A. Baird).** Mast cell activation on patterned lipid bilayers of subcellular dimensions. *Langmuir* 19:1599–605, 2003.

Maggio-Hall, L. A., P. C. Dorrestein, J. C. Escalante-Semerena, and **T. P. Begley.** Formation of the dimethylbenzimidazole ligand of coenzyme B12 under physiological conditions by a facile oxidative cascade. *Org. Lett.* 5(13):2211–3, 2003.

Sacks G. L., and **J. T. Brenna.** High-precision position-specific isotope analysis of 13C/12C in leucine and methionine analogues. *Anal. Chem.* 75(20):5495–503, 2003.

Corso, T. N., and **J. T. Brenna.** High-precision, position-specific isotope analysis. *Proc. Natl. Acad. Sci. U.S.A.* 94(4):1049–53, 1997.

Ramachandran, S., and **R. A. Cerione.** Stabilization of an intermediate activation state for transducin by a fluorescent GTP analogue. *Biochemistry* 43:8778–86, 2004.

Wu, W. J., S. Tu, and **R. A. Cerione.** Activated Cdc42 sequesters c-Cbl and prevents EGF receptor degradation. *Cell* 114:715–25, 2003.

Chan, G. K.-L., and M. Head-Gordon. Exact solution (within a triple-zeta, double polarization basis set) of the electronic Schrodinger equation for water. *J. Chem. Phys.* 118 (19):8551–4, 2003.

Chan, G. K.-L., and M. Head-Gordon. Highly correlated calculations with a polynomial cost algorithm: A study of the density matrix renormalization group. *J. Chem. Phys.* 116(11):4462–76, 2002.

Chen, P., S. I. Gorelsky, S. Ghosh, and E. I. Solomon. N_2O reduction by the m_4-sulfide bridged tetranuclear Cu_Z cluster active site. *Angew. Chem. Int. Ed.* 43:4132–40, 2004.

Pool, J. A., E. Lobkovsky, and **P. J. Chirik.** Hydrogenation and cleavage of dinitrogen to ammonia with a well-defined zirconium complex. *Nature* 427:527–30, 2004.

Bart, S. C., and **P. J. Chirik.** Selective, catalytic carbon-carbon bond activiation and functionalization promoted by late transition metal catalysts. *J. Am. Chem. Soc.* 125:886–7, 2003.

Coates, G. W., and D. R. Moore. Discrete metal-based catalysts for the copolymerization of CO_2 and epoxides: Discovery, reactivity, optimization, and mechanism. *Angew. Chem. Int. Ed.* 43(48):6618–39, 2004.

Byrne, C. M., S. D. Allen, E. B. Lobkovsky, and **G. W. Coates.** Alternating copolymerization of limonene oxide and carbon dioxide. *J. Am. Chem. Soc.* 126(37):11404–5, 2004.

Baird, E. J., D. Holowka, **G. W. Coates,** and **B. A. Baird.** Highly effective poly(ethylene glycol) ligands for specific inhibition of cell activation by IgE receptors. *Biochemistry* 42:12739, 2003.

Briggs, T. F., et al. **(D. B. Collum).** Structural and rate studies of the 1,2-additions of lithium phenylacetylide to lithiated quinazolinones: Influence of mixed aggregates on the reaction mechanism. *J. Am. Chem. Soc.* 126:6291, 2004.

Zhao, P., and **D. B. Collum.** Lithium hexamethyldisilazide/triethylamine-mediated ketone enolization: Remarkable rate accelerations stemming from a dimer-based mechanism. *J. Am. Chem. Soc.* 125:4008, 2003.

Kang, S. A., P. J. Marjavaara, and **B. R. Crane.** Electron transfer between cytochrome c and cytochrome c peroxidase in single crystals. *J. Am. Chem. Soc.* 126(35):10836–7, 2004.

Park, S., et al. **(B. R. Crane).** In different organisms the mode of interaction between two signaling proteins is not necessarily conserved. *Proc. Natl. Acad. Sci. U.S.A.* 101:11646–51, 2004.

Hinrichs, R. Z., J. J. Schroden, and **H. F. Davis.** Competition between C-C and C-H insertion in prototype transition metal-hydrocarbon reactions. *J. Am. Chem. Soc.* 125:861, 2003.

Lin, C., M. F. Witinski, and **H. F. Davis.** Oxygen atom Rydberg time-of-flight spectroscopy—ORTOF. *J. Chem. Phys.* 119:251, 2003.

Casado-Rivera, E., et al. **(F. J. DiSalvo** and **H. D. Abruña).** Electrocatalytic activity of ordered intermetallic phases for fuel cell applications. *J. Am. Chem. Soc.* 126(12):4043–9, 2004.

Anand R., A. A. Hoskins, J. Stubbe, and **S. E. Ealick.** Domain organization of *Salmonella typhimurium* formylglycinamide ribonucleotide amidotransferase revealed by X-ray crystallography. *Biochemistry* 43:10328–42, 2004.

Zhang, Y., S. E. Cottet, and **S. E. Ealick.** Structure of *E. coli* AMP nucleosidase reveals similarity to nucleoside phosphorylases. *Structure* 12:1383–94, 2004.

Schroeder, T. W., A. M. Lam, P. F. Ma, and **J. R. Engstrom.** The effects of atomic hydrogen on the selective area growth of Si and Si_{1-x}Gex thin films on Si and SiO_2 surfaces: Inhibition, nucleation and growth. *J. Vac. Sci. Technol., A* 22:578–93, 2004.

Ma, P. F., T. W. Schroeder, and **J. R. Engstrom.** Nucleation of copper on TiN and SiO_2 from the reaction of hexafluoroacetylacetonate copper(I) trimethylvinylsilane. *Appl. Phys. Lett.* 80:2604–6, 2002.

Noid, W. G., **G. S. Ezra,** and **R. F. Loring.** Semiclassical calculation of the vibrational echo. *J. Chem. Phys.* 120:1491–9, 2004.

Arango, C. A., W. W. Kennerly, and **G. S. Ezra.** Quantum monodromy for diatomic molecules in combined electrostatic and pulsed nonresonant laser fields. *Chem. Phys. Lett.* 392:486–92, 2004.

Fay, R. C. *Chemistry,* 4th ed. Upper Saddle River, N.J.: Prentice Hall, 2004. With McMurry.

Fay, R. C. Stereochemistry and molecular rearrangements of some six-, seven-, and eight-coordinate chelates of early transition metals. *Coord. Chem. Rev.* 154:99–124, 1996.

Gau, H.-M., and **R. C. Fay.** NMR studies of inversion and dithiophosphate methyl group exchange in dialkoxybis(O,O'-dimethyl dithiophosphato)titanium (IV) complexes. Evidence for a bond-rupture mechanism. *Inorg. Chem.* 29:4974, 1990.

Freed, J. H., P. P. Borbat, J. H. Davis, and S. E. Butcher. Measurement of large distances in biomolecules using double-quantum-filtered refocused electron-spin-echoes. *J. Am. Chem. Soc.* 126:7746–7, 2004.

Freed, J. H., A. Blank, C. Dunnam, and P. Borbat. High-resolution electron spin resonance microscopy. *J. Magn. Res.* 165:116–27, 2003.

Wang, Q., Q. Xia, and **B. Ganem.** A general synthesis of 2-substituted-5-aminooxazoles: Building blocks for multifunctional heterocycles. *Tetrahedron Lett.* 44:6825, 2003.

Hamilton, D. S., et al. **(B. Ganem).** Mechanism of the glutathionyl transferase-catalyzed conversion of the antitumor 2-crotonyloxy-2-cycloalkenones to the GSH adducts. *J. Am. Chem. Soc.* 125:15049, 2003.

Tyner, K. M., S. Schiffman, and **E. P. Giannelis.** Nanobiohybrids as delivery vehicles for camptothecin. *J. Controlled Release* 95:501, 2004.

Muthu, M. M. M., E. Hackett, and **E. P. Giannelis.** From nanocomposite to nanogel polymer electrolytes. *J. Mater. Chem.* 13:1, 2003.

Hess, G. P. Rapid chemical reaction techniques developed for use in investigations of membrane-bound proteins (neurotransmitter receptors). *Biophys. Chem.* 100:493–506, 2003.

Hess, G. P., et al. Mechanism-based discovery of ligands that prevent inhibition of the nicotinic acetylcholine receptor by cocaine and MK-801. *Proc. Natl. Acad. Sci. U.S.A.* 97:13895, 2000.

Garcia, S. P., H. Bao, and **M. A. Hines.** The effects of diffusional processes on crystal etching: Kinematic theory extended to two dimensions. *J. Phys. Chem. B* 108:606271, 2004.

Hines, M. A. In search of perfection: Understanding the highly defect-selective chemistry of anisotropic etching. *Annu. Rev. Phys. Chem.* 54:29, 2003.

Papoian, G., and **R. Hoffmann.** Hypervalent bonding in one, two, and three dimensions: Extending the Zintl-Klemm concept to nonclassical electron-rich networks. *Angew. Chem.* 39:2408–48, 2000.

Konecny, R., and **R. Hoffmann.** Metal fragments and extended arrays on a Si(100)-(2x1) surface: I. Theoretical study of Ml_n complexation to Si(100). *J. Am. Chem. Soc.* 121(34):7918–24, 1999.

Slinker, J., et al. **(P. L. Houston** and **H. D. Abruña).** Electroluminescence in transition metal complexes. *Chem. Commun.* 2392–9, 2003.

Geiser, J. D., et al. **(P. L. Houston).** The vibrational distribution of 02(X 3Sigmag-) produced in the photodissociation of ozone between 226 and 240 and at 266 nm. *J. Chem. Phys.* 112:1279, 2000.

Cornell University

Lee, S., Mallik, A. B., and D. C. Fredrickson. Dipolar-dipolar interactions and the crystal packing of nitriles, ketones, aldehydes, and C(sp2)-F groups. *Cryst. Growth Des.* 4(2):279–90, 2004.

Lee, S., and **R. Hoffmann.** Bcc and Fcc transition metals and alloys: A central role for the Jahn-Teller effect in explaining their ideal and distorted structures. *J. Am. Chem. Soc.* 124:4811–23, 2002.

Lin, H., H. Tao, and V. W. Cornish. Directed evolution of a glycosynthase via chemical complementation. *J. Am. Chem. Soc.* 126:15051, 2004.

Lin, H., M. A. Fischbach, D. R. Liu, and C. T. Walsh. In vitro characterization of salmochelin and enterobactin trilactone hydrolases IroD, IroE, and Fes. *J. Am. Chem. Soc.* 127:11075, 2005.

Merchant, K. A., et al. **(R. F. Loring).** Myoglobin-CO substate structures and dynamics: Spectrally resolved stimulated vibrational echoes and molecular dynamics simulations. *J. Am. Chem. Soc.* 125:13804, 2003.

Kempf, J. G., and **J. A. Marohn.** Nanoscale Fourier-transform imaging with magnetic resonance force microscopy. *Phys. Rev. Lett.* 90(8):087601-1, 2003.

Silveira, W. R., and **J. A. Marohn.** A vertical inertial coarse approach for variable temperature scanned probe microscopy. *Rev. Sci. Instrum.* 74(1):267, 2003.

Dorrestein, P. C., et al. **(F. W. McLafferty** and **T. P. Begley).** The biosynthesis of the thiazole phosphate moiety of thiamin (vitamin B1): The early steps catalyzed by thiazole synthase. *J. Am. Chem. Soc.* 126(10):3091–6, 2004.

Xu, G., et al. **(F. W. McLafferty** and **H. A. Scheraga).** Simultaneous characterization of the reductive unfolding pathways of RNase B isoforms by top-down mass spectrometry. *Chem. Biol.* 11:517–24, 2004.

Breuker, K., and **F. W. McLafferty.** Native electron capture dissociation for the structural characterization of noncovalent interactions in native cytochrome c. *Angew. Chem. Int. Ed.* 42:4900–4, 2003.

Jung, H. M., K. E. Price, and **D. T. McQuade.** Nanoparticle catalysts via polymerization of reverse micelles. *J. Am. Chem. Soc.* 125:5351–5, 2003.

Broadwater, S. J., M. K. Hickey, and **D. T. McQuade.** A new sensor platform based on exciton migration in small-molecule thin films. *J. Am. Chem. Soc.* 125:11154–5, 2003.

Smedley, S. R., et al. **(J. Meinwald).** Mayolenes: Labile defensive lipids from the glandular hairs of a caterpillar (pieris rapae). *Proc. Natl. Acad. Sci. U.S.A.* 99:6822–7, 2002.

Meinwald, J. Understanding the chemistry of chemical communication: Are we there yet? *Proc. Natl. Acad. Sci. U.S.A.* 100(2):14514–6, 2003.

Gaul, C., **J. T. Njarðarson,** and S. J. Danishefsky. The total synthesis of (+)-migrastatin. *J. Am. Chem. Soc.* 125:6042, 2003.

Biswas, K., et al. **(J. T. Njarðarson).** Highly concise routes to epothilones: The total synthesis and evaluation of epothilone 490. *J. Am. Chem. Soc.* 124:9825, 2002.

Du, P., et al. **(C. K. Ober).** Phase selective chemistry in block copolymer thin films. *Adv. Mater.* 16(12):953–7, 2004.

Yu, T., and **C. K. Ober** et al. Three-dimensional microfabrication in a chemically amplified positive system using 2-photon lithography. *Adv. Mater.* 15(6):517–21, 2003.

Park, J., et al. **(H. D. Abruña).** Coulomb blockade and the Kondo effect in single-atom transistors. *Nature* 417:722–5, 2002.

Ahn, Y., J. Dunning, and **J. Park.** Scanning photocurrent imaging and electronic band studies in silicon nanowire field effect transistors. *Nano Lett.* 5:1367, 2005.

Vila, J. A., D. R. Ripoll, and **H. A. Scheraga.** Atomically detailed folding simulation of the B domain of staphylococcal protein A from random structures. *Proc. Natl. Acad. Sci. U.S.A.* 100:14812–6, 2003.

Rathore, O., and **D. Y. Sogah.** Self-assembly of beta-sheets into nanostructures by poly(alanine) segments incorporated in multiblock copolymers inspired by spider silk. *J. Am. Chem. Soc.* 123(22):5231–9, 2001.

Lee, J. Y., A. R. C. Baljon, **D. Y. Sogah,** and **R. L. Loring.** Molecular dynamics study of the intercalation of diblock copolymers into layered silicates. *J. Chem. Phys.* 112:9112–9, 2000.

Potuzak, J. S., and **D. S. Tan.** Synthesis of *C*1-alkyl and *C*1-acylglycals from glycals using a *B*-alkyl Suzuki-Miyaura cross coupling approach. *Tetrahedron Lett.* 45:1797–801, 2004.

Tan, D. S., G. B. Dudley, and S. J. Danishefsky. Synthesis of the functionalized tricyclic skeleton of guanacastepene A: A tandem epoxide-opening b-elimination/Knoevenagel cyclization. *Angew. Chem. Int. Ed.* 41:2185–8, 2002.

Usher, D. A. Before antisense. *Antisense Nucl. Acid Drug Dev.* 7:445, 1997.

Harris, M., and **D. A. Usher.** A new amide-linked polynucleotide analog. *Origins Life Evol. Biosphere* 26:398, 1996.

Djikaev, Y., and **B. Widom.** Geometric view of the thermodynamics of adsorption at a line of three-phase contact. *J. Chem. Phys.* 121(12):5602–10, 2004.

Widom, B., P. Bhimalapuram, and K. Koga. The hydrophobic effect. *Phys. Chem. Chem. Phys.* 5:3085, 2003.

Cho, B.-K., A. Jain, S. M. Bruner, and **U. Wiesner.** Mesophase structure-mechanical and ionic transport correlations in extended amphiphilic dendrons. *Science* 305, 2004.

Templin, M., et al. **(U. Wiesner).** Organically modified aluminosilicate mesostructures from block copolymer phases. *Science* 278:1795–8, 1997.

Wilcox, C. F., and E. N. Farley. Dicyclooctyl[1,2,3,4-def:1'2'3'4'-jkl]-biphenylene: Benzenoid atropism in a highly antiaromatic polycycle. *J. Am. Chem. Soc.* 105:7191, 1983.

Wilcox, C. F. A topological definition of resonance energy. *Croat. Chim. Acta* 47:87, 1975.

Sydora, O. L., **P. T. Wolczanski,** and E. B. Lobkovsky. Ferrous wheels, ellipse $[(^{t}Bu_3SiS)FeX]_n$ and cube $[(^{t}Bu_3SiS)Fe(CCSi^{t}Bu_3)]_4$. *Angew. Chem. Int. Ed.* 42:2685–7, 2003.

Veige, A. S., et al. **(P. T. Wolczanski).** Symmetry and geometry considerations of atom transfer: Deoxygenation of $(silox)_3WNO$ and R_3PO (R = Me, Ph, ^{t}Bu) by $(silox)_3M$ (M = V, NbL (L = PMe_3, 4-picolene), Ta; silox = $^{t}Bu_3SiO$). *Inorg. Chem.* 42:6204–24, 2003.

Zax, D. B., et al. Variation of mechanical properties with amino acid content in the silk of *Nephila clavipes*. *Biomacromolecules* 5:732–8, 2004.

Bailey, M. S., et al. **(D. B. Zax** and **F. J. DiSalvo).** $Ca_6[Cr_2N_6]H$: The first quaternary nitride-hydride. *Inorg. Chem.* 42:5572–8, 2003.

DUQUESNE UNIVERSITY

School of Natural and Environmental Sciences
Department of Chemistry and Biochemistry

Program of Study

The Department of Chemistry and Biochemistry offers a program of graduate study in chemistry and biochemistry leading to the Ph.D. and M.S. degrees.

Graduate students begin laboratory research during their first semester in residence and participate in two semester-long research rotations in the laboratories of two different investigators during their first year. First-year students enroll in several short, intensive courses that emphasize applied research skills. Students typically do not enroll in any other courses during the first year, allowing complete focus upon research during this period. At the end of the first year, students are advanced to Ph.D. candidacy based upon successful completion and defense of their research rotation projects. Academic requirements during the second and subsequent years are determined by the student's dissertation committee and are designed on an individual basis. Candidates for the Ph.D. degree are required to submit and defend an original research proposal in an area unrelated to their dissertation research. The department sponsors a weekly research colloquium series that features speakers from academia, industry, and government.

For the M.S. degree, a minimum of 30 semester hours of combined course and research credits are required.

Research Facilities

The Department of Chemistry and Biochemistry is housed in the Richard King Mellon Hall of Science, an award-winning laboratory designed by Mies van der Rohe. Spectroscopic capability within the department includes multinuclear 500- and 300-MHz NMRs, GC/MS, LC/MS, laser Raman, FT-IR, UV/visible, fluorescence, and atomic absorption spectroscopies. Separations instrumentation includes ultra-high-speed and high-speed centrifuges, gas chromatographs, an ion chromatographic system, capillary electrophoresis, and HPLCs. An electrochemical instrumentation laboratory, a robotics and automation facility, a computer-controlled chemical microwave system, a clean laboratory facility, a single-crystal X-ray diffraction facility, and ICP-MS capabilities are available for research. Modeling and computing have SGI supercomputers available within the Center for Computational Sciences.

Financial Aid

A number of teaching and research assistantships are available for Ph.D. students. For 2006–07, annual stipends are $19,415, plus tuition remission. Several prestigious fellowships for graduate studies are offered by the Bayer School.

Cost of Study

Tuition and fees in 2006–07 are $822 per credit. Scholarships provide tuition remission for teaching and research assistants, as described above.

Living and Housing Costs

Off-campus housing is available within easy walking or commuting distance of the University. Living costs for off-campus housing are very reasonable compared with those in other urban areas of the United States.

Student Group

Duquesne University has a total enrollment of more than 10,000 students in its ten schools. With 150 graduate students and 43 full-time faculty members in the graduate programs in the Bayer School of Natural and Environmental Sciences, the University offers students a highly personalized learning and advisement environment.

Location

Duquesne University is located on a bluff overlooking the city of Pittsburgh. This location offers ready access to the many cultural, social, and entertainment attractions of the city. Within walking distance of the campus are Heinz Hall for the Performing Arts (home of the symphony, opera, ballet, theater, and other musical and cultural institutions), the Mellon Arena (center for indoor sporting events and various exhibitions, concerts, and conventions), Heinz Field and PNC Park (for outdoor sporting events), and Market Square (entertainment and nightlife center). The libraries, museums, art galleries, and music hall of the Carnegie Institute in the Oakland area are easily accessible by public transportation (routes pass immediately adjacent to the campus) or by private automobile. As the third-largest center for corporate headquarters and one of the twenty largest metropolitan areas in the United States, Pittsburgh also offers many professional career opportunities for its residents.

The University

Founded in 1878 by the Fathers and Brothers of the Congregation of the Holy Ghost, Duquesne University has provided the opportunity for an education to students from many backgrounds, without regard to sex, race, creed, color, or national or ethnic origins. In the past twenty-five years, the University has undergone a dramatic physical transformation, from a makeshift physical plant occupying approximately 12 acres to a modern, highly functional educational facility that is located on its own 40-acre hilltop overlooking downtown Pittsburgh.

Applying

Applications for admission to graduate study with financial aid should be submitted no later than February 1 for the academic year beginning in the following September. Applications for admission without financial aid may be made up to one month prior to the beginning of the term in which the student desires to begin graduate work. All applications require official transcripts of previous undergraduate and graduate work and three letters of recommendation from faculty members who are familiar with the applicant's past academic progress. Application forms are available by writing to or calling the Office of the Dean, Bayer School of Natural and Environmental Sciences, Department of Chemistry and Biochemistry.

Correspondence and Information

Graduate Programs
Bayer School of Natural and Environmental Sciences
100 Mellon Hall
Duquesne University
Pittsburgh, Pennsylvania 15282

Phone: 412-396-4900
Fax: 412-396-4881
E-mail: gradinfo@duq.edu
Web site: http://www.science.duq.edu

Duquesne University

THE FACULTY

Jennifer Aitken, Assistant Professor; Ph.D., Michigan State. Solid-state inorganic materials chemistry: elucidation of new crystal structures, synthesis and study of novel solid-state materials with potential use in optical and electronic technologies, crystal growth.

Partha Basu, Associate Professor; Ph.D., Jadavpur (India), 1991. Inorganic chemistry: synthesis, structure, reactivity, and magnetic interactions of biological and model molecules.

Bruce D. Beaver, Associate Professor; Ph.D., Massachusetts, 1984. Organic chemistry: oxygenation of organic molecules, development of new antioxidants, oxidative degradation of petroleum products, chemistry of wine making.

Charles Dameron, Associate Professor; Ph.D., Texas A&M, 1987. Biochemistry: metals in biology, understanding how metals are chaperoned by proteins and exchanged between proteins.

Jeffrey D. Evanseck, Professor; Ph.D., UCLA, 1990. Theoretical and computational chemistry: quantum and classical simulations coupled with experiment, energy landscapes of biomolecules, novel ionic liquids for catalysis, influence of solvent on organic reaction mechanism, supramolecular complexation for nanotechnology.

Fraser F. Fleming, Associate Professor; Ph.D., British Columbia, 1990. Organic chemistry: application of the chemistry of α,β-unsaturated nitriles to the synthesis of anti-AIDS and anticancer drugs; synthesis of natural products.

Ellen Gawalt, Assistant Professor; Ph.D., Princeton. Bioorganic and materials chemistry: chemical modification of metal oxide surfaces used in biomaterials and reaction mechanisms of interfacial reactions.

Mitchell E. Johnson, Associate Professor; Ph.D., Massachusetts Amherst, 1993. Analytical chemistry: trace analysis of molecular species, fluorescence spectroscopy, high-speed separations, biochemical analysis.

Shahed Khan, Associate Professor; Ph.D., Flinders (Australia), 1977. Physical chemistry: electrochemistry, photoelectrochemistry, solar energy conversion by thin-film organic and inorganic semiconductors, electrocatalytic biosensors, electrosynthesis of conducting polymers, electrochemical surface modification, theory of electron transfer reactions in condensed medium, effect of solvent dynamics on electrochemical electron transfer reactions.

H. M. Kingston, Professor; Ph.D., American, 1978. Analytical and environmental chemistry: microwave chemistry application, environmental methods and instrument development, speciated analysis, ICP-MS clean-room chemistry, chromatography, laboratory automation.

Jeffry D. Madura, Professor and Chair; Ph.D., Purdue, 1985. Theoretical physical chemistry: computational chemistry and biophysics, classical simulations of biomolecules, Poisson-Boltzmann electrostatics coupled to molecular dynamics, simulation of proteins at ice/water interface, simulation of biomolecular diffusion-controlled rate constants, quantum mechanical calculation of small molecules.

Mihaela Rita Mihailescu, Assistant Professor; Ph.D., Wesleyan, 2001. Molecular biology and biochemistry; protein–nucleic acid interactions studied by NMR, fluorescence spectroscopy, and other biophysical techniques.

Tomislav Pintauer, Assistant Professor; Ph.D., Carnegie Mellon, 2002. Chemistry; inorganic, organometallic, and polymer chemistry, with emphasis on homogeneous and heterogeneous catalysis.

David W. Seybert, Professor and Dean; Ph.D., Cornell, 1976. Biochemistry: lipid peroxidation in biomembranes and lipoproteins, antioxidants and inhibition of LDL oxidation, mechanism and regulation of cytochrome P450 catalyzed steroid hydroxylations.

Omar W. Steward, Professor Emeritus; Ph.D., Penn State, 1957. Inorganic chemistry: synthesis and structural studies of carboxylato metal complexes by X-ray diffraction; magneto-structural studies of transition metal complexes with organosilicon ligands; organosilicon and organogermanium compounds; structure-reactivity studies.

Julian Talbot, Associate Professor; Ph.D., Southampton (England), 1985. Theoretical physical chemistry: statistical mechanics, Monte Carlo and molecular dynamic simulation of classical systems, theory of adsorption kinetics and equilibria, biomolecules at interfaces, gases in porous solids.

Theodore J. Weismann, Adjunct Professor; Ph.D., Duquesne, 1956. Physical chemistry: mass spectrometry, ion optics, free radical reactions, organoboron chemistry, geochemistry, stable isotope MS, petroleum source and characterization.

Stephanie Wetzel, Assistant Professor; Ph.D., American, 2001. Analytical chemistry, instrumental methods, forensic chemistry, synthetic polymer characterization by mass spectrometry.

Richard King Mellon Hall of Science, which houses the Department of Chemistry and Biochemistry.

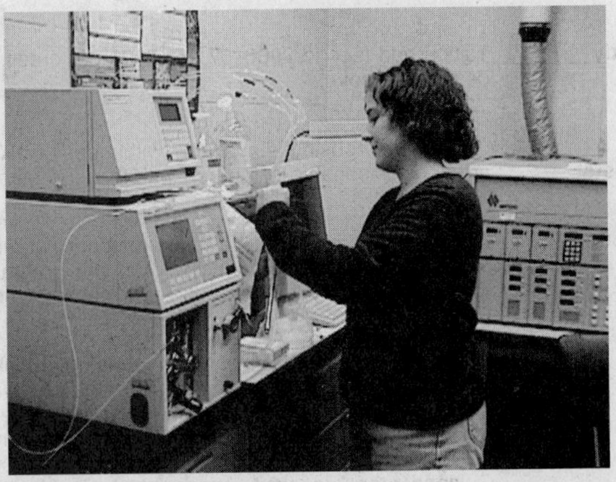

Research laboratory in Mellon Hall.

A student at work at Duquesne University.

SELECTED PUBLICATIONS

Takas, N. J., and **J. A. Aitken.** Phase transitions and second-harmonic generation in sodium monothiophosphate. *Inorg. Chem.* 45:2779–81, 2006.

Jayasekera, B., **J. A. Aitken,** M. J. Heeg, and S. L. Brock. Towards an arsenic analog of Hittorf's phosphorus: Mixed pnictogen chains in $Cu_2P_{1.8}As_{1.2}I_2$. *Inorg. Chem.* 42:658–60, 2002.

Aitken, J. A., M. Evain, L. Iordanidis, and M. G. Kanatzidis. $NaCeP_2Se_6$, $Cu_{0.4}Ce_{1.2}P_2Se_6$, $Ce_{1.33}P_2Se_6$ and the incommensurately modulated, $AgCeP_2Se_6$: New selenophosphates featuring the ethane-like $[P_2Se_6]^{4-}$ anion. *Inorg. Chem.* 41:180–91, 2002.

Aitken, J. A., P. Larson, S. D. Mahanti, and M. G. Kanatzidis. Li_2PbGeS_4 and Li_2EuGeS_4: Polar chalcopyrites with a severe tetragonal compression. *Chem. Mater.* 13:4714–21, 2001.

Aitken, J. A., and M. G. Kanatzidis. New information on the Na-Ti-Se ternary system. *Z. Naturforsch. B* 56:49–56, 2001.

Aitken, J. A., and M. G. Kanatzidis. α-$Na_6Pb_3(PS_4)_4$, a noncentrosymmetric thiophosphate with the novel saucer-shaped $[Pb_3(PS_4)_4]^{6-}$ cluster, and its metastable, 3-dimensionally polymerized allotrope β-$Na_6Pb_3(PS_4)_4$. *Inorg. Chem.* 40:2938–9, 2001.

Aitken, J. A., C. Canlas, D. P. Weliky, and M. G. Kanatzidis. $[P_2S_{10}]^{4-}$: A novel polythiophosphate anion containing a tetrasulfide fragment. *Inorg. Chem.* 40:6496–8, 2001.

Basu, P., et al. Donor atom dependent geometric isomers in mononuclear oxo-molybdenum^{V} complexes: Implications for coordinated endogenous ligation in molybdoenzymes. *Inorg. Chem.* 42:5999–6007, 2003.

Afkar, E., and **P. Basu** et al. The respiratory arsenate reductase from a haloalkaliphilic bacterium *Bacillus selenitireducens* strain MLS10. *FEMS Microbiol. Lett.* 226:107–12, 2003.

Basu, P., V. N. Nemykin, and R. S. Sengar. Electronic properties of para-substituted thiophenols and disulfides from ^{13}C NMR spectroscopy and ab initio calculations: Relations to the Hammett parameters and atomic charges. *New J. Chem.* 27:1115–23, 2003.

Basu, P., and V. N. Nemykin. Comparative theoretical investigation of the vertical excitation energies and the electronic structure of $[Mo^{V}OCl_4]^-$: Influence of basis set and geometry. *Inorg. Chem.* 42:4046–56, 2003.

Basu, P., J. F. Stolz, and M. T. Smith. A coordination chemist's view of the active sites of mononuclear molybdenum enzymes. *Curr. Sci.* 84:1412–8, 2003.

Beaver, B., et. al. High heat sink jet fuels, part 1: Development of potential oxidative and pyrolytic additives for JP-8. *Energy Fuels* 20(4):1639–46, 2006.

Beaver, B., L. Gao, C. Burgess-Clifford, and M. Sobkowiak. On the mechanisms of formation of thermal oxidative deposits in jet fuels. Is a unified mechanism possible for both storage and thermal oxidative deposit formation for middle distillate fuels? *Energy Fuels* 19:1574–9, 2005.

Beaver, B., et al. Development of oxygen scavenger additives for future jet fuels. A role for electron-transfer-initiated oxygenation (ETIO) of 1,2,5-trimethylpyrrole? *Energy Fuels* 14:441–7, 2000.

Cobine, P., C. E. Jones, and **C. T. Dameron.** Role of zinc(II) in the copper regulated protein CopY. *J. Inorg. Biochem.* 88:192–6, 2002.

Harrison, M. D., C. E. Jones, M. Solioz, and **C. T. Dameron.** Intracellular copper routing: The role of copper chaperones. *Trends Biol. Sci.* 25:29–32, 2000.

Cobine, P., et al. **(C. T. Dameron.)** Stoichiometry of the complex formation between copper(I) and the N-terminal domain of the Menkes protein. *Biochemistry* 39:6857–63, 2000.

Joseph, M. A., et al. **(J. D. Evanseck)**. Secondary anchor substitutions in an HLA-A*0201-restricted T-cell epitope derived from Her-2/neu. *Molec. Immun.,* in press.

Mueller Stein, S. A., et al. **(J. D. Evanseck)** Principal components analysis: A review of its application on molecular dynamics data. *Ann. Rep. Comp. Chem.,* in press.

Seybert, D. S., **J. D. Evanseck,** and J. S. Doctor. Integrated biological and chemical laboratory experiences for enhanced education, research opportunities, and career development. *CUR Q.* 26:104–8, 2006.

Henriksen, B. S., et al. **(J. D. Evanseck)**. Computational and conformational evaluation of FTase alternative substrates: Insight into a novel enzyme binding pocket. *J. Chem. Inf. Model.* 45(4):1047–52, 2005.

Ruben, E. A., M. S. Chapman, and **J. D. Evanseck.** Generalized anomeric interpretation of the high energy N-P bond in N-methyl-N'-phosphorylguanidine: Breakdown of opposing resonance theory. *J. Am. Chem. Soc.* 127:17789–98, 2005.

Evanseck, J. D., and S. M. Firestine. Status of research-based experiences for first- and second-year undergraduate students. *Ann. Rep. Comp. Chem.* 1:205–14, 2005.

DeChancie, J., A. Acevedo, and **J. D. Evanseck.** Density functional theory determination of an axial gateway to explain the rate and *endo* selectivity enhancement of Diels-Alder reactions by Bis(oxazoline)-Cu(II). *J. Am. Chem. Soc.* 126:6043, 2004.

Loccisano, A. E., et al. **(J. D. Evanseck.)** Enhanced sampling by multiple molecular dynamics trajectories: Carbonmonoxy myoglobin 10 micros A0$\rightarrow$A$_{1-3}$ transition from ten 400 picosecond simulations. *J. Mol. Graphics Modeling* 22(5):369, 2004.

Acevedo, O., and **J. D. Evanseck.** Transition structure models of organic reactions in ionic liquids. In *Ionic Liquids as Green Solvents,* pp. 174–90, eds. R. D. Rogers and K. R. Seddon. ACS Symposium Series (Ionic Liquids), 2003.

Acevedo, O., and **J. D. Evanseck.** The effect of solvent on a Lewis acid catalyzed Diels-Alder reaction using computed an experimental kinetic isotope effects. *Org. Lett.* 649, 2003.

Macias, A. T., J. E. Norton, and **J. D. Evanseck.** Impact of multiple cation-π interactions upon calix[4]arene substrate binding and specificity. *J. Am. Chem. Soc.* 123:2351, 2003.

Fleming, F. F., and B. C. Shook. 1-Oxo-2-cyclohexenyl-2-carbonitrile. *Org. Syn.* 77:254–60, 2001.

Fleming, F. F., Q. Wang, and **O. W. Steward.** Hydroxylated α,β-unsaturated nitriles: Stereoselective synthesis. *J. Org. Chem.* 66:2171–4, 2001.

Fleming, F. F., and P. Iyer. Flood prevention by recirculating condenser cooling water. *J. Chem. Educ.* 78:946, 2001.

Fleming, F. F., and B. C. Shook. α,β-unsaturated nitriles: Preparative MgO elimination. *Tetrahedron Lett.* 41:8847–51, 2000.

Fleming, F. F., Q. Wang, and **O. W. Steward.** γ-hydroxy unsaturated nitriles: Chelation-controlled conjugate additions. *Org. Lett.* 2:1477–9, 2000.

Raman, A., M. Dubey, I. Gouzman, and **E. S. Gawalt.** Formation of self-assembled monolayers of alkylphosphonic acid on native oxide surface of SS316L. *Langmuir* 22(15):6469–72, 2006.

Gawalt, E. S., et al. Bonding organics to Ti alloys: Facilitating human osteoblast attachment and spreading on surgical implant materials. *Langmuir* 19:200–4, 2003.

Houseman, B. T., **E. S. Gawalt,** and M. Mrksich. Maleimide-functionalized self-assembled monolayers for the preparation of peptide and carbohydrate biochips. *Langmuir* 19:1522–31, 2003.

Schwartz, J., et al. **(E. S. Gawalt.)** Cell attachment and spreading on metal implant materials. *Mater. Sci. Eng., C* 23:395–400, 2003.

Gawalt, E. S., et al. Enhanced bonding of organometallics to titanium via a titanium (III) phosphate interface. *Langmuir* 17:6743–5, 2001.

Gawalt, E. S., M. J. Avaltroni, N. Koch, and J. Schwartz. Self-assembly and bonding of alkanephosphonic acids on the native oxide surface of titanium. *Langmuir* 17:5736–8, 2001.

Sultana, T., and **M. E. Johnson.** Sample preparation and gas chromatography of primary fatty acid. *Amides' J. Chromatogr. B* 1101:278–85, 2006.

Johnson, M. E., and T. S. Carpenter. The use of solid phase supports for derivatization in chromatography and spectroscopy. *Appl. Spectrosc.* 40:1–22, 2005.

Johnson, M. E. Research instrumentation used in education. Liquid chromatography-mass spectrometry experiences in upperclass laboratory settings. In *Invention and Impact: Building Excellence in Undergraduate Science, Technology, Engineering, and Mathematics (STEM) Education,* pp. 99–104, eds. S. Cunningham and Y. S. George. Washington, D.C.: American Association for the Advancement of Science, 2004.

Johnson, M. E., and J. P. Landers. Fundamentals and practice for ultrasensitive laser-induced fluorescence detection in microanalytical systems. *Electrophoresis* 25:3513–27, 2004.

Merkler, D. J., et al. **(M. E. Johnson.)** Oleic acid derived metabolites in mouse neuroblastoma N18TG2 cells. *Biochem.* 43:12667–74, 2004.

Carpenter, T., et al. **(M. E. Johnson.)** Liquid chromatographic separation of fatty acid amides and N-acyl glycines for biological assays. *J. Chromatogr. B Biomed. Appl.* 809:15–21, 2004.

Stokes, J. C., and **M. E. Johnson.** Resolution in sub-micrometer particle separations by capillary electrophoresis. *Microchem. J.* 76:121–9, 2004.

Gallaher, D. L., and **M. E. Johnson.** Nonaqueous capillary electrophoresis of fatty acids derivatized with a near-infrared fluorophore. *Anal. Chem.* 72:2080–6, 2000.

Gee, A. J., L. Groen, and **M. E. Johnson.** Ion trap mass spectrometry of trimethylsilylamides following gas chromatography. *J. Mass Spectrom.* 35:305–10, 2000.

Shah, J. M., and **S. U. M. Khan.** Detection of glucose by electroreduction at a semiconductor electrode: An implantable non-enzymatic glucose sensor. In *Chemical and Biological Sensors and Analytical Methods II,* vol. 2001–18, pp. 259–71, eds. M. Butler, P. Vanysek, and Y. Yamazoe. Pennington: Electrochemical Society, 2001.

Sultana, T., and **S. U. M. Khan.** Photoelectrichemical splitting of water on nanocrystalline n-TiO_2 thin film and quantum wire electrodes. In *Quantum Confinement VI: Nanostructured Materials and Devices,* vol. 2001–19, pp. 9–19, eds. M. Cahay et al. Pennington: Electrochemical Society, 2001.

Boylan, H. M., R. D. Cain, and **H. M. Kingston.** A new method to assess mercury emissions: A study of three coal-fired electric generating power stations. *Air Waste Manage. Assoc.* 53:1318–25, 2003.

Bhandari, S., **H. M. Kingston,** and G. M. M. Rahman. Synthesis and characterization of isotopically enriched methylmercury ($CH_3{}^{201}Hg^+$). *Appl. Organometal. Chem.* 17:913–20, 2003.

Gazmuri, R. J., et al. **(H. M. Kingston.)** Myocardial protection during ventricular fibrillation by interventions that limit proton-driven sarcolemmal sodium influx, American Heart Association. *J. Lab. Clin. Med.* 137(1):43–55, 2001.

Duquesne University

Han, Y., **H. M. Kingston,** R. C. Richter, and C. Pirola. Dual-vessel integrated microwave sample decomposition and digest evaporation for trace element analysis of silicon material by ICP-MS: Design and application. *Anal. Chem.* 73(6):1106–11, 2001.

Richter, R. C., L. Dirk, and **H. M. Kingston.** Microwave enhanced chemistry: Standardizing sample preparation. *Anal. Chem.* 73(1):30A–7A, 2001.

Huo, D., and **H. M. Kingston.** Correction of species transformations in the analysis of Cr(VI) in solid environmental samples using speciated isotope dilution mass spectrometry. *Anal. Chem.* 72(20):5047–54, 2000.

Boylan, H. M., T. A. Ronning, R. L. DeGroot, and **H. M. Kingston.** Field analysis using Method 7473: Minimizing the cost of mercury cleanup. *Environ. Test. Anal.,* 2000.

Link, D. D., and **H. M. Kingston.** Use of microwave-assisted evaporation for the complete recovery of volatile species of inorganic trace analytes. *Anal. Chem.* 72(13):2908–13, 2000.

Richter, R. C., D. D. Link, and **H. M. Kingston.** On demand production of high-purity acids in the analytical laboratory. *Spectroscopy,* 2000.

Zhou, Z., M. Bates, and **J. D. Madura.** Structure modeling, ligand binding, and binding affinity calculation (LR-MM-PBSA) of human heparanase for inhibition and drug design. *Proteins: Struct., Funct., Bioinform.,* in press.

Krouskop, P. E., P. C. Garrison, P. C. Gedeon, and **J. D. Madura.** A novel hybrid simulation for study of multiscale phenomena. *Mol. Simulation,* in press.

Krouskop, P., **J. D. Madura,** D. Paschek, and A. Krukau. Solubility of simple, nonpolar compounds in TIP4P-Ew. *J. Chem. Phys.* 124(1):16102, 2006.

Munshi, R., et al. **(J. D. Madura).** An introduction to simulation and visualization of biological systems at multiple scales: A summer training program for interdisciplinary research *Biotechnol. Prog.* 22(1):179–85, 2006.

Zhou Z., M. Madrid, J. D. Evanseck, and **J. D. Madura.** Effect of a bound non-nucleoside RT inhibitor on the dynamics of wild-type and mutant HIV-1 reverse transcriptase. *J. Amer. Chem. Soc.* 127:17253–60, 2005.

Dalal, P., et al. **(J. D. Madura.)** Molecular dynamics simulation studies of the effect of phosphocitrate on crystal-induced membranolysis. *Biophys. J.* 89(4):2251–7, 2005.

Berberich, J. A., et al. **(J. D. Madura.)** A stable three-enzyme creatinine biosensor. 1. Impact of structure, function, and environment on PEGylated and immobilized sarcosine oxidase. *Acta Biomaterialia* 1:173–81, 2005.

Owens, J. W., M. B. Perry, and **D. W. Seybert.** Reactions of nitric oxide with cobaltous tetraphenylporphyrin and phthalocyanines. *Inorg. Chim. Acta* 277:1–7, 1998.

Hanlon, M. C., and **D. W. Seybert.** The pH dependence of lipid peroxidation using water-soluble azo initiators. *Free Radical Biol. Med.* 23:712–9, 1997.

Warburton, R. J., and **D. W. Seybert.** Structural and functional characterization of bovine adrenodoxin reductase by limited proteolysis. *Biochim. Biophys. Acta* 1246: 39–46, 1995.

Yaukey, T. S., **O. W. Steward,** and S.-C. Chang. Manganese(II) triphenylacetate hydratrate: A manganese(II) complex with a chain structure. *Acta Crystallogr.* C54: 1081–3, 1998.

Muto, Y., et al. **(O. W. Steward.)** Characterization of dimeric copper(II) trichoroacetate complexes by electron spin resonance, infrared, and electronic reflectance spectra: Correlation of spectral parameters with molecular geometry. *Bull. Chem. Soc. Jpn.* 70:1573, 1997.

Steward, O. W., et al. Structural and magnetic studies of dimeric copper(II) 2,2-diphenylpropanoato and triphenylacetato complexes with oxygen donor ligands: The cage geometry of dimeric α-phenyl substituted copper(II) carboxylates. *Bull. Chem. Soc. Jpn.* 69:34123–7, 1996.

Talbot, J. Analysis of adsorption selectively in a one-dimensional model system. *AIChE J.* 43:2471–8, 1997.

Van Tassel, P. R., **J. Talbot,** G. Tarjus, and P. Viot. A distribution function analysis of the structure of depleted particle configurations. *Phys. Rev. E* 56:1299R, 1997.

Choi, H. S., and **J. Talbot.** Effect of diffusion and convection on the flux of depositing particles near preadsorbed particle. *Korean J. Chem. Eng.* 14:117–24, 1997.

Talbot, J. Molecular thermodynamics of binary mixture adsorption. *J. Chem. Phys.* 196:4696–706, 1997.

Van Tassel, P., **J. Talbot,** G. Tarjus, and P. Viot. The kinetics of irreversible adsorption with particle conformational change. *Phys. Rev. E* 53:785, 1996.

Talbot, J. Time-dependent desorption: A memory function approach. *Adsorption* 2:89, 1996.

Pintauer, T. Bis(2,2'-Bipyridine-k2N,N')Copper(I)trifluoromethanesulfonate. *Acta Crystallogr. Section E* E62:m620, 2006.

Pintauer, T. Synthesis, characterization, and the role of counterion in stabilizing trigonal pyramidal copper(I)/2,2 bipyridine complexes containing electron-poor methyl acrylate. *J. Organomet. Chem.,* in press.

Wetzel, S.J., C. M. Guttman, K. M. Flynn, and J. J. Filliban. A statistical analysis of the optimization of the MALDI-TOF-MS. *J. Am. Soc. Mass Spectrom.* 17:246–52, 2006.

Guttman, C.M., and **S. J. Wetzel** et al. Matrix-assisted laser desorption/ionization time-of-flight mass spectrometry interlaboratory comparison of mixtures of polystyrene with different end groups: Statistical analysis of mass fractions and mass moments. *Anal. Chem.* 77:4539–48, 2005.

Wetzel, S. J., C. M. Guttman, and K. M. Flynn. The influence of electrospray deposition in MALDI-MS sample preparation for synthetic polymers. *Rapid Commun. Mass Spectrom.* 18:1139–46, 2004.

Lin-Gibson S., et al. **(S. J. Wetzel.)** Synthesis and characterization of PEG dimethacrylates and their hydrogels, *Biomacromolecules* 5(4):1280–87, 2004.

Wetzel, S.J., C. M. Guttman, and J. E. Girard. Influence of laser power and matrix on the molecular mass distribution of synthetic polymers obtained by MALDI-TOF-MS. *Int. J. Mass Spectrom.* 238(3):215–25, 2004.

Wallace, W. E., C. M. Guttman, **S. J. Wetzel,** and S. D. Hanton. Mass spectrometry of synthetic-polymer mixtures workshop. *Rapid Commun. Mass Spectrom.* 18:518–21, 2004.

FLORIDA STATE UNIVERSITY

Department of Chemistry and Biochemistry

Programs of Study

Graduate education in the Department of Chemistry and Biochemistry, leading to the M.S. and Ph.D. degrees, can be pursued in both traditional and contemporary fields. Five areas—biochemistry and analytical, inorganic, organic, and physical chemistry—provide fundamental curricula in their disciplines, and a variety of research-oriented programs are available in each of these traditional areas. In addition, opportunities for chemical research at the interfaces with physics, biology, and materials science are available in research groups participating in the Materials Research and Technology Center, the Institute of Molecular Biophysics, the School of Computational Science and Information Technology (CSIT), and the National High Magnetic Field Laboratory. Interdisciplinary programs leading to advanced degrees in chemical physics and in molecular biophysics are offered in cooperation with the Department of Physics and the Department of Biological Science. A list of chemistry faculty members and their research interests is given in this Close-Up.

Ph.D. candidates who perform satisfactorily on entrance examinations may immediately begin advanced course work and research. Although Ph.D. programs of study are structured to meet individual needs and vary among the divisions, most programs incorporate seven to twelve 1-semester courses at the graduate level. Mastery of the material in the area of interest is demonstrated by passing either cumulative examinations or comprehensive examinations, depending on the area of specialization. A thesis or dissertation is required in all but the courses-only option of the master's degree program. The presence of about 40 postdoctoral and visiting faculty researchers, in addition to the low faculty-student ratio, permits a high level of student-scientist interaction.

Research Facilities

Departmental research operations are housed mainly in the interconnected Dittmer Laboratory of Chemistry building and Institute of Molecular Biophysics building. Several adjacent structures serve other departmental teaching functions. Major items of research equipment include 150-, 200-, 270-, 300-, 400-, and 500-Mz NMR spectrometers; SQUID magnetometers for characterization of temperature and field-dependent magnetic behavior; several medium- and high-resolution mass spectrometers with electron impact (EI), chemical ionization (CI), electrospray (ESI), and matrix-assisted laser desorption (MALDI) ionization capabilities and GC or LC interfaces; a scanning probe microscope; an X-ray fluorescence unit; a variable temperature powder X-ray diffractometer and automated Bruker SMART CCD single crystal X-ray diffractometer with variable temperature capability and complete in-house solution and refinement capabilities; and a bioanalytical facility containing gas chromatographs, analytical and preparative liquid chromatographs, automated DNA synthesizers, peptide synthesizers, and protein sequencing equipment. The department has excellently staffed glassworking, machine, electronics, photographic/computer graphics, and woodworking shops in support of the teaching and research programs.

The departmental computing needs are met by a variety of workstations and microcomputers that are linked with the Internet. Several members of the chemistry faculty are members of the of the CSIT School and have access to two Silicon Graphics Power Challenge supercomputers on the Florida State University (FSU) campus as well as to several supercomputers around the world. CSIT maintains two high-performance computer environments—a Connection Machine CM-2 and a large cluster of super-workstations providing more than 8.2 gigaflops of computing power and more than 460 gigabytes of disk storage. The department is a sponsoring academic unit for the National High Magnetic Field Laboratory (NHMFL) and has access to the facilities and instrumentation of the NHMFL.

The University's Strozier Library, which includes the Dirac Science Library, houses 1.2 million volumes and maintains 12,650 active journal subscriptions. The Dirac Science Library is located adjacent to the Dittmer Laboratory of Chemistry.

Financial Aid

In addition to providing teaching and research assistantships, the department offers several special fellowships on a competitive basis. Nearly all graduate students are supported by one of these programs. Twelve-month teaching assistantships with stipends of $19,000 can be augmented by special fellowships ranging up to $3000. Competitive fellowships on the University and College levels are also available.

Cost of Study

Tuition for in-state residents was $2753.40 per semester and for out-of-state residents was $10,330.20 per semester in 2005–06. Most of these tuition costs are normally waived for teaching and research assistants, but the number of waivers available each year is determined in part by legislative appropriation. In addition to tuition, nonwaivable fees were $487–$827 per semester and the cost of health insurance was $107 per month.

Living and Housing Costs

Single students sharing a 2-person room in graduate apartment housing paid $397 per month in 2005–06, in addition to utilities and telephone service. The University also operates an apartment complex of 795 units for single and married students. Rents ranged from $370 to $589 per month plus utilities for furnished one- to three-bedroom apartments. Off-campus accommodations began at about $450 per month.

Student Group

The graduate enrollment in the Department of Chemistry and Biochemistry is 144 students. Although students come from all parts of the United States and numerous other countries, many are native to the eastern half of the United States.

Location

Metropolitan Tallahassee has a population of about 330,000 and is recognized for the scenic beauty of its rolling hills, abundant trees and flowers, and seasonal changes. As the capital of Florida, the city is host to many cultural affairs, including symphony, theater, and dance groups. Its location 30 minutes from the Gulf Coast, the area's many lakes, and an average annual temperature of 68°F make the region eminently suitable for a variety of year-round outdoor activities.

The University

Florida State University was founded in 1851 and is one of eleven members of the State University System of Florida. Currently, enrollment is 38,886, including 7,456 graduate students. The University's rapid climb to prominence began with the adoption of an emphasis on graduate studies in 1947. The first doctoral degree was conferred in 1952. In addition to the College of Arts and Sciences, the University is composed of the Colleges of Business, Communication, Criminology and Criminal Justice, Education, Engineering, Human Sciences, Information, Law, Medicine, Music, Social Sciences, Social Work, and Theatre and Visual Arts. Several departments, including chemistry, enjoy international recognition.

Applying

Application for admission may be made at any time; however, initial inquiries concerning assistantships, especially fellowships, should be made nine to twelve months prior to the anticipated enrollment date. Later requests are considered as funds are available for the following fall semester. Requests for forms, detailed requirements, and other information should be directed to the address listed in this Close-Up.

Correspondence and Information

Director of Graduate Admissions
Department of Chemistry and Biochemistry
Florida State University
Tallahassee, Florida 32306-4390

Phone: 850-644-1897
888-525-9286 (toll-free in the United States)
Fax: 850-644-0465
E-mail: gradinfo@chem.fsu.edu
Web site: http://www.chem.fsu.edu (Department of Chemistry)
http://admissions.fsu.edu (the University)

Florida State University

THE FACULTY AND THEIR RESEARCH

Igor V. Alabugin, Assistant Professor; Ph.D., Moscow State, 1995. New reactions for DNA cleavage, pH-activated anticancer agents, radical cyclization cascades in synthesis of conjugated polymers, molecular imprinting and supramolecular control of photochemical reactivity, stereoelectronic effects, improper hydrogen bonding.
Michael Blaber, Associate Professor; Ph.D., California, Irvine, 1990. Protein structure, function, folding, evolution, and engineering; X-ray crystallography and biophysics; serine protease structure and function.
Rafael Bruschweiler, George Matthew Edgar Professor of Chemistry and Biochemistry and Associate Director of Biophysics at the National High Magnetic Field Laboratory; Ph.D., ETH Zurich, 1991. Protein dynamics and structure by NMR spectroscopy and computer simulations, covariance spectroscopy, NMR ensemble computing.
Michael S. Chapman, Professor; Ph.D., UCLA, 1987. Structure-function characterization of enzymes and viruses by X-ray crystallography.
William T. Cooper, Professor and Director, Terrestrial Waters Institute; Ph.D., Indiana, 1981. Environmental chemistry of organic compounds in surface and ground waters, organic geochemistry of recent sediments, FT-ICR mass spectrometry, solid-state ^{13}C NMR spectroscopy; chromatography and capillary electrophoresis.
Timothy A. Cross, Earl Frieden Professor and National High Magnetic Field Lab NMR Program Director; Ph.D., Pennsylvania, 1981. Structural biology and structural genomics of membrane proteins, using solid-state and solution nuclear magnetic resonance and a combination of technologies.
Naresh Dalal, Professor and Chairperson; Ph.D., British Columbia, 1971. Materials science, solid-state chemistry, synthesis and characterization of magnetic and ferroelectric compounds, nanochemistry and quantum computation, high field EPR and NMR, scanning probe microscopy, spectroscopic studies of single crystals.
John G. Dorsey, Katherine Blood Hoffman Professor and Editor, *Journal of Chromatography A;* Ph.D., Cincinnati, 1979. Analytical separations, especially liquid chromatography; retention mechanisms; stationary phase design and synthesis; electrochromatography; quantitative structure-retention relationships; theory and technique development.
Ralph C. Dougherty, Professor; Ph.D., Chicago, 1963. Structure of water and aqueous solutions, superconductivity, tunneling in chemical reactions, mesoscale (10 km) atmospheric chemical modeling, electrolytic hydrogen production.
Gregory B. Dudley, Assistant Professor; Ph.D., MIT, 2000. Natural products synthesis and application of synthesis to medicinal chemistry research, new strategies and tactics for organic chemistry.
Thomas M. Fischer, Associate Professor; Ph.D., Mainz (Germany), 1992. Dynamic phase behavior of soft matter, experimental and theoretical studies of two-dimensional systems, biomimetic self-organization.
Robert L. Fulton, Professor; Ph.D., Harvard, 1964. Development and application of techniques for the description of the behavior of electrons in molecules and in intermolecular complexes, theories of relaxation phenomena, theories of linear and nonlinear dielectric properties and their relation to molecular motion.
Penny J. Gilmer, Professor; Ph.D., Berkeley, 1972; D.Sc.Ed., Curtin (Australia), 2004. Biochemistry, chemistry education, science education, scientific research for teachers, collaboration and technology, education of future science teachers, progression of science graduate students.
Kenneth A. Goldsby, Associate Professor; Ph.D., North Carolina at Chapel Hill, 1983. Redox reactions of transition metal complexes, proton-coupled electron transfer, surface-modified electrodes and self-assembled monolayers, electrochemistry.
Nancy L. Greenbaum, Associate Professor; Ph.D., Pennsylvania, 1984. Structure and function of RNA-RNA, RNA-metal, and RNA-protein interactions studied by multidimensional NMR and other spectroscopic techniques.
Werner Herz, R. O. Lawton Distinguished Professor Emeritus; Ph.D., Colorado, 1947. Organic chemistry, natural products isolation and structure, chemical taxonomy.
Edwin F. Hilinski, Associate Professor; Ph.D., Yale, 1982. Mechanistic studies of photochemical and thermal reactions of organic compounds in solution, time-resolved laser spectroscopy, photoinduced phenomena in solids.
Robert A. Holton, Matthew Suffness Professor; Ph.D., Florida State, 1971. Synthetic organic, organometallic, and bioorganic chemistry; total synthesis of natural products.
Michael Kasha, University Professor; Ph.D., Berkeley, 1945. Chemical physics, molecular spectroscopy; flavonol spectroscopy and biomechanisms; solar-planetary geophysical interactions; new designs of musical string instruments and audio-spectral determination.
Marie E. Krafft, Martin A. Schwartz Professor; Ph.D., Virginia Tech, 1983. Synthetic organic and organometallic chemistry, natural products synthesis.
Harold Kroto, Eppes Professor of Chemistry; Ph.D., Sheffield, 1964. Controlled assembly and physical properties of ID and ZD nanostructures, global educational outreach (GEO) using Internet broadcasting technology.
Susan E. Latturner, Assistant Professor; Ph.D., California, Santa Barbara, 2000. Use of molten salts and metals for the synthesis of new inorganic materials, growth of subvalent clusters in zeolites by ion-exchange and doping, crystallography and electronic properties of inorganic solids.
Hong Li, Assistant Professor; Ph.D., Rochester, 1994. X-ray crystallography, molecular principles of protein and RNA interactions, gene expression and regulation.
Bruno Linder, Professor Emeritus of Chemistry; Ph.D., UCLA, 1955. Theoretical physical chemistry, intermolecular forces, thermodynamics and statistical mechanics, effect of van der Waals forces on spectral intensities, statistical mechanical approach to the thermodynamic functions in the unfolding of biomolecules, effect of dispersion interaction on the work function of metals.
Timothy M. Logan, Associate Professor; Ph.D., Chicago, 1991. Structural and biophysical studies of proteins and protein-ligand complexes, glycoprotein structural biology, multidimensional NMR spectroscopy.
Alan G. Marshall, Kasha Professor and National High Magnetic Field Laboratory ICR Program Director; Ph.D., Stanford, 1970. Fourier transform ion cyclotron resonance mass spectrometry: instrumentation and technique development; applications to protein structure and posttranslational modifications, biomarkers, mapping of contact interfaces in biomacromolecular assemblies, and petroleomics.
Brian G. Miller, Assistant Professor; Ph.D., North Carolina at Chapel Hill, 2001. Enzymology, protein structure-function, enzyme evolution, bacterial genetics, magnetotactic bacteria.
Hugh Nymeyer, Assistant Professor; Ph.D., California, San Diego, 2001. Computational biophysics, protein folding and dynamics, protein-membrane interactions.
Michael Roper, Assistant Professor; Ph.D., Florida, 2003. Proteomes, metabolites, bioanalytics, microfluidics, nanotechnology.
Sanford A. Safron, Professor; Ph.D., Harvard, 1969. He atom-surface scattering experiments; dynamics of crystalline surfaces and interfaces; structure, dynamics, and growth of ultrathin films.
Jack Saltiel, Professor; Ph.D., Caltech, 1964. Photochemistry of organic molecules, elucidation of the mechanisms of photochemical reactions by chemical and spectroscopic means.
Qing-Xiang Amy Sang, Associate Professor; Ph.D., Georgetown, 1990. Protein chemistry, enzymology, molecular biology, and biochemistry of metalloproteinases; biochemical basis of angiogenesis; molecular carcinogenesis and mechanisms of cancer metastasis; biochemistry of cardiovascular diseases and stroke; proteomics.
Joseph B. Schlenoff, Mandelkern Professor of Polymer Science; Ph.D., Massachusetts Amherst, 1987. Polymer science, ultrathin films, charged and water-soluble polymers, surface chemistry, polymers in separation.
Oliver Steinbock, Associate Professor; Ph.D., Göttingen (Germany), 1993. Kinetics, experimental and theoretical studies of nonequilibrium systems, chemical self-organization.
Albert E. Stiegman, Associate Professor; Ph.D., Columbia, 1984. Synthesis of inorganic and hybrid organic/inorganic materials and nanocomposite structures; optical, dielectric, catalytic, and photochemical processes in solids; spectroscopy and photophysics of reactive metal centers in glasses.
Andre Striegel, Assistant Professor; Ph.D., New Orleans, 1996. Separation science of natural and synthetic macromolecules; structure-property relations of polymers; light scattering, viscometry, and polymer science; fundamental separation science.
Geoffrey F. Strouse, Associate Professor; Ph.D., North Carolina at Chapel Hill, 1993. Materials science with a principle focus on nanometer-scale materials, including novel synthetic methodology, development of analytical methods (XAFS, mass spectroscopy, magnetism, optical, vibrational), and biomaterial integration.
Kenneth D. Weston, Assistant Professor; Ph.D., California, Santa Barbara, 1998. Developing and applying single-molecule fluorescence spectroscopy techniques and microfluidics in a complimentary fashion to study a broad range of topics in biology, chemistry, and physics.
Wei Yang, Assistant Professor; Ph.D., SUNY at Stony Brook, 2001. Free-energy simulations, quantum mechanical calculations, computational protein design, computational biophysics/biochemistry.
Armen Zakarian, Assistant Professor; Ph.D., Florida State, 2002. Organic chemistry: synthetic methodology, natural product synthesis, bioorganic chemistry; development of new transformations for organic synthesis involving transition metals.
Lei Zhu, Assistant Professor; Ph.D., NYU, 2003. DNA nanotechnology, molecular sensing and logic applications, replicable chemical systems, supramolecular chemistry.

HARVARD UNIVERSITY

Department of Chemistry and Chemical Biology

Program of Study

The Department of Chemistry and Chemical Biology offers a program of study that leads to the degree of Doctor of Philosophy in chemistry in the special fields of biological, inorganic, organic, and physical chemistry. An interdepartmental Ph.D. program in chemical physics is also available. Upon entering the program, students formulate a plan of study in consultation with a Graduate Advisory Committee. Students must obtain honor grades in four advanced half courses (five for chemical physics). The course work is usually expected to be completed by the end of the second term of residence. All students must present and defend a research proposal in their second year of residence. Although the curriculum for the Ph.D. degree includes the course, research proposal, and oral defense requirements, the majority of the graduate student's time and energy is devoted to original investigations in a chosen field of research. Students are expected to join a research group in their second term of residence, but no later than the third. The Ph.D. dissertation is based on independent scholarly research, which, upon conclusion, is defended in an oral examination before a Ph.D. committee. The preparation of a satisfactory thesis normally requires at least four years of full-time research.

Research Facilities

The facilities of the Department of Chemistry and Chemical Biology are housed in five buildings in the Cabot Science Complex, with the adjacent Science Center providing major undergraduate lecture and laboratory areas. Three centers of research provide a central location for the following research instruments: for NMR research, one Bruker Avance 700i, one Varian 600-MHz NMR, three Varian 500-MHz NMRs, two Varian 400-MHz NMRs, one Varian 300-MHz NMR, and one Bruker ESP 300 EPR spectrometer; for mass spectroscopy, a JEOL-SX102A mass spectrometer, a Micromass LCT Platform II mass spectrometer equipped with APCI ionization, a Waters Q-Tof micro MS/MS mass spectrometer equipped with both electrospray and APCI ionization, an Agilent 6890/5973 GC-MS (gas chromatography–mass spectrometer), an Applied Biosystems MALDI (matrix-assisted laser desorption ionization time-of-flight mass spectrometer); for X-ray crystallography, two Siemens X-ray diffractometers, both with SMART area-detection systems; and a Nicolet 7000 Fourier transform infrared spectrometer. Computing in the Department is done mostly on workstations in individual research groups, with more than 1,200 devices linked by a Department-wide network. The Department, along with the Materials Research Laboratories at Harvard and MIT, operates and manages a Surface Sciences Center.

Financial Aid

The Department of Chemistry and Chemical Biology meets the financial needs of its graduate students through Departmental scholarships, Departmental fellowships, teaching fellowships, research assistantships, and independent outside fellowships. Financial support is awarded on a twelve-month basis, enabling students to pursue their research throughout the year. Tuition is afforded to all graduate students in good standing for the tenure of the Ph.D. program.

Cost of Study

As stated in the Financial Aid section, tuition is waived for all Ph.D. students in good standing.

Living and Housing Costs

Dormitory rooms for single students are available, with costs (excluding meals) that ranged from $4921 for a single room to $7729 for a two-room suite in 2005–06. Married and single students may apply for apartments managed by Harvard Planning and Real Estate. The monthly costs are studio apartment, $743–$1265; one-bedroom apartment, $1110–$1634; two-bedroom apartment, $1203–$2433; and three-bedroom apartment, $1728–$4132. There are also many privately owned apartments nearby and within commuting distance.

Student Group

The Graduate School of Arts and Sciences (GSAS) has an enrollment of about 3,500 graduate students. There are nearly 200 students in the Department of Chemistry and Chemical Biology, 30 percent of whom are international students.

Student Outcomes

In 2002, 4 percent of the Ph.D. recipients entered positions in academia, 34 percent accepted permanent positions in industry, 49 percent conducted postdoctoral research before accepting permanent positions in academia or industry, and 12 percent pursued other directions.

Location

Cambridge, a city of 101,355, is just minutes away from Boston. It is a scientific and intellectual center, teeming with activities in all areas of creativity and study. The Cambridge/Boston area is a major cultural center, with its many public and university museums, theaters, symphony, and numerous private, special interest, and historical collections and performances. New England abounds in possibilities for recreational pursuits, from camping, hiking, and skiing in the mountains of New Hampshire and Vermont to swimming and sailing on the seashores of Cape Cod and Maine.

The University

Harvard College was established in 1636, and its charter, which still guides the University, was granted in 1650. An early brochure, published in 1643, justified the College's existence: "To advance Learning and perpetuate it to Posterity...." Today, Harvard University, with its network of graduate and professional schools, occupies a noteworthy position in the academic world, and the Department of Chemistry and Chemical Biology offers an educational program in keeping with the University's long-standing record of achievement.

Applying

Applications for admission to study for the Ph.D. degree in chemistry may be obtained from GSAS and are accepted from students who have received a bachelor's degree or equivalent. The application process should begin during the summer or fall of the year preceding desired entrance. Completed application forms and supporting materials should be returned to the GSAS Admissions Office by December 8, though this date may vary slightly from year to year.

Correspondence and Information

Graduate Admissions Office
Department of Chemistry and Chemical Biology
Harvard University
12 Oxford Street
Cambridge, Massachusetts 02138
Phone: 617-496-3208
E-mail: admissions@chemistry.harvard.edu
Web site: http://www.chem.harvard.edu

Harvard University

THE FACULTY AND THEIR RESEARCH

James G. Anderson, Professor; Ph.D. (physical chemistry), Colorado, 1970. Chemical reactivity of radical-molecule systems, molecular orbital analysis of barrier height control, coupling of chemistry, radiation and climate in the earth system, photochemistry of planetary atmospheres, in situ detection of radicals in troposphere and stratosphere.

Alan Aspuru-Guzik, Assistant Professor; Ph.D. (physical chemistry), Berkeley, 2004. Theoretical physical chemistry, quantum computation and its application to chemistry problems, development of electronic structure methods for atoms and molecules: Density functional theory and quantum Monte Carlo, theoretical understanding and design of renewable energy materials.

David A. Evans, Professor; Ph.D. (organic chemistry), Caltech, 1967. Organic synthesis, organometallic chemistry, asymmetric synthesis.

Cynthia M. Friend, Professor; Ph.D. (physical chemistry), Berkeley, 1981. Physical chemistry of surface phenomena, materials chemistry and catalysis, electron spectroscopies and chemical techniques applied to the understanding of complex surface reactions, relating chemical processes to electronic structure on surfaces.

Roy Gerald Gordon, Professor; Ph.D. (physical chemistry), Harvard, 1964. Intermolecular forces, transport processes and molecular motion, theory of crystal structures and phase transitions, kinetics of crystal growth, solar energy, chemical vapor deposition, synthesis of inorganic precursors to new materials, thin films and their applications to microelectronics and solar cells.

Eric J. Heller, Professor; Ph.D. (chemical physics), Harvard, 1973. Few-body quantum mechanics, scattering theory, and quantum chaos; physics of semiconductor devices, ultracold molecular collisions, and nonadiabatic l interactions in molecules and gases.

Richard H. Holm, Professor; Ph.D. (inorganic and bioinorganic chemistry), MIT, 1959. Synthetic, structural, electronic, and reactivity properties of transition-element compounds; structure and function of metallobiomolecules.

Eric N. Jacobsen, Professor; Ph.D. (organic chemistry), Berkeley, 1986. Mechanistic and synthetic organic chemistry; development of new synthetic methods, with emphasis on asymmetric catalysis; physical-organic studies of reactivity and recognition phenomena in homogeneous catalysis; stereoselective synthesis of natural products.

Daniel Kahne, Professor; Ph.D. (organic chemistry), Columbia, 1986. Synthetic organic chemistry and its applications to problems in chemistry and biology.

Charles M. Lieber, Professor; Ph.D. (physical chemistry), Stanford, 1985. Chemistry and physics of materials, with an emphasis on nanoscale systems; rational synthesis of new nanoscale building blocks and nanostructured solids; development of methodologies for hierarchical assembly of nanoscale building blocks into complex and functional systems; investigation of fundamental electronic, optical, and optoelectronic properties of nanoscale materials; design and development of nanoelectronics and nanophotonic systems, with emphasis on electrically based biological detection, digital and quantum computing, and photonic systems.

David Liu, Professor; Ph.D. (organic chemistry and chemical biology), Berkeley, 1999. Organic chemistry and chemical biology of molecular evolution, nucleic acid–templated organic synthesis, reaction discovery, protein and nucleic acid evolution and engineering, synthetic polymer evolution; generally, effective molarity-based approaches to controlling reactivity and evolution-based approaches to the discovery of functional synthetic and biological molecules.

Gavin MacBeath, Associate Professor; Ph.D. (organic chemistry and chemical biology), Scripps, 1997. Interdisciplinary, combining proteomics, organic chemistry, and the development of array-based technology to reveal how groups of proteins function as networks inside the cell.

Andrew G. Myers, Professor; Ph.D. (organic chemistry), Harvard, 1985. Synthesis and study of complex organic molecules of importance in biology and human medicine.

Hongkun Park, Professor; Ph.D. (physical chemistry and chemical physics), Stanford, 1996. Physics and chemistry of nanostructured materials; electron transport individual molecules, inorganic clusters, nanowires, and nanotubes; single-molecule optoelectronics; synthesis and characterization of transition-metal-oxide and chalcogenide nanostructures with novel electronic and magnetic properties.

Tobias Ritter, Assistant Professor; Ph.D. (organic chemistry), Swiss Federal Institute of Technology, 2004. Synthetic organic and organometallic chemistry, development of new synthetic methods based on transition metal catalysis, steroselective synthesis of biologically active natural and unnatural products.

Alan Saghatelian, Assistant Professor; Ph.D. (organic chemistry), California, San Diego (Scripps), 2002. Development and application of global metabolite profiling (metabolomics) as a general discovery tool for chemical biology.

Stuart L. Schreiber, Professor; Ph.D. (organic synthesis), Harvard, 1981. Development and application of diversity-oriented organic synthesis to cell circuitry and genomic medicine.

Matthew D. Shair, Professor; Ph.D. (synthetic chemistry and chemical biology), Columbia, 1995. Synthesis of small molecules that have interesting biological functions and elucidation of their cellular mechanisms; development of organic synthesis.

Eugene I. Shakhnovich, Professor; Ph.D. (physical chemistry), Moscow, 1984. Theoretical biomolecular science, including protein folding, theory of molecular evolution, structural bioinformatics, rational drug design, populational genomics, and other systems, including complex polymers, spin glasses, etc.

Gregory L. Verdine, Professor; Ph.D. (organic chemistry, chemical biology, structural biology), Columbia, 1986. DNA repair; transcriptional control, chemistry for the conversion of peptides to ligands having cellular activity.

George M. Whitesides, Professor; Ph.D. (organic chemistry), Caltech, 1964. Physical organic chemistry, materials science, biophysics, complexity, surface science, microfluidics, self-assembly, microtechnology and nanotechnology, cell-surface biochemistry.

X. Sunney Xie, Professor; Ph.D. (physical chemistry), California, San Diego, 1990. Biophysical chemistry, single-molecule spectroscopy and dynamics, developments of new approaches for molecular and cellular imaging.

Xiaowei Zhuang, Professor; Ph.D. (physics), Berkeley, 1996. Biophysical chemistry, single-molecule biophysics, fluorescence microscopy and spectroscopy, microscopic and nanoscopic imaging of biomolecular and cellular systems.

Affiliate Member of the Department of Chemistry and Chemical Biology

Jon Clardy, Professor of Biological Chemistry and Molecular Pharmacology. Chemical Biology: Discovery of biologically active small molecules using DNA-based approaches, discovery of biologically active small molecules using high-throughput screening and chemical analysis, insight into how small molecules function as carriers of biological information.

Suzanne Walker, Professor of Microbiology and Molecular Genetics. Chemical Biology: Synthetic organic chemistry applied to the study of biochemical molecules, enzymology, mechanism of action of antibiotics.

Christopher T. Walsh, Hamilton Kuhn Professor of Biological Chemistry and Molecular Pharmacology. Chemical Biology: Molecular basis of biological catalysis with focus on the structure and function of enzymes, biosynthesis and mechanism of action of antibiotics and bacterial siderophores.

KENT STATE UNIVERSITY

KENT STATE UNIVERSITY

Department of Chemistry

Programs of Study The Department of Chemistry offers programs leading to the Master of Science (M.S.) and Doctor of Philosophy (Ph.D.) degrees in the divisions of analytical, inorganic, organic, and physical chemistry and biochemistry. Many faculty members also have research interests in the specialty areas of liquid crystals, materials and spectroscopy, separations, and surface science. A variety of interdisciplinary and collaborative projects are available, in addition to interdisciplinary doctoral programs in chemical physics and molecular and cellular biology.

Graduate students are required to complete a program of core courses in their area of specialization and at least one (for M.S. candidates) or two (for Ph.D. candidates) courses in other areas of chemistry. In addition to these courses, students may choose from a wide variety of electives. The program thus gives students considerable flexibility in curriculum design. At the end of the second year, doctoral candidates must pass a written examination in their field of specialization and present, and subsequently defend, an original research proposal for their dissertation. Students normally complete their doctoral program after four years.

Research Facilities Research laboratories are located primarily in Williams Hall and the adjoining Science Research Laboratory. In addition, facilities in the Liquid Crystal Institute, housed in the Materials Science Building, are available to chemistry students. Williams Hall houses two large lecture halls, classrooms, undergraduate and research laboratories, the research laboratories of the Separation Science Consortium, the Chemistry-Physics Library, chemical stockrooms, and glass and electronics shops. A machine shop, which is jointly operated with the physics department, is located in nearby Smith Hall. Spectrometers include 500-MHz and 300-MHz high-resolution NMR instruments; several FT-IR spectrometers, including a Bruker Equinox system; photon-counting fluorometer; circular dichroism; FPLC, UV/visible spectrometers, and cell culture; an ion-trap GC-MS with MS/MS(m) capability; AA/AE equipment; a EDX-700 energy dispersive X-ray spectrometer; a Bruker Biospin Avance 400-MHz digital NMR spectrometer; and a Shimadzu EDX700 X-ray fluorescence spectrometer. The X-ray facility includes a Siemens D5000 Powder diffractometer and a Bruker AXS CCD instrument for single-crystal structural elucidation. Equipment available in specialty areas includes an EMX-A EPR spectrometer system, a microwave spectrometer, an LCQ-Electrospray mass spectrometer with MS/MS(m) capability, a phosphorimager, microcal VP DSC and ITC calorimeters, Bruker Vector 33 FTIR-NIR, Cary Eclipse fluorescence spectrophotometer, Bruker Esquire 300plus MS with Agilent HP1100 HPLC, MALDI-TOF MS and LC-MS, a Cary 5 UV-VIS-NIR spectrophotometer, ThermoFinnigan Polaris Q115W GC-MS, a BAS electrochemical analyzer, various preparative centrifuges, a molecular dynamic Typhoon 8600 imaging system, and PCR and DNA sequencing facilities. Additional equipment available to the Separation Science Consortium includes a 400-MHz high-resolution and solids multinuclear NMR spectrometer, thermal analysis and gas adsorption equipment, and a variety of HPLC and GC instruments. There are a wide range of computer facilities available at Kent State, from microcomputers to supercomputers. Individual research groups in the Department of Chemistry maintain a variety of computer systems, including PCs and workstations. The Department has advanced molecular modeling facilities, including Cerius, Felix, Hyperchem, InsightII/Discover, Macromodel, and Spartan packages for modeling surfaces and interfaces, polymers, proteins, and nucleic acids, as well as facilities for performing ab initio calculations of molecular properties and molecular dynamics. University Computer Services maintains a number of microcomputer labs on campus as well as an IBM 4381-R24 mainframe computer that runs the VM/CMS operating system. High-performance computing is available at the Ohio Supercomputer Center, which maintains Cray T94, Cray T3E, IBM SP2, and SGI Origin 2000 supercomputers. The abstracting and indexing service maintains an extensive collection of books in chemistry and physics. There is also online access to a variety of chemical databases, including the Chemical Abstracts Service.

Financial Aid Graduate students are supported through teaching and research assistantships and University fellowships. Students in good academic standing are guaranteed appointments for periods of 4½ years (Ph.D. candidates) or 2½ years (M.S. candidates). Stipends for 2005–06 ranged from $16,000 (M.S.) to $17,000 (Ph.D.) for a twelve-month appointment. An $810 credit is made toward the University's health insurance plan. Renewable merit fellowships providing an additional $2500 per year are available to outstanding applicants.

Cost of Study Graduate tuition and fees for the 2005–06 academic year were $6192, for which a tuition scholarship was provided to students in good academic standing.

Living and Housing Costs Rooms in the graduate hall of residence are $2020 to $2705 per semester; married students' apartments may be rented for $660 to $690 per month (all utilities included). Information concerning off-campus housing may be obtained from the University housing office. Costs vary widely, but apartments typically rent for $500 to $600 per month.

Student Group Graduate students in chemistry currently number about 45. There are approximately 20,000 students enrolled at the main campus of Kent State University; 8,000 additional students attend the seven regional campuses.

Location Kent, a city of about 28,000, is located 35 miles southeast of Cleveland and 12 miles east of Akron in a peaceful suburban setting. Kent offers the cultural advantages of a major metropolitan complex as well as the relaxed pace of semirural living. There are a number of theater and art groups at the University and in the community. Blossom Music Center, the summer home of the Cleveland Orchestra and the site of Kent State's cooperative programs in art, music, and theater, is only 15 miles from the main campus. The Akron and Cleveland art museums are also within easy reach of the campus. There are a wide variety of recreational facilities available on the campus and within the local area, including West Branch State Park and the Cuyahoga Valley National Recreation Area. Opportunities for outdoor activities such as summer sports, ice skating, swimming, and downhill and cross-country skiing abound.

The University Established in 1910, Kent State University is one of Ohio's largest state universities. The campus contains 820 acres of wooded hillsides plus an airport and an eighteen-hole golf course. There are approximately 100 buildings on the main campus. Bachelor's, master's, and doctoral degrees are offered in more than thirty subject areas. The faculty numbers approximately 800.

Applying Forms for admission to the graduate programs are available on request from the Graduate Coordinator. There is no formal deadline for admission, but graduate assistantships are normally awarded by May for the following fall. Applicants requesting assistantships should apply by March 1.

Correspondence and Information

Graduate Coordinator
Department of Chemistry
Kent State University
Kent, Ohio 44242

Phone: 330-672-2032
Fax: 330-672-3816
E-mail: chemgc@kent.edu
Web site: http://www.kent.edu/chemistry

Kent State University

THE FACULTY AND THEIR RESEARCH

Soumitra Basu, Assistant Professor; Ph.D., Thomas Jefferson, 1996. Biochemistry: molecular modulation of RNA function, anticancer therapeutics using RNAi, alternative translation modes with implications for tumor angiogenesis, chemical modification of RNA, toxicoribonomics.

Nicola E. Brasch, Assistant Professor; Ph.D., Otago (New Zealand), 1994. Bioinorganic and medicinal chemistry; vitamin B_{12} and the B_{12}-dependent enzyme reactions; vanadium chemistry; inorganic drug delivery systems; synthesis, kinetics, and mechanism.

Bansidhar Datta, Associate Professor; Ph.D., Nebraska–Lincoln, 1989. Biochemistry and molecular biology: mechanism of protein synthesis initiation in mammals; studies of posttranslational modifications, such as O-glycosylation and phosphorylation of translational regulator, p67; molecular cloning of translational regulatory proteins; studies of the evolutionary origins of the regulatory/structural domains in p67.

Arne Gericke, Assistant Professor; Dr.rer.nat., Hamburg (Germany), 1994. Biophysical chemistry: characterization of lipid-mediated protein functions; infrared spectroscopy, fluorescence, and calorimetric measurements.

Edwin S. Gould, University Professor; Ph.D., UCLA, 1950. Inorganic chemistry: mechanisms of inorganic redox reactions; catalysis of redox reactions by organic species; electron-transfer reactions of flavin-related systems; reactions of cobalt, chromium, vanadium, titanium, europium, uranium, ruthenium, indium, peroxynitrous acid, and trioxodinitrate; reactions of water-soluble radical species.

Roger B. Gregory, Professor; Ph.D., Sheffield (England), 1980. Biochemistry: protein conformational dynamics; the characterization of dynamically distinct substructures in proteins; protein hydration; protein glass transition behavior and its role in protein function, stability, and folding; development and application of high-sensitivity methods for protein characterization, including protein-protein interactions and protein chemical modifications.

Songping D. Huang, Associate Professor; Ph.D., Michigan State, 1993. Inorganic chemistry: molecule-based magnetic and nonlinear optical materials, organic conductors and superconductors, novel microporous and mesoporous materials, synthesis and crystal growth of metal oxides and chalcogenides.

Mietek Jaroniec, Professor; Ph.D., Lublin (Poland), 1976. Physical/analytical/materials chemistry: adsorption and chromatography at the gas/solid and liquid/solid interface; synthesis and modification of adsorbents, catalysts, and chromatographic packings with tailored surface and structural properties; self-assembled organic-inorganic nanomaterials, such as ordered mesoporous silica, periodic mesoporous organo-silica, and other inorganic oxides; ordered mesoporous carbons synthesized via templating and imprinting methods; characterization of nanoporous materials by using adsorption, thermal analysis, chromatography, and other techniques.

Anatoly K. Khitrin, Associate Professor; Ph.D., Institute of Chemical Physics, Russian Academy of Sciences, 1985. Physical chemistry: NMR techniques, theory of magnetic resonance, material science, quantum computing and microimaging.

Kenneth K. Laali, Professor; Ph.D., Manchester (England), 1977. Organic, physical organic, and organometallic/heteroatom chemistry: modeling biological electrophiles from polycylic aromatic hydrocarbons (PAHs), PAH-chemistry, structure/reactivity relationships, correlation between DNA-binding and PAH-carbocation stability and structure, hetero-PAHs and their onium salts as DNA intercalating agents, host-guest chemistry, functional materials in particular conducting organic molecules based on cyclophane chemistry, green chemistry, organofluorine chemistry.

Paul Sampson, Professor; Ph.D., Birmingham (England), 1983. Synthetic organic chemistry: development of new synthetic methods; synthetic (stereoselective) organofluorine chemistry, with applications to the synthesis of fluorinated liquid crystals and carbohydrate analogs; development of new organometallic synthons as building blocks for organic synthesis.

Alexander J. Seed, Associate Professor; Ph.D., Hull (England), 1995. Organic chemistry, design, synthesis, and physical characterization of liquid crystals; ferroelectric, antiferroelectric, and high-twisting power materials for optical applications; new heterocyclic synthetic methodology.

Diane Stroup, Assistant Professor; Ph.D., Ohio State, 1992. Biochemistry: control of mammalian gene expression by regulation of transcriptional and posttranscriptional processes, study of nuclear hormone receptors and signal transduction events.

Yuriy V. Tolmachev, Assistant Professor; Ph.D., Case Western Reserve, 1999. Analytical chemistry: electrochemistry, fuel cells, electrocatalysis, X-ray spectroscopy, microfabrication, nanofabrication.

Chun-che Tsai, Professor; Ph.D., Indiana, 1968. Biochemistry: interaction of drugs with nucleic acids; structure and activity of anticancer drugs, antiviral agents, antibiotic drugs, and interferon inducers; structure and biological function relationships; X-ray diffraction; quantitative structure-activity relationships (QSAR); molecular and drug design.

Michael J. Tubergen, Associate Professor; Ph.D., Chicago, 1991. Physical chemistry: high-resolution microwave spectroscopy for molecular structure determination of hydrogen-bonded complexes and biological molecules.

Robert J. Twieg, Professor; Ph.D., Berkeley, 1976. Organic chemistry and materials science: development of organic and polymeric materials with novel electronic and optoelectronic properties, including nonlinear optical chromophores, photorefractive chromophores, organic semiconductors, fluorescent tags, and liquid crystals, with emphasis on applications and durability issues.

Frederick G. Walz, Professor; Ph.D., SUNY Downstate Medical Center, 1966. Biochemistry: site-directed and random-combinatorial mutagenesis and in vitro recombination methods used in studies of ribonuclease T_1; investigation, catalytic perfection, alternate substrate recognition, and catalytic mechanism at the active site and the functional role of enzyme subsites.

John L. West, Professor; Vice President, Research; and Dean, Graduate Studies; Ph.D., Carnegie Mellon, 1980. Materials science: liquid crystal polymer formulations for display applications, basic studies of liquid crystal alignment.

SELECTED PUBLICATIONS

Basu, S., and S. A. Strobel. Identification of specific monovalent metal ion binding sites within RNA. *Methods* 122:264–75, 2001.

Basu, S., A. Szewczak, M. Cocco, and S. A. Strobel. Direct detection of specific monovalent metal binding to a DNA G-quartet by ^{205}T1 NMR. *J. Am. Chem. Soc.* 122:3240–1, 2000.

Basu, S., and S. A. Strobel. Thiophilic metal ion rescue of phosphorothioate interference within the Tetrahymena ribozyme P4-P6 domain. *RNA* 5:1399–407, 1999.

Fry, F. H., et al. **(N. E. Brasch).** Characterization of novel vanadium(III)/acetate clusters formed in aqueous solution. *Inorg. Chem.* 44:5197, 2005.

Xia, L., A. G. Cregan, L. A. Berben, and **N. E. Brasch.** Studies on the formation of glutathionylcobalamin: Any free intracellular aquacobalamin is likely to be rapidly and irreversibly converted to glutathionylcobalamin. *Inorg. Chem.* 43:6848, 2004.

Hamza, M. S. A., A. G. Cregan, **N. E. Brasch,** and R. van Eldik. Mechanistic insight from activation parameters for the reaction between coenzymes B_{12} and cyanide: Further evidence that heterolytic Co-C bond cleavage is solvent-assisted. *Dalton Trans.* 596, 2003.

Brasch, N. E., A. G. Cregan, and M. L. Vanselow. Studies on the mechanism of the reaction between 5'-deoxyadenosylcobinamide and cyanide. *Dalton Trans.* 1287, 2002.

Datta, B., R. Datta, A. Majumdar, and A. Ghosh. The stability of eukaryotic initiation factor 2-associated glycoprotein, p67 increases during skeletal muscle differentiation and that inhibits the phosphorylation of extracellular signal-regulated kinases 1 and 2. *Exp. Cell. Res.* 303:174–82, 2005.

Datta, B., A. Majumdar, R. Datta, and R. Balusu. Treatment of cells with the angiogenic inhibitor fumagillin results in increased stability of eukaryotic initiation factor 2-associated glycoprotein, p67 and that inhibits the phosphorylation of extracellular signal-regulated kinases. *Biochemistry* 43:14821–31, 2004.

Datta, B., R. Datta, A. Ghosh, and A. Majumdar. Eukaryotic initiation factor 2-associated glycoprotein, p67, shows differential effects on the activity of certain kinases during serum-starved conditions. *Arch. Biochem. Biophys.* 427:68–78, 2004.

Wu, F. J., and **A. Gericke** et al. Domain structure and molecular conformation in annexin V/1,2-dimyristoyl-sn-glycero-3-phosphate/Ca2+ aqueous monolayers: A Brewster angle microscopy/infrared reflection-absorption spectroscopy study. *Biophys. J.* 74:3273–81, 1998.

Gericke, A., C. R. Flach, and R. Mendelsohn. Structure and orientation of lung surfactant SP-C and L. Alpha dipalmitoylphosphatidylcholine in aqueous monolayers. *Biophys. J.* 73:492–9, 1997.

Mendelsohn, R., J. W. Brauner, and **A. Gericke.** External infrared reflection-absorption spectrometry of monolayers at the air/water interface. *Ann. Rev. Phys. Chem.* 46:305–34, 1995.

Yang, Z., and **E. S. Gould.** Electron transfer. 160. Reductions by aquatitanium(II). *Dalton* 1781, 2005.

Yang, Z., and **E. S. Gould.** Electron transfer. 159. Reactions of trisoxalatocobaltate(III) with two-electron reductants. *Dalton* 3601, 2004.

Yang, Z., and **E. S. Gould.** Electron transfer. 158. Reactions of octacyanomolybdate(V) and octacyanotungstate(V) with s^2 metal-ion reducing centers. *Dalton* 1858, 2004.

Roh, J. H., et al. **(R. B. Gregory).** Onsets of anharmonicity in protein dynamics. *Phys. Rev. Lett.* 95:038101, 2005.

Gregory, R. B. Protein hydration and glass transitions. In *The Role of Water in Foods,* pp. 55–99, ed. D. Reid. New York: Chapman-Hall, 1997.

Gregory, R. B. Protein hydration and glass transition behavior. In *Protein-Solvent Interactions,* pp. 191–264, ed. **R. B. Gregory.** New York: Marcel Dekker, Inc., 1995.

Ranasinghe, M. I., et al. **(S. D. Huang).** Temperature dependence of excitation energy transport in a benzene branching molecular system. *Chem. Phys. Lett.* 383(3–4):411–7, 2004.

Lu, Z. K., et al. **(S. D. Huang).** Copper-catalyzed amination of aromatic halides with 2-N,N-dimethylaminoethanol as solvent. *Tetrahedron Lett.* 44(33):6289–92, 2003.

Olkhovyk, O., and **M. Jaroniec.** Periodic mesoporous organosilica with large heterocyclic bridging groups. *J. Am. Chem. Soc.* 127:60–1, 2005.

Yoon, S. B., et al. **(M. Jaroniec).** Graphitized pitch-based carbons with ordered nanopores synthesized by using colloidal crystals as templates. *J. Am. Chem. Soc.* 127:4188–9, 2005.

Kowalczyk, P., and **M. Jaroniec** et al. An improvement of the Derjaguin-Broekhoff-de Boer Theory for capillary condensation/evaporation of nitrogen in mesoposous systems and its implications for pore size analysis of MCM-41 silicas and related materials. *Langmuir* 21:1827–33, 2005.

Li, Z., and **M. Jaroniec.** Colloid-imprinted carbons as stationary phases for reverse phase liquid chromatography. *Anal. Chem.* 76:5479–85, 2004.

Lee, J.-S., and **A. K. Khitrin.** Pseudopure state of a twelve-spin system. *J. Chem. Phys.* 122:041101-1/3, 2005.

Lee, J.-S., and **A. K. Khitrin.** Experimental demonstration of quantum state expansion in a cluster of dipolar-coupled nuclear spins. *Phys. Rev. Lett.* 94:150504-1/4, 2005.

Lee, J.-S., and **A. K. Khitrin.** Stimulated wave of polarization in a one-dimensional Ising chain. *Phys. Rev. A* 71:062338-1/5, 2005.

Okazaki, T., and **K. K. Laali.** Probing the intermediates of halogen addition to alkynes: Bridged halonium versus open halovinyl cation; a theoretical study. *J. Org. Chem.* 70(23):9139–46, 2005.

Laali, K. K., S. Hupertz, A. G. Temu, and S. E. Galembeck. Electrospray mass spectrometric and DFT study of substituent effects in Ag^+ complexation to polycyclic aromatic hydrocarbons (PAHs). *Org. Biomol. Chem.* 2319–26, 2005.

Borosky, G. L., and **K. K. Laali.** Theoretical study of aza-polycyclic aromatic hydrocarbons (Aza-PAHs), modeling carbocations from oxidized metabolites and their covalent adducts with representative nucleophiles. *Org. Biomol. Chem.* 3:1180–8, 2005.

Laali, K. K., et al. **(S. D. Huang).** 1-triflato-3,5,7-trimethyl-1,3,5,7-tetrasilaadamantane and 1,3-bis-triflato-5,7-dimethyl-1,3,5,7-tetrasilaadamantane; synthesis, complexation study and X-ray structure of 1-hydroxy-3,5,7-trimethyl-1,3,5,7-tetrasilaadamantane. *J. Organomet. Chem.* 658(1–2):141–6, 2002.

Chumachenko, N., **P. Sampson,** A. D. Hunter, and M. Zeller. β-acyloxysulfonyl tethers for intramolecular Diels-Alder cycloaddition reactions. *Org. Lett.* 7:3203–6, 2005.

Novikov, Y. Y., and **P. Sampson.** 1-bromo-1-lithioethene: A practical reagent for the efficient preparation of 2-bromo-1-alken-3-ols. *Org. Lett.* 5:2263–6, 2003.

Kiryanov, A. A., **P. Sampson,** and **A. J. Seed.** Synthesis of 2-alkoxy-substituted thiophenes 1.3-thiazoles and related W-heterocycles via Lawesson's reagent-mediated cyclization under microwave irradiation: Applications for liquid crystal synthesis. *J. Org. Chem.* 66(23):7925–9, 2001.

Seed, A. J., J. W. Doane, A. Khan, and M. E. Walzh. A new high twisting power material for use as a single asymmetric dopant in cholesteric displays with a temperature independence of the helical twisting power. *Mol. Cryst. Liq. Cryst.* 410:201–8, 2004.

Seed, A. J., G. J. Cross, K. J. Toyne, and J. W. Goodby. Novel, highly polarizable thiophene derivatives for use in nonlinear optical applications. *Liq. Cryst.* 30(9):1089–107, 2003.

Hirst, L. S., et al. **(A. J. Seed).** Interlayer structures of the chiral smectic liquid crystal phases revealed by resonant X-ray scattering. *Phys. Rev. E* 65(4):article number 041705, 2002.

Kiryanov, A. A., **A. J. Seed,** and **P. Sampson.** Ring fluorinated thiophenes: Applications to liquid crystal synthesis. *Tetrahedron Lett.* 42:8797–800, 2001.

Stroup, D., and J. R. Ramasaran. Cholesterol 7 alpha-hydroxylase is phosphorylated at multiple amino acids. *Biochem. Biophys. Res. Commun.* 329:957–65, 2005.

Stroup, D. Kinase/phosphatease regulation of CYP7A1. *Frontiers in Bioscience* (invited review), 2005.

Chen, W., et al. **(D. Stroup).** Bile acid repression of human sterol 27-hydroxylase and cholesterol 7alpha-hydroxylase. *J. Lipid Res.* 42:1402–12, 2001.

Kent State University

Chiang, J. Y. L., R. Kimmel, C. Weinberger, and **D. Stroup.** Farnesoid X receptor responds to bile acids and represses cholesterol 7alpha-hydroxylase gene (CPY7A1) transcription. *J. Biol. Chem.* 275:10918–24, 2000.

Sapozhnikov, M. V., I. S. Aronson, W.-K. Kwok, and **Y. V. Tolmachev.** Field assisted self-assembly of conducting microchains. *Phys. Rev. Lett.* 93:084502/1–4, 2004.

Tolmachev, Y. V., et al. In situ X-ray surface scattering observation of long-range-ordered (√19x√19) R 23.4°-13CO structure on Pt (111) in aqueous electrolytes. *Electrochem. Solid-State Lett.* 7:3/E23–6, 2004.

Rhee, C. K., et al. **(Y. Tolmachev).** Osmium nanoislands spontaneously deposited on a Pt(111) electrode: The XPS, STM and GIF-XAS study. *J. Electroanal. Chem.* 554–555:367–78, 2003.

Durand, P. J., R. Pasari, J. W. Baker, and **C.-c. Tsai.** An efficient algorithm for similarity analysis of molecules. *Internet J. Chem.* 2(17):1–16, 1999.

Lesniewski, M. L., et al. **(C.-c. Tsai).** QSAR studies of antiviral agents using structure-activity maps. *Internet J. Chem.* 2(7):1–59, 1999.

Parakulam, R. R., M. L. Lesniewski, K. J. Taylor-McCabe, and **C.-c. Tsai.** QSAR studies of antiviral agents using molecular similarity analysis and structure-activity maps. *SAR QSAR Environ. Res.* 10:179–206, 1999.

Craig, N. C., et al. **(M. J. Tubergen).** Equilibrium structures for the cis and trans isomers of 1,2-difluoroethylene and the cis,trans isomer of 1,4-difluorobutadiene. *Int. J. Quantum Chem.* 95:837–52, 2003.

Tubergen, M. J., C. R. Torok, and R. J. Lavrich. Effect of solvent on molecular conformation: Microwave spectra and structures of 2-aminoethanol van der Waals complexes. *J. Chem. Phys.* 119:8397–403, 2003.

Lavrich, R. J., et al. **(M. J. Tubergen).** Experimental studies of peptide bonds: Identification of the C_7^{eq} conformation of the alanine dipeptide analog N-acetyl-alanine n'-methylamide from torsion-rotation interactions. *J. Chem. Phys.* 118:1253–65, 2003.

Lavrich, R. J., C. R. Torok, and **M. J. Tubergen.** Effect of the bulky side chain on the backbone structure of the amino and derivative valinamide. *J. Phys. Chem. A* 106:8013–8, 2002.

Willets, K. A., et al. **(R. J. Twieg).** Nonlinear optical chromophores as nonoscale emitters for single-molecule spectroscopy. *Acc. Chem. Res.* 38:549–56, 2005.

Lu, Z., and **R. J. Twieg.** A mild and practical amination of unactivated halothiophenes. *Tetrahedron* 61:903–18, 2005.

Lu, Z., and **R. J. Twieg.** Copper catalyzed amination of aromatic halides in aqueous media with 2-dimethylaminoethanol ligand. *Tetrahedron Lett.* 46:2997–3001, 2005.

Chitester, B. J., and **F. G. Walz Jr.** Kinetic studies of guanine recognition and a phosphate group subsite on ribonuclease T_1 using substitution mutants at Glu46 and Lys41. *Arch. Biochem. Biophys.* 406:73–7, 2002.

Kumar, K., and **F. G. Walz Jr.** Probing functional perfection in substructures of ribonuclease T_1: Double random mutagenesis involving Asn43, Asn44, and Glu46 in the guanine binding loop. *Biochemistry* 40:3748–57, 2001.

Arni, R. K., et al. **(F. G. Walz Jr.).** Three-dimensional structure of ribonuclease T_1 complexed with an isosteric analogue of GpU: Alternate substrate binding modes and catalysis. *Biochemistry* 38:2452–61, 1999.

West, J. L., G. Zhang, Y. Reznikov, and A. Glushchenko. Fast birefringent mode of stressed liquid crystal. *Appl. Phys. Lett.* 86:031111, 2005.

Reznikov, Y., et al. **(J. L. West).** Ferroelectric nematic suspension. *Appl. Phys. Lett.* 82:1917–9, 2003.

West, J. L., et al. Drag on particles in a nematic suspension by a moving nematic-isotropic interface. *Phys. Rev. E* 66:012702, 2002.

NEW YORK UNIVERSITY

Department of Chemistry

Programs of Study

The Department of Chemistry at NYU offers programs leading to the degrees of Master of Science (M.S.) and Doctor of Philosophy (Ph.D.) in chemistry. A major focus of the Department is in the study of molecules in living systems; faculty members and students in all subdisciplines of chemistry at NYU work on problems relevant to the molecules of life. The Department is highly interdisciplinary, including both traditional areas of chemistry and areas involving biophysics, chemical biology, computational chemistry, materials, and nanoscience. The Department has a relatively small faculty, which promotes outstanding interactions among students and faculty and staff members. The major requirements for the doctoral degree in chemistry are the successful completion of an original research project and the presentation and defense of the Ph.D. thesis.

A tightly coupled series of requirements and exams has been designed to advance students to Ph.D. candidacy status and to continue to educate them throughout their graduate work. These steps train, challenge, and broaden students in preparation for excellence in research and an independent scientific career. Entering graduate students meet with the Director of Graduate Studies to plan an academic course schedule that best suits each individual's background and career goals. The Ph.D. students are required to take six courses (three per semester) during their first year. During the first year, students also work in different laboratories for one semester each. This research rotation is expected to take roughly half of the student's time, with an oral presentation at the end of each rotation that is attended by other students and members of the lab. The research rotation provides valuable exposure to the diversity of the program and allows the student to gain firsthand experience for choosing a thesis laboratory. After the first year, students concentrate on their research. Students generally finish the program in approximately five years. In addition to research and course work, most students gain teaching experience. The Department hosts a weekly colloquium series as well as regular seminars in several disciplines.

Research Facilities

A new shared instrumentation facility in the chemistry department includes MALDI-TOF and electrospray mass spectrometers; several GC-MS and LC-MS systems; 300-, 400-, and 500-MHz NMR spectrometers; polarimeter; circular dichroism spectropolarimeter; FT-IR, UV-Vis, and fluorescence spectrophotometers; phosphorimager; and polarizing microscope. The facility is managed by a full-time staff member who has earned a Ph.D. in NMR spectroscopy. Additional NMR facilities include a 400-MHz wide-bore spectrometer for solids research and several very-high-field NMR instruments located at the nearby New York State Center for Structural Biology.

Other equipment in the Department that is available to researchers includes a scanning tunneling microscope with a high-vacuum specimen-coating device, an atomic force microscope, a state-of-the-art laser laboratory for fast time-resolved studies of chemical processes, peptide, DNA and other robotic synthesizers, photoacoustic and photothermal beam deflection spectrometers, microcalorimeters, and a highly automated, high-performance liquid chromatography system (LC-MS) with electrospray mass spectrometry, UV-Vis, and fluorescence detectors. Mass spectrometry facilities are complemented by extensive resources available at the Protein Analysis Facility located at the NYU School of Medicine.

Computing facilities are extensive within research groups, in the department, at the Scientific Visualization Center (located in the Courant Institute of Mathematics), and at the University in general. Bobst Library is one of the largest open-stack research libraries in the nation.

Financial Aid

All students admitted to the Ph.D. program receive financial support in the form of teaching assistantships, research assistantships, or University fellowships. Students usually receive support for the duration of their studies. The basic stipend in 2005–06 for eleven months was $24,000.

Cost of Study

Research and teaching appointments carry a waiver of tuition, which amounted to $23,304 for the 2004–05 academic year, and of registration fees, which were estimated at $1740 per academic year.

Living and Housing Costs

University housing for graduate students is limited. It consists mainly of shared studio apartments and shared suites in residence halls within walking distance of the University. University housing rents in the 2004–05 academic year ranged from $10,250 to $17,234. Students may apply for subsidized housing ($825 per month for 2004–05) for their first year only.

Student Group

The Graduate School of Arts and Science has an enrollment of 4,000 graduate students; about 70 are pursuing advanced degrees in chemistry. They represent a wide diversity of ethnic and national groups; more than a third are women. Upon receiving their Ph.D.'s, about 10 percent of recent graduates entered positions in academia, and the others were approximately equally divided between those who accepted industrial employment and those who elected to gain postdoctoral research experience before accepting permanent positions.

Location

Greenwich Village, the home of the University, has long been famous for its contributions to the fine arts, literature, and drama and for its personalized, smaller-scale, European style of living. It is one of the most desirable places to live in the city. New York City is the business, cultural, artistic, and financial center of the nation, and its extraordinary resources enrich both the academic programs and the experience of living at NYU.

The University

New York University, a private university, awarded its first doctorate in chemistry in 1866. Ten years later, the American Chemical Society was founded in the original University building, and the head of the chemistry department, John W. Draper, assumed the presidency.

Applying

Application forms may be obtained by writing to the address below. Students beginning graduate study are accepted only for September admission. Applicants are expected to submit scores on the GRE General Test and the Subject Test in chemistry or a related discipline. Students whose native language is not English must submit a score on the Test of English as a Foreign Language (TOEFL). The application deadline is December 15. Applicants are invited to visit the University but are advised to contact the Department beforehand to arrange an appointment.

Correspondence and Information

Director of Graduate Programs
Department of Chemistry
New York University
New York, New York 10003
Phone: 212-998-8400
Fax: 212-260-7905
E-mail: chem.web@nyu.edu
Web site: http://www.nyu.edu/pages/chemistry/

New York University

THE FACULTY AND THEIR RESEARCH

Paramjit S. Arora, Assistant Professor; Ph.D., California, Irvine, 1999. Bioorganic chemistry, chemical biology, molecular recognition.
Zlatko Bacic, Professor; Ph.D., Utah, 1982. Theoretical chemistry: spectra and dynamics of highly vibrationally excited floppy molecules, dissociation dynamics of rare-gas clusters in collisions with solid surfaces.
Henry C. Brenner, Associate Professor; Ph.D., Chicago, 1972. Physical chemistry: optical and magnetic resonance studies of molecular solids and biological systems, energy transfer and luminescence in condensed phases.
James W. Canary, Professor; Ph.D., UCLA, 1988. Organic, bioorganic, and bioinorganic chemistry; chiral materials; nanoscience; molecular imaging; supramolecular chemistry.
Young-Tae Chang, Assistant Professor; Ph.D., Pohang University of Science and Technology (Korea), 1996. Combinatorial and bioorganic chemistry.
John S. Evans, Associate Professor; Ph.D., Caltech, 1993. Biomolecular materials: solution and solid-state NMR structure and dynamics of biomineralization and structural proteins, computational and molecular modeling of proteins and biomaterials.
Paul J. Gans, Professor; Ph.D., Case Tech, 1959. Theoretical chemistry: determination of conformational and thermodynamic properties of macromolecules by Monte Carlo simulation.
Nicholas E. Geacintov, Professor; Ph.D., Syracuse, 1961. Physical and biophysical chemistry: interaction of polycyclic aromatic carcinogens with nucleic acids, laser studies of fluorescence mechanisms and photoinduced electron transfer.
Alexej Jerschow, Assistant Professor; Ph.D., Linz (Austria), 1997. NMR spectroscopy, imaging, and microscopy; theory and applications in material sciences, biophysics, and quantum computation.
Neville R. Kallenbach, Professor; Ph.D., Yale, 1961. Biophysical chemistry of proteins and nucleic acids: structure, sequence, and site selectivity in DNA-drug interactions, protein folding, model helix and beta sheet structures.
Kent Kirshenbaum, Assistant Professor; Ph.D., California, San Francisco, 1999. Bioorganic and biophysical chemistry: artificial proteins, biomimetic heteropolymers, biomolecular conformational rearrangements.
Edward J. McNelis, Professor; Ph.D., Columbia, 1960. Organic chemistry: oxidation as a route to synthetically useful substances, novel organometallic catalysts.
Johannes P. M. Schelvis, Assistant Professor; Ph.D., Leiden (Netherlands), 1995. Biophysical chemistry: steady-state and time-resolved vibrational and optical spectroscopy of biological systems, structure-function relationship in proteins, and enzyme catalysis.
Tamar Schlick, Professor; Ph.D., NYU, 1987. Molecular mechanics and dynamics, computational and structural biology, nucleic acid structure, nucleic acid and protein interaction.
David I. Schuster, Professor; Ph.D., Caltech, 1961. Synthesis of functionalized fullerenes; supramolecular complexes of fullerenes and carbon nanotubes; photoinduced electron transfer in porphyrin/fullerene hybrids, including dyads, rotaxanes, catenanes, and molecular wires; inhibition of HIV-1 protease by fullerenes.
Nadrian C. Seeman, Professor; Ph.D., Pittsburgh, 1970. Biophysical chemistry: structural DNA nanotechnology, structural chemistry of recombination, catenated and knotted DNA topologies, DNA-based computation, crystallography.
Robert Shapiro, Professor; Ph.D., Harvard, 1959. Organic and bioorganic chemistry: effects of mutagens on the structure and function of nucleic acids.
Mark Tuckerman, Assistant Professor; Ph.D., Columbia, 1993. Theoretical chemistry: ab initio molecular dynamic simulations and statistical mechanics.
Alexander Vologodskii, Research Professor; Ph.D., Moscow Physical Technical Institute, 1975. Statistical mechanical properties of DNA; supercoiling, catenanes, knots; effect of supercoiling on DNA-protein interaction.
Marc A. Walters, Associate Professor; Ph.D., Princeton, 1981. Inorganic chemistry: kinetics and energetics of metal thiolate and selenolate formation, synthesis of complexes for the uptake of many-electron equivalents.
Stephen Wilson, Professor; Ph.D., Rice, 1972. Organic chemistry: total synthesis of natural products, new synthetic methodology, synthesis of enzyme mimics.
John Z. H. Zhang, Professor; Ph.D., Houston, 1987. Theory: study of molecular collision dynamics and chemical reactions in the gas phase and on surfaces; methods for quantum energy calculation of protein and macromolecules; protein-drug interaction and free energy study.
Yingkai Zhang, Assistant Professor; Ph.D., Duke, 2000. Computational biochemistry and biophysics: multiscale modeling of biological systems, enzyme catalysis and regulation, DNA damage and repair, biomolecular recognition.

New York University

SELECTED PUBLICATIONS

Angelo, N. G., and **P. S. Arora.** Nonpeptidic foldamers from amino acids: Synthesis and characterization of 1,3-substituted triazole oligomers. *J. Am. Chem. Soc.* 127:17134–5, 2005.

Wang, D., W. Liao, and **P. S. Arora.** Enhanced metabolic stability and protein-binding properties of artificial alpha-helices derived from a hydrogen-bond surrogate: Application to Bcl-xL. *Angew. Chem. Int. Ed.* 44:6525–9, 2005.

Chapman, R. N., G. Dimartino, and **P. S. Arora.** A highly stable short alpha-helix constrained by a main chain hydrogen bond surrogate. *J. Am. Chem. Soc.* 126:12252–3, 2004.

Jiang, H., et al. **(Z. Bacic).** $(^3HCl)_2$ and $(HF)_2$ in small helium clusters: Quantum solvation of hydrogen-bonded dimmers. *J. Chem. Phys.* 123: 224313, 2005.

Jiang, H., M. Xu, J. M. Hutson, and **Z. Bacic.** ArnHF van der Walls clusters revisited: II. Energetics and HF vibrational frequency shifts from diffusion Monte Carlo calculations on additive and nonadditive potential-energy surfaces for n=1–12. *J. Chem. Phys.* 123:054305, 2005.

Jiang, H., and **Z. Bacic**. HF in clusters of molecular hydrogen: I. Size evolution of quantum solvation by parahydrogen molecules. *J. Chem. Phys.* 122:244306, 2005.

Dourandin, A., **H. C. Brenner,** and M. Pope. New broad spectrum auto-correlator for ultrashort light pulses based on multiphoton photoemission. *Rev. Sci. Instrum.* 71:1589–94, 2000.

Tringali, A. E., S. K. Kim, and **H. C. Brenner.** ODMR and fluorescence studies of pyrene solubilized in anionic and cationic micelles. *J. Lumin.* 81:85–100, 1999.

Tringali, A. E., and **H. C. Brenner.** Spin-lattice relaxation and ODMR linenarrowing of the photoexcited triplet state of pyrene in polycrystalline Shpol'skii hosts and glassy matrices. *Chem. Phys.* 226, 1998.

Royzen, M., Z. Dai, and **J. W. Canary.** Ratiometric displacement approach to Cu(II) sensing by fluorescence. *J. Am. Chem. Soc.* 127:1612–3, 2005.

Zhang, J., K. Siu, C. H. Lin, and **J. W. Canary.** Conformational dynamics of Cu(I) complexes of tripodal ligands: Steric control of molecular motion. *New J. Chem.* 29:1147–51, 2005.

Holmes, A. E., S. A. Simpson, and **J. W. Canary.** Stereodynamic coordination complexes: Dependence of exciton-coupled circular dichroism spectra on molecular conformation and shape. *Monatsh. Chem.* 136:461–75, 2005.

Walsh, D. P., and **Y. T. Chang.** Chemical genetics. *Chem. Rev.,* in press.

Snyder, J. R., et al. **(Y. T. Chang).** Dissection of melanogenesis with small molecules identifies prohibitin as a regulator. *Chem. Biol.* 12:477–84, 2005.

Li, Q., et al. **(Y. T. Chang).** Solid-phase synthesis of styryl dye library and its application to amyloid sensors. *Angew. Chem., Int. Ed. Engl.* 43:6331–5, 2004.

Collino, S., I. W. Kim, and **J. S. Evans.** Identification of an "acidic" C-terminal mineral modification sequence from the mollusk shell protein asprich. *Cryst. Growth Design,* in press.

Kim, I. W., S. Collino, D. E. Morse, and **J. S. Evans.** A crystal-modulating protein from molluscan nacre that limits the growth of calcite in vitro. *Cryst. Growth Design* 6:839–42, 2006.

Kim, I. W, M. R. Darragh, C. Orme, and **J. S. Evans.** Molecular "tuning" of crystal growth by nacre-associated polypeptides. *Cryst. Growth Design* 5:5–10, 2006.

Gans, P. J. Review of *Statistical Mechanics: Fundamentals and Modern Applications. J. Am. Chem. Soc.* 120:11026, 1998.

Lyu, P. C., **P. J. Gans,** and **N. R. Kallenbach.** Energetic contribution of solvent-exposed ion pairs to alpha-helix structure. *J. Mol. Biol.* 223:343–50, 1992.

Gans, P. J., et al. **(N. R. Kallenbach).** The helix-coil transition in heterogeneous peptides with specific side-chain interactions: Theory and comparison with CD spectral data. *Biopolymers* 31:1605–14, 1991.

Misiaszek, R., et al. **(N. E. Geacintov).** Combination of nitrogen dioxide radicals with 8-oxo-7,8-dihydroguanine and guanine radicals in DNA: Oxidation and nitration end-products. *J. Am. Chem. Soc.* 127:2191–200, 2005.

Hsu, G. W., et al. **(N. E. Geacintov).** Structure of a high-fidelity DNA polymerase bound to a benzo[a]pyrene adduct that blocks replication. *J. Biol. Chem.* 280:3764–70, 2005.

Zhang, N., et al. **(N. E. Geacintov).** Methylation of cytosine at C5 in a CpG sequence context causes a conformational switch of a benzo[a]pyrene diol epoxide-N2-guanine adduct in DNA from a minor groove alignment to intercalation with base displacement. *J. Mol. Biol.* 346:951–65, 2005.

Müller, N., and **A. Jerschow.** Nuclear spin noise imaging. *Proc. Natl. Acad. Sci.,* in press.

Choy, J., W. Ling, and **A. Jerschow**. Selective detection of ordered sodium signals via the central transition. *J. Magn. Reson.* 180:105–9, 2006.

Shi, Z., K. Chen, L. Zhigang, and **N. R. Kallenbach.** Conformation of the backbone in unfolded proteins. *Chem. Rev.,* in press.

Lee-Huang, S., et al. **(N. R. Kallenbach).** Structural and functional modeling of human lysozyme reveals a unique nonapeptide HL9 with anti-HIV activity. *Biochemistry* 44:4648–55, 2005.

Chen, K., Z. Liu, Z. Shi, and **N. R. Kallenbach.** Neighbor effect on PPII conformation in alanine peptides. *J. Am. Chem. Soc.* 127:10146–7, 2005.

Spencer, T., B. Yoo, and **K. Kirshenbaum.** Purification and modification of fullerene C60 in the undergraduate laboratory. *J. Chem. Ed.,* in press.

Holub, J. M., H. Jang, and **K. Kirshenbaum.** Clickity-click: Highly functionalized peptoid oligomers generated by sequential azide/alkyne [3+2] cycloadditions on solid-phase support. *Org. Biomol. Chem.* 4:1497–502, 2006.

Yoo, B., and **K. Kirshenbaum.** Protease-mediated ligation of abiotic oligomers. *J. Am. Chem. Soc.* 127:17132–3, 2005.

Blandino, M., and **E. McNelis.** Rearrangements of haloalykynol derivatives of glucofuranose. *Org. Lett.* 4:3387–90, 2002.

McNelis, E., and M. Blandino. A method for estimating tetrahedral bond angles. *New J. Chem.* 25:772–4, 2001.

Djuardi, E., and **E. McNelis.** Furo[3,4-b]furan formations from alkynols of xylose. *Tetrahedron Lett.* 40:7193–6, 1999.

Del Federico, E., et al. **(J. Schelvis** and **A. Jerschow).** Insight into framework destruction in ultramarine pigments. *Inorg. Chem.* 45:1270–6, 2006.

Gindt, Y. M., et al. **(J. P. M. Schelvis).** Substrate binding modulates the reduction potential of DNA photolyase. *J. Am. Chem. Soc.* 127:10472–3, 2005.

Gurudas, U., and **J. P. M. Schelvis.** Resonance Raman spectroscopy of the neutral radical Trp306 in DNA photolyase. *J. Am. Chem. Soc.* 126:12788–9, 2004.

Schelvis, J. P. M., et al. Resonance Raman and UV-vis spectroscopic characterization of the complex of photolyase with UV-damaged DNA. *J. Phys. Chem. B* 107:12352–62, 2003.

Gevertz, J., et al. **(T. Schlick).** In vitro RNA random pools are not structurally diverse: A computational analysis. *RNA* 11:853–63, 2005.

Radhakrishnan, R., and **T. Schlick.** Orchestration of cooperative events in DNA synthesis and repair mechanism unraveled by transition path sampling of DNA polymerases beta's closing. *Proc. Natl. Acad. Sci. U.S.A.* 101:5970–5, 2004.

Schlick, T. *Molecular Modeling and Simulation: An Interdisciplinary Guide.* New York: Springer-Verlag, 2002.

Schuster, D. I., K. Li, and D. M. Guldi. Porphyrin-fullerene photosynthetic model systems with rotaxane and catenane architectures. *C. R. Chimie*, in press.

Schuster, D. I., et al. Synthesis and photophysics of porphyrin-fullerene donor-acceptor dyads with conformationally flexible linkers. *Tetrahedron* 62:1928–36, 2006.

Vail, S. A., et al. **(D. I. Schuster).** Energy and electron transfer in polyacetylene-linked zinc-porphyrin-[60]fullerene molecular wires. *Chem. Eur. J.* 11: 3375–80, 2005.

Seeman, N. C. From genes to machines: DNA nanomechanical devices. *Trends Biochem. Sci.* 30:119–235, 2005.

New York University

Mathieu, F., et al. **(N. C. Seeman).** Six-helix bundles designed from DNA. *Nano Lett.* 5:661–5, 2005.

Liao, S., and **N. C. Seeman.** Translation of DNA signals into polymer assembly instructions. *Science* 306:2072–4, 2004.

Shapiro, R. Small-molecule interactions were central to the origin of life. *Q. Rev. Biol.* 81:105–25, 2006.

Zhang, L., **R. Shapiro,** and S. Broyde. Molecular dynamics of a food carcinogen-DNA adduct in a replicative DNA polymerase suggest hindered nucleotide incorporation and extension. *Chem. Res. Toxicol.* 18:1347–63, 2005.

Jia, L., et al. **(R. Shapiro** and **N. E. Geacintov).** Structural and thermodynamic features of spiroiminodihydantoin-damaged DNA duplexes. *Biochemistry* 44:13342–53, 2005.

Corminboeuf C., P. Hu, **M. E. Tuckerman,** and **Y. Zhang.** Unexpected deacetylation mechanism suggested by a density functional theory QM/MM study of histone-deacetylase-like protein. *J. Am. Chem. Soc.* 128:4530–1, 2006.

Tuckerman, M. E., A. Chandra, and D. Marx. Structure and dynamics of OH-(aq). *Acc. Chem. Res.* 39:151, 2006.

Minary, P., and **M. E. Tuckerman.** Reaction mechanism of cis-1,3-butadiene addition to the Si(100)-2x1 surface. *J. Am. Chem. Soc.* 127:1110, 2005.

Minary, P., **M. E. Tuckerman,** and G. J. Martyna. Long-time molecular dynamics for enhanced conformational sampling in biomolecular systems. *Phys. Rev. Lett.* 93:150201, 2004.

Vologodskii, A. Energy transformation in biological molecular motors. *Phys. Life Rev.,* in press.

Vologodskii, A. Simulation of equilibrium and dynamic properties of large DNA molecules. In *Computational Studies of DNA and RNA,* eds. F. Lankas and J. Sponer. Springer, in press.

Du, Q., et al. **(A. Vologodskii).** Cyclization of short DNA fragments and bending fluctuations of the double helix. *Proc. Natl. Acad. Sci. U.S.A.* 102:5397–402, 2005.

Walters, M. A., et al. Amide-ligand hydrogen bonding in reverse micelles. *Inorg. Chem.* 44:1172–4, 2005.

Walters, M. A., C. L. Roche, A. L. Rheingold, and S. W. Kassel. N-H···S hydrogen bonds in a ferredoxin model. *Inorg. Chem.* 44(11):3777–9, 2005.

Dey, A., et al. **(M. A. Walters).** Sulfur K-edge XAS and DFT calculations on $[Fe_4S_4]^{2+}$ clusters: Effects of H-bonding and structural distortion on covalency and spin topology. *Inorg. Chem.* 44(23):8349–54, 2005.

Schapira, M., et al. **(S. R. Wilson).** Discovery of diverse thyroid hormone receptor antagonists by high-throughput docking. *Proc. Natl. Acad. Sci. U.S.A.* 100(12):7354–9, 2003.

Gharbi, N., et al. **(S. R. Wilson).** Chromatographic separation and identification of a water-soluble dendritic methano[60]fullerene octadeca-acid. *Anal. Chem.* 75(16):4217–22, 2003.

Wilson, S. R., et al. Synthesis and photophysics of a linear noncovalently linked porphyrin–fullerene dyad. *J. Chem. Soc., Chem. Commun.* 226–7, 2003.

Cui, Q., M. L. Wang, and **J. Z. H. Zhang.** Effect of entrance channel topology on reaction dynamics: $O(^3P)+CH_3 \rightarrow CH_3 + OH^{11}$. *Chem. Phys. Lett.* 410:115–9, 2005.

He, X., and **J. Z. H. Zhang.** A new method for direct calculation of total energy of protein. *J. Chem. Phys.* 122:031103, 2005.

X. He, et al. **(J. Z. H. Zhang).** Quantum computational analysis for drug resistance of HIV-1 reverse transcriptase to nevirapine through point mutations. *Proteins: Struct., Funct., Bioinformatics* 61:423–32, 2005.

Hu, P., and **Y. Zhang.** Catalytic mechanism and product specificity of the histone lysine methyltransferase SET7/9. An ab initio QM/MM-FE study with multiple initial structures. *J. Am. Chem. Soc.* 128:1272–8, 2006.

Zhang, Y. Improved pseudobonds for combined ab initio quantum mechanical/molecular mechanical (QM/MM) methods. *J. Chem. Phys.* 122:024114, 2005.

Northeastern
UNIVERSITY

NORTHEASTERN UNIVERSITY

Department of Chemistry and Chemical Biology
Graduate Programs in Chemistry

Programs of Study The Department of Chemistry and Chemical Biology offers thesis-based M.S. and Ph.D. degrees with research concentrations in a wide variety of experimental and theoretical fields, including bioanalysis, bioorganic and medicinal chemistry, biophysical chemistry, computational biology, nanomaterials, and proteomics. Those students who are unable to pursue their graduate degree full-time or who are employed full-time in industry may opt for the nonthesis M.S. program, for which the program of study is largely course work that can be completed on weekday evenings.

Research Facilities Research facilities in the Department of Chemistry and Chemical Biology are located in three buildings on the main Northeastern campus in Boston: Hurtig Hall (departmental headquarters); Mugar Hall, and the Egan Science and Engineering Center. The department is closely allied with the Barnett Institute of Chemical Analysis and the Center for Drug Discovery. The department is equipped with a wide variety of instrumentation: four NMR instruments (from 300 MHz to 700 MHz); extensive and variously interfaced mass spectrometry facilities, including several triple-quad mass spectrometers, ion trap and time-of-flight mass spectrometers, and several GC-MS instruments; a molecular modeling facility with Silicon Graphics Octane workstations; Raman and FT-IR spectometers, including confocal, imaging, and single point microspectrometers; TGA and DSC instruments; Mössbauer spectrometers; potentiostats, galvanostats, and other electrochemical equipment; AES and AAS spectrometers; a high resolution field emission SEM-EDAX; a peptide sequencer; UV-vis spectrometers and a steady-state spectrofluorometer; 9 and 220 GHz ESR spectrometers; and extensive bioanalytical facilities, including HPLC and CE systems, SDS-PAGE and IEF flat bed electrophoresis, SDS-CGE, CIEF, CITP, CEC/PEC/TLC, Elisa, RIA, immuno-CE, affinity-CE, and PCR systems.

Research in the areas of analytical chemistry, bioanalysis, and chemical biology includes separation science, with an emphasis on capillary electrophoresis and HPLC applied to biological systems; DNA sequencing, peptide mapping, and protein characterization; combinatorial chemistry and DNA biomarkers; LC- and CZE-mass spectrometry; microscale separations; spectroscopic and electrochemical studies of metalloproteins; cellular responses to genotoxic stress; physical/optical methods for medical diagnosis; mass spectrometric studies of protein conformation; and proteomic analysis of biological fluids and tissues.

Topics of research in organic and medicinal chemistry include the design, synthesis, and characterization of anticancer drugs; sequence-specific DNA cleaving and crosslinking agents; enantioselective catalysts; drug-membrane interactions; endocannabinoid systems in drug discovery; RNA analogs with nonphosphorus internucleoside linkages; design of enzyme inhibitors; selective derivatization of proteins; theoretical studies of structure and properties; dynamic NMR; and natural products.

Departmental research in physical, materials, and inorganic chemistry includes development of computational methods for identifying enzyme-active sites; solid-state chemistry, spectroscopy, and electrochemistry of materials for electrochemical energy conversion and storage (fuel cells and batteries); catalysts and catalyst supports for electrochemical and fuel-processing reactors; surface analysis; high-field ESR; spin-label studies of polymer and protein dynamics; synthesis and characterization of structurally engineered materials of nanosized dimensions possessing an array of unique properties; humic substances, synthesis of organometallic complexes as precursors for advanced materials, and Mössbauer spectrometry applied to solid-state chemistry and biological systems.

Financial Aid Full-time graduate students in the Department of Chemistry and Chemical Biology are ordinarily supported by a teaching or research assistantship that includes a stipend, tuition remission, and a discount on a health insurance plan. The stipend for 2006–07 is approximately $16,968 for the fall and spring semesters together. Including 2007 summer support, the minimum total stipend will be $22,300, or up to $25,534 for those with full summer assistantship support. These assistantships require a maximum of 20 hours of work per week and are only available for students studying in the full-time graduate program. A few higher-paying fellowships are also available. There are also a limited number of tuition assistantships that provide partial- or full-tuition remission and require 10 hours of work per week.

Cost of Study The tuition rate for 2006–07 is $930 per semester hour, but tuition is waived for students holding RAs or TAs. The annual health and accident insurance fee of $1915 that is required for all full-time students is reduced by 40 percent for TAs and RAs. Student fees are $128 per semester, and a one-time international student fee of $100 is charged.

Living and Housing Costs Most graduate students live in private apartments or share living expenses with roommates. On-campus housing for graduate students is very limited and is granted on a space-available basis. Students should visit http://www.housing.neu.edu for more information on housing options.

Student Group In 2005–06, 100 graduate students were enrolled in the Department, 62 on a full-time basis. Overall, more than 14,700 undergraduate and 4,800 graduate and professional students were enrolled at the University.

Student Outcomes The majority of graduates find employment in various high-tech industries across the United States, especially in biotech and pharmaceutical firms. Ph.D. graduates are also employed by academic institutions in teaching and research.

Location Northeastern University is set in the heart of the ultimate college town—Boston. A high-energy hub of cultural, educational, and social activity, Boston is home to more than 300,000 college students from around the country and the world. The city is alive with people of every race, ethnicity, political persuasion, and religion. Within walking distance of Northeastern are the world-renowned Museum of Fine Arts, Symphony Hall, and stylish Newbury Street, with great shopping and dining.

The University Northeastern University is a world leader in practice-oriented education and is recognized for its expert faculty members and first-rate academic and research facilities. Northeastern has six undergraduate colleges, eight graduate and professional schools, two part-time undergraduate divisions, and an extensive variety of research institutes and divisions. Northeastern's graduate programs offer both professional and research degrees at the master's or doctoral level.

Applying Applications for admission to the full-time M.S. and Ph.D. programs are only accepted for the fall semester each year. Review of all applications begins in early December. Admission with offers of financial support is offered on a rolling basis, so it is best to apply early.

Students applying to the Ph.D. program are expected to have an earned B.A., B.S., or M.S. in chemistry or to have completed equivalent course work with an overall GPA of 3.0 or better at an accredited college or university.

The application form is available online at http://www.applyweb.com/aw?neuga. Students must have a valid credit card in order to complete the online application. A traditional paper-based application may be requested from the department. Required materials include the application, fees, transcripts, and three letters of recommendation. Submission of GRE scores is strongly recommended. Transcripts of all previous undergraduate and graduate course work are required. Letters of recommendation should be completed by persons acquainted with the applicant's academic and personal qualifications.

Correspondence and Information

Department of Chemistry and Chemical Biology
102 Hurtig Hall
Northeastern University
360 Huntington Avenue
Boston, Massachusetts 02115

Phone: 617-373-2822
Fax: 617-373-8795
E-mail: chemistry-grad-info@neu.edu
Web site: http://www.chem.neu.edu

Northeastern University

THE FACULTY AND THEIR RESEARCH

Penny J. Beuning, Assistant Professor; Ph.D., Minnesota. Cellular response to genotoxic stress, Y family of DNA polymerases, clamp proteins.

David E. Budil, Associate Professor; Ph.D., Chicago. Electrostatic mapping of protein and polymer surfaces by ESR, molecular dynamics–based simulation of ESR spectra, high-field ESR studies of the photosynthetic reaction center.

Geoffrey Davies, Professor; Ph.D., D.Sc., Birmingham (England). Isolation, purification, properties, and structures of humic acids (HAs), the brown polymers responsible for water retention, metal binding, and solute adsorption in soils and sediments.

Max Diem, Professor; Ph.D., Toledo. Physical/optical methods for medical diagnosis, infrared and Raman microspectroscopy.

John R. Engen, Associate Professor; Ph.D., Nebraska. Protein conformation via mass spectroscopy and deuterium exchange, Src-family of tyrosine kinases, viral protein conformation.

David A. Forsyth, Professor and Graduate Coordinator; Ph.D., Berkeley. Organic structure and activity through computational modeling, multidimensional and dynamic NMR methods.

Bill C. Giessen, Professor; Sc.D., Göttingen (Germany). Econometric analysis of usage, supply, and pricing patterns of chemicals and materials using pattern recognition and other advanced computer methods.

Thomas R. Gilbert, Associate Professor; Ph.D., MIT. Chemical education, curriculum development in analytical chemistry.

William S. Hancock, Professor; Ph.D., Adelaide (Australia). Disease mechanisms and discovery of potential therapeutic agents by proteomic analysis of biological fluids and tissue samples.

Robert N. Hanson, Professor; Ph.D., Berkeley. Application of contemporary organic chemistry to the design, synthesis, and characterization of biologically active small molecules; synthetic techniques, including the utilization of palladium-catalyzed coupling of organoboranes and stannanes, introduction of isotopes for radio imaging, and solid-phase and solution combinatorial chemistry.

Graham Jones, Professor and Chair; Ph.D., Imperial College (London). Enediyne antitumor antibiotics, enantioselective catalyst design, enantioselective catalyst design, clinical chemistry.

Barry L. Karger, Professor; Ph.D., Cornell. High-performance DNA sequencing and mutational analysis using capillary electrophoresis, proteome analysis using capillary array MALDI/TOF MS, microfabricated devices coupled online to electrospray MS.

Rein U. Kirss, Associate Professor; Ph.D., Wisconsin–Madison. Application of organometallic compounds to organic and materials synthesis with characterization by NMR, electrochemistry, and Mössbauer spectroscopy.

Ira S. Krull, Associate Professor; Ph.D., NYU. Derivatization of phospholipids and proteins for improved separation, detection, and quantitation; HPCE for protein, peptide, and synthetic organic polymer resolutions with improved detection; immunodetection and immunoanalysis in HPLC/HPCE for proteins/peptides (prions).

Philip W. LeQuesne, Professor; Ph.D., Auckland (New Zealand). Chemistry of natural products, development of analogues of natural wound-healing compounds.

P. A. Mabrouk, Professor; Ph.D., MIT. Nonaqueous enzymology, green chemistry, electrochemistry in supercritical fluids and redox liquids, novel methods of synthesizing conducting polymers, chemical education.

Alexandros Makriyannis, Professor and Director for Drug Discovery; Ph.D., Kansas. Design and synthesis of cannainergic drugs, interactions of drugs with membranes.

Sanjeev Mukerjee, Associate Professor; Ph.D., Texas A&M. Design and testing of novel membranes for PEM fuel cells, electrosynthesis and sensors (such as immunoselective biosensors), materials characterization using synchrotron-based in situ spectroscopic methods.

Mary Jo Ondrechen, Professor; Ph.D., Northwestern. Theoretical and computational chemistry and chemical biology, prediction of the functional roles of gene products (proteins), modeling of enzyme-substrate interactions, rational design of materials, nonlinear optical properties of polymers, modeling for proteomics applications.

William M. Reiff, Professor; Ph.D., Syracuse. The application of Mössbauer spectroscopy to the characterization of the magnetic properties of the organometallic solid state.

Eriks Rozners, Assistant Professor; Ph.D., Riga Technical (Latvia). The chemistry and biochemistry of nucleic acids, with a focus on elucidation of RNA structure and function; synthesis and biophysical exploration of RNA analogs having amides as internucleoside linkages.

Eugene Smotkin, Professor; Ph.D., Texas at Austin. Catalysts and catalyst supports for electrochemical and fuel processing reactors, development of in-situ surface and bulk analysis.

Paul Vouros, Professor; Ph.D., MIT. Mass spectrometry and its applications to bioorganic analysis, analysis of DNA adducts through development of techniques for the trace-level detection and characterization of modified nucleosides or nucleotides as a result of the covalent binding of carcinogens with DNA, application of LC-MS and GC-MS to study the metabolic pathways of vitamin D, characterization and sequencing of oligosaccharides by ion trap MS.

Philip M. Warner, Professor; Ph.D., UCLA. Application of modern theoretical methods to the study of organic chemistry and materials, e.g., chemical behavior of enediyne antibiotics and study of the rearrangements of organometallic cyclobutadiene complexes.

Sunny Zhou, Associate Professor; Ph.D., Scripps. Design of enzyme inhibitors, proteomic analysis of protein post-translation modifications, selective derivatization of proteins.

PENNSTATE

THE PENNSYLVANIA STATE UNIVERSITY

Eberly College of Science
Department of Chemistry

Programs of Study

An integral part of the research community, graduate students in the Department of Chemistry interact frequently with faculty members, a distinguishing feature of the program. The program builds each student's academic career on a solid foundation of interdisciplinary research, an important facet of the chemical sciences. The Department offers a Ph.D. in chemistry. In the first year, each student selects a faculty adviser who heads a committee that guides the student's research development. Students may choose from a wide range of subject areas, including analytical, biological, inorganic, organic, physical, and theoretical chemistry and chemical physics. Courses in related fields, such as biochemistry, biophysics, environmental chemistry, materials, polymers, molecular dynamics, nanotechnology, and neuroscience may also be taken. The Ph.D. program requires six graduate-level courses, the first of which combines writing assignments with scientific-writing tutorials to sharpen the student's communication and critical-thinking skills. A second-year writing class focuses these skills on proposal writing and scientific creativity. To hone their public-speaking abilities, students present a Departmental seminar during their second year. Formal course work, which is tailored to a student's individual needs and special interests, normally takes about twelve to eighteen months to complete, after which students concentrate on research. Students must pass an oral comprehensive exam in which they describe their research progress and defend a brief original research proposal. The Ph.D. program culminates in the oral presentation and defense of a dissertation on the student's research.

The Department also offers students opportunities to interact with scientists outside the classroom. The colloquium program attracts highly renowned scientists to the campus to present their latest work, and the annual lecture series brings distinguished visitors to the campus from all over the world. The Cooperative Education Program links students directly to industry through on-site research experience in industrial laboratories.

Research Facilities

The Department of Chemistry provides a variety of resources and support services, maintaining a pleasant and modern physical environment. The Department office, a purchasing office, the stockroom, and six floors of research laboratories are located in the new Chemistry Building. The new building also houses the NMR and mass spectrometry facilities. Whitmore Laboratory houses the undergraduate laboratories and the glass-blowing shop. Wartik Laboratory houses the Biotechnology Institute and some of the biochemical research laboratories. The Materials Research Institute houses major instrumentation for materials research. Each research group has the extensive instrumentation required for modern research in chemistry. Some major instruments, however, are located in facilities that can be shared by all research groups in the Department. These include NMR, ESR, and mass spectrometry equipment; X-ray diffraction facilities; computational facilities; biologically related instrumentation; and equipment for time-resolved laser spectroscopy. The NMR and mass spectrometry laboratories have full-time professional operators, but graduate students who frequently use them are encouraged to learn to operate the instruments themselves.

Financial Aid

The Department offers many teaching and research assistantships, which usually require 20 hours of work a week. Because teaching is integral to the student's academic experience, teaching assistantships are offered to first- and second-year students. Students can obtain research assistantships after the spring of their first year, which offers them more time to focus on their research. Some fellowships are available, and they provide between $2000 and $4000 more than the usual stipend.

Cost of Study

All appointments include a grant-in-aid that covers tuition. Both fellowships and assistantships support virtually all chemistry graduate students. Medical insurance is available at a minimal cost. For more information, students should visit http://www.sa.psu.edu/uhs/currentstudents/gafellows.cfm.

Living and Housing Costs

Although most graduate students prefer off-campus housing (Web site: http://www.sa.psu.edu/ocl/), with both furnished and unfurnished accommodations available in and near State College (Web site: http://www.statecollege.com), the University offers accommodations for both single and married students (Web site: http://www.hfs.psu.edu/housing/).

Student Group

Currently, the Department has 240 full-time students. Each year, about 45 new students are admitted.

Student Outcomes

Penn State graduate students are highly successful in attaining employment in major industries and academic laboratories across the United States.

Location

Penn State sits near the tree-lined ridges of the Appalachians in State College, a friendly, modern city in a region with a population of 72,000. The area provides excellent outdoor recreational activities. Pittsburgh, Philadelphia, Baltimore, and Washington, D.C., lie within a reasonable drive (3 to 4 hours). Air, rail, and bus transportation is available.

The Department

Penn State's tradition of excellence in chemistry continues with the Department's current faculty members and students. One faculty member was recently elected to the National Academy of Sciences and 1 to the American Academy of Arts and Sciences. Graduate students have received such prestigious awards as the Nobel Signature Award of the ACS, the Henkel Award in Colloid and Surface Chemistry, and the Eli Lilly Award for Women Scientists.

Applying

Application requirements include a completed one-page chemistry application (which can be downloaded), a one-page Graduate School Application, a personal statement (minimum length, one page), three letters of recommendation, GRE General Test scores, an official undergraduate transcript, and an official graduate transcript, if applicable. The GRE Subject Test in chemistry is recommended but not required. International students must also submit their TOEFL and TSE score reports. No applications are considered for review until all materials have been submitted. Photocopies of the Educational Testing Service exams (GRE, TOEFL, TSE) suffice until the originals are received. The Admissions Committee does not review extra materials, such as publications or curricula vitae, enclosed with applications. U.S. students must apply by January 31 and international students by December 15. All application materials should be submitted to the address listed in the contact section.

Correspondence and Information

Dana Coval-Dinant
Graduate Admissions
Department of Chemistry
104 Chemistry Building
The Pennsylvania State University

Phone: 814-865-1383
877-688-4234 (toll-free)
Fax: 814-865-3226
E-mail: dmc6@psu.edu
Web site: http://www.chem.psu.edu

The Pennsylvania State University

THE FACULTY AND THEIR RESEARCH

David L. Allara, Professor of Chemistry and Professor of Materials Science; Ph.D., UCLA, 1964. Interface chemistry and technology. The effect of local environment on molecular conduction: Isolated molecule versus self-assembled monolayer. *Nanoletters* 5:61–5, 2005 (with Selzer et al.).

Harry R. Allcock, Evan Pugh Professor; Ph.D., London, 1956. Polymer chemistry and materials synthesis. Influence of reaction parameters on the living cationic polymerization of phosphoranimines to polyphosphazenes. *Macromolecules* 34:748–54, 2001 (with Reeves, de Denus, and Crane).

James B. Anderson, Evan Pugh Professor; Ph.D., Princeton, 1963. Quantum chemistry by Monte Carlo methods. Monte Carlo methods in electronic structure for large systems. *Ann. Rev. Phys. Chem.* 51:501–26, 2000 (with Luechow).

John B. Asbury, Assistant Professor; Ph.D., Emory, 2001. Ultrafast multidimensional vibrational spectroscopy of molecular transformations during chemical reaction, studies on mechansim of light-induced defect formation in hydrogenated amorphous silicon photovoltaics.

John V. Badding, Associate Professor; Ph.D., Berkeley, 1989. Solid-state and materials chemistry. Pressure tuning in the chemical search for improved thermoelectric materials: NdxCe3-xPt3Sb4. *Chem. Mater.* 12(1):197–201, 2000 (with Meng et al.).

Alan J. Benesi, Lecturer and Director, NMR Facility; Ph.D., Berkeley, 1975. Multinuclear NMR spectroscopy of liquids and solids. Multiple-rotor-cycle QPASS pulse sequences: Separation of quadrupolar spinning sidebands with an application to 139La NMR. *J. Magn. Reson.* 138:320–5, 1999 (with Aurentz, Vogt, and Mueller).

Stephen J. Benkovic, Evan Pugh Professor and Eberly Chair; Ph.D., Cornell, 1963. Chemistry related to biological systems. A perspective on enzyme catalysis. *Science* 301(5637):1196–202, 2003 (with Hammes-Schiffer).

Philip C. Bevilacqua, Associate Professor; Ph.D., Rochester, 1993. Biological chemistry. Mechanistic considerations for general acid-base catalysis by RNA: Revisiting the mechanism of the hairpin ribozyme. *Biochemistry* 42:2259–65, 2003.

A. Welford Castleman Jr., Evan Pugh Professor of Chemistry and Physics and Eberly Distinguished Chair in Science; Ph.D., Polytechnic of Brooklyn, 1969. Nanoscale science. Clusters: Structure, energetics, and dynamics of intermediate states of matter. *J. Phys. Chem.* 100:12911–44, 1996 (with K. H. Bowen Jr.).

Andrew G. Ewing, Professor of Chemistry and Neural and Behavioral Science and J. Lloyd Huck Chair in Natural Sciences; Ph.D., Indiana, 1983. Single-cell neurochemistry. Mass spectrometric imaging of highly curved membranes during *Tetrahymena* mating. *Science* 305:71–3, 2004 (with Ostrowski, VanBell, and Winograd).

Ken S. Feldman, Professor; Ph.D., Stanford, 1984. Allenyl azide cycloaddition chemistry. Synthesis of pyrrolidine-containing bicycles and tricycles via the possible intermediacy of azatrimethylenemethane species. *J. Am. Chem. Soc.* 127:4590–1, 2005 (with Iyer).

Raymond L. Funk, Professor; Ph.D., Berkeley, 1978. Total synthesis of natural products. Preparation of 2-alkyl- and 2-acylpropenals from 5-(trifluoromethanesulfonyloxy)-4H-1,3 dioxin: A versatile acrolein a-cation synthon. *Tetrahedron* 56:10275, 2000 (with Fearnley and Gregg).

Barbara J. Garrison, Shapiro Professor; Ph.D., Berkeley, 1975. Microscopic insights into sputtering of Ag{111} induced by C60 and Ga bombardment of Ag{111}. *J. Phys. Chem. B* 108:7831–8, 2004 (with Postawa et al.).

Michael T. Green, Assistant Professor; Ph.D., Chicago, 1998. Using a mixture of theory and experiment to investigate the factors that determine enzymatic reactivity. The structure and spin coupling of catalase compound I: A study of non-covalent effects. *J. Am. Chem. Soc.* 123:9218–9, 2001.

Sharon Hammes-Schiffer, Professor; Ph.D., Stanford, 1993. Effect of mutation on enzyme motion in dihydrofolate reductase, *J. Am. Chem. Soc.* 125:3745–50, 2003 (with Watney and Agarwal).

Christine D. Keating, Assistant Professor; Ph.D., Penn State, 1997. Cytomimetic chemistry. Aqueous phase separation in giant vesicles. *J. Am. Chem. Soc.* 124:13374–5, 2002 (with Helfrich et al.).

Juliette T. J. Lecomte, Associate Professor; Ph.D., Carnegie-Mellon, 1982. The solution structure of the recombinant hemoglobin from the cyanobacterium *Synechocystis* sp. PCC 6803 in its hemichrome state. *J. Mol. Biol.* 324:1015–29, 2002 (with Falzone, Vu, and Scott).

Thomas E. Mallouk, DuPont Professor of Materials Chemistry and Physics; Ph.D., Berkeley, 1983. Chemistry of nanoscale inorganic materials. Autonomous movement of striped nanorods. *J. Am. Chem. Soc.*. 126:13424–31, 2004 (with Paxton et al.).

Mark Maroncelli, Professor; Ph.D., Berkeley, 1984. Solvation and solvent effects on chemical reactions. Solvent dependence of the spectra and kinetics of the LE→CT reaction in three alkylaminobenzonitriles. *J. Phys. Chem. B* 109:1563–85, 2005.

Pshemak Maslak, Associate Professor; Ph.D., Kentucky, 1982. Physical organic chemistry of organic molecular materials and devices. On the molecular and electronic structure of spiro-ketones and half-molecule models. *J. Phys. Chem. A* 106:10622–9, 2002 (with Galasso et al.).

Karl T. Mueller, Associate Professor; Ph.D., Berkeley, 1991. New experimental techniques in NMR spectroscopy for the study of solids. F-19 MAS NMR quantification of accessible hydroxyl sites on fiberglass surfaces. *J. Am. Chem. Soc.* 125:2378–9, 2003 (with Fry, Tsomaia, and Pantano).

Blake R. Peterson, Associate Professor; Ph.D., UCLA, 1994. Bioorganic chemistry, medicinal chemistry, and chemical biology. Synthetic mimics of small mammalian cell surface receptors. *J. Am. Chem. Soc.* 126:16379–86, 2004 (with Boonyarattanakalin et al.).

Ayusman Sen, Professor and Department Head; Ph.D., Chicago, 1978. Synthetic and mechanistic organotransition metal chemistry, homogeneous and heterogeneous catalysis, environmental chemistry, polymer chemistry, nanotechnology. Living/controlled copolymerization of acrylates with non-activated alkenes. *J. Polym. Sci., A: Polym. Chem.* 42:6175, 2004 (with Liu).

Erin D. Sheets, Assistant Professor; Ph.D., North Carolina at Chapel Hill, 1997. Single molecule dynamics in living cells and development of nanofabricated tools for cell biology. Quantitative analysis of the fluorescence properties of intrinsically fluorescent proteins in living cells. *Biophys. J.* 85:2566–80, 2003 (with Hess, Wagenknecht-Wiesner, and Heikal).

Steven M. Weinreb, Russell and Mildred Marker Professor of Natural Products Chemistry; Ph.D., Rochester, 1967. Synthesis of natural products. A new enantioselective approach to total synthesis of the securinega alkaloids: Application to (-)-norsecurinine and phyllanthine. *Angew. Chem. Int. Ed.* 39:237, 2000 (with Han et al.).

Paul S. Weiss, Professor; Ph.D., Berkeley, 1986. Surface chemistry and physics. Molecular rulers for scaling down nanostructures. *Science* 291:1019, 2001 (with Hatzor).

Mary Elizabeth Williams, Assistant Professor; Ph.D., North Carolina at Chapel Hill, 1999. Analytical chemistry of hybrid inorganic/organic materials. Artificial oligopeptide scaffolds for stoichiometric metal binding. *J. Am. Chem. Soc.*, in press (with Gilmartin et al.).

Nicholas Winograd, Evan Pugh Professor; Ph.D., Case Western Reserve, 1970. Surface chemistry studies with ion beams and femtosecond pulse lasers. The interaction of vapor-deposited Al atoms with CO_2H groups at the surface of a self-assembled alkanethiolate monolayer on gold. *J. Phys. Chem.* 104(14):3267, 2000 (with Fisher et al.).

Xumu Zhang, Professor; Ph.D., Stanford, 1992. Synthetic organic, inorganic, and organometallic chemistry. Rh-catalyzed enzyme cycloisomerization. *J. Am. Chem. Soc.* 122:6490, 2000 (with Cao and Wang).

PRINCETON UNIVERSITY

Department of Chemistry

Programs of Study

The Department of Chemistry offers a program of study leading to the degree of Doctor of Philosophy. The graduate program emphasizes research, and students enter a research group by the end of the first semester. Students are required to take six graduate courses in chemistry and allied areas, satisfying at least four of ten areas of distribution, and are expected to participate in the active lecture and seminar programs throughout their graduate careers.

Early in the second year, students take a general examination that consists of an oral defense of a thesis-related subject. Upon satisfactory performance in the general examination, students advance to candidacy for the degree of Doctor of Philosophy in chemistry. The degree is awarded primarily on the basis of a thesis describing original research in one of the areas of chemistry. The normal length of the entire Ph.D. program is four to five years.

Programs of graduate study in neuroscience, molecular biophysics, and materials are also offered, in cooperation with other science departments at Princeton University.

Research Facilities

Research is conducted in Frick Chemical Laboratory and the adjoining Hoyt Laboratory. NMR, IR, ESR, FT-IR, atomic absorption, UV-visible, CD, and vacuum UV spectrometers and departmental computers are available to students. In addition, high-resolution FT-NMR spectrometers and mass spectrometers are run by operators for any research group. There is a wide variety of equipment in individual research groups, including lasers of many kinds, high-resolution spectrographs, molecular-beam instrumentation, a microwave spectrometer, computers, gas chromatographs, and ultrahigh-vacuum systems for surface studies. The department has an electronics shop, a machine shop, a student machine shop, and a glassblower for designing and building equipment. Extensive shop facilities are available on campus, and there is a large supercomputing center.

Financial Aid

All admitted students receive tuition plus a maintenance allowance, typically in the form of assistantships in instruction or research. In 2004–05, students earned $18,200 to $21,050 during the academic year for approximately 20 hours per week of work, plus a summer stipend of $4500. First-year graduate students are not required to work; they receive fellowship funds that allow them to concentrate on course work.

Cost of Study

See Financial Aid section.

Living and Housing Costs

Rooms at the Graduate College cost from $2894 to $5289 for the 2004–05 academic year of thirty-five weeks. Several meal plans are available, priced from $2813 to $4282. University apartments for married students currently rent for $626 to $1452 per month. Accommodations are also available in the surrounding community.

Student Group

The total number of graduate students in chemistry is currently about 130. Postdoctoral students number about 70. A wide variety of academic, ethnic, and national backgrounds are represented among these students.

Location

Princeton University and the surrounding community together provide an ideal environment for learning and research. From the point of view of a chemist at the University, the engineering, physics, mathematics, and molecular biology departments, as well as the plasma physics lab on the Forrestal campus, provide valuable associates, supplementary facilities, and sources of special knowledge. Many corporations have located their research laboratories near Princeton, leading to fruitful collaborations, seminars and lecture series, and employment opportunities after graduation.

Because of the nature of the institutions located here, the small community of Princeton has a very high proportion of professional people. To satisfy the needs of this unusual community, the intellectual and cultural activities approach the number and variety ordinarily found only in large cities, but with the advantage that everything is within walking distance. There are many film series, a resident repertory theater, orchestras, ballet, and chamber music and choral groups. Scientific seminars and other symposia bring prominent visitors from every field of endeavor.

Princeton's picturesque and rural countryside provides a pleasant area for work and recreation, yet New York City and Philadelphia are each only about an hour away.

The University

Princeton University was founded in 1746 as the College of New Jersey. At its 150th anniversary in 1896, the trustees changed the name to Princeton University. The Graduate School was organized in 1901 and has since won international recognition in mathematics, the natural sciences, philosophy, and the humanities.

Applying

Application instructions, including an online application, are available at http://gso.princeton.edu/admission/e2/index.html. The application deadlines are December 1 for applicants who currently do not reside in the United States, Canada, or Mexico and December 30 for those who do. All applications must be accompanied by an application fee, which is discounted for online and early applications.

Admission consideration is given to all candidates without regard to race, color, national origin, religion, sex, or handicap.

Correspondence and Information

Sallie Dunner
Graduate Administrator
Department of Chemistry
Frick Laboratory
Princeton University
Washington Road
Princeton, New Jersey 08544
Phone: 609-258-5502
E-mail: sdunner@princeton.edu
Web site: http://www.princeton.edu/~chemdept

Princeton University

THE FACULTY AND THEIR RESEARCH

Although all major areas of chemistry are represented by the faculty members, the department is small and housed in Frick Chemical Laboratory and the adjoining Hoyt Laboratory so that fruitful collaborations may develop. The following list briefly indicates the areas of interest of each professor.

S. L. Bernasek. Chemical physics of surfaces: basic studies of chemisorption and reaction on well-characterized transition-metal surfaces using electron diffraction and electron spectroscopy, surface reaction dynamics, heterogeneous catalysis, electronic materials.
S. Bernhard. Synthesis and characterization of transition metal complex–based materials for optoelectronics (light-emitting devices, photovoltaics), redox polymers, chiral metal complexes.
A. B. Bocarsly. Inorganic photochemistry, photoelectrochemistry, chemically modified electrodes, electrocatalysis, sensors, applications to solar energy conversion, fuel cells, materials chemistry.
R. Car. Chemical physics and materials science; electronic structure theory and ab initio molecular dynamics; computer modeling and simulation of solids, liquids, disordered systems, and molecular structures; structural phase transitions and chemical reactions.
J. Carey. Biophysical chemistry: protein and nucleic acid structure, function, and interactions; protein folding and stability.
R. Cava. Synthesis, structure, and physical property characterization of new superconductors, magnetic materials, transparent electronic conductors, dielectrics, thermoelectrics, and correlated electron systems.
G. C. Dismukes. Biochemistry and inorganic chemistry, molecular evolution of metalloenzymes, catalysis, photosynthetic water splitting.
J. T. Groves. Organic and inorganic chemistry: synthetic and mechanistic studies of reactions of biological interest, metalloenzymes, siderophores, membrane self-assembly, transition-metal redox catalysis, asymmetric catalysis, and biological oxidations.
M. Hecht. Biochemistry: sequence determinants of protein structure and design of novel proteins, molecular causes of Alzheimer's disease.
M. Jones Jr. Reactions and spin states of carbenes, arynes, twisted p systems, and other reactive intermediates; carborane chemistry.
C. Lee. Organic chemistry: synthetic organic chemistry, organometallic chemistry, bioorganic chemistry, synthetic methodologies, asymmetric catalysis.
K. K. Lehmann. Experimental and theoretical molecular spectroscopy, including spectroscopy in quantum liquids and solids, intramolecular dynamics, and development of spectroscopic tools for trace-species detection, particularly using cavity-enhanced methods.
R. A. Pascal Jr. Aromatic compounds of unusual structure, organic materials chemistry, enzymatic reaction mechanisms and inhibitors.
J. Rabinowitz. Biophysical chemistry: biochemical kinetics, cellular metabolism, chemical basis of complex biological processes.
H. Rabitz. Physical chemistry: atomic and molecular collisions, theory of chemical reactions and chemical kinetics, biodynamics phenomena, heterogeneous phenomena, control of molecular motion.
W. Richter. Magnetic resonance imaging (MRI) and in vivo NMR spectroscopy, investigation of brain function in children and adults during mental tasks, physiological correlates of learning and intelligence, origin and interpretation of the BOLD signal, integration of complementary methodologies (electroencephalography, optical imaging, and NMR spectroscopy), analysis of 4-dimensional imaging data by wavelet methods and independent component analysis (ICA).
C. E. Schutt. Structural biology, crystallography of actin-binding proteins, theories of muscle contraction, structure of viral proteins, structural neurobiology, architectonics.
J. Schwartz. Organometallic chemistry, applications to organic and materials synthesis, surface and interface organometallic chemistry.
G. Scoles. Chemical physics: laser spectroscopy; chemical dynamics and cluster studies with molecular beams; structure, dynamics, and spectroscopic properties of organic overlayers adsorbed on crystal surfaces; nanomaterials and molecular manipulation on surfaces; protein-surface interactions; biosensors.
A. Selloni. Computational physics and chemistry and modeling of materials; structural, electronic, and dynamic properties of semiconductor and oxide surfaces; chemisorption and surface reactions.
M. F. Semmelhack. Organic synthesis and organometallic chemistry: development of synthesis methodology with transition-metal reagents, synthesis of unusual ring systems in natural and unnatural molecules, functional models of the enediyne toxins.
Z. G. Soos. Chemical physics: electronic states of π-molecular crystals and conjugated polymers, paramagnetic and charge-transfer excitons, linear chain crystals, energy transfer, electronic polarization and hopping transport.
E. J. Sorensen. Organic chemistry: chemical synthesis of bioactive natural products and molecular probes for biological research, bioinspired strategies for chemical synthesis, architectural self-constructions and novel methods for synthesis.
T. G. Spiro. Biological structure and dynamics from spectroscopic probes, role of metals in biology, environmental chemistry.
E. I. Stiefel. Bioinorganic chemistry; the metabolism of iron in marine environments; ferritin and bacterioferritin; hydrogenase, nitrogenase, and molybdenum enzymes; synthetic inorganic chemistry for biological models and new technology.
S. Torquato. Statistical mechanics and materials science; theory and computer simulations of disordered heterogeneous materials, liquids, amorphous solids, and biological materials; optimization in materials science; simulations of peptide binding; modeling the growth of tumors.
W. S. Warren. Laser spectroscopy in gaseous and solid phases, coherence effects, multiphoton processes, nuclear magnetic resonance.

Associated Faculty
F. M. Hughson, Department of Molecular Biology. Structural biology of neurotransmitter release, intracellular trafficking, and bacterial quorum sensing.
F. M. M. Morel, Department of Geosciences. Environmental chemistry, metals and metalloproteins in the environment, biogeochemistry.
S. M. Myneni, Department of Geosciences. Environmental chemistry, ion hydration and complexation, interfacial chemistry, X-ray spectroscopy.
Y. Shi, Department of Molecular Biology. Biophysics and structural biology, biochemistry and cancer biology.
J. Stock, Department of Molecular Biology. Receptor-mediated signal transduction, role of protein methylation in degenerative diseases such as Alzheimer's.

Frick Laboratory, Princeton University.

SELECTED PUBLICATIONS

Cai, Y., and **S. L. Bernasek.** Scanning tunneling microscopy of chiral pair self assembled monolayers. In *Encyclopedia of Nanoscience and Nanotechnology,* p. 3305. New York: Marcel Dekker, 2004.

Cai, Y., and **S. L. Bernasek.** Chiral pair monolayer adsorption of iodine-substituted octadecanol molecules on graphite. *J. Am. Chem. Soc.* 125:1655, 2003.

Slinker, J. D., et al. **(S. Bernhard).** Efficient yellow electroluminescence from a single layer of a cyclometalated iridium complex. *J. Am. Chem. Soc.* 126(9):2763–7, 2004.

Slinker, J. D., et al. **(S. Bernhard).** Solid-state electroluminescent devices based on transition metal complexes. *Chem. Commun.* 19:2392–9, 2003.

Barron, J. A., et al. **(S. Bernhard).** Electroluminescence in ruthenium(Ii) dendrimers. *J. Phys. Chem. A* 107(40):8130–3, 2003.

Bernhard, S., J. I. Goldsmith, K. Takada, and H. D. Abruna. Iron(Ii) and copper(I) coordination polymers: Electrochromic materials with and without chiroptical properties. *Inorg. Chem.* 42(14):4389–93, 2003.

Barron, J. A., et al. **(S. Bernhard).** Photophysics and redox behavior of chiral transition metal polymers. *Inorg. Chem.* 42(5):1448–55, 2003.

Zhu, S., and **A. B. Bocarsly.** Spin-coated cyanogels: A new approach to the synthesis of nanoscopic metal alloy particles. In *Encyclopedia of Nanoscience and Nanotechnology,* pp. 3667–74. New York: Marcel Dekker, 2004.

Watson, D. F., et al. **(A. B. Bocarsly).** Femtosecond pump-probe spectroscopy of trinuclear transition metal mixed-valence complexes. *J. Phys. Chem. A* 108:3261–7, 2004.

Yang, C., et al. **(A. B. Bocarsly).** A comparison of physical properties and fuel cell performance of nafion and zirconium phosphate/nafion composite membranes. *J. Membrane Sci.* 237:145–61, 2004.

Haataja, M., D. J. Srolovitz, and **A. B. Bocarsly.** Morphological stability during electrodeposition I: Steady states and stability analysis. *J. Electrochem. Soc.* 150 (10):C699–707, 2003.

Haataja, M., D. J. Srolovitz, and **A. B. Bocarsly.** Morphological stability during electrodeposition II: Additive effects. *J. Electrochem. Soc.* 150(10):C708–16, 2003.

Deshpande, R. S., et al. **(A. B. Bocarsly).** Morphology and gas adsorption properties of the palladium-cobalt based cyanogels. *Chem. Mater.* 15:4239–46, 2003.

Giannozzi, P., **R. Car,** and **G. Scoles.** Oxygen adsorption on graphite and nanotubes. *J. Chem. Phys.* 118:1003, 2003.

Han, S., et al. **(R. Car).** Interatomic potential for vanadium suitable for radiation damage simulations. *J. Appl. Phys.* 93:3328, 2003.

Zepeda-Ruiz, L. A., et al. **(R. Car).** Molecular dynamics study of the threshold displacement energy in vanadium. *Phys. Rev. B* 67:134114, 2003.

Savage, T., et al. **(R. Car).** Photoinduced oxydation of carbon nanotubes. *J. Phys.: Condens. Matter* 15:5915, 2003.

Sharma, M., Y. D. Wu, and **R. Car.** Ab initio molecular dynamics with maximally localized Wannier functions. *Int. J. Quantum Chem.* 95:821, 2003.

Lawson, C. L., et al. **(J. Carey).** *E. coli* trp repressor forms a domain-swapped array in aqueous alcohol. *Structure* 12:1098–107, 2004.

Szwajkajzer, D., and **J. Carey.** RaPID plots: Affinity and mechanism at a glance. *Biacore J.* 3(1):19, 2003.

McWhirter, A., and **J. Carey.** T-cell receptor recognition of peptide antigen/MHC complexes. *Biacore J.* 3(1):18, 2003.

Cava, R. J., H. W. Zandbergen, and K. Inumaru. The substitutional chemistry of MgB_2. *Physica C* 385:8, 2003.

Yu, A., et al. **(R. J. Cava).** Observation of a low-symmetry crystal structure for superconducting $MgCNi_3$ by Ni K-edge X-ray absorption measurements. *Phys. Rev. B* 67:064509, 2003.

Foo, M. L., et al. **(R. J. Cava).** Chemical instability of the cobalt oxyhydrate superconductor under ambient conditions. *Solid State Commun.* 127:33, 2003.

Mao, Z. Q., et al. **(R. J. Cava).** Experimental determination of superconducting parameters for the intermetallic perovskite superconductor $MgCNi_3$. *Phys. Rev. B* 67:094502, 2003.

Khalifah, P., D. A. Huse, and **R. J. Cava.** Magnetic behavior of $La_7Ru_3O_{18}$. *J. Phys.: Condens. Matter* 15:1, 2003.

Rogado, N., et al. **(R. J. Cava).** β-$Cu_3V_2O_8$: Magnetic ordering in a spin-1/2 Kagome-staircase lattice. *J. Phys.: Condens. Matter* 15:907, 2003.

Baranov, S., et al. **(G. C. Dismukes).** Bicarbonate is a native cofactor for assembly of the manganese cluster of the photosynthetic water oxidizing complex: Kinetics of reconstitution of O_2 evolution by photoactivation. *Biochemistry* 43:2070–9, 2004.

Maneiro, M., et al. **(G. C. Dismukes).** Kinetics of proton-coupled electron transfer in the manganese-oxo "cubane" core complexes containing the $[Mn_4O_4]^{6+}$ and $[Mn_4O_4]^{7+}$ core types. *Proc. Nat. Acad. Sci. U.S.A.* 100(7):3703–12, 2003.

Carrell, T. G., E. Bourles, M. Lin, and **G. C. Dismukes.** Transition from hydrogen atom to hydride abstraction by manganese-oxo cubanes: Contrasting chemistry of $Mn_4O_4(O_2PPh_2)_6$ vs. $[Mn_4O_4(O_2PPh_2)_6]^+$ in the formation of $Mn_4O_3(OH)(O_2PPh_2)_6$. *Inorg. Chem.* 42:2849-58, 2003.

Sanford, M. S., and **J. T. Groves.** Anti-Markovnikov hydrofunctionalization of olefins mediated by rhodium porphyrins. *Angew. Chim.* 116:598–600, 2004.

Puranik, M., et al. **(J. T. Groves** and **T. G. Spiro).** Dynamics of carbon monoxide binding to CooA. *J. Biol. Chem.* 279:21096–108, 2004.

Phillips-McNaughton, K., and **J. T. Groves.** Zinc-coordination oligomers of phenanthrolinylporphyrins. *Org. Lett.* 5(11):1829–32, 2003.

Naidu, B. V., et al. **(J. T. Groves).** Enhanced peroxynitrite decomposition protects against experimental obliterative bronchiolitis. *Exp. Mol. Pathol.* 75(1):12–7, 2003.

Pacher, P., et al. **(J. T. Groves).** Role of superoxide, NO and peroxynitrite in doxorubicin-induced cardiac dysfunction. *FASEB J.* 17(4):A229, Part 1 Suppl., 2003.

Soriano, F. G., et al. **(J. T. Groves).** Role of peroxynitrite in the pathogenesis of endotoxic and septic shock in rodents. *FASEB J.* 17(5):A1069–70, Part 2 Suppl., 2003.

Wei, Y., and **M. H. Hecht.** Enzyme-like proteins from an unselected library of designed amino acid sequences. *Protein Eng. Design Selection* 17:67–75, 2004.

Wei, Y., et al. **(M. H. Hecht).** Stably folded de novo proteins from a designed combinatorial library. *Protein Sci.* 12:92–102, 2003.

Moffet, D. A., J. Foley, and **M. H. Hecht.** Midpoint reduction potentials and heme binding stoichiometries of de novo proteins from designed combinatorial libraries. *Biophys. Chem.* 105:231–9, 2003.

Wei, Y., et al. **(M. H. Hecht).** Solution structure of a de novo protein from a designed combinatorial library. *Proc. Natl. Acad. Sci. U.S.A.* 100:13270–3, 2003.

Oka, T., D. Ungar, **F. M. Hughson,** and M. Krieger. The COG and COPI complexes interact to control the abundance of GEARs: A subset of golgi integral membrane proteins. *Mol. Biol. Cell* 15:2423–35, 2004.

Ungar, D., and **F. M. Hughson.** SNARE protein structure and function [review]. *Annu. Rev. Cell Dev. Biol.* 19:493–517, 2003.

Kim, H., H. Men, and **C. Lee.** Stereoselective palladium-catalyzed O-glycosylation using glycals. *J. Am. Chem. Soc.* 126:1336–7, 2004.

Çarçabal, P., R. Schmied, **K. K. Lehmann,** and **G. Scoles.** Helium nanodroplet isolation spectroscopy of perylene and its complexes with molecular oxygen. *J. Chem. Phys.* 120:6792–3, 2004.

Callegari, A., et al. **(K. K. Lehmann** and **G. Scoles).** Intramolecular vibrational relaxation in aromatic molecules II: An experimental and computational study of pyrrole and triazine. *Mol. Phys.* 101:551–68, 2003.

Shaked, Y., A. B. Kustka, **F. M. M. Morel,** and Y. Erel. Simultaneous determination of iron reduction and uptake by phytoplankton. *Limnol. Oceanogr.: Methods* 2:137–45, 2004.

Kilway, K. V., et al. **(R. A. Pascal Jr.).** Unexpected conversion of a polycyclic thiophene to a macrocyclic anhydride. *Tetrahedron* 60:2433–8, 2004.

Shen, X. F., D. M. Ho, and **R. A. Pascal Jr.** Synthesis of polyphenylene dendrimers related to "cubic graphite". *J. Am. Chem. Soc.* 126:5798–805, 2004.

Shen, X., D. M. Ho, and **R. A. Pascal Jr.** Synthesis and structure of a polyphenylene macrocycle related to "cubic graphite". *Org. Lett.* 5:369–71, 2003.

Feng, X.-J., and **H. Rabitz.** Optimal identification of biochemical reaction networks. *Biophys. J.* 86:1270–81, 2004.

Alis, O. F., et al. **(H. Rabitz).** On the inversion of quantum mechanical systems: Determining the amount and type of data for a unique solution. *J. Math. Chem.* 35:65–78, 2004.

Turinici, G., V. Ramakrishna, B. Li, and **H. Rabitz.** Optimal discrimination of multiple quantum systems: Controllability analysis. *J. Phys. A* 37:273–82, 2004.

Rabitz, H., M. Hsieh, and C. Rosenthal. Quantum optimally controlled transition landscapes. *Science* 303:998, 2004.

Cheong, B.-S., and **H. Rabitz.** Revealing the roles of Hamiltonian matrix coupling in bound state quantum systems. *J. Chem. Phys.* 120:6874, 2004.

Xu, R., et al. **(H. Rabitz).** Optimal control of quantum non-Markovian dissipation: Reduced Liouville-space theory. *J. Chem. Phys.* 120:6600, 2004.

Nyåkern-Meazza, M., K. Narayan, **C. E. Schutt,** and U. Lindberg. Tropomyosin and gelsolin cooperate in controlling the microfilament system. *J. Biol. Chem.* 277:28774–9, 2003.

Danahy, M. P., et al. **(J. Schwartz).** Self-assembled monolayers of α,ω-diphosphonic acids on Ti enable complete or spatially controlled surface derivatization. *Langmuir* 20, 2004 (online).

Midwood, K. M., et al. **(J. Schwartz).** Rapid and efficient bonding of biomolecules to the native oxide surface of silicon. *Langmuir* 20, 2004 (online).

Gawalt, E. S., et al. **(J. Schwartz).** Bonding organics to Ti alloys: Facilitating human osteoblast attachment and spreading on surgical implant materials. *Langmuir* 19:200–4, 2003.

Schwartz, J., et al. Cell attachment and spreading on metal implant materials. *Mater. Sci. Eng. C* 23:395–400, 2003.

Schwartz, J., et al. **(S. L. Bernasek).** Controlling the work function of indium tin oxide: Differentiating dipolar from local surface effects. II. *Synth. Met.* 138:223–7, 2003.

Koch, N., C. Chan, A. Kahn, and **J. Schwartz.** Lack of thermodynamic equilibrium in conjugated organic molecular thin films. *Phys. Rev. B* 67:195330, 2003.

Suo, Z., Y. F. Gao, and **G. Scoles.** Nanoscale domain stability in organic monolayers on metals. *J. Appl. Mech.* 71:24–31, 2004.

Princeton University

Case, M., et al. **(G. Scoles).** Using nanografting to achieve directed assembly of de novo designed metalloproteins on gold. *Nano Lett.* 3(4):425–9, 2003.

Casalis, L., et al. **(G. Scoles).** Hyperthermal molecular beam deposition of highly ordered organic thin films. *Phys. Rev. Lett.* 90(20):206101-1–4, 2003.

Bracco, G., and **G. Scoles.** Study of the interaction potential between a He atom and a SAM of decanethiols *J. Chem. Phys.* 119:6277, 2003.

Schreiber, F., M. G. Gerstenberg, H. Dosch, and **G. Scoles.** Melting point enhancement of a SAM induced by a van der Waals bound capping layer. *Langmuir* 19:10004–6, 2003.

Ruiz, R., et al. **(G. Scoles).** Dynamic scaling, island size distribution and morphology in the aggregation regime of sub-monolayer pentacene films. *Phys. Rev. Lett.* 91: 136102, 2003.

De Renzi, V., et al. **(A. Selloni).** Ordered (3x4) high-density phase of methylthiolate on Au(111). *J. Phys. Chem. B* 108:16, 2004.

De Angelis, F., S. Fantacci, and **A. Selloni.** Time-dependent density functional theory study of the absorption spectrum of [Ru(4,4'-COOH-2,2'-bpy)_2(NCS)_2] in water solution: Influence of the pH. *Chem. Phys. Lett.* 389:204, 2004.

Vittadini, A., and **A. Selloni.** Periodic density functional theory studies of vanadia-titania catalysts: Structure and stability of the oxidized monolayer. *J. Phys. Chem. B* 108:7337, 2004.

Tilocca, A., and **A. Selloni.** Structure and reactivity of water layers on defect-free and defective anatase TiO_2(101) surfaces. *J. Phys. Chem. B* 108:4743, 2004.

Di Felice, R., and **A. Selloni.** Adsorption modes of cysteine on Au(111): Thiolate, amino-thiolate, disulfide. *J. Chem. Phys.* 120:4906, 2004 (also in *Virtual J. Biol. Phys. Res.*, 2004).

Wu, X., **A. Selloni,** and S. K. Nayak. First principles study of CO oxidation on TiO_2(110): The role of surface oxygen vacancies. *J. Chem. Phys.* 120:4512, 2004.

Semmelhack, M. F., and R. J. Hooley. Palladium-catalyzed hydrostannylations of highly hindered acetyolenes in hexane. *Tetrahedron Lett.* 44:5737–9, 2003.

Semmelhack, M. F., L. Wu, **R. A. Pascal Jr.,** and D. M. Ho. Conformational control in activation of an enediyne. *J. Am. Chem. Soc.* 125:10496–7, 2003.

Shiozaki, E. N., L. Gu, N. Yan, and **Y. Shi.** Structure of the BRCT repeats of BRCA1 bound to a BACH1 phosphopeptide: Implications for signaling. *Mol. Cell* 14:405–12, 2004.

Frederick, J. P., et al. **(Y. Shi).** Transforming growth factor β-mediated transcriptional repression of c-myc is dependent on direct binding of Smad3 to a novel repressive Smad binding element. *Mol. Cell. Biol.* 24(6):2546–59, 2004.

Huh, J. R., et al. **(Y. Shi).** Multiple apoptotic caspase cascades are required in non-apoptotic roles for *Drosophila spermatid* individualization. *PLoS Biol.* 2:43–53, 2004.

Soos, Z. G., and E. V. Tsiper. Polarization energies, transport gap and charge transfer states of organic molecular crystals. *Macromol. Symp.* 212:1–12, 2004.

Girlando, A., A. Painelli, S. A. Bewick, and **Z. G. Soos.** Charge fluctuations and electron-phonon coupling in organic charge-transfer salts with neutral-ionic and Peierls transitions. *Synth. Met.* 141:129–38, 2004.

Soos, Z. G., S. A. Bewick, A. Peri, and A. Painelli. Dielectric response of modified Hubbard models with neutral-ionic and Peierls transitions. *J. Chem. Phys.* 120: 6712, 2004.

Sin, J. M., and **Z. G. Soos.** Hopping transport in molecularly doped polymers: Positional disorder, donor packing and rate law. *Recent Res. Dev. Chem. Phys.* 3: 563–83, 2003.

Massino, M., A. Girlando, and **Z. G. Soos.** Evidence for a soft mode in the temperature induced neutral-ionic transition of TTF-CA. *Chem. Phys. Lett.* 369:428–33, 2003.

Tsiper, E. V., and **Z. G. Soos.** Electronic polarization in pentacene crystals and thin films. *Phys. Rev. B* 68:085301, 2003.

Anderson, E. A., E. J. Alexanian, and **E. J. Sorensen.** A synthesis of the furanosteroidal antibiotic viridian. *Angew. Chem. Int. Ed.* 43:1998–2001, 2004.

Adam, G. C., C. D. Vanderwal, **E. J. Sorensen,** and B. F. Cravatt. (-)-FR182877 is a potent and selective inhibitor of carboxylesterase-1. *Angew. Chem. Int. Ed.* 42:5480–4, 2003.

Vosburg, D. A., S. Weiler, and **E. J. Sorensen.** Concise stereocontrolled routes to fumagillol, fumagillin, and TNP-470. *Chirality* 15:156–66, 2003.

Vanderwal, C. D., D. A. Vosburg, S. Weiler, and **E. J. Sorensen.** An enantioselective synthesis of FR182877 provides a chemical rationalization of its structure and affords multigram quantities of its direct precursor. *J. Am. Chem. Soc.* 125:5393–407, 2003.

Sorensen, E. J. Architectural self-construction in nature and chemical synthesis. *Bioorg. Med. Chem.* 11:3225–8, 2003.

Tamiya, J., and **E. J. Sorensen.** A spontaneous bicyclization facilitates a synthesis of (-)-hispidospermidin. *Tetrahedron* 59:6921–32, 2003.

Jarzecki, A. A., A. D. Anbar, and **T. G. Spiro.** DFT analysis of $Fe(H_2O)_6^{3+}$ and $Fe(H_2O)_6^{2+}$ structure and vibrations: Implications for isotope fractionation. *J. Phys. Chem. A* 108:2726–32, 2004.

Coyle, C. M., et al. **(T. G. Spiro).** Activation mechanism of the CO sensor CooA: Mutational and resonance Raman spectroscopic studies. *J. Biol. Chem.* 278:35384–93, 2003.

Wang, Y., and **T. G. Spiro.** Vibrational and electronic couplings in ultraviolet resonance Raman spectra of cyclic peptides. *Biophys. Chem.* 105:461–70, 2003.

Stiefel, E. I., ed. Dithiolene chemistry: Synthesis, properties, and applications. In *Progress in Inorganic Chemistry,* vol. 52, pp. 1–738. John Wiley and Sons, 2004.

Beswick, C., J. M. Schulman, and **E. I. Stiefel.** Structures and structural trends in homoleptic dithiolene complexes. In dithiolene chemistry: synthesis, properties and applications. In *Progress in Inorganic Chemistry,* vol. 52, pp. 55–110. John Wiley and Sons, 2004.

Horvath, I., et al. **(E. I. Stiefel),** eds. *Encyclopedia of Catalysis.* Hoboken, N.J.: John Wiley and Sons, 2003.

Wolanin, P. M., and **J. B. Stock.** Transmembrane signaling and the regulation of histidine kinase activity. In *Histidine Kinases in Signal Transduction,* pp. 73–122, eds. M. Inouye and R. Dutta. New York: Academic Press, 2003.

Webre, D., P. M. Wolanin, and **J. B. Stock.** Bacterial chemotaxis. *Curr. Biol.* 13(2):R47–9, 2003.

Park, S., et al. **(J. B. Stock).** From the cover: Influence of topology on bacterial social interaction. *Proc. Natl. Acad. Sci. U.S.A.* 100:13910–5, 2003.

Wolanin, P. M., D. Webre, and **J. B. Stock.** Mechanism of phosphate activity in the chemotaxis response regulator CheY. *Biochemistry* 42(47):14075–82, 2003.

Donev, A., **S. Torquato,** F. H. Stillinger, and R. Connelly. A linear programming algorithm to test for jamming in hard-sphere packings. *J. Comput. Phys.* 197:139, 2004.

Donev, A., and **S. Torquato.** Energy-efficient actuation in infinite lattice structures. *J. Mech. Phys. Solids* 51:1459, 2003.

Crawford, J., **S. Torquato,** and F. H. Stillinger. Aspects of correlation function realizability. *J. Chem. Phys.* 119:7065, 2003.

Torquato, S., S. Hyun, and A. Donev. Optimal design of manufacturable three-dimensional composites with multifunctional characteristics. *J. Appl. Phys.* 94: 5748, 2003.

Hyun, S., A. M. Karlsson, **S. Torquato,** and A. G. Evans. Simulated properties of kagome and tetragonal truss core panels. *Int. J. Sol. Struct.* 40:6989, 2003.

Bosacchi, B., et al. **(W. S. Warren** and **H. Rabitz).** Computational intelligence in bacterial sport detection and identification. *Proc. SPIE* 5200:31–45, 2004.

RENSSELAER POLYTECHNIC INSTITUTE

Department of Chemistry and Chemical Biology

Programs of Study

Rensselaer's Department of Chemistry and Chemical Biology provides courses and programs of study that reflect the central role of chemistry in the science and technology of tomorrow. The Department offers Master of Science and Doctor of Philosophy degree programs; both programs require research and a thesis.

Research, the heart of graduate study, is conducted by groups of students and postdoctoral research associates under the direction of individual faculty members. During the first semester of study, graduate students are encouraged to join one of these research groups and begin their thesis research. Research groups vary widely in size and organization, but all are characterized by close interaction and exchange of ideas among faculty members and students. Faculty members devote substantial time to their students not only as advisers but as active collaborators in research.

The Department's graduate program maintains well-developed research programs not only in the traditional areas of analytical, inorganic, organic, and physical chemistry but also in the interdisciplinary fields of polymer chemistry, biochemistry, and medicinal chemistry. There are extensive interactions among chemistry, chemical engineering, and materials science in the area of polymers/bio/nano/materials and collaborative programs with the Departments of Biology, Computer Science, Physics, and Mathematical Science and the School of Engineering and the Center for Integrated Electronics. These and off-campus collaborations, which include Albany Medical College, the University at Albany, and the New York State Wadsworth Laboratories, provide essential connections between chemistry and other areas vital to modern society.

Graduate students typically begin their studies with graduate courses in analytical chemistry, biochemistry, inorganic chemistry, organic chemistry, and/or physical chemistry. The courses that are required depend on the student's background, area of interest, and performance in entering placement exams. In addition, in consultation with the adviser, students may select a number of specialized advanced-level courses in chemistry as well as courses that meet their needs that are offered in other departments. Each student plans a program with his or her adviser to meet individual professional goals.

The M.S. degree requires 30 credit hours beyond the B.S. and the Ph.D., 90 credit hours beyond the B.S. The M.S. degree is not a prerequisite for the Ph.D. degree. No overall courses are required for the Ph.D.; requirements are determined for each student based on the results of initial placement exams, the student's background, and proposed area of research. Normally, students are expected to have graduate-level understanding in at least four of the areas of analytical chemistry, biochemistry, inorganic chemistry, organic chemistry, and physical chemistry, as shown by examination results or by taking courses in those areas. At least 21 course credits (15 at the 6000 level or above) are required for the M.S. degree; there is no fixed number required for the Ph.D. Teaching assistants are required to attend a 1-credit Teaching Seminar, and all graduate students are expected to participate in the Departmental seminar program.

Most Ph.D. students complete all their course work by the end of the third semester of residence, and thereafter concentrate on their research. Each doctoral candidate takes a written qualifying exam or a set of cumulative exams, depending on the division, and an oral candidacy exam, the format of which is set by the student's doctoral committee. The final and by far the most important requirement for the Ph.D. is successful completion of a research problem that culminates in the writing and defense of the dissertation.

The requirements for the M.S. are parallel to, but less demanding than, those for the Ph.D.; the major difference is the research problem is more focused. There is no qualifying or candidacy examination or oral thesis defense.

Research Facilities

The discovery of new scientific concepts and technologies, especially in emerging interdisciplinary fields, is the lifeblood of Rensselaer's culture and a core goal for faculty and staff members and students. The Department sustains well-developed research programs and facilities and conducts research in interdisciplinary centers on campus through collaborative programs with other departments.

A variety of modern instruments are available in individual laboratories and in the Department's Major Instrumentation Center, which provides state-of-the-art equipment for nuclear magnetic resonance (both solution and solid state) and other techniques. This equipment, serviced and operated by a professional staff, is available to all researchers in the Department.

The central Mass Spectroscopy Facility includes GC-MS, MALDI- TOF for macromolecular analysis, and LC-MS (ion trap) equipment. Other instruments available for research include NIR, visible, UV, fluorescence, atomic absorption, surface plasma resonance, and FTIR spectrophotometers; G.C. and HPLC equipment; electrochemical equipment; ESR spectrometers; DSC, DTA, TGA, and TMA instruments for thermal studies; and X-ray fluorescence and diffraction instruments.

A molecular modeling laboratory contains computer workstations and a variety of sophisticated computer programs for molecular modeling, conformational analysis, energy calculation, and synthesis design.

Advances in the generation, mining, and analysis of chemical information is crucial to the development of new drug therapies and to modern methods of bioinformatics and molecular medicine. The Rensselaer Exploratory Center for Cheminformatics Research (RECCR) emphasizes the central role of cheminformatics in modern biotechnology efforts, molecular design projects, and bioinformatics programs. The center brings together and stimulates collaborative pilot projects among a constantly-evolving nucleus of experts in cheminformatics-related fields ranging from methods of encoding and capturing molecular information to machine learning and data-mining techniques to predictive model development, validation, interpretation, and utilization.

The Center for Biotechnology and Interdisciplinary Studies houses faculty members and researchers engaged in interdisciplinary research and hosts world-class programs and symposia. It exemplifies a new research paradigm, as no department offices reside in the building; rather, it is occupied by researchers and their laboratories. The core research facilities within the center contain laboratories for molecular biology, analytical biochemistry, microbiology, imaging, histology, tissue and cell culture, proteomics, and scientific computing and visualization. The center contains an 800-MHz nuclear magnetic resonance (NMR) spectrometer and the computing and visualization infrastructure needed to model molecular structure at the atomic level.

The Materials Research Center (MRC) was constructed under the first facilities grant awarded by NASA on September 25, 1962. According to James E. Webb, NASA administrator, the purpose of the grant program was "to house interdisciplinary activities in space-related sciences and technology to universities which are making substantial contributions to the national space program." Twenty-five laboratories, designed in part by faculty members, provide facilities for powder metallurgy, polymer, ceramics, ultrasonics, cryogenics, corrosion, and other materials research.

The New York Center for Studies on the Origins of Life promotes education and research in all relevant fields of biology, chemistry, Earth sciences, and astrophysics. This center is sponsored by the NASA Office of Space Science.

The New York State Center for Polymer Synthesis, dedicated in 1998, is the synthesis of a history that has impacted several generations of polymer scientists and a radical interdisciplinary approach to faculty-student research. The facility houses advanced technology for the discovery, scale-up, processing, and evaluation of unique polymers needed by many industries. The center's focus is grounded in three areas: ground-breaking research, corporate and government partnerships, and undergraduate and graduate education.

Financial Aid

Financial aid is available in the forms of teaching and research assistantships and fellowships, which include tuition scholarships and stipends. Rensselaer assistantships and university, corporate, or national fellowships fund many of Rensselaer's full-time graduate students. Outstanding students may qualify for university-sponsored Rensselaer Graduate Fellowship Awards, which carry a minimum stipend of $20,000 and a full tuition and fees scholarship. All fellowship awards are calendar-year awards for full-time graduate students. Summer support is also available in many departments. Low-interest, deferred-repayment graduate loans are available to U.S. citizens with demonstrated need.

Cost of Study

Full-time graduate tuition for the 2006–07 academic year is $32,600. Other costs (estimated living expenses, insurance, etc.) are projected to be about $12,400. Therefore, the cost of attendance for full-time graduate study is approximately $45,000. Part-time study and cohort programs are priced differently. Students should contact Rensselaer for specific cost information related to the program they wish to study.

Living and Housing Costs

Graduate students at Rensselaer may choose from a variety of housing options. On campus, students can select one of the many residence halls, and there are abundant options off campus as well, many within easy walking distance.

Student Group

There are 1,234 graduate students, of whom 30 percent are women, 90 percent are full-time, and 69 percent study at the doctoral level.

Student Outcomes

Rensselaer's graduate students are hired in a variety of industries and sectors of the economy and by private and public organizations, the government, and institutions of higher education. Starting salaries average $63,262 for master's degree recipients.

Location

Located just 10 miles northeast of Albany, New York State's capital city, Rensselaer's historic 275-acre campus sits on a hill overlooking the city of Troy, New York, and the Hudson River. The area offers a relaxed lifestyle with many cultural and recreational opportunities, with easy access to both the high-energy metropolitan centers of the Northeast—such as Boston, New York City, and Montreal, Canada—and the quiet beauty of the neighboring Adirondack Mountains.

The Institute

Recognized as a leader in interactive learning and interdisciplinary research, Rensselaer continues a tradition of excellence and technological innovation dating back to 1824. More than 100 graduate programs in more than fifty disciplines attract top students, researchers, and professors. The discovery of new scientific concepts and technologies, especially in emerging interdisciplinary fields, is the lifeblood of Rensselaer's culture and a core goal for the faculty, staff, and students. Fueled by significant support from government, industry, and private donors, Rensselaer provides a world-class education in an environment tailored to the individual.

Applying

The admission deadline for the fall semester is January 1. Basic admission requirements are the submission of a completed application form (available online), the required application fee, a statement of background and goals, official transcripts, official GRE scores, TOEFL or IELTS scores (if applicable), and two recommendations. In addition, the GRE Subject Test in either biochemistry or chemistry is strongly recommended.

Correspondence and Information

Department of Chemistry and Chemical Biology
120 Cogswell
110 8th Street
Rensselaer Polytechnic Institute
Troy, New York 12180

Phone: 518-276-6456
E-mail: mcgrae@rpi.edu
Web site: http://www.rpi.edu/dept/chem/index.html

Rensselaer Polytechnic Institute

THE FACULTY AND THEIR RESEARCH

Analytical and Bioanalytical Chemistry

Yvonne A. Akpalu, Assistant Professor; Ph.D. (polymer science and engineering), Massachusetts Amherst. Integrated light, X-ray, neutron scattering, and spectroscopy for characterizing macromolecular structure, aggregation, and dynamics. (akpaly@rpi.edu; http://www.rpi.edu/dept/chem/chem_faculty/profiles/akpalu.html)

Gerald M. Korenowski, Professor; Ph.D., Cornell. Interfacial studies; nonlinear optics; spectroscopy; surface imaging, optical materials, in situ laser spectroscopy as environmental probes. (koreng@rpi.edu; http://www.rpi.edu/dept/chem/chem_faculty/profiles/korenowski.html)

Linda B. McGown, Professor and Department Chair; Ph.D. (chemistry), Washington (Seattle). Fluorescence lifetime analysis, fluorescence probe techniques, genomics, proteomics, capillary electrochromatography, biomaterials and environmental analysis. (mcgowl@rpi.edu; http://www.rpi.edu/dept/chem/chem_faculty/profiles/mcgown.html)

Julie Stenken, Associate Professor; Ph.D. (bioanalytical chemistry), Kansas. Enhancing microdialysis sampling relative recovery, quantitation of the foreign-body response, oxidative stress measurement of the highly reactive hydroxyl radical, online analytical methods for anesthesia monitoring. (stenkj@rpi.edu; http://www.rpi.edu/dept/chem/chem_faculty/profiles/stenken.html)

Chemical Biology and Biochemistry

Wilfredo Colón, Associate Professor; Ph.D. (chemistry), Texas A&M. Molecular recognition in protein oligomerization and aggregation, protein folding, biophysical mechanism of amyloid formation, protein folding defects in human diseases. (colonw@rpi.edu; http://www.rpi.edu/dept/chem/chem_faculty/profiles/colon.html)

James Kempf, Assistant Professor; Ph.D., Caltech. Application of nuclear magnetic resonance techniques to the study of dynamic processes in proteins. (kempfj2@rpi.edu; http://www.rpi.edu/dept/chem/chem_faculty/profiles/kempf.html)

Robert J. Linhardt, Professor, Department of Chemistry and Chemical Biology; Ann and John H. Broadbent, Jr. '59 Senior Constellation Professor of Biocatalysis and Metabolic Engineering; and Professor, Department of Biology; Ph.D. (organic chemistry), Johns Hopkins. Synthetic carbohydrate chemistry, biochemistry and structural biochemistry, biophysical chemistry. (linhar@rpi.edu; http://www.rpi.edu/dept/chem/chem_faculty/profiles/linhardt.html)

Linda B. McGown, Professor and Department Chair; Ph.D. (chemistry), Washington (Seattle). Fluorescence lifetime analysis, fluorescence probe techniques, genomics, proteomics, capillary electrochromatography, biomaterials and environmental analysis. (mcgowl@rpi.edu; http://www.rpi.edu/dept/chem/chem_faculty/profiles/mcgown.html)

Julie Stenken, Associate Professor; Ph.D. (bioanalytical chemistry), Kansas. Enhancing microdialysis sampling relative recovery, quantitation of the foreign-body response, oxidative stress measurement of the highly reactive hydroxyl radical, online analytical methods for anesthesia monitoring. (stenkj@rpi.edu; http://www.rpi.edu/dept/chem/chem_faculty/profiles/stenken.html)

Inorganic and Organometallic Chemistry

Ronald A. Bailey, Professor; Ph.D., McGill. Coordination chemistry, rhenium chemistry. (bailer@rpi.edu; http://www.rpi.edu/dept/chem/chem_faculty/profiles/baily.html)

Alan R. Cutler, Professor; Ph.D. (organic and transition organometallic chemistry), Brandeis. Carbonylation chemistry, catalytic hydrosilation and hydrometalation, CO_2 fixation. (cutlea@rpi.edu; http://www.rpi.edu/dept/chem/chem_faculty/profiles/cutler.html)

Leonard V. Interrante, Professor; Ph.D. (inorganic chemistry), Illinois at Urbana-Champaign. Synthesis and study of new inorganic polymers, polymer precursors to sic and other ceramic materials, chemical vapor deposition of inorganic thin films, fabrication and study of ceramic composites from organometallic precursors. (interl@rpi.edu; http://www.rpi.edu/dept/chem/chem_faculty/profiles/interrante.html)

Organic and Medicinal Chemistry

Brian C. Benicewicz, Professor and Director of the New York State Center for Polymer Synthesis; Ph.D. (polymer chemistry), Connecticut. High-temperature fuel cell membranes, polymer synthesis, conducting polymers, liquid crystalline polymers and thermosets. (benice@rpi.edu; http://www.rpi.edu/dept/chem/chem_faculty/profiles/benicewicz.html)

Alan R. Cutler, Professor; Ph.D. (organic and transition organometallic chemistry), Brandeis. Carbonylation chemistry, catalytic hydrosilation and hydrometalation, CO_2 fixation. (cutlea@rpi.edu; http://www.rpi.edu/dept/chem/chem_faculty/profiles/cutler.html)

James V. Crivello, Professor; Ph.D. (organic chemistry), Notre Dame. New polymer forming reaction, initiators, block polymers, metal-catalyzed ring-opening polymerizations. (crivej@rpi.edu; http://www.rpi.edu/dept/chem/chem_faculty/profiles/crivello.html)

Robert J. Linhardt, Professor, Department of Chemistry and Chemical Biology; Ann and John H. Broadbent, Jr. '59 Senior Constellation Professor of Biocatalysis and Metabolic Engineering; and Professor, Department of Biology; Ph.D. (organic chemistry), Johns Hopkins. Synthetic carbohydrate chemistry, biochemistry and structural biochemistry, biophysical chemistry. (linhar@rpi.edu; http://www.rpi.edu/dept/chem/chem_faculty/profiles/linhardt.html)

James A. Moore, Professor; Ph.D. (organic-polymer chemistry), Polytechnic of Brooklyn. Vapor deposition polymerization, dendrimer polyelectrolytes, novel polymers and polymerizations. (moorej@rpi.edu; http://www.rpi.edu/dept/chem/chem_faculty/profiles/moore.html)

Mark P. Wentland, Professor; Ph.D. (synthetic organic chemistry), Rice. Anti-cocaine medications, identification of novel analgesic targets. (wentmp@rpi.edu; http://www.rpi.edu/~wentmp/; http://www.rpi.edu/dept/chem/chem_faculty/profiles/wentland.html)

Physical and Computational Chemistry

Yvonne A. Akpalu, Assistant Professor; Ph.D. (polymer science and engineering), Massachusetts Amherst. Integrated light, X-ray, and neutron scattering for characterizing polymer structure in solution and thin films; solution, gel, and membrane structure of polymer exchange membranes for fuel cells; structure-property relationships for semicrystalline polymers, ion-containing polymers, polymer blends, and nanocomposites. (akpaly@rpi.edu; http://www.rpi.edu/dept/chem/chem_faculty/profiles/akpalu.html)

Curt M. Breneman, Professor and Director, Rensselaer Exploratory Center for Cheminformatics Research; Ph.D. (organic chemistry), California, Santa Barbara. Ab initio computational chemistry, rapid construction of molecular electron density distributions: RECON/TAE transferable atom equivalent (TAE) modeling, machine learning in high-throughput screening and data mining of molecular databases, electron density-based qsar and qspr descriptor computation, automated drug discovery methods and "materials by design," molecular recognition. (brenec@rpi.edu; http://www.rpi.edu/dept/chem/chem_faculty/profiles/breneman.html)

Charles W. Gillies, Associate Professor; Ph.D. (chemistry), Michigan. Van der Waals complexes, transient chemical species. (gillic@rpi.edu; http://www.rpi.edu/dept/chem/faculty/gillies/gillies.html)

James Kempf, Assistant Professor; Ph.D., Caltech. Application of nuclear magnetic resonance techniques to the study of dynamic processes in proteins. (kempfj2@rpi.edu; http://www.rpi.edu/dept/chem/chem_faculty/profiles/kempf.html)

Gerald M. Korenowski, Professor; Ph.D., Cornell. Interfacial studies; nonlinear optics; spectroscopy; surface imaging, optical materials, in situ laser spectroscopy as environmental probes. (koreng@rpi.edu; http://www.rpi.edu/dept/chem/chem_faculty/profiles/korenowski.html)

Chang Y. Ryu, Associate Professor; Ph.D. (chemical engineering), Minnesota. Macromolecular separation and adsorption, block copolymer self-assembly in solution and thin films, polymer-carbon nanotube hybrid materials. (ryuc@rpi.edu; http://www.rpi.edu/~ryuc/; http://www.rpi.edu/dept/chem/chem_faculty/profiles/ryu.html)

Joseph T. Warden, Professor; Ph.D. (physical chemistry), Minnesota. ESR spectroscopy, biophysical chemistry. (wardej@rpi.edu; http://www.rpi.edu/dept/chem/chem_faculty/profiles/warden.html)

Mark P. Wentland, Professor; Ph.D. (synthetic organic chemistry), Rice. Anti-cocaine medications, identification of novel analgesic targets. (wentmp@rpi.edu; http://www.rpi.edu/~wentmp/; http://www.rpi.edu/dept/chem/chem_faculty/profiles/wentland.html)

Heribert Wiedemeier, Research Professor; D.Sc., Münster (Germany). Solid-state and high-temperature chemistry, crystal growth of electronic materials on earth and in space, thermodynamic and kinetic studies of reactions, structural and defect studies on solids. (wiedeh@rpi.edu; http://www.rpi.edu/dept/chem/chem_faculty/profiles/wiedemeier.html)

Polymer and Materials Chemistry

Yvonne A. Akpalu, Assistant Professor; Ph.D. (polymer science and engineering), Massachusetts Amherst. Integrated light, X-ray neutron scattering for characterizing polymer structure in solution and thin films; solution, gel, and membrane structure of polymer exchange membranes for fuel cells; structure-property relationships for semicrystalline polymers, ion-containing polymers, polymer blends, and nanocomposites. (akpaly@rpi.edu; http://www.rpi.edu/dept/chem/chem_faculty/profiles/akpalu.html)

Brian C. Benicewicz, Professor and Director of the New York State Center for Polymer Synthesis; Ph.D. (polymer chemistry), Connecticut. High-temperature fuel cell membranes, polymer synthesis, conducting polymers, liquid crystalline polymers and thermosets. (benice@rpi.edu; http://www.rpi.edu/dept/chem/chem_faculty/profiles/benicewicz.html)

James V. Crivello, Professor; Ph.D. (organic chemistry), Notre Dame. New polymer forming reaction, initiators, block polymers, metal-catalyzed ring-opening polymerizations. (crivej@rpi.edu; http://www.rpi.edu/dept/chem/chem_faculty/profiles/crivello.html)

James A. Moore, Professor; Ph.D. (organic-polymer chemistry), Polytechnic of Brooklyn. Vapor deposition polymerization, dendrimer polyelectrolytes, novel polymers and polymerizations. (moorej@rpi.edu; http://www.rpi.edu/dept/chem/chem_faculty/profiles/moore.html)

Chang Y. Ryu, Associate Professor; Ph.D. (chemical engineering), Minnesota. Macromolecular separation and adsorption, block copolymer self-assembly in solution and thin films, polymer-carbon nanotube hybrid materials. (ryuc@rpi.edu; http://www.rpi.edu/~ryuc/; http://www.rpi.edu/dept/chem/chem_faculty/profiles/ryu.html)

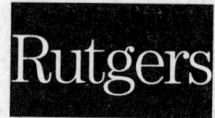

Rutgers

RUTGERS, THE STATE UNIVERSITY OF NEW JERSEY, NEW BRUNSWICK/PISCATAWAY

Graduate Program in Chemistry and Chemical Biology

Programs of Study

The Graduate Program in Chemistry and Chemical Biology at Rutgers in New Brunswick offers programs leading to the degrees of Master of Science and Doctor of Philosophy.

The principal requirement for the Ph.D. degree is the completion and successful oral defense of a thesis based on original research. A wide variety of research specializations are available in the traditional areas of chemistry—analytical, inorganic, organic, and physical—as well as in related areas and subdisciplines, including biological, bioinorganic, bioorganic, and biophysical chemistry; chemical physics; theoretical chemistry; and solid-state and surface chemistry.

The M.S. degree may be taken with or without a research thesis. The principal requirements are completion of 30 credits of graduate courses, a passing grade on the master's examination, and a master's essay or thesis. When the thesis option is chosen, 6 of the 30 credits may be in research.

Research Facilities

The research facilities of the program, located in the Wright-Rieman chemistry laboratories on the Busch campus, include a comprehensive chemistry library and glassworking, electronics, and machine shops. Research instruments of particular note include 800-, 600-, 500-, 400-, and 300-MHz NMR spectrometers; ESR spectrometers; single-crystal and powder X-ray diffractometers; multiwire area detectors for macromolecular structure determination; laser flash photolysis systems; a temperature-programmable ORD-CD spectropolarimeter; automated peptide and DNA synthesizers; a SQUID magnetometer; ultrahigh-vacuum surface analysis systems; scanning tunneling and atomic force microscopes; a helium-atom scattering apparatus; molecular beam and supersonic jet apparatuses; GC/quadrupole and ICP mass spectrometers; and extensive laser instrumentation, crystal-growing facilities, and calorimetric equipment. Computing facilities in the Wright-Rieman Laboratories include four multiprocessor servers, more than fifty graphics workstations, a ninety-six-processor cluster of Linux-based workstations, video animation equipment, and an assortment of personal computers, X-terminals, and laser and color printers.

Financial Aid

Full-time Ph.D. students receive financial assistance in the form of fellowships, research assistantships, teaching assistantships, or a combination of these. Financial assistance for entering students ranges from approximately $19,000 to $22,000 plus tuition remission for a calendar-year appointment. This includes the J. R. L. Morgan fellowships awarded annually to outstanding applicants.

Cost of Study

In 2004–05, the full-time tuition (remitted for assistantship and fellowship recipients) was $4834 per semester for New Jersey residents and $7185 per semester for out-of-state residents. There was a fee for full-time students of $585 per semester. All of these fees are subject to change for the next academic year.

Living and Housing Costs

A furnished double room in the University residence halls or apartments rented for $5898 to $6774 per person for the 2004–05 calendar year. Married student apartments rented for $787 to $998 per month. Current information may be obtained from the Department of Housing (732-445-2215), which also has information on private housing in the New Brunswick area.

Student Group

Total University enrollment is more than 48,000. Graduate and professional enrollment is approximately 13,500, of whom about 8,500 are in the Graduate School. Enrollment of graduate students in chemistry totals approximately 130. Of these, about three fourths are full-time students. Students come from all parts of the United States as well as from other countries. In addition, there are approximately 50 postdoctoral research associates in residence.

Location

New Brunswick, with a population of about 49,000, is located in central New Jersey, roughly midway between New York City and Philadelphia. The cultural facilities of these two cities are easily accessible by automobile or regularly scheduled bus and train service. Within a 1½-hour drive of New Brunswick are the recreational areas of the Pocono Mountains of Pennsylvania and the beaches of the New Jersey shore. The University also offers a rich program of cultural, recreational, and social activities.

The University

Graduate instruction and research in chemistry and chemical biology are conducted on the University's Busch campus, which has a rural-suburban environment and is a few minutes' drive from downtown New Brunswick. On the same campus, within walking distance, are the Hill Center for the Mathematical Sciences (home of the University's computer center), the Library of Science and Medicine, the physics and biology laboratories, the Waksman Institute of Microbiology, the Robert Wood Johnson Medical School, the Center for Advanced Biotechnology and Medicine, and the School of Engineering. The University provides a free shuttle-bus service between the Busch and New Brunswick campuses.

Applying

Applications for assistantships and fellowships should be made at the same time as applications to the Graduate School. All information describing current research programs may be obtained at the program's Web address. Admission consideration is open to all qualified candidates without regard to race, color, national origin, religion, sex, or handicap.

Correspondence and Information

Vice Chair for the Graduate Program
Department of Chemistry and Chemical Biology
Wright-Rieman Laboratories
Rutgers, The State University of New Jersey, New Brunswick/Piscataway
610 Taylor Road
Piscataway, New Jersey 08854-8087
Phone: 732-445-3223
E-mail: gradexec@rutchem.rutgers.edu
Web site: http://rutchem.rutgers.edu

Rutgers, The State University of New Jersey, New Brunswick/Piscataway

THE FACULTY AND THEIR RESEARCH

Stephen Anderson, Associate Professor. Proteases and protease inhibitors, protein folding, molecular recognition, and structural bioinformatics.
Georgia A. Arbuckle-Keil, Professor. Synthesis and properties of conducting polymers, quartz crystal microbalance study of electroactive surfaces, dynamic infrared linear dichroism (DIRLD).
Edward Arnold, Professor. Crystallographic studies of human viruses and viral proteins; molecular design, including drugs and vaccines; polymerase structure.
Jean Baum, Professor. Structural studies of proteins by nuclear magnetic resonance techniques, protein folding.
Helen M. Berman, Board of Governors Professor. Structural biology, structural nucleic acids, bioinformatics, protein–nucleic acid interaction.
Robert S. Boikess, Professor. Chemical education.
John G. Brennan, Professor. Molecular and solid-state inorganic chemistry; lanthanide chalcogenides and pnictides; molecular approaches to semiconductor thin films, optical fibers, and nanometer-sized clusters.
Kenneth J. Breslauer, Linus C. Pauling Professor. Characterization of the molecular interactions that control biopolymer structure and stability, drug-binding affinity and specificity, nucleic acid–based diagnostics and therapeutics.
Kieron Burke, Professor. Density functional theory in quantum chemistry and solid-state physics, nanoscience, surface science, proteins, and atomic and molecular structure.
Edward Castner Jr., Associate Professor. Ultrafast molecular dynamics and photoreactions in solution, hydrophilic polymer micelles and hydrogels, room-temperature ionic liquids.
Yves J. Chabal, Professor. Surface and interface chemistry of electronic and photonic materials and nanomaterials; in situ infrared spectroscopy of surface etching, thin-film growth, and solid/biological interfaces.
Kuang-Yu Chen, Professor. Hypusine formation and function of protein eIF5A, screening for anti-inflammatory and antitumor nutraceuticals, small-molecule interference of cell senescence, stress and gene expression.
Martha A. Cotter, Professor and Vice Chair, Graduate Program. Theoretical investigations of liquid crystals and micellar solutions, phase transitions in simple model systems, theory of liquids.
Richard H. Ebright, Professor and HHMI Investigator. Protein-DNA interaction; protein-protein interaction; structure, function, and regulation of transcription initiation complexes; single-molecule microscopy and nanomanipulation.
Eric L. Garfunkel, Professor. Surface chemistry, ultrathin films and interfaces for nanoelectronics, nanomaterials and technology, molecular electronics, organic-inorganic interfaces.
Alan S. Goldman, Professor. Organometallic chemistry: homogeneous catalysis, reactions, and mechanisms; catalytic functionalization of hydrocarbons.
Martha Greenblatt, Board of Governors Professor. Solid-state inorganic chemistry, low-dimensional transition-metal oxides and chalcogenides, high-T_c superconductors, CMR materials, solid electrolytes and electrodes for fuel cells and sensors.
Gene S. Hall, Professor. Applications of Raman microscopy, ICP-MS, XRF, and FT-IR in biological and environmental samples.
Gregory F. Herzog, Professor. Origin and evolution of meteorites, cosmogenic radioisotopes by accelerator mass spectrometry; stable isotopes by conventional forms of mass spectrometry.
Jane Hinch, Associate Professor. Molecular beam–surface interactions, surface diffractive techniques, scanning tunneling microscopy.
Stephan S. Isied, Professor. Bioinorganic and physical inorganic chemistry, photoinduced electron transfer in proteins, electron mediation by peptides with secondary structures, hydrogen bonding and other molecular recognition sites.
Leslie S. Jimenez, Associate Professor. Synthesis and characterization of analogs of antitumor antibiotics, total synthesis of natural products.
Roger A. Jones, Professor and Chair. Nucleoside and nucleic acid synthesis, including specifically labeled and modified nucleosides to probe ligand–nucleic acid interactions of both DNA and RNA.
Spencer Knapp, Professor. Total synthesis of natural products; design and synthesis of enzyme models and inhibitors and of complex ligands, new synthetic methods.
Joachim Kohn, Professor. Development of structurally new polymers as biomaterials for medical applications, tissue engineering, drug delivery, studies on the interactions of cells with artificial surfaces.
John Krenos, Professor. Chemical physics, energy transfer in hyperthermal collisions and collisions involving electronically excited reactants.
Karsten Krogh-Jespersen, Professor. Computational studies of molecular electronic structure, excited electronic states, solvation effects on photophysical properties, computational inorganic chemistry, catalysis.
Jeehiun Katherine Lee, Associate Professor. Experimental and theoretical studies of biological and organic reactivity, recognition, and catalysis; DNA duplex stability; mass spectrometry.
Ronald M. Levy, Board of Governors Professor. Biophysical chemistry, chemical physics of liquids, computational biology, protein structure, dynamics and protein folding.
Jing Li, Professor. Inorganic and materials chemistry: design, synthesis, crystal structure, and properties of solid-state inorganic materials; hybrid nanostructures; microporous and nanoporous metal organic frameworks.
Richard D. Ludescher, Professor. Food science; protein structure, dynamics, and function; luminescence spectroscopy; biophysics of amorphous solids.
Theodore E. Madey, State of New Jersey Professor of Surface Science. Structure and reactivity of surfaces and ultrathin films, electron-and photon-induced surface processes.
Gaetano T. Montelione, Professor. Protein NMR spectroscopy, molecular recognition, rational drug design, and structural bioinformatics.
Robert A. Moss, Louis P. Hammett Professor. Chemistry of reactive intermediates: carbenes, carbocations, diazirines; laser flash photolysis and fast kinetics.
Wilma K. Olson, Mary I. Bunting Professor. Theoretical and computational studies of nucleic acid conformation, properties, and interactions.
Daniel S. Pilch, Associate Professor. Mechanism of action of topoisomerase poisoning drugs, structure and energetics of specific RNA recognition by drugs and proteins.
Joseph A. Potenza, University Professor. Molecular structure, X-ray diffraction, magnetic resonance.
Laurence S. Romsted, Professor. Theory of micellar effects on reaction rates and equilibria, ion binding at aqueous interfaces, organic reaction mechanisms, antioxidant and dediazoniation chemistries.
Heinz D. Roth, Professor. Electron transfer–induced chemistry, physical organic chemistry of reactive intermediates, nuclear spin polarization, electron spin resonance, zeolite-induced chemistry, history of chemistry.
Daniel Seidel, Assistant Professor. Synthesis and synthetic methodology; enantioselective catalysis, using organocomponent and multicomponent catalysts; reaction development.
Stanley Stein, Adjunct Professor. Methods development in protein analysis, synthesis of biologically active peptides and oligonucleotides.
Ann Stock, Professor and HHMI Investigator. Structure and function of signal transduction proteins.
David Talaga, Assistant Professor. Protein folding and conformational dynamics, single molecule studies of inorganic and biological polymers, vibrational spectroscopy of inorganic complexes.
John W. Taylor, Associate Professor. Bioactive peptide design and synthesis, multicyclic peptides, peptide conformation, protein engineering, peptide ligand-receptor interactions, peptide-based HIV vaccines.
Kathryn E. Uhrich, Associate Professor. Synthesis and characterization of novel polymers for drug delivery, preparation and analysis of micropatterned polymer surfaces for cell growth.
Ralf Warmuth, Associate Professor. Design and synthesis of conformationally constrained peptides and of ion channels, host-guest chemistry, molecular container chemistry, strained organic molecules and reactive intermediates.
Lawrence J. Williams, Assistant Professor. Organic chemistry: natural product total synthesis, synthetic methods development, reaction discovery, molecular design.

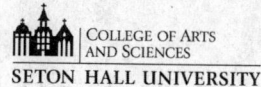

COLLEGE OF ARTS AND SCIENCES
SETON HALL UNIVERSITY

SETON HALL UNIVERSITY

Department of Chemistry and Biochemistry

Programs of Study

The Department of Chemistry and Biochemistry offers four programs leading to the Master of Science (M.S.) degree: Plan A (with thesis) and Plans B, C, and D (without thesis). In all of the programs, course work is selected by the student in consultation with an adviser. For students desiring research participation, the course work is augmented by the investigation of a research problem in collaboration with a faculty member. When the student completes the research project and the results are incorporated into a thesis, the M.S. degree is awarded. A nonthesis M.S. degree program (Plan B or C) based only on course study is also available. A flexible program combining graduate course work in chemistry and business administration (Plan D) is available for those who want advanced training directed toward the management aspects of chemical and pharmaceutical companies. Active participation in the Department's seminars is an important part of all programs leading to the M.S. degree in chemistry.

The principal requirement for the Ph.D. degree is the investigation of an original research project. A student may undertake a research project with any faculty member, irrespective of his or her area of specialization. Faculty members conduct state-of-the-art research in diverse areas, such as gas and liquid chromatography, surface science, computational chemistry, biophysical chemistry, carbohydrate synthesis, and catalysis. A broad range of research choices is available, and most groups participate in interdisciplinary research within the department and with collaborators in other departments and institutions. In addition, the student must meet the departmental requirements for course work, residency, the cumulative exams, the oral matriculation exam, and the seminar.

These graduate programs are unique in that they can all be completed on a part-time basis, with classes and seminars at night or on weekends, to serve the needs of busy working professionals.

Research Facilities

The science building, McNulty Hall, contains chemical research laboratories, instrument laboratories, computer facilities, a stockroom, and a machine shop. Labs are well equipped with modern instrumentation, including NMR spectrometers, gas chromatographs, high-performance liquid chromatographs, GC/MS and LC/MS, FT-IR, UV-Visible, fluorescence and atomic absorption spectrometers, and CD-ORD spectrometers. Advanced computing facilities include an Indigo 2 Workstation for molecular modeling and Pentium personal computers in the Department's Student Computing facility. Students also have access to a well-equipped local area network, the Internet, and Risc 6000 computers. The Center for Applied Catalysis was established at Seton Hall University in 1997 to assist industrial clients in developing catalytic processes for commercially important reactions. The center is well equipped to undertake catalyst evaluations and optimization of catalytic reactions in either a batch process or a continuous mode. Various pieces of equipment are available for use in reactions run from atmospheric pressure to 1500 psig (100 bar). These reactors range in size from 40 mL to 300 mL. These small-scale reactors provide an economy of scale that is important when a large number of reactions are being run.

The Walsh Library, a state-of-the-art 155,000-square-foot building, houses 500,000 titles, 1,875 current periodicals, and an extensive collection of microform and other nonprint items that include videotapes, CD-ROM music, and other electronic media.

Financial Aid

The Department offers a number of teaching assistantships for the academic year that are usually extended for the summer term to provide teaching and research support. A number of research fellowships also are available to students at the beginning of their second year of graduate study. In addition, the Reverend Owen Garrigan Graduate Biochemistry Supplemental Award is given to an incoming student who plans to do graduate work in the biochemistry area. This award is available each year to supplement the stipend of 1 or more full-time biochemistry graduate students. Seton Hall University is one of the beneficiaries of the Clare Booth Luce Fund, which supports women in science. Research fellowships for women pursuing graduate study are available on a competitive basis.

Cost of Study

Tuition is $743 per credit. Fees total about $300 per semester.

Living and Housing Costs

Seton Hall maintains a limited supply of housing to accommodate graduate students. On-campus room and board total about $9000 a year. Housing and living costs in South Orange and surrounding towns are comparable to most suburban cities, with studio and one-bedroom apartments renting for $650 to $900 per month. Off-campus listings are available through the Department of Housing and Residence Life.

Student Group

The Department currently enrolls about 100 graduate students on both a full- and part-time basis. Approximately 25 full-time students hail from a variety of national and international backgrounds, while the part-time students are generally industrial scientists working in the tri-state area. This provides a unique, daily, and direct interaction between students and working industrial scientists.

Location

Seton Hall is located on 58 acres in the village of South Orange, New Jersey, a suburban residential area 14 miles southwest of New York City. The town center is a 10-minute walk from the campus and features bookstores, coffee shops, and restaurants. The heart of midtown Manhattan is about 25 minutes away; students can take advantage of everything this exciting city has to offer while still living in a suburban area.

The University

Founded in 1856, Seton Hall is a private coeducational Catholic institution—the nation's oldest diocesan institution of higher education in the United States. With a total enrollment of about 10,000, including approximately 4,500 graduate students, the University comprises nine colleges and schools. Seton Hall is accredited by the Middle States Association of Colleges and Schools. Through the incorporation of technology into the curriculum, the College of Arts and Sciences seeks to enhance and enliven the learning environment. Rooted in tradition yet looking to the future, the College offers a rich set of opportunities for intellectual discovery. Graduate students are guided by scholars and specialists toward the mastery of academic and professional areas.

Applying

In addition to the general University requirements for admission, the Department strongly recommends that all applicants have completed a minimum of 30 credits in chemistry, including a two-semester course in physical chemistry, a one-year course in physics, and mathematics through differential and integral calculus. Students must submit the completed application, the $50 application fee, three letters of recommendation from individuals who can evaluate the applicant's scientific ability, and transcripts from all previously attended universities or colleges. Applicants whose native language is not English must submit TOEFL score results (minimum score of 620 on the paper-based test). The deadlines for fall and spring admission are July 1 and November 1, respectively. Applications are processed on a rolling basis.

Correspondence and Information

Nicholas H. Snow, Chair
Department of Chemistry and Biochemistry
Seton Hall University
400 South Orange Avenue
South Orange, New Jersey 07079

Phone: 973-761-9430
Fax: 973-761-9453
E-mail: artsci@shu.edu
Web site: http://artsci.shu.edu/graduateprograms/chem.htm

Seton Hall University

THE FACULTY AND THEIR RESEARCH

Robert Augustine, Professor Emeritus of Organic Chemistry and Executive Director, Center for Applied Catalysis; Ph.D., Columbia, 1957. Heterogeneous catalysis in organic synthesis, enantioselective catalysis.

Rev. Al Celiano, Professor Emeritus of Physical Chemistry; Ph.D., Fordham, 1959. Kinetics and mechanisms of substitution reactions at pentacoordinated platinum (II).

Alexander Fadeev, Associate Professor of Surface Chemistry and Materials Science; Ph.D., Moscow, 1990. Study of molecular mechanisms of wettability and adsorption, self-assembled and covalently attached organic monolayers, development of methods for chemical modification of solid surfaces. Reaction of organosilicon hydrides with solid surfaces: An example of surface-catalyzed self-assembly. *J. Am. Chem. Soc.* 126(24):7595, 2004. With Helmy and Wenslow.

James Hanson, Associate Professor of Organic Chemistry; Ph.D., Caltech, 1990. Organic and polymer synthesis and photochemistry, photochemical acid and base generation, novel polymer structures, dendritic polymers, molecular imprinting, microlithography. Photophysical studies of pyrene focused poly(aryl ether) monodendrons: Quenching and excimer formation. *Polym. Mater. Sci. Eng.* 80:68, 1999. With Khan and Murphy Jr.

Yuri Kazakevich, Assistant Professor of Analytical Chemistry; Ph.D., Moscow State, 1982. Physical studies of retention in liquid chromatography and the solution of unique analytical problems using chemical separations. Wetting in hydrophobic nanochannels: A challenge of classical capillarity. *J. Am. Chem. Soc.* 127(36):12446–7, 2005. With Helmy, Ni, and Fadeev.

Stephen Kelty, Associate Professor of Physical Chemistry and Graduate Chair; Ph.D., Harvard, 1993. Characterization of surface electronic and structural properties of heterogeneous catalysts, structure-function relationship of chemically active surfaces, scanning probe microscopy, novel properties of inorganic colloidal particles. Investigation of human hair fibers using lateral force microscopy. *Scanning* 23:337–45, 2001. With McMullen.

Joseph Maloy, Associate Professor of Analytical Chemistry; Ph.D., Texas at Austin, 1970. Electroanalytical methods, computer modeling of high-energy density batteries, digital simulation of electrochemical and chromatographic phenomena, data-acquisition techniques. Short-term chronoamperometric screening of chlorpromazine-package interactions. *J. Pharm. Sci.* 87(9):1130–7, 1998. With Sarsfield.

Cecilia Marzabadi, Assistant Professor of Organic Chemistry; Ph.D., Missouri–St. Louis, 1994. Synthetic organic and carbohydrate chemistry, solid-phase synthesis of modified cyclodextrins, synthesis of septanose sugars and hydroxymethyl C-glycosides, preparation of novel nucleoside analogs. Chair-chair conformational equilibria in silyloxycyclohexanes and their dependence on the substituents on silicon. The wider roles of eclipsing, of 1,3-repulsive steric interactions, and of attractive steric interactions. *J. Am. Chem. Soc.* 125(49):15163–73, 2003. With Anderson et al.

W. Rorer Murphy, Professor of Inorganic Chemistry and Undergraduate Chair; Ph.D., North Carolina at Chapel Hill, 1984. Fluorescent probes for study of dendritic polymers and nucleic acids; binding of metal complexes to nucleic acids; synthesis, photochemistry, and electrochemistry of novel transition metal complexes; development of organized molecular structures. A review and assessment of potential sources of ethnic differences in drug responsiveness. *J. Clin. Pharmacol.* 43(9):943–67, 2003. With Bjornsson et al.

Richard Sheardy, Professor of Biochemistry; Ph.D., Florida, 1979. Oligonucleotide synthesis; synthesis of DNA binding molecules; structural and thermodynamical characterizations of DNA oligomers via CD spectropolarimetry, UV-Vis Spectroscopy, NMR spectroscopy, and gel electrophoresis; small molecule–DNA interactions. The influence of sequence context and length on the kinetics of DNA duplex formation from complementary hairpins possessing (CNG) repeats. *J. Am. Chem. Soc.* 127(15):5581–5, 2005.

Nicholas Snow, Professor of Analytical Chemistry and Chair; Ph.D., Virginia Tech, 1992. Chemical separations; GC and HPLC; GC/MS; sampling for chromatography; analysis of trace organic compounds from inorganic, aqueous, and biological matrices; molecular imprinted polymers as stationary phases. Stir-bar sorptive extraction and thermal desorption–ion mobility spectrometry for the determination of trinitrotoluene and l,3,5-trinitro-l,3,5-triazine in water samples. *J. Chromatogr. A.*, online, 2005. With Lokhnauth.

John Sowa, Associate Professor of Organic Chemistry; Ph.D., Iowa State, 1991. Organic and organometallic chemistry, synthetic and mechanistic studies of homogeneous, achiral, and chiral catalytic reactions. Kinetic study on the epimerization of trityloxymethyl butyrolactol by liquid chromatography. *J. Chromatogr. A* 1043(2):171–5, 2004. With Li, Thompson, Clausen, and Dowling.

George Turner, Assistant Professor of Chemistry and Biochemistry; Ph.D., Duke, 1989. Molecular mechanism of function for the integral membrane proteins involved in communication cascades, particularly those involving signal and solute transduction. Downstream coding region determinants of bacterio-opsin, muscarinic acetylcholine receptor and α_{2b}-adrenergic receptor expression in *Halobacterium salinarum. Biochim. Biophys. Acta* 1610(1):109–23, 2002. With Bartus et al.

SOUTHERN ILLINOIS UNIVERSITY CARBONDALE

Department of Chemistry and Biochemistry
Doctoral Program

Program of Study

The Department of Chemistry and Biochemistry offers a Ph.D. program in chemistry with specializations in analytical chemistry, biochemistry, inorganic chemistry, organic chemistry, physical chemistry, and materials chemistry. The Department also has an interdisciplinary focus in the fields of materials and biological chemistry and employs several faculty members whose research interests overlap in these areas.

The doctoral degree in chemistry is a research degree. To be awarded this degree, the student must, to the satisfaction of the graduate committee, demonstrate the ability to conduct original and independent research within some area of chemistry and must make an original contribution to science. Candidates must also successfully complete cumulative exams, required graduate course work, and meet other requirements of the Department and Graduate School. In addition, the Ph.D. candidate must write and defend an original research proposal.

Research Facilities

The Department's research activities are supported by a full spectrum of modern chemical instruments and support facilities. Research shops for electronics, machining, fine instruments, and electron and atomic force microscopy provide essential services for many research projects. Major equipment available to the Department includes three Varian Inova NMR spectrometers (300, 400, 500 MHz) numerous FTIRs, an ABI 4700 MALDI TOF/TOF, a Bruker Microflex MALDI reTOF, a Bruker HPLC-ESI HCT, a Thermofinnigan GCMS, and a Varian Ultramass ICP-MS, as well as a class 100 clean room. More routine equipment, such as UV-VIS absorbance, ICP-AES, HPLC, CE, GC, and AA is commonly available. Other major equipment items, such as gel permeation chromatography, DSC, TGA, polarized optical microscopy, modern electrochemical instrumentation, and materials characterization equipment are associated with specific research groups. In addition, a multitude of personal computers connected to a campuswide network, including a twenty-station computer laboratory, are also located within the Department.

Financial Aid

Successful candidates are offered an opening in the Department's graduate program and an assistantship at the current stipend rate for an academic year. Assuming graduate school approval, the assistantship includes a tuition waiver. Student fees are not considered a part of tuition. Incoming graduate students are usually teaching assistants for the first year, with approximately 6 to 9 contact hours per week as laboratory instructors. Some grading, proctoring, and laboratory preparation may also be assigned. By the end of the first semester, graduate students are expected to join a research group and become research assistants. Assuming satisfactory performance, assistantships are renewable.

Fellowships for outstanding students, as well as dedicated fellowships for minority students, are available through the Department and the University.

Cost of Study

In-state graduate tuition is $243 per credit hour in 2006–07. Out-of-state tuition is 2.5 times the in-state tuition rate ($607.50 per credit hour). Graduate students with at least a 25 percent appointment as a graduate assistant receive a tuition waiver. Fees vary from $441.62 (1 credit hour) to $987.30 (12 credit hours).

Living and Housing Costs

For married couples, students with families, and single graduate students, the University has 589 efficiency and one-, two-, and three-bedroom apartments that rent for $438 to $505 per month in 2006–07. Residence halls for single graduate students are also available, as are accessible residence hall rooms and apartments for students with disabilities.

Student Group

The Southern Illinois University Carbondale (SIUC) campus has more than 21,000 students, approximately 4,500 of whom are enrolled in graduate programs with more than 1,000 international graduate students representing over 100 countries. The Department of Chemistry and Biochemistry has a total of approximately 45 graduate students enrolled in the Ph.D. and master's programs.

Student Outcomes

The Department has a distinguished list of alumni making substantial impact at places such as the University of California at Berkeley, the University of Florida, Kansas State, DuPont, Monsanto, Genentech, BASF, and Bristol-Meyers Squibb.

Location

SIUC is 350 miles south of Chicago and 100 miles southeast of St. Louis. Nestled in rolling hills bordered by the Ohio and Mississippi Rivers and enhanced by a mild climate, the area has state parks, national forests and wildlife refuges, and large lakes for outdoor recreation. Cultural offerings include theater, opera, concerts, art exhibits, and cinema. Educational facilities for the families of students are excellent.

The University

Southern Illinois University Carbondale is a comprehensive public university with a variety of general and professional education programs. The University offers associate, bachelor's, master's, and doctoral degrees; the J.D. degree; and the M.D. degree. The University is fully accredited by the North Central Association of Colleges and Schools. The Graduate School has an essential role in the development and coordination of graduate instruction and research programs. The Graduate Council has academic responsibility for determining graduate standards, recommending new graduate programs and research centers, and establishing policies to facilitate the research effort.

Applying

Applications for admission may be submitted at any time online through the Graduate School or directly to the Department, and students may start in the fall or, in special cases, in the spring. While there are no specific deadlines in the Department, domestic applications should be completed as early in January as possible for the following fall term if the applicant is interested in nomination for fellowships or scholarships offered outside the Department. International applications have the best opportunity for consideration in the Department if completed applications are on file at least six months in advance of the requested date of admission. TOEFL and TSE scores are required for international applicants. GRE scores are required for all applicants.

Further information and application packets can be obtained directly from the Department by mail, e-mail, or printed from the Department's Web site at http://www.chem.siu.edu/.

Correspondence and Information

Graduate Admissions Chair
c/o Graduate Admissions Secretary
Chemistry-MC 4409
Southern Illinois University Carbondale
Carbondale, Illinois 62901
E-mail: chemistry@chem.siu.edu
Web site: http://www.chem.siu.edu/

Southern Illinois University Carbondale

THE FACULTY AND THEIR RESEARCH

Mark J. Bausch, Associate Professor; Ph.D., Northwestern 1982. Organic chemistry, radical anion basicities, radical acidities, stability of organic cations.
Roger E. Beyler, Professor Emeritus; Ph.D., Illinois, 1949.
Albert L. Caskey, Associate Professor Emeritus; Ph.D., Iowa State 1961.
Bakul Dave, Associate Professor; Ph.D., Houston, 1993. Inorganic/materials chemistry, inorganic and organic nanocomposites, sol-gel based materials, bioinorganic chemistry.
Joe M. Davis, Professor; Ph.D., Utah, 1985. Analytical chemistry, mass transport, separations, statistics, electrokinetic separations.
Daniel J. Dyer, Associate Professor; Ph.D., Colorado, 1996. Organic/materials chemistry, design and synthesis of organic materials and polymers.
Yong Gao, Associate Professor; Ph.D., Alberta, 1998. Organic/materials/biological chemistry, design and synthesis of nanomaterials for biological applications.
Qingfeng Ge, Assistant Professor; Ph.D., Tianjin University (China), 1991. Theoretical studies of electrocatalytic processes in fuel cells and biological systems.
Boyd Goodson, Assistant Professor; Ph.D., Berkeley, 1999. Physical chemistry, optical/nuclear double resonance spectroscopy.
John C. Guyon, Professor Emeritus; Ph.D., Purdue, 1961.
Herbert I. Hadler, Professor Emeritus; Ph.D., Wisconsin, 1952.
Kara Huff Hartz, Assistant Professor; Ph.D., Purdue, 2002. Analytical chemistry, environmental nanoparticle sampling and analysis, filter collection, chromatography, mass spectrometry.
Conrad C. Hinckley, Professor Emeritus; Ph.D., Texas, 1964.
Gary Kinsel, Professor; Ph.D., Texas at Arlington, 1995. Analytical chemistry, laser mass spectrometry for biomolecular analysis.
Punit Kohli, Assistant Professor; Ph.D., Michigan State, 2000. Organic/materials chemistry, self-assembly of nanotubes and nanoparticles using molecular recognition principles, highly selective biological-tailored nanotube membranes for transport studies.
John A. Koropchak, Professor and Vice Chancellor for Research; Ph.D., Georgia, 1980. Analytical chemistry, atomic spectroscopy, metal speciation, separations detection, condensation nucleation light scattering detection, single molecule detection, capillary separations.
David F. Koster, Professor Emeritus; Ph.D., Texas A&M, 1965.
Brian Lee, Assistant Professor; Ph.D., Maryland, 1997. Biochemistry, biomolecular NMR spectroscopy to study the interactions and structures of proteins and nucleic acids that participate in translational regulation.
Matthew E. McCarroll, Assistant Professor; Ph.D., Idaho, 1998. Analytical chemistry, florescence spectroscopy, chiral and molecular recognition, organized media, stationary-phase development, capillary electrophoresis, development of florescence sensors.
Cal Y. Meyers, Professor Emeritus; Ph.D., Illinois, 1951.
Gabriela Pérez-Alvarado, Assistant Professor; Ph.D., Maryland, 1995. Biochemistry; application of NMR spectroscopy, molecular biological, biochemical, and biophysical techniques to characterize the structure, dynamics, and molecular interactions of multidomain proteins involved in normal cell regulation and cancer metastasis.
C. David Schmulbach, Professor Emeritus; Ph.D., Illinois, 1958.
Gerard V. Smith, Professor Emeritus; Ph.D., Arkansas, 1959.
Luke T. Tolley, Assistant Professor; Ph.D., North Carolina, 2001. Analytical chemistry, chromatography, capillary electrophoresis, mass spectrometry, intercellular signaling biomarkers.
Russell F. Trimble, Professor Emeritus; Ph.D., MIT, 1950.
James Tyrrell, Professor and Associate Dean; Ph.D., Glasgow, 1963. Physical chemistry, computational chemistry, transition states, reaction surfaces.
Lori Vermeulen, Associate Professor and Chair; Ph.D., Princeton, 1994. Inorganic/organic/materials chemistry, solid state chemistry, drug delivery, polymers.
Lichang Wang, Assistant Professor; Ph.D., Copenhagen, 1993. Physical/materials chemistry, computational/theoretical, material and catalytic properties of transition metal nanoparticles, hydrogen bonding network in self-assembled monolayers, activities of enzymes in biochemical reactions.
Ling Zang, Assistant Professor; Ph.D., Chinese Academy of Sciences, 1995. Analytical/physical/materials chemistry, nanoscale imaging and spectroscopy, nanostructure assembling and patterning, nanodevices for fluorescence sensing and probing.

STANFORD UNIVERSITY

Department of Chemistry

Program of Study

The Department of Chemistry strives for excellence in education and research. Only candidates for the Ph.D. degree are accepted. The department has a relatively small faculty, which promotes outstanding interactions between students and faculty and staff members. The faculty has achieved broad national and international recognition for its outstanding research contributions, and more than a third of its members belong to the National Academy of Sciences. The graduate program is based strongly on research, and students enter a research group by the end of the winter quarter of their first year. Students are also expected to complete a rigorous set of core courses in various areas of chemistry in their first year and to complement these courses later by studying upper-level subjects of their choice. Placement examinations are administered early in the fall quarter of the first year in inorganic, organic, and physical chemistry, with optional exams in biophysical chemistry and chemical physics. Students with deficiencies in undergraduate training in these areas are identified and work with the faculty to make them up. No other departmental examinations or orals are required of students progressing toward the Ph.D. degree. Much of the department's instruction is informal and includes diverse and active seminar programs, group meetings, and discussions with visiting scholars and with colleagues in other departments of the University. Stanford Ph.D. recipients are particularly well prepared for advanced scientific and technological study, and chemistry graduates typically accept positions on highly regarded university faculties or enter a wide variety of positions in industry.

Research Facilities

The department occupies six buildings with approximately 200,000 square feet of research space. The department has a strong commitment to obtaining and maintaining state-of-the-art instrumentation for analysis and spectroscopy. Equipment available includes 200-, 300-, 400-, 500-, 600- and 800-MHz NMR spectrometers; X-ray crystallography facilities; ultrafast absorption and fluorescence spectroscopy facilities; ultrahigh-resolution laser spectroscopy facilities; ion cyclotron resonance facilities; dynamic light-scattering spectroscopy facilities; tissue culture facilities; electrochemical systems; ultrahigh-vacuum facilities for surface analysis, including ESCA, Auger, EELS, LEED, and UPS; laser-Raman facilities; a superconducting magnetometer; and a GC/MS. The University has fully staffed machine and glass shops available for use by the department. Extensive computing capabilities are available in all research groups, and these are supplemented by a department computer network and a regional computational facility. The department and library maintain licenses for many of the most popular software packages for the chemical and life sciences. Additional major instrumentation and expertise are available in the Stanford Nanofabrication Facility, the Stanford Synchrotron Radiation Laboratory, the Stanford University Mass Spectrometry facility, the Stanford Magnetic Resonance Laboratory, and the Laboratory for Advanced Materials.

Financial Aid

Financial support of graduate students is provided in the form of teaching assistantships, research assistantships, and fellowships. All graduate students in good standing receive full financial support (tuition and stipend) for the duration of their graduate studies. The stipend for incoming first-year graduate students is $27,150; the amount of the stipend is adjusted annually to allow for inflation. Typical appointments involve teaching assistantships in the first year and research assistantships in subsequent years. Supplements are provided to holders of outside fellowships.

Cost of Study

Holders of teaching assistantships or research assistantships pay no academic tuition. In 2006–07, 10-unit tuition is $28,720 for the year (four quarters).

Living and Housing Costs

Both University-owned and private-owned housing accommodations are available. Due to the residential nature of the surrounding area, it is not uncommon for several graduate students to share in a house rental. Escondido Village, an apartment development on campus, provides one- to three-bedroom apartments for married students and single parents. Rains Houses and Lyman provide additional apartment residences on campus. The average monthly expenditure for rent is $550 for on-campus housing.

Student Group

The total enrollment at Stanford University is 18,836, of which there are 7,800 graduate students. The Department of Chemistry has 220 graduate students in its Ph.D. program.

Location

Stanford University is located in Palo Alto, a community of 61,200 about 35 miles south of San Francisco. Extensive cultural and recreational opportunities are available at the University and in surrounding areas, as well as in San Francisco. To the west lie the Santa Cruz Mountains and the Pacific Ocean and to the east, the Sierra Nevada range with its many national parks, hiking and skiing trails, and redwood forests.

The University

Stanford is a private university that was founded in 1885 and ranks in the top few for academic excellence in physical and natural sciences, liberal arts, humanities, and engineering. The campus occupies 8,800 acres of land, of which 5,200 acres are in general academic use. In all disciplines, the University has a primary commitment to excellence in education and research.

Applying

Admission to the chemistry department is by competitive application. Applications are available from the World Wide Web and should be filed before January 1 for admission in the fall quarter. All applicants are required to submit GRE scores from the verbal, quantitative, and analytical tests and the Subject Test in chemistry, as well as transcripts and three letters of recommendation. Applicants are notified of admission decisions before March 15. In unusual circumstances, late applications or a deferred enrollment are considered.

Correspondence and Information

Graduate Admissions Committee
Department of Chemistry
Stanford University
Stanford, California 94305-5080
Phone: 650-723-1525
E-mail: chem.admissions@stanford.edu
Web site: http://www.stanford.edu/dept/chemistry

Stanford University

THE FACULTY AND THEIR RESEARCH

Hans C. Andersen, Professor; Ph.D., MIT, 1966. Physical chemistry: statistical mechanics, theories of the structure and dynamics of liquids, computer simulation methods, glass transition.
Steven G. Boxer, Professor; Ph.D., Chicago, 1976. Physical and biophysical chemistry: structure, function, dynamics, and electrostatics in proteins and membranes; spectroscopy; photosynthesis; GFP; membrane biotechnology; cell-cell interactions.
John I. Brauman, Professor; Ph.D., Berkeley, 1963. Organic and physical chemistry: structure and reactivity of ions in the gas phase, photochemistry and spectroscopy of gas phase ions, electron photodetachment spectroscopy, electron affinities, reaction mechanisms.
Christopher E. D. Chidsey, Associate Professor; Ph.D., Stanford, 1983. Physical chemistry: molecular electronics, nanowire patterning, interfacial electron transfer, monomolecular films, electrocatalysis by discrete metal complexes.
James P. Collman, Professor Emeritus; Ph.D., Illinois, 1958. Inorganic, organic, and organometallic chemistry: synthetic analogues of the active sites in hemoproteins, homogeneous catalysis, multielectron redox catalysts, multiple metal-metal bonds.
Hongjie Dai, Professor; Ph.D., Harvard, 1994. Physical and materials chemistry: condensed-matter physics; materials science; biophysics; carbon nanotube and semiconducting nanowire synthesis; electrical, mechanical, electromechanical, and electrochemical characterizations; molecular electronics; nanobiotechnology, nanosensors and nanomaterial intracellular transporters.
Carl Djerassi, Professor Emeritus; Ph.D., Wisconsin–Madison, 1945. Organic chemistry: chemistry of steroids, terpenes, and alkaloids, with major emphasis on marine sources; application of chiroptical methods—especially circular dichroism and magnetic circular dichroism—to organic and biochemical problems; organic chemical applications of mass spectrometry; use of computer artificial-intelligence techniques in structure elucidation of organic compounds. Author of novels in genre of science-in-fiction and of science-in-theater plays.
Justin Du Bois, Associate Professor; Ph.D., Caltech, 1997. Organic chemistry: reaction development and transition metal catalysis, natural product synthesis, molecular recognition, ion channel physiology.
Michael D. Fayer, Professor; Ph.D., Berkeley, 1974. Physical chemistry and biophysical chemistry: dynamics in molecular condensed phases; laser spectroscopy; ultrafast nonlinear techniques; infrared and visible studies of dynamics and intermolecular interactions in hydrogen-bonding liquids, supercooled liquids, liquid crystals, supercritical fluids and proteins, and other biological systems.
Keith O. Hodgson, Professor; Ph.D., Berkeley, 1972. Inorganic, biophysical, and structural chemistry: chemistry and structure of metal sites in biomolecules, molecular and crystal structure analysis, protein crystallography, extended X-ray absorption fine-structure spectroscopy.
Wray H. Huestis, Professor; Ph.D., Caltech, 1972. Biophysical chemistry: chemistry of cell-surface receptors, membrane-mediated control mechanisms, biochemical studies of membrane protein complexes in situ, magnetic resonance studies of conformation and function in soluble proteins and protein-lipid complexes, viral fusion mechanisms, drug delivery.
Chaitan Khosla, Professor; Ph.D., Caltech, 1990. Biological chemistry: mechanism and engineering of polyketide biosynthesis; biology of new polydetides; chemistry and biology of Celiac Sprue.
Jennifer J. Kohler, Assistant Professor; Ph.D., Yale, 2000. Organic chemistry: bioorganic chemistry, chemical biology, glycobiology.
Eric T. Kool, Professor; Ph.D., Columbia, 1988. Organic, biological, and biophysical chemistry: synthetic mimics of nucleic acid structures, mechanistic studies of DNA replication and DNA repair, fluorescence methods for detecting and imaging RNA in cells, combinatorial biosensor discovery, design of new biological pathways.
Harden M. McConnell, Professor Emeritus; Ph.D., Caltech, 1951. Physical chemistry, biophysics, immunology: membrane biophysics, with emphasis on immunology and the activity of cholesterol in membranes.
W. E. Moerner, Professor; Ph.D., Cornell, 1982. Physical chemistry: individual molecules in solids, polymers, and proteins probed by far-field and near-field optical spectroscopy; single-molecule biophysics; nanophotonics; quantum optics of single molecules; chemistry of optical materials.
Vijay Pande, Assistant Professor; Ph.D., MIT, 1995. Physical chemistry and biophysics: theoretical models and computer simulations to examine the equilibrium and nonequilibrium statistical mechanics of biological molecules; thermodynamics and kinetics of protein folding, RNA folding, and protein design.
Robert Pecora, Professor; Ph.D., Columbia, 1962. Physical chemistry: statistical mechanics of fluids and macromolecules, molecular motions in fluids, light-scattering spectroscopy of liquids, macromolecules and biological systems.
John Ross, Professor Emeritus; Ph.D., MIT, 1951. Physical chemistry: experimental and theoretical studies of chemical kinetics, chemical instabilities, oscillatory reactions, strategies of determining complex reaction mechanisms, chemical computations, thermodynamics and fluctuations of systems far from equilibrium.
Edward I. Solomon, Professor; Ph.D., Princeton, 1972. Physical, inorganic, and bioinorganic chemistry: inorganic spectroscopy and ligand field, molecular orbital, and density functional theory; active sites; spectral and magnetic studies on bioinorganic systems directed toward understanding the geometric and electronic structural origins of their activity; structure/function correlations; development of new spectroscopic methods in bioinorganic chemistry.
T. Daniel P. Stack, Associate Professor; Ph.D., Harvard, 1988. Inorganic and organic chemistry: bioinspired oxidation catalysis.
Barry M. Trost, Professor; Ph.D., MIT, 1965. Organic, organometallic, and bioorganic chemistry: new synthetic methods, natural product synthesis and structure determinations, insect chemistry, potentially antiaromatic unsaturated hydrocarbons, chemistry of ylides.
Robert M. Waymouth, Professor; Ph.D., Caltech, 1987. Inorganic, organometallic, and polymer chemistry: mechanistic and synthetic chemistry of the early transition elements, mechanisms of olefin polymerization, design of new polymerization catalysts.
Paul A. Wender, Professor; Ph.D., Yale, 1973. Organic, organometallic, bioorganic, and medicinal chemistry: synthesis of biologically active compounds, synthetic methods, computer drug design, drug mechanisms, drug delivery.
Dmitry V. Yandulov, Assistant Professor; Ph.D., Indiana, 2000. Inorganic chemistry: redox catalysis, bioinorganic modeling, organometallic catalysis, catalytic fluorination.
Richard N. Zare, Professor; Ph.D., Harvard, 1964. Physical and analytical chemistry, chemical physics: application of lasers to chemical problems, molecular structure and molecular reaction dynamics.

Courtesy Faculty
Stacey F. Bent, Professor of Chemical Engineering; Ph.D., Stanford, 1992. Semiconductor processing and surface reactivity, surface functionalization, atomic layer deposition, electronic materials, biological interfaces.
James K. Chen, Assistant Professor of Molecular Pharmacology; Ph.D., Harvard, 1999. Mechanistic studies of embryonic signaling pathways, modulation of embryonic and oncogenic processes by small molecule probes, chemical approaches to zebrafish development.
Karlene Cimprich, Assistant Professor of Molecular Pharmacology; Ph.D., Harvard, 1994. Use of chemical and biochemical approaches to understand and control the DNA damage–induced cell cycle checkpoints and signal transduction cascades that allow the cell to detect and respond to DNA damage.
Curtis W. Frank, Professor of Chemical Engineering; Ph.D., Illinois, 1972. Polymer physics: dependence of polymer chain configuration on interactions with its environment.
Daniel Herschlag, Professor of Biochemistry; Ph.D., Brandeis, 1988. Biological, bioorganic, and biophysical chemistry: chemical and physical principles of biological catalysis elucidated through study of RNA and protein catalysis, RNA folding kinetics and thermodynamics, systems analysis of RNA processing elucidated via global microarray analysis.
Thomas J. Wandless, Assistant Professor of Molecular Pharmacology; Ph.D., Harvard, 1993. Organic and biological chemistry: design and synthesis of molecules that regulate specific biological processes in both cultured cells and in animals.

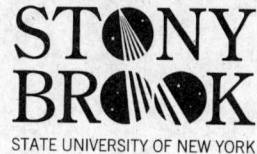

STONY BROOK
STATE UNIVERSITY OF NEW YORK

STONY BROOK UNIVERSITY, STATE UNIVERSITY OF NEW YORK

Department of Chemistry

Program of Study

The Department of Chemistry at Stony Brook offers courses of study leading to the M.S. and Ph.D. degrees. The emphasis of the graduate program is on research. Upon arrival at Stony Brook, each new student meets with a faculty adviser to choose appropriate course work. During the first semester, new students get to know the faculty members through courses, Departmental activities, and a research seminar in which the faculty members describe their research programs. Students choose a faculty research adviser at the end of the first semester, and many begin their research projects in the second semester.

The faculty members at Stony Brook have diverse backgrounds and research interests, with a significant emphasis on interdisciplinary research. Students have access to research projects in the broad range of chemistry fields, including biological, inorganic, organic, and physical chemistry, as well as studies at the interfaces of chemistry and other disciplines (materials science, physics, molecular biology, pharmacology, earth and space science, computer science, and others). Several faculty members have joint appointments with other University departments or with Brookhaven National Laboratory. Students have the option of pursuing a degree with a concentration in chemical biology or chemical physics, if desired. In addition, nearby Brookhaven National Laboratory and Cold Spring Harbor Laboratory provide many opportunities for collaborative research.

The graduate program is individually tailored to each student, with students finishing in four to six years. Graduates pursue careers both in industry and in academia, including undergraduate and research institutions.

Research Facilities

Graduate students at Stony Brook have access to excellent research facilities. The Chemistry Building is a modern, seven-story, 170,000-square-foot structure designed for research and advanced teaching. Student and faculty offices overlook the densely wooded campus and Long Island Sound.

In-house services include the machine shop and electronics shop, as well as a mass spectrometry facility with GC/MS, FAB, and TOF-MS capabilities. The NMR facility provides equipment for standard analytical characterization as well as sophisticated solution and solid-state experiments and includes 250-, 300-, 500-, and 600-MHz instruments.

The Chemistry Library, located in the Chemistry Building, maintains subscriptions to 240 journals, with online access to ninety-two, and provides 24-hour access to databases such as Beilstein Commander, SciFinder, and the Science Citation Index.

Financial Aid

All students in good standing are fully supported as either teaching or research assistants. Twelve-month stipends for 2006–07 are $22,000. Students teach in their first year and then receive research assistantships during the summer. Advanced graduate students receive similar support, most often via research assistantships.

Cost of Study

Tuition is waived for all students in good standing. Students are responsible for fees of approximately $600 per year, as well as subsidized health insurance premiums of approximately $125 per year.

Living and Housing Costs

Housing is available both on and off campus. Rates for graduate apartments on campus range from $400 to $950 per month, depending on the size of the apartment and the number of residents. For students living off campus, the estimate for total living expenses is $17,000 per year.

Student Group

The chemistry graduate program includes 130 students, of whom approximately 40 percent are women. Students in the program come from across the U.S. and a dozen other countries. The Department values a diverse student body, united in their excitement about chemistry.

Student Outcomes

Graduates from Stony Brook's Department of Chemistry go on to top industrial and academic positions. Recent graduates have taken research positions at companies such as Merck, Pfizer, and Duracell; postdoctoral positions at research institutions such as Berkeley, Harvard, Oxford, and Sloan-Kettering; and faculty positions at schools such as Manhattan College, the University of North Carolina, Florida State University, and City College, CUNY.

Location

Located near the historic village of Stony Brook, the University lies approximately 60 miles east of Manhattan on the wooded north shore of Long Island, convenient to both New York City's cultural life and Suffolk County's tranquil, recreational countryside and seashores. Long Island offers spectacular beaches, excellent fishing, and some of the East Coast's best wineries.

The University

Stony Brook is a relatively young university, founded in 1957. In part because of close contacts with Cold Spring Harbor and Brookhaven National Labs, the University has quickly grown to become one of the country's major research universities. Stony Brook has an enrollment of about 22,000 students, including 14,000 undergraduates and 8,000 graduate and professional students. The Department of Chemistry is part of the College of Arts and Science; the University also includes a medical school and colleges of engineering, health sciences, management, and marine sciences.

Applying

The Department of Chemistry uses a rolling admissions system. Most applications are received in December and January for admission in the fall semester. Applications that arrive after February 1 are considered, if possible. Information on downloading an application or applying directly online can be found at http://www.sunysb.edu/chemistry/gradprogram.

Correspondence and Information

For additional information and application forms, students should contact:

Department of Chemistry
Stony Brook University, State University of New York
Stony Brook, New York 11794-3400

Phone: 631-632-7880
Fax: 631-632-7960
E-mail: chemistry@notes.cc.sunysb.edu
Web site: http://www.sunysb.edu/chemistry

Stony Brook University, State University of New York

THE FACULTY AND THEIR RESEARCH

John M. Alexander, Professor; Ph.D., MIT, 1956. Nuclear chemistry.
Elizabeth M. Boon, Assistant Professor; Ph.D., Caltech, 2002. Biochemistry, bioinorganic chemistry, chemical biology.
Isaac S. Carrico, Assistant Professor; Ph.D., Caltech, 2003. Biochemistry, chemical biology.
Benjamin Chu, Distinguished Professor; Ph.D., Cornell, 1959. Physical chemistry, polymer physics, materials science.
Dale G. Drueckhammer, Professor; Ph.D., Texas A&M, 1987. Bioorganic chemistry.
Frank W. Fowler, Professor; Ph.D., Colorado, 1968. Synthetic chemistry.
Nancy S. Goroff, Associate Professor; Ph.D., UCLA, 1994. Organic molecules and materials.
Clare P. Grey, Professor; D.Phil., Oxford, 1991. Inorganic chemistry, solid-state chemistry and solid-state NMR.
David M. Hanson, Professor; Ph.D., Caltech, 1968. Physical chemistry, soft X-ray photochemistry.
Benjamin S. Hsiao, Professor; Ph.D., Connecticut, 1987. Physical chemistry, polymer physics, materials science.
Jiangyong Jia, Assistant Professor; Ph.D., Stony Brook, SUNY, 2003. Nuclear chemistry.
Francis Johnson, Professor (also with Pharmacological Sciences, School of Medicine); Ph.D., Glasgow (Scotland), 1954. Organic chemistry, medicinal chemistry.
Philip M. Johnson, Professor; Ph.D., Cornell, 1967. Physical chemistry, molecular spectroscopy and photophysics.
Robert C. Kerber, Professor; Ph.D., Purdue, 1965. Organic and organometallic chemistry.
Stephen A. Koch, Professor; Ph.D., MIT, 1975. Inorganic chemistry.
Roy A. Lacey, Professor; Ph.D., SUNY at Stony Brook, 1987. Nuclear chemistry.
Joseph W. Lauher, Professor; Ph.D., Northwestern, 1974. Structural chemistry.
Erwin London, Professor (also with Biochemistry and Cell Biology); Ph.D., Cornell, 1980. Structural biology, membrane protein structure.
Andreas Mayr, Professor; Ph.D., Munich, 1978. Inorganic and organometallic chemistry.
Michelle Millar, Associate Professor; Ph.D., MIT, 1975. Inorganic chemistry.
Iwao Ojima, Distinguished Professor; Ph.D., Tokyo, 1973. Synthetic, organometallic, and medicinal chemistry.
John B. Parise, Professor (also with Geosciences); Ph.D., James Cook (Australia), 1981. Inorganic chemistry.
Kathlyn A. Parker, Professor; Ph.D., Stanford, 1971. Organic chemistry.
Daniel Raleigh, Professor; Ph.D., MIT, 1988. Structural biology.
Nicole S. Sampson, Professor; Ph.D., Berkeley, 1990. Biological chemistry.
Orlando D. Schärer, Associate Professor (also with Pharmacological Sciences, School of Medicine); Ph.D., Harvard, 1986. Chemical biology of mammalian DNA repair.
Carlos Simmerling, Associate Professor; Ph.D., Illinois at Chicago, 1994. Computational structural biology.
George Stell, Distinguished Professor; Ph.D., NYU, 1961. Statistical mechanics.
Peter J. Tonge, Professor; Ph.D., Birmingham (England), 1986. Biological chemistry.
Jin Wang, Assistant Professor; Ph.D., Illinois, 1991. Physics and chemistry of biomolecules.
Michael White, Professor and Chair (also with Brookhaven National Lab); Ph.D., Berkeley, 1979. Physical chemistry, dynamics at surfaces.
Arnold Wishnia, Associate Professor; Ph.D., NYU, 1957. Biochemistry.
Stanislaus S. Wong, Assistant Professor (also with Brookhaven National Lab); Ph.D., Harvard, 1999. Physical chemistry, materials science, biophysics.

Adjunct Faculty
Joanna Fowler, Adjunct Professor (Brookhaven National Lab); Ph.D., Colorado, 1967. Biochemical affects of drugs, aging and selected diseases on the brain.
Fernando Raineri, Adjunct Assistant Professor; Ph.D., Buenos Aires, 1987. Theoretical chemistry.
Robert C. Rizzo, Adjunct Assistant Professor (Department of Applied Math and Statistics); Ph.D., Yale, 2001. Computational structural biology.
Jose R. Rodriguez, Adjunct Professor (Brookhaven National Laboratory); Ph.D., Indiana, 1988. Catalysis.
Trevor J. Sears, Adjunct Professor (Brookhaven National Lab); Ph.D., Southampton (England), 1979. Physical chemistry.

TEXAS A&M UNIVERSITY

Department of Chemistry

Programs of Study

The Department of Chemistry offers a Ph.D. degree with programs of study in traditional areas of chemistry as well as in atmospheric, biological, catalytic, environmental, materials and surface science, nuclear, polymer, solid-state, spectroscopy, and theoretical chemistry. A nonthesis M.S. (emphasis in chemical education) degree is also available.

Graduate students pursuing the Ph.D. degree select a research supervisor and formulate a plan of study during their first semester. The majority of a student's time is spent on independent research. Students present and defend a research proposal in their third year. Upon conclusion of their research, a dissertation suitable for publication is defended before their faculty advisory committee. The average time required to complete the Ph.D. degree is four to five years.

Research Facilities

The chemistry complex has 224,000 square feet of new or recently renovated space for teaching and research in four contiguous buildings, with major institutes housed in three other buildings. It maintains professionally staffed laboratories for high-resolution mass spectrometry; solution NMR; solid-state NMR; CCD-equipped, single-crystal, and powder X-ray diffractometers; a SQUID magnetometer; and departmental computing. Departmental instrumentation includes ESCA, SIMS, Auger, and other surface-science instruments; a PerSeptive Biosystems, high-performance MALDI-TOF; two Extrel FTMS 2001 systems; an XPS; and a variety of EPR, ENDOR infrared, Raman, UV-visible, fluorescence, atomic absorption, gamma-ray, and photoelectron spectrometers. Other campus facilities include the Nuclear Science Center (1-MW reactor) and the Cyclotron Institute, which includes a superconducting cyclotron. In addition, there are a number of specialized facilities, including the Center for Biological Nuclear Magnetic Resonance, the Laboratory for Molecular Structure and Bonding, the Center for Chemical Characterization and Analysis, the Laboratory for Biological Mass Spectrometry, the Laboratory for Protein Chemistry, the Laboratory for Molecular Simulation, the Center for Integrated Microchemical Systems, and the Center for Catalysis and Surface Science.

The Evans Library houses 1.6 million volumes and maintains subscriptions to approximately 8,000 scientific and technical journals.

Financial Aid

Graduate students in good standing receive full financial support for the duration of their studies. The 2005–06 stipend for twelve-month research or teaching assistantships was $21,000. All graduate assistants receive the same health-care benefits as faculty and staff members. Additional fellowships are available for outstanding applicants.

Cost of Study

Tuition and all mandatory fees are paid for up to five years of graduate study for domestic students in good standing pursuing Ph.D. degrees.

Living and Housing Costs

The cost of living in the area is low: 93 percent of the national average. University apartments are available; applications for them should be made early. Their costs range from $290 to $452 per month. Private apartments and houses for rent are available close to the campus, with prices ranging from $350 to $750 per month.

Student Group

Students in the program come from all fifty states and a dozen other countries.

Location

As a university town, College Station has a high proportion of professional people and enjoys many of the advantages of a cosmopolitan center without the disadvantages of a congested urban environment. The crime rate is very low, and students feel safe on campus. There are many film series, a symphony, chamber music, and choral groups. College Station is situated in the middle of a triangle formed by Dallas, Houston, and Austin, and the symphonies, ballets, sporting events, museums, and concerts of these cities are within easy day-trip distance.

Mild, sunny winters make the region eminently suitable for year-round activities, from fishing and hiking in the beautiful piney woods of eastern Texas to boating, bicycling, and camping in the Texas hill country. There are more than 100 state parks within a day's drive of College Station.

The University and The Department

Texas A&M University was founded in 1876 as the state's first public institution of higher education. The University's enrollment includes approximately 44,000 students studying for degrees in ten academic colleges, of whom about 7,500 are in graduate or professional programs. Vigorous research programs in biochemistry, engineering, physics, mathematics, medicine, and veterinary medicine provide chemists with supplementary facilities and intellectual resources.

The Department of Chemistry is among the top ten in the country of those at public universities and is tenth in the nation in spending on chemical research and development. The internationally known faculty members include a National Medal of Science awardee, holders of international medals in a variety of chemistry subdisciplines, and members of both the National Academy of Sciences and Royal Society. The 45 members of the graduate faculty generated approximately 450 publications and more than $14 million in external grant funding in 2002. More than 100 research fellows and visiting scientists and a graduate student body of about 240 support their efforts. Though the department is large, most research groups have 3 to 10 students and thus provide an intensive, personalized learning environment.

Applying

There is no application fee for domestic applicants. Online application forms and a more detailed description of requirements are available at the department's Web site. Admission decisions are made on a continuous basis beginning in December. Departmental fellowship awards are made in February for the next academic year. Domestic applications for fall admission should arrive by April 1 for preferential consideration. International applications must be received by January 31. Applications and all supporting material should be filed no later than six weeks prior to the opening of the preferred semester of entrance.

Correspondence and Information

Graduate Student Office
Department of Chemistry
Texas A&M University
P.O. Box 30012
College Station, Texas 77842-3012
Phone: 979-845-5345
800-334-1082 (toll-free)
Fax: 979-845-5211
E-mail: gradmail@mail.chem.tamu.edu
Web site: http://www.chem.tamu.edu

Texas A&M University

THE FACULTY AND THEIR RESEARCH

David Barondeau, Assistant Professor; Ph.D., Texas A&M, 1996. Bioinorganic chemistry.
James Batteas, Associate Professor; Ph.D., Berkeley, 1995. Physical/analytical chemistry.
David Bergbreiter, Professor; Ph.D., MIT, 1974. Organic chemistry.
John Bevan, Professor; Ph.D., London, 1975. Physical chemistry.
Kevin Burgess, Professor; Ph.D., Cambridge, 1983. Organic chemistry.
Abraham Clearfield, Professor; Ph.D., Rutgers, 1954. Inorganic chemistry.
Brian Connell, Assistant Professor; Ph.D., Harvard, 2002. Organic chemistry.
F. Albert Cotton, Distinguished Professor and Doherty-Welch Chair; Ph.D., Harvard, 1955. Inorganic chemistry.
Paul Cremer, Professor; Ph.D., Berkeley, 1996. Surface science.
Donald J. Darensbourg, Professor; Ph.D., Illinois at Urbana-Champaign, 1968. Organometallic chemistry.
Marcetta Y. Darensbourg, Professor; Ph.D., Illinois at Urbana-Champaign, 1967. Organometallic chemistry.
Kim R. Dunbar, Professor; Ph.D., Purdue, 1984. Inorganic chemistry.
John P. Fackler Jr., Distinguished Professor; Ph.D., MIT, 1960. Inorganic chemistry.
Paul F. Fitzpatrick, Professor; Ph.D., Michigan, 1981. Biochemistry.
Francois Gabbai, Associate Professor; Ph.D., Texas at Austin, 1994. Inorganic chemistry.
Yi-Qin Gao, Assistant Professor; Ph.D., Caltech, 2001. Theoretical chemistry.
D. Wayne Goodman, Distinguished Professor and Robert A. Welch Chair; Ph.D., Texas at Austin, 1974. Physical chemistry.
Michael B. Hall, Professor; Ph.D., Wisconsin–Madison, 1971. Inorganic chemistry.
Kenn E. Harding, Professor; Ph.D., Stanford, 1968. Organic chemistry.
Christian Hilty, Assistant Professor; Ph.D., Swiss Federal Institute of Technology, 2004. Biophysical chemistry.
John L. Hogg, Professor; Ph.D., Kansas, 1974. Bioorganic chemistry.
Timothy Hughbanks, Professor; Ph.D., Cornell, 1983. Inorganic chemistry.
Arthur Johnson, Professor and Wehner-Welch Chair; Ph.D., Oregon, 1973. Biochemistry.
Jaan Laane, Professor; Ph.D., MIT, 1967. Physical chemistry.
Paul Lindahl, Professor; Ph.D., MIT, 1985. Inorganic chemistry.
Robert R. Lucchese, Professor; Ph.D., Caltech, 1982. Theoretical chemistry.
Jack H. Lunsford, Distinguished Professor Emeritus; Ph.D., Rice, 1962. Physical chemistry.
Ronald D. Macfarlane, Professor; Ph.D., Carnegie Tech, 1959. Bioanalytical chemistry.
Stephen A. Miller, Assistant Professor; Ph.D., Caltech, 1999. Organic chemistry.
Joseph B. Natowitz, Distinguished Professor; Ph.D., Pittsburgh, 1965. Nuclear chemistry.
Simon North, Associate Professor; Ph.D., Berkeley, 1995. Physical chemistry.
Frank M. Raushel, Professor; Ph.D., Wisconsin–Madison, 1976. Biochemistry.
Daniel Romo, Professor; Ph.D., Colorado State, 1991. Organic chemistry.
Michael P. Rosynek, Professor and Associate Head; Ph.D., Rice, 1972. Physical chemistry.
Marvin W. Rowe, Professor; Ph.D., Arkansas, 1966. Analytical cosmochemistry.
David H. Russell, Professor; Ph.D., Nebraska–Lincoln, 1978. Analytical chemistry.
James C. Sacchettini, Professor; Ph.D., Washington (St. Louis), 1987. Biochemistry.
Raymond E. Schaak, Assistant Professor; Ph.D., Penn State, 2001. Inorganic chemistry.
Emile A. Schweikert, Professor and Head; Ph.D., Paris IV (Sorbonne), 1964. Activation analysis and analytical chemistry.
A. Ian Scott, Distinguished Professor, Derek Barton Professor, and Robert A. Welch Chair; Ph.D., Glasgow, 1952. Organic chemistry, biochemistry.
Eric E. Simanek, Associate Professor; Ph.D., Harvard, 1996. Organic chemistry.
Daniel A. Singleton, Professor; Ph.D., Minnesota, 1986. Organic chemistry.
Dong Hee Son, Assistant Professor; Ph.D., Texas at Austin, 2002. Physical/analytical/materials chemistry.
Manual P. Soriaga, Professor; Ph.D., Hawaii, 1978. Analytical chemistry.
Gyula Vigh, Professor; Ph.D., Veszperm (Hungary), 1975. Analytical chemistry.
Coran Watanabe, Assistant Professor; Ph.D., Johns Hopkins, 1998. Biological chemistry.
Rand L. Watson, Professor; Ph.D., Berkeley, 1966. Nuclear chemistry.
Robert D. Wells, Professor; Ph.D., Pittsburgh, 1964. Biochemistry.
Danny L. Yeager, Professor; Ph.D., Caltech, 1975. Theoretical chemistry.
Sherry J. Yennello, Professor; Ph.D., Indiana, 1990. Nuclear chemistry.

TUFTS UNIVERSITY

Department of Chemistry

Programs of Study

The Department of Chemistry offers graduate programs leading to the degrees of professional Master of Science and Doctor of Philosophy in the fields of analytical, inorganic, organic, and physical chemistry and in the subdisciplinary areas of bioorganic, environmental, and materials science and chemistry-biotechnology. Programs of study may involve collaborations in related science departments or in the Sackler School of Graduate Biomedical Sciences, the Tufts–New England Medical Center, or Tufts Biotechnology Center.

Entering graduate students meet with the Graduate Committee to plan an academic course schedule best suited to the student's background and career goals. New students must complete four graduate courses, one course in each of the four chemistry disciplines (analytical, inorganic, organic, and physical) by the end of the third academic semester. This course of study is to ensure that by the end of the third semester each student has a firm foundation in the fundamentals of chemistry. A research advisory committee, consisting of the thesis adviser and two other faculty members, then directs the student's course and research program.

A Ph.D. candidate must complete a minimum of six formal courses (exclusive of research) and present a departmental seminar. The student must also present an additional independent study topic, successfully defend a research proposal, and complete a dissertation reporting significant research of publishable quality. Additional course work may be required at the discretion of the research adviser.

The professional master's program is very flexible in order to accommodate individual goals. Each student must pass eight graduate-level courses, at least six of which must be formal classroom instruction. Up to half of the courses may be taken outside the Department of Chemistry in related fields. A thesis may or may not be required, depending on the importance of a thesis for the candidate's career plans.

Research Facilities

Research is carried out in the Pearson and Michael Laboratories, combined facilities of 66,000 square feet. A wide array of modern instrumentation necessary for cutting-edge research in chemistry is available for general use by graduate students, including FT-NMR, FT-IR, UV-Vis, AA, AES, and fluorescence spectrometers; scanning probe (STM, AFM) and scanning electron microscopes; GC-MS and MALDI-TOF mass spectrometers; analytical and preparative gas and liquid chromotography equipment; pulsed and CW laser systems for analytical and physical measurement; computerized electrochemical instrumentation; UHV surface analysis equipment; a fermentor, incubator, and coldroom; and professionally staffed electronics and machine shops. Complementary instrumentation is available at other Tufts facilities, including the Sackler School of Graduate Biomedical Sciences, the Tufts–New England Medical Center, the Science and Technology Center, and the School of Nutrition. All laboratories, classrooms, and offices are wired for high-speed Internet connections, with access to most online scientific journals.

Financial Aid

To help students whose records indicate scholarly promise, a variety of financial awards and work opportunities are available. All Ph.D. candidates receive twelve months of financial support, which is guaranteed for five years. Students must remain in good academic standing and make steady progress toward the Ph.D. degree. The 2004–05 twelve-month minimum stipend was $22,800 and was derived from teaching or research assistantships. Graduate students are supported by departmental fellowships during the first summer, enabling them to devote full-time to research. Supplemental fellowship awards are also available for outstanding students. Stipends are reviewed annually.

Cost of Study

All Ph.D. candidates receive tuition scholarships.

Living and Housing Costs

Most graduate students attending Tufts University live off campus in moderately priced apartments in the surrounding metropolitan area. Meal plans are available.

Student Group

At present the total enrollment in all divisions is about 7,000 students, including approximately 2,000 graduate and professional students.

Location

There are a variety of local restaurants and entertainment options close to the Tufts campus. The Boston area offers an excellent environment for the pursuit of academic interests. Due to the high density of world-famous institutions, many distinguished chemists visit the Boston area to present seminars and to confer with colleagues. All graduate students may obtain Boston library consortium privileges, enabling them to use the library facilities of other local universities. Researchers at Tufts have been able to take advantage of various facilities available at other local universities such as MIT and Harvard. The cultural offerings of the Boston area are some of the finest in the world, including the Boston Symphony, the Museum of Fine Arts, the Museum of Science, and the New England Aquarium, as well as innumerable chamber groups, performing groups, and theaters showing first-run and major international films.

The University

Since its designation as Tufts College in 1852, Tufts University has grown to include seven primary divisions: Arts and Sciences; the Fletcher School of Law and Diplomacy; the Schools of Medicine, Dental Medicine, and Veterinary Medicine; the Sackler School of Graduate Biomedical Sciences; and the School of Nutrition.

Applying

Application materials should be submitted directly to the Graduate School Office by February 15 for September enrollment and by October 15 for January enrollment. Applications received after these dates are considered on a space-available basis. All U.S. applicants must submit their test scores on the General Test of the Graduate Record Examinations (GRE) and are strongly encouraged to take the Subject Test in chemistry. International students must submit scores on both the General Test and Subject Test in chemistry, the results of the Test of English as a Foreign Language (TOEFL), and the Test of Spoken English (not available after September 2005 as the TOEFL will then include a test of speaking proficiency). Applicants are urged to take the appropriate tests in October or December. Application forms for admission and support may be obtained directly from the Graduate School of Arts and Sciences or the chemistry department. Applicants are encouraged to visit the University; an appointment can be arranged with the department beforehand. Tufts University is an equal opportunity institution.

Correspondence and Information

Graduate Committee Chair
Department of Chemistry
Tufts University
Medford, Massachusetts 02155

Phone: 617-627-3441
Fax: 617-627-3443
E-mail: chemgradinfo@tufts.edu
Web site: http://chem.tufts.edu

Tufts University

THE FACULTY AND THEIR RESEARCH

Marc d'Alarcao, Ph.D., Illinois. Synthesis and evaluation of compounds of biological interest, especially inositol-containing carbohydrates and nucleosides.
Robert R. Dewald, Ph.D., Michigan State. Mechanistic studies of metal-ammonia reductions, chemistry of metal metalides and nonaqueous solvents.
Terry E. Haas, Ph.D., MIT. Physical inorganic chemistry; structure and electronic structure and optical, electronic, and transport properties of thin films, solids, and thin-film devices; synthesis and characterization of thin films; X-ray crystallography.
Jonathan E. Kenny, Ph.D., Chicago. In situ detection of contaminants in soil and groundwater using laser-induced fluorescence, Raman spectroscopies, and fiber optics; fluorescence fingerprinting of natural waters; multidimensional fluorescence characterization of sediments.
Samuel P. Kounaves, Ph.D., Geneva (Switzerland). Fundamental questions in planetary science using techniques of analytical chemistry and especially electrochemically based sensors; use of in situ autonomous chemical analysis systems to study Martian geochemistry and possible biology, in both the regolith (soil) and polar ice caps, and to investigate the subglacial oceans on Jupiter's moon Europa.
Krishna Kumar, Ph.D., Brown. Organic and bioorganic chemistry, self-assembling and self-organizing systems, peptide and protein design, combinatorial chemistry using dynamic libraries, studies into the origin of exon-intron gene structure of modern-day enzymes.
David H. Lee, Ph.D., Scripps Research Institute. Biomaterials, self-assembly, bionanotechnology.
Albert Robbat Jr., Ph.D., Penn State. Development of analytical methods for hazardous-waste site field investigations—gas, liquid, and supercritical chromatographies and a mobile mass spectrometer; PCBs in marine life, ocean water, and sediment; PAHs in hazardous-waste incinerator and coal combustion emissions; volatile and semivolatile organics in soil, ground, and surface water; electron transfer mechanisms and rate measurements in biological systems.
Elena V. Rybak-Akimova, Ph.D., Kiev. Coordination, supramolecular, and bioinorganic chemistry; synthetic macrocyclic transition metal complexes; molecular tweezers; redox catalysis and enzyme mimics; dioxygen binding and activation; spatiotemporal self-organization (oscillating reactions and pattern formation) in chemical reactions; molecular modeling of macrocyclic supramolecular aggregates.
Mary J. Shultz, Ph.D., MIT. Development of methods for probing liquid, gas/liquid, solid/liquid, or gas/solid interface; probing dynamical processes at high-vapor-pressure interfaces; mechanism of heterogeneous photochemistry of transition metal oxides; heterogeneous processes in atmospheric chemistry, including ozone depletion nanomaterials.
Robert D. Stolow, Ph.D., Illinois. NMR and computational studies of conformational equilibria in solution and in the gas phase; the influence of electrostatic interactions among polar groups upon conformational energies.
E. Charles Sykes. Ph.D., Cambridge. Investigating chemical systems by high-resolution scanning tunneling microscopy (STM).
Arthur L. Utz, Ph.D., Wisconsin. Dynamics of gas-surface reactions relevant to materials and catalytic chemistry; supersonic molecular beam, laser excitation, and ultrahigh vacuum techniques; laser-induced chemistry at surfaces; mechanisms for heterogeneous catalysis; vibrational and translational energy as synthetic tools in materials chemistry.
David R. Walt, Ph.D., SUNY at Stony Brook. Chemical sensors, microfabrication, genechips, micromaterials, nanomaterials.

UNIVERSITY OF CALIFORNIA, SAN DIEGO

Department of Chemistry and Biochemistry

Program of Study

The goal of the program is to prepare students for careers in science as researchers and educators by expanding their knowledge of chemistry while developing their ability for critical analysis, creativity, and independent study. Research opportunities are comprehensive and interdisciplinary, spanning biochemistry; bioinformatics; biophysics; inorganic, organic, physical, analytical, computational, and theoretical chemistry; surface and materials chemistry; and atmospheric and environmental chemistry. During the first year, students take courses, begin their teaching apprenticeships, choose a research adviser, and embark on their thesis research; students whose native language is not English must pass an English proficiency examination. In the second year, there is an oral examination, which includes critical discussion of a recent research article. In the third year, students advance to candidacy for the doctorate by defending the topic, preliminary findings, and future research plans for their dissertation. Subsequent years focus on thesis research and writing the dissertation. Most students graduate during their fifth year.

At the University of California, San Diego (UCSD), chemists and biochemists are part of a thriving community that stretches across the campus and out into research institutions throughout the San Diego area, uniting researchers in substantive interactions and collaborations. Seminars are presented weekly in biochemistry and inorganic, organic, and physical chemistry. Interdisciplinary programs in nonlinear science, materials science, biophysics, bioinformatics, and atmospheric and planetary chemistry also hold regular seminars.

Research Facilities

State-of-the-art facilities include a national laboratory for protein crystal structure determination; high-field nuclear magnetic resonance (NMR) instruments; the Natural Sciences Graphics Laboratory, which provides high-end graphic workstation resources; and laser spectroscopic equipment. Buildings specially designed for chemical research, a computational center, a laboratory fabrication and construction facility, and machine, glass, and electronic shops are part of the high-quality research support system.

The UCSD library collection is one of the largest in the country, with superior computerized reference and research services. Access to the facilities at the Scripps Institute of Oceanography, the San Diego Supercomputer Center, the Salk Institute, the Scripps Research Institute, and a thriving technological park, all within blocks of the campus, make the overall scope of available research facilities among the best in the world.

Financial Aid

All students who remain in good academic standing are provided year-round support packages of a stipend plus fees and tuition. Support comes from a variety of sources, including teaching assistantships, research assistantships, fellowships, and awards. Special fellowships, such as the GAANN, Urey, Cota Robles, and San Diego fellowships, are available to outstanding students. Students are strongly encouraged to apply for outside fellowships, and the Department supplements such awards. The twelve-month stipend is adjusted annually; for 2005–06, it is $22,400. Entering students receive $600 in relocation allowance. Emergency short-term as well as long-term loan programs are administered by the UCSD Student Financial Services office.

Cost of Study

Registration fees and tuition (paid by the Department) are $6365 and $18,202, respectively, per year. Premiums for a primary health-care program, which covers most major medical expenses and a portion of dental fees, are covered by registration fees. The Student Health Center treats minor illnesses and injuries. Optional health and dental coverage for dependents is at the student's expense.

Living and Housing Costs

The University Housing Service operates more than 1,000 apartments for families, couples, and single graduate students. There are several apartment complexes near the campus at higher rents; rental sharing is a common way to reduce the expense. The Off Campus Student Housing Office, located on campus, maintains extensive current rental and rental-share opportunities.

Student Group

Students are drawn from the top ranks of U.S. and international colleges and universities. There are 3,300 graduate students on the general campus and at the Scripps Institute of Oceanography and 500 in the School of Medicine. Within the Department, there are 250 graduate students and 900 undergraduate majors. The graduate student population reflects diversity in culture, gender, and ethnicity.

Student Outcomes

Graduates typically obtain jobs in academia (55 percent) or in the chemical industry (45 percent). Many take postdoctoral research positions in academic institutions or national laboratories that lead to future academic or industrial careers. The Departmental Industrial Relations Office assists students with placement in industrial positions and with networking with industry.

Location

The campus sits on 1,200 acres of eucalyptus groves near the Pacific Ocean. Surrounding the campus is La Jolla, a picturesque community of boutiques, bistros, and businesses. Seven miles south of the campus is San Diego, with its world-acclaimed zoo, museums, and theaters. The Laguna Mountains and the Anza-Borrego Desert are within a 2-hour drive east of the campus.

The University

UCSD, a comparatively young university, has already achieved widespread recognition, ranking fifth in federal funding for research and development and in the top ten of all doctoral degree–granting institutions in a study conducted by the National Research Council. Recently, the Institute for Scientific Information ranked UCSD's Department of Chemistry and Biochemistry as third in the nation for "High Impact U.S. Universities, 1994–98." Programs span the arts and humanities, engineering, international studies, and the social, natural, and physical sciences. The intellectual climate is enhanced by a variety of social, educational, professional, political, religious, and recreational opportunities and services.

Applying

Application packets include a completed UCSD application form, a statement of purpose, official transcripts from previous colleges, three letters of recommendation, GRE scores (general and advanced chemistry or biochemistry), and a TOEFL score (for noncitizens only; a minimum score of 550 on the paper-based test or 220 on the computer-based test is required). Research experience should be described. Copies of or references to any publications should be included with the application. Applications received by January 15 receive the highest priority.

Correspondence and Information

Graduate Admissions Coordinator
Department of Chemistry and Biochemistry 0301
University of California, San Diego
9500 Gilman Drive
La Jolla, California 92093-0301
Phone: 858-534-6870
Fax: 858-534-7687
E-mail: gradinfo@chem.ucsd.edu
Web site: http://www-chem.ucsd.edu/

University of California, San Diego

THE FACULTY AND THEIR RESEARCH

Timothy S. Baker, Ph.D., UCLA. Biochemistry.
Michael D. Burkart, Ph.D., Scripps Research Institute. Biological chemistry, chemo-enzymatic synthesis, natural product biosynthesis. Focus on antibiotic design, synthesis, biosynthesis.
Seth Cohen, Ph.D., Berkeley. Bioinorganic and coordination chemistry; metalloregulatory proteins, zinc proteinase inhibitors, and supramolecular materials proteins); design and synthesis of zinc proteinase inhibitors, synthesis of materials containing coordination clusters as building blocks.
Robert E. Continetti, Ph.D., Berkeley. Dissociation dynamics of transient species, three-body reaction dynamics, novel mass-spectrometric methods.
John E. Crowell, Ph.D., Berkeley. Materials chemistry, surface kinetics of metals/semiconductors, CVD, photo-induced deposition, thin-film spectroscopy.
Edward A. Dennis, Ph.D., Harvard. Biochemistry: phospholipase A2, signal transduction in macrophages, mechanism, prostaglandin regulation, mass spec of lipids and proteins.
Jack Dixon, Ph.D., California, Santa Barbara. Biochemistry.
Daniel J. Donoghue, Ph.D., MIT. Signal transduction, human cancer, receptor tyrosine kinase, cell cycle, tumor suppressor, oncogenesis.
Marye Anne Fox, Chancellor; Ph.D., Dartmouth. Organic chemistry: physical organic chemistry, photochemistry, nanoscience of organized arrays.
Gourisankar Ghosh, Ph.D., Yeshiva (Einstein). Biochemistry and biophysics: transcription; signaling; pre-mRNA splicing; mRNA transport; protein/DNA/RNA interactions.
Partho Ghosh, Ph.D., California, San Francisco. Mechanisms of bacterial and protozoan pathogenesis, host response against infectious microbes.
Sergio Guazzotti, Ph.D., California, Riverside. Chemical education.
David N. Hendrickson, Ph.D., Berkeley. Inorganic chemistry, materials chemistry, single-molecule magnets, dynamics of transition metal complexes.
Thomas C. Hermann, Ph.D., Ludwig Maximilians (Germany). Organic chemistry.
Alexander Hoffman, Ph.D., Rockefeller. Biochemistry: signaling, transcription, computational network, stress and immune responses, apoptosis, proliferation.
Patricia A. Jennings, Ph.D., Penn State. Biophysical chemistry: protein structure, dynamics, and folding; 2-, 3-, and 4-D NMR; equilibrium/kinetic-fluorescence; circular dichroism.
Simpson Joseph, Ph.D., Vermont. Biochemistry and biophysics: ribosome structure, function and dynamics; discovery of novel antibiotics.
Judy E. Kim, Ph.D., Berkeley. Biophysical chemistry: spectroscopic studies of membrane protein folding and dynamics.
Yoshihisa Kobayashi, Ph.D., Tokyo (Japan). Natural product synthesis, new reaction and catalyst, heterocycles synthesis, elucidation of stereostructure.
Elizabeth A. Komives, Ph.D., California, San Francisco. Structure, function, dynamics, and thermodynamics of protein-protein interactions: NMR, mass spectrometry and kinetics.
Clifford P. Kubiak, Chair; Ph.D., Rochester. Inorganic chemistry: electron transfer, organometallic chemistry. Nanoscience: molecular electronics, nanosensors.
Andrew C. Kummel, Ph.D., Stanford. STM/STS of gate oxides on compound semiconductors and adsorbates on organic semiconductor.
Katja Lindenberg, Ph.D., Cornell. Theoretical chemical physics: nonequilibrium statistical mechanics, stochastic processes, nonlinear phenomena, condensed-matter theory.
Douglas Magde, Ph.D., Cornell. Experimental physical chemistry: photochemistry and photobiophysics, picosecond and femtosecond lasers.
J. Andrew McCammon, Ph.D., Harvard. Statistical mechanics and computational chemistry, with applications to biological systems.
Karsten Meyer, Ph.D., Mulheim (Max-Planck). Inorganic chemistry: bridges coordination chemistry with the fields of organometallic and actinide chemistry.
Mario J. Molina, Ph.D., Berkeley. Atmospheric chemistry; environmental chemistry.
Terunaga Nakagawa, Ph.D., Tokyo. Biochemistry.
K. C. Nicolaou, Ph.D., University College, London. Total synthesis and chemical biology of natural and designed molecules.
Joseph M. O'Connor, Ph.D., Wisconsin–Madison. Organotransition metal; organic, bioorganometallic, and inorganic chemistry.
Hans Oesterreicher, Ph.D., Vienna. Solid-state science: magnetic information storage, superconductivity.
Stanley Opella, Ph.D., Stanford. NMR structural studies of proteins in membranes and other supramolecular assemblies.
Charles L. Perrin, Ph.D., Harvard. Physical-organic chemistry: stereoelectronic effects, hydrogen bonding, isotope effects, ionic solvation.
Kimberly A. Prather, Ph.D., California, Davis. Environmental, analytical chemistry: gas/particle processes of tropospheric significance; mass spectrometry; laser-based techniques.
Arnold Rheingold, Ph.D., Maryland. Inorganic chemistry: small-molecule crystallography, synthesis of transition metal/p-block clusters.
Michael J. Sailor, Ph.D., Northwestern. Nanomaterials: porous silicon, chemical and biological sensors, biomaterials, electrochemistry.
Barbara A. Sawrey, Ph.D., California, San Diego, and San Diego State. Chemical education: development of computer-based multimedia to assist student learning of complex scientific processes and concepts.
Amitabha Sinha, Ph.D., MIT. Experimental physical chemistry: photochemistry, laser spectroscopy, reaction dynamics of vibrationally excited molecules.
Susan S. Taylor, Ph.D., Johns Hopkins. Protein kinases/signal transduction: structure/function and localization; biophysics; crystallography; NMR, fluorescence/FRET.
F. Akif Tezcan, Ph.D., Caltech. Inorganic chemistry.
Emmanuel A. Theodorakis, Ph.D., Paris XI (South). Synthetic, medicinal, bioorganic, and biological chemistry; methods and strategies in natural products chemistry.
Mark H. Thiemens, Ph.D., Florida State. Atmospheric chemistry: physical chemistry of isotope effects, solar system information.
Yitzhak Tor, Ph.D., Weizmann (Israel). Ligand-nucleic acid interactions, metal-containing materials, new emissive molecules.
William C. Trogler, Ph.D., Caltech. Inorganic chemistry: polymer chemistry, nanotechnology applied to chemical and environmental sensing.
Roger Y. Tsien, Ph.D., Cambridge. Chemical biology; design, synthesis, and application of molecular probes of biological function.
Robert H. Tukey, Ph.D., Iowa. Environmental toxicology: the role of environmental and chemical toxicants on gene expression.
Peter L. van der Geer, Ph.D., Amsterdam. Biochemistry: molecular, biological, and biochemical analysis of signal transduction downstream of normally and malignantly activated protein-tyrosine kinases.
Michael S. VanNieuwenhze, Ph.D., Indiana. Chemical biology, synthetic methods, natural products synthesis, solid-phase synthesis, carbohydrate chemistry.
Wei Wang, Ph.D., California, San Francisco. Inference of gene regulatory networks and determination of protein specificity.
John H. Weare, Ph.D., Johns Hopkins. Physical chemistry: calculations of the dynamics of complex systems, theoretical geochemistry.
Haim Weizman, Ph.D., Weizmann (Israel). Organic chemistry.
John C. Wheeler, Ph.D., Cornell. Physical chemistry: calculations of the dynamics of complex systems, theoretical geochemistry.
James K. Whitesell, Ph.D., Harvard. Organic chemistry.
Peter Wolynes, Ph.D., Harvard. Theoretical chemical physics, protein folding and function, glasses and stochastic cell biology.
Jerry Yang, Ph.D., Columbia. Bioorganic chemistry: molecular self-assembly, materials chemistry, bionanotechnology.

Adjunct Faculty
Kim Baldridge, Ph.D., North Dakota State; Senyon Choe, Ph.D., Berkeley; Daniel F. Harvey, Ph.D., Yale; John E. Johnson, Ph.D., Iowa State; Joseph P. Noel, Ph.D., Ohio State; Leslie E. Orgel, Ph.D., Oxford; Shankar Subramaniam, Ph.D., Indian Institute of Technology (Kanpur).

UCSF
University of California
San Francisco

UNIVERSITY OF CALIFORNIA, SAN FRANCISCO

Chemistry and Chemical Biology Graduate Program

Program of Study

The Ph.D. program in chemistry and chemical biology (CCB) provides students with a broad and rigorous background in modern chemistry that includes training in molecular thermodynamics, bioorganic chemistry, computational chemistry, and structural biology. The program is distinctive in its orientation toward the study of molecules in living systems. It is further distinguished by providing integrated training in the sciences related to chemical biology: integrated both with respect to the levels of structure (atomic, molecular, cellular) and with respect to the traditional disciplines of chemistry and biology in the setting of a health science campus. The training objectives for students of the program are met through course work, laboratory rotations, and activities of the program such as journal clubs and research presentations and through thesis research in a specific laboratory.

The CCB curriculum includes a series of core required courses, including molecular thermodynamics, reaction mechanisms, physical organic chemistry, and chemical biology. In addition to these required courses, students must also take two electives. During the first year, students work in three separate laboratories for one quarter each. The rotation is expected to take roughly half of the student's time, with an oral presentation at the end of each rotation that is attended by other students and members of the lab. This provides valuable exposure to the diversity of the program and allows the student to gain firsthand experience for choosing a thesis laboratory. Students generally finish the program in approximately five years.

Research Facilities

The faculty has excellent facilities for carrying out research in chemistry and chemical biology. These include a National Bio-organic, Biomedical Mass Spectrometry Laboratory; a National Research Computer Graphics Laboratory; a Nuclear Magnetic Resonance Laboratory for both high-resolution NMR and in vivo spectroscopy and imaging; a high-resolution X-ray crystallography laboratory; and numerous research support laboratories in addition to individual faculty laboratories.

Financial Aid

All doctoral candidates receive funding as research assistants, teaching assistants, fellows, or trainees at a standardized level of $26,000 for 2006–07.

Cost of Study

The graduate program covers annual fees for its students ($8899 in 2005–06). Nonresident tuition was an additional $14,694 in 2005–06. Nonresident tuition is paid for the first year for U.S. citizens and permanent residents. After the first year, students are expected to have established California residency.

Living and Housing Costs

The Mission Bay housing complex of 431 newly built apartments opened in fall 2005. The four high-rise buildings offer studios and one-, two-, three-, and four-bedroom units. With views of the city and the bay, the complex is located at the main plaza entrance to the new campus on 3rd Street at 13th Street. Within walking distance are SBC Park, Muni and Cal trains, supermarkets, restaurants, bookstores, banks and retail services. The Campus Housing Office maintains listings of current community-supplied rental information concerning vacant off-campus houses, apartments, rooms, and various types of shared housing.

Student Group

Approximately 55 Ph.D. students are enrolled full-time and receive full financial support. A wide range of ethnic, academic, and national backgrounds are represented.

Student Outcomes

A need exists for a large number of faculty members qualified in scientific research and teaching, and the need for qualified chemists and chemical biologists in the biotechnology and pharmaceutical industries has never been higher. Such fields as patent law and government also have job openings in this area. There is a strong demand for Ph.D. scientists capable of research and teaching in this cross-disciplinary field of chemistry and chemical biology.

Location

The Parnassus Heights campus is located on a hill in a residential area of San Francisco. San Francisco has a diverse collection of ethnic and social traditions, lifestyles, and community groups. An abundance of visual and performing arts events, including opera, symphony, theater, museums, galleries, dance, music, and arts organizations, are available, as are neighborhood street fairs and sporting events.

The Chemistry and Chemical Biology Graduate Program occupies space in Genentech Hall at the University of California, San Francisco's (UCSF) new Mission Bay campus. In addition to Genentech Hall, two more life science research buildings, Rock Hall and the newly constructed QB3 building, are now occupied. At the Mission Bay campus, UCSF has built a campus community center with a state-of-the-art fitness center and pub cafe, parking structures, and a housing complex for students and postdocs. By the time the campus is complete, it will have 2 million square feet of space in twenty new buildings.

The University

The University of California, San Francisco, one of ten campuses of a statewide university system, is the only UC campus dedicated exclusively to the health sciences. The campus is home to professional schools in dentistry, medicine, nursing, and pharmacy and graduate programs in the biological, biomedical, and behavioral sciences. UCSF encompasses several major sites in San Francisco.

Applying

A formal application is required of all persons seeking admission to the CCB program. A baccalaureate degree is required. Other required materials include official transcripts of previous college work with a minimum 3.0 GPA, official GRE General Test scores and a GRE Subject Test score from tests taken within the last five years, a statement of purpose, letters of recommendation, a program application and Graduate Division application, and a $60 fee. Completed forms must be submitted by January 5. Students are admitted for the fall quarter only.

Correspondence and Information

Graduate Program Administrator
Chemistry and Chemical Biology Graduate Program
University of California, San Francisco
600 16th Street, Room 522
San Francisco, California 94143-2280

Phone: 415-476-1914
Fax: 415-514-1546
E-mail: ccb@picasso.ucsf.edu
Web site: http://www.ucsf.edu/ccb/

University of California, San Francisco

THE FACULTY AND THEIR RESEARCH

Dave Agard, Professor of Biochemistry and Biophysics and of Pharmaceutical Chemistry; Ph.D., Caltech. Steroid receptor structure-function, protein folding and centrosome structure.

Patricia Babbitt, Associate Professor of Biopharmaceutical Sciences and of Pharmaceutical Chemistry; Ph.D., California, San Francisco. Protein bioinformatics and protein engineering.

Carolina Bertozzi, Professor of Chemistry and Molecular Cell Biology (University of California, Berkeley, and Howard Hughes Medical Institute) and of Cellular and Molecular Pharmacology; Ph.D., Berkeley. Molecular basis of cell surface interactions.

Frances Brodsky, Professor of Biopharmaceutical Sciences and of Pharmaceutical Chemistry; D.Phil., Oxford. Molecular mechanisms of intracellular membrane traffic.

Alma Burlingame, Professor of Pharmaceutical Chemistry; Ph.D., MIT. Protein machines, posttranslational dynamics and mass spectrometry.

M. Almira Correia, Professor of Cellular and Molecular Pharmacology, Medicine, Pharmacy, and Pharmaceutical Chemistry; Ph.D., Minnesota. Structure-function relationships and regulation of hepatic hemoproteins.

Charles Craik, Professor of Pharmaceutical Chemistry, Cellular and Molecular Pharmacology, and Biochemistry and Biophysics; Ph.D., Columbia. Chemical biology of proteolysis.

Joe DeRisi, Assistant Professor of Biochemistry and Biophysics; Ph.D., Stanford. DNA microarrays.

Marc Diamond, Assistant Professor of Neurology and of Cellular and Molecular Pharmacology; M.D., California, San Francisco. Regulating protein conformation to treat human disease.

Ken Dill, Professor of Pharmaceutical Chemistry and of Biopharmaceutical Sciences; Ph.D., California, San Diego. Statistical mechanics of biomolecules.

Jonathan Ellman, Associate Professor of Chemistry (University of California, Berkeley) and Cellular and Molecular Pharmacology; Ph.D., Harvard. Design, synthesis, and evaluation of small-molecule libraries.

Pamela England, Assistant Professor of Pharmaceutical Chemistry and of Cellular and Molecular Pharmacology; Ph.D., MIT. Chemical neurobiology: ion channel structure-function, synaptic plasticity.

Tom Ferrin, Professor of Pharmaceutical Chemistry; Ph.D., California, San Francisco. Macromolecular structure and function through use of computational algorithms and interactive three-dimensional computer graphics.

Robert Fletterick, Professor of Biochemistry and Biophysics and of Cellular and Molecular Pharmacology; Ph.D., Cornell. Protein structure and function.

Alan Frankel, Professor of Biochemistry and of Cellular and Molecular Pharmacology; Ph.D., Johns Hopkins. RNA-protein recognition.

Bradford Gibson, Professor of Pharmaceutical Chemistry; Ph.D., MIT. Coupling mass spectroscopy to structural biology.

John Gross, Assistant Professor of Pharmaceutical Chemistry; Ph.D., MIT. Translational control of gene expression, RNA turnover.

Holly Ingraham, Associate Professor of Physiology, Obstetrics, Gynecology, and Reproductive Sciences; Ph.D., California, San Diego. Gene expression and cellular signaling in reproductive development.

Matthew Jacobson, Assistant Professor of Pharmaceutical Chemistry; Ph.D., MIT. Physical chemistry–based approaches to predictive protein modeling.

Tom James, Professor and Chair of Pharmaceutical Chemistry and Professor of Radiology; Ph.D., Wisconsin. Three-dimensional structure determination by NMR.

David Julius, Professor of Cellular and Molecular Pharmacology; Ph.D., Berkeley. Molecular mechanisms of neurotransmitter action.

Stephen Kahl, Professor of Pharmaceutical Chemistry; Ph.D., Indiana. Synthesis and evaluation of cancer chemotherapeutics.

Andrew Krutchinsky, Assistant Professor of Pharmaceutical Chemistry; Ph.D., Manitoba. Biological mass spectrometry.

Wendell Lim, Associate Professor of Cellular and Molecular Pharmacology and of Biochemistry and Biophysics; Ph.D., MIT. Mitosis and cell motility.

Michael Marletta, Professor of Chemistry and Biochemistry and of Molecular Biology (University of California, Berkeley) and Cellular and Molecular Pharmacology; Ph.D., California, San Francisco. Nitric oxide signaling and mechanisms of catalysis.

James McKerrow, Professor of Pathology, Medicine, and Pharmaceutical Chemistry; M.D., SUNY at Stony Brook. Protease structure and biology, protease inhibitor development.

Susan Miller, Associate Professor of Pharmaceutical Chemistry; Ph.D., Berkeley. Enzyme mechanisms and regulation.

Dan Minor, Assistant Professor of Biochemistry and Biophysics and Investigator, Cardiovascular Research Institute; Ph.D., MIT. Molecular structure and mechanism of ion channel action, development of novel molecules for ion channel regulation.

Geeta Narlikar, Assistant Professor of Biochemistry and of Biophysics; Ph.D., Stanford. Mechanisms of chromatin remodeling.

Norman Oppenheimer, Professor of Pharmaceutical Chemistry; Ph.D., California, San Diego. Enzymology and function of dehydrogenases and glycosidases.

Paul Ortiz de Montellano, Professor of Pharmaceutical Chemistry and of Cellular and Molecular Pharmacology; Ph.D., Harvard. Chemical biology of hemoproteins.

Andrej Sali, Professor of Biopharmaceutical Sciences and of Pharmaceutical Chemistry; Ph.D., London. Structure and function of proteins.

Richard Shafer, Professor of Pharmaceutical Chemistry; Ph.D., Harvard. Nucleic acid structure and interactions.

Martin Shetlar, Professor of Pharmaceutical Chemistry; Ph.D., Berkeley. Protein-DNA interactions.

Brian Shoichet, Associate Professor of Cellular and Molecular Pharmacology; Ph.D., California, San Francisco. Structure-based inhibitor discovery.

Kevan Shokat, Professor of Cellular and Molecular Pharmacology; Ph.D., Berkeley. Deciphering cellular protein kinase cascades.

Robert Stroud, Professor of Biochemistry and Biophysics and of Pharmaceutical Chemistry; Ph.D., London. Cellular signaling and molecular mechanisms.

Francis Szoka, Professor of Biopharmaceutical Sciences and of Pharmaceutical Chemistry; Ph.D., SUNY at Buffalo. Gene therapy and macromolecular drug-delivery systems.

Jack Taunton, Assistant Professor of Cellular and Molecular Pharmacology; Ph.D., Harvard. Biochemical mechanisms of regulating the actin cytoskeleton.

Ronald Vale, Professor and Chairman of Cellular and Molecular Pharmacology and Professor of Biochemistry and Biophysics; Ph.D., Stanford. Microtube-based motility.

Christopher Voigt, Assistant Professor of Pharmaceutical Chemistry; Ph.D., Caltech. Design and evolution of complex gene circuits.

Peter Walter, Professor of Biochemistry and Biophysics; Ph.D., Rockefeller. Intracellular signaling, regulated mRNA splicing, protein sorting.

Ching Chung Wang, Professor of Pharmaceutical Chemistry; Ph.D., Berkeley. Molecular approaches to combating infectious diseases.

Jonathan Weissman, Associate Professor of Cellular and Molecular Pharmacology and of Biochemistry and Biophysics; Ph.D., MIT. Protein folding in vivo.

James Wells, Professor of Cellular and Molecular Pharmacology and of Pharmaceutical Chemistry; Ph.D., Washington. Discovery and design of small molecules that trigger or modulate cellular processes in inflammation and cancer.

Keith Yamamoto, Professor of Cellular and Molecular Pharmacology and of Biochemistry and Biophysics; Ph.D., Princeton. Signaling and regulation by intracellular receptors.

UMBC

AN HONORS UNIVERSITY IN MARYLAND

UNIVERSITY OF MARYLAND, BALTIMORE COUNTY

Department of Chemistry and Biochemistry

Programs of Study

The Department of Chemistry and Biochemistry at the University of Maryland, Baltimore County (UMBC), offers programs of graduate study in chemistry and biochemistry leading to the M.S. and Ph.D. degrees. The department also participates in a biotechnology M.S. program (applied molecular biology) in collaboration with the Department of Biological Sciences, an interdisciplinary Ph.D. program in molecular and cellular biology, an Interface Program that provides cross-disciplinary experience in chemistry and biology, and the Meyerhoff Program in Biomedical Sciences. The department's graduate programs in biochemistry and in molecular and cellular biology benefit from being part of larger intercampus joint programs in those fields, sponsored in conjunction with departments at the medical, dental, and pharmacy schools of the downtown Baltimore campus.

Upon entering the graduate program, both M.S. and Ph.D. students are required to take a set of placement examinations that are designed to test their proficiency at the senior undergraduate level and to indicate any areas of deficiency. Under the guidance of an advisory committee, they next complete a group of courses, constituting a core curriculum, which has been designed to bring them to a minimum level of proficiency in each of the major areas of chemistry. In addition, they are expected to take specialized courses in their field of interest. To fulfill the course requirements normally requires one to two years, depending upon the student's initial level of proficiency, as demonstrated by the placement examinations.

To qualify for the degree, M.S. students must pass a comprehensive examination in their major field. Thesis research must be approved by a thesis committee, which also administers an oral examination based on the thesis research. Completing the course requirements, passing the comprehensive and oral examinations, and gaining approval of the thesis constitute fulfillment of the M.S. degree requirements. Candidates for the M.S. also have the option of substituting additional course work for the thesis. Graduate students in the Ph.D. program are expected to pass comprehensive examinations, prepare and defend an acceptable research proposal, present a dissertation based upon original research, and pass an oral defense.

The principal areas of thesis research for both the M.S. and Ph.D. degrees include analytical chemistry; biochemistry; bioinorganic chemistry; enzymology; mass spectrometry; models for enzymic reactions; organic mechanisms; organic synthesis; protein and nucleic acid chemistry; theoretical, physical, bioanalytical, biophysical, and carbohydrate chemistry; and chemistry at the biology interface.

Research Facilities

Extensive facilities are available for cutting-edge research. The specialized research instrumentation available includes calorimetry, chromatography, stopped-flow and temperature-jump kinetics, nanosecond laser flash photolysis and nuclear magnetic resonance spectroscopy (one 200-, one 300-, one 500-, two 600-, and one 800-MHz instruments), electron spin resonance spectroscopy, circular dichroism, X-ray diffraction, infrared spectroscopy, laser fluorescence spectroscopy, atomic absorption, gas chromatography–mass spectrometry, and Fourier-transform/ion cyclotron resonance mass spectrometry, as well as extensive molecular modeling computer facilities. Also located in the department is a Center for Structural Biochemistry, which specializes in the structural analysis of biological molecules (e.g., biopolymers, peptides, and glycoproteins). In addition to a laser desorption mass spectrometer and two 500-MHz and 600-MHz NMRs, the center houses one of the few tandem mass spectrometers located in academic institutions worldwide. The Howard Hughes Medical Investigator Suite houses the second 600-MHz and the 800-MHz NMRs, which are used for high-dimensional studies of HIV proteins, metallobiomolecules, and macromolecular interactions. The main University library contains more than 2,500 volumes of chemistry texts and subscribes to 150 chemistry and biochemistry periodicals.

Financial Aid

Financial aid packages with twelve-month stipends of $17,000 to $21,000, health insurance, and tuition remission are offered to qualified students. Enhanced stipend fellowships are also available. Approximately 80 percent of the students are receiving some form of financial aid.

Cost of Study

Tuition in 2004–05 was $395 per credit hour for Maryland residents and $606 per credit hour for nonresidents. Fees were $82 per credit hour. New students were charged a one-time orientation fee of $75.

Living and Housing Costs

There are a limited number of on-campus dormitory rooms available for graduate students. Most graduate students are housed in apartments or rooming houses in the nearby communities of Arbutus and Catonsville. A single graduate student can expect living and educational expenses of approximately $14,000 to $16,000 per year.

Student Group

In fall 2004, the department's graduate students included 34 men and 43 women.

Student Outcomes

All of the students plan careers in chemistry and biochemistry in either teaching or research and in associated regulatory and financial areas.

Location

The University has a scenic location on the periphery of the Baltimore metropolitan area. Downtown Baltimore can be reached in 15 minutes by car, while Washington is an hour away. Both Baltimore and Washington have very extensive cultural facilities, including eight major universities, a number of museums and art galleries of international reputation, two major symphony orchestras, and numerous theaters.

The University

UMBC was established in 1966 on a 500-acre campus. The Department of Chemistry and Biochemistry of the University of Maryland Graduate School, Baltimore, is located at the UMBC campus. The University has about 11,700 students, who are drawn primarily from Maryland, although an increasing number have enrolled from other states and countries. The undergraduates are predominantly interested in professional or business careers. Because a high percentage of the students are the first members of their families to attend college, UMBC has a very different atmosphere from that encountered in older institutions and has made a particular effort to attract minority students, who now account for about 36 percent of the undergraduate student body.

Applying

Applications should include an academic transcript, three references, and the results of the General Test of the Graduate Record Examinations.

Correspondence and Information

Graduate Program Director
Department of Chemistry and Biochemistry
University of Maryland, Baltimore County
1000 Hilltop Circle
Baltimore, Maryland 21250

Phone: 866-PhD-UMBC (743-8622; toll-free for chemistry and biochemistry only)
E-mail: chemgrad@umbc.edu
Web site: http://www.umbc.edu/chem-biochem

University of Maryland, Baltimore County

THE FACULTY AND THEIR RESEARCH

Bradley R. Arnold, Assistant Professor; Ph.D., Utah, 1991; postdoctoral studies at the National Research Council of Canada, Ottawa, and the Center for Photoinduced Charge Transfer, University of Rochester. Physical chemistry, application of time-resolved polarized spectroscopy.

C. Allen Bush, Professor; Ph.D., Berkeley, 1965; postdoctoral studies at Cornell. Biophysical chemistry: conformation and dynamics of carbohydrates, glycoproteins, glycopeptides, and polysaccharides by NMR spectroscopy, computer modeling, and circular dichroism.

Donald Creighton, Professor; Ph.D., UCLA, 1972; postdoctoral studies at the Institute for Cancer Research. Enzyme mechanisms and protein structure, studies on sulfhydryl proteases and glyoxalase enzymes.

Brian M. Cullum, Assistant Professor; Ph.D., South Carolina, 1998; postdoctoral studies at Oak Ridge National Laboratory. Analytical chemistry, development of optical sensors and optical sensing techniques for biomedical and environmental research.

Dan Fabris, Assistant Professor; Ph.D., Padua (Italy), 1989; postdoctoral studies at the National Research Council, Area of Research of Padua (Italy) and the University of Maryland, Baltimore County. Bioanalytical and biomedical applications of mass spectrometry, nucleic acid adducts, and protein–nucleic acid interactions.

James C. Fishbein, Professor; Ph.D., Brandeis, 1985; postdoctoral studies at Toronto. Mechanisms of organic reactions in aqueous solutions; generation and study of reactive intermediates, particularly those involved in nitrosamine and nitrosamide carcinogenesis.

Colin W. Garvie, Assistant Professor; Ph.D., Leeds (England), 1997; postdoctoral studies at Howard Hughes Medical Institute, Johns Hopkins University School of Medicine. Structural studies of the redox regulation of circadian clock proteins, structural studies of the MHC II enhanceosome.

Susan K. Gregurick, Assistant Professor; Ph.D., Maryland, College Park, 1994; postdoctoral studies at Maryland Biotechnology Institute. Development of evolutionary algorithms for use in intricate chemical problems.

Ramachandra S. Hosmane, Professor; Ph.D., South Florida, 1978; postdoctoral studies at Illinois. Organic synthesis; biomedicinal chemistry, with applications in antiviral and anticancer therapy; biomedical technology, with applications in artificial blood.

Richard L. Karpel, Professor; Ph.D., Brandeis, 1970; postdoctoral studies at Princeton. Interactions of helix destabilizing proteins with nucleic acids and the involvement of such proteins in various aspects of RNA function, metal ion–nucleic acid interactions.

Lisa A. Kelly, Assistant Professor; Ph.D., Bowling Green, 1993; postdoctoral studies at Brookhaven National Laboratory. Photoredox initiated chemical bond cleavage in biological and model systems.

William R. LaCourse, Associate Professor; Ph.D., Northeastern, 1987; postdoctoral studies at Ames Laboratory (USDOE) and Iowa State. Pulsed electrochemical detection techniques for bioanalytical separations.

Joel F. Liebman, Professor; Ph.D., Princeton, 1970; postdoctoral studies at Cambridge and the National Institute of Standards and Technology. Energetics of organic molecules, especially considerations of strain and aromaticity; gaseous ions; noble gas, fluorine, boron, and silicon chemistry; mathematical and quantum chemistry.

Wuyuan Lu, Affiliate Assistant Professor; Ph.D., Purdue, 1994; postdoctoral studies at the Scripps Research Institute and the University of Chicago. Structural and functional relationships of maternin and D-peptides as receptor antagonists and enzyme inhibitors.

Ralph M. Pollack, Professor; Ph.D., Berkeley, 1968; postdoctoral studies at Northwestern. Enzyme reactions, model systems for enzyme mechanisms, organic reaction mechanisms.

Katherine L. Seley-Radtke, Associate Professor; Ph.D., Auburn, 1996; postdoctoral studies at Auburn. Discovery, design, and synthesis of nucleoside/nucleotide and heterocyclic enzyme inhibitors for use as medicinal agents, with chemotherapeutic emphasis in the areas of anticancer, antiviral, antibiotic, and antiparasitic targets.

Paul J. Smith, Assistant Professor; Ph.D., Pittsburgh, 1993; postdoctoral studies at Johns Hopkins. Bioorganic and physical organic chemistry, host-guest chemistry, DNA structure and DNA binding by small molecules.

Michael F. Summers, Professor and Howard Hughes Associate Medical Investigator; Ph.D., Emory, 1984; postdoctoral studies at the Center for Drugs and Biologics, FDA, Bethesda, Maryland. Elucidation of structural, dynamic, and thermodynamic features of metallobiomolecules utilizing advanced two-dimensional and multinuclear NMR methods.

Veronika Szalai, Assistant Professor; Ph.D., Yale, 1988; postdoctoral studies at North Carolina, Chapel Hill. Spectroscopic characterization of biomolecular assemblies containing paramagnetic transition metals or spin labels.

James S. Vincent, Associate Professor; Ph.D., Harvard, 1963; postdoctoral studies at Caltech. Infrared and Raman spectroscopy of phospholipid membrane systems, magnetic spectroscopy of transition-metal complexes.

Dale L. Whalen, Professor; Ph.D., Berkeley, 1965; postdoctoral studies at UCLA. Reactions of carcinogenic polycyclic aromatic hydrocarbon epoxides, organic reaction mechanisms.

SELECTED PUBLICATIONS

Levy, D., and **B. R. Arnold.** Influence of localized excited states on the transition moment direction of charge-transfer complex absorptions. *J. Phys. Chem. A* 109:2113–9, 2005.

Levy, D., and **B. R. Arnold.** Analysis of charge-transfer absorption and emission spectra on an absolute scale: Evaluation of free energies, matrix elements, and reorganization energies. *J. Phys. Chem. A* 109:8572–8, 2005.

Levy, D., and **B. R. Arnold.** The nature of the rapid relaxation of excited charge-transfer complexes. *J. Am. Chem. Soc.* 126:10727–31, 2004.

Bae, S. Y., and **B. R. Arnold.** Characterization of solvatochromic probes: Simulation of merocyanine 540 absorption spectra in binary solvent mixtures and pure solvent systems. *J. Phys. Org. Chem.* 17:187–93, 2004.

Yoshida, Y., S. Ganguly, **C. A. Bush,** and J. O. Cisar. Carbohydrate engineering of the recognition motifs in streptococcal coaggregation receptor polysaccharides. *Mol. Microbiol.* 58:244–56, 2005.

Adeyeye, J., et al. **(C. A. Bush).** Conformation of the hexasaccharide repeating subunit from the *Vibrio cholerae* O139 capsular polysaccharide. *Biochemistry* 42:3979–88, 2003.

Stroop, C. J. M., **C. A. Bush,** R. L. Marple, and **W. R. LaCourse.** Carbohydrate analysis of bacterial polysaccharides by high-pH anion-exchange chromatography and on-line polarimetric determination of absolute configuration. *Anal. Biochem.* 303:176–85, 2002.

Azurmendi, H. F., and **C. A. Bush.** Tracking alignment from the moment of inertia tensor (TRAMITE) of biomolecules in neutral dilute liquid crystal solutions. *J. Am. Chem. Soc.* 124:2426–7, 2002.

Joseph, E., B. Ganem, J. L. Eiseman, and **D. J. Creighton.** Selective inhibition of MCF-7^{piGST} breast tumors using glutathione transferase-derived 2-methylene-cycloalkenones. *J. Med. Chem.* 48:6549–52, 2005.

Zheng, Z.-B., et al. **(D. J. Creighton).** N-(2-hydroxypropyl)methacrylamide copolymers of a glutathione (GSH)-activated glyoxalase I inhibitor and DNA alkylating agent: Synthesis, reaction kinetics with GSH, and in vitro antitumor activities. *Bioconj. Chem.* 16(5):598–607, 2005.

Akoachere, M., et al. **(D. J. Creighton).** Characterization of the glyoxalases of the malarial parasite *Plasmodium falciparum* and comparisons with their human counterparts. *Biol. Chem.* 386:41–52, 2005.

Hamilton, D. S., et al. **(D. J. Creighton).** Mechanism of the glutathione transferase-catalyzed conversion of antitumor 2-crotonyloxymethyl-2-cycloalkenones to GSH adducts. *J. Am. Chem. Soc.* 125:15049–58, 2003.

Li, H., and **B. M. Cullum.** Multilayer enhancements from continuous silver films based surface-enhanced Raman scattering (SERS) substrates. *Appl. Spectrosc.* 58(4):410–7, 2005.

Hankus, M., G. Gibson, N. Chandrasekharan, and **B. M. Cullum.** Surface enhanced Raman scattering (SERS)–Nanoimaging probes for biological analysis. *Proc. SPIE* 5588:106–17, 2005.

Cullum, B. M. Optical nanosensors and nanobiosensors. In *Encyclopedia of Nanoscience and Nanotechnology,* pp. 2757–68. New York: Marcel Dekker, Inc., 2004.

Chandrasekharan, N., B. Gonzales, and **B. M. Cullum.** Non-resonant multiphoton photoacoustic spectroscopy for non-invasive subsurface chemical diagnostics. *Appl. Spectrosc.* 58(11):1325–33, 2004.

Akinsiku, O., E. T. Yu, and **D. Fabris.** Mass spectrometric investigation of protein alkylation by the RNA footprinting probe kethoxal. *J. Mass Spectrom.* 40:1372–81, 2005.

Richter, S., **D. Fabris,** S. Moro, and M. Palumbo. Dissecting the reactivity of clerocidin towards common buffer systems by means of selected drug analogues. *Chem. Res. Toxicol.* 18:35–40, 2005.

Yu, E. T., Q. Zhang, and **D. Fabris.** Untying the FIV frameshifting pseudoknot structure by MS3D. *J. Mol. Biol.* 345:69–80, 2005.

Fabris, D. Mass spectrometric approaches for the investigation of dynamic processes in condensed phase. *Mass Spectrom. Rev.* 24:30–54, 2005.

Lu, X., J. M. Heilman, P. Blans, and **J. C. Fishbein.** The structure of DNA dictates purine atom site selectivity in alkylation by primary diazonium ions. *Chem. Res. Toxicol.* 18:1462–70, 2005.

Velayutham, M., et al. **(J. C. Fishbein).** Glutathione mediated formation of oxygen free radicals by the major metabolite of Oltipraz. *Chem. Res. Toxicol.* 18(6):970–5, 2005.

Velayutham, M., F. A. Villamena, **J. C. Fishbein,** and J. L. Zweier. Cancer chemopreventive Oltipraz generates superoxide anion radical. *Arch. Biochem. Biophys.* 435:83–8, 2005.

Maier, H., et al. **(C. W. Garvie).** Requirements for selective recruitment of Ets proteins and activation of mb-1/Ig-alpha gene transcription by Pax-5 (BSAP). *Nucleic Acids Res.* 31:5483–9, 2003.

He, Y. Y., and **C. W. Garvie** et al. Structural and functional studies of an intermediate on the pathway to operator binding by *Escherichia coli* MetJ. *J. Mol. Biol.* 320:39–53, 2002.

Garvie, C. W., M. A. Pufall, B. J. Graves, and C. Wolberger. Structural analysis of the autoinhibition of Ets-1 and its role in protein partnerships. *J. Biol. Chem.* 277:45529–36, 2002.

Garvie, C. W., J. Hagman, and C. Wolberger. Structural studies of Ets-1/Pax5 complex formation on DNA. *Mol. Cell* 8(6):1267–76, 2001.

Zhou, Z., A. Deyheim, S. Krueger, and **S. K. Gregurick.** LORES: Low resolution shape program for the calculation of small-angle scattering profiles for biological macromolecules in solution. *Comput. Phys. Commun.* 170(2):186–204, 2005.

Ling, Y., M. Ascano, P. R. Robinson, and **S. K. Gregurick.** Experimental and computational studies of the desensitization process in the bovine rhodopsin-arrestin complex. *Biophys. J.* 86(4):2445–54, 2004.

Krueger, S., and **S. K. Gregurick** et al. Entropic nature of the interaction between promoter bound CRP mutants and RNA polymerase. *Biochemistry* 47(7):1958–68, 2003.

Krueger, S., **S. K. Gregurick,** J. Zondlo, and E. Eisenstein. Interaction of GroEL and GroEL/GroES complex with a non-native subtilisin variant: A small angle neutron scattering study. *J. Struct. Biol.* 141(3):240–58, 2003.

Ujjinamatada, R. K., and **R. S. Hosmane.** Selective functional group transformation using guanidine: The conversion of an ester group into an amide in vinylogous ester-aldehydes. *Tetrahedron Lett.* 46:6005–9, 2005.

Reayi, A., and **R. S. Hosmane.** Inhibition of adenosine deaminase by novel 5:7-fused heterocycles containing the imidazo[4,5-e][1,2,4]triazepine ring system: A structure-activity relationship study. *J. Med. Chem.* 47:1044–50, 2004.

Gunther, S., et al. **(R. S. Hosmane).** Applications of real-time PCR for testing antiviral compounds against lassa virus, SARS corona virus, and ebola virus in vitro. *Antiviral Res.* 63:209–15, 2004.

Roach, T. A., V. W. Macdonald, and **R. S. Hosmane.** A novel site-directed affinity reagent for cross-linking human hemoglobin: Bis[2-(4-phosphonooxy phenoxy)carbonylethyl]phosphinic acid (BPPCEP). *J. Med. Chem.* 47:5847–59, 2004.

Pant, K., **R. L. Karpel,** I. Rouzina, and M. C. Williams. Salt dependent binding of T4 gene 32 protein to single- and double-stranded DNA: Single molecule force spectroscopy measurements. *J. Mol. Biol.* 349:317–30, 2005.

Pant, K., **R. L. Karpel,** I. Rouzina, and M. C. Williams. Mechanical measurement of single molecule binding rates: Kinetics of DNA helix-destabilization by T4 gene 32 protein. *J. Mol. Biol.* 336:851–70, 2004.

Karpel, R. L. LAST motifs and SMART domains in gene 32 protein: An unfolding story of autoregulation? *IUBMB Life* 53:161–6, 2002.

Urbaneja, M. A., M. Wu, J. R. Casas-Finet, and **R. L. Karpel.** HIV-1 nucleocapsid protein as a nucleic acid chaperone: Spectroscopic study of its helix-destabilizing properties, structural binding specificity, and annealing activity. *J. Mol. Biol.* 318:749–64, 2002.

Abraham, B., S. McMasters, M. A. Mullan, and **L. A. Kelly.** Reactivities of carboxyalkyl-substituted 1,4,5,8-naphthalene diimides in aqueous solution. *J. Am. Chem. Soc.* 126:4293–300, 2004.

Chandrasekharan, N., and **L. A. Kelly.** Progress towards fluorescent molecular thermometers. In *Reviews in Fluorescence 2004,* pp. 21–40, ed. C. D. Geddes. 2004.

Abraham, B., and **L. A. Kelly.** Photo-oxidation of amino acids and proteins mediated by novel 1,8-naphthalimide derivatives. *J. Phys. Chem. B* 107:12534–41, 2003.

Chandrasekharan, N., and **L. A. Kelly.** A dual fluorescence temperature sensor based on perylene/exciplex interconversion. *J. Am. Chem. Soc.* 123:9898–9, 2001.

University of Maryland, Baltimore County

Perrino, F. W., et al. **(W. R. LaCourse** and **J. C. Fishbein).** Polymerization past the N2-isopropylguanine and the N6-isopropyladenine DNA lesions with the translesion synthesis DNA polymerases η and ι and the replicative DNA polymerase α. *Chem. Res. Toxicol.* 18:1451–61, 2005.

Marple, R. L., and **W. R. LaCourse.** Application of photoassisted electrochemical detection to explosive-containing environmental samples. *Anal. Chem.* 6709–14, 2005.

LaCourse, W. R., and S. J. Modi. Microelectrode applications of pulsed electrochemical detection. *Electroanalysis* 17(3):1141–52, 2005.

Marple, R. L., and **W. R. LaCourse.** A platform for on-site environmental analysis of explosives using HPLC with UV absorbance and photo-assisted electrochemical detection. *Talanta* 66(3):581–90, 2005.

Gelhaus, S. L., and **W. R. LaCourse.** Separation of modified 2'deoxyoligonucleotides using ion-pairing reversed phase HPLC. *J. Chromatogr. B* 820(2):157–63, 2005.

Liebman, J. F., and M. Ponikvar. Ion selective electrode determination of free versus total fluoride ion in simple and fluoroligand coordinated hexafluoropnictate (PnF_6–, Pn = P, As, Sb, Bi) salts. *Struct. Chem.* 16:516–28, 2005.

Jenkins, H. D. B., and **J. F. Liebman.** Volume of solid state ions and their estimation. *Inorg. Chem.* 44:6359–72, 2005.

Roux, M. V., et al. **(J. F. Liebman).** Cubane, cuneane and their carboxylates: A calorimetric, crystallographic, calculational and conceptual coinvestigation. *J. Org. Chem.* 70:5461–70, 2005.

Matos, M. A. R., et al. **(J. F. Liebman).** Thermochemistry of diphenic anhydride. A combined experimental and theoretical study. *Mol. Phys.* 103:1885–94, 2005.

Houck, W. J., and **R. M. Pollack.** Temperature effects on the catalytic activity of the D38E mutant of 3-oxo-Δ5-steroid isomerase: Favorable enthalpies and entropies of activation relative to the nonenzymatic reaction catalyzed by acetate ion. *J. Am. Chem. Soc.* 126:16416–25, 2004.

Pollack, R. M. Enzymatic mechanisms for catalysis of enolization: Ketosteroid isomerase. *Bioorg. Chem.* 32:341–53, 2004.

Houck, W. J., and **R. M. Pollack.** Activation enthalpies and entropies for the microscopic rate constants of acetate-catalyzed isomerization of 5-androstene-3,17-dione. *J. Am. Chem. Soc.* 125:10206–12, 2003.

Thornburg, L. D., Y. R. Goldfeder, T. C. Wilde, and **R. M. Pollack.** Selective catalysis of elementary steps by Asp-99 and Tyr-14 of 3-oxo-Δ^5-steroid isomerase. *J. Am. Chem. Soc.* 123:9912–3, 2001.

Quirk, S., and **K. L. Seley.** Identification of catalytic amino acids in the human GTP fucose pyrophosphorylase active site. *Biochemistry* 44:13172–8, 2005.

Quirk, S., and **K. L. Seley.** Substrate discrimination by the human GTP-fucose pyrophosphorylase. *Biochemistry* 44:10854–63, 2005.

Seley, K. L., S. Salim, L. Zhang, and P. I. O'Daniel. 'Molecular chameleons'. Design and synthesis of a new series of flexible nucleosides. *J. Org. Chem.* 70:1612–9, 2005.

Polak, M., **K. L. Seley,** and J. Plavec. Fleximers: Conformational properties of novel shaped nucleosides—Fleximers. *J. Am. Chem. Soc.* 126:8159–66, 2004.

Roux, M. V., **P. J. Smith,** and **J. F. Liebman.** Paradigms and paradoxes: Thoughts on the enthalpy of formation of guanidine and its monosubstituted derivatives. *Struct. Chem.* 16:73–5, 2005.

Hauser, S. L., and **P. J. Smith.** Induced-fit binding of an aryl phosphate by a macrobicyclic dicationic cyclodextrin derivative. *J. Phys. Org. Chem.* 18:473–6, 2005.

Wang, B. B., N. Maghami, V. L. Goodlin, and **P. J. Smith.** Critical structural motif for the catalytic inhibition of human topoisomerase II by UK-1 and analogs. *Bioorg. Med. Chem. Lett.* 14:3221–6, 2004.

Wang, B. B., and **P. J. Smith.** Synthesis of a terbenzimidazole topoisomerase I poison via iterative borinate ester couplings. *Tetrahedron Lett.* 44:8967–9, 2003.

Wu, L-Q., et al. **(P. J. Smith).** Utilizing renewable resources to create functional polymers: Chitosan-based associative thickener. *Environ. Sci. Tech.* 36:3446–54, 2002.

Zhou, J., J. K. McAllen, Y. Tailor, and **M. F. Summers.** High affinity nucleocapsid protein binding to the μΨ RNA packaging signal of Rous sarcoma virus. *J. Mol. Biol.* 349:976–88, 2005.

D'Souza, V. D., and **M. F. Summers.** Structural basis for packaging the dimeric genome of Moloney murine leukemia virus. *Nature* 431:586–90, 2004.

D'Souza, V. D., A. Dey, D. Habib, and **M. F. Summers.** NMR structure of the 101-nucleotide core encapsidation signal of the Moloney murine leukemia virus. *J. Mol. Biol.* 337:427–42, 2004.

Tang, C., et al. **(M. F. Summers).** Entropic switch regulates myristate exposure in the HIV-1 matrix protein. *Proc. Natl. Acad. Sci. U.S.A.* 101:517–22, 2004.

Karr, J. W., H. Akintoye, L. J. Kaupp, and **V. A. Szalai.** N-terminal deletions modify the Cu^{2+} binding site in amyloid-β. *Biochemistry* 44:5478–87, 2005.

Karr, J. W., L. J. Kaupp, and **V. A. Szalai.** Amyloid-β binds copper (II) in a mononuclear metal ion binding site. *J. Am. Chem. Soc.* 126:3534–8, 2004.

Keating, L. R., and **V. A. Szalai.** Parallel stranded guanine quadruplex interactions with a cationic copper porphyrin. *Biochemistry* 43:15891–900, 2004.

Szalai, V. A., M. J. Singer, and H. H. Thorp. Site-specific probing of oxidative reactivity and telomerase function using 7,8-dihydro-8-oxoguanine in telomeric DNA. *J. Am. Chem. Soc.* 124:1625–31, 2002.

Ukachukwu, V. C., R. S. Mohan, and **D. L. Whalen.** Kinetic deuterium isotope effects on the reactions of 2-(4-methoxyphenyl)oxirane in water solutions. *ARKIVOC (iii)*:45–58, 2005.

Doan, L., H. Yagi, D. M. Jerina, and **D. L. Whalen.** Synthesis and hydrolysis of a cis-chlorohydrin derived from a benzo[a]pyrene 7,8-diol 9,10-epoxide. *J. Org. Chem.* 69:8012–7, 2004.

Sampson, K., A. Paik, B. Duvall, and **D. L. Whalen.** Transition state effects in acid-catalyzed aryl epoxide hydrolyses. *J. Org. Chem.* 69:5204–11, 2004.

Doan, L., H. Yagi, D. M. Jerina, and **D. L. Whalen.** Chloride ion-catalyzed conformational inversion of carbocation intermediates in the hydrolysis of a benzo[a]pyrene 7,8-diol 9,10-epoxide. *J. Am. Chem. Soc.* 124(48):14382–7, 2002.

UNIVERSITY OF PITTSBURGH

Department of Chemistry

Programs of Study

The department offers programs of study leading to the M.S. and Ph.D. degrees in analytical, biological, inorganic, organic, and physical chemistry and in chemical physics. Interdisciplinary research is currently conducted in the areas of surface science, natural product synthesis, biological chemistry, bioanalytical chemistry, combinatorial chemistry, laser spectroscopy, materials science, electrochemistry, nanoscience, organometallic chemistry, and computational and theoretical chemistry. Both advanced degree programs involve original research and course work. Other requirements include a comprehensive examination, a thesis, a seminar, and, for the Ph.D. candidate, a proposal. For the typical Ph.D. candidate, this process takes four to five years. Representative of current research activities in the department in analytical chemistry are techniques in electroanalytical chemistry, in vivo electrochemistry, UV resonance Raman spectroscopy, microseparations and nanoseparations, sensors and selective extraction, NMR, EPR, mass spectrometry, and vibrational circular dichroism. Fields of research in biological chemistry include structural dynamics of biological systems, design of soluble membrane proteins, neurochemistry, and molecular design and recognition. In inorganic chemistry, studies are being conducted on organotransition metal complexes, redox reactions, complexes of biological interest, transition metal polymers, and optoelectronic materials; in organic chemistry, on reaction mechanisms, ion transport, total synthesis, drug design, natural products synthesis, bioorganic chemistry, synthetic methodology, organometallics, enzyme mechanisms, and physical-organic chemistry. Research areas in physical chemistry include Raman, photoelectron, Auger, NMR, EPR, infrared, and mass spectroscopy; electron-stimulated desorption ion angular distribution (ESDIAD); condensed phase spectroscopy; high-resolution laser spectroscopy; molecular spectroscopy; electron and molecular beam scattering; electronic emission spectroscopy; and catalysis. Theoretical fields of research include electronic structure, reaction mechanisms, electron transfer theory, quantum mechanics, and new material design. Research on computer applications to chemistry is under way in a variety of areas.

Research Facilities

The Department of Chemistry is housed in two buildings, a fifteen-story and a three-story complex, containing a vast array of modern research instruments and in-house machine, electronics, and glassblowing shops. The Chemistry Library is a spacious 6,000-square-foot facility that contains more than 30,000 monographs and bound periodicals and more than 200 maintained journal subscriptions. These and many related journals, as well as search capabilities, are available online for free. Three other chemistry libraries are nearby. In 2002, the Department of Chemistry received a five-year, $9.6-million grant from the National Institute of General Medical Sciences (NIGMS, a subdivision of NIH) to build one of the nation's first Centers for Excellence in Chemical Methods and Library Development. Shared departmental instrumentation includes four 300-MHz NMRs, one 500-MHz NMR, and one 600-MHz NMR with LC-NMR and MAS capabilities; three high-resolution mass spectrometers; an LC/MS, a triple quadrupole MS, and four low-resolution mass spectrometers; a light-scattering instrument; a circular dichroism spectrophotometer; a spectropolarimeter; X-ray systems—single crystal, powder, small angle scattering, and fluorescence; a scanning electron microscope; a vibrating sample magnetometer; several FT-IR and UV-VIS spectrophotometers; and computer and workstation clusters.

Financial Aid

Seventy-five teaching assistantships and teaching fellowships are available. The former provided $20,332 in 2005–06 for the three trimesters of the year; the fellowships (awarded to superior students after their first year) carried an annual stipend of $21,150. Most advanced students are supported by research assistantships and fellowships, which paid up to $1937 per month. All teaching assistantships, fellowships, and research assistantships include a full scholarship that covers all tuition, fees, and medical insurance. Special Kaufman Fellowships provide up to an additional $4000 award to truly outstanding incoming Ph.D. candidates. In addition, Bayer Fellowships, Ashe Fellowships, and Sunoco Fellowships provide salary supplements to research or teaching assistantships that range from $2000 to $5000 annually. In some cases, these supplements may be used to begin research in the summer prior to the formal initiation of graduate study.

Cost of Study

All graduate assistants and fellows receive full-tuition scholarships. Tuition and fees for full-time study in 2005 were $12,776 per term for out-of-state students and $6867 for state residents.

Living and Housing Costs

Most graduate students prefer private housing, which is available in a wide range of apartments and rooms in areas of Pittsburgh near the campus. The University maintains a housing office to assist students seeking off-campus housing. Living costs compare favorably with other urban areas.

Student Group

The University enrolls about 17,000 students, including about 9,500 graduate and professional school students. Most parts of the United States and many other countries are represented. Almost 200 full-time graduate chemistry students are supported by the various sources listed under Financial Aid. The University is coeducational in all schools and divisions; more than one third of the graduate chemistry students are women. An honorary chemistry society promotes a social program for all faculty members and graduate students in the department.

Location

Deservedly, Pittsburgh is currently ranked "among the most livable cities in the United States" by Rand McNally. It is recognized for its outstanding blend of cultural, educational, and technological resources. Pittsburgh's famous Golden Triangle is enclosed by the Allegheny and Monongahela Rivers, which meet at the Point in downtown Pittsburgh to form the Ohio River. Pittsburgh has enjoyed a dynamic renaissance in the last few years. The city's cultural resources include the Pittsburgh Ballet, Opera Company, Symphony Orchestra, Civic Light Opera, and Public Theatre and the Three Rivers Shakespeare Festival. Many outdoor activities, such as rock climbing, rafting, sailing, skiing, and hunting, are also available within a 50-mile radius.

The University

The University of Pittsburgh, founded in 1787, is the oldest school west of the Allegheny Mountains. Although privately endowed and controlled, the University is state related to permit lower tuition rates for Pennsylvania residents and to provide a steady source of funds for all of its programs. Attracting more than $310 million in sponsored research annually, the University has continued to increase in stature.

Applying

Applications for September admission and assistantships should be made prior to February 1. However, special cases may be considered throughout the year. A background that includes a B.S. degree in chemistry, with courses in mathematics through integral calculus, is preferred. GRE scores, including the chemistry Subject Test, are required for fellowship consideration (see above). For admission, the General Test of the GRE is required, and the chemistry Subject Test is suggested. International applicants must submit TOEFL results and GRE scores.

Correspondence and Information

Graduate Admissions
Department of Chemistry
University of Pittsburgh
Pittsburgh, Pennsylvania 15260

Phone: 412-624-8501
E-mail: gradadm@pitt.edu
Web site: http://www.chem.pitt.edu

University of Pittsburgh

THE FACULTY AND THEIR RESEARCH

S. Amemiya, Assistant Professor; Ph.D., Tokyo, 1998. Analytical chemistry: electrochemical sensors, scanning electrochemical microscopy, ion and electron transfer at interfaces, liquid-liquid interfaces, biomembranes, molecular recognition, ion channel.

S. A. Asher, Professor; Ph.D., Berkeley, 1977. Analytical and physical chemistry: resonance Raman spectroscopy, biophysical chemistry, material science, protein folding, nanoscale and mesoscale smart materials, heme proteins, photonic crystals.

K. Brummond, Associate Professor; Ph.D., Penn State, 1991. Organic chemistry: organometallic chemistry, synthesis of natural products, solid phase synthesis.

T. M. Chapman, Associate Professor; Ph.D., Polytechnic of Brooklyn, 1965. Organic chemistry: new polymers of uncommon architecture, dendritic polymers, polymer surfactants and emulsifiers, tissue engineering, controlled drug delivery, gene transfer, electron transfer in dendritic polymers.

R. D. Coalson, Professor of Chemistry and Physics; Ph.D., Harvard, 1984. Physical chemistry: quantum theory of rate processes, optical spectroscopy, computational techniques for quantum dynamics, structure and energetics of macroions in solution; design of optical waveguides and photonic bandgap structures; laser control of condensed-phase electron transfer; theoretical/computational approaches to the transport of ions and polymers through biological (protein) pores.

T. Cohen, Professor Emeritus; Ph.D., USC, 1955. Organic chemistry: new synthetic methods, particularly those involving organometallics, most often of main group elements; synthesis of natural products using the new synthetic methods; mechanistic studies.

N. J. Cooper, Professor and Dean, School of Arts and Sciences; D.Phil., Oxford, 1976. Inorganic chemistry: synthetic and mechanistic inorganic and organometallic chemistry, transition metal chemistry of carbon dioxide, synthesis of highly reduced complexes containing metals in negative oxidation states, use of highly reduced metal centers to activate arenas and polyaromatic hydrocarbons, synthesis and reactivity of cationic alkylidene complexes of transition metals, organometallic photochemistry.

D. P. Curran, Bayer Professor and Distinguished Service Professor; Ph.D., Rochester, 1979. Organic chemistry: natural products total synthesis and new synthetic methodology, synthesis via free-radical reactions, fluorous chemistry, combinatorial chemistry.

P. Floreancig, Associate Professor; Ph.D., Stanford, 1997. Organic chemistry: total synthesis of natural products and bioactive analogs, methodology development, electron transfer chemistry.

M. F. Golde, Associate Professor; Ph.D., Cambridge, 1972. Physical chemistry: kinetic and spectroscopic studies of mechanisms of formation and removal of electronically excited atoms and small molecules, ion-electron recombination and similar species.

J. J. Grabowski, Associate Professor; Ph.D., Colorado, 1983. Physical-organic chemistry: reactive intermediates; reaction mechanisms; novel uses of mass spectrometry and photochemistry for organic, analytical, or environmental chemistry; novel uses of the World Wide Web for chemical education.

K. D. Jordan, Professor; Ph.D., MIT, 1974. Physical chemistry: theoretical studies of the electronic structure of molecules, electron-induced chemistry, computer simulations, hydrogen-bonded clusters, chemical reactions at semiconductor and carbon nanotube surfaces, parallel computational methods.

K. Koide, Assistant Professor; Ph.D., California, San Diego, 1997. Organic chemistry and chemical biology: natural product synthesis, diversity-oriented synthesis, RNA imaging.

T. Y. Meyer, Associate Professor; Ph.D., Iowa, 1991. Inorganic and polymer chemistry: organometallic chemistry, application of transition metal catalysis to polymer synthesis, polymers with functionalized backbones, polymeric materials for biological applications.

A. C. Michael, Associate Professor; Ph.D., Emory, 1987. Analytical chemistry: new microsensor technologies for neurochemical monitoring in the central nervous system; investigations of the chemical aspects of brain disorders such as Parkinson's disease, schizophrenia, and substance abuse; quantitative aspects of in vivo chemical measurements.

S. G. Nelson, Associate Professor; Ph.D., Rochester, 1991. Organic chemistry: natural products total synthesis, new synthetic methods, asymmetric catalysis and organometallic chemistry.

S. Petoud, Assistant Professor, Ph.D., Lausanne, 1997. Design, synthesis, and investigation of luminescent lanthanide complexes and luminescent material for application in biology, biotechnology, medical diagnostics, genomic and proteomic, and flat color display technologies.

D. W. Pratt, Professor; Ph.D., Berkeley, 1967. Physical chemistry: molecular structure and dynamics, as revealed by high-resolution laser and magnetic resonance spectroscopy, in the gas phase and in the condensed phase; optical trapping, nucleation, and imaging using focused laser beams; science education, especially for nonscience students.

S. K. Saxena, Assistant Professor, Ph.D., Cornell, 1997. Analytical, biophysical, and physical chemistry: two-dimensional Fourier-transform electron spin resonance, conformational dynamics, self-assembly and global folding patterns in membrane-associated protein complexes.

C. E. Schafmeister, Assistant Professor; Ph.D., San Francisco, 1997. Organic and biological chemistry: molecular building block design, macromolecular design, ligand design, protein-ligand interactions, engineering of soluble membrane proteins.

P. E. Siska, Professor; Ph.D., Harvard, 1970. Physical chemistry: crossed molecular beam and theoretical studies of intermolecular forces and chemical reaction dynamics, scattering and reactions of excited atoms and ions.

M. M. Spence, Professor; Ph.D., Berkeley, 2002. Physical chemistry: membrane proteins, antimicrobial peptides, peptide neurotoxins, lipid structure and dynamics, liquid- and solid-state NMR.

A. Star, Assistant Professor; Ph.D., Tel Aviv, 2000. Physical organic chemistry: carbon nanotubes, nanoelectronic chemical and biological sensors; supramolecular and dynamic chemistry.

D. H. Waldeck, Professor and Department Chair; Ph.D., Chicago, 1983. Analytical and physical chemistry: ultrafast spectroscopy, electrochemistry, homogenous and heterogenous electron transfer, solvation; electron tunneling; bioelectrochemistry; biophysics; nanoscience; molecular electronics.

S. G. Weber, Professor; Ph.D., McGill, 1979. Analytical chemistry: microcolumn HPLC and sensitive detection for bioanalysis, sensors and selective extraction, screening and molecular diversity, and electrochemistry.

C. S. Wilcox, Professor; Ph.D., Caltech, 1979. Organic chemistry: diffusion-reaction processes, precipitons and separation methods for parallel synthesis and combinatorial chemistry, chemical synthesis, molecular recognition and the molecular torsion balance, self-assembling materials.

P. Wipf, University Professor and Director, Center for Combinatorial Chemistry; Ph.D., Zurich, 1987. Organic chemistry: total synthesis of natural products; organometallic, heterocyclic, and combinatorial chemistry.

J. T. Yates Jr., Mellon Chair Professor and Director, Surface Science Center; Ph.D., MIT, 1960. Surface science: kinetics of surface processes; vibrational spectroscopy of surface species; electronic spectroscopy of surfaces; catalytic and surface chemistry on model clusters, oxides, and single crystals; semiconductor surfaces; scanning tunneling microscopy; nanoscience; photochemistry of surfaces.

UNIVERSITY OF PUERTO RICO, RÍO PIEDRAS

Department of Chemistry

Programs of Study

The Department of Chemistry of the University of Puerto Rico, Río Piedras campus, offers programs of study that lead to the M.S. and Ph.D. degrees. Both degree programs require the presentation and oral defense of a thesis. Students should have knowledge of Spanish and/or English. All students select research advisers and begin research by the end of the first year. Areas of concentration are biochemistry and analytical, inorganic, organic, and physical chemistry. An interdepartmental doctoral program in chemical physics is also offered.

M.S. students normally complete the thesis and the required 21 semester credits in courses during the second year. Students pursuing the Ph.D. degree must pass qualifying examinations in three major areas by the end of the first year, after which they must present two research proposals; one proposal is related to the student research project. Forty-two semester credits in courses and 24 in research are required for the Ph.D. One year as a teaching assistant is required for all students.

Research Facilities

The Facundo Bueso Building houses the Office of the Graduate Program and all the laboratories, classrooms, and services used by the program. Extensive renovation and modernization of the laboratories has been recently completed. Major equipment and instruments necessary for research are available, including lasers, a single photon-counting spectrofluorimeter, two X-ray diffractometers, and high-resolution FT-NMR (300 and 500 MHz), FT-IR, and GC-MS spectrometers. There is a wide variety of equipment in individual research laboratories, including electrochemical analyzers, a quartz-crystal microbalance, UV-Vis spectrophotometers, DSC, HPLC/GC, a fluorometer, and computers. The Department has electronics, scientific illustration, machine, and glassblowing shops. Facilities to support research include the Materials Science and Surface Characterization, Biotesting, and Computational Chemistry Centers. The Science Library is in a new building of the College of Natural Sciences.

Financial Aid

The research program of the Department is supported by funds from the University and from government and industrial grants. Both teaching and research assistantships are available from these sources. Support for students from Latin America may also be obtained from foundations and from the OAS. Support of $1000 to $1200 per month plus remission of tuition and fees is available, depending on the qualifications of the student.

Cost of Study

For 2005–06, tuition and fees for students who were not teaching or research assistants were $75 per credit hour for residents. There is a nonresident fee, which was about $1750 per semester for international students or equal to the nonresident fee charged by the state university in the state where the student resides.

Living and Housing Costs

University housing is very limited, but private apartments and rooms are available in the University area. Housing costs vary from $150 to $300 per month per student.

Student Group

A total of 135 students (local and international) are enrolled in the graduate programs in chemistry, a number that permits careful supervision of each student's progress and needs. Postdoctorals and visiting scientists also participate in the program.

Location

The University is located in a residential suburb of San Juan, the capital and cultural center of Puerto Rico. The numerous historical sites and the carefully restored buildings of Old San Juan give the city a highly individual character. With its perennially pleasant climate, excellent beaches, and convenient transportation to North and South America and all the Caribbean, San Juan is the center of tourism in the Caribbean, and its residents enjoy a wide variety of entertainment and recreational facilities.

The University

The University of Puerto Rico, which was founded in 1903, is supported by the Commonwealth of Puerto Rico. Río Piedras, the oldest and largest campus, includes the Colleges of Natural Sciences, Social Sciences, Humanities, General Studies, Law, Education, and Business Administration; a School of Architecture; a graduate School of Planning; and a School of Library Science. The Medical School is located near the Río Piedras campus and the Engineering School is at the Mayagüez campus of the University. Facilities at the 288-acre Río Piedras campus are rapidly being expanded and remodeled. The student body of the Río Piedras campus consists of 22,000 full-time students in eight colleges.

Applying

The application for admission, including scores on the chemistry Subject Test of the Graduate Record Examinations, should be addressed to the Department of Chemistry and returned no later than the first week of February. The application for financial assistance also should be sent to the Department of Chemistry.

Correspondence and Information

Graduate Program Coordinator
Department of Chemistry
University of Puerto Rico, Río Piedras
P.O. Box 23346
San Juan, Puerto Rico 00931-3346
Phone: 787-764-0000 (Ext. 2445, 4818, or 4791)
Fax: 787-756-8242
E-mail: nmcarballeira@uprrp.edu
Web site: http://graduados.uprrp.edu/cnquimica

University of Puerto Rico, Río Piedras

THE FACULTY AND THEIR RESEARCH

Rafael Arce, Professor; Ph.D., Wisconsin, 1971. Physical chemistry (photochemistry): photochemistry and photophysics of purine bases, dinucleotides and polynucleotides, photocross-linking reactions, heterogeneous photochemistry of polycyclic aromatic hydrocarbons, mechanisms of photodestruction of adsorbed pollutants.

Carlos R. Cabrera, Professor; Ph.D., Cornell, 1987. Analytical chemistry photoelectrochemistry, electrocatalysis, surface analysis, molecular sensors, fuel cell catalyst design.

Néstor Carballeira, Professor and Graduate Program Coordinator; Ph.D., Würzburg (Germany), 1983. Organic chemistry: isolation, characterization, and total synthesis of novel marine natural products, with emphasis on lipids; new antimycobacterial and antifungal agents.

Jorge L. Colón, Associate Professor; Ph.D., Texas A&M, 1989. Inorganic and bioinorganic chemistry: photophysics and photochemistry of transition metal complexes in microheterogeneous environments, photocatalytic studies of inorganic layered materials, bioinorganic chemistry.

Fernando A. González, Professor; Ph.D., Cornell, 1989. Biochemistry and molecular biology: signal transduction by purinoceptors in mammalian cells, modulation of signal transduction pathways by natural products.

Kai Griebenow, Associate Professor; Ph.D., Max Planck Institute (Mühlheim), 1992. Biochemistry: structure of proteins in unusual environments, protein stability and formulation, nonaqueous enzymology, FT-IR spectroscopy.

Ana R. Guadalupe, Professor; Ph.D., Cornell, 1987. Analytical chemistry–electrochemistry, chemical sensors and biosensors, electrocatalysis and electron transfer reactions in redox enzymes.

Yasuyuki Ishikawa, Professor; Ph.D., Iowa, 1976. Theoretical chemistry: relativistic quantum chemistry, theoretical study of photodissociation dynamics, direct molecular dynamics study of chemical reactions in clusters.

Reginald Morales, Professor; Ph.D., Rutgers, 1976. Biochemistry: phospholipid organization on cell surfaces, phospholipases as probes of membrane structure, structure-function relationships of lipid analogues, phospholipid synthesis and analysis.

José A. Prieto, Professor; Ph.D., Puerto Rico, 1982. Organic chemistry: synthetic methodology, chemistry of epoxides, development of new reagents and the synthesis of biologically active molecules, organosilanes in synthesis.

Edwin Quiñones, Professor; Ph.D., Puerto Rico, 1986. Physical chemistry: dynamics of highly excited electronic states, laser-induced processes in van der Waals clusters, Doppler spectroscopy, magnetic field effects in small molecules, protein folding, spectroscopy of proteins immobilized in sol-gel materials, single-molecule spectroscopy of biomolecules.

Raphael G. Raptis, Associate Professor; Ph.D., Texas A&M, 1988. Inorganic chemistry: chemical, electrochemical, and structural studies of oligonuclear transition-metal pyrazolato complexes; bioinorganic chemistry; iron-based MRI contrast agents; nanoscale materials; molecular precursors of metal nanoparticles; magnetostructural correlations.

José M. Rivera, Assistant Professor; Ph.D., MIT, 2000. Organic chemistry: supramolecular chemistry, molecular recognition, organic synthesis, nanotechnology, bioorganic chemistry, medicinal chemistry.

Abimael D. Rodríguez, Professor; Ph.D., Johns Hopkins, 1983. Organic chemistry: isolation, characterization, and synthesis of marine natural products.

Osvaldo Rosario, Professor; Ph.D., Puerto Rico, 1978. Analytical chemistry: development of methods for the analysis of environmental pollutants, analysis of air pollutants, gas chromatography–mass spectrometry, artifacts during sampling.

Eugene S. Smotkin, Associate Professor; Ph.D., Texas at Austin, 1989. Physical chemistry, electrochemistry, vibrational spectroscopy, catalysis and fuel cells, mass spectroscopy, combinatorial chemistry, in situ X-ray adsorption spectroscopy.

John A. Soderquist, Professor; Ph.D., Colorado, 1977. Organic chemistry: organometallics in organic synthesis, natural product synthesis, organosilicon and organotin chemistry; organoboranes in synthesis, catalytic processes, asymmetric synthesis.

Brad R. Weiner, Professor and Dean; Ph.D., California, Davis, 1986. Physical chemistry: laser photochemistry and molecular reaction dynamics in the gas phase, mechanisms of laser ablation.

University of Puerto Rico, Río Piedras

SELECTED PUBLICATIONS

Crespo-Hernández, C. E., and **R. Arce.** Formamidopyrimidines as major products in the low- and high- intensity UV irradiation of guanine derivatives. *J. Photochem. Photobiol. B* 73:167, 2004.

Sotero, P., and **R. Arce.** Surface and adsorbates effects on the photochemistry and photophysics of adsorbed perylene on unactivated silica gel and alumina. *J. Photochem. Photobiol. A* 167:191–9, 2004.

Negrón-Encarnación, I., and **R. Arce.** Acridine species adsorbed on models of atmospheric particulate matter and their role in the photodegradation mechanisms under N_2 or O_2 atmospheres. *Polycyclic Aromat. Compd.* 24:6007–16, 2004.

Crespo-Hernández, C., **R. Arce,** and **Y. Ishikawa** et al. Ab initio ionization energy-thresholds of DNA and RNA bases in gas phase and in aqueous solution. *J. Phys. Chem. A* 108:6373–7, 2004.

Crespo-Hernández, C., **R. Arce,** and **E. Quiñones.** Magnetic field enhanced photoionization of 6-methylpurine. *Chem. Phys. Lett.* 382:661–4, 2003.

Diaz-Morales, R. R., et al. **(C. R. Cabrera** and **E. S. Smotkin).** XRD and XPS analysis of as-prepared and conditioned DMFC array membrane electrode assemblies. *J. Electrochem. Soc.* 151:A1314–8, 2004.

Rodriguez-Nieto, F. J., T. Y. Morante-Catacora, and **C. R. Cabrera.** Sequential and simultaneous electrodeposition of Pt-Ru electrocatalysts on a HOPG substrate and the electro-oxidation of methanol in aqueous sulfuric acid. *J. Electroanal. Chem.* 571:15–26, 2004.

Brito, R., R. Tremont, and **C. R. Cabrera.** Electron transfer kinetics across derivatized self-assembled monolayers on platinum: A cyclic voltammetry and electrochemical impedance spectroscopy study. *J. Electroanal. Chem.* 574:15–22, 2004.

Medina, J. R., G. Cruz, **C. R. Cabrera,** and **J. A. Soderquist.** New direct ^{11}B NMR-based analysis of organoboranes through their potassium borohydrides. *J. Org. Chem.* 68:4631–42, 2003.

Carballeira, N. M., C. Miranda, E. A. Orellano, and **F. González.** Synthesis of a novel series of 2-methylsulfanyl fatty acids and their toxicity on the human K-562 and U-937 leukemia cell lines. *Lipids* 40:1063–8, 2005.

Carballeira, N. M., D. Sanabria, and K. Parang. Total synthesis and further scrutiny of the *in vitro*-antifungal activity of 6-nonadecynoic acid. *Arch. Pharm.* 338:441–3, 2005.

Carballeira, N. M., R. O'Neill, and K. Parang. Racemic and optically active 2-methoxy-4-oxatetradecanoic acids: Novel synthetic fatty acids with selective antifungal properties. *Chem. Phys. Lipids* 136:47–54, 2005.

Carballeira, N. M., D. Sanabria, N. L. Ayala, and C. Cruz. A stereoselective synthesis for the (5Z,9Z)-14-methyl-5,9-pentadecadienoic acid and its mono-unsaturated analog (Z)-14-methyl-9-pentadecenoic acid. *Tetrahedron Lett.* 45:3761–3, 2004.

Carballeira, N. M., D. Oritz, K. Parang, and S. Sardari. Total synthesis and in vitro antifungal activity of (±)-2-methoxytetradecanoic acid. *Arch. Pharm.* 337:152–5, 2004.

Marti, A. A., and **J. L. Colón.** Direct ion exchange of tris(2,2'-bipyridl)ruthenium (ii) into a zirconium phosphate type framework. *Inorg. Chem.* 42:2830–2, 2003.

Langen, R., and **J. L. Colón** et al. Electron tunneling in proteins: Role of the intervening medium. *J. Biol. Inorg. Chem.* 1:221–5, 1996.

Navarro, A. M., et al. **(J. L. Colón).** Control of carbon monoxide–binding states and dynamics in hemoglobin I of *Lucina pectinata* by nearby aromatic residues. *Inorg. Chim. Acta* 243:161–6, 1996.

Casimiro, D. R., et al. **(J. L. Colón).** Electron transfer in ruthenium/zinc porphyrin derivatives of recombinant human myoglobins. Analysis of tunneling pathways in myoglobin and cytochrome c. *J. Am. Chem. Soc.* 115:1485–9, 1993.

Colón, J. L., and C. R. Martin. Luminescence probe studies of ionomers 3. Distribution of decay rate constants for tris(2,2'-bipyridyl)ruthenium(ii) in nafion membranes. *Langmuir* 9:1066–70, 1993.

Trujillo, C. A., et al. **(F. A. González).** Inhibition mechanism of the recombinant rat P2X2 receptor in glial cells by suramin and TNP-ATP. *Biochemistry* 45:221–33, 2006.

Flores, R. V., et al. **(F. A. González).** Agonist-induced phosphorylation and desensitization of the P2Y2 nucleotide receptor. *Mol. Cell. Biochem.* 280:35–45, 2005.

Tulapurkar, M. E., et al. **(F. A. González).** Endocytosis mechanism of P2Y2 nucleotide receptor tagged with green fluorescent protein: Clathrin and actin cytoskeleton dependence. *Cell. Mol. Life Sci.* 62:1388–99, 2005.

Bagchi, S., et al. **(F. A. González).** The P2Y2 nucleotide receptor interacts with aV integrins to activate GO and induce cell migration. *J. Biol. Chem.* 280:39050–7, 2005.

González, F. A., et al. Mechanisms for inhibition of P2 receptors signaling in neural cells. *Mol. Neurobiol.* 31:65–79, 2005.

Eker, F., and **K. Griebenow** et al. Tripeptides with ionizable side chains adopt a perturbed polyproline II structure in water. *Biochemistry* 43:613–21, 2004.

Schweitzer-Stenner, R., et al. **(K. Griebenow).** The conformation of tetraalanine in water determined by polarized Raman, FT-IR, and VCD spectroscopy. *J. Am. Chem. Soc.* 126:2768–76, 2004.

Eker, F., **K. Griebenow,** and R. Schweitzer-Stenner. A $\beta_{1\text{-}28}$ fragment of the amyloid peptide predominantly adopts a polyproline II conformation in an acidic solution. *Biochemistry* 43:6893–8, 2004.

Eker, F., and **K. Griebenow** et al. Preferred peptide backbone conformations in the unfolded state revealed by the structure analysis of alanine-based (AXA) tripeptides in aqueous solution. *Proc. Natl. Acad. Sci. U.S.A.* 101:10054–9, 2004.

Castellanos, I. J., and **K. Griebenow.** Improved α-chymotrypsin stability upon encapsulation in PLGA microspheres by solvent replacement. *Pharm. Res.* 20:1873–80, 2003.

Rosario-Canales, M., **A. R. Guadalupe,** L. F. Fonseca, and O. Resto. Physicochemical characterization of porous silicon surfaces etched in salt solutions of varying compositions and pH. *Mater. Res. Soc. Symp. Proc.* 762:779–84, 2003.

Zhiqin, J., H. Songping, and **A. R. Guadalupe.** Synthesis, X-ray structures, spectroscopic and electrochemical properties of ruthenium (II) complex containing 2,2'-bipyrimidine. *Inorg. Chim. Acta* 305:127–34, 2000.

Guo, Y., and **A. R. Guadalupe.** Chemical-derived Prussian blue sol-gel composite thin films. *Chem. Mater.* 11(1):135–40, 1999.

Landau, A., E. Eliav, **Y. Ishikawa,** and U. Kaldor. Mixed-sector intermediate Hamiltonian Fock-space coupled cluster approach. *J. Chem. Phys.* 121:6634–9, 2004.

Vilkas, M. J., and **Y. Ishikawa.** Relativistic many-body perturbation calculations on extreme ultraviolet and soft-X-ray transition energies in siliconlike iron. *Phys. Rev. A: At. Mol. Opt. Phys.* 69:062503/1–15, 2004.

Ishikawa, Y., et al. Direct ab initio molecular dynamics study of CH_3^+ + benzene. *Chem. Phys. Lett.* 396:16–20, 2004.

Ishikawa, Y., M. S. Liao, and **C. R. Cabrera.** A theory-guided design of bimetallic nanoparticle catalysts for fuel cell applications. *Theor. Comput. Chem.* 15:325–65, 2004.

Ikegami, T., N. Kurita, H. Sekino, and **Y. Ishikawa.** Mechanism of cis-to-trans isomerization of azobenzene. Direct MD study. *J. Phys. Chem. A* 107:4555, 2003.

Kirchstetter, T. W., T. Novakov, **R. Morales,** and **O. Rosario.** Differences in the volatility of organic aerosols in unpolluted tropical and polluted continental atmospheres. *J. Geophys. Res. (Atmos.)* 105:26547–54, 2000.

Torres, G., W. Torres, and **J. A. Prieto.** Microwave-assisted epoxidation of hindered homoallylic alcohols using $VO(acac)_2$: Application to polypropionate synthesis. *Tetrahedron* 60:10245–51, 2004.

Arbelo, D. O., L. Castro-Rosario, and **J. A. Prieto.** Efficient hydroxyl inversion in polypropionates via cesium carboxylates. *Synthetic Commun.* 33:3211–23, 2003.

Montes, I., **J. A. Prieto,** and M. García. Using molecular modeling in the organic chemistry course for majors. *Chem. Educator* 7:293–6, 2002.

Arbelo, D. O., and **J. A. Prieto.** A new epoxide-inversion methodology mediated by cesium propionate. *Tetrahedron Lett.* 43:4111–4, 2002.

Arias, L., D. Arbelo, A. Alzérreca, and **J. A. Prieto.** Synthesis of functionalized enamines from lithium alkyl phenyl sulfones and N-carbo-*tert*-butoxy lactams. *J. Heterocyclic Chem.* 38:29–33, 2001.

University of Puerto Rico, Río Piedras

Cox, O., and **J. A. Prieto** et al. Synthesis and complete 1H and 13C assignments of thiazolo[3,2-a]quinolinium derivatives. *J. Heterocyclic Chem.* 36:937–42, 1999.

Dixit, A. A., et al. **(E. Quiñones).** Dissociation of sulfur dioxide by ultraviolet absorption between 224 and 232 nm. *J. Phys. Chem.* 109:1770–5, 2005.

Fuentes, L., J. Oyola, M. Fernandez, and **E. Quiñones.** Conformational changes in azurin from *Pseudomonas aeruginosa* induced through chemical and physical protocols. *Biophys. J.* 87:1873–80, 2004.

Makarov, V. I., and **E. Quiñones.** Magnetic field quenching of individual rotational levels of the ~A^1A_u, $2v_3$' state of acetylene. *J. Chem. Phys.* 118:87–92, 2003.

Lei, Y., V. I. Makarov, C. Conde, and **E. Quiñones.** Laser-initiated processes within $(SO_2)_m(NO)_n$ weakly bound clusters. *J. Chem. Phys.* 295:131–6, 2003.

Demadis, K. D., et al. **(R. G. Raptis).** Alkali earth metal triphosphonates: Inorganic-organic polymeric hybrids from dication-dianion association. *Cryst. Growth Des.* 6:836–8, 2006.

Demadis, K. D., **R. G. Raptis,** and P. Baran. Chemistry of organophosphonate scale growth inhibitors: 2. Structural aspects of 2-phosphonobutane-1,2,4-tricarboxylic acid monohydrate (PBTC.H_2O). *Bioinorg. Chem. Appl.* 3:119–24, 2005.

Mezei, G., J. E. McGrady, and **R. G. Raptis.** First structural characterization of a delocalized, mixed-valent, triangular Cu_3^{7+} species: Chemical and electrochemical oxidation of a $Cu^{II}_3(m_3\text{-}O)$ pyrazolate and electronic structure of the oxidation product. *Inorg. Chem.* 44:7271–3, 2005.

Demadis, K. D., C. Mantzaridis, **R. G. Raptis,** and G. Mezei. Metal-organophosphonate inorganic-organic hybrids: Crystal structure and anticorrosion effects of zinc hexamethylenediaminetetrakis(methylenephosphonate) on carbon steels. *Inorg. Chem.* 44:4469–71, 2005.

Miras, H. N., and **R. G. Raptis** et al. A novel series of vanadium-sulfite polyoxometalates: Synthesis, structural, and physical studies. *Chem. Eur. J.* 11:2295–306, 2005.

Gubala, V., J. E. Betancourt, and **J. M. Rivera.** Expanding the Hoogsteen Edge of 2'-deoxyguanosine: Consequences for G-quadruplex formation. *Org. Lett.* 6:4735–8, 2004.

Rivera, J. M., T. Martín, and J. Rebek Jr. Chiral softballs: Synthesis and molecular recognition properties. *J. Am. Chem. Soc.* 123:5213–20, 2001.

Rivera, J. M., S. L. Craig, T. Martín, and J. Rebek Jr. Chiral guests and their ghosts in reversibly assembled hosts. *Angew. Chem. Int. Ed.* 39:2130–2, 2000.

Rivera, J. M., and J. Rebek Jr. Chiral space in a unimolecular capsule. *J. Am. Chem. Soc.* 122:7811–2, 2000.

Marrero, J., et al. **(A. D. Rodríguez** and **R. G. Raptis).** Bielschowskysin, a gorgonian-derived biologically active diterpene with an unprecedented carbon skeleton. *Org. Lett.* 6:1661–4, 2004.

Marrero, J., **A. D. Rodríguez,** P. Baran, and **R. G. Raptis.** Ciereszkolide: Isolation and structure characterization of a novel rearranged cembrane from the Caribbean sea plume *Pseudopterogorgia kallos. Eur. J. Org. Chem.* 3909–12, 2004.

Rodríguez, I. I., et al. **(A. D. Rodríguez).** New pseudopterosin and seco-pseudopterosin diterpene glycosides from two Colombian isolates of *Pseudopterogorgia elisabethae* and their diverse biological activities. *J. Nat. Prod.* 67:1672–80, 2004.

Marrero, J., **A. D. Rodríguez,** P. Baran, and **R. G. Raptis.** Isolation and characterization of kallosin A: A novel rearranged pseudopterane diterpenoid from the Caribbean Sea plume *Pseudopterogorgia kallos* (Bielschowsky). *J. Org. Chem.* 68:4977–9, 2003.

Ospina, C. A., **A. D. Rodríguez,** E. Ortega-Barria, and T. L. Capson. Briarellins J-P and polyanthellin A: New eunicellin-based diterpenes from the gorgonian coral *Briareum polyanthes* and their antimalarial activity. *J. Nat. Prod.* 66:357–63, 2003.

Colón, I., D. Caro, and **O. Rosario.** Analysis of exogenous compounds in the serum of young Puerto Rican girls with premature thelarche. *Environ. Health Perspect.*, in press.

Reyes, D. R., **O. Rosario,** J. F. Rodríguez, and B. D. Jiménez. Toxic evaluation of organic extracts from airborne particulate matter in Puerto Rico. *Environ. Health Perspect.*, in press.

Mayol, O., et al. **(O. Rosario** and **R. Morales).** Chemical characterization of submicron, organic aerosols in the tropical trade winds in the Caribbean, using gas chromatograph/mass spectrometry. *Atmos. Environ.* 35:1735–45, 2001.

Rodríguez, J. F., J. L. Rodríguez, J. Santana, and **O. Rosario.** Simultaneous quantitation of intracellular zidovudine and lamivudine-triphosphate in HIV-infected individuals. *Antimicrob. Agents Chemother.* 44:3097, 2000.

Nayar, A., et al. **(E. S. Smotkin).** High speed laser activated membrane introduction mass spectrometric evaluation of bulk methylcyclohexane dehydrogenation catalysts. *Appl. Surf. Sci.* 223:118–23, 2004.

Smotkin, E. S. Fuel cells operating in the "gap" temperature regime. *Prepr. Symp.—Am. Chem. Soc., Div. Fuel Chem.* 48:887–8, 2003.

Smotkin, E. S., and R. R. Diaz-Morales. New electrocatalysts by combinatorial methods. *Annu. Rev. Mater. Res.* 33:557–79, 2003.

Mallouk, T. E., and **E. S. Smotkin.** Combinatorial catalyst development methods. In *Handbook of Fuel Cells—Fundamentals, Technology and Applications,* vol. 2, pp. 334–47, eds. W. Vielstich, A. Lamm, and H. A. Gasteiger. Chichester, England: John Wiley & Sons, Ltd., 2003.

Hernandez, E., E. Canales, E. Gonzalez, and **J. A. Soderquist.** Asymmetric synthesis with the robust and versatile 10-substituted 9-borabicyclo[3.3.2]decanes: Homoallylic amines from aldimines. *Pure Appl. Chem.*, in press.

Lai, C., and **J. A. Soderquist.** Non-racemic homopropargylic alcohols via asymmetric allenylboration with the robust and versatile 10-tms-9-borabicyclo[3.3.2]decanes. *Org. Lett.* 7:799–803, 2005.

Burgos, C. H., E. Canales, K. Matos, and **J. A. Soderquist.** Asymmetric allyl- and crotylboration with the robust, versatile and recyclable 10-tms-9-borabicyclo[3.3.2]decanes. *J. Am. Chem. Soc.* 127:8044–51, 2005.

Canales, E., K. G. Prasad, and **J. A. Soderquist.** B-allyl-10-ph-9-borabicyclo[3.3.2]decanes: Strategically designed for the asymmetric allylboration of ketones. *J. Am. Chem. Soc.* 127:11553–72, 2005.

Hernandez, E., and **J. A. Soderquist.** Non-racemic α-allenyl carbinols from asymmetric propargylation with the 10-trimethylsilyl-9-borabicyclo[3.3.2]decanes. *Org. Lett.* 7:5397–400, 2005.

Weiner, B. R., and G. Morell. Effects of heavy-ion radiation on the electron field emission properties of sulfur-doped nanocomposite carbon films. *Diamond Relat. Mater.* 13:221–5, 2004.

Feng, P. X., and **B. R. Weiner.** High abundance of metastable helium atoms for diagnostic applications. *J. Phys. B: At., Mol. Opt. Phys.* 37:1–5, 2004.

Gupta, S., G. Morell, and **B. R. Weiner.** Role of H in hot-wire deposited a-Si:H films revisited: Optical characterization and modeling. *J. Non-Cryst. Solids* 343:131–42, 2004.

Gupta, S., **B. R. Weiner,** and G. Morell. Ex situ spectroscopic ellipsometry investigations of chemical vapor deposited nanocomposite carbon thin films. *Thin Solid Films* 455:422–8, 2004.

Gupta, S., G. Morell, and **B. R. Weiner.** Electron field-emission mechanism in nanostructured carbon films: A quest. *J. Appl. Phys.* 95:8314–20, 2004.

THE UNIVERSITY OF TOLEDO

THE UNIVERSITY OF TOLEDO

Department of Chemistry

Programs of Study

The University of Toledo Department of Chemistry offers graduate programs leading to the M.S. and Ph.D. degrees. A wide range of research topics is available for study within the subdisciplines of analytical, biological, inorganic, materials, organic, and physical chemistry. Collaborative interactions with the departments of biology, medicinal and biological chemistry, physics, and chemical engineering as well as various industrial partners enhance the research opportunities and interactions for students.

The major purpose of the graduate program is to educate graduate students to participate in the advancing field of chemistry as a professional chemist in industry or academia. This education involves specialized course work, teaching classes as a teaching assistant, attending seminars and scientific meetings, and completing research projects that contribute to the growth of scientific knowledge. The main requirements for the M.S. degree are the satisfactory completion of course work and the research leading to the successful defense of a written thesis. Students typically complete an M.S. degree in two to three years. The Ph.D. program is described in three stages. The first stage involves course work to provide a foundation for chemical research. During the second stage, students undertake research toward their dissertation and take a comprehensive examination for admission to Ph.D. candidacy. The final stage is devoted to research for completion and defense of the dissertation. Ph.D. students also present a literature seminar during this stage. Ph.D. students usually complete these requirements in four to five years.

Research Facilities

The chemistry research facilities are housed in the Bowman-Oddy/Wolfe Hall complex. Wolfe Hall, which opened in 1998 adjacent to the Bowman-Oddy Laboratories, is a 165,000-square-foot research and teaching facility.

The department maintains an extensive array of modern scientific instrumentation for research. This includes the College of Arts and Sciences Instrumentation Center, which houses more than $5 million of instrumentation, and the Ohio Crystallography Consortium, which distinguishes the department as the center of excellence in Ohio for crystallographic research. Single-crystal X-ray analysis for small molecules and biological macromolecules, as well as powders, is routine. In addition to a wide range of specialized equipment found in individual research laboratories, the department has a dedicated NMR laboratory with 200-, 400-, and 600-MHz NMR spectrometers for solution samples and a 200-MHz NMR spectrometer for solids. A separations laboratory includes an LC-MS, three GC-MS, two GCs, and two LC systems for isocratic and gradient capabilities. Hands-on use by students is encouraged and is integrated into graduate course work. Department and college instrumentation specialists are available for training, consultation, and equipment maintenance.

Financial Aid

Most full-time chemistry graduate students receive some financial support. Fellowships and teaching and research assistantships, which include a stipend and a tuition waiver, are available for qualified students on a competitive basis.

The out-of-state tuition surcharge normally charged to out-of-state and international students is waived for students whose permanent address is within one of the following Michigan counties: Hillsdale, Lenawee, Macomb, Oakland, Washtenaw, and Wayne. In addition the University of Toledo offers an out-of-state tuition surcharge waiver to cities and regions that are a part of the Sister Cities Agreement. These regions include Toledo, Spain; Londrina, Brazil; Qinhuangdao, China; Csongrad County, Hungary; Delmenhorst, Germany; Toyohashi, Japan; Tanga, Tanzania; Bekaa Valley, Lebanon; and Poznan, Poland.

The University of Toledo Graduate College offers a variety of memorial and minority scholarship awards including the Ronald E. McNair Postbaccalaureate Achievement Scholarship, the Graduate Minority Assistantship Award, and two full University Fellowships.

Cost of Study

The graduate tuition rate for the 2006–07 academic year is $390.05 per semester credit hour for in-state students. For nonresidents, the out-of-state surcharge is $367.15 per semester credit hour. Additional fees are required and include the general fee, technology fee, and mandatory insurance.

Living and Housing Costs

The University of Toledo has a diverse offering of student housing options including suite-style and traditional residential halls. Housing is offered to graduate students through Residence Life or contracted individually by the student. Affordable, high-quality off-campus apartment-style housing within walking distance of campus is abundant.

Student Group

There are approximately 20,000 students at the University of Toledo. About 4,000 are graduate and professional students. The University has a rich diversity of student organizations. Students join groups that are organized around common cultural, religious, athletic, and educational interests.

Student Outcomes

Graduates of the chemistry program are very successful obtaining high-profile and rewarding positions in industry and academia. Recent graduates have accepted industrial positions at companies such as Pfizer, Merck, Procter and Gamble, Eli-Lilly, DuPont, and Millennium Pharmaceuticals. Graduates have accepted faculty or postdoctoral research positions at numerous colleges and universities throughout the U.S. and the world.

Location

The University of Toledo has several campus sites in the city of Toledo. Most chemistry graduate students take classes on the main campus, which is located in suburban western Toledo. With a population of more than 330,000, Toledo is the fiftieth-largest city in the United States. It is located on the western shores of Lake Erie within a 2-hour drive of Cleveland and Detroit.

The University and The Department

The University of Toledo was founded by Jessup W. Scott in 1872 as a municipal institution and became part of the state of Ohio's system of higher education in 1967. On July 1, 2006, the University of Toledo merged with the Medical University of Ohio becoming one of only seventeen American universities to offer professional and graduate academic programs in medicine, law, pharmacy, nursing, health sciences, engineering, and business.

The Department of Chemistry is a strong department and a leading producer of M.S. and Ph.D. students in the sciences at the University of Toledo. External funding for research averages $100,000 per faculty member per year.

Applying

Applications should be submitted directly to the Graduate Admissions Committee. An application can be obtained online at the department Web site. A complete application includes the application form, three letters of recommendation, a copy of all previous undergraduate or graduate transcripts, and scores from the aptitude portion of the Graduate Record Examinations (GRE). International students are also required to submit scores from the TOEFL examination. Applications should be completed by February 1 for fall admission and merit fellowship consideration, although later applications are considered depending on the availability of assistantship funds.

Correspondence and Information

Mark R. Mason, Ph.D.
Chair, Graduate Admissions Committee
Department of Chemistry
Mail Stop 602
The University of Toledo
Toledo, Ohio 43606-3390
Phone: 419-530-2100
Fax: 419-530-4033
E-mail: utchem@uoft02.utoledo.edu
Web site: http://www.chem.utoledo.edu

The University of Toledo Graduate College
3240 University Hall, MS 933
The University of Toledo
2801 West Bancroft Street
Toledo, Ohio 43606
Phone: 419-530-4723
E-mail: grdsch@utnet.utoledo.edu
Web site: http://www.gradschool.utoledo.edu

The University of Toledo

THE FACULTY AND THEIR RESEARCH

Jared L. Anderson, Assistant Professor; Ph.D., Iowa State, 2005. Analytical: application of analytical instrumentation and methodology to topics in bioanalytical, environmental, and green chemistry.
Bruce A. Averill, Distinguished University Professor; Ph.D., MIT, 1973. Inorganic and Biochemistry: synthetic models for metalloproteins, spectroscopic and mechanistic studies of metalloenzymes.
Terry P. Bigioni, Assistant Professor; Ph.D., Georgia Tech, 1999. Physical: nanocrystals and nanocomposite materials, non-covalent self-assembly.
Eric W. Findsen, Associate Professor; Ph.D., New Mexico, 1986. Biophysical and Physical: Raman and time-resolved Raman spectroscopy of metalloporphyrins and protein systems.
Max O. Funk, Professor; Ph.D., Duke, 1975. Bioorganic and Biochemistry: investigations of lipoxygenase structure and mechanism of action, enzymology.
Dean M. Giolando, Professor; Ph.D., Illinois, 1987. Inorganic and Organometallic: precursors to solid-state materials; thin-film deposition; photovoltaics.
Xiche Hu, Associate Professor; Ph.D., Wayne State, 1991. Computational and Biophysical: quantum chemical analysis of protein-ligand interactions in molecular recognition and structure-based drug design.
Xuefei Huang, Associate Professor; Ph.D., Columbia, 1999. Organic and Bioorganic: synthesis of oligosaccharides, assembly of oligosaccharide libraries, study of carbohydrate-protein interactions and conformational change of proteins.
Richard A. Hudson, Professor; Ph.D., Chicago, 1966. Bioorganic and Medicinal: design of active site–directed agents.
Andrew D. Jorgensen, Associate Professor; Ph.D., Illinois at Chicago, 1976. Physical and Chemical Education: classroom communication in large lecture environments, factors that influence student performance.
Jon R. Kirchhoff, Professor; Ph.D., Purdue, 1985. Analytical: development and applications of microelectrode devices and sensor systems for bioanalysis and capillary electrophoresis detection, spectroelectrochemical methods.
Yun-Ming Lin, Assistant Professor; Ph.D., Notre Dame, 2000. Organic and Chemical Biology: synthetic methodology development, syntheses and mechanistic studies of biologically active small molecules.
Cora Lind, Assistant Professor; Ph.D., Georgia Tech, 2001. Inorganic and Materials: synthesis and characterization of new solids, nonhydrolytic sol-gel chemistry, negative thermal expansion materials, composites, powder XRD, Rietveld.
Mark R. Mason, Associate Professor; Ph.D., Iowa State, 1991. Inorganic and Organometallic: synthesis, characterization, and reactivity of molecules and materials with catalytic applications; group 13 chemistry.
Timothy C. Mueser, Associate Professor; Ph.D., Nebraska, 1989. Biochemistry and Crystallography: X-ray crystallography and protein chemistry of DNA replication and DNA repair multiprotein complexes.
A. Alan Pinkerton, Professor and Chair; Ph.D., Alberta, 1971. X-ray Crystallography: nonroutine structure determination, charge density analysis, new techniques.
Donald R. Ronning, Assistant Professor; Ph.D., Texas A&M, 2001. Biochemistry and Crystallography: X-ray crystallography of protein/DNA complexes representing intermediates along the DNA recombination pathways.
Joseph A. R. Schmidt, Assistant Professor; Ph.D., Berkeley, 2002. Inorganic and Organometallic: design, synthesis, and characterization of homogeneous catalysts for the functionalization of alkenes and alkynes, hydroamination, hydrosilylation, and olefin metathesis reaction chemistry.
James Slama, Professor; Ph.D., Berkeley, 1977. Medicinal and Drug Design: mechanism-based enzyme inhibitors and their application to biochemical and pharmacological problems.
Steven J. Sucheck, Assistant Professor; Ph.D., Virginia, 1998. Organic and Bioorganic: synthesis and study of nucleic acid-interactive small molecules, glycoconjugates, and biologically active natural products.
Ronald E. Viola, Professor; Ph.D., Penn State, 1976. Biochemistry and Enzymology: mechanistic studies of enzyme catalyzed reactions, mapping of enzyme-active sites, introduction of unnatural amino acids, high-resolution structural studies.

The University of Toledo main campus.

Wolfe Hall.

SELECTED PUBLICATIONS

Anderson, J. L., et. al. Structure and properties of high stability geminal dicationic ionic liquids. *J. Am. Chem. Soc.* 127:593–604, 2005.

Anderson, J. L., and D. W. Armstrong. High-stability ionic liquids. A new class of stationary phases for gas chromatography. *Anal. Chem.* 75:4851–8, 2003.

Anderson, J. L., et. al. Surfactant solvation effects and micelle formation in ionic liquids. *Chem. Commun.* 19:2444–5, 2003.

Anderson, J. L., et al. Separation of racemic sulfoxides and sulfinate esters on four derivatized cyclodextrin chiral stationary phases using capillary gas chromatography. *J. Chromatogr. A* 946:197–208, 2002.

Anderson, J. L., J. Ding, T. Welton, and D. W. Armstrong. Characterizing ionic liquids on the basis of multiple solvation interactions. *J. Am. Chem. Soc.* 124:14247–54, 2002.

Funhoff, E. G., Y. Wang, G. Andersson, and **B. A. Averill.** Substrate positioning by His92 is an important factor in catalysis by purple acid phosphatase. *FEBS J.* 272:2968–77, 2005.

Funhoff, E. G., M. Bollen, and **B. A. Averill.** The Fe(III)Zn(II) form of recombinant human purple acid phosphatase is not activated by proteolysis. *J. Inorg. Biochem.* 99:521–9, 2004.

Averill, B. A. Dimetal hydrolases. In *Comprehensive Coordination Chemistry,* 2nd ed., vol. 8 *(Bio-coordination Chemistry),* pp. 641–76, eds. L. Que Jr. and W. B. Tolman. 2004.

Dikiy, A., E. G. Funhoff, **B. A. Averill**, and S. Ciurli. New insights into the mechanism of purple acid phosphatase through 1H NMR spectrometry. *J. Am. Chem. Soc.* 124:13974–5, 2002.

Funhoff, E. G., et al. **(B. A. Averill).** Mutational analysis of the interaction between active site residues and the loop region in mammalian purple acid phosphatases. *Biochemistry* 40:11614–22, 2001.

Blanchette, F., and **T. P. Bigioni.** Partial coalescence of drops on liquid interfaces. *Nature Phys.* 2:254–7, 2006.

Bigioni, T. P., et al. Kinetically-driven self assembly of highly-ordered nanoparticle monolayers. *Nature Materials* 5:265–70, 2006.

Tran, T. B., et al. (**T. P. Bigioni**). Multiple cotunneling in large quantum dot arrays. *Phys. Rev. Lett.* 95:076806/1–4, 2005.

Bigioni, T. P., et al. Scanning tunneling microscopy determination of single nanocrystal core sizes via correlation with mass spectrometry. *J. Phys. Chem. B* 108:3772–6, 2004.

Larsen, R. W., P. E. Wheeler, and **E. W. Findsen.** Ligand photolysis and recombination of Fe(II) protoporphyrin IX complexes in tetramethylene sulfoxide. *Inorg. Chim. Acta* 319:1–7, 2001.

Vahedi-Faridi, A., et al. **(M. O. Funk).** Interaction between non-heme iron of lipoxygenases and cumene hydroperoxide: Basis for enzyme activation, inactivation, and inhibition. *J. Am. Chem. Soc.* 126:2006–15, 2004.

Brault, P. A., et al. **(M. O. Funk).** Protein micelles from lipoxygenase-3. *Biomacromolecules* 3:649–54, 2002.

Kariapper, M. S. T., W. R. Dunham, and **M. O. Funk.** Iron extraction from soybean lipoxygenase 3 and reconstitution of catalytic activity from the apoenzyme. *Biochem. Biophys. Res. Commun.* 284:563–7, 2001.

Skrzypczak-Jankun, E., et al. **(M. O. Funk).** Three-dimensional structure of a purple lipoxygenase. *J. Am. Chem. Soc.* 123:10814–20, 2001.

Roussillon, Y., and **D. M. Giolando** et al. Blocking thin-film nonuniformities: Photovoltaic self-healing. *Appl. Phys. Lett.* 84:616–8, 2004.

Rosen, T., K. Kirschbaum, and **D. M. Giolando.** Homoleptic phenolate complexes of zirconium(IV): Syntheses and structural characterization of the first six coordinate complexes. *J. Chem. Soc., Dalton Trans.* 120–5, 2003.

Giolando, D. M., and **J. R. Kirchhoff** et al. Chemical vapor deposition of alumina on carbon and silicon carbide microfiber substrates for microelectrode development. *Chem. Vap. Deposition* 8:93–8, 2002.

Bozon, J. P., **D. M. Giolando,** and **J. R. Kirchhoff.** Development of metal-based microelectrode sensor platforms by chemical vapor deposition. *Electroanalysis* 13:911–6, 2001.

Mao, L. S., Y. L. Wang, Y. M. Liu, and **X. C. Hu.** Molecular determinants for ATP-binding in proteins: A data mining and quantum chemical analysis. *J. Mol. Biol.* 336:787–807, 2004.

Wang, Y. L., L. S. Mao, and **X. C. Hu.** Insight into the structural role of carotenoids in the photosystem I: A quantum chemical analysis. *Biophys. J.* 86:3097–111, 2004.

Mao, L. S., Y. L. Wang, Y. M. Liu, and **X. C. Hu.** Multiple intermolecular interaction modes of positively charged residues with adenine in ATP-binding proteins. *J. Am. Chem. Soc.* 125:14216–7, 2003.

Wang, Y., and **X. Hu.** A quantum chemistry study of binding carotenoids in the bacterial light-harvesting complexes. *J. Am. Chem. Soc.* 124:8445–51, 2002.

Wang, Y., and **X. Hu.** Quantum chemical study of pi-pi stacking interactions of the bacteriochlorophyll dimer in the photosynthetic reaction center of *rhodobacter sphaeroides. J. Chem. Phys.* 117:1–4, 2002.

Wang, Y., **X. Huang,** L.-H. Zhang, and X.-S. Ye. A four-component one-pot synthesis of alpha-Gal pentasaccharide. *Org. Lett.* 6:4415–7, 2004.

Huang, L., Z. Wang, and **X. Huang.** One-pot oligosaccharide synthesis: Reactivity tuning by post-synthetic modification of aglycon. *Chem. Commun.* 1960–1, 2004.

Huang, X., L. Huang, H. Wang, and X.-S. Ye. Iterative one-pot oligosaccharide synthesis. *Angew. Chem. Int. Ed.* 43:5221–4, 2004.

Jing, Y., and **X. Huang.** Fluorous thiols in oligosaccharide synthesis. *Tetrahedron Lett.* 45:4615–8, 2004.

Huang, X., et al. Absolute configurational assignments of secondary amines by CD-sensitive dimeric zinc porphyrin host. *J. Am. Chem. Soc.* 124:10320–35, 2002.

Fouchard, D., L. M. V. Tillekeratne, and **R. A. Hudson.** An alternative one-pot benzimidazole imidazole synthesis. *Synthesis* 17–20, 2005.

Nacario, R., et al. **(R. A. Hudson).** Reductive monoalkylation of aromatic and aliphatic nitro compounds and the corresponding amines with nitriles. *Org. Lett.* 7:471–4, 2005.

Schroeder, M. M., R. J. Belloto Jr., **R. A. Hudson,** and M. M. McInerney. Effects of the antioxidant coenzyme Q_{10} and lipoic acid on interleukin-1 β mediated inhibition of glucose-stimulated insulin release from cultured mouse pancreatic islets. *Immunopharmacol. Immunotoxicol.* 27:109–22, 2005.

Fouchard, D., L. M. V. Tillekeratne, and **R. A. Hudson.** Synthesis of imidazolo analogues of the oxidation-reduction cofactor pyrroloquinoline quinone. *J. Org. Chem.* 69:2626–9, 2004.

Barkhimer, T. V., **J. R. Kirchhoff, R. A. Hudson,** and W. S. Messer Jr. Classification of the mode of inhibition of high-affinity choline uptake using capillary electrophoresis with electrochemical detection at an enzyme-modified microelectrode. *Anal. Biochem.* 339:216-22, 2005.

Liu, Q., et al. **(J. R. Kirchhoff** and **R. A. Hudson).** Chiral separation of highly negatively charged enantiomers by capillary electrophoresis. *J. Chromatogr. A* 1033:349–56, 2004.

Kovalcik, K. **D., J. R. Kirchhoff, D. M. Giolando,** and J. P. Bozon. Copper ring-disk microelectrodes: Fabrication, characterization, and application as amperometric detectors for capillary columns. *Anal. Chim. Acta* 507:237–45, 2004.

Inoue, T., **J. R. Kirchhoff,** and **R. A. Hudson.** Enhanced measurement stability and selectivity for choline and acetylcholine by capillary electrophoresis with electrochemical detection at a covalently linked enzyme modified electrode. *Anal. Chem.* 74:5321–6, 2002.

Barkhimer, T. V., **J. R. Kirchhoff, R. A. Hudson,** and W. S. Messer Jr. Evaluation of the inhibition of choline uptake in synaptosomes by capillary electrophoresis with electrochemical detection. *Electrophoresis* 23:3699–704, 2002.

Braun, P. D., et al. **(Y.-M. Lin).** A bifunctional molecule that displays context-dependent cellular activity. *J. Am. Chem. Soc.* 125:7575–80, 2003.

Lin, Y.-M., M. J. Miller, and U. Möllmann. The remarkable hydrophobic effect of a fatty acid side chain on the microbial growth promoting activity of a synthetic siderophore. *BioMetals* 14:153–7, 2001.

Lin, Y.-M., and M. J. Miller. Oxidation of primary amines to oxaziridines using molecular oxygen (O_2) as the ultimate oxidant. *J. Org. Chem.* 66:8282–5, 2001.

Lin, Y.-M., P. Helquist, and M. J. Miller. Synthesis and biological evaluation of a siderophore-virginiamycin conjugate. *Synthesis* 1510–4, 1999.

Lin, Y.-M., and M. J. Miller. Practical synthesis of hydroxamate-derived siderophore components by indirect oxidation method and syntheses of a DIG-siderophore conjugate and a biotin-siderophore conjugate. *J. Org. Chem.* 64:7451–8, 1999.

Varga, T., et al. **(C. Lind).** In-situ diffraction study of $Sc_2(WO_4)_3$ at up to 10 GPa. *Phys. Rev. B* 71:214106/1–214106/8, 2005.

The University of Toledo

Mittal, R., et al. **(C. Lind).** Negative thermal expansion in cubic $ZrMo_2O_8$: Inelastic neutron scattering and lattice dynamical studies. *Phys. Rev. B* 70:214303/1–214306/1, 2004.

Casado-Rivera, E., et al. **(C. Lind).** Electrocatalytic activity of ordered intermetallic phases for fuel cell applications. *J. Am. Chem. Soc.* 126:4043–9, 2004.

Stevens, R., et al. **(C. Lind).** Heat capacities, third-law entropies and thermodynamic functions of the negative thermal expansion materials, cubic α-ZrW_2O_8 and cubic $ZrMo_2O_8$, from T = (0 to 400) K. *J. Chem. Thermodyn.* 35(6):919–37, 2003.

Lind, C., et al. New high-pressure form of the negative thermal expansion materials zirconium molybdate and hafnium molybdate. *Chem. Mater.* 13:487–90, 2001.

Mason, M. R., et al. Deprotonated diindolylmethanes as dianionic analogues of scorpionate bis(pyrazolyl)borate ligands: Synthesis and structural characterization of representative titanocene and zirconocene complexes. *J. Organomet. Chem.* 690:157–62, 2005.

Mason, M. R., F. A. Beckford, K. Kirschbaum, and B. J. Gorecki. Fortuitous formation of $[(LiCl)_6\{(Me_2NCH_2C_6H_5N)_3P\}_2]$: An amine-ligated hexameric lithium chloride aggregate. *Inorg. Chem. Commun.* 8:331–4, 2005.

Mason, M. R., B. Song, and K. Kirschbaum. Remarkable room-temperature insertion of carbon monoxide into an aluminum–carbon bond of tri-*tert*-butylaluminum. *J. Am. Chem. Soc.* 126:11812–3, 2004.

Mason, M. R., B. N. Fneich, and K. Kirschbaum. Titanium and zirconium amido complexes ligated by 2,2'-di(3-methylindolyl)methanes: Synthesis, characterization, and ethylene polymerization activity. *Inorg. Chem.* 42:6592–4, 2003.

Marteel, A., et al. **(M. R. Mason)**. Supported platinum/tin complexes as catalysts for hydroformylation of 1-hexene in supercritical carbon dioxide. *Catal. Commun.* 4:309–14, 2003.

Senger, A. B., and **T. C. Mueser.** Rapid preparation of custom grid screens for protein crystal growth optimization. *J. Appl. Crystallogr.*, in press.

Collins, B. K., et al. **(T. C. Mueser).** A preliminary solubility screen used to improve crystallization trials. Crystallization and preliminary X-ray structure determination of *Aeropyrum pernix* flap endonuclease-1. *Acta Crystallogr. D, Biol. Crystallogr.* 60:1674–8, 2004.

Jones, C. E., **T. C. Mueser,** and N. G. Nossal. Bacteriophage T4 32 protein is required for helicase-dependent leading strand synthesis when the helicase is loaded by the T4 59 helicase-loading protein. *J. Biol. Chem.* 279:12067–75, 2004.

Jones, C. E., et al. **(T. C. Mueser).** Mutations of bacteriophage T4 59 helicase loader defective in fork binding and interactions with T4 32 single-stranded DNA binding protein. *J. Biol. Chem.* 279:25721–8, 2004.

Jones, C. E., and **T. C. Mueser** et al. Bacteriophage T4 gene 41 helicase and gene 59 helicase loading protein: A versatile couple with roles in replication and recombination. *Proc. Natl. Acad. Sci. Colloquium* 98:8312–8, 2001.

Chinte, U., et al. **(A. A. Pinkerton).** Sample size: An important parameter in flash-cooling macromolecular crystallization solutions. *J. Appl. Crystallogr.* 38:412–9, 2005.

Zhurova, E. A., et al. **(A. A. Pinkerton).** Electronic energy distributions in energetic materials: NTO and the biguanidinium dinitramides. *J. Phys. Chem. B.* 108:20173–9, 2004.

Bolotina, N. B., M. J. Hardie, R. L. Speer, and **A. A. Pinkerton.** Energetic materials: Variable temperature crystal structures of g and e-HNIW polymorphs. *J. Appl. Crystallogr.* 37:808–14, 2004.

Chen, Y.-S., et al. **(A. A. Pinkerton).** A standard local coordinate system for multipole refinements of the estrogen core structure. *J. Appl. Crystallogr.* 36:1464–6, 2003.

Ritchie, J. P., E. A. Zhurova, A. Martin, and **A. A. Pinkerton.** Dinitramide ion: Robust molecular charge topology accompanies an enhanced dipole moment in its ammonium salt. *J. Phys. Chem. B* 107:14576–89, 2003.

Ronning, D. R., et al. The carboxy-terminal portion of TnsC activates the Tn7 transposase through a specific interaction with TnsA. *EMBO J.* 23:2972–81, 2004.

Ronning, D. R., et al. Mycobacterium tuberculosis antigen 85A and 85C structures confirm binding orientation and conserved substrate specificity. *J. Biol. Chem.* 279:36771–7, 2004.

Hickman A. B., et al. **(D. R. Ronning).** The nuclease domain of adeno-associated virus rep coordinates replication initiation using two distinct DNA recognition interfaces. *Mol. Cell.* 13:403–14, 2004.

Sharma V., et al. **(D. R. Ronning).** Crystal structure of *Mycobacterium tuberculosis* SecA, a preprotein translocating ATPase. *Proc. Natl. Acad. Sci. U. S. A.* 100:2243–8, 2003.

Hickman A. B., **D. R. Ronning,** R. M. Kotin, and F. Dyda. Structural unity among viral origin binding proteins: Crystal structure of the nuclease domain of adeno-associated virus rep. *Mol. Cell* 10:327–37, 2002.

Getzler, Y. D. Y. L., **J. A. R. Schmidt,** and G. W. Coates. Synthesis of an epoxide carbonylation catalyst: Exploration of contemporary chemistry for advanced undergraduates. *J. Chem. Educ.* 82:621–4, 2005.

Schmidt, J. A. R., V. Mahadevan, Y. D. Y. L. Getzler, and G. W. Coates. A readily synthesized and highly active epoxide carbonylation catalyst based on a chromium porphyrin framework: Expanding the range of available b-lactones. *Org. Lett.* 6:373–6, 2004.

Schmidt, J. A. R., G. R. Giesbrecht, C. Cui, and J. Arnold. Anionic triazacyclononanes: New supporting ligands in main group and transition metal organometallic chemistry. *Chem. Commun.* 1025–33, 2003.

Schmidt, J. A. R., and J. Arnold. First-row transition metal complexes of sterically-hindered amidinates. *J. Chem. Soc., Dalton Trans.* 3454–61, 2002.

Giesbrecht, G. R., et al. **(J. A. R. Schmidt).** Divalent lanthanide metal complexes of a triazacyclononane-functionalized tetramethylcyclopentadienyl ligand: X-ray crystal structures of $[C_5Me_4SiMe_2(iPr_2\text{-tacn})]LnI$ (Ln = Sm, Yb; tacn = 1,4-diisopropyl-1,4,7-triazacyclononane). *Organometallics* 21:3841–4, 2002.

Liang, C.-H., et al. **(S. J. Sucheck).** Structure–activity relationships of bivalent aminoglycosides and evaluation of their microbiological activities. *Bioorg. Med. Chem. Lett.* 15:2123–8, 2005.

Romero, A., et al. **(S. J. Sucheck).** An efficient entry to new sugar modified ketolide antibiotics. *Tetrahedron Lett.* 46:1483–7, 2005.

Yao, S., et al. **(S. J. Sucheck).** Glyco-optimization of aminoglycosides: New aminoglycosides as novel anti-infective agents. *Bioorg. Med. Chem. Lett.* 14:3733–8, 2004.

Lee, L. V., et al. **(S. J. Sucheck).** Inhibition of the proteolytic activity of anthrax lethal factor by aminoglycosides. *J. Am. Chem. Soc.* 126:4774–5, 2004.

Agnelli, F., et al. **(S. J. Sucheck).** Dimeric aminoglycosides as antibiotics. *Angew. Chem. Int. Ed. Engl.* 43:1562–6, 2004.

Elder, I., et al. **(R. E. Viola).** Activation of carbonic anhydrase II by active-site incorporation of histidine analogs. *Arch. Biochem. Biophys.* 421:283–9, 2004.

Han, S. and **R. E. Viola.** Splicing of unnatural amino acids into proteins: A peptide model study. *Prot. Pept. Lett.* 11:107–14, 2004.

Blanco, J., et al. **(R. E. Viola).** The role of substrate binding groups in the mechanism of aspartate-*b*-semialdehyde dehydrogenase. *Acta Crystallogr.* D60:1388–95, 2004.

Cama, E., et al. **(R. E. Viola).** Inhibitor coordination interactions in the binuclear manganese cluster of arginase. *Biochemistry* 43:8987–99, 2004.

Blanco, J., R. A. Moore, and **R. E. Viola.** Capture of an intermediate in the catalytic cycle of L-aspartate-β-semialdehyde dehydrogenase. *Proc. Natl. Acad. Sci. U.S.A.* 100:12613–7, 2003.

UNIVERSITY OF VIRGINIA

Department of Chemistry

Program of Study

The Department of Chemistry at the University of Virginia (UVA) combines outstanding physical facilities with a close-knit community of scholars who have received some of the most distinguished awards in their fields. The goal of its graduate program is to provide students with a broad, fundamental background in chemistry and develop an ability to make significant contributions in their chosen fields. With this in mind, emphasis is placed on research that contributes to the fundamental body of knowledge.

The Ph.D. program requires completion of 72 credits with a grade of B or higher, as well as a written dissertation. Upon entering the program, students are required to take qualifying examinations in four of the following six areas: analytical, biological, inorganic, organic, and physical chemistry and chemical physics. The results of these examinations are used to plan the first-year course program, which includes six courses plus the selection of a research project and adviser and the initiation of a research program that is intended to generate new fundamental chemical knowledge and methodology.

During the second year, students take an oral examination that includes a critique of a recent scientific paper and a summary of their thesis research problem and the progress made to date, which may be assessed by the Research Advisory Committee. This is followed by a written dissertation and oral defense of the dissertation. It is expected that students should complete the program in five years.

Research Facilities

The Barksdale Chemistry Library, along with the science, engineering, and physics libraries, contain more than 20,000 volumes on chemistry-related subjects, plus up-to-date sets of chemical journals, books, and online databases. Alderman Library has more than 1.6 million books and extensive collections of manuscripts, maps, prints, and microfilms. The Department also has a number of facilities and instruments that complement classroom learning. Five nuclear magnetic resonance spectrometers are available for routine use in a wide range of experiments. The Molecular Structure Laboratory consists of an X-ray diffraction facility and a molecular modeling laboratory and has access to several large computing systems. The Center for Nanoscopic Materials Design provides extensive facilities for creating, imaging, and spectroscopically analyzing nanoscale and nanopatterned materials. The Ultrafast Laser Laboratory conducts a wide range of multidisciplinary research involving lasers and houses a versatile collection of high-power lasers. The Bioanalytical Microchip Fabrication Facility is used to design and fabricate analytical microchip devices with the goal of revolutionizing conventional analytical techniques.

Financial Aid

Full financial support, including tuition, stipend, and health insurance, is offered to graduate students who are in good academic standing. Financial aid for beginning graduate students may take the form of a teaching assistantship or fellowship. Teaching assistants normally work less than 12 hours per week in laboratory instruction, supervision, and consultation with students. Some teaching assistants are awarded supplementary fellowship support in recognition of their undergraduate records. Although teaching assistantships are available beyond the first year, most students ultimately receive support from research grants or fellowships, funded through private endowments or from University, state, or federal sources.

Cost of Study

In 2005–06, full-time graduate tuition (9 or more hours) was $4876.50 per semester for Virginia residents and $10,176.50 for nonresidents. All students also pay an additional $28.50 in fees. Tuition for research is $1324 and $1332.50, respectively, plus an additional $13.50 in fees.

Living and Housing Costs

Single students may live on campus at a cost of $3930 for a single room, $3600 for a double room, or $3630 for two bedrooms. Housing for married students is available for $657 to $814 per month, depending on size. Board plans range from $585 per semester for fifty meals to $1550 for unlimited meals. Many students live off campus, paying $650 to $850 per month for a one-bedroom apartment or $700 to $950 for two bedrooms.

Student Group

Approximately 120 students are enrolled in the graduate program each year, most with undergraduate degrees in chemistry or course work in related areas.

Student Outcomes

Graduates of the program fill a number of positions in academia and business. Many become professors in chemistry departments at colleges and universities around the world, while others work as chemists and scientists at pharmaceutical companies and government agencies. Recent graduates are working at GlaxoSmithKline, Abbott Laboratories, Westinghouse, Merck, Eli Lilly, Dupont, Roche, the Defense Intelligence Agency, University of Virginia, University of Vermont, and College of William and Mary.

Location

Charlottesville is often listed among *Money* magazine's 100 best places to live in the U.S. and is the most highly ranked city in Virginia. It is close to historic homes, including Thomas Jefferson's Monticello, but it also supports a variety of businesses and services. Charlottesville hosts the annual Virginia Festival of American Film, and it has the most newspaper readers and computer owners per capita in America.

The University

The University of Virginia was born in 1825 from founder Thomas Jefferson's idea of a state-supported college. Since then, it has grown to include 19,000 students and 1,900 full-time faculty members. UVA's curriculum offers more than fifty bachelor's degrees, ninety master's degrees, and fifty-five doctoral degrees, all of which benefit from the University's interdisciplinary focus. UVA is ranked by *U.S. News & World Report* as the nation's top public university, and it has the highest academic standards of any state university in the U.S.

Applying

Applicants of the graduate program are required to submit an application for admission, which includes a personal essay, two original copies of transcripts showing a GPA of B or better, two letters of recommendation, and GRE General and Subject Test scores. International students for whom English is a second language must submit TOEFL and TSE scores. The deadline for application is February 1 for fall admission. Applications can be submitted online or mailed to the University of Virginia's Graduate Admissions Office.

Correspondence and Information

Department of Chemistry
University of Virginia
McCormick Road
P.O. Box 400319
Charlottesville, Virginia 22904-4319
Phone: 434-924-3344
Fax: 434-924-3710
E-mail: chem@virginia.edu
Web site: http://www.virginia.edu/chem

University of Virginia

THE FACULTY AND THEIR RESEARCH

Ralph O. Allen, Professor; Ph.D., Wisconsin, 1970. Trace analysis; environmental, archaeological, and forensic applications.
W. Lester S. Andrews, Professor; Ph.D., Berkeley, 1966. Spectroscopy and photochemistry of matrix isolated species.
Robert G. Bryant, Professor; Ph.D., Stanford, 1969. Nuclear magnetic resonance in chemistry.
John H. Bushweller, Associate Professor; Ph.D., Berkeley, 1989. Structural and functional basis for oncogenesis.
David S. Cafiso, Professor; Ph.D., Berkeley, 1979. Biophysical chemistry, membrane electrostatics, membrane protein structure.
Jason Chruma, Assistant Professor; Ph.D., Pennsylvania, 2003. Synthetic, medicinal, and bioorganic chemistry: synthesis of biologically active natural products, synthesis of unnatural amino acids and conformationally constrained peptides, modulation of protein-protein interactions.
James N. Demas, Professor; Ph.D., New Mexico, 1970. Photochemistry and photophysics of transition-metal complexes.
Sergei A. Egorov, Associate Professor; Ph.D., Wisconsin–Madison, 1996. Theoretical physical chemistry, statistical mechanics, quantum dynamics in condensed phases.
Cassandra L. Fraser, Professor and Cavaliers' Distinguished Teaching Professor; Ph.D., Chicago, 1993. Polymeric metal complexes, biomaterials, bioinspired design, multifunctional nanoscale assemblies.
H. Mario Geysen, Alfred Burger Professor; Ph.D., Melbourne, 1976. Combinatorial chemistry.
Charles M. Grisham, Professor; Ph.D., Minnesota, 1973. Biophysical chemistry, magnetic resonance spectroscopy of complex biological structures.
W. Dean Harman, Professor; Ph.D., Stanford, 1987. Inorganic chemistry, organometallic chemistry, synthetic organic methods activation of aromatic molecules.
Ian Harrison, Professor and Chair; Ph.D., Toronto, 1987. Surface chemistry: adsorbate photochemistry, reaction kinetics and dynamics; single-molecule chemistry and nanotechnology; laser interactions with matter.
Sidney M. Hecht, John W. Mallet Professor; Ph.D., Illinois, 1970. Organic and biological chemistry, synthesis of bleomycin group antibiotics.
Donald F. Hunt, University Professor; Ph.D., Massachusetts, 1967. Analytical biochemistry, developing new methods and instrumentation for structural characterization of proteins and their posttranslational modifications.
James P. Landers, Professor; Ph.D., Guelph, 1988. Biological, bioanalytical, and clinical chemistry.
Kevin Lehmann, Professor; Ph.D., Harvard, 1983. Experimental and theoretical molecular spectroscopy, studies of molecular dynamics in the gas phase and superfluid helium, development of ultrasensitive spectroscopic methods with applications in trace-gas detection.
Timothy L. Macdonald, Professor and Chair; Ph. D., Columbia, 1975. Bioorganic and synthetic organic chemistry.
James A. Marshall, Thomas Jefferson Professor; Ph.D., Michigan, 1960. Organic chemistry, synthesis of organic natural products with possible medicinal applications, development of synthesis methodology.
Glenn J. McGarvey, Associate Professor; Ph.D., California, Davis, 1977. Synthetic organic chemistry, structural glycobiological chemistry.
Matthew Neurock, Professor; Ph.D., Delaware, 1992. Molecular modeling, computational heterogeneous catalysis, kinetics of complex reaction systems.
Brooks H. Pate, Professor and William R. Kenan Jr. Professor; Ph.D., Princeton, 1992. High-resolution infrared spectroscopy studies of complex vibrational dynamics in molecules.
Lin Pu, Professor; Ph.D., California, San Diego, 1990. Organic, polymer and organometallic chemistry, asymmetric catalysis, chiral sensors.
Frederick S. Richardson, Commonwealth Professor; Ph.D., Princeton, 1966. Atomic and molecular spectroscopy, chirality-dependent molecular dynamics and recognition, lanthanide chemistry and spectroscopy, optical studies of biomolecular systems.
Richard J. Sundberg, Professor; Ph.D., Minnesota, 1962. Synthesis, reactivity, and biological activity of heterocyclic compounds.
Carl Trindle, Professor; Ph.D., Tufts, 1967. Theoretical chemistry.
B. Jill Venton, Assistant Professor; Ph.D., North Carolina at Chapel Hill, 2002. Bioanalytical and neurochemistry, development and characterization of analytical techniques to measure neurochemical changes.

UTAH STATE UNIVERSITY

Department of Chemistry and Biochemistry

Programs of Study

Utah State University offers graduate degree programs leading to the M.S. and Ph.D. in chemistry and biochemistry. Both the M.S. and Ph.D. degree programs require the completion of an original research project and the defense of a thesis (dissertation). The M.S. degree is not a prerequisite for the Ph.D. degree. A diversity of research topics in the Department allows incoming students to pursue specific interests and contributes to the enrichment of course work.

Upon completion of interviews with faculty members and/or laboratory rotations, entering students choose an adviser and initiate an appropriate research project to be the basis of the dissertation.

Research Facilities

The Department currently occupies two adjoining buildings, Maeser Laboratory and Widtsoe Hall. Widtsoe Hall is a 75,000-square-foot facility that houses state-of-the-art teaching and research laboratories.

Departmental research instrumentation includes a Bruker ARX-400-MHz and a JEOL ECX-300 MHz NMR, both with broadband multinuclear capabilities and a Bruker ESP-300 X-band EPR. The Analytical Sciences Laboratory, equipped by a generous donation from the Shimadzu Corporation, contains a GC-MS, a fast-scan UV-VIS spectrophotometer, and a gradient HPLC system with autosampler and UV-VIS and fluorescence detection. Individual research laboratories within the Department have additional extensive instrumentation, including small-molecule and protein X-ray diffractometers, pulsed-laser spectrometers, MALDI time-of-flight mass spectrometers, FTIR and UV-VIS spectrophotometers, stopped-flow kinetics equipment, and a microcalorimeter.

The Department has an electronics shop staffed by a full-time professional. A wood/machine shop is housed within the Department, as is a storeroom with an extensive inventory of glassware, chemicals, and other supplies for research and teaching.

Financial Aid

All incoming students are supported by teaching assistantships, research assistantships, or fellowships. Assistantships for incoming students include a stipend ($17,000 for 2005–06), $800 toward the purchase of health insurance, and a tuition waiver. Highly qualified applicants are considered for the Willard L. Eccles Science Foundation Fellowship, which provides three years of support at $18,000 per year, or for the Presidential Fellowship, which provides $15,000 for one year. Both fellowships may be supplemented with a half-time teaching assistantship position and a stipend. The Department strongly encourages students with strong academic records to apply for these fellowships. Assistance is provided in obtaining and submitting the appropriate forms.

Cost of Study

Out-of-state tuition for an academic year is $13,879. A full-tuition waiver is included in the support package for all graduate students.

Living and Housing Costs

Costs for housing in the immediate vicinity of the University range from $250 to $500 per month, depending on whether a student prefers to live alone or in shared housing.

Student Group

The Department's current graduate student population consists of 44 students, 41 of whom are pursuing the Ph.D. and 3 of whom are pursuing the M.S. degree. Women make up 38 percent of the total, and 59 percent are international students.

Location

Utah State University is located in the city of Logan in northern Utah's Cache Valley in the heart of the Rocky Mountains. The local area offers many outdoor activities, including camping, hiking, boating, fishing, and a whole spectrum of winter activities; students can take advantage of the "greatest snow on earth."

The University

Utah State University was founded in 1888 as Utah's land-grant college. The University has an international reputation for research and teaching, is recognized as a Carnegie Foundation Research I institution, and ranks seventh in research expenditures among land-grant universities. The University also offers cultural and athletic programs that feature regional, national, and international participants. With an on-campus enrollment of nearly 20,000, the University population is made up of students from all fifty states and sixty-nine other countries.

Applying

There are no application deadlines; however, submission of application materials by April 15 is strongly encouraged for fall admission. U.S. applicants may also apply for spring admission. To be considered for admission to the graduate program, students must submit the following material: official transcripts, three letters of recommendation, and official GRE General Test scores. International students must also include official TOEFL scores. For consideration in the Willard L. Eccles Foundation Science Fellowship competition, applications should be completed by March.

Correspondence and Information

Graduate Admissions Committee Chair
Department of Chemistry and Biochemistry
Utah State University
Logan, Utah 84322-0300
Phone: 435-797-1618
Fax: 435-797-3390
E-mail: chemgrad@cc.usu.edu
Web site: http://www.chem.usu.edu/faculty/Recruiting/index.html

Utah State University

THE FACULTY AND THEIR RESEARCH

Ann E. Aust, Trustee Professor; Ph.D., Michigan State, 1975. Role of iron, glutathione, and nitric oxide in asbestos-induced mutagenesis and carcinogenesis. (Biochemistry)

Steven D. Aust, Professor; Ph.D., Illinois, 1965. Structure-function relationships of fungal enzymes for the metabolism of environmental pollutants and the loading of iron into ferritin by ceruloplasmin. (Biochemistry)

Lisa M. Berreau, Associate Professor; Ph.D., Iowa State, 1994. Synthetic inorganic and bioinorganic chemistry. (Inorganic Chemistry)

Stephen E. Bialkowski, Professor; Ph.D., Utah, 1978. Applications of novel optical methods for chemical analysis. (Analytical Chemistry)

Alexander I. Boldyrev, Associate Professor; Ph.D./Dr.Sci., Novosibirsk (Russia), 1974. Theory of nonstoichiometric molecules, clusters, and materials; superalkalies; superhalogens; high-spin molecules. (Physical Chemistry)

Robert S. Brown, Associate Professor; Ph.D., Virginia Tech, 1983. Fundamental and applied aspects of mass spectrometry (particularly TOF-MS and FT-MS), application of mass spectrometry to various chemical problems, with particular emphasis on biochemical systems. (Analytical Chemistry)

Cheng-Wei Tom Chang, Assistant Professor; Ph.D., Washington (St. Louis), 1997. Development of novel antibiotics, parallel synthesis of unusual sugars, combinatorial synthesis of polyketide and aminoglycoside antibiotic mimics, combinatorial synthesis of fluoroalkylthiophosphonate as enzyme inhibitors. (Organic Chemistry)

Bradley S. Davidson, Associate Professor; Ph.D., Cornell, 1989. Structural chemistry and synthesis of marine and microbial natural products. (Organic Chemistry)

Scott A. Ensign, Professor; Ph.D., Wisconsin–Madison, 1991. Microbial metabolism of aliphatic hydrocarbons, biochemistry, and enzymology of enzymes involved in hydrocarbon metabolism. (Biochemistry)

David Farrelly, Professor; Ph.D., Manchester (England), 1980. Theory of intramolecular energy flow, chaotic dynamics, and diffusion quantum Monte Carlo studies of quantum clusters. (Physical Chemistry)

Alvan C. Hengge, Professor; Ph.D., Cincinnati, 1987. Mechanistic organic and bioorganic chemistry, molecular mechanisms of chemical reactions, particularly the details of enzymatic catalysis. (Organic Chemistry)

Joan M. Hevel, Hansen Assistant Professor; Ph.D., Michigan, 1993. Molecular mechanisms of cellular communication, with emphasis on protein structure and biochemistry. (Biochemistry)

Richard C. Holz, Professor; Ph.D., Penn State, 1989. Hydrolytic chemistry of dinuclear centers, structure-function relationships of metalloproteins. (Inorganic Chemistry)

John L. Hubbard, Associate Professor; Ph.D., Arizona, 1982. Structure/bonding/reactivity relationships in transition metal catalysis. (Inorganic Chemistry)

Sean J. Johnson, Hansen Assistant Professor; Ph.D., Duke, 2003. Protein-nucleic acid interactions, X-ray crystallography, structure and mechanism of proteins and macromolecular assemblies central to replication and transcriptional regulation. (Biochemistry)

Tapas Kar, Research Assistant Professor; Ph.D., Indian Institute of Technology (Kharagpur), 1989. Doped fullerenes and nanotubes, functionalized nanotubes, hydrogen and dihydrogen bonds, Li-nano-battery, electronic and optical properties. (Computational Chemistry)

Vernon D. Parker, Professor; Ph.D., Stanford, 1964. Reactive intermediate chemistry, electron transfer reactions. (Organic Chemistry)

Steve Scheiner, Professor and Department Head; Ph.D., Harvard, 1976. Fundamental properties of hydrogen bonds, proton transfer reactions, electronic properties of materials. (Computational Chemistry)

Lance C. Seefeldt, Professor; Ph.D., California, Riverside, 1989. Structure and function studies of metalloproteins, functions of MgATP hydrolysis in electron transfer and substrate reduction in the enzyme nitrogenase. (Biochemistry)

Philip J. Silva, Assistant Professor; Ph.D., California, Riverside, 2000. Chemical analysis of aerosol particles and study of their surface chemistry, development of mass spectrometry methods for analysis of environmental systems. (Analytical Chemistry)

VILLANOVA UNIVERSITY

College of Liberal Arts and Sciences
Department of Chemistry

Programs of Study

The Department of Chemistry at Villanova offers the Master of Science (M.S.) degree in all traditional areas of chemistry. The degree can be earned full-time or part-time. Part-time students may choose from thesis and nonthesis options. The thesis option requires the successful completion of six courses and a research project culminated by a written thesis. The nonthesis option requires the completion of ten courses and a seminar course based on work experience. All students are required to take three core courses in either analytical, biological, inorganic, organic, or physical chemistry, followed by elective courses. Comprehensive exams are required, and may be taken anytime after the completion of four courses. Thirty credits are required for the Master of Science degree.

Research Facilities

In 1999, Villanova completed a $35-million expansion renovation to Mendel Science Center, resulting in a state-of-the-art teaching and research facility. The department is well equipped with instrumentation. Two FT-NMR spectrometers (both 300 MHz), including a new Varian Mercury instrument with an MAS probe for solids analysis, are available. Other instrumentation includes an HP GC–mass spectrometer, a Siemens single crystal diffractometer, DSC, TG, ultracentrifugation, and polarimetry. Spectroscopy is performed with several FT-IRs and UV-visible spectrophotometers, CE, and fluorescence. Chromatographs include several GC and LC instruments, along with CE. The Department Computational Chemistry Lab holds an IBM RISC station along with a Linux cluster and a Silicon Graphics O2 workstation.

Financial Aid

Most full-time graduate students in the department hold teaching or research assistantships of $14,500 for nine months plus full tuition remission. Limited research and teaching fellowships are available for summer months.

Cost of Study

The tuition for graduate chemistry at Villanova was $600 per credit hour in 2005–06, with a general fee of $30 per semester.

Living and Housing Costs

Although on-campus housing is not available, ample apartments/rooms are available in the suburban neighborhoods surrounding Villanova. The Office of Residence Life offers assistance by providing rental lists to students. Living expenses for a single student are estimated at $13,000 per year.

Student Group

There are 40 graduate students (16 full-time) and 14 tenure-track faculty members in the department. About 40 percent of the students are women and 10 percent are international.

Student Outcomes

Some graduates are employed by the nine major pharmaceutical firms in the Philadelphia area. Others choose employment at smaller companies, and some continue their studies at Ph.D. programs in chemistry and related areas.

Location

Located in a safe, suburban community 12 miles west of Philadelphia, the picturesque 254-acre campus features sixty buildings. Bryn Mawr and Haverford Colleges are nearby, and major Universities in Philadelphia (Penn, Temple, Drexel, and others) are easily accessible by public transportation. Philadelphia supports many cultural opportunities including theater, opera, symphony concerts, and ballet, as well as professional sports teams of every variety.

The University

Founded in 1842 by the priests and brothers of the Order of St. Augustine, Villanova University is the oldest and largest Catholic university in the Commonwealth of Pennsylvania. The University's commitment to love and service is reflected in the Latin words of its seal, which translate into truth, unity, and love.

Applying

An application form, with full instructions, is available on the Web at http://www.gradartsci.villanova.edu. The application fee is $50. Application deadlines are August 1 (fall), December 1 (spring), and May 1 (summer). Applicants who wish to be considered for assistantships should submit their application by March 1 (fall) and October 1 (spring) for priority evaluation. The GRE General Test is required of all students; the TOEFL is required of international applicants whose native language is not English. The most important criterion for admission is a sincere desire to study chemistry. Applications from second career, older, and other nontraditional chemistry students are encouraged.

Correspondence and Information

Graduate Chairperson
Department of Chemistry
Villanova University
800 Lancaster Avenue
Villanova, Pennsylvania 19085-1699
Phone: 610-519-4840
Fax: 610-519-7167
E-mail: chemistrygrad@villanova.edu
Web site: http://www.chemistry.villanova.edu

Villanova University

THE FACULTY AND THEIR RESEARCH

Temer S. Ahmadi, Ph.D., UCLA. Materials/physical chemistry: synthesis and optical properties of metal-polymer, metal-semiconductor, and luminescent semiconductor nanomaterials.

Joseph W. Bausch, Ph.D., USC. Organic and computational chemistry: synthetic and computational studies of electron-deficient clusters, carborane synthesis, structure prediction.

Carol A. Bessel, Ph.D., SUNY at Buffalo. Inorganic and materials chemistry: carbon nanofibers, electrochemistry, catalysis in supercritical fluids, octahedral ruthenium complexes.

Eduard G. Casillas, Ph.D., Johns Hopkins. Organic chemistry: natural product synthesis, synthesis of antagonists for plant/fungal secondary metabolic pathways, terpene biomimetic synthesis.

Robert M. Giuliano, Ph.D., Virginia. Organic chemistry: carbohydrate chemistry, synthesis of vinyl glycosides and carbohydrate vinyl ethers, branched-chain carbohydrates, nitrosugars.

Amanda M. Grannas, Ph.D., Purdue. Analytical/environmental chemistry: Photochemical degradation of environmental pollutants in surface waters, photo chemistry of organics in snow and ice, redox chemistry of soil and sediments, Arctic climate change.

W. Scott Kassel, Ph.D., Florida. Inorganic chemistry: solid-base catalysis, X-ray diffraction, synthesis of chiral pyrrolidine transition metal complexes as enantioselective catalysts.

Anthony F. Lagalante, Ph.D., Colorado. Analytical/environmental chemistry: environmental/food/agricultural applications of solid phase microextraction (SPME), high-pressure spectroscopy in supercritical fluids used as "green" solvents.

Christine A. Martey-Ochola, Ph.D., Lehigh. Research: effect of cigarette smoke and airborne nanoparticles on normal human lung cells, synthesis and characterization of novel drug-polymer conjugates.

Brian K. Ohta, Ph.D., California, San Diego. Organic chemistry: NMR spectroscopy, intermediate characterization in photosensitized oxidation reactions, hydrogen bond asymmetry.

Jennifer B. Palenchar, Ph.D., Delaware. Research: enzymology of bacterial aspartate kinase, transcriptional initiation complexes in *Trypanosoma brucei.*

Barry S. Selinsky, Ph.D., SUNY at Buffalo. Biochemistry: membrane biophysics, structural analysis of membrane proteins, membrane-active antibiotics, anticoagulants.

Deanna L. Zubris, Ph.D., Caltech. Inorganic chemistry: synthesis of organometallic complexes as polymerization catalysts, mechanistic studies.

WESLEYAN UNIVERSITY

Department of Chemistry

Program of Study

The Department of Chemistry offers a program of study leading to the Ph.D. degree. Students are awarded this degree upon demonstration of creativity and scholarly achievement. This demands intensive specialization in one field of chemistry as well as broad knowledge of related areas. The department provides coverage of physical, organic, inorganic, bioorganic, and biophysical chemistry.

The first year of graduate study contains much of the required course work, although most students also choose a research adviser and begin a research program at the beginning of the second semester. Students are expected to demonstrate knowledge of five core areas of chemistry, either by taking the appropriate course or by passing a placement examination. In addition, students take advanced courses in their area of specialization. Classes are small (5–10 students) and emphasize interaction and discussion. Student seminar presentations are also emphasized. Election of interdisciplinary programs in chemical physics and molecular biophysics in conjunction with the Departments of Physics and Molecular Biology and Biochemistry, respectively, is also possible. Students are admitted to Ph.D. candidacy, generally in the second year, by demonstrating proficiency in the core course curriculum, passing a specified number of regularly scheduled progress exams, demonstrating an aptitude for original research, and defending a research proposal. The progress and development of a student is monitored throughout by a 3-member faculty advisory committee. The Ph.D. program, culminating in the completion of a Ph.D. thesis, is normally completed within four to five years. Two semesters of teaching in undergraduate courses is required, where the load is, on average, about 5 hours per week during the academic year. This requirement is normally met in the first year.

Research Facilities

The Department of Chemistry is housed in an air-conditioned building, sharing space with the Departments of Biology and Molecular Biology and Biochemistry. Major items of equipment include a Hewlett-Packard GC (5890) mass spectrometer, two Perkin-Elmer M1600 FT-IR, a Fluoromax-2 spectrofluorimeter, a stop-flow reactions kinetics system, a picosecond CW mode-locked Nd:YAG laser with cavity-dumped dye laser, a JASCO J810 spectropolarimeter equipped with a Peltier temperature controller, a Beckman DU650 spectrophotometer equipped with a Peltier temperature controller, a Johnson-Matthey magnetic susceptibility balance, a variety of gas and liquid chromatographs, a Perkin-Elmer LS-50 spectrofluorimeter, a Hitachi F2000 spectrofluorimeter, three Hitachi U-2000 UV/VIS, three HP diode array UV/VIS, a Perkin-Elmer 241 polarimeter, a Hewlett-Packard 5972 GC mass spectrometer, a Photon Technologies LS-100 luminescence lifetime apparatus, a Jobin Yvon 1500 high-resolution optical spectrometer, two high-throughput molecular jet machines, a pulsed-jet Fabry-Perot Fourier-transform microwave spectrometer, and a Storm R40 Imagequant gel and blot analysis system. The NMR facilities consist of a Varian Gemini two-channel, broadband Mercury Vx with gradients, a two-channel Unity Plus 400 with gradients, and a three-channel Unity Inova 500 with gradients. The Department of Chemistry depends heavily on computers. Among other systems, the department has a Sun IPX workstation, a Sun UltraSPARC workstation, an IBM RISK6000 3CT workstation, a Silicon Graphics Octane2 workstation, a Silicon Graphics Indigo 2 10000 workstation, and two IBM RD6000 workstations. Ethernet connects all department computers to the University computer and to the Internet.

Financial Aid

All students receive a twelve-month stipend, which, for 2005–06, was $20,541. In the first year, this stipend derives from a teaching assistantship. In later academic years, students may be supported by research assistantships where funds are available or by further teaching assistantships.

Cost of Study

Tuition for 2005–06 was $4122 per course credit, but remission of this is granted to all holders of teaching and research assistantships.

Living and Housing Costs

Most graduate students, both single and married, live in houses administered and maintained by the University, with rents ranging between $500 and $750 per month.

Student Group

The student body at Wesleyan is composed of some 2,800 undergraduates and 204 graduate students. Of the latter, most are in the sciences, with 40 divided between men and women in the Department of Chemistry. Most graduates obtain industrial positions, although some choose academic careers, normally after postdoctoral experience in each case.

Student Outcomes

All Ph.D. graduates in the last two years have gone on to postdoctoral fellowships at major universities such as Harvard, Yale, and California Institute of Technology. Most earlier graduates are now on college faculties or have research positions in the chemical industry.

Location

Middletown is a small city on the west bank of the Connecticut River, 15 miles south of Hartford, the state capital. New Haven is 24 miles to the southwest; New York City and Boston are 2 hours away by automobile. Middletown's population of 50,000 is spread over an area of 43 square miles, much of which is rural. Although Wesleyan is the primary source of cultural activity in Middletown, the city is not a "college town" but serves as a busy commercial center for the region between Hartford and the coast. Water sports, skiing, hiking, and other outdoor activities can be enjoyed in the hills, lakes, and river nearby.

The University

For more than 150 years, Wesleyan University has been identified with the highest aspirations and achievements of private liberal arts higher education. Wesleyan's commitment to the sciences dates from the founding of the University, when natural sciences and modern languages were placed on an equal footing with traditional classical studies. In order to maintain and strengthen this commitment, graduate programs leading to the Ph.D. degree in the sciences were established in the late 1960s. The program in chemistry was designed to be small, distinctive, and personal, emphasizing research, acquisition of a broad knowledge of advanced chemistry, and creative thinking.

Applying

By and large, a rolling admissions policy is in place, although applicants seeking admission in September are advised to submit applications (no application fee) as early as possible in the calendar year. Three letters of recommendation are required, and applicants are required to take the Graduate Record Examinations. Students whose native language is not English should take the TOEFL. Applicants are strongly encouraged to visit the University after arrangements are made with the department.

Correspondence and Information

Ms. Roslyn N. Carrier-Brault
Administrative Assistant
Department of Chemistry
Hall-Atwater Laboratories
Wesleyan University
Middletown, Connecticut 06459-0001
Phone: 860-685-2210
Fax: 860-685-2211
E-mail: chemistry@wesleyan.edu
Web site: http://www.wesleyan.edu/chem/

Wesleyan University

THE FACULTY AND THEIR RESEARCH

Anne M. Baranger, Associate Professor; Ph.D., Berkeley, 1993. Bioorganic chemistry: mechanism of RNA folding, molecular origins of RNA–protein complex affinity and specificity.

David L. Beveridge, Professor; Ph.D., Cincinnati, 1965. Theoretical physical chemistry and molecular biophysics: statistical thermodynamics and computer simulation studies of hydrated biological molecules, structure and motions of nucleic acid, environmental effects on conformational stability, organization of water in crystal hydrates.

Philip H. Bolton, Professor; Ph.D., California, San Diego, 1976. Biochemistry and physical chemistry: NMR and modeling studies of duplex DNA; the structure of DNA containing abasic sites and other damaged DNA; studies on aptamer, telomere, and triplet repeat DNA; development of NMR methodology.

Joseph W. Bruno, Professor; Ph.D., Northwestern, 1983. Inorganic and organometallic chemistry: synthetic and mechanistic studies of transition-metal compounds; organometallic photochemistry, metal-mediated reactions of unsaturated organics.

Michael A. Calter, Associate Professor; Ph.D., Harvard, 1993. Synthetic organic chemistry, particularly in the area of asymmetric catalysis.

Michael J. Frisch, Visiting Scholar; Ph.D., Carnegie-Mellon, 1983. Theoretical chemistry: method development and applications to problems of current interest.

Albert J. Fry, Professor; Ph.D., Wisconsin, 1963. Organic chemistry: mechanisms of organic electrode processes, development of synthetically useful organic electrochemical reactions.

Joseph L. Knee, Professor; Ph.D., SUNY at Stony Brook, 1983. Chemical physics: investigation of ultrafast energy redistribution in molecules using picosecond laser techniques, emphasis on isolated molecule processes including unimolecular photodissociation reaction rates.

Stewart E. Novick, Professor; Ph.D., Harvard, 1973. Physical chemistry: pulsed-jet Fabry-Perot Fourier transform microwave spectroscopy, structure and dynamics of weakly bound complexes, high-resolution spectroscopy of radicals important in the interstellar medium.

George A. Petersson, Professor; Ph.D., Caltech, 1970. Theoretical chemistry: development of improved methods for electronic structure calculations, with applications to small molecular systems and chemical reactions.

Rex F. Pratt, Professor; Ph.D., Melbourne (Australia), 1969. Bioorganic chemistry: enzyme mechanisms and inhibitor design, beta-lactam antibiotics and beta-lactamases, protein chemistry.

Wallace C. Pringle, Professor; Ph.D., MIT, 1966. Physical chemistry: spectroscopic studies of internal interactions in small molecules, collision-induced spectra, environmental chemistry.

Irina M. Russu, Professor; Ph.D., Pittsburgh, 1979. Biochemistry and molecular biophysics: structure and dynamics of nucleic acids, allosteric mechanisms in human hemoglobin, nuclear magnetic resonance spectroscopy.

T. David Westmoreland, Associate Professor; Ph.D., North Carolina, 1985. Inorganic and bioinorganic chemistry: electronic structure and mechanism in molybdenum-containing enzymes, EPR spectroscopy of transition-metal complexes, fundamental aspects of atom transfer reactions in solution.

Peter S. Wharton, Professor Emeritus; Ph.D., Yale, 1959. Organic chemistry.

The Hall-Atwater Laboratory, which houses the Department of Chemistry.

SELECTED PUBLICATIONS

Yan, Z., and **A. M. Baranger.** Binding of an aminoacridine derivative to a tetralopp RNA. *Bioorg. Med. Chem. Lett.* 14:5889, 2004.

Zhao, Y., and **A. M. Baranger.** Design of an adenosine analog that selectively improves the affinity of a mutant U1A protein for RNA. *J. Am. Chem. Soc.* 125:2480, 2003.

Tuite, J. B., J. C. Shields, and **A. M. Baranger.** Substitution of an essential adenine in the U1A-RNA complex with a non-polar isostere. *Nucleic Acids Res.* 30:5269, 2002.

Gayle, A. Y., and **A. M. Baranger.** Inhibition of the U1A-RNA complex by an aminoacridine derivative. *Bioorg. Med. Chem. Lett.* 12:2839, 2002.

Shiels, J. C., et al. **(A. M. Baranger).** Investigation of a conserved stacking interaction in target site recognition by the U1A protein. *Nucleic Acids Res.* 30:550, 2002.

Shiels, J. C., et al. **(A. M. Baranger).** RNA-DNA hybrids containing damaged DNA are substrates for RNase H. *Bioorg. Med. Chem. Lett.* 11:2623, 2001.

Blakaj, D. M., et al. **(A. M. Baranger** and **D. L. Beveridge).** Molecular dynamics and thermodynamics of a protein-RNA complex: Mutation of a conserved aromatic residue modifies stacking interactions and structural adaptation in the U1A-stem loop 2 RNA complex. *J. Am. Chem. Soc.* 123:2548, 2001.

Luchansky, S., et al. **(A. M. Baranger).** Contribution of RNA conformation to the stability of a high-affinity RNA-protein complex. *J. Am. Chem. Soc.* 122:7130, 2000.

Nolan, S. J., et al. **(A. M. Baranger).** Recognition of an essential adenine at a protein-RNA interface: Comparison of the contributions of hydrogen bonds and a stacking interaction. *J. Am. Chem. Soc.* 121:8951, 1999.

Beveridge, D. L., et al. Molecular dynamics of DNA and protein-DNA complexes: Progress on sequence effects, conformational stability, axis curvature, and structural bioinformatics. *Nucleic Acids: Curvature Deformation* 884:13–64, 2004.

Beveridge, D. L., et al. Molecular dynamics simulations of DNA curvature and flexibility: Helix phasing and premelting. *Biopolymers* 73(3):380–403, 2004.

Byun, K. S., and **D. L. Beveridge.** Molecular dynamics simulation of papilloma virus E2 DNA sequences: Dynamical models for oligonucleotide structures in solution. *Biopolymers* 73(3):369–79, 2004.

Ponomarev, S. Y., K. M. Thayer, and **D. L. Beveridge.** Ion motions in molecular dynamics simulations on DNA. *Proc. Natl. Acad. Sci. U.S.A.* 101(41):14771–5, 2004.

Beveridge, D. L., et al. Molecular dynamics simulations of the 136 unique tetranucleotide sequences of DNA oligonucleotides: I. Research design and results on d(CpG) steps. *Biophys. J.* 87:3799–813, 2004.

Arthanari, H., et al. **(D. L. Beveridge).** Assessment of the molecular dynamics structure of DNA in solution based on calculated and observed NMR NOESY volumes and dihedral angles from scalar coupling constants. *Biopolymers* 68(1):3–15, 2003.

Liu, Y., and **D. L. Beveridge.** Exploratory studies of ab initio protein structure prediction: Multiple copy simulated annealing, AMBER energy functions, and a generalized born/solvent accessibility solvation model. *Proteins* 46(1):128–46, 2002.

Pitici, F., **D. L. Beveridge,** and **A. M. Baranger.** Molecular dynamics simulation studies of induced fit and conformational capture in U1A-RNA binding: Do molecular substates code for specificity? *Biopolymers* 65(6):424–35, 2002.

Thayer, K. M., and **D. L. Beveridge.** Hidden Markov models from molecular dynamics simulations on DNA. *Proc. Natl. Acad. Sci. U.S.A.* 99(13):8642–7, 2002.

Kombo, D. C., et al. **(D. L. Beveridge).** Calculation of the affinity of the λ-repressor-operator complex based on free energy component analysis. *Mol. Simulation* 28:187–211, 2002.

Jayaram, B., et al. **(D. L. Beveridge).** Free-energy component analysis of 40 protein-DNA complexes: A consensus view on the thermodynamics of binding at the molecular level. *J. Comput. Chem.* 23:1, 2002.

McConnell, K. J., and **D. L. Beveridge.** Molecular dynamics simulations of B'-DNA: Sequence effects on A-tract bending and bendability. *J. Mol. Biol.* 314:23, 2001.

Kombo, D. C., et al. **(D. L. Beveridge).** Molecular dynamics simulation reveals sequence-intrinsic and protein-induced geometrical features of the OL1 DNA operator. *Biopolymers* 59:205, 2001.

Liu, Y., and **D. L. Beveridge.** A refined prediction method for gel retardation of DNA oligonucleotides from dinucleotide step parameters: Reconciliation of DNA bending models with crystal structure data. *J. Biomol. Struct. Dyn.* 18:505, 2001.

Rujan, I. N., J. C. Meleney, and **P. H. Bolton.** Vertebrate telomere repeat DNAs favor external loop propeller quadruplex structures in the presence of high concentrations of potassium. *Nucleic Acids Res.* 33:2022–31, 2005.

Arthanari, H., K. Wojtuszewski, I. Mukerji, and **P. H. Bolton.** Effects of HU binding on the equilibrium cyclization of mismatched, curved, and normal DNA. *Biophysics* 86:1625–31, 2004.

Arthanari, H., and **P. H. Bolton.** Did quadruplex DNA play a role in the evolution of the eukaryotic linear chromosome? *Mini Rev. Med. Chem.* 3(1):1–9, 2003.

Arthanari, H., and **P. H. Bolton.** Functional and dysfunctional roles of quadruplex DNA. *Chem. Biol.* 8:221, 2001.

Marathias, V. M., et al. **(P. H. Bolton).** Flexibility and curvature of duplex DNA containing mismatched sites as a function of temperature. *Biochemistry* 39:153, 2000.

Jerkovic, B., and **P. H. Bolton.** The curvature of dA tracts is temperature dependent [in process citation]. *Biochemistry* 39:12121, 2000.

Marathias, V. M., et al. **(P. H. Bolton).** 6-Thioguanine alters the structure and stability of duplex DNA and inhibits quadruplex DNA formation. *Nucleic Acids Res.* 27:2860, 1999.

Ellis, W. W., et al. **(J. W. Bruno).** Hydricities of BzNADH, C5H5Mo(PMe3)(CO)2H, and C5Me5Mo(PMe3)(CO)2H in acetonitrile. *Am. Chem. Soc.* 126:2738–43, 2004.

Smith, A. R., **J. W. Bruno,** and S. D. Pastor. Sterically-congested bisphosphite ligands for the catalytic hydrosilation of ketones. *Phosphorus, Sulfur Silicon Relat. Elem.* 177:479–85, 2002.

Albert, D. F., et al. **(J. W. Bruno).** Supercritical methanol drying as a convenient route to phenolic-furfural aerogels. *J. Non-Cryst. Solids* 296:1, 2001.

Michalczyk, L., et al. **(J. W. Bruno).** Chelating aryloxide ligands in the synthesis of titanium, niobium, and tantalum compounds: Electrochemical studies and styrene polymerization activities. *Organometallics* 20:5547, 2001.

Sarker, N., and **J. W. Bruno.** Thermodynamic studies of hydride transfer for a series of niobium and tantalum compounds. *Organometallics* 20:51, 2001.

Bruno, J. W., and X. J. Li. Use of niobium(III) and niobium(V) compounds in catalytic imine metathesis under mild conditions. *Organometallics* 19:4672, 2000.

Kerr, M. E., et al. **(J. W. Bruno).** Hydride and proton transfer reactions of niobium-bound ligands. Synthetic and thermodynamic studies of ketene, enacyl, and vinylketene complexes. *Organometallics* 19:901, 2000.

Sarker, N., and **J. W. Bruno.** Thermodynamic and kinetic studies of hydride transfer for a series of molybdenum and tungsten hydrides. *J. Am. Chem. Soc.* 120:2174, 1999.

Calter, M. A., O. A. Tretyak, and C. Flaschenriem. Formation of disubstituted β-lactones using bifunctional catalysis. *Org. Lett.* 7:1809–12, 2005.

Calter, M. A., and J. G. Zhou. Synthesis of the C_1-C_{27} portion of the aplyronines. *Tetrahedron Lett.* 45:4847–50, 2004.

Calter, M. A., W. Song, and J. G. Zhou. Catalytic, asymmetric synthesis and diastereoselective aldol reactions of dipropionate equivalents *J. Org. Chem.* 69:1270–5, 2004.

Calter, M. A., R. K. Orr, and W. Song. Catalytic, asymmetric preparation of ketene dimers from acid chlorides. *Org. Lett.* 5:4745–8, 2003.

Calter, M. A., and W. Liao. First total synthesis of a natural product containing a chiral, beta-diketone: Synthesis and stereochemical reassignment of siphonarienedione and siphonarienolone. *J. Am. Chem. Soc.* 124:13127–9, 2002.

Calter, M. A., and X. Guo. Synthesis of the C_{21}-C_{34} segment of the aplyronines using the dimer of methylketene. *Tetrahedron* 58:7093–100, 2002.

Calter, M. A., C. Zhu, and R. J. Lachicotte. Rapid synthesis of the 7-deoxy zaragozic acid core. *Org. Lett.* 4:209–12, 2002.

Calter, M. A, and C. Zhu. The scope and diastereoselectivity of the "interrupted" Feist-Benary reaction. *Org. Lett.* 4:205–8, 2002.

Calter, M. A., W. Liao, and J. A. Struss. Catalytic, asymmetric synthesis of siphonarienal. *J. Org. Chem.* 66:7500–4, 2001.

Calter, M. A., X. Guo, and W. Liao. One-pot, catalytic, asymmetric synthesis of polypropionates. *Org. Lett.* 3:1499–501, 2001.

Vreven, T., and **M. J. Frisch** et al. Geometry optimization with QM/MM methods II: Explicit quadratic coupling. *Mol. Phys.*, in press.

Fermann, J. T., et al. **(M. J. Frisch).** Modeling proton transfer in zeolites: Convergence behavior of embedded and constrained cluster of calculations. *J. Comput. Theor. Chem.*, in press.

Cammi, R., et al. **(M. J. Frisch).** Second-order Møller-Plesset second derivatives for the polarizable continuum model: Theoretical bases and application to solvent effects in electrophilic bromination of ethylene. *Theor. Chem. Acc.* 111(2–6):66–77, 2004.

Rega, N., et al. **(M. J. Frisch).** Hybrid ab-initio/empirical molecular dynamics: Combining the ONIOM scheme with the atom-centered density matrix propagation (ADMP) approach. *J. Phys. Chem. B* 108(13):4210–20, 2004.

Scalmani, G., et al. **(M. J. Frisch).** Achieving linear scaling computational cost for the polarizable continuum model of salvation. *Theor. Chem. Acc.* 111(2–6):90–100, 2004.

Wiberg, K. B., et al. **(M. J. Frisch).** Optical activity of 1-butene, butane and related hydrocarbons. *J. Phys. Chem. A* 108(1):32–8, 2004.

Stephens, P. J., et al. **(M. J. Frisch).** Determination of the absolute configuration of [3_2](1,4)barrelenophanedicarbonitrile using concerted time-dependent density functional theory calculations of optical rotation and electronic circular dichroism. *J. Am. Chem. Soc.* 126(24): 7514–21, 2004.

Toyota, K., et al. **(M. J. Frisch).** Singularity-free analytical energy gradients for the SAC/SAC-CI method: Coupled perturbed minimum orbital-deformation (CPMOD) approach. *Chem. Phys. Lett.* 367(5–6):730–6, 2003.

Li, X., et al. **(M. J. Frisch).** Density matrix search using direct inversion in the iterative subspace as a linear scaling alternative to diagonalization in electronic structure calculations. *J. Chem. Phys.* 119(15):7651–8, 2003.

Fry, A. J. A computational study of solution effects on the disproportionation of electrochemically generated polycyclic aromatic hydrocarbon radical anions. *Tetrahedron*, in press.

Gordon, P. E., **A. J. Fry,** and L. D. Hicks. Further studies on the reduction of benzylic alcohols by hypophosphorous acid/iodine. *ARKIVOC* 6:393, 2005.

Fry, A. J. Disproportionation of arene radical anions is driven overwhelmingly by solvation, not ion pairing. *Electrochemistry Commun.* 7:602, 2005.

Hicks, L. D., **A. J. Fry,** and V. C. Kurzweil. Ab initio computation of electron affinities of substituted benzalacetophenones (chalcones). A new approach to substituent effects in organic electrochemistry. *Electrochim. Acta* 50:1039, 2004.

Fry, A. J. Electrochemical processes (review). In *Kirk-Othmer Encyclopedia of Technology,* 10th ed., 2004.

Halas, S. M., K. Okyne, and **A. J. Fry.** Anodic oxidation of negatively substituted stilbenes. *Electrochim. Acta* 48:1837, 2003.

Fry, A. J. Strong ion-pairing effects in a room temperature ionic liquid. *J. Electroanal. Chem.* 546:35, 2003.

Kaimakliotis, C., and **A. J. Fry.** Novel desilylation of alpha-dimethylsilyl esters by electrochemically generated superoxide ion. *Tetrahedron Lett.* 44:5859, 2003.

Fry, A. J., and C. Kaimakliotis. Electrochemical oxidation and reduction of alpha-dimethylsilyl esters: A novel silicon gamma-effect. In *Mechanistic and Synthetic Aspects of Organic and Biological Electrochemistry,* pp. 77–80, eds. D. G. Peters, J. Simonet, and H. Tanaka. Pennington, N.J.: The Electrochemical Society, Inc., 2003.

Kaimakliotis, C., and **A. J. Fry.** Anodic oxidation of methyl a-dimethylsilydihydrocinnamate: A novel silicon gamma-aryl effect. *J. Org. Chem.* 68:9893, 2003.

Kaimakliotis, C., H. Arthanari, and **A. J. Fry.** Synthesis, NMR spectroscopy, and conformational analysis of *alpha*-dimethylsilyl esters. *J. Organomet. Chem.* 671:126, 2003.

Taylor, K., K. Miura, F. Akinfaderin, and **A. J. Fry.** Reactions cascade in the anodic oxidation of benzyl silanes in methanol. *J. Electrochem. Soc.* 150:D85, 2003.

Fry, A. J., et al. Pinacol reduction-*cum*-rearrangement. A re-examination of the reduction of aryl alkyl ketones by zinc-aluminum chloride. *Tetrahedron Lett.* 43:4391, 2002.

Fry, A. J., et al. Reduction of diaryl alkenes by hypophosphorous acid-iodine in acetic acid. *Tetrahedron* 58:4411, 2002.

Basu, S., and **J. L. Knee.** Vibrational dynamics of 9-fluorenemethanol using infrared-ultraviolet double resonance spectroscopy. *J. Chem. Phys.* 120:5631, 2004.

Pitts, J. D., et al. **(J. L. Knee).** 3-Ethylindole electronic spectroscopy: S_1 and cation torsional potential surfaces. *J. Chem. Phys.* 113:1857, 2000.

Wesleyan University

Basu, S., and **J. L. Knee.** Conformational studies of the neutral and cation of several substituted fluorenes. *J. Electron Spectrosc. Relat. Phenom.* 112:209, 2000.

Pitts, J. D., and **J. L. Knee.** Structure and dynamics of 9-ethylfluorene-Ar_n van der Waals complexes. *J. Chem. Phys.* 110:3389, 1998.

Pitts, J. D., et al. **(J. L. Knee).** Conformational energy and dynamics of 9-ethylfluorene. *J. Chem. Phys.* 110:3378, 1998.

Pitts, J. D., and **J. L. Knee.** Electronic spectroscopy and dynamics of the monomer and Ar_n clusters of 9-phenylfluorene. *J. Chem. Phys.* 109:7113, 1998.

Pitts, J. D., and **J. L. Knee.** Dynamics of vibronically excited fluorene-Ar_n (n=4-5) cluster. *J. Chem. Phys.* 108:9632, 1998.

Zhang, X., et al. **(J. L. Knee).** Neutral and cation spectroscopy of fluorene-Ar_n clusters. *J. Chem. Phys.* 107:8239, 1997.

Lin, W., L. Kang, and **S. E. Novick.** The microwave spectrum of HGeCl. *J. Mol. Spectrosc.* 230:93–8, 2005.

Kang, L., and **S. E. Novick.** The microwave spectrum of cyanophosphine, H_2PCN. *J. Mol. Spectrosc.* 225:66, 2005.

Lehmann, K. K., **S. E. Novick,** R. W. Field, and A. J. Merer. William A. Klemperer: An appreciation. *J. Mol. Spectrosc.* 222:1, 2003.

Subramanian, R., **S. E. Novick,** and R. K. Bohn. Torsional analysis of 2-butynol. *J. Mol. Spectrosc.* 222:57, 2003.

Kang, L., et al. **(S. E. Novick).** Rotational spectra of argon acetone: A two-top internally rotating complex. *J. Mol. Spectrosc.* 213:122, 2002.

Kang, L., and **S. E. Novick.** Microwave spectra of four new perfluoromethyl polyyne chains, trifluoropentadiyne, CF_3-C≡C-C≡C-H, trifluoroheptatriyne, CF_3-C≡C-C≡C-C≡C-H, tetrafluoropentadiyne, CF_3-C≡C-C≡C-F, and trifluoromethylcyanoacetylene, CF_3-C≡C-C≡N. *J. Phys. Chem. A* 106:3749, 2002.

Lin, W., et al. **(S. E. Novick).** Hyperfine interactions in HSiCl. *J. Phys. Chem. A* 106:7703, 2002.

Chen, W., et al. **(S. E. Novick).** Microwave spectroscopy of the methylpolyynes $CH_3(C{\equiv}C)_6H$ and $CH_3(C{\equiv}C)_7H$. *J. Mol. Spectrosc.* 196:335, 1999.

Munrow, M. R., et al. **(S. E. Novick).** Determination of the structure of the argon cyclobutanone van der Waals complex. *J. Phys. Chem.* 103:2256, 1999.

Chen, W., et al. **(S. E. Novick).** Microwave spectroscopy of the 2,4-pentadiynyl radical: $H_2CCCCCH$. *J. Chem. Phys.* 109:10190, 1998.

Chen, W., et al. **(S. E. Novick).** Microwave spectra of the methylcyanopolynes $CH_3 (C{\equiv}C)_n CN$(n=2,3,4,5). *J. Mol. Spectrosc.* 192:1, 1998.

Chen, W., et al. **(S. E. Novick).** Laboratory detection of a new carbon radical: H_2CCCCN. *Astrophys. J.* 492:849, 1998.

Petersson, G. A., S. Zhong, J. A. Montgomery Jr., and **M. J. Frisch.** On the optimization of Gaussian basis sets. *J. Chem. Phys.* 118:1101, 2003.

Nimlos, M. R. et al. **(G. A. Petersson).** Photoelectron spectroscopy of $HCCN^-$ and $HCNC^-$ reveals the quasilinear triplet carbenes, HCCN and HCNC. *J. Chem. Phys.* 117:4323, 2002.

Austin, A. J., et al. **(G. A. Petersson).** An overlap criterion for selection of core orbitals. *Theor. Chem. Acc.* 107:180, 2002.

Petersson, G. A. Complete basis set models for chemical reactivity: From the helium atom to enzyme kinetics. In *Theoretical Thermochemistry,* ed. J. Cioslowski. Kluwer Academic Publishers, 2001.

Montgomery, J. A., Jr., et al. **(G. A. Petersson).** A complete basis set model chemistry. VII. Use of the minimum population localization method. *J. Chem. Phys.* 112:6532, 2000.

Petersson, G. A., and M. J. Frisch. A journey from generalized valence bond theory to the full CI complete basis set limit. *J. Phys. Chem.* 104:2183, 2000.

Petersson, G. A. Perspective on: "The activated complex in chemical reactions" by Henry Eyring [*J. Chem. Phys.* 3:107 (1935)]. *Theor. Chem. Acc.* 103:190, 2000.

Montgomery, J. A., Jr., et al. **(G. A. Petersson).** A complete basis set model chemistry. VI. Use of density functional geometries and frequencies. *J. Chem. Phys.* 110:2822, 1999.

Perumal, S. K., and **R. F. Pratt.** Ketophosph(on)ates—a new lead to inhibitors of beta-lactamases. *FASEB J.* 19:A862, 2005.

Adediran, S. A., et al. **(R. F. Pratt).** The D-methyl group in ß-lactamase evolution: Evidence from the Y221G and GC1 mutants of the class C ß-lactamase of *Enterobacter cloacae* P99. *Biochemistry* 44:7543, 2005.

Silvaggi, N. R., et al. **(R. F. Pratt).** Crystal structures of complexes between the R61 DD-peptidase and peptidoglycan-mimetic ß-lactams: A non-covalent complex with a "perfect Penicillin." *J. Mol. Biol.* 345:521, 2005.

Nukaga, M., et al. **(R. F. Pratt).** Hydrolysis of third-generation cephalosporins by class C ß-lactamases: Structures of a transition state analog of cefotaxime in wild-type and extended spectrum enzymes. *J. Biol. Chem.* 279:9344, 2004.

Kumar, S., S. A. Adediran, M. Nukaga, and **R. F. Pratt.** Kinetics of turnover of cefotaxime by the *Enterobacter cloacae* P99 and GCl ß-lactamases: Two free enzyme forms of the P99 ß-lactamase detected by a combination of pre- and post-steady state kinetics. *Biochemistry* 43:2664, 2004.

Ahn, Y. M., and **R. F. Pratt.** Kinetic and structural consequences of the leaving group in substrates of a class C ß-lactamase. *Bioorg. Med. Chem.* 12:1537, 2004.

Silvaggi, N. R., et al. **(R. F. Pratt).** Toward better antibiotics: Crystallographic studies of a novel class of DD-peptidase/ß-lactamase inhibitors. *Biochemistry* 43:7046, 2004.

Josephine, H. R., I. Kumar, and **R. F. Pratt.** The perfect Penicillin? Inhibition of a bacterial DD-peptidase by peptidoglycan-mimetic ß-lactams. *J. Am. Chem. Soc.* 126:8122, 2004.

Nagarajan, R., and **R. F. Pratt.** Thermodynamic evaluation of a covalently bonded transition state analogue inhibitor: Inhibition of ß-lactamases by phosphonates. *Biochemistry* 43:9664, 2004.

Adediran, S. A., et al. **(R. F. Pratt).** Benzopyranones with retro-amide side chains as (inhibitory) ß-lactamase substrates. *Bioorg. Med. Chem. Lett.* 14:5117, 2004.

Nagarajan, R., and **R. F. Pratt.** Synthesis and evaluation of new substrate analogues of the Streptomyces R61 DD-peptidase: Dissection of a specific ligand. *J. Org. Chem.* 69:7472, 2004.

Cabaret, D., S. A. Adediran, **R. F. Pratt,** and M. Wakselman. New substrates for ß-lactam-recognizing enzymes: Aryl malonamates. *Biochemistry* 42:6719, 2003.

Anderson, J. W., et al. **(R. F. Pratt).** On the substrate specificity of bacterial DD-peptidases: Evidence from two series of peptidoglycan-mimetic peptides. *Biochem. J.* 373:949, 2003.

Silvaggi, N. R., et al. **(R. F. Pratt).** The crystal structure of phosphonate-inhibited D-Ala-D-Ala peptidase reveals an analogue of a tetrahedral transition state. *Biochemistry* 42:1199, 2003.

Kaur, K., S. A. Adediran, M. J. K. Lan, and **R. F. Pratt.** Inhibition of ß-lactamases by monocyclic acyl phosph(on)ates. *Biochemistry* 42:1429, 2003.

McDonough, M. A., et al. **(R. F. Pratt).** Structures of two kinetic intermediates reveal species specificity of Penicillin-binding proteins. *J. Mol. Biol.* 322:111, 2002.

Bell, J. H., and **R. F. Pratt.** Mechanism of inhibition of the ß-lactamase of *Enterobacter cloacae* P99 by 1:1 complexes of vanadate with hydroxamic acids. *Biochemistry* 41:4329, 2002.

Bell, J. H., and **R. F. Pratt.** Formation and structure of 1:1 complexes between aryl hydroxamic acids and vanadate at neutral pH. *Inorg. Chem.* 41:2747, 2002.

Pratt, R. F. Functional evolution of the β-lactamase active site. *J. Chem. Soc. Perkin Trans. II* 851, 2002.

Kaur, K., et al. **(R. F. Pratt).** Mechanism of inhibition of the class C β-lactamase of *Enterobacter cloacae* P99 by cyclic acyl phosph(on)ates: Rescue by return. *J. Am. Chem. Soc.* 123:10436, 2001.

Morrison, M. J., et al. **(R. F. Pratt).** Inverse acyl phosph(on)ates: Substrates or inhibitors of β-lactam-recognizing enzymes? *Bioorg. Chem.* 29:271, 2001.

Kumar, S., et al. **(R. F. Pratt).** Design, synthesis and evaluation of α-ketoheterocycles as class C β-lactamase inhibitors. *Bioorg. Med. Chem.* 9:2035, 2001.

Bebrone, C., et al. **(R. F. Pratt).** CENTA as a chromogenic substrate for studying β-lactamases. *Antimicrob. Agents Chemother.* 45:1868, 2001.

Adediran, S. A., et al. **(R. F. Pratt).** The synthesis and evaluation of benzofuranones as β-lactamase substrates. *Bioorg. Med. Chem.* 9:1175, 2001.

Kaur, K., and **R. F. Pratt.** Mechanism of reaction of acyl phosph(on)ates with the β-lactamase of *Enterobacter cloacae* P99. *Biochemistry* 40:4610, 2001.

Cabaret, D., et al. **(R. F. Pratt).** Synthesis, hydrolysis, and evaluation of 3-acylamino-3,4-dihydro-2-oxo-2*H*-1,3-benzoxazinecarboxylic acids and linear azadepsipeptides as potential substrates/inhibitors of β-lactam-recognizing enzymes. *Eur. J. Org. Chem.* 141, 2001.

Subramanian, R., J. M. Szarko, **W. C. Pringle,** and **S. E. Novick.** Rotational spectrum, nuclear quadrupole coupling constants, and structure of six isotopomers of the argon-chlorocyclobutane van der Waals complex. *J. Mol. Struct.* 742:165–72, 2005.

Munrow, M. R., et al. **(W. C. Pringle).** Determination of the structure of the argon cyclobutanone van der Waals complex. *J. Phys. Chem.* 103:2256, 1999.

Pringle, W. C., et al. Collision induced far infrared spectrum of cyclopropane. *Mol. Phys.* 62:669, 1987.

Pringle, W. C., et al. Analysis of collision induced far infrared spectrum of ethylene. *Mol. Phys.* 62:661, 1987.

Coman, D., and **I. M. Russu.** Base pair opening in three DNA-unwinding elements. *J. Biol. Chem.* 280:20216, 2005.

Russu, I. M. Probing site-specific energetics in proteins and nucleic acids by hydrogen exchange and NMR spectroscopy. *Methods Enzymol.* 379:152–75, 2004.

Coman, D., and **I. M. Russu.** Site-resolved stabilization a DNA triple helix by magnesium ions. *Nucleic Acids Res.* 32:878–83, 2004.

Chen, C., and **I. M. Russu.** Sequence-dependence of the energetics of opening of AT base pairs in DNA. *Biophys. J.* 87:2545–51, 2004.

Coman, D., and **I. M. Russu.** Probing hydrogen bonding in a DNA triple helix using protium-deuterium fractionation factors. *J. Am. Chem. Soc.* 125:6626–7, 2003.

Rujan, I. N., and **I. M. Russu.** Allosteric effects of chloride ions at the "interfaces of human hemoglobin." *Proteins Struct. Funct. Genet.* 49:413–49, 2002.

Jiang, L., and **I. M. Russu.** Internal dynamics in a DNA triple helix probed by ^1H-^{15}N NMR spectroscopy. *Biophys. J.* 82:3181–5, 2002.

Coman, D., and **I. M. Russu.** Site-resolved energetics in DNA triple helices containing GTA and TCG triads. *Biochemistry* 41:4407–14, 2002.

Mihailescu, M. R., et al. **(I. M. Russu).** Allosteric free energy changes at the $\alpha1\beta2$ interface of human hemoglobin probed by proton exchange of Trpβ37. *Proteins Struct. Funct. Genet.* 44:73, 2001.

Mihailescu, M. R., and **I. M. Russu.** A signature of the T→P transition in human hemoglobin. *Proc. Natl. Acad. Sci. U.S.A.* 98:3773, 2001.

Russu, I. M., and C. Fronticelli. Structural design of hemoglobin blood substitute. Invited paper for *Blood Substitutes.* New York: Academic Press, 2001.

Michalczyk, R., and **I. M. Russu.** Rotational dynamics of adenine amino groups in a DNA double helix. *Biophys. J.* 76:2679, 1999.

Michalczyk, R., and **I. M. Russu.** Studies of the dynamics of adenine protons in DNA by ^{15}N-labeling and heteronuclear NMR spectroscopy. In *Proceedings of the 10th Conversation.* Albany, N.Y.: SUNY, 1998.

Shea, T. M., and **T. D. Westmoreland.** Electronic effects on the rates of coupled two-electron/halide self-exchange reactions of substituted ruthenocenes. *Inorg. Chem.* 39:1573, 2000.

Holmer, S. A., et al. **(T. D. Westmoreland).** A new irreversibly inhibited form of xanthine oxidase from ethylisonitrile. *J. Inorg. Biochem.* 66:63, 1997.

Swann, J., and **T. D. Westmoreland.** Density functional calculations of *g* values and molybdenum hyperfine coupling constants for a series of molybdenum(V) oxyhalide anions. *Inorg. Chem.* 36:5348, 1997.

Nipales, N. S., and **T. D. Westmoreland.** Correlation of EPR parameters with electronic structure in the homologous series of low symmetry complexes Tp^*MoOX_2 (Tp^* = tris(3,5-dimethylpyrazol-1-yl)borate; X = F, Cl, Br). *Inorg. Chem.* 36:756, 1997.

WORCESTER POLYTECHNIC INSTITUTE

Department of Chemistry and Biochemistry

Program of Study

The Department of Chemistry and Biochemistry at Worcester Polytechnic Institute (WPI) offers M.S. degrees in chemistry and biochemistry and a Ph.D. degree in chemistry. The Department provides state-of-the-art infrastructure, equipment, and resources to perform scientific research at the highest level. The diverse learning environment of the Department promotes an easy exchange of ideas, access to all the necessary resources, and personalized training of future scientists. Frequent participation in national and international meetings provides students with a broader scientific perspective, networking opportunities, and the possibility to plan their postgraduation professional path.

Research is conducted in several broad areas, including plant biochemistry and biophysics, molecular engineering and synthesis, and nanotechnology. In biochemistry and biophysics, active research involves heavy metal transport and metal homeostasis of both plants and bacteria, plant pathogen interactions, enzyme structure and function, and regulation of plant development by light. In the molecular design and synthesis area, there is active research on topics encompassing supramolecular materials, photovoltaic materials, polymorphism in pharmaceutical drugs, spectroscopy of heterocyclic molecules, photophysical properties of cumulenes, and host-guest chemistry. Research in nanotechnology encompasses such projects as photonic and nonlinear optical materials, nanoporous and microporous crystals of organic and coordination compounds, and molecular interactions at surfaces.

Each student selects a research adviser no later than the end of the first term (seven weeks) of residence, and research is started by the beginning of the second term. The M.S. degree in chemistry or biochemistry requires 30 semester hours of credit, of which at least 6 or more must be thesis research and the remainder in approved independent studies and courses. An M.S. candidate submits a thesis based on research conducted under the direction of a faculty member during tenure at WPI. The thesis must be approved by the faculty adviser and the Department chair. Before formal admission to the doctoral program, Ph.D. candidates must take the qualifying examination in their field of specialization. At the end of the first semester of the second year of residence, the student must submit a written and oral progress report on completed research to the Department. A committee of 3 faculty members, including the research adviser, then recommends whether the student should complete an M.S. degree or be formally admitted to the Ph.D. program. Presentation and defense of the Ph.D. dissertation concludes the Ph.D. studies.

Research Facilities

The Department of Chemistry and Biochemistry is located in Goddard Hall, which has 20,000 square feet of research laboratories, shops, and instrument laboratories. Department facilities and instrumentation in individual research laboratories include 200 and 400 MHz FT-NMR, GC-MS, GC, HPLC, capillary electrophoresis, a differential scanning calorimeter (DSC), thermogravometric analysis (TGA), polarizing optical stereomicroscope, FT-IR, UV-VIS absorption, fluorescence and phosphorescence spectroscopy, cyclic voltammetry, impedence spectroscopy, ellipsometer, quartz-crystal microbalance, grazing incidence IR, atomic-force microscopic (AFM), and other surface-related facilities. Additional equipment in the biochemistry area includes centrifuges, ultracentrifuges, PCR, a phosphor-imager, a scintillation counter, FPLC, bacteria and eukaryotic cell cultures, and plant-growth facilities. The Department is exceptionally well set up with computer facilities and is also networked to the University's computer infrastructure.

Financial Aid

Graduate assistantships are available for teaching or research. For the academic year 2006–07, teaching assistantships carry a stipend of $1667 per month, and research assistantship stipends varied between $1667 and $2500 per month. Both assistantships provide remission of tuition for up to 20 credits. Additional assistance may be available for the summer. U.S. citizens with exceptional qualifications are encouraged to apply for the Robert F. Goddard Fellowship. This prestigious award offers full-time tuition for one year (up to 20 credits) and a stipend of $1667 per month for twelve months. Other fellowship opportunities are also available. Information may be found online at: http://www.grad.wpi.edu/Financial/.

Cost of Study

Graduate tuition for the 2006–07 academic year is $997 per credit hour. There are nominal extra charges for the thesis, health insurance, and other fees.

Living and Housing Costs

On-campus graduate student housing is limited to a space-available basis. There is no on-campus housing for married students. Apartments and rooms in private homes near the campus are available at varying costs. For further information and apartment listings, students should check the Residential Services Office Web site at http://www.wpi.edu/Admin/RSO/Offcampus/.

Student Group

Worcester Polytechnic Institute has a student body of 3,869, of whom 837 are full- or part-time graduate students. Most states and nearly fifty countries are represented.

Location

The university is located on an 80-acre campus in a residential section of Worcester, Massachusetts. The city, the second largest in New England, has many colleges and an unusual variety of cultural opportunities. Located three blocks from the campus, the nationally famous Worcester Art Museum contains one of the finest permanent collections in the country and offers many special activities of interest to students. The community also provides outstanding programs in music and theater. The DCU Center offers rock concerts and semiprofessional athletic events. Easily reached for recreation are Boston and Cape Cod to the east and the Berkshires to the west, and good skiing is nearby to the north. Complete athletic and recreational facilities and a program of concerts and special events are available on the campus to graduate students.

The Institute

Worcester Polytechnic Institute, founded in 1865, is the third-oldest independent university of engineering and science in the United States. Graduate study has been a part of the Institute's activity for more than 100 years. Classes are small and provide for close student-faculty relationships. Graduate students frequently interact in research with undergraduates participating in WPI's innovative project-based program of education.

Applying

Applicants must have a B.S. degree with demonstrated proficiency in chemistry or biochemistry. Applicants must submit WPI application forms, official college transcript(s) from all undergraduate and graduate institutions attended, three letters of recommendation, a $70 application fee (waived for WPI alumni), and scores from the GRE General Test. International students whose primary language is not English must also submit proof of English language proficiency. WPI accepts either the Test of English as a Foreign Language (TOEFL) or the International English Language Testing System (IELTS). A paper-based TOEFL score of at least 550 (213 on the computer-based test or 79–80 on the Internet-based test) or an IELTS overall band score of 6.5 (no band score below 6.0) is required for admission.

To be considered for funding (assistantships and fellowships), complete applications must be on file by January 15 for fall admission and October 15 for spring admission. Files completed after those deadlines are reviewed on a rolling basis and may not be considered for funding. Some fellowships require an additional application. Prospective students should visit http://www.grad.wpi.edu/Financial/ for more information. Inquiries should be directed to the Department of Chemistry and Biochemistry or to the Office of Graduate Studies and Enrollment at grad_studies@wpi.edu.

Correspondence and Information

Department of Chemistry and Biochemistry
Worcester Polytechnic Institute
100 Institute Road
Worcester, Massachusetts 01609-2280

Phone: 508-831-5326
Fax: 508-831-5933
E-mail: chem-biochem@wpi.edu
Web site: http://www.wpi.edu/Academics/Depts/Chemistry/Graduate

Worcester Polytechnic Institute

THE FACULTY AND THEIR RESEARCH

José M. Argüello, Associate Professor; Ph.D., National of Rió Cuarto (Argentina). The mechanism of ion transport by P-type ATPases.
Identification of the transmembrane metal binding site in Cu+ transporting PIB-type ATPases. *J. Biol. Chem.* 279:54802–7, 2004. With Mandal, Yang, and Kertesz.

Robert E. Connors, Professor; Ph.D., Northeastern. The spectroscopy and photophysics of organic and inorganic systems.
Electronic absorption and fluorescence properties of 2,5-diarylidene-cyclopentanones. *J. Phys. Chem. A* 107:7684, 2003. With Ucak-Astarlioglu.

James P. Dittami, Professor and Department Head; Ph.D., Rensselaer. Synthetic organic chemistry, organic photochemistry, and heterocyclic chemistry, specifically the development of novel chemical reactions with application to the synthesis of biologically active natural products.

John C. MacDonald, Assistant Professor; Ph.D., Minnesota. Supramolecular assembly of molecules in bulk molecular crystals and in multilayer thin films on surfaces, the crystallization of drugs and proteins on chemically modified surfaces as a means to control nucleation and the incidence of polymorphism.
Fabrication of complex crystals using kinetic control, chemical additives, and epitaxial growth. *Chem. Mater.* 16:4916–27, 2004. With Luo and Palmore.

William Grant McGimpsey, Professor; Ph.D., Queen's. The practical applications of molecular nanotechnology and sensors.
Modeling the oxidative degradation of ultrahigh molecular weight polyethylene. *J. Appl. Polym. Sci.* 87:814, 2003. With Medhekar, Thompson, and Wang.

James W. Pavlik, Professor and Interim Department Head; Ph.D., George Washington. Understanding the phototransposition chemistry of aromatic heterocycles.
Synthesis and spectroscopic properties of some dideuterated cyanopyridines. *J. Heterocyclic Chem.* 42:73, 2005. With Laohhasurayotin.

Alfred A. Scala, Professor; Ph.D., Polytechnic. Fundamental reaction dynamics, specifically the relationship between thermodynamic and kinetic control of reaction products and fundamental transition state theory.

Venkat R. Thalladi, Assistant Professor; Ph.D., Hyderabad (India). Polymorphism in pharmaceutical drugs, organic octupolar nonlinear optical materials, self-assembly approaches to nanofabrication, crystals of crystals—photonic materials at the micron and submicron length-scales, chemical applications of microfluidics.
Dissections: Self-assembled aggregates that spontaneously reconfigure their structures when their environment changes. *J. Am. Chem. Soc.* 124:14508–9, 2002. With Mao, Wolfe, S. Whitesides, and G. M. Whitesides.

Kristin K. Wobbe, Associate Professor; Ph.D., Harvard. The molecular interactions that determine the outcome of plant/pathogen interactions.
The mevalonate-independent pathway is expressed in transformed roots of *Artemisia annua* and regulated by light and culture age. *Vitro Cell. Dev. Biol. Plant* 38:581–8, 2002. With Souret and Weathers.

YALE UNIVERSITY

Department of Chemistry

Program of Study

The Department of Chemistry offers programs of study leading to the Ph.D. in chemistry. The Department is organized around four principal areas of teaching and research—biophysical, inorganic, organic, and physical and theoretical chemistry. Courses are chosen according to the student's background and area of research interest. To be admitted to candidacy, a student must receive at least two term grades of honors, pass either three cumulative examinations and one oral examination (organic and inorganic students) or two oral examinations (physical and biophysical students) by the end of the second year of study, and submit a thesis prospectus no later than the end of the third year of study. Remaining degree requirements include completing eight cumulative examinations (organic and inorganic students), a written thesis describing the research, and an oral defense of the thesis. In addition, all students are required to teach two afternoons per week (or the equivalent) for at least two semesters.

To assist students in choosing a research group, faculty members present a series of brief seminars about their research in the early fall. Incoming students then select three or four research groups of interest and for a period of three (nonbiophysical students) or ten (biophysical students) weeks participate in these groups' activities. These rotations provide informal and easy interactions with several research groups and enable the incoming student to make an informed choice of a research group.

Research Facilities

The world-class research effort at Yale is supported by numerous professionally staffed facilities, many of which are available to students 24 hours a day. Located in the center of the Sterling Chemistry Laboratory and open to students 24 hours a day, the Sterling Chemistry Library houses a large collection of periodicals and reference books as well as traditional and electronic resources for searching the chemical literature. In addition, a very large collection spanning most of the physical and biological sciences is located less than 100 yards away in the Kline Science Library. The Chemical Instrumentation Center, which is situated in the heart of the Sterling Chemistry Laboratory, is a comprehensive facility containing EPR and GC mass spectrometers as well as a vast array of medium- and high-frequency NMR spectrometers. The center also maintains a collection of instruments for optical spectroscopy that includes FT-IR, Raman, CD, UV-visible, fluorescence, and polarimetric and atomic adsorption instruments. Under the direction of Chris Incarvito, the Chemistry X-Ray Diffraction Center contains state-of-the-art X-ray diffraction equipment with area detection capabilities. These instruments provide rapid and unambiguous structural information for crystalline compounds of moderate size. Located adjacent to the chemical instrumentation center, the Keck High-Field Magnetic Resonance Laboratory provides access to very high field strengths (800 MHz proton) and supports both solution and solid-state NMR methods. Established in conjunction with the Department of Applied Physics, the Yale Ultrafast Kinetics Laboratory enables students to perform experiments requiring tunable optical pulses having less than 100-femtosecond duration (roughly the period of a molecular vibration). Although the decreasing cost of computers has enabled individual research groups to develop their own computational resources, the chemistry department maintains an array of Silicon Graphics and VAX workstations as well as several Macintosh/PC systems in the Chemistry Computer Center for the general use of students. The Glass Shop provides access to custom-made scientific equipment as well as rapid repair of existing apparatus. The Sterling Student Machine Shop is designed to provide students with training in modern materials and instrument fabrication techniques. After completing a shop course, students have full access to the shop facilities for work on routine or small, specialized projects. The Chemistry Stockroom maintains an extensive inventory of chemicals and supplies. In addition, the stockroom staff serves as liaison between the students and the vast array of commercial vendors of research equipment and supplies.

Financial Aid

Students who are accepted for admission normally receive health insurance, full tuition, and a twelve-month stipend ($25,333 in 2006–07). Students are generally expected to serve as teaching fellows for at least two semesters during their time at Yale.

Cost of Study

The cost of study and health care is provided for all students for the duration of their enrollment.

Living and Housing Costs

Single rooms and apartments are available on campus for rents that range from $350 to $800 per month. A wide variety of off-campus housing is also available in all price ranges in neighborhoods immediately adjacent to Yale as well as in suburban, rural, and shoreline communities adjacent to New Haven.

Student Group

There are 113 full-time students in the chemistry department; 50 are women and 50 are international.

Location

Yale University is situated in New Haven, a small (population 130,000), historic New England city located on Long Island Sound. New Haven has an active cultural life and is situated only 1½ hours from New York City and 2½ hours from Boston, both of which are easily visited by car or train. Although part of a major metropolitan and cultural center, New Haven is surrounded by picturesque communities and scenic rural areas that offer easily accessible opportunities for outdoor activities and housing.

The University

Yale University was founded in 1701 and has grown to become a large and diverse campus consisting of Yale College, the Graduate School of Arts and Sciences, the School of Medicine, and nine other professional schools. The University has a total enrollment of more than 11,000 students, nearly half of whom are studying at the graduate level.

Applying

Applicants are expected to have completed a standard undergraduate chemistry major that includes a year of elementary organic chemistry with a laboratory and a year of elementary physical chemistry. Other majors are acceptable if these requirements are met. Undergraduate research experience is strongly recommended. Students must submit the completed application, the application fee, official transcripts, and scores from both the GRE General Test and the Chemistry Advanced Test. Students whose native language is not English are required to take the Test of English as a Foreign Language and the Test of Spoken English. The application deadline for the fall semester is January 2.

Correspondence and Information

Gary Brudvig, Chair
Department of Chemistry
Yale University
225 Prospect Street
P.O. Box 208107
New Haven, Connecticut 06520-8107
Phone: 203-432-3915
Fax: 203-432-6144
E-mail: chemistry.admissions@yale.edu
Web site: http://www.chem.yale.edu

Yale University

THE FACULTY AND THEIR RESEARCH

Sidney Altman, Sterling Professor of Molecular Cellular and Developmental Biology and Professor of Chemistry, Biophysical Chemistry, and Organic Chemistry; Ph.D., Colorado, 1967. Function and structure of ribonuclease P in both bacteria and human cells. Characterization of RNase P from *Thermotoga maritima*. *Nucl. Acids Res.* 29(4):880–5, 2001 (with Paul and Lazarev).

Victor S. Batista, Associate Professor of Biophysical Chemistry and of Physical and Theoretical Chemistry; Ph.D., Boston University, 1997. Development of rigorous and practical methods for simulations of quantum processes in complex systems and applications studies of photochemical processes in proteins, semiconductor materials, and systems of environmental interest. Model study of coherent-control of the femtosecond primary event of vision. *J. Phys. Chem. B* 108:6745, 2004 (with Flores).

Gary Brudvig, Professor and Chair of Chemistry and Professor of Molecular Biophysics and Biochemistry, Biophysical Chemistry, and Inorganic Chemistry; Ph.D., Caltech, 1981. Molecular basis for energy transduction in plant photosynthesis. Pulsed high-frequency EPR study on the location of carotenoid and chlorophyll cation radicals in photosystem II. *J. Am. Chem. Soc.* 125:5005–14, 2003 (with Lakshmi et al.).

Robert H. Crabtree, Professor of Inorganic Chemistry and Organic Chemistry; Ph.D., Sussex, 1973. Inorganic, organometallic, and bioinorganic chemistry; design and synthesis of inorganic, coordination, or organometallic molecules with unusual structures, useful catalytic properties, or bioinorganic relevance; molecular recognition in homogeneous catalysis; computational and physicochemical insights from collaborative work. Dimer-of-dimers model for the oxygen-evolving complex of photosystem II: Synthesis and properties of $[Mn^{IV}_4O_5(terpy)_4(H_2O)_2](ClO_4)_6$. *J. Am. Chem. Soc.* 126, 2004 (with Chen, Faller, and Brudvig).

Craig Crews, Associate Professor of Chemistry; Pharmacology; Molecular, Cell, and Developmental Biology; Organic Chemistry; and Bioorganic Chemistry; Ph.D., Harvard, 1993. Development of drug candidates into clinically useful therapeutic agents. Chemical genetic control of protein levels: Selective in vivo targeted degradation. *J. Am. Chem. Soc.* 126(12):3748–54, 2004 (with Schneekloth et al.).

R. James Cross Jr., Professor of Physical Chemistry; Ph.D., Harvard, 1966. Study of chemical systems under unusual conditions. Kr extended X-ray absorption fine structure study of endohedral Kr@C60. *J. Phys. Chem. B* 108:3191, 2004 (with Ito et al.).

Jack W. Faller, Professor of Chemistry, Inorganic Chemistry, and Organic Chemistry; Ph.D., MIT, 1967. Understanding how transition metal complexes and organometallic compounds can be used to control the stereochemistry of organic reactions. Kinetic resolution and unusual regioselectivity in palladium-catalyzed allylic alkylations with a chiral P,S ligand. *Org. Lett.* 6:1301–4, 2004 (with Wilt and Parr).

Gary W. Haller, Henry Prentiss Becton Professor of Engineering and Applied Science and Professor of Chemistry, Physical Chemistry, and Inorganic Chemistry; Ph.D., Northwestern, 1966. Understanding and rationalizing heterogeneous catalytic activity and selectivity in terms of surface or deduced site structure. Pore curvature effect on the stability of co-MCM-41 and the formation of size-controllable subnanometer co-clusters. *J. Phys. Chem. B* 108, 2004 (with Lim et al.).

Andrew D. Hamilton, Benjamin Silliman Professor of Organic Chemistry and Biophysical Chemistry; Ph.D., Cambridge, 1980. Molecular recognition. FTI-2153 inhibits bipolar spindle formation and chromosome alignment and causes prometaphase accumulation during mitosis of human lung cancer cells. *J. Biol. Chem.* 276: 16161–7, 2001 (with Crespo, Ohkanda, and Sebti).

John F. Hartwig, Irénée P. duPont Professor of Chemistry, Inorganic Chemistry, Organic Chemistry, and Organic Synthesis; Ph.D., Berkeley, 1990. Discovering, developing, and understanding the mechanisms of new transition metal–catalyzed reactions. Synthesis, structure, and reductive elimination chemistry of three-coordinate arylpalladium amido complexes. *J. Am. Chem. Soc.* 126:5344–5, 2004 (with Yamashita).

Francesco Iachello, J. W. Gibbs Professor of Physics and Chemistry and of Theoretical Physical Chemistry; Ph.D., MIT, 1969. Developing and exploiting advanced mathematical (algebraic) methods as a means of attacking problems of current interest in chemistry. A novel algebraic scheme for describing non-rigid molecules. *Chem. Phys. Lett.* 375:309, 2003 (with Perez-Bernal and Vaccaro).

Mark A. Johnson, Arthur T. Kemp Professor of Chemistry and Physical Chemistry; Ph.D., Stanford, 1983. Cooperative mechanics that act to split a proton free from a water molecule or a dissolved acid. Spectral signatures of hydrated proton vibrations in water clusters. *Science* 308:1765–9, 2005 (with Headrick et al.).

William L. Jorgensen, Whitehead Professor of Chemistry, Organic Chemistry, Theoretical Chemistry, and Biophysical Chemistry; Ph.D., Harvard, 1975. Application of computational methods to solve problems concerning structure and reactivity for biomolecular and organic systems. Polypeptide folding using Monte Carlo sampling, concerted rotation, and continuum solvation. *J. Am. Chem. Soc.* 126:1849–57, 2004 (with Ulmschneider).

J. Patrick Loria, Associate Professor of Chemistry, Biophysical Chemistry, and Physical Chemistry; Ph.D., Notre Dame, 1997. Understanding how the dynamic and structural properties of proteins correlate with their function. Multiple time-scale backbone dynamics of homologous thermophilic and mesophilic ribonuclease HI enzymes. *J. Mol. Biol.* 339:855–71, 2004 (with Butterwick et al.).

J. Michael McBride, Richard M. Colgate Professor of Organic Chemistry; Ph.D., Harvard, 1967. Using physical-organic chemistry to understand and control the chemical reactivity and physical properties of organic solids. Using crystal optics to demonstrate single-layer localization of a solid-state chain reaction. *Helv. Chim. Acta* 83:2352–62, 2000 (with Pate).

Glenn C. Micalizio, Assistant Professor of Organic Chemistry and Inorganic Chemistry; Ph.D., Michigan, 2001. Natural product total synthesis and development of new synthetic methods that enable rapid access to molecular complexity and structural diversity. An alkynylboronic ester annulation: Development of synthetic methods for application to diversity-oriented organic synthesis. *Angew. Chem. Int. Ed.* 41:3272, 2002 (with Schreiber).

Scott J. Miller, Professor of Organic Chemistry; Ph.D., Harvard, 1994. Development of new methods for the synthesis and derivatization of complex molecules, involving reaction design, development, and application. Thiazolylalanine-derived catalysts for enantioselective intermolecular aldehyde-imine cross-couplings. *J. Am. Chem. Soc.* 127:1654–5, 2005 (with Mennen, Gipson, and Kim).

Peter B. Moore, Sterling Professor of Chemistry and Professor of Molecular Biophysics and Biochemistry and of Biophysical Chemistry; Ph.D., Harvard, 1966. Structures of RNAs and the complexes they form with proteins, relationship between RNA structure and biological function. The structure of a ribosomal protein S8/spc operon mRNA complex. *RNA* 10:954–64, 2004 (with Merianos and Wang).

Lynne J. Regan, Professor of Molecular Biophysics and Biochemistry and Chemistry, Biophysical Chemistry, and Organic Chemistry; Ph.D., MIT, 1987. How a protein's primary sequence specifies its three-dimensional structure. Antiparallel leucine zipper-directed protein reassembly: Application to the green fluorescent protein. *J. Am. Chem. Soc.* 122:5658–9, 2000 (with Ghosh and Hamilton).

Martin Saunders, Professor of Organic Chemistry; Ph.D., Harvard, 1956. Buckminsterfullerene and its larger relatives. Insertion of helium and molecular hydrogen through the orifice of an open fullerene. *Angew. Chem. Int. Ed.* 40:1543–6, 2001 (with Rubin et al.).

Alanna Schepartz, Milton Harris, '29, Ph.D., Professor of Chemistry and of Molecular Cellular and Developmental Biology, Organic Chemistry, Chemical Biology, and Biophysical Chemistry; Ph.D., Columbia, 1987. Chemical biology of protein-protein and protein-DNA interactions inside the cell. Specific recognition of hDM2 by β–peptide helices. *J. Am. Chem. Soc.* 126:9468, 2004 (with Kritzer, Lear, and Hodsdon).

Charles A. Schmuttenmaer, Professor of Physical Chemistry, Biophysical Chemistry, Biochemistry, Materials Chemistry, and Nanochemistry; Ph.D., Berkeley, 1991. The far-infrared (FIR) region of the spectrum in time-resolved studies. Exploring dynamics in the far-infrared with terahertz spectroscopy. *Chem. Rev.* 104:1759–79, 2004.

Dieter Söll, Sterling Professor of Molecular Biophysics and Biochemistry and Professor of Chemistry, Organic Chemistry, and Biophysical Chemistry; Ph.D., Stuttgart, 1962. The diverse roles of transfer RNA in various biological systems. Aminoacyl-tRNA synthetases, the genetic code, and the evolutionary process. *Microbiol. Mol. Biol. Rev.* 64:202–36, 2000 (with Woese, Olsen, and Ibba).

Thomas A. Steitz, Sterling Professor of Molecular Biophysics and Biochemistry, Professor of Chemistry, and Investigator of Biophysical Chemistry at the Howard Hughes Medical Institute; Ph.D., Harvard, 1966. Understanding, in terms of atomic structures, the molecular and chemical mechanisms by which proteins and nucleic acids achieve their biological functions. The structural mechanism of translocation and helicase activity in T7 RNA polymerase. *Cell* 116:393–404, 2004 (with Yin).

Scott Strobel, Professor of Molecular Biophysics and Biochemistry and Chemistry, Biochemistry, Biophysical Chemistry, and Bioorganic Chemistry; Ph.D., Caltech, 1992. How RNA catalyzes biologically essential chemical reactions, including protein synthesis and RNA splicing. Crystal structure of a self-splicing group I intron with both exons. *Nature* 430:45–50, 2004 (with Adams, Stahley, Kosek, and Wang).

John C. Tully, Sterling Professor of Chemistry and Professor of Physics and Applied Physics and of Physical and Theoretical Chemistry; Ph.D., Chicago, 1968. Understanding, at the molecular level, of dynamical processes such as energy transfer and chemical reaction at surfaces in condensed phases and in biological environments. Efficient thermal rate constant calculation for rare event systems. *J. Chem. Phys.* 118:1085, 2003 (with Corcelli and Rahman).

Patrick H. Vaccaro, Professor of Physical Chemistry; Ph.D., MIT, 1986. Fundamental properties of molecular systems, as revealed through detailed interpretation of spectroscopic measurements and related optical phenomena. Temperature dependence of optical rotation: α-pinene, β-pinene, pinane, camphene, camphor, and fenchone. *J. Phys. Chem. A* 108(26):5559–63, 2004 (with Wiberg, Wang, and Murphy).

Ann M. Valentine, Assistant Professor of Inorganic and Biophysical Chemistry; Ph.D., MIT, 1998. Study of the molecular transactions of inorganic elements in biological systems. Titanium: Inorganic and coordination chemistry. In *Encyclopedia of Inorganic Chemistry*, 2nd ed., ed. King. Chichester, UK: John Wiley and Sons, 2004.

John L. Wood, Professor of Organic Chemistry; Ph.D., Pennsylvania, 1991. Natural product synthesis and methods development. Total synthesis of epoxysorbicillinol. *J. Am. Chem. Soc.* 125:4022, 2001 (with Thompson, Yusuff, and Pflum).

Kurt W. Zilm, Professor of Chemistry and Chemical Engineering, Physical Chemistry, and Biophysical Chemistry; Ph.D., Utah, 1981. Development of high-resolution solid-state NMR techniques for probing the structure and physical chemistry of macromolecules in nanocrystalline form. RF homogeneity in high field solid-state NMR probes. *J. Magn. Reson.* 171:314–23, 2004 (with Paulson and Martin).

Section 3
Geosciences

This section contains a directory of institutions offering graduate work in geosciences, followed by in-depth entries submitted by institutions that chose to prepare detailed program descriptions. Additional information about programs listed in the directory but not augmented by an in-depth entry may be obtained by writing directly to the dean of a graduate school or chair of a department at the address given in the directory.

For programs offering related work, see all other areas in this book. In Book 2, see Geography; in Book 3, see Biological and Biomedical Sciences, Biophysics, and Botany and Plant Biology; in Book 5, see Aerospace/Aeronautical Engineering; Agricultural Engineering and Bioengineering; Civil and Environmental Engineering; Energy and Power Engineering (Nuclear Engineering); Engineering and Applied Sciences; Geological, Mineral/Mining, and Petroleum Engineering; and Mechanical Engineering and Mechanics.

CONTENTS

Program Directories

Announcements

Close-Ups

See also:

Geochemistry

California Institute of Technology, Division of Geological and Planetary Sciences, Pasadena, CA 91125-0001. Offers cosmochemistry (PhD); geobiology (PhD); geochemistry (MS, PhD); geology (MS, PhD); geophysics (MS, PhD); planetary science (MS, PhD). Part-time programs available. *Degree requirements:* For doctorate, thesis/dissertation. *Entrance requirements:* For doctorate, GRE General Test. Additional exam requirements/recommendations for international students: Required—TOEFL. Electronic applications accepted. *Faculty research:* Astronomy, evolution of anaerobic respiratory processes, structural geology and tectonics, theoretical and numerical seismology, global biogeochemical cycles.

California State University, Fullerton, Graduate Studies, College of Natural Science and Mathematics, Department of Chemistry and Biochemistry, Fullerton, CA 92834-9480. Offers analytical chemistry (MS); biochemistry (MS); geochemistry (MS); inorganic chemistry (MS); organic chemistry (MS); physical chemistry (MS). Part-time programs available. *Students:* 19 full-time (6 women), 20 part-time (9 women); includes 20 minority (1 African American, 13 Asian Americans or Pacific Islanders, 6 Hispanic Americans), 8 international. Average age 28. 36 applicants, 47% accepted, 4 enrolled. In 2005, 11 degrees awarded. *Degree requirements:* For master's, thesis, departmental qualifying exam. *Entrance requirements:* For master's, minimum GPA of 2.5 in last 60 units of course work, major in chemistry or related field. Application fee: $55. *Expenses:* Tuition, area resident: Part-time $2,270 per year. Tuition, state resident: full-time $2,572; part-time $339 per unit. Tuition, nonresident: full-time $339; part-time $339 per unit. International tuition: $339 full-time. *Financial support:* Teaching assistantships, career-related internships or fieldwork, Federal Work-Study, institutionally sponsored loans, and scholarships/grants available. Support available to part-time students. Financial award application deadline: 3/1. *Unit head:* Dr. Maria Linder, Chair, 714-278-3621. *Application contact:* Dr. Gregory Williams, Adviser, 714-278-2170.

Colorado School of Mines, Graduate School, Department of Chemistry and Geochemistry and Department of Geology and Geological Engineering, Program in Geochemistry, Golden, CO 80401-1887. Offers MS, PhD. Part-time programs available. *Students:* 10 full-time (5 women), 3 part-time (1 woman). 12 applicants, 83% accepted, 5 enrolled. In 2005, 2 master's awarded. *Degree requirements:* For master's, thesis/dissertation; for doctorate, thesis/dissertation, comprehensive exam. *Entrance requirements:* For master's and doctorate, GRE General Test. Additional exam requirements/recommendations for international students: Required—TOEFL (minimum score 550 paper-based; 213 computer-based). *Application deadline:* For fall admission, 1/1 priority date for domestic students, 1/1 priority date for international students; for spring admission, 9/1 priority date for domestic students, 9/1 priority date for international students. Application fee: $50. Electronic applications accepted. *Expenses:* Tuition, state resident: full-time $7,240; part-time $362 per credit hour. Tuition, nonresident: full-time $19,840; part-time $992 per credit hour. Required fees: $895. *Financial support:* In 2005–06, 7 students received support, including 1 fellowship with full tuition reimbursement available (averaging $9,600 per year), 2 research assistantships with full tuition reimbursements available (averaging $9,600 per year), 4 teaching assistantships with full tuition reimbursements available (averaging $9,600 per year); scholarships/grants and unspecified assistantships also available. *Faculty research:* Geochemical analysis, organic geochemistry, hydrochemical systems, environmental microbiology, process control programming. *Unit head:* John Curtis, Professor, 303-273-3887, Fax: 303-273-3859, E-mail: jbcurtis@mines.edu.

Colorado School of Mines, Graduate School, Department of Geology and Geological Engineering, Golden, CO 80401-1887. Offers engineering geology (Diploma); exploration geosciences (Diploma); geochemistry (MS, PhD); geological engineering (ME, MS, PhD, Diploma); geology (MS, PhD); hydrogeology (Diploma). Part-time programs available. *Faculty:* 28 full-time (7 women), 15 part-time/adjunct (4 women). *Students:* 72 full-time (26 women), 29 part-time (5 women); includes 3 minority (1 American Indian/Alaska Native, 2 Hispanic Americans), 25 international. 108 applicants, 60% accepted, 24 enrolled. In 2005, 18 master's, 4 doctorates awarded. *Degree requirements:* For master's, thesis/dissertation; for doctorate, thesis/dissertation, comprehensive exam. *Entrance requirements:* For master's and doctorate, GRE General Test; for Diploma, GRE General Test, minimum GPA of 3.0. Additional exam requirements/recommendations for international students: Required—TOEFL (minimum score 550 paper-based; 213 computer-based). *Application deadline:* For fall admission, 1/1 for domestic students, 1/1 for international students; for spring admission, 9/1 for domestic students, 9/1 for international students. Application fee: $50. Electronic applications accepted. *Expenses:* Tuition, state resident: full-time $7,240; part-time $362 per credit hour. Tuition, nonresident: full-time $19,840; part-time $992 per credit hour. Required fees: $895. *Financial support:* In 2005–06, 4 fellowships with full tuition reimbursements (averaging $9,600 per year), 14 research assistantships with full tuition reimbursements (averaging $9,600 per year), 11 teaching assistantships with full tuition reimbursements (averaging $9,600 per year) were awarded; scholarships/grants, health care benefits, and unspecified assistantships also available. Financial award applicants required to submit FAFSA. *Faculty research:* Predictive sediment modeling, petrophysics, aquifer-contaminant flow modeling, water-rock interactions, geotechnical engineering. Total annual research expenditures: $2.1 million. *Unit head:* Dr. Murray W. Hitzman, Head, 303-384-2127, Fax: 303-273-3859, E-mail: mhitzman@mines.edu. *Application contact:* Marilyn Schwinger, Administrative Assistant, 303-273-3800, Fax: 303-273-3859, E-mail: mschwing@mines.edu.

Columbia University, Graduate School of Arts and Sciences, Division of Natural Sciences, Department of Earth and Environmental Sciences, New York, NY 10027. Offers geochemistry (M Phil, MA, PhD); geodetic sciences (M Phil, MA, PhD); geophysics (M Phil, MA, PhD); oceanography (M Phil, MA, PhD). *Faculty:* 21 full-time, 19 part-time/adjunct. *Students:* 78 full-time (31 women), 6 part-time (2 women); includes 4 minority (3 Asian Americans or Pacific Islanders, 1 Hispanic American), 31 international. Average age 32. 115 applicants, 20% accepted. In 2005, 16 master's, 11 doctorates awarded. *Degree requirements:* For master's, thesis or alternative, fieldwork, written exam; for doctorate, one foreign language, thesis/dissertation. *Entrance requirements:* For master's and doctorate, GRE General Test, GRE Subject Test, major in natural or physical science. Additional exam requirements/recommendations for international students: Required—TOEFL. Application fee: $75. *Expenses:* Tuition: Full-time $31,448. Tuition and fees vary according to course level, course load, campus/location and program. *Financial support:* Fellowships, teaching assistantships, Federal Work-Study and institutionally sponsored loans available. Support available to part-time students. Financial award application deadline: 1/5; financial award applicants required to submit FAFSA. *Faculty research:* Structural geology and stratigraphy, petrology, paleontology, rare gas, isotope and aqueous geochemistry. *Unit head:* Dr. Nicholas Christie-Blick, Chair, 845-365-8821, Fax: 845-365-8150.

Cornell University, Graduate School, Graduate Fields of Engineering, Field of Geological Sciences, Ithaca, NY 14853-0001. Offers economic geology (M Eng, MS, PhD); engineering geology (M Eng, MS, PhD); environmental geophysics (M Eng, MS, PhD); general geology (M Eng, MS, PhD); geobiology (M Eng, MS, PhD); geochemistry and isotope geology (M Eng, MS, PhD); geohydrology (M Eng, MS, PhD); geomorphology (M Eng, MS, PhD); geophysics (M Eng, MS, PhD); geotectonics (M Eng, MS, PhD); marine geology (MS, PhD); mineralogy (M Eng, MS, PhD); paleontology (M Eng, MS, PhD); petroleum geology (M Eng, MS, PhD); petrology (M Eng, MS, PhD); planetary geology (M Eng, MS, PhD); Precambrian geology (M Eng, MS, PhD); Quaternary geology (M Eng, MS, PhD); rock mechanics (M Eng, MS, PhD); sedimentology (M Eng, MS, PhD); seismology (M Eng, MS, PhD); stratigraphy (M Eng, MS, PhD); structural geology (M Eng, MS, PhD). *Faculty:* 39 full-time (5 women). *Students:* 29 full-time (15 women); includes 2 minority (both Hispanic Americans), 7 international. 73 applicants, 15% accepted, 6 enrolled. In 2005, 5 master's, 1 doctorate awarded. *Degree requirements:* For master's, thesis (MS); for doctorate, thesis/dissertation, comprehensive exam. *Entrance requirements:* For master's and doctorate, GRE General Test, 3 letters of recommendation. Additional exam requirements/recommendations for international students: Required—TOEFL (minimum score 550 paper-based; 213 computer-based). *Application deadline:* For fall admission, 1/15 for domestic students. Applications are processed on a rolling basis. Application fee: $60. Electronic applications accepted. *Financial support:* In 2005–06, 8 fellowships with full tuition reimbursements, 10 research assistantships with full tuition reimbursements, 7 teaching assistantships with full tuition reimbursements were awarded; institutionally sponsored loans, scholarships/grants, health care benefits, tuition waivers (full and partial), and unspecified assistantships also available. Financial award applicants required to submit FAFSA. *Faculty research:* Geophysics, structural geology, petrology, geochemistry, geodynamics. *Unit head:* Director of Graduate Studies, 607-255-5466, Fax: 607-254-4780. *Application contact:* Graduate Field Assistant, 607-255-5466, Fax: 607-254-4780, E-mail: gradprog@geology.cornell.edu.

Georgia Institute of Technology, Graduate Studies and Research, College of Sciences, School of Earth and Atmospheric Sciences, Atlanta, GA 30332-0001. Offers atmospheric chemistry and air pollution (MS, PhD); atmospheric dynamics and climate (MS, PhD); geochemistry (MS, PhD); hydrologic cycle (MS, PhD); ocean sciences (MS, PhD); solid-earth and environmental geophysics (MS, PhD). Part-time programs available. Terminal master's awarded for partial completion of doctoral program. *Degree requirements:* For master's, thesis or alternative; for doctorate, thesis/dissertation, comprehensive exam. *Entrance requirements:* For master's, GRE, minimum GPA of 3.0; for doctorate, GRE General Test, minimum GPA of 2.7. Additional exam requirements/recommendations for international students: Required—TOEFL (minimum score 550 paper-based; 213 computer-based). *Faculty research:* Geophysics, atmospheric chemistry, atmospheric dynamics, seismology.

See Close-Up on page 227.

Indiana University Bloomington, Graduate School, College of Arts and Sciences, Department of Geological Sciences, Bloomington, IN 47405-7000. Offers biogeochemistry (MS, PhD); economic geology (MS, PhD); geobiology (MS, PhD); geophysics, structural geology and tectonics (MS, PhD); hydrogeology (MS, PhD); mineralogy (MS, PhD); stratigraphy and sedimentology (MS, PhD). PhD offered through the University Graduate School. Part-time programs available. *Faculty:* 14 full-time (1 woman). *Students:* 30 full-time (13 women), 12 part-time (5 women); includes 5 minority (2 African Americans, 1 American Indian/Alaska Native, 1 Asian American or Pacific Islander, 1 Hispanic American), 21 international. Average age 29. In 2005, 7 master's, 2 doctorates awarded. Terminal master's awarded for partial completion of doctoral program. *Degree requirements:* For master's, one foreign language, thesis or alternative; for doctorate, thesis/dissertation. *Entrance requirements:* For master's and doctorate, GRE General Test. Additional exam requirements/recommendations for international students: Required—TOEFL. *Application deadline:* For fall admission, 1/15 priority date for domestic students, 12/15 priority date for international students; for spring admission, 9/1 priority date for domestic students, 9/1 priority date for international students. Applications are processed on a rolling basis. Application fee: $50 ($60 for international students). *Expenses:* Tuition, state resident: full-time $5,437; part-time $227 per credit hour. Tuition, nonresident: full-time $15,836; part-time $660 per credit hour. Required fees: $821. Tuition and fees vary according to campus/location and program. *Financial support:* Fellowships with tuition reimbursements, research assistantships with full and partial tuition reimbursements, teaching assistantships with tuition reimbursements, career-related internships or fieldwork, Federal Work-Study, and institutionally sponsored loans available. Financial award application deadline: 2/15. *Faculty research:* Geophysics, geochemistry, hydrogeology, igneous and metamorphic petrology and clay mineralogy. Total annual research expenditures: $289,139. *Unit head:* Mark Person, Graduate Advisor, 812-855-7214. *Application contact:* Mary Iverson, Secretary, Committee for Graduate Studies, 812-855-7214, Fax: 812-855-7899, E-mail: geograd@indiana.edu.

Massachusetts Institute of Technology, School of Science, Department of Earth, Atmospheric, and Planetary Sciences, Cambridge, MA 02139-4307. Offers atmospheric chemistry (PhD, Sc D); atmospheric science (SM, PhD, Sc D); climate physics and chemistry (PhD, Sc D); earth and planetary sciences (SM); geochemistry (PhD, Sc D); geology (PhD, Sc D); geophysics (PhD, Sc D); marine geology and geophysics (SM); oceanography (SM, PhD, Sc D); planetary sciences (PhD, Sc D). *Faculty:* 38 full-time (3 women). *Students:* 165 full-time (63 women); includes 12 minority (7 Asian Americans or Pacific Islanders, 5 Hispanic Americans), 51 international. Average age 27. 179 applicants, 34% accepted, 31 enrolled. In 2005, 2 master's, 15 doctorates awarded. Terminal master's awarded for partial completion of doctoral program. *Degree requirements:* For master's, thesis/dissertation; for doctorate, thesis/dissertation, comprehensive exam. *Entrance requirements:* For master's, GRE General Test; for doctorate, GRE General Test, GRE Subject Test (chemistry or physics for planetary science program). Additional exam requirements/recommendations for international students: Required—TOEFL (minimum score 577 paper-based; 233 computer-based). *Application deadline:* For fall admission, 1/5 for domestic students, 1/5 for international students; for spring admission, 11/1 for domestic students, 11/1 for international students. Application fee: $70. Electronic applications accepted. *Expenses:* Tuition: Full-time $32,100. Required fees: $200. Part-time tuition and fees vary according to course load. *Financial support:* In 2005–06, 115 students received support, including 36 fellowships with tuition reimbursements available (averaging $26,951 per year), 60 research assistantships with tuition reimbursements available (averaging $24,520 per year), 18 teaching assistantships with tuition reimbursements available (averaging $22,773 per year); Federal Work-Study, institutionally sponsored loans, scholarships/grants, health care benefits, and unspecified assistantships also available. *Faculty research:* Evolution of main features of the planetary system; origin, composition, structure, and state of the atmospheres, oceans, surfaces, and interiors of planets; dynamics of planets and satellite motions. Total annual research expenditures: $18.6 million. *Unit head:* Prof. Maria Zuber, Department Head, 617-253-6397, E-mail: zuber@mit.edu. *Application contact:* EAPS Education Office.

McMaster University, School of Graduate Studies, Faculty of Science, School of Geography and Earth Sciences, Hamilton, ON L8S 4M2, Canada. Offers geochemistry (PhD); geology (M Sc, PhD); human geography (MA, PhD); physical geography (M Sc, PhD). Part-time programs available. Terminal master's awarded for partial completion of doctoral program. *Degree requirements:* For master's, thesis/dissertation; for doctorate, thesis/dissertation, comprehensive exam. *Entrance requirements:* For master's, minimum B+ average. Additional exam requirements/recommendations for international students: Required—TOEFL (minimum score 550 paper-based; 213 computer-based).

Montana Tech of The University of Montana, Graduate School, Geoscience Program, Butte, MT 59701-8997. Offers geochemistry (MS); geological engineering (MS); geology (MS); geophysical engineering (MS); hydrogeological engineering (MS); hydrogeology (MS). Part-time programs available. *Faculty:* 17 full-time (3 women). *Students:* 8 full-time (4 women), 3 part-time (1 woman). 15 applicants, 40% accepted, 2 enrolled. In 2005, 5 degrees awarded. *Degree requirements:* For master's, thesis (for some programs), comprehensive exam (for some programs), registration. *Entrance requirements:* For master's, GRE General Test, minimum GPA of 3.0. Additional exam requirements/recommendations for international students: Required—TOEFL (minimum score 525 paper-based; 195 computer-based). *Application deadline:* For fall admission, 4/1 priority date for domestic students, 3/1 priority date for international students; for spring admission, 10/1 priority date for domestic students, 7/1 priority date for international students. Applications are processed on a rolling basis. Application fee: $30. *Expenses:* Tuition, area resident: Full-time $5,133. Tuition, state resident: full-time $5,333. Tuition, nonresident: full-time $15,746. *Financial support:* In 2005–06, 8 students received support, including 3 research assistantships with partial tuition reimbursements available (averaging $3,667 per year), 6 teaching assistantships with partial tuition reimbursements available (averaging $6,667 per year); career-related internships or fieldwork, tuition waivers (full and partial), and unspecified assistantships also available. Financial award application deadline: 4/1; financial award applicants required to submit FAFSA. *Faculty research:*

Water resource development, seismic processing, petroleum reservoir characterization, environmental geochemistry, molecular modeling, magmatic and hydrothermal ore deposits. Total annual research expenditures:$542,109. *Unit head:* Dr. Diane Wolfgram, Department Head, 406-496-4353, Fax: 406-496-4260, E-mail: dwolfgram@mtech.edu. *Application contact:* Cindy Dunstan, Administrator, Graduate School, 406-496-4304, Fax: 406-496-4710, E-mail: cdunstan@mtech.edu.

New Mexico Institute of Mining and Technology, Graduate Studies, Department of Earth and Environmental Science, Program in Geology and Geochemistry, Socorro, NM 87801. Offers geochemistry (MS, PhD); geology (MS, PhD). *Degree requirements:* For master's, thesis optional; for doctorate, thesis/dissertation. *Entrance requirements:* For master's, GRE General Test; for doctorate, GRE General Test, GRE Subject Test. Additional exam requirements/recommendations for international students: Required—TOEFL (minimum score 540 paper-based; 207 computer-based). Electronic applications accepted. *Faculty research:* Care and karst topography, soil/water chemistry and properties, geochemistry of ore deposits.

Ohio University, Graduate Studies, College of Arts and Sciences, Department of Geological Sciences, Athens, OH 45701-2979. Offers environmental geochemistry (MS); environmental geology (MS); environmental/hydrology (MS); geology (MS); geology education (MS); geomorphology/surficial processes (MS); geophysics (MS); hydrogeology (MS); sedimentology (MS); structure/tectonics (MS). Part-time programs available. *Faculty:* 10 full-time (4 women), 4 part-time/adjunct (1 woman). *Students:* 22 full-time (6 women), 3 part-time (2 women); includes 1 minority (Hispanic American), 4 international. Average age 23. 15 applicants, 67% accepted, 8 enrolled. In 2005, 7 degrees awarded. *Degree requirements:* For master's, thesis, thesis proposal defense and thesis defense. *Entrance requirements:* Additional exam requirements/recommendations for international students: Required—TOEFL (minimum score 550 paper-based; 217 computer-based). *Application deadline:* For fall admission, 2/1 priority date for domestic students, 1/1 priority date for international students. Application fee: $45. Electronic applications accepted. *Financial support:* In 2005–06, 18 students received support, including 3 research assistantships with full tuition reimbursements available (averaging $11,900 per year), 13 teaching assistantships with full tuition reimbursements available (averaging $11,900 per year); institutionally sponsored loans, scholarships/grants, tuition waivers (full), and unspecified assistantships also available. Financial award application deadline: 2/1. *Faculty research:* Geoscience education, tectonics, flurial geomorphology, invertebrate paleontology, mine/hydrology. Total annual research expenditures: $649,020. *Unit head:* Dr. David Kidder, Chair, 740-593-1101, Fax: 740-593-0486, E-mail: kidder@ohio.edu. *Application contact:* Dr. David Schneider, Graduate Chair, 740-593-1101, Fax: 740-593-0486, E-mail: schneidd@ohio.edu.

Rensselaer Polytechnic Institute, Graduate School, School of Science, Department of Earth and Environmental Sciences, Troy, NY 12180-3590. Offers environmental chemistry (MS, PhD); geochemistry (MS, PhD); geology (MS, PhD); geophysics (MS, PhD); petrology (MS, PhD). Part-time programs available. Terminal master's awarded for partial completion of doctoral program. *Degree requirements:* For master's, thesis (for some programs), comprehensive exam; for doctorate, thesis/dissertation, comprehensive exam. *Entrance requirements:* For master's and doctorate, GRE General Test. Additional exam requirements/recommendations for international students: Required—TOEFL. Electronic applications accepted. *Expenses:* Tuition: Full-time $31,000; part-time $1,320 per credit. Required fees: $1,623. *Faculty research:* Mantel geochemistry, contaminant geochemistry, seismology, GPS geodesy, remote sensing petrology.

See Close-Up on page 661.

University of California, Los Angeles, Graduate Division, College of Letters and Science, Department of Earth and Space Sciences, Program in Geochemistry, Los Angeles, CA 90095. Offers MS, PhD. *Degree requirements:* For master's, comprehensive exams or thesis; for doctorate, thesis/dissertation, oral and written qualifying exams. *Entrance requirements:* For master's, GRE General Test, minimum GPA of 3.0; for doctorate, GRE General Test, minimum undergraduate GPA of 3.0. Electronic applications accepted.

University of Hawaii at Manoa, Graduate Division, School of Ocean and Earth Science and Technology, Department of Geology and Geophysics, Honolulu, HI 96822. Offers high-pressure geophysics and geochemistry (MS, PhD); hydrogeology and engineering geology (MS, PhD); marine geology and geophysics (MS, PhD); planetary geosciences and remote sensing (MS, PhD); seismology and solid-earth geophysics (MS, PhD); volcanology, petrology, and geochemistry (MS, PhD). *Faculty:* 65 full-time (12 women), 19 part-time/adjunct (1 woman). *Students:* 50 full-time (23 women), 8 part-time (1 woman); includes 2 minority (1 African American, 1 Asian American or Pacific Islander), 15 international. Average age 29. 89 applicants, 18% accepted, 12 enrolled. In 2005, 9 master's, 6 doctorates awarded. Terminal master's awarded for partial completion of doctoral program. *Median time to degree:* Of those who began their doctoral program in fall 1997, 100% received their degree in 8 years or less. *Degree requirements:* For master's, thesis/dissertation; for doctorate, thesis/dissertation, comprehensive exam. *Entrance requirements:* For master's and doctorate, GRE General Test, minimum GPA of 3.0. Additional exam requirements/recommendations for international students: Required—TOEFL. *Application deadline:* For fall admission, 1/15 for domestic students, 1/1 for international students; for spring admission, 9/1 for domestic students, 8/15 for international students. Application fee: $50. *Expenses:* Tuition, state resident: full-time $8,400; part-time $200 per credit hour. Tuition, nonresident: full-time $11,088; part-time $462 per credit hour. Tuition and fees vary according to program. *Financial support:* In 2005–06, 41 research assistantships (averaging $19,366 per year), 5 teaching assistantships (averaging $18,198 per year) were awarded. *Unit head:* Dr. Charles Fletcher, Chair, 808-956-8763, Fax: 808-956-5512, E-mail: fletcher@soest.hawaii.edu. *Application contact:* Dr. Charles Fletcher, Chair, 808-956-8763, Fax: 808-956-5512, E-mail: fletcher@soest.hawaii.edu.

University of Illinois at Chicago, Graduate College, College of Liberal Arts and Sciences, Department of Earth and Environmental Sciences, Chicago, IL 60607-7128. Offers crystallography (MS, PhD); environmental geology (MS, PhD); geochemistry (MS, PhD); geology (MS, PhD); geomorphology (MS, PhD); geophysics (MS, PhD); geotechnical engineering and geosciences (PhD); hydrogeology (MS, PhD); low-temperature and organic geochemistry (MS, PhD); mineralogy (MS, PhD); paleoclimatology (MS, PhD); paleontology (MS, PhD); petrology (MS, PhD); quaternary geology (MS, PhD); sedimentology (MS, PhD); water resources (MS, PhD). *Degree requirements:* For master's and doctorate, thesis/dissertation. *Entrance requirements:* For master's and doctorate, GRE General Test, minimum GPA of 2.75. Additional exam requirements/recommendations for international students: Required—TOEFL. Electronic applications accepted.

University of Illinois at Urbana–Champaign, Graduate College, College of Liberal Arts and Sciences, Department of Geology, Champaign, IL 61820. Offers earth sciences (MS, PhD); geochemistry (MS, PhD); geology (MS, PhD); geophysics (MS, PhD). *Faculty:* 13 full-time (3 women). *Students:* 32 full-time (13 women), 4 part-time; includes 1 minority (American Indian/Alaska Native), 16 international. Average age 26. 44 applicants, 39% accepted, 13 enrolled. In 2005, 4 master's, 5 doctorates awarded. Terminal master's awarded for partial completion of doctoral program. *Degree requirements:* For master's and doctorate, thesis/dissertation. *Entrance requirements:* For master's and doctorate, GRE General Test, minimum GPA of 3.0. Additional exam requirements/recommendations for international students: Required—TOEFL. *Application deadline:* For fall admission, 2/22 for domestic students; for spring admission, 10/16 for domestic students. Applications are processed on a rolling basis. Application fee: $50 ($60 for international students). *Financial support:* In 2005–06, 8 fellowships, 14 research assistantships, 18 teaching assistantships were awarded; Federal Work-Study and tuition waivers (full and partial) also available. Financial award application deadline: 2/15. *Faculty research:* Hydrogeology, structure/tectonics, mineral science. *Unit head:* Stephen Marshak, Head, 217-333-3542, Fax: 217-244-4996, E-mail: smarshak@uiuc.edu. *Application contact:* Barbara Elmore, Graduate Admissions Secretary, 217-333-3542, Fax: 217-244-4996, E-mail: belmore@uiuc.edu.

University of Michigan, Horace H. Rackham School of Graduate Studies, College of Literature, Science, and the Arts, Department of Geological Sciences, Program in Oceanography: Marine Geology and Geochemistry, Ann Arbor, MI 48109. Offers MS, PhD. Terminal master's awarded for partial completion of doctoral program. *Degree requirements:* For master's, thesis; for doctorate, thesis/dissertation, oral defense of dissertation, preliminary exam. *Entrance requirements:* For master's and doctorate, GRE General Test. Electronic applications accepted. *Expenses:* Tuition, state resident: full-time $14,082; part-time $894 per credit hour. Tuition, nonresident: full-time $28,500; part-time $1,675 per credit hour. Required fees: $189; $189 per unit. *Faculty research:* Paleoceanography, paleolimnology, marine geochemistry, seismic stratigraphy.

University of Missouri–Rolla, Graduate School, School of Materials, Energy, and Earth Resources, Department of Geological Sciences and Engineering, Program in Geology and Geophysics, Rolla, MO 65409-0910. Offers geochemistry (MS, PhD); geology (MS, PhD); geophysics (MS, PhD); groundwater and environmental geology (MS, PhD). Part-time programs available. *Degree requirements:* For master's and doctorate, thesis/dissertation. *Entrance requirements:* For master's, GRE General Test, GRE Subject Test, minimum GPA of 3.0 in last 4 semesters; for doctorate, GRE General Test, GRE Subject Test. Additional exam requirements/recommendations for international students: Required—TOEFL. Electronic applications accepted. *Faculty research:* Economic geology, geophysical modeling, seismic wave analysis.

University of Nevada, Reno, Graduate School, College of Science, Mackay School of Earth Sciences and Engineering, Department of Geological Sciences, Reno, NV 89557. Offers geochemistry (MS, PhD); geological engineering (MS, Geol E); geology (MS, PhD); geophysics (MS, PhD). *Degree requirements:* For master's, thesis optional; for doctorate, one foreign language, thesis/dissertation. *Entrance requirements:* For master's, GRE General Test, GRE Subject Test, minimum GPA of 2.75; for doctorate, GRE General Test, GRE Subject Test, minimum GPA of 3.0. Additional exam requirements/recommendations for international students: Required—TOEFL. *Faculty research:* Hydrothermal ore deposits, metamorphic and igneous petrogenesis, sedimentary rock record of earth history, field and petrographic investigation of magnetism, rock fracture mechanics.

University of New Hampshire, Graduate School, College of Engineering and Physical Sciences, Department of Earth Sciences, Durham, NH 03824. Offers earth sciences (MS), including geochemical, geology, ocean mapping, oceanography; hydrology (MS). *Faculty:* 29 full-time. *Students:* 17 full-time (5 women), 24 part-time (10 women); includes 2 minority (1 African American, 1 Asian American or Pacific Islander), 6 international. Average age 31. 37 applicants, 84% accepted, 10 enrolled. In 2005, 10 master's awarded. *Degree requirements:* For master's, thesis. *Entrance requirements:* For master's, GRE General Test. Additional exam requirements/recommendations for international students: Required—TOEFL (minimum score 550 paper-based; 213 computer-based); Recommended—TSE. *Application deadline:* For fall admission, 4/1 priority date for domestic students, 4/1 priority date for international students. For winter admission, 12/1 for domestic students. Applications are processed on a rolling basis. Application fee: $60. Electronic applications accepted. *Expenses:* Tuition, state resident: full-time $8,010; part-time $445 per credit hour. Tuition, nonresident: full-time $19,730; part-time $810 per credit hour. Required fees: $322 per semester. Tuition and fees vary according to course load and program. *Financial support:* In 2005–06, 1 fellowship, 9 research assistantships, 9 teaching assistantships were awarded; career-related internships or fieldwork, Federal Work-Study, scholarships/grants, and tuition waivers (full and partial) also available. Support available to part-time students. Financial award application deadline: 2/15. *Unit head:* Dr. Matt Davis, Chairperson, 603-862-1718, E-mail: earth.sciences@unh.edu. *Application contact:* Nancy Gauthier, Administrative Assistant, 603-862-1720, E-mail: earth.sciences@unh.edu.

Washington State University, Graduate School, College of Sciences, Department of Geology, Pullman, WA 99164. Offers hydrology (MS, PhD); mineralology-petrology geochemistry (MS); minerology-petrology geochemistry (PhD); sedimentology-stratogeography (MS, PhD); structural geology-tectonics (MS, PhD). *Faculty:* 12. *Students:* 22 full-time (8 women), 4 part-time (1 woman), 5 international. Average age 29. 39 applicants, 49% accepted, 7 enrolled. In 2005, 5 master's, 1 doctorate awarded. *Degree requirements:* For master's, thesis, oral exam; for doctorate, one foreign language, thesis/dissertation, oral exam, written exam. *Entrance requirements:* For master's and doctorate, GRE General Test, minimum GPA of 3.0, 3 letters of recommendation. Additional exam requirements/recommendations for international students: Required—TOEFL (minimum score 560 paper-based; 220 computer-based). *Application deadline:* For fall admission, 2/1 priority date for domestic students, 2/1 priority date for international students; for spring admission, 9/1 priority date for domestic students, 7/1 priority date for international students. Applications are processed on a rolling basis. Application fee: $35. Electronic applications accepted. *Expenses:* Tuition, state resident: full-time $6,295; part-time $336 per credit. Tuition, nonresident: full-time $15,949; part-time $819 per credit. Required fees: $933. Part-time tuition and fees vary according to campus/location and program. *Financial support:* In 2005–06, 4 fellowships (averaging $2,700 per year), 5 research assistantships with full and partial tuition reimbursements (averaging $14,200 per year), 18 teaching assistantships with full and partial tuition reimbursements (averaging $13,831 per year) were awarded; career-related internships or fieldwork, Federal Work-Study, institutionally sponsored loans, and scholarships/grants also available. Financial award application deadline: 2/1; financial award applicants required to submit FAFSA. *Faculty research:* Genesis of ore deposits, geohydrology of the Pacific Northwest, geochemistry and petrology of plateau basalts. Total annual research expenditures: $379,240. *Unit head:* Dr. Peter Larson, Chair, 509-335-3009, Fax: 509-335-7816, E-mail: plarson@wsu.edu. *Application contact:* E-mail: geology@wsu.edu.

Washington University in St. Louis, Graduate School of Arts and Sciences, Department of Earth and Planetary Sciences, St. Louis, MO 63130-4899. Offers earth and planetary sciences (MA); geochemistry (PhD); geology (MA, PhD); geophysics (PhD); planetary sciences (PhD). Terminal master's awarded for partial completion of doctoral program. *Degree requirements:* For master's and doctorate, thesis/dissertation. *Entrance requirements:* For master's and doctorate, GRE General Test. Electronic applications accepted.

Woods Hole Oceanographic Institution, MIT/WHOI Joint Program in Oceanography/Applied Ocean Science and Engineering, Woods Hole, MA 02543-1541. Offers applied ocean sciences (PhD); biological oceanography (PhD, Sc D); chemical oceanography (PhD, Sc D); civil and environmental and oceanographic engineering (PhD); electrical and oceanographic engineering (PhD); geochemistry (PhD); geophysics (PhD); marine biology (PhD); marine geochemistry (PhD, Sc D); marine geology (PhD, Sc D); marine geophysics (PhD); mechanical and oceanographic engineering (PhD); ocean engineering (PhD); oceanographic engineering (M Eng, MS, PhD, Sc D, Eng); paleoceanography (PhD); physical oceanography (PhD, Sc D). MS, PhD, and Sc D offered jointly with MIT. Terminal master's awarded for partial completion of doctoral program. *Degree requirements:* For master's and Eng, thesis (for some programs);

Geochemistry

Woods Hole Oceanographic Institution (continued)
for doctorate, thesis/dissertation. *Entrance requirements:* For master's, GRE General Test; for doctorate, GRE General Test, GRE Subject Test. Additional exam requirements/recommendations for international students: Required—TOEFL. Electronic applications accepted.

See Close-Up on page 259.

Wright State University, School of Graduate Studies, College of Science and Mathematics, Department of Geological Sciences, Program in Geological Sciences, Dayton, OH 45435. Offers environmental geochemistry (MS); environmental geology (MS); environmental sciences (MS); geological sciences (MS); geophysics (MS); hydrogeology (MS); petroleum geology (MS). Part-time programs available. *Degree requirements:* For master's, thesis. *Entrance requirements:* Additional exam requirements/recommendations for international students: Required—TOEFL.

Yale University, Graduate School of Arts and Sciences, Department of Geology and Geophysics, New Haven, CT 06520. Offers geochemistry (PhD); geophysics (PhD); meteorology (PhD); mineralogy and crystallography (PhD); oceanography (PhD); paleoecology (PhD); paleontology and stratigraphy (PhD); petrology (PhD); structural geology (PhD). *Degree requirements:* For doctorate, thesis/dissertation. *Entrance requirements:* For doctorate, GRE General Test. Additional exam requirements/recommendations for international students: Required—TOEFL.

Geodetic Sciences

Columbia University, Graduate School of Arts and Sciences, Division of Natural Sciences, Department of Earth and Environmental Sciences, New York, NY 10027. Offers geochemistry (M Phil, MA, PhD); geodetic sciences (M Phil, MA, PhD); geophysics (M Phil, MA, PhD); oceanography (M Phil, MA, PhD). *Faculty:* 21 full-time, 19 part-time/adjunct. *Students:* 78 full-time (31 women), 6 part-time (2 women); includes 4 minority (3 Asian Americans or Pacific Islanders, 1 Hispanic American), 31 international. Average age 32. 115 applicants, 20% accepted. In 2005, 16 master's, 11 doctorates awarded. *Degree requirements:* For master's, thesis or alternative, fieldwork, written exam; for doctorate, one foreign language, thesis/dissertation. *Entrance requirements:* For master's and doctorate, GRE General Test, GRE Subject Test, major in natural or physical science. Additional exam requirements/recommendations for international students: Required—TOEFL. Application fee: $75. *Expenses:* Tuition: Full-time $31,448. Tuition and fees vary according to course level, course load, campus/location and program. *Financial support:* Fellowships, teaching assistantships, Federal Work-Study and institutionally sponsored loans available. Support available to part-time students. Financial award application deadline: 1/5; financial award applicants required to submit FAFSA. *Faculty research:* Structural geology and stratigraphy, petrology, paleontology, rare gas, isotope and aqueous geochemistry. *Unit head:* Dr. Nicholas Christie-Blick, Chair, 845-365-8821, Fax: 845-365-8150.

George Mason University, College of Science, Fairfax, VA 22030. Offers bioinformatics (MS, PhD); climate dynamics (PhD); computational sciences (MS); computational sciences and informatics (PhD); computational social science (PhD); computational techniques and applications (Certificate); earth systems and geoinformation science (PhD); earth systems science (MS); nanotechnology and nanoscience (Certificate); neuroscience (PhD); physical sciences (PhD); remote sensing and earth image processing (Certificate). Part-time and evening/weekend programs available. *Degree requirements:* For doctorate, thesis/dissertation, comprehensive exam, registration. *Entrance requirements:* For master's and doctorate, GRE General Test, minimum GPA of 3.0 in last 60 hours. Additional exam requirements/recommendations for international students: Required—TOEFL. Electronic applications accepted. *Expenses:* Tuition, area resident: Full-time $5,244; part-time $219 per credit. Tuition, state resident: part-time $651 per credit. Tuition, nonresident: full-time $15,636. Required fees: $1,524; $65 per credit. *Faculty research:* Space sciences and astrophysics, fluid dynamics, materials modeling and simulation, bioinformatics, global changes and statistics.

See Close-Up on page 501.

The Ohio State University, Graduate School, College of Engineering, Program in Geodetic Science and Surveying, Columbus, OH 43210. Offers MS, PhD. *Degree requirements:* For master's, thesis optional; for doctorate, thesis/dissertation. *Entrance requirements:* For master's and doctorate, GRE General Test. Additional exam requirements/recommendations for international students: Recommended—TOEFL (minimum score 600 paper-based; 250 computer-based). Electronic applications accepted. *Faculty research:* Photogrammetry, cartography, geodesy, land information systems.

Université Laval, Faculty of Forestry and Geomatics, Department of Geomatics Sciences, Programs in Geomatics Sciences, Québec, QC G1K 7P4, Canada. Offers M Sc, PhD. Terminal master's awarded for partial completion of doctoral program. *Degree requirements:* For master's, thesis (for some programs); for doctorate, thesis/dissertation, comprehensive exam. *Entrance requirements:* For master's and doctorate, knowledge of French and English. Electronic applications accepted.

University of New Brunswick Fredericton, School of Graduate Studies, Faculty of Engineering, Department of Geodesy and Geomatics, Fredericton, NB E3B 5A3, Canada. Offers land information management (Diploma); mapping, charting and geodesy (Diploma); surveying engineering (M Eng, M Sc E, PhD). Part-time programs available. *Degree requirements:* For master's, thesis; for doctorate, thesis/dissertation, qualifying exam. *Entrance requirements:* For master's and doctorate, minimum GPA of 3.0. Additional exam requirements/recommendations for international students: Required—TOEFL, TWE.

Geology

Acadia University, Faculty of Pure and Applied Science, Department of Geology, Wolfville, NS B4P 2R6, Canada. Offers M Sc. *Faculty:* 6 full-time (2 women). *Students:* 8 full-time (3 women), 1 (woman) part-time. Average age 24. 3 applicants, 100% accepted, 3 enrolled. In 2005, 4 degrees awarded. *Degree requirements:* For master's, thesis. *Entrance requirements:* For master's, bachelor of science in geology or equivalent. Additional exam requirements/recommendations for international students: Required—TOEFL (minimum score 580 paper-based; 237 computer-based). *Application deadline:* For fall admission, 2/1 priority date for domestic students, 2/1 priority date for international students. Applications are processed on a rolling basis. Application fee: $50. *Financial support:* In 2005–06, 6 students received support, including 5 teaching assistantships (averaging $8,500 per year); scholarships/grants also available. Financial award application deadline: 2/1. *Faculty research:* Igneous, metamorphic, and Quaternary geology; stratigraphy; remote sensing; sentimentology. Total annual research expenditures: $318,013. *Unit head:* Dr. Robert Raeside, Head, 902-585-1208, Fax: 902-585-1816, E-mail: geology@acadiau.ca. *Application contact:* Dr. Sandra Barr, Graduate Coordinator, 902-585-1340, Fax: 902-585-1816, E-mail: sandra.barr@acadiau.ca.

American University of Beirut, Graduate Programs, Faculty of Arts and Sciences, Beirut, Lebanon. Offers anthropology (MA); Arabic language and literature (MA); archaeology (MA); biology (MS); business administration (MBA); chemistry (MS); computer science (MS); economics (MA); education (MA); English language (MA); English literature (MA); environmental policy planning (MSES); finance and banking (MFB); financial economics (MFE); geology (MS); history (MA); mathematics (MS); Middle Eastern studies (MA); philosophy (MA); physics (MS); political studies (MA); psychology (MA); public administration (MA); sociology (MA). *Degree requirements:* For master's, one foreign language, thesis (for some programs), comprehensive exam, registration. *Entrance requirements:* For master's, GRE, letter of recommendation.

Auburn University, Graduate School, College of Sciences and Mathematics, Department of Geology and Geography, Auburn University, AL 36849. Offers MS. Part-time programs available. *Faculty:* 13 full-time (1 woman). *Students:* 19; includes 3 minority (1 American Indian/Alaska Native, 1 Asian American or Pacific Islander, 1 Hispanic American), 1 international. 9 applicants, 100% accepted, 3 enrolled. In 2005, 6 degrees awarded. *Degree requirements:* For master's, computer language or Geographic Information Systems, field camp. *Entrance requirements:* For master's, GRE General Test. *Application deadline:* For fall admission, 7/7 for domestic students; for spring admission, 11/24 for domestic students. Applications are processed on a rolling basis. Application fee: $25 ($50 for international students). Electronic applications accepted. *Financial support:* Research assistantships, teaching assistantships, Federal Work-Study available. Support available to part-time students. Financial award application deadline: 3/15. *Faculty research:* Empirical magma dynamics and melt migration, ore mineralogy, role of terrestrial plant biomass in deposition, metamorphic petrology and isotope geochemistry, reef development, crinoid taphology. *Unit head:* Dr. Charles E. Savrda, Chair, 334-844-4282. *Application contact:* Dr. Stephen L. McFarland, Acting Dean of the Graduate School, 334-844-4700.

Announcement: The master's program offers a broad-based curriculum that takes advantage of Auburn's location on the boundary between the Appalachian front and the Gulf Coastal Plain. Low student-faculty ratio results in small class size, close relationships between students and faculty members, and relaxed and informal atmosphere. Research opportunities include hydrology and aqueous and hydrothermal geochemistry; economic geology, petrochemistry, and tectonics of the southern Appalachians; environmental geophysics; coastal-plain stratigraphy/sedimentology; and taphonomy and paleoecology of invertebrate assemblages. Current investigations extend beyond the region to other areas in the US as well as the Bahamas, Scandinavia, and Himalayan foreland basins. For additional information, visit the Web site: http://www.auburn.edu/academic/science_math/geology/docs.

Ball State University, Graduate School, College of Sciences and Humanities, Department of Geology, Muncie, IN 47306-1099. Offers MA, MS. *Faculty:* 6. *Students:* 3 full-time (2 women), 2 part-time; includes 2 minority (1 Asian American or Pacific Islander, 1 Hispanic American). Average age 39. 2 applicants, 100% accepted, 1 enrolled. In 2005, 4 degrees awarded. *Degree requirements:* For master's, thesis (for some programs). *Entrance requirements:* For master's, GRE General Test. Application fee: $25 ($35 for international students). *Expenses:* Tuition, state resident: full-time $6,246. Tuition, nonresident: full-time $16,006. *Financial support:* In 2005–06, 5 teaching assistantships with full tuition reimbursements (averaging $8,924 per year) were awarded; research assistantships, career-related internships or fieldwork also available. Financial award application deadline: 3/1. *Faculty research:* Environmental geology, geophysics, stratigraphy. *Unit head:* Dr. Jeffry Grigsby, Chairman, 765-285-8270, Fax: 765-285-8265. *Application contact:* Scott Rice-Snow, Director of Graduate Programs, 765-285-8270, Fax: 765-285-8265, E-mail: ricesnow@bsu.edu.

Baylor University, Graduate School, College of Arts and Sciences, Department of Geology, Waco, TX 76798. Offers earth science (MA); geology (MS, PhD). *Faculty:* 12 full-time (1 woman). *Students:* 14 full-time (7 women), 4 part-time (2 women); includes 1 minority (Hispanic American), 1 international. In 2005, 10 master's, 1 doctorate awarded. *Degree requirements:* For master's and doctorate, thesis/dissertation. *Entrance requirements:* For master's and doctorate, GRE General Test. *Application deadline:* For fall admission, 3/15 for domestic students. Applications are processed on a rolling basis. Application fee: $25. *Financial support:* In 2005–06, 18 teaching assistantships were awarded; Federal Work-Study and institutionally sponsored loans also available. *Faculty research:* Petroleum geology, geophysics, engineering geology, hydrogeology. *Unit head:* Dr. Stacy Atchley, Graduate Program Director, 254-710-2196, Fax: 254-710-2673, E-mail: stacy_atchley@baylor.edu. *Application contact:* Suzanne Keener, Administrative Assistant, 254-710-3588, Fax: 254-710-3870.

Boise State University, Graduate College, College of Arts and Sciences, Department of Geosciences, Program in Geology, Boise, ID 83725-0399. Offers MS. Part-time programs available. *Degree requirements:* For master's, thesis. *Entrance requirements:* For master's, GRE General Test, BS in related field, minimum GPA of 3.0. Electronic applications accepted.

Boston College, Graduate School of Arts and Sciences, Department of Geology and Geophysics, Chestnut Hill, MA 02467-3800. Offers MS, MBA/MS. *Students:* 13 full-time (5 women), 8 part-time (4 women), 3 international. 14 applicants, 86% accepted, 9 enrolled. In 2005, 4 degrees awarded. *Degree requirements:* For master's, thesis. *Entrance requirements:* For master's, GRE General Test, GRE Subject Test. Additional exam requirements/recommendations for international students: Required—TOEFL (minimum score 550 paper-based; 213 computer-based). *Application deadline:* For fall admission, 1/15 for domestic students. Application fee: $70. Electronic applications accepted. *Financial support:* Research assistantships with full tuition reimbursements, teaching assistantships with full tuition reimbursements, Federal Work-Study available. Support available to part-time students. Financial award application deadline: 3/1; financial award applicants required to submit FAFSA. *Faculty research:* Coastal and marine geology, experimental sedimentology, geomagnetism, igneous petrology, paleontology.

Unit head: Dr. Alan Kafka, Chairperson, 617-552-3650. *Application contact:* Dr. John Ebel, Graduate Program Director, 617-552-3640, E-mail: john.ebel@bc.edu.

See Close-Up on page 223.

Bowling Green State University, Graduate College, College of Arts and Sciences, Department of Geology, Bowling Green, OH 43403. Offers MS. Part-time programs available. *Faculty:* 12 full-time (2 women), 8 part-time/adjunct (1 woman). *Students:* 23 full-time (5 women), 2 part-time, 4 international. Average age 26. 34 applicants, 91% accepted, 7 enrolled. In 2005, 8 degrees awarded. *Degree requirements:* For master's, thesis. *Entrance requirements:* For master's, GRE General Test. Additional exam requirements/recommendations for international students: Required—TOEFL. *Application deadline:* For fall admission, 3/1 for domestic students; for spring admission, 11/1 for domestic students. Application fee: $30. Electronic applications accepted. *Financial support:* In 2005–06, 2 research assistantships with full tuition reimbursements (averaging $10,777 per year), 15 teaching assistantships with full tuition reimbursements (averaging $15,432 per year) were awarded; career-related internships or fieldwork, institutionally sponsored loans, tuition waivers (full), and unspecified assistantships also available. Financial award applicants required to submit FAFSA. *Faculty research:* Remote sensing, environmental geology, geological information systems, structural geology, geochemistry. *Unit head:* Dr. Charles Onasch, Chair, 419-372-7197. *Application contact:* Dr. Sheila Roberts, Graduate Coordinator, 419-372-0354.

Brigham Young University, Graduate Studies, College of Physical and Mathematical Sciences, Department of Geology, Provo, UT 84602-1001. Offers MS. *Faculty:* 13 full-time (0 women). *Students:* 30 full-time (12 women); includes 1 minority (African American) Average age 23. 14 applicants, 71% accepted, 7 enrolled. In 2005, 16 degrees awarded. *Degree requirements:* For master's, thesis. *Entrance requirements:* For master's, GRE General Test, minimum GPA of 3.0 in last 60 hours of course work. Additional exam requirements/recommendations for international students: Required—TOEFL. *Application deadline:* For fall admission, 2/1 priority date for domestic students, 2/1 priority date for international students. For winter admission, 9/15 for domestic students. Applications are processed on a rolling basis. Application fee: $50. *Financial support:* In 2005–06, 20 students received support, including 1 research assistantship (averaging $12,000 per year), 7 teaching assistantships (averaging $12,000 per year); fellowships, career-related internships or fieldwork, institutionally sponsored loans, scholarships/grants, and tuition waivers (partial) also available. Financial award application deadline: 2/1. *Faculty research:* Regional tectonics, hydrogeochemistry, crystal chemistry and crystallography, stratigraphy, environmental geophysics. Total annual research expenditures: $88,051. *Unit head:* Dr. Jeffrey D. Keith, Chairman, 801-422-2189, Fax: 801-422-0267, E-mail: jeffrey_keith@byu.edu. *Application contact:* Dr. Michael J. Dorais, Graduate Coordinator, 801-422-1347, Fax: 801-422-0267, E-mail: michael_dorais@byu.edu.

Brooklyn College of the City University of New York, Division of Graduate Studies, Department of Geology, Brooklyn, NY 11210-2889. Offers applied geology (MA); geology (MA, PhD). The department offers courses at Brooklyn College that are creditable toward the CUNY doctoral degree (with permission of the executive officer of the doctoral program). Evening/weekend programs available. Terminal master's awarded for partial completion of doctoral program. *Degree requirements:* For master's, qualifying exams. *Entrance requirements:* For master's, GRE or qualifying exam, bachelor's degree in geology or equivalent, fieldwork, 2 letters of recommendation; for doctorate, GRE. Additional exam requirements/recommendations for international students: Required—TOEFL. *Faculty research:* Geochemistry, petrology, tectonophysics, hydrogeology, sedimentary geology, environmental geology.

California Institute of Technology, Division of Geological and Planetary Sciences, Pasadena, CA 91125-0001. Offers cosmochemistry (PhD); geobiology (PhD); geochemistry (MS, PhD); geology (MS, PhD); geophysics (MS, PhD); planetary science (MS, PhD). Part-time programs available. *Degree requirements:* For doctorate, thesis/dissertation. *Entrance requirements:* For doctorate, GRE General Test. Additional exam requirements/recommendations for international students: Required—TOEFL. Electronic applications accepted. *Faculty research:* Astronomy, evolution of anaerobic respiratory processes, structural geology and tectonics, theoretical and numerical seismology, global biogeochemical cycles.

California State University, Bakersfield, Division of Graduate Studies and Research, School of Natural Sciences and Mathematics, Program in Geology, Bakersfield, CA 93311-1022. Offers geology (MS); hydrology (MS). Part-time and evening/weekend programs available. *Faculty:* 5 full-time (1 woman). *Students:* 2 full-time (1 woman), 7 part-time (3 women); includes 2 minority (1 Asian American or Pacific Islander, 1 Hispanic American). Average age 33. In 2005, 4 degrees awarded. *Degree requirements:* For master's, thesis. *Entrance requirements:* For master's, GRE General Test, BS in geology. *Application deadline:* Applications are processed on a rolling basis. Application fee: $55. *Financial support:* Career-related internships or fieldwork and institutionally sponsored loans available. *Unit head:* Dr. Dirk Baron, Graduate Coordinator, 661-664-3044, Fax: 661-664-2040, E-mail: dbaron@csub.edu.

California State University, Chico, Graduate School, College of Natural Sciences, Department of Geological and Environmental Sciences, Chico, CA 95929-0205. Offers environmental science (MS); geosciences (MS), including hydrology/hydrogeology. *Degree requirements:* For master's, thesis, competency exam. *Entrance requirements:* For master's, GRE General Test. Additional exam requirements/recommendations for international students: Required—TOEFL (minimum score 550 paper-based; 213 computer-based). Electronic applications accepted.

California State University, East Bay, Academic Programs and Graduate Studies, College of Science, Department of Geological Sciences, Hayward, CA 94542-3000. Offers geology (MS). Evening/weekend programs available. *Students:* 5. 3 applicants, 0% accepted. *Degree requirements:* For master's, thesis. *Entrance requirements:* For master's, GRE, minimum GPA of 2.75 in field, 2.5 overall. Additional exam requirements/recommendations for international students: Required—TOEFL (minimum score 550 paper-based; 213 computer-based). *Application deadline:* For fall admission, 5/31 for domestic students, 4/30 for international students. For winter admission, 9/30 for domestic students. Applications are processed on a rolling basis. Application fee: $55. Electronic applications accepted. *Financial support:* Career-related internships or fieldwork, Federal Work-Study, and institutionally sponsored loans available. Support available to part-time students. Financial award application deadline: 3/2. *Unit head:* Dr. Dietz Warnke, Chair, 510-885-4716, E-mail: dietz.warnke@csueastbay.edu. *Application contact:* Deborah Baker, Associate Director, 510-885-3286, Fax: 510-885-4777, E-mail: deborah.baker@csueastbay.edu.

California State University, Fresno, Division of Graduate Studies, College of Science and Mathematics, Department of Earth and Environmental Sciences, Fresno, CA 93740-8027. Offers geology (MS). Part-time programs available. *Degree requirements:* For master's, thesis. *Entrance requirements:* For master's, GRE General Test, undergraduate geology degree, minimum GPA of 2.7. Additional exam requirements/recommendations for international students: Required—TOEFL. Electronic applications accepted. *Faculty research:* Water drainage, pollution, cartography, creek restoration, nitrate contamination.

California State University, Fullerton, Graduate Studies, College of Natural Science and Mathematics, Department of Geological Sciences, Fullerton, CA 92834-9480. Offers MS. Part-time programs available. *Students:* 1 full-time (0 women), 16 part-time (8 women); includes 4 minority (1 Asian American or Pacific Islander, 3 Hispanic Americans). Average age 38. 10 applicants, 70% accepted, 5 enrolled. In 2005, 3 degrees awarded. *Degree requirements:* For master's, thesis. *Entrance requirements:* For master's, bachelor's degree in geology, minimum GPA of 3.0 in geology courses. Application fee: $55. *Expenses:* Tuition, nonresident: part-time $339 per unit. *Unit head:* Dr. Diane Clemens-Knott, Chair, 714-278-3882. *Application contact:* Dr. Brady Rhodes, Adviser, 714-278-2942.

California State University, Long Beach, Graduate Studies, College of Natural Sciences and Mathematics, Department of Geological Sciences, Long Beach, CA 90840. Offers geology/geosciences (MS). Part-time programs available. *Faculty:* 8 full-time (3 women), 2 part-time/adjunct (0 women). *Students:* 2 full-time (both women), 6 part-time (4 women). Average age 37. 7 applicants, 71% accepted, 3 enrolled. In 2005, 1 degree awarded. *Degree requirements:* For master's, thesis. *Entrance requirements:* For master's, GRE General Test. *Application deadline:* For fall admission, 7/1 for domestic students; for spring admission, 12/1 for domestic students. Applications are processed on a rolling basis. Application fee: $55. Electronic applications accepted. *Expenses:* Tuition, nonresident: part-time $339 per semester hour. *Financial support:* Research assistantships, teaching assistantships, Federal Work-Study, institutionally sponsored loans, and scholarships/grants available. Financial award application deadline: 3/2. *Faculty research:* Paleontology, geophysics, structural geology, organic geochemistry, sedimentary geology. *Unit head:* Dr. Stanley C. Finney, Chair, 562-985-4809, Fax: 562-985-8638, E-mail: scfinney@csulb.edu. *Application contact:* Dr. R. Dan Francis, Graduate Coordinator, 562-985-4929, Fax: 562-985-8638, E-mail: rfrancis@csulb.edu.

California State University, Los Angeles, Graduate Studies, College of Natural and Social Sciences, Department of Geological Sciences, Los Angeles, CA 90032-8530. Offers MS. Part-time and evening/weekend programs available. *Faculty:* 6 full-time (0 women), 2 part-time/adjunct (0 women). *Students:* 3 full-time (1 woman), 11 part-time (5 women); includes 1 minority (Asian American or Pacific Islander) In 2005, 2 degrees awarded. *Degree requirements:* For master's, comprehensive exam or thesis. *Entrance requirements:* Additional exam requirements/recommendations for international students: Required—TOEFL. *Application deadline:* For fall admission, 6/30 for domestic students; for spring admission, 2/1 for domestic students. Applications are processed on a rolling basis. Application fee: $55. *Financial support:* Federal Work-Study available. Support available to part-time students. Financial award application deadline: 3/1. *Unit head:* Dr. Pedro Ramirez, Chair, 323-343-2400, Fax: 323-343-5609.

California State University, Northridge, Graduate Studies, College of Science and Mathematics, Department of Geological Sciences, Northridge, CA 91330. Offers MS. Part-time and evening/weekend programs available. *Degree requirements:* For master's, thesis. *Entrance requirements:* For master's, GRE General Test, minimum GPA of 2.75. Additional exam requirements/recommendations for international students: Required—TOEFL. *Faculty research:* Petrology of California Miocene volcanics, sedimentology of California Miocene formations, Eocene gastropods, structure of White/Inyo Mountains, seismology of Californian and Mexican earthquakes.

Case Western Reserve University, School of Graduate Studies, Department of Geological Sciences, Cleveland, OH 44106. Offers MS, PhD. Part-time programs available. Terminal master's awarded for partial completion of doctoral program. *Degree requirements:* For master's, thesis or alternative; for doctorate, thesis/dissertation. *Entrance requirements:* For master's and doctorate, GRE General Test, GRE Subject Test. Additional exam requirements/recommendations for international students: Required—TOEFL. *Faculty research:* Geochemistry, hydrology, geochronology, paleoclimates, geomorphology.

Central Washington University, Graduate Studies, Research and Continuing Education, College of the Sciences, Department of Geological Sciences, Ellensburg, WA 98926. Offers MS. Part-time programs available. *Faculty:* 14 full-time (5 women). *Students:* 11 full-time (5 women), 3 part-time. 5 applicants, 100% accepted, 5 enrolled. In 2005, 5 degrees awarded. *Degree requirements:* For master's, thesis. *Entrance requirements:* For master's, GRE General Test, minimum GPA of 3.0. Additional exam requirements/recommendations for international students: Required—TOEFL (minimum score 550 paper-based; 213 computer-based). *Application deadline:* For fall admission, 4/1 for domestic students. For winter admission, 10/1 for domestic students; for spring admission, 1/1 for domestic students. Applications are processed on a rolling basis. Application fee: $50. *Expenses:* Tuition, state resident: full-time $1,968; part-time $197 per credit. Tuition, nonresident: full-time $4,320; part-time $432 per credit. Required fees: $623. Tuition and fees vary according to degree level. *Financial support:* In 2005–06, 4 research assistantships with partial tuition reimbursements (averaging $8,100 per year), 9 teaching assistantships with partial tuition reimbursements (averaging $8,100 per year) were awarded; career-related internships or fieldwork and Federal Work-Study also available. Financial award application deadline: 3/1; financial award applicants required to submit FAFSA. *Unit head:* Dr. Lisa Ely, Chair, 509-963-2701. *Application contact:* Justine Eason, Admissions Program Coordinator, 509-963-3103, Fax: 509-963-1799, E-mail: masters@cwu.edu.

Colorado School of Mines, Graduate School, Department of Geology and Geological Engineering, Golden, CO 80401-1887. Offers engineering geology (Diploma); exploration geosciences (Diploma); geochemistry (MS, PhD); geological engineering (ME, MS, PhD, Diploma); geology (MS, PhD); hydrogeology (Diploma). Part-time programs available. *Faculty:* 28 full-time (7 women), 15 part-time/adjunct (4 women). *Students:* 72 full-time (26 women), 29 part-time (5 women); includes 3 minority (1 American Indian/Alaska Native, 2 Hispanic Americans), 25 international. 108 applicants, 60% accepted, 24 enrolled. In 2005, 18 master's, 4 doctorates awarded. *Degree requirements:* For master's, thesis/dissertation; for doctorate, thesis/dissertation, comprehensive exam. *Entrance requirements:* For master's and doctorate, GRE General Test; for Diploma, GRE General Test, minimum GPA of 3.0. Additional exam requirements/recommendations for international students: Required—TOEFL (minimum score 550 paper-based; 213 computer-based). *Application deadline:* For fall admission, 1/1 for domestic students, 1/1 for international students; for spring admission, 9/1 for domestic students, 9/1 for international students. Application fee: $50. Electronic applications accepted. *Expenses:* Tuition, state resident: full-time $7,240; part-time $362 per credit hour. Tuition, nonresident: full-time $19,840; part-time $992 per credit hour. Required fees: $895. *Financial support:* In 2005–06, 4 fellowships with full tuition reimbursements (averaging $9,600 per year), 14 research assistantships with full tuition reimbursements (averaging $9,600 per year), 11 teaching assistantships with full tuition reimbursements (averaging $9,600 per year) were awarded; scholarships/grants, health care benefits, and unspecified assistantships also available. Financial award applicants required to submit FAFSA. *Faculty research:* Predictive sediment modeling, petrophysics, aquifer-contaminant flow modeling, water-rock interactions, geotechnical engineering. Total annual research expenditures: $2.1 million. *Unit head:* Dr. Murray W. Hitzman, Head, 303-384-2127, Fax: 303-273-3859, E-mail: mhitzman@mines.edu. *Application contact:* Marilyn Schwinger, Administrative Assistant, 303-273-3800, Fax: 303-273-3859, E-mail: mschwing@mines.edu.

Colorado State University, Graduate School, Warner College of Natural Resources, Department of Geosciences, Fort Collins, CO 80523-0015. Offers earth sciences (PhD); geology (MS), including geomorphology, geophysics, hydrogeology, petrology/geochemistry and economic geology, sedimentology, structural geology; geosciences (PhD); wateshed (PhD). Part-time programs available. *Faculty:* 8 full-time (3 women). *Students:* 17 full-time (11 women), 22 part-time (7 women); includes 1 minority (American Indian/Alaska Native), 2 international. Average age 32. 28 applicants, 61% accepted, 7 enrolled. In 2005, 3 master's, 1 doctorate awarded. *Degree requirements:* For master's, thesis/dissertation, registration; for doctorate, thesis/dissertation, comprehensive exam, registration. *Entrance requirements:* For master's and doctorate, GRE General Test, minimum GPA of 3.0. Additional exam requirements/recommendations for international students: Required—TOEFL (minimum score 550 paper-based; 213 computer-based). *Application deadline:* For fall admission, 2/1 priority date for domestic students, 2/1 priority date for international students. Applications are processed on a rolling basis. Application fee: $50. Electronic applications accepted. *Expenses:* Tuition, state resident: full-time $3,690; part-time $205 per credit. Tuition, nonresident: full-time $14,958; part-time $831 per credit. Required fees: $1,061. *Financial support:* In 2005–06, 6 fellowships (averaging $9,000 per year), 13 research assistantships with partial tuition reimbursements (averaging $16,000 per year), 9 teaching assistantships with full tuition reimbursements (averaging $16,300 per year) were awarded; career-related internships or fieldwork, Federal Work-Study, institutionally sponsored loans, scholarships/grants, and traineeships also available. Financial award application deadline: 2/15. *Faculty research:* Snow, surface, and groundwater hydrology; fluvial geomorphology; geographic information systems; geochemistry; bedrock geology. Total annual research expenditures: $1.3 million. *Unit head:* Dr. Judith L. Hannah, Head, 970-491-5662, Fax: 970-491-6307, E-mail: jhannah@cnr.colostate.edu. *Application contact:*

Geology

Colorado State University (continued)

Sharyl Pierson, Administrative Assistant, 970-491-5662, Fax: 970-491-6307, E-mail: sharyl@cnr.colostate.edu.

Cornell University, Graduate School, Graduate Fields of Engineering, Field of Geological Sciences, Ithaca, NY 14853-0001. Offers economic geology (M Eng, MS, PhD); engineering geology (M Eng, MS, PhD); environmental geophysics (M Eng, MS, PhD); general geology (M Eng, MS, PhD); geobiology (M Eng, MS, PhD); geochemistry and isotope geology (M Eng, MS, PhD); geohydrology (M Eng, MS, PhD); geomorphology (M Eng, MS, PhD); geophysics (M Eng, MS, PhD); geotectonics (M Eng, MS, PhD); marine geology (MS, PhD); mineralogy (M Eng, MS, PhD); paleontology (M Eng, MS, PhD); petroleum geology (M Eng, MS, PhD); petrology (M Eng, MS, PhD); planetary geology (M Eng, MS, PhD); Precambrian geology (M Eng, MS, PhD); Quaternary geology (M Eng, MS, PhD); rock mechanics (M Eng, MS, PhD); sedimentology (M Eng, MS, PhD); seismology (M Eng, MS, PhD); stratigraphy (M Eng, MS, PhD); structural geology (M Eng, MS, PhD). *Faculty:* 39 full-time (5 women). *Students:* 29 full-time (15 women); includes 2 minority (both Hispanic Americans), 7 international. 73 applicants, 15% accepted, 6 enrolled. In 2005, 5 master's, 1 doctorate awarded. *Degree requirements:* For master's, thesis (MS); for doctorate, thesis/dissertation, comprehensive exam. *Entrance requirements:* For master's and doctorate, GRE General Test, 3 letters of recommendation. Additional exam requirements/recommendations for international students: Required—TOEFL (minimum score 550 paper-based; 213 computer-based). *Application deadline:* For fall admission, 1/15 for domestic students. Applications are processed on a rolling basis. Application fee: $60. Electronic applications accepted. *Financial support:* In 2005–06, 8 fellowships with full tuition reimbursements, 10 research assistantships with full tuition reimbursements, 7 teaching assistantships with full tuition reimbursements were awarded; institutionally sponsored loans, scholarships/grants, health care benefits, tuition waivers (full and partial), and unspecified assistantships also available. Financial award applicants required to submit FAFSA. *Faculty research:* Geophysics, structural geology, petrology, geochemistry, geodynamics. *Unit head:* Director of Graduate Studies, 607-255-5466, Fax: 607-254-4780. *Application contact:* Graduate Field Assistant, 607-255-5466, Fax: 607-254-4780, E-mail: gradprog@geology.cornell.edu.

Duke University, Graduate School, Department of Earth and Ocean Sciences (Geology), Durham, NC 27708-0586. Offers MS, PhD. Part-time programs available. *Faculty:* 18 full-time. *Students:* 22 full-time (10 women), 4 international. 45 applicants, 31% accepted, 5 enrolled. In 2005, 2 master's, 3 doctorates awarded. Terminal master's awarded for partial completion of doctoral program. *Degree requirements:* For master's and doctorate, thesis/dissertation. *Entrance requirements:* For master's and doctorate, GRE General Test. Additional exam requirements/recommendations for international students: Required—IELT (preferred) or TOEFL. *Application deadline:* For fall admission, 12/31 for domestic students, 12/31 for international students; for spring admission, 11/1 for domestic students. Application fee: $75. Electronic applications accepted. *Financial support:* Fellowships, research assistantships, teaching assistantships, Federal Work-Study available. Financial award application deadline:12/31. *Unit head:* Alan Boudreau, Director of Graduate Studies, 919-684-5646, Fax: 919-684-5833, E-mail: dcgooch@duke.edu.

East Carolina University, Graduate School, Thomas Harriot College of Arts and Sciences, Department of Geology, Greenville, NC 27858-4353. Offers MS. Part-time programs available. *Faculty:* 13 full-time (2 women). *Students:* 17 full-time (8 women), 11 part-time (5 women); includes 1 minority (Hispanic American) Average age 25. 3 applicants, 67% accepted, 1 enrolled. In 2005, 4 degrees awarded. *Degree requirements:* For master's, one foreign language, thesis, comprehensive exam. *Entrance requirements:* For master's, GRE General Test. Additional exam requirements/recommendations for international students: Required—TOEFL. *Application deadline:* For fall admission, 6/1 for domestic students; for spring admission, 10/15 for domestic students. Applications are processed on a rolling basis. Application fee: $50. *Expenses:* Tuition, state resident: full-time $2,516. Tuition, nonresident: full-time $12,832. *Financial support:* Research assistantships with partial tuition reimbursements, teaching assistantships with partial tuition reimbursements available. Support available to part-time students. Financial award application deadline: 6/1. *Unit head:* Dr. Stephen Culver, Chair, 252-328-6360, Fax: 252-328-4391, E-mail: culvers@ecu.edu. *Application contact:* Dean of Graduate School, 252-328-6012, Fax: 252-328-6071, E-mail: gradschool@ecu.edu.

Eastern Kentucky University, The Graduate School, College of Arts and Sciences, Department of Earth Sciences, Richmond, KY 40475-3102. Offers geology (MS, PhD). Part-time programs available. *Degree requirements:* For master's, thesis. *Entrance requirements:* For master's, GRE General Test, minimum GPA of 2.5. *Faculty research:* Hydrogeology, sedimentary geology, geochemistry, environmental geology, tectonics.

Florida Atlantic University, Charles E. Schmidt College of Science, Department of Geosciences, Program in Geology, Boca Raton, FL 33431-0991. Offers MS. Part-time programs available. *Faculty:* 6 full-time (0 women). *Students:* 6 full-time (2 women), 9 part-time (6 women), 1 international. Average age 31. 4 applicants, 75% accepted, 3 enrolled. In 2005, 7 degrees awarded. *Degree requirements:* For master's, thesis (for some programs). *Entrance requirements:* For master's, GRE General Test, minimum GPA of 3.0. *Application deadline:* For fall admission, 6/1 for domestic students; for spring admission, 10/15 for domestic students. Applications are processed on a rolling basis. Application fee: $30. Electronic applications accepted. *Expenses:* Tuition, state resident: full-time $4,394; part-time $244 per credit. Tuition, nonresident: full-time $16,441; part-time $912 per credit. *Financial support:* In 2005–06, 2 research assistantships with partial tuition reimbursements (averaging $9,100 per year), 5 teaching assistantships with partial tuition reimbursements (averaging $9,100 per year) were awarded; Federal Work-Study also available. *Faculty research:* Paleontology, beach erosion, stratigraphy, hydrogeology, environmental geology. Total annual research expenditures:$150,000. *Application contact:* Dr. David Warburton, Graduate Coordinator, 561-297-3250, Fax: 561-297-2745, E-mail: warburto@fau.edu.

Florida State University, Graduate Studies, College of Arts and Sciences, Department of Geological Sciences, Tallahassee, FL 32306. Offers MS, PhD. *Faculty:* 14 full-time (2 women), 2 part-time/adjunct (0 women). *Students:* 25 full-time (14 women), 12 part-time (6 women), 11 international. Average age 27. *Degree requirements:* For master's and doctorate, thesis/dissertation. *Entrance requirements:* For master's and doctorate, GRE General Test, minimum GPA of 3.0. Additional exam requirements/recommendations for international students: Required—TOEFL. *Application deadline:* For fall admission, 3/1 for domestic students; for spring admission, 11/1 priority date for domestic students. Applications are processed on a rolling basis. Application fee: $30. Electronic applications accepted. *Financial support:* In 2005–06, 12 students received support; fellowships, research assistantships, teaching assistantships, career-related internships or fieldwork and Federal Work-Study available. Financial award application deadline: 2/7; financial award applicants required to submit FAFSA. *Faculty research:* Appalachian and collisional tectonics, surface and groundwater hydrogeology, micropaleontology, isotope and trace element geochemistry, coastal and estuarine studies. Total annual research expenditures: $2.3 million. *Unit head:* Dr. A. Leroy Odom, Chairman, 850-644-3743, Fax: 850-644-4214, E-mail: odom@magnet.fsu.edu. *Application contact:* Tami Karl, Program Assistant, 850-644-5861, Fax: 850-644-4214, E-mail: karl@gly.fsu.edu.

Fort Hays State University, Graduate School, College of Arts and Sciences, Department of Geosciences, Program in Geology, Hays, KS 67601-4099. Offers MS. *Faculty:* 8 full-time (0 women). *Students:* 5 full-time (0 women), 7 part-time (1 woman); includes 2 minority (both Asian Americans or Pacific Islanders) Average age 27. 7 applicants, 57% accepted. In 2005, 1 degree awarded. *Degree requirements:* For master's, thesis, comprehensive exam. *Entrance requirements:* For master's, GRE General Test. Additional exam requirements/recommendations for international students: Required—TOEFL (minimum score 550 paper-based; 213 computer-based). *Application deadline:* For fall admission, 7/1 for domestic students. Applications are processed on a rolling basis. Application fee: $30 ($35 for international students). Electronic applications accepted. *Expenses:* Tuition, state resident: part-time $141 per credit. Tuition, nonresident: part-time $371 per credit. *Financial support:* In 2005–06, 5 teaching assistantships with tuition reimbursements (averaging $5,000 per year) were awarded; research assistantships, career-related internships or fieldwork and institutionally sponsored loans also available. Support available to part-time students. *Faculty research:* Cretaceous and late Cenozoic stratigraphy, sedimentation, paleontology. *Unit head:* Dr. John Heinrichs, Chair, Department of Geosciences, 785-628-5389, E-mail: jheinric@fhsu.edu.

The George Washington University, Columbian College of Arts and Sciences, Department of Environmental Studies, Washington, DC 20052. Offers geology (MS, PhD); geosciences (MS, PhD); hominid paleobiology (MS, PhD). Part-time and evening/weekend programs available. Terminal master's awarded for partial completion of doctoral program. *Degree requirements:* For master's, thesis or alternative, comprehensive exam; for doctorate, thesis/dissertation, general exam. *Entrance requirements:* For master's, GRE General Test, bachelor's degree in field, interview, minimum GPA of 3.0; for doctorate, GRE General Test, interview, minimum GPA of 3.0. Additional exam requirements/recommendations for international students: Required—TOEFL (minimum score 550 paper-based; 213 computer-based). *Faculty research:* Engineering geology.

Georgia State University, College of Arts and Sciences, Department of Geosciences, Atlanta, GA 30303-3083. Offers earth science—hydrology (MS), including earth science—environmental management, earth science—GIS; geographic information systems (Certificate); geography (MA); geology (MS). Part-time and evening/weekend programs available. *Degree requirements:* For master's, one foreign language, thesis or alternative, comprehensive exam (for some programs), registration. *Entrance requirements:* For master's, GRE General Test, minimum GPA of 2.75. Additional exam requirements/recommendations for international students: Required—TOEFL. Electronic applications accepted. *Expenses:* Tuition, state resident: full-time $4,368; part-time $182 per term. Tuition, nonresident: full-time $8,732; part-time $728 per term. Required fees: $46 per hour. *Faculty research:* Clay mineralogy, metamorphism, fracture analysis, carbonates, groundwater.

ICR Graduate School, Graduate Programs, Santee, CA 92071. Offers astro/geophysics (MS); biology (MS); geology (MS); science education (MS). Part-time programs available. *Faculty:* 5 full-time (1 woman), 1 part-time/adjunct (0 women). *Students:* 6 full-time (2 women), 6 part-time (3 women). Average age 45. In 2005, 4 degrees awarded. *Degree requirements:* For master's, thesis (for some programs), comprehensive exam (for some programs). *Entrance requirements:* For master's, minimum undergraduate GPA of 3.0, bachelor's degree in science or science education. *Application deadline:* Applications are processed on a rolling basis. Application fee: $30. *Expenses:* Tuition: Part-time $150 per semester hour. *Faculty research:* Age of the earth, limits of variation, catastrophe, optimum methods for teaching. Total annual research expenditures: $200,000. *Unit head:* Dr. Kenneth B. Cumming, Dean, 619-448-0900, Fax: 619-448-3469. *Application contact:* Dr. Jack Kriege, Registrar, 619-448-0900 Ext. 6016, Fax: 619-448-3469, E-mail: jkriege@icr.org.

Idaho State University, Office of Graduate Studies, College of Arts and Sciences, Department of Geosciences, Pocatello, ID 83209. Offers geographic information science (MS); geology (MNS, MS); geophysics/hydrology (MS); geotechnology (Postbaccalaureate Certificate). Part-time programs available. Postbaccalaureate distance learning degree programs offered. *Degree requirements:* For master's, thesis, comprehensive exam, registration (for some programs); for Postbaccalaureate Certificate, thesis optional. *Entrance requirements:* For master's and Postbaccalaureate Certificate, GRE General Test, 3 letters of recommendation. Additional exam requirements/recommendations for international students: Required—TOEFL (minimum score 550 paper-based; 213 computer-based). *Faculty research:* Structural geography, stratigraphy, geochemistry, remote sensing, geomorphology.

Indiana University Bloomington, Graduate School, College of Arts and Sciences, Department of Geological Sciences, Bloomington, IN 47405-7000. Offers biogeochemistry (MS, PhD); economic geology (MS, PhD); geobiology (MS, PhD); geophysics, structural geology and tectonics (MS, PhD); hydrogeology (MS, PhD); mineralogy (MS, PhD); stratigraphy and sedimentology (MS, PhD). PhD offered through the University Graduate School. Part-time programs available. *Faculty:* 14 full-time (1 woman). *Students:* 30 full-time (13 women), 12 part-time (5 women); includes 5 minority (2 African Americans, 1 American Indian/Alaska Native, 1 Asian American or Pacific Islander, 1 Hispanic American), 21 international. Average age 29. In 2005, 7 master's, 2 doctorates awarded. Terminal master's awarded for partial completion of doctoral program. *Degree requirements:* For master's, one foreign language, thesis or alternative; for doctorate, thesis/dissertation. *Entrance requirements:* For master's and doctorate, GRE General Test. Additional exam requirements/recommendations for international students: Required—TOEFL. *Application deadline:* For fall admission, 1/15 priority date for domestic students, 12/15 priority date for international students; for spring admission, 9/1 priority date for domestic students, 9/1 priority date for international students. Applications are processed on a rolling basis. Application fee: $50 ($60 for international students). *Expenses:* Tuition, state resident: full-time $5,437; part-time $227 per credit hour. Tuition, nonresident: full-time $15,836; part-time $660 per credit hour. Required fees: $821. Tuition and fees vary according to campus/location and program. *Financial support:* Fellowships with tuition reimbursements, research assistantships with full and partial tuition reimbursements, teaching assistantships with tuition reimbursements, career-related internships or fieldwork, Federal Work-Study, and institutionally sponsored loans available. Financial award application deadline: 2/15. *Faculty research:* Geophysics, geochemistry, hydrogeology, igneous and metamorphic petrology and clay minerology. Total annual research expenditures: $289,139. *Unit head:* Mark Person, Graduate Advisor, 812-855-7214. *Application contact:* Mary Iverson, Secretary, Committee for Graduate Studies, 812-855-7214, Fax: 812-855-7899, E-mail: geograd@indiana.edu.

Indiana University–Purdue University Indianapolis, School of Science, Department of Geology, Indianapolis, IN 46202-3272. Offers MS. Part-time and evening/weekend programs available. *Faculty:* 8 full-time (2 women). *Students:* 4 full-time (2 women), 2 part-time (both women), 1 international. Average age 24. In 2005, 4 degrees awarded. *Degree requirements:* For master's, thesis (for some programs). *Entrance requirements:* For master's, GRE General Test, minimum GPA of 3.0. Application fee: $50 ($60 for international students). *Expenses:* Tuition, state resident: full-time $5,159; part-time $215 per credit hour. Tuition, nonresident: full-time $14,890; part-time $620 per credit hour. Required fees: $614. Tuition and fees vary according to campus/location and program. *Financial support:* Fellowships with full tuition reimbursements, research assistantships with full tuition reimbursements, teaching assistantships with full tuition reimbursements, scholarships/grants available. Financial award application deadline: 3/1. *Faculty research:* Wetland hydrology, groundwater contamination, soils, sedimentology, sediment chemistry. *Unit head:* Gabriel Filippelli, Chair, 317-274-7484, Fax: 317-274-7966. *Application contact:* Lenore P. Tedesco, Associate Professor, 317-274-7484, Fax: 317-274-7966, E-mail: ltedesco@iupui.edu.

Iowa State University of Science and Technology, Graduate College, College of Liberal Arts and Sciences, Department of Geological and Atmospheric Sciences, Ames, IA 50011. Offers earth science (MS, PhD); geology (MS, PhD); meteorology (MS, PhD); water resources (MS, PhD). *Faculty:* 15 full-time, 2 part-time/adjunct. *Students:* 32 full-time (12 women), 2 part-time; includes 1 minority (African American), 15 international. 40 applicants, 60% accepted, 13 enrolled. In 2005, 11 master's awarded. *Degree requirements:* For master's, thesis (for some programs); for doctorate, thesis/dissertation. *Entrance requirements:* For master's and doctorate, GRE General Test. Additional exam requirements/recommendations for international students: Required—TOEFL (paper score 530; computer score 197) or IELTS (score 6.0). *Application deadline:* For fall admission, 1/1 for domestic students. Applications are processed on a rolling basis. Application fee: $30 ($70 for international students). Electronic applications accepted. *Expenses:* Tuition, state resident: full-time $6,410. Tuition, nonresident: full-time $16,422. Tuition and fees vary according to program. *Financial support:* In 2005–06, 21 research assistantships with full and partial tuition reimbursements (averaging $13,545 per year), 9 teaching assistantships with full and partial tuition reimbursements (averaging $13,507 per year) were awarded; fellowships, scholarships/grants, health care benefits, and unspecified assistantships also available. *Unit head:* Dr. Carl E. Jacobson, Chair, 515-294-4477.

★ **Kansas State University,** Graduate School, College of Arts and Sciences, Department of Geology, Manhattan, KS 66506. Offers MS. *Faculty:* 7 full-time (2 women), 12 part-time/adjunct (2 women). *Students:* 15 full-time (6 women), 4 part-time (1 woman); includes 2 minority (both African Americans), 5 international. Average age 24. 11 applicants, 73% accepted, 8 enrolled. In 2005, 4 degrees awarded. *Degree requirements:* For master's, thesis. *Entrance requirements:* For master's, GRE General Test, GRE Subject Test. Additional exam requirements/recommendations for international students: Required—TOEFL. *Application deadline:* For fall admission, 3/15 for domestic students, 2/1 for international students; for spring admission, 10/1 for domestic students, 8/1 for international students. Applications are processed on a rolling basis. Application fee: $30 ($55 for international students). Electronic applications accepted. *Expenses:* Tuition, state resident: full-time $5,160; part-time $215. Tuition, nonresident: full-time $12,816; part-time $534. Required fees: $564. *Financial support:* In 2005–06, 3 research assistantships (averaging $11,098 per year), 4 teaching assistantships with full tuition reimbursements (averaging $9,362 per year) were awarded; career-related internships or fieldwork, Federal Work-Study, institutionally sponsored loans, and scholarships/grants also available. Support available to part-time students. Financial award application deadline: 3/1; financial award applicants required to submit FAFSA. *Faculty research:* Seismology/tectonics, sedimentology and paleobiology, quarternary geology, orogenesis, earth science education. Total annual research expenditures:$183,107. *Unit head:* Mary Hubbard, Head, 785-532-2245, Fax: 785-532-5159, E-mail: mhub@ksu.edu. *Application contact:* Charles G. Oviatt, Director, 785-532-2245, Fax: 785-532-5159, E-mail: ioviatt@ksu.edu.

Kent State University, College of Arts and Sciences, Department of Geology, Kent, OH 44242-0001. Offers MS, PhD. *Degree requirements:* For master's, thesis; for doctorate, one foreign language, thesis/dissertation. *Entrance requirements:* For master's, minimum GPA of 2.75; for doctorate, GRE General Test, GRE Subject Test, minimum GPA of 3.0. Additional exam requirements/recommendations for international students: Required—TOEFL (minimum score 575 paper-based; 232 computer-based). Electronic applications accepted. *Faculty research:* Groundwater, surface water, engineering geology, paleontology, structural geology.

Lakehead University, Graduate Studies, Department of Geology, Thunder Bay, ON P7B 5E1, Canada. Offers M Sc. Part-time and evening/weekend programs available. *Degree requirements:* For master's, thesis, department seminar, oral exam. *Entrance requirements:* For master's, minimum B average, honours bachelors degree in geology. Additional exam requirements/recommendations for international students: Required—TOEFL. *Faculty research:* Rock physics, sedimentology, mineralogy and economic geology, geochemistry, petrology of alkaline rocks.

Laurentian University, School of Graduate Studies and Research, Programme in Geology (Earth Sciences), Sudbury, ON P3E 2C6, Canada. Offers M Sc. Part-time programs available. *Degree requirements:* For master's, thesis. *Entrance requirements:* For master's, honors degree with second class or better. *Faculty research:* Localization and metallogenesis of Ni-Cu-(PGE) sulfide mineralization in the Thompson Nickel Belt, mapping lithology and ore-grade and monitoring dissolved organic carbon in lakes using remote sensing, global reefs, volcanic effects on VMS deposits.

Lehigh University, College of Arts and Sciences, Department of Earth and Environmental Sciences, Bethlehem, PA 18015-3094. Offers MS, PhD. *Faculty:* 12 full-time (1 woman), 1 (woman) part-time/adjunct. *Students:* 24 full-time (10 women), 5 part-time (1 woman), 4 international. Average age 26. 30 applicants, 27% accepted, 2 enrolled. In 2005, 3 master's, 1 doctorate awarded. Terminal master's awarded for partial completion of doctoral program. *Degree requirements:* For master's, thesis, registration; for doctorate, thesis/dissertation, language at the discretion of the PhD committee, comprehensive exam, registration. *Entrance requirements:* For master's and doctorate, GRE General Test, 2 letters of recommendation. Additional exam requirements/recommendations for international students: Required—TOEFL. *Application deadline:* For fall admission, 1/15 for domestic students; for spring admission, 10/15 priority date for domestic students. Applications are processed on a rolling basis. Application fee: $60. *Financial support:* In 2005–06, 5 fellowships with full tuition reimbursements (averaging $13,670 per year), 16 research assistantships with full tuition reimbursements (averaging $13,670 per year), 10 teaching assistantships with full tuition reimbursements (averaging $13,670 per year) were awarded; Federal Work-Study, institutionally sponsored loans, and tuition waivers (full and partial) also available. Financial award application deadline: 1/15. *Faculty research:* Tectonics, surficial processes, aquatic ecology. Total annual research expenditures: $1.5 million. *Unit head:* Dr. Peter K. Zeitler, Chairman, 610-758-3660 Ext. 3671, Fax: 610-758-3677, E-mail: pkz0@lehigh.edu. *Application contact:* Dr. Gray E. Bebout, Graduate Coordinator, 610-758-3660 Ext. 5831, Fax: 610-758-3677, E-mail: geb0@lehigh.edu.

Louisiana State University and Agricultural and Mechanical College, Graduate School, College of Basic Sciences, Department of Geology and Geophysics, Baton Rouge, LA 70803. Offers MS, PhD. *Faculty:* 22 full-time (6 women). *Students:* 38 full-time (21 women), 4 part-time (2 women); includes 6 minority (3 African Americans, 1 Asian American or Pacific Islander, 2 Hispanic Americans), 7 international. Average age 29. 40 applicants, 48% accepted, 21 enrolled. In 2005, 6 master's, 4 doctorates awarded. Terminal master's awarded for partial completion of doctoral program. *Degree requirements:* For master's and doctorate, thesis/dissertation. *Entrance requirements:* For master's and doctorate, GRE General Test, minimum GPA of 3.0. Additional exam requirements/recommendations for international students: Required—TOEFL (minimum score 550 paper-based; 213 computer-based). *Application deadline:* For fall admission, 1/25 priority date for domestic students, 5/15 priority date for international students. Applications are processed on a rolling basis. Application fee: $25. Electronic applications accepted. *Financial support:* In 2005–06, 33 students received support, including 5 fellowships with full tuition reimbursements available (averaging $15,833 per year), 9 research assistantships with full and partial tuition reimbursements available (averaging $21,577 per year), 19 teaching assistantships with full and partial tuition reimbursements available (averaging $14,855 per year); career-related internships or fieldwork, Federal Work-Study, institutionally sponsored loans, tuition waivers (full and partial), and unspecified assistantships also available. Financial award application deadline: 3/15; financial award applicants required to submit FAFSA. *Faculty research:* Geophysics, geochemistry of sediments, isotope geochemistry, igneous and metamorphic petrology, micropaleontology. Total annual research expenditures: $1.1 million. *Unit head:* Dr. Laurie Anderson, Chair, 225-578-3353, Fax: 225-578-2302, E-mail: landerson@geol.lsu.edu. *Application contact:* Jeffrey Nunn, Graduate Coordinator, 225-578-6657, E-mail: jeff@geol.lsu.edu.

Massachusetts Institute of Technology, School of Science, Department of Earth, Atmospheric, and Planetary Sciences, Cambridge, MA 02139-4307. Offers atmospheric chemistry (PhD, Sc D); atmospheric science (SM, PhD, Sc D); climate physics and chemistry (PhD, Sc D); earth and planetary sciences (SM); geochemistry (PhD, Sc D); geology (PhD, Sc D); geophysics (PhD, Sc D); marine geology and geophysics (SM); oceanography (SM, PhD, Sc D); planetary sciences (PhD, Sc D). *Faculty:* 38 full-time (3 women). *Students:* 165 full-time (63 women); includes 12 minority (7 Asian Americans or Pacific Islanders, 5 Hispanic Americans), 51 international. Average age 27. 179 applicants, 34% accepted, 31 enrolled. In 2005, 2 master's, 15 doctorates awarded. Terminal master's awarded for partial completion of doctoral program. *Degree requirements:* For master's, thesis/dissertation; for doctorate, thesis/dissertation, comprehensive exam. *Entrance requirements:* For master's, GRE General Test; for doctorate, GRE General Test, GRE Subject Test (chemistry or physics for planetary science program). Additional exam requirements/recommendations for international students: Required—TOEFL (minimum score 577 paper-based; 233 computer-based). *Application deadline:* For fall admission, 1/5 for domestic students, 1/5 for international students; for spring admission, 11/1 for domestic students, 11/1 for international students. Application fee: $70. Electronic applications accepted. *Expenses:* Tuition: Full-time $32,100. Required fees: $200. Part-time tuition and fees vary according to course load. *Financial support:* In 2005–06, 115 students received support, including 36 fellowships with tuition reimbursements available (averaging $26,951 per year), 60 research assistantships with tuition reimbursements available (averaging $24,520 per year), 18 teaching assistantships with tuition reimbursements available (averaging $22,773 per year); Federal Work-Study, institutionally sponsored loans, scholarships/grants, health care benefits, and unspecified assistantships also available. *Faculty research:* Evolution of main features of the planetary system; origin, composition, structure, and state of the atmospheres, oceans, surfaces, and interiors of planets; dynamics of planets and satellite motions. Total annual research expenditures: $18.6 million. *Unit head:* Prof. Maria Zuber, Department Head, 617-253-6397, E-mail: zuber@mit.edu. *Application contact:* EAPS Education Office.

McMaster University, School of Graduate Studies, Faculty of Science, School of Geography and Earth Sciences, Hamilton, ON L8S 4M2, Canada. Offers geochemistry (PhD); geology (M Sc, PhD); human geography (MA, PhD); physical geography (M Sc, PhD). Part-time programs available. Terminal master's awarded for partial completion of doctoral program. *Degree requirements:* For master's, thesis/dissertation; for doctorate, thesis/dissertation, comprehensive exam. *Entrance requirements:* For master's, minimum B+ average. Additional exam requirements/recommendations for international students: Required—TOEFL (minimum score 550 paper-based; 213 computer-based).

Memorial University of Newfoundland, School of Graduate Studies, Department of Earth Sciences, St. John's, NL A1C 5S7, Canada. Offers geology (M Sc, PhD); geophysics (M Sc, PhD). Part-time programs available. *Students:* 53 full-time (27 women), 6 part-time, 9 international. 23 applicants, 70% accepted, 16 enrolled. In 2005, 10 master's, 2 doctorates awarded. *Degree requirements:* For master's, thesis; for doctorate, thesis/dissertation, oral thesis defense; PhD entry evaluation, comprehensive exam. *Entrance requirements:* For master's, honors B Sc; for doctorate, M Sc. *Application deadline:* Applications are processed on a rolling basis. Application fee: $40 Canadian dollars. Electronic applications accepted. *Expenses:* Tuition: Part-time $733 per term. Tuition and fees vary according to degree level and program. *Financial support:* Fellowships, research assistantships, teaching assistantships available. *Faculty research:* Geochemistry, sedimentology, paleoceanography and global change, mineral deposits, petroleum geology, hydrology. *Unit head:* Dr. John Hanchar, Head, 709-737-2334, Fax: 709-737-2589, E-mail: head@esd.mun.ca. *Application contact:* Raymund Patzold, Graduate Officer, 709-737-4464, Fax: 709-737-2589, E-mail: rpatzold@esd.mun.ca.

Miami University, Graduate School, College of Arts and Sciences, Department of Geology, Oxford, OH 45056. Offers MA, MS, PhD. Part-time programs available. *Degree requirements:* For master's, thesis (for some programs), final exam; for doctorate, thesis/dissertation, final exams, comprehensive exam. *Entrance requirements:* For master's, GRE General Test, GRE Subject Test, minimum undergraduate GPA of 3.0 during previous 2 years or 2.75 overall; for doctorate, GRE General Test, GRE Subject Test, minimum GPA of 2.75 (undergraduate) or 3.0 (graduate). Additional exam requirements/recommendations for international students: Required—TOEFL (minimum score 550 paper-based; 213 computer-based), TWE (minimum score 4). Electronic applications accepted.

Michigan State University, The Graduate School, College of Natural Science, Department of Geological Sciences, East Lansing, MI 48824. Offers environmental geosciences (MS, PhD); environmental geosciences-environmental toxicology (PhD); geological sciences (MS, PhD). *Faculty:* 11 full-time (1 woman). *Students:* 20 full-time (12 women), 4 part-time (1 woman); includes 2 minority (both Asian Americans or Pacific Islanders), 5 international. Average age 28. 37 applicants, 59% accepted. In 2005, 8 master's, 2 doctorates awarded. *Degree requirements:* For master's, thesis for those without prior thesis work; for doctorate, thesis/dissertation, registration. *Entrance requirements:* For master's, GRE General Test, minimum GPA of 3.0, course work in geoscience, 3 letters of recommendation; for doctorate, GRE General Test, 3 letters of recommendation. Additional exam requirements/recommendations for international students: Required—TOEFL (minimum score 550 paper-based; 213 computer-based), Michigan State Univeristy ELT (85), Michigan ELAB (83). *Application deadline:* For fall admission, 12/27 for domestic students. Application fee: $50. Electronic applications accepted. *Expenses:* Tuition, state resident: part-time $330 per credit hour. Tuition, nonresident: part-time $685 per credit hour. Tuition and fees vary according to program. *Financial support:* In 2005–06, 10 fellowships with tuition reimbursements (averaging $3,479 per year), 4 research assistantships with tuition reimbursements (averaging $12,710 per year), 13 teaching assistantships with tuition reimbursements (averaging $12,997 per year) were awarded; Federal Work-Study, scholarships/grants, and unspecified assistantships also available. *Faculty research:* Water in the environment, global and biological change, crystal dynamics. Total annual research expenditures: $1.1 million. *Unit head:* Dr. Ralph E. Taggart, Chairperson, 517-355-4626, Fax: 517-353-8787, E-mail: taggart@msu.edu. *Application contact:* Information Contact, 517-355-4626, Fax: 517-353-8787, E-mail: geosci@msu.edu.

Michigan Technological University, Graduate School, College of Engineering, Department of Geological and Mining Engineering and Sciences, Program in Geology, Houghton, MI 49931-1295. Offers MS, PhD. Part-time programs available. *Faculty:* 12 full-time (1 woman), 10 part-time/adjunct (2 women). *Students:* 25 full-time (17 women), 5 part-time (4 women); includes 5 minority (1 African American, 4 Hispanic Americans), 3 international. Average age 29. 25 applicants, 76% accepted, 10 enrolled. In 2005, 2 master's, 3 doctorates awarded. Terminal master's awarded for partial completion of doctoral program. *Median time to degree:* Of those who began their doctoral program in fall 1997, 50% received their degree in 8 years or less. *Degree requirements:* For master's, comprehensive exam, registration; for doctorate, thesis/dissertation, comprehensive exam, registration. *Entrance requirements:* Additional exam requirements/recommendations for international students: Required—TOEFL (minimum score 550 paper-based; 213 computer-based). *Application deadline:* For fall admission, 3/15 for domestic students. Applications are processed on a rolling basis. Application fee: $40 ($45 for international students). Electronic applications accepted. *Expenses:* Tuition, nonresident: full-time $11,232; part-time $468 per credit. Required fees: $754; $377 per semester. Full-time tuition and fees vary according to course load, degree level and program. *Financial support:* In 2005–06, 19 students received support, including 4 fellowships with full tuition reimbursements available (averaging $9,542 per year), 13 research assistantships with full tuition reimbursements available (averaging $9,542 per year), 2 teaching assistantships with full tuition reimbursements available (averaging $9,542 per year); career-related internships or fieldwork, Federal Work-Study, scholarships/grants, health care benefits, tuition waivers (partial), unspecified assistantships, and co-op also available. Financial award applicants required to submit FAFSA. *Application contact:* Amie S. Ledgerwood, Secretary 5, 906-487-2531, Fax: 906-487-3371, E-mail: asledger@mtu.edu.

Missouri State University, Graduate College, College of Natural and Applied Sciences, Department of Geography, Geology, and Planning, Springfield, MO 65804-0094. Offers geography, geology and planning (MNAS); geospatial sciences (MS); secondary education (MS Ed), including earth science, geography. Part-time and evening/weekend programs available. *Faculty:* 20 full-time (3 women), 1 part-time/adjunct (0 women). *Students:* 13 full-time (5 women), 8 part-time (3 women). Average age 32. 14 applicants, 79% accepted, 5 enrolled. In 2005, 7 degrees awarded. *Degree requirements:* For master's, thesis (for some programs), comprehensive exam. *Entrance requirements:* For master's, GRE General Test (MS, MNAS), minimum undergraduate GPA of 3.0 (MS, MNAS), 9-12 teacher certification (MS Ed). Additional exam requirements/recommendations for international students: Required—TOEFL (minimum score 550 paper-based; 213 computer-based), IELT (minimum score 6). *Application deadline:* For fall admission, 7/20 for domestic students; for spring admission, 12/20 priority date for domestic students. Applications are processed on a rolling basis. Application fee: $30. Electronic applications accepted. *Expenses:* Tuition, state resident: full-time $3,402; part-time $189 per credit. Tuition, nonresident: full-time $6,804; part-time $378 per credit. Required fees: $207 per semester. Part-time tuition and fees vary according to course level, course load and program. *Financial support:* In 2005–06, 4 research assistantships with full tuition reimbursements (averaging $8,750 per year), 6 teaching assistantships with full tuition reimbursements (averaging $6,575 per year) were awarded; career-related internships or fieldwork, Federal Work-

Geology

Missouri State University (continued)

Study, scholarships/grants, and unspecified assistantships also available. Financial award application deadline: 3/31; financial award applicants required to submit FAFSA. *Faculty research:* Water resources, small town planning, recreation and open space planning. *Unit head:* Dr. Tom Plymate, Acting Head, 417-836-5800, Fax: 417-836-6934, E-mail: tomplymate@missouristate.edu. *Application contact:* Dr. Robert T. Pavlowsky, Graduate Adviser, 417-836-8473, Fax: 417-836-6006, E-mail: bobpavlowsky@missouristate.edu.

Montana Tech of The University of Montana, Graduate School, Geoscience Program, Butte, MT 59701-8997. Offers geochemistry (MS); geological engineering (MS); geology (MS); geophysical engineering (MS); hydrogeological engineering (MS); hydrogeology (MS). Part-time programs available. *Faculty:* 17 full-time (3 women). *Students:* 8 full-time (4 women), 3 part-time (1 woman). 15 applicants, 60% accepted, 2 enrolled. In 2005, 5 degrees awarded. *Degree requirements:* For master's, thesis (for some programs), comprehensive exam (for some programs), registration. *Entrance requirements:* For master's, GRE General Test, minimum GPA of 3.0. Additional exam requirements/recommendations for international students: Required—TOEFL (minimum score 525 paper-based; 195 computer-based). *Application deadline:* For fall admission, 4/1 priority date for domestic students, 3/1 priority date for international students; for spring admission, 10/1 priority date for domestic students, 7/1 priority date for international students. Applications are processed on a rolling basis. Application fee: $30. *Expenses:* Tuition, area resident: Full-time $5,133. Tuition, state resident: full-time $5,333. Tuition, nonresident: full-time $15,746. *Financial support:* In 2005–06, 8 students received support, including 3 research assistantships with partial tuition reimbursements available (averaging $3,667 per year), 6 teaching assistantships with partial tuition reimbursements available (averaging $6,667 per year); career-related internships or fieldwork, tuition waivers (full and partial), and unspecified assistantships also available. Financial award application deadline: 4/1; financial award applicants required to submit FAFSA. *Faculty research:* Water resource development, seismic processing, petroleum reservoir characterization, environmental geochemistry, molecular modeling, magmatic and hydrothermal ore deposits. Total annual research expenditures:$542,109. *Unit head:* Dr. Diane Wolfgram, Department Head, 406-496-4353, Fax: 406-496-4260, E-mail: dwolfgram@mtech.edu. *Application contact:* Cindy Dunstan, Administrator, Graduate School, 406-496-4304, Fax: 406-496-4710, E-mail: cdunstan@mtech.edu.

New Mexico Institute of Mining and Technology, Graduate Studies, Department of Earth and Environmental Science, Program in Geology and Geochemistry, Socorro, NM 87801. Offers geochemistry (MS, PhD); geology (MS, PhD). *Degree requirements:* For master's, thesis optional; for doctorate, thesis/dissertation. *Entrance requirements:* For master's, GRE General Test; for doctorate, GRE General Test, GRE Subject Test. Additional exam requirements/recommendations for international students: Required—TOEFL (minimum score 540 paper-based; 207 computer-based). Electronic applications accepted. *Faculty research:* Care and karst topography, soil/water chemistry and properties, geochemistry of ore deposits.

New Mexico State University, Graduate School, College of Arts and Sciences, Department of Geological Sciences, Las Cruces, NM 88003-8001. Offers MS. Part-time programs available. *Faculty:* 6 full-time (2 women), 1 part-time/adjunct (0 women). *Students:* 18 full-time (9 women), 5 part-time (3 women); includes 1 minority (Hispanic American), 1 international. Average age 30. 19 applicants, 89% accepted, 6 enrolled. In 2005, 7 degrees awarded. *Degree requirements:* For master's, thesis. *Entrance requirements:* For master's, GRE General Test, BS in geology or the equivalent. Additional exam requirements/recommendations for international students: Required—TOEFL. *Application deadline:* For fall admission, 7/1 priority date for domestic students, 7/1 priority date for international students; for spring admission, 11/1 priority date for domestic students, 11/1 priority date for international students. Applications are processed on a rolling basis. Application fee: $30 ($50 for international students). Electronic applications accepted. *Expenses:* Tuition, state resident: full-time $3,156; part-time $175 per credit. Tuition, nonresident: full-time $12,510; part-time $565 per credit. Required fees: $1,050. *Financial support:* Fellowships with partial tuition reimbursements, research assistantships with tuition reimbursements, teaching assistantships with partial tuition reimbursements, career-related internships or fieldwork, Federal Work-Study, institutionally sponsored loans, scholarships/grants, health care benefits, and unspecified assistantships available. Support available to part-time students. Financial award application deadline: 2/15. *Faculty research:* Geochemistry, tectonics, sedimentology, stratigraphy, igneous petrology. *Unit head:* Dr. Timothy Lawton, Head, 505-646-2708, Fax: 505-646-1056, E-mail: tlawton@nmsu.edu. *Application contact:* Dr. Katherine A. Giles, Professor, 505-646-2033, Fax: 505-646-1056, E-mail: kgiles@nmsu.edu.

Northern Arizona University, Graduate College, College of Engineering and Natural Science, Department of Geology, Program in Geology, Flagstaff, AZ 86011. Offers MS. *Degree requirements:* For master's, thesis.

Northern Arizona University, Graduate College, College of Engineering and Natural Science, Program in Quaternary Sciences, Flagstaff, AZ 86011. Offers MS. *Degree requirements:* For master's, thesis. *Faculty research:* Sandbar stability in the Grand Canyon; Stone Age site excavation in South Africa; neogene reptile and mammal evolution; mammoths of Hot Springs, South Dakota; Quaternary science of national parks on Colorado Plateau.

Northern Illinois University, Graduate School, College of Liberal Arts and Sciences, Department of Geology and Environmental Geosciences, De Kalb, IL 60115-2854. Offers geology (MS, PhD). Part-time programs available. *Faculty:* 11 full-time (1 woman), 1 (woman) part-time/adjunct. *Students:* 26 full-time (10 women), 14 part-time (8 women), 8 international. Average age 31. 22 applicants, 68% accepted, 12 enrolled. In 2005, 2 master's, 2 doctorates awarded. Terminal master's awarded for partial completion of doctoral program. *Degree requirements:* For master's, research seminar, thesis optional; for doctorate, thesis/dissertation, candidacy exam, dissertation defense, internship, research seminar. *Entrance requirements:* For master's, GRE General Test, bachelor's degree in engineering or science, minimum GPA of 2.75; for doctorate, GRE General Test, bachelor's or master's degree in engineering or science, minimum graduate GPA of 3.2. Additional exam requirements/recommendations for international students: Required—TOEFL (minimum score 550 paper-based; 213 computer-based). *Application deadline:* For fall admission, 6/1 for domestic students, 5/1 for international students; for spring admission, 11/1 for domestic students, 10/1 for international students. Applications are processed on a rolling basis. Application fee: $30. Electronic applications accepted. *Expenses:* Tuition, state resident: full-time $4,565; part-time $191 per credit hour. Tuition, nonresident: full-time $9,129; part-time $382 per credit hour. *Financial support:* In 2005–06, 22 teaching assistantships with full tuition reimbursements were awarded; fellowships with full tuition reimbursements, research assistantships with full tuition reimbursements, career-related internships or fieldwork, Federal Work-Study, scholarships/grants, tuition waivers (full), and unspecified assistantships also available. Support available to part-time students. Financial award applicants required to submit FAFSA. *Faculty research:* Micropaleontology, environmental geochemistry, glacial geology, igneous petrology, statistical analyses of fracture networks. *Unit head:* Dr. Jonathan Berg, Chair, 815-753-1943, Fax: 815-753-1945, E-mail: jon@geol.niu.edu. *Application contact:* Dr. James Walker, Director of Graduate Studies, 815-753-7936, E-mail: jim@geol.niu.edu.

Northwestern University, The Graduate School, Judd A. and Marjorie Weinberg College of Arts and Sciences, Department of Geological Sciences, Evanston, IL 60208. Offers MS, PhD. Admissions and degrees offered through The Graduate School. Part-time programs available. *Faculty:* 10 full-time (3 women). *Students:* 18 full-time (11 women); includes 2 minority (both Hispanic Americans), 5 international. Average age 25. 12 applicants, 58% accepted, 4 enrolled. In 2005, 1 master's, 2 doctorates awarded. *Median time to degree:* Of those who began their doctoral program in fall 1997, 33% received their degree in 8 years or less. *Degree requirements:* For doctorate, thesis/dissertation. *Entrance requirements:* For master's and doctorate, GRE General Test. Additional exam requirements/recommendations for international students: Required—TOEFL. *Application deadline:* For fall admission, 1/30 for domestic students. Application fee: $60 ($75 for international students). Electronic applications accepted. *Financial support:* In 2005–06, 15 students received support, including 2 fellowships with full tuition reimbursements available (averaging $12,600 per year), 2 research assistantships with full tuition reimbursements available (averaging $13,419 per year), 6 teaching assistantships with full tuition reimbursements available (averaging $13,419 per year); career-related internships or fieldwork, Federal Work-Study, institutionally sponsored loans, scholarships/grants, tuition waivers (full and partial), and research awards also available. Financial award application deadline: 1/31; financial award applicants required to submit FAFSA. Total annual research expenditures: $710,913. *Unit head:* Donna M. Jurdy, Chair, 847-491-7163, Fax: 847-491-8060, E-mail: chair@earth.northwestern.edu. *Application contact:* Susan Van der Lee, Admission Officer, 847-491-8183, Fax: 847-491-8060, E-mail: geodept@eath.northwestern.edu.

The Ohio State University, Graduate School, College of Mathematical and Physical Sciences, Department of Geological Sciences, Columbus, OH 43210. Offers MS, PhD. *Degree requirements:* For master's, thesis; for doctorate, one foreign language, thesis/dissertation. *Entrance requirements:* For master's and doctorate, GRE General Test. Additional exam requirements/recommendations for international students: Required—TOEFL (minimum score 600 paper-based; 250 computer-based); Recommended—TSE. Electronic applications accepted.

Ohio University, Graduate Studies, College of Arts and Sciences, Department of Geological Sciences, Athens, OH 45701-2979. Offers environmental geochemistry (MS); environmental geology (MS); environmental/hydrology (MS); geology (MS); geology education (MS); geomorphology/surficial processes (MS); geophysics (MS); hydrogeology (MS); sedimentology (MS); structure/tectonics (MS). Part-time programs available. *Faculty:* 10 full-time (4 women), 4 part-time/adjunct (1 woman). *Students:* 22 full-time (6 women), 3 part-time (2 women); includes 1 minority (Hispanic American), 4 international. Average age 23. 15 applicants, 67% accepted, 8 enrolled. In 2005, 7 degrees awarded. *Degree requirements:* For master's, thesis, thesis proposal defense and thesis defense. *Entrance requirements:* Additional exam requirements/recommendations for international students: Required—TOEFL (minimum score 550 paper-based; 217 computer-based). *Application deadline:* For fall admission, 2/1 priority date for domestic students, 1/1 priority date for international students. Application fee: $45. Electronic applications accepted. *Financial support:* In 2005–06, 18 students received support, including 3 research assistantships with full tuition reimbursements available (averaging $11,900 per year), 13 teaching assistantships with full tuition reimbursements available (averaging $11,900 per year); institutionally sponsored loans, scholarships/grants, tuition waivers (full), and unspecified assistantships also available. Financial award application deadline: 2/1. *Faculty research:* Geoscience education, tectonics, flurial geomorphology, invertebrate paleontology, mine/hydrology. Total annual research expenditures: $649,020. *Unit head:* Dr. David Kidder, Chair, 740-593-1101, Fax: 740-593-0486, E-mail: kidder@ohio.edu. *Application contact:* Dr. David Schneider, Graduate Chair, 740-593-1101, Fax: 740-593-0486, E-mail: schneidd@ohio.edu.

Oklahoma State University, College of Arts and Sciences, School of Geology, Stillwater, OK 74078. Offers MS. *Faculty:* 10 full-time (2 women). *Students:* 22 full-time (4 women), 14 part-time (6 women); includes 2 minority (1 American Indian/Alaska Native, 1 Asian American or Pacific Islander), 6 international. Average age 28. 24 applicants, 71% accepted, 13 enrolled. In 2005, 7 degrees awarded. *Degree requirements:* For master's, thesis. *Entrance requirements:* For master's, minimum GPA of 3.0. Additional exam requirements/recommendations for international students: Required—TOEFL. *Application deadline:* For fall admission, 6/1 priority date for domestic students, 3/1 priority date for international students. Applications are processed on a rolling basis. Application fee: $40 ($75 for international students). Electronic applications accepted. *Expenses:* Tuition, state resident: full-time $4,253; part-time $139 per credit hour. Tuition, nonresident: full-time $12,569; part-time $485 per credit hour. Required fees: $43 per credit hour. One-time fee: $20 part-time. Tuition and fees vary according to course load and program. *Financial support:* In 2005–06, 5 research assistantships (averaging $11,424 per year), 20 teaching assistantships (averaging $12,664 per year) were awarded; career-related internships or fieldwork, Federal Work-Study, scholarships/grants, health care benefits, tuition waivers (partial), and unspecified assistantships also available. Support available to part-time students. Financial award application deadline: 3/1. *Faculty research:* Groundwater hydrology, petroleum geology. *Unit head:* Dr. Jay Gregg, Head, 405-744-6358, E-mail: jay.gregg@okstate.edu.

Oregon State University, Graduate School, College of Science, Department of Geosciences, Program in Geology, Corvallis, OR 97331. Offers MA, MAIS, MS, PhD. Part-time programs available. *Students:* 28 full-time (11 women), 4 part-time (1 woman); includes 1 minority (Hispanic American), 1 international. Average age 28. In 2005, 5 master's, 3 doctorates awarded. Terminal master's awarded for partial completion of doctoral program. *Degree requirements:* For master's, variable foreign language requirement, thesis; for doctorate, one foreign language, thesis/dissertation. *Entrance requirements:* For master's and doctorate, GRE General Test, GRE Subject Test, minimum GPA of 3.0 in last 90 hours. Additional exam requirements/recommendations for international students: Required—TOEFL. *Application deadline:* For fall admission, 2/1 for domestic students. Applications are processed on a rolling basis. Application fee: $50. *Expenses:* Tuition, area resident: Part-time $301 per credit. Tuition, state resident: full-time $8,139; part-time $501 per credit. Tuition, nonresident: full-time $14,376; part-time $532 per credit. Required fees: $1,266. *Financial support:* Fellowships, research assistantships, teaching assistantships, Federal Work-Study and institutionally sponsored loans available. Support available to part-time students. Financial award application deadline: 2/1. *Faculty research:* Hydrogeology, geomorphology, ocean geology, geochemistry, earthquake geology. *Unit head:* Dr. Roger L. Nielsen, Director, 541-737-1201, Fax: 541-737-1200, E-mail: mielsen@oce.orst.edu. *Application contact:* Joanne VanGeest, Graduate Admissions Coordinator, 541-737-1204, Fax: 541-737-1200, E-mail: vangeesj@geo.orst.edu.

Portland State University, Graduate Studies, College of Liberal Arts and Sciences, Department of Geology, Portland, OR 97207-0751. Offers environmental sciences and resources (PhD); geology (MA, MS); science/geology (MAT, MST). Part-time programs available. *Faculty:* 9 full-time (2 women). *Students:* 14 full-time (8 women), 12 part-time (6 women). Average age 29. 19 applicants, 74% accepted, 5 enrolled. In 2005, 5 degrees awarded. *Degree requirements:* For master's, thesis, field comprehensive; for doctorate, thesis/dissertation, 2 years of residency. *Entrance requirements:* For master's, GRE General Test, GRE Subject Test, BA/BS in geology, minimum GPA of 3.0 in upper-division course work or 2.75 overall. Additional exam requirements/recommendations for international students: Required—TOEFL (minimum score 550 paper-based; 213 computer-based). *Application deadline:* 1/31 for domestic students, 1/31 for international students. Applications are processed on a rolling basis. Application fee: $50. *Expenses:* Tuition, state resident: full-time $6,648; part-time $231 per credit. Tuition, nonresident: full-time $11,319; part-time $231 per credit. Required fees: $686; $67 per credit. *Financial support:* In 2005–06, 4 research assistantships with full tuition reimbursements (averaging $14,951 per year), 6 teaching assistantships with full tuition reimbursements (averaging $10,001 per year) were awarded; career-related internships or fieldwork, Federal Work-Study, scholarships/grants, and unspecified assistantships also available. Support available to part-time students. Financial award application deadline: 3/1; financial award applicants required to submit FAFSA. *Faculty research:* Sediment transport, volcanic environmental geology, coastal and fluvial processes. Total annual research expenditures: $1.6 million. *Unit head:* Dr. Michael L. Cummings, Head, 503-725-3022, Fax: 503-725-3025. *Application contact:* Nancy Eriksson, Office Coordinator, 503-725-3022, Fax: 503-725-3025, E-mail: erikssonn@pdx.edu.

Princeton University, Graduate School, Department of Geosciences, Princeton, NJ 08544-1019. Offers atmospheric and oceanic sciences (PhD); geological and geophysical sciences (PhD). *Degree requirements:* For doctorate, one foreign language, thesis/dissertation. *Entrance requirements:* For doctorate, GRE General Test. Additional exam requirements/recommendations for international students: Required—TOEFL (minimum score 600 paper-based; 250 computer-based). Electronic applications accepted. *Faculty research:* Biogeochemistry, climate science, earth history, regional geology and tectonics, solid-earth geophysics.

Queens College of the City University of New York, Division of Graduate Studies, Mathematics and Natural Sciences Division, School of Earth and Environmental Sciences, Flushing, NY 11367-1597. Offers MA. Part-time and evening/weekend programs available. *Faculty:* 14 full-time (4 women). *Students:* 12 applicants, 100% accepted, 8 enrolled. In 2005, 1 degree awarded. *Degree requirements:* For master's, thesis, comprehensive exam. *Entrance requirements:* For master's, GRE, previous course work in calculus, physics, and chemistry; minimum GPA of 3.0. Additional exam requirements/recommendations for international students: Required—TOEFL. *Application deadline:* For fall admission, 4/1 for domestic students; for spring admission, 11/1 for domestic students. Applications are processed on a rolling basis. Application fee: $125. *Expenses:* Tuition, state resident: part-time $270 per credit. Tuition, nonresident: part-time $500 per credit. Required fees: $112 per year. *Financial support:* Career-related internships or fieldwork, Federal Work-Study, institutionally sponsored loans, tuition waivers (partial), unspecified assistantships, and adjunct lectureships available. Support available to part-time students. Financial award application deadline: 4/1; financial award applicants required to submit FAFSA. *Faculty research:* Sedimentology/stratigraphy, paleontology, field petrology. *Unit head:* Dr. Daniel Habib, Chairperson, 718-997-3300, E-mail: daniel_habib@qc.edu. *Application contact:* Dr. Hannes Brueckner, Graduate Adviser, 718-997-3300, E-mail: hannes_brueckner@qc.edu.

Queen's University at Kingston, School of Graduate Studies and Research, Faculty of Arts and Sciences, Department of Geological Sciences and Geological Engineering, Kingston, ON K7L 3N6, Canada. Offers M Sc, M Sc Eng, PhD. Part-time programs available. *Degree requirements:* For master's, thesis (for some programs); for doctorate, thesis/dissertation, comprehensive exam. *Entrance requirements:* Additional exam requirements/recommendations for international students: Required—TOEFL. *Faculty research:* Geochemistry, sedimentology, geophysics, economic geology, structural geology.

Rensselaer Polytechnic Institute, Graduate School, School of Science, Department of Earth and Environmental Sciences, Program in Geology, Troy, NY 12180-3590. Offers MS, PhD. Part-time programs available. Terminal master's awarded for partial completion of doctoral program. *Degree requirements:* For master's, thesis (for some programs), comprehensive exam; for doctorate, thesis/dissertation, comprehensive exam. *Entrance requirements:* For master's and doctorate, GRE General Test. Additional exam requirements/recommendations for international students: Required—TOEFL. Electronic applications accepted. *Expenses:* Tuition: Full-time $31,000; part-time $1,320 per credit. Required fees: $1,623. *Faculty research:* Geochemistry, petrology, geophysics, environmental geochemistry, planetary geology.

Rush University, College of Nursing, Department of Community and Mental Health Nursing, Chicago, IL 60612-3832. Offers community and mental health nursing (DN Sc, DNP); family nurse practitioner (MSN, Post-Master's Certificate); psychiatric clinical specialist (MSN); psychiatric nurse practitioner—adult (MSN); psychiatric nurse practitioner—family (MSN); psychiatric-mental health clinical specialist (Post-Master's Certificate); psychiatric-mental health nurse practitioner (Post-Master's Certificate); public health nursing (MSN). *Accreditation:* AACN. Part-time programs available. Postbaccalaureate distance learning degree programs offered (minimal on-campus study). *Faculty:* 26. *Students:* 5 full-time, 48 part-time; includes 11 minority (6 African Americans, 4 Asian Americans or Pacific Islanders, 1 Hispanic American). Average age 35. 30 applicants, 93% accepted, 27 enrolled. In 2005, 1 master's, 13 doctorates awarded. Terminal master's awarded for partial completion of doctoral program. *Degree requirements:* For master's, Capstone project; for doctorate, thesis/dissertation, DNP leadership project. *Entrance requirements:* For master's, GRE General Test (waived if nursing GPA is above 3.0 or cumulative GPA is above 3.25), interview; for doctorate, GRE General Test, interview, course work in statistics (DN Sc). *Application deadline:* For fall admission, 7/1 for domestic students. For winter admission, 11/1 for domestic students; for spring admission, 1/15 for domestic students. Applications are processed on a rolling basis. Application fee: $40. Electronic applications accepted. *Financial support:* In 2005–06, 11 students received support; teaching assistantships with tuition reimbursements available, Federal Work-Study, institutionally sponsored loans, scholarships/grants, and traineeships available. Support available to part-time students. Financial award applicants required to submit FAFSA. *Faculty research:* Immigrant mental health, de-escalation strategies, caregiver interventions, parent-teacher training, restraint use. *Unit head:* Dr. Julia Cowell, Chairperson, 312-942-7117. *Application contact:* Hicela Castruita Woods, Director, College Admissions Services, 312-942-7100, Fax: 312-942-2219, E-mail: hicela_castruita@rush.edu.

Rutgers, The State University of New Jersey, Newark, Graduate School, Program in Environmental Geology, Newark, NJ 07102. Offers MS. Part-time and evening/weekend programs available. *Faculty:* 6 full-time (1 woman), 1 part-time/adjunct (0 women). *Students:* 1 full-time (0 women), 8 part-time. 5 applicants, 80% accepted, 2 enrolled. In 2005, 1 degree awarded. *Degree requirements:* For master's, thesis optional. *Entrance requirements:* For master's, GRE General Test, minimum B average. *Application deadline:* For fall admission, 6/1 for domestic students; for spring admission, 12/1 for domestic students. Application fee: $50. Electronic applications accepted. *Expenses:* Tuition, state resident: full-time $10,440; part-time $435 per credit. Tuition, nonresident: full-time $15,520; part-time $637 per credit. *Faculty research:* Environmental geology, plate tectonics, geoarchaeology, geophysics, mineralogy-petrology. Total annual research expenditures: $124,000. *Unit head:* Dr. Alex Gates, Program Coordinator and Adviser, 973-353-5034, Fax: 973-353-5100, E-mail: agates@andromeda.rutgers.edu.

Rutgers, The State University of New Jersey, New Brunswick/Piscataway, Graduate School, Program in Geological Sciences, New Brunswick, NJ 08901-1281. Offers MS, PhD. Part-time programs available. *Faculty:* 40 full-time, 3 part-time/adjunct. *Students:* 17 full-time (8 women), 9 part-time (4 women); includes 1 minority (African American), 3 international. Average age 32. 30 applicants, 33% accepted, 6 enrolled. In 2005, 2 degrees awarded. *Degree requirements:* For master's, thesis/dissertation; for doctorate, thesis/dissertation, comprehensive exam. *Entrance requirements:* For master's and doctorate, GRE General Test, GRE Subject Test (recommended). *Application deadline:* For fall admission, 5/1 for domestic students; for spring admission, 2/15 for domestic students. Applications are processed on a rolling basis. Application fee: $50. Electronic applications accepted. *Expenses:* Tuition, state resident: full-time $10,440; part-time $435 per credit. Tuition, nonresident: full-time $15,520; part-time $647 per credit. Required fees: $129 per credit. Tuition and fees vary according to program. *Financial support:* In 2005–06, 13 students received support, including 5 fellowships with full tuition reimbursements available (averaging $18,000 per year), 2 research assistantships with full tuition reimbursements available (averaging $15,500 per year), 5 teaching assistantships with full tuition reimbursements available (averaging $16,988 per year); Federal Work-Study and scholarships/grants also available. Financial award application deadline: 3/1; financial award applicants required to submit FAFSA. *Faculty research:* Stratigraphy and basins analysis; volcanology and geochemistry; quaternary studies; structure and geophysics; marine science, biogeochemistry and paleoceanography. Total annual research expenditures: $1 million. *Unit head:* Kenneth G. Miller, Chair, 732-445-3622, Fax: 732-445-3374, E-mail: kgm@rci.rutgers.edu. *Application contact:* Carl Swisher, Graduate Director, 732-445-5363, Fax: 732-445-3374, E-mail: cswisherci.rutgers.edu.

St. Francis Xavier University, Graduate Studies, Department of Earth Sciences, Antigonish, NS B2G 2W5, Canada. Offers M Sc. *Faculty:* 2 full-time (0 women). *Students:* 2 full-time (1 woman). *Degree requirements:* For master's, thesis, registration. *Entrance requirements:* Additional exam requirements/recommendations for international students: Required—TOEFL (minimum score 580 paper-based; 236 computer-based). Application fee: $40. *Faculty research:* Environmental earth sciences, global change tectonics, paleoclimatology, crustal fluids. Total annual research expenditures: $300,000. *Unit head:* Dr. Brendan Murphy, Professor, 902-867-2481, Fax: 902-867-5153, E-mail: bmurphy@stfx.ca.

San Diego State University, Graduate and Research Affairs, College of Sciences, Department of Geological Sciences, San Diego, CA 92182. Offers MS. Part-time programs available. *Students:* 16 full-time (8 women), 13 part-time (8 women); includes 4 minority (all Hispanic Americans) Average age 27. 24 applicants, 67% accepted, 10 enrolled. In 2005, 7 degrees awarded. *Degree requirements:* For master's, thesis. *Entrance requirements:* For master's, GRE General Test, bachelor's degree in related field, 2 letters of reference. Additional exam requirements/recommendations for international students: Required—TOEFL. *Application deadline:* For fall admission, 5/1 for domestic students, 5/1 for international students; for spring admission, 11/1 for domestic students, 10/1 for international students. Applications are processed on a rolling basis. Application fee: $55. Electronic applications accepted. *Financial support:* Fellowships, research assistantships, teaching assistantships, unspecified assistantships available. Financial award applicants required to submit FAFSA. *Faculty research:* Earthquakes, hydrology, meteorological analysis and tomography studies. Total annual research expenditures: $965,830. *Unit head:* Gary Girty, Chair, 619-594-5586, Fax: 619-594-4372, E-mail: ggirty@geology.sdsu.edu. *Application contact:* Kathryn Thorbjarnarson, Graduate Coordinator, 619-594-1392, Fax: 619-594-4372, E-mail: thorbjarnarson@geology.sdsu.edu.

San Jose State University, Graduate Studies and Research, College of Science, Department of Geology, San Jose, CA 95192-0001. Offers MS. *Students:* 4 full-time (2 women), 8 part-time (7 women). Average age 33. 6 applicants, 83% accepted, 0 enrolled. In 2005, 2 degrees awarded. *Degree requirements:* For master's, thesis. *Entrance requirements:* For master's, GRE. *Application deadline:* For fall admission, 6/29 for domestic students; for spring admission, 11/30 for domestic students. Applications are processed on a rolling basis. Application fee: $59. Electronic applications accepted. *Expenses:* Tuition, nonresident: part-time $339 per unit. Required fees: $1,286 per semester. Tuition and fees vary according to course load and degree level. *Financial support:* Teaching assistantships, Federal Work-Study available. Financial award applicants required to submit FAFSA. *Unit head:* John Williams, Chair, 408-924-5050, Fax: 408-924-5053. *Application contact:* Dr. Robert Miller, Graduate Adviser, 408-924-5025.

South Dakota School of Mines and Technology, Graduate Division, College of Engineering, Department of Geology and Geological Engineering, Rapid City, SD 57701-3995. Offers geology and geological engineering (MS, PhD); paleontology (MS). Part-time programs available. *Faculty:* 7 full-time (0 women). *Students:* 13 full-time (3 women), 12 part-time (4 women), 3 international. Average age 33. In 2005, 10 degrees awarded. *Degree requirements:* For master's and doctorate, thesis/dissertation. *Entrance requirements:* For master's and doctorate, GRE General Test, GRE Subject Test. Additional exam requirements/recommendations for international students: Required—TOEFL, TWE. *Application deadline:* For fall admission, 7/1 priority date for domestic students, 4/1 priority date for international students; for spring admission, 11/1 for domestic students, 9/1 for international students. Applications are processed on a rolling basis. Application fee: $35. Electronic applications accepted. *Expenses:* Tuition, area resident: Part-time $116 per credit hour. Tuition, state resident: full-time $2,084. Tuition, nonresident: full-time $6,146; part-time $341 per credit hour. Required fees: $1,805; $100 per credit hour. *Financial support:* In 2005–06, 4 fellowships (averaging $1,800 per year), 4 research assistantships with partial tuition reimbursements (averaging $7,047 per year), 8 teaching assistantships with partial tuition reimbursements (averaging $5,286 per year) were awarded; Federal Work-Study and institutionally sponsored loans also available. Support available to part-time students. Financial award application deadline: 5/15. *Faculty research:* Contaminants in soil, nitrate leaching, environmental changes, fracture formations, greenhouse effect. Total annual research expenditures:$18,865. *Unit head:* Dr. Arden Davis, Dean, 605-394-2461. *Application contact:* Jeannette R. Nilson, Program Assistant-Research and Graduate Education, 800-454-8162 Ext. 1206, Fax: 605-394-5360, E-mail: graduate_admissions@silver.sdsmt.edu.

Southern Illinois University Carbondale, Graduate School, College of Science, Department of Geology, Carbondale, IL 62901-4701. Offers environmental resources and policy (PhD); geology (MS, PhD). *Faculty:* 12 full-time (0 women). *Students:* 5 full-time (2 women), 19 part-time (3 women); includes 1 minority (Asian American or Pacific Islander), 5 international. Average age 25. 23 applicants, 35% accepted, 7 enrolled. In 2005, 6 degrees awarded. *Degree requirements:* For master's, thesis; for doctorate, one foreign language, thesis/dissertation. *Entrance requirements:* For master's, GRE, minimum GPA of 2.7; for doctorate, GRE General Test, minimum GPA of 3.25. Additional exam requirements/recommendations for international students: Required—TOEFL. *Application deadline:* For fall admission, 2/15 for domestic students. Applications are processed on a rolling basis. Application fee: $20. *Financial support:* In 2005–06, 17 students received support; fellowships with full tuition reimbursements available, research assistantships with full tuition reimbursements available, teaching assistantships with full tuition reimbursements available, Federal Work-Study, institutionally sponsored loans, and tuition waivers (full) available. Support available to part-time students. Total annual research expenditures:$720,000. *Unit head:* Steven Esling, Chair, 618-453-3351, Fax: 618-453-7393.

Southern Methodist University, Dedman College, Department of Geological Sciences, Program in Geology, Dallas, TX 75275. Offers MS, PhD. Part-time programs available. *Degree requirements:* For master's and doctorate, thesis/dissertation, qualifying exam. *Entrance requirements:* For master's and doctorate, GRE General Test, minimum GPA of 3.0, letters of recommendation. Additional exam requirements/recommendations for international students: Required—TOEFL; Recommended—TSE. *Faculty research:* Geothermal, paleontology, environmental, stable isotope geochemistry.

State University of New York at Binghamton, Graduate School, School of Arts and Sciences, Department of Geological Sciences, Binghamton, NY 13902-6000. Offers MA, PhD. Part-time programs available. Terminal master's awarded for partial completion of doctoral program. *Degree requirements:* For master's, thesis or alternative; for doctorate, variable foreign language requirement, thesis/dissertation, departmental qualifying exam. *Entrance requirements:* For master's and doctorate, GRE General Test, GRE Subject Test. Additional exam requirements/recommendations for international students: Required—TOEFL. Electronic applications accepted.

State University of New York at Buffalo, Graduate School, College of Arts and Sciences, Department of Geology, Buffalo, NY 14260. Offers MA, MS, PhD. Part-time programs available. *Faculty:* 9 full-time (2 women), 2 part-time/adjunct (0 women). *Students:* 40 full-time (20 women), 9 part-time (3 women), 9 international. Average age 30. 58 applicants, 57% accepted, 6 enrolled. In 2005, 11 master's awarded. *Degree requirements:* For master's, thesis (for some programs), project, thesis, or exam; for doctorate, thesis/dissertation, dissertation defense. *Entrance requirements:* For master's and doctorate, GRE General Test. Additional exam requirements/recommendations for international students: Required—TOEFL (minimum score 550 paper-based; 213 computer-based); Recommended—TSE. *Application deadline:* For fall admission, 2/1 priority date for domestic students, 2/1 priority date for international students; for spring admission, 10/1 priority date for domestic students, 10/1 priority date for international students. Applications are processed on a rolling basis. Application fee: $35. Electronic applications accepted. *Financial support:* In 2005–06, fellowships with full tuition reimbursements (averaging $4,000 per year), 7 research assistantships with full tuition reimbursements (averaging $12,000 per year), 15 teaching assistantships with full tuition reimbursements (averaging $11,800 per year) were awarded; Federal Work-Study, scholarships/grants, health care benefits, and unspecified assistantships also available. Financial award application deadline: 3/1; financial award applicants required to submit FAFSA. *Faculty research:* Environmental geophysics, hydrogeology, geochemistry, fractured rocks, volcanology. Total annual research expenditures: $1.2 million. *Unit head:* Dr. Charles E. Mitchell, Professor and Chair, 716-645-6800 Ext. 6100, Fax: 716-645-3999, E-mail: geology@acsu.buffalo.edu. *Application contact:* Dr. Robert D. Jacobi, Director of Graduate Studies, 716-645-6800 Ext. 2468, Fax: 716-645-3999, E-mail: rdjacobi@acsu.buffalo.edu.

State University of New York at New Paltz, Graduate School, School of Science and Engineering, Department of Geological Sciences, New Paltz, NY 12561. Offers MA, MAT, MS Ed. Part-time and evening/weekend programs available. *Faculty:* 5 full-time (0 women), 1 (woman) part-time/adjunct. *Students:* Average age 38. *Degree requirements:* For master's, thesis, comprehensive exam. *Entrance requirements:* For master's, GRE General Test, minimum

Geology

State University of New York at New Paltz (continued)
GPA of 3.0. Additional exam requirements/recommendations for international students: Required—TOEFL (minimum score 550 paper-based). *Application deadline:* For fall admission, 3/1 priority date for domestic students, 3/1 priority date for international students; for spring admission, 10/1 for domestic students, 10/1 for international students. Applications are processed on a rolling basis. Application fee: $50. *Expenses:* Tuition, state resident: full-time $3,450; part-time $288 per credit hour. Tuition, nonresident: full-time $3,550; part-time $455 per credit hour. Required fees: $27 per credit. $130 per semester. *Financial support:* Federal Work-Study and institutionally sponsored loans available. *Unit head:* Dr. Frederick Vollmer, Chairman, 845-257-3760, E-mail: vollmerf@newpaltz.edu.

Stephen F. Austin State University, Graduate School, College of Sciences and Mathematics, Department of Geology, Nacogdoches, TX 75962. Offers MS, MSNS. *Faculty:* 6 full-time (0 women). *Students:* 10 full-time (4 women), 5 part-time (2 women); includes 2 minority (1 African American, 1 Hispanic American). Average age 23. 7 applicants, 100% accepted. In 2005, 4 degrees awarded. *Degree requirements:* For master's, comprehensive exam. *Entrance requirements:* For master's, GRE General Test, minimum GPA of 2.8 in last 60 hours, 2.5 overall. Additional exam requirements/recommendations for international students: Required—TOEFL. *Application deadline:* For fall admission, 8/1 for domestic students; for spring admission, 12/15 for domestic students. Applications are processed on a rolling basis. Application fee: $0 ($50 for international students). *Expenses:* Tuition, state resident: full-time $2,628; part-time $146 per credit hour. Tuition, nonresident: full-time $7,596; part-time $422 per credit hour. Required fees:$900; $170. *Financial support:* In 2005–06, 5 teaching assistantships (averaging $8,100 per year) were awarded; Federal Work-Study and unspecified assistantships also available. Financial award application deadline: 3/1. *Faculty research:* Stratigraphy of Kaibab limestone, Utah; structure of Ouachita Mountains, Arkansas; groundwater chemistry of Carrizo Sand, Texas. *Unit head:* Dr. R. LaRell Nielson, Chair, 936-468-3701, E-mail: lnielson@sfasu.edu. *Application contact:* Dr. R. LaRell Nielson, Director of Graduate Program, 936-468-2248.

Sul Ross State University, School of Arts and Sciences, Department of Geology and Chemistry, Alpine, TX 79832. Offers MS. Part-time programs available. *Degree requirements:* For master's, thesis optional. *Entrance requirements:* For master's, GRE General Test, minimum GPA of 2.5 in last 60 hours of undergraduate work.

Announcement: Program stresses integrated field and laboratory research. The University is situated in an area of diverse and well-exposed geology. Research equipment includes GIS, GPS, XRF, XRD, AA, and CL. Current faculty research in environmental geology, volcanology, trace-element geochemistry, paleontology, carbonate depositional environments, arid-region hydrogeology, and remote sensing. URL: http://www.sulross.edu/pages/1007.asp

Syracuse University, Graduate School, College of Arts and Sciences, Department of Geology, Syracuse, NY 13244. Offers MA, MS, PhD. Part-time programs available. Postbaccalaureate distance learning degree programs offered. *Students:* 14 full-time (6 women), 2 part-time (both women). 13 applicants, 15% accepted, 1 enrolled. *Degree requirements:* For master's, thesis (for some programs), research tool; for doctorate, thesis/dissertation, 2 research tools. *Entrance requirements:* For master's and doctorate, GRE General Test, GRE Subject Test. Additional exam requirements/recommendations for international students: Required—TOEFL. *Application deadline:* For fall admission, 12/10 for domestic students. Application fee: $65. Electronic applications accepted. *Financial support:* Fellowships with full tuition reimbursements, research assistantships with full tuition reimbursements, teaching assistantships with full and partial tuition reimbursements, tuition waivers (partial) available. *Unit head:* Dr. Scott Samson, Chair, 315-443-2672, Fax: 315-443-3363, E-mail: sdsamson@syr.edu. *Application contact:* Bonnie Windey, Information Contact, 315-443-2672, E-mail: bgwindey@syr.edu.

Temple University, Graduate School, College of Science and Technology, Department of Geology, Philadelphia, PA 19122-6096. Offers MS. *Faculty:* 7 full-time (1 woman). *Students:* 5 full-time (2 women), 7 part-time (4 women), 1 international. 7 applicants, 57% accepted, 3 enrolled. In 2005, 3 degrees awarded. *Degree requirements:* For master's, thesis, qualifying exam. *Entrance requirements:* For master's, GRE General Test, minimum GPA of 3.0. Additional exam requirements/recommendations for international students: Required—TOEFL (minimum score 620 paper-based; 260 computer-based). *Application deadline:* For fall admission, 2/1 for domestic students, 12/15 for international students; for spring admission, 10/1 for domestic students, 8/1 for international students. Application fee: $50. Electronic applications accepted. *Expenses:* Tuition, state resident: full-time $8,694; part-time $483 per credit. Tuition, nonresident: full-time $12,672; part-time $704 per credit. Required fees: $500; $122 per semester. Tuition and fees vary according to course level, campus/location and program. *Financial support:* Fellowships, research assistantships with full tuition reimbursements, teaching assistantships with full tuition reimbursements, scholarships/grants available. Financial award application deadline: 1/15; financial award applicants required to submit FAFSA. *Faculty research:* Hydrolic modeling, environmental geochemistry and geophysics, paleosas, cyclic stratigraphy, materials research. Total annual research expenditures: $50,000. *Unit head:* Dr. David Grandstaff, Chair, 215-204-8228, Fax: 215-204-3496, E-mail: grand@temple.edu.

Texas A&M University, College of Geosciences, Department of Geology and Geophysics, College Station, TX 77843. Offers geology (MS, PhD); geophysics (MS, PhD). *Faculty:* 24 full-time (2 women), 1 part-time/adjunct (0 women). *Students:* 79 full-time (27 women), 22 part-time (5 women); includes 6 minority (1 African American, 1 American Indian/Alaska Native, 1 Asian American or Pacific Islander, 3 Hispanic Americans), 45 international. Average age 31. 76 applicants, 46% accepted, 16 enrolled. In 2005, 16 master's, 6 doctorates awarded. *Degree requirements:* For master's and doctorate, thesis/dissertation. *Entrance requirements:* For master's and doctorate, GRE General Test. Additional exam requirements/recommendations for international students: Required—TOEFL. *Application deadline:* For fall admission, 3/1 priority date for domestic students, 1/15 priority date for international students; for spring admission, 10/1 priority date for domestic students, 8/15 priority date for international students. Applications are processed on a rolling basis. Application fee: $50 ($75 for international students). Electronic applications accepted. *Expenses:* Tuition, state resident: full-time $4,488; part-time $187 per credit hour. Tuition, nonresident: full-time $11,112; part-time $463 per credit hour. Required fees:$1,974. *Financial support:* In 2005–06, fellowships with partial tuition reimbursements (averaging $1,000 per year), research assistantships with partial tuition reimbursements (averaging $11,925 per year), teaching assistantships with partial tuition reimbursements (averaging $11,925 per year) were awarded; Federal Work-Study, institutionally sponsored loans, scholarships/grants, tuition waivers (partial), and unspecified assistantships also available. Financial award application deadline: 3/1; financial award applicants required to submit FAFSA. *Faculty research:* Environmental and engineering geology and geophysics, petroleum geology, tectonophysics, geochemistry. *Unit head:* Dr. Rick Carlson, Head, 979-845-2451, Fax: 979-845-6162. *Application contact:* Robert K. Popp, Graduate Adviser, 979-845-2451, Fax: 979-845-6162, E-mail: popp@geo.tamu.edu.

Texas A&M University–Kingsville, College of Graduate Studies, College of Arts and Sciences, Department of Geosciences, Kingsville, TX 78363. Offers applied geology (MS). Part-time and evening/weekend programs available. *Degree requirements:* For master's, thesis, comprehensive exam. *Entrance requirements:* For master's, GRE General Test, minimum GPA of 3.0. Additional exam requirements/recommendations for international students: Required—TOEFL. *Faculty research:* Stratigraphy and sedimentology of modern coastal sediments, sandstone diagnosis, vertebrate paleontology, structural geology.

Texas Christian University, College of Science and Engineering, Department of Geology, Fort Worth, TX 76129-0002. Offers MS. Part-time and evening/weekend programs available. *Degree requirements:* For master's, thesis, preliminary exam. *Entrance requirements:* For master's, GRE General Test. Additional exam requirements/recommendations for international students: Required—TOEFL. *Application deadline:* For fall admission, 3/1 for domestic students; for spring admission, 12/1 for domestic students. Applications are processed on a rolling basis. Application fee: $0. *Expenses:* Tuition: Part-time $740 per credit hour. *Financial support:* Teaching assistantships, unspecified assistantships available. Financial award application deadline: 3/1. *Unit head:* Dr. Richard Hanson, Chairperson, 817-257-7270, E-mail: r.hanson@tcu.edu.

Tulane University, Graduate School, Department of Earth and Environmental Sciences, New Orleans, LA 70118-5669. Offers geology (MS, PhD); paleontology (PhD). *Degree requirements:* For master's, one foreign language, thesis or alternative; for doctorate, one foreign language, thesis/dissertation. *Entrance requirements:* For master's, GRE General Test, minimum B average in undergraduate course work; for doctorate, GRE General Test. Additional exam requirements/recommendations for international students: Required—TOEFL; Recommended—TSE. Electronic applications accepted. *Faculty research:* Sedimentation, isotopes, biogeochemistry, marine geology, structural geology.

See Close-Up on page 233.

Université du Québec à Montréal, Graduate Programs, Program in Earth Sciences, Montreal, QC H3C 3P8, Canada. Offers geology-research (M Sc); mineral resources (PhD); non-renewable resources (DESS). Part-time programs available. Terminal master's awarded for partial completion of doctoral program. *Degree requirements:* For master's, thesis (for some programs); for doctorate, thesis/dissertation. *Entrance requirements:* For master's, appropriate bachelor's degree or equivalent, proficiency in French. *Faculty research:* Economic geology, structural geology, geochemistry, Quaternary geology, isotopic geochemistry.

Université Laval, Faculty of Sciences and Engineering, Department of Geology and Geological Engineering, Québec, QC G1K 7P4, Canada. Offers earth sciences (M Sc, PhD), including earth sciences, environmental technologies (M Sc); geology (M Sc, PhD). Terminal master's awarded for partial completion of doctoral program. *Degree requirements:* For master's, thesis (for some programs); for doctorate, thesis/dissertation, comprehensive exam. *Entrance requirements:* For master's and doctorate, knowledge of French. Electronic applications accepted. *Faculty research:* Engineering, economics, regional geology.

University at Albany, State University of New York, College of Arts and Sciences, Department of Earth and Atmospheric Sciences, Albany, NY 12222-0001. Offers atmospheric science (MS, PhD); geology (MS, PhD). *Students:* 38 full-time (10 women), 10 part-time (5 women). Average age 29. In 2005, 5 master's, 4 doctorates awarded. *Degree requirements:* For master's, one foreign language, thesis, comprehensive exam; for doctorate, 2 foreign languages, thesis/dissertation, oral exams, comprehensive exam. *Entrance requirements:* For master's and doctorate, GRE General Test. Additional exam requirements/recommendations for international students: Required—TOEFL (minimum score 550 paper-based; 213 computer-based). *Application deadline:* For fall admission, 6/1 for domestic students, 5/1 for international students; for spring admission, 11/1 for domestic students, 11/11 for international students. Applications are processed on a rolling basis. Application fee: $60. Electronic applications accepted. *Financial support:* Fellowships, research assistantships, teaching assistantships, minority assistantships available. Financial award application deadline:3/1. *Faculty research:* Environmental geochemistry, tectonics, mesoscale meteorology, atmospheric chemistry. *Unit head:* Dr. Vincent Idone, Chair, 518-442-4466. *Application contact:* William Kidd, Graduate Program Director, Geology.

The University of Akron, Graduate School, Buchtel College of Arts and Sciences, Department of Geology, Akron, OH 44325. Offers earth science (MS); environmental (MS); geology (MS); geophysics (MS). Part-time programs available. *Faculty:* 10 full-time (2 women). *Students:* 17 full-time (3 women), 4 international. Average age 29. 20 applicants, 85% accepted, 5 enrolled. In 2005, 6 degrees awarded. *Degree requirements:* For master's, thesis, seminar, proficiency exam, comprehensive exam. *Entrance requirements:* For master's, minimum GPA of 2.75. Additional exam requirements/recommendations for international students: Required—TOEFL (minimum score 550 paper-based; 213 computer-based). *Application deadline:* For fall admission, 3/1 for domestic students. Applications are processed on a rolling basis. Application fee: $30 ($40 for international students). *Expenses:* Tuition, state resident: full-time $5,816; part-time $323 per credit. Tuition, nonresident: full-time $9,976; part-time $554 per credit. Required fees: $794; $43 per credit. $12 per term. Tuition and fees vary according to course load, degree level and program. *Financial support:* In 2005–06, 13 teaching assistantships with full tuition reimbursements were awarded; research assistantships with full tuition reimbursements, Federal Work-Study and tuition waivers (full) also available. *Faculty research:* Broad-range geology, petrology (sedimentary, igneous, metamorphic, and clay), geochemistry, geophysics. Total annual research expenditures:$396,632. *Unit head:* Dr. John Szabo, Chair, 330-972-8039, E-mail: jszabo@uakron.edu. *Application contact:* Dr. LaVerne Friberg, Director of Graduate Studies, 330-972-8046.

The University of Alabama, Graduate School, College of Arts and Sciences, Department of Geological Sciences, Tuscaloosa, AL 35487. Offers MS, PhD. *Faculty:* 12 full-time (2 women). *Students:* 28 full-time (8 women), 8 part-time (1 woman); includes 3 minority (all Hispanic Americans), 15 international. Average age 28. 13 applicants, 46% accepted, 3 enrolled. In 2005, 6 master's, 3 doctorates awarded. *Median time to degree:* Of those who began their doctoral program in fall 1997, 33% received their degree in 8 years or less. *Degree requirements:* For master's and doctorate, thesis/dissertation. *Entrance requirements:* For master's and doctorate, GRE General Test, minimum GPA of 3.0. Additional exam requirements/recommendations for international students: Required—TOEFL, TSE. *Application deadline:* For fall admission, 3/15 for domestic students; for spring admission, 8/31 priority date for domestic students. Applications are processed on a rolling basis. Application fee: $25. Electronic applications accepted. *Expenses:* Tuition, area resident: Full-time $2,432. Tuition, nonresident: full-time $6,758. *Financial support:* In 2005–06, 4 fellowships with full tuition reimbursements (averaging $11,000 per year), 14 research assistantships with full tuition reimbursements (averaging $10,269 per year), 13 teaching assistantships with full tuition reimbursements (averaging $10,269 per year) were awarded; career-related internships or fieldwork, Federal Work-Study, and institutionally sponsored loans also available. Financial award application deadline: 3/15. *Faculty research:* Structure, petrology, stratigraphy, geochemistry, hydrogeology, geophysics. Total annual research expenditures: $1.6 million. *Unit head:* Dr. Harold H. Stowell, Chairperson, 205-348-5098, Fax: 205-348-0818, E-mail: hstowell@wgs.geo.ua.edu. *Application contact:* Dr. Andrew Mark Goodliffe, Assistant Professor of Geological Science, 205-348-7167, Fax: 205-348-0818, E-mail: amg@ua.edu.

University of Alaska Fairbanks, College of Natural Sciences and Mathematics, Department of Geology and Geophysics, Fairbanks, AK 99775-7520. Offers geology (MS, PhD); geophysics (MS, PhD). Part-time programs available. *Faculty:* 15 full-time (2 women), 1 (woman) part-time/adjunct. *Students:* 53 full-time (27 women), 18 part-time (8 women); includes 4 minority (1 American Indian/Alaska Native, 1 Asian American or Pacific Islander, 2 Hispanic Americans), 10 international. Average age 29. 62 applicants, 48% accepted, 7 enrolled. In 2005, 13 master's, 4 doctorates awarded. Terminal master's awarded for partial completion of doctoral program. *Degree requirements:* For master's and doctorate, thesis/dissertation, comprehensive exam, registration. *Entrance requirements:* For master's and doctorate, GRE General Test. Additional exam requirements/recommendations for international students: Required—TOEFL (minimum score 550 paper-based; 213 computer-based). *Application deadline:* For fall admission, 1/15 priority date for domestic students, 1/15 priority date for international students; for spring admission, 12/1 for domestic students, 9/1 for international students. Applications are processed on a rolling basis. Application fee: $50. Electronic applications accepted. *Expenses:* Tuition, state resident: full-time $4,392; part-time $244 per credit. Tuition, nonresident: full-time $8,964; part-time $498 per credit. Required fees: $800; $5 per credit. $48 per contact hour. Tuition and fees vary according to course level, course load, campus/location and reciprocity agreements. *Financial support:* In 2005–06, 39 research assistantships with tuition reimbursements (averaging $9,816 per year), 4 teaching assistantships with tuition reimbursements (averaging $9,562 per year) were awarded; fellowships with tuition reimbursements, Federal Work-Study, scholarships/grants, and unspecified assistantships also available. Financial award application deadline: 1/15; financial award applicants required to submit FAFSA. *Faculty research:* Glacial surging, Alaska as geologic fragments, natural zeolites, seismology, volcanology. *Unit head:* Dr. John C. Eichelberger, Co-Chair, 907-474-7565, Fax: 907-474-5163, E-mail: geology@uaf.edu.

University of Arkansas, Graduate School, J. William Fulbright College of Arts and Sciences, Department of Geosciences, Program in Geology, Fayetteville, AR 72701-1201. Offers MS. Part-time programs available. *Students:* 16 full-time (5 women), 3 part-time (1 woman), 1 international. 9 applicants, 44% accepted. In 2005, 7 degrees awarded. *Degree requirements:* For master's, thesis. Application fee: $40 ($50 for international students). *Financial support:* In 2005–06, 1 research assistantship, 12 teaching assistantships were awarded; career-related internships or fieldwork and Federal Work-Study also available. Support available to part-time students. Financial award application deadline: 4/1; financial award applicants required to submit FAFSA. *Unit head:* Doy Zachry, Graduate Coordinator, 479-575-3355, Fax: 479-575-3469, E-mail: dzachry@uark.edu.

The University of British Columbia, Faculty of Graduate Studies, Faculty of Science, Department of Earth and Ocean Sciences, Vancouver, BC V6T 1Z1, Canada. Offers atmospheric science (M Sc, PhD); geological engineering (M Eng, MA Sc, PhD); geological sciences (M Sc, PhD); geophysics (M Sc, MA Sc, PhD); oceanography (M Sc, PhD). *Faculty:* 44 full-time (7 women), 17 part-time/adjunct (1 woman). *Students:* 155 full-time (49 women), 1 (woman) part-time. Average age 30. 96 applicants, 50% accepted, 30 enrolled. In 2005, 11 master's, 6 doctorates awarded. *Degree requirements:* For master's, thesis (for some programs); for doctorate, thesis/dissertation, comprehensive exam. *Entrance requirements:* Additional exam requirements/recommendations for international students: Required—TOEFL (minimum score 600 paper-based; 250 computer-based). *Application deadline:* For fall admission, 2/1 for domestic students, 1/1 for international students. For winter admission, 7/1 for domestic students. Applications are processed on a rolling basis. Application fee: $90 Canadian dollars ($150 Canadian dollars for international students). Electronic applications accepted. *Financial support:* In 2005–06, fellowships (averaging $16,000 per year), research assistantships (averaging $13,000 per year), teaching assistantships (averaging $5,000 per year) were awarded; Federal Work-Study, institutionally sponsored loans, scholarships/grants, tuition waivers (full and partial), and unspecified assistantships also available. *Unit head:* Dr. Paul L. Smith, Head, 604-822-6456, Fax: 604-822-6088, E-mail: head@eos.ubc.ca. *Application contact:* Alex Allen, Graduate Secretary, 604-822-2713, Fax: 604-822-6088, E-mail: aallen@eos.ubc.ca.

University of Calgary, Faculty of Graduate Studies, Faculty of Science, Department of Geology and Geophysics, Calgary, AB T2N 1N4, Canada. Offers geology (M Sc, PhD); geophysics (M Sc, PhD). Part-time programs available. *Faculty:* 22 full-time (3 women), 5 part-time/adjunct (0 women). *Students:* 107 full-time (37 women). 104 applicants, 30% accepted, 26 enrolled. In 2005, 17 master's, 5 doctorates awarded. Terminal master's awarded for partial completion of doctoral program. *Degree requirements:* For master's, thesis; for doctorate, thesis/dissertation, candidacy exam. *Entrance requirements:* For master's, B Sc; for doctorate, honors B Sc or M Sc. Additional exam requirements/recommendations for international students: Required—TOEFL. *Application deadline:* For fall admission, 2/1 priority date for domestic students, 2/1 priority date for international students. For winter admission, 9/1 for domestic students. Applications are processed on a rolling basis. Application fee: $60. Electronic applications accepted. *Financial support:* In 2005–06, 50 students received support, including 9 fellowships, 15 teaching assistantships; career-related internships or fieldwork, institutionally sponsored loans, and scholarships/grants also available. Financial award application deadline: 2/1. *Faculty research:* Geochemistry, petrology, paleontology, stratigraphy, exploration and solid-earth geophysics. *Unit head:* Larry R. Lines, Head, 403-220-8863, Fax: 403-284-0074, E-mail: lrlines@ucalgary.ca. *Application contact:* Cathy H. Hubbell, Graduate Program Administrator, 403-220-3254, Fax: 403-284-0074, E-mail: geosciencegrad@ucalgary.ca.

University of California, Berkeley, Graduate Division, College of Letters and Science, Department of Earth and Planetary Science, Division of Geology, Berkeley, CA 94720-1500. Offers MA, MS, PhD. Terminal master's awarded for partial completion of doctoral program. *Degree requirements:* For master's, oral exam (MA), thesis (MS); for doctorate, thesis/dissertation, candidacy exams, comprehensive exam. *Entrance requirements:* For master's and doctorate, GRE General Test, minimum GPA of 3.0. Additional exam requirements/recommendations for international students: Required—TOEFL. *Faculty research:* Tectonics, environmental geology, economic geology, mineralogy, geochemistry.

University of California, Davis, Graduate Studies, Program in Geology, Davis, CA 95616. Offers MS, PhD. *Faculty:* 32 full-time. *Students:* 44 full-time (25 women); includes 1 minority (Hispanic American), 7 international. Average age 28. 99 applicants, 22% accepted, 11 enrolled. In 2005, 3 master's, 3 doctorates awarded. Terminal master's awarded for partial completion of doctoral program. *Median time to degree:* Of those who began their doctoral program in fall 1997, 0% received their degree in 8 years or less. *Degree requirements:* For master's and doctorate, thesis/dissertation. *Entrance requirements:* For master's and doctorate, GRE General Test, GRE Subject Test, minimum GPA of 3.0. Additional exam requirements/recommendations for international students: Required—TOEFL (minimum score 550 paper-based; 213 computer-based). *Application deadline:* For fall admission, 1/15 for domestic students, 1/15 for international students. Application fee: $60. Electronic applications accepted. *Financial support:* In 2005–06, 13 fellowships with full and partial tuition reimbursements (averaging $9,532 per year), 19 research assistantships with full and partial tuition reimbursements (averaging $12,701 per year), 11 teaching assistantships with partial tuition reimbursements (averaging $13,711 per year) were awarded; Federal Work-Study, institutionally sponsored loans, scholarships/grants, and unspecified assistantships also available. Financial award application deadline: 1/15; financial award applicants required to submit FAFSA. *Faculty research:* Petrology, paleontology, geophysics, sedimentology, structure/tectonics. *Unit head:* Louise Kellogg, Chair, 530-754-6673, E-mail: kellogg@geology.ucdavis.edu. *Application contact:* Marlene Belz, Administrative Assistant, 530-752-9100, Fax: 530-752-0951, E-mail: belz@geology.ucdavis.edu.

University of California, Los Angeles, Graduate Division, College of Letters and Science, Department of Earth and Space Sciences, Program in Geology, Los Angeles, CA 90095. Offers MS, PhD. *Degree requirements:* For master's, comprehensive exams or thesis; for doctorate, thesis/dissertation, oral and written qualifying exams. *Entrance requirements:* For master's, GRE General Test, minimum GPA of 3.0; for doctorate, GRE General Test, minimum undergraduate GPA of 3.0. Electronic applications accepted.

University of California, Riverside, Graduate Division, Department of Earth Sciences, Riverside, CA 92521-0102. Offers geological sciences (MS, PhD). *Faculty:* 13 full-time (1 woman). *Students:* 26 full-time (7 women); includes 1 minority (Hispanic American), 4 international. Average age 31. In 2005, 5 master's, 2 doctorates awarded. Terminal master's awarded for partial completion of doctoral program. *Median time to degree:* Of those who began their doctoral program in fall 1997, 100% received their degree in 8 years or less. *Degree requirements:* For master's, thesis, final oral exam; for doctorate, thesis/dissertation, qualifying exams, final oral exam. *Entrance requirements:* For master's and doctorate, GRE General Test, minimum GPA of 3.2. Additional exam requirements/recommendations for international students: Required—TOEFL (minimum score 550 paper-based; 213 computer-based); Recommended—TSE (minimum score 50). *Application deadline:* For fall admission, 5/1 for domestic students, 2/1 for international students. For winter admission, 9/1 for domestic students; for spring admission, 12/1 for domestic students. Applications are processed on a rolling basis. Application fee: $60 ($75 for international students). Electronic applications accepted. *Expenses:* Tuition, nonresident: full-time $14,694. Full-time tuition and fees vary according to program. *Financial support:* In 2005–06, 17 students received support; fellowships with full and partial tuition reimbursements available, research assistantships with full and partial tuition reimbursements available, teaching assistantships with full and partial tuition reimbursements available, career-related internships or fieldwork, Federal Work-Study, institutionally sponsored loans, health care benefits, tuition waivers (full and partial), and unspecified assistantships available. Financial award application deadline: 1/5; financial award applicants required to submit FAFSA. *Faculty research:* Applied and solid earth geophysics, tectonic geomorphology, fluid-rock interaction, paleobiology-ecology, sedimentary-geochemistry. *Unit head:* Dr. Michael O. Woodburne, Chair, 951-827-5028, Fax: 951-827-4324, E-mail: michael.woodburne@ucr.edu. *Application contact:* John Herring, Graduate Program Assistant, 951-827-3435, Fax: 951-827-4324, E-mail: geology@ucr.edu.

Announcement: The geological sciences program offers research opportunities with a strong field-based curriculum in astrobiology, paleobiology, paleoecology and developmental evolution, Neoproterozoic geobiology, sedimentary geochemistry, stable-isotope geochemistry, sequence stratigraphy, quantitative stratigraphy, experimental tectonophysics, heat flow, earthquake physics and modeling, mineral deposits, hydrothermal geochemistry, and groundwater resources and hydrogeology. Deadline for funding: January 5.

University of California, Santa Barbara, Graduate Division, College of Letters and Sciences, Division of Mathematics, Life, and Physical Sciences, Department of Earth Science, Santa Barbara, CA 93106. Offers geological sciences (MS, PhD); geophysics (MS). *Faculty:* 37 full-time (4 women). *Students:* 57 full-time (25 women); includes 2 minority (1 Asian American or Pacific Islander, 1 Hispanic American), 6 international. 87 applicants, 24% accepted, 10 enrolled. In 2005, 7 master's, 5 doctorates awarded. Terminal master's awarded for partial completion of doctoral program. *Degree requirements:* For master's, thesis, comprehensive exam, registration; for doctorate, thesis/dissertation, dissertation defense, comprehensive exam, registration. *Entrance requirements:* For master's and doctorate, GRE General Test. Additional exam requirements/recommendations for international students: Required—TOEFL (minimum score 550 paper-based; 213 computer-based). *Application deadline:* For fall admission, 1/1 for domestic students, 1/1 for international students. Application fee: $60. Electronic applications accepted. *Financial support:* In 2005–06, 7 students received support, including 8 fellowships (averaging $6,000 per year), 50 teaching assistantships with full and partial tuition reimbursements available (averaging $7,500 per year); research assistantships with full and partial tuition reimbursements available, career-related internships or fieldwork, Federal Work-Study, scholarships/grants, traineeships, health care benefits, and unspecified assistantships also available. Financial award application deadline: 1/1; financial award applicants required to submit FAFSA. *Faculty research:* Geochemistry, geophysics, tectonics, seismology, petrology. *Unit head:* Dr. James Mattinson, Chair, 805-893-3219, E-mail: mattison@geol.ucsb.edu. *Application contact:* Samuel Rifkin, Graduate Program Assistant, 805-893-3329, Fax: 805-893-2314, E-mail: rifkin@geol.ucsb.edu.

University of Cincinnati, Division of Research and Advanced Studies, McMicken College of Arts and Sciences, Department of Geology, Cincinnati, OH 45221. Offers MS, PhD. Part-time programs available. *Degree requirements:* For master's, thesis/dissertation; for doctorate, thesis/dissertation, comprehensive exam. *Entrance requirements:* For master's and doctorate, GRE General Test, 1 year of course work in physics, chemistry, and calculus. Additional exam requirements/recommendations for international students: Required—TOEFL. Electronic applications accepted. *Faculty research:* Paleobiology, sequence stratigraphy, earth systems history, quaternary, groundwater.

University of Colorado at Boulder, Graduate School, College of Arts and Sciences, Department of Geological Sciences, Boulder, CO 80309. Offers geology (MS, PhD); geophysics (PhD). *Faculty:* 27 full-time (7 women). *Students:* 52 full-time (24 women), 15 part-time (7 women); includes 6 minority (1 Asian American or Pacific Islander, 5 Hispanic Americans), 6 international. Average age 30. 22 applicants, 82% accepted. In 2005, 8 master's, 3 doctorates awarded. Terminal master's awarded for partial completion of doctoral program. *Degree requirements:* For master's and doctorate, thesis/dissertation, comprehensive exam. *Entrance requirements:* For master's and doctorate, GRE General Test, minimum GPA of 2.75. *Application deadline:* For fall admission, 1/15 priority date for domestic students, 12/1 priority date for international students. Application fee: $50 ($60 for international students). *Financial support:* In 2005–06, fellowships with full tuition reimbursements (averaging $9,707 per year), research assistantships with full tuition reimbursements (averaging $13,873 per year), teaching assistantships with full tuition reimbursements (averaging $12,684 per year) were awarded; Federal Work-Study, institutionally sponsored loans, scholarships/grants, and tuition waivers (full) also available. Financial award application deadline: 1/15. *Faculty research:* Sedimentology, stratigraphy, economic geology of mineral deposits, fossil fuels, hydrogeology and water resources. Total annual research expenditures: $5.1 million. *Unit head:* Mary Kraus, Chair, 303-492-2330, Fax: 303-492-2606, E-mail: mary.kraus@colorado.edu. *Application contact:* Graduate Secretary, 303-492-2607, Fax: 303-492-2606, E-mail: geolinfo@colorado.edu.

University of Connecticut, Graduate School, College of Liberal Arts and Sciences, Center for Integrative Geosciences, Storrs, CT 06269. Offers geological sciences (MS, PhD). *Faculty:* 9 full-time (1 woman). *Students:* 11 full-time (7 women), 8 part-time (5 women); includes 1 minority (Hispanic American), 1 international. Average age 33. 12 applicants, 42% accepted, 5 enrolled. In 2005, 6 degrees awarded. *Degree requirements:* For doctorate, thesis/dissertation. *Entrance requirements:* For master's and doctorate, GRE General Test. Additional exam requirements/recommendations for international students: Required—TOEFL (minimum score 550 paper-based; 213 computer-based). *Application deadline:* For fall admission, 2/1 priority date for domestic students, 2/1 priority date for international students; for spring admission, 11/1 for domestic students, 10/1 for international students. Applications are processed on a rolling basis. Application fee: $55. Electronic applications accepted. *Expenses:* Tuition, state resident: part-time $444 per credit hour. Tuition, nonresident: part-time $1,154 per credit hour. Tuition and fees vary according to course load. *Financial support:* In 2005–06, 3 research assistantships with full tuition reimbursements, 4 teaching assistantships with full tuition reimbursements were awarded; fellowships, Federal Work-Study, scholarships/grants, health care benefits, and unspecified assistantships also available. Financial award application deadline: 2/1. *Unit head:* Raymond Joesten, Head, 860-486-4434, Fax: 860-486-1383, E-mail: raymond.joesten@uconn.edu. *Application contact:* Timothy Byrne, Chairperson, 860-486-1388, Fax: 860-486-1383, E-mail: tim.byrne@uconn.edu.

University of Delaware, College of Arts and Sciences, Department of Geology, Newark, DE 19716. Offers MS, PhD. Part-time programs available. *Faculty:* 7 full-time (1 woman), 1 part-time/adjunct (0 women). *Students:* 18 full-time (8 women), 2 part-time; includes 1 minority (Hispanic American), 3 international. Average age 31. 10 applicants, 70% accepted, 5 enrolled. In 2005, 3 master's, 1 doctorate awarded. *Degree requirements:* For master's and doctorate, thesis/dissertation. *Entrance requirements:* For master's and doctorate, GRE General Test. Additional exam requirements/recommendations for international students: Required—TOEFL (minimum score 600 paper-based). *Application deadline:* For fall admission, 7/1 for domestic students. Application fee: $60. Electronic applications accepted. *Financial support:* In 2005–06, 16 students received support, including 1 fellowship with full tuition reimbursement available (averaging $18,000 per year), 7 research assistantships with full tuition reimbursements available (averaging $15,000 per year), 5 teaching assistantships with full tuition reimbursements available (averaging $10,500 per year); Federal Work-Study, institutionally sponsored loans, scholarships/grants, and tuition waivers (full and partial) also available. Financial award application deadline: 3/15; financial award applicants required to submit FAFSA. *Faculty research:* Coastal plain mollusk geochemistry, taxonomy of marsh forams, coastal and marine geology, geomorphology, geophysics. Total annual research expenditures: $190,528. *Unit head:* Dr. James E. Pizzuto, Chair, 302-831-2710, Fax: 302-831-4158, E-mail: pizzuto@udel.edu. *Application contact:* Dr. Susan McGeary, Associate Professor, 302-831-8174, Fax: 302-831-4158, E-mail: smcgeary@udel.edu.

Announcement: With allied faculty in engineering and marine studies, the department emphasizes coastal and marine geology and geophysics, coastal plain geology, geochronology and geochemistry, and fluvial geomorphology and watershed dynamics. Available equipment includes ground-penetrating radar, a high-resolution seismic system, CHIRP sonar, GPS, X-ray diffractometer, chromatographs, research vessels, and a total station.

Geology

University of Florida, Graduate School, College of Liberal Arts and Sciences, Department of Geological Sciences, Gainesville, FL 32611. Offers geology (MS, MST, PhD). *Faculty:* 17 full-time (2 women). *Students:* 43 (22 women); includes 1 minority (Hispanic American) 8 international. In 2005, 8 master's, 5 doctorates awarded. Terminal master's awarded for partial completion of doctoral program. *Degree requirements:* For master's, thesis (for some programs); for doctorate, one foreign language, thesis/dissertation. *Entrance requirements:* For master's and doctorate, GRE General Test, GRE Subject Test, minimum GPA of 3.0. Additional exam requirements/recommendations for international students: Required—TOEFL (minimum score 550 paper-based; 213 computer-based). *Application deadline:* For fall admission, 6/1 for domestic students; for spring admission, 10/1 priority date for domestic students. Applications are processed on a rolling basis. Application fee: $30. Electronic applications accepted. *Expenses:* Tuition, state resident: full-time $6,234. Tuition, nonresident: full-time $21,359. Tuition and fees vary according to program. *Financial support:* In 2005–06, fellowships with full tuition reimbursements (averaging $15,000 per year), 7 research assistantships with full tuition reimbursements (averaging $11,931 per year), 21 teaching assistantships with full tuition reimbursements (averaging $11,868 per year) were awarded; career-related internships or fieldwork, Federal Work-Study, institutionally sponsored loans, and scholarships/grants also available. Support available to part-time students. Financial award application deadline: 3/1. *Faculty research:* Paleoclimatology, tectonophysics, petrochemistry, marine geology, geochemistry, hydrology. Total annual research expenditures: $1.5 million. *Unit head:* Dr. Paul Mueller, Chair, 352-392-2231, Fax: 352-392-9294, E-mail: mueller@geology.ufl.edu. *Application contact:* Dr. Michael R. Perfit, Graduate Coordinator, 352-392-2128, Fax: 352-392-9294, E-mail: perfit@geology.ufl.edu.

University of Georgia, Graduate School, College of Arts and Sciences, Department of Geology, Athens, GA 30602. Offers MS, PhD. *Faculty:* 16 full-time (3 women), 1 part-time/adjunct (0 women). *Students:* 33 full-time, 4 part-time; includes 2 minority (both African Americans), 4 international. 32 applicants, 72% accepted, 8 enrolled. In 2005, 2 degrees awarded. *Degree requirements:* For master's, thesis; for doctorate, one foreign language, thesis/dissertation. *Entrance requirements:* For master's and doctorate, GRE General Test. *Application deadline:* For fall admission, 7/1 for domestic students; for spring admission, 11/15 for domestic students. Application fee: $50. Electronic applications accepted. *Financial support:* Fellowships, research assistantships, teaching assistantships, unspecified assistantships available. *Unit head:* Dr. Susan Goldstein, Head, 706-542-2397, Fax: 706-542-2425, E-mail: sgoldst@gly.uga.edu. *Application contact:* Dr. Alberto Patino-Douce, Graduate Coordinator, 706-542-2394, Fax: 706-542-2425, E-mail: alpatino@uga.edu.

University of Hawaii at Manoa, Graduate Division, School of Ocean and Earth Science and Technology, Department of Geology and Geophysics, Honolulu, HI 96822. Offers high-pressure geophysics and geochemistry (MS, PhD); hydrogeology and engineering geology (MS, PhD); marine geology and geophysics (MS, PhD); planetary geosciences and remote sensing (MS, PhD); seismology and solid-earth geophysics (MS, PhD); volcanology, petrology, and geochemistry (MS, PhD). *Faculty:* 65 full-time (12 women), 19 part-time/adjunct (1 woman). *Students:* 50 full-time (23 women), 8 part-time (1 woman); includes 2 minority (1 African American, 1 Asian American or Pacific Islander), 15 international. Average age 29. 89 applicants, 18% accepted, 12 enrolled. In 2005, 9 master's, 6 doctorates awarded. Terminal master's awarded for partial completion of doctoral program. *Median time to degree:* Of those who began their doctoral program in fall 1997, 100% received their degree in 8 years or less. *Degree requirements:* For master's, thesis/dissertation; for doctorate, thesis/dissertation, comprehensive exam. *Entrance requirements:* For master's and doctorate, GRE General Test, minimum GPA of 3.0. Additional exam requirements/recommendations for international students: Required—TOEFL. *Application deadline:* For fall admission, 1/15 for domestic students, 1/1 for international students; for spring admission, 9/1 for domestic students, 8/15 for international students. Application fee: $50. *Expenses:* Tuition, state resident: full-time $8,400; part-time $200 per credit hour. Tuition, nonresident: full-time $11,088; part-time $462 per credit hour. Tuition and fees vary according to program. *Financial support:* In 2005–06, 41 research assistantships (averaging $19,366 per year), 5 teaching assistantships (averaging $18,198 per year) were awarded. *Unit head:* Dr. Charles Fletcher, Chair, 808-956-8763, Fax: 808-956-5512, E-mail: fletcher@soest.hawaii.edu. *Application contact:* Dr. Charles Fletcher, Chair, 808-956-8763, Fax: 808-956-5512, E-mail: fletcher@soest.hawaii.edu.

University of Houston, College of Natural Sciences and Mathematics, Department of Geosciences, Houston, TX 77204. Offers geology (MA, MS, PhD); geophysics (MA, MS, PhD). Part-time and evening/weekend programs available. *Faculty:* 18 full-time (3 women), 4 part-time/adjunct (0 women). *Students:* 70 full-time (24 women), 52 part-time (17 women); includes 12 minority (4 African Americans, 1 American Indian/Alaska Native, 2 Asian Americans or Pacific Islanders, 5 Hispanic Americans), 65 international. Average age 33. 73 applicants, 85% accepted, 32 enrolled. In 2005, 10 master's, 5 doctorates awarded. *Degree requirements:* For master's, thesis (for some programs); for doctorate, one foreign language, thesis/dissertation. *Entrance requirements:* For master's and doctorate, GRE General Test. Additional exam requirements/recommendations for international students: Required—TOEFL. *Application deadline:* For fall admission, 9/20 for domestic students; for spring admission, 12/4 for domestic students. Applications are processed on a rolling basis. Application fee: $0 ($75 for international students). *Financial support:* In 2005–06, 7 fellowships with full tuition reimbursements (averaging $16,800 per year), 27 research assistantships with full tuition reimbursements (averaging $12,750 per year), 28 teaching assistantships with full tuition reimbursements (averaging $12,750 per year) were awarded; career-related internships or fieldwork, Federal Work-Study, institutionally sponsored loans, scholarships/grants, health care benefits, and unspecified assistantships also available. Support available to part-time students. Financial award application deadline: 3/10. *Faculty research:* Seismic and solid earth geophysics, tectonics, environmental hydrochemistry, carbonates, micropaleontology, structure and tectonics, petroleum geology. *Unit head:* Dr. John Casey, Chairman, 713-743-3399, Fax: 713-748-7906, E-mail: jfcasey@uh.edu. *Application contact:* Dr. Charlotte Sullivan, Graduate Adviser, 713-743-3396, Fax: 713-748-7906, E-mail: esulliva@bayou.uh.edu.

University of Idaho, College of Graduate Studies, College of Science, Department of Geological Sciences, Program in Geology, Moscow, ID 83844-2282. Offers MS, PhD. In 2005, 1 degree awarded. *Entrance requirements:* For master's, minimum GPA of 2.8. *Application deadline:* For fall admission, 8/1 for domestic students; for spring admission, 12/15 for domestic students. Application fee: $55 ($60 for international students). *Expenses:* Tuition, nonresident: full-time $8,770; part-time $130 per credit. Required fees: $4,508; $217 per credit. *Financial support:* Application deadline: 2/15. *Unit head:* Dr. Dennis Geist, Head, Department of Geological Sciences, 208-885-6491.

University of Illinois at Chicago, Graduate College, College of Liberal Arts and Sciences, Department of Earth and Environmental Sciences, Chicago, IL 60607-7128. Offers crystallography (MS, PhD); environmental geology (MS, PhD); geochemistry (MS, PhD); geology (MS, PhD); geomorphology (MS, PhD); geophysics (MS, PhD); geotechnical engineering and geosciences (PhD); hydrogeology (MS, PhD); low-temperature and organic geochemistry (MS, PhD); mineralogy (MS, PhD); paleoclimatology (MS, PhD); paleontology (MS, PhD); petrology (MS, PhD); quaternary geology (MS, PhD); sedimentology (MS, PhD); water resources (MS, PhD). *Degree requirements:* For master's and doctorate, thesis/dissertation. *Entrance requirements:* For master's and doctorate, GRE General Test, minimum GPA of 2.75. Additional exam requirements/recommendations for international students: Required—TOEFL. Electronic applications accepted.

University of Illinois at Urbana–Champaign, Graduate College, College of Liberal Arts and Sciences, Department of Geology, Champaign, IL 61820. Offers earth sciences (MS, PhD); geochemistry (MS, PhD); geology (MS, PhD); geophysics (MS, PhD). *Faculty:* 13 full-time (3 women). *Students:* 32 full-time (13 women), 4 part-time; includes 1 minority (American Indian/Alaska Native), 16 international. Average age 26. 44 applicants, 39% accepted, 13 enrolled. In 2005, 4 master's, 5 doctorates awarded. Terminal master's awarded for partial completion of doctoral program. *Degree requirements:* For master's and doctorate, thesis/dissertation. *Entrance requirements:* For master's and doctorate, GRE General Test, minimum GPA of 3.0. Additional exam requirements/recommendations for international students: Required—TOEFL. *Application deadline:* For fall admission, 2/22 for domestic students; for spring admission, 10/16 for domestic students. Applications are processed on a rolling basis. Application fee: $50 ($60 for international students). *Financial support:* In 2005–06, 8 fellowships, 14 research assistantships, 18 teaching assistantships were awarded; Federal Work-Study and tuition waivers (full and partial) also available. Financial award application deadline: 2/15. *Faculty research:* Hydrogeology, structure/tectonics, mineral science. *Unit head:* Stephen Marshak, Head, 217-333-3542, Fax: 217-244-4996, E-mail: smarshak@uiuc.edu. *Application contact:* Barbara Elmore, Graduate Admissions Secretary, 217-333-3542, Fax: 217-244-4996, E-mail: belmore@uiuc.edu.

University of Kansas, Graduate School, College of Liberal Arts and Sciences, Department of Geology, Lawrence, KS 66045. Offers MS, PhD. *Faculty:* 22. *Students:* 44 full-time (18 women), 17 part-time (6 women); includes 3 minority (all Hispanic Americans), 13 international. Average age 28. 43 applicants, 51% accepted. In 2005, 8 master's, 1 doctorate awarded. *Degree requirements:* For master's, thesis or alternative; for doctorate, thesis/dissertation, comprehensive exam. *Entrance requirements:* For master's and doctorate, GRE General Test, 3 letters of recommendation. Additional exam requirements/recommendations for international students: Required—TOEFL, IELT (minimum score 6). *Application deadline:* For fall admission, 2/1 priority date for domestic students, 2/1 priority date for international students; for spring admission, 10/31 priority date for domestic students, 10/31 priority date for international students. Applications are processed on a rolling basis. Application fee: $55 ($60 for international students). Electronic applications accepted. *Expenses:* Tuition, state resident: full-time $4,859. Tuition, nonresident: full-time $1,200. Required fees: $589. Tuition and fees vary according to program. *Financial support:* Fellowships with full and partial tuition reimbursements, research assistantships with full and partial tuition reimbursements, teaching assistantships with full and partial tuition reimbursements available. Financial award application deadline: 2/1. *Faculty research:* Sedimentology, paleontology, tectonics, geophysics, hyrdogeology. *Unit head:* Robert H Goldstein, Chair, 785-864-4974, E-mail: gold@ku.edu. *Application contact:* Yolanda G. Davis, Graduate Coordinator, 785-864-4974, Fax: 785-864-5276, E-mail: yolanda@ku.edu.

University of Kentucky, Graduate School, Graduate School Programs from the College of Arts and Sciences, Program in Geology, Lexington, KY 40506-0032. Offers MS, PhD. *Faculty:* 12 full-time (2 women), 11 part-time/adjunct (0 women). *Students:* 26 full-time (10 women), 5 part-time (3 women), 4 international. Average age 30. 26 applicants, 46% accepted, 7 enrolled. In 2005, 2 master's, 1 doctorate awarded. *Median time to degree:* Of those who began their doctoral program in fall 1997, 64.3% received their degree in 8 years or less. *Degree requirements:* For master's and doctorate, thesis/dissertation, comprehensive exam. *Entrance requirements:* For master's, GRE General Test, minimum undergraduate GPA of 2.5; for doctorate, GRE General Test, minimum graduate GPA of 3.0. Additional exam requirements/recommendations for international students: Required—TOEFL (minimum score 550 paper-based; 213 computer-based). *Application deadline:* For fall admission, 7/17 priority date for domestic students, 2/1 priority date for international students; for spring admission, 12/13 priority date for domestic students, 6/15 priority date for international students. Applications are processed on a rolling basis. Application fee: $40 ($55 for international students). Electronic applications accepted. *Expenses:* Tuition, state resident: full-time $3,159; part-time $331 per credit hour. Tuition, nonresident: full-time $6,984; part-time $756 per credit hour. Tuition and fees vary according to course load, degree level and program. *Financial support:* In 2005–06, 24 students received support, including 1 fellowship with full tuition reimbursement available, 9 research assistantships with full tuition reimbursements available (averaging $11,338 per year), 14 teaching assistantships with full tuition reimbursements available (averaging $11,338 per year); Federal Work-Study, institutionally sponsored loans, scholarships/grants, traineeships, health care benefits, tuition waivers (partial), and unspecified assistantships also available. Support available to part-time students. Financial award application deadline: 3/15. *Faculty research:* Structure tectonics, geophysics, stratigraphy, hydrogeology, coal geology. *Unit head:* Dr. Alan Fryar, Director of Graduate Studies, 859-257-4392, Fax: 859-323-1938, E-mail: alan.fryar@uky.edu. *Application contact:* Dr. Brian Jackson, Senior Associate Dean, 859-257-8176, Fax: 859-323-1928, E-mail: lance.brunner@uky.edu.

University of Louisiana at Lafayette, Graduate School, College of Sciences, Department of Geology, Lafayette, LA 70504. Offers MS. Part-time programs available. *Faculty:* 5 full-time (1 woman), 1 part-time/adjunct (0 women). *Students:* 10 full-time (2 women), 4 part-time (1 woman). Average age 28. 15 applicants, 67% accepted, 5 enrolled. In 2005, 4 degrees awarded. *Degree requirements:* For master's, thesis, comprehensive exam, registration. *Entrance requirements:* For master's, GRE General Test, minimum GPA of 2.75. Additional exam requirements/recommendations for international students: Required—TOEFL (minimum score 550 paper-based; 213 computer-based). *Application deadline:* For fall admission, 5/15 for domestic students, 5/15 for international students; for spring admission, 10/1 for domestic students, 10/1 for international students. Applications are processed on a rolling basis. Application fee: $20 ($30 for international students). Electronic applications accepted. *Expenses:* Tuition, state resident: full-time $3,330; part-time $93 per credit hour. Tuition, nonresident: full-time $9,510; part-time $350 per semester. International tuition: $9,646 full-time. *Financial support:* In 2005–06, 6 research assistantships with full tuition reimbursements (averaging $6,417 per year) were awarded; teaching assistantships, Federal Work-Study, tuition waivers (full and partial), and unspecified assistantships also available. Financial award application deadline: 5/1. *Faculty research:* Aquifer contamination, coastal erosion, geochemistry of peat, petroleum geology and geophysics, remote sensing and geographic information systems applications. *Unit head:* Dr. Carl Richeter, Head, 337-482-6468, Fax: 337-482-5723, E-mail: crichter@louisiana.edu. *Application contact:* Dr. Brian Lock, Graduate Coordinator, 337-482-6823, Fax: 337-482-5723, E-mail: block@louisiana.edu.

University of Maine, Graduate School, Climate Change Institute, Orono, ME 04469. Offers MS. Part-time programs available. *Students:* 9 full-time (3 women), 1 (woman) part-time, 2 international. Average age 33. 12 applicants, 42% accepted, 1 enrolled. In 2005, 1 master's awarded. *Degree requirements:* For master's, thesis. *Entrance requirements:* For master's, GRE General Test. Additional exam requirements/recommendations for international students: Required—TOEFL. *Application deadline:* For fall admission, 2/1 for domestic students. Applications are processed on a rolling basis. Application fee: $50. Electronic applications accepted. *Financial support:* In 2005–06, 8 research assistantships with tuition reimbursements (averaging $14,800 per year) were awarded; tuition waivers (full and partial) also available. Financial award application deadline: 3/1. *Faculty research:* Geology, glacial geology, anthropology, climate. *Unit head:* Dr. Paul Mayewski, Director, 207-581-3019, Fax: 207-581-1203. *Application contact:* Scott G. Delcourt, Associate Dean of the Graduate School, 207-581-3219, Fax: 207-581-3232, E-mail: graduate@maine.edu.

University of Maine, Graduate School, College of Natural Sciences, Forestry, and Agriculture, Department of Earth Sciences, Orono, ME 04469. Offers MS, PhD. Part-time programs available. *Students:* 28 full-time (13 women), 16 part-time (10 women), 6 international. Average age 30. 31 applicants, 77% accepted, 9 enrolled. In 2005, 5 master's, 2 doctorates awarded. *Degree requirements:* For master's, thesis; for doctorate, one foreign language, thesis/dissertation. *Entrance requirements:* For master's and doctorate, GRE General Test. Additional exam requirements/recommendations for international students: Required—TOEFL. *Application deadline:* For fall admission, 2/1 for domestic students. Applications are processed on a rolling basis. Application fee: $50. Electronic applications accepted. *Financial support:* In 2005–06, 5 research assistantships with tuition reimbursements (averaging $14,800 per year), 5 teaching assistantships with tuition reimbursements (averaging $10,100 per year) were awarded; Federal Work-Study, institutionally sponsored loans, and tuition waivers (full and partial) also available. Financial award application deadline: 3/1. *Faculty research:* Appalachian bedrock geology, Quaternary studies, marine geology. *Unit head:* Dr. Daniel Belknap, Chair, 207-581-2152, Fax: 207-581-2202. *Application contact:* Scott G. Delcourt, Associate Dean of the Graduate School, 207-581-3219, Fax: 207-581-3232, E-mail: graduate@maine.edu.

University of Manitoba, Faculty of Graduate Studies, Faculty of Environment, Earth and Resources, Department of Geological Sciences, Winnipeg, MB R3T 2N2, Canada. Offers geology (M Sc, PhD); geophysics (M Sc, PhD). *Degree requirements:* For master's and doctorate, thesis/dissertation. *Entrance requirements:* For master's and doctorate, GRE General Test, GRE Subject Test (geology), minimum GPA of 3.0. Additional exam requirements/recommendations for international students: Required—TOEFL.

University of Maryland, College Park, Graduate Studies, College of Computer, Mathematical and Physical Sciences, Department of Geology, College Park, MD 20742. Offers MS, PhD. *Faculty:* 25 full-time (4 women), 3 part-time/adjunct (1 woman). *Students:* 20 full-time (11 women), 1 (woman) part-time; includes 3 minority (1 African American, 2 Hispanic Americans), 4 international. 33 applicants, 30% accepted, 5 enrolled. In 2005, 4 degrees awarded. *Degree requirements:* For master's, thesis, oral defense; for doctorate, thesis/dissertation. *Entrance requirements:* For master's, GRE General Test, minimum GPA of 3.0, 3 letters of recommendation; for doctorate, GRE General Test, 3 letters of recommendation. Additional exam requirements/recommendations for international students: Required—TOEFL. *Application deadline:* For fall admission, 3/15 for domestic students, 2/1 for international students. Applications are processed on a rolling basis. Application fee: $60. Electronic applications accepted. *Financial support:* In 2005–06, 7 fellowships with full tuition reimbursements (averaging $9,380 per year), 1 research assistantship with tuition reimbursement (averaging $20,800 per year), 17 teaching assistantships with tuition reimbursements (averaging $18,491 per year) were awarded; Federal Work-Study and scholarships/grants also available. Support available to part-time students. Financial award application deadline: 2/15; financial award applicants required to submit FAFSA. *Faculty research:* Metamorphic petrogenesis, phase equilibria, wetland hydrology, glacial geology, origin and evolution of the earth's crust . Total annual research expenditures: $1.5 million. *Unit head:* Dr. Michael Brown, Chairman, 301-405-4065, Fax: 301-314-9661. *Application contact:* Dean of Graduate School, 301-405-4190, Fax: 301-314-9305.

University of Memphis, Graduate School, College of Arts and Sciences, Department of Earth Sciences, Memphis, TN 38152. Offers earth sciences (PhD); geography (MA, MS); geology (MS); geophysics (MS). Part-time programs available. *Faculty:* 23 full-time (3 women), 3 part-time/adjunct (1 woman). *Students:* Average age 29. 13 applicants, 46% accepted. In 2005, 5 master's, 1 doctorate awarded. Terminal master's awarded for partial completion of doctoral program. *Degree requirements:* For master's, thesis, seminar presentation, comprehensive exam; for doctorate, thesis/dissertation. *Entrance requirements:* For master's and doctorate, GRE General Test. Additional exam requirements/recommendations for international students: Required—TOEFL. *Application deadline:* For fall admission, 8/1 for domestic students; for spring admission, 12/1 for domestic students. Applications are processed on a rolling basis. Application fee: $25 ($50 for international students). Electronic applications accepted. *Financial support:* In 2005–06, 34 students received support, including 2 fellowships with full tuition reimbursements available, 10 research assistantships with full tuition reimbursements available, 11 teaching assistantships with full tuition reimbursements available (averaging $10,000 per year) Financial award application deadline: 1/15. *Faculty research:* Hazards, active tectonics, geophysics, hydrology and water resources, spatial analysis. Total annual research expenditures: $1.8 million. *Unit head:* Dr. Mervin J. Bartholomew, Chair, 901-678-1613, Fax: 901-678-4467, E-mail: jbrthlm1@memphis.edu. *Application contact:* Dr. George Swihart, Coordinator of Graduate Studies, 901-678-2606, Fax: 901-678-2178, E-mail: gswihart@memphis.edu.

University of Michigan, Horace H. Rackham School of Graduate Studies, College of Literature, Science, and the Arts, Department of Geological Sciences, Ann Arbor, MI 48109. Offers geology (MS, PhD); mineralogy (MS, PhD); oceanography: marine geology and geochemistry (MS, PhD). Terminal master's awarded for partial completion of doctoral program. *Degree requirements:* For master's, thesis; for doctorate, thesis/dissertation, oral defense of dissertation, preliminary exam. *Entrance requirements:* For master's and doctorate, GRE General Test. Electronic applications accepted. *Expenses:* Tuition, state resident: full-time $14,082; part-time $894 per credit hour. Tuition, nonresident: full-time $28,500; part-time $1,675 per credit hour. Required fees: $189; $189 per unit. *Faculty research:* Isotope geochemistry, paleoclimatology, mineral physics, tectonics, paleontology.

University of Minnesota, Duluth, Graduate School, College of Science and Engineering, Department of Geological Sciences, Duluth, MN 55812-2496. Offers MS, PhD. Part-time programs available. *Faculty:* 10 full-time (3 women), 3 part-time/adjunct (0 women). *Students:* 15 full-time (7 women), 13 part-time (6 women); includes 1 minority (Hispanic American) Average age 25. 25 applicants, 40% accepted, 6 enrolled. In 2005, 3 master's, 2 doctorates awarded. *Degree requirements:* For master's, thesis, final oral exam, written and oral research proposal. *Entrance requirements:* For master's, GRE General Test, minimum GPA of 3.0. Additional exam requirements/recommendations for international students: Required—TOEFL (minimum score 550 paper-based; 213 computer-based). *Application deadline:* For fall admission, 7/15 for domestic students, 7/15 for international students; for spring admission, 11/15 for domestic students, 11/15 for international students. Applications are processed on a rolling basis. Application fee: $55 ($75 for international students). *Financial support:* In 2005–06, research assistantships with full and partial tuition reimbursements (averaging $11,825 per year), teaching assistantships with full and partial tuition reimbursements (averaging $11,825 per year) were awarded; career-related internships or fieldwork, health care benefits, tuition waivers (partial), and unspecified assistantships also available. Support available to part-time students. Financial award application deadline: 1/5. *Faculty research:* Surface processes, tectonics, planetary geology, paleoclimate, petrology. Total annual research expenditures: $375,000. *Unit head:* Dr. Vicki Hansen, Director of Graduate Studies, 218-726-6211, Fax: 218-726-8275, E-mail: vhansen@d.umn.edu.

University of Minnesota, Twin Cities Campus, Graduate School, Institute of Technology, Department of Geology and Geophysics, Minneapolis, MN 55455-0213. Offers geology (MS, PhD); geophysics (MS, PhD). Terminal master's awarded for partial completion of doctoral program. *Degree requirements:* For master's, thesis optional; for doctorate, thesis/dissertation. *Entrance requirements:* For master's and doctorate, GRE General Test. Additional exam requirements/recommendations for international students: Required—TOEFL. *Expenses:* Tuition, state resident: full-time $8,748; part-time $729 per credit. Tuition, nonresident: full-time $15,848; part-time $1,321 per credit. Full-time tuition and fees vary according to class time, course load, program and reciprocity agreements. *Faculty research:* Hydrogeology, geochemistry, structural geology, sedimentology.

University of Missouri–Columbia, Graduate School, College of Arts and Sciences, Department of Geological Sciences, Columbia, MO 65211. Offers MS, PhD. *Faculty:* 13 full-time (2 women), 1 part-time/adjunct (0 women). *Students:* 16 full-time (6 women), 10 part-time (5 women), 8 international. In 2005, 6 master's, 1 doctorate awarded. *Degree requirements:* For master's, thesis; for doctorate, variable foreign language requirement, thesis/dissertation. *Entrance requirements:* For master's and doctorate, GRE General Test, minimum GPA of 3.0. *Application deadline:* For fall admission, 2/15 for domestic students. Applications are processed on a rolling basis. Application fee: $45 ($60 for international students). *Financial support:* Fellowships, research assistantships, teaching assistantships, institutionally sponsored loans available. *Unit head:* Dr. Robert L. Bauer, Director of Graduate Studies, 573-882-3759, E-mail: bauerr@missouri.edu.

University of Missouri–Kansas City, College of Arts and Sciences, Department of Geosciences, Kansas City, MO 64110-2499. Offers environmental and urban geosciences (MS); geosciences (PhD). PhD offered through the School of Graduate Studies. Part-time programs available. *Faculty:* 10 full-time (2 women), 4 part-time/adjunct (0 women). *Students:* 3 full-time (2 women), 4 part-time; includes 1 minority (African American), 2 international. Average age 32. 8 applicants, 75% accepted, 3 enrolled. *Degree requirements:* For master's and doctorate, thesis/dissertation. *Entrance requirements:* For master's, GRE General Test, minimum GPA of 3.0; for doctorate, qualifying exam. Additional exam requirements/recommendations for international students: Required—TOEFL. *Application deadline:* For fall admission, 3/15 priority date for domestic students, 3/15 priority date for international students. Applications are processed on a rolling basis. Application fee: $35 ($50 for international students). Electronic applications accepted. *Expenses:* Tuition, state resident: full-time $4,738; part-time $263 per credit hour. Tuition, nonresident: full-time $12,235; part-time $679 per credit hour. Required fees: $582. Tuition and fees vary according to course load, program and student level. *Financial support:* In 2005–06, research assistantships with partial tuition reimbursements (averaging $11,000 per year), teaching assistantships with partial tuition reimbursements (averaging $10,500 per year) were awarded; fellowships, Federal Work-Study, institutionally sponsored loans, and tuition waivers (full and partial) also available. Support available to part-time students. Financial award application deadline: 3/15. *Faculty research:* Neotectonics and applied geophysics, environmental geosciences, urban geoscience, geoinformatics-remote sensing, atmospheric research. Total annual research expenditures:$90,000. *Unit head:* Dr. Syed Hasan, Chair, 816-235-2976, Fax: 816-235-5535. *Application contact:* Dr. James Murowchick, Associate Professor, 816-235-2979, Fax: 816-235-5535, E-mail: murowchickj@umkc.edu.

University of Missouri–Rolla, Graduate School, School of Materials, Energy, and Earth Resources, Department of Geological Sciences and Engineering, Program in Geology and Geophysics, Rolla, MO 65409-0910. Offers geochemistry (MS, PhD); geology (MS, PhD); geophysics (MS, PhD); groundwater and environmental geology (MS, PhD). Part-time programs available. *Degree requirements:* For master's and doctorate, thesis/dissertation. *Entrance requirements:* For master's, GRE General Test, GRE Subject Test, minimum GPA of 3.0 in last 4 semesters; for doctorate, GRE General Test, GRE Subject Test. Additional exam requirements/recommendations for international students: Required—TOEFL. Electronic applications accepted. *Faculty research:* Economic geology, geophysical modeling, seismic wave analysis.

The University of Montana, Graduate School, College of Arts and Sciences, Department of Geology, Missoula, MT 59812-0002. Offers applied geoscience (PhD); geology (MS, PhD). *Faculty:* 12 full-time (1 woman), 7 part-time/adjunct (1 woman). *Students:* 30 full-time (10 women), 17 part-time (7 women); includes 2 minority (1 American Indian/Alaska Native, 1 Asian American or Pacific Islander), 1 international. 41 applicants, 63% accepted, 11 enrolled. In 2005, 15 degrees awarded. *Degree requirements:* For doctorate, thesis/dissertation. *Entrance requirements:* For master's and doctorate, GRE General Test. Additional exam requirements/recommendations for international students: Required—TOEFL (minimum score 525 paper-based; 197 computer-based). *Application deadline:* For fall admission, 2/15 for domestic students. Application fee: $45. *Expenses:* Tuition, state resident: part-time $267 per credit. Tuition, nonresident: part-time $665 per credit. Part-time tuition and fees vary according to course load and degree level. *Financial support:* In 2005–06, 9 teaching assistantships with full tuition reimbursements (averaging $9,000 per year) were awarded; Federal Work-Study and unspecified assistantships also available. Financial award application deadline: 3/1; financial award applicants required to submit FAFSA. *Faculty research:* Environmental geoscience, regional structure and tectonics, groundwater geology, petrology, mineral deposits. Total annual research expenditures: $533,844. *Unit head:* Dr. Steven D. Sheriff, Chair, 406-243-6560, Fax: 406-243-4028, E-mail: gl_sds@selway.umt.edu. *Application contact:* Dr. Graham Thompson, Graduate Coordinator, 406-243-4953, E-mail: gl_grt@selway.umt.edu.

University of Nevada, Reno, Graduate School, College of Science, Mackay School of Earth Sciences and Engineering, Department of Geological Sciences, Reno, NV 89557. Offers geochemistry (MS, PhD); geological engineering (MS, Geol E); geology (MS, PhD); geophysics (MS, PhD). *Degree requirements:* For master's, thesis optional; for doctorate, one foreign language, thesis/dissertation. *Entrance requirements:* For master's, GRE General Test, GRE Subject Test, minimum GPA of 2.75; for doctorate, GRE General Test, GRE Subject Test, minimum GPA of 3.0. Additional exam requirements/recommendations for international students: Required—TOEFL. *Faculty research:* Hydrothermal ore deposits, metamorphic and igneous petrogenesis, sedimentary rock record of earth history, field and petrographic investigation of magnetism, rock fracture mechanics.

University of New Brunswick Fredericton, School of Graduate Studies, Faculty of Science, Department of Geology, Fredericton, NB E3B 5A3, Canada. Offers M Sc, PhD. Part-time programs available. *Degree requirements:* For master's and doctorate, thesis/dissertation. *Entrance requirements:* For master's and doctorate, minimum GPA of 3.0. Additional exam requirements/recommendations for international students: Required—TOEFL, TWE. Electronic applications accepted. *Faculty research:* Hydrogeology, glacial geology, petrology, paleontology, planetary geology.

University of New Hampshire, Graduate School, College of Engineering and Physical Sciences, Department of Earth Sciences, Durham, NH 03824. Offers earth sciences (MS), including geochemical, geology, ocean mapping, oceanography; hydrology (MS). *Faculty:* 29 full-time. *Students:* 17 full-time (5 women), 24 part-time (10 women); includes 2 minority (1 African American, 1 Asian American or Pacific Islander), 6 international. Average age 31. 37 applicants, 84% accepted, 10 enrolled. In 2005, 10 master's awarded. *Degree requirements:* For master's, thesis. *Entrance requirements:* For master's, GRE General Test. Additional exam requirements/recommendations for international students: Required—TOEFL (minimum score 550 paper-based; 213 computer-based); Recommended—TSE. *Application deadline:* For fall admission, 4/1 priority date for domestic students, 4/1 priority date for international students. For winter admission, 12/1 for domestic students. Applications are processed on a rolling basis. Application fee: $60. Electronic applications accepted. *Expenses:* Tuition, state resident: full-time $8,010; part-time $445 per credit hour. Tuition, nonresident: full-time $19,730; part-time $810 per credit hour. Required fees: $322 per semester. Tuition and fees vary according to course load and program. *Financial support:* In 2005–06, 1 fellowship, 9 research assistantships, 9 teaching assistantships were awarded; career-related internships or fieldwork, Federal Work-Study, scholarships/grants, and tuition waivers (full and partial) also available. Support available to part-time students. Financial award application deadline: 2/15. *Unit head:* Dr. Matt Davis, Chairperson, 603-862-1718, E-mail: earth.sciences@unh.edu. *Application contact:* Nancy Gauthier, Administrative Assistant, 603-862-1720, E-mail: earth.sciences@unh.edu.

University of New Orleans, Graduate School, College of Sciences, Department of Geology and Geophysics, New Orleans, LA 70148. Offers MS. Evening/weekend programs available. *Degree requirements:* For master's, thesis. *Entrance requirements:* For master's, GRE General Test. Additional exam requirements/recommendations for international students: Required—TOEFL (minimum score 550 paper-based; 213 computer-based). Electronic applications accepted. *Faculty research:* Continental margin structure and seismology, burial diagenesis of siliclastic sediments, tectonics at convergent plate margins, continental shelf sediment stability, early diagenesis of carbonates.

The University of North Carolina at Chapel Hill, Graduate School, College of Arts and Sciences, Department of Geological Sciences, Chapel Hill, NC 27599. Offers MS, PhD. *Degree requirements:* For master's, thesis, comprehensive exam; for doctorate, one foreign language, thesis/dissertation, comprehensive exam. *Entrance requirements:* For master's and doctorate, GRE General Test, minimum GPA of 3.0. Electronic applications accepted. *Faculty research:* Paleoceanography, igneous petrology, paleontology, geophysics, structural geology.

The University of North Carolina Wilmington, College of Arts and Sciences, Department of Earth Sciences, Wilmington, NC 28403-3297. Offers geology (MS); marine science (MS). *Faculty:* 20 full-time (4 women), 1 (woman) part-time/adjunct. *Students:* 11 full-time (7 women), 57 part-time (38 women); includes 1 minority (Hispanic American), 1 international. Average age 35. 42 applicants, 48% accepted, 17 enrolled. In 2005, 26 degrees awarded. *Degree requirements:* For master's, thesis, comprehensive exam. *Entrance requirements:* For master's, GRE General Test, GRE Subject Test, minimum B average in undergraduate major and basic courses for prerequisite to geology. *Application deadline:* For fall admission, 2/15 for domestic students. Applications are processed on a rolling basis. Application fee: $45. *Financial support:* In 2005–06, 9 teaching assistantships were awarded; career-related internships or fieldwork

Geology

The University of North Carolina Wilmington (continued)

and Federal Work-Study also available. Support available to part-time students. Financial award application deadline: 3/15. *Unit head:* Dr. Richard A. Laws, Chair, 910-962-4125, Fax: 910-962-7077. *Application contact:* Dr. Robert D. Roer, Dean, Graduate School, 910-962-4117, Fax: 910-962-3787, E-mail: roer@uncw.edu.

See Close-Up on page 237.

University of North Dakota, Graduate School, School of Engineering and Mines, Department of Geology, Grand Forks, ND 58202. Offers MA, MS, PhD. *Faculty:* 11 full-time (0 women). *Students:* 11 applicants, 45% accepted, 4 enrolled. In 2005, 1 degree awarded. *Degree requirements:* For master's, thesis, final exam; for doctorate, one foreign language, thesis/dissertation, final exam, comprehensive exam. *Entrance requirements:* For master's and doctorate, GRE General Test, minimum GPA of 3.0. Additional exam requirements/recommendations for international students: Required—TOEFL (minimum score 550 paper-based; 213 computer-based). *Application deadline:* For fall admission, 2/15 priority date for domestic students, 2/15 priority date for international students; for spring admission, 10/15 priority date for domestic students, 10/15 priority date for international students. Applications are processed on a rolling basis. Application fee: $35. Electronic applications accepted. *Financial support:* In 2005–06, 10 students received support, including 3 research assistantships with full tuition reimbursements available (averaging $11,921 per year), 6 teaching assistantships with full tuition reimbursements available (averaging $9,979 per year); fellowships, career-related internships or fieldwork, Federal Work-Study, institutionally sponsored loans, scholarships/grants, and tuition waivers (full and partial) also available. Support available to part-time students. Financial award application deadline: 3/15; financial award applicants required to submit FAFSA. *Faculty research:* Hydrogeology, environmental geology, geological engineering, sedimentology, geomorphology. *Unit head:* Dr. Philip J. Gerla, Graduate Director, 701-777-3305, Fax: 701-777-4449, E-mail: phil.gerla@mail.und.nodak.edu. *Application contact:* Linda Baeza, Admissions Officer, 701-777-2945, Fax: 701-777-3619, E-mail: gradschool@mail.und.nodak.edu.

University of Oklahoma, Graduate College, College of Earth and Energy, School of Geology and Geophysics, Program in Geology, Norman, OK 73019-0390. Offers MS, PhD. Part-time programs available. *Students:* 44 full-time (23 women), 15 part-time (5 women); includes 3 minority (1 African American, 1 Asian American or Pacific Islander, 1 Hispanic American), 29 international. 24 applicants, 54% accepted, 8 enrolled. In 2005, 11 master's, 2 doctorates awarded. Terminal master's awarded for partial completion of doctoral program. *Degree requirements:* For master's, thesis, comprehensive exam; for doctorate, one foreign language, thesis/dissertation, general exam. *Entrance requirements:* For master's, GRE General Test, bachelor's degree in geology; for doctorate, GRE General Test. Additional exam requirements/recommendations for international students: Required—TOEFL (minimum score 550 paper-based; 213 computer-based). *Application deadline:* For fall admission, 2/1 priority date for domestic students, 4/1 priority date for international students; for spring admission, 9/1 for domestic students, 9/1 for international students. Applications are processed on a rolling basis. Application fee: $40 ($90 for international students). *Expenses:* Tuition, state resident: full-time $3,029; part-time $126 per credit hour. Tuition, nonresident: full-time $10,807; part-time $450 per credit hour. Required fees: $1,231; $44 per credit hour. Tuition and fees vary according to course load and program. *Financial support:* In 2005–06, 13 students received support, including 3 fellowships (averaging $1,500 per year); research assistantships with partial tuition reimbursements available, teaching assistantships with partial tuition reimbursements available, career-related internships or fieldwork, scholarships/grants, tuition waivers (partial), and unspecified assistantships also available. Financial award application deadline: 2/1; financial award applicants required to submit FAFSA. *Faculty research:* Petroleum geology, geochemistry; sedimentology; structural geology and paleontology, paleobotany. *Application contact:* Donna S. Mullins, Coordinator of Administrative Student Services, 405-325-3255, Fax: 405-325-3140, E-mail: dsmullins@ou.edu.

University of Oregon, Graduate School, College of Arts and Sciences, Department of Geological Sciences, Eugene, OR 97403. Offers MA, MS, PhD. *Faculty:* 13 full-time (2 women), 4 part-time/adjunct (2 women). *Students:* 29 full-time (18 women), 1 part-time; includes 2 minority (1 Asian American or Pacific Islander, 1 Hispanic American), 7 international. 58 applicants, 31% accepted. In 2005, 6 master's, 11 doctorates awarded. *Degree requirements:* For master's, foreign language (MA). *Entrance requirements:* For master's and doctorate, GRE General Test, GRE Subject Test. *Application deadline:* For fall admission, 2/1 for domestic students. Application fee: $50. *Financial support:* In 2005–06, 22 teaching assistantships were awarded; research assistantships, career-related internships or fieldwork and Federal Work-Study also available. Financial award application deadline:2/1. *Unit head:* Dana Johnston, Head, 541-346-5588. *Application contact:* Pat Kallunki, Admissions Contact, 541-346-4573, E-mail: pkallunk@uoregon.edu.

University of Pennsylvania, School of Arts and Sciences, Graduate Group in Geology, Philadelphia, PA 19104. Offers MS, PhD. Part-time programs available. *Degree requirements:* For master's and doctorate, one foreign language, thesis/dissertation. *Entrance requirements:* For master's and doctorate, GRE General Test. Additional exam requirements/recommendations for international students: Required—TOEFL. Electronic applications accepted. *Faculty research:* Isotope geochemistry, regional tectonics, environmental geology, metamorphic and igneous petrology, paleontology.

University of Pittsburgh, School of Arts and Sciences, Department of Geology and Planetary Science, Pittsburgh, PA 15260. Offers geographical information systems (PM Sc); geology and planetary science (MS, PhD). Part-time programs available. *Faculty:* 12 full-time (1 woman), 2 part-time/adjunct (0 women). *Students:* 20 full-time (6 women), 12 part-time (4 women); includes 2 minority (1 Asian American or Pacific Islander, 1 Hispanic American), 2 international. Average age 30. 43 applicants, 21% accepted, 7 enrolled. In 2005, 5 master's, 1 doctorate awarded. *Median time to degree:* Of those who began their doctoral program in fall 1997, 100% received their degree in 8 years or less. *Degree requirements:* For master's, thesis, oral thesis defense; for doctorate, thesis/dissertation, oral dissertation defense. *Entrance requirements:* For master's and doctorate, GRE General Test. Additional exam requirements/recommendations for international students: Required—TOEFL. *Application deadline:* For fall admission, 8/1 priority date for domestic students, 4/30 priority date for international students. For winter admission, 12/1 for domestic students; for spring admission, 4/1 for domestic students. Applications are processed on a rolling basis. Application fee: $50. Electronic applications accepted. *Expenses:* Tuition, state resident: full-time $13,194; part-time $537 per credit. Tuition, nonresident: full-time $25,012; part-time $1,026 per credit. Required fees: $700; $164 per term. Tuition and fees vary according to campus/location and program. *Financial support:* In 2005–06, 19 students received support, including fellowships with tuition reimbursements available (averaging $13,690 per year), 7 research assistantships with tuition reimbursements available (averaging $13,200 per year), 12 teaching assistantships with tuition reimbursements available (averaging $13,555 per year); career-related internships or fieldwork, Federal Work-Study, institutionally sponsored loans, scholarships/grants, and tuition waivers (full and partial) also available. Support available to part-time students. Financial award application deadline: 2/1; financial award applicants required to submit FAFSA. *Faculty research:* Geographical information systems, hydrology, low temperature geochemistry, radiogenic isotopes, volcanology. Total annual research expenditures: $745,848. *Unit head:* Dr. Brian W Stewart, Chair, 412-624-8783, Fax: 412-624-3914, E-mail: bstewart@pitt.edu. *Application contact:* Dr. Thomas Anderson, Graduate Adviser, 412-624-8870, Fax: 412-624-3914, E-mail: taco@pitt.edu.

University of Puerto Rico, Mayagüez Campus, Graduate Studies, College of Arts and Sciences, Department of Geology, Mayagüez, PR 00681-9000. Offers MS. Part-time programs available. *Faculty:* 8 full-time (1 woman). *Students:* 6 full-time (5 women), 13 part-time (7 women); includes 17 minority (all Hispanic Americans), 2 international. 8 applicants, 75% accepted, 6 enrolled. *Degree requirements:* For master's, thesis, comprehensive exam. *Entrance requirements:* For master's, GRE. *Application deadline:* For fall admission, 2/15 for domestic students; for spring admission, 9/15 for domestic students. Applications are processed on a rolling basis. Application fee: $20. *Expenses:* Tuition, state resident: full-time $900; part-time $100 per credit. International tuition: $4,655 full-time. Part-time tuition and fees vary according to course level and course load. *Financial support:* In 2005–06, 7 students received support, including fellowships (averaging $1,500 per year), 5 research assistantships (averaging $1,200 per year), 2 teaching assistantships (averaging $987 per year) *Faculty research:* Seismology, applied geophysics, GIS, environmental remote sensing, petrology. Total annual research expenditures: $16,250. *Unit head:* Dr. Johannes Schellekens, Director, 787-265-3845.

University of Regina, Faculty of Graduate Studies and Research, Faculty of Science, Department of Geology, Regina, SK S4S 0A2, Canada. Offers M Sc, PhD. PhD program offered on a special case basis. *Faculty:* 7 full-time (2 women), 5 part-time/adjunct (0 women). *Students:* 13 full-time (5 women), 7 part-time (2 women). 5 applicants, 20% accepted. In 2005, 1 master's, 1 doctorate awarded. *Degree requirements:* For master's and doctorate, thesis/dissertation, registration. *Entrance requirements:* Additional exam requirements/recommendations for international students: Required—TOEFL (minimum score 580 paper-based; 237 computer-based). *Application deadline:* Applications are processed on a rolling basis. Application fee: $60 ($100 for international students). *Financial support:* In 2005–06, 2 fellowships (averaging $14,886 per year), 1 research assistantship (averaging $12,750 per year), 13 teaching assistantships (averaging $13,501 per year) were awarded; scholarships/grants also available. Financial award application deadline: 6/15. *Faculty research:* Geological and planetary science, petrology, mineralogy and economic geology. *Unit head:* Dr. Janis Dale, Head, 306-585-4840, Fax: 306-585-5433, E-mail: janis.dale@uregina.ca. *Application contact:* Dr. Ian Coulson, Associate Professor and Graduate Student Coordinator, 306-585-4184.

University of Rochester, The College, Arts and Sciences, Department of Earth and Environmental Sciences, Rochester, NY 14627-0250. Offers geological sciences (MS, PhD). *Degree requirements:* For doctorate, thesis/dissertation, qualifying exam. *Entrance requirements:* For master's and doctorate, GRE General Test. Additional exam requirements/recommendations for international students: Required—TOEFL.

See Close-Up on page 239.

University of Saskatchewan, College of Graduate Studies and Research, College of Arts and Sciences and College of Engineering, Department of Geological Sciences, Saskatoon, SK S7N 5A2, Canada. Offers M Sc, PhD, Diploma. *Degree requirements:* For master's and doctorate, thesis/dissertation, registration. *Entrance requirements:* Additional exam requirements/recommendations for international students: Required—TOEFL.

University of South Carolina, The Graduate School, College of Arts and Sciences, Department of Geological Sciences, Columbia, SC 29208. Offers environmental geoscience (PMS); geological sciences (MS, PhD). Terminal master's awarded for partial completion of doctoral program. *Degree requirements:* For master's, thesis; for doctorate, thesis/dissertation, published paper, comprehensive exam. *Entrance requirements:* For master's and doctorate, GRE General Test. Additional exam requirements/recommendations for international students: Required—TOEFL. Electronic applications accepted. *Faculty research:* Environmental geology, tectonics, petrology, coastal processes, paleoclimatology.

University of Southern Mississippi, Graduate School, College of Science and Technology, Department of Geology, Hattiesburg, MS 39406-0001. Offers MS. Part-time programs available. *Degree requirements:* For master's, thesis, comprehensive exam. *Entrance requirements:* For master's, GRE General Test, BS in geology, minimum GPA of 2.75 in last 60 hours. Additional exam requirements/recommendations for international students: Required—TOEFL. *Faculty research:* Volcanic rocks and associated minerals, marine stratigraphy and seismology, hydrology, micropaleontology, isotope geology.

University of South Florida, College of Graduate Studies, College of Arts and Sciences, Department of Geology, Tampa, FL 33620-9951. Offers MS, PhD. Part-time programs available. *Faculty:* 12 full-time (3 women). *Students:* 29 full-time (18 women), 21 part-time (9 women); includes 4 minority (1 African American, 1 Asian American or Pacific Islander, 2 Hispanic Americans), 15 international. 45 applicants, 51% accepted, 14 enrolled. In 2005, 6 degrees awarded. *Degree requirements:* For master's, thesis (for some programs); for doctorate, thesis/dissertation. *Entrance requirements:* For master's, GRE General Test, minimum GPA of 3.0 in last 60 hours of course work; for doctorate, GRE General Test. *Application deadline:* For fall admission, 2/15 for domestic students; for spring admission, 10/15 for domestic students. Application fee: $30. Electronic applications accepted. *Financial support:* Fellowships with partial tuition reimbursements, research assistantships with partial tuition reimbursements, teaching assistantships with partial tuition reimbursements, institutionally sponsored loans and scholarships/grants available. Financial award application deadline: 6/30; financial award applicants required to submit FAFSA. *Faculty research:* Coastal geology, environmental geology and hydrogeology, paleontology and geochemistry. *Unit head:* Chuck Connor, Chairperson, 813-974-0325, Fax: 813-974-2654. *Application contact:* Dr. Peter Harries, Graduate Coordinator, 813-974-4974, Fax: 813-974-2654, E-mail: harries@chuma.cas.usf.edu.

The University of Tennessee, Graduate School, College of Arts and Sciences, Department of Geological Sciences, Knoxville, TN 37996. Offers geology (MS, PhD). Part-time programs available. *Degree requirements:* For master's, thesis; for doctorate, one foreign language, thesis/dissertation. *Entrance requirements:* For master's and doctorate, GRE General Test, minimum GPA of 2.7. Additional exam requirements/recommendations for international students: Required—TOEFL. Electronic applications accepted.

The University of Texas at Arlington, Graduate School, College of Science, Department of Geology, Arlington, TX 76019. Offers environmental science (MS, PhD); geology (MS); math: geoscience (PhD). Part-time and evening/weekend programs available. *Faculty:* 4 full-time (0 women), 2 part-time/adjunct (0 women). *Students:* 5 full-time (2 women), 7 part-time (4 women); includes 1 minority (African American), 2 international. 5 applicants, 100% accepted, 2 enrolled. In 2005, 2 master's awarded. Terminal master's awarded for partial completion of doctoral program. *Degree requirements:* For master's, thesis optional; for doctorate, thesis/dissertation, comprehensive exam. *Entrance requirements:* For master's, GRE General Test. Additional exam requirements/recommendations for international students: Required—TOEFL (minimum score 550 paper-based; 213 computer-based). *Application deadline:* For fall admission, 6/16 for domestic students. Applications are processed on a rolling basis. Application fee: $35 ($50 for international students). Electronic applications accepted. *Expenses:* Tuition, state resident: full-time $3,350. Tuition, nonresident: full-time $8,318. International tuition: $8,448 full-time. Required fees: $1,277. Full-time tuition and fees vary according to course level and program. *Financial support:* In 2005–06, 7 students received support, including 4 fellowships (averaging $1,000 per year), 7 teaching assistantships (averaging $14,700 per year); career-related internships or fieldwork, Federal Work-Study, institutionally sponsored loans, scholarships/grants, health care benefits, and unspecified assistantships also available. Financial award application deadline: 6/1; financial award applicants required to submit FAFSA. *Faculty research:* Hydrology, aqueous geochemistry, biostratigraphy, structural geology, petroleum geology. Total annual research expenditures: $250,000. *Unit head:* Dr. John S. Wickham, Chair, 817-272-2987, Fax: 817-272-2628, E-mail: wickham@uta.edu. *Application contact:* Dr. William L. Balsam, Graduate Adviser, 817-272-2987, Fax: 817-272-2628, E-mail: balsam@uta.edu.

The University of Texas at Austin, Graduate School, College of Natural Sciences, Department of Geological Sciences, Austin, TX 78712-1111. Offers MA, MS, PhD. Part-time programs available. *Degree requirements:* For master's, report (MA), thesis (MS); for doctorate, thesis/dissertation. *Entrance requirements:* For master's and doctorate, GRE General Test. Electronic applications accepted. *Faculty research:* Sedimentary geology, geophysics, hydrogeology, structure/tectonics, vertebrate paleontology.

The University of Texas at El Paso, Graduate School, College of Science, Department of Geological Sciences, El Paso, TX 79968-0001. Offers geological sciences (MS, PhD); geophysics (MS). Part-time and evening/weekend programs available. *Degree requirements:* For master's, thesis; for doctorate, one foreign language, thesis/dissertation. *Entrance requirements:* For master's, GRE, minimum GPA of 3.0, BS in geology or equivalent; for doctorate, GRE, minimum GPA of 3.0, MS in geology or equivalent. Additional exam requirements/recommendations for international students: Required—TOEFL. Electronic applications accepted.

The University of Texas at San Antonio, College of Sciences, Department of Earth and Environmental Sciences, San Antonio, TX 78249-0617. Offers environmental science and engineering (PhD); environmental sciences (MS); geology (MS). *Degree requirements:* For master's, thesis optional; for doctorate, thesis/dissertation, comprehensive exam, registration. *Entrance requirements:* For master's, GRE General Test, minimum GPA of 3.0 in last 60 hours; for doctorate, GRE, resumé, 3 letters of recommendation. Additional exam requirements/recommendations for international students: Required—TOEFL (minimum score 500 paper-based; 173 computer-based). Electronic applications accepted.

The University of Texas of the Permian Basin, Office of Graduate Studies, College of Arts and Sciences, Department of Sciences and Mathematics, Program in Geology, Odessa, TX 79762-0001. Offers MS. *Degree requirements:* For master's, thesis or alternative, comprehensive exam, registration. *Entrance requirements:* For master's, GRE General Test. Additional exam requirements/recommendations for international students: Required—TOEFL (minimum score 550 paper-based; 213 computer-based).

The University of Toledo, Graduate School, College of Arts and Sciences, Department of Earth, Ecological and Environmental Sciences, Toledo, OH 43606-3390. Offers biology (ecology track) (MS, PhD); geology (MS), including earth surface processes, general geology. Part-time programs available. *Faculty:* 28. *Students:* 8 full-time (6 women), 2 part-time (both women). Average age 31. 6 applicants, 83% accepted, 3 enrolled. In 2005, 2 degrees awarded. *Degree requirements:* For master's, thesis. *Entrance requirements:* For master's, GRE General Test. Additional exam requirements/recommendations for international students: Required—TOEFL. *Application deadline:* For fall admission, 8/1 for domestic students. Applications are processed on a rolling basis. Application fee: $45. Electronic applications accepted. *Expenses:* Tuition, area resident: Part-time $308 per credit hour. Tuition, state resident: full-time $3,312. Tuition, nonresident: full-time $6,616; part-time $735 per credit hour. *Financial support:* In 2005–06, 2 research assistantships (averaging $14,000 per year), 30 teaching assistantships (averaging $11,724 per year) were awarded; Federal Work-Study, institutionally sponsored loans, and tuition waivers (full) also available. Support available to part-time students. Financial award application deadline: 4/1; financial award applicants required to submit FAFSA. *Faculty research:* Environmental geochemistry, geophysics, petrology and mineralogy, paleontology, geohydrology. *Unit head:* Dr. Michael Phillips, Chair, 419-530-4572, Fax: 419-530-4421, E-mail: michael.phillips@utoledo.edu. *Application contact:* Johan Gottgens, 419-530-8451, E-mail: john.gottgens@utoledo.edu.

University of Toronto, School of Graduate Studies, Physical Sciences Division, Department of Geology, Toronto, ON M5S 1A1, Canada. Offers M Sc, MA Sc, PhD. Part-time programs available. *Degree requirements:* For master's, thesis (for some programs); for doctorate, thesis/dissertation. *Entrance requirements:* For master's, B Sc or BA Sc, or equivalent; minimum B average; letters of reference; for doctorate, M Sc or equivalent, minimum B+ average, letters of reference.

University of Tulsa, Graduate School, College of Business Administration and College of Engineering and Natural Sciences, Department of Engineering and Technology Management, Tulsa, OK 74104-3189. Offers chemical engineering (METM); computer science (METM); electrical engineering (METM); geological science (METM); mathematics (METM); mechanical engineering (METM); petroleum engineering (METM). Part-time and evening/weekend programs available. *Students:* 3 full-time (0 women), 2 part-time (1 woman); includes 3 minority (1 American Indian/Alaska Native, 2 Hispanic Americans), 1 international. Average age 28. 9 applicants, 33% accepted, 2 enrolled. In 2005, 1 degree awarded. *Entrance requirements:* For master's, GRE General Test or GMAT. Additional exam requirements/recommendations for international students: Required—TOEFL (minimum score 575 paper-based; 231 computer-based). *Application deadline:* Applications are processed on a rolling basis. Application fee: $30. Electronic applications accepted. *Expenses:* Tuition: Full-time $12,132; part-time $674 per credit hour. Required fees: $60; $3 per credit hour. *Financial support:* Fellowships, research assistantships with full and partial tuition reimbursements, teaching assistantships, Federal Work-Study, scholarships/grants, tuition waivers (full and partial), and unspecified assistantships available. Support available to part-time students. Financial award application deadline: 2/1; financial award applicants required to submit FAFSA. *Unit head:* Ron Cooper, Director of Graduate Business Studies, 918-631-2680, Fax: 918-631-2142, E-mail: ron-cooper@utulsa.edu.

University of Utah, The Graduate School, College of Mines and Earth Sciences, Department of Geology and Geophysics, Salt Lake City, UT 84112-1107. Offers environmental engineering (ME, MS, PhD); geological engineering (ME, MS, PhD); geology (MS, PhD); geophysics (MS, PhD). *Faculty:* 22 full-time (4 women), 3 part-time/adjunct (0 women). *Students:* 45 full-time (13 women), 20 part-time (7 women), 14 international. Average age 31. 81 applicants, 37% accepted, 16 enrolled. In 2005, 10 master's, 3 doctorates awarded. Terminal master's awarded for partial completion of doctoral program. *Median time to degree:* Of those who began their doctoral program in fall 1997, 33% received their degree in 8 years or less. *Degree requirements:* For master's, thesis, comprehensive exam; for doctorate, one foreign language, thesis/dissertation, qualifying exam. *Entrance requirements:* For master's and doctorate, GRE General Test, minimum GPA of 3.25. Additional exam requirements/recommendations for international students: Required—TOEFL (minimum score 500 paper-based; 173 computer-based). *Application deadline:* For fall admission, 1/15 priority date for domestic students, 1/15 priority date for international students. Applications are processed on a rolling basis. Application fee: $45 ($65 for international students). Electronic applications accepted. *Expenses:* Tuition, state resident: full-time $2,932; part-time $2,212 per term. Tuition, nonresident: full-time $10,350; part-time $7,812 per term. Required fees: $590; $516 per term. Tuition and fees vary according to course load and program. *Financial support:* In 2005–06, 7 fellowships with full tuition reimbursements (averaging $30,000 per year), 22 research assistantships with full tuition reimbursements (averaging $29,200 per year), 11 teaching assistantships with full tuition reimbursements (averaging $12,183 per year) were awarded; career-related internships or fieldwork, institutionally sponsored loans, scholarships/grants, tuition waivers, unspecified assistantships, and stipends also available. Financial award application deadline: 1/15; financial award applicants required to submit FAFSA. *Faculty research:* Igneous, metamorphic, and sedimentary petrology; ore deposits; aqueous geochemistry; isotope geochemistry; heat flow. Total annual research expenditures: $2.3 million. *Unit head:* Dr. Marjorie A. Chan, Chair, 801-581-7162, Fax: 801-581-7065, E-mail: gg_chair@mines.utah.edu. *Application contact:* Dr. Richard D Jarrad, Director of Graduate Studies, 801-581-6553, Fax: 801-581-7065, E-mail: jarrad@earth.utah.edu.

University of Vermont, Graduate College, College of Arts and Sciences, Department of Geology, Burlington, VT 05405. Offers MS. *Students:* 9 (5 women); includes 1 minority (African American) 2 international. 15 applicants, 27% accepted, 4 enrolled. In 2005, 5 degrees awarded. *Degree requirements:* For master's, thesis. *Entrance requirements:* For master's, GRE General Test. Additional exam requirements/recommendations for international students: Required—TOEFL (minimum score 550 paper-based; 213 computer-based). *Application deadline:* For fall admission, 2/15 for domestic students. Applications are processed on a rolling basis. Application fee: $40. Electronic applications accepted. *Expenses:* Tuition, area resident: Part-time $410 per credit hour. Tuition, nonresident: part-time $1,034 per credit hour. *Financial support:* Research assistantships, teaching assistantships available. Financial award application deadline: 3/1. *Faculty research:* Mineralogy, lake sediments, structural geology. *Unit head:* Dr. Char Mehrtens, Chairperson, 802-656-3396. *Application contact:* Dr. T. Rushmer, Coordinator, 802-656-3396.

University of Washington, Graduate School, College of Arts and Sciences, Department of Earth and Space Sciences, Seattle, WA 98195. Offers geology (MS, PhD); geophysics (MS, PhD). *Degree requirements:* For master's, thesis or alternative, departmental qualifying exam, final exam; for doctorate, thesis/dissertation, departmental qualifying exam, general and final exams. *Entrance requirements:* For master's and doctorate, GRE General Test, minimum GPA of 3.0. Additional exam requirements/recommendations for international students: Required—TOEFL (minimum score 580 paper-based). Electronic applications accepted.

The University of Western Ontario, Faculty of Graduate Studies, Physical Sciences Division, Department of Earth Sciences, London, ON N6A 5B8, Canada. Offers geology (M Sc, PhD); geology and environmental science (M Sc, PhD); geophysics (M Sc, PhD); geophysics and environmental science (M Sc, PhD). *Degree requirements:* For master's, thesis, registration; for doctorate, thesis/dissertation, qualifying exam. *Entrance requirements:* For master's, honors in B Sc; for doctorate, M Sc. Additional exam requirements/recommendations for international students: Required—TOEFL. *Faculty research:* Geophysics, geochemistry, paleontology, sedimentology/stratigraphy, glaciology/quaternary.

University of Wisconsin–Madison, Graduate School, College of Letters and Science, Department of Geology and Geophysics, Program in Geology, Madison, WI 53706-1380. Offers MS, PhD. *Degree requirements:* For master's, thesis; for doctorate, one foreign language, thesis/dissertation. *Entrance requirements:* For master's and doctorate, GRE General Test.

University of Wisconsin–Milwaukee, Graduate School, College of Letters and Sciences, Department of Geosciences, Milwaukee, WI 53201-0413. Offers geological sciences (MS, PhD). *Faculty:* 10 full-time (2 women). *Students:* 6 full-time (5 women), 6 part-time (3 women). 10 applicants, 50% accepted, 4 enrolled. In 2005, 4 degrees awarded. *Degree requirements:* For master's, thesis; for doctorate, one foreign language, thesis/dissertation. *Entrance requirements:* For master's and doctorate, GRE General Test. *Application deadline:* For fall admission, 1/1 for domestic students; for spring admission, 9/1 for domestic students. Applications are processed on a rolling basis. Application fee: $45 ($75 for international students). *Expenses:* Tuition, area resident: Part-time $716 per credit. Tuition, state resident: part-time $776 per credit. Tuition, nonresident: part-time $1,614. Required fees: $229 per term. Tuition and fees vary according to course load and program. *Financial support:* In 2005–06, 8 teaching assistantships were awarded; fellowships, research assistantships, career-related internships or fieldwork and unspecified assistantships also available. Support available to part-time students. Financial award application deadline: 4/15. *Unit head:* Douglas Cherkauer, Representative, 414-229-4562, Fax: 414-229-5452, E-mail: aquadoc@uwm.edu.

University of Wyoming, Graduate School, College of Arts and Sciences, Department of Geology and Geophysics, Laramie, WY 82070. Offers geology (MS, PhD); geophysics (MS, PhD). Part-time programs available. *Faculty:* 17 full-time (2 women), 3 part-time/adjunct (2 women). *Students:* 24 full-time (13 women), 28 part-time (8 women); includes 2 minority (1 Asian American or Pacific Islander, 1 Hispanic American), 6 international. 21 applicants, 67% accepted. In 2005, 12 master's, 2 doctorates awarded. *Degree requirements:* For master's and doctorate, variable foreign language requirement, thesis/dissertation. *Entrance requirements:* For master's and doctorate, GRE General Test, minimum GPA of 3.0. *Application deadline:* For fall admission, 1/15 for domestic students, 1/15 for international students. Applications are processed on a rolling basis. Application fee: $50. *Expenses:* Tuition, state resident: full-time $3,720; part-time $155 per credit hour. Tuition, nonresident: full-time $10,704; part-time $446 per credit hour. Required fees: $666; $162 per semester. Tuition and fees vary according to course load and program. *Financial support:* Fellowships, research assistantships, teaching assistantships, career-related internships or fieldwork, Federal Work-Study, and institutionally sponsored loans available. Financial award application deadline: 3/1. *Faculty research:* Geochemistry and petroleum geology, tectonics and sedimentation, geomorphology and remote sensing, igneous and metamorphic petrology, structure, geohydrology. *Unit head:* Dr. Arthur W. Snoke, Head, 307-766-3386. *Application contact:* Sondra S. Cawley, Admissions Coordinator, 307-766-3389, Fax: 307-766-6679, E-mail: acadcoord.geol@uwyo.edu.

Utah State University, School of Graduate Studies, College of Science, Department of Geology, Logan, UT 84322. Offers MS. *Faculty:* 7 full-time (0 women), 2 part-time/adjunct (both women). *Students:* 33 full-time (7 women), 12 part-time (1 woman). Average age 24. 30 applicants, 63% accepted, 15 enrolled. In 2005, 2 degrees awarded. *Degree requirements:* For master's, thesis. *Entrance requirements:* For master's, GRE General Test, minimum GPA of 3.0. Additional exam requirements/recommendations for international students: Required—TOEFL. *Application deadline:* For fall admission, 2/15 for domestic students; for spring admission, 10/15 for domestic students. Applications are processed on a rolling basis. Application fee: $50 ($60 for international students). *Financial support:* In 2005–06, 1 fellowship with partial tuition reimbursement (averaging $18,000 per year), 3 research assistantships with partial tuition reimbursements (averaging $11,500 per year), 9 teaching assistantships with partial tuition reimbursements (averaging $11,500 per year) were awarded; career-related internships or fieldwork, Federal Work-Study, and institutionally sponsored loans also available. Financial award application deadline: 2/15. *Faculty research:* Sedimentary geology, structural geology, regional tectonics, hydrogeology petrology. Total annual research expenditures: $400,000. *Unit head:* Dr. John W. Shervais, Head, 435-797-1274, Fax: 435-797-1588, E-mail: geology@cc.usu.edu. *Application contact:* Dr. W. David Liddell, Program Director, 435-797-1261, Fax: 435-797-1588, E-mail: davel@cc.usu.edu.

Virginia Polytechnic Institute and State University, Graduate School, College of Science, Department of Geosciences, Blacksburg, VA 24061. Offers geological sciences (MS, PhD); geophysics (MS, PhD). *Faculty:* 18 full-time (2 women), 1 (woman) part-time/adjunct. *Students:* 55 full-time (20 women), 2 part-time (1 woman); includes 2 minority (1 American Indian/Alaska Native, 1 Asian American or Pacific Islander), 18 international. Average age 27. 55 applicants, 36% accepted, 11 enrolled. In 2005, 9 master's, 1 doctorate awarded. *Entrance requirements:* For master's and doctorate, GRE General Test. Additional exam requirements/recommendations for international students: Required—TOEFL (minimum score 550 paper-based; 213 computer-based). *Application deadline:* Applications are processed on a rolling basis. Application fee: $45. Electronic applications accepted. *Expenses:* Tuition, state resident: full-time $6,558; part-time $364 per credit. Tuition, nonresident: full-time $11,296; part-time $628 per credit. Required fees: $1,419; $468 per credit. $234 per term. *Financial support:* In 2005–06, 8 fellowships with full tuition reimbursements (averaging $15,121 per year), 24 research assistantships with full tuition reimbursements (averaging $16,689 per year), 27 teaching assistantships with full tuition reimbursements (averaging $13,902 per year) were awarded; career-related internships or fieldwork, Federal Work-Study, scholarships/grants, tuition waivers (full), and unspecified assistantships also available. Financial award application deadline: 4/1. *Faculty research:* Paleontology/geobiology, active tectonics, geomorphology, mineralogy/crystallography, mineral physics. *Unit head:* Dr. Robert Tracy, Head, 540-231-6521, Fax: 540-231-3386. *Application contact:* Connie Lowe, Student Program Coordinator, 540-231-8824, Fax: 540-231-3386, E-mail: clowe@vt.edu.

Washington State University, Graduate School, College of Sciences, Department of Geology, Pullman, WA 99164. Offers hydrology (MS, PhD); mineralology-petrology geochemistry (MS); minerology-petrology geochemistry (PhD); sedimentology-stratogeography (MS, PhD); structural geology-tectonics (MS, PhD). *Faculty:* 12. *Students:* 22 full-time (8 women), 4 part-time (1 woman), 5 international. Average age 29. 39 applicants, 49% accepted, 7 enrolled. In 2005, 5 master's, 1 doctorate awarded. *Degree requirements:* For master's, thesis, oral exam; for doctorate, one foreign language, thesis/dissertation, oral exam, written exam. *Entrance requirements:* For master's and doctorate, GRE General Test, minimum GPA of 3.0, 3 letters of recommendation. Additional exam requirements/recommendations for international students: Required—TOEFL (minimum score 560 paper-based; 220 computer-based). *Application deadline:* For fall admission, 2/1 priority date for domestic students, 2/1 priority date for international students; for spring admission, 9/1 priority date for domestic students, 7/1 priority date for international students. Applications are processed on a rolling basis. Application fee: $35. Electronic applications accepted. *Expenses:* Tuition, state resident: full-time $6,295;

Geology

Washington State University (continued)
part-time $336 per credit. Tuition, nonresident: full-time $15,949; part-time $819 per credit. Required fees: $933. Part-time tuition and fees vary according to campus/location and program. *Financial support:* In 2005–06, 4 fellowships (averaging $2,700 per year), 5 research assistantships with full and partial tuition reimbursements (averaging $14,200 per year), 18 teaching assistantships with full and partial tuition reimbursements (averaging $13,831 per year) were awarded; career-related internships or fieldwork, Federal Work-Study, institutionally sponsored loans, and scholarships/grants also available. Financial award application deadline: 2/1; financial award applicants required to submit FAFSA. *Faculty research:* Genesis of ore deposits, geohydrology of the Pacific Northwest, geochemistry and petrology of plateau basalts. Total annual research expenditures: $379,240. *Unit head:* Dr. Peter Larson, Chair, 509-335-3009, Fax: 509-335-7816, E-mail: plarson@wsu.edu. *Application contact:* E-mail: geology@wsu.edu.

Washington University in St. Louis, Graduate School of Arts and Sciences, Department of Earth and Planetary Sciences, St. Louis, MO 63130-4899. Offers earth and planetary sciences (MA); geochemistry (PhD); geology (MA, PhD); geophysics (PhD); planetary sciences (PhD). Terminal master's awarded for partial completion of doctoral program. *Degree requirements:* For master's and doctorate, thesis/dissertation. *Entrance requirements:* For master's and doctorate, GRE General Test. Electronic applications accepted.

Wayne State University, Graduate School, College of Liberal Arts and Sciences, Department of Geology, Detroit, MI 48202. Offers MA, MS. *Faculty:* 2 full-time (0 women). *Students:* 3 full-time (1 woman), 3 part-time; includes 1 minority (African American) Average age 35. 3 applicants, 67% accepted, 2 enrolled. In 2005, 1 degree awarded. *Degree requirements:* For master's, thesis. *Entrance requirements:* For master's, GRE General Test. Additional exam requirements/recommendations for international students: Required—TOEFL (minimum score 550 paper-based; 213 computer-based); Recommended—TWE (minimum score 6). *Application deadline:* For fall admission, 7/1 for domestic students, 6/1 for international students. Applications are processed on a rolling basis. Application fee: $30 ($50 for international students). Electronic applications accepted. *Expenses:* Tuition, state resident: part-time $338 per credit hour. Tuition, nonresident: part-time $746 per credit hour. Required fees: $24 per credit hour. Full-time tuition and fees vary according to program. *Financial support:* In 2005–06, 2 teaching assistantships with tuition reimbursements (averaging $15,749 per year) were awarded *Faculty research:* Geologic history of SW US, heavy metal contamination of soils, role of colloids in the removal of particle reactive radionuclides and contaminants, environmental radioactivity and geochronology, light stable isotope geochemistry and mineral ore formation. Total annual research expenditures: $84,779. *Unit head:* James D. Tucker, Chair, 313-577-2783, Fax: 313-577-6891, E-mail: ao1754@wayne.edu. *Application contact:* Jeffrey Howard, Graduate Director, 313-577-3258, E-mail: aa2675@wayne.edu.

West Chester University of Pennsylvania, Graduate Studies, College of Arts and Sciences, Department of Geology and Astronomy, West Chester, PA 19383. Offers physical science (MA). Part-time and evening/weekend programs available. *Students:* 3 full-time (0 women), 10 part-time (3 women); includes 1 African American. Average age 30. 7 applicants, 86% accepted, 3 enrolled. In 2005, 6 degrees awarded. *Degree requirements:* For master's, thesis optional. *Entrance requirements:* For master's, GRE General Test, interview. *Application deadline:* For fall admission, 4/15 for domestic students; for spring admission, 10/15 for domestic students. Applications are processed on a rolling basis. Application fee: $35. *Expenses:* Tuition, state resident: full-time $2,944; part-time $327 per credit. Tuition, nonresident: full-time $4,711; part-time $523 per credit. Required fees: $54 per semester. *Financial support:* In 2005–06, 1 research assistantship with full tuition reimbursement (averaging $5,000 per year) was awarded; unspecified assistantships also available. Support available to part-time students. Financial award application deadline: 2/15; financial award applicants required to submit FAFSA. *Faculty research:* Developing and using a meteorological data station. *Unit head:* Dr. Gil Wiswall, Chair, 610-436-2727, E-mail: gwiswall@wcupa.edu. *Application contact:* Dr. Steven Good, Information Contact, 610-436-2203, E-mail: sgood@wcupa.edu.

Western Kentucky University, Graduate Studies, Ogden College of Science and Engineering, Department of Geography and Geology, Bowling Green, KY 42101-3576. Offers MAE, MS. *Faculty:* 9 full-time (1 woman). *Students:* 25 full-time (15 women), 17 part-time (7 women); includes 3 minority (1 African American, 1 American Indian/Alaska Native, 1 Hispanic American), 6 international. Average age 31. 14 applicants, 64% accepted, 5 enrolled. In 2005, 5 master's awarded. *Degree requirements:* For master's, thesis or alternative, comprehensive exam. *Entrance requirements:* For master's, GRE General Test, minimum GPA of 2.75. Additional exam requirements/recommendations for international students: Required—TOEFL (minimum score 555 paper-based; 213 computer-based). *Application deadline:* For fall admission, 7/1 priority date for domestic students, 5/15 priority date for international students; for spring admission, 11/1 for domestic students, 9/15 for international students. Applications are processed on a rolling basis. Application fee: $35. *Expenses:* Tuition, state resident: full-time $5,816; part-time $299 per credit hour. Tuition, nonresident: full-time $6,356; part-time $326 per credit hour. *Financial support:* In 2005–06, 7 students received support, including 7 research assistantships with partial tuition reimbursements available (averaging $9,000 per year); teaching assistantships, Federal Work-Study, institutionally sponsored loans, unspecified assistantships, and service awards also available. Support available to part-time students. Financial award application deadline: 4/1; financial award applicants required to submit FAFSA. *Faculty research:* Hydroclimatology, electronic data sets, groundwater, sinkhole liquification potential, meteorological analysis. Total annual research expenditures: $38,185. *Unit head:* Dr. David J. Keeling, Head, 270-745-4555, E-mail: david.keeling@wku.edu.

Western Michigan University, Graduate College, College of Arts and Sciences, Department of Geosciences, Geosciences Department, Kalamazoo, MI 49008-5202. Offers MS, PhD. *Degree requirements:* For master's, oral exam; for doctorate, thesis/dissertation, oral exam. *Entrance requirements:* For master's and doctorate, GRE General Test.

Western Washington University, Graduate School, College of Sciences and Technology, Department of Geology, Bellingham, WA 98225-5996. Offers MS. Part-time programs available. *Faculty:* 12. *Students:* 12 full-time (4 women), 12 part-time (7 women). 35 applicants, 57% accepted, 9 enrolled. In 2005, 3 degrees awarded. *Degree requirements:* For master's, thesis. *Entrance requirements:* For master's, GRE General Test, minimum GPA of 3.0 in last 60 semester hours or last 90 quarter hours. Additional exam requirements/recommendations for international students: Required—TOEFL (minimum score 567 paper-based; 227 computer-based). *Application deadline:* For fall admission, 1/31 for domestic students. For winter admission, 10/1 for domestic students; for spring admission, 2/1 for domestic students. Applications are processed on a rolling basis. Application fee: $50. *Expenses:* Tuition, area resident: Part-time $188 per credit. Tuition, state resident: full-time $5,628; part-time $539 per credit. Tuition, nonresident: full-time $16,176. Required fees: $624. *Financial support:* In 2005–06, 9 teaching assistantships with partial tuition reimbursements (averaging $9,816 per year) were awarded; career-related internships or fieldwork, Federal Work-Study, institutionally sponsored loans, scholarships/grants, tuition waivers (partial), and unspecified assistantships also available. Support available to part-time students. Financial award application deadline: 2/15; financial award applicants required to submit FAFSA. *Faculty research:* Structure/tectonics; sedimentary, glacial and quaternary geomorphology; paleomagnetism; igneous and metamorphic petrology; paleontology. *Unit head:* Dr. Scott Babcock, Chair, 360-650-3592. *Application contact:* Chris Sutton, Graduate Coordinator, 360-650-3581.

West Virginia University, Eberly College of Arts and Sciences, Department of Geology and Geography, Program in Geology, Morgantown, WV 26506. Offers geomorphology (MS, PhD); geophysics (MS, PhD); hydrogeology (MS); hydrology (PhD); paleontology (MS, PhD); petrology (MS, PhD); stratigraphy (MS, PhD); structure (MS, PhD). Part-time programs available. *Students:* 26 full-time (7 women), 17 part-time (7 women); includes 2 minority (1 American Indian/Alaska Native, 1 Asian American or Pacific Islander), 12 international. Average age 30. 34 applicants, 29% accepted. In 2005, 6 master's, 1 doctorate awarded. Terminal master's awarded for partial completion of doctoral program. *Degree requirements:* For master's, thesis/dissertation; for doctorate, thesis/dissertation, comprehensive exam. *Entrance requirements:* For master's, GRE General Test, GRE Subject Test, minimum GPA of 2.5; for doctorate, GRE General Test, GRE Subject Test, minimum GPA of 3.3. Additional exam requirements/recommendations for international students: Required—TOEFL. *Application deadline:* For fall admission, 2/1 for domestic students; for spring admission, 10/1 for domestic students. Applications are processed on a rolling basis. Application fee: $45. *Expenses:* Tuition, state resident: full-time $4,582; part-time $258 per credit hour. Tuition, nonresident: full-time $1,382; part-time $741 per credit hour. *Financial support:* In 2005–06, 5 research assistantships, 18 teaching assistantships were awarded; career-related internships or fieldwork, Federal Work-Study, institutionally sponsored loans, and tuition waivers (full and partial) also available. Financial award application deadline: 2/1; financial award applicants required to submit FAFSA. *Unit head:* Dr. Thomas H. Wilson, Associate Chair, 304-293-5603 Ext. 4316, Fax: 304-293-6522, E-mail: tom.wilson@mail.wvu.edu. *Application contact:* Dr. Joe Donovan, Associate Professor, 304-293-5603 Ext. 4308, Fax: 304-293-6522, E-mail: joe.donovan@mail.wvu.edu.

Wichita State University, Graduate School, Fairmount College of Liberal Arts and Sciences, Department of Geology, Wichita, KS 67260. Offers MS. Part-time programs available. *Degree requirements:* For master's, thesis. *Entrance requirements:* For master's, GRE General Test. Additional exam requirements/recommendations for international students: Required—TOEFL. Electronic applications accepted. *Faculty research:* Midcontinent and Permian basin stratigraphy studies, recent sediments of Belize and Florida, image analysis of sediments and porosity.

Wright State University, School of Graduate Studies, College of Science and Mathematics, Department of Geological Sciences, Program in Geological Sciences, Dayton, OH 45435. Offers environmental geochemistry (MS); environmental geology (MS); environmental sciences (MS); geological sciences (MS); geophysics (MS); hydrogeology (MS); petroleum geology (MS). Part-time programs available. *Degree requirements:* For master's, thesis. *Entrance requirements:* Additional exam requirements/recommendations for international students: Required—TOEFL.

Yale University, Graduate School of Arts and Sciences, Department of Geology and Geophysics, New Haven, CT 06520. Offers geochemistry (PhD); geophysics (PhD); meteorology (PhD); mineralogy and crystallography (PhD); oceanography (PhD); paleoecology (PhD); paleontology and stratigraphy (PhD); petrology (PhD); structural geology (PhD). *Degree requirements:* For doctorate, thesis/dissertation. *Entrance requirements:* For doctorate, GRE General Test. Additional exam requirements/recommendations for international students: Required—TOEFL.

Geophysics

Boise State University, Graduate College, College of Arts and Sciences, Department of Geosciences, Master's Program in Geophysics, Boise, ID 83725-0399. Offers MS. Part-time programs available. *Degree requirements:* For master's, thesis. *Entrance requirements:* For master's, GRE General Test, minimum GPA of 3.0, BS in related field. Additional exam requirements/recommendations for international students: Required—TOEFL. Electronic applications accepted. *Faculty research:* Shallow seismic profile, seismic hazard, tectonics, hazardous waste disposal.

Boise State University, Graduate College, College of Arts and Sciences, Department of Geosciences, Program in Geophysics, Boise, ID 83725-0399. Offers PhD. Part-time programs available. *Degree requirements:* For doctorate, thesis/dissertation, comprehensive exam. *Entrance requirements:* For doctorate, GRE General Test. Electronic applications accepted.

Boston College, Graduate School of Arts and Sciences, Department of Geology and Geophysics, Chestnut Hill, MA 02467-3800. Offers MS, MBA/MS. *Students:* 13 full-time (5 women), 8 part-time (4 women), 3 international. 14 applicants, 86% accepted, 9 enrolled. In 2005, 4 degrees awarded. *Degree requirements:* For master's, thesis. *Entrance requirements:* For master's, GRE General Test, GRE Subject Test. Additional exam requirements/recommendations for international students: Required—TOEFL (minimum score 550 paper-based; 213 computer-based). *Application deadline:* For fall admission, 1/15 for domestic students. Application fee: $70. Electronic applications accepted. *Financial support:* Research assistantships with full tuition reimbursements, teaching assistantships with full tuition reimbursements, Federal Work-Study available. Support available to part-time students. Financial award application deadline: 3/1; financial award applicants required to submit FAFSA. *Faculty research:* Coastal and marine geology, experimental sedimentology, geomagnetism, igneous petrology, paleontology. *Unit head:* Dr. Alan Kafka, Chairperson, 617-552-3650. *Application contact:* Dr. John Ebel, Graduate Program Director, 617-552-3640, E-mail: john.ebel@bc.edu.

See Close-Up on page 223.

California Institute of Technology, Division of Geological and Planetary Sciences, Pasadena, CA 91125-0001. Offers cosmochemistry (PhD); geobiology (PhD); geochemistry (MS, PhD); geology (MS, PhD); geophysics (MS, PhD); planetary science (MS, PhD). Part-time programs available. *Degree requirements:* For doctorate, thesis/dissertation. *Entrance requirements:* For doctorate, GRE General Test. Additional exam requirements/recommendations for international students: Required—TOEFL. Electronic applications accepted. *Faculty research:* Astronomy, evolution of anaerobic respiratory processes, structural geology and tectonics, theoretical and numerical seismology, global biogeochemical cycles.

Colorado School of Mines, Graduate School, Department of Geophysics, Golden, CO 80401-1887. Offers geophysical engineering (ME, MS, PhD); geophysics (MS, PhD, Diploma). Part-time programs available. *Faculty:* 14 full-time (2 women), 8 part-time/adjunct (0 women). *Students:* 44 full-time (15 women), 7 part-time (1 woman); includes 1 minority (African American), 24 international. 54 applicants, 70% accepted, 12 enrolled. In 2005, 13 master's, 4 doctorates awarded. *Degree requirements:* For master's, thesis; for doctorate, one foreign language, thesis/dissertation, oral exams, comprehensive exam. *Entrance requirements:* For master's, doctorate, and Diploma, GRE General Test. Additional exam requirements/recommendations for international students: Required—TOEFL (minimum score 550 paper-based; 213 computer-based). *Application deadline:* For fall admission, 1/1 for domestic students, 1/1 for international students; for spring admission, 9/1 for domestic students, 9/1 for international students. Application fee: $50. Electronic applications accepted. *Expenses:* Tuition, state resident: full-time $7,240; part-time $362 per credit hour. Tuition, nonresident: full-time $19,840; part-time $992 per credit hour. Required fees: $895. *Financial support:* In 2005–06, 1 fellowship with full tuition reimbursement (averaging $9,600 per year), 26 research assistantships with full tuition reimbursements (averaging $9,600 per year), 4 teaching assistantships with full tuition reimbursements (averaging $9,600 per year) were awarded; scholarships/grants, health care benefits, and unspecified assistantships also available. Financial award applicants required to submit FAFSA. *Faculty research:* Seismic exploration, gravity and geomagnetic fields, electrical mapping and sounding, bore hole measurements, environmental physics. Total annual research expenditures: $4.2 million. *Unit head:* Dr. Terence K. Young, Head, 303-273-3454, Fax: 303-273-3478, E-mail: tkyoung@mine.edu. *Application contact:* Sara Summers, Program Assistant, 303-273-3935, Fax: 303-273-3478, E-mail: ssummers@mines.edu.

Colorado State University, Graduate School, Warner College of Natural Resources, Department of Geosciences, Fort Collins, CO 80523-0015. Offers earth sciences (PhD); geology (MS), including geomorphology, geophysics, hydrogeology, petrology/geochemistry and economic geology, sedimentology, structural geology; geosciences (PhD); wateshed (PhD). Part-time programs available. *Faculty:* 8 full-time (3 women). *Students:* 17 full-time (11 women), 22 part-time (7 women); includes 1 minority (American Indian/Alaska Native), 2 international. Average age 32. 28 applicants, 61% accepted, 7 enrolled. In 2005, 3 master's, 1 doctorate awarded. *Degree requirements:* For master's, thesis/dissertation, registration; for doctorate, thesis/dissertation, comprehensive exam, registration. *Entrance requirements:* For master's and doctorate, GRE General Test, minimum GPA of 3.0. Additional exam requirements/recommendations for international students: Required—TOEFL (minimum score 550 paper-based; 213 computer-based). *Application deadline:* For fall admission, 2/1 priority date for domestic students, 2/1 priority date for international students. Applications are processed on a rolling basis. Application fee: $50. Electronic applications accepted. *Expenses:* Tuition, state resident: full-time $3,690; part-time $205 per credit. Tuition, nonresident: full-time $14,958; part-time $831 per credit. Required fees: $1,061. *Financial support:* In 2005–06, 6 fellowships (averaging $9,000 per year), 13 research assistantships with partial tuition reimbursements (averaging $16,000 per year), 9 teaching assistantships with full tuition reimbursements (averaging $16,300 per year) were awarded; career-related internships or fieldwork, Federal Work-Study, institutionally sponsored loans, scholarships/grants, and traineeships also available. Financial award application deadline: 2/15. *Faculty research:* Snow, surface, and groundwater hydrology; fluvial geomorphology; geographic information systems; geochemistry; bedrock geology. Total annual research expenditures: $1.3 million. *Unit head:* Dr. Judith L. Hannah, Head, 970-491-5662, Fax: 970-491-6307, E-mail: jhannah@cnr.colostate.edu. *Application contact:* Sharyl Pierson, Administrative Assistant, 970-491-5662, Fax: 970-491-6307, E-mail: sharyl@cnr.colostate.edu.

Columbia University, Graduate School of Arts and Sciences, Division of Natural Sciences, Department of Earth and Environmental Sciences, New York, NY 10027. Offers geochemistry (M Phil, MA, PhD); geodetic sciences (M Phil, MA, PhD); geophysics (M Phil, MA, PhD); oceanography (M Phil, MA, PhD). *Faculty:* 21 full-time, 19 part-time/adjunct. *Students:* 78 full-time (31 women), 6 part-time (2 women); includes 4 minority (3 Asian Americans or Pacific Islanders, 1 Hispanic American), 31 international. Average age 32. 115 applicants, 20% accepted. In 2005, 16 master's, 11 doctorates awarded. *Degree requirements:* For master's, thesis or alternative, fieldwork, written exam; for doctorate, one foreign language, thesis/dissertation. *Entrance requirements:* For master's and doctorate, GRE General Test, GRE Subject Test, major in natural or physical science. Additional exam requirements/recommendations for international students: Required—TOEFL. Application fee: $75. *Expenses:* Tuition: Full-time $31,448. Tuition and fees vary according to course level, course load, campus/location and program. *Financial support:* Fellowships, teaching assistantships, Federal Work-Study and institutionally sponsored loans available. Support available to part-time students. Financial award application deadline: 1/5; financial award applicants required to submit FAFSA. *Faculty research:* Structural geology and stratigraphy, petrology, paleontology, rare gas, isotope and aqueous geochemistry. *Unit head:* Dr. Nicholas Christie-Blick, Chair, 845-365-8821, Fax: 845-365-8150.

Cornell University, Graduate School, Graduate Fields of Engineering, Field of Geological Sciences, Ithaca, NY 14853-0001. Offers economic geology (M Eng, MS, PhD); engineering geology (M Eng, MS, PhD); environmental geophysics (M Eng, MS, PhD); general geology (M Eng, MS, PhD); geobiology (M Eng, MS, PhD); geochemistry and isotope geology (M Eng, MS, PhD); geohydrology (M Eng, MS, PhD); geomorphology (M Eng, MS, PhD); geophysics (M Eng, MS, PhD); geotectonics (M Eng, MS, PhD); marine geology (MS, PhD); mineralogy (M Eng, MS, PhD); paleontology (M Eng, MS, PhD); petroleum geology (M Eng, MS, PhD); petrology (M Eng, MS, PhD); planetary geology (M Eng, MS, PhD); Precambrian geology (M Eng, MS, PhD); Quaternary geology (M Eng, MS, PhD); rock mechanics (M Eng, MS, PhD); sedimentology (M Eng, MS, PhD); seismology (M Eng, MS, PhD); stratigraphy (M Eng, MS, PhD); structural geology (M Eng, MS, PhD). *Faculty:* 39 full-time (5 women). *Students:* 29 full-time (15 women); includes 2 minority (both Hispanic Americans), 7 international. 73 applicants, 15% accepted, 6 enrolled. In 2005, 5 master's, 1 doctorate awarded. *Degree requirements:* For master's, thesis (MS); for doctorate, thesis/dissertation, comprehensive exam. *Entrance requirements:* For master's and doctorate, GRE General Test, 3 letters of recommendation. Additional exam requirements/recommendations for international students: Required—TOEFL (minimum score 550 paper-based; 213 computer-based). *Application deadline:* For fall admission, 1/15 for domestic students. Applications are processed on a rolling basis. Application fee: $60. Electronic applications accepted. *Financial support:* In 2005–06, 8 fellowships with full tuition reimbursements, 10 research assistantships with full tuition reimbursements, 7 teaching assistantships with full tuition reimbursements were awarded; institutionally sponsored loans, scholarships/grants, health care benefits, tuition waivers (full and partial), and unspecified assistantships also available. Financial award applicants required to submit FAFSA. *Faculty research:* Geophysics, structural geology, petrology, geochemistry, geodynamics. *Unit head:* Director of Graduate Studies, 607-255-5466, Fax: 607-254-4780. *Application contact:* Graduate Field Assistant, 607-255-5466, Fax: 607-254-4780, E-mail: gradprog@geology.cornell.edu.

Florida State University, Graduate Studies, College of Arts and Sciences, Interdisciplinary Program in Geophysical Fluid Dynamics, Tallahassee, FL 32306. Offers PhD. *Faculty:* 26 full-time (3 women). *Students:* 8 full-time (1 woman), 5 international. Average age 30. 6 applicants, 33% accepted, 1 enrolled. *Degree requirements:* For doctorate, thesis/dissertation, departmental qualifying exam. *Entrance requirements:* For doctorate, GRE General Test, GRE Subject Test, minimum GPA of 3.0. Additional exam requirements/recommendations for international students: Required—TOEFL. *Application deadline:* For fall admission, 12/30 for domestic students. Application fee: $30. *Financial support:* In 2005–06, 1 fellowship (averaging $18,000 per year), 2 research assistantships (averaging $17,500 per year) were awarded; unspecified assistantships also available. Financial award applicants required to submit FAFSA. *Faculty research:* Hurricane dynamics, topography, convection, air-sea interaction, wave-mean flow interaction, numerical models. Total annual research expenditures: $662,000. *Unit head:* Dr. Carol A. Clayson, Director, 850-644-2488, Fax: 850-644-8972, E-mail: clayson@met.fsu.edu.

Georgia Institute of Technology, Graduate Studies and Research, College of Sciences, School of Earth and Atmospheric Sciences, Atlanta, GA 30332-0001. Offers atmospheric chemistry and air pollution (MS, PhD); atmospheric dynamics and climate (MS, PhD); geochemistry (MS, PhD); hydrologic cycle (MS, PhD); ocean sciences (MS, PhD); solid-earth and environmental geophysics (MS, PhD). Part-time programs available. Terminal master's awarded for partial completion of doctoral program. *Degree requirements:* For master's, thesis or alternative; for doctorate, thesis/dissertation, comprehensive exam. *Entrance requirements:* For master's, GRE, minimum GPA of 3.0; for doctorate, GRE General Test, minimum GPA of 2.7. Additional exam requirements/recommendations for international students: Required—TOEFL (minimum score 550 paper-based; 213 computer-based). *Faculty research:* Geophysics, atmospheric chemistry, atmospheric dynamics, seismology.

See Close-Up on page 227.

ICR Graduate School, Graduate Programs, Santee, CA 92071. Offers astro/geophysics (MS); biology (MS); geology (MS); science education (MS). Part-time programs available. *Faculty:* 5 full-time (1 woman), 1 part-time/adjunct (0 women). *Students:* 6 full-time (2 women), 6 part-time (3 women). Average age 45. In 2005, 4 degrees awarded. *Degree requirements:* For master's, thesis (for some programs), comprehensive exam (for some programs). *Entrance requirements:* For master's, minimum undergraduate GPA of 3.0, bachelor's degree in science or science education. *Application deadline:* Applications are processed on a rolling basis. Application fee: $30. *Expenses:* Tuition: Part-time $150 per semester hour. *Faculty research:* Age of the earth, limits of variation, catastrophe, optimum methods for teaching. Total annual research expenditures: $200,000. *Unit head:* Dr. Kenneth B. Cumming, Dean, 619-448-0900, Fax: 619-448-3469. *Application contact:* Dr. Jack Kriege, Registrar, 619-448-0900 Ext. 6016, Fax: 619-448-3469, E-mail: jkriege@icr.org.

Idaho State University, Office of Graduate Studies, College of Arts and Sciences, Department of Geosciences, Pocatello, ID 83209. Offers geographic information science (MS); geology (MNS, MS); geophysics/hydrology (MS); geotechnology (Postbaccalaureate Certificate). Part-time programs available. Postbaccalaureate distance learning degree programs offered. *Degree requirements:* For master's, thesis, comprehensive exam, registration (for some programs); for Postbaccalaureate Certificate, thesis optional. *Entrance requirements:* For master's and Postbaccalaureate Certificate, GRE General Test, 3 letters of recommendation. Additional exam requirements/recommendations for international students: Required—TOEFL (minimum score 550 paper-based; 213 computer-based). *Faculty research:* Structural geography, stratigraphy, geochemistry, remote sensing, geomorphology.

Indiana University Bloomington, Graduate School, College of Arts and Sciences, Department of Geological Sciences, Bloomington, IN 47405-7000. Offers biogeochemistry (MS, PhD); economic geology (MS, PhD); geobiology (MS, PhD); geophysics, structural geology and tectonics (MS, PhD); hydrogeology (MS, PhD); mineralogy (MS, PhD); stratigraphy and sedimentology (MS, PhD). PhD offered through the University Graduate School. Part-time programs available. *Faculty:* 14 full-time (1 woman). *Students:* 30 full-time (13 women), 12 part-time (5 women); includes 5 minority (2 African Americans, 1 American Indian/Alaska Native, 1 Asian American or Pacific Islander, 1 Hispanic American), 21 international. Average age 29. In 2005, 7 master's, 2 doctorates awarded. Terminal master's awarded for partial completion of doctoral program. *Degree requirements:* For master's, one foreign language, thesis or alternative; for doctorate, thesis/dissertation. *Entrance requirements:* For master's and doctorate, GRE General Test. Additional exam requirements/recommendations for international students: Required—TOEFL. *Application deadline:* For fall admission, 1/15 priority date for domestic students, 12/15 priority date for international students; for spring admission, 9/1 priority date for domestic students, 9/1 priority date for international students. Applications are processed on a rolling basis. Application fee: $50 ($60 for international students). *Expenses:* Tuition, state resident: full-time $5,437; part-time $227 per credit hour. Tuition, nonresident: full-time $15,836; part-time $660 per credit hour. Required fees: $821. Tuition and fees vary according to campus/location and program. *Financial support:* Fellowships with tuition reimbursements, research assistantships with full and partial tuition reimbursements, teaching assistantships with tuition reimbursements, career-related internships or fieldwork, Federal Work-Study, and institutionally sponsored loans available. Financial award application deadline: 2/15. *Faculty research:* Geophysics, geochemistry, hydrogeology, igneous and metamorphic petrology and clay minerology. Total annual research expenditures: $289,139. *Unit head:* Mark Person, Graduate Advisor, 812-855-7214. *Application contact:* Mary Iverson, Secretary, Committee for Graduate Studies, 812-855-7214, Fax: 812-855-7899, E-mail: geograd@indiana.edu.

Louisiana State University and Agricultural and Mechanical College, Graduate School, College of Basic Sciences, Department of Geology and Geophysics, Baton Rouge, LA 70803. Offers MS, PhD. *Faculty:* 22 full-time (6 women). *Students:* 38 full-time (21 women), 4 part-time (2 women); includes 6 minority (3 African Americans, 1 Asian American or Pacific Islander, 2 Hispanic Americans), 7 international. Average age 29. 40 applicants, 48% accepted, 21 enrolled. In 2005, 6 master's, 4 doctorates awarded. Terminal master's awarded for partial completion of doctoral program. *Degree requirements:* For master's and doctorate, thesis/dissertation. *Entrance requirements:* For master's and doctorate, GRE General Test, minimum GPA of 3.0. Additional exam requirements/recommendations for international students: Required—TOEFL (minimum score 550 paper-based; 213 computer-based). *Application deadline:* For fall admission, 1/25 priority date for domestic students, 5/15 priority date for international students. Applications are processed on a rolling basis. Application fee: $25. Electronic applications accepted. *Financial support:* In 2005–06, 33 students received support, including 5 fellowships with full tuition reimbursements available (averaging $15,833 per year), 9 research assistantships with full and partial tuition reimbursements available (averaging $21,577 per year), 19 teaching assistantships with full and partial tuition reimbursements available (averaging $14,855 per year); career-related internships or fieldwork, Federal Work-Study, institutionally sponsored loans, tuition waivers (full and partial), and unspecified assistantships also available. Financial award application deadline: 3/15; financial award applicants required to submit FAFSA. *Faculty research:* Geophysics, geochemistry of sediments, isotope geochemistry, igneous and metamorphic petrology, micropaleontology. Total annual research expenditures: $1.1 million. *Unit head:* Dr. Laurie Anderson, Chair, 225-578-3353, Fax: 225-578-2302, E-mail: landerson@geol.lsu.edu. *Application contact:* Jeffrey Nunn, Graduate Coordinator, 225-578-6657, E-mail: jeff@geol.lsu.edu.

Massachusetts Institute of Technology, School of Science, Department of Earth, Atmospheric, and Planetary Sciences, Cambridge, MA 02139-4307. Offers atmospheric chemistry (PhD, Sc D); atmospheric science (SM, PhD, Sc D); climate physics and chemistry (PhD, Sc D); earth and planetary sciences (SM); geochemistry (PhD, Sc D); geology (PhD, Sc D); geophysics (PhD, Sc D); marine geology and geophysics (SM); oceanography (SM, PhD, Sc D); planetary sciences (PhD, Sc D). *Faculty:* 38 full-time (3 women). *Students:* 165 full-time (63 women); includes 12 minority (7 Asian Americans or Pacific Islanders, 5 Hispanic Americans), 51 international. Average age 27. 179 applicants, 34% accepted, 31 enrolled. In 2005, 2 master's, 15 doctorates awarded. Terminal master's awarded for partial completion of doctoral program. *Degree requirements:* For master's, thesis/dissertation; for doctorate, thesis/dissertation, comprehensive exam. *Entrance requirements:* For master's, GRE General Test; for doctorate, GRE General Test, GRE Subject Test (chemistry or physics for planetary science program). Additional exam requirements/recommendations for international students: Required—TOEFL (minimum score 577 paper-based; 233 computer-based). *Application deadline:* For fall admission, 1/5 for domestic students, 1/5 for international students; for spring admission, 11/1 for domestic students, 11/1 for international students. Application fee: $70. Electronic applications accepted. *Expenses:* Tuition: Full-time $32,100. Required fees: $200. Part-time tuition and fees vary according to course load. *Financial support:* In 2005–06, 115 students received support, including 36 fellowships with tuition reimbursements available (averaging $26,951 per year), 60 research assistantships with tuition reimbursements available (averaging $24,520 per year), 18 teaching assistantships with tuition reimbursements available (averaging $22,773 per year); Federal Work-Study, institutionally sponsored loans, scholarships/grants, health care benefits, and unspecified assistantships also available. *Faculty research:* Evolution of main features of the planetary system; origin, composition, structure, and state of the atmospheres, oceans, surfaces, and interiors of planets; dynamics of planets and satellite motions. Total annual research expenditures: $18.6 million. *Unit head:* Prof. Maria Zuber, Department Head, 617-253-6397, E-mail: zuber@mit.edu. *Application contact:* EAPS Education Office.

Memorial University of Newfoundland, School of Graduate Studies, Department of Earth Sciences, St. John's, NL A1C 5S7, Canada. Offers geology (M Sc, PhD); geophysics (M Sc, PhD). Part-time programs available. *Students:* 53 full-time (27 women), 6 part-time, 9 international. 23 applicants, 70% accepted, 16 enrolled. In 2005, 10 master's, 2 doctorates awarded. *Degree requirements:* For master's, thesis; for doctorate, thesis/dissertation, oral thesis defense; PhD entry evaluation, comprehensive exam. *Entrance requirements:* For master's, honors B Sc; for doctorate, M Sc. *Application deadline:* Applications are processed

Geophysics

Memorial University of Newfoundland (continued)
on a rolling basis. Application fee: $40 Canadian dollars. Electronic applications accepted. *Expenses:* Tuition: Part-time $733 per term. Tuition and fees vary according to degree level and program. *Financial support:* Fellowships, research assistantships, teaching assistantships available. *Faculty research:* Geochemistry, sedimentology, paleoceanography and global change, mineral deposits, petroleum geology, hydrology. *Unit head:* Dr. John Hanchar, Head, 709-737-2334, Fax: 709-737-2589, E-mail: head@esd.mun.ca. *Application contact:* Raymund Patzold, Graduate Officer, 709-737-4464, Fax: 709-737-2589, E-mail: rpatzold@esd.mun.ca.

Michigan Technological University, Graduate School, College of Engineering, Department of Geological and Mining Engineering and Sciences, Program in Geophysics, Houghton, MI 49931-1295. Offers MS. Part-time programs available. *Faculty:* 12 full-time (1 woman), 10 part-time/adjunct (2 women). *Students:* 4 full-time. Average age 25. 7 applicants, 43% accepted, 1 enrolled. In 2005, 2 degrees awarded. *Degree requirements:* For master's, comprehensive exam, registration. *Entrance requirements:* Additional exam requirements/recommendations for international students: Required—TOEFL (minimum score 550 paper-based; 213 computer-based). *Application deadline:* For fall admission, 3/15 for domestic students. Applications are processed on a rolling basis. Application fee: $40 ($45 for international students). Electronic applications accepted. *Expenses:* Tuition, nonresident: full-time $11,232; part-time $468 per credit. Required fees: $754; $377 per semester. Full-time tuition and fees vary according to course load, degree level and program. *Financial support:* In 2005–06, 3 students received support, including 2 fellowships with full tuition reimbursements available (averaging $9,542 per year), 1 research assistantship with full tuition reimbursement available (averaging $9,542 per year), teaching assistantships with full tuition reimbursements available (averaging $9,542 per year); career-related internships or fieldwork, Federal Work-Study, scholarships/grants, health care benefits, tuition waivers (partial), unspecified assistantships, and co-op also available. Financial award applicants required to submit FAFSA. *Application contact:* Amie S. Ledgerwood, Secretary 5, 906-487-2531, Fax: 906-487-3371, E-mail: asledger@mtu.edu.

New Mexico Institute of Mining and Technology, Graduate Studies, Department of Earth and Environmental Science, Program in Geophysics, Socorro, NM 87801. Offers MS, PhD. *Degree requirements:* For master's, thesis optional; for doctorate, thesis/dissertation. *Entrance requirements:* For master's, GRE General Test; for doctorate, GRE General Test, GRE Subject Test. Additional exam requirements/recommendations for international students: Required—TOEFL (minimum score 540 paper-based; 207 computer-based). *Faculty research:* Earthquake and volcanic seismology, subduction zone tectonics, network seismology, physical properties of sediments in fault zones.

Ohio University, Graduate Studies, College of Arts and Sciences, Department of Geological Sciences, Athens, OH 45701-2979. Offers environmental geochemistry (MS); environmental geology (MS); environmental/hydrology (MS); geology (MS); geology education (MS); geomorphology/surficial processes (MS); geophysics (MS); hydrogeology (MS); sedimentology (MS); structure/tectonics (MS). Part-time programs available. *Faculty:* 10 full-time (4 women), 4 part-time/adjunct (1 woman). *Students:* 22 full-time (6 women), 3 part-time (2 women); includes 1 minority (Hispanic American), 4 international. Average age 23. 15 applicants, 67% accepted, 8 enrolled. In 2005, 7 degrees awarded. *Degree requirements:* For master's, thesis, thesis proposal defense and thesis defense. *Entrance requirements:* Additional exam requirements/recommendations for international students: Required—TOEFL (minimum score 550 paper-based; 217 computer-based). *Application deadline:* For fall admission, 2/1 priority date for domestic students, 1/1 priority date for international students. Application fee: $45. Electronic applications accepted. *Financial support:* In 2005–06, 18 students received support, including 3 research assistantships with full tuition reimbursements available (averaging $11,900 per year), 13 teaching assistantships with full tuition reimbursements available (averaging $11,900 per year); institutionally sponsored loans, scholarships/grants, tuition waivers (full), and unspecified assistantships also available. Financial award application deadline: 2/1. *Faculty research:* Geoscience education, tectonics, flurial geomorphology, invertebrate paleontology, mine/hydrology. Total annual research expenditures: $649,020. *Unit head:* Dr. David Kidder, Chair, 740-593-1101, Fax: 740-593-0486, E-mail: kidder@ohio.edu. *Application contact:* Dr. David Schneider, Graduate Chair, 740-593-1101, Fax: 740-593-0486, E-mail: schneidd@ohio.edu.

Oregon State University, Graduate School, College of Oceanic and Atmospheric Sciences, Program in Geophysics, Corvallis, OR 97331. Offers MA, MS, PhD. *Students:* 1 full-time (0 women). Terminal master's awarded for partial completion of doctoral program. *Degree requirements:* For master's, thesis optional; for doctorate, thesis/dissertation. *Entrance requirements:* For master's and doctorate, GRE General Test, minimum GPA of 3.0 in last 90 hours. Additional exam requirements/recommendations for international students: Required—TOEFL. *Application deadline:* For fall admission, 2/1 for domestic students. Applications are processed on a rolling basis. Application fee: $50. *Expenses:* Tuition, area resident: Part-time $301 per credit. Tuition, state resident: full-time $8,139; part-time $501 per credit. Tuition, nonresident: full-time $14,376; part-time $532 per credit. Required fees: $1,266. *Financial support:* Fellowships, research assistantships, teaching assistantships, career-related internships or fieldwork, Federal Work-Study, and institutionally sponsored loans available. Support available to part-time students. Financial award application deadline: 2/1. *Faculty research:* Seismic waves; gravitational, geothermal, and electromagnetic fields; rock magnetism; paleomagnetism. *Unit head:* Irma Delson, Assistant Director, Student Services, 541-737-5189, Fax: 541-737-2064, E-mail: student_adviser@oce.orst.edu.

Princeton University, Graduate School, Department of Geosciences, Princeton, NJ 08544-1019. Offers atmospheric and oceanic sciences (PhD); geological and geophysical sciences (PhD). *Degree requirements:* For doctorate, one foreign language, thesis/dissertation. *Entrance requirements:* For doctorate, GRE General Test. Additional exam requirements/recommendations for international students: Required—TOEFL (minimum score 600 paper-based; 250 computer-based). Electronic applications accepted. *Faculty research:* Biogeochemistry, climate science, earth history, regional geology and tectonics, solid-earth geophysics.

Rensselaer Polytechnic Institute, Graduate School, School of Science, Department of Earth and Environmental Sciences, Troy, NY 12180-3590. Offers environmental chemistry (MS, PhD); geochemistry (MS, PhD); geology (MS, PhD); geophysics (MS, PhD); petrology (MS, PhD). Part-time programs available. Terminal master's awarded for partial completion of doctoral program. *Degree requirements:* For master's, thesis (for some programs), comprehensive exam; for doctorate, thesis/dissertation, comprehensive exam. *Entrance requirements:* For master's and doctorate, GRE General Test. Additional exam requirements/recommendations for international students: Required—TOEFL. Electronic applications accepted. *Expenses:* Tuition: Full-time $31,000; part-time $1,320 per credit. Required fees: $1,623. *Faculty research:* Mantel geochemistry, contaminant geochemistry, seismology, GPS geodesy, remote sensing petrology.

See Close-Up on page 661.

Rice University, Graduate Programs, Wiess School of Natural Sciences, Professional Master's Program in Subsurface Geosciences, Houston, TX 77251-1892. Offers geophysics (MS). Part-time programs available. *Degree requirements:* For master's, internship. *Entrance requirements:* For master's, GRE, letters of recommendation (4). Additional exam requirements/recommendations for international students: Required—TOEFL (minimum score 600 paper-based; 250 computer-based). Electronic applications accepted. *Faculty research:* Seismology, geodynamics, wave propogation, bio-geochemistry, remote sensing.

Saint Louis University, Graduate School, College of Arts and Sciences and Graduate School, Department of Earth and Atmospheric Sciences, St. Louis, MO 63103-2097. Offers geophysics (PhD); geoscience (MS); meteorology (M Pr Met, MS-R, PhD). Part-time programs available. *Faculty:* 17 full-time (0 women), 2 part-time/adjunct (0 women). *Students:* 30 full-time (11 women), 4 part-time (2 women); includes 2 minority (1 African American, 1 Hispanic American), 16 international. Average age 28. 35 applicants, 86% accepted, 14 enrolled. In 2005, 4 master's, 3 doctorates awarded. *Degree requirements:* For master's, thesis (for some programs), comprehensive oral exam; for doctorate, thesis/dissertation, preliminary exams. *Entrance requirements:* For master's and doctorate, GRE General Test, letters of recommendation, resumé. Additional exam requirements/recommendations for international students: Required—TOEFL (minimum score 550 paper-based; 213 computer-based). *Application deadline:* For fall admission, 7/1 for domestic students, 7/1 for international students; for spring admission, 11/1 for domestic students, 11/1 for international students. Applications are processed on a rolling basis. Application fee: $40. *Expenses:* Tuition: Part-time $760 per credit hour. Required fees: $55 per semester. *Financial support:* In 2005–06, 25 students received support, including 8 research assistantships with full tuition reimbursements available (averaging $16,000 per year), 7 teaching assistantships with full tuition reimbursements available (averaging $15,500 per year); health care benefits and unspecified assistantships also available. Financial award application deadline: 6/1; financial award applicants required to submit FAFSA. *Faculty research:* Structural geology, mesoscale meteorology and severe storms, weather and climate change prediction. Total annual research expenditures: $800,000. *Unit head:* Dr. Bill Dannevik, Interim Chairperson, 314-977-3115, Fax: 314-911-3117, E-mail: dannevik@slu.edu. *Application contact:* Gary Behrman, Associate Dean of the Graduate School, 314-977-3827, E-mail: behrmang@slu.edu.

Southern Methodist University, Dedman College, Department of Geological Sciences, Program in Exploration Geophysics, Dallas, TX 75275. Offers MS. Part-time programs available. *Degree requirements:* For master's, qualifying exam, thesis optional. *Entrance requirements:* For master's, GRE General Test, minimum GPA of 3.0, letters of recommendation. Additional exam requirements/recommendations for international students: Required—TOEFL; Recommended—TSE. *Faculty research:* Geothermal energy, seismology.

Southern Methodist University, Dedman College, Department of Geological Sciences, Program in Geophysics, Dallas, TX 75275. Offers applied geophysics (MS); geophysics (MS, PhD). Part-time programs available. *Degree requirements:* For master's, thesis (for some programs), qualifying exam; for doctorate, thesis/dissertation, qualifying exam. *Entrance requirements:* For master's and doctorate, GRE General Test, minimum GPA of 3.0, letters of recommendation. Additional exam requirements/recommendations for international students: Required—TOEFL; Recommended—TSE. *Faculty research:* Seismology, heat flow, tectonics.

Stanford University, School of Earth Sciences, Department of Geophysics, Stanford, CA 94305-9991. Offers MS, PhD. Terminal master's awarded for partial completion of doctoral program. *Degree requirements:* For master's and doctorate, thesis/dissertation. *Entrance requirements:* For master's and doctorate, GRE General Test. Additional exam requirements/recommendations for international students: Required—TOEFL. Electronic applications accepted.

Texas A&M University, College of Geosciences, Department of Geology and Geophysics, College Station, TX 77843. Offers geology (MS, PhD); geophysics (MS, PhD). *Faculty:* 24 full-time (2 women), 1 part-time/adjunct (0 women). *Students:* 79 full-time (27 women), 22 part-time (5 women); includes 6 minority (1 African American, 1 American Indian/Alaska Native, 1 Asian American or Pacific Islander, 3 Hispanic Americans), 45 international. Average age 31. 76 applicants, 46% accepted, 16 enrolled. In 2005, 16 master's, 6 doctorates awarded. *Degree requirements:* For master's and doctorate, thesis/dissertation. *Entrance requirements:* For master's and doctorate, GRE General Test. Additional exam requirements/recommendations for international students: Required—TOEFL. *Application deadline:* For fall admission, 3/1 priority date for domestic students, 1/15 priority date for international students; for spring admission, 10/1 priority date for domestic students, 8/15 priority date for international students. Applications are processed on a rolling basis. Application fee: $50 ($75 for international students). Electronic applications accepted. *Expenses:* Tuition, state resident: full-time $4,488; part-time $187 per credit hour. Tuition, nonresident: full-time $11,112; part-time $463 per credit hour. Required fees:$1,974. *Financial support:* In 2005–06, fellowships with partial tuition reimbursements (averaging $1,000 per year), research assistantships with partial tuition reimbursements (averaging $11,925 per year), teaching assistantships with partial tuition reimbursements (averaging $11,925 per year) were awarded; Federal Work-Study, institutionally sponsored loans, scholarships/grants, tuition waivers (partial), and unspecified assistantships also available. Financial award application deadline: 3/1; financial award applicants required to submit FAFSA. *Faculty research:* Environmental and engineering geology and geophysics, petroleum geology, tectonophysics, geochemistry. *Unit head:* Dr. Rick Carlson, Head, 979-845-2451, Fax: 979-845-6162. *Application contact:* Robert K. Popp, Graduate Adviser, 979-845-2451, Fax: 979-845-6162, E-mail: popp@geo.tamu.edu.

The University of Akron, Graduate School, Buchtel College of Arts and Sciences, Department of Geology, Program in Geophysics, Akron, OH 44325. Offers MS. *Students:* 1 full-time (0 women), 1 international. Average age 27. 1 applicant, 0% accepted, 0 enrolled. In 2005, 2 degrees awarded. *Degree requirements:* For master's, thesis, seminar, comprehensive exam. *Entrance requirements:* For master's, minimum GPA of 2.75. Additional exam requirements/recommendations for international students: Required—TOEFL (minimum score 550 paper-based; 213 computer-based), Michigan English Language Assessment Battery. *Application deadline:* For fall admission, 3/1 for domestic students. Applications are processed on a rolling basis. Application fee: $30 ($40 for international students). Electronic applications accepted. *Expenses:* Tuition, state resident: full-time $5,816; part-time $323 per credit. Tuition, nonresident: full-time $9,976; part-time $554 per credit. Required fees: $794; $43 per credit. $12 per term. Tuition and fees vary according to course load, degree level and program. *Unit head:* Dr. LaVerne Friberg, Director of Graduate Studies, 330-972-8064.

University of Alaska Fairbanks, College of Natural Sciences and Mathematics, Department of Geology and Geophysics, Fairbanks, AK 99775-7520. Offers geology (MS, PhD); geophysics (MS, PhD). Part-time programs available. *Faculty:* 15 full-time (2 women), 1 (woman) part-time/adjunct. *Students:* 53 full-time (27 women), 18 part-time (8 women); includes 4 minority (1 American Indian/Alaska Native, 1 Asian American or Pacific Islander, 2 Hispanic Americans), 10 international. Average age 29. 62 applicants, 48% accepted, 7 enrolled. In 2005, 13 master's, 4 doctorates awarded. Terminal master's awarded for partial completion of doctoral program. *Degree requirements:* For master's and doctorate, thesis/dissertation, comprehensive exam, registration. *Entrance requirements:* For master's and doctorate, GRE General Test. Additional exam requirements/recommendations for international students: Required—TOEFL (minimum score 550 paper-based; 213 computer-based). *Application deadline:* For fall admission, 1/15 priority date for domestic students, 1/15 priority date for international students; for spring admission, 12/1 for domestic students, 9/1 for international students. Applications are processed on a rolling basis. Application fee: $50. Electronic applications accepted. *Expenses:* Tuition, state resident: full-time $4,392; part-time $244 per credit. Tuition, nonresident: full-time $8,964; part-time $498 per credit. Required fees: $800; $5 per credit. $48 per contact hour. Tuition and fees vary according to course level, course load, campus/location and reciprocity agreements. *Financial support:* In 2005–06, 39 research assistantships with tuition reimbursements (averaging $9,816 per year), 4 teaching assistantships with tuition reimbursements (averaging $9,562 per year) were awarded; fellowships with tuition reimbursements, Federal Work-Study, scholarships/grants, and unspecified assistantships also available. Financial award application deadline: 1/15; financial award applicants required to submit FAFSA. *Faculty research:* Glacial surging, Alaska as geologic fragments, natural zeolites, seismology, volcanology. *Unit head:* Dr. John C. Eichelberger, Co-Chair, 907-474-7565, Fax: 907-474-5163, E-mail: geology@uaf.edu.

University of Alberta, Faculty of Graduate Studies and Research, Department of Physics, Edmonton, AB T6G 2E1, Canada. Offers astrophysics (M Sc, PhD); condensed matter (M Sc, PhD); geophysics (M Sc, PhD); medical physics (M Sc, PhD); subatomic physics (M Sc, PhD). *Faculty:* 36 full-time (3 women), 7 part-time/adjunct (0 women). *Students:* 56 full-time (6 women), 16 part-time (2 women), 25 international. 85 applicants, 35% accepted. In 2005, 7 master's, 10 doctorates awarded. *Degree requirements:* For master's and doctorate, thesis/dissertation. *Entrance requirements:* For master's and doctorate, minimum GPA of 7.0 on a 9.0

scale. Additional exam requirements/recommendations for international students: Required—TOEFL. *Application deadline:* For fall admission, 2/15 for domestic students. Applications are processed on a rolling basis. Tuition and fees charges are reported in Canadian dollars. *Expenses:* Tuition, state resident: part-time $562 Canadian dollars per term. Tuition, nonresident: full-time $3,375 Canadian dollars. Required fees: $573 Canadian dollars; $84 Canadian dollars per term. *Financial support:* In 2005–06, 45 students received support, including 6 fellowships with partial tuition reimbursements available, 40 teaching assistantships; research assistantships, career-related internships or fieldwork, institutionally sponsored loans, and scholarships/grants also available. Financial award application deadline: 2/15. *Faculty research:* Cosmology, astroparticle physics, high-intermediate energy, magnetism, superconductivity. Total annual research expenditures: $3.1 million. *Unit head:* Dr. R. Marchand, Associate Chair, 780-492-1072, E-mail: assoc-chair@phys.ualberta.ca. *Application contact:* Lynn Chandler, Program Advisor, 780-492-1072, Fax: 780-492-0714, E-mail: lynn@phys.ualberta.ca.

The University of British Columbia, Faculty of Graduate Studies, Faculty of Science, Department of Earth and Ocean Sciences, Vancouver, BC V6T 1Z1, Canada. Offers atmospheric science (M Sc, PhD); geological engineering (M Eng, MA Sc, PhD); geological sciences (M Sc, PhD); geophysics (M Sc, MA Sc, PhD); oceanography (M Sc, PhD). *Faculty:* 44 full-time (7 women), 17 part-time/adjunct (1 woman). *Students:* 155 full-time (49 women), 1 (woman) part-time. Average age 30. 96 applicants, 50% accepted, 30 enrolled. In 2005, 11 master's, 6 doctorates awarded. *Degree requirements:* For master's, thesis (for some programs); for doctorate, thesis/dissertation, comprehensive exam. *Entrance requirements:* Additional exam requirements/recommendations for international students: Required—TOEFL (minimum score 600 paper-based; 250 computer-based). *Application deadline:* For fall admission, 2/1 for domestic students, 1/1 for international students. For winter admission, 7/1 for domestic students. Applications are processed on a rolling basis. Application fee: $90 Canadian dollars ($150 Canadian dollars for international students). Electronic applications accepted. *Financial support:* In 2005–06, fellowships (averaging $16,000 per year), research assistantships (averaging $13,000 per year), teaching assistantships (averaging $5,000 per year) were awarded; Federal Work-Study, institutionally sponsored loans, scholarships/grants, tuition waivers (full and partial), and unspecified assistantships also available. *Unit head:* Dr. Paul L. Smith, Head, 604-822-6456, Fax: 604-822-6088, E-mail: head@eos.ubc.ca. *Application contact:* Alex Allen, Graduate Secretary, 604-822-2713, Fax: 604-822-6088, E-mail: aallen@eos.ubc.ca.

University of Calgary, Faculty of Graduate Studies, Faculty of Science, Department of Geology and Geophysics, Calgary, AB T2N 1N4, Canada. Offers geology (M Sc, PhD); geophysics (M Sc, PhD). Part-time programs available. *Faculty:* 22 full-time (3 women), 5 part-time/adjunct (0 women). *Students:* 107 full-time (37 women). 104 applicants, 30% accepted, 26 enrolled. In 2005, 17 master's, 5 doctorates awarded. Terminal master's awarded for partial completion of doctoral program. *Degree requirements:* For master's, thesis; for doctorate, thesis/dissertation, candidacy exam. *Entrance requirements:* For master's, B Sc; for doctorate, honors B Sc or M Sc. Additional exam requirements/recommendations for international students: Required—TOEFL. *Application deadline:* For fall admission, 2/1 priority date for domestic students, 2/1 priority date for international students. For winter admission, 9/1 for domestic students. Applications are processed on a rolling basis. Application fee: $60. Electronic applications accepted. *Financial support:* In 2005–06, 50 students received support, including 9 fellowships, 15 teaching assistantships; career-related internships or fieldwork, institutionally sponsored loans, and scholarships/grants also available. Financial award application deadline: 2/1. *Faculty research:* Geochemistry, petrology, paleontology, stratigraphy, exploration and solid-earth geophysics. *Unit head:* Larry R. Lines, Head, 403-220-8863, Fax: 403-284-0074, E-mail: lrlines@ucalgary.ca. *Application contact:* Cathy H. Hubbell, Graduate Program Administrator, 403-220-3254, Fax: 403-284-0074, E-mail: geosciencegrad@ucalgary.ca.

University of California, Berkeley, Graduate Division, College of Letters and Science, Department of Earth and Planetary Science, Division of Geophysics, Berkeley, CA 94720-1500. Offers MA, MS, PhD. Terminal master's awarded for partial completion of doctoral program. *Degree requirements:* For master's, oral exam; for doctorate, thesis/dissertation, candidacy exams, comprehensive exam. *Entrance requirements:* For master's and doctorate, GRE General Test, minimum GPA of 3.0. *Faculty research:* High-pressure geophysics and seismology.

University of California, Los Angeles, Graduate Division, College of Letters and Science, Department of Earth and Space Sciences, Program in Geophysics and Space Physics, Los Angeles, CA 90095. Offers MS, PhD. *Degree requirements:* For master's, comprehensive exams or thesis; for doctorate, thesis/dissertation, oral and written qualifying exams. *Entrance requirements:* For master's, GRE General Test, minimum GPA of 3.0; for doctorate, GRE General Test, minimum undergraduate GPA of 3.0. Electronic applications accepted.

University of California, Santa Barbara, Graduate Division, College of Letters and Sciences, Division of Mathematics, Life, and Physical Sciences, Department of Earth Science, Santa Barbara, CA 93106. Offers geological sciences (MS, PhD); geophysics (MS). *Faculty:* 37 full-time (4 women). *Students:* 57 full-time (25 women); includes 2 minority (1 Asian American or Pacific Islander, 1 Hispanic American), 6 international. 87 applicants, 24% accepted, 10 enrolled. In 2005, 7 master's, 5 doctorates awarded. Terminal master's awarded for partial completion of doctoral program. *Degree requirements:* For master's, thesis, comprehensive exam, registration; for doctorate, thesis/dissertation, dissertation defense, comprehensive exam, registration. *Entrance requirements:* For master's and doctorate, GRE General Test. Additional exam requirements/recommendations for international students: Required—TOEFL (minimum score 550 paper-based; 213 computer-based). *Application deadline:* For fall admission, 1/1 for domestic students, 1/1 for international students. Application fee: $60. Electronic applications accepted. *Financial support:* In 2005–06, 7 students received support, including 8 fellowships (averaging $6,000 per year), 50 teaching assistantships with full and partial tuition reimbursements available (averaging $7,500 per year); research assistantships with full and partial tuition reimbursements available, career-related internships or fieldwork, Federal Work-Study, scholarships/grants, traineeships, health care benefits, and unspecified assistantships also available. Financial award application deadline: 1/1; financial award applicants required to submit FAFSA. *Faculty research:* Geochemistry, geophysics, tectonics, seismology, petrology. *Unit head:* Dr. James Mattinson, Chair, 805-893-3219, E-mail: mattison@geol.ucsb.edu. *Application contact:* Samuel Rifkin, Graduate Program Assistant, 805-893-3329, Fax: 805-893-2314, E-mail: rifkin@geol.ucsb.edu.

University of Chicago, Division of the Physical Sciences, Department of the Geophysical Sciences, Chicago, IL 60637-1513. Offers atmospheric sciences (SM, PhD); earth sciences (SM, PhD); paleobiology (PhD); planetary and space sciences (SM, PhD). *Faculty:* 24 full-time (3 women). *Students:* 35 full-time (15 women); includes 1 minority (Hispanic American), 11 international. Average age 29. 52 applicants, 29% accepted. In 2005, 2 master's, 3 doctorates awarded. Terminal master's awarded for partial completion of doctoral program. *Entrance requirements:* For master's and doctorate, GRE General Test. Additional exam requirements/recommendations for international students: Required—TOEFL, IELT. *Application deadline:* For fall admission, 1/15 for domestic students, 1/15 for international students. Application fee: $55. Electronic applications accepted. *Financial support:* In 2005–06, 32 students received support, including research assistantships with full tuition reimbursements available (averaging $18,675 per year), teaching assistantships with full tuition reimbursements available (averaging $19,098 per year); fellowships, Federal Work-Study, institutionally sponsored loans, scholarships/grants, tuition waivers (partial), and unspecified assistantships also available. Financial award application deadline: 1/15. *Faculty research:* Climatology, evolutionary paleontology, petrology, geochemistry, oceanic sciences. *Unit head:* Dr. David Rowley, Chairman, 773-702-8102, Fax: 773-702-9505. *Application contact:* David J. Leslie, Graduate Student Services Coordinator, 773-702-8180, Fax: 773-702-9505, E-mail: info@geosci.uchicago.edu.

University of Colorado at Boulder, Graduate School, College of Arts and Sciences, Department of Geological Sciences, Boulder, CO 80309. Offers geology (MS, PhD); geophysics (PhD). *Faculty:* 27 full-time (7 women). *Students:* 52 full-time (24 women), 15 part-time (7 women); includes 6 minority (1 Asian American or Pacific Islander, 5 Hispanic Americans), 6 international. Average age 30. 22 applicants, 82% accepted. In 2005, 8 master's, 3 doctorates awarded. Terminal master's awarded for partial completion of doctoral program. *Degree requirements:* For master's and doctorate, thesis/dissertation, comprehensive exam. *Entrance requirements:* For master's and doctorate, GRE General Test, minimum GPA of 2.75. *Application deadline:* For fall admission, 1/15 priority date for domestic students, 12/1 priority date for international students. Application fee: $50 ($60 for international students). *Financial support:* In 2005–06, fellowships with full tuition reimbursements (averaging $9,707 per year), research assistantships with full tuition reimbursements (averaging $13,873 per year), teaching assistantships with full tuition reimbursements (averaging $12,684 per year) were awarded; Federal Work-Study, institutionally sponsored loans, scholarships/grants, and tuition waivers (full) also available. Financial award application deadline: 1/15. *Faculty research:* Sedimentology, stratigraphy, economic geology of mineral deposits, fossil fuels, hydrogeology and water resources. Total annual research expenditures: $5.1 million. *Unit head:* Mary Kraus, Chair, 303-492-2330, Fax: 303-492-2606, E-mail: mary.kraus@colorado.edu. *Application contact:* Graduate Secretary, 303-492-2607, Fax: 303-492-2606, E-mail: geolinfo@colorado.edu.

University of Colorado at Boulder, Graduate School, College of Arts and Sciences, Department of Physics, Boulder, CO 80309. Offers chemical physics (PhD); geophysics (PhD); liquid crystal science and technology (PhD); mathematical physics (PhD); medical physics (PhD); optical sciences and engineering (PhD); physics (MS, PhD). *Faculty:* 39 full-time (3 women). *Students:* 157 full-time (37 women), 47 part-time (3 women); includes 10 minority (2 African Americans, 1 American Indian/Alaska Native, 4 Asian Americans or Pacific Islanders, 3 Hispanic Americans), 46 international. Average age 27. 81 applicants, 93% accepted. In 2005, 10 master's, 16 doctorates awarded. Terminal master's awarded for partial completion of doctoral program. *Degree requirements:* For master's, thesis or alternative, comprehensive exam; for doctorate, thesis/dissertation, comprehensive exam. *Entrance requirements:* For master's and doctorate, GRE General Test, GRE Subject Test, minimum undergraduate GPA of 3.0. Additional exam requirements/recommendations for international students: Required—TOEFL. *Application deadline:* For fall admission, 1/15 priority date for domestic students, 1/15 priority date for international students. Applications are processed on a rolling basis. Application fee: $50 ($60 for international students). Electronic applications accepted. *Financial support:* In 2005–06, fellowships with full tuition reimbursements (averaging $4,376 per year), research assistantships with full tuition reimbursements (averaging $15,185 per year), teaching assistantships with full tuition reimbursements (averaging $13,786 per year) were awarded; scholarships/grants also available. Financial award application deadline: 1/15. *Faculty research:* Atomic and molecular physics, nuclear physics, condensed matter, elementary particle physics, laser or optical physics. Total annual research expenditures: $24.1 million. *Unit head:* John Cumalat, Chair, 303-492-6952, Fax: 303-492-3352, E-mail: jcumalat@pizero.colorado.edu. *Application contact:* Graduate Program Assistant, 303-492-6954, Fax: 303-492-3352, E-mail: phys@bogart.colorado.edu.

University of Hawaii at Manoa, Graduate Division, School of Ocean and Earth Science and Technology, Department of Geology and Geophysics, Honolulu, HI 96822. Offers high-pressure geophysics and geochemistry (MS, PhD); hydrogeology and engineering geology (MS, PhD); marine geology and geophysics (MS, PhD); planetary geosciences and remote sensing (MS, PhD); seismology and solid-earth geophysics (MS, PhD); volcanology, petrology, and geochemistry (MS, PhD). *Faculty:* 65 full-time (12 women), 19 part-time/adjunct (1 woman). *Students:* 50 full-time (23 women), 8 part-time (1 woman); includes 2 minority (1 African American, 1 Asian American or Pacific Islander), 15 international. Average age 29. 89 applicants, 18% accepted, 12 enrolled. In 2005, 9 master's, 6 doctorates awarded. Terminal master's awarded for partial completion of doctoral program. *Median time to degree:* Of those who began their doctoral program in fall 1997, 100% received their degree in 8 years or less. *Degree requirements:* For master's, thesis/dissertation; for doctorate, thesis/dissertation, comprehensive exam. *Entrance requirements:* For master's and doctorate, GRE General Test, minimum GPA of 3.0. Additional exam requirements/recommendations for international students: Required—TOEFL. *Application deadline:* For fall admission, 1/15 for domestic students, 1/1 for international students; for spring admission, 9/1 for domestic students, 8/15 for international students. Application fee: $50. *Expenses:* Tuition, state resident: full-time $8,400; part-time $200 per credit hour. Tuition, nonresident: full-time $11,088; part-time $462 per credit hour. Tuition and fees vary according to program. *Financial support:* In 2005–06, 41 research assistantships (averaging $19,366 per year), 5 teaching assistantships (averaging $18,198 per year) were awarded. *Unit head:* Dr. Charles Fletcher, Chair, 808-956-8763, Fax: 808-956-5512, E-mail: fletcher@soest.hawaii.edu. *Application contact:* Dr. Charles Fletcher, Chair, 808-956-8763, Fax: 808-956-5512, E-mail: fletcher@soest.hawaii.edu.

University of Houston, College of Natural Sciences and Mathematics, Department of Geosciences, Houston, TX 77204. Offers geology (MA, MS, PhD); geophysics (MA, MS, PhD). Part-time and evening/weekend programs available. *Faculty:* 18 full-time (3 women), 4 part-time/adjunct (0 women). *Students:* 70 full-time (24 women), 52 part-time (17 women); includes 12 minority (4 African Americans, 1 American Indian/Alaska Native, 2 Asian Americans or Pacific Islanders, 5 Hispanic Americans), 65 international. Average age 33. 73 applicants, 85% accepted, 32 enrolled. In 2005, 10 master's, 5 doctorates awarded. *Degree requirements:* For master's, thesis (for some programs); for doctorate, one foreign language, thesis/dissertation. *Entrance requirements:* For master's and doctorate, GRE General Test. Additional exam requirements/recommendations for international students: Required—TOEFL. *Application deadline:* For fall admission, 9/20 for domestic students; for spring admission, 12/4 for domestic students. Applications are processed on a rolling basis. Application fee: $0 ($75 for international students). *Financial support:* In 2005–06, 7 fellowships with full tuition reimbursements (averaging $16,800 per year), 27 research assistantships with full tuition reimbursements (averaging $12,750 per year), 28 teaching assistantships with full tuition reimbursements (averaging $12,750 per year) were awarded; career-related internships or fieldwork, Federal Work-Study, institutionally sponsored loans, scholarships/grants, health care benefits, and unspecified assistantships also available. Support available to part-time students. Financial award application deadline: 3/10. *Faculty research:* Seismic and solid earth geophysics, tectonics, environmental hydrochemistry, carbonates, micropaleontology, structure and tectonics, petroleum geology. *Unit head:* Dr. John Casey, Chairman, 713-743-3399, Fax: 713-748-7906, E-mail: jfcasey@uh.edu. *Application contact:* Dr. Charlotte Sullivan, Graduate Adviser, 713-743-3396, Fax: 713-748-7906, E-mail: esulliva@bayou.uh.edu.

University of Illinois at Chicago, Graduate College, College of Liberal Arts and Sciences, Department of Earth and Environmental Sciences, Chicago, IL 60607-7128. Offers crystallography (MS, PhD); environmental geology (MS, PhD); geochemistry (MS, PhD); geology (MS, PhD); geomorphology (MS, PhD); geophysics (MS, PhD); geotechnical engineering and geosciences (PhD); hydrogeology (MS, PhD); low-temperature and organic geochemistry (MS, PhD); mineralogy (MS, PhD); paleoclimatology (MS, PhD); paleontology (MS, PhD); petrology (MS, PhD); quaternary geology (MS, PhD); sedimentology (MS, PhD); water resources (MS, PhD). *Degree requirements:* For master's and doctorate, thesis/dissertation. *Entrance requirements:* For master's and doctorate, GRE General Test, minimum GPA of 2.75. Additional exam requirements/recommendations for international students: Required—TOEFL. Electronic applications accepted.

University of Illinois at Urbana–Champaign, Graduate College, College of Liberal Arts and Sciences, Department of Geology, Champaign, IL 61820. Offers earth sciences (MS, PhD); geochemistry (MS, PhD); geology (MS, PhD); geophysics (MS, PhD). *Faculty:* 13 full-time (3 women). *Students:* 32 full-time (13 women), 4 part-time; includes 1 minority (American Indian/Alaska Native), 16 international. Average age 26. 44 applicants, 39% accepted, 13 enrolled. In 2005, 4 master's, 5 doctorates awarded. Terminal master's awarded for partial completion of doctoral program. *Degree requirements:* For master's and doctorate, thesis/dissertation. *Entrance requirements:* For master's and doctorate, GRE General Test, minimum GPA of 3.0. Additional exam requirements/recommendations for international students: Required—TOEFL. *Application deadline:* For fall admission, 2/22 for domestic students; for spring admission, 10/16 for domestic students. Applications are processed on a rolling basis. Application fee: $50 ($60 for

Geophysics

University of Illinois at Urbana–Champaign (continued)
international students). *Financial support:* In 2005–06, 8 fellowships, 14 research assistantships, 18 teaching assistantships were awarded; Federal Work-Study and tuition waivers (full and partial) also available. Financial award application deadline: 2/15. *Faculty research:* Hydrogeology, structure/tectonics, mineral science. *Unit head:* Stephen Marshak, Head, 217-333-3542, Fax: 217-244-4996, E-mail: smarshak@uiuc.edu. *Application contact:* Barbara Elmore, Graduate Admissions Secretary, 217-333-3542, Fax: 217-244-4996, E-mail: belmore@uiuc.edu.

University of Manitoba, Faculty of Graduate Studies, Faculty of Environment, Earth and Resources, Department of Geological Sciences, Winnipeg, MB R3T 2N2, Canada. Offers geology (M Sc, PhD); geophysics (M Sc, PhD). *Degree requirements:* For master's and doctorate, thesis/dissertation. *Entrance requirements:* For master's and doctorate, GRE General Test, GRE Subject Test (geology), minimum GPA of 3.0. Additional exam requirements/recommendations for international students: Required—TOEFL.

University of Memphis, Graduate School, College of Arts and Sciences, Department of Earth Sciences, Memphis, TN 38152. Offers earth sciences (PhD); geography (MA, MS); geology (MS); geophysics (MS). Part-time programs available. *Faculty:* 23 full-time (3 women), 3 part-time/adjunct (1 woman). *Students:* Average age 29. 13 applicants, 46% accepted. In 2005, 5 master's, 1 doctorate awarded. Terminal master's awarded for partial completion of doctoral program. *Degree requirements:* For master's, thesis, seminar presentation, comprehensive exam; for doctorate, thesis/dissertation. *Entrance requirements:* For master's and doctorate, GRE General Test. Additional exam requirements/recommendations for international students: Required—TOEFL. *Application deadline:* For fall admission, 8/1 for domestic students; for spring admission, 12/1 for domestic students. Applications are processed on a rolling basis. Application fee: $25 ($50 for international students). Electronic applications accepted. *Financial support:* In 2005–06, 34 students received support, including 2 fellowships with full tuition reimbursements available, 10 research assistantships with full tuition reimbursements available, 11 teaching assistantships with full tuition reimbursements available (averaging $10,000 per year) Financial award application deadline: 1/15. *Faculty research:* Hazards, active tectonics, geophysics, hydrology and water resources, spatial analysis. Total annual research expenditures: $1.8 million. *Unit head:* Dr. Mervin J. Bartholomew, Chair, 901-678-1613, Fax: 901-678-4467, E-mail: jbrthlm1@memphis.edu. *Application contact:* Dr. George Swihart, Coordinator of Graduate Studies, 901-678-2606, Fax: 901-678-2178, E-mail: gswihart@memphis.edu.

University of Miami, Graduate School, Rosenstiel School of Marine and Atmospheric Science, Division of Marine Geology and Geophysics, Coral Gables, FL 33124. Offers MS, PhD. *Faculty:* 14 full-time (2 women), 14 part-time/adjunct (4 women). *Students:* 26 full-time (13 women), 8 international. Average age 27. 23 applicants, 26% accepted, 4 enrolled. In 2005, 1 master's, 4 doctorates awarded. Terminal master's awarded for partial completion of doctoral program. *Median time to degree:* Of those who began their doctoral program in fall 1997, 100% received their degree in 8 years or less. *Degree requirements:* For master's and doctorate, thesis/dissertation, comprehensive exam, registration. *Entrance requirements:* For master's and doctorate, GRE General Test. Additional exam requirements/recommendations for international students: Required—TOEFL (minimum score 550 paper-based; 213 computer-based). *Application deadline:* For fall admission, 1/1 priority date for domestic students, 1/1 priority date for international students. Applications are processed on a rolling basis. Application fee: $50. Electronic applications accepted. *Financial support:* In 2005–06, 23 students received support, including 4 fellowships with tuition reimbursements available (averaging $22,380 per year), 15 research assistantships with tuition reimbursements available (averaging $22,380 per year), 4 teaching assistantships with tuition reimbursements available (averaging $22,380 per year); institutionally sponsored loans also available. Financial award application deadline: 3/1; financial award applicants required to submit FAFSA. *Faculty research:* Carbonate sedimentology, low-temperature geochemistry, paleoceanography, geodesy and tectonics. *Unit head:* Dr. Gregor Eberli, Chairperson, 305-421-4678, Fax: 305-421-4632, E-mail: geberli@rsmas.miami.edu. *Application contact:* Dr. Larry Peterson, Associate Dean, 305-421-4155, Fax: 305-421-4771, E-mail: gso@rsmas.miami.edu.

University of Minnesota, Twin Cities Campus, Graduate School, Institute of Technology, Department of Geology and Geophysics, Minneapolis, MN 55455-0213. Offers geology (MS, PhD); geophysics (MS, PhD). Terminal master's awarded for partial completion of doctoral program. *Degree requirements:* For master's, thesis optional; for doctorate, thesis/dissertation. *Entrance requirements:* For master's and doctorate, GRE General Test. Additional exam requirements/recommendations for international students: Required—TOEFL. *Expenses:* Tuition, state resident: full-time $8,748; part-time $729 per credit. Tuition, nonresident: full-time $15,848; part-time $1,321 per credit. Full-time tuition and fees vary according to class time, course load, program and reciprocity agreements. *Faculty research:* Hydrogeology, geochemistry, structural geology, sedimentology.

University of Missouri–Rolla, Graduate School, School of Materials, Energy, and Earth Resources, Department of Geological Sciences and Engineering, Program in Geology and Geophysics, Rolla, MO 65409-0910. Offers geochemistry (MS, PhD); geology (MS, PhD); geophysics (MS, PhD); groundwater and environmental geology (MS, PhD). Part-time programs available. *Degree requirements:* For master's and doctorate, thesis/dissertation. *Entrance requirements:* For master's, GRE General Test, GRE Subject Test, minimum GPA of 3.0 in last 4 semesters; for doctorate, GRE General Test, GRE Subject Test. Additional exam requirements/recommendations for international students: Required—TOEFL. Electronic applications accepted. *Faculty research:* Economic geology, geophysical modeling, seismic wave analysis.

University of Nevada, Reno, Graduate School, College of Science, Mackay School of Earth Sciences and Engineering, Department of Geological Sciences, Reno, NV 89557. Offers geochemistry (MS, PhD); geological engineering (MS, Geol E); geology (MS, PhD); geophysics (MS, PhD). *Degree requirements:* For master's, thesis optional; for doctorate, one foreign language, thesis/dissertation. *Entrance requirements:* For master's, GRE General Test, GRE Subject Test, minimum GPA of 2.75; for doctorate, GRE General Test, GRE Subject Test, minimum GPA of 3.0. Additional exam requirements/recommendations for international students: Required—TOEFL. *Faculty research:* Hydrothermal ore deposits, metamorphic and igneous petrogenesis, sedimentary rock record of earth history, field and petrographic investigation of magnetism, rock fracture mechanics.

University of New Orleans, Graduate School, College of Sciences, Department of Geology and Geophysics, New Orleans, LA 70148. Offers MS. Evening/weekend programs available. *Degree requirements:* For master's, thesis. *Entrance requirements:* For master's, GRE General Test. Additional exam requirements/recommendations for international students: Required—TOEFL (minimum score 550 paper-based; 213 computer-based). Electronic applications accepted. *Faculty research:* Continental margin structure and seismology, burial diagenesis of siliclastic sediments, tectonics at convergent plate margins, continental shelf sediment stability, early diagenesis of carbonates.

University of Oklahoma, Graduate College, College of Earth and Energy, School of Geology and Geophysics, Program in Geophysics, Norman, OK 73019-0390. Offers MS. Part-time programs available. *Students:* 7 full-time (3 women), 7 part-time (4 women), 7 international. 8 applicants, 50% accepted, 1 enrolled. In 2005, 7 degrees awarded. *Degree requirements:* For master's, thesis, comprehensive exam. *Entrance requirements:* For master's, GRE General Test. Additional exam requirements/recommendations for international students: Required—TOEFL (minimum score 550 paper-based; 213 computer-based). *Application deadline:* For fall admission, 2/1 priority date for domestic students, 4/1 priority date for international students; for spring admission, 9/1 for domestic students, 9/1 for international students. Applications are processed on a rolling basis. Application fee: $40 ($90 for international students). *Expenses:* Tuition, state resident: full-time $3,029; part-time $126 per credit hour. Tuition, nonresident: full-time $10,807; part-time $450 per credit hour. Required fees: $1,231; $44 per credit hour. Tuition and fees vary according to course load and program. *Financial support:* In 2005–06, 4 students received support; research assistantships with partial tuition reimbursements available, teaching assistantships with partial tuition reimbursements available, career-related internships or fieldwork, scholarships/grants, tuition waivers (partial), unspecified assistantships, and full-time employment possible during degree study on an individual basis available. Financial award application deadline: 2/1; financial award applicants required to submit FAFSA. *Faculty research:* Near surface geophysics, exploration geophysics, reservoir, characterization, basin modeling, lithospheric structure and techfonics. *Application contact:* Donna S. Mullins, Coordinator of Administrative Student Services, 405-325-3255, Fax: 405-325-3140, E-mail: dsmullins@ou.edu.

The University of Texas at El Paso, Graduate School, College of Science, Department of Geological Sciences, Program in Geophysics, El Paso, TX 79968-0001. Offers MS. Part-time and evening/weekend programs available. *Degree requirements:* For master's, thesis. *Entrance requirements:* For master's, GRE, minimum GPA of 3.0, BS in geology or equivalent. Additional exam requirements/recommendations for international students: Required—TOEFL. Electronic applications accepted.

University of Utah, The Graduate School, College of Mines and Earth Sciences, Department of Geology and Geophysics, Salt Lake City, UT 84112-1107. Offers environmental engineering (ME, MS, PhD); geological engineering (ME, MS, PhD); geology (MS, PhD); geophysics (MS, PhD). *Faculty:* 22 full-time (4 women), 3 part-time/adjunct (0 women). *Students:* 45 full-time (13 women), 20 part-time (7 women), 14 international. Average age 31. 81 applicants, 37% accepted, 16 enrolled. In 2005, 10 master's, 3 doctorates awarded. Terminal master's awarded for partial completion of doctoral program. *Median time to degree:* Of those who began their doctoral program in fall 1997, 33% received their degree in 8 years or less. *Degree requirements:* For master's, thesis, comprehensive exam; for doctorate, one foreign language, thesis/dissertation, qualifying exam. *Entrance requirements:* For master's and doctorate, GRE General Test, minimum GPA of 3.25. Additional exam requirements/recommendations for international students: Required—TOEFL (minimum score 500 paper-based; 173 computer-based). *Application deadline:* For fall admission, 1/15 priority date for domestic students, 1/15 priority date for international students. Applications are processed on a rolling basis. Application fee: $45 ($65 for international students). Electronic applications accepted. *Expenses:* Tuition, state resident: full-time $2,932; part-time $2,212 per term. Tuition, nonresident: full-time $10,350; part-time $7,812 per term. Required fees: $590; $516 per term. Tuition and fees vary according to course load and program. *Financial support:* In 2005–06, 7 fellowships with full tuition reimbursements (averaging $30,000 per year), 22 research assistantships with full tuition reimbursements (averaging $29,200 per year), 11 teaching assistantships with full tuition reimbursements (averaging $12,183 per year) were awarded; career-related internships or fieldwork, institutionally sponsored loans, scholarships/grants, tuition waivers, unspecified assistantships, and stipends also available. Financial award application deadline: 1/15; financial award applicants required to submit FAFSA. *Faculty research:* Igneous, metamorphic, and sedimentary petrology; ore deposits; aqueous geochemistry; isotope geochemistry; heat flow. Total annual research expenditures: $2.3 million. *Unit head:* Dr. Marjorie A. Chan, Chair, 801-581-7162, Fax: 801-581-7065, E-mail: gg_chair@mines.utah.edu. *Application contact:* Dr. Richard D Jarrad, Director of Graduate Studies, 801-581-6553, Fax: 801-581-7065, E-mail: jarrad@earth.utah.edu.

University of Victoria, Faculty of Graduate Studies, Faculty of Science, Department of Physics and Astronomy, Victoria, BC V8W 2Y2, Canada. Offers astronomy and astrophysics (M Sc, PhD); condensed matter physics (M Sc, PhD); experimental particle physics (M Sc, PhD); medical physics (M Sc, PhD); ocean physics (PhD); ocean physics and geophysics (M Sc); theoretical physics (M Sc, PhD). *Faculty:* 16 full-time (0 women), 13 part-time/adjunct (1 woman). *Students:* 45 full-time, 8 international. Average age 25. 66 applicants, 45% accepted, 10 enrolled. In 2005, 4 master's, 1 doctorate awarded. *Median time to degree:* Of those who began their doctoral program in fall 1997, 100% received their degree in 8 years or less. *Degree requirements:* For master's, thesis, registration; for doctorate, thesis/dissertation, Candidacy exam, comprehensive exam, registration. *Entrance requirements:* For master's and doctorate, GRE. Additional exam requirements/recommendations for international students: Required—TOEFL (minimum score 575 paper-based; 233 computer-based), IELT (minimum score 7). *Application deadline:* For fall admission, 5/31 priority date for domestic students, 12/15 priority date for international students. Applications are processed on a rolling basis. Application fee: $75 ($125 for international students). Electronic applications accepted. Tuition and fees charges are reported in Canadian dollars. *Expenses:* Tuition, area resident: Full-time $4,492 Canadian dollars; part-time $749 Canadian dollars per term. International tuition: $5,346 Canadian dollars full-time. Required fees: $4,492 Canadian dollars; $749 Canadian dollars per term. Tuition and fees vary according to course load, campus/location and program. *Financial support:* In 2005–06, 3 students received support; fellowships, research assistantships, teaching assistantships, career-related internships or fieldwork, institutionally sponsored loans, and awards available. Financial award application deadline: 2/15. *Faculty research:* Old stellar populations; observational cosmology and large scale structure; cp violation; atlas. *Unit head:* Dr. J. Michael Ronev, Chair, 250-721-7698, Fax: 250-721-7715, E-mail: chair@phys.uvic.ca. *Application contact:* Dr. Chris J. Pritchet, Graduate Adviser, 250-721-7744, Fax: 250-721-7715, E-mail: pritchet@uvic.ca.

University of Washington, Graduate School, College of Arts and Sciences, Department of Earth and Space Sciences, Seattle, WA 98195. Offers geology (MS, PhD); geophysics (MS, PhD). *Degree requirements:* For master's, thesis or alternative, departmental qualifying exam, final exam; for doctorate, thesis/dissertation, departmental qualifying exam, general and final exams. *Entrance requirements:* For master's and doctorate, GRE General Test, minimum GPA of 3.0. Additional exam requirements/recommendations for international students: Required—TOEFL (minimum score 580 paper-based). Electronic applications accepted.

The University of Western Ontario, Faculty of Graduate Studies, Physical Sciences Division, Department of Earth Sciences, London, ON N6A 5B8, Canada. Offers geology (M Sc, PhD); geology and environmental science (M Sc, PhD); geophysics (M Sc, PhD); geophysics and environmental science (M Sc, PhD). *Degree requirements:* For master's, thesis, registration; for doctorate, thesis/dissertation, qualifying exam. *Entrance requirements:* For master's, honors in B Sc; for doctorate, M Sc. Additional exam requirements/recommendations for international students: Required—TOEFL. *Faculty research:* Geophysics, geochemistry, paleontology, sedimentology/stratigraphy, glaciology/quaternary.

University of Wisconsin–Madison, Graduate School, College of Letters and Science, Department of Geology and Geophysics, Program in Geophysics, Madison, WI 53706-1380. Offers MS, PhD. *Degree requirements:* For master's, thesis; for doctorate, one foreign language, thesis/dissertation. *Entrance requirements:* For master's and doctorate, GRE General Test.

University of Wyoming, Graduate School, College of Arts and Sciences, Department of Geology and Geophysics, Laramie, WY 82070. Offers geology (MS, PhD); geophysics (MS, PhD). Part-time programs available. *Faculty:* 17 full-time (2 women), 3 part-time/adjunct (2 women). *Students:* 24 full-time (13 women), 28 part-time (8 women); includes 2 minority (1 Asian American or Pacific Islander, 1 Hispanic American), 6 international. 21 applicants, 67% accepted. In 2005, 12 master's, 2 doctorates awarded. *Degree requirements:* For master's and doctorate, variable foreign language requirement, thesis/dissertation. *Entrance requirements:* For master's and doctorate, GRE General Test, minimum GPA of 3.0. *Application deadline:* For fall admission, 1/15 for domestic students, 1/15 for international students. Applications are processed on a rolling basis. Application fee: $50. *Expenses:* Tuition, state resident: full-time $3,720; part-time $155 per credit hour. Tuition, nonresident: full-time $10,704; part-time $446 per credit hour. Required fees: $666; $162 per semester. Tuition and fees vary according to course load and program. *Financial support:* Fellowships, research assistantships, teaching assistantships, career-related internships or fieldwork, Federal Work-Study, and institutionally sponsored loans available. Financial award application deadline: 3/1. *Faculty research:* Geochemistry and petroleum geology, tectonics and sedimentation, geomorphology and remote sensing, igneous and metamorphic petrology, structure, geohydrology. *Unit head:* Dr. Arthur W. Snoke, Head, 307-766-3386. *Application contact:* Sondra S. Cawley, Admissions Coordinator, 307-766-3389, Fax: 307-766-6679, E-mail: acadcoord.geol@uwyo.edu.

Virginia Polytechnic Institute and State University, Graduate School, College of Science, Department of Geosciences, Blacksburg, VA 24061. Offers geological sciences (MS, PhD); geophysics (MS, PhD). *Faculty:* 18 full-time (2 women), 1 (woman) part-time/adjunct. *Students:* 55 full-time (20 women), 2 part-time (1 woman); includes 2 minority (1 American Indian/Alaska Native, 1 Asian American or Pacific Islander), 18 international. Average age 27. 55 applicants, 36% accepted, 11 enrolled. In 2005, 9 master's, 1 doctorate awarded. *Entrance requirements:* For master's and doctorate, GRE General Test. Additional exam requirements/recommendations for international students: Required—TOEFL (minimum score 550 paper-based; 213 computer-based). *Application deadline:* Applications are processed on a rolling basis. Application fee: $45. Electronic applications accepted. *Expenses:* Tuition, state resident: full-time $6,558; part-time $364 per credit. Tuition, nonresident: full-time $11,296; part-time $628 per credit. Required fees: $1,419; $468 per credit. $234 per term. *Financial support:* In 2005–06, 8 fellowships with full tuition reimbursements (averaging $15,121 per year), 24 research assistantships with full tuition reimbursements (averaging $16,689 per year), 27 teaching assistantships with full tuition reimbursements (averaging $13,902 per year) were awarded; career-related internships or fieldwork, Federal Work-Study, scholarships/grants, tuition waivers (full), and unspecified assistantships also available. Financial award application deadline: 4/1. *Faculty research:* Paleontology/geobiology, active tectonics, geomorphology, mineralogy/crystallography, mineral physics. *Unit head:* Dr. Robert Tracy, Head, 540-231-6521, Fax: 540-231-3386. *Application contact:* Connie Lowe, Student Program Coordinator, 540-231-8824, Fax: 540-231-3386, E-mail: clowe@vt.edu.

Washington University in St. Louis, Graduate School of Arts and Sciences, Department of Earth and Planetary Sciences, St. Louis, MO 63130-4899. Offers earth and planetary sciences (MA); geochemistry (PhD); geology (MA, PhD); geophysics (PhD); planetary sciences (PhD). Terminal master's awarded for partial completion of doctoral program. *Degree requirements:* For master's and doctorate, thesis/dissertation. *Entrance requirements:* For master's and doctorate, GRE General Test. Electronic applications accepted.

West Virginia University, Eberly College of Arts and Sciences, Department of Geology and Geography, Program in Geology, Morgantown, WV 26506. Offers geomorphology (MS, PhD); geophysics (MS, PhD); hydrogeology (MS); hydrology (PhD); paleontology (MS, PhD); petrology (MS, PhD); stratigraphy (MS, PhD); structure (MS, PhD). Part-time programs available. *Students:* 26 full-time (7 women), 17 part-time (7 women); includes 2 minority (1 American Indian/Alaska Native, 1 Asian American or Pacific Islander), 12 international. Average age 30. 34 applicants, 29% accepted. In 2005, 6 master's, 1 doctorate awarded. Terminal master's awarded for partial completion of doctoral program. *Degree requirements:* For master's, thesis/dissertation; for doctorate, thesis/dissertation, comprehensive exam. *Entrance requirements:* For master's, GRE General Test, GRE Subject Test, minimum GPA of 2.5; for doctorate, GRE General Test, GRE Subject Test, minimum GPA of 3.3. Additional exam requirements/recommendations for international students: Required—TOEFL. *Application deadline:* For fall admission, 2/1 for domestic students; for spring admission, 10/1 for domestic students. Applications are processed on a rolling basis. Application fee: $45. *Expenses:* Tuition, state resident: full-time $4,582; part-time $258 per credit hour. Tuition, nonresident: full-time $1,382; part-time $741 per credit hour. *Financial support:* In 2005–06, 5 research assistantships, 18 teaching assistantships were awarded; career-related internships or fieldwork, Federal Work-Study, institutionally sponsored loans, and tuition waivers (full and partial) also available. Financial award application deadline: 2/1; financial award applicants required to submit FAFSA. *Unit head:* Dr. Thomas H. Wilson, Associate Chair, 304-293-5603 Ext. 4316, Fax: 304-293-6522, E-mail: tom.wilson@mail.wvu.edu. *Application contact:* Dr. Joe Donovan, Associate Professor, 304-293-5603 Ext. 4308, Fax: 304-293-6522, E-mail: joe.donovan@mail.wvu.edu.

Woods Hole Oceanographic Institution, MIT/WHOI Joint Program in Oceanography/Applied Ocean Science and Engineering, Woods Hole, MA 02543-1541. Offers applied ocean sciences (PhD); biological oceanography (PhD, Sc D); chemical oceanography (PhD, Sc D); civil and environmental and oceanographic engineering (PhD); electrical and oceanographic engineering (PhD); geochemistry (PhD); geophysics (PhD); marine biology (PhD); marine geochemistry (PhD, Sc D); marine geology (PhD, Sc D); marine geophysics (PhD); mechanical and oceanographic engineering (PhD); ocean engineering (PhD); oceanographic engineering (M Eng, MS, PhD, Sc D, Eng); paleoceanography (PhD); physical oceanography (PhD, Sc D). MS, PhD, and Sc D offered jointly with MIT. Terminal master's awarded for partial completion of doctoral program. *Degree requirements:* For master's and Eng, thesis (for some programs); for doctorate, thesis/dissertation. *Entrance requirements:* For master's, GRE General Test; for doctorate, GRE General Test, GRE Subject Test. Additional exam requirements/recommendations for international students: Required—TOEFL. Electronic applications accepted.

See Close-Up on page 259.

Wright State University, School of Graduate Studies, College of Science and Mathematics, Department of Geological Sciences, Program in Geological Sciences, Dayton, OH 45435. Offers environmental geochemistry (MS); environmental geology (MS); environmental sciences (MS); geological sciences (MS); geophysics (MS); hydrogeology (MS); petroleum geology (MS). Part-time programs available. *Degree requirements:* For master's, thesis. *Entrance requirements:* Additional exam requirements/recommendations for international students: Required—TOEFL.

Yale University, Graduate School of Arts and Sciences, Department of Geology and Geophysics, New Haven, CT 06520. Offers geochemistry (PhD); geophysics (PhD); meteorology (PhD); mineralogy and crystallography (PhD); oceanography (PhD); paleoecology (PhD); paleontology and stratigraphy (PhD); petrology (PhD); structural geology (PhD). *Degree requirements:* For doctorate, thesis/dissertation. *Entrance requirements:* For doctorate, GRE General Test. Additional exam requirements/recommendations for international students: Required—TOEFL.

Geosciences

Arizona State University, Division of Graduate Studies, College of Liberal Arts and Sciences, Division of Natural Sciences and Mathematics, Department of Geological Sciences, Tempe, AZ 85287. Offers geological engineering (MS, PhD); natural science (MNS). *Degree requirements:* For master's, thesis or alternative; for doctorate, thesis/dissertation. *Entrance requirements:* For master's and doctorate, GRE.

Ball State University, Graduate School, College of Sciences and Humanities, Department of Geography, Muncie, IN 47306-1099. Offers earth sciences (MA). *Faculty:* 8. *Students:* 6 full-time (2 women), 5 part-time (2 women), 4 international. Average age 29. 10 applicants, 70% accepted, 4 enrolled. In 2005, 4 degrees awarded. Application fee: $25 ($35 for international students). *Expenses:* Tuition, state resident: full-time $6,246. Tuition, nonresident: full-time $16,006. *Financial support:* In 2005–06, 9 teaching assistantships with full tuition reimbursements (averaging $8,400 per year) were awarded; research assistantships with full tuition reimbursements Financial award application deadline: 3/1. *Faculty research:* Remote sensing, tourism and recreation, Latin American urbanization. *Unit head:* Dr. Gopalan Venugopal, Chairman, 765-285-1776.

Baylor University, Graduate School, College of Arts and Sciences, Department of Geology, Waco, TX 76798. Offers earth science (MA); geology (MS, PhD). *Faculty:* 12 full-time (1 woman). *Students:* 14 full-time (7 women), 4 part-time (2 women); includes 1 minority (Hispanic American), 1 international. In 2005, 10 master's, 1 doctorate awarded. *Degree requirements:* For master's and doctorate, thesis/dissertation. *Entrance requirements:* For master's and doctorate, GRE General Test. *Application deadline:* For fall admission, 3/15 for domestic students. Applications are processed on a rolling basis. Application fee: $25. *Financial support:* In 2005–06, 18 teaching assistantships were awarded; Federal Work-Study and institutionally sponsored loans also available. *Faculty research:* Petroleum geology, geophysics, engineering geology, hydrogeology. *Unit head:* Dr. Stacy Atchley, Graduate Program Director, 254-710-2196, Fax: 254-710-2673, E-mail: stacy_atchley@baylor.edu. *Application contact:* Suzanne Keener, Administrative Assistant, 254-710-3588, Fax: 254-710-3870.

Boise State University, Graduate College, College of Arts and Sciences, Department of Geosciences, Program in Earth Science, Boise, ID 83725-0399. Offers MS. Part-time programs available. *Degree requirements:* For master's, thesis. *Entrance requirements:* For master's, GRE General Test, minimum GPA of 3.0, BS in related field. Electronic applications accepted.

Boston University, Graduate School of Arts and Sciences, Department of Earth Sciences, Boston, MA 02215. Offers MA, PhD. *Students:* 23 full-time (16 women); includes 1 minority (Asian American or Pacific Islander), 5 international. Average age 29. 45 applicants, 33% accepted, 8 enrolled. In 2005, 4 master's, 1 doctorate awarded. Terminal master's awarded for partial completion of doctoral program. *Degree requirements:* For master's and doctorate, one foreign language, thesis/dissertation, comprehensive exam, registration. *Entrance requirements:* For master's and doctorate, GRE General Test, 3 letters of recommendation. Additional exam requirements/recommendations for international students: Required—TOEFL (minimum score 550 paper-based; 213 computer-based). *Application deadline:* For fall admission, 7/1 for domestic students, 7/1 for international students; for spring admission, 11/15 for domestic students, 11/15 for international students. Application fee: $60. *Expenses:* Tuition: Full-time $31,530; part-time $985 per credit. Required fees: $316; $40 per semester. Tuition and fees vary according to course level and program. *Financial support:* In 2005–06, 1 fellowship with full tuition reimbursement (averaging $16,500 per year), 10 research assistantships with tuition reimbursements (averaging $16,000 per year), 7 teaching assistantships with full tuition reimbursements (averaging $16,000 per year) were awarded; Federal Work-Study and unspecified assistantships also available. Support available to part-time students. Financial award application deadline: 1/15; financial award applicants required to submit FAFSA. *Unit head:* Dr. Guido Salvucci, Chairman, 617-353-4213, Fax: 617-353-3290, E-mail: gdsalvuc@bu.edu. *Application contact:* Katie Bowes, Senior Program Coordinator, 617-353-2529, Fax: 617-353-3290, E-mail: kbowes@bu.edu.

Brock University, Graduate Studies, Faculty of Mathematics and Science, Program in Earth Sciences, St. Catharines, ON L2S 3A1, Canada. Offers M Sc. Part-time programs available. *Faculty:* 10 full-time (1 woman), 1 part-time/adjunct (0 women). *Students:* 2 full-time (both women). 10 applicants, 10% accepted. In 2005, 2 degrees awarded. *Degree requirements:* For master's, thesis. *Entrance requirements:* For master's, honors B Sc in geology. Additional exam requirements/recommendations for international students: Required—TOEFL. *Application deadline:* Applications are processed on a rolling basis. Application fee: $75. Electronic applications accepted. *Financial support:* Fellowships, research assistantships, teaching assistantships, career-related internships or fieldwork, scholarships/grants, and unspecified assistantships available. Support available to part-time students. Financial award application deadline: 4/30. *Faculty research:* Clastic sedimentology, environmental geology, geochemistry, micropaleontology, structural geology. *Unit head:* Dr. Uwe Brand, Graduate Program Director, 905-688-5550 Ext. 3529, Fax: 905-682-9020, E-mail: uwe.brand@brocku.ca. *Application contact:* Dr. Uwe Brand, Graduate Program Director, 905-688-5550 Ext. 3529, Fax: 905-682-9020, E-mail: uwe.brand@brocku.ca.

Brown University, Graduate School, Department of Geological Sciences, Providence, RI 02912. Offers MA, Sc M, PhD. *Degree requirements:* For doctorate, thesis/dissertation, 1 semester of teaching experience, preliminary exam. *Faculty research:* Geochemistry, mineral kinetics, igneous and metamorphic petrology, tectonophysics including geophysics and structural geology, paleoclimatology, paleoceanography, sedimentation, planetary geology.

California State University, Long Beach, Graduate Studies, College of Natural Sciences and Mathematics, Department of Geological Sciences, Long Beach, CA 90840. Offers geology/geosciences (MS). Part-time programs available. *Faculty:* 8 full-time (3 women), 2 part-time/adjunct (0 women). *Students:* 2 full-time (both women), 6 part-time (4 women). Average age 37. 7 applicants, 71% accepted, 3 enrolled. In 2005, 1 degree awarded. *Degree requirements:* For master's, thesis. *Entrance requirements:* For master's, GRE General Test. *Application deadline:* For fall admission, 7/1 for domestic students; for spring admission, 12/1 for domestic students. Applications are processed on a rolling basis. Application fee: $55. Electronic applications accepted. *Expenses:* Tuition, nonresident: part-time $339 per semester hour. *Financial support:* Research assistantships, teaching assistantships, Federal Work-Study, institutionally sponsored loans, and scholarships/grants available. Financial award application deadline: 3/2. *Faculty research:* Paleontology, geophysics, structural geology, organic geochemistry, sedimentary geology. *Unit head:* Dr. Stanley C. Finney, Chair, 562-985-4809, Fax: 562-985-8638, E-mail: scfinney@csulb.edu. *Application contact:* Dr. R. Dan Francis, Graduate Coordinator, 562-985-4929, Fax: 562-985-8638, E-mail: rfrancis@csulb.edu.

California University of Pennsylvania, School of Graduate Studies, School of Science and Technology, Program in Earth Science, California, PA 15419-1394. Offers MS. Part-time and evening/weekend programs available. *Students:* 2 full-time (1 woman), 7 part-time (2 women). Average age 34. In 2005, 3 degrees awarded. *Degree requirements:* For master's, thesis optional. *Entrance requirements:* For master's, MAT, minimum GPA of 2.5, teaching certificate (M Ed). Additional exam requirements/recommendations for international students: Required—TOEFL. *Application deadline:* Applications are processed on a rolling basis. Application fee: $25. Electronic applications accepted. *Financial support:* Tuition waivers (full) and unspecified assistantships available. Financial award applicants required to submit FAFSA. *Unit head:* Dr. Chad Kauffman, Coordinator, 724-938-4130, E-mail: moses@cup.edu.

Carleton University, Faculty of Graduate Studies, Faculty of Science, Department of Earth Sciences, Ottawa, ON K1S 5B6, Canada. Offers M Sc, PhD. *Degree requirements:* For master's, thesis/dissertation, seminar; for doctorate, thesis/dissertation, seminar, comprehensive exam. *Entrance requirements:* For master's, honors degree in science; for doctorate, M Sc. Additional exam requirements/recommendations for international students: Required—TOEFL. Application fee: $75 Canadian dollars. *Financial support:* Fellowships, research assistantships, teaching assistantships, institutionally sponsored loans, scholarships/grants, and unspecified assistantships available. *Faculty research:* Resource geology, geophysics, basin analysis, lithosphere dynamics. *Unit head:* Claudia Schroeder-Adams, Chair, 613-520-2600 Ext. 1852, Fax: 613-520-4490, E-mail: earth_sciences@carleton.ca. *Application contact:* Shelia Thayer, Graduate Administrator, 613-520-2600 Ext. 8769, Fax: 613-520-2569, E-mail: earth_science@carleton.ca.

Geosciences

Case Western Reserve University, School of Graduate Studies, Department of Geological Sciences, Cleveland, OH 44106. Offers MS, PhD. Part-time programs available. Terminal master's awarded for partial completion of doctoral program. *Degree requirements:* For master's, thesis or alternative; for doctorate, thesis/dissertation. *Entrance requirements:* For master's and doctorate, GRE General Test, GRE Subject Test. Additional exam requirements/recommendations for international students: Required—TOEFL. *Faculty research:* Geochemistry, hydrology, geochronology, paleoclimates, geomorphology.

Central Connecticut State University, School of Graduate Studies, School of Arts and Sciences, Department of Physics and Earth Science, New Britain, CT 06050-4010. Offers earth science (MS); physics (MS). Part-time and evening/weekend programs available. *Faculty:* 11 full-time (3 women), 13 part-time/adjunct (2 women). *Students:* 7 full-time (3 women), 30 part-time (23 women); includes 2 minority (1 Asian American or Pacific Islander, 1 Hispanic American), 1 international. Average age 32. 14 applicants, 64% accepted, 4 enrolled. In 2005, 12 master's awarded. *Degree requirements:* For master's, thesis or alternative, comprehensive exam. *Entrance requirements:* For master's, minimum GPA of 2.7. Additional exam requirements/recommendations for international students: Required—TOEFL. *Application deadline:* For fall admission, 7/1 for domestic students; for spring admission, 12/1 for domestic students. Applications are processed on a rolling basis. Application fee: $50. Electronic applications accepted. *Expenses:* Tuition, area resident: Full-time $3,780. Tuition, state resident: full-time $5,670; part-time $362 per credit. Tuition, nonresident: full-time $10,530; part-time $362 per credit. Required fees: $3,064. One-time fee: $62 part-time. Tuition and fees vary according to degree level and program. *Financial support:* In 2005–06, 1 student received support, including 1 research assistantship; career-related internships or fieldwork, Federal Work-Study, scholarships/grants, and unspecified assistantships also available. Support available to part-time students. Financial award application deadline: 3/1; financial award applicants required to submit FAFSA. *Faculty research:* Elementary/secondary science education, particle and solid states, weather patterns, planetary studies. *Unit head:* Dr. Ali Antar, Chair, 860-832-2930.

City College of the City University of New York, Graduate School, College of Liberal Arts and Science, Division of Science, Department of Earth and Atmospheric Sciences, New York, NY 10031-9198. Offers earth and environmental science (PhD); earth systems science (MA). *Students:* 10 applicants, 70% accepted, 5 enrolled. In 2005, 2 degrees awarded. *Degree requirements:* For master's, thesis, comprehensive exam. *Entrance requirements:* For master's, GRE, appropriate bachelor's degree. Additional exam requirements/recommendations for international students: Required—TOEFL (minimum score 500 paper-based; 173 computer-based). *Application deadline:* For fall admission, 5/1 for domestic students; for spring admission, 11/15 for domestic students. Application fee: $125. *Financial support:* Fellowships, career-related internships or fieldwork available. *Faculty research:* Water resources, high-temperature geochemistry, sedimentary basin analysis, tectonics. *Unit head:* Jeffrey Steiner, Chair, 212-650-6984, Fax: 212-650-6473, E-mail: steiner@sci.ccny.cuny.edu.

Colorado School of Mines, Graduate School, Department of Geology and Geological Engineering, Golden, CO 80401-1887. Offers engineering geology (Diploma); exploration geosciences (Diploma); geochemistry (MS, PhD); geological engineering (ME, MS, PhD, Diploma); geology (MS, PhD); hydrogeology (Diploma). Part-time programs available. *Faculty:* 28 full-time (7 women), 15 part-time/adjunct (4 women). *Students:* 72 full-time (26 women), 29 part-time (5 women); includes 3 minority (1 American Indian/Alaska Native, 2 Hispanic Americans), 25 international. 108 applicants, 60% accepted, 24 enrolled. In 2005, 18 master's, 4 doctorates awarded. *Degree requirements:* For master's, thesis/dissertation; for doctorate, thesis/dissertation, comprehensive exam. *Entrance requirements:* For master's and doctorate, GRE General Test; for Diploma, GRE General Test, minimum GPA of 3.0. Additional exam requirements/recommendations for international students: Required—TOEFL (minimum score 550 paper-based; 213 computer-based). *Application deadline:* For fall admission, 1/1 for domestic students, 1/1 for international students; for spring admission, 9/1 for domestic students, 9/1 for international students. Application fee: $50. Electronic applications accepted. *Expenses:* Tuition, state resident: full-time $7,240; part-time $362 per credit hour. Tuition, nonresident: full-time $19,840; part-time $992 per credit hour. Required fees: $895. *Financial support:* In 2005–06, 4 fellowships with full tuition reimbursements (averaging $9,600 per year), 14 research assistantships with full tuition reimbursements (averaging $9,600 per year), 11 teaching assistantships with full tuition reimbursements (averaging $9,600 per year) were awarded; scholarships/grants, health care benefits, and unspecified assistantships also available. Financial award applicants required to submit FAFSA. *Faculty research:* Predictive sediment modeling, petrophysics, aquifer-contaminant flow modeling, water-rock interactions, geotechnical engineering. Total annual research expenditures: $2.1 million. *Unit head:* Dr. Murray W. Hitzman, Head, 303-384-2127, Fax: 303-273-3859, E-mail: mhitzman@mines.edu. *Application contact:* Marilyn Schwinger, Administrative Assistant, 303-273-3800, Fax: 303-273-3859, E-mail: mschwing@mines.edu.

Colorado State University, Graduate School, Warner College of Natural Resources, Department of Geosciences, Fort Collins, CO 80523-0015. Offers earth sciences (PhD); geology (MS), including geomorphology, geophysics, hydrogeology, petrology/geochemistry and economic geology, sedimentology, structural geology; geosciences (PhD); wateshed (PhD). Part-time programs available. *Faculty:* 8 full-time (3 women). *Students:* 17 full-time (11 women), 22 part-time (7 women); includes 1 minority (American Indian/Alaska Native), 2 international. Average age 32. 28 applicants, 61% accepted, 7 enrolled. In 2005, 3 master's, 1 doctorate awarded. *Degree requirements:* For master's, thesis/dissertation, registration; for doctorate, thesis/dissertation, comprehensive exam, registration. *Entrance requirements:* For master's and doctorate, GRE General Test, minimum GPA of 3.0. Additional exam requirements/recommendations for international students: Required—TOEFL (minimum score 550 paper-based; 213 computer-based). *Application deadline:* For fall admission, 2/1 priority date for domestic students, 2/1 priority date for international students. Applications are processed on a rolling basis. Application fee: $50. Electronic applications accepted. *Expenses:* Tuition, state resident: full-time $3,690; part-time $205 per credit. Tuition, nonresident: full-time $14,958; part-time $831 per credit. Required fees: $1,061. *Financial support:* In 2005–06, 6 fellowships (averaging $9,000 per year), 13 research assistantships with partial tuition reimbursements (averaging $16,000 per year), 9 teaching assistantships with full tuition reimbursements (averaging $16,300 per year) were awarded; career-related internships or fieldwork, Federal Work-Study, institutionally sponsored loans, scholarships/grants, and traineeships also available. Financial award application deadline: 2/15. *Faculty research:* Snow, surface, and groundwater hydrology; fluvial geomorphology; geographic information systems; geochemistry; bedrock geology. Total annual research expenditures: $1.3 million. *Unit head:* Dr. Judith L. Hannah, Head, 970-491-5662, Fax: 970-491-6307, E-mail: jhannah@cnr.colostate.edu. *Application contact:* Sharyl Pierson, Administrative Assistant, 970-491-5662, Fax: 970-491-6307, E-mail: sharyl@cnr.colostate.edu.

Columbia University, Graduate School of Arts and Sciences, Division of Natural Sciences, Department of Earth and Environmental Sciences, New York, NY 10027. Offers geochemistry (M Phil, MA, PhD); geodetic sciences (M Phil, MA, PhD); geophysics (M Phil, MA, PhD); oceanography (M Phil, MA, PhD). *Faculty:* 21 full-time, 19 part-time/adjunct. *Students:* 78 full-time (31 women), 6 part-time (2 women); includes 4 minority (3 Asian Americans or Pacific Islanders, 1 Hispanic American), 31 international. Average age 32. 115 applicants, 20% accepted. In 2005, 16 master's, 11 doctorates awarded. *Degree requirements:* For master's, thesis or alternative, fieldwork, written exam; for doctorate, one foreign language, thesis/dissertation. *Entrance requirements:* For master's and doctorate, GRE General Test, GRE Subject Test, major in natural or physical science. Additional exam requirements/recommendations for international students: Required—TOEFL. Application fee: $75. *Expenses:* Tuition: Full-time $31,448. Tuition and fees vary according to course level, course load, campus/location and program. *Financial support:* Fellowships, teaching assistantships, Federal Work-Study and institutionally sponsored loans available. Support available to part-time students. Financial award application deadline: 1/5; financial award applicants required to submit FAFSA. *Faculty research:* Structural geology and stratigraphy, petrology, paleontology, rare gas, isotope and aqueous geochemistry. *Unit head:* Dr. Nicholas Christie-Blick, Chair, 845-365-8821, Fax: 845-365-8150.

Cornell University, Graduate School, Graduate Fields of Engineering, Field of Geological Sciences, Ithaca, NY 14853-0001. Offers economic geology (M Eng, MS, PhD); engineering geology (M Eng, MS, PhD); environmental geophysics (M Eng, MS, PhD); general geology (M Eng, MS, PhD); geobiology (M Eng, MS, PhD); geochemistry and isotope geology (M Eng, MS, PhD); geohydrology (M Eng, MS, PhD); geomorphology (M Eng, MS, PhD); geophysics (M Eng, MS, PhD); geotectonics (M Eng, MS, PhD); marine geology (MS, PhD); mineralogy (M Eng, MS, PhD); paleontology (M Eng, MS, PhD); petroleum geology (M Eng, MS, PhD); petrology (M Eng, MS, PhD); planetary geology (M Eng, MS, PhD); Precambrian geology (M Eng, MS, PhD); Quaternary geology (M Eng, MS, PhD); rock mechanics (M Eng, MS, PhD); sedimentology (M Eng, MS, PhD); seismology (M Eng, MS, PhD); stratigraphy (M Eng, MS, PhD); structural geology (M Eng, MS, PhD). *Faculty:* 39 full-time (5 women). *Students:* 29 full-time (15 women); includes 2 minority (both Hispanic Americans), 7 international. 73 applicants, 15% accepted, 6 enrolled. In 2005, 5 master's, 1 doctorate awarded. *Degree requirements:* For master's, thesis (MS); for doctorate, thesis/dissertation, comprehensive exam. *Entrance requirements:* For master's and doctorate, GRE General Test, 3 letters of recommendation. Additional exam requirements/recommendations for international students: Required—TOEFL (minimum score 550 paper-based; 213 computer-based). *Application deadline:* For fall admission, 1/15 for domestic students. Applications are processed on a rolling basis. Application fee: $60. Electronic applications accepted. *Financial support:* In 2005–06, 8 fellowships with full tuition reimbursements, 10 research assistantships with full tuition reimbursements, 7 teaching assistantships with full tuition reimbursements were awarded; institutionally sponsored loans, scholarships/grants, health care benefits, tuition waivers (full and partial), and unspecified assistantships also available. Financial award applicants required to submit FAFSA. *Faculty research:* Geophysics, structural geology, petrology, geochemistry, geodynamics. *Unit head:* Director of Graduate Studies, 607-255-5466, Fax: 607-254-4780. *Application contact:* Graduate Field Assistant, 607-255-5466, Fax: 607-254-4780, E-mail: gradprog@geology.cornell.edu.

Dalhousie University, Faculty of Graduate Studies, College of Arts and Science, Faculty of Science, Department of Earth Sciences, Halifax, NS B3H 4R2, Canada. Offers M Sc, PhD. Part-time programs available. *Degree requirements:* For master's and doctorate, one foreign language, thesis/dissertation. *Entrance requirements:* For doctorate, M Sc. Additional exam requirements/recommendations for international students: Required—TOEFL. *Faculty research:* Marine geology and geophysics, Appalachian and Grenville geology, micropaleontology, geodynamics and structural geology, geochronology.

Dartmouth College, School of Arts and Sciences, Department of Earth Sciences, Hanover, NH 03755. Offers MS, PhD. *Faculty:* 11 full-time (2 women), 2 part-time/adjunct (1 woman). *Students:* 24 full-time (10 women); includes 2 minority (1 Asian American or Pacific Islander, 1 Hispanic American), 3 international. Average age 27. 36 applicants, 28% accepted, 7 enrolled. In 2005, 5 degrees awarded. Terminal master's awarded for partial completion of doctoral program. *Degree requirements:* For master's and doctorate, thesis/dissertation. *Entrance requirements:* For master's and doctorate, GRE General Test, GRE Subject Test. Additional exam requirements/recommendations for international students: Required—TOEFL. *Application deadline:* For fall admission, 1/15 for domestic students. Application fee: $15. *Expenses:* Tuition: Full-time $31,770. *Financial support:* In 2005–06, 24 students received support, including fellowships with full tuition reimbursements available (averaging $21,000 per year), research assistantships with full tuition reimbursements available (averaging $21,000 per year); career-related internships or fieldwork, Federal Work-Study, institutionally sponsored loans, scholarships/grants, tuition waivers (full), and unspecified assistantships also available. *Faculty research:* Geochemistry, remote sensing, geophysics, hydrology, economic geology. Total annual research expenditures: $450,189. *Unit head:* Dr. Xiahong Feng, Chair, 603-646-2373. *Application contact:* Jodi Davi, Departmental Administration, 603-646-2373, Fax: 603-646-3922, E-mail: jodi.davi@dartmouth.edu.

Emporia State University, School of Graduate Studies, College of Liberal Arts and Sciences, Department of Physical Sciences, Emporia, KS 66801-5087. Offers chemistry (MS); earth science (MS); physical science (MS); physics (MS). *Faculty:* 15 full-time (2 women), 1 (woman) part-time/adjunct. *Students:* 5 full-time (1 woman), 19 part-time (5 women); includes 1 minority (African American) 8 applicants, 88% accepted, 5 enrolled. In 2005, 6 degrees awarded. *Degree requirements:* For master's, comprehensive exam or thesis. *Entrance requirements:* For master's, physical science qualifying exam, appropriate undergraduate degree. Additional exam requirements/recommendations for international students: Required—TOEFL. *Application deadline:* For fall admission, 8/15 for domestic students. Applications are processed on a rolling basis. Application fee: $30 ($75 for international students). Electronic applications accepted. *Expenses:* Tuition, state resident: full-time $2,890; part-time $132 per credit. Tuition, nonresident: full-time $9,258; part-time $422 per credit. Required fees: $626; $41 per credit. Tuition and fees vary according to degree level. *Financial support:* In 2005–06, research assistantships with full tuition reimbursements (averaging $6,492 per year), 8 teaching assistantships with full tuition reimbursements (averaging $6,492 per year) were awarded; Federal Work-Study, institutionally sponsored loans, health care benefits, and unspecified assistantships also available. Financial award application deadline: 3/15; financial award applicants required to submit FAFSA. *Faculty research:* Bredigite, larnite, and dicalcium silicates–Marble Canyon. *Unit head:* Dr. DeWayne Backhus, Chair, 620-341-5330, Fax: 620-341-6055, E-mail: dbackhus@emporia.edu.

Florida International University, College of Arts and Sciences, Department of Earth Sciences, Miami, FL 33199. Offers MS, PhD. Part-time and evening/weekend programs available. *Faculty:* 12 full-time (2 women). *Students:* 22 full-time (8 women), 7 part-time (4 women); includes 6 minority (1 African American, 1 Asian American or Pacific Islander, 4 Hispanic Americans), 14 international. Average age 32. 118 applicants, 5% accepted, 4 enrolled. In 2005, 2 degrees awarded. *Degree requirements:* For master's, one foreign language, thesis; for doctorate, thesis/dissertation. *Entrance requirements:* For master's and doctorate, GRE General Test, 3 letters of recommendation. Additional exam requirements/recommendations for international students: Required—TOEFL. *Application deadline:* For fall admission, 4/1 for domestic students; for spring admission, 10/1 for domestic students. Applications are processed on a rolling basis. Application fee: $25. *Expenses:* Tuition, area resident: Part-time $239 per credit. Tuition, state resident: full-time $4,294; part-time $869 per credit. Tuition, nonresident: full-time $15,641. Required fees: $252; $126 per term. Tuition and fees vary according to program. *Financial support:* Research assistantships, teaching assistantships available. Financial award application deadline:4/1. *Faculty research:* Determination of dispersivity and hydraulic conductivity in the Biscayne Aquifer. *Unit head:* Dr. Bradford Clement, Chairperson, 305-348-3085, Fax: 305-348-3877, E-mail: bradford.clement@fiu.edu.

The George Washington University, Columbian College of Arts and Sciences, Department of Environmental Studies, Program in Geosciences, Washington, DC 20052. Offers MS, PhD. *Degree requirements:* For master's, thesis or alternative, comprehensive exam. *Entrance requirements:* For master's and doctorate, GRE General Test, bachelor's degree in field, minimum GPA of 3.0. Additional exam requirements/recommendations for international students: Required—TOEFL (minimum score 550 paper-based; 213 computer-based). Electronic applications accepted.

Georgia Institute of Technology, Graduate Studies and Research, College of Sciences, School of Earth and Atmospheric Sciences, Atlanta, GA 30332-0001. Offers atmospheric chemistry and air pollution (MS, PhD); atmospheric dynamics and climate (MS, PhD); geochemistry (MS, PhD); hydrologic cycle (MS, PhD); ocean sciences (MS, PhD); solid-earth and environmental geophysics (MS, PhD). Part-time programs available. Terminal master's awarded for partial completion of doctoral program. *Degree requirements:* For master's, thesis or alternative; for doctorate, thesis/dissertation, comprehensive exam. *Entrance requirements:*

For master's, GRE, minimum GPA of 3.0; for doctorate, GRE General Test, minimum GPA of 2.7. Additional exam requirements/recommendations for international students: Required—TOEFL (minimum score 550 paper-based; 213 computer-based). *Faculty research:* Geophysics, atmospheric chemistry, atmospheric dynamics, seismology.

See Close-Up on page 227.

Georgia State University, College of Arts and Sciences, Department of Geosciences, Atlanta, GA 30303-3083. Offers earth science—hydrology (MS), including earth science—environmental management, earth science—GIS; geographic information systems (Certificate); geography (MA); geology (MS). Part-time and evening/weekend programs available. *Degree requirements:* For master's, one foreign language, thesis or alternative, comprehensive exam (for some programs), registration. *Entrance requirements:* For master's, GRE General Test, minimum GPA of 2.75. Additional exam requirements/recommendations for international students: Required—TOEFL. Electronic applications accepted. *Expenses:* Tuition, state resident: full-time $4,368; part-time $182 per term. Tuition, nonresident: full-time $8,732; part-time $728 per term. Required fees: $46 per hour. *Faculty research:* Clay mineralogy, metamorphism, fracture analysis, carbonates, groundwater.

Graduate School and University Center of the City University of New York, Graduate Studies, Program in Earth and Environmental Sciences, New York, NY 10016-4039. Offers PhD. *Faculty:* 36 full-time (5 women). *Students:* 58 full-time (22 women), 9 part-time (4 women); includes 9 minority (3 African Americans, 1 Asian American or Pacific Islander, 5 Hispanic Americans), 17 international. Average age 39. 24 applicants, 63% accepted, 9 enrolled. In 2005, 3 degrees awarded. *Degree requirements:* For doctorate, one foreign language, thesis/dissertation, comprehensive exam. *Entrance requirements:* For doctorate, GRE General Test. Additional exam requirements/recommendations for international students: Required—TOEFL. *Application deadline:* For fall admission, 4/15 for domestic students. Application fee: $125. Electronic applications accepted. *Financial support:* In 2005–06, 28 fellowships, 2 research assistantships, 1 teaching assistantship were awarded; career-related internships or fieldwork, Federal Work-Study, institutionally sponsored loans, and tuition waivers (full and partial) also available. Financial award application deadline: 2/1; financial award applicants required to submit FAFSA. *Unit head:* Dr. Yehuda Klein, Executive Officer, 212-817-8241, Fax: 212-817-1513.

Harvard University, Graduate School of Arts and Sciences, Department of Earth and Planetary Sciences, Cambridge, MA 02138. Offers AM, PhD. *Faculty:* 27 full-time (3 women). *Students:* 51 full-time (25 women); includes 15 minority (11 Asian Americans or Pacific Islanders, 4 Hispanic Americans). 103 applicants, 18% accepted, 11 enrolled. In 2005, 11 master's, 4 doctorates awarded. Terminal master's awarded for partial completion of doctoral program. *Degree requirements:* For master's, registration; for doctorate, thesis/dissertation, comprehensive exam, registration. *Entrance requirements:* For doctorate, GRE General Test. Additional exam requirements/recommendations for international students: Required—TOEFL. *Application deadline:* For fall admission, 1/2 for domestic students. Application fee: $85. Electronic applications accepted. *Expenses:* Tuition: Full-time $28,752. Full-time tuition and fees vary according to program and student level. *Financial support:* In 2005–06, 50 students received support; fellowships with full tuition reimbursements available, research assistantships, teaching assistantships, scholarships/grants and health care benefits available. *Faculty research:* Economic geography, geochemistry, geophysics, mineralogy, crystallography. *Unit head:* Ventatesh Narayanamurti, Dean, 617-495-5829, Fax: 617-495-5264, E-mail: venky@deas.harvard.edu. *Application contact:* Office of Admissions and Financial Aid, 617-495-5315.

Hunter College of the City University of New York, Graduate School, School of Arts and Sciences, Department of Geography, New York, NY 10021-5085. Offers analytical geography (MA); earth system science (MA); environmental and social issues (MA); geographic information science (Certificate); geographic information systems (MA); teaching earth science (MA). Part-time and evening/weekend programs available. *Faculty:* 15 full-time (3 women), 4 part-time/adjunct (1 woman). *Students:* 3 full-time (1 woman), 33 part-time (15 women); includes 4 minority (2 African Americans, 2 Hispanic Americans). Average age 33. 14 applicants, 71% accepted, 8 enrolled. In 2005, 14 degrees awarded. *Degree requirements:* For master's, comprehensive exam or thesis. *Entrance requirements:* For master's, GRE General Test, minimum B average in major, minimum B- average overall, 18 credits of course work in geography, 2 letters of recommendation; for Certificate, minimum of B average in major, B- overall. Additional exam requirements/recommendations for international students: Required—TOEFL. *Application deadline:* For fall admission, 4/1 for domestic students; for spring admission, 11/1 for domestic students. Applications are processed on a rolling basis. Application fee: $125. *Expenses:* Tuition, state resident: full-time $6,400; part-time $270 per credit. Tuition, nonresident: part-time $500 per credit. International tuition: $12,000 full-time. Required fees: $50 per term. Part-time tuition and fees vary according to course load and program. *Financial support:* In 2005–06, 1 fellowship (averaging $3,000 per year), 2 research assistantships (averaging $10,000 per year), 10 teaching assistantships (averaging $6,000 per year) were awarded; career-related internships or fieldwork, Federal Work-Study, institutionally sponsored loans, and unspecified assistantships also available. Financial award application deadline: 3/1. *Faculty research:* Urban geography, economic geography, geographic information science, demographic methods, climate change. *Unit head:* Prof. Marianna Pavlovskaya, Chair, 212-772-5320, Fax: 212-772-5268, E-mail: mpavlov@geo.hunter.cuny.edu. *Application contact:* Prof. Marianna Pavlovskaya, Graduate Adviser, 212-772-5320, Fax: 212-772-5268, E-mail: mpavlov@geo.hunter.cuny.edu.

Hunter College of the City University of New York, Graduate School, School of Education, Programs in Secondary Education, New York, NY 10021-5085. Offers biology education (MA); chemistry education (MA); earth science (MA); English education (MA); French education (MA); Italian education (MA); mathematics education (MA); physics education (MA); social studies education (MA); Spanish education (MA). *Accreditation:* NCATE. *Students:* 17 full-time (11 women), 255 part-time (158 women); includes 52 minority (16 African Americans, 17 Asian Americans or Pacific Islanders, 19 Hispanic Americans). Average age 33. 197 applicants, 51% accepted, 57 enrolled. In 2005, 39 degrees awarded. *Degree requirements:* For master's, thesis. *Application deadline:* For fall admission, 4/1 for domestic students, 2/1 for international students; for spring admission, 11/1 for domestic students, 9/1 for international students. Applications are processed on a rolling basis. Application fee: $125. *Expenses:* Tuition, state resident: full-time $6,400; part-time $270 per credit. Tuition, nonresident: part-time $500 per credit. International tuition: $12,000 full-time. Required fees: $50 per term. Part-time tuition and fees vary according to course load and program. *Financial support:* Fellowships, tuition waivers (full and partial) available. Support available to part-time students. *Unit head:* Dr. Kate Garret, Coordinator, 212-772-5049, E-mail: kgarret@hunter.cuny.edu. *Application contact:* William Zlata, Director for Graduate Admissions, 212-772-4482, Fax: 212-650-3336, E-mail: admissions@hunter.cuny.edu.

Idaho State University, Office of Graduate Studies, College of Arts and Sciences, Department of Geosciences, Pocatello, ID 83209. Offers geographic information science (MS); geology (MNS, MS); geophysics/hydrology (MS); geotechnology (Postbaccalaureate Certificate). Part-time programs available. Postbaccalaureate distance learning degree programs offered. *Degree requirements:* For master's, thesis, comprehensive exam, registration (for some programs); for Postbaccalaureate Certificate, thesis optional. *Entrance requirements:* For master's and Postbaccalaureate Certificate, GRE General Test, 3 letters of recommendation. Additional exam requirements/recommendations for international students: Required—TOEFL (minimum score 550 paper-based; 213 computer-based). *Faculty research:* Structural geography, stratigraphy, geochemistry, remote sensing, geomorphology.

Indiana State University, School of Graduate Studies, College of Arts and Sciences, Department of Geography, Geology and Anthropology, Terre Haute, IN 47809-1401. Offers earth sciences (MA, MS); economic geography (PhD); geography (MA); geology (MS); physical geography (PhD). *Faculty:* 21 full-time (5 women), 3 part-time/adjunct (1 woman). *Students:* 25 full-time (6 women), 16 part-time (6 women); includes 5 minority (1 African American, 4 Asian Americans or Pacific Islanders), 11 international. Average age 31. 22 applicants, 86% accepted, 7 enrolled. In 2005, 8 master's, 1 doctorate awarded. *Degree requirements:* For master's, thesis or alternative, registration (for some programs); for doctorate, thesis/dissertation, departmental qualifying exam, comprehensive exam, registration. *Entrance requirements:* For doctorate, GRE General Test. Additional exam requirements/recommendations for international students: Required—TOEFL (minimum score 550 paper-based). *Application deadline:* For fall admission, 7/1 for domestic students; for spring admission, 11/1 priority date for domestic students. Applications are processed on a rolling basis. Application fee: $45. Electronic applications accepted. *Expenses:* Tuition, state resident: full-time $6,288; part-time $262 per credit hour. Tuition, nonresident: full-time $12,504; part-time $521 per credit hour. *Financial support:* In 2005–06, 13 teaching assistantships with partial tuition reimbursements (averaging $9,147 per year) were awarded; research assistantships with partial tuition reimbursements, tuition waivers (partial) also available. Financial award application deadline: 3/1; financial award applicants required to submit FAFSA. *Unit head:* Dr. Susan Berta, Chairperson, 812-237-2261.

Indiana University Bloomington, Graduate School, College of Arts and Sciences, Department of Geological Sciences, Bloomington, IN 47405-7000. Offers biogeochemistry (MS, PhD); economic geology (MS, PhD); geobiology (MS, PhD); geophysics, structural geology and tectonics (MS, PhD); hydrogeology (MS, PhD); mineralogy (MS, PhD); stratigraphy and sedimentology (MS, PhD). PhD offered through the University Graduate School. Part-time programs available. *Faculty:* 14 full-time (1 woman). *Students:* 30 full-time (13 women), 12 part-time (5 women); includes 5 minority (2 African Americans, 1 American Indian/Alaska Native, 1 Asian American or Pacific Islander, 1 Hispanic American), 21 international. Average age 29. In 2005, 7 master's, 2 doctorates awarded. Terminal master's awarded for partial completion of doctoral program. *Degree requirements:* For master's, one foreign language, thesis or alternative; for doctorate, thesis/dissertation. *Entrance requirements:* For master's and doctorate, GRE General Test. Additional exam requirements/recommendations for international students: Required—TOEFL. *Application deadline:* For fall admission, 1/15 priority date for domestic students, 12/15 priority date for international students; for spring admission, 9/1 priority date for domestic students, 9/1 priority date for international students. Applications are processed on a rolling basis. Application fee: $50 ($60 for international students). *Expenses:* Tuition, state resident: full-time $5,437; part-time $227 per credit hour. Tuition, nonresident: full-time $15,836; part-time $660 per credit hour. Required fees: $821. Tuition and fees vary according to campus/location and program. *Financial support:* Fellowships with tuition reimbursements, research assistantships with full and partial tuition reimbursements, teaching assistantships with tuition reimbursements, career-related internships or fieldwork, Federal Work-Study, and institutionally sponsored loans available. Financial award application deadline: 2/15. *Faculty research:* Geophysics, geochemistry, hydrogeology, igneous and metamorphic petrology and clay minerology. Total annual research expenditures: $289,139. *Unit head:* Mark Person, Graduate Advisor, 812-855-7214. *Application contact:* Mary Iverson, Secretary, Committee for Graduate Studies, 812-855-7214, Fax: 812-855-7899, E-mail: geograd@indiana.edu.

Iowa State University of Science and Technology, Graduate College, College of Liberal Arts and Sciences, Department of Geological and Atmospheric Sciences, Ames, IA 50011. Offers earth science (MS, PhD); geology (MS, PhD); meteorology (MS, PhD); water resources (MS, PhD). *Faculty:* 15 full-time, 2 part-time/adjunct. *Students:* 32 full-time (12 women), 2 part-time; includes 1 minority (African American), 15 international. 40 applicants, 60% accepted, 13 enrolled. In 2005, 11 master's awarded. *Degree requirements:* For master's, thesis (for some programs); for doctorate, thesis/dissertation. *Entrance requirements:* For master's and doctorate, GRE General Test. Additional exam requirements/recommendations for international students: Required—TOEFL (paper score 530; computer score 197) or IELTS (score 6.0). *Application deadline:* For fall admission, 1/1 for domestic students. Applications are processed on a rolling basis. Application fee: $30 ($70 for international students). Electronic applications accepted. *Expenses:* Tuition, state resident: full-time $6,410. Tuition, nonresident: full-time $16,422. Tuition and fees vary according to program. *Financial support:* In 2005–06, 21 research assistantships with full and partial tuition reimbursements (averaging $13,545 per year), 9 teaching assistantships with full and partial tuition reimbursements (averaging $13,507 per year) were awarded; fellowships, scholarships/grants, health care benefits, and unspecified assistantships also available. *Unit head:* Dr. Carl E. Jacobson, Chair, 515-294-4477.

Lehigh University, College of Arts and Sciences, Department of Earth and Environmental Sciences, Bethlehem, PA 18015-3094. Offers MS, PhD. *Faculty:* 12 full-time (1 woman), 1 (woman) part-time/adjunct. *Students:* 24 full-time (10 women), 5 part-time (1 woman), 4 international. Average age 26. 30 applicants, 27% accepted, 2 enrolled. In 2005, 3 master's, 1 doctorate awarded. Terminal master's awarded for partial completion of doctoral program. *Degree requirements:* For master's, thesis, registration; for doctorate, thesis/dissertation, language at the discretion of the PhD committee, comprehensive exam, registration. *Entrance requirements:* For master's and doctorate, GRE General Test, 2 letters of recommendation. Additional exam requirements/recommendations for international students: Required—TOEFL. *Application deadline:* For fall admission, 1/15 for domestic students; for spring admission, 10/15 priority date for domestic students. Applications are processed on a rolling basis. Application fee: $60. *Financial support:* In 2005–06, 5 fellowships with full tuition reimbursements (averaging $13,670 per year), 16 research assistantships with full tuition reimbursements (averaging $13,670 per year), 10 teaching assistantships with full tuition reimbursements (averaging $13,670 per year) were awarded; Federal Work-Study, institutionally sponsored loans, and tuition waivers (full and partial) also available. Financial award application deadline: 1/15. *Faculty research:* Tectonics, surficial processes, aquatic ecology. Total annual research expenditures: $1.5 million. *Unit head:* Dr. Peter K. Zeitler, Chairman, 610-758-3660 Ext. 3671, Fax: 610-758-3677, E-mail: pkz0@lehigh.edu. *Application contact:* Dr. Gray E. Bebout, Graduate Coordinator, 610-758-3660 Ext. 5831, Fax: 610-758-3677, E-mail: geb0@lehigh.edu.

Loma Linda University, School of Science and Technology, Department of Biological and Earth Sciences, Loma Linda, CA 92350. Offers MS, PhD. *Faculty:* 8 full-time, 7 part-time/adjunct. *Students:* 15 full-time (7 women), 11 part-time (2 women); includes 1 American Indian/Alaska Native, 1 Asian American or Pacific Islander, 3 Hispanic Americans, 1 international. *Unit head:* Dr. Ron L. Carter, Coordinator, 909-824-4530.

Massachusetts Institute of Technology, School of Science, Department of Earth, Atmospheric, and Planetary Sciences, Cambridge, MA 02139-4307. Offers atmospheric chemistry (PhD, Sc D); atmospheric science (SM, PhD, Sc D); climate physics and chemistry (PhD, Sc D); earth and planetary sciences (SM); geochemistry (PhD, Sc D); geology (PhD, Sc D); geophysics (PhD, Sc D); marine geology and geophysics (SM); oceanography (SM, PhD, Sc D); planetary sciences (PhD, Sc D). *Faculty:* 38 full-time (3 women). *Students:* 165 full-time (63 women); includes 12 minority (7 Asian Americans or Pacific Islanders, 5 Hispanic Americans), 51 international. Average age 27. 179 applicants, 34% accepted, 31 enrolled. In 2005, 2 master's, 15 doctorates awarded. Terminal master's awarded for partial completion of doctoral program. *Degree requirements:* For master's, thesis/dissertation; for doctorate, thesis/dissertation, comprehensive exam. *Entrance requirements:* For master's, GRE General Test; for doctorate, GRE General Test, GRE Subject Test (chemistry or physics for planetary science program). Additional exam requirements/recommendations for international students: Required—TOEFL (minimum score 577 paper-based; 233 computer-based). *Application deadline:* For fall admission, 1/5 for domestic students, 1/5 for international students; for spring admission, 11/1 for domestic students, 11/1 for international students. Application fee: $70. Electronic applications accepted. *Expenses:* Tuition: Full-time $32,100. Required fees: $200. Part-time tuition and fees vary according to course load. *Financial support:* In 2005–06, 115 students received support, including 36 fellowships with tuition reimbursements available (averaging $26,951 per year), 60 research assistantships with tuition reimbursements available (averaging $24,520 per year), 18 teaching assistantships with tuition reimbursements available (averaging $22,773 per year); Federal Work-Study, institutionally sponsored loans, scholarships/grants, health care benefits, and unspecified assistantships also available. *Faculty research:* Evolution of main features of the planetary system; origin, composition, structure, and state of the

Geosciences

Massachusetts Institute of Technology (continued)
atmospheres, oceans, surfaces, and interiors of planets; dynamics of planets and satellite motions. Total annual research expenditures: $18.6 million. *Unit head:* Prof. Maria Zuber, Department Head, 617-253-6397, E-mail: zuber@mit.edu. *Application contact:* EAPS Education Office.

McGill University, Faculty of Graduate and Postdoctoral Studies, Faculty of Science, Department of Earth and Planetary Sciences, Montréal, QC H3A 2T5, Canada. Offers M Sc, PhD. *Degree requirements:* For master's, thesis/dissertation, registration; for doctorate, thesis/dissertation, comprehensive exam, registration. *Entrance requirements:* For master's, minimum GPA of 3.0; academic background in geology, geophysics, chemistry, or physics; for doctorate, B Sc or M Sc. Additional exam requirements/recommendations for international students: Required—TOEFL (minimum score 550 paper-based; 213 computer-based), IELT (minimum score 7). Electronic applications accepted. *Faculty research:* Geochemistry, sedimentary petrology, igneous petrology, theoretical geophysics, economic geology.

McMaster University, School of Graduate Studies, Faculty of Science, School of Geography and Earth Sciences, Hamilton, ON L8S 4M2, Canada. Offers geochemistry (PhD); geology (M Sc, PhD); human geography (MA, PhD); physical geography (M Sc, PhD). Part-time programs available. Terminal master's awarded for partial completion of doctoral program. *Degree requirements:* For master's, thesis/dissertation; for doctorate, thesis/dissertation, comprehensive exam. *Entrance requirements:* For master's, minimum B+ average. Additional exam requirements/recommendations for international students: Required—TOEFL (minimum score 550 paper-based; 213 computer-based).

Memorial University of Newfoundland, School of Graduate Studies, Department of Earth Sciences, St. John's, NL A1C 5S7, Canada. Offers geology (M Sc, PhD); geophysics (M Sc, PhD). Part-time programs available. *Students:* 53 full-time (27 women), 6 part-time, 9 international. 23 applicants, 70% accepted, 16 enrolled. In 2005, 10 master's, 2 doctorates awarded. *Degree requirements:* For master's, thesis; for doctorate, thesis/dissertation, oral thesis defense; PhD entry evaluation, comprehensive exam. *Entrance requirements:* For master's, honors B Sc; for doctorate, M Sc. *Application deadline:* Applications are processed on a rolling basis. Application fee: $40 Canadian dollars. Electronic applications accepted. *Expenses:* Tuition: Part-time $733 per term. Tuition and fees vary according to degree level and program. *Financial support:* Fellowships, research assistantships, teaching assistantships available. *Faculty research:* Geochemistry, sedimentology, paleoceanography and global change, mineral deposits, petroleum geology, hydrology. *Unit head:* Dr. John Hanchar, Head, 709-737-2334, Fax: 709-737-2589, E-mail: head@esd.mun.ca. *Application contact:* Raymund Patzold, Graduate Officer, 709-737-4464, Fax: 709-737-2589, E-mail: rpatzold@esd.mun.ca.

Michigan State University, The Graduate School, College of Natural Science, Department of Geological Sciences, East Lansing, MI 48824. Offers environmental geosciences (MS, PhD); environmental geosciences-environmental toxicology (PhD); geological sciences (MS, PhD). *Faculty:* 11 full-time (1 woman). *Students:* 20 full-time (12 women), 4 part-time (1 woman); includes 2 minority (both Asian Americans or Pacific Islanders), 5 international. Average age 28. 37 applicants, 59% accepted. In 2005, 8 master's, 2 doctorates awarded. *Degree requirements:* For master's, thesis for those without prior thesis work; for doctorate, thesis/dissertation, registration. *Entrance requirements:* For master's, GRE General Test, minimum GPA of 3.0, course work in geoscience, 3 letters of recommendation; for doctorate, GRE General Test, 3 letters of recommendation. Additional exam requirements/recommendations for international students: Required—TOEFL (minimum score 550 paper-based; 213 computer-based), Michigan State Univeristy ELT (85), Michigan ELAB (83). *Application deadline:* For fall admission, 12/27 for domestic students. Application fee: $50. Electronic applications accepted. *Expenses:* Tuition, state resident: part-time $330 per credit hour. Tuition, nonresident: part-time $685 per credit hour. Tuition and fees vary according to program. *Financial support:* In 2005–06, 10 fellowships with tuition reimbursements (averaging $3,479 per year), 4 research assistantships with tuition reimbursements (averaging $12,710 per year), 13 teaching assistantships with tuition reimbursements (averaging $12,997 per year) were awarded; Federal Work-Study, scholarships/grants, and unspecified assistantships also available. *Faculty research:* Water in the environment, global and biological change, crystal dynamics. Total annual research expenditures: $1.1 million. *Unit head:* Dr. Ralph E. Taggart, Chairperson, 517-355-4626, Fax: 517-353-8787, E-mail: taggart@msu.edu. *Application contact:* Information Contact, 517-355-4626, Fax: 517-353-8787, E-mail: geosci@msu.edu.

Middle Tennessee State University, College of Graduate Studies, College of Liberal Arts, Department of Geosciences, Murfreesboro, TN 37132. Offers Graduate Certificate. *Entrance requirements:* Additional exam requirements/recommendations for international students: Required—TOEFL (minimum score 525 paper-based; 195 computer-based).

Millersville University of Pennsylvania, Graduate School, School of Science and Mathematics, Millersville, PA 17551-0302. Offers biology (MS); earth science (MS); mathematics (M Ed); nursing (MSN). Part-time and evening/weekend programs available. *Faculty:* 44 full-time (17 women), 13 part-time/adjunct (11 women). *Students:* 5 full-time (4 women), 32 part-time (29 women). Average age 37. 8 applicants, 88% accepted, 3 enrolled. In 2005, 22 degrees awarded. *Degree requirements:* For master's, thesis optional. *Entrance requirements:* For master's, GRE. Additional exam requirements/recommendations for international students: Required—TOEFL (minimum score 500 paper-based; 183 computer-based). *Application deadline:* For fall admission, 3/1 priority date for domestic students, 3/1 priority date for international students; for winter and spring admission, 10/1; for winter admission, 10/1 for domestic students. Applications are processed on a rolling basis. Application fee: $35. *Expenses:* Tuition, state resident: full-time $5,888; part-time $327 per credit. Tuition, nonresident: full-time $9,422; part-time $523 per credit. Required fees: $1,216; $60 per credit. Tuition and fees vary according to course load. *Financial support:* In 2005–06, 4 students received support, including 4 research assistantships with full tuition reimbursements available (averaging $4,000 per year); career-related internships or fieldwork, Federal Work-Study, institutionally sponsored loans, and unspecified assistantships also available. Support available to part-time students. Financial award application deadline: 3/15; financial award applicants required to submit FAFSA. *Unit head:* Dr. Edward C. Shane, Dean, 717-872-3407, Fax: 717-872-3985. *Application contact:* Dr. Victor S. DeSantis, Dean of Graduate Studies, 717-872-3099, Fax: 717-871-2022, E-mail: victor.desantis@millersville.edu.

Mississippi State University, College of Arts and Sciences, Department of Geosciences, Mississippi State, MS 39762. Offers MS. *Faculty:* 19 full-time (3 women), 1 part-time/adjunct (0 women). *Students:* 52 full-time (16 women), 332 part-time (162 women); includes 28 minority (10 African Americans, 3 American Indian/Alaska Native, 3 Asian Americans or Pacific Islanders, 12 Hispanic Americans), 7 international. Average age 37. 160 applicants, 94% accepted, 137 enrolled. In 2005, 132 degrees awarded. *Degree requirements:* For master's, thesis (for some programs), comprehensive oral or written exam. *Entrance requirements:* For master's, minimum QPA of 2.75. Additional exam requirements/recommendations for international students: Required—TOEFL. *Application deadline:* For fall admission, 7/1 for domestic students; for spring admission, 11/1 for domestic students. Applications are processed on a rolling basis. Application fee: $30. *Expenses:* Tuition, state resident: full-time $4,312; part-time $240 per hour. Tuition, nonresident: full-time $9,772; part-time $543 per hour. International tuition: $10,102 full-time. Tuition and fees vary according to course load. *Financial support:* In 2005–06, 3 students received support, including 21 teaching assistantships with full tuition reimbursements available (averaging $7,629 per year); research assistantships with full tuition reimbursements available, Federal Work-Study, institutionally sponsored loans, tuition waivers (partial), and unspecified assistantships also available. Financial award application deadline: 4/1; financial award applicants required to submit FAFSA. *Faculty research:* Climatology, hydrogeology, sedimentology, meteorology. Total annual research expenditures: $39,695. *Unit head:* Dr. Darrel Schmitz, Interim Head, 662-325-3915, Fax: 662-325-9423, E-mail: schmitz@geosci.msstate.edu. *Application contact:* Philip G. Bonfanti, Director of Admissions, 662-325-4104, Fax: 662-325-8872, E-mail: admit@msstate.edu.

Missouri State University, Graduate College, College of Natural and Applied Sciences, Department of Geography, Geology, and Planning, Springfield, MO 65804-0094. Offers geography, geology and planning (MNAS); geospatial sciences (MS); secondary education (MS Ed), including earth science, geography. Part-time and evening/weekend programs available. *Faculty:* 20 full-time (3 women), 1 part-time/adjunct (0 women). *Students:* 13 full-time (5 women), 8 part-time (3 women). Average age 32. 14 applicants, 79% accepted, 5 enrolled. In 2005, 7 degrees awarded. *Degree requirements:* For master's, thesis (for some programs), comprehensive exam. *Entrance requirements:* For master's, GRE General Test (MS, MNAS), minimum undergraduate GPA of 3.0 (MS, MNAS), 9-12 teacher certification (MS Ed). Additional exam requirements/recommendations for international students: Required—TOEFL (minimum score 550 paper-based; 213 computer-based), IELT (minimum score 6). *Application deadline:* For fall admission, 7/20 for domestic students; for spring admission, 12/20 priority date for domestic students. Applications are processed on a rolling basis. Application fee: $30. Electronic applications accepted. *Expenses:* Tuition, state resident: full-time $3,402; part-time $189 per credit. Tuition, nonresident: full-time $6,804; part-time $378 per credit. Required fees: $207 per semester. Part-time tuition and fees vary according to course level, course load and program. *Financial support:* In 2005–06, 4 research assistantships with full tuition reimbursements (averaging $8,750 per year), 6 teaching assistantships with full tuition reimbursements (averaging $6,575 per year) were awarded; career-related internships or fieldwork, Federal Work-Study, scholarships/grants, and unspecified assistantships also available. Financial award application deadline: 3/31; financial award applicants required to submit FAFSA. *Faculty research:* Water resources, small town planning, recreation and open space planning. *Unit head:* Dr. Tom Plymate, Acting Head, 417-836-5800, Fax: 417-836-6934, E-mail: tomplymate@missouristate.edu. *Application contact:* Dr. Robert T. Pavlowsky, Graduate Adviser, 417-836-8473, Fax: 417-836-6006, E-mail: bobpavlowsky@missouristate.edu.

Montana State University, College of Graduate Studies, College of Letters and Science, Department of Earth Sciences, Bozeman, MT 59717. Offers MS, PhD. Part-time programs available. *Faculty:* 14 full-time (2 women), 2 part-time/adjunct (1 woman). *Students:* 16 full-time (9 women), 26 part-time (16 women); includes 2 minority (1 American Indian/Alaska Native, 1 Hispanic American), 2 international. Average age 29. 58 applicants, 28% accepted, 15 enrolled. In 2005, 8 degrees awarded. *Degree requirements:* For master's, thesis (for some programs), comprehensive exam, registration; for doctorate, thesis/dissertation, comprehensive exam, registration. *Entrance requirements:* For master's and doctorate, GRE General Test. Additional exam requirements/recommendations for international students: Required—TOEFL (minimum score 550 paper-based; 213 computer-based). *Application deadline:* For fall admission, 7/15 priority date for domestic students, 5/15 priority date for international students; for spring admission, 12/1 priority date for domestic students, 10/1 priority date for international students. Applications are processed on a rolling basis. Application fee: $30. Electronic applications accepted. *Expenses:* Tuition, state resident: full-time $4,132. Tuition, nonresident: full-time $1,132. *Financial support:* In 2005–06, 30 students received support, including 12 research assistantships (averaging $7,200 per year), 15 teaching assistantships with full tuition reimbursements available (averaging $10,000 per year); fellowships, scholarships/grants and tuition waivers (full and partial) also available. Financial award application deadline: 3/1; financial award applicants required to submit FAFSA. *Faculty research:* Arc magnetism, digital education resources in microbial ecology, cretaceous terrestrial record of Montana, climatic and ecohydrologic variability, Holocene fire-climate-vegetation changes. Total annual research expenditures:$796,239. *Unit head:* Dr. David Lageson, Department Head, 406-994-3331, Fax: 406-994-6923, E-mail: lageson@montana.edu.

Montana Tech of The University of Montana, Graduate School, Geoscience Program, Butte, MT 59701-8997. Offers geochemistry (MS); geological engineering (MS); geology (MS); geophysical engineering (MS); hydrogeological engineering (MS); hydrogeology (MS). Part-time programs available. *Faculty:* 17 full-time (3 women). *Students:* 8 full-time (4 women), 3 part-time (1 woman). 15 applicants, 60% accepted, 2 enrolled. In 2005, 5 degrees awarded. *Degree requirements:* For master's, thesis (for some programs), comprehensive exam (for some programs), registration. *Entrance requirements:* For master's, GRE General Test, minimum GPA of 3.0. Additional exam requirements/recommendations for international students: Required—TOEFL (minimum score 525 paper-based; 195 computer-based). *Application deadline:* For fall admission, 4/1 priority date for domestic students, 3/1 priority date for international students; for spring admission, 10/1 priority date for domestic students, 7/1 priority date for international students. Applications are processed on a rolling basis. Application fee: $30. *Expenses:* Tuition, area resident: Full-time $5,133. Tuition, state resident: full-time $5,333. Tuition, nonresident: full-time $15,746. *Financial support:* In 2005–06, 8 students received support, including 3 research assistantships with partial tuition reimbursements available (averaging $3,667 per year), 6 teaching assistantships with partial tuition reimbursements available (averaging $6,667 per year); career-related internships or fieldwork, tuition waivers (full and partial), and unspecified assistantships also available. Financial award application deadline: 4/1; financial award applicants required to submit FAFSA. *Faculty research:* Water resource development, seismic processing, petroleum reservoir characterization, environmental geochemistry, molecular modeling, magmatic and hydrothermal ore deposits. Total annual research expenditures:$542,109. *Unit head:* Dr. Diane Wolfgram, Department Head, 406-496-4353, Fax: 406-496-4260, E-mail: dwolfgram@mtech.edu. *Application contact:* Cindy Dunstan, Administrator, Graduate School, 406-496-4304, Fax: 406-496-4710, E-mail: cdunstan@mtech.edu.

Montclair State University, The Graduate School, College of Science and Mathematics, Department of Earth and Environmental Studies, Montclair, NJ 07043-1624. Offers environmental management (MA, D Env M); environmental studies (MS), including environmental education, environmental health, environmental management, environmental science; geoscience (MS, Certificate), including geoscience (MS), water resource management (Certificate). Part-time and evening/weekend programs available. *Faculty:* 17 full-time (3 women), 9 part-time/adjunct (2 women). *Students:* 21 full-time (13 women), 48 part-time (20 women); includes 14 minority (8 African Americans, 3 Asian Americans or Pacific Islanders, 3 Hispanic Americans), 1 international. 32 applicants, 78% accepted, 16 enrolled. In 2005, 19 master's, 3 other advanced degrees awarded. *Degree requirements:* For master's, thesis or alternative, comprehensive exam; for doctorate, thesis/dissertation. *Entrance requirements:* For master's, GRE General Test, 2 letters of recommendation. Additional exam requirements/recommendations for international students: Required—TOEFL (minimum score 83 computer-based). *Application deadline:* Applications are processed on a rolling basis. Application fee: $60. Electronic applications accepted. *Expenses:* Tuition: Full-time $3,001; part-time $409 per credit. Required fees: $56 per credit. Tuition and fees vary according to course load, degree level and program. *Financial support:* In 2005–06, 14 research assistantships with full tuition reimbursements were awarded; Federal Work-Study, scholarships/grants, and unspecified assistantships also available. Support available to part-time students. Financial award application deadline: 3/1; financial award applicants required to submit FAFSA. *Faculty research:* Antarctica, carbon pools, contaminated sediments, wetlands. Total annual research expenditures: $127,880. *Unit head:* Dr. Gregory Pope, Chairperson, 973-655-7385. *Application contact:* Dr. Harbans Singh, Adviser, 973-655-7383.

Murray State University, College of Science, Engineering and Technology, Department of Geosciences, Murray, KY 42071-0009. Offers MA, MS. Part-time programs available. *Degree requirements:* For master's, thesis (for some programs). *Entrance requirements:* For master's, GRE General Test. Additional exam requirements/recommendations for international students: Required—TOEFL.

New Mexico Institute of Mining and Technology, Graduate Studies, Department of Earth and Environmental Science, Socorro, NM 87801. Offers geology and geochemistry (MS, PhD), including geochemistry, geology; geophysics (MS, PhD); hydrology (MS, PhD). *Degree requirements:* For master's, thesis optional; for doctorate, thesis/dissertation. *Entrance requirements:* For master's, GRE General Test; for doctorate, GRE General Test, GRE Subject Test. Additional exam requirements/recommendations for international students: Required—

TOEFL. *Faculty research:* Seismology, geochemistry, caves and karst topography, hydrology, volcanology.

North Carolina Central University, Division of Academic Affairs, College of Arts and Sciences, Department of Earth Sciences, Durham, NC 27707-3129. Offers MS. *Degree requirements:* For master's, one foreign language, comprehensive exam. *Entrance requirements:* For master's, GRE, minimum GPA of 3.0 in major, 2.5 overall. Additional exam requirements/recommendations for international students: Required—TOEFL.

North Carolina State University, Graduate School, College of Physical and Mathematical Sciences, Department of Marine, Earth, and Atmospheric Sciences, Raleigh, NC 27695. Offers marine, earth, and atmospheric sciences (MS, PhD); meteorology (MS, PhD); oceanography (MS, PhD). Terminal master's awarded for partial completion of doctoral program. *Degree requirements:* For master's, thesis (for some programs), final oral exam; for doctorate, thesis/dissertation, final oral exam, preliminary oral and written exams, comprehensive exam, registration. *Entrance requirements:* For master's, GRE General Test, minimum GPA of 3.0; for doctorate, GRE General Test, GRE Subject Test (for disciplines in biological oceanography and geology), minimum GPA of 3.0. Additional exam requirements/recommendations for international students: Required—TOEFL (minimum score 550 paper-based). Electronic applications accepted. *Faculty research:* Boundary layer and air quality meteorology; climate and mesoscale dynamics; biological, chemical, geological, and physical oceanography; hard rock, soft rock, environmental, and paleo geology.

See Close-Up on page 229.

Northeastern Illinois University, Graduate College, College of Arts and Sciences, Department of Earth Science, Program in Earth Science, Chicago, IL 60625-4699. Offers MS. Part-time and evening/weekend programs available. *Degree requirements:* For master's, oral presentation, thesis optional. *Entrance requirements:* For master's, 15 undergraduate hours in earth science, 8 undergraduate hours in chemistry and physics, minimum GPA of 2.75. *Faculty research:* Coastal engineering, Paleozoic and Precambrian tectonics and volcanology, ravine erosion control, well head protection delineation, genesis and evolution of basaltic magma.

Northern Arizona University, Graduate College, College of Engineering and Natural Science, Department of Geology, Program in Earth Science, Flagstaff, AZ 86011. Offers MAT, MS.

Northwestern University, The Graduate School, Judd A. and Marjorie Weinberg College of Arts and Sciences, Department of Geological Sciences, Evanston, IL 60208. Offers MS, PhD. Admissions and degrees offered through The Graduate School. Part-time programs available. *Faculty:* 10 full-time (3 women). *Students:* 18 full-time (11 women); includes 2 minority (both Hispanic Americans), 5 international. Average age 25. 12 applicants, 58% accepted, 4 enrolled. In 2005, 1 master's, 2 doctorates awarded. *Median time to degree:* Of those who began their doctoral program in fall 1997, 33% received their degree in 8 years or less. *Degree requirements:* For doctorate, thesis/dissertation. *Entrance requirements:* For master's and doctorate, GRE General Test. Additional exam requirements/recommendations for international students: Required—TOEFL. *Application deadline:* For fall admission, 1/30 for domestic students. Application fee: $60 ($75 for international students). Electronic applications accepted. *Financial support:* In 2005–06, 15 students received support, including 2 fellowships with full tuition reimbursements available (averaging $12,600 per year), 2 research assistantships with full tuition reimbursements available (averaging $13,419 per year), 6 teaching assistantships with full tuition reimbursements available (averaging $13,419 per year); career-related internships or fieldwork, Federal Work-Study, institutionally sponsored loans, scholarships/grants, tuition waivers (full and partial), and research awards also available. Financial award application deadline: 1/31; financial award applicants required to submit FAFSA. Total annual research expenditures: $710,913. *Unit head:* Donna M. Jurdy, Chair, 847-491-7163, Fax: 847-491-8060, E-mail: chair@earth.northwestern.edu. *Application contact:* Susan Van der Lee, Admission Officer, 847-491-8183, Fax: 847-491-8060, E-mail: geodept@eath.northwestern.edu.

Oregon State University, Graduate School, College of Science, Department of Geosciences, Corvallis, OR 97331. Offers geography (MA, MAIS, MS, PhD); geology (MA, MAIS, MS, PhD). Part-time programs available. *Faculty:* 15 full-time (3 women), 8 part-time/adjunct (4 women). *Students:* 67 full-time (28 women), 10 part-time (3 women); includes 5 minority (1 African American, 2 Asian Americans or Pacific Islanders, 2 Hispanic Americans), 4 international. Average age 31. In 2005, 16 master's, 4 doctorates awarded. Terminal master's awarded for partial completion of doctoral program. *Degree requirements:* For doctorate, one foreign language, thesis/dissertation. *Entrance requirements:* For master's and doctorate, GRE General Test, GRE Subject Test, minimum GPA of 3.0 in last 90 hours. Additional exam requirements/recommendations for international students: Required—TOEFL. *Application deadline:* For fall admission, 2/1 for domestic students. Applications are processed on a rolling basis. Application fee: $50. *Expenses:* Tuition, area resident: Part-time $301 per credit. Tuition, state resident: full-time $8,139; part-time $501 per credit. Tuition, nonresident: full-time $14,376; part-time $532 per credit. Required fees: $1,266. *Financial support:* Fellowships, research assistantships, teaching assistantships, career-related internships or fieldwork, Federal Work-Study, and institutionally sponsored loans available. Support available to part-time students. Financial award application deadline: 2/1. *Unit head:* Dr. Roger L. Nielsen, Director, 541-737-1201, Fax: 541-737-1200, E-mail: rnielsen@oce.orst.edu. *Application contact:* Joanne VanGeest, Graduate Admissions Coordinator, 541-737-1204, Fax: 541-737-1200, E-mail: vangeesj@geo.orst.edu.

The Pennsylvania State University University Park Campus, Graduate School, College of Earth and Mineral Sciences, Department of Geosciences, State College, University Park, PA 16802-1503. Offers astrobiology (PhD); earth sciences (M Ed); geosciences (MS, PhD). *Students:* 98 full-time (41 women), 5 part-time (1 woman); includes 7 minority (1 African American, 2 American Indian/Alaska Native, 1 Asian American or Pacific Islander, 3 Hispanic Americans), 19 international. *Entrance requirements:* For master's and doctorate, GRE General Test. Additional exam requirements/recommendations for international students: Required—TOEFL. *Expenses:* Tuition, state resident: full-time $12,518; part-time $522 per credit. Tuition, nonresident: full-time $23,004; part-time $959 per credit. Required fees: $484. Tuition and fees vary according to course load, campus/location and program. *Unit head:* Dr. Timothy J. Bralower, Professor of Geosciences, 814-863-8177, Fax: 814-863-7823.

Princeton University, Graduate School, Department of Geosciences, Princeton, NJ 08544-1019. Offers atmospheric and oceanic sciences (PhD); geological and geophysical sciences (PhD). *Degree requirements:* For doctorate, one foreign language, thesis/dissertation. *Entrance requirements:* For doctorate, GRE General Test. Additional exam requirements/recommendations for international students: Required—TOEFL (minimum score 600 paper-based; 250 computer-based). Electronic applications accepted. *Faculty research:* Biogeochemistry, climate science, earth history, regional geology and tectonics, solid-earth geophysics.

Purdue University, Graduate School, School of Science, Department of Earth and Atmospheric Sciences, West Lafayette, IN 47907. Offers MS, PhD. *Faculty:* 33 full-time (4 women). *Students:* 50 full-time (17 women), 6 part-time (3 women); includes 3 minority (1 African American, 2 Hispanic Americans), 17 international. Average age 29. 64 applicants, 30% accepted, 14 enrolled. In 2005, 7 master's, 3 doctorates awarded. *Degree requirements:* For master's, thesis; for doctorate, one foreign language, thesis/dissertation. *Entrance requirements:* For master's and doctorate, GRE General Test. Additional exam requirements/recommendations for international students: Required—TOEFL. *Application deadline:* For fall admission, 5/1 for domestic students, 2/1 for international students; for spring admission, 9/1 for domestic students, 7/1 for international students. Applications are processed on a rolling basis. Application fee: $55. Electronic applications accepted. *Financial support:* In 2005–06, 10 fellowships with partial tuition reimbursements (averaging $15,133 per year), 8 research assistantships with partial tuition reimbursements (averaging $14,400 per year), 26 teaching assistantships with partial tuition reimbursements (averaging $14,400 per year) were awarded. Support available to part-time students. Financial award application deadline: 3/1; financial award applicants required to submit FAFSA. *Faculty research:* Geology, geophysics, hydrogeology, paleoclimatology, environmental science. *Unit head:* Dr. Jonathan M Harbor, Head, 765-494-4753, Fax: 765-496-1210. *Application contact:* Kathy Kincade, Graduate Secretary, 765-494-5984, Fax: 765-496-1210, E-mail: kkincade@purdue.edu.

Rensselaer Polytechnic Institute, Graduate School, School of Science, Department of Earth and Environmental Sciences, Troy, NY 12180-3590. Offers environmental chemistry (MS, PhD); geochemistry (MS, PhD); geology (MS, PhD); geophysics (MS, PhD); petrology (MS, PhD). Part-time programs available. Terminal master's awarded for partial completion of doctoral program. *Degree requirements:* For master's, thesis (for some programs), comprehensive exam; for doctorate, thesis/dissertation, comprehensive exam. *Entrance requirements:* For master's and doctorate, GRE General Test. Additional exam requirements/recommendations for international students: Required—TOEFL. Electronic applications accepted. *Expenses:* Tuition: Full-time $31,000; part-time $1,320 per credit. Required fees: $1,623. *Faculty research:* Mantel geochemistry, contaminant geochemistry, seismology, GPS geodesy, remote sensing petrology.

See Close-Up on page 661.

Rice University, Graduate Programs, Wiess School of Natural Sciences, Department of Earth Science, Houston, TX 77251-1892. Offers MA, PhD. *Degree requirements:* For master's and doctorate, thesis/dissertation. *Entrance requirements:* For master's and doctorate, GRE General Test, minimum GPA of 3.0. Additional exam requirements/recommendations for international students: Required—TOEFL. Electronic applications accepted. *Faculty research:* Marine geology/paleocenography; stratigraphy/sedimentology; petrology/geochemistry; seismology/computational geophysics; structure/tectonics.

Rice University, Graduate Programs, Wiess School of Natural Sciences, Professional Master's Program in Subsurface Geosciences, Houston, TX 77251-1892. Offers geophysics (MS). Part-time programs available. *Degree requirements:* For master's, internship. *Entrance requirements:* For master's, GRE, letters of recommendation (4). Additional exam requirements/recommendations for international students: Required—TOEFL (minimum score 600 paper-based; 250 computer-based). Electronic applications accepted. *Faculty research:* Seismology, geodynamics, wave propogation, bio-geochemistry, remote sensing.

St. Francis Xavier University, Graduate Studies, Department of Earth Sciences, Antigonish, NS B2G 2W5, Canada. Offers M Sc. *Faculty:* 2 full-time (0 women). *Students:* 2 full-time (1 woman). *Degree requirements:* For master's, thesis, registration. *Entrance requirements:* Additional exam requirements/recommendations for international students: Required—TOEFL (minimum score 580 paper-based; 236 computer-based). Application fee: $40. *Faculty research:* Environmental earth sciences, global change tectonics, paleoclimatology, crustal fluids. Total annual research expenditures: $300,000. *Unit head:* Dr. Brendan Murphy, Professor, 902-867-2481, Fax: 902-867-5153, E-mail: bmurphy@stfx.ca.

Saint Louis University, Graduate School, College of Arts and Sciences and Graduate School, Department of Earth and Atmospheric Sciences, St. Louis, MO 63103-2097. Offers geophysics (PhD); geoscience (MS); meteorology (M Pr Met, MS-R, PhD). Part-time programs available. *Faculty:* 17 full-time (0 women), 2 part-time/adjunct (0 women). *Students:* 30 full-time (11 women), 4 part-time (2 women); includes 2 minority (1 African American, 1 Hispanic American), 16 international. Average age 28. 35 applicants, 86% accepted, 14 enrolled. In 2005, 4 master's, 3 doctorates awarded. *Degree requirements:* For master's, thesis (for some programs), comprehensive oral exam; for doctorate, thesis/dissertation, preliminary exams. *Entrance requirements:* For master's and doctorate, GRE General Test, letters of recommendation, resumé. Additional exam requirements/recommendations for international students: Required—TOEFL (minimum score 550 paper-based; 213 computer-based). *Application deadline:* For fall admission, 7/1 for domestic students, 7/1 for international students; for spring admission, 11/1 for domestic students, 11/1 for international students. Applications are processed on a rolling basis. Application fee: $40. *Expenses:* Tuition: Part-time $760 per credit hour. Required fees: $55 per semester. *Financial support:* In 2005–06, 25 students received support, including 8 research assistantships with full tuition reimbursements available (averaging $16,000 per year), 7 teaching assistantships with full tuition reimbursements available (averaging $15,500 per year); health care benefits and unspecified assistantships also available. Financial award application deadline: 6/1; financial award applicants required to submit FAFSA. *Faculty research:* Structural geology, mesoscale meteorology and severe storms, weather and climate change prediction. Total annual research expenditures: $800,000. *Unit head:* Dr. Bill Dannevik, Interim Chairperson, 314-977-3115, Fax: 314-911-3117, E-mail: dannevik@slu.edu. *Application contact:* Gary Behrman, Associate Dean of the Graduate School, 314-977-3827, E-mail: behrmang@slu.edu.

San Francisco State University, Division of Graduate Studies, College of Science and Engineering, Department of Geosciences, San Francisco, CA 94132-1722. Offers applied geosciences (MS). *Entrance requirements:* For master's, minimum GPA of 2.5 in last 60 units.

Simon Fraser University, Graduate Studies, Faculty of Science, Department of Earth Sciences, Burnaby, BC V5A 1S6, Canada. Offers M Sc. Part-time programs available. *Degree requirements:* For master's, thesis. *Entrance requirements:* For master's, minimum GPA of 3.0. Additional exam requirements/recommendations for international students: Required—TOEFL or IELTS. Electronic applications accepted. *Faculty research:* Earth surface processes, environmental geoscience, surficial and Quaternary geology, sedimentology.

Stanford University, School of Earth Sciences, Department of Geological and Environmental Sciences, Stanford, CA 94305-9991. Offers MS, PhD, Eng. Terminal master's awarded for partial completion of doctoral program. *Degree requirements:* For masters, doctorate, and Eng, thesis/dissertation. *Entrance requirements:* For master's, doctorate, and Eng, GRE General Test. Additional exam requirements/recommendations for international students: Required—TOEFL. Electronic applications accepted.

Stanford University, School of Earth Sciences, Earth Systems Program, Stanford, CA 94305-9991. Offers MS. Students admitted at the undergraduate level. Electronic applications accepted.

State University of New York College at Oneonta, Graduate Studies, Department of Earth Sciences, Oneonta, NY 13820-4015. Offers MA. Part-time and evening/weekend programs available. *Students:* 9. 3 applicants, 67% accepted, 2 enrolled. *Degree requirements:* For master's, thesis. *Entrance requirements:* For master's, GRE General Test. *Application deadline:* For fall admission, 3/25 for domestic students; for spring admission, 10/1 priority date for domestic students. Applications are processed on a rolling basis. Application fee: $50. *Expenses:* Tuition, area resident: Full-time $6,900. Tuition, nonresident: full-time $10,920. *Financial support:* Fellowships available. *Unit head:* Dr. James Ebert, Chair, 607-436-3065, E-mail: ebertj@oneonta.edu.

Stony Brook University, State University of New York, Graduate School, College of Arts and Sciences, Department of Geosciences, Stony Brook, NY 11794. Offers earth and space science (MS, PhD); earth science (MAT). MAT offered through the School of Professional Development and Continuing Studies. *Faculty:* 17 full-time (2 women), 1 part-time/adjunct (0 women). *Students:* 43 full-time (16 women), 6 part-time (3 women); includes 2 minority (1 African American, 1 Hispanic American), 14 international. Average age 29. 45 applicants, 40% accepted. In 2005, 10 master's, 3 doctorates awarded. Terminal master's awarded for partial completion of doctoral program. *Degree requirements:* For master's, thesis or alternative; for doctorate, thesis/dissertation. *Entrance requirements:* For master's and doctorate, GRE General Test, minimum GPA of 3.0. Additional exam requirements/recommendations for international students: Required—TOEFL. *Application deadline:* For fall admission, 1/15 for domestic students. Application fee: $50. *Expenses:* Tuition, area resident: Part-time $288. Tuition, state resident: full-time $6,900. Tuition, nonresident: full-time $10,920; part-time $455. Required fees: $704. *Financial support:* In 2005–06, 1 fellowship, 29 research assistantships, 7 teaching assistant-

Geosciences

Stony Brook University, State University of New York (continued)

ships were awarded. *Faculty research:* Astronomy, theoretical and observational astrophysics, paleontology, petrology, crystallography. Total annual research expenditures: $5.7 million. *Unit head:* Dr. Teng-Fong Wong, Chair, 631-632-8194, Fax: 631-632-6900. *Application contact:* Dr. John Parise, Director, 631-632-8200, Fax: 631-632-8240, E-mail: jparise@notes.cc.sunysb.edu.

Announcement: Many versed in the sciences say that rocks and their formations give clues to the basis of life on Earth and how it evolved. This is a key point of study for the Department of Geosciences, which has the facilities available to both faculty and students to continue to unlock the mysteries of this segment of nature.

See Close-Up on page 231.

Texas A&M University–Commerce, Graduate School, College of Arts and Sciences, Department of Biological and Earth Sciences, Commerce, TX 75429-3011. Offers M Ed, MS. *Faculty:* 7 full-time (2 women), 2 part-time/adjunct (0 women). *Students:* 10 full-time (5 women), 13 part-time (10 women); includes 7 minority (2 African Americans, 4 Asian Americans or Pacific Islanders, 1 Hispanic American), 3 international. Average age 36. In 2005, 3 degrees awarded. *Degree requirements:* For master's, thesis (for some programs), comprehensive exam. *Entrance requirements:* For master's, GRE General Test. *Application deadline:* For fall admission, 6/1 for domestic students; for spring admission, 11/1 priority date for domestic students. Applications are processed on a rolling basis. Application fee: $0 ($25 for international students). Electronic applications accepted. *Financial support:* In 2005–06, research assistantships (averaging $7,875 per year), teaching assistantships (averaging $7,875 per year) were awarded; Federal Work-Study, institutionally sponsored loans, and scholarships/grants also available. Financial award application deadline: 5/1; financial award applicants required to submit FAFSA. *Faculty research:* Microbiology, botany, environmental science, birds. Total annual research expenditures: $3,000. *Unit head:* Dr. Don R. Lee, Interim Head, 903-886-5378, Fax: 903-886-5997. *Application contact:* Tammi Thompson, Graduate Admissions Adviser, 843-886-5167, Fax: 843-886-5165, E-mail: tammi_thompson@tamu-commerce.edu.

Texas Christian University, College of Science and Engineering, Department of Biology, Program in Environmental Sciences, Fort Worth, TX 76129-0002. Offers earth sciences (MS); ecology (MS). Part-time and evening/weekend programs available. *Degree requirements:* For master's, thesis optional. *Entrance requirements:* For master's, GRE General Test, GRE Subject Test, 1 year course work in biology and chemistry; 1 semester course work in calculus, government, and physical geology. Additional exam requirements/recommendations for international students: Required—TOEFL. *Application deadline:* For fall admission, 3/1 for domestic students; for spring admission, 12/1 for domestic students. Applications are processed on a rolling basis. Application fee: $0. *Expenses:* Tuition: Part-time $740 per credit hour. *Financial support:* Unspecified assistantships available. Financial award application deadline: 3/1. *Unit head:* Dr. Mike Slattery, Director, 817-257-7506. *Application contact:* Dr. Bonnie Melhart, Associate Dean, College of Science and Engineering, E-mail: b.melhart@tcu.edu.

Texas Tech University, Graduate School, College of Arts and Sciences, Department of Geosciences, Lubbock, TX 79409. Offers atmospheric sciences (MS); geoscience (MS, PhD). Part-time programs available. *Faculty:* 15 full-time (2 women). *Students:* 27 full-time (11 women), 10 part-time (1 woman); includes 3 minority (1 African American, 2 Hispanic Americans), 4 international. Average age 31. 52 applicants, 56% accepted, 11 enrolled. In 2005, 12 master's, 2 doctorates awarded. *Degree requirements:* For master's and doctorate, thesis/dissertation. *Entrance requirements:* For master's and doctorate, GRE General Test. Additional exam requirements/recommendations for international students: Required—TOEFL (minimum score 550 paper-based; 213 computer-based). *Application deadline:* Applications are processed on a rolling basis. Application fee: $50 ($60 for international students). Electronic applications accepted. *Expenses:* Tuition, state resident: full-time $4,296. Tuition, nonresident: full-time $10,920. Required fees: $1,992. Tuition and fees vary according to program. *Financial support:* In 2005–06, 21 students received support, including 3 research assistantships with partial tuition reimbursements available (averaging $13,851 per year), 23 teaching assistantships with partial tuition reimbursements available (averaging $15,434 per year); Federal Work-Study and institutionally sponsored loans also available. Support available to part-time students. Financial award application deadline: 4/15; financial award applicants required to submit FAFSA. *Faculty research:* Ophiolites and oceanic lower crust; petroleum geology; tectonics and arc magnetism; aqueous and environmental geochemistry; near-ground high wind phenomenon (hurricanes and severe storms). Total annual research expenditures:$346,436. *Unit head:* Dr. James Barrick, Chairman, 806-742-3107, Fax: 806-742-0100, E-mail: jim.barrick@ttu.edu. *Application contact:* Dr. Moira Ridley, Graduate Adviser, 806-742-3102, Fax: 806-724-0100, E-mail: moira.ridley@ttu.edu.

Université du Québec à Chicoutimi, Graduate Programs, Program in Earth Sciences, Chicoutimi, QC G7H 2B1, Canada. Offers M Sc A. Part-time programs available. *Degree requirements:* For master's, thesis. *Entrance requirements:* For master's, appropriate bachelor's degree, proficiency in French.

Université du Québec à Montréal, Graduate Programs, Program in Earth Sciences, Montreal, QC H3C 3P8, Canada. Offers geology-research (M Sc); mineral resources (PhD); non-renewable resources (DESS). Part-time programs available. Terminal master's awarded for partial completion of doctoral program. *Degree requirements:* For master's, thesis (for some programs); for doctorate, thesis/dissertation. *Entrance requirements:* For master's, appropriate bachelor's degree or equivalent, proficiency in French. *Faculty research:* Economic geology, structural geology, geochemistry, Quaternary geology, isotopic geochemistry.

Université du Québec, Institut National de la Recherche Scientifique, Graduate Programs, Research Center—Water, Earth and Environment, Québec, QC G1K 9A9, Canada. Offers earth sciences (M Sc, PhD); earth sciences-environmental technologies (M Sc); water sciences (MA, PhD). Part-time programs available. *Faculty:* 38. *Students:* 172 full-time (80 women), 10 part-time (3 women), 44 international. Average age 29. In 2005, 37 master's, 10 doctorates awarded. *Degree requirements:* For master's, thesis optional; for doctorate, thesis/dissertation. *Entrance requirements:* For master's, appropriate bachelor's degree, proficiency in French; for doctorate, appropriate master's degree, proficiency in French. *Application deadline:* For fall admission, 3/30 for domestic students, 3/30 for international students. For winter admission, 11/1 for domestic students. Application fee: $30. *Financial support:* Fellowships, research assistantships, teaching assistantships available. *Faculty research:* Land use, impacts of climate change, adaptation to climate change, integrated management of resources (mineral and water). *Unit head:* Jean Pierre Villeneuve, Director, 418-654-2575, Fax: 418-654-2615, E-mail: jp_villeneuve@ete.inrs.ca. *Application contact:* Michel Barbeau, Registrar, 418-654-2518, Fax: 418-654-3858, E-mail: michel.barbeau@adm.inrs.ca.

Université Laval, Faculty of Sciences and Engineering, Department of Geology and Geological Engineering, Programs in Earth Sciences, Québec, QC G1K 7P4, Canada. Offers earth sciences (M Sc, PhD); environmental technologies (M Sc). Offered jointly with INRS-Géressources. Terminal master's awarded for partial completion of doctoral program. *Degree requirements:* For master's, thesis (for some programs); for doctorate, thesis/dissertation, comprehensive exam. *Entrance requirements:* For master's and doctorate, knowledge of French. Electronic applications accepted.

University at Albany, State University of New York, College of Arts and Sciences, Department of Earth and Atmospheric Sciences, Albany, NY 12222-0001. Offers atmospheric science (MS, PhD); geology (MS, PhD). *Students:* 38 full-time (10 women), 10 part-time (5 women). Average age 29. In 2005, 5 master's, 4 doctorates awarded. *Degree requirements:* For master's, one foreign language, thesis, comprehensive exam; for doctorate, 2 foreign languages, thesis/dissertation, oral exams, comprehensive exam. *Entrance requirements:* For master's and doctorate, GRE General Test. Additional exam requirements/recommendations for international students: Required—TOEFL (minimum score 550 paper-based; 213 computer-based). *Application deadline:* For fall admission, 6/1 for domestic students, 5/1 for international students; for spring admission, 11/1 for domestic students, 11/11 for international students. Applications are processed on a rolling basis. Application fee: $60. Electronic applications accepted. *Financial support:* Fellowships, research assistantships, teaching assistantships, minority assistantships available. Financial award application deadline:3/1. *Faculty research:* Environmental geochemistry, tectonics, mesoscale meteorology, atmospheric chemistry. *Unit head:* Dr. Vincent Idone, Chair, 518-442-4466. *Application contact:* William Kidd, Graduate Program Director, Geology.

The University of Akron, Graduate School, Buchtel College of Arts and Sciences, Department of Geology, Program in Earth Science, Akron, OH 44325. Offers MS. *Students:* 4 full-time (2 women). *Degree requirements:* For master's, thesis, seminar, comprehensive exam. *Entrance requirements:* For master's, minimum GPA of 2.75. Additional exam requirements/recommendations for international students: Required—TOEFL (minimum score 550 paper-based; 213 computer-based). *Application deadline:* For fall admission, 3/1 for domestic students. Applications are processed on a rolling basis. Application fee: $30 ($40 for international students). Electronic applications accepted. *Expenses:* Tuition, state resident: full-time $5,816; part-time $323 per credit. Tuition, nonresident: full-time $9,976; part-time $554 per credit. Required fees: $794; $43 per credit. $12 per term. Tuition and fees vary according to course load, degree level and program. *Unit head:* Dr. LaVerne Friberg, Director of Graduate Studies, 330-972-8046.

University of Alberta, Faculty of Graduate Studies and Research, Department of Earth and Atmospheric Sciences, Edmonton, AB T6G 2E1, Canada. Offers M Sc, MA, PhD. *Faculty:* 38 full-time (4 women), 15 part-time/adjunct (0 women). *Students:* 19 full-time (6 women). Average age 30. 88 applicants, 43% accepted. In 2005, 14 master's, 7 doctorates awarded. *Degree requirements:* For master's and doctorate, thesis/dissertation, residency. *Entrance requirements:* For master's, B Sc, minimum GPA of 6.5; for doctorate, M Sc. Additional exam requirements/recommendations for international students: Required—TOEFL or Michigan English Language Assessment Battery. *Application deadline:* Applications are processed on a rolling basis. Electronic applications accepted. Tuition and fees charges are reported in Canadian dollars. *Expenses:* Tuition, state resident: part-time $562 Canadian dollars per term. Tuition, nonresident: full-time $3,375 Canadian dollars. Required fees: $573 Canadian dollars; $84 Canadian dollars per term. *Financial support:* In 2005–06, 10 fellowships, 15 research assistantships were awarded; teaching assistantships, scholarships/grants and unspecified assistantships also available. *Faculty research:* Geology, human geography, physical geography, meteorology. Total annual research expenditures: $10 million. *Unit head:* Dr. Martin J. Sharp, Chair, 403-492-3329, Fax: 403-492-7598, E-mail: martin.sharp@ualberta.ca.

The University of Arizona, Graduate College, College of Science, Department of Geosciences, Tucson, AZ 85721. Offers MS, PhD. Part-time programs available. *Degree requirements:* For master's, thesis or prepublication; for doctorate, one foreign language, thesis/dissertation, comprehensive exam. *Entrance requirements:* For master's and doctorate, GRE General Test. Additional exam requirements/recommendations for international students: Required—TOEFL. *Faculty research:* Tectonics, geophysics, geochemistry/petrology, economic geology, Quaternary studies, stratigraphy/paleontology.

University of California, Irvine, Office of Graduate Studies, School of Physical Sciences, Department of Earth System Science, Irvine, CA 92697. Offers MS, PhD. *Degree requirements:* For doctorate, thesis/dissertation. *Entrance requirements:* For master's and doctorate, GRE General Test, GRE Subject Test, minimum GPA of 3.0. Additional exam requirements/recommendations for international students: Required—TOEFL (minimum score 550 paper-based; 213 computer-based). Electronic applications accepted. *Faculty research:* Atmospheric chemistry, climate change, isotope biogeochemistry, global environmental chemistry.

University of California, Los Angeles, Graduate Division, College of Letters and Science, Department of Earth and Space Sciences, Los Angeles, CA 90095. Offers geochemistry (MS, PhD); geology (MS, PhD); geophysics and space physics (MS, PhD). *Degree requirements:* For master's, comprehensive exams or thesis; for doctorate, thesis/dissertation, oral and written qualifying exams. *Entrance requirements:* For master's, GRE General Test, minimum GPA of 3.0; for doctorate, GRE General Test, minimum undergraduate GPA of 3.0. Electronic applications accepted.

University of California, San Diego, Graduate Studies and Research, Scripps Institution of Oceanography, La Jolla, CA 92093. Offers earth sciences (PhD); marine biodiversity and conservation (MAS); marine biology (PhD); oceanography (PhD). Postbaccalaureate distance learning degree programs offered (minimal on-campus study). *Faculty:* 97. *Students:* 243 (126 women); includes 21 minority (2 African Americans, 2 American Indian/Alaska Native, 11 Asian Americans or Pacific Islanders, 6 Hispanic Americans) 63 international. 311 applicants, 24% accepted, 37 enrolled. In 2005, 9 master's, 25 doctorates awarded. *Median time to degree:* Of those who began their doctoral program in fall 1997, 100% received their degree in 8 years or less. *Entrance requirements:* For doctorate, GRE General Test, GRE Subject Test. Additional exam requirements/recommendations for international students: Required—TOEFL (minimum score 550 paper-based; 213 computer-based). *Application deadline:* For fall admission, 1/4 for domestic students. Application fee: $60 ($80 for international students). Electronic applications accepted. *Financial support:* Fellowships, research assistantships, health care benefits available. *Unit head:* Myrl C. Hendershott, Chair, 858-534-3206, E-mail: siodept@sio.ucsd.edu. *Application contact:* Dawn Huffman, Graduate Coordinator, 858-534-3206.

University of California, Santa Barbara, Graduate Division, College of Letters and Sciences, Division of Mathematics, Life, and Physical Sciences, Department of Earth Science, Santa Barbara, CA 93106. Offers geological sciences (MS, PhD); geophysics (MS). *Faculty:* 37 full-time (4 women). *Students:* 57 full-time (25 women); includes 2 minority (1 Asian American or Pacific Islander, 1 Hispanic American), 6 international. 87 applicants, 24% accepted, 10 enrolled. In 2005, 7 master's, 5 doctorates awarded. Terminal master's awarded for partial completion of doctoral program. *Degree requirements:* For master's, thesis, comprehensive exam, registration; for doctorate, thesis/dissertation, dissertation defense, comprehensive exam, registration. *Entrance requirements:* For master's and doctorate, GRE General Test. Additional exam requirements/recommendations for international students: Required—TOEFL (minimum score 550 paper-based; 213 computer-based). *Application deadline:* For fall admission, 1/1 for domestic students, 1/1 for international students. Application fee: $60. Electronic applications accepted. *Financial support:* In 2005–06, 7 students received support, including 8 fellowships (averaging $6,000 per year), 50 teaching assistantships with full and partial tuition reimbursements available (averaging $7,500 per year); research assistantships with full and partial tuition reimbursements available, career-related internships or fieldwork, Federal Work-Study, scholarships/grants, traineeships, health care benefits, and unspecified assistantships also available. Financial award application deadline: 1/1; financial award applicants required to submit FAFSA. *Faculty research:* Geochemistry, geophysics, tectonics, seismology, petrology. *Unit head:* Dr. James Mattinson, Chair, 805-893-3219, E-mail: mattison@geol.ucsb.edu. *Application contact:* Samuel Rifkin, Graduate Program Assistant, 805-893-3329, Fax: 805-893-2314, E-mail: rifkin@geol.ucsb.edu.

University of California, Santa Cruz, Division of Graduate Studies, Division of Physical and Biological Sciences, Program in Earth Sciences, Santa Cruz, CA 95064. Offers MS, PhD. *Faculty:* 21 full-time (4 women). *Students:* 54 (29 women); includes 7 minority (3 Asian Americans or Pacific Islanders, 4 Hispanic Americans) 3 international. 121 applicants, 31% accepted, 22 enrolled. In 2005, 8 master's, 5 doctorates awarded. *Degree requirements:* For master's, thesis; for doctorate, one foreign language, thesis/dissertation, qualifying exam. *Entrance requirements:* For master's and doctorate, GRE General Test, GRE Subject Test. *Application deadline:* For fall admission, 1/15 for domestic students. Application fee: $60. *Expenses:* Tuition, nonresident: full-time $14,694. *Financial support:* Fellowships, research assistantships, teaching assistantships, career-related internships or fieldwork, Federal Work-Study, and institutionally sponsored loans available. Financial award application deadline: 1/15. *Faculty research:* Evolution of continental margins and orogenic belts, geologic processes occurring at plate boundaries, deep-sea sediment diagenesis, paleoecology, hydrogeology. *Unit head:* Elise Knittle, Chairperson, 831-459-3164. *Application contact:* Judy L. Glass, Reporting Analyst for Graduate Admissions, 831-459-5906, Fax: 831-459-4843, E-mail: jlglass@ucsc.edu.

University of Chicago, Division of the Physical Sciences, Department of the Geophysical Sciences, Chicago, IL 60637-1513. Offers atmospheric sciences (SM, PhD); earth sciences (SM, PhD); paleobiology (PhD); planetary and space sciences (SM, PhD). *Faculty:* 24 full-time (3 women). *Students:* 35 full-time (15 women); includes 1 minority (Hispanic American), 11 international. Average age 29. 52 applicants, 29% accepted. In 2005, 2 master's, 3 doctorates awarded. Terminal master's awarded for partial completion of doctoral program. *Entrance requirements:* For master's and doctorate, GRE General Test. Additional exam requirements/recommendations for international students: Required—TOEFL, IELT. *Application deadline:* For fall admission, 1/15 for domestic students, 1/15 for international students. Application fee: $55. Electronic applications accepted. *Financial support:* In 2005–06, 32 students received support, including research assistantships with full tuition reimbursements available (averaging $18,675 per year), teaching assistantships with full tuition reimbursements available (averaging $19,098 per year); fellowships, Federal Work-Study, institutionally sponsored loans, scholarships/grants, tuition waivers (partial), and unspecified assistantships also available. Financial award application deadline: 1/15. *Faculty research:* Climatology, evolutionary paleontology, petrology, geochemistry, oceanic sciences. *Unit head:* Dr. David Rowley, Chairman, 773-702-8102, Fax: 773-702-9505. *Application contact:* David J. Leslie, Graduate Student Services Coordinator, 773-702-8180, Fax: 773-702-9505, E-mail: info@geosci.uchicago.edu.

University of Florida, Graduate School, College of Liberal Arts and Sciences, Department of Geological Sciences, Gainesville, FL 32611. Offers geology (MS, MST, PhD). *Faculty:* 17 full-time (2 women). *Students:* 43 (22 women); includes 1 minority (Hispanic American) 8 international. In 2005, 8 master's, 5 doctorates awarded. Terminal master's awarded for partial completion of doctoral program. *Degree requirements:* For master's, thesis (for some programs); for doctorate, one foreign language, thesis/dissertation. *Entrance requirements:* For master's and doctorate, GRE General Test, GRE Subject Test, minimum GPA of 3.0. Additional exam requirements/recommendations for international students: Required—TOEFL (minimum score 550 paper-based; 213 computer-based). *Application deadline:* For fall admission, 6/1 for domestic students; for spring admission, 10/1 priority date for domestic students. Applications are processed on a rolling basis. Application fee: $30. Electronic applications accepted. *Expenses:* Tuition, state resident: full-time $6,234. Tuition, nonresident: full-time $21,359. Tuition and fees vary according to program. *Financial support:* In 2005–06, fellowships with full tuition reimbursements (averaging $15,000 per year), 7 research assistantships with full tuition reimbursements (averaging $11,931 per year), 21 teaching assistantships with full tuition reimbursements (averaging $11,868 per year) were awarded; career-related internships or fieldwork, Federal Work-Study, institutionally sponsored loans, and scholarships/grants also available. Support available to part-time students. Financial award application deadline: 3/1. *Faculty research:* Paleoclimatology, tectonophysics, petrochemistry, marine geology, geochemistry, hydrology. Total annual research expenditures: $1.5 million. *Unit head:* Dr. Paul Mueller, Chair, 352-392-2231, Fax: 352-392-9294, E-mail: mueller@geology.ufl.edu. *Application contact:* Dr. Michael R. Perfit, Graduate Coordinator, 352-392-2128, Fax: 352-392-9294, E-mail: perfit@geology.ufl.edu.

University of Illinois at Chicago, Graduate College, College of Liberal Arts and Sciences, Department of Earth and Environmental Sciences, Chicago, IL 60607-7128. Offers crystallography (MS, PhD); environmental geology (MS, PhD); geochemistry (MS, PhD); geology (MS, PhD); geomorphology (MS, PhD); geophysics (MS, PhD); geotechnical engineering and geosciences (PhD); hydrogeology (MS, PhD); low-temperature and organic geochemistry (MS, PhD); mineralogy (MS, PhD); paleoclimatology (MS, PhD); paleontology (MS, PhD); petrology (MS, PhD); quaternary geology (MS, PhD); sedimentology (MS, PhD); water resources (MS, PhD). *Degree requirements:* For master's and doctorate, thesis/dissertation. *Entrance requirements:* For master's and doctorate, GRE General Test, minimum GPA of 2.75. Additional exam requirements/recommendations for international students: Required—TOEFL. Electronic applications accepted.

University of Illinois at Urbana–Champaign, Graduate College, College of Liberal Arts and Sciences, Department of Geology, Champaign, IL 61820. Offers earth sciences (MS, PhD); geochemistry (MS, PhD); geology (MS, PhD); geophysics (MS, PhD). *Faculty:* 13 full-time (3 women). *Students:* 32 full-time (13 women), 4 part-time; includes 1 minority (American Indian/Alaska Native), 16 international. Average age 26. 44 applicants, 39% accepted, 13 enrolled. In 2005, 4 master's, 5 doctorates awarded. Terminal master's awarded for partial completion of doctoral program. *Degree requirements:* For master's and doctorate, thesis/dissertation. *Entrance requirements:* For master's and doctorate, GRE General Test, minimum GPA of 3.0. Additional exam requirements/recommendations for international students: Required—TOEFL. *Application deadline:* For fall admission, 2/22 for domestic students; for spring admission, 10/16 for domestic students. Applications are processed on a rolling basis. Application fee: $50 ($60 for international students). *Financial support:* In 2005–06, 8 fellowships, 14 research assistantships, 18 teaching assistantships were awarded; Federal Work-Study and tuition waivers (full and partial) also available. Financial award application deadline: 2/15. *Faculty research:* Hydrogeology, structure/tectonics, mineral science. *Unit head:* Stephen Marshak, Head, 217-333-3542, Fax: 217-244-4996, E-mail: smarshak@uiuc.edu. *Application contact:* Barbara Elmore, Graduate Admissions Secretary, 217-333-3542, Fax: 217-244-4996, E-mail: belmore@uiuc.edu.

The University of Iowa, Graduate College, College of Liberal Arts and Sciences, Department of Geoscience, Iowa City, IA 52242-1316. Offers MS, PhD. *Faculty:* 16 full-time, 7 part-time/adjunct. *Students:* 23 full-time (14 women), 23 part-time (8 women); includes 1 minority (Hispanic American), 5 international. 33 applicants, 64% accepted, 11 enrolled. In 2005, 4 master's, 1 doctorate awarded. *Degree requirements:* For master's, exam, thesis optional; for doctorate, thesis/dissertation, comprehensive exam, registration. *Entrance requirements:* For master's and doctorate, GRE General Test, minimum GPA of 3.0. Additional exam requirements/recommendations for international students: Required—TOEFL (minimum score 550 paper-based; 213 computer-based). *Application deadline:* For fall admission, 1/15 for domestic students, 1/15 for international students; for spring admission, 11/1 for domestic students. Application fee: $60 ($85 for international students). Electronic applications accepted. *Expenses:* Tuition, state resident: part-time $1,882 per term. Tuition, nonresident: full-time $17,338; part-time $4,907 per term. Tuition and fees vary according to course load and program. *Financial support:* In 2005–06, 11 research assistantships with partial tuition reimbursements, 27 teaching assistantships with partial tuition reimbursements were awarded; fellowships Financial award applicants required to submit FAFSA. *Unit head:* C. T. Foster, Chair, 319-335-1820, Fax: 319-335-1821.

University of Louisiana at Monroe, Graduate Studies and Research, College of Arts and Sciences, Department of Geosciences, Monroe, LA 71209-0001. Offers MS. *Degree requirements:* For master's, thesis. *Entrance requirements:* For master's, GRE General Test, minimum GPA of 2.8 during previous 2 years or 3.0 in 21 hours of geosciences. *Faculty research:* Sedimentology, environmental hydrology, planetary geosciences, micropaleontology.

University of Maine, Graduate School, College of Natural Sciences, Forestry, and Agriculture, Department of Earth Sciences, Orono, ME 04469. Offers MS, PhD. Part-time programs available. *Students:* 28 full-time (13 women), 16 part-time (10 women), 6 international. Average age 30. 31 applicants, 77% accepted, 9 enrolled. In 2005, 5 master's, 2 doctorates awarded. *Degree requirements:* For master's, thesis; for doctorate, one foreign language, thesis/dissertation. *Entrance requirements:* For master's and doctorate, GRE General Test. Additional exam requirements/recommendations for international students: Required—TOEFL. *Application deadline:* For fall admission, 2/1 for domestic students. Applications are processed on a rolling basis. Application fee: $50. Electronic applications accepted. *Financial support:* In 2005–06, 5 research assistantships with tuition reimbursements (averaging $14,800 per year), 5 teaching assistantships with tuition reimbursements (averaging $10,100 per year) were awarded; Federal Work-Study, institutionally sponsored loans, and tuition waivers (full and partial) also available. Financial award application deadline: 3/1. *Faculty research:* Appalachian bedrock geology, Quaternary studies, marine geology. *Unit head:* Dr. Daniel Belknap, Chair, 207-581-2152, Fax: 207-581-2202. *Application contact:* Scott G. Delcourt, Associate Dean of the Graduate School, 207-581-3219, Fax: 207-581-3232, E-mail: graduate@maine.edu.

University of Massachusetts Amherst, Graduate School, College of Natural Sciences and Mathematics, Department of Geosciences, Program in Geosciences, Amherst, MA 01003. Offers MS, PhD. Postbaccalaureate distance learning degree programs offered. *Students:* 30 full-time (18 women), 25 part-time (8 women); includes 4 minority (3 Asian Americans or Pacific Islanders, 1 Hispanic American), 6 international. Average age 32. 70 applicants, 59% accepted, 14 enrolled. In 2005, 5 master's, 3 doctorates awarded. *Degree requirements:* For doctorate, one foreign language, thesis/dissertation. *Entrance requirements:* For doctorate, GRE General Test. Additional exam requirements/recommendations for international students: Required—TOEFL (minimum score 530 paper-based; 197 computer-based). *Application deadline:* For fall admission, 2/1 for domestic students, 2/1 for international students; for spring admission, 10/1 for domestic students, 10/1 for international students. Applications are processed on a rolling basis. Application fee: $40 ($65 for international students). Electronic applications accepted. *Expenses:* Tuition, state resident: part-time $110 per credit. Tuition, nonresident: part-time $414 per credit. Required fees: $2,824 per term. One-time fee: $250 part-time. Full-time tuition and fees vary according to course load, campus/location, program and reciprocity agreements. *Financial support:* Fellowships with full tuition reimbursements, research assistantships with full tuition reimbursements, teaching assistantships with full tuition reimbursements, career-related internships or fieldwork, Federal Work-Study, scholarships/grants, traineeships, and unspecified assistantships available. Support available to part-time students. Financial award application deadline: 2/1. *Unit head:* Dr. Richard Yuretich, Director, 413-545-2286.

University of Memphis, Graduate School, College of Arts and Sciences, Department of Earth Sciences, Memphis, TN 38152. Offers earth sciences (PhD); geography (MA, MS); geology (MS); geophysics (MS). Part-time programs available. *Faculty:* 23 full-time (3 women), 3 part-time/adjunct (1 woman). *Students:* Average age 29. 13 applicants, 46% accepted. In 2005, 5 master's, 1 doctorate awarded. Terminal master's awarded for partial completion of doctoral program. *Degree requirements:* For master's, thesis, seminar presentation, comprehensive exam; for doctorate, thesis/dissertation. *Entrance requirements:* For master's and doctorate, GRE General Test. Additional exam requirements/recommendations for international students: Required—TOEFL. *Application deadline:* For fall admission, 8/1 for domestic students; for spring admission, 12/1 for domestic students. Applications are processed on a rolling basis. Application fee: $25 ($50 for international students). Electronic applications accepted. *Financial support:* In 2005–06, 34 students received support, including 2 fellowships with full tuition reimbursements available, 10 research assistantships with full tuition reimbursements available, 11 teaching assistantships with full tuition reimbursements available (averaging $10,000 per year) Financial award application deadline: 1/15. *Faculty research:* Hazards, active tectonics, geophysics, hydrology and water resources, spatial analysis. Total annual research expenditures: $1.8 million. *Unit head:* Dr. Mervin J. Bartholomew, Chair, 901-678-1613, Fax: 901-678-4467, E-mail: jbrthlm1@memphis.edu. *Application contact:* Dr. George Swihart, Coordinator of Graduate Studies, 901-678-2606, Fax: 901-678-2178, E-mail: gswihart@memphis.edu.

University of Missouri–Kansas City, College of Arts and Sciences, Department of Geosciences, Kansas City, MO 64110-2499. Offers environmental and urban geosciences (MS); geosciences (PhD). PhD offered through the School of Graduate Studies. Part-time programs available. *Faculty:* 10 full-time (2 women), 4 part-time/adjunct (0 women). *Students:* 3 full-time (2 women), 4 part-time; includes 1 minority (African American), 2 international. Average age 32. 8 applicants, 75% accepted, 3 enrolled. *Degree requirements:* For master's and doctorate, thesis/dissertation. *Entrance requirements:* For master's, GRE General Test, minimum GPA of 3.0; for doctorate, qualifying exam. Additional exam requirements/recommendations for international students: Required—TOEFL. *Application deadline:* For fall admission, 3/15 priority date for domestic students, 3/15 priority date for international students. Applications are processed on a rolling basis. Application fee: $35 ($50 for international students). Electronic applications accepted. *Expenses:* Tuition, state resident: full-time $4,738; part-time $263 per credit hour. Tuition, nonresident: full-time $12,235; part-time $679 per credit hour. Required fees: $582. Tuition and fees vary according to course load, program and student level. *Financial support:* In 2005–06, research assistantships with partial tuition reimbursements (averaging $11,000 per year), teaching assistantships with partial tuition reimbursements (averaging $10,500 per year) were awarded; fellowships, Federal Work-Study, institutionally sponsored loans, and tuition waivers (full and partial) also available. Support available to part-time students. Financial award application deadline: 3/15. *Faculty research:* Neotectonics and applied geophysics, environmental geosciences, urban geoscience, geoinformatics-remote sensing, atmospheric research. Total annual research expenditures:$90,000. *Unit head:* Dr. Syed Hasan, Chair, 816-235-2976, Fax: 816-235-5535. *Application contact:* Dr. James Murowchick, Associate Professor, 816-235-2979, Fax: 816-235-5535, E-mail: murowchickj@umkc.edu.

The University of Montana, Graduate School, College of Arts and Sciences, Department of Geology, Missoula, MT 59812-0002. Offers applied geoscience (PhD); geology (MS, PhD). *Faculty:* 12 full-time (1 woman), 7 part-time/adjunct (1 woman). *Students:* 30 full-time (10 women), 17 part-time (7 women); includes 2 minority (1 American Indian/Alaska Native, 1 Asian American or Pacific Islander), 1 international. 41 applicants, 63% accepted, 11 enrolled. In 2005, 15 degrees awarded. *Degree requirements:* For doctorate, thesis/dissertation. *Entrance requirements:* For master's and doctorate, GRE General Test. Additional exam requirements/recommendations for international students: Required—TOEFL (minimum score 525 paper-based; 197 computer-based). *Application deadline:* For fall admission, 2/15 for domestic students. Application fee: $45. *Expenses:* Tuition, state resident: part-time $267 per credit. Tuition, nonresident: part-time $665 per credit. Part-time tuition and fees vary according to course load and degree level. *Financial support:* In 2005–06, 9 teaching assistantships with full tuition reimbursements (averaging $9,000 per year) were awarded; Federal Work-Study and unspecified assistantships also available. Financial award application deadline: 3/1; financial award applicants required to submit FAFSA. *Faculty research:* Environmental geoscience, regional structure and tectonics, groundwater geology, petrology, mineral deposits. Total annual research expenditures: $533,844. *Unit head:* Dr. Steven D. Sheriff, Chair, 406-243-6560, Fax: 406-243-4028, E-mail: gl_sds@selway.umt.edu. *Application contact:* Dr. Graham Thompson, Graduate Coordinator, 406-243-4953, E-mail: gl_grt@selway.umt.edu.

University of Nebraska–Lincoln, Graduate College, College of Arts and Sciences, Department of Geosciences, Lincoln, NE 68588. Offers MS, PhD. *Degree requirements:* For master's, departmental qualifying exam, thesis optional; for doctorate, thesis/dissertation, departmental qualifying exams, comprehensive exam. *Entrance requirements:* For master's and doctorate, GRE General Test. Additional exam requirements/recommendations for international students: Required—TOEFL (minimum score 550 paper-based; 213 computer-based). Electronic applications accepted. *Faculty research:* Hydrogeology, sedimentology, environmental geology, vertebrate paleontology.

University of Nevada, Las Vegas, Graduate College, College of Science, Department of Geoscience, Las Vegas, NV 89154-9900. Offers MS, PhD. Part-time programs available. *Faculty:* 21 full-time (6 women), 10 part-time/adjunct (2 women). *Students:* 24 full-time (14 women), 19 part-time (10 women); includes 5 minority (2 Asian Americans or Pacific Islanders, 3 Hispanic Americans), 3 international. 29 applicants, 48% accepted, 9 enrolled. In 2005, 8 degrees awarded. *Degree requirements:* For master's and doctorate, thesis/dissertation, comprehensive exam. *Entrance requirements:* For master's and doctorate, GRE General Test, minimum GPA of 3.0. Additional exam requirements/recommendations for international students: Required—TOEFL (minimum score 550 paper-based; 213 computer-based). *Application*

Geosciences

University of Nevada, Las Vegas (continued)

deadline: For fall admission, 2/1 for domestic students, 2/1 for international students; for spring admission, 1/1 for domestic students, 10/1 for international students. Application fee: $60 ($75 for international students). Electronic applications accepted. *Expenses:* Tuition, state resident: part-time $150 per credit. Tuition, nonresident: part-time $315 per credit. Tuition and fees vary according to course load, program and reciprocity agreements. *Financial support:* In 2005–06, 4 research assistantships with partial tuition reimbursements (averaging $11,000 per year), 14 teaching assistantships with partial tuition reimbursements (averaging $10,000 per year) were awarded; career-related internships or fieldwork, Federal Work-Study, institutionally sponsored loans, scholarships/grants, health care benefits, and unspecified assistantships also available. Support available to part-time students. Financial award application deadline: 3/1. *Unit head:* Dr. Wanda Taylor, Chair, 702-895-3262. *Application contact:* Graduate College Admissions Evaluator, 702-895-3320, Fax: 702-895-4180, E-mail: gradcollege@unlv.edu.

University of New Hampshire, Graduate School, College of Engineering and Physical Sciences, Department of Earth Sciences, Durham, NH 03824. Offers earth sciences (MS), including geochemical, geology, ocean mapping, oceanography; hydrology (MS). *Faculty:* 29 full-time. *Students:* 17 full-time (5 women), 24 part-time (10 women); includes 2 minority (1 African American, 1 Asian American or Pacific Islander), 6 international. Average age 31. 37 applicants, 84% accepted, 10 enrolled. In 2005, 10 master's awarded. *Degree requirements:* For master's, thesis. *Entrance requirements:* For master's, GRE General Test. Additional exam requirements/recommendations for international students: Required—TOEFL (minimum score 550 paper-based; 213 computer-based); Recommended—TSE. *Application deadline:* For fall admission, 4/1 priority date for domestic students, 4/1 priority date for international students. For winter admission, 12/1 for domestic students. Applications are processed on a rolling basis. Application fee: $60. Electronic applications accepted. *Expenses:* Tuition, state resident: full-time $8,010; part-time $445 per credit hour. Tuition, nonresident: full-time $19,730; part-time $810 per credit hour. Required fees: $322 per semester. Tuition and fees vary according to course load and program. *Financial support:* In 2005–06, 1 fellowship, 9 research assistantships, 9 teaching assistantships were awarded; career-related internships or fieldwork, Federal Work-Study, scholarships/grants, and tuition waivers (full and partial) also available. Support available to part-time students. Financial award application deadline: 2/15. *Unit head:* Dr. Matt Davis, Chairperson, 603-862-1718, E-mail: earth.sciences@unh.edu. *Application contact:* Nancy Gauthier, Administrative Assistant, 603-862-1720, E-mail: earth.sciences@unh.edu.

University of New Mexico, Graduate School, College of Arts and Sciences, Department of Earth and Planetary Sciences, Albuquerque, NM 87131-2039. Offers MS, PhD. *Faculty:* 20 full-time (5 women), 7 part-time/adjunct (2 women). *Students:* 39 full-time (26 women), 12 part-time (3 women); includes 3 minority (all Hispanic Americans), 3 international. Average age 29. 69 applicants, 20% accepted, 8 enrolled. In 2005, 17 master's, 2 doctorates awarded. *Degree requirements:* For master's and doctorate, thesis/dissertation, comprehensive exam, registration. *Entrance requirements:* For master's and doctorate, GRE General Test. Additional exam requirements/recommendations for international students: Required—TOEFL. *Application deadline:* For fall admission, 1/31 for domestic students; for spring admission, 11/1 priority date for domestic students. Electronic applications accepted. *Expenses:* Tuition, nonresident: full-time $3,388; part-time $238 per credit hour. Required fees: $385 per term. Tuition and fees vary according to course load and program. *Financial support:* In 2005–06, 3 fellowships with full tuition reimbursements (averaging $15,000 per year), 18 research assistantships with full tuition reimbursements (averaging $14,050 per year), 13 teaching assistantships with full tuition reimbursements (averaging $12,050 per year) were awarded; scholarships/grants and health care benefits also available. Financial award application deadline: 3/1; financial award applicants required to submit FAFSA. *Faculty research:* Geochemistry, meteoritics, tectonics, igcodynamics, climate and surface processes. Total annual research expenditures: $2.5 million. *Unit head:* Dr. Les D. McFadden, Chair, 505-277-4204, Fax: 505-277-8843, E-mail: lmcfadnm@unm.edu. *Application contact:* Cindy Jaramillo, Administrative Assistant II, 505-277-1635, Fax: 505-277-8843, E-mail: epsdept@unm.edu.

The University of North Carolina at Charlotte, Graduate School, College of Arts and Sciences, Department of Geography and Earth Sciences, Program in Earth Sciences, Charlotte, NC 28223-0001. Offers MS. Part-time and evening/weekend programs available. *Students:* 10 full-time (4 women), 10 part-time (7 women), 1 international. Average age 31. 7 applicants, 71% accepted, 4 enrolled. In 2005, 5 degrees awarded. *Degree requirements:* For master's, thesis optional. *Entrance requirements:* For master's, GRE General Test, minimum GPA of 3.0 in science major. Additional exam requirements/recommendations for international students: Required—TOEFL (minimum score 557 paper-based; 220 computer-based). *Application deadline:* For fall admission, 7/1 for domestic students, 5/1 for international students; for spring admission, 11/1 for domestic students, 10/1 for international students. Applications are processed on a rolling basis. Application fee: $55. Electronic applications accepted. *Expenses:* Tuition, state resident: full-time $2,504; part-time $157 per credit. Tuition, nonresident: full-time $12,711; part-time $794 per credit. Required fees: $1,424; $89 per credit. Tuition and fees vary according to course load and program. *Financial support:* In 2005–06, 7 research assistantships (averaging $5,487 per year), 9 teaching assistantships (averaging $6,648 per year) were awarded; fellowships Financial award application deadline: 4/1; financial award applicants required to submit FAFSA. *Faculty research:* Environmental geology; trace element geochemistry; geomorphology; hydrogeology; mineralogy and petrology. *Unit head:* Dr. John F. Bender, Coordinator, 704-687-4251, Fax: 704-687-3182, E-mail: jfbender@email.uncc.edu. *Application contact:* Kathy B. Giddings, Director of Graduate Admissions, 704-687-3366, Fax: 704-687-3279, E-mail: gradadm@email.uncc.edu.

The University of North Carolina Wilmington, College of Arts and Sciences, Department of Earth Sciences, Wilmington, NC 28403-3297. Offers geology (MS); marine science (MS). *Faculty:* 20 full-time (4 women), 1 (woman) part-time/adjunct. *Students:* 11 full-time (7 women), 57 part-time (38 women); includes 1 minority (Hispanic American), 1 international. Average age 35. 42 applicants, 48% accepted, 17 enrolled. In 2005, 26 degrees awarded. *Degree requirements:* For master's, thesis, comprehensive exam. *Entrance requirements:* For master's, GRE General Test, GRE Subject Test, minimum B average in undergraduate major and basic courses for prerequisite to geology. *Application deadline:* For fall admission, 2/15 for domestic students. Applications are processed on a rolling basis. Application fee: $45. *Financial support:* In 2005–06, 9 teaching assistantships were awarded; career-related internships or fieldwork and Federal Work-Study also available. Support available to part-time students. Financial award application deadline: 3/15. *Unit head:* Dr. Richard A. Laws, Chair, 910-962-4125, Fax: 910-962-7077. *Application contact:* Dr. Robert D. Roer, Dean, Graduate School, 910-962-4117, Fax: 910-962-3787, E-mail: roer@uncw.edu.

See Close-Up on page 237.

University of North Dakota, Graduate School, Program in Earth System Science and Policy, Grand Forks, ND 58202. Offers MEM, MS, PhD. *Faculty:* 7 full-time (1 woman). *Students:* 12 applicants, 50% accepted, 5 enrolled. *Entrance requirements:* For master's and doctorate, GRE General Test, minimum GPA of 3.0. Additional exam requirements/recommendations for international students: Required—TOEFL (minimum score 550 paper-based; 213 computer-based). *Application deadline:* For fall admission, 2/15 priority date for domestic students, 2/15 priority date for international students; for spring admission, 10/15 priority date for domestic students, 10/15 priority date for international students. Applications are processed on a rolling basis. Application fee: $35. Electronic applications accepted. *Financial support:* In 2005–06, 4 research assistantships with full tuition reimbursements (averaging $11,715 per year) were awarded *Unit head:* Dr. Rodney Hanley, Graduate Director, 701-777-3909, E-mail: rshanley@umac.org. *Application contact:* Linda Baeza, Admissions Officer, 701-777-2945, Fax: 701-777-3619, E-mail: gradschool@mail.und.nodak.edu.

University of Northern Colorado, Graduate School, College of Natural and Health Sciences, School of Chemistry, Earth Sciences and Physics, Program in Earth Sciences, Greeley, CO 80639. Offers MA. Part-time programs available. *Faculty:* 4 full-time (0 women). *Students:* 5 full-time (2 women), 4 part-time (2 women). Average age 35. 3 applicants, 67% accepted, 1 enrolled. In 2005, 7 degrees awarded. *Degree requirements:* For master's, comprehensive exam. *Entrance requirements:* For master's, GRE General Test, 3 letters of recommendation. *Application deadline:* Applications are processed on a rolling basis. Application fee: $50 ($60 for international students). Electronic applications accepted. *Expenses:* Tuition, state resident: full-time $4,968; part-time $207 per credit hour. Tuition, nonresident: full-time $14,688; part-time $612 per credit hour. Required fees: $645; $32 per credit hour. *Financial support:* In 2005–06, 2 research assistantships (averaging $8,266 per year), 4 teaching assistantships (averaging $14,603 per year) were awarded; fellowships, unspecified assistantships also available. Financial award application deadline: 3/1; financial award applicants required to submit FAFSA. *Unit head:* Dr. William Nesse, Program Coordinator, 970-351-2647.

University of Notre Dame, Graduate School, College of Engineering, Department of Civil Engineering and Geological Sciences, Notre Dame, IN 46556. Offers bioengineering (MS Bio E); civil engineering (MSCE); civil engineering and geological sciences (PhD); environmental engineering (MS Env E); geological sciences (MS). *Faculty:* 14 full-time (3 women), 1 part-time/adjunct (0 women). *Students:* 48 full-time (24 women), 2 part-time (1 woman); includes 4 minority (2 African Americans, 1 Asian American or Pacific Islander, 1 Hispanic American), 8 international. 58 applicants, 26% accepted, 9 enrolled. In 2005, 5 master's, 8 doctorates awarded. Terminal master's awarded for partial completion of doctoral program. *Median time to degree:* Of those who began their doctoral program in fall 1997, 54% received their degree in 8 years or less. *Degree requirements:* For master's, comprehensive exam; for doctorate, thesis/dissertation. *Entrance requirements:* For master's and doctorate, GRE General Test. Additional exam requirements/recommendations for international students: Required—TOEFL. *Application deadline:* For fall admission, 2/1 for domestic students; for spring admission, 10/15 for domestic students. Applications are processed on a rolling basis. Application fee: $50. Electronic applications accepted. *Financial support:* In 2005–06, 8 fellowships with full tuition reimbursements (averaging $22,000 per year), 21 research assistantships with full tuition reimbursements (averaging $15,250 per year), 16 teaching assistantships with full tuition reimbursements (averaging $16,000 per year) were awarded; tuition waivers (full) also available. Financial award application deadline:2/1. *Faculty research:* Environmental modeling, biological-waste treatment, petrology, environmental geology, geochemistry. Total annual research expenditures: $2.7 million. *Unit head:* Dr. Yahya C. Kurama, Director of Graduate Studies, 574-631-8227, Fax: 574-631-9236, E-mail: cegeos@nd.edu. *Application contact:* Dr. Terrence J. Akai, Director of Graduate Admissions, 574-631-7706, Fax: 574-631-4183, E-mail: gradad@nd.edu.

University of Ottawa, Faculty of Graduate and Postdoctoral Studies, Faculty of Science, Ottawa-Carleton Geoscience Centre, Ottawa, ON K1N 6N5, Canada. Offers earth sciences (M Sc, PhD). *Faculty:* 27 full-time (12 women), 4 part-time/adjunct (2 women). *Students:* 33 full-time, 5 part-time. 21 applicants, 81% accepted, 12 enrolled. In 2005, 4 master's, 3 doctorates awarded. *Degree requirements:* For master's, thesis/dissertation, seminar; for doctorate, thesis/dissertation, seminar, comprehensive exam. *Entrance requirements:* For master's, honors B Sc degree or equivalent, minimum B average; for doctorate, honors B Sc with minimum B average or M Sc with minimum B+ average. *Application deadline:* For fall admission, 3/1 for domestic students, 2/15 for international students. For winter admission, 11/15 for domestic students; for spring admission, 4/1 for domestic students. Application fee: $75. Electronic applications accepted. *Expenses:* Tuition: Part-time $260 per credit. Tuition and fees vary according to course load and program. *Financial support:* Fellowships, research assistantships with full tuition reimbursements, teaching assistantships with full tuition reimbursements, career-related internships or fieldwork, Federal Work-Study, scholarships/grants, traineeships, tuition waivers (full and partial), and unspecified assistantships available. Financial award application deadline: 2/15. *Faculty research:* Environmental geoscience, geochemistry/petrology, geomatics/geomathematics, mineral resource studies. *Unit head:* Dr. Andre Desrochers, Chair, 613-562-5800 Ext. 6854, Fax: 613-562-5192, E-mail: estchair@science.uottawa.ca. *Application contact:* Lise Maisonneuve, Graduate Studies Administrator, 613-562-5800 Ext. 6335, Fax: 613-562-5486, E-mail: lise@science.uottawa.ca.

University of Rhode Island, Graduate School, College of the Environment and Life Sciences, Department of Geosciences, Kingston, RI 02881. Offers MS. *Degree requirements:* For master's, thesis optional. *Application deadline:* For fall admission, 4/15 for domestic students. Applications are processed on a rolling basis. Application fee: $35. *Expenses:* Tuition, state resident: full-time $5,522; part-time $307 per credit. Tuition, nonresident: full-time $15,992; part-time $888 per credit. Required fees: $1,786; $73 per credit. One-time fee: $80 part-time. *Unit head:* Dr. Daniel Murray, Chairperson, 401-874-2197.

University of Rochester, The College, Arts and Sciences, Department of Earth and Environmental Sciences, Rochester, NY 14627-0250. Offers geological sciences (MS, PhD). *Degree requirements:* For doctorate, thesis/dissertation, qualifying exam. *Entrance requirements:* For master's and doctorate, GRE General Test. Additional exam requirements/recommendations for international students: Required—TOEFL.

See Close-Up on page 239.

University of San Diego, Hahn School of Nursing and Health Sciences, San Diego, CA 92110-2492. Offers adult nurse practitioner (MSN, Post Master's Certificate); case management for vulnerable populations (MSN); family nurse practitioner (MSN, Post Master's Certificate); health care systems (MSN); nursing science (PhD); pediatric nurse practitioner (MSN, Post Master's Certificate). *Accreditation:* AACN. Part-time and evening/weekend programs available. *Faculty:* 18 full-time (15 women), 16 part-time/adjunct (15 women). *Students:* 94 full-time (82 women), 126 part-time (116 women); includes 38 minority (8 African Americans, 1 American Indian/Alaska Native, 19 Asian Americans or Pacific Islanders, 10 Hispanic Americans), 5 international. Average age 38. 259 applicants, 51% accepted, 85 enrolled. In 2005, 23 master's, 9 doctorates awarded. *Degree requirements:* For doctorate, thesis/dissertation, administrative residency. *Entrance requirements:* For master's, GRE General Test, BSN (required for all except master's entry program), minimum GPA of 3.0, current California RN licensure; for doctorate, GRE General Test, minimum GPA of 3.5, MSN, current California RN licensure. Additional exam requirements/recommendations for international students: Required—TOEFL, TWE. *Application deadline:* For fall admission, 5/1 for domestic students; for spring admission, 11/1 priority date for domestic students. Applications are processed on a rolling basis. Application fee: $45. Electronic applications accepted. *Financial support:* In 2005–06, 27 fellowships with partial tuition reimbursements (averaging $4,831 per year) were awarded; institutionally sponsored loans, scholarships/grants, traineeships, tuition waivers (partial), and graduate work program also available. Support available to part-time students. Financial award application deadline: 5/1; financial award applicants required to submit FAFSA. *Faculty research:* Health promotion, decision making, psychogeriatric nursing, historical nursing, leadership behavior. *Unit head:* Dr. Sally Hardin, Dean, 619-260-4550, Fax: 619-260-6814. *Application contact:* Stephen Pultz, Director of Admissions, 619-260-4524, Fax: 619-260-4158, E-mail: grads@sandiego.edu.

University of South Carolina, The Graduate School, College of Arts and Sciences, Department of Geological Sciences, Columbia, SC 29208. Offers environmental geoscience (PMS); geological sciences (MS, PhD). Terminal master's awarded for partial completion of doctoral program. *Degree requirements:* For master's, thesis; for doctorate, thesis/dissertation, published paper, comprehensive exam. *Entrance requirements:* For master's and doctorate, GRE General Test. Additional exam requirements/recommendations for international students: Required—TOEFL. Electronic applications accepted. *Faculty research:* Environmental geology, tectonics, petrology, coastal processes, paleoclimatology.

University of Southern California, Graduate School, College of Letters, Arts and Sciences, Department of Earth Sciences, Los Angeles, CA 90089. Offers MS, PhD. *Degree requirements:* For master's and doctorate, thesis/dissertation. *Entrance requirements:* For master's and doctorate, GRE General Test. *Expenses:* Tuition: Full-time $25,416; part-time $1,059 per unit. Required fees: $484; $484 per year. Tuition and fees vary according to course load and program.

The University of Texas at Arlington, Graduate School, College of Science, Department of Geology, Arlington, TX 76019. Offers environmental science (MS, PhD); geology (MS); math: geoscience (PhD). Part-time and evening/weekend programs available. *Faculty:* 4 full-time (0 women), 2 part-time/adjunct (0 women). *Students:* 5 full-time (2 women), 7 part-time (4 women); includes 1 minority (African American), 2 international. 5 applicants, 100% accepted, 2 enrolled. In 2005, 2 master's awarded. Terminal master's awarded for partial completion of doctoral program. *Degree requirements:* For master's, thesis optional; for doctorate, thesis/dissertation, comprehensive exam. *Entrance requirements:* For master's, GRE General Test. Additional exam requirements/recommendations for international students: Required—TOEFL (minimum score 550 paper-based; 213 computer-based). *Application deadline:* For fall admission, 6/16 for domestic students. Applications are processed on a rolling basis. Application fee: $35 ($50 for international students). Electronic applications accepted. *Expenses:* Tuition, state resident: full-time $3,350. Tuition, nonresident: full-time $8,318. International tuition: $8,448 full-time. Required fees: $1,277. Full-time tuition and fees vary according to course level and program. *Financial support:* In 2005–06, 7 students received support, including 4 fellowships (averaging $1,000 per year), 7 teaching assistantships (averaging $14,700 per year); career-related internships or fieldwork, Federal Work-Study, institutionally sponsored loans, scholarships/grants, health care benefits, and unspecified assistantships also available. Financial award application deadline: 6/1; financial award applicants required to submit FAFSA. *Faculty research:* Hydrology, aqueous geochemistry, biostratigraphy, structural geology, petroleum geology. Total annual research expenditures: $250,000. *Unit head:* Dr. John S. Wickham, Chair, 817-272-2987, Fax: 817-272-2628, E-mail: wickham@uta.edu. *Application contact:* Dr. William L. Balsam, Graduate Adviser, 817-272-2987, Fax: 817-272-2628, E-mail: balsam@uta.edu.

The University of Texas at Arlington, Graduate School, College of Science, Program in Environmental and Earth Sciences, Arlington, TX 76019. Offers MS, PhD. *Students:* 2 full-time (1 woman), 1 (woman) part-time. 4 applicants, 100% accepted, 1 enrolled. In 2005, 2 master's awarded. *Expenses:* Tuition, state resident: full-time $3,350. Tuition, nonresident: full-time $8,318. International tuition: $8,448 full-time. Required fees: $1,277. Full-time tuition and fees vary according to course level and program. *Application contact:* Dr. Robert F. McMahon, Director, 817-272-3492, Fax: 817-272-3511, E-mail: r.mcmahon@uta.edu.

The University of Texas at Austin, Graduate School, College of Natural Sciences, Department of Geological Sciences, Austin, TX 78712-1111. Offers MA, MS, PhD. Part-time programs available. *Degree requirements:* For master's, report (MA), thesis (MS); for doctorate, thesis/dissertation. *Entrance requirements:* For master's and doctorate, GRE General Test. Electronic applications accepted. *Faculty research:* Sedimentary geology, geophysics, hydrogeology, structure/tectonics, vertebrate paleontology.

The University of Texas at Dallas, School of Natural Sciences and Mathematics, Program in Geosciences, Richardson, TX 75083-0688. Offers MS, PhD. Part-time and evening/weekend programs available. *Faculty:* 11 full-time (0 women), 1 part-time/adjunct (0 women). *Students:* 30 full-time (8 women), 21 part-time (11 women); includes 4 minority (2 African Americans, 1 American Indian/Alaska Native, 1 Asian American or Pacific Islander), 25 international. Average age 33. 40 applicants, 50% accepted, 9 enrolled. In 2005, 4 master's, 2 doctorates awarded. *Degree requirements:* For master's, thesis optional; for doctorate, thesis/dissertation. *Entrance requirements:* For master's and doctorate, GRE General Test, minimum GPA of 3.0 in upper-level course work in field. Additional exam requirements/recommendations for international students: Required—TOEFL (minimum score 550 paper-based; 213 computer-based). *Application deadline:* For fall admission, 7/15 for domestic students; for spring admission, 11/15 for domestic students. Applications are processed on a rolling basis. Application fee: $50 ($100 for international students). Electronic applications accepted. *Expenses:* Tuition, state resident: full-time $5,450; part-time $303 per credit. Tuition, nonresident: full-time $12,648; part-time $703 per credit. Tuition and fees vary according to program. *Financial support:* In 2005–06, 12 research assistantships with tuition reimbursements (averaging $10,235 per year), 11 teaching assistantships with tuition reimbursements (averaging $9,760 per year) were awarded; fellowships, career-related internships or fieldwork, Federal Work-Study, institutionally sponsored loans, and scholarships/grants also available. Support available to part-time students. Financial award application deadline: 4/30; financial award applicants required to submit FAFSA. *Faculty research:* Hydrology, organic geochemistry, tectonic structures, seismic characteristics, digital geologic mapping. Total annual research expenditures: $867,780. *Unit head:* Dr. Robert Stern, Head, 972-883-2401, Fax: 972-883-2537. *Application contact:* Dr. Carlos Aiken, Graduate Adviser, 972-883-2450, Fax: 972-883-2537, E-mail: aiken@utdallas.edu.

The University of Toledo, Graduate School, College of Arts and Sciences, Department of Earth, Ecological and Environmental Sciences, Toledo, OH 43606-3390. Offers biology (ecology track) (MS, PhD); geology (MS), including earth surface processes, general geology. Part-time programs available. *Faculty:* 28. *Students:* 8 full-time (6 women), 2 part-time (both women). Average age 31. 6 applicants, 83% accepted, 3 enrolled. In 2005, 2 degrees awarded. *Degree requirements:* For master's, thesis. *Entrance requirements:* For master's, GRE General Test. Additional exam requirements/recommendations for international students: Required—TOEFL. *Application deadline:* For fall admission, 8/1 for domestic students. Applications are processed on a rolling basis. Application fee: $45. Electronic applications accepted. *Expenses:* Tuition, area resident: Part-time $308 per credit hour. Tuition, state resident: full-time $3,312. Tuition, nonresident: full-time $6,616; part-time $735 per credit hour. *Financial support:* In 2005–06, 2 research assistantships (averaging $14,000 per year), 30 teaching assistantships (averaging $11,724 per year) were awarded; Federal Work-Study, institutionally sponsored loans, and tuition waivers (full) also available. Support available to part-time students. Financial award application deadline: 4/1; financial award applicants required to submit FAFSA. *Faculty research:* Environmental geochemistry, geophysics, petrology and mineralogy, paleontology, geohydrology. *Unit head:* Dr. Michael Phillips, Chair, 419-530-4572, Fax: 419-530-4421, E-mail: michael.phillips@utoledo.edu. *Application contact:* Johan Gottgens, 419-530-8451, E-mail: john.gottgens@utoledo.edu.

University of Tulsa, Graduate School, College of Engineering and Natural Sciences, Department of Geosciences, Tulsa, OK 74104-3189. Offers MS, PhD. Part-time programs available. *Faculty:* 7 full-time (1 woman). *Students:* 9 full-time (3 women), 3 part-time (2 women), 7 international. Average age 37. 9 applicants, 44% accepted, 2 enrolled. In 2005, 5 master's awarded. *Degree requirements:* For master's, thesis (for some programs); for doctorate, thesis/dissertation, comprehensive exam. *Entrance requirements:* For master's and doctorate, GRE General Test. Additional exam requirements/recommendations for international students: Required—TOEFL (minimum score 550 paper-based; 213 computer-based), IELT (minimum score 6). *Application deadline:* Applications are processed on a rolling basis. Application fee: $30. Electronic applications accepted. *Expenses:* Tuition: Full-time $12,132; part-time $674 per credit hour. Required fees: $60; $3 per credit hour. *Financial support:* In 2005–06, 5 students received support, including 6 teaching assistantships with full and partial tuition reimbursements available (averaging $11,110 per year); fellowships with full and partial tuition reimbursements available, research assistantships with full and partial tuition reimbursements available, career-related internships or fieldwork, scholarships/grants, tuition waivers (full and partial), and unspecified assistantships also available. Support available to part-time students. Financial award application deadline: 2/1; financial award applicants required to submit FAFSA. *Faculty research:* Petroleum geology, carbonate and marine geology, exploration geophysics, structural geology. Total annual research expenditures: $34,952. *Unit head:* Dr. Bryan Tapp, Chairperson, 918-631-3018, Fax: 918-631-2091, E-mail: jbt@utulsa.edu. *Application contact:* Dr. Peter J. Michael, Adviser, 918-631-3017, Fax: 918-631-2156, E-mail: pjm@utulsa.edu.

University of Victoria, Faculty of Graduate Studies, Faculty of Science, School of Earth and Ocean Sciences, Victoria, BC V8W 2Y2, Canada. Offers M Sc, PhD. Part-time programs available. *Faculty:* 15 full-time, 41 part-time/adjunct. *Students:* 70, 18 international. Average age 24. 137 applicants, 66% accepted, 15 enrolled. In 2005, 5 master's, 4 doctorates awarded. *Degree requirements:* For master's, thesis, registration; for doctorate, thesis/dissertation, Candidacy exam. *Entrance requirements:* For master's and doctorate, GRE (75th percentile). Additional exam requirements/recommendations for international students: Required—TOEFL (minimum score 575 paper-based; 233 computer-based), IELT (minimum score 7). *Application deadline:* For fall admission, 2/15 priority date for domestic students, 12/15 priority date for international students. Applications are processed on a rolling basis. Application fee: $75 ($125 for international students). Electronic applications accepted. Tuition and fees charges are reported in Canadian dollars. *Expenses:* Tuition, area resident: Full-time $4,492 Canadian dollars; part-time $749 Canadian dollars per term. International tuition: $5,346 Canadian dollars full-time. Required fees: $4,492 Canadian dollars; $749 Canadian dollars per term. Tuition and fees vary according to course load, campus/location and program. *Financial support:* In 2005–06, 16 fellowships, 22 research assistantships, 25 teaching assistantships were awarded; career-related internships or fieldwork, institutionally sponsored loans, and awards also available. Financial award application deadline: 2/15. *Faculty research:* Climate modeling, geology. *Unit head:* Dr. Kathryn Gillis, Director, 250-472-4023, Fax: 250-721-6200, E-mail: kgillis@uvic.ca. *Application contact:* Dr. Andrew Weaver, Graduate Adviser, 250-472-4006, Fax: 250-721-6200, E-mail: weaver@ocean.seas.uvic.ca.

University of Waterloo, Graduate Studies, Faculty of Science, Department of Earth Sciences, Waterloo, ON N2L 3G1, Canada. Offers M Sc, PhD. Part-time programs available. *Faculty:* 42 full-time (7 women), 32 part-time/adjunct (2 women). *Students:* 79 full-time (37 women), 13 part-time (4 women). 53 applicants, 47% accepted, 21 enrolled. In 2005, 17 master's, 3 doctorates awarded. *Degree requirements:* For master's, research paper or thesis; for doctorate, thesis/dissertation, comprehensive exam, registration. *Entrance requirements:* For master's, GRE, honors degree, minimum B average; for doctorate, GRE, master's degree, minimum B average. Additional exam requirements/recommendations for international students: Required—TOEFL, TWE. *Application deadline:* For fall admission, 8/1 for domestic students. Applications are processed on a rolling basis. Application fee: $75 Canadian dollars. Electronic applications accepted. *Financial support:* Research assistantships, teaching assistantships, career-related internships or fieldwork and institutionally sponsored loans available. *Faculty research:* Environmental geology, soil physics. *Unit head:* Dr. J. F. Barker, Chair, 519-888-4567 Ext. 2103, Fax: 519-746-7484, E-mail: jfbarker@uwaterloo.ca. *Application contact:* S. Fisher, Administrative Graduate Coordinator, 519-888-4567 Ext. 5836, Fax: 519-746-7484, E-mail: sfisher@sciborg.uwaterloo.ca.

The University of Western Ontario, Faculty of Graduate Studies, Physical Sciences Division, Department of Earth Sciences, London, ON N6A 5B8, Canada. Offers geology (M Sc, PhD); geology and environmental science (M Sc, PhD); geophysics (M Sc, PhD); geophysics and environmental science (M Sc, PhD). *Degree requirements:* For master's, thesis, registration; for doctorate, thesis/dissertation, qualifying exam. *Entrance requirements:* For master's, honors in B Sc; for doctorate, M Sc. Additional exam requirements/recommendations for international students: Required—TOEFL. *Faculty research:* Geophysics, geochemistry, paleontology, sedimentology/stratigraphy, glaciology/quaternary.

University of Windsor, Faculty of Graduate Studies and Research, Faculty of Science, Department of Earth Sciences, Windsor, ON N9B 3P4, Canada. Offers M Sc, PhD. Part-time programs available. *Faculty:* 12 full-time (1 woman), 2 part-time/adjunct (0 women). *Students:* 18 full-time (4 women), 4 part-time (3 women). 22 applicants, 27% accepted. In 2005, 4 degrees awarded. *Degree requirements:* For master's, thesis/dissertation; for doctorate, thesis/dissertation, comprehensive exam. *Entrance requirements:* For master's, minimum B average; for doctorate, minimum B average, copies of publication abstract. Additional exam requirements/recommendations for international students: Required—TOEFL (minimum score 560 paper-based; 220 computer-based). *Application deadline:* For fall admission, 7/1 for domestic students. For winter admission, 11/1 for domestic students; for spring admission, 3/1 for domestic students. Applications are processed on a rolling basis. Application fee: $55. *Financial support:* In 2005–06, 17 teaching assistantships (averaging $8,456 per year) were awarded; Federal Work-Study, scholarships/grants, tuition waivers (full and partial), unspecified assistantships, and bursaries also available. Financial award application deadline: 2/15. *Faculty research:* Aqueous geochemistry and hydrothermal processes, igneous petrochemistry, radiogenic isotopes, radiometric age-dating, diagenetic and sedimentary geochemistry. *Unit head:* Dr. Ihsan Al-Aasm, Head, 519-253-3000 Ext. 2494, Fax: 519-971-7081, E-mail: earth@uwindsor.ca. *Application contact:* Applicant Services, 519-253-3000 Ext. 6459, Fax: 519-971-3653, E-mail: gradadmit@uwindsor.ca.

Virginia Polytechnic Institute and State University, Graduate School, College of Science, Department of Geosciences, Blacksburg, VA 24061. Offers geological sciences (MS, PhD); geophysics (MS, PhD). *Faculty:* 18 full-time (2 women), 1 (woman) part-time/adjunct. *Students:* 55 full-time (20 women), 2 part-time (1 woman); includes 2 minority (1 American Indian/Alaska Native, 1 Asian American or Pacific Islander), 18 international. Average age 27. 55 applicants, 36% accepted, 11 enrolled. In 2005, 9 master's, 1 doctorate awarded. *Entrance requirements:* For master's and doctorate, GRE General Test. Additional exam requirements/recommendations for international students: Required—TOEFL (minimum score 550 paper-based; 213 computer-based). *Application deadline:* Applications are processed on a rolling basis. Application fee: $45. Electronic applications accepted. *Expenses:* Tuition, state resident: full-time $6,558; part-time $364 per credit. Tuition, nonresident: full-time $11,296; part-time $628 per credit. Required fees: $1,419; $468 per credit. $234 per term. *Financial support:* In 2005–06, 8 fellowships with full tuition reimbursements (averaging $15,121 per year), 24 research assistantships with full tuition reimbursements (averaging $16,689 per year), 27 teaching assistantships with full tuition reimbursements (averaging $13,902 per year) were awarded; career-related internships or fieldwork, Federal Work-Study, scholarships/grants, tuition waivers (full), and unspecified assistantships also available. Financial award application deadline: 4/1. *Faculty research:* Paleontology/geobiology, active tectonics, geomorphology, mineralogy/crystallography, mineral physics. *Unit head:* Dr. Robert Tracy, Head, 540-231-6521, Fax: 540-231-3386. *Application contact:* Connie Lowe, Student Program Coordinator, 540-231-8824, Fax: 540-231-3386, E-mail: clowe@vt.edu.

Washington State University Tri-Cities, Graduate Programs, Program in Environmental Science, Richland, WA 99352-1671. Offers applied environmental science (MS); atmospheric science (MS); earth science (MS); environmental and occupational health science (MS); environmental regulatory compliance (MS); environmental science (PhD); environmental toxicology and risk assessment (MS); water resource science (MS). Part-time programs available. *Faculty:* 1 full-time (0 women), 53 part-time/adjunct. *Students:* 4 full-time (3 women), 22 part-time (10 women); includes 1 Asian American or Pacific Islander, 2 Hispanic Americans. Average age 41. 11 applicants, 55% accepted, 6 enrolled. In 2005, 1 degree awarded. *Degree requirements:* For master's, oral exam, thesis optional. *Entrance requirements:* For master's, GRE General Test, minimum GPA of 3.0, 3 letters of recommendation. Additional exam requirements/recommendations for international students: Required—TOEFL (minimum score 550 paper-based; 213 computer-based). *Application deadline:* For fall admission, 2/1 priority date for domestic students, 3/1 priority date for international students; for spring admission, 9/1 priority date for domestic students, 7/1 priority date for international students. Application fee: $35. *Expenses:* Tuition, state resident: full-time $6,295; part-time $336 per credit. Tuition, nonresident: full-time $15,949; part-time $819 per credit. Required fees: $429. Full-time tuition and fees vary according to campus/location and program. Part-time tuition and fees vary according to course load and program. *Financial support:* In 2005–06, 8 students received support, including 1 fellowship (averaging $2,200 per year); research assistantships with full and partial tuition reimbursements available, teaching assistantships with full and partial tuition reimbursements available, Federal Work-Study, scholarships/grants, health care benefits, and unspecified assistantships also available. *Faculty research:* Radiation ecology, cytogenetics. *Unit head:* Dr. Gene Schreckhise, Associate Dean/Coordinator, 509-372-7323, E-mail: gschreck@wsu.edu.

Geosciences

Washington University in St. Louis, Graduate School of Arts and Sciences, Department of Earth and Planetary Sciences, St. Louis, MO 63130-4899. Offers earth and planetary sciences (MA); geochemistry (PhD); geology (MA, PhD); geophysics (PhD); planetary sciences (PhD). Terminal master's awarded for partial completion of doctoral program. *Degree requirements:* For master's and doctorate, thesis/dissertation. *Entrance requirements:* For master's and doctorate, GRE General Test. Electronic applications accepted.

Wesleyan University, Graduate Programs, Department of Earth Sciences, Middletown, CT 06459-0260. Offers MA. *Faculty:* 8 full-time (3 women). *Students:* 6 full-time (5 women), 1 international. Average age 25. In 2005, 4 degrees awarded. *Degree requirements:* For master's, thesis. *Entrance requirements:* For master's, GRE General Test, GRE Subject Test. *Application deadline:* For fall admission, 3/1 for domestic students. Applications are processed on a rolling basis. Application fee: $0. *Expenses:* Tuition: Full-time $24,732. One-time fee: $20 full-time. *Financial support:* Teaching assistantships, tuition waivers (partial) available. *Faculty research:* Tectonics, volcanology, stratigraphy, coastal processes, geochemistry. *Unit head:* Dr. Johan C. Vasekamp, Chair, 860-685-2248. *Application contact:* Ginny Harris, Administrative Assistant, 860-685-2244, E-mail: vharris@wesleyan.edu.

Western Connecticut State University, Division of Graduate Studies, School of Arts and Sciences, Department of Physics, Astronomy and Meteorology, Danbury, CT 06810-6885. Offers earth and planetary sciences (MA). Part-time and evening/weekend programs available. *Degree requirements:* For master's, thesis. *Entrance requirements:* For master's, minimum GPA of 2.5.

Western Michigan University, Graduate College, College of Arts and Sciences, Department of Geosciences, Program in Earth Science, Kalamazoo, MI 49008-5202. Offers MS. *Degree requirements:* For master's, thesis or alternative, oral exam. *Entrance requirements:* For master's, GRE General Test.

Yale University, Graduate School of Arts and Sciences, Department of Geology and Geophysics, New Haven, CT 06520. Offers geochemistry (PhD); geophysics (PhD); meteorology (PhD); mineralogy and crystallography (PhD); oceanography (PhD); paleoecology (PhD); paleontology and stratigraphy (PhD); petrology (PhD); structural geology (PhD). *Degree requirements:* For doctorate, thesis/dissertation. *Entrance requirements:* For doctorate, GRE General Test. Additional exam requirements/recommendations for international students: Required—TOEFL.

York University, Faculty of Graduate Studies, Faculty of Pure and Applied Science, Program in Earth and Space Science, Toronto, ON M3J 1P3, Canada. Offers M Sc, PhD. Part-time and evening/weekend programs available. *Faculty:* 46 full-time (4 women), 25 part-time/adjunct (3 women). *Students:* 51 full-time (15 women), 15 part-time (1 woman). 41 applicants. In 2005, 10 master's, 1 doctorate awarded. *Degree requirements:* For master's, thesis or alternative, registration; for doctorate, thesis/dissertation, registration. *Application deadline:* Applications are processed on a rolling basis. Application fee: $80. Electronic applications accepted. *Expenses:* Tuition, state resident: full-time $3,190; part-time $798 per term. International tuition: $7,515 full-time. Required fees: $217. Tuition and fees vary according to program. *Financial support:* In 2005–06, fellowships (averaging $3,060 per year), research assistantships (averaging $10,259 per year), teaching assistantships (averaging $7,493 per year) were awarded; fee bursaries also available. *Unit head:* Peter Taylor, Director, 416-736-5247.

Hydrogeology

California State University, Chico, Graduate School, College of Natural Sciences, Department of Geological and Environmental Sciences, Program in Geosciences, Chico, CA 95929-0722. Offers hydrology/hydrogeology (MS). *Degree requirements:* For master's, thesis, oral exam. *Entrance requirements:* For master's, GRE General Test. Additional exam requirements/recommendations for international students: Required—TOEFL (minimum score 550 paper-based; 213 computer-based). Electronic applications accepted.

Clemson University, Graduate School, College of Engineering and Science, Department of Geological Sciences, Program in Hydrogeology, Clemson, SC 29634. Offers MS. *Students:* 12 full-time (5 women), 3 part-time. 12 applicants, 75% accepted, 4 enrolled. In 2005, 9 degrees awarded. *Degree requirements:* For master's, thesis optional. *Entrance requirements:* For master's, GRE General Test, minimum GPA of 3.0 during previous 2 years. Additional exam requirements/recommendations for international students: Required—TOEFL. *Application deadline:* For fall admission, 6/1 for domestic students, 4/15 for international students. Application fee: $50. Electronic applications accepted. *Financial support:* Fellowships, research assistantships, teaching assistantships, career-related internships or fieldwork and institutionally sponsored loans available. Support available to part-time students. Financial award application deadline: 6/1; financial award applicants required to submit FAFSA. *Faculty research:* Groundwater, geology, environmental geology, geochemistry, remediation, stratigraphy. Total annual research expenditures: $670,000. *Unit head:* Dr. Jim Castle, Coordinator, 864-656-5015, E-mail: jcastle@clemson.edu.

See Close-Up on page 225.

Colorado School of Mines, Graduate School, Department of Geology and Geological Engineering, Golden, CO 80401-1887. Offers engineering geology (Diploma); exploration geosciences (Diploma); geochemistry (MS, PhD); geological engineering (ME, MS, PhD, Diploma); geology (MS, PhD); hydrogeology (Diploma). Part-time programs available. *Faculty:* 28 full-time (7 women), 15 part-time/adjunct (4 women). *Students:* 72 full-time (26 women), 29 part-time (5 women); includes 3 minority (1 American Indian/Alaska Native, 2 Hispanic Americans), 25 international. 108 applicants, 60% accepted, 24 enrolled. In 2005, 18 master's, 4 doctorates awarded. *Degree requirements:* For master's, thesis/dissertation; for doctorate, thesis/dissertation, comprehensive exam. *Entrance requirements:* For master's and doctorate, GRE General Test; for Diploma, GRE General Test, minimum GPA of 3.0. Additional exam requirements/recommendations for international students: Required—TOEFL (minimum score 550 paper-based; 213 computer-based). *Application deadline:* For fall admission, 1/1 for domestic students, 1/1 for international students; for spring admission, 9/1 for domestic students, 9/1 for international students. Application fee: $50. Electronic applications accepted. *Expenses:* Tuition, state resident: full-time $7,240; part-time $362 per credit hour. Tuition, nonresident: full-time $19,840; part-time $992 per credit hour. Required fees: $895. *Financial support:* In 2005–06, 4 fellowships with full tuition reimbursements (averaging $9,600 per year), 14 research assistantships with full tuition reimbursements (averaging $9,600 per year), 11 teaching assistantships with full tuition reimbursements (averaging $9,600 per year) were awarded; scholarships/grants, health care benefits, and unspecified assistantships also available. Financial award applicants required to submit FAFSA. *Faculty research:* Predictive sediment modeling, petrophysics, aquifer-contaminant flow modeling, water-rock interactions, geotechnical engineering. Total annual research expenditures: $2.1 million. *Unit head:* Dr. Murray W. Hitzman, Head, 303-384-2127, Fax: 303-273-3859, E-mail: mhitzman@mines.edu. *Application contact:* Marilyn Schwinger, Administrative Assistant, 303-273-3800, Fax: 303-273-3859, E-mail: mschwing@mines.edu.

Colorado State University, Graduate School, Warner College of Natural Resources, Department of Geosciences, Fort Collins, CO 80523-0015. Offers earth sciences (PhD); geology (MS), including geomorphology, geophysics, hydrogeology, petrology/geochemistry and economic geology, sedimentology, structural geology; geosciences (PhD); wateshed (PhD). Part-time programs available. *Faculty:* 8 full-time (3 women). *Students:* 17 full-time (11 women), 22 part-time (7 women); includes 1 minority (American Indian/Alaska Native), 2 international. Average age 32. 28 applicants, 61% accepted, 7 enrolled. In 2005, 3 master's, 1 doctorate awarded. *Degree requirements:* For master's, thesis/dissertation, registration; for doctorate, thesis/dissertation, comprehensive exam, registration. *Entrance requirements:* For master's and doctorate, GRE General Test, minimum GPA of 3.0. Additional exam requirements/recommendations for international students: Required—TOEFL (minimum score 550 paper-based; 213 computer-based). *Application deadline:* For fall admission, 2/1 priority date for domestic students, 2/1 priority date for international students. Applications are processed on a rolling basis. Application fee: $50. Electronic applications accepted. *Expenses:* Tuition, state resident: full-time $3,690; part-time $205 per credit. Tuition, nonresident: full-time $14,958; part-time $831 per credit. Required fees: $1,061. *Financial support:* In 2005–06, 6 fellowships (averaging $9,000 per year), 13 research assistantships with partial tuition reimbursements (averaging $16,000 per year), 9 teaching assistantships with full tuition reimbursements (averaging $16,300 per year) were awarded; career-related internships or fieldwork, Federal Work-Study, institutionally sponsored loans, scholarships/grants, and traineeships also available. Financial award application deadline: 2/15. *Faculty research:* Snow, surface, and groundwater hydrology; fluvial geomorphology; geographic information systems; geochemistry; bedrock geology. Total annual research expenditures: $1.3 million. *Unit head:* Dr. Judith L. Hannah, Head, 970-491-5662, Fax: 970-491-6307, E-mail: jhannah@cnr.colostate.edu. *Application contact:* Sharyl Pierson, Administrative Assistant, 970-491-5662, Fax: 970-491-6307, E-mail: sharyl@cnr.colostate.edu.

Indiana University Bloomington, Graduate School, College of Arts and Sciences, Department of Geological Sciences, Bloomington, IN 47405-7000. Offers biogeochemistry (MS, PhD); economic geology (MS, PhD); geobiology (MS, PhD); geophysics, structural geology and tectonics (MS, PhD); hydrogeology (MS, PhD); mineralogy (MS, PhD); stratigraphy and sedimentology (MS, PhD). PhD offered through the University Graduate School. Part-time programs available. *Faculty:* 14 full-time (1 woman). *Students:* 30 full-time (13 women), 12 part-time (5 women); includes 5 minority (2 African Americans, 1 American Indian/Alaska Native, 1 Asian American or Pacific Islander, 1 Hispanic American), 21 international. Average age 29. In 2005, 7 master's, 2 doctorates awarded. Terminal master's awarded for partial completion of doctoral program. *Degree requirements:* For master's, one foreign language, thesis or alternative; for doctorate, thesis/dissertation. *Entrance requirements:* For master's and doctorate, GRE General Test. Additional exam requirements/recommendations for international students: Required—TOEFL. *Application deadline:* For fall admission, 1/15 priority date for domestic students, 12/15 priority date for international students; for spring admission, 9/1 priority date for domestic students, 9/1 priority date for international students. Applications are processed on a rolling basis. Application fee: $50 ($60 for international students). *Expenses:* Tuition, state resident: full-time $5,437; part-time $227 per credit hour. Tuition, nonresident: full-time $15,836; part-time $660 per credit hour. Required fees: $821. Tuition and fees vary according to campus/location and program. *Financial support:* Fellowships with tuition reimbursements, research assistantships with full and partial tuition reimbursements, teaching assistantships with tuition reimbursements, career-related internships or fieldwork, Federal Work-Study, and institutionally sponsored loans available. Financial award application deadline: 2/15. *Faculty research:* Geophysics, geochemistry, hydrogeology, igneous and metamorphic petrology and clay minerology. Total annual research expenditures: $289,139. *Unit head:* Mark Person, Graduate Advisor, 812-855-7214. *Application contact:* Mary Iverson, Secretary, Committee for Graduate Studies, 812-855-7214, Fax: 812-855-7899, E-mail: geograd@indiana.edu.

Montana Tech of The University of Montana, Graduate School, Geoscience Program, Butte, MT 59701-8997. Offers geochemistry (MS); geological engineering (MS); geology (MS); geophysical engineering (MS); hydrogeological engineering (MS); hydrogeology (MS). Part-time programs available. *Faculty:* 17 full-time (3 women). *Students:* 8 full-time (4 women), 3 part-time (1 woman). 15 applicants, 60% accepted, 2 enrolled. In 2005, 5 degrees awarded. *Degree requirements:* For master's, thesis (for some programs), comprehensive exam (for some programs), registration. *Entrance requirements:* For master's, GRE General Test, minimum GPA of 3.0. Additional exam requirements/recommendations for international students: Required—TOEFL (minimum score 525 paper-based; 195 computer-based). *Application deadline:* For fall admission, 4/1 priority date for domestic students, 3/1 priority date for international students; for spring admission, 10/1 priority date for domestic students, 7/1 priority date for international students. Applications are processed on a rolling basis. Application fee: $30. *Expenses:* Tuition, area resident: Full-time $5,133. Tuition, state resident: full-time $5,333. Tuition, nonresident: full-time $15,746. *Financial support:* In 2005–06, 8 students received support, including 3 research assistantships with partial tuition reimbursements available (averaging $3,667 per year), 6 teaching assistantships with partial tuition reimbursements available (averaging $6,667 per year); career-related internships or fieldwork, tuition waivers (full and partial), and unspecified assistantships also available. Financial award application deadline: 4/1; financial award applicants required to submit FAFSA. *Faculty research:* Water resource development, seismic processing, petroleum reservoir characterization, environmental geochemistry, molecular modeling, magmatic and hydrothermal ore deposits. Total annual research expenditures:$542,109. *Unit head:* Dr. Diane Wolfgram, Department Head, 406-496-4353, Fax: 406-496-4260, E-mail: dwolfgram@mtech.edu. *Application contact:* Cindy Dunstan, Administrator, Graduate School, 406-496-4304, Fax: 406-496-4710, E-mail: cdunstan@mtech.edu.

Ohio University, Graduate Studies, College of Arts and Sciences, Department of Geological Sciences, Athens, OH 45701-2979. Offers environmental geochemistry (MS); environmental geology (MS); environmental/hydrology (MS); geology (MS); geology education (MS); geomorphology/surficial processes (MS); geophysics (MS); hydrogeology (MS); sedimentology (MS); structure/tectonics (MS). Part-time programs available. *Faculty:* 10 full-time (4 women), 4 part-time/adjunct (1 woman). *Students:* 22 full-time (6 women), 3 part-time (2 women); includes 1 minority (Hispanic American), 4 international. Average age 23. 15 applicants, 67% accepted, 8 enrolled. In 2005, 7 degrees awarded. *Degree requirements:* For master's, thesis, thesis proposal defense and thesis defense. *Entrance requirements:* Additional exam requirements/recommendations for international students: Required—TOEFL (minimum score 550 paper-based; 217 computer-based). *Application deadline:* For fall admission, 2/1 priority date for domestic students, 1/1 priority date for international students. Application fee: $45. Electronic applications accepted. *Financial support:* In 2005–06, 18 students received support, including 3 research assistantships with full tuition reimbursements available (averaging $11,900 per year), 13 teaching assistantships with full tuition reimbursements available (averaging $11,900 per year); institutionally sponsored loans, scholarships/grants, tuition waivers (full), and unspecified assistantships also available. Financial award application deadline: 2/1. *Faculty research:* Geoscience education, tectonics, fluvial geomorphology, invertebrate paleontology, mine/hydrology. Total annual research expenditures: $649,020. *Unit head:* Dr. David Kidder, Chair, 740-593-1101, Fax: 740-593-0486, E-mail: kidder@ohio.edu. *Application contact:* Dr. David Schneider, Graduate Chair, 740-593-1101, Fax: 740-593-0486, E-mail: schneidd@ohio.edu.

University of Hawaii at Manoa, Graduate Division, School of Ocean and Earth Science and Technology, Department of Geology and Geophysics, Honolulu, HI 96822. Offers high-pressure geophysics and geochemistry (MS, PhD); hydrogeology and engineering geology (MS, PhD); marine geology and geophysics (MS, PhD); planetary geosciences and remote sensing (MS, PhD); seismology and solid-earth geophysics (MS, PhD); volcanology, petrology, and geochemistry (MS, PhD). *Faculty:* 65 full-time (12 women), 19 part-time/adjunct (1 woman). *Students:* 50 full-time (23 women), 8 part-time (1 woman); includes 2 minority (1 African American, 1 Asian American or Pacific Islander), 15 international. Average age 29. 89 applicants, 18% accepted, 12 enrolled. In 2005, 9 master's, 6 doctorates awarded. Terminal master's awarded for partial completion of doctoral program. *Median time to degree:* Of those who began their doctoral program in fall 1997, 100% received their degree in 8 years or less. *Degree requirements:* For master's, thesis/dissertation; for doctorate, thesis/dissertation, comprehensive exam. *Entrance requirements:* For master's and doctorate, GRE General Test, minimum GPA of 3.0. Additional exam requirements/recommendations for international students: Required—TOEFL. *Application deadline:* For fall admission, 1/15 for domestic students, 1/1 for international students; for spring admission, 9/1 for domestic students, 8/15 for international students. Application fee: $50. *Expenses:* Tuition, state resident: full-time $8,400; part-time $200 per credit hour. Tuition, nonresident: full-time $11,088; part-time $462 per credit hour. Tuition and fees vary according to program. *Financial support:* In 2005–06, 41 research assistantships (averaging $19,366 per year), 5 teaching assistantships (averaging $18,198 per year) were awarded. *Unit head:* Dr. Charles Fletcher, Chair, 808-956-8763, Fax: 808-956-5512, E-mail: fletcher@soest.hawaii.edu. *Application contact:* Dr. Charles Fletcher, Chair, 808-956-8763, Fax: 808-956-5512, E-mail: fletcher@soest.hawaii.edu.

University of Illinois at Chicago, Graduate College, College of Liberal Arts and Sciences, Department of Earth and Environmental Sciences, Chicago, IL 60607-7128. Offers crystallography (MS, PhD); environmental geology (MS, PhD); geochemistry (MS, PhD); geology (MS, PhD); geomorphology (MS, PhD); geophysics (MS, PhD); geotechnical engineering and geosciences (PhD); hydrogeology (MS, PhD); low-temperature and organic geochemistry (MS, PhD); mineralogy (MS, PhD); paleoclimatology (MS, PhD); paleontology (MS, PhD); petrology (MS, PhD); quaternary geology (MS, PhD); sedimentology (MS, PhD); water resources (MS, PhD). *Degree requirements:* For master's and doctorate, thesis/dissertation. *Entrance requirements:* For master's and doctorate, GRE General Test, minimum GPA of 2.75. Additional exam requirements/recommendations for international students: Required—TOEFL. Electronic applications accepted.

University of Nevada, Reno, Graduate School, College of Science, Graduate Program in Hydrologic Sciences, Reno, NV 89557. Offers hydrogeology (MS, PhD); hydrology (MS, PhD). Offered through the M. C. Fleischmann College of Agriculture, the College of Engineering, the Mackay School of Mines, and the Desert Research Institute. Part-time programs available. Terminal master's awarded for partial completion of doctoral program. *Degree requirements:* For master's, thesis optional; for doctorate, thesis/dissertation. *Entrance requirements:* For master's and doctorate, GRE General Test, minimum GPA of 3.0. Additional exam requirements/recommendations for international students: Required—TOEFL. *Faculty research:* Groundwater, water resources, surface water, soil science.

West Virginia University, Eberly College of Arts and Sciences, Department of Geology and Geography, Program in Geology, Morgantown, WV 26506. Offers geomorphology (MS, PhD); geophysics (MS, PhD); hydrogeology (MS); hydrology (PhD); paleontology (MS, PhD); petrology (MS, PhD); stratigraphy (MS, PhD); structure (MS, PhD). Part-time programs available. *Students:* 26 full-time (7 women), 17 part-time (7 women); includes 2 minority (1 American Indian/Alaska Native, 1 Asian American or Pacific Islander), 12 international. Average age 30. 34 applicants, 29% accepted. In 2005, 6 master's, 1 doctorate awarded. Terminal master's awarded for partial completion of doctoral program. *Degree requirements:* For master's, thesis/dissertation; for doctorate, thesis/dissertation, comprehensive exam. *Entrance requirements:* For master's, GRE General Test, GRE Subject Test, minimum GPA of 2.5; for doctorate, GRE General Test, GRE Subject Test, minimum GPA of 3.3. Additional exam requirements/recommendations for international students: Required—TOEFL. *Application deadline:* For fall admission, 2/1 for domestic students; for spring admission, 10/1 for domestic students. Applications are processed on a rolling basis. Application fee: $45. *Expenses:* Tuition, state resident: full-time $4,582; part-time $258 per credit hour. Tuition, nonresident: full-time $1,382; part-time $741 per credit hour. *Financial support:* In 2005–06, 5 research assistantships, 18 teaching assistantships were awarded; career-related internships or fieldwork, Federal Work-Study, institutionally sponsored loans, and tuition waivers (full and partial) also available. Financial award application deadline: 2/1; financial award applicants required to submit FAFSA. *Unit head:* Dr. Thomas H. Wilson, Associate Chair, 304-293-5603 Ext. 4316, Fax: 304-293-6522, E-mail: tom.wilson@mail.wvu.edu. *Application contact:* Dr. Joe Donovan, Associate Professor, 304-293-5603 Ext. 4308, Fax: 304-293-6522, E-mail: joe.donovan@mail.wvu.edu.

Wright State University, School of Graduate Studies, College of Science and Mathematics, Department of Geological Sciences, Program in Geological Sciences, Dayton, OH 45435. Offers environmental geochemistry (MS); environmental geology (MS); environmental sciences (MS); geological sciences (MS); geophysics (MS); hydrogeology (MS); petroleum geology (MS). Part-time programs available. *Degree requirements:* For master's, thesis. *Entrance requirements:* Additional exam requirements/recommendations for international students: Required—TOEFL.

Hydrology

Auburn University, Graduate School, College of Engineering, Department of Civil Engineering, Auburn University, AL 36849. Offers construction engineering and management (MCE, MS, PhD); environmental engineering (MCE, MS, PhD); geotechnical/materials engineering (MCE, MS, PhD); hydraulics/hydrology (MCE, MS, PhD); structural engineering (MCE, MS, PhD); transportation engineering (MCE, MS, PhD). Part-time programs available. *Faculty:* 23 full-time (2 women). *Students:* 48 full-time (7 women), 29 part-time (6 women); includes 5 minority (3 African Americans, 2 Hispanic Americans), 29 international. 111 applicants, 55% accepted, 21 enrolled. In 2005, 24 master's, 2 doctorates awarded. *Degree requirements:* For master's, project (MCE), thesis (MS); for doctorate, thesis/dissertation, comprehensive exam. *Entrance requirements:* For master's and doctorate, GRE General Test. *Application deadline:* For fall admission, 7/7 for domestic students; for spring admission, 11/24 for domestic students. Applications are processed on a rolling basis. Application fee: $25 ($50 for international students). Electronic applications accepted. *Financial support:* Fellowships, research assistantships, teaching assistantships, Federal Work-Study available. Support available to part-time students. Financial award application deadline: 3/15. *Unit head:* Dr. J. Michael Stallings, Head, 334-844-4320. *Application contact:* Dr. Stephen L. McFarland, Acting Dean of the Graduate School, 334-844-4700.

California State University, Bakersfield, Division of Graduate Studies and Research, School of Natural Sciences and Mathematics, Program in Geology, Bakersfield, CA 93311-1022. Offers geology (MS); hydrology (MS). Part-time and evening/weekend programs available. *Faculty:* 5 full-time (1 woman). *Students:* 2 full-time (1 woman), 7 part-time (3 women); includes 2 minority (1 Asian American or Pacific Islander, 1 Hispanic American). Average age 33. In 2005, 4 degrees awarded. *Degree requirements:* For master's, thesis. *Entrance requirements:* For master's, GRE General Test, BS in geology. *Application deadline:* Applications are processed on a rolling basis. Application fee: $55. *Financial support:* Career-related internships or fieldwork and institutionally sponsored loans available. *Unit head:* Dr. Dirk Baron, Graduate Coordinator, 661-664-3044, Fax: 661-664-2040, E-mail: dbaron@csub.edu.

California State University, Chico, Graduate School, College of Natural Sciences, Department of Geological and Environmental Sciences, Program in Geosciences, Chico, CA 95929-0722. Offers hydrology/hydrogeology (MS). *Degree requirements:* For master's, thesis, oral exam. *Entrance requirements:* For master's, GRE General Test. Additional exam requirements/recommendations for international students: Required—TOEFL (minimum score 550 paper-based; 213 computer-based). Electronic applications accepted.

Colorado State University, Graduate School, College of Engineering, Department of Civil Engineering, Fort Collins, CO 80523-0015. Offers environmental engineering (ME, MS, PhD); geotechnical engineering (ME); hydraulics and wind engineering (MS, PhD); infrastructure engineering (ME); irrigation engineering (MS, PhD); structural and geotechnical engineering (MS, PhD); structural engineering (ME); water resources (ME); water resources planning and management (MS, PhD); water resources, hydrologic and environmental sciences (MS, PhD). Part-time programs available. Postbaccalaureate distance learning degree programs offered (no on-campus study). *Faculty:* 29 full-time (2 women), 2 part-time/adjunct (0 women). *Students:* 80 full-time (28 women), 123 part-time (27 women); includes 6 minority (4 Asian Americans or Pacific Islanders, 2 Hispanic Americans), 71 international. Average age 31. 145 applicants, 79% accepted, 41 enrolled. In 2005, 33 master's, 15 doctorates awarded. Terminal master's awarded for partial completion of doctoral program. *Median time to degree:* Of those who began their doctoral program in fall 1997, 50% received their degree in 8 years or less. *Degree requirements:* For master's, thesis or alternative; for doctorate, thesis/dissertation. *Entrance requirements:* For master's and doctorate, GRE General Test, minimum GPA of 3.0. Additional exam requirements/recommendations for international students: Required—TOEFL. *Application deadline:* For fall admission, 4/1 priority date for domestic students, 3/1 priority date for international students; for spring admission, 8/1 priority date for domestic students, 8/1 priority date for international students. Applications are processed on a rolling basis. Application fee: $50. Electronic applications accepted. *Expenses:* Tuition, state resident: full-time $3,690; part-time $205 per credit. Tuition, nonresident: full-time $14,958; part-time $831 per credit. Required fees: $1,061. *Financial support:* In 2005–06, 10 fellowships (averaging $13,725 per year), 11 research assistantships with tuition reimbursements (averaging $18,000 per year), 56 teaching assistantships (averaging $13,950 per year) were awarded; Federal Work-Study, institutionally sponsored loans, and traineeships also available. Financial award application deadline: 2/15. *Faculty research:* Hydraulics, hydrology, water resources, infrastructure, environmental engineering. Total annual research expenditures: $8.5 million. *Unit head:* Luis Garcia, Interim Head, 970-491-5049, Fax: 970-491-7727. *Application contact:* Kathy Stencel, Student Advisor, 970-491-5844, Fax: 970-491-7727, E-mail: kstencel@colostate.edu.

Colorado State University, Graduate School, Warner College of Natural Resources, Department of Forest, Rangeland, and Watershed Stewardship, Fort Collins, CO 80523-0015. Offers forest sciences (MS, PhD); natural resource stewardship (MNRS); rangeland ecosystem science (MS, PhD); watershed science (MS). Part-time programs available. *Faculty:* 27 full-time (5 women), 1 part-time/adjunct (0 women). *Students:* 58 full-time (23 women), 90 part-time (39 women); includes 11 minority (1 African American, 2 American Indian/Alaska Native, 1 Asian American or Pacific Islander, 7 Hispanic Americans), 10 international. Average age 33. 101 applicants, 35% accepted, 30 enrolled. In 2005, 18 master's, 5 doctorates awarded. Terminal master's awarded for partial completion of doctoral program. *Degree requirements:* For master's, thesis optional; for doctorate, thesis/dissertation, comprehensive exam, registration. *Entrance requirements:* For master's and doctorate, GRE General Test, minimum GPA of 3.0. Additional exam requirements/recommendations for international students: Required—TOEFL. *Application deadline:* For fall admission, 4/1 for domestic students; for spring admission, 9/1 priority date for domestic students. Applications are processed on a rolling basis. Application fee: $50. Electronic applications accepted. *Expenses:* Tuition, state resident: full-time $3,690; part-time $205 per credit. Tuition, nonresident: full-time $14,958; part-time $831 per credit. Required fees: $1,061. *Financial support:* Career-related internships or fieldwork, Federal Work-Study, institutionally sponsored loans, and traineeships available. Financial award application deadline: 5/1. *Faculty research:* Ecology, natural resource management, hydrology, restoration, human dimensions. Total annual research expenditures: $4.1 million. *Unit head:* Dr. N. Thompson Hobbs, Head, 970-491-6911, Fax: 970-491-6754, E-mail: nthobbs@warnercnr@colostate.edu. *Application contact:* Graduate Coordinator, 970-491-4994, Fax: 970-491-6754.

Cornell University, Graduate School, Graduate Fields of Engineering, Field of Civil and Environmental Engineering, Ithaca, NY 14853-0001. Offers engineering management (M Eng, MS, PhD); environmental engineering (M Eng, MS, PhD); environmental fluid mechanics and hydrology (M Eng, MS, PhD); environmental systems engineering (M Eng, MS, PhD); geotechnical engineering (M Eng, MS, PhD); remote sensing (M Eng, MS, PhD); structural engineering (M Eng, MS, PhD); structural mechanics (M Eng, MS); transportation engineering (MS, PhD); transportation systems engineering (M Eng); water resource systems (M Eng, MS, PhD). *Faculty:* 40 full-time (6 women). *Students:* 378 applicants, 36% accepted, 73 enrolled. In 2005, 51 master's, 22 doctorates awarded. Terminal master's awarded for partial completion of doctoral program. *Degree requirements:* For master's, thesis (MS); for doctorate, thesis/dissertation, comprehensive exam. *Entrance requirements:* For master's and doctorate, GRE General Test (recommended), 2 letters of recommendation. Additional exam requirements/recommendations for international students: Required—TOEFL (minimum score 600 paper-based; 250 computer-based). *Application deadline:* For fall admission, 1/15 for domestic students; for spring admission, 10/15 for domestic students. Application fee: $60. Electronic applications accepted. *Financial support:* In 2005–06, 64 students received support, including 28 fellowships with full tuition reimbursements available, 20 research assistantships with full tuition reimbursements available, 16 teaching assistantships with full tuition reimbursements available; institutionally sponsored loans, scholarships/grants, health care benefits, tuition waivers (full and partial), and unspecified assistantships also available. Financial award applicants required to submit FAFSA. *Faculty research:* Environmental engineering, geotechnical engineering remote sensing, environmental fluid mechanics and hydrology, structural engineering. *Unit head:* Director of Graduate Studies, 607-255-7560, Fax: 607-255-9004. *Application contact:* Graduate Field Assistant, 607-255-7560, Fax: 607-255-9004, E-mail: cee_grad@cornell.edu.

Cornell University, Graduate School, Graduate Fields of Engineering, Field of Geological Sciences, Ithaca, NY 14853-0001. Offers economic geology (M Eng, MS, PhD); engineering geology (M Eng, MS, PhD); environmental geophysics (M Eng, MS, PhD); general geology (M Eng, MS, PhD); geobiology (M Eng, MS, PhD); geochemistry and isotope geology (M Eng, MS, PhD); geohydrology (M Eng, MS, PhD); geomorphology (M Eng, MS, PhD); geophysics (M Eng, MS, PhD); geotectonics (M Eng, MS, PhD); marine geology (MS, PhD); mineralogy (M Eng, MS, PhD); paleontology (M Eng, MS, PhD); petroleum geology (M Eng, MS, PhD); petrology (M Eng, MS, PhD); planetary geology (M Eng, MS, PhD); Precambrian geology (M Eng, MS, PhD); Quaternary geology (M Eng, MS, PhD); rock mechanics (M Eng, MS, PhD); sedimentology (M Eng, MS, PhD); seismology (M Eng, MS, PhD); stratigraphy (M Eng, MS, PhD); structural geology (M Eng, MS, PhD). *Faculty:* 39 full-time (5 women). *Students:* 29 full-time (15 women); includes 2 minority (both Hispanic Americans), 7 international. 73

Hydrology

Cornell University (continued)
applicants, 15% accepted, 6 enrolled. In 2005, 5 master's, 1 doctorate awarded. *Degree requirements:* For master's, thesis (MS); for doctorate, thesis/dissertation, comprehensive exam. *Entrance requirements:* For master's and doctorate, GRE General Test, 3 letters of recommendation. Additional exam requirements/recommendations for international students: Required—TOEFL (minimum score 550 paper-based; 213 computer-based). *Application deadline:* For fall admission, 1/15 for domestic students. Applications are processed on a rolling basis. Application fee: $60. Electronic applications accepted. *Financial support:* In 2005–06, 8 fellowships with full tuition reimbursements, 10 research assistantships with full tuition reimbursements, 7 teaching assistantships with full tuition reimbursements were awarded; institutionally sponsored loans, scholarships/grants, health care benefits, tuition waivers (full and partial), and unspecified assistantships also available. Financial award applicants required to submit FAFSA. *Faculty research:* Geophysics, structural geology, petrology, geochemistry, geodynamics. *Unit head:* Director of Graduate Studies, 607-255-5466, Fax: 607-254-4780. *Application contact:* Graduate Field Assistant, 607-255-5466, Fax: 607-254-4780, E-mail: gradprog@geology.cornell.edu.

Georgia Institute of Technology, Graduate Studies and Research, College of Sciences, School of Earth and Atmospheric Sciences, Atlanta, GA 30332-0001. Offers atmospheric chemistry and air pollution (MS, PhD); atmospheric dynamics and climate (MS, PhD); geochemistry (MS, PhD); hydrologic cycle (MS, PhD); ocean sciences (MS, PhD); solid-earth and environmental geophysics (MS, PhD). Part-time programs available. Terminal master's awarded for partial completion of doctoral program. *Degree requirements:* For master's, thesis or alternative; for doctorate, thesis/dissertation, comprehensive exam. *Entrance requirements:* For master's, GRE, minimum GPA of 3.0; for doctorate, GRE General Test, minimum GPA of 2.7. Additional exam requirements/recommendations for international students: Required—TOEFL (minimum score 550 paper-based; 213 computer-based). *Faculty research:* Geophysics, atmospheric chemistry, atmospheric dynamics, seismology.

See Close-Up on page 227.

Georgia State University, College of Arts and Sciences, Department of Geosciences, Atlanta, GA 30303-3083. Offers earth science—hydrology (MS), including earth science—environmental management, earth science—GIS; geographic information systems (Certificate); geography (MA); geology (MS). Part-time and evening/weekend programs available. *Degree requirements:* For master's, one foreign language, thesis or alternative, comprehensive exam (for some programs), registration. *Entrance requirements:* For master's, GRE General Test, minimum GPA of 2.75. Additional exam requirements/recommendations for international students: Required—TOEFL. Electronic applications accepted. *Expenses:* Tuition, state resident: full-time $4,368; part-time $182 per term. Tuition, nonresident: full-time $8,732; part-time $728 per term. Required fees: $46 per hour. *Faculty research:* Clay mineralogy, metamorphism, fracture analysis, carbonates, groundwater.

Idaho State University, Office of Graduate Studies, College of Arts and Sciences, Department of Geosciences, Pocatello, ID 83209. Offers geographic information science (MS); geology (MNS, MS); geophysics/hydrology (MS); geotechnology (Postbaccalaureate Certificate). Part-time programs available. Postbaccalaureate distance learning degree programs offered. *Degree requirements:* For master's, thesis, comprehensive exam, registration (for some programs); for Postbaccalaureate Certificate, thesis optional. *Entrance requirements:* For master's and Postbaccalaureate Certificate, GRE General Test, 3 letters of recommendation. Additional exam requirements/recommendations for international students: Required—TOEFL (minimum score 550 paper-based; 213 computer-based). *Faculty research:* Structural geography, stratigraphy, geochemistry, remote sensing, geomorphology.

Illinois State University, Graduate School, College of Arts and Sciences, Department of Geography-Geology, Normal, IL 61790-2200. Offers geohydrology (MS). *Faculty:* 8 full-time (2 women), 1 part-time/adjunct (0 women). *Students:* 13 full-time (5 women), 3 part-time (1 woman); includes 2 minority (1 Asian American or Pacific Islander, 1 Hispanic American), 2 international. 14 applicants, 86% accepted. In 2005, 9 degrees awarded. *Entrance requirements:* For master's, GRE General Test. *Application deadline:* Applications are processed on a rolling basis. Application fee: $30. *Expenses:* Tuition, state resident: full-time $3,060; part-time $170 per credit hour. Tuition, nonresident: full-time $6,390; part-time $355 per credit hour. Required fees: $1,411; $47 per credit hour. *Financial support:* In 2005–06, 2 research assistantships (averaging $9,225 per year), 9 teaching assistantships (averaging $8,505 per year) were awarded; unspecified assistantships also available. Financial award application deadline: 4/1. *Faculty research:* Revised structural, stratigraphic, and hydrogeologic model of the Troy Grove Gas Storage Field; delineating abandoned underground coal mines in Illinois using shallow seismic techniques; geologic mapping of the Streator North and Tonica 7.5 minute Quadrangles. *Unit head:* Dr. David Malone, Chairperson.

Massachusetts Institute of Technology, School of Engineering, Department of Civil and Environmental Engineering, Cambridge, MA 02139-4307. Offers biological oceanography (PhD, Sc D); chemical oceanography (PhD, Sc D); civil and environmental engineering (M Eng, SM, PhD, Sc D, CE); civil and environmental systems (PhD, Sc D); civil engineering (PhD, Sc D); coastal engineering (PhD, Sc D); construction engineering and management (PhD, Sc D); environmental biology (PhD, Sc D); environmental chemistry (PhD, Sc D); environmental engineering (PhD, Sc D); environmental fluid mechanics (PhD, Sc D); geotechnical and geoenvironmental engineering (PhD, Sc D); hydrology (PhD, Sc D); information technology (PhD, Sc D); oceanographic engineering (PhD, Sc D); structures and materials (PhD, Sc D); transportation (PhD, Sc D). *Faculty:* 34 full-time (3 women). *Students:* 179 full-time (66 women), 1 part-time; includes 15 minority (1 African American, 9 Asian Americans or Pacific Islanders, 5 Hispanic Americans), 116 international. Average age 26. 374 applicants, 38% accepted, 71 enrolled. In 2005, 76 master's, 27 doctorates, 1 other advanced degree awarded. *Degree requirements:* For master's and CE, thesis/dissertation; for doctorate, thesis/dissertation, comprehensive exam. *Entrance requirements:* For master's and doctorate, GRE General Test. Additional exam requirements/recommendations for international students: Required—TOEFL (minimum score 577 paper-based; 233 computer-based). *Application deadline:* For fall admission, 1/2 for domestic students, 1/2 for international students. Application fee: $70. Electronic applications accepted. *Expenses:* Tuition: Full-time $32,100. Required fees: $200. Part-time tuition and fees vary according to course load. *Financial support:* In 2005–06, 134 students received support, including 38 fellowships with tuition reimbursements available (averaging $23,116 per year), 86 research assistantships with tuition reimbursements available (averaging $22,765 per year), 13 teaching assistantships with tuition reimbursements available (averaging $19,735 per year); career-related internships or fieldwork, Federal Work-Study, institutionally sponsored loans, scholarships/grants, health care benefits, and unspecified assistantships also available. Total annual research expenditures: $12.4 million. *Unit head:* Prof. Patrick Jaillet, Department Head, 617-452-3379, Fax: 617-452-3294, E-mail: jaillet@mit.edu. *Application contact:* Graduate Admissions, 617-253-7119, Fax: 617-258-6775, E-mail: cee-admissions@mit.edu.

New Mexico Institute of Mining and Technology, Graduate Studies, Department of Earth and Environmental Science, Program in Hydrology, Socorro, NM 87801. Offers MS, PhD. *Degree requirements:* For master's and doctorate, thesis/dissertation. *Entrance requirements:* For master's, GRE General Test; for doctorate, GRE General Test, GRE Subject Test. Additional exam requirements/recommendations for international students: Required—TOEFL (minimum score 540 paper-based; 207 computer-based). *Faculty research:* Surface and subsurface hydrology, numerical simulation, stochastic hydrology, water quality, modeling.

State University of New York College of Environmental Science and Forestry, Faculty of Forest and Natural Resources Management, Syracuse, NY 13210-2779. Offers environmental and natural resource policy (MS, PhD); environmental and natural resources policy (MPS); forest management and operations (MF); forestry ecosystems science and applications (MPS, MS, PhD); natural resources management (MPS, MS, PhD); quantitative methods and management in forest science (MPS, MS, PhD); recreation and resource management (MPS, MS, PhD); watershed management and forest hydrology (MPS, MS, PhD). *Faculty:* 30 full-time (7 women), 1 (woman) part-time/adjunct. *Students:* 43 full-time (18 women), 26 part-time (12 women); includes 1 minority (Hispanic American), 13 international. Average age 31. 38 applicants, 55% accepted, 12 enrolled. In 2005, 13 master's, 4 doctorates awarded. *Degree requirements:* For master's, thesis (for some programs), registration; for doctorate, thesis/dissertation, comprehensive exam, registration. *Entrance requirements:* For master's and doctorate, GRE General Test, minimum GPA of 3.0. Additional exam requirements/recommendations for international students: Required—TOEFL (minimum score 550 paper-based; 213 computer-based). *Application deadline:* For fall admission, 2/1 priority date for domestic students, 2/1 priority date for international students; for spring admission, 11/1 priority date for domestic students, 11/1 priority date for international students. Applications are processed on a rolling basis. Application fee: $60. *Expenses:* Tuition, area resident: Full-time $6,900; part-time $288 per credit. Tuition, nonresident: full-time $10,920; part-time $455 per credit. Required fees: $395; $32 per credit. $20 per term. One-time fee: $145. *Financial support:* In 2005–06, 43 students received support, including 8 fellowships with full and partial tuition reimbursements available (averaging $9,446 per year), 20 research assistantships with full and partial tuition reimbursements available (averaging $10,000 per year), 11 teaching assistantships with full and partial tuition reimbursements available (averaging $9,446 per year); career-related internships or fieldwork, Federal Work-Study, institutionally sponsored loans, scholarships/grants, health care benefits, and unspecified assistantships also available. Financial award application deadline: 6/30; financial award applicants required to submit FAFSA. *Faculty research:* Silviculture recreation management, tree improvement, operations management, economics. Total annual research expenditures: $2.1 million. *Unit head:* Dr. Chad P. Dawson, Chair, 315-470-6536, Fax: 315-470-6535, E-mail: cpdawson@esf.edu. *Application contact:* Dr. Dudley J. Raynal, Dean, Instruction and Graduate Studies, 315-470-6599, Fax: 315-470-6978, E-mail: esfgrad@esf.edu.

Texas A&M University, Interdisciplinary Program in Water Management and Hydrological Sciences, College Station, TX 77843. Offers MS, PhD. *Expenses:* Tuition, state resident: full-time $4,488; part-time $187 per credit hour. Tuition, nonresident: full-time $11,112; part-time $463 per credit hour. Required fees: $1,974. *Unit head:* Ronald A. Kaiser, Co-Chair, 979-845-5303, Fax: 979-845-0446, E-mail: rkaiser@tamu.edu.

Université du Québec, Institut National de la Recherche Scientifique, Graduate Programs, Research Center—Water, Earth and Environment, Québec, QC G1K 9A9, Canada. Offers earth sciences (M Sc, PhD); earth sciences-environmental technologies (M Sc); water sciences (MA, PhD). Part-time programs available. *Faculty:* 38. *Students:* 172 full-time (80 women), 10 part-time (3 women), 44 international. Average age 29. In 2005, 37 master's, 10 doctorates awarded. *Degree requirements:* For master's, thesis optional; for doctorate, thesis/dissertation. *Entrance requirements:* For master's, appropriate bachelor's degree, proficiency in French; for doctorate, appropriate master's degree, proficiency in French. *Application deadline:* For fall admission, 3/30 for domestic students, 3/30 for international students. For winter admission, 11/1 for domestic students. Application fee: $30. *Financial support:* Fellowships, research assistantships, teaching assistantships available. *Faculty research:* Land use, impacts of climate change, adaptation to climate change, integrated management of resources (mineral and water). *Unit head:* Jean Pierre Villeneuve, Director, 418-654-2575, Fax: 418-654-2615, E-mail: jp_villeneuve@ete.inrs.ca. *Application contact:* Michel Barbeau, Registrar, 418-654-2518, Fax: 418-654-3858, E-mail: michel.barbeau@adm.inrs.ca.

The University of Arizona, Graduate College, College of Engineering, Department of Hydrology and Water Resources, Tucson, AZ 85721. Offers hydrology (MS, PhD); water resources engineering (M Eng). Part-time programs available. *Degree requirements:* For master's and doctorate, thesis/dissertation. *Entrance requirements:* For master's, GRE General Test, minimum undergraduate GPA of 3.0; for doctorate, GRE General Test, minimum undergraduate GPA of 3.2, 3.4 graduate. Additional exam requirements/recommendations for international students: Required—TOEFL. *Faculty research:* Subsurface and surface hydrology, hydrometeorology/climatology, applied remote sensing, water resource systems, environmental hydrology and water quality.

University of California, Davis, Graduate Studies, Graduate Group in Hydrologic Sciences, Davis, CA 95616. Offers MS, PhD. *Faculty:* 48 full-time. *Students:* 19 full-time (8 women); includes 2 minority (both Asian Americans or Pacific Islanders), 2 international. Average age 32. 42 applicants, 12% accepted, 4 enrolled. In 2005, 2 master's, 6 doctorates awarded. Terminal master's awarded for partial completion of doctoral program. *Median time to degree:* Of those who began their doctoral program in fall 1997, 60% received their degree in 8 years or less. *Degree requirements:* For master's, thesis (for some programs), comprehensive exam (for some programs); for doctorate, thesis/dissertation. *Entrance requirements:* For master's, GRE General Test, minimum GPA of 3.0; for doctorate, GRE. Additional exam requirements/recommendations for international students: Required—TOEFL (minimum score 550 paper-based; 213 computer-based). *Application deadline:* For fall admission, 4/1 priority date for domestic students, 3/1 priority date for international students. Application fee: $60. Electronic applications accepted. *Financial support:* In 2005–06, 14 students received support, including 3 fellowships with full and partial tuition reimbursements available (averaging $10,989 per year), 5 research assistantships with full and partial tuition reimbursements available (averaging $11,172 per year), 2 teaching assistantships with partial tuition reimbursements available (averaging $15,082 per year); career-related internships or fieldwork, Federal Work-Study, institutionally sponsored loans, scholarships/grants, tuition waivers (full and partial), and unspecified assistantships also available. Financial award application deadline: 1/15; financial award applicants required to submit FAFSA. *Faculty research:* Pollutant transport in surface and subsurface waters, subsurface heterogeneity, micrometeorology evaporation, biodegradation. *Unit head:* Dr. Mark Grismer, Chair, 530-752-3243, Fax: 530-752-5262, E-mail: megrismer@ucdavis.edu. *Application contact:* Merlyn Potters, Graduate Staff Adviser, 530-752-1669, Fax: 530-752-1552, E-mail: lawradvising@ucdavis.edu.

University of California, Los Angeles, Graduate Division, Henry Samueli School of Engineering and Applied Science, Department of Civil and Environmental Engineering, Los Angeles, CA 90095. Offers environmental engineering (MS, PhD); geotechnical engineering (MS, PhD); hydrology and water resources engineering (MS); structures (MS, PhD), including structural mechanics and earthquake engineering; water resource systems engineering (PhD). *Faculty:* 13 full-time (2 women). *Students:* 111 full-time (36 women); includes 28 minority (2 African Americans, 21 Asian Americans or Pacific Islanders, 5 Hispanic Americans), 47 international. In 2005, 36 master's, 14 doctorates awarded. *Degree requirements:* For master's, comprehensive exam or thesis; for doctorate, thesis/dissertation, qualifying exams. *Entrance requirements:* For master's, GRE General Test, minimum GPA of 3.0; for doctorate, GRE General Test, minimum GPA of 3.25. Additional exam requirements/recommendations for international students: Required—TOEFL (minimum score 560 paper-based; 220 computer-based). *Application deadline:* For fall admission, 1/15 for domestic students. Application fee: $60 ($80 for international students). Electronic applications accepted. *Financial support:* In 2005–06, 87 fellowships, 98 research assistantships, 35 teaching assistantships were awarded; Federal Work-Study, institutionally sponsored loans, and tuition waivers (full and partial) also available. Financial award application deadline: 12/15; financial award applicants required to submit FAFSA. Total annual research expenditures: $2.8 million. *Unit head:* Dr. William Yeh, Chair, 310-825-2300. *Application contact:* Maida Bassili, Graduate Affairs Officer, 310-825-1851, Fax: 310-206-2222, E-mail: maida@ea.ucla.edu.

University of Idaho, College of Graduate Studies, College of Science, Department of Geological Sciences, Program in Hydrology, Moscow, ID 83844-2282. Offers MS. *Students:* 6 full-time (2 women), 13 part-time (7 women); includes 1 minority (Asian American or Pacific Islander) In 2005, 3 degrees awarded. *Entrance requirements:* For master's, minimum GPA of 2.8. *Application deadline:* For fall admission, 8/1 for domestic students; for spring admission, 12/15 for domestic students. Application fee: $55 ($60 for international students). *Expenses:* Tuition, nonresident: full-time $8,770; part-time $130 per credit. Required fees: $4,508; $217 per credit. *Financial support:* Application deadline: 2/15. *Unit head:* Dr. Dennis Geist, Head, Department of Geological Sciences, 208-885-6491.

University of Missouri–Rolla, Graduate School, School of Engineering, Department of Civil, Architectural, and Environmental Engineering, Program in Hydrology and Hydraulic Engineering, Rolla, MO 65409-0910. Offers MS, DE, PhD. *Degree requirements:* For master's, thesis or alternative; for doctorate, thesis/dissertation. *Entrance requirements:* For master's and doctorate, GRE General Test, minimum GPA of 3.0. Additional exam requirements/recommendations for international students: Required—TOEFL.

University of Nevada, Reno, Graduate School, College of Science, Graduate Program in Hydrologic Sciences, Reno, NV 89557. Offers hydrogeology (MS, PhD); hydrology (MS, PhD). Offered through the M. C. Fleischmann College of Agriculture, the College of Engineering, the Mackay School of Mines, and the Desert Research Institute. Part-time programs available. Terminal master's awarded for partial completion of doctoral program. *Degree requirements:* For master's, thesis optional; for doctorate, thesis/dissertation. *Entrance requirements:* For master's and doctorate, GRE General Test, minimum GPA of 3.0. Additional exam requirements/recommendations for international students: Required—TOEFL. *Faculty research:* Groundwater, water resources, surface water, soil science.

University of New Brunswick Fredericton, School of Graduate Studies, Faculty of Engineering, Department of Civil Engineering, Fredericton, NB E3B 5A3, Canada. Offers construction engineering and management (M Eng, M Sc E, PhD); environmental engineering (M Eng, M Sc E, PhD); geotechnical engineering (M Eng, M Sc E, PhD); groundwater/hydrology (M Eng, M Sc E, PhD); materials (M Eng, M Sc E, PhD); pavements (M Eng, M Sc E, PhD); structures (M Eng, M Sc E, PhD); transportation (M Eng, M Sc E, PhD). Part-time programs available. *Degree requirements:* For master's, thesis; for doctorate, thesis/dissertation, qualifying exam. *Entrance requirements:* For master's and doctorate, minimum GPA of 3.0. Additional exam requirements/recommendations for international students: Required—TOEFL, TWE. *Faculty research:* Steel and masonry structures, traffic engineering, highway safety, centrifuge modeling, transport and fate of reactive contaminants, durability of marine concrete.

University of New Hampshire, Graduate School, College of Engineering and Physical Sciences, Department of Earth Sciences, Durham, NH 03824. Offers earth sciences (MS), including geochemical, geology, ocean mapping, oceanography; hydrology (MS). *Faculty:* 29 full-time. *Students:* 17 full-time (5 women), 24 part-time (10 women); includes 2 minority (1 African American, 1 Asian American or Pacific Islander), 6 international. Average age 31. 37 applicants, 84% accepted, 10 enrolled. In 2005, 10 master's awarded. *Degree requirements:* For master's, thesis. *Entrance requirements:* For master's, GRE General Test. Additional exam requirements/recommendations for international students: Required—TOEFL (minimum score 550 paper-based; 213 computer-based); Recommended—TSE. *Application deadline:* For fall admission, 4/1 priority date for domestic students, 4/1 priority date for international students. For winter admission, 12/1 for domestic students. Applications are processed on a rolling basis. Application fee: $60. Electronic applications accepted. *Expenses:* Tuition, state resident: full-time $8,010; part-time $445 per credit hour. Tuition, nonresident: full-time $19,730; part-time $810 per credit hour. Required fees: $322 per semester. Tuition and fees vary according to course load and program. *Financial support:* In 2005–06, 1 fellowship, 9 research assistantships, 9 teaching assistantships were awarded; career-related internships or fieldwork, Federal Work-Study, scholarships/grants, and tuition waivers (full and partial) also available. Support available to part-time students. Financial award application deadline: 2/15. *Unit head:* Dr. Matt Davis, Chairperson, 603-862-1718, E-mail: earth.sciences@unh.edu. *Application contact:* Nancy Gauthier, Administrative Assistant, 603-862-1720, E-mail: earth.sciences@unh.edu.

University of Southern Mississippi, Graduate School, College of Science and Technology, Department of Marine Science, Stennis Space Center, MS 39529. Offers hydrographic science (MS); marine science (MS, PhD). Part-time programs available. *Faculty:* 10 full-time (1 woman), 1 part-time/adjunct (0 women). *Students:* 31 full-time (16 women), 6 part-time (1 woman); includes 3 minority (2 African Americans, 1 Hispanic American), 9 international. Average age 32. 17 applicants, 88% accepted, 15 enrolled. In 2005, 14 master's, 1 doctorate awarded. *Degree requirements:* For master's, thesis, oral qualifying exam (marine science), comprehensive exam; for doctorate, 2 foreign languages, thesis/dissertation, oral qualifying exam, comprehensive exam. *Entrance requirements:* For master's, GRE General Test, minimum GPA of 3.0; for doctorate, GRE General Test, minimum GPA of 3.0 (undergraduate), 3.5 (graduate). Additional exam requirements/recommendations for international students: Required—TOEFL. *Application deadline:* For fall admission, 3/1 for domestic students; for spring admission, 12/13 for domestic students. Applications are processed on a rolling basis. Application fee: $0 ($25 for international students). Electronic applications accepted. *Financial support:* In 2005–06, 4 students received support, including research assistantships with full tuition reimbursements available (averaging $16,800 per year), teaching assistantships with full tuition reimbursements available (averaging $16,800 per year); Federal Work-Study and institutionally sponsored loans also available. Financial award application deadline: 3/15. *Faculty research:* Chemical, biological, physical, and geological marine science; remote sensing; bio-optics; numerical modeling; hydrography. Total annual research expenditures: $2.1 million. *Unit head:* Dr. Steven E. Lohrenz, Chair, 228-688-3177, Fax: 228-688-1121, E-mail: marine.science@usm.edu.

University of Washington, Graduate School, College of Forest Resources, Seattle, WA 98195. Offers forest economics (MS, PhD); forest ecosystem analysis (MS, PhD); forest engineering/forest hydrology (MS, PhD); forest products marketing (MS, PhD); forest soils (MS, PhD); paper science and engineering (MS, PhD); quantitative resource management (MS, PhD); silviculture (MFR); silviculture and forest protection (MS, PhD); social sciences (MS, PhD); urban horticulture (MFR, MS, PhD); wildlife science (MS, PhD). *Degree requirements:* For master's, thesis (for some programs), registration; for doctorate, thesis/dissertation, comprehensive exam (for some programs), registration. *Entrance requirements:* For master's and doctorate, GRE, minimum GPA of 3.0. Additional exam requirements/recommendations for international students: Required—TOEFL. Electronic applications accepted. *Faculty research:* Ecosystem analysis, silviculture and forest protection, paper science and engineering, environmental horticulture and urban forestry, natural resource policy and economics.

Washington State University, Graduate School, College of Sciences, Department of Geology, Pullman, WA 99164. Offers hydrology (MS, PhD); mineralology-petrology geochemistry (MS); minerology-petrology geochemistry (PhD); sedimentology-stratogeography (MS, PhD); structural geology-tectonics (MS, PhD). *Faculty:* 12. *Students:* 22 full-time (8 women), 4 part-time (1 woman), 5 international. Average age 29. 39 applicants, 49% accepted, 7 enrolled. In 2005, 5 master's, 1 doctorate awarded. *Degree requirements:* For master's, thesis, oral exam; for doctorate, one foreign language, thesis/dissertation, oral exam, written exam. *Entrance requirements:* For master's and doctorate, GRE General Test, minimum GPA of 3.0, 3 letters of recommendation. Additional exam requirements/recommendations for international students: Required—TOEFL (minimum score 560 paper-based; 220 computer-based). *Application deadline:* For fall admission, 2/1 priority date for domestic students, 2/1 priority date for international students; for spring admission, 9/1 priority date for domestic students, 7/1 priority date for international students. Applications are processed on a rolling basis. Application fee: $35. Electronic applications accepted. *Expenses:* Tuition, state resident: full-time $6,295; part-time $336 per credit. Tuition, nonresident: full-time $15,949; part-time $819 per credit. Required fees: $933. Part-time tuition and fees vary according to campus/location and program. *Financial support:* In 2005–06, 4 fellowships (averaging $2,700 per year), 5 research assistantships with full and partial tuition reimbursements (averaging $14,200 per year), 18 teaching assistantships with full and partial tuition reimbursements (averaging $13,831 per year) were awarded; career-related internships or fieldwork, Federal Work-Study, institutionally sponsored loans, and scholarships/grants also available. Financial award application deadline: 2/1; financial award applicants required to submit FAFSA. *Faculty research:* Genesis of ore deposits, geohydrology of the Pacific Northwest, geochemistry and petrology of plateau basalts. Total annual research expenditures: $379,240. *Unit head:* Dr. Peter Larson, Chair, 509-335-3009, Fax: 509-335-7816, E-mail: plarson@wsu.edu. *Application contact:* E-mail: geology@wsu.edu.

West Virginia University, Eberly College of Arts and Sciences, Department of Geology and Geography, Program in Geology, Morgantown, WV 26506. Offers geomorphology (MS, PhD); geophysics (MS, PhD); hydrogeology (MS); hydrology (PhD); paleontology (MS, PhD); petrology (MS, PhD); stratigraphy (MS, PhD); structure (MS, PhD). Part-time programs available. *Students:* 26 full-time (7 women), 17 part-time (7 women); includes 2 minority (1 American Indian/Alaska Native, 1 Asian American or Pacific Islander), 12 international. Average age 30. 34 applicants, 29% accepted. In 2005, 6 master's, 1 doctorate awarded. Terminal master's awarded for partial completion of doctoral program. *Degree requirements:* For master's, thesis/dissertation; for doctorate, thesis/dissertation, comprehensive exam. *Entrance requirements:* For master's, GRE General Test, GRE Subject Test, minimum GPA of 2.5; for doctorate, GRE General Test, GRE Subject Test, minimum GPA of 3.3. Additional exam requirements/recommendations for international students: Required—TOEFL. *Application deadline:* For fall admission, 2/1 for domestic students; for spring admission, 10/1 for domestic students. Applications are processed on a rolling basis. Application fee: $45. *Expenses:* Tuition, state resident: full-time $4,582; part-time $258 per credit hour. Tuition, nonresident: full-time $1,382; part-time $741 per credit hour. *Financial support:* In 2005–06, 5 research assistantships, 18 teaching assistantships were awarded; career-related internships or fieldwork, Federal Work-Study, institutionally sponsored loans, and tuition waivers (full and partial) also available. Financial award application deadline: 2/1; financial award applicants required to submit FAFSA. *Unit head:* Dr. Thomas H. Wilson, Associate Chair, 304-293-5603 Ext. 4316, Fax: 304-293-6522, E-mail: tom.wilson@mail.wvu.edu. *Application contact:* Dr. Joe Donovan, Associate Professor, 304-293-5603 Ext. 4308, Fax: 304-293-6522, E-mail: joe.donovan@mail.wvu.edu.

Limnology

Baylor University, Graduate School, College of Arts and Sciences, Department of Biology, Waco, TX 76798. Offers biology (MA, MS, PhD); environmental biology (MS); limnology (MSL). Part-time programs available. *Faculty:* 13 full-time (3 women). *Students:* 21 full-time (6 women), 16 part-time (8 women); includes 2 minority (1 African American, 1 Hispanic American), 7 international. In 2005, 4 master's, 2 doctorates awarded. *Degree requirements:* For master's, thesis (for some programs); for doctorate, thesis/dissertation. *Entrance requirements:* For master's and doctorate, GRE General Test. *Application deadline:* For fall admission, 1/31 for domestic students. Applications are processed on a rolling basis. Application fee: $25. *Financial support:* Teaching assistantships, career-related internships or fieldwork, Federal Work-Study, institutionally sponsored loans, tuition waivers (full and partial) also available. Support available to part-time students. Financial award application deadline:2/28. *Faculty research:* Terrestrial ecology, aquatic ecology, genetics. *Unit head:* Dr. Ken Wilkins, Graduate Program Director, 254-710-2911, Fax: 254-710-2969, E-mail: ken_wilkins@baylor.edu. *Application contact:* Sandy Tighe, Administrative Assistant, 254-710-2911, Fax: 254-710-2969, E-mail: sandy_tighe@baylor.edu.

Cornell University, Graduate School, Graduate Fields of Agriculture and Life Sciences, Field of Ecology and Evolutionary Biology, Ithaca, NY 14853-0001. Offers ecology (PhD), including animal ecology, applied ecology, biogeochemistry, community and ecosystem ecology, limnology, oceanography, physiological ecology, plant ecology, population ecology, theoretical ecology, vertebrate zoology; evolutionary biology (PhD), including ecological genetics, paleobiology, population biology, systematics. *Faculty:* 62 full-time (11 women). *Students:* 61 full-time (33 women); includes 6 minority (1 African American, 3 Asian Americans or Pacific Islanders, 2 Hispanic Americans), 10 international. 107 applicants, 15% accepted, 12 enrolled. In 2005, 7 doctorates awarded. *Degree requirements:* For doctorate, thesis/dissertation, 2 semesters of teaching experience, comprehensive exam. *Entrance requirements:* For doctorate, GRE General Test, GRE Subject Test (biology), 2 letters of recommendation. Additional exam requirements/recommendations for international students: Required—TOEFL (minimum score 550 paper-based; 213 computer-based). *Application deadline:* For fall admission, 12/15 for domestic students. Application fee: $60. Electronic applications accepted. *Financial support:* In 2005–06, 59 students received support, including 25 fellowships with full tuition reimbursements available, 10 research assistantships with full tuition reimbursements available, 24 teaching assistantships with full tuition reimbursements available; institutionally sponsored loans, scholarships/grants, health care benefits, tuition waivers (full and partial), and unspecified assistantships also available. Financial award applicants required to submit FAFSA. *Faculty research:* Population and organismal biology, population and evolutionary genetics, systematics and macroevolution, biochemistry, conservation biology. *Unit head:* Director of Graduate Studies, 607-254-4230. *Application contact:* Graduate Field Assistant, 607-254-4230, E-mail: eeb_grad_req@cornell.edu.

University of Alaska Fairbanks, School of Fisheries and Ocean Sciences, Department of Marine Sciences and Limnology, Fairbanks, AK 99775-7520. Offers marine biology (MS, PhD); oceanography (MS, PhD), including biological oceanography (PhD), chemical oceanography (PhD), fisheries (PhD), geological oceanography (PhD), physical oceanography (PhD). Part-time programs available. Terminal master's awarded for partial completion of doctoral program. *Degree requirements:* For master's and doctorate, thesis/dissertation, comprehensive exam, registration. *Entrance requirements:* For master's and doctorate, GRE General Test. Additional exam requirements/recommendations for international students: Required—TOEFL. Electronic applications accepted. *Expenses:* Tuition, state resident: full-time $4,392; part-time $244 per credit. Tuition, nonresident: full-time $8,964; part-time $498 per credit. Required fees: $800; $5 per credit. $48 per contact hour. Tuition and fees vary according to course level, course load, campus/location and reciprocity agreements. *Faculty research:* Seafood science and nutrition, sustainable harvesting, chemical oceanography, marine biology, physical oceanography.

University of Florida, Graduate School, College of Agricultural and Life Sciences, Department of Fisheries and Aquatic Sciences, Gainesville, FL 32611. Offers MFAS, MS, PhD. *Faculty:* 14 full-time (2 women), 1 part-time/adjunct (0 women). *Students:* 50 (22 women); includes 2 minority (both Asian Americans or Pacific Islanders) 2 international. 29 applicants, 31% accepted. In 2005, 10 master's, 2 doctorates awarded. *Degree requirements:* For master's, thesis optional; for doctorate, thesis/dissertation. *Entrance requirements:* For master's and doctorate, GRE General Test, minimum GPA of 3.0. Additional exam requirements/recommendations for international students: Required—TOEFL. *Application deadline:* For fall admission, 6/1 for domestic students. Applications are processed on a rolling basis. Application fee: $20. Electronic applications accepted. *Expenses:* Tuition, state resident: full-

Limnology

University of Florida (continued)
time $6,234. Tuition, nonresident: full-time $21,359. Tuition and fees vary according to program. *Financial support:* In 2005–06, 22 research assistantships (averaging $9,546 per year) were awarded; fellowships, unspecified assistantships also available. *Unit head:* Dr. Karl Havens, Chair, 352-392-9617, Fax: 352-392-3672, E-mail: khavens@ifas.ufl.edu. *Application contact:* Dr. Chuck Chichra, Graduate Coordinator, 352-392-9617 Ext. 249, Fax: 352-392-3672, E-mail: fish@ifas.ufl.edu.

University of Wisconsin–Madison, Graduate School, College of Engineering, Program in Limnology and Marine Science, Madison, WI 53706-1380. Offers MS, PhD. Terminal master's awarded for partial completion of doctoral program. *Degree requirements:* For master's and doctorate, thesis/dissertation. *Entrance requirements:* For master's and doctorate, GRE General Test. Additional exam requirements/recommendations for international students: Required—TOEFL. Electronic applications accepted. *Faculty research:* Lake ecosystems, ecosystem modeling, geochemistry, physiological ecology, chemical limnology.

William Paterson University of New Jersey, College of Science and Health, Department of Biology, General Biology Program, Wayne, NJ 07470-8420. Offers general biology (MA); limnology and terrestrial ecology (MA); molecular biology (MA); physiology (MA). Part-time and evening/weekend programs available. *Students:* 2 full-time (1 woman), 13 part-time (9 women); includes 4 Hispanic Americans. In 2005, 4 degrees awarded. *Degree requirements:* For master's, independent study or thesis. *Entrance requirements:* For master's, GRE General Test, minimum GPA of 2.75. *Application deadline:* Applications are processed on a rolling basis. Application fee: $50. Electronic applications accepted. *Expenses:* Tuition, state resident: full-time $476. Tuition, nonresident: full-time $717. *Financial support:* Research assistantships, career-related internships or fieldwork and unspecified assistantships available. Financial award application deadline: 4/1; financial award applicants required to submit FAFSA. *Application contact:* Danielle Liautaud, Assistant Director, 973-720-3579, Fax: 973-720-2035, E-mail: liautaudd@wpunj.edu.

Marine Geology

Cornell University, Graduate School, Graduate Fields of Engineering, Field of Geological Sciences, Ithaca, NY 14853-0001. Offers economic geology (M Eng, MS, PhD); engineering geology (M Eng, MS, PhD); environmental geophysics (M Eng, MS, PhD); general geology (M Eng, MS, PhD); geobiology (M Eng, MS, PhD); geochemistry and isotope geology (M Eng, MS, PhD); geohydrology (M Eng, MS, PhD); geomorphology (M Eng, MS, PhD); geophysics (M Eng, MS, PhD); geotectonics (M Eng, MS, PhD); marine geology (MS, PhD); mineralogy (M Eng, MS, PhD); paleontology (M Eng, MS, PhD); petroleum geology (M Eng, MS, PhD); petrology (M Eng, MS, PhD); planetary geology (M Eng, MS, PhD); Precambrian geology (M Eng, MS, PhD); Quaternary geology (M Eng, MS, PhD); rock mechanics (M Eng, MS, PhD); sedimentology (M Eng, MS, PhD); seismology (M Eng, MS, PhD); stratigraphy (M Eng, MS, PhD); structural geology (M Eng, MS, PhD). *Faculty:* 39 full-time (5 women). *Students:* 29 full-time (15 women); includes 2 minority (both Hispanic Americans), 7 international. 73 applicants, 15% accepted, 6 enrolled. In 2005, 5 master's, 1 doctorate awarded. *Degree requirements:* For master's, thesis (MS); for doctorate, thesis/dissertation, comprehensive exam. *Entrance requirements:* For master's and doctorate, GRE General Test, 3 letters of recommendation. Additional exam requirements/recommendations for international students: Required—TOEFL (minimum score 550 paper-based; 213 computer-based). *Application deadline:* For fall admission, 1/15 for domestic students. Applications are processed on a rolling basis. Application fee: $60. Electronic applications accepted. *Financial support:* In 2005–06, 8 fellowships with full tuition reimbursements, 10 research assistantships with full tuition reimbursements, 7 teaching assistantships with full tuition reimbursements were awarded; institutionally sponsored loans, scholarships/grants, health care benefits, tuition waivers (full and partial), and unspecified assistantships also available. Financial award applicants required to submit FAFSA. *Faculty research:* Geophysics, structural geology, petrology, geochemistry, geodynamics. *Unit head:* Director of Graduate Studies, 607-255-5466, Fax: 607-254-4780. *Application contact:* Graduate Field Assistant, 607-255-5466, Fax: 607-254-4780, E-mail: gradprog@geology.cornell.edu.

Massachusetts Institute of Technology, School of Science, Department of Earth, Atmospheric, and Planetary Sciences, Cambridge, MA 02139-4307. Offers atmospheric chemistry (PhD, Sc D); atmospheric science (SM, PhD, Sc D); climate physics and chemistry (PhD, Sc D); earth and planetary sciences (SM); geochemistry (PhD, Sc D); geology (PhD, Sc D); geophysics (PhD, Sc D); marine geology and geophysics (SM); oceanography (SM, PhD, Sc D); planetary sciences (PhD, Sc D). *Faculty:* 38 full-time (3 women). *Students:* 165 full-time (63 women); includes 12 minority (7 Asian Americans or Pacific Islanders, 5 Hispanic Americans), 51 international. Average age 27. 179 applicants, 34% accepted, 31 enrolled. In 2005, 2 master's, 15 doctorates awarded. Terminal master's awarded for partial completion of doctoral program. *Degree requirements:* For master's, thesis/dissertation; for doctorate, thesis/dissertation, comprehensive exam. *Entrance requirements:* For master's, GRE General Test; for doctorate, GRE General Test, GRE Subject Test (chemistry or physics for planetary science program). Additional exam requirements/recommendations for international students: Required—TOEFL (minimum score 577 paper-based; 233 computer-based). *Application deadline:* For fall admission, 1/5 for domestic students, 1/5 for international students; for spring admission, 11/1 for domestic students, 11/1 for international students. Application fee: $70. Electronic applications accepted. *Expenses:* Tuition: Full-time $32,100. Required fees: $200. Part-time tuition and fees vary according to course load. *Financial support:* In 2005–06, 115 students received support, including 36 fellowships with tuition reimbursements available (averaging $26,951 per year), 60 research assistantships with tuition reimbursements available (averaging $24,520 per year), 18 teaching assistantships with tuition reimbursements available (averaging $22,773 per year); Federal Work-Study, institutionally sponsored loans, scholarships/grants, health care benefits, and unspecified assistantships also available. *Faculty research:* Evolution of main features of the planetary system; origin, composition, structure, and state of the atmospheres, oceans, surfaces, and interiors of planets; dynamics of planets and satellite motions. Total annual research expenditures: $18.6 million. *Unit head:* Prof. Maria Zuber, Department Head, 617-253-6397, E-mail: zuber@mit.edu. *Application contact:* EAPS Education Office.

University of Delaware, College of Marine Studies, Newark, DE 19716. Offers geology (MS, PhD); marine management (MMM); marine policy (MS); marine studies (MMP, MS, PhD); oceanography (MS, PhD). *Faculty:* 41 full-time (4 women). *Students:* 115 full-time (57 women), 5 part-time (2 women); includes 7 minority (2 African Americans, 4 Asian Americans or Pacific Islanders, 1 Hispanic American), 30 international. Average age 29. 127 applicants, 35% accepted, 24 enrolled. In 2005, 12 master's, 8 doctorates awarded. *Degree requirements:* For master's and doctorate, thesis/dissertation. *Entrance requirements:* For master's and doctorate, GRE General Test. Additional exam requirements/recommendations for international students: Required—TOEFL. *Application deadline:* For fall admission, 3/1 for domestic students; for spring admission, 10/1 for domestic students. Applications are processed on a rolling basis. Application fee: $60. Electronic applications accepted. *Financial support:* In 2005–06, 78 students received support, including 14 fellowships with full tuition reimbursements available (averaging $19,000 per year), 62 research assistantships with full tuition reimbursements available (averaging $19,000 per year), 2 teaching assistantships with full tuition reimbursements available (averaging $19,000 per year); career-related internships or fieldwork, Federal Work-Study, and tuition waivers (full and partial) also available. Financial award application deadline: 3/1. *Faculty research:* Marine biology and biochemistry, oceanography, marine policy, physical ocean science and engineering, ocean engineering. Total annual research expenditures: $10.5 million. *Unit head:* Dr. Nancy Targett, Dean, 302-831-2841. *Application contact:* Lisa Perelli, Coordinator, 302-645-4226, E-mail: lperelli@udel.edu.

University of Hawaii at Manoa, Graduate Division, School of Ocean and Earth Science and Technology, Department of Geology and Geophysics, Honolulu, HI 96822. Offers high-pressure geophysics and geochemistry (MS, PhD); hydrogeology and engineering geology (MS, PhD); marine geology and geophysics (MS, PhD); planetary geosciences and remote sensing (MS, PhD); seismology and solid-earth geophysics (MS, PhD); volcanology, petrology, and geochemistry (MS, PhD). *Faculty:* 65 full-time (12 women), 19 part-time/adjunct (1 woman). *Students:* 50 full-time (23 women), 8 part-time (1 woman); includes 2 minority (1 African American, 1 Asian American or Pacific Islander), 15 international. Average age 29. 89 applicants, 18% accepted, 12 enrolled. In 2005, 9 master's, 6 doctorates awarded. Terminal master's awarded for partial completion of doctoral program. *Median time to degree:* Of those who began their doctoral program in fall 1997, 100% received their degree in 8 years or less. *Degree requirements:* For master's, thesis/dissertation; for doctorate, thesis/dissertation, comprehensive exam. *Entrance requirements:* For master's and doctorate, GRE General Test, minimum GPA of 3.0. Additional exam requirements/recommendations for international students: Required—TOEFL. *Application deadline:* For fall admission, 1/15 for domestic students, 1/1 for international students; for spring admission, 9/1 for domestic students, 8/15 for international students. Application fee: $50. *Expenses:* Tuition, state resident: full-time $8,400; part-time $200 per credit hour. Tuition, nonresident: full-time $11,088; part-time $462 per credit hour. Tuition and fees vary according to program. *Financial support:* In 2005–06, 41 research assistantships (averaging $19,366 per year), 5 teaching assistantships (averaging $18,198 per year) were awarded. *Unit head:* Dr. Charles Fletcher, Chair, 808-956-8763, Fax: 808-956-5512, E-mail: fletcher@soest.hawaii.edu. *Application contact:* Dr. Charles Fletcher, Chair, 808-956-8763, Fax: 808-956-5512, E-mail: fletcher@soest.hawaii.edu.

University of Miami, Graduate School, Rosenstiel School of Marine and Atmospheric Science, Division of Marine Geology and Geophysics, Coral Gables, FL 33124. Offers MS, PhD. *Faculty:* 14 full-time (2 women), 14 part-time/adjunct (4 women). *Students:* 26 full-time (13 women), 8 international. Average age 27. 23 applicants, 26% accepted, 4 enrolled. In 2005, 1 master's, 4 doctorates awarded. Terminal master's awarded for partial completion of doctoral program. *Median time to degree:* Of those who began their doctoral program in fall 1997, 100% received their degree in 8 years or less. *Degree requirements:* For master's and doctorate, thesis/dissertation, comprehensive exam, registration. *Entrance requirements:* For master's and doctorate, GRE General Test. Additional exam requirements/recommendations for international students: Required—TOEFL (minimum score 550 paper-based; 213 computer-based). *Application deadline:* For fall admission, 1/1 priority date for domestic students, 1/1 priority date for international students. Applications are processed on a rolling basis. Application fee: $50. Electronic applications accepted. *Financial support:* In 2005–06, 23 students received support, including 4 fellowships with tuition reimbursements available (averaging $22,380 per year), 15 research assistantships with tuition reimbursements available (averaging $22,380 per year), 4 teaching assistantships with tuition reimbursements available (averaging $22,380 per year); institutionally sponsored loans also available. Financial award application deadline: 3/1; financial award applicants required to submit FAFSA. *Faculty research:* Carbonate sedimentology, low-temperature geochemistry, paleoceanography, geodesy and tectonics. *Unit head:* Dr. Gregor Eberli, Chairperson, 305-421-4678, Fax: 305-421-4632, E-mail: geberli@rsmas.miami.edu. *Application contact:* Dr. Larry Peterson, Associate Dean, 305-421-4155, Fax: 305-421-4771, E-mail: gso@rsmas.miami.edu.

University of Michigan, Horace H. Rackham School of Graduate Studies, College of Literature, Science, and the Arts, Department of Geological Sciences, Program in Oceanography: Marine Geology and Geochemistry, Ann Arbor, MI 48109. Offers MS, PhD. Terminal master's awarded for partial completion of doctoral program. *Degree requirements:* For master's, thesis; for doctorate, thesis/dissertation, oral defense of dissertation, preliminary exam. *Entrance requirements:* For master's and doctorate, GRE General Test. Electronic applications accepted. *Expenses:* Tuition, state resident: full-time $14,082; part-time $894 per credit hour. Tuition, nonresident: full-time $28,500; part-time $1,675 per credit hour. Required fees: $189; $189 per unit. *Faculty research:* Paleoceanography, paleolimnology, marine geochemistry, seismic stratigraphy.

University of Washington, Graduate School, College of Ocean and Fishery Sciences, School of Oceanography, Seattle, WA 98195. Offers biological oceanography (MS, PhD); chemical oceanography (MS, PhD); marine geology and geophysics (MS, PhD); physical oceanography (MS, PhD). Terminal master's awarded for partial completion of doctoral program. *Degree requirements:* For master's, research project; for doctorate, thesis/dissertation. *Entrance requirements:* For master's and doctorate, GRE General Test, minimum GPA of 3.0. Additional exam requirements/recommendations for international students: Required—TOEFL. Electronic applications accepted. *Faculty research:* Global climate change, hydrothermal vent systems, marine microbiology, marine and freshwater biogeochemistry, biological-physical interactions.

Woods Hole Oceanographic Institution, MIT/WHOI Joint Program in Oceanography/Applied Ocean Science and Engineering, Woods Hole, MA 02543-1541. Offers applied ocean sciences (PhD); biological oceanography (PhD, Sc D); chemical oceanography (PhD, Sc D); civil and environmental and oceanographic engineering (PhD); electrical and oceanographic engineering (PhD); geochemistry (PhD); geophysics (PhD); marine biology (PhD); marine geochemistry (PhD, Sc D); marine geology (PhD, Sc D); marine geophysics (PhD); mechanical and oceanographic engineering (PhD); ocean engineering (PhD); oceanographic engineering (M Eng, MS, PhD, Sc D, Eng); paleoceanography (PhD); physical oceanography (PhD, Sc D). MS, PhD, and Sc D offered jointly with MIT. Terminal master's awarded for partial completion of doctoral program. *Degree requirements:* For master's and Eng, thesis (for some programs); for doctorate, thesis/dissertation. *Entrance requirements:* For master's, GRE General Test; for doctorate, GRE General Test, GRE Subject Test. Additional exam requirements/recommendations for international students: Required—TOEFL. Electronic applications accepted.

See Close-Up on page 259.

Mineralogy

Cornell University, Graduate School, Graduate Fields of Engineering, Field of Geological Sciences, Ithaca, NY 14853-0001. Offers economic geology (M Eng, MS, PhD); engineering geology (M Eng, MS, PhD); environmental geophysics (M Eng, MS, PhD); general geology (M Eng, MS, PhD); geobiology (M Eng, MS, PhD); geochemistry and isotope geology (M Eng, MS, PhD); geohydrology (M Eng, MS, PhD); geomorphology (M Eng, MS, PhD); geophysics (M Eng, MS, PhD); geotectonics (M Eng, MS, PhD); marine geology (MS, PhD); mineralogy (M Eng, MS, PhD); paleontology (M Eng, MS, PhD); petroleum geology (M Eng, MS, PhD); petrology (M Eng, MS, PhD); planetary geology (M Eng, MS, PhD); Precambrian geology (M Eng, MS, PhD); Quaternary geology (M Eng, MS, PhD); rock mechanics (M Eng, MS, PhD); sedimentology (M Eng, MS, PhD); seismology (M Eng, MS, PhD); stratigraphy (M Eng, MS, PhD); structural geology (M Eng, MS, PhD). *Faculty:* 39 full-time (5 women). *Students:* 29 full-time (15 women); includes 2 minority (both Hispanic Americans), 7 international. 73 applicants, 15% accepted, 6 enrolled. In 2005, 5 master's, 1 doctorate awarded. *Degree requirements:* For master's, thesis (MS); for doctorate, thesis/dissertation, comprehensive exam. *Entrance requirements:* For master's and doctorate, GRE General Test, 3 letters of recommendation. Additional exam requirements/recommendations for international students: Required—TOEFL (minimum score 550 paper-based; 213 computer-based). *Application deadline:* For fall admission, 1/15 for domestic students. Applications are processed on a rolling basis. Application fee: $60. Electronic applications accepted. *Financial support:* In 2005–06, 8 fellowships with full tuition reimbursements, 10 research assistantships with full tuition reimbursements, 7 teaching assistantships with full tuition reimbursements were awarded; institutionally sponsored loans, scholarships/grants, health care benefits, tuition waivers (full and partial), and unspecified assistantships also available. Financial award applicants required to submit FAFSA. *Faculty research:* Geophysics, structural geology, petrology, geochemistry, geodynamics. *Unit head:* Director of Graduate Studies, 607-255-5466, Fax: 607-254-4780. *Application contact:* Graduate Field Assistant, 607-255-5466, Fax: 607-254-4780, E-mail: gradprog@geology.cornell.edu.

Indiana University Bloomington, Graduate School, College of Arts and Sciences, Department of Geological Sciences, Bloomington, IN 47405-7000. Offers biogeochemistry (MS, PhD); economic geology (MS, PhD); geobiology (MS, PhD); geophysics, structural geology and tectonics (MS, PhD); hydrogeology (MS, PhD); mineralogy (MS, PhD); stratigraphy and sedimentology (MS, PhD). PhD offered through the University Graduate School. Part-time programs available. *Faculty:* 14 full-time (1 woman). *Students:* 30 full-time (13 women), 12 part-time (5 women); includes 5 minority (2 African Americans, 1 American Indian/Alaska Native, 1 Asian American or Pacific Islander, 1 Hispanic American), 21 international. Average age 29. In 2005, 7 master's, 2 doctorates awarded. Terminal master's awarded for partial completion of doctoral program. *Degree requirements:* For master's, one foreign language, thesis or alternative; for doctorate, thesis/dissertation. *Entrance requirements:* For master's and doctorate, GRE General Test. Additional exam requirements/recommendations for international students: Required—TOEFL. *Application deadline:* For fall admission, 1/15 priority date for domestic students, 12/15 priority date for international students; for spring admission, 9/1 priority date for domestic students, 9/1 priority date for international students. Applications are processed on a rolling basis. Application fee: $50 ($60 for international students). *Expenses:* Tuition, state resident: full-time $5,437; part-time $227 per credit hour. Tuition, nonresident: full-time $15,836; part-time $660 per credit hour. Required fees: $821. Tuition and fees vary according to campus/location and program. *Financial support:* Fellowships with tuition reimbursements, research assistantships with full and partial tuition reimbursements, teaching assistantships with tuition reimbursements, career-related internships or fieldwork, Federal Work-Study, and institutionally sponsored loans available. Financial award application deadline: 2/15. *Faculty research:* Geophysics, geochemistry, hydrogeology, igneous and metamorphic petrology and clay mineralogy. Total annual research expenditures: $289,139. *Unit head:* Mark Person, Graduate Advisor, 812-855-7214. *Application contact:* Mary Iverson, Secretary, Committee for Graduate Studies, 812-855-7214, Fax: 812-855-7899, E-mail: geograd@indiana.edu.

Université du Québec à Chicoutimi, Graduate Programs, Program in Mineral Resources, Chicoutimi, QC G7H 2B1, Canada. Offers PhD. Part-time programs available. *Degree requirements:* For doctorate, thesis/dissertation. *Entrance requirements:* For doctorate, appropriate master's degree, proficiency in French.

Université du Québec à Montréal, Graduate Programs, Program in Earth Sciences, Montreal, QC H3C 3P8, Canada. Offers geology-research (M Sc); mineral resources (PhD); nonrenewable resources (DESS). Part-time programs available. Terminal master's awarded for partial completion of doctoral program. *Degree requirements:* For master's, thesis (for some programs); for doctorate, thesis/dissertation. *Entrance requirements:* For master's, appropriate bachelor's degree or equivalent, proficiency in French. *Faculty research:* Economic geology, structural geology, geochemistry, Quaternary geology, isotopic geochemistry.

Université du Québec à Montréal, Graduate Programs, Program in Mineral Resources, Montréal, QC H3C 3P8, Canada. Offers PhD. Part-time programs available. *Degree requirements:* For doctorate, thesis/dissertation. *Entrance requirements:* For doctorate, appropriate master's degree or equivalent and proficiency in French.

University of Illinois at Chicago, Graduate College, College of Liberal Arts and Sciences, Department of Earth and Environmental Sciences, Chicago, IL 60607-7128. Offers crystallography (MS, PhD); environmental geology (MS, PhD); geochemistry (MS, PhD); geology (MS, PhD); geomorphology (MS, PhD); geophysics (MS, PhD); geotechnical engineering and geosciences (PhD); hydrogeology (MS, PhD); low-temperature and organic geochemistry (MS, PhD); mineralogy (MS, PhD); paleoclimatology (MS, PhD); paleontology (MS, PhD); petrology (MS, PhD); quaternary geology (MS, PhD); sedimentology (MS, PhD); water resources (MS, PhD). *Degree requirements:* For master's and doctorate, thesis/dissertation. *Entrance requirements:* For master's and doctorate, GRE General Test, minimum GPA of 2.75. Additional exam requirements/recommendations for international students: Required—TOEFL. Electronic applications accepted.

University of Michigan, Horace H. Rackham School of Graduate Studies, College of Literature, Science, and the Arts, Department of Geological Sciences, Ann Arbor, MI 48109. Offers geology (MS, PhD); mineralogy (MS, PhD); oceanography: marine geology and geochemistry (MS, PhD). Terminal master's awarded for partial completion of doctoral program. *Degree requirements:* For master's, thesis; for doctorate, thesis/dissertation, oral defense of dissertation, preliminary exam. *Entrance requirements:* For master's and doctorate, GRE General Test. Electronic applications accepted. *Expenses:* Tuition, state resident: full-time $14,082; part-time $894 per credit hour. Tuition, nonresident: full-time $28,500; part-time $1,675 per credit hour. Required fees: $189; $189 per unit. *Faculty research:* Isotope geochemistry, paleoclimatology, mineral physics, tectonics, paleontology.

Yale University, Graduate School of Arts and Sciences, Department of Geology and Geophysics, New Haven, CT 06520. Offers geochemistry (PhD); geophysics (PhD); meteorology (PhD); mineralogy and crystallography (PhD); oceanography (PhD); paleoecology (PhD); paleontology and stratigraphy (PhD); petrology (PhD); structural geology (PhD). *Degree requirements:* For doctorate, thesis/dissertation. *Entrance requirements:* For doctorate, GRE General Test. Additional exam requirements/recommendations for international students: Required—TOEFL.

Paleontology

Cornell University, Graduate School, Graduate Fields of Engineering, Field of Geological Sciences, Ithaca, NY 14853-0001. Offers economic geology (M Eng, MS, PhD); engineering geology (M Eng, MS, PhD); environmental geophysics (M Eng, MS, PhD); general geology (M Eng, MS, PhD); geobiology (M Eng, MS, PhD); geochemistry and isotope geology (M Eng, MS, PhD); geohydrology (M Eng, MS, PhD); geomorphology (M Eng, MS, PhD); geophysics (M Eng, MS, PhD); geotectonics (M Eng, MS, PhD); marine geology (MS, PhD); mineralogy (M Eng, MS, PhD); paleontology (M Eng, MS, PhD); petroleum geology (M Eng, MS, PhD); petrology (M Eng, MS, PhD); planetary geology (M Eng, MS, PhD); Precambrian geology (M Eng, MS, PhD); Quaternary geology (M Eng, MS, PhD); rock mechanics (M Eng, MS, PhD); sedimentology (M Eng, MS, PhD); seismology (M Eng, MS, PhD); stratigraphy (M Eng, MS, PhD); structural geology (M Eng, MS, PhD). *Faculty:* 39 full-time (5 women). *Students:* 29 full-time (15 women); includes 2 minority (both Hispanic Americans), 7 international. 73 applicants, 15% accepted, 6 enrolled. In 2005, 5 master's, 1 doctorate awarded. *Degree requirements:* For master's, thesis (MS); for doctorate, thesis/dissertation, comprehensive exam. *Entrance requirements:* For master's and doctorate, GRE General Test, 3 letters of recommendation. Additional exam requirements/recommendations for international students: Required—TOEFL (minimum score 550 paper-based; 213 computer-based). *Application deadline:* For fall admission, 1/15 for domestic students. Applications are processed on a rolling basis. Application fee: $60. Electronic applications accepted. *Financial support:* In 2005–06, 8 fellowships with full tuition reimbursements, 10 research assistantships with full tuition reimbursements, 7 teaching assistantships with full tuition reimbursements were awarded; institutionally sponsored loans, scholarships/grants, health care benefits, tuition waivers (full and partial), and unspecified assistantships also available. Financial award applicants required to submit FAFSA. *Faculty research:* Geophysics, structural geology, petrology, geochemistry, geodynamics. *Unit head:* Director of Graduate Studies, 607-255-5466, Fax: 607-254-4780. *Application contact:* Graduate Field Assistant, 607-255-5466, Fax: 607-254-4780, E-mail: gradprog@geology.cornell.edu.

Duke University, Graduate School, Department of Biological Anthropology and Anatomy, Durham, NC 27710. Offers cellular and molecular biology (PhD); gross anatomy and physical anthropology (PhD), including comparative morphology of human and non-human primates, primate social behavior, vertebrate paleontology; neuroanatomy (PhD). *Faculty:* 6 full-time. *Students:* 14 full-time (9 women); includes 1 minority (African American), 2 international. 39 applicants, 18% accepted, 3 enrolled. In 2005, 3 doctorates awarded. *Degree requirements:* For doctorate, one foreign language, thesis/dissertation. *Entrance requirements:* For doctorate, GRE General Test. Additional exam requirements/recommendations for international students: Required—IELT (preferred) or TOEFL. *Application deadline:* For fall admission, 12/31 for domestic students, 12/31 for international students. Application fee: $75. Electronic applications accepted. *Financial support:* Fellowships, teaching assistantships, Federal Work-Study available. Financial award application deadline: 12/31. *Unit head:* Daniel Schmitt, Director of Graduate Studies, 919-684-5664, Fax: 919-684-8034, E-mail: l.squires@baa.mc.duke.edu.

South Dakota School of Mines and Technology, Graduate Division, College of Engineering, Department of Geology and Geological Engineering, Program in Paleontology, Rapid City, SD 57701-3995. Offers MS. Part-time programs available. *Faculty:* 3 part-time/adjunct (1 woman). *Students:* 1 full-time (0 women), 2 part-time (1 woman). Average age 32. In 2005, 3 degrees awarded. *Degree requirements:* For master's, thesis. *Entrance requirements:* For master's, GRE General Test, GRE Subject Test. Additional exam requirements/recommendations for international students: Required—TOEFL, TWE. *Application deadline:* For fall admission, 7/1 priority date for domestic students, 4/1 priority date for international students; for spring admission, 11/1 for domestic students, 9/1 for international students. Applications are processed on a rolling basis. Application fee: $35. Electronic applications accepted. *Expenses:* Tuition, area resident: Part-time $116 per credit hour. Tuition, state resident: full-time $2,084. Tuition, nonresident: full-time $6,146; part-time $341 per credit hour. Required fees: $1,805; $100 per credit hour. *Financial support:* In 2005–06, 1 fellowship (averaging $2,000 per year), 3 research assistantships with partial tuition reimbursements (averaging $12,349 per year), 4 teaching assistantships with partial tuition reimbursements (averaging $2,940 per year) were awarded; Federal Work-Study and institutionally sponsored loans also available. Support available to part-time students. Financial award application deadline: 5/15. *Faculty research:* Cretaceous vertebrates, Miocene vertebrates, Oligocene vertebrates. *Unit head:* Dr. Gale Bishop, Dean, 605-394-2467. *Application contact:* Jeannette R. Nilson, Program Assistant-Research and Graduate Education, 800-454-8162 Ext. 1206, Fax: 605-394-5360, E-mail: graduate_admissions@silver.sdsmt.edu.

Tulane University, Graduate School, Department of Earth and Environmental Sciences, New Orleans, LA 70118-5669. Offers geology (MS, PhD); paleontology (PhD). *Degree requirements:* For master's, one foreign language, thesis or alternative; for doctorate, one foreign language, thesis/dissertation. *Entrance requirements:* For master's, GRE General Test, minimum B average in undergraduate course work; for doctorate, GRE General Test. Additional exam requirements/recommendations for international students: Required—TOEFL; Recommended—TSE. Electronic applications accepted. *Faculty research:* Sedimentation, isotopes, biogeochemistry, marine geology, structural geology.

See Close-Up on page 233.

University of Chicago, Division of the Physical Sciences, Department of the Geophysical Sciences, Chicago, IL 60637-1513. Offers atmospheric sciences (SM, PhD); earth sciences (SM, PhD); paleobiology (PhD); planetary and space sciences (SM, PhD). *Faculty:* 24 full-time (3 women). *Students:* 35 full-time (15 women); includes 1 minority (Hispanic American), 11 international. Average age 29. 52 applicants, 29% accepted. In 2005, 2 master's, 3 doctorates awarded. Terminal master's awarded for partial completion of doctoral program. *Entrance requirements:* For master's and doctorate, GRE General Test. Additional exam requirements/recommendations for international students: Required—TOEFL, IELT. *Application deadline:* For fall admission, 1/15 for domestic students, 1/15 for international students. Application fee: $55. Electronic applications accepted. *Financial support:* In 2005–06, 32 students received support, including research assistantships with full tuition reimbursements available (averaging $18,675 per year), teaching assistantships with full tuition reimbursements available (averaging $19,098 per year); fellowships, Federal Work-Study, institutionally sponsored loans, scholarships/grants, tuition waivers (partial), and unspecified assistantships also available. Financial award application deadline: 1/15. *Faculty research:* Climatology, evolutionary paleontology, petrology, geochemistry, oceanic sciences. *Unit head:* Dr. David Rowley, Chairman, 773-702-8102, Fax: 773-702-9505. *Application contact:* David J. Leslie,

Paleontology

University of Chicago (continued)
Graduate Student Services Coordinator, 773-702-8180, Fax: 773-702-9505, E-mail: info@geosci.uchicago.edu.

University of Illinois at Chicago, Graduate College, College of Liberal Arts and Sciences, Department of Earth and Environmental Sciences, Chicago, IL 60607-7128. Offers crystallography (MS, PhD); environmental geology (MS, PhD); geochemistry (MS, PhD); geology (MS, PhD); geomorphology (MS, PhD); geophysics (MS, PhD); geotechnical engineering and geosciences (PhD); hydrogeology (MS, PhD); low-temperature and organic geochemistry (MS, PhD); mineralogy (MS, PhD); paleoclimatology (MS, PhD); paleontology (MS, PhD); petrology (MS, PhD); quaternary geology (MS, PhD); sedimentology (MS, PhD); water resources (MS, PhD). *Degree requirements:* For master's and doctorate, thesis/dissertation. *Entrance requirements:* For master's and doctorate, GRE General Test, minimum GPA of 2.75. Additional exam requirements/recommendations for international students: Required—TOEFL. Electronic applications accepted.

West Virginia University, Eberly College of Arts and Sciences, Department of Geology and Geography, Program in Geology, Morgantown, WV 26506. Offers geomorphology (MS, PhD); geophysics (MS, PhD); hydrogeology (MS); hydrology (PhD); paleontology (MS, PhD); petrology (MS, PhD); stratigraphy (MS, PhD); structure (MS, PhD). Part-time programs available. *Students:* 26 full-time (7 women), 17 part-time (7 women); includes 2 minority (1 American Indian/Alaska Native, 1 Asian American or Pacific Islander), 12 international. Average age 30. 34 applicants, 29% accepted. In 2005, 6 master's, 1 doctorate awarded. Terminal master's awarded for partial completion of doctoral program. *Degree requirements:* For master's, thesis/dissertation; for doctorate, thesis/dissertation, comprehensive exam. *Entrance requirements:* For master's, GRE General Test, GRE Subject Test, minimum GPA of 2.5; for doctorate, GRE General Test, GRE Subject Test, minimum GPA of 3.3. Additional exam requirements/recommendations for international students: Required—TOEFL. *Application deadline:* For fall admission, 2/1 for domestic students; for spring admission, 10/1 for domestic students. Applications are processed on a rolling basis. Application fee: $45. *Expenses:* Tuition, state resident: full-time $4,582; part-time $258 per credit hour. Tuition, nonresident: full-time $1,382; part-time $741 per credit hour. *Financial support:* In 2005–06, 5 research assistantships, 18 teaching assistantships were awarded; career-related internships or fieldwork, Federal Work-Study, institutionally sponsored loans, and tuition waivers (full and partial) also available. Financial award application deadline: 2/1; financial award applicants required to submit FAFSA. *Unit head:* Dr. Thomas H. Wilson, Associate Chair, 304-293-5603 Ext. 4316, Fax: 304-293-6522, E-mail: tom.wilson@mail.wvu.edu. *Application contact:* Dr. Joe Donovan, Associate Professor, 304-293-5603 Ext. 4308, Fax: 304-293-6522, E-mail: joe.donovan@mail.wvu.edu.

Yale University, Graduate School of Arts and Sciences, Department of Geology and Geophysics, New Haven, CT 06520. Offers geochemistry (PhD); geophysics (PhD); meteorology (PhD); mineralogy and crystallography (PhD); oceanography (PhD); paleoecology (PhD); paleontology and stratigraphy (PhD); petrology (PhD); structural geology (PhD). *Degree requirements:* For doctorate, thesis/dissertation. *Entrance requirements:* For doctorate, GRE General Test. Additional exam requirements/recommendations for international students: Required—TOEFL.

Planetary and Space Sciences

Air Force Institute of Technology, Graduate School of Engineering and Management, Department of Operational Sciences, Dayton, OH 45433-7765. Offers logistics management (MS); operations research (MS, PhD); space operations (MS). Part-time programs available. *Faculty:* 23 full-time (1 woman), 1 part-time/adjunct (0 women). *Students:* 165 full-time (22 women), 1 part-time. Average age 33. In 2005, 68 master's, 4 doctorates awarded. *Degree requirements:* For master's and doctorate, thesis/dissertation. *Entrance requirements:* For doctorate, GRE General Test, minimum GPA of 3.0, must be U.S. citizen. *Application deadline:* For fall admission, 3/1 for domestic students. Applications are processed on a rolling basis. *Financial support:* Fellowships, research assistantships with tuition reimbursements, scholarships/grants available. Financial award application deadline: 3/15. *Faculty research:* Optimization, simulation, combat modeling and analysis, reliability and maintainability, resource scheduling. Total annual research expenditures: $1.1 million. *Unit head:* Lt. Col. Jeffery D. Weir, Head, 937-255-2549, Fax: 937-656-4943, E-mail: jeffery.weir@afit.edu. *Application contact:* Lt. Col. John O. Miller, Information Contact, 937-255-3636 Ext. 4326, E-mail: john.miller@afit.edu.

California Institute of Technology, Division of Geological and Planetary Sciences, Pasadena, CA 91125-0001. Offers cosmochemistry (PhD); geobiology (PhD); geochemistry (MS, PhD); geology (MS, PhD); geophysics (MS, PhD); planetary science (MS, PhD). Part-time programs available. *Degree requirements:* For doctorate, thesis/dissertation. *Entrance requirements:* For doctorate, GRE General Test. Additional exam requirements/recommendations for international students: Required—TOEFL. Electronic applications accepted. *Faculty research:* Astronomy, evolution of anaerobic respiratory processes, structural geology and tectonics, theoretical and numerical seismology, global biogeochemical cycles.

Columbia University, Graduate School of Arts and Sciences, Division of Natural Sciences, Program in Atmospheric and Planetary Science, New York, NY 10027. Offers M Phil, PhD. Offered jointly through the Departments of Geological Sciences, Astronomy, and Physics and in cooperation with NASA Goddard Space Flight Center's Institute for Space Studies. *Degree requirements:* For doctorate, variable foreign language requirement, thesis/dissertation. *Entrance requirements:* For doctorate, GRE General Test, GRE Subject Test, previous course work in mathematics and physics. Additional exam requirements/recommendations for international students: Required—TOEFL. Application fee: $75. *Expenses:* Tuition: Full-time $31,448. Tuition and fees vary according to course level, course load, campus/location and program. *Financial support:* Available to part-time students. Application deadline: 1/5; *Faculty research:* Climate, weather prediction. *Unit head:* Wallace S. Broecker, Chair, 914-365-8413, Fax: 845-365-8169.

See Close-Up on page 289.

Cornell University, Graduate School, Graduate Fields of Arts and Sciences, Field of Astronomy and Space Sciences, Ithaca, NY 14853-0001. Offers astronomy (PhD); astrophysics (PhD); general space sciences (PhD); infrared astronomy (PhD); planetary studies (PhD); radio astronomy (PhD); radiophysics (PhD); theoretical astrophysics (PhD). *Faculty:* 38 full-time (3 women). *Students:* 31 full-time (10 women); includes 1 minority (Asian American or Pacific Islander), 11 international. 110 applicants, 21% accepted, 7 enrolled. In 2005, 6 doctorates awarded. *Degree requirements:* For doctorate, thesis/dissertation, comprehensive exam. *Entrance requirements:* For doctorate, GRE General Test, GRE Subject Test (physics), 3 letters of recommendation. Additional exam requirements/recommendations for international students: Required—TOEFL (minimum score 600 paper-based; 250 computer-based). *Application deadline:* For fall admission, 1/15 for domestic students. Application fee: $60. Electronic applications accepted. *Financial support:* In 2005–06, 31 students received support, including 3 fellowships with full tuition reimbursements available, 19 research assistantships with full tuition reimbursements available, 9 teaching assistantships with full tuition reimbursements available; institutionally sponsored loans, scholarships/grants, health care benefits, tuition waivers (full and partial), and unspecified assistantships also available. Financial award applicants required to submit FAFSA. *Faculty research:* Observational astrophysics, planetary sciences, cosmology, instrumentation, gravitational astrophysics. *Unit head:* Director of Graduate Studies, 607-255-4341. *Application contact:* Graduate Field Assistant, 607-255-4341, E-mail: oconnor@astro.cornell.edu.

Embry-Riddle Aeronautical University, Daytona Beach Campus Graduate Program, Department of Physical Sciences, Daytona Beach, FL 32114-3900. Offers space science (MS Sp C); space science (engineering physics) (MS). Part-time and evening/weekend programs available. *Students:* 15 full-time (5 women), 2 part-time; includes 3 minority (1 African American, 1 Asian American or Pacific Islander, 1 Hispanic American), 2 international. Average age 25. 15 applicants, 67% accepted, 6 enrolled. In 2005, 8 degrees awarded. *Application deadline:* For fall admission, 8/1 for domestic students; for spring admission, 12/1 priority date for domestic students. Applications are processed on a rolling basis. Application fee: $50. *Expenses:* Tuition: Full-time $11,700; part-time $955 per credit. Required fees: $730. *Financial support:* In 2005–06, 16 students received support, including 4 research assistantships with full and partial tuition reimbursements available (averaging $6,044 per year), 8 teaching assistantships with full and partial tuition reimbursements available (averaging $6,044 per year); career-related internships or fieldwork also available. Financial award application deadline: 4/15; financial award applicants required to submit FAFSA. *Unit head:* Dr. Peter Erdman, Program Coordinator, 386-226-6712, Fax: 386-226-6621, E-mail: erdmanp@erau.edu. *Application contact:* Alicia Richardson, Associate Director, International and Graduate Admissions, 800-388-3728, Fax: 386-226-7070, E-mail: graduate.admissions@erau.edu.

Florida Institute of Technology, Graduate Programs, College of Science, Department of Physics and Space Sciences, Melbourne, FL 32901-6975. Offers physics (MS, PhD); space science (MS, PhD). Part-time programs available. *Faculty:* 9 full-time (0 women). *Students:* 18 full-time (3 women), 13 part-time (4 women); includes 3 minority (1 African American, 2 Hispanic Americans), 10 international. Average age 31. 49 applicants, 35% accepted, 8 enrolled. In 2005, 2 master's, 3 doctorates awarded. Terminal master's awarded for partial completion of doctoral program. *Degree requirements:* For master's, oral exam, thesis optional; for doctorate, one foreign language, thesis/dissertation, publication in referred journal, seminar on dissertation research, comprehensive exam, registration. *Entrance requirements:* For master's, GRE General Test, GRE Subject Test, minimum GPA of 3.0, proficiency in a computer language, resumé, 3 letters of recommendation, vector analysis; for doctorate, GRE General Test, GRE Subject Test, minimum GPA of 3.2, resumé, 3 letters of recommendation, proficiency in a computer program language. Additional exam requirements/recommendations for international students: Required—TOEFL (minimum score 550 paper-based; 213 computer-based). *Application deadline:* Applications are processed on a rolling basis. Application fee: $50. Electronic applications accepted. *Expenses:* Tuition: Part-time $825 per credit. *Financial support:* In 2005–06, 15 students received support, including 3 research assistantships with full and partial tuition reimbursements available (averaging $10,630 per year), 12 teaching assistantships with full and partial tuition reimbursements available (averaging $10,890 per year); career-related internships or fieldwork and tuition remissions also available. Financial award application deadline: 3/1; financial award applicants required to submit FAFSA. *Faculty research:* Lasers, semiconductors, magnetism, quantum devices, high energy physics. Total annual research expenditures: $1.4 million. *Unit head:* Dr. Laszlo A. Baksay, Department Head, 321-674-7367, Fax: 321-674-7482, E-mail: baksay@fit.edu. *Application contact:* Carolyn P. Farrior, Director of Graduate Admissions, 321-674-7118, Fax: 321-723-9468, E-mail: cfarrior@fit.edu.

Harvard University, Graduate School of Arts and Sciences, Department of Earth and Planetary Sciences, Cambridge, MA 02138. Offers AM, PhD. *Faculty:* 27 full-time (3 women). *Students:* 51 full-time (25 women); includes 15 minority (11 Asian Americans or Pacific Islanders, 4 Hispanic Americans). 103 applicants, 18% accepted, 11 enrolled. In 2005, 11 master's, 4 doctorates awarded. Terminal master's awarded for partial completion of doctoral program. *Degree requirements:* For master's, registration; for doctorate, thesis/dissertation, comprehensive exam, registration. *Entrance requirements:* For doctorate, GRE General Test. Additional exam requirements/recommendations for international students: Required—TOEFL. *Application deadline:* For fall admission, 1/2 for domestic students. Application fee: $85. Electronic applications accepted. *Expenses:* Tuition: Full-time $28,752. Full-time tuition and fees vary according to program and student level. *Financial support:* In 2005–06, 50 students received support; fellowships with full tuition reimbursements available, research assistantships, teaching assistantships, scholarships/grants and health care benefits available. *Faculty research:* Economic geography, geochemistry, geophysics, mineralogy, crystallography. *Unit head:* Ventatesh Narayanamurti, Dean, 617-495-5829, Fax: 617-495-5264, E-mail: venky@deas.harvard.edu. *Application contact:* Office of Admissions and Financial Aid, 617-495-5315.

Massachusetts Institute of Technology, School of Science, Department of Earth, Atmospheric, and Planetary Sciences, Cambridge, MA 02139-4307. Offers atmospheric chemistry (PhD, Sc D); atmospheric science (SM, PhD, Sc D); climate physics and chemistry (PhD, Sc D); earth and planetary sciences (SM); geochemistry (PhD, Sc D); geology (PhD, Sc D); geophysics (PhD, Sc D); marine geology and geophysics (SM); oceanography (SM, PhD, Sc D); planetary sciences (PhD, Sc D). *Faculty:* 38 full-time (3 women). *Students:* 165 full-time (63 women); includes 12 minority (7 Asian Americans or Pacific Islanders, 5 Hispanic Americans), 51 international. Average age 27. 179 applicants, 34% accepted, 31 enrolled. In 2005, 2 master's, 15 doctorates awarded. Terminal master's awarded for partial completion of doctoral program. *Degree requirements:* For master's, thesis/dissertation; for doctorate, thesis/dissertation, comprehensive exam. *Entrance requirements:* For master's, GRE General Test; for doctorate, GRE General Test, GRE Subject Test (chemistry or physics for planetary science program). Additional exam requirements/recommendations for international students: Required—TOEFL (minimum score 577 paper-based; 233 computer-based). *Application deadline:* For fall admission, 1/5 for domestic students, 1/5 for international students; for spring admission, 11/1 for domestic students, 11/1 for international students. Application fee: $70. Electronic applications accepted. *Expenses:* Tuition: Full-time $32,100. Required fees: $200. Part-time tuition and fees vary according to course load. *Financial support:* In 2005–06, 115 students received support, including 36 fellowships with tuition reimbursements available (averaging $26,951 per year), 60 research assistantships with tuition reimbursements available (averaging $24,520 per year), 18 teaching assistantships with tuition reimbursements available (averaging $22,773 per year); Federal Work-Study, institutionally sponsored loans, scholarships/grants, health care benefits, and unspecified assistantships also available. *Faculty research:* Evolution of main features of the planetary system; origin, composition, structure, and state of the atmospheres, oceans, surfaces, and interiors of planets; dynamics of planets and satellite motions. Total annual research expenditures: $18.6 million. *Unit head:* Prof. Maria Zuber, Department Head, 617-253-6397, E-mail: zuber@mit.edu. *Application contact:* EAPS Education Office.

McGill University, Faculty of Graduate and Postdoctoral Studies, Faculty of Science, Department of Earth and Planetary Sciences, Montréal, QC H3A 2T5, Canada. Offers M Sc, PhD. *Degree requirements:* For master's, thesis/dissertation, registration; for doctorate, thesis/dissertation, comprehensive exam, registration. *Entrance requirements:* For master's, minimum GPA of 3.0; academic background in geology, geophysics, chemistry, or physics; for doctorate, B Sc or M Sc. Additional exam requirements/recommendations for international students: Required—TOEFL (minimum score 550 paper-based; 213 computer-based), IELT (minimum score

7). Electronic applications accepted. *Faculty research:* Geochemistry, sedimentary petrology, igneous petrology, theoretical geophysics, economic geology.

Stony Brook University, State University of New York, Graduate School, College of Arts and Sciences, Department of Physics and Astronomy, Program in Astronomy, Stony Brook, NY 11794. Offers earth and space sciences (MS, PhD). *Degree requirements:* For master's, thesis or alternative; for doctorate, thesis/dissertation. *Entrance requirements:* For master's and doctorate, GRE General Test, minimum GPA of 3.0. Additional exam requirements/recommendations for international students: Required—TOEFL. *Application deadline:* For fall admission, 1/15 for domestic students. Application fee: $50. *Expenses:* Tuition, area resident: Part-time $288. Tuition, state resident: full-time $6,900. Tuition, nonresident: full-time $10,920; part-time $455. Required fees: $704. *Financial support:* Fellowships, research assistantships, teaching assistantships available. Financial award application deadline: 2/1. *Application contact:* Dr. Peter Stephens, Director, 631-632-8279, Fax: 631-632-8176, E-mail: pstephens@notes.cc.sunysb.edu.

The University of Arizona, Graduate College, College of Science, Department of Planetary Sciences/Lunar and Planetary Laboratory, Tucson, AZ 85721. Offers MS, PhD. *Degree requirements:* For master's, thesis (for some programs); for doctorate, one foreign language, thesis/dissertation. *Entrance requirements:* For master's and doctorate, GRE General Test, GRE Subject Test. Additional exam requirements/recommendations for international students: Required—TOEFL. *Faculty research:* Cosmochemistry, planetary geology, astronomy, space physics, planetary physics.

See Close-Up on page 235.

University of California, Los Angeles, Graduate Division, College of Letters and Science, Department of Earth and Space Sciences, Los Angeles, CA 90095. Offers geochemistry (MS, PhD); geology (MS, PhD); geophysics and space physics (MS, PhD). *Degree requirements:* For master's, comprehensive exams or thesis; for doctorate, thesis/dissertation, oral and written qualifying exams. *Entrance requirements:* For master's, GRE General Test, minimum GPA of 3.0; for doctorate, GRE General Test, minimum undergraduate GPA of 3.0. Electronic applications accepted.

University of Chicago, Division of the Physical Sciences, Department of the Geophysical Sciences, Chicago, IL 60637-1513. Offers atmospheric sciences (SM, PhD); earth sciences (SM, PhD); paleobiology (PhD); planetary and space sciences (SM, PhD). *Faculty:* 24 full-time (3 women). *Students:* 35 full-time (15 women); includes 1 minority (Hispanic American), 11 international. Average age 29. 52 applicants, 29% accepted. In 2005, 2 master's, 3 doctorates awarded. Terminal master's awarded for partial completion of doctoral program. *Entrance requirements:* For master's and doctorate, GRE General Test. Additional exam requirements/recommendations for international students: Required—TOEFL, IELT. *Application deadline:* For fall admission, 1/15 for domestic students, 1/15 for international students. Application fee: $55. Electronic applications accepted. *Financial support:* In 2005–06, 32 students received support, including research assistantships with full tuition reimbursements available (averaging $18,675 per year), teaching assistantships with full tuition reimbursements available (averaging $19,098 per year); fellowships, Federal Work-Study, institutionally sponsored loans, scholarships/grants, tuition waivers (partial), and unspecified assistantships also available. Financial award application deadline: 1/15. *Faculty research:* Climatology, evolutionary paleontology, petrology, geochemistry, oceanic sciences. *Unit head:* Dr. David Rowley, Chairman, 773-702-8102, Fax: 773-702-9505. *Application contact:* David J. Leslie, Graduate Student Services Coordinator, 773-702-8180, Fax: 773-702-9505, E-mail: info@geosci.uchicago.edu.

University of Hawaii at Manoa, Graduate Division, School of Ocean and Earth Science and Technology, Department of Geology and Geophysics, Honolulu, HI 96822. Offers high-pressure geophysics and geochemistry (MS, PhD); hydrogeology and engineering geology (MS, PhD); marine geology and geophysics (MS, PhD); planetary geosciences and remote sensing (MS, PhD); seismology and solid-earth geophysics (MS, PhD); volcanology, petrology, and geochemistry (MS, PhD). *Faculty:* 65 full-time (12 women), 19 part-time/adjunct (1 woman). *Students:* 50 full-time (23 women), 8 part-time (1 woman); includes 2 minority (1 African American, 1 Asian American or Pacific Islander), 15 international. Average age 29. 89 applicants, 18% accepted, 12 enrolled. In 2005, 9 master's, 6 doctorates awarded. Terminal master's awarded for partial completion of doctoral program. *Median time to degree:* Of those who began their doctoral program in fall 1997, 100% received their degree in 8 years or less. *Degree requirements:* For master's, thesis/dissertation; for doctorate, thesis/dissertation, comprehensive exam. *Entrance requirements:* For master's and doctorate, GRE General Test, minimum GPA of 3.0. Additional exam requirements/recommendations for international students: Required—TOEFL. *Application deadline:* For fall admission, 1/15 for domestic students, 1/1 for international students; for spring admission, 9/1 for domestic students, 8/15 for international students. Application fee: $50. *Expenses:* Tuition, state resident: full-time $8,400; part-time $200 per credit hour. Tuition, nonresident: full-time $11,088; part-time $462 per credit hour. Tuition and fees vary according to program. *Financial support:* In 2005–06, 41 research assistantships (averaging $19,366 per year), 5 teaching assistantships (averaging $18,198 per year) were awarded. *Unit head:* Dr. Charles Fletcher, Chair, 808-956-8763, Fax: 808-956-5512, E-mail: fletcher@soest.hawaii.edu. *Application contact:* Dr. Charles Fletcher, Chair, 808-956-8763, Fax: 808-956-5512, E-mail: fletcher@soest.hawaii.edu.

University of Michigan, Horace H. Rackham School of Graduate Studies, College of Engineering, Department of Atmospheric, Oceanic, and Space Sciences, Ann Arbor, MI 48109. Offers atmospheric (MS); atmospheric and space sceinces (PhD); geoscience and remote sensing (PhD); space and planetary sciences (PhD); space engineering (M Eng); space sciences (MS). Part-time programs available. *Faculty:* 34 full-time (6 women). *Students:* 53 full-time (21 women), 1 part-time; includes 5 minority (1 American Indian/Alaska Native, 3 Asian Americans or Pacific Islanders, 1 Hispanic American), 22 international. Average age 24. 53 applicants, 28% accepted, 10 enrolled. In 2005, 22 master's, 2 doctorates awarded. Terminal master's awarded for partial completion of doctoral program. *Degree requirements:* For master's, thesis (for some programs); for doctorate, thesis/dissertation, oral defense of dissertation, preliminary exams. *Entrance requirements:* For master's and doctorate, GRE General Test. Additional exam requirements/recommendations for international students: Required—TOEFL. *Application deadline:* For fall admission, 1/15 priority date for domestic students, 1/15 priority date for international students. Applications are processed on a rolling basis. Application fee: $60 ($75 for international students). Electronic applications accepted. *Financial support:* In 2005–06, 3 fellowships with tuition reimbursements (averaging $16,163 per year), 42 research assistantships with tuition reimbursements (averaging $15,159 per year), 3 teaching assistantships with tuition reimbursements (averaging $14,396 per year) were awarded; career-related internships or fieldwork, Federal Work-Study, institutionally sponsored loans, and health care benefits also available. Support available to part-time students. Financial award application deadline: 3/15; financial award applicants required to submit FAFSA. *Faculty research:* Modeling of atmospheric and aerosol chemistry, radiative transfer, remote sensing, atmospheric dynamics, space weather modeling. Total annual research expenditures: $18 million. *Unit head:* Tamas Gombosi, Chair, 734-764-7222, Fax: 734-615-4645, E-mail: tamas@umich.edu. *Application contact:* Margaret Reid, Student Services Associate, 734-936-0482, Fax: 734-763-0437, E-mail: aoss.um@umich.edu.

University of New Mexico, Graduate School, College of Arts and Sciences, Department of Earth and Planetary Sciences, Albuquerque, NM 87131-2039. Offers MS, PhD. *Faculty:* 20 full-time (5 women), 7 part-time/adjunct (2 women). *Students:* 39 full-time (26 women), 12 part-time (3 women); includes 3 minority (all Hispanic Americans), 3 international. Average age 29. 69 applicants, 20% accepted, 8 enrolled. In 2005, 17 master's, 2 doctorates awarded. *Degree requirements:* For master's and doctorate, thesis/dissertation, comprehensive exam, registration. *Entrance requirements:* For master's and doctorate, GRE General Test. Additional exam requirements/recommendations for international students: Required—TOEFL. *Application deadline:* For fall admission, 1/31 for domestic students; for spring admission, 11/1 priority date for domestic students. Electronic applications accepted. *Expenses:* Tuition, nonresident: full-time $3,388; part-time $238 per credit hour. Required fees: $385 per term. Tuition and fees vary according to course load and program. *Financial support:* In 2005–06, 3 fellowships with full tuition reimbursements (averaging $15,000 per year), 18 research assistantships with full tuition reimbursements (averaging $14,050 per year), 13 teaching assistantships with full tuition reimbursements (averaging $12,050 per year) were awarded; scholarships/grants and health care benefits also available. Financial award application deadline: 3/1; financial award applicants required to submit FAFSA. *Faculty research:* Geochemistry, meteoritics, tectonics, igcodynamics, climate and surface processes. Total annual research expenditures: $2.5 million. *Unit head:* Dr. Les D. McFadden, Chair, 505-277-4204, Fax: 505-277-8843, E-mail: lmcfadnm@unm.edu. *Application contact:* Cindy Jaramillo, Administrative Assistant II, 505-277-1635, Fax: 505-277-8843, E-mail: epsdept@unm.edu.

University of North Dakota, Graduate School, John D. Odegard School of Aerospace Sciences, Space Studies Program, Grand Forks, ND 58202. Offers MS. Part-time programs available. Postbaccalaureate distance learning degree programs offered (minimal on-campus study). *Faculty:* 7 full-time (0 women). *Students:* 5 full-time (3 women), 95 part-time (24 women). 43 applicants, 63% accepted, 19 enrolled. In 2005, 25 degrees awarded. *Degree requirements:* For master's, thesis or alternative, comprehensive exam. *Entrance requirements:* For master's, minimum GPA of 3.0. Additional exam requirements/recommendations for international students: Required—TOEFL (minimum score 550 paper-based; 213 computer-based). *Application deadline:* For fall admission, 2/15 priority date for domestic students, 2/15 priority date for international students; for spring admission, 10/15 priority date for domestic students, 10/15 priority date for international students. Applications are processed on a rolling basis. Application fee: $35. Electronic applications accepted. *Financial support:* In 2005–06, 13 research assistantships with full tuition reimbursements (averaging $7,151 per year), 1 teaching assistantship (averaging $5,206 per year) were awarded; fellowships, career-related internships or fieldwork, Federal Work-Study, institutionally sponsored loans, scholarships/grants, tuition waivers (full and partial), and unspecified assistantships also available. Support available to part-time students. Financial award application deadline: 3/15; financial award applicants required to submit FAFSA. *Faculty research:* Earth-approaching asteroids, international remote sensing statutes, Mercury fly-by design, origin of meteorites, craters on Venus. *Unit head:* Dr. Eligar Sadeh, Graduate Director, 701-777-3462, Fax: 701-777-3711, E-mail: sadeh@aero.und.edu. *Application contact:* Brenda Halle, Admissions Specialist, 701-777-2947, Fax: 701-777-3619, E-mail: brendahalle@mail.und.edu.

University of Pittsburgh, School of Arts and Sciences, Department of Geology and Planetary Science, Pittsburgh, PA 15260. Offers geographical information systems (PM Sc); geology and planetary science (MS, PhD). Part-time programs available. *Faculty:* 12 full-time (1 woman), 2 part-time/adjunct (0 women). *Students:* 20 full-time (6 women), 12 part-time (4 women); includes 2 minority (1 Asian American or Pacific Islander, 1 Hispanic American), 2 international. Average age 30. 43 applicants, 21% accepted, 7 enrolled. In 2005, 5 master's, 1 doctorate awarded. *Median time to degree:* Of those who began their doctoral program in fall 1997, 100% received their degree in 8 years or less. *Degree requirements:* For master's, thesis, oral thesis defense; for doctorate, thesis/dissertation, oral dissertation defense. *Entrance requirements:* For master's and doctorate, GRE General Test. Additional exam requirements/recommendations for international students: Required—TOEFL. *Application deadline:* For fall admission, 8/1 priority date for domestic students, 4/30 priority date for international students. For winter admission, 12/1 for domestic students; for spring admission, 4/1 for domestic students. Applications are processed on a rolling basis. Application fee: $50. Electronic applications accepted. *Expenses:* Tuition, state resident: full-time $13,194; part-time $537 per credit. Tuition, nonresident: full-time $25,012; part-time $1,026 per credit. Required fees: $700; $164 per term. Tuition and fees vary according to campus/location and program. *Financial support:* In 2005–06, 19 students received support, including fellowships with tuition reimbursements available (averaging $13,690 per year), 7 research assistantships with tuition reimbursements available (averaging $13,200 per year), 12 teaching assistantships with tuition reimbursements available (averaging $13,555 per year); career-related internships or fieldwork, Federal Work-Study, institutionally sponsored loans, scholarships/grants, and tuition waivers (full and partial) also available. Support available to part-time students. Financial award application deadline: 2/1; financial award applicants required to submit FAFSA. *Faculty research:* Geographical information systems, hydrology, low temperature geochemistry, radiogenic isotopes, volcanology. Total annual research expenditures: $745,848. *Unit head:* Dr. Brian W Stewart, Chair, 412-624-8783, Fax: 412-624-3914, E-mail: bstewart@pitt.edu. *Application contact:* Dr. Thomas Anderson, Graduate Adviser, 412-624-8870, Fax: 412-624-3914, E-mail: taco@pitt.edu.

Washington University in St. Louis, Graduate School of Arts and Sciences, Department of Earth and Planetary Sciences, St. Louis, MO 63130-4899. Offers earth and planetary sciences (MA); geochemistry (PhD); geology (MA, PhD); geophysics (PhD); planetary sciences (PhD). Terminal master's awarded for partial completion of doctoral program. *Degree requirements:* For master's and doctorate, thesis/dissertation. *Entrance requirements:* For master's and doctorate, GRE General Test. Electronic applications accepted.

Western Connecticut State University, Division of Graduate Studies, School of Arts and Sciences, Department of Physics, Astronomy and Meteorology, Danbury, CT 06810-6885. Offers earth and planetary sciences (MA). Part-time and evening/weekend programs available. *Degree requirements:* For master's, thesis. *Entrance requirements:* For master's, minimum GPA of 2.5.

York University, Faculty of Graduate Studies, Faculty of Pure and Applied Science, Program in Earth and Space Science, Toronto, ON M3J 1P3, Canada. Offers M Sc, PhD. Part-time and evening/weekend programs available. *Faculty:* 46 full-time (4 women), 25 part-time/adjunct (3 women). *Students:* 51 full-time (15 women), 15 part-time (1 woman). 41 applicants. In 2005, 10 master's, 1 doctorate awarded. *Degree requirements:* For master's, thesis or alternative, registration; for doctorate, thesis/dissertation, registration. *Application deadline:* Applications are processed on a rolling basis. Application fee: $80. Electronic applications accepted. *Expenses:* Tuition, state resident: full-time $3,190; part-time $798 per term. International tuition: $7,515 full-time. Required fees: $217. Tuition and fees vary according to program. *Financial support:* In 2005–06, fellowships (averaging $3,060 per year), research assistantships (averaging $10,259 per year), teaching assistantships (averaging $7,493 per year) were awarded; fee bursaries also available. *Unit head:* Peter Taylor, Director, 416-736-5247.

BOSTON COLLEGE

Graduate School of Arts and Sciences
Department of Geology and Geophysics

Programs of Study

The Department of Geology and Geophysics at Boston College offers research and educational programs that focus on understanding the Earth's complex systems and on the evolution of the Earth over its billions of years of geologic history. The academic programs cover a broad range of topics, including coastline and estuarine processes, earthquake and exploration seismology, environmental geology and geophysics, process geomorphology, groundwater hydrology, igneous and metamorphic petrology and geochemistry, physical sedimentation, plate tectonics, and structural geology.

The department offers Master of Science (M.S.) degrees in geology or geophysics. Students are encouraged to obtain broad backgrounds by taking courses in geology, geophysics, and environmental areas as well as the other sciences and mathematics. Multidisciplinary preparation is particularly useful for students seeking future employment in industry. There are approximately 20 graduate students in the department. The program prepares students for careers as geoscientists in industry, oil exploration, or government service or for continued studies toward a Ph.D. degree.

In conjunction with the Carroll Graduate School of Management at Boston College, the Department of Geology and Geophysics also offers a combined M.S./M.B.A. degree program. This program prepares students for careers in industrial or financial geoscience management, including such areas as the environmental and petroleum industries, natural hazard assessment, and natural resource evaluation and investment.

Students enjoy close working relationships with faculty members while being able to undertake research using the most modern scientific equipment available. The program stresses a strong background in the earth sciences as well as the ability to carry out research. The department offers a number of teaching and research assistantships.

Research Facilities

The Department of Geology and Geophysics is housed in Devlin Hall and has additional research facilities at Weston Observatory, a geophysical research laboratory located in Weston, Massachusetts, approximately 13 miles west of downtown Boston. The observatory houses seismic instruments for the World-Wide Standardized Seismic Network and for the New England Seismic Network. Observatory staff members monitor the Northeast United States for seismic activity and disseminate information about any earthquakes that are recorded.

Research in the department covers a broad range of topics, including coastal and estuarine processes, physical sedimentation, earthquake and exploration seismology, geomorphology, structural geology, igneous and metamorphic petrology and geochemistry, global change geochemistry, interpretative tectonics, groundwater hydrology, and environmental geology and geophysics.

Financial Aid

Boston College offers the following stipends and scholarships to academically and financially qualified students: graduate assistantships, teaching fellowships, research assistantships, tuition scholarships, teaching assistantships, and university fellowships. Students are routinely considered for financial aid by the department in which they hope to study; no additional application is necessary. The amounts of the awards and the number of years for which they are awarded vary by department. Research grants are also available on a year-by-year basis and vary among academic disciplines.

Student loans and work-study programs are also available. The Boston College Office of Student Services administers and awards need-based federal financial aid programs, including Federal Subsidized and Unsubsidized Stafford Loans, Perkins Loans, and Federal Work Study.

Cost of Study

Tuition for the Graduate School of Arts and Sciences was $990 per credit for the 2005–06 academic year. Student activity fees ranged from $25 to $50 per semester, depending on the number of credits taken. The cost of books varies each semester. Additional fees may apply.

Living and Housing Costs

Boston College does not currently offer on-campus graduate housing. However, rental housing is plentiful in Newton as well as in surrounding cities and towns. There are many different types of housing available, ranging from one-room rentals in large Victorian homes, to triple-decker brownstones and apartment high-rises. Allston-Brighton and Jamaica Plain are among the nearby Boston neighborhoods that attract students from many colleges and universities because of their diverse communities and relative affordability. Graduate students also have found Newton, Brookline, Waltham, Watertown, and Boston's West Roxbury neighborhood attractive places to live. Boston College's Off-Campus Housing Office is available to assist in the housing search. The office maintains an extensive database of available rental listings, roommates, and helpful local realtors.

Student Group

In 2005–06 there were approximately 20 students pursuing master's degrees in the geology and geophysics program. The students come from throughout the United States and five countries. The relatively small size of the program fosters strong relationships among the students and faculty members. Faculty members look for students who have a strong analytical mind, enjoy research, and have a keen interest in the natural sciences.

Student Outcomes

Students who graduate with a master's degree in geology and geophysics go on to pursue rewarding careers in a variety of corporate, government, and nonprofit agencies. Employment opportunities have included some of the world's largest oil and gas companies, such as ExxonMobil and Texaco, the U.S. Army Corps of Engineers, American Airlines, the New Jersey Water Supply Authority, Weston and Sampson, Shaw Environmental Infrastructure, and many other corporations.

Location

The Graduate School of Arts and Sciences is located on the Chestnut Hill campus of Boston College, approximately 5 miles west of the city of Boston, Massachusetts. Boston offers students the opportunity to experience one of the oldest cities in the U.S., with museums, a symphony orchestra, and world championship professional basketball, baseball, ice hockey, and football teams. The city of Boston also offers a wide variety of shopping, dining, and cultural experiences—all located on the beautiful Boston Harbor and Charles River.

The College and The School

Founded in 1863, Boston College is one of the oldest Jesuit-sponsored universities in the United States. It has professional and graduate schools, doctoral programs, research institutes, community service programs, an excellent faculty, and rich resources of libraries, research equipment, computers, and other facilities. A coeducational university, it has an enrollment of 9,000 undergraduate and 4,700 graduate students representing nearly 100 countries. Boston College confers degrees in more than fifty fields of study through its eleven schools and colleges. It has more than 600 full-time faculty members committed to both teaching and research.

The Graduate School of Arts and Sciences offers programs of study in the humanities, social sciences, and natural sciences, leading to the degrees of Doctor of Philosophy (Ph.D.), Master of Arts (M.A.), and Master of Science (M.S.). The Graduate School may also admit students as Special Students, those not seeking a degree but who are interested in pursuing course work at the graduate level for personal or professional edification. The Graduate School of Arts and Sciences operates on a semester calendar, with the fall semester running from late August until mid-December and the spring semester running from late January until late May.

Applying

The deadline for applying for the fall semester is January 15. Applicants are encouraged to apply early. Students applying for a spot in the Graduate School of Arts and Sciences must hold a bachelor's degree in the natural sciences or a related field of study from an accredited institution. Students must submit the following materials as part of the application packet: the Boston College application form, a $70 nonrefundable application fee, official undergraduate transcripts, abstract of courses form, GRE scores, and three letters of recommendation.

Correspondence and Information

Graduate School of Arts and Sciences
221 McGuinn Hall
Boston College
140 Commonwealth Avenue
Chestnut Hill, Massachusetts 02467

Phone: 617-552-3640
Fax: 617-552-2462
E-mail: gsasinfo@bc.edu
Web site: http://www.bc.edu/schools/cas/geo

Boston College

THE FACULTY AND THEIR RESEARCH

Emanuel G. Bombolakis, Professor; Ph.D., MIT. Structural geology, rock mechanics.
Peter M. Dillon, Lecturer; M.S., Boston College. Geology, geochemistry, GIS modeling.
John E. Ebel, Professor; Ph.D., Caltech. Earthquake seismology, exploration seismology, theoretical seismology.
Kenneth Galli, Lecturer; Ph.D., Massachusetts Amherst. Regional geology, sedimentology.
J. Christopher Hepburn, Professor; Ph.D., Harvard. Regional geology and tectonics, metamorphic petrology.
Judith Hepburn, Lecturer; Ph.D., Harvard. Environmental geoscience.
Rudolph Hon, Associate Professor; Ph.D., MIT. Solid earth and environmental chemistry, GIS.
Alan Kafka, Associate Professor and Department Chair; Ph.D., SUNY at Stony Brook. Seismology, earthquake hazards, science education, science and public policy.
Gail C. Kineke, Associate Professor; Ph.D., Washington (Seattle). Coastal and estuarine processes, marine sediment transport.
Randolph J. Martin III, Adjunct Professor; Ph.D., MIT. Rock mechanics, experimental geophysics.
James W. Skehan, S.J., Professor Emeritus; M.Div., Ph.L., Weston; Ph.D., Harvard. Assembly and dispersal of supercontinents.
Noah P. Snyder, Assistant Professor; Ph.D., MIT. Earth surface processes, river erosion, transport and deposition.
Paul Strother, Adjunct Research Professor; Ph.D., Harvard. Paleobotany and palynology, Precambrian paleobiology, origin of land plants.
Alfredo Urzua, Adjunct Professor; Ph.D., MIT. Groundwater flow, geotechnical engineering, risk assessment for geologic hazards.
Dale Weiss, Lecturer; M.S., Boston College. Geology, geophysics, environmental hydrology.

CLEMSON UNIVERSITY

CLEMSON UNIVERSITY

Master of Science in Hydrogeology

Program of Study

Clemson's graduate program in geological sciences, which offers a master's degree in hydrogeology, is best known for its outstanding reputation in hydrogeology and related areas such as groundwater modeling, remediation, stratigraphy, and sedimentology. Research opportunities for graduate students include contaminant transport and numerical modeling, constructed wetlands and sediments, environmental geochemistry, geophysics, geoscience education, geostatistics, groundwater and hydraulic fracturing, hydrology, petrology and mineralogy, and structural geology and mapping. Research and courses are enhanced by close ties with the Department of Environmental Engineering and Science.

Students typically take three courses during each semester of their first year. A six-week summer field camp in hydrogeology is offered during the summer. The second year of graduate study focuses on research, with completion of all degree requirements expected within approximately two years of beginning graduate study.

Graduate-level courses taught by faculty members in the Department of Geological Sciences include applied geophysics, invertebrate paleontology, geomorphology, geohydrology, stratigraphy, GIS applications in geology, field geophysics, geostatistics, advanced stratigraphy, aquifer characterization, ground water modeling, subsurface remediation modeling, analytical methods for hydrogeology, hydrogeophysics, environmental geochemistry, environmental sedimentology, aquifer systems, and hydrogeology of fractured aquifers. Additional, related courses are offered by the Department of Environmental Engineering and Science.

Research Facilities

The Department of Geological Sciences is housed in Brackett Hall, which is a modern classroom and laboratory building. The department maintains laboratories for geochemistry, mineralogy-petrology, hydraulic properties, cartography/GIS, and thin-section preparation. Instruments for geochemical analysis include a Perkin Elmer Zeeman/5100 PC Atomic Absorption Spectrometer and an HP 6890 Gas Chromatograph with an FID and autosampler. A CME 45 drill rig with diamond bit coredrill and geoprobe sampler is available for well installation. Various pumps, transducers, and data acquisition systems are used for aquifer testing. The department maintains field equipment for full-scale hydraulic fracturing experiments. Geophysical field instruments include a Pulse Ekko 100 low-frequency GPR, Pulse Ekko 1000 high-frequency GPR, EM-34 electromagnetic ground conductivity meter, fluxgate magnetometer, portable gamma-ray scintillometer, and borehole gamma, resistivity, temperature, and caliper tools.

Financial Aid

Graduate research assistantships, teaching assistantships, and fellowships are available. Assistantships may begin in the fall or spring semester and sometimes in the summer. Various types of traineeships, scholarships, loans, work-study programs, job opportunities for spouses, and aid for international students are also available. Through a special program within the Department of Geological Sciences, internships with local companies are available for qualified students.

Cost of Study

Tuition for 2006–07 is $4643 per semester for in-state students and $9255 per semester for nonresidents. Off-campus rates are $535 per hour for in-state students and $918 per hour for nonresidents. Graduate assistants pay a flat fee of $1079 per semester and $348 per summer session. Graduate fellows pay South Carolina resident fees.

Living and Housing Costs

On-campus housing is available; for information, students should visit http://www.housing.clemson.edu. The cost of living in Clemson is quite low compared to the national average; students who choose to live off campus typically spend $300 to $400 per month for rent, depending on location, amenities, and roommates.

Student Group

The program has approximately 21 students. Thirty-three percent are women, 81 percent are full-time students, and 10 percent are international students.

Student Outcomes

Graduates of this program are in high demand to fill positions in the environmental industry and other areas such as petroleum, education, and government. An active alumni network contributes to job placement.

Location

Clemson is a small, beautiful college town near the Blue Ridge Mountains and Lake Hartwell. The Upstate is one of the country's fastest-growing areas and is an important part of the I-85 corridor, a multistate area along Interstate 85 that runs from metro Atlanta to Richmond, Virginia, and encompasses Charlotte, North Carolina, and North Carolina's Research Triangle. Atlanta and Charlotte are each a 2-hour drive away. Many financial institutions and other industries have national or a major presence in the Upstate, including Wachovia, Bank of America, BMW, Bon Secours St. Francis Health System, Bosch North America, Bowater, Charter Communications, Ernst and Young, Fluor Corporation, IBM, Microsoft, and Michelin of North America.

The University

Clemson is classified by the Carnegie Foundation as Doctoral/Research University–Extensive, a category comprising less than 4 percent of all universities in America. The University's mission is to fulfill the covenant between its founder and the people of South Carolina to establish a "high seminary of learning" through its responsibilities of teaching, research, and extended public service. The University has identified eight areas of academic emphasis that create collaborations that, in turn, help fulfill the University's mission.

Applying

Minimum requirements for admission include a four-year bachelor's degree and a previous academic record of high quality. Majors in all areas of science and engineering are encouraged to apply. Applicants may apply on the Web at http://www.grad.clemson.edu/p_apply.html. Applications with a $50 nonrefundable fee should be received no later than five weeks prior to registration. Every required item in support of the application must be on file by that date. Students are advised to contact the department for the deadlines of the program of proposed study.

Correspondence and Information

Dr. James W. Castle, Graduate Coordinator
Department of Geological Sciences
School of the Environment
340 Brackett Hall
Clemson University
Clemson, South Carolina 29634-0919
Phone: 864-656-5015
Fax: 864-656-1041
E-mail: jcastle@clemson.edu
Web site: http://www.ces.clemson.edu/geology/hydro/

Clemson University

THE FACULTY AND THEIR RESEARCH

James W. Castle, Associate Professor; Ph.D., Illinois. Geology.
Raymond A. Christopher, Associate Professor; Ph.D., LSU. Geology.
Alan W. Elzerman, Academic Program Director; Ph.D., Wisconsin. Water chemistry.
Ronald W. Falta Jr., Professor; Ph.D., Berkeley. Mineral engineering.
Robert A. Fjeld, Named Professor; Ph.D., Penn State. Nuclear engineering.
David L. Freedman, Associate Professor; Ph.D., Cornell. Environmental engineering.
Tanju Karanfil, Associate Professor; Ph.D., Michigan. Environmental engineering.
Cindy M. Lee, Associate Professor; Ph.D., Colorado School of Mines. Geochemistry.
Fred J. Molz III, Named Professor; Ph.D., Stanford. Hydrology.
Stephen M. J. Moysey, Assistant Professor; Ph.D., Stanford. Geophysics.
Lawrence C. Murdoch, Associate Professor; Ph.D., Cincinnati. Geology.
James D. Navratil, Professor; Ph.D., Colorado-Boulder. Analytical chemistry.
Thomas J. Overcamp, Professor; Ph.D., MIT. Mechanical engineering.
Mark A. Schlautman, Associate Professor; Ph.D., Caltech. Environmental engineering science.
John R. Wagner, Professor; Ph.D., South Carolina. Geology.
Richard D. Warner, Professor; Ph.D., Stanford. Geology.
Yanru Yang, Assistant Professor; Ph.D., Tsinghua University (China). Environmental engineering.

GEORGIA INSTITUTE OF TECHNOLOGY

School of Earth and Atmospheric Sciences

Programs of Study

The School of Earth and Atmospheric Sciences (EAS) offers graduate programs in the geosciences leading to the degrees of Master of Science (M.S.) and Doctor of Philosophy (Ph.D.) in six areas of specialization: atmospheric chemistry and air pollution, atmospheric dynamics and climate, geochemistry, solid earth and environmental geophysics, ocean sciences, and hydrologic cycle.

The core curricula in each area of specialization are designed to provide students from diverse academic backgrounds with a common introduction to fundamental chemical and physical principles. More advanced courses are also available to introduce students to current academic and research topics. Doctoral students pursue their thesis research upon successful completion of the comprehensive examination, which consists of a written original research paper or proposal and an oral examination that covers the paper and fundamental principles within the student's area of specialization.

In addition to the required courses in a student's area of specialization, doctoral candidates complete 9 credit hours of study in an academic minor. This can be satisfied in another discipline within the School or in other academic units at Georgia Tech, such as in chemistry and biochemistry, physics, mathematics, public policy, computer sciences, or environmental engineering. EAS students can also participate in a certificate program in geohydrology, which is based on educational criteria of the American Institute of Hydrology and is administered by the School of Civil and Environmental Engineering. To accomplish this, students supplement their graduate program of study with a specified set of engineering and EAS courses. Also, marine science research may be carried out in cooperation with the Skidaway Institute of Oceanography. Students conduct their thesis research at Skidaway after completing course work at Georgia Tech.

Research Facilities

The School is well equipped with a wide variety of computational, laboratory, and field measurement research tools. Computational facilities include a large array of high-performance workstations, personal computers, and data servers that are used to analyze, simulate, and predict different components of the earth system, including global climate, regional air quality, and oceanic hydrothermal systems.

Several chromatographs, spectrophotometers, and various elemental analyzers are available for analytical measurements of chemical constituents in solid, liquid, and gaseous samples. In addition, there are several mass spectrometers equipped for the measurement of isotopic ratios in different types of samples. Interaction of EAS with the Departments of Biology and Civil and Environmental Engineering has resulted in the recent opening of an interdisciplinary instrumental facility to quantify biomolecules and natural substrates involved in geomicrobial processes.

For field studies, equipment includes a reverse osmosis system to separate natural organic matter from aquatic environments; several benthic landers and underwater instruments for in situ measurements in marine and freshwater environments; ground-penetrating radar and electromagnetic conductivity meters to determine the resistivity of soils and sedimentary environments; magnetometers and gravimeters to measure the magnetic properties of the earth; seismometers, geophones, and seismographs to study earthquakes; and a variety of chemical instruments to collect and analyze the composition of aerosols from airplanes. These instruments are used in research projects that include understanding and quantifying biogeochemical processes in the water column and sediments of rivers, lakes, coastal marine environments, and the open oceans. They are also used in studies aimed at understanding and quantifying the formation and reactivity of aerosols in polluted and pristine environments or to identify past environmental events recorded in ice at the earth's poles or in deep-sea sediments.

Finally, the School maintains several facilities for environmental monitoring. The geophysics program maintains a seismic network in Georgia to study earthquakes in the region. The atmospheric chemistry program maintains mobile and fixed-site sampling facilities, including state-of-the-art chemical ionization mass spectrometers and laser spectrometers for detection of trace-gas species and a variety of meteorological and analytical instruments for detailed studies of chemical processes. Several field stations are also accessible for environmental studies on the Georgia coast, and a research vessel is available for oceanographic research with scientists at the Skidaway Institute of Oceanography.

Financial Aid

Graduate research and teaching assistantships are available to applicants with outstanding records and high research potential. Research and teaching assistants receive a tuition waiver plus a twelve-month stipend of $21,840. President's Fellowships and President's Minority Fellowships are awarded to qualified matriculants on a competitive basis. These fellowships provide stipend supplements of $5500 and are renewable for up to four years. The Institute also participates in a number of fellowship and traineeship programs sponsored by federal agencies. Traineeships associated with specific programs, such as water resources planning and management, are also available through the Environmental Resources Center.

Cost of Study

Nonresident tuition is estimated to be $9605 per semester in 2006–07. Additional information regarding tuition waivers for graduate assistants is described above. The 2006–07 matriculation fees for graduate assistants are estimated to be $555 per semester.

Living and Housing Costs

Room and board costs for individual graduate students were about $5920 per semester for 2005–06. Contemporary on-campus graduate student housing is available as well as private off-campus housing.

Student Group

There are currently about 90 graduate students in the School, representing a diverse body of academic, ethnic, and national backgrounds. Successful applicants typically have degrees in the physical, chemical, or biological sciences or in engineering, as well as a keen desire to understand the chemistry and physics of the natural environment.

Location

Georgia Tech is located on a 400-acre campus in the heart of midtown Atlanta, a modern, cosmopolitan city with a variety of cultural, historical, and outdoor attractions. The city benefits from a moderate climate, which permits a broad range of year-round outdoor activities. Additional information on Atlanta can be found on the Web at http://www.accessatlanta.com.

The Institute

Georgia Tech was founded in 1888 and is a member of the University System of Georgia. The Institute has a tradition of excellence as a center of technological research and education, with a strong focus on interdisciplinary activities. The School of Earth and Atmospheric Sciences is the cornerstone of a new campus building that fosters interdisciplinary research in environmental sciences and technology.

Applying

Application information is available from the Graduate Admissions Committee from the address listed in the Correspondence and Information section. Prospective applicants are also encouraged to directly contact faculty members with whom their interests best coincide. Applicants are required to submit scores from the General Test of the Graduate Record Examinations. Minimum TOEFL scores of 550 (paper-based) or 213 (computer-based) are required of all international applicants whose native language is not English. To ensure full consideration of available fellowships and assistantships, completed applications for the fall term should be received by January 15.

Correspondence and Information

EAS Graduate Admissions Committee
School of Earth and Atmospheric Sciences
Georgia Institute of Technology
Atlanta, Georgia 30332-0340

Phone: 404-894-3893
Fax: 404-894-5638
E-mail: gradinfo@eas.gatech.edu
Web site: http://www.eas.gatech.edu/

Georgia Institute of Technology

THE FACULTY AND THEIR RESEARCH

Michael H. Bergin, Associate Professor; Ph.D., Carnegie Mellon, 1994. Atmospheric chemistry, atmospheric aerosols, climate impacts.
Robert X. Black, Associate Professor; Ph.D., MIT, 1990. Atmospheric climate dynamics, diagnostic methods, model validation.
George Chimonas, Professor; Ph.D., Sussex (England), 1965. Atmospheric dynamics, waves, turbulence, and stability.
Kim M. Cobb, Assistant Professor; Ph.D., California, San Diego (Scripps), 2002. Tropical Pacific climate change (past and present), carbonate geochemistry, multiproxy approaches.
Derek M. Cunnold, Professor; Ph.D., Cornell, 1965. Atmospheric dynamics, remote sensing and modeling of trace gases.
Judith A. Curry, Professor and Chair; Ph.D., Chicago, 1982. Climate, remote sensing, atmospheric modeling, air-sea interactions.
Robert E. Dickinson, Professor, GRA/Georgia Power Chair, and Member of the National Academy of Sciences and the National Academy of Engineering; Ph.D., MIT, 1966. Climate dynamics and modeling, land-atmosphere interactions.
Emanuele Di Lorenzo, Assistant Professor; Ph.D., California, San Diego (Scripps), 2003. Ocean and climate dynamics, ocean modeling and data assimilation.
Rong Fu, Associate Professor; Ph.D., Columbia, 1991. Climate, atmospheric hydrological processes, remote sensing.
L. Gregory Huey, Associate Professor; Ph.D., Wisconsin–Madison, 1992. Atmospheric chemistry, chemical kinetics, trace-gas measurements.
Ellery Ingall, Associate Professor; Ph.D., Yale, 1991. Marine biogeochemistry; carbon, nitrogen, and phosphorus cycling.
Daniel Lizarralde, Assistant Professor; Ph.D., MIT (Woods Hole), 1997. Geophysics, lithospheric evolution, continental margins, marine gas hydrates, seismology.
L. Timothy Long, Professor; Ph.D., Oregon State, 1968. Intraplate seismotectonics, surface wave imaging in environmental geophysics, earthquakes in education.
Robert P. Lowell, Professor; Ph.D., Oregon State, 1972. Marine geophysics, magmatic and hydrothermal processes.
Jean Lynch-Stieglitz, Associate Professor; Ph.D., Columbia, 1995. Paleooceanography, paleoclimatology, stable isotope geochemistry.
Athanasios Nenes, Assistant Professor; Ph.D., Caltech, 2002. Atmospheric aerosols, clouds, and climate; cloud microphysical processes.
Andrew Newman, Assistant Professor; Ph.D., Northwestern, 2000. Active deformation and brittle failure of the earth's lithosphere in seismic and volcanic provinces.
Carolyn D. Ruppel, Associate Professor; Ph.D., MIT, 1992. Environmental geophysics, physical hydrology, methane gas hydrates.
Irina M. Sokolik, Professor; Candidate of Science (Ph.D. equivalent), Russian Academy of Sciences, 1989. Radiation, remote sensing, aerosols.
Andrew Stack, Assistant Professor; Ph.D., Wyoming, 2002. Mineral surface reactions and how they affect groundwater and soil chemistry.
Marc Stieglitz, Associate Professor; Ph.D., Columbia, 1995. Surface hydrology, watershed dynamics, land surface–climate interactions.
Martial Taillefert, Assistant Professor; Ph.D., Northwestern, 1997. Aqueous inorganic geochemistry, chemical oceanography, geomicrobiology, metal cycling in aquatic systems.
David Tan, Associate Professor; Ph.D., Cornell, 1994. Atmospheric chemistry, tropospheric photochemistry and trace gases.
Yuhang Wang, Associate Professor; Ph.D., Harvard, 1997. Atmospheric chemistry, chemical modeling and forecasting.
Rodney J. Weber, Associate Professor; Ph.D., Minnesota, 1995. Atmospheric chemistry, aerosol measurements and formation.
Peter J. Webster, Professor; Ph.D., MIT, 1972. Atmospheric and ocean dynamics, ocean-atmosphere interaction, wave propagation, prediction and decision theory.
Paul H. Wine, Professor; Ph.D., Florida State, 1974. Atmospheric chemistry, photochemical kinetics.

NORTH CAROLINA STATE UNIVERSITY

Department of Marine, Earth and Atmospheric Sciences

Programs of Study

The department offers M.S. and Ph.D. degrees with majors in oceanography, meteorology, and geology.

In oceanography, students specialize in biological, chemical, geological, or physical oceanography. In the biological area, research topics are in benthic, plankton, or invertebrate physiological ecology. Research in the chemical area concentrates on the study of organic and inorganic processes in estuarine, coastal, and deep-sea environments. Emphasis in geological oceanography is in sedimentology and micropaleontology. In physical oceanography, research topics include the study of the dynamics of estuarine, shelf, slope, and deep-sea waters.

In meteorology, research topics exist in modeling and parameterizing the planetary atmospheric boundary layer, in physically and theoretically modeling dispersion over complex terrain, in air-sea interaction, in atmospheric chemistry, and in climate dynamics. Other research areas are those of cloud-aerosol interaction, cloud chemistry and acid rain deposition, plant-atmosphere interaction, severe localized storm systems, and mesoscale phenomena and processes related to East Coast fronts and cyclones.

In earth science, research topics are in the areas of hard-rock and soft-rock geology, hydrogeology, tectonics, sedimentary geochemistry, and paleoecology. In hard-rock geology the emphasis is on igneous and metamorphic petrology. Soft-rock studies span both recent and ancient detrital deposits, including economic deposits, with field-based studies of facies relationships and associated depositional environments.

The M.S. degree program requires 30 semester credit hours of course work, a research thesis, and a final oral examination. A nonthesis option is available to students on leave from government or industry. The Ph.D. program requires at least 54 credit hours beyond the M.S. degree, as well as a thesis, preliminary written and oral examinations, and a dissertation defense.

Research Facilities

Jordan Hall, the department's home, is a modern structure dedicated to research in natural resources which has been specially designed to accommodate department research laboratories. Modern facilities currently exist in all program areas. Students have access to the million-volume D. H. Hill Library and the University Computing Center resources, which link the department to local, national, and international networks. The department operates a facility for ocean/atmosphere modeling and visualization; Nextlab, with seventeen networked Sun SPARCstations; and a general computer facility with twelve networked Sun SPARCstations. Other specialized departmental equipment includes a Quorum Communications HRTP satellite receiver, a Finnigan MAT 251 Ratio Mass Spectrometer, a McIDAS workstation, an electron microprobe, an X-ray diffractometer, and an atomic absorption spectrometer. Elsewhere on campus, students have access to electron microscopes, ion microprobes, and a nuclear reactor for neutron activation analyses. The department is a member of the Duke/UNC consortium, which operates the 131-foot R/V *Cape Hatteras,* a vessel used for both educational and research cruises. The department participates in the operation of the Center for Marine Science and Technology, a coastal facility in Morehead City, North Carolina. The department is a member of the University Corporation for Atmospheric Research, which provides access to the computing and observational systems of the National Center for Atmospheric Research.

Financial Aid

A number of teaching and research assistantships are available on a competitive basis. The stipends for 2005–06, for 20 hours of service per week, were $1555 per month on a nine-month basis for teaching assistants and a twelve-month basis for research assistants. Students on assistantships received paid health insurance and had tuition waived; they were responsible only for in-state fees of $580 per semester.

Cost of Study

Tuition and fees for 2005–06 for a full course load of 9 or more credits were $2429 per semester for in-state students and $8453 per semester for out-of-state students. U.S. citizens may be able to establish North Carolina residence after one year and then be eligible for in-state tuition rates.

Living and Housing Costs

The University has graduate dormitory rooms that cost about $1700 per semester. Married student housing is available at King Village for about $500 per month for a one-bedroom apartment. Off-campus housing is available starting at about $600 per month.

Student Group

University enrollment is approximately 27,200, with an undergraduate enrollment of 18,700, a graduate enrollment of 4,400, and a continuing-education enrollment of 4,100. The department has approximately 225 undergraduate majors and 100 graduate students, with 25 in marine science, 25 in earth science, and 50 in atmospheric science. Presently, 22 of the graduate students are women.

Location

Raleigh, a modern growing city with a population of more than 200,000 in a metropolitan area of more than 1 million, is situated in rolling terrain in the eastern Piedmont near the upper Coastal Plain. Raleigh, the state capital, is one vertex of the Research Triangle area, with Durham and Chapel Hill the other vertices. Numerous colleges and industrial and government laboratories are located in the Triangle area, which each year also attracts some of the world's foremost symphony orchestras and ballet companies. Located within a 3-hour drive of the campus are both the seashore and the mountains, which offer many opportunities for skiing, hiking, swimming, boating, and fishing.

The University

North Carolina State University, the state's land-grant and chief technological institution, recently celebrated its centennial year. A graduate faculty of 1,400 and more than 100 major buildings are located on the 623-acre main campus. The 780-acre Centennial Campus, which has just been acquired adjacent to the main campus, ensures room for future expansion. The University is organized into nine colleges plus the Graduate School. The department is one of six in the College of Physical and Mathematical Sciences.

Applying

For fall admission, the completed application form, transcripts, recommendation forms, GRE scores, and TOEFL scores (for international students) should be received no later than February 1 to ensure full consideration for assistantship support. Applications for summer and spring admission are also considered, but assistantship support is less likely.

Correspondence and Information

Graduate Administrator
Department of Marine, Earth and Atmospheric Sciences
North Carolina State University
Box 8208
Raleigh, North Carolina 27695-8208
Phone: 919-515-7837
E-mail: janowitz@ncsu.edu
Web site: http://www.meas.ncsu.edu/

North Carolina State University

THE FACULTY AND THEIR RESEARCH

Anantha Aiyyer, Assistant Professor; Ph.D., SUNY at Albany, 2003. Atmospheric dynamics.
Viney Aneja, Professor; Ph.D., North Carolina State, 1977. Atmospheric chemistry.
S. Pal Arya, Professor; Ph.D., Colorado State, 1968. Micrometeorology, atmospheric turbulence and diffusion, air-sea interaction.
Neal E. Blair, Professor; Ph.D., Stanford, 1980. Chemical oceanography, biogeochemistry, organic geochemistry.
DelWayne R. Bohnenstiehl, Assistant Professor; Ph.D., Columbia, 2002. Marine geology and geophysics.
Julia Clark, Assistant Professor; Ph.D., Yale, 2002. Paleoecology, theropod relationships.
Cynthia Cudaback, Assistant Professor; Ph.D., Washington (Seattle), 1998. Physical oceanography of coastal and estuarine systems.
Jerry M. Davis, Professor Emeritus; Ph.D., Ohio State, 1971. Agricultural meteorology, climatology, statistical meteorology, planetary boundary layer.
David J. DeMaster, Professor; Ph.D., Yale, 1979. Marine geochemistry and radio chemistry in the nearshore and deep-sea environments.
David B. Eggleston, Professor; Ph.D., William and Mary, 1991. Marine benthic ecology, epifauna.
Ronald V. Fodor, Professor; Ph.D., New Mexico, 1972. Igneous petrology, volcanoes, meteorites.
John C. Fountain, Professor and Head; Ph.D., California, Santa Barbara, 1975. Geochemistry, contaminant hydrogeology.
David P. Genereux, Associate Professor; Ph.D., MIT, 1991. Hydrogeology.
Rouying He, Associate Professor; Ph.D., South Florida, 2002. Physical oceanography of estuaries, shelf and slope.
James Hibbard, Professor; Ph.D., Cornell, 1988. Structural geology.
Gerald S. Janowitz, Professor and Graduate Administrator; Ph.D., Johns Hopkins, 1967. Geophysical fluid mechanics, continental shelf and ocean circulation.
Daniel L. Kamykowski, Professor; Ph.D., California, San Diego, 1973. Effects of physical factors on phytoplankton behavior and physiology, global plant nutrient distributions.
Michael M. Kimberley, Associate Professor; Ph.D., Princeton, 1974. Sedimentary geochemistry, sedimentary ore deposits, chemistry of natural and polluted water.
Gary M. Lackmann, Associate Professor; Ph.D., SUNY at Albany, 1995. Synoptic and mesoscale meteorology.
Elana L. Leithold, Associate Professor; Ph.D., Washington (Seattle), 1987. Nearshore and shelf sedimentation and stratigraphy, sediment transport.
Yuh Lang Lin, Professor; Ph.D., Yale, 1984. Modeling of mesoscale atmospheric dynamics.
Jingpu Liu, Assistant Professor; Ph.D., William and Mary, 2001. Geological oceanography and geomorphology.
Nicholas Meskhidze, Assistant Professor; Ph.D., Georgia Tech, 2003. Atmospheric chemistry.
Matthew Parker, Assistant Professor; Ph.D., Colorado State. 2002. Dynamics and thermodynamics of mesoscale convective systems and severe local storms and mesoscale numerical modeling.
Leonard J. Pietrafesa, Professor; Ph.D., Washington (Seattle), 1973. Estuarine and continental margin physical processes, seismology.
Sethu S. Raman, Professor; Ph.D., Colorado State, 1972. Air-sea interactions, boundary layer meteorology and air pollution.
Allen J. Riordan, Associate Professor Emeritus; Ph.D., Wisconsin, 1977. Satellite meteorology, Antarctic meteorology.
Mary Schweitzer, Assistant Professor; Ph.D., Montana State, 1995. Vertebrate paleontology.
Frederick Semazzi, Professor; Ph.D., Nairobi, 1983. Climate dynamics.
Ping Tung Shaw, Associate Professor; Ph.D., Woods Hole/MIT, 1982. Shelf-slope physical oceanography and Lagrangian analysis.
William J. Showers, Associate Professor; Ph.D., Hawaii, 1982. Stable-isotope geochemistry, paleoceanography, micropaleontology, environmental monitoring, geoarchaeology.
Edward F. Stoddard, Associate Professor Emeritus; Ph.D., UCLA, 1976. Metamorphic petrology, silicate mineralogy, Piedmont geology.
Carrie Thomas, Visiting Assistant Professor; Ph.D., North Carolina State, 1998. Biogeochemistry, animal-sediment interaction.
Donna L. Wolcott, Associate Professor; Ph.D., Berkeley, 1972. Physiological ecology of terrestrial crabs.
Thomas G. Wolcott, Professor Emeritus; Ph.D., Berkeley, 1971. Physiological ecology of marine invertebrates, biotelemetry.
Lian Xie, Professor; Ph.D., Miami (Florida), 1992. Air-sea interaction processes.
Sandra E. Yuter, Assistant Professor; Ph.D., Washington (Seattle), 1996. Deep convection over tropical oceans, orographic precipitation, and marine stratocumulus.
Yang Zhang, Assistant Professor; Ph.D., Iowa, 1994. Air quality modeling.

Jordan Hall, home of the department.

STONY BROOK UNIVERSITY, STATE UNIVERSITY OF NEW YORK

Department of Geosciences
M.A. and Ph.D. in Geosciences
M.A.T. in Earth Science

Programs of Study

The Department offers programs leading to the Master of Arts in Teaching (M.A.T.) and M.S. and Ph.D. degrees in the geosciences. The M.A.T. in earth science is a nonthesis degree for which all requirements can be completed in three semesters. The M.S. degree with a concentration in hydrogeology is a nonthesis M.S., with most courses offered at times appropriate for working professionals. The M.S. degree in geosciences with thesis is typically not a terminal degree. Many students seeking Ph.D. candidacy first earn an M.S. Students become candidates for the Ph.D. in geosciences by completing preparatory work leading to completion of the Ph.D. preliminary examination.

The nonthesis M.S. with a concentration in hydrogeology requires a total of 30 credits. Of these, at least 21 credits must be in the required and approved courses and at least 6 credits must be in approved research. In addition to formal course work, the curriculum for the M.S. with a concentration in hydrogeology includes a minimum of 6 credits of research. The M.S. in geosciences with thesis is typically a nonterminal degree completed by some students before seeking Ph.D. candidacy. All requirements for the M.S. degree must be completed within a period of three years after entry. There are no residence or language requirements.

Research Facilities

The Department of Geosciences occupies the Earth and Space Science (ESS) Building, a modern, well-equipped building that houses extensive experimental and analytical labs, faculty and graduate student offices, numerous computers and workstations, a machine shop, an electronics support group, and the Geosciences Resource Room. The Mineral Physics Institute (MPI), the Consortium for Materials Properties Research (COMPRES), the Center for Environmental Molecular Science (CEMS), the Long Island Groundwater Research Institute (LIGRI), the Marine Sciences Research Center (MSRC), and nearby Brookhaven National Laboratory offer additional support and laboratory facilities for graduate student research. In particular, the National Synchrotron Light Source (NSLS) at Brookhaven, only 20 miles away, offers unparalleled opportunities for faculty members and graduate students to perform unique experiments requiring high-intensity X-rays. COMPRES serves the high-pressure research community through enabling support at national facilities, such as the Brookhaven National Lab, Argon National Lab, and Lawrence Berkeley National Lab. The Center for Environmental Molecular Science, funded by the National Science Foundation, has its main office in the Department of Geosciences and involves several faculty members in geosciences and related departments, as well as scientists at nearby Brookhaven National Laboratory. The center focuses on geochemical aspects of environmental contaminants and includes research, education, and outreach activities.

Financial Aid

Because Stony Brook is committed to attracting high-quality students, the Graduate School provides two competitive fellowships for U.S. citizens and permanent residents. Graduate Council fellowships are for outstanding doctoral candidates studying in any discipline, and the W. Burghardt Turner Fellowships target outstanding African-American, Hispanic-American, and Native American students entering either a doctoral or master's degree program. For doctoral students, both fellowships provide an annual stipend of at least $15,600 for up to five years as well as a full-tuition scholarship. For master's students, the Turner Fellowship provides an annual stipend of $10,000 for up to two years, along with a full-tuition scholarship. Health insurance subsidies are also provided within a scale depending on the size of the fellow's dependent family. Departments and degree programs award approximately 900 teaching and graduate assistantships and approximately 600 research assistantships on an annual basis. Full assistantships carry a stipend amount that usually ranges from $11,655 to $20,000, depending on the department.

Cost of Study

In 2006–07, full-time tuition is $3450 per semester for state residents and $5460 per semester for nonresidents. Part-time tuition is $288 per credit hour for residents and $455 per credit hour for nonresidents, plus fees.

Living and Housing Costs

University apartments range in cost from approximately $316 per month to approximately $1126 per month, depending on the size of the unit. Off-campus housing options include furnished rooms to rent and houses and/or apartments to share that can be rented for $350 to $650 per month.

Location

Stony Brook's campus is approximately 50 miles east of Manhattan on the north shore of Long Island. The cultural offerings of New York City and Suffolk County's countryside and seashore are conveniently located nearby. Cold Spring Harbor Laboratories and Brookhaven National Laboratories are easily accessible from, and have close relationships with, the University.

The University

The University, established in 1957, achieved national stature within a generation. Founded at Oyster Bay, Long Island, the school moved to its present location in 1962. Stony Brook has grown to encompass more than 110 buildings on 1,100 acres. There are more than 1,568 faculty members, and the annual budget is more than $805 million. The Graduate Student Organization oversees the spending of the student activity fee for graduate student campus events. International students find the additional four-week Summer Institute in American Living very helpful. The Intensive English Center offers classes in English as a second language. The Career Development Office assists with career planning and has information on permanent full-time employment. Disabled Student Services has a Resource Center that offers placement testing, tutoring, vocational assessment, and psychological counseling. The Counseling Center provides individual, group, family, and marital counseling and psychotherapy. Day-care services are provided in four on-campus facilities. The Writing Center offers tutoring in all phases of writing.

Applying

Applicants are judged on the basis of distinguished undergraduate records (and graduate records, if applicable), thorough preparation for advanced study and research in the field of interest, candid appraisals from those familiar with the applicant's academic/professional work, potential for graduate study, and a clearly defined statement of purpose and scholarly interest germane to the program. Applicants are required to have a bachelor's degree in one of the earth and space sciences, biology, chemistry, physics, mathematics, or engineering; a minimum average of B for all undergraduate course work; and at least a B average for all courses in the sciences. In some cases, students not meeting the degree and GPA requirements are admitted on a provisional basis. Results of the Graduate Record Examinations (GRE) are required. Students should submit admission and financial aid applications by January 15 for the fall semester. Students are admitted for the spring semester only under special circumstances.

Correspondence and Information

Lianxing Wen
Department of Geosciences
Stony Brook University, State University of New York
Stony Brook, New York 11794-4433
Phone: 631-632-8554
Fax: 631-632-8240
E-mail: lbudd@notes.cc.sunysb.edu
Web site: http://pbisotopes.ess.sunysb.edu/geo/

Stony Brook University, State University of New York

THE FACULTY AND THEIR RESEARCH

Robert C. Aller, Distinguished Service Professor; Ph.D., Yale, 1977. Marine geochemistry, early marine diagenesis.
Henry J. Bokuniewicz, Professor; Ph.D., Yale, 1976. Marine geophysics.
Bruce Brownawell, Associate Professor; Ph.D., MIT (Woods Hole), 1986. Biogeochemistry, environmental chemistry, diagenesis.
Jiuhua Chen, Research Assistant Professor; Ph.D., Institute of Materials Structure Science, KEK (Japan), 1994. Mineral physics, mantle petrology, application of synchrotron radiation to earth sciences.
J. Kirk Cochran, Professor; Ph.D., Yale, 1979. Marine geochemistry, use of radionuclides as geochemical tracers, diagenesis of marine sediments.
Daniel M. Davis, Professor; Ph.D., MIT, 1983. Geomechanical modeling of active margins, shallow surface geophysics.
Catherine A. Forster, Associate Professor; Ph.D., Pennsylvania, 1990. Vertebrate paleontology, systematics, functional morphology.
Tibor Gasparik, Research Associate Professor; Ph.D., SUNY at Stony Brook, 1981. Experimental petrology and mineral physics.
Marvin Geller, Professor; Ph.D., MIT, 1969. Atmospheric dynamics, upper atmosphere, climate variability, aeronomy, physical oceanography.
Gilbert N. Hanson, Distinguished Service Professor; Ph.D., Minnesota, 1964. Isotope and trace-element geochemistry, geochronology.
Garman Harbottle, Professor; Ph.D., Columbia, 1949. Nuclear chemistry, archeology.
William Holt, Professor; Ph.D., Arizona, 1989. Tectonophysics, kinematics and dynamics of large-scale deformation of the earth's crust and upper mantle.
David W. Krause, Professor; Ph.D., Michigan, 1982. Vertebrate paleontology; mammalian evolution, including primates.
Baosheng Li, Research Assistant Professor; Ph.D., SUNY at Stony Brook, 1996. Mineral physics, elasticity of minerals, high-pressure research.
Robert C. Liebermann, Distinguished Service Professor; Ph.D., Columbia, 1969. Mineral physics, solid earth geophysics.
Donald H. Lindsley, Distinguished Professor; Ph.D., Johns Hopkins, 1961. Geochemistry, petrology.
Scott M. McLennan, Professor; Ph.D., Australian National, 1981. Geochemistry, planetary science, crustal evolution, sedimentary petrology.
Hanna Nekvasil, Professor; Ph.D., Penn State, 1986. Silicate melt/crystal equilibria, Mars volcanism, thermodynamics of silicate systems.
Maureen O'Leary, Assistant Professor; Ph.D., Johns Hopkins, 1997. Vertebrate paleontology, phylogenetic systematics, mammalian evolution.
John B. Parise, Professor; Ph.D., James Cook (Australia), 1981. Crystal structure-property relations, solid-state synthesis.
Brian L. Phillips, Assistant Professor; Ph.D., Illinois at Urbana-Champaign, 1990. Mineralogy and low-temperature geochemistry, structure/reactivity relationships in aqueous fluids and mineral/fluid interfaces.
Troy Rasbury, Assistant Professor; Ph.D., SUNY at Stony Brook, 1998. Sedimentary geology and geochemistry, geochronology.
Richard J. Reeder, Professor; Ph.D., Berkeley, 1980. Geochemistry and mineralogy relating to near-earth's surface processes.
Martin A. A. Schoonen, Professor; Ph.D., Penn State, 1989. Geochemistry of sulfur and sulfides, hydrogeochemistry, catalysis.
Michael T. Vaughan, Research Associate Professor; Ph.D., SUNY at Stony Brook, 1979. Experimental geophysics, crystallography, synchrotron X-ray studies.
Donald J. Weidner, Distinguished Professor; Ph.D., MIT, 1972. Mineral physics and seismology.
Lianxing Wen, Assistant Professor; Ph.D., Caltech, 1998. Seismology, geodynamics, global geophysics.

TULANE UNIVERSITY

Department of Earth and Environmental Sciences

Programs of Study

The Department of Earth and Environmental Sciences (EES) offers graduate programs leading to the degrees of Master of Science in broad areas of geology and paleontology and Doctor of Philosophy. Two master's degree programs are available: the principal one requires 24 semester hours of graduate course work and successful completion, presentation, and defense of a thesis that reflects individual research accomplishments. A second, nonthesis, program requires 36 semester hours of course work. The Ph.D. program requires 48 semester hours of course work, oral and written examinations, and an original contribution in the form of a written dissertation suitable for publication in a learned journal. Areas of research in EES include sedimentary geochemistry, organic geochemistry, global and biogeochemical cycles, theoretical geochemistry, sedimentary geology, process-oriented sedimentology, isotope geochemistry, global climate change, coastal and marine geology, environmental geology, structural geology, subsurface geology, neotectonics, igneous petrology, volcanology, and paleontology. Special emphasis is given to geology of the Gulf Coast region and Latin America.

Research Facilities

The Department's research facilities, partially supported by a Departmental endowment, include a scanning electron microscope with an energy-dispersive X-ray system, an X-ray fluorescence spectrometer, an electron microprobe, an X-ray diffractometer, a cathodoluminescence microscope, an ICP spectrometer, a wet chemistry laboratory, and a computer laboratory, including a 3-D seismic interpretation laboratory. In addition, single-crystal X-ray diffraction equipment, a transmission electron microscope with an energy-dispersive X-ray system, a high-resolution optical microscope, and other equipment are available in a coordinated instrumentation facility. A variety of field instrumentation is also available for use on the University's 60-foot research vessel, the R/V *Eugenie*.

Financial Aid

Graduate teaching and research assistantships are available to all qualified students and provide nine-month stipends, including Departmental supplements, that average $17,500. All assistantships and fellowships are accompanied by a tax-free full tuition scholarship. Funding to support research activities during the summer months is available each year.

Cost of Study

Full-time tuition for 2003–04 was $14,250 per semester plus a $300 University fee. Tuition on a part-time basis was $1583 per credit hour plus fees.

Living and Housing Costs

A limited amount of University housing is available for graduate students. Most graduate students choose to live off campus, where costs vary greatly depending on the type of accommodation selected. A cost-of-living figure of $750 per month is quoted to international graduate students for purposes of entry.

Student Group

Tulane currently enrolls 8,750 full-time and 2,320 part-time students. Of these, approximately 800 are registered in the Graduate School. In recent years, graduate students have come to Tulane from more than 380 colleges and universities, from all fifty states, and from thirty-six other countries.

The Department seeks to admit 4 to 6 students per year. There are currently 23 students in residence. Graduate students, in coordination with a member of the faculty and with Departmental support, organize a program of speakers.

Location

Tulane's eleven colleges and schools, with the exception of the medical divisions, are located on 100 acres in a residential area of New Orleans. New Orleans' mild climate, many parks, and proximity to the Gulf Coast provide opportunities for a wide variety of outdoor activities. The city's many art galleries and museums offer regularly scheduled exhibits throughout the year. New Orleans is famous for its French Quarter, Mardi Gras, Creole cuisine, and jazz.

The University and The Department

Tulane is a private nonsectarian university offering a wide range of undergraduate, graduate, and professional courses of study for men and women. The University's history dates from 1834, when it was founded as the Medical College of Louisiana. Graduate work was first offered in 1883. In 1884, the University was organized under its present form of administration and renamed for Paul Tulane, a wealthy New Orleans merchant who endowed it generously. Tulane is a member of the American Association of Universities, a group of fifty-six major North American research universities. It is among the top twenty-five private universities in the amount of outside support received for research each year.

The Department of Earth and Environmental Sciences is in the Liberal Arts and Sciences division of the University, which has strong programs in biology, chemistry, mathematics, and physics as well as in earth and environmental sciences. Graduate students in EES are encouraged to enroll in appropriate courses in one or more of these disciplines. Cross-enrollment with the School of Engineering is also available, as are environmental courses offered by the School of Law. EES faculty members play a leadership role in the Tulane Center for River-Ocean Studies (CeROS), which serves as a nexus for multidisciplinary research and education activities related to major world rivers (including the Mississippi) and their role in global change and impact on coastal environments.

Applying

For those requesting financial aid, the application deadline is February 1. Students should write to the Dean of the Graduate School for application forms, or they can download the application forms from the Web site http://www.tulane.edu/~gradprog/. The Graduate School will not forward the application to the Department for consideration for admission until all of the following documents, plus the $45 application fee, have been received: a completed application form, three completed recommendation forms, official transcripts of all undergraduate and graduate work, and official results of the Graduate Record Examinations General Test, taken within the past five years. International applicants for admission must present satisfactory evidence of competence in English by submitting a score of at least 220 on the TSE (Test of Spoken English) or, if this test is not available, a minimum score of 600 on the TOEFL. Admission is based on academic accomplishments and promise. Admission preference is given to students applying to the Ph.D. program. Tulane is an affirmative action/equal employment opportunity institution.

Correspondence and Information

For application forms and admission:
Dean of the Graduate School
Tulane University
New Orleans, Louisiana 70118
Phone: 504-865-5100
Web site: http://www.tulane.edu/~gradprog/

For specific information regarding programs:
Director of Graduate Studies
Department of Earth and Environmental Sciences
Tulane University
New Orleans, Louisiana 70118
Phone: 504-865-5198
E-mail: cdillon@tulane.edu
Web site: http://www.tulane.edu/~eens/

Tulane University

THE FACULTY AND THEIR RESEARCH

Mead A. Allison, Ph.D., SUNY at Stony Brook, 1993. Continental margin sedimentology, high-resolution geophysics of river-ocean margins, marine sediment transport.
Thomas S. Bianchi, Ph.D., Maryland, 1987. Organic geochemistry, organic carbon cycling in coastal and wetland environments, molecular biomarkers as tracers of organic carbon inputs to land-margin ecosystems, chemical biomarkers as paleoindicators of carbon cycling in past environments.
Nancye H. Dawers, Ph.D., Columbia, 1997. Fault growth and interaction, studies of fault scaling relations, neotectonics, basin analysis and syntectonic stratigraphy using 3-D seismic data.
George C. Flowers, Ph.D., Berkeley, 1979. Theoretical geochemistry, sedimentary geochemistry, and environmental geochemistry of estuarine sediments.
Suzanne F. Leclair, Ph.D., SUNY at Binghamton, 2000. Process-oriented sedimentology, bedform development and sediment transport, morphological and sedimentological response of rivers to neotectonism, interactions between depositional systems, paleoenvironmental reconstruction.
Franco Marcantonio, Ph.D., Columbia, 1994. Isotope geochemistry, global environmental change.
Brent A. McKee, Ph.D., North Carolina State, 1986. Sedimentary geochemistry, geochemical cycling in river-ocean margins, lakes and anoxic environments, use of radioisotopes as tools to quantify rates of environmental processes.
Stephen A. Nelson, Ph.D., Berkeley, 1979. Igneous petrology: petrologic studies of volcanoes; relationships between volcanism and tectonism, particularly in Mexico; volcanic hazards studies; mechanisms of explosive volcanism; thermodynamic modeling of silicate systems; fluid mechanical processes in magmatic systems.
Ronald L. Parsley, Ph.D., Cincinnati, 1969. Paleontology: paleobiology, paleoecology, and evolution of lower Paleozoic primitive Echinodermata; Paleozoic faunas in general.

RECENT PUBLICATIONS

Allison, M. A., S. R. Khan, S. L. Goodbred, and S. A. Kuehl. Stratigraphic evolution of the late Holocene Ganges-Brahmaputra lower delta plain. *Sedimentary Geol.* 155:317–42, 2003.
Allison, M. A., and C. F. Neill. Accumulation rates and stratigraphic character of the modern Atchafalaya River prodelta, Louisiana. *Trans. Gulf Coast Assoc. Geol. Soc.* 52:1031–40, 2002.
Allison, M. A., M. T. Lee, A. S. Ogston, and R. C. Aller. Origin of mudbanks along the northeast coast of South America. *Mar. Geol.* 163:241–56, 2000.
Chen, N., **T. S. Bianchi,** and J. M. Bland. Novel decomposition products of chlorophyll-α in continental shelf (Louisiana shelf) sediments: Formation and transformation of carotenol chlorin esters. *Geochim. Cosmochim. Acta* 67:2027–42, 2003.
Bianchi, T. S., S. Mitra, and **B. McKee.** Sources of terrestrially-derived carbon in the Lower Mississippi River and Louisiana shelf; implications for differential sedimentation and transport at the coastal margin. *Mar. Chem.* 77:211–23, 2002.
Bianchi, T. S., et al. Cyanobacterial blooms in the Baltic Sea: Natural or human-induced? *Limnol. Oceanogr.* 45(3):716–26, 2000.
Mitra, S., **T. S. Bianchi,** L. Guo, and P. H. Santschi. Sources and transport of terrestrially-derived organic matter in the Chesapeake Bay and Middle Atlantic Bight. *Geochim. Cosmochim. Acta* 64:3547–57, 2000.
Dawers, N. H. and J. R. Underhill. The role of fault interaction and linkage in controlling synrift stratigraphic sequences: Late Jurassic, Statfjord East area, northern North Sea. *AAPG Bull.* 84:45–64, 2000.
Dawers, N. H., et al. Controls on Late Jurassic, subtle sand distribution in the Tampen area, northern North Sea. In *Petroleum Geology of NW Europe: Proceedings of the 5th Conference,* vol. 2, pp. 827–38, eds. A. J. Fleet and S. A. R. Boldy. London: The Geological Society, 1999.
Flowers, G. C. Environmental sedimentology of the Pontchartrain Estuary. *Trans. Gulf Coast Assoc. Geol. Soc.* 40:237–50, 1990.
Leclair, S. F. Preservation of cross-strata due to migration of subaqueous dunes: An experimental investigation. *Sedimentology* 49:1157–80, 2002.
Leclair, S. F., and J. S. Bridge. Quantitative interpretation of sedimentary structures formed by river dunes. *J. Sedimentary Res.* 71(5):714–7, 2001.
Marcantonio, F., A. Zimmerman, Y. Xu, and E. Canuel. A record of eastern U.S. atmospheric Pb emissions in Chesapeake Bay sediments: Stable Pb isotopes as a chronological tool. *Mar. Chem.* 77:123–321, 2002.
Marcantonio, F., et al. Sediment focusing in the central equatorial Pacific Ocean. *Paleoceanography* 16:260–7, 2001.
McKee, B. A., et al. **(M. A. Allison** and **T. S. Bianchi).** Transport and transformation of dissolved and particulate materials on continental margins influenced by major rivers: Benthic boundary layer and seabed processes. *Continental Shelf Res.,* in press.
O'Reilly, C. M., et al. **(B. A. McKee).** Climate change decreases aquatic ecosystem productivity of Lake Tanganyika, East Africa. *Nature* 424:766–8, 2003.
McKee, B. A. and M. Baskaran. Sedimentary processes in Gulf of Mexico estuaries: Inputs and dynamics. In *Biogeochemistry of Gulf of Mexico Estuaries,* eds. **T. Bianchi,** J. Pennock, and R. Twilley. New York: John Wiley and Sons, 1998.
McKee, B. A., P. W. Swarzenski, and J. G. Booth. The flux of uranium isotopes from river-dominated shelf sediments. In *Geochemistry of the Earth's Surface,* pp. 85–91, ed. S. H. Bottrell. Leeds: University of Leeds Press, 1996.
Nelson, S. A., E. Gonzalez-Caver, and T. K. Kyser. Constraints on the origin of alkaline and calc-alkaline magmas from the Tuxtla Volcanic Field, Veracruz, Mexico. *Contr. Mineral. Petrol.* 122:191-211, 1995.
Nelson, S. A., and E. Gonzalez-Caver. Geology and K-Ar dating of the Tuxtla volcanic field, Veracruz, Mexico. *Bull. Volcanol.* 55:85–96, 1992.
Verma, S. P., and **S. A. Nelson.** Isotopic and trace-element constraints on the origin and evolution of calc-alkaline and alkaline magmas in the northwestern portion of the Mexican Volcanic Belt. *J. Geophys. Res.* 94:4531–44, 1989.
Parsley, R. L. Community setting and functional morphology of Echinoshpaerites infaustus (Fistuliporita: Echinodermata) from the Ordovician of Bohemia, the Czech Republic. *Vestnik (Bull. Czech Geol. Survey)* 73:252–65, 1998.
Parsley, R. L. The echinoderm classes Stylophora and Homoiostelea: Non-Calcichordata. *Paleontol. Soc. Pap.* 3:225–48, 1997.

RECENT THESIS AND DISSERTATION TOPICS

"Dolomitization and Evolution of the Puerto Rico North Coast Confined Aquifer System," Wilson R. Ramirez Martinez (2000).
"Geology and Petrology of Sierra Las Navajas, Hidalgo Mexico, A Pliocene Peralkaline Rhyolite Volcano in the Mexican Volcanic Belt," Alyson Lighthart (2001).
"Anthropogenic and Natural Controls on Shoreface Evolution Along Galveston Island, Texas," Bethany K. Robb (2003).
"Partitioning of Metals Throughout a Winter Storm-Generated Fluid Mud Event, Atchafalaya Shelf, Louisiana," F. Ryan Clark (2003).

THE UNIVERSITY OF ARIZONA

Department of Planetary Sciences / Lunar and Planetary Laboratory

Program of Study

The graduate program prepares students for careers in solar system research. For this interdisciplinary enterprise, the department maintains faculty expertise in the important areas of planetary science. Through a combination of core courses, minor requirements, and interaction with faculty and research personnel, students are provided with a comprehensive education in modern planetary science. The program is oriented toward granting the Ph.D., although M.S. degrees are now awarded as well.

Upon admission, a student is assigned an adviser in his or her general scientific area. Students advance to Ph.D. candidacy by passing an oral preliminary examination after completing the required major and minor course work. The examination is normally taken two years after matriculation. Students typically complete their dissertations and receive the Ph.D. three to four years later.

Because of the low student-faculty ratio, students receive close supervision and guidance. Dissertation areas include, but are not limited to, observational planetary astronomy; physics of the sun and interplanetary medium; observational, experimental, and theoretical studies of planetary atmospheres, surfaces, and interiors; studies of the interstellar medium and the origin of the solar system; and the geology and chemistry of the surfaces and interiors of solar system bodies.

Research Facilities

The Lunar and Planetary Laboratory (LPL) and the Department of Planetary Sciences function as a single unit to carry out solar system research and education. The department and laboratory are housed in the Gerard P. Kuiper Space Sciences and the Gould-Simpson Buildings on the campus. Neighboring facilities include the Tucson headquarters of the National Optical Astronomy Observatory, the National Radio Astronomy Observatory, Steward Observatory, the Optical Sciences Center, the Flandrau Planetarium, the Department of Geosciences, and the Planetary Sciences Institute.

The facilities of the University observatories are available to all researchers in LPL. These include the multiple-mirror telescope as well as numerous midsize and smaller telescopes. For cosmochemical research, LPL operates a scanning electron microprobe, high-temperature and high-pressure apparatus for rock-melting experiments, a noble gas mass spectrometry laboratory, and a radiochemistry separation facility for neutron activation analysis; these are used for studying meteorites, lunar samples, and terrestrial analogues. Also available in LPL are well-equipped electronics, machine, and photo shops as well as a graphic arts facility.

The Space Imagery Center at the LPL is one of several regional facilities supported by NASA as a repository for spacecraft images and maps of planets and satellites. The Planetary Image Research Laboratory is a modern remote sensing and image processing center for analysis of astronomical and spacecraft data. The Laboratory maintains an extensive computer network; various research groups maintain specialized computer systems for particular applications. University central computing facilities include a variety of systems and network facilities as well as several superminicomputers.

Financial Aid

Most planetary sciences graduate students receive graduate research assistantships for the academic year. These assistantships normally require 20 hours of work per week on a sponsored research project. For the nine-month academic year, such assistantships pay $13,940. For students who pass their preliminary examination and are advanced to Ph.D. candidacy, the pay increases to $15,508. In addition, most students work full-time (40 hours per week) on research projects during the summer term, earning $6444 more.

Cost of Study

For 2005–06, fees for Arizona residents taking 1–6 units were $253 per unit. For 7 or more units, the cost was $2424 per semester. Out-of-state students who do not have a research or teaching assistantship are also charged tuition, but tuition scholarships are frequently available.

Living and Housing Costs

Typical costs for off-campus housing, food, and entertainment for a single graduate student total about $700 to $950 per month. Housing is generally inexpensive and plentiful.

Student Group

In 2005–06, there were 37 graduate students enrolled in the planetary science program. Most came from undergraduate or M.S. programs in chemistry, physics, geology, and astronomy, and some were employed for several years prior to entering graduate school.

Location

Tucson is located in the Sonoran Desert, about 100 kilometers north of the Mexican border. The climate is dry and warm; hiking, mountain climbing, horseback riding, swimming, golf, and tennis are popular year-round activities.

The University

The University is a state-supported institution with an enrollment of approximately 35,700 and ranks in the top twenty research universities. In addition to the planetary sciences program, major research efforts and graduate programs exist in chemistry, engineering, physics, astronomy, optical sciences, and geosciences. LPL interacts closely with these groups.

Applying

Completed application forms, three letters of reference, and GRE scores must be received by January 15 in order to receive full consideration. All applicants are required to submit GRE General Test scores as well as the Subject Test score in a physical science or other relevant area.

Correspondence and Information

Graduate Admissions Secretary
Lunar and Planetary Laboratory
Kuiper Space Sciences Building
The University of Arizona
1629 East University Boulevard
Tucson, Arizona 85721-0092

Phone: 520-621-6954
E-mail: acad_info@lpl.arizona.edu
Web site: http://www.lpl.arizona.edu/

The University of Arizona

THE FACULTY AND THEIR RESEARCH

Victor R. Baker, Regents Professor; Ph.D., Colorado, 1971. Planetary surfaces, geomorphology.
William V. Boynton, Professor; Ph.D., Carnegie Mellon, 1971. Neutron-activation analysis, cosmochemistry.
Lyle A. Broadfoot, Senior Research Scientist; Ph.D., Saskatchewan, 1963. Planetary atmospheres, ultraviolet spectroscopy.
Robert H. Brown, Professor; Ph.D., Hawaii, 1982. Ground-based, space-based, and theoretical studies of surfaces and satellites in the outer solar system.
Alexander J. Dessler, Senior Research Scientist Emeritus; Ph.D., Duke, 1956. Magnetospheric and space-plasma physics.
Michael J. Drake, Professor, Head, and Director; Ph.D., Oregon, 1972. Lunar samples, meteorites.
Uwe Fink, Professor; Ph.D., Penn State, 1965. Planetary atmospheres, infrared spectroscopy.
Tom Gehrels, Professor; Ph.D., Chicago, 1956. Polarimetry, asteroids, comets.
Joe Giacalone, Assistant Professor; Ph.D., Kansas, 1991. Solar and heliospheric physics, astrophysics.
Richard J. Greenberg, Professor; Ph.D., MIT, 1972. Celestial mechanics, studies of planetary accumulation, satellite and ring dynamics.
Caitlin Griffith, Associate Professor; Ph.D., SUNY at Stony Brook, 1991. Infrared astronomy, evolution of planetary atmospheres, radiative transfer in planetary atmospheres.
Jay Holberg, Senior Research Scientist; Ph.D., Berkeley, 1974. Far-ultraviolet spectra, planetary rings and atmospheres.
Lon L. Hood, Senior Research Scientist; Ph.D., UCLA, 1979. Geophysics, space physics.
William B. Hubbard, Professor; Ph.D., Berkeley, 1967. High-pressure theory, planetary interiors.
Donald M. Hunten, Regents Professor Emeritus; Ph.D., McGill, 1950. Earth and planetary atmospheres, aeronomy.
J. R. Jokipii, Regents Professor; Ph.D., Caltech, 1965. Cosmic rays, solar wind, plasma.
Jozsef Kota, Senior Research Scientist; Ph.D., Budapest (Hungary), 1980. Theoretical space physics, space weather.
David A. Kring, Associate Professor; Ph.D., Harvard, 1989. Cosmochemistry.
Emil R. Kursinski Jr., Associate Professor; Ph.D., Caltech, 1997. Study of climate dynamics using remote sensing techniques, atmospheric physics, weather and climate of Mars.
Harold P. Larson, Professor; Ph.D., Purdue, 1967. Planetary atmospheres, infrared spectroscopy.
Dante Lauretta, Assistant Professor; Ph.D., Washington (St. Louis), 1997. Origin and chemical evolution of the solar system.
Larry A. Lebofsky, Senior Research Scientist; Ph.D., MIT, 1974. Infrared photometry, asteroidal observations.
John S. Lewis, Professor; Ph.D., California, San Diego, 1968. Cosmochemistry.
Ralph D. Lorenz, Assistant Research Scientist; Ph.D., Kent (England), 1994. Planetary climate, aerospace systems.
Jonathan I. Lunine, Professor; Ph.D., Caltech, 1985. Theoretical planetary physics, condensed-matter studies, structure of planets.
Renu Malhotra, Associate Professor; Ph.D., Cornell, 1988. Solar system dynamics.
Alfred McEwen, Associate Professor; Ph.D., Arizona State, 1988. Planetary geology and image processing, multispectral studies of many bodies, volcanism on Io, Calderas in Guatemala, mass movements of Earth and Mars, Copernican craters on the moon, remote sensing.
Robert S. McMillan, Associate Research Scientist; Ph.D., Texas at Austin, 1977. Doppler spectroscopy asteroid-detection survey.
H. Jay Melosh, Professor; Ph.D., Caltech, 1972. Planetary rings.
George H. Rieke, Professor; Ph.D., Harvard, 1969. Infrared astronomy.
Elizabeth Roemer, Professor Emerita; Ph.D., Berkeley, 1955. Comets, minor planets, astrometry.
Bill R. Sandel, Senior Research Scientist; Ph.D., Rice, 1972. Astrophysical plasmas, planetary atmospheres.
Adam Showman, Assistant Professor; Ph.D., Caltech, 1998. Dynamics and evolution of planetary atmospheres.
Peter H. Smith, Senior Research Scientist; M.S., Arizona, 1977. Optical sciences and radiative transfer in planetary atmospheres.
Charles P. Sonett, Regents Professor Emeritus; Ph.D., UCLA, 1954. Planetary physics, solar wind.
Robert G. Strom, Professor Emeritus; M.S., Stanford, 1957. Extraterrestrial geology.
Timothy D. Swindle, Professor; Ph.D., Washington (St. Louis), 1986. Cosmochemistry, noble gas studies of meteorites.
Martin G. Tomasko, Research Professor; Ph.D., Princeton, 1969. Planetary atmospheres, radiative transfer theory.
Elizabeth P. Turtle, Assistant Research Scientist; Ph.D., Arizona, 1998. Modeling and observations of impact craters and tectonic processes.
Roger Yelle, Professor; Ph.D., Wisconsin–Madison, 1984. Atmospheres and icy surfaces in this solar system, atmospheres of extrasolar planets.

Space Sciences Building, which houses the Department of Planetary Sciences and the Lunar and Planetary Laboratory.

THE UNIVERSITY OF NORTH CAROLINA WILMINGTON

Department of Geography and Geology
Master of Science in Geology

Program of Study

The Department of Geography and Geology at the University of North Carolina Wilmington (UNCW) offers a program of study leading to the Master of Science (M.S.) degree in geology. The program develops professional geologists who are capable of conducting research in geology through the broadly based study of modern geological processes and their ancient analogs. The program has six areas of research emphasis: surface and subsurface hydrology; oceanography, marine geology, and geophysics; coastal and estuarine processes; basin analysis and paleontology; hard-rock petrogenesis and structural analysis; and preparation for professional certification.

Students may choose from a thesis or a nonthesis option, both of which provide a foundation for employment in the environmental fields, mineral and energy industries, and government agencies. The thesis option prepares students for advanced study leading to the doctoral degree, while the nonthesis option prepares students for professional licensure in geology. The specific goals of the program are to provide advanced research and educational opportunities in the geological sciences and to prepare geologists for solving contemporary geologic problems. Thesis-track students must complete 30 semester hours of graduate study, at least 18 of which must be in geology. Each student must pass a comprehensive oral examination based on prior course work and must present and defend a thesis based on original research. The nonthesis program requires at least 36 semester hours of graduate credit, including a maximum of 3 credits for an internship or final project, 3 for seminars, and 6 of directed studies. Each nonthesis student must complete a core curriculum. Nonthesis students must take a written comprehensive examination upon successful completion of the required core course work (other than the final project or internship). Nonthesis candidates must prepare and present a scholarly paper or report prior to graduation. A final seminar is required.

Research Facilities

The Department of Geography and Geology is located in DeLoach Hall, which has classrooms, faculty offices, teaching and research laboratories, computer laboratories, and instrument rooms. The Department is well equipped with modern instruments for geologic and geographic studies. The program uses computers extensively for both laboratory and classroom instruction. The Department maintains state-of-the-art spatial analysis and cartography laboratories that are used by earth sciences students for data analysis, report writing, geographic information systems, digital image processing, and programming. Instructors use microcomputers equipped with advanced multimedia systems for classroom presentations. Approximately one third of the faculty members have offices and research laboratories at the Center for Marine Science, which is located 7 miles southeast of the main UNCW campus on the Atlantic intercoastal waterway. This modern facility provides space for other related agencies and supports marine research projects conducted in the coastal region of North Carolina, in the southeast region of the United States, and in other locations as required. Research laboratories include the Coastal and Marine Geophysics Laboratory, the Coastal Geology Laboratory, the Coastal Hydrology and Sedimentology Laboratory, the Invertebrate Paleontology Laboratory, the Laboratory for Applied Climate Research, the Marine Micropaleontology Laboratory, the Petrology Preparation Laboratory, the Soils Analysis Laboratory, and the Spatial Analysis Laboratory.

Financial Aid

The Department of Geography and Geology offers teaching assistantships, which are typically worth $9500, paid over a ten-month period. Teaching assistantships are awarded based on evaluation of the student's grade point average, scores on the Graduate Record Examinations, letters of recommendation, and research experience. Teaching assistantships are typically awarded for one academic year. Students may normally expect to have an assistantship for a second year; however, this is based on continued availability of funding and the student's standing in the program. Research assistantships are offered through individual faculty members who have funds available from research grants or contracts. Students should contact the Graduate School for a complete listing of current scholarships.

Cost of Study

Tuition is $279 per hour for North Carolina residents and $1508 for nonresidents. Fees vary depending on the number of hours taken. In 2005–06, a North Carolina resident paid a total of $1998 in tuition and fees for 9 hours; nonresidents paid $6916. The majority of students awarded teaching or research assistantships also receive tuition remission.

Living and Housing Costs

There are a variety of rental properties available in the Wilmington area—both apartments and single-family housing. Prices vary considerably, but one-bedroom units are available for approximately $500 to $750 per month. The University housing office maintains an online off-campus housing ad system.

Student Group

There are 27 students currently enrolled in the program. More than one third of the students are women, and many are from out of state. The University has an undergraduate population of approximately 11,650 and a graduate student population of approximately 950.

Location

The University of North Carolina Wilmington is located in the southeastern region of North Carolina, near the historic district of Wilmington, 4 miles from the Cape Fear River and 5 miles west of Wrightsville Beach and the Atlantic Ocean. Moss-laden oaks, towering pines, and stately brick buildings make up the arboretum campus on Highway 132 (College Road). By car, UNCW is accessible via I-40 and U.S. Highways 421, 17, and 74/76. USAirways and ASA Delta offer flights into Wilmington International Airport, which is only minutes from the campus.

The University and The Department

The University of North Carolina Wilmington is an active learning community that uniquely combines a small college's commitment to excellence in teaching with a research university's opportunities for student involvement in significant faculty scholarship. At UNCW, students are afforded opportunities to learn through collaborative scholarly activities with world-class faculty members at a level that rivals exclusive research institutions of similar size. The University is committed to service as both an obligation and an opportunity to improve the quality of life of the institution and the region.

The Department of Geography and Geology at the University of North Carolina Wilmington offers programs of study in geology that prepare the student for advanced studies or for teaching, research, and technical careers. The 21 faculty members of the Department of Geography and Geology have collective expertise in many areas of geology and geography. Research areas represented in the Department include marine and coastal geology, oceanography, geochemistry, marine and coastal geophysics, stratigraphy and basin analysis, environmental and resource geology, hydrogeology, geomorphology, mineralogy and petrology, paleontology and paleoecology, structural geology and tectonics, geographic information systems, climatology, spatial analysis, and historical and cultural geography.

Applying

Students must hold a bachelor's degree in any of the biological, earth, physical, or mathematical sciences from an accredited college or university in the United States (or its equivalent in an international institution based on a four-year program) and have a strong overall academic record, with a B average or better in the basic courses that are prerequisite to geology and satisfactory scores on the Graduate Record Examinations. All students must have completed two semesters each of chemistry, physics, and calculus and have working knowledge of physical and historical geology. Applicants must submit to the UNCW Graduate School an application for graduate admission, including the statement of interest and degree option form; official transcripts of all college work completed (undergraduate and graduate); official scores on the Graduate Record Examinations (verbal, quantitative, and analytical); and three letters of recommendation from individuals in professionally relevant fields. The application deadline is April 15 for the fall semester and November 15 for the spring semester. The application deadline for students who are interested in financial aid is February 15 for the fall semester and October 15 for the spring semester.

Correspondence and Information

Nancy Grindlay, Graduate Program Coordinator
Department of Geography and Geology
The University of North Carolina Wilmington
601 South College Road
Wilmington, North Carolina 28403-5944
Phone: 910-962-2352
Fax: 910-962-2410
E-mail: grindlayn@uncw.edu
Web site: http://www.uncw.edu/earsci/

The University of North Carolina Wilmington

THE FACULTY AND THEIR RESEARCH

Lewis Abrams, Associate Professor of Geology; Ph.D., Rhode Island, 1992. Investigating subsurface features, both submarine and terrestrial, using a variety of geophysical instruments, including seismic and multibeam bathymetry systems along with gravimeters, magnetometers, and ground-penetrating radar.

Frank Ainsley, Professor of Geography; Ph.D., North Carolina at Chapel Hill, 1977. Chattel houses and settlements in Barbados, tourism images and perceptions of Barbados, vernacular architecture of alpine Europe, rural settlement patterns in Norway, immigrant farm colonies in southeastern North Carolina.

Robert Argenbright, Associate Professor of Geography; Ph.D., Berkeley, 1990. Historical geography of the Soviet Union, particularly the periods of the Russian Civil War and World War II; current changes in the urban structure and pattern of land use in Moscow.

Michael Benedetti, Associate Professor of Geography; Ph.D., Wisconsin–Madison, 2000. Postglacial evolution of the upper Mississippi River, soil erosion and sedimentation in agricultural watersheds, influence of floods on sediment transport in large rivers.

David Blake, Associate Professor of Geology; Ph.D., Washington State, 1991. Study of the tectonic evolution of the North American craton from the Late Proterozoic to the Mesozoic era in the eastern Piedmont of North Carolina and west-central Idaho.

William Cleary, Professor of Geology; Ph.D., South Carolina, 1972. Coastal management–oriented investigations involving inlet- and hurricane-related shoreline changes, modification of inlet systems for sand resources for nourishment purposes.

James Dockal, Professor of Geology; Ph.D., Iowa State, 1980. Petrography of carbonate and carbonate-associated rocks in high-grade metamorphic terrains of North Carolina and Colorado; structural analysis of the Blue Ridge to Piedmont transition in Henderson and Transylvania Counties, North Carolina.

Doug Gamble, Associate Professor of Geography; Ph.D., Georgia, 1997. Assessment of the spatial variability in rainfall on San Salvador Island, Bahamas; linking of weather patterns to Wilmington rainwater chemistry; development of a coastal climatology for the U.S. East Coast in collaboration with the NOAA Coastal Services Center.

Nancy Grindlay, Professor of Geology and Graduate Program Coordinator; Ph.D., Rhode Island, 1991. Morphology, structure, and tectonic evolution of submarine plate boundaries; seabed classification techniques; sediment transport and mass wasting along continental and insular margins.

Joanne Halls, Associate Professor of Geography and Director of the Spatial Analysis Lab; Ph.D., South Carolina, 1996. Watershed modeling; coastal land-use development; integration of temporal ecological models with GIS; Internet GIS, with emphasis on underwater GIS and near-shore coastal GIS applications; spatiotemporal analysis of species distribution patterns.

W. Burleigh Harris, Professor of Geology; Ph.D., North Carolina, 1975. Surface and subsurface spatial and temporal distribution of depositional sequences of the southeastern Atlantic and Gulf Coastal Plains, differentiating the impact of eustasy from tectonism, geochronology, Rb-Sr and K-Ar dating of glauconite and Sr isotopes to elucidate sequence age.

Eric Henry, Assistant Professor of Geology; Ph.D., Arizona, 2001. Groundwater flow and contaminant transport, vadose-zone processes, coastal hydrology.

Elizabeth Hines, Associate Professor of Geography; Ph.D., Louisiana. The changing national perceptions of Southern cultural characteristics, modern hate-crime perpetrators and victims and changing hate-crimes statutes, North Carolina's gold rush and the immigration of Cornish miners to the United States and other mining areas.

John Huntsman, Associate Professor of Geology; Ph.D., Bryn Mawr, 1978. Fabrics and deformation mechanisms in crystalline thrust faults in the Blue Ridge of northwestern North Carolina, the transition from the Blue Ridge to the Valley and Ridge in northwestern North Carolina and northeastern Tennessee, inventories and implications of brittle structures in the crystalline and sedimentary rocks of the Blue Ridge, Piedmont, and Coastal Plain regions of North Carolina.

Patricia Kelley, Professor of Geology; Ph.D., Harvard, 1979. The tempo and mode of evolution, the role of biological factors such as predation in evolution, predator-prey coevolution and escalation, mass extinction and recovery of mollusk faunas.

Richard Laws, Professor of Geology and Department Chair; Ph.D., Berkeley, 1983. Biostratigraphy and paleoecology; taxonomy, biostratigraphy, and paleoecology of marine diatoms and calcareous nanofossils; distribution and ecology of deep-living, benthic marine diatoms.

Lynn Leonard, Professor of Geology; Ph.D., South Florida, 1994. Biogeochemical fingerprinting of phosphorus sources in anthropogenically impacted watersheds, sediment dispersal in tidal marshes of the Chesapeake Bay (and other SEUS marsh systems), marsh restoration using dredge spoil material, sediment transport during storms on the midcontinental shelf, effect of river dredging on tidal amplitude and marsh sedimentation in the Cape Fear River.

Roger Shew, Lecturer of Geology; M.S., North Carolina at Chapel Hill, 1979; M.Sci.Ed., Houston, 1996. Reservoir characterization and modeling (seismic to pore scale) of siliciclastic (in particular deep-water, deltaic, and coastal) and carbonate depositional environments, developing integrated science modules for earth and environmental science education.

Michael Smith, Associate Professor of Geology; Ph.D., Washington (St. Louis), 1990. Study of mid-Proterozoic crustal rocks in west Greenland, 2.2-Ma-old volcanic rocks of the upper Yellowstone Valley, the occurrence and petrology of cordierite-orthoamphibole rocks, ceramic petrology and the determination of the provenance of historic and prehistoric ceramic shards, technological exchange and innovation in the history of the gold mining rush(es) in the southeastern United States from 1799 to 1864, the sometimes-erratic development of the North Carolina geological surveys.

Paul Thayer, Professor of Geology; Ph.D., North Carolina at Chapel Hill, 1967. Sedimentology, petrology, and provenance of sediment and sedimentary rocks; geology and hydrology of nuclear repository sites throughout the United States.

Craig Tobias, Assistant Professor of Geology; Ph.D., William and Mary, 1999. Removal of groundwater nitrogen loads by coastal salt marshes, constructing estuary-scale nitrogen budgets, examining coupled carbon and oxygen dynamics in aquatic systems, developing stable-isotope techniques to measure reaction rates in the field across hydraulically active interfaces.

UNIVERSITY OF ROCHESTER

UNIVERSITY OF ROCHESTER

Department of Earth and Environmental Sciences

Programs of Study

The Department of Earth and Environmental Sciences offers programs of study leading to the degrees of M.S. and Ph.D. in geological sciences. These programs provide classroom, laboratory, and field instruction as well as research experience to prepare students for successful careers in academia and industry. The Department faculty members conduct active research in paleomagnetism and geophysics, solid earth geochemistry, noble gas geochemistry, cosmogenic isotope geochemistry, environmental geochemistry, paleoclimatology and paleoaltimetry, paleoceanography, sedimentary geology, stratigraphy, structural geology, tectonics, and geodynamics.

All graduate students are expected to take a combination of courses designed to provide an in-depth understanding of their area of specialization as well as general expertise in geological sciences. This curricular program is designed individually for each student in consultation with the student's research adviser and thesis committee. To ensure that candidates for the M.S. and Ph.D. obtain teaching experience, all students are required to aid in instruction for at least one term.

Research Facilities

The Department of Earth and Environmental Sciences has state-of-the-art instrumentation to complement field-based research programs. Instruments include the following: a multicollector, magnetic-sector ICP-MS, a quadrupole ICP-MS, and a thermal ionization mass spectrometer, which is used to determine trace-metal contents and isotopic compositions of geological, environmental, and biological materials. A rare gas mass spectrometer is used for high-precision He, Ne, and Ar isotopic measurements. A gas source mass spectrometer is available for analyzing light stable isotopes of H, C, O, and N. The instrument is used to analyze carbonates, fossil teeth, organic matter, and water for paleoenvironmental, climate, and ecology studies. A DC SQUID superconducting rock magnetometer with a high-resolution coil configuration is used for paleomagnetic and paleointensity investigations of single minerals. An alternating gradient force magnetometer, a high-speed automatic spinner magnetometer, an automated magnetic susceptibility system (allowing low- and high-temperature measurements), and a range of demagnetization devices allow further detailed rock magnetic analyses. A structural geology laboratory includes research microscopes for photomicrography, semi-automated point counting stage, Image Pro Plus image analysis system for petrofabric studies, a cold-cathode luminoscope, and a digitizer and plotter. Equipment is also available for X-ray diffractometry, UV-Vis spectrophotometry, and ion chromatography. Sample preparation is carried out in the Department's rock-cutting and thin-section facilities and in clean labs and wet chemistry labs, which also house a Frantz isodynamic separator and a microwave digestion system. PC, Macintosh, and Sun workstations are used for data processing and general-purpose computing.

Financial Aid

Students enrolled full-time in the Ph.D. program are generally awarded a full-tuition scholarship as well as a competitive stipend. The stipend is in the form of a teaching or research assistantship. All students are expected to serve as teaching assistants for at least one year as part of their training. A number of merit-based and minority representation scholarships are also available for qualified students.

Cost of Study

Generally, students admitted to the Ph.D. program receive a full-tuition scholarship.

Living and Housing Costs

Monthly rents for on-campus facilities for married and unmarried graduate students range from $391 for a furnished sleeping room with a shared bath and kitchen to $692 for a three-bedroom town house. The Community Living Program provides a variety of referral and apartment-hunting services for members of the University who prefer to live elsewhere in the community.

Student Group

Currently, the Department has 11 full-time graduate students and 4 postdoctoral research associates of varying ages, nationalities, and experiences. Of this number, 7 are international students and 6 are women.

Location

Rochester is located in upstate New York, south of Lake Ontario. The metropolitan area includes nearly a million people and high-tech corporations such as Eastman Kodak Company and Xerox Corporation. The city and eight area colleges support theater, dance, film series, art galleries, museums, and musical activities centering on the Eastman School of Music. Rochester is home to professional teams in baseball, hockey, lacrosse, and soccer. The Finger Lakes, Adirondack Mountains, and other nearby wilderness areas offer many recreational opportunities.

The University

The University of Rochester, which was founded in 1850, is a private, coeducational institution with some 1,000 full-time faculty members, 2,500 full-time graduate students, 1,100 part-time graduate students, and 4,700 full-time undergraduates.

Applying

Graduate students are generally expected to have a strong background in geoscience and a broad knowledge of other sciences and mathematics. However, because of the multidisciplinary nature of research in the Department, special consideration is given to students with strong backgrounds in particular areas of science (especially chemistry, biology, physics, engineering, and material science) even if they have only a modest background in geoscience. Applications from qualified women and members of minority groups are strongly encouraged.

Students are required to submit scores from the Graduate Record Examinations (GRE). The Test of English as a Foreign Language (TOEFL) is required of all applicants whose native language is not English. Applications should be submitted by February 1 to guarantee consideration for September enrollment.

Correspondence and Information

Director of Graduate Studies
Department of Earth and Environmental Sciences
University of Rochester
227 Hutchison Hall
Rochester, New York 14627
Phone: 585-275-5713
Fax: 585-244-5689
E-mail: ees@earth.rochester.edu
Web site: http://www.earth.rochester.edu

University of Rochester

THE FACULTY AND THEIR RESEARCH

Asish R. Basu, Professor; Ph.D., California, Davis, 1975. Applications of trace elements and radiogenic isotopes in understanding the age, origin, and evolution of the crust-mantle system of rocks; geochemistry of igneous, metamorphic, and sedimentary rocks; plume volcanism and continental flood basalt genesis; groundwater flux of Sr to the oceans and the marine Sr-isotopic record; meteorite impacts and mass extinctions; slab devolatilization and material transfer in the mantle wedge.

William Chaisson, Adjunct Assistant Professor; Ph.D., Massachusetts Amherst, 1996. Assessing variations in the character of the Gulf Stream during the especially cold MIS 12 to the especially warm MIS 11, documenting fluctuations in the path of the Gulf Stream/North Atlantic Current near the Northwest Corner during the Holocene, reconstructing the Pliocene development of the present surface hydrography in the equatorial Pacific, reconstructing Holocene climate change based on the sedimentary record of the Finger Lakes.

Cynthia Ebinger, Professor; Ph.D., MIT/Woods Hole Oceanographic Institute, 1988. Mechanical properties of continental and oceanic lithosphere at divergent plate boundaries, interaction between lithospheric and asthenospheric processes at the transition between continental and oceanic rifting. Both geophysical and geological data are used to determine the relative importance of processes in these regional tectonic studies. Analytical and numerical models of lithospheric deformation processes allow one to assess the primary factors controlling a process and to identify the key observations needed to differentiate between models. Specific research themes are the initiation and evolution of along-axis rift segmentation and the modification of continental lithosphere by magmatic processes. Research necessarily includes studies of pre-existing crust and upper mantle structure and rheology. Current research focuses on measurement and modeling of ongoing earthquake and volcanic processes along the divergent plate boundary in East Africa, with particular focus on the Afar Depression. Ongoing research also includes the use of regional gravity-topography data sets to test and refine spectral and time-frequency models of flexural rigidity in continental and oceanic settings.

Udo Fehn, Professor; Ph.D., Munich Technical, 1973. Origin and movement of fluids in the crust, application of cosmogenic isotopes for tracing and dating of fluids, origin of volatiles in volcanic and geothermal fluids, recycling of marine sediments in island arcs, formation of gas hydrates in marine and terrestrial settings, sources of fluids associated with oil and gas reservoirs, tracing of anthropogenic radioisotopes in the environment.

Carmala Garzione, Associate Professor; Ph.D., Arizona, 2000. The study of sedimentary basin evolution and related mountain belts; fieldwork emphasizes reconstructing paleogeography and basin evolution from provenance, facies, and paleocurrent information; application of geochemical and petrologic provenance techniques to understand mountain belt unroofing history and stable isotopes as indicators of paleoelevation and/or paleoenvironment; recent work in the Tibetan Plateau and Bolivian Altiplano.

Gautam Mitra, Professor; Ph.D., Johns Hopkins, 1977. Field-based structural studies in the North American Cordillera (Montana-Idaho-Wyoming-Utah), the Central and Southern Appalachians, the Kumaon and Darjeeling Himalaya, and the Scottish Caledonides aimed at deciphering the tectonic evolution of mountain belts; finite strain and strain history analysis aimed at understanding large-scale deformational patterns in mountain belts; microstructural and textural studies to determine deformation mechanisms in rocks and its implications for large-scale kinematics and mechanics of deformation.

Robert J. Poreda, Professor; Ph.D., California, San Diego, 1983. Groundwater flow models and the tritium/^{3}He dating of young groundwater, cosmic ray–produced ^{3}He and its application to geological problems, use of isotopic and geochemical tracers to study regional flow systems and the discharge of groundwater to the coastal ocean, understanding the structure and composition of the subcontinental and oceanic mantles through the study of the rare gases and other volatiles, origin of the volatile components in back-arc basalts (Lau Basin and the Mariana Trough) and island arc lavas (Alaska-Aleutian Arc and the Philippines), chemical and isotopic tracers of extraterrestrial impacts.

John A. Tarduno, Professor and Chair; Ph.D., Stanford, 1987. Application of paleomagnetism to geodynamics, geomagnetism, and environmental change; studies of plate and hot spot motion and true polar wander; use of single crystals to determine geomagnetic paleointensity and growth of the inner core; magnetostratigraphic studies of Cretaceous climate change; distribution and properties of biogenic magnetite; field studies in California, India, South Africa, New Zealand, and the high Arctic; marine geological work in the Pacific, including ocean drilling of Hawaiian-Emperor seamounts.

SELECTED PUBLICATIONS

Chakrabarti, R., and **A. R. Basu.** Trace element and isotopic evidence for Archean basement in the Lonar crater impact Breccia, Deccan Volcanic Province. *Earth Planet. Sci. Lett.*, in press.

Basu, A. R., et al. **(R. J. Poreda).** Chondritic meteorite fragments associated with the Permian-Triassic boundary in Antarctica. *Science* 302(5649):1388–92, 2003.

Basu, A. R., et al. **(R. J. Poreda).** Large ground water Sr flux to the oceans from the Bengal Basin and the Marine Sr isotope record. *Science* 293:5534, 2470–3, 2001.

Basu, A. R., and S. R. Hart, eds. Earth processes: Reading the Isotopic Code. *Geophys. Monogr.*, vol. 95. Washington, D.C.: American Geophysical Union, 1996.

Basu, A. R., and **R. J. Poreda** et al. High ^{3}He plume origin and temporal-spatial evolution of the Siberian flood basalts. *Science* 269:822–5, 1995.

Chaisson, W. P., M.-S. Poli, and R. C. Thunell. Gulf Stream and western boundary undercurrent variations during MIS 10-12 at Site 1056, Blake Bahama Outer Ridge. *Mar. Geol.* 189:79–105, 2002.

Gruetzner, J., et al. **(W. P. Chaisson).** Astronomical Age models for Pleistocene drift sediments from the western North Atlantic (ODP Sites 1055 to 1063). *Mar. Geol.* 189:5–23, 2002.

Chaisson, W. P., and A. C. Ravelo. Pliocene development of the east-west hydrographic gradient in the equatorial Pacific. *Paleoceanography* 15:497–505, 2000.

Chaisson, W. P., and S. L. D'Hondt. Neogene planktonic foraminifer biostratigraphy at site 999: Western Caribbean Sea. In *Proceedings of the Ocean Drilling Program, Scientific Results,* vol. 165, pp. 19–56, eds. H. Sigurdsson et al. College Station, Tex.: Ocean Drilling Program, 2000.

Schneider, D. A., J. Backman, **W. P. Chaisson,** and I. Raffi. Miocene calibration for calcareous microfossils from low-latitude Ocean Drilling Program sites and the Jamaican Conundrum. *GSA Bull.* 109(9):1073–9, 1997.

Wright, T. C., and **C. Ebinger** et al. Magma-fed along-axis segmentation in a nascent oceanic rift—the 2005 Dabbahu rifting event, Afar. *Nature,* in press.

Keir, D., and **C. Ebinger** et al. Strain accommodation by magmatism and faulting at continental breakup: Sismicity of the northern Ethiopian rift. *J. Geophys. Res.*, 2006.

Wolfenden, E., and **C. Ebinger** et al. Evolution of the southern Red Sea rift: Birth of a magmatic margin. *Bull. Geol. Soc. Am.* 117:846–64, 2005.

Kendall, J.-M., et al. **(C. Ebinger).** Magma-assisted rifting in Ethiopia. *Nature* 433:146–8, 2005.

Tiberi, C., and **C. Ebinger** et al. Inverse models of gravity data from the Red Sea-Gulf of Aden-Ethiopian rift triple junction zone. *Geophys. J. Int.* 163:775–87, 2005.

Sleep, N., **C. Ebinger,** and M. Kendall. Deflection of mantle plume material by cratonic keels. In *Early Earth,* vol. 199, pp. 135–50, eds. C. M. Fowler, **C. Ebinger,** and C. J. Hawkesworth. Geological Society of London Special Publication, 2002.

Fehn, U., Z. Lu, and H. Tomaru. ^{129}I/I ratios and halogen concentrations in pore waters of the Hydrate Ridge and their relevance for the origin of gas hydrates: A progress report. *Proc. ODP, Sci. Results* 204:MS 204SR-107, 2006.

Hurwitz, R., H. Mariner, **U. Fehn,** and G. T. Snyder. Systematics of halogen elements and their radioisotopes in thermal springs of the Cascade Range, Central Oregon, Western USA. *Earth Planet. Sci. Lett.* 235:700–14, 2005.

Fehn, U., and G. T. Snyder. Residence times and source ages of deep crustal fluids: Interpretation of ^{129}I and ^{36}Cl results from the KTB-VB drill site, Germany. *Geofluids* 5:42–51, 2005.

Fehn, U., et al. Iodine dating of pore waters associated with gas hydrates in the Nankai area, Japan. *Geology* 31:521–4, 2003.

Fehn, U., G. Snyder, and P. K. Egeberg. Dating of pore waters with ^{129}I: Relevance for the origin of marine gas hydrates. *Science* 289:2332–5, 2000.

Garzione, C. N., P. Molnar, J. C. Libarkin, and B. MacFadden. Rapid late Miocene rise of the Andean plateau: Evidence for removal of mantle lithosphere. *Earth Planet. Sci. Lett.* 241:543–56, 2006.

Ghosh, P., **C. N. Garzione,** and J. Eiler. Paleothermometry of Altiplano paleosols: Implications for Late Miocene surface uplift of the Andean plateau. *Science* 311:511–5, 2006.

Garzione, C. N., M. Ikari, and **A. Basu.** Source of Oligocene to Pliocene sedimentary rocks in the Linxia Basin in NE Tibet from Nd isotopes: Implications for tectonic forcing of climate. *Geol. Soc. Am. Bull.* 117:1156–66, 2005.

Garzione, C. N., J. Quade, P. G. DeCelles, and N. B. English. Predicting paleoelevation of Tibet and the Himalaya from δ^{18} O vs. altitude gradients of meteoric water across the Nepal Himalaya. *Earth Planet. Sci. Lett.* 183:215–29, 2000.

Garzione, C. N., et al. High times on the Tibetan Plateau: Paleoelevation of the Thakkhola Graben, Nepal. *Geology* 28:339–42, 2000.

Kwon, S., and **G. Mitra.** Three-dimensional kinematic history at an oblique ramp, Leamington zone, Sevier belt, Utah. *J. Struct. Geol.* 28:474–93, 2006.

Ismat, Z., and **G. Mitra.** Folding by cataclastic flow: Evolution of controlling factors during deformation. *J. Struct. Geol.* 27:2181–203, 2005.

Ismat, Z., and **G. Mitra.** Fold-thrust belt evolution expresses in an internal thrust sheet, Sevier Orogen: The role of cataclastic flow. *Geol. Soc. Am. Bull.* 116:764–82, 2005.

Kwon, S., and **G. Mitra.** Three-dimensional FE modeling of a thin-skinned FTB wedge: Insights from the example of the Provo salient, Sevier FTB, Utah. *Geology* 32:561–4, 2004.

University of Rochester

Strine, M., and **G. Mitra.** Preliminary kinematic data from a salient-recess pair along the Moine thrust, NW Scotland. In *Salients in Mountain Belts,* vol. 383, pp. 87–107, eds. A. Sussman and A. Weil. Geological Society of America Special Paper, 2004.

Mitra, G., and Z. Ismat. Microfracturing associated with reactivated fault zones and shear zones: What it can tell us about deformation history. In *The Nature and Tectonic Significance of Fault Zone Weakening,* vol. 186, pp. 113–40. Geological Society of London Special Publication, 2001.

Poreda, R. J., A. G. Hunt, K. Welch, and W. B. Lyons. Chemical and isotopic evolution of Lake Bonney, Taylor Valley: Timing of Late Holocene climate change in Antarctica. *Aquatic Geochem.* 10:353–71, 2004.

Dowling, C. B., **R. J. Poreda,** A. G. Hunt, and A. E. Carey. Ground water discharge and nitrate flux to the Gulf of Mexico. *Ground Water* 42:401–17, 2004.

Poreda, R. J., and L. Becker. Fullerenes and interplanetary dust at the Permian-Triassic boundary. *Astrobiology* 3:75–90, 2003.

Snyder, G., **R. J. Poreda,** A. G. Hunt, and **U. Fehn.** Sources of nitrogen and methane in Central American geothermal settings: Noble gases and ^{129}I evidence of crustal magmatic sources. *Geochem. Geophys. Geosyst.* 4, 2003 (DOI 10.1029/2002GC000363).

Dowling, C. B., **R. J. Poreda,** and **A. R. Basu** et al. Geochemical study of arsenic release mechanisms in the Bengal Basin groundwater. *Water Resour. Res.* 38:1173–90, 2002.

Tarduno, J. A., R. D. Cottrell, and A. V. Smirnov. The paleomagnetism of single silicate crystals: Recording geomagnetic field strength during mixed polarity intervals, superchrons and inner core growth. *Rev. Geophys.* RG1002, 2006.

Smirnov, A. V., and **J. A. Tarduno.** Secular variation of the late Archean–Early Proterozoic geodynamo. *Geophys. Res. Lett.* 31: L16607, 2004.

Tarduno, J. A., et al. The Emperor Seamounts: Southward motion of the Hawaiian hotspot plume in the Earth's mantle. *Science* 301:1064–9, 2003.

Tarduno, J. A., R. D. Cottrell, and A. V. Smirnov. The Cretaceous Superchron geodynamo: Observations near the tangent cylinder. *Proc. Natl. Acad. Sci. U.S.A.* 99:14020–5, 2002.

Tarduno, J. A., and A. V. Smirnov. Stability of the Earth with respect to the spin axis for the last 130 million years. *Earth Planet. Sci. Lett.* 184:549–53, 2001.

Tarduno, J. A., R. D. Cottrell, and A. V. Smirnov. High geomagnetic field intensity during the mid-Cretaceous from Thellier analyses of single plagioclase crystals. *Science* 291:1779–83, 2001.

Section 4
Marine Sciences and Oceanography

This section contains a directory of institutions offering graduate work in marine sciences and oceanography, followed by in-depth entries submitted by institutions that chose to prepare detailed program descriptions. Additional information about programs listed in the directory but not augmented by an in-depth entry may be obtained by writing directly to the dean of a graduate school or chair of a department at the address given in the directory.

For programs offering related work, see also in this book Chemistry, Geosciences, Meteorology and Atmospheric Sciences, and Physics. In Book 3, see Biological and Biomedical Sciences; Ecology, Environmental Biology, and Evolutionary Biology; and Marine Biology; and in Book 5, see Civil and Environmental Engineering, Engineering and Applied Sciences, and Ocean Engineering.

CONTENTS

Program Directories

Announcements

Cross-Discipline Announcement

Close-Ups

See also:

Marine Sciences

American University, College of Arts and Sciences, Department of Biology, Environmental Science Program, Washington, DC 20016-8001. Offers environmental science (MS); marine science (MS). *Students:* 4 full-time (3 women), 1 (woman) part-time; includes 1 minority (Hispanic American), 1 international. Average age 28. In 2005, 6 degrees awarded. *Degree requirements:* For master's, thesis or alternative, comprehensive exam. *Entrance requirements:* For master's, GRE General Test, GRE Subject Test, minimum GPA of 3.0. Additional exam requirements/recommendations for international students: Required—TOEFL. *Application deadline:* For fall admission, 2/1 for domestic students; for spring admission, 10/1 for domestic students. Application fee: $50. *Expenses:* Tuition: Full-time $17,802; part-time $989 per credit. Required fees: $380. *Financial support:* Research assistantships, teaching assistantships available. Financial award application deadline: 2/1. *Unit head:* Dr. Kiho Kim, Director, 202-885-2181, Fax: 202-885-2181.

California State University, East Bay, Academic Programs and Graduate Studies, College of Science, Department of Biological Sciences, Moss Landing Marine Laboratory, Hayward, CA 94542-3000. Offers MS. *Degree requirements:* For master's, thesis. *Entrance requirements:* For master's, GRE Subject Test, minimum GPA of 3.0 in field, 2.75 overall. Application fee: $55. *Financial support:* Application deadline: 3/2. *Unit head:* Dr. Kenneth H Coale, Director, 832-632-4400, E-mail: coale@mlml.calstate.edu. *Application contact:* Deborah Baker, Associate Director, 510-885-3286, Fax: 510-885-4777, E-mail: deborah.baker@csueastbay.edu.

California State University, Fresno, Division of Graduate Studies, College of Science and Mathematics, Program in Marine Sciences, Fresno, CA 93740-8027. Offers MS. Part-time programs available. Postbaccalaureate distance learning degree programs offered. *Degree requirements:* For master's, thesis. *Entrance requirements:* For master's, GRE General Test, minimum GPA of 3.0. Additional exam requirements/recommendations for international students: Required—TOEFL. Electronic applications accepted. *Faculty research:* Wetlands ecology, land/water conservation, water irrigation.

California State University, Monterey Bay, College of Science, Media Arts and Technology, Program in Marine Studies, Seaside, CA 93955-8001. Offers marine science (MS). Program offered in conjunction with Moss Landing Marine Laboratories. Part-time programs available. *Degree requirements:* For master's, thesis, thesis defense. Electronic applications accepted. *Faculty research:* Remote sensing microbiology trace elements; chemistry ecology of birds, mammals, turtles and fish; invasive species; marine phycology.

California State University, Sacramento, Graduate Studies, College of Natural Sciences and Mathematics, Department of Biological Sciences, Sacramento, CA 95819-6048. Offers biological sciences (MA, MS); immunohematology (MS); marine science (MS). Part-time programs available. *Students:* 22 full-time (11 women), 53 part-time (37 women); includes 16 minority (3 African Americans, 7 Asian Americans or Pacific Islanders, 6 Hispanic Americans), 4 international. Average age 30. 79 applicants, 42% accepted, 24 enrolled. *Degree requirements:* For master's, thesis, writing proficiency exam. *Entrance requirements:* For master's, bachelor's degree in biology or equivalent, minimum GPA of 3.0 in biology, minimum overall GPA of 2.75 during last 2 years of course work. Additional exam requirements/recommendations for international students: Required—TOEFL. *Application deadline:* Applications are processed on a rolling basis. Application fee: $55. Electronic applications accepted. *Expenses:* Tuition, nonresident: part-time $339 per unit. Required fees: $276 per semester hour. *Financial support:* Research assistantships, teaching assistantships, career-related internships or fieldwork and Federal Work-Study available. Support available to part-time students. Financial award application deadline: 3/1. *Unit head:* Dr. Nick Ewing, Chair, 916-278-6535, Fax: 916-278-6993.

Coastal Carolina University, College of Natural and Applied Sciences, Conway, SC 29528-6054. Offers coastal marine and wetland studies (MS). Part-time and evening/weekend programs available. *Faculty:* 11 full-time (2 women). *Students:* 13 full-time (6 women), 10 part-time (6 women). Average age 27. In 2005, 1 degree awarded. *Degree requirements:* For master's, thesis. *Entrance requirements:* For master's, GRE, 2 letters of recommendation. Additional exam requirements/recommendations for international students: Required—TOEFL. *Application deadline:* For fall admission, 8/15 for domestic students. Applications are processed on a rolling basis. Application fee: $45. Electronic applications accepted. *Expenses:* Tuition, state resident: full-time $7,160; part-time $295 per credit hour. Tuition, nonresident: full-time $15,920; part-time $660 per credit hour. *Financial support:* Fellowships, research assistantships, unspecified assistantships available. Support available to part-time students. Financial award application deadline: 4/1; financial award applicants required to submit FAFSA. *Unit head:* Dr. Joan F. Piroch, Interim Dean, 843-349-2271, Fax: 843-349-2545, E-mail: pirochj@coastal.edu. *Application contact:* Dr. Judy W. Vogt, Vice President, Enrollment Services, 843-349-2037, Fax: 843-349-2127, E-mail: jvogt@coastal.edu.

The College of William and Mary, School of Marine Science, Williamsburg, VA 23187-8795. Offers MS, PhD. *Faculty:* 53 full-time (10 women), 1 part-time/adjunct (0 women). *Students:* 117 full-time (59 women), 5 part-time (2 women); includes 6 African Americans, 2 Asian Americans or Pacific Islanders, 3 Hispanic Americans, 16 international. 112 applicants, 30% accepted, 26 enrolled. In 2005, 16 master's, 8 doctorates awarded. *Degree requirements:* For master's, thesis, qualifying exam defense; for doctorate, thesis/dissertation, qualifiying exam defense. *Entrance requirements:* For master's, GRE, appropriate bachelor's degree; for doctorate, GRE, appropriate bachelor's/master's degree. Additional exam requirements/recommendations for international students: Required—TOEFL. *Application deadline:* For fall admission, 1/15 for domestic students, 1/15 for international students. Application fee: $50. *Expenses:* Tuition, state resident: full-time $5,828; part-time $245 per credit. Tuition, nonresident: full-time $17,980; part-time $685 per credit. Required fees: $3,051. Tuition and fees vary according to program. *Financial support:* In 2005–06, fellowships with full tuition reimbursements (averaging $17,200 per year), research assistantships (averaging $16,400 per year) were awarded; teaching assistantships, career-related internships or fieldwork, Federal Work-Study, health care benefits, and unspecified assistantships also available. Support available to part-time students. Financial award application deadline: 6/15; financial award applicants required to submit FAFSA. Total annual research expenditures: $20.6 million. *Application contact:* Sue Presson, Graduate School Registrar.

The College of William and Mary, School of Marine Science/Virginia Institute of Marine Science, Gloucester Point, VA 23062. Offers MS, PhD. *Degree requirements:* For master's and doctorate, thesis/dissertation, defense, qualifying exam. *Entrance requirements:* For master's, GRE, appropriate bachelor's degree; for doctorate, GRE, appropriate master's degree. Additional exam requirements/recommendations for international students: Required—TOEFL. Electronic applications accepted. *Expenses:* Tuition, state resident: full-time $5,828; part-time $245 per credit. Tuition, nonresident: full-time $17,980; part-time $685 per credit. Required fees: $3,051. Tuition and fees vary according to program. *Faculty research:* Physical, biological, geological, and chemical oceanography; marine fisheries science; resource management.

Cornell University, Graduate School, Graduate Fields of Agriculture and Life Sciences, Field of Natural Resources, Ithaca, NY 14853-0001. Offers aquatic science (MPS, MS, PhD); environmental management (MPS); fishery science (MPS, MS, PhD); forest science (MPS, MS, PhD); resource policy and management (MPS, MS, PhD); wildlife science (MPS, MS, PhD). *Faculty:* 62 full-time (9 women). *Students:* 62 full-time (33 women); includes 9 minority (1 African American, 2 American Indian/Alaska Native, 4 Asian Americans or Pacific Islanders, 2 Hispanic Americans), 14 international. 60 applicants, 23% accepted, 12 enrolled. In 2005, 11 master's, 5 doctorates awarded. *Degree requirements:* For master's, thesis (MS), project paper (MPS); for doctorate, thesis/dissertation, comprehensive exam. *Entrance requirements:* For master's and doctorate, GRE General Test, 2 letters of recommendation. Additional exam requirements/recommendations for international students: Required—TOEFL (minimum score 550 paper-based; 213 computer-based). *Application deadline:* For spring admission, 10/30 for domestic students. Applications are processed on a rolling basis. Application fee: $60. Electronic applications accepted. *Financial support:* In 2005–06, 49 students received support, including 15 fellowships with full tuition reimbursements available, 16 research assistantships with full tuition reimbursements available, 18 teaching assistantships with full tuition reimbursements available; institutionally sponsored loans, scholarships/grants, health care benefits, tuition waivers (full and partial), and unspecified assistantships also available. Financial award applicants required to submit FAFSA. *Faculty research:* Ecosystem-level dynamics, systems modeling, conservation biology/management, resource management's human dimensions, biogeochemistry. *Unit head:* Director of Graduate Studies, 607-255-2807, Fax: 607-255-0349. *Application contact:* Graduate Field Assistant, 607-255-2807, Fax: 607-255-0349, E-mail: nrgrad@cornell.edu.

Duke University, Nicholas School of the Environment and Earth Sciences, Durham, NC 27708-0328. Offers coastal environmental management (MEM); environmental economics and policy (MEM); environmental health and security (MEM); forest resource management (MF); global environmental change (MEM); resource ecology (MEM); water and air resources (MEM). *Accreditation:* SAF (one or more programs are accredited). Part-time programs available. *Degree requirements:* For master's, thesis, registration. *Entrance requirements:* For master's, GRE General Test, previous course work in biology or ecology, calculus, statistics, and microeconomics; computer familiarity with word processing and data analysis. Additional exam requirements/recommendations for international students: Required—TOEFL (minimum score 550 paper-based; 213 computer-based). Electronic applications accepted. Expenses: Contact institution. *Faculty research:* Ecosystem management, conservation ecology, earth systems, risk assessment.

Announcement: Two-year professional program provides excellent preparation for careers in coastal environmental management. Core courses emphasize coastal sedimentary and biological processes, ecology, economics, policy, and quantitative analytical methods. Program includes courses on the main Durham campus and at the Duke University Marine Laboratory located at Beaufort, North Carolina. PhD program also available.

See Close-Up on page 709.

Florida Institute of Technology, Graduate Programs, College of Engineering, Department of Marine and Environmental Systems, Program in Oceanography, Melbourne, FL 32901-6975. Offers biological oceanography (MS); chemical oceanography (MS); coastal zone management (MS); geological oceanography (MS); oceanography (PhD); physical oceanography (MS). Part-time programs available. *Students:* Average age 30.Terminal master's awarded for partial completion of doctoral program. *Degree requirements:* For master's, thesis (for some programs); for doctorate, one foreign language, thesis/dissertation, departmental qualifying exams, comprehensive exam. *Entrance requirements:* For master's, GRE General Test, minimum GPA of 3.0; for doctorate, GRE General Test, minimum GPA of 3.3, resumé. *Application deadline:* Applications are processed on a rolling basis. Electronic applications accepted. *Expenses:* Tuition: Part-time $825 per credit. *Financial support:* Research assistantships with full and partial tuition reimbursements, teaching assistantships with full and partial tuition reimbursements, career-related internships or fieldwork and tuition remissions available. Financial award application deadline: 3/1; financial award applicants required to submit FAFSA. *Faculty research:* Marine geochemistry, ecosystem dynamics, coastal processes, marine pollution, environmental modeling. Total annual research expenditures: $938,395. *Unit head:* Dr. Dean R. Norris, Chair, 321-674-7377, Fax: 321-674-7212, E-mail: norris@fit.edu. *Application contact:* Carolyn P. Farrior, Director of Graduate Admissions, 321-674-7118, Fax: 321-723-9468, E-mail: cfarrior@fit.edu.

See Close-Ups on pages 253 and 653.

Memorial University of Newfoundland, School of Graduate Studies, Interdisciplinary Program in Marine Studies, St. John's, NL A1C 5S7, Canada. Offers fisheries resource management (MMS, Advanced Diploma). Part-time programs available. *Students:* 10 full-time (5 women), 12 part-time (3 women), 3 international. 9 applicants, 100% accepted, 4 enrolled. In 2005, 4 degrees awarded. *Degree requirements:* For master's, report. *Entrance requirements:* For master's and Advanced Diploma, high 2nd class degree from a recognized university. *Application deadline:* For fall admission, 4/30 for domestic students, 4/30 for international students. Application fee: $40 Canadian dollars. *Expenses:* Tuition: Part-time $733 per term. Tuition and fees vary according to degree level and program. *Financial support:* Fellowships, research assistantships, teaching assistantships available. *Faculty research:* Biological, ecological and oceanographic aspects of world fisheries; economics; political science; sociology. *Unit head:* Dr. Peter Fisher, Chair, 709-778-0356, Fax: 709-778-0346, E-mail: peter.fisher@mi.mun.ca. *Application contact:* Nancy Smith, Program Support, 709-778-0522, E-mail: nancy.smith@mi.mun.ca.

Murray State University, College of Science, Engineering and Technology, Department of Water Science, Murray, KY 42071-0009. Offers MS. Part-time programs available.

North Carolina State University, Graduate School, College of Physical and Mathematical Sciences, Department of Marine, Earth, and Atmospheric Sciences, Raleigh, NC 27695. Offers marine, earth, and atmospheric sciences (MS, PhD); meteorology (MS, PhD); oceanography (MS, PhD). Terminal master's awarded for partial completion of doctoral program. *Degree requirements:* For master's, thesis (for some programs), final oral exam; for doctorate, thesis/dissertation, final oral exam, preliminary oral and written exams, comprehensive exam, registration. *Entrance requirements:* For master's, GRE General Test, minimum GPA of 3.0; for doctorate, GRE General Test, GRE Subject Test (for disciplines in biological oceanography and geology), minimum GPA of 3.0. Additional exam requirements/recommendations for international students: Required—TOEFL (minimum score 550 paper-based). Electronic applications accepted. *Faculty research:* Boundary layer and air quality meteorology; climate and mesoscale dynamics; biological, chemical, geological, and physical oceanography; hard rock, soft rock, environmental, and paleo geology.

See Close-Up on page 229.

Nova Southeastern University, Oceanographic Center, Program in Coastal-Zone Management, Fort Lauderdale, FL 33314-7796. Offers MS. *Faculty:* 15 full-time (1 woman), 5 part-time/adjunct (0 women). *Students:* 18 applicants, 94% accepted, 14 enrolled. *Entrance requirements:* For master's, GRE. Additional exam requirements/recommendations for international students: Required—TOEFL (minimum score 550 paper-based). *Application deadline:* Applications are processed on a rolling basis. Application fee: $50. *Financial support:* Career-related internships or fieldwork, Federal Work-Study, scholarships/grants, and unspecified assistantships available. Financial award applicants required to submit FAFSA. *Unit head:* Dr. Andrew Rogerson, Associate Dean, Director of Graduate Programs, 954-262-3600, Fax: 954-262-4020, E-mail: arogerso@nsu.nova.edu.

See Close-Up on page 261.

Nova Southeastern University, Oceanographic Center, Program in Marine Environmental Science, Fort Lauderdale, FL 33314-7796. Offers MS. *Faculty:* 15 full-time (1 woman), 5 part-time/adjunct (0 women). *Students:* 14 applicants, 79% accepted, 5 enrolled. In 2005, 1 degree awarded. *Degree requirements:* For master's, thesis. *Entrance requirements:* For master's, GRE. Additional exam requirements/recommendations for international students: Required—TOEFL (minimum score 550 paper-based). *Application deadline:* Applications are processed on a rolling basis. Application fee: $50. *Application contact:* Dr. Andrew Rogerson, Associate Dean, Director of Graduate Programs, 954-262-3600, Fax: 954-262-4020, E-mail: arogerso@nsu.nova.edu.

Oregon State University, Graduate School, College of Oceanic and Atmospheric Sciences, Program in Marine Resource Management, Corvallis, OR 97331. Offers MA, MS. *Students:* 25

full-time (15 women), 8 part-time (5 women); includes 1 minority (African American), 6 international. Average age 30. In 2005, 13 degrees awarded. *Degree requirements:* For master's, thesis optional. *Entrance requirements:* For master's, GRE General Test, minimum GPA of 3.0 in last 90 hours of course work. Additional exam requirements/recommendations for international students: Required—TOEFL. *Application deadline:* For fall admission, 2/1 for domestic students. Applications are processed on a rolling basis. Application fee: $50. *Expenses:* Tuition, area resident: Part-time $301 per credit. Tuition, state resident: full-time $8,139; part-time $501 per credit. Tuition, nonresident: full-time $14,376; part-time $532 per credit. Required fees: $1,266. *Financial support:* Fellowships, research assistantships, teaching assistantships, career-related internships or fieldwork, Federal Work-Study, and institutionally sponsored loans available. Support available to part-time students. Financial award application deadline: 2/1. *Faculty research:* Ocean and coastal resources, fisheries resources, marine pollution, marine recreation and tourism. *Unit head:* Dr. Robert Allen, Assistant Director, 541-737-1339, Fax: 541-737-2064. *Application contact:* Irma Delson, Assistant Director, Student Services, 541-737-5189, Fax: 541-737-2064, E-mail: student_adviser@oce.orst.edu.

San Francisco State University, Division of Graduate Studies, College of Science and Engineering, Department of Biology, Program in Marine Science, San Francisco, CA 94132-1722. Offers MS. *Entrance requirements:* For master's, minimum GPA of 2.5 in last 60 units.

See Close-Up on page 267.

San Jose State University, Graduate Studies and Research, College of Science, Program in Marine Science, San Jose, CA 95192-0001. Offers MS. *Students:* 1 full-time (0 women), 20 part-time (17 women); includes 1 minority (Hispanic American) Average age 28. 17 applicants, 35% accepted, 5 enrolled. In 2005, 1 degree awarded. *Degree requirements:* For master's, thesis, qualifying exam. *Entrance requirements:* For master's, GRE. *Application deadline:* For fall admission, 6/29 for domestic students; for spring admission, 11/30 for domestic students. Applications are processed on a rolling basis. Application fee: $59. Electronic applications accepted. *Expenses:* Tuition, nonresident: part-time $339 per unit. Required fees: $1,286 per semester. Tuition and fees vary according to course load and degree level. *Financial support:* Teaching assistantships, career-related internships or fieldwork available. Support available to part-time students. Financial award applicants required to submit FAFSA. *Faculty research:* Physical oceanography, marine geology, ecology, ichthyology, invertebrate zoology. *Unit head:* Dr. Kenneth Coale, Director, 831-771-4400, Fax: 831-753-2826.

Savannah State University, Program in Marine Science, Savannah, GA 31404. Offers MS. Part-time programs available. *Students:* 1 full-time (0 women), 12 part-time (4 women); includes 5 African Americans. In 2005, 6 degrees awarded. *Entrance requirements:* For master's, GRE General Test. *Application deadline:* For fall admission, 7/31 for domestic students; for spring admission, 12/31 for domestic students. Applications are processed on a rolling basis. Application fee: $20. Electronic applications accepted. *Unit head:* Dr. Matthew Gilligan, Coordinator, 912-356-2809, E-mail: gilligann@savstate.edu.

Stony Brook University, State University of New York, Graduate School, Marine Sciences Research Center, Program in Marine and Atmospheric Sciences, Stony Brook, NY 11794. Offers MS, PhD. Evening/weekend programs available. *Faculty:* 37 full-time (7 women). *Students:* 113 full-time (63 women), 6 part-time (4 women); includes 11 minority (2 African Americans, 1 American Indian/Alaska Native, 2 Asian Americans or Pacific Islanders, 6 Hispanic Americans), 45 international. Average age 27. 86 applicants, 52% accepted. In 2005, 15 master's, 3 doctorates awarded. *Degree requirements:* For doctorate, one foreign language, thesis/dissertation, comprehensive exam. *Entrance requirements:* For doctorate, GRE General Test, minimum graduate GPA of 3.0. Additional exam requirements/recommendations for international students: Required—TOEFL. *Application deadline:* For fall admission, 1/15 for domestic students. Application fee: $50. *Expenses:* Tuition, area resident: Part-time $288. Tuition, state resident: full-time $6,900. Tuition, nonresident: full-time $10,920; part-time $455. Required fees: $704. *Financial support:* In 2005–06, 26 fellowships, 55 research assistantships, 31 teaching assistantships were awarded; career-related internships or fieldwork also available. Total annual research expenditures: $9.3 million. *Application contact:* Dr. Glen Lopez, Acting Director, 631-632-8660, Fax: 631-632-8200, E-mail: glopez@notes.cc.sunysb.edu.

Announcement: Located on Long Island's North Shore, 50 miles from New York City, the Marine Sciences Research Center and the Institute for Terrestrial and Planetary Atmospheres offer close-by opportunities for coastal research along a gradient from urban to pristine, as well as for research in many regions of the globe.

See Close-Up on page 269.

Texas A&M University at Galveston, Department of Marine Sciences, Galveston, TX 77553-1675. Offers marine resources management (MMRM). *Faculty:* 10 full-time (2 women). *Students:* 9 full-time (6 women), 32 part-time (24 women); includes 4 minority (1 African American, 1 Asian American or Pacific Islander, 2 Hispanic Americans), 3 international. Average age 27. 15 applicants, 93% accepted, 12 enrolled. In 2005, 9 degrees awarded. *Entrance requirements:* For master's, GRE, course work in economics. Additional exam requirements/recommendations for international students: Required—TOEFL (minimum score 550 paper-based; 213 computer-based). *Application deadline:* Applications are processed on a rolling basis. Application fee: $50 ($75 for international students). Electronic applications accepted. *Expenses:* Tuition, state resident: part-time $175 per credit. Tuition, nonresident: full-time $3,141; part-time $451 per credit. International tuition: $8,109 full-time. Required fees: $907. *Financial support:* In 2005–06, 14 students received support, including 1 research assistantship, 2 teaching assistantships; scholarships/grants, health care benefits, and unspecified assistantships also available. Financial award application deadline: 4/1; financial award applicants required to submit FAFSA. *Faculty research:* Biogeochemistry, physical oceanography, theoretical chemistry, marine policy. Total annual research expenditures: $3.1 million. *Unit head:* Dr. William Seitz, Head, 409-740-4515, Fax: 409-740-4429, E-mail: seitzw@tamug.edu. *Application contact:* Dr. Frederick C. Schlemmer, Associate Professor/ Graduate Advisor, 409-740-4518, Fax: 409-740-4429, E-mail: schlemme@tamug.edu.

University of Alaska Fairbanks, School of Fisheries and Ocean Sciences, Department of Marine Sciences and Limnology, Fairbanks, AK 99775-7520. Offers marine biology (MS, PhD); oceanography (MS, PhD), including biological oceanography (PhD), chemical oceanography (PhD), fisheries (PhD), geological oceanography (PhD), physical oceanography (PhD). Part-time programs available. Terminal master's awarded for partial completion of doctoral program. *Degree requirements:* For master's and doctorate, thesis/dissertation, comprehensive exam, registration. *Entrance requirements:* For master's and doctorate, GRE General Test. Additional exam requirements/recommendations for international students: Required—TOEFL. Electronic applications accepted. *Expenses:* Tuition, state resident: full-time $4,392; part-time $244 per credit. Tuition, nonresident: full-time $8,964; part-time $498 per credit. Required fees: $800; $5 per credit. $48 per contact hour. Tuition and fees vary according to course level, course load, campus/location and reciprocity agreements. *Faculty research:* Seafood science and nutrition, sustainable harvesting, chemical oceanography, marine biology, physical oceanography.

The University of British Columbia, Faculty of Graduate Studies, Faculty of Science, Department of Earth and Ocean Sciences, Vancouver, BC V6T 1Z1, Canada. Offers atmospheric science (M Sc, PhD); geological engineering (M Eng, MA Sc, PhD); geological sciences (M Sc, PhD); geophysics (M Sc, MA Sc, PhD); oceanography (M Sc, PhD). *Faculty:* 44 full-time (7 women), 17 part-time/adjunct (1 woman). *Students:* 155 full-time (49 women), 1 (woman) part-time. Average age 30. 96 applicants, 50% accepted, 30 enrolled. In 2005, 11 master's, 6 doctorates awarded. *Degree requirements:* For master's, thesis (for some programs); for doctorate, thesis/dissertation, comprehensive exam. *Entrance requirements:* Additional exam requirements/recommendations for international students: Required—TOEFL (minimum score 600 paper-based; 250 computer-based). *Application deadline:* For fall admission, 2/1 for domestic students, 1/1 for international students. For winter admission, 7/1 for domestic students. Applications are processed on a rolling basis. Application fee: $90 Canadian dollars ($150 Canadian dollars for international students). Electronic applications accepted. *Financial support:* In 2005–06, fellowships (averaging $16,000 per year), research assistantships (averaging $13,000 per year), teaching assistantships (averaging $5,000 per year) were awarded; Federal Work-Study, institutionally sponsored loans, scholarships/grants, tuition waivers (full and partial), and unspecified assistantships also available. *Unit head:* Dr. Paul L. Smith, Head, 604-822-6456, Fax: 604-822-6088, E-mail: head@eos.ubc.ca. *Application contact:* Alex Allen, Graduate Secretary, 604-822-2713, Fax: 604-822-6088, E-mail: aallen@eos.ubc.ca.

University of California, San Diego, Graduate Studies and Research, Scripps Institution of Oceanography, Program in Marine Biodiversity and Conservation, La Jolla, CA 92093. Offers MAS.

University of California, Santa Barbara, Graduate Division, College of Letters and Sciences, Division of Mathematics, Life, and Physical Sciences, Interdepartmental Program in Marine Science, Santa Barbara, CA 93106. Offers MS, PhD. *Faculty:* 42 full-time (11 women). *Students:* 41 full-time (21 women); includes 4 minority (1 African American, 2 Asian Americans or Pacific Islanders, 1 Hispanic American). Average age 28. 54 applicants, 19% accepted, 6 enrolled. In 2005, 4 master's, 3 doctorates awarded. *Median time to degree:* Of those who began their doctoral program in fall 1997, 100% received their degree in 8 years or less. *Degree requirements:* For master's, thesis, 39 units; for doctorate, thesis/dissertation, 31 units, comprehensive exam, registration. *Entrance requirements:* For master's and doctorate, GRE. Additional exam requirements/recommendations for international students: Required—TOEFL (minimum score 650 paper-based; 213 computer-based). *Application deadline:* For fall admission, 12/15 for domestic students, 12/15 for international students. Application fee: $60. Electronic applications accepted. *Financial support:* In 2005–06, 1 fellowship with full tuition reimbursement (averaging $18,000 per year), 28 research assistantships with full tuition reimbursements (averaging $15,600 per year), 4 teaching assistantships with full tuition reimbursements (averaging $15,000 per year) were awarded; career-related internships or fieldwork, Federal Work-Study, institutionally sponsored loans, scholarships/grants, health care benefits, tuition waivers (full and partial), and unspecified assistantships also available. Support available to part-time students. Financial award application deadline: 12/15; financial award applicants required to submit FAFSA. *Faculty research:* Ocean carbon cycling, paleooceanography, physiology of marine organisms, bio-optical oceanography, biological oceanography. *Unit head:* Dr. Mark Brzezinski, Chair, 805-893-8605, E-mail: brzezins@lifesci.ucsb.edu. *Application contact:* Melanie Fujii-Abe, Graduate Program Assistant, 805-893-8162, Fax: 805-893-4724, E-mail: abe@lifesci.ucsb.edu.

University of California, Santa Cruz, Division of Graduate Studies, Division of Physical and Biological Sciences, Program in Ocean Sciences, Santa Cruz, CA 95064. Offers MS, PhD. *Faculty:* 9 full-time (4 women). *Students:* 39 (24 women); includes 6 minority (2 Asian Americans or Pacific Islanders, 4 Hispanic Americans) 2 international. 42 applicants, 31% accepted, 7 enrolled. In 2005, 1 master's, 3 doctorates awarded. *Degree requirements:* For doctorate, one foreign language, thesis/dissertation. *Entrance requirements:* For doctorate, GRE General Test, GRE Subject Test. *Application deadline:* For fall admission, 1/1 for domestic students. Application fee: $60. *Expenses:* Tuition, nonresident: full-time $14,694. *Unit head:* Ken Bruland, Chairperson, 831-459-4736. *Application contact:* Judy L. Glass, Reporting Analyst for Graduate Admissions, 831-459-5906, Fax: 831-459-4843, E-mail: jlglass@ucsc.edu.

University of Connecticut, Graduate School, College of Liberal Arts and Sciences, Department of Marine Sciences, Field of Oceanography, Storrs, CT 06269. Offers MS, PhD. *Faculty:* 26 full-time (7 women). *Students:* 36 full-time (24 women), 8 part-time (5 women); includes 1 minority (Hispanic American), 14 international. Average age 29. 38 applicants, 58% accepted, 14 enrolled. In 2005, 3 master's, 3 doctorates awarded. Terminal master's awarded for partial completion of doctoral program. *Degree requirements:* For master's, comprehensive exam; for doctorate, thesis/dissertation. *Entrance requirements:* For master's and doctorate, GRE General Test, GRE Subject Test. Additional exam requirements/recommendations for international students: Required—TOEFL (minimum score 550 paper-based; 213 computer-based). *Application deadline:* For fall admission, 2/1 priority date for domestic students, 2/1 priority date for international students; for spring admission, 11/1 for domestic students, 10/1 for international students. Applications are processed on a rolling basis. Application fee: $55. Electronic applications accepted. *Expenses:* Tuition, state resident: part-time $444 per credit hour. Tuition, nonresident: part-time $1,154 per credit hour. Tuition and fees vary according to course load. *Financial support:* In 2005–06, 33 research assistantships with full tuition reimbursements, 2 teaching assistantships with full tuition reimbursements were awarded; fellowships, Federal Work-Study, scholarships/grants, health care benefits, and unspecified assistantships also available. Financial award application deadline: 2/1; financial award applicants required to submit FAFSA. *Unit head:* Robert B. Whitlatch, Head, 860-445-3467, Fax: 860-405-9153, E-mail: robert.whitlatch@uconn.edu. *Application contact:* Barbara Mahoney, Administrative Assistant, 860-405-9151, Fax: 860-405-9153, E-mail: mscadm03@uconnvm.uconn.edu.

University of Delaware, College of Marine Studies, Newark, DE 19716. Offers geology (MS, PhD); marine management (MMM); marine policy (MS); marine studies (MMP, MS, PhD); oceanography (MS, PhD). *Faculty:* 41 full-time (4 women). *Students:* 115 full-time (57 women), 5 part-time (2 women); includes 7 minority (2 African Americans, 4 Asian Americans or Pacific Islanders, 1 Hispanic American), 30 international. Average age 29. 127 applicants, 35% accepted, 24 enrolled. In 2005, 12 master's, 8 doctorates awarded. *Degree requirements:* For master's and doctorate, thesis/dissertation. *Entrance requirements:* For master's and doctorate, GRE General Test. Additional exam requirements/recommendations for international students: Required—TOEFL. *Application deadline:* For fall admission, 3/1 for domestic students; for spring admission, 10/1 for domestic students. Applications are processed on a rolling basis. Application fee: $60. Electronic applications accepted. *Financial support:* In 2005–06, 78 students received support, including 14 fellowships with full tuition reimbursements available (averaging $19,000 per year), 62 research assistantships with full tuition reimbursements available (averaging $19,000 per year), 2 teaching assistantships with full tuition reimbursements available (averaging $19,000 per year); career-related internships or fieldwork, Federal Work-Study, and tuition waivers (full and partial) also available. Financial award application deadline: 3/1. *Faculty research:* Marine biology and biochemistry, oceanography, marine policy, physical ocean science and engineering, ocean engineering. Total annual research expenditures: $10.5 million. *Unit head:* Dr. Nancy Targett, Dean, 302-831-2841. *Application contact:* Lisa Perelli, Coordinator, 302-645-4226, E-mail: lperelli@udel.edu.

University of Florida, Graduate School, College of Agricultural and Life Sciences, Department of Fisheries and Aquatic Sciences, Gainesville, FL 32611. Offers MFAS, MS, PhD. *Faculty:* 14 full-time (2 women), 1 part-time/adjunct (0 women). *Students:* 50 (22 women); includes 2 minority (both Asian Americans or Pacific Islanders) 2 international. 29 applicants, 31% accepted. In 2005, 10 master's, 2 doctorates awarded. *Degree requirements:* For master's, thesis optional; for doctorate, thesis/dissertation. *Entrance requirements:* For master's and doctorate, GRE General Test, minimum GPA of 3.0. Additional exam requirements/recommendations for international students: Required—TOEFL. *Application deadline:* For fall admission, 6/1 for domestic students. Applications are processed on a rolling basis. Application fee: $20. Electronic applications accepted. *Expenses:* Tuition, state resident: full-time $6,234. Tuition, nonresident: full-time $21,359. Tuition and fees vary according to program. *Financial support:* In 2005–06, 22 research assistantships (averaging $9,546 per year) were awarded; fellowships, unspecified assistantships also available. *Unit head:* Dr. Karl Havens, Chair, 352-392-9617, Fax: 352-392-3672, E-mail: khavens@ifas.ufl.edu. *Application contact:* Dr. Chuck Chichra, Graduate Coordinator, 352-392-9617 Ext. 249, Fax: 352-392-3672, E-mail: fish@ifas.ufl.edu.

University of Georgia, Graduate School, College of Arts and Sciences, Department of Marine Sciences, Athens, GA 30602. Offers MS, PhD. *Faculty:* 14 full-time (5 women). *Students:* 36 full-time, 3 part-time; includes 2 minority (1 Asian American or Pacific Islander, 1 Hispanic American), 8 international. Average age 28. 40 applicants, 30% accepted, 10 enrolled. In 2005, 2 master's, 3 doctorates awarded. *Degree requirements:* For master's, thesis; for

Marine Sciences

University of Georgia (continued)

doctorate, thesis/dissertation, teaching experience, field research experience, comprehensive exam. *Entrance requirements:* For master's and doctorate, GRE General Test. Additional exam requirements/recommendations for international students: Required—TOEFL. *Application deadline:* For fall admission, 2/1 priority date for domestic students, 2/1 priority date for international students; for spring admission, 10/15 priority date for domestic students, 9/1 priority date for international students. Applications are processed on a rolling basis. Application fee: $50. Electronic applications accepted. *Financial support:* In 2005–06, 9 fellowships with full tuition reimbursements (averaging $20,000 per year), 21 research assistantships with full tuition reimbursements (averaging $18,000 per year), 11 teaching assistantships with full tuition reimbursements (averaging $18,000 per year) were awarded. *Faculty research:* Microbial ecology, biogeochemistry, polar biology, coastal ecology, coastal circulation. *Unit head:* Dr. James T. Hollibaugh, Director, 706-542-3016, Fax: 706-542-5888, E-mail: aquadoc@uga.edu. *Application contact:* Dr. Mary Ann Moran, Graduate Coordinator, 706-542-6481, Fax: 706-542-5888, E-mail: mmoran@uga.edu.

University of Maine, Graduate School, College of Natural Sciences, Forestry, and Agriculture, School of Marine Sciences, Orono, ME 04469. Offers marine biology (MS, PhD); marine policy (MS); oceanography (MS, PhD). Part-time programs available. *Faculty:* 28. *Students:* 47 full-time (27 women), 27 part-time (21 women); includes 2 minority (both Asian Americans or Pacific Islanders), 13 international. Average age 28. 67 applicants, 18% accepted, 10 enrolled. *Degree requirements:* For master's and doctorate, thesis/dissertation. *Entrance requirements:* For master's and doctorate, GRE General Test. Additional exam requirements/recommendations for international students: Required—TOEFL. *Application deadline:* For fall admission, 2/1 for domestic students. Applications are processed on a rolling basis. Application fee: $50. Electronic applications accepted. *Financial support:* In 2005–06, fellowships with tuition reimbursements (averaging $30,000 per year), research assistantships with tuition reimbursements (averaging $16,500 per year), teaching assistantships with tuition reimbursements (averaging $16,000 per year) were awarded; career-related internships or fieldwork, Federal Work-Study, and tuition waivers (full and partial) also available. Support available to part-time students. Financial award application deadline:3/1. *Faculty research:* Coastal processes, microbial ecology, crustacean systematics. *Unit head:* Dr. David Townsend, Director, 207-581-4367, Fax: 207-581-4388. *Application contact:* Scott G. Delcourt, Associate Dean of the Graduate School, 207-581-3219, Fax: 207-581-3232, E-mail: graduate@maine.edu.

University of Maryland, Graduate School, Program in Marine-Estuarine-Environmental Sciences, Baltimore, MD 21201. Offers MS, PhD. An intercampus, interdisciplinary program. Part-time programs available. *Faculty:* 8. *Students:* 2 full-time (0 women), 1 (woman) part-time, 1 international. 1 applicant, 100% accepted, 1 enrolled.Terminal master's awarded for partial completion of doctoral program. *Degree requirements:* For master's, thesis; for doctorate, thesis/dissertation, proposal defense, comprehensive exam. *Entrance requirements:* For master's and doctorate, GRE General Test, minimum GPA of 3.0. Additional exam requirements/recommendations for international students: Required—TOEFL. *Application deadline:* For fall admission, 2/1 for domestic students; for spring admission, 9/1 for domestic students. Applications are processed on a rolling basis. Application fee: $50. Electronic applications accepted. *Expenses:* Tuition, state resident: full-time $8,079; part-time $409 per credit hour. Tuition, nonresident: full-time $18,384; part-time $731 per credit hour. Required fees: $695; $10 per credit hour. Tuition and fees vary according to degree level and program. *Financial support:* Research assistantships with tuition reimbursements, teaching assistantships with tuition reimbursements, scholarships/grants and unspecified assistantships available. *Unit head:* Dr. Kennedy T. Paynter, Director, 301-405-6938, Fax: 301-314-4139, E-mail: mees@mees.umd.edu.

See Close-Up on page 271.

University of Maryland, Baltimore County, Graduate School, College of Natural Sciences and Mathematics, Department of Biological Sciences, Program in Marine-Estuarine-Environmental Sciences, Baltimore, MD 21250. Offers MS, PhD. Part-time programs available. *Faculty:* 17. *Students:* 9 full-time (6 women); includes 2 minority (both African Americans), 3 international. 16 applicants, 25% accepted, 2 enrolled. *Degree requirements:* For master's, thesis; for doctorate, thesis/dissertation, proposal defense, comprehensive exam (for some programs). *Entrance requirements:* For master's and doctorate, GRE General Test, minimum GPA of 3.0. Additional exam requirements/recommendations for international students: Required—TOEFL. *Application deadline:* For fall admission, 2/1 for domestic students, 1/1 for international students; for spring admission, 9/1 for domestic students. Applications are processed on a rolling basis. Application fee: $50. Electronic applications accepted. *Expenses:* Tuition, state resident: part-time $395 per credit. Tuition, nonresident: part-time $652 per credit. Required fees: $82 per credit. Tuition and fees vary according to course load, program and reciprocity agreements. *Financial support:* In 2005–06, 11 students received support, including 1 fellowship with tuition reimbursement available (averaging $22,500 per year), research assistantships with tuition reimbursements available (averaging $21,000 per year), teaching assistantships with tuition reimbursements available (averaging $20,000 per year); career-related internships or fieldwork, scholarships/grants, and unspecified assistantships also available. Financial award application deadline: 1/1. *Unit head:* Dr. Kennedy T. Paynter, Director, 301-405-6938, Fax: 301-314-4139, E-mail: mees@mees.umd.edu. *Application contact:* Dr. Thomas Cronin, Graduate Program Director, 410-455-3669, Fax: 410-455-3875, E-mail: biograd@umbc.edu.

University of Maryland, College Park, Graduate Studies, College of Chemical and Life Sciences, Program in Marine-Estuarine-Environmental Sciences, College Park, MD 20742. Offers MS, PhD. An intercampus, interdisciplinary program. Part-time programs available. *Faculty:* 135. *Students:* 145 (73 women); includes 6 minority (1 African American, 1 American Indian/Alaska Native, 2 Asian Americans or Pacific Islanders, 2 Hispanic Americans) 29 international. 118 applicants, 22% accepted, 15 enrolled. In 2005, 22 master's, 10 doctorates awarded. Terminal master's awarded for partial completion of doctoral program. *Degree requirements:* For master's, thesis, oral defense; for doctorate, thesis/dissertation, proposal defense, comprehensive exam. *Entrance requirements:* For master's and doctorate, GRE General Test, minimum GPA of 3.0. Additional exam requirements/recommendations for international students: Required—TOEFL. *Application deadline:* For fall admission, 2/1 for domestic students, 2/1 for international students; for spring admission, 9/1 for domestic students, 6/1 for international students. Applications are processed on a rolling basis. Application fee: $60. Electronic applications accepted. *Financial support:* In 2005–06, 9 teaching assistantships with full tuition reimbursements were awarded; fellowships with full tuition reimbursements, research assistantships with full tuition reimbursements, Federal Work-Study, scholarships/grants, traineeships, health care benefits, and unspecified assistantships also available. Financial award application deadline: 1/1; financial award applicants required to submit FAFSA. *Faculty research:* Marine and estuarine organisms, terrestrial and freshwater ecology, remote environmental sensing. *Unit head:* Dr. Kennedy T. Paynter, Director, 301-405-6938, Fax: 301-314-4139, E-mail: mees@mees.umd.edu.

University of Maryland Eastern Shore, Graduate Programs, Department of Natural Sciences, Program in Marine-Estuarine-Environmental Sciences, Princess Anne, MD 21853-1299. Offers MS, PhD. Part-time programs available. *Faculty:* 30. *Students:* 36 (18 women); includes 12 minority (10 African Americans, 1 Asian American or Pacific Islander, 1 Hispanic American) 14 international. 28 applicants, 57% accepted, 13 enrolled. In 2005, 5 master's awarded. *Degree requirements:* For master's, thesis; for doctorate, thesis/dissertation, proposal defense, comprehensive exam. *Entrance requirements:* For master's and doctorate, GRE General Test, minimum GPA of 3.0. Additional exam requirements/recommendations for international students: Required—TOEFL. *Application deadline:* For fall admission, 2/1 for domestic students; for spring admission, 9/1 for domestic students. Applications are processed on a rolling basis. Application fee: $30. Electronic applications accepted. *Expenses:* Tuition, area resident: Part-time $216. Tuition, nonresident: part-time $392. Required fees: $40. *Financial support:* In 2005–06, 30 students received support; fellowships with tuition reimbursements available, research assistantships with tuition reimbursements available, teaching assistantships with tuition reimbursements available, career-related internships or fieldwork, scholarships/grants, and unspecified assistantships available. Support available to part-time students. Financial award application deadline: 1/1. *Unit head:* Dr. Kennedy T. Paynter, Director, 301-405-6938, Fax: 301-314-4139, E-mail: mees@mees.umd.edu.

University of Massachusetts Amherst, Graduate School, Interdisciplinary Programs, Program in Marine Science and Technology, Amherst, MA 01003. Offers MS. Part-time programs available. *Students:* 1 (woman) full-time, 2 part-time (both women), 1 international. Average age 27. *Entrance requirements:* Additional exam requirements/recommendations for international students: Required—TOEFL (minimum score 530 paper-based; 197 computer-based). *Application deadline:* For fall admission, 1/15 for domestic students. *Expenses:* Tuition, state resident: part-time $110 per credit. Tuition, nonresident: part-time $414 per credit. Required fees: $2,824 per term. One-time fee: $250 part-time. Full-time tuition and fees vary according to course load, campus/location, program and reciprocity agreements. *Financial support:* In 2005–06, 3 research assistantships (averaging $3,867 per year) were awarded *Unit head:* Dr. Kevin McGarigal, Director, 413-545-2666, Fax: 413-545-4358.

University of Massachusetts Boston, Office of Graduate Studies and Research, College of Science and Mathematics, Department of Environmental, Coastal and Ocean Sciences, Track in Environmental, Coastal and Ocean Sciences, Boston, MA 02125-3393. Offers PhD. Part-time and evening/weekend programs available. *Degree requirements:* For doctorate, thesis/dissertation, oral exams, comprehensive exam. *Entrance requirements:* For doctorate, GRE General Test, minimum GPA of 2.75. *Faculty research:* Conservation genetics, anthropogenic and natural influences on community structures of coral reef factors, geographical variation in mitochondrial DNA, protein chemistry and enzymology pertaining to insect cuticle.

University of Massachusetts Dartmouth, Graduate School, School of Marine Science and Technology, Program in Marine Science and Technology, North Dartmouth, MA 02747-2300. Offers MS, PhD. *Faculty:* 9 full-time, 1 part-time/adjunct. *Students:* 10 full-time (3 women), 8 part-time (4 women), 10 international. Average age 26. 17 applicants, 47% accepted, 3 enrolled. In 2005, 1 degree awarded. *Degree requirements:* For master's, thesis or alternative; for doctorate, thesis/dissertation, comprehensive exam. *Entrance requirements:* For master's and doctorate, GRE, minimum GPA of 3.0. Additional exam requirements/recommendations for international students: Required—TOEFL (minimum score 600 paper-based; 213 computer-based). *Application deadline:* For fall admission, 4/20 priority date for domestic students, 2/20 priority date for international students. Applications are processed on a rolling basis. Application fee: $35 ($55 for international students). Electronic applications accepted. *Expenses:* Tuition, state resident: full-time $2,071; part-time $86 per credit. Tuition, nonresident: full-time $8,099; part-time $337 per credit. Required fees: $9,437; $393 per credit. *Financial support:* In 2005–06, 15 research assistantships with full tuition reimbursements (averaging $11,745 per year) were awarded; unspecified assistantships also available. Financial award application deadline: 3/1; financial award applicants required to submit FAFSA. *Faculty research:* Northeast water quality, oceanography, osmerus mordax, shellfish analysis, marsh and estuaries. Total annual research expenditures: $9.5 million. *Unit head:* Dr. Avijit Gangopadhyay, Director, 508-910-6330, E-mail: avijit@umassd.edu. *Application contact:* Carol Novo, Graduate Admissions Officer, 508-999-8604, Fax: 508-999-8183, E-mail: graduate@umassd.edu.

University of Miami, Graduate School, Rosenstiel School of Marine and Atmospheric Science, Division of Applied Marine Physics, Coral Gables, FL 33124. Offers applied marine physics (MS, PhD), including coastal ocean dynamics, underwater acoustics and geoacoustics (PhD), wave surface dynamics and air-sea interaction (PhD). Part-time programs available. *Faculty:* 17 full-time (1 woman), 13 part-time/adjunct (2 women). *Students:* 16 full-time (7 women); includes 1 minority (Hispanic American), 11 international. Average age 30. 10 applicants, 50% accepted, 2 enrolled. In 2005, 2 master's awarded. Terminal master's awarded for partial completion of doctoral program. *Median time to degree:* Of those who began their doctoral program in fall 19967 100% received their degree in 8 years or less. *Degree requirements:* For master's and doctorate, thesis/dissertation, comprehensive exam, registration. *Entrance requirements:* For master's and doctorate, GRE General Test. Additional exam requirements/recommendations for international students: Required—TOEFL (minimum score 550 paper-based; 213 computer-based). *Application deadline:* For fall admission, 1/1 priority date for domestic students, 1/1 priority date for international students. Applications are processed on a rolling basis. Application fee: $50. Electronic applications accepted. *Financial support:* In 2005–06, 15 students received support, including 2 fellowships with tuition reimbursements available (averaging $22,380 per year), 13 research assistantships with tuition reimbursements available (averaging $22,380 per year); teaching assistantships with tuition reimbursements available, Federal Work-Study, institutionally sponsored loans, scholarships/grants, and unspecified assistantships also available. Financial award application deadline: 3/1; financial award applicants required to submit FAFSA. Total annual research expenditures: $2.5 million. *Unit head:* Dr. Mark Donelan, Chair, 305-421-4640, E-mail: mdonelan@rsmas.miami.edu. *Application contact:* Dr. Larry Peterson, Associate Dean, 305-421-4155, Fax: 305-421-4771, E-mail: gso@rsmas.miami.edu.

University of Miami, Graduate School, Rosenstiel School of Marine and Atmospheric Science, Division of Marine and Atmospheric Chemistry, Coral Gables, FL 33124. Offers MS, PhD. *Faculty:* 16 full-time (2 women), 14 part-time/adjunct (2 women). *Students:* 12 full-time (5 women); includes 1 minority (Hispanic American), 6 international. Average age 25. 14 applicants, 29% accepted, 1 enrolled.Terminal master's awarded for partial completion of doctoral program. *Degree requirements:* For master's and doctorate, thesis/dissertation, comprehensive exam, registration. *Entrance requirements:* For master's and doctorate, GRE General Test. Additional exam requirements/recommendations for international students: Required—TOEFL (minimum score 550 paper-based; 213 computer-based). *Application deadline:* For fall admission, 1/1 priority date for domestic students, 1/1 priority date for international students. Applications are processed on a rolling basis. Application fee: $50. Electronic applications accepted. *Financial support:* In 2005–06, 12 students received support, including 1 fellowship with tuition reimbursement available (averaging $22,380 per year), 11 research assistantships with tuition reimbursements available (averaging $22,380 per year); teaching assistantships with tuition reimbursements available, Federal Work-Study, institutionally sponsored loans, scholarships/grants, and unspecified assistantships also available. Financial award application deadline: 3/1; financial award applicants required to submit FAFSA. *Faculty research:* Global change issues, chemistry of marine waters and marine atmosphere. *Unit head:* Dr. Dennis Hansell, Chairperson, 305-421-4922, Fax: 305-421-4689, E-mail: dhansell@rsmas.miami.edu. *Application contact:* Dr. Larry Peterson, Associate Dean, 305-421-4155, Fax: 305-421-4771, E-mail: gso@rsmas.miami.edu.

University of New England, College of Arts and Sciences, Programs in Professional Science, Biddeford, ME 04005-9526. Offers applied biosciences (MS); marine sciences (MS). *Faculty:* 5 full-time (4 women). *Students:* 5 full-time (3 women); includes 1 minority (Asian American or Pacific Islander), 2 international. Average age 25. 5 applicants, 100% accepted, 5 enrolled. Application fee: $40. *Expenses:* Contact institution. *Unit head:* Lawrence Fritz, Chair, Department of Biological Sciences, E-mail: lfritz@une.edu. *Application contact:* Robert Pecchia, Associate Dean of Admissions, 207-283-0171 Ext. 2297, Fax: 207-602-5900, E-mail: admissions@une.edu.

The University of North Carolina at Chapel Hill, Graduate School, College of Arts and Sciences, Department of Marine Sciences, Chapel Hill, NC 27599. Offers MS, PhD. *Degree requirements:* For master's and doctorate, thesis/dissertation, comprehensive exam. *Entrance requirements:* For master's and doctorate, GRE General Test, GRE Subject Test, minimum GPA of 3.0.

The University of North Carolina at Chapel Hill, Graduate School, School of Public Health, Department of Environmental Sciences and Engineering, Chapel Hill, NC 27599. Offers air, radiation and industrial hygiene (MPH, MS, MSEE, MSPH, PhD); aquatic and atmospheric sciences (MPH, MS, MSPH, PhD); environmental engineering (MPH, MS, MSEE, MSPH,

PhD); environmental health sciences (MPH, MS, MSPH, PhD); environmental management and policy (MPH, MS, MSPH, PhD). *Faculty:* 33 full-time (3 women), 35 part-time/adjunct. *Students:* 141 full-time (74 women); includes 37 minority (10 African Americans, 25 Asian Americans or Pacific Islanders, 2 Hispanic Americans). Average age 27. 216 applicants, 37% accepted, 29 enrolled. In 2005, 14 master's, 11 doctorates awarded. Terminal master's awarded for partial completion of doctoral program. *Median time to degree:* Of those who began their doctoral program in fall 1997, 100% received their degree in 8 years or less. *Degree requirements:* For master's, thesis (for some programs), research paper, comprehensive exam, registration; for doctorate, thesis/dissertation, comprehensive exam, registration. *Entrance requirements:* For master's and doctorate, GRE General Test, minimum GPA of 3.0. Additional exam requirements/recommendations for international students: Required—TOEFL. *Application deadline:* For fall admission, 1/1 priority date for domestic students, 1/1 priority date for international students; for spring admission, 9/15 for domestic students. Applications are processed on a rolling basis. Application fee: $70. Electronic applications accepted. *Financial support:* In 2005–06, 134 students received support, including 36 fellowships with tuition reimbursements available (averaging $6,358 per year), 86 research assistantships with tuition reimbursements available (averaging $6,197 per year), 12 teaching assistantships with tuition reimbursements available (averaging $6,729 per year); career-related internships or fieldwork, Federal Work-Study, traineeships, health care benefits, and unspecified assistantships also available. Support available to part-time students. Financial award application deadline: 1/1; financial award applicants required to submit FAFSA. *Faculty research:* Air, radiation and industrial hygiene, aquatic and atmospheric sciences, environmental health sciences, environmental management and policy, water resources engineering. Total annual research expenditures: $9.6 million. *Unit head:* Dr. Don Fox, Interim Chair, 919-966-1024, Fax: 919-966-7911, E-mail: don_fox@unc.edu. *Application contact:* Jack Whaley, Registrar, 919-966-3844, Fax: 919-966-7911, E-mail: jack_whaley@unc.edu.

The University of North Carolina Wilmington, College of Arts and Sciences, Department of Earth Sciences, Wilmington, NC 28403-3297. Offers geology (MS); marine science (MS). *Faculty:* 20 full-time (4 women), 1 (woman) part-time/adjunct. *Students:* 11 full-time (7 women), 57 part-time (38 women); includes 1 minority (Hispanic American), 1 international. Average age 35. 42 applicants, 48% accepted, 17 enrolled. In 2005, 26 degrees awarded. *Degree requirements:* For master's, thesis, comprehensive exam. *Entrance requirements:* For master's, GRE General Test, GRE Subject Test, minimum B average in undergraduate major and basic courses for prerequisite to geology. *Application deadline:* For fall admission, 2/15 for domestic students. Applications are processed on a rolling basis. Application fee: $45. *Financial support:* In 2005–06, 9 teaching assistantships were awarded; career-related internships or fieldwork and Federal Work-Study also available. Support available to part-time students. Financial award application deadline: 3/15. *Unit head:* Dr. Richard A. Laws, Chair, 910-962-4125, Fax: 910-962-7077. *Application contact:* Dr. Robert D. Roer, Dean, Graduate School, 910-962-4117, Fax: 910-962-3787, E-mail: roer@uncw.edu.

See Close-Up on page 237.

University of Puerto Rico, Mayagüez Campus, Graduate Studies, College of Arts and Sciences, Department of Marine Sciences, Mayagüez, PR 00681-9000. Offers biological oceanography (MMS, PhD); chemical oceanography (MMS, PhD); geological oceanography (MMS, PhD); physical oceanography (MMS, PhD). *Faculty:* 25. *Students:* 18 full-time (16 women), 32 part-time (20 women); includes 46 minority (all Hispanic Americans), 4 international. 29 applicants, 59% accepted, 7 enrolled. In 2005, 6 master's, 2 doctorates awarded. *Degree requirements:* For master's, one foreign language, thesis, departmental and comprehensive final exams; for doctorate, one foreign language, thesis/dissertation, qualifying, comprehensive, and final exams. *Application deadline:* For fall admission, 2/15 for domestic students; for spring admission, 9/15 for domestic students. Applications are processed on a rolling basis. Application fee: $20. *Expenses:* Tuition, state resident: full-time $900; part-time $100 per credit. International tuition: $4,655 full-time. Part-time tuition and fees vary according to course level and course load. *Financial support:* In 2005–06, 49 students received support, including 1 fellowship (averaging $1,500 per year), 39 research assistantships (averaging $1,200 per year), 9 teaching assistantships (averaging $987 per year); Federal Work-Study and institutionally sponsored loans also available. *Faculty research:* Marine botany, ecology, chemistry, and parasitology; fisheries; ichthyology; aquaculture. Total annual research expenditures: $173,037. *Unit head:* Dr. Nilda Aponte, Director, 787-265-3838, E-mail: naponte@uprm.edu.

See Close-Up on page 273.

University of Rhode Island, Graduate School, College of the Environment and Life Sciences, Department of Fisheries, Animal and Veterinary Science, Kingston, RI 02881. Offers animal health and disease (MS); animal science (MS); aquaculture (MS); aquatic pathology (MS); environmental sciences (PhD), including animal science, aquacultural science, aquatic pathology, fisheries science; fisheries (MS). In 2005, 6 degrees awarded. *Application deadline:* For fall admission, 4/15 for domestic students. Applications are processed on a rolling basis. Application fee: $35. *Expenses:* Tuition, state resident: full-time $5,522; part-time $307 per credit. Tuition, nonresident: full-time $15,992; part-time $888 per credit. Required fees: $1,786; $73 per credit. One-time fee: $80 part-time. *Unit head:* Dr. David Bengtson, Chairperson, 401-874-2688.

University of San Diego, College of Arts and Sciences, Program in Marine and Environmental Studies, San Diego, CA 92110-2492. Offers marine science (MS). Part-time programs available. *Faculty:* 6 full-time (2 women). *Students:* 4 full-time (2 women), 15 part-time (7 women), 1 international. Average age 26. 17 applicants, 59% accepted, 4 enrolled. In 2005, 4 degrees awarded. *Entrance requirements:* For master's, GRE General Test, minimum GPA of 3.0, undergraduate major in science. Additional exam requirements/recommendations for international students: Required—TOEFL (minimum score 580 paper-based; 237 computer-based), TWE. *Application deadline:* For fall admission, 4/1 for domestic students. Applications are processed on a rolling basis. Application fee: $45. Electronic applications accepted. *Financial support:* Career-related internships or fieldwork, Federal Work-Study, institutionally sponsored loans, tuition waivers (partial), and unspecified assistantships available. Support available to part-time students. Financial award application deadline: 5/1; financial award applicants required to submit FAFSA. *Faculty research:* Marine ecology; paleoclimatology; geochemistry; functional morphology; marine zoology of mammals, birds and turtles. *Unit head:* Dr. Hugh I. Ellis, Director, 619-260-4075, Fax: 619-260-6804, E-mail: ellis@sandiego.edu. *Application contact:* Stephen Pultz, Director of Admissions, 619-260-4524, Fax: 619-260-4158, E-mail: grads@sandiego.edu.

University of South Alabama, Graduate School, College of Arts and Sciences, Department of Marine Sciences, Mobile, AL 36688-0002. Offers MS, PhD. *Faculty:* 4 full-time (0 women). *Students:* 33 full-time (16 women), 11 part-time (3 women), 9 international. 26 applicants, 19% accepted, 5 enrolled. In 2005, 5 master's, 2 doctorates awarded. *Degree requirements:* For master's, comprehensive exam; for doctorate, one foreign language, thesis/dissertation, research project, comprehensive exam. *Entrance requirements:* For master's, GRE, minimum GPA of 3.0. *Application deadline:* For fall admission, 9/1 for domestic students. Applications are processed on a rolling basis. Application fee: $25. *Expenses:* Tuition, state resident: full-time $4,008. Tuition, nonresident: full-time $8,016. Required fees: $692. *Financial support:* Fellowships, research assistantships available. Financial award application deadline: 4/1. *Unit head:* Dr. Robert Shipp, Chair, 251-460-7136.

University of South Carolina, The Graduate School, College of Arts and Sciences, Marine Science Program, Columbia, SC 29208. Offers MS, PhD. *Degree requirements:* For master's and doctorate, thesis/dissertation. *Entrance requirements:* For master's and doctorate, GRE General Test. Additional exam requirements/recommendations for international students: Required—TOEFL (minimum score 570 paper-based; 230 computer-based). Electronic applications accepted. *Faculty research:* Biological, chemical, geological, and physical oceanography; policy.

University of Southern California, Graduate School, College of Letters, Arts and Sciences, Department of Biological Sciences, Program in Marine Environmental Biology, Los Angeles, CA 90089. Offers PhD. *Degree requirements:* For doctorate, thesis/dissertation. *Entrance requirements:* For doctorate, GRE General Test. Additional exam requirements/recommendations for international students: Required—TOEFL. *Expenses:* Tuition: Full-time $25,416; part-time $1,059 per unit. Required fees: $484; $484 per year. Tuition and fees vary according to course load and program. *Faculty research:* Microbial ecology, physiology of larval development, biological community structure, Cambrian radiation.

University of Southern Mississippi, Graduate School, College of Science and Technology, Department of Coastal Sciences, Hattiesburg, MS 39406-0001. Offers MS, PhD. Part-time programs available. *Degree requirements:* For master's and doctorate, thesis/dissertation, comprehensive exam. *Entrance requirements:* For master's, GRE General Test, minimum GPA of 3.0; for doctorate, GRE General Test, minimum undergraduate GPA of 3.0, graduate 3.5. Additional exam requirements/recommendations for international students: Required—TOEFL. Electronic applications accepted.

University of Southern Mississippi, Graduate School, College of Science and Technology, Department of Marine Science, Stennis Space Center, MS 39529. Offers hydrographic science (MS); marine science (MS, PhD). Part-time programs available. *Faculty:* 10 full-time (1 woman), 1 part-time/adjunct (0 women). *Students:* 31 full-time (16 women), 6 part-time (1 woman); includes 3 minority (2 African Americans, 1 Hispanic American), 9 international. Average age 32. 17 applicants, 88% accepted, 15 enrolled. In 2005, 14 master's, 1 doctorate awarded. *Degree requirements:* For master's, thesis, oral qualifying exam (marine science), comprehensive exam; for doctorate, 2 foreign languages, thesis/dissertation, oral qualifying exam, comprehensive exam. *Entrance requirements:* For master's, GRE General Test, minimum GPA of 3.0; for doctorate, GRE General Test, minimum GPA of 3.0 (undergraduate), 3.5 (graduate). Additional exam requirements/recommendations for international students: Required—TOEFL. *Application deadline:* For fall admission, 3/1 for domestic students; for spring admission, 12/13 for domestic students. Applications are processed on a rolling basis. Application fee: $0 ($25 for international students). Electronic applications accepted. *Financial support:* In 2005–06, 4 students received support, including research assistantships with full tuition reimbursements available (averaging $16,800 per year), teaching assistantships with full tuition reimbursements available (averaging $16,800 per year); Federal Work-Study and institutionally sponsored loans also available. Financial award application deadline: 3/15. *Faculty research:* Chemical, biological, physical, and geological marine science; remote sensing; bio-optics; numerical modeling; hydrography. Total annual research expenditures: $2.1 million. *Unit head:* Dr. Steven E. Lohrenz, Chair, 228-688-3177, Fax: 228-688-1121, E-mail: marine.science@usm.edu.

University of South Florida, College of Graduate Studies, College of Marine Science, St. Petersburg, FL 33701-5016. Offers MS, PhD. Part-time and evening/weekend programs available. *Faculty:* 29 full-time (5 women). *Students:* 77 full-time (51 women), 44 part-time (29 women); includes 4 minority (2 African Americans, 2 Hispanic Americans), 37 international. Average age 31. 83 applicants, 36% accepted, 17 enrolled. In 2005, 4 master's, 1 doctorate awarded. *Degree requirements:* For master's, thesis; for doctorate, thesis/dissertation, proficiency foreign language and relevant skill directly related to area of study. *Entrance requirements:* For master's and doctorate, GRE General Test, minimum GPA of 3.0 in last 60 hours. Additional exam requirements/recommendations for international students: Required—TOEFL. *Application deadline:* For fall admission, 3/1 for domestic students; for spring admission, 10/1 for domestic students. Applications are processed on a rolling basis. Application fee: $30. *Financial support:* Fellowships with partial tuition reimbursements, research assistantships with partial tuition reimbursements, teaching assistantships with partial tuition reimbursements available. *Faculty research:* Trace metal chemistry, water quality, organic and isotopic geochemistry, physical chemistry, nutrient chemistry. *Unit head:* Dr. Peter R. Betzer, Dean, 727-553-1130, Fax: 727-553-1189, E-mail: pbetzer@marine.usf.edu. *Application contact:* Dr. Edward VanVleet, Coordinator, 727-553-1165, Fax: 727-553-1189, E-mail: advisor@marine.usf.edu.

The University of Texas at Austin, Graduate School, College of Natural Sciences, Department of Marine Science, Austin, TX 78712-1111. Offers MS, PhD. *Degree requirements:* For master's and doctorate, thesis/dissertation. *Entrance requirements:* For master's and doctorate, GRE General Test. Additional exam requirements/recommendations for international students: Required—TOEFL.

See Close-Up on page 275.

University of Wisconsin–La Crosse, Office of University Graduate Studies, College of Science and Health, Department of Biology, La Crosse, WI 54601-3742. Offers aquatic sciences (MS); biology (MS); cellular and molecular biology (MS); clinical microbiology (MS); microbiology (MS); nurse anesthesia (MS); physiology (MS). *Accreditation:* AANA/CANAEP. Part-time programs available. *Faculty:* 18 full-time (5 women), 1 part-time/adjunct (0 women). *Students:* 23 full-time (10 women), 52 part-time (25 women); includes 5 minority (1 American Indian/Alaska Native, 3 Asian Americans or Pacific Islanders, 1 Hispanic American), 3 international. Average age 26. 61 applicants, 44% accepted, 23 enrolled. In 2005, 14 degrees awarded. *Degree requirements:* For master's, thesis, comprehensive exam, registration. *Entrance requirements:* For master's, GRE General Test, minimum GPA of 2.85. Additional exam requirements/recommendations for international students: Required—TOEFL (minimum score 550 paper-based; 213 computer-based). *Application deadline:* For fall admission, 3/1 for domestic students. Applications are processed on a rolling basis. Application fee: $45. Electronic applications accepted. *Expenses:* Tuition, state resident: part-time $354 per credit. Tuition, nonresident: part-time $943 per credit. Tuition and fees vary according to course load, program and reciprocity agreements. *Financial support:* In 2005–06, 10 students received support, including 4 research assistantships with partial tuition reimbursements available (averaging $10,000 per year), 10 teaching assistantships with partial tuition reimbursements available (averaging $9,600 per year); career-related internships or fieldwork, Federal Work-Study, health care benefits, unspecified assistantships, and grant-funded positions also available. Support available to part-time students. Financial award application deadline: 3/15; financial award applicants required to submit FAFSA. *Faculty research:* Cell and molecular biology, physiology, environmental sciences, mycology, biomedical general. Total annual research expenditures:$700,000. *Unit head:* Dr. Tom Volk, Program Director, 608-785-6972, Fax: 608-785-6959, E-mail: volk.thom@uwlax.edu. *Application contact:* Kathryn Kiefer, Associate Director of Admissions, 608-785-8939, E-mail: admissions@uwlax.edu.

University of Wisconsin–Madison, Graduate School, College of Letters and Science, Department of Atmospheric and Oceanic Sciences, Madison, WI 53706-1380. Offers MS, PhD. Part-time programs available. *Degree requirements:* For master's, thesis (for some programs); for doctorate, thesis/dissertation. *Entrance requirements:* For master's and doctorate, GRE General Test, minimum GPA of 3.0; previous course work in chemistry, mathematics, and physics. Electronic applications accepted. *Faculty research:* Satellite meteorology, weather systems, global climate change, numerical modeling, atmosphere-ocean interaction.

Oceanography

Columbia University, Graduate School of Arts and Sciences, Division of Natural Sciences, Department of Earth and Environmental Sciences, New York, NY 10027. Offers geochemistry (M Phil, MA, PhD); geodetic sciences (M Phil, MA, PhD); geophysics (M Phil, MA, PhD); oceanography (M Phil, MA, PhD). *Faculty:* 21 full-time, 19 part-time/adjunct. *Students:* 78 full-time (31 women), 6 part-time (2 women); includes 4 minority (3 Asian Americans or Pacific Islanders, 1 Hispanic American), 31 international. Average age 32. 115 applicants, 20% accepted. In 2005, 16 master's, 11 doctorates awarded. *Degree requirements:* For master's, thesis or alternative, fieldwork, written exam; for doctorate, one foreign language, thesis/dissertation. *Entrance requirements:* For master's and doctorate, GRE General Test, GRE Subject Test, major in natural or physical science. Additional exam requirements/recommendations for international students: Required—TOEFL. Application fee: $75. *Expenses:* Tuition: Full-time $31,448. Tuition and fees vary according to course level, course load, campus/location and program. *Financial support:* Fellowships, teaching assistantships, Federal Work-Study and institutionally sponsored loans available. Support available to part-time students. Financial award application deadline: 1/5; financial award applicants required to submit FAFSA. *Faculty research:* Structural geology and stratigraphy, petrology, paleontology, rare gas, isotope and aqueous geochemistry. *Unit head:* Dr. Nicholas Christie-Blick, Chair, 845-365-8821, Fax: 845-365-8150.

Cornell University, Graduate School, Graduate Fields of Agriculture and Life Sciences, Field of Ecology and Evolutionary Biology, Ithaca, NY 14853-0001. Offers ecology (PhD), including animal ecology, applied ecology, biogeochemistry, community and ecosystem ecology, limnology, oceanography, physiological ecology, plant ecology, population ecology, theoretical ecology, vertebrate zoology; evolutionary biology (PhD), including ecological genetics, paleobiology, population biology, systematics. *Faculty:* 62 full-time (11 women). *Students:* 61 full-time (33 women); includes 6 minority (1 African American, 3 Asian Americans or Pacific Islanders, 2 Hispanic Americans), 10 international. 107 applicants, 15% accepted, 12 enrolled. In 2005, 7 doctorates awarded. *Degree requirements:* For doctorate, thesis/dissertation, 2 semesters of teaching experience, comprehensive exam. *Entrance requirements:* For doctorate, GRE General Test, GRE Subject Test (biology), 2 letters of recommendation. Additional exam requirements/recommendations for international students: Required—TOEFL (minimum score 550 paper-based; 213 computer-based). *Application deadline:* For fall admission, 12/15 for domestic students. Application fee: $60. Electronic applications accepted. *Financial support:* In 2005–06, 59 students received support, including 25 fellowships with full tuition reimbursements available, 10 research assistantships with full tuition reimbursements available, 24 teaching assistantships with full tuition reimbursements available; institutionally sponsored loans, scholarships/grants, health care benefits, tuition waivers (full and partial), and unspecified assistantships also available. Financial award applicants required to submit FAFSA. *Faculty research:* Population and organismal biology, population and evolutionary genetics, systematics and macroevolution, biochemistry, conservation biology. *Unit head:* Director of Graduate Studies, 607-254-4230. *Application contact:* Graduate Field Assistant, 607-254-4230, E-mail: eeb_grad_req@cornell.edu.

Dalhousie University, Faculty of Graduate Studies, College of Arts and Science, Faculty of Science, Department of Oceanography, Halifax, NS B3H 4R2, Canada. Offers M Sc, PhD. Part-time programs available. *Degree requirements:* For master's and doctorate, thesis/dissertation. *Entrance requirements:* Additional exam requirements/recommendations for international students: Required—TOEFL. *Faculty research:* Biological and physical oceanography, chemical and geological oceanography, atmospheric sciences.

Florida Institute of Technology, Graduate Programs, College of Engineering, Department of Marine and Environmental Systems, Program in Oceanography, Melbourne, FL 32901-6975. Offers biological oceanography (MS); chemical oceanography (MS); coastal zone management (MS); geological oceanography (MS); oceanography (PhD); physical oceanography (MS). Part-time programs available. *Students:* Average age 30.Terminal master's awarded for partial completion of doctoral program. *Degree requirements:* For master's, thesis (for some programs); for doctorate, one foreign language, thesis/dissertation, departmental qualifying exams, comprehensive exam. *Entrance requirements:* For master's, GRE General Test, minimum GPA of 3.0; for doctorate, GRE General Test, minimum GPA of 3.3, resumé. *Application deadline:* Applications are processed on a rolling basis. Electronic applications accepted. *Expenses:* Tuition: Part-time $825 per credit. *Financial support:* Research assistantships with full and partial tuition reimbursements, teaching assistantships with full and partial tuition reimbursements, career-related internships or fieldwork and tuition remissions available. Financial award application deadline: 3/1; financial award applicants required to submit FAFSA. *Faculty research:* Marine geochemistry, ecosystem dynamics, coastal processes, marine pollution, environmental modeling. Total annual research expenditures: $938,395. *Unit head:* Dr. Dean R. Norris, Chair, 321-674-7377, Fax: 321-674-7212, E-mail: norris@fit.edu. *Application contact:* Carolyn P. Farrior, Director of Graduate Admissions, 321-674-7118, Fax: 321-723-9468, E-mail: cfarrior@fit.edu.

See Close-Ups on pages 253 and 653.

Florida State University, Graduate Studies, College of Arts and Sciences, Department of Oceanography, Tallahassee, FL 32306. Offers MS, PhD. *Faculty:* 20 full-time (2 women), 4 part-time/adjunct (0 women). *Students:* 51 full-time (22 women), 19 international. Average age 28. 51 applicants, 27% accepted, 12 enrolled. In 2005, 5 master's, 6 doctorates awarded. *Median time to degree:* Of those who began their doctoral program in fall 1997, 100% received their degree in 8 years or less. *Degree requirements:* For master's, thesis/dissertation; for doctorate, thesis/dissertation, comprehensive exam. *Entrance requirements:* For master's and doctorate, GRE General Test. Additional exam requirements/recommendations for international students: Required—TOEFL (minimum score 550 paper-based; 213 computer-based). *Application deadline:* For fall admission, 2/1 for domestic students; for spring admission, 7/1 priority date for domestic students. Applications are processed on a rolling basis. Application fee: $30. Electronic applications accepted. *Financial support:* In 2005–06, 1 fellowship with full tuition reimbursement, 39 research assistantships with full tuition reimbursements, 3 teaching assistantships with full tuition reimbursements were awarded. Financial award application deadline: 2/1; financial award applicants required to submit FAFSA. *Faculty research:* Trace metals in seawater, currents and waves, modeling, benthic ecology, marine biogeochemistry. Total annual research expenditures: $3.8 million. *Unit head:* Dr. William K Dewar, Chair, 850-644-6700, Fax: 850-644-2581, E-mail: dewar@ocean.fsu.edu. *Application contact:* Michaela Lupiani, Academic Coordinator, 850-644-6700, Fax: 850-644-2581, E-mail: admissions@ocean.fsu.edu.

See Close-Up on page 257.

Louisiana State University and Agricultural and Mechanical College, Graduate School, School of the Coast and Environment, Department of Oceanography and Coastal Sciences, Baton Rouge, LA 70803. Offers MS, PhD. *Faculty:* 32 full-time (2 women). *Students:* 51 full-time (30 women), 15 part-time (4 women); includes 1 African American, 1 Hispanic American, 17 international. Average age 30. 27 applicants, 37% accepted, 17 enrolled. In 2005, 10 master's, 7 doctorates awarded. *Degree requirements:* For master's, thesis (for some programs); for doctorate, one foreign language, thesis/dissertation. *Entrance requirements:* For master's, GRE General Test, minimum GPA of 3.0; for doctorate, GRE General Test, MA or MS, minimum GPA of 3.0. Additional exam requirements/recommendations for international students: Required—TOEFL (minimum score 550 paper-based; 213 computer-based). *Application deadline:* For fall admission, 1/25 priority date for domestic students, 5/15 priority date for international students. Applications are processed on a rolling basis. Application fee: $25. *Financial support:* In 2005–06, 54 students received support, including 7 fellowships (averaging $20,200 per year), 39 research assistantships with full and partial tuition reimbursements available (averaging $19,760 per year), 3 teaching assistantships with full and partial tuition reimbursements available (averaging $12,750 per year); Federal Work-Study, institutionally sponsored loans, scholarships/grants, tuition waivers (full and partial), and unspecified assistantships also available. Support available to part-time students. Financial award applicants required to submit FAFSA. *Faculty research:* Management and development of estuarine and coastal areas and resources; physical, chemical, geological, and biological research. Total annual research expenditures:$88,496. *Unit head:* Dr. Lawrence Rouse, Chair, 225-578-2453, Fax: 225-578-6307, E-mail: lrouse@lsu.edu. *Application contact:* Dr. Masamichi Inoue, Graduate Adviser, 225-578-6308, Fax: 225-578-6307, E-mail: coiino@lsu.edu.

Massachusetts Institute of Technology, School of Engineering, Department of Civil and Environmental Engineering, Cambridge, MA 02139-4307. Offers biological oceanography (PhD, Sc D); chemical oceanography (PhD, Sc D); civil and environmental engineering (M Eng, SM, PhD, Sc D, CE); civil and environmental systems (PhD, Sc D); civil engineering (PhD, Sc D); coastal engineering (PhD, Sc D); construction engineering and management (PhD, Sc D); environmental biology (PhD, Sc D); environmental chemistry (PhD, Sc D); environmental engineering (PhD, Sc D); environmental fluid mechanics (PhD, Sc D); geotechnical and geoenvironmental engineering (PhD, Sc D); hydrology (PhD, Sc D); information technology (PhD, Sc D); oceanographic engineering (PhD, Sc D); structures and materials (PhD, Sc D); transportation (PhD, Sc D). *Faculty:* 34 full-time (3 women). *Students:* 179 full-time (66 women), 1 part-time; includes 15 minority (1 African American, 9 Asian Americans or Pacific Islanders, 5 Hispanic Americans), 116 international. Average age 26. 374 applicants, 38% accepted, 71 enrolled. In 2005, 76 master's, 27 doctorates, 1 other advanced degree awarded. *Degree requirements:* For master's and CE, thesis/dissertation; for doctorate, thesis/dissertation, comprehensive exam. *Entrance requirements:* For master's and doctorate, GRE General Test. Additional exam requirements/recommendations for international students: Required—TOEFL (minimum score 577 paper-based; 233 computer-based). *Application deadline:* For fall admission, 1/2 for domestic students, 1/2 for international students. Application fee: $70. Electronic applications accepted. *Expenses:* Tuition: Full-time $32,100. Required fees: $200. Part-time tuition and fees vary according to course load. *Financial support:* In 2005–06, 134 students received support, including 38 fellowships with tuition reimbursements available (averaging $23,116 per year), 86 research assistantships with tuition reimbursements available (averaging $22,765 per year), 13 teaching assistantships with tuition reimbursements available (averaging $19,735 per year); career-related internships or fieldwork, Federal Work-Study, institutionally sponsored loans, scholarships/grants, health care benefits, and unspecified assistantships also available. Total annual research expenditures: $12.4 million. *Unit head:* Prof. Patrick Jaillet, Department Head, 617-452-3379, Fax: 617-452-3294, E-mail: jaillet@mit.edu. *Application contact:* Graduate Admissions, 617-253-7119, Fax: 617-258-6775, E-mail: cee-admissions@mit.edu.

See Close-Up on page 259.

Massachusetts Institute of Technology, School of Science, Department of Biology, Cambridge, MA 02139-4307. Offers biological oceanography (PhD); biology (PhD). *Faculty:* 52 full-time (11 women). *Students:* 247 full-time (129 women); includes 50 minority (3 African Americans, 1 American Indian/Alaska Native, 34 Asian Americans or Pacific Islanders, 12 Hispanic Americans), 31 international. Average age 27. 550 applicants, 17% accepted, 29 enrolled. In 2005, 35 doctorates awarded. *Degree requirements:* For doctorate, thesis/dissertation, comprehensive exam. *Entrance requirements:* For doctorate, GRE General Test. Additional exam requirements/recommendations for international students: Required—TOEFL (minimum score 577 paper-based; 233 computer-based). *Application deadline:* For fall admission, 12/15 for domestic students, 12/15 for international students. Application fee: $70. Electronic applications accepted. *Expenses:* Tuition: Full-time $32,100. Required fees: $200. Part-time tuition and fees vary according to course load. *Financial support:* In 2005–06, 214 students received support, including 107 fellowships with tuition reimbursements available (averaging $26,164 per year), 114 research assistantships with tuition reimbursements available (averaging $26,424 per year); teaching assistantships, Federal Work-Study, institutionally sponsored loans, scholarships/grants, traineeships, health care benefits, and unspecified assistantships also available. *Faculty research:* DNA recombination, replication, and repair; transcription and gene regulation; signal transduction; cell cycle; neuronal cell fate. Total annual research expenditures: $95.4 million. *Unit head:* Prof. Chris Kaiser, Department Head, 617-253-4701, Fax: 617-253-8699. *Application contact:* Biology Education Office, 617-253-3717, Fax: 617-258-9329, E-mail: gradbio@mit.edu.

Massachusetts Institute of Technology, School of Science, Department of Earth, Atmospheric, and Planetary Sciences, Cambridge, MA 02139-4307. Offers atmospheric chemistry (PhD, Sc D); atmospheric science (SM, PhD, Sc D); climate physics and chemistry (PhD, Sc D); earth and planetary sciences (SM); geochemistry (PhD, Sc D); geology (PhD, Sc D); geophysics (PhD, Sc D); marine geology and geophysics (SM); oceanography (SM, PhD, Sc D); planetary sciences (PhD, Sc D). *Faculty:* 38 full-time (3 women). *Students:* 165 full-time (63 women); includes 12 minority (7 Asian Americans or Pacific Islanders, 5 Hispanic Americans), 51 international. Average age 27. 179 applicants, 34% accepted, 31 enrolled. In 2005, 2 master's, 15 doctorates awarded. Terminal master's awarded for partial completion of doctoral program. *Degree requirements:* For master's, thesis/dissertation; for doctorate, thesis/dissertation, comprehensive exam. *Entrance requirements:* For master's, GRE General Test; for doctorate, GRE General Test, GRE Subject Test (chemistry or physics for planetary science program). Additional exam requirements/recommendations for international students: Required—TOEFL (minimum score 577 paper-based; 233 computer-based). *Application deadline:* For fall admission, 1/5 for domestic students, 1/5 for international students; for spring admission, 11/1 for domestic students, 11/1 for international students. Application fee: $70. Electronic applications accepted. *Expenses:* Tuition: Full-time $32,100. Required fees: $200. Part-time tuition and fees vary according to course load. *Financial support:* In 2005–06, 115 students received support, including 36 fellowships with tuition reimbursements available (averaging $26,951 per year), 60 research assistantships with tuition reimbursements available (averaging $24,520 per year), 18 teaching assistantships with tuition reimbursements available (averaging $22,773 per year); Federal Work-Study, institutionally sponsored loans, scholarships/grants, health care benefits, and unspecified assistantships also available. *Faculty research:* Evolution of main features of the planetary system; origin, composition, structure, and state of the atmospheres, oceans, surfaces, and interiors of planets; dynamics of planets and satellite motions. Total annual research expenditures: $18.6 million. *Unit head:* Prof. Maria Zuber, Department Head, 617-253-6397, E-mail: zuber@mit.edu. *Application contact:* EAPS Education Office.

McGill University, Faculty of Graduate and Postdoctoral Studies, Faculty of Science, Department of Atmospheric and Oceanic Sciences, Montréal, QC H3A 2T5, Canada. Offers atmospheric science (M Sc, PhD); physical oceanography (M Sc, PhD). Terminal master's awarded for partial completion of doctoral program. *Degree requirements:* For master's, thesis/dissertation, registration; for doctorate, thesis/dissertation, comprehensive exam, registration. *Entrance requirements:* For master's, GRE General Test, minimum GPA of 3.2 during last 2 years of full-time study or 3.0 overall; B Sc or equivalent in biochemistry or related disciplines (biology, chemistry, physics, physiology, microbiology); for doctorate, GRE General Test, master's degree in meteorology or related field. Additional exam requirements/recommendations for international students: Required—TOEFL, IELT (minimum score 7). *Faculty research:* Dynamic meteorology and climate dynamics, synoptic and mesoscale meteorology, radar meteorology, atmospheric chemistry.

Memorial University of Newfoundland, School of Graduate Studies, Department of Physics and Physical Oceanography, St. John's, NL A1C 5S7, Canada. Offers atomic and molecular physics (M Sc, PhD); condensed matter physics (M Sc, PhD); physical oceanography (M Sc, PhD); physics (M Sc). Part-time programs available. *Students:* 24 full-time (10 women), 1 part-time, 16 international. 18 applicants, 56% accepted, 7 enrolled. In 2005, 9 master's, 2 doctorates awarded. *Degree requirements:* For master's, thesis, seminar presentation on thesis topic; for doctorate, thesis/dissertation, oral defense of thesis, comprehensive exam. *Entrance requirements:* For master's, honors B Sc or equivalent; for doctorate, M Sc or equivalent. *Application deadline:* Applications are processed on a rolling basis. Application fee: $40 Canadian dollars. Electronic applications accepted. *Expenses:* Tuition: Part-time $733 per term. Tuition and fees vary according to degree level and program. *Financial support:* Fellowships, research assistantships, teaching assistantships available. *Faculty research:* Experiment and theory in atomic and molecular physics, condensed matter physics, physical oceanography, theoretical geophysics and applied nuclear physics. *Unit head:* Dr. John Whitehead, Head, 709-737-8737, Fax: 709-737-8739, E-mail: johnw@physics.mun.ca. *Application contact:* Dr. Mike Morrow, Graduate Officer, 709-737-4361, Fax: 709-737-8739, E-mail: myke@physics.mun.ca.

Naval Postgraduate School, Graduate Programs, Department of Oceanography, Monterey, CA 93943. Offers MS, PhD. Program only open to commissioned officers of the United States and friendly nations and selected United States federal civilian employees. Part-time programs available. *Students:* 12 full-time. In 2005, 5 degrees awarded. *Degree requirements:* For master's, thesis; for doctorate, one foreign language, thesis/dissertation. *Unit head:* Dr. Roland W. Garwood, Chairman, 831-656-3260, E-mail: garwood@nps.navy.mil. *Application contact:* Tracy Hammond, Acting Director of Admissions, 831-656-3059, Fax: 831-656-2891, E-mail: thammond@bps.navy.mil.

Naval Postgraduate School, Graduate Programs, Program in Undersea Warfare, Monterey, CA 93943. Offers applied science (MS); electrical engineering (MS); engineering acoustics (MS); operations research (MS); physical oceanography (MS). Program only open to commissioned officers of the United States and friendly nations and selected United States federal civilian employees. Part-time programs available. *Students:* 12 full-time. *Degree requirements:* For master's, thesis. *Unit head:* Dr. Clyde Scandrett, Academic Group Chairman, 831-656-2654. *Application contact:* Tracy Hammond, Acting Director of Admissions, 831-656-3059, Fax: 831-656-2891, E-mail: thammond@bps.navy.mil.

North Carolina State University, Graduate School, College of Physical and Mathematical Sciences, Department of Marine, Earth, and Atmospheric Sciences, Raleigh, NC 27695. Offers marine, earth, and atmospheric sciences (MS, PhD); meteorology (MS, PhD); oceanography (MS, PhD). Terminal master's awarded for partial completion of doctoral program. *Degree requirements:* For master's, thesis (for some programs), final oral exam; for doctorate, thesis/dissertation, final oral exam, preliminary oral and written exams, comprehensive exam, registration. *Entrance requirements:* For master's, GRE General Test, minimum GPA of 3.0; for doctorate, GRE General Test, GRE Subject Test (for disciplines in biological oceanography and geology), minimum GPA of 3.0. Additional exam requirements/recommendations for international students: Required—TOEFL (minimum score 550 paper-based). Electronic applications accepted. *Faculty research:* Boundary layer and air quality meteorology; climate and mesoscale dynamics; biological, chemical, geological, and physical oceanography; hard rock, soft rock, environmental, and paleo geology.

See Close-Up on page 229.

Nova Southeastern University, Oceanographic Center, Program in Marine Biology and Oceanography, Fort Lauderdale, FL 33314-7796. Offers marine biology (PhD); oceanography (PhD). *Faculty:* 15 full-time (1 woman), 5 part-time/adjunct (0 women). *Students:* 4 applicants, 75% accepted, 3 enrolled. *Degree requirements:* For doctorate, thesis/dissertation, comprehensive exam. *Entrance requirements:* For doctorate, GRE, master's degree. Application fee: $50. *Application contact:* Dr. Andrew Rogerson, Associate Dean, Director of Graduate Programs, 954-262-3600, Fax: 954-262-4020, E-mail: arogerso@nsu.nova.edu.

See Close-Up on page 261.

Nova Southeastern University, Oceanographic Center, Program in Physical Oceanography, Fort Lauderdale, FL 33314-7796. Offers MS. *Faculty:* 15 full-time (1 woman), 5 part-time/adjunct (0 women). *Students:* 6 applicants, 67% accepted, 1 enrolled. *Degree requirements:* For master's, thesis. *Entrance requirements:* For master's, GRE, 1 year course work in calculus. Additional exam requirements/recommendations for international students: Required—TOEFL (minimum score 550 paper-based). *Application deadline:* Applications are processed on a rolling basis. Application fee: $50. *Application contact:* Dr. Andrew Rogerson, Associate Dean, Director of Graduate Programs, 954-262-3600, Fax: 954-262-4020, E-mail: arogerso@nsu.nova.edu.

Old Dominion University, College of Sciences, Programs in Oceanography, Norfolk, VA 23529. Offers ocean and earth sciences (MS); oceanography (PhD). Part-time programs available. *Faculty:* 21 full-time (3 women), 14 part-time/adjunct (3 women). *Students:* 18 full-time (10 women), 27 part-time (11 women); includes 4 minority (1 African American, 1 Asian American or Pacific Islander, 2 Hispanic Americans), 12 international. Average age 30. 23 applicants, 65% accepted, 10 enrolled. In 2005, 3 master's, 8 doctorates awarded. Terminal master's awarded for partial completion of doctoral program. *Degree requirements:* For master's, 10 days of ship time, thesis optional; for doctorate, thesis/dissertation, 10 days of ship time, comprehensive exam. *Entrance requirements:* For master's, GRE General Test, minimum GPA of 3.0 in major, 2.7 overall; for doctorate, GRE General Test. Additional exam requirements/recommendations for international students: Required—TOEFL (minimum score 550 paper-based; 213 computer-based). *Application deadline:* For fall admission, 2/15 priority date for domestic students, 2/15 priority date for international students. Applications are processed on a rolling basis. Application fee: $40. Electronic applications accepted. *Expenses:* Tuition, state resident: part-time $263 per credit hour. Tuition, nonresident: part-time $661 per credit hour. Required fees: $39 per semester. Part-time tuition and fees vary according to campus/location. *Financial support:* In 2005–06, 2 fellowships with full tuition reimbursements (averaging $20,000 per year), 28 research assistantships with full tuition reimbursements (averaging $18,000 per year), 14 teaching assistantships with full tuition reimbursements (averaging $15,000 per year) were awarded; career-related internships or fieldwork, scholarships/grants, and unspecified assistantships also available. Support available to part-time students. Financial award application deadline: 2/15; financial award applicants required to submit FAFSA. *Faculty research:* Biological, chemical, geological and physical oceanography. Total annual research expenditures: $3.3 million. *Unit head:* Dr. Fred Dobbs, Graduate Program Director, 757-683-4285, Fax: 757-683-5303, E-mail: oceangpd@odu.edu.

Oregon State University, Graduate School, College of Oceanic and Atmospheric Sciences, Program in Oceanography, Corvallis, OR 97331. Offers MA, MS, PhD. *Students:* 49 full-time (18 women), 7 part-time (6 women); includes 2 minority (1 American Indian/Alaska Native, 1 Hispanic American), 9 international. Average age 30. In 2005, 3 master's, 6 doctorates awarded. Terminal master's awarded for partial completion of doctoral program. *Degree requirements:* For master's, thesis optional; for doctorate, thesis/dissertation. *Entrance requirements:* For master's and doctorate, GRE General Test, minimum GPA of 3.0 in last 90 hours of course work. Additional exam requirements/recommendations for international students: Required—TOEFL. *Application deadline:* For fall admission, 2/1 for domestic students. Applications are processed on a rolling basis. Application fee: $50. *Expenses:* Tuition, area resident: Part-time $301 per credit. Tuition, state resident: full-time $8,139; part-time $501 per credit. Tuition, nonresident: full-time $14,376; part-time $532 per credit. Required fees: $1,266. *Financial support:* Fellowships, research assistantships, teaching assistantships, career-related internships or fieldwork, Federal Work-Study, and institutionally sponsored loans available. Support available to part-time students. Financial award application deadline: 2/1. *Faculty research:* Biological, chemical, geological, and physical oceanography. *Unit head:* Irma Delson, Assistant Director, Student Services, 541-737-5189, Fax: 541-737-2064, E-mail: student_adviser@oce.orst.edu.

Princeton University, Graduate School, Department of Geosciences, Program in Atmospheric and Oceanic Sciences, Princeton, NJ 08544-1019. Offers PhD. *Degree requirements:* For doctorate, one foreign language, thesis/dissertation. *Entrance requirements:* For doctorate, GRE General Test, GRE Subject Test. Additional exam requirements/recommendations for international students: Required—TOEFL (minimum score 600 paper-based; 250 computer-based). Electronic applications accepted. *Faculty research:* Climate dynamics, middle atmosphere dynamics and chemistry, oceanic circulation, marine geochemistry, numerical modeling.

Rutgers, The State University of New Jersey, New Brunswick/Piscataway, Graduate School, Program in Oceanography, New Brunswick, NJ 08901-1281. Offers MS, PhD. *Faculty:* 43 full-time, 1 part-time/adjunct. *Students:* 30 full-time (16 women); includes 2 minority (both Hispanic Americans), 7 international. Average age 28. 41 applicants, 20% accepted, 3 enrolled. In 2005, 1 master's, 2 doctorates awarded. Terminal master's awarded for partial completion of doctoral program. *Median time to degree:* Of those who began their doctoral program in fall 1997, 100% received their degree in 8 years or less. *Degree requirements:* For master's, thesis/dissertation; for doctorate, thesis/dissertation, comprehensive exam. *Entrance requirements:* For master's and doctorate, GRE General Test, 1 year course work in calculus, physics, chemistry. Additional exam requirements/recommendations for international students: Required—TOEFL. *Application deadline:* For fall admission, 2/1 for domestic students; for spring admission, 11/1 for domestic students. Applications are processed on a rolling basis. Application fee: $50. Electronic applications accepted. *Expenses:* Tuition, state resident: full-time $10,440; part-time $435 per credit. Tuition, nonresident: full-time $15,520; part-time $647 per credit. Required fees: $129 per credit. Tuition and fees vary according to program. *Financial support:* In 2005–06, 28 students received support, including 2 fellowships with full tuition reimbursements available (averaging $20,000 per year), 24 research assistantships with full tuition reimbursements available (averaging $19,367 per year), 2 teaching assistantships with full tuition reimbursements available (averaging $19,367 per year); career-related internships or fieldwork, institutionally sponsored loans, health care benefits, and unspecified assistantships also available. Financial award application deadline:3/1. *Faculty research:* Coastal observations and modeling, estuarine ecology/fish/benthos, geochemistry, deep sea ecology/hydrothermal vents, molecular biology applications. Total annual research expenditures: $14.7 million. *Unit head:* Dr. Paul Falkowski, Director, 732-932-6555 Ext. 370, Fax: 732-932-8578, E-mail: falko@marine.rutgers.edu. *Application contact:* Sarah N. Kasule, Administrative Assistant, 732-932-6555 Ext. 500, Fax: 732-932-8578.

See Close-Up on page 263.

Texas A&M University, College of Geosciences, Department of Oceanography, College Station, TX 77843. Offers MS, PhD. *Faculty:* 19 full-time (3 women), 2 part-time/adjunct (0 women). *Students:* 57 full-time (27 women), 9 part-time (5 women); includes 3 minority (1 Asian American or Pacific Islander, 2 Hispanic Americans), 20 international. Average age 28. 39 applicants, 54% accepted, 11 enrolled. In 2005, 11 master's, 8 doctorates awarded. *Degree requirements:* For master's and doctorate, thesis/dissertation. *Entrance requirements:* For master's and doctorate, GRE General Test. Additional exam requirements/recommendations for international students: Required—TOEFL. *Application deadline:* For fall admission, 1/15 for domestic students; for spring admission, 10/1 for domestic students. Applications are processed on a rolling basis. Application fee: $50 ($75 for international students). Electronic applications accepted. *Expenses:* Tuition, state resident: full-time $4,488; part-time $187 per credit hour. Tuition, nonresident: full-time $11,112; part-time $463 per credit hour. Required fees:$1,974. *Financial support:* In 2005–06, fellowships with partial tuition reimbursements (averaging $18,000 per year), research assistantships with partial tuition reimbursements (averaging $18,000 per year), teaching assistantships with partial tuition reimbursements (averaging $18,000 per year) were awarded; Federal Work-Study, scholarships/grants, and tuition waivers (partial) also available. Financial award application deadline: 1/15. *Faculty research:* Ocean circulation, climate studies, coastal and shelf dynamics, marine phytoplankton, stable isotope geochemistry. *Unit head:* Dr. Wilford D. Gardner, Head, 979-845-7211, Fax: 979-845-6331, E-mail: wgardner@ocean.tamu.edu. *Application contact:* Donna Dunlap, Academic Advisor, 979-845-7412, Fax: 979-845-6331.

Université du Québec à Rimouski, Graduate Programs, Program in Oceanography, Rimouski, QC G5L 3A1, Canada. Offers M Sc, PhD. Part-time programs available. *Students:* 97 full-time (56 women), 2 part-time (1 woman), 31 international. 50 applicants, 74% accepted. In 2005, 11 master's, 6 doctorates awarded. *Degree requirements:* For master's and doctorate, thesis/dissertation. *Entrance requirements:* For master's, appropriate bachelor's degree, proficiency in French; for doctorate, appropriate master's degree, proficiency in French. *Application deadline:* For fall admission, 5/1 for domestic students. Application fee: $30. Tuition charges are reported in Canadian dollars. *Expenses:* Tuition, state resident: full-time $2,000 Canadian dollars. Tuition, nonresident: full-time $9,000 Canadian dollars. Tuition and fees vary according to course load and program. *Financial support:* Fellowships, research assistantships, teaching assistantships available. *Unit head:* Jean-Francois Dumais, Director, 418-724-1770, Fax: 418-724-1525, E-mail: jean-francois_dumais@uqar.ca. *Application contact:* Marc Berube, Office of Admissions, 418-724-1433, Fax: 418-724-1525, E-mail: marc_berube@uqar.ca.

Université Laval, Faculty of Sciences and Engineering, Program in Oceanography, Québec, QC G1K 7P4, Canada. Offers PhD. *Degree requirements:* For doctorate, thesis/dissertation, comprehensive exam. *Entrance requirements:* For doctorate, knowledge of French, knowledge of English. Additional exam requirements/recommendations for international students: Required—TOEFL. Electronic applications accepted.

University of Alaska Fairbanks, School of Fisheries and Ocean Sciences, Department of Marine Sciences and Limnology, Fairbanks, AK 99775-7520. Offers marine biology (MS, PhD); oceanography (MS, PhD), including biological oceanography (PhD), chemical oceanography (PhD), fisheries (PhD), geological oceanography (PhD), physical oceanography (PhD). Part-time programs available. Terminal master's awarded for partial completion of doctoral program. *Degree requirements:* For master's and doctorate, thesis/dissertation, comprehensive exam, registration. *Entrance requirements:* For master's and doctorate, GRE General Test. Additional exam requirements/recommendations for international students: Required—TOEFL. Electronic applications accepted. *Expenses:* Tuition, state resident: full-time $4,392; part-time $244 per credit. Tuition, nonresident: full-time $8,964; part-time $498 per credit. Required fees: $800; $5 per credit. $48 per contact hour. Tuition and fees vary according to course level, course load, campus/location and reciprocity agreements. *Faculty research:* Seafood science and nutrition, sustainable harvesting, chemical oceanography, marine biology, physical oceanography.

The University of British Columbia, Faculty of Graduate Studies, Faculty of Science, Department of Earth and Ocean Sciences, Vancouver, BC V6T 1Z1, Canada. Offers atmospheric science (M Sc, PhD); geological engineering (M Eng, MA Sc, PhD); geological sciences (M Sc, PhD); geophysics (M Sc, MA Sc, PhD); oceanography (M Sc, PhD). *Faculty:* 44 full-time (7 women), 17 part-time/adjunct (1 woman). *Students:* 155 full-time (49 women), 1 (woman) part-time. Average age 30. 96 applicants, 50% accepted, 30 enrolled. In 2005, 11 master's, 6 doctorates awarded. *Degree requirements:* For master's, thesis (for some programs); for doctorate, thesis/dissertation, comprehensive exam. *Entrance requirements:* Additional exam requirements/recommendations for international students: Required—TOEFL (minimum score 600 paper-based; 250 computer-based). *Application deadline:* For fall admission, 2/1 for domestic students, 1/1 for international students. For winter admission, 7/1 for domestic students. Applications are processed on a rolling basis. Application fee: $90 Canadian dollars ($150 Canadian dollars for international students). Electronic applications accepted. *Financial support:* In 2005–06, fellowships (averaging $16,000 per year), research assistantships (averaging $13,000 per year), teaching assistantships (averaging $5,000 per year) were awarded; Federal Work-Study, institutionally sponsored loans, scholarships/grants, tuition waivers (full and partial), and unspecified assistantships also available. *Unit head:* Dr. Paul L. Smith, Head, 604-822-6456, Fax: 604-822-6088, E-mail: head@eos.ubc.ca. *Application contact:* Alex Allen, Graduate Secretary, 604-822-2713, Fax: 604-822-6088, E-mail: aallen@eos.ubc.ca.

University of California, San Diego, Graduate Studies and Research, Scripps Institution of Oceanography, La Jolla, CA 92093. Offers earth sciences (PhD); marine biodiversity and conservation (MAS); marine biology (PhD); oceanography (PhD). Postbaccalaureate distance learning degree programs offered (minimal on-campus study). *Faculty:* 97. *Students:* 243 (126 women); includes 21 minority (2 African Americans, 2 American Indian/Alaska Native, 11 Asian

Oceanography

University of California, San Diego (continued)
Americans or Pacific Islanders, 6 Hispanic Americans) 63 international. 311 applicants, 24% accepted, 37 enrolled. In 2005, 9 master's, 25 doctorates awarded. *Median time to degree:* Of those who began their doctoral program in fall 1997, 100% received their degree in 8 years or less. *Entrance requirements:* For doctorate, GRE General Test, GRE Subject Test. Additional exam requirements/recommendations for international students: Required—TOEFL (minimum score 550 paper-based; 213 computer-based). *Application deadline:* For fall admission, 1/4 for domestic students. Application fee: $60 ($80 for international students). Electronic applications accepted. *Financial support:* Fellowships, research assistantships, health care benefits available. *Unit head:* Myrl C. Hendershott, Chair, 858-534-3206, E-mail: siodept@sio.ucsd.edu. *Application contact:* Dawn Huffman, Graduate Coordinator, 858-534-3206.

University of Colorado at Boulder, Graduate School, College of Arts and Sciences, Department of Atmospheric and Oceanic Sciences, Boulder, CO 80309. Offers MS, PhD. *Faculty:* 10 full-time (2 women). *Entrance requirements:* For master's, minimum undergraduate GPA of 3.0. *Application deadline:* For fall admission, 2/1 for domestic students, 12/1 for international students. *Faculty research:* Large-scale dynamics of the ocean and the atmosphere, air-sea interaction, radiative transfer and remote sensing of the ocean and the atmosphere, sea ice and its role in climate. Total annual research expenditures: $2 million. *Unit head:* Brian Toon, Chair, 303-492-1534, Fax: 303-492-3524, E-mail: brian.toon@lasp.colorado.edu. *Application contact:* Graduate Secretary, 303-492-6633, Fax: 303-492-3524, E-mail: paosasst@colorado.edu.

University of Connecticut, Graduate School, College of Liberal Arts and Sciences, Department of Marine Sciences, Field of Oceanography, Storrs, CT 06269. Offers MS, PhD. *Faculty:* 26 full-time (7 women). *Students:* 36 full-time (24 women), 8 part-time (5 women); includes 1 minority (Hispanic American), 14 international. Average age 29. 38 applicants, 58% accepted, 14 enrolled. In 2005, 3 master's, 3 doctorates awarded. Terminal master's awarded for partial completion of doctoral program. *Degree requirements:* For master's, comprehensive exam; for doctorate, thesis/dissertation. *Entrance requirements:* For master's and doctorate, GRE General Test, GRE Subject Test. Additional exam requirements/recommendations for international students: Required—TOEFL (minimum score 550 paper-based; 213 computer-based). *Application deadline:* For fall admission, 2/1 priority date for domestic students, 2/1 priority date for international students; for spring admission, 11/1 for domestic students, 10/1 for international students. Applications are processed on a rolling basis. Application fee: $55. Electronic applications accepted. *Expenses:* Tuition, state resident: part-time $444 per credit hour. Tuition, nonresident: part-time $1,154 per credit hour. Tuition and fees vary according to course load. *Financial support:* In 2005–06, 33 research assistantships with full tuition reimbursements, 2 teaching assistantships with full tuition reimbursements were awarded; fellowships, Federal Work-Study, scholarships/grants, health care benefits, and unspecified assistantships also available. Financial award application deadline: 2/1; financial award applicants required to submit FAFSA. *Unit head:* Robert B. Whitlatch, Head, 860-445-3467, Fax: 860-405-9153, E-mail: robert.whitlatch@uconn.edu. *Application contact:* Barbara Mahoney, Administrative Assistant, 860-405-9151, Fax: 860-405-9153, E-mail: mscadm03@uconnvm.uconn.edu.

University of Delaware, College of Marine Studies, Newark, DE 19716. Offers geology (MS, PhD); marine management (MMM); marine policy (MS); marine studies (MMP, MS, PhD); oceanography (MS, PhD). *Faculty:* 41 full-time (4 women). *Students:* 115 full-time (57 women), 5 part-time (2 women); includes 7 minority (2 African Americans, 4 Asian Americans or Pacific Islanders, 1 Hispanic American), 30 international. Average age 29. 127 applicants, 35% accepted, 24 enrolled. In 2005, 12 master's, 8 doctorates awarded. *Degree requirements:* For master's and doctorate, thesis/dissertation. *Entrance requirements:* For master's and doctorate, GRE General Test. Additional exam requirements/recommendations for international students: Required—TOEFL. *Application deadline:* For fall admission, 3/1 for domestic students; for spring admission, 10/1 for domestic students. Applications are processed on a rolling basis. Application fee: $60. Electronic applications accepted. *Financial support:* In 2005–06, 78 students received support, including 14 fellowships with full tuition reimbursements available (averaging $19,000 per year), 62 research assistantships with full tuition reimbursements available (averaging $19,000 per year), 2 teaching assistantships with full tuition reimbursements available (averaging $19,000 per year); career-related internships or fieldwork, Federal Work-Study, and tuition waivers (full and partial) also available. Financial award application deadline: 3/1. *Faculty research:* Marine biology and biochemistry, oceanography, marine policy, physical ocean science and engineering, ocean engineering. Total annual research expenditures: $10.5 million. *Unit head:* Dr. Nancy Targett, Dean, 302-831-2841. *Application contact:* Lisa Perelli, Coordinator, 302-645-4226, E-mail: lperelli@udel.edu.

University of Georgia, Graduate School, College of Arts and Sciences, Department of Marine Sciences, Athens, GA 30602. Offers MS, PhD. *Faculty:* 14 full-time (5 women). *Students:* 36 full-time, 3 part-time; includes 2 minority (1 Asian American or Pacific Islander, 1 Hispanic American), 8 international. Average age 28. 40 applicants, 30% accepted, 10 enrolled. In 2005, 2 master's, 3 doctorates awarded. *Degree requirements:* For master's, thesis; for doctorate, thesis/dissertation, teaching experience, field research experience, comprehensive exam. *Entrance requirements:* For master's and doctorate, GRE General Test. Additional exam requirements/recommendations for international students: Required—TOEFL. *Application deadline:* For fall admission, 2/1 priority date for domestic students, 2/1 priority date for international students; for spring admission, 10/15 priority date for domestic students, 9/1 priority date for international students. Applications are processed on a rolling basis. Application fee: $50. Electronic applications accepted. *Financial support:* In 2005–06, 9 fellowships with full tuition reimbursements (averaging $20,000 per year), 21 research assistantships with full tuition reimbursements (averaging $18,000 per year), 11 teaching assistantships with full tuition reimbursements (averaging $18,000 per year) were awarded. *Faculty research:* Microbial ecology, biogeochemistry, polar biology, coastal ecology, coastal circulation. *Unit head:* Dr. James T. Hollibaugh, Director, 706-542-3016, Fax: 706-542-5888, E-mail: aquadoc@uga.edu. *Application contact:* Dr. Mary Ann Moran, Graduate Coordinator, 706-542-6481, Fax: 706-542-5888, E-mail: mmoran@uga.edu.

University of Hawaii at Manoa, Graduate Division, School of Ocean and Earth Science and Technology, Department of Oceanography, Honolulu, HI 96822. Offers MS, PhD. Part-time programs available. *Faculty:* 52 full-time (6 women), 7 part-time/adjunct (0 women). *Students:* 62 full-time (31 women), 5 part-time (all women); includes 8 minority (7 Asian Americans or Pacific Islanders, 1 Hispanic American), 25 international. Average age 32. 141 applicants, 21% accepted, 19 enrolled. In 2005, 8 master's, 2 doctorates awarded. Terminal master's awarded for partial completion of doctoral program. *Median time to degree:* Of those who began their doctoral program in fall 1997, 38% received their degree in 8 years or less. *Degree requirements:* For master's, thesis, field experience; for doctorate, one foreign language, thesis/dissertation, field experience. *Entrance requirements:* For master's and doctorate, GRE. *Application deadline:* For fall admission, 2/1 for domestic students, 1/15 for international students; for spring admission, 9/1 for domestic students, 8/1 for international students. Application fee: $50. *Expenses:* Tuition, state resident: full-time $8,400; part-time $200 per credit hour. Tuition, nonresident: full-time $11,088; part-time $462 per credit hour. Tuition and fees vary according to program. *Financial support:* In 2005–06, 45 research assistantships (averaging $18,555 per year), 14 teaching assistantships (averaging $18,198 per year) were awarded; fellowships, career-related internships or fieldwork, institutionally sponsored loans, and tuition waivers (full and partial) also available. Financial award applicants required to submit FAFSA. *Faculty research:* Physical oceanography, marine chemistry, biological oceanography, atmospheric chemistry, marine geology. Total annual research expenditures: $6.6 million. *Unit head:* Dr. Lorenz Magaard, Chairperson, 808-956-2913, Fax: 808-956-5035, E-mail: lorenz@hawaii.edu. *Application contact:* Dr. Lorenz Magaard, Chairperson, 808-956-2913, Fax: 808-956-5035, E-mail: lorenz@hawaii.edu.

University of Maine, Graduate School, College of Natural Sciences, Forestry, and Agriculture, School of Marine Sciences, Program in Oceanography, Orono, ME 04469. Offers MS, PhD. Part-time programs available. *Students:* 19 full-time (12 women), 6 part-time (5 women); includes 1 minority (Asian American or Pacific Islander), 6 international. Average age 28. 16 applicants, 50% accepted, 7 enrolled. In 2005, 4 master's awarded. *Degree requirements:* For master's and doctorate, thesis/dissertation. *Entrance requirements:* For master's and doctorate, GRE General Test. Additional exam requirements/recommendations for international students: Required—TOEFL. *Application deadline:* For fall admission, 2/1 for domestic students. Applications are processed on a rolling basis. Application fee: $50. Electronic applications accepted. *Financial support:* Fellowships with tuition reimbursements, research assistantships with tuition reimbursements, teaching assistantships with tuition reimbursements, career-related internships or fieldwork, Federal Work-Study, and tuition waivers (full and partial) available. Support available to part-time students. Financial award application deadline: 3/1. *Faculty research:* Coastal processes, microbial ecology, crustacean systematics. *Unit head:* Nuijie Xoe, Coordinator, 207-581-4318 Ext.245. *Application contact:* Scott G. Delcourt, Associate Dean of the Graduate School, 207-581-3219, Fax: 207-581-3232, E-mail: graduate@maine.edu.

University of Maryland, College Park, Graduate Studies, College of Computer, Mathematical and Physical Sciences, Department of Atmospheric and Oceanic Science, College Park, MD 20742. Offers MS, PhD. Part-time and evening/weekend programs available. Postbaccalaureate distance learning degree programs offered. *Faculty:* 34 full-time (5 women), 3 part-time/adjunct (1 woman). *Students:* 52 full-time (19 women), 7 part-time (3 women); includes 6 minority (3 Asian Americans or Pacific Islanders, 3 Hispanic Americans), 29 international. 82 applicants, 41% accepted, 14 enrolled. In 2005, 7 master's, 5 doctorates awarded. Terminal master's awarded for partial completion of doctoral program. *Median time to degree:* Of those who began their doctoral program in fall 1997, 25% received their degree in 8 years or less. *Degree requirements:* For master's, scholarly paper, written and oral exams; for doctorate, thesis/dissertation, exam. *Entrance requirements:* For master's, GRE General Test, GRE, background in mathematics, experience in scientific computer languages, 3 letters of recommendation; for doctorate, GRE General Test. *Application deadline:* For fall admission, 5/15 for domestic students, 2/1 for international students; for spring admission, 10/15 for domestic students, 6/1 for international students. Applications are processed on a rolling basis. Application fee: $60. Electronic applications accepted. *Financial support:* In 2005–06, 5 fellowships with full tuition reimbursements (averaging $3,722 per year), 8 research assistantships with tuition reimbursements (averaging $20,246 per year), 31 teaching assistantships with tuition reimbursements (averaging $19,081 per year) were awarded; Federal Work-Study and scholarships/grants also available. Support available to part-time students. Financial award applicants required to submit FAFSA. *Faculty research:* Weather, atmospheric chemistry, air pollution, global change, radiation. Total annual research expenditures: $3.9 million. *Unit head:* Dr. Russ Dickerson, Chairman, 301-405-5364, Fax: 301-314-9482. *Application contact:* Dean of Graduate School, 301-405-4190, Fax: 301-314-9305.

University of Miami, Graduate School, Rosenstiel School of Marine and Atmospheric Science, Division of Meteorology and Physical Oceanography, Coral Gables, FL 33124. Offers meteorology (PhD); physical oceanography (MS, PhD). *Faculty:* 30 full-time (5 women), 21 part-time/adjunct (3 women). *Students:* 34 full-time (7 women), 2 part-time; includes 1 minority (Hispanic American), 17 international. Average age 27. 34 applicants, 41% accepted, 8 enrolled. In 2005, 4 master's, 2 doctorates awarded. Terminal master's awarded for partial completion of doctoral program. *Median time to degree:* Of those who began their doctoral program in fall 1997, 50% received their degree in 8 years or less. *Degree requirements:* For master's and doctorate, thesis/dissertation, comprehensive exam, registration. *Entrance requirements:* For master's and doctorate, GRE General Test. Additional exam requirements/recommendations for international students: Required—TOEFL (minimum score 550 paper-based; 213 computer-based). *Application deadline:* For fall admission, 1/1 priority date for domestic students, 1/1 priority date for international students. Applications are processed on a rolling basis. Application fee: $50. Electronic applications accepted. *Financial support:* In 2005–06, 31 students received support, including 5 fellowships with tuition reimbursements available (averaging $22,380 per year), 21 research assistantships with tuition reimbursements available (averaging $22,380 per year), 2 teaching assistantships with tuition reimbursements available (averaging $22,380 per year); institutionally sponsored loans and scholarships/grants also available. Financial award application deadline: 3/1; financial award applicants required to submit FAFSA. *Unit head:* Dr. William Johns, Chairperson, 305-421-4057, E-mail: wjohns@rsmas.miami.edu. *Application contact:* Dr. Larry Peterson, Associate Dean, 305-421-4155, Fax: 305-421-4771, E-mail: gso@rsmas.miami.edu.

University of Michigan, Horace H. Rackham School of Graduate Studies, College of Literature, Science, and the Arts, Department of Geological Sciences, Program in Oceanography: Marine Geology and Geochemistry, Ann Arbor, MI 48109. Offers MS, PhD. Terminal master's awarded for partial completion of doctoral program. *Degree requirements:* For master's, thesis; for doctorate, thesis/dissertation, oral defense of dissertation, preliminary exam. *Entrance requirements:* For master's and doctorate, GRE General Test. Electronic applications accepted. *Expenses:* Tuition, state resident: full-time $14,082; part-time $894 per credit hour. Tuition, nonresident: full-time $28,500; part-time $1,675 per credit hour. Required fees: $189; $189 per unit. *Faculty research:* Paleoceanography, paleolimnology, marine geochemistry, seismic stratigraphy.

University of New Hampshire, Graduate School, College of Engineering and Physical Sciences, Department of Earth Sciences, Durham, NH 03824. Offers earth sciences (MS), including geochemical, geology, ocean mapping, oceanography; hydrology (MS). *Faculty:* 29 full-time. *Students:* 17 full-time (5 women), 24 part-time (10 women); includes 2 minority (1 African American, 1 Asian American or Pacific Islander), 6 international. Average age 31. 37 applicants, 84% accepted, 10 enrolled. In 2005, 10 master's awarded. *Degree requirements:* For master's, thesis. *Entrance requirements:* For master's, GRE General Test. Additional exam requirements/recommendations for international students: Required—TOEFL (minimum score 550 paper-based; 213 computer-based); Recommended—TSE. *Application deadline:* For fall admission, 4/1 priority date for domestic students, 4/1 priority date for international students. For winter admission, 12/1 for domestic students. Applications are processed on a rolling basis. Application fee: $60. Electronic applications accepted. *Expenses:* Tuition, state resident: full-time $8,010; part-time $445 per credit hour. Tuition, nonresident: full-time $19,730; part-time $810 per credit hour. Required fees: $322 per semester. Tuition and fees vary according to course load and program. *Financial support:* In 2005–06, 1 fellowship, 9 research assistantships, 9 teaching assistantships were awarded; career-related internships or fieldwork, Federal Work-Study, scholarships/grants, and tuition waivers (full and partial) also available. Support available to part-time students. Financial award application deadline: 2/15. *Unit head:* Dr. Matt Davis, Chairperson, 603-862-1718, E-mail: earth.sciences@unh.edu. *Application contact:* Nancy Gauthier, Administrative Assistant, 603-862-1720, E-mail: earth.sciences@unh.edu.

University of New Hampshire, Graduate School, College of Engineering and Physical Sciences, Department of Ocean Engineering, Durham, NH 03824. Offers ocean engineering (MS, PhD); ocean mapping (MS). *Faculty:* 5 full-time. *Students:* 7 full-time (0 women), 6 part-time (1 woman); includes 1 minority (Asian American or Pacific Islander), 4 international. Average age 32. 8 applicants, 75% accepted, 3 enrolled. In 2005, 3 degrees awarded. *Degree requirements:* For master's, thesis. *Entrance requirements:* Additional exam requirements/recommendations for international students: Required—TOEFL (minimum score 550 paper-based; 213 computer-based); Recommended—TSE. *Application deadline:* For fall admission, 4/1 for domestic students. For winter admission, 12/1 for domestic students. Applications are processed on a rolling basis. Application fee: $60. Electronic applications accepted. *Expenses:* Tuition, state resident: full-time $8,010; part-time $445 per credit hour. Tuition, nonresident: full-time $19,730; part-time $810 per credit hour. Required fees: $322 per semester. Tuition and fees vary according to course load and program. *Financial support:* In 2005–06, 1 fellowship, 3 research assistantships were awarded; teaching assistantships, Federal Work-Study, scholarships/grants, and tuition waivers (full and partial) also available. Support available to part-time students. Financial award application deadline:2/15. *Unit head:* Dr. Kenneth Baldwin,

Chairperson, 603-862-1898. *Application contact:* Jennifer Bedsole, Information Contact, 603-862-0672, E-mail: ocean.engineering@unh.edu.

University of Puerto Rico, Mayagüez Campus, Graduate Studies, College of Arts and Sciences, Department of Marine Sciences, Mayagüez, PR 00681-9000. Offers biological oceanography (MMS, PhD); chemical oceanography (MMS, PhD); geological oceanography (MMS, PhD); physical oceanography (MMS, PhD). *Faculty:* 25. *Students:* 18 full-time (16 women), 32 part-time (20 women); includes 46 minority (all Hispanic Americans), 4 international. 29 applicants, 59% accepted, 7 enrolled. In 2005, 6 master's, 2 doctorates awarded. *Degree requirements:* For master's, one foreign language, thesis, departmental and comprehensive final exams; for doctorate, one foreign language, thesis/dissertation, qualifying, comprehensive, and final exams. *Application deadline:* For fall admission, 2/15 for domestic students; for spring admission, 9/15 for domestic students. Applications are processed on a rolling basis. Application fee: $20. *Expenses:* Tuition, state resident: full-time $900; part-time $100 per credit. International tuition: $4,655 full-time. Part-time tuition and fees vary according to course level and course load. *Financial support:* In 2005–06, 49 students received support, including 1 fellowship (averaging $1,500 per year), 39 research assistantships (averaging $1,200 per year), 9 teaching assistantships (averaging $987 per year); Federal Work-Study and institutionally sponsored loans also available. *Faculty research:* Marine botany, ecology, chemistry, and parasitology; fisheries; ichthyology; aquaculture. Total annual research expenditures: $173,037. *Unit head:* Dr. Nilda Aponte, Director, 787-265-3838, E-mail: naponte@uprm.edu.

See Close-Up on page 273.

University of Rhode Island, Graduate School, Graduate School of Oceanography, Kingston, RI 02881. Offers MO, MS, PhD. In 2005, 4 master's, 10 doctorates awarded. *Application deadline:* For fall admission, 4/15 for domestic students. Applications are processed on a rolling basis. Application fee: $35. *Expenses:* Tuition, state resident: full-time $5,522; part-time $307 per credit. Tuition, nonresident: full-time $15,992; part-time $888 per credit. Required fees: $1,786; $73 per credit. One-time fee: $80 part-time. *Unit head:* David Farmer, Dean, 401-874-6222.

University of Southern California, Graduate School, College of Letters, Arts and Sciences, Department of Biological Sciences, Program in Marine Environmental Biology, Los Angeles, CA 90089. Offers PhD. *Degree requirements:* For doctorate, thesis/dissertation. *Entrance requirements:* For doctorate, GRE General Test. Additional exam requirements/recommendations for international students: Required—TOEFL. *Expenses:* Tuition: Full-time $25,416; part-time $1,059 per unit. Required fees: $484; $484 per year. Tuition and fees vary according to course load and program. *Faculty research:* Microbial ecology, physiology of larval development, biological community structure, Cambrian radiation.

University of South Florida, College of Graduate Studies, College of Marine Science, St. Petersburg, FL 33701-5016. Offers MS, PhD. Part-time and evening/weekend programs available. *Faculty:* 29 full-time (5 women). *Students:* 77 full-time (51 women), 44 part-time (29 women); includes 4 minority (2 African Americans, 2 Hispanic Americans), 37 international. Average age 31. 83 applicants, 36% accepted, 17 enrolled. In 2005, 4 master's, 1 doctorate awarded. *Degree requirements:* For master's, thesis; for doctorate, thesis/dissertation, proficiency foreign language and relevant skill directly related to area of study. *Entrance requirements:* For master's and doctorate, GRE General Test, minimum GPA of 3.0 in last 60 hours. Additional exam requirements/recommendations for international students: Required—TOEFL. *Application deadline:* For fall admission, 3/1 for domestic students; for spring admission, 10/1 for domestic students. Applications are processed on a rolling basis. Application fee: $30. *Financial support:* Fellowships with partial tuition reimbursements, research assistantships with partial tuition reimbursements, teaching assistantships with partial tuition reimbursements available. *Faculty research:* Trace metal chemistry, water quality, organic and isotopic geochemistry, physical chemistry, nutrient chemistry. *Unit head:* Dr. Peter R. Betzer, Dean, 727-553-1130, Fax: 727-553-1189, E-mail: pbetzer@marine.usf.edu. *Application contact:* Dr. Edward VanVleet, Coordinator, 727-553-1165, Fax: 727-553-1189, E-mail: advisor@marine.usf.edu.

University of Victoria, Faculty of Graduate Studies, Faculty of Science, School of Earth and Ocean Sciences, Victoria, BC V8W 2Y2, Canada. Offers M Sc, PhD. Part-time programs available. *Faculty:* 15 full-time, 41 part-time/adjunct. *Students:* 70, 18 international. Average age 24. 137 applicants, 66% accepted, 15 enrolled. In 2005, 5 master's, 4 doctorates awarded. *Degree requirements:* For master's, thesis, registration; for doctorate, thesis/dissertation, candidacy exam. *Entrance requirements:* For master's and doctorate, GRE (75th percentile). Additional exam requirements/recommendations for international students: Required—TOEFL (minimum score 575 paper-based; 233 computer-based), IELT (minimum score 7). *Application deadline:* For fall admission, 2/15 priority date for domestic students, 12/15 priority date for international students. Applications are processed on a rolling basis. Application fee: $75 ($125 for international students). Electronic applications accepted. Tuition and fees charges are reported in Canadian dollars. *Expenses:* Tuition, area resident: Full-time $4,492 Canadian dollars; part-time $749 Canadian dollars per term. International tuition: $5,346 Canadian dollars full-time. Required fees: $4,492 Canadian dollars; $749 Canadian dollars per term. Tuition and fees vary according to course load, campus/location and program. *Financial support:* In 2005–06, 16 fellowships, 22 research assistantships, 25 teaching assistantships were awarded; career-related internships or fieldwork, institutionally sponsored loans, and awards also available. Financial award application deadline: 2/15. *Faculty research:* Climate modeling, geology. *Unit head:* Dr. Kathryn Gillis, Director, 250-472-4023, Fax: 250-721-6200, E-mail: kgillis@uvic.ca. *Application contact:* Dr. Andrew Weaver, Graduate Adviser, 250-472-4006, Fax: 250-721-6200, E-mail: weaver@ocean.seas.uvic.ca.

University of Washington, Graduate School, College of Ocean and Fishery Sciences, School of Oceanography, Programs in Biological Oceanography, Seattle, WA 98195. Offers MS, PhD. Terminal master's awarded for partial completion of doctoral program. *Degree requirements:* For master's, research project; for doctorate, thesis/dissertation. *Entrance requirements:* For master's and doctorate, GRE General Test, minimum GPA of 3.0. Additional exam requirements/recommendations for international students: Required—TOEFL. Electronic applications accepted. *Faculty research:* Immunological techniques, thermophilic and archae-bacteria in hydrothermal systems, remote sensing, astrobiology.

University of Wisconsin–Madison, Graduate School, College of Engineering, Program in Limnology and Marine Science, Madison, WI 53706-1380. Offers MS, PhD. Terminal master's awarded for partial completion of doctoral program. *Degree requirements:* For master's and doctorate, thesis/dissertation. *Entrance requirements:* For master's and doctorate, GRE General Test. Additional exam requirements/recommendations for international students: Required—TOEFL. Electronic applications accepted. *Faculty research:* Lake ecosystems, ecosystem modeling, geochemistry, physiological ecology, chemical limnology.

University of Wisconsin–Madison, Graduate School, College of Letters and Science, Department of Atmospheric and Oceanic Sciences, Madison, WI 53706-1380. Offers MS, PhD. Part-time programs available. *Degree requirements:* For master's, thesis (for some programs); for doctorate, thesis/dissertation. *Entrance requirements:* For master's and doctorate, GRE General Test, minimum GPA of 3.0; previous course work in chemistry, mathematics, and physics. Electronic applications accepted. *Faculty research:* Satellite meteorology, weather systems, global climate change, numerical modeling, atmosphere-ocean interaction.

Woods Hole Oceanographic Institution, MIT/WHOI Joint Program in Oceanography/Applied Ocean Science and Engineering, Woods Hole, MA 02543-1541. Offers applied ocean sciences (PhD); biological oceanography (PhD, Sc D); chemical oceanography (PhD, Sc D); civil and environmental and oceanographic engineering (PhD); electrical and oceanographic engineering (PhD); geochemistry (PhD); geophysics (PhD); marine biology (PhD); marine geochemistry (PhD, Sc D); marine geology (PhD, Sc D); marine geophysics (PhD); mechanical and oceanographic engineering (PhD); ocean engineering (PhD); oceanographic engineering (M Eng, MS, PhD, Sc D, Eng); paleoceanography (PhD); physical oceanography (PhD, Sc D). MS, PhD, and Sc D offered jointly with MIT. Terminal master's awarded for partial completion of doctoral program. *Degree requirements:* For master's and Eng, thesis (for some programs); for doctorate, thesis/dissertation. *Entrance requirements:* For master's, GRE General Test; for doctorate, GRE General Test, GRE Subject Test. Additional exam requirements/recommendations for international students: Required—TOEFL. Electronic applications accepted.

See Close-Up on page 259.

Yale University, Graduate School of Arts and Sciences, Department of Geology and Geophysics, New Haven, CT 06520. Offers geochemistry (PhD); geophysics (PhD); meteorology (PhD); mineralogy and crystallography (PhD); oceanography (PhD); paleoecology (PhD); paleontology and stratigraphy (PhD); petrology (PhD); structural geology (PhD). *Degree requirements:* For doctorate, thesis/dissertation. *Entrance requirements:* For doctorate, GRE General Test. Additional exam requirements/recommendations for international students: Required—TOEFL.

Cross-Discipline Announcement

Florida Institute of Technology, Graduate Programs, College of Science, Department of Biological Sciences, Melbourne, FL 32901-6975.

MS and PhD in marine biology with research emphases in mollusks, echinoderms, coral reef fishes, and manatees and community studies of lagoonal, mangrove, and reef systems. Applications range from physiological and ecological systematics to biochemical and molecular biological studies in these areas.

FLORIDA INSTITUTE OF TECHNOLOGY

Department of Marine and Environmental Systems Programs in Oceanography and Coastal Zone Management

Programs of Study

Florida Institute of Technology offers programs of research and study options in the fields of biological, chemical, geological, and physical oceanography; marine meteorology; and environmental and marine chemistry that lead to M.S. and Ph.D. degrees in oceanography. An M.S. in oceanography with an option in coastal zone management is also offered. Those students interested in the graduate program in ocean engineering should consult the program description in Book 5 of these guides.

Research Facilities

Florida Institute of Technology is conveniently located on the Indian River Lagoon, a major east-central Florida estuarine system recently designated an Estuary of National Significance. Marine and environmental laboratories and field research stations are located on the lagoon and at an oceanfront marine research facility. Marine operations, located just 5 minutes from the campus, house a fleet of small outboard-powered craft and medium-sized work boats. These boats are available to students and faculty members for teaching and research in the freshwater tributaries and the Indian River Lagoon. In addition, the university operates the 60-foot Research Vessel *Delphinus,* which is berthed at Port Canaveral. With its own captain and crew, requisite marine and oceanographic cranes, winches, state-of-the-art sampling equipment, instrumentation, and laboratories, the vessel is the focal point of both marine and estuarine research in the region. The ship can accommodate a scientific team and crew for periods of seven to ten days. The *Delphinus* conducts short research and teaching cruises throughout the year and teaching trips to the Atlantic Ocean each summer.

Florida Tech's oceanfront marine research facility, the Vero Beach Marine Laboratory, located at Vero Beach just 40 minutes from the campus, provides facilities, including flowing seawater from the Atlantic Ocean, to support research in such areas as aquaculture, biofouling, and corrosion. There is also a permanent research platform, centrally located in the Indian River Lagoon, to support marine research projects. On the campus, the departmental teaching and research facilities include separate laboratories for biological, chemical, physical, geological, and instrumentation investigations. In addition, high-pressure, hydroacoustics, fluid dynamics, and GIS/remote sensing facilities are available in the department. An electron microscope is also available for research work.

About an hour from campus is the Harbor Branch Oceanographic Institution; scientists and engineers there pursue their own research and development activities and interact with Florida Tech students and faculty members on projects of mutual interest.

The Biological Oceanography Laboratory is fully equipped for research on plankton, benthos, and fishes of coastal and estuarine ecosystems. Collection gear; analytical equipment, including a flow-through fluorometer; and a controlled environment room are available for student and research use. Areas of research have included toxic algal blooms, seagrass ecology, and artificial and natural reef communities.

The Chemical Oceanography Laboratory is equipped to do both routine and research-level operations on open ocean and coastal lagoonal waters. Major and minor nutrients, heavy-metal contaminants, and biological pollutants can be quantitatively determined. Analytical methods available include gas and liquid chromatography, infrared and visible light spectrophotometry, and atomic absorption spectrometry.

The Physical Oceanography and the Surf Mechanics laboratories support graduate research in ocean waves, coastal processes, circulation, and pollutant transport. In addition, CTD and XBT systems, ADCP and other current meters, tide and wind recorders, salinometers, wave-height gauges, side-scan sonar, and other oceanographic instruments are available for field work. A remote sensing and optics lab provides capabilities for analyzing ocean color data and collecting in situ hydrologic optics data.

The Marine Geology and Geophysics Laboratory is used to study near-shore sedimentation and stratigraphy. The lab equipment includes a state-of-the-art computerized rapid sediment analyzer, a magnetic heavy-mineral separator, and computer-assisted sieve systems.

Financial Aid

Graduate teaching, research assistantships, and endowed fellowships are available to qualified students. For 2006–07, financial support ranges from approximately $9000 to $16,000, including stipend and tuition, per academic year for approximately half-time duties. Stipend-only assistantships are sometimes awarded for less time commitment. Most coastal zone management students receive support through internship appointments.

Cost of Study

In 2006–07, tuition is $900 per graduate semester credit hour.

Living and Housing Costs

Room and board on campus cost approximately $3000 per semester in 2006–07. On-campus housing (dormitories and apartments) is available for full-time single and married graduate students, but priority for dormitory rooms is given to undergraduate students. Many apartment complexes and rental houses are available near the campus.

Student Group

The College of Engineering has 450 graduate students. Oceanography currently has approximately 25 graduate and 40 undergraduate students.

Student Outcomes

Graduates have gone on to careers with such institutions as NOAA, EPA, Florida Water Management Districts, Western Geophysical, Naval Oceanographic Office, Digicon, and with county and state agencies.

Location

The campus is located in Melbourne, on Florida's east coast. It is an area, located 4 miles from the Atlantic Ocean beaches, with a year-round subtropical climate. The area's economy is supported by a well-balanced mix of industries in electronics, aviation, light manufacturing, optics, communications, agriculture, and tourism. Many industries support activities at the Kennedy Space Center.

The Institute

Florida Institute of Technology is a distinctive, independent university, founded in 1958 by a group of scientists and engineers to fulfill a need for specialized advanced educational opportunities on the Space Coast of Florida. Florida Tech is the only independent technological university in the Southeast. Supported by both industry and the community, Florida Tech is the recipient of many research grants and contracts, a number of which provide financial support for graduate students.

Applying

Forms and instructions for applying for admission and assistantships are sent on request. Admission is possible at the beginning of any semester, but admission in the fall semester is recommended. It is advantageous to apply early.

Correspondence and Information

Dr. John G. Windsor Jr., Program Chairman
Oceanography Program
Florida Institute of Technology
Melbourne, Florida 32901-6975
Phone: 321-674-8096
Fax: 321-674-7212
E-mail: dmes@fit.edu
Web site: http://www.fit.edu/AcadRes/dmes

Graduate Admissions Office
Florida Institute of Technology
Melbourne, Florida 32901-6975
Phone: 321-674-8027
800-944-4348 (toll-free)
Fax: 321-723-9468
E-mail: grad-admissions@fit.edu
Web site: http://www.fit.edu/grad

Florida Institute of Technology

THE FACULTY AND THEIR RESEARCH

Charles R. Bostater Jr., Associate Professor; Ph.D., Delaware. Remote sensing, hydrologic optics, particle dynamics in estuaries, modeling of toxic substances, physical oceanography of coastal waters, environmental modeling.

Iver W. Duedall, Professor Emeritus; Ph.D., Dalhousie. Chemical oceanography, physical chemistry of seawater, geochemistry, marine pollution, ocean management.

Lee E. Harris, Associate Professor; Ph.D., Florida Atlantic; PE. Coastal engineering, coastal structures, beach erosion and control, physical oceanography.

Elizabeth A. Irlandi, Associate Professor; Ph.D., North Carolina. Landscape ecology in aquatic environments, seagrass ecosystems, coastal zone management.

Kevin B. Johnson, Assistant Professor; Ph.D., Oregon. Zooplankton ecology, predator-prey interactions, metamorphosis, larval transport and settlement, larval behavior.

George A. Maul, Professor; Ph.D., Miami (Florida). Physical oceanography, marine meteorology, climate and sea level change, satellite oceanography, earth system science, tsunamis and coastal hazards.

Dean R. Norris, Professor Emeritus; Ph.D., Texas A&M. Taxonomy and ecology of marine phytoplankton, particularly dinoflagellates; ecology and life cycles of toxic dinoflagellates.

Geoffrey W. J. Swain, Professor; Ph.D., Southampton. Materials corrosion, biofouling, offshore technology, ship operations.

Eric D. Thosteson, Assistant Professor; Ph.D., Florida; PE. Coastal and nearshore engineering, coastal processes, wave mechanics, sediment transport.

John H. Trefry, Professor; Ph.D., Texas A&M. Trace metal geochemistry and pollution, geochemistry of rivers, global chemical cycles, deep-sea hydrothermal systems.

John G. Windsor Jr., Professor; Ph.D., William and Mary. Trace organic analysis, organic chemistry, sediment-sea interaction, air-sea interaction, mass spectrometry, hazardous/toxic substance research.

Gary Zarillo, Professor; Ph.D., Georgia; PG. Sediment transport and morphodynamics, tidal inlet–barrier dynamics, numerical modeling of inlet hydrodynamics.

Adjunct Faculty

Diane D. Barile, M.S., Florida Tech. Environmental planning, environmental policy.
D. E. DeFreese, Ph.D., Florida Tech. Marine biology.
R. Grant Gilmore, Ph.D., Florida Tech. Bioacoustics, biological oceanography.
M. Dennis Hanisak, Ph.D., Rhode Island. Biological oceanography.
Brian E. LaPointe, Ph.D., South Florida. Riverine and estuarine systems.
Francis J. Merceret, Ph.D., Johns Hopkins. Atmospheric physics, spacecraft meterorology.
Donald T. Resio, Ph.D., Virginia. Ocean waves, inlet dynamics, physical oceanography.
Ned P. Smith, Ph.D., Wisconsin. Physical oceanography, marine observations, marine meteorology.
Robert W. Virnstein, Ph.D., William and Mary. Limnology.
Craig M. Young, Ph.D., Alberta. Benthic ecology.

SELECTED PUBLICATIONS

Bostater, C. Remote sensing methods using aircraft and ships for estimating optimal bands and coefficients related to ecosystem responses. *Int. Soc. Opt. Eng. (SPIE)* 1930:1051–62, 1992.

Bostater, C. Mathematical techniques for spectral discrimination between corals, seagrasses, bottom and water types using high spectral resolution reflectance signatures. In *ISSSR, Spectral Sensing Research,* pp. 526–36, 1992.

Duedall, I. W., and **G. A. Maul.** Demography of coastal populations. In *Encyclopedia of Coastal Science,* pp. 368–74, ed. M. L. Schwartz. Dordrecht: Springer, 2005.

Dickenson, C., and **I. W. Duedall.** "An Analysis of the International Maritime Organization—London Convention Annual Ocean Dumping Reports." Chapter in *Water Encyclopedia; Oceanography; Meteorology; Physics and Chemistry; Water Law; and Water History, Art, and Culture,* pp. 144–9, eds. J. H. Lehr and J. Keeley. New York: John Wiley and Sons.

Williams, J., and **I. W. Duedall.** *Florida Hurricanes and Tropical Storms, 1871–2001,* 167 pp. University of Florida Press, 2002.

Shieh, C. S., and **I. W. Duedall.** Disposal of wastes at sea in tropical areas. In *Pollution in Tropical Aquatic Systems,* pp. 218–29, eds. D. W. Connell and D. W. Hawker. Boca Raton: CRC Press, 1992.

Shieh, C. S., and **I. W. Duedall.** Cd and Pb in waste-to-energy residues. *Chem. Ecol.* 6:247–58, 1992.

Duedall, I. W., and M. A. Champ. Artificial reefs: Emerging science and technology. *Oceanus* 34:94–101, 1991.

Duedall, I. W. A brief history of ocean disposal. *Oceanus* 33(2):29–38, 1990.

Harris, L. E., and J. W. Sample. The evolution of multi-celled sand-filled geosynthetic systems for coastal protection and surfing enhancement. *J. Coastal Res.*, special issue, in press.

Harris, L. E., and Woodring. Artificial reefs for submerged and subaerial habitat protection, mitigation and restoration. In *Gulf and Caribbean Fisheries Institute (GCFI),* pp. 386–95, 2001.

Zadikoff, Covello, **L. E. Harris,** and Skornick. Concrete tetrahedrons and sand-filled geotextile containers: New technologies for shoreline stabilization. *J. Coastal Res.*, special issue, 26:261–8, 1998.

Harris, L. E., Childress, Winder, and Perry. Real-time wave data collection system at Sebastian Inlet, Florida. In *Fifth International Workshop on Wave Hindcasting and Forecasting,* pp. 146–53. Ontario: Environment Canada, 1998.

Smith, J. T., **L. E. Harris,** and J. Tabar. *Preliminary Evaluation of the Vero Beach, FL Prefabricated Submerged Breakwater.* Tallahassee, Fla.: FSBPA, 1998.

Zadikovv, Covello, **L. E. Harris,** and Skornick. Concrete tetrahedrons and sand-filled geotextile containers: New technologies for shoreline stabilization. In *Proceedings of the International Coastal Symposium (ICS98).* J. Coastal Res., special issue, 26:261–8, 1998.

Harris, L. E. Dredged material used in sand-filled containers for scour and erosion control. In *Dredging 94,* American Society of Civil Engineers (ASCE), 1994.

Bishop, M. J., et al. **(E. A. Irlandi).** Spatio-temporal patterns in the mortality of bay scallop recruits in North Carolina: Investigation of a life-history anomaly. *J. Exp. Mar. Biol. Ecol.* 315:127–46, 2005.

Irlandi, E. A., B.A. Orlando and P. D. Biber. Drift algae-epiphyte-seagrass interactions in a subtropical *Thalassia testudinum* meadow. *Mar. Ecol. Prog. Ser.* 279:81–91, 2004.

Irlandi, E. A., B. A. Orlando, and W. Cropper Jr. Short-term effects of nutrient addition on growth and biomass of *Thalassia testudinum* in Biscayne Bay, FL. *Fla. Scientist* 67:18–26, 2003.

Irlandi, E. A., et al. The influence of freshwater runoff on biomass, morphometrics, and production of *Thalassia testudinum. Aquat. Botany* 1536:1–12, 2001.

Chambers, P. A., R. E. DeWreede, **E. A. Irlandi,** and H. Vandermeulen. Management issues in aquatic macrophyte ecology: A Canadian perspective. *Can. J. Botany,* 77:471–87, 1999.

Irlandi, E. A., B. A. Orlando, and W. G. Ambrose Jr. The effect of habitat patch size on growth and survival of juvenile bay scallops *(Argopecten irradians). J. Exp. Marine Biol. Ecol.* 235:21–43, 1999.

Irlandi, E. A. Seagrass patch size and survivorship of an infaunal bivalve. *Oikos* 78:511–8, 1997.

Irlandi, E. A., S. Macia, and J. Serafy. Salinity reduction from freshwater canal discharge: Effects on mortality and feeding of an urchin *(Lytechinus variegatus)* and a gastropod *(Lithopoma tectum). Bull. Mar. Sci.* 61:869–79, 1997.

Irlandi, E. A. The effects of seagrass patch size and energy regime on growth of a suspension-feeding bivalve. *J. Mar. Res.* 54:161–85, 1996.

Irlandi, E. A., and M. E. Mehlich. The effect of tissue cropping and disturbance by browsing fishes on growth of two species of suspension-feeding bivalves. *J. Exp. Mar. Biol. Ecol.* 197:279–93, 1996.

Johnson, K. B., and R. B. Forward Jr., Larval photoresponses of the polyclad flatworm *Maritigrella crozieri* (Platyhelminthes, Polycladida). *J. Exp. Mar. Biol. Ecol.* 282:103–12, 2003.

Johnson, K. B., and A. L. Shanks. Low rates of predation on planktonic marine invertebrate larvae. *Mar. Ecol. Prog. Ser.* 248:125–39, 2003.

Smith, D. L., and **K. B. Johnson.** *A Guide to Marine Coastal Plankton and Marine Invertebrate Larvae,* 2nd ed. Dubuque, Iowa: Kendall/Hunt, 2001.

Johnson, K. B., and A. L. Shanks. The importance of prey densities and background plankton in studies of predation on invertebrate larvae. *Mar. Ecol. Prog. Ser.* 158:293–6, 1997.

Maul, G. A. Wave climate. In *Encyclopedia of Coastal Science,* ed. M. Schwartz. Dordrecht: Kluwer Academic Publishers, pp: 1049–52, 2005.

Maul, G. A. Ocean wind system. In *Interdisciplinary Encyclopedia of Marine Sciences,* 3 vols., ed. J. Nybakken. Danbury, Conn.: Grolier Academic Reference, pp. 400–6, 2003.

Maul, G. A., A. M. Davis, and J. W. Simmons. Seawater temperature trends at USA tide gauge sites. *Geophys. Res. Lett.* 28(20):3935–7, 2001.

Pugh, D. T., and **G. A. Maul.** Coastal sea level prediction for climate change. In *Coastal Ocean Prediction,* no. 56, pp. 377–404, ed. C. N. K. Mooers. Washington: American Geophysical Union, Coastal and Estuarine Studies, 1999.

Mooers, C. N. K., and **G. A. Maul.** Intra-Americas sea circulation. In *The Sea,* chap. 7, pp. 183–208, eds. A. R. Robinson and K. H. Brink, 1998.

Maul, G. A. *Small Islands: Marine Science and Sustainable Development.* Washington: American Geophysical Union, Coastal and Estuarine Studies, number 51, 1996.

Maul, G. A. *Climatic Changes in the Intra-Americas Sea.* United Nations Environment Programme. London: Edward Arnold Publishers, 1993.

Maul, G. A., and D. M. Martin. Sea level rise at Key West, Florida, 1846–1992: America's longest instrument record? *Geophys. Res. Lett.* 20(18):1955–59, 1993.

Hanson, D. V., and **G. A. Maul.** Anticyclonic current rings in the eastern tropical Pacific Ocean. *J. Geophys. Res.* 96(C4):6965–79, 1991.

Hanson, K., **G. A. Maul,** and T. R. Karl. Are atmospheric greenhouse effects apparent in the climatic record of the contiguous U.S. (1895–1987)? *Geophys. Res. Lett.* 16(1):49–52, 1989.

Maul, G. A. *Introduction to Satellite Oceanography.* Dordrecht/Boston/Lancaster: Martinus Nijhoff Publishers, 1985.

Swain, G. W. J., J. Griffith, D. Bultman, and H. Vincent. Barnacle adhesion measurements for the field evaluation of candidate anti-fouling surfaces. *Biofouling* 6:105–14, 1992.

Swain, G. W. J., and E. Muller. Oxygen concentration cells and corrosion in a seawater aquarium. *Corrosion* 92(394), 1992.

Swain, G. W. J., and J. Patrick-Maxwell. The effect of biofouling on the performance of Al-Zn-Hg anodes. *Corrosion* 46(3):256–60, 1990.

Swain, G. W. J., and W. Thomason. Cathodic protection and the use of copper anti-fouling systems on fixed offshore steel structures. Presented at Offshore Mechanics and Arctic Engineering 9th International Conference, Houston, February 1990.

Thosteson, E. D., D. M. Hanes, and S. L. Schonfield. Design of a littoral sedimentation processes monitoring system. *J. Oceanic Eng.*, in press.

Florida Institute of Technology

Hanes, D. M., and **E. D. Thosteson** et al. Field observations of small scale sediment suspension. In *Proceedings of the 26th International Conference on Coastal Engineering.* Copenhagen, Denmark: ASCE, 1998.

Thosteson, E. D., and D. M. Hanes. The time lag between fluid forcing and wave generated sediment suspension. *AGU Fall Meeting,* 1998.

Thosteson, E. D., and D. M. Hanes. A simplified method for determining sediment size and concentration from multiple frequency acoustic backscatter measurements. *J. Acoust. Soc. Am.* 104(2):820, 1998.

Rember, R. D., and **J. H. Trefry.** Increased concentrations of dissolved trace metals and organic carbon during snowmelt in rivers of the Alaskan Arctic. *Geochim. Cosmochim. Acta* 68:477–89, 2004.

Trefry, J. H., R. D. Rember, R. P. Trocine, and J. S. Brown. Trace metals in sediments near offshore oil exploration and production sites in the Alaskan Arctic. *Environ. Geol.* 45:149–60, 2003.

Kang, W.-J., **J. H. Trefry,** T. A. Nelsen, and H. R. Wanless. Direct atmospheric inputs versus runoff fluxes of mercury to the lower Everglades and Florida Bay. *Environ. Sci. Technol.* 34(19):4058–63, 2000.

Trefry, J. H., R. P. Trocine, K. L. Naito, and S. Metz. Assessing the potential for enhanced bioaccumulation of heavy metals from produced water discharges to the Gulf of Mexico. In *Produced Water: Environmental Issues and Mitigation Technologies,* pp. 339–54. New York: Plenum Press, 1996.

Trefry, J. H., et al. Transport of particulate organic carbon by the Mississippi River and its fate in the Gulf of Mexico. *Estuaries* 17:839–49, 1994.

Grguric, G., **J. H. Trefry,** and J. J. Keaffaber. Reactions of bromine and chlorine in ozonated artificial seawater systems. *Water Res.* 28:1087–94, 1994.

Feely, R. A., et al. **(J. H. Trefry).** Composition and sedimentation of hydrothermal plume particles from North Cleft segment, Juan de Fuca Ridge. *J. Geophys. Res.* 99:4985–5006, 1994.

Trefry, J. H., et al. Trace metals in hydrothermal solutions from cleft segment on the Southern Juan de Fuca Ridge, *J. Geophys. Res.* 99:4925–35, 1994.

Trefry, J. H., and S. Metz. Role of hydrothermal precipitates in the geochemical cycling of vanadium. *Nature* 342:531–3, 1989.

Trefry, J. H., S. Metz, R. P. Trocine, and T. A. Nelsen. A decline in lead transport by the Mississippi River. *Science* 230:439–41, 1985.

Frease, R. A., and **J. G. Windsor Jr.** Behavior of selected polycyclic aromatic hydrocarbons associated with a stabilized oil and coal ash reef. *Mar. Pollut. Bull.* 22:15–19, 1991.

Windsor, J. G., Jr. Fate and transport of oil, dispersants, and dispersed oil in the Florida coastal environment. Oil Spill Dispersant Research Program: Technical Advisory Group Workshop, University of Florida, Gainesville, Florida, April 25–26, 1991.

Windsor, J. G., Jr., and **L. E. Harris.** SEEAS—Science and Engineering Education at Sea. *MTS '90, Marine Technology Society,* Washington, D.C., September 1990.

Holm, S. E., and **J. G. Windsor Jr.** Exposure assessment of sewage treatment plant effluent by a selected chemical marker method. *Arch. Environ. Contam. Toxicol.* 19:674–79, 1990.

Windsor, J. G., Jr. Marine field projects: An established, unique undergraduate curriculum in ocean science. Presented at the 200th National Meeting of the American Chemical Society, August 26–31, 1990.

Zarillo, G. A., and T. S. Bacchus. Application of seismic profile measurements to sand source studies. In *Handbook of Geophysical Exploration at Sea,* 2nd ed., *Hard Minerals,* pp. 241–58, ed. R. A. Geyer. Boca Raton: CRC Press, 1992.

Zarillo, G. A., and J. Liu. Resolving components of the upper shoreface of a wave-dominated coast using empirical orthogonal functions. *Mar. Geol.* 82:169–86, 1988.

Zarillo, G. A., and M. J. Park. Prediction of sediment transport in a tide-dominated environment using a numerical model. *J. Coastal Res.* 3:429–44, 1987.

FLORIDA STATE UNIVERSITY

Department of Oceanography

Programs of Study

A graduate program in oceanography has existed at Florida State University since 1949, first in an interdisciplinary institute and since 1966 in a department within the College of Arts and Sciences. The Department of Oceanography, which offers both the M.S. and Ph.D. degrees in oceanography with specializations in physical, biological, or chemical oceanography, is the center for marine studies at the University. Additional marine and environmental research is conducted by the Departments of Biological Sciences, Chemistry, Geology, Mathematics, Meteorology, Physics, and Statistics, as well as the Geophysical Fluid Dynamics Institute and the Institute of Molecular Biophysics. Both formal and informal cooperative efforts between these science departments and the Department of Oceanography have flourished for years.

The M.S. degree program requires the completion of 33 semester hours of course work and a thesis covering an original research topic. Students pursuing the Ph.D. degree must complete 18 semester hours of formal course work beyond the master's degree course requirements and perform original research leading to a dissertation that makes a contribution to the science of oceanography.

The first year of graduate study is generally concerned with required course work and examinations. A supervisory committee, chosen by the individual student, directs the examinations and supervises the student's progress. Under its direction, the student begins thesis research as soon as possible. There is no foreign language requirement for either the M.S. or the Ph.D.

Research Facilities

Oceanography department headquarters, offices, and laboratories are located in the Oceanography-Statistics Building in the science area of the campus. Some of the laboratories currently in operation are for water quality analysis, organic geochemistry, trace-element analysis, radiochemistry, microbial ecology, phytoplankton ecology, numerical modeling, and fluid dynamics. The department also has the benefit of a fully equipped machine shop and a current-meter facility with state-of-the-art instruments. The Florida State University Marine Laboratory on the Gulf of Mexico is located at Turkey Point near Carrabelle, about 45 miles southwest of Tallahassee. The R/V *Bellows,* a 65-foot research vessel, is shared by FSU and other campuses of the State University System of Florida.

Departmental facilities are augmented by those in other FSU departments and institutes, such as the Van de Graaff accelerator in the physics department, the Antarctic Marine Geology Research Facility and Core Library in the geology department, the Geophysical Fluid Dynamics Institute laboratories, the Electron Microscopy Laboratory in the biological sciences department, the Statistical Consulting Center in the statistics department, and the IBM RS/6000 SP supercomputer in the School of Computational Science and Information Technology.

The research activities of the faculty members are heavily supported by federal funding, and these programs involve fieldwork, often at sea, all over the world. Faculty members and students have worked aboard a great many of the major research vessels of the U.S. fleet; because this kind of active collaboration works so well, it has not seemed necessary for the University to have its own major research vessel.

Financial Aid

Fellowships and teaching and research assistantships are available on a competitive basis. The Department of Oceanography fellowship carries a $21,500 stipend per calendar year. University fellowships pay $18,000 per academic year. Research and teaching assistantships range from $16,180 to $19,180 for the year, including summer. In addition, out-of-state tuition waivers are available for assistantship and fellowship recipients. Currently, most of the full-time students in the department receive financial assistance.

Cost of Study

Tuition for 2006–07 is $235.45 per credit hour for Florida residents; out-of-state students paid $866.85 per credit hour. The normal course load is 9 to 12 credit hours per semester for research and teaching assistants receiving tuition waivers.

Living and Housing Costs

A double room for single students in the graduate dormitories on campus costs about $400 per month (including utilities and local telephone). Married student housing at Alumni Village costs $345 to $589 per month. There are many off-campus apartment complexes in Tallahassee, with shared rents beginning at about $300 per month.

Student Group

The department currently has 49 full-time graduate students enrolled in the program (17 M.S., 32 Ph.D.). The students come from all areas of the country, with the greatest number representing the Northeast, South, and Midwest. Fifteen of the students are women. During the last five years, twenty-seven M.S. and twenty-three Ph.D. degrees were awarded in oceanography. Graduates have taken positions in federal and state agencies, universities, and private companies.

Location

Florida State University is located in Tallahassee, the state capital. Although it is among the nation's fastest-growing cities, Tallahassee has managed to preserve its natural beauty. The northern Florida location has a landscape and climate that are substantially different from those of southern Florida. Heavy forest covers much of the area, with the giant live oak being the chief tree of the clay hills. Five large lakes and the nearby Gulf of Mexico offer numerous recreational opportunities. Life in Tallahassee has been described as a combination of the ambience of traditional southern living with the bustle of a modern capital city.

The University

Florida State University is a public coeducational institution founded in 1851. Current enrollment is more than 38,000. The University has great diversity in its cultural offerings and is rich in traditions. It has outstanding science departments and excellent schools and departments in law, music, theater, and religion.

Applying

Applications should be submitted as early as possible in the academic year prior to anticipated enrollment. The deadline for applications for fall semester enrollment is mid-February. Each prospective candidate must have a bachelor's degree, with a major pertinent to the student's chosen specialty area in oceanography. Minimum undergraduate preparation must include one year of calculus, chemistry, and physics. A minimum undergraduate GPA of 3.0 and a minimum GRE General Test score of 1100 (combined verbal and quantitative scores) are required. The average undergraduate GPA and the average GRE score of currently enrolled students are 3.45 and 1230 (verbal and quantitative), respectively.

Correspondence and Information

Academic Coordinator
Department of Oceanography
Florida State University
Tallahassee, Florida 32306-4320

Phone: 850-644-6700
Fax: 850-644-2581
E-mail: admissions@ocean.fsu.edu
Web site: http://ocean.fsu.edu/

Florida State University

THE FACULTY AND THEIR RESEARCH

William C. Burnett, Professor of Oceanography; Ph.D., Hawaii. Uranium-series isotopes and geochemistry of authigenic minerals of the seafloor, elemental composition of suspended material from estuaries and the deep ocean, environmental studies.
Jeffrey P. Chanton, Professor of Oceanography; Ph.D., North Carolina at Chapel Hill. Major element cycling, light stable isotopes, methane production and transport, coastal biogeochemical processes.
Allan J. Clarke, Professor of Oceanography; Ph.D., Cambridge. Climate dynamics, coastal oceanography, equatorial dynamics, tides. Fellow, American Meteorological Society.
William K. Dewar, Professor of Oceanography; Ph.D., MIT. Gulf Stream ring dynamics, general circulation theory, intermediate- and large-scale interaction, mixed layer processes.
Thorsten Dittmar, Professor of Oceanography; Ph.D., Bremen, (West Germany). Marine biogeochemistry, molecular tracer techniques, major element cycling in coastal zones (mangroves) and polar oceans (Arctic Ocean, Antarctica).
Phillip Froelich, Francis Epps Professor of Oceanography; Ph.D., Rhode Island. Marine geochemistry, paleoceanography, paleoclimatology, global biogeochemical dynamics.
Markus H. Huettel, Professor of Oceanography; Ph.D., Kiel (Germany). Benthic ecology, biogeochemistry, transport mechanisms in sediments, effect of boundary layer flows on benthic processes.
Richard L. Iverson, Professor of Oceanography; Ph.D., Oregon State. Physiology and ecology of marine phytoplankton.
Joel E. Kostka, Associate Professor of Oceanography; Ph.D., Delaware. Microbial ecology and biogeochemistry, carbon and nutrient cycling in coastal marine environments, bacteria-mineral interactions.
Ruby E. Krishnamurti, Professor of Oceanography; Ph.D., UCLA. Ocean circulation, atmospheric convection, bioconvection, stability and transition to turbulence. Fellow, American Meteorological Society.
William M. Landing, Professor of Oceanography; Ph.D., California, Santa Cruz. Biogeochemistry of trace elements in the oceans, with emphasis on the effects of biological and inorganic processes on dissolved/particulate fractionation.
Nancy H. Marcus, Professor and Chair, Department of Oceanography; Ph.D., Yale. Population biology and genetics of marine zooplankton dormancy, photoperiodism, biological rhythms. Fellow, AAAS.
Doron Nof, Professor of Oceanography; Ph.D., Wisconsin–Madison. Fluid motions in the ocean, dynamics of equatorial outflows and formation of eddies, geostrophic adjustment in sea straits and estuaries, generation of oceanfronts. Fellow, American Meteorological Society.
Douglas P. Nowacek, Professor of Oceanography; Ph.D., MIT. Behavioral ecology and bioacoustics of marine mammals, including foraging behavior, response of marine mammals to anthropogenic noise, controlled exposure experiments or playbacks, and tagging.
James J. O'Brien, Professor of Oceanography; Ph.D., Texas A&M. Modeling of coastal upwelling and equatorial circulation, upper oceanfronts, climate scale fluctuations. Fellow, American Meteorological Society; Recipient, Sverdrup Gold Medal in Air-Sea Interactions.
Louis St. Laurent, Assistant Professor of Oceanography; Ph.D., MIT. Small-scale mixing processes and turbulence and double diffusion; buoyancy forcing and secondary circulation; internal waves and internal tides; flow over rough topography.
Kevin G. Speer, Professor of Oceanography; Ph.D., MIT. Deep-ocean circulation, observations and dynamics; water-mass formation; thermohaline flow; hydrothermal sources and circulation.
Melvin E. Stern, Professor of Oceanography; Ph.D., MIT. Theory of ocean circulation, salt fingers. Fellow, American Geophysical Union; Member, National Academy of Sciences.
David Thistle, Professor of Oceanography; Ph.D., California, San Diego (Scripps). Ecology of sediment communities, meiofauna ecology, deep-sea biology, crustacean systematics. Fellow, AAAS.
Georges L. Weatherly, Professor of Oceanography; Ph.D., Nova. Deep-ocean circulation and near-bottom currents.

Professors Emeritus
Ya Hsueh, Professor of Oceanography; Ph.D., Johns Hopkins. Variabilities in sea level, coastal currents, water density in continental shelf waters.
Wilton Sturges III, Professor of Oceanography; Ph.D., Johns Hopkins. Ocean currents.
John W. Winchester, Professor of Oceanography; Ph.D., MIT. Atmospheric chemistry, trace-element and aerosol-particle analysis.

AFFILIATED FACULTY
These faculty members are important in the academic program of students in the Department of Oceanography. This list includes faculty members from other departments who interact regularly with the department's faculty members and students.

Lawrence G. Abele, Professor of Biological Sciences and Provost; Ph.D., Miami (Florida). Ecology, community biology, systematics of decapod crustaceans.
David Balkwill, Professor of Medical Sciences; Ph.D., Penn State. Environmental microbiology.
Steven L. Blumsack, Associate Professor of Mathematics; Ph.D., MIT. Theory of rotating fluids.
William F. Herrnkind, Professor of Biological Sciences; Ph.D., Miami (Florida). Behavior and migration of marine animals.
Louis N. Howard, Professor of Mathematics; Ph.D., Princeton. Theory of rotating and stratified flows, hydrodynamic stability, bifurcation theory, geophysical fluid dynamics, chemical waves, biological oscillations. Member, National Academy of Sciences.
Christopher Hunter, Professor of Mathematics; Ph.D., Cambridge. Dynamics of fluids and stellar systems.
Charles L. Jordan, Professor of Meteorology; Ph.D., Chicago. Synoptic meteorology.
Michael Kasha, Professor of Chemistry, Institute of Molecular Biophysics; Ph.D., Berkeley. Molecular electronic spectroscopy, molecular quantum mechanics. Member, National Academy of Sciences.
Robley J. Light, Professor of Chemistry; Ph.D., Duke. Biosynthesis, metabolism, and structure of lipids and related compounds.
Robert J. Livingston, Professor of Biological Sciences; Ph.D., Miami (Florida). Estuarine ecology, aquatic pollution biology.
John K. Osmond, Professor of Geology; Ph.D., Wisconsin–Madison. Uranium-series isotopes.
Paul C. Ragland, Professor of Geology; Ph.D., Rice. Petrology, geochemistry.
William F. Tanner, Professor of Geology; Ph.D., Oklahoma. Fluid dynamics, paleoceanography, sedimentology, structural geology.
Thomas J. Vickers, Professor of Chemistry; Ph.D., Florida. Spectroscopic techniques for chemical analysis of trace elements relating to human health.
Sherwood W. Wise, Professor of Geology; Ph.D., Illinois. Micropaleontology, marine geology, diagenesis of pelagic sediments, biomineralization.

MASSACHUSETTS INSTITUTE OF TECHNOLOGY/ WOODS HOLE OCEANOGRAPHIC INSTITUTION

Joint Program in Oceanography/Applied Ocean Science and Engineering

Program of Study

The Massachusetts Institute of Technology (MIT) and the Woods Hole Oceanographic Institution (WHOI) offer joint doctoral and professional degrees in oceanography and in applied ocean science and engineering. The Joint Program leads to a single degree awarded by both institutions. Graduate study in oceanography encompasses virtually all of the basic sciences as they apply to the marine environment: physics, chemistry, geochemistry, geology, geophysics, and biology. Oceanographic engineering allows for concentration in the major engineering fields of civil, mechanical, electrical, and ocean engineering. The graduate programs are administered by joint MIT/WHOI committees drawn from the faculty and staff members of both institutions. The Joint Program involves several departments at MIT: Earth, Atmospheric, and Planetary Sciences and Biology in the School of Science and Civil and Environmental Engineering, Electrical Engineering and Computer Science, and Mechanical Engineering in the School of Engineering. WHOI departments are Physical Oceanography, Biology, Marine Chemistry and Geochemistry, Geology and Geophysics, and Applied Ocean Physics and Engineering. Upon admission, students register in the appropriate MIT department and at WHOI simultaneously and are assigned academic advisers at each institution. The usual steps to a doctoral degree are entering the program the summer preceding the first academic year and working in a laboratory at WHOI, following an individually designed program in preparation for a general (qualifying) examination to be taken before the third year, submitting a dissertation of significant original theoretical or experimental research, and conducting a public oral defense of the thesis. The guideline for time to achieve the doctoral degree is about five years from the bachelor's degree. Students entering with a master's degree in the field may need less time. Each student is expected to become familiar with the principal areas of oceanography in addition to demonstrating a thorough knowledge of at least one major field. Subjects, seminars, and opportunities for research participation are offered at both MIT and WHOI. Courses and seminars are supplemented by cross-registration privileges with Harvard, Brown, and the Boston University Marine Program. Students also have the opportunity to participate in oceanographic cruises during graduate study.

Research Facilities

A broad spectrum of equipment and facilities is available. The wide-ranging deep-sea research vessels at WHOI include *Oceanus, Atlantis,* and *Knorr.* In addition, the deep-diving submersible *ALVIN,* which is carried on *Atlantis,* is operated by WHOI, as are several smaller coastal vessels. Both MIT and WHOI utilize the latest developments in computer technology, from personal computers to large multiuser access systems. Videoconferencing between MIT and WHOI provides interactive transmission for classes, and a high-speed data link for research is provided. Broad-based engineering design and support shop facilities (machining, electronics) are available at both MIT and WHOI. There are more than twelve libraries at MIT containing more than 2 million volumes. Cooperative arrangements with other libraries in the Boston area provide students with access to substantial research collections. WHOI library facilities are shared with the Marine Biological Laboratory and are supplemented by collections of the Northeast Fisheries Center and the U.S. Geological Survey, all located in Woods Hole.

Financial Aid

Research assistantships are available to most entering graduate students in the Joint Program and are usually awarded on a full-year basis. Such awards, as well as a few special fellowships, cover full tuition and provide a stipend adjusted periodically to current living expenses. For the 2006–07 academic year, the stipend is approximately $25,450 per year.

Cost of Study

Because tuition and stipend are usually paid for students in good standing, the main costs to the student are for medical insurance, books, and supplies.

Living and Housing Costs

Place of residence is determined by the student's selected program of study and research interests. Graduate students traditionally live off campus at both MIT and WHOI, although there is some graduate housing at both campuses. Housing and living costs tend to be expensive in both the Cambridge and Woods Hole areas, although reasonable housing is available.

Student Group

There are 139 graduate students enrolled in the Joint Program. They are divided among the five disciplines as follows: physical oceanography (28), marine geology and geophysics (22), biological oceanography (35), chemical oceanography (23), and applied ocean science and engineering (31).

Student Outcomes

Graduates of the program are employed in a various number of areas. Fifty-four percent are employed in academic/universities, 11 percent are in civilian government, 15 percent are employed in the private sector, 9 percent are in the military service, and 11 percent are employed in a variety of other areas. Recent graduates are employed by Lamont Doherty Earth Observatory, Monterey Bay Aquarium Research Institute, George Mason University, USGS, Oregon State University, Shell, Exxon/Mobil, University of Hawaii, University of Chicago, Harvard University, Computer Motion, and Mote Marine Lab.

Location

MIT's 146-acre campus extends more than a mile along the Cambridge side of the Charles River, overlooking downtown Boston. Metropolitan Boston offers diverse recreational and cultural opportunities. WHOI is located on two campuses in the village of Woods Hole on Cape Cod, about 80 miles southeast of Boston, and offers excellent access to the sea.

The Institute and The Institution

MIT is an independent, coeducational, privately endowed university. It is broadly organized into five academic schools: Architecture and Planning, Engineering, Humanities and Social Sciences, Management, and Science. Within these schools there are twenty-two academic departments. Total enrollment is approximately 9,800 divided almost evenly between undergraduate and graduate students. The MIT faculty numbers approximately 1,000 with a total teaching staff of 1,940. WHOI is the largest independent, unaffiliated oceanographic institution and research fleet operator in the world. There is a staff of approximately 900 scientists, engineers, technicians, research vessel crews, and support personnel organized into five research and academic departments, two centers, and four ocean institutes.

Applying

Application for admission to the Joint Program is made on the MIT graduate school application forms. Complete application files, including college transcripts, three letters of recommendation, test scores, and the application fee should be filed no later than January 15 for admission beginning in June or September. The General Test and one Subject Test of the GRE are required. The TOEFL is required of all international students whose schooling has not been predominantly in English.

Admission is offered on a competitive basis to those who appear most likely to benefit from the Joint Program. Notification of admission decisions are sent out in early March.

Correspondence and Information

Academic Programs Office
Woods Hole Oceanographic Institution MS #31
360 Wood Hole Road, Clark 223
Woods Hole, Massachusetts 02543-1546
Phone: 508-289-2219
Web site: http://web.mit.edu/mit-whoi/www

MIT/WHOI Joint Program Office
Room 54-911
Massachusetts Institute of Technology
77 Massachusetts Avenue
Cambridge, Massachusetts 02139
Phone: 617-253-7544
Web site: http://web.mit.edu/mit-whoi/www

Massachusetts Institute of Technology / Woods Hole Oceanographic Institution

THE DEPARTMENTS AND THEIR RESEARCH

Oceanography and applied ocean science and engineering are fields that naturally lead to interdisciplinary research. In the following departmental program sections, brief descriptions of the research areas covered by the faculty members are given. It is quite common for students to pursue research problems that cross the disciplinary lines of the given departments.

BIOLOGICAL OCEANOGRAPHY

Patrick Jaillet, Ph.D., Head, Civil and Environmental Engineering Department, MIT.
Chris Kaiser, Ph.D., Head, Biology Department, MIT.
Judith E. McDowell, Ph.D., Chair, Biology Department, WHOI.

Phytoplankton and zooplankton ecology; regulation of primary and secondary production; population biology; natural history and biology of oceanic fishes; comparative physiology; biochemical toxicology in marine species; biochemical and physiological adaptations; toxic algae and red tides; theoretical and experimental population ecology; estuarine and salt marsh ecology; ecology of deep and coastal benthos; microbial ecology and biochemistry; development and reproductive biology of marine invertebrates; larval dispersal mechanisms; behavior of marine mammals; symbiotic relationships; biogeochemistry of aquatic systems; biodegradation of aquatic contaminants; cell biology; molecular biology and evolution; synthesis, shape, and structure of macromolecules; cellular and molecular immunology; gene expression.

CHEMICAL OCEANOGRAPHY

Patrick Jaillet, Ph.D., Head, Civil and Environmental Engineering Department, MIT.
Maria Zuber, Ph.D., Head, Department of Earth, Atmospheric, and Planetary Sciences, MIT.
Ken O. Buesseler, Ph.D., Chair, Marine Chemistry and Geochemistry Department, WHOI.

Water columns (open and coastal oceans, estuaries, rivers): organic and inorganic cycles of particulate and dissolved carbon, oxygen, nitrogen, phosphorous, sulfur, and trace metals (redox transformations, rare earths); stable and radioisotopic tracers; noble gases; air-sea exchange; remote sensing and modeling; environmental quality; oil and gas geochemistry; colloids and particle-reactive tracers; weathering. Sedimentary geochemistry: major and minor elements, radionuclides and their paleoceanographic applications, diagenesis and preservation of organic matter, modeling.

Seawater-basalt interactions: major and trace elements, stable isotopes, solid phases and hydrothermal solutions, laboratory experiments, modeling.

MARINE GEOLOGY AND GEOPHYSICS

Maria Zuber, Ph.D., Head, Department of Earth, Atmospheric, and Planetary Sciences, MIT.
Susan E. Humphris, Ph.D., Chair, Geology and Geophysics Department, WHOI.

Micropaleontological biostratigraphy, planktonic and benthonic foraminifera, paleoceanography, paleobiogeography, benthic boundary-layer processes, paleocirculation and paleoecology, igneous petrology and volcanic processes, crustal structure and tectonics, marine magnetic anomalies, heat flow of the ocean floor, upper mantle petrology, seismic stratigraphy, fractionation processes of stable isotopes and stable isotope stratigraphy, metamorphosis of high-strain zones, gravity, observational and theoretical reflection and refraction seismology, earthquake seismology, relative and absolute plate motions, coastal processes, marine sedimentation.

OCEANOGRAPHIC ENGINEERING

Patrick Jaillet, Ph.D., Head, Civil and Environmental Engineering Department, MIT.
John V. Guttag, Ph.D., Head, Electrical Engineering and Computer Sciences Department, MIT.
Rohan Abeyaratne, Ph.D., Head, Mechanical Engineering Department, MIT.
James F. Lynch, Ph.D., Chairman, Applied Ocean Physics and Engineering Department, WHOI.

Optical instrumentation, laser velocimetry; volcanic, tectonic, and hydrothermal processes; deep submergence systems (imaging, control, robotics); underwater acoustics (acoustic tomography, scattering, remote sensing, Arctic acoustics, bottom acoustics propagation through the ocean interior and sediments, array design); buoy and mooring engineering; ocean instrumentation; signal processing theory; fluid dynamics, sediment transport, nearshore processes, bottom boundary layer and mixed layer dynamics; turbulence, wave prediction, numerical modeling; seismic profiling; data acquisition and communication systems, microprocessor-based instrumentation; fiber optics; sonar systems; marsh ecology; ship design; offshore structures; material science; groundwater flow.

PHYSICAL OCEANOGRAPHY

Maria Zuber, Ph.D., Head, Department of Earth, Atmospheric, and Planetary Sciences, MIT.
Nelson Hogg, Ph.D., Chair, Physical Oceanography Department, WHOI.
General circulation: distribution of tracer fields, models of idealized gyres, abyssal circulation, heat transport.
Air-sea interaction: water mass transportation, upper ocean response to atmospheric forcing, equatorial ocean circulation.
Shelf dynamics: coastal upwelling and fronts, coastal-trapped waves, deep ocean-shelf exchange.
Mesoscale processes: Gulf Stream Rings, oceanic fronts, barotropic and baroclinic instability, eddy-mean interactions.
Small-scale processes: double diffusion, intrusion, internal waves, convection.

NOVA SOUTHEASTERN UNIVERSITY

Oceanographic Center
Programs in Marine and Coastal Studies

Programs of Study

The Oceanographic Center through the Institute of Marine and Coastal Studies offers the M.S. degree in coastal zone management, marine biology, marine environmental sciences, and physical oceanography; joint M.S. degrees in the aforementioned areas (e.g., marine biology/coastal zone management); and the Ph.D. degree in oceanography/marine biology. The M.S. in coastal zone management is also offered in a distance learning format with a Capstone Review Track (as listed below). The M.S. and Ph.D. programs contain a common core of five courses encompassing the major disciplines of oceanography. Specialty and tutorial courses in each program provide depth. The Oceanographic Center operates on a quarter-term system with twelve-week courses.

Classes for the M.S. programs meet one evening per week in a 3-hour session. Capstone Review and Thesis tracks are offered. The Capstone Review Track requires a minimum of 45 credits, which includes thirteen 3-credit courses and a 6-credit paper. The paper is usually an extended literature review of an approved subject, which the student defends before the Advisory Committee. The Thesis Track requires a minimum of 39 credits, including ten 3-credit courses and at least 9 credits of master's thesis research. The number of research credits depends upon the time needed to complete the thesis research. The thesis is formally defended before the committee. All students admitted to the program are placed in the Capstone Review Track. To enter the Thesis Track, students must have approval of the major professor and complete an approved thesis proposal.

The joint M.S. degree ecompasses two of the following: marine biology/coastal zone management, marine environmental science, and physical oceanography. It requires a minimum of 51 to 57 credits, depending upon the student's track: Capstone Review or Thesis.

The Ph.D. program consists of upper-level course work and original research on a selected topic of importance in the ocean sciences. Requirements include general core courses, electives, and tutorial studies with the major professor. The Ph.D. degree requires a minimum of 90 credits beyond the baccalaureate; at least 48 credits must consist of dissertation research and at least 42 credits must consist of upper-level course work. The student must successfully complete the Ph.D. comprehensive examination and defend the dissertation before Oceanographic Center faculty members. Students are expected to complete the Ph.D. program in nine years or less, a minimum of three years of which must be in residence.

The Oceanographic Center also offers a distance learning graduate certificate in coastal studies. Enrollment in the certificate program is designed for those who do not wish to enroll in the full M.S. graduate program of study at this time. The flexible format of the certificate program makes it ideal for working professionals and college graduates in a variety of fields related to the coastal zone. Distance courses bring the learning to the student, whether online, by CD-ROM, or through written materials. The graduate certificate in coastal studies is awarded upon successful completion of four of the Oceanographic Center's graduate distance learning courses. Successful completion of the graduate certificate program awards the equivalent of 12 graduate credits. Pending full graduate acceptance, these credits may be applied towards the full online M.S. in coastal zone management. Any of the Center's distance learning courses may also be taken individually at an undergraduate or general interest (audit) level.

In conjunction with other schools at Nova Southeastern University (NSU), the Center also offers a one-year distance M.S. in environmental education, and a distance M.A. in cross-disciplinary studies with specialization in environment.

Research Facilities

The Center is composed of three main buildings, several modulars, and one houseboat. The two-story houseboat contains a student center and ten student offices. The main buildings contain a conference room, two classrooms with digital projection capability, a warehouse bay staging area, an electron microscopy laboratory, a darkroom, a machine shop, a carpentry shop, an electronics laboratory, a computer center with student microlabs, a wetlab/classroom, a coral workshop, a filtered seawater facility, eight working biology laboratories, and twenty-four additional offices.

The William Springer Richardson Library contains 3,000 books as well as 100 active and thirty-three inactive periodicals. There is also a large selection of newsletters and magazines. A joint-use library facility is maintained on the main campus in Davie, Florida.

The computer department consists of two student microlabs with Windows PCs. Also available are networked HP laser printers, a networked color laser printer, flatbed scanners with imaging software, and a large format color poster printer. The Center also operates a LAN consisting of approximately 130 PCs for student and faculty and staff member use that is connected to the Internet via a T-1 link. Wireless internet connectivity is available everywhere on campus grounds.

Financial Aid

There is limited financial aid available in the form of undergraduate laboratory teaching assistantships and graduate research assistantships. The Office of Student Financial Aid helps students finance tuition, fees, books, and other costs, drawing on a variety of public and private aid programs. For more information, students should call 800-541-6682 Ext. 7411 (toll-free).

Cost of Study

In 2005–06, tuition costs were $595 per credit hour for students enrolled in the M.S. programs and $4074 per term for students enrolled in the Ph.D. program.

Living and Housing Costs

For housing information, students should call 800-541-6682 Ext. 7052 (toll-free). Numerous apartments, condominiums, and other rental housing are available in Hollywood, Dania Beach, and Ft. Lauderdale.

Student Group

There are 150 students enrolled in the M.S. programs and 8 students enrolled in the Ph.D. program.

Student Outcomes

M.S. graduates find positions in city, county, and state governments or private industry, including consulting companies. Graduates also go on for further education and enter Ph.D. programs.

Location

The Center is located in Dania Beach, Florida, just south of Ft. Lauderdale, on a 10-acre site on the ocean side of Port Everglades and is easily accessible from I-95 and the Ft. Lauderdale airport. The Center has a 1-acre boat basin, and its location affords immediate access to the Gulf Stream and the open sea, the Florida Straits, and the Bahamas Banks.

The University

Nova Southeastern University was chartered by the State of Florida in 1964 and currently, with nearly 23,000 students, is the largest independent university in Florida. The main campus is located on 227 acres in Davie, Florida, near Ft. Lauderdale.

Applying

When applying, students must submit an application form, application fee, transcripts from other schools attended, GRE scores, and letters of recommendation. Applicants interested in the M.S. program in marine biology should hold a bachelor's degree in biology, oceanography, or a closely related field, including science education. Due to the discipline's diversity, applicants with any undergraduate major are considered for admission into the M.S. program in coastal zone management or marine environmental science. However, a science major is most useful, and a science background is essential. For the M.S. in physical oceanography degree in computer science, engineering, mathematics, and physics is appropriate. Applicants for the distance graduate certificate must supply a copy of their undergraduate degree, in any major.

Correspondence and Information

Oceanographic Center
Institute of Marine and Coastal Studies
Nova Southeastern University
8000 North Ocean Drive

Phone: 954-262-3600
Fax: 954-262-4020
E-mail: imcs@nsu.nova.edu
Web site: http://www.nova.edu/ocean/

Nova Southeastern University

THE FACULTY AND THEIR RESEARCH

The Oceanographic Center pursues studies and investigations in biological, observational, and theoretical oceanography. Research interests include modeling of large-scale ocean circulation, coastal dynamics, ocean-atmosphere coupling, surface gravity waves, biological oceanography, chemical oceanography, coral reef assessment, Pleistocene and Holocene sea level changes, benthic ecology, marine biodiversity, calcification of invertebrates, marine fisheries, molecular ecology and evolution, wetlands ecology, marine microbiology, and nutrient dynamics. Regions of interest include not only Florida's coastal waters and the continental shelf/slope waters of the southeastern United States, but also the waters of the Caribbean Sea, the Gulf of Mexico, and the Antarctic, Atlantic, Indian, and Pacific oceans.

Professors
Richard E. Dodge: Coral reefs and reef-building corals, effects of pollution and past climatic changes.
Charles Messing: Systematics of crinoids and macroinvertebrate communities.
Andrew Rogerson: Ecology of eukaryotic microbes (the protists) in the cycling of carbon and nutrients in coastal waters, particularly the amoeboid protozoa.
Richard Spieler: Fish chronobiology, artificial reefs, and habitat assessment.
James Thomas: Marine biodiversity, invertebrate systematics.

Associate Professors
Patricia Blackwelder: Calcification and distribution of marine microfauna, a historical record of the past.
Curtis Burney: Dissolved nutrients and marine microbes, especially bacteria.
Veljko Dragojlovic: Isolation, characterization, and synthesis of natural products.
Joshua Feingold: Coral reef ecology.
Edward Keith: Structure, function, and evolution of milk and tear proteins; physiological ecology of terrestrial and marine mammals; molecular phylogenetics and evolution of marine mammals.
Bernhard Riegl: Coral reefs, spatial ecology, remote sensing, hydrographic surveying.
Mahmood Shivji: Conservation biology, biodiversity, evolution, molecular ecology, and population biology.
Alexander Soloviev: Measurement and modeling of near-surface turbulence and air/sea exchange.

Assistant Professors
Sean Kennan: Physical oceanography, large and mesoscale ocean circulation, tropical and equatorial dynamics, physical forcing of marine ecosystems.
Alexander Yankovsky: Wind- and buoyancy-driven currents on the continental shelf and slope, their mesoscale variability, and adjustment to realistic shelf topography.

Adjunct Professors
Barry Barker: Environmental studies and conservation of natural resources.
Brion Blackwelder: Coastal law.
Gregory Booten: Molecular genetics.
Nancy Craig: Coastal marine systems.
Jane Dougan: Distance education coordinator.
O. P. Dwivedi: Cross-cultural, scientific, and spiritual perspectives of water.
Randy Edwards: Marine and estuarine ecology and fish ecology.
Ruth Ewing: Diseases in marine mammals.
Mark Farber: Statistical analysis.
Nancy Gassman: Marine biology, coastal zone management, and marine environmental sciences internships.
David Gilliam: Coral reef assessment.
Patrick Hardigan: Statistics.
Vladimir Kosmynin: Geomorphology and ecology of coral reefs.
Phillip Light: Fish assemblage structure.
Frank Mazzotti: Conservation planning, landscape ecology, impacts of human activities on fish and wildlife resources.
Donald McCorquodale: Water pollution indicators and testing.
Esther Peters: Aquatic ecotoxicology.
Brian Polkinghorn: Conflict resolution.
Sam Purkis: Remote sensing techniques to monitor coral reef systems.
Keith Ronald: Marine mammals.
Scott Schatz: Fungal and protozoan pathogens.
Steffen Schmidt: Coastal policy, international integrated coastal zone management.
Michael Stanhope: Evolutionary ecology, molecular population genetics.
Bernardo Vargas-Angel: Coral reef ecology.
William Venezia: Ocean engineering.
Alan Watson: Nature interpretation.
Brad Weatherbee: Marine environmental physiology and fish ecology.
Scott White: Eco-terrorism.

The Oceanographic Center.

Rutgers

RUTGERS, THE STATE UNIVERSITY OF NEW JERSEY, NEW BRUNSWICK/PISCATAWAY

Institute of Marine and Coastal Sciences
Graduate Program in Oceanography

Programs of Study

The Graduate Program in Oceanography at the Institute of Marine and Coastal Sciences is dedicated to an interdisciplinary course of study that encompasses all aspects of oceanography, including the biology, chemistry, physics, and geology of the oceans. The basic goal of the program is to provide a graduate student with a rigorous, quantitative understanding of ocean processes; an environment that encourages critical thinking; and opportunities for hands-on experience in both the laboratory and the field. A broad range of research opportunities are available, including real-time studies of the coastal ocean using advanced underwater instrumentation, molecular genetics and evolution of marine organisms, biogeochemical cycles, fisheries and fish behavior, coastal geomorphology, organism-sediment interactions, and estuarine processes. The course of study is tailored to meet the needs of the individual student.

Candidates with a baccalaureate degree may apply for either the Doctor of Philosophy (Ph.D.) or Master of Science (M.S.) degree program. The Ph.D. degree requires the completion of 72 credit hours of course work and research beyond the baccalaureate degree and the writing and defense of a dissertation resulting from the candidate's independent, original research in oceanography. The M.S. degree requires the completion of 30 credit hours of course work and research beyond the baccalaureate degree and the writing and defense of a thesis. All students are required to complete a program of core courses in oceanography; additional courses are chosen by students in consultation with their major professors and program committees. Ph.D. students must pass a written and oral qualifying examination upon completion of their course work.

Research Facilities

The Institute of Marine and Coastal Sciences is housed in a modern research building that includes flow-through seawater, analytical chemistry, remote sensing, and ocean modeling laboratories. Major equipment includes seawater annular and racetrack flumes; two satellite receiving stations; a network of small (PCs), medium (UNIX-based workstations), and large (multiprocessor and Beowulf-type) computer platforms; a variety of mass spectrometers; a state-of-the-art molecular biological lab; and coral culturing facilities. The Rutgers University Marine Field Station, located at the northern entrance to Great Bay, is the site of a large tract of pristine marsh and a major estuary that retains most of its natural characteristics. Great Bay connects with adjoining bays and has direct access to the Atlantic Ocean. An extensive program of long-term oceanographic and ecosystems research is under way at the station. The Haskin Shellfish Research Laboratory, located on the Delaware Bay, includes facilities for research on fisheries, aquaculture, biology, and ecology. Wet and dry laboratories and a research hatchery support molecular genetics, cytogenetics, microbiology, cell culture, histopathology, population/fishery modeling, animal husbandry, phytoplankton culture, physiology, and general ecology.

Financial Aid

Graduate assistantships are available from sponsored research grants and contracts awarded to the faculty. In addition, a limited number of state-supported teaching assistantships and fellowships are available each year. All assistantships and fellowships include a stipend of $19,367 per calendar year and full tuition remission. Virtually all full-time students who are accepted receive financial aid.

Cost of Study

Tuition for 2005–06, for a full course load of 12 or more credits, was $5220 per semester for New Jersey residents and $7760.40 for out-of-state residents.

Living and Housing Costs

Graduate students traditionally live off campus. University housing is also available in dormitory/apartment-style accommodations that cost $6084 per academic year. A variety of meal plans that average $3460 per academic year are also available.

Student Group

Currently, there are 30 full-time students in the program, 16 women and 14 men. Twenty-five are pursuing the Ph.D. degree. Due to the highly interdisciplinary nature of much of the research conducted at the Institute, students share a unique rapport with each other and faculty members.

Student Outcomes

The graduate program was established in 1994 and has granted sixteen Ph.D. degrees and eighteen M.S. degrees; several M.S. students continued their studies toward a Ph.D. at Rutgers.

Location

The Institute of Marine and Coastal Sciences is located on Rutgers University's Cook Campus in New Brunswick, New Jersey. A wealth of cultural and recreational opportunities are nearby. Several accomplished repertory companies are housed in New Brunswick and nearby Princeton. The major metropolitan areas of New York City and Philadelphia are only short (1 hour) train rides away. The world-famous New Jersey shore, with its beaches, swimming, and fishing, is readily accessible.

The University and The Institute

Rutgers, The State University of New Jersey, with 50,000 undergraduate and graduate students, traces its origins back to 1766 when it was chartered as Queen's College, the eighth institution of higher learning founded in the colonies. The Institute of Marine and Coastal Sciences was established in 1989 to develop research programs in marine and coastal sciences and to provide a center for the education of marine scientists. The main building is located on the Cook College Campus in New Brunswick, with the Rutgers Marine Field Station in Tuckerton, the Pinelands Research Station in New Lisbon, and the Haskin Shellfish Research Laboratory in Port Norris.

Applying

Applicants to the program are expected to have an undergraduate degree in either mathematics, science, or engineering. Scores on the GRE General Test are required. International students must show proficiency in English. Application deadlines are February 1 for admission the following fall semester and November 1 for admission the following spring semester. Early submission is encouraged, especially for students seeking financial aid. Applicants are strongly encouraged to contact faculty members with complementary interests before and during the application process.

Correspondence and Information

Graduate Program in Oceanography
Institute of Marine and Coastal Sciences
Rutgers, The State University of New Jersey
71 Dudley Road
New Brunswick, New Jersey 08901-8521
Phone: 732-932-6555 Ext. 500
Fax: 732-932-8578
E-mail: gpo@imcs.rutgers.edu
Web site: http://marine.rutgers.edu/gpo/GradProg.html

Rutgers, The State University of New Jersey, New Brunswick/Piscataway

THE FACULTY AND THEIR RESEARCH

Kenneth W. Able, Professor; Ph.D., William and Mary. Life history, ecology, and behavior of fishes.
James W. Ammerman, Associate Research Professor; Ph.D., California, San Diego (Scripps). Aquatic microbial ecology and biogeochemistry, phosphorus cycling, microbial enzymes, development of automated assay methods.
Gail M. Ashley, Professor; Ph.D., British Columbia. Sedimentology, geomorphology, environmental ecology, modern processes.
Karen G. Bemis, Research Associate; Ph.D., Rutgers. Marine hydrothermal systems, plume behavior, volcanology, visualization.
Kay D. Bidle, Assistant Professor; Ph.D., California, San Diego (Scripps). Molecular evolution and ecology, marine microbial ecology, biogeochemistry.
David Bushek, Assistant Professor; Ph.D., Rutgers. Shellfish ecology, aquaculture, and host-parasite interactions.
Robert J. Chant, Assistant Professor; Ph.D., SUNY at Stony Brook. Observations and numerical modeling of estuarine and coastal processes.
Colomban de Vargas, Assistant Professor; Ph.D., Geneva (Switzerland). Molecular ecology and evolution of unicellular organisms in the ocean.
Richard H. Dunk, Adjunct Professor; Ph.D., Rutgers. Meteorology, air-sea interactions, sea breezes.
Paul G. Falkowski, Professor; Ph.D., British Columbia. Biological oceanography: photosynthesis and biogeochemical cycles, application of molecular and biophysical techniques to the marine environment.
Katja Fennel, Assistant Professor; Ph.D., Rostock (Germany). Ecological and biogeochemical modeling, data assimilation.
Jennifer A. Francis, Associate Research Professor; Ph.D., Washington (Seattle). Satellite remote sensing of polar regions, air/ice/ocean transfer, Arctic climate and polar meteorology.
Scott M. Glenn, Professor; Sc.D., MIT/Woods Hole Oceanographic Institution. Physical oceanography, satellite remote sensing.
J. Frederick Grassle, Professor; Ph.D., Duke. Marine ecology, oceanography.
Judith P. Grassle, Professor; Ph.D., Duke. Population genetics, marine benthic ecology.
Thomas M. Grothues, Assistant Research Professor; Ph.D., SUNY at Stony Brook. Fish recruitment, dispersal, migration, and habitat use.
Ximing Guo, Associate Professor; Ph.D., Washington (Seattle). Molluscan genetics and genomics, aquaculture.
Dale B. Haidvogel, Professor; Ph.D., MIT/Woods Hole Oceanographic Institution. Physical oceanography, numerical ocean circulation modeling.
Michael J. Kennish, Associate Research Professor; Ph.D., Rutgers. Marine geology, estuarine and marine ecology, marine pollution.
Lee J. Kerkhof, Associate Professor; Ph.D., California, San Diego (Scripps). Marine microbiology–molecular biology, microbial population dynamics.
Julia C. Levin, Assistant Research Professor; Ph.D., Columbia. Ocean modeling, data assimilation, computational fluid dynamics.
Richard A. Lutz, Professor; Ph.D., Maine. Marine ecology and paleoecology, shellfish ecology, biology of deep-sea hydrothermal vents.
George R. McGhee, Professor; Ph.D., Rochester. Marine paleoecology, evolutionary theory, theoretical morphology, mass extinction.
James R. Miller, Professor; Ph.D., Maryland. Air-sea interactions, remote sensing, climate modeling, earth system science.
Kenneth G. Miller Sr., Professor; Ph.D., MIT/Woods Hole Oceanographic Institution. Cenozoic stratigraphy and paleoceanography; integrated biostratigraphy, isotope stratigraphy, magnetostratigraphy, and seismic stratigraphy.
Michael R. Muller, Associate Professor; Ph.D., Brown. Fluid mechanics, internal gravity waves and thermals.
Karl F. Nordstrom, Professor; Ph.D., Rutgers. Geomorphology, sedimentology.
Richard K. Olsson, Professor Emeritus of Geological Sciences; Ph.D., Princeton. Foraminiferal paleoecology and paleobathymetry, planktonic foraminiferal biostratigraphy and phylogeny of K/T boundary, sequence stratigraphy of passive margins.
Eric N. Powell, Professor; Ph.D., North Carolina. Shellfish biology/modeling, carbonate preservation, reproductive biology, fisheries management.
Norbert P. Psuty, Professor Emeritus; Ph.D., LSU. Coastal geomorphology, coastal dune evolution, sea-level rise, coastal zone management.
John A. Quinlan, Assistant Professor; Ph.D., North Carolina at Chapel Hill. Fisheries oceanography, management, biophysical interactions, modeling, acoustics.
John R. Reinfelder, Associate Professor; Ph.D., SUNY at Stony Brook. Trace-element cycling and bioavailability, inorganic carbon acquisition in marine phytoplankton.
Peter A. Rona, Professor; Ph.D., Yale. Seafloor hydrothermal systems, ocean ridge processes, geology of Atlantic continental margins, genesis of seafloor mineral and energy resources.
Yair Rosenthal, Associate Professor; Ph.D., MIT/Woods Hole Oceanographic Institution. Quaternary paleoceanography, trace metal and isotope biogeochemistry, estuarine and coastal geochemistry.
Oscar M. E. Schofield, Associate Professor; Ph.D., California, Santa Barbara. Marine phytoplankton ecology, hydrological optics and remote sensing, integrated ocean observatories.
Sybil P. Seitzinger, Visiting Professor; Ph.D., Rhode Island. Nutrient dynamics in marine, freshwater, and terrestrial ecosystems.
Robert E. Sheridan, Professor; Ph.D., Columbia. Geology and geophysics of the Atlantic continental margin.
Robert M. Sherrell, Associate Professor; Ph.D., MIT/Woods Hole Oceanographic Institution. Trace metals in the oceanic water column, metal-biota interactions, environmental chemistry in the present and past.
Elisabeth L. Sikes, Associate Research Professor; Ph.D., MIT/Woods Hole Oceanographic Institution. Paleoceanography, marine organic geochemistry.
Peter Smouse, Professor; Ph.D., North Carolina State. Population genetics of marine and terrestrial organisms.
Gary L. Taghon, Associate Professor; Ph.D., Washington (Seattle). Marine benthic ecology.
Christopher G. Uchrin, Professor; Ph.D., Michigan. Mathematical modeling of contaminant transport in surface and ground waters.
Dana E. Veron, Assistant Professor; Ph.D., California, San Diego (Scripps). Cloud-aerosol-radiation interactions, climate modeling, remote sensing.
Costantino Vetriani, Assistant Professor; Ph.D., Rome (Italy). Deep-sea microbiology, thermophiles, microbial adaptations to extreme environments.
Michael P. Weinstein, Visiting Professor; Ph.D., Florida State. Coastal ecology, habitat utilization (nekton) secondary production, restoration ecology, ecological engineering.
John L. Wilkin, Assistant Professor; Ph.D., MIT/Woods Hole Oceanographic Institution. Physical oceanography, coastal dynamics, coupled physical/biological modeling.

Rutgers, The State University of New Jersey, New Brunswick/Piscataway

SELECTED PUBLICATIONS

Able, K. W., D. Nemerson, and **T. M. Grothues**. Evaluating salt marsh restoration in Delaware Bay: Continued analysis of fish response at former salt hay farms. *Estuaries* 27(1):58–69, 2004.

Currin, C. A., et al. **(K. W. Able** and **M. P. Weinstein).** Determination of food web support and trophic position of the mummichog, *Fundulus heteroclitus,* in New Jersey smooth cordgrass *(Spartina alterniflora),* common reed *(Phragmites australis),* and restored salt marshes. *Estuaries* 26(2B):495–510, 2003.

Minello, T. J., **K. W. Able, M. P. Weinstein,** and C. Hays. Salt marsh nurseries for nekton: Testing hypotheses on density, growth and survival through meta-analysis. *Mar. Ecol. Prog. Ser.* 246:39–59, 2003.

Able, K. W. Measures of juvenile fish habitat quality: Examples from a national estuarine research reserve. In *Fish Habitat: Essential Fish Habitat and Rehabilitation,* pp. 134–47, ed. L. R. Benaka. American Fisheries Society Symp. 22, Bethesda, Maryland, 1999.

Able, K. W., and M. P. Fahay. *The First Year in the Life of Estuarine Fishes in the Middle Atlantic Bight.* New Brunswick, N. J.: Rutgers University Press, 1998.

Lomas, M. W., A. Swain, R. Shelton, and **J. W. Ammerman.** Taxonomic variability of phosphorus stress in Sargasso Sea phytoplankton. *Limnol. Oceanogr.* 49:2303–10, 2004.

Ammerman, J. W. Phosphorus cycling in aquatic environments: Role of bacteria. In *The Encyclopedia of Environmental Microbiology,* vol. 5, pp. 2448–53, ed. G. Bitton. John Wiley & Sons, Inc., 2002.

Ashley, G. M., and N. D. Smith. Marine sedimentation at a subpolar calving ice margin, Antarctic, Peninsula. *Geo. Soc. Am. Bull.* 112(5):657–67, 2000.

Smith, N. D., and **G. M. Ashley.** A study of brash ice in the proximal marine zone of a sub-polar tidewater glacier. *Mar. Geol.* 133:75–87, 1996.

Santilli, K., et al. **(K. Bemis** and **P. Rona).** Generating realistic images from hydrothermal plume data. *IEEE Visualization 2004 Proc.* 91–8, 2004.

Bemis, K. G., and **P. A. Rona** et al. A comparison of black smoker hydrothermal plume behavior at Monolith Vent and at Clam Acres vent field: Dependence on source configuration. *Mar. Geophys. Res.* 23:81–96, 2002.

Bidle, K. D., and **P. G. Falkowski.** Cell death in planktonic, photosynthetic microorganisms. *Nat. Rev. Microbiol.* 2:643–55, 2004.

Bidle, K. D., et al. Diminished efficiency of the oceanic silica pump by bacterially-mediated silica dissolution. *Limnol. Oceanogr.* 48:1855–68, 2003.

Ford, S. E., M. M. Chintala, and **D. Bushek.** Comparison of in vitro cultured and wild-type *Perkinsus marinus* I. Pathogen virulence. *Dis. Aquat. Org.* 51:187–201, 2002.

Dame, R., and **D. Bushek** et al. Ecosystem response to bivalve density reduction: Management implications. *Aquat. Ecol.* 36:51–65, 2002.

Chant, R. J., S. Glenn, and J. Kohut. Flow reversals during upwelling conditions on the New Jersey inner shelf. *J. Geophys. Res.* 109:C12, C12S03, doi:10.1029/2003JC001941, 2004.

Chant, R. J. Secondary flows in a region of flow curvature: Relationship with tidal forcing and river discharge. *J. Geophys. Res.* doi: 10.1029/2001JC001082, 2002.

de Vargas, C., A. Garcia-Saez, L. K. Medlin, and H. Thierstein. Super-species in the calcareous plankton. In *Coccolithophores: From Molecular Processes to Global Impact,* pp. 271–98, eds. H. R. Thierstein and J. R. Young. New York: Springer-Verlag, in press.

de Vargas, C., et al. Molecular evidence of cryptic speciation in planktonic foraminifera and their relation to oceanic provinces. *Proc. Natl. Acad. Sci. U.S.A.* 96:2864–8, 1999.

Ratcliff, M., R. Petersen, **R. Dunk,** and J. DeToro. Comparison of wind tunnel and ISDM model simulations of sea breeze fumigations. *Annual AWMA Meeting,* Atlanta, Georgia, 1996.

Peterson, R., B. Cochran, and **R. Dunk.** Wind tunnel determined building heights for modeling combustion turbines with ISC. *74th AMS Meeting, 8th Conference on the Applications of Air Pollution Meteorology,* Nashville, Tennessee, 1994.

Falkowski, P. G. The ocean's invisible forest. *Sci. Am.* 287:38–45, 2002.

Falkowski, P. G., and J. Raven. *Aquatic Photosynthesis,* p. 375. Oxford: Blackwell, 1997.

Fennel, K., M. Follows, and **P. G. Falkowski.** The coevolution of the nitrogen, carbon, and oxygen cycles in the proterozoic ocean. *Am. J. Sci.* 305:526–45, 2005.

Fennel, K., and E. Boss. Subsurface maxima of phytoplankton and chlorophyll: Steady-state solutions from a simple model. *Limnol. Oceanogr.* 48(4):1521–34, 2003.

Fennel, K., et al. Impacts of iron control on phytoplankton production in the modern and glacial Southern Ocean. *Deep-Sea Res. II* 50:833–51, 2003.

Serreze, M. C., and **J. A. Francis**. The Arctic amplification debate. *Climate Dynamics,* in press.

Francis, J. A., E. Hunter, J. Key, and X. Wang. Clues to variability in Arctic minimum sea ice extent. *Geophys. Res. Lett.* 32:L21501, doi:10.1029/2005GL024376.

Francis, J. A., E. Hunter, and C.-Z. Zou. Arctic tropospheric winds derived from TOVS satellite retrievals. *J. Climate* 18:2270–85, 2005.

Glenn, S. M., et al. Biogeochemical impact of summertime coastal upwelling on the New Jersey shelf. *J. Geophys. Res.* 109:C12S02, doi:10.1029/2003JC002265, 2004.

Glenn, S. M., et al. The expanding role of ocean color optics in the changing field of operational oceanography. *Oceanography* 17:86–95, 2004.

Zhang, Y., and **J. F. Grassle.** A portal for the Ocean Biogeographic Information System. *Oceanol. Acta* 25:193–7, 2003.

Snelgrove, P. V. R., et al. **(J. F. Grassle** and **J. P. Grassle).** The role of colonization in establishing patterns of community composition and diversity in shallow-water sedimentary communities. *J. Mar. Res.* 59:813–30, 2001.

Ma, H., and **J. P. Grassle.** Invertebrate larval availability during summer upwelling and downwelling on the inner continental shelf off New Jersey. *J. Mar. Res.* 62:837–65, 2004.

Weissberger, E. J., and **J. P. Grassle.** Settlement, first-year growth, and mortality of surfclams *Spisula solidissima. Est. Coastal Shelf Sci.* 56:669–84, 2003.

Grothues, T. M., K. W. Able, J. McDonnell, and M. Sisak. An estuarine observatory for real-time telemetry of migrant macrofauna: Design, performance, and constraints. *Oceanogr. Limnol.: Methods* 3:275–89, 2005.

Grothues, T. M. and **K. W. Able.** Response of juvenile fish assemblages in tidal marsh creeks during treatment for phragmites removal. *Estuaries* 26(2B):563–73, 2003.

Grothues, T. M., and **K. W. Able.** Discerning vegetation and environmental correlates with subtidal marsh fish assemblage dynamics during Phragmites eradication efforts: Interannual trend measures. *Estuaries* 26(2B):547–86, 2003.

Grothues, T. M., et al. Flux of larval fish around Cape Hatteras. *Limnol. Oceanogr.* 47(1):165–75, 2002.

Grothues, T. M. and R. K. Cowen. Larval fish assemblages and water mass history in a major faunal transition zone. *Cont. Shelf Res.* 19:1171-1198, 1999.

Tanguy, A., **X. Guo,** and S. E. Ford. Discovery of genes expressed in response to *Perkinsus marinus* challenge in eastern *(Crassostrea virginica)* and Pacific *(C. gigas)* oysters. *Gene* 338:121–31, 2004.

Yu, Z., and **X. Guo.** Genetic linkage map of the eastern oyster *Crassostrea virginica* Gmelin. *Biol. Bull.* 204:327–38, 2003.

Iskandarani, M., **D. B. Haidvogel,** and **J. Levin.** A three-dimensional spectral element model for the solution of the hydrostatic primitive equations. *J. Computational Phys.* 186(2):397–425, 2003.

Haidvogel, D. B., and A. Beckmann. *Numerical Ocean Circulation Modeling,* p. 318. London: Imperial College Press, 1999.

Kennish, M. J., Trends of PCBs in New Jersey's estuarine and coastal marine environments: A literature review and update of findings. *Bull. N.J. Acad. Sci.* 50:1–15, 2005.

Kennish, M. J., et al. Benthic macrofaunal community structure along a well-defined salinity gradient in the Mullica River–Great Bay Estuary. *J. Coastal Res.* SI45:209–26, 2004.

Kennish, M. J., et al. Side-scan sonar imaging of subtidal benthic habitats in the Mullica River–Great Bay estuarine system. *J. Coastal Res.* SI45:227–40, 2004.

Corredor, J., et al. **(L. Kerkhof).** Geochemical rate-RNA integration study: Ribulose 1,5 bisphosphate carboxylase/oxygenase gene transcription and photosynthetic capacity of planktonic photoautotrophs. *Appl. Environ. Microbiol.* 70:5459–68, 2004.

Iskandarani, M., **J. Levin,** B.-J. Choi, and **D. B. Haidvogel.** Comparison of advection schemes for high-order h-p finite element and finite volume methods. *Ocean Modeling* 10:233–52, 2005.

Rutgers, The State University of New Jersey, New Brunswick/Piscataway

Choi, B.-J., M. Iskandarani, **J. Levin**, and **D. B. Haidvogel.** A spectral finite volume method for the shallow water equations. *Mon. Wea. Rev.* 132:1777–91, 2004.

Levin, J., M. Iskandarani, and **D. Haidvogel.** A nonconforming spectral element ocean model. *Int. J. Numer. Methods Fluids* 34(6):495–525, 2000.

Levin, J., M. Iskandarani, and **D. Haidvogel.** A spectral filtering procedure for eddy-resolving simulations with a spectral element ocean model. *J. Comput. Phys.* 13(1):130–54, 1997.

Lutz, R. A. Dawn in the deep. *Natl. Geograph.* 203(2):92–103, 2003.

Lutz, R. A., T. A. Shank, and R. Evans. Life after death in the deep-sea. *Am. Scientist* 89:422–31, 2001.

McGhee, G. R. Jr. *Theoretical Morphology.* New York: Columbia University Press, 1999.

McGhee, G. R. Jr. *The Late Devonian Mass Extinction.* New York: Columbia University Press, 1996.

Chen, Y., **J. R. Miller,** and **J. A. Francis** et al. Observed and modeled relationships among Arctic climate variables. *J. Geophys. Res.* 108(D24): 4799, doi:10.1029/2003JD003824, 2003.

Miller, J. R., and G. L. Russell. Projected impact of climate change on the energy budget of the Arctic Ocean by a global climate model. *J. Climate* 15(21):3028–42, 2002.

Miller, J. R., G. L. Russell, and G. Caliri. Continental scale river flow in climate models. *J. Climate* 7:914–28, 1994.

Miller, K. G., et al. A chronology of Late Cretaceous sequences and sea-level history: Glacioeustasy during the Greenhouse World. *Geology* 31(7):585–8, 2003.

Miller, K. G. The role of ODP in understanding the causes and effects of global sea-level change, accomplishments and opportunities of the ODP. *JOIDES J.* 28(1):23–8, 2002.

Miller, J. R., G. L. Russell, and G. Caliri. Continental scale river flow in climate models. *J. Climate* 7:914–28, 1994

Nordstrom, K. F., N. L. Jackson, J. R. Allen, and D. J. Sherman. Longshore sediment transport rates on a microtidal estuarine beach. *J. Waterway, Port, Coastal, Ocean Eng.* 129:1–4, 2003.

Nordstrom, K. F. *Beaches and Dunes of Developed Coasts.* Cambridge: Cambridge University Press, 2000.

Olsson, R. K., and **K. G. Miller** et al. Sequence stratigraphy and sea-level change across the Cretaceous-Tertiary boundary on the New Jersey passive margin. *Geo. Soc. Am. Spec. Paper* 336:97–108, 2002.

Powell, E. N., E. A. Bochenek, J. M. Klinck, and E. E. Hofmann. Influence of short-term variations in food on survival of *Crassostrea gigas* larvae: A modeling study. *J. Mar. Res.* 62:117–52, 2004.

Powell, E. N., A. J. Bonner, B. Muller, and E. A. Bochenek. Assessment of the effectiveness of scup bycatch-reduction regulations in the *Loligo* squid fishery. *J. Environ. Manage.* 71:155–67, 2004.

M. L. Martinez and **N. P. Psuty,** eds. *Coastal Dunes: Ecology and Conservation, Ecological Studies 171* Berlin: Springer-Verlag, 2004.

Psuty, N. P., and D. D. Ofiara. *Coastal Hazard Management: Lessons and Future Directions from New Jersey.* New Brunswick, N. J.: Rutgers University Press, 2002.

Quinlan, J. A., B. O. Blanton, T. J. Miller, and F. E. Werner. From spawning grounds to the estuary: Using linked individual-based and hydrodynamic models to interpret patterns and processes in the oceanic phase of Atlantic menhaden life history. *Fish. Oceanogr.* 8(2):224–46, 1999.

Reinfelder, J. R., A. J. Milligan, and F. M. M. Morel. The role of the C4 pathway in carbon accumulation and fixation in a marine diatom. *Plant Phys.* 135:2106–11, 2004.

Schaefer, J. K., et al. **(J. R. Reinfelder).** Role of the bacterial organomercury lyase (MerB) in controlling methylmercury accumulation in mercury-contaminated natural waters. *Environ. Sci. Technol.* 38:4304–11, 2004.

Rona, P. A. Resources of the sea floor. *Science* 299:673–4, 2003.

Rona, P. A., K. G. Bemis, D. Silver, and C. D. Jones. Acoustic imaging, visualization, and quantification of buoyant hydrothermal plumes in the ocean *Mar. Geophys. Res.* 23:147–68, 2002.

Rosenthal, Y., D. W. Oppo, and B. K. Linsley. The amplitude and phasing of climate change during the last deglaciation in the Sulu Sea, western equatorial Pacific. *Geophys. Res. Lett.* 30(8):1428 doi:10.1029/2002GL016612, 2003.

Lear, C. H., **Y. Rosenthal,** and J. D. Wright. The closing of a seaway: Ocean water masses and global climate change. *Earth Planet. Sci. Lett.* 210:425–36, 2003.

Schofield, O., et al. Inverting inherent optical signatures in the nearshore coastal waters at the Long Term Ecosystem Observatory. *J. Geophys. Res.* 109:C12S04, doi: 10.1029/2003JC002071, 2004.

Schofield, O., et al. Watercolors in the coastal zone: What can we see? *Oceanography* 17:28–37, 2004.

Seitzinger, S. P., et al. Molecular-level chemical characterizastion and biovailability of dissolved organic matter in stream water using ESI mass spectrometry. *Limnol. Oceanogr.* 50(1):1–12, 2005.

Seitzinger, S. P., et al. Global patterns of dissolved inorganic and particulate nitrogen inputs to coastal systems: Recent conditions and future projections. *Estuaries* 25(4b):640–55, 2002.

Cullen, J. T., et al. **(R. M. Sherrell).** Effect of iron limitation on the cadmium to phosphorus ratio of natural phytoplankton assemblages from the Southern Ocean. *Limnol. Oceanogr.* 48:1079–87, 2003.

Sherrell, R. M., M. P. Field, and G. Ravizza. Uptake and fractionation of rare earth elements on hydrothermal plume particles at 9 45N, East Pacific Rise. *Geochim. Cosmochim. Acta* 63:1709–22, 1999.

Sikes, E. L., T. O'Leary, S. D. Nodder, and J. K. Volkman. Alkenone temperature records and biomarker flux at the subtropical front on the Chatham Rise SW Pacific Ocean. *Deep Sea Res.* 52(5):721–48, 2005.

Sikes, E. L., C. R. Samson, T. P. Guilderson, and W. R. Howard. Old radiocarbon ages in the sowthwest Pacific at 11,900 years ago and the last glaciation. *Nature* 405:555–9, 2000.

Smouse, P. E., R. J. Dyer, R. D. Westfall, and V. L. Sork. Two-generation analysis of pollen flow across a landscape. I. Male gamete heterogeneity among females. *Evolution* 55:260–71, 2001.

Smouse, P. E., T. R. Meagher, and C. J. Kobak. Parentage analysis in *Chamaelirium luteum* (L.): Why do some males have disproportionate reproductive contributions? *J. Evol. Biol.* 12:1056–68, 1999.

Reimers, C. E., et al. **(G. L. Taghon).** In situ measurements of advective solute transport in permeable shelf sands. *Continental Shelf Res.* 24:183–201, 2004.

Linton, D. L., and **G. L. Taghon.** Feeding, growth, and fecundity of *Abarenicola pacifica* in relation to sediment organic concentration. *J. Exp. Mar. Biol. Ecol.* 2254:85–107, 2000.

Park, S. S., Y. Na, and **C. G. Uchrin.** An Oxygen Equivalent Model for water quality dynamics in a macrophyte dominated river. *Ecol. Modell.* 168:1–12, 2003.

Park, S. S., J. W. Park, **C. G. Uchrin,** and M. A. Cheney. A Micelle Inhibition Model for the availability of PAHs in aquatic systems. *Environ. Toxicol. Chem.* 21:2737–41, 2002.

Veron, D. E., and R. C. J. Somerville. Radiative transfer through broken cloud fields: An application of stochastic theory. *J. Geophys. Res.* 109, doi:10.1029/2004JD004524, 2004.

Feingold, G., W. L. Eberhard, **D. E. Veron,** and M. Previdi. First measurement of the Twomey indirect effect using ground-based remote sensors. *Geophys. Res. Lett.* 30(6):1287, doi:10.1029/2002GL016633, 2003.

Voordeckers, J. W., V. Starovoytov, and **C. Vetriani.** Caminibacter mediatlanticus sp.nov., a thermophilic, chemolithoautotrophic, nitrate ammonifying bacterium isolated from a deep-sea hydrothermal vent on the Mid-Atlantic Ridge. *Int. J. Syst. Evol. Microbiol.* 55:773–9, 2005.

Vetriani, C., et al. **(R. A. Lutz).** Mercury adaptation among bacteria from a deep-sea hydrothermal vent. *Appl. Environ. Microbiol.* 71:220–66, 2005.

Vetriani, C., et al. **(R. A. Lutz).**. Thermovibrio ammonificans sp.nov., a thermophilic, chemolithotrophic, nitrate ammonifying bacterium from deep-sea hydrothermal vents. *Int. J. Syst. Evol. Microbiol.* 54:175–81, 2004.

Wilkin, J. L., M. M. Bowen, and W. J. Emery. Mapping mesoscale currents by optimal interpolation of satellite radiometer and altimeter data. *Ocean Dynamics* 52:95–103, 2002.

Griffin, D., and **J. L. Wilkin,** et al. Ocean currents and the larval phase of Australian western rock lobster, *Panulirus cygnus. Mar. Freshwater Res.* 52:1187–200, 2001.

SAN FRANCISCO STATE UNIVERSITY

Romberg Tiburon Center for Environmental Studies
Program in Marine Science

Program of Study

The Master of Science in biology with a concentration in marine biology, the Master of Science in chemistry, and the Master of Science in applied geoscience are offered by San Francisco State University (SFSU) through the Departments of Biology, Chemistry and Biochemistry, and Geosciences on the main campus, with study based at its program at the Romberg Tiburon Center for Environmental Studies (RTC).

Programs in this area reflect the fact that marine science is a meeting place for all the biological sciences and the physical sciences. Graduate work is conducted in conjunction with a research professor or faculty member at RTC and follows specific program areas, such as ecology or oceanography.

RTC scientists train and support students in their laboratories, out in the field, and through collaborations with fellow scientists at universities, institutions, and environmental agencies around the state and nation. Students also take courses at RTC and on the main campus to fulfill their graduate degree requirements. Courses offered at RTC encompass the subject areas of benthic ecology, biological oceanography, marine conservation biology, marine microbial ecology, molecular tools in marine biology, fisheries biology, physical oceanography, remote sensing and GIS, and restoration and wetlands ecology, in addition to others.

Research conducted at RTC has contributed significantly to the existing body of knowledge on estuarine environments—from a fundamental understanding of estuaries to applied science and management methodologies—and has been widely published in a variety of scientific journals and texts. RTC is known throughout the academic community as a major center for scientific research, in addition to serving as a local and regional resource for studies and information on estuarine and marine environments.

Research Facilities

The Romberg Tiburon Center is the only academic research facility situated on San Francisco Bay, the largest estuary on the west coast of the United States. The Center's mission is to conduct fundamental scientific research and to educate and train the next generation of scientists. RTC scientists pursue their research in their laboratories at the Center, at field sites around the world, and through collaborations with colleagues at other universities and institutions.

Financial Aid

Graduate students have a number of financial aid opportunities available to them. These include graduate teaching fellowships, state university grants, University-administered scholarships, Graduate Assistance in Area of National Need (GAANN) in biology and chemistry, MBRS-RISE (Research Initiative for Scientific Enhancement), M.A.-M.S./Ph.D. Bridge to the Future, NSF Graduate K–12 Grant, Postbaccalaureate Research Education Program (PREP), Graduate Equity Fellowships, Federal Work-Study, and Federal Direct Student Loans. Students may obtain more information on the University's financial aid Web site at http://www.sfsu.edu/~seo.

Cost of Study

Per-semester fees for California residents include a state university fee of $819 for up to 6 units ($1410 for more than 6 units) and local fees of $273, for a total registration fee per semester of $1092 for up to 6 units ($1683 for more than 6 units). Further information is available on the Web at http://www.sfsu.edu/~bursar/Feepayment/Current/schedule.htm#undergrad.

Living and Housing Costs

Currently, there is no on-campus housing at RTC, but housing is available at the University Park Apartments (Web site: http://www.sfsu.edu/~housing/). In addition, local houses, studios, and apartments are available off campus. Rent ranges from $600 to more than $2000 per month. Further housing information is available on the Web at http://www.craigslist.org.

Location

The Romberg Tiburon Center is situated on a breathtaking 34-acre parcel of bayfront property located just outside the town of Tiburon, California. RTC's location is an ideal setting for scientists to conduct their research, much of which focuses on understanding the natural forces at work in the San Francisco Bay and its surrounding wetland environments. RTC research scientists are also studying natural phenomena in the open ocean, both close to home along California's coastline and in remote locations such as the Ross Sea (Antarctica) and the Equatorial Pacific. The Center is a close-knit community of scientists and students working together to fulfill RTC's mission of education and research.

The University

Founded in 1899 as the San Francisco State Normal School, a two-year teaching college, San Francisco State University has undergone five name changes in its history. The main campus is situated on 102 acres in the southwest corner of San Francisco. It awards bachelor's degrees in 116 areas of specialization and master's degrees in ninety-five areas. The University also offers a Ph.D. and an Ed.D. in education, with a concentration in special education. In addition, the school offers twenty-seven credential programs and thirty-four certificate programs.

Applying

Applicants must have the equivalent of the Bachelor of Arts in biology at San Francisco State University, supplemented with additional training in marine biology that indicates the capability to pursue graduate study in this field.

The Division of Graduate Studies can best and most quickly process an application if it is submitted in the same packet with the completed California State University application form, two sets of official transcripts from each postsecondary institution attended (in a sealed envelope, as issued by the school), GRE results, two letters of support, a personal statement, and the processing fee of $55. The admission deadline is January 16 for the fall semester. Graduate students are required to have a minimum 2.75 GPA in undergraduate study. Additional materials may be required, including a TOEFL score sent directly by ETS, a financial statement, and a medical insurance agreement.

Correspondence and Information

Romberg Tiburon Center for Environmental Studies
3152 Paradise Drive
Tiburon, California 94920
Phone: 415-338-6063
Fax: 415-435-7120
Web site: http://www.rtc.sfsu.edu

SFSU–Graduate Studies, ADM 254
San Francisco State University
1600 Holloway Avenue
San Francisco, California 94132
Phone: 415-338-2234
Fax: 415-405-0340
Web site: http://www.sfsu.edu

San Francisco State University

THE FACULTY AND THEIR RESEARCH

Alissa J. Arp, Marine Ecological Physiologist. How organisms cope with hypoxia and toxic conditions in estuaries and on the ocean floor.
Roger Bland, Physicist. Underwater acoustical monitoring, using sonar signals to measure water temperature and current speed circulation patterns in San Francisco Bay.
Katharyn E. Boyer, Wetland and Coastal Community Ecologist. Role of species interactions in ecosystem functioning, particularly in restoration settings, and the effects of nutrients and other perturbations on wetland communities.
Edward J. Carpenter, Biological Oceanographer. Ecology of marine phytoplankton, particularly cyanobacteria, and the factors affecting the significance of nitrogen fixation in the sea.
William P. Cochlan, Marine Microbial Ecologist/Biological Oceanographer. Physiology and ecology of phytoplankton and bacteria, including harmful algal blooms (HABs); quantifying the nitrogenous nutrition of microorganisms in the sea from coastal, oceanic, and polar environments.
C. Sarah Cohen, Ecological/Evolutionary Biologist and Population Geneticist. Connectivity of marine populations; human impacts on aquatic systems; larval biology; speciation in marine systems; immunity in marine organisms; genetic responses to pollution; recognition and mating systems in aquatic organisms.
Richard C. Dugdale, Biological Oceanographer. Distributions and effects of nutrients on oceanic productivity in coastal and equatorial upwelling areas, using isotopes and remote sensing.
Patricia G. Foschi, Remote Sensing Specialist and Physical Geographer. Integration of remote sensing, GIS, and artificial intelligence for environmental applications; development of automated systems for detecting invasive vegetation in wetlands.
Newell Garfield, Physical Oceanographer. Oceanic circulation in coastal regions and over continental margins, using remote sensing and free-drifting buoy technologies.
Wim Kimmerer, Biological Oceanographer. Growth and predation processes in zooplankton; computer modeling of ecological systems; analysis of human impacts on estuarine and marine ecosystems.
Tomoko Komada, Biogeochemist. Studies the dynamics of nonliving organic matter in freshwater and marine systems, with a focus on the factors affecting the long-term organic carbon cycle.
Jaime C. Kooser, Resource Geographer. Management of San Francisco Bay National Estuarine Research Reserve, with a focus on tidal marsh restoration; using science to inform coastal resource management decisions; relationship between land use and water quality.
Dale Robinson, Phytoplankton Ecologist and Physiologist. Changes in ocean productivity and photosynthesis that result from variations in the physical environment; remote sensing, field, and laboratory studies.
Jonathan Stillman, Marine Ecological Physiologist. Thermal physiology of intertidal-zone organisms from ecological to molecular levels; mechanistic bases of thermal limits and thermal ranges; linkages between thermal phenotype and responses to climate change.
Drew Talley, Biological Oceanographer and Research Coordinator of the San Francisco Bay NERR. Studies the influence of habitat connectivity on wetland and coastal community structure and function, focusing on conservation and restoration importance.
Frances P. Wilkerson, Marine Biologist. Nutrient flux in symbiotic associations between invertebrates and algae; response of phytoplankton to upwelled nutrients and eutrophication.

Representative Research Projects

Nitrogen fixation in the tropical Atlantic and Pacific Oceans and the role of iron and physical factors in affecting rates of carbon and nitrogen fixation.
Effects of wetland restoration stage on fishes, invertebrates, primary production, and nutrients in tidal wetlands of the San Francisco Bay delta.
Molecular approaches to population variation in estuarine fish in relation to environmental stress gradients.
ECOHAB PNW: the ecology and oceanography of the toxic diatom *Pseudonitzschia* in the northeast Pacific Ocean.
SFBEAMS: a collaborative study of long-term water quality monitoring of San Francisco Bay.
CoOP WEST: a multidisciplinary study of upwelling responses over the California shelf north of Point Reyes.
Experimental evaluation of restoration techniques for eelgrass *(Zostera marina)* in San Francisco Bay.
COCMP: a California statewide effort to monitor ocean surface currents.
Thermal physiology of marine invertebrates at organismal and molecular levels.
Functional genomics of porcelain crabs.
netBEAMS: technology development to use cell phone technology and open-source software to control an array of water-quality instruments throughout San Francisco Bay.
Investigating mating probability as a limiting factor for plankton populations at low densities.
Assessing changes in the pelagic food web of the San Francisco Estuary.
Polar macromolecular lipids in marine sediments and their role in the oceanic export of fossil carbon.
Seagrass restoration genetics in San Francisco Bay.
Genetic and behavioral variation in invasive tunicates.
Connectivity and larval dispersal in lobster populations in southern Massachusetts.
Conservation genetics of Bermuda killifish.
The effects of upwelling along the Pacific Equator on biological productivity and the large biogeochemical cycles of the ocean and atmosphere.
The effect of phytoplankton species composition on rates of carbon sequestration in the Southern Ocean.
Understanding the importance of diatoms in coupling nitrogen and carbon cycles in coastal and estuarine ecosystems, using biogeochemical and molecular approaches.
Role of microzooplankton in planktonic food webs of the San Francisco estuary.
Researching the introduction of exotic species in San Francisco Bay and their impact on its estuarine food webs.
Community ecology of restored and natural wetlands of San Francisco Bay.
Monitoring the aquatic plant *Egeria densa* in the Sacramento–San Joaquin Delta and developing an automated system for detecting *Egeria* in scan-digitized color-infrared aerial photography.
Ecology, evolution, and conservation of California native plants and their pollinators.

STONY BROOK
STATE UNIVERSITY OF NEW YORK

STONY BROOK UNIVERSITY, STATE UNIVERSITY OF NEW YORK

Marine Sciences Research Center and the Institute for Terrestrial and Planetary Atmospheres

Program of Study

The Marine Sciences Research Center (MSRC) and the Institute for Terrestrial and Planetary Atmospheres (ITPA) offer a Master of Science (M.S.) degree in marine and atmospheric sciences, with specializations in marine sciences and atmospheric sciences. MSRC and ITPA are top-rated research centers that maintain a collegial, cooperative atmosphere among students and faculty members.

A typical program of study begins with a coordinated set of core courses covering fundamental principles of marine sciences or atmospheric sciences. At the same time, students are encouraged to join an ongoing research activity to help identify an area of specialization. All M.S. students complete a thesis project that usually results in a published manuscript. The M.S. program can be completed in two years, although some students take as long as three to finish.

Oceanographic research and teaching in marine sciences focuses on the collaborative, interdisciplinary study of oceanographic processes—combining biological, chemical, geological, and physical approaches to examine a wide range of issues, including biogeochemical cycling, fate and effects of contaminants, coastal environmental health, habitat destruction, and living marine resources. An important focus of the Center is regional coastal problems, such as coastal habitat alteration, diseases of marine animals, harmful algal blooms, and effects of contaminants on biota. Equally important are studies of diverse problems throughout the world; faculty members at MSRC are currently carrying out research in the Caribbean Sea, the Mediterranean Sea, Papua New Guinea, Bangladesh, and the Canadian Arctic, among other locations. A wide variety of approaches are employed, including shipboard sampling, remote sensing, field and laboratory experiments, laboratory analyses, and computer modeling and simulation. More information about marine research at MSRC can be found on the Web at http://msrc.sunysb.edu.

The Institute for Terrestrial and Planetary Atmospheres teaches students how to apply their knowledge of mathematics, physics, and chemistry to increase understanding of the atmospheres of Earth and other planets. Completion of the degree program requires a thorough understanding of the principles of atmospheric science coupled with the ability to apply that knowledge to significant problems. Research is conducted at various temporal and spatial scales, from the daily evolution of the atmospheric state—weather—to longer-scale climate variabilities, including those associated with El Niño and global warming. Comprehensive data sets from satellites, field experiments, laboratory measurements, and meteorological observations are analyzed in the context of global three-dimensional weather and climate models and simplified conceptual models. A key goal is to achieve better understanding of the physical bases of numerical weather forecasting and climate prediction, including the size and timing of future greenhouse warming. More information can be found at http://atmos.msrc.sunysb.edu.

Research Facilities

Facilities at the Marine Sciences Research Center are modern and comprehensive and support a wide range of oceanographic research. Major shared-use facilities include the R/V *Seawolf,* a 24-meter research ship; a running seawater laboratory; a multibeam echosounder for detailed seabed mapping; a laser ablation/inductively coupled plasma/mass spectrometer (LA-ICP-MS); a liquid chromatograph/time of flight/mass spectrometer (LC/TOF/MS); and an analytical facility for CHN and nutrient analyses. A "clean" laboratory and environmental incubators are also available. Individual research laboratories contain facilities for a wide range of biological, chemical, physical, and geological research. The Center is also home to the Marine Animal Disease Laboratory, a diagnostic and research center equipped with facilities needed to culture and evaluate marine pathogens and host responses.

The Institute for Terrestrial and Planetary Atmospheres' computer facilities include several high-end multiple-processor Alpha UNIX stations, plus a large network of PC, UNIX, Linux, and Macintosh computers; printers; graphics terminals; and hard-copy plotters. The Institute maintains a comprehensive system to display real-time weather data, satellite measurements, and numerical model products. It has a state-of-the-art weather laboratory and a remote-sensing laboratory for students to use. The spectroscopy laboratory has infrared (grating) spectrometers, low-temperature absorption cells, a tunable diode laser spectrometer, and a Fourier-transform spectrometer. A stable-isotope mass spectrometer is maintained in the atmospheric isotope laboratory. Students have access to millimeter-wavelength remote-sensing equipment, developed at Stony Brook, and to data from NASA missions.

Financial Aid

Teaching assistantships of $12,592 are provided to most incoming students for the 2006–07 academic year, after which students are generally supported on research assistantships that range, depending on status, from $18,000 to $23,000 for the calendar year. Students receiving support also receive tuition scholarships that completely cover the cost of tuition.

Cost of Study

Tuition for the 2005–06 academic year was $6900 for residents of New York State and $10,500 for nonresidents. Miscellaneous fees, such as insurance and activity fees, total about $526 annually. As described in the Financial Aid section, tuition scholarships are available.

Living and Housing Costs

In 2005–06, the estimated cost of on-campus room and board was $7458. Off-campus rentals are also available and are preferred by most students.

Student Group

Including Ph.D. students, approximately 95 students are engaged in research in marine science, and about 25 graduate students are in atmospheric sciences.

Location

Stony Brook is located about 50 miles east of Manhattan on the wooded North Shore of Long Island. It is convenient to New York City's cultural life and Suffolk County's recreational countryside and seashores. Long Island's hundreds of miles of magnificent coastline attract many swimming, boating, and fishing enthusiasts from around the world.

The University

Established forty years ago as New York's comprehensive state university center for Long Island and metropolitan New York, Stony Brook offers excellent programs in a broad spectrum of academic subjects. The University conducts major research and public service projects. Over the past decade, externally funded support for Stony Brook's research programs has grown faster than that of any other university in the United States and now exceeds $125 million per year. The University's internationally renowned faculty members teach courses from the undergraduate to the doctoral level to more than 22,000 students. More than 100 undergraduate and graduate departmental and interdisciplinary majors are offered. Extensive resources and expert support services help foster intellectual and personal growth.

Applying

Students applying for graduate study in marine sciences should have a B.S. degree in biology, chemistry, geology, physics, or a related discipline. Students applying for atmospheric sciences should have a B.S. degree in physics, chemistry, mathematics, engineering, or atmospheric science. Before applying, applicants should contact program faculty members whose research is of primary interest to them. Applications for September admission should be received by January 15 to ensure consideration for the widest range of support opportunities.

Students may request an application online at http://www.msrc.sunysb.edu/pages/gradapp.html and apply online at http://www.grad.sunysb.edu/applying/applying.htm.

Correspondence and Information

Carol Dovi
Marine Sciences Research Center
Stony Brook University, State University of New York
Stony Brook, New York 11794-5000
Phone: 631-632-8681
Fax: 631-632-8820
E-mail: cdovi@notes.cc.sunysb.edu

Gina Gartin
Institute for Terrestrial and Planetary Atmospheres
Stony Brook University, State University of New York
Stony Brook, New York 11794-5000
Phone: 631-632-8009
Fax: 631-632.6251
E-mail: ggartin@notes.cc.sunysb.edu

Stony Brook University, State University of New York

THE FACULTY AND THEIR RESEARCH

Marine Sciences

Bassem Allam, Assistant Professor; Ph.D., Western Brittany (France), 1998. Pathology and immunology of marine bivalves.
Josephine Y. Aller, Research Professor; Ph.D., USC, 1975. Marine benthic ecology, invertebrate zoology, marine microbiology, biogeochemistry.
Robert C. Aller, Distinguished Professor; Ph.D., Yale, 1977. Marine geochemistry, marine animal-sediment relations.
Robert A. Armstrong, Associate Professor; Ph.D., Minnesota, 1975. Mathematical modeling in marine ecology and biogeochemistry.
David Black, Assistant Professor; Ph.D., Miami (Florida). 1998. Paleoceanography.
Henry J. Bokuniewicz, Professor; Ph.D., Yale, 1976. Nearshore transport processes, coastal groundwater hydrology, coastal sedimentation, marine geophysics.
Malcolm J. Bowman, Distinguished Service Professor; Ph.D., Saskatchewan, 1970. Coastal ocean and estuarine dynamics.
Bruce J. Brownawell, Associate Professor; Ph.D., MIT (Woods Hole), 1986. Biogeochemistry of organic pollutants in seawater and groundwater.
Robert M. Cerrato, Associate Professor; Ph.D., Yale, 1980. Benthic ecology, population and community dynamics.
J. Kirk Cochran, Professor; Ph.D., Yale, 1979. Marine geochemistry, use of radionuclides as geochemical tracers, diagenesis of marine sediments.
Jackie Collier, Assistant Professor; Ph.D., Stanford, 1994. Phytoplankton ecology, physiology, and molecular genetics.
David O. Conover, Professor and Dean, Marine Sciences Research Center; Ph.D., Massachusetts, 1982. Ecology of fish, fisheries biology.
Alistair Dove, Assistant Research Professor; Ph.D., Queensland (Australia), 1999. Pathology, taxonomy, life cycles/ecology.
Mark Fast, Assistant Professor; Ph.D., Dalhousie, 2005. Host-pathogen interactions in marine fish.
Nicholas S. Fisher, Distinguished Professor; Ph.D., SUNY at Stony Brook, 1974. Marine phytoplankton physiology and ecology, biogeochemistry of metals, marine pollution.
Charles N. Flagg, Research Professor; Ph.D., MIT (Woods Hole), 1977. Continental shelf dynamics, biophysical interactions and climate change effects on coastal systems.
Roger D. Flood, Professor; Ph.D., MIT (Woods Hole), 1978. Marine geology, sediment dynamics, continental margin sedimentation.
Michael G. Frisk, Assistant Professor; Ph.D., Maryland, College Park, 2004. Fish ecology, population modeling and life history theory.
Christopher J. Gobler, Associate Professor; Ph.D., SUNY at Stony Brook, 1999. Phytoplankton, harmful algal blooms, estuarine ecology, aquatic biogeochemistry.
Steven L. Goodbred Jr., Adjunct Assistant Professor; Ph.D., William and Mary, 1999. Coastal marine sedimentology, Quaternary development of continental margins, salt-marsh processes.
Paul F. Kemp, Research Professor; Ph.D., Oregon State, 1985. Growth and activity of marine microbes, benthic-pelagic interactions, molecular ecology of marine bacteria.
Cindy Lee, Distinguished Professor; Ph.D., California, San Diego (Scripps), 1975. Ocean carbon cycle, marine geochemistry of organic compounds, nitrogen-cycle biochemistry, biomineralization.
Darcy J. Lonsdale, Associate Professor; Ph.D., Maryland, 1979. Ecology and physiology of marine zooplankton, food web dynamics of estuarine plankton, impacts of harmful algal blooms.
Glenn R. Lopez, Professor; Ph.D., SUNY at Stony Brook, 1976. Marine benthic ecology, animal-sediment interactions, contaminant uptake.
Kamazima M. M. Lwiza, Associate Professor; Ph.D., Wales, 1990. Structure and dynamics of shelf seas, remote-sensing oceanography.
Anne E. McElroy, Associate Professor; Ph.D., MIT (Woods Hole), 1985. Toxicology of aquatic organisms, contaminant bioaccumulation, estrogenicity of organic contaminants.
Stephan B. Munch, Assistant Professor; Ph.D., SUNY at Stony Brook, 2002. Evolutionary ecology of growth and life history traits, applied population dynamics modeling.
Bradley J. Peterson, Assistant Professor; Ph.D., South Alabama, 1998. Community ecology of seagrass-dominated ecosystems.
Sergey A. Piontkovski, Associate Research Professor; Ph.D., National Academy of Sciences of Ukraine (Institute of Biology of the Southern Seas), 1978; Ph.D., Moscow State (Russia), 2004. Physical-biological coupling in coastal and oceanic ecosystems, spatial-temporal structure of marine plankton communities.
Frank J. Roethel, Lecturer; Ph.D., SUNY at Stony Brook, 1982. Environmental chemistry, municipal solid-waste management impacts.
Sergio A. Sañudo-Wilhelmy, Professor; Ph.D., California, Santa Cruz, 1993. Geochemical cycles of trace elements, marine pollution.
Mary I. Scranton, Professor; Ph.D., MIT (Woods Hole), 1977. Marine geochemistry, biological-chemical interactions in seawater.
R. Lawrence Swanson, Adjunct Professor; Ph.D., Oregon State, 1971. Recycling and reuse of waste materials, waste management.
Gordon T. Taylor, Professor; Ph.D., USC, 1983. Marine microbiology; interests in microbial ecology, plankton trophodynamics, and marine biofouling.
Dong-Ping Wang, Professor; Ph.D., Miami (Florida), 1975. Coastal ocean dynamics.
Joseph D. Warren, Assistant Professor; Ph.D., MIT (Woods Hole). 2001. Acoustical oceanography, zooplankton behavior and ecology.
Robert E. Wilson, Associate Professor; Ph.D., Johns Hopkins, 1974. Estuarine and coastal ocean dynamics.
Peter M. J. Woodhead, Adjunct Professor; B.Sc.Hon., Durham (England), 1953. Behavior and physiology of fish, coral reef ecology, ocean energy conversion systems.

Atmospheric Sciences

Robert D. Cess, Distinguished Professor Emeritus; Ph.D., Pittsburgh, 1959. Radiative transfer and climate modeling, greenhouse effect, intercomparison of global climate models.
Edmund K. M. Chang, Associate Professor; Ph.D., Princeton, 1993. Atmospheric dynamics and diagnoses, climate dynamics, synoptic meteorology.
Brian A. Colle, Associate Professor; Ph.D., Washington (Seattle), 1997. Synaptic meteorology, weather forecasting, mesoscale modeling.
Robert L. de Zafra, Professor Emeritus (Department of Physics, with joint appointment in Marine Sciences Research Center); Ph.D., Maryland, 1958. Monitoring and detection of trace gases in the terrestrial stratosphere, changes in the ozone layer, remote-sensing instrumentation.
Marvin A. Geller, Professor; Ph.D., MIT, 1969. Atmospheric dynamics, stratosphere dynamics and transport, climate dynamics.
Sultan Hameed, Professor; Ph.D., Manchester (England), 1968. Analysis of climate change using observational data and climate models, interannual variations in climate, climate predictability.
Daniel Knopf, Assistant Professor; Ph.D., Swiss Federal Institute of Technology, 2003. Atmospheric chemistry, heterogeneous chemical processes, chemical ionization mass spectrometry, organic aerosols.
John E. Mak, Associate Professor; Ph.D., California, San Diego (Scripps), 1992. Atmospheric chemistry, biosphere-atmosphere interactions, trace gas isotope chemistry, marine geochemistry, mass spectrometry.
Nicole Reimer, Assistant Professor; Ph.D., Karlsruhe (Germany), 2002. Cloud microphysics, aerosol physics and chemistry.
Prasad Varanasi, Professor; Ph.D., California, San Diego (Scripps), 1967. Infrared spectroscopic measurements in support of NASA's space missions, atmospheric remote sensing, greenhouse effect and climate research, molecular physics at low temperatures.
Duane E. Waliser, Adjunct Associate Professor; Ph.D., California, San Diego,1992. Ocean-atmosphere interactions, tropical climate dynamics.
Minghua Zhang, Professor and Director of ITPA; Ph.D., Academia Sinica (China), 1987. Atmospheric dynamics and climate modeling.

UNIVERSITY OF MARYLAND

Graduate Program in Marine-Estuarine-Environmental Sciences

Program of Study

The specific objective of the all-University Graduate Program in Marine-Estuarine-Environmental Sciences (MEES) is the training of qualified graduate students, working toward the M.S. or Ph.D. degree, who have research interests in fields of study that involve interactions between biological systems and physical or chemical systems in the marine, estuarine, or terrestrial environments. The program comprises six Areas of Specialization (AOS): Oceanography, Environmental Chemistry (and toxicology), Ecology, Environmental Molecular Biology/Biotechnology, Fisheries Science, and Environmental Science. Students work with their advisory committee to develop a customized course of study based on research interests and previous experience.

All students must demonstrate competence in statistics. Each student is required to complete a thesis or dissertation reporting the results of an original investigation. The research problem is selected and pursued under the guidance of the student's adviser and advisory committee.

Research Facilities

Students may conduct their research either in the laboratories and facilities of the College Park (UMCP), Baltimore (UMB), Baltimore County (UMBC), or Eastern Shore (UMES) campuses or in one of the laboratories of the University of Maryland Center for Environmental Science (UMCES): Chesapeake Biological Laboratory (CBL) at Solomons, Maryland; the Horn Point Laboratory (HPL) in Cambridge, Maryland; and the Appalachian Laboratory (AL) in Frostburg, Maryland; or at the Center of Marine Biotechnology (COMB) in Baltimore, Maryland. CBL and HPL are located on the Chesapeake Bay. They include excellent facilities for the culture of estuarine organisms. The laboratories are provided with running salt water, which may be heated or cooled and may be filtered. Berthed at CBL are the University's research vessels, which range from the 65-foot *Aquarius* to a variety of smaller vessels for various specialized uses. At HPL, there are extensive marshes, intertidal areas, oyster reefs, tidal creeks, and rock jetties. AL, located in the mountains of western Maryland, specializes in terrestrial and freshwater ecology.

Specialized laboratory facilities for environmental research are located on the campuses. These facilities provide space for microbiology, biotechnology, water chemistry, and cellular, molecular, and organismal biology, as well as specialized facilities for the rearing and maintenance of both terrestrial and aquatic organisms of all kinds. There are extensive facilities for remote sensing of the environment. Extensive field sites for environmental research are available through the University's agricultural programs and through cooperation with many other organizations in the state.

Financial Aid

University fellowships, research assistantships and traineeships, and teaching assistantships are available. In general, aid provides for full living and educational expenses. Some partial assistance may also be available. Research support from federal, state, and private sources often provides opportunities for additional student support through either research assistantships or part-time employment on research projects.

Cost of Study

In 2005–06, tuition for graduate students was $803 for Maryland residents and $1323 for nonresidents for each credit hour. In addition, stipulated fees ranged from $270 to $460 per semester for each student. However, financial aid typically covers most of these expenses.

Living and Housing Costs

Commercial housing is plentiful in the area around the campuses. For students who are working at HPL or CBL, limited dormitory-type housing is available on site. Minimum living expenses for a year's study at College Park or in the Baltimore area are about $14,000, exclusive of tuition and fees. Costs are lower at the UMES campus.

Student Group

About 210 students are enrolled in the program. They come from a variety of academic backgrounds. There are a number of international students. About 50 percent of the students are in the doctoral program, and 50 percent are working toward the M.S. Some of the master's students expect to continue toward the doctorate. While most of the students are biologists, some come with undergraduate majors in chemistry, biochemistry, geology, economics, political science, or engineering. The program encourages and accommodates such diversity in its students.

Location

The MEES program is offered on campuses of the University at College Park, Baltimore, Baltimore County, and Eastern Shore and at the UMCES laboratories and COMB. Students normally enroll on the campus where their adviser is located. Of particular relevance for the MEES program is the University's location near Chesapeake Bay, one of the world's most important estuarine systems, which in many aspects serves as the program's principal laboratory resource.

The University

The University of Maryland is the state's land-grant and sea-grant university. It has comprehensive programs at both the undergraduate and graduate levels on the campuses at College Park, Baltimore County, and Eastern Shore. Programs in the health sciences and the professions are located in Baltimore. There are approximately 8,400 graduate students at College Park, 800 at Baltimore, 300 at Baltimore County, and 75 at Eastern Shore.

Applying

Applications for admission in the fall semester must be filed by February 1; however, to be considered for financial support, it is better to apply by December 1. Some students will be admitted for the semester starting in January, for which the deadline is September 1. Applicants must submit an official application to the University of Maryland Graduate School, along with official transcripts of all previous collegiate work, three letters of recommendation, and scores on the General Test (aptitude) of the Graduate Record Examinations. It is particularly important that a student articulate clearly in the application a statement of goals and objectives pertaining to their future work in the field. Because of the interdisciplinary and interdepartmental nature of the program, only students for whom a specific adviser is identified in advance can be admitted. Prior communication with individual members of the faculty is encouraged.

Correspondence and Information

Graduate Program in Marine-Estuarine-Environmental Sciences
0105 Cole
University of Maryland
College Park, Maryland 20742

Phone: 301-405-6938
Fax: 301-314-4139
E-mail: mees@mees.umd.edu
Web site: http://www.mees.umd.edu

University of Maryland

THE FACULTY AND THEIR RESEARCH

Baltimore Campus. Da-Wei Gong: molecular and cell biology of energy metabolism. Raymond T. Jones: pathophysiology of elasmobranch and teleost fishes. Andrew S. Kane: environmental pathology, toxicology, and husbandry of aquatic and marine organisms, with emphasis on Chesapeake Bay fauna and captive fish species. Henry N. Williams: ecology of the bacterial predator, *Bdellovibrio,* in the Chesapeake Bay.

Baltimore County Campus. Brian P. Bradley: zooplankton physiology and genetics. C. Allen Bush: environmental molecular biology, molecular structure determination. Thomas W. Cronin: vision in marine animals. Erle C. Ellis: landscape ecology, biogeochemistry, sustainable resource management. Upal Ghosh: experimental investigation, design, and modeling of physiochemical and biological processes that affect water quality. Jin Ping (Jack) Gwo: theoretical and computational aspects of subsurface, multispecies solute and microbial transport. Raymond Hoff: optical properties of aerosols and gases in the atmosphere, pathways and fates of toxic organic and elemental chemicals in the environment. Jeffrey Leips: evolution of life history traits, specifically focused on understanding how the genetic architecture underlying these traits guides and constrains their evolutionary responses to natural selection. Andrew J. Miller: surface-water hydrology and fluvial geomorphology, effects of human activities on watershed hydrology and river channels. Robert R. Provine: fish and waterfowl behavioral ecology. Brian E. Reed: sorption of organics/inorganics, surface chemistry, water and wastewater treatment, soil and site remediation. Youngsinn Sohn: applications of GIS and digital image analysis for addressing environmental problems, monitoring, and mapping. Philip G. Sokolove: endogenous rhythms and neuroendocrinology. Christopher Swan: benthic evolution and ecology, community ecology, limnology, systems, biostatistics. Carl Weber: benthic ecology, systems ecology. Claire Welty: fundamental understanding of transport processes in aquifiers, mathematical modeling of groundwater flow.

College Park Campus. Lowell W. Adams: wildlife biology, ecology, and management. Andrew H. Baldwin: wetland ecology, plant community ecology of coastal marshes and mangroves. Jennifer Becker: microbial communities that biodegrade xenobiotics, bioremediation of contaminated groundwater systems, anaerobic biological treatment processes for waste streams. Amy Brown: toxicology, epidemiology, effects of pesticides on human health. Kaye L. Brubaker: physical hydrology, numerical modeling, stream and estuary water-quality modeling. James Carton: physical oceanography, ocean modeling, atmosphere/ocean interactions. James Dietz: mammalian ecology and conservation. Jocelyne DiRuggiero: study of DNA repair mechanisms, genome plasticity and lateral gene transfer in hyperthermophilic Archae using functional genomics. Irwin Forseth: plant ecology and physiology. Oliver J. Hao: waste management and environmental engineering. Matthew P. Hare: population and conservation genetics of marine organisms, invasion biology, phylogeography. Robert L. Hill: soil runoff, nonpoint source pollution in soil systems. David W. Inouye: terrestrial ecology, especially plant-animal interactions. Patrick Kangas: modeling and measuring of whole ecosystems with emphasis on management and ecology. Michael S. Kearney: pollen analytical investigations of tidal marsh sediments. William O. Lamp: crop protection from arthropods through integration of crop management practices with arthropod/plant interactions, development of non-pesticide management tactics. Marla McIntosh: sludge utilization in woodlands, genetic diversity of food crops. Bahram Momem: applied statistics. Judd O. Nelson: environmental toxicology of pesticides. Mary Ann Ottinger: effect of toxic substances on avian reproduction. Margaret Palmer: stream and estuarine ecology and hydrodynamics. Michael Paolisso: applied anthropology, environment and pollution, international and rural development. Kennedy T. Paynter Jr.: physiology and biochemistry of estuarine organisms, oyster reef restoration. Karen Prestegaard: watershed and wetland hydrology. Marjorie Reaka-Kudla: zoogeography, symbiosis, and behavior of marine crustaceans. Miranda Schreurs: comparative environmental and energy politics in northeast Asia and Europe. Adel Shirmohammadi: impact of agricultural pest management practices on water quality. Eugene B. Small: estuarine and marine protozoology. Joseph H. Soares Jr.: waterfowl nutrition, calcification, vitamin and mineral metabolism. Daniel E. Terlizzi: plant aquaculture, phycology. David Tilley: ecological engineering, industrial ecology, ecological decision making for sustainable development. Alba Torrents: organic pollutants, soil/water interface. Ray R. Weil: disturbed-land revegetation, land application of organic wastes. Ronald M. Weiner: environmental bacteriology. Richard Weismiller: agriculture and natural resources, remote sensing. L. Curry Woods: aquaculture, larviculture.

Eastern Shore Campus. Isoken Aighewi: soil-water pollution, environmental soil chemistry. Eugene L. Bass: algal toxins, acclimatization of animals to environmental variables. Dwayne W. Boucaud: biodiversity of the microbes of the coastal bays of Maryland and northern Virginia, DNA fingerprinting to identify benthic bacterial species. Carolyn B. Brooks: microbial insecticides: symbiotic nitrogen fixation. Robert B. Dadson: soybean breeding, insect resistance, biological nitrogen fixation. Joseph Dodoo: coal technology, kinetic studies of coal pyrolysis. Gian C. Gupta: environmental chemistry, soil science, water and wastewater recycling. Youssef Hafez: nutrition, effects of processing on bioavailability of peptides. Thomas Handwerker: small-scale alternative crops. Jeannine M. Harter-Dennis: roasters chicken nutrition, reduction of fat. George E. Heath: food safety and drug residues, pharmokinetics. Ali B. Ishaque: marine ecotoxicology; behavior, transport, distribution, and fate of chemical stressors in marine environments. Jagmohan Joshi: plant breeding and genetics. Gerald E. Kananen: analytical instrumentation, environmental pollutants. Eric P. May: responses of fish to injurious agents, markers of population health. Madhumi Mitra: paleontology, paleocology, and paleoenvironmental studies of cretaceous-quaternary sediments of Atlantic coastal plain. Abhijit Naghaudhuri: Integration of advanced technologies of mechatronics in the fields of precision agriculture, environmental, marine, and geosciences. Joseph Okoh: carbon reaction chemistry. Salina Parveen: genotypic and phenotypic methods for detecting sources of fecal pollution in aquatic environments. Douglas E. Ruby: population ecology and behavior of reptiles. Jeurel Singleton: aquatic and terrestrial invertebrate community ecology and population dynamics; population dynamics of phyto- and zooplankton communities in reservoirs, lakes, and fish rearing facilities. Yan Waguespeck: spectroscopic studies of temperature- and gas pressure–induced chemical changes.

Appalachian Laboratory. Mark S. Castro: atmosphere-biosphere interactions. Katharina Engelhardt: effects of species richness on wetland ecosystem functioning and services, community and ecosystem ecology. Keith N. Eshleman: watershed and wetlands hydrology and hydrobiogeochemistry. Robert H. Gardner: landscape ecology, ecosystem modeling. J. Edward Gates: behavioral ecology of vertebrates, habitat analysis and evaluation. Robert H. Hilderbrand: ecology and conservation biology of running waters, watershed and stream habitat restoration, linking landscapes and populations, dynamic modeling of watersheds. John L. Hoogland: vertebrate behavioral ecology, evolutionary biology of mammals. Sujay S. Kaushal: fate and transport of pollutants, biochemistry, limnology, organic geochemistry and environmental history. Kenneth R. McKaye: evolution, behavior, and community ecology of fishes. Raymond P. Morgan: pollution ecology, fisheries genetics. Louis Pitelka: plant ecology, including population biology and ecosystem dynamics. Cathlyn D. Stylinski: environmental science education and scientific inquiry in precollege classrooms.

Chesapeake Biological Laboratory. Robert Anderson: biochemical toxicology, effects of stress on marine invertebrate immunology. Joel Baker: behavior of organic contaminants in marine/estuarine systems. Walter R. Boynton: nutrient cycling in estuarine systems, food-web studies. H. Rodger Harvey: sources and fates of organic compounds in aquatic environments. Edward D. Houde: fishery science, population dynamics, ecology of the larval stage. Roberta L. Marinelli: benthic ecology, animal-sediment interaction, benthic larval recruitment, modeling benthic processes. Thomas J. Miller: fish ecology, population dynamics. Carys L. Mitchelmore: molecular, biochemical, and cellular responses of aquatic organisms to inorganic and organic pollutants. Margaret Palmer: stream and estuarine ecology and hydrodynamics. Kennedy T. Paynter Jr.: physiology and biochemistry of estuarine organisms, oyster reef restoration. Christopher L. Rowe: physiological, population and community responses to sublethal levels of pollutants. David H. Secor: fisheries ecology, demographics, migration. Marcelino Suzuki: marine microbial ecology, application of molecular approaches to the study of aquatic microbes. Mario Tamburri: coastal sensor development, ecosystem monitoring. Kenneth Tenore: bioenergetics of detritus-based food chains, nutrition of marine invertebrates. Robert E. Ulanowicz: estuarine food-chain dynamics, hydrological-biological modeling. David A. Wright: comparative physiology of marine and estuarine animals, inorganic pollutants.

Horn Point Laboratory. William Boicourt: physical oceanography, continental shelf and estuarine circulation. Shenn-yu Chao: physical oceanography, continental shelf and slope circulation, western boundary currents. Louis A. Codispoti: chemical oceanography, oceanic nitrogen cycle. Victoria J. Coles: observation and modeling of seasonal to climate-scale variability in ocean circulation. Jeffrey C. Cornwell: nutrient, metal, and sulfur cycling in estuaries and wetlands. Byron C. Crump: microbial ecology, bacterial and Achaeal diversity, organic matter and nutrient cycling. William C. Dennison: coastal ecosystem ecology, ecophysiology of marine plants. Thomas R. Fisher Jr.: nitrogen cycles in Atlantic coastal plain estuaries, nutrient cycling in tropical lakes. Patricia M. Glibert: phytoplankton and microplankton ecology, nitrogen cycling, photosynthesis. Lawrence W. Harding: biological oceanography, phytoplankton physiology and ecology. Raleigh R. Hood: phytoplankton production and light response, modeling of primary production. Todd M. Kana: phytoplankton physiology. W. Michael Kemp: systems ecology, watershed nutrient budgets, submerged aquatic vegetation. Victor S. Kennedy: ecology and dynamics of benthic communities, particularly bivalves. Evamaria W. Koch: ecology of submerged aquatic vegetation and coastal seagrass ecosystems. Andrew Lazur: food and baitfish culture, integration of aquaculture with agriculture for nutrient reduction. Ming Li: geophysical fluid dynamics, ocean mixing processes, numerical modeling, biological/physical interactions, marine pollution. Thomas Malone: phytoplankton ecology and nutrient cycling. Donald W. Meritt: oyster aquaculture and restoration. Laura Murray: wetlands, seagrass ecology. Roger I. E. Newell: physiological and behavioral adaptations of invertebrates, especially bivalve mollusks. Elizabeth W. North: biological-physical interactions hydrodynamics and particle trajectory modeling, ichthyoplankton and zooplankton ecology. Judith O'Neil: cyanobacteria ecophysiology, plankton trophodynamics. Michael R. Roman: zooplankton ecology, plankton food-chain energetics, detrital food chains. Lawrence P. Sanford: physical oceanography, geophysical boundary layers, turbulence and mixing processes. J. Court Stevenson: marsh ecology, nutrient loading in coastal watersheds. Diane Stoecker: role of heterotrophic and mixotrophic protists in food webs. William Van Heukelem: behavior of crab and oyster larvae as related to dispersal.

Center of Marine Biotechnology. Hafiz Ahmed: biological roles of galectins in early embryo development and immune function, structure-function studies of galectins. Robert M. Belas Jr.: sensory transduction and genetic regulation of gram-negative bacteria. Feng Chen: bacterio- and phyto-plankton production, biomass and growth in aquatic environments; ecological interaction among marine viruses, bacteria, and phytoplankton; phylogenetic relationship and co-evolution among marine microorganisms. J. Sook Chung: response of crustaceans to the neurotransmitters, neurohormones, hormone, and phermones that regulate critical life cycle events. Shiladitya DasSarma: halophilic archael genomes; structure, function, and evolution of genomes. Shao-Jun (Jim) Du: cellular and molecular mechanisms controlling differentiation of muscle and nerve cells during embryogenesis. John D. Hansen: genetic organization of the rainbow trout MHC and its inducibility during viral infection. Russell T. Hill: natural products from marine microorganisms; actinomycete molecular biology and ecology, use in bioremediation. Anwarul Huq: isolation, identification, and characterization of enteric bacterial agents using conventional, immunological, and genetic methods. Rosemary Jagus: developmental regulation of gene expression in sea urchin embryos. Zeev Pancer: vertebrate adaptive immunity, antigen receptors of jawless vertebrates. Allen Place: biochemical adaptations in marine organisms. Frank T. Robb: genetics of thermophilic marine bacteria. Harold J. Schreier: adaptation of microorganisms to extreme environments, biochemistry and molecular biology of Archaea. Kevin R. Sowers: molecular genetics and adaptation of anaerobic archaebacteria. John M. Trant: reproductive physiology, molecular endocrinology. Gerardo Vasta: cellular nonself recognition and cell-cell interactions. Yonathan Zohar: physiology and endocrinology of fish reproduction.

UNIVERSITY OF PUERTO RICO, MAYAGÜEZ CAMPUS

Department of Marine Sciences

Programs of Study

The Department of Marine Sciences is a graduate department offering instruction leading to the degrees of Master of Marine Sciences (M.M.S.) and Doctor of Philosophy (Ph.D.) in marine sciences. The primary aim of the Department is to train marine scientists for careers in teaching, research, and management of marine resources. Students specialize in biological, chemical, geological, or physical oceanography; fisheries biology; or aquaculture or through core courses and electives. Much of the teaching and research is carried out at the marine station, 22 miles south of Mayagüez, but students are able to elect courses in other departments and have access to all the facilities at the main campus.

A minimum of 35 semester hours of credit in approved graduate courses is required for the M.M.S. degree; 72 for the Ph.D. Courses in the Department of Marine Sciences are taught in English and Spanish. Because Puerto Rico has a Spanish culture and the University is bilingual, candidates are expected to gain a functional knowledge of Spanish as well as English before finishing their degrees. Further requirements for the M.M.S. are residence of at least one academic year, passing a Departmental comprehensive examination, completing a satisfactory thesis, and passing a thesis defense examination. For the Ph.D. degree, residence of at least two years, passing a qualifying examination, subsequently passing a comprehensive examination, completing a satisfactory dissertation, and passing a final examination in defense of the dissertation are required. Prospective students can visit the Web page at http://cima.uprm.edu.

Research Facilities

Modern teaching and sophisticated research facilities are available both on campus and at the field station. A department library specializing in marine science publications is located at the main campus. The field station is on 18-acre Magueyes Island within a protected embayment off La Parguera, 22 miles from Mayagüez. In addition to classroom and laboratory facilities, the marine station has indoor and outdoor aquaria and tanks with running seawater and three museums containing reference collections of fish, invertebrates, and algae. Boats include a 51-foot Thompson trawler, a 35-foot diesel Downeast, and a number of medium and small open boats. Research facilities for warm-water aquaculture include some 8 acres of earthen ponds, two hatcheries and numerous concrete tanks, plastic pools, and aquarium facilities with running water available for controlled environmental studies.

Financial Aid

Some graduate students receive tuition waivers and stipends for their work as teaching or research assistants. Student support is also available through research grants awarded to faculty members and through the University of Puerto Rico, Mayagüez (UPRM), Financial Aid Office.

Cost of Study

Residents carrying a full program (9–12 credits) paid $100 per credit hour in 2005–06. General fees of approximately $100 per semester are added to these costs. Resident status may be established in one year. Nonresidents, including aliens, pay $3500 per year, except for students from U.S. institutions having reciprocal tuition-reduction agreements with Puerto Rico (a list is available on request). Students must have an accepted health insurance or must acquire the health insurance available from the University.

Living and Housing Costs

Apartments and houses can be found in Mayagüez, San Germán, Lajas, and La Parguera for roughly $250 to $450 per month. Single rooms may be obtained for less. Single students in the Department frequently share apartments.

Student Group

Total enrollment at the Mayagüez campus of the University of Puerto Rico is about 12,500. The Department's enrollment is 80–120 graduate students.

Location

Mayagüez, the third-largest city in Puerto Rico, has a population of 200,000. It is a seaport on the west coast of the island. The economy of the city centers largely on shipping, commercial fishing, light industry, and the University. San Germán and Lajas, where much of the University community lives, are 10 and 18 miles south of Mayagüez, respectively. The main campus of Inter-American University is located in San Germán. An increasing number of concerts, art exhibits, and other cultural activities are arranged in Mayagüez, Lajas, and San Germán, although San Juan, a 3-hour drive from Mayagüez, remains the principal island center for cultural events (for example, the yearly Casals Festival, drama, and art as well as other activities sponsored by various civic entities and the Institute of Puerto Rican Culture). Many of these events are held at the Center for Fine Arts. In addition, repertory theaters are becoming very popular in the metropolitan area.

The University and The Department

The University of Puerto Rico, Mayagüez (http://www.uprm.edu), had its beginning in 1911 as the College of Agriculture, an extension of the University in Río Piedras. In 1912 the name was changed to the College of Agriculture and Mechanic Arts. Following a general reform of the University in 1942, the college became a regular campus of the University under a vice-chancellor and, in 1966, an autonomous campus with its own chancellor. The marine sciences program began in 1954 with the establishment of the Institute of Marine Biology. A master's degree program in marine biology was initiated in 1963. In 1968, the institute became the Department of Marine Sciences, a graduate department, with its own academic staff. The doctoral program began in 1972. Research remains an important function of the Department.

Applying

Application forms are available from the Graduate School and at http://grad.uprm.edu. An undergraduate science degree is required. The applicant should have had at least basic courses in biology, chemistry, physics, geology, and mathematics through calculus. An engineering degree may be acceptable under some circumstances. Applications should be submitted before February 15 for the fall semester and before September 15 for the spring semester. A late fee is applied to applications received after the set deadlines.

Correspondence and Information

Director, Graduate School
Box 9020
University of Puerto Rico
Mayagüez, Puerto Rico 00681-9020
Web site: http://grad.uprm.edu

Director, Department of Marine Sciences
Box 9013
University of Puerto Rico
Mayagüez, Puerto Rico 00681-9013
E-mail: director@cima.uprm.edu
Web site: http://cima.uprm.edu

University of Puerto Rico, Mayagüez Campus

THE FACULTY AND THEIR RESEARCH

Dallas E. Alston, Professor; Ph.D. (invertebrate aquaculture), Auburn. Culture of invertebrate organisms.
Nilda E. Aponte, Professor and Director of the Department; Ph.D. (marine botany), Puerto Rico, Mayagüez. Taxonomy, morphology, and life history of marine algae.
Richard S. Appeldoorn, Professor; Ph.D. (fisheries biology), Rhode Island. Fisheries biology.
Roy A. Armstrong, Professor; Ph.D. (biooptical oceanography), Puerto Rico, Mayagüez. Remote sensing, water optics.
David L. Ballantine, Professor; Ph.D. (marine botany), Puerto Rico, Mayagüez. Taxonomy and ecology of marine algae.
Jorge E. Corredor, Professor; Ph.D. (chemical oceanography), Miami (Florida). Chemical oceanography, pollution.
Dannie A. Hensley, Professor; Ph.D. (ichthyology), South Florida. Systematics and ecology of fishes.
Aurelio Mercado Irizarry, Professor; M.S. (physical oceanography), Miami (Florida). Geophysical fluid dynamics.
John M. Kubaryk, Professor and Associate Director of the Department; Ph.D. (seafood technology), Auburn. Seafood technology, aquatic nutrition, water quality.
José M. López-Díaz, Professor; Ph.D. (environmental chemistry), Texas at Dallas. Water pollution control.
Ricardo Cortés Maldonado, Professor; M.S. (aquaculture), Puerto Rico, Mayagüez. Aquaculture.
Ernesto Otero Morales, Associate Investigator; Ph.D. (microbial ecology), Georgia. Microbial utilization of organic matter, microbial diversity and ecological application of stable isotopes.
Julio Morell, Investigator; M.S. (chemical oceanography), Puerto Rico, Mayagüez. Biogeochemistry and environmental chemistry.
Govind S. Nadathur, Professor; Ph.D. (molecular microbiology), Gujarat (India). Genetics and biotechnology of marine organisms.
Jorge R. García Sais, Investigator; Ph.D. (biological oceanography), Rhode Island. Zooplankton ecology.
Nikolaos V. Schizas, Associate Professor; Ph.D. (invertebrate zoology, molecular evolution), South Carolina. Molecular invertebrate zoology.
Wilford Schmidt, Associate Professor; Ph. D. (physical oceanography, coastal processes), California, San Diego (Scripps). Nearshore and coastal physical processes, instrument development.
Thomas R. Tosteson, Professor; Ph.D. (physiology), Pennsylvania. Marine physiology and pharmacology.
Ernesto Weil, Professor; Ph.D. (zoology), Texas at Austin. Coral systematics, ecology and evolution, coral reef ecology.
Ernest H. Williams, Professor; Ph.D. (parasitology), Auburn. Systematics and culture of parasites of fishes.
Amos Winter, Professor; Ph.D. (paleoceanography), Hebrew (Jerusalem). Paleoceanography.
Paul Yoshioka, Professor; Ph.D. (marine ecology), California, San Diego. Marine ecology.
Baqar R. Zaidi, Investigator; Ph.D. (marine microbiology and physiology), Puerto Rico, Mayagüez. Marine physiology, bioremediation.

Field station at Magueyes Island, La Parguera, Puerto Rico.

THE UNIVERSITY OF TEXAS AT AUSTIN

Department of Marine Science

Programs of Study

The Department of Marine Science offers research opportunities and course work leading to the M.S. and Ph.D. degrees in marine science. Graduate students usually begin their academic program with course work on the Austin campus and move to the University of Texas Marine Science Institute at Port Aransas for specialized advanced courses and thesis or dissertation research. Core courses are required in several subdisciplines, including marine ecosystem dynamics, marine biogeochemistry, and adaptations to the marine environment. Areas of research available in Port Aransas include physiology and ecology of marine organisms, biological oceanography, geochemistry, marine environmental quality, coastal processes, and mariculture.

Research Facilities

The Marine Science Institute in Port Aransas is located near Corpus Christi and provides opportunities to study living organisms in the laboratory and under field conditions. A wide variety of environments are readily accessible, such as the pass connecting Corpus Christi Bay with the Gulf of Mexico, the continental shelf, and many bays and estuaries, including brackish estuaries and the hypersaline Laguna Madre. There are outside open and covered seawater tanks, a pier lab with running seawater, a reference collection of most of the plants and animals of the area, and controlled-environment chambers. Vessels include the *R/V Longhorn,* a 105-foot research vessel with navigation and laboratory capabilities for most research projects; the *R/V Katy,* a 57-foot vessel with dredge and trawl equipment for collection of specimens; and several smaller boats. A remote terminal is linked with the computation center in Austin. Laboratories are equipped to study animal physiology, toxicology, bacterial and algal physiology, bacterial and algal ecology, fish ecology, marine phycology, mariculture, sea grass ecology and physiology, and geochemistry. Research is under way in benthic ecology, biological oceanography, fish behavior, invertebrate biology, phytoplankton ecology, and taxonomy of marine organisms. The institute also provides teaching facilities in Port Aransas, including upper-division and graduate course offerings during the summer. Facilities are available on the Austin campus for research in marine sedimentology and in marine mineral deposits, including genesis, exploration, and recovery.

Financial Aid

Research and teaching assistantships are available through graduate advisers or the department chairman. E. J. Lund Fellowships and Scholarships for research at the Marine Science Institute are awarded annually.

Cost of Study

In 2005–06, tuition and required fees for Texas residents and any students holding an assistantship were approximately $2900 per semester for 9 credit hours. Nonresident tuition and required fees for 9 credit hours totaled $5000 per semester.

Living and Housing Costs

In Port Aransas, furnished University apartments are available for students at approximately $410–$555 per month plus utilities. Non-University housing off campus costs approximately $600–$800 per month. For the Department of Marine Science's Summer Program, dormitory and dining facilities are also available to registered students.

In Austin, University dormitories and apartments, furnished and unfurnished, are available. Rooms and apartments are conveniently situated near the campus. There is also a shuttle bus service.

Student Group

The enrollment of the University of Texas at Austin is more than 50,000, including approximately 12,000 graduate students. The College of Natural Sciences has about 1,500 students. An average of 25 graduate students reside at the Marine Science Institute in Port Aransas.

Location

Austin is the state capital, with a population of approximately 550,000. Cultural events sponsored by the University are abundant, and there are many recreational facilities available. Port Aransas is a small coastal town approximately 200 miles south of Austin, where the Gulf of Mexico and surrounding bays and estuaries provide excellent boating, fishing, and swimming.

The University

The University of Texas at Austin was founded in 1883 and is part of the University of Texas System. The Department of Marine Science is in the College of Natural Sciences.

Applying

Prospective students must apply to both the Director of Admissions of the Graduate School and the Graduate Studies Committee of the Department of Marine Science in order to be considered for admission to the department. Application forms may be obtained from the Graduate School and from the department online. Only admission applications completed by January 1 can be considered for fellowship or teaching assistantship awards.

Correspondence and Information

Graduate Adviser
Marine Science Institute
The University of Texas at Austin
750 Channel View Drive
Port Aransas, Texas 78373-5015
Phone: 361-749-6721
E-mail: gradinfo@utmsi.utexas.edu
Web site: http://www.utmsi.utexas.edu

Chairman
Department of Marine Science
The University of Texas at Austin
750 Channel View Drive
Port Aransas, Texas 78373-5015
Phone: 361-749-6721

The University of Texas at Austin

THE FACULTY AND THEIR RESEARCH

Edward J. Buskey, Professor, Port Aransas; Ph.D., Rhode Island, 1983. Behavior and sensory perception of zooplankton, bioluminescence of marine plankton, predator-prey interactions, role of zooplankton grazers in phytoplankton bloom dynamics.

Kenneth H. Dunton, Professor, Port Aransas; Ph.D., Alaska, Fairbanks, 1985. Transport and fate of carbon and nitrogen into estuarine food webs from adjacent watersheds, benthic-pelagic coupling and role of sediment biogeochemical processes, biology of high-latitude kelps, use of stable isotopes in the elucidation of trophic structure and carbon flow in aquatic food webs, application of nitrogen isotopes as tracers of anthropogenic N in estuarine and coastal habitats.

Henrietta N. Edmonds, Associate Professor, Port Aransas; Ph.D., MIT and Woods Hole Oceanographic Institution, 1997. Natural and anthropogenic radionuclides as tracers of ocean biogeochemistry and circulation, Arctic and northern North Atlantic oceanography, Arctic Ocean hydrothermal systems, scavenging and particle cycling in hydrothermal plumes, marine geochemistry of ^{230}Th and ^{231}Pa, groundwater inputs to Texas bays and estuaries.

Deana L. Erdner, Assistant Professor, Port Aransas; Ph.D., MIT and Woods Hole Oceanographic Institution, 1997. Harmful algae; marine dinoflagellates; physiology, biochemistry, and genetics of nutrient stress; population biology; molecular methods for cell identification and enumeration.

Lee A. Fuiman, Professor, Port Aransas (also Professor in the School of Biological Sciences–Integrative Biology); Ph.D., Michigan, 1983. Behavioral, sensory, and developmental ecology of aquatic animals; development of fish larvae.

Wayne S. Gardner, Professor, Port Aransas; Ph.D., Wisconsin, 1971. Nitrogen dynamics in water column and sediments, nutrient-organism interactions in coastal ecosystems.

G. Joan Holt, Professor, Port Aransas; Ph.D., Texas A&M, 1976. Physiological ecology of larval fish and biochemical measures of adaptation, studies of larval fish in estuarine and oceanic nursery grounds, growth and development in controlled culture, larval fish feeding and nutrition and condition, tropical reef fish, marine aquaculture.

Izhar A. Khan, Assistant Professor, Port Aransas; Ph.D., Banaras Hindu (India), 1990. Reproductive physiology and neuroendocrinology of marine fishes; neuroendocrine control of puberty, sexual maturation, and spawning; induced breeding of captive broodstock fishes; mechanisms of neuroendocrine toxicity of environmental chemicals.

James W. McClelland, Assistant Professor, Port Aransas; Ph.D., Boston University, 1998. Effects of human activity on water, carbon, and nutrient fluxes from land to sea; responses of estuarine and coastal food webs to changes in land-derived resources; use of stable isotopes and other natural tracers to follow water and waterborne constituents across the land-sea interface.

Dong-Ha Min, Assistant Professor, Port Aransas; Ph.D., California, San Diego, 1999. Dissolved oxygen and anthropogenic halocarbon compounds as tracers of large-scale ocean circulation and ventilation processes, temporal changes in the Southern Ocean and the North Atlantic Ocean, rapid climate change responses in the marginal seas and coastal environments.

Paul A. Montagna, Professor, Port Aransas; Ph.D., South Carolina, 1983. Benthic invertebrate community ecology, ecosystems ecology, environmental studies, resource management.

B. Scott Nunez, Assistant Professor, Port Aransas; Ph.D., LSU, 1996. Regulation of steroid hormone action; molecular endocrinology of stress; comparative endocrinology of osmoregulation; biology of sharks, skates, and rays.

Tamara K. Pease, Assistant Professor, Port Aransas; Ph.D., North Carolina at Chapel Hill, 2000. Sources, behavior, and cycling of organic compounds in the marine and estuarine environments; use of specific organic compounds to elucidate the sources and processes that control the distributions of organic matter; biogeochemistry; molecular stable and radioisotope organic geochemistry.

Peter Thomas, Professor, Port Aransas (also Professor in the School of Biological Sciences-Integrative Biology); Ph.D., Leicester (England), 1977. Environmental and neuroendocrine control of reproduction; molecular mechanisms of steroid hormone action; cloning and characterization of steroid nuclear and membrane receptors; gonadal and gamete physiology; applications of endocrinology in fish culture; environmental and reproductive toxicology of marine fishes, especially mechanisms of endocrine disruption and molecular biomarkers of reproductive impairment in fish populations exposed to pollutants and hypoxia.

Tracy A. Villareal, Associate Professor, Port Aransas; Ph.D., Rhode Island, 1989. Ecology of harmful algal blooms, ciguatera-causing dinoflagellates, ecology of large oceanic phytoplankton.

The University of Texas at Austin

SELECTED PUBLICATIONS

Clarke, R. D., and **E. J. Buskey.** Effects of water motion and prey behavior on zooplankton capture by two coral reef fishes. *Mar. Biol.* 146:1145–55, 2005.

Buskey, E. J. Behavioral characteristics of copepods that affect their suitability as food for larval fishes. In *Copepods in Aquaculture,* pp. 91–106, eds. C.-S. Lee, P. J. O'Bryen, and N. H. Marcus. Blackwell Publishing, 2005.

Buskey, E. J., C. J. Hyatt, and C. L. Speekmann. Trophic interactions within the planktonic food web in mangrove channels of Twin Cays, Belize, Central America. *Atoll Res. Bull.* 529:1–22, 2004.

Buskey, E. J. Behavioral adaptations of the cubozoan medusae *Tripedalia cystophora* for feeding on copepod swarms. *Mar. Biol.* 142:225–32, 2003.

Buskey, E. J., and D. K. Hartline. High speed video analysis of the escape responses of the copepod *Acartia tonsa* to shadows. *Biol. Bull.* 204:28–37, 2003.

Dunton, K. H., et al. Multi-decadal synthesis of benthic-pelagic coupling in the western arctic: Role of cross-shelf advective processes. *Deep-Sea Res.,* in press.

Alexander, H. D., and **K. H. Dunton.** Treated wastewater effluent as an alternative freshwater source in a hypersaline salt marsh: Impacts on salinity, inorganic nitrogen, and emergent vegetation. *J. Coast. Res.,* in press.

Dunton, K. H. $\delta^{15}C$ and $\delta^{13}C$ measurements of Antarctic peninsular fauna: Trophic relationships and assimilation of benthic seaweeds. *Am. Zool.* 41(1):99–112, 2001.

Dunton, K. H., and S. V. Schonberg. The benthic faunal assemblage of the Boulder Patch kelp community. In *The Natural History of an Arctic Oil Field,* pp. 338–59, eds. J. C. Truett and S. R. Johnson. Academic Press, 2000.

Lee, K.-S., and **K. H. Dunton.** Inorganic nitrogen acquisition in the seagrass *Thalassia testudinum:* Development of a whole-plant nitrogen budget. *Limnol. Oceanogr.* 44(5):1204–15, 1999.

Edmonds, H. N., S. B. Moran, H. Cheng, and R. L. Edwards. ^{230}Th and ^{231}Pa in the Arctic Ocean: Implications for particle fluxes and basin-scale Th/Pa fractionation. *Earth Planet. Sci. Lett.* 227:155–67, 2004.

Baker, E. T., and **H. N. Edmonds** et al. Hydrothermal venting in magma deserts: The ultraslow-spreading Gakkel and Southwest Indian Ridges. *Geochem. Geophys. Geosyst.* 5(8):Q08002, doi: 10.1029/2004GC000712, 2004.

Severmann, S., et al. **(H. N. Edmonds).** Origin of the Fe isotope composition of the oceans as inferred from the Rainbow vent site, Mid-Atlantic Ridge, 36°14'N. *Earth Planet. Sci. Lett.* 225(1–2):63–76, 2004.

Edmonds, H. N., and C. R. German. Particle geochemistry in the Rainbow hydrothermal plume, Mid-Atlantic Ridge. *Geochim. Cosmochim. Acta* 68(4): 759–72, 2004.

Edmonds, H. N., et al. Discovery of abundant hydrothermal venting on the ultraslow-spreading Gakkel ridge in the Arctic Ocean. *Nature* 421:252–6, 2003.

Dyhrman, S. T., and **D. L. Erdner** et al. Molecular quantification of toxic *Alexandrium* fundyense in the Gulf of Maine using a newly developed PCR-based assay. *Harmful Algae,* in press.

Hackett, J. D., D. M. Anderson, **D. L. Erdner,** and D. Bhattacharya. Dinoflagellates: A remarkable evolutionary experiment. *Am. J. Botany* 91:1523–34, 2004.

Lindell, D., and **D. Erdner** et al. The nitrogen stress response of *Prochlorococcus* strain PCC 9511 (oxyphotobacteria) involves contrasting regulation of ntcA and amtl. *J. Phycol.* 38(6):1113–24, 2002.

Beeson, K. E., and **D. L. Erdner** et al. Differentiation of plasmids in marine diazotroph assemblages determined by randomly amplified polymorphic DNA analysis. *Microbiology* 148:179–89, 2002.

Erdner, D. L., and D. M. Anderson. Ferredoxin and flavodoxin as biochemical indicators of iron limitation during open-ocean iron enrichment. *Limnol. Oceanogr.* 44(7):1609–15, 1999.

Alvarez, M. C., and **L. A. Fuiman.** Environmental levels of atrazine and its degradation products impair survival skills and growth of red drum larvae. *Aquat. Toxicol.* 74:229–41, 2005.

Fuiman, L. A., J. H. Cowan Jr., M. E. Smith, and J. P. O'Neal. Behavior and recruitment success in fish larvae: Variation with growth rate and the batch effect. *Can. J. Fish. Aquat. Sci.* 62:1337–49, 2005.

Smith, M. E., and **L. A. Fuiman.** Behavioral performance of wild-caught and laboratory-reared red drum *Sciaenops ocellatus* (Linnaeus) larvae. *J. Exp. Mar. Biol. Ecol.* 302:17–33, 2004.

Fuiman, L. A., D. M. Higgs, and K. R. Poling. Changing structure and function of the ear and lateral line system of fishes during development. *Am. Fish. Soc. Symp.* 40:117–44, 2004.

Williams, T. M., **L. A. Fuiman,** M. Horning, and R. W. Davis. The cost of foraging by a marine predator, the Weddell seal *(Leptonychotes weddellii):* Pricing by the stroke. *J. Exp. Biol.* 207:973–82, 2004.

Jochem, F. J., M. J. McCarthy, and **W. S. Gardner.** Microbial ammonium cycling in the Mississippi River plume during the drought spring of 2000. *J. Plankton Res.* 1265–75, 2004.

Gardner, W. S., et al. The distribution and dynamics of nitrogen and microbial plankton in southern Lake Michigan during spring transition 1999–2000. *J. Geophys. Res.* 109:CO3007, 2004.

An, S., and **W. S. Gardner.** Dissimilatory nitrate reduction to ammonium (DNRA) as a nitrogen link, versus denitrification as a sink in a shallow estuary (Laguna Madre/Baffin Bay, Texas). *Mar. Ecol. Prog. Ser.* 237:41–50, 2002.

An, S., **W. S. Gardner,** and T. Kana. Simultaneous measurement of denitrification and nitrogen fixation using isotope pairing with membrane inlet mass spectrometry analysis. *Appl. Environ. Microbiol.* 67:1171–8, 2001.

Gardner, W. S., et al. Nitrogen cycling rates and light effects in tropical Lake Maracaibo, Venezuela. *Limnol. Oceanogr.* 43(8):1814–25, 1998.

Faulk, C. K., and **G. J. Holt.** Advances in rearing cobia *Rachycentron canadum* larvae in recirculating aquaculture systems: Live prey enrichment and greenwater culture. *Aquaculture* 249:231–3, 2005.

Holt, G. J. Research on culturing the early life history stages of marine ornamental species. In *Marine Ornamental Species: Collection, Culture and Conservation,* pp. 252–4, eds. J. C. Cato and C. L. Brown. Iowa State Press, 2003.

Applebaum, S. L., and **G. J. Holt.** The digestive protease, chymotrypsin, as an indicator of nutritional condition in larval red drum *(Sciaenops ocellatus). Mar. Biol.* 142:1159–67, 2003.

Holt, G. J. Ecophysiology, growth, and development of larvae and juveniles for aquaculture. *Fish. Sci.* 68(1):867–71, 2002.

Holt, G. J., and S. A. Holt. Effects of variable salinity on reproduction and early life stages of spotted seatrout. In *Biology of Spotted Seatrout,* pp. 135–45, ed. S. A. Bartone. Boca Raton, Fla.: CRC Marine Biology Series, CRC Press, 2002.

Mohamed, J. S., et al. **(I. A. Khan).** Isolation, cloning, and expression of three prepro-GnRH mRNAs in Atlantic croaker brain and pituitary. *J. Comp. Neurol.* 488:384–95, 2005.

Khan, I. A., and **P. Thomas.** Vitamin E co-treatment reduces Aroclor 1254-induced impairment of reproductive neuroendocrine function in Atlantic croaker. *Mar. Environ. Res.* 58:333–6, 2004.

Khan, I. A., and **P. Thomas.** Disruption of neuroendocrine control of luteinizing hormone secretion in Atlantic croaker by Aroclor 1254 involves inhibition of hypothalamic tryptophan hydroxylase activity. *Biol. Reprod.* 64:955–64, 2001.

Khan, I. A., and **P. Thomas.** GABA exerts stimulatory and inhibitory influences on gonadotropin II secretion in the Atlantic croaker *(Micropogonias undulatus). Neuroendocrinology* 69:261–8, 1999.

Khan, I. A., et al. Gonadal stage-dependent effects of gonadal steroids on gonadotropin II secretion in the Atlantic croaker *(Micropogonias undulatus). Biol. Reprod.* 61:834–41, 1999.

Cooper, L., et al. **(J. McClelland).** Linkage among runoff, dissolved organic carbon, and the stable oxygen isotope composition of seawater and other water mass indicators in the Arctic Ocean. *J. Geophys. Res.* 110:G02013, doi: 1029/2005JG000031, 2005.

McClelland, J. W., R. M. Holmes, B. J. Peterson, and M. Stieglitz. Increasing river discharge in the Eurasian Arctic: Consideration of dams,

The University of Texas at Austin

permafrost thaw, and fires as potential agents of change. *J. Geophys. Res.* 109:D18102, doi: 10.1029/2004JD004583, 2004.

McClelland, J. W., C. M. Holl, and J. P. Montoya. Attributing low $\delta^{15}N$ values of zooplankton to an N_2-fixing source in the tropical North Atlantic: Insights provided by stable isotope ratios of amino acids. *Deep-Sea Res. I* 50:849–61, 2003.

Peterson, B. J., et al. **(J. W. McClelland).** Increasing river discharge to the Arctic Ocean. *Science* 298:2171–3, 2002.

Valiela, I., M. Geist, **J. McClelland,** and G. Tomasky. Nitrogen loadings from watersheds to estuaries: Verification of the Waquoit Bay Nitrogen Loading Model. *Biogeochemistry* 49:277–93, 2000.

Min, D.-H., and K. Keller. Errors in estimated temporal tracer trends due to changes in the historical observation network: A case study of oxygen trends in the Southern Ocean. *Ocean Polar Res.* 27(2):189–95, 2005.

Min, D.-H., and M. J. Warner. Basin-wide circulation and ventilation study in the East Sea (Sea of Japan) using chlorofluorocarbon tracers. *Deep-Sea Res. II* 52:1580–616, 2005.

Talley, L. D., et al. **(D.-H. Min).** Atlas of Japan (East) Sea hydrographic properties in summer, 1999. *Prog. Ocean.* 61:277–348, 2004.

Min, D.-H., J. L. Bullister, and R. F. Weiss. Anomalous chlorofluorocarbons in the Southern California Borderland Basin. *Geophys. Res. Lett.* 29(20):1955, doi: 10.1029/2002GL015408, 2002.

Min, D.-H., J. L. Bullister, and R. F. Weiss. Constant ventilation age of thermocline water in the eastern subtropical North Pacific Ocean from chlorofluorocarbon measurements over a 12-year period. *Geophys. Res. Lett.* 27:3909–12, 2000.

Applebaum, S., **P. A. Montagna,** and C. Ritter. Status and trends of dissolved oxygen in Corpus Christi Bay, Texas, U.S.A. *Environ. Monit. Assess.* 107:297–311, 2005.

Montagna, P. A., M. Alber, P. Doering, and M. S. Connor. Freshwater inflow: Science, policy, management. *Estuaries* 25:1243–5, 2002.

Montagna, P. A., R. D. Kalke, and C. Ritter. Effect of restored freshwater inflow on macrofauna and meiofauna in upper Rincon Bayou, Texas, USA. *Estuaries* 25:1436–47, 2002.

Montagna, P. A., S. C. Jarvis, and M. C. Kennicutt II. Distinguishing between contaminant and reef effects on meiofauna near offshore hydrocarbon platforms in the Gulf of Mexico. *Can. J. Fish. Aquat. Sci.* 59:1584–92, 2002.

Montagna, P. A., and J. Li. Modeling contaminant effects on deposit feeding nematodes near Gulf of Mexico production platforms. *Ecol. Model.* 98:151–62, 1997.

Nunez, B. S., A. E. Evans, P. M. Piermarini, and S. A. Applebaum. Cloning and characterization of cDNAs encoding steroidogenic acute regulatory protein (StAR) from freshwater stingrays (Potamotrygon sp). *Mol. Endocrinol.* 35:557–69, 2005.

Nunez, B. S., et al. Interaction between the interferon signaling pathway and the human glucocorticoid receptor gene 1A promoter. *Endocrinology* 146(3):1449–57, 2005.

Nunez, B. S., and W. V. Vedeckis. Monitoring nuclear receptor function. In *Receptors: New Objectives, New Techniques,* pp. 233–56, eds. S. C. Stanford and R. W. Horton. Oxford: Oxford University Press, 2001.

Nunez, B. S., and J. M. Trant. Regulation of interrenal steroidogenesis in the Atlantic stingray *(Dasyatis sabina). J. Exp. Zool.* 284:517–25, 1999.

Nunez, B. S., and J. M. Trant. Molecular biology and enzymology of elasmobranch 3β-hydroxysteroid dehydrogenase. *Fish Physiol. Biochem.* 19(4):293–304, 1998.

Wakeham, S. G., A. P. McNichol, J. Kostka, and **T. K. Pease.** Natural-abundance radiocarbon as a tracer of assimilation of petroleum carbon by bacteria in salt marsh sediments. *Geochim. Cosmochim. Acta,* in press.

Muri, G., S. G. Wakeham, **T. K. Pease,** and J. Faganeli. Evaluation of lipid biomarkers as indicators of changes in organic matter delivery to sediments from Lake Planina, a remote mountain lake in NW Slovenia. *Org. Geochem.* 35:1083–93, 2004.

Wakeham, S. G., **T. K. Pease,** and R. Benner. Hydroxy fatty acids in marine dissolved organic matter as indicators of bacterial membrane material. *Org. Geochem.* 34:857–68, 2003.

Hee, C., **T. K. Pease,** M. J. Alpern, and C. S. Martens. DOC production and consumption in anoxic marine sediments: A pulsed-tracer experiment. *Limnol. Oceanogr.* 46:1908–20, 2002.

Pease, T. K., et al. Simulated degradation of glyceryl ethers by hydrous and flash pyrolysis. *Org. Geochem.* 29:979–88, 1998.

Thomas, P., and **I. A. Khan.** Disruption of nongenomic steroid actions on gametes and serotonergic pathways controlling reproductive neuroendocrine function by environmental chemicals. In *Endocrine Disruptors: Effects on Male and Female Reproductive Systems,* vol. 2, pp. 3–45, ed. R. K. Naz. Boca Raton: CRC Press, 2005.

Thomas, P., Y. Pang, E. J. Filardo, and J. Dong. Identity of an estrogen membrane receptor coupled to a G-protein in human breast cancer cells. *Endocrinology* 146:624–32, 2005.

Thomas, P., and K. Doughty. Disruption of rapid, nongenomic steroid actions by environmental chemicals: Interference with progestin stimulation of sperm motility in Atlantic croaker. *Environ. Sci. Technol.* 38:328–32, 2004.

Thomas, P. Nongenomic steroid actions initiated at the cell surface: Lessons from studies in fish. *Fish. Biochem. Physiol.* 28:3–12, 2004.

Zhu, Y., et al. **(P. Thomas).** Cloning, expression, and characterization of a membrane progestin receptor and evidence it is an intermediary in meiotic maturation of fish oocytes. *Proc. Natl. Acad. Sci. U.S.A.* 100(5):2231–6, 2003.

Zhu. U., J. Bond, and **P. Thomas.** Identification, classification, and partial characterization of genes in humans and other vertebrates homologous to a fish membrane progestin receptor. *Proc. Natl. Acad. Sci. U.S.A.* 100:2237–42, 2003.

Pilskaln, C. H., and **T. A. Villareal** et al. High concentrations of marine snow and diatom algal mats in the North Pacific Subtropical Gyre: Implications for carbon and nitrogen cycles in the oligotrophic ocean. *Deep-Sea Res. I* 52:2315–32, 2005.

Biegalski, S. R., and **T. A. Villareal.** Correlations between atmospheric aerosol trace element concentrations and red tide at Port Aransas, Texas, on the Gulf of Mexico. *J. Radioanal. Nucl. Chem.* 263(3):767–72, 2005.

Villareal, T. A. Active fluorescence in the giant diatom *Ethmodiscus* (Bacillariophyceae) documented using single cell PAM fluorometry. *J. Phycol.* 40(6):1052–61, 2004.

Montoya, J. P., et al. **(T. A. Villareal).** High rates of N2-fixation by unicellular diazotrophs in the oligotrophic Pacific Ocean. *Nature* 430:1027–31, 2004.

Villareal, T. A., and E. J. Carpenter. Buoyancy regulation and the potential for vertical migration in the oceanic cyanobacterium *Trichodesmium. Microb. Ecol.* 45:1–10, 2003.

Section 5
Meteorology and Atmospheric Sciences

This section contains a directory of institutions offering graduate work in meteorology and atmospheric sciences, followed by in-depth entries submitted by institutions that chose to prepare detailed program descriptions. Additional information about programs listed in the directory but not augmented by an in-depth entry may be obtained by writing directly to the dean of a graduate school or chair of a department at the address given in the directory.

For programs offering related work, see also in this book Astronomy and Astrophysics, Geosciences, Marine Sciences and Oceanography, and Physics. In Book 3, see Biological and Biomedical Sciences and Biophysics; and in Book 5, see Aerospace/Aeronautical Engineering, Civil and Environmental Engineering, Engineering and Applied Sciences, and Mechanical Engineering and Mechanics.

CONTENTS

Program Directories

Announcements

Close-Ups

See also:

Atmospheric Sciences

City College of the City University of New York, Graduate School, College of Liberal Arts and Science, Division of Science, Department of Earth and Atmospheric Sciences, New York, NY 10031-9198. Offers earth and environmental science (PhD); earth systems science (MA). *Students:* 10 applicants, 70% accepted, 5 enrolled. In 2005, 2 degrees awarded. *Degree requirements:* For master's, thesis, comprehensive exam. *Entrance requirements:* For master's, GRE, appropriate bachelor's degree. Additional exam requirements/recommendations for international students: Required—TOEFL (minimum score 500 paper-based; 173 computer-based). *Application deadline:* For fall admission, 5/1 for domestic students; for spring admission, 11/15 for domestic students. Application fee: $125. *Financial support:* Fellowships, career-related internships or fieldwork available. *Faculty research:* Water resources, high-temperature geochemistry, sedimentary basin analysis, tectonics. *Unit head:* Jeffrey Steiner, Chair, 212-650-6984, Fax: 212-650-6473, E-mail: steiner@sci.ccny.cuny.edu.

Clemson University, Graduate School, College of Engineering and Science, Department of Physics and Astronomy, Program in Physics, Clemson, SC 29634. Offers astronomy and astrophysics (MS, PhD); atmospheric physics (MS, PhD); biophysics (MS, PhD). Part-time programs available. *Students:* 53 full-time (15 women), 2 part-time; includes 2 minority (1 Asian American or Pacific Islander, 1 Hispanic American), 19 international. 46 applicants, 41% accepted, 10 enrolled. In 2005, 5 master's, 4 doctorates awarded. Terminal master's awarded for partial completion of doctoral program. *Degree requirements:* For master's, thesis or alternative; for doctorate, thesis/dissertation. *Entrance requirements:* For master's and doctorate, GRE General Test. Additional exam requirements/recommendations for international students: Required—TOEFL. *Application deadline:* For fall admission, 2/15 for domestic students. Applications are processed on a rolling basis. Application fee: $50. *Financial support:* Fellowships, research assistantships, teaching assistantships available. Financial award application deadline: 6/1; financial award applicants required to submit FAFSA. *Faculty research:* Radiation physics, solid-state physics, nuclear physics, radar and lidar studies of atmosphere. *Unit head:* Dr. Brad Myer, Head, 864-656-5320. *Application contact:* Dr. Miguel Larsen, Coordinator, 864-656-5309, Fax: 864-656-0805, E-mail: mlarsen@clemson.edu.

See Close-Ups on pages 331 and 333.

Colorado State University, Graduate School, College of Engineering, Department of Atmospheric Science, Fort Collins, CO 80523-0015. Offers MS, PhD. Part-time programs available. *Faculty:* 13 full-time (1 woman). *Students:* 73 full-time (32 women), 24 part-time (5 women); includes 7 minority (2 African Americans, 1 American Indian/Alaska Native, 1 Asian American or Pacific Islander, 3 Hispanic Americans), 13 international. Average age 28. 105 applicants, 28% accepted, 19 enrolled. In 2005, 18 master's, 7 doctorates awarded. *Median time to degree:* Of those who began their doctoral program in fall 1997, 61% received their degree in 8 years or less. *Degree requirements:* For master's, thesis or alternative; for doctorate, thesis/dissertation, preliminary exam. *Entrance requirements:* For master's and doctorate, GRE General Test, minimum GPA of 3.0. Additional exam requirements/recommendations for international students: Required—TOEFL (minimum score 550 paper-based; 213 computer-based). *Application deadline:* For fall admission, 2/1 priority date for domestic students, 2/1 priority date for international students; for spring admission, 9/15 priority date for domestic students, 9/15 priority date for international students. Applications are processed on a rolling basis. Application fee: $50. Electronic applications accepted. *Expenses:* Tuition, state resident: full-time $3,690; part-time $205 per credit. Tuition, nonresident: full-time $14,958; part-time $831 per credit. Required fees: $1,061. *Financial support:* In 2005–06, 14 fellowships (averaging $15,624 per year), 7 research assistantships with full tuition reimbursements (averaging $5,160 per year), 65 teaching assistantships (averaging $18,756 per year) were awarded; traineeships also available. Financial award application deadline:4/15. *Faculty research:* Global circulation and climate, atmospheric chemistry, radiation and remote sensing, marine meteorology, mesoscale meteorology. Total annual research expenditures: $14.2 million. *Unit head:* Dr. Steven A. Rutledge, Head, 970-491-8360, Fax: 970-491-8449, E-mail: rutledge@atmos.colostate.edu. *Application contact:* Dr. David Thompson, Student Counselor, 970-491-8360, Fax: 970-491-8449, E-mail: davet@atmos.colostate.edu.

Columbia University, Graduate School of Arts and Sciences, Division of Natural Sciences, Program in Atmospheric and Planetary Science, New York, NY 10027. Offers M Phil, PhD. Offered jointly through the Departments of Geological Sciences, Astronomy, and Physics and in cooperation with NASA Goddard Space Flight Center's Institute for Space Studies. *Degree requirements:* For doctorate, variable foreign language requirement, thesis/dissertation. *Entrance requirements:* For doctorate, GRE General Test, GRE Subject Test, previous course work in mathematics and physics. Additional exam requirements/recommendations for international students: Required—TOEFL. Application fee: $75. *Expenses:* Tuition: Full-time $31,448. Tuition and fees vary according to course level, course load, campus/location and program. *Financial support:* Available to part-time students. Application deadline: 1/5; *Faculty research:* Climate, weather prediction. *Unit head:* Wallace S. Broecker, Chair, 914-365-8413, Fax: 845-365-8169.

See Close-Up on page 289.

Cornell University, Graduate School, Graduate Fields of Agriculture and Life Sciences, Field of Atmospheric Science, Ithaca, NY 14853-0001. Offers MS, PhD. *Faculty:* 16 full-time (2 women). *Students:* 8 full-time (3 women), 4 international. 27 applicants, 19% accepted, 3 enrolled. In 2005, 1 degree awarded. *Degree requirements:* For master's, thesis/dissertation; for doctorate, thesis/dissertation, comprehensive exam. *Entrance requirements:* For master's and doctorate, GRE General Test, 2 letters of recommendation. Additional exam requirements/recommendations for international students: Required—TOEFL (minimum score 550 paper-based; 213 computer-based). *Application deadline:* For fall admission, 2/1 for domestic students; for spring admission, 8/1 priority date for domestic students. Application fee: $60. Electronic applications accepted. *Financial support:* In 2005–06, 8 students received support, including 1 fellowship with full tuition reimbursement available, 7 research assistantships with full tuition reimbursements available; teaching assistantships with full tuition reimbursements available, institutionally sponsored loans, traineeships, health care benefits, tuition waivers (full and partial), and unspecified assistantships also available. Financial award applicants required to submit FAFSA. *Faculty research:* Applied climatology, climate dynamics, statistical meteorology/climatology, synoptic meteorology, upper atmospheric science. *Unit head:* Director of Graduate Studies, 607-255-3034, Fax: 607-255-2106, E-mail: atmscigradfield@cornell.edu. *Application contact:* Graduate Field Assistant, 607-255-3034, Fax: 607-255-2106, E-mail: atmscigradfield@cornell.edu.

Creighton University, Graduate School, College of Arts and Sciences, Program in Atmospheric Sciences, Omaha, NE 68178-0001. Offers MS. *Faculty:* 3 full-time, 10 part-time/adjunct. *Students:* 6 full-time (1 woman); includes 1 minority (Asian American or Pacific Islander) In 2005, 2 degrees awarded. *Degree requirements:* For master's, thesis. *Entrance requirements:* For master's, GRE General Test. Additional exam requirements/recommendations for international students: Required—TOEFL. *Application deadline:* For fall admission, 3/1 for domestic students. Applications are processed on a rolling basis. Application fee: $40. *Unit head:* Dr. Art Douglas, Chair, 402-280-5759. *Application contact:* Dr. Barbara J. Braden, Dean, 402-280-2870, Fax: 402-280-5762, E-mail: bbraden@creighton.edu.

George Mason University, College of Science, Fairfax, VA 22030. Offers bioinformatics (MS, PhD); climate dynamics (PhD); computational sciences (MS); computational sciences and informatics (PhD); computational social science (PhD); computational techniques and applications (Certificate); earth systems and geoinformation science (PhD); earth systems science (MS); nanotechnology and nanoscience (Certificate); neuroscience (PhD); physical sciences (PhD); remote sensing and earth image processing (Certificate). Part-time and evening/weekend programs available. *Degree requirements:* For doctorate, thesis/dissertation, comprehensive exam, registration. *Entrance requirements:* For master's and doctorate, GRE General Test, minimum GPA of 3.0 in last 60 hours. Additional exam requirements/recommendations for international students: Required—TOEFL. Electronic applications accepted. *Expenses:* Tuition, area resident: Full-time $5,244; part-time $219 per credit. Tuition, state resident: part-time $651 per credit. Tuition, nonresident: full-time $15,636. Required fees: $1,524; $65 per credit. *Faculty research:* Space sciences and astrophysics, fluid dynamics, materials modeling and simulation, bioinformatics, global changes and statistics.

See Close-Up on page 501.

Georgia Institute of Technology, Graduate Studies and Research, College of Sciences, School of Earth and Atmospheric Sciences, Atlanta, GA 30332-0001. Offers atmospheric chemistry and air pollution (MS, PhD); atmospheric dynamics and climate (MS, PhD); geochemistry (MS, PhD); hydrologic cycle (MS, PhD); ocean sciences (MS, PhD); solid-earth and environmental geophysics (MS, PhD). Part-time programs available. Terminal master's awarded for partial completion of doctoral program. *Degree requirements:* For master's, thesis or alternative; for doctorate, thesis/dissertation, comprehensive exam. *Entrance requirements:* For master's, GRE, minimum GPA of 3.0; for doctorate, GRE General Test, minimum GPA of 2.7. Additional exam requirements/recommendations for international students: Required—TOEFL (minimum score 550 paper-based; 213 computer-based). *Faculty research:* Geophysics, atmospheric chemistry, atmospheric dynamics, seismology.

See Close-Up on page 227.

Howard University, Graduate School of Arts and Sciences, Department of Chemistry, Washington, DC 20059-0002. Offers analytical chemistry (MS, PhD); atmospheric (MS, PhD); biochemistry (MS, PhD); environmental (MS, PhD); inorganic chemistry (MS, PhD); organic chemistry (MS, PhD); physical chemistry (MS, PhD); polymer chemistry (MS, PhD). Part-time programs available. *Degree requirements:* For master's, one foreign language, thesis, teaching experience, comprehensive exam, registration; for doctorate, 2 foreign languages, thesis/dissertation, teaching experience, comprehensive exam, registration. *Entrance requirements:* For master's, GRE General Test, minimum GPA of 2.7; for doctorate, GRE General Test, minimum GPA of 3.0. *Faculty research:* Stratospheric aerosols, liquid crystals, polymer coatings, terrestrial and extraterrestrial atmospheres, amidogen reaction.

Howard University, Graduate School of Arts and Sciences and School of Engineering and Computer Science, Program in Atmospheric Sciences, Washington, DC 20059-0002. Offers MS, PhD. Part-time programs available. Terminal master's awarded for partial completion of doctoral program. *Degree requirements:* For master's, thesis, comprehensive exam; for doctorate, one foreign language, thesis/dissertation, comprehensive exam. *Entrance requirements:* For master's, GRE General Test, minimum GPA of 3.0; for doctorate, GRE General Test, minimum GPA of 3.2. Additional exam requirements/recommendations for international students: Required—TOEFL (minimum score 550 paper-based; 213 computer-based). *Faculty research:* Atmospheric chemistry, climate, ionospheric physics, gravity waves, aerosols, extraterrestrial atmospheres, turbulence.

Massachusetts Institute of Technology, School of Science, Department of Earth, Atmospheric, and Planetary Sciences, Cambridge, MA 02139-4307. Offers atmospheric chemistry (PhD, Sc D); atmospheric science (SM, PhD, Sc D); climate physics and chemistry (PhD, Sc D); earth and planetary sciences (SM); geochemistry (PhD, Sc D); geology (PhD, Sc D); geophysics (PhD, Sc D); marine geology and geophysics (SM); oceanography (SM, PhD, Sc D); planetary sciences (PhD, Sc D). *Faculty:* 38 full-time (3 women). *Students:* 165 full-time (63 women); includes 12 minority (7 Asian Americans or Pacific Islanders, 5 Hispanic Americans), 51 international. Average age 27. 179 applicants, 34% accepted, 31 enrolled. In 2005, 2 master's, 15 doctorates awarded. Terminal master's awarded for partial completion of doctoral program. *Degree requirements:* For master's, thesis/dissertation; for doctorate, thesis/dissertation, comprehensive exam. *Entrance requirements:* For master's, GRE General Test; for doctorate, GRE General Test, GRE Subject Test (chemistry or physics for planetary science program). Additional exam requirements/recommendations for international students: Required—TOEFL (minimum score 577 paper-based; 233 computer-based). *Application deadline:* For fall admission, 1/5 for domestic students, 1/5 for international students; for spring admission, 11/1 for domestic students, 11/1 for international students. Application fee: $70. Electronic applications accepted. *Expenses:* Tuition: Full-time $32,100. Required fees: $200. Part-time tuition and fees vary according to course load. *Financial support:* In 2005–06, 115 students received support, including 36 fellowships with tuition reimbursements available (averaging $26,951 per year), 60 research assistantships with tuition reimbursements available (averaging $24,520 per year), 18 teaching assistantships with tuition reimbursements available (averaging $22,773 per year); Federal Work-Study, institutionally sponsored loans, scholarships/grants, health care benefits, and unspecified assistantships also available. *Faculty research:* Evolution of main features of the planetary system; origin, composition, structure, and state of the atmospheres, oceans, surfaces, and interiors of planets; dynamics of planets and satellite motions. Total annual research expenditures: $18.6 million. *Unit head:* Prof. Maria Zuber, Department Head, 617-253-6397, E-mail: zuber@mit.edu. *Application contact:* EAPS Education Office.

McGill University, Faculty of Graduate and Postdoctoral Studies, Faculty of Science, Department of Atmospheric and Oceanic Sciences, Montréal, QC H3A 2T5, Canada. Offers atmospheric science (M Sc, PhD); physical oceanography (M Sc, PhD). Terminal master's awarded for partial completion of doctoral program. *Degree requirements:* For master's, thesis/dissertation, registration; for doctorate, thesis/dissertation, comprehensive exam, registration. *Entrance requirements:* For master's, GRE General Test, minimum GPA of 3.2 during last 2 years of full-time study or 3.0 overall; B Sc or equivalent in biochemistry or related disciplines (biology, chemistry, physics, physiology, microbiology); for doctorate, GRE General Test, master's degree in meteorology or related field. Additional exam requirements/recommendations for international students: Required—TOEFL, IELT (minimum score 7). *Faculty research:* Dynamic meteorology and climate dynamics, synoptic and mesoscale meteorology, radar meteorology, atmospheric chemistry.

New Mexico Institute of Mining and Technology, Graduate Studies, Department of Physics, Socorro, NM 87801. Offers astrophysics (MS, PhD); atmospheric physics (MS, PhD); instrumentation (MS); mathematical physics (PhD). *Degree requirements:* For master's, thesis optional; for doctorate, thesis/dissertation. *Entrance requirements:* For master's, GRE General Test; for doctorate, GRE General Test, GRE Subject Test. Additional exam requirements/recommendations for international students: Required—TOEFL (minimum score 540 paper-based; 207 computer-based). *Faculty research:* Cloud physics, stellar and extragalactic processes.

North Carolina State University, Graduate School, College of Physical and Mathematical Sciences, Department of Marine, Earth, and Atmospheric Sciences, Raleigh, NC 27695. Offers marine, earth, and atmospheric sciences (MS, PhD); meteorology (MS, PhD); oceanography (MS, PhD). Terminal master's awarded for partial completion of doctoral program. *Degree requirements:* For master's, thesis (for some programs), final oral exam; for doctorate, thesis/dissertation, final oral exam, preliminary oral and written exams, comprehensive exam, registration. *Entrance requirements:* For master's, GRE General Test, minimum GPA of 3.0; for doctorate, GRE General Test, GRE Subject Test (for disciplines in biological oceanography and geology), minimum GPA of 3.0. Additional exam requirements/recommendations for international students: Required—TOEFL (minimum score 550 paper-based). Electronic applications accepted. *Faculty research:* Boundary layer and air quality meteorology; climate and mesoscale dynamics; biological, chemical, geological, and physical oceanography; hard rock, soft rock, environmental, and paleo geology.

See Close-Up on page 229.

The Ohio State University, Graduate School, College of Social and Behavioral Sciences, Department of Geography, Program in Atmospheric Sciences, Columbus, OH 43210. Offers MS, PhD. *Degree requirements:* For master's and doctorate, thesis/dissertation. *Entrance requirements:* For master's and doctorate, GRE General Test. Additional exam requirements/recommendations for international students: Required—TOEFL (minimum score 600 paper-based; 250 computer-based). Electronic applications accepted. *Faculty research:* Climatology, aeronomy, solar-terrestrial physics, air environment.

Oregon State University, Graduate School, College of Oceanic and Atmospheric Sciences, Program in Atmospheric Sciences, Corvallis, OR 97331. Offers MA, MS, PhD. *Students:* 8 full-time (2 women), 1 part-time, 3 international. Average age 37. In 2005, 2 doctorates awarded. Terminal master's awarded for partial completion of doctoral program. *Degree requirements:* For master's, variable foreign language requirement, thesis, qualifying exams; for doctorate, thesis/dissertation, qualifying exams. *Entrance requirements:* For master's and doctorate, GRE General Test, minimum GPA of 3.0 in last 90 hours of course work. Additional exam requirements/recommendations for international students: Required—TOEFL. *Application deadline:* For fall admission, 2/1 for domestic students. Applications are processed on a rolling basis. Application fee: $50. *Expenses:* Tuition, area resident: Part-time $301 per credit. Tuition, state resident: full-time $8,139; part-time $501 per credit. Tuition, nonresident: full-time $14,376; part-time $532 per credit. Required fees: $1,266. *Financial support:* Fellowships, research assistantships, teaching assistantships, career-related internships or fieldwork, Federal Work-Study, and institutionally sponsored loans available. Support available to part-time students. Financial award application deadline: 2/1. *Faculty research:* Planetary atmospheres, boundary layer dynamics, climate, statistical meteorology, satellite meteorology, atmospheric chemistry. *Unit head:* Dr. Mike Unsworth, Director, 541-737-5428, Fax: 541-737-2540, E-mail: unswortm@oce.orst.edu. *Application contact:* Irma Delson, Assistant Director, Student Services, 541-737-5189, Fax: 541-737-2064, E-mail: student_adviser@oce.orst.edu.

Princeton University, Graduate School, Department of Geosciences, Program in Atmospheric and Oceanic Sciences, Princeton, NJ 08544-1019. Offers PhD. *Degree requirements:* For doctorate, one foreign language, thesis/dissertation. *Entrance requirements:* For doctorate, GRE General Test, GRE Subject Test. Additional exam requirements/recommendations for international students: Required—TOEFL (minimum score 600 paper-based; 250 computer-based). Electronic applications accepted. *Faculty research:* Climate dynamics, middle atmosphere dynamics and chemistry, oceanic circulation, marine geochemistry, numerical modeling.

Purdue University, Graduate School, School of Science, Department of Earth and Atmospheric Sciences, West Lafayette, IN 47907. Offers MS, PhD. *Faculty:* 33 full-time (4 women). *Students:* 50 full-time (17 women), 6 part-time (3 women); includes 3 minority (1 African American, 2 Hispanic Americans), 17 international. Average age 29. 64 applicants, 30% accepted, 14 enrolled. In 2005, 7 master's, 3 doctorates awarded. *Degree requirements:* For master's, thesis; for doctorate, one foreign language, thesis/dissertation. *Entrance requirements:* For master's and doctorate, GRE General Test. Additional exam requirements/recommendations for international students: Required—TOEFL. *Application deadline:* For fall admission, 5/1 for domestic students, 2/1 for international students; for spring admission, 9/1 for domestic students, 7/1 for international students. Applications are processed on a rolling basis. Application fee: $55. Electronic applications accepted. *Financial support:* In 2005–06, 10 fellowships with partial tuition reimbursements (averaging $15,133 per year), 8 research assistantships with partial tuition reimbursements (averaging $14,400 per year), 26 teaching assistantships with partial tuition reimbursements (averaging $14,400 per year) were awarded. Support available to part-time students. Financial award application deadline: 3/1; financial award applicants required to submit FAFSA. *Faculty research:* Geology, geophysics, hydrogeology, paleoclimatology, environmental science. *Unit head:* Dr. Jonathan M Harbor, Head, 765-494-4753, Fax: 765-496-1210. *Application contact:* Kathy Kincade, Graduate Secretary, 765-494-5984, Fax: 765-496-1210, E-mail: kkincade@purdue.edu.

Rutgers, The State University of New Jersey, New Brunswick/Piscataway, Graduate School, Program in Environmental Sciences, New Brunswick, NJ 08901-1281. Offers air resources (MS, PhD); aquatic biology (MS, PhD); aquatic chemistry (MS, PhD); atmospheric science (MS, PhD); chemistry and physics of aerosol and hydrosol systems (MS, PhD); environmental chemistry (MS, PhD); environmental microbiology (MS, PhD); environmental toxicology (PhD); exposure assessment (PhD); fate and effects of pollutants (MS, PhD); pollution prevention and control (MS, PhD); water and wastewater treatment (MS, PhD); water resources (MS, PhD). *Faculty:* 81 full-time, 7 part-time/adjunct. *Students:* 49 full-time (27 women), 48 part-time (19 women); includes 10 minority (3 African Americans, 6 Asian Americans or Pacific Islanders, 1 Hispanic American), 24 international. Average age 32. 79 applicants, 41% accepted, 15 enrolled. In 2005, 8 master's, 10 doctorates awarded. Terminal master's awarded for partial completion of doctoral program. *Degree requirements:* For master's, thesis or alternative, oral final exam, comprehensive exam; for doctorate, thesis/dissertation, thesis defense, qualifying exam, comprehensive exam. *Entrance requirements:* For master's and doctorate, GRE General Test. Additional exam requirements/recommendations for international students: Required—TOEFL. *Application deadline:* For fall admission, 3/1 for domestic students; for spring admission, 11/1 for domestic students. Applications are processed on a rolling basis. Application fee: $50. Electronic applications accepted. *Expenses:* Tuition, state resident: full-time $10,440; part-time $435 per credit. Tuition, nonresident: full-time $15,520; part-time $647 per credit. Required fees: $129 per credit. Tuition and fees vary according to program. *Financial support:* In 2005–06, 10 fellowships with full tuition reimbursements (averaging $21,887 per year), 34 research assistantships with full tuition reimbursements (averaging $19,367 per year), 3 teaching assistantships with full tuition reimbursements (averaging $17,583 per year) were awarded; career-related internships or fieldwork and Federal Work-Study also available. Financial award application deadline: 1/15; financial award applicants required to submit FAFSA. *Faculty research:* Atmospheric sciences; biological waste treatment; contaminant fate and transport; exposure assessment; air, soil and water quality. Total annual research expenditures: $5.7 million. *Unit head:* John Reinfelder, Director, 732-932-8013, Fax: 732-932-8644, E-mail: reinfelder@envsci.rutgers.edu. *Application contact:* Dr. Paul J. Lioy, Graduate Admissions Committee, 732-932-0150, Fax: 732-445-0116, E-mail: plioy@eohsi.rutgers.edu.

South Dakota School of Mines and Technology, Graduate Division, College of Science and Letters, Department of Atmospheric Sciences, Rapid City, SD 57701-3995. Offers MS. Part-time programs available. *Faculty:* 6 part-time/adjunct (0 women). *Students:* 7 full-time (2 women), 7 part-time (1 woman). In 2005, 2 master's awarded. *Degree requirements:* For master's, thesis/dissertation. *Entrance requirements:* Additional exam requirements/recommendations for international students: Required—TOEFL, TWE. *Application deadline:* For fall admission, 7/1 priority date for domestic students, 4/1 priority date for international students; for spring admission, 11/1 for domestic students, 9/1 for international students. Applications are processed on a rolling basis. Application fee: $35. Electronic applications accepted. *Expenses:* Tuition, area resident: Part-time $116 per credit hour. Tuition, state resident: full-time $2,084. Tuition, nonresident: full-time $6,146; part-time $341 per credit hour. Required fees: $1,805; $100 per credit hour. *Financial support:* In 2005–06, 1 fellowship (averaging $2,500 per year), 10 research assistantships with partial tuition reimbursements (averaging $15,919 per year) were awarded; teaching assistantships with partial tuition reimbursements, Federal Work-Study and institutionally sponsored loans also available. Support available to part-time students. Financial award application deadline: 5/15. *Faculty research:* Hailstorm observations and numerical modeling, microbursts and lightning, radiative transfer, remote sensing. Total annual research expenditures: $1.3 million. *Unit head:* Dr. Andrew Detwiler, Chair, 605-394-2291. *Application contact:* Jeannette R. Nilson, Program Assistant-Research and Graduate Education, 800-454-8162 Ext. 1206, Fax: 605-394-5360, E-mail: graduate_admissions@silver.sdsmt.edu.

South Dakota School of Mines and Technology, Graduate Division, College of Science and Letters, Joint PhD Program in Atmospheric, Environmental, and Water Resources, Rapid City, SD 57701-3995. Offers PhD. *Faculty:* 9 full-time (0 women), 1 (woman) part-time/adjunct. *Students:* 4 full-time (2 women), 7 part-time (2 women); includes 1 minority (American Indian/Alaska Native), 1 international. In 2005, 2 degrees awarded. *Degree requirements:* For doctorate, thesis/dissertation. *Entrance requirements:* For doctorate, GRE General Test, GRE Subject Test. Additional exam requirements/recommendations for international students: Required—TOEFL, TWE. *Application deadline:* For fall admission, 7/1 priority date for domestic students, 4/1 priority date for international students; for spring admission, 11/1 for domestic students, 9/1 for international students. Applications are processed on a rolling basis. Application fee: $35. Electronic applications accepted. *Expenses:* Tuition, area resident: Part-time $116 per credit hour. Tuition, state resident: full-time $2,084. Tuition, nonresident: full-time $6,146; part-time $341 per credit hour. Required fees: $1,805; $100 per credit hour. *Financial support:* In 2005–06, 3 fellowships (averaging $4,000 per year), 4 teaching assistantships with partial tuition reimbursements (averaging $6,774 per year) were awarded; research assistantships with partial tuition reimbursements *Unit head:* Dr. Andrew Detwiler, Chair, 605-394-2291. *Application contact:* Jeannette R. Nilson, Program Assistant-Research and Graduate Education, 800-454-8162 Ext. 1206, Fax: 605-394-5360, E-mail: graduate_admissions@silver.sdsmt.edu.

South Dakota State University, Graduate School, College of Engineering, Joint PhD Program in Atmospheric, Environmental, and Water Resources, Brookings, SD 57007. Offers PhD. Postbaccalaureate distance learning degree programs offered (minimal on-campus study). *Degree requirements:* For doctorate, thesis/dissertation, preliminary oral and written exams. *Entrance requirements:* Additional exam requirements/recommendations for international students: Required—TOEFL (minimum score 525 paper-based). Expenses: Contact institution.

Stony Brook University, State University of New York, Graduate School, Institute for Terrestrial and Planetary Atmospheres, Stony Brook, NY 11794. Offers PhD. *Application deadline:* For fall admission, 3/1 for domestic students. Application fee: $50. *Expenses:* Tuition, area resident: Part-time $288. Tuition, state resident: full-time $6,900. Tuition, nonresident: full-time $10,920; part-time $455. Required fees: $704. *Financial support:* Fellowships available. *Unit head:* Minghua Zhang, Director, 631-632-8318.

See Close-Up on page 291.

Stony Brook University, State University of New York, Graduate School, Marine Sciences Research Center, Program in Marine and Atmospheric Sciences, Stony Brook, NY 11794. Offers MS, PhD. Evening/weekend programs available. *Faculty:* 37 full-time (7 women). *Students:* 113 full-time (63 women), 6 part-time (4 women); includes 11 minority (2 African Americans, 1 American Indian/Alaska Native, 2 Asian Americans or Pacific Islanders, 6 Hispanic Americans), 45 international. Average age 27. 86 applicants, 52% accepted. In 2005, 15 master's, 3 doctorates awarded. *Degree requirements:* For doctorate, one foreign language, thesis/dissertation, comprehensive exam. *Entrance requirements:* For doctorate, GRE General Test, minimum graduate GPA of 3.0. Additional exam requirements/recommendations for international students: Required—TOEFL. *Application deadline:* For fall admission, 1/15 for domestic students. Application fee: $50. *Expenses:* Tuition, area resident: Part-time $288. Tuition, state resident: full-time $6,900. Tuition, nonresident: full-time $10,920; part-time $455. Required fees: $704. *Financial support:* In 2005–06, 26 fellowships, 55 research assistantships, 31 teaching assistantships were awarded; career-related internships or fieldwork also available. Total annual research expenditures: $9.3 million. *Application contact:* Dr. Glen Lopez, Acting Director, 631-632-8660, Fax: 631-632-8200, E-mail: glopez@notes.cc.sunysb.edu.

See Close-Up on page 269.

Texas Tech University, Graduate School, College of Arts and Sciences, Department of Geosciences, Lubbock, TX 79409. Offers atmospheric sciences (MS); geoscience (MS, PhD). Part-time programs available. *Faculty:* 15 full-time (2 women). *Students:* 27 full-time (11 women), 10 part-time (1 woman); includes 3 minority (1 African American, 2 Hispanic Americans), 4 international. Average age 31. 52 applicants, 56% accepted, 11 enrolled. In 2005, 12 master's, 2 doctorates awarded. *Degree requirements:* For master's and doctorate, thesis/dissertation. *Entrance requirements:* For master's and doctorate, GRE General Test. Additional exam requirements/recommendations for international students: Required—TOEFL (minimum score 550 paper-based; 213 computer-based). *Application deadline:* Applications are processed on a rolling basis. Application fee: $50 ($60 for international students). Electronic applications accepted. *Expenses:* Tuition, state resident: full-time $4,296. Tuition, nonresident: full-time $10,920. Required fees: $1,992. Tuition and fees vary according to program. *Financial support:* In 2005–06, 21 students received support, including 3 research assistantships with partial tuition reimbursements available (averaging $13,851 per year), 23 teaching assistantships with partial tuition reimbursements available (averaging $15,434 per year); Federal Work-Study and institutionally sponsored loans also available. Support available to part-time students. Financial award application deadline: 4/15; financial award applicants required to submit FAFSA. *Faculty research:* Ophiolites and oceanic lower crust; petroleum geology; tectonics and arc magnetism; aqueous and environmental geochemistry; near-ground high wind phenomenon (hurricanes and severe storms). Total annual research expenditures:$346,436. *Unit head:* Dr. James Barrick, Chairman, 806-742-3107, Fax: 806-742-0100, E-mail: jim.barrick@ttu.edu. *Application contact:* Dr. Moira Ridley, Graduate Adviser, 806-742-3102, Fax: 806-724-0100, E-mail: moira.ridley@ttu.edu.

Université du Québec à Montréal, Graduate Programs, Programs in Atmospheric Sciences and Meteorology, Montréal, QC H3C 3P8, Canada. Offers atmospheric sciences (M Sc); meteorology (PhD, Diploma). Part-time programs available. *Degree requirements:* For master's, thesis. *Entrance requirements:* For master's and Diploma, appropriate bachelor's degree or equivalent and proficiency in French; for doctorate, appropriate master's degree or equivalent and proficiency in French.

University at Albany, State University of New York, College of Arts and Sciences, Department of Earth and Atmospheric Sciences, Albany, NY 12222-0001. Offers atmospheric science (MS, PhD); geology (MS, PhD). *Students:* 38 full-time (10 women), 10 part-time (5 women). Average age 29. In 2005, 5 master's, 4 doctorates awarded. *Degree requirements:* For master's, one foreign language, thesis, comprehensive exam; for doctorate, 2 foreign languages, thesis/dissertation, oral exams, comprehensive exam. *Entrance requirements:* For master's and doctorate, GRE General Test. Additional exam requirements/recommendations for international students: Required—TOEFL (minimum score 550 paper-based; 213 computer-based). *Application deadline:* For fall admission, 6/1 for domestic students, 5/1 for international students; for spring admission, 11/1 for domestic students, 11/11 for international students. Applications are processed on a rolling basis. Application fee: $60. Electronic applications accepted. *Financial support:* Fellowships, research assistantships, teaching assistantships, minority assistantships available. Financial award application deadline:3/1. *Faculty research:* Environmental geochemistry, tectonics, mesoscale meteorology, atmospheric chemistry. *Unit head:* Dr. Vincent Idone, Chair, 518-442-4466. *Application contact:* William Kidd, Graduate Program Director, Geology.

The University of Alabama in Huntsville, School of Graduate Studies, College of Science, Department of Atmospheric and Environmental Science, Huntsville, AL 35899. Offers MS, PhD. Part-time and evening/weekend programs available. *Faculty:* 7 full-time (0 women), 3 part-time/adjunct (0 women). *Students:* 26 full-time (6 women), 11 part-time (5 women); includes 2 minority (both African Americans), 11 international. Average age 29. 21 applicants, 95% accepted, 10 enrolled. In 2005, 9 master's, 2 doctorates awarded. *Degree requirements:* For master's, thesis or alternative, oral and written exams, comprehensive exam, registration; for doctorate, thesis/dissertation, oral and written exams, comprehensive exam, registration. *Entrance requirements:* For master's and doctorate, GRE General Test, minimum GPA of 3.0. Additional exam requirements/recommendations for international students: Required—TOEFL (minimum score 550 paper-based; 213 computer-based). *Application deadline:* For fall admission, 5/30 for domestic students; for spring admission, 10/10 priority date for domestic students, 7/10 priority date for international students. Applications are processed on a rolling basis. Application fee: $40. *Expenses:* Tuition, state resident: full-time $5,866; part-time $244 per credit hour. Tuition, nonresident: full-time $12,060; part-time $500 per credit hour. Tuition and

Atmospheric Sciences

The University of Alabama in Huntsville (continued)

fees vary according to course load. *Financial support:* In 2005–06, 24 students received support, including 22 research assistantships with full and partial tuition reimbursements available (averaging $12,777 per year), 1 teaching assistantship with full and partial tuition reimbursement available (averaging $13,500 per year); fellowships with full and partial tuition reimbursements available, career-related internships or fieldwork, Federal Work-Study, institutionally sponsored loans, scholarships/grants, health care benefits, tuition waivers (full and partial), and unspecified assistantships also available. Support available to part-time students. Financial award application deadline: 4/1; financial award applicants required to submit FAFSA. Total annual research expenditures:$225,474. *Unit head:* Dr. Ronald Welch, Chair, 256-961-7754, Fax: 256-961-7755, E-mail: ron.welch@atmos.uah.edu.

University of Alaska Fairbanks, College of Natural Sciences and Mathematics, Department of Physics, Fairbanks, AK 99775-7520. Offers atmospheric science (MS, PhD); computational physics (MS); general physics (MS); physics (MAT, PhD); space physics (PhD). Part-time programs available. *Faculty:* 14 full-time (1 woman). *Students:* 24 full-time (5 women), 2 part-time; includes 2 minority (1 African American, 1 Hispanic American), 4 international. Average age 29. 16 applicants, 56% accepted, 3 enrolled. In 2005, 4 master's, 1 doctorate awarded. Terminal master's awarded for partial completion of doctoral program. *Degree requirements:* For master's, thesis or alternative, comprehensive exam, registration; for doctorate, thesis/dissertation, comprehensive exam, registration. *Entrance requirements:* For master's, GRE General Test, BS in physics; for doctorate, GRE General Test. Additional exam requirements/recommendations for international students: Required—TOEFL (minimum score 550 paper-based; 213 computer-based). *Application deadline:* For fall admission, 3/1 for domestic students, 3/1 for international students. Applications are processed on a rolling basis. Application fee: $50. Electronic applications accepted. *Expenses:* Tuition, state resident: full-time $4,392; part-time $244 per credit. Tuition, nonresident: full-time $8,964; part-time $498 per credit. Required fees: $800; $5 per credit. $48 per contact hour. Tuition and fees vary according to course level, course load, campus/location and reciprocity agreements. *Financial support:* In 2005–06, 14 research assistantships with tuition reimbursements (averaging $10,507 per year), 4 teaching assistantships with tuition reimbursements (averaging $11,027 per year) were awarded; fellowships with tuition reimbursements, Federal Work-Study, scholarships/grants, and unspecified assistantships also available. Financial award applicants required to submit FAFSA. *Faculty research:* Atmospheric and ionospheric radar studies, space plasma theory, magnetospheric dynamics, space weather and auroral studies, turbulence and complex systems. *Unit head:* John D. Craven, Chair, 907-474-7339, Fax: 907-474-6130, E-mail: physics@uaf.edu.

University of Alaska Fairbanks, College of Natural Sciences and Mathematics, Program in Atmospheric Science, Fairbanks, AK 99775-7520. Offers MS, PhD. Part-time programs available. *Faculty:* 3 full-time (2 women). *Students:* 11 full-time (4 women), 2 part-time (1 woman); includes 2 minority (1 African American, 1 Asian American or Pacific Islander), 5 international. Average age 31. 23 applicants, 70% accepted, 2 enrolled. In 2005, 1 master's, 2 doctorates awarded. Terminal master's awarded for partial completion of doctoral program. *Degree requirements:* For master's, thesis or alternative, comprehensive exam, registration; for doctorate, thesis/dissertation, comprehensive exam, registration. *Entrance requirements:* For master's, GRE General Test. Additional exam requirements/recommendations for international students: Required—TOEFL (minimum score 550 paper-based; 213 computer-based). *Application deadline:* For fall admission, 3/1 for domestic students, 3/1 for international students; for spring admission, 12/1 for domestic students, 9/1 for international students. Applications are processed on a rolling basis. Application fee: $50. Electronic applications accepted. *Expenses:* Tuition, state resident: full-time $4,392; part-time $244 per credit. Tuition, nonresident: full-time $8,964; part-time $498 per credit. Required fees: $800; $5 per credit. $48 per contact hour. Tuition and fees vary according to course level, course load, campus/location and reciprocity agreements. *Financial support:* In 2005–06, 12 research assistantships with tuition reimbursements (averaging $11,685 per year) were awarded; fellowships with tuition reimbursements, teaching assistantships with tuition reimbursements, Federal Work-Study and scholarships/grants also available. Financial award applicants required to submit FAFSA. *Faculty research:* Sea ice, climate modeling, atmospheric chemistry, global change, cloud and aerosol physics. *Unit head:* Dr. Kenneth Sassen, Program Chair, 907-474-7845, Fax: 907-474-7290, E-mail: ksassen@gi.alaska.edu.

The University of Arizona, Graduate College, College of Science, Department of Atmospheric Sciences, Tucson, AZ 85721. Offers MS, PhD. *Degree requirements:* For master's, thesis or alternative, registration; for doctorate, thesis/dissertation, comprehensive exam, registration. *Entrance requirements:* For master's and doctorate, GRE General Test. Additional exam requirements/recommendations for international students: Required—TOEFL. Electronic applications accepted. *Faculty research:* Climate dynamics, radiative transfer and remote sensing, atmospheric chemistry, atmosphere dynamics, atmospheric electricity.

The University of British Columbia, Faculty of Graduate Studies, Faculty of Science, Department of Earth and Ocean Sciences, Vancouver, BC V6T 1Z1, Canada. Offers atmospheric science (M Sc, PhD); geological engineering (M Eng, MA Sc, PhD); geological sciences (M Sc, PhD); geophysics (M Sc, MA Sc, PhD); oceanography (M Sc, PhD). *Faculty:* 44 full-time (7 women), 17 part-time/adjunct (1 woman). *Students:* 155 full-time (49 women), 1 (woman) part-time. Average age 30. 96 applicants, 50% accepted, 30 enrolled. In 2005, 11 master's, 6 doctorates awarded. *Degree requirements:* For master's, thesis (for some programs); for doctorate, thesis/dissertation, comprehensive exam. *Entrance requirements:* Additional exam requirements/recommendations for international students: Required—TOEFL (minimum score 600 paper-based; 250 computer-based). *Application deadline:* For fall admission, 2/1 for domestic students, 1/1 for international students. For winter admission, 7/1 for domestic students. Applications are processed on a rolling basis. Application fee: $90 Canadian dollars ($150 Canadian dollars for international students). Electronic applications accepted. *Financial support:* In 2005–06, fellowships (averaging $16,000 per year), research assistantships (averaging $13,000 per year), teaching assistantships (averaging $5,000 per year) were awarded; Federal Work-Study, institutionally sponsored loans, scholarships/grants, tuition waivers (full and partial), and unspecified assistantships also available. *Unit head:* Dr. Paul L. Smith, Head, 604-822-6456, Fax: 604-822-6088, E-mail: head@eos.ubc.ca. *Application contact:* Alex Allen, Graduate Secretary, 604-822-2713, Fax: 604-822-6088, E-mail: aallen@eos.ubc.ca.

University of California, Davis, Graduate Studies, Graduate Group in Atmospheric Sciences, Davis, CA 95616. Offers MS, PhD. *Faculty:* 17 full-time. *Students:* 21 full-time (7 women); includes 2 minority (1 American Indian/Alaska Native, 1 Hispanic American), 7 international. Average age 28. 29 applicants, 34% accepted, 4 enrolled. In 2005, 1 master's awarded. *Median time to degree:* Of those who began their doctoral program in fall 1997, 20% received their degree in 8 years or less. *Degree requirements:* For master's, thesis (for some programs), comprehensive exam or thesis, comprehensive exam (for some programs); for doctorate, thesis/dissertation, 3 part qualifying exam. *Entrance requirements:* For master's and doctorate, GRE General Test, minimum GPA of 3.0. Additional exam requirements/recommendations for international students: Required—TOEFL (minimum score 550 paper-based; 213 computer-based). *Application deadline:* For fall admission, 1/15 for domestic students, 1/15 for international students. Application fee: $60. Electronic applications accepted. *Financial support:* In 2005–06, 19 students received support, including 1 fellowship with full and partial tuition reimbursement available (averaging $12,060 per year), 9 research assistantships with full and partial tuition reimbursements available (averaging $12,422 per year), 3 teaching assistantships with partial tuition reimbursements available (averaging $15,083 per year); career-related internships or fieldwork, Federal Work-Study, institutionally sponsored loans, scholarships/grants, tuition waivers (full and partial), and unspecified assistantships also available. Financial award application deadline: 1/15; financial award applicants required to submit FAFSA. *Faculty research:* Air quality, biometeorology, climate dynamics, boundary layer large-scale dynamics. Total annual research expenditures: $1.2 million. *Unit head:* Richard Snyder, Graduate Chair, 530-752-4628, E-mail: rlsnyder@ucdavis.edu. *Application contact:* Merlyn Potters, Graduate Staff Adviser, 530-752-1669, Fax: 530-752-1552, E-mail: lawradvising@ucdavis.edu.

University of California, Los Angeles, Graduate Division, College of Letters and Science, Department of Atmospheric Sciences, Los Angeles, CA 90095. Offers MS, PhD. *Degree requirements:* For master's, comprehensive exam or thesis; for doctorate, thesis/dissertation, oral and written qualifying exams. *Entrance requirements:* For master's, GRE General Test, minimum GPA of 3.0; for doctorate, GRE General Test, minimum undergraduate GPA of 3.0. Electronic applications accepted.

University of Chicago, Division of the Physical Sciences, Department of the Geophysical Sciences, Chicago, IL 60637-1513. Offers atmospheric sciences (SM, PhD); earth sciences (SM, PhD); paleobiology (PhD); planetary and space sciences (SM, PhD). *Faculty:* 24 full-time (3 women). *Students:* 35 full-time (15 women); includes 1 minority (Hispanic American), 11 international. Average age 29. 52 applicants, 29% accepted. In 2005, 2 master's, 3 doctorates awarded. Terminal master's awarded for partial completion of doctoral program. *Entrance requirements:* For master's and doctorate, GRE General Test. Additional exam requirements/recommendations for international students: Required—TOEFL, IELT. *Application deadline:* For fall admission, 1/15 for domestic students, 1/15 for international students. Application fee: $55. Electronic applications accepted. *Financial support:* In 2005–06, 32 students received support, including research assistantships with full tuition reimbursements available (averaging $18,675 per year), teaching assistantships with full tuition reimbursements available (averaging $19,098 per year); fellowships, Federal Work-Study, institutionally sponsored loans, scholarships/grants, tuition waivers (partial), and unspecified assistantships also available. Financial award application deadline: 1/15. *Faculty research:* Climatology, evolutionary paleontology, petrology, geochemistry, oceanic sciences. *Unit head:* Dr. David Rowley, Chairman, 773-702-8102, Fax: 773-702-9505. *Application contact:* David J. Leslie, Graduate Student Services Coordinator, 773-702-8180, Fax: 773-702-9505, E-mail: info@geosci.uchicago.edu.

University of Colorado at Boulder, Graduate School, College of Arts and Sciences, Department of Atmospheric and Oceanic Sciences, Boulder, CO 80309. Offers MS, PhD. *Faculty:* 10 full-time (2 women). *Entrance requirements:* For master's, minimum undergraduate GPA of 3.0. *Application deadline:* For fall admission, 2/1 for domestic students, 12/1 for international students. *Faculty research:* Large-scale dynamics of the ocean and the atmosphere, air-sea interaction, radiative transfer and remote sensing of the ocean and the atmosphere, sea ice and its role in climate. Total annual research expenditures: $2 million. *Unit head:* Brian Toon, Chair, 303-492-1534, Fax: 303-492-3524, E-mail: brian.toon@lasp.colorado.edu. *Application contact:* Graduate Secretary, 303-492-6633, Fax: 303-492-3524, E-mail: paosasst@colorado.edu.

University of Delaware, College of Arts and Sciences, Department of Geography, Program in Climatology, Newark, DE 19716. Offers PhD. *Faculty:* 12 full-time (3 women), 3 part-time/adjunct (2 women). *Students:* 8 full-time (5 women), 2 international. Average age 24. 5 applicants, 100% accepted, 2 enrolled. In 2005, 2 doctorates awarded. *Degree requirements:* For doctorate, thesis/dissertation. *Entrance requirements:* For doctorate, GRE General Test. Additional exam requirements/recommendations for international students: Required—TOEFL. *Application deadline:* For fall admission, 2/1 for domestic students, 2/1 for international students. Application fee: $60. Electronic applications accepted. *Financial support:* In 2005–06, 6 students received support, including 1 fellowship with full tuition reimbursement available (averaging $13,100 per year), 4 research assistantships with full tuition reimbursements available (averaging $13,100 per year), 1 teaching assistantship with full tuition reimbursement available (averaging $13,100 per year) Financial award application deadline: 2/1. *Faculty research:* Physical and applied climatology, synoptic climatology, glaciology, hydroclimatology, cryospheric studies. Total annual research expenditures: $700,000. *Application contact:* Janice Spry, Assistant to the Chair, 302-831-8998, Fax: 302-831-6654, E-mail: jspry@udel.edu.

University of Guelph, Graduate Program Services, Ontario Agricultural College, Department of Land Resource Science, Guelph, ON N1G 2W1, Canada. Offers atmospheric science (M Sc, PhD); environmental and agricultural earth sciences (M Sc, PhD); land resources management (M Sc, PhD); soil science (M Sc, PhD). Part-time programs available. *Faculty:* 19 full-time (5 women), 5 part-time/adjunct (1 woman). *Students:* 47 full-time (20 women), 3 part-time; includes 9 minority (1 African American, 6 Asian Americans or Pacific Islanders, 2 Hispanic Americans), 2 international. Average age 28. 25 applicants, 24% accepted. In 2005, 4 master's, 3 doctorates awarded. *Degree requirements:* For master's and doctorate, thesis/dissertation. *Entrance requirements:* For master's, minimum B- average during previous 2 years of course work; for doctorate, minimum B average during previous 2 years of course work. Additional exam requirements/recommendations for international students: Required—TOEFL (minimum score 550 paper-based; 213 computer-based). *Application deadline:* For fall admission, 7/1 priority date for domestic students, 5/1 priority date for international students. For winter admission, 10/1 for domestic students; for spring admission, 3/1 for domestic students. Applications are processed on a rolling basis. Application fee: $75 Canadian dollars. Electronic applications accepted. *Financial support:* In 2005–06, 30 students received support, including 40 research assistantships (averaging $16,500 Canadian dollars per year), 15 teaching assistantships (averaging $3,800 Canadian dollars per year); fellowships, scholarships/grants also available. *Faculty research:* Soil science, environmental earth science, land resource management. Total annual research expenditures: $2.1 million Canadian dollars. *Unit head:* Dr. S. Hilts, Chair, 519-824-4120 Ext. 52447, Fax: 519-824-5730, E-mail: shilts@uoguelph.ca. *Application contact:* Dr. B. Hale, Graduate Coordinator, 519-824-4120 Ext. 53434, Fax: 519-824-5730, E-mail: bhale@uoguelph.ca.

University of Illinois at Urbana–Champaign, Graduate College, College of Liberal Arts and Sciences, Department of Atmospheric Science, Champaign, IL 61820. Offers MS, PhD. *Faculty:* 10 full-time (1 woman). *Students:* 36 full-time (10 women), 1 (woman) part-time; includes 1 minority (Asian American or Pacific Islander), 16 international. 69 applicants, 7% accepted, 5 enrolled. In 2005, 5 master's, 6 doctorates awarded. *Degree requirements:* For master's and doctorate, thesis/dissertation. *Entrance requirements:* For master's and doctorate, GRE General Test, minimum GPA of 3.0. *Application deadline:* For fall admission, 2/15 for domestic students. Applications are processed on a rolling basis. Application fee: $50 ($60 for international students). Electronic applications accepted. *Financial support:* In 2005–06, 2 fellowships, 29 research assistantships, 9 teaching assistantships were awarded; tuition waivers (full and partial) also available. Financial award application deadline: 2/15. *Unit head:* Donald Wuebbles, Head, 217-244-1568, Fax: 217-244-4393, E-mail: wuebbles@uiuc.edu. *Application contact:* Peggy Cook, Office Administrator, 217-333-2046, Fax: 217-244-4393, E-mail: pjcook@uiuc.edu.

University of Maryland, Baltimore County, Graduate School, College of Natural Sciences and Mathematics, Department of Physics, Program in Atmospheric Physics, Baltimore, MD 21250. Offers MS, PhD. *Expenses:* Tuition, state resident: part-time $395 per credit. Tuition, nonresident: part-time $652 per credit. Required fees: $82 per credit. Tuition and fees vary according to course load, program and reciprocity agreements. *Application contact:* Dr. Wallace McMillan, Director, 410-455-6315, E-mail: mcmillan@umbc.edu.

University of Michigan, Horace H. Rackham School of Graduate Studies, College of Engineering, Department of Atmospheric, Oceanic, and Space Sciences, Ann Arbor, MI 48109. Offers atmospheric (MS); atmospheric and space sciences (PhD); geoscience and remote sensing (PhD); space and planetary sciences (PhD); space engineering (M Eng); space sciences (MS). Part-time programs available. *Faculty:* 34 full-time (6 women). *Students:* 53 full-time (21 women), 1 part-time; includes 5 minority (1 American Indian/Alaska Native, 3 Asian Americans or Pacific Islanders, 1 Hispanic American), 22 international. Average age 24. 53 applicants, 28% accepted, 10 enrolled. In 2005, 22 master's, 2 doctorates awarded. Terminal master's awarded for partial completion of doctoral program. *Degree requirements:* For master's, thesis (for some programs); for doctorate, thesis/dissertation, oral defense of dissertation, preliminary exams. *Entrance requirements:* For master's and doctorate, GRE General Test. Additional exam requirements/recommendations for international students: Required—TOEFL.

Application deadline: For fall admission, 1/15 priority date for domestic students, 1/15 priority date for international students. Applications are processed on a rolling basis. Application fee: $60 ($75 for international students). Electronic applications accepted. *Financial support:* In 2005–06, 3 fellowships with tuition reimbursements (averaging $16,163 per year), 42 research assistantships with tuition reimbursements (averaging $15,159 per year), 3 teaching assistantships with tuition reimbursements (averaging $14,396 per year) were awarded; career-related internships or fieldwork, Federal Work-Study, institutionally sponsored loans, and health care benefits also available. Support available to part-time students. Financial award application deadline: 3/15; financial award applicants required to submit FAFSA. *Faculty research:* Modeling of atmospheric and aerosol chemistry, radiative transfer, remote sensing, atmospheric dynamics, space weather modeling. Total annual research expenditures: $18 million. *Unit head:* Tamas Gombosi, Chair, 734-764-7222, Fax: 734-615-4645, E-mail: tamas@umich.edu. *Application contact:* Margaret Reid, Student Services Associate, 734-936-0482, Fax: 734-763-0437, E-mail: aoss.um@umich.edu.

University of Missouri–Columbia, Graduate School, School of Natural Resources, Department of Soil, Environmental, and Atmospheric Sciences, Columbia, MO 65211. Offers atmospheric science (MS, PhD); soil science (MS, PhD). *Faculty:* 8 full-time (0 women). *Students:* 19 full-time (7 women), 10 part-time (3 women); includes 3 minority (1 African American, 1 Asian American or Pacific Islander, 1 Hispanic American), 9 international. In 2005, 10 master's, 5 doctorates awarded. *Degree requirements:* For doctorate, thesis/dissertation. *Entrance requirements:* For master's and doctorate, GRE General Test, minimum GPA of 3.0. *Application deadline:* Applications are processed on a rolling basis. Application fee: $45 ($60 for international students). *Financial support:* Fellowships, research assistantships, teaching assistantships, institutionally sponsored loans and scholarships/grants available. *Unit head:* Dr. Anthony Lupo, Director of Graduate Studies, 573-884-1638.

University of Nevada, Reno, Graduate School, College of Science, Interdisciplinary Program in Atmospheric Sciences, Reno, NV 89557. Offers MS, PhD. *Entrance requirements:* For master's, GRE (recommended), minimum GPA of 2.75; for doctorate, GRE (recommended), minimum GPA of 3.0. Additional exam requirements/recommendations for international students: Required—TOEFL.

University of New Hampshire, Climate Change Research Center., Durham, NH 03824.

The University of North Carolina at Chapel Hill, Graduate School, School of Public Health, Department of Environmental Sciences and Engineering, Chapel Hill, NC 27599. Offers air, radiation and industrial hygiene (MPH, MS, MSEE, MSPH, PhD); aquatic and atmospheric sciences (MPH, MS, MSPH, PhD); environmental engineering (MPH, MS, MSEE, MSPH, PhD); environmental health sciences (MPH, MS, MSPH, PhD); environmental management and policy (MPH, MS, MSPH, PhD). *Faculty:* 33 full-time (3 women), 35 part-time/adjunct. *Students:* 141 full-time (74 women); includes 37 minority (10 African Americans, 25 Asian Americans or Pacific Islanders, 2 Hispanic Americans). Average age 27. 216 applicants, 37% accepted, 29 enrolled. In 2005, 14 master's, 11 doctorates awarded. Terminal master's awarded for partial completion of doctoral program. *Median time to degree:* Of those who began their doctoral program in fall 1997, 100% received their degree in 8 years or less. *Degree requirements:* For master's, thesis (for some programs), research paper, comprehensive exam, registration; for doctorate, thesis/dissertation, comprehensive exam, registration. *Entrance requirements:* For master's and doctorate, GRE General Test, minimum GPA of 3.0. Additional exam requirements/recommendations for international students: Required—TOEFL. *Application deadline:* For fall admission, 1/1 priority date for domestic students, 1/1 priority date for international students; for spring admission, 9/15 for domestic students. Applications are processed on a rolling basis. Application fee: $70. Electronic applications accepted. *Financial support:* In 2005–06, 134 students received support, including 36 fellowships with tuition reimbursements available (averaging $6,358 per year), 86 research assistantships with tuition reimbursements available (averaging $6,197 per year), 12 teaching assistantships with tuition reimbursements available (averaging $6,729 per year); career-related internships or fieldwork, Federal Work-Study, traineeships, health care benefits, and unspecified assistantships also available. Support available to part-time students. Financial award application deadline: 1/1; financial award applicants required to submit FAFSA. *Faculty research:* Air, radiation and industrial hygiene, aquatic and atmospheric sciences, environmental health sciences, environmental management and policy, water resources engineering. Total annual research expenditures: $9.6 million. *Unit head:* , Dr. Don Fox, Interim Chair, 919-966-1024, Fax: 919-966-7911, E-mail: don_fox@unc.edu. *Application contact:* Jack Whaley, Registrar, 919-966-3844, Fax: 919-966-7911, E-mail: jack_whaley@unc.edu.

University of North Dakota, Graduate School, John D. Odegard School of Aerospace Sciences, Department of Atmospheric Sciences, Grand Forks, ND 58202. Offers MS. *Faculty:* 8 full-time (0 women). *Students:* 15 applicants, 33% accepted, 4 enrolled. In 2005, 6 degrees awarded. *Degree requirements:* For master's, thesis or alternative, comprehensive exam. *Entrance requirements:* For master's, GRE General Test, minimum GPA of 3.0. Additional exam requirements/recommendations for international students: Required—TOEFL (minimum score 550 paper-based; 213 computer-based). *Application deadline:* For fall admission, 2/15 priority date for domestic students, 2/15 priority date for international students; for spring admission, 10/15 priority date for domestic students, 10/15 priority date for international students. Applications are processed on a rolling basis. Application fee: $35. Electronic applications accepted. *Financial support:* In 2005–06, 5 students received support, including 12 research assistantships with full tuition reimbursements available (averaging $9,945 per year), 3 teaching assistantships with full tuition reimbursements available (averaging $10,413 per year); Federal Work-Study, institutionally sponsored loans, scholarships/grants, and tuition waivers (full and partial) also available. Support available to part-time students. Financial award application deadline: 3/15; financial award applicants required to submit FAFSA. *Unit head:* Dr. Leon F. Osborne, Graduate Director, 701-777-2184, Fax: 701-777-5032. *Application contact:* Brenda Halle, Admissions Specialist, 701-777-2947, Fax: 701-777-3619, E-mail: brendahalle@mail.und.edu.

University of Washington, Graduate School, College of Arts and Sciences, Department of Atmospheric Sciences, Seattle, WA 98195. Offers MS, PhD. *Degree requirements:* For master's, thesis; for doctorate, thesis/dissertation, qualifying exam. *Entrance requirements:* For master's and doctorate, GRE General Test, minimum GPA of 3.0. Additional exam requirements/recommendations for international students: Required—TOEFL. *Faculty research:* Climate change, synoptic and mesoscale meteorology, atmospheric chemistry, cloud physics, dynamics of the atmosphere.

University of Wisconsin–Madison, Graduate School, College of Letters and Science, Department of Atmospheric and Oceanic Sciences, Madison, WI 53706-1380. Offers MS, PhD. Part-time programs available. *Degree requirements:* For master's, thesis (for some programs); for doctorate, thesis/dissertation. *Entrance requirements:* For master's and doctorate, GRE General Test, minimum GPA of 3.0; previous course work in chemistry, mathematics, and physics. Electronic applications accepted. *Faculty research:* Satellite meteorology, weather systems, global climate change, numerical modeling, atmosphere-ocean interaction.

University of Wyoming, Graduate School, College of Engineering, Department of Atmospheric Science, Laramie, WY 82070. Offers MS, PhD. Postbaccalaureate distance learning degree programs offered (minimal on-campus study). *Faculty:* 9 full-time (0 women). *Students:* 16 full-time (6 women), 6 part-time (4 women), 10 international. Average age 26. 23 applicants, 43% accepted, 7 enrolled. In 2005, 4 master's, 1 doctorate awarded. Terminal master's awarded for partial completion of doctoral program. *Degree requirements:* For master's and doctorate, thesis/dissertation. *Entrance requirements:* For master's and doctorate, GRE General Test, minimum GPA of 3.0. Additional exam requirements/recommendations for international students: Required—TOEFL. *Application deadline:* For fall admission, 4/15 priority date for domestic students, 4/15 priority date for international students. Applications are processed on a rolling basis. Application fee: $50. Electronic applications accepted. *Expenses:* Contact institution. *Financial support:* In 2005–06, 21 research assistantships with full tuition reimbursements (averaging $17,181 per year) were awarded; career-related internships or fieldwork, Federal Work-Study, and institutionally sponsored loans also available. Support available to part-time students. Financial award application deadline: 3/1. *Faculty research:* Cloud and precipitation processes, mesoscale dynamics, weather modification, winter storms, aircraft instrumentation. Total annual research expenditures: $3.3 million. *Unit head:* Dr. Alfred R. Rodi, Head, 307-766-4945, Fax: 307-766-2635, E-mail: rodi@uwyo.edu. *Application contact:* Susan R. Allen, Graduate Coordinator, 307-766-5352, Fax: 307-766-2635, E-mail: sallen@uwyo.edu.

Washington State University Tri-Cities, Graduate Programs, Program in Environmental Science, Richland, WA 99352-1671. Offers applied environmental science (MS); atmospheric science (MS); earth science (MS); environmental and occupational health science (MS); environmental regulatory compliance (MS); environmental science (PhD); environmental toxicology and risk assessment (MS); water resource science (MS). Part-time programs available. *Faculty:* 1 full-time (0 women), 53 part-time/adjunct. *Students:* 4 full-time (3 women), 22 part-time (10 women); includes 1 Asian American or Pacific Islander, 2 Hispanic Americans. Average age 41. 11 applicants, 55% accepted, 6 enrolled. In 2005, 1 degree awarded. *Degree requirements:* For master's, oral exam, thesis optional. *Entrance requirements:* For master's, GRE General Test, minimum GPA of 3.0, 3 letters of recommendation. Additional exam requirements/recommendations for international students: Required—TOEFL (minimum score 550 paper-based; 213 computer-based). *Application deadline:* For fall admission, 2/1 priority date for domestic students, 3/1 priority date for international students; for spring admission, 9/1 priority date for domestic students, 7/1 priority date for international students. Application fee: $35. *Expenses:* Tuition, state resident: full-time $6,295; part-time $336 per credit. Tuition, nonresident: full-time $15,949; part-time $819 per credit. Required fees: $429. Full-time tuition and fees vary according to campus/location and program. Part-time tuition and fees vary according to course load and program. *Financial support:* In 2005–06, 8 students received support, including 1 fellowship (averaging $2,200 per year); research assistantships with full and partial tuition reimbursements available, teaching assistantships with full and partial tuition reimbursements available, Federal Work-Study, scholarships/grants, health care benefits, and unspecified assistantships also available. *Faculty research:* Radiation ecology, cytogenetics. *Unit head:* Dr. Gene Schreckhise, Associate Dean/Coordinator, 509-372-7323, E-mail: gschreck@wsu.edu.

Meteorology

Columbia University, Graduate School of Arts and Sciences, Program in Climate and Society, New York, NY 10027. Offers MA. Application fee: $75. *Expenses:* Tuition: Full-time $31,448. Tuition and fees vary according to course level, course load, campus/location and program. *Unit head:* Mark A. Cane, Chair, 212-854-8344, Fax: 845-365-8736, E-mail: mcane@ideo.columbia.edu.

Announcement: Columbia's new 12-month M.A. Program in Climate and Society trains professionals, researchers, and students to understand and cope with the impacts of climate variability and climate change on society and the environment, with an emphasis on the problems of developing societies. The curriculum combines elements of established programs in earth sciences, earth engineering, international relations, political science, sociology, and economics with unique classes in interdisciplinary applications specially designed for the program's students.

See Close-Up on page 287.

Florida Institute of Technology, Graduate Programs, College of Engineering, Department of Marine and Environmental Systems, Melbourne, FL 32901-6975. Offers environmental resource management (MS); environmental science (MS, PhD); meteorology (MS); ocean engineering (MS, PhD); oceanography (MS, PhD), including biological oceanography (MS), chemical oceanography (MS), coastal zone management (MS), geological oceanography (MS), oceanography (PhD), physical oceanography (MS). Part-time programs available. *Faculty:* 11 full-time (1 woman). *Students:* 40 full-time (15 women), 20 part-time (13 women); includes 2 minority (both Hispanic Americans), 15 international. Average age 29. 101 applicants, 50% accepted, 15 enrolled. In 2005, 16 master's awarded. Terminal master's awarded for partial completion of doctoral program. *Degree requirements:* For master's, thesis, comprehensive exam, registration; for doctorate, one foreign language, thesis/dissertation, attendance of graduate seminar, internships (oceanography and environmental science), publications, comprehensive exam, registration. *Entrance requirements:* For master's, GRE General Test (environmental science), 3 letters of recommendation, minimum GPA of 3.0; for doctorate, GRE General Test (oceanography and environmental science), resumé, 3 letters of recommendation, minimum GPA of 3.2. Additional exam requirements/recommendations for international students: Required—TOEFL (minimum score 550 paper-based; 213 computer-based). *Application deadline:* Applications are processed on a rolling basis. Application fee: $50. Electronic applications accepted. *Expenses:* Tuition: Part-time $825 per credit. *Financial support:* In 2005–06, 18 students received support, including 1 fellowship with full and partial tuition reimbursement available (averaging $1,064 per year), 8 research assistantships with full and partial tuition reimbursements available (averaging $4,116 per year), 9 teaching assistantships with full and partial tuition reimbursements available (averaging $6,867 per year); career-related internships or fieldwork and tuition remissions also available. Financial award application deadline: 3/1; financial award applicants required to submit FAFSA. *Faculty research:* Environmental modeling, coastal processes, exploring marine pollution, marine geophysics, remote sensing . Total annual research expenditures: $1 million. *Unit head:* Dr. George Maul, Department Head, 321-674-7453, Fax: 321-674-7212, E-mail: gmaul@fit.edu. *Application contact:* Carolyn P. Farrior, Director of Graduate Admissions, 321-674-7118, Fax: 321-723-9468, E-mail: cfarrior@fit.edu.

See Close-Up on page 653.

Florida State University, Graduate Studies, College of Arts and Sciences, Department of Meteorology, Tallahassee, FL 32306. Offers MS, PhD. *Faculty:* 18 full-time (3 women). *Students:* 81 full-time (28 women), 3 part-time (1 woman); includes 5 minority (1 African American, 4 Hispanic Americans), 11 international. Average age 27. 91 applicants, 66% accepted, 25 enrolled. In 2005, 11 master's, 4 doctorates awarded. Terminal master's awarded for partial completion of doctoral program. *Median time to degree:* Of those who began their doctoral program in fall 1997, 90% received their degree in 8 years or less. *Degree requirements:* For master's, thesis optional; for doctorate, thesis/dissertation, comprehensive exam. *Entrance requirements:* For master's, GRE General Test (1000 combined on verbal and quantitative

Meteorology

Florida State University (continued)

portions), minimum GPA of 3.0 in upper division work; for doctorate, GRE General Test, minimum GPA of 3.0. Additional exam requirements/recommendations for international students: Required—TOEFL (minimum score 550 paper-based; 213 computer-based). *Application deadline:* For fall admission, 2/15 priority date for domestic students, 2/1 priority date for international students; for spring admission, 11/1 for domestic students, 6/30 for international students. Applications are processed on a rolling basis. Application fee: $30. *Financial support:* In 2005–06, 76 students received support, including fellowships with partial tuition reimbursements available (averaging $18,000 per year), 65 research assistantships with partial tuition reimbursements available (averaging $18,500 per year), 12 teaching assistantships with partial tuition reimbursements available (averaging $18,500 per year); career-related internships or fieldwork, scholarships/grants, and unspecified assistantships also available. *Faculty research:* Physical, dynamic, and synoptic meteorology; climatology. Total annual research expenditures: $5.3 million. *Unit head:* Dr. Robert G. Ellingson, Chairman, 850-644-6205, Fax: 850-644-9642, E-mail: bobe@met.fsu.edu. *Application contact:* Marc Unger, Academic Coordinator, 850-644-8580, Fax: 850-644-9642, E-mail: ungerm@met.fsu.edu.

Iowa State University of Science and Technology, Graduate College, College of Liberal Arts and Sciences, Department of Geological and Atmospheric Sciences, Ames, IA 50011. Offers earth science (MS, PhD); geology (MS, PhD); meteorology (MS, PhD); water resources (MS, PhD). *Faculty:* 15 full-time, 2 part-time/adjunct. *Students:* 32 full-time (12 women), 2 part-time; includes 1 minority (African American), 15 international. 40 applicants, 60% accepted, 13 enrolled. In 2005, 11 master's awarded. *Degree requirements:* For master's, thesis (for some programs); for doctorate, thesis/dissertation. *Entrance requirements:* For master's and doctorate, GRE General Test. Additional exam requirements/recommendations for international students: Required—TOEFL (paper score 530; computer score 197) or IELTS (score 6.0). *Application deadline:* For fall admission, 1/1 for domestic students. Applications are processed on a rolling basis. Application fee: $30 ($70 for international students). Electronic applications accepted. *Expenses:* Tuition, state resident: full-time $6,410. Tuition, nonresident: full-time $16,422. Tuition and fees vary according to program. *Financial support:* In 2005–06, 21 research assistantships with full and partial tuition reimbursements (averaging $13,545 per year), 9 teaching assistantships with full and partial tuition reimbursements (averaging $13,507 per year) were awarded; fellowships, scholarships/grants, health care benefits, and unspecified assistantships also available. *Unit head:* Dr. Carl E. Jacobson, Chair, 515-294-4477.

McGill University, Faculty of Graduate and Postdoctoral Studies, Faculty of Agricultural and Environmental Sciences, Department of Natural Resource Sciences, Montréal, QC H3A 2T5, Canada. Offers agrometeorology (M Sc, PhD); entomology (M Sc, PhD); forest science (M Sc, PhD); microbiology (M Sc, PhD); neotropical environment (M Sc, PhD); soil science (M Sc, PhD); wildlife biology (M Sc, PhD). *Degree requirements:* For master's and doctorate, thesis/dissertation, registration. *Entrance requirements:* For master's, minimum GPA of 3.0 or 3.2 in the last 2 years of university study. Additional exam requirements/recommendations for international students: Required—TOEFL (minimum score 550 paper-based; 213 computer-based), IELT (minimum score 7). Electronic applications accepted. *Faculty research:* Toxicology, reproductive physiology, parasites, wildlife management, genetics.

Naval Postgraduate School, Graduate Programs, Department of Meteorology, Monterey, CA 93943. Offers MS, PhD. Program only open to commissioned officers of the United States and friendly nations and selected United States federal civilian employees. Part-time programs available. *Students:* 38 full-time. In 2005, 25 master's, 2 doctorates awarded. *Degree requirements:* For master's, thesis; for doctorate, one foreign language, thesis/dissertation. *Unit head:* Dr. Carlyle H. Wash, Chairman, 831-656-2517. *Application contact:* Tracy Hammond, Acting Director of Admissions, 831-656-3059, Fax: 831-656-2891, E-mail: thammond@bps.navy.mil.

North Carolina State University, Graduate School, College of Physical and Mathematical Sciences, Department of Marine, Earth, and Atmospheric Sciences, Raleigh, NC 27695. Offers marine, earth, and atmospheric sciences (MS, PhD); meteorology (MS, PhD); oceanography (MS, PhD). Terminal master's awarded for partial completion of doctoral program. *Degree requirements:* For master's, thesis (for some programs), final oral exam; for doctorate, thesis/dissertation, final oral exam, preliminary oral and written exams, comprehensive exam, registration. *Entrance requirements:* For master's, GRE General Test, minimum GPA of 3.0; for doctorate, GRE General Test, GRE Subject Test (for disciplines in biological oceanography and geology), minimum GPA of 3.0. Additional exam requirements/recommendations for international students: Required—TOEFL (minimum score 550 paper-based). Electronic applications accepted. *Faculty research:* Boundary layer and air quality meteorology; climate and mesoscale dynamics; biological, chemical, geological, and physical oceanography; hard rock, soft rock, environmental, and paleo geology.

See Close-Up on page 229.

The Pennsylvania State University University Park Campus, Graduate School, College of Earth and Mineral Sciences, Department of Meteorology, State College, University Park, PA 16802-1503. Offers MS, PhD. *Students:* 58 full-time (17 women), 1 part-time; includes 4 minority (3 African Americans, 1 Asian American or Pacific Islander), 14 international. *Entrance requirements:* For master's and doctorate, GRE General Test. *Expenses:* Tuition, state resident: full-time $12,518; part-time $522 per credit. Tuition, nonresident: full-time $23,004; part-time $959 per credit. Required fees: $484. Tuition and fees vary according to course load, campus/location and program. *Unit head:* Dr. William H. Brune, Head, 814-865-3286, Fax: 814-865-3663, E-mail: whb2@psu.edu. *Application contact:* Dr. William H. Brune, Head, 814-865-3286, Fax: 814-865-3663, E-mail: whb2@psu.edu.

Plymouth State University, College of Graduate Studies, Graduate Studies in Education, Program in Science, Plymouth, NH 03264-1595. Offers applied meteorology (MS); environmental science and policy (MS); science education (MS). *Students:* 1 (woman) full-time, 18 part-time (12 women). 19 applicants, 100% accepted. *Unit head:* Dr. Steve Kahl, Director of the Center for the Environment, E-mail: jskahl@plymouth.edu.

Saint Louis University, Graduate School, College of Arts and Sciences and Graduate School, Department of Earth and Atmospheric Sciences, St. Louis, MO 63103-2097. Offers geophysics (PhD); geoscience (MS); meteorology (M Pr Met, MS-R, PhD). Part-time programs available. *Faculty:* 17 full-time (0 women), 2 part-time/adjunct (0 women). *Students:* 30 full-time (11 women), 4 part-time (2 women); includes 2 minority (1 African American, 1 Hispanic American), 16 international. Average age 28. 35 applicants, 86% accepted, 14 enrolled. In 2005, 4 master's, 3 doctorates awarded. *Degree requirements:* For master's, thesis (for some programs), comprehensive oral exam; for doctorate, thesis/dissertation, preliminary exams. *Entrance requirements:* For master's and doctorate, GRE General Test, letters of recommendation, resumé. Additional exam requirements/recommendations for international students: Required—TOEFL (minimum score 550 paper-based; 213 computer-based). *Application deadline:* For fall admission, 7/1 for domestic students, 7/1 for international students; for spring admission, 11/1 for domestic students, 11/1 for international students. Applications are processed on a rolling basis. Application fee: $40. *Expenses:* Tuition: Part-time $760 per credit hour. Required fees: $55 per semester. *Financial support:* In 2005–06, 25 students received support, including 8 research assistantships with full tuition reimbursements available (averaging $16,000 per year), 7 teaching assistantships with full tuition reimbursements available (averaging $15,500 per year); health care benefits and unspecified assistantships also available. Financial award application deadline: 6/1; financial award applicants required to submit FAFSA. *Faculty research:* Structural geology, mesoscale meteorology and severe storms, weather and climate change prediction. Total annual research expenditures: $800,000. *Unit head:* Dr. Bill Dannevik, Interim Chairperson, 314-977-3115, Fax: 314-911-3117, E-mail: dannevik@slu.edu. *Application contact:* Gary Behrman, Associate Dean of the Graduate School, 314-977-3827, E-mail: behrmang@slu.edu.

San Jose State University, Graduate Studies and Research, College of Science, Department of Meteorology, San Jose, CA 95192-0001. Offers MS. *Students:* 3 full-time (1 woman), 4 part-time (1 woman); includes 1 minority (Asian American or Pacific Islander), 3 international. Average age 37. 7 applicants, 57% accepted, 0 enrolled. In 2005, 2 degrees awarded. *Degree requirements:* For master's, thesis or alternative. *Entrance requirements:* For master's, GRE. *Application deadline:* For fall admission, 6/29 for domestic students; for spring admission, 11/30 for domestic students. Applications are processed on a rolling basis. Application fee: $59. Electronic applications accepted. *Expenses:* Tuition, nonresident: part-time $339 per unit. Required fees: $1,286 per semester. Tuition and fees vary according to course load and degree level. *Financial support:* Applicants required to submit FAFSA. *Unit head:* Allison Bridger, Chair, 408-924-5200, Fax: 408-924-5191.

Texas A&M University, College of Geosciences, Department of Atmospheric Sciences, College Station, TX 77843. Offers MS, PhD. *Faculty:* 6 full-time (0 women), 1 part-time/adjunct (0 women). *Students:* 51 full-time (17 women), 8 part-time (3 women); includes 2 minority (1 African American, 1 Asian American or Pacific Islander), 20 international. Average age 28. 40 applicants, 90% accepted, 10 enrolled. In 2005, 9 master's, 5 doctorates awarded. *Degree requirements:* For master's and doctorate, thesis/dissertation. *Entrance requirements:* For master's and doctorate, GRE General Test. Additional exam requirements/recommendations for international students: Required—TOEFL. *Application deadline:* For fall admission, 3/1 for domestic students; for spring admission, 10/1 for domestic students. Applications are processed on a rolling basis. Application fee: $50 ($75 for international students). Electronic applications accepted. *Expenses:* Tuition, state resident: full-time $4,488; part-time $187 per credit hour. Tuition, nonresident: full-time $11,112; part-time $463 per credit hour. Required fees:$1,974. *Financial support:* In 2005–06, fellowships (averaging $16,500 per year), research assistantships with tuition reimbursements (averaging $15,000 per year), teaching assistantships (averaging $15,000 per year) were awarded; career-related internships or fieldwork, institutionally sponsored loans, scholarships/grants, and tuition waivers (partial) also available. Financial award application deadline: 3/1; financial award applicants required to submit FAFSA. *Faculty research:* Radar- and satellite-rainfall relationships, mesoscale dynamics and numerical modeling, climatology. *Unit head:* Dr. Richard Orville, Head, 979-845-7671, Fax: 979-862-4466. *Application contact:* Patricia Price, Academic Advisor, 979-845-7688, Fax: 979-862-4466, E-mail: pprice@ariel.net.tamu.edu.

Université du Québec à Montréal, Graduate Programs, Programs in Atmospheric Sciences and Meteorology, Montréal, QC H3C 3P8, Canada. Offers atmospheric sciences (M Sc); meteorology (PhD, Diploma). Part-time programs available. *Degree requirements:* For master's, thesis. *Entrance requirements:* For master's and Diploma, appropriate bachelor's degree or equivalent and proficiency in French; for doctorate, appropriate master's degree or equivalent and proficiency in French.

University of Hawaii at Manoa, Graduate Division, School of Ocean and Earth Science and Technology, Department of Meteorology, Honolulu, HI 96822. Offers MS, PhD. Part-time programs available. *Faculty:* 17 full-time (1 woman), 7 part-time/adjunct (1 woman). *Students:* 33 full-time (12 women), 1 part-time; includes 4 minority (all Hispanic Americans), 19 international. Average age 31. 46 applicants, 15% accepted, 4 enrolled. In 2005, 6 master's, 4 doctorates awarded. *Median time to degree:* Of those who began their doctoral program in fall 1997, 100% received their degree in 8 years or less. *Degree requirements:* For master's and doctorate, thesis/dissertation. *Entrance requirements:* For master's and doctorate, GRE General Test. *Application deadline:* For fall admission, 3/1 for domestic students, 1/15 for international students; for spring admission, 9/1 for domestic students, 8/1 for international students. Application fee: $50. *Expenses:* Tuition, state resident: full-time $8,400; part-time $200 per credit hour. Tuition, nonresident: full-time $11,088; part-time $462 per credit hour. Tuition and fees vary according to program. *Financial support:* In 2005–06, 24 research assistantships (averaging $17,825 per year), 3 teaching assistantships (averaging $15,760 per year) were awarded; fellowships, Federal Work-Study and tuition waivers (full) also available. *Faculty research:* Tropical cyclones, air-sea interactions, mesoscale meteorology, intraseasonal oscillations, tropical climate. Total annual research expenditures: $1.2 million. *Unit head:* Dr. Kevin Hamilton, Chairperson, 808-956-7476, Fax: 808-956-2877, E-mail: tas@soest.hawaii.edu. *Application contact:* Yi-Leng Chen, Graduate Chairperson, 808-956-8775, Fax: 808-956-2877, E-mail: yileng@hawaii.edu.

University of Maryland, College Park, Graduate Studies, College of Computer, Mathematical and Physical Sciences, Department of Atmospheric and Oceanic Science, College Park, MD 20742. Offers MS, PhD. Part-time and evening/weekend programs available. Postbaccalaureate distance learning degree programs offered. *Faculty:* 34 full-time (5 women), 3 part-time/adjunct (1 woman). *Students:* 52 full-time (19 women), 7 part-time (3 women); includes 6 minority (3 Asian Americans or Pacific Islanders, 3 Hispanic Americans), 29 international. 82 applicants, 41% accepted, 14 enrolled. In 2005, 7 master's, 5 doctorates awarded. Terminal master's awarded for partial completion of doctoral program. *Median time to degree:* Of those who began their doctoral program in fall 1997, 25% received their degree in 8 years or less. *Degree requirements:* For master's, scholarly paper, written and oral exams; for doctorate, thesis/dissertation, exam. *Entrance requirements:* For master's, GRE General Test, GRE, background in mathematics, experience in scientific computer languages, 3 letters of recommendation; for doctorate, GRE General Test. *Application deadline:* For fall admission, 5/15 for domestic students, 2/1 for international students; for spring admission, 10/15 for domestic students, 6/1 for international students. Applications are processed on a rolling basis. Application fee: $60. Electronic applications accepted. *Financial support:* In 2005–06, 5 fellowships with full tuition reimbursements (averaging $3,722 per year), 8 research assistantships with tuition reimbursements (averaging $20,246 per year), 31 teaching assistantships with tuition reimbursements (averaging $19,081 per year) were awarded; Federal Work-Study and scholarships/grants also available. Support available to part-time students. Financial award applicants required to submit FAFSA. *Faculty research:* Weather, atmospheric chemistry, air pollution, global change, radiation. Total annual research expenditures: $3.9 million. *Unit head:* Dr. Russ Dickerson, Chairman, 301-405-5364, Fax: 301-314-9482. *Application contact:* Dean of Graduate School, 301-405-4190, Fax: 301-314-9305.

University of Miami, Graduate School, Rosenstiel School of Marine and Atmospheric Science, Division of Meteorology and Physical Oceanography, Coral Gables, FL 33124. Offers meteorology (PhD); physical oceanography (MS, PhD). *Faculty:* 30 full-time (5 women), 21 part-time/adjunct (3 women). *Students:* 34 full-time (7 women), 2 part-time; includes 1 minority (Hispanic American), 17 international. Average age 27. 34 applicants, 41% accepted, 8 enrolled. In 2005, 4 master's, 2 doctorates awarded. Terminal master's awarded for partial completion of doctoral program. *Median time to degree:* Of those who began their doctoral program in fall 1997, 50% received their degree in 8 years or less. *Degree requirements:* For master's and doctorate, thesis/dissertation, comprehensive exam, registration. *Entrance requirements:* For master's and doctorate, GRE General Test. Additional exam requirements/recommendations for international students: Required—TOEFL (minimum score 550 paper-based; 213 computer-based). *Application deadline:* For fall admission, 1/1 priority date for domestic students, 1/1 priority date for international students. Applications are processed on a rolling basis. Application fee: $50. Electronic applications accepted. *Financial support:* In 2005–06, 31 students received support, including 5 fellowships with tuition reimbursements available (averaging $22,380 per year), 21 research assistantships with tuition reimbursements available (averaging $22,380 per year), 2 teaching assistantships with tuition reimbursements available (averaging $22,380 per year); institutionally sponsored loans and scholarships/grants also available. Financial award application deadline: 3/1; financial award applicants required to submit FAFSA. *Unit head:* Dr. William Johns, Chairperson, 305-421-4057, E-mail: wjohns@rsmas.miami.edu. *Application contact:* Dr. Larry Peterson, Associate Dean, 305-421-4155, Fax: 305-421-4771, E-mail: gso@rsmas.miami.edu.

Announcement: The Division is engaged in education and research in the physical processes governing the motion and composition of the ocean and atmosphere through observational, diagnostic, modeling, and theoretical explorations. Research apprenticeships are integral to the Division's education program. Students receive tuition waivers and financial aid as research assistantships or fellowships.

University of Oklahoma, Graduate College, College of Atmospheric and Geographic Sciences, School of Meteorology, Norman, OK 73019-0390. Offers M Pr Met, MS Metr, PhD. Part-time programs available. *Faculty:* 43 full-time (3 women), 2 part-time/adjunct (0 women). *Students:* 67 full-time (22 women), 18 part-time (4 women); includes 3 minority (1 African American, 1 Asian American or Pacific Islander, 1 Hispanic American), 18 international. 60 applicants, 40% accepted, 19 enrolled. In 2005, 13 master's, 5 doctorates awarded. *Degree requirements:* For master's, thesis or alternative, comprehensive exam; for doctorate, one foreign language, thesis/dissertation, departmental qualifying exam. *Entrance requirements:* For master's, GRE, bachelor's degree in related area; for doctorate, GRE. Additional exam requirements/recommendations for international students: Required—TOEFL (minimum score 600 paper-based). *Application deadline:* For fall admission, 2/1 priority date for domestic students, 4/1 priority date for international students; for spring admission, 11/1 for domestic students, 9/1 for international students. Applications are processed on a rolling basis. Application fee: $40 ($90 for international students). *Expenses:* Tuition, state resident: full-time $3,029; part-time $126 per credit hour. Tuition, nonresident: full-time $10,807; part-time $450 per credit hour. Required fees: $1,231; $44 per credit hour. Tuition and fees vary according to course load and program. *Financial support:* In 2005–06, 8 fellowships with full tuition reimbursements (averaging $5,000 per year), 58 research assistantships (averaging $14,650 per year), 13 teaching assistantships with partial tuition reimbursements (averaging $14,306 per year) were awarded; career-related internships or fieldwork, institutionally sponsored loans, scholarships/grants, health care benefits, tuition waivers (partial), and unspecified assistantships also available. Financial award application deadline: 2/1; financial award applicants required to submit FAFSA. *Faculty research:* Radar meteorology, synoptic and dynamic meteorology, mesoscale meteorology, numerical weather prediction, regional and global climate. Total annual research expenditures: $743,452. *Unit head:* Dr. Frederick H. Carr, Director, 405-325-6561, Fax: 405-325-7689, E-mail: fcarr@ou.edu. *Application contact:* Celia Jones, Coordinator, Academic Student Services, 405-325-6571, Fax: 405-325-7689, E-mail: cjones@ou.edu.

University of Utah, The Graduate School, College of Mines and Earth Sciences, Department of Meteorology, Salt Lake City, UT 84112-1107. Offers MS, PhD. Part-time programs available. *Faculty:* 9 full-time (1 woman). *Students:* 25 full-time (9 women), 6 part-time (2 women), 12 international. Average age 28. 32 applicants, 44% accepted, 10 enrolled. In 2005, 3 master's, 4 doctorates awarded. Terminal master's awarded for partial completion of doctoral program. *Median time to degree:* Of those who began their doctoral program in fall 1997, 33% received their degree in 8 years or less. *Degree requirements:* For master's, thesis optional; for doctorate, thesis/dissertation. *Entrance requirements:* For master's and doctorate, GRE General Test, minimum GPA of 3.0, 3 letters of reference. Additional exam requirements/recommendations for international students: Required—TOEFL (minimum score 500 paper-based; 173 computer-based). *Application deadline:* For fall admission, 12/31 priority date for domestic students, 12/31 priority date for international students. Applications are processed on a rolling basis. Application fee: $45 ($65 for international students). Electronic applications accepted. *Expenses:* Tuition, state resident: full-time $2,932; part-time $2,212 per term. Tuition, nonresident: full-time $10,350; part-time $7,812 per term. Required fees: $590; $516 per term. Tuition and fees vary according to course load and program. *Financial support:* In 2005–06, 31 students received support, including 2 fellowships, 25 research assistantships (averaging $2,000 per year), 4 teaching assistantships; unspecified assistantships also available. Financial award application deadline: 2/15; financial award applicants required to submit FAFSA. *Faculty research:* Clouds, aerosols, and climate; numerical weather prediction; mountain weather and climate; tropical convection and storms; and climate variability and change. Total annual research expenditures: $2.2 million. *Unit head:* Dr. W James Steenburgh, Chair, 801-585-9482, Fax: 801-585-3681, E-mail: jimsteen@met.utah.edu. *Application contact:* Kathy Roberts, Executive Secretary, 801-581-6136, Fax: 801-585-3681, E-mail: kroberts@met.utah.edu.

Utah State University, School of Graduate Studies, College of Agriculture, Department of Plants, Soils, and Biometeorology, Logan, UT 84322. Offers biometeorology (MS, PhD); ecology (MS, PhD); plant science (MS, PhD); soil science (MS, PhD). Part-time programs available. *Faculty:* 31 full-time (4 women), 13 part-time/adjunct (0 women). *Students:* 117 full-time (58 women), 16 part-time (5 women), 14 international. Average age 26. 25 applicants, 80% accepted, 19 enrolled. In 2005, 8 master's, 1 doctorate awarded. Terminal master's awarded for partial completion of doctoral program. *Median time to degree:* Of those who began their doctoral program in fall 1997, 100% received their degree in 8 years or less. *Degree requirements:* For master's and doctorate, thesis/dissertation. *Entrance requirements:* For master's, GRE General Test, BS in plant, soil, atmospheric science, or related field; minimum GPA of 3.0; for doctorate, GRE General Test, minimum GPA of 3.0. Additional exam requirements/recommendations for international students: Required—TOEFL. *Application deadline:* For fall admission, 6/15 priority date for domestic students, 3/15 priority date for international students; for spring admission, 10/15 priority date for domestic students, 9/15 priority date for international students. Applications are processed on a rolling basis. Application fee: $50 ($60 for international students). Electronic applications accepted. *Financial support:* In 2005–06, 23 research assistantships with partial tuition reimbursements (averaging $15,000 per year) were awarded; Federal Work-Study, institutionally sponsored loans, and tuition waivers (full) also available. Support available to part-time students. Financial award application deadline: 3/1. *Faculty research:* Biotechnology and genomics, plant physiology and biology, nutrient and water efficient landscapes, physical-chemical-biological processes in soil, environmental biophysics and climate. Total annual research expenditures: $4.5 million. *Unit head:* Dr. Larry A. Rupp, Head, 435-797-2099, Fax: 435-797-3376, E-mail: larryr@ext.usu.edu. *Application contact:* Dr. Paul G. Johnson, Graduate Program Coordinator, 435-797-7039, Fax: 435-797-3376, E-mail: paul.johnson@usu.edu.

Yale University, Graduate School of Arts and Sciences, Department of Geology and Geophysics, New Haven, CT 06520. Offers geochemistry (PhD); geophysics (PhD); meteorology (PhD); mineralogy and crystallography (PhD); oceanography (PhD); paleoecology (PhD); paleontology and stratigraphy (PhD); petrology (PhD); structural geology (PhD). *Degree requirements:* For doctorate, thesis/dissertation. *Entrance requirements:* For doctorate, GRE General Test. Additional exam requirements/recommendations for international students: Required—TOEFL.

COLUMBIA UNIVERSITY

Master of Arts Program in Climate and Society

Program of Study

Recent research has generated a wealth of new knowledge about long-term climate change, shorter-term climate variability, and their socioeconomic impacts. Decision makers need clear and reliable guidance on impending climate shocks as well as practical information and tools to deal with their consequences. Building on improved scientific understanding of climate and improved coping mechanisms, Columbia University is training a new generation of academics and professionals at the nexus of social science, climate science, and public policy.

The twelve-month Master of Arts (M.A.) Program in Climate and Society at Columbia University is designed to train professionals and academics to understand and cope with the impacts of climate variability and climate change on society and the environment. This rigorous program emphasizes the problems of developing societies. Columbia is at the forefront of research on climate and climate applications, and is supported by an extensive network of research units and faculty members. Drawing on the superb educational and research facilities of Columbia University, the M.A. Program in Climate and Society combines elements of established programs in earth sciences, earth engineering, international relations, political science, sociology, and economics with unique classes in interdisciplinary applications especially designed for students in this program.

A set of tailor-made core courses is designed to provide a scientific basis for inquiry and to stress interdisciplinary problem solving. The core modules include dynamics of climate variability and change, regional climate and climate impacts, quantitative models of climate-sensitive natural and human systems, and the integrative seminar: Managing Climate Variability and Adapting to Climate Change. A professional development seminar and the choice of a summer internship or a research thesis complete the required core curriculum.

Research Facilities

The M.A. Program in Climate and Society uses an interdisciplinary approach in combination with other departments and institutions at Columbia. The staff and facilities overlap with the Lamont-Doherty Earth Observatory (LDEO) research campus. The program also draws on the School of International and Public Affairs (SIPA), the Henry Krumb School of Mines, the Department of Earth and Environmental Engineering, the Earth Engineering Center, the Earth Institute at Columbia University, the International Research Institute for Climate and Society (IRI), and the NASA Goddard Institute for Space Studies.

Financial Aid

Applicants from the United States are typically eligible for Federal Stafford Student loans. All interested U.S. applicants must complete the Free Application for Federal Student Aid (FAFSA), which is available online at http://www.fafsa.ed.gov. Following admission, all students who submitted a FAFSA are advised by the school of their eligibility for federal aid via an award letter with the forms and materials needed to apply for the Stafford loans. Additional financial support in the form of private loans can help to meet the full cost of tuition. Students enrolled in the program are eligible to have prior federal student loans deferred during their term of study.

Cost of Study

Costs for 2006–07 are as follows: three semesters of tuition (total tuition for the entire M.A. program), $37,462; fees, including basic health insurance, $2800; modest living expenses, $15,000; personal expenses, $5000; books and supplies, $1500. The total estimated cost: $61,762.

Living and Housing Costs

Every effort is made to find housing for full-time entering students in Columbia University–owned buildings in the Morningside Heights and Washington Heights neighborhoods, but housing is not guaranteed. Applications must be received in a timely manner to be considered. University Apartment Housing (UAH) consists of apartments and dormitory-style suites located within walking distance of the campus. Most students in the program find housing off campus in the neighborhoods of New York City.

Student Outcomes

At the end of twelve months, graduates of the program will be prepared to address environmental issues from positions in government, business, and nongovernmental organizations. Some may wish to continue their academic careers in the social or natural sciences.

Location

Columbia University is located in the heart of the borough of Manhattan in the City of New York. With a population of more than 8 million, New York City boasts many of the world's most recognized sites. It is host to the United Nations, the financial industry, a wealth of museums and cultural institutions, and leading universities and research institutions. The city's international character and diverse immigrant populations contribute to a thriving environment for study and exploration.

The University

Columbia University was founded in 1754 as King's College by royal charter of King George II of England. It is the oldest institution of higher learning in the state of New York and the fifth oldest in the United States. From its beginnings in a schoolhouse in lower Manhattan, the University has grown to encompass two principal campuses: the historic, neoclassical campus in the Morningside Heights neighborhood and the modern Medical Center in Washington Heights. Today, Columbia is one of the top academic and research institutions in the world, conducting research in medicine, science, the arts, and the humanities. It includes three undergraduate schools, thirteen graduate and professional schools, and a School of Continuing Education. Sixty-four Nobel laureates have taught or studied at Columbia. Each year, the faculty of approximately 4,000 teaches more than 23,000 students from more than 150 countries.

Applying

Applicants should have completed a bachelor's degree in physical sciences, engineering, social sciences, or planning and policy studies. Work experience in a related field is desirable. Students are selected for admission based on their academic background and related work experiences. The General Test of the Graduate Record Examinations (GRE) is required. The Test of English as a Foreign Language (TOEFL) or International English Language Testing System (IELTS) exam is required for nonnative speakers of English.

The University does not discriminate on the basis of race, sex, religion, sexual orientation, or national or ethnic origin.

Correspondence and Information

M.A. Program in Climate and Society
554 Schermerhorn Extension, MC 5505
Columbia University
1200 Amsterdam Avenue
New York, New York 10027
Phone: 212-854-9896
Fax: 212-854-7975
E-mail: climatesociety@ei.columbia.edu
Web site: http://www.columbia.edu/climatesociety

Graduate School of Arts and Sciences
107 Low Memorial Library
Columbia University
535 West 116th Street, MC 4304
New York, New York 10027
Phone: 212-854-4737
Fax: 212-854-2863
E-mail: gsas-admit@columbia.edu
Web site: http://www.columbia.edu/cu/gsas

Columbia University

AFFILIATED FACULTY AND RESEARCHERS

Tony Barnston, Head of Forecast Operations, International Research Institute for Climate Prediction (IRI); Ph.D. Improvement of IRI's forecast operation, maximizing accuracy of forecasts, streamlining and automating forecasting process.

Volker Berghahn, Seth Low Professor of History; Ph.D., London, 1964. Modern German history, European-American relations.

Sylvie Le Blancq, Assistant Professor of Clinical Environmental Health Sciences in the Mailman School of Public Health at Columbia University and the Center for Environmental Research and Conservation; Ph.D., London, 1983. Study of *Cryptosporidium parvum* and *Giardia lamblia.*

Wallace S. Broecker, Newberry Professor of Earth and Environmental Sciences. Climate systems, especially as they involve the role of oceans in climate change.

Coralie Bryant, Director, Program in Economic and Political Development, School of International and Public Affairs; Ph.D., London School of Economics. Capacity building for poverty reduction, institutional development, development management, policy analysis.

Mark Cane, G. Unger Vetlesen Professor of Earth and Climate Sciences and Program Director.

Steven A. Cohen, Director, Master of Public Administration Program in Environmental Science and Policy, School of International and Public Affairs; Director, Executive Master of Public Administration Program; and Director, Office of Educational Programs of the Earth Institute at Columbia University; Ph.D., SUNY at Buffalo, 1979. Environmental policy management, energy policy, nuclear waste policy, urban policy, total quality management, management effectiveness.

Anthony Del Genio, Adjunct Professor, Goddard Institute for Space Studies; Ph.D., UCLA, 1978. Investigating long-term global climate change—how terrestrial climate system will respond to increasing concentrations of greenhouse gases over the next few decades.

David Downie, Associate Director, M.A. Program in Climate and Society, and Director, Global Roundtable on Climate Change, Columbia University; Ph.D., North Carolina at Chapel Hill, 1996. Environmental politics, international environmental politics.

Richard G. Fairbanks, Professor, Department of Earth and Environmental Sciences, and Senior Scientist, Lamont-Doherty Earth Observatory of Columbia University; Ph.D., Brown. Using geochemical tracers to document modulations of North Atlantic Deep Water (NADW) in the Pleistocene.

Dana R. Fisher, Assistant Professor, Department of Sociology; Ph.D., Wisconsin–Madison, 2001. Exploring ways that civil society participates in political processes on local, national, and international levels.

Lisa Goddard, Research Scientist, International Research Institute (IRI) for Climate Prediction; Ph.D., Princeton. Climate dynamics and potential predictability; assessing climate prediction tools; advancing strategies for research, development, and implementation of climate forecasts.

Geoffrey Heal, Paul Garrett Professor of Public Policy and Business Responsibility, Columbia Business School; Ph.D., Cambridge. Studying ways of controlling the impact of economic activity on the environment and ways of valuing economic services provided by environmental assets.

Patrick L. Kinney, Associate Professor of Public Health, Environmental Health Sciences Department; Ph.D., Harvard, 1986. Epidemiologic research addressing the respiratory health impacts of air pollution, with particular focus on the impacts of motor vehicle pollution in urban areas, in context of suburban sprawl.

David H. Krantz, Professor, Departments of Psychology and Statistics; Ph.D., Pennsylvania, 1964. Problem solving, especially decision making, induction, and math education; applications of axiomatic measurement theory; perception, especially psychophysics and color vision.

Upmanu Lall, Professor, Department of Earth and Environmental Engineering, Senior Research Scientist, International Research Institute for Climate Prediction; Ph.D., Texas, 1981. Hydroclimate modeling, spatial data analysis and visualization, time series analysis and forecasting.

Douglas G. Martinson, Senior Research Scientist, Climate Modeling and Diagnostics Group, Lamont-Doherty Earth Observatory, Ph.D. Elucidating understanding of processes controlling climatically relevant polar characteristics through complex interactions between ocean, sea ice, and atmospheric boundary layer.

Mary Northridge, Associate Professor, Department of Sociomedical Sciences, Mailman School of Public Health at Columbia University; Ph.D., Columbia. Urbanism and public health.

Dorothy Peteet, Adjunct Senior Research Scientist, NASA/Goddard Institute for Space Studies/Lamont-Doherty Earth Observatory; Ph.D., NYU, 1983. Utilizing general circulation modeling (GCM) to perform climate sensitivity tests in order to better understand mechanisms and causes of climatic change.

Alexander S. Pfaff, Associate Research Scientist, International Research Institute for Climate Prediction, and Director, Center on Globalization and Sustainable Development, the Earth Institute at Columbia University; Ph.D., MIT, 1995. Environmental and natural resource economics, environment and development policy, applied microeconomics and policy.

Thomas Pogge, Associate Professor, Department of Philosophy; Ph.D., Harvard, 1983. Global justice, social and political philosophy, ethics, moral philosophy.

David Rind, Adjunct Professor, Department of Earth and Environmental Sciences; Climate Modeler, NASA Goddard Institute for Space Studies; Ph.D., Columbia, 1976. Climate modeling, paleoclimate studies, and atmospheric dynamics.

Cynthia Rosenzweig, Adjunct Professor, Barnard College, Adjunct Senior Research Scientist, Earth Institute of Columbia University, and Research Scientist, Goddard Institute for Space Studies; Ph.D., Massachusetts, 1991. Climate variability and change in relation to agriculture at the regional, national, and global scales.

Anji Seth, Associate Research Scientist, Climate Monitoring and Dissemination, International Research Institute for Climate Prediction; Ph.D., Michigan. Role of land surface processes on climate simulations and predictability of seasonal rainfall using regional climate models.

Christopher Small, Associate Research Scientist, Marine Geology and Geophysics, Lamont-Doherty Earth Observatory; Ph.D., Scripps Research Institute, 1993. Urban remote sensing; marine gravity field; continental physiography, climate, and human population; hypsographic demography; global volcanism and human population; urban vegetation.

Shiv Someshwar, Research Scientist, International Research Institute for Climate Prediction; Ph.D., UCLA, 1995. Reducing livelihood vulnerability to climate variability.

Awash Teklehaimanot, Senior Staff Member of World Health Organization (WHO/Geneva), Director of Malaria Program at Columbia, and member of the Task Force on Malaria for the U.N. Millennium Project; Ph.D., Purdue. Infectious disease epidemiology.

Mingfang Ting, Associate Director, M.A. Program in Climate and Society, and Doherty Senior Research Scientist, Lamont-Doherty Earth Observatory; Ph.D., Princeton, 1990. Impact of global climate change on regional scales and teleconnection dynamics, modeling and diagnostics of climatological and anomalous stationary waves and impact of sea surface temperatures on global climate, dynamics of droughts and floods circulation for U.S. and North American monsoon system.

M. Neil Ward, Head of Forecast Development, International Research Institute for Climate Prediction; Ph.D. Important link between forecast products and user applications.

Duncan Watts, Associate Professor, Department of Sociology; Ph.D., Cornell. Mathematical and computational modeling of complex networks as applied to problems in social network theory, contagion, computation, and theory of the firm.

Elke U. Weber, Professor of Psychology and Management; Ph.D., Harvard, 1984. Behavioral models of judgment and decision making under risk and uncertainty; psychologically appropriate ways to measure and model individual and cultural differences in risk taking, specifically in risky financial situations; environmental decision making and policy.

Paige West, Assistant Professor, Department of Anthropology at Barnard College; Ph.D., Rutgers, 2000. Linkages between environmental conservation and international development, environmental ethics, anthropology of science, the imagination, American environmentalism, ethnography of nature, critical analysis of conservation and development interventions.

COLUMBIA UNIVERSITY/ NASA GODDARD SPACE FLIGHT CENTER'S INSTITUTE FOR SPACE STUDIES

Atmospheric and Planetary Science Program

Program of Study

The Departments of Earth and Environmental Sciences and Applied Physics and Applied Mathematics jointly offer a graduate program in atmospheric and planetary science leading to the Ph.D. degree. Four to six years are generally required to complete the Ph.D., including the earning of M.A. and M.Phil. degrees. Applicants should have a strong background in physics and mathematics, including advanced undergraduate courses in mechanics, electromagnetism, advanced calculus, and differential equations.

The program is conducted in cooperation with the NASA Goddard Space Flight Center's Institute for Space Studies, which is adjacent to Columbia University. Members of the Institute hold adjunct faculty appointments, offer courses, and supervise the research of graduate students in the program. The Institute holds colloquia and scientific conferences in which the University community participates. Opportunities for visiting scientists to conduct research at the Institute are provided by postdoctoral research programs administered by the National Academy of Sciences–National Research Council and Columbia and supported by NASA.

Research at the Institute focuses on broad studies of natural and anthropogenic global changes. Areas of study include global climate, earth observations, biogeochemical cycles, planetary atmospheres, and related interdisciplinary studies. The global climate involves basic research on climatic variations and climate processes, including the development of global numerical models to study the climate effects of increasing carbon dioxide and other trace gases, aerosols, solar variability, and changing surface conditions. The earth observations program entails research in the retrieval of cloud, aerosol, and surface radiative properties from global satellite radiance data to further understanding of their effects on climate. Biogeochemical cycles research utilizes three-dimensional models to study in the distribution of trace gases in the troposphere and stratosphere and to examine the role of the biosphere in the global carbon cycle. The planetary atmospheres program includes comparative modeling of radiative transfer and dynamics applied to Venus, Titan, Mars, and the Jovian planets; participation in spacecraft experiments; and analysis of ground-based observations. Interdisciplinary research includes studies of turbulence, solar system formation, and astrobiology.

Research Facilities

The Institute operates a modern general-purpose scientific computing facility consisting of a Compaq ES45 with 32 processors operating on Tru 64 UNIX; a 96-processor SGI Origin 3000 server; an 8-processor SGI Power Challenge development server; eighty workstations, including sixty IBM RS/6000 and twenty SGI IRIX; and PCs, Macs, and peripheral equipment. Spyglass, AVS, IDL, NCAR graphics, and in-house software permit interactive processing, display, and analysis of satellite imagery and other digital data. The Institute is the Global Processing Center for the International Satellite Cloud Climatology Project, which uses satellite observations to create a multidecadal record of cloud and surface variations. Institute personnel frequently collaborate with scientists at the Goddard Space Flight Center in Greenbelt, Maryland. Close research ties also exist with the Lamont-Doherty Earth Observatory of Columbia University, especially in the areas of geochemistry, oceanography, and paleoclimate studies. All facilities, including the Institute's library containing approximately 17,000 volumes, are made available to students in the program.

Financial Aid

Research assistantships are available to most students in the program. Graduate assistantships in 2006–07 carry a twelve-month stipend of approximately $2150 per month and include a tuition waiver and payment of fees.

Cost of Study

Tuition and fees for 2006–07 are estimated at $34,878. As noted above, tuition and fees are paid for graduate students holding research assistantships.

Living and Housing Costs

Limited on-campus housing is available for single and married graduate students on 350-day contracts. Rates range from $8028 for a double room to $12,132 for a single room. Studios, suites, and one-bedroom apartments range from $9528 to $16,872. Most students live off campus, many of them in apartments owned and operated by the University within a few blocks of the campus.

Student Group

Of the 23,000 students at Columbia, 3,500 are students in the Graduate School of Arts and Sciences. Currently, there are 8 Columbia students at the Institute for Space Studies, all of whom are Ph.D. candidates in the Atmospheric and Planetary Science Program. There are also 22 University research appointments at the Institute.

Location

Columbia University is located in the Morningside Heights section of Manhattan in New York City. New York's climate is moderate, with average maximum and minimum temperatures of 85 and 69 degrees in July and 40 and 28 in January. New York is one of the top cultural centers in the United States and, as such, provides unrivaled opportunities for attending concerts, operas, and plays and for visiting world-renowned art, scientific, and historical museums. Student discount tickets for many musical and dramatic performances are available in the Graduate Student Lounge. A comprehensive network of public transportation alleviates the need for keeping an automobile in the city. The superb beaches of Long Island, including the Fire Island National Seashore, are within easy driving distance, as are the numerous ski slopes, state parks, and other mountain recreational areas of upstate New York and southern New England.

The University

Columbia University, founded in 1754 by royal charter of King George II of England, is a member of the Ivy League. It is the oldest institution of higher learning in New York State and the fifth-oldest in the United States. It consists of sixteen separate schools and colleges with more than 3,200 full-time faculty members.

Applying

To enter the program, an application must be submitted to one of the participating departments. For students applying for September admission, completed forms should be received by December 1 for the Department of Earth and Environmental Sciences, and by December 15 for the Department of Applied Physics and Applied Mathematics.

Correspondence and Information

Dr. Anthony D. Del Genio
Atmospheric and Planetary Science Program
Armstrong Hall—GISS
Columbia University
2880 Broadway
New York, New York 10025
Phone: 212-678-5588
E-mail: adelgenio@giss.nasa.gov

Columbia University/NASA Goddard Space Flight Center's Institute for Space Studies

THE INSTITUTE STAFF AND THEIR RESEARCH

Brian Cairns, Ph.D., Rochester, 1992. Radiative transfer, remote sensing, statistical physics.
Vittorio M. Canuto, Ph.D., Turin (Italy), 1960. Theory of fully developed turbulence, analytical models for large-scale turbulence and their applications to geophysics and astrophysics.
Barbara E. Carlson, Ph.D., SUNY at Stony Brook, 1984. Radiative transfer in planetary atmospheres, remote sensing and cloud modeling of Earth and Jovian planets.
Mark A. Chandler, Ph.D., Columbia, 1992. Paleoclimate reconstruction and modeling, role of oceans in climate change.
Anthony D. Del Genio, Ph.D., UCLA, 1978. Dynamics of planetary atmospheres, parameterization of clouds and cumulus convection, climate change, general circulation.
Leonard M. Druyan, Ph.D., NYU, 1971. Tropical climate, African climate, Sahel drought, regional climate remodeling.
Timothy M. Hall, Ph.D., Cornell, 1991. Atmosphere and ocean transport processes, atmospheric chemistry, ocean carbon, modeling and interpretation of observations.
James E. Hansen, Head of the Institute for Space Studies; Ph.D., Iowa, 1967. Remote sensing of Earth and planetary atmospheres, global modeling of climate processes and climate sensitivity.
Nancy Y. Kiang, Ph.D., Berkeley, 2002. Interaction between terrestrial ecosystems and the atmosphere, biogeochemistry, plant ecophysiology, micrometeorology, photosynthesis, mathematical modeling, extensions to astrobiology.
Andrew A. Lacis, Ph.D., Iowa, 1970. Radiative transfer, climate modeling, remote sensing of Earth and planetary atmospheres.
Ron L. Miller, Ph.D., MIT, 1990. Tropical climate, coupled ocean-atmosphere dynamics, interannual and decadal variability.
Michael I. Mishchenko, Ph.D., Ukrainian Academy of Sciences, 1987. Radiative transfer, electromagnetic scattering, remote sensing of Earth and planetary atmospheres.
Jan Perlwitz, Ph.D., Hamburg, 1997. Soil dust aerosol modeling, effect of tropospheric aerosols on Earth's past, present, and future climate.
Dorothy M. Peteet, Ph.D., NYU, 1983. Paleoclimatology, palynology, ecology, botany.
David H. Rind, Ph.D., Columbia, 1976. Atmospheric and climate dynamics, stratospheric modeling and remote sensing.
Cynthia Rosenzweig, Ph.D., Massachusetts at Amherst, 1991. Parameterization of ground hydrology and biosphere, impacts of climate change on agriculture.
Gary L. Russell, Ph.D., Columbia, 1976. Numerical methods, general circulation modeling.
Gavin A. Schmidt, Ph.D., London, 1994. Physical oceanography, paleoclimate, coupled atmosphere-ocean general circulation models.
Drew T. Shindell, Ph.D., SUNY at Stony Brook, 1995. Atmospheric chemistry and climate change.
Larry D. Travis, Associate Chief of the Institute for Space Studies; Ph.D., Penn State, 1971. Remote sensing of Earth and planetary atmospheres, radiative transfer, numerical modeling.

The Institute hosts conferences and workshops that bring scientists together to discuss relevant dynamics, radiation, and chemistry issues. Conferences and workshops on satellite long-term climate monitoring, tropospheric aerosols, and astrobiology have been held recently.

Fossil pollen and spores obtained by coring swamp sediments are used by Institute scientists to document ancient climate changes.

Graduate students can use a variety of computing platforms to process and analyze satellite and surface remote sensing data and to conduct and view the results of global climate model simulations.

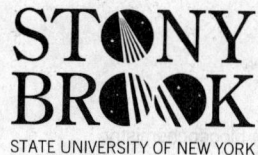

STONY BROOK
STATE UNIVERSITY OF NEW YORK

STONY BROOK UNIVERSITY, STATE UNIVERSITY OF NEW YORK

Marine Sciences Research Center and the Institute for Terrestrial and Planetary Atmospheres

Programs of Study The Marine Sciences Research Center (MSRC) and the Institute for Terrestrial and Planetary Atmospheres (ITPA) offer a Ph.D. degree in Marine and Atmospheric Sciences, with specializations in marine sciences and atmospheric sciences. MSRC and ITPA are top-rated research centers that maintain a collegial, cooperative atmosphere among students and faculty members.

A typical program of study begins with a coordinated set of core courses covering fundamental principles of marine sciences or atmospheric sciences. At the same time, students are encouraged to join an ongoing research activity to help identify an area of specialization. Approximately five years are required to complete the Ph.D.

Oceanographic research and teaching in marine sciences focus on collaborative, interdisciplinary study of oceanographic processes, combining biological, chemical, geological, and physical oceanographic approaches, to examine a wide range of issues, including biogeochemical cycling, fate and effects of contaminants, coastal environmental health, habitat destruction, and living marine resources. An important focus of the Center is regional coastal problems, such as coastal habitat alteration, diseases of marine animals, harmful algal blooms, and effects of contaminants on biota. Equally important are studies of diverse problems throughout the world; faculty members at MSRC are currently carrying out research in the Caribbean Sea, the Mediterranean Sea, Papua New Guinea, Bangladesh, and the Canadian Arctic, among others. A wide variety of approaches are employed, including shipboard sampling, remote sensing, field and laboratory experiments, laboratory analyses, and computer modeling and simulation. More information can be found at http://msrc.sunysb.edu.

The Institute for Terrestrial and Planetary Atmospheres teaches students how to apply their knowledge of mathematics, physics, and chemistry to increase understanding of the atmospheres of Earth and other planets. Completion of the degree program requires a thorough understanding of principles of atmospheric science, coupled with the ability to apply that knowledge to significant problems. Research is conducted at various temporal and spatial scales, from the daily evolution of the atmospheric state—weather—to longer-scale climate variabilities, including those associated with El Niño and global warming. Comprehensive data sets from satellites, field experiments, laboratory measurements, and meteorological observations are analyzed in the context of global three-dimensional weather and climate models and simplified conceptual models. A key goal is to achieve better understanding of the physical bases of numerical weather forecasting and climate prediction, including the size and timing of future greenhouse warming. More information can be found at http://atmos.msrc.sunysb.edu.

Research Facilities Facilities at the Marine Sciences Research Center are modern and comprehensive and support a wide range of oceanographic research. Major shared-use facilities include the R/V *Seawolf,* a 24-meter research ship; a running seawater laboratory; a multibeam echosounder for detailed seabed mapping; a laser ablation/inductively coupled plasma/mass spectrometer (LA-ICP-MS); a liquid chromatograph/time of flight/mass spectrometer (LC/TOF/MS); and an analytical facility for CHN and nutrient analyses. A "clean" laboratory and environmental incubators are also available. Individual research laboratories contain facilities for a wide range of biological, chemical, physical, and geological research. The Center is also the home of the Marine Animal Disease Laboratory, a diagnostic and research center equipped with facilities needed to culture and evaluate marine pathogens and host responses.

The Institute for Terrestrial and Planetary Atmospheres' computer facilities include several high-end multiple processor alpha UNIX stations, plus a large network of PC/UNIX/LINUX/Mac computers, printers, graphics terminals, and hard-copy plotters. The Institute maintains a comprehensive system to display real-time weather data, satellite measurements, and numerical model products. It has a state-of-the-art weather laboratory and remote sensing laboratory. The spectroscopy laboratories have infrared (grating) spectrometers, low-temperature absorption cells, a tunable diode laser spectrometer, and a high-resolution Fourier-transform spectrometer. A stable isotope mass spectrometer is maintained in the atmospheric isotope laboratory. Students have access to millimeter-wavelength remote-sensing equipment, developed at Stony Brook, and to data from NASA missions.

Financial Aid For the 2006–07 academic year, teaching assistantships of $12,592 are provided to most incoming students, after which students are generally supported on research assistantships that range from $18,000 to $23,000 for the calendar year, depending on status. Additional stipend support of $2000 to $3000 is also sometimes awarded to first-year Ph.D. students on a competitive basis. Students receiving support also receive tuition scholarship that completely cover the cost of tuition.

Cost of Study Tuition for 2005–06 was $6900 for residents of New York State and $10,500 for nonresidents. Miscellaneous fees, such as insurance and activity fees, totaled about $526 annually. As described in the Financial Aid section, tuition scholarships are available.

Living and Housing Costs In 2005–06, estimated costs of room and board on campus were $7458. Off-campus rentals are also available and are preferred by most students.

Student Group Approximately 95 students are engaged in research in marine sciences, and about 25 graduate students are in atmospheric sciences.

Location Stony Brook is located about 50 miles east of Manhattan on the wooded North Shore of Long Island, convenient to New York City's cultural life and Suffolk County's recreational countryside and seashores. Long Island's hundreds of miles of magnificent coastline attract many swimming, boating, and fishing enthusiasts from around the world.

The University Established forty years ago as New York's comprehensive State University Center for Long Island and metropolitan New York, Stony Brook offers excellent programs in a broad spectrum of academic subjects. The University conducts major research and public service projects. Over the past decade, externally funded support for Stony Brook's research programs has grown faster than that of any other university in the United States and now exceeds $125 million per year. The University's renowned faculty members teach courses from the undergraduate to the doctoral level to more than 22,000 students. More than 100 undergraduate and graduate departmental and interdisciplinary majors are offered.

Applying Students applying for graduate study in marine sciences should have a B.S. in biology, chemistry, geology, physics, or a related discipline. Students applying for atmospheric sciences should have a B.S. in physics, chemistry, mathematics, engineering, or atmospheric science. Before applying, applicants should contact program faculty members whose research is of primary interest to them. Applications for September admission should be received by January 15 to ensure consideration for the widest range of support opportunities. Students may request an online application at http://www.msrc.sunysb.edu/pages/gradapp.html and apply online at http://www.grad.sunysb.edu/applying/applying.htm.

Correspondence and Information

Carol Dovi
Marine Sciences Research Center
Stony Brook University, State University of New York
Stony Brook, New York 11794-5000
Phone: 631-632-8681
Fax: 631-632-8820
E-mail: cdovi@notes.cc.sunysb.edu

Gina Gartin
Institute for Terrestrial and Planetary Atmospheres
Stony Brook University, State University of New York
Stony Brook, New York 11794-5000
Phone: 631-632-8009
Fax: 631-632-6251
E-mail: ggartin@notes.cc.sunysb.edu

Stony Brook University, State University of New York

THE FACULTY AND THEIR RESEARCH

MARINE SCIENCES

Bassem Allam, Assistant Professor; Ph.D., Western Brittany (France), 1998. Pathology and immunology of marine bivalves.
Josephine Y. Aller, Research Professor; Ph.D., USC, 1975. Marine benthic ecology, invertebrate zoology, marine microbiology, biogeochemistry.
Robert C. Aller, Distinguished Professor; Ph.D., Yale, 1977. Marine geochemistry, marine animal-sediment relations.
Robert A. Armstrong, Associate Professor; Ph.D., Minnesota, 1975. Mathematical modeling in marine ecology and biogeochemistry.
David Black, Assistant Professor; Ph.D., Miami, 1998. Paleoceanography.
Henry J. Bokuniewicz, Professor; Ph.D., Yale, 1976. Nearshore transport processes, coastal groundwater hydrology, coastal sedimentation, marine geophysics.
Malcolm J. Bowman, Distinguished Service Professor; Ph.D., Saskatchewan, 1970. Coastal ocean and estuarine dynamics.
Bruce J. Brownawell, Associate Professor; Ph.D., MIT (Woods Hole), 1986. Biogeochemistry of organic pollutants in seawater and groundwater.
Robert M. Cerrato, Associate Professor; Ph.D., Yale, 1980. Benthic ecology, population and community dynamics.
J. Kirk Cochran, Professor; Ph.D., Yale, 1979. Marine geochemistry, use of radionuclides as geochemical tracers; diagenesis of marine sediments.
Jackie Collier, Assistant Professor; Ph.D., Stanford, 1994. Phytoplankton ecology, physiology, and molecular genetics.
David O. Conover, Professor and Dean, MSRC; Ph.D., Massachusetts, 1982. Ecology of fish, fisheries biology.
Alistair Dove, Adjunct Assistant Professor; Ph.D., Queensland (Australia), 1999. Pathology, taxonomy, life cycles/ecology.
Mark Fast, Assistant Professor; Ph.D., Dalhousie (Canada), 2005. Host-pathogen interactions in marine fish.
Nicholas S. Fisher, Distinguished Professor; Ph.D., SUNY at Stony Brook, 1974. Marine phytoplankton physiology and ecology, biogeochemistry of metals, marine pollution.
Charles N. Flagg, Research Professor; Ph.D., MIT (Woods Hole), 1977. Continental shelf dynamics, bio-physical interactions, climate change effects on coastal systems.
Roger D. Flood, Professor; Ph.D., MIT (Woods Hole), 1978. Marine geology, sediment dynamics, continental margin sedimentation.
Michael G. Frisk, Assistant Professor; Ph.D., Maryland, College Park, 2004. Fish ecology, population modeling, life history theory.
Christopher J. Gobler, Associate Professor; Ph.D., SUNY at Stony Brook, 1999. Phytoplankton, harmful algal blooms, estuarine ecology, aquatic biogeochemistry.
Steven L. Goodbred Jr., Adjunct Assistant Professor; Ph.D., William and Mary, 1999. Coastal marine sedimentology, quaternary development of continental margins, salt-marsh processes.
Paul F. Kemp, Research Professor; Ph.D., Oregon State, 1985. Growth and activity of marine microbes, benthic-pelagic interactions, molecular ecology of marine bacteria.
Cindy Lee, Distinguished Professor; Ph.D., California, San Diego (Scripps), 1975. Ocean carbon cycle, marine geochemistry of organic compounds, nitrogen-cycle biochemistry, biomineralization.
Darcy J. Lonsdale, Associate Professor; Ph.D., Maryland, 1979. Ecology and physiology of marine zooplankton, food web dynamics of estuarine plankton, impacts of harmful algal blooms.
Glenn R. Lopez, Professor; Ph.D., SUNY at Stony Brook, 1976. Marine benthic ecology, animal-sediment interactions, contaminant uptake.
Kamazima M. M. Lwiza, Associate Professor; Ph.D., Wales (United Kingdom), 1990. Structure and dynamics of shelf-seas and remote sensing oceanography.
Anne E. McElroy, Associate Professor; Ph.D., MIT (Woods Hole), 1985. Toxicology of aquatic organisms, contaminant bioaccumulation, estrogenicity of organic contaminants.
Stephan B. Munch, Assistant Professor; Ph.D., SUNY at Stony Brook, 2002. Evolutionary ecology of growth and life history traits, applied population dynamics modeling.
Bradley J. Peterson, Assistant Professor; Ph.D., South Alabama, 1998. Community ecology of seagrass-dominated ecosystems.
Sergey A. Piontkovski, Associate Research Professor; Ph.D., Ukraine, 1978; Ph.D., Moscow, 2004. Physical-biological coupling in coastal and oceanic ecosystems, spatial-temporal structure of marine plankton communities.
Frank J. Roethel, Lecturer; Ph.D., SUNY at Stony Brook, 1982. Environmental chemistry, municipal solid waste management impacts.
Sergio A. Sañudo-Wilhelmy, Associate Professor; Ph.D., California, Santa Cruz, 1993. Geochemical cycles of trace elements, marine pollution.
Mary I. Scranton, Professor; Ph.D., MIT (Woods Hole), 1977. Marine geochemistry, biological-chemical interactions in seawater.
R. Lawrence Swanson, Adjunct Professor; Ph.D., Oregon State, 1971. Recycling and reuse of waste materials, waste management.
Gordon T. Taylor, Professor; Ph.D., USC, 1983. Marine microbiology; interests in microbial ecology, plankton trophodynamics, and marine biofouling.
Joseph D. Warren, Assistant Professor; Ph.D., MIT (Woods Hole); 2001. Acoustical oceanography, zooplankton behavior and ecology.
Dong-Ping Wang, Professor; Ph.D., Miami, 1975. Coastal ocean dynamics.
Robert E. Wilson, Associate Professor; Ph.D., Johns Hopkins, 1974. Estuarine and coastal ocean dynamics.
Peter M. J. Woodhead, Adjunct Professor; B.Sc.Hon. 1 Cl., Durham (England), 1953. Behavior and physiology of fish, coral reef ecology, ocean energy conversion systems.

ATMOSPHERIC SCIENCES

Robert D. Cess, Distinguished Professor Emeritus; Ph.D., Pittsburgh, 1959. Radiative transfer and climate modeling, greenhouse effect, intercomparison of global climate models.
Edmund K. M. Chang, Associate Professor; Ph.D., Princeton, 1993. Atmospheric dynamics and diagnoses, climate dynamics, synoptic meteorology.
Brian A. Colle, Associate Professor; Ph.D., Washington (Seattle), 1997. Synaptic meteorology, weather forecasting, mesoscale modeling.
Robert L. de Zafra, Professor Emeritus (Department of Physics with joint appointment in Marine Sciences Research Center); Ph.D., Maryland, 1958. Monitoring and detection of trace gases in the terrestrial stratosphere, changes in the ozone layer, remote-sensing instrumentation.
Marvin A. Geller, Professor; Ph.D., MIT, 1969. Atmospheric dynamics, stratosphere dynamics and transport, climate dynamics.
Sultan Hameed, Professor; Ph.D., Manchester (England), 1968. Analysis of climate change using observational data and climate models, interannual variations in climate, climate predictability.
Daniel Knopf, Assistant Professor; Ph.D., Swiss Federal Institute of Technology, 2003. Atmospheric chemistry, heterogeneous chemical processes, chemical ionization mass spectrometry, organic aerosols.
John E. Mak, Associate Professor; Ph.D., California, San Diego (Scripps), 1992. Atmospheric chemistry, biosphere-atmosphere interactions, trace gas isotope chemistry, marine geochemistry, mass spectrometry.
Nicole Reimer, Assistant Professor; Ph.D., Karlsruhe (Germany), 2002. Cloud microphysics, aerosol physics and chemistry.
Prasad Varanasi, Professor; Ph.D., California, San Diego (Scripps), 1967. Infrared spectroscopic measurements in support of NASA's space missions, atmospheric remote sensing, greenhouse effect and climate research, molecular physics at low temperatures.
Duane E. Waliser, Adjunct Associate Professor; Ph.D., California, San Diego, 1992. Ocean-atmosphere interactions, tropical climate dynamics.
Minghua Zhang, Professor and Director, ITPA; Ph.D., Academia Sinica (China), 1987. Atmospheric dynamics and climate modeling.

Section 6
Physics

This section contains a directory of institutions offering graduate work in physics, followed by in-depth entries submitted by institutions that chose to prepare detailed program descriptions. Additional information about programs listed in the directory but not augmented by an in-depth entry may be obtained by writing directly to the dean of a graduate school or chair of a department at the address given in the directory.

For programs offering related work, see all other areas in this book. In Book 3, see Biological and Biomedical Sciences and Biophysics; in Book 5, see Aerospace/Aeronautical Engineering, Electrical and Computer Engineering, Energy and Power Engineering (Nuclear Engineering), Engineering and Applied Sciences, Engineering Physics, Materials Sciences and Engineering, and Mechanical Engineering and Mechanics; and in Book 6, see Allied Health and Optometry and Vision Sciences.

CONTENTS

Program Directories

Announcements

Cross-Discipline Announcement

Close-Ups

See also:

Acoustics

The Catholic University of America, School of Engineering, Department of Mechanical Engineering, Washington, DC 20064. Offers design (D Engr, PhD); design and robotics (MME, D Engr, PhD); fluid mechanics and thermal science (MME, D Engr, PhD); mechanical design (MME); ocean and structural acoustics (MME, MS Engr, PhD). Part-time and evening/weekend programs available. *Faculty:* 6 full-time (0 women), 2 part-time/adjunct (0 women). *Students:* Average age 39. 12 applicants, 33% accepted, 0 enrolled. In 2005, 3 master's, 2 doctorates awarded. *Degree requirements:* For master's, thesis or alternative; for doctorate, thesis/dissertation, oral exams, comprehensive exam. *Entrance requirements:* For master's and doctorate, 3 letters of recommendation. Additional exam requirements/recommendations for international students: Required—TOEFL (minimum score 550 paper-based; 213 computer-based). *Application deadline:* For fall admission, 2/1 for domestic students; for spring admission, 11/15 priority date for domestic students. Applications are processed on a rolling basis. Application fee: $55. Electronic applications accepted. *Expenses: Contact institution.* Part-time tuition and fees vary according to course load and program. *Financial support:* Research assistantships, teaching assistantships, career-related internships or fieldwork, Federal Work-Study, scholarships/grants, tuition waivers (full and partial), and unspecified assistantships available. Support available to part-time students. Financial award application deadline: 2/1; financial award applicants required to submit FAFSA. *Faculty research:* Automated engineering. *Unit head:* Dr. J. Steven Brown, Chair, 202-319-5170, Fax: 202-319-5173, E-mail: brownjs@cua.edu.

The Pennsylvania State University University Park Campus, Graduate School, Intercollege Graduate Programs and College of Engineering, Intercollege Graduate Program in Acoustics, State College, University Park, PA 16802-1503. Offers M Eng, MS, PhD. Postbaccalaureate distance learning degree programs offered (minimal on-campus study). *Students:* 49 full-time (12 women), 6 part-time (1 woman); includes 5 minority (1 African American, 3 Asian Americans or Pacific Islanders, 1 Hispanic American), 12 international. *Degree requirements:* For doctorate, thesis/dissertation. *Entrance requirements:* For master's and doctorate, GRE General Test. Application fee: $45. *Expenses:* Tuition, state resident: full-time $12,518; part-time $522 per credit. Tuition, nonresident: full-time $23,004; part-time $959 per credit. Required fees: $484. Tuition and fees vary according to course load, campus/location and program. *Unit head:* Dr. Anthony Atchley, Head, 814-865-6364, Fax: 814-865-7595, E-mail: atchley@psu.edu.

Applied Physics

Air Force Institute of Technology, Graduate School of Engineering and Management, Department of Engineering Physics, Dayton, OH 45433-7765. Offers applied physics (MS, PhD); electro-optics (MS, PhD); materials science (PhD); nuclear engineering (MS, PhD); space physics (MS). Part-time programs available. *Faculty:* 20 full-time (1 woman). *Students:* 86 full-time (9 women), 4 part-time (all women). Average age 31. In 2005, 36 master's, 4 doctorates awarded. *Degree requirements:* For master's and doctorate, thesis/dissertation. *Entrance requirements:* For master's and doctorate, GRE General Test, minimum GPA of 3.0, U.S. citizenship. *Application deadline:* For spring admission, 3/1 for domestic students. Applications are processed on a rolling basis. Application fee: $0. *Financial support:* Fellowships with full tuition reimbursements, research assistantships with full and partial tuition reimbursements, scholarships/grants and unspecified assistantships available. Support available to part-time students. Financial award application deadline:3/15. *Faculty research:* High-energy lasers, space physics, nuclear weapon effects, semiconductor physics. Total annual research expenditures: $2.1 million. *Unit head:* Dr. Robert L. Hengehold, Head, 937-255-3636 Ext. 4502, Fax: 937-255-2921, E-mail: robert.hengehold@afit.edu. *Application contact:* Dr. David E. Weeks, Associate Professor of Physics, 937-255-3636 Ext. 4561, Fax: 937-255-2921, E-mail: david.weeks@afit.edu.

Alabama Agricultural and Mechanical University, School of Graduate Studies, School of Arts and Sciences, Department of Natural and Physical Sciences, Huntsville, AL 35811. Offers biology (MS); physics (MS, PhD), including applied physics (PhD), materials science (PhD), optics (PhD), physics (MS). Part-time and evening/weekend programs available. *Degree requirements:* For doctorate, thesis/dissertation. *Entrance requirements:* For master's and doctorate, GRE General Test. Electronic applications accepted.

Appalachian State University, Cratis D. Williams Graduate School, College of Arts and Sciences, Department of Physics and Astronomy, Boone, NC 28608. Offers applied physics (MS). *Faculty:* 10 full-time (1 woman). *Students:* 10 full-time (1 woman), 2 part-time. 7 applicants, 86% accepted, 4 enrolled. In 2005, 3 degrees awarded. *Degree requirements:* For master's, thesis optional. *Entrance requirements:* For master's, GRE General Test. Additional exam requirements/recommendations for international students: Required—TOEFL (minimum score 570 paper-based; 230 computer-based). *Application deadline:* For fall admission, 7/1 for domestic students, 1/1 for international students; for spring admission, 11/1 for domestic students, 6/1 for international students. Applications are processed on a rolling basis. Application fee: $45. *Expenses:* Tuition, state resident: full-time $2,593. Tuition, nonresident: full-time $12,176. Required fees: $1,726. *Financial support:* In 2005–06, 8 teaching assistantships with tuition reimbursements (averaging $8,125 per year) were awarded; fellowships, research assistantships with tuition reimbursements, career-related internships or fieldwork, Federal Work-Study, scholarships/grants, and unspecified assistantships also available. Support available to part-time students. Financial award application deadline: 7/1; financial award applicants required to submit FAFSA. *Faculty research:* Raman spectroscopy, applied electrostatics, scanning tunneling microscope/atomic force microscope (STM/AFM), stellar spectroscopy and photometry, surface physics, remote sensing. Total annual research expenditures: $132,812. *Unit head:* Dr. Anthony Calamai, Chairperson, 828-262-3090, E-mail: calamai@appstate.edu. *Application contact:* Dr. Sid Clements, Director, 828-262-2447, E-mail: clementsjs@appstate.edu.

Brooklyn College of the City University of New York, Division of Graduate Studies, Department of Physics, Brooklyn, NY 11210-2889. Offers applied physics (MA); physics (MA, PhD). The department is a full participant in the PhD program; it offers a complete sequence of courses that are creditable toward the CUNY doctoral degree, and a wide range of research opportunities in fulfillment of the doctoral dissertation requirements for that degree. Part-time programs available. Terminal master's awarded for partial completion of doctoral program. *Degree requirements:* For master's, comprehensive exam. *Entrance requirements:* For master's, GRE, 2 letters of recommendation; for doctorate, GRE. Additional exam requirements/recommendations for international students: Required—TOEFL.

California Institute of Technology, Division of Engineering and Applied Science, Option in Applied Physics, Pasadena, CA 91125-0001. Offers MS, PhD. *Faculty:* 13 full-time (0 women). *Students:* 64 full-time (12 women); includes 6 minority (1 African American, 5 Asian Americans or Pacific Islanders), 27 international. 102 applicants, 21% accepted, 5 enrolled. In 2005, 12 master's, 3 doctorates awarded. *Degree requirements:* For doctorate, thesis/dissertation. *Application deadline:* For fall admission, 1/15 for domestic students. Application fee: $0. Electronic applications accepted. *Financial support:* In 2005–06, 18 research assistantships were awarded; fellowships, teaching assistantships *Faculty research:* Solid-state electronics, quantum electronics, plasmas, linear and nonlinear laser optics, electromagnetic theory. *Unit head:* Dr. Kerry Vahala, Representative, 626-395-2144.

Christopher Newport University, Graduate Studies, Department of Physics, Computer Science, and Engineering, Newport News, VA 23606-2998. Offers applied physics and computer science (MS). Part-time and evening/weekend programs available. *Degree requirements:* For master's, thesis or alternative, comprehensive exam. *Entrance requirements:* For master's, GRE General Test, minimum GPA of 3.0. Electronic applications accepted. *Faculty research:* Advanced programming methodologies, experimental nuclear physics, computer architecture, semiconductor nanophysics, laser and optical fiber sensors.

Colorado School of Mines, Graduate School, Department of Physics, Golden, CO 80401-1887. Offers applied physics (PhD); physics (MS). Part-time programs available. *Faculty:* 16 full-time (0 women), 12 part-time/adjunct (2 women). *Students:* 28 full-time (7 women), 3 part-time; includes 2 minority (1 Asian American or Pacific Islander, 1 Hispanic American), 4 international. 29 applicants, 31% accepted, 8 enrolled. In 2005, 3 master's, 7 doctorates awarded. *Degree requirements:* For master's, thesis/dissertation; for doctorate, thesis/dissertation, comprehensive exam. *Entrance requirements:* For master's and doctorate, GRE General Test, GRE Subject Test. Additional exam requirements/recommendations for international students: Required—TOEFL (minimum score 550 paper-based; 213 computer-based). *Application deadline:* For fall admission, 1/1 priority date for domestic students, 1/1 priority date for international students; for spring admission, 9/1 priority date for domestic students, 9/1 priority date for international students. Application fee: $50. Electronic applications accepted. *Expenses:* Tuition, state resident: full-time $7,240; part-time $362 per credit hour. Tuition, nonresident: full-time $19,840; part-time $992 per credit hour. Required fees: $895. *Financial support:* In 2005–06, 8 students received support, including fellowships with full tuition reimbursements available (averaging $9,600 per year), 15 research assistantships with full tuition reimbursements available (averaging $9,600 per year), 10 teaching assistantships with full tuition reimbursements available (averaging $9,600 per year); scholarships/grants, health care benefits, and unspecified assistantships also available. Financial award applicants required to submit FAFSA. *Faculty research:* Light scattering, low-energy nuclear physics, high fusion plasma diagnostics, laser operations, mathematical physics. Total annual research expenditures: $5.3 million. *Unit head:* Dr. James A. McNeil, Head, 303-273-3844, Fax: 303-273-3919, E-mail: jamcneil@mine.edu. *Application contact:* Jeff Squier, Professor, 303-384-2385, Fax: 303-273-3919, E-mail: jsquier@mines.edu.

Columbia University, Fu Foundation School of Engineering and Applied Science, Department of Applied Physics and Applied Mathematics, New York, NY 10027. Offers applied physics (MS, PhD), including applied mathematics (PhD), optical physics (PhD), plasma physics (PhD), solid state physics (PhD); applied physics and applied mathematics (Eng Sc D); materials science and engineering (MS, Eng Sc D, PhD); medical physics (MS); minerals engineering and materials science (Eng Sc D, PhD, Engr). Part-time programs available. *Faculty:* 30 full-time (2 women), 15 part-time/adjunct (1 woman). *Students:* 94 full-time (26 women), 30 part-time (5 women); includes 17 minority (1 American Indian/Alaska Native, 12 Asian Americans or Pacific Islanders, 4 Hispanic Americans), 53 international. Average age 24. 294 applicants, 22% accepted, 38 enrolled. In 2005, 33 master's, 5 doctorates awarded. Terminal master's awarded for partial completion of doctoral program. *Degree requirements:* For master's, English Proficiency Test; for doctorate, thesis/dissertation, qualifying exam, English Proficiency Test; for Engr, English Profiency Test. *Entrance requirements:* For master's and doctorate, GRE General Test, GRE Subject Test (strongly recommended). Additional exam requirements/recommendations for international students: Required—TOEFL. *Application deadline:* For fall admission, 12/15 priority date for domestic students, 12/15 priority date for international students; for spring admission, 10/1 priority date for domestic students, 10/1 priority date for international students. Application fee: $45. Electronic applications accepted. *Expenses:* Tuition: Full-time $31,448. Tuition and fees vary according to course level, course load, campus/location and program. *Financial support:* In 2005–06, 8 fellowships with full tuition reimbursements, 56 research assistantships with full tuition reimbursements (averaging $24,750 per year), 16 teaching assistantships with full tuition reimbursements (averaging $24,750 per year) were awarded; Federal Work-Study and unspecified assistantships also available. Financial award application deadline: 12/15; financial award applicants required to submit FAFSA. *Faculty research:* Plasma physics, applied mathematics, solid-state and optical physics, atmospheric, oceanic and earth physics, materials science and engineering. *Unit head:* Dr. Michael E. Mauel, Professor and Chairman, 212-854-4457, E-mail: seasinfo.apam@columbia.edu. *Application contact:* Marlene Arbo, Department Administrator, 212-854-4458, Fax: 212-854-8257, E-mail: seasinfo.apam@columbia.edu.

See Close-Up on page 335.

Cornell University, Graduate School, Graduate Fields of Engineering, Field of Applied Physics, Ithaca, NY 14853-0001. Offers applied physics (PhD); engineering physics (M Eng). *Faculty:* 57 full-time (4 women). *Students:* 84 full-time (15 women); includes 13 minority (1 American Indian/Alaska Native, 11 Asian Americans or Pacific Islanders, 1 Hispanic American), 33 international. 121 applicants, 25% accepted, 13 enrolled. In 2005, 16 master's, 9 doctorates awarded. *Degree requirements:* For doctorate, thesis/dissertation, written exams, comprehensive exam. *Entrance requirements:* For master's, GRE General Test, 3 letters of recommendation; for doctorate, GRE General Test, GRE Subject Test (physics), GRE writing assessment, 3 letters of recommendation. Additional exam requirements/recommendations for international students: Required—TOEFL (minimum score 600 paper-based; 250 computer-based). *Application deadline:* For fall admission, 1/15 for domestic students. Application fee: $60. Electronic applications accepted. *Financial support:* In 2005–06, 76 students received support, including 13 fellowships with full tuition reimbursements available, 56 research assistantships with full tuition reimbursements available, 7 teaching assistantships with full tuition reimbursements available (averaging $76 per year); institutionally sponsored loans, scholarships/grants, health care benefits, tuition waivers (full and partial), and unspecified assistantships also available. *Faculty research:* Quantum and nonlinear optics, plasma physics, solid state physics, condensed matter physics and nanotechnology, electron and x-ray spectroscopy. *Unit head:* Graduate Faculty Representative, 607-255-0638. *Application contact:* Graduate Field Assistant, 607-255-0638, E-mail: aep_info@cornell.edu.

DePaul University, College of Liberal Arts and Sciences, Department of Physics, Chicago, IL 60604-2287. Offers applied physics (MS). Part-time and evening/weekend programs available. *Faculty:* 7 full-time (1 woman), 3 part-time/adjunct (0 women). *Students:* 6 full-time (1 woman), 4 part-time (1 woman); includes 1 minority (African American), 2 international. Average age 23. 12 applicants, 42% accepted, 3 enrolled. In 2005, 3 degrees awarded. *Degree requirements:* For master's, thesis, oral exams. *Entrance requirements:* For master's, 2 letters of recommendation, BA in physics or closely related field. Additional exam requirements/recommendations for international students: Required—TOEFL. *Application deadline:* For fall admission, 6/1 priority date for domestic students, 6/1 priority date for international students. Application fee: $25. Electronic applications accepted. *Financial support:* In 2005–06, teaching assistantships with full tuition reimbursements (averaging $9,500 per year); tuition waivers (full) also available. *Faculty research:* Optics, solid-state physics, comology, atomic physics, nuclear physics. Total annual research expenditures: $54,000. *Unit head:* Dr. Christopher G. Goedde, Chairman, 773-325-7330, Fax: 773-325-7334, E-mail: egoedde@condor.depaul.edu. *Application contact:* Dr. Jesus Pando, Associate Professor, 773-325-7330, Fax: 773-325-7334.

George Mason University, College of Arts and Sciences, Department of Physics, Fairfax, VA 22030. Offers applied and engineering physics (MS). *Faculty:* 23 full-time (8 women), 3 part-time/adjunct (0 women). *Students:* 2 full-time (0 women), 18 part-time (5 women); includes 2 minority (1 American Indian/Alaska Native, 1 Hispanic American), 1 international. Average age 28. 16 applicants, 94% accepted, 9 enrolled. In 2005, 13 degrees awarded. *Degree requirements:* For master's, thesis optional. *Entrance requirements:* For master's, minimum GPA of 2.75 in last 60 hours of course work. *Application deadline:* For fall admission, 5/1 for domestic students; for spring admission, 11/1 for domestic students. Electronic applications accepted. *Expenses:* Tuition, area resident: Full-time $5,244; part-time $219 per credit. Tuition, state resident: part-time $651 per credit. Tuition, nonresident: full-time $15,636. Required fees: $1,524; $65 per credit. *Financial support:* Research assistantships, teaching assistantships available. Support available to part-time students. Financial award application deadline: 3/1; financial award applicants required to submit FAFSA. *Unit head:* Dr. Maria Dworzecka, Chairman, 703-993-1280, Fax: 703-993-1269, E-mail: mdworzecka@gmu.edu. *Application contact:* Dr. Paul So, Information Contact, 703-993-1280, E-mail: physics@gmu.edu.

Harvard University, Graduate School of Arts and Sciences, Department of Physics, Cambridge, MA 02138. Offers experimental physics (PhD); medical engineering/medical physics (PhD), including applied physics, engineering sciences, physics; theoretical physics (PhD). *Students:* 174. *Degree requirements:* For doctorate, thesis/dissertation, final exams, laboratory experience. *Entrance requirements:* For doctorate, GRE General Test, GRE Subject Test. Additional exam requirements/recommendations for international students: Required—TOEFL. *Application deadline:* For fall admission, 12/14 for domestic students. Application fee: $60. *Expenses:* Tuition: Full-time $28,752. Full-time tuition and fees vary according to program and student level. *Financial support:* Fellowships, research assistantships, teaching assistantships, career-related internships or fieldwork, Federal Work-Study, and institutionally sponsored loans available. Financial award application deadline: 12/30. *Faculty research:* Particle physics, condensed matter physics, atomic physics. *Unit head:* Sheila Ferguson, Administrator, 617-495-4327. *Application contact:* Office of Admissions and Financial Aid, 617-495-5315.

Harvard University, Graduate School of Arts and Sciences, Division of Engineering and Applied Sciences, Cambridge, MA 02138. Offers applied mathematics (ME, SM, PhD); applied physics (ME, SM, PhD); computer science (ME, SM, PhD); computing technology (PhD); engineering science (ME); engineering sciences (SM, PhD). Part-time programs available. *Faculty:* 75 full-time (5 women), 12 part-time/adjunct (3 women). *Students:* 277 full-time (65 women), 11 part-time (1 woman); includes 40 minority (5 African Americans, 3 American Indian/Alaska Native, 26 Asian Americans or Pacific Islanders, 6 Hispanic Americans), 126 international. 1,197 applicants, 14% accepted, 86 enrolled. In 2005, 68 master's, 19 doctorates awarded. Terminal master's awarded for partial completion of doctoral program. *Median time to degree:* Of those who began their doctoral program in fall 1997, 92% received their degree in 8 years or less. *Degree requirements:* For master's, thesis optional; for doctorate, thesis/dissertation, comprehensive exam, registration. *Entrance requirements:* For master's and doctorate, GRE General Test, GRE Subject Test (recommended), 3 letters of recommendation. Additional exam requirements/recommendations for international students: Required—TOEFL (minimum score 550 paper-based; 213 computer-based). *Application deadline:* For fall admission, 12/15 priority date for domestic students, 12/15 priority date for international students. For winter admission, 1/2 for domestic students. Application fee: $95. Electronic applications accepted. *Expenses:* Tuition: Full-time $28,752. Full-time tuition and fees vary according to program and student level. *Financial support:* In 2005–06, 55 fellowships with full tuition reimbursements (averaging $19,350 per year), 187 research assistantships (averaging $31,856 per year), 112 teaching assistantships (averaging $5,138 per year) were awarded; Federal Work-Study, institutionally sponsored loans, and traineeships also available. *Faculty research:* Applied mathematics, applied physics, computer science & electrical engineering, environmental engineering, mechanical and biomedical engineering. Total annual research expenditures: $33.5 million. *Unit head:* Betsey Cogswell, Dean, 617-495-5829, Fax: 617-495-5264, E-mail: venky@deas.harvard.edu. *Application contact:* Office of Admissions and Financial Aid, 617-495-5315, E-mail: admissions@deas.harvard.edu.

Iowa State University of Science and Technology, Graduate College, College of Liberal Arts and Sciences, Department of Physics and Astronomy, Ames, IA 50011. Offers applied physics (MS, PhD); astrophysics (MS, PhD); condensed matter physics (MS, PhD); high energy physics (MS, PhD); nuclear physics (MS, PhD); physics (MS, PhD). Part-time programs available. *Faculty:* 44 full-time, 3 part-time/adjunct. *Students:* 79 full-time (14 women), 6 part-time; includes 1 minority (Asian American or Pacific Islander), 56 international. 161 applicants, 34% accepted, 23 enrolled. In 2005, 5 master's, 7 doctorates awarded. Terminal master's awarded for partial completion of doctoral program. *Degree requirements:* For master's, thesis (for some programs); for doctorate, thesis/dissertation. *Entrance requirements:* For master's and doctorate, GRE General Test, GRE Subject Test (physics). Additional exam requirements/recommendations for international students: Required—TOEFL (paper score 550; computer score 213) or IELTS (score 6.5). *Application deadline:* For fall admission, 2/15 priority date for domestic students, 2/15 priority date for international students; for spring admission, 10/15 for domestic students, 10/15 for international students. Applications are processed on a rolling basis. Application fee: $30 ($70 for international students). Electronic applications accepted. *Expenses:* Tuition, state resident: full-time $6,410. Tuition, nonresident: full-time $16,422. Tuition and fees vary according to program. *Financial support:* In 2005–06, 48 research assistantships with full tuition reimbursements (averaging $14,626 per year), 30 teaching assistantships with full tuition reimbursements (averaging $14,533 per year) were awarded; fellowships, Federal Work-Study, institutionally sponsored loans, scholarships/grants, health care benefits, and unspecified assistantships also available. Support available to part-time students. Financial award application deadline: 2/15. *Faculty research:* Condensed-matter physics, including superconductivity and new materials; high-energy and nuclear physics; astronomy and astrophysics; atmospheric and environmental physics. Total annual research expenditures: $8.8 million. *Unit head:* Dr. Eli Rosenberg, Chair, 515-294-5441, Fax: 515-294-6027, E-mail: phys_astro@iastate.edu. *Application contact:* Dr. Steven Kawaler, Director of Graduate Education, 515-294-9728, E-mail: phys_astro@iastate.edu.

The Johns Hopkins University, G. W. C. Whiting School of Engineering, Engineering and Applied Science Programs for Professionals, Department of Applied Physics, Baltimore, MD 21218-2699. Offers MS. Part-time and evening/weekend programs available. In 2005, 15 degrees awarded. *Degree requirements:* For master's, registration. *Application deadline:* Applications are processed on a rolling basis. Application fee: $70. Electronic applications accepted. *Expenses:* Tuition: Full-time $30,960. Tuition and fees vary according to degree level and program. *Financial support:* Institutionally sponsored loans available. *Application contact:* Doug Schiller, Assistant Director of Registration, 410-540-2960, Fax: 410-579-8049, E-mail: schiller@jhu.edu.

Laurentian University, School of Graduate Studies and Research, Programme in Physics and Astronomy, Sudbury, ON P3E 2C6, Canada. Offers M Sc. Part-time programs available. *Degree requirements:* For master's, thesis or alternative. *Entrance requirements:* For master's, honors degree with second class or better. *Faculty research:* Solar neutrino physics and astrophysics, applied acoustics and ultrasonics, powder science and technology, solid state physics, theoretical physics.

Naval Postgraduate School, Graduate Programs, Department of Physics, Monterey, CA 93943. Offers applied physics (MS); engineering acoustics (MS); physics (MS, PhD). Program only open to commissioned officers of the United States and friendly nations and selected United States federal civilian employees. Part-time programs available. *Students:* 59 full-time. In 2005, 28 master's, 1 doctorate awarded. *Degree requirements:* For master's, thesis; for doctorate, one foreign language, thesis/dissertation. *Unit head:* Dr. James H. Luscombe, Chairman, 831-656-2896, E-mail: luscombe@nps.navy.mil. *Application contact:* Tracy Hammond, Acting Director of Admissions, 831-656-3059, Fax: 831-656-2891, E-mail: thammond@bps.navy.mil.

New Jersey Institute of Technology, Office of Graduate Studies, College of Science and Liberal Arts, Department of Physics, Program in Applied Physics, Newark, NJ 07102. Offers MS, PhD. Part-time and evening/weekend programs available. *Students:* 29 full-time (7 women), 4 part-time (1 woman); includes 6 minority (3 Asian Americans or Pacific Islanders, 3 Hispanic Americans), 21 international. Average age 32. 31 applicants, 39% accepted, 3 enrolled. In 2005, 1 master's, 4 doctorates awarded. Terminal master's awarded for partial completion of doctoral program. *Degree requirements:* For master's, thesis; for doctorate, thesis/dissertation, residency. *Entrance requirements:* For master's, GRE General Test; for doctorate, GRE General Test, minimum graduate GPA of 3.5. Additional exam requirements/recommendations for international students: Required—TOEFL (minimum score 550 paper-based; 213 computer-based). *Application deadline:* For fall admission, 6/5 for domestic students; for spring admission, 10/15 for domestic students. Applications are processed on a rolling basis. Application fee: $60. Electronic applications accepted. *Expenses:* Tuition, state resident: full-time $9,620; part-time $520 per credit. Tuition, nonresident: full-time $13,542; part-time $715 per credit. Required fees: $78; $54 per credit. $78 per year. Tuition and fees vary according to course load. *Financial support:* Fellowships with full and partial tuition reimbursements, research assistantships with full and partial tuition reimbursements, teaching assistantships with full and partial tuition reimbursements, career-related internships or fieldwork, Federal Work-Study, institutionally sponsored loans, and unspecified assistantships available. Financial award application deadline: 3/15. *Unit head:* Dr. Ken K. Chin, Director, 973-596-3297, E-mail: chin@njit.edu. *Application contact:* Kathryn Kelly, Director of Admissions, 973-596-3300, Fax: 973-596-3461, E-mail: admissions@njit.edu.

Northern Arizona University, Graduate College, College of Engineering and Natural Science, Department of Physics and Astronomy, Flagstaff, AZ 86011. Offers applied physics (MS); physical science (MAT). Part-time programs available. *Degree requirements:* For master's, thesis optional. *Entrance requirements:* For master's, GRE.

Pittsburg State University, Graduate School, College of Arts and Sciences, Department of Physics, Pittsburg, KS 66762. Offers applied physics (MS); physics (MS); professional physics (MS). *Students:* 3. *Degree requirements:* For master's, thesis or alternative. Application fee: $70 ($60 for international students). *Expenses:* Tuition, state resident: full-time $2,015; part-time $170 per credit hour. Tuition, nonresident: full-time $4,953; part-time $415 per credit hour. Tuition and fees vary according to course load, campus/location and program. *Financial support:* Teaching assistantships, career-related internships or fieldwork and Federal Work-Study available. *Unit head:* Dr. Charles Blatchley, Chairperson, 620-235-4398. *Application contact:* Marvene Darraugh, Administrative Officer, 620-235-4220, Fax: 620-235-4219, E-mail: mdarraug@pittstate.edu.

Princeton University, Graduate School, Department of Mechanical and Aerospace Engineering, Princeton, NJ 08544. Offers applied physics (M Eng, MSE, PhD); computational methods (M Eng, MSE); dynamics and control systems (M Eng, MSE, PhD); energy and environmental policy (M Eng, MSE, PhD); energy conversion, propulsion, and combustion (M Eng, MSE, PhD); flight science and technology (M Eng, MSE, PhD); fluid mechanics (M Eng, MSE, PhD). Part-time programs available. *Faculty:* 22 full-time (3 women). *Students:* 76 full-time (13 women); includes 5 minority (1 African American, 1 Asian American or Pacific Islander, 3 Hispanic Americans), 43 international. Average age 24. 252 applicants, 20% accepted, 20 enrolled. In 2005, 7 master's, 8 doctorates awarded. Terminal master's awarded for partial completion of doctoral program. *Degree requirements:* For master's, thesis/dissertation; for doctorate, thesis/dissertation, comprehensive exam. *Entrance requirements:* For master's and doctorate, GRE General Test. Additional exam requirements/recommendations for international students: Required—IELT. *Application deadline:* For fall admission, 12/31 for domestic students, 12/1 for international students. Application fee: $105. Electronic applications accepted. *Financial support:* In 2005–06, 12 fellowships with full tuition reimbursements (averaging $8,800 per year), 36 research assistantships with full tuition reimbursements (averaging $27,461 per year), 9 teaching assistantships with full tuition reimbursements (averaging $21,641 per year) were awarded; Federal Work-Study and institutionally sponsored loans also available. Financial award application deadline: 1/2. Total annual research expenditures: $6.2 million. *Unit head:* Prof. Luigi Martinelli, Director of Graduate Studies, 609-258-6652, Fax: 609-258-1918, E-mail: gigi@princeton.edu. *Application contact:* Janice Hueng, Director of Graduate Admissions, 609-258-3034, Fax: 609-258-6180, E-mail: gsadmit@princeton.edu.

Rensselaer Polytechnic Institute, Graduate School, School of Science, Department of Physics, Applied Physics and Astronomy, Troy, NY 12180-3590. Offers physics (MS, PhD). *Faculty:* 24 full-time (3 women), 3 part-time/adjunct (0 women). *Students:* 59 full-time (11 women); includes 36 minority (all Asian Americans or Pacific Islanders) Average age 28. 90 applicants, 28% accepted. In 2005, 11 master's, 4 doctorates awarded. *Degree requirements:* For doctorate, thesis/dissertation. *Entrance requirements:* For master's and doctorate, GRE General Test, GRE Subject Test. Additional exam requirements/recommendations for international students: Required—TOEFL (minimum score 600 paper-based; 250 computer-based). *Application deadline:* For fall admission, 1/15 priority date for domestic students, 1/15 priority date for international students; for spring admission, 8/15 priority date for domestic students, 8/15 priority date for international students. Applications are processed on a rolling basis. Application fee: $75. Electronic applications accepted. *Expenses:* Tuition: Full-time $31,000; part-time $1,320 per credit. Required fees: $1,623. *Financial support:* In 2005–06, 10 fellowships with tuition reimbursements (averaging $25,000 per year), 27 research assistantships with tuition reimbursements (averaging $18,700 per year), 16 teaching assistantships with tuition reimbursements (averaging $19,000 per year) were awarded; career-related internships or fieldwork and institutionally sponsored loans also available. Financial award application deadline: 2/1. *Faculty research:* Astrophysics, condensed matter, nuclear physics, optics, physics education. Total annual research expenditures: $4.5 million. *Unit head:* Dr. G. C. Wang, Chair, 518-276-8387, Fax: 518-276-6680, E-mail: wangg@rpi.edu. *Application contact:* Dr. Toh-Ming Lu, Chair, Graduate Recruitment Committee, 518-276-8391, Fax: 518-276-6680, E-mail: mcquade@rpi.edu.

See Close-Up on page 363.

Rice University, Rice Quantum Institute, Houston, TX 77251-1892. Offers MS, PhD. *Degree requirements:* For master's and doctorate, thesis/dissertation. *Entrance requirements:* For master's and doctorate, GRE General Test, GRE Subject Test (physics), minimum GPA of 3.0. Additional exam requirements/recommendations for international students: Required—TOEFL (minimum score 600 paper-based; 250 computer-based). Electronic applications accepted. *Faculty research:* Nanotechnology, solid state materials, atomic physics, thin films.

Rutgers, The State University of New Jersey, Newark, Graduate School, Program in Applied Physics, Newark, NJ 07102. Offers MS, PhD. *Faculty:* 29 full-time (1 woman), 5 part-time/adjunct (0 women). *Students:* 3 full-time (1 woman), 2 part-time (1 woman); includes 1 minority (Asian American or Pacific Islander) 17 applicants, 47% accepted, 2 enrolled. In 2005, 2 doctorates awarded. *Entrance requirements:* For master's and doctorate, GRE. Additional exam requirements/recommendations for international students: Required—TOEFL. *Application deadline:* For fall admission, 7/1 for domestic students; for spring admission, 12/1 for domestic students. Application fee: $50. *Expenses:* Tuition, state resident: full-time $10,440; part-time $435 per credit. Tuition, nonresident: full-time $15,520; part-time $637 per credit. *Financial support:* In 2005–06, 3 teaching assistantships with full tuition reimbursements (averaging $16,988 per year) were awarded *Unit head:* Zhen Wu, Program Coordinator, 973-353-1311, E-mail: zwu@andromeda.rutgers.edu. *Application contact:* Elizabeth Wheeler, Administrative Assistant, 201-973-1312, E-mail: ewheeler@andromeda.rutgers.edu.

Applied Physics

Southern Illinois University Carbondale, Graduate School, College of Science, Department of Physics, Carbondale, IL 62901-4701. Offers MS, PhD. *Faculty:* 9 full-time (0 women). *Students:* 8 full-time (2 women), 9 part-time (1 woman); includes 1 minority (African American), 11 international. 25 applicants, 48% accepted, 2 enrolled. In 2005, 6 degrees awarded. *Degree requirements:* For master's, one foreign language, thesis. *Entrance requirements:* For master's, minimum GPA of 2.7. Additional exam requirements/recommendations for international students: Required—TOEFL. *Application deadline:* Applications are processed on a rolling basis. Application fee: $20. *Financial support:* In 2005–06, 1 fellowship with full tuition reimbursement, 9 teaching assistantships with full tuition reimbursements were awarded; research assistantships with full tuition reimbursements, career-related internships or fieldwork, Federal Work-Study, institutionally sponsored loans, and tuition waivers (full) also available. Support available to part-time students. Financial award application deadline: 2/15. *Faculty research:* Atomic, molecular, nuclear, and mathematical physics; statistical mechanics; solid-state and low-temperature physics; rheology; material science. Total annual research expenditures: $773,352. *Unit head:* Dr. Aldo Migone, Chairperson, 618-453-1054. *Application contact:* Graduate Admissions Committee, 618-453-2643.

Announcement: This program provides many research opportunities within the department, and a low student-faculty ratio allows students to work closely with faculty members. Areas of research include experimental (IR spectroscopy, thin films, magnetism, gas adsorption, ellipsometry), theoretical (condensed-matter theory, statistical physics, quantum computing, simulation of materials), and applied physics (novel materials from agricultural byproducts, coal and coal ash composites, permanent magnetic materials, chemical nanosensors, gas storage, superhard coatings). Assistantships for the academic year are offered to every selected applicant.

See Close-Up on page 365.

Stanford University, School of Humanities and Sciences, Department of Applied Physics, Stanford, CA 94305-9991. Offers MS, PhD. Terminal master's awarded for partial completion of doctoral program. *Degree requirements:* For doctorate, thesis/dissertation. *Entrance requirements:* For master's and doctorate, GRE General Test, GRE Subject Test. Additional exam requirements/recommendations for international students: Required—TOEFL. Electronic applications accepted.

State University of New York at Binghamton, Graduate School, School of Arts and Sciences, Department of Physics, Applied Physics, and Astronomy, Binghamton, NY 13902-6000. Offers applied physics (MS); physics (MA, MS). *Degree requirements:* For master's, thesis or alternative. *Entrance requirements:* For master's, GRE General Test, GRE Subject Test. Additional exam requirements/recommendations for international students: Required—TOEFL. Electronic applications accepted.

Texas A&M University, College of Science, Department of Physics, College Station, TX 77843. Offers applied physics (PhD); physics (MS, PhD). *Faculty:* 35 full-time (1 woman). *Students:* 140 full-time (18 women), 10 part-time (2 women); includes 19 minority (3 African Americans, 2 American Indian/Alaska Native, 6 Asian Americans or Pacific Islanders, 8 Hispanic Americans), 90 international. 117 applicants, 51% accepted, 43 enrolled. In 2005, 15 master's, 6 doctorates awarded. Terminal master's awarded for partial completion of doctoral program. *Degree requirements:* For master's, thesis (for some programs), registration; for doctorate, thesis/dissertation, registration. *Entrance requirements:* For master's and doctorate, GRE General Test, GRE Subject Test. Additional exam requirements/recommendations for international students: Required—TOEFL. *Application deadline:* For fall admission, 3/1 for domestic students; for spring admission, 8/1 for domestic students. Application fee: $50 ($75 for international students). Electronic applications accepted. *Expenses:* Tuition, state resident: full-time $4,488; part-time $187 per credit hour. Tuition, nonresident: full-time $11,112; part-time $463 per credit hour. Required fees:$1,974. *Financial support:* In 2005–06, research assistantships (averaging $16,200 per year), teaching assistantships (averaging $16,200 per year) were awarded; fellowships Financial award application deadline: 3/1; financial award applicants required to submit FAFSA. *Faculty research:* Condensed-matter, atomic/molecular, high-energy, and nuclear physics; quantum optics. *Unit head:* Dr. Edward S. Fry, Head, 979-845-7717, Fax: 979-845-2590, E-mail: fry@physics.tamu.edu. *Application contact:* Dr. George W. Kattawar, Professor, 979-845-1180, Fax: 979-845-2590, E-mail: kattawar@physics.tamu.edu.

Texas Tech University, Graduate School, College of Arts and Sciences, Department of Physics, Lubbock, TX 79409. Offers applied physics (MS); physics (MS, PhD). Part-time programs available. *Faculty:* 17 full-time (1 woman). *Students:* 34 full-time (4 women), 6 part-time (1 woman); includes 4 minority (all Hispanic Americans), 18 international. Average age 29. 26 applicants, 62% accepted, 2 enrolled. In 2005, 7 master's, 2 doctorates awarded. *Degree requirements:* For master's and doctorate, variable foreign language requirement, thesis/dissertation. *Entrance requirements:* For master's and doctorate, GRE General Test. Additional exam requirements/recommendations for international students: Required—TOEFL (minimum score 550 paper-based; 213 computer-based). *Application deadline:* Applications are processed on a rolling basis. Application fee: $50 ($60 for international students). Electronic applications accepted. *Expenses:* Tuition, state resident: full-time $4,296. Tuition, nonresident: full-time $10,920. Required fees: $1,992. Tuition and fees vary according to program. *Financial support:* In 2005–06, 15 students received support, including 6 research assistantships with partial tuition reimbursements available (averaging $16,110 per year), 25 teaching assistantships with partial tuition reimbursements available (averaging $15,608 per year); career-related internships or fieldwork, Federal Work-Study, and institutionally sponsored loans also available. Support available to part-time students. Financial award application deadline: 4/15; financial award applicants required to submit FAFSA. *Faculty research:* Molecular spectroscopy of biological membranes, thin films and semiconductor characterization, muon spin rotation defect, characterization of semiconductors, nanotechnology of magnetic materials, theory of impurities and complexes in semiconductors. Total annual research expenditures: $1.1 million. *Unit head:* Dr. Lynn L. Hatfield, Chair, 806-742-3767, Fax: 806-742-1182, E-mail: lynn.hatfield@ttu.edu. *Application contact:* Dr. Wallace L. Glab, Graduate Recruiter, 806-742-3767, Fax: 806-742-1182, E-mail: wallace.glab@ttu.edu.

The University of Arizona, Graduate College, College of Science, Department of Physics, Applied and Industrial Physics, Professional Program, Tucson, AZ 85721. Offers MS. Part-time programs available. *Degree requirements:* For master's, thesis or alternative, internship, colloquium, business courses. *Entrance requirements:* For master's, GRE General Test, 3 letters of recommendation. *Faculty research:* Nanotechnology, optics, medical imaging, high energy physics, biophysics.

University of Arkansas, Graduate School, J. William Fulbright College of Arts and Sciences, Department of Physics, Program in Applied Physics, Fayetteville, AR 72701-1201. Offers MS. *Students:* 2 full-time (both women), 1 international. *Degree requirements:* For master's, thesis. Application fee: $40 ($50 for international students). *Financial support:* Teaching assistantships available. Financial award applicants required to submit FAFSA. *Application contact:* Dr. Raj Gupta, Graduate Coordinator, 479-575-5933, E-mail: rgupta@uark.edu.

University of California, San Diego, Graduate Studies and Research, Department of Electrical and Computer Engineering, La Jolla, CA 92093. Offers applied ocean science (MS, PhD); applied physics (MS, PhD); communication theory and systems (MS, PhD); computer engineering (MS, PhD); electrical engineering (M Eng); electronic circuits and systems (MS, PhD); intelligent systems, robotics and control (MS, PhD); photonics (MS, PhD); signal and image processing (MS, PhD). MS only offered to students who have been admitted to the PhD program. *Entrance requirements:* For master's and doctorate, GRE General Test. Electronic applications accepted.

University of Maryland, Baltimore County, Graduate School, College of Natural Sciences and Mathematics, Department of Physics, Program in Applied Physics, Baltimore, MD 21250. Offers astrophysics (PhD); optics (MS, PhD); quantum optics (PhD); solid state physics (MS, PhD). *Expenses:* Tuition, state resident: part-time $395 per credit. Tuition, nonresident: part-time $652 per credit. Required fees: $82 per credit. Tuition and fees vary according to course load, program and reciprocity agreements. *Application contact:* Dr. Terrance L Worchesky, Director, 410-455-6779, Fax: 410-455-1072, E-mail: workchesk@umbc.edu.

University of Massachusetts Boston, Office of Graduate Studies and Research, College of Science and Mathematics, Program in Applied Physics, Boston, MA 02125-3393. Offers MS. Part-time and evening/weekend programs available. *Degree requirements:* For master's, thesis optional. *Entrance requirements:* For master's, minimum GPA of 2.75. *Faculty research:* Experimental laser research, nonlinear optics, experimental and theoretical solid state physics, semiconductor devices, opto-electronics.

University of Massachusetts Lowell, Graduate School, College of Arts and Sciences, Department of Physics and Applied Physics, Program in Applied Physics, Lowell, MA 01854-2881. Offers applied mechanics (PhD); applied physics (MS, PhD), including optical sciences (MS). Terminal master's awarded for partial completion of doctoral program. *Degree requirements:* For master's, thesis; for doctorate, 2 foreign languages, thesis/dissertation. *Entrance requirements:* For master's and doctorate, GRE General Test.

University of Michigan, Horace H. Rackham School of Graduate Studies, College of Engineering and College of Literature, Science, and the Arts, Interdepartmental Program in Applied Physics, Ann Arbor, MI 48109. Offers PhD. *Faculty:* 98 full-time (9 women), 1 part-time/adjunct (0 women). *Students:* 62 full-time (17 women); includes 26 minority (9 African Americans, 14 Asian Americans or Pacific Islanders, 3 Hispanic Americans). Average age 23. 115 applicants, 8% accepted. In 2005, 4 degrees awarded. *Median time to degree:* Of those who began their doctoral program in fall 1997, 86% received their degree in 8 years or less. *Degree requirements:* For doctorate, oral defense of dissertation, preliminary and qualifying exams. *Entrance requirements:* For doctorate, GRE General Test. Additional exam requirements/recommendations for international students: Required—TOEFL. *Application deadline:* For fall admission, 1/15 for domestic students, 1/15 for international students. Applications are processed on a rolling basis. Application fee: $60 ($75 for international students). Electronic applications accepted. *Financial support:* In 2005–06, 18 fellowships with full tuition reimbursements (averaging $21,000 per year), 44 research assistantships with full tuition reimbursements (averaging $21,000 per year) were awarded; teaching assistantships with full tuition reimbursements, traineeships, health care benefits, and unspecified assistantships also available. Financial award application deadline: 1/15; financial award applicants required to submit FAFSA. *Faculty research:* Optical sciences, materials research, quantum structures, medical imaging, environment and science policy. Total annual research expenditures: $1.1 million. *Unit head:* Bradford Orr, Director, 734-936-0653, Fax: 734-764-2193, E-mail: orr@umich.edu. *Application contact:* Charles N. Sutton, Program Assistant, 734-764-4595, Fax: 734-764-2193, E-mail: csutton@umich.edu.

University of Missouri–St. Louis, College of Arts and Sciences, Department of Physics and Astronomy, St. Louis, MO 63121. Offers applied physics (MS); astrophysics (MS); physics (PhD). Part-time and evening/weekend programs available. *Faculty:* 13. *Students:* 10 full-time (2 women), 15 part-time (1 woman); includes 1 minority (Asian American or Pacific Islander), 6 international. Average age 30. In 2005, 3 master's awarded. Terminal master's awarded for partial completion of doctoral program. *Degree requirements:* For master's, thesis optional; for doctorate, thesis/dissertation. *Entrance requirements:* For master's, 2 letters of recommendation; for doctorate, GRE General Test, 2 letters of recommendation. Additional exam requirements/recommendations for international students: Required—TOEFL (minimum score 550 paper-based; 213 computer-based). *Application deadline:* For fall admission, 7/1 for domestic students; for spring admission, 12/1 priority date for domestic students. Applications are processed on a rolling basis. Application fee: $35 ($40 for international students). Electronic applications accepted. *Expenses:* Tuition, state resident: part-time $263 per credit hour. Tuition, nonresident: part-time $680 per credit hour. Required fees: $53 per credit hour. Tuition and fees vary according to program. *Financial support:* In 2005–06, 3 research assistantships with full and partial tuition reimbursements (averaging $13,333 per year), 11 teaching assistantships with full and partial tuition reimbursements (averaging $11,575 per year) were awarded; fellowships with full tuition reimbursements, career-related internships or fieldwork also available. *Faculty research:* Biophysics, atomic physics, nonlinear dynamics, materials science. *Unit head:* Dr. Ricardo Flores, Director of Graduate Studies, 314-516-5931, Fax: 314-516-6152, E-mail: flores@jinx.umsl.edu. *Application contact:* 314-516-5458, Fax: 314-516-5310, E-mail: gradadm@umsl.edu.

The University of North Carolina at Charlotte, Graduate School, College of Arts and Sciences, Department of Physics and Optical Science, Charlotte, NC 28223-0001. Offers applied physics (MS); optical science and engineering (MS, PhD). *Faculty:* 13 full-time (3 women). *Students:* 26 full-time (5 women), 9 part-time (1 woman); includes 4 minority (2 African Americans, 2 Asian Americans or Pacific Islanders), 17 international. Average age 29. 30 applicants, 87% accepted, 70 enrolled. In 2005, 3 master's awarded. *Degree requirements:* For master's, thesis optional. *Entrance requirements:* For master's, GRE General Test, minimum GPA of 3.0 during previous 2 years, 2.75 overall. Additional exam requirements/recommendations for international students: Required—TOEFL (minimum score 557 paper-based; 220 computer-based). *Application deadline:* For fall admission, 7/15 for domestic students, 5/1 for international students; for spring admission, 11/15 for domestic students, 10/1 for international students. Applications are processed on a rolling basis. Application fee: $55. Electronic applications accepted. *Expenses:* Tuition, state resident: full-time $2,504; part-time $157 per credit. Tuition, nonresident: full-time $12,711; part-time $794 per credit. Required fees: $1,424; $89 per credit. Tuition and fees vary according to course load and program. *Financial support:* In 2005–06, 46 research assistantships (averaging $6,612 per year), 39 teaching assistantships (averaging $10,000 per year) were awarded; fellowships, career-related internships or fieldwork, Federal Work-Study, institutionally sponsored loans, scholarships/grants, and unspecified assistantships also available. Support available to part-time students. Financial award application deadline: 4/1; financial award applicants required to submit FAFSA. *Faculty research:* Optics, lasers, microscopy, fibers, astrophysics. *Unit head:* Dr. Faramarz Farahi, Chair, 704-687-2537, Fax: 704-687-3160, E-mail: ffarahi@uncc.edu. *Application contact:* Kathy B. Giddings, Director of Graduate Admissions, 704-687-3366, Fax: 704-687-3279, E-mail: gradadm@email.uncc.edu.

University of South Florida, College of Graduate Studies, College of Arts and Sciences, Department of Physics, Tampa, FL 33620-9951. Offers applied physics (PhD); physics (MS). Part-time programs available. *Faculty:* 15 full-time (1 woman). *Students:* 51 full-time (15 women), 5 part-time; includes 4 minority (2 African Americans, 1 Asian American or Pacific Islander, 1 Hispanic American), 28 international. 39 applicants, 72% accepted, 15 enrolled. In 2005, 2 degrees awarded. *Degree requirements:* For master's, thesis optional; for doctorate, 2 foreign languages, thesis/dissertation. *Entrance requirements:* For master's, GRE General Test, minimum GPA of 3.0 in last 60 hours of course work; for doctorate, GRE General Test, minimum graduate GPA of 3.2. Additional exam requirements/recommendations for international students: Required—TOEFL (minimum score 550 paper-based), TSE (minimum score 50). *Application deadline:* For fall admission, 6/1 for domestic students; for spring admission, 10/1 for domestic students. Applications are processed on a rolling basis. Application fee: $30. Electronic applications accepted. *Financial support:* Fellowships with full tuition reimbursements, research assistantships with full tuition reimbursements, teaching assistantships with full tuition reimbursements, career-related internships or fieldwork, scholarships/grants, and unspecified assistantships available. *Faculty research:* Laser, medical, and solid-state physics. *Unit head:* Dr. Sarath Witanachchi, Director of Graduate Studies, 813-974-2789, Fax: 813-974-5813, E-mail: switanac@cas.usf.edu. *Application contact:* Evelyne Keeton-Williams, Program Assistant, 813-974-2871, Fax: 813-974-5813, E-mail: ekeeton@cas.usf.edu.

University of Washington, Graduate School, College of Arts and Sciences, Department of Physics, Seattle, WA 98195. Offers MS, PhD. Part-time and evening/weekend programs available. Terminal master's awarded for partial completion of doctoral program. *Degree requirements:*

For doctorate, thesis/dissertation. *Entrance requirements:* For master's, GRE; for doctorate, GRE General Test, GRE Subject Test. Additional exam requirements/recommendations for international students: Required—TOEFL. Electronic applications accepted. *Faculty research:* Astro-, atomic, condensed-matter, nuclear, and particle physics; physics education.

Virginia Commonwealth University, Graduate School, College of Humanities and Sciences, Department of Physics, Richmond, VA 23284-9005. Offers applied physics (MS); physics (MS). Part-time programs available. *Faculty:* 7 full-time (3 women). *Students:* 7 full-time (4 women), 2 part-time. 8 applicants, 100% accepted. In 2005, 5 degrees awarded. *Degree requirements:* For master's, thesis optional. *Entrance requirements:* For master's, GRE. *Application deadline:* For fall admission, 8/1 for domestic students; for spring admission, 12/1 for domestic students. Applications are processed on a rolling basis. Application fee: $50. *Expenses:* Tuition, state resident: full-time $3,185; part-time $405 per credit. Tuition, nonresident: full-time $7,952; part-time $940 per credit. Required fees: $751 per semester hour. Tuition and fees vary according to course load and program. *Financial support:* Fellowships, teaching assistantships, Federal Work-Study, institutionally sponsored loans, and tuition waivers (full and partial) available. Support available to part-time students. *Faculty research:* Condensed-matter theory and experimentation, electronic instrumentation, relativity. *Unit head:* Dr. Robert H. Gowdy, Chair, 804-828-1821, Fax: 804-828-7073, E-mail: rhgowdy@vcu.edu. *Application contact:* Dr. Alison Baski, Graduate Program Director, 804-828-8295, Fax: 804-828-7073, E-mail: aabaski@vcu.edu.

Virginia Polytechnic Institute and State University, Graduate School, College of Science, Department of Physics, Blacksburg, VA 24061. Offers applied physics (MS, PhD); physics (MS, PhD). *Faculty:* 22 full-time (3 women). *Students:* 46 full-time (8 women), 7 part-time (2 women); includes 4 minority (1 African American, 2 Asian Americans or Pacific Islanders, 1 Hispanic American), 21 international. Average age 27. 65 applicants, 22% accepted, 11 enrolled. In 2005, 6 master's, 5 doctorates awarded. *Entrance requirements:* For master's and doctorate, GRE Subject Test. Additional exam requirements/recommendations for international students: Required—TOEFL (minimum score 550 paper-based; 213 computer-based). *Application deadline:* Applications are processed on a rolling basis. Application fee: $45. Electronic applications accepted. *Expenses:* Tuition, state resident: full-time $6,558; part-time $364 per credit. Tuition, nonresident: full-time $11,296; part-time $628 per credit. Required fees: $1,419; $468 per credit. $234 per term. *Financial support:* In 2005–06, 2 fellowships with full tuition reimbursements (averaging $6,000 per year), 14 research assistantships with full tuition reimbursements (averaging $16,689 per year), 26 teaching assistantships with full tuition reimbursements (averaging $13,902 per year) were awarded; career-related internships or fieldwork, Federal Work-Study, scholarships/grants, and unspecified assistantships also available. Financial award application deadline: 4/1. *Faculty research:* Condensed matter, particle physics, theoretical and experimental astrophysics, biophysics, mathematical physics. *Unit head:* Dr. Royce Zia, Head, 540-231-5767, Fax: 540-231-7511. *Application contact:* Chris C. Thomas, Graduate Program Coordinator, 540-231-8728, Fax: 540-231-7511, E-mail: chris.thomas@vt.edu.

West Virginia University, Eberly College of Arts and Sciences, Department of Physics, Morgantown, WV 26506. Offers applied physics (MS, PhD); astrophysics (MS, PhD); chemical physics (MS, PhD); condensed matter physics (MS, PhD); elementary particle physics (MS, PhD); materials physics (MS, PhD); plasma physics (MS, PhD); solid state physics (MS, PhD); statistical physics (MS, PhD); theoretical physics (MS, PhD). *Faculty:* 17 full-time (3 women), 1 part-time/adjunct (0 women). *Students:* 44 full-time (10 women), 4 part-time (1 woman), 32 international. Average age 28. 60 applicants, 20% accepted. In 2005, 2 master's, 3 doctorates awarded. Terminal master's awarded for partial completion of doctoral program. *Degree requirements:* For master's, thesis or alternative, qualifying exam; for doctorate, thesis/dissertation, qualifying exam. *Entrance requirements:* For master's and doctorate, GRE General Test, GRE Subject Test, minimum GPA of 3.0. Additional exam requirements/recommendations for international students: Required—TOEFL. *Application deadline:* For fall admission, 2/15 for domestic students. Applications are processed on a rolling basis. Application fee: $50. *Expenses:* Tuition, state resident: full-time $4,582; part-time $258 per credit hour. Tuition, nonresident: full-time $1,382; part-time $741 per credit hour. *Financial support:* In 2005–06, 30 research assistantships with full and partial tuition reimbursements (averaging $18,000 per year), 10 teaching assistantships with full and partial tuition reimbursements (averaging $16,000 per year) were awarded; fellowships, Federal Work-Study, institutionally sponsored loans, and tuition waivers (full and partial) also available. Financial award application deadline: 2/1; financial award applicants required to submit FAFSA. *Faculty research:* Experimental and theoretical condensed-matter, plasma, high-energy theory, nonlinear dynamics, space physics. Total annual research expenditures: $3.3 million. *Unit head:* Dr. Earl E. Scime, Chair, 304-293-3422 Ext. 1437, Fax: 304-293-5732, E-mail: earl.scime@mail.wvu.edu.

Yale University, Graduate School of Arts and Sciences, Programs in Engineering and Applied Science, Department of Applied Physics, New Haven, CT 06520. Offers MS, PhD. Terminal master's awarded for partial completion of doctoral program. *Degree requirements:* For doctorate, thesis/dissertation, area exam. *Entrance requirements:* For master's and doctorate, GRE General Test. Additional exam requirements/recommendations for international students: Required—TOEFL.

Chemical Physics

Columbia University, Graduate School of Arts and Sciences, Division of Natural Sciences, Department of Chemistry, Program in Chemical Physics, New York, NY 10027. Offers M Phil, PhD. *Students:* 9 full-time (3 women), 4 international. Average age 26. 21 applicants, 24% accepted. *Entrance requirements:* For master's, GRE General Test, GRE Subject Test. Additional exam requirements/recommendations for international students: Required—TOEFL. Application fee: $75. *Expenses:* Tuition: Full-time $31,448. Tuition and fees vary according to course level, course load, campus/location and program. *Financial support:* Fellowships, teaching assistantships available. Support available to part-time students. Financial award application deadline: 1/5; financial award applicants required to submit FAFSA. *Unit head:* Philip Pechukas, Head, 212-854-4231, Fax: 212-932-1289.

Cornell University, Graduate School, Graduate Fields of Arts and Sciences, Field of Chemistry and Chemical Biology, Ithaca, NY 14853-0001. Offers analytical chemistry (PhD); bio-organic chemistry (PhD); biophysical chemistry (PhD); chemical biology (PhD); chemical physics (PhD); inorganic chemistry (PhD); materials chemistry (PhD); organic chemistry (PhD); organometallic chemistry (PhD); physical chemistry (PhD); polymer chemistry (PhD); theoretical chemistry (PhD). *Faculty:* 44 full-time (2 women). *Students:* 190 full-time (73 women); includes 23 minority (4 African Americans, 8 Asian Americans or Pacific Islanders, 11 Hispanic Americans), 65 international. 339 applicants, 35% accepted, 49 enrolled. In 2005, 23 doctorates awarded. *Degree requirements:* For doctorate, thesis/dissertation, comprehensive exam. *Entrance requirements:* For doctorate, GRE General Test, GRE Subject Test (chemistry), 3 letters of recommendation. Additional exam requirements/recommendations for international students: Required—TOEFL (minimum score 600 paper-based; 250 computer-based). *Application deadline:* For fall admission, 1/10 for domestic students. Application fee: $60. Electronic applications accepted. *Financial support:* In 2005–06, 185 students received support, including 33 fellowships with full tuition reimbursements available, 85 research assistantships with full tuition reimbursements available, 67 teaching assistantships with full tuition reimbursements available; institutionally sponsored loans, scholarships/grants, health care benefits, tuition waivers (full and partial), and unspecified assistantships also available. Financial award applicants required to submit FAFSA. *Faculty research:* Analytical, organic, inorganic, physical, materials, chemical biology. *Unit head:* Director of Graduate Studies, 607-255-4139, Fax: 607-255-4137. *Application contact:* Graduate Field Assistant, 607-255-4139, Fax: 607-255-4137, E-mail: chemgrad@cornell.edu.

See Close-Up on page 109.

Florida State University, Graduate Studies, College of Arts and Sciences, Department of Chemistry and Biochemistry and Department of Physics, Program in Chemical Physics, Tallahassee, FL 32306. Offers MS, PhD. *Faculty:* 17 full-time (0 women). *Students:* 3 full-time (1 woman). 1 applicant, 100% accepted, 1 enrolled. *Degree requirements:* For master's, cumulative and diagnostic exams; for doctorate, thesis/dissertation, cumulative and diagnostic exams. *Entrance requirements:* For master's and doctorate, GRE General Test, minimum B average in undergraduate course work. Additional exam requirements/recommendations for international students: Required—TOEFL (minimum score 515 paper-based; 213 computer-based). *Application deadline:* For fall admission, 4/15 for domestic students. Applications are processed on a rolling basis. Application fee: $30. Electronic applications accepted. *Financial support:* In 2005–06, 1 research assistantship with tuition reimbursement (averaging $19,000 per year), 2 teaching assistantships with tuition reimbursements (averaging $19,000 per year) were awarded; career-related internships or fieldwork, institutionally sponsored loans, tuition waivers (full), and unspecified assistantships also available. Financial award application deadline: 2/15; financial award applicants required to submit FAFSA. *Faculty research:* Theoretical and experimental research in molecular and solid-state physics and chemistry, statistical mechanics. *Application contact:* Dr. Oliver Steinbock, Chair, Graduate Admissions Committee, 888-525-9286, Fax: 850-644-8281, E-mail: gradinfo@chem.fsu.edu.

See Close-Up on page 117.

Georgetown University, Graduate School of Arts and Sciences, Department of Chemistry, Washington, DC 20057. Offers analytical chemistry (MS, PhD); biochemistry (MS, PhD); chemical physics (MS, PhD); inorganic chemistry (MS, PhD); organic chemistry (MS, PhD); physical chemistry (MS, PhD); theoretical chemistry (MS, PhD). Terminal master's awarded for partial completion of doctoral program. *Degree requirements:* For master's, thesis (for some programs), qualifying exam; for doctorate, thesis/dissertation, comprehensive exam. *Entrance requirements:* For master's and doctorate, GRE General Test. Additional exam requirements/recommendations for international students: Required—TOEFL.

Harvard University, Graduate School of Arts and Sciences, Committee on Chemical Physics, Cambridge, MA 02138. Offers chemical physics (PhD). *Students:* 2 full-time (0 women); includes 1 minority (Hispanic American) *Degree requirements:* For doctorate, one foreign language, thesis/dissertation, cumulative exams. *Entrance requirements:* For doctorate, GRE General Test, GRE Subject Test. Additional exam requirements/recommendations for international students: Required—TOEFL. *Application deadline:* For fall admission, 1/1 for domestic students. Application fee: $60. *Expenses:* Tuition: Full-time $28,752. Full-time tuition and fees vary according to program and student level. *Financial support:* Fellowships, research assistantships, teaching assistantships, career-related internships or fieldwork, Federal Work-Study, and institutionally sponsored loans available. Financial award application deadline: 12/30. *Unit head:* Betsey Cogswell, Administrator, 617-495-5497, Fax: 617-495-5264. *Application contact:* Department of Chemistry and Chemical Biology, 617-496-3208.

Kent State University, College of Arts and Sciences, Chemical Physics Interdisciplinary Program, Kent, OH 44242-0001. Offers MS, PhD. Offered in cooperation with the Departments of Chemistry, Mathematics and Computer Science, and Physics and the Liquid Crystal Institute. Terminal master's awarded for partial completion of doctoral program. *Degree requirements:* For master's, thesis; for doctorate, thesis/dissertation, candidacy exam. *Entrance requirements:* For master's and doctorate, GRE. Additional exam requirements/recommendations for international students: Required—TOEFL (minimum score 525 paper-based; 197 computer-based). Electronic applications accepted.

Marquette University, Graduate School, College of Arts and Sciences, Department of Chemistry, Milwaukee, WI 53201-1881. Offers analytical chemistry (MS, PhD); bioanalytical chemistry (MS, PhD); biophysical chemistry (MS, PhD); chemical physics (MS, PhD); inorganic chemistry (MS, PhD); organic chemistry (MS, PhD); physical chemistry (MS, PhD). Part-time programs available. Terminal master's awarded for partial completion of doctoral program. *Degree requirements:* For master's, comprehensive exam; for doctorate, thesis/dissertation, cumulative exams. *Entrance requirements:* For master's and doctorate, GRE Subject Test. Additional exam requirements/recommendations for international students: Required—TOEFL. *Faculty research:* Inorganic complexes, laser Raman spectroscopy, organic synthesis, chemical dynamics, biophysiology.

McMaster University, School of Graduate Studies, Faculty of Science, Department of Chemistry, Hamilton, ON L8S 4M2, Canada. Offers analytical chemistry (M Sc, PhD); chemical physics (M Sc, PhD); chemistry (M Sc, PhD); inorganic chemistry (M Sc, PhD); organic chemistry (M Sc, PhD); physical chemistry (M Sc, PhD); polymer chemistry (M Sc, PhD). Part-time programs available. Terminal master's awarded for partial completion of doctoral program. *Degree requirements:* For master's, thesis/dissertation; for doctorate, thesis/dissertation, comprehensive exam. *Entrance requirements:* For master's, minimum B+ average. Additional exam requirements/recommendations for international students: Required—TOEFL (minimum score 550 paper-based; 213 computer-based).

Michigan State University, The Graduate School, College of Natural Science, Department of Chemistry, East Lansing, MI 48824. Offers chemical physics (PhD); chemistry (MS, PhD); chemistry-environmental toxicology (PhD); computational chemistry (MS). *Faculty:* 35 full-time (4 women). *Students:* 221 full-time (100 women), 4 part-time (2 women); includes 13 minority (4 African Americans, 1 American Indian/Alaska Native, 6 Asian Americans or Pacific Islanders, 2 Hispanic Americans), 129 international. Average age 27. 138 applicants, 66% accepted. In 2005, 9 master's, 19 doctorates awarded. *Degree requirements:* For master's, oral defense of thesis, thesis optional; for doctorate, thesis/dissertation, oral defense of dissertation, comprehensive exam. *Entrance requirements:* For master's, GRE General Test, bachelor's degree in chemistry; course work in chemistry, physics, and calculus; 3 letters of recommendation; for doctorate, GRE General Test, minimum GPA of 3.0; bachelor's or master's degree in chemistry; coursework in chemistry, physics, and calculus; 3 letters of recommendation. Additional exam requirements/recommendations for international students: Required—TOEFL (minimum score 550 paper-based; 213 computer-based), Michigan State University ELT (85), Michigan ELAB (83). *Application deadline:* For fall admission, 12/27 for domestic students. Application fee: $50. Electronic applications accepted. *Expenses:* Tuition, state resident: part-time $330 per credit hour. Tuition, nonresident: part-time $685 per credit hour. Tuition and fees vary according to program. *Financial support:* In 2005–06, 85 fellowships with tuition reimbursements (averaging $3,132 per year), 71 research assistantships with tuition reimbursements (averaging $15,305 per year), 138 teaching assistantships with tuition reimbursements (averaging $15,041 per year) were awarded; scholarships/grants and unspecified assistantships also available. *Faculty research:* Analytical chemistry, inorganic and organic chemistry,

Chemical Physics

Michigan State University (continued)
nuclear chemistry, physical chemistry, theoretical and computational chemistry. Total annual research expenditures: $8.7 million. *Unit head:* Dr. John L. McCracken, Chairperson, 517-355-9715 Ext. 346, Fax: 517-353-1793, E-mail: chair@chemistry.msu.edu. *Application contact:* Deborah Roper, Graduate Admissions Secretary, 517-355-9715 Ext. 362, Fax: 517-353-1793, E-mail: gradoff@crm.msu.edu.

The Ohio State University, Graduate School, College of Mathematical and Physical Sciences, Program in Chemical Physics, Columbus, OH 43210. Offers MS, PhD. *Degree requirements:* For master's, thesis optional; for doctorate, thesis/dissertation. *Entrance requirements:* For master's and doctorate, GRE General Test, GRE Subject Test (chemistry or physics). Additional exam requirements/recommendations for international students: Recommended—TOEFL (minimum score 600 paper-based; 250 computer-based). Electronic applications accepted.

Princeton University, Graduate School, Department of Physics, Princeton, NJ 08544-1019. Offers applied and computational mathematics (PhD); mathematical physics (PhD); physics (PhD); physics and chemical physics (PhD). *Degree requirements:* For doctorate, thesis/dissertation, qualifying exam. *Entrance requirements:* For doctorate, GRE General Test, GRE Subject Test. Additional exam requirements/recommendations for international students: Required—TOEFL (minimum score 600 paper-based; 250 computer-based). Electronic applications accepted.

Simon Fraser University, Graduate Studies, Faculty of Science, Department of Chemistry, Burnaby, BC V5A 1S6, Canada. Offers chemical physics (M Sc, PhD); chemistry (M Sc, PhD). *Degree requirements:* For master's and doctorate, thesis/dissertation. *Entrance requirements:* For master's, minimum GPA of 3.0; for doctorate, minimum GPA of 3.5. Additional exam requirements/recommendations for international students: Required—TOEFL or IELTS. *Faculty research:* Organic chemistry, nuclear chemistry, physical chemistry, inorganic chemistry, theoretical chemistry.

Simon Fraser University, Graduate Studies, Faculty of Science, Department of Physics, Burnaby, BC V5A 1S6, Canada. Offers biophysics (M Sc, PhD); chemical physics (M Sc, PhD); physics (M Sc, PhD). *Degree requirements:* For master's and doctorate, thesis/dissertation. *Entrance requirements:* For master's, minimum GPA of 3.0; for doctorate, minimum GPA of 3.5. Additional exam requirements/recommendations for international students: Required—TOEFL or IELTS. *Faculty research:* Solid-state physics, magnetism, energy research, superconductivity, nuclear physics.

University of Colorado at Boulder, Graduate School, College of Arts and Sciences, Department of Physics, Boulder, CO 80309. Offers chemical physics (PhD); geophysics (PhD); liquid crystal science and technology (PhD); mathematical physics (PhD); medical physics (PhD); optical sciences and engineering (PhD); physics (MS, PhD). *Faculty:* 39 full-time (3 women). *Students:* 157 full-time (37 women), 47 part-time (3 women); includes 10 minority (2 African Americans, 1 American Indian/Alaska Native, 4 Asian Americans or Pacific Islanders, 3 Hispanic Americans), 46 international. Average age 27. 81 applicants, 93% accepted. In 2005, 10 master's, 16 doctorates awarded. Terminal master's awarded for partial completion of doctoral program. *Degree requirements:* For master's, thesis or alternative, comprehensive exam; for doctorate, thesis/dissertation, comprehensive exam. *Entrance requirements:* For master's and doctorate, GRE General Test, GRE Subject Test, minimum undergraduate GPA of 3.0. Additional exam requirements/recommendations for international students: Required—TOEFL. *Application deadline:* For fall admission, 1/15 priority date for domestic students, 1/15 priority date for international students. Applications are processed on a rolling basis. Application fee: $50 ($60 for international students). Electronic applications accepted. *Financial support:* In 2005–06, fellowships with full tuition reimbursements (averaging $4,376 per year), research assistantships with full tuition reimbursements (averaging $15,185 per year), teaching assistantships with full tuition reimbursements (averaging $13,786 per year) were awarded; scholarships/grants also available. Financial award application deadline: 1/15. *Faculty research:* Atomic and molecular physics, nuclear physics, condensed matter, elementary particle physics, laser or optical physics. Total annual research expenditures: $24.1 million. *Unit head:* John Cumalat, Chair, 303-492-6952, Fax: 303-492-3352, E-mail: jcumalat@pizero.colorado.edu. *Application contact:* Graduate Program Assistant, 303-492-6954, Fax: 303-492-3352, E-mail: phys@bogart.colorado.edu.

University of Louisville, Graduate School, College of Arts and Sciences, Department of Chemistry, Louisville, KY 40292-0001. Offers analytical chemistry (MS, PhD); biochemistry (MS, PhD); chemical physics (PhD); inorganic chemistry (MS, PhD); organic chemistry (MS, PhD); physical chemistry (MS, PhD). *Students:* 37 full-time (16 women), 16 part-time (7 women); includes 3 minority (2 African Americans, 1 Hispanic American), 24 international. Average age 28. In 2005, 1 master's, 8 doctorates awarded. *Degree requirements:* For master's, thesis/dissertation; for doctorate, thesis/dissertation, comprehensive exam. *Entrance requirements:* For master's and doctorate, GRE General Test. Additional exam requirements/recommendations for international students: Required—TOEFL. *Application deadline:* Applications are processed on a rolling basis. Application fee: $50. *Expenses:* Tuition, state resident: full-time $3,003; part-time $334 per credit hour. Tuition, nonresident: full-time $8,277; part-time $920 per credit hour. Tuition and fees vary according to course load, degree level and program. *Financial support:* In 2005–06, 33 teaching assistantships with tuition reimbursements were awarded; fellowships, research assistantships *Unit head:* Dr. George R. Pack, Chair, 502-852-6798, Fax: 502-852-8149, E-mail: george.pack@louisville.edu.

University of Maryland, College Park, Graduate Studies, College of Computer, Mathematical and Physical Sciences, Institute for Physical Science and Technology, Program in Chemical Physics, College Park, MD 20742. Offers MS, PhD. Part-time and evening/weekend programs available. *Students:* 33 full-time (9 women), 6 part-time (2 women); includes 3 minority (2 Asian Americans or Pacific Islanders, 1 Hispanic American), 24 international. 18 applicants, 61% accepted, 4 enrolled. In 2005, 2 doctorates awarded. Terminal master's awarded for partial completion of doctoral program. *Median time to degree:* Of those who began their doctoral program in fall 1997, 57% received their degree in 8 years or less. *Degree requirements:* For master's, paper, qualifying exam, thesis optional; for doctorate, thesis/dissertation, seminars. *Entrance requirements:* For master's, GRE General Test, GRE Subject Test (chemistry, math or physics), minimum GPA of 3.3, 3 letters of recommendation; for doctorate, GRE Subject Test (chemistry, math, or physics), GRE General Test, minimum GPA of 3.3, 3 letters of recommendation. *Application deadline:* For fall admission, 5/15 for domestic students, 2/1 for international students; for spring admission, 10/15 for domestic students, 6/1 for international students. Applications are processed on a rolling basis. Application fee: $60. Electronic applications accepted. *Financial support:* In 2005–06, 14 fellowships (averaging $5,741 per year) were awarded; research assistantships, teaching assistantships, Federal Work-Study and scholarships/grants also available. Financial award applicants required to submit FAFSA. *Faculty research:* Discrete molecules and gases; dynamic phenomena; thermodynamics, statistical mechanical theory and quantum mechanical theory; atmospheric physics; biophysics. *Unit head:* Dr. Michael Coplan, Director, 301-405-4780, Fax: 301-314-9396. *Application contact:* Dean of Graduate School, 301-405-4190, Fax: 301-314-9305.

University of Nevada, Reno, Graduate School, College of Science, Department of Chemical Physics, Reno, NV 89557. Offers PhD. *Entrance requirements:* For doctorate, GRE, minimum GPA of 3.0. Additional exam requirements/recommendations for international students: Required—TOEFL.

University of Southern California, Graduate School, College of Letters, Arts and Sciences, Department of Chemistry, Program in Chemical Physics, Los Angeles, CA 90089. Offers PhD. *Faculty:* 30 full-time (3 women). *Students:* Average age 25. *Degree requirements:* For doctorate, thesis/dissertation, qualifying exam. *Entrance requirements:* For doctorate, GRE General Test. *Application deadline:* For fall admission, 3/1 for domestic students. Applications are processed on a rolling basis. Application fee: $0. *Expenses:* Tuition: Full-time $25,416; part-time $1,059 per unit. Required fees: $484; $484 per year. Tuition and fees vary according to course load and program. *Financial support:* In 2005–06, fellowships (averaging $22,000 per year), research assistantships with tuition reimbursements (averaging $22,000 per year), teaching assistantships (averaging $22,000 per year) were awarded; Federal Work-Study, institutionally sponsored loans, scholarships/grants, and health care benefits also available. Financial award application deadline: 3/1. *Faculty research:* Inorganic chemistry, polymer chemistry, theoretical chemistry. *Application contact:* Heather Connor, Graduate Advisor, 213-740-6855, Fax: 213-740-2701, E-mail: hconnor@usc.edu.

The University of Tennessee, Graduate School, College of Arts and Sciences, Department of Chemistry, Knoxville, TN 37996. Offers analytical chemistry (MS, PhD); chemical physics (PhD); environmental chemistry (MS, PhD); inorganic chemistry (MS, PhD); organic chemistry (MS, PhD); physical chemistry (MS, PhD); polymer chemistry (MS, PhD); theoretical chemistry (PhD). Part-time programs available. Terminal master's awarded for partial completion of doctoral program. *Degree requirements:* For master's and doctorate, thesis/dissertation. *Entrance requirements:* For master's and doctorate, GRE General Test, minimum GPA of 2.7. Additional exam requirements/recommendations for international students: Required—TOEFL. Electronic applications accepted.

University of Utah, The Graduate School, College of Science, Department of Chemistry, Salt Lake City, UT 84112-1107. Offers chemical physics (PhD); chemistry (M Phil, MA, MS, PhD); science teacher education (MS). Part-time programs available. Postbaccalaureate distance learning degree programs offered. *Faculty:* 27 full-time (4 women), 2 part-time/adjunct (0 women). *Students:* 151 full-time (66 women), 21 part-time (6 women); includes 9 minority (1 African American, 4 Asian Americans or Pacific Islanders, 4 Hispanic Americans), 76 international. Average age 28. 42 applicants, 79% accepted, 30 enrolled. In 2005, 14 master's, 22 doctorates awarded. Terminal master's awarded for partial completion of doctoral program. *Median time to degree:* Of those who began their doctoral program in fall 1997, 100% received their degree in 8 years or less. *Degree requirements:* For master's, 26 hours coursework, 10 hours research, thesis optional; for doctorate, thesis/dissertation, 18 hours coursework, 14 hours research. *Entrance requirements:* For master's and doctorate, GRE General Test, minimum GPA of 3.0. Additional exam requirements/recommendations for international students: Required—TOEFL (minimum score 620 paper-based; 260 computer-based), TSE. *Application deadline:* For fall admission, 7/1 for domestic students, 4/1 for international studentsFor winter admission, 3/15 for domestic students; for spring admission, 11/1 for domestic students. Applications are processed on a rolling basis. Application fee: $45 ($65 for international students). Electronic applications accepted. *Expenses:* Tuition, area resident: Full-time $2,932; part-time $2,212 per term. Tuition, state resident: full-time $2,932; part-time $2,212 per term. Tuition, nonresident: full-time $10,350; part-time $7,812 per term. International tuition: $10,350 full-time. Required fees: $590; $516 per term. Tuition and fees vary according to course load and program. *Financial support:* In 2005–06, 172 students received support, including research assistantships with tuition reimbursements available (averaging $21,500 per year), teaching assistantships with tuition reimbursements available (averaging $20,500 per year); fellowships with tuition reimbursements available, scholarships/grants and tuition waivers (full) also available. Financial award application deadline: 7/1; financial award applicants required to submit FAFSA. *Faculty research:* Biological, theoretical, inorganic, organic, and physical-analytical chemistry. Total annual research expenditures: $11.1 million. *Unit head:* Peter B. Armentrout, Chair, 801-581-6681, Fax: 801-581-8433, E-mail: armentrout@chemistry.utah.edu. *Application contact:* Jo Hoovey, Graduate Coordinator, 801-581-4393, Fax: 801-581-5408, E-mail: jhoovey@chem.utah.edu.

University of Utah, The Graduate School, College of Science, Department of Physics, Salt Lake City, UT 84112-1107. Offers chemical physics (PhD); physics (MA, MS, PhD). Part-time programs available. *Faculty:* 31 full-time (1 woman), 1 part-time/adjunct (0 women). *Students:* 83 full-time (15 women), 18 part-time (3 women); includes 4 minority (2 Asian Americans or Pacific Islanders, 2 Hispanic Americans), 46 international. Average age 29. 73 applicants, 34% accepted, 24 enrolled. In 2005, 9 master's, 7 doctorates awarded. Terminal master's awarded for partial completion of doctoral program. *Degree requirements:* For master's, thesis or alternative, teaching experience, comprehensive exam (for some programs); for doctorate, thesis/dissertation, departmental qualifying exam, comprehensive exam. *Entrance requirements:* For master's and doctorate, GRE General Test, GRE Subject Test, minimum GPA of 3.0. Additional exam requirements/recommendations for international students: Required—TOEFL (minimum score 500 paper-based; 173 computer-based). *Application deadline:* For fall admission, 2/1 for domestic students. Applications are processed on a rolling basis. Application fee: $45 ($65 for international students). Electronic applications accepted. *Expenses:* Tuition, state resident: full-time $2,932; part-time $2,212 per term. Tuition, nonresident: full-time $10,350; part-time $7,812 per term. Required fees: $590; $516 per term. Tuition and fees vary according to course load and program. *Financial support:* Fellowships, research assistantships with full and partial tuition reimbursements, teaching assistantships with full and partial tuition reimbursements, Federal Work-Study, institutionally sponsored loans, and scholarships/grants available. Financial award application deadline: 2/15; financial award applicants required to submit FAFSA. *Faculty research:* High-energy, cosmic-ray, astrophysics, medical physics, condensed matter, relativity applied physics. Total annual research expenditures: $5.8 million. *Unit head:* Dr. Pierre Sokolsky, Chair, 801-581-6901, Fax: 801-581-4801, E-mail: ps@cosmic.utah.edu. *Application contact:* Jackie Hadley, Graduate Secretary, 801-581-6861, Fax: 801-581-4801, E-mail: jackie@physics.utah.edu.

University of Utah, The Graduate School, College of Science, Interdepartmental Program in Chemical Physics, Salt Lake City, UT 84112-1107. Offers PhD. *Students:* 1 full-time (0 women), 6 part-time (4 women). *Entrance requirements:* For doctorate, minimum undergraduate GPA of 3.0. Additional exam requirements/recommendations for international students: Required—TOEFL (minimum score 500 paper-based; 173 computer-based). *Application deadline:* For fall admission, 4/1 for domestic students, 4/1 for international students; for spring admission, 11/1 for domestic students, 11/1 for international students. Application fee: $45 ($65 for international students). *Expenses:* Tuition, state resident: full-time $2,932; part-time $2,212 per term. Tuition, nonresident: full-time $10,350; part-time $7,812 per term. Required fees: $590; $516 per term. Tuition and fees vary according to course load and program. *Financial support:* Applicants required to submit FAFSA. *Unit head:* Peter J. Stang, Dean, 801-581-6958, Fax: 801-585-3169, E-mail: stang@chemistry.utah.edu. *Application contact:* Information Contact, 801-581-6958, E-mail: office@science.utah.edu.

Virginia Commonwealth University, Graduate School, College of Humanities and Sciences, Department of Chemistry, Richmond, VA 23284-9005. Offers analytical (MS, PhD); chemical physics (PhD); inorganic (MS, PhD); organic (MS, PhD); physical (MS, PhD). Part-time programs available. *Faculty:* 18 full-time (6 women). *Students:* 36 full-time (13 women), 11 part-time (1 woman); includes 13 minority (6 African Americans, 1 American Indian/Alaska Native, 5 Asian Americans or Pacific Islanders, 1 Hispanic American), 13 international. 47 applicants, 74% accepted. In 2005, 2 master's, 8 doctorates awarded. Terminal master's awarded for partial completion of doctoral program. *Degree requirements:* For master's, thesis; for doctorate, thesis/dissertation, comprehensive cumulative exams, research proposal. *Entrance requirements:* For master's, GRE General Test, 30 undergraduate credits in chemistry; for doctorate, GRE General Test. *Application deadline:* For fall admission, 3/15 for domestic students; for spring admission, 11/15 for domestic students. Applications are processed on a rolling basis. Application fee: $50. *Expenses:* Tuition, state resident: full-time $3,185; part-time $405 per credit. Tuition, nonresident: full-time $7,952; part-time $940 per credit. Required fees: $751 per semester hour. Tuition and fees vary according to course load and program. *Financial support:* Fellowships, research assistantships, teaching assistantships, career-related internships or fieldwork and institutionally sponsored loans available. Support available to part-time students. Financial award application deadline: 7/1. *Faculty research:* Physical, organic, inorganic, analytical, and polymer chemistry; chemical physics. *Unit head:* Dr. Fred M. Hawkridge, Chair, 804-828-1298, Fax: 804-828-8599, E-mail: fmhawkri@vcu.edu. *Application contact:* Dr.

M. Samy El-Shall, Chair, Graduate Recruiting and Admissions Committee, 804-828-3518, E-mail: mselshal@vcu.edu.

Wesleyan University, Graduate Programs, Department of Chemistry, Program in Chemical Physics, Middletown, CT 06459-0260. Offers MA, PhD. *Faculty:* 4 full-time (0 women). Terminal master's awarded for partial completion of doctoral program. *Degree requirements:* For master's and doctorate, one foreign language, thesis/dissertation. *Entrance requirements:* For master's, GRE General Test, GRE Subject Test; for doctorate, GRE Subject Test, BA or BS in chemistry or physics. *Application deadline:* For fall admission, 3/1 for domestic students. Applications are processed on a rolling basis. Application fee: $0. *Expenses:* Tuition: Full-time $24,732. One-time fee: $20 full-time. *Faculty research:* Spectroscopy, photochemistry, reactive collisions, surface physics, quantum theory. *Unit head:* Dr. Suzanne O'Connell, Chair, 860-685-2248, E-mail: soconnell@wesleyan.edu. *Application contact:* Gloria Augeri, Administrative Assistant, 860-685-2244, E-mail: gaugeri@wesleyan.edu.

See Close-Up on page 177.

West Virginia University, Eberly College of Arts and Sciences, Department of Physics, Morgantown, WV 26506. Offers applied physics (MS, PhD); astrophysics (MS, PhD); chemical physics (MS, PhD); condensed matter physics (MS, PhD); elementary particle physics (MS, PhD); materials physics (MS, PhD); plasma physics (MS, PhD); solid state physics (MS, PhD); statistical physics (MS, PhD); theoretical physics (MS, PhD). *Faculty:* 17 full-time (3 women), 1 part-time/adjunct (0 women). *Students:* 44 full-time (10 women), 4 part-time (1 woman), 32 international. Average age 28. 60 applicants, 20% accepted. In 2005, 2 master's, 3 doctorates awarded. Terminal master's awarded for partial completion of doctoral program. *Degree requirements:* For master's, thesis or alternative, qualifying exam; for doctorate, thesis/dissertation, qualifying exam. *Entrance requirements:* For master's and doctorate, GRE General Test, GRE Subject Test, minimum GPA of 3.0. Additional exam requirements/recommendations for international students: Required—TOEFL. *Application deadline:* For fall admission, 2/15 for domestic students. Applications are processed on a rolling basis. Application fee: $50. *Expenses:* Tuition, state resident: full-time $4,582; part-time $258 per credit hour. Tuition, nonresident: full-time $1,382; part-time $741 per credit hour. *Financial support:* In 2005–06, 30 research assistantships with full and partial tuition reimbursements (averaging $18,000 per year), 10 teaching assistantships with full and partial tuition reimbursements (averaging $16,000 per year) were awarded; fellowships, Federal Work-Study, institutionally sponsored loans, and tuition waivers (full and partial) also available. Financial award application deadline: 2/1; financial award applicants required to submit FAFSA. *Faculty research:* Experimental and theoretical condensed-matter, plasma, high-energy theory, nonlinear dynamics, space physics. Total annual research expenditures: $3.3 million. *Unit head:* Dr. Earl E. Scime, Chair, 304-293-3422 Ext. 1437, Fax: 304-293-5732, E-mail: earl.scime@mail.wvu.edu.

Condensed Matter Physics

Cleveland State University, College of Graduate Studies, College of Science, Department of Physics, Cleveland, OH 44115. Offers applied optics (MS); condensed matter physics (MS); medical physics (MS). Part-time and evening/weekend programs available. *Faculty:* 5 full-time (1 woman), 3 part-time/adjunct (0 women). *Students:* 3 full-time (1 woman), 16 part-time (6 women); includes 4 minority (2 African Americans, 1 American Indian/Alaska Native, 1 Hispanic American), 1 international. Average age 31. 14 applicants, 43% accepted, 5 enrolled. In 2005, 3 degrees awarded. *Degree requirements:* For master's, exit project. *Entrance requirements:* For master's, undergraduate degree in engineering, physics, chemistry or mathematics. Additional exam requirements/recommendations for international students: Required—TOEFL (minimum score 525 paper-based; 197 computer-based), GRE. *Application deadline:* For fall admission, 7/15 priority date for domestic students, 7/15 priority date for international students. Applications are processed on a rolling basis. Application fee: $30. Electronic applications accepted. *Expenses:* Tuition, state resident: full-time $10,700. Tuition, nonresident: full-time $14,628. Tuition and fees vary according to program. *Financial support:* In 2005–06, 1 research assistantship with full and partial tuition reimbursement (averaging $5,666 per year) was awarded; fellowships with tuition reimbursements, teaching assistantships, tuition waivers (full) also available. *Faculty research:* Statistical mechanics of phase transitions, low-temperature and solid-state physics, superconductivity, theoretical light scattering. Total annual research expenditures: $350,000. *Unit head:* Dr. Miron Kaufman, Chairperson, 216-687-2436, Fax: 216-523-7268, E-mail: m.kaufman@csuohio.edu. *Application contact:* Dr. James A. Lock, Director, 216-687-2425, Fax: 216-523-7268, E-mail: j.lock@csuohio.edu.

Emory University, Graduate School of Arts and Sciences, Department of Physics, Atlanta, GA 30322-1100. Offers biophysics (PhD); condensed matter physics (PhD); non-linear physics (PhD); radiological physics (PhD); soft condensed matter physics (PhD); solid-state physics (PhD); statistical physics (PhD). *Faculty:* 19 full-time (2 women). *Students:* 17 full-time (4 women); includes 2 minority (1 African American, 1 Hispanic American), 13 international. Average age 24. 40 applicants, 13% accepted. *Degree requirements:* For doctorate, thesis/dissertation, qualifier proposal (PhD qualification). *Entrance requirements:* For doctorate, GRE General Test, minimum GPA of 3.0. Additional exam requirements/recommendations for international students: Required—TOEFL (minimum score 600 paper-based). *Application deadline:* For fall admission, 1/3 priority date for domestic students, 1/3 priority date for international students. Application fee: $50. Electronic applications accepted. *Expenses:* Tuition: Full-time $14,400. Required fees: $217. *Financial support:* In 2005–06, 17 students received support, including 6 fellowships (averaging $20,000 per year); institutionally sponsored loans, scholarships/grants, health care benefits, and tuition waivers (full) also available. Financial award application deadline: 1/3; financial award applicants required to submit FAFSA. *Faculty research:* Experimental studies of the structure and function of metalloproteins, soft condensed matter, granular materials, biophotonics and fluorescence correlation spectroscopy, single molecule studies of DNA-protein systems. Total annual research expenditures: $1.5 million. *Unit head:* Dr. Raymond DuVarney, Chair, 404-727-4296, Fax: 404-727-0873, E-mail: phsrcd@physics.emory.edu. *Application contact:* Dr. Kurt Warncke, Director of Graduate Studies, 404-727-2975, Fax: 404-727-0873, E-mail: kwarncke@physics.emory.edu.

Iowa State University of Science and Technology, Graduate College, College of Liberal Arts and Sciences, Department of Physics and Astronomy, Ames, IA 50011. Offers applied physics (MS, PhD); astrophysics (MS, PhD); condensed matter physics (MS, PhD); high energy physics (MS, PhD); nuclear physics (MS, PhD); physics (MS, PhD). Part-time programs available. *Faculty:* 44 full-time, 3 part-time/adjunct. *Students:* 79 full-time (14 women), 6 part-time; includes 1 minority (Asian American or Pacific Islander), 56 international. 161 applicants, 34% accepted, 23 enrolled. In 2005, 5 master's, 7 doctorates awarded. Terminal master's awarded for partial completion of doctoral program. *Degree requirements:* For master's, thesis (for some programs); for doctorate, thesis/dissertation. *Entrance requirements:* For master's and doctorate, GRE General Test, GRE Subject Test (physics). Additional exam requirements/recommendations for international students: Required—TOEFL (paper score 550; computer score 213) or IELTS (score 6.5). *Application deadline:* For fall admission, 2/15 priority date for domestic students, 2/15 priority date for international students; for spring admission, 10/15 for domestic students, 10/15 for international students. Applications are processed on a rolling basis. Application fee: $30 ($70 for international students). Electronic applications accepted. *Expenses:* Tuition, state resident: full-time $6,410. Tuition, nonresident: full-time $16,422. Tuition and fees vary according to program. *Financial support:* In 2005–06, 48 research assistantships with full tuition reimbursements (averaging $14,626 per year), 30 teaching assistantships with full tuition reimbursements (averaging $14,533 per year) were awarded; fellowships, Federal Work-Study, institutionally sponsored loans, scholarships/grants, health care benefits, and unspecified assistantships also available. Support available to part-time students. Financial award application deadline: 2/15. *Faculty research:* Condensed-matter physics, including superconductivity and new materials; high-energy and nuclear physics; astronomy and astrophysics; atmospheric and environmental physics. Total annual research expenditures: $8.8 million. *Unit head:* Dr. Eli Rosenberg, Chair, 515-294-5441, Fax: 515-294-6027, E-mail: phys_astro@iastate.edu. *Application contact:* Dr. Steven Kawaler, Director of Graduate Education, 515-294-9728, E-mail: phys_astro@iastate.edu.

Memorial University of Newfoundland, School of Graduate Studies, Department of Physics and Physical Oceanography, St. John's, NL A1C 5S7, Canada. Offers atomic and molecular physics (M Sc, PhD); condensed matter physics (M Sc, PhD); physical oceanography (M Sc, PhD); physics (M Sc). Part-time programs available. *Students:* 24 full-time (10 women), 1 part-time, 16 international. 18 applicants, 56% accepted, 7 enrolled. In 2005, 9 master's, 2 doctorates awarded. *Degree requirements:* For master's, thesis, seminar presentation on thesis topic; for doctorate, thesis/dissertation, oral defense of thesis, comprehensive exam. *Entrance requirements:* For master's, honors B Sc or equivalent; for doctorate, M Sc or equivalent. *Application deadline:* Applications are processed on a rolling basis. Application fee: $40 Canadian dollars. Electronic applications accepted. *Expenses:* Tuition: Part-time $733 per term. Tuition and fees vary according to degree level and program. *Financial support:* Fellowships, research assistantships, teaching assistantships available. *Faculty research:* Experiment and theory in atomic and molecular physics, condensed matter physics, physical oceanography, theoretical geophysics and applied nuclear physics. *Unit head:* Dr. John Whitehead, Head, 709-737-8737, Fax: 709-737-8739, E-mail: johnw@physics.mun.ca. *Application contact:* Dr. Mike Morrow, Graduate Officer, 709-737-4361, Fax: 709-737-8739, E-mail: myke@physics.mun.ca.

Rutgers, The State University of New Jersey, New Brunswick/Piscataway, Graduate School, Program in Physics and Astronomy, New Brunswick, NJ 08901-1281. Offers astronomy (MS, PhD); biophysics (PhD); condensed matter physics (MS, PhD); elementary particle physics (MS, PhD); intermediate energy nuclear physics (MS); nuclear physics (MS, PhD); physics (MST); surface science (PhD); theoretical physics (MS, PhD). Part-time programs available. *Faculty:* 92 full-time. *Students:* 105 full-time (18 women), 5 part-time (3 women); includes 6 minority (2 African Americans, 4 Asian Americans or Pacific Islanders), 54 international. Average age 27. 258 applicants, 19% accepted, 23 enrolled. In 2005, 10 master's, 14 doctorates awarded. Terminal master's awarded for partial completion of doctoral program. *Degree requirements:* For master's, thesis or alternative, comprehensive exam; for doctorate, thesis/dissertation, comprehensive exam. *Entrance requirements:* For master's and doctorate, GRE General Test, GRE Subject Test. Additional exam requirements/recommendations for international students: Required—TOEFL, TSE. *Application deadline:* For fall admission, 1/2 priority date for domestic students, 1/2 priority date for international students; for spring admission, 11/1 for domestic students, 11/1 for international students. Applications are processed on a rolling basis. Application fee: $50. Electronic applications accepted. *Expenses:* Tuition, state resident: full-time $10,440; part-time $435 per credit. Tuition, nonresident: full-time $15,520; part-time $647 per credit. Required fees: $129 per credit. Tuition and fees vary according to program. *Financial support:* In 2005–06, 19 fellowships with full tuition reimbursements (averaging $21,000 per year), 40 research assistantships with full tuition reimbursements (averaging $18,088 per year), 36 teaching assistantships with full tuition reimbursements (averaging $18,088 per year) were awarded; health care benefits and unspecified assistantships also available. Financial award application deadline: 1/2; financial award applicants required to submit FAFSA. *Faculty research:* Astronomy, high energy, condensed matter, surface, nuclear physics. Total annual research expenditures: $7.8 million. *Unit head:* Ronald Ransome, Director, 732-445-2516, Fax: 732-445-4343, E-mail: ransome@physics.rutgers.edu. *Application contact:* Shirley Hinds, Administrative Assistant, 732-445-2502, Fax: 732-445-4343, E-mail: graduate@physics.rutgers.edu.

University of Alberta, Faculty of Graduate Studies and Research, Department of Physics, Edmonton, AB T6G 2E1, Canada. Offers astrophysics (M Sc, PhD); condensed matter (M Sc, PhD); geophysics (M Sc, PhD); medical physics (M Sc, PhD); subatomic physics (M Sc, PhD). *Faculty:* 36 full-time (3 women), 7 part-time/adjunct (0 women). *Students:* 56 full-time (6 women), 16 part-time (2 women), 25 international. 85 applicants, 35% accepted. In 2005, 7 master's, 10 doctorates awarded. *Degree requirements:* For master's and doctorate, thesis/dissertation. *Entrance requirements:* For master's and doctorate, minimum GPA of 7.0 on a 9.0 scale. Additional exam requirements/recommendations for international students: Required—TOEFL. *Application deadline:* For fall admission, 2/15 for domestic students. Applications are processed on a rolling basis. Tuition and fees charges are reported in Canadian dollars. *Expenses:* Tuition, state resident: part-time $562 Canadian dollars per term. Tuition, nonresident: full-time $3,375 Canadian dollars. Required fees: $573 Canadian dollars; $84 Canadian dollars per term. *Financial support:* In 2005–06, 45 students received support, including 6 fellowships with partial tuition reimbursements available, 40 teaching assistantships; research assistantships, career-related internships or fieldwork, institutionally sponsored loans, and scholarships/grants also available. Financial award application deadline: 2/15. *Faculty research:* Cosmology, astroparticle physics, high-intermediate energy, magnetism, superconductivity. Total annual research expenditures: $3.1 million. *Unit head:* Dr. R. Marchand, Associate Chair, 780-492-1072, E-mail: assoc-chair@phys.ualberta.ca. *Application contact:* Lynn Chandler, Program Advisor, 780-492-1072, Fax: 780-492-0714, E-mail: lynn@phys.ualberta.ca.

University of Victoria, Faculty of Graduate Studies, Faculty of Science, Department of Physics and Astronomy, Victoria, BC V8W 2Y2, Canada. Offers astronomy and astrophysics (M Sc, PhD); condensed matter physics (M Sc, PhD); experimental particle physics (M Sc, PhD); medical physics (M Sc, PhD); ocean physics (PhD); ocean physics and geophysics (M Sc); theoretical physics (M Sc, PhD). *Faculty:* 16 full-time (0 women), 13 part-time/adjunct (1 woman). *Students:* 45 full-time, 8 international. Average age 25. 66 applicants, 45% accepted, 10 enrolled. In 2005, 4 master's, 1 doctorate awarded. *Median time to degree:* Of those who began their doctoral program in fall 1997, 100% received their degree in 8 years or less. *Degree requirements:* For master's, thesis, registration; for doctorate, thesis/dissertation, Candidacy exam, comprehensive exam, registration. *Entrance requirements:* For master's and doctorate, GRE. Additional exam requirements/recommendations for international students: Required—TOEFL (minimum score 575 paper-based; 233 computer-based), IELT (minimum score 7). *Application deadline:* For fall admission, 5/31 priority date for domestic students, 12/15 priority date for international students. Applications are processed on a rolling basis. Application fee: $75 ($125 for international students). Electronic applications accepted. Tuition and fees charges are reported in Canadian dollars. *Expenses:* Tuition, area resident: Full-time $4,492 Canadian dollars; part-time $749 Canadian dollars per term. International tuition: $5,346 Canadian dollars full-time. Required fees: $4,492 Canadian dollars; $749 Canadian dollars per term. Tuition and fees vary according to course load, campus/location and program. *Financial support:* In 2005–06, 3 students received support; fellowships, research assistantships, teaching assistantships, career-related internships or fieldwork, institutionally sponsored loans, and awards available. Financial award application deadline: 2/15. *Faculty research:* Old

Condensed Matter Physics

University of Victoria (continued)
stellar populations; observational cosmology and large scale structure; cp violation; atlas. *Unit head:* Dr. J. Michael Ronev, Chair, 250-721-7698, Fax: 250-721-7715, E-mail: chair@phys.uvic.ca. *Application contact:* Dr. Chris J. Pritchet, Graduate Adviser, 250-721-7744, Fax: 250-721-7715, E-mail: pritchet@uvic.ca.

West Virginia University, Eberly College of Arts and Sciences, Department of Physics, Morgantown, WV 26506. Offers applied physics (MS, PhD); astrophysics (MS, PhD); chemical physics (MS, PhD); condensed matter physics (MS, PhD); elementary particle physics (MS, PhD); materials physics (MS, PhD); plasma physics (MS, PhD); solid state physics (MS, PhD); statistical physics (MS, PhD); theoretical physics (MS, PhD). *Faculty:* 17 full-time (3 women), 1 part-time/adjunct (0 women). *Students:* 44 full-time (10 women), 4 part-time (1 woman), 32 international. Average age 28. 60 applicants, 20% accepted. In 2005, 2 master's, 3 doctorates awarded. Terminal master's awarded for partial completion of doctoral program. *Degree requirements:* For master's, thesis or alternative, qualifying exam; for doctorate, thesis/dissertation, qualifying exam. *Entrance requirements:* For master's and doctorate, GRE General Test, GRE Subject Test, minimum GPA of 3.0. Additional exam requirements/recommendations for international students: Required—TOEFL. *Application deadline:* For fall admission, 2/15 for domestic students. Applications are processed on a rolling basis. Application fee: $50. *Expenses:* Tuition, state resident: full-time $4,582; part-time $258 per credit hour. Tuition, nonresident: full-time $1,382; part-time $741 per credit hour. *Financial support:* In 2005–06, 30 research assistantships with full and partial tuition reimbursements (averaging $18,000 per year), 10 teaching assistantships with full and partial tuition reimbursements (averaging $16,000 per year) were awarded; fellowships, Federal Work-Study, institutionally sponsored loans, and tuition waivers (full and partial) also available. Financial award application deadline: 2/1; financial award applicants required to submit FAFSA. *Faculty research:* Experimental and theoretical condensed-matter, plasma, high-energy theory, nonlinear dynamics, space physics. Total annual research expenditures: $3.3 million. *Unit head:* Dr. Earl E. Scime, Chair, 304-293-3422 Ext. 1437, Fax: 304-293-5732, E-mail: earl.scime@mail.wvu.edu.

Mathematical Physics

New Mexico Institute of Mining and Technology, Graduate Studies, Department of Physics, Socorro, NM 87801. Offers astrophysics (MS, PhD); atmospheric physics (MS, PhD); instrumentation (MS); mathematical physics (PhD). *Degree requirements:* For master's, thesis optional; for doctorate, thesis/dissertation. *Entrance requirements:* For master's, GRE General Test; for doctorate, GRE General Test, GRE Subject Test. Additional exam requirements/recommendations for international students: Required—TOEFL (minimum score 540 paper-based; 207 computer-based). *Faculty research:* Cloud physics, stellar and extra-galactic processes.

Princeton University, Graduate School, Department of Mathematics, Princeton, NJ 08544-1019. Offers applied and computational mathematics (PhD); mathematical physics (PhD); mathematics (PhD). *Degree requirements:* For doctorate, 2 foreign languages, thesis/dissertation. *Entrance requirements:* For doctorate, GRE General Test, GRE Subject Test. Additional exam requirements/recommendations for international students: Required—TOEFL (minimum score 600 paper-based; 250 computer-based). Electronic applications accepted.

Princeton University, Graduate School, Department of Physics, Princeton, NJ 08544-1019. Offers applied and computational mathematics (PhD); mathematical physics (PhD); physics (PhD); physics and chemical physics (PhD). *Degree requirements:* For doctorate, thesis/dissertation, qualifying exam. *Entrance requirements:* For doctorate, GRE General Test, GRE Subject Test. Additional exam requirements/recommendations for international students: Required—TOEFL (minimum score 600 paper-based; 250 computer-based). Electronic applications accepted.

University of Alberta, Faculty of Graduate Studies and Research, Department of Mathematical and Statistical Sciences, Edmonton, AB T6G 2E1, Canada. Offers applied mathematics (M Sc, PhD); biostatistics (M Sc); mathematical finance (M Sc, PhD); mathematical physics (M Sc, PhD); mathematics (M Sc, PhD); statistics (M Sc, PhD, Postgraduate Diploma). Part-time programs available. *Faculty:* 48 full-time (4 women). *Students:* 112 full-time (41 women), 5 part-time. Average age 24. 776 applicants, 5% accepted, 34 enrolled. In 2005, 12 master's, 10 doctorates awarded. Terminal master's awarded for partial completion of doctoral program. *Median time to degree:* Of those who began their doctoral program in fall 1997, 100% received their degree in 8 years or less. *Degree requirements:* For master's, thesis (for some programs); for doctorate, thesis/dissertation, comprehensive exam. *Entrance requirements:* Additional exam requirements/recommendations for international students: Required—TOEFL (minimum score 580 paper-based; 237 computer-based). *Application deadline:* For fall admission, 3/1 for domestic students, 2/1 for international students. Applications are processed on a rolling basis. Application fee: $0. Electronic applications accepted. Tuition and fees charges are reported in Canadian dollars. *Expenses:* Tuition, state resident: part-time $562 Canadian dollars per term. Tuition, nonresident: full-time $3,375 Canadian dollars. Required fees: $573 Canadian dollars; $84 Canadian dollars per term. *Financial support:* In 2005–06, 51 research assistantships, 88 teaching assistantships with full and partial tuition reimbursements were awarded; scholarships/grants also available. Financial award application deadline:5/1. *Faculty research:* Classical and functional analysis, algebra, differential equations, geometry. *Unit head:* Dr. Anthony To-Ming Lau, Chair, 403-492-5141, E-mail: tlau@math.ualberta.ca. *Application contact:* Dr. Yau Shu Wong, Associate Chair, Graduate Studies, 403-492-5799, Fax: 403-492-6828, E-mail: gradstudies@math.ualberta.ca.

University of Colorado at Boulder, Graduate School, College of Arts and Sciences, Department of Physics, Boulder, CO 80309. Offers chemical physics (PhD); geophysics (PhD); liquid crystal science and technology (PhD); mathematical physics (PhD); medical physics (PhD); optical sciences and engineering (PhD); physics (MS, PhD). *Faculty:* 39 full-time (3 women). *Students:* 157 full-time (37 women), 47 part-time (3 women); includes 10 minority (2 African Americans, 1 American Indian/Alaska Native, 4 Asian Americans or Pacific Islanders, 3 Hispanic Americans), 46 international. Average age 27. 81 applicants, 93% accepted. In 2005, 10 master's, 16 doctorates awarded. Terminal master's awarded for partial completion of doctoral program. *Degree requirements:* For master's, thesis or alternative, comprehensive exam; for doctorate, thesis/dissertation, comprehensive exam. *Entrance requirements:* For master's and doctorate, GRE General Test, GRE Subject Test, minimum undergraduate GPA of 3.0. Additional exam requirements/recommendations for international students: Required—TOEFL. *Application deadline:* For fall admission, 1/15 priority date for domestic students, 1/15 priority date for international students. Applications are processed on a rolling basis. Application fee: $50 ($60 for international students). Electronic applications accepted. *Financial support:* In 2005–06, fellowships with full tuition reimbursements (averaging $4,376 per year), research assistantships with full tuition reimbursements (averaging $15,185 per year), teaching assistantships with full tuition reimbursements (averaging $13,786 per year) were awarded; scholarships/grants also available. Financial award application deadline: 1/15. *Faculty research:* Atomic and molecular physics, nuclear physics, condensed matter, elementary particle physics, laser or optical physics. Total annual research expenditures: $24.1 million. *Unit head:* John Cumalat, Chair, 303-492-6952, Fax: 303-492-3352, E-mail: jcumalat@pizero.colorado.edu. *Application contact:* Graduate Program Assistant, 303-492-6954, Fax: 303-492-3352, E-mail: phys@bogart.colorado.edu.

Virginia Polytechnic Institute and State University, Graduate School, College of Science, Department of Mathematics, Blacksburg, VA 24061. Offers applied mathematics (MS, PhD); mathematical physics (MS, PhD); pure mathematics (MS, PhD). *Faculty:* 69 full-time (20 women). *Students:* 57 full-time (18 women), 5 part-time (2 women); includes 5 minority (2 African Americans, 1 Asian American or Pacific Islander, 2 Hispanic Americans), 27 international. Average age 28. 108 applicants, 20% accepted, 15 enrolled. In 2005, 20 master's, 8 doctorates awarded. *Entrance requirements:* For master's and doctorate, GRE. Additional exam requirements/recommendations for international students: Required—TOEFL (minimum score 550 paper-based; 213 computer-based). *Application deadline:* Applications are processed on a rolling basis. Application fee: $45. Electronic applications accepted. *Expenses:* Tuition, state resident: full-time $6,558; part-time $364 per credit. Tuition, nonresident: full-time $11,296; part-time $628 per credit. Required fees: $1,419; $468 per credit. $234 per term. *Financial support:* In 2005–06, 1 fellowship with full tuition reimbursement (averaging $1,481 per year), 1 research assistantship with full tuition reimbursement (averaging $16,689 per year), 44 teaching assistantships with full tuition reimbursements (averaging $13,902 per year) were awarded; career-related internships or fieldwork, Federal Work-Study, scholarships/grants, and unspecified assistantships also available. *Faculty research:* Differential equations, operator theory, numerical analysis, algebra, control theory. *Unit head:* Dr. John Rossi, Head, 540-231-6536, Fax: 540-231-5960, E-mail: rossi@math.vt.edu. *Application contact:* Hannah Swiger, Information Contact, 540-231-6537, Fax: 540-231-5960, E-mail: hsswiger@math.vt.edu.

Optical Sciences

Air Force Institute of Technology, Graduate School of Engineering and Management, Department of Electrical and Computer Engineering, Dayton, OH 45433-7765. Offers computer engineering (MS, PhD); computer systems/science (MS); electrical engineering (MS, PhD); electro-optics (MS, PhD). *Accreditation:* ABET (one or more programs are accredited). Part-time programs available. *Faculty:* 34 full-time (0 women), 4 part-time/adjunct (0 women). *Students:* 201 full-time (9 women), 8 part-time (1 woman). Average age 31. In 2005, 64 master's, 6 doctorates awarded. *Degree requirements:* For master's and doctorate, thesis/dissertation. *Entrance requirements:* For master's and doctorate, GRE General Test, minimum GPA of 3.0, U.S. citizenship. *Application deadline:* For fall admission, 3/1 for domestic students. Applications are processed on a rolling basis. Application fee: $0. *Financial support:* Fellowships with full and partial tuition reimbursements, research assistantships with full and partial tuition reimbursements available. Financial award application deadline: 3/15. *Faculty research:* Remote sensing, information survivability, microelectronics, computer networks, artificial intelligence. Total annual research expenditures: $3.4 million. *Unit head:* Dr. Nathaniel J. Davis, Head, 937-255-2024, Fax: 937-656-4055, E-mail: nathaniel.davis@afit.edu.

Air Force Institute of Technology, Graduate School of Engineering and Management, Department of Engineering Physics, Dayton, OH 45433-7765. Offers applied physics (MS, PhD); electro-optics (MS, PhD); materials science (PhD); nuclear engineering (MS, PhD); space physics (MS). Part-time programs available. *Faculty:* 20 full-time (1 woman). *Students:* 86 full-time (9 women), 4 part-time (all women). Average age 31. In 2005, 36 master's, 4 doctorates awarded. *Degree requirements:* For master's and doctorate, thesis/dissertation. *Entrance requirements:* For master's and doctorate, GRE General Test, minimum GPA of 3.0, U.S. citizenship. *Application deadline:* For spring admission, 3/1 for domestic students. Applications are processed on a rolling basis. Application fee: $0. *Financial support:* Fellowships with full tuition reimbursements, research assistantships with full and partial tuition reimbursements, scholarships/grants and unspecified assistantships available. Support available to part-time students. Financial award application deadline:3/15. *Faculty research:* High-energy lasers, space physics, nuclear weapon effects, semiconductor physics. Total annual research expenditures: $2.1 million. *Unit head:* Dr. Robert L. Hengehold, Head, 937-255-3636 Ext. 4502, Fax: 937-255-2921, E-mail: robert.hengehold@afit.edu. *Application contact:* Dr. David E. Weeks, Associate Professor of Physics, 937-255-3636 Ext. 4561, Fax: 937-255-2921, E-mail: david.weeks@afit.edu.

Alabama Agricultural and Mechanical University, School of Graduate Studies, School of Arts and Sciences, Department of Natural and Physical Sciences, Huntsville, AL 35811. Offers biology (MS); physics (MS, PhD), including applied physics (PhD), materials science (PhD), optics (PhD), physics (MS). Part-time and evening/weekend programs available. *Degree requirements:* For doctorate, thesis/dissertation. *Entrance requirements:* For master's and doctorate, GRE General Test. Electronic applications accepted.

Cleveland State University, College of Graduate Studies, College of Science, Department of Physics, Cleveland, OH 44115. Offers applied optics (MS); condensed matter physics (MS); medical physics (MS). Part-time and evening/weekend programs available. *Faculty:* 5 full-time (1 woman), 3 part-time/adjunct (0 women). *Students:* 3 full-time (1 woman), 16 part-time (6 women); includes 4 minority (2 African Americans, 1 American Indian/Alaska Native, 1 Hispanic American), 1 international. Average age 31. 14 applicants, 43% accepted, 5 enrolled. In 2005, 3 degrees awarded. *Degree requirements:* For master's, exit project. *Entrance requirements:* For master's, undergraduate degree in engineering, physics, chemistry or mathematics. Additional exam requirements/recommendations for international students: Required—TOEFL (minimum score 525 paper-based; 197 computer-based), GRE. *Application deadline:* For fall admission, 7/15 priority date for domestic students, 7/15 priority date for international students. Applications are processed on a rolling basis. Application fee: $30. Electronic applications accepted. *Expenses:* Tuition, state resident: full-time $10,700. Tuition, nonresident: full-time $14,628. Tuition and fees vary according to program. *Financial support:* In 2005–06, 1 research assistantship with full and partial tuition reimbursement (averaging $5,666 per year) was awarded; fellowships with tuition reimbursements, teaching assistantships, tuition waivers (full) also available. *Faculty research:* Statistical mechanics of phase transitions, low-temperature and solid-state physics, superconductivity, theoretical light scattering. Total annual research expenditures: $350,000. *Unit head:* Dr. Miron Kaufman, Chairperson, 216-687-2436, Fax: 216-523-7268, E-mail: m.kaufman@csuohio.edu. *Application contact:* Dr. James A. Lock, Director, 216-687-2425, Fax: 216-523-7268, E-mail: j.lock@csuohio.edu.

Columbia University, Fu Foundation School of Engineering and Applied Science, Department of Applied Physics and Applied Mathematics, New York, NY 10027. Offers applied physics (MS, PhD), including applied mathematics (PhD), optical physics (PhD), plasma physics (PhD), solid state physics (PhD); applied physics and applied mathematics (Eng Sc D); materials science and engineering (MS, Eng Sc D, PhD); medical physics (MS); minerals engineering and materials science (Eng Sc D, PhD, Engr). Part-time programs available. *Faculty:* 30 full-time (2 women), 15 part-time/adjunct (1 woman). *Students:* 94 full-time (26 women), 30 part-time (5 women); includes 17 minority (1 American Indian/Alaska Native, 12 Asian Americans or Pacific Islanders, 4 Hispanic Americans), 53 international. Average age 24. 294 applicants, 22% accepted, 38 enrolled. In 2005, 33 master's, 5 doctorates awarded. Terminal master's awarded for partial completion of doctoral program. *Degree requirements:* For master's, English Proficiency Test; for doctorate, thesis/dissertation, qualifying exam, English Proficiency Test; for Engr, English Profiency Test. *Entrance requirements:* For master's and doctorate, GRE General Test, GRE Subject Test (strongly recommended). Additional exam requirements/recommendations for international students: Required—TOEFL. *Application deadline:* For fall admission, 12/15 priority date for domestic students, 12/15 priority date for international students; for spring admission, 10/1 priority date for domestic students, 10/1 priority date for international students. Application fee: $45. Electronic applications accepted. *Expenses:* Tuition: Full-time $31,448. Tuition and fees vary according to course level, course load, campus/location and program. *Financial support:* In 2005–06, 8 fellowships with full tuition reimbursements, 56 research assistantships with full tuition reimbursements (averaging $24,750 per year), 16 teaching assistantships with full tuition reimbursements (averaging $24,750 per year) were awarded; Federal Work-Study and unspecified assistantships also available. Financial award application deadline: 12/15; financial award applicants required to submit FAFSA. *Faculty research:* Plasma physics, applied mathematics, solid-state and optical physics, atmospheric, oceanic and earth physics, materials science and engineering. *Unit head:* Dr. Michael E. Mauel, Professor and Chairman, 212-854-4457, E-mail: seasinfo.apam@columbia.edu. *Application contact:* Marlene Arbo, Department Administrator, 212-854-4458, Fax: 212-854-8257, E-mail: seasinfo.apam@columbia.edu.

See Close-Up on page 335.

École Polytechnique de Montréal, Graduate Programs, Department of Engineering Physics, Montréal, QC H3C 3A7, Canada. Offers optical engineering (M Eng, M Sc A, PhD); solid-state physics and engineering (M Eng, M Sc A, PhD). Part-time programs available. *Degree requirements:* For master's and doctorate, one foreign language, thesis/dissertation. *Entrance requirements:* For master's, minimum GPA of 2.75; for doctorate, minimum GPA of 3.0. *Faculty research:* Optics, thin-film physics, laser spectroscopy, plasmas, photonic devices.

Norfolk State University, School of Graduate Studies, School of Science and Technology, Program in Optical Engineering, Norfolk, VA 23504. Offers MS.

The Ohio State University, College of Optometry and Graduate School, Program in Vision Science, Columbus, OH 43210. Offers MS, PhD, OD/MS. *Faculty:* 19 full-time (6 women), 3 part-time/adjunct (0 women). *Students:* 36 full-time (17 women). In 2005, 9 master's, 2 doctorates awarded. *Degree requirements:* For master's and doctorate, thesis/dissertation. *Entrance requirements:* For master's and doctorate, GRE General Test. *Application deadline:* Applications are processed on a rolling basis. Application fee: $40 ($50 for international students). Electronic applications accepted. *Financial support:* In 2005–06, fellowships with tuition reimbursements (averaging $42,000 per year), research assistantships with full tuition reimbursements (averaging $12,000 per year), teaching assistantships with full tuition reimbursements (averaging $22,000 per year) were awarded; Federal Work-Study, scholarships/grants, traineeships, and unspecified assistantships also available. Financial award application deadline: 2/1; financial award applicants required to submit FAFSA. *Faculty research:* Ocular development, myopia, cornea, refractive error, quality of life. Total annual research expenditures: $9 million. *Unit head:* Dr. Karla Zadnik, Associate Dean for Research and Graduate Studies, 614-292-6603, Fax: 614-292-4705, E-mail: zadnik.4@osu.edu. *Application contact:* Dr. Ronald Jones, Graduate Chair, 614-292-3246, Fax: 614-292-7493, E-mail: grad@optometry.osu.edu.

Rochester Institute of Technology, Graduate Enrollment Services, College of Science, Center for Imaging Science, Rochester, NY 14623-5603. Offers MS, PhD. *Students:* 61 full-time (17 women), 35 part-time (6 women); includes 8 minority (2 African Americans, 3 Asian Americans or Pacific Islanders, 3 Hispanic Americans), 34 international. 57 applicants, 63% accepted, 26 enrolled. In 2005, 17 master's, 4 doctorates awarded. *Degree requirements:* For master's, thesis. *Entrance requirements:* For master's, GRE General Test, minimum GPA of 3.0. Additional exam requirements/recommendations for international students: Required—TOEFL. *Application deadline:* For fall admission, 2/15 for domestic students. Applications are processed on a rolling basis. Application fee: $50. Electronic applications accepted. *Expenses:* Tuition: Full-time $25,392; part-time $713 per credit. Required fees: $183; $61 per term. *Financial support:* Research assistantships, teaching assistantships available. *Unit head:* Dr. Stefi Baum, Director, 585-475-6220, E-mail: sabpci@rit.edu.

Rose-Hulman Institute of Technology, Faculty of Engineering and Applied Sciences, Department of Physics and Optical Engineering, Terre Haute, IN 47803-3999. Offers optical engineering (MS). Part-time programs available. *Faculty:* 14 full-time (2 women), 2 part-time/adjunct (0 women). *Students:* 8 full-time (3 women), 3 part-time; includes 1 minority (Asian American or Pacific Islander), 5 international. Average age 26. 8 applicants, 100% accepted, 7 enrolled. In 2005, 2 degrees awarded. *Degree requirements:* For master's, thesis. *Entrance requirements:* For master's, GRE, minimum GPA of 3.0. Additional exam requirements/recommendations for international students: Required—TOEFL (minimum score 550 paper-based; 210 computer-based). *Application deadline:* For fall admission, 2/1 for domestic students. Applications are processed on a rolling basis. Application fee: $0. *Expenses:* Tuition: Full-time $26,658; part-time $768 per credit hour. Part-time tuition and fees vary according to course load. *Financial support:* In 2005–06, 4 students received support; fellowships with full and partial tuition reimbursements available, research assistantships with full and partial tuition reimbursements available, teaching assistantships, institutionally sponsored loans, scholarships/grants, and tuition waivers (full and partial) available. Financial award application deadline: 2/1. *Faculty research:* Optical instrument design and prototypes, photorefractive phenomena, speckle techniques, holography, fiber-optic sensors. Total annual research expenditures: $659,969. *Unit head:* Dr. Charles Joenathan, Chairman, 812-877-8494, Fax: 812-877-8023, E-mail: charles.joenathan@rose-hulman.edu. *Application contact:* Dr. Daniel J. Moore, Associate Dean of the Faculty, 812-877-8110, Fax: 812-877-8061, E-mail: daniel.j.moore@rose-hulman.edu.

The University of Alabama in Huntsville, School of Graduate Studies, Interdisciplinary Program in Optical Science and Engineering, Huntsville, AL 35899. Offers PhD. Part-time programs available. *Faculty:* 1 full-time (0 women). *Students:* 20 full-time (7 women), 9 part-time; includes 1 minority (African American), 17 international. Average age 30. 9 applicants, 100% accepted, 8 enrolled. In 2005, 5 degrees awarded. *Degree requirements:* For doctorate, thesis/dissertation, written and oral exams, comprehensive exam, registration. *Entrance requirements:* For doctorate, GRE General Test (1600 preferred), minimum GPA of 3.0, BS in physical science or engineering. Additional exam requirements/recommendations for international students: Required—TOEFL (minimum score 550 paper-based; 213 computer-based). *Application deadline:* For fall admission, 5/30 for domestic students; for spring admission, 10/10 priority date for domestic students, 7/10 priority date for international students. Applications are processed on a rolling basis. Application fee: $40. *Expenses:* Tuition, state resident: full-time $5,866; part-time $244 per credit hour. Tuition, nonresident: full-time $12,060; part-time $500 per credit hour. Tuition and fees vary according to course load. *Financial support:* In 2005–06, 19 students received support, including 13 research assistantships with full and partial tuition reimbursements available (averaging $11,617 per year), 6 teaching assistantships with full and partial tuition reimbursements available (averaging $9,900 per year); fellowships with full and partial tuition reimbursements available, career-related internships or fieldwork, Federal Work-Study, institutionally sponsored loans, scholarships/grants, health care benefits, tuition waivers (full and partial), and unspecified assistantships also available. Support available to part-time students. Financial award application deadline: 4/1; financial award applicants required to submit FAFSA. *Faculty research:* Laser technology, holography, optical communications, medical image processing, computer design. *Unit head:* Dr. John Dimmock, Director, 256-824-2512, Fax: 256-824-6803, E-mail: dimmockj@uah.edu.

The University of Arizona, Optical Sciences Center, Tucson, AZ 85721. Offers MS, PhD. Part-time programs available. *Degree requirements:* For master's, thesis (for some programs), exam; for doctorate, thesis/dissertation, oral and written exams. *Entrance requirements:* For master's and doctorate, GRE General Test, GRE Subject Test. Additional exam requirements/recommendations for international students: Required—TOEFL. *Faculty research:* Medical optics, medical imaging, optical data storage, optical bistability, nonlinear optical effects.

University of Central Florida, College of Optics and Photonics, Orlando, FL 32816. Offers optics (MS, PhD). Part-time and evening/weekend programs available. *Faculty:* 21 full-time (1 woman), 4 part-time/adjunct (2 women). *Students:* 114 full-time (14 women), 10 part-time (3 women); includes 10 minority (1 African American, 5 Asian Americans or Pacific Islanders, 4 Hispanic Americans), 63 international. Average age 28. 122 applicants, 70% accepted, 23 enrolled. In 2005, 16 master's, 13 doctorates awarded. *Degree requirements:* For master's, thesis or alternative; for doctorate, thesis/dissertation, departmental qualifying exam, candidacy exam. *Entrance requirements:* For master's, GRE General Test, minimum GPA of 3.0 in last 60 hours; for doctorate, GRE General Test, minimum GPA of 3.5 in last 60 hours. Additional exam requirements/recommendations for international students: Required—TOEFL. *Application deadline:* For fall admission, 2/1 for domestic students; for spring admission, 12/1 for domestic students. Application fee: $30. Electronic applications accepted. *Expenses:* Tuition, state resident: full-time $5,788. Tuition, nonresident: full-time $21,927. Required fees: $241 per credit hour. *Financial support:* In 2005–06, fellowships with partial tuition reimbursements (averaging $4,700 per year), research assistantships with partial tuition reimbursements (averaging $11,300 per year) were awarded; teaching assistantships with partial tuition reimbursements, career-related internships or fieldwork, Federal Work-Study, institutionally sponsored loans, tuition waivers (partial), and unspecified assistantships also available. Financial award application deadline: 3/1; financial award applicants required to submit FAFSA. *Unit head:* Dr. Eric W. Van Stryland, Dean and Director, 407-823-6835, E-mail: cwvs@mail.creol.ucf.edu. *Application contact:* Dr. David J. Hagan, Coordinator, 407-823-6817, E-mail: dhagan@creol.ucf.edu.

University of Colorado at Boulder, Graduate School, College of Arts and Sciences, Department of Physics, Boulder, CO 80309. Offers chemical physics (PhD); geophysics (PhD); liquid crystal science and technology (PhD); mathematical physics (PhD); medical physics (PhD); optical sciences and engineering (PhD); physics (MS, PhD). *Faculty:* 39 full-time (3 women). *Students:* 157 full-time (37 women), 47 part-time (3 women); includes 10 minority (2 African Americans, 1 American Indian/Alaska Native, 4 Asian Americans or Pacific Islanders, 3 Hispanic Americans), 46 international. Average age 27. 81 applicants, 93% accepted. In 2005, 10 master's, 16 doctorates awarded. Terminal master's awarded for partial completion of doctoral program. *Degree requirements:* For master's, thesis or alternative, comprehensive exam; for doctorate, thesis/dissertation, comprehensive exam. *Entrance requirements:* For master's and doctorate, GRE General Test, GRE Subject Test, minimum undergraduate GPA of 3.0. Additional exam requirements/recommendations for international students: Required—TOEFL. *Application deadline:* For fall admission, 1/15 priority date for domestic students, 1/15 priority date for international students. Applications are processed on a rolling basis. Application fee: $50 ($60 for international students). Electronic applications accepted. *Financial support:* In 2005–06, fellowships with full tuition reimbursements (averaging $4,376 per year), research assistantships with full tuition reimbursements (averaging $15,185 per year), teaching assistantships with full tuition reimbursements (averaging $13,786 per year) were awarded; scholarships/grants also available. Financial award application deadline: 1/15. *Faculty research:* Atomic and molecular physics, nuclear physics, condensed matter, elementary particle physics, laser or optical physics. Total annual research expenditures: $24.1 million. *Unit head:* John Cumalat, Chair, 303-492-6952, Fax: 303-492-3352, E-mail: jcumalat@pizero.colorado.edu. *Application contact:* Graduate Program Assistant, 303-492-6954, Fax: 303-492-3352, E-mail: phys@bogart.colorado.edu.

University of Dayton, Graduate School, School of Engineering, Program in Electro-Optics, Dayton, OH 45469-1300. Offers MSEO, PhD. Part-time and evening/weekend programs available. *Faculty:* 4 full-time (0 women), 14 part-time/adjunct (0 women). *Students:* 21 full-time (4 women), 8 part-time (1 woman); includes 3 minority (2 Asian Americans or Pacific Islanders, 1 Hispanic American), 10 international. Average age 24. 48 applicants, 58% accepted, 4 enrolled. In 2005, 3 master's, 2 doctorates awarded. *Degree requirements:* For master's, thesis; for doctorate, thesis/dissertation, departmental qualifying exam. *Entrance requirements:* Additional exam requirements/recommendations for international students: Required—TOEFL (minimum score 550 paper-based; 213 computer-based). *Application deadline:* For fall admission, 8/1 for domestic students, 3/1 for international students. Applications are processed on a rolling basis. Application fee: $0. Electronic applications accepted. *Expenses:* Tuition: Part-time $567 per credit hour. Required fees: $25 per term. Tuition and fees vary according to degree level and program. *Financial support:* In 2005–06, 22 students received support, including 23 research assistantships with full tuition reimbursements available (averaging $10,365 per year), 4 teaching assistantships with full tuition reimbursements available (averaging $7,025 per year); fellowships, career-related internships or fieldwork, institutionally sponsored loans, health care benefits, and unspecified assistantships also available. Financial award application deadline: 2/1; financial award applicants required to submit FAFSA. *Faculty research:* Fiber optics, optical materials, computational optics, holography, laser diagnostics. Total annual research expenditures: $485,840. *Unit head:* Dr. Joseph W. Haus, Director, 937-229-2797, Fax: 937-229-2097, E-mail: jhaus@notes.udayton.edu. *Application contact:* E. Eavers.

University of Maryland, Baltimore County, Graduate School, College of Natural Sciences and Mathematics, Department of Physics, Program in Applied Physics, Baltimore, MD 21250. Offers astrophysics (PhD); optics (MS, PhD); quantum optics (PhD); solid state physics (MS, PhD). *Expenses:* Tuition, state resident: part-time $395 per credit. Tuition, nonresident: part-time $652 per credit. Required fees: $82 per credit. Tuition and fees vary according to course load, program and reciprocity agreements. *Application contact:* Dr. Terrance L Worchesky, Director, 410-455-6779, Fax: 410-455-1072, E-mail: workchesk@umbc.edu.

University of Massachusetts Lowell, Graduate School, College of Arts and Sciences, Department of Physics and Applied Physics, Program in Applied Physics, Lowell, MA 01854-2881. Offers applied mechanics (PhD); applied physics (MS, PhD), including optical sciences (MS). Terminal master's awarded for partial completion of doctoral program. *Degree requirements:* For master's, thesis; for doctorate, 2 foreign languages, thesis/dissertation. *Entrance requirements:* For master's and doctorate, GRE General Test.

University of New Mexico, Graduate School, College of Arts and Sciences, Department of Physics and Astronomy, Albuquerque, NM 87131-2039. Offers biomedical physics (MS, PhD); optical science and engineering (MS); optical sciences and engineering (PhD); physics (MS, PhD). *Faculty:* 35 full-time (4 women), 11 part-time/adjunct (1 woman). *Students:* 89 full-time (16 women), 15 part-time (2 women); includes 4 minority (2 Asian Americans or Pacific Islanders, 2 Hispanic Americans), 50 international. Average age 29. 99 applicants, 37% accepted, 18 enrolled. In 2005, 12 master's, 9 doctorates awarded. *Degree requirements:* For master's, thesis (for some programs), comprehensive exam (for some programs); for doctorate, thesis/dissertation, comprehensive exam. *Entrance requirements:* Additional exam requirements/recommendations for international students: Required—TOEFL (minimum score 610 paper-based; 253 computer-based). *Application deadline:* For fall admission, 2/1 for

Optical Sciences

University of New Mexico (continued)

domestic students; for spring admission, 8/1 for domestic students. Application fee: $40. Electronic applications accepted. *Expenses:* Tuition, nonresident: full-time $3,388; part-time $238 per credit hour. Required fees: $385 per term. Tuition and fees vary according to course load and program. *Financial support:* In 2005–06, 15 students received support, including research assistantships with full tuition reimbursements available (averaging $20,000 per year), teaching assistantships with full tuition reimbursements available (averaging $13,500 per year); fellowships with full tuition reimbursements available, career-related internships or fieldwork, scholarships/grants, health care benefits, and unspecified assistantships also available. Financial award application deadline: 2/1; financial award applicants required to submit FAFSA. *Faculty research:* High-energy and particle physics, optical and laser sciences, condensed matter, nuclear and particle physics, surface physics, biomedical physics, quantum information, subatomic physics, nonlinear science, general relativity. Total annual research expenditures: $6.1 million. *Unit head:* Dr. Bernd Bassalleck, Chair, 505-277-1517, Fax: 505-277-1520, E-mail: bassek@unm.edu. *Application contact:* Mary De Witt, Program Advisement Coordinator, 505-277-1514, Fax: 505-277-1514, E-mail: mdewitt@unm.edu.

University of New Mexico, Graduate School, School of Engineering, Department of Electrical and Computer Engineering, Albuquerque, NM 87131-2039. Offers electrical engineering (MS); engineering (PhD); optical sciences (PhD). *Faculty:* 30 full-time (2 women), 18 part-time/adjunct (1 woman). *Students:* 165 full-time (35 women), 88 part-time (16 women); includes 31 minority (1 African American, 2 American Indian/Alaska Native, 8 Asian Americans or Pacific Islanders, 20 Hispanic Americans), 118 international. Average age 30. 291 applicants, 30% accepted, 56 enrolled. In 2005, 4 master's, 15 doctorates awarded. *Degree requirements:* For master's, thesis (for some programs); for doctorate, thesis/dissertation. *Entrance requirements:* For master's, GRE General Test, minimum GPA of 3.0; for doctorate, GRE General Test, minimum GPA of 3.5. *Application deadline:* For fall admission, 7/30 for domestic students; for spring admission, 11/30 for domestic students. Application fee: $40. Electronic applications accepted. *Expenses:* Tuition, nonresident: full-time $3,388; part-time $238 per credit hour. Required fees: $385 per term. Tuition and fees vary according to course load and program. *Financial support:* In 2005–06, 22 students received support, including fellowships (averaging $4,000 per year), research assistantships (averaging $9,625 per year), teaching assistantships (averaging $5,400 per year); scholarships/grants and unspecified assistantships also available. Financial award application deadline: 3/1; financial award applicants required to submit FAFSA. *Faculty research:* Applied electromagnetics, high performance computing, wireless communications, optoelectronics, control systems. Total annual research expenditures: $3.1 million. *Unit head:* Dr. Christos Christodoulou, Chair, 505-277-2436, Fax: 505-277-1439, E-mail: christos@eece.unm.edu. *Application contact:* Maryellen Missik, Graduate Coordinator, 505-277-2600, Fax: 505-277-1439, E-mail: maryellen@eece.unm.edu.

The University of North Carolina at Charlotte, Graduate School, College of Arts and Sciences, Department of Physics and Optical Science, Program in Optical Science and Engineering, Charlotte, NC 28223-0001. Offers MS, PhD. Part-time programs available. *Students:* 21 full-time (4 women), 7 part-time (1 woman); includes 3 minority (1 African American, 2 Asian Americans or Pacific Islanders), 17 international. Average age 30. 24 applicants, 88% accepted, 6 enrolled. In 2005, 1 master's awarded. *Degree requirements:* For master's and doctorate, thesis/dissertation. *Entrance requirements:* For master's, GRE, minimum GPA of 3.0; for doctorate, GRE, minimum GPA of 3.2 in major, 3.0 overall. Additional exam requirements/recommendations for international students: Required—TOEFL (minimum score 557 paper-based; 220 computer-based). *Application deadline:* For fall admission, 7/15 for domestic students, 5/1 for international students; for spring admission, 11/15 for domestic students, 10/1 for international students. Applications are processed on a rolling basis. Application fee: $55. Electronic applications accepted. *Expenses:* Tuition, state resident: full-time $2,504; part-time $157 per credit. Tuition, nonresident: full-time $12,711; part-time $794 per credit. Required fees: $1,424; $89 per credit. Tuition and fees vary according to course load and program. *Financial support:* In 2005–06, 2 fellowships (averaging $20,000 per year), 33 research assistantships (averaging $7,136 per year) were awarded; teaching assistantships, career-related internships or fieldwork, Federal Work-Study, institutionally sponsored loans, scholarships/grants, and unspecified assistantships also available. Support available to part-time students. Financial award application deadline: 4/1; financial award applicants required to submit FAFSA. *Unit head:* Dr. Robert K. Tyson, Graduate Coordinator, 704-687-3399, Fax: 704-687-3160, E-mail: rtyson@email.uncc.edu. *Application contact:* Kathy B. Giddings, Director of Graduate Admissions, 704-687-3366, Fax: 704-687-3279, E-mail: gradadm@email.uncc.edu.

University of Rochester, The College, School of Engineering and Applied Sciences, Institute of Optics, Rochester, NY 14627-0250. Offers MS, PhD. Terminal master's awarded for partial completion of doctoral program. *Degree requirements:* For master's, comprehensive exam; for doctorate, thesis/dissertation, preliminary and qualifying exams. *Entrance requirements:* For master's and doctorate, GRE. Additional exam requirements/recommendations for international students: Required—TOEFL.

Photonics

Boston University, College of Engineering, Department of Electrical and Computer Engineering, Boston, MA 02215. Offers computer engineering (PhD); computer systems engineering (MS); electrical engineering (MS, PhD); photonics (MS); systems engineering (PhD). Part-time programs available. *Faculty:* 41 full-time (4 women). *Students:* 165 full-time (31 women), 21 part-time (2 women); includes 20 minority (4 African Americans, 12 Asian Americans or Pacific Islanders, 4 Hispanic Americans), 98 international. Average age 24. 566 applicants, 31% accepted, 67 enrolled. In 2005, 82 master's, 11 doctorates awarded. Terminal master's awarded for partial completion of doctoral program. *Degree requirements:* For master's, thesis optional; for doctorate, thesis/dissertation, comprehensive exam, registration. *Entrance requirements:* For master's and doctorate, GRE General Test. Additional exam requirements/recommendations for international students: Required—TOEFL (550 paper, 213 computer) or IELT (84). *Application deadline:* For fall admission, 4/1 for domestic students, 4/1 for international students; for spring admission, 10/1 for domestic students, 10/1 for international students. Applications are processed on a rolling basis. Application fee: $70. Electronic applications accepted. *Expenses:* Tuition: Full-time $31,530; part-time $985 per credit. Required fees: $316; $40 per semester. Tuition and fees vary according to course level and program. *Financial support:* In 2005–06, 139 students received support, including 6 fellowships with full tuition reimbursements available (averaging $24,000 per year), 76 research assistantships with full tuition reimbursements available (averaging $16,000 per year), 21 teaching assistantships with full tuition reimbursements available (averaging $16,000 per year); career-related internships or fieldwork, Federal Work-Study, institutionally sponsored loans, scholarships/grants, traineeships, and health care benefits also available. Financial award application deadline: 1/15; financial award applicants required to submit FAFSA. *Faculty research:* Signal and image processing, solid state materials, subsurface imaging, photonics, sensor networks. Total annual research expenditures: $17.4 million. *Unit head:* Dr. Bahaa Saleh, Chairman, 617-353-7176, Fax: 617-353-6440, E-mail: besaleh@bu.edu. *Application contact:* Cheryl Kelley, Director of Graduate Programs, 617-353-9760, Fax: 617-353-0259, E-mail: enggrad@bu.edu.

Lehigh University, College of Arts and Sciences, Department of Physics, Bethlehem, PA 18015-3094. Offers photonics (MS); physics (MS, PhD); polymer science (MS, PhD). Part-time programs available. *Faculty:* 18 full-time (0 women), 3 part-time/adjunct (0 women). *Students:* 39 full-time (7 women), 3 part-time (1 woman); includes 2 minority (1 African American, 1 Hispanic American), 17 international. 61 applicants, 23% accepted, 11 enrolled. In 2005, 8 master's, 7 doctorates awarded. Terminal master's awarded for partial completion of doctoral program. *Degree requirements:* For master's, research project; for doctorate, thesis/dissertation, exam. *Entrance requirements:* For doctorate, GRE General Test. Additional exam requirements/recommendations for international students: Required—TOEFL (minimum score 600 paper-based; 235 computer-based); Recommended—TSE. *Application deadline:* For fall admission, 7/15 for domestic students; for spring admission, 1/15 priority date for domestic students. Applications are processed on a rolling basis. Application fee: $60. Electronic applications accepted. *Financial support:* In 2005–06, 6 fellowships with tuition reimbursements (averaging $18,240 per year), 8 research assistantships with tuition reimbursements (averaging $18,240 per year), 14 teaching assistantships with tuition reimbursements (averaging $18,240 per year) were awarded; Federal Work-Study and institutionally sponsored loans also available. Financial award application deadline:1/15. *Faculty research:* Condensed matter physics; atomic, molecular and optical physics; plasma physics; complex fluids; computational physics. Total annual research expenditures: $2.3 million. *Unit head:* Dr. Michael J. Stavola, Chair, 610-758-3903, Fax: 610-758-5730, E-mail: mjsa@lehigh.edu. *Application contact:* Dr. Volkmar Dierolf, Graduate Admissions Officer, 610-758-3915, Fax: 610-758-5730, E-mail: vod2@lehigh.edu.

Lehigh University, P.C. Rossin College of Engineering and Applied Science, Department of Electrical and Computer Engineering, Bethlehem, PA 18015-3094. Offers electrical engineering (M Eng, MS, PhD); photonics (MS); wireless network engineering (MS). Part-time programs available. *Faculty:* 21 full-time (3 women). *Students:* 86 full-time (18 women), 21 part-time (5 women); includes 3 minority (2 Asian Americans or Pacific Islanders, 1 Hispanic American), 83 international. Average age 25. 214 applicants, 71% accepted, 32 enrolled. In 2005, 21 master's, 7 doctorates awarded. *Degree requirements:* For master's, oral presentation of thesis, thesis optional; for doctorate, thesis/dissertation, qualifying, general, and oral exams. *Entrance requirements:* For master's, GRE General Test, minimum GPA of 3.0; for doctorate, GRE General Test, minimum GPA of 3.25. Additional exam requirements/recommendations for international students: Required—TOEFL (minimum score 550 paper-based; 213 computer-based). *Application deadline:* For fall admission, 4/1 for domestic students; for spring admission, 11/1 for domestic students. Applications are processed on a rolling basis. Application fee: $60. Electronic applications accepted. *Financial support:* In 2005–06, 3 fellowships with full tuition reimbursements (averaging $18,600 per year), 44 research assistantships with full and partial tuition reimbursements (averaging $18,300 per year), 8 teaching assistantships with full tuition reimbursements (averaging $18,300 per year) were awarded. Financial award application deadline: 1/15. *Unit head:* Dr. Filbert J. Bartoli, Interim Chair, 610-758-4069, Fax: 610-758-6279, E-mail: dmb4@lehigh.edu. *Application contact:* Brianne Clapp, Graduate Coordinator, 610-758-4072, Fax: 610-758-6279, E-mail: brc3@lehigh.edu.

Lehigh University, P.C. Rossin College of Engineering and Applied Science, Department of Materials Science and Engineering, Bethlehem, PA 18015-3094. Offers materials science and engineering (M Eng, MS, PhD); photonics (MS); polymer science/engineering (MS, PhD). Part-time programs available. *Faculty:* 14 full-time (1 woman), 1 part-time/adjunct (0 women). *Students:* 31 full-time (3 women), 9 part-time (2 women), 12 international. 57 applicants, 19% accepted, 2 enrolled. In 2005, 6 master's, 7 doctorates awarded. *Degree requirements:* For master's and doctorate, thesis/dissertation. *Entrance requirements:* For master's and doctorate, GRE General Test, minimum GPA of 3.0. Additional exam requirements/recommendations for international students: Required—TOEFL. *Application deadline:* For fall admission, 2/15 for domestic students; for spring admission, 10/1 priority date for domestic students. Applications are processed on a rolling basis. Application fee: $60. *Financial support:* In 2005–06, 5 fellowships with full and partial tuition reimbursements (averaging $20,160 per year), 29 research assistantships with full tuition reimbursements (averaging $19,680 per year), 8 teaching assistantships with full and partial tuition reimbursements (averaging $18,300 per year) were awarded; scholarships/grants also available. Financial award application deadline: 1/15. *Faculty research:* Metals, ceramics, crystals, polymers, fatigue crack propagation. Total annual research expenditures: $5.4 million. *Unit head:* Dr. G. Slade Cargill, Chairperson, 610-758-4207, Fax: 610-758-4244, E-mail: gsc3@lehigh.edu. *Application contact:* Maxine C. Mattie, Graduate Administrative Coordinator, 610-758-4222, Fax: 610-758-4244, E-mail: mcm1@lehigh.edu.

Oklahoma State University, College of Arts and Sciences, Department of Physics, Stillwater, OK 74078. Offers photonics (MS, PhD); physics (MS, PhD). *Faculty:* 35 full-time (7 women), 2 part-time/adjunct (0 women). *Students:* 13 full-time (1 woman), 28 part-time (4 women); includes 3 minority (1 American Indian/Alaska Native, 2 Asian Americans or Pacific Islanders), 21 international. Average age 29. 40 applicants, 48% accepted, 10 enrolled. In 2005, 5 master's, 4 doctorates awarded. *Degree requirements:* For master's, thesis, thesis or report; for doctorate, thesis/dissertation, oral defense of dissertation, preliminary exam, qualifying exam. *Entrance requirements:* Additional exam requirements/recommendations for international students: Required—TOEFL. *Application deadline:* For fall admission, 3/15 priority date for domestic students, 3/1 priority date for international students. Applications are processed on a rolling basis. Application fee: $40 ($75 for international students). Electronic applications accepted. *Expenses:* Tuition, state resident: full-time $4,253; part-time $139 per credit hour. Tuition, nonresident: full-time $12,569; part-time $485 per credit hour. Required fees: $43 per credit hour. One-time fee: $20 part-time. Tuition and fees vary according to course load and program. *Financial support:* In 2005–06, 27 research assistantships (averaging $16,754 per year), 31 teaching assistantships with partial tuition reimbursements (averaging $15,743 per year) were awarded; Federal Work-Study, traineeships, health care benefits, tuition waivers (partial), and unspecified assistantships also available. Support available to part-time students. Financial award application deadline: 3/1. *Faculty research:* Lasers and photonics, non-linear optical materials, turbulence, structure and function of biological membranes, particle theory. *Unit head:* Dr. James Wicksted, Head, 405-744-5796. *Application contact:* Dr. Paul A. Westhaus, Graduate Coordinator, 405-744-5815, E-mail: paul.westhaus@okstate.edu.

Oklahoma State University, Graduate College, Interdisciplinary Program in Photonics, Stillwater, OK 74078. Offers biophotonics (MS, PhD). *Degree requirements:* For master's, thesis or report; for doctorate, thesis/dissertation. *Entrance requirements:* For master's, bachelor's degree in physics, chemistry, electrical engineering, or a related field. *Application deadline:* For fall admission, 3/15 for domestic students. Applications are processed on a rolling basis. Application fee: $25 ($50 for international students). Electronic applications accepted. *Expenses:* Tuition, state resident: full-time $4,253; part-time $139 per credit hour. Tuition, nonresident: full-time $12,569; part-time $485 per credit hour. Required fees: $43 per credit hour. One-time fee: $20 part-time. Tuition and fees vary according to course load and program. *Financial support:* Traineeships available. Financial award application deadline: 3/15. *Faculty research:* Nanostructure quantum well semiconductor growth, characterizations for UV-blue photonics applications, interaction of light with biological materials at the tissue, cellular and molecular levels. *Unit head:* Dr. Paul A. Westhaus, Graduate Coordinator, 405-744-5815, E-mail: paul.westhaus@okstate.edu. *Application contact:* Information Contact, 405-744-5815, Fax: 405-744-6406, E-mail: physpaw@okstate.edu.

Princeton University, Center for Photonic and Optoelectronic Materials (POEM), Princeton, NJ 08544-1019. Offers PhD.

Stevens Institute of Technology, Graduate School, Charles V. Schaefer Jr. School of Engineering, Interdisciplinary Program in Microelectronics and Photonics, Hoboken, NJ 07030. Offers Certificate. *Expenses:* Tuition: Part-time $920 per credit hour. Tuition and fees vary according to program. *Unit head:* , Dr. George Korfiatis, Dean, Charles V. Schaefer Jr. School of Engineering, 201-216-5263.

University of Arkansas, Graduate School, Interdisciplinary Program in Microelectronics and Photonics, Fayetteville, AR 72701-1201. Offers MS, PhD. *Students:* 33 full-time (4 women), 20 part-time (7 women); includes 5 minority (4 African Americans, 1 Hispanic American), 31 international. 31 applicants, 48% accepted. In 2005, 7 master's, 8 doctorates awarded. *Degree requirements:* For doctorate, thesis/dissertation. Application fee: $40 ($50 for international students). *Financial support:* In 2005–06, 3 fellowships with tuition reimbursements, 9 research assistantships, 2 teaching assistantships were awarded. Financial award application deadline: 4/1; financial award applicants required to submit FAFSA. *Unit head:* Dr. Ken Vickers, Head, 479-575-2875, Fax: 479-575-4580, E-mail: vickers@uark.edu.

University of California, San Diego, Graduate Studies and Research, Department of Electrical and Computer Engineering, La Jolla, CA 92093. Offers applied ocean science (MS, PhD); applied physics (MS, PhD); communication theory and systems (MS, PhD); computer engineering (MS, PhD); electrical engineering (M Eng); electronic circuits and systems (MS, PhD); intelligent systems, robotics and control (MS, PhD); photonics (MS, PhD); signal and image processing (MS, PhD). MS only offered to students who have been admitted to the PhD program. *Entrance requirements:* For master's and doctorate, GRE General Test. Electronic applications accepted.

University of Central Florida, College of Optics and Photonics, Orlando, FL 32816. Offers optics (MS, PhD). Part-time and evening/weekend programs available. *Faculty:* 21 full-time (1 woman), 4 part-time/adjunct (2 women). *Students:* 114 full-time (14 women), 10 part-time (3 women); includes 10 minority (1 African American, 5 Asian Americans or Pacific Islanders, 4 Hispanic Americans), 63 international. Average age 28. 122 applicants, 70% accepted, 23 enrolled. In 2005, 16 master's, 13 doctorates awarded. *Degree requirements:* For master's, thesis or alternative; for doctorate, thesis/dissertation, departmental qualifying exam, candidacy exam. *Entrance requirements:* For master's, GRE General Test, minimum GPA of 3.0 in last 60 hours; for doctorate, GRE General Test, minimum GPA of 3.5 in last 60 hours. Additional exam requirements/recommendations for international students: Required—TOEFL. *Application deadline:* For fall admission, 2/1 for domestic students; for spring admission, 12/1 for domestic students. Application fee: $30. Electronic applications accepted. *Expenses:* Tuition, state resident: full-time $5,788. Tuition, nonresident: full-time $21,927. Required fees: $241 per credit hour. *Financial support:* In 2005–06, fellowships with partial tuition reimbursements (averaging $4,700 per year), research assistantships with partial tuition reimbursements (averaging $11,300 per year) were awarded; teaching assistantships with partial tuition reimbursements, career-related internships or fieldwork, Federal Work-Study, institutionally sponsored loans, tuition waivers (partial), and unspecified assistantships also available. Financial award application deadline: 3/1; financial award applicants required to submit FAFSA. *Unit head:* Dr. Eric W. Van Stryland, Dean and Director, 407-823-6835, E-mail: cwvs@mail.creol.ucf.edu. *Application contact:* Dr. David J. Hagan, Coordinator, 407-823-6817, E-mail: dhagan@creol.ucf.edu.

Physics

Alabama Agricultural and Mechanical University, School of Graduate Studies, School of Arts and Sciences, Department of Natural and Physical Sciences, Huntsville, AL 35811. Offers biology (MS); physics (MS, PhD), including applied physics (PhD), materials science (PhD), optics (PhD), physics (MS). Part-time and evening/weekend programs available. *Degree requirements:* For doctorate, thesis/dissertation. *Entrance requirements:* For master's and doctorate, GRE General Test. Electronic applications accepted.

American University, College of Arts and Sciences, Department of Computer Science, Audio Technology, and Physics, Program in Applied Science, Washington, DC 20016-8001. Offers MS. Part-time and evening/weekend programs available. *Students:* Average age 39. *Application deadline:* For fall admission, 2/1 for domestic students; for spring admission, 10/1 for domestic students. Applications are processed on a rolling basis. Application fee: $50. *Expenses:* Tuition: Full-time $17,802; part-time $989 per credit. Required fees: $380. *Financial support:* Fellowships with full tuition reimbursements, teaching assistantships, career-related internships or fieldwork, Federal Work-Study, institutionally sponsored loans, and unspecified assistantships available. Financial award application deadline: 2/1. *Faculty research:* Artificial intelligence, database systems, software engineering, expert systems.

American University of Beirut, Graduate Programs, Faculty of Arts and Sciences, Beirut, Lebanon. Offers anthropology (MA); Arabic language and literature (MA); archaeology (MA); biology (MS); business administration (MBA); chemistry (MS); computer science (MS); economics (MA); education (MA); English language (MA); English literature (MA); environmental policy planning (MSES); finance and banking (MFB); financial economics (MFE); geology (MS); history (MA); mathematics (MS); Middle Eastern studies (MA); philosophy (MA); physics (MS); political studies (MA); psychology (MA); public administration (MA); sociology (MA). *Degree requirements:* For master's, one foreign language, thesis (for some programs), comprehensive exam, registration. *Entrance requirements:* For master's, GRE, letter of recommendation.

Arizona State University, Division of Graduate Studies, College of Liberal Arts and Sciences, Division of Natural Sciences and Mathematics, Department of Physics and Astronomy, Tempe, AZ 85287. Offers MNS, MS, PhD. *Degree requirements:* For master's, thesis, oral and written exams; for doctorate, thesis/dissertation. *Entrance requirements:* For master's and doctorate, GRE.

Auburn University, Graduate School, College of Sciences and Mathematics, Department of Physics, Auburn University, AL 36849. Offers MS, PhD. Part-time programs available. *Faculty:* 21 full-time (1 woman). *Students:* 21 full-time (6 women), 15 part-time; includes 1 minority (Hispanic American), 23 international. 20 applicants, 85% accepted, 8 enrolled. In 2005, 3 master's, 2 doctorates awarded. *Degree requirements:* For doctorate, thesis/dissertation, oral and written exams. *Entrance requirements:* For master's and doctorate, GRE General Test. *Application deadline:* For fall admission, 7/7 for domestic students; for spring admission, 11/24 for domestic students. Applications are processed on a rolling basis. Application fee: $25 ($50 for international students). Electronic applications accepted. *Financial support:* Research assistantships, teaching assistantships, career-related internships or fieldwork and Federal Work-Study available. Support available to part-time students. Financial award application deadline: 3/15. *Faculty research:* Atomic/radiative physics, plasma physics, condensed matter physics, space physics, nonlinear dynamics. *Unit head:* Dr. Joe D. Perez, Head, 334-844-4264. *Application contact:* Dr. Stephen L. McFarland, Acting Dean of the Graduate School, 334-844-4700.

See Close-Up on page 325.

Ball State University, Graduate School, College of Sciences and Humanities, Department of Physics and Astronomy, Program in Physics, Muncie, IN 47306-1099. Offers MA, MS. *Faculty:* 14. *Students:* 7 full-time (1 woman), 8 part-time (2 women), 4 international. Average age 26. 13 applicants, 77% accepted, 5 enrolled. In 2005, 10 degrees awarded. *Entrance requirements:* For master's, GRE General Test. Application fee: $25 ($35 for international students). *Expenses:* Tuition, state resident: full-time $6,246. Tuition, nonresident: full-time $16,006. *Financial support:* Research assistantships with full tuition reimbursements, teaching assistantships with full tuition reimbursements available. Financial award application deadline: 3/1. *Faculty research:* Solar energy, particle physics, atomic spectroscopy.

Baylor University, Graduate School, College of Arts and Sciences, Department of Physics, Waco, TX 76798. Offers MA, MS, PhD. *Students:* 21 full-time (4 women), 1 part-time; includes 1 minority (Hispanic American), 14 international. In 2005, 3 master's, 3 doctorates awarded. *Degree requirements:* For master's, thesis or alternative; for doctorate, one foreign language, thesis/dissertation. *Entrance requirements:* For master's and doctorate, GRE General Test. *Application deadline:* Applications are processed on a rolling basis. Application fee: $25. *Financial support:* Fellowships, teaching assistantships, Federal Work-Study and institutionally sponsored loans available. *Unit head:* Dr. Dwight Russell, Graduate Program Director, 254-710-3938, Fax: 254-710-5083, E-mail: dwight_russell@baylor.edu. *Application contact:* Suzanne Keener, Administrative Assistant, 254-710-3588, Fax: 254-710-3870.

Boston College, Graduate School of Arts and Sciences, Department of Physics, Chestnut Hill, MA 02467-3800. Offers MS, PhD. *Students:* 42 full-time (5 women); includes 4 minority (all Asian Americans or Pacific Islanders), 30 international. 124 applicants, 6% accepted, 7 enrolled. In 2005, 6 doctorates awarded. Terminal master's awarded for partial completion of doctoral program. *Degree requirements:* For master's, thesis (for some programs); for doctorate, thesis/dissertation. *Entrance requirements:* For master's and doctorate, GRE General Test, GRE Subject Test. Additional exam requirements/recommendations for international students: Required—TOEFL (minimum score 550 paper-based; 213 computer-based). *Application deadline:* For fall admission, 1/15 for domestic students. Application fee: $70. Electronic applications accepted. *Financial support:* Fellowships with full tuition reimbursements, research assistantships with full tuition reimbursements, teaching assistantships with full tuition reimbursements, Federal Work-Study and scholarships/grants available. Support available to part-time students. Financial award application deadline: 3/1; financial award applicants required to submit FAFSA. *Faculty research:* Atmospheric/space physics, astrophysics, atomic and molecular physics, fusion and plasmas, solid-state physics. *Unit head:* Dr. Kevin Bedell, Chairperson, 617-552-3576, E-mail: kevin.bedell@bc.edu. *Application contact:* Dr. Rein Uritam, Graduate Program Director, 617-552-3576, E-mail: rein.uritam@bc.edu.

Boston University, Graduate School of Arts and Sciences, Department of Physics, Boston, MA 02215. Offers MA, PhD. *Students:* 109 full-time (17 women), 1 part-time; includes 6 minority (1 African American, 4 Asian Americans or Pacific Islanders, 1 Hispanic American), 62 international. Average age 27. 265 applicants, 29% accepted, 26 enrolled. In 2005, 8 master's, 13 doctorates awarded. Terminal master's awarded for partial completion of doctoral program. *Degree requirements:* For master's, one foreign language, thesis or alternative, comprehensive exam, registration; for doctorate, one foreign language, thesis/dissertation, comprehensive exam, registration. *Entrance requirements:* For master's and doctorate, GRE General Test, GRE Subject Test. Additional exam requirements/recommendations for international students: Required—TOEFL (minimum score 600 paper-based; 250 computer-based). *Application deadline:* For fall admission, 1/15 for domestic students, 1/15 for international students; for spring admission, 11/1 for domestic students, 11/1 for international students. Application fee: $60. *Expenses:* Tuition: Full-time $31,530; part-time $985 per credit. Required fees: $316; $40 per semester. Tuition and fees vary according to course level and program. *Financial support:* In 2005–06, 104 students received support, including 2 fellowships with full tuition reimbursements available (averaging $16,500 per year), 70 research assistantships (averaging $16,000 per year), 30 teaching assistantships with full tuition reimbursements available (averaging $16,000 per year); Federal Work-Study and scholarships/grants also available. Support available to part-time students. Financial award application deadline: 1/15; financial award applicants required to submit FAFSA. *Unit head:* Dr. Bennett Goldberg, Acting Chairman, 617-353-5789, Fax: 617-353-9393, E-mail: goldberg@bu.edu. *Application contact:* Mirtha M. Cabello, Administrative Coordinator, 617-353-2623, Fax: 617-353-9393, E-mail: cabello@bu.edu.

See Close-Up on page 327.

Bowling Green State University, Graduate College, College of Arts and Sciences, Department of Physics and Astronomy, Bowling Green, OH 43403. Offers physics (MAT, MS); physics and astronomy (MAT). *Faculty:* 7 full-time (0 women), 1 part-time/adjunct (0 women). *Students:* 8 full-time (1 woman), 10 part-time (7 women); includes 1 Asian American or Pacific Islander, 4 international. Average age 34. 25 applicants, 36% accepted, 3 enrolled. In 2005, 10 degrees awarded. *Degree requirements:* For master's, thesis or alternative. *Entrance requirements:* For master's, GRE General Test. Additional exam requirements/recommendations for international students: Required—TOEFL. Application fee: $30. Electronic applications accepted. *Financial support:* In 2005–06, 10 teaching assistantships with full tuition reimbursements (averaging $13,214 per year) were awarded; research assistantships with full tuition reimbursements, career-related internships or fieldwork, institutionally sponsored loans, and unspecified assistantships also available. Financial award applicants required to submit FAFSA. *Faculty research:* Computational physics, solid-state physics, materials science, theoretical physics. *Unit head:* Dr. John Laird, Chair, 419-372-7244. *Application contact:* Dr. Lewis Fulcher, Graduate Coordinator, 419-372-2635.

Brandeis University, Graduate School of Arts and Sciences, Department of Physics, Waltham, MA 02454-9110. Offers MS, PhD. Part-time programs available. *Faculty:* 11 full-time (1 woman). *Students:* 30 full-time (4 women); includes 10 minority (all Asian Americans or Pacific Islanders), 8 international. Average age 23. 104 applicants, 21% accepted, 7 enrolled. In 2005, 8 master's, 2 doctorates awarded. Terminal master's awarded for partial completion of doctoral program. *Degree requirements:* For master's, qualifying exam, 1 year in residence, 6 semester courses numbered above 160; for doctorate, thesis/dissertation, advanced exam, 9 semester courses above 160. *Entrance requirements:* For doctorate, GRE General Test, GRE Subject Test, resumé, 2 letters of recommendation (3rd suggested). Additional exam requirements/recommendations for international students: Required—TOEFL (minimum score 600 paper-based; 250 computer-based). *Application deadline:* For fall admission, 1/15 for domestic students. Application fee: $55. Electronic applications accepted. *Financial support:* In 2005–06, 17 students received support, including 15 fellowships with full tuition reimbursements available (averaging $20,000 per year), 15 research assistantships with full tuition reimbursements available (averaging $20,000 per year); scholarships/grants and tuition waivers (full) also available. Financial award application deadline: 1/15. *Faculty research:* Theoretical physics, experimental physics, astrophysics, computational neuroscience, condensed matter, high energy physics. Total annual research expenditures: $3.1 million. *Unit head:* Dr. Bulbul Chakraborty, Chair, 781-736-2843, Fax: 781-736-2915, E-mail: bulbul@brandeis.edu. *Application contact:* Chairman, Graduate Admissions Committee, 781-736-2870, Fax: 781-736-2915, E-mail: physics1@brandeis.edu.

Physics

Brigham Young University, Graduate Studies, College of Physical and Mathematical Sciences, Department of Physics and Astronomy, Provo, UT 84602-1001. Offers physics (MS, PhD); physics and astronomy (PhD). Part-time programs available. *Faculty:* 31 full-time (0 women). *Students:* 38 full-time (7 women), 3 part-time (1 woman); includes 3 minority (1 American Indian/Alaska Native, 2 Hispanic Americans), 9 international. Average age 28. 22 applicants, 68% accepted, 11 enrolled. In 2005, 14 master's, 1 doctorate awarded. Terminal master's awarded for partial completion of doctoral program. *Median time to degree:* Of those who began their doctoral program in fall 1997, 100% received their degree in 8 years or less. *Degree requirements:* For master's, thesis/dissertation, registration; for doctorate, thesis/dissertation, comprehensive exam, registration. *Entrance requirements:* For master's and doctorate, GRE Subject Test in physics, minimum GPA of 3.0 in last 60 hours. Additional exam requirements/recommendations for international students: Required—TOEFL (minimum score 550 paper-based; 213 computer-based). *Application deadline:* For fall admission, 1/15 priority date for domestic students, 1/15 priority date for international students. Application fee: $50. Electronic applications accepted. *Financial support:* In 2005–06, 2 fellowships with full tuition reimbursements (averaging $18,000 per year), 10 research assistantships with full tuition reimbursements (averaging $18,000 per year), 25 teaching assistantships with full tuition reimbursements (averaging $16,000 per year) were awarded; career-related internships or fieldwork, institutionally sponsored loans, and tuition waivers (partial) also available. Support available to part-time students. Financial award application deadline: 1/15. *Faculty research:* Acoustics; astrophysics; atomic, molecular, and optical physics; plasma; theoretical and mathematical physics. Total annual research expenditures:$994,000. *Unit head:* Dr. Scott D. Sommerfeldt, Chair, 801-422-2205, Fax: 801-422-0553, E-mail: scott_sommerfeldt@byu.edu. *Application contact:* Dr. Ross L. Spencer, Graduate Coordinator, 801-422-2341, Fax: 801-422-0553, E-mail: ross_spencer@byu.edu.

Brock University, Graduate Studies, Faculty of Mathematics and Science, Program in Physics, St. Catharines, ON L2S 3A1, Canada. Offers M Sc. Part-time programs available. *Faculty:* 12 full-time (2 women), 1 part-time/adjunct (0 women). *Students:* 9 full-time (3 women), 8 international. 10 applicants, 20% accepted. In 2005, 3 degrees awarded. *Degree requirements:* For master's, thesis. *Entrance requirements:* For master's, honors B Sc in physics. Additional exam requirements/recommendations for international students: Required—TOEFL. *Application deadline:* Applications are processed on a rolling basis. Application fee: $75. Electronic applications accepted. *Financial support:* Fellowships, research assistantships, teaching assistantships, career-related internships or fieldwork, scholarships/grants, and unspecified assistantships available. Support available to part-time students. *Faculty research:* Quantum physics, optical properties, non-crystalline materials, condensed matter physics, biophysics. *Unit head:* Graduate Program Director, 905-688-5550 Ext. 3877, Fax: 905-682-9020, E-mail: reedyk@brocku.ca. *Application contact:* Graduate Program Director, 905-688-5550 Ext. 3877, Fax: 905-682-9020, E-mail: reedyk@brocku.ca.

Brooklyn College of the City University of New York, Division of Graduate Studies, Department of Physics, Brooklyn, NY 11210-2889. Offers applied physics (MA); physics (MA, PhD). The department is a full participant in the PhD program; it offers a complete sequence of courses that are creditable toward the CUNY doctoral degree, and a wide range of research opportunities in fulfillment of the doctoral dissertation requirements for that degree. Part-time programs available. Terminal master's awarded for partial completion of doctoral program. *Degree requirements:* For master's, comprehensive exam. *Entrance requirements:* For master's, GRE, 2 letters of recommendation; for doctorate, GRE. Additional exam requirements/recommendations for international students: Required—TOEFL.

Brown University, Graduate School, Department of Physics, Providence, RI 02912. Offers Sc M, PhD. *Degree requirements:* For doctorate, thesis/dissertation, qualifying and oral exams.

Bryn Mawr College, Graduate School of Arts and Sciences, Department of Physics, Bryn Mawr, PA 19010-2899. Offers MA, PhD. *Students:* 2 full-time (0 women), 2 part-time. 4 applicants, 50% accepted, 2 enrolled. *Degree requirements:* For master's and doctorate, one foreign language, thesis/dissertation. *Entrance requirements:* For master's and doctorate, GRE General Test, GRE Subject Test. Additional exam requirements/recommendations for international students: Required—TOEFL (minimum score 600 paper-based; 250 computer-based). *Application deadline:* For fall admission, 1/13 for domestic students, 1/13 for international students. Application fee: $30. *Financial support:* In 2005–06, 3 teaching assistantships with partial tuition reimbursements were awarded; research assistantships with full tuition reimbursements, Federal Work-Study, scholarships/grants, tuition waivers (partial), and tuition awards also available. Support available to part-time students. Financial award application deadline: 1/13. *Unit head:* Dr. Liz McCormack, Chair, 610-526-5358. *Application contact:* Graduate School of Arts and Sciences, 610-526-5072.

California Institute of Technology, Division of Physics, Mathematics and Astronomy, Department of Physics, Pasadena, CA 91125-0001. Offers PhD. *Degree requirements:* For doctorate, thesis/dissertation, candidacy and final exams. *Entrance requirements:* For doctorate, GRE General Test, GRE Subject Test. Additional exam requirements/recommendations for international students: Required—TOEFL. *Faculty research:* High-energy physics, nuclear physics, condensed-matter physics, theoretical physics and astrophysics, gravity physics.

California State University, Fresno, Division of Graduate Studies, College of Science and Mathematics, Department of Physics, Fresno, CA 93740-8027. Offers MS. Part-time programs available. *Degree requirements:* For master's, thesis or alternative. *Entrance requirements:* For master's, GRE General Test, minimum GPA of 2.5. Additional exam requirements/recommendations for international students: Required—TOEFL. Electronic applications accepted. *Faculty research:* Energy, astronomy, silicon vertex detector, neuroimaging, particle physics.

California State University, Fullerton, Graduate Studies, College of Natural Science and Mathematics, Department of Physics, Fullerton, CA 92834-9480. Offers MA. *Students:* 2 full-time (0 women), 7 part-time (3 women); includes 4 minority (1 Asian American or Pacific Islander, 3 Hispanic Americans), 1 international. Average age 31. 8 applicants, 38% accepted, 1 enrolled. In 2005, 5 degrees awarded. Application fee: $55. *Expenses:* Tuition, nonresident: part-time $339 per unit. *Financial support:* Scholarships/grants available. Financial award application deadline: 3/1. *Unit head:* Dr. Roger Nanes, Chair, 714-278-3366.

California State University, Long Beach, Graduate Studies, College of Natural Sciences and Mathematics, Department of Physics and Astronomy, Long Beach, CA 90840. Offers metals physics (MS); physics (MS). Part-time programs available. *Faculty:* 13 full-time (1 woman). *Students:* 7 full-time (0 women), 18 part-time (4 women); includes 10 minority (2 African Americans, 4 Asian Americans or Pacific Islanders, 4 Hispanic Americans). Average age 31. 17 applicants, 82% accepted, 10 enrolled. In 2005, 3 degrees awarded. *Degree requirements:* For master's, comprehensive exam or thesis. *Application deadline:* For fall admission, 7/1 for domestic students; for spring admission, 12/1 for domestic students. Applications are processed on a rolling basis. Application fee: $55. Electronic applications accepted. *Expenses:* Tuition, nonresident: part-time $339 per semester hour. *Financial support:* Federal Work-Study, institutionally sponsored loans, and scholarships/grants available. Financial award application deadline: 3/2. *Faculty research:* Musical acoustics, modern optics, neutrino physics, quantum gravity, atomic physics. *Unit head:* Dr. Patrick Kenealy, Chair, 562-985-4924, Fax: 562-985-7924, E-mail: kenealyp@csulb.edu. *Application contact:* Information Contact, 562-985-4924, Fax: 562-985-7924.

California State University, Los Angeles, Graduate Studies, College of Natural and Social Sciences, Department of Physics and Astronomy, Los Angeles, CA 90032-8530. Offers physics (MS). Part-time and evening/weekend programs available. *Faculty:* 4 full-time (1 woman), 1 part-time/adjunct (0 women). *Students:* 3 full-time (0 women), 11 part-time; includes 8 minority (2 African Americans, 3 Asian Americans or Pacific Islanders, 3 Hispanic Americans). In 2005, 1 degree awarded. *Degree requirements:* For master's, comprehensive exam or thesis. *Entrance requirements:* Additional exam requirements/recommendations for international students: Required—TOEFL. *Application deadline:* For fall admission, 6/30 for domestic students; for spring admission, 2/1 for domestic students. Applications are processed on a rolling basis. Application fee: $55. *Financial support:* Federal Work-Study available. Support available to part-time students. Financial award application deadline: 3/1. *Faculty research:* Intermediate energy, nuclear physics, condensed-matter physics, biophysics. *Unit head:* Dr. Edward Rezayi, Chair, 323-343-2100, Fax: 323-343-2497.

California State University, Northridge, Graduate Studies, College of Science and Mathematics, Department of Physics and Astronomy, Northridge, CA 91330. Offers physics (MS). Part-time and evening/weekend programs available. *Degree requirements:* For master's, thesis optional. *Entrance requirements:* For master's, GRE General Test or minimum GPA of 3.0. Additional exam requirements/recommendations for international students: Required—TOEFL.

Carleton University, Faculty of Graduate Studies, Faculty of Science, Department of Physics, Ottawa, ON K1S 5B6, Canada. Offers M Sc, PhD. *Degree requirements:* For master's, seminar, thesis optional; for doctorate, thesis/dissertation, seminar, comprehensive exam. *Entrance requirements:* For master's, honors degree in science; for doctorate, M Sc. Additional exam requirements/recommendations for international students: Required—TOEFL. *Application deadline:* Applications are processed on a rolling basis. Application fee: $75 Canadian dollars. *Financial support:* Fellowships, research assistantships, teaching assistantships, institutionally sponsored loans, scholarships/grants, and unspecified assistantships available. *Faculty research:* Experimental and theoretical elementary particle physics, medical physics. *Unit head:* Paul Johns, Chair, 613-520-2600 Ext. 4317, Fax: 613-520-4061, E-mail: physics@carleton.ca. *Application contact:* Stephen Goafrey, Associate Chair, Graduate Studies, 613-520-2600 Ext. 4386, Fax: 613-520-4061, E-mail: grad_supervisor@physics.carleton.ca.

Carnegie Mellon University, Mellon College of Science, Department of Physics, Pittsburgh, PA 15213-3891. Offers PhD. *Degree requirements:* For doctorate, thesis/dissertation, qualifying exam. *Entrance requirements:* For doctorate, GRE General Test, GRE Subject Test. Additional exam requirements/recommendations for international students: Required—TOEFL. Electronic applications accepted. *Faculty research:* Astrophysics, condensed matter physics, biological physics, medium energy and nuclear physics, high-energy physics.

Case Western Reserve University, School of Graduate Studies, Department of Physics, Cleveland, OH 44106. Offers MS, PhD. Part-time programs available. Terminal master's awarded for partial completion of doctoral program. *Degree requirements:* For master's, exam; for doctorate, thesis/dissertation, qualifying exam, topical exam. *Entrance requirements:* Additional exam requirements/recommendations for international students: Required—TOEFL. *Faculty research:* Condensed-matter physics, imaging physics, nonlinear optics, high-energy physics, cosmology and astrophysics.

The Catholic University of America, School of Arts and Sciences, Department of Physics, Washington, DC 20064. Offers MS, PhD. Part-time programs available. *Faculty:* 10 full-time (0 women). *Students:* 9 full-time (2 women), 24 part-time (1 woman); includes 1 African American, 2 Hispanic Americans, 7 international. Average age 34. 15 applicants, 60% accepted, 4 enrolled. In 2005, 4 master's, 1 doctorate awarded. Terminal master's awarded for partial completion of doctoral program. *Degree requirements:* For master's, thesis or alternative, comprehensive exam; for doctorate, thesis/dissertation, comprehensive exam. *Entrance requirements:* For master's and doctorate, GRE General Test, 3 letters of recommendation. Additional exam requirements/recommendations for international students: Required—TOEFL (minimum score 580 paper-based; 237 computer-based). *Application deadline:* For fall admission, 2/1 for domestic students; for spring admission, 11/5 priority date for domestic students. Applications are processed on a rolling basis. Application fee: $55. Electronic applications accepted. *Expenses:* Tuition: Full-time $24,800; part-time $940 per credit. Required fees: $1,090; $285 per term. Part-time tuition and fees vary according to course load and program. *Financial support:* Fellowships, research assistantships, teaching assistantships, career-related internships or fieldwork, Federal Work-Study, scholarships/grants, tuition waivers (full and partial), and unspecified assistantships available. Support available to part-time students. Financial award application deadline: 2/1; financial award applicants required to submit FAFSA. *Faculty research:* Condensed-matter physics, intermediate-energy physics, astrophysics, biophysics. *Unit head:* Dr. Charles Montrose, Chair, 202-319-5347, Fax: 202-319-4448, E-mail: montrose@cua.edu. *Application contact:* Christine Mica, Director, University Admissions, 202-319-5305, Fax: 202-319-6533, E-mail: cua-admissions@cua.edu.

Central Connecticut State University, School of Graduate Studies, School of Arts and Sciences, Department of Physics and Earth Science, New Britain, CT 06050-4010. Offers earth science (MS); physics (MS). Part-time and evening/weekend programs available. *Faculty:* 11 full-time (3 women), 13 part-time/adjunct (2 women). *Students:* 7 full-time (3 women), 30 part-time (23 women); includes 2 minority (1 Asian American or Pacific Islander, 1 Hispanic American), 1 international. Average age 32. 14 applicants, 64% accepted, 4 enrolled. In 2005, 12 master's awarded. *Degree requirements:* For master's, thesis or alternative, comprehensive exam. *Entrance requirements:* For master's, minimum GPA of 2.7. Additional exam requirements/recommendations for international students: Required—TOEFL. *Application deadline:* For fall admission, 7/1 for domestic students; for spring admission, 12/1 for domestic students. Applications are processed on a rolling basis. Application fee: $50. Electronic applications accepted. *Expenses:* Tuition, area resident: Full-time $3,780. Tuition, state resident: full-time $5,670; part-time $362 per credit. Tuition, nonresident: full-time $10,530; part-time $362 per credit. Required fees: $3,064. One-time fee: $62 part-time. Tuition and fees vary according to degree level and program. *Financial support:* In 2005–06, 1 student received support, including 1 research assistantship; career-related internships or fieldwork, Federal Work-Study, scholarships/grants, and unspecified assistantships also available. Support available to part-time students. Financial award application deadline: 3/1; financial award applicants required to submit FAFSA. *Faculty research:* Elementary/secondary science education, particle and solid states, weather patterns, planetary studies. *Unit head:* Dr. Ali Antar, Chair, 860-832-2930.

Central Michigan University, College of Graduate Studies, College of Science and Technology, Department of Physics, Mount Pleasant, MI 48859. Offers MS. *Faculty:* 14 full-time (0 women). *Students:* 3 full-time (0 women), 10 part-time (3 women). Average age 28. In 2005, 1 degree awarded. *Degree requirements:* For master's, thesis or alternative, registration. *Entrance requirements:* For master's, GRE, bachelor's degree in physics, minimum GPA of 2.6. Additional exam requirements/recommendations for international students: Required—TOEFL. *Application deadline:* Applications are processed on a rolling basis. Application fee: $35 ($45 for international students). *Expenses:* Tuition, area resident: Part-time $325 per credit hour. Tuition, state resident: part-time $603 per credit hour. Tuition and fees vary according to degree level and reciprocity agreements. *Financial support:* In 2005–06, 6 research assistantships with tuition reimbursements, 5 teaching assistantships with tuition reimbursements were awarded; fellowships with tuition reimbursements, career-related internships or fieldwork and Federal Work-Study also available. Financial award application deadline: 3/7. *Faculty research:* Polymer physics, laser spectroscopy, observational astronomy, nuclear physics, thin films. *Unit head:* Dr. Stanley Hirschi, Chairperson, 989-774-3321, Fax: 989-774-2697, E-mail: stanley.hirschi@cmich.edu.

Christopher Newport University, Graduate Studies, Department of Physics, Computer Science, and Engineering, Newport News, VA 23606-2998. Offers applied physics and computer science (MS). Part-time and evening/weekend programs available. *Degree requirements:* For master's, thesis or alternative, comprehensive exam. *Entrance requirements:* For master's, GRE General Test, minimum GPA of 3.0. Electronic applications accepted. *Faculty research:* Advanced programming methodologies, experimental nuclear physics, computer architecture, semiconductor nanophysics, laser and optical fiber sensors.

City College of the City University of New York, Graduate School, College of Liberal Arts and Science, Division of Science, Department of Physics, New York, NY 10031-9198. Offers MA,

PhD. *Students:* 2 full-time (1 woman), 2 part-time; includes 3 minority (1 Asian American or Pacific Islander, 2 Hispanic Americans), 1 international. 8 applicants, 88% accepted, 1 enrolled. In 2005, 3 degrees awarded. Terminal master's awarded for partial completion of doctoral program. *Degree requirements:* For master's, comprehensive exam; for doctorate, thesis/dissertation. *Entrance requirements:* For master's and doctorate, GRE. Additional exam requirements/recommendations for international students: Required—TOEFL (minimum score 500 paper-based; 173 computer-based). *Application deadline:* For fall admission, 5/1 for domestic students; for spring admission, 11/1 for domestic students. Application fee: $125. *Financial support:* Fellowships available. *Unit head:* Michael Lubell, Chair, 212-650-6832, Fax: 212-650-6940, E-mail: lubell@sci.ccny.cuny.edu. *Application contact:* Timothy Boyer, MA Advisor, 212-650-5584, Fax: 212-650-6940, E-mail: boyer@sci.ccny.cuny.edu.

See Close-Up on page 329.

Clark Atlanta University, School of Arts and Sciences, Department of Physics, Atlanta, GA 30314. Offers MS. Part-time programs available. *Degree requirements:* For master's, one foreign language, thesis. *Entrance requirements:* For master's, GRE General Test, minimum GPA of 2.5. *Faculty research:* Fusion energy, investigations of nonlinear differential equations, difference schemes, collisions in dense plasma.

Clarkson University, Graduate School, School of Arts and Sciences, Department of Physics, Potsdam, NY 13699. Offers MS, PhD. Part-time programs available. *Faculty:* 6 full-time (0 women), 2 part-time/adjunct (0 women). *Students:* 16 full-time (1 woman), 11 international. Average age 27. 37 applicants, 41% accepted. In 2005, 7 master's, 1 doctorate awarded. *Median time to degree:* Of those who began their doctoral program in fall 1997, 100% received their degree in 8 years or less. *Degree requirements:* For doctorate, thesis/dissertation, departmental qualifying exam. *Entrance requirements:* For master's, GRE. Additional exam requirements/recommendations for international students: Required—TOEFL. *Application deadline:* For fall admission, 5/15 for domestic students; for spring admission, 10/15 priority date for domestic students. Applications are processed on a rolling basis. Application fee: $25 ($35 for international students). Electronic applications accepted. *Expenses:* Tuition: Full-time $20,160; part-time $840 per hour. Required fees: $215. *Financial support:* In 2005–06, 6 research assistantships (averaging $19,032 per year), 11 teaching assistantships (averaging $19,032 per year) were awarded; fellowships, scholarships/grants and tuition waivers (partial) also available. *Faculty research:* Statistical physics, surface sidence optics, quantum computing, colloids, biophysics. Total annual research expenditures: $686,211. *Unit head:* Dr. Phillip A. Christiansen, Division Head, 315-268-6669, Fax: 315-268-2308, E-mail: tony.collins@clarkson.edu. *Application contact:* Donna Brockway, Graduate Admissions International Advisor/Assistant to the Provost, 315-268-6447, Fax: 315-268-7994, E-mail: brockway@clarkson.edu.

Clark University, Graduate School, Department of Physics, Worcester, MA 01610-1477. Offers MA, PhD. Part-time programs available. *Faculty:* 6 full-time (0 women), 2 part-time/adjunct (0 women). *Students:* 12 full-time (4 women), 2 part-time (1 woman), 8 international. Average age 27. 45 applicants, 11% accepted, 3 enrolled. In 2005, 4 doctorates awarded. Terminal master's awarded for partial completion of doctoral program. *Degree requirements:* For master's, thesis or alternative; for doctorate, one foreign language, thesis/dissertation. *Entrance requirements:* Additional exam requirements/recommendations for international students: Required—TOEFL. *Application deadline:* For fall admission, 2/15 for domestic students. Application fee: $50. *Expenses:* Tuition: Full-time $29,300. Required fees: $30. *Financial support:* In 2005–06, fellowships with full and partial tuition reimbursements (averaging $17,250 per year), 8 research assistantships with full tuition reimbursements (averaging $17,250 per year), 5 teaching assistantships with full tuition reimbursements (averaging $17,250 per year) were awarded; Federal Work-Study and tuition waivers (full and partial) also available. Financial award application deadline: 4/1. *Faculty research:* Statistical and thermal physics, magnetic properties of materials, computer simulation, particle diffusion. Total annual research expenditures: $336,000. *Unit head:* Dr. Chris Landee, Chair, 508-793-7169. *Application contact:* Sujata Davis, Department Secretary, 508-793-7169, Fax: 508-793-8861, E-mail: sdavis1@clarku.edu.

Clemson University, Graduate School, College of Engineering and Science, Department of Physics and Astronomy, Program in Physics, Clemson, SC 29634. Offers astronomy and astrophysics (MS, PhD); atmospheric physics (MS, PhD); biophysics (MS, PhD). Part-time programs available. *Students:* 53 full-time (15 women), 2 part-time; includes 2 minority (1 Asian American or Pacific Islander, 1 Hispanic American), 19 international. 46 applicants, 41% accepted, 10 enrolled. In 2005, 5 master's, 4 doctorates awarded. Terminal master's awarded for partial completion of doctoral program. *Degree requirements:* For master's, thesis or alternative; for doctorate, thesis/dissertation. *Entrance requirements:* For master's and doctorate, GRE General Test. Additional exam requirements/recommendations for international students: Required—TOEFL. *Application deadline:* For fall admission, 2/15 for domestic students. Applications are processed on a rolling basis. Application fee: $50. *Financial support:* Fellowships, research assistantships, teaching assistantships available. Financial award application deadline: 6/1; financial award applicants required to submit FAFSA. *Faculty research:* Radiation physics, solid-state physics, nuclear physics, radar and lidar studies of atmosphere. *Unit head:* Dr. Brad Myer, Head, 864-656-5320. *Application contact:* Dr. Miguel Larsen, Coordinator, 864-656-5309, Fax: 864-656-0805, E-mail: mlarsen@clemson.edu.

See Close-Ups on pages 331 and 333.

Cleveland State University, College of Graduate Studies, College of Science, Department of Physics, Cleveland, OH 44115. Offers applied optics (MS); condensed matter physics (MS); medical physics (MS). Part-time and evening/weekend programs available. *Faculty:* 5 full-time (1 woman), 3 part-time/adjunct (0 women). *Students:* 3 full-time (1 woman), 16 part-time (6 women); includes 4 minority (2 African Americans, 1 American Indian/Alaska Native, 1 Hispanic American), 1 international. Average age 31. 14 applicants, 43% accepted, 5 enrolled. In 2005, 3 degrees awarded. *Degree requirements:* For master's, exit project. *Entrance requirements:* For master's, undergraduate degree in engineering, physics, chemistry or mathematics. Additional exam requirements/recommendations for international students: Required—TOEFL (minimum score 525 paper-based; 197 computer-based), GRE. *Application deadline:* For fall admission, 7/15 priority date for domestic students, 7/15 priority date for international students. Applications are processed on a rolling basis. Application fee: $30. Electronic applications accepted. *Expenses:* Tuition, state resident: full-time $10,700. Tuition, nonresident: full-time $14,628. Tuition and fees vary according to program. *Financial support:* In 2005–06, 1 research assistantship with full and partial tuition reimbursement (averaging $5,666 per year) was awarded; fellowships with tuition reimbursements, teaching assistantships, tuition waivers (full) also available. *Faculty research:* Statistical mechanics of phase transitions, low-temperature and solid-state physics, superconductivity, theoretical light scattering. Total annual research expenditures: $350,000. *Unit head:* Dr. Miron Kaufman, Chairperson, 216-687-2436, Fax: 216-523-7268, E-mail: m.kaufman@csuohio.edu. *Application contact:* Dr. James A. Lock, Director, 216-687-2425, Fax: 216-523-7268, E-mail: j.lock@csuohio.edu.

College of Staten Island of the City University of New York, Graduate Programs, Program in Physics, Staten Island, NY 10314-6600. Offers PhD. *Expenses:* Tuition, area resident: Full-time $3,200; part-time $270 per credit. Tuition, nonresident: full-time $500; part-time $500 per credit. Required fees: $328; $101 per semester. *Faculty research:* Worm algorithm and diagrammatic Monte Carlo for strongly correlated and condensed matter systems, strongly interacting quantum phases of atomic gases, nucleon—nucleon Bremsstrahlung processes in the electromagnetic sectors, carbon nanowires on diamond substrates, photoluminescence and Raman studies of Xe-ion implanted diamond. Total annual research expenditures:$74,820. *Unit head:* Syed Rizui, Chairperson, 718-982-2825, E-mail: rizui@mail.csi.cuny.edu. *Application contact:* Emmanuel Esperance, Deputy Director of Office of Recruitment and Admissions, 718-982-2259, Fax: 718-982-2500, E-mail: admissions@mail.csi.cuny.edu.

The College of William and Mary, Faculty of Arts and Sciences, Department of Physics, Williamsburg, VA 23187-8795. Offers MS, PhD. *Faculty:* 26 full-time (2 women). *Students:* 55 full-time (17 women); includes 1 African American, 2 Asian Americans or Pacific Islanders, 2 Hispanic Americans, 14 international. Average age 26. 72 applicants, 15% accepted, 11 enrolled. In 2005, 9 master's, 5 doctorates awarded. Terminal master's awarded for partial completion of doctoral program. *Degree requirements:* For master's, comprehensive exam; for doctorate, thesis/dissertation, final exams, comprehensive exam. *Entrance requirements:* For master's and doctorate, GRE General Test, GRE Subject Test, minimum GPA of 2.5. Additional exam requirements/recommendations for international students: Required—TOEFL. *Application deadline:* For fall admission, 2/1 priority date for domestic students, 2/1 priority date for international students. Applications are processed on a rolling basis. Application fee: $45. Electronic applications accepted. *Expenses:* Tuition, state resident: full-time $5,828; part-time $245 per credit. Tuition, nonresident: full-time $17,980; part-time $685 per credit. Required fees: $3,051. Tuition and fees vary according to program. *Financial support:* In 2005–06, 48 students received support, including 33 research assistantships with full tuition reimbursements available (averaging $14,250 per year), 22 teaching assistantships with full tuition reimbursements available (averaging $14,250 per year); career-related internships or fieldwork, health care benefits, and unspecified assistantships also available. *Faculty research:* Nuclear/particle, condensed-matter, atomic, and plasma physics; accelerator physics; molecular/optical physics; computational/nonlinear physics. Total annual research expenditures: $5.7 million. *Unit head:* Dr. Keith Griffioen, Chair, 757-221-3500, Fax: 757-221-3540. *Application contact:* Dr. Marc Sher, Chair of Admissions, 757-221-3538, Fax: 757-221-3540, E-mail: grad@physics.wm.edu.

Colorado School of Mines, Graduate School, Department of Physics, Golden, CO 80401-1887. Offers applied physics (PhD); physics (MS). Part-time programs available. *Faculty:* 16 full-time (0 women), 12 part-time/adjunct (2 women). *Students:* 28 full-time (7 women), 3 part-time; includes 2 minority (1 Asian American or Pacific Islander, 1 Hispanic American), 4 international. 29 applicants, 31% accepted, 8 enrolled. In 2005, 3 master's, 7 doctorates awarded. *Degree requirements:* For master's, thesis/dissertation; for doctorate, thesis/dissertation, comprehensive exam. *Entrance requirements:* For master's and doctorate, GRE General Test, GRE Subject Test. Additional exam requirements/recommendations for international students: Required—TOEFL (minimum score 550 paper-based; 213 computer-based). *Application deadline:* For fall admission, 1/1 priority date for domestic students, 1/1 priority date for international students; for spring admission, 9/1 priority date for domestic students, 9/1 priority date for international students. Application fee: $50. Electronic applications accepted. *Expenses:* Tuition, state resident: full-time $7,240; part-time $362 per credit hour. Tuition, nonresident: full-time $19,840; part-time $992 per credit hour. Required fees: $895. *Financial support:* In 2005–06, 8 students received support, including fellowships with full tuition reimbursements available (averaging $9,600 per year), 15 research assistantships with full tuition reimbursements available (averaging $9,600 per year), 10 teaching assistantships with full tuition reimbursements available (averaging $9,600 per year); scholarships/grants, health care benefits, and unspecified assistantships also available. Financial award applicants required to submit FAFSA. *Faculty research:* Light scattering, low-energy nuclear physics, high fusion plasma diagnostics, laser operations, mathematical physics. Total annual research expenditures: $5.3 million. *Unit head:* Dr. James A. McNeil, Head, 303-273-3844, Fax: 303-273-3919, E-mail: jamcneil@mine.edu. *Application contact:* Jeff Squier, Professor, 303-384-2385, Fax: 303-273-3919, E-mail: jsquier@mines.edu.

Colorado State University, Graduate School, College of Natural Sciences, Department of Physics, Fort Collins, CO 80523-0015. Offers MS, PhD. Part-time programs available. *Faculty:* 19 full-time (1 woman). *Students:* 21 full-time (4 women), 27 part-time (7 women); includes 2 minority (both African Americans), 20 international. Average age 28. 81 applicants, 33% accepted, 14 enrolled. In 2005, 3 master's, 3 doctorates awarded. Terminal master's awarded for partial completion of doctoral program. *Degree requirements:* For master's, thesis (for some programs); for doctorate, thesis/dissertation. *Entrance requirements:* For master's and doctorate, GRE General Test or GRE Subject Test in physics, minimum GPA of 3.0. Additional exam requirements/recommendations for international students: Required—TOEFL. *Application deadline:* For fall admission, 2/15 priority date for domestic students, 2/15 priority date for international students. Applications are processed on a rolling basis. Application fee: $50. Electronic applications accepted. *Expenses:* Tuition, state resident: full-time $3,690; part-time $205 per credit. Tuition, nonresident: full-time $14,958; part-time $831 per credit. Required fees: $1,061. *Financial support:* In 2005–06, 46 students received support, including 5 fellowships (averaging $1,600 per year), 19 research assistantships with full tuition reimbursements available (averaging $13,680 per year), 22 teaching assistantships with full tuition reimbursements available (averaging $13,680 per year); career-related internships or fieldwork, Federal Work-Study, and traineeships also available. Financial award application deadline: 2/15. *Faculty research:* Experimental condensed-matter physics, laser spectroscopy, optics, theoretical condensed-matter physics, particle physics. Total annual research expenditures: $2.5 million. *Unit head:* Hans D. Hochheimer, Chair, 970-491-6206, Fax: 970-491-7947, E-mail: dieter@lamar.colostate.edu. *Application contact:* Sandy Demlow, Secretary, Graduate Admissions Committee, 970-491-6207, Fax: 970-491-7947, E-mail: demlow@lamar.colostate.edu.

Columbia University, Graduate School of Arts and Sciences, Division of Natural Sciences, Department of Physics, New York, NY 10027. Offers philosophical foundations of physics (MA); physics (M Phil, PhD). *Faculty:* 23 full-time, 5 part-time/adjunct. *Students:* 90 full-time (10 women), 1 (woman) part-time. Average age 27. 199 applicants, 28% accepted. In 2005, 4 master's, 11 doctorates awarded. *Degree requirements:* For doctorate, thesis/dissertation. *Entrance requirements:* For master's and doctorate, GRE General Test, GRE Subject Test, 3 years of course work in physics. Additional exam requirements/recommendations for international students: Required—TOEFL. Application fee: $75. *Expenses:* Tuition: Full-time $31,448. Tuition and fees vary according to course level, course load, campus/location and program. *Financial support:* Fellowships, teaching assistantships, Federal Work-Study and institutionally sponsored loans available. Support available to part-time students. Financial award application deadline: 1/5; financial award applicants required to submit FAFSA. *Faculty research:* Theoretical physics; astrophysics; low-, medium-, and high-energy physics. *Unit head:* Erick K. Weinberg, Chair, 212-854-4508, Fax: 212-854-3379, E-mail: ejw@phys.columbia.edu.

Cornell University, Graduate School, Graduate Fields of Arts and Sciences, Field of Physics, Ithaca, NY 14853-0001. Offers experimental physics (MS, PhD); physics (MS, PhD); theoretical physics (MS, PhD). *Faculty:* 70 full-time (7 women). *Students:* 429 applicants, 15% accepted, 24 enrolled. In 2005, 40 master's, 18 doctorates awarded. *Degree requirements:* For doctorate, thesis/dissertation, comprehensive exam. *Entrance requirements:* For doctorate, GRE General Test, GRE Subject Test (physics), 3 letters of recommendation. Additional exam requirements/recommendations for international students: Required—TOEFL (minimum score 550 paper-based; 213 computer-based). *Application deadline:* For fall admission, 1/3 for domestic students. Application fee: $60. Electronic applications accepted. *Financial support:* In 2005–06, 175 students received support, including 28 fellowships with full tuition reimbursements available, 95 research assistantships with full tuition reimbursements available, 52 teaching assistantships with full tuition reimbursements available; institutionally sponsored loans, scholarships/grants, health care benefits, tuition waivers (full and partial), and unspecified assistantships also available. Financial award applicants required to submit FAFSA. *Faculty research:* Experimental condensed matter physics, theoretical condensed matter physics, experimental high energy particle physics, theoretical particle physics and field theory, theoretical astrophysics. *Unit head:* Director of Graduate Studies, 607-255-7561. *Application contact:* Graduate Field Assistant, 607-255-7561, E-mail: physics-grad-adm@cornell.edu.

Creighton University, Graduate School, College of Arts and Sciences, Program in Physics, Omaha, NE 68178-0001. Offers MS. *Faculty:* 3 full-time. *Students:* 8 full-time (3 women), 2 international. In 2005, 4 degrees awarded. *Degree requirements:* For master's, one foreign language, thesis or alternative. *Entrance requirements:* For master's, GRE General Test, GRE Subject Test. Additional exam requirements/recommendations for international students: Required—TOEFL. *Application deadline:* For fall admission, 3/1 for domestic students. Applications are processed on a rolling basis. Application fee: $40. *Unit head:* Dr. Sam Cipolla, Chair,

Physics

Creighton University (continued)

402-280-2133. *Application contact:* Dr. Barbara J. Braden, Dean, 402-280-2870, Fax: 402-280-5762, E-mail: bbraden@creighton.edu.

Dalhousie University, Faculty of Graduate Studies, College of Arts and Science, Faculty of Science, Department of Physics, Halifax, NS B3H 4R2, Canada. Offers M Sc, PhD. *Degree requirements:* For master's and doctorate, thesis/dissertation. *Entrance requirements:* Additional exam requirements/recommendations for international students: Required—TOEFL. *Faculty research:* Applied, experimental, and solid-state physics.

Dartmouth College, School of Arts and Sciences, Department of Physics and Astronomy, Hanover, NH 03755. Offers MS, PhD. *Faculty:* 23 full-time (6 women), 3 part-time/adjunct (1 woman). *Students:* 42 full-time (13 women); includes 1 minority (African American), 18 international. Average age 26. 117 applicants, 23% accepted, 10 enrolled. In 2005, 3 master's, 5 doctorates awarded. Terminal master's awarded for partial completion of doctoral program. *Degree requirements:* For master's and doctorate, thesis/dissertation. *Entrance requirements:* For master's and doctorate, GRE General Test, GRE Subject Test. Additional exam requirements/recommendations for international students: Required—TOEFL. *Application deadline:* For fall admission, 2/1 for domestic students. Application fee: $15. *Expenses:* Tuition: Full-time $31,770. *Financial support:* In 2005–06, 43 students received support, including fellowships with full tuition reimbursements available (averaging $21,000 per year), research assistantships with full tuition reimbursements available (averaging $21,000 per year); Federal Work-Study, institutionally sponsored loans, scholarships/grants, and tuition waivers (full) also available. *Faculty research:* Matter physics, plasma and beam physics, space physics, astronomy, cosmology. Total annual research expenditures: $3.8 million. *Unit head:* Robert Caldwell, Chair, Graduate Admissions, 603-646-2742, Fax: 603-646-1446, E-mail: robert.caldwell@dartmouth.edu. *Application contact:* Jean Blandin, Administrative Assistant, 603-646-2854, Fax: 603-646-1446, E-mail: jean.blandin@dartmouth.edu.

Delaware State University, Graduate Programs, Department of Physics, Dover, DE 19901-2277. Offers physics (MS); physics teaching (MS). Part-time and evening/weekend programs available. *Entrance requirements:* For master's, minimum GPA of 3.0 in major, 2.75 overall. Electronic applications accepted. *Faculty research:* Thermal properties of solids, nuclear physics, radiation damage in solids.

DePaul University, College of Liberal Arts and Sciences, Department of Physics, Chicago, IL 60604-2287. Offers applied physics (MS). Part-time and evening/weekend programs available. *Faculty:* 7 full-time (1 woman), 3 part-time/adjunct (0 women). *Students:* 6 full-time (1 woman), 4 part-time (1 woman); includes 1 minority (African American), 2 international. Average age 23. 12 applicants, 42% accepted, 3 enrolled. In 2005, 3 degrees awarded. *Degree requirements:* For master's, thesis, oral exams. *Entrance requirements:* For master's, 2 letters of recommendation, BA in physics or closely related field. Additional exam requirements/recommendations for international students: Required—TOEFL. *Application deadline:* For fall admission, 6/1 priority date for domestic students, 6/1 priority date for international students. Application fee: $25. Electronic applications accepted. *Financial support:* In 2005–06, teaching assistantships with full tuition reimbursements (averaging $9,500 per year); tuition waivers (full) also available. *Faculty research:* Optics, solid-state physics, comology, atomic physics, nuclear physics. Total annual research expenditures: $54,000. *Unit head:* Dr. Christopher G. Goedde, Chairman, 773-325-7330, Fax: 773-325-7334, E-mail: egoedde@condor.depaul.edu. *Application contact:* Dr. Jesus Pando, Associate Professor, 773-325-7330, Fax: 773-325-7334.

Drexel University, College of Arts and Sciences, Physics Program, Philadelphia, PA 19104-2875. Offers MS, PhD. Terminal master's awarded for partial completion of doctoral program. *Degree requirements:* For doctorate, thesis/dissertation. *Entrance requirements:* For master's and doctorate, GRE. Additional exam requirements/recommendations for international students: Required—TOEFL. Electronic applications accepted. *Faculty research:* Nuclear structure, mesoscale meteorology, numerical astrophysics, numerical weather prediction, earth energy radiation budget.

Duke University, Graduate School, Department of Medical Physics, Durham, NC 27708-0586. Offers MS, PhD. *Faculty:* 42 full-time. *Students:* 22 full-time (7 women); includes 2 minority (1 African American, 1 Asian American or Pacific Islander), 6 international. 61 applicants, 56% accepted, 22 enrolled. *Entrance requirements:* Additional exam requirements/recommendations for international students: Required—IELT (preferred) or TOEFL. *Application deadline:* For fall admission, 2/28 for domestic students. Electronic applications accepted. *Unit head:* Ehsan Samei, Director, 919-684-7852, Fax: 919-684-7122, E-mail: olga.baranova@duke.edu.

Duke University, Graduate School, Department of Physics, Durham, NC 27708-0586. Offers PhD. Part-time programs available. *Faculty:* 36 full-time. *Students:* 79 full-time (13 women); includes 4 minority (1 African American, 3 Asian Americans or Pacific Islanders), 40 international. 200 applicants, 21% accepted, 15 enrolled. In 2005, 7 doctorates awarded. *Degree requirements:* For doctorate, thesis/dissertation. *Entrance requirements:* For doctorate, GRE General Test, GRE Subject Test. Additional exam requirements/recommendations for international students: Required—IELT (preferred) or TOEFL. *Application deadline:* For fall admission, 12/31 for domestic students. Application fee: $75. *Financial support:* Fellowships, research assistantships, teaching assistantships, Federal Work-Study available. Financial award application deadline: 12/31. *Unit head:* Dr. Roxanne Springer, Director of Graduate Studies, 919-660-2676, Fax: 919-660-2525, E-mail: donna@phy.duke.edu.

East Carolina University, Graduate School, Thomas Harriot College of Arts and Sciences, Department of Physics, Greenville, NC 27858-4353. Offers applied and biomedical physics (MS); medical physics (MS); physics (PhD). Part-time programs available. *Faculty:* 19 full-time (2 women). *Students:* 25 full-time (8 women), 10 part-time (1 woman); includes 6 minority (4 African Americans, 1 American Indian/Alaska Native, 1 Asian American or Pacific Islander), 7 international. Average age 30. 11 applicants, 36% accepted, 4 enrolled. In 2005, 4 master's, 5 doctorates awarded. *Degree requirements:* For master's, one foreign language, comprehensive exam. *Entrance requirements:* For master's, GRE General Test. Additional exam requirements/recommendations for international students: Required—TOEFL. *Application deadline:* Applications are processed on a rolling basis. Application fee: $50. *Expenses:* Tuition, state resident: full-time $2,516. Tuition, nonresident: full-time $12,832. *Financial support:* Research assistantships with partial tuition reimbursements, teaching assistantships with partial tuition reimbursements, Federal Work-Study available. Support available to part-time students. Financial award application deadline: 6/1. *Unit head:* Dr. John Sutherland, Chair, 252-328-6739, Fax: 252-328-6314, E-mail: sutherlandj@ecu.edu. *Application contact:* Dean of Graduate School, 252-328-6012, Fax: 252-328-6071, E-mail: gradschool@ecu.edu.

Announcement: The graduate program in the Department of Physics offers PhD degrees in biomedical physics and MS degrees in physics with options in applied physics and medical physics to satisfy the career goals of most physics students. This program draws on faculty members in ECU's physical and medical science departments to meet the need for highly trained scientists who can integrate knowledge of the physical sciences with biomedical research. More information at http://www.ecu.edu/physics/grad.htm or e-mail physics@mail.ecu.edu.

Eastern Michigan University, Graduate School, College of Arts and Sciences, Department of Physics and Astronomy, Program in Physics, Ypsilanti, MI 48197. Offers MS. *Entrance requirements:* Additional exam requirements/recommendations for international students: Required—TOEFL. *Application deadline:* For fall admission, 5/15 priority date for domestic students, 5/1 priority date for international students. For winter admission, 10/15 for domestic students; for spring admission, 3/15 for domestic students. Applications are processed on a rolling basis. Application fee: $35. *Expenses:* Tuition, state resident: full-time $7,838; part-time $327 per credit hour. Tuition, nonresident: full-time $15,770; part-time $657 per credit hour. Required fees: $33 per credit hour. $40 per term. Tuition and fees vary according to course level, course load and degree level. *Financial support:* Fellowships, teaching assistantships available. Support available to part-time students. Financial award applicants required to submit FAFSA. *Unit head:* Dr. Marshall Thomsen, Coordinator, 734-487-4144.

Emory University, Graduate School of Arts and Sciences, Department of Physics, Atlanta, GA 30322-1100. Offers biophysics (PhD); condensed matter physics (PhD); non-linear physics (PhD); radiological physics (PhD); soft condensed matter physics (PhD); solid-state physics (PhD); statistical physics (PhD). *Faculty:* 19 full-time (2 women). *Students:* 17 full-time (4 women); includes 2 minority (1 African American, 1 Hispanic American), 13 international. Average age 24. 40 applicants, 13% accepted. *Degree requirements:* For doctorate, thesis/dissertation, qualifier proposal (PhD qualification). *Entrance requirements:* For doctorate, GRE General Test, minimum GPA of 3.0. Additional exam requirements/recommendations for international students: Required—TOEFL (minimum score 600 paper-based). *Application deadline:* For fall admission, 1/3 priority date for domestic students, 1/3 priority date for international students. Application fee: $50. Electronic applications accepted. *Expenses:* Tuition: Full-time $14,400. Required fees: $217. *Financial support:* In 2005–06, 17 students received support, including 6 fellowships (averaging $20,000 per year); institutionally sponsored loans, scholarships/grants, health care benefits, and tuition waivers (full) also available. Financial award application deadline: 1/3; financial award applicants required to submit FAFSA. *Faculty research:* Experimental studies of the structure and function of metalloproteins, soft condensed matter, granular materials, biophotonics and fluorescence correlation spectroscopy, single molecule studies of DNA-protein systems. Total annual research expenditures: $1.5 million. *Unit head:* Dr. Raymond DuVarney, Chair, 404-727-4296, Fax: 404-727-0873, E-mail: phsrcd@physics.emory.edu. *Application contact:* Dr. Kurt Warncke, Director of Graduate Studies, 404-727-2975, Fax: 404-727-0873, E-mail: kwarncke@physics.emory.edu.

Emporia State University, School of Graduate Studies, College of Liberal Arts and Sciences, Department of Physical Sciences, Emporia, KS 66801-5087. Offers chemistry (MS); earth science (MS); physical science (MS); physics (MS). *Faculty:* 15 full-time (2 women), 1 (woman) part-time/adjunct. *Students:* 5 full-time (1 woman), 19 part-time (5 women); includes 1 minority (African American) 8 applicants, 88% accepted, 5 enrolled. In 2005, 6 degrees awarded. *Degree requirements:* For master's, comprehensive exam or thesis. *Entrance requirements:* For master's, physical science qualifying exam, appropriate undergraduate degree. Additional exam requirements/recommendations for international students: Required—TOEFL. *Application deadline:* For fall admission, 8/15 for domestic students. Applications are processed on a rolling basis. Application fee: $30 ($75 for international students). Electronic applications accepted. *Expenses:* Tuition, state resident: full-time $2,890; part-time $132 per credit. Tuition, nonresident: full-time $9,258; part-time $422 per credit. Required fees: $626; $41 per credit. Tuition and fees vary according to degree level. *Financial support:* In 2005–06, research assistantships with full tuition reimbursements (averaging $6,492 per year), 8 teaching assistantships with full tuition reimbursements (averaging $6,492 per year) were awarded; Federal Work-Study, institutionally sponsored loans, health care benefits, and unspecified assistantships also available. Financial award application deadline: 3/15; financial award applicants required to submit FAFSA. *Faculty research:* Bredigite, larnite, and dicalcium silicates–Marble Canyon. *Unit head:* Dr. DeWayne Backhus, Chair, 620-341-5330, Fax: 620-341-6055, E-mail: dbackhus@emporia.edu.

Fisk University, Graduate Programs, Department of Physics, Nashville, TN 37208-3051. Offers MA. *Degree requirements:* For master's, thesis. *Entrance requirements:* For master's, GRE General Test, GRE Subject Test, minimum GPA of 3.0. *Faculty research:* Molecular physics, astrophysics, surface physics, nanobase materials, optical processing.

Florida Agricultural and Mechanical University, Division of Graduate Studies, Research, and Continuing Education, College of Arts and Sciences, Department of Physics, Tallahassee, FL 32307-3200. Offers MS, PhD. *Degree requirements:* For master's, thesis optional; for doctorate, thesis/dissertation, comprehensive exam. *Entrance requirements:* For master's, GRE General Test, minimum GPA of 3.0; for doctorate, GRE General Test, minimum GPA of 3.0, letters of recommendation (2). Additional exam requirements/recommendations for international students: Required—TOEFL (minimum score 550 paper-based). *Faculty research:* Plasma physics, quantum mechanics, condensed matter physics, astrophysics, laser ablation.

Florida Atlantic University, Charles E. Schmidt College of Science, Department of Physics, Boca Raton, FL 33431-0991. Offers MS, MST, PhD. Part-time programs available. *Faculty:* 7 full-time (1 woman), 2 part-time/adjunct (0 women). *Students:* 21 full-time (6 women), 5 part-time (1 woman); includes 4 minority (1 African American, 1 Asian American or Pacific Islander, 2 Hispanic Americans), 13 international. Average age 35. 7 applicants, 71% accepted, 5 enrolled. In 2005, 5 degrees awarded. *Median time to degree:* Of those who began their doctoral program in fall 1997, 90% received their degree in 8 years or less. *Degree requirements:* For master's, thesis (for some programs); for doctorate, thesis/dissertation. *Entrance requirements:* For master's, GRE General Test, minimum GPA of 3.0; for doctorate, GRE General Test. Additional exam requirements/recommendations for international students: Required—TOEFL (minimum score 500 paper-based; 173 computer-based). *Application deadline:* For fall admission, 7/1 for domestic students, 2/15 for international students; for spring admission, 11/1 for domestic students, 8/15 for international students. Applications are processed on a rolling basis. Application fee: $30. *Expenses:* Tuition, state resident: full-time $4,394; part-time $244 per credit. Tuition, nonresident: full-time $16,441; part-time $912 per credit. *Financial support:* In 2005–06, 3 research assistantships with tuition reimbursements (averaging $17,372 per year), 18 teaching assistantships with tuition reimbursements (averaging $17,372 per year) were awarded; fellowships, Federal Work-Study and unspecified assistantships also available. *Faculty research:* Astrophysics, spectroscopy, mathematical physics, theory of metals, superconductivity. Total annual research expenditures:$123,700. *Unit head:* Dr. Warner Miller, Chair, 561-297-3382, Fax: 561-297-2662, E-mail: wam@physics.fau.edu.

Florida Institute of Technology, Graduate Programs, College of Science, Department of Physics and Space Sciences, Melbourne, FL 32901-6975. Offers physics (MS, PhD); space science (MS, PhD). Part-time programs available. *Faculty:* 9 full-time (0 women). *Students:* 18 full-time (3 women), 13 part-time (4 women); includes 3 minority (1 African American, 2 Hispanic Americans), 10 international. Average age 31. 49 applicants, 35% accepted, 8 enrolled. In 2005, 2 master's, 3 doctorates awarded. Terminal master's awarded for partial completion of doctoral program. *Degree requirements:* For master's, oral exam, thesis optional; for doctorate, one foreign language, thesis/dissertation, publication in referred journal, seminar on dissertation research, comprehensive exam, registration. *Entrance requirements:* For master's, GRE General Test, GRE Subject Test, minimum GPA of 3.0, proficiency in a computer language, resumé, 3 letters of recommendation, vector analysis; for doctorate, GRE General Test, GRE Subject Test, minimum GPA of 3.2, resumé, 3 letters of recommendation, proficiency in a computer program language. Additional exam requirements/recommendations for international students: Required—TOEFL (minimum score 550 paper-based; 213 computer-based). *Application deadline:* Applications are processed on a rolling basis. Application fee: $50. Electronic applications accepted. *Expenses:* Tuition: Part-time $825 per credit. *Financial support:* In 2005–06, 15 students received support, including 3 research assistantships with full and partial tuition reimbursements available (averaging $10,630 per year), 12 teaching assistantships with full and partial tuition reimbursements available (averaging $10,890 per year); career-related internships or fieldwork and tuition remissions also available. Financial award application deadline: 3/1; financial award applicants required to submit FAFSA. *Faculty research:* Lasers, semiconductors, magnetism, quantum devices, high energy physics. Total annual research expenditures: $1.4 million. *Unit head:* Dr. Laszlo A. Baksay, Department Head, 321-674-7367, Fax: 321-674-7482, E-mail: baksay@fit.edu. *Application contact:* Carolyn P. Farrior, Director of Graduate Admissions, 321-674-7118, Fax: 321-723-9468, E-mail: cfarrior@fit.edu.

Florida International University, College of Arts and Sciences, Department of Physics, Miami, FL 33199. Offers MS, PhD. Part-time and evening/weekend programs available. *Faculty:* 19 full-time (2 women). *Students:* 30 full-time (7 women), 5 part-time (1 woman); includes 15 minority (3 African Americans, 1 Asian American or Pacific Islander, 11 Hispanic Americans), 16 international. Average age 30. 33 applicants, 55% accepted, 9 enrolled. In 2005, 4 master's, 1 doctorate awarded. *Degree requirements:* For master's and doctorate, one foreign language, thesis/dissertation. *Entrance requirements:* For master's and doctorate, GRE General Test. Additional exam requirements/recommendations for international students: Required—TOEFL. *Application deadline:* For fall admission, 4/1 for domestic students; for spring admission, 10/1 for domestic students. Applications are processed on a rolling basis. Application fee: $25. *Expenses:* Tuition, area resident: Part-time $239 per credit. Tuition, state resident: full-time $4,294; part-time $869 per credit. Tuition, nonresident: full-time $15,641. Required fees: $252; $126 per term. Tuition and fees vary according to program. *Financial support:* Application deadline: 4/1. *Faculty research:* Molecular collision processes (molecular beams), biophysical optics. *Unit head:* Dr. Walter Van Hamme, Chairperson, 305-348-2605, Fax: 305-348-3053, E-mail: walter.vanhamme@fiu.edu.

Florida State University, Graduate Studies, College of Arts and Sciences, Department of Physics, Tallahassee, FL 32306. Offers MS, PhD. *Faculty:* 43 full-time (4 women). *Students:* 129 full-time (19 women); includes 55 minority (1 African American, 49 Asian Americans or Pacific Islanders, 5 Hispanic Americans). Average age 28. 254 applicants, 9% accepted, 23 enrolled. In 2005, 21 master's, 8 doctorates awarded. *Median time to degree:* Of those who began their doctoral program in fall 1997, 42% received their degree in 8 years or less. *Degree requirements:* For doctorate, thesis/dissertation, comprehensive exam. *Entrance requirements:* For master's and doctorate, GRE General Test, minimum GPA of 3.0. Additional exam requirements/recommendations for international students: Required—TOEFL (minimum score 550 paper-based; 213 computer-based). *Application deadline:* For fall admission, 2/15 for domestic students, 1/15 for international students. Applications are processed on a rolling basis. Application fee: $30. Electronic applications accepted. *Financial support:* In 2005–06, 129 students received support, including 84 research assistantships with full tuition reimbursements available (averaging $17,000 per year), 45 teaching assistantships with full tuition reimbursements available (averaging $17,000 per year); career-related internships or fieldwork and Federal Work-Study also available. Financial award application deadline: 2/15; financial award applicants required to submit FAFSA. *Faculty research:* High energy physics, computational physics, biophysics, condensed matter physics, nuclear physics. Total annual research expenditures: $4 million. *Unit head:* Dr. David H. Van Winkle, Chairman, 850-644-2867, Fax: 850-644-2338, E-mail: rip@phy.fsu.edu. *Application contact:* Sherry Ann Tointigh, Program Assistant, 850-644-4473, Fax: 850-644-8630, E-mail: graduate@phy.fsu.edu.

Announcement: Extensive research opportunities exist in theoretical and experimental physics in the areas of atomic and molecular physics, condensed-matter, high-energy, and nuclear physics. This research makes use of extensive computers and instrumentation at the National High-Magnetic Field Laboratory at FSU. Each full-time graduate student has an assistantship.

See Close-Up on page 339.

George Mason University, College of Arts and Sciences, Department of Physics, Fairfax, VA 22030. Offers applied and engineering physics (MS). *Faculty:* 23 full-time (8 women), 3 part-time/adjunct (0 women). *Students:* 2 full-time (0 women), 18 part-time (5 women); includes 2 minority (1 American Indian/Alaska Native, 1 Hispanic American), 1 international. Average age 28. 16 applicants, 94% accepted, 9 enrolled. In 2005, 13 degrees awarded. *Degree requirements:* For master's, thesis optional. *Entrance requirements:* For master's, minimum GPA of 2.75 in last 60 hours of course work. *Application deadline:* For fall admission, 5/1 for domestic students; for spring admission, 11/1 for domestic students. Electronic applications accepted. *Expenses:* Tuition, area resident: Full-time $5,244; part-time $219 per credit. Tuition, state resident: part-time $651 per credit. Tuition, nonresident: full-time $15,636. Required fees: $1,524; $65 per credit. *Financial support:* Research assistantships, teaching assistantships available. Support available to part-time students. Financial award application deadline: 3/1; financial award applicants required to submit FAFSA. *Unit head:* Dr. Maria Dworzecka, Chairman, 703-993-1280, Fax: 703-993-1269, E-mail: mdworzecka@gmu.edu. *Application contact:* Dr. Paul So, Information Contact, 703-993-1280, E-mail: physics@gmu.edu.

The George Washington University, Columbian College of Arts and Sciences, Department of Physics, Washington, DC 20052. Offers MA, PhD. Part-time and evening/weekend programs available. *Degree requirements:* For doctorate, thesis/dissertation, general exam. *Entrance requirements:* For master's and doctorate, GRE General Test, minimum GPA of 3.0. Additional exam requirements/recommendations for international students: Required—TOEFL (minimum score 550 paper-based; 213 computer-based). Electronic applications accepted.

Georgia Institute of Technology, Graduate Studies and Research, College of Sciences, School of Physics, Atlanta, GA 30332-0001. Offers MS, PhD. Part-time programs available. Terminal master's awarded for partial completion of doctoral program. *Degree requirements:* For doctorate, thesis/dissertation, comprehensive exam. *Entrance requirements:* For master's, GRE General Test, GRE Subject Test, minimum GPA of 3.0; for doctorate, GRE General Test, GRE Subject Test, minimum GPA of 3.4. Additional exam requirements/recommendations for international students: Required—TOEFL. Electronic applications accepted. *Faculty research:* Atomic and molecular physics, chemical physics, condensed matter, optics, nonlinear physics and chaos.

Georgia State University, College of Arts and Sciences, Department of Physics and Astronomy, Program in Physics, Atlanta, GA 30303-3083. Offers MS, PhD. Part-time and evening/weekend programs available. *Faculty:* 17 full-time (0 women), 2 part-time/adjunct (both women). *Students:* 29 full-time (6 women), 1 part-time; includes 1 African American, 23 Asian Americans or Pacific Islanders. 64 applicants, 16% accepted, 10 enrolled. In 2005, 9 master's, 3 doctorates awarded. Terminal master's awarded for partial completion of doctoral program. *Degree requirements:* For master's, one foreign language, thesis or alternative, exam; for doctorate, 2 foreign languages, thesis/dissertation, exam. *Entrance requirements:* For master's and doctorate, GRE General Test. Additional exam requirements/recommendations for international students: Required—TOEFL. *Application deadline:* For fall admission, 7/1 for domestic students, 3/1 for international students; for spring admission, 11/15 for domestic students, 11/15 for international students. Applications are processed on a rolling basis. Application fee: $50. Electronic applications accepted. *Expenses:* Tuition, state resident: full-time $4,368; part-time $182 per term. Tuition, nonresident: full-time $8,732; part-time $728 per term. Required fees: $46 per hour. *Financial support:* In 2005–06, fellowships with tuition reimbursements (averaging $22,000 per year), research assistantships with tuition reimbursements (averaging $17,500 per year), teaching assistantships with tuition reimbursements (averaging $16,500 per year) were awarded; Federal Work-Study, institutionally sponsored loans, tuition waivers (full), and unspecified assistantships also available. Financial award application deadline: 5/1; financial award applicants required to submit FAFSA. *Faculty research:* Biophysics; nuclear, condensed-matter, and atomic physics; astrophysics. Total annual research expenditures: $630,700. *Unit head:* Dr. Unil Perera, Faculty Advisor, 404-651-2279, Fax: 404-651-1427, E-mail: uperera@gsu.edu. *Application contact:* Dr. Unil Perera, Faculty Advisor, 404-651-2279, Fax: 404-651-1427, E-mail: uperera@gsu.edu.

Graduate School and University Center of the City University of New York, Graduate Studies, Program in Physics, New York, NY 10016-4039. Offers PhD. *Faculty:* 105 full-time (3 women). *Students:* 85 full-time (17 women), 3 part-time; includes 6 minority (1 African American, 2 American Indian/Alaska Native, 1 Asian American or Pacific Islander, 2 Hispanic Americans), 61 international. Average age 27. 41 applicants, 54% accepted, 19 enrolled. In 2005, 10 degrees awarded. *Degree requirements:* For doctorate, thesis/dissertation. *Entrance requirements:* For doctorate, GRE General Test. Additional exam requirements/recommendations for international students: Required—TOEFL. *Application deadline:* For fall admission, 4/15 for domestic students. Application fee: $125. Electronic applications accepted. *Financial support:* In 2005–06, 52 students received support, including 51 fellowships, 1 teaching assistantship; research assistantships, career-related internships or fieldwork, Federal Work-Study, institutionally sponsored loans, and tuition waivers (full and partial) also available. Financial award application deadline: 2/1; financial award applicants required to submit FAFSA. *Faculty research:* Condensed-matter, particle, nuclear, and atomic physics. *Unit head:* Dr. Sultan Catto, Executive Officer, 212-817-8651, Fax: 212-817-1531, E-mail: scatto@gc.cuny.edu.

Hampton University, Graduate College, Department of Physics, Hampton, VA 23668. Offers MS, PhD. Part-time and evening/weekend programs available. Terminal master's awarded for partial completion of doctoral program. *Degree requirements:* For master's, thesis optional; for doctorate, thesis/dissertation, oral defense, qualifying exam. *Entrance requirements:* For master's, GRE General Test; for doctorate, GRE General Test, minimum GPA of 3.0 or master's degree in physics or related field. *Faculty research:* Laser optics, remote sensing.

Harvard University, Graduate School of Arts and Sciences, Department of Physics, Cambridge, MA 02138. Offers experimental physics (PhD); medical engineering/medical physics (PhD), including applied physics, engineering sciences, physics; theoretical physics (PhD). *Students:* 174. *Degree requirements:* For doctorate, thesis/dissertation, final exams, laboratory experience. *Entrance requirements:* For doctorate, GRE General Test, GRE Subject Test. Additional exam requirements/recommendations for international students: Required—TOEFL. *Application deadline:* For fall admission, 12/14 for domestic students. Application fee: $60. *Expenses:* Tuition: Full-time $28,752. Full-time tuition and fees vary according to program and student level. *Financial support:* Fellowships, research assistantships, teaching assistantships, career-related internships or fieldwork, Federal Work-Study, and institutionally sponsored loans available. Financial award application deadline: 12/30. *Faculty research:* Particle physics, condensed matter physics, atomic physics. *Unit head:* Sheila Ferguson, Administrator, 617-495-4327. *Application contact:* Office of Admissions and Financial Aid, 617-495-5315.

Howard University, Graduate School of Arts and Sciences, Department of Physics and Astronomy, Washington, DC 20059-0002. Offers physics (MS, PhD). *Degree requirements:* For master's, thesis (for some programs), comprehensive exam (for some programs); for doctorate, thesis/dissertation, departmental qualifying exam, final comprehensive exam, comprehensive exam. *Entrance requirements:* For master's, GRE General Test, bachelor's degree in physics or related field, minimum GPA of 3.0; for doctorate, GRE General Test, bachelor's or master's degree in physics or related field, minimum GPA of 3.0. Additional exam requirements/recommendations for international students: Required—TOEFL (minimum score 550 paper-based; 213 computer-based). *Faculty research:* Atmospheric physics, spectroscopy and optical physics, high energy physics, condensed matter.

Hunter College of the City University of New York, Graduate School, School of Arts and Sciences, Department of Physics, New York, NY 10021-5085. Offers MA, PhD. Part-time programs available. *Faculty:* 5 full-time (1 woman). *Students:* Average age 31. 1 applicant, 100% accepted, 0 enrolled. In 2005, 3 degrees awarded. Terminal master's awarded for partial completion of doctoral program. *Degree requirements:* For master's, comprehensive exam or thesis. *Entrance requirements:* For master's, minimum 36 credits of course work in mathematics and physics. Additional exam requirements/recommendations for international students: Required—TOEFL. *Application deadline:* For fall admission, 4/1 for domestic students, 2/1 for international students; for spring admission, 11/1 for domestic students, 9/1 for international students. Application fee: $125. *Expenses:* Tuition, state resident: full-time $6,400; part-time $270 per credit. Tuition, nonresident: part-time $500 per credit. International tuition: $12,000 full-time. Required fees: $50 per term. Part-time tuition and fees vary according to course load and program. *Financial support:* In 2005–06, research assistantships (averaging $20,000 per year), teaching assistantships (averaging $9,000 per year) were awarded; Federal Work-Study, scholarships/grants, and tuition waivers (partial) also available. Support available to part-time students. *Faculty research:* Experimental and theoretical quantum optics, experimental and theoretical condensed matter, mathematical physics. *Unit head:* Ying-Chin Chen, Chairperson, 212-650-4526, Fax: 212-772-5390, E-mail: y.c.chen@hunter.cuny.edu. *Application contact:* William Zlata, Director for Graduate Admissions, 212-772-4482, Fax: 212-650-3336, E-mail: admissions@hunter.cuny.edu.

Idaho State University, Office of Graduate Studies, College of Arts and Sciences, Department of Physics, Pocatello, ID 83209. Offers MNS, MS. Part-time programs available. *Degree requirements:* For master's, thesis (for some programs), comprehensive exam, registration. *Entrance requirements:* For master's, GRE General Test, 3 letters of recommendation, BS or BA in physics, teaching certificate (MNS). Additional exam requirements/recommendations for international students: Required—TOEFL (minimum score 550 paper-based; 213 computer-based). *Faculty research:* Ion beam applications, low-energy nuclear physics, relativity and cosmology, observational astronomy.

See Close-Up on page 341.

Illinois Institute of Technology, Graduate College, College of Science and Letters, Department of Biological, Chemical and Physical Sciences, Physics Division, Chicago, IL 60616-3793. Offers health physics (MHP); physics (MS, PhD). Part-time programs available. Postbaccalaureate distance learning degree programs offered. Terminal master's awarded for partial completion of doctoral program. *Degree requirements:* For master's, thesis (for some programs), comprehensive exam; for doctorate, thesis/dissertation, comprehensive exam. *Entrance requirements:* For master's and doctorate, GRE General Test, minimum undergraduate GPA of 3.0. Additional exam requirements/recommendations for international students: Required—TOEFL (minimum score 550 paper-based; 213 computer-based). Electronic applications accepted. *Faculty research:* Biophysics, condensed matter physics, high energy physics, surface physics, theoretical physics.

Indiana University Bloomington, Graduate School, College of Arts and Sciences, Department of Physics, Bloomington, IN 47405-7000. Offers MAT, MS, PhD. PhD offered through the University Graduate School. Part-time programs available. *Faculty:* 35 full-time (3 women). *Students:* 63 full-time (18 women), 21 part-time (3 women); includes 2 minority (1 Asian American or Pacific Islander, 1 Hispanic American), 52 international. Average age 28. In 2005, 18 master's, 9 doctorates awarded. Terminal master's awarded for partial completion of doctoral program. *Degree requirements:* For master's, qualifying exam; for doctorate, thesis/dissertation, qualifying exam. *Entrance requirements:* For master's and doctorate, GRE General Test, GRE Subject Test (physics). Additional exam requirements/recommendations for international students: Required—TOEFL. *Application deadline:* For fall admission, 1/15 priority date for domestic students, 12/15 priority date for international students; for spring admission, 9/1 for domestic students, 9/1 for international students. Application fee: $50 ($60 for international students). *Expenses:* Tuition, state resident: full-time $5,437; part-time $227 per credit hour. Tuition, nonresident: full-time $15,836; part-time $660 per credit hour. Required fees: $821. Tuition and fees vary according to campus/location and program. *Financial support:* Research assistantships with partial tuition reimbursements, teaching assistantships with partial tuition reimbursements, career-related internships or fieldwork available. Financial award application deadline:2/1. *Unit head:* Dr. James Musser, Chair, 812-855-1247. *Application contact:* June Dizer, Student Affairs Administrator, 812-855-3973, E-mail: gradphys@indiana.edu.

Announcement: Theoretical and experimental nuclear, particle, condensed matter, accelerator, astrophysics, biophysics, biocomplexity, and chemical physics. IUCF electron-cooled storage ring; nuclear theory center, supercomputers. Low-temperature clean room, cryostats, photolithograph, STM, UHV analysis, 14T magnet, X-ray diffraction, squid magnetometer, and microwave facilities. Experiments at national and international laboratories. Financial support available.

See Close-Up on page 343.

Indiana University of Pennsylvania, Graduate School and Research, College of Natural Sciences and Mathematics, Department of Physics, Program in Physics, Indiana, PA 15705-1087. Offers MA, MS. Part-time programs available. *Degree requirements:* For master's, thesis (for some programs), comprehensive exam (for some programs).

Physics

Indiana University–Purdue University Indianapolis, School of Science, Department of Physics, Indianapolis, IN 46202-2896. Offers MS, PhD. Part-time programs available. *Faculty:* 4 full-time (0 women). *Students:* 6 full-time (1 woman), 1 (woman) part-time; includes 2 minority (1 African American, 1 Hispanic American), 3 international. Average age 29. In 2005, 1 degree awarded. Terminal master's awarded for partial completion of doctoral program. *Degree requirements:* For master's, thesis optional; for doctorate, thesis/dissertation. *Entrance requirements:* Additional exam requirements/recommendations for international students: Required—TOEFL. *Application deadline:* For fall admission, 3/1 for domestic students. Applications are processed on a rolling basis. Application fee: $50 ($60 for international students). *Expenses:* Tuition, state resident: full-time $5,159; part-time $215 per credit hour. Tuition, nonresident: full-time $14,890; part-time $620 per credit hour. Required fees: $614. Tuition and fees vary according to campus/location and program. *Financial support:* Fellowships with full tuition reimbursements, research assistantships with full tuition reimbursements, teaching assistantships with full tuition reimbursements, Federal Work-Study, institutionally sponsored loans, and tuition waivers (full and partial) available. Support available to part-time students. Financial award application deadline: 3/1. *Faculty research:* Magnetic resonance, photosynthesis, optical physics, biophysics, physics of materials. *Unit head:* Guantam Vemuri, Chair, 317-274-6900, E-mail: gnamuri@iupui.edu. *Application contact:* Z. Ou, Chair, Graduate Committee, 317-274-2125, Fax: 317-274-2393, E-mail: zou@iupui.edu.

Iowa State University of Science and Technology, Graduate College, College of Liberal Arts and Sciences, Department of Physics and Astronomy, Ames, IA 50011. Offers applied physics (MS, PhD); astrophysics (MS, PhD); condensed matter physics (MS, PhD); high energy physics (MS, PhD); nuclear physics (MS, PhD); physics (MS, PhD). Part-time programs available. *Faculty:* 44 full-time, 3 part-time/adjunct. *Students:* 79 full-time (14 women), 6 part-time; includes 1 minority (Asian American or Pacific Islander), 56 international. 161 applicants, 34% accepted, 23 enrolled. In 2005, 5 master's, 7 doctorates awarded. Terminal master's awarded for partial completion of doctoral program. *Degree requirements:* For master's, thesis (for some programs); for doctorate, thesis/dissertation. *Entrance requirements:* For master's and doctorate, GRE General Test, GRE Subject Test (physics). Additional exam requirements/recommendations for international students: Required—TOEFL (paper score 550; computer score 213) or IELTS (score 6.5). *Application deadline:* For fall admission, 2/15 priority date for domestic students, 2/15 priority date for international students; for spring admission, 10/15 for domestic students, 10/15 for international students. Applications are processed on a rolling basis. Application fee: $30 ($70 for international students). Electronic applications accepted. *Expenses:* Tuition, state resident: full-time $6,410. Tuition, nonresident: full-time $16,422. Tuition and fees vary according to program. *Financial support:* In 2005–06, 48 research assistantships with full tuition reimbursements (averaging $14,626 per year), 30 teaching assistantships with full tuition reimbursements (averaging $14,533 per year) were awarded; fellowships, Federal Work-Study, institutionally sponsored loans, scholarships/grants, health care benefits, and unspecified assistantships also available. Support available to part-time students. Financial award application deadline: 2/15. *Faculty research:* Condensed-matter physics, including superconductivity and new materials; high-energy and nuclear physics; astronomy and astrophysics; atmospheric and environmental physics. Total annual research expenditures: $8.8 million. *Unit head:* Dr. Eli Rosenberg, Chair, 515-294-5441, Fax: 515-294-6027, E-mail: phys_astro@iastate.edu. *Application contact:* Dr. Steven Kawaler, Director of Graduate Education, 515-294-9728, E-mail: phys_astro@iastate.edu.

John Carroll University, Graduate School, Department of Physics, University Heights, OH 44118-4581. Offers MS. Part-time programs available. *Degree requirements:* For master's, essay or thesis. *Entrance requirements:* For master's, bachelor's degree in electrical engineering or physics. *Faculty research:* Fiber optics, ultrasonics, atomic force microscopy, computational materials science, transport properties.

The Johns Hopkins University, Zanvyl Krieger School of Arts and Sciences, Henry A. Rowland Department of Physics and Astronomy, Baltimore, MD 21218-2699. Offers astronomy (PhD); physics (PhD). *Faculty:* 34 full-time (3 women), 17 part-time/adjunct (2 women). *Students:* 98 full-time (23 women), 1 part-time; includes 9 minority (1 African American, 1 American Indian/Alaska Native, 5 Asian Americans or Pacific Islanders, 2 Hispanic Americans), 49 international. Average age 28. 267 applicants, 22% accepted, 21 enrolled. *Degree requirements:* For doctorate, thesis/dissertation, comprehensive exam, registration. *Entrance requirements:* For doctorate, GRE General Test, GRE Subject Test. Additional exam requirements/recommendations for international students: Required—TOEFL (minimum score 600 paper-based; 250 computer-based). *Application deadline:* For fall admission, 1/15 for domestic students, 1/15 for international students. Application fee: $60. Electronic applications accepted. *Expenses:* Tuition: Full-time $30,960. Tuition and fees vary according to degree level and program. *Financial support:* In 2005–06, 9 fellowships with tuition reimbursements (averaging $2,500 per year), 53 research assistantships with full tuition reimbursements (averaging $20,000 per year), 43 teaching assistantships with full tuition reimbursements (averaging $15,000 per year) were awarded; career-related internships or fieldwork, Federal Work-Study, and institutionally sponsored loans also available. Financial award application deadline: 4/15; financial award applicants required to submit FAFSA. *Faculty research:* High-energy physics, condensed-matter astrophysics, particle and experimental physics, plasma physics. Total annual research expenditures: $26.2 million. *Unit head:* Dr. Jonathan A. Bagger, Chair, 410-516-7346, Fax: 410-516-7239, E-mail: bagger@jhu.edu. *Application contact:* Carmelita D. King, Academic Affairs Administrator, 410-516-7344, Fax: 410-516-7239, E-mail: jazzy@pha.jhu.edu.

See Close-Up on page 347.

Kansas State University, Graduate School, College of Arts and Sciences, Department of Physics, Manhattan, KS 66506. Offers MS, PhD. *Faculty:* 27 full-time (3 women), 2 part-time/adjunct (0 women). *Students:* 61 full-time (18 women); includes 3 minority (1 African American, 2 Hispanic Americans), 42 international. 77 applicants, 34% accepted, 13 enrolled. In 2005, 4 master's, 7 doctorates awarded. Terminal master's awarded for partial completion of doctoral program. *Degree requirements:* For master's, thesis; for doctorate, one foreign language, thesis/dissertation, preliminary exams. *Entrance requirements:* For master's, GRE Subject Test, BS degree in physics, minimum GPA of 3.0 in math, chemistry or engineering; for doctorate, GRE Subject Test. Additional exam requirements/recommendations for international students: Required—TOEFL (minimum score 550 paper-based; 213 computer-based). *Application deadline:* For fall admission, 2/1 for domestic students, 2/1 for international students; for spring admission, 10/1 for domestic students, 8/1 for international students. Applications are processed on a rolling basis. Application fee: $0 ($25 for international students). Electronic applications accepted. *Expenses:* Tuition, state resident: full-time $5,160; part-time $215. Tuition, nonresident: full-time $12,816; part-time $534. Required fees: $564. *Financial support:* In 2005–06, 48 research assistantships with partial tuition reimbursements (averaging $16,072 per year), 17 teaching assistantships with full tuition reimbursements (averaging $12,327 per year) were awarded; fellowships, career-related internships or fieldwork, Federal Work-Study, institutionally sponsored loans, and scholarships/grants also available. Support available to part-time students. Financial award application deadline: 3/1; financial award applicants required to submit FAFSA. *Faculty research:* Physics education, atomic physics, condensed matter physics, high energy physics, cosmology. Total annual research expenditures: $6.9 million. *Unit head:* Dean Zollman, Head, 785-532-1619, Fax: 785-532-6806, E-mail: dzollman@phys.ksu.edu. *Application contact:* Brett Esry, Director, 785-532-1630, Fax: 785-532-6808, E-mail: esry@phys.ksu.edu.

Kent State University, College of Arts and Sciences, Department of Physics, Kent, OH 44242-0001. Offers MA, MS, PhD. Terminal master's awarded for partial completion of doctoral program. *Degree requirements:* For master's, thesis/dissertation, registration; for doctorate, thesis/dissertation, comprehensive exam, registration. *Entrance requirements:* For master's and doctorate, GRE, minimum GPA of 3.0. Additional exam requirements/recommendations for international students: Required—TOEFL. Electronic applications accepted. *Faculty research:* Correlated electron materials physics, liquid crystals, complex fluids, computational biophysics, QCD-Hadranphysics.

See Close-Up on page 349.

Lakehead University, Graduate Studies, Department of Physics, Thunder Bay, ON P7B 5E1, Canada. Offers M Sc. *Degree requirements:* For master's, thesis or alternative. *Entrance requirements:* For master's, minimum B average. Additional exam requirements/recommendations for international students: Required—TOEFL. *Faculty research:* Absorbed water, radiation reaction, superlattices and quantum well structures, polaron interactions.

Lehigh University, College of Arts and Sciences, Department of Physics, Bethlehem, PA 18015-3094. Offers photonics (MS); physics (MS, PhD); polymer science (MS, PhD). Part-time programs available. *Faculty:* 18 full-time (0 women), 3 part-time/adjunct (0 women). *Students:* 39 full-time (7 women), 3 part-time (1 woman); includes 2 minority (1 African American, 1 Hispanic American), 17 international. 61 applicants, 23% accepted, 11 enrolled. In 2005, 8 master's, 7 doctorates awarded. Terminal master's awarded for partial completion of doctoral program. *Degree requirements:* For master's, research project; for doctorate, thesis/dissertation, exam. *Entrance requirements:* For doctorate, GRE General Test. Additional exam requirements/recommendations for international students: Required—TOEFL (minimum score 600 paper-based; 235 computer-based); Recommended—TSE. *Application deadline:* For fall admission, 7/15 for domestic students; for spring admission, 1/15 priority date for domestic students. Applications are processed on a rolling basis. Application fee: $60. Electronic applications accepted. *Financial support:* In 2005–06, 6 fellowships with tuition reimbursements (averaging $18,240 per year), 8 research assistantships with tuition reimbursements (averaging $18,240 per year), 14 teaching assistantships with tuition reimbursements (averaging $18,240 per year) were awarded; Federal Work-Study and institutionally sponsored loans also available. Financial award application deadline:1/15. *Faculty research:* Condensed matter physics; atomic, molecular and optical physics; plasma physics; complex fluids; computational physics. Total annual research expenditures: $2.3 million. *Unit head:* Dr. Michael J. Stavola, Chair, 610-758-3903, Fax: 610-758-5730, E-mail: mjsa@lehigh.edu. *Application contact:* Dr. Volkmar Dierolf, Graduate Admissions Officer, 610-758-3915, Fax: 610-758-5730, E-mail: vod2@lehigh.edu.

Louisiana State University and Agricultural and Mechanical College, Graduate School, College of Basic Sciences, Department of Physics and Astronomy, Baton Rouge, LA 70803. Offers astronomy (PhD); astrophysics (PhD); physics (MS, PhD). *Faculty:* 45 full-time (1 woman). *Students:* 72 full-time (8 women), 4 part-time (1 woman); includes 2 African Americans, 2 Asian Americans or Pacific Islanders, 1 Hispanic American, 33 international. Average age 28. 99 applicants, 23% accepted, 76 enrolled. In 2005, 6 master's, 4 doctorates awarded. Terminal master's awarded for partial completion of doctoral program. *Degree requirements:* For master's, thesis or alternative; for doctorate, thesis/dissertation. *Entrance requirements:* For master's and doctorate, GRE General Test, minimum GPA of 3.0. Additional exam requirements/recommendations for international students: Required—TOEFL (minimum score 550 paper-based; 213 computer-based). *Application deadline:* For fall admission, 1/25 priority date for domestic students, 5/15 priority date for international students. Applications are processed on a rolling basis. Application fee: $25. Electronic applications accepted. *Financial support:* In 2005–06, 72 students received support, including 3 fellowships with full tuition reimbursements available (averaging $15,000 per year), 40 research assistantships with full and partial tuition reimbursements available (averaging $16,240 per year), 27 teaching assistantships with full and partial tuition reimbursements available (averaging $17,652 per year); institutionally sponsored loans, tuition waivers (full and partial), and unspecified assistantships also available. Financial award application deadline: 3/15; financial award applicants required to submit FAFSA. *Faculty research:* Experimental and theoretical atomic, nuclear, particle, cosmic-ray, low-temperature, and condensed-matter physics. Total annual research expenditures: $5.8 million. *Unit head:* Dr. Roger McNeil, Chair, 225-578-2261, Fax: 225-578-5855, E-mail: mcneil@phys.lsu.edu. *Application contact:* Dr. James Matthews, Graduate Adviser, 225-578-8598, Fax: 225-578-5855, E-mail: jmatth5@lsu.edu.

Louisiana Tech University, Graduate School, College of Engineering and Science, Department of Physics, Ruston, LA 71272. Offers applied computational analysis and modeling (PhD); physics (MS). Part-time programs available. *Degree requirements:* For master's, thesis or alternative; for doctorate, thesis/dissertation. *Entrance requirements:* For master's, GRE General Test, minimum GPA of 3.0 in last 60 hours. Additional exam requirements/recommendations for international students: Required—TOEFL. *Faculty research:* Experimental high energy physics, laser/optics, computational physics, quantum gravity.

Marshall University, Academic Affairs Division, Graduate College, College of Science, Department of Physical Science and Physics, Huntington, WV 25755. Offers physical science (MS). *Faculty:* 6 full-time (0 women), 1 part-time/adjunct (0 women). *Students:* 18 full-time (8 women), 7 part-time (2 women); includes 1 minority (African American), 3 international. Average age 29. In 2005, 12 degrees awarded. *Degree requirements:* For master's, thesis optional. *Entrance requirements:* For master's, GRE General Test. *Unit head:* Dr. Nicola Orsini, Chairperson, 304-696-2756, E-mail: orsini@marshall.edu. *Application contact:* Information Contact, 304-746-1900, Fax: 304-746-1902, E-mail: services@marshall.edu.

Massachusetts Institute of Technology, School of Science, Department of Physics, Cambridge, MA 02139-4307. Offers PhD. *Faculty:* 70 full-time (5 women). *Students:* 229 full-time (30 women); includes 13 minority (1 African American, 9 Asian Americans or Pacific Islanders, 3 Hispanic Americans), 112 international. Average age 27. 663 applicants, 12% accepted, 27 enrolled. In 2005, 36 doctorates awarded. *Degree requirements:* For doctorate, thesis/dissertation, comprehensive exam. *Entrance requirements:* For doctorate, GRE General Test, GRE Subject Test in physics. Additional exam requirements/recommendations for international students: Required—TOEFL (minimum score 600 paper-based; 250 computer-based). *Application deadline:* For fall admission, 1/1 for domestic students, 1/1 for international students; for spring admission, 11/1 for domestic students, 11/1 for international students. Application fee: $70. Electronic applications accepted. *Expenses:* Tuition: Full-time $32,100. Required fees: $200. Part-time tuition and fees vary according to course load. *Financial support:* In 2005–06, 225 students received support, including 35 fellowships with tuition reimbursements available (averaging $22,885 per year), 153 research assistantships with tuition reimbursements available (averaging $24,422 per year), 34 teaching assistantships with tuition reimbursements available (averaging $25,361 per year); career-related internships or fieldwork, Federal Work-Study, institutionally sponsored loans, scholarships/grants, health care benefits, and unspecified assistantships also available. *Faculty research:* Particle/QCD physics, condensed matter physics, atomic physics, astrophysics, string theory. Total annual research expenditures: $64.2 million. *Unit head:* Prof. Marc A. Kastner, Department Head, 617-253-4801, Fax: 617-253-8554, E-mail: mkastner@mit.edu. *Application contact:* Department of Physics, 617-253-9703, Fax: 617-258-8319, E-mail: physics-grad@mit.edu.

McGill University, Faculty of Graduate and Postdoctoral Studies, Faculty of Science, Department of Physics, Montréal, QC H3A 2T5, Canada. Offers M Sc, PhD. Terminal master's awarded for partial completion of doctoral program. *Degree requirements:* For master's and doctorate, thesis/dissertation, registration. *Entrance requirements:* For master's, minimum GPA of 3.0, B.Sc. in physics; for doctorate, M.Sc. in physics. Additional exam requirements/recommendations for international students: Required—TOEFL (minimum score 550 paper-based; 213 computer-based), IELT (minimum score 7). *Faculty research:* High-energy, condensed-matter, and nuclear physics; biophysics; mathphysics; geophysics/atmospheric physics; astrophysics.

McMaster University, School of Graduate Studies, Faculty of Science, Department of Physics and Astronomy, Hamilton, ON L8S 4M2, Canada. Offers astrophysics (PhD); physics (PhD). Part-time programs available. *Degree requirements:* For doctorate, thesis/dissertation, comprehensive exam. *Entrance requirements:* For doctorate, minimum B+ average. Additional exam requirements/recommendations for international students: Required—TOEFL (minimum score 550 paper-based; 213 computer-based). *Faculty research:* Condensed matter, astrophysics, nuclear, medical, nonlinear dynamics.

Memorial University of Newfoundland, School of Graduate Studies, Department of Physics and Physical Oceanography, St. John's, NL A1C 5S7, Canada. Offers atomic and molecular physics (M Sc, PhD); condensed matter physics (M Sc, PhD); physical oceanography (M Sc, PhD); physics (M Sc). Part-time programs available. *Students:* 24 full-time (10 women), 1 part-time, 16 international. 18 applicants, 56% accepted, 7 enrolled. In 2005, 9 master's, 2 doctorates awarded. *Degree requirements:* For master's, thesis, seminar presentation on thesis topic; for doctorate, thesis/dissertation, oral defense of thesis, comprehensive exam. *Entrance requirements:* For master's, honors B Sc or equivalent; for doctorate, M Sc or equivalent. *Application deadline:* Applications are processed on a rolling basis. Application fee: $40 Canadian dollars. Electronic applications accepted. *Expenses:* Tuition: Part-time $733 per term. Tuition and fees vary according to degree level and program. *Financial support:* Fellowships, research assistantships, teaching assistantships available. *Faculty research:* Experiment and theory in atomic and molecular physics, condensed matter physics, physical oceanography, theoretical geophysics and applied nuclear physics. *Unit head:* Dr. John Whitehead, Head, 709-737-8737, Fax: 709-737-8739, E-mail: johnw@physics.mun.ca. *Application contact:* Dr. Mike Morrow, Graduate Officer, 709-737-4361, Fax: 709-737-8739, E-mail: myke@physics.mun.ca.

Miami University, Graduate School, College of Arts and Sciences, Department of Physics, Oxford, OH 45056. Offers MAT, MS. Part-time programs available. *Degree requirements:* For master's, final exam. *Entrance requirements:* For master's, minimum undergraduate GPA of 3.0 during previous 2 years or 2.75 overall. Additional exam requirements/recommendations for international students: Required—TOEFL (minimum score 550 paper-based; 213 computer-based), TWE (minimum score 4). Electronic applications accepted.

Michigan State University, The Graduate School, College of Natural Science, Department of Physics and Astronomy, East Lansing, MI 48824. Offers astrophysics and astronomy (MS, PhD); physics (MS, PhD). *Faculty:* 52 full-time (3 women). *Students:* 135 full-time (23 women), 6 part-time (3 women); includes 6 minority (2 American Indian/Alaska Native, 3 Asian Americans or Pacific Islanders, 1 Hispanic American), 70 international. Average age 28. 252 applicants, 8% accepted. In 2005, 18 master's, 19 doctorates awarded. *Degree requirements:* For master's, qualifying exam, thesis optional; for doctorate, thesis/dissertation, qualifying exam, comprehensive exam. *Entrance requirements:* For master's, GRE General Test (recommended), minimum GPA of 3.0 in science/math courses, coursework equivalent to a major in physics or astronomy, 3 letters of recommendation; for doctorate, GRE General Test (recommended), minimum GPA of 3.0 in science/math courses, coursework equivalent to a major in physics or astronomy, 3 letters of recommendation, research experience (recommended). Additional exam requirements/recommendations for international students: Required—TOEFL (minimum score 550 paper-based; 213 computer-based), Michigan State University ELT (85), Michigan ELAB (83). *Application deadline:* For fall admission, 12/27 for domestic students; for spring admission, 9/30 for domestic students. Application fee: $50. Electronic applications accepted. *Expenses:* Tuition, state resident: part-time $330 per credit hour. Tuition, nonresident: part-time $685 per credit hour. Tuition and fees vary according to program. *Financial support:* In 2005–06, 20 fellowships with tuition reimbursements (averaging $8,526 per year), 88 research assistantships with tuition reimbursements (averaging $14,281 per year), 40 teaching assistantships with tuition reimbursements (averaging $12,680 per year) were awarded; scholarships/grants and unspecified assistantships also available. *Faculty research:* Nuclear and accelerator physics, high energy physics, condensed matter physics, biophysics, astrophysics and astronomy. Total annual research expenditures: $6.4 million. *Unit head:* Dr. Wolfgang W. Bauer, Chairperson, 517-355-9200 Ext. 2015, Fax: 517-355-4500, E-mail: bauer@pa.msu.edu. *Application contact:* Debbie Simmons, Graduate Secretary, 517-355-9200 Ext. 2032, Fax: 517-353-4500, E-mail: grd_chair@pa.msu.edu.

Michigan Technological University, Graduate School, College of Sciences and Arts, Department of Physics, Houghton, MI 49931-1295. Offers engineering physics (PhD); physics (MS, PhD). Part-time programs available. *Faculty:* 20 full-time (1 woman), 1 part-time/adjunct (0 women). *Students:* 38 full-time (9 women), 29 international. Average age 28. 47 applicants, 32% accepted, 11 enrolled. In 2005, 5 master's, 2 doctorates awarded. Terminal master's awarded for partial completion of doctoral program. *Median time to degree:* Of those who began their doctoral program in fall 1997, 17% received their degree in 8 years or less. *Degree requirements:* For master's, thesis (for some programs), comprehensive exam (for some programs), registration; for doctorate, thesis/dissertation, preliminary exam (research proposal), comprehensive exam, registration. *Entrance requirements:* For master's and doctorate, BS in physics or related discipline. Additional exam requirements/recommendations for international students: Required—TOEFL (minimum score 570 paper-based; 230 computer-based). *Application deadline:* For fall admission, 3/1 for domestic students. Applications are processed on a rolling basis. Application fee: $40 ($45 for international students). Electronic applications accepted. *Expenses:* Tuition, nonresident: full-time $11,232; part-time $468 per credit. Required fees: $754; $377 per semester. Full-time tuition and fees vary according to course load, degree level and program. *Financial support:* In 2005–06, 37 students received support, including 1 fellowship with full tuition reimbursement available (averaging $9,542 per year), 19 research assistantships with full tuition reimbursements available (averaging $9,542 per year), 17 teaching assistantships with full tuition reimbursements available (averaging $9,542 per year); career-related internships or fieldwork, Federal Work-Study, scholarships/grants, health care benefits, tuition waivers (partial), unspecified assistantships, and co-op also available. Financial award applicants required to submit FAFSA. *Faculty research:* Atmospheric physics, astrophysics, biophysics, materials physics, atomic/molecular physics. Total annual research expenditures: $1.3 million. *Unit head:* Dr. Ravindra Pandey, Chair, 906-487-2831, Fax: 906-487-2933, E-mail: pandey@mtu.edu. *Application contact:* Kathleen S. Wollan, Secretary, 906-487-2086, Fax: 906-487-2933, E-mail: kswollan@mtu.edu.

Minnesota State University Mankato, College of Graduate Studies, College of Science, Engineering and Technology, Department of Physics and Astronomy, Mankato, MN 56001. Offers physics (MS); physics and astronomy (MT). *Students:* 3 full-time (1 woman), 2 part-time. Average age 34. In 2005, 2 degrees awarded. *Degree requirements:* For master's, one foreign language, thesis or alternative, comprehensive exam. *Entrance requirements:* For master's, minimum GPA of 3.0 during previous 2 years. Additional exam requirements/recommendations for international students: Required—TOEFL. *Application deadline:* For fall admission, 7/1 for domestic students; for spring admission, 11/1 for domestic students. Applications are processed on a rolling basis. Application fee: $40. Electronic applications accepted. *Expenses:* Tuition, state resident: part-time $243 per credit. Tuition, nonresident: part-time $400 per credit. Required fees: $30 per credit. *Financial support:* Research assistantships, teaching assistantships with full tuition reimbursements, Federal Work-Study and unspecified assistantships available. Support available to part-time students. Financial award application deadline: 3/15; financial award applicants required to submit FAFSA. *Unit head:* Dr. Mark Pickar, Chairperson, 507-389-5743. *Application contact:* 507-389-2321, E-mail: grad@mnsu.edu.

Mississippi State University, College of Arts and Sciences, Department of Physics and Astronomy, Mississippi State, MS 39762. Offers engineering physics (PhD); physics (MS). Part-time programs available. *Faculty:* 17 full-time (2 women), 4 part-time/adjunct (1 woman). *Students:* 3 full-time (0 women), 2 part-time; includes 1 minority (African American), 1 international. Average age 28. 10 applicants, 70% accepted, 5 enrolled. In 2005, 4 degrees awarded. *Degree requirements:* For master's and doctorate, thesis/dissertation, comprehensive oral or written exam. *Entrance requirements:* Additional exam requirements/recommendations for international students: Required—TOEFL, TSE. *Application deadline:* For fall admission, 7/1 for domestic students; for spring admission, 11/1 priority date for domestic students. Applications are processed on a rolling basis. Application fee: $30. Electronic applications accepted. *Expenses:* Tuition, state resident: full-time $4,312; part-time $240 per hour. Tuition, nonresident: full-time $9,772; part-time $543 per hour. International tuition: $10,102 full-time. Tuition and fees vary according to course load. *Financial support:* In 2005–06, 8 teaching assistantships with full tuition reimbursements (averaging $9,509 per year) were awarded; research assistantships with full tuition reimbursements, Federal Work-Study, institutionally sponsored loans, and unspecified assistantships also available. Financial award application deadline: 3/15; financial award applicants required to submit FAFSA. *Faculty research:* Atomic/molecular spectroscopy, theoretical optics, gamma-ray astronomy, experimental nuclear physics, computational physics. Total annual research expenditures: $3.5 million. *Unit head:* Dr. Mark A. Novotny, Head, 662-325-2806, Fax: 662-325-8898, E-mail: man40@ra.msstate.edu. *Application contact:* Philip G. Bonfanti, Director of Admissions, 662-325-4104, Fax: 662-325-8872, E-mail: admit@msstate.edu.

Montana State University, College of Graduate Studies, College of Letters and Science, Department of Physics, Bozeman, MT 59717. Offers MS, PhD. Part-time programs available. *Faculty:* 17 full-time (4 women), 3 part-time/adjunct (2 women). *Students:* 46 full-time (11 women), 7 part-time (2 women); includes 1 minority (Asian American or Pacific Islander), 15 international. Average age 28. 42 applicants, 40% accepted, 11 enrolled. In 2005, 11 master's, 3 doctorates awarded. *Degree requirements:* For master's, thesis (for some programs), comprehensive exam, registration; for doctorate, thesis/dissertation, comprehensive exam, registration. *Entrance requirements:* For master's and doctorate, GRE General Test. Additional exam requirements/recommendations for international students: Required—TOEFL (minimum score 550 paper-based; 213 computer-based). *Application deadline:* For fall admission, 7/15 priority date for domestic students, 5/15 priority date for international students; for spring admission, 12/1 priority date for domestic students, 10/1 priority date for international students. Applications are processed on a rolling basis. Application fee: $30. Electronic applications accepted. *Expenses:* Tuition, state resident: full-time $4,132. Tuition, nonresident: full-time $1,132. *Financial support:* In 2005–06, 51 students received support, including 4 fellowships with full tuition reimbursements available (averaging $18,000 per year), 30 research assistantships with full tuition reimbursements available (averaging $18,000 per year), 17 teaching assistantships with full tuition reimbursements available (averaging $18,000 per year); scholarships/grants, traineeships, health care benefits, and unspecified assistantships also available. Support available to part-time students. Financial award application deadline: 3/1; financial award applicants required to submit FAFSA. *Faculty research:* Astrophysics and relativity and cosmology, condensed matter physics, optics, solar physics, physics education. Total annual research expenditures: $8.2 million. *Unit head:* Dr. William Hiscock, Head, 406-994-3614, Fax: 406-994-4452, E-mail: hiscock@physics.montana.edu.

Morgan State University, School of Graduate Studies, School of Computer, Mathematical, and Natural Sciences, Interdisciplinary Program in Science, Baltimore, MD 21251. Offers biology (MS); chemistry (MS); physics (MS). *Students:* 9 (2 women); includes 4 minority (all African Americans) 4 international. In 2005, 5 degrees awarded. *Degree requirements:* For master's, thesis, oral defense of thesis, comprehensive exam. *Entrance requirements:* For master's, GRE General Test, minimum GPA of 2.5. *Application deadline:* For fall admission, 2/1 for domestic students; for spring admission, 10/1 priority date for domestic students. Applications are processed on a rolling basis. Application fee: $0. *Expenses:* Tuition, state resident: part-time $272 per credit. Tuition, nonresident: part-time $478 per credit. Required fees: $58 per credit. *Financial support:* Fellowships, research assistantships, career-related internships or fieldwork, Federal Work-Study, institutionally sponsored loans, scholarships/grants, health care benefits, and unspecified assistantships available. Support available to part-time students. *Unit head:* Dr. Juarine Stewart, Dean, 443-885-4515, Fax: 443-885-8215. *Application contact:* Dr. James E. Waller, Admissions and Program Officer, 443-885-3185, Fax: 443-885-8226, E-mail: jwaller@moae.morgan.edu.

Naval Postgraduate School, Graduate Programs, Department of Physics, Monterey, CA 93943. Offers applied physics (MS); engineering acoustics (MS); physics (MS, PhD). Program only open to commissioned officers of the United States and friendly nations and selected United States federal civilian employees. Part-time programs available. *Students:* 59 full-time. In 2005, 28 master's, 1 doctorate awarded. *Degree requirements:* For master's, thesis; for doctorate, one foreign language, thesis/dissertation. *Unit head:* Dr. James H. Luscombe, Chairman, 831-656-2896, E-mail: luscombe@nps.navy.mil. *Application contact:* Tracy Hammond, Acting Director of Admissions, 831-656-3059, Fax: 831-656-2891, E-mail: thammond@bps.navy.mil.

New Mexico Institute of Mining and Technology, Graduate Studies, Department of Physics, Socorro, NM 87801. Offers astrophysics (MS, PhD); atmospheric physics (MS, PhD); instrumentation (MS); mathematical physics (PhD). *Degree requirements:* For master's, thesis optional; for doctorate, thesis/dissertation. *Entrance requirements:* For master's, GRE General Test; for doctorate, GRE General Test, GRE Subject Test. Additional exam requirements/recommendations for international students: Required—TOEFL (minimum score 540 paper-based; 207 computer-based). *Faculty research:* Cloud physics, stellar and extragalactic processes.

New Mexico State University, Graduate School, College of Arts and Sciences, Department of Physics, Las Cruces, NM 88003-8001. Offers MS, PhD. Part-time programs available. *Faculty:* 8 full-time (0 women), 8 part-time/adjunct (1 woman). *Students:* 41 full-time (5 women), 3 part-time (1 woman); includes 7 minority (1 African American, 1 Asian American or Pacific Islander, 5 Hispanic Americans), 25 international. Average age 31. 26 applicants, 58% accepted, 8 enrolled. In 2005, 5 master's, 2 doctorates awarded. Terminal master's awarded for partial completion of doctoral program. *Median time to degree:* Of those who began their doctoral program in fall 1997, 100% received their degree in 8 years or less. *Degree requirements:* For master's, thesis optional; for doctorate, thesis/dissertation, comprehensive exam. *Entrance requirements:* For master's and doctorate, GRE General Test, GRE Subject Test. Additional exam requirements/recommendations for international students: Required—TOEFL. *Application deadline:* For fall admission, 3/1 priority date for domestic students, 3/1 priority date for international students; for spring admission, 10/1 priority date for domestic students, 10/1 priority date for international students. Applications are processed on a rolling basis. Application fee: $30 ($50 for international students). Electronic applications accepted. *Expenses:* Tuition, state resident: full-time $3,156; part-time $175 per credit. Tuition, nonresident: full-time $12,510; part-time $565 per credit. Required fees: $1,050. *Financial support:* In 2005–06, 2 fellowships, 15 research assistantships, 21 teaching assistantships were awarded. Financial award application deadline: 3/15. *Faculty research:* Nuclear and particle physics, optics, materials science, geophysics, physics education, atmospheric physics. *Unit head:* Dr. Gary S. Kyle, Head, 505-646-3831, Fax: 505-646-1934, E-mail: kyle@nmsu.edu. *Application contact:* Dr. Matthias Burkhardt, Information Contact, 505-646-1928, Fax: 505-646-1934, E-mail: physics@nmsu.edu.

New York University, Graduate School of Arts and Science, Department of Physics, New York, NY 10012-1019. Offers MS, PhD. Part-time programs available. *Faculty:* 25 full-time (1 woman), 5 part-time/adjunct. *Students:* 43 full-time (5 women); includes 1 minority (Hispanic American), 33 international. Average age 27. 178 applicants, 11% accepted, 8 enrolled. In 2005, 11 degrees awarded. Terminal master's awarded for partial completion of doctoral program. *Degree requirements:* For master's, thesis (for some programs); for doctorate, one foreign language, thesis/dissertation, research seminar, teaching experience. *Entrance requirements:* For master's, GRE General Test, GRE Subject Test, bachelor's degree in physics; for doctorate, GRE General Test, GRE Subject Test. Additional exam requirements/recommendations for international students: Required—TOEFL. *Application deadline:* For fall admission, 1/4 for domestic students. Application fee: $80. *Financial support:* Fellowships with tuition reimbursements, research assistantships with tuition reimbursements, teaching assistantships with tuition reimbursements, Federal Work-Study, institutionally sponsored loans, scholarships/grants, health care benefits, and unspecified assistantships available. Financial award application deadline: 1/4; financial award applicants required to submit FAFSA. *Faculty research:* Atomic physics, elementary particles and fields, astrophysics, condensed-matter physics, neuromagnetism. *Unit head:* David Grier, Chairman, 212-998-7700, Fax: 212-995-4016, E-mail: dgphys@nyu.

Physics

New York University (continued)
edu. *Application contact:* Andrew Kent, Director of Graduate Studies, 212-998-7700, Fax: 212-995-4016, E-mail: dgsphys@nyu.edu.

North Carolina State University, Graduate School, College of Physical and Mathematical Sciences, Department of Physics, Raleigh, NC 27695. Offers MS, PhD. Part-time programs available. Terminal master's awarded for partial completion of doctoral program. *Degree requirements:* For master's, thesis (for some programs); for doctorate, thesis/dissertation. *Entrance requirements:* For master's and doctorate, GRE General Test, GRE Subject Test. Electronic applications accepted. *Faculty research:* Astrophysics, optics, physics education, biophysics, geophysics.

North Dakota State University, The Graduate School, College of Science and Mathematics, Department of Physics, Fargo, ND 58105. Offers MS, PhD. Part-time programs available. *Faculty:* 3 full-time. *Students:* 6 full-time (0 women); includes 5 minority (all Asian Americans or Pacific Islanders) Average age 25. 10 applicants, 0% accepted.Terminal master's awarded for partial completion of doctoral program. *Degree requirements:* For master's, thesis/dissertation; for doctorate, thesis/dissertation, comprehensive exam. *Entrance requirements:* Additional exam requirements/recommendations for international students: Required—TOEFL (minimum score 550 paper-based; 215 computer-based). *Application deadline:* For fall admission, 3/1 priority date for domestic students, 5/1 priority date for international students; for spring admission, 9/1 priority date for domestic students, 9/1 priority date for international students. Applications are processed on a rolling basis. Application fee: $45 ($60 for international students). *Financial support:* In 2005–06, 2 students received support, including 2 research assistantships with tuition reimbursements available (averaging $16,000 per year), teaching assistantships with tuition reimbursements available (averaging $12,000 per year); career-related internships or fieldwork, scholarships/grants, and unspecified assistantships also available. Support available to part-time students. Financial award application deadline: 4/15; financial award applicants required to submit FAFSA. *Faculty research:* Biophysics; condensed matter; surface physics; general relativity, gravitation, and space physics; nonlinear physics. Total annual research expenditures: $305,839. *Unit head:* Dr. Daniel Kroll, Chair, 701-231-8974, Fax: 701-231-7088. *Application contact:* Dr. Alexander Wagner, Graduate Advisory Committee Chair, 701-231-9582, Fax: 701-231-7088, E-mail: alexander.wagner@ndsu.edu.

Northeastern University, College of Arts and Sciences, Department of Physics, Boston, MA 02115-5096. Offers MS, PhD. Part-time programs available. Terminal master's awarded for partial completion of doctoral program. *Degree requirements:* For master's, thesis optional; for doctorate, thesis/dissertation. *Entrance requirements:* Additional exam requirements/recommendations for international students: Required—TOEFL. *Faculty research:* High-energy theory and experimentation, astrophysics, biophysics, condensed-matter theory and experimentation.

See Close-Up on page 351.

Northern Illinois University, Graduate School, College of Liberal Arts and Sciences, Department of Physics, De Kalb, IL 60115-2854. Offers MS, PhD. Part-time programs available. *Faculty:* 18 full-time (3 women), 3 part-time/adjunct (0 women). *Students:* 28 full-time (4 women), 28 part-time (4 women); includes 1 minority (African American), 12 international. Average age 32. 37 applicants, 59% accepted, 14 enrolled. In 2005, 7 master's, 1 doctorate awarded. Terminal master's awarded for partial completion of doctoral program. *Degree requirements:* For master's, thesis or alternative, research seminar, comprehensive exam; for doctorate, thesis/dissertation, candidacy exam, dissertation defense, research seminar. *Entrance requirements:* For master's, GRE General Test, minimum GPA of 2.75; for doctorate, GRE General Test, GRE Subject Test (physics), bachelor's degree in physics or related field, minimum undergraduate GPA of 2.75, minimum graduate GPA of 3.2. Additional exam requirements/recommendations for international students: Required—TOEFL (minimum score 550 paper-based; 213 computer-based). *Application deadline:* For fall admission, 6/1 for domestic students, 5/1 for international students; for spring admission, 11/1 for domestic students, 10/1 for international students. Applications are processed on a rolling basis. Application fee: $30. Electronic applications accepted. *Expenses:* Tuition, state resident: full-time $4,565; part-time $191 per credit hour. Tuition, nonresident: full-time $9,129; part-time $382 per credit hour. *Financial support:* In 2005–06, 18 research assistantships with full tuition reimbursements, 19 teaching assistantships with full tuition reimbursements were awarded; fellowships with full tuition reimbursements, career-related internships or fieldwork, Federal Work-Study, scholarships/grants, and unspecified assistantships also available. Support available to part-time students. Financial award applicants required to submit FAFSA. *Faculty research:* Band-structure interpolation schemes, nonlinear procession beams, Mossbauer spectroscopy, beam physics. *Unit head:* Dr. Susan Mini, Chair, 815-753-6470, Fax: 815-753-8565, E-mail: mini@niuhep.physics.niu.edu. *Application contact:* Dr. David Hedin, Director of Graduate Studies, 815-753-6483, E-mail: hedin@niu.edu.

Northwestern University, The Graduate School, Judd A. and Marjorie Weinberg College of Arts and Sciences, Department of Physics and Astronomy, Evanston, IL 60208. Offers astrophysics (PhD); physics (MS, PhD). Admissions and degrees offered through The Graduate School. *Faculty:* 30 full-time (4 women), 2 part-time/adjunct (0 women). *Students:* 113 (26 women). Average age 24. 140 applicants, 21% accepted, 12 enrolled. In 2005, 4 master's, 10 doctorates awarded. *Median time to degree:* Of those who began their doctoral program in fall 1997, 50% received their degree in 8 years or less. *Degree requirements:* For doctorate, thesis/dissertation, qualifying exam. *Entrance requirements:* For doctorate, GRE General Test, GRE Subject Test. Additional exam requirements/recommendations for international students: Required—TOEFL. Application fee: $60 ($75 for international students). *Financial support:* In 2005–06, 57 students received support, including 9 fellowships with full tuition reimbursements available (averaging $12,906 per year), 28 research assistantships with partial tuition reimbursements available (averaging $18,732 per year), 20 teaching assistantships with full tuition reimbursements available (averaging $13,329 per year); career-related internships or fieldwork, Federal Work-Study, and institutionally sponsored loans also available. Financial award application deadline: 1/15; financial award applicants required to submit FAFSA. *Faculty research:* Nuclear and particle physics, condensed-matter physics, nonlinear physics, astrophysics. Total annual research expenditures: $6.3 million. *Unit head:* Melvin Ulmer, Chair, 847-491-5633, Fax: 847-491-9982, E-mail: physics-astronomy@northwestern.edu. *Application contact:* Mayda Velasco, Admission Officer, 847-467-7099, Fax: 847-491-9982, E-mail: physics-astronomy@northwestern.edu.

Oakland University, Graduate Study and Lifelong Learning, College of Arts and Sciences, Department of Physics, Rochester, MI 48309-4401. Offers medical physics (PhD); physics (MS). *Faculty:* 3 full-time (0 women). *Students:* 15 full-time (6 women), 1 part-time; includes 1 minority (Asian American or Pacific Islander), 6 international. Average age 35. 10 applicants, 80% accepted, 4 enrolled. In 2005, 4 master's, 2 doctorates awarded. *Degree requirements:* For doctorate, thesis/dissertation. *Entrance requirements:* For master's, minimum GPA of 3.0 for unconditional admission; for doctorate, GRE Subject Test, GRE General Test, minimum GPA of 3.0 for unconditional admission. Additional exam requirements/recommendations for international students: Required—TOEFL (minimum score 550 paper-based; 213 computer-based). *Application deadline:* For fall admission, 7/15 priority date for domestic students, 5/1 priority date for international students. For winter admission, 12/1 for domestic students; for spring admission, 3/15 for domestic students. Applications are processed on a rolling basis. Application fee: $30. Electronic applications accepted. *Expenses:* Contact institution. *Financial support:* Fellowships, career-related internships or fieldwork, Federal Work-Study, institutionally sponsored loans, and tuition waivers (full) available. Financial award application deadline: 3/1; financial award applicants required to submit FAFSA. *Faculty research:* Multifunctional ferromagnetic ferroelectric heterostructures, elastic and plastic deformation in binary alloy crystalization, magnoelectric interactions and microwave signal processors, quantitative molecular imagings of articular cartilage. Total annual research expenditures: $348,451. *Unit head:* Dr. Andrei N. Slavin, Chair, 248-370-2352, Fax: 248-370-3401, E-mail: slavin@oakland.edu.

The Ohio State University, Graduate School, College of Mathematical and Physical Sciences, Department of Physics, Columbus, OH 43210. Offers MS, PhD. *Degree requirements:* For master's, thesis optional; for doctorate, thesis/dissertation. *Entrance requirements:* For master's and doctorate, GRE General Test, GRE Subject Test. Additional exam requirements/recommendations for international students: Required—TOEFL (minimum score 600 paper-based; 250 computer-based). Electronic applications accepted.

Ohio University, Graduate Studies, College of Arts and Sciences, Department of Physics and Astronomy, Athens, OH 45701-2979. Offers astronomy (MS, PhD); physics (MS, PhD). Part-time programs available. *Faculty:* 29 full-time (3 women), 1 (woman) part-time/adjunct. *Students:* 67 full-time (21 women), 57 international. Average age 24. 95 applicants, 42% accepted, 19 enrolled. In 2005, 12 master's, 12 doctorates awarded. Terminal master's awarded for partial completion of doctoral program. *Median time to degree:* Of those who began their doctoral program in fall 1997, 100% received their degree in 8 years or less. *Degree requirements:* For master's, thesis or alternative; for doctorate, thesis/dissertation, comprehensive exam. *Entrance requirements:* For master's and doctorate, minimum GPA of 3.0. Additional exam requirements/recommendations for international students: Required—TOEFL (minimum score 600 paper-based; 250 computer-based), IELT (minimum score 7), TWE (minimum score 4). *Application deadline:* For fall admission, 4/1 priority date for domestic students, 4/1 priority date for international students. Applications are processed on a rolling basis. Application fee: $45. Electronic applications accepted. *Financial support:* In 2005–06, 2 fellowships with full tuition reimbursements (averaging $9,400 per year), 31 research assistantships with full tuition reimbursements (averaging $19,000 per year), 33 teaching assistantships with full tuition reimbursements (averaging $15,600 per year) were awarded; scholarships/grants and unspecified assistantships also available. Financial award application deadline: 4/1. *Faculty research:* Nuclear physics, condensed-matter physics, nonlinear systems, acoustics, astrophysics. Total annual research expenditures: $3.2 million. *Unit head:* Dr. Joseph Shields, Chair, 740-593-0336, Fax: 740-593-6433, E-mail: shields@helios.phy.ohiou.edu. *Application contact:* Dr. Daniel S. Carman, Graduate Admissions Chair, 740-593-3964, Fax: 740-593-0433, E-mail: gradapp@phy.ohiou.edu.

See Close-Up on page 355.

Oklahoma State University, College of Arts and Sciences, Department of Physics, Stillwater, OK 74078. Offers photonics (MS, PhD); physics (MS, PhD). *Faculty:* 35 full-time (7 women), 2 part-time/adjunct (0 women). *Students:* 13 full-time (1 woman), 28 part-time (4 women); includes 3 minority (1 American Indian/Alaska Native, 2 Asian Americans or Pacific Islanders), 21 international. Average age 29. 40 applicants, 48% accepted, 10 enrolled. In 2005, 5 master's, 4 doctorates awarded. *Degree requirements:* For master's, thesis, thesis or report; for doctorate, thesis/dissertation, oral defense of dissertation, preliminary exam, qualifying exam. *Entrance requirements:* Additional exam requirements/recommendations for international students: Required—TOEFL. *Application deadline:* For fall admission, 3/15 priority date for domestic students, 3/1 priority date for international students. Applications are processed on a rolling basis. Application fee: $40 ($75 for international students). Electronic applications accepted. *Expenses:* Tuition, state resident: full-time $4,253; part-time $139 per credit hour. Tuition, nonresident: full-time $12,569; part-time $485 per credit hour. Required fees: $43 per credit hour. One-time fee: $20 part-time. Tuition and fees vary according to course load and program. *Financial support:* In 2005–06, 27 research assistantships (averaging $16,754 per year), 31 teaching assistantships with partial tuition reimbursements (averaging $15,743 per year) were awarded; Federal Work-Study, traineeships, health care benefits, tuition waivers (partial), and unspecified assistantships also available. Support available to part-time students. Financial award application deadline: 3/1. *Faculty research:* Lasers and photonics, non-linear optical materials, turbulence, structure and function of biological membranes, particle theory. *Unit head:* Dr. James Wicksted, Head, 405-744-5796. *Application contact:* Dr. Paul A. Westhaus, Graduate Coordinator, 405-744-5815, E-mail: paul.westhaus@okstate.edu.

Old Dominion University, College of Sciences, Program in Physics, Norfolk, VA 23529. Offers MS, PhD. *Faculty:* 20 full-time (2 women), 20 part-time/adjunct (3 women). *Students:* 10 full-time (2 women), 32 part-time (9 women), 27 international. Average age 30. 40 applicants, 33% accepted, 7 enrolled. In 2005, 15 master's awarded. Terminal master's awarded for partial completion of doctoral program. *Median time to degree:* Of those who began their doctoral program in fall 1997, 100% received their degree in 8 years or less. *Degree requirements:* For master's, thesis optional; for doctorate, thesis/dissertation, comprehensive exam. *Entrance requirements:* For master's, BS in physics or related field, minimum GPA of 3.0 in major; for doctorate, GRE General Test, minimum GPA of 3.0. Additional exam requirements/recommendations for international students: Required—TOEFL (minimum score 217 paper-based). *Application deadline:* For fall admission, 7/1 for domestic students, 7/1 for international students. Applications are processed on a rolling basis. Application fee: $40. Electronic applications accepted. *Expenses:* Tuition, state resident: part-time $263 per credit hour. Tuition, nonresident: part-time $661 per credit hour. Required fees: $39 per semester. Part-time tuition and fees vary according to campus/location. *Financial support:* In 2005–06, 4 research assistantships with full tuition reimbursements (averaging $19,500 per year), 8 teaching assistantships with full tuition reimbursements (averaging $19,500 per year) were awarded; fellowships, career-related internships or fieldwork, scholarships/grants, and tuition waivers (partial) also available. Support available to part-time students. Financial award application deadline: 2/15; financial award applicants required to submit FAFSA. *Faculty research:* Nuclear and particle physics, atomic physics, condensed-matter physics, plasma physics, ultra-cold physics. Total annual research expenditures: $1.4 million. *Unit head:* Dr. Larry Weinstein, Graduate Program Director, 757-683-5803, Fax: 757-683-3038, E-mail: physgpd@odu.edu.

Oregon State University, Graduate School, College of Science, Department of Physics, Corvallis, OR 97331. Offers MA, MS, PhD. Part-time programs available. *Faculty:* 12 full-time (3 women), 5 part-time/adjunct (1 woman). *Students:* 30 full-time (6 women), 2 part-time (1 woman); includes 2 minority (1 Asian American or Pacific Islander, 1 Hispanic American), 5 international. Average age 30. In 2005, 5 master's, 7 doctorates awarded. Terminal master's awarded for partial completion of doctoral program. *Degree requirements:* For master's, qualifying exam, thesis optional; for doctorate, thesis/dissertation, qualifying exam. *Entrance requirements:* For master's and doctorate, minimum GPA of 3.0 in last 90 hours. Additional exam requirements/recommendations for international students: Required—TOEFL. *Application deadline:* For fall admission, 3/1 for domestic students. Application fee: $50. *Expenses:* Tuition, area resident: Part-time $301 per credit. Tuition, state resident: full-time $8,139; part-time $501 per credit. Tuition, nonresident: full-time $14,376; part-time $532 per credit. Required fees: $1,266. *Financial support:* Fellowships, research assistantships, teaching assistantships, Federal Work-Study and institutionally sponsored loans available. Support available to part-time students. Financial award application deadline: 2/1. *Unit head:* Dr. Henri J.F. Jansen, Chair, 541-737-1668, Fax: 541-737-1683, E-mail: chair@physics.orst.edu.

Announcement: Solid-state research: transparent conductors, defects in semiconductors, magnetic anisotropy theory, magnetic semiconductors, superlattices. Nuclear/particle research: nucleon structure, superstring theory. Optics research: nonlinear optics, surface physics, photonics, laser cooling and trapping, atomic interferometry, terahertz spectroscopy, ultrafast and single molecule spectroscopy. Computational physics: electromagnetism in composite media, nanostructured materials, magnetic anisotropy. Teaching apprentice program for future college teachers. PhysTEC site.

The Pennsylvania State University University Park Campus, Graduate School, Eberly College of Science, Department of Physics, State College, University Park, PA 16802-1503. Offers M Ed, MS, D Ed, PhD. *Students:* 110 full-time (13 women), 1 part-time; includes 6 minority (1 Asian American or Pacific Islander, 5 Hispanic Americans), 63 international. *Entrance requirements:* For master's and doctorate, GRE General Test. Application fee: $45. *Expenses:* Tuition, state resident: full-time $12,518; part-time $522 per credit. Tuition, nonresident: full-time $23,004; part-time $959 per credit. Required fees: $484. Tuition and fees vary according to course load, campus/location and program. *Unit head:* Dr. Jayanth R. Banavar, Head, 814-863-1089, Fax: 814-865-0978, E-mail: jrb16@psu.edu.

See Close-Up on page 359.

Pittsburg State University, Graduate School, College of Arts and Sciences, Department of Physics, Pittsburg, KS 66762. Offers applied physics (MS); physics (MS); professional physics (MS). *Students:* 3. *Degree requirements:* For master's, thesis or alternative. Application fee: $70 ($60 for international students). *Expenses:* Tuition, state resident: full-time $2,015; part-time $170 per credit hour. Tuition, nonresident: full-time $4,953; part-time $415 per credit hour. Tuition and fees vary according to course load, campus/location and program. *Financial support:* Teaching assistantships, career-related internships or fieldwork and Federal Work-Study available. *Unit head:* Dr. Charles Blatchley, Chairperson, 620-235-4398. *Application contact:* Marvene Darraugh, Administrative Officer, 620-235-4220, Fax: 620-235-4219, E-mail: mdarraug@pittstate.edu.

Polytechnic University, Brooklyn Campus, Program in Physics, Brooklyn, NY 11201-2990. Offers MS, PhD. Part-time and evening/weekend programs available. *Degree requirements:* For master's, thesis (for some programs), comprehensive exam (for some programs), registration; for doctorate, thesis/dissertation, comprehensive exam, registration. *Entrance requirements:* For master's, BA in physics; for doctorate, departmental qualifying exam, BS in physics. Additional exam requirements/recommendations for international students: Required—TOEFL (minimum score 550 paper-based; 213 computer-based); Recommended—IELT (minimum score 7). *Application deadline:* For fall admission, 7/15 priority date for domestic students, 4/1 priority date for international students; for spring admission, 12/15 priority date for domestic students, 10/1 priority date for international students. Applications are processed on a rolling basis. Application fee: $55. Electronic applications accepted. *Expenses:* Tuition: Part-time $950 per unit. Required fees: $330 per semester. *Financial support:* Fellowships, research assistantships, teaching assistantships, institutionally sponsored loans available. Support available to part-time students. Financial award applicants required to submit FAFSA. *Faculty research:* Combining microdroplets, UHV cryogenic scanning, tunneling, surface spectroscopy of a single aerosol particle. Total annual research expenditures: $294,623. *Unit head:* Dr. Edward Wolf, Head, 718-260-3629, E-mail: ewolf@poly.edu.

Portland State University, Graduate Studies, College of Liberal Arts and Sciences, Department of Physics, Portland, OR 97207-0751. Offers MA, MS, PhD. Part-time programs available. *Faculty:* 13 full-time (1 woman), 4 part-time/adjunct (1 woman). *Students:* 18 full-time (3 women), 6 part-time (1 woman); includes 3 minority (1 Asian American or Pacific Islander, 2 Hispanic Americans), 10 international. Average age 28. 16 applicants, 69% accepted, 8 enrolled. In 2005, 6 degrees awarded. *Degree requirements:* For master's, variable foreign language requirement, thesis, oral exam; for doctorate, thesis/dissertation. *Entrance requirements:* For master's, GRE General Test, minimum GPA of 3.0 in upper-division course work or 2.75 overall, 2 letters of recommendation. Additional exam requirements/recommendations for international students: Required—TOEFL (minimum score 550 paper-based; 213 computer-based). *Application deadline:* For fall admission, 4/1 priority date for domestic students, 3/1 priority date for international students. For winter admission, 9/1 for domestic students; for spring admission, 11/1 for domestic students. Applications are processed on a rolling basis. Application fee: $50. *Expenses:* Tuition, state resident: full-time $6,648; part-time $231 per credit. Tuition, nonresident: full-time $11,319; part-time $231 per credit. Required fees: $686; $67 per credit. *Financial support:* In 2005–06, 1 research assistantship with full tuition reimbursement (averaging $15,012 per year), 15 teaching assistantships with full tuition reimbursements (averaging $13,500 per year) were awarded; career-related internships or fieldwork, Federal Work-Study, and unspecified assistantships also available. Support available to part-time students. Financial award application deadline: 3/1; financial award applicants required to submit FAFSA. *Faculty research:* Statistical physics, membrane biophysics, low-temperature physics, electron microscopy, atmospheric physics. Total annual research expenditures: $1.5 million. *Unit head:* Dr. Eric Bodegom, Chair, 503-725-3812, Fax: 503-725-3888, E-mail: bodegom@pdx.edu. *Application contact:* Peter Leung, Coordinator, 503-725-3812, Fax: 503-725-3888.

Princeton University, Graduate School, Department of Physics, Princeton, NJ 08544-1019. Offers applied and computational mathematics (PhD); mathematical physics (PhD); physics (PhD); physics and chemical physics (PhD). *Degree requirements:* For doctorate, thesis/dissertation, qualifying exam. *Entrance requirements:* For doctorate, GRE General Test, GRE Subject Test. Additional exam requirements/recommendations for international students: Required—TOEFL (minimum score 600 paper-based; 250 computer-based). Electronic applications accepted.

Purdue University, Graduate School, School of Science, Department of Physics, West Lafayette, IN 47907. Offers MS, PhD. Part-time programs available. *Faculty:* 44 full-time (3 women), 2 part-time/adjunct (0 women). *Students:* 93 full-time (21 women), 45 part-time (11 women); includes 5 minority (3 African Americans, 2 Asian Americans or Pacific Islanders), 93 international. Average age 28. 251 applicants, 16% accepted, 22 enrolled. In 2005, 18 master's, 12 doctorates awarded. Terminal master's awarded for partial completion of doctoral program. *Degree requirements:* For master's, qualifying exam; for doctorate, thesis/dissertation, qualifying exam. *Entrance requirements:* For master's and doctorate, GRE General Test, GRE Subject Test (physics). Additional exam requirements/recommendations for international students: Required—TOEFL. *Application deadline:* For fall admission, 2/1 for domestic students, 2/1 for international students. Applications are processed on a rolling basis. Application fee: $55. Electronic applications accepted. *Financial support:* In 2005–06, 4 fellowships with partial tuition reimbursements (averaging $15,000 per year), 30 research assistantships with partial tuition reimbursements (averaging $16,500 per year), 59 teaching assistantships with partial tuition reimbursements (averaging $14,700 per year) were awarded. Support available to part-time students. Financial award application deadline: 2/1; financial award applicants required to submit FAFSA. *Faculty research:* Solid-state, elementary particle, and nuclear physics; biological physics; acoustics; astrophysics. *Unit head:* Dr. A. S. Hirsch, Head, 765-494-3000, Fax: 765-494-0706. *Application contact:* Sangita Handa, Graduate Coordinator, 765-494-5383, Fax: 765-494-0706, E-mail: shanda@purdue.edu.

Queens College of the City University of New York, Division of Graduate Studies, Mathematics and Natural Sciences Division, Department of Physics, Flushing, NY 11367-1597. Offers MA, PhD. Part-time and evening/weekend programs available. *Faculty:* 11 full-time (1 woman). *Students:* 2 full-time (1 woman), 1 (woman) part-time. 6 applicants, 67% accepted, 2 enrolled. In 2005, 4 degrees awarded. *Degree requirements:* For master's, comprehensive exam. *Entrance requirements:* For master's, previous course work in calculus, minimum GPA of 3.0. Additional exam requirements/recommendations for international students: Required—TOEFL. *Application deadline:* For fall admission, 4/1 for domestic students; for spring admission, 11/1 for domestic students. Applications are processed on a rolling basis. Application fee: $125. *Expenses:* Tuition, state resident: part-time $270 per credit. Tuition, nonresident: part-time $500 per credit. Required fees: $112 per year. *Financial support:* Career-related internships or fieldwork, Federal Work-Study, institutionally sponsored loans, and tuition waivers (partial) available. Support available to part-time students. Financial award application deadline: 4/1; financial award applicants required to submit FAFSA. *Faculty research:* Solid-state physics, low temperature physics, elementary particles and fields. *Unit head:* Dr. Alexander Lisyansky, Chairperson, 718-997-3350, E-mail: alexander_lisyansky@qc.edu. *Application contact:* Dr. J. Marion Dickey, Graduate Adviser, 718-997-3350.

Queen's University at Kingston, School of Graduate Studies and Research, Faculty of Arts and Sciences, Department of Physics, Kingston, ON K7L 3N6, Canada. Offers M Sc, M Sc Eng, PhD. Part-time programs available. *Degree requirements:* For master's, thesis/dissertation; for doctorate, thesis/dissertation, comprehensive exam. *Entrance requirements:* For master's, first or upper second class honours in Physics; for doctorate, M Sc or M Sc Eng. Additional exam requirements/recommendations for international students: Required—TOEFL (minimum score 550 paper-based; 213 computer-based). *Faculty research:* Theoretical physics, astronomy and astrophysics, subatomic, condensed matter, applied and engineering.

Rensselaer Polytechnic Institute, Graduate School, School of Science, Department of Physics, Applied Physics and Astronomy, Troy, NY 12180-3590. Offers physics (MS, PhD). *Faculty:* 24 full-time (3 women), 3 part-time/adjunct (0 women). *Students:* 59 full-time (11 women); includes 36 minority (all Asian Americans or Pacific Islanders) Average age 28. 90 applicants, 28% accepted. In 2005, 11 master's, 4 doctorates awarded. *Degree requirements:* For doctorate, thesis/dissertation. *Entrance requirements:* For master's and doctorate, GRE General Test, GRE Subject Test. Additional exam requirements/recommendations for international students: Required—TOEFL (minimum score 600 paper-based; 250 computer-based). *Application deadline:* For fall admission, 1/15 priority date for domestic students, 1/15 priority date for international students; for spring admission, 8/15 priority date for domestic students, 8/15 priority date for international students. Applications are processed on a rolling basis. Application fee: $75. Electronic applications accepted. *Expenses:* Tuition: Full-time $31,000; part-time $1,320 per credit. Required fees: $1,623. *Financial support:* In 2005–06, 10 fellowships with tuition reimbursements (averaging $25,000 per year), 27 research assistantships with tuition reimbursements (averaging $18,700 per year), 16 teaching assistantships with tuition reimbursements (averaging $19,000 per year) were awarded; career-related internships or fieldwork and institutionally sponsored loans also available. Financial award application deadline: 2/1. *Faculty research:* Astrophysics, condensed matter, nuclear physics, optics, physics education. Total annual research expenditures: $4.5 million. *Unit head:* Dr. G. C. Wang, Chair, 518-276-8387, Fax: 518-276-6680, E-mail: wangg@rpi.edu. *Application contact:* Dr. Toh-Ming Lu, Chair, Graduate Recruitment Committee, 518-276-8391, Fax: 518-276-6680, E-mail: mcquade@rpi.edu.

See Close-Up on page 363.

Rice University, Graduate Programs, Wiess School of Natural Sciences, Department of Physics and Astronomy, Houston, TX 77251-1892. Offers physics (MA); physics and astronomy (MS, PhD). *Degree requirements:* For master's and doctorate, thesis/dissertation. *Entrance requirements:* For master's and doctorate, GRE General Test, GRE Subject Test (physics), minimum GPA of 3.0. Additional exam requirements/recommendations for international students: Required—TOEFL (minimum score 600 paper-based; 250 computer-based). Electronic applications accepted. *Faculty research:* Atomic, solid-state, and molecular physics; biophysics; medium- and high-energy physics, magnetospheric physics, planetary atmospheres, astrophysics.

Rice University, Graduate Programs, Wiess School of Natural Sciences, Professional Master's Program in Nanoscale Physics, Houston, TX 77251-1892. Offers MS. *Degree requirements:* For master's, internship. *Entrance requirements:* For master's, GRE General Test, bachelor's in physics and related field, letters of recommendation (4). Additional exam requirements/recommendations for international students: Required—TOEFL. Electronic applications accepted. *Faculty research:* Atomic, molecular, and applied physics, surface and condensed matter physics.

Royal Military College of Canada, Division of Graduate Studies and Research, Science Division, Department of Physics, Kingston, ON K7K 7B4, Canada. Offers M Sc. *Degree requirements:* For master's, thesis, registration. Electronic applications accepted.

Rutgers, The State University of New Jersey, New Brunswick/Piscataway, Graduate School, Program in Physics and Astronomy, New Brunswick, NJ 08901-1281. Offers astronomy (MS, PhD); biophysics (PhD); condensed matter physics (MS, PhD); elementary particle physics (MS, PhD); intermediate energy nuclear physics (MS); nuclear physics (MS, PhD); physics (MST); surface science (PhD); theoretical physics (MS, PhD). Part-time programs available. *Faculty:* 92 full-time. *Students:* 105 full-time (18 women), 5 part-time (3 women); includes 6 minority (2 African Americans, 4 Asian Americans or Pacific Islanders), 54 international. Average age 27. 258 applicants, 19% accepted, 23 enrolled. In 2005, 10 master's, 14 doctorates awarded. Terminal master's awarded for partial completion of doctoral program. *Degree requirements:* For master's, thesis or alternative, comprehensive exam; for doctorate, thesis/dissertation, comprehensive exam. *Entrance requirements:* For master's and doctorate, GRE General Test, GRE Subject Test. Additional exam requirements/recommendations for international students: Required—TOEFL, TSE. *Application deadline:* For fall admission, 1/2 priority date for domestic students, 1/2 priority date for international students; for spring admission, 11/1 for domestic students, 11/1 for international students. Applications are processed on a rolling basis. Application fee: $50. Electronic applications accepted. *Expenses:* Tuition, state resident: full-time $10,440; part-time $435 per credit. Tuition, nonresident: full-time $15,520; part-time $647 per credit. Required fees: $129 per credit. Tuition and fees vary according to program. *Financial support:* In 2005–06, 19 fellowships with full tuition reimbursements (averaging $21,000 per year), 40 research assistantships with full tuition reimbursements (averaging $18,088 per year), 36 teaching assistantships with full tuition reimbursements (averaging $18,088 per year) were awarded; health care benefits and unspecified assistantships also available. Financial award application deadline: 1/2; financial award applicants required to submit FAFSA. *Faculty research:* Astronomy, high energy, condensed matter, surface, nuclear physics. Total annual research expenditures: $7.8 million. *Unit head:* Ronald Ransome, Director, 732-445-2516, Fax: 732-445-4343, E-mail: ransome@physics.rutgers.edu. *Application contact:* Shirley Hinds, Administrative Assistant, 732-445-2502, Fax: 732-445-4343, E-mail: graduate@physics.rutgers.edu.

St. Francis Xavier University, Graduate Studies, Department of Physics, Antigonish, NS B2G 2W5, Canada. Offers M Sc. *Faculty:* 6 full-time (0 women), 2 part-time/adjunct (0 women). *Degree requirements:* For master's, thesis, registration. *Entrance requirements:* For master's, minimum B average in undergraduate course work, honors degree in physics or related area. Additional exam requirements/recommendations for international students: Required—TOEFL (minimum score 580 paper-based; 236 computer-based). *Application deadline:* For fall admission, 9/1 for domestic students. Applications are processed on a rolling basis. Application fee: $40. *Faculty research:* Atomic and molecular spectroscopy, quantum theory, many body theory, mathematical physics, phase transitions. Total annual research expenditures: $600,000. *Unit head:* Dr. Douglas L. Hunter, Chair, 902-867-2104, Fax: 902-867-2414, E-mail: dhunter@stfx.ca. *Application contact:* 902-867-2219, Fax: 902-867-2329, E-mail: admit@stfx.ca.

San Diego State University, Graduate and Research Affairs, College of Sciences, Department of Physics, Program in Physics, San Diego, CA 92182. Offers MA, MS. Part-time programs available. *Students:* 4 full-time (0 women), 26 part-time (2 women); includes 10 minority (3 Asian Americans or Pacific Islanders, 7 Hispanic Americans), 2 international. 28 applicants, 68% accepted, 13 enrolled. In 2005, 2 degrees awarded. *Degree requirements:* For master's, thesis, oral exam. *Entrance requirements:* For master's, GRE General Test, GRE Physics Subject Test, 2 letters of recommendation. Additional exam requirements/recommendations for international students: Required—TOEFL. *Application deadline:* For fall admission, 5/1 for domestic students, 5/1 for international students; for spring admission, 11/1 for domestic students, 10/1 for international students. Applications are processed on a rolling basis. Application fee: $55. Electronic applications accepted. *Financial support:* Teaching assistantships, career-related internships or fieldwork and unspecified assistantships available. Financial award applicants required to submit FAFSA. *Unit head:* Dr. Saul Oseroff, Chair, 619-594-5146, Fax: 619-594-5485, E-mail: soseroff@sciences.sdsu.edu. *Application contact:* Calvin Johnson, Graduate Advisor, 619-594-1284, E-mail: cjohnson@sciences.sdsu.edu.

San Francisco State University, Division of Graduate Studies, College of Science and Engineering, Department of Physics and Astronomy, San Francisco, CA 94132-1722. Offers physics (MS). Part-time programs available. *Degree requirements:* For master's, thesis, registration. *Entrance requirements:* For master's, minimum GPA of 2.5 in last 60 units. Additional exam requirements/recommendations for international students: Required—TOEFL (minimum score 550 paper-based; 213 computer-based). Electronic applications accepted. *Faculty research:* Quark search, thin-films, dark matter detection, search for planetary systems, low temperature.

Physics

San Jose State University, Graduate Studies and Research, College of Science, Department of Physics, San Jose, CA 95192-0001. Offers computational physics (MS); physics (MS). Part-time and evening/weekend programs available. *Students:* 10 full-time (3 women), 17 part-time (8 women); includes 8 minority (6 Asian Americans or Pacific Islanders, 2 Hispanic Americans), 3 international. Average age 35. 20 applicants, 80% accepted, 10 enrolled. In 2005, 5 degrees awarded. *Degree requirements:* For master's, thesis optional. *Entrance requirements:* For master's, GRE. *Application deadline:* For fall admission, 6/29 for domestic students; for spring admission, 11/30 for domestic students. Applications are processed on a rolling basis. Application fee: $59. Electronic applications accepted. *Expenses:* Tuition, nonresident: part-time $339 per unit. Required fees: $1,286 per semester. Tuition and fees vary according to course load and degree level. *Financial support:* In 2005–06, 7 teaching assistantships were awarded; career-related internships or fieldwork, Federal Work-Study, and institutionally sponsored loans also available. Support available to part-time students. Financial award application deadline: 3/1; financial award applicants required to submit FAFSA. *Faculty research:* Astrophysics, atmospheric physics, elementary particles, dislocation theory, general relativity. *Unit head:* Kiu Mars Parnin, Chair, 408-924-5210, Fax: 408-924-2917. *Application contact:* Dr. Karamjeet Arya, Graduate Adviser, 408-924-5267.

Simon Fraser University, Graduate Studies, Faculty of Science, Department of Physics, Burnaby, BC V5A 1S6, Canada. Offers biophysics (M Sc, PhD); chemical physics (M Sc, PhD); physics (M Sc, PhD). *Degree requirements:* For master's and doctorate, thesis/dissertation. *Entrance requirements:* For master's, minimum GPA of 3.0; for doctorate, minimum GPA of 3.5. Additional exam requirements/recommendations for international students: Required—TOEFL or IELTS. *Faculty research:* Solid-state physics, magnetism, energy research, superconductivity, nuclear physics.

South Dakota School of Mines and Technology, Graduate Division, College of Engineering, Doctoral Program in Materials Engineering and Science, Rapid City, SD 57701-3995. Offers chemical engineering (PhD); chemistry (PhD); civil engineering (PhD); electrical engineering (PhD); mechanical engineering (PhD); metallurgical engineering (PhD); physics (PhD). Part-time programs available. *Faculty:* 6 full-time (0 women), 1 part-time/adjunct (0 women). *Students:* 3 full-time (0 women), 4 part-time, 2 international. In 2005, 5 degrees awarded. *Degree requirements:* For doctorate, thesis/dissertation. *Entrance requirements:* For doctorate, minimum graduate GPA of 3.0. Additional exam requirements/recommendations for international students: Required—TOEFL, TWE. *Application deadline:* For fall admission, 7/1 priority date for domestic students, 4/1 priority date for international students; for spring admission, 11/1 for domestic students, 9/1 for international students. Applications are processed on a rolling basis. Application fee: $35. Electronic applications accepted. *Expenses:* Tuition, area resident: Part-time $116 per credit hour. Tuition, state resident: full-time $2,084. Tuition, nonresident: full-time $6,146; part-time $341 per credit hour. Required fees: $1,805; $100 per credit hour. *Financial support:* In 2005–06, 1 fellowship (averaging $1,000 per year), 6 research assistantships with partial tuition reimbursements (averaging $15,390 per year), teaching assistantships with partial tuition reimbursements (averaging $4,482 per year) were awarded; Federal Work-Study and institutionally sponsored loans also available. Support available to part-time students. Financial award application deadline: 5/15. *Faculty research:* Thermophysical properties of solids, development of multiphase materials and composites, concrete technology, electronic polymer materials. *Unit head:* Dr. Duane C. Hrncir, Dean, 605-394-1237. *Application contact:* Jeannette R. Nilson, Program Assistant-Research and Graduate Education, 800-454-8162 Ext. 1206, Fax: 605-394-5360, E-mail: graduate_admissions@silver.sdsmt.edu.

South Dakota School of Mines and Technology, Graduate Division, College of Engineering, Master's Program in Materials Engineering and Science, Rapid City, SD 57701-3995. Offers chemistry (MS); metallurgical engineering (MS); physics (MS). *Faculty:* 6 full-time (0 women), 1 part-time/adjunct (0 women). *Students:* 9 full-time (1 woman), 3 part-time; includes 1 minority (Hispanic American), 6 international. Average age 26. In 2005, 4 degrees awarded. *Entrance requirements:* For master's, GRE General Test. Additional exam requirements/recommendations for international students: Required—TOEFL, TWE. *Application deadline:* For fall admission, 7/1 priority date for domestic students, 4/1 priority date for international students; for spring admission, 11/1 for domestic students, 9/1 for international students. Applications are processed on a rolling basis. Application fee: $35. Electronic applications accepted. *Expenses:* Tuition, area resident: Part-time $116 per credit hour. Tuition, state resident: full-time $2,084. Tuition, nonresident: full-time $6,146; part-time $341 per credit hour. Required fees: $1,805; $100 per credit hour. *Financial support:* In 2005–06, 15 research assistantships with partial tuition reimbursements (averaging $11,400 per year), 11 teaching assistantships with partial tuition reimbursements (averaging $4,063 per year) were awarded; fellowships Financial award application deadline: 5/15. *Unit head:* Dr. Daniel Heglund, 605-394-1241. *Application contact:* Jeannette R. Nilson, Program Assistant-Research and Graduate Education, 800-454-8162 Ext. 1206, Fax: 605-394-5360, E-mail: graduate_admissions@silver.sdsmt.edu.

South Dakota State University, Graduate School, College of Engineering, Department of Physics, Brookings, SD 57007. Offers MS. *Degree requirements:* For master's, thesis, oral exam. *Entrance requirements:* Additional exam requirements/recommendations for international students: Required—TOEFL. *Faculty research:* Materials science, astrophysics, remote sensing and atmospheric corrections, theoretical and computational physics, applied physics.

Southern Illinois University Carbondale, Graduate School, College of Science, Department of Physics, Carbondale, IL 62901-4701. Offers MS, PhD. *Faculty:* 9 full-time (0 women). *Students:* 8 full-time (2 women), 9 part-time (1 woman); includes 1 minority (African American), 11 international. 25 applicants, 48% accepted, 2 enrolled. In 2005, 6 degrees awarded. *Degree requirements:* For master's, one foreign language, thesis. *Entrance requirements:* For master's, minimum GPA of 2.7. Additional exam requirements/recommendations for international students: Required—TOEFL. *Application deadline:* Applications are processed on a rolling basis. Application fee: $20. *Financial support:* In 2005–06, 1 fellowship with full tuition reimbursement, 9 teaching assistantships with full tuition reimbursements were awarded; research assistantships with full tuition reimbursements, career-related internships or fieldwork, Federal Work-Study, institutionally sponsored loans, and tuition waivers (full) also available. Support available to part-time students. Financial award application deadline: 2/15. *Faculty research:* Atomic, molecular, nuclear, and mathematical physics; statistical mechanics; solid-state and low-temperature physics; rheology; material science. Total annual research expenditures: $773,352. *Unit head:* Dr. Aldo Migone, Chairperson, 618-453-1054. *Application contact:* Graduate Admissions Committee, 618-453-2643.

See Close-Up on page 365.

Southern Illinois University Edwardsville, Graduate Studies and Research, College of Arts and Sciences, Department of Physics, Edwardsville, IL 62026-0001. Offers MS. Part-time programs available. *Students:* 2 full-time (0 women), 4 part-time (1 woman), 3 international. Average age 33. 9 applicants, 67% accepted. In 2005, 1 degree awarded. *Degree requirements:* For master's, thesis or alternative, final exam. *Entrance requirements:* Additional exam requirements/recommendations for international students: Required—TOEFL. *Application deadline:* For fall admission, 7/21 for domestic students, 6/1 for international students; for spring admission, 12/8 for domestic students, 10/1 for international students. Application fee: $30. Electronic applications accepted. *Expenses:* Tuition, state resident: part-time $190 per semester hour. Tuition, nonresident: part-time $380 per semester hour. Tuition and fees vary according to course load, reciprocity agreements and student level. *Financial support:* In 2005–06, 3 teaching assistantships with full tuition reimbursements were awarded; fellowships with full tuition reimbursements, research assistantships with full tuition reimbursements, Federal Work-Study, institutionally sponsored loans, and unspecified assistantships also available. Support available to part-time students. Financial award application deadline: 3/1. *Unit head:* Dr. Kimberly Shaw, Chair, 618-650-5326, E-mail: kshaw@siue.edu.

Southern Methodist University, Dedman College, Department of Physics, Dallas, TX 75275. Offers MS, PhD. Part-time programs available. *Faculty:* 10 full-time (0 women). *Students:* 9 full-time (3 women), 3 part-time (1 woman), 7 international. Average age 29. 30 applicants, 27% accepted, 6 enrolled. In 2005, 1 doctorate awarded. Terminal master's awarded for partial completion of doctoral program. *Median time to degree:* Of those who began their doctoral program in fall 1997, 100% received their degree in 8 years or less. *Degree requirements:* For master's, oral exam, thesis optional; for doctorate, thesis/dissertation, written exam. *Entrance requirements:* For master's, GRE General Test, GRE Subject Test in physics, minimum GPA of 3.0; for doctorate, GRE General Test, GRE Subject Test (physics), minimum GPA of 3.0. Additional exam requirements/recommendations for international students: Required—TOEFL. *Application deadline:* For fall admission, 2/1 priority date for domestic students, 2/1 priority date for international students. Application fee: $60. Electronic applications accepted. *Financial support:* In 2005–06, 2 research assistantships with full tuition reimbursements (averaging $19,500 per year), 6 teaching assistantships with full tuition reimbursements (averaging $17,500 per year) were awarded; health care benefits and tuition waivers (partial) also available. Financial award application deadline: 2/1; financial award applicants required to submit FAFSA. *Faculty research:* Particle physics, cosmology, astrophysics, mathematics physics, computational physics. Total annual research expenditures:$1 million. *Unit head:* Prof. Fredrick Olness, Head, 214-768-2495, Fax: 214-768-4095, E-mail: olness@smu.edu. *Application contact:* Prof. Jingbo Ye, Director of Graduate Recruitment, Fax: 214-768-4095.

Southern University and Agricultural and Mechanical College, Graduate School, College of Sciences, Department of Physics, Baton Rouge, LA 70813. Offers MS. *Faculty:* 13 full-time (0 women), 1 part-time/adjunct (0 women). *Students:* 2 full-time (1 woman), 1 part-time; all minorities (all African Americans) Average age 26. 6 applicants, 83% accepted, 0 enrolled. In 2005, 2 degrees awarded. *Degree requirements:* For master's, thesis. *Entrance requirements:* For master's, GMAT or GRE General Test. Additional exam requirements/recommendations for international students: Required—TOEFL (minimum score 525 paper-based; 193 computer-based). *Application deadline:* For fall admission, 4/15 priority date for domestic students, 4/15 priority date for international students; for spring admission, 11/1 for domestic students, 11/1 for international students. Applications are processed on a rolling basis. Application fee: $25. *Financial support:* In 2005–06, fellowships (averaging $15,000 per year), research assistantships with full tuition reimbursements (averaging $12,000 per year), teaching assistantships with partial tuition reimbursements (averaging $12,000 per year) were awarded. Financial award application deadline: 4/15. *Faculty research:* Piezoelectric materials and devices, predictive ab-instio calculations, high energy physics, surface growth studies, semiconductor and intermetallics. Total annual research expenditures: $2.1 million. *Unit head:* Dr. Stephen C. McGuire, Chair, 225-771-4130 Ext. 12, Fax: 225-771-2310, E-mail: mcguire@grant.phys.subr.edu. *Application contact:* Dr. Ali R. Fazely, Professor, 225-771-3070, Fax: 225-771-4341, E-mail: fazely@phys.subr.edu.

Stanford University, School of Humanities and Sciences, Department of Physics, Stanford, CA 94305-9991. Offers PhD. *Degree requirements:* For doctorate, thesis/dissertation, oral exam, qualifying exam. *Entrance requirements:* For doctorate, GRE General Test, GRE Subject Test. Additional exam requirements/recommendations for international students: Required—TOEFL. Electronic applications accepted.

State University of New York at Binghamton, Graduate School, School of Arts and Sciences, Department of Physics, Applied Physics, and Astronomy, Binghamton, NY 13902-6000. Offers applied physics (MS); physics (MA, MS). *Degree requirements:* For master's, thesis or alternative. *Entrance requirements:* For master's, GRE General Test, GRE Subject Test. Additional exam requirements/recommendations for international students: Required—TOEFL. Electronic applications accepted.

State University of New York at Buffalo, Graduate School, College of Arts and Sciences, Department of Physics, Buffalo, NY 14260. Offers MS, PhD. Part-time programs available. *Faculty:* 23 full-time (3 women). *Students:* 86 full-time (17 women), 2 part-time; includes 3 minority (2 Asian Americans or Pacific Islanders, 1 Hispanic American), 54 international. Average age 29. 166 applicants, 31% accepted, 21 enrolled. In 2005, 3 master's, 7 doctorates awarded. Terminal master's awarded for partial completion of doctoral program. *Median time to degree:* Of those who began their doctoral program in fall 1997, 33% received their degree in 8 years or less. *Degree requirements:* For master's, thesis, qualifying exam; for doctorate, thesis/dissertation, qualifying exams. *Entrance requirements:* For master's and doctorate, GRE General Test, letters of recommendation. Additional exam requirements/recommendations for international students: Required—TOEFL (minimum score 550 paper-based; 213 computer-based). *Application deadline:* For fall admission, 2/1 priority date for domestic students, 2/1 priority date for international students; for spring admission, 10/1 priority date for domestic students, 10/1 priority date for international students. Applications are processed on a rolling basis. Application fee: $35. Electronic applications accepted. *Financial support:* In 2005–06, 92 students received support, including 11 fellowships with full tuition reimbursements available (averaging $20,300 per year), 42 research assistantships with full tuition reimbursements available (averaging $18,000 per year), 39 teaching assistantships with full tuition reimbursements available (averaging $14,300 per year); Federal Work-Study, institutionally sponsored loans, scholarships/grants, health care benefits, and unspecified assistantships also available. Financial award application deadline: 3/1; financial award applicants required to submit FAFSA. *Faculty research:* Condensed-matter physics (experimental and theoretical), cosmology (theoretical), high energy and particle physics (experimental and theoretical), computational physics, medical physics, materials physics. Total annual research expenditures: $1.8 million. *Unit head:* Dr. Frank Gasparini, Chairman, 716-645-2017 Ext. 126, Fax: 716-645-2507, E-mail: fmg@buffalo.edu. *Application contact:* Dr. John Ho, Director of Graduate Studies, 716-645-2017 Ext. 128, Fax: 716-645-2507, E-mail: proho@buffalo.edu.

Stephen F. Austin State University, Graduate School, College of Sciences and Mathematics, Department of Physics and Astronomy, Nacogdoches, TX 75962. Offers physics (MS). Part-time programs available. *Faculty:* 7 full-time (0 women), 1 part-time/adjunct (0 women). *Students:* 5 full-time (1 woman), 1 part-time; includes 2 minority (both Hispanic Americans), 1 international. Average age 30. 5 applicants, 100% accepted. *Degree requirements:* For master's, comprehensive exam. *Entrance requirements:* For master's, GRE General Test, minimum GPA of 2.8 in last 60 hours, 2.5 overall. Additional exam requirements/recommendations for international students: Required—TOEFL. *Application deadline:* For fall admission, 8/1 for domestic students; for spring admission, 12/15 for domestic students. Applications are processed on a rolling basis. Application fee: $0 ($50 for international students). *Expenses:* Tuition, state resident: full-time $2,628; part-time $146 per credit hour. Tuition, nonresident: full-time $7,596; part-time $422 per credit hour. Required fees: $900; $170. *Financial support:* In 2005–06, 5 teaching assistantships (averaging $8,100 per year) were awarded; Federal Work-Study, institutionally sponsored loans, and unspecified assistantships also available. Financial award application deadline: 3/1. *Faculty research:* Low-temperature physics, x-ray spectroscopy and metallic glasses, infrared spectroscopy. *Unit head:* Dr. Harry D. Downing, Chair, 936-468-3001.

Stevens Institute of Technology, Graduate School, Arthur E. Imperatore School of Sciences and Arts, Department of Physics and Engineering Physics, Hoboken, NJ 07030. Offers applied optics (Certificate); engineering physics (M Eng); physics (MS, PhD); surface physics (Certificate). Part-time and evening/weekend programs available. *Students:* 26 full-time (6 women), 28 part-time (5 women); includes 4 minority (1 African American, 3 Hispanic Americans), 26 international. Average age 25. 44 applicants, 100% accepted.Terminal master's awarded for partial completion of doctoral program. *Degree requirements:* For master's, thesis optional; for doctorate, thesis/dissertation. *Entrance requirements:* For master's and doctorate, GRE. Additional exam requirements/recommendations for international students: Required—TOEFL. *Application deadline:* Applications are processed on a rolling basis. Application fee: $50. Electronic applications accepted. *Expenses:* Tuition: Part-time $920 per credit hour. Tuition and fees vary according to program. *Financial support:* Fellowships, research assistantships, teaching assistantships, Federal Work-Study and institutionally sponsored loans available. *Faculty research:* Laser spectroscopy, physical kinetics, semiconductor-device physics, condensed-matter theory. *Unit head:* Dr. Kurt Becker, Director, 201-216-5671. *Application* *contact:* H. L. Cui, Chairman, Graduate Committee, 201-216-5637, Fax: 201-216-5638, E-mail: hcui@stevens-tech.edu.

Stony Brook University, State University of New York, Graduate School, College of Arts and Sciences, Department of Physics and Astronomy, Program in Physics, Stony Brook, NY 11794. Offers MA, MAT, MS, PhD. *Students:* 194 full-time (37 women); includes 11 minority (1 African American, 5 Asian Americans or Pacific Islanders, 5 Hispanic Americans), 104 international. *Degree requirements:* For doctorate, one foreign language, thesis/dissertation. *Entrance requirements:* For master's and doctorate, GRE General Test. Additional exam requirements/recommendations for international students: Required—TOEFL. *Application deadline:* For fall admission, 1/15 for domestic students. Application fee: $50. *Expenses:* Tuition, area resident: Part-time $288. Tuition, state resident: full-time $6,900. Tuition, nonresident: full-time $10,920; part-time $455. Required fees: $704. *Financial support:* Fellowships, research assistantships, teaching assistantships available. Financial award application deadline: 2/1. *Application contact:* Dr. Peter Stephens, Director, 631-632-8279, Fax: 631-632-8176, E-mail: pstephens@notes.cc.sunysb.edu.

Announcement: Graduate students in the department engage in a broad range of experimental, observational, and theoretical research activities both on campus and at major facilities, including the nearby Brookhaven National Laboratory. Students work with leaders in their fields and enjoy personal attention in the development of their education.

See Close-Up on page 367.

Syracuse University, Graduate School, College of Arts and Sciences, Department of Physics, Syracuse, NY 13244. Offers MS, PhD. Part-time programs available. *Students:* 46 full-time (14 women), 7 part-time (1 woman); includes 3 minority (1 African American, 1 Asian American or Pacific Islander, 1 Hispanic American), 39 international. 111 applicants, 17% accepted, 6 enrolled.Terminal master's awarded for partial completion of doctoral program. *Degree requirements:* For master's, thesis or alternative; for doctorate, thesis/dissertation. *Entrance requirements:* For master's and doctorate, GRE General Test, GRE Subject Test. Additional exam requirements/recommendations for international students: Required—TOEFL. *Application deadline:* For fall admission, 1/10 for domestic students. Applications are processed on a rolling basis. Application fee: $65. Electronic applications accepted. *Financial support:* Fellowships with full tuition reimbursements, research assistantships with full and partial tuition reimbursements, teaching assistantships with full and partial tuition reimbursements, tuition waivers (partial) available. *Unit head:* Dr. Edward Lipson, Chair, 315-443-5690, Fax: 315-443-9103, E-mail: edlipson@syr.edu. *Application contact:* Joseph Schecter, Graduate Program Director, 315-443-5968, E-mail: jmschech@syr.edu.

Temple University, Graduate School, College of Science and Technology, Department of Physics, Philadelphia, PA 19122-6096. Offers MA, PhD. *Faculty:* 12 full-time (2 women). *Students:* 9 full-time (1 woman), 20 part-time (5 women), 18 international. 41 applicants, 32% accepted, 8 enrolled. In 2005, 5 master's, 1 doctorate awarded. Terminal master's awarded for partial completion of doctoral program. *Degree requirements:* For master's, thesis or alternative, comprehensive exam; for doctorate, thesis/dissertation, 2 comprehensive exams. *Entrance requirements:* For master's and doctorate, GRE General Test, minimum GPA of 3.0. Additional exam requirements/recommendations for international students: Required—TOEFL (minimum score 575 paper-based; 230 computer-based). *Application deadline:* For fall admission, 7/15 for domestic students, 12/15 for international students; for spring admission, 11/15 for domestic students, 8/1 for international students. Applications are processed on a rolling basis. Application fee: $50. Electronic applications accepted. *Expenses:* Tuition, state resident: full-time $8,694; part-time $483 per credit. Tuition, nonresident: full-time $12,672; part-time $704 per credit. Required fees: $500; $122 per semester. Tuition and fees vary according to course level, campus/location and program. *Financial support:* Fellowships, research assistantships, teaching assistantships, tuition waivers (full and partial) available. Financial award application deadline: 1/15; financial award applicants required to submit FAFSA. *Faculty research:* Laser-based molecular spectroscopy, elementary particle physics, statistical mechanics, solid-state physics. *Unit head:* Dr. C. Jeff Martoff, Chair, 215-204-3877, Fax: 215-204-5652, E-mail: martoff@temple.edu.

See Close-Up on page 369.

Texas A&M International University, Office of Graduate Studies and Research, College of Arts and Sciences, Department of Mathematical and Physical Science, Laredo, TX 78041-1900. Offers MAIS. *Faculty:* 2 full-time (0 women). *Students:* Average age 29. 7 applicants, 100% accepted, 4 enrolled. In 2005, 1 degree awarded. *Entrance requirements:* For master's, GRE General Test. Additional exam requirements/recommendations for international students: Required—TOEFL (minimum score 550 paper-based; 213 computer-based). *Application deadline:* For fall admission, 7/15 for domestic students; for spring admission, 11/12 for domestic students. Applications are processed on a rolling basis. Application fee: $25. *Expenses:* Tuition, state resident: full-time $1,580. Tuition, nonresident: full-time $5,432. Required fees: $3,808. *Financial support:* In 2005–06, 1 student received support. Application deadline: 11/1. *Unit head:* Dr. Chen Snung, Chair, 956-326-2567, Fax: 956-326-2439, E-mail: csung@tamiu.edu. *Application contact:* Rosie Espinoza, Director of Admissions, 956-326-2200, Fax: 956-326-2199, E-mail: enroll@tamiu.edu.

Texas A&M University, College of Science, Department of Physics, College Station, TX 77843. Offers applied physics (PhD); physics (MS, PhD). *Faculty:* 35 full-time (1 woman). *Students:* 140 full-time (18 women), 10 part-time (2 women); includes 19 minority (3 African Americans, 2 American Indian/Alaska Native, 6 Asian Americans or Pacific Islanders, 8 Hispanic Americans), 90 international. 117 applicants, 51% accepted, 43 enrolled. In 2005, 15 master's, 6 doctorates awarded. Terminal master's awarded for partial completion of doctoral program. *Degree requirements:* For master's, thesis (for some programs), registration; for doctorate, thesis/dissertation, registration. *Entrance requirements:* For master's and doctorate, GRE General Test, GRE Subject Test. Additional exam requirements/recommendations for international students: Required—TOEFL. *Application deadline:* For fall admission, 3/1 for domestic students; for spring admission, 8/1 for domestic students. Application fee: $50 ($75 for international students). Electronic applications accepted. *Expenses:* Tuition, state resident: full-time $4,488; part-time $187 per credit hour. Tuition, nonresident: full-time $11,112; part-time $463 per credit hour. Required fees:$1,974. *Financial support:* In 2005–06, research assistantships (averaging $16,200 per year), teaching assistantships (averaging $16,200 per year) were awarded; fellowships Financial award application deadline: 3/1; financial award applicants required to submit FAFSA. *Faculty research:* Condensed-matter, atomic/molecular, high-energy, and nuclear physics; quantum optics. *Unit head:* Dr. Edward S. Fry, Head, 979-845-7717, Fax: 979-845-2590, E-mail: fry@physics.tamu.edu. *Application contact:* Dr. George W. Kattawar, Professor, 979-845-1180, Fax: 979-845-2590, E-mail: kattawar@physics.tamu.edu.

Texas A&M University–Commerce, Graduate School, College of Arts and Sciences, Department of Physics, Commerce, TX 75429-3011. Offers M Ed, MS. Part-time programs available. *Faculty:* 6 full-time (1 woman). *Students:* 2 full-time (0 women), 4 part-time. Average age 36. In 2005, 3 degrees awarded. *Degree requirements:* For master's, thesis (for some programs), comprehensive exam. *Entrance requirements:* For master's, GRE General Test. *Application deadline:* For fall admission, 6/1 for domestic students; for spring admission, 11/1 priority date for domestic students. Applications are processed on a rolling basis. Application fee: $0 ($25 for international students). Electronic applications accepted. *Financial support:* In 2005–06, research assistantships (averaging $7,875 per year), teaching assistantships (averaging $7,875 per year) were awarded; Federal Work-Study, institutionally sponsored loans, and scholarships/grants also available. Financial award application deadline: 5/1; financial award applicants required to submit FAFSA. Total annual research expenditures: $7,345. *Unit head:* Dr. Ben Doughty, Head, 903-886-5488, Fax: 903-886-5480, E-mail: ben_doughty@tamu-commerce.edu. *Application contact:* Tammi Thompson, Graduate Admissions Adviser, 843-886-5167, Fax: 843-886-5165, E-mail: tammi_thompson@tamu-commerce.edu.

Texas Christian University, College of Science and Engineering, Department of Physics and Astronomy, Fort Worth, TX 76129-0002. Offers physics (MA, MS, PhD), including astrophysics (PhD), business (PhD), physics (PhD). Part-time and evening/weekend programs available. *Degree requirements:* For doctorate, thesis/dissertation, qualifying exams. *Entrance requirements:* For doctorate, GRE General Test. Additional exam requirements/recommendations for international students: Required—TOEFL. *Application deadline:* For fall admission, 3/1 for domestic students; for spring admission, 12/1 for domestic students. Applications are processed on a rolling basis. Application fee: $0. *Expenses:* Tuition: Part-time $740 per credit hour. *Financial support:* Fellowships, teaching assistantships available. Financial award application deadline: 3/1. *Unit head:* Dr. T W Zerda, Chairperson, 817-257-7375. *Application contact:* Dr. Bonnie Melhart, Associate Dean, College of Science and Engineering, E-mail: b.melhart@tcu.edu.

Texas State University-San Marcos, Graduate School, College of Science, Department of Physics, San Marcos, TX 78666. Offers MS. Part-time programs available. *Faculty:* 5 full-time (1 woman), 1 part-time/adjunct (0 women). *Students:* 10 full-time (1 woman), 6 part-time (1 woman); includes 2 minority (both Hispanic Americans), 1 international. Average age 27. 10 applicants, 90% accepted, 8 enrolled. In 2005, 2 degrees awarded. *Degree requirements:* For master's, thesis (for some programs), comprehensive exam. *Entrance requirements:* For master's, GRE General Test, minimum GPA of 2.75 in last 60 hours of course work. Additional exam requirements/recommendations for international students: Required—TOEFL. *Application deadline:* For fall admission, 6/15 priority date for domestic students, 6/1 priority date for international students; for spring admission, 10/15 priority date for domestic students, 10/1 priority date for international students. Applications are processed on a rolling basis. Application fee: $40 ($90 for international students). *Expenses:* Tuition, area resident: Part-time $116 per credit. Tuition, state resident: full-time $3,168; part-time $176 per credit. Tuition, nonresident: full-time $8,136; part-time $452 per credit. Required fees: $1,112; $74 per credit. Full-time tuition and fees vary according to course load. *Financial support:* In 2005–06, 11 students received support, including 1 research assistantship (averaging $4,932 per year), 10 teaching assistantships (averaging $4,932 per year); career-related internships or fieldwork, Federal Work-Study, and institutionally sponsored loans also available. Support available to part-time students. Financial award application deadline: 4/1; financial award applicants required to submit FAFSA. *Faculty research:* High-temperature superconductors, historical astronomy, general relativity. *Unit head:* Dr. Vedaraman Sriraman, Interim Chair, 512-245-2131, Fax: 512-245-8233, E-mail: vs04@txstate.edu.

Texas Tech University, Graduate School, College of Arts and Sciences, Department of Physics, Lubbock, TX 79409. Offers applied physics (MS); physics (MS, PhD). Part-time programs available. *Faculty:* 17 full-time (1 woman). *Students:* 34 full-time (4 women), 6 part-time (1 woman); includes 4 minority (all Hispanic Americans), 18 international. Average age 29. 26 applicants, 62% accepted, 2 enrolled. In 2005, 7 master's, 2 doctorates awarded. *Degree requirements:* For master's and doctorate, variable foreign language requirement, thesis/dissertation. *Entrance requirements:* For master's and doctorate, GRE General Test. Additional exam requirements/recommendations for international students: Required—TOEFL (minimum score 550 paper-based; 213 computer-based). *Application deadline:* Applications are processed on a rolling basis. Application fee: $50 ($60 for international students). Electronic applications accepted. *Expenses:* Tuition, state resident: full-time $4,296. Tuition, nonresident: full-time $10,920. Required fees: $1,992. Tuition and fees vary according to program. *Financial support:* In 2005–06, 15 students received support, including 6 research assistantships with partial tuition reimbursements available (averaging $16,110 per year), 25 teaching assistantships with partial tuition reimbursements available (averaging $15,608 per year); career-related internships or fieldwork, Federal Work-Study, and institutionally sponsored loans also available. Support available to part-time students. Financial award application deadline: 4/15; financial award applicants required to submit FAFSA. *Faculty research:* Molecular spectroscopy of biological membranes, thin films and semiconductor characterization, muon spin rotation defect, characterization of semiconductors, nanotechnology of magnetic materials, theory of impurities and complexes in semiconductors. Total annual research expenditures: $1.1 million. *Unit head:* Dr. Lynn L. Hatfield, Chair, 806-742-3767, Fax: 806-742-1182, E-mail: lynn.hatfield@ttu.edu. *Application contact:* Dr. Wallace L. Glab, Graduate Recruiter, 806-742-3767, Fax: 806-742-1182, E-mail: wallace.glab@ttu.edu.

Trent University, Graduate Studies, Program in Applications of Modeling in the Natural and Social Sciences, Department of Physics, Peterborough, ON K9J 7B8, Canada. Offers M Sc. Part-time programs available. *Degree requirements:* For master's, thesis. *Entrance requirements:* For master's, honours degree. *Faculty research:* Radiation physics, chemical physics.

Tufts University, Graduate School of Arts and Sciences, Department of Physics and Astronomy, Medford, MA 02155. Offers physics (MS, PhD). *Faculty:* 19 full-time, 1 part-time/adjunct. *Students:* 32 (12 women); includes 1 minority (Hispanic American) 13 international. 99 applicants, 9% accepted, 5 enrolled. In 2005, 6 master's, 3 doctorates awarded. Terminal master's awarded for partial completion of doctoral program. *Degree requirements:* For master's, thesis optional; for doctorate, thesis/dissertation. *Entrance requirements:* For master's and doctorate, GRE General Test. Additional exam requirements/recommendations for international students: Required—TOEFL (minimum score 550 paper-based; 213 computer-based). *Application deadline:* For fall admission, 2/15 for domestic students, 12/30 for international students; for spring admission, 10/15 for domestic students, 9/15 for international students. Applications are processed on a rolling basis. Application fee: $65. Electronic applications accepted. *Expenses:* Tuition: Full-time $32,360. Tuition and fees vary according to program. *Financial support:* Research assistantships with full and partial tuition reimbursements, teaching assistantships with full and partial tuition reimbursements, Federal Work-Study, scholarships/grants, and tuition waivers (partial) available. Financial award application deadline: 2/15; financial award applicants required to submit FAFSA. *Unit head:* William Oliver, Chair, 617-627-3029. *Application contact:* Dr. Krzysztof Sliwa, Information Contact, 617-627-3029.

Tulane University, Graduate School, Department of Physics, New Orleans, LA 70118-5669. Offers MS, PhD. *Degree requirements:* For master's, thesis or alternative; for doctorate, thesis/dissertation. *Entrance requirements:* For master's, GRE General Test, minimum B average in undergraduate course work; for doctorate, GRE General Test. Additional exam requirements/recommendations for international students: Required—TOEFL; Recommended—TSE. Electronic applications accepted. *Faculty research:* Surface physics, condensed-matter experiment, condensed-matter theory, nuclear theory, polymers.

Université de Moncton, Faculty of Science, Department of Physics and Astronomy, Moncton, NB E1A 3E9, Canada. Offers M Sc. Part-time programs available. *Degree requirements:* For master's, thesis. *Entrance requirements:* For master's, proficiency in French. Electronic applications accepted. *Faculty research:* Thin films, optical properties, solar selective surfaces, microgravity and photonic materials.

Université de Montréal, Faculty of Graduate Studies, Faculty of Arts and Sciences, Department of Physics, Montréal, QC H3C 3J7, Canada. Offers M Sc, PhD. *Faculty:* 36 full-time (3 women), 9 part-time/adjunct (0 women). *Students:* 132 full-time (30 women). 90 applicants, 30% accepted, 26 enrolled. In 2005, 29 master's, 8 doctorates awarded. *Degree requirements:* For doctorate, thesis/dissertation, general exam. *Application deadline:* For fall and spring admission, 2/1. For winter admission, 11/1 for domestic students. Electronic applications accepted. *Financial support:* Fellowships, research assistantships, teaching assistantships available. *Faculty research:* Astronomy; biophysics; solid-state, plasma, and nuclear physics. Total annual research expenditures: $5.8 million. *Unit head:* Laurent J. Lewis, Chairman, 514-343-6669, Fax: 514-343-2071. *Application contact:* Louise La Fortune, Student Files Management Technician, 514-343-6667, Fax: 514 343 2071.

Université de Sherbrooke, Faculty of Sciences, Department of Physics, Sherbrooke, QC J1K 2R1, Canada. Offers M Sc, PhD. *Faculty:* 11 full-time (1 woman), 2 part-time/adjunct (0 women). *Students:* 32 full-time (5 women). 43 applicants, 14% accepted. In 2005, 6 master's, 2 doctorates awarded. *Degree requirements:* For master's, thesis/dissertation; for doctorate, thesis/dissertation, comprehensive exam. *Entrance requirements:* For doctorate, master's degree.

Physics

Université de Sherbrooke (continued)

Application deadline: For fall admission, 6/30 for domestic students. Applications are processed on a rolling basis. Application fee: $50. Electronic applications accepted. *Financial support:* Fellowships, research assistantships, teaching assistantships available. *Faculty research:* Solid-state physics, quantum computing. *Unit head:* Dr. Denis Morris, Chairman, 819-821-7055, Fax: 819-821-8046, E-mail: denis.morris@usherbrooke.ca.

Université Laval, Faculty of Sciences and Engineering, Department of Physics, Physical Engineering, and Optics, Programs in Physics, Québec, QC G1K 7P4, Canada. Offers M Sc, PhD. Terminal master's awarded for partial completion of doctoral program. *Degree requirements:* For master's, thesis/dissertation; for doctorate, thesis/dissertation, comprehensive exam. *Entrance requirements:* For master's and doctorate, knowledge of French, comprehension of written English. Electronic applications accepted.

University at Albany, State University of New York, College of Arts and Sciences, Department of Physics, Albany, NY 12222-0001. Offers MS, PhD. Evening/weekend programs available. *Students:* 21 full-time (3 women), 18 part-time (6 women). Average age 33. In 2005, 13 master's, 7 doctorates awarded. *Degree requirements:* For master's, one foreign language; for doctorate, one foreign language, thesis/dissertation. *Entrance requirements:* Additional exam requirements/recommendations for international students: Required—TOEFL (minimum score 550 paper-based; 213 computer-based). *Application deadline:* For fall admission, 6/15 for domestic students, 5/1 for international students. Applications are processed on a rolling basis. Application fee: $60. Electronic applications accepted. *Financial support:* Fellowships, research assistantships, teaching assistantships, minority assistantships available. Financial award application deadline: 6/15. *Faculty research:* Condensed-matter physics, high-energy physics, applied physics, electronic materials, theoretical particle physics. *Unit head:* Mohammed Sajjad Alam, Chair, 518-442-4500.

The University of Akron, Graduate School, Buchtel College of Arts and Sciences, Department of Physics, Akron, OH 44325. Offers MS. Part-time and evening/weekend programs available. *Faculty:* 10 full-time (1 woman). *Students:* 23 full-time (8 women), 1 part-time, 10 international. Average age 27. 17 applicants, 65% accepted, 5 enrolled. In 2005, 1 degree awarded. *Degree requirements:* For master's, thesis or written exam or formal report, thesis optional. *Entrance requirements:* For master's, minimum GPA of 2.75. Additional exam requirements/recommendations for international students: Required—TOEFL (minimum score 550 paper-based; 213 computer-based), Michigan English Language Assessment Battery. *Application deadline:* For fall admission, 8/15 for domestic students. Applications are processed on a rolling basis. Application fee: $30 ($40 for international students). Electronic applications accepted. *Expenses:* Tuition, state resident: full-time $5,816; part-time $323 per credit. Tuition, nonresident: full-time $9,976; part-time $554 per credit. Required fees: $794; $43 per credit. $12 per term. Tuition and fees vary according to course load, degree level and program. *Financial support:* In 2005–06, 3 research assistantships with full tuition reimbursements, 17 teaching assistantships with full tuition reimbursements were awarded; tuition waivers (full) also available. *Faculty research:* Polymer physics, statistical physics, NMR, electron tunneling, solid-state physics. Total annual research expenditures:$381,147. *Unit head:* Dr. Robert Mallik, Chair, 330-972-7145, E-mail: rmallik@uakron.edu. *Application contact:* Dr. Jutta Luettmer-Strathman, Head, 330-972-8029, E-mail: jutta@uakron.edu.

The University of Alabama, Graduate School, College of Arts and Sciences, Department of Physics and Astronomy, Tuscaloosa, AL 35487. Offers physics (MS, PhD). *Faculty:* 18 full-time (0 women). *Students:* 28 full-time (7 women), 11 part-time (2 women); includes 12 minority (11 African Americans, 1 Hispanic American), 27 international. Average age 29. 42 applicants, 36% accepted, 5 enrolled. In 2005, 3 master's, 3 doctorates awarded. Terminal master's awarded for partial completion of doctoral program. *Median time to degree:* Of those who began their doctoral program in fall 1997, 50% received their degree in 8 years or less. *Degree requirements:* For master's, oral exam, thesis optional; for doctorate, thesis/dissertation, oral and written exams. *Entrance requirements:* For master's and doctorate, GRE General Test or GRE Subject Test, minimum GPA of 3.0. Additional exam requirements/recommendations for international students: Required—TOEFL. *Application deadline:* For fall admission, 7/6 for domestic students; for spring admission, 11/22 for domestic students. Applications are processed on a rolling basis. Application fee: $25. Electronic applications accepted. *Expenses:* Tuition, area resident: Full-time $2,432. Tuition, nonresident: full-time $6,758. *Financial support:* In 2005–06, 2 fellowships with full tuition reimbursements (averaging $15,000 per year), 19 research assistantships with full tuition reimbursements (averaging $13,000 per year), 16 teaching assistantships with full tuition reimbursements (averaging $12,500 per year) were awarded; career-related internships or fieldwork and institutionally sponsored loans also available. Financial award application deadline: 4/1. *Faculty research:* Condensed-matter, high-energy physics; optics; molecular spectroscopy; astrophysics. Total annual research expenditures: $708,400. *Unit head:* Dr. Stanley T. Jones, Chairman and Professor, 205-348-5050, Fax: 205-348-5051, E-mail: stjones@bama.ua.edu.

The University of Alabama at Birmingham, School of Natural Sciences and Mathematics, Department of Physics, Birmingham, AL 35294. Offers MS, PhD. *Students:* 27 full-time (7 women), 1 (woman) part-time; includes 5 minority (4 African Americans, 1 Hispanic American), 11 international. 31 applicants, 81% accepted. In 2005, 4 master's, 5 doctorates awarded. Terminal master's awarded for partial completion of doctoral program. *Degree requirements:* For master's, thesis optional; for doctorate, thesis/dissertation. *Entrance requirements:* For master's and doctorate, GRE General Test, minimum GPA of 3.0. Additional exam requirements/recommendations for international students: Required—TOEFL. *Application deadline:* Applications are processed on a rolling basis. Application fee: $35 ($60 for international students). Electronic applications accepted. *Expenses:* Tuition, state resident: part-time $170 per credit hour. Tuition, nonresident: full-time $4,612; part-time $425 per credit hour. International tuition: $10,732 full-time. Required fees: $11 per credit hour. $124 per term. Tuition and fees vary according to course load, degree level and program. *Financial support:* In 2005–06, 9 fellowships with full tuition reimbursements (averaging $16,898 per year), 4 research assistantships (averaging $15,170 per year), 8 teaching assistantships with full tuition reimbursements (averaging $12,825 per year) were awarded; career-related internships or fieldwork, Federal Work-Study, institutionally sponsored loans, scholarships/grants, traineeships, and unspecified assistantships also available. Support available to part-time students. Financial award application deadline: 4/15; financial award applicants required to submit FAFSA. *Faculty research:* Laser physics, space physics, optics, biophysics, material physics. *Unit head:* Dr. David L. Shealy, Chair, 205-934-4736, Fax: 205-934-8042.

The University of Alabama in Huntsville, School of Graduate Studies, College of Science, Department of Physics, Huntsville, AL 35899. Offers MS, PhD. Part-time and evening/weekend programs available. *Faculty:* 12 full-time (0 women), 1 part-time/adjunct (0 women). *Students:* 22 full-time (4 women), 13 part-time (5 women); includes 1 minority (American Indian/Alaska Native), 9 international. Average age 30. 22 applicants, 95% accepted, 12 enrolled. In 2005, 8 master's, 3 doctorates awarded. *Degree requirements:* For master's, thesis or alternative, oral and written exams, comprehensive exam, registration; for doctorate, thesis/dissertation, oral and written exams, comprehensive exam, registration. *Entrance requirements:* For master's and doctorate, GRE General Test, minimum GPA of 3.0. Additional exam requirements/recommendations for international students: Required—TOEFL (minimum score 550 paper-based; 213 computer-based). *Application deadline:* For fall admission, 5/30 for domestic students; for spring admission, 10/10 priority date for domestic students, 7/10 priority date for international students. Applications are processed on a rolling basis. Application fee: $40. *Expenses:* Tuition, state resident: full-time $5,866; part-time $244 per credit hour. Tuition, nonresident: full-time $12,060; part-time $500 per credit hour. Tuition and fees vary according to course load. *Financial support:* In 2005–06, 23 students received support, including 13 research assistantships with full and partial tuition reimbursements available (averaging $10,411 per year), 10 teaching assistantships with full and partial tuition reimbursements available (averaging $9,405 per year); fellowships with full and partial tuition reimbursements available, career-related internships or fieldwork, Federal Work-Study, institutionally sponsored loans, scholarships/grants, health care benefits, tuition waivers (full and partial), and unspecified assistantships also available. Support available to part-time students. Financial award application deadline: 4/1; financial award applicants required to submit FAFSA. *Faculty research:* Space sciences, solid state/materials, optics/quantum electronics, astrophysics, crystal growth. Total annual research expenditures: $338,954. *Unit head:* Dr. James Miller, Chair, 256-824-2482, Fax: 256-824-6873, E-mail: hillmanl@email.uah.edu.

University of Alaska Fairbanks, College of Natural Sciences and Mathematics, Department of Physics, Fairbanks, AK 99775-7520. Offers atmospheric science (MS, PhD); computational physics (MS); general physics (MS); physics (MAT, PhD); space physics (PhD). Part-time programs available. *Faculty:* 14 full-time (1 woman). *Students:* 24 full-time (5 women), 2 part-time; includes 2 minority (1 African American, 1 Hispanic American), 4 international. Average age 29. 16 applicants, 56% accepted, 3 enrolled. In 2005, 4 master's, 1 doctorate awarded. Terminal master's awarded for partial completion of doctoral program. *Degree requirements:* For master's, thesis or alternative, comprehensive exam, registration; for doctorate, thesis/dissertation, comprehensive exam, registration. *Entrance requirements:* For master's, GRE General Test, BS in physics; for doctorate, GRE General Test. Additional exam requirements/recommendations for international students: Required—TOEFL (minimum score 550 paper-based; 213 computer-based). *Application deadline:* For fall admission, 3/1 for domestic students, 3/1 for international students. Applications are processed on a rolling basis. Application fee: $50. Electronic applications accepted. *Expenses:* Tuition, state resident: full-time $4,392; part-time $244 per credit. Tuition, nonresident: full-time $8,964; part-time $498 per credit. Required fees: $800; $5 per credit. $48 per contact hour. Tuition and fees vary according to course level, course load, campus/location and reciprocity agreements. *Financial support:* In 2005–06, 14 research assistantships with tuition reimbursements (averaging $10,507 per year), 4 teaching assistantships with tuition reimbursements (averaging $11,027 per year) were awarded; fellowships with tuition reimbursements, Federal Work-Study, scholarships/grants, and unspecified assistantships also available. Financial award applicants required to submit FAFSA. *Faculty research:* Atmospheric and ionospheric radar studies, space plasma theory, magnetospheric dynamics, space weather and auroral studies, turbulence and complex systems. *Unit head:* John D. Craven, Chair, 907-474-7339, Fax: 907-474-6130, E-mail: physics@uaf.edu.

University of Alberta, Faculty of Graduate Studies and Research, Department of Physics, Edmonton, AB T6G 2E1, Canada. Offers astrophysics (M Sc, PhD); condensed matter (M Sc, PhD); geophysics (M Sc, PhD); medical physics (M Sc, PhD); subatomic physics (M Sc, PhD). *Faculty:* 36 full-time (3 women), 7 part-time/adjunct (0 women). *Students:* 56 full-time (6 women), 16 part-time (2 women), 25 international. 85 applicants, 35% accepted. In 2005, 7 master's, 10 doctorates awarded. *Degree requirements:* For master's and doctorate, thesis/dissertation. *Entrance requirements:* For master's and doctorate, minimum GPA of 7.0 on a 9.0 scale. Additional exam requirements/recommendations for international students: Required—TOEFL. *Application deadline:* For fall admission, 2/15 for domestic students. Applications are processed on a rolling basis. Tuition and fees charges are reported in Canadian dollars. *Expenses:* Tuition, state resident: part-time $562 Canadian dollars per term. Tuition, nonresident: full-time $3,375 Canadian dollars. Required fees: $573 Canadian dollars; $84 Canadian dollars per term. *Financial support:* In 2005–06, 45 students received support, including 6 fellowships with partial tuition reimbursements available, 40 teaching assistantships; research assistantships, career-related internships or fieldwork, institutionally sponsored loans, and scholarships/grants also available. Financial award application deadline: 2/15. *Faculty research:* Cosmology, astroparticle physics, high-intermediate energy, magnetism, superconductivity. Total annual research expenditures: $3.1 million. *Unit head:* Dr. R. Marchand, Associate Chair, 780-492-1072, E-mail: assoc-chair@phys.ualberta.ca. *Application contact:* Lynn Chandler, Program Advisor, 780-492-1072, Fax: 780-492-0714, E-mail: lynn@phys.ualberta.ca.

The University of Arizona, Graduate College, College of Science, Department of Physics, Tucson, AZ 85721. Offers M Ed, MS, PhD. Part-time programs available. Terminal master's awarded for partial completion of doctoral program. *Degree requirements:* For master's, thesis optional; for doctorate, thesis/dissertation, comprehensive exam. *Entrance requirements:* For master's and doctorate, GRE General Test, GRE Subject Test, minimum GPA of 3.0. Additional exam requirements/recommendations for international students: Required—TOEFL; Recommended—TSE. *Faculty research:* Astrophysics; high-energy, condensed-matter, atomic and molecular physics; optics.

University of Arkansas, Graduate School, J. William Fulbright College of Arts and Sciences, Department of Physics, Fayetteville, AR 72701-1201. Offers applied physics (MS); physics (MS, PhD); physics education (MA). *Students:* 21 full-time (7 women), 9 part-time (1 woman), 7 international. 55 applicants, 25% accepted. In 2005, 5 master's, 5 doctorates awarded. *Degree requirements:* For master's and doctorate, thesis/dissertation. Application fee: $40 ($50 for international students). *Financial support:* In 2005–06, 5 fellowships with tuition reimbursements, 9 research assistantships, 13 teaching assistantships were awarded; career-related internships or fieldwork and Federal Work-Study also available. Support available to part-time students. Financial award application deadline: 4/1; financial award applicants required to submit FAFSA. *Unit head:* Dr. Lin Oliver, Departmental Chairperson, 479-575-2506, Fax: 479-575-4580, E-mail: woliver@uark.edu. *Application contact:* Dr. Raj Gupta, Graduate Coordinator, 479-575-5933, E-mail: rgupta@uark.edu.

The University of British Columbia, Faculty of Graduate Studies, Faculty of Science, Program in Physics, Vancouver, BC V6T 1Z1, Canada. Offers engineering physics (MA Sc); physics (M Sc, PhD). *Faculty:* 55. *Students:* 97. Average age 24. 163 applicants, 37% accepted, 33 enrolled. In 2005, 15 master's, 16 doctorates awarded. *Degree requirements:* For master's, thesis/dissertation; for doctorate, thesis/dissertation, comprehensive exam. *Entrance requirements:* For master's, GRE General Test, honors degree; for doctorate, GRE General Test, master's degree. Additional exam requirements/recommendations for international students: Required—TOEFL. *Application deadline:* For fall admission, 3/2 for domestic students, 2/1 for international students. Applications are processed on a rolling basis. Application fee: $60. *Financial support:* Fellowships, research assistantships, teaching assistantships, career-related internships or fieldwork and Federal Work-Study available. *Faculty research:* Applied physics, astrophysics, condensed matter, plasma physics, subatomic physics, astronomy. Total annual research expenditures:$3 million. *Unit head:* Dr. Jeff Young, Head, 604-822-3150, E-mail: young@physics.ubc.ca. *Application contact:* Oliva Dela Cruz-Lordero, Graduate Program Coordinator, 604-822-4245, E-mail: oliva.dela.cruz-cordero@ubc.ca.

University of Calgary, Faculty of Graduate Studies, Faculty of Science, Department of Physics and Astronomy, Calgary, AB T2N 1N4, Canada. Offers M Sc, PhD. Part-time programs available. *Faculty:* 22 full-time (1 woman), 12 part-time/adjunct (0 women). *Students:* 56 full-time (15 women), 4 part-time (2 women). Average age 28. 75 applicants, 100 enrolled. In 2005, 7 master's, 2 doctorates awarded. *Degree requirements:* For master's, thesis; for doctorate, thesis/dissertation, oral candidacy exam, written qualifying exam. *Entrance requirements:* For master's and doctorate, GRE General Test, GRE Subject Test. Additional exam requirements/recommendations for international students: Required—TOEFL (minimum score 550 paper-based; 213 computer-based). *Application deadline:* For fall admission, 3/1 for domestic students, 3/1 for international students. For winter admission, 7/1 for domestic students. Applications are processed on a rolling basis. Application fee: $130. Electronic applications accepted. *Financial support:* Fellowships with full and partial tuition reimbursements, research assistantships, teaching assistantships, institutionally sponsored loans available. Financial award application deadline:2/1. *Faculty research:* Astronomy and astrophysics, mass spectrometry, atmospheric physics, space physics, medical physics. Total annual research expenditures: $4.6 million. *Unit head:* Dr. A. R. Taylor, Head, 403-220-5385, Fax: 403-289-3331, E-mail: russ@ras.ucalgary.ca. *Application contact:* Dr. R. I. Thompson, Chairman, Graduate Affairs, 403-220-5407, Fax: 403-289-3331, E-mail: gradinfo@ucalgary.ca.

University of California, Berkeley, Graduate Division, College of Letters and Science, Department of Physics, Berkeley, CA 94720-1500. Offers PhD. *Degree requirements:* For doctorate, thesis/dissertation, qualifying exam. *Entrance requirements:* For doctorate, GRE General Test, GRE Subject Test, minimum GPA of 3.0. Additional exam requirements/recommendations for international students: Required—TOEFL (minimum score 570 paper-based; 230 computer-based); Recommended—TSE (minimum score 50). *Faculty research:* Astrophysics (experimental and theoretical), condensed matter physics (experimental and theoretical), particle physics (experimental and theoretical), atomic/molecular/botical physics, biophysics and complex systems.

University of California, Davis, Graduate Studies, Program in Physics, Davis, CA 95616. Offers MS, PhD. *Faculty:* 53 full-time. *Students:* 135 full-time (27 women); includes 17 minority (1 African American, 13 Asian Americans or Pacific Islanders, 3 Hispanic Americans), 25 international. Average age 29. 199 applicants, 40% accepted, 31 enrolled. In 2005, 4 master's, 9 doctorates awarded. Terminal master's awarded for partial completion of doctoral program. *Median time to degree:* Of those who began their doctoral program in fall 1997, 66.7% received their degree in 8 years or less. *Degree requirements:* For master's, thesis (for some programs), comprehensive exam (for some programs); for doctorate, thesis/dissertation. *Entrance requirements:* For master's and doctorate, GRE General Test, GRE Subject Test, minimum GPA of 3.0. Additional exam requirements/recommendations for international students: Required—TOEFL (minimum score 550 paper-based; 213 computer-based). *Application deadline:* For fall admission, 4/1 for domestic students, 3/1 for international students. Application fee: $60. Electronic applications accepted. *Financial support:* In 2005–06, 125 students received support, including 13 fellowships with full and partial tuition reimbursements available (averaging $14,638 per year), 32 research assistantships with full and partial tuition reimbursements available (averaging $15,986 per year), 79 teaching assistantships with partial tuition reimbursements available (averaging $15,253 per year); Federal Work-Study, institutionally sponsored loans, scholarships/grants, and tuition waivers (full and partial) also available. Financial award application deadline: 1/15; financial award applicants required to submit FAFSA. *Faculty research:* Astrophysics, condensed-matter physics, nuclear physics, particle physics, quantum optics. *Unit head:* Shirley Chiang, Chair, 530-752-8538, E-mail: chiang@physics.ucdavis.edu. *Application contact:* Kristi Case, Administrative Assistant, 530-752-1501, E-mail: kacase@ucdavis.edu.

University of California, Irvine, Office of Graduate Studies, School of Physical Sciences, Department of Physics and Astronomy, Irvine, CA 92697. Offers physics (MS, PhD). Terminal master's awarded for partial completion of doctoral program. *Degree requirements:* For doctorate, thesis/dissertation. *Entrance requirements:* For master's and doctorate, GRE General Test, GRE Subject Test, minimum GPA of 3.0. Additional exam requirements/recommendations for international students: Required—TOEFL (minimum score 550 paper-based; 213 computer-based). Electronic applications accepted. *Faculty research:* Condensed-matter physics, plasma physics, astrophysics, particle physics, chemical and materials physics.

University of California, Los Angeles, Graduate Division, College of Letters and Science, Department of Physics and Astronomy, Program in Physics, Los Angeles, CA 90095. Offers physics (MS, PhD); physics education (MAT). MAT admits only applicants whose objective is PhD. *Degree requirements:* For master's, comprehensive exam or thesis; for doctorate, thesis/dissertation, oral and written qualifying exams. *Entrance requirements:* For master's, GRE General Test, GRE Subject Test (physics), minimum GPA of 3.0; for doctorate, GRE General Test, GRE Subject Test (physics), minimum undergraduate GPA of 3.0. Electronic applications accepted.

Announcement: Strong, broad research and graduate student programs in both experimental and theoretical physics: condensed matter, low temperature, plasma, astrophysics, biophysics, high energy, intermediate energy, nuclear physics. Approximately 60 faculty members, 110 graduate students; 20 PhDs per year. Strong research funding. Financial support for essentially all graduate students. Near ocean in attractive West Los Angeles. http://www.physics.ucla.edu.

University of California, Riverside, Graduate Division, Department of Physics, Riverside, CA 92521-0102. Offers MS, PhD. Part-time programs available. *Faculty:* 31 full-time (7 women). *Students:* 72 full-time (16 women); includes 11 minority (1 African American, 4 Asian Americans or Pacific Islanders, 6 Hispanic Americans), 36 international. Average age 29. In 2005, 7 master's, 6 doctorates awarded. Terminal master's awarded for partial completion of doctoral program. *Degree requirements:* For master's, comprehensive exams or thesis; for doctorate, thesis/dissertation, qualifying exams. *Entrance requirements:* For master's and doctorate, GRE General Test, minimum GPA of 3.2. Additional exam requirements/recommendations for international students: Required—TOEFL (minimum score 550 paper-based; 213 computer-based); Recommended—TSE (minimum score 50). *Application deadline:* For fall admission, 5/1 for domestic students, 2/1 for international students. For winter admission, 9/1 for domestic students; for spring admission, 12/1 for domestic students. Applications are processed on a rolling basis. Application fee: $60 ($75 for international students). Electronic applications accepted. *Expenses:* Tuition, nonresident: full-time $14,694. Full-time tuition and fees vary according to program. *Financial support:* In 2005–06, fellowships (averaging $12,000 per year); research assistantships, teaching assistantships, career-related internships or fieldwork, Federal Work-Study, institutionally sponsored loans, scholarships/grants, health care benefits, and unspecified assistantships also available. Financial award application deadline: 1/5; financial award applicants required to submit FAFSA. *Faculty research:* Laser physics and surface science, elementary particle and heavy ion physics, plasma physics, optical physics, astrophysics. *Application contact:* Pat Brooks, Graduate Program Assistant, 951-827-5332, Fax: 951-827-4529, E-mail: gophysics@ucr.edu.

University of California, San Diego, Graduate Studies and Research, Department of Physics, La Jolla, CA 92093. Offers biophysics (MS, PhD); physics (MS, PhD); physics/materials physics (MS). *Degree requirements:* For doctorate, thesis/dissertation. *Entrance requirements:* For master's and doctorate, GRE General Test, GRE Subject Test. Additional exam requirements/recommendations for international students: Required—TOEFL. Electronic applications accepted.

University of California, Santa Barbara, Graduate Division, College of Letters and Sciences, Division of Mathematics, Life, and Physical Sciences, Department of Physics, Santa Barbara, CA 93106. Offers PhD. *Students:* 149 full-time (21 women); includes 14 minority (10 Asian Americans or Pacific Islanders, 4 Hispanic Americans), 28 international. Average age 25. 535 applicants, 20% accepted, 37 enrolled. In 2005, 10 doctorates awarded. *Degree requirements:* For doctorate, thesis/dissertation, comprehensive exam, registration. *Entrance requirements:* For doctorate, GRE General Test, GRE Subject Test in physics, bachelor's degree in physics or other related field. Additional exam requirements/recommendations for international students: Required—TOEFL (minimum score 550 paper-based; 213 computer-based). *Application deadline:* For fall admission, 12/15 for domestic students, 12/15 for international students. Application fee: $60. *Financial support:* In 2005–06, 149 students received support, including 73 research assistantships with partial tuition reimbursements available (averaging $19,542 per year), 51 teaching assistantships with partial tuition reimbursements available (averaging $15,083 per year); fellowships, Federal Work-Study, health care benefits, tuition waivers (full and partial), and unspecified assistantships also available. Financial award application deadline: 12/15; financial award applicants required to submit FAFSA. *Faculty research:* Astrophysics, biophysics, condensed matter physics, high energy physics, gravity and relativity. Total annual research expenditures: $368.5 million. *Unit head:* Prof. Mark Srednicki, Chair, 805-893-2165, E-mail: mark@vulcan2.physics.vesb.edu. *Application contact:* Prof. David Berenstein, Admissions Committee Chair, 805-893-6120, Fax: 805-893-3307, E-mail: dberens@physics.ucsb.edu.

University of California, Santa Cruz, Division of Graduate Studies, Division of Physical and Biological Sciences, Program in Physics, Santa Cruz, CA 95064. Offers MS, PhD. *Faculty:* 19 full-time (1 woman). *Students:* 56 full-time (18 women); includes 8 minority (1 African American, 4 Asian Americans or Pacific Islanders, 3 Hispanic Americans), 4 international. 151 applicants, 31% accepted, 12 enrolled. In 2005, 10 master's, 5 doctorates awarded. *Degree requirements:* For master's, thesis; for doctorate, one foreign language, thesis/dissertation, qualifying exam. *Entrance requirements:* For master's and doctorate, GRE General Test, GRE Subject Test. *Application deadline:* For fall admission, 1/15 for domestic students. Application fee: $60. *Expenses:* Tuition, nonresident: full-time $14,694. *Financial support:* Fellowships, research assistantships, teaching assistantships, career-related internships or fieldwork, Federal Work-Study, and institutionally sponsored loans available. Financial award application deadline: 1/15. *Faculty research:* Theoretical and experimental high-energy physics, theoretical and experimental solid-state physics, critical phenomena, theoretical fluid dynamics, experimental biophysics. *Unit head:* Dr. David Dorfan, Chair, 831-459-2327. *Application contact:* Judy L. Glass, Reporting Analyst for Graduate Admissions, 831-459-5906, Fax: 831-459-4843, E-mail: jlglass@ucsc.edu.

University of Central Florida, College of Sciences, Department of Physics, Orlando, FL 32816. Offers MS, PhD. Part-time and evening/weekend programs available. *Faculty:* 25 full-time (6 women), 2 part-time/adjunct (0 women). *Students:* 57 full-time (10 women), 3 part-time; includes 9 minority (6 African Americans, 1 Asian American or Pacific Islander, 2 Hispanic Americans), 23 international. Average age 30. 56 applicants, 48% accepted, 19 enrolled. In 2005, 7 master's, 2 doctorates awarded. *Degree requirements:* For master's, thesis or alternative; for doctorate, thesis/dissertation, candidacy and qualifying exams. *Entrance requirements:* For master's, GRE General Test, minimum GPA of 3.0 in last 60 hours of course work; for doctorate, GRE General Test, GRE Subject Test, minimum GPA of 3.0 in last 60 hours or master's qualifying exam. Additional exam requirements/recommendations for international students: Required—TOEFL. *Application deadline:* For fall admission, 2/15 for domestic students. Application fee: $30. Electronic applications accepted. *Expenses:* Tuition, state resident: full-time $5,788. Tuition, nonresident: full-time $21,927. Required fees: $241 per credit hour. *Financial support:* In 2005–06, 9 fellowships with partial tuition reimbursements (averaging $6,144 per year), 23 research assistantships with partial tuition reimbursements (averaging $10,900 per year), 25 teaching assistantships with partial tuition reimbursements (averaging $10,700 per year) were awarded; career-related internships or fieldwork, Federal Work-Study, institutionally sponsored loans, tuition waivers (partial), and unspecified assistantships also available. Financial award application deadline: 3/1; financial award applicants required to submit FAFSA. *Faculty research:* Atomic-molecular physics, condensed-matter physics, biophysics of proteins, laser physics. *Unit head:* Dr. Ralph A. Llewellyn, Interim Chair, 407-823-5785, E-mail: ral@pysics.ucf.edu. *Application contact:* Dr. Eduardo Mucciolo, Coordinator, 407-823-5208, Fax: 407-823-5112, E-mail: graduate@physics.ucf.edu.

University of Central Oklahoma, College of Graduate Studies and Research, College of Mathematics and Science, Department of Physics and Engineering, Edmond, OK 73034-5209. Offers MS. Part-time programs available. *Faculty:* 4 full-time (0 women). *Students:* 4 full-time (2 women), 7 part-time (1 woman); includes 2 minority (1 Asian American or Pacific Islander, 1 Hispanic American), 4 international. Average age 33. 2 applicants, 100% accepted. In 2005, 1 degree awarded. *Degree requirements:* For master's, thesis optional. *Entrance requirements:* For master's, 24 hours of course work in physics. Additional exam requirements/recommendations for international students: Required—TOEFL (minimum score 550 paper-based; 213 computer-based). *Application deadline:* Applications are processed on a rolling basis. Application fee: $25. Electronic applications accepted. *Expenses:* Tuition, state resident: full-time $2,988; part-time $125 per credit hour. Tuition, nonresident: full-time $4,728; part-time $197 per credit hour. Required fees: $716; $16 per credit hour. *Financial support:* Unspecified assistantships available. Financial award application deadline: 3/31; financial award applicants required to submit FAFSA. *Faculty research:* Acoustics, solid-state physics/optical properties, molecular dynamics, nuclear physics, crystallography. *Unit head:* Dr. Ronald Miller, Chairperson, 405-974-5461, Fax: 405-974-3824, E-mail: physics@ucok.edu.

University of Chicago, Division of the Physical Sciences, Department of Physics, Chicago, IL 60637-1513. Offers PhD. *Faculty:* 37 full-time (3 women), 7 part-time/adjunct (0 women). *Students:* 136 full-time (19 women); includes 8 minority (7 Asian Americans or Pacific Islanders, 1 Hispanic American), 71 international. Average age 25. 379 applicants, 20% accepted, 22 enrolled. In 2005, 16 degrees awarded. *Median time to degree:* Of those who began their doctoral program in fall 1997, 70% received their degree in 8 years or less. *Degree requirements:* For doctorate, thesis/dissertation, submission of a paper to a refereed professional journal, comprehensive exam. *Entrance requirements:* For doctorate, GRE General Test, GRE Subject Test. Additional exam requirements/recommendations for international students: Required—TOEFL (minimum score 600 paper-based; 250 computer-based). *Application deadline:* For fall admission, 12/28 for domestic students, 12/28 for international students. Applications are processed on a rolling basis. Application fee: $55. Electronic applications accepted. *Financial support:* In 2005–06, 23 fellowships with full tuition reimbursements (averaging $26,500 per year), 78 research assistantships with full tuition reimbursements (averaging $22,500 per year), 35 teaching assistantships with full tuition reimbursements (averaging $16,447 per year) were awarded; institutionally sponsored loans, scholarships/grants, health care benefits, tuition waivers (partial), and unspecified assistantships also available. Financial award application deadline: 12/28. *Faculty research:* Astrophysics, particle physics, condensed-matter physics, statistical physics, relativity. Total annual research expenditures: $20.3 million. *Unit head:* Robert Wald, Chair, 773-702-7006, Fax: 773-702-2045, E-mail: rmwa@midway.uchicago.edu. *Application contact:* Nobuko B. McNeill, Assistant to the Chairman for Graduate Affairs, 773-702-7007, Fax: 773-702-2045, E-mail: n-mcneill@uchicago.edu.

University of Chicago, Division of the Physical Sciences, Program in the Physical Sciences, Chicago, IL 60637-1513. Offers MS. Part-time programs available. *Students:* 4 full-time (1 woman). Average age 24. 4 applicants, 75% accepted, 1 enrolled. *Degree requirements:* For master's, thesis. *Entrance requirements:* For master's, GRE. Additional exam requirements/recommendations for international students: Required—TOEFL. *Application deadline:* For fall admission, 2/28 priority date for domestic students, 2/28 priority date for international students. Applications are processed on a rolling basis. Application fee: $55. *Financial support:* In 2005–06, 4 students received support; fellowships with partial tuition reimbursements available, research assistantships, teaching assistantships available. Financial award application deadline: 2/28; financial award applicants required to submit FAFSA. *Unit head:* Robert Wald, Chair, 773-702-7006, Fax: 773-702-2045, E-mail: rmwa@midway.uchicago.edu. *Application contact:* Richard Hefley, Dean of Students, 773-702-8789.

University of Cincinnati, Division of Research and Advanced Studies, McMicken College of Arts and Sciences, Department of Physics, Cincinnati, OH 45221. Offers MS, PhD. Terminal master's awarded for partial completion of doctoral program. *Degree requirements:* For master's, thesis optional; for doctorate, thesis/dissertation. *Entrance requirements:* For master's and doctorate, GRE General Test, GRE Subject Test. Additional exam requirements/recommendations for international students: Required—TOEFL (minimum score 540 paper-based; 207 computer-based). Electronic applications accepted. *Faculty research:* Condensed matter physics, experimental particle physics, theoretical high energy physics, astronomy and astrophysics, computational physics.

University of Colorado at Boulder, Graduate School, College of Arts and Sciences, Department of Physics, Boulder, CO 80309. Offers chemical physics (PhD); geophysics (PhD); liquid crystal science and technology (PhD); mathematical physics (PhD); medical physics (PhD); optical sciences and engineering (PhD); physics (MS, PhD). *Faculty:* 39 full-time (3 women). *Students:* 157 full-time (37 women), 47 part-time (3 women); includes 10 minority (2 African Americans, 1 American Indian/Alaska Native, 4 Asian Americans or Pacific Islanders, 3 Hispanic Americans), 46 international. Average age 27. 81 applicants, 93% accepted. In 2005, 10 master's, 16 doctorates awarded. Terminal master's awarded for partial completion of doctoral program. *Degree requirements:* For master's, thesis or alternative, comprehensive exam; for doctorate, thesis/dissertation, comprehensive exam. *Entrance requirements:* For master's and doctorate, GRE General Test, GRE Subject Test, minimum undergraduate GPA of 3.0. Additional exam requirements/recommendations for international students: Required—TOEFL. *Application deadline:* For fall admission, 1/15 priority date for domestic students, 1/15 priority date for international students. Applications are processed on a rolling basis. Application fee: $50 ($60 for international students). Electronic applications accepted. *Financial support:* In 2005–06, fellowships with full tuition reimbursements (averaging $4,376 per year),

Physics

University of Colorado at Boulder (continued)
research assistantships with full tuition reimbursements (averaging $15,185 per year), teaching assistantships with full tuition reimbursements (averaging $13,786 per year) were awarded; scholarships/grants also available. Financial award application deadline: 1/15. *Faculty research:* Atomic and molecular physics, nuclear physics, condensed matter, elementary particle physics, laser or optical physics. Total annual research expenditures: $24.1 million. *Unit head:* John Cumalat, Chair, 303-492-6952, Fax: 303-492-3352, E-mail: jcumalat@pizero.colorado.edu. *Application contact:* Graduate Program Assistant, 303-492-6954, Fax: 303-492-3352, E-mail: phys@bogart.colorado.edu.

University of Connecticut, Graduate School, College of Liberal Arts and Sciences, Department of Physics, Field of Physics, Storrs, CT 06269. Offers MS, PhD. *Faculty:* 37 full-time (3 women). *Students:* 74 full-time (14 women), 2 part-time; includes 1 minority (Asian American or Pacific Islander), 47 international. Average age 28. 82 applicants, 17% accepted, 14 enrolled. In 2005, 7 master's, 4 doctorates awarded. Terminal master's awarded for partial completion of doctoral program. *Degree requirements:* For master's, comprehensive exam; for doctorate, thesis/dissertation. *Entrance requirements:* For master's and doctorate, GRE General Test, GRE Subject Test. Additional exam requirements/recommendations for international students: Required—TOEFL (minimum score 550 paper-based; 213 computer-based). *Application deadline:* For fall admission, 2/1 priority date for domestic students, 2/1 priority date for international students; for spring admission, 11/1 for domestic students, 10/1 for international students. Applications are processed on a rolling basis. Application fee: $55. Electronic applications accepted. *Expenses:* Tuition, state resident: part-time $444 per credit hour. Tuition, nonresident: part-time $1,154 per credit hour. Tuition and fees vary according to course load. *Financial support:* In 2005–06, 45 research assistantships with full tuition reimbursements, 26 teaching assistantships with full tuition reimbursements were awarded; fellowships, Federal Work-Study, scholarships/grants, health care benefits, and unspecified assistantships also available. Financial award application deadline: 2/1; financial award applicants required to submit FAFSA. *Unit head:* Gerald Dunne, Chairperson, 860-486-4978, E-mail: gerald.dunne@uconn.edu. *Application contact:* Lorraine Smurra, Administrative Assistant, 860-486-0449, Fax: 860-486-3346, E-mail: physadm@uconnvm.uconn.edu.

University of Delaware, College of Arts and Sciences, Department of Physics and Astronomy, Newark, DE 19716. Offers MS, PhD. Part-time programs available. *Faculty:* 33 full-time (2 women), 6 part-time/adjunct (1 woman). *Students:* 90 full-time (21 women); includes 4 minority (2 African Americans, 1 Asian American or Pacific Islander, 1 Hispanic American), 68 international. Average age 26. 123 applicants, 33% accepted, 20 enrolled. In 2005, 3 master's, 1 doctorate awarded. Terminal master's awarded for partial completion of doctoral program. *Degree requirements:* For master's and doctorate, thesis/dissertation. *Entrance requirements:* For master's and doctorate, GRE General Test, GRE Subject Test. Additional exam requirements/recommendations for international students: Required—TOEFL (minimum score 600 paper-based; 250 computer-based). *Application deadline:* For fall admission, 2/1 for domestic students. Application fee: $60. Electronic applications accepted. *Financial support:* In 2005–06, 82 students received support, including 2 fellowships with full tuition reimbursements available (averaging $19,000 per year), 27 research assistantships with full tuition reimbursements available (averaging $19,000 per year), 25 teaching assistantships with full tuition reimbursements available (averaging $19,000 per year); career-related internships or fieldwork, Federal Work-Study, institutionally sponsored loans, and corporate sponsorships also available. Financial award application deadline: 3/1. *Faculty research:* Magnetoresistance and magnetic materials, ultrafast optical phenomena, superfluidity, elementary particle physics, stellar atmospheres and interiors. Total annual research expenditures: $6.9 million. *Unit head:* Dr. George Hadjipanayis, Chair, 302-831-3361. *Application contact:* Dr. Norbert Mulders, Information Contact, 302-831-1995, E-mail: grad.physics@udel.edu.

University of Denver, Faculty of Natural Sciences and Mathematics, Department of Physics and Astronomy, Denver, CO 80208. Offers MS, PhD. Part-time programs available. *Faculty:* 5 full-time (0 women). *Students:* 4 full-time (3 women), 5 part-time (2 women), 4 international. 17 applicants, 71% accepted. In 2005, 2 master's awarded. Terminal master's awarded for partial completion of doctoral program. *Degree requirements:* For master's, thesis optional; for doctorate, thesis/dissertation. *Entrance requirements:* For master's and doctorate, GRE General Test, diagnostic exam. Additional exam requirements/recommendations for international students: Required—TOEFL, TSE. *Application deadline:* Applications are processed on a rolling basis. Application fee: $45. *Expenses:* Tuition: Full-time $27,756; part-time $771 per credit. Required fees: $174. *Financial support:* In 2005–06, 2 research assistantships with full and partial tuition reimbursements (averaging $16,407 per year), 6 teaching assistantships with full and partial tuition reimbursements (averaging $16,407 per year) were awarded; career-related internships or fieldwork, Federal Work-Study, institutionally sponsored loans, and scholarships/grants also available. Support available to part-time students. Financial award application deadline: 3/1; financial award applicants required to submit FAFSA. *Faculty research:* Atomic and molecular beams and collisions, infrared astronomy, acoustic emission from stressed solids. Total annual research expenditures:$751,000. *Unit head:* Dr. Herschel Neumann, Chair, 303-871-3544. *Application contact:* Information Contact, 303-871-2238, E-mail: bstephen@du.edu.

University of Florida, Graduate School, College of Liberal Arts and Sciences, Department of Physics, Gainesville, FL 32611. Offers MS, MST, PhD. *Faculty:* 57 full-time (4 women), 2 part-time/adjunct (0 women). *Students:* 134 (21 women); includes 4 minority (2 Asian Americans or Pacific Islanders, 2 Hispanic Americans) 84 international. In 2005, 17 master's, 16 doctorates awarded. *Degree requirements:* For master's, variable foreign language requirement, thesis (for some programs); for doctorate, one foreign language, thesis/dissertation. *Entrance requirements:* For master's and doctorate, GRE General Test, minimum GPA of 3.0. Additional exam requirements/recommendations for international students: Required—TOEFL (minimum score 550 paper-based; 213 computer-based). *Application deadline:* For fall admission, 6/1 for domestic students. Applications are processed on a rolling basis. Application fee: $30. Electronic applications accepted. *Expenses:* Tuition, state resident: full-time $6,234. Tuition, nonresident: full-time $21,359. Tuition and fees vary according to program. *Financial support:* In 2005–06, fellowships with tuition reimbursements (averaging $20,000 per year), 67 research assistantships with tuition reimbursements (averaging $18,393 per year), 49 teaching assistantships with tuition reimbursements (averaging $19,760 per year) were awarded; unspecified assistantships also available. *Faculty research:* Astrophysics, condensed-matter physics, elementary particle physics, statistical mechanics, quantum theory. *Unit head:* Dr. Alan Dorsey, Chair, 352-392-0521, Fax: 352-392-0524, E-mail: chair@phys.ufl.edu. *Application contact:* Dr. Mark W. Meisel, Coordinator, 352-392-0521, Fax: 352-392-0524, E-mail: meisel@phys.ufl.edu.

University of Georgia, Graduate School, College of Arts and Sciences, Department of Physics and Astronomy, Athens, GA 30602. Offers physics (MS, PhD). *Faculty:* 24 full-time (2 women). *Students:* 54 full-time, 3 part-time; includes 6 minority (4 African Americans, 1 American Indian/Alaska Native, 1 Asian American or Pacific Islander), 27 international. 120 applicants, 23% accepted, 17 enrolled. In 2005, 4 master's, 2 doctorates awarded. *Degree requirements:* For master's, thesis; for doctorate, one foreign language, thesis/dissertation. *Entrance requirements:* For master's and doctorate, GRE General Test. *Application deadline:* For fall admission, 7/1 for domestic students; for spring admission, 11/15 for domestic students. Application fee: $50. Electronic applications accepted. *Financial support:* Fellowships, research assistantships, teaching assistantships, unspecified assistantships available. *Unit head:* Dr. Heinz-Bernd Schüttler, Head, 706-542-2485, Fax: 706-542-2492, E-mail: hbs@physast.uga.edu. *Application contact:* Dr. F. Todd Baker, Graduate Coordinator, 706-542-0979, Fax: 706-542-2492, E-mail: tbaker@physast.uga.edu.

University of Guelph, Graduate Program Services, College of Physical and Engineering Science, Guelph-Waterloo Physics Institute, Guelph, ON N1G 2W1, Canada. Offers M Sc, PhD. Part-time programs available. *Faculty:* 49 full-time (8 women), 29 part-time/adjunct (4 women). *Students:* 90 full-time (16 women), 6 part-time. 100 applicants, 57% accepted, 30 enrolled. In 2005, 10 master's, 7 doctorates awarded. *Degree requirements:* For master's, thesis or alternative, project or thesis; for doctorate, thesis/dissertation, comprehensive exam, registration. *Entrance requirements:* For master's, GRE Subject Test, minimum B average for honors degree; for doctorate, GRE Subject Test, minimum B average. Additional exam requirements/recommendations for international students: Required—TOEFL (minimum score 550 paper-based; 213 computer-based), TWE (minimum score 4). *Application deadline:* For fall admission, 6/1 priority date for domestic students, 1/31 priority date for international students. For winter admission, 10/1 for domestic students; for spring admission, 2/1 for domestic students. Applications are processed on a rolling basis. Application fee: $75. *Financial support:* In 2005–06, research assistantships (averaging $11,481 per year), teaching assistantships (averaging $9,558 per year) were awarded; fellowships, career-related internships or fieldwork, scholarships/grants, and unspecified assistantships also available. *Faculty research:* Condensed matter and material physics, quantum computing, astrophysics and gravitation, industrial and applied physics, subatomic physics. Total annual research expenditures: $4 million. *Unit head:* Dr. Jamie Forrest, Director, 519-888-4567 Ext. 7598, Fax: 519-746-8115, E-mail: gwp@sciborg.uwaterloo.ca. *Application contact:* Margaret M. O'Neill, Administrative Assistant, 519-838-4567 Ext. 7598, Fax: 519-746-8115, E-mail: jwp@saborg.uwaterloo.ca.

University of Hawaii at Manoa, Graduate Division, Colleges of Arts and Sciences, College of Natural Sciences, Department of Physics and Astronomy, Honolulu, HI 96822. Offers astronomy (MS, PhD); physics (MS, PhD). *Faculty:* 66 full-time (5 women), 4 part-time/adjunct (1 woman). *Students:* 64 full-time (17 women), 3 part-time; includes 9 minority (1 African American, 6 Asian Americans or Pacific Islanders, 2 Hispanic Americans), 21 international. 176 applicants, 23% accepted, 12 enrolled. In 2005, 11 master's, 5 doctorates awarded. *Median time to degree:* Of those who began their doctoral program in fall 1997, 100% received their degree in 8 years or less. *Degree requirements:* For master's, qualifying exam or thesis; for doctorate, thesis/dissertation, oral comprehensive and qualifying exams. *Entrance requirements:* For master's and doctorate, GRE General Test, GRE Subject Test. Application fee: $50. *Expenses:* Tuition, state resident: full-time $8,400; part-time $200 per credit hour. Tuition, nonresident: full-time $11,088; part-time $462 per credit hour. Tuition and fees vary according to program. *Financial support:* In 2005–06, 43 research assistantships, 19 teaching assistantships were awarded. *Unit head:* Dr. Michael Peters, Chairperson, 808-956-7087, Fax: 808-956-7107, E-mail: mwp@phys.hawaii.edu. *Application contact:* Dr. Joshua Barnes, Graduate Chair, Astronomy, 808-956-8138, Fax: 808-956-4604, E-mail: barnes@ifa.hawaii.edu.

University of Houston, College of Natural Sciences and Mathematics, Department of Physics, Houston, TX 77204. Offers MA, MS, PhD. Part-time programs available. *Faculty:* 20 full-time (1 woman), 2 part-time/adjunct (0 women). *Students:* 85 full-time (18 women), 12 part-time (2 women); includes 6 minority (3 Asian Americans or Pacific Islanders, 3 Hispanic Americans), 69 international. Average age 29. 20 applicants, 90% accepted, 17 enrolled. In 2005, 9 master's, 12 doctorates awarded. Terminal master's awarded for partial completion of doctoral program. *Degree requirements:* For doctorate, thesis/dissertation. *Entrance requirements:* For master's and doctorate, GRE General Test. Additional exam requirements/recommendations for international students: Required—TOEFL. *Application deadline:* For fall admission, 7/20 for domestic students; for spring admission, 11/20 for domestic students. Applications are processed on a rolling basis. Application fee: $0 ($75 for international students). *Financial support:* In 2005–06, 80 research assistantships with full tuition reimbursements (averaging $15,150 per year), 17 teaching assistantships with full tuition reimbursements (averaging $15,150 per year) were awarded; fellowships with full tuition reimbursements, career-related internships or fieldwork, Federal Work-Study, institutionally sponsored loans, scholarships/grants, health care benefits, and unspecified assistantships also available. Support available to part-time students. Financial award application deadline:3/10. *Faculty research:* Condensed-matter, particle physics, high-temperature superconductivity, material/space physics, chaos. *Unit head:* Dr. Lawrence Pinsky, Chairman, 713-743-3552, Fax: 713-743-3589, E-mail: pinsky@uh.edu. *Application contact:* Advising Assistant, 713-743-3550, Fax: 713-743-3589.

University of Idaho, College of Graduate Studies, College of Science, Department of Physics, Moscow, ID 83844-2282. Offers physics (MS, PhD); physics education (MAT). *Students:* 20 full-time (3 women), 13 international. Average age 28. In 2005, 1 master's, 1 doctorate awarded. *Degree requirements:* For master's and doctorate, thesis/dissertation. *Entrance requirements:* For master's, GRE, minimum GPA of 2.8; for doctorate, GRE, minimum undergraduate GPA of 2.8, 3.0 graduate. *Application deadline:* For fall admission, 8/1 for domestic students; for spring admission, 12/15 for domestic students. Application fee: $55 ($60 for international students). *Expenses:* Tuition, nonresident: full-time $8,770; part-time $130 per credit. Required fees: $4,508; $217 per credit. *Financial support:* Research assistantships, teaching assistantships available. Financial award application deadline: 2/15. *Unit head:* Dr. Rupert Machleidt, Interim Chair, 208-885-6380.

University of Illinois at Chicago, Graduate College, College of Liberal Arts and Sciences, Department of Physics, Chicago, IL 60607-7128. Offers MS, PhD. Terminal master's awarded for partial completion of doctoral program. *Degree requirements:* For doctorate, thesis/dissertation. *Entrance requirements:* For master's and doctorate, GRE General Test, minimum GPA of 3.0. Additional exam requirements/recommendations for international students: Required—TOEFL. Electronic applications accepted. *Faculty research:* High-energy, laser, and solid-state physics.

University of Illinois at Urbana–Champaign, Graduate College, College of Engineering, Department of Physics, Champaign, IL 61820. Offers MS, PhD. *Faculty:* 60 full-time (6 women), 5 part-time/adjunct (2 women). *Students:* 269 full-time (33 women), 26 part-time (5 women); includes 17 minority (2 African Americans, 13 Asian Americans or Pacific Islanders, 2 Hispanic Americans), 123 international. 490 applicants, 18% accepted, 46 enrolled. In 2005, 30 master's, 30 doctorates awarded. *Degree requirements:* For doctorate, thesis/dissertation, departmental qualifying exam. *Entrance requirements:* For master's, GRE, minimum GPA of 3.0. *Application deadline:* Applications are processed on a rolling basis. Application fee: $50 ($60 for international students). Electronic applications accepted. *Financial support:* In 2005–06, 28 fellowships, 173 research assistantships, 154 teaching assistantships were awarded. Financial award application deadline: 2/15. *Unit head:* Jeremiah D. Sullivan, Head, 217-333-3760, Fax: 217-244-4293, E-mail: jdsacdis@uiuc.edu. *Application contact:* Wendy Wimmer, Secretary, 217-333-3645, Fax: 217-244-5073, E-mail: wwimmer@uiuc.edu.

The University of Iowa, Graduate College, College of Liberal Arts and Sciences, Department of Physics and Astronomy, Program in Physics, Iowa City, IA 52242-1316. Offers MS, PhD. *Students:* 31 full-time (7 women), 31 part-time (6 women); includes 3 minority (all Hispanic Americans), 32 international. 160 applicants, 15% accepted, 8 enrolled. In 2005, 5 master's, 5 doctorates awarded. *Degree requirements:* For master's, exam, thesis optional; for doctorate, thesis/dissertation, comprehensive exam, registration. *Entrance requirements:* For master's and doctorate, GRE General Test, GRE Subject Test, minimum GPA of 3.0. Additional exam requirements/recommendations for international students: Required—TOEFL (minimum score 550 paper-based; 213 computer-based). *Application deadline:* For fall admission, 2/1 priority date for domestic students, 2/1 priority date for international students. Application fee: $60 ($85 for international students). Electronic applications accepted. *Expenses:* Tuition, state resident: part-time $1,882 per term. Tuition, nonresident: full-time $17,338; part-time $4,907 per term. Tuition and fees vary according to course load and program. *Financial support:* In 2005–06, 1 fellowship, 32 research assistantships with partial tuition reimbursements, 24 teaching assistantships with partial tuition reimbursements were awarded. Financial award applicants required to submit FAFSA. *Unit head:* Thomas Boggess, Chair, Department of Physics and Astronomy, 319-335-1688, Fax: 319-335-1753.

University of Kansas, Graduate School, College of Liberal Arts and Sciences, Department of Physics and Astronomy, Lawrence, KS 66045. Offers computational physics and astronomy (MS); physics (MS, PhD). *Faculty:* 30. *Students:* 43 full-time (12 women), 9 part-time; includes 2 minority (both Hispanic Americans), 29 international. Average age 29. 62 applicants, 31% accepted. In 2005, 4 master's, 3 doctorates awarded. *Degree requirements:* For master's, thesis (for some programs); for doctorate, thesis/dissertation, comprehensive exam. *Entrance* *requirements:* Additional exam requirements/recommendations for international students: Required—TOEFL, TWE; Recommended—TSE. *Application deadline:* For fall admission, 3/1 priority date for domestic students, 3/1 priority date for international students; for spring admission, 10/1 priority date for domestic students, 10/1 priority date for international students. Applications are processed on a rolling basis. Application fee: $55 ($60 for international students). Electronic applications accepted. *Expenses:* Tuition, state resident: full-time $4,859. Tuition, nonresident: full-time $1,200. Required fees: $589. Tuition and fees vary according to program. *Financial support:* Fellowships with tuition reimbursements, research assistantships with full and partial tuition reimbursements, teaching assistantships with full and partial tuition reimbursements available. Financial award application deadline: 3/1. *Faculty research:* Condensed-matter, cosmology, elementary particles, nuclear physics, space physics. *Unit head:* Dr. Stephen J. Sanders, Chair, 785-864-4626, Fax: 785-864-5262. *Application contact:* Patricia Marvin, Graduate Admission Specialist, 785-864-4626, Fax: 785-864-5262, E-mail: physics@ku.edu.

See Close-Up on page 371.

University of Kentucky, Graduate School, Graduate School Programs from the College of Arts and Sciences, Program in Physics and Astronomy, Lexington, KY 40506-0032. Offers MS, PhD. *Faculty:* 26 full-time (2 women). *Students:* 53 full-time (16 women), 4 part-time (1 woman); includes 1 minority (Asian American or Pacific Islander), 35 international. 120 applicants, 27% accepted, 25 enrolled. In 2005, 5 master's, 2 doctorates awarded. *Median time to degree:* Of those who began their doctoral program in fall 1997, 54% received their degree in 8 years or less. *Degree requirements:* For master's, thesis optional; for doctorate, thesis/dissertation, comprehensive exam. *Entrance requirements:* For master's, GRE General Test, minimum undergraduate GPA of 2.5; for doctorate, GRE General Test, minimum graduate GPA of 3.0. Additional exam requirements/recommendations for international students: Required—TOEFL (minimum score 550 paper-based; 213 computer-based). *Application deadline:* For fall admission, 7/17 priority date for domestic students, 2/1 priority date for international students; for spring admission, 12/13 priority date for domestic students, 6/15 priority date for international students. Applications are processed on a rolling basis. Application fee: $40 ($55 for international students). Electronic applications accepted. *Expenses:* Tuition, state resident: full-time $3,159; part-time $331 per credit hour. Tuition, nonresident: full-time $6,984; part-time $756 per credit hour. Tuition and fees vary according to course load, degree level and program. *Financial support:* In 2005–06, 3 fellowships with full tuition reimbursements, 21 research assistantships with full tuition reimbursements (averaging $13,674 per year), 32 teaching assistantships with full tuition reimbursements (averaging $13,674 per year) were awarded; Federal Work-Study, institutionally sponsored loans, scholarships/grants, traineeships, health care benefits, tuition waivers (partial), and unspecified assistantships also available. Support available to part-time students. Financial award application deadline:3/15. *Faculty research:* Astrophysics, active galactic nuclei, and radio astronomy; Rydbert atoms, and electron scattering; TOF spectroscopy, hyperon interactions and muons; particle theory, lattice gauge theory, quark, and skyrmion models. Total annual research expenditures: $3.1 million. *Unit head:* Dr. Thomas Troland, Director of Graduate Studies, 859-257-8620, Fax: 859-323-2846, E-mail: troland@asta.pa.uky.edu. *Application contact:* Dr. Brian Jackson, Senior Associate Dean, 859-257-8176, Fax: 859-323-1928, E-mail: lance.brunner@uky.edu.

See Close-Up on page 373.

University of Lethbridge, School of Graduate Studies, Lethbridge, AB T1K 3M4, Canada. Offers accounting (MScM); addictions counseling (M Sc); agricultural biotechnology (M Sc); agricultural studies (M Sc, MA); anthropology (MA); archaeology (MA); art (MA); biochemistry (M Sc); biological sciences (M Sc); biomolecular science (PhD); biosystems and biodiversity (PhD); Canadian studies (MA); chemistry (M Sc); computer science (M Sc); computer science and geographical information science (M Sc); counseling psychology (M Ed); dramatic arts (MA); earth, space, and physical science (PhD); economics (MA); educational leadership (M Ed); English (MA); environmental science (M Sc); evolution and behavior (PhD); exercise science (M Sc); finance (MScM); French (MA); French/German (MA); French/Spanish (MA); general education (M Ed); general management (MScM); geography (M Sc, MA); German (MA); health sciences (M Sc, MA); history (MA); human resource management and labour relations (MScM); individualized multidisciplinary (M Sc, MA); information systems (MScM); international management (MScM); kinesiology (M Sc, MA); management (M Sc, MA); marketing (MScM); mathematics (M Sc); music (MA); Native American studies (MA); neuroscience (M Sc, PhD); new media (MA); nursing (M Sc); philosophy (MA); physics (M Sc); policy and strategy (MScM); political science (MA); psychology (M Sc, MA); religious studies (MA); sociology (MA); theoretical and computational science (PhD); urban and regional studies (MA). Part-time and evening/weekend programs available. *Faculty:* 250. *Students:* 193 full-time, 145 part-time. 35 applicants, 100% accepted, 35 enrolled. In 2005, 40 degrees awarded. *Degree requirements:* For doctorate, thesis/dissertation, comprehensive exam. *Entrance requirements:* For master's, GMAT (M Sc management), bachelor's degree in related field, minimum GPA of 3.0 during previous 20 graded semester courses, 2 years teaching or related experience (M Ed); for doctorate, master's degree, minimum graduate GPA of 3.5. Additional exam requirements/recommendations for international students: Required—TOEFL. Application fee: $60 Canadian dollars. *Expenses:* Tuition, nonresident: part-time $531 per course. Required fees: $83 per year. Tuition and fees vary according to degree level and program. *Financial support:* Fellowships, research assistantships, teaching assistantships, scholarships/grants, health care benefits, and unspecified assistantships available. *Faculty research:* Movement and brain plasticity, gibberellin physiology, photosynthesis, carbon cycling, molecular properties of main-group ring components. *Unit head:* Dr. Shamsul Alam, Dean, 403-329-2121, Fax: 403-329-2097, E-mail: inquiries@uleth.ca. *Application contact:* Kathy Schrage, Administrative Assistant, Office of the Academic Vice President, 403-329-2121, Fax: 403-329-2097, E-mail: inquiries@uleth.ca.

University of Louisiana at Lafayette, Graduate School, College of Sciences, Department of Physics, Lafayette, LA 70504. Offers MS. Part-time programs available. *Faculty:* 8 full-time (1 woman). *Students:* 4 full-time (2 women), 2 international. Average age 31. 9 applicants, 33% accepted, 2 enrolled. In 2005, 5 degrees awarded. *Degree requirements:* For master's, thesis, registration. *Entrance requirements:* For master's, GRE General Test, minimum GPA of 2.75. Additional exam requirements/recommendations for international students: Required—TOEFL (minimum score 550 paper-based; 213 computer-based). *Application deadline:* For fall admission, 5/15 for domestic students, 5/15 for international students; for spring admission, 10/1 for domestic students, 10/1 for international students. Applications are processed on a rolling basis. Application fee: $20 ($30 for international students). Electronic applications accepted. *Expenses:* Tuition, state resident: full-time $3,330; part-time $93 per credit hour. Tuition, nonresident: full-time $9,510; part-time $350 per semester. International tuition: $9,646 full-time. *Financial support:* In 2005–06, 1 fellowship with full tuition reimbursement (averaging $14,500 per year), 3 research assistantships with full tuition reimbursements (averaging $10,000 per year) were awarded; Federal Work-Study and unspecified assistantships also available. Financial award application deadline:5/1. *Faculty research:* Environmental physics, geophysics, astrophysics, acoustics, atomic physics. *Unit head:* Dr. John Meriwether, Head, 337-482-6691, Fax: 337-482-6699, E-mail: meriwether@louisiana.edu. *Application contact:* Dr. Daniel Whitmire, Graduate Coordinator, 337-482-6185, Fax: 337-482-6699, E-mail: whitmire@louisiana.edu.

University of Louisville, Graduate School, College of Arts and Sciences, Department of Physics, Louisville, KY 40292-0001. Offers MS. *Students:* 12 full-time (2 women), 5 part-time (1 woman), 5 international. Average age 31. In 2005, 8 degrees awarded. *Degree requirements:* For master's, thesis optional. *Entrance requirements:* For master's, GRE General Test. *Application deadline:* Applications are processed on a rolling basis. Application fee: $50. *Expenses:* Tuition, state resident: full-time $3,003; part-time $334 per credit hour. Tuition, nonresident: full-time $8,277; part-time $920 per credit hour. Tuition and fees vary according to course load, degree level and program. *Financial support:* In 2005–06, 12 teaching assistantships with tuition reimbursements (averaging $13,000 per year) were awarded *Unit head:* Dr. David Brown, Acting Chair, 502-852-6790, Fax: 502-852-0742, E-mail: brown@louisville.edu.

University of Maine, Graduate School, College of Liberal Arts and Sciences, Department of Physics and Astronomy, Orono, ME 04469. Offers engineering physics (M Eng); physics (MS, PhD). *Faculty:* 16. *Students:* 36 full-time (9 women), 8 part-time; includes 2 minority (1 African American, 1 American Indian/Alaska Native), 5 international. Average age 30. 27 applicants, 41% accepted, 10 enrolled. In 2005, 1 master's awarded. Terminal master's awarded for partial completion of doctoral program. *Degree requirements:* For doctorate, thesis/dissertation. *Entrance requirements:* For master's, GRE General Test, GRE Subject Test; for doctorate, GRE General Test. Additional exam requirements/recommendations for international students: Required—TOEFL. *Application deadline:* For fall admission, 2/1 for domestic students. Applications are processed on a rolling basis. Application fee: $50. Electronic applications accepted. *Financial support:* In 2005–06, 5 research assistantships with tuition reimbursements (averaging $19,500 per year), 15 teaching assistantships with tuition reimbursements (averaging $12,000 per year) were awarded; fellowships with tuition reimbursements, tuition waivers (full and partial) also available. Financial award application deadline: 3/1. *Faculty research:* Solid-state physics, fluids, biophysics, plasma physics, surface physics. *Unit head:* Dr. David Batuski, Chair, 207-581-1039, Fax: 207-581-3410. *Application contact:* Scott G. Delcourt, Associate Dean of the Graduate School, 207-581-3219, Fax: 207-581-3232, E-mail: graduate@maine.edu.

University of Manitoba, Faculty of Graduate Studies, Faculty of Science, Department of Physics, Winnipeg, MB R3T 2N2, Canada. Offers M Sc, PhD. *Degree requirements:* For master's, thesis; for doctorate, one foreign language, thesis/dissertation.

University of Maryland, Baltimore County, Graduate School, College of Natural Sciences and Mathematics, Department of Physics, Baltimore, MD 21250. Offers applied physics (MS, PhD), including astrophysics (PhD), optics, quantum optics (PhD), solid state physics; atmospheric physics (MS, PhD). Part-time programs available. *Faculty:* 19 full-time (3 women), 26 part-time/adjunct (0 women). *Students:* 44 full-time (12 women), 4 part-time (1 woman); includes 6 minority (all African Americans), 16 international. 40 applicants, 33% accepted, 10 enrolled. In 2005, 5 master's, 3 doctorates awarded. Terminal master's awarded for partial completion of doctoral program. *Degree requirements:* For master's, thesis optional; for doctorate, thesis/dissertation, comprehensive exam. *Entrance requirements:* For master's and doctorate, GRE General Test, GRE Subject Test, minimum GPA of 3.0. Additional exam requirements/recommendations for international students: Required—TOEFL. *Application deadline:* For fall admission, 7/31 for domestic students; for spring admission, 12/31 priority date for domestic students. Applications are processed on a rolling basis. Application fee: $50. Electronic applications accepted. *Expenses:* Tuition, state resident: part-time $395 per credit. Tuition, nonresident: part-time $652 per credit. Required fees: $82 per credit. Tuition and fees vary according to course load, program and reciprocity agreements. *Financial support:* In 2005–06, 42 students received support, including 5 fellowships with tuition reimbursements available (averaging $30,000 per year), 22 research assistantships with full tuition reimbursements available (averaging $21,000 per year), 14 teaching assistantships with full tuition reimbursements available (averaging $20,000 per year); scholarships/grants, health care benefits, and unspecified assistantships also available. Financial award application deadline:3/1. *Faculty research:* Optics, solid state physics, astrophysics, atmospheric physics, quantum optics. Total annual research expenditures: $6.8 million. *Unit head:* Dr. L Michael Summers, Chairman, 410-455-2513, Fax: 410-455-1072. *Application contact:* Dr. Lazlo Takacs, Graduate Admissions, 410-455-2513, Fax: 410-455-1072, E-mail: physics@umbc.edu.

University of Maryland, College Park, Graduate Studies, College of Computer, Mathematical and Physical Sciences, Department of Physics, College Park, MD 20742. Offers MS, PhD. Part-time and evening/weekend programs available. *Faculty:* 140 full-time (16 women), 26 part-time/adjunct (2 women). *Students:* 210 full-time (26 women), 10 part-time (1 woman); includes 12 minority (4 African Americans, 1 American Indian/Alaska Native, 6 Asian Americans or Pacific Islanders, 1 Hispanic American), 86 international. 634 applicants, 18% accepted, 40 enrolled. In 2005, 2 master's, 29 doctorates awarded. Terminal master's awarded for partial completion of doctoral program. *Median time to degree:* Of those who began their doctoral program in fall 1997, 39% received their degree in 8 years or less. *Degree requirements:* For master's, thesis optional; for doctorate, thesis/dissertation. *Entrance requirements:* For master's, GRE General Test, GRE Advanced Subject Test in physics, minimum GPA of 3.0, 3 letters of recommendation; for doctorate, GRE General Test, GRE Advanced Subject Test in physics, 3 letters of recommendation. *Application deadline:* For fall admission, 2/1 for domestic students, 2/1 for international students. Applications are processed on a rolling basis. Application fee: $60. Electronic applications accepted. *Financial support:* In 2005–06, 30 fellowships with full tuition reimbursements (averaging $12,410 per year), 123 research assistantships with tuition reimbursements (averaging $19,351 per year), 54 teaching assistantships with tuition reimbursements (averaging $14,005 per year) were awarded; Federal Work-Study and scholarships/grants also available. Support available to part-time students. Financial award applicants required to submit FAFSA. *Faculty research:* Astrometeorology, superconductivity, particle astrophysics, plasma physics, elementary particle theory. Total annual research expenditures: $17.9 million. *Unit head:* Dr. Jordan A. Goodman, Chair, 301-405-5946, Fax: 301-405-0327. *Application contact:* Linda O'Hara, Administrative Assistant, 301-405-5982.

Announcement: Maryland, a leading recipient of federal physics research grants, has more than 30 programs and institutes, including the nation's #1 ranked nonlinear dynamics/chaos program and a major new AMO program. Others include condensed matter, particle astrophysics, and centers in superconductivity, plasma, and materials science. Contact Linda O'Hara, 301-405-5982; lohara@physics.umd.edu; www.physics.umd.edu.

See Close-Up on page 375.

University of Massachusetts Amherst, Graduate School, College of Natural Sciences and Mathematics, Department of Physics, Amherst, MA 01003. Offers MS, PhD. Part-time programs available. *Faculty:* 40 full-time (1 woman). *Students:* 65 full-time (9 women), 1 part-time; includes 4 minority (1 African American, 3 Hispanic Americans), 45 international. Average age 27. 153 applicants, 19% accepted, 8 enrolled. In 2005, 16 master's, 5 doctorates awarded. Terminal master's awarded for partial completion of doctoral program. *Degree requirements:* For doctorate, thesis/dissertation. *Entrance requirements:* For master's and doctorate, GRE General Test, GRE Subject Test. Additional exam requirements/recommendations for international students: Required—TOEFL (minimum score 530 paper-based; 197 computer-based). *Application deadline:* For fall admission, 2/1 priority date for domestic students, 2/1 priority date for international students; for spring admission, 10/1 for domestic students, 10/1 for international students. Applications are processed on a rolling basis. Application fee: $40 ($65 for international students). Electronic applications accepted. *Expenses:* Tuition, state resident: part-time $110 per credit. Tuition, nonresident: part-time $414 per credit. Required fees: $2,824 per term. One-time fee: $250 part-time. Full-time tuition and fees vary according to course load, campus/location, program and reciprocity agreements. *Financial support:* In 2005–06, fellowships with full tuition reimbursements (averaging $625 per year), research assistantships with full tuition reimbursements (averaging $13,021 per year), teaching assistantships with full tuition reimbursements (averaging $10,648 per year) were awarded; career-related internships or fieldwork, Federal Work-Study, scholarships/grants, traineeships, and unspecified assistantships also available. Support available to part-time students. Financial award application deadline: 2/1. *Unit head:* Dr. Jonathan Machta, Director, 413-545-2545, Fax: 413-545-0648, E-mail: machta@physics.umass.edu. *Application contact:* Dr. Donald Candela, Chair, Admissions Committee, 413-545-2407, E-mail: candela@phast.umass.edu.

University of Massachusetts Dartmouth, Graduate School, College of Engineering, Department of Physics, North Dartmouth, MA 02747-2300. Offers MS. Part-time programs available. *Faculty:* 12 full-time (2 women), 2 part-time/adjunct (1 woman). *Students:* 17 full-time (4 women), 3 part-time (1 woman), 12 international. Average age 24. 15 applicants, 87% accepted, 10 enrolled. In 2005, 3 degrees awarded. *Degree requirements:* For master's, thesis or alternative. *Entrance requirements:* For master's, GRE General Test. Additional exam

Physics

University of Massachusetts Dartmouth (continued)
requirements/recommendations for international students: Required—TOEFL (minimum score 500 paper-based). *Application deadline:* For fall admission, 4/20 priority date for domestic students, 2/20 priority date for international students; for spring admission, 11/15 priority date for domestic students, 9/15 priority date for international students. Applications are processed on a rolling basis. Application fee: $35 ($55 for international students). Electronic applications accepted. *Expenses:* Tuition, state resident: full-time $2,071; part-time $86 per credit. Tuition, nonresident: full-time $8,099; part-time $337 per credit. Required fees: $9,437; $393 per credit. *Financial support:* In 2005–06, 12 research assistantships with full tuition reimbursements (averaging $11,495 per year), 8 teaching assistantships with full tuition reimbursements (averaging $7,800 per year) were awarded; Federal Work-Study and unspecified assistantships also available. Support available to part-time students. Financial award application deadline: 3/1; financial award applicants required to submit FAFSA. *Faculty research:* Real-time physics dissemination, four-dimensional symmetry principle, physical and biological variability of Northeast Massachusetts coast, light atoms and molecules interaction with photons, property fluxes in the North Atlantic. Total annual research expenditures: $201,000. *Unit head:* Elliot Horch, Director, 508-999-8360, Fax: 508-999-9115, E-mail: ehorch@umassd.edu. *Application contact:* Carol Novo, Graduate Admissions Officer, 508-999-8604, Fax: 508-999-8183, E-mail: graduate@umassd.edu.

University of Massachusetts Lowell, Graduate School, College of Arts and Sciences, Department of Physics and Applied Physics, Program in Physics, Lowell, MA 01854-2881. Offers MS, PhD. *Degree requirements:* For master's, thesis; for doctorate, 2 foreign languages, thesis/dissertation. *Entrance requirements:* For master's and doctorate, GRE General Test.

University of Memphis, Graduate School, College of Arts and Sciences, Department of Physics, Memphis, TN 38152. Offers MS. Part-time programs available. *Faculty:* 10 full-time (1 woman), 3 part-time/adjunct (0 women). *Students:* 6 full-time (1 woman), 2 part-time; includes 1 minority (African American), 3 international. Average age 28. *Degree requirements:* For master's, thesis or alternative, comprehensive exam. *Entrance requirements:* For master's, GRE General Test or MAT, 20 undergraduate hours of course work in physics. *Application deadline:* For fall admission, 8/1 for domestic students; for spring admission, 12/1 for domestic students. Applications are processed on a rolling basis. Application fee: $25 ($50 for international students). Electronic applications accepted. *Financial support:* In 2005–06, research assistantships (averaging $7,000 per year), teaching assistantships (averaging $5,150 per year) were awarded; Federal Work-Study and institutionally sponsored loans also available. Financial award application deadline: 5/1; financial award applicants required to submit CSS PROFILE. *Faculty research:* Solid-state physics, materials science, biophysics, astrophysics, physics education. Total annual research expenditures: $422,716. *Unit head:* Dr. M. Shah Jahan, Chairman, 901-678-2620, Fax: 901-678-4733, E-mail: mjahan@memphis.edu. *Application contact:* Dr. Sanjay Mishra, Coordinator of Graduate Studies, 901-678-2410, Fax: 901-678-4733, E-mail: srmishra@memphis.edu.

University of Miami, Graduate School, College of Arts and Sciences, Department of Physics, Coral Gables, FL 33124. Offers MS, PhD. *Faculty:* 16 full-time (0 women), 1 (woman) part-time/adjunct. *Students:* 25 full-time (6 women); includes 1 minority (Hispanic American), 22 international. Average age 26. 59 applicants, 12% accepted, 5 enrolled. In 2005, 4 degrees awarded. Terminal master's awarded for partial completion of doctoral program. *Degree requirements:* For master's, comprehensive exam, registration; for doctorate, thesis/dissertation, comprehensive exam, registration. *Entrance requirements:* For master's and doctorate, GRE General Test, GRE Subject Test. Additional exam requirements/recommendations for international students: Required—TOEFL (minimum score 550 paper-based; 213 computer-based). *Application deadline:* For fall admission, 3/1 priority date for domestic students, 2/1 priority date for international students. Applications are processed on a rolling basis. Application fee: $50. Electronic applications accepted. *Financial support:* In 2005–06, 1 fellowship with tuition reimbursement (averaging $18,000 per year), 3 research assistantships with tuition reimbursements (averaging $18,000 per year), 21 teaching assistantships with tuition reimbursements (averaging $18,000 per year) were awarded. Financial award application deadline: 2/1; financial award applicants required to submit FAFSA. *Faculty research:* High-energy theory, marine and atmospheric optics, plasma physics, solid-state physics. Total annual research expenditures:$780,565. *Unit head:* Dr. George Alexandrakis, Chairman, 305-284-2323 Ext. 9, Fax: 305-284-4222, E-mail: alexandrakis@physics.miami.edu. *Application contact:* Dr. Josef Ashkenazi, Chairman, Graduate Recruitment Committee, 305-284-2323 Ext. 3, Fax: 305-284-4222, E-mail: ashkenazi@physics.miami.edu.

See Close-Up on page 379.

University of Michigan, Horace H. Rackham School of Graduate Studies, College of Literature, Science, and the Arts, Department of Physics, Ann Arbor, MI 48109. Offers MS, PhD. *Faculty:* 60 full-time (7 women). *Students:* 142 full-time (32 women); includes 8 minority (2 African Americans, 2 Asian Americans or Pacific Islanders, 4 Hispanic Americans), 65 international. Average age 29. 393 applicants, 13% accepted, 16 enrolled. In 2005, 26 master's, 18 doctorates awarded. Terminal master's awarded for partial completion of doctoral program. *Degree requirements:* For doctorate, oral defense of dissertation, preliminary exam. *Entrance requirements:* For master's and doctorate, GRE General Test. Additional exam requirements/recommendations for international students: Required—TOEFL (minimum score 560 paper-based; 220 computer-based). *Application deadline:* For fall admission, 1/15 for domestic students, 12/8 for international students. Application fee: $60 ($75 for international students). Electronic applications accepted. *Expenses:* Tuition, state resident: full-time $14,082; part-time $894 per credit hour. Tuition, nonresident: full-time $28,500; part-time $1,675 per credit hour. Required fees: $189; $189 per unit. *Financial support:* In 2005–06, fellowships with full tuition reimbursements (averaging $24,300 per year), research assistantships with full tuition reimbursements (averaging $21,489 per year), teaching assistantships with full tuition reimbursements (averaging $21,489 per year) were awarded. *Faculty research:* Elementary particle, solid-state, atomic, and molecular physics (theoretical and experimental). Total annual research expenditures: $14.2 million. *Unit head:* Dr. Myron Campbell, Chair, 734-764-4437. *Application contact:* Kimberly A. Smith, Graduate Coordinator, 734-936-0658, Fax: 734-763-9694, E-mail: physics.inquiries@umich.edu.

See Close-Up on page 383.

University of Minnesota, Duluth, Graduate School, College of Science and Engineering, Department of Physics, Duluth, MN 55812-2496. Offers MS. Part-time programs available. *Faculty:* 6 full-time (0 women), 2 part-time/adjunct (0 women). *Students:* 7 full-time (0 women), 2 international. Average age 27. 7 applicants, 71% accepted, 4 enrolled. In 2005, 2 degrees awarded. *Degree requirements:* For master's, final oral exam, thesis optional. *Entrance requirements:* For master's, minimum GPA of 3.0. Additional exam requirements/recommendations for international students: Required—TOEFL (minimum score 550 paper-based; 213 computer-based). *Application deadline:* For fall admission, 7/15 for domestic students, 7/15 for international students; for spring admission, 11/1 for domestic students, 11/1 for international students. Applications are processed on a rolling basis. Application fee: $55 ($75 for international students). *Financial support:* In 2005–06, 7 students received support, including 1 research assistantship (averaging $11,895 per year), 6 teaching assistantships with full tuition reimbursements available (averaging $11,895 per year); Federal Work-Study, institutionally sponsored loans, scholarships/grants, and health care benefits also available. Support available to part-time students. Financial award application deadline: 3/15. *Faculty research:* Computational physics, solid–state physics, physical oceanography, high energy neutrinos, quantum chromodynamics. Total annual research expenditures: $70,000. *Unit head:* Dr. John R. Hiller, Head, 218-726-7594, Fax: 218-726-6942, E-mail: jhiller@d.umn.edu. *Application contact:* Prof. Jonathan Maps, Director of Graduate Studies, 218-726-8125, Fax: 218-726-6942, E-mail: jmaps@d.umn.edu.

Announcement: MS in physics: concentrations in particle physics, condensed matter, and physical oceanography. Thesis projects provide experience in computational physics, instrumentation, optics, and data analysis. Current research areas include neutrinos, quark models, scanned probe microscopy, remote sensing and environmental optics, and observation and modeling of circulation of coastal shelves, estuaries, and lakes.

University of Minnesota, Twin Cities Campus, Graduate School, Institute of Technology, School of Physics and Astronomy, Department of Physics, Minneapolis, MN 55455-0213. Offers MS, PhD. Part-time programs available. *Degree requirements:* For master's and doctorate, thesis/dissertation. *Entrance requirements:* For master's and doctorate, GRE General Test, GRE Subject Test. *Expenses:* Tuition, state resident: full-time $8,748; part-time $729 per credit. Tuition, nonresident: full-time $15,848; part-time $1,321 per credit. Full-time tuition and fees vary according to class time, course load, program and reciprocity agreements. *Faculty research:* Condensed matter, elementary particle, space, nuclear and atomic physics.

University of Mississippi, Graduate School, College of Liberal Arts, Department of Physics and Astronomy, Oxford, University, MS 38677. Offers physics (MA, MS, PhD). *Faculty:* 17 full-time (0 women), 1 part-time/adjunct (0 women). *Students:* 22 full-time (3 women), 1 (woman) part-time, 10 international. 28 applicants, 43% accepted, 8 enrolled. In 2005, 3 master's awarded. *Degree requirements:* For master's, thesis (for some programs); for doctorate, thesis/dissertation. *Entrance requirements:* For master's, GRE General Test, minimum GPA of 3.0; for doctorate, GRE General Test. Additional exam requirements/recommendations for international students: Required—TOEFL. *Application deadline:* For fall admission, 4/1 for domestic students; for spring admission, 10/1 for domestic students. Applications are processed on a rolling basis. Application fee: $25. Electronic applications accepted. *Expenses:* Tuition, state resident: full-time $4,320; part-time $240 per credit hour. Tuition, nonresident: full-time $9,744; part-time $301 per credit hour. Tuition and fees vary according to program. *Financial support:* Scholarships/grants available. Financial award application deadline: 3/1; financial award applicants required to submit FAFSA. *Unit head:* Thomas C. Marshall, Chairman, 662-915-5325, Fax: 662-915-5045, E-mail: physics@phy.olemiss.edu.

University of Missouri–Columbia, Graduate School, College of Arts and Sciences, Department of Physics and Astronomy, Columbia, MO 65211. Offers MS, PhD. *Faculty:* 23 full-time (5 women). *Students:* 33 full-time (6 women), 10 part-time (2 women), 26 international. In 2005, 4 master's, 4 doctorates awarded. Terminal master's awarded for partial completion of doctoral program. *Degree requirements:* For doctorate, one foreign language, thesis/dissertation. *Entrance requirements:* For master's and doctorate, GRE General Test, minimum GPA of 3.0. *Application deadline:* For fall admission, 4/15 for domestic students. Applications are processed on a rolling basis. Application fee: $45 ($60 for international students). *Financial support:* Research assistantships, teaching assistantships, institutionally sponsored loans available. *Unit head:* Dr. H. R. Chandrasekhar, Director of Graduate Studies, 573-882-6086, E-mail: chandra@missouri.edu.

University of Missouri–Kansas City, College of Arts and Sciences, Department of Physics, Kansas City, MO 64110-2499. Offers MS, PhD. PhD offered through the School of Graduate Studies. Part-time and evening/weekend programs available. *Faculty:* 10 full-time (1 woman). *Students:* 5 full-time (0 women), 7 part-time (1 woman), 4 international. Average age 29. 7 applicants, 86% accepted, 6 enrolled. In 2005, 2 degrees awarded. Terminal master's awarded for partial completion of doctoral program. *Degree requirements:* For master's, thesis optional; for doctorate, thesis/dissertation, comprehensive exam. *Entrance requirements:* For doctorate, GRE General Test. Additional exam requirements/recommendations for international students: Required—TOEFL. *Application deadline:* For fall admission, 6/1 for domestic students; for spring admission, 11/1 priority date for domestic students. Applications are processed on a rolling basis. Application fee: $35 ($50 for international students). Electronic applications accepted. *Expenses:* Tuition, state resident: full-time $4,738; part-time $263 per credit hour. Tuition, nonresident: full-time $12,235; part-time $679 per credit hour. Required fees: $582. Tuition and fees vary according to course load, program and student level. *Financial support:* In 2005–06, fellowships with full tuition reimbursements (averaging $12,500 per year), research assistantships with full and partial tuition reimbursements (averaging $11,000 per year), teaching assistantships with full and partial tuition reimbursements (averaging $11,000 per year) were awarded; Federal Work-Study, institutionally sponsored loans, and tuition waivers (full and partial) also available. Support available to part-time students. Financial award application deadline: 4/1. *Faculty research:* Surface physics, material science, statistical mechanics, computational physics, relativity and quantum theory. Total annual research expenditures: $550,000. *Unit head:* Dr. Michael Kruger, Chairperson, 816-235-1604, E-mail: krugerm@umkc.edu. *Application contact:* Da Ming Zhu, Principal Graduate Adviser, 816-235-5326, Fax: 816-235-5221, E-mail: zhud@umkc.edu.

University of Missouri–Rolla, Graduate School, College of Arts and Sciences, Department of Physics, Rolla, MO 65409-0910. Offers MS, PhD.

University of Missouri–St. Louis, College of Arts and Sciences, Department of Physics and Astronomy, St. Louis, MO 63121. Offers applied physics (MS); astrophysics (MS); physics (PhD). Part-time and evening/weekend programs available. *Faculty:* 13. *Students:* 10 full-time (2 women), 15 part-time (1 woman); includes 1 minority (Asian American or Pacific Islander), 6 international. Average age 30. In 2005, 3 master's awarded. Terminal master's awarded for partial completion of doctoral program. *Degree requirements:* For master's, thesis optional; for doctorate, thesis/dissertation. *Entrance requirements:* For master's, 2 letters of recommendation; for doctorate, GRE General Test, 2 letters of recommendation. Additional exam requirements/recommendations for international students: Required—TOEFL (minimum score 550 paper-based; 213 computer-based). *Application deadline:* For fall admission, 7/1 for domestic students; for spring admission, 12/1 priority date for domestic students. Applications are processed on a rolling basis. Application fee: $35 ($40 for international students). Electronic applications accepted. *Expenses:* Tuition, state resident: part-time $263 per credit hour. Tuition, nonresident: part-time $680 per credit hour. Required fees: $53 per credit hour. Tuition and fees vary according to program. *Financial support:* In 2005–06, 3 research assistantships with full and partial tuition reimbursements (averaging $13,333 per year), 11 teaching assistantships with full and partial tuition reimbursements (averaging $11,575 per year) were awarded; fellowships with full tuition reimbursements, career-related internships or fieldwork also available. *Faculty research:* Biophysics, atomic physics, nonlinear dynamics, materials science. *Unit head:* Dr. Ricardo Flores, Director of Graduate Studies, 314-516-5931, Fax: 314-516-6152, E-mail: flores@jinx.umsl.edu. *Application contact:* 314-516-5458, Fax: 314-516-5310, E-mail: gradadm@umsl.edu.

University of Nebraska–Lincoln, Graduate College, College of Arts and Sciences, Department of Physics and Astronomy, Lincoln, NE 68588. Offers astronomy (MS, PhD); physics (MS, PhD). *Degree requirements:* For master's, thesis optional; for doctorate, thesis/dissertation, comprehensive exam. *Entrance requirements:* For master's and doctorate, GRE General Test. Additional exam requirements/recommendations for international students: Required—TOEFL (minimum score 550 paper-based; 213 computer-based). Electronic applications accepted. *Faculty research:* Electromagnetics of solids and thin films, photoionization, ion collisions with atoms, molecules and surfaces, nanostructures.

University of Nevada, Las Vegas, Graduate College, College of Science, Department of Physics, Las Vegas, NV 89154-9900. Offers MS, PhD. Part-time programs available. *Faculty:* 18 full-time (3 women), 1 part-time/adjunct (0 women). *Students:* 11 full-time (2 women), 6 part-time; includes 1 minority (Hispanic American), 6 international. 11 applicants, 45% accepted, 3 enrolled. *Degree requirements:* For master's, thesis, oral exam; for doctorate, thesis/dissertation, comprehensive exam. *Entrance requirements:* For master's, GRE General Test, GRE Subject Test, minimum GPA of 3.0 during previous 2 years, 2.75 overall; for doctorate, GRE General Test, GRE Subject Test, minimum GPA of 3.25 during previous 2 years, 3.0 overall. Additional exam requirements/recommendations for international students: Required—TOEFL (minimum score 550 paper-based; 213 computer-based). *Application deadline:* For fall admission, 6/15 for domestic students, 5/1 for international students; for spring admission, 11/15 for domestic students, 10/1 for international students. Application fee: $60 ($75 for international students). Electronic applications accepted. *Expenses:* Tuition, state resident:

part-time $150 per credit. Tuition, nonresident: part-time $315 per credit. Tuition and fees vary according to course load, program and reciprocity agreements. *Financial support:* In 2005–06, 3 research assistantships with partial tuition reimbursements (averaging $12,000 per year), 11 teaching assistantships with partial tuition reimbursements (averaging $11,000 per year) were awarded; career-related internships or fieldwork, Federal Work-Study, institutionally sponsored loans, scholarships/grants, health care benefits, and unspecified assistantships also available. Support available to part-time students. Financial award application deadline: 3/1. *Faculty research:* Laser (atomic, molecular, and optical) physics, astronomy/astrophysics, condensed-matter physics. *Unit head:* Dr. James Selser, Chair, 702-895-3084. *Application contact:* Graduate College Admissions Evaluator, 702-895-3320, Fax: 702-895-4180, E-mail: gradcollege@unlv.edu.

University of Nevada, Reno, Graduate School, College of Science, Department of Physics, Reno, NV 89557. Offers MS, PhD. Terminal master's awarded for partial completion of doctoral program. *Degree requirements:* For master's, thesis optional; for doctorate, thesis/dissertation. *Entrance requirements:* For master's, GRE General Test, minimum GPA of 2.75; for doctorate, GRE General Test, minimum GPA of 3.0. Additional exam requirements/recommendations for international students: Required—TOEFL. *Faculty research:* Atomic and molecular physics.

University of New Brunswick Fredericton, School of Graduate Studies, Faculty of Science, Department of Physics, Fredericton, NB E3B 5A3, Canada. Offers M Sc, PhD. Part-time programs available. *Degree requirements:* For master's and doctorate, thesis/dissertation. *Entrance requirements:* For master's and doctorate, minimum GPA of 3.0. Additional exam requirements/recommendations for international students: Required—TOEFL, TWE.

University of New Hampshire, Graduate School, College of Engineering and Physical Sciences, Department of Physics, Durham, NH 03824. Offers MS, PhD. *Faculty:* 29 full-time. *Students:* 28 full-time (8 women), 14 part-time (3 women), 22 international. Average age 27. 52 applicants, 35% accepted, 9 enrolled. In 2005, 4 master's, 3 doctorates awarded. Terminal master's awarded for partial completion of doctoral program. *Degree requirements:* For master's, thesis or alternative; for doctorate, thesis/dissertation. *Entrance requirements:* For master's and doctorate, GRE General Test. Additional exam requirements/recommendations for international students: Required—TOEFL (minimum score 550 paper-based; 213 computer-based); Recommended—TSE. *Application deadline:* For fall admission, 4/1 priority date for domestic students, 4/1 priority date for international students. For winter admission, 12/1 for domestic students. Applications are processed on a rolling basis. Application fee: $60. Electronic applications accepted. *Expenses:* Tuition, state resident: full-time $8,010; part-time $445 per credit hour. Tuition, nonresident: full-time $19,730; part-time $810 per credit hour. Required fees: $322 per semester. Tuition and fees vary according to course load and program. *Financial support:* In 2005–06, 1 fellowship, 27 research assistantships, 11 teaching assistantships were awarded; Federal Work-Study, scholarships/grants, and tuition waivers (full and partial) also available. Support available to part-time students. Financial award application deadline: 2/15. *Faculty research:* Astrophysics and space physics, nuclear physics, atomic and molecular physics, nonlinear dynamical systems. *Unit head:* Dr. Dawn Meredith, Chairperson, 603-862-2063. *Application contact:* Katie Makem, Administrative Assistant, 603-862-2669, E-mail: physics.grad.info@unh.edu.

University of New Mexico, Graduate School, College of Arts and Sciences, Department of Physics and Astronomy, Albuquerque, NM 87131-2039. Offers biomedical physics (MS, PhD); optical science and engineering (MS); optical sciences and engineering (PhD); physics (MS, PhD). *Faculty:* 35 full-time (4 women), 11 part-time/adjunct (1 woman). *Students:* 89 full-time (16 women), 15 part-time (2 women); includes 4 minority (2 Asian Americans or Pacific Islanders, 2 Hispanic Americans), 50 international. Average age 29. 99 applicants, 37% accepted, 18 enrolled. In 2005, 12 master's, 9 doctorates awarded. *Degree requirements:* For master's, thesis (for some programs), comprehensive exam (for some programs); for doctorate, thesis/dissertation, comprehensive exam. *Entrance requirements:* Additional exam requirements/recommendations for international students: Required—TOEFL (minimum score 610 paper-based; 253 computer-based). *Application deadline:* For fall admission, 2/1 for domestic students; for spring admission, 8/1 for domestic students. Application fee: $40. Electronic applications accepted. *Expenses:* Tuition, nonresident: full-time $3,388; part-time $238 per credit hour. Required fees: $385 per term. Tuition and fees vary according to course load and program. *Financial support:* In 2005–06, 15 students received support, including research assistantships with full tuition reimbursements available (averaging $20,000 per year), teaching assistantships with full tuition reimbursements available (averaging $13,500 per year); fellowships with full tuition reimbursements available, career-related internships or fieldwork, scholarships/grants, health care benefits, and unspecified assistantships also available. Financial award application deadline: 2/1; financial award applicants required to submit FAFSA. *Faculty research:* High-energy and particle physics, optical and laser sciences, condensed matter, nuclear and particle physics, surface physics, biomedical physics, quantum information, subatomic physics, nonlinear science, general relativity. Total annual research expenditures: $6.1 million. *Unit head:* Dr. Bernd Bassalleck, Chair, 505-277-1517, Fax: 505-277-1520, E-mail: bassek@unm.edu. *Application contact:* Mary De Witt, Program Advisement Coordinator, 505-277-1514, Fax: 505-277-1514, E-mail: mdewitt@unm.edu.

University of New Orleans, Graduate School, College of Sciences, Department of Physics, New Orleans, LA 70148. Offers MS, PhD. Part-time and evening/weekend programs available. *Degree requirements:* For master's, thesis (for some programs). *Entrance requirements:* For master's, GRE General Test. Additional exam requirements/recommendations for international students: Required—TOEFL (minimum score 550 paper-based; 213 computer-based). Electronic applications accepted. *Faculty research:* Underwater acoustics, applied electromagnetics, experimental atomic beams, digital signal processing, astrophysics.

The University of North Carolina at Chapel Hill, Graduate School, College of Arts and Sciences, Department of Physics and Astronomy, Chapel Hill, NC 27599. Offers physics (MS, PhD). Terminal master's awarded for partial completion of doctoral program. *Degree requirements:* For master's, comprehensive exam, registration; for doctorate, thesis/dissertation, comprehensive exam, registration. *Entrance requirements:* For master's and doctorate, GRE General Test, minimum GPA of 3.0. Electronic applications accepted. *Faculty research:* Observational astronomy, fullerenes, polarized beams, nanotubes, nucleosynthesis in stars and supernovae, superstring theory, ballistic transport in semiconductors, gravitation.

University of North Dakota, Graduate School, College of Arts and Sciences, Department of Physics, Grand Forks, ND 58202. Offers MS, PhD. *Faculty:* 10 full-time (1 woman). *Students:* 23 applicants, 22% accepted, 1 enrolled. In 2005, 7 degrees awarded. *Degree requirements:* For master's, thesis/dissertation, final exam; for doctorate, thesis/dissertation, final exam, comprehensive exam. *Entrance requirements:* For master's, minimum GPA of 3.0; for doctorate, minimum GPA of 3.5. Additional exam requirements/recommendations for international students: Required—TOEFL (minimum score 550 paper-based; 213 computer-based). *Application deadline:* For fall admission, 2/15 priority date for domestic students, 2/15 priority date for international students; for spring admission, 10/15 priority date for domestic students, 10/15 priority date for international students. Applications are processed on a rolling basis. Application fee: $35. Electronic applications accepted. *Financial support:* In 2005–06, 7 students received support, including 7 research assistantships with full tuition reimbursements available (averaging $11,648 per year), 8 teaching assistantships with full tuition reimbursements available (averaging $9,946 per year); fellowships, Federal Work-Study, institutionally sponsored loans, scholarships/grants, and tuition waivers (full and partial) also available. Support available to part-time students. Financial award application deadline: 3/15; financial award applicants required to submit FAFSA. *Faculty research:* Solid state physics, atomic and molecular physics, astrophysics, health physics. *Unit head:* Dr. Kanishka Marasinghe, Graduate Director, 701-777-2911, Fax: 701-777-3523, E-mail: k.macasinghe@und.nodak.edu. *Application contact:* Brenda Halle, Admissions Specialist, 701-777-2947, Fax: 701-777-3619, E-mail: brendahalle@mail.und.edu.

University of North Texas, Robert B. Toulouse School of Graduate Studies, College of Arts and Sciences, Department of Physics, Denton, TX 76203. Offers MA, MS, PhD. *Faculty:* 19 full-time (1 woman). *Students:* 31 full-time (5 women), 21 part-time (3 women); includes 4 minority (1 African American, 1 Asian American or Pacific Islander, 2 Hispanic Americans), 27 international. Average age 32. 24 applicants, 38% accepted, 2 enrolled. In 2005, 5 master's, 3 doctorates awarded. Terminal master's awarded for partial completion of doctoral program. *Degree requirements:* For master's, thesis or problems; for doctorate, one foreign language, thesis/dissertation, comprehensive exam. *Entrance requirements:* For master's and doctorate, GRE General Test. Additional exam requirements/recommendations for international students: Recommended—TOEFL (minimum score 550 paper-based; 213 computer-based). *Application deadline:* For fall admission, 7/15 for domestic students. Applications are processed on a rolling basis. Application fee: $50 ($75 for international students). Electronic applications accepted. *Expenses:* Tuition, state resident: full-time $3,258; part-time $181 per semester hour. Tuition, nonresident: full-time $8,226; part-time $451 per semester hour. Required fees: $1,219; $68 per semester hour. *Financial support:* Fellowships, research assistantships, teaching assistantships available. *Faculty research:* Accelerator physics, chaos. *Unit head:* Dr. Floyd D. McDaniel, Chair, 940-565-2630, Fax: 940-565-2515, E-mail: fmcdaniel@unt.edu. *Application contact:* Dr. Duncan Weathers, Graduate Adviser, 940-565-2630, Fax: 940-565-2515, E-mail: weathers@unt.edu.

University of Notre Dame, Graduate School, College of Science, Department of Physics, Notre Dame, IN 46556. Offers PhD. *Faculty:* 41 full-time (5 women), 5 part-time/adjunct (1 woman). *Students:* 93 full-time (23 women), 1 part-time; includes 3 minority (2 Asian Americans or Pacific Islanders, 1 Hispanic American), 52 international. 141 applicants, 35% accepted, 24 enrolled. In 2005, 14 doctorates awarded. *Median time to degree:* Of those who began their doctoral program in fall 1997, 64% received their degree in 8 years or less. *Degree requirements:* For doctorate, thesis/dissertation. *Entrance requirements:* For doctorate, GRE General Test, GRE Subject Test. Additional exam requirements/recommendations for international students: Required—TOEFL. *Application deadline:* For fall admission, 2/1 for domestic students; for spring admission, 10/15 for domestic students. Applications are processed on a rolling basis. Application fee: $25. Electronic applications accepted. *Financial support:* In 2005–06, 94 students received support, including 4 fellowships with full tuition reimbursements available (averaging $22,000 per year), 28 research assistantships with full tuition reimbursements available (averaging $15,250 per year), 61 teaching assistantships with full tuition reimbursements available (averaging $16,000 per year); tuition waivers (full) also available. Financial award application deadline: 2/1. *Faculty research:* High energy, nuclear, atomic, condensed-matter physics; astrophysics; biophysics. Total annual research expenditures: $6.8 million. *Unit head:* Dr. Kathie Newman, Director of Graduate Studies, 574-631-6387, Fax: 574-631-5952. *Application contact:* Dr. Terrence J. Akai, Director of Graduate Admissions, 574-631-7706, Fax: 574-631-4183, E-mail: gradad@nd.edu.

See Close-Up on page 385.

University of Oklahoma, Graduate College, College of Arts and Sciences, Department of Physics and Astronomy, Norman, OK 73019-0390. Offers astrophysics (MS, PhD); physics (MS, PhD). Part-time programs available. *Faculty:* 29 full-time (4 women), 1 part-time/adjunct (0 women). *Students:* 51 full-time (15 women), 4 part-time (1 woman); includes 3 minority (2 African Americans, 1 Asian American or Pacific Islander), 27 international. 7 applicants, 86% accepted, 5 enrolled. In 2005, 4 master's, 8 doctorates awarded. Terminal master's awarded for partial completion of doctoral program. *Degree requirements:* For master's, thesis or alternative, departmental qualifying exam; for doctorate, thesis/dissertation, comprehensive, departmental qualifying, oral, and written exams. *Entrance requirements:* For master's and doctorate, GRE General Test, GRE Subject Test, 3 letters of recommendation. Additional exam requirements/recommendations for international students: Required—TOEFL (minimum score 600 paper-based; 250 computer-based). *Application deadline:* For fall admission, 3/1 for domestic students. Application fee: $40 ($90 for international students). *Expenses:* Tuition, state resident: full-time $3,029; part-time $126 per credit hour. Tuition, nonresident: full-time $10,807; part-time $450 per credit hour. Required fees: $1,231; $44 per credit hour. Tuition and fees vary according to course load and program. *Financial support:* In 2005–06, 9 students received support, including 2 fellowships with full tuition reimbursements available (averaging $4,000 per year), 23 research assistantships with partial tuition reimbursements available (averaging $13,909 per year), 34 teaching assistantships with partial tuition reimbursements available (averaging $13,748 per year); Federal Work-Study, scholarships/grants, health care benefits, tuition waivers (full), and unspecified assistantships also available. Financial award application deadline: 3/1; financial award applicants required to submit FAFSA. *Faculty research:* Atomic, molecular, and chemical physics; high energy physics; solid state and applied physics; astrophysics. Total annual research expenditures: $4.6 million. *Unit head:* Dr. Ryan Doezema, Chair, 405-325-3961, Fax: 405-325-7557, E-mail: rdoezema@ou.edu. *Application contact:* Debbie Barnhill, Graduate Studies Coordinator, 405-325-3961, Fax: 405-325-7557, E-mail: dbarnhill@ou.edu.

University of Oregon, Graduate School, College of Arts and Sciences, Department of Physics, Eugene, OR 97403. Offers MA, MS, PhD. *Faculty:* 36 full-time (2 women), 11 part-time/adjunct (3 women). *Students:* 91 full-time (16 women), 3 part-time; includes 1 minority (Asian American or Pacific Islander), 38 international. 55 applicants, 49% accepted. In 2005, 10 master's, 14 doctorates awarded. Terminal master's awarded for partial completion of doctoral program. *Degree requirements:* For doctorate, thesis/dissertation. *Entrance requirements:* For master's and doctorate, GRE General Test, GRE Subject Test, minimum GPA of 3.0. Additional exam requirements/recommendations for international students: Required—TOEFL. *Application deadline:* For fall admission, 3/15 for domestic students. Applications are processed on a rolling basis. Application fee: $50. *Financial support:* In 2005–06, 56 teaching assistantships were awarded; Federal Work-Study, institutionally sponsored loans, and traineeships also available. Financial award application deadline: 2/15. *Faculty research:* Solid-state and chemical physics, optical physics, elementary particle physics, astrophysics, atomic and molecular physics. *Unit head:* Dr. Davison E. Soper, Head, 541-346-5826. *Application contact:* Jani Levy, Admissions Contact, 541-346-4751, Fax: 541-346-4787, E-mail: jlevy@uoregon.edu.

University of Ottawa, Faculty of Graduate and Postdoctoral Studies, Faculty of Science, Ottawa-Carleton Institute for Physics, Ottawa, ON K1N 6N5, Canada. Offers M Sc, PhD. *Faculty:* 30 full-time (2 women). *Students:* 45 full-time, 8 part-time. 43 applicants, 42% accepted, 10 enrolled. In 2005, 5 master's, 2 doctorates awarded. *Degree requirements:* For master's, thesis or alternative; for doctorate, thesis/dissertation, seminar, comprehensive exam. *Entrance requirements:* For master's, honors B Sc degree or equivalent, minimum B average; for doctorate, M Sc, minimum B+ average. *Application deadline:* For fall admission, 3/1 for domestic students, 2/15 for international students. For winter admission, 11/15 for domestic students; for spring admission, 4/1 for domestic students. Application fee: $75. Electronic applications accepted. *Expenses:* Tuition: Part-time $260 per credit. Tuition and fees vary according to course load and program. *Financial support:* Fellowships, research assistantships with full tuition reimbursements, teaching assistantships with full tuition reimbursements, career-related internships or fieldwork, Federal Work-Study, scholarships/grants, traineeships, tuition waivers (full and partial), and unspecified assistantships available. Financial award application deadline: 2/15. *Faculty research:* Condensed matter physics and statistical physics (CMS), subatomic physics (SAP), medical physics (Med). *Unit head:* Dr. Richard Hodgson, Chair, 613-520-7546, Fax: 613-520-7539. *Application contact:* Lise Maisonneuve, Graduate Studies Administrator, 613-562-5800 Ext. 6335, Fax: 613-562-5486, E-mail: lise@science.uottawa.ca.

University of Pennsylvania, School of Arts and Sciences, Graduate Group in Physics and Astronomy, Philadelphia, PA 19104. Offers medical physics (MS); physics (PhD). Part-time programs available. *Degree requirements:* For doctorate, thesis/dissertation, oral, preliminary, and final exams. *Entrance requirements:* For doctorate, GRE General Test, GRE Subject Test (recommended). Additional exam requirements/recommendations for international students:

Physics

University of Pennsylvania (continued)

Required—TOEFL; Recommended—TSE. Electronic applications accepted. *Faculty research:* Astrophysics, condensed matter experiment, condensed matter theory, particle experiment, particle theory.

University of Pittsburgh, School of Arts and Sciences, Department of Physics and Astronomy, Pittsburgh, PA 15260. Offers physics (MS, PhD). *Faculty:* 40 full-time (4 women). *Students:* 85 full-time (15 women), 5 part-time; includes 4 minority (1 African American, 3 Asian Americans or Pacific Islanders), 59 international. Average age 20. 284 applicants, 22% accepted, 18 enrolled. In 2005, 16 master's, 7 doctorates awarded. Terminal master's awarded for partial completion of doctoral program. *Median time to degree:* Of those who began their doctoral program in fall 1997, 57% received their degree in 8 years or less. *Degree requirements:* For master's, thesis optional; for doctorate, thesis/dissertation, teaching present seminar. *Entrance requirements:* For master's and doctorate, GRE General Test, GRE Subject Test, minimum QPA of 3.0. Additional exam requirements/recommendations for international students: Required—TOEFL (minimum score 550 paper-based; 213 computer-based), IELT or iBT (internet-Based TOEFL. *Application deadline:* For fall admission, 1/31 priority date for domestic students, 1/31 priority date for international students. Applications are processed on a rolling basis. Application fee: $0 ($50 for international students). Electronic applications accepted. *Expenses:* Tuition, state resident: full-time $13,194; part-time $537 per credit. Tuition, nonresident: full-time $25,012; part-time $1,026 per credit. Required fees: $700; $164 per term. Tuition and fees vary according to campus/location and program. *Financial support:* In 2005–06, 7 fellowships with full tuition reimbursements, 45 research assistantships with full tuition reimbursements (averaging $20,332 per year), 29 teaching assistantships with full tuition reimbursements (averaging $20,332 per year) were awarded; scholarships/grants, health care benefits, and unspecified assistantships also available. Financial award application deadline:1/31. *Faculty research:* Astrophysics and cosmology employing ground-and space-based telescopes; particle astrophysics; atomic, molecular, and optical physics; condensed-matter and solid-state physics; particle physics. Total annual research expenditures: $6.6 million. *Unit head:* Dr. David Turnshek, Chairman, 412-624-9000, Fax: 412-624-9163, E-mail: turnshek@pitt.edu. *Application contact:* Dr. Anthony Duncan, Admissions, 412-624-9000, Fax: 412-624-9163, E-mail: tony@dectony.phyast.pitt.edu.

See Close-Up on page 389.

University of Puerto Rico, Mayagüez Campus, Graduate Studies, College of Arts and Sciences, Department of Physics, Mayagüez, PR 00681-9000. Offers MS. Part-time programs available. *Faculty:* 24. *Students:* 7 full-time (5 women), 19 part-time (4 women); includes 21 minority (all Hispanic Americans), 5 international. 8 applicants, 50% accepted, 2 enrolled. In 2005, 2 degrees awarded. *Degree requirements:* For master's, thesis, comprehensive exam. *Application deadline:* For fall admission, 2/15 for domestic students; for spring admission, 8/15 for domestic students. Applications are processed on a rolling basis. Application fee: $20. *Expenses:* Tuition, state resident: full-time $900; part-time $100 per credit. International tuition: $4,655 full-time. Part-time tuition and fees vary according to course level and course load. *Financial support:* In 2005–06, 24 students received support, including 2 fellowships (averaging $1,500 per year), 6 research assistantships (averaging $1,200 per year), 16 teaching assistantships (averaging $987 per year); Federal Work-Study and institutionally sponsored loans also available. *Faculty research:* Atomic and molecular physics, nuclear physics, nonlinear thermostatics, fluid dynamics, molecular spectroscopy. Total annual research expenditures: $45,668. *Unit head:* Dr. Hector Jimenez, Director, 787-265-3844.

University of Puerto Rico, Río Piedras, College of Natural Sciences, Department of Physics, San Juan, PR 00931-3300. Offers chemical physics (PhD); physics (MS). Part-time and evening/weekend programs available. *Students:* 61 full-time (21 women), 6 part-time (1 woman); includes 64 minority (10 Asian Americans or Pacific Islanders, 54 Hispanic Americans). Average age 29. In 2005, 1 master's, 1 doctorate awarded. *Degree requirements:* For master's and doctorate, one foreign language, thesis/dissertation, comprehensive exam. *Entrance requirements:* For master's, GRE, EXADEP, interview, minimum GPA of 3.0, letter of recommendation; for doctorate, GRE, master's degree, minimum GPA of 3.0, letter of recommendation. Additional exam requirements/recommendations for international students: Required—TOEFL. *Application deadline:* For fall admission, 2/1 for domestic students, 2/1 for international students. Application fee: $17. *Expenses:* Tuition, state resident: part-time $100 per credit. Tuition, nonresident: part-time $294 per credit. Required fees: $72 per term. *Financial support:* Fellowships, research assistantships, teaching assistantships, Federal Work-Study, institutionally sponsored loans, and tuition waivers (partial) available. Financial award application deadline: 5/31. *Faculty research:* Energy transfer process through Van der Vacqs interactions, study of the photodissociation of ketene. *Unit head:* Luis F. Fonseca, Coordinator of Doctoral Program, 787-764-0000 Ext. 4773, Fax: 787-764-4063.

University of Regina, Faculty of Graduate Studies and Research, Faculty of Science, Department of Physics, Regina, SK S4S 0A2, Canada. Offers M Sc, PhD. *Faculty:* 9 full-time (0 women), 3 part-time/adjunct (0 women). *Students:* 4 full-time (1 woman), 2 part-time (1 woman). 5 applicants, 0% accepted. In 2005, 1 master's, 1 doctorate awarded. Terminal master's awarded for partial completion of doctoral program. *Degree requirements:* For master's and doctorate, thesis/dissertation, registration. *Entrance requirements:* For master's, GRE (recommended for foreign applicants), honors degree in physics or engineering physics; for doctorate, GRE (recommended for foreign applicants), M Sc or equivalent. Additional exam requirements/recommendations for international students: Required—TOEFL (minimum score 580 paper-based; 237 computer-based). *Application deadline:* For fall admission, 5/15 for domestic students. For winter admission, 8/15 for domestic students. Applications are processed on a rolling basis. Application fee: $60 ($100 for international students). *Financial support:* In 2005–06, 3 fellowships (averaging $14,886 per year), research assistantships (averaging $12,750 per year), 2 teaching assistantships (averaging $13,501 per year) were awarded; career-related internships or fieldwork and scholarships/grants also available. Financial award application deadline:6/15. *Faculty research:* Experimental and theoretical subatomic physics. Total annual research expenditures: $2.1 million. *Unit head:* Dr. Zisis Papandreaou, Head, 306-585-5378, Fax: 306-585-5659, E-mail: zisis@uregina.ca. *Application contact:* Dr. Garth Huber, Graduate Coordinator, 306-585-4240, Fax: 306-585-5659, E-mail: grad@phys.uregina.ca.

University of Rhode Island, Graduate School, College of Arts and Sciences, Department of Physics, Kingston, RI 02881. Offers MS, PhD. In 2005, 1 master's, 6 doctorates awarded. *Application deadline:* For fall admission, 4/15 for domestic students. Applications are processed on a rolling basis. Application fee: $35. *Expenses:* Tuition, state resident: full-time $5,522; part-time $307 per credit. Tuition, nonresident: full-time $15,992; part-time $888 per credit. Required fees: $1,786; $73 per credit. One-time fee: $80 part-time. *Unit head:* Jan Northby, Chair, 401-874-2074.

University of Rochester, The College, Arts and Sciences, Department of Physics and Astronomy, Rochester, NY 14627-0250. Offers physics (MA, MS, PhD); physics and astronomy (PhD). Part-time programs available. Terminal master's awarded for partial completion of doctoral program. *Degree requirements:* For master's, thesis (for some programs), comprehensive exam; for doctorate, thesis/dissertation, qualifying exam, comprehensive exam. *Entrance requirements:* For master's and doctorate, GRE General Test. Additional exam requirements/recommendations for international students: Required—TOEFL.

See Close-Up on page 391.

University of Saskatchewan, College of Graduate Studies and Research, College of Arts and Sciences, Department of Physics and Engineering Physics, Saskatoon, SK S7N 5A2, Canada. Offers M Sc, PhD. *Degree requirements:* For master's and doctorate, thesis/dissertation, registration. *Entrance requirements:* Additional exam requirements/recommendations for international students: Required—TOEFL.

University of South Carolina, The Graduate School, College of Arts and Sciences, Department of Physics and Astronomy, Columbia, SC 29208. Offers IMA, MAT, MS, PSM, PhD. IMA and MAT offered in cooperation with the College of Education. Part-time programs available. Terminal master's awarded for partial completion of doctoral program. *Degree requirements:* For master's, thesis, comprehensive exam, registration; for doctorate, one foreign language, thesis/dissertation, comprehensive exam, registration. *Entrance requirements:* For master's and doctorate, GRE General Test, GRE Subject Test. Additional exam requirements/recommendations for international students: Required—TOEFL (minimum score 570 paper-based; 230 computer-based). Electronic applications accepted. *Faculty research:* Condensed matter, intermediate-energy nuclear physics, foundations of quantum mechanics, astronomy/astrophysics.

University of Southern California, Graduate School, College of Letters, Arts and Sciences, Department of Physics and Astronomy, Los Angeles, CA 90089. Offers physics (MA, MS, PhD). Part-time programs available. Terminal master's awarded for partial completion of doctoral program. *Degree requirements:* For master's, thesis (for some programs); for doctorate, thesis/dissertation. *Entrance requirements:* For master's and doctorate, GRE General Test, GRE Subject Test. *Expenses:* Tuition: Full-time $25,416; part-time $1,059 per unit. Required fees: $484; $484 per year. Tuition and fees vary according to course load and program. *Faculty research:* Space physics, laser physics, high-energy particle theory, condensed matter physics, atomic and molecular physics.

University of Southern Mississippi, Graduate School, College of Science and Technology, Department of Physics and Astronomy, Hattiesburg, MS 39406-0001. Offers physics (MS). *Degree requirements:* For master's, thesis, comprehensive exam. *Entrance requirements:* For master's, GRE General Test, minimum GPA of 2.75 in last 60 hours. Additional exam requirements/recommendations for international students: Required—TOEFL. *Faculty research:* Polymers, atomic physics, fluid mechanics, liquid crystals, refractory materials.

University of South Florida, College of Graduate Studies, College of Arts and Sciences, Department of Physics, Tampa, FL 33620-9951. Offers applied physics (PhD); physics (MS). Part-time programs available. *Faculty:* 15 full-time (1 woman). *Students:* 51 full-time (15 women), 5 part-time; includes 4 minority (2 African Americans, 1 Asian American or Pacific Islander, 1 Hispanic American), 28 international. 39 applicants, 72% accepted, 15 enrolled. In 2005, 2 degrees awarded. *Degree requirements:* For master's, thesis optional; for doctorate, 2 foreign languages, thesis/dissertation. *Entrance requirements:* For master's, GRE General Test, minimum GPA of 3.0 in last 60 hours of course work; for doctorate, GRE General Test, minimum graduate GPA of 3.2. Additional exam requirements/recommendations for international students: Required—TOEFL (minimum score 550 paper-based), TSE (minimum score 50). *Application deadline:* For fall admission, 6/1 for domestic students; for spring admission, 10/1 for domestic students. Applications are processed on a rolling basis. Application fee: $30. Electronic applications accepted. *Financial support:* Fellowships with full tuition reimbursements, research assistantships with full tuition reimbursements, teaching assistantships with full tuition reimbursements, career-related internships or fieldwork, scholarships/grants, and unspecified assistantships available. *Faculty research:* Laser, medical, and solid-state physics. *Unit head:* Dr. Sarath Witanachchi, Director of Graduate Studies, 813-974-2789, Fax: 813-974-5813, E-mail: switanac@cas.usf.edu. *Application contact:* Evelyne Keeton-Williams, Program Assistant, 813-974-2871, Fax: 813-974-5813, E-mail: ekeeton@cas.usf.edu.

The University of Tennessee, Graduate School, College of Arts and Sciences, Department of Physics and Astronomy, Knoxville, TN 37996. Offers physics (MS, PhD). Part-time programs available. *Degree requirements:* For master's, thesis or alternative; for doctorate, thesis/dissertation. *Entrance requirements:* For master's and doctorate, minimum GPA of 2.7. Additional exam requirements/recommendations for international students: Required—TOEFL. Electronic applications accepted.

The University of Tennessee Space Institute, Graduate Programs, Program in Physics, Tullahoma, TN 37388-9700. Offers MS, PhD. *Faculty:* 5 full-time (1 woman). *Students:* 8 full-time (1 woman), 2 part-time, 2 international. 14 applicants, 57% accepted, 2 enrolled. In 2005, 2 degrees awarded. *Degree requirements:* For master's, thesis (for some programs); for doctorate, one foreign language, thesis/dissertation. *Entrance requirements:* For master's and doctorate, GRE General Test, GRE Subject Test. *Application deadline:* Applications are processed on a rolling basis. Application fee: $35. *Financial support:* Fellowships with full and partial tuition reimbursements, research assistantships with full tuition reimbursements, career-related internships or fieldwork, Federal Work-Study, tuition waivers (full and partial), and unspecified assistantships available. Financial award applicants required to submit FAFSA. *Unit head:* Dr. Horace Crater, Degree Program Chairman, 931-393-7469, Fax: 931-393-7444, E-mail: hcrater@utsi.edu. *Application contact:* Callie Taylor, Coordinator II, 931-393-7432, Fax: 931-393-7346, E-mail: ctaylor@utsi.edu.

The University of Texas at Arlington, Graduate School, College of Science, Department of Physics, Arlington, TX 76019. Offers physics (MS); physics and applied physics (PhD). Part-time programs available. *Faculty:* 5 full-time (0 women). *Students:* 31 full-time (10 women), 6 part-time (1 woman); includes 2 minority (both Asian Americans or Pacific Islanders), 22 international. 13 applicants, 85% accepted, 9 enrolled. In 2005, 8 master's awarded. Terminal master's awarded for partial completion of doctoral program. *Median time to degree:* Of those who began their doctoral program in fall 1997, 95% received their degree in 8 years or less. *Degree requirements:* For master's, thesis optional; for doctorate, thesis/dissertation, internship or substitute, comprehensive exam. *Entrance requirements:* For master's, GRE General Test, minimum GPA of 3.0 in last 60 hours of course work; for doctorate, GRE General Test, minimum GPA of 3.0 in last 60 hours of course work, 30 hours graduate course work in physics. Additional exam requirements/recommendations for international students: Required—TOEFL (minimum score 550 paper-based; 213 computer-based). *Application deadline:* For fall admission, 6/16 for domestic students. Applications are processed on a rolling basis. Application fee: $35 ($50 for international students). *Expenses:* Tuition, state resident: full-time $3,350. Tuition, nonresident: full-time $8,318. International tuition: $8,448 full-time. Required fees: $1,277. Full-time tuition and fees vary according to course level and program. *Financial support:* In 2005–06, 25 students received support, including 4 fellowships (averaging $1,000 per year), research assistantships (averaging $18,000 per year), 11 teaching assistantships (averaging $18,000 per year); career-related internships or fieldwork, Federal Work-Study, institutionally sponsored loans, scholarships/grants, health care benefits, tuition waivers (full and partial), and unspecified assistantships also available. Support available to part-time students. Financial award application deadline: 6/1; financial award applicants required to submit FAFSA. *Faculty research:* Particle physics, astrophysics, condensed matter theory and experiment. Total annual research expenditures: $1.5 million. *Unit head:* Dr. James Horowitz, Chair, 807-272-2266, Fax: 817-272-3637. *Application contact:* Dr. Ciming Zhang, Graduate Advisor, 817-272-2266, Fax: 817-272-3637, E-mail: zhang@uta.edu.

The University of Texas at Austin, Graduate School, College of Natural Sciences, Department of Physics, Austin, TX 78712-1111. Offers MA, MS, PhD. *Degree requirements:* For master's and doctorate, thesis/dissertation, registration. *Entrance requirements:* For master's and doctorate, GRE General Test, GRE Subject Test (physics). Electronic applications accepted.

The University of Texas at Brownsville, Graduate Studies, College of Science, Mathematics and Technology, Brownsville, TX 78520-4991. Offers biological sciences (MS, MSIS); mathematics (MS); physics (MS). Part-time and evening/weekend programs available. *Faculty:* 51 full-time (10 women). *Students:* 34 (16 women); includes 24 minority (all Hispanic Americans) 7 applicants. In 2005, 9 degrees awarded. *Degree requirements:* For master's, thesis optional. *Entrance requirements:* For master's, GRE General Test. Additional exam requirements/recommendations for international students: Required—TOEFL. *Application deadline:* For fall admission, 7/1 for domestic students; for spring admission, 12/1 priority date for domestic students. Applications are processed on a rolling basis. Application fee: $30. *Financial support:* Federal Work-Study, scholarships/grants, and tuition waivers (partial) available. Support available to part-time students. Financial award application deadline: 4/3; financial award applicants required to submit FAFSA. *Faculty research:* Fish, insects, barrier islands, algae, curlits. *Unit head:* Terry Jay Phillips, Interim Dean, 956-882-6701, Fax: 956-882-8988. *Application contact:* Irma C. Hernandez, Information Contact, 956-882-7787, Fax: 956-882-7279, E-mail: irma.c.hernandez@utb.edu.

The University of Texas at Dallas, School of Natural Sciences and Mathematics, Program in Physics, Richardson, TX 75083-0688. Offers applied physics (MS); physics (PhD). Part-time and evening/weekend programs available. *Faculty:* 15 full-time (0 women). *Students:* 41 full-time (11 women), 28 part-time (5 women); includes 9 minority (1 African American, 4 Asian Americans or Pacific Islanders, 4 Hispanic Americans), 22 international. Average age 31. 64 applicants, 69% accepted, 20 enrolled. In 2005, 12 master's, 10 doctorates awarded. *Degree requirements:* For master's, industrial internship, thesis optional; for doctorate, thesis/dissertation, publishable paper. *Entrance requirements:* For master's and doctorate, GRE General Test, minimum GPA of 3.0 in upper-level coursework in field. Additional exam requirements/recommendations for international students: Required—TOEFL (minimum score 550 paper-based; 213 computer-based). *Application deadline:* For fall admission, 7/15 for domestic students; for spring admission, 11/15 for domestic students. Applications are processed on a rolling basis. Application fee: $50 ($100 for international students). Electronic applications accepted. *Expenses:* Tuition, state resident: full-time $5,450; part-time $303 per credit. Tuition, nonresident: full-time $12,648; part-time $703 per credit. Tuition and fees vary according to program. *Financial support:* In 2005–06, 16 research assistantships with tuition reimbursements (averaging $11,602 per year), 17 teaching assistantships with tuition reimbursements (averaging $9,410 per year) were awarded; fellowships, career-related internships or fieldwork, Federal Work-Study, institutionally sponsored loans, scholarships/grants, and unspecified assistantships also available. Support available to part-time students. Financial award application deadline: 4/30; financial award applicants required to submit FAFSA. *Faculty research:* Atomic, molecular, atmospheric, chemical, solid-state, and space physics; optics and quantum electronics; relativity and astrophysics; high-energy particles. Total annual research expenditures: $7.1 million. *Unit head:* Dr. Roderick A. Heelis, Department Head, 972-883-2822, Fax: 972-883-2848, E-mail: heelis@utdallas.edu. *Application contact:* Dr. Roy Chaney, Graduate Advisor, 972-883-2887, Fax: 972-883-2848, E-mail: chaneyr@utdallas.edu.

The University of Texas at El Paso, Graduate School, College of Science, Department of Physics, El Paso, TX 79968-0001. Offers MS. Part-time and evening/weekend programs available. *Degree requirements:* For master's, thesis. *Entrance requirements:* For master's, GRE General Test, minimum GPA of 3.0. Additional exam requirements/recommendations for international students: Required—TOEFL. Electronic applications accepted.

The University of Toledo, Graduate School, College of Arts and Sciences, Department of Physics and Astronomy, Toledo, OH 43606-3390. Offers physics (MS, PhD). *Faculty:* 23. *Students:* 71 full-time (16 women), 7 part-time; includes 5 minority (2 African Americans, 2 Asian Americans or Pacific Islanders, 1 Hispanic American), 34 international. Average age 28. 59 applicants, 29% accepted, 10 enrolled. In 2005, 2 degrees awarded. *Degree requirements:* For master's, thesis; for doctorate, thesis/dissertation, departmental qualifying exam. *Entrance requirements:* For master's and doctorate, GRE General Test, GRE Subject Test. Additional exam requirements/recommendations for international students: Required—TOEFL. *Application deadline:* For fall admission, 5/31 for domestic students. Applications are processed on a rolling basis. Application fee: $45. Electronic applications accepted. *Expenses:* Tuition, area resident: Part-time $308 per credit hour. Tuition, state resident: full-time $3,312. Tuition, nonresident: full-time $6,616; part-time $735 per credit hour. *Financial support:* In 2005–06, 2 research assistantships (averaging $8,000 per year), 30 teaching assistantships (averaging $14,267 per year) were awarded; Federal Work-Study, institutionally sponsored loans, and tuition waivers (full) also available. Support available to part-time students. Financial award application deadline: 4/1; financial award applicants required to submit FAFSA. *Faculty research:* Atomic physics, solid-state physics, materials science, astrophysics. *Unit head:* Dr. Alan Compaan, Chair, 419-530-4906, Fax: 419-530-2723, E-mail: adc@physics.utoledo.edu. *Application contact:* Nancy Morrison, Information Contact, 419-530-2659, Fax: 419-530-2723, E-mail: ndm@astro.utoledo.edu.

See Close-Up on page 393.

University of Toronto, School of Graduate Studies, Physical Sciences Division, Department of Physics, Toronto, ON M5S 1A1, Canada. Offers M Sc, PhD. *Degree requirements:* For master's, thesis optional; for doctorate, thesis/dissertation. *Entrance requirements:* For master's, minimum B+ average in an honors physics program or equivalent, 2 letters of reference; for doctorate, M Sc degree in physics or a related field, 2 letters of reference.

University of Utah, The Graduate School, College of Science, Department of Physics, Salt Lake City, UT 84112-1107. Offers chemical physics (PhD); physics (MA, MS, PhD). Part-time programs available. *Faculty:* 31 full-time (1 woman), 1 part-time/adjunct (0 women). *Students:* 83 full-time (15 women), 18 part-time (3 women); includes 4 minority (2 Asian Americans or Pacific Islanders, 2 Hispanic Americans), 46 international. Average age 29. 73 applicants, 34% accepted, 24 enrolled. In 2005, 9 master's, 7 doctorates awarded. Terminal master's awarded for partial completion of doctoral program. *Degree requirements:* For master's, thesis or alternative, teaching experience, comprehensive exam (for some programs); for doctorate, thesis/dissertation, departmental qualifying exam, comprehensive exam. *Entrance requirements:* For master's and doctorate, GRE General Test, GRE Subject Test, minimum GPA of 3.0. Additional exam requirements/recommendations for international students: Required—TOEFL (minimum score 500 paper-based; 173 computer-based). *Application deadline:* For fall admission, 2/1 for domestic students. Applications are processed on a rolling basis. Application fee: $45 ($65 for international students). Electronic applications accepted. *Expenses:* Tuition, state resident: full-time $2,932; part-time $2,212 per term. Tuition, nonresident: full-time $10,350; part-time $7,812 per term. Required fees: $590; $516 per term. Tuition and fees vary according to course load and program. *Financial support:* Fellowships, research assistantships with full and partial tuition reimbursements, teaching assistantships with full and partial tuition reimbursements, Federal Work-Study, institutionally sponsored loans, and scholarships/grants available. Financial award application deadline: 2/15; financial award applicants required to submit FAFSA. *Faculty research:* High-energy, cosmic-ray, astrophysics, medical physics, condensed matter, relativity applied physics. Total annual research expenditures: $5.8 million. *Unit head:* Dr. Pierre Sokolsky, Chair, 801-581-6901, Fax: 801-581-4801, E-mail: ps@cosmic.utah.edu. *Application contact:* Jackie Hadley, Graduate Secretary, 801-581-6861, Fax: 801-581-4801, E-mail: jackie@physics.utah.edu.

University of Vermont, Graduate College, College of Arts and Sciences, Department of Physics, Burlington, VT 05405. Offers MS. *Students:* 5, 1 international. 6 applicants, 33% accepted, 2 enrolled. *Entrance requirements:* For master's, GRE General Test. Additional exam requirements/recommendations for international students: Required—TOEFL (minimum score 550 paper-based; 213 computer-based). *Application deadline:* For fall admission, 4/1 for domestic students. Applications are processed on a rolling basis. Application fee: $40. *Expenses:* Tuition, area resident: Part-time $410 per credit hour. Tuition, nonresident: part-time $1,034 per credit hour. *Financial support:* Fellowships, research assistantships, teaching assistantships available. Financial award application deadline:3/1. *Unit head:* Dr. J. Wu, Chairperson, 802-656-2644. *Application contact:* K. Spartalian, Coordinator, 802-656-2644.

University of Victoria, Faculty of Graduate Studies, Faculty of Science, Department of Physics and Astronomy, Victoria, BC V8W 2Y2, Canada. Offers astronomy and astrophysics (M Sc, PhD); condensed matter physics (M Sc, PhD); experimental particle physics (M Sc, PhD); medical physics (M Sc, PhD); ocean physics (PhD); ocean physics and geophysics (M Sc); theoretical physics (M Sc, PhD). *Faculty:* 16 full-time (0 women), 13 part-time/adjunct (1 woman). *Students:* 45 full-time, 8 international. Average age 25. 66 applicants, 45% accepted, 10 enrolled. In 2005, 4 master's, 1 doctorate awarded. *Median time to degree:* Of those who began their doctoral program in fall 1997, 100% received their degree in 8 years or less. *Degree requirements:* For master's, thesis, registration; for doctorate, thesis/dissertation, Candidacy exam, comprehensive exam, registration. *Entrance requirements:* For master's and doctorate, GRE. Additional exam requirements/recommendations for international students: Required—TOEFL (minimum score 575 paper-based; 233 computer-based), IELT (minimum score 7). *Application deadline:* For fall admission, 5/31 priority date for domestic students, 12/15 priority date for international students. Applications are processed on a rolling basis. Application fee: $75 ($125 for international students). Electronic applications accepted. Tuition and fees charges are reported in Canadian dollars. *Expenses:* Tuition, area resident: Full-time $4,492 Canadian dollars; part-time $749 Canadian dollars per term. International tuition: $5,346 Canadian dollars full-time. Required fees: $4,492 Canadian dollars; $749 Canadian dollars per term. Tuition and fees vary according to course load, campus/location and program. *Financial support:* In 2005–06, 3 students received support; fellowships, research assistantships, teaching assistantships, career-related internships or fieldwork, institutionally sponsored loans, and awards available. Financial award application deadline: 2/15. *Faculty research:* Old stellar populations; observational cosmology and large scale structure; cp violation; atlas. *Unit head:* Dr. J. Michael Ronev, Chair, 250-721-7698, Fax: 250-721-7715, E-mail: chair@phys.uvic.ca. *Application contact:* Dr. Chris J. Pritchet, Graduate Adviser, 250-721-7744, Fax: 250-721-7715, E-mail: pritchet@uvic.ca.

University of Virginia, College and Graduate School of Arts and Sciences, Department of Physics, Charlottesville, VA 22903. Offers physics (MA, MS, PhD); physics education (MA). *Faculty:* 29 full-time (0 women). *Students:* 86 full-time (15 women); includes 2 minority (1 Asian American or Pacific Islander, 1 Hispanic American), 37 international. Average age 26. 191 applicants, 21% accepted, 12 enrolled. In 2005, 17 master's, 9 doctorates awarded. *Degree requirements:* For master's and doctorate, thesis/dissertation. *Entrance requirements:* For master's and doctorate, GRE General Test, GRE Subject Test. *Application deadline:* Applications are processed on a rolling basis. Application fee: $40. Electronic applications accepted. *Expenses:* Tuition, state resident: full-time $7,731. Tuition, nonresident: full-time $18,672. Required fees: $1,479. Full-time tuition and fees vary according to degree level and program. *Financial support:* Applicants required to submit FAFSA. *Unit head:* Dinko Poucanic, Chairman, 434-924-3781, Fax: 434-924-4576, E-mail: phys-chair@physics.virginia.edu. *Application contact:* Peter C. Brunjes, Associate Dean for Graduate Programs and Research, 434-924-7184, Fax: 434-924-6737, E-mail: grad-a-s@virginia.edu.

University of Washington, Graduate School, College of Arts and Sciences, Department of Physics, Seattle, WA 98195. Offers MS, PhD. Part-time and evening/weekend programs available. Terminal master's awarded for partial completion of doctoral program. *Degree requirements:* For doctorate, thesis/dissertation. *Entrance requirements:* For master's, GRE; for doctorate, GRE General Test, GRE Subject Test. Additional exam requirements/recommendations for international students: Required—TOEFL. Electronic applications accepted. *Faculty research:* Astro-, atomic, condensed-matter, nuclear, and particle physics; physics education.

University of Waterloo, Graduate Studies, Faculty of Science, Guelph-Waterloo Physics Institute, Waterloo, ON N2L 3G1, Canada. Offers M Sc, PhD. Part-time programs available. *Faculty:* 35 full-time (3 women), 32 part-time/adjunct (1 woman). *Students:* 52 full-time (8 women), 6 part-time (2 women). 54 applicants, 39% accepted, 12 enrolled. In 2005, 7 master's, 3 doctorates awarded. *Degree requirements:* For master's, project or thesis; for doctorate, thesis/dissertation, registration. *Entrance requirements:* For master's, GRE Subject Test, honors degree, minimum B average; for doctorate, GRE Subject Test, master's degree, minimum B average. Additional exam requirements/recommendations for international students: Required—TOEFL, TWE. *Application deadline:* For fall admission, 7/1 for domestic students. For winter admission, 11/1 for domestic students; for spring admission, 3/1 for domestic students. Applications are processed on a rolling basis. Application fee: $75 Canadian dollars. Electronic applications accepted. *Financial support:* Research assistantships, teaching assistantships, career-related internships or fieldwork, scholarships/grants, and unspecified assistantships available. *Faculty research:* Condensed-matter and materials physics; industrial and applied physics; subatomic physics; astrophysics and gravitation; atomic, molecular, and optical physics. *Unit head:* Dr. Robert Mann, Director, 519-888-4567 Ext. 6285, Fax: 519-746-8115, E-mail: gwp@scimail.uwaterloo.ca. *Application contact:* M. M. O'Neill, Administrative Assistant, 519-888-4567 Ext. 6874, Fax: 519-746-8115, E-mail: gwp@scimail.uwaterloo.ca.

The University of Western Ontario, Faculty of Graduate Studies, Physical Sciences Division, Department of Applied Mathematics, London, ON N6A 5B8, Canada. Offers applied mathematics (M Sc, PhD); theoretical physics (PhD). *Degree requirements:* For master's, thesis or alternative; for doctorate, thesis/dissertation, comprehensive exam. *Entrance requirements:* For master's and doctorate, minimum B average. *Faculty research:* Fluid dynamics, mathematical and computational methods, theoretical physics.

The University of Western Ontario, Faculty of Graduate Studies, Physical Sciences Division, Department of Physics and Astronomy, Program in Physics, London, ON N6A 5B8, Canada. Offers M Sc, PhD. Terminal master's awarded for partial completion of doctoral program. *Degree requirements:* For master's, thesis/dissertation; for doctorate, thesis/dissertation, comprehensive exam. *Entrance requirements:* For master's, GRE Physics Test, honors B Sc degree, minimum B average (Canadian), A- (international); for doctorate, minimum B average (Canadian), A- (international). Additional exam requirements/recommendations for international students: Required—TOEFL (minimum score 580 paper-based; 237 computer-based). *Faculty research:* Condensed-matter and surface science, space and atmospheric physics, atomic and molecular physics, medical physics, theoretical physics.

University of Windsor, Faculty of Graduate Studies and Research, Faculty of Science, Department of Physics, Windsor, ON N9B 3P4, Canada. Offers M Sc, PhD. Part-time programs available. *Faculty:* 13 full-time (2 women), 2 part-time/adjunct (0 women). *Students:* 31 full-time (4 women), 1 part-time. 20 applicants, 50% accepted. In 2005, 2 degrees awarded. *Degree requirements:* For master's, thesis or alternative; for doctorate, thesis/dissertation. *Entrance requirements:* For master's, GRE, minimum B average; for doctorate, GRE General Test, master's degree. Additional exam requirements/recommendations for international students: Required—TOEFL (minimum score 560 paper-based; 220 computer-based), GRE Advanced Physics Test. *Application deadline:* For fall admission, 7/1 for domestic students. For winter admission, 11/1 for domestic students; for spring admission, 3/1 for domestic students. Applications are processed on a rolling basis. Application fee: $55. Electronic applications accepted. *Financial support:* In 2005–06, 24 teaching assistantships (averaging $8,956 per year) were awarded; research assistantships, Federal Work-Study, scholarships/grants, tuition waivers (full and partial), unspecified assistantships, and bursaries also available. Financial award application deadline: 2/15. *Faculty research:* Electrodynamics, plasma physics, atomic structure/particles, spectroscopy, quantum mechanics. *Unit head:* Dr. Gordon W. Drake, Head, 519-253-3000 Ext. 2647, Fax: 519-973-7075, E-mail: gdrake@uwindsor.ca. *Application contact:* Marlene Bezaire, Graduate Secretary, 519-253-3000 Ext. 3520, Fax: 519-971-7098, E-mail: spsgrad@uwindsor.ca.

University of Wisconsin–Madison, Graduate School, College of Letters and Science, Department of Physics, Madison, WI 53706-1380. Offers MA, MS, PhD. Terminal master's awarded for partial completion of doctoral program. *Degree requirements:* For master's, thesis (for some programs), qualifying exam, thesis (MS); for doctorate, thesis/dissertation, preliminary and qualifying exams. *Entrance requirements:* For master's and doctorate, GRE, minimum of 3.0. Additional exam requirements/recommendations for international students: Required—TOEFL. Electronic applications accepted. *Faculty research:* Atomic, physics, condensed matter, astrophysics, particles and fields.

University of Wisconsin–Milwaukee, Graduate School, College of Letters and Sciences, Department of Physics, Milwaukee, WI 53201-0413. Offers MS, PhD. *Faculty:* 19 full-time (2 women). *Students:* 22 full-time (4 women), 11 part-time (1 woman), 21 international. 51 applicants, 27% accepted, 6 enrolled. In 2005, 2 master's, 3 doctorates awarded. *Degree requirements:* For master's, thesis or alternative; for doctorate, one foreign language, thesis/dissertation. *Entrance requirements:* For master's and doctorate, GRE General Test. *Application deadline:* For fall admission, 1/1 for domestic students; for spring admission, 9/1 for domestic students. Applications are processed on a rolling basis. Application fee: $45 ($75 for

Physics

University of Wisconsin–Milwaukee (continued)

international students). *Expenses:* Tuition, area resident: Part-time $716 per credit. Tuition, state resident: part-time $776 per credit. Tuition, nonresident: part-time $1,614. Required fees: $229 per term. Tuition and fees vary according to course load and program. *Financial support:* In 2005–06, 6 research assistantships, 28 teaching assistantships were awarded; fellowships, career-related internships or fieldwork and unspecified assistantships also available. Support available to part-time students. Financial award application deadline: 4/15. *Unit head:* Richard Sorbello, Representative, 414-229-6266, Fax: 414-229-4474, E-mail: sorbello@uwm.edu.

Utah State University, School of Graduate Studies, College of Science, Department of Physics, Logan, UT 84322. Offers MS, PhD. Part-time programs available. *Faculty:* 25 full-time (1 woman), 16 part-time/adjunct (1 woman). *Students:* 68 full-time (8 women), 9 part-time (1 woman); includes 5 minority (all Hispanic Americans), 31 international. Average age 34. 28 applicants, 79% accepted, 16 enrolled. In 2005, 2 master's, 2 doctorates awarded. Terminal master's awarded for partial completion of doctoral program. *Degree requirements:* For master's, thesis/dissertation; for doctorate, thesis/dissertation, comprehensive exam. *Entrance requirements:* For master's and doctorate, GRE General Test, minimum GPA of 3.0. Additional exam requirements/recommendations for international students: Required—TOEFL (minimum score 550 paper-based; 213 computer-based). *Application deadline:* For fall admission, 6/15 priority date for domestic students, 5/15 priority date for international students; for spring admission, 10/15 for domestic students, 9/15 for international students. Applications are processed on a rolling basis. Application fee: $50 ($60 for international students). Electronic applications accepted. *Financial support:* In 2005–06, 3 fellowships with partial tuition reimbursements (averaging $12,000 per year), 7 research assistantships with partial tuition reimbursements (averaging $14,000 per year), 10 teaching assistantships with partial tuition reimbursements (averaging $10,500 per year) were awarded; Federal Work-Study and institutionally sponsored loans also available. Support available to part-time students. Financial award application deadline: 3/1. *Faculty research:* Upper-atmosphere physics, relativity, gravitational magnetism, particle physics, nanotechnology. Total annual research expenditures: $1.6 million. *Unit head:* Dr. Jan J. Sojka, Head, 435-797-2848, Fax: 435-797-2492. *Application contact:* Dr. David Peak, Assistant Head, 435-797-2884, Fax: 435-797-2492, E-mail: physics@cc.usu.edu.

Vanderbilt University, Graduate School, Department of Physics and Astronomy, Nashville, TN 37240-1001. Offers astronomy (MS); physics (MA, MAT, MS, PhD). *Faculty:* 65 full-time (3 women). *Students:* 61 full-time (14 women), 2 part-time; includes 3 minority (2 African Americans, 1 Hispanic American), 31 international. 153 applicants, 20% accepted, 22 enrolled. In 2005, 3 master's, 5 doctorates awarded. *Degree requirements:* For master's, thesis; for doctorate, thesis/dissertation, final and qualifying exams. *Entrance requirements:* For master's, GRE General Test; for doctorate, GRE General Test, GRE Subject Test. *Application deadline:* For fall admission, 1/15 for domestic students, 1/15 for international students. Application fee: $0. Electronic applications accepted. *Expenses:* Tuition: Full-time $15,396; part-time $1,283 per semester hour. Required fees: $2,202; $1,101 per semester. One-time fee: $30. Tuition and fees vary according to course load, program and student level. *Financial support:* Fellowships with full and partial tuition reimbursements, research assistantships with full tuition reimbursements, teaching assistantships with full tuition reimbursements, career-related internships or fieldwork, Federal Work-Study, and institutionally sponsored loans available. Financial award application deadline: 1/15. *Faculty research:* Experimental and theoretical physics, free electron laser, living-state physics, heavy-ion physics, nuclear structure. *Unit head:* Robert J. Scherrer, Chair, 615-322-2828, Fax: 615-343-7263. *Application contact:* Charles F. Maguire, Director of Graduate Studies, 615-322-2828, Fax: 615-343-7263, E-mail: charles.f.maguire@vanderbilt.edu.

Virginia Commonwealth University, Graduate School, College of Humanities and Sciences, Department of Physics, Richmond, VA 23284-9005. Offers applied physics (MS); physics (MS). Part-time programs available. *Faculty:* 7 full-time (3 women). *Students:* 7 full-time (4 women), 2 part-time. 8 applicants, 100% accepted. In 2005, 5 degrees awarded. *Degree requirements:* For master's, thesis optional. *Entrance requirements:* For master's, GRE. *Application deadline:* For fall admission, 8/1 for domestic students; for spring admission, 12/1 for domestic students. Applications are processed on a rolling basis. Application fee: $50. *Expenses:* Tuition, state resident: full-time $3,185; part-time $405 per credit. Tuition, nonresident: full-time $7,952; part-time $940 per credit. Required fees: $751 per semester hour. Tuition and fees vary according to course load and program. *Financial support:* Fellowships, teaching assistantships, Federal Work-Study, institutionally sponsored loans, and tuition waivers (full and partial) available. Support available to part-time students. *Faculty research:* Condensed-matter theory and experimentation, electronic instrumentation, relativity. *Unit head:* Dr. Robert H. Gowdy, Chair, 804-828-1821, Fax: 804-828-7073, E-mail: rhgowdy@vcu.edu. *Application contact:* Dr. Alison Baski, Graduate Program Director, 804-828-8295, Fax: 804-828-7073, E-mail: aabaski@vcu.edu.

Virginia Polytechnic Institute and State University, Graduate School, College of Science, Department of Physics, Blacksburg, VA 24061. Offers applied physics (MS, PhD); physics (MS, PhD). *Faculty:* 22 full-time (3 women). *Students:* 46 full-time (8 women), 7 part-time (2 women); includes 4 minority (1 African American, 2 Asian Americans or Pacific Islanders, 1 Hispanic American), 21 international. Average age 27. 65 applicants, 22% accepted, 11 enrolled. In 2005, 6 master's, 5 doctorates awarded. *Entrance requirements:* For master's and doctorate, GRE Subject Test. Additional exam requirements/recommendations for international students: Required—TOEFL (minimum score 550 paper-based; 213 computer-based). *Application deadline:* Applications are processed on a rolling basis. Application fee: $45. Electronic applications accepted. *Expenses:* Tuition, state resident: full-time $6,558; part-time $364 per credit. Tuition, nonresident: full-time $11,296; part-time $628 per credit. Required fees: $1,419; $468 per credit. $234 per term. *Financial support:* In 2005–06, 2 fellowships with full tuition reimbursements (averaging $6,000 per year), 14 research assistantships with full tuition reimbursements (averaging $16,689 per year), 26 teaching assistantships with full tuition reimbursements (averaging $13,902 per year) were awarded; career-related internships or fieldwork, Federal Work-Study, scholarships/grants, and unspecified assistantships also available. Financial award application deadline: 4/1. *Faculty research:* Condensed matter, particle physics, theoretical and experimental astrophysics, biophysics, mathematical physics. *Unit head:* Dr. Royce Zia, Head, 540-231-5767, Fax: 540-231-7511. *Application contact:* Chris C. Thomas, Graduate Program Coordinator, 540-231-8728, Fax: 540-231-7511, E-mail: chris.thomas@vt.edu.

Virginia State University, School of Graduate Studies, Research, and Outreach, School of Engineering, Science and Technology, Department of Chemistry and Physics, Petersburg, VA 23806-0001. Offers physics (MS). *Degree requirements:* For master's, one foreign language, thesis. *Entrance requirements:* For master's, GRE General Test.

Wake Forest University, Graduate School, Department of Physics, Winston-Salem, NC 27109. Offers MS, PhD. Part-time programs available. *Faculty:* 18 full-time (2 women), 3 part-time/adjunct (0 women). *Students:* 16 full-time (4 women), 1 part-time, 5 international. Average age 28. 62 applicants, 10% accepted, 3 enrolled. In 2005, 1 master's, 1 doctorate awarded. *Degree requirements:* For master's, one foreign language, thesis, comprehensive exam, registration; for doctorate, 2 foreign languages, thesis/dissertation, comprehensive exam, registration. *Entrance requirements:* For master's and doctorate, GRE General Test. Additional exam requirements/recommendations for international students: Required—TOEFL (minimum score 213 computer-based). *Application deadline:* For fall admission, 1/15 for domestic students, 1/15 for international students. Application fee: $45. Electronic applications accepted. *Financial support:* In 2005–06, 17 students received support, including 1 fellowship with full tuition reimbursement available (averaging $19,500 per year), 4 research assistantships with full tuition reimbursements available (averaging $16,000 per year), 11 teaching assistantships with full tuition reimbursements available (averaging $16,000 per year); scholarships/grants and tuition waivers (full and partial) also available. Support available to part-time students. Financial award application deadline: 1/15; financial award applicants required to submit FAFSA. *Unit head:* Dr. Keith Bonin, Director, 336-758-4962, Fax: 336-758-6142, E-mail: bonin@wfu.edu.

See Close-Up on page 397.

Washington State University, Graduate School, College of Sciences, Department of Physics and Astronomy, Pullman, WA 99164. Offers physics (MS, PhD). *Faculty:* 22. *Students:* 41 full-time (10 women), 3 part-time; includes 2 minority (1 Asian American or Pacific Islander, 1 Hispanic American), 21 international. Average age 28. 80 applicants, 21% accepted, 11 enrolled. In 2005, 7 master's, 2 doctorates awarded. Terminal master's awarded for partial completion of doctoral program. *Degree requirements:* For master's, oral exam; for doctorate, thesis/dissertation, oral exam, written exam. *Entrance requirements:* For master's and doctorate, GRE General Test, GRE Subject Test, minimum GPA of 3.0, 3 letters of recommendation. Additional exam requirements/recommendations for international students: Required—TOEFL (minimum score 550 paper-based; 214 computer-based). *Application deadline:* For fall admission, 2/1 priority date for domestic students, 3/1 priority date for international students; for spring admission, 9/1 priority date for domestic students, 7/1 priority date for international students. Applications are processed on a rolling basis. Application fee: $35. Electronic applications accepted. *Expenses:* Tuition, state resident: full-time $6,295; part-time $336 per credit. Tuition, nonresident: full-time $15,949; part-time $819 per credit. Required fees: $933. Part-time tuition and fees vary according to campus/location and program. *Financial support:* In 2005–06, 41 students received support, including 1 fellowship with full and partial tuition reimbursement available (averaging $5,000 per year), 15 research assistantships with full and partial tuition reimbursements available (averaging $14,523 per year), 25 teaching assistantships with full and partial tuition reimbursements available (averaging $14,002 per year); Federal Work-Study and institutionally sponsored loans also available. Financial award application deadline: 3/1; financial award applicants required to submit FAFSA. *Faculty research:* Linear and nonlinear acoustics and optics, shock wave dynamics, solid-state physics, surface physics, high-pressure and semi conductor physics. Total annual research expenditures: $2.3 million. *Unit head:* Dr. Steven Tomsovic, Chair, 509-335-9532, E-mail: physics@wsu.edu. *Application contact:* Sabreen Yamini Dodson, Graduate Coordinator, 509-335-9532, Fax: 509-335-7816, E-mail: sabreen@wsu.edu.

Washington University in St. Louis, Graduate School of Arts and Sciences, Department of Physics, St. Louis, MO 63130-4899. Offers MA, PhD. Terminal master's awarded for partial completion of doctoral program. *Degree requirements:* For master's, thesis or alternative; for doctorate, thesis/dissertation. *Entrance requirements:* For master's and doctorate, GRE General Test. Electronic applications accepted.

See Close-Up on page 401.

Wayne State University, Graduate School, College of Liberal Arts and Sciences, Department of Physics and Astronomy, Detroit, MI 48202. Offers physics (MA, MS, PhD). *Faculty:* 19 full-time (1 woman). *Students:* 37 full-time (7 women), 3 part-time; includes 2 minority (both Asian Americans or Pacific Islanders), 20 international. Average age 31. 42 applicants, 21% accepted, 8 enrolled. In 2005, 7 master's, 1 doctorate awarded. *Degree requirements:* For doctorate, thesis/dissertation. *Entrance requirements:* Additional exam requirements/recommendations for international students: Required—TOEFL (minimum score 550 paper-based; 213 computer-based); Recommended—TWE (minimum score 6). *Application deadline:* For fall admission, 7/1 for domestic students, 6/1 for international students. Applications are processed on a rolling basis. Application fee: $30 ($50 for international students). Electronic applications accepted. *Expenses:* Tuition, state resident: part-time $338 per credit hour. Tuition, nonresident: part-time $746 per credit hour. Required fees: $24 per credit hour. Full-time tuition and fees vary according to program. *Financial support:* In 2005–06, 1 fellowship with tuition reimbursement, 11 research assistantships with tuition reimbursements (averaging $16,025 per year), 17 teaching assistantships with tuition reimbursements (averaging $14,997 per year) were awarded; Federal Work-Study also available. Financial award application deadline: 7/1. *Faculty research:* High energy particle physics, relativistic heavy ion physics, theoretical physics, positron and atomic physics, condensed matter and nano-scale physics. Total annual research expenditures: $1.3 million. *Unit head:* Ratna Naik, Chair, 313-577-2756, Fax: 313-577-3932, E-mail: naik@physics.wayne.edu. *Application contact:* Jogindra Wadehra, Graduate Director, 313-577-2740, E-mail: wadehra@physics.wayne.edu.

Wesleyan University, Graduate Programs, Department of Physics, Middletown, CT 06459-0260. Offers MA, PhD. *Faculty:* 9 full-time (1 woman). *Students:* 17 full-time (4 women); includes 1 minority (Hispanic American), 8 international. Average age 25. In 2005, 2 master's awarded. Terminal master's awarded for partial completion of doctoral program. *Degree requirements:* For master's and doctorate, thesis/dissertation. *Entrance requirements:* For master's, GRE General Test, GRE Subject Test; for doctorate, GRE Subject Test. *Application deadline:* For fall admission, 3/1 for domestic students. Applications are processed on a rolling basis. Application fee: $0. *Expenses:* Tuition: Full-time $24,732. One-time fee: $20 full-time. *Financial support:* Teaching assistantships, institutionally sponsored loans and tuition waivers (full) available. *Faculty research:* Low-temperature physics, magnetic resonance, atomic collisions, laser spectroscopy, surface physics. *Unit head:* Dr. Lutz Huwel, Chairman, 860-685-2052, E-mail: lhuwel@wesleyan.edu. *Application contact:* Anna Milardo, Information Contact, 860-685-2030, Fax: 860-685-2031, E-mail: amilardo@wesleyan.edu.

Western Illinois University, School of Graduate Studies, College of Arts and Sciences, Department of Physics, Macomb, IL 61455-1390. Offers MS. Part-time programs available. *Students:* 11 full-time (1 woman), 8 international. Average age 29. 12 applicants, 67% accepted. In 2005, 5 degrees awarded. *Degree requirements:* For master's, thesis or alternative. *Entrance requirements:* Additional exam requirements/recommendations for international students: Required—TOEFL (minimum score 500 paper-based; 173 computer-based). *Application deadline:* Applications are processed on a rolling basis. Application fee: $30. Electronic applications accepted. *Expenses:* Tuition, state resident: full-time $3,599; part-time $200 per semester hour. Tuition, nonresident: full-time $7,198; part-time $400 per semester hour. Required fees: $890; $49 per semester hour. Tuition and fees vary according to campus/location. *Financial support:* In 2005–06, 10 students received support, including 10 research assistantships with full tuition reimbursements available (averaging $6,288 per year) Financial award applicants required to submit FAFSA. *Unit head:* Dr. Vivian Incera, Chairperson, 309-298-1538. *Application contact:* Dr. Barbara Baily, Director of Graduate Studies/Associate Provost, 309-298-1806, Fax: 309-298-2345, E-mail: grad-office@wiu.edu.

Western Michigan University, Graduate College, College of Arts and Sciences, Department of Physics, Kalamazoo, MI 49008-5202. Offers MA, PhD. *Degree requirements:* For master's, thesis; for doctorate, thesis/dissertation, oral exam. *Entrance requirements:* For doctorate, GRE General Test.

West Virginia University, Eberly College of Arts and Sciences, Department of Physics, Morgantown, WV 26506. Offers applied physics (MS, PhD); astrophysics (MS, PhD); chemical physics (MS, PhD); condensed matter physics (MS, PhD); elementary particle physics (MS, PhD); materials physics (MS, PhD); plasma physics (MS, PhD); solid state physics (MS, PhD); statistical physics (MS, PhD); theoretical physics (MS, PhD). *Faculty:* 17 full-time (3 women), 1 part-time/adjunct (0 women). *Students:* 44 full-time (10 women), 4 part-time (1 woman), 32 international. Average age 28. 60 applicants, 20% accepted. In 2005, 2 master's, 3 doctorates awarded. Terminal master's awarded for partial completion of doctoral program. *Degree requirements:* For master's, thesis or alternative, qualifying exam; for doctorate, thesis/dissertation, qualifying exam. *Entrance requirements:* For master's and doctorate, GRE General Test, GRE Subject Test, minimum GPA of 3.0. Additional exam requirements/recommendations for international students: Required—TOEFL. *Application deadline:* For fall admission, 2/15 for domestic students. Applications are processed on a rolling basis. Application fee: $50. *Expenses:* Tuition, state resident: full-time $4,582; part-time $258 per credit hour. Tuition, nonresident: full-time $1,382; part-time $741 per credit hour. *Financial support:* In 2005–06, 30 research assistantships with full and partial tuition reimbursements (averaging $18,000 per year), 10 teaching assistantships with full and partial tuition reimbursements (averaging $16,000 per year) were awarded; fellowships, Federal Work-Study, institutionally sponsored loans, and tuition waivers (full and partial) also available. Financial award application deadline: 2/1; financial award applicants required to submit FAFSA. *Faculty research:* Experimental and theoretical condensed-matter, plasma, high-energy theory, nonlinear dynamics, space physics. Total annual research expenditures: $3.3 million. *Unit head:* Dr. Earl E. Scime, Chair, 304-293-3422 Ext. 1437, Fax: 304-293-5732, E-mail: earl.scime@mail.wvu.edu.

Wichita State University, Graduate School, Fairmount College of Liberal Arts and Sciences, Department of Physics, Wichita, KS 67260. Offers MS. Part-time programs available. *Degree requirements:* For master's, qualifying exam, thesis optional. *Entrance requirements:* For master's, GRE. Additional exam requirements/recommendations for international students: Required—TOEFL. Electronic applications accepted. *Faculty research:* Condensed matter experiment and theory, low-mass stellar atmospheres.

Worcester Polytechnic Institute, Graduate Studies and Enrollment, Department of Physics, Worcester, MA 01609-2280. Offers MS, PhD. *Faculty:* 13 full-time (2 women). *Students:* 21 full-time (4 women), 1 (woman) part-time; includes 2 minority (1 African American, 1 Asian American or Pacific Islander), 8 international. 29 applicants, 59% accepted, 11 enrolled. In 2005, 3 master's, 4 doctorates awarded. *Degree requirements:* For master's, thesis/dissertation; for doctorate, thesis/dissertation, comprehensive exam. *Entrance requirements:* For master's and doctorate, 3 letters of recommendation. Additional exam requirements/recommendations for international students: Required—TOEFL (minimum score 550 paper-based; 213 computer-based). *Application deadline:* For fall admission, 1/15 for domestic students; for spring admission, 10/15 priority date for domestic students. Applications are processed on a rolling basis. Application fee: $70. Electronic applications accepted. *Expenses:* Tuition: Part-time $997 per credit hour. *Financial support:* In 2005–06, 22 students received support, including 9 research assistantships with full tuition reimbursements available, 13 teaching assistantships with full tuition reimbursements available; fellowships with full tuition reimbursements available, career-related internships or fieldwork, institutionally sponsored loans, scholarships/grants, and unspecified assistantships also available. Financial award application deadline: 1/15. *Faculty research:* Chemical and biochemical physics, materials research, classical and quantum optics, relativity, solid state physics. Total annual research expenditures:$396,838. *Unit head:* Dr. Germano S. Iannacchione, Interim Head, 508-831-5365, Fax: 508-831-5886, E-mail: gsiannac@wpi.edu. *Application contact:* Dr. Nancy A Burnham, Graduate Coordinator, 508-831-5365, Fax: 508-831-5886, E-mail: nab@wpi.edu.

See Close-Up on page 403.

Wright State University, School of Graduate Studies, College of Science and Mathematics, Department of Physics, Program in Physics, Dayton, OH 45435. Offers medical physics (MS); physics (MS). Part-time and evening/weekend programs available. *Degree requirements:* For master's, thesis. *Entrance requirements:* Additional exam requirements/recommendations for international students: Required—TOEFL. *Faculty research:* Solid-state physics, optics, geophysics.

Yale University, Graduate School of Arts and Sciences, Department of Physics, New Haven, CT 06520. Offers PhD. *Degree requirements:* For doctorate, thesis/dissertation. *Entrance requirements:* For doctorate, GRE General Test, GRE Subject Test.

See Close-Up on page 405.

York University, Faculty of Graduate Studies, Faculty of Pure and Applied Science, Program in Physics and Astronomy, Toronto, ON M3J 1P3, Canada. Offers M Sc, PhD. Part-time and evening/weekend programs available. *Faculty:* 51 full-time (5 women), 2 part-time/adjunct (0 women). *Students:* 42 full-time (8 women), 4 part-time (1 woman). 54 applicants, 20% accepted, 10 enrolled. In 2005, 4 master's, 2 doctorates awarded. *Degree requirements:* For master's, thesis or alternative, registration; for doctorate, thesis/dissertation, comprehensive exam, registration. *Application deadline:* Applications are processed on a rolling basis. Application fee: $80. Electronic applications accepted. *Expenses:* Tuition, state resident: full-time $3,190; part-time $798 per term. International tuition: $7,515 full-time. Required fees: $217. Tuition and fees vary according to program. *Financial support:* In 2005–06, fellowships (averaging $11,388 per year), research assistantships (averaging $11,216 per year), teaching assistantships (averaging $8,084 per year) were awarded; fee bursaries also available. *Unit head:* Helen Freedhoff, Director, 416-736-5249.

Plasma Physics

Columbia University, Fu Foundation School of Engineering and Applied Science, Department of Applied Physics and Applied Mathematics, New York, NY 10027. Offers applied physics (MS, PhD), including applied mathematics (PhD), optical physics (PhD), plasma physics (PhD), solid state physics (PhD); applied physics and applied mathematics (Eng Sc D); materials science and engineering (MS, Eng Sc D, PhD); medical physics (MS); minerals engineering and materials science (Eng Sc D, PhD, Engr). Part-time programs available. *Faculty:* 30 full-time (2 women), 15 part-time/adjunct (1 woman). *Students:* 94 full-time (26 women), 30 part-time (5 women); includes 17 minority (1 American Indian/Alaska Native, 12 Asian Americans or Pacific Islanders, 4 Hispanic Americans), 53 international. Average age 24. 294 applicants, 22% accepted, 38 enrolled. In 2005, 33 master's, 5 doctorates awarded. Terminal master's awarded for partial completion of doctoral program. *Degree requirements:* For master's, English Proficiency Test; for doctorate, thesis/dissertation, qualifying exam, English Proficiency Test; for Engr, English Profiency Test. *Entrance requirements:* For master's and doctorate, GRE General Test, GRE Subject Test (strongly recommended). Additional exam requirements/recommendations for international students: Required—TOEFL. *Application deadline:* For fall admission, 12/15 priority date for domestic students, 12/15 priority date for international students; for spring admission, 10/1 priority date for domestic students, 10/1 priority date for international students. Application fee: $45. Electronic applications accepted. *Expenses:* Tuition: Full-time $31,448. Tuition and fees vary according to course level, course load, campus/location and program. *Financial support:* In 2005–06, 8 fellowships with full tuition reimbursements, 56 research assistantships with full tuition reimbursements (averaging $24,750 per year), 16 teaching assistantships with full tuition reimbursements (averaging $24,750 per year) were awarded; Federal Work-Study and unspecified assistantships also available. Financial award application deadline: 12/15; financial award applicants required to submit FAFSA. *Faculty research:* Plasma physics, applied mathematics, solid-state and optical physics, atmospheric, oceanic and earth physics, materials science and engineering. *Unit head:* Dr. Michael E. Mauel, Professor and Chairman, 212-854-4457, E-mail: seasinfo.apam@columbia.edu. *Application contact:* Marlene Arbo, Department Administrator, 212-854-4458, Fax: 212-854-8257, E-mail: seasinfo.apam@columbia.edu.

See Close-Up on page 335.

Massachusetts Institute of Technology, School of Engineering, Department of Aeronautics and Astronautics, Cambridge, MA 02139-4307. Offers aeroacoustics (PhD, Sc D); aerodynamics (PhD, Sc D); aeroelasticity (PhD, Sc D); aeronautics and astronautics (SM, PhD, Sc D, EAA); aerospace systems (PhD, Sc D); aircraft propulsion (PhD, Sc D); astrodynamics (PhD, Sc D); biomedical engineering (PhD, Sc D); computational fluid dynamics (PhD, Sc D); computer systems (PhD, Sc D); dynamics energy conversion (PhD, Sc D); estimation and control (PhD, Sc D); flight transportation (PhD, Sc D); fluid mechanics (PhD, Sc D); gas turbine structures (PhD, Sc D); gas turbines (PhD, Sc D); humans and automation (PhD, Sc D); instrumentation (PhD, Sc D); materials engineering (PhD, Sc D); navigation and control systems (PhD, Sc D); physics of fluids (PhD, Sc D); plasma physics (PhD, Sc D); space propulsion (PhD, Sc D); structural dynamics (PhD, Sc D); structures technology (PhD, Sc D); vehicle design (PhD, Sc D). *Faculty:* 34 full-time (7 women), 1 part-time/adjunct (0 women). *Students:* 222 full-time (55 women), 1 (woman) part-time; includes 34 minority (5 African Americans, 1 American Indian/Alaska Native, 24 Asian Americans or Pacific Islanders, 4 Hispanic Americans), 89 international. Average age 26. 335 applicants, 50% accepted, 74 enrolled. In 2005, 75 master's, 17 doctorates awarded. *Degree requirements:* For master's, thesis/dissertation; for doctorate, thesis/dissertation, comprehensive exam. *Entrance requirements:* For master's and doctorate, GRE General Test. Additional exam requirements/recommendations for international students: Required—TOEFL (minimum score 600 paper-based; 250 computer-based). *Application deadline:* For fall admission, 12/15 for domestic students, 12/15 for international students; for spring admission, 10/1 for domestic students, 10/1 for international students. Application fee: $70. Electronic applications accepted. *Expenses:* Tuition: Full-time $32,100. Required fees: $200. Part-time tuition and fees vary according to course load. *Financial support:* In 2005–06, 36 fellowships with tuition reimbursements (averaging $19,224 per year), 146 research assistantships with tuition reimbursements (averaging $22,516 per year), 11 teaching assistantships with tuition reimbursements (averaging $24,532 per year) were awarded; Federal Work-Study, institutionally sponsored loans, scholarships/grants, health care benefits, and unspecified assistantships also available. Total annual research expenditures: $23.1 million. *Unit head:* Prof. Wesley L. Harris, Department Head, 617-253-0911, E-mail: weslhar@mit.edu. *Application contact:* Information Contact, 617-253-0043, Fax: 617-253-0823, E-mail: aa-gradadmit@mit.edu.

Princeton University, Graduate School, Department of Astrophysical Sciences, Program in Plasma Physics, Princeton, NJ 08544-1019. Offers PhD. *Degree requirements:* For doctorate, thesis/dissertation. *Entrance requirements:* For doctorate, GRE General Test, GRE Subject Test. Additional exam requirements/recommendations for international students: Required—TOEFL (minimum score 600 paper-based; 250 computer-based). *Faculty research:* Magnetic fusion energy research, plasma physics, x-ray laser studies.

University of Colorado at Boulder, Graduate School, College of Arts and Sciences, Department of Astrophysical and Planetary Sciences, Boulder, CO 80309. Offers astrophysics (MS, PhD); planetary science (MS, PhD). *Faculty:* 19 full-time (3 women). *Students:* 70 full-time (34 women), 29 part-time (6 women); includes 6 minority (4 Asian Americans or Pacific Islanders, 2 Hispanic Americans), 9 international. Average age 29. 59 applicants, 100% accepted. In 2005, 16 master's, 14 doctorates awarded. Terminal master's awarded for partial completion of doctoral program. *Degree requirements:* For master's, thesis or alternative, comprehensive exam; for doctorate, one foreign language, thesis/dissertation. *Entrance requirements:* For master's, GRE General Test, GRE Subject Test, minimum undergraduate GPA of 3.0; for doctorate, GRE General Test, GRE Subject Test. *Application deadline:* For fall admission, 1/15 priority date for domestic students, 12/1 priority date for international students. Applications are processed on a rolling basis. Application fee: $50 ($60 for international students). *Financial support:* In 2005–06, fellowships (averaging $5,307 per year), research assistantships (averaging $15,770 per year), teaching assistantships (averaging $14,900 per year) were awarded; tuition waivers (full) also available. Support available to part-time students. Financial award application deadline: 2/1. *Faculty research:* Stellar and extragalactic astrophysics cosmology, space astronomy, planetary science. Total annual research expenditures: $8.1 million. *Unit head:* James Green, Chair, 303-492-8915, Fax: 303-492-3822, E-mail: jgreen@casa.colorado.edu. *Application contact:* Graduate Program Assistant, 303-492-8914, Fax: 303-492-3822, E-mail: apsgradsec@colorado.edu.

West Virginia University, Eberly College of Arts and Sciences, Department of Physics, Morgantown, WV 26506. Offers applied physics (MS, PhD); astrophysics (MS, PhD); chemical physics (MS, PhD); condensed matter physics (MS, PhD); elementary particle physics (MS, PhD); materials physics (MS, PhD); plasma physics (MS, PhD); solid state physics (MS, PhD); statistical physics (MS, PhD); theoretical physics (MS, PhD). *Faculty:* 17 full-time (3 women), 1 part-time/adjunct (0 women). *Students:* 44 full-time (10 women), 4 part-time (1 woman), 32 international. Average age 28. 60 applicants, 20% accepted. In 2005, 2 master's, 3 doctorates awarded. Terminal master's awarded for partial completion of doctoral program. *Degree requirements:* For master's, thesis or alternative, qualifying exam; for doctorate, thesis/dissertation, qualifying exam. *Entrance requirements:* For master's and doctorate, GRE General Test, GRE Subject Test, minimum GPA of 3.0. Additional exam requirements/recommendations for international students: Required—TOEFL. *Application deadline:* For fall admission, 2/15 for domestic students. Applications are processed on a rolling basis. Application fee: $50. *Expenses:* Tuition, state resident: full-time $4,582; part-time $258 per credit hour. Tuition, nonresident: full-time $1,382; part-time $741 per credit hour. *Financial support:* In 2005–06, 30 research assistantships with full and partial tuition reimbursements (averaging $18,000 per year), 10 teaching assistantships with full and partial tuition reimbursements (averaging $16,000 per year) were awarded; fellowships, Federal Work-Study, institutionally sponsored loans, and tuition waivers (full and partial) also available. Financial award application deadline: 2/1; financial award applicants required to submit FAFSA. *Faculty research:* Experimental and theoretical condensed-matter, plasma, high-energy theory, nonlinear dynamics, space physics. Total annual research expenditures: $3.3 million. *Unit head:* Dr. Earl E. Scime, Chair, 304-293-3422 Ext. 1437, Fax: 304-293-5732, E-mail: earl.scime@mail.wvu.edu.

Theoretical Physics

Cornell University, Graduate School, Graduate Fields of Arts and Sciences, Field of Physics, Ithaca, NY 14853-0001. Offers experimental physics (MS, PhD); physics (MS, PhD); theoretical physics (MS, PhD). *Faculty:* 70 full-time (7 women). *Students:* 429 applicants, 15% accepted, 24 enrolled. In 2005, 40 master's, 18 doctorates awarded. *Degree requirements:* For doctorate, thesis/dissertation, comprehensive exam. *Entrance requirements:* For doctorate, GRE General Test, GRE Subject Test (physics), 3 letters of recommendation. Additional exam requirements/recommendations for international students: Required—TOEFL (minimum score 550 paper-based; 213 computer-based). *Application deadline:* For fall admission, 1/3 for domestic students. Application fee: $60. Electronic applications accepted. *Financial support:* In 2005–06, 175 students received support, including 28 fellowships with full tuition reimbursements available, 95 research assistantships with full tuition reimbursements available, 52 teaching assistantships with full tuition reimbursements available; institutionally sponsored loans, scholarships/grants, health care benefits, tuition waivers (full and partial), and unspecified assistantships also available. Financial award applicants required to submit FAFSA. *Faculty research:* Experimental condensed matter physics, theoretical condensed matter physics, experimental high energy particle physics, theoretical particle physics and field theory, theoretical astrophysics. *Unit head:* Director of Graduate Studies, 607-255-7561. *Application contact:* Graduate Field Assistant, 607-255-7561, E-mail: physics-grad-adm@cornell.edu.

Harvard University, Graduate School of Arts and Sciences, Department of Physics, Cambridge, MA 02138. Offers experimental physics (PhD); medical engineering/medical physics (PhD), including applied physics, engineering sciences, physics; theoretical physics (PhD). *Students:* 174. *Degree requirements:* For doctorate, thesis/dissertation, final exams, laboratory experience. *Entrance requirements:* For doctorate, GRE General Test, GRE Subject Test. Additional exam requirements/recommendations for international students: Required—TOEFL. *Application deadline:* For fall admission, 12/14 for domestic students. Application fee: $60. *Expenses:* Tuition: Full-time $28,752. Full-time tuition and fees vary according to program and student level. *Financial support:* Fellowships, research assistantships, teaching assistantships, career-related internships or fieldwork, Federal Work-Study, and institutionally sponsored loans available. Financial award application deadline: 12/30. *Faculty research:* Particle physics, condensed matter physics, atomic physics. *Unit head:* Sheila Ferguson, Administrator, 617-495-4327. *Application contact:* Office of Admissions and Financial Aid, 617-495-5315.

Rutgers, The State University of New Jersey, New Brunswick/Piscataway, Graduate School, Program in Physics and Astronomy, New Brunswick, NJ 08901-1281. Offers astronomy (MS, PhD); biophysics (PhD); condensed matter physics (MS, PhD); elementary particle physics (MS, PhD); intermediate energy nuclear physics (MS); nuclear physics (MS, PhD); physics (MST); surface science (PhD); theoretical physics (MS, PhD). Part-time programs available. *Faculty:* 92 full-time. *Students:* 105 full-time (18 women), 5 part-time (3 women); includes 6 minority (2 African Americans, 4 Asian Americans or Pacific Islanders), 54 international. Average age 27. 258 applicants, 19% accepted, 23 enrolled. In 2005, 10 master's, 14 doctorates awarded. Terminal master's awarded for partial completion of doctoral program. *Degree requirements:* For master's, thesis or alternative, comprehensive exam; for doctorate, thesis/dissertation, comprehensive exam. *Entrance requirements:* For master's and doctorate, GRE General Test, GRE Subject Test. Additional exam requirements/recommendations for international students: Required—TOEFL, TSE. *Application deadline:* For fall admission, 1/2 priority date for domestic students, 1/2 priority date for international students; for spring admission, 11/1 for domestic students, 11/1 for international students. Applications are processed on a rolling basis. Application fee: $50. Electronic applications accepted. *Expenses:* Tuition, state resident: full-time $10,440; part-time $435 per credit. Tuition, nonresident: full-time $15,520; part-time $647 per credit. Required fees: $129 per credit. Tuition and fees vary according to program. *Financial support:* In 2005–06, 19 fellowships with full tuition reimbursements (averaging $21,000 per year), 40 research assistantships with full tuition reimbursements (averaging $18,088 per year), 36 teaching assistantships with full tuition reimbursements (averaging $18,088 per year) were awarded; health care benefits and unspecified assistantships also available. Financial award application deadline: 1/2; financial award applicants required to submit FAFSA. *Faculty research:* Astronomy, high energy, condensed matter, surface, nuclear physics. Total annual research expenditures: $7.8 million. *Unit head:* Ronald Ransome, Director, 732-445-2516, Fax: 732-445-4343, E-mail: ransome@physics.rutgers.edu. *Application contact:* Shirley Hinds, Administrative Assistant, 732-445-2502, Fax: 732-445-4343, E-mail: graduate@physics.rutgers.edu.

St. John's University, St. John's College of Liberal Arts and Sciences, Institute of Asian Studies, Queens, NY 11439. Offers Asian and African cultural studies (Adv C); Asian studies (Adv C); Chinese studies (MA, Adv C); East Asian culture studies (Adv C); East Asian studies (MA). Part-time and evening/weekend programs available. *Faculty:* 1 (woman) full-time, 7 part-time/adjunct (5 women). *Students:* 6 full-time (2 women), 7 part-time (5 women); includes 1 minority (Hispanic American), 6 international. Average age 28. 12 applicants, 100% accepted, 8 enrolled. In 2005, 5 degrees awarded. *Degree requirements:* For master's, one foreign language, comprehensive exam. *Entrance requirements:* For master's, 18 hours of course work in the field, minimum GPA of 3.0. Additional exam requirements/recommendations for international students: Required—TOEFL (minimum score 500 paper-based; 173 computer-based). *Application deadline:* For fall admission, 5/1 priority date for domestic students, 5/1 priority date for international students; for spring admission, 11/1 priority date for domestic students, 11/1 priority date for international students. Applications are processed on a rolling basis. Application fee: $40. Electronic applications accepted. *Expenses:* Tuition: Full-time $8,760; part-time $730 per credit. Required fees: $250; $125 per term. Tuition and fees vary according to program. *Financial support:* Research assistantships, scholarships/grants available. Support available to part-time students. Financial award application deadline: 3/1; financial award applicants required to submit FAFSA. *Faculty research:* East Asian philosophy and religion, Chinese language and literature, Japanese language, modern Japan, Chinese art and history. *Unit head:* Dr. Bernadette Li, Chair, 718-990-1657, E-mail: lib@stjohns.edu. *Application contact:* Matthew Whelan, Director, Office of Admissions, 718-990-2000, Fax: 718-990-2096, E-mail: admissions@stjohns.edu.

University of Victoria, Faculty of Graduate Studies, Faculty of Science, Department of Physics and Astronomy, Victoria, BC V8W 2Y2, Canada. Offers astronomy and astrophysics (M Sc, PhD); condensed matter physics (M Sc, PhD); experimental particle physics (M Sc, PhD); medical physics (M Sc, PhD); ocean physics (PhD); ocean physics and geophysics (M Sc); theoretical physics (M Sc, PhD). *Faculty:* 16 full-time (0 women), 13 part-time/adjunct (1 woman). *Students:* 45 full-time, 8 international. Average age 25. 66 applicants, 45% accepted, 10 enrolled. In 2005, 4 master's, 1 doctorate awarded. *Median time to degree:* Of those who began their doctoral program in fall 1997, 100% received their degree in 8 years or less. *Degree requirements:* For master's, thesis, registration; for doctorate, thesis/dissertation, Candidacy exam, comprehensive exam, registration. *Entrance requirements:* For master's and doctorate, GRE. Additional exam requirements/recommendations for international students: Required—TOEFL (minimum score 575 paper-based; 233 computer-based), IELT (minimum score 7). *Application deadline:* For fall admission, 5/31 priority date for domestic students, 12/15 priority date for international students. Applications are processed on a rolling basis. Application fee: $75 ($125 for international students). Electronic applications accepted. Tuition and fees charges are reported in Canadian dollars. *Expenses:* Tuition, area resident: Full-time $4,492 Canadian dollars; part-time $749 Canadian dollars per term. International tuition: $5,346 Canadian dollars full-time. Required fees: $4,492 Canadian dollars; $749 Canadian dollars per term. Tuition and fees vary according to course load, campus/location and program. *Financial support:* In 2005–06, 3 students received support; fellowships, research assistantships, teaching assistantships, career-related internships or fieldwork, institutionally sponsored loans, and awards available. Financial award application deadline: 2/15. *Faculty research:* Old stellar populations; observational cosmology and large scale structure; cp violation; atlas. *Unit head:* Dr. J. Michael Ronev, Chair, 250-721-7698, Fax: 250-721-7715, E-mail: chair@phys.uvic.ca. *Application contact:* Dr. Chris J. Pritchet, Graduate Adviser, 250-721-7744, Fax: 250-721-7715, E-mail: pritchet@uvic.ca.

West Virginia University, Eberly College of Arts and Sciences, Department of Physics, Morgantown, WV 26506. Offers applied physics (MS, PhD); astrophysics (MS, PhD); chemical physics (MS, PhD); condensed matter physics (MS, PhD); elementary particle physics (MS, PhD); materials physics (MS, PhD); plasma physics (MS, PhD); solid state physics (MS, PhD); statistical physics (MS, PhD); theoretical physics (MS, PhD). *Faculty:* 17 full-time (3 women), 1 part-time/adjunct (0 women). *Students:* 44 full-time (10 women), 4 part-time (1 woman), 32 international. Average age 28. 60 applicants, 20% accepted. In 2005, 2 master's, 3 doctorates awarded. Terminal master's awarded for partial completion of doctoral program. *Degree requirements:* For master's, thesis or alternative, qualifying exam; for doctorate, thesis/dissertation, qualifying exam. *Entrance requirements:* For master's and doctorate, GRE General Test, GRE Subject Test, minimum GPA of 3.0. Additional exam requirements/recommendations for international students: Required—TOEFL. *Application deadline:* For fall admission, 2/15 for domestic students. Applications are processed on a rolling basis. Application fee: $50. *Expenses:* Tuition, state resident: full-time $4,582; part-time $258 per credit hour. Tuition, nonresident: full-time $1,382; part-time $741 per credit hour. *Financial support:* In 2005–06, 30 research assistantships with full and partial tuition reimbursements (averaging $18,000 per year), 10 teaching assistantships with full and partial tuition reimbursements (averaging $16,000 per year) were awarded; fellowships, Federal Work-Study, institutionally sponsored loans, and tuition waivers (full and partial) also available. Financial award application deadline: 2/1; financial award applicants required to submit FAFSA. *Faculty research:* Experimental and theoretical condensed-matter, plasma, high-energy theory, nonlinear dynamics, space physics. Total annual research expenditures: $3.3 million. *Unit head:* Dr. Earl E. Scime, Chair, 304-293-3422 Ext. 1437, Fax: 304-293-5732, E-mail: earl.scime@mail.wvu.edu.

Cross-Discipline Announcement

Dartmouth College, Thayer School of Engineering, Hanover, NH 03755.

Thayer School offers MS and PhD programs in applied sciences. The interdisciplinary character of the institution, modern laboratories, computing facilities, and active collaborations with other departments provide unique opportunities for study and research in space plasma physics, nonlinear optics, electromagnetism, molecular materials, image and signal processing, fluid mechanics, and oceanography.

AUBURN UNIVERSITY

Department of Physics

Program of Study

The Auburn University Department of Physics offers the Ph.D. degree in physics to students who complete at least 60 semester hours (30 hours of graded course work in graduate-level physics), pass a written and oral general doctoral examination, and successfully defend a research dissertation. The Ph.D. degree program takes approximately five years to complete. The student's research is in one of the areas in which the department has active research groups: plasma physics, especially magnetically confined plasmas with applications to the development of fusion energy; condensed-matter physics, especially semiconductors for microelectronic applications; atomic and molecular physics; dusty plasmas; and space physics, especially in the earth's magnetosphere, with applications to space weather.

An M.S. degree with a thesis or nonthesis option is also offered. With the nonthesis option, 30 hours (21 hours of graduate-level physics) are required, as is an acceptable grade on a written examination. Students who elect the thesis option take similar courses but do a thesis instead of a written examination. The nonthesis M.S. option takes about two years to complete, and the thesis option takes about three years.

Research Facilities

The Department of Physics offices and research laboratories are housed in the Allison Lab (37,000 square feet) and the Leach Science Center (46,000 square feet). Major research equipment includes the Accelerator Lab with a new 2-megavolt Tandem Ion Accelerator; the Compact Toroidal Hybrid, a magnetic fusion device; the Scanning Tunneling Electron Microscope Lab; a molecular beam epitaxy facility; the Epitaxial Growth Laboratory; the Surface Science Laboratory; Laboratories for Plasma Physics; and a 96-processor Beowulf cluster for parallel processing. Physics faculty members and students also collaborate with scientists in the Space Research Institute and the College of Engineering, expanding the facilities available for developing new knowledge.

Auburn's libraries, which ranked third among more than 300 of the nation's top colleges and universities according to a poll taken by the *Princeton Review* for its guide to the best universities, are also available to students.

Financial Aid

Students admitted to graduate study in physics are offered teaching assistantships, with an annual stipend of $18,500 and full tuition remission. Students pay only a $200 registration fee per semester. As students progress toward their degrees, they usually become research assistants with similar monetary remuneration.

Auburn physics graduate students also often compete successfully for special fellowships from various government agencies.

Cost of Study

Tuition is covered for all teaching and research assistants. Only a $200 registration fee per semester is required.

Living and Housing Costs

Official estimates of living expenses are less than $14,000 per year. Actual costs can be lower.

Student Group

The Department of Physics has 23 full-time faculty members, between 30 and 40 undergraduate physics majors, and 30 to 40 graduate students. Physics faculty members and students, both men and women, come from all areas of the United States as well as from Africa, Asia, Europe, and South America.

Student Outcomes

Students who finish with a Ph.D. go on to jobs in academia, postdoctoral research, teaching, government research labs, and industrial research. Students who finish with an M.S. go on to jobs teaching in junior colleges or doing research at government and industrial labs or continue their education.

Location

Auburn, a city of 45,000 people, is beautiful, convenient, and friendly and offers easy living. It is located 125 miles southwest of Atlanta and 60 miles east of Montgomery, the state capital. Auburn is surrounded by farms and woodlands. The name comes from the line, "Sweet Auburn, loveliest village of the Plain," in Oliver Goldsmith's poem *The Deserted Village.*

The University and The Department

The main campus enrollment is more than 21,500, with approximately 1,200 full-time faculty members. Large enough to provide an enriched educational and cultural environment for students from more than thirty countries around the world, Auburn retains the charm and civility of the New South. Yet, the bright lights of Atlanta are only about 2 hours away by car.

The Department of Physics, with its 1:2 faculty-student ratio, provides a nurturing environment in which the student is treated as an individual and also has the opportunity to experience the joy of discovery in world-class research groups.

Applying

To apply for graduate admission and financial assistance, students should go to the department's Web site and apply directly to the Department of Physics. After completing the online form, students should send their transcripts, GRE scores, and letters of recommendation to the department. No application fee is required if application is made in this manner.

Correspondence and Information

To submit application materials:
Graduate Admissions
Department of Physics
206 Allison Lab
Auburn University, Alabama 36849

For information and other correspondence:
Professor J. D. Perez, Head
Department of Physics
206 Allison Lab
Auburn University, Alabama 36849
Phone: 334-844-4264
Fax: 334-844-4613
E-mail: perez@physics.auburn.edu
Web site: http://www.physics.auburn.edu/

Auburn University

THE FACULTY AND THEIR RESEARCH

Robert F. Boivin, Assistant Professor; Ph.D., Quebec. Experimental plasma physics and propulsion.
Michael J. Bozack, Professor; Ph.D., Oregon. Experimental surface science and semiconductor physics.
An-Ban Chen, Professor; Ph.D., William and Mary. Theoretical condensed-matter physics.
Eugene J. Clothiaux, Professor; Ph.D., New Mexico State. Spectroscopy.
Jianjun Dong, Associate Professor; Ph.D., Ohio. Theoretical condensed-matter physics and computational physics.
Albert T. Fromhold, Professor; Ph.D., Cornell. Condensed-matter physics.
Junichiro Fukai, Associate Professor; Ph.D., Tennessee. Fundamentals of electricity and magnetism.
James D. Hanson, Professor; Ph.D., Maryland. Theoretical plasma physics and fusion science.
Satoshi Hinata, Professor; Ph.D., Illinois. Theoretical space physics and solar physics.
Stephen F. Knowlton, Professor; Ph.D., MIT. Experimental plasma physics and fusion science.
Allen L. Landers, Assistant Professor; Ph.D., Kansas State. Experimental atomic physics.
Yu Lin, Associate Professor; Ph.D., Alaska. Theoretical space physics and magnetospheric physics.
Eugene Oks, Professor; Ph.D., Moscow Physical Technical Institute. Theoretical atomic and molecular physics and econophysics.
Minseo Park, Assistant Professor; Ph.D., North Carolina. Experimental condensed-matter physics.
Joseph D. Perez, Professor and Department Head; Ph.D., Maryland. Theoretical space (magnetospheric) and plasma physics.
Michael S. Pindzola, Professor; Ph.D., Virginia. Theoretical atomic and molecular physics and computational physics.
Francis J. Robicheaux, Professor; Ph.D., Chicago. Theoretical atomic and molecular physics.
Marllin L. Simon, Associate Professor; Ph.D., Missouri. Physics education.
D. Gary Swanson, Professor; Ph.D., Caltech. Theoretical plasma physics.
Edward Thomas Jr., Associate Professor; Ph.D., Auburn. Experimental plasma physics and dusty plasmas.
Chin-Che Tin, Professor; Ph.D., Alberta. Experimental condensed-matter physics and epitaxial growth.
Jean-Marie P. Wersinger, Associate Professor; Ph.D., Lausanne. Remote sensing and theoretical plasma physics.
John R. Williams, Professor; Ph.D., North Carolina State. Experimental condensed-matter physics, semiconductors, and low-energy accelerator applications.

BOSTON UNIVERSITY

Department of Physics

Programs of Study

The Department of Physics offers programs leading to the Ph.D., with an optional M.A. degree in physics. The department offers research opportunities in experimental high-energy and medium-energy physics, particle astrophysics, theoretical particle physics and cosmology, biological physics, experimental condensed-matter physics, theoretical condensed-matter physics, polymer physics, and statistical physics.

The M.A. degree requires the completion of eight semester courses, passed with a grade of B– or better; evidence of having successfully completed undergraduate courses in a modern language or passing the departmental language exam; and achieving a passing grade on the departmental comprehensive exam or the completion of a master's thesis. The requirements for a master's degree may be satisfied as part of the Ph.D. degree program. Each student must satisfy a residency requirement of a minimum of two consecutive semesters of full-time graduate study at Boston University.

The Ph.D. requires the completion of eight semester courses beyond the M.A. degree, passed with a grade of B– or better; passing of the departmental language exam; an honors grade on the departmental comprehensive exam; passing of an oral exam; and the completion of a dissertation and a dissertation defense. The dissertation must exhibit an original contribution to the field. Each student must satisfy a residency requirement of a minimum of two consecutive semesters of full-time graduate study at Boston University. The time it takes to obtain a Ph.D. degree is approximately 5½ years, although students have obtained their degree in as short a time as four years and as long as eight.

Research Facilities

The Department of Physics is part of Boston University's $250-million Science and Engineering Complex, centrally located on the main Charles River Campus. Condensed-matter physics facilities include electronic and mechanical nanostructure fabrication and measurement, metastable-helium-atom probes of surface spin order and dynamics, photoemission and soft X-ray fluorescence probes of electronic structure in novel materials, X-ray diffractometers, and the optics and transport of electrons at high fields and low temperatures. Biological physics and polymer physics labs include dynamical light scattering, Raman and Brillouin scattering, and infrared and far-infrared absorption spectroscopy as well as modern facilities for genetically manipulating biomolecules. Physicists at the Center for Photonics develop and use near-field scanning optical and infrared microscopy, ultrafast infrared spectroscopy, entangled photons for quantum information processing and entangled photon microscopy, and a full complement of molecular beam epitaxy and device processing facilities, the latter primarily with InGaAl-nitride wide-band-gap semiconductor materials and devices. The high-energy physics labs include facilities for the design, production, and testing of key components of various particle detectors. Collaborations include the D0 experiment at Fermilab; the ATLAS and CMS experiments at CERN; the MuLan experiment at PSI, Switzerland; and the Super-Kamiokande experiment in Kamioka, Japan, including the K2K and JHF neutrino accelerator projects. For computation, workstations are networked to two major departmental SGI servers as well as computer clusters provided by the University for general student use. In addition, students have access to the University's high-end computational resources, which include IBM p690 servers with 112 processors (580 Gflops), an IBM p655 system with forty-eight processors (210 Gflops), an IBM Linux cluster with fifty-two dual-processor compute nodes and twenty-four display nodes, and advanced visualization facilities.

Financial Aid

Through a combination of teaching fellowships, research assistantships, and University fellowships, the department provides stipends and full-tuition scholarships for essentially all students. The standard stipend for teaching fellows and research assistants is expected to be $24,000 per calendar year plus student medical insurance.

Cost of Study

Tuition and fees are provided for as already described. Books and supplies cost an additional $400.

Living and Housing Costs

There is limited graduate student housing available on the Boston University campus at approximately $10,000 per year for room and board. However, students generally rent apartments in the Boston area. The cost of apartments varies widely, depending on the area.

Student Group

Currently, the department has 108 graduate students engaged in work toward the Ph.D. and M.A. degrees, and it prides itself on the close contact maintained between students and faculty members.

Student Outcomes

Recent Ph.D. recipients from the Department of Physics have been awarded the Wigner Fellowship at Oak Ridge, National Research Council Postdoctoral Fellowships, and the IBM Supercomputer Research Award, among others. Other graduates have gone on to permanent positions at Bell Laboratories, NEC Corporation, NASA, and NIST and to tenured faculty positions at major universities.

Location

Boston University is located in Boston, Massachusetts, which is a major metropolitan center of cultural, scholarly, scientific, and technological activity. Besides Boston University, there are many major academic institutions in the area. Seminars and colloquia are announced in a Boston Area Physics Calendar.

The University and The Department

Boston University is a private urban university with a faculty of 3,813 members and a student population of 31,697. The University consists of fifteen schools and colleges. The Department of Physics is part of the College of Arts and Sciences and the Graduate School. The department has a young and active faculty of 36 full-time members and has experienced significant growth in recent years. Among the recent additions to the faculty is Nobel laureate Sheldon Glashow.

Applying

The application deadlines are January 15 for fall admission and November 1 for spring admission. Application information and forms are available online at http://physics.bu.edu/grad.html. For admission to the graduate programs, a bachelor's degree in physics or astronomy is required. Exceptional candidates from other fields are considered. Official test results of the Graduate Record Examinations (GRE) (General Test and Subject Test in Physics) are required. The minimum acceptable score for admission is dependent on the applicant's overall record. Official results of the Test of English as a Foreign Language (TOEFL) are required of all applicants whose native language is not English. The minimum score requirement is 250 (computer-based test) or 600 (paper-based test).

Correspondence and Information

Chair, Graduate Admissions Committee
Department of Physics
Boston University
590 Commonwealth Avenue
Boston, Massachusetts 02215
Phone: 617-353-2623
E-mail: dept@physics.bu.edu
Web site: http://physics.bu.edu/

Boston University

THE FACULTY AND THEIR RESEARCH

Steven Ahlen, Ph.D., Berkeley, 1976. Experimental particle physics and astrophysics, ATLAS.
Rama Bansil, Ph.D., Rochester, 1974. Biological physics, polymers.
Irving Bigio, joint appointment with the College of Engineering; Ph.D., Michigan, 1974. Biomedical and biological physics.
Kenneth Brecher, joint appointment with the Department of Astronomy; Ph.D., MIT, 1969. Theoretical astrophysics, relativity, cosmology.
John Butler, Ph.D., Stanford, 1986. Experimental high-energy physics, D0.
David Campbell, joint appointment with the College of Engineering; Ph.D., Cambridge, 1970. Theoretical physics and applied mathematics.
Robert Carey, Ph.D., Harvard, 1989. Experimental high-energy physics, muon g-2.
Antonio H. Castro Neto, Ph.D., Illinois, 1994. Condensed-matter theory.
Claudio Chamon, Ph.D., MIT, 1996. Condensed-matter theory.
Bernard Chasan, Emeritus; Ph.D., Cornell, 1961. Biological physics.
Andrew G. Cohen, Ph.D., Harvard, 1986. Elementary particle physics.
Charles Delisi, joint appointment with the College of Engineering; Ph.D., NYU, 1969. Elementary particle theory.
Alvaro DeRújula, joint appointment with CERN; Ph.D., Madrid, 1968. Theoretical particle physics, phenomenology.
Andrew G. Duffy, Ph.D., Queen's at Kingston, 1995. Physics education research.
Dean S. Edmonds Jr., Emeritus; Ph.D., MIT, 1958. Electronics and instrumentation.
Maged El-Batanouny, Ph.D., California, Davis, 1978. Surface physics, solitons.
Shyamsunder Erramilli, Ph.D., Illinois, 1986. Biological physics.
Evan Evans, joint appointment with the College of Engineering; Ph.D., California, San Diego, 1970. Biological physics.
Roscoe Giles, joint appointment with the College of Engineering; Ph.D., Stanford. Theoretical condensed matter.
Sheldon Glashow, Ph.D., Harvard, 1958. Theoretical particle physics.
Bennett Goldberg, Ph.D., Brown, 1987. Condensed-matter physics.
Ulrich Heintz, Ph.D., SUNY at Stony Brook, 1991. Experimental high-energy physics, D0.
Edward Kearns, Ph.D., Harvard, 1990. Neutrino physics and particle astrophysics, Super-Kamiokande.
William Klein, Ph.D., Temple, 1972. Condensed-matter theory.
Emanuel Katz, Ph.D., MIT, 2001. High-energy theory.
Kenneth D. Lane, Ph.D., Johns Hopkins, 1970. Theoretical high-energy physics.
Karl Ludwig, Ph.D., Stanford, 1986. Experimental condensed-matter physics.
Jerome Mertz, joint appointment with the College of Engineering; Ph.D., Paris VI (joint with California, Santa Barbara), 1991. Biological physics.
James Miller, Ph.D., Carnegie Mellon, 1974. Intermediate- and high-energy experimental physics, muon g-2.
Pritiraj Mohanty, Ph.D., Maryland, College Park, 1998. Experimental condensed-matter physics.
Theodore Moustakas, joint appointment with the College of Engineering; Ph.D., Columbia, 1974. Synthetic novel materials.
Meenakshi Narain, Ph.D., SUNY at Stony Brook, 1991. Experimental high-energy physics, D0.
So-Young Pi, Ph.D., SUNY at Stony Brook, 1974. Field theory, theoretical elementary particle physics.
Anatoli Polkovnikov, Ph.D., Yale, 2003. Condensed-matter theory.
Claudio Rebbi, Ph.D., Turin (Italy), 1967. Theoretical physics, lattice quantum chromodynamics, computational physics.
Sidney Redner, Ph.D., MIT, 1977. Statistical physics, condensed-matter theory.
B. Lee Roberts, Ph.D., William and Mary, 1974. Intermediate- and high-energy experimental physics, muon g-2, CP violation.
James Rohlf, Ph.D., Caltech, 1980. Experimental particle physics, hadron collider physics, CMS.
Kenneth Rothschild, Ph.D., MIT, 1973. Biophysics, molecular electronics, physics of vision.
Anders Sandvik, Ph.D., California, Santa Barbara, 1993. Condensed-matter computational physics.
Martin Schmaltz, Ph.D., California, San Diego, 1995. Theoretical particle physics.
William J. Skocpol, Ph.D., Harvard, 1974. Experimental condensed-matter physics.
Kevin E. Smith, Ph.D., Yale, 1988. Experimental condensed-matter physics.
John Stachel, Emeritus, Curator of Einstein papers in the United States; Ph.D., Stevens, 1952. General relativity, foundations of relativistic space-time theories.
H. Eugene Stanley, Ph.D., Harvard, 1967. Phase transitions, scaling, polymer physics, fractals and chaos.
James L. Stone, Ph.D., Michigan, 1976. Experimental particle physics and astrophysics, neutrinos, proton decay, Super-Kamiokande.
Lawrence R. Sulak, Ph.D., Princeton, 1970. Experimental particle physics, proton decay, monopoles, muon g-2, neutrinos.
Ophelia Tsui, Ph.D., Princeton, 1996. Soft condensed-matter physics.
J. Scott Whitaker, Ph.D., Berkeley, 1976. Experimental colliding-beam physics, supersymmetric particle searches.
Charles R. Willis, Emeritus; Ph.D., Syracuse, 1957. Biophysics, nonlinear physics, statistical physics.
George O. Zimmerman, Emeritus; Ph.D., Yale, 1963. Low-temperature physics, magnetism.

Research Faculty and Staff

Mi Kyung Hong, Ph.D., Illinois, 1988. Experimental biophysics.
James Shank, Ph.D., Berkeley, 1988. High-energy physics.
Christopher Walter, Ph.D., Caltech, 1996. High-energy physics.

CITY COLLEGE OF THE CITY UNIVERSITY OF NEW YORK

Department of Physics

Programs of Study

The Department of Physics offers students the opportunity for study and research leading to the degrees of Doctor of Philosophy (Ph.D.) and Master of Arts (M.A.).

Students in the Ph.D. program usually take a year of graduate courses before the first qualifying examination, although some advanced students take the examination after half a year of course work or even upon entering the program. The examination tests classical mechanics and electromagnetism, quantum theory, and general undergraduate physics. Students entering the biophysics subspecialty are allowed to substitute a biophysics examination for classical mechanics. Sixty credits of course work are normally required for the Ph.D. degree program; advanced students with an M.A. degree can usually transfer 30 credits of previous graduate work. In addition, arrangements are always made so that advanced students meet course requirements by working at their appropriate level in connection with their anticipated thesis research. After passing the qualifying examination, students choose faculty mentors for their thesis research. When student and mentor feel confident of the area of thesis research, the student takes an oral second examination before an appropriately chosen thesis committee. During this examination, the student describes the proposed research and demonstrates familiarity with the physics in the area of research. When students complete their original research, they defend a written thesis before their thesis committee at a final thesis defense.

Students in the M.A. program normally take the qualifying examination after 1 or 1½ years, when they have completed the necessary course work. Students who pass the qualifying examination are often admitted to the Ph.D. program. Students who do not pass the qualifying examination but show satisfactory performance at the master's level are awarded a master's degree when they have completed 30 credits of course work. The M.A. program normally requires 1½ years to complete.

Research Facilities

The physics department is housed on three floors (about 70,000 square feet) of the thirteen-story Marshak Science Building, which also houses the other CCNY science departments. FT-IR, X-ray diffraction, UV-visible spectrometers, ultrafast laser instrumentation in picosecond and femtosecond regimes, and departmental computers are available to students. In addition, high-resolution FT-NMR spectrometers and mass spectrometers are run by operators for any research group. A wide variety of equipment is used by individual research groups, including lasers of many kinds, molecular beam instrumentation, a microwave spectrometer, computers, ultrahigh-vacuum systems for surface studies, two He3-He4 dilution refrigerators, a SQUID-based magnetometer, e-beam evaporators, crystal growing equipment, Raman spectrometers, ultrafast time-resolving instrumentation, and atomic beam systems. The department has an electronics shop, a machine shop, a student machine shop, and a glassblower available for designing and building equipment. The Institute for Ultrafast Spectroscopy and Lasers has eight laboratories in the Science Building and the Engineering Building. The New York State Center for Advanced Technology in Ultrafast Photonic Materials and Applications focuses on photonics research with commercial applications.

Financial Aid

Students accepted into the Ph.D. program are normally offered financial support by the Department of Physics. The support is in the form of fellowships and/or research assistantships, for a total stipend of $16,000 (taxable) per year, plus tuition. The exact amount depends on the student's progress in the program, tuition costs, and need. Some New York State residents are also eligible for other stipends or awards. More advanced students are generally awarded research assistantships.

Cost of Study

Tuition for fall 2005 was $6720 for an entering student ($2860 for New York residents), $3983 for an intermediate-level student ($1793 for New York residents), and $1423 for an advanced student ($710 for New York residents).

Living and Housing Costs

There is new but very limited on-campus housing available at City College. For more information, students should visit http://www.ccny.towers.com. Graduate student housing is available for some students and is run by the City University of New York in midtown Manhattan. Many students live in rooms and apartments throughout New York City, paying $550–$750 per person per month.

Student Group

The total number of graduate students in the physics department is currently about 35. There are about 15 postdoctoral assistants. A wide variety of academic, ethnic, and national backgrounds are represented among these students.

Location

The City College is located in an urban setting in the upper part of Manhattan. The College is part of the City University of New York, which includes eighteen campuses—among them Brooklyn, Hunter, and Queens colleges. Physics research at these other branches of the City University complements that at City College. The College is near many other institutions in the New York metropolitan area, including Columbia University, Rockefeller University, and Polytechnic University of New York, and has cooperative arrangements with Brookhaven National Laboratory on Long Island. A number of world-famous industrial research laboratories are near New York City, including AT&T Bell Laboratories, IBM's Thomas J. Watson Laboratory, RCA's David Sarnoff Laboratory, and the Exxon Research Center.

New York City is a major cultural, artistic, communications, medical, and scientific center with numerous resources and opportunities. The city is also a focus of international travel, and visiting scientists often come to City College as part of their itinerary in the United States.

The College

The City College of the City University of New York is the lineal descendant of the Free Academy of New York City, founded in 1847. City College is the oldest and best-known component of the City University of New York.

Applying

Information and application forms can be obtained from the Department of Physics at the address below. An application fee of $125 must accompany the application, with the exception of international students with financial difficulties, for whom the fee can be deferred until registration.

Correspondence and Information

Chairman
Graduate Admissions Committee
Department of Physics
City College of the City University of New York
New York, New York 10031

Fax: 212-650-6940
E-mail: physdept@sci.ccny.cuny.edu
Web site: http://www.sci.ccny.cuny.edu/physics/

City College of the City University of New York

THE FACULTY AND THEIR RESEARCH

Adolf A. Abrahamson, Professor; Ph.D., NYU. Atomic and nuclear structure, properties of superheavy elements.
Robert R. Alfano, Distinguished Professor; Ph.D., NYU. Ultrafast picosecond and femtosecond laser spectroscopy applied to physical and biological systems: nonlinear optics, optical imaging, medical applications of photonics, laser development.
Joseph L. Birman, Distinguished Professor; Ph.D., Columbia. Theoretical physics: condensed-matter theory; symmetry and symmetry breaking and restoration; optical response of matter, including nonlinear response and response of strongly correlated electronic systems (quantum Hall systems); microscopic theory of high-Tc superconductors; many-body theory, including use of quantum deformed algebras.
Timothy Boyer, Professor; Ph.D., Harvard. Connections between classical and quantum theories: zero-point radiation, stochastic electrodynamics, van der Waals forces, classical electromagnetism.
Ngee-Pong Chang, Professor; Ph.D., Columbia. Unification and dynamical symmetry breaking: origin of mass and chirality, quark-gluon plasma and handedness of the early universe, neutrino mass oscillations.
Victor Chung, Professor; Ph.D., Berkeley. Administration, physics instruction.
Harold Falk, Professor; Ph.D., Washington (Seattle). Statistical mechanics, especially exact results for spin-systems: discrete-time, nonlinear, and stochastic models.
Swapan K. Gayen, Associate Professor; Ph.D., Connecticut. Optical biomedical imaging, tunable solid-state lasers, spectroscopy of impurity ions in solids, ultrafast laser spectroscopy, near-field scanning optical spectroscopy.
Joel Gersten, Professor; Ph.D., Columbia. Solid-state theory: interactions involving small solid-state particles or solid-state surfaces, sonoluminescence.
Daniel M. Greenberger, Professor; Ph.D., Illinois. Fundamental problems in quantum mechanics: the neutron interferometer, coherence in and interpretation of quantum theory, relativistic considerations.
Marilyn Gunner, Professor; Ph.D., Pennsylvania. Experimental and theoretical biophysics: proteins in electron and proton transfer reactions, time-resolved spectroscopic measurements in photosynthesis.
Michio Kaku, Professor; Ph.D., Berkeley. Superstring theory, supersymmetry, supergravity, string field theory, quantum gravity, quantum chromodynamics.
Ronald Koder, Assistant Professor; Ph.D., Johns Hopkins. Computational protein design and nuclear magnetic resonance to test understanding of the fundamental interactions underlying protein folding as well as create new proteins for use in cancer therapies, explosives biosensing and the bioremediation of hazardous wastes.
Joel Koplik, Professor; Ph.D., Berkeley. Molecular dynamics of microscopic fluid flow: transport in disordered systems, superfluid vortex dynamics, pattern selection in nonequilibrium growth processes.
Matthias Lenzner, Associate Professor; Ph.D., Schiller (Germany). Application of ultrafast lasers in biomedical optics, machining, and imaging; femtosecond UV solid-state lasers; quantum cryptography.
Michael S. Lubell, Professor; Ph.D., Yale. Photon-atom interactions, synchrotron radiation studies, polarized electron physics, two-electron systems, science and technology policy.
Herman Makse, Associate Professor; Ph.D., Boston University. Condensed-matter physics, granular materials, nonlinear elasticity, Edwards thermodynamics and jamming, discrete element modeling, interface roughening, porous media, dynamics of urban populations.
Carlos A. Meriles, Assistant Professor; Ph.D., Córdoba (Argentina). Nobel magnetic resonance methods and instruments, hyperpolarization and ultrasensitive detection, optical NMR, low/zero field spectroscopy and imaging, applications to semiconductors and spintronics.
Vangal N. Muthukumar, Associate Professor; Ph.D., Indian Institute of Mathematical Sciences, Madras. Theoretical condensed matter, superconductivity, magnetism, transport phenomena, oxides and the physics of strong correlation.
V. Parameswaran Nair, Professor; Ph.D., Syracuse. Mathematical and topological aspects of quantum field theory: skyrmions, quantum breaking of classical symmetries, conformal field theory, black holes, quantum chromodynamics, interaction of anyons.
Vladimir Petricevic, Professor; Ph.D., CUNY. Growth of solid-state laser materials, laser development, photonics, spectroscopy of ions in solids, ultrafast phenomena.
Alexios Polchronakos, Professor; Ph.D., Caltech. Quantum field theory, mathematical physics.
Myriam P. Sarachik, Distinguished Professor and Chair; Ph.D., Columbia. Low-temperature studies of metal-insulator transitions, Anderson localization, disordered systems, strongly correlated systems; mesoscopic tunneling of magnetization, molecular magnets.
David Schmeltzer, Professor; D.Sc., Technion (Israel). Many-body physics of strongly correlated fermions: Fermi and non-Fermi liquids, Luttinger liquids, fractional quantum Hall effect, renormalization group, bosonization; metal-insulator transition, persistent currents; high-Tc superconductivity.
Mark Shattuck, Associate Professor; Ph.D., Duke. Soft condensed matter, granular media, pattern formation, nonlinear dynamics.
Frederick W. Smith, Professor; Ph.D., Brown. Deposition and characterization of semiconducting and dielectric thin films; modeling of local atomic bonding in amorphous films; chemical vapor deposition of diamond.
Richard N. Steinberg, Professor; Ph.D., Yale. Physics education research.
Jiufeng, J. Tu, Assistant Professor; Ph.D., Cornell. Optical studies of correlated systems and nanosystems, infrared and Raman studies of superconductors and nanosystems.
Sergey A. Vitkalov, Assistant Professor; Ph.D., Russian Academy of Sciences. Experimental condensed-matter physics, dynamical properties of low-dimensional quantum systems.

Professors Emeriti
Michael E. Arons, Joseph Aschner, Alvin Bachman, Arthur Bierman, Robert Callender, Herman Z. Cummins, Erich Erlbach, Paul Harris, Hiram Hart, Martin Kramer, Robert M. Lea, S. J. Lindenbaum, Harry Lustig, William Miller, Marvin Mittleman, Leonard Roellig, David Shelupsky, Harry Soodak, Peter Tea, Martin Tiersten, Chi Yuan.

CLEMSON UNIVERSITY

CLEMSON UNIVERSITY

M.S. in Physics and Astronomy

Program of Study

The M.S. in physics and astronomy program has two options: a thesis option and a nonthesis option. Most students take the thesis option, which requires 24 hours of course work in physics and astronomy, with at least 18 hours at the 800 level or above, plus at least 6 hours of PHYS 891, Research in Physics and Astronomy, which culminates in the writing of a thesis submitted to the Graduate School. The nonthesis option requires 30 hours of course work plus at least 6 hours of PHYS 890, Directed Studies in Physics and Astronomy, which leads to a written paper describing those studies. Both options require a final examination (defense). Students with a normal physics B.S. background are expected to complete the M.S. program in 2 to 2½ years.

Research Facilities

More than $3 million in equipment exists in the department, including tunneling and atomic force microscopes, X-ray equipment, and low-temperature cryostats. Extensive crystal growing equipment for bulk materials, epitaxial materials, and carbon and other nanomaterials is available. Extensive characterization equipment under computer control is available. Two multiparallel computers (Beowulf) are used by the Astrophysics, Biophysics, and Solid State Physics groups. Raman and photoluminescence spectrometers, optical tweezers, and optical absorption apparatus are in place. Telescopes in Arizona and Hawaii are accessible to departmental astronomers.

The Clemson University Planetarium is located in 112 Kinard Laboratory of Physics. The planetarium is equipped with a 24-foot dome; a model A3P Spitz Laboratories, Inc., star projector; two slide projectors; a full sound system; and an online computer with projector.

Financial Aid

Graduate teaching assistantships are the most common type of financial support for incoming graduate students. The duties usually entail teaching two or three undergraduate labs per semester, plus some grading for large courses. The stipend is $13,000 per calendar year with the expectation that students take courses or do research during the summer when they have no specific duties.

Research assistantships may pay a somewhat higher rate than teaching assistantships and are available through research grants and contracts held by faculty members in the department. These are awarded based on availability and the qualifications of the recipients.

R. C. Edwards Fellowships and Alumni Fellowships are awarded by the Graduate School. The department nominates candidates who compete with other nominees throughout the University. These fellowships pay $5000 per academic year in addition to an assistantship granted by the department.

Cost of Study

Tuition for 2006–07 is $4643 per semester for in-state students and $9255 per semester for nonresidents. Off-campus rates are $535 per hour for in-state students and $918 per hour for nonresidents. Graduate assistants pay a flat fee of $1079 per semester and $348 per summer session. Graduate fellows pay South Carolina resident fees.

Living and Housing Costs

On-campus housing is available. For more information, students should visit http://www.housing.clemson.edu. The cost of living in Clemson is quite low compared to the national average. Students who choose to live off campus typically spend $300–$400 per month for rent, depending on such things as location, amenities, and roommates.

Student Group

There are approximately 20 students in the M.S. program. Of those, 75 percent are men, all are full-time students, and 40 percent are international students.

Student Outcomes

Graduates of Clemson's M.S. program have gone on to excel in positions at NASA, national labs, and in various types of industry. Many go on to pursue Ph.D.'s in top-ranked programs. Recent graduates have included one Clemson physicist who now works on Wall Street and one top graduate in general relativity theory who is now the premier network analyst for a power company.

Location

Clemson is a small, beautiful college town near the Blue Ridge Mountains and Lake Hartwell. The Upstate is one of the country's fastest-growing areas and is an important part of the I-85 corridor, a multistate area along Interstate 85 that runs from metropolitan Atlanta to Richmond, Virginia, and encompasses Charlotte, North Carolina, and North Carolina's Research Triangle. Atlanta and Charlotte are each a 2-hour drive away. Many financial institutions and other industries have a national or major presence in the Upstate, including Wachovia, Bank of America, BMW, Bon Secours St. Francis Health System, Bosch North America, Bowater, Charter Communications, Ernst & Young, Fluor Corporation, IBM, Microsoft, Michelin of North America, and many others.

The University

Clemson is classified by the Carnegie Foundation as Doctoral/Research University–Extensive, a category comprising less than 4 percent of all universities in America. The University's mission is to fulfill the covenant between its founder and the people of South Carolina to establish a "high seminary of learning" through its responsibilities of teaching, research, and extended public service. The University has identified eight areas of academic emphasis that create collaborations that, in turn, help fulfill the University's mission.

Applying

Applicants to the M.S. in Physics and Astronomy program must have top scores on the GRE and strong letters of recommendation. Applicants may apply on the Web at http://www.grad.clemson.edu/p_apply.html. Applications with a $50 nonrefundable fee should be received no later than five weeks prior to registration. Every required item in support of the application must be on file by that date. Students are advised to contact the department for the deadlines of the program of proposed study.

Correspondence and Information

Wendy May, Student Coordinator
118 Kinard Laboratory
Department of Physics and Astronomy
Clemson University
Clemson, South Carolina 29634-0978
Phone: 864-656-3418
E-mail: wmay@clemson.edu
Web site: http://physicsnt.clemson.edu

Dr. Miguel Larsen, Graduate Acceptance Committee Chair
Phone: 864-656-5309
E-mail: mlarson@clemson.edu

Clemson University

THE FACULTY AND THEIR RESEARCH

Peter A. Barnes, Department Chair; Ph.D., Waterloo, 1969. Engineering physics.
Donald D. Clayton, Professor; Ph.D., Caltech, 1962. Physics.
Murray S. Daw, Named Professor; Ph.D., Caltech, 1981. Physics.
Phillip J. Flower, Associate Professor; Ph.D., Washington (Seattle), 1976. Astronomy.
Dieter H. Hartmann, Associate Professor; Ph.D., California, Santa Cruz, 1989. Astronomy and astrophysics.
Huabei Jiang, Professor; Ph.D., Dartmouth, 1995. Physics.
Pu-Chun Ke, Assistant Professor; Ph.D., Victoria (Australia), 2000. Physics.
Jeremy King, Associate Professor; Ph.D., Hawaii, 1993. Astronomy.
Lyndon L. Larcom, Professor; Ph.D., Pittsburgh, 1968. Biophysics.
Miguel F. Larsen, Professor; Ph.D., Cornell, 1979. Atmospheric science.
Gerald A. Lehmacher, Assistant Professor; Ph.D., Bonn (Germany), 1993. Physics.
Mark D. Leising, Professor; Ph.D., Rice, 1987. Physics and astronomy.
Joseph R. Manson, Professor; Ph.D., Virginia, 1969. Physics.
Domnita Catalina Marinescu, Assistant Professor; Ph.D., Purdue, 1996. Physics.
Peter J. McNulty, Professor; Ph.D., SUNY, 1965. Physics.
John W. Meriwether Jr., Professor; Ph.D., Maryland, 1970. Chemical physics.
Bradley S. Meyer, Professor; Ph.D., Chicago, 1989. Astrophysics.
J. Bruce Rafert, Professor; Ph.D., Florida, 1979. Physics.
Apparao M. Rao, Professor; Ph.D., Kentucky, 1989. Physics.
Chad E. Sosolik, Assistant Professor; Ph.D., Cornell, 2001. Physics.
Terry M. Tritt, Professor; Ph.D., Clemson, 1985. Physics.

CLEMSON UNIVERSITY

CLEMSON UNIVERSITY

Ph.D. in Physics and Astronomy

Program of Study

The doctoral program in physics and astronomy simply requires students to gain admission to Ph.D. candidacy and write and defend a dissertation. Admission to candidacy requires a degree in physics or closely related subject, passing three written qualifying examinations, and passing an oral qualifying examination

There are no specific course requirements for obtaining a Doctor of Philosophy in physics at Clemson. The requirements are that the student successfully passes the following in the five-year period prior to graduation: the Ph.D. qualifying exam, defense of the dissertation, and approval of the dissertation by the Graduate School.

Research Facilities

More than $3 million in equipment exists in the department, including tunneling and atomic force microscopes, X-ray equipment, and low-temperature cryostats. Extensive crystal growing equipment for bulk materials, epitaxial materials, and carbon and other nanomaterials is available. Extensive characterization equipment under computer control is available. Two multiparallel computers (Beowulf) are used by the Astrophysics, Biophysics, and Solid State Physics groups. Raman and photoluminescence spectrometers, optical tweezers, and optical absorption apparatus are in place. Telescopes in Arizona and Hawaii are accessible to departmental astronomers.

The Clemson University Planetarium is located in 112 Kinard Laboratory of Physics. The planetarium is equipped with a 24-foot dome; a model A3P Spitz Laboratories, Inc., star projector; two slide projectors; a full sound system; and an online computer with projector.

Financial Aid

Graduate teaching assistantships are the most common type of financial support for incoming graduate students. The duties usually entail teaching two or three undergraduate labs per semester, plus some grading for large courses. The stipend is $13,000 per calendar year with the expectation that students take courses or do research during the summer when they have no specific duties.

Research assistantships may pay a somewhat higher rate than teaching assistantships and are available through research grants and contracts held by faculty members in the department. These are awarded based on availability and the qualifications of the recipients.

R. C. Edwards Fellowships and Alumni Fellowships are awarded by the Graduate School. The department nominates candidates who compete with other nominees throughout the University. These fellowships pay $5000 per academic year in addition to an assistantship granted by the department.

Cost of Study

Tuition for 2006–07 is $4643 per semester for in-state students and $9255 per semester for nonresidents. Off-campus rates are $535 per hour for in-state students and $918 per hour for nonresidents. Graduate assistants pay a flat fee of $1079 per semester and $348 per summer session. Graduate fellows pay South Carolina resident fees.

Living and Housing Costs

On-campus housing is available. For more information, students should visit http://www.housing.clemson.edu. The cost of living in Clemson is quite low compared to the national average. Students who choose to live off campus typically spend $300–$400 per month for rent, depending on such things as location, amenities, and roommates.

Student Group

There are approximately 33 students in the Ph.D. program. Of those, 73 percent are men, 85 percent are full-time students, and 39 percent are international students.

Student Outcomes

Graduates of Clemson's Ph.D. program have gone on to excel in positions at NASA, national labs, and in various types of industry. Recent graduates have included one Clemson physicist who now works on Wall Street and one top graduate in general relativity theory who is now the premier network analyst for a power company.

Location

Clemson is a small, beautiful college town near the Blue Ridge Mountains and Lake Hartwell. The Upstate is one of the country's fastest-growing areas and is an important part of the I-85 corridor, a multistate area along Interstate 85 that runs from metropolitan Atlanta to Richmond, Virginia, and encompasses Charlotte, North Carolina, and North Carolina's Research Triangle. Atlanta and Charlotte are each a 2-hour drive away. Many financial institutions and other industries have a national or major presence in the Upstate, including Wachovia, Bank of America, BMW, Bon Secours St. Francis Health System, Bosch North America, Bowater, Charter Communications, Ernst & Young, Fluor Corporation, IBM, Microsoft, Michelin of North America, and many others.

The University

Clemson is classified by the Carnegie Foundation as Doctoral/Research University–Extensive, a category comprising less than 4 percent of all universities in America. The University's mission is to fulfill the covenant between its founder and the people of South Carolina to establish a "high seminary of learning" through its responsibilities of teaching, research, and extended public service. The University has identified eight areas of academic emphasis that create collaborations that, in turn, help fulfill the University's mission.

Applying

Applicants may apply on the Web at http://www.grad.clemson.edu/p_apply.html. Applications with a $50 nonrefundable fee should be received no later than five weeks prior to registration. Every required item in support of the application must be on file by that date. Students are advised to contact the department for the deadlines of the program of proposed study.

Correspondence and Information

Wendy May, Student Coordinator
118 Kinard Laboratory
Department of Physics and Astronomy
Clemson University
Clemson, South Carolina 29634-0978
Phone: 864-656-3418
E-mail: wmay@clemson.edu
Web site: http://physicsnt.clemson.edu

Dr. Miguel Larsen, Graduate Acceptance Committee Chair
Phone: 864-656-5309
E-mail: mlarson@clemson.edu

Clemson University

THE FACULTY AND THEIR RESEARCH

Peter A. Barnes, Department Chair; Ph.D., Waterloo, 1969. Engineering physics.
Donald D. Clayton, Professor; Ph.D., Caltech, 1962. Physics.
Murray S. Daw, Named Professor; Ph.D., Caltech, 1981. Physics.
Phillip J. Flower, Associate Professor; Ph.D., Washington (Seattle), 1976. Astronomy.
Dieter H. Hartmann, Associate Professor; Ph.D., California, Santa Cruz, 1989. Astronomy and astrophysics.
Huabei Jiang, Professor; Ph.D., Dartmouth, 1995. Physics.
Pu-Chun Ke, Assistant Professor; Ph.D., Victoria (Australia), 2000. Physics.
Jeremy King, Associate Professor; Ph.D., Hawaii, 1993. Astronomy.
Lyndon L. Larcom, Professor; Ph.D., Pittsburgh, 1968. Biophysics.
Miguel F. Larsen, Professor; Ph.D., Cornell, 1979. Atmospheric science.
Gerald A. Lehmacher, Assistant Professor; Ph.D., Bonn (Germany), 1993. Physics.
Mark D. Leising, Professor; Ph.D., Rice, 1987. Physics and astronomy.
Joseph R. Manson, Professor; Ph.D., Virginia, 1969. Physics.
Domnita Catalina Marinescu, Assistant Professor; Ph.D., Purdue, 1996. Physics.
Peter J. McNulty, Professor; Ph.D., SUNY, 1965. Physics.
John W. Meriwether Jr., Professor; Ph.D., Maryland, 1970. Chemical physics.
Bradley S. Meyer, Professor; Ph.D., Chicago, 1989. Astrophysics.
J. Bruce Rafert, Professor; Ph.D., Florida, 1979. Physics.
Apparao M. Rao, Professor; Ph.D., Kentucky, 1989. Physics.
Chad E. Sosolik, Assistant Professor; Ph.D., Cornell, 2001. Physics.
Terry M. Tritt, Professor; Ph.D., Clemson, 1985. Physics.

COLUMBIA UNIVERSITY

Department of Applied Physics and Applied Mathematics

Programs of Study

The Department of Applied Physics and Applied Mathematics offers graduate study leading to the degrees of Master of Science (M.S), Doctor of Engineering Science (Eng.Sc.D.), and Doctor of Philosophy (Ph.D.).

The following fields of research (topics of emphasis in parentheses) are available for doctoral study: theoretical and experimental plasma physics (fusion and space plasmas), applied mathematics (analysis of partial differential equations, large-scale scientific computing, nonlinear dynamics, inverse problems, geophysical/geological fluid dynamics, and biomathematics), solid-state physics (semiconductor, surface, and low-dimensional physics), optical and laser physics (free-electron lasers and laser interactions with matter), nuclear science (medical applications), earth science (atmosphere, ocean, and climate science and geophysics), and materials science and engineering (thin films; nanomaterials; electronic, optical, and magnetic materials; and mechanical response of materials). Successful completion of 30 points (semester hours) or more of approved graduate course work beyond the master's degree is required for the doctoral degree. Candidates must pass written and oral qualifying exams and successfully defend an approved dissertation based on original research. For the M.S. degree, candidates must successfully complete a minimum of 30 points of credit of approved graduate course work at Columbia. A 35-point M.S. degree in medical physics is offered in collaboration with faculty members from the College of Physicians and Surgeons. It prepares students for careers in medical physics and provides preparation for the ABMP certification exam.

Research Facilities

Research equipment in the Plasma Physics Laboratory includes a toroidal high-beta tokamak for basic and applied research, a steady-state plasma experiment using a linear magnetic mirror, a large laboratory collisionless terrella used to investigate space plasma physics, and a stellarator for nonneutral and antimatter plasma research. The plasma physics group is jointly operating a new plasma confinement experiment, LDX, with MIT, incorporating a levitated superconducting ring. The plasma physics group is also actively involved in the NSTX experiment at the Princeton Plasma Physics Laboratory and on the DIII-D Tokamak at General Atomics in San Diego and is part of the U.S. national effort on the ITER project. Research equipment in the solid-state physics and quantum electronics laboratories includes extensive laser and spectroscopy facilities, a microfabrication laboratory, ultra high-vacuum surface preparation and analysis chambers, direct laser writing stations, a molecular beam epitaxy machine, picosecond and femtosecond lasers, and diamond anvil cells. Research is also conducted in the shared characterization laboratories and clean room operated by the NSF Materials Research Science and Engineering Center and the NSF Nanoscale Science and Engineering Center. Materials science and engineering facilities include transmission and scanning electron microscopes, scanning-tunneling microscopes and atomic force microscopes, X-ray diffractometers, ellipsometer, X-ray photoelectron spectrometer, laser processing equipment, and mechanical testing equipment. Magnetic and electrical measurement characterization equipment is also available.

There are research opportunities in medical physics at the Columbia–Presbyterian Medical Center, as well as at other medical institutes, employing state-of-the-art medical diagnostic imaging and treatment equipment.

The Applied Mathematics Division is closely linked with the Lamont Doherty Earth Observatory (LDEO), with 5 faculty members sharing appointments in the Department of Earth and Environmental Sciences and with the NASA Goddard Institute for Space Studies (GISS). There are also close ties with Columbia's Center for Computational Biology and Biomathemetics (C2B2) and Columbia's Center for Computational Learning Systems (CLASS).

The Department maintains an extensive network of workstations and desktop computers. The research of the plasma physics group is supported by a dedicated data acquisition/data analysis system. Computational researchers have local access to Columbia's 256-processor Linux cluster, to the 10-Tflop/s Columbia-Brookhaven QCDOC computer, and to IBM SP and Cray X-1 systems at the National Center for Atmospheric Research and the Lawrence Berkeley and Oak Ridge National Laboratories, in the 3 Tflop/s to 10 Tflop/s performance range.

Financial Aid

Financial support is awarded on a competitive basis in the form of assistantships that provide a stipend, a tuition allowance, and medical fees. For 2006–07, the stipend for teaching assistants is $20,063 for nine months; for research assistants, the stipend is $26,750 for twelve months.

Cost of Study

For 2006–07, full-time tuition for the academic year is $33,660; for part-time study, the cost is $1122 per credit. In addition to medical fees (approximately $2150), annual fees are $325–$425.

Living and Housing Costs

The cost of on-campus, single-student housing (dormitories, suites, and apartments) ranges from $3000 to $4500 per term; married student accommodations range from $1000 to $1600 per month. For the single student, a minimum of $20,000 should be allowed for board, room, and personal expenses for the academic year.

Student Group

Approximately 23,000 students attend the fifteen schools and colleges of Columbia University; more than half are graduate students. On average, the Department has 120 graduate and 90 undergraduate students. The student population has a diverse and international character. Admission is highly competitive; in 2005–06, 22 percent of the application pool of 274 was admitted.

Student Outcomes

Recent Ph.D. recipients have found employment as postdoctoral research scientists at universities in the United States and abroad and as staff members in advanced technology industries and at national laboratories. Some have secured college-level faculty positions. Most M.S. graduates continue studying for the doctorate; a few go on to medical school or law school. M.S. graduates from the program in medical physics have secured positions in hospital departments of radiology and nuclear medicine or have entered doctoral programs.

Location

The 32-acre campus is situated in Morningside Heights on the Upper West Side of Manhattan. This location, 15 minutes from the heart of New York City, allows Columbia to be an integral part of the city while maintaining the character of a unique neighborhood. Cultural, recreational, and athletic opportunities abound at city museums, libraries, concert halls, theaters, restaurants, stadiums, parks, and beaches.

The University and The Department

With extensive resources and an outstanding faculty, Columbia University has played an eminent role in American education since its founding in 1754. The Department of Applied Physics and Applied Mathematics, a department at the forefront of interdisciplinary research and teaching, was established in 1978 as part of the Graduate School of Arts and Sciences and the Fu Foundation School of Engineering and Applied Sciences. The Graduate Program in Materials Science and Engineering joined the Department in fall 2000.

Applying

For fall admission, applications should be submitted as follows: December 15 for doctoral, doctoral-track, and all financial aid applicants; applications for Master of Science, part-time, and nondegree candidates are reviewed on a rolling basis. Scores from the GRE General Test are required; GRE Subject Test scores are strongly urged. TOEFL scores are required for students from non-English-speaking countries.

Correspondence and Information

Chairman, Graduate Admissions Committee
200 S. W. Mudd Building, MC 4701
Columbia University
New York, New York 10027

Phone: 212-854-4457
E-mail: seasinfo.apam@columbia.edu
Web site: http://www.apam.columbia.edu

Columbia University

THE FACULTY AND THEIR RESEARCH

In the Department of Applied Physics and Applied Mathematics, theoretical and experimental research is conducted by 30 full-time faculty members, 13 adjunct professors, and 56 research scientists. Areas of research include applied mathematics, earth/atmosphere/ocean science, biomathematics, biophysics, numerical analysis, inverse problems, space physics, surface physics, condensed-matter physics, electromagnetism, materials science, nanoscience, medical physics, optical and laser physics, plasma physics, and fusion energy science.

William Bailey, Assistant Professor; Ph.D., Stanford, 1999. Nanoscale magnetic films and heterostructures, materials issues in spin-polarized transport, materials engineering of magnetic dynamics.

Guillaume Bal, Associate Professor; Ph.D., Paris, 1997. Applied mathematics, wave propagation in random media and applications to time reversal, inverse problems with applications to medical imaging and earth science.

Harish S. Bhat, Assistant Professor; Ph.D., Caltech, 2005. Applied mathematics, Hamiltonian/Lagrangian dynamical systems, partial differential equations and numerical analysis with applications to wave phenomena in both discrete and continuous media.

Allen H. Boozer, Professor; Ph.D., Cornell, 1970. Plasma theory, theory of magnetic confinement for fusion energy, nonlinear dynamics.

Mark A. Cane, Professor (joint with Earth and Environmental Sciences); Ph.D., MIT, 1975. Climate dynamics, physical oceanography, geophysical fluid dynamics, computational fluid dynamics.

Siu-Wai Chan, Professor; Sc.D., MIT, 1985. Nanoparticles, electronic ceramics, grain boundaries and interfaces, oxide thin films.

C. K. Chu, Professor Emeritus; Ph.D., NYU (Courant), 1959. Applied mathematics.

Anthony Del Genio, Adjunct Professor (NASA Goddard Institute for Space Studies); Ph.D., UCLA, 1978. Dynamics of planetary atmospheres, parameterization of cloud and cumulus convection, climate change, general circulation.

Morton B. Friedman, Professor (joint with Civil Engineering); D.Sc., NYU, 1953. Applied mathematics and mechanics, numerical analysis, parallel computing.

Irving P. Herman, Professor; Ph.D., MIT, 1977. Nanocrystals, optical spectroscopy of nanostructured materials, laser diagnostics of thin-film processing, physics of solids at high pressure, plasma processing of materials.

James Im, Professor; Ph.D., MIT, 1985. Laser-induced crystallization of thin films, phase transformations and nucleation in condensed systems.

David E. Keyes, Professor; Ph.D., Harvard, 1984. Applied and computational mathematics for PDEs, computational science, parallel numerical algorithms, parallel performance analysis, PDE-constrained optimization.

Thomas C. Marshall, Professor; Ph.D., Illinois, 1960. Accelerator concepts, relativistic beams and radiation, free-electron lasers.

Michael E. Mauel, Professor; Sc.D., MIT, 1983. Plasma physics, waves and instabilities, fusion and equilibrium control; space physics; plasma processing.

Gerald A. Navratil, Professor; Ph.D., Wisconsin–Madison, 1976. Plasma physics, plasma diagnostics, fusion energy science.

Gertrude Neumark, Professor; Ph.D., Columbia, 1979. Materials science and physics of semiconductors, with emphasis on optical and electrical properties of wide bandgap semiconductors and their light-emitting devices.

I. Cevdet Noyan, Professor; Ph.D., Northwestern, 1984. Characterization and modeling of mechanical and micromechanical deformation, residual stress analysis and nondestructive testing, X-ray and neutron diffraction, microdiffrication analysis.

Stephen O'Brien, Associate Professor; D.Phil., Oxford, 1998. Inorganic materials science, synthesis of novel nanocrystals, molecule-based nanoscale design.

Richard M. Osgood, Professor (joint with Electrical Engineering); Ph.D., MIT, 1973. Nanoscale optical and electronic phenomena (experimental and computational), femtosecond lasers and laser probing, low-dimensional physics, integrated optics, nanofabrication and materials growth.

Thomas S. Pedersen, Associate Professor; Ph.D., MIT, 2000. Plasma physics, magnetic confinement, fusion energy, nonneutral plasmas, positron-electron plasmas, plasma turbulence.

Aron Pinczuk, Professor; Ph.D., Pennsylvania, 1969. Spectroscopy of semiconductors and insulators, quantum structures and interfaces, electrons in systems of reduced dimensions, electron quantum fluids.

Lorenzo M. Polvani, Professor; Ph.D., MIT, 1988. Atmospheric, oceanic, and planetary science; geophysical fluid dynamics; computational fluid mechanics.

Malvin A. Ruderman, Professor (joint with Physics); Ph.D., Caltech, 1951. Theoretical astrophysics, neutron stars, pulsars, early universe, cosmic gamma rays.

Christopher H. Scholz, Professor (joint with Earth and Environmental Sciences); Ph. D., MIT, 1967. Experimental and theoretical rock mechanics, especially friction, fracture, and hydraulic transport properties; nonlinear systems; mechanics of earthquakes and faulting.

Amiya K. Sen, Professor (joint with Electrical Engineering); Ph.D., Columbia, 1963. Plasma physics, fluctuations and anomalous transport in plasmas, control of plasma instabilities.

Adam Sobel, Associate Professor; Ph.D., MIT, 1998. Atmospheric science, geophysical fluid dynamics, tropical meteorology, climate dynamics.

Marc Spiegelman, Associate Professor; Ph.D., Cambridge, 1989. Coupled fluid/solid mechanics, reactive fluid flow, solid earth and magma dynamics, scientific computation/modeling.

Horst Stormer, Professor; Ph.D., Stuttgart, 1977. Semiconductors, electronic transport, lower-dimensional physics, transport in nanostructures.

Wen I. Wang, Professor (joint with Electrical Engineering); Ph.D., Cornell, 1981. Heterostructure devices and physics, materials properties, molecular beam epitaxy.

Michael I. Weinstein, Professor; Ph.D., NYU (Courant), 1982. Applied mathematics; nonlinear partial differential equations; analysis and dynamical systems; waves in nonlinear, inhomogenous, and random media; multiscale phenomena; applications to nonlinear optics; earth science; fluid dynamics; mathematical physics.

Chris H. Wiggins, Assistant Professor; Ph.D., Princeton, 1998. Applied mathematics, mathematical biology, biopolymer dynamics, soft condensed matter, genetic networks and network inference, machine learning.

Cheng-Shie Wuu, Associate Professor (Public Health, Environmental Health Sciences, and Applied Physics); Ph.D., Kansas, 1985. Microdosimetry, biophysical modeling, dosimetry of brachytherapy, gel dosimetry, second cancers induced by radiotherapy.

The Schapiro Center for Engineering and Physical Science Research; to the right, the Seeley W. Mudd Building, home of the Fu Foundation School of Engineering and Applied Science.

Faculty, research staff, and students of the Plasma Physics Laboratory in front of the Tokamak, HBT-EP.

Low Memorial Library and grounds.

SELECTED PUBLICATIONS

Scheck, C., L. Cheng, and **W. E. Bailey.** Low Ghz loss in sputtered epitaxial Fe. *Appl. Phys. Lett.*, in press.

Bailey, W. E., S. E. Russek, X.-G. Zhang, and W. H. Butler. Experimental separability of channeling GMR in Co/Cu/Co. *Phys. Rev. B* 72:012409, 2005.

Bailey, W. E., et al. Precessional dynamics of elemental moments in a ferromagnetic alloy. *Phys. Rev. B* 70:172403, 2004.

Reidy, S., L. Cheng, and **W. E. Bailey.** Dopants for independent control of precessional frequency and damping in $Ni_{81}Fe_{19}$(50 nm). *Appl. Phys. Lett.* 82(8): 1254–6, 2003.

Bal, G., and L. Ryzhik. Time reversal and refocusing in random media. *SIAM J. Appl. Math.* 63(5):1475–98, 2003.

Bal, G. Transport through diffusive and nondiffusive regions, embedded objects, and clear layers. *SIAM J. Appl. Math.* 62(5):1677–97, 2002.

Bal, G., G. Papanicolaou, and L. Ryzhik. Radiative transport limit for the random Schroedinger equation. *Nonlinearity* 15:513–29, 2002.

Bal, G. Inverse problems for homogeneous transport equations, Parts I and II. *Inverse Problems* 16:997–1028, 2000.

Boozer, A. H. Density limit for electron plasmas confined by magnetic surfaces. *Phys. Plasmas* 12:104502, 2005.

Boozer, A. H. Plasma effects on the location of the outermost magnetic surface. *Phys. Plasmas* 12:092504, 2005.

Maslovsky, D. A., and **A. H. Boozer.** Effective plasma inductance computation. *Phys. Plasmas* 12:042108, 2005.

Boozer, A. H. Physics of magnetically confined plasmas. *Rev. Modern Phys.* 76: 1071–141, 2004.

Mann, M. E., et al. **(M. A. Cane).** Volcanic and solar forcing of El Niño over the past 1000 years. *J. Climate,* in press.

Cane, M. A. The evolution of El Niño, past and future. *Earth Planet. Sci. Lett.* 104:1–10, 2005.

Chen, D., et al. **(M. A. Cane).** Predictability of El Niño over the past 148 years. *Nature* 428:733–6, 2004.

Cane, M. A., and P. Molnar. Closing of the Indonesian Seaway as a precursor to East African aridification around 3 to 4 million years ago. *Nature* 411:157–62, 2001.

Mei, L., and **S.-W. Chan.** Enthalpy and entropy of twin boundaries in superconducting $YBa_2Cu_3O_{7-x}$. *J. Appl. Phys.* 98:033908–16, 2005.

Tang, J., et al. **(S.-W. Chan).** Martensitic phase transformation of isolated HfO_2, ZrO_2, and $Hf_xZr_{1-x}O_2$ (0<x<1) nanocrystals. *Adv. Mater.* 15:1595–602, 2005.

Zhang, F., et al. **(S.-W. Chan).** Cerium oxidation state in ceria nanoparticles using X-ray photoelectron spectroscopy and X-ray absorption near-edge spectroscopy. *Surf. Sci.* 563:74–82, 2004.

Zhang, F., et al. **(S.-W. Chan).** Nanoparticles: Size, size distribution and shape. *J. Appl. Phys.* 95:4319–26, 2004.

Chefter, J. G., **C. K. Chu,** and E. E. Keyes. Domain decomposition for shallow water equations. In *Contemporary Mathematics, Proceedings of the 7th International Conference on Domain Decomposition Methods in Science and Engineering,* October 1993.

Yin, F. L., I. Y. Fung, and **C. K. Chu.** Equilibrium response of ocean deep-water circulation to variations in Ekman pumping and deep-water sources. *J. Phys. Oceanogr.* 22:1129, 1992.

Chaiken, J., **C. K. Chu,** M. Tabor, and Q. M. Tan. Lagrangian turbulence in Stokes flow. *Phys. Fluids* 30:687, 1987.

Chu, C. K., L. W. Xiang, and Y. Baransky. Solitary waves generated by boundary motion. *Comm. Pure Appl. Math.* 36:495, 1983.

Del Genio, A. D., W. Kovari, M.-S. Yao, and J. Jonas. Cumulus microphysics and climate sensitivity. *J. Clim.* 18:2376–87, 2005.

Del Genio, A. D., A. B. Wolf, and M.-S. Yao. Evaluation of regional cloud feedbacks using single-column models. *J. Geophys. Res.* 110:D15S13, doi:10.1029/2004JD005011, 2005.

Porco, C. C., et al. **(A. D. Del Genio).** Cassini imaging science: Initial results on Saturn's atmosphere. *Science* 307:1243–7, 2005.

Yao, M.-S., and **A. D. Del Genio.** Effects of cloud parameterization on the simulation of climate changes in the GISS GCM, Part II: Sea surface temperature and cloud feedbacks. *J. Climate* 15:2491, 2002.

Banerjee, S., et al. **(I. P. Herman).** Raman microprobe analysis of elastic strain and fracture in electrophoretically deposited CdSe nanocrystal films. *Nano Lett.* 6:175–80, 2006.

Robinson, R. D., et al. **(I. P. Herman).** Raman scattering in $Hf_xZr_{1-x}O_2$ nanoparticles. *Phys. Rev. B* 71:115408, 2005.

Islam, M. A., and **I. P. Herman.** Electrodeposition of patterned CdSe nanocrystal films using thermally charged nanocrystals. *Appl. Phys. Lett.* 80:3823, 2002.

Herman, I. P. *Optical Diagnostics for Thin Film Processing.* San Diego, Calif.: Academic Press, 1996.

Leonard, J. P., and **J. S. Im.** Stochastic modeling of solid nucleation in supercooled liquids. *Appl. Phys. Lett.* 78:3454–6, 2001.

Im, J. S., V. V. Gupta, and M. A. Crowder. On determining the relevance of athermal nucleation in rapidly quenched liquids. *Appl. Phys. Lett.* 72:662,1998.

Sposili, R. S., and **J. S. Im.** Sequential lateral solidification of thin silicon films on SiO_2. *Appl. Phys. Lett.* 69:2864, 1996.

Im, J. S., H. J. Kim, and M. O. Thompson. Phase transformation mechanisms involved in excimer laser crystallization of amorphous silicon films. *Appl. Phys. Lett.* 63:1969–71, 1993.

Knoll, D. A., and **D. E. Keyes.** Jacobian-free Newton-Krylov methods: A survey of approaches and application. *J. Comp. Phys.* 193:357, 2004.

Keyes, D. E., ed. *A Science-Based Case for Large-Scale Simulation.* U.S. Department of Energy Office of Science, 2003, http://www.pnl.gov/scales.

Coffey, T. S., et al. **(D. E. Keyes).** Pseudotransient continuation and differential-algebraic equations. *SIAM J. Sci. Comp.* 25:553–69, 2003.

Keyes, D. E. Domain decomposition methods in the mainstream of computational science. In *Proceedings of the 14th International Conference on Domain Decomposition Methods,* pp. 79–93. Mexico City: UNAM Press, 2003.

Shchelkunov, S. V., and **T. C. Marshall** et al. Experimental observation of constructive superposition of wake fields generated by electron bunches in a dielectric-lined waveguide. *Phys. Rev. Sci. Tech.* 9:011301, 2006.

Wang, C., **T. C. Marshall,** V. P. Yakovlev, and J. L. Hirshfield. Rectangular dielectric-lined two-beam accelerator structure. In *Particle Accelerator Conference Proceedings,* May 2005.

Shchelkunov, S. V., **T. C. Marshall,** J. L. Hirshfield, and M. A. LaPointe. Nondestructive diagnostic for electron bunch length in accelerators using the wake field radiation spectrum. *Phys. Rev. Sci. Tech.* 8:062801, 2005.

Wang, C., et al. **(T. C. Marshall).** Strong wake fields generated by a train of femtosecond bunches in a planar dielectric microstructure. *Phys. Rev. Special Top.–Accelerators Beams* 7:05130, 2004.

Levitt, B., D. Maslovsky, and **M. E. Mauel.** Observation of centrifugally driven interchange instabilities in a plasma confined by a magnetic dipole. *Phys. Rev. Lett.* 94:175002, 2005.

Shilov, M., et al. **(M. E. Mauel).** Dynamical plasma response of resistive wall modes to changing magnetic perturbations. *Phys. Plasmas* 11:2573–9, 2004.

Kesner, J., et al. **(M. E. Mauel).** Helium-catalyzed D-D fusion in a levitated dipole. *Nucl. Fusion* 44:193–203, 2004.

Maslovsky, D., B. Levitt, and **M. E. Mauel.** Observation of nonlinear frequency-sweeping suppression with rf diffusion. *Phys. Rev. Lett.* 90:185001, 2003.

Reimerdes, H., et al. **(G. A. Navratil).** Measurement of resistive wall mode stability in rotating high-beta DIII-D plasmas. *Nucl. Fusion* 45:368, 2005.

LaHaye, R. J., et al. **(G. A. Navratil).** Scaling of the critical plasma rotation for stabilization of the n+1 RWM in DIII-D. *Nucl. Fusion* 44:1197, 2004.

Garofalo, A. M., et al. **(G. A. Navratil).** Sustained rotational stabilization of DIII-D plasmas above the no-wall beta limit. *Phys. Plasmas* 9:1997, 2002.

Bialek, James, et al. **(G. A. Navratil).** Modeling of active control of external MHD instabilities. *Phys. Plasmas* 8:2170, 2001.

Gu, Y., et al. **(G. F. Neumark).** Determination of size and composition of optically active CdZnSe/ZnBeSe quantum dots. *Appl. Phys. Lett.* 83:3779, 2003.

Kuskovsky, I., **G. F. Neumark,** V. N. Bondarer, and P. V. Pikhitsa. Decay dynamics in disordered systems: Application to heavily doped semiconductors. *Phys. Rev. Lett.* 80:2413, 1998.

Neumark, G. F. Defects In wide-bandgap II-VI crystals. *Mater. Sci. Eng. Rep. R* 21:(1), 1997.

Neumark, G. F. Wide-bandgap light-emitting device materials and doping problems. *Mater. Lett.* 30:131, 1997 (published as materials update).

Hanfei, Y., and **I. C. Noyan.** Dynamical diffraction artifacts in Laue microdiffraction images. *J. Appl. Phys.* 98:073527, 2005.

Murray, C. E., et al. **(I. C. Noyan).** High-resolution strain mapping in heteroepitaxial thin-film features. *J. Appl. Phys.* 98:013504, 2005.

Murray, C. E., C. C. Goldsmith, and **I. C. Noyan.** Spatially transient stress effects in thin films by X-ray diffraction. *Powder Diffr.* 20:112, 2005.

Noyan, I. C., C. E. Murray, J. S. Chey, and C. C. Goldsmith. Finite size effects in stress analysis of interconnect structures. *Appl. Phys. Lett.* 85:724, 2004.

Turro, N. J., et al. **(S. O'Brien).** Spectroscopic probe of the surface of iron oxide nanocrystals. *Nano Lett.* 2(4):325–8, 2002.

O'Brien, S., et al. Synthesis and characterization of nanocrystals of barium titanate, towards a generalized synthesis of oxide nanoparticles. *J. Am. Chem. Soc.* 123: 12085–6, 2001.

O'Brien, S., et al. Time-resolved in situ X-ray powder diffraction study of the formation of mesoporous silicates. *Chem. Mater.* 11:1822–32, 1999.

Zhang, S., et al. **(R. M. Osgood Jr.).** Experimental demonstration of near-infrared negative-index metamaterials. *Phys. Rev. Lett.* 95:137404, 2005.

Song, Z., J. Hrbek, and **R. M. Osgood Jr.** Formation of TiO_2 nanoparticles by reactive-layer-assisted deposition and characterization by XPS and STM. *Nano Lett.* 5:1357, 2005.

Smadici, S., and **R. M. Osgood Jr.** Image-state electron scattering on flat Ag/Pt (111) and stepped Ag/Pt(997) surfaces. *Phys. Rev. B* 71:165424, 2004.

Djukic, D., et al. **(R. M. Osgood Jr.).** Low-voltage planar-waveguide electrooptic prism scanner in crystal-ion-sliced thin-film $LiNbO_3$. *Opt. Expr.* 12:6159, 2004.

Columbia University

Pedersen, T. S., et al. Confinement of plasmas of arbitrary neutrality in a stellarator. *Phys. Plasmas* 11:2377, 2004.

Pedersen, T. S., et al. The Columbia nonneutral torus: A new experiment to confine nonneutral and positron-electron plasmas in a stellarator. *Fusion Sci. Technol.* 46:200, 2004.

Pedersen, T. S., et al. Prospects for the creation of positron-electron plasmas in a nonneutral stellarator. *J. Phys. B* 36:1029, 2003.

Pedersen, T. S., and **A. H. Boozer.** Confinement of nonneutral plasmas on magnetic surfaces. *Phys. Rev. Lett.* 88:205002, 2002.

He, R., et al. **(A. Pinczuk).** Extrinsic optical recombination in pentacene single crystals: Evidence of gap states. *Appl. Phys. Lett.* 87(21):211117, 2005.

Hirjibehedin, C. F., et al. **(A. Pinczuk).** Splitting of long-wavelength modes of the fractional quantum hall liquid at nu=1/3. *Phys. Rev. Lett.* 95(6):066803, 2005.

Dujovne, I., and **A. Pinczuk** et al. Composite-fermion spin excitations as nu approaches 1/2: Interactions in the fermi sea. *Phys. Review Lett.* 95(5):056808, 2005.

He, R., et al. **(A. Pinczuk).** Resonant Raman scattering in nanoscale pentacene films. *Appl. Phys. Lett.* 84:987–9, 2004.

Polvani, L. M., et al. Numerically converged solutions of the global primitive equations for testing the dynamical core of atmospheric GCMs. *Monthly Weather Rev.* 11:2539–52, 2004.

Polvani, L. M., and P. J. Kushner. Tropospheric response to stratospheric perturbations in a relatively simple general circulation model. *Geophys. Res. Lett.* 29, 2002.

Cho, J. Y.-K., and **L. M. Polvani.** The morphogenesis of bands and zonal winds with the atmospheres of the giant outer plants. *Science* 273:335–7, 1996.

Polvani, L. M., et al. Simple dynamical models of Neptune's Great Dark Spot. *Science* 249:1393–8, 1990.

Ruderman, M., L. Tao, and W. Kluzniak. A central engine for cosmic gamma-ray burst sources. *Astrophys. J.* 542:243, 2000.

Ruderman, M., K. Chen, and T. Zhu. Millisecond pulsar alignment: PSR 0437–47. *Astrophys. J.* 493:397, 1998.

Ruderman, M., K. Chen, and T. Zhu. Neutron star magnetic field evolution, crust movement and glitches. *Astrophys. J.* 492:267, 1998.

Ruderman, M., F. Wang, J. Halpern, and T. Zhu. Models for X-ray emission from isolated pulsars. *Astrophys. J.* 498:373, 1998.

Spyropoulos, C., **C. H. Scholz,** and B. E. Shaw. Transition regimes for growing crack populations. *Phys. Rev. E* 65:056105, 2002.

Shaw, B. E., and **C. H. Scholz.** Slip-length scaling for earthquakes: Observations and theory and implications for earthquake physics. *Geophys. Res. Lett.* 28:2995–8, 2001.

Scholz, C. H. Evidence of a strong San Andreas fault. *Geology* 28:163–6, 2000.

Spyropoulos, C., W. J. Griffith, **C. H. Scholz,** and B. E. Shaw. Experimental evidence for different strain regimes of crack populations in a clay model. *Geophys. Res. Lett.* 26:1081–4, 1999.

Sen, A. K. Feedback control of kink and tearing modes via novel ECH modulation. *Plasma Phys. Controlled Fusion* 92:L41, 2004.

Sokolov, V., and **A. K. Sen.** A new paradigm for the isotope scaling of plasma transport paradox. *Phys. Rev. Lett.* 92:165002–11, 2004.

Sokolov, V., and **A. K. Sen.** Experimental investigation of isotope scaling of anomalous ion thermal transport. *Phys. Plasmas* 10:3174, 2003.

Sokolov V., and **A. K. Sen.** Experimental study of isotope scaling of ion thermal transport. *Phys. Rev. Lett.* 89:09001, 2002.

Sobel, A. H., and H. Gildor. A simple model of SST hot spots. *J. Climate* 16:3978–92, 2003.

Sobel, A. H., J. Nilsson, and L. M. Polvani. The weak temperature gradient approximation and balanced tropical moisture waves. *J. Atmos. Sci.* 58:3650–65, 2001.

Sobel, A. H., and C. S. Bretherton. Modeling tropical precipitation in a single column. *J. Climate* 13:4378–92, 2000.

Sobel, A. H., and R. A. Plumb. Quantitative diagnostics of mixing in a shallow-water model of the stratosphere. *J. Atmos. Sci.* 56:2811–29, 1999.

Spiegelman, M., and R. Katz. A semi-Lagrangian Crank-Nicolson algorithm for the numerical solution of advection-diffusion problems. *Geochem. Geophys. Geosyst.* , in press.

Spiegelman, M. Linear analysis of melt band formation by simple shear. *Geochem. Geophys. Geosyst.* 4(9):8615, 2003.

Spiegelman, M., and P. B. Kelemen. Extreme chemical variability as a consequence of channelized melt transport. *Geochem. Geophys. Geosyst.* 4(7):1055, 2003.

Spiegelman, M., P. B. Kelemen, and E. Aharonov. Causes and consequences of flow organization during melt transport: The reaction infiltration instability in compactible media. *J. Geophys. Res.* 106(B2):2061–77, 2001.

Tan, Y.-W., et al. **(H. L. Stormer).** Measurements of the density-dependent many-body electron mass in two-dimensional GaAs/AlGaAs heterostructures. *Phys. Rev. Lett.* 94:016405, 2005.

Pan, W., et al. **(H. L. Stormer).** Fractional quantum Hall effect of composite fermions. *Phys. Rev. Lett.* 90:16801, 2003.

Zhu, J., et al. **(H. L. Stormer).** Spin susceptibility of an ultralow density, two-dimensional electron gas system. *Phys. Rev. Lett.* 90:056805, 2003.

Syed, S., et al. **(H. L. Stormer).** Large splitting of the cyclotron-resonance line in AlGaN/GaN heterostructures. *Phys. Rev. B* 67:241304, 2003.

Stormer, H. L. The fractional quantum Hall effect. *Rev. Mod. Phys.* 71:875, 1999.

Katz, J., Y. Zhang, and **W. I. Wang.** Normal incidence intervalence subband absorption in GaSb quantum well enhanced by coupling to InAs conduction band. *Appl. Phys. Lett.* 62:609–11, 1993.

Katz, J., Y. Zhang, and **W. I. Wang.** Normal incidence infrared absorption in AlAs/AlGaAs x-valley multiquantum wells. *Appl. Phys. Lett.* 61:1697–9, 1992.

Li, X., K. F. Longenbach, Y. Wang, and **W. I. Wang.** High breakdown voltage AlSbAs/InAs n-channel field effect transistors. *IEEE Electron Device Lett.* 13:192–4, 1992.

Golowich, S. E., and **M. I. Weinstein.** Scattering resonances of microstructures and homogenization theory. *SIAM J. Multiscale Modeling Simulation* 3(3):477–521, 2005.

Soffer, A., and **M. I. Weinstein.** Theory of nonlinear dispersive waves and selection of the ground state. *Phys. Rev. Lett.* 95:213905, 2005.

Goodman, R. H., R. E. Slusher, and **M. I. Weinstein.** Stopping light on a defect. *J. Opt. Soc. B: Opt. Phys.* 19(7):1635–52, 2002.

Soffer, A., and **M. I. Weinstein.** Resonances, radiation damping, and instability of Hamiltonian nonlinear waves. *Inventiones Mathematicae* 136:9–74, 1999.

Koster, D. A., **C. H. Wiggins,** and N. H. Dekker. Multiple events on single molecules: Unbiased estimation in single-molecule biophysics. *Proc. Natl. Acad. Sci.* 103(6): 1750–5, 2006.

Ziv, E., M. Middendorf, and **C. H. Wiggins.** Information-theoretic approach to network modularity. *J. Phys. Rev. E* 71(4 Pt 2):046117, 2005.

Middendorf, M., and E. Ziv, and **C. H. Wiggins.** Inferring network mechanisms: The *Drosophila melanogaster* protein interaction network. *Proc. Natl. Acad. Sci.* 102 (9):3192-7, 2005.

Middendorf, M., et al. **(C. H. Wiggins).** Predicting genetic regulatory response using classification. *Bioinformatics* 20(1), 2004; *Proceedings of the Twelfth International Conference on Intelligent Systems for Molecular Biology,* ISMB, 2004.

Wuu, C. S., et al. Dosimetry study of Re-188 liquid balloon for intravascular brachytherapy using polymer gel dosimeters and laser-beam optical CT scanner. *Med. Phys.* 30(2):132–7, 2003.

Wuu, C. S., et al. Dosimetric and volumetric criteria for selecting a source activity and/or a source type (I-125 or Pd-103) in the presence of irregular seed placement in permanent prostate implants. *Int. J. Radiat. Oncol. Biol. Phys.* 47:815–20, 2000.

Wuu, C. S., et al. Microdosimetric evaluation of relative biological effectiveness for Pd-103, I-125, Am-241 and Ir-192 brachytherapy sources. *Int. J. Radiat. Oncol. Biol. Phys.* 36:689–97, 1996.

Wuu, C. S., and M. Zaider. A mathematical description of sublethal damage repair and interaction for continuous low-dose rate irradiation. *Radiat. Prot. Dosim.* 52: 211–5, 1994.

FLORIDA STATE UNIVERSITY

Department of Physics

Programs of Study

The Department of Physics at Florida State University (FSU) offers programs of study that lead to the M.S. and Ph.D. degrees. The Department has approximately 45 teaching faculty members, including Nobel Laureate Professor Robert Schrieffer, and another 40 Ph.D. physicists engaged in a variety of research programs. The graduate program has 132 students and almost all hold research or teaching assistantships. The programs of study include experimental and theoretical atomic, condensed-matter, high-energy, materials science, and nuclear physics. Two University institutes have major physics research components—the Material Science and Technology Center (MARTECH) for condensed-matter physics and the National High Magnetic Field Laboratory (NHMFL) for research on materials using very high magnetic fields.

The Department offers both course work only and thesis-type M.S. degrees. Five-year B.S./M.S. programs in computational physics and physics education have been introduced. Students studying for the Ph.D. degree are also required to pass a written qualifying examination on classical mechanics, electricity magnetism, quantum mechanics, thermodynamics and statistical mechanics, and modern physics. Within six months of passing the written qualifying examination, students should pass an oral examination on the subject of the student's prospective research. The only formal course requirement is to take three advanced topics courses and a course in field theory.

Research Facilities

The Department occupies three adjacent buildings: an eight-story Physics Research Building, the Leroy Collins Research Laboratory Building, and an undergraduate physics classroom and laboratory building. Extensive experimental facilities include a 9.5-MV Super FN Tandem Van de Graaff accelerator with superconducting post-accelerator, a precision Penning trap mass spectrometer, a detector development laboratory for high-energy particle detectors, high-resolution Fourier-transform IR spectrometers, an ion implantation facility, instrumentation for research at liquid helium temperature and thin-film preparation, UHV (including surface analysis, molecular beam epitaxy, and atomic cluster facilities), facilities for high- and low-temperature superconductivity, small-angle and standard X-ray diffractometry, crystal-growth facilities and ion implantation facility, scanning electron and tunneling microscopy, image analysis, quasielastic light scattering, polarized electron energy loss spectroscopy, a He-atom beam crystal surface scattering apparatus, and a unique aerosol physics-electron irradiation system.

In addition to in-house facilities, ongoing experiments use accelerator and other research equipment at Fermilab, Bates, Brookhaven, Oak Ridge, Thomas Jefferson National Accelerator Facility (TJNAF), and CERN. Computational facilities at FSU include an IBM Multiprocessor Supercomputer, a state-of-the-art visualization lab, and a 120 CPU Beowulf cluster in the Department. Extensive networking facilities provide access to computers on and off campus. More information on individual faculty member research can be found on the Department's Web site at http://www.physics.fsu.edu.

Financial Aid

The Department offers teaching and research assistantships and fellowships. The fellowships include several that are designed to help develop promising young minority physicists. The assistantship stipend is $17,000 for twelve months, with a workload equivalent to 6 contact hours in an elementary laboratory. In general, summer assistantships are provided for all students. Students are teaching assistants during the first academic year but most are supported by research assistantships during and after their first summer.

Cost of Study

All tuition and fees for Florida residents were covered by the Department in 2005–06. The additional charge for out-of-state tuition is waived for assistants and fellows.

Living and Housing Costs

Apartments and houses are readily available in Tallahassee. A typical one-bedroom unfurnished apartment within walking distance of the physics building rents for $450 per month. The University has married student housing with rents that in 2004–05 ranged from $300 to $550 per month for a one-bedroom apartment to $390 to $500 per month for a two or three-bedroom apartment. National surveys show that the cost of living in Tallahassee is 10 to 15 percent lower than that in most areas of the United States.

Student Group

Florida State University is a comprehensive university with a total of 35,345 students, of whom 6,605 are graduate or professional students. The Department of Physics has 132 graduate students. Students entering with a B.S. degree in physics typically attain the Ph.D. within 5½ years.

Location

Tallahassee is the capital city of the state of Florida. Its population is about 185,000. Many employment opportunities exist for students' spouses in Tallahassee. Students can live in relatively rural surroundings and still be only 20 minutes from the University. Extensive sports facilities and active city leagues exist in the city. Graduate students' fees cover membership in a state-of-the-art, on-campus recreation center. Because of the mild winter climate, people in this region tend to be outdoor-oriented. The Gulf of Mexico is about 30 miles from campus.

The University and The Department

The presentations of the Schools of Fine Arts and Music provide cultural opportunities that are usually available only in much larger cities. The University Symphony, the Flying High Circus, and other theater and music groups give students the opportunity to participate in many activities in addition to their physics studies. FSU has active programs in intercollegiate and intramural sports.

Recent major additions in the FSU Science Center have been an interdisciplinary Materials Sciences and Technology Center and the National High Magnetic Field Laboratory. The NHMFL houses the world's highest field D.C. magnets, making FSU one of the principal centers for magnetic research. In addition to the teaching faculty at the Department of Physics, there are 8 research faculty members at the NHMFL.

Applying

Assistantship decisions are based on a student's transcript, GRE General Test scores, TOEFL scores, statement of purpose, and three letters of reference. The deadline for completed applications to be on file with the physics department is January 15 for international students and February 15 for U.S. citizens. Application forms can be printed from the Department's Web site at http://www.physics.fsu.edu.

Correspondence and Information

Professor Jorge Piekarewicz
Graduate Physics Program
Department of Physics
Florida State University
Tallahassee, Florida 32306-4350

Phone: 850-644-4473
Fax: 850-644-8630
E-mail: graduate@phy.fsu.edu
Web site: http://www.physics.fsu.edu

Florida State University

THE FACULTY AND THEIR RESEARCH

Todd Adams, Assistant Professor; Ph.D., Notre Dame, 1997. Experimental high-energy physics, particle physics, supersymmetry.
Howard Baer, Professor; Ph.D., Wisconsin–Madison, 1984. Theoretical physics: elementary particle.
Bernd Berg, Professor; Ph.D., Free University of Berlin, 1977. Theoretical physics: lattice gauge theory, computational physics.
Susan K. Blessing, Associate Professor; Ph.D., Indiana, 1989. Experimental high energy physics: elementary particle physics.
Gregory S. Boebinger, Professor and Director, National High Magnetic Field Laboratory; Ph.D., MIT, 1986. Magnetism, experimental condensed matter physics; correlated electron systems.
Nicholas Bonesteel, Professor; Ph.D., Cornell, 1991. Theoretical physics: condensed-matter physics, many-body theory, magnetism, Quantum Hall Effect.
James S. Brooks, Professor; Ph.D., Oregon, 1973. Experimental physics: low temperature, high–magnetic field condensed matter, organic conductor, quantum fluid physics.
Jianming Cao, Associate Professor; Ph.D., Rochester, 1996. Experimental condensed-matter physics, ultrafast dynamics probed by lasers.
Simon C. Capstick, Associate Professor; Ph.D., Toronto, 1986. Theoretical physics: hadronic and nuclear physics.
Irinel Chiorescu, Assistant Professor; Ph.D., CNRS Grenoble (France), 2000. Experimental condensed matter physics, magnetic flux qubits.
Volker Crede, Assistant Professor; Ph.D., Bonn (Germany), 2000. Experimental nuclear physics, quark matter.
Paul Cottle, Professor; Ph.D., Yale, 1986. Experimental: heavy-ion nuclear physics, teacher preparation.
Vladimir Dobrosavljivic, Professor; Ph.D, Brown, 1988. Theoretical condensed-matter physics, disordered systems and glasses, metal-insulator transitions.
Dennis Duke, Professor; Ph.D., Iowa State, 1974. Theoretical physics: elementary particle physics, computational physics.
Paul M. Eugenio, Assistant Professor; Ph.D., Massachusetts Amherst, 1998. Experimental nuclear/particle physics, search for new mesons.
Marcia Fenley, Assistant Professor; Ph.D., Rutgers, 1991. Computational biophysics, electrostatics in macromolecules.
Yuri Gershstein, Assistant Professor; Ph.D., Moscow, 1996. Experimental high-energy physics.
Vasken Hagopian, Professor; Ph.D., Pennsylvania, 1963. Experimental physics: elementary particle physics.
Linda Hirst, Assistant Professor; Ph.D., Manchester (England), 2001. Experimental soft condensed matter physics, biophysics.
Peter Hoeflich, Associate Professor; Ph.D., Heidelberg, 1986. Computational and nuclear astrophysics, supernovae explosions, cosmology.
Kirby Kemper, Professor; Ph.D., Indiana, 1968. Experimental physics: polarization studies in heavy-ion reactions.
David M. Lind, Associate Professor; Ph.D., Rice, 1986. Experimental condensed matter physics, magnetic superlattices.
Efstratios Manousakis, Professor; Ph.D., Illinois at Urbana-Champaign, 1985. Theoretical physics: condensed-matter physics, many-body theory, superconductivity.
H. K. Ng, Associate Professor; Ph.D., McMaster, 1984. Experimental physics: far-infrared spectroscopy, superconductivity, highly correlated electron systems, spectroscopy in high-magnetic fields.
Joseph F. Owens, Professor; Ph.D., Tufts, 1973. Theoretical physics: elementary particle theory.
Jorge Piekarewicz, Professor; Ph.D., Pennsylvania, 1985. Theoretical physics, interface between nuclear and particle physics.
Harrison B. Prosper, Professor; Ph.D., Manchester (England), 1980. Experimental high energy physics: particle physics, computational physics.
Laura Reina, Associate Professor; Ph.D., Trieste, 1992. Theoretical high energy physics: elementary particle physics.
Per Arne Rikvold, Professor; Ph.D., Temple, 1983. Theoretical condensed-matter physics: statistical physics, surface and interface science.
Mark A. Riley, Professor; Ph.D., Liverpool, 1985. Experimental physics: nuclear structure physics.
Winston Roberts, Professor; Ph.D., Guelph, 1988. Theoretical physics: nuclear and particle physics, hadronic physics.
Grigory Rogachev, Assistant Professor; Ph.D., Kurchatov Institute (Moscow), 1999. Experimental nuclear physics: nucleosynthesis.
Pedro Schlottmann, Professor; Ph.D., Technical University of Munich, 1973. Theoretical physics; high-T_c condensed-matter physics, superconductors; heavy fermions, magnetism.
Shahid A. Shaheen, Associate Professor; Ph.D., Ruhr-Bochum, 1985. Experimental condensed-matter physics: permanent magnets, superconductivity, magnetism, materials science.
Samuel L. Tabor, Professor; Ph.D., Stanford, 1972. Experimental physics: high-spin states in nuclei, nuclei far from stability.
Oskar Vafek, Associate Professor; Ph.D., Johns Hopkins, 2003. Theoretical condensed-matter physics: quantum phase transitions, superconductivity.
David Van Winkle, Professor and Chairman of the Department; Ph.D., Colorado, 1984. Experimental condensed-matter physics: liquid crystals, colloids, macromolecules, teacher preparation.
Alexander Volya, Assistant Professor; Ph.D., Michigan State, 2000. Theoretical nuclear physics, nuclear structure models.
Stephan von Molnar, Professor; Ph.D., California, Riverside, 1965. Experimental physics: correlation effects in electronic systems, magnetic semiconductors, magnetic nanostructures.
Horst Wahl, Professor; Ph.D., Vienna, 1969. Experimental physics: elementary particle physics.
Christopher Wiebe, Assistant Professor; Ph.D., McMaster, 2002. Experimental condensed-matter physics: highly correlated electron systems, geometrically frustrated materials, superconductivity.
Ingo Wiedenhöver, Assistant Professor; Ph.D., Cologne (Germany), 1995. Nuclear experimental physics, complete gamma spectroscopy of 127 xe.
Peng Xiong, Associate Professor; Ph.D., Brown, 1994. Experimental condensed-matter physics: magnetism and superconductivity, soft/hard hybrid nanostructures.
Kun Yang, Associate Professor; Ph.D., Indiana, 1994. Theoretical physics: condensed matter, computational physics.
Huan-Xiang Zhou, Professor; Ph.D., Drexel, 1988. Computational and experimental biophysics; protein stability folding; and protein-protein interactions.

RESEARCH ACTIVITIES

Theoretical

Computational Biophysics. Electrostatics and dynamics of biomolecules in aqueous environments (Zhou and Fenley).
Condensed Matter. Many-body theory of magnetism, magnetic properties of solids; high-temperature superconductivity; heavy fermions, adsorption, phase transitions, numerical simulations, quantum information theory (Bonesteel, Dobrosavljevic, Manousakis, Rikvold, Schlottmann, Schrieffer, Yang).
Elementary Particles and Fields. Strong and electroweak interaction phenomenology in high-energy particle physics (Owens, Reina). Lattice gauge theory and numerical simulations of various physical systems, computational quantum gravity (Baer, Berg, Duke).
Nuclear Theory. Nuclear structure and studies emphasizing transitions (Volya).
Structure and Electromagnetic interactions of baryons and nuclei in Quard Model (Capstick, Piekarewicz).

Experimental

Atomic and Molecular Physics. Precision atomic measurements using Penning trap (Myers).
Condensed Matter. Biomolecular ordering; nano/biophysics; liquid crystals; gels; spintronics, hard magnetic materials; surface physics; sub-picosecond spectroscopy; low- and high-temperature superconductivity; highly correlated electron systems; organic crystals; quantum qubits (Boebinger, Brooks, Cao, Chiorescu, Hirst, Lind, Ng, Shaheen, von Molnar, Van Winkle, Xiong).
Elementary Particles and Fields. Hadron spectroscopy; collider physics, strong and electroweak interactions in high-energy particle physics (Adams, Blessing, Gershtein, S. Hagopian, V. Hagopian, Prosper, Wahl).
Nuclear Physics. Heavy-ion reactions and radioactive beams; heavy-ion fusion and fragmentation studies; properties of nuclear systems at high angular momentum and far from stability; laser-induced polarization; octupole structure in nuclei; electron scattering at TJNAF; light-ion nuclear spectroscopy; alpha, beta, and gamma spectroscopy; relativistic heavy-ion reactions (Cottle, Crede, Eugenio, Frawley, Kemper, Riley, Rogachev, Tabor, Wiedenhoever).

IDAHO STATE UNIVERSITY

Department of Physics

Programs of Study

The Doctor of Philosophy degree in engineering and applied science is offered jointly by the College of Engineering and the Department of Physics at Idaho State University (ISU). Research areas emphasized are radiation science, accelerator applications, applied nuclear physics, and health physics. All applicants must meet ISU Graduate School admission requirements for doctoral programs. In addition, applicants must have attained a master's degree in engineering, physics, or a closely related field. To attain a degree in this program, a student must demonstrate scholarly achievement and an ability in independent investigation. The program normally requires three years of full-time study beyond the master's degree, including research and preparation of the dissertation.

Master of Science degrees are offered in physics, health physics, and natural science. The Master of Science degree in physics is a thesis program that requires 30 credits, 15 of which are required 600-level courses. The Master of Science (health physics emphasis) is a thesis program that prepares students for radiation protection careers leading to upper technical and management levels. This program requires 30 credits, 15 of which must be at the 600 course level. The Master of Natural Science degree is a nonthesis option available for those planning a teaching career in primary or secondary education. The program requires a minimum of 30 credits, 22 of which must be in residence. A final oral examination is required.

Research Facilities

Research is conducted in the Particle Beam Laboratory (PBL), the Idaho Accelerator Center (IAC), the Environmental Monitoring Laboratory (EML), and the Environmental Assessment Laboratory (EAL), with emphases on experimental low-energy nuclear physics, health physics, accelerator-produced radiation effects, and ion beam analysis of materials. The laboratories house a 2-MeV Van de Graaff and ten electron LINACs with energies from 4 to 30 MeV. The 30-MeV high-current traveling wave electron LINAC has a pulse width adjustable from a few microseconds down to 12 picoseconds. Lab space doubled in summer 2004 with the completion of two additions to the Idaho Accelerator Center. The 9-MeV, 10-kA pulsed power spiral line accelerator is scheduled to be installed in the 2004 academic year. The EML consists of a complete wet lab with two fume hoods, tritium enrichment capabilities to achieve MDC of 10 to 15 pCi/L, 3 ICB-controlled gamma spectrometry systems that include two 23 percent relative efficiency p-type high-purity germanium detectors and one extended-range 50 percent relative efficiency n-type beryllium windowed high-purity germanium detector, one Wallac 14145 low-background liquid scintillation counter, and one 5-inch automatic low-background gas proportional counter. The EAL provides a Beckman LS5000TA liquid scintillation counter, Canberra 2404 proportional counter, and fout high-purity intrinsic germanium detectors along with an SEM. There are collaborative research projects with Sandia National Laboratory (Albuquerque); Idaho National Environmental and Engineering Laboratory; Los Alamos; CEBAF; TUNL; DFEL; the state of Idaho; Positron Systems, Inc.; and Stoller, Inc.

Financial Aid

Graduate assistantships in 2001–02 carried stipends of approximately $10,000 per academic year, plus waiver of fees and tuition. Summer support of $3000 to $4000 is generally available. Research assistantships of up to $18,000 are also available. A University work-study program allows eligible students to work up to 20 hours per week, and student loans are available. Spouses of students may obtain local or campus employment. Assistance for graduate research has been received from a variety of sources, including the National Science Foundation, the Department of Energy, the Idaho State Board of Education, Bechtel BWXT Idaho, the Stoller Corporation, and NSF EpSCOR. Most full-time graduate students receive financial aid.

Cost of Study

In 2004–05, registration costs were $2520 per semester for residents of Idaho and $6060 for nonresidents. Tuition and fees for students on teaching or research assistantships are usually waived. Books cost between $200 and $300 per semester.

Living and Housing Costs

Room and board for single students living on campus cost approximately $3000 per year. Apartments for married students rent for approximately $460 per month (including utilities).

Student Group

The University's total enrollment is approximately 13,000 students. The 1,600 graduate students come from almost every state in the U.S. and from many other countries.

Location

Pocatello (population 52,000) is situated at the edge of the Snake River plain in southeastern Idaho. There are mountains on three sides of town. The climate is pleasant, with an average winter temperature of 28 degrees and summer temperatures seldom exceeding 95 degrees. The climate is dry (annual precipitation of about 12 inches), with many days of sunshine. ISU, the region's premier nuclear research institution, is located within easy driving distance of world-class skiing (Sun Valley, Jackson Hole, Salt Lake City), several national parks, monuments, and wilderness areas and some of the country's best hunting, fishing, kayaking, white-water rafting, climbing, and backpacking.

The University

Idaho State University is composed of the Colleges of Education, Business, Pharmacy, Arts and Sciences, Engineering, and Health-Related Professions. The history of the institution goes back to 1901, but a full four-year curriculum was not begun until 1947. Master's degree programs were initiated in 1958 and those leading to the Ph.D. in 1969. The University is accredited by the Northwest Association of Colleges and Schools. Graduate classes are small enough to permit the faculty members to give students individual attention. Semesters run from late August to mid-December and from mid-January to mid-May. Summer sessions run from mid-May to early August.

Applying

Application forms for admission to graduate study may be downloaded from the Web site (listed below) or obtained from the graduate admissions office or by writing to the Department of Physics. Applications are accepted at any time, but applications for assistantships must be submitted by March 1. GRE General Test scores must be submitted as part of the applications for admission and for financial support.

Correspondence and Information

Department of Physics
Box 8106
Idaho State University
Pocatello, Idaho 83209

Phone: 208-282-2350
Fax: 208-282-4649
E-mail: office@physics.isu.edu
Web site: http://www.physics.isu.edu

Idaho State University

THE FACULTY AND THEIR RESEARCH

Wendland Beezhold, Research Professor (Physics); Ph.D., Washington (Seattle), 1969. Semiconductor physics, radiation physics, electron and particle beam accelerators, nanotechnologies, and modeling and simulation of radiation response of electronics and materials.

Richard R. Brey, Associate Professor (Health Physics); Ph.D., Purdue, 1994.

Philip L. Cole, Associate Professor (Physics); Ph.D., Purdue, 1991. Baryon spectroscopy, exotic and hybrid meson spectroscopy, radiation applications, active interrogation techniques with photon probes, coherent bremsstrahlung.

Thomas F. Gesell, Professor (Physics); Ph.D., Tennessee, 1971. Environmental radiation and radionuclides.

Martin H. Hackworth, Senior Lecturer–Lab Supervisor (Physics); M.S., Eastern Kentucky, 1992. Acoustics and meteorology.

J. Frank Harmon, Professor (Physics); Ph.D., Wyoming, 1969. Director of Idaho Accelerator Center, which is studying the use of nuclear physics in a wide range of applied activities.

Alan W. Hunt, Research Assistant Professor (Physics); Ph.D., Harvard, 2000. Wide band-gap, semiconductors, positron production, isometric nuclei production.

Kara J. Keeter, Associate Professor (Physics); Ph.D., Duke, 1990. Experimental low- and medium-energy nuclear physics and astrophysics: photonuclear reactions, few-body systems and neutrinoless double beta decay, investigating theories ranging from QCD and CPT to massive neutrinos.

John M. Knox, Professor (Ion Beam Analysis); Ph.D., Wyoming, 1981. Ion beam analysis of materials: RBS, PIXE, NRA, ERD; proton microbeam analysis; neutron elastic recoil detection of hydrogen isotopes.

Ernest B. Nieschmidt, Visiting Associate Professor (Physics); M.S., San Diego State, 1961. Nuclear physics, detectors, laser isotope separation, sonoluminiscense.

Steven L. Shropshire, Associate Professor (Physics); Ph.D., Washington State, 1991. Teacher training, physics education, defects and diffusion in solids, nuclear spectroscopies applied to materials science.

Eddie Tatar, Assistant Professor (Experimental Physics); Ph.D., Notre Dame, 2000. Quantum interference phenomena in particle physics and optics, study of exotic mesons, partial wave analysis techniques, lie groups and representation theory, supersymmetry, weak interactions and neutrino physics.

Douglas P. Wells, Associate Professor (Experimental Nuclear Physics) and Chair; Ph.D., Illinois, 1990. Applied accelerator physics, photonuclear physics, health physics.

Postdoctoral Fellows

Mohamed Reda (Nuclear Engineering), Ph.D., Alexandria (Egypt), 1988. Monte Carlo simulation, interaction of charged particles with solids, sputtering phenomena, inelastic effects in low-temperature plasma, design and operation of control systems for electric networks.

Farida Selim (Physics), Ph.D., Alexandria (Egypt) and Harvard, 1999. Atomic physics, positron physics.

Jagoda Mary Urban-Klaehn (Physics), Ph.D., Texas Christian, 1998. Application of physical methods (X-ray, gamma rays, positron annihilation spectroscopy, accelerator techniques, scanning electron microscopy, and other spectroscopic techniques) to the investigation of the internal physical structure of materials.

Professors Emeriti

Barry R. Parker, Ph.D., Utah State, 1968.
Joseph E. Price, Ph.D., Rice, 1959.
Stanley H. Vegors, Ph.D., Illinois, 1955.

INDIANA UNIVERSITY BLOOMINGTON

Department of Physics

Programs of Study

Physics research at Indiana University (IU) is conducted in the subfields of nuclear physics, accelerator physics, biophysics, particle physics, condensed-matter physics, astrophysics, and mathematical physics. M.S., M.A.T., and Ph.D. degrees are offered. A Ph.D. program in biophysics has been established with the addition of several new faculty members. Some areas of specialization in biophysics include neuroscience, experimental and theoretical models of development, networks, pattern formation, and cell signaling. An interdisciplinary scientific computing minor and an M.S. in beam physics and technology are offered.

M.S. candidates must complete 30 credit hours of graduate work (including a minimum of 20 hours in physics) and either pass a written comprehensive exam or, for some programs, complete a thesis. The M.A.T. requires 20 hours in physics and an additional 16 hours in mathematics, astronomy, chemistry, computer science, and education.

To obtain the Ph.D., the candidate must demonstrate an ability to do research by carrying out an investigation and presenting a publishable thesis. The requirements for the Ph.D. include a minimum of 90 hours of graduate credit that consists of course work, supervised reading, and research. A qualifying exam is required no later than one year after arrival; two attempts at the exam are allowed. The great majority of students at IU pass the qualifying exam and are soon involved in thesis research. The final oral exam is conducted by the candidate's doctoral committee and consists of questions on the major and minor fields of work as well as on the thesis.

Research Facilities

The Indiana University Cyclotron Facility/Nuclear Theory Center is a national facility for nuclear, condensed-matter, and medical physics research. It consists of a cyclotron and a low-energy neutron source and extensive support facilities. Other experiments in nuclear, particle and accelerator physics, and astrophysics are conducted at Fermilab, Thomas Jefferson Lab, Brookhaven, NIST, Los Alamos, CERN, and CIDA. Local facilities include high-bay assembly areas, machine shops, electronic design facilities, a large open-bore superconducting magnet for balloon studies, a high-vacuum sputtering system, X-ray diffractometers, ultrahigh-vacuum surface analysis systems, a scanning tunneling microscope, a class 1000 clean room with photolithographic equipment, a 14-Tesla superconducting magnet with pumped ^{3}He and dilution refrigerator inserts, and several standard cryostats for transport measurements from DC to 20 GHz. Computing facilities include numerous workstations and Linux clusters, a 600-CPU IBM SP, a 64-CPU Sun E 10000, an AVIDD cluster, and a data visualization cave.

Financial Aid

Teaching assistantships carried stipends of at least $15,000 for the ten-month 2005–06 academic year. Research assistantship stipends averaged $20,000 for twelve months. Teaching and research positions are also available that pay at least $2150 for the two summer months. Over the last ten years, 98 percent of the students who finished the Ph.D. received full financial support throughout their graduate careers.

Cost of Study

In 2005–06, fees per credit hour for in-state graduate students were $226.55; for out-of-state graduate students, $659.85. Teaching and research assistants ordinarily pay fees of approximately $600 per semester.

Living and Housing Costs

Indiana University's Residential Programs and Services (RPS) Office offers a variety of housing and meal plans, which include many options, from single dormitory rooms to four-bedroom apartments. Rates may include room, utilities (including local telephone service), cable TV, and Ethernet connections. For detailed information, including rates, students may visit the RPS Web site at http://www.rps.indiana.edu.

Student Group

Indiana University is a large institution, with 36,200 students enrolled at the Bloomington campus, including 7,549 graduate and professional school students. In fall 2005, there were 87 graduate students in physics, almost all of whom received full financial support.

Student Outcomes

IU physics Ph.D. graduates are currently employed by national laboratories such as FNAL, BNL, SLAC, LANL, and LBNL; by universities such as Duke, Columbia, Ohio, and Rice; and by companies such as Lucent, Intel, Microsoft, Motorola, JPMorgan, Worldcom, and Battelle. Recent postdoctoral students have obtained positions at Harvard, Johns Hopkins, MIT, NIST, Northwestern University, and the EPA.

Location

Bloomington is located in the picturesque hills of southern Indiana, 50 miles south of Indianapolis, the state capital. It is close to five state parks, two state forests, and the state's largest lake. It has consistently been chosen in national rankings as having a high quality of life.

The University

Indiana University is the oldest state university west of the Allegheny Mountains. It was founded in 1820 and has been a pioneer in higher education in the Midwest. It is widely recognized for the beauty of its campus and for the diversity and high quality of its graduate programs in the arts, humanities, and sciences. The campus provides numerous facilities for all types of sports. The School of Music presents concerts and opera. Lectures, dramatic and musical productions, ballet, drama, and concerts are presented by the Auditorium and the University Theatre.

Applying

The priority deadline for assistantship and fellowship applications for the fall semester is December 1 for international applicants and January 15 for U.S. citizens. For further information, students should visit the Department's Web site (http://www.physics.indiana.edu), or they may write or call the Deparment. Applications are submitted online using a link on the Department's Web site or at http://www.gradapp.indiana.edu.

Correspondence and Information

Chairperson
Department of Physics
Indiana University
Bloomington, Indiana 47405-4201
Phone: 812-855-1247

Graduate Admissions Committee
Department of Physics
Indiana University
Bloomington, Indiana 47405-4201
Phone: 812-855-3973
E-mail: gradphys@indiana.edu
Web site: http://physics.indiana.edu

Indiana University Bloomington

THE FACULTY AND THEIR RESEARCH

Professors Emeriti
Ethan D. Alyea, Ph.D., Caltech, 1962. Astrophysics (experimental).
Robert D. Bent, Ph.D., Rice, 1954. Nuclear physics (experimental).
Bennet B. Brabson, Ph.D., MIT, 1966. Elementary particle physics (experimental).
John M. Cameron, Ph.D., UCLA, 1967. Nuclear physics (experimental).
Ray R. Crittenden, Ph.D., Wisconsin, 1960. Elementary particle physics (experimental).
Charles Goodman, Ph.D., Rochester, 1959. Nuclear physics (experimental).
Richard R. Hake, Ph.D., Illinois, 1955. Condensed-matter physics (experimental).
Richard M. Heinz, Ph.D., Michigan, 1964. Astrophysics (experimental).
Archibald W. Hendry, Ph.D., Glasgow, 1962. Elementary particle physics (theory).
Andrew A. Lenard, Ph.D., Iowa, 1953. Theoretical physics, mathematical physics.
Don B. Lichtenberg, Ph.D., Illinois, 1955. Elementary particle physics (theory).
Malcolm Macfarlane, Ph.D., Rochester, 1959. Nuclear physics (theory).
Hugh J. Martin, Ph.D., Caltech, 1956. Elementary particle physics (experimental).
Daniel W. Miller, Ph.D., Wisconsin, 1951. Nuclear physics (experimental).
Roger G. Newton, Distinguished Professor Emeritus; Ph.D., Harvard, 1953. Theoretical and mathematical physics.
Robert E. Pollock, Distinguished Professor; Ph.D., Princeton, 1963. Nuclear physics (experimental).
Peter Schwandt, Ph.D., Wisconsin, 1967. Nuclear physics (experimental).
James C. Swihart, Ph.D., Purdue, 1955. Condensed-matter physics (theory).
George E. Walker, Ph.D., Case Tech, 1966. Nuclear physics (theory).
John G. Wills, Ph.D., Washington (Seattle), 1963. Nuclear physics (theory).

Professors
Andrew D. Bacher, Ph.D., Caltech, 1967. Nuclear physics (experimental).
David V. Baxter, Ph.D., Caltech, 1984. Condensed-matter physics (experimental).
John L. Challifour, Ph.D., Cambridge, 1963. Theoretical and mathematical physics.
Rob de Ruyter van Steveninck, Ph.D., Groningen (Netherlands), 1986. Biophysics and soft condensed-matter physics (experimental).
Alex R. Dzierba, Ph.D., Notre Dame, 1969. Elementary particle physics (experimental).
Herbert A. Fertig, Ph.D., Harvard, 1988. Condensed-matter physics (theory).
G. C. Fox, Ph.D., Cambridge, 1967. Computational physics.
James A. Glazier, Ph.D., Chicago, 1989. Biophysics and soft condensed-matter physics (experimental).
Steven A. Gottlieb, Ph.D., Princeton, 1978. Elementary particle physics (theory).
Charles J. Horowitz, Ph.D., Stanford, 1981. Nuclear physics (theory).
Larry L. Kesmodel, Ph.D., Texas, 1974. Condensed-matter physics (experimental).
V. Alan Kostelecký, Ph.D., Yale, 1982. Elementary particle physics (theory).
S. Y. Lee, Ph.D., SUNY at Stony Brook, 1972. Accelerator physics (experimental).
J. Timothy Londergan, D.Phil., Oxford, 1969. Nuclear physics (theory).
Hans Otto Meyer, Ph.D., Basel, 1970. Nuclear physics (experimental).
James A. Musser, Ph.D., Berkeley, 1984. High-energy astrophysics (experimental).
Hermann Nann, Ph.D., Frankfurt, 1967. Intermediate-energy nuclear physics (experimental).
Harold Ogren, Ph.D., Cornell, 1970. Elementary particle physics (experimental).
Catherine Olmer, Ph.D., Yale, 1976. Intermediate-energy nuclear physics (experimental).
Roger Pynn, Ph.D., Cambridge, 1969. Condensed-matter physics (experimental).
William L. Schaich, Ph.D., Cornell, 1970. Condensed-matter physics (theory).
Brian D. Serot, Ph.D., Stanford, 1979. Nuclear physics (theory).
William M. Snow, Ph.D., Harvard, 1990. Nuclear physics (experimental).
Paul E. Sokol, Ph.D., Ohio State, 1981. Condensed-matter physics (experimental).
Kumble R. Subbaswamy, Ph.D., Indiana, 1976. Condensed-matter physics (theory).
Richard J. Van Kooten, Ph.D., Stanford, 1990. Elementary particle physics (experimental).
Steven E. Vigdor, Ph.D., Wisconsin, 1973. Nuclear physics (experimental).
Scott W. Wissink, Ph.D., Stanford, 1986. Nuclear physics (experimental).
Andrej Zieminski, Ph.D., Warsaw, 1971. Elementary particle physics (experimental).

Associate Professors
Michael S. Berger, Ph.D., Berkeley, 1991. Elementary particle physics (theory).
John P. Carini, Ph.D., Chicago, 1988. Condensed-matter physics (experimental).
Harold G. Evans, Ph.D., UCLA, 1991. Elementary particle physics (experimental).
Adam P. Szczepaniak, Ph.D., Washington (Seattle), 1990. Nuclear physics (theory).
Jon Urheim, Ph.D., Pennsylvania, 1990. High-energy astrophysics (experimental).

Assistant Professors
John Beggs, Ph.D., Yale, 1998. Biophysics and soft condensed-matter physics (experimental).
Dobrin Bossev, Ph.D., Kyoto (Japan), 1999. Condensed-matter physics (experimental).
Mark H. Hess, Ph.D., MIT, 2004. Accelerator physics (experimental).
Chen-Yu Liu, Ph.D., Princeton, 2002. Nuclear physics (experimental).
Mark D. Messier, Ph.D., Boston University, 1999. High-energy astrophysics (experimental).
Sima Setayeshgar, Ph.D., MIT, 1998. Biophysics and soft condensed-matter physics (theoretical).
Matthew R. Shepherd, Ph.D., Cornell, 2005. Elementary particle physics (experimental).
Rex Tayloe, Ph.D., Illinois, 1995. Nuclear physics (experimental).

Indiana University Bloomington

SELECTED PUBLICATIONS

Betker, A. C., et al. **(A. D. Bacher, C. Olmer** and **S.W. Wissink).** Reaction mechanism for natural parity (p,p′) transitions in ^{10}B. *Phys. Rev. C* 71:064607, 2005.

Fujita, Y., et al. **(A. D. Bacher).** Isospin symmetry-structure study at new high-resolution course of RCNP.*Nucl. Phys. A* 687:311, 2001.

Lisantti, J., et al. **(A. D. Bacher, C. Olmer** and **S.W. Wissink).** Neutron transition densities for low lying states in ^{58}Ni obtained by using 200~MeV inelastic proton scattering. *Phys. Rev. C* 58:2217, 1998.

Baxter, D. V., et al. **(J. M. Cameron, H. O. Meyer,** and **W. M. Snow).** LENS—A pulsed neutron source for education and research. *Nucl. Instr. Meth. A* 542:28, 2005.

Chipara, J., et al **(D. V. Baxter).** Polymer magnetic investigations of titanium-doped gamma iron oxides dispersed in polymers. *Sci. B: Polymer Phys.* 43:3423, 2005.

Baxter, D. V., et al. **(J. M. Cameron, H. O. Meyer** and **W. M. Snow).** Status of the low energy neutron source at Indiana University. *Nucl. Instr. Meth. B* 241:209, 2005.

Ruzmetov, D., et al **(D. V. Baxter).** High-temperature hall effect in $Ga_{1-x}Mn_xAs$. *Phys. Rev. B* 69:155207, 2004.

Haldeman, C., and **J. M. Beggs.** Critical branching captures activity in living neural networks and maximizes the number of metastable states. *Phys. Rev. Lett.* 94:058101, 2005.

Beggs, J. M., and D. Plenz. Neuronal avalanches are diverse and precise activity patterns that are stable for many hours in cortical slice cultures. *J. Neurosci.* 24 (22):5216, 2004.

Beggs, J. M., and D. Plenz. Neuronal avalanches in neocortical circuits. *J. Neurosci.* 23(35):11167, 2003.

Berger, M. S., and B. Zerbe. Signals for low scale gravity in the process $\gamma\gamma \rightarrow ZZ$. *Phys. Rev. D* 72:95007, 2005.

Berger, M. S. Higgs sector radiative corrections and s-channel production. *Phys. Rev. Lett.* 87:131801, 2001.

Berger, M. S. Abelian family symmetries and leptogenesis. *Phys. Rev. D* 62: 013007, 2000.

Bossev, D. P., M. Matsumoto, and M. Nakahara. Effect of mixed counterions on the micelle structure of perfluorinated anionic surfactants. In *Mixed Surfactant Systems,* 2nd ed., eds. K Ogino and M Abe. New York: Marcel Dekker, 2005.

Prabhu, V. M., E. J. Amis, **D. P. Bossev,** and N. Rosov. Counterion associative behavior with flexible polyelectrolytes. *J. Chem. Phys.* 121:4424, 2004.

Paliwal, A., D. Asthagiri, **D. P. Bossev,** and M. E. Paulaitis. Pressure denaturation of staphylococcal nuclease studied by neutron small-angle scattering and molecular simulation. *Biophys. J.* 87:3479, 2004.

Lewis, R. M., and **J. P. Carini.** Frequency scaling of microwave conductivity in the integer quantum Hall effect minima. *Phys. Rev. B* 64:073310, 2001.

Lee, H.-L., et al. **(J. P. Carini** and **D. V. Baxter).** Quantum-critical conductivity scaling for a metal-insulator transition. *Science* 287:633, 2000.

Lee, H.-L., et al. **(J. P. Carini** and **D. V. Baxter).** Temperature-frequency scaling in niobium-silicon near the metal-insulator transition. *Phys. Rev. Lett.* 78:4261–4, 1998.

Challifour, J. L., and J. P. Clancy. A path space formula for gauss vectors in chern-simons quantum electrodynamics. *J. Math. Phys.* 40:5318, 1999.

Gregor, T., et al. **(R. de Ruyter van Steveninck).** Diffusion and scaling during early embryonic pattern formation. *Proc. Natl. Acad. Sci.* 102:18403, 2005.

Simmons, P. J., and **R. de Ruyter van Steveninck.** Reliability of signal transfer at a tonically transmitting, graded potential synapse of the locust ocellar pathway. *J. Neurosci.* 25:7529, 2005.

Nemenman, I., W. Bialek, and **R. de Ruyter van Steveninck.** Entropy and information in neural spike trains: Progress on the sampling problem. *Phys. Rev. E* 69:056111, 2004.

Dzierba, A. R., et al. Comment on: The evidence for a pentaquark signal and kinematic reflections. *Phys. Rev. D* 71:098502, 2005.

Denisov, S., et al. **(A. R. Dzierba).** Systematic studies of timing characteristics for 2m long scintillation counters. *Nucl. Instr. Methods A* 525:183, 2004.

Denisov, S., et al. **(A. R. Dzierba).** Studies of magnetic shielding for phototubes. Nuclear instruments and methods. *Physics Research Section A: Accelerators, Spectrometers, Detectors, and Associated Equipment* 553:467, 2004.

Eugenio, P., et al. **(A. R. Dzierba).** Observation of a new $J^{PC} = 1^{+-}$ isoscalar state in the reaction $\pi^- p \rightarrow \omega\eta n$ at 18-GeV/C. *Phys. Lett. B* 497:190, 2001.

Abazov, V. M., et al. **(H. Evans, R. Van Kooten** and **A. Zieminski).** (D0 Collaboration). Measurement of the t-tbar production cross-section in p-pbar collisions at sqrt(s) = 1.96 TeV in dilepton final states. *Phys. Lett. B* 626:55, 2005.

Abazov, V. M., et al. **(H. Evans, R. Van Kooten** and **A. Zieminski).** (D0 Collaboration). A search for the flavor-changing neutral current decay B_s^0 -> mu+ mu- in p-pbar collisions at sqrt(s) = 1.96 TeV. *Phys. Rev. Lett.* 94:042001, 2005.

Abazov, V. M., et al. **(H. Evans, R. Van Kooten** and **A. Zieminski).** Measurement of the lifetime difference in the B_s system. *Phys. Rev. Lett.* 95:171801, 2005.

Majumdar, K., and **H. A. Fertig.** Deconfinement and phase diagram of bosons in a linear optical lattice with a particle reservoir. *Phys. Rev. Lett.* 94:220402, 2005.

Li, M., **H. A. Fertig,** R. Cote, and H. Yi. Dynamical conductivity of pinned quantum Hall stripes. *Phys. Rev. B* 71:155312, 2005.

Zhang, W., and **H. A. Fertig.** Vortices and dissipation in a bilayer thin film superconductor. *Phys. Rev. B* 71:224514, 2005.

Fox, G. C., et al. Building messaging substrates for web and grid applications. *Philosophical Transactions of the Royal Society: Mathematical, Physical and Engineering Sciences* 363(1833):1757–73, 2005.

Grant, L. B., et al. **(G. C. Fox).** A web-service based universal approach to heterogeneous fault databases. *Computing Sci. Engineering* 7(4):51–7. (http://ieeexplore.ieee.org/search/wrapper.jsp?arnumber=1463136)

Berman, F., **G. C. Fox,** and A. J. G. Hey, eds. *Grid Computing: Making the Global Infrastructure a Reality.* Chicester, England: John Wiley & Sons, 2003.

Roeland, H., M. Merks, and **J. A. Glazier.** A cell-centered approach to developmental biology. *Physica A* 352:113, 2005.

Chaturvedi, R., et al. **(J. A. Glazier).** On multiscale approaches to three-dimensional modeling of morphogenesis. *J. Royal Soc. Interface* 2:237, 2005.

Cickovski, T. M., et al **(J. A. Glazier).** A framework for three-dimensional simulation of morphogenesis. *IEEE/ACM Transactions on Computational Biology and Bioinformatics* 2:273, 2005.

Gottlieb, S., et al. QCD thermodynamics with three flavors of improved staggered quarks. *Phys. Rev. D* 71:034504, 2005.

Gottlieb, S., et al. Semileptonic decays of D mesons in three-flavor lattice QCD. *Phys. Rev. Lett.* 94:011601, 2005.

Gottlieb, S., et al. Charmed meson decay constants in three-flavor lattice QCD. *Phys. Rev. Lett.* 95:122002, 2005.

Ambrosio, M., et al. **(R. Heinz** and **J. Musser).** (MACRO Collaboration). Measurements of atmospheric muon neutrino oscillations, global analysis of the data collected with MACRO detector. *European Phys. J. C* 36:323, 2004.

Aglietta, M., et al. **(R. Heinz** and **J. Musser).** (MACRO Collaboration). The cosmic ray proton, helium and CNO fluxes in the 100 TeV energy region from TeV muons and EAS atmospheric Cherenkov Light Observations of MACRO and EAS-TOP. *Astropart. Phys.* 21:223, 2004.

Aglietta, M., et al. **(R. Heinz** and **J. Musser).** (MACRO Collaboration). The cosmic ray primary composition between 10^{15} and 10^{16} eV from extensive air showers electromagnetic and TeV muon data. *Astropart. Phys.* 20:641, 2004.

Kesar, A. S., **M. Hess,** S. E. Korbly, and R. J. Temkin. Time and frequency-domain models for Smith-Purcell radiation from a two-dimensional charge moving above a finite length grating. *Phys. Rev. E* 71:016501, 2005.

Hess, M., and C. Chen. Space-charge limit for a finite-size bunched beam in a circular conducting pipe. *Phys. Rev. ST Accel. Beams* 7:092002, 2004.

Hess, M., and C. Chen. Equilibrium and confinement of bunched annular beams. *Phys. Plasmas* 9:1422, 2002.

Horowitz, C. J., M. A. Perez, D. K. Berry, and J. Piekarewicz. Dynamical response of the nuclear pasta in neutron star crusts. *Phys. Rev. C* 72:035801, 2005.

Cooper, E. D., and **C. J. Horowitz.** The vector analyzing power in elastic electron-nucleus scattering. *Phys. Rev. C* 72:034602, 2005.

Horowitz, C. J. Parity violation in astrophysics. *European Phys. J. A* 24:167, 2005.

Kesmodel, L. L. High-resolution electron energy loss spectroscopy. In *Encyclopedia of Surface and Colloid Science,* ed. A.T. Hubbard. New York: Marcell Dekker, 2002.

Jungwirthova, I., and **L. L. Kesmodel.** Benzene formation from Acetylene on Pd(111): A HREELS Study. *Surface Sci. Lett.* 470A:39, 2000.

Kesmodel, L. L, et al. High resolution electron energy loss spectroscopy of polymer surfaces. *Surface Sci.* 429:L475, 1999.

Kostelecký, V. A., R. Bluhm, C. Lane, and N. Russell. Clock-comparison tests of Lorentz and CPT symmetry in space. *Phys. Rev. Lett.,* 2002.

Kostelecký, V. A., and R. Jackiw. Radioactively induced CPT and Lorentz violation in electrodynamics. *Phys. Rev. Lett.* 82:3572, 1999.

Kostelecký, V. A., and **R. Van Kooten.** Bounding CPT violation in the neutral ∃ system. *Phys. Rev. D* 54:5585, 1996.

Huang, X., **S. Y. Lee,** E. Prebys, and R. Tomlin. Application of independent component analysis to the Fermilab booster. *Phys. Rev. Special Topics in Accelerators and Beams* 8:064001, 2005.

Cousineau, C., et al. **(S. Y. Lee).** Envelope and particles instabilities of space charge dominated beams in synchrotrons. *Phys. Rev. Special Top.: Accelerators Beams* 6:034205, 2003.

Hahn, H., et al. **(S. Y. Lee).** The RHIC design overview. *Nucl. Instrum. Methods Phys. Res., Sect. A* 499:245, 2003.

Hill, R. E., and **C.-Y. Liu.** Temperature dependent neutron scattering cross sections for polyethylene. *Nucl. Instr. Meth. A* 538:686, 2005.

Liu, C.-Y., and S. K. Lamoreaux. A new search for a permanent electric dipole moment of the electron in a solid state system. *Mod. Phys. Lett. A* 19:1235, 2004.

Londergan, J. T., and A. W. Thomas. Implications of current constraints on Parton charge symmetry. *J. Phys. G Nucl. Part. Phys.* 31:1151, 2005.

Londergan, J. T., D. P. Murdock, and A. W. Thomas. Experimental tests of charge symmetry violation in Parton distributions. *Phys. Rev. D* 72:036010, 2005.

Ashie, Y., et al. **(M. D. Messier).** (Super-Kamiokande Collaboration) A measurement of atmospheric neutrino oscillation parameters by super Kamiokande I. *Phys. Rev. D* 71:112005, 2005.

Indiana University Bloomington

Smy, M. B., et al. **(M. D. Messier).** (Super-Kamiokande Collaboration) Precise measurement of the solar neutrino day/night and seasonal variation in Super-Kamiokande-1. *Phys. Rev. D* 69:011104, 2004.

Liu, D. W., et al. **(M. D. Messier).** (Super-Kamiokande Collaboration) Limits on the neutrino magnetic moment using 1496 days of Super-Kamiokande-1 solar neutrino data. *Phys. Rev. Lett.* 93:021802, 2004.

Meyer, H. O. et al. **(R. Pollock).** Axial observables in $\sim d \sim p$ breakup and the three-nucleon force. *Phys. Rev. Lett.* 93(11):112502-(1/4), 2004.

Meyer, H. O., et al. Faddeev calculations of breakup reactions with realistic experimental constraints. *Few-Body Systems* 34:259, 2004.

Meyer, H. O. et al. **(R. Pollock).** Complete set of polarization observables in pp $\rightarrow$ pppi0 close to threshold. *Phys. Rev. C* 63:064002, 2001.

Rhodes, J., et al. **(J. Musser).** (SNAP Collaboration) Weak lensing from space I: Instrumentation and survey strategy. *Astroparticle Phys.* 20:377, 2004.

Beatty, J. J., et al. **(J. Musser).** (HEAT Collaboration) New measurement of the cosmic ray positron fraction from 5 to 15 GeV. *Phys. Rev. Lett.* 93:241102, 2004.

Gericke, M. T., et al. **(H. Nann** and **W. M. Snow).** A current mode detector array for γ-ray asymmetry measurements. *Nucl. Instrum. Methods A* 540:328, 2005.

Mitchel, G., et al. **(H. Nann** and **W. M. Snow).** A measurement of parity-violating gamma-ray asymmetries in polarized cold neutron capture on ^{35}Cl, ^{113}Cd, and ^{139}La. *Nucl. Instr. Methods A* 521:268, 2004.

Newton, R. Preface (IX-X); Three-dimensional direct scattering theory (686–701); Application of the Marchenko method to three-dimensional inverse scattering (742–53). In *Scattering,* eds., R. Pike and P. Sabatier. London: Academic Press, 2002.

Newton, R. *Thinking About Physics.* Princeton University Press, 2000.

Newton, R. *The truth of science.* Harvard University Press, 1997.

Akesson, T., et al. **(H. O. Ogren).** (ATLAS TRT Collaboration) Status of design and construction of the transition radiation tracker (TRT) for the Atlas experiment at the LHC. *Nucl. Instrum. Methods. A* A522:131, 2004.

Akesson, T., et al. **(H. O. Ogren).** (ATLAS TRT Collaboration) Atlas transition radiation tracker test-beam results. *Nucl. Instrum. Methods A* 522:50, 2004.

Akesson, T., et al. **(H. O. Ogren).** (ATLAS TRT Collaboration) Operation of the atlas transition radiation tracker under very high irradiation at the CERN LHC. *Nucl. Instrum. Methods A* 522:25, 2004.

Hirst, L. S., R. Bruinsma, **R. Pynn,** and C. R Safinya. Hierarchical self-assembly of actin bundle networks: gels with surface protein skin-layers. *J. Chem. Phys.* 123:104902, 2005.

Pynn, R., et al. Neutron spin echo scattering angle measurement (SESAME). *Rev. Sci. Instrum.* 76: 053902, 2005.

Hirst, L. S., et al. **(R. Pynn).** Microchannel systems in titanium and silicon for structural and mechanical studies of aligned protein self-assemblies. *Langmuir* 21:12109, 2005.

Schaich, W. L., et al. Optical near-field of multipolar plasmons of rod-shaped gold nanoparticles. *Europhys. Lett.* 69:538, 2005.

Schaich, W. L., et al. Three-dimensional mapping of the light intensity transmitted through nanoaperatures. *Nano Lett.* 5:1227, 2005.

Schaich, W. L., et al. Optical trapping with integrated near-field apertures. *J. Phys. Chem. B* 108:13607, 2004.

Serot, B. D. Building atomic nuclei with the Dirac equation. *Int. J. Mod. Phys. A* 19:107, 2004.

Furnstahl, R. J., and **(B. D. Serot).** Parameter counting in relativistic mean-field models. *Nucl. Phys. A* 671:447, 2000.

Mueller, H., and **B. D. Serot.** Relativistic mean-field theory and the high-density nuclear equation of state. *Nucl. Phys. A* 606:508–37, 1996.

Setayeshgar, S., et al. Application of coarse integration to bacterial chemotaxis. *Multiscale Modeling Simulation* 4(1):301, 2005.

Bialek, W., and **S. Setayeshgar.** Physical limits to biochemical signaling. *Proc. Natl. Acad. Sci. U.S.A.* 102(29):10040, 2005.

Setayeshgar, S., and A. J. Bernoff. Scroll waves in the presence of slowly varying anisotropy with application to the heart. *Phys. Rev. Lett.* 88:028101, 2002.

Adam, N. E., et al. **(M. Shepherd).** (CLEO Collaboration) Observation of 1^-0^- final states from ψ(2S) decays and e^+e^- annihilation. *Phys. Rev. Lett.* 94:012005, 2005.

Dodds, S., et al. **(M. Shepherd).** (CLEO Collaboration) Search for X(3872) in $\gamma\gamma$ fusion and ISR at CLEO.*Phys. Rev. Lett.* 94:032004, 2005.

Adams, G. S., et al. **(M. Shepherd).** (CLEO Collaboration) Measurement of the muonic branching fractions of the narrow Y resonances. *Phys. Rev. Lett.* 94:012001, 2005.

Nico, J. S., et al. **(W. M. Snow).** Measurement of the neutron lifetime by counting trapped protons in a cold neutron beam. *Phys. Rev. C* 71:055502, 2005.

Hussey, D. S., et al. **(W. M. Snow).** A polarized 3He compression system using metastability-exchange optical pumping. *Rev. Sci. Inst.* 76:053503, 2005.

Gentile, T. R., at al. **(W. M. Snow).** Polarized 3He spin filters in neutron scattering. *Physica B* 356:96, 2005.

Poplawski, N., **A. P. Szczepaniak,** and **J. T. Londergan.** Towards a relativistic description of exotic meson decays. *Phys. Rev. D* 71:016004, 2005.

Poplawski, N., **A. P. Szczepaniak,** and **J. T. Londergan.** Final state interactions in decays of the exotic meson Pi_1. *Phys. Rev. D* 71:056003, 2005.

Szczurek, A., and **A. P. Szczepaniak.** Diffractive photoproduction of opposite-charge pseudoscalar meson pairs at high energies. *Phys. Rev. D* 71:054005, 2005.

Auerbach, L. B., et al. **(R. Tayloe).** Tests of Lorentz violation in anti-nu/mu to anti-nu/e oscillations. *Phys. Rev. D* 72:076004, 2005.

Auerbach, L. B., et al. **(R. Tayloe).** Search for pi0 to num anti-numu decay in LSND. *Phys. Rev. Lett.* 92:091801, 2004.

Auerbach, L. B., et al. **(R. Tayloe).** Measurements of charged current reactions of ν_e on ^{12}C. *Phys. Rev. C* 64:065501, 2001.

Arms, K., et al. **(J. Urheim).** (CLEO Collaboration) Study of tau decays to four-hadron final states with Kaons. *Phys. Rev. Lett.* 94:241802, 2005.

Cronin-Hennessy, D., et al. **(J. Urheim).** (CLEO Collaboration) Observation of the hadronic transitions chi_b(1,2,) $\rightarrow$ omega upsilon(1S). *Phys. Rev. Lett.* 92, 222002, 2004.

Coan, T. E., et al. **(J. Urheim).** (CLEO Collaboration) Wess-Zumino current and the structure of the decay tau $\rightarrow$ K- K+pi- nu_tau 2). *Phys. Rev. Lett.* 92, 232001, 2004.

Sarsour, M., et al. **(S. E. Vigdor)** and **S. W. Wissink).** Measurement of the absolute np scattering differential cross section at 194 MeV. *Phys. Rev. Lett.* 94:082303, 2005.

Adams, J., et al. **(S. E. Vigdor** and **S. W. Wissink).** (STAR Collaboration). Experimental and theoretical challenges in the search for the quark-gluon plasma: The STAR Collaboration's Critical Assessment of the Evidence from RHIC Collisions. *Nucl. Phys. A* 757:102, 2005.

Adams, J., et al. **(S. E. Vigdor** and **S. W. Wissink).** (STAR Collaboration). Distributions of charged hadrons associated with high transverse momentum particles in pp and Au+Au collisions at $\sqrt{S_{NN}}$ = 200 GeV. *Phys. Rev. Lett.* 95: 152301, 2005.

Opper, A. K., et al. **(S. W. Wissink, A. D. Bacher,** and **C. Olmer).** Measurements of the spin observables $D_{NN'}$, P, and A_y in elastic proton scattering from ^{12}C and ^{16}O at 198 MeV. *Phys. Rev. C* 63:34614, 2001.

Wissink, S. W. et al. **(A. D. Bacher).** Spin transfer in *pp* elastic scattering at 198 MeV: Implications for the *BNN* coupling constant. *Phys. Rev. Lett.* 83:4498, 1999.

THE JOHNS HOPKINS UNIVERSITY

Henry A. Rowland Department of Physics and Astronomy

Program of Study

The Department offers a broad program for graduate and postdoctoral study in physics and astronomy in which intermediate, advanced, and specialized courses are offered. These courses and student research, begun as soon as possible, form the basis of the Ph.D. program. Considerable flexibility is available in each student's program, which is shaped to individual needs by recommendation from faculty and staff advisers. Students may choose to specialize in either physics or astrophysics, with a full curriculum of graduate courses available in both areas. In addition to required courses, candidates take written and oral preliminary examinations. Written examinations, covering intermediate-level material, must be passed by the end of the third semester. These exams are followed by an intermediate-level oral examination in the second year. A comprehensive oral examination is taken at the beginning of full-time research (usually in the third year). After completion of the student's research, an oral defense of the thesis is required. During residence, some teaching is usually required. Only those students who expect to complete the Ph.D. are admitted.

Research Facilities

The high-energy physics group has facilities for constructing the electronics and detectors needed in experiments and also has independent computing capabilities that allow full analyses of data. Nuclear physics equipment includes facilities for relativistic heavy-ion collision studies. Facilities for condensed matter physics include systems for molecular beam epitaxy, He^3-He^4 dilution refrigeration, high-rate sputtering, ultrahigh-vacuum thin-film deposition, automatic X-ray diffraction, scanning electron microscopy, X-ray fluorescence, LEED/Auger spectroscopy, SQUID and vibrating-sample magnetometry, ferromagnetic resonance, magnetooptics, neutron scattering, four-circle X-ray diffractometry, optical and electron-beam lithography, and dielectric susceptibility. For atomic, molecular, and plasma physics, facilities include high-resolution and very sensitive spectrometers for measurements of infrared to ultraviolet wavelengths, a high-precision X-ray spectrometer, extensive spectroscopic facilities, and lasers. The astrophysics group maintains a calibration and test facility for testing instrumentation for rocket and space flights. Computer facilities in the Department include a large number of Sun, DEC, SGI, HP, and Intel-based workstations. These machines support a wide range of functions, including data reduction, image processing, and simulation of physical processes. All are networked to universities, national laboratories, and supercomputer facilities throughout the world, and Hopkins is part of Internet2 and VBNS. The Johns Hopkins University is the home of the Space Telescope Science Institute, is a partner in the Sloan Digital Sky Survey, and owns a share of the ARC 3.5 meter optical/infrared telescope. The University and its partners manage a space astronomy mission (FUSE) that was launched in 1999. The Materials Research Science and Engineering Center (MRSEC) at Hopkins is one of twenty-four centers funded by the National Science Foundation to confront major challenges in the field of materials research. Facilities at the following laboratories and observatories are also frequently used: Brookhaven National Laboratory, Stanford Linear Accelerator Center, Fermi National Accelerator Laboratory, CERN, the University's own Applied Physics Laboratory, National Institute of Standards and Technology, Lawrence Berkeley Laboratory, Francis Bitter National Magnet Laboratory, Lawrence Livermore National Laboratory, the White Sands Missile Range, Kitt Peak National Observatory, Cerro Tololo Interamerican Observatory, the Very Large Array of the National Radio Astronomy Observatory, the Las Campanas Observatory, NASA's Goddard Space Flight Center and Space Telescope Science Institute, Anglo-Australian Observatory, Gemini Observatories, Chandra and XMM-Neutron X-ray Observatories, Argonne National Laboratory, NIST Center for Neutron Research, ISIS Facility, and Rutherford Appleton Laboratory.

Financial Aid

Various tuition fellowships are usually awarded to all full-time Ph.D. candidates. Nonservice University fellowships and teaching assistantships offer a minimum of $15,000 (plus full tuition remission) for the nine-month academic year in 2006–07. Summer research assistantships may be available at approximately $5170. Holders of teaching assistantships must assist in teaching general physics and introductory courses. This experience is useful for students interested in a college teaching career. In addition to teaching assistantships, research assistantships that pay $20,676 annually (plus full tuition remission) are also available for graduate students. These assistantships are awarded on the basis of experience, merit, and academic performance. (These awards are not usually given to first-year students unless they have special experience.)

Cost of Study

Tuition is $33,900 for the 2006–07 academic year; however, full tuition support is given to all Ph.D. candidates as part of the financial package, which also includes either a full teaching or research assistantship. A one-time matriculation fee of $500 is required at registration.

Living and Housing Costs

In general, graduate students live in apartments or rent private homes in residential areas near the University. In 2005–06, rates for unfurnished and furnished apartments varied from $500 to $950 per month. A campus housing office assists students in finding rooms and apartments in the surrounding residential area.

Student Group

The University's Homewood Campus (the Schools of Arts and Sciences and of Engineering) had 4,273 undergraduates and 1,625 graduate students in 2005–06. There were 98 graduate students in physics and astronomy; all received financial support of some kind. Admission to graduate study in the Department is highly competitive. An average of 20 new students are admitted each year; the majority enroll directly from college.

Location

Located in the northern section of Baltimore, the University is adjacent to one of the finest residential areas of the city, while most of the cultural activities of the large metropolitan area are but minutes away.

The University

The concept of graduate study came into being in America with the founding of the Johns Hopkins University in 1876. From the beginning, the hallmark of the University has been one of creative scholarship.

Applying

Requirements for admission after completion of the bachelor's or master's degree are transcripts of previous academic work, letters of recommendation, and GRE scores, including the General Test and the Subject Test in physics. International students whose native language is not English must submit their scores on the Test of English as a Foreign Language (TOEFL). Students are admitted only in September. Applications and all supporting materials must be received by January 15. The application fee is $60, but it is temporarily waived for students with either financial need or foreign exchange problems. Application materials may be obtained on the Web at the address listed in this description.

Correspondence and Information

Graduate Admissions
Henry A. Rowland Department of Physics and Astronomy
Bloomberg Center, Room 366
The Johns Hopkins University
3400 North Charles Street
Baltimore, Maryland 21218-2686
Phone: 410-516-7344
Fax: 410-516-7239
E-mail: admissions@pha.jhu.edu
Web site: http://www.pha.jhu.edu

The Johns Hopkins University

THE FACULTY AND THEIR RESEARCH

N. Peter Armitage, Assistant Professor. Experimental condensed-matter physics.
Jonathan A. Bagger, Krieger Eisenhower Professor and Chair. Theoretical high-energy physics.
Bruce A. Barnett, Professor. Experimental high-energy physics.
Steven Beckwith, Professor. Infrared astronomy.
Charles L. Bennett, Professor. Experimental cosmology.
William P. Blair, Research Professor. Astrophysics, shockwaves, spectroscopy of plasmas.
Barry J. Blumenfeld, Professor. Experimental high-energy physics.
Collin Broholm, Professor. Experimental condensed-matter physics.
Chia-Ling Chien, Jacob L. Hain Professor and Director, Materials Research Science and Engineering Center. Condensed-matter physics, artificially structured solids.
Chih-Yung Chien, Professor. Experimental high-energy physics.
Gabor Domokos, Professor. Theoretical high-energy physics, astroparticle physics.
Adam Falk, Professor and Dean, Krieger School of Arts and Sciences. Theoretical high-energy physics.
Gordon Feldman, Professor Emeritus. Quantum field theory, theory of elementary particles.
Paul D. Feldman, Professor. Astrophysics, spectroscopy, space physics, planetary and cometary atmospheres.
Michael Finkenthal, Research Professor. Plasma physics.
Holland Ford, Professor. Stellar dynamics, evolution of galaxies, active galactic nuclei, astronomical instrumentation.
Thomas Fulton, Professor Emeritus. Quantum electrodynamics, high-energy particle physics, atomic theory.
Riccardo Giacconi, University Professor. Astrophysics.
Andrei Gritsan, Assistant Professor. Experimental high-energy physics.
Timothy Heckman, Professor and Director, Center for Astrophysical Sciences. Galaxy evolution, starburst galaxies, active galactic nuclei.
Richard C. Henry, Professor and Director, Maryland Space Grant Consortium. Astronomy, astrophysics.
Brian R. Judd, Gerhard H. Dieke Professor Emeritus. Theoretical atomic and molecular physics, group theory, solid-state theory.
David Kaplan, Assistant Professor. Theoretical high-energy physics.
Chung W. Kim, Professor Emeritus. Theory of elementary particles, nuclear theory, cosmology.
Susan Kövesi-Domokos, Professor. Theoretical high-energy physics, astroparticle physics.
Julian H. Krolik, Professor. Theoretical astrophysics.
Yung Keun Lee, Professor Emeritus. Nuclear physics.
Robert Leheny, Assistant Professor. Experimental condensed-matter physics.
Petar Maksimovic, Assistant Professor. Experimental high-energy physics.
Nina Markovic, Assistant Professor. Experimental condensed-matter physics.
H. Warren Moos, Gerhard H. Dieke Professor. Astrophysics, plasma physics.
Charles Mattias Mountain, Professor and Director, Space Telescope Science Institute. Star formation in galaxies, capabilities of "second-generation telescope."
David A. Neufeld, Professor and Director, Theoretical Interdisciplinary Physics and Astrophysics Center. Theoretical astrophysics, interstellar medium, astrophysical masers, submillimeter astronomy.
Colin A. Norman, Professor. Theoretical astrophysics.
Aihud Pevsner, Jacob L. Hain Professor Emeritus. High-energy physics.
Daniel H. Reich, Professor. Experimental condensed-matter physics.
Adam Riess, Professor. Astrophysics.
Mark O. Robbins, Professor. Theoretical condensed-matter physics.
Raman Sundrum, Professor. Theoretical particle physics, including the physics of extra space-time dimensions, supersymmetry and non-perturbative phenomena.
Morris Swartz, Professor. Experimental high-energy physics.
Alexander S. Szalay, Alumni Centennial Professor. Theoretical astrophysics, galaxy formation.
Oleg Tchernyshyov, Assistant Professor. Theoretical condensed-matter physics, frustrated and quantum magnets, superconducting cuprates.
Zlatko Tesanovic, Professor. Theoretical condensed-matter physics.
Ethan T. Vishniac, Professor. Theoretical astrophysics.
J. C. Walker, Professor Emeritus. Condensed-matter physics, thin films and surfaces, nuclear physics.
Rosemary F. G. Wyse, Professor. Astrophysics, galaxy formation and evolution.

Adjunct and Visiting Appointments
Ronald J. Allen, Adjunct Professor, Space Telescope Science Institute. Spiral structure of galaxies, interstellar medium, radio and optical imaging.
Michael Fall, Adjunct Professor, Space Telescope Science Institute. Astrophysics.
Henry Ferguson, Adjunct Associate Professor, Space Telescope Science Institute. Observational cosmology, galaxy evolution, dwarf galaxies, space astronomy instrumentation.
Michael G. Hauser, Adjunct Professor, Space Telescope Science Institute. Cosmology, especially infrared background radiation.
Gerard A. Kriss, Adjunct Associate Professor, Space Telescope Science Institute. Astrophysics, observations of active galactic nuclei and clusters of galaxies.
Belita Kroiller, Adjunct Professor, UFRJ. Condensed-matter theory.
Mario Livio, Adjunct Associate Professor, Space Telescope Science Institute. Theoretical astrophysics, accretion onto white dwarfs, neutron stars and black holes, novae and supernovae.
Markus Luty, Adjunct Assistant Professor, University of Maryland, College Park. Elementary particle theory.
Bruce Margon, Adjunct Professor, Space Telescope Science Institute. High-energy astrophysics, space astronomy.
Antonella Nota, Adjunct Professor, Space Telescope Science Institute. Astronomy.
Ethan Schreier, Adjunct Professor and President, AUI. Astrophysics, active galaxies and jets.
Mark Stiles, Adjunct Professor, NIST. Condensed-matter experiment.
Roeland van der Marel, Adjunct Associate Professor, Space Telescope Science Institute. Black holes, cluster of galaxies, dark halos, galaxy structure and dynamics.
Kimberly Weaver, Adjunct Assistant Professor, NASA Goddard Space Flight Center. High-energy astrophysics.
Robert Williams, Adjunct Professor, Space Telescope Science Institute. Novae, emission line analysis.

Joint Appointments
Jack Morava, Professor, Mathematics. Algebraic topology, mathematical physics.
Darrell F. Strobel, Professor, Earth and Planetary Sciences. Planetary atmospheres, astrophysics.

RESEARCH ACTIVITIES

Astrophysics. Observational programs include the use of ground-based optical and radio telescopes, analysis of archival data from previous space experiments, new research with existing satellites and sounding rockets, and space experiments. There is extensive laboratory work on detectors and instrument development for ultraviolet and optical astronomy. Research is concentrated in the following areas of astrophysics: cosmology, active galactic nuclei and quasars, galaxies and galaxy dynamics, stellar populations, the interstellar medium, comets and planetary atmospheres, and diffuse ultraviolet background studies.

Atomic Physics. Research in this area includes theoretical work on the electronic structure of atoms and molecules.

Condensed-Matter Physics. Research programs involve studies of very thin magnetic films; interfaces and surfaces; amorphous materials; conducting, superconducting, and magnetic properties of artificially structured materials; nanocrystals of metals and alloys; low-dimensional quantum magnets; highly correlated electron systems; high-T_c superconductors; complex fluids; glass-forming systems; and nonequilibrium phenomena. Techniques involve SQUID magnetometry, X-ray diffraction, atomic force and magnetic force microscopy, neutron scattering, various cryogenic techniques, DC and AC conductivity, LEED and Auger spectroscopies, ferromagnetic resonance, and vibrating-sample magnetometry and dielectric susceptibility.

High-Energy Physics. Current programs involve the study of strong, electromagnetic, and weak interactions. Experiments currently in progress are being performed at the Tevatron $p\bar{p}$ collider at Fermilab, at LEP and SPS, in CERN in Switzerland, and an experiment CMS at LHC in CERN. Data analysis is in progress at these facilities and at the Homewood Campus. Facilities for the construction and testing of particle detectors and associated electronics are available.

Plasma Spectroscopy. Extreme ultraviolet/soft X-ray diagnostic instrumentation is used to study high-temperature plasma devices in controlled thermonuclear research.

Relativistic Heavy-Ion and Medium-Energy Nuclear Physics. The heavy-ion physics program includes the study of quark gluon plasma at the RHIC collider with the STAR and the BRAMS detectors at the Brookhaven National Laboratory.

Theoretical Physics. Areas of current research include particle physics, condensed-matter physics, molecular and atomic structure, quantum optics, and astrophysics. The particle theory group currently conducts research in supersymmetric theories, heavy quark theory, and astroparticle physics. The condensed-matter theory group studies superconductivity, quantum Hall effect, magnetism, quantum critical phenomena, and various forms of nonequilibrium and growth phenomena. Members of the theory group specializing in different areas maintain close contact with each other and with the experimental groups.

KENT STATE UNIVERSITY

Department of Physics

Programs of Study

The Department of Physics offers a diverse program of graduate study and research leading to the Master of Arts (M.A.), Master of Science (M.S.), and Doctor of Philosophy (Ph.D.) degrees. Major areas of research include experimental and theoretical research in biophysics, condensed matter, and nuclear physics. Condensed matter includes emphases on novel electron systems, high-T_c superconductors, and liquid crystal physics. The high-energy nuclear physics research area emphasizes quark-gluon structure of hadrons and hot hadronic matter produced in heavy-ion collisions. Biophysics research encompasses protein conformation studies and simulation of cellular-level systems. The Department has ongoing collaborations with the Departments of Biology, Chemistry, Chemical Physics, and Computer Science, through which students may pursue interdisciplinary research opportunities.

A student typically takes core courses during the first year of study. The M.A. and M.S. degrees require 32 semester credit hours. The M.S. degree requires a thesis. The Ph.D. degree requires courses, seminars, and dissertation research. Doctoral students normally pass the candidacy examination in their second year. The average time to the completion of the Ph.D. degree is less than the six-year national average.

Research Facilities

The Department of Physics has extensive and modern facilities supporting condensed-matter research in nonlinear optics, electrooptics, atomic-force microscopy, nuclear magnetic resonance, high-resolution synchrotron X-ray scattering, light scattering, microcalorimetry, millikelvin refrigeration, SQUID magnetometry, magnetoresistance, and the Hall effect. The experimental nuclear physics group is well equipped with state-of-the-art equipment, including large-volume ultrafast neutron detectors and neutron polarimeters developed by Kent researchers. All research programs receive specialized support from an in-house machine shop, an electronics shop, and a materials synthesis facility. Researchers also use the X-ray diffraction sector at the Advanced Photon Source of Argonne National Laboratory and the NIST Center for Neutron Research in Gaithersburg, Maryland. The KSU Center for Nuclear Research pursues its mission to support, enhance, and promote nuclear and particle physics with research at accelerator facilities at TJNAF and RHIC. The Liquid Crystal Institute (LCI) at Kent State University (KSU) is an internationally known research and technology center for liquid crystal science. Many of the Ph.D. researchers employed in the liquid crystal and flat-panel display industry worldwide have gained education and experience at KSU.

Financial Aid

In 2006–07, graduate appointments carry a stipend of $16,000 (over nine months), a full-tuition scholarship, and partial health insurance coverage. Prorated summer appointments as teaching or research assistants are usually available. It is possible to enter the program at midyear.

Cost of Study

All graduate appointees receive a full-tuition scholarship covering their cost of study. The 2006–07 tuition is $12,232 for in-state residents pursuing full-time study and $21,164 for out-of-state and international students.

Living and Housing Costs

The cost of living in Kent is well below the national average. Both on-campus and off-campus living accommodations are available. On-campus housing costs per month are $525 for a 2-person double, $703 for a deluxe single, and $760 for a 2-person semi-suite. Furnished one- and two-bedroom apartments for married students are available in the attractive University-owned Allerton Apartments. Rates per month are $660 for a one-bedroom and $690 for a two-bedroom apartment. All utilities (including cable and Internet access), excluding phone, are included in the costs. Further information is available online at http://www.res.kent.edu/newres/. Reasonably priced off-campus rental housing also can be easily found. The Campus Bus Service provides free transportation on campus and to the surrounding area.

Student Group

Of the approximately 24,000 students on the Kent campus, about 5,000 are graduate students. The physics department has about 60 graduate students from more than twelve countries. Kent State University emphasizes diversity, and the Department attracts students from traditionally underrepresented groups, including minorities and women. Students appreciate the diversity offered by the high percentage of women faculty members (25 percent) and the international composition of the faculty.

Student Outcomes

Students find rewarding positions in academic, government, and industrial institutions. KSU physics graduates enter initial employment in permanent positions at a higher rate than national norms. Graduates go on to careers in both academic and nonacademic areas. Presently, alumni include professors, presidents and vice presidents of companies, managers, directors, researchers at national laboratories, and consultants.

Location

Kent, which is nicknamed "Tree City," is a beautiful city of 30,000 residents located in northeastern Ohio, within a 50-minute drive of the metropolitan areas of Cleveland, Akron, Canton, and Youngstown. Its geographic location offers excellent job opportunities for students and their spouses. Downtown Cleveland is home to the Cleveland Symphony Orchestra, major-league sports (Cleveland Indians, Browns, Cavaliers, Crunch), and world-class museums (Museum of Art, Museum of Natural History, Rock and Roll Hall of Fame and Museum) as well as the Great Lakes Science Center. Visitors to Akron enjoy the Akron Art Museum, the National Inventors Hall of Fame, and the Akron Aeros baseball team. Nearby Canton has attractions such as the Pro Football Hall of Fame and the U.S. First Ladies exhibit in the McKinley Museum. The Cuyahoga Valley National Recreation Area provides excellent recreational opportunities for every season. Blossom Music Center, the Cleveland Orchestra, and the Porthouse Theatre are among many excellent cultural opportunities nearby.

The University and The Department

Kent State University offers degree programs ranging from undergraduate degrees in creative and performing arts to graduate degrees in the sciences. Kent State is ranked among the nation's seventy-seven public research universities demonstrating high research activity by the Carnegie Foundation for the Advancement of Teaching. A research library of exceptional quality provides access to Ohio academic and public libraries through OhioLink. It houses more than 2 million volumes in its collection and offers access to extensive electronic journals and databases.

The physics department has offices, classrooms, and laboratories in Smith Hall and the adjacent Science Research Building. There are 20 regular faculty members, whose per capita research funding is approximately $2 million per year from federal and state agencies and the private sector.

Applying

Graduate study may be initiated during any term, including summer. Links to online application forms are available at the Department's Web site. Paper forms for admission and financial assistance may be obtained by writing to the Department of Physics.

Correspondence and Information

Dr. Gerassimos G. Petratos, Chair
Department of Physics
Kent State University
Kent, Ohio 44242-0001
Phone: 330-672-2246
Fax: 330-672-2959
E-mail: gradprogram@physics.kent.edu
Web site: http://physics.kent.edu

Kent State University

THE FACULTY AND THEIR RESEARCH

David W. Allender, Professor; Ph.D., Illinois at Urbana-Champaign, 1975. Theoretical physics of condensed matter, liquid crystals, and superconductivity.
Carmen C. Almasan, Professor; Ph.D., South Carolina, 1989. Experimental condensed-matter physics, superconductivity, correlated electron systems, magnetism, low-temperature physics.
Bryon D. Anderson, Professor; Ph.D., Case Western Reserve, 1972. Experimental nuclear physics, nuclear force, nucleon structure.
Brett D. Ellman, Associate Professor; Ph.D., Chicago, 1992. Superconductivity, organic semiconductor, conduction mechanics in insulators and semiconductors, phonon physics, disordered magnets.
George Fai, Professor and Director of the Center for Nuclear Research; Ph.D., Eötvös Loránd (Budapest), 1974. Theoretical nuclear physics, relativistic nuclear collisions.
Daniele Finotello, Professor and Acting Dean of Research and Graduate Studies; Ph.D., SUNY at Buffalo, 1985. Nuclear magnetic resonance and calorimetry studies of thermotropic and lyotropic liquid crystals.
James T. Gleeson, Professor; Ph.D., Kent State, 1991. Nonequilibrium dynamics and pattern formation.
A. Mina T. Katramatou, Assistant Professor; Ph.D., American, 1988. Physics education, nuclear particle physics.
Declan Keane, Professor; Ph.D., University College (Dublin), 1981. Relativistic nuclear collisions.
Satyendra Kumar, Professor; Ph.D., Illinois at Urbana-Champaign, 1981. Liquid crystal structure, phase transition, electrooptical effects, complex fluids, biophysics, nanostructured materials, carbon nanotube composites.
Michael A. Lee, Professor and Graduate Coordinator; Ph.D., Northwestern, 1977. Condensed-matter theory, biophysics, computational physics.
D. Mark Manley, Professor; Ph.D., Wyoming, 1981. Experimental/phenomenological hadronic physics, baryon spectroscopy.
Elizabeth K. Mann, Associate Professor; Ph.D., Paris VI (Curie), 1992. Experimental soft-matter physics, surface physics, complex fluids.
Spyridon Margetis, Associate Professor; Ph.D., Frankfurt (Germany), 1990. Experimental high-energy nuclear physics.
Gerassimos Petratos, Professor and Chair; Ph.D., American, 1988. Experimental nuclear/particle physics.
John J. Portman, Assistant Professor; Ph.D., Illinois at Urbana-Champaign, 2000. Theoretical biological physics.
Khandker F. Quader, Professor; Ph.D., SUNY at Stony Brook, 1983. Theoretical condensed-matter physics, superconductivity, strongly correlated systems, low-temperature physics.
Almut Schroeder, Associate Professor; Ph.D., Karlsruhe (Germany), 1991. Experimental condensed-matter physics, strongly correlated electron physics, quantum phase transitions.
Samuel N. Sprunt Jr., Professor; Ph.D., MIT, 1989. Experimental liquid crystal physics, phase transitions, quasi-elastic light scattering, confocal microscopy.
Peter C. Tandy, Professor and Acting Vice President for Research; Ph.D., Flinders (Australia), 1973. Theoretical nuclear and particle physics.

FACULTY IN RELATED DISCIPLINES

Philip J. Bos, Professor, Chemical Physics Interdisciplinary Program, and Associate Director, Liquid Crystal Institute; Ph.D., Kent State, 1978. Liquid crystal applications.
Paul Farrell, Professor, Computer Science; Ph.D., Dublin, 1983. Numerical computation and analysis.
Arne Gericke, Assistant Professor, Biophysical Chemistry; Dr.rer.nat., Hamburg (Germany), 1994. Lipid-mediated protein functions/lytropic liquid crystals.
J. D. (David) Glass, Professor, Biological Sciences; Ph.D., Wesleyan, 1982. Neural regulation of the mammalian biological clock.
Robert T. Heath, Professor, Biological Sciences and Director of the Water Resources Research Institute; Ph.D., USC, 1968. Biochemical limnology, phosphorus dynamics in aquatic ecosystems, planktonic biochemistry, physiological ecology.
Jack R. Kelly, Professor, Chemical Physics Interdisciplinary Program; Ph.D., Clarkson, 1979. Electrooptic and dielectric properties of liquid crystals.
Oleg Lavrentovich, Professor, Chemical Physics Interdisciplinary Program, and Director, Liquid Crystal Institute; Ph.D., 1984, D.Sc., 1990, Ukrainian Academy of Sciences. Defects in liquid crystals, electrooptics of smectic liquid crystals, physics of liquid crystalline dispersions.
Peter Palffy-Muhoray, Professor and Associate Director, Liquid Crystal Institute, Chemical Physics Interdisciplinary Program; Ph.D., British Columbia, 1977. Nonlinear optics, pattern formation in liquid crystals.
Diane Stroup, Associate Professor, Chemistry; Ph.D., Ohio State, 1992. Molecular basis for regulation of gene expression.

ACTIVE RESEARCH TOPICS

Correlated Electron Physics

Theoretical investigation of fluctuation phenomena and mechanisms of superconductivity.
Study of systems displaying strong electronic correlations, e.g., unconventional superconductivity and magnetism, Bose-Einstein condensation, colossal magnetoresistance, non-Fermi liquid states, quantum phase transitions, quantum fluids, physics of novel semiconductors, and role of disorder and dimensionality.
Experimental studies utilize low-temperature and high-field electron/phonon transport, thermodynamic, neutron scattering, and magnetic techniques.
Theoretical studies employing techniques of many-particle quantum statistical mechanics, e.g., diagrammatic crossing-symmetric and functional integral methods, quantum transport equation, Fermi liquid theory, homotopic topology and elements of critical phenomena.
Systems studied include high-T_c cuprate, heavy fermion, and magnetic superconductors, manganites, rare-earth magnets, organic crystalline and liquid crystalline semiconductors.

Condensed-Matter Physics and Related Applications

Pattern formation in liquid crystals and complex fluids.
Phase transitions in liquid crystals in bulk and finite size effects.
High-precision heat capacity measurements of phase transitions in liquid crystals.
Surface effects on liquid crystals.
High-resolution synchrotron X-ray diffraction and small-angle neutron scattering studies of structure and critical phenomena in complex fluids; thermotropic, polymer, lyotropic, and banana liquid crystals; nanostructured systems; and carbon nanotube composites.
Neutron scattering studies of magnetic systems.
Liquid crystal display physics; ferroelectric liquid crystal displays, optical beam modulating, and steering devices.
Dynamics, thermodynamics, and microrheology in Langmuir films.

Biological Physics

Computer simulation models of cellular-level biological processes.
Protein conformation studies.
Experimental and theoretical investigations of model membrane systems.

Computational Physics

Quantum Monte Carlo and many-body physics, parallel scientific algorithms, molecular dynamics of proteins, reaction diffusion and Monte Carlo applied to biological processes.

Nuclear and Particle Physics

Experimental and theoretical investigation of the nuclear matter equation of state.
Signals of the quark-gluon phase in relativistic nuclear collisions.
Nonperturbative quantum field theory modeling of the quark-gluon structure of hadrons and strong interaction processes and matter, electroweak decays and form factors of hadrons.
Measurements to probe the charge and magnetic structure of the neutron.
Electromagnetic form factors of deuterium and helium.
Studies of nucleon resonances with electromagnetic and hadronic probes.
Few-nucleon reaction studies.
Spin-transfer studies in nuclear reactions.

Northeastern
UNIVERSITY

NORTHEASTERN UNIVERSITY

Department of Physics

Programs of Study The Department offers a full-time program leading to the Ph.D. and full-time and part-time evening programs leading to the M.S. Requirements for the Ph.D. include 45 semester hours of course work, a written qualifying examination, a thesis describing the results of independent research, and a final oral examination. Students may pursue basic research in elementary particle physics, condensed-matter physics, and molecular biophysics or in interdisciplinary areas such as materials science, surface sciences, chemical physics, biophysics, and applied engineering physics. They also may carry out cooperative research at technologically advanced industrial, governmental, and national and international laboratories and at medical research institutions in the Boston area. Requirements for the M.S. are 32 semester hours of credit, up to 8 of which may be transfer credit, if approved. There is no language requirement for any of the degrees.

Research Facilities The Department is housed in the Dana Research Center, with optics and condensed-matter labs in the Egan Research Center. There are ample modern research laboratories, Department and student machine shops, an electronics shop, conference and seminar rooms, and faculty and graduate student offices. The Egan Center provides a direct interface with materials researchers in chemistry and engineering and includes extensive meeting space in the Technology Transfer Center. In 1999, the Department received a $1.2-million NSF grant to establish the Advanced Scientific Computing Center (ASCC) in the Dana Research Center, which has been expanded and upgraded since. In addition to the research they do at campus facilities, faculty members and graduate students also work at research centers located in the United States and Europe. High-energy physics experiments are under way at Fermilab in Batavia, Illinois, and at CERN, Geneva, Switzerland. High-magnetic-field experiments are in progress at the National High-Field Magnet Laboratory in Tallahassee, Florida, and Los Alamos National Laboratory, New Mexico. Several groups use the synchrotron facilities at Brookhaven National Laboratory, Long Island, New York, and Argonne National Laboratory, Argonne, Illinois, and many faculty members have flourishing collaborations with scientists in Europe, Asia, and South America.

Financial Aid Northeastern awards financial aid through the Federal Perkins Loan, Federal Work-Study, and Federal Stafford Student Loan Programs and through minority fellowships, including G. E. Fellowships and Martin Luther King, Jr. Scholarships. The Graduate School offers teaching and research assistantships that include tuition remission and a stipend (currently $16,475 for two semesters) and require 20 hours of work per week. Tuition assistantships provide tuition remission and require 10 hours of work per week. The Department's Lawrence Award Program honors students with Excellence in Teaching Awards, Academic Excellence Awards, and a Speaker's Prize.

Cost of Study Tuition for the 2006–07 academic year is $930 per semester hour. Books and supplies cost about $875 per year. Tuition charges are made for Ph.D. thesis and continuation. Other charges include the Student Center fee and health insurance fee ($1915), which are required of all full-time students.

Living and Housing Costs On-campus housing for graduate students is limited and granted on a space-available basis. For more information about on- and off-campus housing options, students may visit http://www.housing.neu.edu. A public transportation system serves the greater Boston area, and there are subway and bus services that are convenient to the University.

Student Group In fall 2005, 23,385 students were enrolled at the University, representing a wide variety of academic, professional, geographic, and cultural backgrounds. The Department enrolled 59 full-time students, of whom 98 percent received some form of financial support. A small number of students were enrolled in the part-time, evening M.S. program. The Department awards roughly seven Ph.D. degrees and five M.S. degrees per year. Most graduates have continued to pursue research careers, either in academic institutions as postdoctoral fellows or in industrial, medical, or government laboratories.

Location Boston, Massachusetts, offers a rich cultural and intellectual history and is the premier educational center of the country, with more than thirty-five colleges in the city region. Cultural offerings, including several world-class museums, a bevy of art galleries, and the Boston Symphony, are diverse, and the city is home to people of every race, ethnicity, political persuasion, and religion. Boston also offers world-class restaurants and a range of outdoor activities and is steeped in New England tradition.

The University and The Department Founded in 1898, Northeastern University is a privately endowed, nonsectarian institution of higher learning. Located in heart of Boston, Massachusetts, Northeastern is a world leader in cooperative education and recognized for its expert faculty and first-rate academic and research facilities. It offers a variety of curricula through seven undergraduate colleges, nine graduate and professional schools, two part-time undergraduate divisions, a number of continuing education programs, an extensive research division, and several institutes. The Department offers opportunities for students to work on a wide range of groundbreaking research programs with an internationally recognized faculty whose goal is to provide an effective education to students with varied backgrounds.

Applying Although there is no absolute deadline for applying, completed applications should be received by February 1 to secure priority consideration for September acceptance, especially if financial assistance is sought. Scores on the GRE General Test and the Subject Test in physics are required. The latter is given considerable weight in the admissions and assistantship awarding process. For international students, a TOEFL or IELTS score is required for admission.

Correspondence and Information

Graduate Coordinator
Department of Physics
111 Dana Hall
Northeastern University
360 Huntington Avenue
Boston, Massachusetts 02115
Phone: 617-373-2902
Fax: 617-373-2943
E-mail: gradphysics@neu.edu
Web site: http://www.physics.neu.edu

Northeastern University

THE FACULTY AND THEIR RESEARCH

Professors

Ronald Aaron (Emeritus), Ph.D., Pennsylvania, 1961. Medical physics.
Petros Argyres (Emeritus), Ph.D., Berkeley, 1954. Condensed-matter theory.
Arun Bansil, Ph.D., Harvard, 1974. Condensed-matter theory.
Paul M. Champion, Ph.D., Illinois at Urbana-Champaign, 1975. Biological and medical physics.
David A. Garelick, Ph.D., MIT, 1963. Medical physics.
Michael J. Glaubman (Emeritus), Ph.D., Illinois, 1953. High-energy experimental physics.
Haim Goldberg, Ph.D., MIT, 1963. Particle theory.
Donald Heiman, Ph.D., California, Irvine, 1975. Condensed-matter experimental physics.
Jorge V. José, D.Sc., National of Mexico, 1976. Condensed-matter theory.
Alain Karma, Ph.D., California, Santa Barbara, 1986. Condensed-matter theory.
Sergy Kravchenko, Ph.D., Institute of Solid State Physics (Chernogolovka), 1988. Condensed-matter experimental physics.
Robert P. Lowndes, Interim Chair of the Department of Physics; Ph.D., London, 1966. Condensed-matter experimental physics.
Bertram J. Malenka (Emeritus), Ph.D., Harvard, 1951. Particle theory.
Robert S. Markiewicz, Ph.D., Berkeley, 1975. Condensed-matter experimental physics.
Pran Nath, Ph.D., Stanford, 1964. Particle theory.
Clive H. Perry (Emeritus), Ph.D., London, 1960. Condensed-matter experimental physics.
Stephen Reucroft, Ph.D., Liverpool, 1969. High-energy experimental physics.
Eugene J. Saletan (Emeritus), Ph.D., Princeton, 1962. High-energy experimental physics.
Carl A. Shiffman (Emeritus), D.Phil., Oxford, 1956. Medical physics.
Jeffrey B. Sokoloff, Ph.D., MIT, 1967. Condensed-matter theory.
Srinivas Sridhar, Ph.D., Caltech, 1983. Condensed-matter experimental physics.
Yogendra N. Srivastava, Ph.D., Indiana, 1964. Particle theory.
Tomasz Taylor, Ph.D., Warsaw, 1981. Particle theory.
Michael T. Vaughn, Ph.D., Purdue, 1960. Particle theory.
Eberhard von Goeler (Emeritus), Ph.D., Illinois, 1961. High-energy experimental physics.
Allan Widom, Ph.D., Cornell, 1967. Condensed-matter theory.
Fa-Yueh Wu, Ph.D., Washington (St. Louis), 1963. Condensed-matter theory.

Associate Professors

George Alverson, Graduate Coordinator; Ph.D., Illinois at Urbana-Champaign, 1979. High-energy experimental physics.
Nathan Israeloff, Ph.D., Illinois at Urbana-Champaign, 1990. Condensed-matter experimental physics.
J. Timothy Sage, Ph.D., Illinois at Urbana-Champaign, 1986. Molecular biophysics.
John D. Swain, Ph.D., Toronto, 1990. High-energy experimental physics.
Darien Wood, Ph.D., Berkeley, 1987. High-energy experimental physics.

Assistant Professors

Emanuela Barberis, Ph.D., California, Santa Cruz, 1996. High-energy experimental physics.
Latika Menon, Ph.D., Tata (Bombay), 1998. Nanoscaled materials.
Armen Stepanyants, Ph.D., Rhode Island, 1999. Condensed-matter theory.
Mark C. Williams, Ph.D., Minnesota, 1998. Molecular biophysics.

Research Associates

Luis Anchordoqui, Ph.D., National University of La Plata (Argentina), 1998: high-energy physics. Bernardo Barbiellini, Ph.D., Geneva, 1991: condensed-matter theory. Christopher Daly, Ph.D., Northeastern, 1996: condensed-matter physics. Gavin Hesketh, Ph.D., Manchester (England), 2003: high-energy experimental physics. Stanislaw Kaprzyk, Ph.D., Academy of Metallurgy (Krakow), 1981: condensed-matter theory. Matti Lindroos, Ph.D., Tampere (Finland), 1979: condensed-matter theory. Wentao Lu, Ph.D., Northeastern, 2001: condensed-matter theory. Micah McCauley, Ph.D., Colorado State, 2001: laser physics, biophysics. Jorge H. Moromisato, Ph.D., Northeastern, 1971: high-energy experimental physics. Pantanjali V. Parimi, Ph.D., Hyderabad (India), 1998: condensed-matter physics. Thomas Paul, Ph.D., Johns Hopkins, 1994: high-energy experimental physics. Antonio J. Pons-Rivero, Ph.D., Barcelona, 2004: statistical physics, nonlinear dynamics. Seppo Sahrakorpi, Ph.D., Tampere (Finland), 2001: condensed-matter theory. Yohannes Shiferaw, Ph.D., Pittsburgh, 2001: condensed-matter theory. Xiong Ye, Ph.D., Northeastern, 2003: biological physics. Anchi Yu, Ph.D., Beijing, 1999: physical chemistry and laser spectroscopy.

Adjunct Professors

George Tze Yung Chen, Ph.D., Brown, 1972: biomedical physics. Graham Farmelo, Ph.D., Liverpool, 1977: high-energy experimental physics. Howard Fenker, Ph.D., Vanderbilt, 1978: high-energy experimental physics. Wolfhard Kern, Ph.D., Bonn (Germany), 1958: high-energy experimental physics and education. Peter Mijnarends, Ph.D., Delft University of Technology, 1969: condensed-matter theory. C. Robert Morgan, Ph.D., MIT, 1969: condensed-matter theory. Fabio Sauli, Ph.D., Trieste (Italy), 1963: high-energy experimental physics.

RESEARCH ACTIVITIES

Experimental Biological and Medical Physics. The group probes the structure and function of macromolecules, metalloproteins, and protein complexes. Specific research areas include electron transport, macromolecular structure, enzyme catalysis, and ligand binding and protein dynamics, using quasi-elastic scattering; transient absorption spectroscopy; Raman, FT-IR, and fluorescence spectroscopy; femtosecond coherence spectroscopy; nuclear resonance vibrational spectroscopy; single-molecule-force spectroscopy using optical tweezers; measurements of human balance; and novel imaging technologies.

Experimental Condensed-Matter Physics. Research activities focus on high-temperature superconductors (HTSC), semiconductors, and magnetic materials. HTSC research includes fundamental studies of order parameter symmetry and vortex dynamics; flux-lattice melting; Josephson-junction arrays; low-field HTSC magnets; linear and nonlinear electrodynamics of HTSCs; electromagnetic response of HTSCs at far-infrared, microwave, and radio frequencies; growth and characterization of new HTSC ceramics and single crystals; and factors limiting critical currents. Research on semiconductors includes correlated electron and quantum Hall effects, 2-D metal-insulator transition and electron solids, magnetooptical spectroscopy of nanostructures and quantum layers, and molecular-beam epitaxy (MBE) crystal growth. Other areas under investigation are electromagnetic and quantum chaos, left-handed metamaterials, and Raman, FT-IR, mesoscopic systems, noise, scanning probe microscopy, and nanoscale properties of materials.

Experimental High-Energy Physics. The group is working on two major collider experiments: the D-Zero experiment at Fermi National Laboratory outside Chicago and the CMS experiment now under construction at CERN (the European Laboratory for Particle Physics in Geneva, Switzerland). These are frontier experiments probing the electroweak and strong interactions at the highest energy scales and represent a phased program of research keeping the group at the cutting edge of the experimental investigation of the structure of matter and the forces by which it interacts.

Particle Physics. The group has begun an active program in particle astrophysics and is involved in the construction of the Pierre Auger Cosmic Ray Observatory in Argentina, which aims to elucidate the origin and nature of the highest-energy cosmic rays. The group has a strong history of doing not just straight experimental particle physics but also the related phenomenology and keeping an eye out for creative spin-offs of its research with applications in other fields.

Theoretical Condensed-Matter Physics. Research topics include transport theory, quantum chaos, Fermi liquid theory, charge density waves, and dense dipolar suspensions and theory of Josephson junctions, catalytic properties of alloys, transport in nanostructures, structural phase transitions in DNA, nanotribology (atomic-level friction), electronic structure of disordered materials, magnetism, ferrites, Fermiology of HTSCs, Van Hove scenario and stripes in HTSCs, exact and rigorous results in statistical mechanics, localization and percolation in order-disorder phase transitions, positron annihilation and photoemission spectroscopy, and nonlinear dynamics and pattern formation.

Theoretical Elementary Particle Physics. Fundamental research includes the study of unified models based on supersymmetry and superstrings; unified gauge theories in the TeV range and precision calculations within and beyond the Standard Model; particle physics in the early universe; proton stability and neutrino masses; electroweak anomaly in the observed asymmetry of the baryon number, gravitational theory and quantum gravity, Kaluza-Klein theories and large-radius compactification, and computer simulations of topological structures in field theory; and finite temperature effects in quantum chromodynamics.

SELECTED PUBLICATIONS

Rutkove, S. B., **R. Aaron,** and **C. A. Shiffman.** Localized bioimpedance analysis in the evaluation of neuromuscular disease. *Muscle Nerve* 25:390, 2002.

Aaron, R., and **C. A. Shiffman.** Localized muscle impedance measurements. In *Skeletal Muscle: Pathology, Diagnosis and Management of Disease,* chap. 45, eds. V. R. Preedy and V. J. Peters. London: Greenwich Medical Media, 2002.

Aaron, R., and **C. A. Shiffman.** Using localized impedance measurements to study muscle changes resulting from injury and disease. *Ann. N.Y. Acad. Sci.* 904: 171, 2000.

Alverson, G., et al. Iguana architecture, framework and toolkit for interactive graphics. In *2003 Conference for Computing in High-Energy Nuclear Physics (CHEP-03) EconfC0303241:MOLT008,* e-Print Archive: cs.se/0306042, 2003.

Alverson, G., et al. The IGUANA Interactive Graphics Toolkit with examples from CMS and DØ. In *Proceedings of CHEP 2001,* ed. H. Chen. Beijing, China, 2001.

Alverson, G. (The HEPVis 2001 Group). Summary of the HEPVis '01 workshop. In *CHEP 2001.* Beijing, China, 2001.

Alverson, G., et al. Coherent and non-invasive open analysis architecture and framework with applications in CMS. In *Proceedings of CHEP 2001,* ed. H. S. Chen. Beijing, China, 2001.

Alverson, G., I. Gaponenko, and L. Taylor. The CMS IGUANA (Interactive Graphical User Analysis) Project. In *CHEP 2000.* Padova, Italy, 2000.

Bansil, A., et al. Angle-resolved photoemission spectra, electronic structure and spin-dependent scattering in Ni_{1-x}Fe_x permalloys. *Phys. Rev. B* 65:075106, 2002.

Bansil, A., et al. Electron momentum density in Cu_{0.9}A1_. *Appl. Phys. A,* 2002.

Bansil, A., M. Lindroos, and S. Sahrakorpi. Matrix element effects in angle-resolved photoemission from Bi2212: Energy and polarization dependencies, final state spectrum, spectral signatures of specific transitions and related issues. *Phys. Rev. B* 65:054514, 2002.

Bansil, A., and B. Barbiellini. Electron momentum density and Compton profile in disordered alloys. *J. Phys. Chem. Solids* 62:2191, 2001.

Bansil, A., and B. Barbiellini. Electron momentum distribution in A1 and A1_Li_. *J. Phys. Chem. Solids* 62:2223, 2001.

Barberis, E., S. Reucroft, and **D. Wood** et al. (D-Zero Collaboration). Observation of diffractively produced W and Z bosons in antiproton-proton collisions at s** (1/2) = 1800-Gev. *Phys. Lett. B* 574:169–79, 2003.

Barberis, E., S. Reucroft, and **D. Wood** et al. (D-Zero Collaboration). Search for large extra dimensions in the monojet E(T) channel at D0. *Phys. Rev. Lett.* 90: 251802, 2003.

Barberis, E., S. Reucroft, and **D. Wood** et al. (D-Zero Collaboration). Top anti-top production cross-section in proton-antiproton collisions at s** (1/2) = 1.8-Tev. *Phys. Rev. D* 67:012004, 2003.

Barberis, E., S. Reucroft, and **D. Wood** et al. (D-Zero Collaboration). Subjet multiplicity of gluon and quark jets reconstructed with the K(T) algorithm in P anti-P collisions. *Phys. Rev. D* 65:052008, 2002.

Barberis, E., S. Reucroft, and **D. Wood** et al. (D-Zero Collaboration). The inclusive jet cross-section in P anti-P collisions at S**(1/2) = 1.8-Tev using the K(T) algorithm. *Phys. Lett. B* 525:211, 2002.

Rosca, F., et al. **(P. M. Champion).** Investigations of heme protein absorption lineshapes, vibrational relaxation, and resonance Raman scattering on ultrafast timescales. *J. Phys. Chem.* 107:8156, 2003.

Berezhna, S., **P. M. Champion,** and H. Wohlrab. Resonance Raman investigations of cytochrome c conformational change upon interaction with the membranes of intact and Ca2+ exposed mitochondria. *Biochemistry* 42:6149, 2003.

Unno, M., et al. **(P. M. Champion).** Effects of complex formation of cytochrome P450 with putidaredoxin: Evidence for protein-specific interactions involving the proximal thiolate ligand. *J. Biol. Chem.* 277:2547, 2002.

Rosca, F., et al. **(P. M. Champion).** Investigations of anharmonic low-frequency oscillations in heme proteins. *J. Phys. Chem. A* 106:3540, 2002.

Ye, X., A. Demidov, and **P. M. Champion.** Measurements of the photodissociation quantum yields of MbNO and MbO2 and the vibrational relaxation of the six-coordinate heme species. *J. Am. Chem. Soc.* 124:5914, 2002.

Garelick, D. A., A. Widom, M. Harris, and R. Koleva. Posture sway and the transition rate for a fall. *Physica A* 293:605, 2001.

Garelick, D. A. (Co-Leader), et al. Balance evaluation with an ultrasonic measuring system. In *Proceedings of the January 1994 Meeting jointly sponsored by the Association of Academic Physiatrists and the American Academy of Physical Medicine and Rehabilitation,* 1994.

Anchordoqui, L., and **H. Goldberg.** Black hole chromosphere at the LHC. *Phys. Rev. D* 67:064010, 2003.

Anchordoqui, L., **H. Goldberg,** and D. F. Torres. Anisotropy at the end of the cosmic ray spectrum? *Phys. Rev. D* 67:123006, 2003.

Anchordoqui, L., and **H. Goldberg.** Time variation of the fine structure constant driven by quintessence. *Phys. Rev. D* 68:083513, 2003.

Anchordoqui, L., J. L. Feng, **H. Goldberg,** and A. D. Shapere. Updated limits on TeV-scale gravity from absence of neutrino cosmic ray showers mediated by black holes. *Phys. Rev. D* 68:104025, 2003.

Heiman, D., and **C. H. Perry.** Magneto-optics of semiconductors. In *High Magnetic Fields: Science and Technology; Theory and Experimentation,* vol. 2, pp. 47–72, eds. F. Herlach and N. Miura. World Scientific, 2003.

Heiman, D., and **C. H. Perry.** Magneto-optics. In *Characterization of Reduced Dimensional Semiconductor Microstructures,* ed. F. H. Pollak. In series *Optoelectronic Properties of Semiconductor Quantum Wells and Superlattices.* Gordon and Breach Publishers, 2002.

Heiman, D., and **C. H. Perry.** Magneto-optics of semiconductors. In *Physics of High Magnetic Fields and their Applications,* eds. F. Herlach and N. Miura. World Scientific, 2002.

Okamura, H., et al. **(D. Heiman).** Inhibited recombination of charged magnetoexcitons. *Phys. Rev. B Rapid Commun.* 58:R15985, 1998.

Okamura, H., et al. **(D. Heiman).** Inhibited recombination of negatively-charged excitons in GaAs quantum wells at high magnetic fields. *Physica B* 470:256–8, 1998.

Israeloff, N. E., and T. S. Grigera. Numerical study of aging in coupled two-level systems. *Philos. Mag. B* 82:313, 2001.

Vidal, R. E., and **N. E. Israeloff.** Direct observation of molecular cooperativity near the glass transition. *Nature* 408:659, 2000.

Grigera, T. S., and **N. E. Israeloff.** Observation of fluctuation dissipation violations in a structural glass. *Phys. Rev. Lett.* 83:5038, 1999.

Walther, L. E., et al. **(N. E. Israeloff).** Atomic force measurement of low frequency dielectric noise. *Appl. Phys. Lett.* 72:3223, 1998.

Gongora, T, A., **J. V. José,** and S. Schaffner. Classical solutions of an electron in magnetized wedge billiards. *Phys. Rev. E* 66:047201, 2002.

José, J. V., and M. V. José. Thermodynamic distributions of heterogeneous receptor populations. In *Drug Receptor Thermodynamics: Introduction and Applications,* p. 593, ed. R. Raffa. Chichester, Sussex, England: J. Wiley and Sons, Ltd., 2001.

Tiesinga, P. H. E., et al. **(J. V. José).** Computational model of carbachol-induced delta, theta, and gamma oscillations in the hippocampus. *Hippocampus* 11:25, 2001.

Gibbons, F., et al. **(J. V. José).** A dynamical model of kinesin-microtubule motility assays. *Biophys. J.* 80:2515, 2001.

José, J. V., and E. Saletan. *Classical Mechanics: A Contemporary Approach.* Cambridge University Press, 1998. Third edition, 2002.

Echebarria, B., and **A. Karma.** Instability and spatiotemporal dynamics of alternans in paced cardiac tissue. *Phys. Rev. Lett.* 88:208101, 2002.

Karma, A., H. Levine, and D. Kessler. Phase-field model of mode-III dynamic fracture. *Phys. Rev. Lett.* 87:045501, 2001.

Karma, A. Phase-field formulation for quantitative modeling of alloy solidification. *Phys. Rev. Lett.* 87:115701, 2001.

Erlebacher, J., et al. **(A. Karma).** Evolution of nanoporosity in dealloying. *Nature* 410:450, 2001.

Plapp, M., and **A. Karma.** Multiscale random-walk algorithm for simulating interfacial pattern formation. *Phys. Rev. Lett.* 84:1740, 2000.

Karma, A. New paradigm for drug therapies of cardiac fibrillation. *Proc. Natl. Acad. Sci. U.S.A.* 97:5687, 2000.

Kravchenko, S. V., and M. P. Sarachik. Metal-insulator transition in two-dimensional electron systems. *Rep. Prog. Phys.* 67:1, 2004.

Rahimi, M., et al. **(S. V. Kravchenko).** Coherent back-scattering near the two-dimensional metal-insulator transition. *Phys. Rev. Lett.* 91:116402, 2003.

Shashkin, A. A., et al. **(S. V. Kravchenko).** Spin-independent origin of the strongly enhanced effective mass in a dilute 2D electron system. *Phys. Rev. Lett.* 91: 046403, 2003.

Abrahams, E., **S. V. Kravchenko,** and M. P. Sarachik. Metallic behavior and related phenomena in two dimensions. *Rev. Mod. Phys.* 73:251, 2001.

Shashkin, A. A., **S. V. Kravchenko,** V. T. Dolgopolov, and T. M. Klapwijk. Indication of the ferromagnetic instability in a dilute two-dimensional electron system. *Phys. Rev. Lett.* 87:086801, 2001.

Kusko, C., and **R. S. Markiewicz.** White-Scalapino-like stripes in the mean-field Hubbard model. *Phys. Rev. B* 65:041102, 2002.

Markiewicz, R. S., and C. Kusko. Phase separation models for cuprate stripe arrays. *Phys. Rev. B* 65:064520, 2002.

Markiewicz, R. S., and C. Kusko. Flux phase as a dynamic Jahn-Teller phase: Berryonic matter in the cuprates? *Phys. Rev. B* 66:024506, 2002.

Kusko, C., **R. S. Markiewicz,** M. Lindroos, and **A. Bansil.** Fermi surface evolution and collapse of the Mott pseudogap in Nd2-x CexCuO4 delta. *Phys. Rev. B* 66: 140513, 2002.

Markiewicz, R. S., et al. Cluster spin glass distribution functions in La2-xSrxCuO4. *Phys. Rev. B* 054409-1, 2001.

Kusko, C., et al. **(R. S. Markiewicz** and **S. Sridhar).** Anomalous microwave conductivity due to collective transport in the pseupseudogap state of cuprate superconductors. *Phys. Rev. B.,* 2001.

Kusko, C., and **R. S. Markiewicz.** Remnant Fermi surfaces in photoemission. *Phys. Rev. Lett.* 84:963, 2000.

Northeastern University

Markiewicz, R. S., and **M. T. Vaughn.** Stripe disordering transition. In *University of Miami Conference on High Temperature Superconductivity,* Miami, Florida, January 7–13, 1999.

Markiewicz, R. S., C. Kusko, and **M. T. Vaughn.** SO(6)-generalized pseudogap model of the cuprates. In *University of Miami Conference on High Temperature Superconductivity,* Miami, Florida, January 7–13, 1999.

Markiewicz, R. S., and **M. T. Vaughn.** Higher symmetries in condensed matter physics. In *Particles, Strings and Cosmology—PASCOS98,* ed. **P. Nath.** Singapore: World Scientific, 1999.

Nath, P., and R. Syed. Coupling the supersymmetric 210 vector multiplet to matter in SO(10). *Nucl. Phys. B* 676:64–98, 2004.

Nath, P., and T. Ibrahim. Decays of Higgs to b anti-b, tau anti-tau and c anti-c as signatures of supersymmetry and CP phases. *Phys. Rev. D* 68:015008, 2003.

Nath, P., and P. Frampton. MSUGRA celebrates its 20th year. *CERN Cour.* 43N7: 27–8, 2003.

Nath, P., and T. Ibrahim. Supersymmetric QCD and supersymmetric electroweak loop corrections to b, t, and tau masses including the effects of CP phases. *Phys. Rev. D* 67:095003, 2003.

Nath, P., and U. Chattopadhyay. WMAP constraints, SUSY dark matter and implications for the direct detection of SUSY. *Phys. Rev. D* 68:035005, 2003.

Kim, Y., K.-S. Lee, and **C. H. Perry.** Optically detected heavy- and light-hold anti-crossing in GaAs quantum wells under pulsed magnetic fields. *Appl. Phys. Lett.* 84:738, 2004.

Russell, K. J., et al. **(C. H. Perry).** Room temperature electro-optic up conversion via internal photoemission. *Appl. Phys. Lett.* 82:2960, 2003.

Appelbaum, I., et al. **(C. H. Perry).** Ballistic electron emission luminescence. *Appl. Phys. Lett.* 82:4498, 2003.

Kim, Y., and **C. H. Perry** et al. Electron-hole separation studies of the v=1 quantum Hall state in modulation-doped GaAs/AlGaAs single heterojunctions in high magnetic fields. *Phys. Rev. B* 64:195302, 2001.

Reucroft, S., and **J. Swain** et al. (L3 Collaboration). Inclusive Pi0 and K0(S) production in two photon collisions at LEP. *Phys. Lett. B* 524:44–54, 2002.

Reucroft, S., and **J. Swain** et al. (L3 Collaboration). Search for R parity violating decays of supersymmetric particles in E+ E- collisions at LEP. *Phys. Lett. B* 524: 65–80, 2002.

Reucroft, S., and **J. Swain** et al. (L3 Collaboration). Study of the W+ W- gamma process and limits on anomalous quartic gauge boson couplings at LEP. *Phys. Lett. B* 527:29–38, 2002.

Reucroft, S., and **J. Swain** et al. (L3 Collaboration). F(1)(1285) formation in two photon collisions at LEP. *Phys. Lett. B* 526:269–77, 2002.

Reucroft, S., and **J. Swain.** *An Introduction to Science,* 2nd ed. McGraw-Hill/Primis, 2000.

Rai, B. K., et al. **(J. T. Sage).** Direct determination of the complete set of iron normal modes in a porphyrin-imidazole model for carbonmonoxy-heme proteins: [Fe(TPP)(CO)(1-MeIm)]. *J. Am. Chem. Soc.* 125:6927–36, 2003.

Budarz, T. E., et al. **(J. T. Sage).** Determination of the complete set of iron normal modes in the heme model compound Fe III (OEP) Cl from nuclear resonance vibrational spectroscopic data. *J. Phys. Chem. B* 107:11170–7, 2003.

Rai, B. K., et al. **(J. T. Sage).** Iron normal mode dynamics in a porphyrin-imidazole model for deoxyheme proteins. *Phys. Rev. E* 66:051904, 2002.

Parimi, P. V., et al. **(J. B. Sokoloff** and **S. Sridhar).** Negative refraction and left-handed electromagnetism in microwave photonic crystals. *Phys. Rev. Lett.,* in press.

Daly, C., J. Zhang, and **J. B. Sokoloff.** Dry friction due to absorbed molecules. *Phys. Rev. Lett.* 90:246101, 2003.

Daly, C., J. Zhang, and **J. B. Sokoloff.** Friction in the zero sliding velocity limit. *Phys. Rev. E* 68:0661, 2003.

Sokoloff, J. B. Explaining the virtual universal occurrence of static friction. *Phys. Rev. B* 65:115415, 2002.

Sokoloff, J. B. Static friction between elastic solids due to random asperities. *Phys. Rev. Lett.* 86:3312, 2001.

Parimi, P. V., W. T. Lu, P. Vodo, and **S. Sridhar.** Imaging by flat lens using negative refraction. *Nature* 426:404, 2003.

Bishop, A. R., S. R. Shenoy, and **S. Sridhar,** eds. *Intrinsic Multiscale Structure and Dynamics in Novel Electronic Oxides.* World Scientific, 2003.

Lu, W. T., **S. Sridhar,** and M. Zworski. Fractal Weyl laws for chaotic open systems. *Phys. Rev. Lett.* 91:154101, 2003.

Rao, D. M., et al. **(S. Sridhar).** Isospectrality in chaotic billiards. *Phys. Rev. E* 68:26208, 2003.

Grau, A., R. Godbole, G. Pancheri, and **Y. N. Srivastava.** The role of soft gluon radiation in the fall and rise of total cross sections. *Nucl. Phys. B Proc. Suppl.* 126, 2003.

Grau, A., S. Pacetti, G. Pancheri, and **Y. N. Srivastava.** Bloch-Nordsieck resummation for QCD processes. *Nucl. Phys. B. Proc. Suppl.* 126, 2003.

Sivasubramanian, S., **Y. N. Srivastava,** G. Vitiello, and **A. Widom.** Quantum dissipation induced noncommutative geometry. *Phys. Lett. A* 311:97, 2003.

Srivastava, Y. N., and **A. Widom.** Dirac analysis of the muon (g-2) measurements and non-commutative geometry of quantum beams. *J. Phys.* 53:1628, 2003.

Stepanyants, A., G. Tamás, and D. B. Chklovskii. Class-specific features of neuronal wiring. *Neuron* 43:251–9, 2004.

Chklovskii, D. B., and **A. Stepanyants.** Power-law for axon diameters at branch point. *BMC Neurosci.* 4:18, 2003.

Stepanyants, A., P. R. Hof, and D. B. Chklovskii. Geometry and structural plasticity of synaptic connectivity. *Neuron* 34:275–88, 2002.

Stepanyants, A. Diffusion and localization of surface gravity waves over irregular bathymetry. *Phys. Rev. E* 63(3):031202, 2001

Swain, J. Anomalous electroweak couplings of the tau and tau neutrino. In *Proceedings of the Sixth International Workshop on Tau LEPton Physics (TAU2000),* Victoria, B.C., Canada, September 18–21, 2000. *Nucl. Phys. Proc. Suppl.* 98:351, 2001.

Taylor, T. R., and P. Khorsand. Renormalization of boundary fermions and world-volume potentials on D-branes. *Nucl. Phys. B* 611:239, 2001.

Taylor, T. R., and A. Fotopoulos. Remarks on two-loop free energy in N=4 supersymmetric Yang-Mills theory at finite temperature. *Phys. Rev. D* 59:61701, 1999.

Taylor, T. R., et al. Duality in superstring compactifications with magnetic field backgrounds. *Nucl. Phys. B* 511:611, 1998.

Taylor, T. R., I. Antoniadis, and B. Pioline. Calculable e effects. *Nucl. Phys. B* 512:61, 1998.

Vaughn, M. T., and **R. S. Markiewicz.** Classification of the Van Hove scenario as an SO(8) spectrum generating algebra. *Phys. Rev. B Rapid Commun.* 57:14052–5, 1998.

Sivasubramanian, S., **A. Widom,** and **Y. Srivastava.** Equivalent circuit and simulations for the Landau-Khalatnikov model of ferroelectric hysteresis. *IEEE (UFFC)* 50:950, 2003.

Sivasubramanian, S., **A. Widom,** and **Y. N. Srivastava.** Microscopic basis of thermal superradiance. *J. Phys.: Condens. Matter* 15:1109, 2003.

Pant, K., R. L. Karpel, I. Rouzina, and **M. C. Williams.** Mechanical measurement of single molecule binding rates: Kinetics of DNA helix-destabilization by T4 gene 32 protein. *J. Mol. Biol.* 336:851–70, 2004.

Pant, K., R. L. Karpel, and **M. C. Williams.** Kinetic regulation of single DNA molecule denaturation by T4 gene 32 protein structural domains. *J. Mol. Biol.* 327:571–8, 2003.

Williams, M. C., R. J. Gorelick, and K. Musier-Forsyth. Specific zinc finger architecture required for HIV-1 nucleocapsid protein's nucleic acid chaperone function. *Proc. Natl. Acad. Sci. U.S.A.* 99:8614–9, 2002.

Williams, M. C. Optical tweezers: Measuring piconewton forces, in single molecule techniques. In *Biophysics Textbook Online,* ed. in chief L. DeFelice, ed. P. Schwille. Bethesda, Md.: Biophysical Society, 2002.

Williams, M. C., and I. Rouzina. Force spectroscopy of single DNA and RNA molecules. *Curr. Opin. Struct. Biol.* 12:330–6, 2002.

Wu, F. Y., and H. Kunz. The odd eight-vertex model. *J. Stat. Phys.,* in press.

Tzeng, W. J., and **F. Y. Wu.** Dimers on a simple-quartic net with a vacancy. *J. Stat. Phys.* 110:671–89, 2003.

Lee, D. H., and **F. Y. Wu.** Duality relation for frustrated spin models. *Phys. Rev. E* 67:026111, 2003.

Lieb, E. H., and **F. Y. Wu.** The one-dimensional Hubbard model: A reminiscence. *Physica A* 321:1–27, 2003.

Wu, F. Y. Dimers and spanning trees: Some recent results. *Int. J. Mod. Phys. B* 16:1951–61, 2002.

King, C., and **F. Y. Wu.** New correlation relations for the planar Potts model. *J. Stat. Phys.* 107:919–40, 2002.

Wu, F. Y., C. King, and W. T. Lu. On the rooted Tutte polynomial. *Ann. Inst. Fourier, Grenoble* 49:101–12, 1999.

Lu, W. T., and **F. Y. Wu.** Partition function zeroes of a self-dual Ising model. *Physica A* 258:157–70, 1998.

Wu, F. Y. The exact solution of a class of three-dimensional lattice statistical model. In *Proc. 7th Asia Pacific Phys. Conf.,* pp. 20–8, ed. H. Chem. Beijing: Science Press, 1998.

Lu, W. T., and **F. Y. Wu.** On the duality relation for correlation functions of the Potts model. *J. Phys. A* 31:2823, 1998.

OHIO UNIVERSITY

Department of Physics and Astronomy

Programs of Study

The Department of Physics and Astronomy offers graduate study and research programs leading to the Master of Arts, Master of Science, and Doctor of Philosophy degrees. The program of study emphasizes individual needs and interests in addition to essential general requirements of the discipline. Major areas of current research are experimental and theoretical nuclear and intermediate-energy physics, experimental condensed-matter and surface physics, theoretical condensed-matter and statistical physics, nonlinear systems and chaos, biophysics, acoustics, atomic physics, mathematical and computational physics, biological physics, geophysics, astronomy and astrophysics.

A student typically takes core courses (mechanics, math, quantum, electrodynamics) during the first year in preparation for the comprehensive exam that is given at the end of the summer following the first year. Students can usually retake the exam during the winter break of the second year if necessary. The courses in the second year cover more advanced topics. Master's degrees require completion of 45 graduate credits in physics and have both thesis and nonthesis options. Applied master's degrees (e.g., computational physics) are under development. The Ph.D. requirements include passing the comprehensive exam and writing and orally defending the dissertation.

Research Facilities

The physics department occupies two wings of Clippinger Laboratories, a modern, well-equipped research building; the Edwards Accelerator Building, which contains Ohio University's 4.5-MV high-intensity tandem accelerator; and the Surface Science Research Laboratory, which is isolated from mechanical and electrical disturbances. Specialized facilities for measuring structural, thermal, transport, optical, and magnetic properties of condensed matter are available. In addition to research computers in laboratories, students have access to a Beowulf cluster and the Ohio Supercomputer Center, where massively parallel systems (e.g. a CRAY SVI, a SGIB Origin 2000, and an Itanium Cluster) are located.

Financial Aid

Financial aid is available in the form of teaching assistantships (TAs) and research assistantships (RAs). All cover the full cost of tuition plus a stipend from which a quarterly fee of $430 must be paid by the student. Current stipend levels for TAs are $19,000 per year. The stipend levels for RAs are set by the research grant holders but are at or above the level of the TA stipends. TAs require approximately 15 hours per week of laboratory and/or teaching duties. Merit stipends of $20,000 per year are available for outstanding applicants. Special assistantships through the Condensed Matter and Surface Science (CMSS) program are also available.

Cost of Study

Tuition and fees are $2165 per quarter for Ohio residents and $4610 per quarter for out-of-state students. Tuition and fees for part-time students are prorated.

Living and Housing Costs

On-campus rooms for single students are $1265 per quarter, while married student apartments cost from $578 to $707 per month. A number of off-campus apartments and rooms are available at various costs.

Student Group

About 19,800 students study on the main campus of the University, and about 2,700 of these are graduate students. The graduate student enrollment in the physics department ranges from 60 to 70.

Location

Athens is a city of about 25,000, situated in the rolling Appalachian foothills of southeastern Ohio. The surrounding landscape consists of wooded hills rising about the Hocking River valley, and the area offers many outdoor recreational opportunities. Eight state parks lie within easy driving distance of the campus and are popular spots for relaxation. The outstanding intellectual and cultural activities sponsored by this diverse university community are pleasantly blended in Athens with a lively tradition of music and crafts.

The University and The Department

Ohio University, founded in 1804 and the oldest institution of higher education in the Northwest Territory, is a comprehensive university with a wide range of graduate and undergraduate programs. The Ph.D. program in physics began in 1959, and more than 220 doctoral degrees have been awarded. Currently, the department has 33 regular faculty members, and additional part-time faculty and postdoctoral fellows. Sponsored research in the department amounts to approximately $3.8 million per year and comes from NSF, DOE, DOD, ONR, BMDO, NASA, and the state of Ohio. Further information can be found at the department's home page listed below.

Applying

Information on application procedures and downloadable forms can be found at http://www.ohiou.edu/graduate/apps.htm. These materials can also be obtained by writing to the Department of Physics and Astronomy at the address listed below.

Correspondence and Information

Graduate Admissions Chair
Department of Physics and Astronomy
Ohio University
Athens, Ohio 45701
Phone: 740-593-1718
Web site: http://www.phy.ohiou.edu

Ohio University

THE FACULTY AND THEIR RESEARCH

Professors
David A. Drabold, Ph.D., Washington (St. Louis), 1989. Theoretical condensed matter, computational methodology for electronic structure, theory of topologically disordered materials.
Charlotte Elster, Dr.rer.nat, Bonn, 1986. Nuclear and intermediate-energy theory.
Steven M. Grimes, Ph.D., Wisconsin–Madison, 1968. Nuclear physics.
Kenneth H. Hicks, Ph.D., Colorado, 1984; Director, Institute for Nuclear and Particle Physics. Nuclear and intermediate-energy physics.
David C. Ingram, Ph.D., Salford (England), 1980. Atomic collisions in solids, thin films, deposition and analysis.
Martin E. Kordesch, Ph.D., Case Western Reserve, 1984. Surface physics.
Prakash Madappa, Ph.D., Bombay, 1979. Nuclear and particle astrophysics.
Jaccobo Rapaport, Distinguished Professor Emeritus, Ph.D., MIT, 1963. Nuclear physics.
Roger W. Rollins, Ph.D., Cornell, 1967. Solid-state physics, superconductivity, chaotic systems.
Folden B. Stumpf, Professor Emeritus, Ph.D., IIT, 1956. Accoustics, ultrasonics.
Sergio E. Ulloa, Ph.D., SUNY at Buffalo, 1984. Theoretical condensed-matter physics.
Louis E. Wright, Ph.D., Duke, 1966; Chair of the Department. Nuclear theory, electrodynamics, intermediate-energy theory.

Associate Professors
Charles E. Brient, Ph.D., Texas at Austin, 1963. Nuclear physics, surface physics.
Daniel S. Carman, Ph.D., Indiana, 1995. Experimental nuclear and particle physics.
Alexander O. Govorov, Ph.D., Novosibirsk, 1991. Theoretical condensed-matter physics, nanoscience.
Peter Jung, Ph.D., Ulm (Germany), 1985. Nonequilibrium statistical physics, nonlinear stochastic processes, pattern formation.
Brian R. McNamara, Ph.D., Virginia, 1991. Astrophysics, galaxy clusters, and X-ray astronomy.
Allena K. Opper, Ph.D., Indiana Bloomington, 1991. Intermediate-energy physics.
Daniel Phillips, Ph.D., Flinders (Australia), 1995. Theoretical nuclear and particle physics.
Joseph C. Shields, Ph.D., Berkeley, 1991. Astrophysics, interstellar medium, active galactic nuclei.
Thomas S. Statler, Ph.D., Princeton, 1986. Astrophysics, galactic structure and dynamics.
Larry A. Wilen, Ph.D., Princeton, 1986. Experimental acoustics, condensed-matter physics, surface melting.

Assistant Professors
Markus Böttcher, Ph.D., Bonn, 1997. High-energy astrophysics.
Ido Braslovsky, Ph.D., Israel Institute of Technology, 1998. Biophysics.
Carl R. Brune, Ph.D., Caltech, 1994. Experimental nuclear astrophysics.
Horacio E. Castillo, Ph.D., Illinois, 1998. Theoretical condensed-matter physics.
Jean Heremans, Ph.D., Princeton, 1994. Experimental condensed-matter and surface physics.
Saw-Wai Hla, Ph.D., Ljubljana, 1997. Experimental condensed-matter and surface physics, nanoscience.
Mark Lucas, Ph.D., Illinois, 1994. Experimental nuclear physics.
Michael G. Moore, Ph.D., Arizona, 1999. Atomic physics and atom optics.
Alexander Neiman, Ph.D., Saratov State (Russia), 1991. Biophysics, nonlinear dynamics, stochastic processes.
Arthur Smith, Ph.D., Texas at Austin, 1995. Experimental condensed-matter and surface physics.
Victoria Soghomonian, Ph.D., Syracuse, 1995. Experimental chemical physics.
David F. J. Tees, Ph.D., McGill, 1996. Experimental biophysics, nanoscience.

THEORETICAL RESEARCH ACTIVITIES

Astrophysics. Galactic structure, stellar dynamics, and galaxy formation, with emphasis on elliptical galaxies, numerical-studies of disk galaxies, ionization structure of and emission from HII regions, modeling of broad-line emission from active galactic nuclei.

Atomic Physics and Atom Optics. Nonlinear and quantum atom optics; Bose-Einstein condensation and degenerate Fermi gas and matter wave coherence theory.

Biophysics. Computational modeling of complex cellular signaling networks, especially intracellular and intercellular calcium signaling, modeling of neural and glial functions in healthy and epileptic tissue, stochastic modeling of electroreceptors in paddle fish, modeling of the neuronal circuitry of the cat's retina, stochastic and coherence resonance in excitable biologic systems, nanoscale ion channel and receptor clusters, and modeling slow axonal transport.

Condensed Matter and Statistical Physics. Collective electronic excitations, semiconductor superlattices, quantum Hall effect, electronic ballistic transport, phase transitions and critical phenomena, resonance and relaxation in magnetic systems, optical properties, transport theory, electronic states in novel semiconductor nanostructures and heterojunctions, ab initio density functional studies of amorphous and glassy systems, semiclassical and ab initio modeling of growth, and development of efficient algorithms for electronic structure calculations, novel methods for exploring configuration space for complex systems, nonequilibrium dynamics of glassy systems.

Mathematical and Computational Physics. Analytical studies and numerical simulations of nonlinear classical and quantum systems and studies of deterministic chaos; quantum simulations, ab initio calculations, and visualization of many-body and few-body systems in condensed-matter and nuclear physics; numerical methods and algorithmic development for high-performance vector and parallel computers; software development for application of computers in classroom teaching; analytical and algorithmic studies in differential and integral equations, probability theory, and series expansions.

Nuclear and Intermediate Energy Physics. Research in theoretical nuclear and particle physics at Ohio University has as its major component the modeling of processes involving atomic nuclei with mass numbers 1, 2, 3, and 4 in an attempt to reveal aspects of the forces that are at work inside the nucleus by examining data obtained when targets made of hydrogen and helium isotopes are bombarded with photons, electrons, neutrons, and pions. In order to understand the dynamics of the nucleus, theoretical descriptions of these reactions are built and the group's predictions are compared to the experimental results. "Light nuclei" (nuclei containing up to four nucleons) are particularly useful in this regard because once the nuclear dynamics are specified, the Schrodinger equation for these systems can be solved exactly. A recent focus of the group has been the application of effective field theory techniques to such reactions. Using these, and other, theoretical techniques, nucleon-nucleon scattering; nucleon-deuteron scattering at intermediate energies; meson-production in nucleon-nucleon collisions at intermediate energies; relativistic effects in nuclear physics; electron-deuteron scattering; Compton scattering from the proton, deuteron, and Helium-3; pion photo- and electro-production on the proton; and charge-symmetry breaking in nuclear physics have been worked on. Charge-symmetry breaking is of particular current interest, since here nuclear reactions such as the production of neutral pions in deuterium-deuterium collisions reveal aspects of Quantum Chromodynamics associated with the difference between up and down quarks. Lastly, providing reliable predictions for processes of relevance to astrophysics and cosmology are being worked on (e.g., neutron interactions that contribute to supernova and neutron-star cooling). Such projects are relevant to the Ohio University's newly-established research priority on "Structure of the Universe: From Quarks to Superclusters."

EXPERIMENTAL RESEARCH ACTIVITIES

Acoustics. Precision measurements of the properties of thermoacoustic "stack" elements. Novel techniques are used to characterize the properties of stacks of unusual geometry. Nonlinear effects as well as the effect of gas mixtures are explored. The results are applied to the development of lumped element themoacoustic primemovers and refrigerators.

Astrophysics. Spectroscopic observations of stellar motions and stellar populations in elliptical galaxies and evidence for dark matter, ionized gas in galaxies, X-ray studies of galaxy clusters, nuclear physics applied to astrophysics.

Biological Physics. Stochastic modeling of neuronal dynamics in the context of stochastic resonance; modeling of glial processes and cortical calcium waves. In collaboration with Cog-netix/Viatech and the Children's Medical Center in Cincinnati, statistical properties of calcium waves in healthy brain are compared with those of tissue from epileptic foci and glioma. Experimental determination of the response of single cell adhesion molecules to applied forces.

Condensed Matter and Surface Science. Current projects include the fabrication of crystalline and amorphous wide bandgap semiconductor alloys and their characterization via novel electron microscopes and MeV ion-beam techniques; molecular beam epitaxial growth of novel electronic materials, including wide band gap and transition metal nitrides; ultrahigh vacuum scanning tunneling microscopy investigations of semiconductor surfaces; synthesis of nanophase materials from simple precursors and their characterization via X-ray diffraction and nuclear magnetic resonance techniques; experimental transport studies of low-dimensional, nanoscopic, and mesoscopic structures and devices in high-magnetic fields and at lower temperatures; transport phenomena in high-mobility III-V semiconductor heterostructures; and narrow-gap semiconductors (InSb, InAs) for magnetic sensor devices. There are also several projects on hyperthermal beam growth, chemical vapor deposition of thin films, their characterization and fabrication of devices based on wide bandgap semiconductors such as GaN and AlN.

Geophysics. Current studies concern interfacial melting of ice at grain boundaries. Optical techniques are employed to measure the thickness of the melted layer as a function of temperature, mismatch between adjacent grains, and impurity concentration.

Nonlinear Dynamics and Chaos. Nonlinear systems exhibiting deterministic chaos are studied using both experimental and computational methods. Present studies include methods of controlling chaos and techniques of nonlinear analysis of time-series data including the estimation of Lyapunov exponents, noise reduction, forecasting, and control. Some of these techniques are applied to experiments on nonlinear electrical circuits and metal passivation in an electrochemical cell operating in a highly nonlinear regime where spontaneous oscillations and chaos are observed. Numerical and experimental studies include the development of adaptive learning techniques applied to the control of chaotic systems and investigations of spatiotemporal waves in convectively unstable open-flow systems.

Nuclear and Intermediate Energy Physics. Contemporary research in nuclear physics necessarily involves heavy use of specialized accelerator facilities around the world, and collaborations with scientists from many other institutions. Ohio University nuclear physicists play central roles in the study of fundamental symmetries in nuclear reactions (tests of charge symmetry breaking at TRIUMF in Canada), the two- and three-nucleon interaction (np and nd measurements at Los Alamos), weak interactions and spin degrees of freedom (via the charge-exchange reaction at Indiana University Cyclotron Facility), exotic nuclei-far from the line of beta stability (at GANIL and Hahn-Meitner Institute in Europe), and pion photoproduction (at the Laser-Electron Gamma Source (LEGS) at Brookhaven National Laboratory). At higher energy, interests include electronuclear phenomena tests of QCD sum rules (at LEGS) and electroproduction of strange mesons (at the new Thomas Jefferson National Accelerator Facility in Virginia). The high-intensity pulsed beam capability of the Ohio University Tandem Van de Graaff accelerator, with its unique beam swinger magnet and long flight path, is used for high-precision measurements of various nuclear cross-sections and projects in medical physics, materials science, nuclear astrophysics, instrument development, and other projects in applied nuclear physics. The entire research program is supported by the Ohio University Institute for Nuclear and Particle Physics.

Ohio University

SELECTED PUBLICATIONS

Böttcher, M., and A. Reimer. Modeling the multiwavelength spectra and variability of BL Lacertae in 2000. *Astrophys. J.* 609:576, 2004.

Böttcher M., et al. Coordinated multiwavelength observations of BL Lacertae in 2000. *Astrophys. J.* 596:847, 2003.

Böttcher M., R. Mukherjee, and A. Reimer. Predictions of the high-energy emission from BL Lac objects: The case of W Comae. *Astrophys. J.* 581:143, 2002.

Brune, C. R., Alternative parametrization of R-matrix theory. *Phys. Rev. C* 66:044611, 2002.

Bardayan, D. W., et al. **(C. R. Brune).** Destruction of 18F via 18F(p,alpha)150 burning through the Ec.m.=665 keV resonance. *Phys. Rev. C* 63:065802, 2001.

Brune, C. R., W. H. Geist, R. W. Kavanagh, and K. D. Veal. Sub-coulomb alpha transfers on ^{12}C and the $^{12}C(\alpha,\gamma)^{16}O$ S factor. *Phys. Rev. Lett.* 83:4025–8, 1999.

DeVita, R. et al. **(D. S. Carman, K. Hicks, A. Opper,** and **M. Lucas)** (CLAS Collaboration). First measurement of the double spin asymmetry in ep→é pit η in the resonance region. *Phys. Rev. Lett.* 88:082001, 2002.

Joo, K., et al. (**D. S. Carman, K. Hicks, A. Opper,** and **M. Lucas)** (CLAS Collaboration). Q-squared dependence of quadrupole strength in the gamma + p→p+piϕ transition. *Phys. Rev. Lett.* 88:12001, 2002.

Castillo, H. E., C. Chamon, L. F. Cugliandolo, and M. P. Kennett. Heterogeneous aging in spin glasses. *Phys. Rev. Lett.* 88(23):237201, 2002.

Goldbart P. M., **H. E. Castillo,** and A. Zippelius. Randomly crosslinked macromolecular systems: Vulcanization transition to and properties of the amorphous solid state. *Adv. Phys.* 45(5):393–468, 1996.

Dong, J., and **D. A. Drabold.** Atomistic structure of band tail states in amorphous silicon. *Phys. Rev. Lett.* 80:1928, 1998.

Taraskin, S. N., **D. A. Drabold,** and S. R. Elliott. Spatial decay of the single-particle density matrix in insulators: analytic results in two and three dimensions. *Phys. Rev. Lett.* 88:196405, 2002.

Fachruddin, I., **C. Elster,** and W. Glöckle. The Nd break-up process in leading order in a three-dimensional approach. *Phys. Rev. C* 68:054003, 2003.

Liu, H., **C. Elster,** and W. Glöckle. Model study of three-body forces in the three-body bound state. *Few-Body Syst.* 33:241, 2003.

Govorov A. O., and **J. J. Heremans.** Hydrodynamic effects in interacting Fermi electron jets. *Phys. Rev. Lett.* 92:26803, 2004.

Govorov, A. O., et al. Self-induced acoustic transparency in semiconductor quantum films. *Phys. Rev. Lett.* 87:226803, 2001.

Rotter, M., et al. **(A. O. Govorov).** Charge conveyance and nonlinear acoustoelectric phenomena for intense surface acoustic waves on a semiconductor quantum well. *Phys. Rev. Lett.* 82:2171–4, 1999.

H. G. Bohlen, et al. **(S. M. Grimes).** Spectroscopy of particle-hole states of ^{16}C. *Phys. Rev. C* 68:054606, 2003.

F. S. Dietrich, J. D. Andersonn, R. W. Bauer, and **S. M. Grimes.** Wick's limit and a new method for estimating neutron reaction cross sections. *Phys. Rev. C* 68:064608, 2003.

Al-Quraishi, S. I., **S. M. Grimes,** T. N. Massey, and D. A. Resler. Are the level densities for r- and rp- process nuclei different from nearby nuclei in the valley of stability? *Phys. Rev. C* 63:065803, 2001.

Chen, H., and **J. J. Heremans** et al. Ballistic transport in InSb/InAlSb antidot lattices. *Appl. Phys. Lett.,* 2004.

Hartzell B., et al. **(J. J. Heremans** and **V. Soghomonian).** Current-voltage characteristics of diversely disulfide terminated lambda-deoxyribonucleic acid molecules. *J. Appl. Phys.* 94(4):2764–6, 2003.

Hartzell B., et al. **(J. J. Heremans** and **V. Soghomonian).** Comparative current-voltage characteristics of nicked and repaired lambda-DNA. *Appl. Phys. Lett.* 82(26):4800–2, 2003.

Nakano, T., and **K. Hicks.** Discovery of the Strangeness S=+1 Pentaquark. *Mod. Phys. Lett. A* 19:645–57, 2004.

A. W. Thomas, **K. Hicks,** and A. Hosaka. A method to unambiguously determine the parity of the Theta+ Pentaquark. *Prog. Theor. Phys.* 111:291–3, 2004.

Hla, S.-W., and K.-H. Rieder, STM control of chemical reactions: Single molecule synthesis. *Ann. Rev. Phys. Chem.* 54:307–30, 2003.

Hla, S.-W., K.-F. Braun, and K.-H. Rieder. Single-atom manipulation mechanisms during a quantum corral construction. *Phys. Rev. B* 67:201402(R), 2003.

Kang, Y., and **D. C. Ingram.** Properties of amorphous GaNx prepared by ion beam assisted deposition at room temperature. *J. Appl. Phys.* 93:3954, 2003.

Haider, M. B., et al. **(D. C. Ingram** and **A. R. Smith).** Ga/N flux ratio influence on Mn incorporation, surface morphology, and lattice polarity during radio frequency molecular beam epitaxy of (Ga,Mn)N. *J. Appl. Phys.* 93:5274, 2003.

Shuai, J. W., and **P. Jung.** Optimal ion channel clustering for intracellular calcium signaling. *PNAS* 100:506–10, 2003.

Nadkarni, S., and **P. Jung.** Spontaneous oscillations of dressed neurons: A new mechanism for epilepsy? *Phys. Rev. Lett.* 91(26):268101, 2003.

Perjeru, F., X. Bai, and **M. E. Kordesch.** Electronic characterization of n-ScN/p(+) Si heterojunctions. *Appl. Phys. Lett.* 80:995–7, 2002.

Perjeru, F., R. L. Woodin, and **M. E. Kordesch.** Influence of annealing temperature upon deep levels in 6H SiC. *Physica B* 308:695–7, 2001.

Blanpied, G., et al. **(M. Lucas).** The N → Δ transition from simultaneous measurements of $p(\gamma,\pi^{+})$ and $p(\gamma,\gamma)$. *Phys. Rev. Lett.* 79:4337, 1997.

Feldman, G., et al. **(M. Lucas).** Compton scattering, meson-exchange, and the polarizabilities of bound nucleons. *Phys. Rev. C: Nucl. Phys.* 54:2124, 1996.

McNamara, B. R., et al. Chandra X-ray observations of the Hydra A Cluster: An interaction between the radio source and the X ray–emitting gas. *Astrophys. J.,* 534(L):135, 2000.

Harris, D. E., et al. **(B. R. McNamara).** Chandra X ray detection of the radio hot spots of 3C 295. *Astrophys. J.* 530(L):81, 2000.

Moore, M. G., and P. Meystre. Atomic four-wave mixing: Fermioins versus bosons. *Phys. Rev. Lett.* 86:4199, 2001.

Moore, M. G., and P. Meystre. Theory of superradiant scattering of laser light from Bose-Einstein condensates. *Phys. Rev. Lett.* 83:5202, 1999.

Neiman, A., and D. F. Russell. Two distinct types of oscillators in electroreceptors of paddlefish. *J. Neurophys.* 92:492–509, 2004.

Neiman, A., and D. F. Russell. Synchronization of noise-induced bursts in noncoupled sensory neurons. *Phys. Rev. Lett.* 88:138103, 2002.

Opper, A. K., et al. A precision measurement of charge symmetry breaking in np→d pi^ 0. *Phys. Rev. Lett.* 91:212302, 2003.

Opper, A. K., et al. Measurements of the spin observables D_{NN}, P and A_y in inelastic proton scattering from ^{12}C at 200 MeV. *Phys. Rev. C* 63:034614, 2001.

Dreiner, H. K., C. Hanhart, U. Langenfeld, and **D. R. Phillips.** Supernovae and light neutralinos: SN1987A bounds on supersymmetry revisted. *Phys. Rev. D* 68:055004, 2003. (e-Print Archive: hep-ph/0304289)

Ohio University

Pascalutsa, V., an **D. R. Phillips.** Effective theory of the delta(1232) in Compton scattering off the nucleon. *Phys. Rev. C* 67:055202, 2003.

Phillips, D. R. Building light nuclei from neutrons, protons, and pions, 14th Summer School on Understanding the Structure of Hadrons (HADRONS 01), Prague, Czech Republic, 9–13 July 2001. *Czech. J. Phys.* 52:B49, 2002.

Rhode, M. A., **R. W. Rollins,** and H. D. Dewald. On a simple recursive control algorithm automated and applied to an electrochemical experiment. *Chaos* 7:653–63, 1997.

Rhode, M. A., et al. **(R. W. Rollins).** Automated adaptive recursive control of unstable orbits in high-dimensional chaotic systems. *Phys. Rev. E* 54:4880–7, 1996.

Constantin, A., et al. **(J. C. Shields).** Emission-line properties of z > 4 quasars. *Astrophys. J.* 565:50, 2002.

Shields, J. C., et al. Evidence for a black hole and accretion disk in the LINER NGC 4203. *Astrophys. J.,* 534(L):27, 2000.

Smith, A. R, R. Yang, H. Yang, and W. R. L. Lambrecht. Aspects of spin-polarized scanning tunneling microscopy at the atomic scale: Experiment, theory, and simulation. *Surface Sci.* 561(2-3):154, 2004.

Yang, H., **A. R. Smith,** M. Prikhodko, and W. R. L. Lambrecht. Atomic-scale spin-polarized scanning tunneling microscopy applied to Mn3N2 (010). *Phys. Rev. Lett.* 89:226101, 2002.

Smith, A. R., et al. Reconstructions of the GaN(000-1) surface. *Phys. Rev. Lett.* 79:3934, 1997.

Statler, T. S., E. Emsellem, R. F. Peletier, and R. Bacon. Long-lived triaxiality in the dynamically old elliptical galaxy NGC 4365: Limits on chaos and black hole mass. *Mon. Not. R. Astron. Soc.,* in press.

Salow, R. M., and **T. S. Statler.** Self-gravitating eccentric disk models for the double nucleus of M31. *Astrophys. J.,* in press.

Tees, D. F. J., J. T. Woodward, and D. A. Hammer. Reliability theory for recepton-ligand bond dissociation. *J. Chem. Phys.* 114:7483–96, 2001.

Tees, D. F. J., R. E. Waugh, and D. A. Hammer. A microcantilever device to assess the effect of force on the lifetime of selectin-carbohydrate bonds. *Biophys. J.* 80:668–82, 2001.

Destefani, C. F., **S. E. Ulloa,** and G. E. Marques. Spin-orbit coupling and intrinsic spin mixing in quantum dots. *Phys. Rev. B* 69:125302, 2004.

Weichselbaum, A., and **S. E. Ulloa.** Potential landscapes and induced charges near metallic islands in three dimensions. *Phys. Rev. E* 68:056707, 2003.

Petculescu, G., and **L. A. Wilen.** Thermoacoustics in a single pore with an applied temperature gradient. *J. Acoust. Soc. Am.* 106:688, 1999.

Wilen, L. A., and J. G. Dash. Frost heave dynamics at a single crystal interface. *Phys. Rev. Lett.* 74:5076, 1995.

Caia, G. L., V. Pascalutsa, and **L. E. Wright.** Solving potential scattering equations without partial wave decomposition. *Phys. Rev. C* 69:034003, 2004.

Kim, K. S., and **L. E. Wright.** Constraints on medium modifications of nucleon form factors from quasielastic scattering. *Phys. Rev. C* 68:027601, 2003.

PENNSTATE

1855

THE PENNSYLVANIA STATE UNIVERSITY

Department of Physics

Programs of Study

The department is committed to offering an outstanding graduate education in a broad range of fields in experimental and theoretical physics, including condensed-matter physics; elementary particle physics; biological physics; materials science; atomic, molecular, and optical physics; particle astrophysics; and gravitational physics. The department has 48 faculty members and 120 graduate students. All of the faculty members are research active, and one third of the faculty was hired in the last five years. The department is home to major national centers in material science (MRSEC), gravitational wave physics (a Physics Frontier Center), and computational physics (an IGERT). The department is thus very dynamic in research. There is a weekly colloquium series and several series of special lectures from distinguished physicists, in addition to approximately three to five specialized weekly physics seminars.

The graduate program is aimed primarily at the attainment of a Ph.D. degree in physics. An M.S. program is also offered. Upon arrival, each graduate student is appointed a mentoring committee to provide personalized guidance during graduate school. The first year of study covers basic courses in graduate physics. Arriving students with advanced backgrounds may obtain course exemptions, thus effectively becoming second-year students. During the second year, after passing the candidacy exam, students take advanced courses in their area of specialization, form a thesis committee, and choose a research adviser. Completion of all the Ph.D. requirements is typically accomplished in a total of five years. The M.S. degree is typically conferred after one year of research beyond the first-year graduate course work through the submission of a thesis or at the end of two years of course work through a nonthesis (review paper) option.

Research Facilities

The department occupies three adjoining buildings on campus. Extensive high-tech equipment is available for research, including state-of-the-art equipment for the study of ultra-cold atoms; thin-film preparation by sputtering and molecular-beam epitaxy; photoelectron spectrometers; numerous pulsed and continuous lasers; a variety of cryostats operating between 77K and 5mK; atomic-scale microscopes, including scanning tunneling microscopes (STM), field ion microscopes (FIM), and a field emission microscope (FEM); and a low-energy electron diffraction apparatus (LEED). Condensed-matter experimentalists make use of the excellent nanofabrication facilities on campus as well as the professionally staffed departmental machine shop. Experimental high-energy physics research and condensed-matter experiments are also conducted at national facilities such as Brookhaven National Laboratory, CERN, and DESY (Hamburg, Germany). Physics faculty are members of international collaborations such as AMANDA, ICECUBE, and AUGER (experimental particle astrophysics) in Antarctica and South America, as well as LIGO and LISA (gravitational wave observatories). A newly renovated Physical and Math Sciences Library is conveniently located in the same building as the physics department and also provides online access to journals. The department has many networked UNIX, PC, and MAC workstations and various other computer facilities, including Beowulf Linux clusters, immersive virtual reality scientific visualization systems, and a high-speed network used for LIGO data analysis, as well as specially designed rooms for computer-based physics instruction. An SP2 parallel supercomputer is available at the University.

Financial Aid

The department offers incoming students teaching assistantships with full coverage of tuition. The nine-month stipend for the assistantships was approximately $14,445 in 2005–06. Additional summer support of $2000 to $3000 is usually available. Graduate assistants average a total income of $17,375 for twelve months from the assistantship stipend, summer wages, and possible departmental awards. Students from their second year on are commonly supported by the research grants of their advisers through research assistantships. The graduate school and the Eberly College of Science provide a few research fellowships and several supplemental fellowships for qualified students.

Cost of Study

Tuition costs for all incoming graduate students are covered by the department. Tuition in 2005–06 was $5325 per semester (full-time), including a mandatory $190 information technology fee and a $52 activity fee. For 2005, tuition for a normal two-semester load was $10,650.

Living and Housing Costs

There is limited graduate student housing on campus (http://www.hfs.psu.edu/housing). Rentals in State College for a one-bedroom apartment range from $530 to $650 per month. Health-care coverage is offered at a rate of $242.60 per year to graduate assistants.

Student Group

The department typically hosts about 120 graduate students with a variety of ethnic backgrounds and nationalities. The vast majority of students are on teaching or research assistantships. About 50 percent of the students are from the United States.

Location

The University Park campus of Penn State is home to approximately 41,290 students, including approximately 34,825 undergraduates, 6,465 graduate students, and more than 2,900 faculty members. It is located in the municipality of State College, nestled amid the picturesque valleys and wooded mountains of central Pennsylvania. State College has an airport with eighteen departing and eighteen arriving flights per day. The town is within 3½ hours' driving distance of Pittsburgh; Washington, D.C.; and Philadelphia. New York City is 4½ hours away.

The University

Penn State, founded in 1855, is Pennsylvania's land-grant university and has twenty-four campuses throughout the state. Penn State has more than 565,600 living alumni. One in every 122 Americans and one in every 8 Pennsylvanians are graduates of Penn State. The University hosts a legendary football team, and its home turf, Beaver Stadium (capacity 106,537), is the second-largest university stadium in the United States.

Applying

The formal deadline for applications for the fall is April 15, but applications are reviewed beginning in late January until all assistantships are offered. GRE (especially Subject Test in physics) scores are strongly preferred. The TOEFL score is mandatory for students from non-English-speaking countries. Applications sent directly to the physics department do not initially require an application fee.

Correspondence and Information

Chair, Graduate Admissions
104 Davey Laboratory
The Pennsylvania State University
University Park, Pennsylvania 16802
Phone: 814-863-0118
800-876-5348 (toll-free within the United States)
Fax: 814-865-0978
E-mail: graduate-admissions@phys.psu.edu
Web site: http://www.phys.psu.edu

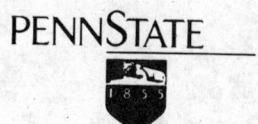

The Pennsylvania State University

THE FACULTY AND THEIR RESEARCH

R. Albert, Assistant Professor; Ph.D., Notre Dame, 2001. Statistical mechanics, network theory, systems biology.
S. H. S. Alexander, Assistant Professor; Ph.D., Brown, 2000. Cosmology, quantum gravity.
J. Anderson, Professor; Ph.D., Princeton, 1963. Quantum chemistry by Monte Carlo methods.
A. Ashtekar, Eberly Professor and Director of Center for Gravitational Physics and Geometry; Ph.D., Chicago, 1974. General relativity, quantum gravity and quantum field theory.
J. R. Banavar, Distinguished Professor and Department Head; Ph.D., Pittsburgh, 1978. Biological physics and statistical mechanics.
M. Bojowald, Assistant Professor; Ph.D., Rhenish-Westphalian Technical (Aachen), 2000. Gravitational physics.
A. W. Castleman Jr., Evan Pugh Professor of Chemistry and Physics and Eberly Distinguished Chair in Science; Ph.D., Polytechnic, 1969. Atomic, molecular, and optical physics; condensed-matter physics; cluster research.
M. H. W. Chan, Evan Pugh Professor and Director of Center for Collective Phenomena in Restricted Geometries; Ph.D., Cornell, 1974. Low-temperature physics.
M. W. Cole, Distinguished Professor; Ph.D., Chicago, 1970. Chemical physics and condensed-matter theory.
J. Collins, Professor; Ph.D., Cambridge, 1974. Perturbative quantum chromodynamics.
S. Coutu, Associate Professor; Ph.D., Caltech, 1993. Experimental studies of high-energy cosmic-ray spectra and composition, cosmic antimatter, highest energy cosmic rays.
D. Cowen, Associate Professor; Ph.D., Wisconsin–Madison, 1990. Astrophysics, particles and fields.
V. H. Crespi, Professor; Ph.D., Berkeley, 1994. Theory of superconducting, transport, electronic, and structural/mechanical properties of novel materials.
P. H. Cutler, Professor Emeritus; Ph.D., Penn State, 1958. Tunneling theory and computer simulation studies of electronic properties of wide band-gap materials (e.g., diamond, GaN, etc.) and their use in thin-film microelectronic devices, theory modeling of electronic coolers.
R. D. Diehl, Professor; Ph.D., Washington (Seattle), 1982. Surface structure and phase transitions.
P. C Eklund, Professor; Ph.D., Purdue, 1974. Fundamental properties and applications of new materials, spectroscopy and thermal/electrical transport.
K. A. Fichthorn, Professor; Ph.D., Michigan, 1989. Condensed-matter simulation and theory.
L. S. Finn, Professor and Director of Center for Gravitational Wave Physics; Ph.D., Caltech, 1987. Detection of gravitational waves, gravitational wave astronomy, relativistic astrophysics, numerical relativity.
N. Freed, Professor and Associate Dean of Eberly College of Science; Ph.D., Case Western Reserve, 1964.
K. E. Gibble, Associate Professor; Ph.D., Colorado at Boulder, 1990. Atomic, molecular, and optical physics.
M. Gunaydin, Professor; Ph.D., Yale, 1973. Superstrings and supergravity.
S. F. Heppelmann, Professor; Ph.D., Minnesota, 1981. Experimental high-energy physics.
J. K. Jain, Erwin W. Mueller Professor of Physics; Ph.D., SUNY at Stony Brook, 1985. Condensed-matter physics and the composite fermion theory of the fractional quantum Hall effect.
D. Jin, Assistant Professor; Ph.D., California, San Diego, 1999. Theory of biological neural networks and computational models of neurobiological functions.
P. Laguna, Professor; Ph.D., Texas at Austin, 1987. Numerical relativity, astronomy.
D. Larson, Professor and Verne M. Williaman Dean of Eberly College of Science; Ph.D., Harvard, 1971. Atomic, molecular, and optical physics.
Q. Li, Professor; Ph.D., Peking, 1989. Magnetic, superconducting, and multifunctional materials and nanostructures.
Y. Liu, Professor; Ph.D., Minnesota, 1991. Experimental condensed-matter and materials physics, superconductivity and physics at nanometer scales.
G. D. Mahan, Distinguished Professor; Ph.D., Berkeley, 1964. Theoretical condensed-matter physics: many-body theory, transport, semiconductor devices.
T. E. Mallouk, DuPont Professor of Materials Chemistry and Physics and Director of the Center for Nanoscale Science; Ph.D., Berkeley, 1983. Materials chemistry and physics.
J. D. Maynard, Distinguished Professor; Ph.D., Princeton, 1974. Quantum and acoustic wave phenomena.
P. Mészáros, Distinguished Professor of Astronomy and Astrophysics and Physics; Ph.D., Berkeley, 1972. Astrophysics, gravitational physics.
A. Mizel, Assistant Professor; Ph.D., Berkeley, 1999. Condensed-matter theory.
I. Mocioiu, Assistant Professor; Ph.D., SUNY at Stony Brook, 2002. High-energy physics and its connections with astrophysics and cosmology.
K. O'Hara, Assistant Professor and Downsbrough Professor; Ph.D., Duke, 2000. Experimental atomic, molecular, and optical physics; condensed-matter physics.
B. Owen, Assistant Professor; Ph.D., Caltech, 1998. Astrophysics, gravitational physics.
R. Penrose, Adjunct Professor; Ph.D., Cambridge, 1957. General relativity.
R. W. Robinett, Professor; Ph.D., Minnesota, 1981. Quantum mechanics.
R. Roiban, Assistant Professor; Ph.D., SUNY at Stony Brook, 2001. String theory, gauge theories, quantum field theory.
N. Samarth, Professor and Director of Center for Materials Physics; Ph.D., Purdue, 1986. Spin transport and coherence in mesoscopic and nanostructured magnetic systems.
P. Schiffer, Professor; Ph.D., Stanford, 1993. Condensed-matter experiment: magnetic materials and granular materials.
D. Shoemaker, Assistant Professor; Ph.D., Texas at Austin, 1999. Numerical relativity and gravitational wave physics.
J. Sofo, Associate Professor and Director of Materials Simulation Center; Ph.D., Instituto Balseiro (Argentina), 1991. Condensed-matter theory, many-body physics, computational physics.
P. Sommers, Professor; Ph.D., Texas at Austin, 1973. High-energy cosmic rays, astrophysics, and general relativity.
M. Strikman, Professor; Ph.D., St. Petersburg Nuclear Physics Institute, 1978. High-energy probes of hadron and nuclear structure.
D. S. Weiss, Associate Professor, Ph.D., Stanford, 1993. Optical lattices.
P. Weiss, Professor; Ph.D., Berkeley, 1986. Surface chemistry and physics.
J. J. Whitmore, Professor; Ph.D., Illinois, 1970. Experimental high-energy physics.
R. F. Willis, Professor; Ph.D., Cambridge, 1967. Experimental ultra-high-vacuum condensed-matter physics: electronic and magnetic behavior.
X. Xi, Professor; Ph.D., Peking, 1987. Materials physics of electronic and photonic thin films.
J. Ye, Assistant Professor; Ph.D., Yale, 1993. Theoretical condensed-matter theory, strongly correlated electron systems.
J. Zhu, Assistant Professor; Ph.D., Columbia, 2003. Experimental condensed-matter physics.

SELECTED PUBLICATIONS

Chaves, M., **R. Albert**, and E. D. Sontag. Robustness and fragility of Boolean models for genetic regulatory networks. *J. Theor. Biol.* 235:431–49, 2005 (http://arxiv.org/abs/q-bio/0501037).

Thadakamalla, H. P., U. N. Raghavan, S. Kumara, and **R. Albert**. Survivability of multiagent-based supply networks: A topological perspective. *IEEE Intell. Syst.* 19(5):24–31, 2004 (http://csdl.computer.org/comp/mags/ex/2004/05/x5024abs.htm).

Albert, I., and **R. Albert**. Conserved network motifs allow protein-protein interaction prediction. *Bioinformatics* 20(18), 2004 (http://arxiv.org/abs/q-bio.MN/0406042).

Alexander, S. H. S. In the realm of the geometric transitions. *Nucl. Phys. B* 704: 231–78, 2005.

Alexander, S. H. S., and L. Smolin. Quantum gravity and inflation. *Phys. Rev. D* 70:044025, 2004.

Alexander, S. H. S., Y. Ling, and L. Smolin. A thermal instability of positive brane cosmological constant in the Randall-Sunfrum Cosmologies. *Phys. Rev. D* 65: 0583503, 2002.

Anderson, J. B., and L. N. Long. Direct Monte Carlo simulation of chemical reaction systems: Prediction of ultrafast detonations. *J. Chem. Phys.* 118:3102–10, 2003.

Riley, K. E., and **J. B. Anderson.** Higher accuracy quantum Monte Carlo calculations of the barrier for the H + H2 reaction. *J. Chem. Phys.* 118:3437–8, 2003.

Sokolova, S., A. Luechow, and **J. B. Anderson.** Energetics of carbon clusters C20 from all-electron quantum Monte Carlo calculations. *Chem. Phys. Lett.* 323:229–33, 2000.

Ashtekar, A., J. Engle, T. Pawlowski, and C. Van Den Broeck. Multipole moments of isolated horizons. *Classical Quantum Gravity* 21:2549–70, 2004.

Ashtekar, A., and J. Lewandowski. Background independent quantum gravity: A status report. *Classical Quantum Gravity* 21:R53–R152, 2004. (Review article commissioned by the British Institute of Physics.)

Ashtekar, A., and B. Krishman. Isolated and dynamical horizons and their applications. *Living Rev. Relativity* 10:1–77, 2004. (Invited review.)

Banavar, J. R., et al. A unified perspective on proteins—a physics approach. *Phys. Rev. E* 70:041905, 2004.

Banavar, J. R., M. Cieplak, and A. Maritan. Lattice tube model of proteins. *Phys. Rev. Lett.* 93: 238101, 2004.

Volkov, I., **J. R. Banavar**, S. P. Hubbell, and A. Maritan. Neutral theory and relative species abundance in ecology, *Nature* 424:1035–37, 2003.

Bojowald, M., and R. Swiderski. Spherically symmetric quantum horizons. *Phys. Rev. D* 71:081501, 2005 (gr-qc/0410147).

Bojowald, M. Inflation from quantum geometry. *Phys. Rev. Lett.* 89:261301, 2002 (gr-qc/0206054).

Bojowald, M. Absence of singularity in loop quantum cosmology. *Phys. Rev. Lett.* 86:5227–30, 2001 (gr-qc/0102069).

Bergeron, D. E., et al. **(A. W. Castleman Jr.).** Aluminum cluster superatoms act as halogens in polyhalide ions and as alkaline earth metals in iodide salt molecules. *Science* 307:231-5, 2005.

Bergeron, D. E., **A. W. Castleman Jr.**, T. Morisato, and S. N. Khanna. Formation of $Al_{13}I^-$: Evidence for the superhalogen character of Al_{13}. *Science* 304:84–7, 2004.

Hurley, S. M., T. E. Dermota, D. P. Hydutsky, and **A. W. Castleman Jr.** The dynamics of acid dissolution in the ground and excited states. *Science* 298:202–4, 2002.

Kim, E., and **M. H. W. Chan.** Observation of superflow in solid helium. *Science* 305:1941–4, 2004.

Kim, E., and **M. H. W. Chan.** Probable observation of a supersolid helium phase. *Nature* 427:225–7, 2004.

Csathy, G., J. D. Reppy, and **M. H. W. Chan.** Substrate-tuned boson localization in superfluid He-4 films. *Phys. Rev. Lett.* 91:235301, 2003.

Taniguchi, J., et al. **(M. W. Cole).** One-dimensional He Fermi fluid formed in nanometer pores of FSM-16. *Phys. Rev. Lett.* 94:065301, 2005.

Ancilotto, F., M. M. Calbi, S. M. Gatica, and **M. W. Cole.** Bose-Einstein condensation of helium and hydrogen inside bundles of carbon nanotubes. *Phys. Rev. B* 70: 165422, 2004.

Trasca, R. A., M. M. Calbi, and **M. W. Cole.** Lattice-gas Monte Carlo study of adsorption in pores. *Phys. Rev. E* 69:011605, 2004.

Collins, J. C., and A. Metz. Universality of soft and collinear factors in hard-scattering factorization. *Phys. Rev. Lett.* 93:252001, 2004.

Collins, J. C., et al. Lorentz invariance: An additional fine-tuning problem. *Phys. Rev. Lett.* 93:191301, 2004.

Collins, J. C. Leading-twist single-transverse-spin asymmetries: Drell-Yan and deep-inelastic scattering. *Phys. Lett. B* 536:43, 2002.

Aglietta, M., et al. **(S. Coutu).** The cosmic ray primary composition between 10^{15} and 10^{16} eV from extensive air showers electromagnetic and TeV muon data. *Astropart. Phys.* 20:641–52, 2004.

Beatty, J. J., et al. **(S. Coutu).** A new measurement of the altitude dependence of the atmospheric muon intensity. *Phys. Rev. D* 70:092005, 2004.

Beatty, J. J., et al. **(S. Coutu).** A new measurement of the cosmic-ray positron fraction from 5 to 15 GeV. *Phys. Rev. Lett.* 93:241102, 2004.

Woschnagg, K., et al. (**D. Cowen** with AMANDA Collaboration). New results from the Antarctic muon and neutrino detector array. *Nucl. Phys. B* 145:319–22, 2005.

Ackermann, M., et al. (**D. Cowen** with AMANDA Collaboration). Search for extraterrestrial point sources of high energy neutrinos with AMANDA-II using data collected in 2000–2002. *Phys. Rev. D* 71:077102, 2005).

Ackermann, M., et al. (**D. Cowen** with AMANDA Collaboration). Flux limits on ultra high energy neutrinos with AMANDA-B10. *Astropart. Phys.* 22:339–53, 2005.

Kolmogorov, A., **V. H. Crespi**, M. H. Schleier-Smith, and J. C. Ellenbogen. Nanotube-substrate interactions: Distinguishing carbon nanotubes by the helical angle. *Phys. Rev. Lett.* 92:085503, 2004.

Han, J. E., O. Gunnarsson, and **V. H. Crespi.** Strong superconductivity with local Jahn-Teller phonons in C60 solids. *Phys. Rev. Lett.* 90:167006, 2003.

Chung, M., et al. **(P. H. Cutler).** Theoretical analysis of a field emission enhanced semiconductor thermoelectric cooler. *Solid State Electronics* 47:1745–51, 2003.

Cutler, P. H., N. M. Miskovsky, N. Kumar, and M. S. Chung. New results on microelectric cooling using the inverse Nottingham effect. *Cold Cathode Proceedings of the Electrochemical Society* 2000–28:98–114, 2001.

Mayer, A., N. Miskovsky, and **P. H. Cutler.** Three-dimensional calculation of field electron energy distribution from open hydrogen-saturated and capped metallic (5,5) carbon nanotubes. *Appl. Phys. Lett.* 79:3338–40, 2001.

McGrath, R., U. Grimm, and **R. D. Diehl.** The forbidden beauty of quasicrystals. *Phys. World* 17:23–7, 2004.

Ferralis, N., and **R. D. Diehl** et al. Low-energy electron diffraction study of Xe adsorption on the 10-fold d-Al-Ni-Co quasicrystal surface. *Phys. Rev. B* 69:075410, 2004.

Ledieu, J., et al. **(R. D. Diehl).** Pseudomorphic growth of a single element quasiperiodic ultrathin film on a quasicrystal substrate. *Phys. Rev. Lett.* 92:135507, 2004.

Romero, H. E., K. Bolton, A. Rosen, and **P. C. Eklund.** Atom collision-induced resistivity of carbon nanotubes. *Science* 307(5706):89–93, 2005.

Chen, G., et al. **(P. C. Eklund).** Anomalous contraction of the C-C bond length in semiconducting carbon nanotubes observed during Cs doping. *Phys. Rev. B* 71: 045408, 2005.

Tanner, D. B., et al. **(P. C. Eklund).** Optical properties of potassium-doped polyacetylene. *Synth. Met.* 141(1–2):75–9, 2004.

Kim, H. -Y., and **K. A. Fichthorn.** Molecular-dynamics simulation of amphiphilic dimers at a liquid-vapor interface. *J. Chem. Phys.* 122:034704, 2005.

Fichthorn, K. A., and M. Scheffler. Nanophysics—a step up to self-assembly. *Nature* 429:617, 2004.

Miron, R. A., and **K. A. Fichthorn.** Multiple-time scale accelerated molecular dynamics: Addressing the small-barrier problem. *Phys. Rev. Lett.* 93:138201, 2004.

Nutzman, P., et. al. **(L. S. Finn).** Gravitational waves from extragalactic inspiraling binaries: Selection effects and expected detection rates. *Astrophys. J.* 212:364–74, 2004.

Finn, L. S., B. Krishnan, and P. J. Sutton. Swift pointing and the association between gamma-ray and gravitational wave bursts. *Astrophys. J.* 607:384–90, 2004.

Dreyer, O., et. al. **(L. S. Finn).** Black-hole spectroscopy: testing general relativity through gravitational wave observations. *Classical Quantum Gravity* 21:787–803, 2004.

Gunaydin, M., and O. Pavlyk. Minimal unitary realizations of exceptional U-duality groups and their subgroups as quasiconformal groups. *J. High Energy Phys.* 501: 019, 2005 (arXiv:hep-th/0409272).

Gunaydin, M., and M. Zagermann. Unified Maxwell-Einstein and Yang-Mills-Einstein supergravity theories in five dimensions. *J. High Energy Phys.* 307:023, 2003 (arXCiv:hep-th/0304109).

Fernando, S., **M. Gunaydin**, and O. Pavlyk. Spectra of PP-wave limits of M-/superstring theory on $AdS_p xS^q$ spaces. *J. High Energy Phys.* 0210:007, 2002 (arXiv: hep-th/0207175).

Scarola, V. W., K. Park, and **J. K. Jain.** Cooper instability of composite Fermions: Pairing from purely repulsive interaction. *Nature,* in press.

Park, J., and **J. K. Jain.** Spontaneous magnetization of composite Fermions. *Phys. Rev. Lett.* 83:5543–6, 1999.

Jain, J. K. Composite Fermion approach for the fractional quantum Hall effect. *Phys. Rev. Lett.* 63:199–202, 1989.

Braccini, V., et al. **(Q. Li).** High-field superconductivity in alloyed MgB_2 thin films. *Phys. Rev. B* 71:012504 .2005.

Pogrebnyakov, A. V., et al. **(Q. Li).** Enhancement of the superconducting transition temperature of MgB_2 by a strain-induced bond-stretching mode softening. *Phys. Rev. Lett.* 93:147006, 2004.

Nelson, K. D., Z. Q. Mao, Y. Maeno, and **Y. Liu.** Direct experimental test of odd-parity superconductivity in Sr_2RuO_4. *Science* 306:1151–4, 2004.

Wang, H., et al. **(Y. Liu).** Metallic contacts with individual Ru nanowires prepared by electrochemical deposition and the suppression of superconductivity in ultrasmall Ru grains. *Appl. Phys. Lett.* 84:5329, 2004.

The Pennsylvania State University

Liu, Y., et al. Destruction of the global phase coherence in ultrathin, doubly connected superconducting cylinders. *Science* 294:2332, 2001.

Mahan, G. D., et al. **(P. C. Eklund).** Optical phonons in polar semiconductor nanowires. *Phys. Rev. B* 68:073402, 2003.

Gupta, R., Q. Xiong, **G. D. Mahan,** and **P. C. Eklund.** Surface optical phonons in gallium phosphide nanowires. *Nano Lett.* 3:1745–50, 2003.

Kovtyukhova, N. I., B. K. Kelly, and **T. E. Mallouk.** Coaxially gated in-wire thin-film transistors made by template assembly. *J. Am. Chem. Soc.* 126:12738–9, 2004.

Lu, Q., et al. **(T. E. Mallouk).** Ordered SBA-15 nanorod arrays inside a porous alumina membrane. *J. Am. Chem. Soc.* 126:8650–1, 2004.

Paxton, W. F., et al. **(T. E. Mallouk).** Autonomous movement of striped nanorods. *J. Am. Chem. Soc.* 126:13424–31, 2004.

Jin H. S., et al. **(J. D. Maynard** and **Q. Li).** Measurements of elastic constants in thin films of colossal magnetoresistance material. *Phys. Rev. Lett.* 90:036103, 2003.

Maynard, J. D. Acoustical analogs of condensed matter problems. *Rev. Mod. Phys.* 73:401–17, 2001.

Maynard, J. D. Resonant ultrasound spectroscopy. *Phys. Today* 49:26–31, 1996.

Alvarez-Muniz, J., and **P. Mészáros.** High energy neutrinos from radio-quiet AGNs. *Phys. Rev. D* 70:123001, 2004 (astro-ph/0409034).

Razzaque, S., **P. Mészáros,** and E. Waxman. TeV neutrinos from core collapse supernovae and hypernovae. *Phys. Rev. Lett.* 93:181101, 2004 (astro-ph/0407064).

Ioka, K., S. Kobayashi, and **P. Mészáros.** Extended GeV-TeV emission around a gamma-ray burst remnant: The case of W49B. *Astrophys. J. Lett.* 613:L171, 2004 (astro-ph/0406555).

Mizel, A., and D. A. Lidar. Three- and four-body interactions in spin-based quantum computation. *Phys. Rev. Lett.* 92:077903, 2004.

Mizel, A. Mimicking time evolution within a quantum ground state: Ground-state quantum computation, cloning, and teleportation. *Phys. Rev. A* 70:012304, 2004.

Mizel, A. Quantum vortex tunneling: Microscopic theory and application to d-wave superconductors (http://arxiv.org/abs/cond-mat/0107530).

Snyder, J., et al. **(A. Mizel** and **P. Schiffer).** Quantum-classical reentrant relaxation crossover in $Dy_2Ti_2O_7$ spin ice. *Phys. Rev. Lett.* 91:107201-1–4, 2003.

Jones, J., **I. Mocioiu,** M. H. Reno, and I. Sarcevic. Tracing very high-energy neutrinos from cosmological distances in ice. *Phys. Rev. D* 69:033004, 2004.

Fuller, G., A. Kusenko, **I. Mocioiu,** and S. Pascoli. Pulsar kicks from a dark-matter sterile neutrino. *Phys. Rev. D* 68:103002, 2003.

Mocioiu, I., and R. Shrock. Matter effects on neutrino oscillations in long baseline experiments. *Phys. Rev. D* 62:053017, 2000.

Yunes, N., W. Tichy, **B. J. Owen,** and B. Bruegmann. Binary black hole initial data from matched asymptotic expansions. gr-qc/0503011.

Owen, B. J. Maximum elastic deformations of compact stars with exotic equations of state. astro-ph/0503399.

Abbott, B., et al. **(B. J. Owen** with LIGO Scientific Collaboration). Limits on gravitational wave emission from selected pulsars using LIGO data. gr-qc/0410007.

Robinett, R. W. Quantum wave packet revivals. *Phys. Rep.* 392, 1–119, 2004 (quant-ph/0401031).

Belloni, M. A., M. A. Doncheski, and **R. W. Robinett.** Wigner quasi-probability distribution for the infinite square well: energy eigenstates and time-dependent wave packets. *Am. J. Phys.* 72:1183–92, 2004 (quant-ph/0312086).

Doncheski, M. A. and **R. W. Robinett.** Wave packet revivals and the energy eigenvalue spectrum of the quantum pendulum. *Ann. J. Phys.* 308:578–98, 2003 (quant-ph/0307079).

Roiban, R., M. Spradlin, and A. Volovich. Dissolving N=4 loop amplitudes into Qcd tree amplitudes. *Phys. Rev. Lett.* 94:102002, 2005 (e-print archive: hep-th/0412265).

Bena, I., J. Polchinski, and **R. Roiban.** Hidden symmetries of the Ads(5) X S**5 superstring. *Phys. Rev. D* 69:046002, 2004 (e-print archive: hep-th/0305116).

Gross, D. J., A. Mikhailov, and **R. Roiban.** A calculation of the plane wave string Hamiltonian from N=4 Superyang-Mills theory. *J. High Energy Phys.* 0305:025, 2003 (e-print archive: hep-th/0208231).

Eid, K. F., et al. **(N. Samarth).** Exchange biasing of the ferromagnetic semiconductor (Ga,Mn)As. *Appl. Phys. Lett.* 85:1556, 2004.

Awschalom, D. D., M. Flatte, and **N. Samarth.** Spintronics. *Sci. Am.* 286:67, 2002.

Gupta, J. A., R. Knobel, **N. Samarth,** and D. D. Awschalom. Ultrafast manipulation of spin coherence. *Science* 292:2458, 2001.

A. H. MacDonald, **P. Schiffer,** and N. Samarth. Ferromagnetic semiconductors: Moving beyond (Ga,Mn)As. *Nature Materials* 4:195–202, 2005.

Stone, M. B., et al. **(P. Schiffer).** Getting to the bottom of a granular medium. *Nature* 427: 503–4, 2004.

Ku, K. C., et al. **(P. Schiffer).** Highly enhanced Curie temperatures in low temperature annealed (Ga,Mn)As epilayers. *Appl. Phys. Lett.* 82:2302–4, 2003.

Pfeiffer, H. P., L. E. Kidder, M. A. Scheel, and **D. Shoemaker.** Initial data for Einstein's equations with superposed gravitational waves. *Phys. Rev. D* 71:024020, 2005.

Fuhr, J. D., A. G. Sal, and **J. O. Sofo.** STM chemical signature of point defects on the MoS2(0001)surface. *Phys. Rev. Lett.* 92:026802, 2004.

Thonhauser, T., T. J. Scheidemantel, and **J. O. Sofo.** Improved thermoelectric devices using bismuth alloys. *Appl. Phys. Lett.* 85:588–90, 2004.

Scheidemantel, T. J., et al. **(J. O. Sofo).** Transport coefficients from first-principles calculations. *Phys. Rev. B* 68:125210, 2003.

Sommers, P. First estimate of the primary cosmic ray energy spectrum above 3 EeV from the Pierre Auger Observatory. For the Pierre Auger Collaboration; http://arxiv.org/abs/astro-ph/0507150.

Frankfurt, L., and **M. Strikman.** Ion induced quark gluon implosion. *Phys. Rev. Lett.* 91:022301, 2003.

Dumitru, A., L. Gerland, and **M. Strikman.** Proton breakup in high-energy p A collisions from perturbative QCD. *Phys. Rev. Lett.* 90:092301, 2003.

Kinoshita, T., T. Wenger, and **D. S. Weiss.** All-optical Bose-Einstein condensation using a compressible crossed dipole trap. *Phys. Rev. A* 71:011602(R), 2005.

Kinoshita, T., T. Wenger, and **D. S. Weiss.** Observation of a one-dimensional Tonks-Girardeau gas. *Science* 305:1125, 2004.

Olshanii, M., and **D. S. Weiss.** Producing Bose condensates using optical lattices. *Phys. Rev. Lett.* 89:090404-1, 2002.

Checkanov, S., et al. **(J. Whitmore).** ZEUS next-to-leading-order QCD analysis of data on deep inelastic scattering, ZEUS collaboration. *Phys. Rev. D* 67:012007, 2003.

Checkanov, S., et al. **(J. Whitmore).** Measurement of high-Q_2 charged current cross sections in e^-p deep inelastic scattering at HERA, ZEUS collaboration. *Phys. Lett. B* 539:197, 2002.

Checkanov, S., et al. **(J. Whitmore).** Exclusive photoproduction of J/Ψ mesons at HERA, ZEUS collaboration. *Eur. Phys. J.* C24:345, 2002.

Willis, R. F., and N. A. R. Janke-Gilman. Distinguishing magnetic moment and magnetic ordering behavior on the Slater-Pauling curve. *Europhys. Lett.* 69:411, 2005.

Altmann, K. N., et al. **(R. F. Willis).** Effect of magnetic doping on electronic states of Ni. *Phys. Rev. Lett.* 87:137201, 2001.

Zhang, R., and **R. F. Willis.** Thickness-dependence of the Curie temperatures of thin magnetic films: role of spin-spin interactions. *Phys. Rev. Lett.* 86:2665, 2001.

Zeng, X., et al. **(X. X. Xi** and **Q. Li).** In situ epitaxial MgB_2 thin films for superconducting electronics. *Nat. Mat.* 1:35, 2002.

Sirenko, A. A., et al. **(X. X. Xi).** Soft-mode hardening in $SrTiO_3$ thin films. *Nature* 404:373, 2000.

Ye, J. Broken symmetry, excitons, gapless modes and topological excitations in trilayer quantum Hall systems. *Phys. Rev. B* 71:125314, 2005 (cond-mat/0402459).

Jeon, G. S., and **J. Ye.** Investigation of trial wavefunction approach to bilayer quantum Hall systems. Phys. Rev. B 71:035348, 2005.

Ye, J. Gauge-invariant Green function in 3+1 dimensional QED and 2+1 dimensional Chern-Simon theory. *J. Phys.: Condens. Matter* 16:4465–76, 2004.

Zhu, J., et al. Spin susceptibility of an ultra-low density two dimensional electron system. *Phys. Rev. Lett.* 90:056803, 2003.

Zhu, J., et al. Density-induced interchange of anisotropy axes at half-filled high Landau levels. *Phys. Rev. Lett.* 88:116803, 2002.

Zhu, J., et al. Hysteresis and spikes in the Quantum Hall Effect. *Phys. Rev. B* 61:R13361, 2000.

RENSSELAER POLYTECHNIC INSTITUTE

Department of Physics, Applied Physics, and Astronomy

Programs of Study

The Department of Physics, Applied Physics, and Astronomy offers graduate programs that lead to the M.S. or the Ph.D. degree in physics. Graduate students develop flexible individual programs of study and research in one or more of the following areas of specialization: astronomy and astrophysics, biological physics, condensed-matter physics, educational physics, optical physics, and experimental particle physics. For graduate students specializing in astronomy and astrophysics, the M.S. degree is available either in astronomy or physics with specialization in astrophysics.

Rensselaer's faculty members are a collaborative community. They take pride in their dedication to teaching—demonstrating a commitment to excellence that always has been a hallmark of Rensselaer's teacher-scholars. As an important part of their education, graduate students collaborate with faculty members to make original research contributions in their area of specialization. The Department conducts both fundamental and applied research, often in collaboration with researchers from other Rensselaer departments, other universities, industry, or the National Laboratories.

The M.S. degree requires 30 credit hours beyond the B.S., and the Ph.D., 90 credit hours beyond the B.S. The M.S degree is not a prerequisite for the Ph.D. degree.

Research Facilities

Students in the Department enjoy the use of world-class facilities. State-of-the-art equipment for the study of condensed-matter physics allows students to use the following experimental techniques: auger electron spectroscopy, high-resolution low-energy electron diffraction, reflection high-energy electron diffraction, atomic force microscopy, scanning tunneling microscopy, X-ray absorption spectroscopy, X-ray crystallography, and ellipsometry. Other experimental facilities used in the Department's programs include accelerators at the University at Albany and the Stanford Synchrotron Radiation Laboratory.

The optical physics facilities allow researchers to perform optical absorption, luminescence, Brillouin scattering, Raman scattering, Rayleigh scattering, photomodulation spectroscopies, photothermal deflection spectroscopy, magneto-optic Kerr effect, and Faraday rotation.

The Hirsch Observatory, located atop the Science Center, houses a Boller and Chivens 16″ Cassegrain telescope, SBIG imaging camera, SBIG spectrograph, and CFW-8 filter wheel. The astrophysics group also makes use of ground-based telescopes at world-class observing sites in Hawaii, Australia, Chile, and South Africa and has access to data from the Hubble Space Telescope, Chandra, and the Infrared Space Observatory as well as Sloan Digital Sky Survey and Two Micron All Sky Survey (2MASS).

Affiliated research centers include the Center for Integrated Electronics (http://www.rpi.edu/dept/cie/), Center for Broadband Data Transport Science and Technology (http://nina.ecse.rpi.edu/shur/BroadBand_overviewShort_files/frame.htm), SRC Center for Advanced Interconnect Systems Technologies (http://www.rpi.edu/dept/cie/caist/), MARCO Interconnect Focus Center (http://www.rpi.edu/dept/cie/fc-ny/), New York Center for Studies of the Origins of Life (http://www.origins.rpi.edu/), and the Center for Terahertz Research (http://www.rpi.edu/terahertz).

Financial Aid

Financial aid is available in the forms of teaching and research assistantships and fellowships, which include tuition scholarships and stipends. Rensselaer assistantships and university, corporate, or national fellowships fund many of Rensselaer's full-time graduate students. Outstanding students may qualify for university-sponsored Rensselaer Graduate Fellowship Awards, which carry a minimum stipend of $20,000 and a full tuition and fees scholarship. All fellowship awards are calendar-year awards for full-time graduate students. Summer support is also available in many departments. Low-interest, deferred-repayment graduate loans are available to U.S. citizens with demonstrated need.

Cost of Study

Full-time graduate tuition for the 2006–07 academic year is $32,600. Other costs (estimated living expenses, insurance, etc.) are projected to be about $12,400. Therefore, the cost of attendance for full-time graduate study is approximately $45,000. Part-time study and cohort programs are priced differently. Students should contact Rensselaer for specific cost information related to the program they wish to study.

Living and Housing Costs

Graduate students at Rensselaer may choose from a variety of housing options. On campus, students can select one of the many residence halls, and there are abundant options off campus as well, many within easy walking distance.

Student Group

There are 1,234 graduate students, of whom 30 percent are women, 90 percent are full-time, and 69 percent study at the doctoral level.

Student Outcomes

Rensselaer's graduate students are hired in a variety of industries and sectors of the economy and by private and public organizations, the government, and institutions of higher education. Starting salaries average $63,262 for master's degree recipients.

Location

Located just 10 miles northeast of Albany, New York State's capital city, Rensselaer's historic 275-acre campus sits on a hill overlooking the city of Troy, New York, and the Hudson River. The area offers a relaxed lifestyle with many cultural and recreational opportunities, with easy access to both the high-energy metropolitan centers of the Northeast—such as Boston, New York City, and Montreal—and the quiet beauty of the neighboring Adirondack Mountains.

The Institute

Recognized as a leader in interactive learning and interdisciplinary research, Rensselaer continues a tradition of excellence and technological innovation dating back to 1824. More than 100 graduate programs in more than fifty disciplines attract top students, researchers, and professors. The discovery of new scientific concepts and technologies, especially in emerging interdisciplinary fields, is the lifeblood of Rensselaer's culture and a core goal for the faculty, staff, and students. Fueled by significant support from government, industry, and private donors, Rensselaer provides a world-class education in an environment tailored to the individual.

Applying

The admission deadline for the fall semester is January 1. Basic admission requirements are the submission of a completed application form (available online), the required application fee ($75), a statement of background and goals, official transcripts, official scores on the GRE General Test and on the Subject Test in physics, TOEFL or IELTS scores (if applicable), and two recommendations.

Correspondence and Information

Department of Physics, Applied Physics, and Astronomy
Rensselaer Polytechnic Institute
110 Eighth Street
Troy, New York 12180-3590

Phone: 518-276-6310
E-mail: physics@rpi.edu
Web site: http://www.rpi.edu

Rensselaer Polytechnic Institute

THE FACULTY AND THEIR RESEARCH

Gary S. Adams, Professor; Ph.D, Indiana. Experimental particle physics: photo reactions, hadron structure, exotic hadrons. (adamsg@rpi.edu)
Gary Bedrosian, Clinical Associate Professor; Ph.D., Caltech. Educational physics, electromagnetic analysis. (bedrog@rpi.edu)
Philip Casabella, Professor and Associate Chair; Ph.D., Brown. Educational physics, nuclear magnetic resonance in solids. (casabp@rpi.edu)
John Cummings, Research Assistant Professor; Ph.D., Rice. Studying fundamental interactions using an electron-positron collider. (cummij@rpi.edu)
Sang-Kee Eah, Assistant Professor; Ph.D., Seoul. Nano-optics of single nanoparticles, synthesis and self-assembly of monodisperse nanoparticles. (eahs@rpi.edu)
Angel E. García, Professor; Ph.D., Cornell. Theoretical and computational aspects of the structure, dynamics, and stability of biological molecules. (angel@rpi.edu)
Timothy Hayes, Professor; Ph.D., Harvard. Atomic-scale structure of materials and the relationship between that structure and electronic or optical properties. (thayes@rpi.edu)
Shirley Ann Jackson, Professor and President, Rensselaer Polytechnic Institute; Ph.D., MIT. Theoretical condensed-matter physics, especially layered systems; physics of opto-electronic materials.
György Korniss, Associate Professor; Ph.D., Virginia Tech. Scalability and synchronization in distributed computer networks. (korniss@rpi.edu)
Valeri Koubarovski, Research Scientist; Ph.D., Institute for High Energy Physics. Experimental high-energy and nuclear physics. (vpk@rpi.edu)
Sabrina Lee, Adjunct Professor; Ph.D., Michigan. Solid-state physics, X-ray crystallography and metallography, high-temperature materials, refractory coatings, thin solid films, electrolytic deposition, sputtering deposition.
Shawn-Yu Lin, Professor; Ph.D., Princeton. Nanophotonics, silicon photonics, quantum optics, integrated optics, photonic lattice structures, emerging energy applications. (sylin@rpi.edu)
James Jian-Qiang Lü, Research Associate Professor; Ph.D., Munich Technical. Micro- and nano-electronics technology from theory and design to materials, devices, processing, and system integration. (luj@rpi.edu)
Toh-Ming Lu, Professor and Associate Director, Center for Integrated Electronics; Ph.D., Wisconsin–Madison. Morphological evolution of films, real-time diffraction during film growth, surface passivation and atomic layer deposition, deposition of novel low and high dielectric constant films. (lut@rpi.edu)
Cynthia McIntyre, Clinical Associate Professor, Chief of Staff, and Associate Vice President for Policy and Planning; Ph.D., MIT. Semiconductor materials, solid-state theory. (crm@rpi.edu)
James Napolitano, Professor; Ph.D., Stanford. Experimental nuclear and particle physics, scientific computation, distributed computing. (napolj@rpi.edu)
Saroj Nayak, Associate Professor; Ph.D., Jawaharlal Nehru (New Delhi). Atomic and electronic structures of matters using ab initio electronic structure calculation methods with classical and quantum molecular dynamics simulations and Monte Carlo methods. (nayaks@rpi.edu)
Heidi Jo Newberg, Associate Professor; Ph.D., Berkeley. Understanding the structure of the galaxy through using A stars as tracers of the galactic halo. (newbeh@rpi.edu)
Peter D. Persans, Professor; Ph.D., Chicago. Optical, electro-optic, electronic, and structural properties of nanocrystalline, quantum dot, and thin-film semiconductor materials; X-ray optics (thin film and capillary); waveguide optical interconnects; nanostructured materials. (persap@rpi.edu)
Wayne G. Roberge, Professor; Ph.D., Harvard. Evolution of ices in the interstellar medium and solar nebula; computer simulations of multifluid, magnetohydrodynamic (MHD) shock waves; analytic and numerical methods for multifluid MHD; physics of interstellar dust. (roberw@rpi.edu)
Leo J. Schowalter, Professor; Ph.D., Illinois at Urbana-Champaign. Molecular beam epitaxy, electron transport at interfaces, optical properties of semiconductors, growth of wide band gap semiconductors. (schowal@rpi.edu)
John Schroeder, Professor; Ph.D., Catholic University. Glass physics, nanoparticle physics. (schroj@rpi.edu)
E. Fred Schubert, Professor; Ph.D., Stuttgart (Germany). Compound semiconductor materials and devices, including epitaxial growth, materials characterization, device processing and fabrication, device design, and device characterization. (EFSchubert@rpi.edu)
Michael S. Shur, Professor and Director, Center for Broadband Data Transport Science and Technology; Ph.D., Dr.Sc., A. F. Ioffe Institute of Physics and Technology (Russia). Plasma wave excitation in submicron field-effect transistors (FET) and related device structures. (shurm@rpi.edu)
Paul Stoler, Professor; Ph.D., Rutgers. Electroproduction from protons of excited baryons and observing their decays via pions and other mesons. (stolep@rpi.edu)
Gwo-Ching Wang, Professor and Chair; Ph.D., Wisconsin–Madison. Nonequilibrium growth and etching of metal and semiconductor films, magnetism of ultrathin magnetic films and dots, transport properties of metallic and magnetic films and nanotubes, fabrication and growth mechanism of sculpture films. (wangg@rpi.edu)
Morris Washington, Clinical Professor and Associate Director, Center for Integrated Electronics; Ph.D., NYU. Photonic and electronic devices. (washim@rpi.edu)
Christian M. Wetzel, Associate Professor; Dr.rer.nat., Munich Technical. Electronic band and defect structure of wide band gap semiconductor materials and devices by means of optical spectroscopy and electronic transport under external perturbation. (wetzel@rpi.edu)
Douglas C. B. Whittet, Professor and Associate Director, New York Center for Studies on the Origins of Life; Ph.D., St. Andrews. Testing hypothesis that organic molecules relevant to the origin of life are ubiquitous to interstellar condensations from which planetary systems are born. (whittd@rpi.edu)
Ingrid Wilke, Assistant Professor; Ph.D., Swiss Federal Institute of Technology. Ultrafast and terahertz spectroscopy. (wilkei@rpi.edu)
Masashi Yamaguchi, Assistant Professor; Ph.D., Hokkaido. Structural and electronic dynamics in condensed matter, THz spectroscopy of advanced materials, THz science and technology. (yamagm@rpi.edu)
Xi-Cheng Zhang, Professor; Ph.D., Brown. Terahertz optics sensing, terahertz optics imaging. (zhangxc@rpi.edu)

ASTRONOMY AND ASTROPHYSICS
Faculty: Heidi Newberg, Wayne Roberge, and Doug Whittet.

Astrobiology and Interstellar Chemistry: Current interest focuses on spectroscopic detection of organic molecules in interstellar dust and gas and their contribution to the organic inventory of protoplanetary disks. Theoretical projects include simulations of protostellar collapse, multifluid magnetohydrodynamic shock waves, and shock chemistry.

Galactic and Intergalactic Astronomy: Research includes studies of how galaxies form and, in particular, how the Milky Way galaxy formed. The process of galaxy formation includes the gravitational effects of dark matter, the hydrodynamics of gas clouds, and energy feedback from formation and death of stars. This research is carried out in conjunction with a large international collaboration called the Sloan Digital Sky Survey (SDSS II).

Astrophysics: Astrophysics uses physics and math to determine why the solar system, stars, and galaxies exist the way they do. Physical characteristics, chemical constitution, light, heat, and atmospheres are included in this study. Recent research includes the synthesis of biogenic compounds by shocks in the solar nebula and the physical and chemical properties of dust grains that carry biogenic elements in interstellar clouds.

BIOLOGICAL PHYSICS
Faculty: Angel E. García, Saroj Nayak, and Ingrid Wilke

Theoretical and Computational Biological Physics: Theoretical and computer simulation studies of the dynamics and statistical mechanics of biological molecules to understand the folding, dynamics, and stability of biological molecules. Research interests include pressure effects on protein stability, the hydrophobic effect, protein aggregation, enzyme catalysis, DNA and RNA structure and dynamics, and protein/membrane interactions.

Cellular Biophysics: Cellular biophysics is represented by a project that measures the motion of single cells in culture by a novel electrical method. The aim of this research is to characterize cells, their interactions with surfaces, and to measure the effects of various chemical agents on cell motility.

CONDENSED-MATTER PHYSICS
Faculty: Sang-Kee Eah, Tim Hayes, György Korniss, James Jian-Qiang Lü, Toh-Ming Lu, Saroj Nayak, Peter Persans, Leo Schowalter, John Schroeder, Michael Shur, Gwo-Ching Wang, Christian Wetzel, Shawn Lin, Masashi Yamaguchi.

Surfaces, Interfaces, and Nanostructures: Experimental and theoretical work on surfaces, interfaces, and nanostructures involves the deposition, growth, and characterization of metals, semiconductors, and insulators. The phenomena studied include homo- and hetero-epitaxy, initial stages of epitaxy, nucleation of thin films, surface phase transitions, and interface (solid-solid and solid-liquid) structure and bonding. Theoretical work also includes applications of statistical physics and large-scale simulations to study the dynamics of natural, artificial, and social systems, including ecological systems, agent-based models, and social networks.

Nanoelectronic Transport: This research enhances understanding of transport in nanostructures. The experimental work includes studies of ballistic electron transport in ultrathin epitaxial multilayers, electrical resistance of metallic films, and plasma wave electronics in high-electron mobility transistors. The electron transport in nanoscale systems (single molecule to atomic wire to carbon nanotube) is studied using the state-of-the-art first principles calculation. The current research includes spin-assisted transport (spintronics) at the nanoscale.

Optical and Electronic Materials: The optical and electronic materials under study include wide band gap semiconductors, photonic crystals, polymers, semiconductor nanoparticle composites, dielectrics, and magnetic thin films.

EDUCATIONAL PHYSICS
Faculty: Gary Bedrosian and Philip Casabella.

Physics Education Group: Students in the Physics Education Group engage in undergraduate research opportunities and earn graduate degrees by conducting research on the learning and teaching of physics. Graduate students in the group actively participate in the development and teaching of departmental courses.

Studio Approach: Rensselaer's Physics Education Group pioneered the studio approach to physics instruction. The defining characteristics of studio physics classes are integrated lecture/laboratory form, a reduction in lecture time, a technology-enhanced learning environment, collaborative group work, and a high level of faculty-student interaction. The studio physics environment employs activities, computer tools, and multimedia materials that allow students to participate in their own learning and to construct their own scientific knowledge.

OPTICAL PHYSICS
Faculty: Sang-Kee Eah, Timothy Hayes, Shawn-Yu Lin, James Jian-Qiang Lü, Peter Persans, John Schroeder, Ingrid Wilke, Jingzhou Xu, and X.-C. Zhang.

Characterization of Materials: The focus is on achieving optical characterization of materials such as nanocrystalline metal and semiconductor particles in glass or in organic materials. Experimental measurements gain further understanding of the optical properties of novel materials.

Optical Interconnects: Research centers on developing and testing polymer and inorganic optical waveguides to address interconnect problems that will arise as computer chips get faster.

Terahertz Pulses: Ultrafast photonics and optoelectronics involve the generation and detection of picosecond and femtosecond electromagnetic pulses. Of particular interest are time-resolved experiments on THz pulses. THz spectroscopy opens up novel opportunities in material characterization and information technology. A current project applies THz pulses for biophotonic imaging. Other projects deal with switching semiconductor devices at THz frequencies.

EXPERIMENTAL PARTICLE PHYSICS
Faculty: Gary Adams, John Cummings, Jim Napolitano, and Paul Stoler.

Beyond the Simple Quark: How these fundamental interactions lead to effects that cannot be understood in terms of simple quark model picture. For example, the group has provided some of the first evidence for exotic mesons and baryons. Exotic mesons suggest that gluons, as well as quarks, manifest themselves in the structure of elementary particles.

Collaborations: Rensselaer has a long tradition of experimental nuclear and particle physics research. In recent years this group has made contributions to collaborations at a number of facilities, including the Wilson Laboratory for Elementary Particle Physics at Cornell University; Jefferson Laboratory in Newport News, Virginia; and the Alternating Gradient Synchrotron at Brookhaven National Laboratory. Most recently, as part of the CLEO collaboration, the role of hadronic structure in the weak decays of D mesons is being investigated, and a conclusive study of glueballs is planned.

Southern™ Illinois University Carbondale

SOUTHERN ILLINOIS UNIVERSITY CARBONDALE

Ph.D. in Applied Physics

Program of Study

The Department of Physics offers graduate work leading to the Ph.D. in applied physics. This unique program, begun in 2005, provides many research opportunities within the department, offering students a wide choice of specific areas of study. A low student-faculty ratio allows students to work closely with faculty members. Areas of research include experimental (IR spectroscopy, thin films, magnetism, gas adsorption, ellipsometry), theoretical (condensed-matter theory, statistical physics, quantum computing, simulation of materials), and applied physics (novel materials from agricultural byproducts, coal and coal ash composites, permanent magnetic materials, chemical nanosensors, gas storage, superhard coatings).

In addition to completing all the requirements set by the Graduate School, Ph.D. students must complete a sequence of required basic core courses, which include classical mechanics, quantum mechanics, electromagnetic theory, statistical mechanics, and solid-state physics. One additional course—in computational physics, scanning electron microscopy, transmission electron microscopy, advanced topics in surface physics, advanced topics in magnetism and magnetic materials, advanced topics in quantum computing, advanced topics in applied physics, advanced topics in the spectroscopy of materials, or advanced topics in the physics of hybrid materials—is also required. Another 9 credit hours of graduate-level elective courses selected from a list approved by the department must be completed next. Starting no later than the third semester in the doctoral program, students must enroll for two consecutive semesters in the Special Projects in Physics course. To be admitted into candidacy, students must pass the qualifying examination, taken no later than during the third semester. No later than six months after admission to candidacy, students must request the appointment of a dissertation committee, to which they must present and describe orally a written dissertation proposal. Students must complete 24 credit hours of Physics 600 (Dissertation). Upon completion, the dissertation committee administers a final oral examination, the dissertation defense.

Research Facilities

The thin-film laboratory includes a sputter coater, a spectroscopic ellipsometer, a wear tester, and a spectrograph. The magnetic properties laboratory includes a SQUID magnetometer, high-temperature ovens, a diamond-cutting saw, assorted other magnetometers, and a variable-temperature electrical-resistance measurement setup. The magnetic film properties laboratory is equipped with a Pulsed Laser Deposition (PLD) setup, an x-ray apparatus, a magneto-optic Kerr effect (MOKE) instrument, and a point contact Andreev reflectivity (PCAR) system. The applied materials and spectroscopy laboratory includes four Fourier-transform infrared (FTIR) spectrometers (one coupled with a high-temperature reactor and another with such attachments as a photoacoustic cell system, a grazing angle, a specular reflectance, three diffuse reflectance systems, an attenuated reflectance, a transmission high-temperature system, transmission low-temperature systems, and various gas cells); an electron paramagnetic resonance spectrometer; a Perkin-Elmer diamond differential scanning calorimeter (DSC) system with low-temperature capabilities, a TGA/DTA system, and a differential thermal analyzer (DTA7) system; two grinding mills; two combustion tube assembly systems; two diamond saws; two 900-watt microwave oven systems; four high-temperature, high-pressure presses for fabricating composites; and stainless-steel dies for fabricating composites of various sizes. The low-temperature laboratory has six automated, variable-temperature, volumetric adsorption setups (one with high-pressure capabilities); a helium mass-spectrometer leak detector; a tube furnace; a vacuum oven; and a setup for AC calorimetry measurements and AC thermal diffusivity measurements. The nanosensor laboratory has chemical vapor deposition (CVD) and physical vapor deposition (PVD) systems for nanowire growth, a high-resolution optical microscopy and microspectroscopy system, a variable pressure and variable temperature probe station for conduction measurements on individual nanostructures, and a variable temperature and variable pressure STM/AFM microscope. Theorists in the department are equipped with individual multinode computer clusters, which enable them to conduct sophisticated simulations and calculations.

Financial Aid

Assistantships for the academic year are offered to every selected applicant. An assistantship guarantees employment and includes a waiver of tuition (except for student fees). Incoming graduate students are teaching assistants or research assistants. In addition to teaching assistantships, other sources of financial assistance include fellowships, scholarships, loans, and work-study programs.

Cost of Study

In-state graduate tuition is $243 per credit hour in 2006–07. Out-of-state tuition is 2.5 times the in-state tuition rate ($607.50 per credit hour). Graduate students with at least a 25 percent appointment as a graduate assistant receive a tuition waiver. Fees vary from $441.62 (1 credit hour) to $987.30 (12 credit hours).

Living and Housing Costs

For married couples, students with families, and single graduate students, the University has 589 efficiency and one-, two-, and three-bedroom apartments that rent for $438 to $505 per month in 2006–07. Residence halls for single graduate students are also available, as are accessible residence hall rooms and apartments for students with disabilities.

Student Group

Of the 10 students currently enrolled in the program, 9 are full-time, and 7 are international students.

Location

Southern Illinois University Carbondale (SIUC) is 350 miles south of Chicago and 100 miles southeast of St. Louis. The scenic main campus occupies 981 acres and includes a wooded area preserved in a natural state, a lake with a beach and swimming area, canoe and boat-rental facilities, a walking (or jogging) trail, and fishing piers. The campus provides an array of cultural activities, including frequent performances by opera, theater, symphony, and dance groups given by both local and traveling performers.

The University and The Department

Since its chartering in 1869, Southern Illinois University Carbondale has grown into a comprehensive university with a student body of approximately 24,000. Supported by the state of Illinois, the University offers a wide variety of undergraduate and graduate programs in liberal arts, sciences, engineering, medicine, and law. The objective of the University is to provide a comprehensive educational program that meets a student's needs. The Department of Physics offers the B.S. and M.S. degrees in physics in addition to the Ph.D. in applied physics. Active student participation in research is encouraged at the undergraduate level and required at the graduate level.

Applying

Candidates should have a minimum GPA of 3.25 for admission into the doctoral program. The GRE is required. Applicants must submit the application to the Graduate School and to the physics department, the professional record form, the $10 application fee, official transcripts from all universities attended, and three letters of reference. Applications are processed on a rolling basis.

Correspondence and Information

Dr. Aldo Migone, Chair
Department of Physics
College of Science
Southern Illinois University Carbondale
Carbondale, Illinois 62901-4701

Phone: 618-453-2643
Fax: 618-453-1056
E-mail: physics@physics.siu.edu
Web site: http://www.physics.siu.edu

Southern Illinois University Carbondale

THE FACULTY AND THEIR RESEARCH

Naushad Ali, Professor; Ph.D., Alberta, 1984. Use of synchrotron radiation in magnetic and superconductivity studies; colossal magnetoresistance; photo-induced magnetization and molecular magnets; permanent magnetic materials; electrical, magnetic, and thermal properties of magnetically ordered rare earth compounds; the study of spin-glass and reentrant magnetic phase transitions; valence fluctuations, heavy fermion, and kondo lattice in $YbSi_2$, $CeSi_x$, $UPt_{1-x}Pd_1$ and like systems; evolution of Mn magnetic moments in RMn_2 (R=Rare Earth $Y_{1-x}R_xMn_2$ systems).
Local structure of Yni2B2C superconductor determined by x-ray adsorption spectroscopy. *Phys. Rev. B* 1:3175, 2000. With Yu, Ignatov, Tischer, Tsvyashchenko, and Foricheva).

Samir Aouadi, Assistant Professor; Ph.D., British Columbia, 1994. Design, fabrication, characterization, and testing of optical and protective coatings, specifically nanocomposite and nanocrystalline single-phase coatings for orthopedic implants and other tribological applications, spectroscopic ellipsometry as a real-time technique for process monitoring and control of composite and multilayer coatings, diffusion barrier coatings for interconnects, and ab-initio density functional theory calculations for materials design and analysis.
Real-time spectroscopic ellipsometry study of ultra-thin diffusion barriers for integrated circuits. *J. Appl. Phys.* 96:3949, 2004. With Shreeman.

Mark Byrd, Assistant Professor; Ph.D., Texas at Austin, 1999. Theoretical quantum computation and quantum error correction.
Universal leakage elimination. *Phys. Rev. A* 71:052301, 2005. With Lidar, Wu, and Zanardi.

Maria de las Mercedes Calbi, Assistant Professor; Ph.D., Buenos Aires, 1999. Gases adsorbed on and inside carbon nanotube bundles (novel phases of matter, adsorption and thermodynamical properties, diffusion behavior, quantum effects, and phase transitions), Bose-Einstein condensation in low-dimensional systems (condensates in optical lattices), forces between nanosized particles (Van der Waals forces between atomic clusters, size and many-body effects, and aggregation phenomena in fluids).
Universal anisotropic condensation transition of gases in nanotube bundles. *J. Low Temp. Phys.* 133:399–406, 2003. With Gatica and Cole.

John D. Cutnell, Professor Emeritus; Ph.D., Wisconsin, 1967.

Frank Gaitan, Assistant Professor; Ph.D., Illinois at Urbana-Champaign, 1992. Quantum computing, condensed-matter theory.
Temporal interferometry: A mechanism for controlling qubit transitions during twisted rapid passage with possible application to quantum computing. *Phys. Rev. A* 68:052314, 2003.

Bruno J. Gruber, Professor Emeritus; Ph.D., Vienna, 1961.

Walter C. Henneberger, Professor Emeritus; Ph.D., Göttingen, 1959.

Richard Holland, Lecturer; Ph.D., Southern Illinois Carbondale, 1999.

Kenneth W. Johnson, Professor Emeritus; Ph.D., Ohio State, 1967.

Andrei Kolmakov, Assistant Professor; Ph.D., Kurchatov Institute and Moscow Institute of Physics and Technology, 1995. Interplay between surface and bulk electronic properties of individual nanostructures (nanowires/nanobelts, clusters, nanoparticles), their assemblies, and nanostructured surfaces (oxides); fabrication-functionalization-characterization of nanostructures and nanostructured surfaces; manufacturing of corresponding nanodevices for sensing, (opto)-electronics, energy conversion, and other applications; development of (spectro)-microscopic techniques and instruments for in situ/in vivo characterization of the functioning nanostructures and nanodevices.
Chemical sensing and catalysis by one-dimensional metal-oxide nanostructures. *Ann. Rev. Mater. Res.* 34:151–80, 2004. With Moskovits.

Ruth Ann Levinson, Lecturer; M.S., Southern Illinois Carbondale, 1996.

Vivak Malhotra, Professor; Ph.D., Kanpur, 1978. Porous materials, nanocomposites, friction materials, biocomposites from agricultural by-products, structural materials, and materials from coal combustion by-products.
Structural, thermal, and thermodynamical behavior of phenolic-inorganic hybrid composites. *MRS Symp. Ser.* 628:6.5.1–6.6.6, 2000. With Amanuel.

F. Bary Malik, Professor Emeritus; Ph.D., Göttingen, 1958.

J. Thomas Masden, Associate Professor; Ph.D., Purdue, 1983. Electrical properties of thin films and wires, giant magnetoresistance in superlattices such as Fe-Cr.
Temperature dependence of the resistance in ultrathin Nb wires. *Phys. Rev. B Condens. Matter* 38(15):10297–301, 1988. With Hussein and Chin.

Aldo D. Migone, Professor and Chair; Ph.D., Penn State, 1984. Thermodynamic studies of phases and phase transitions in systems of reduced dimensionality: binding energies and first-layer phases of gases adsorbed on carbon nanostructures (nanotubes and nanohorns), thermodynamic measurements of molecules adsorbed on well-characterized planar substrates.
Gases do not adsorb in the interstitial channels of SWNT bundles. *Phys. Rev. Lett.* 85:138, 2000. With Talapatra, Zambano, and Weber.

Frank C. Sanders Jr., Associate Professor Emeritus; Ph.D., Texas, 1968.

Mykola Saporoschenko, Professor Emeritus; Ph.D., Washington (St. Louis), 1958.

Shane Stadler, Assistant Professor; Ph.D., Tulane, 1998. Magnetic materials, pulsed-laser deposition of thin films, MOKE measurements, magnetic properties of thin films, spin-polarized materials, x-ray magnetic circular dichroism measurements.

Mesfin Tsige, Assistant Professor; Ph.D., Case Western Reserve, 2001. Modeling polymer systems with molecular dynamic simulations.

Richard E. Watson, Professor Emeritus; Ph.D., Illinois, 1938.

STONY BROOK
STATE UNIVERSITY OF NEW YORK

STONY BROOK UNIVERSITY, STATE UNIVERSITY OF NEW YORK

Department of Physics and Astronomy

Programs of Study

The Department of Physics and Astronomy offers four graduate degrees. Students in the Ph.D. program gain a solid background in the breadth of physics and astronomy and demonstrate their ability to carry out research and overcome new challenges in a specific area of interest. A Master of Arts in Teaching (M.A.T.) provides students who have a B.A. degree in a physical science or engineering with the preparation needed for New York State certification as a secondary school teacher. This program addresses the widely known shortage of well-trained high school physics teachers. A Master of Science (M.S.) program in instrumentation prepares students with undergraduate degrees in physical science, mathematics, or engineering to enter modern technological enterprises such as research labs, industries, and hospitals as professional physicists with expertise in instrumentation.

The program offers both course work and an original instrumentation project in one of the Department's cutting-edge research labs. The Department does not recruit candidates seeking only the Master of Arts (M.A.) in physics degree. However, many students who come to Stony Brook on limited-time exchange programs enroll in the M.A. program. The M.A. may also be awarded to students originally enrolled in the Ph.D. program, either en route to the Ph.D. or as a terminal degree.

Research Facilities

The Department is involved in a wide range of activities in an array of laboratories. Accelerator and beam physics is concerned with the development of novel accelerator concepts, free-electron lasers, and instrumentation. This work is carried out at nearby Brookhaven National Laboratory and includes Stony Brook adjunct faculty members and graduate students.

The astronomy faculty members are interested in many areas of astronomy and astrophysics: extragalactic astronomy and cosmology, including studies of the Hubble Deep Field, which contains the most distant galaxies ever seen; radio- and millimeter-wave studies of molecular clouds and galaxies and the stratosphere; nuclear astrophysics, including studies of supernovae, neutron stars, the equation of state, and merging neutron star binaries; star formation and properties of low-mass (cool) stars observed with IR, optical, and X-radiation; and studies of supergiants and space interferometry. In atmospheric physics, molecular spectroscopy is applied to the study of stratospheric trace gases that regulate the chemical equilibrium of ozone, radiative heating and cooling of the atmosphere, and other physical processes. The work is carried out in collaboration with NASA scientists and with other faculty members in Stony Brook's Institute for Terrestrial and Planetary Atmospheres.

The condensed matter–experimental and device physics group studies fundamental phenomena such as phase transitions and electronic and magnetic properties of materials and does applied research on devices based on superconductor and semiconductor structures, rapid single flux quantum (RSFQ) logic, single-electron tunneling (SET), and powder diffraction. The condensed matter–theory group is interested in the theory of superconductors. The group has played a major role in the theory of the fractional quantum Hall effect and in the development of the ideas of single-electron and single flux quantum device physics. The particle physics–experimental group studies fundamental forces and the constituents of matter at the D0 experiment at the 2-TeV Fermilab proton-antiproton collider and at the ATLAS experiment at the 14-TeV CERN collider. Rare K decays are investigated in the KOPIO experiment in Brookhaven Lab, and neutrino oscillations and proton decay are studied at the SuperKamiokande and K2K experiments in Japan. The C. N. Yang Institute for Theoretical Physics is dedicated to research in fundamental theory, such as the standard model of elementary particles, string theory, supersymmetry, and statistical mechanics. The nuclear and heavy-ion physics–experimental group operates a superconducting linear accelerator for nuclear physics research on campus (including the recently reported trapping of francium atoms) and is engaged in the PHENIX experiment in Brookhaven. Phenix is one of the two experiments on the Relativistic Heavy Ion Collider (RHIC) at Brookhaven. RHIC collides high-energy heavy ions to create matter at extremely large density. The nuclear physics–theory group is working on the theory of hadronic matter (including conditions such as those that are produced at the RHIC) and nuclear astrophysics (such as the theory of supernovae and gamma ray bursters). The atomic, molecular, and optical physics group includes a broad range of activities, including laser spectroscopy and cooling to micro-Kelvin temperatures, interactive control of molecular dynamics, and quantum chaos. The X-ray optics and microscopy group studies X-ray optics and microscopy of biological and materials science applications.

Financial Aid

New assistantships and fellowships provide stipends at a minimum of $18,000 for the calendar year starting in September 2006. All assistants and fellows receive full-tuition scholarships. The Department offers financial support in the form of teaching assistantships or fellowships to essentially every member of the entering class, and all applicants are considered for such support. Awards are renewable as long as good academic standing is maintained. Support from research grants is available for all full-time students in the doctoral program.

Cost of Study

In 2005–06, full-time tuition was $6900 per academic year for state residents and $10,920 per academic year for nonresidents. All assistants and fellows receive full-tuition scholarships. Part-time tuition was $288 per credit hour for residents and $455 per credit hour for nonresidents. Additional charges included an activity fee of $22 and a comprehensive fee of $351.50 per semester.

Living and Housing Costs

University apartments range in cost from approximately $300 per month to approximately $1300 per month, depending on the size of the unit. Off-campus housing options include furnished rooms to rent and houses and/or apartments to share that can be rented for $400 to $1500 per month.

Student Group

Stony Brook's current enrollment is about 22,000 students. There are approximately 7,500 graduate students, and they come from all states in the nation, as well as from some seventy-five countries. International students, both graduate and undergraduate, represent about 10 percent of the total student body. There are about 180 graduate students in the Department.

Student Outcomes

Students often go on to traditional research positions at universities and national laboratories as well as to industrial labs and to other careers, including medical physics and analysis in technological and financial settings. In addition, many graduates fill positions in high school or higher education, while others contribute to research programs in laboratories in the public, private, and educational sectors.

Location

Stony Brook's campus is approximately 60 miles east of Manhattan on the north shore of Long Island. The cultural offerings of New York City and Suffolk County's countryside and seashore are conveniently located nearby. Cold Spring Harbor Laboratories and Brookhaven National Laboratories are easily accessible and have close relationships with the University.

The University

The University, established in 1957, achieved national stature within a generation. Founded at Oyster Bay, Long Island, the school moved to its present location in 1962. Stony Brook has grown to encompass more than 110 buildings on 1,100 acres. There are more than 1,500 faculty members, and the annual budget is more than $800 million. The Graduate Student Organization oversees the spending of the student activity fee for graduate student campus events. The Intensive English Center offers classes in English as a second language. The Career Development Office assists with career planning and has information on permanent full-time employment. Disabled Student Services has a Resource Center that offers placement testing, tutoring, vocational assessment, and psychological counseling. The Counseling Center provides individual, group, family, and marital counseling and psychotherapy. Day-care services are provided in four on-campus facilities. The Writing Center offers tutoring in all phases of writing.

Applying

Information is available at http://graduate.physics.sunysb.edu. This Web page is updated more frequently than any of the printed or PDF documents related to the graduate program. Online application is encouraged. For information related to applications, students should contact Diane Siegel at diane.siegel@sunysb.edu. For specific questions about the Department's academic graduate program, students should contact Pat Peiliker, Assistant Graduate Program Director, at ppeiliker@notes.cc.sunysb.edu or Professor Laszlo Mihaly, Graduate Program Director, at laszlo.mihaly@sunysb.edu.

Correspondence and Information

Peter M. Koch, Chairman
Pam Burris, Assistant to the Chair
Department of Physics and Astronomy
Stony Brook University, State University of New York
Stony Brook, New York 11794-3800

Phone: 631-632-8100
Fax: 631-632-8176
E-mail: graduate.physics@sunysb.edu
Web site: http://www.physics.sunysb.edu/Physics

Stony Brook University, State University of New York

THE FACULTY AND THEIR RESEARCH

Accelerator and Beam Physics

Ilan Ben-Zvi, Adjunct Professor and Head, Brookhaven National Laboratory Accelerator Test Facility; Ph.D. , Weizmann (Israel), 1970. Nuclear physics.

Vladimir Litvinenko, Adjunct Professor; Ph.D., Institute of Nuclear Physics (Russia), 1989. Accelerator physics and free-electron lasers.

Steve Peggs, Adjunct Professor; Ph.D., Cornell, 1981. Accelerator physics.

Astronomy

Aaron Evans, Associate Professor; Ph.D., Hawaii, 1996. Extragalactic astronomy.

Kenneth Lanzetta, Professor; Ph.D., Pittsburgh, 1988. Extragalactic astronomy.

James Lattimer, Professor; Ph.D., Texas at Austin, 1976. Astrophysics.

Deane Peterson, Associate Professor; Ph.D., Harvard, 1968. Stellar and galactic astronomy.

Michal Simon, Professor; Ph.D., Cornell, 1967. Star formation.

Philip Solomon, Distinguished Professor; Ph.D., Wisconsin, 1964. Millimeter-wave astronomy.

Anand Sivaramakrishnan, Adjunct Professor; Ph.D., Texas at Austin, 1983. Astronomy.

F. Douglas Swesty, Research Assistant Professor and Computer System Manager; Ph.D., SUNY at Stony Brook, 1993.

Frederick Walter, Professor; Ph.D., Berkeley, 1981. Astronomy and astrophysics.

Amos Yahil, Professor; Ph.D., Caltech, 1970. Astrophysics.

Michael Zingale, Assistant Professor; Ph.D., Chicago, 2000. Computational astrophysics.

Atmospheric Physics

Robert L. de Zafra, Research Professor; Ph.D., Maryland College Park, 1958. Stratospheric dynamics, development of instrumentation for remote measurement of stratospheric trace gases.

Marvin A. Geller, Adjunct Professor and Professor of Atmospheric Sciences; Ph.D., MIT, 1969. Atmospheric dynamics.

Atomic, Molecular, and Optical Physics and Quantum Electronics

Thomas Bergeman, Adjunct Professor; Ph.D., Harvard, 1971.

Louis DiMauro, Adjunct Professor; Ph.D., Connecticut, 1980. Optics.

Peter M. Koch, Chair, Department of Physics and Astronomy; Ph.D., Yale, 1974. Experimental atomic physics, nonlinear dynamical systems.

John H. Marburger, Professor (on leave), Science Adviser to the President, and Director, Office of Science and Technology Policy; Ph.D., Stanford, 1966. Theoretical nonlinear optics.

Harold J. Metcalf, Distinguished Teaching Professor; Ph. D., Brown, 1967. Atomic physics, level crossing techniques.

Dominik Schneble, Assistant Professor; Ph.D., Konstanz, 2002. Experimental atomic physics, ultracold quantum gases.

Thomas Weinacht, Assistant Professor; Ph.D., Michigan, 2000. Coherent control of molecules and atoms with ultrafast laser pulses.

Biological Physics

Chris Jacobsen, Professor; Ph.D., SUNY at Stony Brook, 1988. X-ray physics.

Janos Kirz, Distinguished Professor; Ph.D., Berkeley, 1963. X-ray physics.

Sergei Maslov, Adjunct Professor; Ph.D., SUNY at Stony Brook, 1996. Condensed matter theory, biological physics.

Jin Wang, Adjunct Professor; Ph.D., Illinois, 1991. Computational biological physics.

Condensed Matter–Experimental and Device Physics

Peter Abbamonte, Adjunct Professor; Ph.D., Illinois, 1999. Experimental condensed-matter physics.

Megan Aronson, Professor; Ph.D., Illinois at Urbana-Champaign, 1982. Experimental condensed matter physics.

Steven Dierker, Professor; Ph.D., Illinois at Urbana-Champaign, 1983. Experimental condensed matter physics.

Vladimir J. Goldman, Professor; Ph.D., Maryland College Park, 1985. Experimental solid-state physics.

Michael Gurvitch, Professor; Ph.D., SUNY at Stony Brook, 1978. Experimental solid-state physics.

Peter Johnson, Adjunct Professor; Ph.D., Warwick (England). Photoelectron spectroscopy of strongly correlated systems.

Chi-Chang Kao, Adjunct Professor; Ph.D., Cornell, 1988. Condensed-matter physics, X-ray physics.

Konstantin K. Likharev, Distinguished Professor; Ph.D., Moscow State, 1969. Solid-state physics and electronics.

James Lukens, Professor; Ph.D., California, San Diego, 1968. Experimental solid-state physics.

Emilio E. Mendez, Professor and Director, Undergraduate Studies, Department of Physics and Astronomy; Ph.D., MIT, 1979. Experimental solid-state physics.

Laszlo Mihaly, Professor and Director, Graduate Studies, Department of Physics and Astronomy; Ph.D., Eötvös Loránd (Budapest), 1977. Experimental solid-state physics.

Vasili Semenov, Research Associate Professor; Ph.D., Moscow State, 1975.

Peter W. Stephens, Professor; Ph.D., MIT, 1978. Experimental solid-state physics.

Sergey Tolpygo, Adjunct Professor; Ph.D., Academy of Sciences (Moscow), 1984. Mesoscopic physics.

Yimei Zhu, Adjunct Professor; Ph.D., Nagoya (Japan), 1987. Materials science, electron microscopy.

Condensed Matter–Theory

Alexandre Abanov, Assistant Professor; Ph.D., Chicago, 1997. Interference effects in strongly correlated electronic systems.

Philip B. Allen, Professor; Ph.D., Berkeley, 1969. Structural and electronic properties of nanoscale systems.

Dmitri V. Averin, Professor; Ph.D., Moscow State, 1987. Mesoscopic systems, quantum computing.

James Davenport, Adjunct Professor; Ph.D., Pennsylvania, 1976. Theoretical condensed matter physics.

Adam Durst, Assistant Professor; Ph.D., MIT, 2002. Condensed-matter theory.

Wei Ku, Adjunct Professor; Ph.D., Tennessee, 2000. Condensed matter theory.

Konstantin K. Likharev, Distinguished Professor; Ph.D., Moscow State, 1969. Solid-state physics and electronics.

Sergei Maslov, Adjunct Professor; Ph.D., SUNY at Stony Brook, 1996. Condensed matter theory, biological physics.

Alexei Tsvelik, Adjunct Professor; Ph.D., Kurchatov Institute of Atomic Energy (Moscow), 1980. Correlated electrons, quantum solids.

C. N. Yang Institute for Theoretical Physics

Gerald E. Brown, Distinguished Professor; Ph.D., Yale, 1950; D.Sc., Birmingham, 1957. Theoretical nuclear physics.

Michael Creutz, Adjunct Professor (Brookhaven National Laboratory); Ph.D., Stanford, 1970. Physics.

Sally Dawson, Adjunct Professor; Ph.D., Harvard, 1981. Theoretical physics, collider phenomenology.

Alfred Goldhaber, Professor; Ph.D., Princeton, 1964. Theoretical physics, nuclear theory, particle physics.

Maria Concepcion Gonzalez-Garcia, Associate Professor; Ph.D., Valencia, 1991. Theoretical physics, neutrino physics.

Vladimir Korepin, Professor; Ph.D., Leningrad, 1977. Mathematical physics, statistical mechanics, condensed matter, exactly solvable models, quantum computing.

Barry McCoy, Distinguished Professor; Ph.D., Harvard, 1967. Statistical mechanics.

Leonardo Rastelli, Assistant Professor; Ph.D., MIT, 2000. Theoretical physics.

Martin Rocek, Professor; Ph.D., Harvard, 1979. Theoretical physics, supersymmetry.

Robert Shrock, Professor; Ph.D., Princeton, 1975. Particle physics, field theory, statistical mechanics.

Warren Siegel, Professor; Ph.D., Berkeley, 1977. Theoretical physics, strings.

Jack Smith, Professor and Deputy Director, C. N. Yang Institute for Theoretical Physics; Ph.D., Edinburgh, 1963. Elementary particle physics.

George Sterman, Professor and Director, C. N. Yang Institute for Theoretical Physics; Ph.D., Maryland College Park, 1974. Theoretical physics, elementary particles.

Peter van Nieuwenhuizen, Distinguished Professor; Ph. D., Utrecht, 1971. Theoretical physics.

William Weisberger, Professor Emeritus; Ph.D., MIT, 1964. Theoretical physics.

Chen Ning Yang, Albert Einstein Professor Emeritus; D.Sc., Princeton; Ph.D., Chicago, 1948. Theoretical physics, field theory, statistical mechanics, particle physics.

High-Energy Physics–Experimental

Roderich Engelmann, Professor; Ph.D., Heidelberg, 1966. Experimental elementary particle physics.

Paul D. Grannis, Distinguished Professor; Ph.D., Berkeley, 1965. Experimental high-energy physics, elementary particle reactions.

John Hobbs, Associate Professor; Ph.D., Chicago, 1991. Experimental high-energy physics.

Chang Kee Jung, Professor; Ph.D., Indiana, 1986. Experimental high-energy physics.

Michael Marx, Professor and Associate Dean, College of Arts and Sciences; Ph.D., MIT, 1974. Experimental high-energy physics.

Robert L. McCarthy, Professor; Ph.D., Berkeley, 1971. Experimental elementary particle physics.

Clark McGrew, Assistant Professor; Ph.D., California, Irvine, 1994. Physics.

Michael Rijssenbeek, Professor; Ph.D., Amsterdam, 1979. Experimental high-energy physics.

R. Dean Schamberger, Senior Scientist; Ph.D., SUNY at Stony Brook, 1977. Experimental high-energy physics.

Helio Takai, Adjunct Professor; Ph.D., Rio de Janeiro, 1986. Experimental particle physics.

Chiaki Yanagisawa, Research Professor; Ph.D., Tokyo, 1981. Experimental high-energy physics.

Nuclear and Heavy-Ion Physics–Experimental

Ralf Averbeck, Research Assistant Professor; Dr.rer.nat. , Justus-Liebig (Germany), 1996.

Abhay Deshpande, Assistant Professor; Ph.D., Yale, 1995. Relativistic heavy-ion physics.

Axel Drees, Professor; Dr.rer.nat., Heidelberg, 1989. Relativistic heavy-ion physics.

Thomas Hemmick, Professor; Ph.D., Rochester, 1989. Experimental relativistic heavy-ion physics.

Barbara Jacak, Professor; Ph.D., Michigan State, 1984. Experimental relativistic heavy-ion physics.

Linwood L. Lee Jr., Professor Emeritus; Ph.D., Yale, 1955. Experimental nuclear structure.

Robert L. McGrath, Professor (on leave); Provost and Executive Vice President, Academic Affairs; and Vice President, Brookhaven Affairs; Ph.D., Iowa, 1965. Heavy-ion reaction studies from low to relativistic energies.

Peter Paul, Distinguished Service Professor; Ph.D., Freiburg, 1959. Experimental nuclear physics.

Gene D. Sprouse, Professor and Director of the Nuclear Structure Laboratory; Ph.D., Stanford, 1968. Laser spectroscopy of radioactive atoms, development of radioactive beams.

Nuclear Physics–Theory

Gerald E. Brown, Distinguished Professor; Ph.D., Yale, 1950; D.Sc., Birmingham, 1957. Theoretical nuclear physics.

Thomas T. S. Kuo, Professor; Ph.D., Pittsburgh, 1964. Nuclear theory.

Edward Shuryak, Distinguished Professor; Ph.D., Novosibirsk Institute of Nuclear Physics, 1974. Theoretical nuclear physics.

Jacobus Verbaarschot, Professor; Ph.D., Utrecht, 1982. Statistical theory of spectra.

Ismail Zahed, Professor; Ph.D., MIT, 1983. Theoretical nuclear physics.

Optics–X-ray Optics and Microscopy

Chris Jacobsen, Professor; Ph.D., SUNY at Stony Brook, 1988. X-ray physics.

Janos Kirz, Distinguished Professor; Ph.D., Berkeley, 1963. High-energy physics.

David Sayre, Adjunct Professor; Ph.D., Oxford, 1951. X-ray physics.

Other Research Areas

Miriam A. Forman, Adjunct Professor; Ph.D., SUNY at Stony Brook, 1972. Cosmic rays in interplanetary space, space physics.

Erlend H. Graf, Associate Professor; Ph.D., Cornell, 1967. Experimental low-temperature physics.

Peter B. Kahn, Professor Emeritus; Ph.D., Northwestern, 1960. Theoretical physics, nonlinear dynamics.

Richard Mould, Professor Emeritus; Ph.D., Yale, 1957. Relativity.

Clifford Swartz, Professor Emeritus; Ph.D., Rochester, 1951. Pedagogy.

TEMPLE UNIVERSITY

of the Commonwealth System of Higher Education

Department of Physics

Programs of Study

The Department offers the M.A. and Ph.D. degrees. The M.A. program requires 24 semester hours of credit. Normally, required courses for the M.A. degree encompass 18 hours; the other 6 semester hours are used for thesis research or for additional courses. The student must also pass the M.A. comprehensive examination in physics. No specific number of graduate credits is required for the Ph.D. degree, but an approved program of graduate courses must be satisfactorily completed. A dissertation and dissertation examination are required. An M.A. degree is not necessary for the Ph.D. degree. The Ph.D. qualifying examination in physics is taken after completion of two years of graduate study. There is a one-year residence requirement for the Ph.D. degree. Students whose native language is not English must pass an examination in spoken and written English. There is no other language requirement for either the M.A. or the Ph.D. degree. Each full-time graduate student is given a desk in one of several student offices. Lecturers from other institutions describe their research activities at a weekly colloquium, and informal discussions with members of the faculty are frequent.

Research Facilities

The Department is housed in Barton Hall, which has "smart" lecture theaters, offices, classrooms, and laboratories. The Physics Department Library contains frequently used journals and books; several thousand additional volumes are located in the Paley Library across the street from Barton Hall. A student shop and a materials preparation facility are available. The University computer facilities are based on a UNIX-cluster-composed Digital Equipment Corporation Alphas, including a high-performance numerical compute-server. The departmental computer facilities include a local area network (LAN) of five Windows XP workstations, and eight host LAN of Linux workstations. The departmental local area networks are connected to a fiber-optic campus backbone through which all University mainframe computer facilities can be reached. High-speed access to the Internet is readily available from all departmental computers. Electronic information retrieval is provided by the Temple University library's Scholars Information System, which subscribes to a wide range of online databases. The research laboratories are conducting a variety of studies on optical hole-burning and multiple quantum well structures; laser-based molecular spectroscopy; low-temperature properties of alloys and intermetallics, including valence fluctuations and heavy fermion behavior; high-temperature superconductivity; Mössbauer spectroscopy; nucleon structure; dark-matter detection; and electrorheology and magnetorheology. The Department also uses outside facilities, including the Los Alamos Meson Physics Facility, the Brookhaven National Laboratory, the Stanford Linear Accelerator Center, the Thomas Jefferson National Accelerator Facility, and the National High Magnetic Laboratory. Theoretical work is being conducted in such areas as elementary particles and their interactions, statistical mechanics, biophysics, general relativity, and condensed-matter theory.

Financial Aid

Aid is available to qualified full-time students in the form of assistantships and fellowships funded by the University and various extramural agencies. All forms of financial aid include a stipend plus tuition. The specific type of aid offered to a particular student depends on the student's qualifications and program of study. Summer support for qualified students is also normally available. Currently, the minimum stipend for the academic year is $14,535, but it can be supplemented up to $18,535. Also, academic year stipends can be supplemented by summer stipends. For students with grant-supported research assistantships, the stipend is much higher.

Cost of Study

The annual tuition for full-time graduate study in 2004–05 was $456 per credit hour for residents of Pennsylvania and $664 per credit hour for nonresidents. Minimal fees are charged for various services, such as microfilming theses.

Living and Housing Costs

Room and board costs for students living on campus were $7522 per year in 2004–05. University-sponsored apartments, both furnished and unfurnished, are also available on the edge of the campus.

Student Group

The Department has 29 full-time graduate students; nearly all are supported by assistantships or fellowships.

Location

Philadelphia is the fifth-largest city in the country, with a metropolitan population of more than 2 million. The city has a world-renowned symphony orchestra, a ballet company, two professional opera companies, and a chamber music society. Besides attracting touring plays, Philadelphia has a professional repertory theater and many amateur troupes. All sports and forms of recreation are easily accessible. The city is world famous for its historic sites and parks and for the eighteenth-century charm that is carefully maintained in the oldest section. The climate is temperate, with an average winter temperature of 33 degrees and an average summer temperature of 75 degrees.

The University

The development of Temple University has been in line with the ideal of "educational opportunity for the able and deserving student of limited means." With a rich heritage of social purpose, Temple seeks to provide the opportunity for high-quality education without regard to a student's race, creed, or station in life. Affiliation with the Commonwealth System of Higher Education underpins Temple's character as a public institution.

Applying

All application material, both for admission and for financial awards, should be received by early March for admission in the fall semester. Notification regarding admission and the awarding of an assistantship is made as soon as the application has been screened.

Correspondence and Information

For program information and all applications:
Graduate Chairman
Department of Physics 009-00
Barton Hall
Temple University
Philadelphia, Pennsylvania 19122-6052
Phone: 215-204-7736
Fax: 215-204-5652
E-mail: physics@temple.edu
Web site: http://www.temple.edu/physics

For general information on graduate programs:
Dean
Graduate School
Temple University
Philadelphia, Pennsylvania 19122

Temple University

THE FACULTY AND THEIR RESEARCH

Atomic, Molecular and Optical Physics

Z. Hasan, Professor; Ph.D., Australian National, 1979. Laser materials, laser spectroscopy of solids.
R. L. Intemann, Professor Emeritus; Ph.D., Stevens, 1964. Theoretical atomic physics, inner-shell processes.
S. Kotochigova, Research Associate Professor; Ph.D., St. Petersburg. Relativistic quantum theory: atomic and molecular applications.
M. Lyyra, Professor; Ph.D., Stockholm, 1984. Laser spectroscopy, molecular coherence effects, laser-atom interactions.
M. Mackie, Research Assistant Professor; Ph.D., Connecticut, 1999. Bose-Einstein condensates.
R. Tao, Professor; Ph.D., Columbia, 1982. Photonic crystals, nonlinear optics.

Condensed-Matter Physics

Z. Hasan, Professor; Ph.D., Australian National, 1979. Optical and magnetooptical properties of solids.
C. L. Lin, Associate Professor; Ph.D., Temple, 1985. Heavy fermions, crystal fields, valence fluctuations, the Kondo effect, high-temperature superconductivity.
T. Mihalisin, Professor; Ph.D., Rochester, 1967. Crystal fields, valence fluctuations and the Kondo effect in magnetic systems.
P. Riseborough, Professor; Ph.D., Imperial College (London), 1977. Theoretical condensed-matter physics and statistical mechanics.
R. Tahir-Kheli, Professor; D.Phil., Oxford, 1962. Theory of magnetism, randomly disordered systems.
R. Tao, Professor; Ph.D., Columbia, 1982. Electrorheological and magnetorheological fluids, self-aggregation of superconducting particles.
T. Yuen, Associate Professor; Ph.D., Temple, 1990. Experimental condensed-matter physics, Mössbauer spectroscopy.

Educational Development Physics

L. Dubeck, Professor; Ph.D., Rutgers, 1965. Development, publication, and testing of precollege science materials.
Z. Dziembowski, Associate Professor; Ph.D., Warsaw, 1975. In-service elementary and secondary teacher training, inquiry-based instruction.
R. B. Weinberg, Professor Emeritus; Ph.D., Columbia, 1963. Teaching physicist.

Elementary Particle Physics and Cosmology

Z. Dziembowski, Associate Professor; Ph.D., Warsaw, 1975. Theoretical particle physics.
J. Franklin, Professor Emeritus; Ph.D., Illinois, 1956. Theoretical particle physics; quark and parton theory, S-matrix theory.
C. J. Martoff, Professor and Chairman; Ph.D., Berkeley, 1980. Experimental particle physics: investigation of weak interactions and development of particle detectors for the study of dark matter, using negative ion drift.
Z.-E. Meziani, Professor; Ph.D., Paris, 1984. Experimental high-energy nuclear physics: investigation of the flavor and spin structure of the nucleon at the Stanford Linear Accelerator Center, search for transition region between nucleon-meson to quark-gluon description of few-body nuclear systems at the Continuous Electron Beam Accelerator Facility.
D. E. Neville, Professor Emeritus; Ph.D., Chicago, 1962. Theoretical particle physics; symmetries and quark models, quantum gravity.

Statistical Physics

T. Burkhardt, Professor; Ph.D., Stanford, 1967. Statistical mechanics and many-body theory.
D. Forster, Professor; Ph.D., Harvard, 1969. Statistical mechanics and many-body theory.
E. Gawlinski, Associate Professor; Ph.D., Boston University, 1983. Statistical mechanics and computational physics.

Barton Hall, the physics building.

The Elementary Particle Physics Laboratory.

UNIVERSITY OF KANSAS

Department of Physics and Astronomy

Programs of Study

The Department of Physics and Astronomy offers programs of study leading to the Ph.D. in physics and the M.S. in physics and computational physics and astronomy.

The master's degree in physics requires 30 hours of advanced courses (up to 6 of which may be transferred from another accredited university) and at least 2 hours of master's research with satisfactory progress. A minimum average of B is required, as is a general examination in physics. The various master's programs differ in their detailed requirements.

The Ph.D. program begins with formal course work (which typically extends through two years for a well-prepared student) and, after admission to candidacy, is followed by Ph.D. research. The required courses include those needed for the M.S. in physics, so it is possible to obtain the M.S. on the way to the Ph.D. degree. Course work should average better than a B. There is no language requirement, but a demonstrated skill in computer programming related to the student's field of study is required. A written preliminary exam and a comprehensive exam are required for admission to candidacy. Following the comprehensive exam, the student may choose a research project from the broad spectrum of experimental and theoretical research areas represented within the Department. These include high-energy particle physics, astrophysics and cosmology, biophysics, astrobiology, space physics, plasma physics, solid-state and condensed-matter physics, nonlinear dynamics, and nuclear physics. After carrying out the research project under the guidance of a faculty member, the student must submit a dissertation showing the results of original research and must defend it in a final oral examination. A minimum of three full academic years of residency is required; the actual time taken to complete the Ph.D. varies considerably.

Research Facilities

Extensive computing facilities exist both in the Department and at the University. Condensed-matter physics facilities include an advanced materials research lab and a quantum electronics lab. These labs are well equipped with thin–film deposition systems, a new scanning electron microscope, a unique UHV multiprobe scanning microscopy system, an X-ray diffractometer, SQUID magnetometers, a 6-mK dilution refrigerator, microwave synthesizers, and a vector network analyzer. A clean room with photo- and electron-beam lithography as well as wafer processing tools is also available for micro- and nanofabrication of solid state devices and circuits. The high-energy physics and nuclear physics groups utilize experimental facilities at various universities and national laboratories as part of collaborative experiments. The Kansas Institute of Theoretical and Computational Science sponsors interdisciplinary research among the Departments of Physics and Anatomy, Mathematics, and Chemistry. The Astrobiology Working Group collaborates with the Biodiversity Research Institute, the Department of Geology, and the Department of Ecology and Evolutionary Biology.

Financial Aid

The principal form of financial aid is the graduate teaching assistantship; most first-year graduate students in the Department have this type of support. A half-time teaching assistantship, which is the usual appointment, carries a nine-month stipend of at least $13,500 plus a 100 percent tuition fee waiver. Summer support is also available. Beginning graduate students may also be considered for graduate school fellowships in a University-wide competition. A few research assistantships are available for qualified first-year students, although the tendency is to award such assistantships to more advanced students.

Cost of Study

Full-time students with private support or with fellowships from sources outside the University paid tuition of $202 per credit hour for graduate-level courses in 2005–06 if they were Kansas residents and $500 per credit hour if they were nonresidents. Typical enrollments range from about 9 to 12 credit hours per semester during the first year. University fees are set by the Board of Regents and are subject to change at any time.

Living and Housing Costs

Room and board are available in University dormitories. The starting cost for the 2006–07 academic year is $3634. There are a limited number of one- and two-bedroom University apartments for married students and their families, and the rent for 2006–07 is $283–$500 per month plus utilities. Many rooms and apartments, both furnished and unfurnished, are available off campus.

Student Group

The University of Kansas has an enrollment of more than 26,000 students, including about 6,000 graduate students. The Department enrolls approximately 45 graduate students drawn from throughout the United States and abroad. Most of these students are supported as either teaching assistants or research assistants.

Location

The University's main campus occupies 1,000 acres on and around Mount Oread in the city of Lawrence, a growing community of 75,000 located among the forested, rolling hills of eastern Kansas. Near Lawrence are four lake resort areas for boating, fishing, and swimming. Metropolitan Kansas City lies about 40 miles east of Lawrence via interstate highway and offers a variety of cultural and recreational activities.

The University

The University of Kansas is a state-supported school founded in 1866. Long known for its commitment to academic excellence, the University considers research an important part of the educational process. In addition to the College of Liberal Arts and Sciences and the Graduate School, the University houses a number of professional schools and programs, which include Engineering, Medicine, Law, Business, Journalism, and many others.

Applying

Online applications should be completed by April 1. Paper applications of international students must be completed by May 1. Domestic applications will be considered through July. For further application information, students should visit http://www.physics.ku.edu/graduate/how.shtml.

Correspondence and Information

For application forms and admission:
Graduate Admissions Officer
Department of Physics and Astronomy
1251 Wescoe Hall Drive, Room 1082
University of Kansas
Lawrence, Kansas 66045-7582
Phone: 785-864-4626
E-mail: physics@ku.edu
Web site: http://www.physics.ku.edu

University of Kansas

THE FACULTY AND THEIR RESEARCH

Raymond G. Ammar, Professor; Ph.D., Chicago, 1959. Experimental high-energy physics.
Barbara J. Anthony-Twarog, Professor; Ph.D., Yale, 1981. Observational astronomy, stellar evolution in open star clusters, CCD and photoelectric photometry, globular clusters.
Thomas P. Armstrong, Professor Emeritus; Ph.D., Iowa, 1966. Space physics, plasma physics.
Scott R. Baird, Adjunct Professor; Ph.D., Washington (Seattle), 1979. Stellar spectroscopy, variable stars.
Philip S. Baringer, Professor and Associate Chairman; Ph.D., Indiana, 1985. Experimental high-energy physics.
Alice L. Bean, Professor; Ph.D., Carnegie Mellon, 1987. Experimental high-energy physics.
Robert C. Bearse, Professor Emeritus; Ph.D., Rice, 1964. Experimental nuclear physics, nuclear safeguards, materials control and accounting, computer database applications.
David Z. Besson, Professor; Ph.D., Rutgers, 1986. Experimental high-energy physics.
Thomas E. Cravens, Professor; Ph.D., Harvard, 1975. Space physics, plasma physics.
John P. Davidson, Professor Emeritus; Ph.D., Washington (St. Louis), 1952. Theoretical nuclear structure physics, atomic physics, astrophysics.
Robin E. P. Davis, Professor; D.Phil., Oxford, 1962. Experimental high-energy physics.
Gisela Dreschhoff, Courtesy Associate Professor; Dr.Sc., Braunschweig Technical (Germany), 1972. Geophysics, energy storage in solids.
Joe R. Eagleman, Professor Emeritus; Ph.D., Missouri, 1963. Atmospheric science.
Jacob Enoch, Associate Professor Emeritus; Ph.D., Wisconsin, 1956. Theoretical physics.
Hume A. Feldman, Associate Professor; Ph.D., SUNY at Stony Brook, 1989. Astrophysics and cosmology.
Christopher J. Fischer, Assistant Professor; Ph.D., Michigan, 2000. Biophysics.
Robert J. Friauf, Professor Emeritus; Ph.D., Chicago, 1953. Experimental condensed-matter physics, diffusion and color centers, molecular dynamics and Monte Carlo simulations.
Paul Goldhammer, Professor Emeritus; Ph.D., Washington (St. Louis), 1956. Theoretical physics, nuclear structure physics, atomic physics.
Siyuan Han, Professor; Ph.D., Iowa State, 1986. Experimental condensed-matter physics.
Ralph W. Krone, Professor Emeritus; Ph.D., Johns Hopkins, 1949. Experimental nuclear physics.
Nowhan Kwak, Professor; Ph.D., Tufts, 1962. Experimental high-energy physics.
Danny Marfatia, Assistant Professor; Ph.D., Wisconsin, 2001. Particle astrophysics.
Carl D. McElwee, Courtesy Professor; Ph.D., Kansas, 1970. Geophysics, magnetic properties of solids.
Douglas W. McKay, Professor; Ph.D., Northwestern, 1968. Theoretical elementary particle physics and particle astrophysics.
Mikhail V. Medvedev, Associate Professor; Ph.D., California, San Diego, 1996. Astrobiology, theoretical astrophysics, space physics and plasma physics.
Adrian L. Melott, Professor; Ph.D., Texas, 1981. Astrobiology, astrophysics and cosmology.
Herman J. Munczek, Professor Emeritus; Ph.D., Buenos Aires, 1958. Theoretical elementary particle physics.
Michael J. Murray, Assistant Professor; Ph.D., Pittsburgh, 1989. Experimental nuclear physics.
Francis W. Prosser, Professor Emeritus; Ph.D., Kansas, 1955. Experimental nuclear physics.
John P. Ralston, Professor; Ph.D., Oregon, 1980. Theoretical elementary particle physics and particle astrophysics.
Stephen J. Sanders, Professor and Department Chairman; Ph.D., Yale, 1977. Experimental nuclear physics.
Richard C. Sapp, Professor Emeritus; Ph.D., Ohio State, 1955. Experimental solid-state physics.
Sergei F. Shandarin, Professor; Ph.D., Moscow Physical Technical Institute, 1971. Astrophysics and cosmology, large-scale structure, nonlinear dynamics, computational physics.
Stephen J. Shawl, Professor; Ph.D., Texas, 1972. Observational astronomy, stellar astronomy, polarization, globular clusters.
Jicong Shi, Associate Professor; Ph.D., Houston, 1991. Theoretical physics, nonlinear dynamics, beam dynamics, accelerator physics.
Don W. Steeples, Courtesy Professor; Ph.D., Stanford, 1975. Geophysics.
Robert Stump, Professor Emeritus; Ph.D., Illinois, 1950. Experimental high-energy physics.
Carsten Timm, Assistant Professor; Ph.D., Hamburg (Germany), 1996. Theoretical condensed matter physics.
Bruce A. Twarog, Professor; Ph.D., Yale, 1980. Observational astronomy, stellar nucleosynthesis, chemical evolution of galaxies, stellar photometry.
Graham W. Wilson, Associate Professor; Ph.D., Lancaster, 1989. Experimental high-energy physics.
Gordon G. Wiseman, Professor Emeritus; Ph.D., Kansas, 1950. Experimental solid-state physics.
Kai-Wai Wong, Professor Emeritus; Ph.D., Northwestern, 1962. Many-body theory, superconductivity, liquid helium.
Judy Z. Wu, Professor; Ph.D., Houston, 1993. Experimental condensed-matter physics, low-temperature physics.

UNIVERSITY OF KENTUCKY

Department of Physics and Astronomy

Programs of Study

The Department of Physics and Astronomy requires dedication to hard work, but the rewards of meeting this challenge can provide membership in one of the great modern adventures of mankind. Research is divided into five areas: astronomy and astrophysics, atomic physics, condensed matter physics, nuclear physics, and particle theory. Within each of these areas, students can complete thesis research projects leading to the Master of Science (M.S.) or Ph.D. degree.

The core curriculum consists of one semester of advanced mechanics, two semesters of advanced electrodynamics, two semesters of quantum mechanics, and one semester of statistical mechanics. The M.S. nonthesis option requires 30 hours of course work. The thesis option requires at least 24 hours, followed by an oral final examination and submission of a written thesis. All thesis students are also required to enroll in two 1-credit courses in which they learn about the faculty research programs.

Prior to qualification for the Ph.D., students must complete at least 36 hours of course work, including two years as a full-time student. They must also complete at least 3 credit hours of a research course, hold a research assistantship for at least one semester, or participate in research during a summer semester. By the fifth semester of study, students should submit a written Ph.D. research plan. When these requirements are met, students take the Qualifying Examination, followed by two semesters of full-time dissertation study. Once the research is complete, students present a thesis defense and then submit their written dissertations. All requirements for the Ph.D. must be completed within five years following the semester in which the student passes the Qualifying Examination.

Research Facilities

The University's Library System, one of the nation's top research libraries, contains more than 2.5 million volumes, 5 million microform units, and 13,000 linear feet of manuscripts and subscribes to over 27,000 periodicals, including more than 270 newspapers from around the world. The collection also includes over 220,000 maps, 9,600 music records and CD's, and more than 2.5 million government documents. The William T. Young Library, the library's central facility, houses 1.2 million volumes and seats 4,000 people. The Chemistry-Physics Library contains more than 54,000 volumes and subscribes to over 400 journals.

The Department currently has 65 computers running Linux, most of which are personal workstations using Intel or AMD CPUs. The Center for Computational Sciences runs a variety of platforms to accommodate the needs of various researchers, including SGI, HP, Sun Microsystems, FreeBSD, and Linux. Its N-Class Supercomputer SDX is ranked among the top 200 Supercomputers in the world, and is among the top ten supercomputers running at academic institutions.

Financial Aid

Nearly all graduate students receive financial support to cover the costs of tuition, books, and room and board in the form of a teaching or research assistantship. Most entering students receive teaching appointments, which they generally hold during their first and second academic years. Students can also make arrangements with faculty members to work as research assistants; these appointments can be made both for the academic year and the summer months. Several fellowships are available to graduate students from either the University or the federal government on a competitive basis. Eligible applicants are automatically considered for these fellowships by the Department.

Cost of Study

In the 2006–07 academic year, full-time tuition and fees are $3518 per semester for residents and $7577 for nonresidents. Part-time tuition is $368.15 per credit for residents and $819.15 for nonresidents. Full-time students are also eligible for health insurance at an additional cost.

Living and Housing Costs

Four on-campus apartment complexes provide housing for graduate students and families. The monthly rent is $500 for an efficiency apartment, $618 for a one-bedroom apartment, or $672 for a two-bedroom unit; rent includes utilities and cable and network connectivity. Meal plans range from $240 per semester for twenty-four meals to $720 for eighty meals. Rent for an off-campus apartment typically costs $400 to $800 per month.

Student Group

Approximately 10 to 15 new graduate students enter the program each year. Roughly half of them are from Kentucky and the surrounding states of West Virginia, Ohio, Indiana, and Tennessee, while the other half come from elsewhere in the U.S. and from around the world. Although students have a wide range of backgrounds, most of them have taken undergraduate courses in mechanics, electricity and magnetism, thermal physics, modern physics, and/or quantum mechanics and have a strong interest in research.

Location

The University is located in Lexington, a metropolitan area of approximately 265,000. Lexington is located in the heart of the famous Bluegrass region of central Kentucky, about 85 miles from both Louisville, Kentucky, and Cincinnati, Ohio. Lexington has numerous theaters, concert halls, and restaurants and more than 400 horse farms. There are many opportunities for outdoor recreation, including a readily accessible network of state parks with opportunities for hiking and canoeing. In addition, the University has recreational facilities and offers a variety of artistic presentations. Lexington's downtown area is within walking distance of the University campus.

The University

The University of Kentucky was established in 1865 as the Agricultural and Mechanical College of the Kentucky University. Today, it is a public, research-extensive, land-grant university with an enrollment of more than 26,000 and an annual budget of $1.4 billion. The University is also Kentucky's ninth-largest organization. The University has more than eighty national rankings for academic excellence, and it ranks tenth in the nation among all universities for the number of start-up companies formed per $10 million in research spending.

Applying

Prospective students must submit the following for admission: a Departmental online preapplication, a completed graduate school application form, a $35 application fee, official transcripts from all colleges/universities attended, official GRE scores in the 50th percentile or higher, three letters of recommendation, and a personal statement of career aims and research interests. The deadline is February 1 for fall admission.

Correspondence and Information

Graduate Admissions Committee
177 Chemistry-Physics Building
University of Kentucky
600 Rose Street
Lexington, Kentucky 40506-0055
Phone: 859-257-6722
Fax: 859-323-2846
E-mail: dgs@pa.uky.edu
Web site: http://www.pa.uky.edu/

University of Kentucky

THE FACULTY AND THEIR RESEARCH

Suketu Bhavsar, Professor; Ph.D. (theoretical astrophysics), Princeton, 1978. Large-scale structure formation, visual perception of large-scale structure, quantifying filamentary structure, first-ranked galaxies in galaxy clusters, computational physics, N-body methods.

Joseph Brill, Professor; Ph.D. (experimental solid state physics), Stanford, 1978. Novel thermal, elastic, and infrared probes of crystals with "low-dimensional" electronic properties, including microcalorimetry and electrooptical and electromechanical measurements of charge-density-wave materials.

Gang Cao, Professor; Ph.D. (experimental condensed matter physics), Temple, 1992. Development of novel materials and synthesis of single crystals and epitaxial thin films of d- and f-electron-based oxides; highly correlated electron systems with an emphasis on magnetic, transport, and thermal properties.

Michael Cavagnero, Professor; Ph.D. (theoretical atomic physics), Chicago, 1987. Energy-transfer processes mediated by weakly bound states of matter, ranging from Rydberg states of atoms or molecules to weakly bound states of diatomic molecules involved in ultracold collisions.

John Christopher, Professor; Ph.D. (physics), Virginia. Investigation, development, and application of improved ways of helping students learn physics and develop scientific reasoning.

John W. D. Connolly, Professor and Director, Center for Computational Sciences; Ph.D., Florida, 1966.

Daniel Dale, Professor; Ph.D. (experimental nuclear physics), Illinois, 1991. Use of electromagnetic probes, photons, and electrons to study mesons and baryons at the quark level; measuring properties of the neutral pi meson as fundamental tests of quantum chromodynamics.

Sumit R. Das, Professor; Ph.D. (theoretical physics), Chicago, 1983. String theory, black-hole physics and gauge theory; manifestations of the holographic principle in string theory; physics of Hawking radiation; noncommutative gauge theories.

Lance DeLong, Professor; Ph.D. (experimental condensed matter physics), California, San Diego, 1977. Rare earth, actinide and transition metal alloys and compounds, organic conductors and high-temperature superconductors; heavy fermion and non-Fermi liquid effects, anomalous upper-critical magnetic fields and "peak effect" in magnetization of superconductors; superconducting thin films patterned with lattices of artificial pinning centers; development of superconducting levitation technologies for applications in bioreactors and stirrers.

Terrence Draper, Professor; Ph.D. (theoretical particle physics), UCLA, 1984. Consequences of the theory of quantum chromodynamics.

Michael Eides, Professor; Ph.D. (theoretical and mathematical physics), Leningrad State, 1977. Theoretical particle physics: quantum field theory, supersymmetry, Standard Model, and beyond; bound-state theory in QED and QCD: high-order corrections to Lamb shift and hyperfine splitting; low energy QCD, spontaneous chiral symmetry breaking, effective chiral Lagrangians.

Moshe Elitzur, Professor; Ph.D. (elementary particles), Weizman (Israel), 1971. Theoretical astrophysics: studies of astronomical masers, infrared dust emission with special emphasis on young stellar objects and winds around evolved stars.

Gary Ferland, Professor; Ph.D. (theoretical astrophysics), Texas at Austin, 1978. Quantitative spectroscopy, quasar emission lines, numerical simulations.

Susan Gardner, Professor; Ph.D. (theoretical nuclear physics), M.I.T., 1988. Discrete symmetries C, P, T—and their violation—in the weak interactions of hadrons; theoretical predictions of hadronic weak decays.

Tim Gorringe, Professor; Ph.D. (experimental nuclear physics), Birmingham (England), 1985. Studies of chiral symmetry and isospin symmetry, investigations of chiral symmetry breaking via the muon capture reaction and isospin symmetry breaking via pion charge exchange reaction.

Howard Grotch, Professor Emeritus; Ph.D. (Cornell), 1967. Quantum electrodynamic, QED corrections to hyperfine splitting, Lamb shift, and g factors.

David Harmin, Professor; Ph.D. (theoretical atomic physics), Chicago, 1981. General interactions of atoms and photons; weakly bound states of diatomic molecules and ultracold collisions; "wave-particle duality," especially intermediate cases; asymmetric two-way interferometers; entanglement and information exchange between two-level atoms and micromaser cavities.

Ivan Horváth, Professor; Ph.D. (theoretical particle physics), Rochester and Brookhaven National Laboratory, 1995. Study of nonperturbative phenomena in particle physics; study of QCD vacuum structure using new methods motivated by recent availability of lattice chiral fermions.

Wolfgang Korsch, Professor; Ph.D. (nuclear physics), Marburg (Germany), 1990. Understanding the structure of nucleons.

Michael Kovash, Professor; Ph.D. (experimental nuclear physics), Ohio State, 1978. Electromagnetic studies of the nucleon, two-nucleon system, and nuclei of special astrophysical interest.

Wasley Krogdahl, Professor Emeritus; Ph.D. (astronomy and astrophysics), Chicago, 1942. Kinematic relativity.

Richard Lamb, Adjunct Professor; Ph.D. (physics), Kentucky, 1963. Observational high-energy astrophysics, analysis of archival satellite data using instruments on the Gamma Ray Observatory.

Nancy A. Levenson, Professor; Ph.D. (astronomy), Berkeley, 1997. Understanding physical processes that shape galaxies and govern their evolution.

Bing An Li, Professor; Ph.D. (theoretical particle physics), Chinese Academy of Sciences, 1968 . Effective theory of hadrons, electroweak theory.

Keh-Fei Liu, Professor; Ph.D. (theoretical nuclear physics), SUNY at Stony Brook, 1975. Lattice QCD: glueballs, nucleon structure, finite density; phenomenology: deep inelastic scattering, hadron spectroscopy.

Keith MacAdam, Professor; Ph.D. (experimental atomic physics), Harvard, 1971. Interaction of highly excited, Rydberg-state atoms with charged particles and fields; alignment and orientation in electron-atom and ion-atom collisions; relationship of quantal, semiclassical, and classical theoretical formulations of the Three-Body Problem applied to atomic collisions.

Alan MacKellar, Professor Emeritus; Ph.D. (physics), Texas A&M, 1966. Application of classical trajectory methods in study of atomic processes in ion-atom collisions; dependence of specific initial (n,l,m) states on the final (n,l,m) distributions for ion-Rydberg atom charge-transfer reactions.

Nicholas Martin, Professor; D.Phil. (experimental atomic physics), Oxford (England), 1978. Electron impact excitation and ionization of atoms, (e,2e) spectroscopy of autoionizing levels.

Marcus McEllistrem, Professor Emeritus; Ph.D. (physics), Wisconsin. Basic nuclear structure studies, specialized experiments in nuclear astrophysics, applied nuclear physics.

Madhu Menon, Professor; Ph.D., Notre Dame. Theoretical predictions of structural and vibrational properties of new materials using quantum mechanical molecular dynamics methods; carbon fullerenes and nanotubes, silicon clathrates, carbon-based nanoscale devices, complex hetero-atomic systems, magnetism in transition metals.

Ganpathy Murthy, Professor; Ph.D. (theoretical condensed matter physics), Yale, 1987. Strongly correlated electron systems, role of disorder in the bilayer quantum Hall systems, states with strong correlations and strong quantum fluctuations in mesoscopic systems, effect of disorder on gauge theories.

Kwok-Wai Ng, Professor; Ph.D. (experimental solid-state physics), Iowa State. Quasiparticle density of states of different superconductors by tunneling methods, angular dependence of the pairing strength, effect of impurity doping on pairing of these electrons.

Alfred Shapere, Professor; Ph.D. (physics), California, Santa Barbara, 1988. String theory, quantum gravity, supersymmetry.

Isaac Shlosman, Professor; Ph.D. (physics), Tel-Aviv, 1986. Galactic dynamics, active galactic nuclei, accretion disks.

Joseph Straley, Professor; Ph.D. (theoretical condensed matter physics), Cornell, 1970. Critical phenomena, electrical behavior of extremely inhomogeneous systems, percolation problem, phase diagram and dynamical behavior of a Josephson array in a magnetic field, one-dimensional many-body quantum systems, phase transitions of the Kosterlitz-Thouless class.

Yuri Sushko, Professor; Ph.D. (physics), Kiev. Strongly correlated electron systems and high-pressure physics, interplay between magnetism and superconductivity in contemporary classes of superconducting materials, superconductor-insulator and metal-insulator transitions in low-dimensional materials.

Thomas H. Troland, Professor; Ph.D. (astrophysics), Berkeley, 1980. Use of radio astronomy techniques to measure interstellar magnetic field strengths.

Jesse Weil, Professor Emeritus; Ph.D. (experimental nuclear physics), Columbia, 1959. Investigation of the level structure of the even-A tin isotopes 116,120,124Sn using the n,n'gamma reaction; evidence of multiphonon vibrational excitations embedded in a background of shell model states.

UNIVERSITY OF MARYLAND, COLLEGE PARK

Department of Physics

Programs of Study

With an exceptional breadth of research programs and numerous outstanding faculty members, the University of Maryland's Department of Physics explores both theoretical and experimental physics.

The department offers a Doctor of Philosophy (Ph.D.) program and a Master of Science (M.S.) program both with and without a thesis. Typically, doctoral candidates pass a written qualifying examination and complete core course work in their first year and pass the Graduate Laboratory in their second year. In their third and fourth years, candidates complete advanced course work (including courses outside their fields of specialization), seminars, and dissertation research before presenting and defending their dissertations.

Students pursuing a master's degree without a thesis complete 30 credits of course work, present a scholarly paper for review by faculty members, and must pass a comprehensive examination. Students pursuing a master's degree with a thesis complete 30 credits of course work and research before taking an oral examination defending their thesis.

The department comprises more than thirty research groups and centers, including Atomic, Molecular, and Optical Physics; Biophysics; Chaos and Non-Linear Dynamics (currently ranked number one in the country); Charged Particles Beam Research; Chemical Physics; Cosmic Ray Physics; Dynamical Systems and Accelerator Theory; Condensed Matter (Theory and Experiment); the East West Space Science Center; Gravitation (Theory and Experiment); Elementary Particle and Quantum Field Theory; High Energy Physics; the Institute for Advanced Computer Studies; the Institute for Physical Science and Technology; the Institute for Research in Electronics and Applied Physics; Materials Research Science and Engineering Center; Mathematical Physics; Center for Multiscale Plasma Dynamics; Nuclear Physics; Particle Astrophysics; the Center for Particle and String Theory; Physics Education; Plasma Physics (Theory and Experiment); Space Physics; the Center for Superconductivity Research; Quantum Electronics and Relativity; and Quarks, Hadrons, and Nuclear Theory.

Research Facilities

The University of Maryland Department of Physics has world-renowned, state-of-the-art facilities, including instrumentation for fabricating nanostructures with characterization down to the atomic scale, ultrahigh-sensitivity electrical and magnetic properties measurement, ion beam research with high-brightness ion beam sources, advanced accelerator applications using charged particle beams, high-power microwave generation, and plasma spectroscopy and diagnostics.

The department possesses a variety of microscopes—a low-energy electron, holographic laser tweezer array, optical, tunneling, microwave, magnetic force, electrostatic force, atomic force, photoemission electron, and SQUID (superconducting quantum interference device) systems—for extensive research in a variety of fields. Through key collaborations, Maryland students, faculty members, and researchers also have access to the facilities of government agencies, private laboratories, and peer institutions. The department has also developed the largest collection of physics lecture demonstrations in the United States and one of the best in the world with two revolving stage auditoriums and 1,500 demonstrations. It also boasts world-class shop facilities including Mechanical Development, Electronic Development, and Engineering and Design.

Financial Aid

Appointments as teaching assistants (compensation taxable) are normally available to all entering students. In addition to tuition remission, students receive a stipend, which was $14,802 for the 2005–06 academic year (20 hours per week) with a $4198 (also taxable) summer stipend for a total calendar year stipend of $19,000. Fellowships, research assistantships, and student loans are available.

Cost of Study

Students in good standing are fully supported for their entire graduate careers, and the majority of entering students are also awarded full tuition remission.

Living and Housing Costs

Housing is available within walking distance to campus. One-bedroom apartments in the area currently rent for $700 to $1100 per month, and efficiency/studios are generally $600 to $900 per month. Trained off-campus housing peer advisers are available for consultation or any questions that students may have (telephone: 301-314-3645).

Student Group

The department's 221 physics graduate students represent a diversity in gender and national, ethnic, and academic backgrounds.

Location

Nine miles from downtown Washington, D.C., and 30 miles from Baltimore and Annapolis, the department is surrounded by the nation's foremost physics associations, a growing corridor of private industry, and some of the world's best government laboratories, including NASA Goddard Space Flight Center, the National Institutes of Health, the Naval Research Laboratory, and the National Institute of Standards and Technology. These neighbors provide students with unparalleled opportunities for work, study, and collaboration.

The University

Established in 1856, the University of Maryland is a large, diverse research university of national stature, highly regarded for its broad base of excellence in both teaching and research. According to *U.S. News & World Report,* the department's doctoral program is ranked fourth among physics departments at public universities and thirteenth overall nationwide.

Applying

Applications are available online at http://www.gradschool.umd.edu. A nonrefundable fee of $60 for U.S. residents and international students must accompany each application. The application deadline for all applicants is February 1. Spring applications are not accepted. Admission consideration is open to all qualified candidates without regard to race, color, national origin, religion, sex, or handicap.

Correspondence and Information

Linda O'Hara
Department of Physics Graduate Secretary
1120 Physics Building
University of Maryland
College Park, Maryland 20742
Phone: 301-405-5982
Fax: 301-405-4061
E-mail: lohara@physics.umd.edu
Web site: http://www.umd.edu

University of Maryland, College Park

THE FACULTY AND THEIR RESEARCH

Professors
Carroll O. Alley Jr., Ph.D., Princeton, 1962: atomic physics, quantum electronics–quantum mechanics, relativistic gravity. J. Robert Anderson, Ph.D., Iowa State, 1963: experimental condensed-matter physics, diluted semiconductors, electronic structures of metals and semimetals. Steven Anlage, Ph.D., Caltech, 1988: superconductivity–electromagnetic properties, proximity effect, near-field microwave microscopy, experimental chaos. Thomas Antonsen, Ph.D., Cornell, 1977: plasma physics. Andrew R. Baden, Ph.D., Berkeley, 1986: experimental high-energy physics, data acquisition, high-performance computing, data analysis. Elizabeth J. Beise, Ph.D., MIT, 1988: experimental nuclear physics–intermediate energy, electron scattering, polarization, few-nucleon and subnucleon systems. Satindar M. Bhagat, Ph.D., Delhi (India), 1956: experimental condensed-matter physics, magnetic resonance, exotic magnetic phases, high-temperature superconductors. Derek A. Boyd, Ph.D., Stevens, 1973: plasma diagnostics, far-infrared spectroscopy, microwave optics. Dieter R. Brill, Ph.D., Princeton, 1959: general relativity and gravitation. Nicholas S. Chant, D.Phil., Oxford (England), 1966: experimental nuclear physics, electron physics. Hsing-Hen Chen, Ph.D., Columbia, 1973: plasma physics, nonlinear dynamical systems. Andrey Chubukov, Ph.D., Moscow State (USSR), 1985: condensed-matter theory. Thomas D. Cohen, Ph.D., Pennsylvania, 1984: nuclear theory, soliton models, chiral symmetry, low-energy models for QCD. Sankar Das Sarma, Distinguished University Professor; Ph.D., Brown, 1979: theoretical condensed matter, many-body theory, semiconductor nanostructures, nonequilibrium statistical mechanics. James F. Drake, Ph.D., UCLA, 1975: plasma physics, magnetic reconnection, tokamak transport. H. Dennis Drew, Ph.D., Cornell, 1967: experimental condensed matter, statistical and thermal physics, semiconductor heterostructures, infrared properties of superconductors. Theodore L. Einstein, Ph.D., Pennsylvania, 1973: condensed-matter theory, surface physics, statistical and thermal physics. Sarah C. Eno, Ph.D., Rochester, 1990: experimental high-energy physics. Michael E. Fisher, Distinguished University Professor and University System of Maryland Regents Professor; Ph.D., King's College (England), 1957: statistical physics, condensed-matter theory, theoretical chemistry, phase transitions and critical phenomena. S. James Gates, John S. Toll Professor of Physics; Ph.D., MIT, 1977: elementary particles–supersymmetry, supergravity, and superstrings. George Gloeckler, Distinguished University Professor; Ph.D., Chicago, 1965: space physics, heliospheric physics. Jordan A. Goodman, Department Chair; Ph.D., Maryland, 1978: particle astrophysics. Oscar W. Greenberg, Ph.D., Princeton, 1956: elementary particles and field theory. Richard L. Greene, Ph.D., Stanford, 1967: experimental condensed matter. James J. Griffin, Ph.D., Princeton, 1956: nuclear theory, nuclear heavy ion physics. Nicholas J. Hadley, Ph.D., Berkeley, 1983: high-energy physics. Douglas C. Hamilton, Ph.D., Chicago, 1977: experimental space physics, magnetospheric physics, solar wind, particle acceleration and transport. David Hammer (joint with Department of Curriculum and Instruction, Science Teaching Center), Ph.D., Berkeley, 1991: physics education—learning and teaching. Adil B. Hassam, Ph.D., Princeton, 1978: plasma physics of the sun, thermonuclear fusion. Wendell T. Hill III, (affiliate), Ph.D., Stanford, 1980: plasma physics of the sun, thermonuclear fusion. Bei-Lok Hu, Ph.D., Princeton, 1972: general relativity, gravitation and cosmology, quantum field theory. Theodore A. Jacobson, Ph.D., Texas at Austin, 1983: gravitation theory, quantum gravity, black hole thermodynamics. Abolhassan Jawahery, Ph.D., Tufts, 1981: high-energy physics with accelerators. Xiangdong Ji, Ph.D., Drexel, 1987: theoretical nuclear physics, quantum chromodynamics, quark and gluon structure of hadrons. James J. Kelly, Ph.D., MIT, 1981: experimental nuclear physics. Young Suh Kim, Ph.D., Princeton, 1961: elementary particles, group theory. Theodore Kirkpatrick, Ph.D., Rockefeller, 1981: theoretical statistical mechanics, condensed matter. Donald N. Langenberg, Chancellor Emeritus, University of Maryland System; Ph.D., Berkeley, 1959: condensed-matter physics. Chuan Sheng Liu, Ph.D., Berkeley, 1968: plasma physics, fusion and space science. Christopher J. Lobb, Ph.D., Harvard, 1980: experimental superconductivity, superconducting devices, mesoscopic systems. Markus A. Luty, Ph.D., Chicago, 1991: particle theory: nonperturbative QCD, dynamical symmetry breaking, nonperturbative supersymmetry, particle cosmology. Howard Milchberg, Ph.D., Princeton, 1985: atomic, molecular, and optical physics. Rabindra N. Mohapatra, Ph.D., Rochester, 1969: elementary particles, quantum field theory, cosmology. Gottlieb Oehrlein (affiliate), Ph.D., SUNY at Albany, 1981. Novel materials, low-temperature plasma science. Luis Orozco, Ph.D., Texas at Austin, 1987: experimental AMO physics, quantum optics, precision measurements and fundamental interactions. Edward Ott, Distinguished University Professor; Ph.D., Polytechnic of Brooklyn, 1967: chaotic dynamics, plasmas. Ho Jung Paik, Ph.D., Stanford, 1974: experimental general relativity, gravitational waves, precision tests of laws of gravity. Dennis Papadopoulos, Ph.D., Maryland, 1968: space plasma physics, lightning, photoconducting plasmas. Robert L. Park, Ph.D., Brown, 1964: experimental condensed-matter physics, surface physics, science policy. William D. Phillips, Distinguished University Professor and Nobel Laureate; Ph.D., MIT, 1976: AMO physics, laser trapping and cooling. Edward F. Redish, Ph.D., MIT, 1968: physics education research. Douglas A. Roberts, Ph.D., UCLA, 1994: experimental high-energy physics with accelerators. Steven Rolston, Ph.D., SUNY at Stony Brook. AMO physics, laser spectroscopy of isotopes. Rajarshi Roy, Ph.D., Rochester, 1981: nonlinear dynamics, laser physics, optical fibers, coherence and stochastic processes. Roald Z. Sagdeev, Ph.D., Institute of Physical Problems (Russia), 1960; D.S., USSR Academy of Sciences, 1962: plasma physics, controlled fusion, space physics, arms control, science policy, global security and environment. Andris Skuja, Ph.D., Berkeley, 1972: experimental particle physics. Katepalli Sreenivasan, Distinguished University Professor; Ph.D., Indian Institute of Science (India), 1975: fluid turbulence, complex fluids, combustion, cryogenic helium and nonlinear dynamics. Stephen J. Wallace, Ph.D., Washington (Seattle), 1971: scattering theory, nucleon-nucleon interactions, relativistic bound states, electron scattering. Frederick C. Wellstood, Ph.D., Berkeley, 1988: superconductivity–high-T_c (YBCO), superconducting quantum interference devices, magnetic microscopy, Coulomb blockade electrometers. Ellen D. Williams, Distinguished University Professor; Ph.D., Caltech, 1982: condensed-matter physics, surface science, scanning tunneling microscopy, statistical mechanics of surfaces. Victor M. Yakovenko, Ph.D., Landau Institute for Theoretical Physics (Russia), 1987: condensed-matter theory, organic and high-T_c superconductors, quantum Hall effect, high magnetic fields. James A. Yorke, Distinguished University Professor; Ph.D., Maryland, 1966. Chaos and nonlinear dynamics.

Associate Professors
Alessandra Buonanno, Ph.D., Pisa (Italy), 1996: gravitational waves. William Dorland, Ph.D., Princeton, 1993: particle astrophysics. Richard F. Ellis, Ph.D., Princeton, 1970: experimental plasma physics, plasma waves and instabilities, microwave and far-infrared diagnostics. Michael Fuhrer, Ph.D., Berkeley, 1998: electronic properties of carbon nanotubes and other nanostructures. Daniel P. Lathrop, Ph.D., Texas at Austin, 1991: nonlinear dynamics and chaos, turbulence, fluid dynamics. Raymond J. Phaneuf (affiliate), Ph.D., Wisconsin–Madison, 1985: materials research. Douglas Roberts, Ph.D., UCLA, 1994. Eun-Suk Seo, Ph.D. LSU, 1991: cosmic ray physics. Gregory W. Sullivan, Ph.D., Illinois, 1990: electroweak physics, standard model, top quark search. Ichiro Takeuchi (affiliate), Ph.D., Maryland, 1996: combinatorial materials synthesis and characterization.

Assistant Professors
Kara Hoffman, Ph.D., Purdue, 1998: particle astrophysics. Arthur LaPorta, Ph.D., California, San Diego, 1996: molecular biophysics. Wolfgang Losert, Ph.D., CUNY, City College, 1998: nonlinear dynamics, pattern formation. Min Ouyang, Ph.D., Harvard, 2001: probing spin physics and chemistry in nanometer scale. Arpita Upadhyaya, Ph.D., Notre Dame, 2000: cellular biophysics.

Research Faculty
Manoj Banerjee, Ph.D., Calcutta, 1956: nuclear chemistry, Baryon structure, Baryon-Meson interactions. Fatiha Benmokhtar, Ph.D., Rutgers, 2004: experimental nuclear physics. Eric Blaufuss, Ph.D., LSU, 2000: particle astrophysics. Gavin Brennen, Ph.D., New Mexico, 2001: general relativity. Herbert Breuer, Ph.D., Heidelberg (Germany), 1976: experimental nuclear physics. Vinh Lam Olivier Buu, Ph.D., CRTBT (France), 1998: condensed matter. Maria Jose Calderon Prieto, Ph.D., Madrid (Spain), 2001: condensed matter. Chia-Cheh Chang, Ph.D., USC, 1968: experimental nuclear physics. Chenhui Chen, Ph.D., Pennsylvania, 2003: high energy. William Cullen, Ph.D., Georgia Tech, 1999: materials science. Douglas Currie, Ph.D., Rochester, 1962: astro metrology. Yoram Dagan, Ph.D., Tel Aviv (Israel), 2001: superconductivity. Mihir Desai, Birmingham, Ph.D., 1995: space physics. Tyce DeYoung, Ph.D., Wisconsin–Madison, 2001: particle astrophysics. Sankar Dhar, Indian Institute of Technology (Kanpur), 1997: superconductivity. J. Robert Dorfman, Ph.D., Johns Hopkins, 1961: statistical and thermal physics, dynamical systems. Alex Dragt, Ph.D., Berkeley, 1963: dynamical systems. Andrew Elby, Ph.D., Berkeley, 1995: physics education. Richard Ferrell, Ph.D., Princeton, 1952: condensed matter. Chad Fertig, Ph.D., Yale, 2002: atomic, molecular, and optical physics. Daniel Freimund, Ph.D., Nebraska–Lincoln, 2003: atomic, molecular, and optical physics. Arnold Glick, Ph.D., Maryland, 1961: condensed matter, statistical and thermal physics. Robert Gluckstern, Ph.D., MIT, 1948: dynamical systems. Tullio Grassi, Padova (Italy), 2000: high-energy physics. Hong Hao, Ph.D., Maryland, 1991: condensed matter. Matthew Hill, Ph.D., Maryland, 2001: space physics. Wouter Hulsbergen, Ph.D., Amsterdam (Netherlands), 2002: high energy. Euyheon Hwang, Ph.D., Maryland, 1996: condensed matter. Fred M. Ipavich, Ph.D., Maryland, 1972: space physics. Masahiro Ishigami, Ph.D., Berkeley, 2004: condensed matter. Hung-Chih Kan, Ph.D., Maryland, 1997: condensed matter. Shrikanta Kanekal, Ph.D., Kansas, 1988: space physics. Richard G. Kellogg, Ph.D., Yale, 1975: high-energy. Eam Khor, Ph.D, Monash (Australia), 1971: condensed matter. Axel Krause, Ph.D., Hamburg (Germany), 2000: elementary particles. Pavel Krotkov, Ph.D., Russian Academy of Sciences L.D. Landau (Russia), 2001: condensed matter. Curtis Lansdell, Ph.D., Texas at Austin, 2002: cosmic-ray physics. Rupert Lewis, Ph.D., Indiana at Bloomington, 2001: superconductivity. Bing Liang, Chinese Academy of Sciences, 1999: superconductivity. Daniela Manoel, Ph.D., UNICAMP (Brazil), 2003: atomic, molecular, and optical physics. Jeremiah Mans, Ph.D., Princeton, 2002: high energy. Glenn Mason, Ph.D., Chicago, 1971: space physics. John Matthews, Ph.D., Maryland, 2002: superconductivity. Colin McCormick, Ph.D., Berkeley, 2003: atomic, molecular, and optical physics. Charles Misner, Ph.D., Chicago, 1971: general relativity. Sergey Mezhenny, Ph.D., Pittsburgh, 2003: condensed matter. Martin Vol Moody, Ph.D., Virginia, 1980: gravitation experiment. Salah Nasri, Ph.D. Syracuse, 2003: elementary particles. Alexander Olivas, Ph.D., Colorado at Boulder, 2004: cosmic-ray physics. Jorge Ovalle, Ph.D., Simon Bolivar-Caracas (Venezuela), 2000: elementary particles. Jogesh Pati, Ph.D., Maryland, 1960: elementary particles. John Paquette, Ph.D., Maryland, 1992: space physics. Kwon Park, Ph.D., SUNY at Stony Brook, 2000: condensed-matter theory. Dmitri Petrovykh, Wisconsin–Madison, 2000: condensed matter. Donald Priour, Ph.D., Princeton, 2000: condensed-matter. Madhav Ranganathan, Ph.D., Stanford, 2003: materials science. Fedor Ratnikov, Ph.D., Institute of Theoretical and Experimental Physics (Russia), 1994: high energy. Philip Roos, Ph.D. MIT, 1964: experimental nuclear physics. Albert Roura, Ph.D., Barcelona, 2002: gravitation theory. Changhyun Ryu, Ph.D., Texas at Austin, 2004: atomic, molecular, and optical physics. Michiel Sanders, Nijmegen (Netherlands), 2001: high energy. Vito Scarola, Ph.D., Penn State, 2002: condensed-matter. Rachel Scherr, Ph.D., Washington (Seattle), 2000: physics education. Don Schmadel, Ph.D., Maryland, 2002: condensed matter. Lingyun Shi, Ph.D., Arizona State, 2003: condensed matter. Kazutomu Shiokawa, Ph.D., Maryland, 1997: gravitation theory. Gabriele Simi, Ph.D., Pisa (Italy), 2001: high energy. Jeffrey Simpson, Ph.D., Maryland, 2004: condensed matter. Andrew Smith, Ph.D., California, Irvine, 1996: particle astrophysics. Andrei Souchkov, Ph.D., Kapitsa Institute for Physical Problems (Russia), 1994: condensed matter. Lowell Swank, Ph.D., Illinois at Urbana–Champaign, 1967: elementary particles. Ferenc Szalma, Ph.D. Szeged (Hungary), 1999: materials science. Lin Tian, Ph.D., MIT, 2002: atomic, molecular, and optical physics. Terrence Toole, Ph.D., American, 2000: high energy. Dusan Turcan: Ph.D., Maryland, 2003: particle astrophysics. Alipasha Vaziri, Ph.D., Vienna (Austria), 2003: atomic, molecular, and optical physics. T. Venky Venkatesan, Ph.D., CUNY, Brooklyn, 1977: superconductivity. Marco Verzocchi, Ph.D., Albert-Ludwigs (Germany), 1998: high energy. Ratnakar Vispute, Ph.D., Poona (India), 1993: superconductivity. Florin Zavaliche, Ph.D., Martin Luther (Germany), 2002: materials science. Ying Zhang, Ph.D. Yale, 2002: condensed matter. Youxiang Zhang, Ph.D., Santa Barbara, 2003: condensed matter. Hui Zhou, Ph.D., Maryland, 2004: condensed matter. Igor Zutic, Ph.D., Minnesota, Twin Cities, 1998: condensed-matter.

University of Maryland, College Park

SELECTED PUBLICATIONS

Alley, C. O., D. Leiter, Y. Mizobuchi, and H. Yilmaz. Energy crisis in astrophysics: Black holes vs. N-body metrics. *Los Alamos E-Print Archive* astro-ph/9906458, 1999.

M. Gorska, **J. R. Anderson** et al. Magnetization of $Sn_{1-x}Gd_xTe$. *Phys. Rev. B* 64:115210, 2001.

Booth, J. C., et al. **(S. M. Anlage).** Large dynamical fluctuations in the microwave conductivity of YBa_2Cu_3O_. *Phys. Rev. Lett.* 77:4438, 1996.

Baden, A. R. Jets and kinematics in hadronic collisions. *Int. J. Mod. Phys. A* 13:1817–45, 1998.

Banerjee, M. K., and J. Milana. Baryon mass splitting in chiral perturbation theory. *Phys. Rev. D: Part. Fields* 54:6451, 1995.

Spayde, D. T., et al. **(E. J. Beise** and **H. Breuer).** The Strange Quark Contribution to the Proton's Magnetic Moment. *Phys. Rev. Lett. B* 79:583, 2004.

Abbott, D., et al. **(E. J. Beise, N. S. Chant,** and **P. G. Roos).** Measurement of Tensor Polarization in Elastic Electron-Deuteron Scattering at Large Momentum Transfer. *Phys. Rev. Lett.* 84:5053, 2000.

Meekins, D. G., **E. J. Beise,** and **N. S. Chant.** Coherent pi^0 photoproduction on the deuteron up to 4 GeV. *Phys. Rev. C* 60:052201, 1999.

Lofland, S. E., et al. **(S. M. Bhagat** and **R. Ramesh).** Magnetic imaging of perovskite thin films by ferromagnetic resonance microscopy—$La_{0.7}Sr_{0.3}MnO_3$. *Appl. Phys. Lett.* V75(12):1947–8, 1999.

Celata, C. M., and **D. A. Boyd**. Cyclotron radiation as a diagnostic tool for tokamak plasmas. *Nucl. Fusion* 17:735, 1977.

Naples, D., et al. **(H. Breuer** and **C. C. Chang).** A dependence of photoproduced dijets. *Phys. Rev. Lett.* 72:2341, 1994.

Brill, D., G. Horowitz, D. Kastor, and J. Traschen. Testing cosmic censorship with black hole collisions. *Phys. Rev. D* 49:840, 1994.

Chen, H. H., and J. E. Lin. Integrability of higher dimensional nonlinear Hamiltonian systems. *J. Math. Phys.,* 1998.

Cohen, T. D. Chiral soliton models, large Nc QCD and the Theta^+ exotic baryon. *Phys. Lett. B* 581:175, 2004.

Das Sarma, S., and D. W. Wang. Many-body renormalization of semiconductor quantum wire excitons: Absorption, gain, binding, and unbinding. *Phys. Rev. Lett.* 84:2010, 2000.

Dorfman, J. R. *An Introduction to Chaos in Nonequilibrium Statistical Mechanics.* Cambridge: Cambridge University Press, 1999.

Jenko, F. and **W. Dorland.** Prediction of significant tokamak turbulence at electron gyroradius scales. *Phys. Rev. Lett.* 89:225001, 2002.

Drake, J. F., et al. Formation of electron holes and particle energization during magnetic reconnection. *Science* 299:873, 2003.

M. Grayson, et al. **(H. D. Drew).** Spectral measurement of the Hall angle response in normal state cuprate superconductors. *Phys. Rev. Lett.* 89:037003, 2002.

Einstein, T. L., H. L. Richards, S. D. Cohen , and O. Pierre-Louis. Terrace-width distributions and step-step repulsions on vicinal surfaces: symmetries, scaling, simplifications, subtleties, and Schrödinger. *Surface Sci.* 493:460–74, 2001.

Austin, M. E., **R. F. Ellis,** R. A. James, and T. C. Luce. Electron temperature measurements from optically gray third harmonic electron cyclotron emission in the DIII-D tokamak. *Phys. Plasmas* 3:10, 3725, 1996.

Carena, M., et al. **(S. Eno).** Searches for supersymmetric particles at the tevatron collider. *Rev. Mod. Phys.* 71:937–81, 1999.

Aqua, J.-N., and **M. E. Fisher.** Ionic criticality: An exactly soluble model. *Phys. Rev. Lett.* 92 135702:1–4, 2004.

Dürkop, T., S. A. Getty, Enrique Cobas, and **M. S. Fuhrer.** Extraordinary mobility in semiconducting carbon nanotubes. *Nano Lett.* 4:35, 2004.

Gates Jr., S. J., and O. Lebedev. Searching for supersymmetry in hadrons. *Phys. Lett. B* 477:216, 2000.

Glick, A., L. E. Henrickson, G. W. Bryant, and D. F. Barbe. Nonequilibrium green's function theory of transport in interacting quantum dots. *Phys. Rev. B* 50:4482, 1994.

Geiss, J., et al. **(G. Gloeckler** and **F. M. Ipavich).** The southern high speed stream: Results from SWICS/Ulysses. *Science,* May 1995.

Gluckstern, R. L., and B. Zotter. Analysis of shielding charged particle beams by thin conductors. *Phys. Rev. ST Accel. Beams* 4:024402, 2001.

Atkins R, et al. **(J. A. Goodman).** TeV gamma-ray survey of the northern hemisphere sky using the Milagro observatory. *Astrophys. J.* 608(2):680–5 Part 1, 2004.

Greenberg, O. W. CPT violation implies violation of Lorentz invariance. *Phys. Rev. Lett.* 89:231602, 2002.

Parkin, S. S. P., et al. **(R. L. Greene).** Superconductivity in a new family of organic conductors. *Phys. Rev. Lett.* 50:270, 1983.

Griffin, J. J. The statistical model of intermediate structure. *Phys. Rev. Lett.* 17:478, 1966.

Abachi, S., and **N. J. Hadley.** Observation of the top quark. *Phys. Rev. Lett.* 74:2632, 1995.

Hamilton, D. C., G. Gloeckler, S. M. Krimigis and L. J. Lanzerotti. Composition of nonthermal ions in the Jovian magnetosphere. *J. Geophys. Res.* 86:8301, 1981.

Hammer, D. Student resources for learning introductory physics. *Am. J. Phys. Phys. Educ. Res. Suppl.* 68(S1):S52–9, 2000.

Hassam, A. B. Reconnection of stressed magnetic fields. *Astrophys. J.* 399:159, 1992.

Hu, B. L., and E. Verdaguer. Stochastic gravity: Theory and applications. *Living Rev. Relativity* 7:3, 2004.

Jacobson, T. Trans-Planckian redshifts and the substance of the space-time river. *Prog. Theor. Phys. Suppl.* 136:1, 1999.

Aubert, B. et al (**A. Jawahery** with the BaBar collaboration) Observation of CP violation in the B^0 meson system. *Phys. Rev. Lett.* 87:091801, 2001.

Ji, X. Gauge-invariant decomposition of nucleon spin and its spin-off. *Phys. Rev. Lett.* 78:610, 1997.

Kelly, J. J. Nucleon knockout by intermediate energy electrons. *Adv. Nucl. Phys.* 23:75, 1996.

Kim, Y. S. Observable gauge transformation in the parton picture. *Phys. Rev. Lett.* 63:348–51, 1989.

Belitz, D., and **T. R. Kirkpatrick**. The Anderson-Mott Transition. *Rev. Mod. Phys.* 66:261–380, 1994.

Zeff, B. W., J. Fineberg, and **D. P. Lathrop**. The dynamics of finite-time singularities: Curvature collapse and jet eruption on a fluid surface. *Nature* 403:401, 2000.

Novakovskii, S. V., **C. S. Liu, R. Z. Sagdeev,** and M. N. Rosenbluth. The radial electric field dynamics in the neoclassical plasmas. *Phys. Plasmas* 4:12, 1997.

Berkley, A. J., et al. **(J. R. Anderson, A. J. Dragt, C. J. Lobb,** and **F. Wellstood).** Entangled macroscopic quantum states in two superconducting qubits. *Science* 300:1548–50, 2003.

Toiya, M., J. Stambaugh, and **W. Losert.** Transient and oscillatory granular shear flow. *Phys. Rev. Lett.* 83: 088001-1, 2004.

Giudice, G. F., **M. A. Luty,** H. Murayama, and R. Rattazzi. Gaugino mass without singlets. *J. High Energy Phys.* 9812:027, 1998.

Mason, G. M., J. E. Mazur, and J. R. Dwyer. 3He enhancements in large solar energetic particle events. *Astrophys. J. Lett.* 525:L133–6, 1999.

Misner, C. W., K. S. Thorne, and J. A. Wheeler. *Gravitation.* San Francisco: W. H. Freeman and Co., 1973.

Mohapatra, R. N., and G. Senjanovic. Spontaneous parity violation and neutrino masses. *Phys. Rev. Lett.* 44:912, 1980.

Hendrey, M., **E. Ott,** and **T. M. Antonsen Jr.** Effect of inhomogeneity on spiral wave dynamics. *Phys. Rev. Lett.* 82:859, 1999.

Ott, E., C. Grebogi, and **J. A. Yorke.** Controlling chaos. *Phys. Rev. Lett.* 64:1196, 1990.

Ouyang, M., et al. Coherent spin transfer between molecularly bridged quantum dots. *Science* 301:1074–8, 2003.

University of Maryland, College Park

Paik, H. J. Superconducting accelerometers, gravitational-wave transducers, and gravity gradiometers. In *SQUID Sensors: Fundamentals, Fabrication and Applications,* ed. H. Weinstock, pp. 569–98. Dordrecht: Kluwer, 1996.

Papadopoulos, K., et al. The physics of substorms as revealed by the ISTP. *Phys. Chem. Earth* 24:1–3, 189–202, 1999.

Park, R. *Voodoo Science: The Road from Foolishness to Fraud.* New York: Oxford University Press, 2000.

Pati, J. C., and A. Salam. Lepton number as the fourth color. *Phys. Rev. D* 10:275, 1974.

Kan, H.-C., et al. **(R. J. Phaneuf).** Transient evolution of surface roughness on patterned GaAs(001) during homoepitaxial growth. *Phys. Rev. Lett.* 92, 146101, 2004.

Fertig, C. D., et al. **(W. D. Phillips).** Strongly inhibited transport of a degenerate 1D Bose gas in a lattice. *Phys. Rev. Lett.* 94:120403, 2005.

Redish, E. F., J. M. Saul, and R. N. Steinberg. Student expectations in introductory physics. *Am. J. Phys.* 66:212, 1998.

CLEO Collaboration et al. **(D. Roberts).** First observation of the decay $\tau^- \to K^- \eta \nu_\tau$. *Phys. Rev. Lett.* 82:281, 1999. hep-ex/9809012.

Rolston, S. L., and **W. D. Phillips.** Nonlinear and quantum atom optics. *Nature* 416:219, 2002.

Roos, P. G., and **N. S. Chant.** Photopion production from polarized nuclear targets. *Phys. Rev. C* 52:2591, 1995.

VanWiggeren, G. D., and **R. Roy**. Communication with chaotic lasers. *Science* 279:1198, 1998.

Mukhin, L., L. Marochnic, and **R. Sagdeev.** Estimates of mass and angular momentum in the Oort cloud. *Science* 242:547–50, 1988.

Seo, E. S., and V. S. Ptuskin. Stochastic reacceleration of cosmic rays in the interstellar medium. *Astrophys. J.* 431:705–14, 1994.

Akrawy, M. Z., **A. Skuja,** and the OPAL Collaboration. Measurement of the Z-zero mass and width with the OPAL detector at LEP. *Phys. Lett. B* 231:530, 1989.

Super-Kamiokande Collaboration and **G. Sullivan**. Measurement of the solar neutrino energy spectrum using neutrino-electron scattering. *Phys. Rev. Lett.* 82:2430, 1999.

Strachan, D. R., et al. **(T. Venkatesan** and **C. J. Lobb).** Do superconductors have zero resistance in a magnetic field? *Phys. Rev. Lett.* 87:067007, 2001.

Wallace, S. J., and N. Devine. Instant two-body equation in Breit frame. *Phys. Rev. C: Nucl. Phys.* 51:3222, 1995.

Xu, H., et al. **(A. J. Dragt, C. J. Lobb,** and **F. C. Wellstood).** Spectroscopy of three-particle entanglement in a macroscopic superconducting circuit. *Phys. Rev. Lett.* 94:027003, 2005.

Thuermer, K., et al. **(E. D. Williams).** Step dynamics in 3D crystal shape relaxation. *Phys. Rev. Lett.* 87:186102-4, 2001.

Kwon, H.-J., and **V. M. Yakovenko.** Spontaneous formation of a pi soliton in a superconducting wire with an odd number of electrons. *Phys. Rev. Lett.* 89:017002, 2002.

UNIVERSITY OF Miami
DEPARTMENT OF PHYSICS

UNIVERSITY OF MIAMI

Department of Physics

Programs of Study

The Department of Physics offers programs leading to the M.S. and Ph.D. degrees, and both thesis and nonthesis M.S. tracks are available. Usually a Ph.D. student devotes most of the first year to basic courses and takes the qualifying exam in the first January following arrival. Students should become involved with a research project by the second year and, after passing the qualifying exam, must present the beginnings of a research project to a committee within six months. This presentation normally turns into a dissertation, but the student is not bound to it and can switch later to another project or even another area of research.

Experimental research in the Department is in the areas of astrophysics, nonlinear phenomena and chaos, optics, optical oceanography, plasmas, and solid-state physics. Theoretical research is in elementary particles, environmental optics, plasmas, nonlinear phenomena and chaos, and solid-state physics. In addition to the research projects, the activities of research groups include seminars where visitors, Department faculty members, and graduate students present results of their research.

Research Facilities

The physics building includes 20,000 square feet of research laboratories and workshops. Major experimental instrumentation includes lasers, a radiometric calibration facility, CCD camera systems for measurements of radiance distribution and the point spread function under water, a very-high-resolution spectroradiometer, instruments for measuring spectral attenuation and scattering under water, an optical and a microwave spectrum analyzer, a UHV pumping station, vibration isolated optical tables, high-speed data acquisition systems, RF power sources, high-resolution video data systems, a transmission resonance spectrometer, cryogenic probes (0.05–330 K), a SQUID magnetometer, a 9-tesla superconducting magnet, thin films deposition systems (evaporator and pulsed excimer laser), high-temperature furnaces, polishing and cutting instruments, a Philips MRD thin-film X-ray diffractometer, and an adiabatic demagnetization refrigerator.

Computing services for the University are provided by the Ungar Computing Center and the Division of Information Technology. These include numerous servers and workstations distributed in various locations, a wireless network for the entire campus, and high-speed wired Internet in most offices, through which research groups have ready access to journals and other archival data, Web-based resources, and national supercomputer facilities. Within the Department there is also the Copernicus Computation Laboratory, with its dedicated servers and a multiteraflop Beowulf cluster.

Financial Aid

Financial support is available in several forms. Research assistantships (RAs) and teaching assistantships (TAs) include a stipend of $18,000 per year (additional stipend for summer teaching may be available) plus tuition for 9 credits per semester. Fellowships, with no teaching duties, include a stipend of $17,000 per year plus tuition for 9 credits per semester. Summer research fellowships are available on a competitive basis.

Cost of Study

In 2006–07, the tuition for one semester of full-time graduate study (9 credits) is $11,520. An additional $220 per semester covers student activity and athletics fees.

Living and Housing Costs

The cost of a single room in the dormitories is $5224 per year (an additional $1841 for meals) per person per semester. Typical rent for off-campus apartments is $700–$800 per month for one bedroom and $800–$1000 per month for two bedrooms. A typical rent for a room in a house off campus is $400–$500 per month, and the cost of living is generally lower than in the urban areas of the northern and western United States.

Student Group

In fall 2005, 26 graduate students were enrolled in the Department. All of them received financial aid in the form of a TA, an RA, a fellowship, or external support (exchange students). The majority of the students were international.

Student Outcomes

Recent Ph.D. graduates have chosen a variety of work environments. Students have gone on to postdoctoral positions at schools such as the University of Chicago and Brown, to government positions in agencies such as NASA and NOAA, and to corporate positions such as with the Beckman Coulter Company.

Location

The Department is housed in the James L. Knight Physics Building on the Coral Gables campus. This campus occupies 260 beautifully landscaped acres in a predominantly residential area. Coral Gables is an affluent suburb of metropolitan Miami, which is the largest urban area in Florida. Downtown Miami is readily accessible by Metrorail train, which has a stop next to the campus. Miami is a center of Latin culture and is the commercial gateway to Latin America. It offers all the amenities of a large and prosperous city, as well as an excellent oceanside climate of warm winters and moderate summers.

The University

The University of Miami was founded in 1925 and is accredited by the Southern Association of Colleges and Schools. Individual programs are accredited by a total of twelve professional agencies. The University of Miami is the largest private university in the Southeast and has a full-time enrollment of more than 14,000, including more than 3,000 graduate students and 2,000 law and medical students. Two colleges and ten schools are located on the main campus in Coral Gables. The University's medical school, the fourth largest in the United States, is situated in Miami's civic center, and the Rosentiel School of Marine and Atmospheric Science is located on Virginia Key. Funded research activities total more than $220 million per year.

Applying

Consideration is given to applicants who have a B.S. degree in physics with a minimum undergraduate GPA of 3.0 (B). The GRE is required. Applicants from non-English-speaking countries must demonstrate proficiency in English via the TOEFL, and the minimum acceptable score for admission is 550.

The application deadline for the fall semester is February 1, and application for financial aid is made at the time of application for admission. Forms should be requested from the Department.

Correspondence and Information

Professor Josef Ashkenazi
Chairman of the Graduate Recruitment Committee
Department of Physics
University of Miami
P.O. Box 248046
Coral Gables, Florida 33124
Phone: 305-284-2323/3
Fax: 305-284-4222
E-mail: ashkenazi@physics.miami.edu

University of Miami

THE FACULTY AND THEIR RESEARCH

George C. Alexandrakis, Professor and Chairman of the Department; Ph.D., Princeton, 1968. Solid-state experiment, transmission resonance, magneto-acoustic propagation in ferromagnetic metals.
Orlando Alvarez, Professor; Ph.D., Harvard, 1979. Theory of elementary particles.
Josef Ashkenazi, Associate Professor; Ph.D., Hebrew (Jerusalem), 1975. Solid-state theory, first-principles band structure methods, many-body physics, high-temperature superconductors.
Stewart E. Barnes, Professor; Ph.D., UCLA, 1972. Solid-state theory, many-body theory, superconductivity and magnetism.
G. Chris Boynton, Research Assistant Professor; Ph.D., Miami (Florida), 1991. Computational MHD and radiative transfer.
Joshua L. Cohn, Professor; Ph.D., Michigan, 1989. Condensed matter, experiment, materials physics, electronic and lattice transport.
Thomas L. Curtright, Professor; Ph.D., Caltech, 1977. Theory of elementary particles.
Massimiliano Galeazzi, Assistant Professor; Ph.D., Genoa, 1999. X-ray astrophysics, study of the interstellar/intergalactic medium and X-ray sources, development of X-ray detectors.
Ghassan Ghandour, Adjunct Professor and Professor Emeritus, Kuwait University; Ph.D., Berkeley, 1974. Theory of elementary particles.
Howard R. Gordon, Professor; Ph.D., Penn State, 1965. Optical oceanography, light scattering, radiative transfer, remote sensing.
Joshua O. Gundersen, Associate Professor; Ph.D., California, Santa Barbara, 1995. Experimental cosmology and astrophysics.
Joseph G. Hirschberg, Professor Emeritus; Ph.D., Wisconsin, 1952. Physical optics, Fabry-Perot interferometry, plasma spectroscopy.
Manuel A. Huerta, Professor; Ph.D., Miami (Florida), 1970. Statistical mechanics, plasma physics, numerical simulations in MHD.
Luca Mezincescu, Professor; Ph.D., Bucharest, 1978. Theory of elementary particles.
James C. Nearing, Associate Professor and Associate Chairman of the Department; Ph.D., Columbia, 1965. Theoretical physics, bifurcation theory in fully nonlinear plasma systems.
Rafael I. Nepomechie, Professor; Ph.D., Chicago, 1982. Theory of elementary particles.
William B. Pardo, Associate Professor; Ph.D., Northwestern, 1957. Experimental physics, plasma physics, nonlinear dynamics.
Arnold Perlmutter, Professor Emeritus; Ph.D., NYU, 1955. Nuclear and particle physics.
Carolyne M. Van Vliet, Adjunct Professor and Professor Emerita, University of Montréal; Ph.D., Free University, Amsterdam, 1956. Equilibrium and nonequilibrium statistical mechanics, stochastic processes, quantum transport in solids.
Kenneth J. Voss, Professor; Ph.D., Texas A&M, 1984. Ocean optics, light scattering, atmospheric optics.
Fulin Zuo, Associate Professor; Ph.D., Ohio State, 1988. Condensed matter, experiment.

RESEARCH ACTIVITIES

Experimental Cosmology and Astrophysics. Studies of the cosmic microwave and infrared background, studies of the interstellar/intergalactic medium and X-ray sources, instrumentation for low-noise RF and mm-wave detectors and telescopes, development of high-resolution cryogenic microcalorimeters and bolometers. (Galeazzi, Gundersen)

Experimental and Theoretical Nonlinear Dynamics. Study of instabilities and chaotic oscillations in systems exhibiting complex dynamical behavior; dripping faucets, electronic circuits, lasers, athletes, inert-gas plasmas at low fractional ionization, phase synchronization and communication with chaos. (Pardo) More information is available through http://ndl.physics.miami.edu.

Experimental Ocean Optics. Light scattering and absorption by marine particulates; instrumentation for measurement of optical properties of the ocean and atmosphere. (Voss)

Experimental Solid-State Physics. Ferromagnetic transmission resonance in metals, spin relaxation, exchange energy, phonon excitation and propagation, nonlinear phenomena. (Alexandrakis)

Transport and magnetic properties of materials at low temperatures; transition metal oxides; high-temperature and organic superconductors and reduced dimensional systems (e.g., layered systems and thin films); electrical and thermal conduction, thermoelectric effects; vortex dynamics, critical currents, quantum tunneling. (Cohn, Zuo)

Theoretical Elementary Particle Physics. Quantum field theory (especially integrable models), supergravity, superstrings. (Alvarez, Curtright, Ghandour, Mezincescu, Nepomechie)

Theoretical Environmental Optics. Radiative transfer, remote determination of ocean chlorophyll concentrations. (Gordon, Boynton)

Theoretical Plasma Physics. Numerical simulations in plasmas and other systems. (Huerta, Nearing, Boynton)

Theoretical Solid-State Physics. Electronic structure of solids, many-body physics, high-temperature superconductivity, magnetism. (Ashkenazi, Barnes)

Linear and nonlinear quantum transport, reduced-dimensionality systems. (Van Vliet)

More information is available through http://www.miami.edu/physics/.

The James L. Knight Physics Building.

SELECTED PUBLICATIONS

Rittenmyer, K. M., **G. C. Alexandrakis,** and P. S. Dubbelday. Detection of fluid velocity and hydroacoustic particle velocity using a temperature autostabilized nonlinear dielectric element (Tandel). *J. Acoust. Soc. Am.* 84:2002, 1988.

Alexandrakis, G. C., and G. Dewar. Electromagnetic generation of 9.4 GHz phonons in Fe and Ni. *J. Appl. Phys.* 55:2467, 1984.

Abeles, J. H., T. R. Carver, and **G. C. Alexandrakis.** Microwave transmission measurement of the critical exponent beta in iron and iron-silicon. *J. Appl. Phys.* 53:8116, 1982.

Alexandrakis, G. C., R. A. B. Devine, and J. H. Abeles. High frequency sound as a probe of exchange energy in nickel. *J. Appl. Phys.* 53:2095, 1982.

Alexandrakis, G. C. Determination of the molecular size and the Avogadro number: A student experiment. *Am. J. Phys.* 46:810, 1978.

Alvarez, O., I. M. Singer, and P. Windey. The supersymmetric σ-model and the geometry of the Weyl-Kac character formula. *Nucl. Phys. B* 373:647, 1992.

Alvarez, O., T. P. Killingback, M. Mangano, and P. Windey. The Dirac-Ramond operator in string theory and loop space index theorems. *Nucl. Phys. Proc. Suppl.* B1A:189, 1987.

Alvarez, O., T. P. Killingback, M. Mangano, and P. Windey. String theory and loop space index theorems. *Comm. Math. Phys.* 111:1, 1987.

Alvarez, O., I. M. Singer, and B. Zumino. Gravitational anomalies and the family's index theorem. *Comm. Math. Phys.* 96:409, 1984.

Alvarez, O. Theory of strings with boundaries: Fluctuations, topology and quantum geometry. *Nucl. Phys. B* 216:125, 1983.

Ashkenazi, J. Stripelike inhomogeneities, coherence, and the physics of the high-T_c cuprates. In *New Challenges in Superconductivity: Experimental Advances and Emerging Theories,* pp. 187–212, eds. **J. Ashkenazi,** et al. **(J. L. Cohn** and **F. Zuo).** Springer, 2005.

Ashkenazi, J., et al. (J. L. Cohn and F. Zuo), eds. *New Challenges in Superconductivity: Experimental Advances and Emerging Theories.* Springer, 2005.

Ashkenazi, J. Stripelike inhomogeneities, spectroscopies, pairing, and coherence in the high-T_c cuprates. *J. Phys. Chem. Solids* 65:1461–72, 2004.

Ashkenazi, J. Stripe fluctuations, carriers, spectroscopies, transport, and BCS-BEC crossover in the high-T_c cuprates. *J. Phys. Chem. Solids* 63:2277–85, 2002.

Ashkenazi, J. Stripes, carriers, and high-T_c in the cuprates. In *High-Temperature Superconductivity, AIP Conference Proceedings,* vol. 483, pp. 12–21, eds. **S. E. Barnes, J. Ashkenazi, J. L. Cohn,** and **F. Zuo.** 1999.

Vacaru, D., and **S. E. Barnes.** A new auxiliary particle method for the Hubbard, t-J and Heisenberg models. *J. Phys.: Condens. Matter* 6:719, 1994.

Nagashpur, M., and **S. E. Barnes.** Nonuniversality in the Kondo effect. *Phys. Rev. Lett.* 69:3824, 1992.

Barnes, S. E. Spinon-holon statistics, and broken statistical symmetry for the t-J and Hubbard models in 2D. In *High Temperature Superconductivity: Physical Properties, Microscopic Theory, and Mechanisms, Proceedings of the University of Miami Workshop on Electronic Structure and Mechanisms of High-Temperature Superconductivity.* Held at Coral Gables, Florida, January 1991; pp. 95–105, eds. **J. Ashkenazi, S. E. Barnes,** and **F. Zuo** et al. New York: Plenum Press, 1992.

Barnes, S. E. Theory of the Jahn-Teller-Kondo effect. *Phys. Rev.* 37:3671, 1988.

Barnes, S. E. Theory of electron paramagnetic resonance of ions in metals. *Adv. Phys.* 30:801–938, 1981.

Boynton, G. C., and **H. R. Gordon.** Irradiance inversion algorithm for estimating the absorption and backscattering coefficients of natural waters: Raman scattering effects. *Appl. Opt.* 39:3012–22, 2000.

Boynton, G. C., and U. Torkelson. Dissipation of nonlinear Alfven waves. *Astron. Astrophys.* 308:299–308, 1996.

Boynton, G. C., M. A. Huerta, and Y. C. Thio. 2-D MHD numerical simulations of EML plasma armatures with ablation. *IEEE Trans. Magn.* 29:751–6, 1993.

Chiorescu, C., J. J. Neumeier, and **J. L. Cohn.** Magnetic inhomogeneity and magnetotransport in electron-doped manganites Ca1-xLaxMnO3 ($0 \leq x \leq 0.10$). *Phys. Rev. B* 73:014406–12, 2006.

Cohn, J. L., C. Chiorescu, and J. J. Neumeier. Polaron transport in the paramagnetic phase of electron-doped manganites. *Phys. Rev. B* 72:024422–8, 2005.

Cohn, J. L., M. Peterca, and J. J. Neumeier. Low-temperature permittivity of insulating perovskite manganites. *Phys. Rev. B* 70:214433–9, 2004.

Neumeier, J. J., and **J. L. Cohn.** Possible signatures of magnetic phase segregation in electron-doped antiferromagnetic CaMnO3. *Phys. Rev. B* 61:14319–22, 2000.

Cohn, J. L., et al. Glasslike heat conduction in high-mobility semiconductors. *Phys. Rev. Lett.* 82:779–82, 1999.

Curtright, T., and **L. Mezincescu.** Biorthogonal quantum systems. *J. Math. Phys.* 47, 2006.

Zachos, C. K., D. B. Fairlie, and **T. L. Curtright,** eds. *Quantum Mechanics in Phase Space, An Overview with Selected Papers; World Scientific Series in 20th Century Physics, Vol. 34,* 2005.

Brugues, J., **T. Curtright,** J. Gomis, and **L. Mezincescu.** Nonrelativistic strings and branes as nonlinear realizations of Galilei groups. *Phys. Lett. B* 594:227–33, 2004.

Curtright, T., and D. Fairlie. Morphing quantum mechanics and fluid dynamics. *J. Phys. A* 36:8885–902, 2003.

Curtright, T. L., and C. Zachos. Classical and quantum Nambu mechanics. *Phys. Rev. D* 68:085001, 2003.

Curtright, T. L., G. I. Ghandour, and C. K. Zachos. Quantum algebra deforming maps, Clebsch-Gordan coefficients, coproducts, U and R matrices. *J. Math. Phys.* 32:676–88, 1991.

Curtright, T. L., G. I. Ghandour, C. B. Thorn, and C. K. Zachos. Trajectories of strings with rigidity. *Phys. Rev. Lett.* 57:799–802, 1986.

Braaten, E., **T. Curtright, G. Ghandour,** and C. B. Thorn. Nonperturbative weak coupling analysis of the Liouville quantum field theory. *Phys. Rev. Lett.* 51:19, 1983.

Galeazzi, M., and D. McCammon. A microcalorimeter and bolometer model. *J. Appl. Phys.* 93:4856, 2003.

McCammon, D., et al. **(M. Galeazzi).** A high-spectral resolution observation of the soft X-ray diffuse background with thermal detectors. *Astrophys. J.* 576:188, 2002.

Galeazzi, M., et al. Limits on the existence of heavy neutrinos in the range 50–1000 eV from the study of the ^{187}Re beta decay. *Phys. Rev. Lett.* 86:1978, 2001.

Galeazzi, M., et al. The end-point energy and half life of the ^{187}Re beta decay. *Phys. Rev. C* 63:014302, 2001.

Gatti, F., et al. **(M. Galeazzi).** Detection of environmental fine structure in the low-energy beta-decay spectrum of ^{187}Re. *Nature* 397:137, 1999.

Davis, E. D., and **G. I. Ghandour.** Implications of invariance of the Hamiltonian under canonical transformations in phase space. *J. Phys. A* 35:5875–91, 2002.

Barnes, T., and **G. I. Ghandour.** Variational treatment of the effective potential and renormalization in Fermi-Bose interacting field theories. *Phys. Rev. D* 22:924, 1980.

Gordon, H. R., et al. Retrieval of coccolithophore calcite concentration from SeaWiFS imagery. *Geophys. Res. Lett.* 28:1587–90, 2001.

Moulin, C., and **H. R. Gordon** et al. Atmospheric correction of ocean color imagery through thick layers of Saharan dust. *Geophys. Res. Lett.* 28:5–8, 2001.

Gordon, H. R., and **G. C. Boynton.** A radiance-irradiance inversion algorithm for estimating the absorption and backscattering coefficients of natural water: Homogenous waters. *Appl. Opt.* 36:2636–41, 1997.

Gordon, H. R., et al. Phytoplankton pigment concentrations in the Middle Atlantic Bight: Comparison of ship determinations and coastal zone color scanner measurements. *Appl. Opt.* 22:20–36, 1983.

Gordon, H. R., D. K. Clark, J. L. Mueller, and W. A. Hovis. Phytoplankton pigments derived from the Nimbus-7 CZCS: Initial comparisons with surface measurements. *Science* 210:63–6, 1980.

Gundersen, J., et al. BLAST—a balloon-borne large-aperture submillimeter telescope. In *AIP Proceedings from the 9th International Conference on Low Temperature Detectors,* vol. 605, pp. 585, eds. F. S. Porter, D. McCammon, **M. Galeazzi,** and C. K. Stahle, 2002.

Barkats, D., et al. **(J. O. Gundersen).** First measurements of the polarization of the cosmic microwave background radiation at small angular scales from CAPMAP. *Astrophys. J.* 619:L127, 2005.

Farese, P. C., et al. **(J. O. Gundersen).** COMPASS: An upper limit on CMB polarization at an angular scale of 20. *Astrophys. J.* 610:625, 2004.

University of Miami

Gundersen, J. O. (KUPID Collaboration). The Ku-band polarization identifier. In *Proceedings of the Workshop on the Cosmic Microwave Background and Its Polarization,* eds. S. Hanany and K. A. Olive. *New Astron. Rev.* 47(11):1097, 2003.

Hedman, M. M., et al. **(J. O. Gundersen).** New limits on the polarized anisotropy of the cosmic microwave background at subdegree angular scales. *Astrophys. J.* 573:L73, 2002.

Hirschberg, J. G., and T. N. Veziroglu. Tidal energy for inexpensive power. Compact disc *Proceedings, 13th World Hydrogen Energy Conference,* Beijing, China, June 11–15, 2000.

Hirschberg, J. G., and E. Kohen. A new spectral method for fluorescence excitation. In *Recent Research Developments in Optical Engineering, Applications of Optical Engineering to the Study of Cellular Pathology,* pp. 91–7, eds. E. Kohen and **J. G. Hirschberg** (Scientific Information Guild). Trivandrum, Kerala (India): Research Signpost, 1999.

Skinner, C. H., et al. **(J. G. Hirschberg).** Contact microscopy with a soft X-ray laser. *J. Microsc.* 159(1):51–60, 1990.

Hirschberg, J. G. A long range microscope phase condensing system. *Appl. Opt.* 29:1409–10, 1990.

Orta, J. A., **M. A. Huerta,** and **G. C. Boynton.** Magnetohydrodynamic shock heating of the solar corona. *Astrophys. J.* 596:646–55, 2003.

Cardelli, E., N. Esposito, and **M. A. Huerta.** Numerical comparison between a differential and an integrated approach in MHD simulations. *IEEE Trans. Magn.* 33:219–24, 1997.

Thio, Y. C., **M. A. Huerta,** and **J. C. Nearing.** On some techniques to achieve ablation-free operation of electromagnetic rail launchers. *IEEE Trans. Magn.* 29, 1993.

Castillo, J. L., and **M. A. Huerta.** Effect of resistivity on the Rayleigh-Taylor instability in an accelerated plasma. *Phys. Rev. E* 48:3849–66, 1993.

Huerta, M. A. Steady detonation waves with losses. *Phys. Fluids* 28:2735–43, 1985.

Huerta, M. A., and J. Magnan. Spatial structures in plasmas with metastable states as bifurcation phenomena. *Phys. Rev. A* 26:539–55, 1982.

Grisaru, M. T., **L. Mezincescu,** and **R. I. Nepomechie.** Direct calculation of boundary S matrix for open Heisenberg chain. *J. Phys. A* 28:1027–45, 1995.

de Vega, H. J., **L. Mezincescu,** and **R. I. Nepomechie.** Scalar kinks. *Int. J. Mod. Phys. B* 8:3473–85, 1994.

Mezincescu, L., and **R. I. Nepomechie.** Analytical Bethe ansatz for quantum-algebra-invariant spin chains. *Nucl. Phys. B* 372:597–621, 1992.

Mezincescu, L., and **R. I. Nepomechie.** Integrability of open spin chains with quantum algebra symmetry. *Int. J. Mod. Phys. A* 6:5231–48, 1991 (Addendum—A7:5657–9, 1992).

Mezincescu, L., and M. Henneaux. A σ model interpretation of Green-Schwarz covariant superstrings. *Phys. Lett.* 152B:340–2, 1985.

Jones, D. R. T., and **L. Mezincescu.** The chiral anomaly and a class of two-loop finite supersymmetric theories. *Phys. Lett.* 138B:293–5, 1984.

Mezincescu, L. On the superfield approach for 0(2) supersymmetry. Preprint *Joint Institute for Nuclear Research (JINR),* Dubna P2-12572 (in Russian), pp. 1–19, 1979.

Nearing, J. C., and **M. A. Huerta.** Skin and heating effects of railgun currents. *IEEE Trans. Magn.* 25:381–6, 1989.

Nepomechie, R. I. Solving the open XXZ spin chain with nondiagonal boundary terms at roots of unity. *Nucl. Phys. B* 622:615–32, 2002.

Nepomechie, R. I. The boundary supersymmetric sine-Gordon model revisited. *Phys. Lett. B* 509:183–8, 2001.

Nepomechie, R. I. Magnetic monopoles from antisymmetric tensor gauge fields. *Phys. Rev. D* 31:1921–4, 1985.

Pardo, W. B., et al. Pacing a chaotic plasma with a music signal. *Phys. Lett. A* 284:259, 2001.

Rosa, E., Jr., and **W. B. Pardo** et al. Phase synchronization of chaos in a plasma discharge tube. *Int. J. Bifurcation Chaos* 10:2551, 2000.

Ticos, C. M., et al. **(W. B. Pardo).** Experimental real-time phase synchronization of a paced chaotic plasma discharge. *Phys. Rev. Lett.* 85:2929, 2000.

Monti, M., and **W. B. Pardo** et al. Color map of Lyapunov exponents of invariant sets. *Int. J. Bifurcation Chaos* 9:1459, 1999.

Walkenstein, J. A., **W. B. Pardo,** M. Monti, and E. Rosa Jr. Chaotic moving striations in inert gas plasmas. *Phys. Lett. A* 261:183, 1999.

Perlmutter, A., et al. Spin analyzing power in p-p elastic scattering at 28 GeV/c. *Phys. Rev. Lett.* 50:802–6, 1983.

Perlmutter, A., et al. Spin-spin forces in 6 GeV/c neutron-proton elastic scattering. *Phys. Rev. Lett.* 43:983–6, 1979.

Friedmann, M., D. Kessler, A. Levy, and **A. Perlmutter.** The coulomb field in $\Sigma\pi$ production by slow K-mesons in complex nuclei. *Nuovo Cimento* 35:355–76, 1965.

Cox, J., and **A. Perlmutter.** A method for the determination of the S matrix for scattering by a tensor potential. *Nuovo Cimento* 37:76–87, 1965.

Van Vliet, C. M. Electronic noise due to multiple trap levels in homogenous solids and in space-charge layers. *J. Appl. Phys.,* 2003.

Van Vliet, C. M., and A. Barrios. Quantum electron transport beyond linear response. *Physica A* 315:493–536, 2002.

Van Vliet, C. M. Random walk and 1/f noise. *Physica A* 303:421–6, 2002.

Guillon, S., P. Vasilopoulos, and **C. M. Van Vliet.** Magnetoconductance of parabolically confined quasi-onedimensional channels. *J. Phys. C: Condens. Matter* 14:803–14, 2002.

Van Vliet, C. M. Quantum transport in solids. In *Advances in Mathematical Sciences, CRM Proceedings and Lecture Notes,* vol. 11, pp. 21–48, ed. L. Vinet. Providence, Rhode Island: American Mathematical Society, 1997.

Voss, K. J., et al. The spectral upwelling radiance distribution in optically shallow waters. *Limnol. Oceanogr.* 48:364–73, 2003.

Voss, K. J., et al. **(H. R. Gordon).** Lidar measurements during Aerosols99. *J. Geophys. Res.* 106:20821–32, 2001.

Voss, K. J., A. Chapin, M. Monti, and H. Zhang. An instrument to measure the bidirectional reflectance distribution function (BRDF) of surfaces. *Appl. Opt.* 39:6197–206, 2000.

Ritter, J. M., and **K. J. Voss.** A new instrument to measure the solar aureole from an unstable platform. *J. Atmos. Ocean. Tech.* 17:1040–7, 2000.

Voss, K. J., and Y. Liu. Polarized radiance distribution measurements of skylight: I. system description and characterization. *Appl. Opt.* 36:6083–94, 1997.

Zhang, P., and **F. Zuo** et al. Irreversible magnetization in nickel nanoparticles. *J. Magn. Magn. Mater.* 225:337–45, 2001.

Zuo, F., et al. Paramagnetic limiting of the upper critical field of the layered organic superconductor k-(BEDT-TTF)$_2$ Cu(NCS)$_2$. *Phys. Rev. B* 61:750–5, 2000.

Zuo, F., X. Su, and W. K. Wu. Magnetic properties of the pre-martensitic transition in Ni_2MnGa alloys. *Phys. Rev. B* 58:11127, 1998.

Zuo, F., J. Schlueter, and J. W. Williams. Mixed state magnetoresistance in organic superconductors k-(ET)$_2$Cu[N(CN)$_2$]Br. *Phys. Rev. B* 54:11973, 1996.

Zuo, F., et al. **(G. C. Alexandrakis).** Anomalous magnetization in single-crystal κ-[bis(ethylenedithiotetrathiafulvalene)]$_2$Cu[N(CN)$_2$]Br superconductors. *Phys. Rev. B* 52:R13126, 1995.

Zuo, F., et al. Josephson decoupling in single crystal $Nd_{1.85}Ce_{0.15}CuO_{4-y}$ superconductors. *Phys. Rev. Lett.* 72:1746–9, 1994.

UNIVERSITY OF MICHIGAN

Department of Physics

Program of Study

The Department of Physics offers a program leading to the Doctor of Philosophy in physics. The Department offers research opportunities in theoretical and experimental fields, including atomic physics, astrophysics, biophysics, optical physics, condensed-matter physics, elementary particle physics, and nuclear physics. The requirements for the Ph.D. are as follows. Students must pass, with a grade of B- or better, nine prescribed graduate physics courses (500 level) or show equivalent competence, 4 credits of cognate courses, one advanced graduate physics course (600 level) on a special topic, and a 4-credit course of supervised nonthesis research. Students must also pass a two-part written qualifying examination on advanced undergraduate material no later than the beginning of their third year. Ph.D. students are expected to attain candidacy by the beginning of their fifth term. Completion of the degree involves writing a thesis based on independent research done under the supervision of a faculty adviser and passing a final oral exam.

Research Facilities

Physics research is focused at the Randall Laboratory/West Hall complex, which includes a 65,000-square-foot laboratory addition. The physics laboratories house state-of-the-art space and facilities that support the Department's research activities. Individual investigators use tools such as atomic scale and positron microscopes, lasers of all sorts (CW, pulsed, Q-switched, mode-locked, ultrafast, frequency-stabilized, ion, dye, solid-state, and diode), dilution refrigerators and cryogenic equipment, laser tweezers, and a Mössbauer spectrometer for the study of active sites in protein. Nuclear and high-energy physics groups use campus laboratories to build and test apparatus used at accelerator facilities around the world. Apparatus for beams of radioactive nuclei, for polarized beams and targets, and for detector facilities used in fixed target and colliding beam experiments are among those developed on the University of Michigan campus. Collaborations in the medical sciences take place at the University Hospital and Medical School. The School of Engineering on the University's North Campus is the site of the Center for Ultrafast Optical Sciences, an NSF science and technology center exploring ultrafast and high-intensity laser science.

Department computer facilities are state-of-the-art, based on a professionally managed, distributed network with powerful workstations and high-speed network connections among department and University computers and to the Internet. Department shop facilities include a well-equipped student shop, a large instrument shop with computerized numerically controlled milling machines, and an electronics shop with custom VLSI circuit design facilities. Other University shop facilities complete the technical support necessary for state-of-the-art research. The University of Michigan libraries house one of the country's largest science libraries and employ modern computerized catalogs, databases, and information retrieval.

Financial Aid

The University's Regents' Fellowships paid a stipend of $23,400, plus full tuition and fees, in 2005–06. Several other merit-based scholarships are available to incoming students. Graduate student instructorships (GSIs) cover a period of eight months and paid a stipend of $14,326 plus tuition in 2005–06; teaching loads usually consist of two to four 2-hour elementary lab sections per week. GSIs are represented by a union. Graduate research assistantship stipends were $21,489 for twelve months in 2005–06. Summer RA appointments are available for most students. A very small number of spring GSI appointments are also available. Students who have at least a one-quarter-time appointment (half a normal appointment) as an RA or GSI are eligible to participate in the University's group health insurance plan.

Cost of Study

For 2005–06, tuition was $7041 per term for full-time in-state residents and $14,250 per term for full-time out-of-state students; candidates paid $4562 per term. Tuition is waived for students with one-quarter-time or more teaching or research assistantships. Most fees are included in the tuition; fees not included total about $95 each semester. Books and supplies cost approximately $400 per term.

Living and Housing Costs

Living costs, including room and board, transportation, and personal needs, are estimated at $15,584 per academic year for a single student with no dependents.

Student Group

The University of Michigan has approximately 39,995 students, of whom 10,986 are graduate students. The Department of Physics has approximately 125 graduate students.

Location

The University is in Ann Arbor, 40 miles west of Detroit in the Huron River Valley, in a beautiful, tree-lined town that combines the charm of a small city with the sophistication of cities many times its size. Regarded as a cultural center of the Midwest, it offers numerous opportunities for recreation and enjoyment.

The University and The Department

The University of Michigan, founded in 1817, is one of the nation's oldest public institutions of higher education. Consistently ranked among the great universities in the world, Michigan has a strong tradition of leadership in the development of the modern American research university—a tradition sustained by the wide-ranging interests and activities of its faculty members and students. Exceptional facilities and programs, both academic and nonacademic, are available.

The Department of Physics has played a leading role in the development of modern physics, with accomplishments ranging from the discovery of spin, the invention of the racetrack synchrotron, and the bubble chamber to the birth of nonlinear optics, the detection of neutrinos from supernova 1987A, and evidence of the top quark.

Applying

Applications for admission in the fall term are due by January 15 of the preceding spring; international students must submit applications by December 8. The GRE General Test is required, and the GRE Subject Test in physics is strongly recommended. For further information on admission requirements, students should write to the Associate Chair for Graduate Studies.

Correspondence and Information

Professor James Wells, Associate Chair for Graduate Studies
Department of Physics
2464 Randall Laboratory of Physics
University of Michigan
450 Church Street
Ann Arbor, Michigan 48109-1040

Phone: 734-936-0658
Web site: http://www.lsa.umich.edu/physics

University of Michigan

THE FACULTY AND THEIR RESEARCH

Fred C. Adams, Professor; Ph.D., Berkeley, 1988. Theoretical astrophysics.
Carl W. Akerlof, Professor; Ph.D., Cornell, 1967. Experimental high-energy physics, astrophysics, cosmic rays.
Ratindranath Akhoury, Professor; Ph.D., SUNY at Stony Brook, 1980. Theoretical high-energy physics.
James W. Allen, Professor; Ph.D., Stanford, 1968. Experimental condensed-matter physics.
Dante E. Amidei, Professor; Ph.D., Berkeley, 1984. Experimental high-energy physics.
Meigan C. Aronson, Professor; Ph.D., Illinois at Urbana-Champaign, 1988. Experimental condensed-matter physics.
Frederick D. Becchetti Jr., Professor; Ph.D., Minnesota, 1969. Experimental nuclear physics.
Paul Berman, Professor; Ph.D., Yale, 1969. Theoretical atomic, molecular, and optical physics.
Michael Bretz, Professor; Ph.D., Washington (Seattle), 1971. Experimental low-temperature physics, condensed-matter physics.
Philip H. Bucksbaum, Otto Laporte Professor of Physics and Associate Director, NSF Center for Ultrafast Optical Science; Ph.D., Berkeley, 1980. Experimental atomic and optical physics.
Myron Campbell, Professor and Department Chair; Ph.D., Yale, 1982. Experimental high-energy physics, elementary particles.
J. Wehrley Chapman, Professor; Ph.D., Duke, 1966. Experimental high-energy physics.
Timothy E. Chupp, Professor; Ph.D., Washington (Seattle), 1983. Experimental atomic physics.
Roy Clarke, Professor and Director, Applied Physics Program; Ph.D., Queen Mary College (London), 1973. Experimental condensed-matter physics.
Steven Dierker, Professor; Ph.D., Illinois at Urbana-Champaign, 1983. Experimental condensed-matter physics.
Luming Duan, Assistant Professor; Ph.D., University of Science and Technology (China), 1994. Quantum information.
August Evrard, Associate Professor; Ph.D., SUNY at Stony Brook, 1980. Theoretical astrophysics.
Stephen Forrest, Professor and Vice President of Research; Ph.D., Michigan, 1979. Photonic materials, devices and systems.
Katherine Freese, Professor; Ph.D., Chicago, 1984. Theoretical astrophysics.
David Gerdes, Associate Professor; Ph.D., Chicago, 1992. High-energy physics.
David W. Gidley, Professor; Ph.D., Michigan, 1979. Experimental physics, atomic physics.
Sharon Glotzer, Associate Professor; Ph.D., Boston University, 1993. Materials science and engineering, macromolecular science and engineering.
Gordon L. Kane, Professor; Ph.D., Illinois, 1963. Theoretical physics, elementary particles.
Alan D. Krisch, Professor; Ph.D., Cornell, 1964. Experimental high-energy physics, elementary particles.
Jean P. Krisch, Professor; Ph.D., Cornell, 1965. Theoretical physics, elementary particles, physics teaching.
Cagliyan Kurdak, Associate Professor; Ph.D., Princeton, 1995. Condensed-matter physics.
Finn Larsen, Assistant Professor; Ph.D., Princeton, 1996. Theoretical high-energy physics.
James T. Liu, Assistant Professor; Ph.D., Princeton, 1991. Theoretical high-energy physics.
Michael J. Longo, Professor; Ph.D., Berkeley, 1961. Experimental high-energy physics, instrumentation.
Wolfgang B. Lorenzon, Associate Professor; Ph.D., Basel (Switzerland), 1988. Experimental high-energy physics.
Timothy McKay, Associate Professor; Ph.D., Chicago, 1992. Astrophysics.
Jens-Christian D. Meiners, Assistant Professor; Ph.D., Constance (Germany), 1997. Experimental biophysics.
Roberto D. Merlin, Professor; Dr.rer.nat., Stuttgart, 1978. Experimental condensed-matter physics, applied physics.
Christopher Monroe, Professor; Ph.D., Colorado at Boulder, 1992. Experimental atomic, molecular, and optical physics.
Samuel Moukouri, Assistant Professor; Ph.D., Paris, 1993. Condensed-matter theory.
Homer A. Neal, Professor, Vice President Emeritus for Research, and Director, Project ATLAS; Ph.D., Michigan, 1966. Experimental high-energy physics.
Mark Newman, Associate Professor; D.Phil., Oxford, 1988. Statistical physics theory.
Franco Nori, Professor; Ph.D., Illinois at Urbana-Champaign, 1987. Condensed-matter theory.
Jennifer Ogilvie, Assistant Professor; Ph.D., Toronto, 2003. Experimental biophysics.
Bradford G. Orr, Professor; Ph.D., Minnesota, 1985. Experimental condensed-matter physics, applied physics.
Leopoldo Pando Zayas, Assistant Professor; Ph.D., Moscow State, 1998. Particle theory.
Jian-Ming Qian, Associate Professor; Ph.D., MIT, 1991. Experimental high-energy physics.
Georg Raithel, Professor; Ph.D., Munich, 1990. Experimental atomic, molecular, and optical physics.
Steven Rand, Associate Professor; Ph.D., Toronto, 1978. Optical physics, applied physics.
David Reis, Assistant Professor; Ph.D., Rochester, 1999. Experimental condensed-matter physics.
J. Keith Riles, Professor; Ph.D., Stanford, 1989. Experimental high-energy physics.
Leonard M. Sander, Professor; Ph.D., Berkeley, 1968. Theoretical physics, condensed matter.
Robert S. Savit, Professor; Ph.D., Stanford, 1973. Theoretical condensed-matter physics, elementary particles.
Roseanne J. Sension, Associate Professor; Ph.D., Berkeley, 1986. Experimental atomic and molecular optical physics, biophysics.
Duncan G. Steel, Professor; Ph.D., Michigan, 1976. Experimental atomic and optical physics.
Gregory Tarlé, Professor; Ph.D., Berkeley, 1978. Experimental astrophysics, nuclear physics.
Rudolf P. Thun, Professor; Ph.D., SUNY at Stony Brook, 1980. Experimental high-energy physics, elementary particles.
Alexei Tkachenko, Assistant Professor; Ph.D., Bar-Ilan, 1998. Theoretical condensed matter and complex systems.
Ctirad Uher, Professor; Ph.D., New South Wales, 1975. Experimental condensed-matter physics, applied physics.
James Wells, Associate Professor; Ph.D., Michigan, 1995. Elementary particle physics.
Y. P. Edward Yao, Professor; Ph.D., Harvard, 1964. Theoretical high-energy physics, elementary particles.
Bing Zhou, Professor; Ph.D., MIT, 1987. High-energy physics.
Michal Zochowski, Assistant Professor; Ph.D., Warsaw, 1995. Biophysics, complex systems, neuroscience.
Jens C. Zorn, Professor; Ph.D., Yale, 1961. Experimental physics, atomic physics.

UNIVERSITY OF NOTRE DAME

Department of Physics

Programs of Study

The Department of Physics graduate program is primarily a doctoral program leading to a Doctor of Philosophy (Ph.D.) degree, and the Department ordinarily does not accept students who intend to complete only a master's degree. Major areas of research include astronomy, astrophysics, atomic physics, biophysics, condensed-matter physics, cosmology, general relativity, high-energy elementary particle physics, nuclear physics, radiation physics, and statistical physics. Interdisciplinary programs are available in radiation physics, biophysics, and chemical physics. Requirements for the Ph.D. include 39 course credit hours, seminars, and research. Students are expected to become actively involved in research during the first year and take a first-year qualifying exam. Both oral and written candidacy examinations are normally completed early in the third year. The candidate must demonstrate the ability to perform research and must show a broad understanding of physics. A thesis is required and must be approved by and defended orally before the student's doctoral committee.

Research Facilities

The Department has excellent research facilities on and off campus. Astronomy/astrophysics research facilities include twenty nights a year at the 1.8-meter Vatican Advanced Technology Telescope (VATT) and ten nights a year at the 2x 8.5-meter Large Binocular Telescope (LBT). Current research is also conducted using a variety of telescopes, including the Hubble Space Telescope (HST), the Keck Telescope, the NASA Infrared Telescope (IRTF), and the Steward and Cerro-Telolo Observatories. An air-shower array located next to the campus is used to study high- (30–300GeV) and ultrahigh-energy (greater than 100TeV) cosmic rays, utilizing position-sensitive proportional wire detectors for precision angle measurements and particle identification. Facilities for accelerator-based atomic physics research include the Atomic Physics Accelerator Lab (APAL) at Notre Dame, which includes a 200 kV heavy-ion accelerator and various vacuum ultraviolet and visible monochromators, high-resolution position-sensitive photon detectors, and Doppler-free laser excitation chambers as well as other tabletop laser excitation systems. Precision measurements in atomic Cs, necessary for interpretation of parity nonconservation experiments, are carried out using Ti-sapphire, dye, and diode laser. Experiments on highly charged ions are also carried out at Argonne National Laboratory (ANL) and at GSI-Darmstadt, Germany. X-ray–atom interactions are also studied at national synchrotron radiation facilities. In biophysics, a 300-Mhz magnetic resonance imager (MRI) is available. The MRI has a vertical superwide-bore 7-Tesla magnet with exchangeable probes (up to 64 mm in diameter) and gradient sets (up to 100 Gauss/cm) for imaging microscopy and biological applications. The facility is equipped for in vivo study of small animals. Condensed-matter physics facilities are available for molecular-beam epitaxy (MBE) of semiconductor films, superlattices, and microstructures and for bulk crystal growth, including a traveling solvent floating zone furnace; low-temperature electron tunneling; microwave, optical, and infrared photoresponse studies of superconductors; resonance studies in ferromagnetic and paramagnetic materials; surface physics; X-ray and fluorescence characterization of solids; low-temperature thermodynamic studies; and optical and far-infrared studies of semiconductors. XAFS and X-ray–scattering experiments are also carried out at the ANL, and neutron diffraction studies are performed at the National Institute of Standards and Technology (NIST). High-energy elementary particle physics research is carried out at the Tevatron Collider at Fermi National Accelerator Laboratory (FNAL), Brookhaven National Laboratory (BNL), Stanford Linear Accelerator (SLAC), and the Large Hadron Collider at the CERN Laboratory in Geneva, Switzerland. On-campus facilities are used for the development of new particle detection systems, including scintillating fiber tracking and tile-fiber calorimeter detectors, and for detector development and instruction for the QuarkNet education and outreach project. Facilities for research in nuclear physics include 1-MV, 4-MV, and 10-MV Van de Graaff accelerators; a multidetector array for gamma-ray spectroscopy; and a dual superconducting solenoid system for radioactive beam studies. Nuclear physics programs are also under way at ANL, the National Superconducting Cyclotron Laboratory (NCSL) at Michigan State University, Oak Ridge National Laboratory, and Los Alamos National Laboratory, as well as several laboratories abroad. Computing facilities include the University's supercomputers plus Departmental computer clusters. Wireless connections are available in all offices, laboratories, residences, and other campus locations. The Department has an extensive research library and state-of-the-art machine shop.

Financial Aid

Graduate teaching assistantships are normally available to all Ph.D. students and for 2006–07 include a minimum nine-month stipend of $16,000, plus payment of tuition and fees. Higher stipends are available for exceptionally well-qualified applicants. Summer support is normally provided, though not guaranteed, from federal and external research funding. Research fellowships are available on a competitive basis. Advanced students often receive support as research assistants.

Cost of Study

Graduate tuition for 2006–07 is $32,800; summer tuition and fees are $364. Payment of tuition and fees is provided in addition to the student stipends.

Living and Housing Costs

Accommodations for single students are available on campus at a cost of $4365 to $4465 for nine months. Accommodations for married students are available near the campus for $465 to $710 (utilities extra) per month. Privately owned rooms and apartments are also for rent near the campus.

Student Group

There are 97 graduate students in the Department of Physics. In 2005–06, the University had an enrollment of 11,417 students, of whom 3,142 were graduate students.

Student Outcomes

The Department has current employment data on most former graduate students. Of 41 students who completed their degrees from 2000 to 2004, 17 percent have accepted academic or research faculty positions; 19 percent are employed by industry in physics and 5 percent by industry in computing; 49 percent now hold postdoctoral positions.

Location

The University is located in South Bend, a city at the northernmost edge of Indiana a few miles from Michigan. Chicago is about 90 miles west, Lake Michigan is about 35 miles northwest, and Indianapolis is about 140 miles south. The population of South Bend is roughly 100,000, but the city merges with surrounding communities to form a greater Michiana (Michigan and Indiana) area that is home to more than 275,000 people. The city's name derives from the bend in the St. Joseph River, which winds through the city and provides various forms of recreation, including a nationally recognized white-water site. South Bend is home to the College Football Hall of Fame; the Studebaker Museum; the South Bend Regional Museum of Art; several theaters, including the historic Morris Performing Arts Center; the South Bend Symphony Orchestra; South Bend Chocolate Factory; Broadway Theater League; and the Silver Hawks minor-league baseball team.

The University

The University of Notre Dame, which was founded in 1842, is a private, independent coeducational school. The 1,250-acre campus offers a spacious setting, with lakes and wooded areas. The intellectual, cultural, and athletic traditions at Notre Dame, coupled with the beauty of the campus, contribute to the University's fine reputation. A new science teaching facility is scheduled to open in summer 2006. The Jordan Hall of Science contains a four-story concourse with new undergraduate lab space for the Departments of Chemistry and Biochemistry, Biology, and Physics. The 201,782-square-foot facility also houses two 250-seat lecture halls, a 150-seat multivisualization room, an observatory, teaching labs, a herbarium, a greenhouse, and Departmental offices for preprofessional studies.

The University offers a variety of cultural and recreational activities, including plays, concerts, and lecture series. The Rolfs Sports Recreation Center provides indoor space for a wide variety of health and recreation activities; tennis and ice skating can be enjoyed year-round. The students and faculty members represent a rich diversity of religious, racial, and ethnic backgrounds.

Applying

Applications are invited from qualified students without regard to sex, race, religion, or national or ethnic origin. Both the General Test and the Subject Test in physics of the GRE and three letters of recommendation are required. Complete applications should be submitted by February 1. Detailed Departmental information is available on the Department's Web site. Applications are available online at http://graduateschool.nd.edu/html/admissions/application_gateway.html.

Correspondence and Information

Chair, Admissions Committee, Department of Physics
225 Nieuwland Science Hall
University of Notre Dame
Notre Dame, Indiana 46556-5670

Phone: 574-631-6386
E-mail: physics@nd.edu
Web site: http://www.physics.nd.edu

University of Notre Dame

THE FACULTY AND THEIR RESEARCH

Ani Aprahamian, Ph.D., Clark, 1986. Experimental nuclear physics: gamma-ray spectroscopy, nuclear masses, lifetimes, astrophysics.
Gerald B. Arnold, Ph.D., UCLA, 1977. Theoretical solid-state physics: magnetism, high-temperature superconductivity.
Dinshaw Balsara, Ph.D., Illinois at Urbana-Champaign, 1990. Theoretical and computational astrophysics.
Albert-László Barabási, Emil T. Hofman Professor; Ph.D., Boston University, 1994. Theoretical physics, statistical mechanics, nonlinear systems, networks, biophysics.
H. Gordon Berry, Ph.D., Wisconsin, 1967. Experimental atomic physics.
Ikaros I. Bigi, Grace-Rupley II Professor; Ph.D., Munich, 1977. Theoretical high-energy physics.
Howard A. Blackstead, Ph.D., Rice, 1967. Experimental physics: solid-state physics, magnetism and acoustics.
Bruce A. Bunker, Ph.D., Washington (Seattle), 1980. Experimental physics: X-ray, UV, and electron spectroscopy of condensed-matter and biological/environmental systems.
Neal M. Cason, Ph.D., Wisconsin, 1964. Experimental physics: high-energy elementary particle physics, particle spectroscopy.
Philippe A. Collon, Ph.D., Vienna, 1999. Experimental nuclear physics: new techniques, AMS.
Malgorzata Dobrowolska-Furdyna, Ph.D., Polish Academy of Sciences, 1979. Experimental solid-state physics.
Morten R. Eskildsen, Ph.D., Copenhagen, 1998. Experimental condensed matter.
Stefan G. Frauendorf, Ph.D., Dresden Technical (Germany), 1971. Theoretical nuclear physics, atomic physics, mesoscopic systems.
Jacek K. Furdyna, Marquez Professor; Ph.D., Northwestern, 1960. Experimental solid-state physics: man-made materials.
Umesh Garg, Ph.D., SUNY at Stony Brook, 1978. Experimental nuclear physics: nuclear structure, giant resonances, gamma-ray spectroscopy, high-spin states.
Peter Garnavich, Ph.D., Washington (Seattle), 1991. Astrophysics, observational cosmology.
Anna Goussiou, Ph.D., Wisconsin–Madison, 1995. Experimental high-energy elementary particle physics.
Michael D. Hildreth, Ph.D., Stanford, 1995. Experimental high-energy elementary particle physics.
J. Christopher Howk, Ph.D., Wisconsin, 1999. Observational astrophysics.
Anthony Hyder, Ph.D., Air Force Tech, 1971. Experimental nuclear physics.
Boldizsár Jankó, Ph.D., Cornell, 1996. Theoretical condensed-matter physics.
Colin Jessop, Ph.D., Harvard, 1994. Experimental high-energy elementary particle physics.
Walter R. Johnson, Freimann Professor of Physics; Ph.D., Michigan, 1957. Theoretical physics: quantum electrodynamics, atomic physics.
James J. Kolata, Ph.D., Michigan State, 1969. Experimental physics: nuclear structure, heavy-ion reactions, radioactive beam physics.
Christopher F. Kolda, Associate Chair and Director of Undergraduate Studies; Ph.D., Michigan, 1995. Theoretical high-energy physics.
A. Eugene Livingston, Ph.D., Alberta, 1974. Experimental physics: atomic physics, spectroscopy of highly ionized atoms.
John M. LoSecco, Ph.D., Harvard, 1976. Experimental and theoretical physics: high-energy elementary particle physics.
Grant J. Mathews, Ph.D., Maryland, 1977. Theoretical astrophysics/cosmology, general relativity.
Kathie E. Newman, Associate Chair and Director of Graduate Studies; Ph.D., Washington (Seattle), 1981. Theoretical physics: statistical mechanics, semiconductors.
Terrence W. Rettig, Ph.D., Indiana, 1976. Observational astronomy: comets, solar system formation, and T Tauri stars.
Randal C. Ruchti, Ph.D., Michigan State, 1973. Experimental physics: high-energy elementary particle physics.
Steven T. Ruggiero, Ph.D., Stanford, 1981. Experimental physics: condensed-matter and low-temperature physics, superconductivity.
Jonathan R. Sapirstein, Ph.D., Stanford, 1979. Theoretical physics: quantum electrodynamics.
Carol E. Tanner, Ph.D., Berkeley, 1985. Experimental physics: atomic physics.
Zoltan Toroczkai, Ph.D., Virginia Tech, 1997. Theoretical condensed-matter physics, biophysics.
Mitchell R. Wayne, Chair, Department of Physics; Ph.D., UCLA, 1985. Experimental high-energy elementary particle physics.
Michael Wiescher, Ph.D., Münster (Germany), 1980. Experimental nuclear physics: nuclear astrophysics.

RESEARCH ACTIVITIES

Theoretical

Astrophysics/Cosmology: inflationary cosmology, primordial nucleosynthesis, cosmic microwave background, galaxy formation and evolution, large-scale structure, neutrino physics, dark matter, stellar evolution and nucleosynthesis, neutron star binaries, gravity waves, gamma-ray bursts, supernovae. (Balsara, Kolda, Mathews, Wiescher, 1 adjunct professor, 1 assistant professional specialist, 5 postdoctoral research associates, 3 research visitors)
Atomic Physics: quantum electrodynamics, weak interactions, atomic many-body theory, photoionization and photoexcitation. (Johnson, Sapirstein, 1 adjunct professor)
Biophysics: bioinformatics, cellular networks, modeling morphogenesis. (Alber, Barabási, Merz, Toroczkai)
Condensed Matter: many-body problem, high-temperature superconductivity, superconductivity and magnetism on the nanoscale, tunneling phenomena, metal-metal interfaces, inhomogeneous and layered superconductors, hopping transport, studies of ordering in semiconductors, magnetic semiconductors. (Arnold, Barabási, Frauendorf, Jankó, Newman, 1 research assistant professor, 3 postdoctoral research assistants, 2 research visitors)
Elementary Particle Physics: formal properties of quantum field theories, supersymmetry, grand unification, spontaneous breaking symmetry, phenomenology of strong and weak processes, rare decays, CP violation, supergravity, extra dimensions, new particles. (Bigi, Kolda, 1 research assistant professor, 1 adjunct associate professor, 1 postdoctoral research assistant)
General Relativity: black holes in a magnetic field, charged black holes, neutron stars, numerical relativity, gravity waves. (Mathews)
Nuclear Physics: many-body problem, nuclear reactions, few-body problem, boson expansions, structure of nuclei with momentum, high angular momentum, exotic proton and neutron numbers. (Frauendorf, Hyder, Mathews, 1 adjunct professor, 1 postdoctoral research associate)
Statistical Mechanics: complex networks, phase transitions, critical phenomena in fluids, networks. (Barabási, Newman)

Experimental

Astrophysics/Astronomy: air shower array measurements of ultrahigh-energy cosmic rays, spectra and images of comets, stellar nuclear reaction rates, high redshift supernovae, cosmological parameters. (Garnavich, Howk, LoSecco, Rettig, Wiescher, 1 research associate professor, 1 research assistant professor, 2 adjunct professors, 1 postdoctoral research associate)
Atomic Physics: atomic structure, parity violation, tests of fundamental symmetries, excitation mechanisms, and radiative decays in neutral and ionized atoms; precision lifetimes. (Berry, Livingston, Tanner, 1 postdoctoral research associate)
Biological Physics and Molecular Environmental Science: structure and function in metalloproteins, metals in biological and geo-environmental systems. (Bunker)
Condensed-Matter Physics: low-temperature physics, superconducting microwave absorption, metal and semiconductor superlattices, magnetism, granular materials, magnetic resonance, magnetoelastic effects, high-temperature superconductivity, optical and far-infrared spectroscopy of semiconductors, crystal growth and MBE of semiconductors, magnetostatic effects, layered superconductors, single-electron tunneling, optical and infrared photoresponse, X-ray absorption spectroscopy and X-ray scattering, condensed-matter systems. (Blackstead, Bunker, Dobrowolska-Furdyna, Eskildsen, Furdyna, Ruggiero, 1 adjunct professor, 1 adjunct assistant professor, 1 research assistant professor, 1 concurrent professor, 2 postdoctoral research associates)
High-Energy Elementary Particle Physics: Fermilab D0 experiment (study of the top quark, bottom quark, W boson, and physics beyond the standard model), BaBar experiment at SLAC (CP violation in the b-quark system), CMS at CERN (search for the Higgs boson). (Cason, Goussiou, Hildreth, Jessop, LoSecco, Ruchti, Wayne, 1 research professor, 1 research assistant professor, 1 adjunct assistant professor, 1 adjunct assistant research professor, 1 professional specialist, 2 postdoctoral research associates, 1 guest assistant professor)
Nuclear Physics: nuclear structure, reaction energies, electromagnetic transitions, gamma-ray spectroscopy, high-spin states, polarized particles, giant resonances, heavy-ion reactions, radioactive beam studies, nuclear astrophysics. (Aprahamian, Collon, Garg, Kolata, Wiescher, 2 research professors, 1 research assistant professor, 1 visiting professor, 3 adjunct professors, 1 professional specialist, 1 assistant professional specialist, 1 research associate, 2 postdoctoral research associates, 1 visiting research professor, 1 visiting scholar)
Radiation Physics: low-energy electrons in condensed media; determination of product yields in the radiolysis of water, aqueous solutions, liquid hydrocarbons, and liquefied rare gases; diffusion-kinetic modeling of transient species produced by ionizing radiation. (LaVerne, Pimblott)

SELECTED PUBLICATIONS

Boutachkov, P., et al. **(A. Aprahamian).** Study of the low spin states of ^{208}Bi through γ–γ spectroscopy. *Nucl. Phys. A* 768:22–42, 2006.

Aprahamian, A., et al. Complete spectroscopy of the ^{162}Dy nucleus. *Nucl. Phys. A* 764:42–78, 2006.

Caamano, M., et al. **(A. Aprahamian).** Isomers in neutron-rich A~190 nuclides from ^{208}Pb fragmentation. *Eur. Phys. J. A—Hadrons Nuclei* 23:2, 201–15, 2005.

Boutachkov, P., et al. **(A. Aprahamian** and **J. J. Kolata).** Doppler shift as a tool for studies of isobaric analog states of neutron-rich nuclei: Application to 7He. *Phys. Rev. Lett.* 95:132502, 2005.

Boutachkov, P., et al. **(A. Aprahamian** and **J. J. Kolata).** Isobaric analog states of neutron-rich nuclei. Doppler shift as a measurement tool for resonance excitation functions. *Eur. Phys. J. A* 25:259–60, Suppl. 1, 2005.

Arnold, G. B., and R. A. Klemm. Theory of coherent c-axis Josephson tunneling between layered superconductors. *Phys. Rev. B* 62:661–70, 2000.

Klemm, R. A., and **G. B. Arnold.** Coherent versus incoherent c-axis Josephson tunneling beyond layered superconductors. *Phys. Rev. B* 58:14203–6, 1998.

MacLow, M. M., **D. S. Balsara,** M. de Avillez, and J. S. Kim. The distribution of pressures in a SN-driven interstellar medium I. Magnetized medium. *Astrophys. J.* 626:864–76, 2005.

Balsara, D. S. Second order accurate schemes for magnetohydrodynamics with divergence-free reconstruction. *Astrophys. J. Suppl.* 151(1):149–84, 2004.

Balsara, D. S., J. S. Kim, M. M. Mac Low, and G. J. Mathews. Amplification of magnetic fields in the multi-phase ISM with supernova-driven turbulence. *Astrophys. J.* 617:339–49, 2004.

Barabási, A.-L. Network theory–The emergence of creative enterprise. *Science* 308:639, 2005.

Barabási, A.-L. Taming complexity. *Nat. Phys.* 1:68–70, 2005.

Barabási, A.-L. *Linked: The New Science of Networks.* Cambridge: Perseus Publishing, 2002.

Berry, G., and B. Lin. A brief and personalized history of doubly-excited states. *Phys. Scr. T* 120:105–16, 2005.

Lin, B., et al. **(H. G. Berry** and **A. E. Livingston).** 1s2s2p^{2}3p ^{6}L-1s2p^{3}3p ^{6}P transitions in O IV, F V and Ne VI. *J. Phys. B: At., Mol. Opt. Phys.* 37:13, 2797–809, 2004.

Bigi, I. I., et al. Four-quark mesons in non-leptonic B decays—Could they resolve some old puzzles? *Phys. Rev. D* 72:114016, 2005.

Bigi, I. I., and A. I. Sanda. A 'known' CP asymmetry in tau decays. *Phys. Lett. B* 625:47, 2005.

Bigi, I. I., and A. I. Sanda. *CP Violation.* Cambridge University Press, 1999.

Yelon, W. B., et al. **(H. A. Blackstead).** Neutron diffraction studies of magnetic and superconducting compounds. *Phys. Status Solidi A* 201:1428–35, 2004.

Dow, J. D., **H. A. Blackstead,** and D. R. Harshman. High-temperature superconductivity: The roles of oxide layers. *Physica C* 364:74–8, 2001.

Blackstead, H. A., et al. Magnetically ordered Cu and Ru in $Ba_2GdRu_{1-u}Cu_uO_6$ and in $Sr_2YRu_{1-u}Cu_uO_6$. *Phys. Rev. B* 63(21): 214412, 1–11, 2001.

Robel, I., et al. **(B. A. Bunker).** Structural changes and catalytic activity of platinum nanoparticles supported on C60 and carbon nanotube films during the operation of direct methanol fuel cells. *Appl. Phys. Lett.* 88:073113-1–3, 2006.

Lahiri, D., et al. **(B. A. Bunker).** EXAFS studies of bimetallic AgPt and AgPd nanorods. *Phys. Scr. T* 15:776–80, 2005.

Lahiri, D., and **B. A. Bunker** et al. Bimetallic Pt-Ag and Pd-Ag nanoparticles. *J. Appl. Phys.* 97:094304, 2005.

Abazov, V. M., et al. **(N. M. Cason, A. Goussiou, M. D. Hildreth, R. Ruchti,** and **M. Wayne)** (Dzero Collaboration). Measurement of the lifetime difference in the B_s system. *Phys. Rev. Lett.* 95:171801, 2005.

Abazov, V. M., et al., **(N. M. Cason, A. Goussiou, M. D. Hildreth, R. Ruchti,** and **M. Wayne)** (Dzero Collaboration). Measurement of semileptonic branching fractions of B mesons to narrow D** states. *Phys. Rev. Lett.* 95:171803, 2005.

Abazov, V. M., et al. **(N. M. Cason, A. Goussiou, M. D. Hildreth, R. Ruchti,** and **M. Wayne)** (Dzero Collaboration). Search for large extra spatial dimensions in dimuon production at DZero. *Phys. Rev. Lett.* 95:161602, 2005.

Abazov, V. M., et al. **(N. M. Cason, A. Goussiou, M. D. Hildreth, R. Ruchti,** and **M. Wayne)** (Dzero Collaboration). Measurement of the $t\bar{t}$ production cross section in $\bar{p}p$ collisions at $\sqrt{s}$ =1.96 TeV in dilepton final states. *Phys. Lett. B* 626:55, 2005.

Abazov, V. M., et al. **(N. M. Cason, A. Goussiou, M. D. Hildreth, R. Ruchti,** and **M. Wayne)** (Dzero Collaboration). First measurement of $\sigma(\bar{p}p \to Z)$ * $Br(Z \to \tau\tau)$ at $\sqrt{s}$ =1.96 TeV. *Phys. Rev. D* 71:072004, 2005.

Abazov, V. M., et al. **(N. M. Cason, A. Goussiou, M. D. Hildreth, R. Ruchti,** and **M. Wayne)** (Dzero Collaboration). Search for Randall-Sundrum gravitons in dilepton and diphoton final states. *Phys. Rev. Lett.* 95:091,801, 2005.

Lu, M., and **N. M. Cason** et al. Exotic meson decay to $\omega\pi^0\pi^-$. *Phys. Rev. Lett.* 94:032002, 2005.

Kuhn, J., and **N. M. Cason** et al. Exotic meson production in the $f_1(1285)\pi^-$ system observed in the reaction $\pi^-p \to \eta\pi^+\pi^-\pi^-p$ at 18 GeV/cf. *Phys. Lett. B* 595:109–17, 2004.

Ivanov, I. E., and **N. M. Cason** et al. Observation of exotic meson production in the reaction $\pi^-p \to \eta'\pi^-p$ at 18 GeV/c. *Phys. Rev. Lett.* 86:3977–80, 2001.

Nassar, H., and **P. Collon** et al. Stellar (n, γ) cross section of ^{62}Ni. *Phys. Rev. Lett.* 94:092504, 2005.

Jiang, C. L., et al. **(P. Collon).** Hindrance of heavy-ion fusion at extreme sub-barrier energies in open-shell colliding systems. *Phys. Rev. C* 71:044613, 2005.

Collon, P., Z.-T. Lu, and W. Kutschera. Tracing noble gas radionuclides in the environment. *Ann. Rev. Nucl. Part. Sci.* 54:39–67, 2004.

Liu, X., et al. **(M. Dobrowolska** and **J. K. Furdyna).** Strain-engineered ferromagnetic $In_{1-x}Mn_xAs$ films with in-plane easy axis. *Appl. Phys. Lett.* 86:112512, 2005.

Liu, X., et al. **(M. Dobrowolska** and **J. K. Furdyna).** Ferromagnetic resonance study of the free-hole contribution to magnetization and magnetic anisotropy in modulation-doped Ga1-xMnxAs/Ga1-yAlyAs: Be. *Phys. Rev. B* 71(3):035307, 2005.

Lee, S., **M. Dobrowolska,** and **J. K. Furdyna.** Effect of spin-dependent Mn2+ internal transitions in CdSe/$Zn_{1-x}Mn_xSe$ magnetic semiconductor quantum dot systems. *Phys. Rev. B* 72(7):075320, 2005.

Cubitt, R., et al. **(M. R. Eskildsen).** Penetration depth anisotropy in MgB_2 single crystals and powders. *J. Phys. Chem. Solids* 67:493–6, 2006.

Jaiswal-Nagar, D., et al. **(M. R. Eskildsen).** dHvA oscillations, upper critical field and the peak effect studies in a single crystal of $LuNi_2B_2C$. *Physica B* 359–361:476–8, 2005.

Eskildsen, M. R., et al. Vortex imaging in magnesium diboride with H $\perp$ c. *Phys. Rev. B* 68:100508(R), 2003.

Frauendorf, S. Symmetries in nuclear structure. *Nucl. Phys. A* 752:203C–12C, 2005.

Afanasjev, A. V., and **S. Frauendorf.** Description of rotating N=Z nuclei in terms of isovector pairing. *Phys. Rev. C* 71:064318, 2005.

Zhu, S. J., et al. **(S. Frauendorf).** Soft ciral vibrations in ^{106}Mo. *Eur. Phys. J. A* 25:459–62, Suppl. 1, 2005.

Bao, J. M., A. V. Bragas, **J. K. Furdyna,** and R. Merlin. Control of spin dynamics with laser pulses: Generation of entangled states of donor-bound electrons in a $Cd_{1-x}Mn_xTe$ quantum well. *Phys. Rev. B* 71:045314, 2005.

Kutrowski, M., et al. **(J. K. Furdyna** and **M. Dobrowolska).** Observation of photoluminescence related to Lomer-Cottrell-like dislocations in ZnSe epilayers grown on in situ cleaved (110)GaAs surfaces. *J. Appl. Phys.* 97:013519, 2005.

Naguleswaran, S., et al. **(U. Garg).** Magnetic and intruder rotational bands in ^{113}In. *Phys. Rev. C* 72:044304, 2005.

Chakraborty, A., et al. **(U. Garg).** Spectroscopy of ^{90}Nb at high spin. *Phys. Rev. C* 72:054309, 2005.

Garg, U. The isoscalar giant dipole resonance: A status report. *Nucl. Phys. A* 731:3, 2004.

Gallagher, J. S., and **P. M. Garnavich** et al. Chemistry and star formation in the host galaxies of type Ia supernovae. *Astrophys. J.* 634:210, 2005.

Krisciunas, K., and **P. M. Garnavich** et al. Hubble Space Telescope observations of nine high-redshift ESSENCE supernovae. *Astron. J.* 130:2453, 2005.

Levan, A., et al. **(P. Garnavich).** GRB 020410: A gamma-ray burst afterglow discovered by its supernova light. *Astrophys. J.* 624:880, 2005.

University of Notre Dame

Kajino, T., et al. **(P. M. Garnavich** and **G. J. Mathews).** Dark matter and dark radiation in brane world cosmology and its observational test in the BBN, CMB and supernovae. *Nucl. Phys. B* 138:82–5, 2005.

Assmann, R., et al. **(M. Hildreth)** (LEP Energy Working Group). Calibration of centre-of-mass energies at LEP 2 for a precise measurement of the W boson mass. *Eur. Phys. J. C* 39:253, 2005.

Hildreth, M. D., et al. (DØ Collaboration). B physics at DØ. *Eur. Phys. J. C* 33:S192–4, 2004.

Abbiendi, G., et al. **(M. Hildreth)** (OPAL Collaboration). Searches for prompt light gravitino signatures in e^+e^- collisions at $\sqrt{s}$ = 189 GeV. *Phys. Lett. B* 501:12, 2001.

Howk, J. C., K. R. Sembach, and B. D. Savage. A method for deriving accurate gas-phase abundances for the multiphase interstellar galactic halo. *Astrophys. J.* 637:333–41, 2006.

Howk, J. C., A. M. Wolfe, and J. X. Prochaska. Cold neutral gas in a z = 4.2 damped Lyman-α system: The fuel for star formation. *Astrophys. J. Lett.* 622:L81, 2005.

Prochaska, J. X., T. M. Tripp, and **J. C. Howk.** Evidence of correlated titanium and deuterium depletion in the galactic interstellar medium. *Astrophys. J. Lett.* 620:L39, 2005.

Berciu, M., T. G. Rappoport, and **B. Jankó.** Manipulating spin and charge in magnetic semiconductors using superconducting vortices. *Nature* 435(7038): 71–5, 2005.

Zarand, G., P. M. Moca, and **B. Jankó.** Scaling theory of magnetoresistance in disordered local moment ferromagnets. *Phys. Rev. Lett.* 94:247202, 2005.

Csontos, M., et al. **(B. Jankó).** Pressure-induced ferromagnetism in (In,Mn)Sb dilute magnetic semiconductor. *Nat. Mater.* 4(6):447–9, 2005.

Aubert, B., et al. **(C. Jessop)** (BaBar Collaboration). Search for the radiative penguin decays B+ $\rightarrow$ rho+ gamma, B0 $\rightarrow$ rho0 gamma, and B0 $\rightarrow$ omega gamma. *Phys. Rev. Lett.* 94:011801, 2005.

Aubert, B., et al. **(C. Jessop)** (BaBar Collaboration). Search for the w-exchange decays B0 $\rightarrow$ D/s(*)- D/s(*)+. *Phys. Rev. D* 72:111101, 2005.

Aubert, B., et al. **(C. Jessop)** (BaBar Collaboration). Measurement of the branching ratios Gamma D/s*+ $\rightarrow$ D/s+ pi0) / Gamma D/s*+ $\rightarrow$ D/s+ gamma) and Gamma (D*0 $\rightarrow$ D0 pi0) / Gamma (D*0 $\rightarrow$ D0 gamma). *Phys. Rev. D* 72:091101, 2005.

Nilson, J., and **W. R. Johnson.** Plasma interferometry and how the bound-electron contribution can bend fringes in unexpected ways. *Appl. Opt.* 44:7295–301, 2005.

Lapierre, A., et al. **(W. R. Johnson).** Relativistic electron correlation, quantum electrodynamics and the lifetime of the P(1s2)(2s2)2p 2Po3/2 level in boronlike argon. *Phys. Rev. Lett.* 95:183001, 2005.

Dzuba, V. A., **W. R. Johnson,** and M. S. Safronova. Calculation of isotope shifts for cesium and francium. *Phys. Rev. A* 72:022503, 1–9, 2005.

Johnson, W. R., H. C. Ho, **C. E. Tanner,** and A. Derevianko. Off-diagonal hyperfine interaction between the $6p_{1/2}$ and $6p_{3/2}$ levels in ^{133}Cs. *Phys. Rev. A* 70:014501, 1–3, 2004.

Kolata, J. J. Direct neutron transfer in the $^{238}U(^{6}He$, fission) reaction near the Coulomb barrier. *Phys. Rev. C* 71:067603, 2005.

Kolda, C. Minimal flavor violation. *J. Korean Phys. Soc.* 45:S381, 2004.

Kane, G., et al. **(C. Kolda).** B(D) $\rightarrow$ ϕK(S) and supersymmetry. *Phys. Rev. D* 70:035015, 2004.

Babu, K., and **C. F. Kolda.** Higgs-mediated $\tau \rightarrow 3\mu$ in the supersymmetric seesaw model. *Phys. Rev. Lett.* 89:241802, 2002.

Kukla, K. W., et al. **(A. E. Livingston** and **H. G. Berry).** Extreme-ultraviolet wavelength and lifetime measurements in highly-ionized krypton. *Can. J. Phys.* 83:1–16, 2005.

Feili, D., et al. **(A. E. Livingston).** $2s^2\ {}^1S_0$–$2s2p\ {}^3P_1$ intercombination transition wavelengths in Be-like Ag^{43+}, Sn^{46+}, and Xe^{50+} ions. *Phys. Scr.* 71:48–51, 2005.

Lu, M., and **J. M. LoSecco** et al. Exotic meson decay to $\omega\pi^0\pi^-$. *Phys. Rev. Lett.* 94:032002, 2005.

LoSecco, J. M. Detector depth dependence of the high energy atmospheric neutrino flux. *Phys. Rev. D* 70:097301, 2004.

Kuhn, J., et al. **(J. M. LoSecco).** Exotic meson production in the f1(1285) pi-system observed in the reaction pi- p $\rightarrow$ eta pi+ pi- pi- p at 18-Ge V/c. *Phys. Lett. B* 595:109, 2004.

Mathews, G. J., T. Kajino, and T. Shima. Big bang nucleosynthesis with a new neutron lifetime. *Phys. Rev. D* 71:021302, 2005.

Wilson, J. R., and **G. J. Mathews.** White dwarfs near black holes: A new paradigm for Type I supernovae. *Astrophys. J.* 610:368–77, 2004.

Vandeworp, E. M., and **K. E. Newman.** Coherent alloy separation: Differences in canonical and grand canonical ensembles. *Phys. Rev. B* 55:14222–9, 1997.

Weidmann, M. R., and **K. E. Newman.** Effects of site correlations on the local structure of strain-relaxed semiconductor alloys. *Phys. Rev. B* 51:4962–81, 1995.

Cohen, R. J., and **K. E. Newman.** Commensurate and incommensurate phases of epitaxial semiconductor antiferromagnets with "built-in" strain. *Phys. Rev. B* 46:14282, 1992.

Rettig, T. W., et al. CO emission and absorption toward V1647 Ori (McNeil's Nebula). *Astrophys. J.* 626:245, 2005.

Rettig, T., et al. Discovery of CO gas in the inner disk of TW Hydrae. *Astrophys. J.* 616, 2004.

Brittain, S., T. Simon, C. Kulesa, and **T. Rettig.** Interstellar H_3^+ line absorption toward LkHα101. *Astrophys. J.* 606:911, 2004.

Clark, A. M., et al. **(S. T. Ruggiero).** Cooling of bulk material by electron-tunneling refrigerators. *Appl. Phys. Lett.* 86:173508, 2005.

Clark, A. M., et al. **(S. T. Ruggiero).** Practical tunneling refrigerator. *Appl. Phys. Lett.* 84:625–7, 2004.

Ruggiero, S. T., et al. Magneto-optic effects in spin-injection devices. *Appl. Phys. Lett.* 82:4599–601, 2003.

Sapirstein, J. Accurate S-state helium wave functions in momentum space. *Phys. Rev. A* 69:042515, 2004.

Sapirstein, J., K. Pachucki, and K. T. Cheng. Radiative corrections to one-photon decays of hydrogenic ions. *Phys. Rev. A* 69:022113, 2004.

Sapirstein, J., and K. T. Cheng. Calculation of radiative corrections to hyperfine splittings in the neutral alkalis. *Phys. Rev. A* 67:022512, 2003.

Gerginov, V., and **C. E. Tanner** et al. High resolution spectroscopy with a femtosecond laser frequency comb. *Opt. Lett.* 30, 2005.

Gerginov, V., and **C. E. Tanner** et al. Optical frequency measurements of 6s $^2S_{1/2}$-6p $^2P_{3/2}$ transitions in a ^{133}Cs atomic beam using a femtosecond laser frequency comb. *Phys. Rev. A* 70:042505, 2004.

Toroczkai, Z., and K. E. Bassler. Network dynamics: Jamming is limited in scale-free systems. *Nature* 428:716, 2004.

Eubank, S., et al. **(Z. Toroczkai).** Controlling epidemics in realistic urban social networks. *Nature* 429:180, 2004.

Tél, T., et al. **(Z. Toroczkai).** Universality in active chaos. *Chaos* 14:72, 2004.

Karakas, A., et al. **(M. Wiescher).** The uncertainties in the ^{22}Ne + alpha-capture reaction rates and the production of the heavy magnesium isotopes in asymptotic giant branch stars of intermediate mass. *Astrophys. J.* 643:471–83, 2006.

Marrone, S., and **M. Wiescher** et al. Measurement of the Sm151(n,γ) cross section from 0.6 eV to 1 MeV via the neutron time-of-flight technique at the CERN n_TOF facility. *Phys. Rev. C* 73:034604, 2006.

Sun, Y., **M. Wiescher, A. Aprahamian,** and J. Fisker. Nuclear structure of the exotic mass region along the rp process path. *Nucl. Phys. A* 758:765c–8c, 2005.

UNIVERSITY OF PITTSBURGH

Department of Physics and Astronomy

Programs of Study

The graduate programs in the Department of Physics and Astronomy are designed primarily for students who wish to obtain the Ph.D. degree, although the M.S. degree is also offered. The Ph.D. program provides high-quality training for students without needlessly emphasizing formal requirements. Upon arrival, each graduate student is appointed a faculty adviser to provide personalized guidance through the core curriculum. A set of basic courses is to be taken by all graduate students unless the equivalent material has been demonstrably mastered in other ways. These basic courses include dynamical systems, quantum mechanics, electromagnetic theory, and statistical physics. More advanced and special-topics courses are offered in a range of areas, including, but not limited to, high-energy and particle physics, condensed matter, statistical and solid-state physics, astrophysics, astronomy, and relativity.

Students have a wide variety of programs from which to choose a thesis topic. University faculty members have active research programs in atomic, molecular, and optical physics; astrophysics and cosmology employing ground- and space-based telescopes; particle astrophysics; condensed-matter and solid-state physics; particle physics; biological physics; chemical physics; quantum information; general relativity; and physics education.

Interdisciplinary research programs may be arranged on a case-by-case basis. There have been physics doctorates awarded for the work done in collaboration with the faculty members in the Department of Biological Sciences, the Department of Chemistry, the Department of Mathematics, the Department of Materials Science, the Departments of Electrical and Chemical Engineering, the Department of Radiological Sciences, and the Department of Radiology in the School of Medicine, among others.

Research Facilities

The department's facilities include the physics library, an electronics shop, a glassblowing shop, a professionally staffed machine shop, and extensive departmental and University computer resources. Departmental students have easy access to the facilities and expertise available at the Pittsburgh Supercomputing Center (PSC). Other facilities include the Allegheny Observatory (for positional astronomy). Experiments in particle physics are carried out at such national and international facilities as Fermi National Laboratory in Chicago and CERN in Switzerland. Similarly, programs are conducted at national and international observatories, such as at Kitt Peak and Mount Hopkins, Arizona; at Cerro Tololo, Las Campanas, and La Silla in Chile; at Apache Point in New Mexico for collection of Sloan Digital Sky Survey data; at Mauna Kea in Hawaii; and on the Hubble Space Telescope, the Chandra X-ray Observatory, and other space observatories.

Financial Aid

Financial aid is normally provided through teaching assistantships during the first year and through research assistantships thereafter. The University provides individual health insurance under the Graduate Student Plan. The department has several fellowships established for entering graduate students. They are awarded on a competitive basis, with all qualified applicants automatically entered into a pool. Some University fellowships are also available and are awarded in a University-wide competition. Students are generally supported throughout their entire graduate career, provided good academic standing is maintained. Teaching and research assistantship appointments carried a stipend of $6777.50 per term in 2005–06, plus a tuition scholarship, bringing the annual stipend to $20,332.50 for students supported throughout the year. Research assistantship appointments may be held in connection with most of the department's research programs.

Cost of Study

For full-time students who are not Pennsylvania residents, tuition and fees per term in 2005–06 were $12,776. Part-time students paid $1026 per credit plus fees. Full-time students who are Pennsylvania residents paid $6867 per term, including fees, and part-time students who are Pennsylvania residents paid $537 per credit plus fees.

Living and Housing Costs

Most University of Pittsburgh students live in rooms or apartments in the Oakland area. The typical cost of rooms or apartments ranges from $340 to $550 per month for housing. Meals range from $300 to $400 per month.

Student Group

The department's graduate student body in 2005–06 consisted of 90 students; 85 students received financial support. These figures are typical of the department's graduate enrollment.

Student Outcomes

Many Ph.D. graduates accept postdoctoral positions at major research universities, often leading to teaching and research positions at outstanding universities in the United States and around the world. Other recent graduates have entered research careers in the private sector. One former graduate received the American Physical Society's Nicholas Metropolis Award for Outstanding Doctoral Thesis Work in Computational Physics.

Location

Pittsburgh is situated in a hilly and wooded region of western Pennsylvania where the Allegheny and Monongahela Rivers join to form the Ohio. The region has a natural beauty. The terrain of western Pennsylvania and nearby West Virginia is excellent for outdoor activities, including cycling, hiking, downhill and cross-country skiing, white-water rafting and kayaking, rock climbing, hunting, and fishing. The University is located about 3 miles east of downtown Pittsburgh in the city's cultural center. Adjacent to the campus are Carnegie Mellon University and the Carnegie, comprising the Museum of Art, the Museum of Natural History, the Carnegie Library, and the Carnegie Music Hall. Schenley Park adjoins the campus; it has picnic areas, playing fields, jogging trails, and an excellent botanical conservatory. The Pittsburgh area has several professional sports teams; for detailed information, students should visit http://www.phyast.pitt.edu/Resources/Visiting/PITT_INFO.htm.

The Department

The department has long been active in research and has trained more than 500 recipients of the Ph.D. degree. Close cooperation exists between this department and the physics department of Carnegie Mellon University; all seminars, colloquia, and courses are shared. The graduate students of both institutions benefit from belonging to one of the largest communities of active physicists in the country. Furthermore, basic research, conducted at the University of Pittsburgh Medical Center and the School of Medicine, provides additional opportunities for research with multidisciplinary perspectives.

Applying

Students who wish to apply for admission or financial aid should apply online and take the GRE, including the Subject Test in physics. Applicants should request that the registrars of their undergraduate and graduate schools send transcripts of their records to the department. Three letters of recommendation are required for admission with aid. Unless English is the applicant's native language, the TOEFL is required, except in cases in which an international applicant has received an advanced degree from a U.S. institution. The application deadline is January 31. Late applications are accepted on the basis of space availability.

Correspondence and Information

Professor H. E. Anthony Duncan
Admissions Committee
Department of Physics and Astronomy
University of Pittsburgh
Pittsburgh, Pennsylvania 15260
Phone: 412-624-9066
Web site: http://www.phyast.pitt.edu/

University of Pittsburgh

THE FACULTY AND THEIR RESEARCH

Joseph Boudreau, Associate Professor; Ph.D., Wisconsin. Experimental particle physics.
Daniel Boyanovsky, Professor; Ph.D., California, Santa Barbara. Condensed-matter physics, particle astrophysics.
Wolfgang J. Choyke, Research Professor; Ph.D., Ohio State. Solid-state physics, defect states in semiconductors, large bandgap spectroscopy.
Russell Clark, Lecturer/Lab Supervisor; Ph.D., LSU.
Rob Coalson, Professor; Ph.D., Harvard. Chemical physics.
Andrew Connolly, Associate Professor; Ph.D., Imperial College (London). Astrophysics, extragalactic astronomy, observational cosmology.
Robert P. Devaty, Associate Professor; Ph.D., Cornell. Experimental solid-state physics.
H. E. Anthony Duncan, Professor and Chairperson of the Admissions Committee; Ph.D., MIT. Theoretical high-energy physics.
Steven A. Dytman, Professor; Ph.D., Carnegie Mellon. Particle physics.
George D. Gatewood, Professor; Ph.D., Pittsburgh. Astronomy, astrometry, search for planetary systems orbiting neighboring stars.
Yadin Y. Goldschmidt, Professor; Ph.D., Hebrew (Jerusalem). Condensed-matter theory, statistical mechanics.
Albert Heberle, Associate Professor; Ph.D., Stuttgart (Germany). Experimental condensed matter.
D. John Hillier, Professor; Ph.D., Australian National. Theoretical and observational astrophysics, computational physics.
David M. Jasnow, Professor and Department Chair; Ph.D., Illinois. Theory of phase transitions, statistical mechanics.
Rainer Johnsen, Professor; Ph.D., Kiel (Germany). Experimental atomic and plasma physics.
Peter F. M. Koehler, Professor and Graduate Director; Ph.D., Rochester. Experimental high-energy physics.
Arthur Kosowsky, Associate Professor; Ph.D., Chicago. Theoretical and experimental cosmology and astrophysics.
Adam Leibovich, Assistant Professor; Ph.D., Caltech. Theoretical elementary particle physics.
Jeremy Levy, Professor; Ph.D., California, Santa Barbara. Experimental condensed matter.
W. Vincent Liu, Assistant Professor; Ph.D., Texas at Austin. Theoretical condensed matter.
James V. Maher, Professor and Provost; Ph.D., Yale. Experimental solid-state physics, critical phenomena, physics of fluids.
James Mueller, Associate Professor and Undergraduate Director; Ph.D., Cornell. Particle physics.
Donna Naples, Associate Professor; Ph.D., Maryland. Experimental high-energy physics.
Vittorio Paolone, Associate Professor; Ph.D., California, Davis. Experimental high-energy physics.
Hrvoje Petek, Professor; Ph.D., Berkeley. Experimental condensed matter.
Sandhya Rao, Research Assistant Professor; Ph.D., Pittsburgh. Astrophysics, extragalactic astronomy, observational cosmology.
Ralph Z. Roskies, Professor; Ph.D., Princeton. Theoretical high-energy physics, use of computer in theoretical physics.
Vladimir Savinov, Assistant Professor; Ph.D., Minnesota. Experimental intermediate-energy physics.
Regina E. Schulte-Ladbeck, Professor and Associate Dean, College of Arts and Sciences; Ph.D., Heidelberg. Astrophysics.
Paul F. Shepard, Professor; Ph.D., Princeton. Experimental high-energy physics.
Chandralekha Singh, Associate Professor; Ph.D., California, Santa Barbara. Physics education research, polymer physics.
David Snoke, Associate Professor; Ph.D., Illinois at Urbana-Champaign. Solid-state experimental.
G. Alec Stewart, Associate Professor and Dean, University Honors College; Ph.D., Washington (Seattle). Experimental solid-state physics.
Eric Swanson, Assistant Professor; Ph.D., Toronto. Theoretical intermediate-energy physics.
Frank Tabakin, Professor; Ph.D., MIT. Theoretical nuclear physics.
David A. Turnshek, Professor; Ph.D., Arizona. Astrophysics, extragalactic astronomy, observational cosmology.
Xiao-Lun Wu, Professor; Ph.D., Cornell. Experimental condensed matter, biophysics.
John T. Yates, Professor; Ph.D., MIT. Physical chemistry.

EMERITUS FACULTY

Elizabeth U. Baranger, Professor; Ph.D., Cornell. Theoretical nuclear physics.
James E. Bayfield, Professor; Ph.D., Yale. Experimental atomic physics and quantum optics.
Manfred A. Biondi, Professor; Ph.D., MIT. Experimental atomic physics and astronomy.
Wilfred W. Cleland, Professor; Ph.D., Yale. Experimental high-energy physics.
Bernard L. Cohen, Professor; Ph.D., Carnegie Mellon. Energy and environment.
Richard M. Drisko, Professor; Ph.D., Carnegie Mellon. Theoretical nuclear physics.
Eugene Engels Jr., Professor; Ph.D., Princeton. Experimental high-energy physics.
Myron P. Garfunkel, Professor; Ph.D., Rutgers. Experimental low-temperature physics, superconductivity.
Edward Gerjuoy, Professor; Ph.D., Berkeley. Theoretical atomic physics.
Walter I. Goldburg, Professor; Ph.D., Duke. Experimental solid-state physics, phase transitions, light scattering, turbulence.
Cyril Hazard, Professor; Ph.D., Manchester. Astrophysics, extragalactic astronomy, observational cosmology.
Allen I. Janis, Professor; Ph.D., Syracuse. General relativity, philosophy of science.
Irving J. Lowe, Professor; Ph.D., Washington (St. Louis). Experimental solid-state physics, nuclear magnetic resonance, nuclear magnetic resonance imaging.
Ezra T. Newman, Professor; Ph.D., Syracuse. General relativity, twistor theory.
Richard H. Pratt, Professor; Ph.D., Chicago. Theoretical atomic and low-energy particle physics, bremsstrahlung, hot plasma processes, photon scattering.
Juerg X. Saladin, Professor; Ph.D., Swiss Federal Institute of Technology. Experimental nuclear physics.
C. Martin Vincent, Professor; Ph.D., Witwatersrand (South Africa). Theoretical intermediate-energy physics.
Raymond S. Willey, Professor; Ph.D., Stanford. Theoretical high-energy physics.
Jeffrey Winicour, Research Professor; Ph.D., Syracuse. General relativity.
Edward C. Zipf, Professor; Ph.D., Johns Hopkins. Experimental atomic and atmospheric physics.

Allen Hall, home to the department.

M51, the Whirlpool Galaxy, a spiral galaxy in the constellation of Canes Venatici.

Collider detector at Fermilab.

UNIVERSITY OF ROCHESTER

Department of Physics and Astronomy

Programs of Study

The Department offers programs of study leading to the Ph.D. degree in physics or physics and astronomy. Students normally earn the M.A. or M.S. degree in physics en route to the Ph.D. The M.A. is awarded after the completion of 30 semester hours of course work and a comprehensive examination; the M.S. degree in physics requires a thesis in addition to the course work. Students are not usually admitted to work toward a master's degree unless they intend to obtain a Ph.D. (exceptions include students in the 3-2 programs in physics and medical physics).

Most candidates for the Ph.D. degree take two years of course work and a written preliminary examination during their second year. Requirements for the Ph.D. include demonstrating competence in quantum mechanics, mathematical methods, electromagnetic theory, and statistical physics, as well as in an advanced area of specialization. A doctoral thesis, based on a significant piece of original research, and a final oral thesis defense are required of all Ph.D. candidates. A typical program of study takes five or six years to complete. A minor is not required, although students are encouraged to broaden their understanding of other subfields of physics or astronomy beyond the area of their thesis research.

The Department provides opportunities for research in observational astronomy, theoretical and laboratory astrophysics, biological and medical physics, and experimental and theoretical areas of condensed-matter physics, chemical physics, cross-disciplinary physics, elementary particle physics, nuclear physics, plasma physics, quantum optics, and atomic, molecular, and optical physics.

Research Facilities

The infrared astronomy group has extensive programs in the development of advanced detector arrays, electronics, and instrumentation for astronomy. The results are in use on several ground-based observatories in Arizona and Hawaii and in space on the NASA Spitzer Space Telescope. The high-energy densities that can be produced by the Omega laser at the Laboratory for Laser Energetics offer unique opportunities for the laboratory study of matter under conditions ordinarily associated with high-energy astrophysical phenomena such as supernova blasts. The atomic molecular and optical physics group offers extensive facilities, including a broad range of ultraviolet, visible, and infrared lasers; ultrahigh-stability CW dye, solid-state, and diode laser systems; ultrafast lasers (femtosecond); and an ultrahigh-power (psec, chirped-pulse, regeneratively amplified) pulsed solid-state laser system. The group's capabilities include sophisticated photon-counting, laser cooling and trapping, atomic beam, and laser physics experimentation. For research in condensed matter, the Department offers a unique magnetooptical spectroscopy lab and an advanced surface science research lab that is equipped with X-ray, ultraviolet, and inverse photoemission spectroscopy; scanning-tunneling, atomic-force, and near-field microscopy; low-energy electron and photoelectron diffraction facilities; and advanced thin-film deposition systems. Research in high-energy and nuclear physics includes projects to develop state-of-the-art detectors for application in high-energy and nuclear physics, which are designed and tested on campus and then assembled and operated at international and national laboratories, including CERN (CMS), Fermilab (Dzero, CDF, NuTeV, and MINERvA), Brookhaven (PHOBOS/RHIC), the Stanford Linear Accelerator Center, Jefferson Laboratory, Lawrence Berkeley Laboratory, Argonne, KEK/JPARC (Japan), the Boulby Mine (UK), and Wilson Lab at Ithaca, New York. The Physics-Optics-Astronomy Library, within the physics building, provides ready access to more than 225 journals. The facilities of the University's Laboratory for Laser Energetics, the Institute of Optics, and the Center for Optoelectronics and Imaging are also available for collaborative efforts.

Financial Aid

In 2006–07, graduate teaching and research assistantships, which require 16 hours of work per week during the academic year, carry stipends of $16,380 for nine months. Additional support is available for participation in summer research. A graduate assistant who also takes part in full-time summer research receives a total of $21,840 for the calendar year. A few special University and departmental fellowships provide stipends of up to $23,840 for the calendar year. In addition, the Department of Education GAANN, Robert E. Marshak Fellowships, Provost Fellowships, and Sproull Fellowships for academic excellence are available to supplement teaching or research assistantships for outstanding students.

Cost of Study

For students with fewer than 90 credit hours of accumulated graduate course work, tuition for the 2006–07 academic year is $32,640. Tuition for more advanced students is $1528–$1632 per academic year. (Currently, all graduate students in the Department receive special awards to cover tuition.) All full-time graduate students are charged a health service fee of $996 per year. The cost of books and supplies is about $600 per year.

Living and Housing Costs

The cost of living in Rochester is among the lowest for metropolitan areas. Supermarkets with moderate prices are located near the University, or meals can be obtained on campus. University-operated housing for either single or married graduate students is available within easy walking distance of the campus. Free shuttle-bus service is available within the University complex. Additional privately owned rooms and apartments are available in the residential areas near the University.

Student Group

There are approximately 120 graduate students in physics and astronomy; about 18 percent are women, and about 28 percent of the students are married. All full-time students receive some form of financial aid. Admission to graduate study is highly competitive, with only about 20 new students admitted each year; about 50 percent have undergraduate degrees from institutions outside the United States.

Location

With approximately 735,500 inhabitants, the Rochester metropolitan area is the third largest in the state. A city with its economy based on high-technology industries, it is located on the southern shore of Lake Ontario. Niagara Falls, the scenic Finger Lakes district, and the rugged Adirondack Mountains are all within a few hours' drive. The Rochester Philharmonic Orchestra and the Rochester Americans ice-hockey team provide two examples of the range of experiences available. Rochester is readily accessible both by air and by car.

The University and The Department

The University of Rochester is a private institution with 4,726 undergraduates, 3,590 graduate students, and 1,264 faculty members. The Department of Physics and Astronomy, one of the largest and strongest departments within the University, has a reputation for excellence in graduate education and research spanning more than fifty years. Many faculty members of the Department have been awarded major fellowships and prizes in recognition of their research accomplishments.

Applying

Catalogs and application information can be obtained from the Department's Web site. Students are admitted only in September, and completed applications should be received by January 15 in order for applicants to be considered for financial aid. Applicants should take the GRE General Test and physics Subject Test in time for scores to arrive by January 15. TOEFL scores are required of international students whose native language is not English.

Correspondence and Information

Graduate Student Counselor
Department of Physics and Astronomy
University of Rochester
Rochester, New York 14627

Phone: 585-275-4356
Web site: http://www.pas.rochester.edu

University of Rochester

THE FACULTY AND THEIR RESEARCH

G. P. Agrawal, Professor; Ph.D., Indian Institute of Technology (New Delhi), 1974. Fiber optics, lasers, optical communications.

R. Betti, Professor; Ph.D., MIT, 1992. Theoretical plasma physics, nuclear and mechanical engineering, computational and plasma physics.

N. P. Bigelow, Lee A. Du Bridge Professor; Ph.D., Cornell, 1989. Experimental and theoretical quantum optics and quantum physics, studies of BEC and laser-cooled and trapped atoms.

E. G. Blackman, Professor; Ph.D., Harvard, 1995. Theoretical astrophysics, astrophysical plasmas and magnetic fields, accretion and ejection phenomena, relativistic and high-energy astrophysics.

M. F. Bocko, Professor; Ph.D., Rochester, 1984. Superconducting electronics, quantum computing, musical acoustics, digital audio technology, sensors.

A. Bodek, Professor and Chair; Ph.D., MIT, 1972. Experimental elementary particle physics, proton-antiproton collisions, QCD and structure functions, neutrino physics, electron scattering, tile-fiber calorimetric detectors.

R. W. Boyd, M. Parker Givens Professor; Ph.D., Berkeley, 1977. Nonlinear optics.

T. G. Castner, Professor Emeritus; Ph.D., Illinois, 1958. Experimental condensed-matter physics, metal insulator transition.

D. Cline, Professor; Ph.D., Manchester (England), 1963. Extreme states of nuclei pairing and shape correlations in nuclei.

E. Conwell, Professor; Ph.D., Chicago, 1948. Theoretical chemical physics, condensed-matter physics, biological physics.

A. Das, Professor; Ph.D., SUNY at Stony Brook, 1977. Theoretical particle physics, finite temperature field theory, integrable systems, phenomenology, noncommutative field theory and string/M theory.

R. Demina, Associate Professor; Ph.D., Northeastern, 1994. Experimental particle physics, proton-antiproton collisions, top and electroweak physics.

D. H. Douglass, Professor; Ph.D., MIT, 1959. Experimental condensed-matter physics, climate change and pollution.

J. H. Eberly, Andrew Carnegie Professor; Ph.D., Stanford, 1962. Theoretical quantum optics, quantum entanglement, cavity QED, atoms in strong laser fields, dark-state optical control theory.

T. Ferbel, Professor; Ph.D., Yale, 1963. Experimental elementary particle physics, studies of the top quark in hadronic collisions.

W. J. Forrest, Professor and Director, C. E. Kenneth Mees Observatory; Ph.D., California, San Diego, 1974. Observational astrophysics, infrared astronomy, stellar and planetary formation, low-mass stars and brown dwarfs, development of infrared detector arrays and instrumentation.

T. H. Foster, Professor; Ph.D., Rochester, 1990. Biological and medical physics.

A. Frank, Professor; Ph.D., Washington (Seattle), 1992. Theoretical astrophysics, astrophysical plasmas, numerical hydrodynamics and magnetohydrodynamics.

H. W. Fulbright, Professor Emeritus; Ph.D., Washington (St. Louis), 1944. Experimental nuclear physics, radio astronomy, phenomenology of strong interactions.

Y. Gao, Professor; Ph.D., Purdue, 1986. Experimental condensed-matter physics, surface physics.

H. E. Gove, Professor Emeritus; Ph.D., MIT, 1950. Experimental nuclear physics, heavy ions, accelerator mass spectrometry.

C. R. Hagen, Professor; Ph.D., MIT, 1962. Theoretical elementary particle physics; quantum field theory, particularly 2+1 dimensional theories.

H. L. Helfer, Professor Emeritus; Ph.D., Chicago, 1953. Theoretical astrophysics and plasma physics, high-energy astrophysics, dark matter in galactic haloes.

J. Howell, Assistant Professor; Ph.D., Penn State, 2000. Experimental quantum optics and quantum physics, quantum cryptography and quantum computation.

R. S. Knox, Professor Emeritus; Ph.D., Rochester, 1958. Theoretical biological physics and condensed-matter physics, energy-balance models of climate.

D. S. Koltun, Professor Emeritus; Ph.D., Princeton, 1961. Theoretical nuclear physics, meson interactions with nuclei, many-body theory, electron scattering.

S. L. Manly, Mercer Brugler Professor; Ph.D., Columbia, 1989. Experimental relativistic heavy-ion physics, experimental elementary particle physics.

R. L. McCrory, Professor and Director of the Laboratory for Laser Energetics; Ph.D., MIT, 1973. Nuclear and mechanical engineering, computational hydrodynamics.

K. S. McFarland, Professor; Ph.D., Chicago, 1994. Experimental elementary particle physics: properties of top quarks, neutrino physics, electroweak unification.

A. C. Melissinos, Professor; Ph.D., MIT, 1958. Experimental particle physics, high-intensity laser-particle interactions, free-electron lasers, searches for relic gravitational radiation.

D. D. Meyerhofer, Professor; Ph.D., Princeton, 1987. Experimental plasma and laser physics, high-energy-density physics and inertial confinement fusion, high-intensity laser-matter interaction experiments, quantum optics.

L. Novotny, Associate Professor; Dr. sci. techn., ETH Zurich, 1996. Nano-optics, nanoscale phenomena, biophysics.

S. Okubo, Professor Emeritus; Ph.D., Rochester, 1958. Theoretical particle physics and mathematical physics, Lie and nonassociative algebras.

L. Orr, Professor; Ph.D., Chicago, 1991. Theoretical elementary particle physics, phenomenology, quantum chromodynamics and electroweak physics.

J. L. Pipher, Professor Emeritus; Ph.D., Cornell, 1971. Observational astrophysics, infrared astronomy, galactic and extragalactic star formation, low-mass stars and brown dwarfs, development of infrared detector arrays and instrumentation.

A. Quillen, Assistant Professor; Ph.D., Caltech, 1993. Observational astrophysics, galactic structure and dynamics, active galactic nuclei, dynamics of planetary and protoplanetary systems.

S. G. Rajeev, Professor; Ph.D., Syracuse, 1984. Theoretical particle physics, nonperturbative quantum field theory applied to strong interactions.

C. Ren, Assistant Professor; Ph.D., Wisconsin–Madison, 1998. Theoretical and computational plasma physics, controlled fusion.

L. Rothberg, Professor; Ph.D., Harvard, 1983. Experimental chemical physics, organic electronics and biomolecular sensing.

M. P. Savedoff, Professor Emeritus; Ph.D., Princeton, 1957. Theoretical astrophysics, stellar interiors, interstellar matter, high-energy astrophysics.

W.-U. Schröder, Professor; Ph.D., Darmstadt (Germany), 1971. Experimental nuclear physics, dynamics of complex nuclear reactions, fundamental properties of nuclear matter, nuclear transmutation, nuclear technology applications.

Y. Shapir, Professor; Ph.D., Tel-Aviv, 1981. Theoretical condensed-matter physics, statistical mechanics, critical phenomena in ordered and disordered systems, fractal growth.

A. Simon, Professor; Ph.D., Rochester, 1950. Theoretical plasma physics, controlled thermonuclear fusion.

P. F. Slattery, Professor; Ph.D., Yale, 1967. Experimental elementary particle physics, investigation of QCD via direct photon production, top quark studies and searches for new phenomena using high-energy colliders.

R. L. Sproull, Professor Emeritus; Ph.D., Cornell, 1943. Experimental condensed-matter physics.

C. R. Stroud Jr., Professor; Ph.D., Washington (St. Louis), 1969. Quantum optics, short-pulse excitation of atoms and molecules, quantum information.

J. A. Tarduno, Professor; Ph.D., Stanford, 1987. Geophysics, geomagnetism and geodynamics, plate tectonics and polar wander, geomagnetic reversals, fine-particle magnetism, planetary astrophysics.

S. L. Teitel, Professor; Ph.D., Cornell, 1981. Statistical and condensed-matter physics.

J. H. Thomas, Professor; Ph.D., Purdue, 1966. Theoretical astrophysics, astrophysical plasmas, astrophysical fluid dynamics and magnetohydrodynamics, solar physics.

E. H. Thorndike, Professor; Ph.D., Harvard, 1960. Experimental elementary particle physics, weak decays of bottom and charm quarks.

P. L. Tipton, Professor; Ph.D., Rochester, 1987. Experimental elementary particle physics, production and decay of top and b quarks in proton-antiproton collisions at 1.8 TeV.

H. M. Van Horn, Professor Emeritus; Ph.D., Cornell, 1965. Theoretical astrophysics, degenerate stars.

D. M. Watson, Professor; Ph.D., Berkeley, 1983. Observational astrophysics, infrared astronomy, stellar and planetary formation, low-mass stars and brown dwarfs, development of infrared detector arrays and instrumentation.

E. Wolf, Wilson Professor; Ph.D., Bristol (England), 1948. Theoretical optics, statistical optics, theory of coherence and polarization, inverse scattering, diffraction tomography.

F. L. H. Wolfs, Professor; Ph.D., Chicago, 1987. Experimental high-energy/nuclear physics, relativistic heavy-ion physics, dark-matter searches.

J. Zhong, Professor; Ph.D., Brown, 1988. Biological and medical physics, advanced medical imaging, novel MRI techniques, physiological properties, biological tissues.

RECENT FACULTY PUBLICATIONS

Usechak, N. G., and **G. P. Agrawal.** A semi-analytic technique for analyzing mode-locked lasers. *Opt. Exp.* 13:2075–81, 2005.

Tscherneck, M., et al. **(N. P. Bigelow).** Creating, detecting and locating ultracold molecules in a surface trap. *Appl. Phys. B* 80(6):639-43, 2005.

Ohki, A., J. L. Habif, M. J. Feldman, and **M. F. Bocko.** Thermal design of superconducting digital circuits for milliKelvin operation. *IEEE Trans. Appl. Superconductivity* 13:078, 2003.

Bodek, A., and U. K. Yang. Modeling neutrino and electron scattering cross-sections in the few GeV region with effective LO PDFs. *AIP Conf. Proc.* 670:110, 2003.

O'Sullivan-Hale, M. N., I. Ali Khan, **R. W. Boyd,** and **J. C. Howell.** Pixel entanglement: Experimental realization of optically entangled d=3 and d=6 qudits. *Phys. Rev. Lett.* 94:220501, 2005.

Das, A., J. Gamboa, J. Lopez-Sarrion, and F. A. Schaposnik. Gauge field theory in the infrared regime. *Phys. Rev. D* 72:107702, 2005.

Demina, R., et al. A quasi-model-independent search for new high pT physics at DZero. *Phys. Rev. Lett.* 86:3712, 2001.

Douglass, D. H., V. Patel, and **R. S. Knox.** Iceland as a heat island. *Geophys. Res. Lett.* 32:L03709, 2005.

Phay, J., et al. **(J. H. Eberly).** Non-sequential double ionization is a completely classical photoelectric effect. *Phys. Rev. Lett.* 94:093002, 2005.

Ferbel, T., et al. (D0 collaboration). Helicity of the W boson in lepton + jets ttbar events. *Phys. Lett. B* 617:1, 2005.

D'Alessio, P., et al. **(W. J. Forrest** and **D. M. Watson).** The truncated disk of CoKu Tau/4. *Astrophys. J.* 621:461, 2005.

Finlay, J. C., and **T. H. Foster.** Recovery of hemoglobin oxygen saturation and intrinsic fluorescence using a forward adjoint model. *Appl. Opt.* 44:1917–33, 2005.

Frank, A., and **E. G. Blackman** et al. A HED laboratory astrophysics testbed comes of age: JET deflection via cross winds. *Astrophys. Space Sci.* 298:107, 2005.

Tanaka, A., L. Yan, N. J. Watkins, and **Y. Gao.** Femtosecond time-resolved two-photon photoemission study of organic semiconductor copper phthalocyanine film. *J. Electron. Spectrosc. Relat. Phenom.* 144:327, 2005.

Litherland, A. E., **H. E. Gove,** R. P. Beukens, and X.-L. Zhao. Low-level carbon-14 measurements and accelerator mass spectrometry. In *Topical Workshop on Low Radioactivity Techniques,* p. 48, eds. B. Cleveland, R. Ford, and M. Chen. American Institute of Physics, 2005.

Helfer, H. L. The local dark matter. In *Progress in Dark Matter Research,* ed. V. J. Blain. Hauppauge, N.Y.: Nova Science Publishers, 2004.

Marshall, F. J., et al. **(R. L. McCrory** and **D. D. Meyerhofer).** Direct-drive, cryogenic target implosions on OMEGA. *Phys. Plasmas* 12:056302, 2005.

McFarland, K. (CDF Collaboration). Measurement of the W boson polarization in top decay at CDF at sqrt(s)=1.8 TeV. *Phys. Rev. D* 71:031101, 2005.

Fitch, M. J., and **A. C. Melissinos** et al. Electro-optic measurement of the wake fields of a relativistic electron beam. *Phys. Rev. Lett.* 87:034801, 2001.

Ignatovich, F. V., and **L. Novotny.** Real-time and background-free detection of nanoscale particles. *Phys. Rev. Lett.* 96:013901, 2006.

Baur, U., A. Juste, **L. H. Orr,** and D. Rainwater. Probing electroweak top quark couplings at hadron colliders. *Phys. Rev. D* 71:054013, 2005.

Quillen, A. C. A wind-driven warping instability in accretion disks. *Astrophys. J.* 563:313, 2001.

Agarwal, A., L. Akant, G. S. Krishnaswami, and **S. J. Rajeev.** Collective potential for large-N Hamiltonian matrix models and free Fisher information. *Int. J. Mod. Phys. A* 18:917, 2003.

Ren, C., et al. Global simulation for laser-driven MeV electrons in fast ignition. *Phys. Rev. Lett.* 93:185003, 2004.

Chen, A., E. H. Chimowitz, S. De, and **Y. Shapir.** Universal dynamic exponent at the liquid-gas transition from molecular dynamics. *Phys. Rev. Lett.* 95:255701, 2005.

Aronstein, D. L., and **C. R. Stroud Jr.** Phase-difference equations: A calculus for quantum revivals. *Laser Phys.* 15:1496, 2005.

Tarduno, J. A., and R. D. Cottrell. Dipole strength and variation of the time-averaged reversing and nonreversing geodynamo based on Thellier analyses of single plagioclase crystals. *J. Geophys. Res.* 110:B11, 2005.

Gotcheva, V., Y. Wang, A. T. J. Wang, and **S. Teitel.** Continuous time Monte Carlo and spatial ordering in driven lattice gases: Application to driven vortices in superconducting networks. *Phys. Rev. B* 72:064505, 2005.

Thomas, J., and N. O. Weiss. Fine structure in sunspots. *Ann. Rev. Astron. Astrophys.* 42:517, 2004.

He, Q., and **E. H. Thorndike** (CLEO collaboration). Search for rare and forbidden decays $D^+ \to h^{\pm}e^{\mp}e^{+}$. *Phys. Rev. Lett.* 95:221802, 2005.

Tipton, P., et al. (The CDF Collaboration). Search for anomalous kinematics in tt-bar dilepton events at CDF II. *Phys. Rev. Lett.* 95:022001, 2005.

Roychowdhury, H., S. Ponomarenko, and **E. Wolf.** Change of polarization of partially coherent electromagnetic beams propagating through the turbulent atmosphere. *J. Mod. Opt.* 52:1611–8, 2005.

Wolfs, F., et al. The Phobos perspective on discoveries at RHIC. *Nucl. Phys. A* 757:28, 2005.

Cai, C., Z. Chen, S. Cai, and **J. Zhong.** A simulation algorithm based on Bloch equations and product operator matrix: Application to dipolar and scalar couplings. *J. Magn. Reson.* 172:242–53, 2005.

THE UNIVERSITY OF
TOLEDO

THE UNIVERSITY OF TOLEDO

Department of Physics and Astronomy

Programs of Study

The University of Toledo (UT) Department of Physics and Astronomy offers M.S., M.S.E. (Master of Science and Education), and Ph.D. degrees in physics with specializations in astronomy and astrophysics, atomic and molecular physics, biophysics, condensed-matter physics and materials science, medical physics, and photonics. The M.S. in physics is a professional master's degree, preparing students for responsible positions in industrial and academic/government research support. It has flexible course requirements, requires a thesis, and usually requires two years of full-time study to complete. A joint Ph.D. in physics/M.S. in electrical engineering is also available. The Ph.D. has a number of required courses and takes five to seven years to complete. Requirements include residence for at least two consecutive semesters, successful completion of a qualifying and a comprehensive examination, completion of a thesis, and successful defense of the thesis.

A major graduate research focus is in experimental and theoretical studies of thin films, especially photovoltaics, magnetic nanostructures, and surface growth. A second major focus is in astronomy/astrophysics, with studies of stellar atmospheres and envelopes, star formation, interstellar matter, and climate on Mars. The atomic and molecular physics focus includes studies of quantum-condensed phases, Rydberg state lifetimes, and accelerator-based optical spectroscopy. The medical and biological physics includes accelerator-based research in radiation oncology and DNA bonding and structure. The plasma physics focus is on the self-consistent kinetic description of low-pressure discharges, especially under external electric and magnetic field influence. The photonics research focuses on the design of optical integrated circuits and waveguides. The department has a collective strength and focus on advanced computational methods in treating astrophysical, atomic, plasma, and materials problems.

Research collaboration on-campus includes chemists and chemical, electrical, and mechanical engineers. Department faculty members serve as the core of UT's Center for Photovoltaic Electricity and Hydrogen recently established by a major state of Ohio grant with matching support from several industrial collaborators.

Research Facilities

Thin-film materials laboratories include high- and ultrahigh-vacuum deposition systems using glow-discharge and hot-wire deposition, sputtering, and MBE, and incorporate in situ spectroscopic ellipsometry. Other materials and device characterization include the magnetooptical Kerr effect, Raman, photoluminescence, AFM/STM, SEM/EDS, quantum efficiency, and current-voltage dependence under solar simulation. Ritter Observatory houses a 1-meter reflecting telescope that is used for studies of variable stellar spectra. Some UT astronomers' research programs are based on observations made at external ground- and space-based facilities. Atomic physics research is done with 300-keV heavy-ion and 80-keV negative-ion accelerators. Lasers are also used for thin-film scribing and thin-film index-of-refraction measurements. Computing facilities include UNIX workstations and three cluster computing systems. Supercomputer access is provided through the Ohio Supercomputing Center via Internet 2.

Financial Aid

Most full-time physics and astronomy students receive some financial support. Fellowships and teaching and research assistantships, which include a stipend and a tuition waiver, are available for qualified students on a competitive basis.

The out-of-state tuition surcharge normally charged to out-of-state and international students is waived for students whose permanent address is within one of the following Michigan counties: Hillsdale, Lenawee, Macomb, Oakland, Washtenaw, and Wayne. In addition, the University of Toledo offers an out-of-state tuition surcharge waiver to cities and regions that are a part of the Sister Cities Agreement. These regions include Toledo, Spain; Londrina, Brazil; Qinhuangdao, China; Csongrad County, Hungary; Delmenhorst, Germany; Toyohashi, Japan; Tanga, Tanzania; Bekaa Valley, Lebanon; and Poznan, Poland. The University of Toledo Graduate College offers a variety of memorial and minority scholarship awards, including the Ronald E. McNair Postbaccalaureate Achievement Scholarship, the Graduate Minority Assistantship Award, and two full University fellowships.

Cost of Study

The graduate tuition rate for the 2006–07 academic year is $390.05 per semester credit hour for in-state students. For nonresidents, the out-of-state surcharge is $367.15 per semester credit hour. Additional fees are required and include the general fee, technology fee, and mandatory insurance.

Living and Housing Costs

The University of Toledo has a diverse offering of student housing options, including suite-style and traditional residential halls. Housing is offered to graduate students through Residence Life or contracted individually by the student. Affordable, high-quality off-campus apartment-style housing within walking distance of campus is abundant.

Student Group

There are approximately 20,000 students at the University of Toledo. About 4,000 are graduate and professional students. The University has a rich diversity of student organizations. Students join groups that are organized around common cultural, religious, athletic, and educational interests

Student Outcomes

Examples of professional situations of recent graduates include postdoctoral fellowships at Vanderbilt, LSU, Johns Hopkins, Malin Space Science Systems, and Space Telescope Science Institute. Several recent photonics and materials science graduates are employed in Silicon Valley and other industrial jobs.

Location

The University of Toledo has several campus sites in the city of Toledo. Most graduate students take classes on the Main campus, which is located in suburban western Toledo. With a population of more than 330,000, Toledo is the fiftieth-largest city in the United States. It is located on the western shores of Lake Erie, within a 2-hour drive of Cleveland and Detroit.

The University and The Department

The University of Toledo was founded by Jessup W. Scott in 1872 as a municipal institution and became part of the state of Ohio's system of higher education in 1967. On July 1, 2006, the University of Toledo merged with the Medical University of Ohio becoming one of only seventeen American universities to offer professional and graduate academic programs in medicine, law, pharmacy, nursing, health sciences, engineering, and business. The Department of Physics and Astronomy is recognized as one of the University's flagship departments. The Department has strong research collaborations with the Department of Chemistry and the College of Engineering through interdisciplinary research in materials science.

Applying

Applications should be submitted to the Graduate School, UH 3240, University of Toledo, Toledo, Ohio 43606-3390, or online at http://gradschool.utoledo.edu/pages/applyonline.asp. The aptitude section of the Graduate Record Examinations is required of international students and of domestic students whose GPA is less than 2.7 on a 4-point scale. Applications for assistantships should be completed six months before intended first enrollment, although later applications can sometimes be accepted.

Correspondence and Information

Nancy D. Morrison
Professor of Astronomy and Chair, Graduate Admissions Committee
Department of Physics and Astronomy
Mail Stop 113
The University of Toledo
Toledo, Ohio 43606
Phone: 419-530-2659
Fax: 419-530-5167 (Ritter Astrophysical Research Center)
E-mail: ndm@physics.utoledo.edu
Web site: http://www.physics.utoledo.edu

The University of Toledo Graduate College
3240 University Hall, MS 933
The University of Toledo
2801 West Bancroft Street
Toledo, Ohio 43606
Phone: 419-530-4723
E-mail: grdsch@utnet.utoledo.edu
Web site: http://www.gradschool.utoledo.edu

The University of Toledo

THE FACULTY AND THEIR RESEARCH

Astronomy

Lawrence Anderson-Huang, Professor; Ph.D., Berkeley, 1977. Stellar atmospheres.
Jon Bjorkman, Associate Professor; Ph.D., Wisconsin, 1992. Theory of stellar envelopes and winds.
Karen Bjorkman, Professor; Ph.D., Colorado, 1989. Circumstellar matter/stellar winds.
Bernard Bopp, Professor; Ph.D., Texas, 1973. Science education.
Steven Federman, Professor; Ph.D., NYU, 1979. Interstellar chemistry.
Philip James, Emeritus Distinguished University Professor; Ph.D., Wisconsin, 1966. Martian climate and weather.
Tom Megeath, Assistant Professor; Ph.D., Cornell, 1993. Star formation and infrared astronomy.
Nancy Morrison, Professor; Ph.D., Hawaii, 1975. Stellar spectroscopy: massive stars.
Adolf Witt, Emeritus Distinguished University Professor; Ph.D., Chicago, 1967. Interstellar dust.

Physics

Jacques Amar, Associate Professor; Ph.D., Temple, 1985. Condensed matter/materials science.
Brian Bagley, Professor; Ph.D., Harvard, 1968. Optics/materials science.
Randy Bohn, Emeritus Professor; Ph.D., Ohio State, 1969. Solid-state physics.
Song Cheng, Associate Professor; Ph.D., Kansas State, 1991. Atomic physics.
Robert Collins, Professor and NEG Endowed Chair in Silicate and Materials Science; Ph.D., Harvard, 1982. Condensed matter/materials science.
Alvin Compaan, Professor and Chair; Ph.D., Chicago, 1971. Condensed-matter physics/materials science.
Larry Curtis, Emeritus Distinguished University Professor; Ph.D., Michigan, 1963. Atomic spectroscopy.
Robert Deck, Emeritus Professor; Ph.D., Notre Dame, 1961. Nonlinear optics.
Xunming Deng, Professor; Ph.D., Chicago, 1990. Materials science/photovoltaics.
David Ellis, Emeritus Professor; Ph.D., Cornell, 1964. Theoretical atomic physics.
Bo Gao, Associate Professor; Ph.D., Nebraska–Lincoln, 1989. Theoretical physics.
Victor Karpov, Professor; Ph.D., Polytechnic (Russia), 1979. Condensed matter/theoretical physics.
Sanjay V. Khare, Assistant Professor; Ph.D., Maryland, 1996. Theoretical condensed matter/materials science.
Thomas Kvale, Professor; Ph.D., Missouri–Rolla, 1984. Experimental atomic physics.
Scott Lee, Professor; Ph.D., Cincinnati, 1983. Biological physics and high-pressure physics.
R. Ale Lukaszew, Associate Professor; Ph.D., Wayne State, 1996. Condensed matter/materials science.
Sylvain Marsillac, Assistant Professor; Ph.D., Nantes (France), 1996. Materials science/photovoltaics.
Richard Schectman, Emeritus Professor; Ph.D., Cornell, 1962. Atomic physics.
Constantine Theodosiou, Professor; Ph.D., Chicago, 1977. Atomic and plasma physics.

Graduate students at the base of the Ritter Observatory telescope preparing for a night of observations.

Graduate student working on one of the two ion accelerators used for research in the department.

Multiple UHV chamber system for plasma-enhanced CVD and sputter deposition of triple-junction, amorphous silicon solar cells.

SELECTED PUBLICATIONS

Shim, Y. and **J. G. Amar**. Growth instability in Cu multilayer films due to fast edge/corner diffusion. *Phys. Rev. B* 73:035423, 2006.

Shim Y., and **J. G. Amar**. Synchronous sublattice algorithm for parallel kinetic Monte Carlo. *Phys. Rev. B* 71:125432, 2005.

Noreyan, A., **J. G. Amar**, and I. Marinescu. Molecular dynamics simulations of nanoindentation of beta-SiC with diamond indenter. *Mater. Sci. Eng., B* 117:235, 2005.

Mirkov, M. G., **B. G. Bagley**, and **R. T. Deck**. Design of multichannel optical splitter without bends. *Fiber Integr. Opt.* 20:241–55, 2001.

Bjorkman, K. S. Spectropolarimetric variability in hot stars: 15 years of monitoring, and what we've learned. In *Astronomical Polarimetry: Current Status and Future Prospects. Proceedings of the ASP Conference*. San Francisco: ASP, in press.

Wisniewski, J. P., **K. S. Bjorkman,** and A. M. Magalhaes. Identifying circumstellar disks in LMC/SMC clusters. In *Astronomical Polarimetry: Current Status and Future Prospects. Proceedings of the ASP Conference*. San Francisco: ASP, in press.

Miroshnichenko, A. S., and **K. S. Bjorkman** et al. Properties of galactic B[e] supergiants. IV. Hen 3-298 and Hen 3-303. *Astron Astrophys.* 436:653, 2005.

Miroshnichenko, A. S., et al. **(K. S. Bjorkman)**. Fundamental parameters and evolutionary state of the Herbig Ae star candidate HD 35929. *Astron. Astrophys.* 427:937, 2004.

Pogodin, M. A., et al. **(K. S. Bjorkman** and **N. D. Morrison)**. A new phase of activity of the Herbig Be star HD 200775 in 2001: Evidence for binarity. *Astron. Astrophys.* 417:715–23, 2004.

Walker, C., et al. **(J. E. Bjorkman)**. The structure of brown dwarf circumstellar disks. *Mon. Not. R. Astron. Soc.* 351:607, 2004.

Carciofi, A. C., **J. E. Bjorkman**, and A. M. Magalhaes. Effects of grain size on the spectral energy distribution of dusty circumstellar envelopes. *Astrophys. J.* 604:238, 2004.

Whitney, B. A., K. Wood, **J. E. Bjorkman**, and M. Cohen. 2-D radiative transfer in protostellar envelopes: II. An evolutionary sequence. *Astrophys. J.* 598:1079, 2003.

Wisniewski, J. P., **K. S. Bjorkman**, and A. M. Magalhaes. Evolution of the inner circumstellar envelope of V838 Monocerotis. *Astrophys. J. Lett.* 598:L43, 2003.

Collins, R. W. Ellipsometry. In *The Optics Encyclopedia*, vol. 1, pp. 609–70, eds. T. G. Brown, et al. Weinheim, Germany: Wiley-VCH Verlag, 2004.

Collins, R. W., et al. Evolution of microstructure and phase in amorphous protocrystalline, and microcrystalline silicon studied by real time spectroscopic ellipsometry. *Sol. Energy Mater. Sol. Cells* 78:143–80, 2003.

Chen, C., I. An, and **R. W. Collins**. Multichannel Mueller matrix ellipsometry for simultaneous real-time measurement of bulk isotropic and surface anisotropic complex dielectric functions of semiconductors. *Phys. Rev. Lett.* 90:217402, 2003.

Gupta, A., V. Parikh, and **A. D. Compaan**. High efficiency ultra-thin sputtered CdTe solar cells. *Sol. Energy Mater. Sol. Cells* 90:2263, 2006.

Liu, X., **A. D. Compaan**, and J. Terry. Cu K-edge EXAFS studies of CdCl2 effects on CdTe solar cells. *Mater. Res. Soc. Symp. Proc.* 865:F4.2, 2005.

Compaan, A. Photovoltaics: Clean electricity for the 21st century. *Am. Phys. Soc. News* p. 6, April 2005.

Roussillon, Y., et al. **(A. D. Compaan** and **V. G. Karpov)**. Blocking thin film nonuniformities: Photovoltaic self-healing. *Appl. Phys. Lett.* 84:616, 2004.

Curtis, L. J., and I. Martinson. Atomic structure. In *Electrostatic Accelerators*, chap. 14, ed. R. Hellborg. Berlin: Springer-Verlag, in press.

Curtis, L. J., and **D. G. Ellis**. Use of the Einstein-Brillouin-Keller action quantization. *Am. J. Phys.* 72, 2004.

Curtis, L. J. *Atomic Structure and Lifetimes: A Conceptual Approach.* Cambridge: Cambridge University Press, 2003.

Curtis, L. J., R. Matulioniene, **D. G. Ellis**, and C. Froese Fischer. A predictive data-based exposition of $5s5p\,^{1,3}P_1$ lifetimes in the Cd isoelectronic sequence. *Phys. Rev. A* 62:52513, 2000.

Deck, R. T., and **J. G. Amar**. Nuclear size corrections to the energy levels of single-electron and –muon atoms. *J. Phys. B* 38:2173, 2005.

Andaloro, R., **R. T. Deck**, and H. J. Simon. Optical interference pattern resulting from excitation of surface mode with diverging beam. *J. Opt. Soc. Am. B* 22:1512, 2005.

Chen, G., et al. **(R. T. Deck** and **B. G. Bagley)**. Design of single wavelength or polarization filter/combiner. *Fiber Integr. Opt.* 24:1, 2005.

Deck, R. T., A. L. Sala, Y. Sikorski, and **B. G. Bagley**. Loss in a rectangular optical waveguide induced by the crossover of a second waveguide. *Opt. Laser Technol.* 34:351–6, 2002.

Deck, R. T., and J. Walker. The connection between spin-statistics. *Physica Scripta* 63:7, 2001.

Deng, X., and E. Schiff. Amorphous silicon based solar cells. In *The Handbook of Photovoltaic Science and Engineering*, eds. A. Luque and S. Hegedus. New Jersey: John Wiley & Sons, Ltd., 2003.

Povolny, H., and **X. Deng**. High rate deposition of amorphous silicon films using HWCVD with coil-shaped filament. *Thin Solid Films* 430:125, 2003.

Miller, E. L., R. E. Rocheleau, and **X. Deng**. Design considerations for a hybrid amorphous silicon photoelectrochemical multijunction cell for hydrogen production. *Int. J. Hydrogen Energy* 28:615–23, 2003.

Froese Fischer, C., and **D. G. Ellis**. Angular integration using symbolic state expansions. *Lithuanian J. Phys.* 44:121, 2004.

Pan, K., et al. **(S. R. Federman)**. Cloud structure and physical conditions in star-forming regions from optical observations. *Astrophys J.* 633:986, 2005.

Federman, S. R., et al. The interstellar rubidium isotope ratio toward ρ Ophiuchi A. *Astrophys. J. Lett.* 603:L105, 2004.

Sheffer, Y. et al. **(S. R. Federman)**. Ultraviolet detection of interstellar $^{12}C^{17}O$ and the CO isotopomeric ratios toward X Per. *Astrophys. J. Lett.* 574:L171, 2002.

Gao, B. Universal properties of Bose systems with van der Waals interaction. *J. Phys. B: At., Mol. Opt. Phys.* 37:L227, 2004.

Fu, H., Y. Wang, and **B. Gao.** Beyond Fermi pseudopotential: A modified GP equation. *Phys. Rev. A* 67:053612, 2003.

Gao, B. Effective potentials for atom-atom interaction at low temperatures. *J. Phys. B: At., Mol. Opt. Phys.* 36:211, 2003.

Benson, J., et al. **(P. B. James)**. A study of seasonal and short period variation of water ice clouds in the Tharsis and Valles Marineris regions of Mars with Mars Global Surveyor. *Icarus* 165:34–52, 2003.

James, P. B., and B. A. Cantor. Atmospheric monitoring of Mars by the Mars Orbiter camera on Mars Global Surveyor. *Adv. Space Res.* 29:121–9, 2002.

Bonev, B. P., **P. B. James, J. E. Bjorkman**, and M.J. Wolff. Regression of the mountains of Mitchel Polar Ice after the onset of a global dust storm on Mars. *Geophys. Res. Lett.* 29:2017, 2002.

Shvydka, D., and **V. G. Karpov**. Power generation in random diode arrays., *Phys. Rev. B* 71:115314, 2005.

Karpov, V. G., D. Shvydka, and Y. Roussillon. E2 phase transitions: Thin-film breakdown and Schottky barrier suppression. *Phys. Rev. B* 70:155332, 2004.

Karpov, V. G. Critical disorder and phase transition in random diode arrays. *Phys. Rev. Lett.* 91:226806, 2003.

Kodambaka, S., **S. V. Khare**, I. Petrov, and J. E. Greene. Two-dimensional island dynamics: Role of step energy anisotropy. *Surf. Sci. Rep.* 60:55, 2006.

Wang, L. L., and **S. V. Khare** et al. Origin of bulk-like structure and bond length disorder of Pt37 and Pt6Ru31 clusters on carbon: Comparison of theory and experiment. *J. Am. Chem. Soc.* 128:131, 2006.

Kodambaka, S., and **S. V. Khare** et al. Dislocation-driven surface dynamics on solids. *Nature* 429:49, 2004.

Covington, A. M., D. Calabrese, J. S. Thompson, and **T. J. Kvale**. Measurement of the electron affinity of lanthanum. *J. Phys. B* 31:L855, 1998.

Kvale, T. J., et al. Single electron detachment cross sections for 5- to 50-keV H- ions incident on helium, neon, and argon atoms. *Phys. Rev. A* 51:1351, 1995.

The University of Toledo

Woods, K. N., **S. A. Lee**, H.-Y. N. Holman, and H. Wiedemann. The effect of solvent dynamics on the low frequency collective excitations of DNA in solution and unoriented films. *J. Chem. Phys.* Article 234706, 124:2006.

Lee, S. A., I. Lawson, L. Lettress, and A. Anderson. Mid-infrared study of deoxycytidine at high pressures: evidence of a phase transition. *J. Biomolec. Struc. Dyn.* 23:677, 2006.

Cooper, R. L., and **S. A. Lee**. Differential scanning calorimetric study of the binding of water of hydration to deoxyadenosine. *J. Biomolec. Struc. Dyn.* 22:375, 2004.

Lukaszew, R. A., et al. Surface morphology structure and magnetic anisotropy in epitaxial Ni Films. *J. Alloys Compd.* 369(1–2):213–6, 2004.

Lukaszew, R. A., Z. Zhang, D. Pearson, and A. Zambano. Epitaxial Ni films, E-beam nano-patterning and BMR. *J. Magn. Magn. Mater.* 272–6:1864, 2004.

Lukaszew, R. A., Z. Zhang, V. Stoica, and R. Clarke. *AIP Conf. Proc.* 696:629, 2003.

Wisniewski, J. P., et al. **(N. D. Morrison** and **K. S. Bjorkman)**. Spectroscopic and spectropolarimetric observations of V838 Monocerotis. *Astrophys. J.* 588:486, 2003.

Barreau, N., **S. Marsillac**, J. C. Bernede, and L. Assmann. Evolution of the band structure of beta -In2S3-3xO3x buffer layer with its oxygen content. *J. Appl. Phys.* 93:5456, 2003.

Marsillac, S., et al. High-efficiency solar cells based on Cu(InAl)Se2 thin films. *Appl. Phys. Lett.* 81:1350, 2002.

Paulson, P. D., et al. **(S. Marsillac)**. CuInAlSe2 thin films and solar cells. *J. Appl. Phys.* 91:10153, 2002.

Rebull, L. M., et al. **(S. T. Megeath)** A correlation between pre-main sequence stellar rotation rates and IRAC excesses in Orion. *Astrophys. J.* 646:297, 2006.

Megeath, S. T., T. L. Wilson, and M. R. Corbin. Hubble space telescope NICMOS imaging of W3 IRS 5: A trapezium in the making? *Astrophys. J.* 622:L141, 2005.

Charbonneau, D., et al. **(S. T. Megeath)**. Detection of thermal emission from an extrasolar planet. *Astrophys. J.* 626:523, 2005.

Schectman, R. M., and **S. R. Federman** et al. Oscillator strengths for ultraviolet transitions in Cl II and Cl III. *Astrophys. J.* 621:1159, 2005.

Kaganovich, I. D., O. V. Polomarov, and **C. E. Theodosiou**. Landau damping and anomalous skin effect in low-pressure gas discharges: Self consistent treatment of collisionless heating. *Phys. Plasmas* 11(5):2399–410, 2004.

Sosov, Y., and **C. E. Theodosiou**. Determination of electric field-dependent effective secondary emission coefficients for He/Xe ions on brass. *J. Appl. Phys.* 95(8):4385–8, 2004.

Sosov Y., and **C. E. Theodosiou**. A well known boundary value problem requires unusual eigenfunctions. *Am. J. Phys.* 72(2):185–9, 2004.

Witt, A. N., et al. The excitation of extended red emission: new constraints on its carrier from Hubble space telescope observations of NGC 7023. *Astrophys. J.* 636:303, 2006.

Vijh, U. P., **A. N. Witt**, and K. D. Gordon. Blue luminescence and the presence of small polycyclic aromatic hydrocarbons in the interstellar medium. *Astrophys. J.* 633:262, 2005.

Vijh, U. P., **A. N. Witt**, and K. D. Gordon. Discovery of blue luminescence in the red rectangle: possible fluorescence from neutral polycyclic aromatic hydrocarbon molecules. *Astrophys. J.* 606:L65, 2004.

WAKE FOREST UNIVERSITY

WAKE FOREST UNIVERSITY

Department of Physics

Programs of Study

The Department of Physics offers graduate programs of study leading to the M.S. and Ph.D. degrees in the fields of condensed matter; biological physics; gravitation; field theory; cosmology; atomic, molecular, and optical physics; particle physics; and physics related to medicine. Study in these fields satisfies the needs of students with differing career plans. A low student-faculty ratio allows close contact with the faculty.

The entering graduate student is expected to have a sound knowledge of undergraduate mechanics, electromagnetism, thermodynamics, and atomic physics. Provision is made for the beginning graduate student to make up deficiencies in these areas.

For the M.S. degree, a student completes 24 hours of course work, submits a thesis based on his or her completed research, and passes an oral examination based on the thesis. Students normally complete the requirements for the M.S. degree in two years.

The principal requirement of the Ph.D. degree is the solution of an important physics problem at the frontier of current knowledge. Students are encouraged to choose a faculty adviser and to begin research during or just after the first year of course work. The course requirements for the Ph.D. are determined by the student's graduate committee and are tailored to meet the needs of the individual student while providing a broad, well-balanced background. Ph.D. students are required to pass a preliminary examination at the end of the first year and defend their dissertation, based on their research, in an oral examination.

Research Facilities

The Department of Physics is located in the Olin Physical Laboratory, a building near the center of the campus. The building has excellent space for teaching, research, and study. Research instrumentation includes a femtosecond Ti:sapphire laser system with regenerative amplifier, a subpicosecond amplified dye laser and streak, an excimer laser, chambers for pulsed-laser deposition and planar magnetron sputtering, an ultrahigh-vacuum surface analysis system, two nanomanipulators (modified Topometrix AFMs; nM-AFM), an inverted optical microscope, a computer-controlled spectrometer, low-temperature facilities for studies to 4K, and X-ray and laser equipment for sample irradiation. Two high-powered Nd:YAG lasers, an Ar+ laser, a pulsed-dye laser, two research-class microscopes, a molecular-beam apparatus, microwave absorption in biosystems, and magnetic resonance imaging facilities are located at the medical school. Excellent computer facilities include a ninety-eight-node Linux cluster connected with a high-speed network and access to national supercomputing resources. The entire campus has wireless Internet accessibility.

The Center for Nanotechnology is a world-class equipment infrastructure for nanosciences. Experimental capabilities within the center include high-resolution microscopy (low temperature, ultrahigh-vacuum tunneling microscopy, near-field microscopy and spectroscopy, vacuum and in situ AFM, scanning electron microscopy, and high-resolution transmission electron microscopy), transport (LHe cryostat with multiple probes, noise spectral analysis, thermopower, and electrical and thermal transport capabilities), optics (Raman spectroscopy, Z-scan and nonlinear transmission, pulse-probe, photoluminescence and electroluminescence, AM1.5g standard for photovoltaic testing, and coloration test beds), and device fabrication (glove box prototyping, evaporators, spinners, synthesis capabilities, and device testing, including time-of-flight techniques). To learn more, prospective students should visit http://www.wfu.edu/nanotech.

Financial Aid

Most students admitted to graduate study in physics receive financial assistance. In 2006–07, teaching assistantships are $18,000 per year, plus a full tuition scholarship ($26,985). Research assistantships are $18,000 to $19,000 per year, plus a full tuition scholarship. In addition, a Dean's Fellowship ($20,000 for twelve months plus tuition) may be offered to exceptional students. During the fall and spring semesters, teaching assistants are expected to work approximately 12 hours per week in introductory laboratory classes or grading undergraduate problem sets.

All matriculating graduate students receive a new IBM notebook computer with extensive software, such as Microsoft Office, Dreamweaver, Adobe Premiere, Photoshop, and Writer, and the current versions of Maple, SigmaPlot, and other software packages that are useful in research.

Cost of Study

The 2006–07 tuition of $26,985 is remitted for all scholarship and assistantship recipients.

Living and Housing Costs

Apartments near the campus, where most graduate students live, rent from about $350 to $500 per month.

Student Group

The total number of graduate students in physics is currently 24. Twelve have teaching assistantships, and the others have either research assistantships or scholarships.

Location

Winston-Salem is a city in northwestern North Carolina with a population of about 200,000. It is located about 200 miles from the Atlantic Ocean, 70 miles from the Blue Ridge Mountains, and 300 miles from Atlanta and Washington, D.C.

The University

Wake Forest University has a rich tradition going back to its founding as a college in 1834. Currently, its enrollment consists of 4,100 undergraduates and 2,400 graduate and professional (law, medicine, and business) students. The main campus is a wooded 300-acre site on the northwest edge of Winston-Salem surrounded by many scenic walking and jogging paths, grassy picnic areas, and a beautiful public garden.

Applying

Men and women who are completing a bachelor's degree or its equivalent are invited to obtain additional information and application materials by writing to the physics department. GRE General Test scores, transcripts, and three letters of recommendation are required. Submission of scores on the GRE Subject Test in Physics is recommended but not required. Students may be admitted in the fall or spring semester. The deadline for fall application is January 15, but late applicants are considered if positions are still available. The deadline for spring application is November 1.

Correspondence and Information

Graduate Program Director
Department of Physics
Wake Forest University
P.O. Box 7507
Winston-Salem, North Carolina 27109
Phone: 336-758-5337
E-mail: gradphy@wfu.edu
Web site: http://www.wfu.edu/physics/Graduate-studies.html

Wake Forest University

THE FACULTY AND THEIR RESEARCH

Paul R. Anderson, Ph.D., California, Santa Barbara. General relativity and quantum field theory in curved space.
Swati Basu, Ph.D., Illinois. Biophysics.
Keith D. Bonin, Ph.D., Maryland. Optics, nanomotors and nanostructures, molecular motors in biophysics.
J. D. Bourland, Ph.D., North Carolina at Chapel Hill. Radiation oncology.
Eric D. Carlson, Ph.D., Harvard. Particle physics and particle astrophysics.
David Carroll, Ph.D., Wesleyan. Nanostructures and nanotechnology.
Forrest Charnock, Ph.D., Wake Forest. Condensed-matter experiment.
Greg Cook, Ph.D., North Carolina at Chapel Hill. General relativity, numerical relativity, black-hole coalescence.
Jacquelyn Fetrow, Ph.D., Penn State. Computational biophysics, computational drug discovery, cheminformatics, systems biology.
Martin Guthold, Ph.D., Oregon. Biophysics and nanotechnology, scanning probe microscopy.
George Holzwarth, Ph.D., Harvard. Biophysics of molecular motors.
Natalie A. W. Holzwarth, Ph.D., Chicago. First-principles computer modeling of electronic and structural properties of materials, surfaces, and defects in crystals.
William C. Kerr, Ph.D., Cornell. Statistical physics, nonlinearity, solitons, and chaos, with applications to structural phase transitions in solids.
Daniel B. Kim-Shapiro, Ph.D., Berkeley. Biological physics, dynamic monitoring of macromolecular structure, kinetics and mechanism of sickle cell hemoglobin depolymerization.
Janna Levin, Ph.D., Virginia. Geophysics.
Jed Macosko, Ph.D., Berkeley. Biophysics of molecular motors and biopolymers.
G. Eric Matthews Jr., Ph.D., North Carolina at Chapel Hill. Point defects in solids and high-temperature superconductors, first-principles structure calculations.
Timothy Miller, Ph.D., Vanderbilt. High-energy nuclear physics.
Mark Roberson, Ph.D., Princeton. Signal processing.
Fred Salsbury, Ph.D., Berkeley. Theoretical and computational biophysics, density functionality theory.
Peter Santago, Adjunct Associate Professor; Ph.D., North Carolina State. Medical physics, magnetic resonance imaging, tomography.
K. Burak Ucer, Ph.D., Rochester. Multiphoton microscopy and imaging, defect dynamics in oxide materials, photoluminescence dynamics of ZnO, sickle cell hemoglobin photolysis dynamics.
Tim Wagner, Ph.D., Maryland. Condensed-matter physics.
Richard T. Williams, Ph.D., Princeton. Femtosecond laser studies of defects and electrons in solids, metamaterials and negative index materials.

Olin Physical Laboratory.

SELECTED PUBLICATIONS

Anderson, P. R., A. Eftekharzadeh, and B. L. Hu. Self-force on a scalar charge in radial infall from Res using the Hadamard-WKB expansion. *Phys. Rev. D* 73:064023, 2006.

Anderson, P. R., C. Molina-Paris, and E. Mottola. Short distance and initial state effects in inflation: Stress tensor and decoherence. *Phys. Rev. D* 72:043515, 2005.

Anderson, P. R., R. Balbinot, and A. Fabbri. Cutoff anti–de Sitter space/conformal-field-theory duality and the quest for Braneworld black holes. *Phys. Rev. Lett.* 94:061301, 2005.

Anderson, P. R., and B. L. Hu. Radiation reaction in Schwarzschild spacetime: Retarded Green's function via Hadamard-WKB expansion. *Phys. Rev. D* 69:064039, 2004.

Huang, K. T., et al. **(S. Basu** and **D. B. Kim-Shapiro).** Lack of allosterically controlled intramolecular transfer of nitric oxide from the heme to cysteine in the β subunit of hemoglobin. *Blood l* 107(7):2602–4, 2006.

Azarov, I., et al. **(S. Basu** and **D. B. Kim-Shapiro).** Nitric oxide scavenging by red blood cells as a function of hematocrit and oxygenation. *J. Biol. Chem.* 280(47):39024–32, 2005.

Bonin, K. D., W. A. Shelton, and T. G. Walker. Nonlinear motion of optically torqued nanorods. *Phys. Rev. E* 71:036204, 2005.

Bonin, K. D., W. A. Shelton, D. Bonessi, and T. G. Walker. Nonlinear motion of rotating glass fibers. In *Optical Trapping and Optical Micromanipulation, Proceedings of SPIE 5930,* pp. 59302b-1–8, eds. K. Dholakia and G. C. Spalding. Bellingham, Wash.: SPIE, 2005.

Bonin, K., A. Shelton, B. Kourmanov, and T. G. Walker. Light torqued nanomotors in a standing wave. In *Clusters and Nano-Assemblies, Physical and Biological Systems,* pp. 257–63, eds. P. Jena, S. N. Khanna, and B. K. Rao. New Jersey: World Scientific, 2005.

Hill, D., M. Plaza, **K. Bonin,** and **G. Holzwarth.** Vesicle transport in PC12 neurites: Forces and saltatory motion. *Eur. Biophys. J.* 33:623–32, 2004. doi: 10.1007/s00249-004-0403-6

Reyes-Reyes, M., et al. **(D. L. Carroll).** Meso-structure formation for enhanced organic photovoltaic cells. *Org. Lett.* 7:5749–52, 2005.

Kim, K., and **D. L. Carroll.** Roles of Au and Ag nanoparticles in efficiency enhancement of poly(3-octylthiophene) / C60 bulk heterojunction photovoltaic devices. *Appl. Phys. Lett.* 87:203113, 2005.

Reyes-Reyes, M., K. Kim, and **D. L. Carroll.** High efficiency photovoltaic devices based on annealed poly(3-hexylthiophene) and 1-(3-methoxycarbonyl)-propyl-1-phenyl-(6,6)C61 blends. *Appl. Phys. Lett.* 87:083506, 2005.

Bao, H., et al. **(D. L. Carroll).** *Colloid Polym. Sci.* 283:653–61, 2005.

Kim, K., et al. **(D. L. Carroll).** Luminescent poly(phenylene ethynylene) coated silica opals. *Langmuir* 21(11):5207–11, 2005.

Charnock, F. T., R. Lopusnik, and T. J. Silva. Pump-probe Faraday rotation magnetometer using two diode lasers. *Rev. Sci. Instrum.* 76:056105, 2005.

Hannam, M. D., and **G. B. Cook.** Conformal thin-sandwich puncture initial data for boosted black holes. *Phys. Rev. D* 71:084023/1–12, 2005.

Cook, G. B., and H. P. Pfeiffer. Excision boundary conditions for black-hole initial data. *Phys. Rev. D* 70:104016/1–24, 2004.

Creamer, T. P., and **J. S. Fetrow.** Rose is a rose is a rose. Especially if you're a George. *Proteins* 63(2):268–72, 2006.

Fetrow, J. S., S. T. Knutson, and M. H. Edgell. Mutations in alpha-helical solvent-exposed sites of eglin c have long-range effects: Evidence from molecular dynamics simulations. *Proteins* 63(2):356–72, 2006.

Allen, E. E., and **J. S. Fetrow** et al. Algebraic dependency models of protein signal transduction networks from time-series data. *J. Theor. Biol.* 238(2):317–30, 2006.

Allen, E. E., **J. Fetrow,** D. J. John, and S. J. Thomas. Heuristic dependency conjectures in proteomic signaling pathways. In *Proceedings of the 43rd Annual Association for Computing Machinery Southeast Conference,* ed. V. A. Clincy. Kennesaw, Ga., 2005.

Mallakin, A. K., V. M. Inoue, and **M. Guthold.** In-situ quantitative analysis of tumor suppressor protein (hDmp1) using a nanomechanical cantilever beam. In *Proceedings of the 20th Biennial Conference on Mechanical Vibration and Noise, ASME International Design Engineering Technical Conferences & Computers and Information in Engineering Conference,* vol. 1, pp. 2599–606, 2005.

Guthold, M., et al. Visualization and mechanical manipulations of individual fibrin fibers suggest that fiber cross-section has fractal dimension 1.3. *Biophys. J.* 87:4226–36, 2004.

Abraham, Y., and **N. A. W. Holzwarth.** A method for calculating electronic structures near surfaces of semi-infinite crystals. *Phys. Rev. B* 73:035412, 2006.

Kerr, W. C., M. J. Rave, and L. A. Turski. Phase-space dynamics of semiclassical spin-½ Bloch electrons. *Phys. Rev. Lett.* 94, 2005.

Kerr, W. C., and A. J. Graham. Nucleation rate of critical droplets on an elastic string in a ϕ^6 potential. *Phys. Rev. B* 70:066103, 2004.

Huang, J., M. Yakubu, **D. B. Kim-Shapiro,** and S. B. King. Rat liver mediated metabolism of hydroxyurea to nitric oxide. *Free Radical Biol. Med.* 40:1675–81, 2006.

Kim-Shapiro, D. B., A. N. Schechter, and M. T. Gladwin. Unraveling the reactions of nitric oxide, nitrite and hemoglobin in physiology and therapeutics. *Arterioscler. Thromb. Vascular Biol.* 26:697–705, 2006.

Gladwin, M. T., et al. **(D. B. Kim-Shapiro).** The emerging biology of the nitrite anion. *Nature Chem. Biol.* 1:308–14, 2005.

Wake Forest University

Ortiz, T. P., et al. **(J. C. Macosko).** Stepping statistics of single HIV-1 reverse transcriptase molecules during DNA polymerization. *J. Phys. Chem. B* 109:16127–31, 2005.

Lu, H., and **J. Macosko** et al. Closing of the fingers domain generates motor forces in the HIV reverse transcriptase. *J. Biol. Chem.* 24(279):54529–32, 2004. doi: 10.1074/jbc.M407193200

Macosko, J. Molecular machines. In *ISCID Encyclopedia of Science and Philosophy,* 2004.

Salsbury, F. R. Analysis of errors in Stil's equation for macromolecular solvation. *Mol. Phys.* 104:1299–309, 2006.

Drotschmann, K., R. P. Topping, J. E. Clodfelter, and **F. R. Salsbury.** Mutations in the nucleotide-binding domain of MutS homologs uncouple cell death from cell survival. *DNA Repair* 3(7):729–42, 2004.

Thomas, L. R., et al. **(F. R. Salsbury).** Direct binding of FADD to the TRAIL receptor DR5 is regulated by the death effector domain of FADD. *J. Biol. Chem.* 279:31, 2004.

Lee, M. S., **F. R. Salsbury,** and C. L. Brooks III. Constant pH-molecular dynamics using continuous titration coordinates. *Proteins: Struct., Funct., Bioinformatics* 56(4):738–52, 2004.

Huff, R. G., et al. **(P. Santago II** and **J. S. Fetrow).** Chemical and structural diversity in cyclooxygenase protein active sites. *Chem. Biodiversity* 2:1533–52, 2005.

Xiong, G., **K. B. Ucer,** and **R. T. Williams** et al. Donor-acceptor pair luminescence of nitrogen-implanted ZnO single crystal. *J. Appl. Phys.* 97:043528-1–4, 2005.

Xiong, G., J. Wilkinson, **K. B. Ucer,** and **R. T. Williams.** Giant oscillator strength of excitons in bulk and nanostructured systems. *J. Lumin.* 112:1–6, 2005.

Qiu, Y., **K. B. Ucer,** and **R. T. Williams.** Formation time of a small electron polaron in LiNbO3: Measurements and interpretation. *Phys. Status Solidi C* 2:232–35, 2005.

Wilkinson, J., **K. B. Ucer,** and **R. T. Williams.** The oscillator strength of extended exciton states and possibility for very fast scintillators. *Nucl. Instrum. Methods Phys. Res., Sect. A* 537:66–70, 2005.

Xiong, G., J. Wilkinson, **K. B. Ucer,** and **R. T. Williams.** Time-of-flight study of bound exciton polariton dispersive propagation in ZnO. *J. Phys.: Condens. Matter* 17:7287–96, 2005.

Wilkinson, J., **K. B. Ucer,** and **R. T. Williams,** Picosecond excitonic luminescence in ZnO and other wide-gap semiconductors. *Radiat. Meas.* 38:501–5, 2004.

Williams, R. T., and K. S. Song. Dynamics induced by electronic transitions: Finite-temperature relaxation of self-trapped excitons to defects in NaCl. *Surf. Sci.* 593:89–101, 2005.

WASHINGTON UNIVERSITY IN ST. LOUIS

Department of Physics

Programs of Study

The Department of Physics at Washington University offers programs leading to the Ph.D. and M.A. degrees. A minimum of 72 hours is required for the Ph.D., including 36 hours in classroom courses. Candidates must also pass a general qualifying examination and an oral defense of the dissertation research. For the M.A., requirements include 30 hours (at least 24 in classroom courses) and a thesis or final examination. Interdisciplinary studies are facilitated by the McDonnell Center for the Space Sciences, a University-wide center that involves the faculties of the Departments of Physics, Earth and Planetary Sciences, Chemistry, and Engineering.

Research Facilities

The Department of Physics maintains extensive research laboratories in its two buildings, Crow Hall and Compton Hall. The McDonnell Center for the Space Sciences has recently installed the NANOSIMS, a first-of-its-kind ion microprobe with greatly enhanced resolution, and has design, fabrication, and testing facilities for experimental astrophysics research. Experimental condensed matter, materials physics, and biophysics laboratories have equipment for electron microscopy, X-ray diffraction, calorimetry, atomic force microscopy, magnetic resonance imaging, and ultrasonic imaging. The department has several Beowulf clusters, and maintains an extensive, heterogeneous network.

Financial Aid

The department awards a number of teaching and research assistantships. For the 2006–07 academic year, nine-month stipends are $17,500 plus tuition remission. Fellowships and scholarships ranged from $22,000 to $30,000 for twelve months plus tuition remission. Stipends of up to $5500 were available for summer support.

Cost of Study

The cost of graduate study at Washington University is comparable to that at other institutions of its type and caliber. For the 2006–07 academic year, tuition for full-time students is $32,800 per year. Physics department teaching and research assistants receive tuition remission in addition to a stipend.

Living and Housing Costs

Moderately priced rental units may be obtained near the University at average monthly rates of $450 to $850 for one to three bedrooms.

Student Group

About half of the approximately 13,580 students at the University are graduate and professional school students. The faculty has 2,995 full-time members. The department has 89 graduate students, 24 of whom are women.

Student Outcomes

Most recent graduates have gone on to postdoctoral research positions at distinguished institutions and laboratories such as Brookhaven National Laboratory, the University of Chicago, the California Institute of Technology, and the Naval Research Laboratory. Former students who completed their postdoctoral research include an assistant professor at Brown University; a staff scientist at MEMC Corporation in St. Peters, Missouri; and a staff scientist at Los Alamos National Laboratory.

Location

The community surrounding the University is both residential and commercial. Entertainment is varied: music lovers can enjoy the St. Louis Symphony, the St. Louis Philharmonic, and many jazz, blues, rock, and dance clubs; theatergoers can attend performances of several repertory groups and musicals at the summer Municipal Opera. Forest Park, which is within walking distance of the campus, contains a golf course, bicycle and running paths, horseback-riding facilities, a fine zoo, the St. Louis Science Center, and the St. Louis Art Museum.

The University

Washington University in St. Louis was founded in 1853 as a private, coeducational institution. In 1904, the University moved to its present 168-acre hilltop campus bordering on the 1,430 acres of Forest Park. Undergraduate programs are offered in arts and science, engineering, business, and fine arts, and graduate programs are offered in all major fields of human inquiry. Twenty Nobel Prize recipients have done all or part of their distinguished work at Washington University. Since 1976, Washington University has placed first in the William Lowell Putnam Mathematical Competition four times and among the top ten all but two times. Graduates often receive such prestigious graduate study awards as Fulbright, Marshall, Beinecke, and Truman scholarships and Mellon, Putnam, National Science Foundation, and NASA graduate fellowships.

Applying

To ensure consideration of a student for admission in September, a completed application, transcript, financial statement, Graduate Record Examinations (GRE) scores, and three letters of recommendation must be received by January 15. Applications are considered at other times as well. Applicants should have had courses in calculus and physics and should have specialized in physics or a related subject in physical science, engineering, or mathematics.

Correspondence and Information

Graduate Admissions
Department of Physics
Washington University in St. Louis
One Brookings Drive
St. Louis, Missouri 63130-4899
Phone: 314-935-6250
E-mail: gradinfo@wuphys.wustl.edu
Web site: http://www.physics.wustl.edu/

Washington University in St. Louis

THE FACULTY AND THEIR RESEARCH

Professors
Carl M. Bender, Ph.D., Harvard, 1969. Theoretical physics, mathematical physics, particle physics.
Claude W. Bernard, Ph.D., Harvard, 1976. Theoretical physics, mathematical physics, particle physics.
Thomas Bernatowicz, Ph.D., Washington (St. Louis), 1980. Mass spectrometry, transmission electron microscopy.
James H. Buckley, Ph.D., Chicago, 1994. High-energy astrophysics.
Anders E. Carlsson, Ph.D., Harvard, 1981. Condensed-matter theory, biophysics.
John W. Clark, Wayman Crow Professor and Department Chair; Ph.D., Washington (St. Louis), 1959. Theoretical physics and astrophysics, many-body theory, biophysics.
Mark S. Conradi, Ph.D., Washington (St. Louis), 1977. Experimental magnetic resonance: lung imaging and solid-state systems.
Ramanath Cowsik, Ph.D., Bombay, 1968. Theoretical astrophysics.
Willem H. Dickhoff, Ph.D., Free University (Amsterdam), 1981. Theoretical physics, many-body theory.
Peter A. Fedders, Emeritus; Ph.D., Harvard, 1965. Solid-state theory.
Michael W. Friedlander, Ph.D., Bristol, 1955. Cosmic rays, astrophysics.
Patrick C. Gibbons, Ph.D., Harvard, 1971. Experimental solid-state physics, electron microscopy, materials science.
Charles M. Hohenberg, Ph.D., Berkeley, 1968. Experimental space science, rare-gas mass spectroscopy.
Martin H. Israel, Ph.D., Caltech, 1968. Cosmic ray astrophysics.
Jonathan I. Katz, Ph.D., Cornell, 1973. Theoretical astrophysics.
Kenneth F. Kelton, Arthur Holly Compton Professor of Physics; Ph.D., Harvard, 1983. Experimental solid-state physics and materials science.
Joseph Klarmann, Emeritus; Ph.D., Rochester, 1958. Cosmic ray astrophysics.
Kazimierz Luszczynski, Emeritus; Ph.D., London, 1959. Solid-state and low-temperature physics, magnetic resonance.
James G. Miller, Albert Gordon Hill Professor of Physics; Ph.D., Washington (St. Louis), 1969. Ultrasonics, biomedical physics, elastic properties of inhomogeneous media.
Richard E. Norberg, Ph.D., Illinois, 1951. Solid-state and low-temperature physics, magnetic resonance.
Michael C. Ogilvie, Ph.D., Brown, 1980. Theoretical physics, mathematical physics, particle physics.
Peter R. Phillips, Emeritus; Ph.D., Stanford, 1961. General relativity and cosmology.
John H. Scandrett, Emeritus; Ph.D., Wisconsin, 1963. Biomedical physics and computer applications.
James S. Schilling, Ph.D., Wisconsin, 1969. Solid-state and high-pressure physics.
J. Ely Shrauner, Emeritus; Ph.D., Chicago, 1963. Theoretical physics, elementary particle theory, high-energy physics, applied physics.
Stuart A. Solin, Charles M. Hohenberg Professor of Physics; Ph.D., Purdue, 1969. Experimental solid-state and materials physics.
Wai-Mo Suen, Ph.D., Caltech, 1985. Theoretical astrophysics, general relativity, cosmology.
Ronald K. Sundfors, Emeritus; Ph.D., Cornell, 1963. Nuclear acoustic resonance, ultrasonics.
Clifford M. Will, James S. McDonnell Professor of Physics; Ph.D., Caltech, 1971. Theoretical astrophysics, general relativity.

Associate Professor
Ralf Wessel, Ph.D., Cambridge, 1992. Experimental biophysics, neuroscience.

Joint Professors
Shankar Sastry, Ph.D., Toronto, 1974. Materials science and metallurgy, alloys and intermetallic compounds. (Department of Mechanical Engineering)
Lee G. Sobotka, Ph.D., Berkeley, 1982. Nuclear physics, heavy ion reactions. (Department of Chemistry)

Assistant Professors
Mark G. Alford, Ph.D., Harvard, 1990. Quantum field theory and particle physics.
Ramki Kalyanaraman, Ph.D., North Carolina State, 1998. Experimental materials science and solid-state physics.
Henric Krawczynski, Ph.D., Hamburg (Germany), 1997. Astrophysics.
Yan Mei Wang, Ph.D., Berkeley, 2002. Experimental biophysics.

Professors (Courtesy)
Donald P. Ames, Ph.D., Wisconsin–Madison, 1949. Materials research, magnetic resonance.
Charles H. Anderson, Ph.D., Harvard, 1962. Biophysics.
Vijai V. Dixit, Ph.D., Purdue, 1972. Theoretical physics.
Elliot L. Elson, Ph.D., Stanford, 1966. Molecular biophysics.
Robert Falster, Ph.D., Stanford, 1983. Experimental materials science, semiconductors.
Solomon L. Linder, Ph.D., Washington (St. Louis), 1955. Electrooptics.
Jeffrey E. Mandula, Ph.D., Harvard, 1966. Theoretical physics, particle physics, mathematical physics.
Manfred L. Ristig, Ph.D., Köln (Germany), 1966. Nuclear theory.
Dmitriy Yablonsky, Ph.D., Ukraine, 1973. Radiation physics.

Associate Professors (Courtesy)
Thomas E. Conturo, M.D./Ph.D., Vanderbilt, 1989. Biophysics, magnetic resonance imaging.
Philip B. Fraundorf, Ph.D., Washington (St. Louis), 1980. Space physics, solid-state physics.
Sandor J. Kovacs, M.D., Miami (Florida), 1979; Ph.D., Caltech, 1977. Cardiology, astrophysics.
Samuel A. Wickline, M.D., Hawaii, 1980. Cardiology.

Assistant Professors (Courtesy)
Gregory L. Comer, Ph.D., North Carolina, 1990. General relativity.
David A. Feinberg, Ph.D., Berkeley, 1982; M.D., Miami (Florida), 1988. Magnetic resonance.
Mary M. Leopold, Ph.D., Washington (St. Louis), 1985. Semiconductor physics.
Craig W. Lincoln, Ph.D., Washington (St. Louis). Astrophysics, general relativity.
Ian Redmount, Ph.D., Caltech, 1984. General relativity.

Research Professors
Robert W. Binns, Ph.D., Colorado State, 1969. Astrophysics, medical and health physics.
Ernst Zinner, Ph.D., Washington (St. Louis), 1972. Experimental space science, extraterrestrial materials.

Research Associate Professor
Daniel J. Leopold, Ph.D., Washington (St. Louis), 1983. Chemical physics.

Research Assistant Professor
Mark R. Holland, Ph.D., Washington (St. Louis), 1989. Ultrasonics, medical physics, biomedical ultrasound.

WORCESTER POLYTECHNIC INSTITUTE

Department of Physics

Program of Study

The Worcester Polytechnic Institute (WPI) physics graduate program prepares students for careers in research that require a high degree of initiative and responsibility. Prospective employers include industrial laboratories, government or nonprofit research centers, and colleges or universities. WPI's physics courses are generally scheduled during the day but with sufficient flexibility to accommodate part-time students. To improve the course offerings and opportunities for graduate students, the Departments of Physics at WPI and Clark University share their graduate courses. Special topics courses in areas of faculty research interest are often available. Research areas include quantum physics (cold atoms, quantum information, and wave-function engineering), optics (photonics, spectroscopy, and lasers), condensed matter (semiconductors, magnetic solids, and nanomechanics), soft condensed matter/complex fluids (polymers, liquid crystals, liquids, and glasses), and physics education.

The Master of Science (M.S.) degree in physics requires 30 semester hours of credit, with 6 or more in thesis research and the remainder in approved courses and independent studies. Although a thesis defense is not required, students nearing completion of the M.S. program are required to present a seminar based on their thesis research. The Doctor of Philosophy (Ph.D.) degree requires 90 credit hours, including 42 in approved courses or directed study, 30 of dissertation research, and the completion and defense of a Ph.D. thesis. Courses taken to satisfy M.S. degree requirements may be counted against the required 42 credits of courses, but completion of a M.S. degree is not required. One year of residency and passage of a qualifying examination are required.

Research Facilities

The physics department has research thrusts in complex fluids (polymer solutions, surfactants and colloids, light scattering, and liquid crystals), condensed matter (wave-function engineering of nanostructures, semiconductor heterostructure laser design and spintronics in diluted magnetic semiconductors, calorimetry, nanomechanics, wetting, and Casimir forces), optics (photonics, lasers, and spectroscopy), physics education research, and quantum physics (cold atoms and Bose-Einstein condensates and quantum information theory).

The IPG Photonics Laboratory consists of state-of-the-art equipment in the area of photonics, including a fiber communication setup, EDFA, diode-laser-testing equipment, diode laser drivers (500 mA, DC-150 kHz modulated), interferometers, an optical spectrum analyzer (600–1700 nm, 20 pm resolution), CCD camera (400–900 nm), a digital storage oscilloscope (500 MHz bandwidth, waveform averaging), He-Ne lasers (red and green, 0.5 mW), diode lasers (670nm and 785nm, 5mW), Si and InGaAs photodetectors (10 MHz), Si photodiode array, and fiber-optic components (cleaver, positioner, coupler). The Atomic Force Microscopy (AFM) Laboratory's principal instrument is an M5 AFM from TM Microscopes. Features include a scan range up to 100 by 100 square microns, a closed-loop scanner for metrology with 5 percent accuracy, 200-mm translation stage accommodating up to 400 by 400 by 25 mm3 samples, on-axis integrated optics (400–1700X), liquid and air operation, up to eight data acquisition channels, most data-acquisition modes, a Windows 98 operating system, intuitive user interface, and sophisticated image processing. The department also houses the Order-Disorder Phenomena Laboratory.

Financial Aid

Ph.D. students are offered financial support in the forms of teaching assistantships, research assistantships, and fellowships. Teaching assistantships range between $13,000 and $14,000 per year plus tuition. GAANN fellowships are $15,000 per year plus full tuition and require students to teach seven weeks per year.

Cost of Study

Graduate tuition for the 2005–06 academic year was $941 per credit hour. There are nominal extra charges for the thesis, health insurance, and other fees.

Living and Housing Costs

On-campus graduate student housing is limited to a space-available basis. There is no on-campus housing for married students. Apartments and rooms in private homes near the campus are available at varying costs.

Student Group

In 2005, there were 11 students that included 4 women and 5 international students.

Location

The university is located on an 80-acre campus in a residential section of Worcester. The city, the second-largest in New England, has many colleges and an unusual variety of cultural opportunities. Located three blocks from the campus, the nationally famous Worcester Art Museum contains one of the finest permanent collections in the country and offers many special activities of interest to students. The community also provides outstanding programs in music and theater. The DCU Center offers rock concerts and semiprofessional athletic events. Easily reached for recreation are Boston and Cape Cod to the east and the Berkshires to the west, and good skiing is nearby to the north. Complete athletic and recreational facilities and a program of concerts and special events are available on campus to graduate students.

The Institute

Worcester Polytechnic Institute, founded in 1865, is the third-oldest independent university of engineering and science in the United States. Graduate study has been a part of the Institute's activity for more than 100 years. Classes are small and provide for close student-faculty relationships. Graduate students frequently interact in research with undergraduates participating in WPI's innovative project-based program of education.

Applying

Applicants with a B.S. in physics are preferred, but those with comparable backgrounds are considered. Students should submit the completed application, the $70 nonrefundable application fee, three letters of recommendation, official transcripts from all undergraduate and graduate institutions attended, and scores from the GRE Subject Test in Physics. International students must also submit TOEFL scores. The application deadlines are February 1 for the fall semester and October 15 for the spring. Applications are processed on a rolling basis.

Correspondence and Information

Graduate Studies and Enrollment
Department of Physics
Worcester Polytechnic Institute
Worcester, Massachusetts 01609-2280
Phone: 508-831-5631
Fax: 508-831-5886
E-mail: physics@wpi.edu
Web site: http://www.wpi.edu/Academics/Depts/Physics/Graduate/

Worcester Polytechnic Institute

THE FACULTY AND THEIR RESEARCH

P. K. Aravind, Professor; Ph.D., Northwestern. Quantum information theory. Decoherence as a measure of entanglement. *Phys. Rev. A* 71:060308, 2005 (with Tolkunov and Privman).

N. A. Burnham, Associate Professor; Ph.D., Colorado at Boulder. Mechanical properties of nanostructures, instrumentation for nanomechanics. Standard-deviation minimization for calibrating the radii of spheres attached to AFM cantilevers. *Rev. Sci. Instrum.* 75:1359–62, 2004 (with Thoreson).

R. Garcia, Assistant Professor; Ph.D., Penn State. Casimir forces, phase transitions, and wetting phenomena. Quartz microbalance study of thick He-4 films near the superfluid transition. *J. Low Temp. Phys.* 134:527, 2004 (with Jordan, Lazzaretti, and Chan).

G. S. Iannacchione, Associate Professor; Ph.D., Kent State. Soft condensed matter physics/complex fluids, liquid crystals, calorimetry, and order-disorder phenomena. Critical linear thermal expansion in the smectic-A phase near the nematic-smectic-A phase transition. *Phys. Rev. E* 70(4):041703–8, 2004 (with Anesta and Garland).

S. N. Jasperson, Professor; Ph.D., Princeton. Optical properties of solids, optical instruments.

T. H. Keil, Professor; Ph.D., Rochester. Solid-state physics, mathematical physics, fluid mechanics.

C. Koleci, Assistant Professor; Ph.D., Yale. Physics education.

L. C. Lew Yan Voon, Associate Professor; Ph.D., Worcester Polytechnic. Band structure theory, optoelectronic properties of nanostructures, acoustics. Influence of aspect ratio on the lowest states of quantum rods. *Nano Lett.* 4:289–92, 2004 (with Lassen, Melnik, and Willatzen).

J. Norbury, Professor and Department Head; Ph.D., Idaho, 1983. Theoretical nuclear and particle physics and cosmology.

G. D. J. Phillies, Professor; D.Sc., MIT. Light-scattering spectroscopy, biochemical physics, polymers. Fourth-order hydrodynamic contribution to the polymer self-diffusion coefficient. *J. Polym. Sci., Part B: Polym. Phys.* 42:1663–70, 2004 (with Merriam).

S. W. Pierson, Associate Professor; Ph.D., Minnesota. Statistical mechanics, High-T superconductors, vortices.

R. S. Quimby, Associate Professor; Ph.D., Wisconsin–Madison. Optical properties of solids, laser spectroscopy, fiber optics. Multiphonon energy gap law in rare-earth doped chalcogenide glass. *J. Non-Cryst. Solids* 320:100, 2003 (with Aitken).

L. R. Ram-Mohan, Professor; Ph.D., Purdue. Field theory, many-body problems, solid-state physics, and finite-element modeling of quantum systems. Finite Element and Boundary Element Applications. In *Quantum Mechanics.* New York: Oxford University Press, 2002.

A. Zozulya, Professor; Ph.D., Russian Academy of Sciences. Nonlinear optics, photo-refractive materials, atom pipes.

YALE UNIVERSITY

Graduate School of Arts and Sciences
Department of Physics

Program of Study

The Department of Physics offers a program of study leading to the Ph.D. degree. To complete the course requirements for the degree, students are expected to take a set of nine term courses. Students normally take three courses during each of their first three semesters. In addition, all students are required to be proficient and familiar with mathematical methods of physics (such as those necessary to master the material covered in the five core courses) and to be proficient and familiar with advanced laboratory techniques. These requirements can be met either by having had sufficiently advanced prior course work or by taking a course offered by the department. All students also attend a seminar during their first term in order to be introduced to the various research efforts and opportunities at Yale. Those who pass their courses with satisfactory grades and who pass the qualifying exam are admitted to doctoral candidacy for the Ph.D. degree. Dissertation research then becomes the primary activity.

The qualifying exam, normally taken at the beginning of the third semester (although it can be taken earlier), is devoted to graduate-level physics, with special attention to material at the level of courses taken during the first two semesters.

Formal association with a dissertation adviser normally begins in the fourth semester after the qualifying examination has been passed. An adviser from a department other than physics can be chosen in consultation with the Director of Graduate Studies, provided that the dissertation topic is deemed suitable for a physics Ph.D.

Approximately eighteen months after passing the qualifying exam, but no later than the end of the fourth year, students take an oral exam centering on a recently published research paper in the field (but not on the topic) of their dissertation research. The final examination is an oral defense of the dissertation. The average time needed to complete all of the Ph.D. requirements has been six years.

Research Facilities

The physics department occupies the Sloane Physics Laboratory, part of the J. W. Gibbs Laboratories, and the Wright Nuclear Structure Laboratory. Research on condensed-matter physics is also done in the Becton Laboratory. Sloane has recently constructed laboratories for research in atomic, molecular, optical, and condensed-matter physics. The theoretical physicists are located in Sloane. The Wright Laboratory contains an Extended Stretch Transuranium (ESTU) 20-megavolt tandem electrostatic accelerator, the most powerful of its kind in the world. The Wright and Gibbs Laboratories house design facilities used in supporting high-energy experiments at Brookhaven National Laboratories, Fermilab, the Stanford Linear Accelerator Center, and the forthcoming Large Hadron Collider. Experiments are also done at European accelerators, and observations have been taken at South American astronomical observatories. In addition to the centralized University computer system, each research group has its own appropriate computing facilities. There are four libraries of major pertinence to physics—Kline Science, Astronomy, Mathematics, and Engineering and Applied Science. Research areas in the Department of Physics include atomic physics and quantum optics, nuclear physics, particle physics, astrophysics, cosmology, condensed-matter physics, quantum information physics, applied physics, and other areas in collaboration with the faculties of engineering and applied science, chemistry, mathematics, geology and geophysics, and astronomy.

Financial Aid

Virtually all entering graduate students in the Department of Physics are offered financial aid for the first three terms in the form of a Yale University Fellowship. This is a combination of stipend, teaching fellowship, tuition, and payment for an Assistantship in Research for the summer following the first year. After the third or fourth semester, when a student has begun dissertation research, full financial support is provided by the student's thesis adviser in the form of an Assistantship in Research. The total support for 2006–07 is $25,500 for twelve months plus full tuition and health and hospitalization coverage. There are also teaching fellowships available to advanced students.

Cost of Study

Tuition and fees are covered for all students who are not supported in full by outside scholarships.

Living and Housing Costs

The rents of dormitory rooms for the 2006–07 academic year range from $3675 for a single room to $5745 for a deluxe single room. Three-bedroom suites, which include a study, are also available. Board plans are offered. The cost for an apartment ranges from $680 to $980 per month. The lease period for graduate housing apartments is usually July 1 through June 30. Off-campus housing in the vicinity of the physics department is plentiful.

Student Group

The total number of students for 2006–07 is 107, all of whom attend full-time. Of these, about 40 percent are international students, and about 18 percent are women. Students with a strong basic undergraduate physics education, together with some research experience, are prime candidates for admission. Advanced commitment to a particular field is not required.

Location

Yale is located in the center of the city of New Haven (population about 125,000; metropolitan area about 400,000). The city offers an unusually wide variety of activities—especially in theater, music, film, fine arts, sports, and international dining. Frequent rail service to New York City and Boston takes less than 2 hours and about 3 hours, respectively.

The University

Chartered in 1701 as the Collegiate School, Yale was named for Elihu Yale, a London merchant who made a modest donation to help the fledgling school. A medical school was added in 1810. The Department of Philosophy and the Arts was organized in 1847, awarding the first three Ph.D. degrees in the United States in 1861 and becoming the Graduate School in 1892. Women were admitted early in the twentieth century to the graduate and professional schools and to Yale College as undergraduates in 1969.

Applying

All applications are submitted online by accessing the following link: http://www.yale.edu/graduateschool/admissions/index.html. Candidates submitting completed applications and supporting materials before January 2, 2007, are considered for admission in fall 2007. Applications must be accompanied by an application fee of $85. Students are required to take the GRE General Test as well as the GRE Subject Test in physics. Those whose native language is not English must also take the TOEFL; the TSE is recommended. Admission consideration is open to all qualified candidates without regard to race, color, national origin, religion, sex, sexual preference, or handicap.

Correspondence and Information

Director of Graduate Studies
Department of Physics
Yale University
P.O. Box 208120
New Haven, Connecticut 06520-8120
Phone: 203-432-3607
Fax: 203-432-6175
E-mail: graduatephysics@yale.edu
Web site: http://www.yale.edu/physics

Yale University

THE FACULTY AND THEIR RESEARCH

Robert K. Adair, Professor Emeritus and Senior Research Scientist; Ph.D., Wisconsin, 1951. Elementary particle physics.
Charles H. Ahn, Associate Professor (joint with Applied Physics); Ph.D., Stanford, 1996. Condensed-matter physics.
Yoram Alhassid, Professor; Ph.D., Hebrew (Jerusalem), 1979. Nuclear theory.
Thomas Appelquist, Professor; Ph.D., Cornell, 1968. Particle theory.
Charles Bailyn, Professor (joint with Astronomy); Ph.D., Harvard, 1987. High-energy astrophysics.
Charles Baltay, Professor; Ph.D., Yale, 1963. Elementary particle physics.
Sean E. Barrett, Professor; Ph.D., Illinois, 1992. Condensed-matter physics.
Cornelius Beausang, Adjunct Professor; Ph.D., SUNY at Stony Brook, 1987. Nuclear physics.
Sidney B. Cahn, Lecturer and Research Scientist; Ph.D., SUNY at Stony Brook, 1997. Experiemental atomic physics.
Helen L. Caines, Assistant Professor; Ph.D., Birmingham (England), 1996. Experimental nuclear physics.
Richard Casten, Professor; Ph.D., Yale, 1967. Nuclear physics.
Richard K. Chang, Professor (joint with Applied Physics); Ph.D., Harvard, 1965. Condensed-matter and laser physics.
Paolo Coppi, Professor (joint with Astronomy); Ph.D., Caltech, 1990. High-energy astrophysics.
David P. DeMille, Professor; Ph.D., Berkeley, 1994. Atomic physics.
Michel Devoret, Professor (joint with Applied Physics); Ph.D., D'Orsay (France), 1982. Applied physics.
Satish Dhawan, Senior Research Scientist; Ph.D., Tsukuba (Japan), 1984. Elementary particle physics.
Eric Dufresne, Assistant Professor (joint with Mechanical Engineering and Chemical Engineering); Ph.D., Chicago, 2000. Experimental soft condensed-matter physics.
Richard Easther, Assistant Professor; Ph.D., Canterbury, 1994. Particle theory and cosmology.
Bonnie Fleming, Assistant Professor; Ph.D., Columbia, 2001. High-energy physics.
Paul A. Fleury, Professor (joint with Engineering and Applied Physics); Ph.D., MIT, 1965. Applied physics.
Steven Furlanetto, Assistant Professor; Ph.D., Harvard, 2003. Theoretical astrophysics.
Moshe Gai, Adjunct Professor; Ph.D., SUNY at Stony Brook, 1980. Nuclear physics.
Colin Gay, Associate Professor; Ph.D., Toronto, 1991. Elementary particle physics.
Steven M. Girvin, Professor (joint with Applied Physics); Ph.D., Princeton, 1977. Theoretical condensed-matter physics.
Walter Goldberger, Assistant Professor; Ph.D., Caltech, 2001. Theoretical particle physics.
Robert D. Grober, Professor (joint with Applied Physics); Ph.D., Maryland, 1991. Condensed-matter physics.
Martin Gutzwiller, Adjunct Professor; Ph.D., Kansas, 1953. Condensed-matter theory.
Jack Harris, Assistant Professor; Ph.D., California, Santa Barbara, 2000. Atomic physics.
John Harris, Professor; Ph.D., SUNY at Stony Brook, 1978. Relativistic heavy-ion physics.
Andreas Heinz, Assistant Professor; Ph.D., GSI Darmstadt (Germany), 1998. Nuclear physics.
Victor E. Henrich, Professor (joint with Applied Physics); Ph.D., Michigan, 1967. Condensed-matter physics.
Jay L. Hirshfield, Adjunct Professor; Ph.D., MIT, 1960. Beam physics.
Francesco Iachello, Professor (joint with Chemistry); Ph.D., MIT, 1969. Nuclear theory.
Stephen Irons, Lecturer; Ph.D., California, Davis, 1996. Condensed-matter physics.
Sohrab Ismail-Beigi, Assistant Professor (joint with Applied Physics); Ph.D., MIT, 2002. Condensed-matter physics.
Karyn Le Hur, Associate Professor; Ph.D., Paris XI (South), 1998. Condensed-matter physics.
Richard D. Majka, Senior Research Scientist; Ph.D., Yale, 1974. Elementary particle physics.
William J. Marciano, Adjunct Professor; Ph.D., NYU, 1957. Particle theory.
Daniel McKinsey, Assistant Professor; Ph.D., Harvard, 2002. Atomic physics.
Simon Mochrie, Professor (joint with Applied Physics); Ph.D., MIT, 1985. Condensed matter.
Vincent E. Moncrief, Professor (joint with Mathematics); Ph.D., Maryland, 1972. Gravitation and cosmology.
Priyamvada Natarajan, Associate Professor (joint with Astronomy); Ph.D., Cambridge, 1998. Astrophysics.
Homer A. Neal Jr., Associate Professor; Ph.D., Stanford, 1995. Experimental elementary particle physics.
Corey O'Hern, Assistant Professor (joint with Mechanical Engineering); Ph.D., Pennsylvania, 1999. Theoretical soft condensed matter.
Peter D. M. Parker, Professor; Ph.D., Caltech, 1963. Experimental nuclear physics and nuclear astrophysics.
Daniel E. Prober, Professor (joint with Applied Physics); Ph.D., Harvard, 1975. Condensed-matter physics.
Nicholas Read, Professor (joint with Applied Physics); Ph.D., London, 1986. Condensed-matter theory.
Vladimir Rokhlin, Professor (joint with Computer Science and Math); Ph.D., Rice, 1983. Scientific computation.
Jack Sandweiss, Professor; Ph.D., Berkeley, 1957. Elementary particle physics.
Michael P. Schmidt, Professor; Ph.D., Yale, 1979. Elementary particle physics.
Robert J. Schoelkopf, Professor (joint with Applied Physics); Ph.D., Caltech, 1995. Experimental condensed-matter physics.
Ramamurti Shankar, Professor (joint with Applied Physics); Ph.D., Berkeley, 1974. Condensed-matter theory and statistical physics.
Witold Skiba, Associate Professor; Ph.D., MIT, 1997. Particle theory.
A. Douglas Stone, Professor (joint with Applied Physics); Ph.D., MIT, 1983. Condensed-matter theory.
Andrew Szymkowiak, Senior Research Scientist; Ph.D., Maryland, 1984. Astrophysics.
John C. Tully, Professor (joint with Chemistry); Ph.D., Chicago, 1968. Theoretical chemical physics.
Thomas Ullrich, Adjunct Professor; Ph.D., Heidelberg, 1994. Relativistic heavy-ion physics.
C. Megan Urry, Professor; Ph.D. Johns Hopkins, 1984. Astrophysics.
Volker Werner, Assistant Professor; Ph.D., Cologne, 2004. Experimental nuclear physics.
John Wettlaufer, Professor (joint with Geophysics); Ph.D., Washington (Seattle), 1991. Geophysics.
Michael E. Zeller, Professor; Ph.D., UCLA, 1968. Elementary particle physics.

ACADEMIC AND PROFESSIONAL PROGRAMS IN MATHEMATICS

Section 7
Mathematical Sciences

This section contains a directory of institutions offering graduate work in mathematical sciences, followed by in-depth entries submitted by institutions that chose to prepare detailed program descriptions. Additional information about programs listed in the directory but not augmented by an in-depth entry may be obtained by writing directly to the dean of a graduate school or chair of a department at the address given in the directory.

For programs offering work in related fields, see all other areas in this book. In Book 2, see Economics and Psychology and Counseling; in Book 3, see Biological and Biomedical Sciences; Biophysics; Genetics, Developmental Biology, and Reproductive Biology; and Pharmacology and Toxicology; in Book 5, see Biomedical Engineering and Biotechnology; Chemical Engineering (Biochemical Engineering); Computer Science and Information Technology; Electrical and Computer Engineering; Engineering and Applied Sciences; and Industrial Engineering; and in Book 6, see Business Administration and Management, Library and Information Studies, and Public Health.

CONTENTS

Program Directories

Announcements

Close-Ups

See also:

Applied Mathematics

Acadia University, Faculty of Pure and Applied Science, Department of Mathematics and Statistics, Wolfville, NS B4P 2R6, Canada. Offers applied mathematics and statistics (M Sc). *Faculty:* 13 full-time (2 women). *Students:* 7 full-time (4 women). *Degree requirements:* For master's, thesis, 4-8 month industry internship. *Entrance requirements:* For master's, honors degree in mathematics, statistics or equivalent. Additional exam requirements/recommendations for international students: Required—TOEFL (minimum score 580 paper-based; 237 computer-based). *Application deadline:* For fall admission, 2/1 priority date for domestic students, 2/1 priority date for international students. Applications are processed on a rolling basis. Application fee: $50. *Financial support:* Career-related internships or fieldwork and unspecified assistantships available. *Faculty research:* Geophysical fluid dynamics, machine scheduling problems, control theory, stochastic optimization, survival analysis. *Unit head:* Dr. Paul Stephenson, Head, 902-585-1382, Fax: 902-585-1074, E-mail: paul.stephenson@acadiau.ca. *Application contact:* Dr. Richard H. Karsten, Professor, 902-585-1608, Fax: 902-585-1074, E-mail: richard.karsten@acadiau.ca.

Air Force Institute of Technology, Graduate School of Engineering and Management, Department of Mathematics and Statistics, Dayton, OH 45433-7765. Offers applied mathematics (MS, PhD). Part-time programs available. *Faculty:* 14 full-time (2 women), 7 part-time/adjunct (0 women). *Students:* 9 full-time (8 women). Average age 29. In 2005, 4 master's, 2 doctorates awarded. *Degree requirements:* For master's and doctorate, thesis/dissertation. *Entrance requirements:* For master's, GRE General Test, minimum GPA of 3.0, must be U.S. citizen or permanent U.S. resident; for doctorate, GRE General Test, minimum GPA of 3.5, must be U.S. citizen or permanent U.S. resident. *Application deadline:* For fall admission, 3/1 for domestic students. Applications are processed on a rolling basis. Application fee: $0. *Financial support:* Fellowships with full and partial tuition reimbursements, research assistantships with full and partial tuition reimbursements, scholarships/grants available. Financial award application deadline: 3/15. *Faculty research:* Electromagnetics, groundwater modeling, nonlinear diffusion, goodness of fit, finite element analysis. Total annual research expenditures: $450,000. *Unit head:* Dr. Alan V. Lair, Head, 937-255-3098, Fax: 937-656-4413, E-mail: alan.lair@afit.edu.

Arizona State University, Division of Graduate Studies, College of Liberal Arts and Sciences, Division of Natural Sciences and Mathematics, Department of Mathematics and Statistics, Tempe, AZ 85287. Offers applied mathematics (MA, PhD); mathematics (MA, MNS, PhD); statistics (MA, PhD). *Degree requirements:* For master's, thesis or alternative; for doctorate, one foreign language, thesis/dissertation. *Entrance requirements:* For master's and doctorate, GRE General Test.

Brown University, Graduate School, Division of Applied Mathematics, Providence, RI 02912. Offers Sc M, PhD. *Degree requirements:* For master's, thesis or alternative; for doctorate, one foreign language, thesis/dissertation, oral exam. *Entrance requirements:* For master's and doctorate, GRE General Test.

See Close-Up on page 481.

California Institute of Technology, Division of Engineering and Applied Science, Option in Applied and Computational Mathematics, Pasadena, CA 91125-0001. Offers MS, PhD. *Faculty:* 6 full-time (0 women). *Students:* 26 full-time (5 women); includes 2 minority (1 Asian American or Pacific Islander, 1 Hispanic American), 15 international. 83 applicants, 8% accepted, 6 enrolled. In 2005, 3 degrees awarded. *Degree requirements:* For doctorate, thesis/dissertation. *Entrance requirements:* For doctorate, GRE Subject Test. *Application deadline:* For fall admission, 1/15 for domestic students. Application fee: $0. Electronic applications accepted. *Financial support:* In 2005–06, 5 research assistantships were awarded; fellowships, teaching assistantships *Faculty research:* Theoretical and computational fluid mechanics, numerical analysis, ordinary and partial differential equations, linear and nonlinear wave propagation, perturbation and asymptotic methods. *Unit head:* Dr. Yizhao Thomas Hou, Executive Officer, 626-395-4546, E-mail: hou@ama.caltech.edu.

California State Polytechnic University, Pomona, Academic Affairs, College of Science, Program in Mathematics, Pomona, CA 91768-2557. Offers applied mathematics (MS); pure mathematics (MS). Part-time programs available. *Students:* 27 full-time (12 women), 27 part-time (7 women); includes 27 minority (1 African American, 18 Asian Americans or Pacific Islanders, 8 Hispanic Americans), 3 international. Average age 29. 45 applicants, 78% accepted, 17 enrolled. In 2005, 10 degrees awarded. *Degree requirements:* For master's, thesis or alternative. *Entrance requirements:* For master's, GRE General Test. *Application deadline:* For fall admission, 5/1 for domestic students. For winter admission, 10/15 for domestic students; for spring admission, 1/20 for domestic students. Applications are processed on a rolling basis. Application fee: $55. Electronic applications accepted. *Expenses:* Tuition, nonresident: full-time $9,021. Required fees: $3,597. *Financial support:* Career-related internships or fieldwork, Federal Work-Study, and institutionally sponsored loans available. Support available to part-time students. Financial award application deadline: 3/2; financial award applicants required to submit FAFSA. *Unit head:* Dr. Michael L. Green, Graduate Coordinator, 909-869-4007.

California State University, Fullerton, Graduate Studies, College of Natural Science and Mathematics, Department of Mathematics, Fullerton, CA 92834-9480. Offers applied mathematics (MA); mathematics (MA); mathematics for secondary school teachers (MA). Part-time programs available. *Students:* 8 full-time (3 women), 68 part-time (39 women); includes 36 minority (1 African American, 1 American Indian/Alaska Native, 19 Asian Americans or Pacific Islanders, 15 Hispanic Americans), 2 international. Average age 32. 78 applicants, 79% accepted, 40 enrolled. In 2005, 25 degrees awarded. *Degree requirements:* For master's, comprehensive exam or project. *Entrance requirements:* For master's, minimum GPA of 2.5 in last 60 units of course work, major in mathematics or related field. Application fee: $55. *Expenses:* Tuition, nonresident: part-time $339 per unit. *Financial support:* Research assistantships, teaching assistantships, career-related internships or fieldwork, Federal Work-Study, institutionally sponsored loans, and scholarships/grants available. Support available to part-time students. Financial award application deadline: 3/1. *Unit head:* Dr. Paul Deland, Chair, 714-278-3631.

California State University, Long Beach, Graduate Studies, College of Engineering, Department of Civil Engineering and Construction Engineering Management, Long Beach, CA 90840. Offers civil engineering (MSCE, CE); engineering (MS); engineering and industrial applied mathematics (PhD); engineering management/industrial management (MS); waste engineering and management (Graduate Certificate). Part-time programs available. *Faculty:* 12 full-time (1 woman), 8 part-time/adjunct (0 women). *Students:* 6 full-time (1 woman), 41 part-time (9 women); includes 25 minority (2 African Americans, 17 Asian Americans or Pacific Islanders, 6 Hispanic Americans), 1 international. Average age 31. 55 applicants, 56% accepted, 16 enrolled. In 2005, 12 degrees awarded. *Degree requirements:* For master's, comprehensive exam or thesis. *Entrance requirements:* Additional exam requirements/recommendations for international students: Required—TOEFL. *Application deadline:* For fall admission, 7/1 for domestic students; for spring admission, 12/1 for domestic students. Application fee: $55. Electronic applications accepted. *Expenses:* Tuition, nonresident: part-time $339 per semester hour. *Financial support:* Career-related internships or fieldwork, Federal Work-Study, institutionally sponsored loans, scholarships/grants, and unspecified assistantships available. Financial award application deadline: 3/2. *Faculty research:* Soils, hydraulics, seismic structures, composite metals, computer-aided manufacturing. *Unit head:* Dr. Hsiaq-Ling Geng Chu, Graduate Adviser, 562-985-5768, Fax: 562-985-2380, E-mail: chu@csulb.edu. *Application contact:* Dr. Chan-Feng (Steve) Tsai, Graduate Adviser, 562-985-5768, Fax: 562-985-2380, E-mail: stsai@csulb.edu.

California State University, Long Beach, Graduate Studies, College of Engineering, Department of Mechanical and Aerospace Engineering, Long Beach, CA 90840. Offers aerospace engineering (MSAE); engineering and industrial applied mathematics (PhD); interdisciplinary engineering (MSE); management engineering (MSE); mechanical engineering (MSME). Part-time programs available. *Faculty:* 18 full-time (1 woman), 14 part-time/adjunct (2 women). *Students:* 15 full-time (0 women), 59 part-time (5 women); includes 48 minority (34 Asian Americans or Pacific Islanders, 14 Hispanic Americans), 1 international. Average age 31. 59 applicants, 69% accepted, 22 enrolled. In 2005, 20 degrees awarded. *Entrance requirements:* Additional exam requirements/recommendations for international students: Required—TOEFL. *Application deadline:* For fall admission, 7/1 for domestic students; for spring admission, 12/1 for domestic students. Application fee: $55. Electronic applications accepted. *Expenses:* Tuition, nonresident: part-time $339 per semester hour. *Financial support:* Career-related internships or fieldwork, Federal Work-Study, institutionally sponsored loans, scholarships/grants, and unspecified assistantships available. Financial award application deadline: 3/2. *Faculty research:* Unsteady turbulent flows, solar energy, energy conversion, CAD/CAM, computer-assisted instruction. *Unit head:* Dr. Hamid Hefazi, Chairman, 562-985-1563, Fax: 562-985-4408, E-mail: hefazi@csulb.edu. *Application contact:* Dr. Hamid Rahai, Graduate Coordinator, 562-985-5132, Fax: 562-985-4408, E-mail: rahai@engr.csulb.edu.

California State University, Long Beach, Graduate Studies, College of Natural Sciences and Mathematics, Department of Mathematics and Statistics, Long Beach, CA 90840. Offers applied mathematics (MA); mathematics (MSEE). Part-time programs available. *Faculty:* 20 full-time (1 woman). *Students:* 47 full-time (22 women), 64 part-time (25 women); includes 62 minority (3 African Americans, 42 Asian Americans or Pacific Islanders, 17 Hispanic Americans), 3 international. Average age 33. 95 applicants, 62% accepted, 44 enrolled. In 2005, 15 degrees awarded. *Degree requirements:* For master's, comprehensive exam or thesis. *Application deadline:* For fall admission, 7/1 for domestic students; for spring admission, 12/1 for domestic students. Applications are processed on a rolling basis. Application fee: $55. Electronic applications accepted. *Expenses:* Tuition, nonresident: part-time $339 per semester hour. *Financial support:* Teaching assistantships, Federal Work-Study, institutionally sponsored loans, scholarships/grants, and traineeships available. Financial award application deadline: 3/2. *Faculty research:* Algebra, functional analysis, partial differential equations, operator theory, numerical analysis. *Unit head:* Dr. Robert A Mena, Chair, 562-985-4721, Fax: 562-985-8227, E-mail: rmena@csulb.edu. *Application contact:* Dr. Ngo Viet, Graduate Coordinator, 562-985-5610, Fax: 562-985-8227, E-mail: viet@csulb.edu.

California State University, Los Angeles, Graduate Studies, College of Natural and Social Sciences, Department of Mathematics, Los Angeles, CA 90032-8530. Offers mathematics (MS), including applied mathematics, mathematics. Part-time and evening/weekend programs available. *Faculty:* 7 full-time (1 woman). *Students:* 18 full-time (7 women), 50 part-time (17 women); includes 51 minority (6 African Americans, 19 Asian Americans or Pacific Islanders, 26 Hispanic Americans). In 2005, 12 degrees awarded. *Degree requirements:* For master's, comprehensive exam or thesis. *Entrance requirements:* For master's, previous course work in mathematics. Additional exam requirements/recommendations for international students: Required—TOEFL. *Application deadline:* For fall admission, 6/30 for domestic students; for spring admission, 2/1 for domestic students. Applications are processed on a rolling basis. Application fee: $55. *Financial support:* Teaching assistantships, Federal Work-Study available. Support available to part-time students. Financial award application deadline: 3/1. *Faculty research:* Group theory, functional analysis, convexity theory, ordered geometry. *Unit head:* Dr. P. K. Subramanian, Chair, 323-343-2150, Fax: 323-343-5071.

Case Western Reserve University, School of Graduate Studies, Department of Mathematics, Cleveland, OH 44106. Offers applied mathematics (MS, PhD); mathematics (MS, PhD). Part-time programs available. Terminal master's awarded for partial completion of doctoral program. *Degree requirements:* For master's, thesis (applied mathematics); for doctorate, one foreign language, thesis/dissertation. *Entrance requirements:* For master's and doctorate, GRE General Test. Additional exam requirements/recommendations for international students: Required—TOEFL. *Faculty research:* Probability theory, differential equations and control theory, differential geometry and topology, Lie groups, functional and harmonic analysis.

Central Missouri State University, The Graduate School, College of Arts and Sciences, Department of Mathematics and Computer Science, Warrensburg, MO 64093. Offers applied mathematics (MS); mathematics (MS). Part-time programs available. *Faculty:* 16 full-time (4 women). *Students:* 1 full-time (0 women), 10 part-time (4 women); includes 3 minority (all African Americans), 1 international. Average age 30. 3 applicants, 100% accepted, 3 enrolled. In 2005, 4 degrees awarded. *Degree requirements:* For master's, thesis (MS); comprehensive exam or thesis (MSE). *Entrance requirements:* For master's, GRE General Test (MSE), bachelor's degree in mathematics, minimum GPA of 3.0 (MS); minimum GPA of 2.75, teaching certificate (MSE). Additional exam requirements/recommendations for international students: Required—TOEFL (minimum score 500 paper-based; 173 computer-based). *Application deadline:* For fall admission, 6/1 priority date for domestic students, 5/1 priority date for international students; for spring admission, 10/1 priority date for domestic students, 10/1 priority date for international students. Applications are processed on a rolling basis. Application fee: $30 ($50 for international students). *Expenses:* Tuition, area resident: Full-time $5,160; part-time $215 per credit hour. Tuition, nonresident: full-time $10,320; part-time $430 per credit hour. Required fees: $336; $14 per credit hour. *Financial support:* In 2005–06, 2 teaching assistantships (averaging $3,750 per year) were awarded; Federal Work-Study, scholarships/grants, unspecified assistantships, and administrative assistantships also available. Support available to part-time students. Financial award application deadline: 3/1; financial award applicants required to submit FAFSA. *Faculty research:* Number theory, graph theory, mathematics education, topology, differential equations. Total annual research expenditures: $800,000. *Unit head:* Dr. Edward W. Davenport, Chair, 660-543-4931, Fax: 660-543-8006, E-mail: davenport@cmsu1.cmsu.edu.

Claremont Graduate University, Graduate Programs, School of Mathematical Sciences, Claremont, CA 91711-6160. Offers computational and systems biology (PhD); computational science (PhD); engineering mathematics (PhD); operations research and statistics (MA, MS); physical applied mathematics (MA, MS); pure mathematics (MA, MS); scientific computing (MA, MS); systems and control theory (MA, MS). Part-time programs available. *Faculty:* 4 full-time (0 women), 6 part-time/adjunct (2 women). *Students:* 54 full-time (13 women), 13 part-time (1 women); includes 20 minority (1 African American, 14 Asian Americans or Pacific Islanders, 5 Hispanic Americans), 16 international. Average age 38. In 2005, 5 master's, 5 doctorates awarded. Terminal master's awarded for partial completion of doctoral program. *Degree requirements:* For doctorate, 2 foreign languages, thesis/dissertation. *Entrance requirements:* For master's and doctorate, GRE General Test. *Application deadline:* For fall admission, 2/15 for domestic students. Applications are processed on a rolling basis. Electronic applications accepted. *Expenses:* Tuition: Full-time $27,902; part-time $1,214 per term. Required fees: $1,600; $800 per term. Tuition and fees vary according to degree level and program. *Financial support:* Fellowships, research assistantships, career-related internships or fieldwork, Federal Work-Study, institutionally sponsored loans, and tuition waivers (full and partial) available. Support available to part-time students. Financial award application deadline: 2/15; financial award applicants required to submit FAFSA. *Unit head:* John Angus, Dean, 909-621-8080, Fax: 909-607-8261, E-mail: john.angus@cgu.edu. *Application contact:* Susan Townzen, Program Coordinator, 909-621-8080, Fax: 909-607-8261, E-mail: susan.n.townzen@cgu.edu.

Clark Atlanta University, School of Arts and Sciences, Department of Mathematical Sciences, Atlanta, GA 30314. Offers applied mathematics (MS); computer science (MS). Part-time programs available. *Degree requirements:* For master's, one foreign language, thesis. *Entrance requirements:* For master's, GRE General Test, minimum GPA of 2.5. *Faculty research:* Numerical methods for operator equations, Ada language development.

Clemson University, Graduate School, College of Engineering and Science, Department of Mathematical Sciences, Clemson, SC 29634. Offers applied and pure mathematics (MS,

PhD); computational mathematics (MS, PhD); operations research (MS, PhD); statistics (MS, PhD). Part-time programs available. *Students:* 76 full-time (27 women), 5 part-time (2 women); includes 4 minority (2 African Americans, 2 Asian Americans or Pacific Islanders), 25 international. Average age 29. 73 applicants, 84% accepted, 24 enrolled. In 2005, 23 master's, 4 doctorates awarded. *Median time to degree:* Of those who began their doctoral program in fall 1997, 100% received their degree in 8 years or less. *Degree requirements:* For master's, final project, thesis optional; for doctorate, thesis/dissertation, qualifying exams. *Entrance requirements:* For master's and doctorate, GRE General Test. Additional exam requirements/recommendations for international students: Required—TOEFL; Recommended—TSE. *Application deadline:* For fall admission, 1/15 priority date for domestic students, 2/15 priority date for international students; for spring admission, 10/1 priority date for domestic students, 9/15 priority date for international students. Applications are processed on a rolling basis. Application fee: $50. Electronic applications accepted. *Financial support:* In 2005–06, 68 students received support; fellowships with partial tuition reimbursements available, research assistantships with partial tuition reimbursements available, teaching assistantships with partial tuition reimbursements available available. Financial award application deadline:4/15. *Faculty research:* Applied and computational analysis, cryptography, discrete mathematics, optimization, statistics. Total annual research expenditures: $1 million. *Unit head:* Robert L. Taylor, Chair, 864-656-5240, Fax: 864-656-5230, E-mail: rtaylo2@clemson.edu. *Application contact:* Dr. K.B. Kulasekera, Graduate Coordinator, 864-656-5231, Fax: 864-656-5230, E-mail: kk@clemson.edu.

See Close-Ups on pages 489 and 491.

Columbia University, Fu Foundation School of Engineering and Applied Science, Department of Applied Physics and Applied Mathematics, New York, NY 10027. Offers applied physics (MS, PhD), including applied mathematics (PhD), optical physics (PhD), plasma physics (PhD), solid state physics (PhD); applied physics and applied mathematics (Eng Sc D); materials science and engineering (MS, Eng Sc D, PhD); medical physics (MS); minerals engineering and materials science (Eng Sc D, PhD, Engr). Part-time programs available. *Faculty:* 30 full-time (2 women), 15 part-time/adjunct (1 woman). *Students:* 94 full-time (26 women), 30 part-time (5 women); includes 17 minority (1 American Indian/Alaska Native, 12 Asian Americans or Pacific Islanders, 4 Hispanic Americans), 53 international. Average age 24. 294 applicants, 22% accepted, 38 enrolled. In 2005, 33 master's, 5 doctorates awarded. Terminal master's awarded for partial completion of doctoral program. *Degree requirements:* For master's, English Proficiency Test; for doctorate, thesis/dissertation, qualifying exam, English Proficiency Test; for Engr, English Profiency Test. *Entrance requirements:* For master's and doctorate, GRE General Test, GRE Subject Test (strongly recommended). Additional exam requirements/recommendations for international students: Required—TOEFL. *Application deadline:* For fall admission, 12/15 priority date for domestic students, 12/15 priority date for international students; for spring admission, 10/1 priority date for domestic students, 10/1 priority date for international students. Application fee: $45. Electronic applications accepted. *Expenses:* Tuition: Full-time $31,448. Tuition and fees vary according to course level, course load, campus/location and program. *Financial support:* In 2005–06, 8 fellowships with full tuition reimbursements, 56 research assistantships with full tuition reimbursements (averaging $24,750 per year), 16 teaching assistantships with full tuition reimbursements (averaging $24,750 per year) were awarded; Federal Work-Study and unspecified assistantships also available. Financial award application deadline: 12/15; financial award applicants required to submit FAFSA. *Faculty research:* Plasma physics, applied mathematics, solid-state and optical physics, atmospheric, oceanic and earth physics, materials science and engineering. *Unit head:* Dr. Michael E. Mauel, Professor and Chairman, 212-854-4457, E-mail: seasinfo.apam@columbia.edu. *Application contact:* Marlene Arbo, Department Administrator, 212-854-4458, Fax: 212-854-8257, E-mail: seasinfo.apam@columbia.edu.

See Close-Up on page 335.

Cornell University, Graduate School, Graduate Fields of Arts and Sciences, Center for Applied Mathematics, Ithaca, NY 14853-0001. Offers PhD. *Faculty:* 98 full-time. *Students:* 134 applicants, 16% accepted, 11 enrolled. In 2005, 6 doctorates awarded. *Degree requirements:* For doctorate, one foreign language, thesis/dissertation, comprehensive exam. *Entrance requirements:* For doctorate, GRE General Test, GRE Subject Test (mathematics, recommended), 3 letters of recommendation. Additional exam requirements/recommendations for international students: Required—TOEFL (minimum score 550 paper-based; 213 computer-based). *Application deadline:* For fall admission, 1/15 for domestic students. Application fee: $60. Electronic applications accepted. *Financial support:* In 2005–06, 32 students received support, including 8 fellowships with full tuition reimbursements available, 6 research assistantships with full tuition reimbursements available, 18 teaching assistantships with full tuition reimbursements available; institutionally sponsored loans, scholarships/grants, health care benefits, tuition waivers (full and partial), and unspecified assistantships also available. Financial award applicants required to submit FAFSA. *Faculty research:* Nonlinear systems and PDE's, numerical methods, signal and image processing, mathematical biology, discrete mathematics and optimization. *Unit head:* Director of Graduate Studies, 607-255-4756, Fax: 607-255-9860. *Application contact:* Graduate Field Assistant, 607-255-4756, Fax: 607-255-9860, E-mail: appliedmath@cornell.edu.

Announcement: The Center for Applied Mathematics is an interdepartmental program with more than 80 faculty members. Students may pursue PhD studies over a broad range of the mathematical sciences and are admitted to the field from a variety of educational backgrounds with strong mathematical components. Students are normally awarded fellowships or teaching or research assistantships.

Cornell University, Graduate School, Graduate Fields of Engineering, Field of Chemical Engineering, Ithaca, NY 14853-0001. Offers advanced materials processing (M Eng, MS, PhD); applied mathematics and computational methods (M Eng, MS, PhD); biochemical engineering (M Eng, MS, PhD); chemical reaction engineering (M Eng, MS, PhD); classical and statistical thermodynamics (M Eng, MS, PhD); fluid dynamics, rheology and biorheology (M Eng, MS, PhD); heat and mass transfer (M Eng, MS, PhD); kinetics and catalysis (M Eng, MS, PhD); polymers (M Eng, MS, PhD); surface science (M Eng, MS, PhD). *Faculty:* 30 full-time (1 woman). *Students:* 266 applicants, 23% accepted, 30 enrolled. In 2005, 21 master's, 11 doctorates awarded. *Degree requirements:* For master's, thesis (MS); for doctorate, thesis/dissertation, comprehensive exam. *Entrance requirements:* For master's and doctorate, GRE General Test, 2 letters of recommendation. Additional exam requirements/recommendations for international students: Required—TOEFL (minimum score 580 paper-based; 237 computer-based). *Application deadline:* For fall admission, 1/15 for domestic students. Application fee: $60. Electronic applications accepted. *Financial support:* In 2005–06, 81 students received support, including 27 fellowships with full tuition reimbursements available, 44 research assistantships with full tuition reimbursements available, 10 teaching assistantships with full tuition reimbursements available; institutionally sponsored loans, scholarships/grants, health care benefits, tuition waivers (full and partial), and unspecified assistantships also available. Financial award applicants required to submit FAFSA. *Faculty research:* Biochemical, biomedical and metabolic engineering; fluid and polymer dynamics; surface science and chemical kinetics; electronics materials; microchemical systems and nanotechnology. *Unit head:* Director of Graduate Studies, 607-255-4550. *Application contact:* Graduate Field Assistant, 607-255-4550, E-mail: dgs@cheme.cornell.edu.

Cornell University, Graduate School, Graduate Fields of Engineering, Field of Operations Research and Industrial Engineering, Ithaca, NY 14853-0001. Offers applied probability and statistics (PhD); manufacturing systems engineering (PhD); mathematical programming (PhD); operations research and industrial engineering (M Eng). *Faculty:* 40 full-time (5 women). *Students:* 451 applicants, 25% accepted, 89 enrolled. In 2005, 87 master's, 4 doctorates awarded. *Degree requirements:* For doctorate, thesis/dissertation, comprehensive exam. *Entrance requirements:* For master's and doctorate, GRE General Test, 3 letters of recommendation. Additional exam requirements/recommendations for international students: Required—TOEFL (minimum score 600 paper-based; 250 computer-based). *Application deadline:* For fall admission, 1/15 for domestic students. Application fee: $60. Electronic applications accepted. *Financial support:* In 2005–06, 13 fellowships with full tuition reimbursements, 9 research assistantships with full tuition reimbursements, 24 teaching assistantships with full tuition reimbursements were awarded; institutionally sponsored loans, scholarships/grants, health care benefits, tuition waivers (full and partial), and unspecified assistantships also available. Financial award applicants required to submit FAFSA. *Faculty research:* Mathematical programming and combinatorial optimization, statistics, stochastic processes, mathematical finance, simulation, manufacturing, and e-commerce. *Unit head:* Director of Graduate Studies, 607-255-9128, Fax: 607-255-9129. *Application contact:* Graduate Field Assistant, 607-255-9128, Fax: 607-255-9129, E-mail: orie@cornell.edu.

Dalhousie University, Faculty of Graduate Studies, DalTech, Faculty of Engineering, Department of Engineering Mathematics, Halifax, NS B3H 4R2, Canada. Offers M Sc, PhD. *Degree requirements:* For master's and doctorate, thesis/dissertation. *Entrance requirements:* Additional exam requirements/recommendations for international students: Required—TOEFL. *Faculty research:* Piecewise regression and robust statistics, random field theory, dynamical systems, wave loads on offshore structures, digital signal processing.

East Carolina University, Graduate School, Thomas Harriot College of Arts and Sciences, Department of Mathematics, Greenville, NC 27858-4353. Offers applied mathematics (MA); mathematics (MA). Part-time and evening/weekend programs available. *Faculty:* 22 full-time (7 women). *Students:* 9 full-time (6 women), 3 part-time (1 woman); includes 3 minority (2 African Americans, 1 American Indian/Alaska Native). Average age 29. 22 applicants, 0% accepted, 0 enrolled. In 2005, 4 degrees awarded. *Degree requirements:* For master's, comprehensive exam. *Entrance requirements:* For master's, GRE General Test, MAT. Additional exam requirements/recommendations for international students: Required—TOEFL. *Application deadline:* For fall admission, 6/1 for domestic students; for spring admission, 10/15 for domestic students. Applications are processed on a rolling basis. Application fee: $50. *Expenses:* Tuition, state resident: full-time $2,516. Tuition, nonresident: full-time $12,832. *Financial support:* Research assistantships with partial tuition reimbursements, teaching assistantships with partial tuition reimbursements available. Financial award application deadline: 6/1. *Unit head:* Dr. John Daughtry, Chair, 252-328-6461, Fax: 252-328-6414, E-mail: daughtryj@ecu.edu. *Application contact:* Dean of Graduate School, 252-328-6012, Fax: 252-328-6071, E-mail: gradschool@ecu.edu.

École Polytechnique de Montréal, Graduate Programs, Department of Mathematics, Montréal, QC H3C 3A7, Canada. Offers mathematical method in CA engineering (M Eng, M Sc A, PhD); operational research (M Eng, M Sc A, PhD). Part-time programs available. *Degree requirements:* For master's and doctorate, one foreign language, thesis/dissertation. *Entrance requirements:* For master's, minimum GPA of 2.75; for doctorate, minimum GPA of 3.0. *Faculty research:* Statistics and probability, fractal analysis, optimization.

Florida Atlantic University, Charles E. Schmidt College of Science, Department of Mathematical Science, Boca Raton, FL 33431-0991. Offers applied mathematics and statistics (MS); mathematics (MS, MST, PhD). Part-time programs available. *Faculty:* 25 full-time (2 women), 3 part-time/adjunct (0 women). *Students:* 46 full-time (17 women), 14 part-time (8 women); includes 7 minority (4 African Americans, 1 Asian American or Pacific Islander, 2 Hispanic Americans), 30 international. Average age 31. 33 applicants, 82% accepted, 20 enrolled. In 2005, 18 degrees awarded. Terminal master's awarded for partial completion of doctoral program. *Degree requirements:* For master's, thesis (for some programs), comprehensive exam (for some programs), registration; for doctorate, thesis/dissertation, comprehensive exam, registration. *Entrance requirements:* For master's and doctorate, GRE General Test, minimum GPA of 3.0. Additional exam requirements/recommendations for international students: Required—TOEFL (minimum score 500 paper-based; 173 computer-based). *Application deadline:* For fall admission, 6/1 priority date for domestic students, 2/15 priority date for international students; for spring admission, 10/20 priority date for domestic students, 8/15 priority date for international students. Applications are processed on a rolling basis. Application fee: $30. Electronic applications accepted. *Expenses:* Tuition, state resident: full-time $4,394; part-time $244 per credit. Tuition, nonresident: full-time $16,441; part-time $912 per credit. *Financial support:* In 2005–06, fellowships with partial tuition reimbursements (averaging $20,000 per year), 20 teaching assistantships with partial tuition reimbursements (averaging $20,000 per year) were awarded; Federal Work-Study also available. Financial award application deadline: 4/1. *Faculty research:* Cryptography, statistics, algebra, analysis, combinatorics. Total annual research expenditures: $550,000. *Unit head:* Dr. Yoram Sagher, Chair, 561-297-3341, Fax: 561-297-2436, E-mail: sagher@fau.edu. *Application contact:* Dr. Heinrich Niederhausen, Graduate Director, 561-297-3237, Fax: 561-297-2436, E-mail: grad@math.fau.edu.

Florida Institute of Technology, Graduate Programs, College of Science, Department of Mathematical Sciences, Melbourne, FL 32901-6975. Offers applied mathematics (MS, PhD); operations research (MS, PhD). Part-time and evening/weekend programs available. *Faculty:* 9 full-time (2 women). *Students:* 27 full-time (5 women), 13 part-time (3 women); includes 8 minority (4 African Americans, 1 Asian American or Pacific Islander, 3 Hispanic Americans), 16 international. Average age 32. 41 applicants, 76% accepted, 13 enrolled. In 2005, 11 master's, 1 doctorate awarded. Terminal master's awarded for partial completion of doctoral program. *Degree requirements:* For master's, thesis (for some programs), comprehensive exam (for some programs), registration; for doctorate, thesis/dissertation, comprehensive exam, registration. *Entrance requirements:* For master's, minimum GPA of 3.0; for doctorate, minimum GPA of 3.2, resumé, 3 letters of recommendation. Additional exam requirements/recommendations for international students: Required—TOEFL (minimum score 550 paper-based; 213 computer-based). *Application deadline:* Applications are processed on a rolling basis. Application fee: $50. Electronic applications accepted. *Expenses:* Tuition: Part-time $825 per credit. *Financial support:* In 2005–06, 17 students received support, including 17 teaching assistantships with full and partial tuition reimbursements available (averaging $8,179 per year); career-related internships or fieldwork and tuition remissions also available. Financial award application deadline: 3/1; financial award applicants required to submit FAFSA. *Faculty research:* Real analysis, ODE, PDE, numerical analysis, statistics, data analysis, combinatorics, artificial intelligence, simulation. Total annual research expenditures:$99,134. *Unit head:* Dr. V. Lakshmikantham, Department Head, 321-674-7412, Fax: 321-674-7412, E-mail: lakshmik@fit.edu. *Application contact:* Carolyn P. Farrior, Director of Graduate Admissions, 321-674-7118, Fax: 321-723-9468, E-mail: cfarrior@fit.edu.

Florida State University, Graduate Studies, College of Arts and Sciences, Department of Mathematics, Tallahassee, FL 32306. Offers applied mathematics (MS, PhD); biomedical mathematics (MS, PhD); financial mathematics (MS, PhD); pure mathematics (MS, PhD). Part-time programs available. *Faculty:* 41 full-time (4 women), 10 part-time/adjunct (6 women). *Students:* 117 full-time (27 women), 5 part-time; includes 20 minority (3 African Americans, 1 American Indian/Alaska Native, 7 Asian Americans or Pacific Islanders, 9 Hispanic Americans), 56 international. Average age 26. 317 applicants, 66% accepted, 41 enrolled. In 2005, 31 master's, 7 doctorates awarded. Terminal master's awarded for partial completion of doctoral program. *Degree requirements:* For master's, thesis optional; for doctorate, thesis/dissertation, preliminary exam. *Entrance requirements:* For master's and doctorate, GRE General Test, minimum GPA of 3.0, 4-year bachelor's degree. Additional exam requirements/recommendations for international students: Required—TOEFL (minimum score 550 paper-based; 213 computer-based). *Application deadline:* For fall admission, 3/1 priority date for domestic students, 1/1 priority date for international students; for spring admission, 8/15 priority date for domestic students, 5/1 priority date for international students. Applications are processed on a rolling basis. Application fee: $30. Electronic applications accepted. *Financial support:* In 2005–06, 100 students received support, including 2 fellowships with full tuition reimbursements available (averaging $18,000 per year), 12 research assistantships with full tuition reimbursements available (averaging $16,000 per year), 79 teaching assistantships with full tuition reimbursements available (averaging $16,000 per year); career-related intern-

Applied Mathematics

Florida State University (continued)

ships or fieldwork, scholarships/grants, and unspecified assistantships also available. Financial award application deadline: 3/1. *Faculty research:* Low-dimensional manifolds, algebra geometry, fluid dynamics, financial mathematics, biomedical mathematics. *Unit head:* Dr. Philip L Bowers, Chairperson, 850-645-3338, Fax: 850-644-4053, E-mail: bowers@math.fsu.edu. *Application contact:* Dr. Eric P. Klassen, Associate Chair for Graduate Studies, 850-644-2202, Fax: 850-644-4053, E-mail: klassen@math.fsu.edu.

The George Washington University, Columbian College of Arts and Sciences, Department of Mathematics, Washington, DC 20052. Offers applied mathematics (MA, MS); pure mathematics (MA, PhD). Part-time and evening/weekend programs available. Terminal master's awarded for partial completion of doctoral program. *Degree requirements:* For master's, comprehensive exam; for doctorate, one foreign language, thesis/dissertation, general exam. *Entrance requirements:* For master's and doctorate, GRE General Test, minimum GPA of 3.0, interview. Additional exam requirements/recommendations for international students: Required—TOEFL (minimum score 550 paper-based; 213 computer-based). Electronic applications accepted.

Georgia Institute of Technology, Graduate Studies and Research, College of Sciences, School of Mathematics, Atlanta, GA 30332-0001. Offers algorithms, combinatorics, and optimization (PhD); applied mathematics (MS); bioinformatics (PhD); mathematics (PhD); quantitative and computational finance (MS); statistics (MS Stat). Terminal master's awarded for partial completion of doctoral program. *Degree requirements:* For master's, thesis or alternative; for doctorate, one foreign language, thesis/dissertation. *Entrance requirements:* For master's, GRE General Test, minimum GPA of 3.0; for doctorate, GRE General Test, GRE Subject Test, minimum GPA of 3.0. Additional exam requirements/recommendations for international students: Required—TOEFL. Electronic applications accepted. *Faculty research:* Dynamical systems, discrete mathematics, probability and statistics, mathematical physics.

Hampton University, Graduate College, Program in Applied Mathematics, Hampton, VA 23668. Offers MS. *Degree requirements:* For master's, thesis optional. *Entrance requirements:* For master's, GRE General Test.

Harvard University, Graduate School of Arts and Sciences, Division of Engineering and Applied Sciences, Cambridge, MA 02138. Offers applied mathematics (ME, SM, PhD); applied physics (ME, SM, PhD); computer science (ME, SM, PhD); computing technology (PhD); engineering science (ME); engineering sciences (SM, PhD). Part-time programs available. *Faculty:* 75 full-time (5 women), 12 part-time/adjunct (3 women). *Students:* 277 full-time (65 women), 11 part-time (1 woman); includes 40 minority (5 African Americans, 3 American Indian/Alaska Native, 26 Asian Americans or Pacific Islanders, 6 Hispanic Americans), 126 international. 1,197 applicants, 14% accepted, 86 enrolled. In 2005, 68 master's, 19 doctorates awarded. Terminal master's awarded for partial completion of doctoral program. *Median time to degree:* Of those who began their doctoral program in fall 1997, 92% received their degree in 8 years or less. *Degree requirements:* For master's, thesis optional; for doctorate, thesis/dissertation, comprehensive exam, registration. *Entrance requirements:* For master's and doctorate, GRE General Test, GRE Subject Test (recommended), 3 letters of recommendation. Additional exam requirements/recommendations for international students: Required—TOEFL (minimum score 550 paper-based; 213 computer-based). *Application deadline:* For fall admission, 12/15 priority date for domestic students, 12/15 priority date for international students. For winter admission, 1/2 for domestic students. Application fee: $95. Electronic applications accepted. *Expenses:* Tuition: Full-time $28,752. Full-time tuition and fees vary according to program and student level. *Financial support:* In 2005–06, 55 fellowships with full tuition reimbursements (averaging $19,350 per year), 187 research assistantships (averaging $31,856 per year), 112 teaching assistantships (averaging $5,138 per year) were awarded; Federal Work-Study, institutionally sponsored loans, and traineeships also available. *Faculty research:* Applied mathematics, applied physics, computer science & electrical engineering, environmental engineering, mechanical and biomedical engineering. Total annual research expenditures: $33.5 million. *Unit head:* Betsey Cogswell, Dean, 617-495-5829, Fax: 617-495-5264, E-mail: venky@deas.harvard.edu. *Application contact:* Office of Admissions and Financial Aid, 617-495-5315, E-mail: admissions@deas.harvard.edu.

Hofstra University, College of Liberal Arts and Sciences, Division of Natural Sciences, Mathematics, Engineering, and Computer Science, Department of Mathematics, Hempstead, NY 11549. Offers applied mathematics (MS); mathematics (MA). Part-time and evening/weekend programs available. *Faculty:* 5 full-time (1 woman), 1 (woman) part-time/adjunct. *Students:* 6 full-time (2 women), 8 part-time (3 women); includes 3 minority (2 African Americans, 1 Hispanic American), 1 international. Average age 28. 13 applicants, 77% accepted, 4 enrolled. In 2005, 5 degrees awarded. *Degree requirements:* For master's, A-MA. *Entrance requirements:* For master's, bachelor's degree with strong background in math. Additional exam requirements/recommendations for international students: Required—TOEFL (minimum score 550 paper-based; 213 computer-based). *Application deadline:* Applications are processed on a rolling basis. Application fee: $60. Electronic applications accepted. *Expenses:* Tuition: Full-time $12,060; part-time $670 per credit. Required fees: $930; $155 per term. Tuition and fees vary according to course load and program. *Financial support:* In 2005–06, 14 students received support, including 2 fellowships (averaging $2,667 per year); tuition waivers (full) also available. Financial award applicants required to submit FAFSA. *Faculty research:* Logic-four valued logic, homotopy theory, combinatonics, statistics, algebraic number theory. *Unit head:* Dr. Maryisa T. Weiss, Chairperson, 516-463-5580, Fax: 516-463-6596, E-mail: matmtw@hofstra.edu. *Application contact:* Carol Drummer, Dean of Graduate Admissions, 516-463-4876, Fax: 516-463-4664, E-mail: gradstudent@hofstra.edu.

Howard University, Graduate School of Arts and Sciences, Department of Mathematics, Washington, DC 20059-0002. Offers applied mathematics (MS, PhD); mathematics (MS, PhD). Part-time programs available. *Faculty:* 25 full-time (5 women). *Students:* 26 full-time (8 women), 4 part-time; includes 22 minority (all African Americans), 7 international. In 2005, 5 master's, 2 doctorates awarded. Terminal master's awarded for partial completion of doctoral program. *Degree requirements:* For master's, thesis or alternative, qualifying exam, comprehensive exam; for doctorate, 2 foreign languages, thesis/dissertation, qualifying exams, comprehensive exam. *Entrance requirements:* For master's, GRE General Test, minimum GPA of 3.0; for doctorate, GRE General Test. Additional exam requirements/recommendations for international students: Required—TOEFL. *Application deadline:* For fall admission, 2/15 priority date for domestic students, 2/15 priority date for international students; for spring admission, 11/1 priority date for domestic students, 11/1 priority date for international students. Applications are processed on a rolling basis. Application fee: $45. Electronic applications accepted. *Financial support:* In 2005–06, 11 students received support, including fellowships with full tuition reimbursements available (averaging $16,000 per year), 2 research assistantships with full tuition reimbursements available (averaging $15,000 per year), 9 teaching assistantships with full tuition reimbursements available (averaging $13,000 per year); institutionally sponsored loans and scholarships/grants also available. Financial award application deadline: 4/1. *Unit head:* Dr. Abdul-Aziz Yzkubu, Chairman, 202-806-6830, Fax: 202-806-6831, E-mail: ayakubu@howard.edu. *Application contact:* Dr. Neil Hindman, Director of Graduate Studies, 202-806-5927, Fax: 202-806-6831, E-mail: nhindman@howard.edu.

Hunter College of the City University of New York, Graduate School, School of Arts and Sciences, Department of Mathematics and Statistics, New York, NY 10021-5085. Offers applied mathematics (MA); mathematics for secondary education (MA); pure mathematics (MA). Part-time and evening/weekend programs available. *Faculty:* 19 full-time (4 women), 2 part-time/adjunct (both women). *Students:* 2 full-time (1 woman), 18 part-time (8 women); includes 7 minority (5 Asian Americans or Pacific Islanders, 2 Hispanic Americans). Average age 34. 14 applicants, 57% accepted, 4 enrolled. In 2005, 17 degrees awarded. *Degree requirements:* For master's, one foreign language, thesis (for some programs), comprehensive exam. *Entrance requirements:* For master's, GRE General Test, 24 credits in mathematics. Additional exam requirements/recommendations for international students: Required—TOEFL. *Application deadline:* For fall admission, 4/1 for domestic students, 2/1 for international students; for spring admission, 11/1 for domestic students, 9/1 for international students. Application fee: $125. *Expenses:* Tuition, state resident: full-time $6,400; part-time $270 per credit. Tuition, nonresident: part-time $500 per credit. International tuition: $12,000 full-time. Required fees: $50 per term. Part-time tuition and fees vary according to course load and program. *Financial support:* Federal Work-Study, institutionally sponsored loans, scholarships/grants, and tuition waivers (partial) available. Support available to part-time students. *Faculty research:* Data analysis, dynamical systems, computer graphics, topology, statistical decision theory. *Unit head:* Ada Peluso, Chairperson, 212-772-5300, Fax: 212-772-4858, E-mail: peluso@math.hunter.cuny.edu. *Application contact:* William Zlata, Director for Graduate Admissions, 212-772-4482, Fax: 212-650-3336, E-mail: admissions@hunter.cuny.edu.

Illinois Institute of Technology, Graduate College, College of Science and Letters, Department of Applied Mathematics, Chicago, IL 60616-3793. Offers applied mathematics (MS, PhD); financial mathematics (MS). Terminal master's awarded for partial completion of doctoral program. *Degree requirements:* For master's, comprehensive exam; for doctorate, thesis/dissertation, comprehensive exam. *Entrance requirements:* For master's, GRE General Test (combined score of 1100), minimum undergraduate GPA of 3.0; for doctorate, GRE General Test (combined score of 1150), minimum undergraduate GPA of 3.5. Additional exam requirements/recommendations for international students: Required—TOEFL (minimum score 550 paper-based; 213 computer-based).

Indiana University Bloomington, Graduate School, College of Arts and Sciences, Department of Mathematics, Bloomington, IN 47405-7000. Offers applied mathematics–numerical analysis (MA, PhD); mathematics education (MAT); probability-statistics (MA, PhD). PhD offered through the University Graduate School. *Faculty:* 36 full-time (1 woman). *Students:* 116 full-time (39 women), 20 part-time (4 women); includes 6 minority (1 African American, 5 Asian Americans or Pacific Islanders), 74 international. Average age 27. In 2005, 21 master's, 5 doctorates awarded. Terminal master's awarded for partial completion of doctoral program. *Degree requirements:* For doctorate, one foreign language, thesis/dissertation. *Entrance requirements:* For master's and doctorate, GRE General Test, GRE Subject Test. Additional exam requirements/recommendations for international students: Required—TOEFL. *Application deadline:* For fall admission, 1/15 priority date for domestic students, 12/15 priority date for international students; for spring admission, 9/1 priority date for domestic students, 9/1 priority date for international students. Applications are processed on a rolling basis. Application fee: $50 ($60 for international students). Electronic applications accepted. *Expenses:* Tuition, state resident: full-time $5,437; part-time $227 per credit hour. Tuition, nonresident: full-time $15,836; part-time $660 per credit hour. Required fees: $821. Tuition and fees vary according to campus/location and program. *Financial support:* Fellowships with full tuition reimbursements, research assistantships, teaching assistantships with full tuition reimbursements, Federal Work-Study available. Support available to part-time students. Financial award application deadline: 2/1. *Faculty research:* Topology, geometry, algebra. *Unit head:* David Hoff, Chair, 812-855-2200. *Application contact:* Misty Cummings, Graduate Secretary, 812-855-2645, Fax: 812-855-0046, E-mail: gradmath@indiana.edu.

Indiana University of Pennsylvania, Graduate School and Research, College of Natural Sciences and Mathematics, Department of Mathematics, Program in Applied Mathematics, Indiana, PA 15705-1087. Offers MS. *Degree requirements:* For master's, thesis optional. *Entrance requirements:* For master's, 2 letters of recommendation. Additional exam requirements/recommendations for international students: Required—TOEFL.

Indiana University–Purdue University Fort Wayne, School of Arts and Sciences, Department of Mathematical Sciences, Fort Wayne, IN 46805-1499. Offers applied mathematics (MS); applied statistics (Certificate); mathematics (MS); operations research (MS). Part-time and evening/weekend programs available. *Faculty:* 20 full-time (4 women). *Students:* 8 full-time (1 woman), 9 part-time (4 women); includes 2 minority (both American Indian/Alaska Native). Average age 33. 10 applicants, 100% accepted, 8 enrolled. In 2005, 5 degrees awarded. *Degree requirements:* For Certificate, completed a calculus and a statistics course. *Entrance requirements:* For master's, minimum GPA of 3.0, major or minor in mathematics. Additional exam requirements/recommendations for international students: Required—TOEFL (minimum score 600 paper-based; 260 computer-based). *Application deadline:* For fall admission, 7/1 for domestic students; for spring admission, 12/1 for domestic students. Applications are processed on a rolling basis. Application fee: $55. *Expenses:* Tuition, state resident: full-time $4,023; part-time $232 per credit. Tuition, nonresident: full-time $9,182; part-time $503 per credit. Required fees: $383. Tuition and fees vary according to course load. *Financial support:* In 2005–06, 10 teaching assistantships with partial tuition reimbursements (averaging $11,700 per year) were awarded; scholarships/grants and unspecified assistantships also available. Support available to part-time students. Financial award application deadline: 3/1; financial award applicants required to submit FAFSA. *Faculty research:* Graph theory, biostatistics, partial differential equations, 11-venn diagrams, random operator equations. Total annual research expenditures: $67,015. *Unit head:* Dr. David A. Legg, Chair, 260-481-6821, Fax: 260-481-6880, E-mail: legg@ipfw.edu. *Application contact:* Dr. W. Douglas Weakley, Director of Graduate Studies, 260-481-6821, Fax: 260-481-6880, E-mail: weakley@ipfw.edu.

Indiana University–Purdue University Indianapolis, School of Science, Department of Mathematical Sciences, Indianapolis, IN 46202-3216. Offers applied mathematics (MS, PhD); applied statistics (MS); mathematics (MS, PhD). Part-time programs available. *Faculty:* 10 full-time (0 women), 1 part-time/adjunct (0 women). *Students:* 12 full-time (4 women), 42 part-time (20 women); includes 3 minority (1 African American, 2 Asian Americans or Pacific Islanders), 16 international. Average age 32. In 2005, 19 degrees awarded. Terminal master's awarded for partial completion of doctoral program. *Degree requirements:* For master's, thesis optional; for doctorate, one foreign language, thesis/dissertation. *Entrance requirements:* For doctorate, GRE. Additional exam requirements/recommendations for international students: Required—TOEFL. *Application deadline:* For fall admission, 2/1 for domestic students. Application fee: $50 ($60 for international students). *Expenses:* Tuition, state resident: full-time $5,159; part-time $215 per credit hour. Tuition, nonresident: full-time $14,890; part-time $620 per credit hour. Required fees: $614. Tuition and fees vary according to campus/location and program. *Financial support:* In 2005–06, 14 students received support; fellowships with tuition reimbursements available, research assistantships with tuition reimbursements available, teaching assistantships with tuition reimbursements available, career-related internships or fieldwork, Federal Work-Study, and tuition waivers (full and partial) available. Financial award application deadline: 3/1. *Faculty research:* Mathematical physics, analysis, operator theory, functional analysis, integrated systems. *Unit head:* Benzion Boukai, Chair, 317-274-6920, Fax: 317-274-3460, E-mail: bboukai@math.iupui.edu. *Application contact:* Joan Morand, Student Services Specialist, 317-274-4127, Fax: 317-274-3460.

See Close-Up on page 503.

Indiana University South Bend, College of Liberal Arts and Sciences, Program in Applied Mathematics and Computer Science, South Bend, IN 46634-7111. Offers MS. *Students:* 9 full-time (4 women), 20 part-time (5 women); includes 1 minority (Hispanic American), 12 international. Average age 29.Application fee: $52 for international students. *Expenses:* Tuition, state resident: full-time $4,222; part-time $176 per credit hour. Tuition, nonresident: full-time $10,286; part-time $429 per credit hour. Required fees: $406. Tuition and fees vary according to campus/location and program. *Unit head:* Dr. James Wolfer, Graduate Director, 574-237-6521, Fax: 574-237-4335, E-mail: amcs@iusb.edu.

Inter American University of Puerto Rico, San Germán Campus, Graduate Studies Center, Graduate Program in Applied Mathematics, San Germán, PR 00683-5008. Offers MA. Part-time and evening/weekend programs available. *Faculty:* 2 full-time. *Students:* 42. In 2005, 4 degrees awarded. *Degree requirements:* For master's, comprehensive exam. *Entrance requirements:* For master's, EXADEP or GRE General Test, minimum GPA of 3.0. *Application deadline:* For fall admission, 4/30 for domestic students; for spring admission, 11/15 for domestic students. Application fee: $31. Master $170/credit Ph.D Bus. Adm. $410/credit Ph.D

(Psychology) $270/credit. *Expenses:* Tuition: Full-time $3,060; part-time $170 per credit. Required fees: $418; $418 per year. *Financial support:* Teaching assistantships, Federal Work-Study and unspecified assistantships available. *Application contact:* Dr. Alvaro Lecompte, Graduate Coordinator, 787-264-1912 Ext. 7358, Fax: 787-892-7510, E-mail: alecompte@sg.inter.edu.

Iowa State University of Science and Technology, Graduate College, College of Liberal Arts and Sciences, Department of Mathematics, Ames, IA 50011. Offers applied mathematics (MS, PhD); mathematics (MS, PhD); school mathematics (MSM). *Faculty:* 48 full-time, 1 part-time/adjunct. *Students:* 60 full-time (16 women), 7 part-time (5 women); includes 1 minority (Asian American or Pacific Islander), 28 international. 67 applicants, 42% accepted, 15 enrolled. In 2005, 13 master's, 9 doctorates awarded. *Degree requirements:* For master's, thesis or alternative; for doctorate, thesis/dissertation. *Entrance requirements:* For master's and doctorate, GRE General Test. Additional exam requirements/recommendations for international students: Required—TOEFL (paper score 550; computer score 213) or IELTS (score 6.5). *Application deadline:* For fall admission, 2/1 priority date for domestic students, 2/1 priority date for international students. Application fee: $30 ($70 for international students). Electronic applications accepted. *Expenses:* Tuition, state resident: full-time $6,410. Tuition, nonresident: full-time $16,422. Tuition and fees vary according to program. *Financial support:* In 2005–06, 3 research assistantships with full and partial tuition reimbursements (averaging $18,610 per year), 51 teaching assistantships with full and partial tuition reimbursements (averaging $16,173 per year) were awarded; fellowships, scholarships/grants, health care benefits, and unspecified assistantships also available. *Unit head:* Dr. Justin R. Peters, Chair, 515-294-1752, Fax: 515-294-5454, E-mail: gradmath@iastate.edu. *Application contact:* Dr. Paul Sacks, Director of Graduate Education, 515-294-8143, E-mail: gradmath@iastate.edu.

The Johns Hopkins University, G. W. C. Whiting School of Engineering, Department of Applied Mathematics and Statistics, Baltimore, MD 21218-2699. Offers discrete mathematics (MA, MSE, PhD); operations research/optimization/decision science (MA, MSE, PhD); statistics/probability/stochastic processes (MA, MSE, PhD). *Faculty:* 15 full-time (1 woman), 3 part-time/adjunct (all women). *Students:* 36 full-time (12 women), 2 part-time (1 woman); includes 2 minority (both Asian Americans or Pacific Islanders), 20 international. Average age 27. 112 applicants, 67% accepted, 22 enrolled. In 2005, 12 master's, 3 doctorates awarded. Terminal master's awarded for partial completion of doctoral program. *Median time to degree:* Of those who began their doctoral program in fall 1997, 40% received their degree in 8 years or less. *Degree requirements:* For master's, thesis (for some programs), registration; for doctorate, thesis/dissertation, oral exam, introductory exam. *Entrance requirements:* For master's and doctorate, GRE General Test, GRE Subject Test. Additional exam requirements/recommendations for international students: Required—TOEFL (minimum score 600 paper-based; 250 computer-based). *Application deadline:* For fall admission, 1/15 priority date for domestic students, 1/15 priority date for international students. Applications are processed on a rolling basis. Application fee: $0. Electronic applications accepted. *Expenses:* Tuition: Full-time $30,960. Tuition and fees vary according to degree level and program. *Financial support:* In 2005–06, 34 students received support, including 1 fellowship with full tuition reimbursement available (averaging $17,500 per year), 3 research assistantships with full tuition reimbursements available (averaging $15,750 per year), 20 teaching assistantships with full tuition reimbursements available (averaging $15,750 per year); Federal Work-Study, institutionally sponsored loans, health care benefits, tuition waivers (partial), and unspecified assistantships also available. Financial award application deadline: 1/15. *Faculty research:* Discrete mathematics, probability, statistics, optimization and operations research, scientific computation. Total annual research expenditures:$587,000. *Unit head:* Dr. Daniel Q. Naiman, Chair, 410-516-7203, Fax: 410-516-7459, E-mail: daniel.naiman@jhu.edu. *Application contact:* Kristin Bechtel, Academic Program Coordinator, 410-516-7198, Fax: 410-516-7459, E-mail: bechtel@ams.jhu.edu.

The Johns Hopkins University, G. W. C. Whiting School of Engineering, Engineering and Applied Science Programs for Professionals, Department of Applied and Computational Mathematics, Baltimore, MD 21218-2699. Offers MS. Part-time and evening/weekend programs available. *Faculty:* 10 part-time/adjunct (2 women). *Students:* Average age 29. In 2005, 18 degrees awarded. *Degree requirements:* For master's, registration. *Entrance requirements:* For master's, GRE General Test. Additional exam requirements/recommendations for international students: Required—TOEFL. *Application deadline:* For fall admission, 1/15 for domestic students. Applications are processed on a rolling basis. Application fee: $70. Electronic applications accepted. *Expenses:* Tuition: Full-time $30,960. Tuition and fees vary according to degree level and program. *Financial support:* Applicants required to submit FAFSA. Total annual research expenditures:$606,210. *Unit head:* Dr. Jim Spall, Program Chair, 410-540-2960, Fax: 410-579-8049. *Application contact:* Doug Schiller, Assistant Director of Registration, 410-540-2960, Fax: 410-579-8049, E-mail: schiller@jhu.edu.

Kent State University, College of Arts and Sciences, Department of Mathematical Sciences, Kent, OH 44242-0001. Offers applied mathematics (MA, MS, PhD); pure mathematics (MA, MS, PhD). Part-time programs available. *Degree requirements:* For master's, thesis optional; for doctorate, one foreign language, thesis/dissertation. Electronic applications accepted. *Faculty research:* Approximation theory, measure theory, ring theory, functional analysis, complex analysis.

Lehigh University, College of Arts and Sciences, Department of Mathematics, Bethlehem, PA 18015-3094. Offers applied mathematics (MS, PhD); mathematics (MS, PhD); statistics (MS). Part-time programs available. *Faculty:* 20 full-time (1 woman), 3 part-time/adjunct (0 women). *Students:* 33 full-time (11 women), 7 part-time (4 women); includes 1 Asian American or Pacific Islander, 10 international. Average age 26. 68 applicants, 71% accepted, 14 enrolled. In 2005, 5 master's, 1 doctorate awarded. Terminal master's awarded for partial completion of doctoral program. *Median time to degree:* Of those who began their doctoral program in fall 1997, 50% received their degree in 8 years or less. *Degree requirements:* For master's, comprehensive exam; for doctorate, one foreign language, thesis/dissertation, qualifying exams, comprehensive exam. *Entrance requirements:* For master's and doctorate, minimum GPA of 3.0. Additional exam requirements/recommendations for international students: Required—TOEFL. *Application deadline:* For fall admission, 7/15 for domestic students; for spring admission, 12/7 priority date for domestic students. Applications are processed on a rolling basis. Application fee: $60. Electronic applications accepted. *Financial support:* In 2005–06, 30 students received support, including 1 fellowship with full tuition reimbursement available (averaging $16,000 per year), 2 research assistantships with full tuition reimbursements available (averaging $14,000 per year), 20 teaching assistantships with full tuition reimbursements available (averaging $13,700 per year); scholarships/grants, tuition waivers (partial), and unspecified assistantships also available. Financial award application deadline: 1/15. *Faculty research:* Probability and statistics, geometry and topology, number theory, algebra, differential equations. *Unit head:* Dr. Steven Weintraub, Chairman, 610-758-3730, Fax: 610-758-3767, E-mail: shw2@lehigh.edu. *Application contact:* Dr. Terry Napier, Graduate Coordinator, 610-758-3755, E-mail: mathgrad@lehigh.edu.

Long Island University, C.W. Post Campus, College of Liberal Arts and Sciences, Department of Mathematics, Brookville, NY 11548-1300. Offers applied mathematics (MS); mathematics education (MS); mathematics for secondary school teachers (MS). Part-time and evening/weekend programs available. *Degree requirements:* For master's, thesis or alternative, oral presentation. *Entrance requirements:* Additional exam requirements/recommendations for international students: Required—TOEFL. Electronic applications accepted. *Faculty research:* Differential geometry, topological groups, general topology, number theory, analysis and statistics, numerical analysis.

McGill University, Faculty of Graduate and Postdoctoral Studies, Faculty of Science, Department of Mathematics and Statistics, Montréal, QC H3A 2T5, Canada. Offers computational science and engineering (M Sc); mathematics (M Sc, MA, PhD), including applied mathematics (M Sc, MA), pure mathematics (M Sc, MA), statistics (M Sc, MA). Part-time programs available. *Degree requirements:* For master's, thesis (for some programs), registration; for doctorate, one foreign language, thesis/dissertation, comprehensive exam, registration. *Entrance requirements:* For master's, minimum GPA of 3.0, Canadian Honours degree in mathematics or closely related field (statistics or applied mathematics). Additional exam requirements/recommendations for international students: Required—TOEFL (minimum score 550 paper-based; 213 computer-based), IELT (minimum score 7). Electronic applications accepted.

Michigan State University, The Graduate School, College of Natural Science, Department of Mathematics, East Lansing, MI 48824. Offers applied mathematics (MS, PhD); industrial mathematics (MS); mathematics (MAT, MS, PhD); mathematics education (PhD). *Faculty:* 58 full-time (13 women). *Students:* 110 full-time (34 women), 6 part-time (2 women); includes 7 minority (2 African Americans, 3 Asian Americans or Pacific Islanders, 2 Hispanic Americans), 71 international. Average age 28. 131 applicants, 18% accepted. In 2005, 18 master's, 15 doctorates awarded. *Degree requirements:* For doctorate, one foreign language, thesis/dissertation, qualifying exam, seminar presentations, comprehensive exam. *Entrance requirements:* For master's, GRE General Test, bachelor's degree in mathematics, physics, or engineering; mathematics course work beyond calculus; 3 letters of recommendation; for doctorate, GRE General Test, minimum GPA of 3.0, MS in mathematics or equivalent, 3 letters of recommendation. Additional exam requirements/recommendations for international students: Required—TOEFL (minimum score 550 paper-based; 213 computer-based), Michigan State University ELT (85), Michigan ELAB (83). *Application deadline:* For fall admission, 2/1 for domestic students. Application fee: $50. Electronic applications accepted. *Expenses:* Tuition, state resident: part-time $330 per credit hour. Tuition, nonresident: part-time $685 per credit hour. Tuition and fees vary according to program. *Financial support:* In 2005–06, 12 fellowships with tuition reimbursements (averaging $4,370 per year), 21 research assistantships with tuition reimbursements (averaging $14,054 per year), 94 teaching assistantships with tuition reimbursements (averaging $13,770 per year) were awarded; scholarships/grants and unspecified assistantships also available. *Faculty research:* Applied and industrial mathematics; analysis; geometry and topology; logic, combinatorics and graph theory; algebra. Total annual research expenditures: $2.6 million. *Unit head:* Dr. Peter W. Bates, Chairperson, 517-355-9681, Fax: 517-432-1562, E-mail: bates@math.msu.edu. *Application contact:* Barbara S. Miller, Graduate Secretary, 517-353-6338, Fax: 517-432-1562, E-mail: bmiller@math.msu.edu.

Montclair State University, The Graduate School, College of Science and Mathematics, Department of Computer Science, Montclair, NJ 07043-1624. Offers applied mathematics (MS); applied statistics (MS); CISCO (Certificate); informatics (MS); object oriented computing (Certificate). Part-time and evening/weekend programs available. *Faculty:* 13 full-time (2 women), 12 part-time/adjunct (3 women). *Students:* 13 full-time (7 women), 22 part-time (3 women); includes 11 minority (2 African Americans, 8 Asian Americans or Pacific Islanders, 1 Hispanic American), 3 international. 18 applicants, 72% accepted, 12 enrolled. In 2005, 13 degrees awarded. *Degree requirements:* For master's, thesis or alternative, comprehensive exam. *Entrance requirements:* For master's, GRE General Test, minimum GPA of 2.67; 15 undergraduate math credits; bachelor's degree in computer science, math, science or engineering; 2 letters of recommendation. Additional exam requirements/recommendations for international students: Required—TOEFL (minimum score 83 computer-based). *Application deadline:* Applications are processed on a rolling basis. Application fee: $60. Electronic applications accepted. *Expenses:* Tuition: Full-time $3,001; part-time $409 per credit. Required fees: $56 per credit. Tuition and fees vary according to course load, degree level and program. *Financial support:* In 2005–06, 4 research assistantships with full tuition reimbursements (averaging $7,000 per year) were awarded; Federal Work-Study, scholarships/grants, and unspecified assistantships also available. Support available to part-time students. Financial award application deadline: 3/1; financial award applicants required to submit FAFSA. *Unit head:* Dr. Dorothy Deremer, Chairperson, 973-655-4166. *Application contact:* Dr. James Benham, Adviser, 973-655-3746, E-mail: benham@pegasus.montclair.edu.

Montclair State University, The Graduate School, College of Science and Mathematics, Department of Mathematics, Montclair, NJ 07043-1624. Offers mathematics (MS), including computer science, mathematics education, pure and applied mathematics, statistics; teaching middle grades math (Certificate). Part-time and evening/weekend programs available. *Faculty:* 29 full-time (10 women), 29 part-time/adjunct (11 women). *Students:* 23 full-time (12 women), 135 part-time (99 women); includes 28 minority (13 African Americans, 10 Asian Americans or Pacific Islanders, 5 Hispanic Americans), 3 international. 77 applicants, 83% accepted, 51 enrolled. In 2005, 9 master's, 26 other advanced degrees awarded. *Degree requirements:* For master's, comprehensive exam. *Entrance requirements:* For master's, GRE General Test, minimum GPA of 2.67, 2 letters of recommendation. Additional exam requirements/recommendations for international students: Required—TOEFL (minimum score 83 computer-based). *Application deadline:* Applications are processed on a rolling basis. Application fee: $60. *Expenses:* Tuition: Full-time $3,001; part-time $409 per credit. Required fees: $56 per credit. Tuition and fees vary according to course load, degree level and program. *Financial support:* In 2005–06, 8 research assistantships with full tuition reimbursements (averaging $7,000 per year) were awarded; Federal Work-Study, scholarships/grants, and unspecified assistantships also available. Support available to part-time students. Financial award application deadline: 3/1; financial award applicants required to submit FAFSA. *Faculty research:* Infectious disease. Total annual research expenditures:$130,000. *Unit head:* Dr. Helen Roberts, Chairperson, 973-655-5132. *Application contact:* Dr. Ted Williamson, Advisor, 973-655-5146, E-mail: williamsont@mail.montclair.edu.

Naval Postgraduate School, Graduate Programs, Department of Mathematics, Monterey, CA 93943. Offers applied mathematics (MS, PhD). Program only open to commissioned officers of the United States and friendly nations and selected United States federal civilian employees. Part-time programs available. *Faculty:* 13. *Students:* 13 full-time. In 2005, 3 master's, 1 doctorate awarded. *Degree requirements:* For master's, thesis; for doctorate, one foreign language, thesis/dissertation. *Unit head:* Prof. Clyde Scandrett, Chairman, 831-656-2027, Fax: 831-656-2355, E-mail: cscand@nps.navy.mil. *Application contact:* Tracy Hammond, Acting Director of Admissions, 831-656-3059, Fax: 831-656-2891, E-mail: thammond@bps.navy.mil.

New Jersey Institute of Technology, Office of Graduate Studies, College of Science and Liberal Arts, Department of Mathematical Science, Program in Applied Mathematics, Newark, NJ 07102. Offers MS. Part-time and evening/weekend programs available. *Students:* 6 full-time (2 women), 3 part-time; includes 6 minority (1 African American, 3 Asian Americans or Pacific Islanders, 2 Hispanic Americans), 1 international. Average age 28. 23 applicants, 43% accepted, 4 enrolled. In 2005, 5 degrees awarded. *Entrance requirements:* For master's, GRE General Test. Additional exam requirements/recommendations for international students: Required—TOEFL (minimum score 550 paper-based; 213 computer-based). *Application deadline:* For fall admission, 6/5 for domestic students; for spring admission, 10/15 for domestic students. Applications are processed on a rolling basis. Application fee: $60. Electronic applications accepted. *Expenses:* Tuition, state resident: full-time $9,620; part-time $520 per credit. Tuition, nonresident: full-time $13,542; part-time $715 per credit. Required fees: $78; $54 per credit. $78 per year. Tuition and fees vary according to course load. *Financial support:* Fellowships with full and partial tuition reimbursements, research assistantships with full and partial tuition reimbursements, teaching assistantships with full and partial tuition reimbursements, career-related internships or fieldwork, Federal Work-Study, institutionally sponsored loans, and unspecified assistantships available. Financial award application deadline: 3/15. *Unit head:* Dr. Demetrius Papageorgious, Director, 973-596-3498, Fax: 973-596-5591, E-mail: demetrius.papageorgiou@njit.edu. *Application contact:* Kathryn Kelly, Director of Admissions, 973-596-3300, Fax: 973-596-3461, E-mail: admissions@njit.edu.

New Mexico Institute of Mining and Technology, Graduate Studies, Department of Mathematics, Socorro, NM 87801. Offers applied math (PhD); mathematics (MS); operations research (MS). *Degree requirements:* For master's, thesis optional; for doctorate, thesis/dissertation. *Entrance requirements:* For master's, GRE General Test. Additional exam requirements/recommendations for international students: Required—TOEFL (minimum score 540 paper-based; 207 computer-based). *Faculty research:* Applied mathematics, differential equations, industrial mathematics, numerical analysis, stochastic processes.

Applied Mathematics

North Carolina State University, Graduate School, College of Physical and Mathematical Sciences, Department of Mathematics, Program in Applied Mathematics, Raleigh, NC 27695. Offers MS, PhD. *Degree requirements:* For master's, thesis (for some programs); for doctorate, thesis/dissertation. *Entrance requirements:* For master's and doctorate, GRE, GRE Subject Test. Electronic applications accepted. *Faculty research:* Biological and physical modeling, numerical analysis, control, stochastic processes, industrial mathematics.

North Dakota State University, The Graduate School, College of Science and Mathematics, Department of Mathematics, Fargo, ND 58105. Offers applied mathematics (MS, PhD); mathematics (MS, PhD). *Faculty:* 15 full-time (1 woman), 3 part-time/adjunct (1 woman). *Students:* 21 full-time (7 women), 3 part-time (1 woman); includes 1 minority (Asian American or Pacific Islander), 8 international. Average age 28. 15 applicants, 67% accepted, 9 enrolled. In 2005, 2 master's awarded. *Degree requirements:* For master's, thesis, comprehensive exam, registration; for doctorate, one foreign language, thesis/dissertation, computer proficiency, comprehensive exam, registration. *Entrance requirements:* For master's and doctorate, GRE General Test. Additional exam requirements/recommendations for international students: Required—TOEFL, IELT. *Application deadline:* For fall admission, 5/1 priority date for domestic students, 5/1 priority date for international students; for spring admission, 8/1 for domestic students, 8/1 for international students. Applications are processed on a rolling basis. Application fee: $45 ($60 for international students). Electronic applications accepted. *Financial support:* In 2005–06, 5 fellowships with full tuition reimbursements (averaging $18,000 per year), 1 research assistantship with tuition reimbursement (averaging $14,000 per year), 17 teaching assistantships with full tuition reimbursements (averaging $9,300 per year) were awarded; Federal Work-Study, institutionally sponsored loans, and tuition waivers (full) also available. Support available to part-time students. Financial award application deadline: 3/31. *Faculty research:* Discrete mathematics, number theory, analysis theory, algebra, applied math. Total annual research expenditures: $57,844. *Unit head:* Dr. Warren Shreve, Chair, 701-231-8171, Fax: 701-231-7598, E-mail: warren.shreve@ndsu.edu. *Application contact:* Dr. Jim Coykendall, Graduate Program Director, 701-231-8079, Fax: 701-231-7598, E-mail: jim.coykendall@ndsu.edu.

Northeastern University, College of Arts and Sciences, Department of Mathematics, Boston, MA 02115-5096. Offers applied mathematics (MS); mathematics (MS, PhD); operations research (MSOR). Part-time and evening/weekend programs available. *Faculty:* 41 full-time (6 women), 13 part-time/adjunct (2 women). *Students:* 34 full-time (9 women), 6 part-time (1 woman). Average age 32. 63 applicants, 46% accepted. In 2005, 7 master's, 3 doctorates awarded. *Degree requirements:* For doctorate, thesis/dissertation, qualifying exams. *Entrance requirements:* For master's and doctorate, GRE Subject Test, GRE General Test. Additional exam requirements/recommendations for international students: Required—TOEFL. *Application deadline:* For fall admission, 2/1 priority date for domestic students, 2/1 priority date for international students. Applications are processed on a rolling basis. Application fee: $50. Electronic applications accepted. *Financial support:* In 2005–06, 26 teaching assistantships with tuition reimbursements (averaging $16,276 per year) were awarded; research assistantships with tuition reimbursements, Federal Work-Study, institutionally sponsored loans, tuition waivers (full and partial), and unspecified assistantships also available. Financial award application deadline: 3/1; financial award applicants required to submit FAFSA. *Faculty research:* Algebra and singularities, combinatorics, topology, probability and statistics, geometric analysis and partial differential equations. *Unit head:* Dr. Robert McOwen, Chairperson, 617-373-2450, Fax: 617-373-5658, E-mail: mathdept@neu.edu. *Application contact:* Gari Horton, Graduate Secretary, 617-373-2454, Fax: 617-373-5658, E-mail: mathdept@neu.edu.

See Close-Up on page 515.

Northwestern University, The Graduate School, Interdepartmental Degree Programs, Program in Mathematical Methods in Social Science, Evanston, IL 60208. Offers MS. *Students:* 8 (5 women) (all international). *Unit head:* William Rogerson, Director, 847-491-8484, Fax: 847-491-7001, E-mail: wrogerson@northwestern.edu. *Application contact:* David Austen-Smith, Admission Officer, 847-491-2626, Fax: 847-491-7001, E-mail: dasm@kellogg.northwestern.edu.

Northwestern University, McCormick School of Engineering and Applied Science, Program in Applied Mathematics, Evanston, IL 60208. Offers MS, PhD. Admissions and degrees offered through The Graduate School. Part-time programs available. *Faculty:* 14 full-time (1 woman). *Students:* 56 (25 women); includes 6 minority (1 American Indian/Alaska Native, 5 Asian Americans or Pacific Islanders) 7 international. Average age 27. 47 applicants, 23% accepted, 9 enrolled. In 2005, 6 master's, 9 doctorates awarded. Terminal master's awarded for partial completion of doctoral program. *Median time to degree:* Of those who began their doctoral program in fall 1997, 100% received their degree in 8 years or less. *Degree requirements:* For master's, thesis or alternative, comprehensive exam, registration; for doctorate, thesis/dissertation, comprehensive exam, registration. *Entrance requirements:* For master's and doctorate, GRE. Additional exam requirements/recommendations for international students: Required—TOEFL. *Application deadline:* For fall admission, 2/1 for domestic students, 2/1 for international students. Application fee: $60 ($75 for international students). Electronic applications accepted. *Financial support:* In 2005–06, 31 students received support, including 8 fellowships with full tuition reimbursements available (averaging $18,000 per year), 5 research assistantships with full tuition reimbursements available (averaging $18,000 per year), 13 teaching assistantships with full tuition reimbursements available (averaging $14,000 per year); career-related internships or fieldwork, Federal Work-Study, institutionally sponsored loans, and scholarships/grants also available. Financial award application deadline: 2/1; financial award applicants required to submit FAFSA. *Faculty research:* Combustion, interfacial phenomena, nonlinear optics, dynamical systems, scientific computation. Total annual research expenditures: $1.6 million. *Unit head:* Michael J. Miksis, Chair, 847-491-5397, Fax: 847-491-2178, E-mail: miksis@northwestern.edu. *Application contact:* Edward Olmstead, Admission Officer, 847-491-5397, Fax: 847-491-2178, E-mail: weo@northwestern.edu.

Oakland University, Graduate Study and Lifelong Learning, College of Arts and Sciences, Department of Mathematics and Statistics, Program in Applied Mathematical Sciences, Rochester, MI 48309-4401. Offers PhD. *Students:* 13 full-time (5 women), 6 part-time (1 woman), 8 international. Average age 32. 8 applicants, 88% accepted. In 2005, 1 degree awarded. Application fee: $30. *Expenses:* Tuition, state resident: full-time $9,192; part-time $383 per credit. Tuition, nonresident: full-time $15,990; part-time $666 per credit. *Unit head:* Dr. Meir Shillor, Coordinator, Graduate Programs, 248-370-3439, Fax: 248-370-4184, E-mail: shillor@oakland.edu.

Oakland University, Graduate Study and Lifelong Learning, College of Arts and Sciences, Department of Mathematics and Statistics, Program in Industrial Applied Mathematics, Rochester, MI 48309-4401. Offers MS. Part-time and evening/weekend programs available. *Students:* Average age 37. 1 applicant, 100% accepted, 1 enrolled. In 2005, 1 degree awarded. *Entrance requirements:* For master's, minimum GPA of 3.0 for unconditional admission. Additional exam requirements/recommendations for international students: Required—TOEFL (minimum score 550 paper-based; 213 computer-based). *Application deadline:* For fall admission, 7/15 priority date for domestic students, 5/1 priority date for international students. For winter admission, 12/1 for domestic students; for spring admission, 3/15 for domestic students. Applications are processed on a rolling basis. Application fee: $30. Electronic applications accepted. *Expenses:* Contact institution. *Financial support:* Federal Work-Study, institutionally sponsored loans, and tuition waivers (full) available. Financial award application deadline: 3/1; financial award applicants required to submit FAFSA. *Unit head:* Dr. Meir Shillor, Coordinator, Graduate Programs, 248-370-3439, Fax: 248-370-4184, E-mail: shillor@oakland.edu.

Oklahoma State University, College of Arts and Sciences, Department of Mathematics, Stillwater, OK 74078. Offers applied mathematics (MS); mathematics (pure and applied) (PhD); mathematics (pure) (MS); mathematics education (MS, PhD). *Faculty:* 35 full-time (5 women), 7 part-time/adjunct (3 women). *Students:* 17 full-time (7 women), 26 part-time (13 women); includes 3 minority (1 American Indian/Alaska Native, 2 Asian Americans or Pacific Islanders), 20 international. Average age 29. 44 applicants, 34% accepted, 9 enrolled. In 2005, 6 master's, 1 doctorate awarded. *Degree requirements:* For master's, report or thesis; for doctorate, one foreign language, thesis/dissertation, comprehensive exam. *Entrance requirements:* For master's, GRE. Additional exam requirements/recommendations for international students: Required—TOEFL. *Application deadline:* For fall admission, 6/1 priority date for domestic students, 3/1 priority date for international students. Applications are processed on a rolling basis. Application fee: $40 ($75 for international students). Electronic applications accepted. *Expenses:* Tuition, state resident: full-time $4,253; part-time $139 per credit hour. Tuition, nonresident: full-time $12,569; part-time $485 per credit hour. Required fees: $43 per credit hour. One-time fee: $20 part-time. Tuition and fees vary according to course load and program. *Financial support:* In 2005–06, 3 research assistantships (averaging $18,752 per year), 37 teaching assistantships (averaging $17,088 per year) were awarded; career-related internships or fieldwork, Federal Work-Study, scholarships/grants, health care benefits, tuition waivers (partial), and unspecified assistantships also available. Support available to part-time students. Financial award application deadline: 3/1. *Unit head:* Dr. Alan Adolphson, Head, 405-744-5688, Fax: 405-744-8275.

See Close-Up on page 517.

The Pennsylvania State University University Park Campus, Graduate School, Eberly College of Science, Department of Mathematics, State College, University Park, PA 16802-1503. Offers mathematics (M Ed, MA, D Ed, PhD), including applied mathematics (MA, PhD). *Students:* 77 full-time (10 women), 2 part-time (1 woman); includes 3 minority (2 Asian Americans or Pacific Islanders, 1 Hispanic American), 62 international. *Entrance requirements:* For master's and doctorate, GRE General Test. Application fee: $45. *Expenses:* Tuition, state resident: full-time $12,518; part-time $522 per credit. Tuition, nonresident: full-time $23,004; part-time $959 per credit. Required fees: $484. Tuition and fees vary according to course load, campus/location and program. *Unit head:* Dr. Nigel D. Higson, Head, 814-865-7527, Fax: 814-865-3735, E-mail: higson@psu.edu.

Princeton University, Graduate School, Department of Chemical Engineering, Princeton, NJ 08544-1019. Offers applied and computational mathematics (PhD); chemical engineering (M Eng, MSE, PhD); plasma science and technology (MSE, PhD); polymer sciences and materials (MSE, PhD). Terminal master's awarded for partial completion of doctoral program. *Degree requirements:* For master's, thesis; for doctorate, thesis/dissertation, general exam. *Entrance requirements:* For master's and doctorate, GRE General Test. Additional exam requirements/recommendations for international students: Required—TOEFL (minimum score 600 paper-based; 250 computer-based). Electronic applications accepted. *Faculty research:* Applied and computational mathematics; bioengineering; fluid mechanics and transport phenomena; materials synthesis, processing, structure and properties; process engineering and science.

Princeton University, Graduate School, Department of Mathematics, Princeton, NJ 08544-1019. Offers applied and computational mathematics (PhD); mathematical physics (PhD); mathematics (PhD). *Degree requirements:* For doctorate, 2 foreign languages, thesis/dissertation. *Entrance requirements:* For doctorate, GRE General Test, GRE Subject Test. Additional exam requirements/recommendations for international students: Required—TOEFL (minimum score 600 paper-based; 250 computer-based). Electronic applications accepted.

Princeton University, Graduate School, Department of Physics, Princeton, NJ 08544-1019. Offers applied and computational mathematics (PhD); mathematical physics (PhD); physics (PhD); physics and chemical physics (PhD). *Degree requirements:* For doctorate, thesis/dissertation, qualifying exam. *Entrance requirements:* For doctorate, GRE General Test, GRE Subject Test. Additional exam requirements/recommendations for international students: Required—TOEFL (minimum score 600 paper-based; 250 computer-based). Electronic applications accepted.

Princeton University, Graduate School, Program in Applied and Computational Mathematics, Princeton, NJ 08544-1019. Offers PhD. *Degree requirements:* For doctorate, thesis/dissertation. *Entrance requirements:* For doctorate, GRE General Test, GRE Subject Test. Additional exam requirements/recommendations for international students: Required—TOEFL (minimum score 600 paper-based; 250 computer-based). Electronic applications accepted.

Announcement: The program in applied and computational mathematics at Princeton is an interdisciplinary PhD program offering a select group of highly qualified students the opportunity to obtain a thorough knowledge of branches of mathematics indispensable for science and engineering applications, including numerical analysis and other computational methods. Before being admitted to a third year of study, students must sustain the general examination. The general examination is designed as a sequence of interviews with assigned professors that begins in the first year and covers 3 areas of applied mathematics. The generals culminate in a seminar on a research topic, usually delivered toward the end of the fourth semester. The doctoral dissertation may consist of a mathematical contribution to some field of science or engineering or the development or analysis of mathematical or computational methods useful for, inspired by, or relevant to science or engineering. Satisfactory completion of the requirements leads to the PhD degree in applied and computational mathematics. For more information, visit http://www.pacm.princeton.edu.

Rensselaer Polytechnic Institute, Graduate School, School of Science, Department of Mathematical Sciences, Program in Applied Mathematics, Troy, NY 12180-3590. Offers MS. Part-time programs available. *Faculty:* 23 full-time (3 women), 4 part-time/adjunct (1 woman). *Students:* 18 full-time (3 women), 3 part-time (1 woman); includes 3 minority (1 African American, 2 Hispanic Americans), 5 international. Average age 22. 15 applicants, 40% accepted, 0 enrolled. In 2005, 7 master's awarded. *Degree requirements:* For master's, registration. *Entrance requirements:* For master's, GRE General Test. Additional exam requirements/recommendations for international students: Required—TOEFL. *Application deadline:* For fall admission, 1/15 for domestic students. Applications are processed on a rolling basis. Application fee: $75. Electronic applications accepted. *Expenses:* Tuition: Full-time $31,000; part-time $1,320 per credit. Required fees:$1,623. *Financial support:* In 2005–06, 6 students received support. Career-related internships or fieldwork and institutionally sponsored loans available. Financial award application deadline: 2/1. *Faculty research:* Mathematical modeling, differential equations, applications of mathematics in science and engineering, operations research, analysis. Total annual research expenditures: $3.2 million. *Application contact:* Dawnmarie Robens, Graduate Student Coordinator, 518-276-6414, Fax: 518-276-4824, E-mail: robensd@rpi.edu.

Rice University, Graduate Programs, George R. Brown School of Engineering, Department of Computational and Applied Mathematics, Houston, TX 77251-1892. Offers MA, MCAM, MCSE, PhD. *Degree requirements:* For master's, thesis (for some programs), comprehensive exam (for some programs), registration; for doctorate, thesis/dissertation, comprehensive exam, registration. *Entrance requirements:* For master's and doctorate, GRE General Test, minimum GPA of 3.0. Additional exam requirements/recommendations for international students: Required—TOEFL (minimum score 600 paper-based; 250 computer-based). Electronic applications accepted. *Faculty research:* Inverse problems, partial differential equations, computer algorithms, computational modeling, optimization theory.

Rochester Institute of Technology, Graduate Enrollment Services, College of Science, Department of Mathematics and Statistics, Rochester, NY 14623-5603. Offers industrial and applied mathematics (MS). *Students:* 10 full-time (3 women), 2 part-time, 1 international. 10 applicants, 80% accepted, 5 enrolled. In 2005, 6 degrees awarded. *Degree requirements:* For master's, thesis. *Entrance requirements:* For master's, GRE General Test, minimum GPA of 3.0. Additional exam requirements/recommendations for international students: Required—TOEFL (minimum score 550 paper-based). *Application deadline:* For fall admission, 3/1 for domestic students. Applications are processed on a rolling basis. Application fee: $50. *Expenses:*

Tuition: Full-time $25,392; part-time $713 per credit. Required fees: $183; $61 per term. *Unit head:* Sophia Maggelakis, Head, 585-475-2498, E-mail: sxmsma@rit.edu.

Rutgers, The State University of New Jersey, New Brunswick/Piscataway, Graduate School, Program in Mathematics, New Brunswick, NJ 08901-1281. Offers applied mathematics (MS, PhD); math finance (MS); mathematics (MS, PhD). Part-time programs available. *Faculty:* 106 full-time. *Students:* 70 full-time (16 women); includes 2 minority (1 Asian American or Pacific Islander, 1 Hispanic American), 32 international. Average age 26. 248 applicants, 15% accepted, 13 enrolled. In 2005, 7 degrees awarded. *Median time to degree:* Of those who began their doctoral program in fall 1997, 88% received their degree in 8 years or less. *Degree requirements:* For doctorate, one foreign language, thesis/dissertation, comprehensive exam. *Entrance requirements:* For master's and doctorate, GRE General Test, GRE Subject Test. Additional exam requirements/recommendations for international students: Required—TOEFL. *Application deadline:* For fall admission, 2/1 for domestic students; for spring admission, 11/1 for domestic students. Application fee: $50. *Expenses:* Tuition, state resident: full-time $10,440; part-time $435 per credit. Tuition, nonresident: full-time $15,520; part-time $647 per credit. Required fees: $129 per credit. Tuition and fees vary according to program. *Financial support:* In 2005–06, 8 fellowships with full tuition reimbursements (averaging $25,347 per year), 10 research assistantships with full tuition reimbursements (averaging $18,307 per year), 53 teaching assistantships with full tuition reimbursements (averaging $19,347 per year) were awarded; health care benefits, tuition waivers (full), and unspecified assistantships also available. Financial award application deadline: 2/1; financial award applicants required to submit FAFSA. *Faculty research:* Logic and set theory, number theory, mathematical physics, control theory, partial differential equations. Total annual research expenditures: $2 million. *Unit head:* Prof. Charles A. Weibel, Director, 732-445-3864, Fax: 732-445-5530, E-mail: grad-director@math.rutgers.edu.

St. John's University, St. John's College of Liberal Arts and Sciences, Department of Mathematics and Computer Science, Queens, NY 11439. Offers algebra (MA); analysis (MA); applied mathematics (MA); computer science (MA); geometry-topology (MA); logic and foundations (MA); probability and statistics (MA). Part-time and evening/weekend programs available. *Faculty:* 19 full-time (2 women), 10 part-time/adjunct (4 women). *Students:* 4 full-time (all women), 5 part-time (3 women); includes 1 minority (Hispanic American), 1 international. Average age 28. 18 applicants, 83% accepted, 6 enrolled. In 2005, 2 degrees awarded. *Degree requirements:* For master's, thesis optional. *Entrance requirements:* For master's, minimum GPA of 3.0. Additional exam requirements/recommendations for international students: Required—TOEFL (minimum score 500 paper-based; 173 computer-based). *Application deadline:* For fall admission, 5/1 priority date for domestic students, 5/1 priority date for international students; for spring admission, 11/1 priority date for domestic students, 11/1 priority date for international students. Applications are processed on a rolling basis. Application fee: $40. Electronic applications accepted. *Expenses:* Tuition: Full-time $8,760; part-time $730 per credit. Required fees: $250; $125 per term. Tuition and fees vary according to program. *Financial support:* Research assistantships, scholarships/grants available. Support available to part-time students. Financial award application deadline: 3/1; financial award applicants required to submit FAFSA. *Faculty research:* Development of a computerized metabolic map. *Unit head:* Dr. Charles Traina, Chair, 718-990-6166, E-mail: trainac@stjohns.edu. *Application contact:* Matthew Whelan, Director, Office of Admissions, 718-990-2000, Fax: 718-990-2096, E-mail: admissions@stjohns.edu.

San Diego State University, Graduate and Research Affairs, College of Sciences, Department of Mathematical Sciences, Program in Applied Mathematics, San Diego, CA 92182. Offers MS. Part-time programs available. *Students:* 19 full-time (4 women), 9 part-time (2 women); includes 3 minority (all Asian Americans or Pacific Islanders), 6 international. Average age 29. 25 applicants, 72% accepted, 8 enrolled. In 2005, 9 degrees awarded. *Degree requirements:* For master's, comprehensive exam. *Entrance requirements:* For master's, GRE General Test. Additional exam requirements/recommendations for international students: Required—TOEFL. *Application deadline:* For fall admission, 5/1 for domestic students, 5/1 for international students; for spring admission, 11/1 for domestic students, 11/1 for international students. Applications are processed on a rolling basis. Application fee: $55. Electronic applications accepted. *Financial support:* Teaching assistantships, unspecified assistantships available. Financial award applicants required to submit FAFSA. *Faculty research:* Modeling, computational fluid dynamics, biomathematics, thermodynamics. *Unit head:* Peter Salamon, Graduate Advisor, 619-594-7204, Fax: 619-594-6746, E-mail: salamon@math.sdsu.edu. *Application contact:* Peter Salamon, Graduate Advisor, 619-594-7204, Fax: 619-594-6746, E-mail: salamon@math.sdsu.edu.

Santa Clara University, School of Engineering, Department of Applied Mathematics, Santa Clara, CA 95053. Offers MSAM. Part-time and evening/weekend programs available. *Students:* 2 full-time (both women), 4 part-time (2 women); includes 1 minority (Asian American or Pacific Islander), 2 international. Average age 35. 3 applicants, 100% accepted, 1 enrolled. In 2005, 3 degrees awarded. *Degree requirements:* For master's, thesis or alternative. *Entrance requirements:* For master's, GRE General Test, minimum GPA of 2.75. Additional exam requirements/recommendations for international students: Required—TOEFL. *Application deadline:* Applications are processed on a rolling basis. Electronic applications accepted. *Expenses:* Tuition: Full-time $34,000; part-time $680 per unit. Tuition and fees vary according to course level, degree level, program and student level. *Financial support:* Fellowships, research assistantships, teaching assistantships, Federal Work-Study, institutionally sponsored loans, and scholarships/grants available. Support available to part-time students. Financial award application deadline: 3/1; financial award applicants required to submit FAFSA. *Unit head:* Dr. Stephen Chiappari, Chair, 408-554-6866.

Simon Fraser University, Graduate Studies, Faculty of Science, Department of Mathematics, Burnaby, BC V5A 1S6, Canada. Offers applied mathematics (M Sc, PhD); pure mathematics (M Sc, PhD); statistics and actuarial science (M Sc, PhD). *Degree requirements:* For master's and doctorate, thesis/dissertation. *Entrance requirements:* For master's, GRE Subject Test, minimum GPA of 3.0; for doctorate, GRE Subject Test, minimum GPA of 3.5. Additional exam requirements/recommendations for international students: Required—TWE or IELTS. *Faculty research:* Semi-groups, number theory, optimization, combinations.

Southern Methodist University, Dedman College, Department of Mathematics, Dallas, TX 75275. Offers computational and applied mathematics (MS, PhD). *Faculty:* 15 full-time (1 woman). *Students:* 22 full-time (11 women), 1 part-time; includes 3 minority (2 Asian Americans or Pacific Islanders, 1 Hispanic American), 9 international. Average age 30. 25 applicants, 32% accepted, 8 enrolled. In 2005, 4 master's, 2 doctorates awarded. *Degree requirements:* For doctorate, thesis/dissertation, oral and written exams. *Entrance requirements:* For master's and doctorate, GRE General Test, minimum GPA of 3.0, 18 undergraduate hours in mathematics beyond first and second year calculus. Additional exam requirements/recommendations for international students: Required—TOEFL. *Application deadline:* For fall admission, 6/30 for domestic students. For winter admission, 11/30 for domestic students. Applications are processed on a rolling basis. Application fee: $60. Electronic applications accepted. *Financial support:* In 2005–06, 7 teaching assistantships with full tuition reimbursements (averaging $15,000 per year) were awarded; career-related internships or fieldwork, scholarships/grants, health care benefits, and tuition waivers (partial) also available. Support available to part-time students. Financial award applicants required to submit FAFSA. *Faculty research:* Numerical analysis, scientific computation, fluid dynamics, software development, differential equations. Total annual research expenditures: $195,000. *Unit head:* Dr. Peter Moore, Chairman, 214-768-2506, Fax: 214-768-2355, E-mail: mathchair@mail.smu.edu. *Application contact:* Dr. Ian Gladwell, Director of Graduate Studies, 214-768-4338, E-mail: math@mail.smu.edu.

Stevens Institute of Technology, Graduate School, Arthur E. Imperatore School of Sciences and Arts, Department of Mathematical Sciences, Program in Applied Mathematics, Hoboken, NJ 07030. Offers MS, PhD. *Students:* 16 full-time (4 women), 8 part-time; includes 2 minority (both Asian Americans or Pacific Islanders), 4 international. 3 applicants, 33% accepted. *Degree requirements:* For master's, thesis optional; for doctorate, one foreign language, thesis/dissertation. *Entrance requirements:* For master's and doctorate, GRE. Additional exam requirements/recommendations for international students: Required—TOEFL. *Application deadline:* Applications are processed on a rolling basis. Application fee: $50. Electronic applications accepted. *Expenses:* Tuition: Part-time $920 per credit hour. Tuition and fees vary according to program. *Application contact:* Dr. Milos Dostal, Professor, 201-216-5426.

Stony Brook University, State University of New York, Graduate School, College of Engineering and Applied Sciences, Department of Applied Mathematics and Statistics, Stony Brook, NY 11794. Offers MS, PhD. *Faculty:* 20 full-time (4 women), 1 part-time/adjunct (0 women). *Students:* 139 full-time (60 women), 15 part-time (4 women); includes 26 minority (7 African Americans, 15 Asian Americans or Pacific Islanders, 4 Hispanic Americans), 97 international. Average age 28. 195 applicants, 84% accepted. In 2005, 34 master's, 8 doctorates awarded. *Degree requirements:* For master's, thesis or alternative; for doctorate, one foreign language, thesis/dissertation, comprehensive exam. *Entrance requirements:* For master's and doctorate, GRE General Test. Additional exam requirements/recommendations for international students: Required—TOEFL. *Application deadline:* For fall admission, 1/15 for domestic students. Application fee: $50. *Expenses:* Tuition, area resident: Part-time $288. Tuition, state resident: full-time $6,900. Tuition, nonresident: full-time $10,920; part-time $455. Required fees: $704. *Financial support:* In 2005–06, 7 fellowships, 29 research assistantships, 33 teaching assistantships were awarded. *Faculty research:* Biostatistics, combinatorial analysis, differential equations, modeling. Total annual research expenditures: $2.8 million. *Unit head:* Dr. J. Glimm, Chairman, 631-632-8360. *Application contact:* Dr. Woo Jong Kim, Director, 631-632-8360, Fax: 631-632-8490, E-mail: wjkim@ccmail.sunysb.edu.

Announcement: The department offers programs leading to the master's and PhD degrees, covering areas in computational applied mathematics, operations research, and statistics. Its faculty comprises 18 departmental and 22 affiliated members, among whom are a National Medal of Science recipient, 2 National Academy of Sciences members, and recipients of other prestigious awards. The research focus is on applied, interdisciplinary problems.

See Close-Up on page 527.

Temple University, Graduate School, College of Science and Technology, Department of Mathematics, Philadelphia, PA 19122-6096. Offers applied mathematics (MA); mathematics (PhD); pure mathematics (MA). Part-time and evening/weekend programs available. *Faculty:* 20 full-time (1 woman). *Students:* 14 full-time (5 women), 20 part-time (5 women); includes 6 minority (2 African Americans, 3 Asian Americans or Pacific Islanders, 1 Hispanic American), 18 international. 35 applicants, 31% accepted, 9 enrolled. In 2005, 5 master's, 6 doctorates awarded. Terminal master's awarded for partial completion of doctoral program. *Degree requirements:* For master's, written exam, thesis optional; for doctorate, 2 foreign languages, thesis/dissertation, oral and written exams. *Entrance requirements:* For master's, GRE General Test, minimum GPA of 3.0; for doctorate, GRE General Test, GRE Subject Test, minimum GPA of 3.0. Additional exam requirements/recommendations for international students: Required—TOEFL (minimum score 575 paper-based; 230 computer-based). *Application deadline:* For fall admission, 2/15 priority date for domestic students, 12/15 priority date for international students; for spring admission, 11/15 priority date for domestic students, 8/1 priority date for international students. Applications are processed on a rolling basis. Application fee: $50. Electronic applications accepted. *Expenses:* Tuition, state resident: full-time $8,694; part-time $483 per credit. Tuition, nonresident: full-time $12,672; part-time $704 per credit. Required fees: $500; $122 per semester. Tuition and fees vary according to course level, campus/location and program. *Financial support:* Fellowships, research assistantships, teaching assistantships, Federal Work-Study and institutionally sponsored loans available. Financial award application deadline: 1/15; financial award applicants required to submit FAFSA. *Faculty research:* Differential geometry, numerical analysis. *Unit head:* Dr. Omar Hijab, Chair, 215-204-4650, Fax: 215-204-6433, E-mail: hijab@temple.edu.

Texas State University-San Marcos, Graduate School, College of Science, Department of Mathematics, Program in Mathematics, San Marcos, TX 78666. Offers MS. *Students:* 28 full-time (11 women), 14 part-time (7 women); includes 14 minority (4 African Americans, 4 Asian Americans or Pacific Islanders, 6 Hispanic Americans), 1 international. Average age 33. 17 applicants, 100% accepted, 12 enrolled. In 2005, 17 degrees awarded. *Entrance requirements:* For master's, GRE, minimum GPA of 2.75 in last 60 hours of undergraduate course work. Additional exam requirements/recommendations for international students: Required—TOEFL. *Application deadline:* For fall admission, 6/15 priority date for domestic students, 6/1 priority date for international students; for spring admission, 10/15 priority date for domestic students, 10/1 priority date for international students. Application fee: $40 ($90 for international students). *Expenses:* Tuition, area resident: Part-time $116 per credit. Tuition, state resident: full-time $3,168; part-time $176 per credit. Tuition, nonresident: full-time $8,136; part-time $452 per credit. Required fees: $1,112; $74 per credit. Full-time tuition and fees vary according to course load. *Financial support:* In 2005–06, 28 students received support. Application deadline: 4/1; *Unit head:* Dr. Maria Acosta, Graduate Adviser, 512-245-2497, E-mail: ma05@txstate.edu.

Towson University, Graduate School, Program in Applied and Industrial Mathematics, Towson, MD 21252-0001. Offers MS. Part-time and evening/weekend programs available. *Students:* 23. 3 applicants, 100% accepted, 2 enrolled. In 2005, 3 degrees awarded. *Degree requirements:* For master's, internships. *Entrance requirements:* For master's, bachelor's degree in mathematics or related field, minimum GPA of 3.0. Additional exam requirements/recommendations for international students: Required—TOEFL (minimum score 550 paper-based). *Application deadline:* Applications are processed on a rolling basis. Application fee: $40. Electronic applications accepted. *Financial support:* In 2005–06, 3 students received support; teaching assistantships with full tuition reimbursements available, unspecified assistantships available. Financial award application deadline: 4/1; financial award applicants required to submit FAFSA. *Faculty research:* Partial differential equations, numerical computations, statistics, probability, game theory. *Unit head:* Dr. Mostafa Aminzadeh, Graduate Program Director, 410-704-2978, Fax: 410-704-4149, E-mail: maminzadeh@towson.edu. *Application contact:* 410-704-2501, Fax: 410-704-4675, E-mail: grads@towson.edu.

Tulane University, Graduate School, Department of Mathematics, New Orleans, LA 70118-5669. Offers applied mathematics (MS); mathematics (MS, PhD); statistics (MS). *Degree requirements:* For master's, thesis (for some programs); for doctorate, thesis/dissertation. *Entrance requirements:* For master's, GRE General Test, minimum B average in undergraduate course work; for doctorate, GRE General Test. Additional exam requirements/recommendations for international students: Required—TOEFL; Recommended—TSE. Electronic applications accepted.

The University of Akron, Graduate School, Buchtel College of Arts and Sciences, Department of Theoretical and Applied Mathematics, Program in Applied Mathematics, Akron, OH 44325. Offers MS. *Students:* 15 full-time (4 women), 2 part-time; includes 2 minority (both African Americans), 2 international. Average age 30. 13 applicants, 77% accepted, 4 enrolled. In 2005, 5 degrees awarded. *Degree requirements:* For master's, seminar and comprehensive exam or thesis, thesis optional. *Entrance requirements:* For master's, minimum GPA of 2.75. Additional exam requirements/recommendations for international students: Required—TOEFL (minimum score 550 paper-based; 213 computer-based), Michigan English Language Assessment Battery. *Application deadline:* For fall admission, 3/1 for domestic students. Applications are processed on a rolling basis. Application fee: $30 ($40 for international students). Electronic applications accepted. *Expenses:* Tuition, state resident: full-time $5,816; part-time $323 per credit. Tuition, nonresident: full-time $9,976; part-time $554 per credit. Required fees: $794; $43 per credit. $12 per term. Tuition and fees vary according to course load, degree level and program. *Unit head:* Dr. Gerald Young, Coordinator, 330-972-5731, E-mail: gwyoung@uakron.edu.

Applied Mathematics

The University of Akron, Graduate School, College of Engineering, Program in Engineering-Applied Mathematics, Akron, OH 44325. Offers PhD. *Students:* 8 full-time (3 women), 2 part-time, 5 international. Average age 33. 4 applicants, 100% accepted, 2 enrolled. *Degree requirements:* For doctorate, one foreign language, thesis/dissertation, candidacy exam, qualifying exam. *Entrance requirements:* For doctorate, GRE. Additional exam requirements/recommendations for international students: Required—TOEFL (minimum score 550 paper-based; 213 computer-based), Michigan English Language Assessment Battery. *Application deadline:* Applications are processed on a rolling basis. Application fee: $30 ($40 for international students). Electronic applications accepted. *Expenses:* Tuition, state resident: full-time $5,816; part-time $323 per credit. Tuition, nonresident: full-time $9,976; part-time $554 per credit. Required fees: $794; $43 per credit. $12 per term. Tuition and fees vary according to course load, degree level and program. *Financial support:* In 2005–06, 2 teaching assistantships were awarded *Unit head:* Dr. Subramaniya Hariharan, Head, 330-972-6580.

The University of Alabama, Graduate School, College of Arts and Sciences, Department of Mathematics, Tuscaloosa, AL 35487. Offers applied mathematics (PhD); mathematics (MA, PhD); pure mathematics (PhD). *Faculty:* 25 full-time (1 woman). *Students:* 31 full-time (11 women), 5 part-time; includes 5 African Americans, 16 international. Average age 31. 40 applicants, 70% accepted, 11 enrolled. In 2005, 83 master's, 6 doctorates awarded. Terminal master's awarded for partial completion of doctoral program. *Median time to degree:* Of those who began their doctoral program in fall 1997, 85% received their degree in 8 years or less. *Degree requirements:* For master's, thesis or alternative; for doctorate, thesis/dissertation. *Entrance requirements:* For master's and doctorate, GRE General Test, minimum GPA of 3.0. Additional exam requirements/recommendations for international students: Required—TOEFL. *Application deadline:* For fall admission, 7/1 for domestic students. Applications are processed on a rolling basis. Application fee: $25. Electronic applications accepted. *Expenses:* Tuition, area resident: Full-time $2,432. Tuition, nonresident: full-time $6,758. *Financial support:* In 2005–06, 27 students received support, including 3 fellowships with full tuition reimbursements available (averaging $14,834 per year), 28 teaching assistantships with full tuition reimbursements available (averaging $11,007 per year); research assistantships with full tuition reimbursements available, Federal Work-Study, institutionally sponsored loans, scholarships/grants, and unspecified assistantships also available. Support available to part-time students. Financial award application deadline: 7/1. *Faculty research:* Analysis, topology, algebra, fluid mechanics and system control theory, optimization, stochastic processes. Total annual research expenditures:$115,100. *Unit head:* Dr. Zhijian Wu, Chairperson and Professor, 205-348-5080, Fax: 205-348-7067, E-mail: zwu@gp.as.ua.edu. *Application contact:* Dr. Martin Evans, Director, Graduate Programs in Mathematics, 205-348-5301, Fax: 205-348-7067, E-mail: mevans@gp.as.ua.edu.

The University of Alabama at Birmingham, School of Natural Sciences and Mathematics, Department of Mathematics, Birmingham, AL 35294. Offers applied mathematics (PhD); mathematics (MS). *Students:* 32 full-time (13 women), 2 part-time; includes 3 minority (1 African American, 1 Asian American or Pacific Islander, 1 Hispanic American), 12 international. 33 applicants, 79% accepted. In 2005, 8 master's, 2 doctorates awarded. Terminal master's awarded for partial completion of doctoral program. *Degree requirements:* For master's, thesis optional; for doctorate, one foreign language, thesis/dissertation, comprehensive exam. *Entrance requirements:* For master's and doctorate, GRE General Test. *Application deadline:* Applications are processed on a rolling basis. Application fee: $35 ($60 for international students). Electronic applications accepted. *Expenses:* Tuition, state resident: part-time $170 per credit hour. Tuition, nonresident: full-time $4,612; part-time $425 per credit hour. International tuition: $10,732 full-time. Required fees: $11 per credit hour. $124 per term. Tuition and fees vary according to course load, degree level and program. *Financial support:* In 2005–06, 18 teaching assistantships with tuition reimbursements (averaging $14,000 per year) were awarded; fellowships, research assistantships, career-related internships or fieldwork, Federal Work-Study, institutionally sponsored loans, tuition waivers (full and partial), and unspecified assistantships also available. Support available to part-time students. Financial award application deadline: 3/31; financial award applicants required to submit FAFSA. *Faculty research:* Differential equations, topology, mathematical physics, dynamic systems. *Unit head:* Dr. Rudi Weikard, Chair, 205-934-2154, Fax: 205-934-9025, E-mail: weikard@uab.edu.

The University of Alabama in Huntsville, School of Graduate Studies, College of Science, Department of Mathematical Sciences, Huntsville, AL 35899. Offers applied mathematics (PhD); mathematics (MA, MS). Part-time and evening/weekend programs available. *Faculty:* 13 full-time (1 woman). *Students:* 20 full-time (11 women), 6 part-time (3 women); includes 5 minority (1 African American, 1 American Indian/Alaska Native, 2 Asian Americans or Pacific Islanders, 1 Hispanic American), 7 international. Average age 32. 19 applicants, 89% accepted, 12 enrolled. In 2005, 7 master's, 1 doctorate awarded. *Degree requirements:* For master's, thesis or alternative, oral and written exams, comprehensive exam, registration; for doctorate, one foreign language, thesis/dissertation, oral and written exams, comprehensive exam, registration. *Entrance requirements:* For master's and doctorate, GRE General Test, minimum GPA of 3.0. Additional exam requirements/recommendations for international students: Required—TOEFL (minimum score 550 paper-based; 213 computer-based). *Application deadline:* For fall admission, 5/30 for domestic students; for spring admission, 10/10 priority date for domestic students, 7/10 priority date for international students. Applications are processed on a rolling basis. Application fee: $40. *Expenses:* Tuition, state resident: full-time $5,866; part-time $244 per credit hour. Tuition, nonresident: full-time $12,060; part-time $500 per credit hour. Tuition and fees vary according to course load. *Financial support:* In 2005–06, 21 students received support, including 4 research assistantships with full and partial tuition reimbursements available (averaging $12,851 per year), 14 teaching assistantships with full and partial tuition reimbursements available (averaging $9,202 per year); fellowships with full and partial tuition reimbursements available, career-related internships or fieldwork, Federal Work-Study, institutionally sponsored loans, scholarships/grants, health care benefits, tuition waivers (full and partial), and unspecified assistantships also available. Support available to part-time students. Financial award application deadline: 4/1; financial award applicants required to submit FAFSA. *Faculty research:* Statistical modeling, stochastic processes, numerical analysis, combinatorics, fracture mechanics. Total annual research expenditures:$306,144. *Unit head:* Dr. Jia Li, Chair, 256-824-6470, Fax: 256-824-6173, E-mail: chair@math.uah.edu.

University of Alberta, Faculty of Graduate Studies and Research, Department of Mathematical and Statistical Sciences, Edmonton, AB T6G 2E1, Canada. Offers applied mathematics (M Sc, PhD); biostatistics (M Sc); mathematical finance (M Sc, PhD); mathematical physics (M Sc, PhD); mathematics (M Sc, PhD); statistics (M Sc, PhD, Postgraduate Diploma). Part-time programs available. *Faculty:* 48 full-time (4 women). *Students:* 112 full-time (41 women), 5 part-time. Average age 24. 776 applicants, 5% accepted, 34 enrolled. In 2005, 12 master's, 10 doctorates awarded. Terminal master's awarded for partial completion of doctoral program. *Median time to degree:* Of those who began their doctoral program in fall 1997, 100% received their degree in 8 years or less. *Degree requirements:* For master's, thesis (for some programs); for doctorate, thesis/dissertation, comprehensive exam. *Entrance requirements:* Additional exam requirements/recommendations for international students: Required—TOEFL (minimum score 580 paper-based; 237 computer-based). *Application deadline:* For fall admission, 3/1 for domestic students, 2/1 for international students. Applications are processed on a rolling basis. Application fee: $0. Electronic applications accepted. Tuition and fees charges are reported in Canadian dollars. *Expenses:* Tuition, state resident: part-time $562 Canadian dollars per term. Tuition, nonresident: full-time $3,375 Canadian dollars. Required fees: $573 Canadian dollars; $84 Canadian dollars per term. *Financial support:* In 2005–06, 51 research assistantships, 88 teaching assistantships with full and partial tuition reimbursements were awarded; scholarships/grants also available. Financial award application deadline:5/1. *Faculty research:* Classical and functional analysis, algebra, differential equations, geometry. *Unit head:* Dr. Anthony To-Ming Lau, Chair, 403-492-5141, E-mail: tlau@math.ualberta.ca. *Application contact:* Dr. Yau Shu Wong, Associate Chair, Graduate Studies, 403-492-5799, Fax: 403-492-6828, E-mail: gradstudies@math.ualberta.ca.

The University of Arizona, Graduate College, Graduate Interdisciplinary Programs, Graduate Interdisciplinary Program in Applied Mathematics, Tucson, AZ 85721. Offers applied mathematics (MS, PhD); mathematical sciences (PMS). Terminal master's awarded for partial completion of doctoral program. *Degree requirements:* For master's, thesis (for some programs), registration; for doctorate, one foreign language, thesis/dissertation, comprehensive exam, registration. *Entrance requirements:* For master's and doctorate, GRE. Additional exam requirements/recommendations for international students: Required—TOEFL (minimum score 575 paper-based; 230 computer-based), TSE (minimum score 45). *Faculty research:* Dynamical systems and chaos, partial differential equations, pattern formation, fluid dynamics and turbulence, scientific computation, mathematical physics, mathematical biology, medical imaging, applied probability and stochastic processes.

Announcement: The Interdisciplinary Program in Applied Mathematics at the University of Arizona offers outstanding research opportunities at the interface of applied mathematics and disciplines in the physical, biological, and engineering sciences. The program has achieved an excellent reputation for interdisciplinary graduate studies leading to MS and PhD degrees in applied mathematics.

University of Arkansas at Little Rock, Graduate School, College of Science and Mathematics, Department of Mathematics and Statistics, Little Rock, AR 72204-1099. Offers applied mathematics (MS), including applied analysis, mathematical statistics. Part-time and evening/weekend programs available. *Degree requirements:* For master's, comprehensive exam. *Entrance requirements:* For master's, GRE General Test, GRE Subject Test, minimum GPA of 2.7, previous course work in advanced mathematics.

The University of British Columbia, Faculty of Graduate Studies, Institute of Applied Mathematics, Vancouver, BC V6T 1Z1, Canada. Offers M Sc, PhD. *Faculty:* 58 full-time (8 women). *Students:* 46 full-time (16 women). In 2005, 9 master's awarded. *Median time to degree:* Of those who began their doctoral program in fall 1997, 100% received their degree in 8 years or less. *Degree requirements:* For master's, thesis (for some programs); for doctorate, thesis/dissertation, comprehensive exam. *Entrance requirements:* For doctorate, master's degree. Additional exam requirements/recommendations for international students: Required—TOEFL. Application fee: $90 ($128 for international students). *Financial support:* In 2005–06, 4 fellowships, 21 research assistantships, 21 teaching assistantships were awarded. *Faculty research:* Applied analysis, optimization, mathematical biology, numerical analysis, fluid mechanics. *Unit head:* Michael J. Ward, Director, 604-822-4584, Fax: 604-822-0550, E-mail: ward@math.ubc.ca.

University of California, Berkeley, Graduate Division, College of Letters and Science, Department of Mathematics, Program in Applied Mathematics, Berkeley, CA 94720-1500. Offers PhD. *Degree requirements:* For doctorate, 2 foreign languages, thesis/dissertation, qualifying exam. *Entrance requirements:* For doctorate, GRE General Test, GRE Subject Test, minimum GPA of 3.0.

University of California, Davis, Graduate Studies, Graduate Group in Applied Mathematics, Davis, CA 95616. Offers MS, PhD. *Faculty:* 72 full-time. *Students:* 47 full-time (14 women); includes 5 minority (1 American Indian/Alaska Native, 4 Asian Americans or Pacific Islanders), 14 international. Average age 28. 49 applicants, 47% accepted, 11 enrolled. In 2005, 1 doctorate awarded. Terminal master's awarded for partial completion of doctoral program. *Median time to degree:* Of those who began their doctoral program in fall 1997, 28.6% received their degree in 8 years or less. *Degree requirements:* For master's, thesis; for doctorate, one foreign language, thesis/dissertation. *Entrance requirements:* For master's, GRE General Test, GRE Subject Test, minimum GPA of 3.0; for doctorate, GRE General Test, GRE Subject Test, master's degree, minimum GPA of 3.0. Additional exam requirements/recommendations for international students: Required—TOEFL (minimum score 550 paper-based; 213 computer-based). *Application deadline:* For fall admission, 1/15 for domestic students. Application fee: $60. Electronic applications accepted. *Financial support:* In 2005–06, 45 students received support, including 18 fellowships with full and partial tuition reimbursements available (averaging $10,438 per year), 8 research assistantships with full and partial tuition reimbursements available (averaging $13,891 per year), 14 teaching assistantships with partial tuition reimbursements available (averaging $15,521 per year); Federal Work-Study, institutionally sponsored loans, scholarships/grants, traineeships, tuition waivers (full and partial), and unspecified assistantships also available. Financial award application deadline: 1/15; financial award applicants required to submit FAFSA. *Faculty research:* Mathematical biology, control and optimization, atmospheric sciences, theoretical chemistry, mathematical physics. *Unit head:* Blake Temple, Graduate Group Chair, 530-752-2214, E-mail: temple@math.ucdavis.edu. *Application contact:* Celia Davis, Administrative Assistant, 530-752-8131, Fax: 530-752-6635, E-mail: studentservices@math.ucdavis.edu.

University of California, San Diego, Graduate Studies and Research, Department of Mathematics, La Jolla, CA 92093. Offers applied mathematics (MA); mathematics (MA, PhD); statistics (MS). *Degree requirements:* For doctorate, thesis/dissertation. *Entrance requirements:* For master's and doctorate, GRE General Test, GRE Subject Test. Electronic applications accepted.

University of California, Santa Barbara, Graduate Division, College of Letters and Sciences, Division of Mathematics, Life, and Physical Sciences, Department of Mathematics, Santa Barbara, CA 93106. Offers applied mathematics (MA); mathematics (MA, PhD). *Faculty:* 32 full-time (2 women). *Students:* 58 full-time (17 women); includes 4 minority (2 Asian Americans or Pacific Islanders, 2 Hispanic Americans), 11 international. Average age 25. 115 applicants, 25% accepted, 5 enrolled. In 2005, 14 master's, 6 doctorates awarded. Terminal master's awarded for partial completion of doctoral program. *Median time to degree:* Of those who began their doctoral program in fall 1997, 100% received their degree in 8 years or less. *Degree requirements:* For master's, thesis (for some programs), comprehensive exam (for some programs), registration; for doctorate, thesis/dissertation, comprehensive exam, registration. *Entrance requirements:* For master's and doctorate, GRE General Test, GRE Subject Test. Additional exam requirements/recommendations for international students: Required—TOEFL (minimum score 575 paper-based; 231 computer-based); Recommended—TSE. *Application deadline:* For fall admission, 1/1 for domestic students, 1/1 for international students. Application fee: $60. Electronic applications accepted. *Financial support:* In 2005–06, 3 students received support, including 2 fellowships with full and partial tuition reimbursements available (averaging $17,000 per year), 45 teaching assistantships with full and partial tuition reimbursements available (averaging $15,000 per year); Federal Work-Study and health care benefits also available. Financial award application deadline: 1/1; financial award applicants required to submit FAFSA. *Faculty research:* Topology, differential geometry, algebra, analysis, applied mathematics. *Unit head:* Thomas Sideris, Chair, 805-893-8340, E-mail: sideris@math.ucsb.edu. *Application contact:* Medina Teel, Graduate Advisor, 805-893-8192, Fax: 805-893-2385, E-mail: teel@math.ucsb.edu.

University of Central Florida, College of Sciences, Department of Mathematics, Orlando, FL 32816. Offers applied mathematics (Certificate); mathematical science (MS); mathematics (PhD). Part-time and evening/weekend programs available. *Faculty:* 45 full-time (9 women), 4 part-time/adjunct (0 women). *Students:* 34 full-time (9 women), 21 part-time (8 women); includes 12 minority (2 African Americans, 3 Asian Americans or Pacific Islanders, 7 Hispanic Americans), 10 international. Average age 33. 42 applicants, 76% accepted, 19 enrolled. In 2005, 14 master's, 3 doctorates awarded. *Degree requirements:* For master's, thesis or alternative; for doctorate, thesis/dissertation, candidacy exam. *Entrance requirements:* For master's, GRE General Test, minimum GPA of 3.0 in last 60 hours; for doctorate, GRE Subject Test, minimum GPA of 3.0 in last 60 hours or master's qualifying exam. Additional exam requirements/recommendations for international students: Required—TOEFL. *Application deadline:* For fall admission, 7/15 for domestic students; for spring admission, 12/1 for domestic students. Application fee: $30. Electronic applications accepted. *Expenses:* Tuition, state resident: full-time $5,788. Tuition, nonresident: full-time $21,927. Required fees: $241 per credit hour. *Financial support:* In 2005–06, 13 fellowships with partial tuition reimbursements (averaging $5,285 per year), 5 research assistantships with partial tuition reimbursements (averaging $11,700 per year), 26 teaching assistantships with partial tuition reimbursements (averaging $12,600 per year) were awarded; career-related internships or fieldwork, Federal Work-Study, institutionally sponsored loans, tuition waivers (partial), and unspecified assistantships also available. Financial award application deadline: 3/1; financial award applicants required to submit FAFSA. *Faculty research:* Applied mathematics, analysis, approximation theory, graph theory, mathematical statistics. *Unit head:* Dr. Zuhair Nashed, Chair, 407-823-0445, Fax: 407-823-6253, E-mail: znashed@mail.ucf.edu. *Application contact:* Dr. Ram Mohapatra, Coordinator, 407-823-5080, Fax: 407-823-6253, E-mail: ramm@pegasus.cc.ucf.edu.

University of Central Oklahoma, College of Graduate Studies and Research, College of Mathematics and Science, Department of Mathematics and Statistics, Edmond, OK 73034-5209. Offers applied mathematical sciences (MS), including computer science, mathematics, mathematics/computer science teaching, statistics. Part-time programs available. *Faculty:* 8 full-time (3 women), 1 part-time/adjunct (0 women). *Students:* 7 full-time (1 woman), 8 part-time (3 women); includes 2 minority (both Asian Americans or Pacific Islanders), 7 international. Average age 29. 14 applicants, 100% accepted. In 2005, 6 degrees awarded. *Degree requirements:* For master's, thesis. *Entrance requirements:* Additional exam requirements/recommendations for international students: Required—TOEFL (minimum score 550 paper-based; 213 computer-based). *Application deadline:* Applications are processed on a rolling basis. Application fee: $25. Electronic applications accepted. *Expenses:* Tuition, state resident: full-time $2,988; part-time $125 per credit hour. Tuition, nonresident: full-time $4,728; part-time $197 per credit hour. Required fees: $716; $16 per credit hour. *Financial support:* Federal Work-Study and unspecified assistantships available. Financial award application deadline: 3/31; financial award applicants required to submit FAFSA. *Faculty research:* Curvature, FAA, math education. *Unit head:* Dr. Chuck Cooper, Chairperson, 405-974-5294. *Application contact:* Dr. James Yates, Adviser, 405-974-5386, Fax: 405-974-3824, E-mail: jyates@aix1.ucok.edu.

University of Chicago, Division of the Physical Sciences, Department of Mathematics, Program in Applied Mathematics, Chicago, IL 60637-1513. Offers SM, PhD. *Degree requirements:* For master's, one foreign language; for doctorate, one foreign language, thesis/dissertation, 2 qualifying exams. *Entrance requirements:* For master's and doctorate, GRE General Test, GRE Subject Test. Additional exam requirements/recommendations for international students: Required—TOEFL (minimum score 600 paper-based; 250 computer-based). *Application deadline:* For fall admission, 1/5 for domestic students. Application fee: $55. Electronic applications accepted. *Financial support:* Fellowships, research assistantships, teaching assistantships available. Financial award application deadline: 1/5. *Faculty research:* Applied analysis, dynamical systems, theoretical biology, math-physics. *Unit head:* Norman R. Lebovitz, Head, 773-702-7329. *Application contact:* Laurie Wail, Graduate Studies Assistant, 773-702-7358, Fax: 773-702-9787, E-mail: lwail@math.uchicago.edu.

University of Cincinnati, Division of Research and Advanced Studies, McMicken College of Arts and Sciences, Department of Mathematical Sciences, Cincinnati, OH 45221. Offers applied mathematics (MS, PhD); mathematics education (MAT); pure mathematics (MS, PhD); statistics (MS, PhD). *Accreditation:* NCATE (one or more programs are accredited). Part-time programs available. Terminal master's awarded for partial completion of doctoral program. *Degree requirements:* For master's, thesis or alternative, comprehensive exam; for doctorate, one foreign language, thesis/dissertation, comprehensive exam. *Entrance requirements:* For master's, GRE, teacher certification (MAT); for doctorate, GRE. Additional exam requirements/recommendations for international students: Required—TOEFL. Electronic applications accepted. *Faculty research:* Algebra, analysis, differential equations, numerical analysis, statistics.

University of Colorado at Boulder, Graduate School, College of Arts and Sciences, Department of Applied Mathematics, Boulder, CO 80309. Offers MS, PhD. Part-time programs available. *Faculty:* 14 full-time (2 women). *Students:* 67 full-time (14 women), 8 part-time (2 women); includes 2 minority (both Hispanic Americans), 17 international. Average age 28. 29 applicants, 100% accepted. In 2005, 21 master's, 7 doctorates awarded. Terminal master's awarded for partial completion of doctoral program. *Degree requirements:* For master's, thesis or alternative, comprehensive exam; for doctorate, one foreign language, thesis/dissertation, comprehensive exam. *Entrance requirements:* For master's, GRE General Test, minimum undergraduate GPA of 2.75; for doctorate, GRE General Test. Additional exam requirements/recommendations for international students: Required—TOEFL. *Application deadline:* For fall admission, 2/15 priority date for domestic students, 12/1 priority date for international students. Applications are processed on a rolling basis. Application fee: $50 ($60 for international students). *Financial support:* In 2005–06, fellowships (averaging $5,264 per year), research assistantships (averaging $14,324 per year), teaching assistantships (averaging $13,688 per year) were awarded; scholarships/grants and traineeships also available. Support available to part-time students. Financial award application deadline: 2/15. *Faculty research:* Non-linear phenomena, computational mathematics, physical applied mathematics, statistics. Total annual research expenditures: $1.8 million. *Unit head:* Jim Curry, Chair, 303-492-0592, Fax: 303-492-4066, E-mail: james.h.curry@colorado.edu. *Application contact:* Graduate Program Assistant, 303-492-4668, Fax: 303-492-4066, E-mail: appm_app@colorado.edu.

University of Colorado at Colorado Springs, Graduate School, College of Engineering and Applied Science, Department of Mathematics, Colorado Springs, CO 80933-7150. Offers applied mathematics (MS). Part-time and evening/weekend programs available. *Faculty:* 6 full-time, 1 part-time/adjunct. *Students:* 11 full-time (2 women), 5 part-time (3 women). Average age 32. In 2005, 3 degrees awarded. *Degree requirements:* For master's, thesis. *Entrance requirements:* For master's, GRE General Test, minimum GPA of 3.0. Additional exam requirements/recommendations for international students: Required—TOEFL. *Application deadline:* For fall admission, 6/15 for domestic students. Application fee: $60 ($75 for international students). *Expenses:* Tuition, state resident: full-time $4,068; part-time $312 per credit hour. Tuition, nonresident: full-time $11,570; part-time $890 per credit hour. Required fees: $339; $19 per credit hour. $177 per term. Tuition and fees vary according to course load, program and reciprocity agreements. *Financial support:* Teaching assistantships available. *Faculty research:* Abelian groups and noncommutative rings, hormone analysis and computer vision, probability and mathematical physics, stochastic dynamics, probability models. Total annual research expenditures: $6,231. *Unit head:* Dr. Rinaldo Schinazi, Chair, 719-262-3920, Fax: 719-262-3605. *Application contact:* Joan Stephens, Director of Graduate Studies, 719-262-3311, Fax: 719-262-3605, E-mail: mathinfo@math.uccs.edu.

University of Colorado at Denver and Health Sciences Center—Downtown Denver Campus, College of Liberal Arts and Sciences, Department of Mathematical Sciences, Denver, CO 80217-3364. Offers applied mathematics (MS, PhD). Part-time and evening/weekend programs available. *Students:* 24 full-time (11 women), 50 part-time (17 women); includes 11 minority (5 Asian Americans or Pacific Islanders, 6 Hispanic Americans), 8 international. Average age 33. 28 applicants, 71% accepted, 15 enrolled. In 2005, 6 master's, 3 doctorates awarded. *Degree requirements:* For master's, thesis optional; for doctorate, thesis/dissertation, comprehensive exam. *Entrance requirements:* For master's, GRE, 30 hours of course work in mathematics, 24 hours of course work in upper division mathematics, minimum GPA of 2.75; for doctorate, GRE, 24 hours of course work in upper division mathematics. Additional exam requirements/recommendations for international students: Required—TOEFL (minimum score 525 paper-based; 197 computer-based). *Application deadline:* For fall admission, 4/1 for domestic students; for spring admission, 11/1 for domestic students. Applications are processed on a rolling basis. Application fee: $50 ($75 for international students). Electronic applications accepted. *Expenses:* Tuition, state resident: part-time $325 per credit hour. Tuition, nonresident: part-time $1,077 per credit hour. Required fees: $145 per credit hour. One-time fee: $115 part-time. Tuition and fees vary according to course level and program. *Financial support:* Fellowships with partial tuition reimbursements, research assistantships with full tuition reimbursements, teaching assistantships with full tuition reimbursements, Federal Work-Study available. Financial award application deadline: 4/1; financial award applicants required to submit FAFSA. *Faculty research:* Computational mathematics, computational biology, discrete mathematics and geometry probability and statistics, optimization. *Unit head:* Prof. Mike S Jacobson, Chair, 303-556-6270, Fax: 303-556-8550, E-mail: msj@math.cudenver.edu. *Application contact:* Marcia Kelly, Program Assistant, 303-556-2341, Fax: 303-556-8550, E-mail: marcia.kelly@cudenver.edu.

University of Connecticut, Graduate School, College of Liberal Arts and Sciences, Department of Mathematics, Field of Applied Financial Mathematics, Storrs, CT 06269. Offers MS. *Faculty:* 43 full-time (5 women). *Students:* 11 full-time (6 women), 10 international. Average age 28. 30 applicants, 57% accepted, 5 enrolled. In 2005, 2 degrees awarded. *Degree requirements:* For master's, comprehensive exam. *Entrance requirements:* Additional exam requirements/recommendations for international students: Required—TOEFL (minimum score 550 paper-based; 213 computer-based). *Application deadline:* For fall admission, 2/1 priority date for domestic students, 2/1 priority date for international students; for spring admission, 11/1 for domestic students, 10/1 for international students. Applications are processed on a rolling basis. Application fee: $55. Electronic applications accepted. *Expenses:* Tuition, state resident: part-time $444 per credit hour. Tuition, nonresident: part-time $1,154 per credit hour. Tuition and fees vary according to course load. *Financial support:* In 2005–06, 1 research assistantship, 7 teaching assistantships were awarded; Federal Work-Study and scholarships/grants also available. Financial award application deadline: 2/1; financial award applicants required to submit FAFSA. *Unit head:* James Bridgeman, Chairperson, 860-486-8382, Fax: 860-486-4283, E-mail: james.bridgeman@uconn.edu. *Application contact:* Sharon McDermott, Administrative Assistant, 860-486-6452, Fax: 860-486-4283, E-mail: gradadm@math.uconn.edu.

University of Dayton, Graduate School, College of Arts and Sciences, Department of Mathematics, Dayton, OH 45469-1300. Offers applied mathematics (MS). Part-time and evening/weekend programs available. *Faculty:* 8 full-time (2 women). *Students:* 4 full-time (3 women), 3 part-time (1 woman); includes 1 minority (African American), 1 international. Average age 25. 48 applicants, 54% accepted, 8 enrolled. In 2005, 3 degrees awarded. *Entrance requirements:* For master's, minimum undergraduate GPA of 2.8. Additional exam requirements/recommendations for international students: Required—TOEFL (minimum score 550 paper-based; 213 computer-based). *Application deadline:* For fall admission, 3/1 priority date for domestic students, 3/1 priority date for international students. Electronic applications accepted. *Expenses:* Tuition: Part-time $567 per credit hour. Required fees: $25 per term. Tuition and fees vary according to degree level and program. *Financial support:* In 2005–06, 7 teaching assistantships with full tuition reimbursements (averaging $9,706 per year) were awarded; institutionally sponsored loans, health care benefits, and unspecified assistantships also available. Financial award applicants required to submit FAFSA. *Faculty research:* Differential equations, integral equations, general topology, measure theory, graph theory. *Unit head:* Dr. Paul W. Eloe, Chair, 937-229-2511, Fax: 937-229-2566, E-mail: paul.eloe@notes.udayton.edu. *Application contact:* E. Eavers.

University of Delaware, College of Arts and Sciences, Department of Mathematical Sciences, Newark, DE 19716. Offers applied mathematics (MS, PhD); mathematics (MS, PhD). Part-time programs available. *Faculty:* 46 full-time (13 women). *Students:* 40 full-time (16 women), 1 part-time; includes 3 minority (1 African American, 1 American Indian/Alaska Native, 1 Asian American or Pacific Islander), 18 international. Average age 25. 111 applicants, 16% accepted, 13 enrolled. In 2005, 11 master's, 2 doctorates awarded. Terminal master's awarded for partial completion of doctoral program. *Degree requirements:* For master's, thesis (for some programs); for doctorate, one foreign language, thesis/dissertation, qualifying exam. *Entrance requirements:* For master's and doctorate, GRE General Test. Additional exam requirements/recommendations for international students: Required—TOEFL. *Application deadline:* For fall admission, 3/1 for domestic students; for spring admission, 12/15 priority date for domestic students. Applications are processed on a rolling basis. Application fee: $60. Electronic applications accepted. *Financial support:* In 2005–06, 5 fellowships with tuition reimbursements (averaging $12,200 per year), 1 research assistantship with tuition reimbursement (averaging $14,000 per year), 35 teaching assistantships with tuition reimbursements (averaging $12,240 per year) were awarded; career-related internships or fieldwork, institutionally sponsored loans, scholarships/grants, and tuition waivers (full and partial) also available. Financial award application deadline: 3/1. *Faculty research:* Scattering theory, inverse problems, fluid dynamics, numerical analysis, combinatorics. Total annual research expenditures: $863,660. *Unit head:* Prof. Peter B. Monk, Interim Chair, 302-831-2711, Fax: 302-831-4511, E-mail: monk@math.udel.edu. *Application contact:* Dr. George C. Hsiao, Graduate Chair, 302-831-1882, Fax: 302-831-4511, E-mail: hsiao@math.udel.edu.

See Close-Up on page 533.

University of Denver, Faculty of Natural Sciences and Mathematics, Department of Mathematics, Denver, CO 80208. Offers applied mathematics (MA, MS); computer science (MS); mathematics (PhD). Part-time programs available. *Faculty:* 9 full-time (3 women). *Students:* 12 applicants, 100% accepted. In 2005, 5 master's, 2 doctorates awarded. Terminal master's awarded for partial completion of doctoral program. *Degree requirements:* For master's, computer language, foreign language, or laboratory experience; for doctorate, one foreign language, thesis/dissertation, oral and written exams. *Entrance requirements:* For master's and doctorate, GRE General Test. Additional exam requirements/recommendations for international students: Required—TOEFL. *Application deadline:* Applications are processed on a rolling basis. Application fee: $45. *Expenses:* Tuition: Full-time $27,756; part-time $771 per credit. Required fees: $174. *Financial support:* In 2005–06, 11 teaching assistantships with full and partial tuition reimbursements (averaging $12,896 per year) were awarded; career-related internships or fieldwork, Federal Work-Study, institutionally sponsored loans, and scholarships/grants also available. Support available to part-time students. Financial award application deadline: 3/1; financial award applicants required to submit FAFSA. *Faculty research:* Real-time software, convex bodies, multidimensional data, parallel computer clusters. *Unit head:* Dr. Rick Ball, Chairperson, 303-871-2821. *Application contact:* 303-871-2911, E-mail: math-info@math.du.edu.

University of Georgia, Graduate School, College of Arts and Sciences, Department of Mathematics, Athens, GA 30602. Offers applied mathematical science (MAMS); mathematics (MA, PhD). *Faculty:* 31 full-time (3 women). *Students:* 48 full-time; includes 3 minority (2 African Americans, 1 Asian American or Pacific Islander), 19 international. 123 applicants, 19% accepted, 15 enrolled. In 2005, 6 master's, 7 doctorates awarded. *Degree requirements:* For master's, one foreign language, thesis (for some programs); for doctorate, 2 foreign languages, thesis/dissertation. *Entrance requirements:* For master's and doctorate, GRE General Test. *Application deadline:* For fall admission, 7/1 for domestic students; for spring admission, 11/15 for domestic students. Application fee: $50. Electronic applications accepted. *Financial support:* Fellowships, research assistantships, teaching assistantships, unspecified assistantships available. *Unit head:* Dr. Joseph H. G. Fu, Graduate Coordinator, 706-542-2643, Fax: 706-542-2573. *Application contact:* Dr. Joseph H. G. Fu, Graduate Coordinator, 706-542-2643, Fax: 706-542-2573.

University of Guelph, Graduate Program Services, College of Physical and Engineering Science, Department of Mathematics and Statistics, Guelph, ON N1G 2W1, Canada. Offers applied mathematics (PhD); applied statistics (PhD); mathematics and statistics (M Sc). Part-time programs available. *Faculty:* 21 full-time (6 women), 4 part-time/adjunct (0 women). *Students:* 40 full-time, 4 part-time. Average age 22. 103 applicants, 25% accepted, 23 enrolled. In 2005, 20 master's, 1 doctorate awarded. *Median time to degree:* Of those who began their doctoral program in fall 1997, 100% received their degree in 8 years or less. *Degree requirements:* For master's, thesis (for some programs); for doctorate, thesis/dissertation. *Entrance requirements:* For master's, minimum B- average during previous 2 years of course work; for doctorate, minimum B average. Additional exam requirements/recommendations for international students: Required—TOEFL (minimum score 550 paper-based; 213 computer-based), IELT (minimum score 7). *Application deadline:* For fall admission, 2/1 for domestic students, 2/1 for international students. Application fee: $75. *Financial support:* In 2005–06, 20 students received support, including 26 research assistantships (averaging $2,300 per year),

Applied Mathematics

University of Guelph (continued)
27 teaching assistantships (averaging $9,000 per year); fellowships, scholarships/grants also available. *Faculty research:* Dynamical systems, mathematical biology, numerical analysis, linear and nonlinear models, reliability and bioassay. *Unit head:* Dr. O. Brian Allen, Chair, 519-824-4120 Ext. 56556, Fax: 519-837-0221, E-mail: chair@uoguelph.ca. *Application contact:* Susan McCormick, Graduate Administrative Assistant, 519-824-4120 Ext. 56553, Fax: 519-837-0221, E-mail: smccormi@uoguelph.ca.

University of Illinois at Chicago, Graduate College, College of Liberal Arts and Sciences, Department of Mathematics, Statistics, and Computer Science, Chicago, IL 60607-7128. Offers applied mathematics (MS, DA, PhD); computer science (MS, DA, PhD); math and information science for the industry (MS); probability and statistics (MS, DA, PhD); pure mathematics (MS, DA, PhD); teaching of mathematics (MST). Part-time programs available. *Degree requirements:* For master's, comprehensive exam; for doctorate, one foreign language, thesis/dissertation. *Entrance requirements:* For master's and doctorate, GRE General Test, minimum GPA of 2.75. Additional exam requirements/recommendations for international students: Required—TOEFL. Electronic applications accepted.

See Close-Up on page 535.

University of Illinois at Urbana–Champaign, Graduate College, College of Liberal Arts and Sciences, Department of Mathematics, Champaign, IL 61820. Offers applied mathematics (MS); mathematics (MS, PhD); teaching of mathematics (MS). *Faculty:* 68 full-time (5 women), 6 part-time/adjunct (1 woman). *Students:* 198 full-time (54 women), 33 part-time (8 women); includes 14 minority (1 African American, 1 American Indian/Alaska Native, 9 Asian Americans or Pacific Islanders, 3 Hispanic Americans), 119 international. 408 applicants, 43% accepted, 52 enrolled. In 2005, 41 master's, 19 doctorates awarded. *Degree requirements:* For doctorate, 2 foreign languages, thesis/dissertation. *Entrance requirements:* For master's, GRE, minimum GPA of 3.0. *Application deadline:* For fall admission, 2/6 for domestic students. Applications are processed on a rolling basis. Application fee: $50 ($60 for international students). Electronic applications accepted. *Financial support:* In 2005–06, 37 fellowships, 37 research assistantships, 146 teaching assistantships were awarded; tuition waivers (full and partial) also available. Financial award application deadline: 2/15. *Unit head:* Daniel Grayson, Chair, 217-333-6209, Fax: 217-333-9576, E-mail: dan@uiuc.edu. *Application contact:* Lori Dick, Administrative Assistant, 217-333-3350, Fax: 217-333-9576, E-mail: ldick@math.uiuc.edu.

The University of Iowa, Graduate College, Program in Applied Mathematical and Computational Sciences, Iowa City, IA 52242-1316. Offers PhD. *Students:* 15 full-time (4 women), 15 part-time (4 women); includes 4 minority (2 African Americans, 2 Hispanic Americans), 15 international. 27 applicants, 22% accepted, 3 enrolled. In 2005, 4 degrees awarded. *Degree requirements:* For doctorate, thesis/dissertation, comprehensive exam, registration. *Entrance requirements:* For doctorate, GRE General Test, minimum GPA of 3.0. Additional exam requirements/recommendations for international students: Required—TOEFL (minimum score 600 paper-based; 250 computer-based). *Application deadline:* For fall admission, 1/15 priority date for domestic students, 1/15 priority date for international students; for spring admission, 10/1 priority date for domestic students. Applications are processed on a rolling basis. Application fee: $60 ($85 for international students). Electronic applications accepted. *Expenses:* Tuition, state resident: part-time $1,882 per term. Tuition, nonresident: full-time $17,338; part-time $4,907 per term. Tuition and fees vary according to course load and program. *Financial support:* In 2005–06, 2 fellowships, 2 research assistantships with partial tuition reimbursements, 27 teaching assistantships with partial tuition reimbursements were awarded. Financial award applicants required to submit FAFSA. *Unit head:* Yi Li, Director, 319-335-0772.

University of Kansas, Graduate School, College of Liberal Arts and Sciences, Department of Mathematics, Lawrence, KS 66045. Offers applied mathematics and statistics (MA, PhD); mathematics (MA, PhD). *Faculty:* 40. *Students:* 67 full-time (23 women), 12 part-time (6 women); includes 4 minority (1 American Indian/Alaska Native, 3 Asian Americans or Pacific Islanders), 31 international. Average age 27. 67 applicants, 73% accepted. In 2005, 17 master's, 2 doctorates awarded. Terminal master's awarded for partial completion of doctoral program. *Degree requirements:* For master's, thesis or alternative; for doctorate, 2 foreign languages, thesis/dissertation, comprehensive exam. *Entrance requirements:* For master's and doctorate, GRE. Additional exam requirements/recommendations for international students: Required—TOEFL. *Application deadline:* For fall admission, 3/1 priority date for domestic students, 3/1 priority date for international students. Applications are processed on a rolling basis. Application fee: $55 ($60 for international students). Electronic applications accepted. *Expenses:* Tuition, state resident: full-time $4,859. Tuition, nonresident: full-time $1,200. Required fees: $589. Tuition and fees vary according to program. *Financial support:* Fellowships, research assistantships with full and partial tuition reimbursements, teaching assistantships with full and partial tuition reimbursements, institutionally sponsored loans available. Support available to part-time students. Financial award application deadline: 2/1. *Faculty research:* Commutative algebra/algebraic geometry, stochastic adaptive control/stochastic processes analysis/harmonic analysis/PDES, numerical analysis/dynamical systems, topology/set theory. *Unit head:* Jack Porter, Chair, 785-864-3651, Fax: 785-864-5255, E-mail: porter@math.ukans.edu. *Application contact:* David Lerner, Graduate Director, 785-864-3651, E-mail: lerner@ukans.edu.

University of Kentucky, Graduate School, Graduate School Programs from the College of Arts and Sciences, Program in Mathematics, Lexington, KY 40506-0032. Offers applied mathematics (MS); mathematics (MA, MS, PhD). *Faculty:* 34 full-time (1 woman), 3 part-time/adjunct (0 women). *Students:* 63 full-time (14 women), 5 part-time (4 women); includes 3 minority (1 Asian American or Pacific Islander, 2 Hispanic Americans), 16 international. Average age 27. 102 applicants, 42% accepted, 22 enrolled. In 2005, 12 master's, 8 doctorates awarded. *Median time to degree:* Of those who began their doctoral program in fall 1997, 84% received their degree in 8 years or less. *Degree requirements:* For master's, thesis optional; for doctorate, one foreign language, thesis/dissertation, comprehensive exam. *Entrance requirements:* For master's, GRE General Test, minimum undergraduate GPA of 2.5; for doctorate, GRE General Test, minimum graduate GPA of 3.0. Additional exam requirements/recommendations for international students: Required—TOEFL (minimum score 550 paper-based; 213 computer-based). *Application deadline:* For fall admission, 7/17 priority date for domestic students, 2/1 priority date for international students; for spring admission, 12/13 priority date for domestic students, 6/15 priority date for international students. Applications are processed on a rolling basis. Application fee: $40 ($55 for international students). Electronic applications accepted. *Expenses:* Tuition, state resident: full-time $3,159; part-time $331 per credit hour. Tuition, nonresident: full-time $6,984; part-time $756 per credit hour. Tuition and fees vary according to course load, degree level and program. *Financial support:* In 2005–06, 12 fellowships with full tuition reimbursements (averaging $3,000 per year), 8 research assistantships with full tuition reimbursements (averaging $12,800 per year), 55 teaching assistantships with full tuition reimbursements (averaging $13,000 per year) were awarded; Federal Work-Study, institutionally sponsored loans, scholarships/grants, traineeships, health care benefits, tuition waivers (partial), and unspecified assistantships also available. Support available to part-time students. Financial award application deadline:3/15. *Faculty research:* Numerical analysis, combinatorics, partial differential equations, algebra and number theory, real and complex analysis. *Unit head:* Dr. Serge C. Ochanine, Director of Graduate Studies, 859-257-8837, Fax: 859-257-3464, E-mail: ochanine@ms.uky.edu. *Application contact:* Dr. Brian Jackson, Senior Associate Dean, 859-257-8176, Fax: 859-323-1928, E-mail: lance.brunner@uky.edu.

University of Louisville, Graduate School, College of Arts and Sciences, Department of Mathematics, Louisville, KY 40292-0001. Offers applied and industrial mathematics (PhD); mathematics (MA). Evening/weekend programs available. *Students:* 32 full-time (8 women), 8 part-time (4 women); includes 2 minority (1 African American, 1 Asian American or Pacific Islander), 9 international. Average age 28. In 2005, 12 degrees awarded. *Degree requirements:* For master's, thesis optional; for doctorate, thesis/dissertation, internship, project, comprehensive exam. *Entrance requirements:* For master's and doctorate, GRE General Test. *Application deadline:* Applications are processed on a rolling basis. Application fee: $50. *Expenses:* Tuition, state resident: full-time $3,003; part-time $334 per credit hour. Tuition, nonresident: full-time $8,277; part-time $920 per credit hour. Tuition and fees vary according to course load, degree level and program. *Financial support:* In 2005–06, 25 teaching assistantships with full tuition reimbursements were awarded *Unit head:* Dr. Thomas Riedel, Chair, 502-852-5974, Fax: 502-852-7132, E-mail: thomas.riedel@louisville.edu. *Application contact:* Dr. Prasana Sahoo, Graduate Studies Director, 502-852-6826, Fax: 502-852-7132, E-mail: sahoo@louisville.edu.

University of Maryland, Baltimore County, Graduate School, College of Natural Sciences and Mathematics, Department of Mathematics and Statistics, Program in Applied Mathematics, Baltimore, MD 21250. Offers MS, PhD. Part-time and evening/weekend programs available. *Faculty:* 17 full-time (2 women). *Students:* 14 full-time (9 women), 18 part-time (5 women); includes 2 minority (1 African American, 1 Asian American or Pacific Islander), 17 international. Average age 28. 25 applicants, 60% accepted, 11 enrolled. In 2005, 1 master's, 3 doctorates awarded. Terminal master's awarded for partial completion of doctoral program. *Median time to degree:* Of those who began their doctoral program in fall 1997, 60% received their degree in 8 years or less. *Degree requirements:* For master's, thesis (for some programs), comprehensive exam (for some programs), registration; for doctorate, thesis/dissertation, comprehensive exam, registration. *Entrance requirements:* For master's and doctorate, GRE General Test, minimum GPA of 3.0. Additional exam requirements/recommendations for international students: Required—TOEFL (minimum score 600 paper-based; 250 computer-based). *Application deadline:* For fall admission, 2/15 priority date for domestic students, 1/1 priority date for international students; for spring admission, 10/15 priority date for domestic students, 9/15 priority date for international students. Applications are processed on a rolling basis. Application fee: $50. Electronic applications accepted. *Expenses:* Tuition, state resident: part-time $395 per credit. Tuition, nonresident: part-time $652 per credit. Required fees: $82 per credit. Tuition and fees vary according to course load, program and reciprocity agreements. *Financial support:* In 2005–06, 19 students received support, including 4 research assistantships with full tuition reimbursements available (averaging $15,000 per year), 19 teaching assistantships with full tuition reimbursements available (averaging $15,000 per year); career-related internships or fieldwork, scholarships/grants, health care benefits, and unspecified assistantships also available. Support available to part-time students. Financial award application deadline: 2/15. *Faculty research:* Numerical analysis and scientific computation, optimization theory and algorithms, differential equations and mathematical modeling, mathematical biology and bioinformatics. Total annual research expenditures: $580,000. *Application contact:* Dr. Muddappa Gowda, Director of Graduate Programs, 410-455-2431, Fax: 410-455-1066, E-mail: gowda@math.umbc.edu.

University of Maryland, College Park, Graduate Studies, College of Computer, Mathematical and Physical Sciences, Department of Mathematics, Applied Mathematics Program, College Park, MD 20742. Offers MS, PhD. Part-time and evening/weekend programs available. *Students:* 90 full-time (22 women), 13 part-time (3 women); includes 15 minority (6 African Americans, 8 Asian Americans or Pacific Islanders, 1 Hispanic American), 42 international. 126 applicants, 26% accepted, 19 enrolled.Terminal master's awarded for partial completion of doctoral program. *Degree requirements:* For master's, seminar, scholarly paper, thesis optional; for doctorate, thesis/dissertation, exams, seminars, comprehensive exam. *Entrance requirements:* For master's and doctorate, GRE General Test, GRE Subject Test, minimum GPA of 3.0, 3 letters of recommendation. *Application deadline:* For fall admission, 5/1 for domestic students, 2/1 for international students; for spring admission, 10/15 for domestic students, 6/1 for international students. Applications are processed on a rolling basis. Application fee: $60. Electronic applications accepted. *Financial support:* In 2005–06, 11 fellowships (averaging $5,192 per year) were awarded; teaching assistantships Financial award applicants required to submit FAFSA. *Unit head:* Dr. David Levermore, Director, 301-405-5127, Fax: 301-314-8027, E-mail: lvrmr@math.umd.edu. *Application contact:* Dean of Graduate School, 301-405-4190, Fax: 301-314-9305.

University of Massachusetts Amherst, Graduate School, College of Natural Sciences and Mathematics, Department of Mathematics and Statistics, Program in Applied Mathematics, Amherst, MA 01003. Offers MS. *Students:* 8 full-time (4 women); includes 1 minority (Hispanic American), 1 international. Average age 28. 25 applicants, 20% accepted, 4 enrolled. In 2005, 9 degrees awarded. *Entrance requirements:* Additional exam requirements/recommendations for international students: Required—TOEFL (minimum score 530 paper-based; 197 computer-based). *Application deadline:* For fall admission, 2/1 priority date for domestic students, 2/1 priority date for international students. Applications are processed on a rolling basis. Application fee: $40 ($65 for international students). Electronic applications accepted. *Expenses:* Tuition, state resident: part-time $110 per credit. Tuition, nonresident: part-time $414 per credit. Required fees: $2,824 per term. One-time fee: $250 part-time. Full-time tuition and fees vary according to course load, campus/location, program and reciprocity agreements. *Financial support:* Fellowships with full tuition reimbursements, research assistantships with full tuition reimbursements, teaching assistantships with full tuition reimbursements, career-related internships or fieldwork, Federal Work-Study, scholarships/grants, traineeships, and unspecified assistantships available. Support available to part-time students. Financial award application deadline: 2/1. *Unit head:* Dr. Ivan Mirkovic, Director, 413-545-2282, Fax: 413-545-1801.

University of Massachusetts Lowell, Graduate School, College of Arts and Sciences, Department of Mathematics, Lowell, MA 01854-2881. Offers applied mathematics (MS); computational mathematics (PhD); mathematics (MS). Part-time programs available. *Entrance requirements:* For master's, GRE General Test.

University of Memphis, Graduate School, College of Arts and Sciences, Department of Mathematical Sciences, Memphis, TN 38152-3420. Offers applied mathematics (MS); applied statistics (PhD); bioinformatics (MS); computer science (PhD); computer sciences (MS); mathematics (MS, PhD); statistics (MS, PhD). Part-time programs available. *Faculty:* 24 full-time (5 women), 3 part-time/adjunct (0 women). *Students:* 105 full-time (38 women), 34 part-time (8 women); includes 8 minority (7 African Americans, 1 Asian American or Pacific Islander), 89 international. Average age 30. 139 applicants, 37% accepted. In 2005, 43 master's, 5 doctorates awarded. Terminal master's awarded for partial completion of doctoral program. *Degree requirements:* For master's, comprehensive exam; for doctorate, one foreign language, thesis/dissertation, oral exams. *Entrance requirements:* For master's and doctorate, GRE General Test, minimum GPA of 2.5. Additional exam requirements/recommendations for international students: Required—TOEFL (minimum score 550 paper-based; 210 computer-based), WES evaluation of transcript. *Application deadline:* For fall admission, 8/1 for domestic students, 5/1 for international students; for spring admission, 12/1 for domestic students, 9/1 for international students. Applications are processed on a rolling basis. Application fee: $25 ($50 for international students). Electronic applications accepted. *Financial support:* In 2005–06, 58 students received support, including fellowships with full tuition reimbursements available (averaging $17,500 per year), 9 research assistantships with full tuition reimbursements available (averaging $9,000 per year), 30 teaching assistantships with full tuition reimbursements available (averaging $9,000 per year); career-related internships or fieldwork, Federal Work-Study, scholarships/grants, unspecified assistantships, and minority scholarships also available. Financial award application deadline: 2/2. *Faculty research:* Combinatorics, ergodic theory, graph theory, Ramsey theory, applied statistics. Total annual research expenditures: $1.5 million. *Unit head:* Dr. James E. Jamison, Chairman, 901-678-2482, Fax: 901-678-2480, E-mail: jjamison@memphis.edu. *Application contact:* Coordinator of Graduate Studies, 901-678-2482, Fax: 901-678-2480, E-mail: dfwilson@memphis.edu.

University of Michigan–Dearborn, College of Arts, Sciences, and Letters, Program in Applied and Computational Mathematics, Dearborn, MI 48128-1491. Offers MS. Part-time and evening/weekend programs available. *Faculty:* 9 full-time (3 women). *Students:* 3 full-time (1 woman), 11 part-time (3 women); includes 3 minority (2 African Americans, 1 Asian American or Pacific Islander). Average age 33. 3 applicants, 100% accepted. In 2005, 4 degrees awarded.

Degree requirements: For master's, thesis or alternative, project. *Entrance requirements:* For master's, 3 letters of recommendation, minimum GPA of 3.0, 2 years course work in math. Additional exam requirements/recommendations for international students: Required—TOEFL (minimum score 560 paper-based; 220 computer-based). *Application deadline:* For fall admission, 8/1 for domestic students. For winter admission, 12/1 for domestic students; for spring admission, 4/1 for domestic students. Applications are processed on a rolling basis. Application fee: $60 ($75 for international students). Electronic applications accepted. *Financial support:* Federal Work-Study and scholarships/grants available. Support available to part-time students. Financial award application deadline: 4/1; financial award applicants required to submit FAFSA. *Faculty research:* Partial differential equations, statistics, discrete optimization, approximation theory, stochastic processes. *Unit head:* Dr. Joan Remski, Director, 313-593-4994, E-mail: remski@umd.umich.edu. *Application contact:* Carol Ligienza, Administrative Coordinator, Case Graduate Programs, 313-593-1183, Fax: 313-583-6498, E-mail: caslgrad@umd.umich.edu.

University of Minnesota, Duluth, Graduate School, College of Science and Engineering, Department of Mathematics and Statistics, Duluth, MN 55812-2496. Offers applied and computational mathematics (MS). Part-time programs available. *Faculty:* 17 full-time (3 women). *Students:* 28 full-time (11 women); includes 2 minority (1 African American, 1 Hispanic American), 11 international. Average age 24. 23 applicants, 74% accepted, 12 enrolled. In 2005, 10 degrees awarded. *Degree requirements:* For master's, thesis or alternative. *Entrance requirements:* For master's, GRE General Test, minimum GPA of 3.0. Additional exam requirements/recommendations for international students: Required—TOEFL (minimum score 550 paper-based; 213 computer-based); Recommended—TWE, TSE. *Application deadline:* For fall admission, 3/1 priority date for domestic students, 3/1 priority date for international students; for spring admission, 11/15 for domestic students, 9/1 for international students. Applications are processed on a rolling basis. Application fee: $55 ($75 for international students). *Financial support:* In 2005–06, 28 students received support, including 6 research assistantships with full tuition reimbursements available (averaging $11,895 per year), 22 teaching assistantships with full tuition reimbursements available (averaging $11,895 per year); fellowships, scholarships/grants, health care benefits, unspecified assistantships, and summer fellowships also available. Financial award application deadline: 3/1. *Faculty research:* Discrete mathematics, diagnostic markers, combinatorics, biostatistics, mathematical modeling and scientific computation. Total annual research expenditures: $113,454. *Unit head:* Dr. Zhuangyi Liu, Director of Graduate Studies, 218-726-7179, Fax: 218-726-8399, E-mail: zliu@d.umn.edu.

Announcement: The program prepares graduates for jobs in industry, government, and teaching, as well as for subsequent PhD studies. Computational facilities are excellent. Faculty research includes graph theory, combinatorics, number theory, scientific computation, dynamical systems, control theory, numerical methods, statistics, biostatistics, and probability.

University of Missouri–Columbia, Graduate School, College of Arts and Sciences, Department of Mathematics, Program in Applied Mathematics, Columbia, MO 65211. Offers MS. *Students:* 10 full-time (4 women), 1 international. In 2005, 8 degrees awarded. *Degree requirements:* For master's, thesis. *Entrance requirements:* For master's, GRE General Test, minimum GPA of 3.0. *Application deadline:* Applications are processed on a rolling basis. Application fee: $45 ($60 for international students). *Financial support:* Fellowships, research assistantships, teaching assistantships, institutionally sponsored loans available. *Unit head:* Dr. Jan Segert, Director of Graduate Studies, Department of Mathematics, 573-882-6953, E-mail: segertj@missouri.edu.

University of Missouri–Rolla, Graduate School, College of Arts and Sciences, Department of Mathematics and Statistics, Program in Applied Mathematics, Rolla, MO 65409-0910. Offers MS. *Degree requirements:* For master's, thesis or alternative. *Entrance requirements:* For master's, GRE General Test, GRE Subject Test. Electronic applications accepted. *Faculty research:* Analysis, differential equations, statistics, topological dynamics.

University of Missouri–St. Louis, College of Arts and Sciences, Department of Mathematics and Computer Science, St. Louis, MO 63121. Offers applied mathematics (MA, PhD); computer science (MS); telecommunications science (Certificate). Part-time and evening/weekend programs available. *Faculty:* 14. *Students:* 28 full-time (11 women), 73 part-time (22 women); includes 19 minority (3 African Americans, 15 Asian Americans or Pacific Islanders, 1 Hispanic American), 23 international. Average age 33. In 2005, 21 master's, 1 doctorate awarded. *Degree requirements:* For master's, thesis optional; for doctorate, thesis/dissertation. *Entrance requirements:* For master's, 2 letters of recommendation; for doctorate, GRE General Test, GRE Subject Test, 3 letters of recommendation. Additional exam requirements/recommendations for international students: Required—TOEFL (minimum score 550 paper-based; 213 computer-based). *Application deadline:* For fall admission, 7/1 for domestic students; for spring admission, 12/1 for domestic students. Applications are processed on a rolling basis. Application fee: $35 ($40 for international students). Electronic applications accepted. *Expenses:* Tuition, state resident: part-time $263 per credit hour. Tuition, nonresident: part-time $680 per credit hour. Required fees: $53 per credit hour. Tuition and fees vary according to program. *Financial support:* In 2005–06, 8 teaching assistantships with full and partial tuition reimbursements (averaging $12,000 per year) were awarded; fellowships with full tuition reimbursements, research assistantships with full tuition reimbursements *Faculty research:* Statistics, algebra, analysis. *Unit head:* Dr. Shiying Zhao, Director of Graduate Studies, 314-516-5741, Fax: 314-516-5400, E-mail: Zhao@arch.cs.umsl.edu. *Application contact:* 314-516-5458, Fax: 314-516-5310, E-mail: gradadm@umsl.edu.

University of Nevada, Las Vegas, Graduate College, College of Science, Department of Mathematical Sciences, Las Vegas, NV 89154-9900. Offers applied mathematics (MS, PhD); applied statistics (MS); computational mathematics (PhD); pure mathematics (MS, PhD); statistics (PhD); teaching mathematics (MS). Part-time programs available. *Faculty:* 29 full-time (6 women), 1 part-time/adjunct (0 women). *Students:* 29 full-time (13 women), 29 part-time (9 women); includes 10 minority (1 African American, 6 Asian Americans or Pacific Islanders, 3 Hispanic Americans), 22 international. 37 applicants, 57% accepted, 11 enrolled. In 2005, 18 degrees awarded. *Degree requirements:* For master's, thesis (for some programs), oral exam, comprehensive exam (for some programs). *Entrance requirements:* For master's, minimum GPA of 3.0 during previous 2 years, 2.75 overall. Additional exam requirements/recommendations for international students: Required—TOEFL (minimum score 550 paper-based; 213 computer-based). *Application deadline:* For fall admission, 6/15 for domestic students, 5/1 for international students; for spring admission, 11/15 for domestic students, 10/1 for international students. Application fee: $60 ($75 for international students). Electronic applications accepted. *Expenses:* Tuition, state resident: part-time $150 per credit. Tuition, nonresident: part-time $315 per credit. Tuition and fees vary according to course load, program and reciprocity agreements. *Financial support:* In 2005–06, 42 teaching assistantships with partial tuition reimbursements (averaging $10,500 per year) were awarded; career-related internships or fieldwork, Federal Work-Study, institutionally sponsored loans, scholarships/grants, health care benefits, and unspecified assistantships also available. Support available to part-time students. Financial award application deadline: 3/1. *Unit head:* Dr. Malwane Ananda, Chair, 702-895-3567. *Application contact:* Graduate College Admissions Evaluator, 702-895-3320, Fax: 702-895-4180, E-mail: gradcollege@unlv.edu.

University of New Hampshire, Graduate School, College of Engineering and Physical Sciences, Department of Mathematics and Statistics, Durham, NH 03824. Offers applied mathematics (MS); mathematics (MS, MST, PhD); mathematics education (PhD); statistics (MS). *Faculty:* 26 full-time. *Students:* 17 full-time (7 women), 24 part-time (13 women), 21 international. Average age 28. 37 applicants, 78% accepted, 5 enrolled. In 2005, 17 master's, 6 doctorates awarded. Terminal master's awarded for partial completion of doctoral program. *Degree requirements:* For doctorate, 2 foreign languages, thesis/dissertation. *Entrance requirements:* Additional exam requirements/recommendations for international students: Required—TOEFL (minimum score 550 paper-based; 213 computer-based); Recommended—TSE. *Application deadline:* For fall admission, 4/1 priority date for domestic students, 4/1 priority date for international students. For winter admission, 12/1 for domestic students. Applications are processed on a rolling basis. Application fee: $60. Electronic applications accepted. *Expenses:* Tuition, state resident: full-time $8,010; part-time $445 per credit hour. Tuition, nonresident: full-time $19,730; part-time $810 per credit hour. Required fees: $322 per semester. Tuition and fees vary according to course load and program. *Financial support:* In 2005–06, 1 fellowship, 2 research assistantships, 23 teaching assistantships were awarded; Federal Work-Study, scholarships/grants, and tuition waivers (full and partial) also available. Support available to part-time students. Financial award application deadline: 2/15. *Faculty research:* Operator theory, complex analysis, algebra, nonlinear dynamics, statistics. *Unit head:* Dr. Eric Grinberg, Chairperson, 603-862-5772. *Application contact:* Jan Jankowski, Administrative Assistant, 603-862-2320, E-mail: jan.jankowski@unh.edu.

The University of North Carolina at Charlotte, Graduate School, College of Arts and Sciences, Department of Mathematics and Statistics, Program in Applied Mathematics, Charlotte, NC 28223-0001. Offers PhD. *Students:* 15 full-time (7 women), 23 part-time (7 women); includes 4 minority (3 African Americans, 1 Asian American or Pacific Islander), 25 international. Average age 31. 14 applicants, 93% accepted, 9 enrolled. In 2005, 7 doctorates awarded. *Degree requirements:* For doctorate, thesis/dissertation. *Entrance requirements:* For doctorate, GRE General Test, minimum GPA of 2.75. Additional exam requirements/recommendations for international students: Required—TOEFL (minimum score 557 paper-based; 220 computer-based). *Application deadline:* For fall admission, 7/15 for domestic students, 5/1 for international students; for spring admission, 11/15 for domestic students, 10/1 for international students. Application fee: $55. Electronic applications accepted. *Expenses:* Tuition, state resident: full-time $2,504; part-time $157 per credit. Tuition, nonresident: full-time $12,711; part-time $794 per credit. Required fees: $1,424; $89 per credit. Tuition and fees vary according to course load and program. *Financial support:* In 2005–06, 2 fellowships (averaging $20,000 per year), 10 teaching assistantships (averaging $14,140 per year) were awarded; research assistantships, career-related internships or fieldwork, Federal Work-Study, institutionally sponsored loans, scholarships/grants, and unspecified assistantships also available. Support available to part-time students. Financial award application deadline: 4/1; financial award applicants required to submit FAFSA. *Unit head:* Dr. Joel D. Avrin, Graduate Coordinator, 704-687-4929, Fax: 704-687-0415, E-mail: jdavrin@email.uncc.edu. *Application contact:* Kathy B. Giddings, Director of Graduate Admissions, 704-687-3366, Fax: 704-687-3279, E-mail: gradadm@email.uncc.edu.

University of Notre Dame, Graduate School, College of Science, Department of Mathematics, Notre Dame, IN 46556. Offers algebra (PhD); algebraic geometry (PhD); applied mathematics (MSAM); complex analysis (PhD); differential geometry (PhD); logic (PhD); partial differential equations (PhD); topology (PhD). *Faculty:* 45 full-time (5 women). *Students:* 48 full-time (14 women); includes 2 minority (1 American Indian/Alaska Native, 1 Asian American or Pacific Islander), 19 international. 117 applicants, 14% accepted, 9 enrolled. In 2005, 6 master's, 7 doctorates awarded. Terminal master's awarded for partial completion of doctoral program. *Median time to degree:* Of those who began their doctoral program in fall 1997, 73% received their degree in 8 years or less. *Degree requirements:* For doctorate, one foreign language, thesis/dissertation, qualifying exam. *Entrance requirements:* For master's and doctorate, GRE General Test, GRE Subject Test. Additional exam requirements/recommendations for international students: Required—TOEFL. *Application deadline:* For fall admission, 2/1 for domestic students. Applications are processed on a rolling basis. Application fee: $50. Electronic applications accepted. *Financial support:* In 2005–06, 47 students received support, including 9 fellowships with full tuition reimbursements available (averaging $22,000 per year), 8 research assistantships with full tuition reimbursements available (averaging $15,250 per year), 26 teaching assistantships with full tuition reimbursements available (averaging $16,000 per year); tuition waivers (full) also available. Financial award application deadline: 2/1. *Faculty research:* Algebra, analysis, geometry/topology, logic, applied math. Total annual research expenditures: $1.5 million. *Unit head:* Dr. Julia Knight, Director of Graduate Studies, 574-631-7484, E-mail: mathgrad@nd.edu. *Application contact:* Dr. Terrence J. Akai, Director of Graduate Admissions, 574-631-7706, Fax: 574-631-4183, E-mail: gradad@nd.edu.

See Close-Up on page 539.

University of Pittsburgh, School of Arts and Sciences, Department of Mathematics, Pittsburgh, PA 15260. Offers applied mathematics (MA, MS); financial mathematics (PMS); mathematics (MA, MS, PhD). Part-time programs available. *Faculty:* 35 full-time (4 women), 4 part-time/adjunct (1 woman). *Students:* 82 full-time (25 women), 12 part-time (5 women); includes 1 minority (Asian American or Pacific Islander), 43 international. 123 applicants, 67% accepted, 28 enrolled. In 2005, 5 master's, 3 doctorates awarded. Terminal master's awarded for partial completion of doctoral program. *Median time to degree:* Of those who began their doctoral program in fall 1997, 100% received their degree in 8 years or less. *Degree requirements:* For master's, thesis (for some programs), comprehensive exam; for doctorate, thesis/dissertation, preliminary exams, comprehensive exam. *Entrance requirements:* For master's and doctorate, GRE General Test, GRE Subject Test (recommended), minimum GPA of 3.0. Additional exam requirements/recommendations for international students: Required—TOEFL (minimum score 550 paper-based; 213 computer-based). *Application deadline:* For fall admission, 1/15 priority date for domestic students, 1/2 priority date for international students; for spring admission, 9/1 priority date for domestic students, 9/1 priority date for international students. Applications are processed on a rolling basis. Application fee: $50. Electronic applications accepted. *Expenses:* Tuition, state resident: full-time $13,194; part-time $537 per credit. Tuition, nonresident: full-time $25,012; part-time $1,026 per credit. Required fees: $700; $164 per term. Tuition and fees vary according to campus/location and program. *Financial support:* In 2005–06, 6 fellowships with full and partial tuition reimbursements (averaging $16,500 per year), 13 research assistantships with full and partial tuition reimbursements (averaging $13,600 per year), 50 teaching assistantships with full and partial tuition reimbursements (averaging $13,555 per year) were awarded; career-related internships or fieldwork, Federal Work-Study, institutionally sponsored loans, scholarships/grants, health care benefits, tuition waivers (partial), and unspecified assistantships also available. Financial award application deadline:1/15. *Faculty research:* Computational math, math biology, math finance, algebra, analysis. Total annual research expenditures: $700,000. *Unit head:* Juan Manfredi, Chairman, 412-624-8307, Fax: 412-624-8697, E-mail: manfredi@pitt.edu. *Application contact:* Molly Williams, Administrator, 412-624-1175, Fax: 412-624-8397, E-mail: mollyw@pitt.edu.

University of Puerto Rico, Mayagüez Campus, Graduate Studies, College of Arts and Sciences, Department of Mathematics, Mayagüez, PR 00681-9000. Offers applied mathematics (MS); computational sciences (MS); pure mathematics (MS); statistics (MS). Part-time programs available. *Faculty:* 34. *Students:* 21 full-time (9 women), 9 part-time (4 women); includes 6 minority (all Hispanic Americans), 24 international. 49 applicants, 61% accepted, 2 enrolled. In 2005, 6 degrees awarded. *Degree requirements:* For master's, one foreign language, comprehensive exam. *Application deadline:* For fall admission, 2/15 for domestic students; for spring admission, 9/15 for domestic students. Applications are processed on a rolling basis. Application fee: $20. *Expenses:* Tuition, state resident: full-time $900; part-time $100 per credit. International tuition: $4,655 full-time. Part-time tuition and fees vary according to course level and course load. *Financial support:* In 2005–06, fellowships (averaging $1,500 per year), research assistantships (averaging $1,200 per year), teaching assistantships (averaging $987 per year) were awarded; Federal Work-Study and institutionally sponsored loans also available. *Faculty research:* Automata theory, linear algebra, logic. Total annual research expenditures: $13,829. *Unit head:* Dr. Luis A. Caceres, Director, 787-832-4040 Ext. 3848.

University of Rhode Island, Graduate School, College of Arts and Sciences, Department of Computer Science and Statistics, Kingston, RI 02881. Offers applied mathematics (PhD), including computer science, statistics; computer science (MS, PhD); digital forensics (Graduate Certificate); statistics (MS). In 2005, 4 degrees awarded. *Degree requirements:* For master's, thesis optional; for doctorate, one foreign language, thesis/dissertation. *Entrance requirements:* For master's, GRE Subject Test. *Application deadline:* For fall admission, 4/15

Applied Mathematics

University of Rhode Island (continued)
for domestic students. Applications are processed on a rolling basis. Application fee: $35. *Expenses:* Tuition, state resident: full-time $5,522; part-time $307 per credit. Tuition, nonresident: full-time $15,992; part-time $888 per credit. Required fees: $1,786; $73 per credit. One-time fee: $80 part-time. *Financial support:* Unspecified assistantships available. *Unit head:* Dr. James Kowalski, Chair, 401-874-2701.

University of Southern California, Graduate School, College of Letters, Arts and Sciences, Department of Mathematics, Program in Applied Mathematics, Los Angeles, CA 90089. Offers MA, MS, PhD. *Degree requirements:* For master's, thesis (for some programs); for doctorate, 2 foreign languages, thesis/dissertation. *Entrance requirements:* For master's and doctorate, GRE General Test. *Expenses:* Tuition: Full-time $25,416; part-time $1,059 per unit. Required fees: $484; $484 per year. Tuition and fees vary according to course load and program.

University of South Florida, College of Graduate Studies, College of Arts and Sciences, Department of Mathematics, Tampa, FL 33620-9951. Offers applied mathematics (PhD); mathematics (MA, PhD). Part-time and evening/weekend programs available. *Faculty:* 25 full-time (2 women), 1 part-time/adjunct (0 women). *Students:* 48 full-time (14 women), 24 part-time (11 women); includes 9 minority (1 African American, 8 Asian Americans or Pacific Islanders), 38 international. 56 applicants, 91% accepted, 20 enrolled. In 2005, 2 degrees awarded. Terminal master's awarded for partial completion of doctoral program. *Degree requirements:* For master's, one foreign language, thesis optional; for doctorate, 2 foreign languages, thesis/dissertation. *Entrance requirements:* For master's, GRE General Test, minimum GPA of 3.0 in mathematics course work (undergraduate), 3.5 (graduate); for doctorate, GRE General Test. Additional exam requirements/recommendations for international students: Required—TOEFL (minimum score 550 paper-based; 213 computer-based). *Application deadline:* For fall admission, 6/1 for domestic students, 2/1 for international students; for spring admission, 10/15 for domestic students, 8/1 for international students. Application fee: $30. Electronic applications accepted. *Financial support:* Teaching assistantships with partial tuition reimbursements, scholarships/grants and unspecified assistantships available. Financial award application deadline: 2/1. *Faculty research:* Approximation theory, differential equations, discrete mathematics, functional analysis topology. *Unit head:* Dr. Marcus McWaters, Chairperson, 813-974-3838, Fax: 813-974-2700, E-mail: marcus@chuma.cas.usf.edu. *Application contact:* Dr. Natasha Jonoska, Graduate Admissions Director, 813-974-9566, Fax: 813-974-2700, E-mail: jonoska@math.usf.edu.

The University of Tennessee, Graduate School, College of Arts and Sciences, Department of Mathematics, Knoxville, TN 37996. Offers applied mathematics (MS); mathematical ecology (PhD); mathematics (M Math, MS, PhD). Part-time programs available. *Degree requirements:* For master's, thesis or alternative; for doctorate, one foreign language, thesis/dissertation. *Entrance requirements:* For master's and doctorate, minimum GPA of 2.7. Additional exam requirements/recommendations for international students: Required—TOEFL. Electronic applications accepted.

The University of Tennessee Space Institute, Graduate Programs, Program in Applied Mathematics, Tullahoma, TN 37388-9700. Offers MS. Part-time programs available. *Faculty:* 2 full-time (0 women), 1 part-time/adjunct (0 women). *Degree requirements:* For master's, thesis (for some programs). *Entrance requirements:* Additional exam requirements/recommendations for international students: Required—TOEFL (minimum score 550 paper-based; 213 computer-based). *Application deadline:* Applications are processed on a rolling basis. Application fee: $35. *Financial support:* Fellowships with full and partial tuition reimbursements, research assistantships with full tuition reimbursements, career-related internships or fieldwork, Federal Work-Study, tuition waivers (partial), and unspecified assistantships available. Financial award applicants required to submit FAFSA. *Unit head:* Dr. Ken Kimble, Degree Program Chairman, 931-393-7484, Fax: 931-393-7542, E-mail: kkimble@utsi.edu. *Application contact:* Callie Taylor, Coordinator II, 931-393-7432, Fax: 931-393-7346, E-mail: ctaylor@utsi.edu.

The University of Texas at Austin, Graduate School, Program in Computational and Applied Mathematics, Austin, TX 78712-1111. Offers MA, PhD. Terminal master's awarded for partial completion of doctoral program. *Degree requirements:* For master's, thesis optional; for doctorate, thesis/dissertation, 3 area qualifying exams. Electronic applications accepted.

The University of Texas at Dallas, School of Natural Sciences and Mathematics, Programs in Mathematical Sciences, Richardson, TX 75083-0688. Offers applied mathematics (MS, PhD); engineering mathematics (MS); mathematical science (MS); statistics (MS, PhD). Part-time and evening/weekend programs available. *Faculty:* 12 full-time (0 women). *Students:* 33 full-time (12 women), 25 part-time (13 women); includes 13 minority (2 African Americans, 6 Asian Americans or Pacific Islanders, 5 Hispanic Americans), 26 international. Average age 30. 73 applicants, 59% accepted, 28 enrolled. In 2005, 19 master's, 8 doctorates awarded. *Degree requirements:* For master's, thesis optional; for doctorate, thesis/dissertation. *Entrance requirements:* For master's, GRE General Test, minimum GPA of 3.0 in upper-level course work in field; for doctorate, GRE General Test, minimum GPA of 3.5 in upper-level course work in field. Additional exam requirements/recommendations for international students: Required—TOEFL (minimum score 550 paper-based; 213 computer-based). *Application deadline:* For fall admission, 7/15 for domestic students; for spring admission, 11/15 for domestic students. Applications are processed on a rolling basis. Application fee: $50 ($100 for international students). Electronic applications accepted. *Expenses:* Tuition, state resident: full-time $5,450; part-time $303 per credit. Tuition, nonresident: full-time $12,648; part-time $703 per credit. Tuition and fees vary according to program. *Financial support:* In 2005–06, 25 teaching assistantships with tuition reimbursements (averaging $9,451 per year) were awarded; fellowships, research assistantships, career-related internships or fieldwork, Federal Work-Study, institutionally sponsored loans, and scholarships/grants also available. Support available to part-time students. Financial award application deadline: 4/30; financial award applicants required to submit FAFSA. *Faculty research:* Statistical methods, control theory, mathematical modeling and analyses of biological and physical systems. Total annual research expenditures: $59,757. *Unit head:* Dr. M. Ali Hooshyar, Head, 972-883-2161, Fax: 972-883-6622, E-mail: utdmath@utdallas.edu. *Application contact:* Dr. Michael Baron, Graduate Advisor, 972-883-6874, Fax: 972-883-6622, E-mail: mbaron@utdallas.edu.

See Close-Up on page 545.

The University of Toledo, Graduate School, College of Arts and Sciences, Department of Mathematics, Toledo, OH 43606-3390. Offers applied mathematics (MS); mathematics (MA, PhD); statistics (MS). Part-time programs available. *Faculty:* 16. *Students:* 57 full-time (24 women), 13 part-time (3 women); includes 1 minority (Asian American or Pacific Islander), 45 international. Average age 29. 65 applicants, 63% accepted, 21 enrolled. In 2005, 16 degrees awarded. *Degree requirements:* For doctorate, 2 foreign languages, thesis/dissertation. *Entrance requirements:* For master's and doctorate, GRE General Test, GRE Subject Test. *Application deadline:* For fall admission, 8/1 for domestic students. Application fee: $45. Electronic applications accepted. *Expenses:* Tuition, area resident: Part-time $308 per credit hour. Tuition, state resident: full-time $3,312. Tuition, nonresident: full-time $6,616; part-time $735 per credit hour. *Financial support:* In 2005–06, 16 research assistantships with full tuition reimbursements (averaging $3,688 per year), 37 teaching assistantships with full tuition reimbursements (averaging $10,807 per year) were awarded; Federal Work-Study and institutionally sponsored loans also available. Support available to part-time students. Financial award application deadline: 4/1; financial award applicants required to submit FAFSA. *Faculty research:* Topology. *Unit head:* Dr. Geoffrey Martin, Chair, 419-530-2569, Fax: 419-530-4720, E-mail: gmartin@math.utoledo.edu. *Application contact:* Dr. Gerard Thompson, Advising Coordinator, 419-530-2568, Fax: 419-530-4720, E-mail: thompson@math.utoledo.edu.

See Close-Up on page 547.

University of Washington, Graduate School, College of Arts and Sciences, Department of Applied Mathematics, Seattle, WA 98195. Offers MS, PhD. Terminal master's awarded for partial completion of doctoral program. *Degree requirements:* For master's, thesis optional; for doctorate, thesis/dissertation. *Entrance requirements:* For master's and doctorate, GRE, minimum GPA of 3.0. Additional exam requirements/recommendations for international students: Required—TOEFL. Electronic applications accepted. *Faculty research:* Mathematical modeling for physical, biological, social, and engineering sciences; development of mathematical methods for analysis, including perturbation, asymptotic, transform, vocational, and numerical methods.

University of Waterloo, Graduate Studies, Faculty of Mathematics, Department of Applied Mathematics, Waterloo, ON N2L 3G1, Canada. Offers M Math, PhD. Part-time programs available. *Faculty:* 20 full-time (4 women), 22 part-time/adjunct (1 woman). *Students:* 28 full-time (7 women). 73 applicants, 22% accepted, 6 enrolled. In 2005, 1 master's, 3 doctorates awarded. *Degree requirements:* For master's, research paper or thesis; for doctorate, thesis/dissertation. *Entrance requirements:* For master's, honors degree in field, minimum B+ average; for doctorate, master's degree, minimum B+ average. Additional exam requirements/recommendations for international students: Required—TOEFL (minimum score 600 paper-based; 250 computer-based), TWE (minimum score 4). *Application deadline:* For fall admission, 3/1 for domestic students. Applications are processed on a rolling basis. Application fee: $75 Canadian dollars. Electronic applications accepted. *Financial support:* Research assistantships, teaching assistantships available. *Faculty research:* Differential equations, quantum theory, statistical mechanics, fluid mechanics, relativity, control theory. *Unit head:* Dr. S. P. Lipshitz, Associate Chair, 519-888-4567 Ext. 6246, Fax: 519-746-4319, E-mail: spl@audiolab.uwaterloo.ca. *Application contact:* Helen A. Warren, Graduate Secretary, 519-888-4567 Ext. 3170, Fax: 519-746-4319, E-mail: amgrad@math.uwaterloo.ca.

The University of Western Ontario, Faculty of Graduate Studies, Physical Sciences Division, Department of Applied Mathematics, London, ON N6A 5B8, Canada. Offers applied mathematics (M Sc, PhD); theoretical physics (PhD). *Degree requirements:* For master's, thesis or alternative; for doctorate, thesis/dissertation, comprehensive exam. *Entrance requirements:* For master's and doctorate, minimum B average. *Faculty research:* Fluid dynamics, mathematical and computational methods, theoretical physics.

Utah State University, School of Graduate Studies, College of Science, Department of Mathematics and Statistics, Logan, UT 84322. Offers industrial mathematics (MS); mathematical sciences (PhD); mathematics (M Math, MS); statistics (MS). Part-time programs available. *Faculty:* 33 full-time (4 women). *Students:* 91 full-time (29 women), 4 part-time (1 woman), 44 international. Average age 29. 41 applicants, 61% accepted, 15 enrolled. In 2005, 9 master's, 3 doctorates awarded. Terminal master's awarded for partial completion of doctoral program. *Degree requirements:* For master's, qualifying exam, thesis optional; for doctorate, one foreign language, thesis/dissertation, comprehensive exam. *Entrance requirements:* For master's and doctorate, GRE General Test, minimum GPA of 3.0. Additional exam requirements/recommendations for international students: Required—TOEFL. *Application deadline:* For fall admission, 6/15 for domestic students; for spring admission, 10/15 for domestic students. Applications are processed on a rolling basis. Application fee: $50 ($60 for international students). *Financial support:* In 2005–06, 1 fellowship with partial tuition reimbursement (averaging $12,000 per year), 17 teaching assistantships with partial tuition reimbursements (averaging $14,500 per year) were awarded; research assistantships with partial tuition reimbursements Support available to part-time students. Financial award application deadline: 4/1. *Faculty research:* Differential equations, computational mathematics, dynamical systems, probability and statistics, pure mathematics. Total annual research expenditures: $212,000. *Unit head:* Dr. Russell C. Thompson, Head, 435-797-2810, Fax: 435-797-1822, E-mail: thompson@math.usu.edu. *Application contact:* Dr. David Richard Cutler, Graduate Chairman, 435-797-2699, Fax: 435-797-1822, E-mail: richard.cutler@usu.edu.

Virginia Commonwealth University, Graduate School, College of Humanities and Sciences, Department of Mathematical Sciences, Program in Applied Mathematics, Richmond, VA 23284-9005. Offers MS. *Students:* 3 full-time (2 women), 1 part-time; includes 1 minority (Asian American or Pacific Islander), 1 international. 7 applicants, 100% accepted. *Entrance requirements:* For master's, GRE General Test, GRE Subject Test. Additional exam requirements/recommendations for international students: Required—TOEFL. *Application deadline:* For fall admission, 7/1 for domestic students; for spring admission, 11/15 for domestic students. Applications are processed on a rolling basis. Application fee: $50. *Expenses:* Tuition, state resident: full-time $3,185; part-time $405 per credit. Tuition, nonresident: full-time $7,952; part-time $940 per credit. Required fees: $751 per semester hour. Tuition and fees vary according to course load and program. *Unit head:* Dr. John Berglond, Head, 804-828-1301 Ext.115. *Application contact:* Dr. James A. Wood, Information Contact, 804-828-1301, E-mail: jawood@vcu.edu.

Virginia Polytechnic Institute and State University, Graduate School, College of Science, Department of Mathematics, Blacksburg, VA 24061. Offers applied mathematics (MS, PhD); mathematical physics (MS, PhD); pure mathematics (MS, PhD). *Faculty:* 69 full-time (20 women). *Students:* 57 full-time (18 women), 5 part-time (2 women); includes 5 minority (2 African Americans, 1 Asian American or Pacific Islander, 2 Hispanic Americans), 27 international. Average age 28. 108 applicants, 20% accepted, 15 enrolled. In 2005, 20 master's, 8 doctorates awarded. *Entrance requirements:* For master's and doctorate, GRE. Additional exam requirements/recommendations for international students: Required—TOEFL (minimum score 550 paper-based; 213 computer-based). *Application deadline:* Applications are processed on a rolling basis. Application fee: $45. Electronic applications accepted. *Expenses:* Tuition, state resident: full-time $6,558; part-time $364 per credit. Tuition, nonresident: full-time $11,296; part-time $628 per credit. Required fees: $1,419; $468 per credit. $234 per term. *Financial support:* In 2005–06, 1 fellowship with full tuition reimbursement (averaging $1,481 per year), 1 research assistantship with full tuition reimbursement (averaging $16,689 per year), 44 teaching assistantships with full tuition reimbursements (averaging $13,902 per year) were awarded; career-related internships or fieldwork, Federal Work-Study, scholarships/grants, and unspecified assistantships also available. *Faculty research:* Differential equations, operator theory, numerical analysis, algebra, control theory. *Unit head:* Dr. John Rossi, Head, 540-231-6536, Fax: 540-231-5960, E-mail: rossi@math.vt.edu. *Application contact:* Hannah Swiger, Information Contact, 540-231-6537, Fax: 540-231-5960, E-mail: hsswiger@math.vt.edu.

Washington State University, Graduate School, College of Sciences, Department of Mathematics, Pullman, WA 99164. Offers applied mathematics (MS, PhD); mathematics teaching (MS, PhD). *Faculty:* 31. *Students:* 34 full-time (15 women), 1 (woman) part-time; includes 3 minority (all Asian Americans or Pacific Islanders), 11 international. Average age 29. 100 applicants, 28% accepted, 14 enrolled. In 2005, 12 master's, 5 doctorates awarded. *Degree requirements:* For master's, oral exam, project; for doctorate, 2 foreign languages, thesis/dissertation, oral exam, written exam. *Entrance requirements:* For master's and doctorate, minimum GPA of 3.0, 3 letters of recommendation. Additional exam requirements/recommendations for international students: Required—TOEFL (minimum score 600 paper-based; 250 computer-based). *Application deadline:* For fall admission, 2/1 for domestic students, 2/1 for international students; for spring admission, 9/1 for domestic students, 7/1 for international students. Applications are processed on a rolling basis. Application fee: $35. Electronic applications accepted. *Expenses:* Tuition, state resident: full-time $6,295; part-time $336 per credit. Tuition, nonresident: full-time $15,949; part-time $819 per credit. Required fees: $933. Part-time tuition and fees vary according to campus/location and program. *Financial support:* In 2005–06, 33 students received support, including 2 fellowships with tuition reimbursements available (averaging $2,500 per year), 3 research assistantships with full and partial tuition reimbursements available (averaging $13,871 per year), 27 teaching assistantships with full and partial tuition reimbursements available (averaging $14,109 per year); career-related internships or fieldwork, Federal Work-Study, institutionally sponsored loans, and tuition waivers (partial) also available. Financial award application deadline: 2/1; financial award applicants required to submit FAFSA. *Faculty research:* Computational mathematics, operations research, modeling in the natural sciences, applied statistics. Total annual research expenditures:$425,725. *Unit head:* Dr. Alan Genz,

Chair, 509-335-4918, Fax: 509-335-1188, E-mail: chair@math.wsu.edu. *Application contact:* Pam Guptill, Coordinator, 509-335-6868, Fax: 509-335-1188, E-mail: pguptill@wsu.edu.

See Close-Up on page 553.

Wayne State University, Graduate School, College of Liberal Arts and Sciences, Department of Mathematics, Program in Applied Mathematics, Detroit, MI 48202. Offers MA, PhD. *Students:* 4 full-time (1 woman), 7 part-time (2 women); includes 4 minority (2 African Americans, 2 Asian Americans or Pacific Islanders). Average age 31. 9 applicants, 67% accepted, 0 enrolled. *Degree requirements:* For doctorate, thesis/dissertation. *Entrance requirements:* Additional exam requirements/recommendations for international students: Required—TOEFL (minimum score 550 paper-based; 213 computer-based); Recommended—TWE (minimum score 6). *Application deadline:* For fall admission, 7/1 for domestic students, 6/1 for international students. Applications are processed on a rolling basis. Application fee: $30 ($50 for international students). Electronic applications accepted. *Expenses:* Tuition, state resident: part-time $338 per credit hour. Tuition, nonresident: part-time $746 per credit hour. Required fees: $24 per credit hour. Full-time tuition and fees vary according to program. *Application contact:* Bert Schreiber, Professor, 313-577-8838, E-mail: bschreiber@wayne.edu.

Western Michigan University, Graduate College, College of Arts and Sciences, Department of Mathematics, Program in Applied Mathematics, Kalamazoo, MI 49008-5202. Offers MS.

West Virginia University, Eberly College of Arts and Sciences, Department of Mathematics, Morgantown, WV 26506. Offers applied mathematics (MS, PhD); discrete mathematics (PhD); interdisciplinary mathematics (MS); mathematics for secondary education (MS); pure mathematics (MS). Part-time programs available. *Faculty:* 28 full-time (1 woman), 11 part-time/adjunct (3 women). *Students:* 33 full-time (15 women), 8 part-time (2 women); includes 3 minority (1 African American, 2 Asian Americans or Pacific Islanders), 23 international. Average age 29. 30 applicants, 100% accepted, 16 enrolled. In 2005, 5 master's, 6 doctorates awarded. Terminal master's awarded for partial completion of doctoral program. *Degree requirements:* For master's, thesis optional; for doctorate, one foreign language, thesis/dissertation, comprehensive exam. *Entrance requirements:* For master's, minimum GPA of 2.5; for doctorate, master's degree in mathematics. Additional exam requirements/recommendations for international students: Required—TOEFL (paper score 550; computer score 213) or IELTS (paper score 6). *Application deadline:* For fall admission, 2/15 priority date for domestic students, 2/15 priority date for international students. Applications are processed on a rolling basis. Application fee: $50. *Expenses:* Tuition, state resident: full-time $4,582; part-time $258 per credit hour. Tuition, nonresident: full-time $1,382; part-time $741 per credit hour. *Financial support:* In 2005–06, 25 students received support, including 6 research assistantships with full tuition reimbursements available (averaging $1,000 per year), 18 teaching assistantships with full tuition reimbursements available (averaging $9,500 per year); Federal Work-Study, institutionally sponsored loans, and tuition waivers (full and partial) also available. Financial award application deadline: 2/15; financial award applicants required to submit FAFSA. *Faculty research:* Combinatorics and graph theory, topology, differential equations, applied and computational mathematics. Total annual research expenditures:$80,423. *Unit head:* Dr. Sherman Riemenschneider, Chair, 304-293-2011 Ext. 2322, Fax: 304-293-3982, E-mail: sherm.riemenschneider@mail.wvu.edu. *Application contact:* Dr. Harvey R. Diamond, Director of Graduate Studies, 304-293-2011 Ext. 2347, Fax: 304-293-3982, E-mail: harvey.diamond@mail.wvu.edu.

Wichita State University, Graduate School, Fairmount College of Liberal Arts and Sciences, Department of Mathematics and Statistics, Wichita, KS 67260. Offers applied mathematics (PhD); mathematics (MS); statistics (MS). Part-time programs available. *Degree requirements:* For master's, thesis optional; for doctorate, thesis/dissertation. *Entrance requirements:* For master's, GRE; for doctorate, GRE Subject Test. Additional exam requirements/recommendations for international students: Required—TOEFL. Electronic applications accepted. *Faculty research:* Partial differential equations, combinatorics, ring theory, minimal surfaces, several complex variables.

Worcester Polytechnic Institute, Graduate Studies and Enrollment, Department of Mathematical Sciences, Worcester, MA 01609-2280. Offers applied mathematics (MS); applied statistics (MS); financial mathematics (MS); industrial mathematics (MS); mathematical sciences (PhD); mathematics (MME). Part-time and evening/weekend programs available. *Faculty:* 30 full-time (3 women), 4 part-time/adjunct (0 women). *Students:* 28 full-time (13 women), 33 part-time (17 women); includes 5 minority (2 African Americans, 3 Asian Americans or Pacific Islanders), 14 international. 77 applicants, 77% accepted, 17 enrolled. In 2005, 15 degrees awarded. *Degree requirements:* For master's, thesis (for some programs); for doctorate, thesis/dissertation, comprehensive exam. *Entrance requirements:* For master's and doctorate, 3 letters of recommendation. Additional exam requirements/recommendations for international students: Required—TOEFL (minimum score 550 paper-based; 213 computer-based). *Application deadline:* For fall admission, 1/15 for domestic students; for spring admission, 10/15 priority date for domestic students. Applications are processed on a rolling basis. Application fee: $70. Electronic applications accepted. *Expenses:* Tuition: Part-time $997 per credit hour. *Financial support:* In 2005–06, 19 students received support, including fellowships with full tuition reimbursements available (averaging $33,246 per year), 5 research assistantships with full and partial tuition reimbursements available, 14 teaching assistantships with full and partial tuition reimbursements available; career-related internships or fieldwork, institutionally sponsored loans, scholarships/grants, and unspecified assistantships also available. Financial award application deadline: 1/15. *Faculty research:* Applied mathematical modeling and analysis, computational mathematics, discrete mathematics, applied and computational statistics, industrial and financial mathematics. Total annual research expenditures: $1.2 million. *Unit head:* Dr. Bogdan Vernescu, Head, 508-831-5241, Fax: 508-831-5824. *Application contact:* Dr. Homer F Walker, Graduate Coordinator, 508-831-6113, Fax: 508-831-5824, E-mail: walker@wpi.edu.

See Close-Up on page 559.

Wright State University, School of Graduate Studies, College of Science and Mathematics, Department of Mathematics and Statistics, Program in Applied Mathematics, Dayton, OH 45435. Offers MS. *Degree requirements:* For master's, comprehensive exam. *Entrance requirements:* For master's, bachelor's degree in mathematics or related field. Additional exam requirements/recommendations for international students: Required—TOEFL. *Faculty research:* Control theory, ordinary differential equations, partial differential equations, numerical analysis, mathematical modeling.

Yale University, Graduate School of Arts and Sciences, Program in Applied Mathematics, New Haven, CT 06520. Offers M Phil, MS, PhD. *Entrance requirements:* For doctorate, GRE General Test.

York University, Faculty of Graduate Studies, Faculty of Arts, Program in Mathematics and Statistics, Toronto, ON M3J 1P3, Canada. Offers industrial and applied mathematics (M Sc); mathematics and statistics (MA, PhD). Part-time programs available. *Faculty:* 57 full-time (13 women), 6 part-time/adjunct (2 women). *Students:* 69 full-time (33 women), 34 part-time (18 women). 300 applicants, 13% accepted, 38 enrolled. In 2005, 28 master's, 5 doctorates awarded. *Degree requirements:* For master's, thesis optional; for doctorate, one foreign language, thesis/dissertation, comprehensive exam, registration. *Application deadline:* For fall admission, 2/1 for domestic students. Application fee: $80. Electronic applications accepted. *Expenses:* Tuition, state resident: full-time $3,190; part-time $798 per term. International tuition: $7,515 full-time. Required fees: $217. Tuition and fees vary according to program. *Financial support:* In 2005–06, fellowships (averaging $8,354 per year), research assistantships (averaging $5,815 per year), teaching assistantships (averaging $9,475 per year) were awarded; tuition waivers (partial) and fee bursaries also available. *Unit head:* Yuehua Wu, Director, 416-736-2100.

Biometrics

Cornell University, Graduate School, Graduate Fields of Agriculture and Life Sciences, Field of Biometry, Ithaca, NY 14853-0001. Offers MS, PhD. *Faculty:* 19 full-time (1 woman). *Students:* 9 full-time (6 women); includes 3 minority (all Hispanic Americans), 4 international. 20 applicants, 5% accepted, 1 enrolled. In 2005, 2 master's, 2 doctorates awarded. Terminal master's awarded for partial completion of doctoral program. *Degree requirements:* For master's, thesis/dissertation; for doctorate, thesis/dissertation, comprehensive exam. *Entrance requirements:* For master's and doctorate, GRE General Test, 2 letters of recommendation. Additional exam requirements/recommendations for international students: Required—TOEFL (minimum score 550 paper-based; 213 computer-based). *Application deadline:* For fall admission, 1/15 for domestic students. Application fee: $60. Electronic applications accepted. *Financial support:* In 2005–06, 8 students received support, including 3 fellowships with full tuition reimbursements available, 2 research assistantships with full tuition reimbursements available, 3 teaching assistantships with full tuition reimbursements available; institutionally sponsored loans, scholarships/grants, health care benefits, tuition waivers (full and partial), and unspecified assistantships also available. Financial award applicants required to submit FAFSA. *Faculty research:* Environmental, agricultural, and biological statistics; biomathematics; modern nonparametric statistics; statistical genetics; computational statistics. *Unit head:* Director of Graduate Studies, 607-255-8066. *Application contact:* Graduate Field Assistant, 607-255-8066, E-mail: bscb@cornell.edu.

Cornell University, Graduate School, Graduate Fields of Industrial and Labor Relations, Field of Statistics, Ithaca, NY 14853-0001. Offers applied statistics (MPS); biometry (MS, PhD); decision theory (MS, PhD); economic and social statistics (MS, PhD); engineering statistics (MS, PhD); experimental design (MS, PhD); mathematical statistics (MS, PhD); probability (MS, PhD); sampling (MS, PhD); statistical computing (MS, PhD); stochastic processes (MS, PhD). Terminal master's awarded for partial completion of doctoral program. *Degree requirements:* For master's, project (MPS), thesis (MS); for doctorate, one foreign language, thesis/dissertation. *Entrance requirements:* For master's, GRE General Test (MS), 2 letters of recommendation (MS and MPS); for doctorate, GRE General Test, 2 letters of recommendation. Additional exam requirements/recommendations for international students: Required—TOEFL (minimum score 550 paper-based; 213 computer-based). Electronic applications accepted. *Faculty research:* Bayesian analysis, survival analysis, nonparametric statistics, stochastic processes, mathematical statistics.

Louisiana State University Health Sciences Center, School of Graduate Studies in New Orleans, Department of Biometry, New Orleans, LA 70112-2223. Offers MPH, MS. Part-time programs available. *Degree requirements:* For master's, thesis, comprehensive exam. *Entrance requirements:* For master's, GRE General Test. Additional exam requirements/recommendations for international students: Required—TOEFL. *Faculty research:* Longitudinal data, repeated measures, missing data, generalized estimating equations, multivariate methods.

Medical University of South Carolina, College of Graduate Studies, Program in Biostatistics, Bioinformatics, and Epidemiology, Charleston, SC 29425-0002. Offers biometrics (MS, PhD); biostatistics (MS, PhD); clinical research (MCR); epidemiology (MCR, PhD). *Faculty:* 26 full-time (9 women). *Students:* 40 full-time (26 women); includes 3 minority (1 African American, 2 Asian Americans or Pacific Islanders), 6 international. Average age 28. 181 applicants, 34% accepted, 43 enrolled. In 2005, 7 master's, 4 doctorates awarded. Terminal master's awarded for partial completion of doctoral program. *Degree requirements:* For master's, thesis, research seminar; for doctorate, thesis/dissertation, teaching and research seminar, oral and written exams. *Entrance requirements:* For master's, GRE General Test; for doctorate, GRE General Test, interview. Additional exam requirements/recommendations for international students: Required—TOEFL (minimum score 600 paper-based; 250 computer-based). *Application deadline:* For fall admission, 1/15 priority date for domestic students, 1/15 priority date for international students. Applications are processed on a rolling basis. Application fee: $0 ($75 for international students). Electronic applications accepted. *Financial support:* In 2005–06, 8 students received support, including fellowships with partial tuition reimbursements available (averaging $21,000 per year); Federal Work-Study and scholarships/grants also available. Support available to part-time students. Financial award application deadline: 3/15; financial award applicants required to submit FAFSA. *Faculty research:* Health disparities, central nervous system injuries, radiation exposure, analysis of clinical trial data, biomedical information. *Unit head:* Dr. Barbara Tilley, Chair, 873-876-1327, Fax: 873-792-6950, E-mail: tilleyb@musc.edu. *Application contact:* Cheryl Brown, 843-792-4620, Fax: 843-792-4645, E-mail: brownche@musc.edu.

North Carolina State University, Graduate School, College of Physical and Mathematical Sciences, Department of Statistics, Program in Biomathematics, Raleigh, NC 27695. Offers biomathematics (M Biomath, MS, PhD); ecology (PhD). Part-time programs available. Terminal master's awarded for partial completion of doctoral program. *Degree requirements:* For master's, thesis (for some programs); for doctorate, thesis/dissertation. *Entrance requirements:* For master's and doctorate, GRE General Test. Additional exam requirements/recommendations for international students: Required—TOEFL. Electronic applications accepted. *Faculty research:* Theory and methods of biological modeling, theoretical biology (genetics, ecology, neurobiology), applied biology (wildlife).

Oregon State University, Graduate School, College of Science, Department of Statistics, Corvallis, OR 97331. Offers applied statistics (MA, MS, PhD); biometry (MA, MS, PhD); environmental statistics (MA, MS, PhD); mathematical statistics (MA, MS, PhD); operations research (MA, MAIS, MS); statistics (MA, MS, PhD). Part-time programs available. *Faculty:* 10 full-time (4 women), 2 part-time/adjunct (0 women). *Students:* 33 full-time (14 women), 3 part-time (1 woman); includes 5 minority (4 Asian Americans or Pacific Islanders, 1 Hispanic American), 12 international. Average age 30. In 2005, 9 master's, 2 doctorates awarded. *Degree requirements:* For master's, consulting experience; for doctorate, thesis/dissertation, consulting experience. *Entrance requirements:* For master's and doctorate, minimum GPA of 3.0 in last 90 hours. Additional exam requirements/recommendations for international students: Required—TOEFL. *Application deadline:* For fall admission, 2/15 for domestic students. Applications are processed on a rolling basis. Application fee: $50. *Expenses:* Tuition, area resident: Part-time $301 per credit. Tuition, state resident: full-time $8,139; part-time $501 per credit. Tuition, nonresident: full-time $14,376; part-time $532 per credit. Required fees: $1,266. *Financial support:* In 2005–06, 8 research assistantships, 19 teaching assistantships were awarded; Federal Work-Study and institutionally sponsored loans also available. Financial award application deadline: 2/15. *Faculty research:* Analysis of enumerative data, nonparametric statistics, asymptotics, experimental design, generalized regression models, linear model theory, reliability theory, survival analysis, wildlife and general survey methodology. *Unit head:* Dr. Robert T. Smythe, Chair, 541-737-3366, Fax: 541-737-3489, E-mail: symthe@stat.orst.

Biometrics

Oregon State University (continued)
edu. *Application contact:* Dr. Daniel W. Schafer, Director of Graduate Studies, 541-737-3366, Fax: 541-737-3489, E-mail: statoff@stat.orst.edu.

San Diego State University, Graduate and Research Affairs, College of Sciences, Department of Biological Sciences, Program in Biostatistics and Biometry, San Diego, CA 92182. Offers PhD. Program offered jointly with the University of California, Davis. *Degree requirements:* For doctorate, thesis/dissertation. *Entrance requirements:* For doctorate, GRE General Test, GRE Subject Test, resumé or curriculum vitae, 3 letters of recommendation. *Application deadline:* For fall admission, 5/1 for domestic students; for spring admission, 11/1 for domestic students, 10/1 for international students. Applications are processed on a rolling basis. Application fee: $55. Electronic applications accepted. *Financial support:* Research assistantships, teaching assistantships, career-related internships or fieldwork, scholarships/grants, and unspecified assistantships available. *Unit head:* Kung-Jong Lui, Graduate Advisor, 619-594-7239, Fax: 619-594-6746, E-mail: kjl@rohan.sdsu.edu. *Application contact:* Kung-Jong Lui, Graduate Advisor, 619-594-7239, Fax: 619-594-6746, E-mail: kjl@rohan.sdsu.edu.

The University of Alabama at Birmingham, School of Public Health, Department of Biostatistics, Birmingham, AL 35294. Offers biomathematics (MS, PhD); biostatistics (MS, PhD). *Students:* 22 full-time (8 women), 3 part-time (1 woman); includes 3 minority (all African Americans), 14 international. 53 applicants, 25% accepted. In 2005, 1 master's, 1 doctorate awarded. *Degree requirements:* For master's, variable foreign language requirement, thesis, fieldwork, research project; for doctorate, variable foreign language requirement, thesis/dissertation, comprehensive exam. *Entrance requirements:* For master's, GRE General Test or MAT, minimum GPA of 3.0; for doctorate, GRE General Test or MAT, MPH or MSPH, minimum GPA of 3.0, interview. *Application deadline:* Applications are processed on a rolling basis. Application fee: $35 ($60 for international students). Electronic applications accepted. *Expenses:* Contact institution. Tuition and fees vary according to course load, degree level and program. *Financial support:* Fellowships, career-related internships or fieldwork available. *Unit head:* Dr. George Howard, Chair, 205-934-4905, Fax: 205-975-2540, E-mail: ghoward@uab.edu. *Application contact:* Nancy O. Pinson, Coordinator of Student Admissions, 205-934-4993, Fax: 205-975-5484.

University of California, Los Angeles, School of Medicine and Graduate Division, Graduate Programs in Medicine, Department of Biomathematics, Los Angeles, CA 90095. Offers biomathematics (MS, PhD); clinical research (MS). *Degree requirements:* For master's, comprehensive exam or thesis; for doctorate, thesis/dissertation, oral and written qualifying exams. *Entrance requirements:* For master's and doctorate, GRE General Test, GRE Subject Test.

See Close-Up on page 529.

University of Nebraska–Lincoln, Graduate College, College of Agricultural Sciences and Natural Resources, Department of Biometry, Lincoln, NE 68588. Offers MS. *Degree requirements:* For master's, thesis optional. *Entrance requirements:* For master's, GRE General Test. Additional exam requirements/recommendations for international students: Required—TOEFL (minimum score 550 paper-based; 213 computer-based). Electronic applications accepted. *Faculty research:* Design of experiments, linear models, spatial variability, statistical modeling and inference, sampling.

University of Southern California, Keck School of Medicine and Graduate School, Graduate Programs in Medicine, Department of Preventive Medicine, Master of Public Health Program, Los Angeles, CA 90089. Offers biometry/epidemiology (MPH); health communication (MPH); health promotion (MPH); preventive nutrition (MPH). *Accreditation:* CEPH. Part-time programs available. *Faculty:* 16 full-time (9 women), 8 part-time/adjunct (2 women). *Students:* 113 full-time (81 women), 10 part-time (8 women); includes 61 minority (9 African Americans, 1 American Indian/Alaska Native, 39 Asian Americans or Pacific Islanders, 12 Hispanic Americans), 21 international. Average age 26. 268 applicants, 67% accepted, 55 enrolled. In 2005, 54 degrees awarded. *Degree requirements:* For master's, practicum, final report, oral presentation. *Entrance requirements:* For master's, GRE General Test, MCAT, GMAT, DAT, minimum GPA of 3.0. Additional exam requirements/recommendations for international students: Required—TOEFL (minimum score 600 paper-based; 250 computer-based). *Application deadline:* For fall admission, 6/1 priority date for domestic students, 6/1 priority date for international students; for spring admission, 11/15 priority date for domestic students, 10/1 priority date for international students. Applications are processed on a rolling basis. Application fee: $65 ($75 for international students). Electronic applications accepted. *Expenses:* Tuition: Full-time $25,416; part-time $1,059 per unit. Required fees: $484; $484 per year. Tuition and fees vary according to course load and program. *Financial support:* In 2005–06, 120 students received support, including 7 research assistantships with full tuition reimbursements available (averaging $24,876 per year), 11 teaching assistantships with partial tuition reimbursements available (averaging $24,876 per year); career-related internships or fieldwork, Federal Work-Study, institutionally sponsored loans, scholarships/grants, health care benefits, unspecified assistantships, and staff tuition remission also available. Support available to part-time students. Financial award application deadline: 2/1; financial award applicants required to submit CSS PROFILE or FAFSA. *Faculty research:* Substance abuse prevention, cancer and heart disease prevention, mass media and health communication research, health promotion, treatment compliance. Total annual research expenditures: $12 million. *Unit head:* Dr. Thomas W. Valente, Director, 626-457-6678, Fax: 626-457-6699, E-mail: tvalente@usc.edu. *Application contact:* Nemesia P. Kelly, Program Specialist, 626-457-6603, Fax: 626-457-6699, E-mail: nkelly@usc.edu.

The University of Texas Health Science Center at Houston, Graduate School of Biomedical Sciences, Program in Biomathematics and Biostatistics, Houston, TX 77225-0036. Offers MS, PhD, MD/PhD. *Faculty:* 31 full-time (6 women). *Students:* 4 full-time (1 woman), 2 international. Average age 25. 37 applicants, 30% accepted, 7 enrolled. In 2005, 3 degrees awarded. Terminal master's awarded for partial completion of doctoral program. *Degree requirements:* For master's and doctorate, thesis/dissertation. *Entrance requirements:* For master's and doctorate, GRE General Test. Additional exam requirements/recommendations for international students: Required—TOEFL, TWE. *Application deadline:* For fall admission, 1/15 for domestic students; for spring admission, 11/1 for domestic students. Applications are processed on a rolling basis. Application fee: $10. Electronic applications accepted. *Financial support:* Fellowships with full tuition reimbursements, research assistantships with full tuition reimbursements, teaching assistantships, institutionally sponsored loans, scholarships/grants, and health care benefits available. Financial award application deadline: 1/15. *Faculty research:* Statistical and mathematical modeling, development of new models for design and analysis of research studies, formulation of mathematical models of biological systems. *Unit head:* Dr. Li Zhang, Director, 713-563-4298, Fax: 713-563-4243, E-mail: zhangli@mdanderson.org. *Application contact:* Dr. Victoria P. Knutson, Assistant Dean of Admissions, 713-500-9860, Fax: 713-500-9877, E-mail: victoria.p.knutson@uth.tmc.edu.

University of Wisconsin–Madison, Graduate School, College of Agricultural and Life Sciences, Biometry Program, Madison, WI 53706-1380. Offers MS. Application fee: $45. *Unit head:* Murray Clayton, Chair, 608-262-1009.

Biostatistics

Arizona State University, Division of Graduate Studies, College of Liberal Arts and Sciences, Department of Biology, Program in Computational, Statistical, and Mathematical Biology, Tempe, AZ 85287. Offers MS, PhD. *Entrance requirements:* Additional exam requirements/recommendations for international students: Required—TOEFL (minimum score 600 paper-based); Recommended—TSE.

Boston University, Graduate School of Arts and Sciences, Program in Biostatistics, Boston, MA 02215. Offers MA, PhD. *Students:* 33 full-time (24 women), 55 part-time (41 women); includes 9 minority (8 Asian Americans or Pacific Islanders, 1 Hispanic American), 26 international. Average age 32. 112 applicants, 38% accepted, 15 enrolled. In 2005, 20 master's, 2 doctorates awarded. Terminal master's awarded for partial completion of doctoral program. *Degree requirements:* For master's, one foreign language, comprehensive exam, registration; for doctorate, one foreign language, thesis/dissertation, comprehensive exam, registration. *Entrance requirements:* For master's and doctorate, GRE General Test, 2 letters of recommendation. Additional exam requirements/recommendations for international students: Required—TOEFL (minimum score 550 paper-based; 213 computer-based). *Application deadline:* For fall admission, 5/1 for domestic students, 5/1 for international students; for spring admission, 10/15 for domestic students, 10/15 for international students. Application fee: $60. *Expenses:* Tuition: Full-time $31,530; part-time $985 per credit. Required fees: $316; $40 per semester. Tuition and fees vary according to course level and program. *Financial support:* In 2005–06, 21 students received support, including 21 research assistantships with full tuition reimbursements available (averaging $15,500 per year); fellowships, teaching assistantships Support available to part-time students. Financial award application deadline: 1/15; financial award applicants required to submit FAFSA. *Unit head:* Ralph D'Agostino, Director, 617-353-2767, Fax: 617-638-4458, E-mail: ralph@bu.edu. *Application contact:* Sharon Milewits, Administrative Assistant, 617-638-5172, Fax: 617-638-4458, E-mail: sharonm@bu.edu.

Boston University, School of Public Health, Biostatistics Department, Boston, MA 02215. Offers MA, MPH, PhD. Application fee: $60. *Expenses:* Tuition: Full-time $31,530; part-time $985 per credit. Required fees: $316; $40 per semester. Tuition and fees vary according to course level and program. *Unit head:* L. Adrienne Cupples, Chairman, 617-638-5176, Fax: 617-638-4458, E-mail: adrienne@bu.edu. *Application contact:* LePhan Quan, Assistant Director of Admissions, 617-638-4640, Fax: 617-638-5299, E-mail: sphadmis@bu.edu.

Brown University, Graduate School, Division of Biology and Medicine, Department of Community Health, Providence, RI 02912. Offers health services research (MS, PhD); public health (MPH); statistical science (MS, PhD), including biostatistics, epidemiology. *Accreditation:* CEPH. *Degree requirements:* For doctorate, thesis/dissertation, preliminary exam. *Entrance requirements:* For master's and doctorate, GRE General Test. Additional exam requirements/recommendations for international students: Required—TOEFL.

Brown University, Graduate School, Division of Biology and Medicine, Department of Community Health, Center for Statistical Science, Program in Biostatistics, Providence, RI 02912. Offers MS, PhD, MD/PhD. *Degree requirements:* For doctorate, thesis/dissertation, preliminary exam. *Entrance requirements:* For master's and doctorate, GRE General Test.

California State University, East Bay, Academic Programs and Graduate Studies, College of Science, Department of Statistics, Hayward, CA 94542-3000. Offers actuarial statistics (MS); biostatistics (MS); computational statistics (MS); mathematical statistics (MS); statistics (MS); theoretical and applied statistics (MS). *Students:* 68. 35 applicants, 97% accepted. In 2005, 2 degrees awarded. *Degree requirements:* For master's, comprehensive exam. *Entrance requirements:* For master's, minimum GPA of 2.5 during previous 2 years of course work. Additional exam requirements/recommendations for international students: Required—TOEFL (minimum score 550 paper-based; 213 computer-based). *Application deadline:* For fall admission, 5/31 for domestic students, 4/30 for international students. For winter admission, 9/30 for domestic students. Application fee: $55. *Financial support:* Federal Work-Study and institutionally sponsored loans available. Support available to part-time students. Financial award application deadline: 3/2. *Unit head:* Dr. Julia Norton, Chair, 510-885-3435, E-mail: julia.norton@csueastbay.edu. *Application contact:* Deborah Baker, Associate Director, 510-885-3286, Fax: 510-885-4777, E-mail: deborah.baker@csueastbay.edu.

Case Western Reserve University, School of Medicine and School of Graduate Studies, Graduate Programs in Medicine, Department of Epidemiology and Biostatistics, Program in Biostatistics, Cleveland, OH 44106. Offers MS, PhD. Part-time programs available. *Faculty:* 7 full-time (2 women), 12 part-time/adjunct (1 woman). *Students:* 16 full-time (8 women), 1 part-time; includes 1 African American, 8 Asian Americans or Pacific Islanders, 1 Hispanic American. Average age 31. 15 applicants, 40% accepted, 3 enrolled. In 2005, 1 doctorate awarded. Terminal master's awarded for partial completion of doctoral program. *Degree requirements:* For master's, thesis, exam/practicum, comprehensive exam; for doctorate, thesis/dissertation, comprehensive exam. *Entrance requirements:* For master's, GRE General Test (MCAT may be substituted), 3 recommendations; for doctorate, GRE General Test, 3 recommendations. Additional exam requirements/recommendations for international students: Required—TOEFL (minimum score 213 paper-based). *Application deadline:* For fall admission, 2/1 for domestic students. Applications are processed on a rolling basis. Application fee: $50. Electronic applications accepted. *Financial support:* In 2005–06, 3 students received support, including fellowships with full tuition reimbursements available (averaging $20,772 per year), 3 research assistantships with full and partial tuition reimbursements available (averaging $20,772 per year); career-related internships or fieldwork, scholarships/grants, tuition waivers (partial), and unspecified assistantships also available. Support available to part-time students. Financial award application deadline: 2/1. *Faculty research:* Survey sampling and statistical computing, generalized linear models, statistical modeling, models in breast cancer survival. Total annual research expenditures:$300,000. *Unit head:* Dr. Sara M Debanne, Acting Director of the Division of Biostatistics, 216-368-3895, Fax: 216-368-3970, E-mail: smd3@case.edu. *Application contact:* Alicia M Boscarello, Graduate Student Coordinator, 216-368-5957, Fax: 216-368-3970, E-mail: amb62@case.edu.

Columbia University, Joseph L. Mailman School of Public Health, Division of Biostatistics, New York, NY 10032. Offers MPH, MS, Dr PH, PhD. PhD offered in cooperation with the Graduate School of Arts and Sciences. Part-time programs available. *Degree requirements:* For doctorate, thesis/dissertation. *Entrance requirements:* For master's, GRE General Test; for doctorate, GRE General Test, MPH or equivalent (Dr PH). Electronic applications accepted. *Expenses:* Tuition: Full-time $31,448. Tuition and fees vary according to course level, course load, campus/location and program. *Faculty research:* Application of statistics in public policy, medical experiments, and legal processing; clinical trial results; statistical methods in epidemiology.

Drexel University, School of Biomedical Engineering, Science and Health Systems, Philadelphia, PA 19104-2875. Offers biomedical engineering (MS, PhD); biomedical science (MS, PhD); biostatistics (MS); clinical/rehabilitation engineering (MS). *Degree requirements:* For doctorate, thesis/dissertation, 1 year of residency, qualifying exam. *Entrance requirements:* For master's, minimum GPA of 3.0; for doctorate, minimum GPA of 3.0, MS. Additional exam requirements/recommendations for international students: Required—TOEFL. Electronic applications accepted. *Faculty research:* Cardiovascular dynamics, diagnostic and therapeutic ultrasound.

Emory University, Graduate School of Arts and Sciences, Department of Biostatistics, Atlanta, GA 30322-1100. Offers biostatistics (MPH, MSPH, PhD); public health informatics (MSPH). *Faculty:* 21 full-time (8 women), 17 part-time/adjunct (5 women). *Students:* 29 full-time (18 women); includes 5 minority (3 African Americans, 2 Asian Americans or Pacific Islanders), 13 international. Average age 29. 86 applicants, 15% accepted, 6 enrolled. In 2005, 4 degrees awarded. *Median time to degree:* Of those who began their doctoral program in fall 1997, 100% received their degree in 8 years or less. *Degree requirements:* For doctorate, thesis/dissertation, comprehensive exam, registration. *Entrance requirements:* For doctorate, GRE General Test. Additional exam requirements/recommendations for international students: Required—TOEFL (minimum score 550 paper-based; 220 computer-based). *Application deadline:* For fall admission, 1/3 priority date for domestic students, 1/3 priority date for international students. Application fee: $50. Electronic applications accepted. *Expenses:* Tuition: Full-time $14,400. Required fees: $217. *Financial support:* In 2005–06, 6 fellowships with full tuition reimbursements (averaging $19,000 per year) were awarded; career-related internships or fieldwork and scholarships/grants also available. Financial award application deadline:1/20. *Faculty research:* Vaccine efficacy, clinical trials, spatial statistics, statistical genetics, neuroimaging. Total annual research expenditures: $6 million. *Unit head:* Dr. Michael H. Kutner, Chair, 404-727-7693, Fax: 404-727-1370, E-mail: mkutner@sph.emory.edu. *Application contact:* Dr. John J. Hanfelt, Director of Graduate Studies, 404-727-2876, Fax: 404-727-1370, E-mail: jhanfel@sph.emory.edu.

See Close-Up on page 497.

Emory University, Rollins School of Public Health, Department of Biostatistics, Atlanta, GA 30322-1100. Offers MPH, MSPH, PhD. Part-time programs available. *Students:* 7 full-time (5 women), 6 part-time (3 women). Average age 27. 33 applicants, 61% accepted, 9 enrolled. In 2005, 4 degrees awarded. *Degree requirements:* For master's, thesis, practicum. *Entrance requirements:* For master's, GRE General Test. Additional exam requirements/recommendations for international students: Required—TOEFL (minimum score 550 paper-based; 213 computer-based). *Application deadline:* For fall admission, 1/15 priority date for domestic students, 1/1 priority date for international students. Application fee: $75. Electronic applications accepted. *Expenses:* Tuition: Full-time $14,400. Required fees: $217. *Financial support:* Fellowships with full and partial tuition reimbursements, career-related internships or fieldwork, Federal Work-Study, institutionally sponsored loans, and scholarships/grants available. Support available to part-time students. Financial award application deadline: 1/15. *Unit head:* Dr. Michael H. Kutner, Chair, 404-727-7693, Fax: 404-727-1370, E-mail: mkutner@sph.emory.edu. *Application contact:* Catherine Strate, Assistant Director of Academic Programs, 404-727-3968, Fax: 404-727-1370, E-mail: cstrate@emory.edu.

See Close-Up on page 497.

Florida State University, Graduate Studies, College of Arts and Sciences, Department of Statistics, Tallahassee, FL 32306. Offers applied statistics (MS); biostatistics (MS); mathematical statistics (MS, PhD). Part-time programs available. *Faculty:* 13 full-time (2 women). *Students:* 47 full-time (20 women), 8 part-time (2 women); includes 4 minority (all African Americans), 38 international. Average age 30. 141 applicants, 35% accepted, 18 enrolled. In 2005, 2 master's, 4 doctorates awarded. Terminal master's awarded for partial completion of doctoral program. *Median time to degree:* Of those who began their doctoral program in fall 1997, 100% received their degree in 8 years or less. *Degree requirements:* For master's, comprehensive exam (mathematical statistics); for doctorate, thesis/dissertation, departmental qualifying exam. *Entrance requirements:* For master's, GRE General Test, previous course work in calculus, minimum GPA of 3.0; for doctorate, GRE General Test, minimum GPA of 3.0, 1 course in linear algebra (preferred), Calculus I-III. Additional exam requirements/recommendations for international students: Required—TOEFL (minimum score 600 paper-based; 250 computer-based). *Application deadline:* For fall admission, 7/1 for domestic students, 5/2 for international students; for spring admission, 11/1 for domestic students, 9/1 for international students. Applications are processed on a rolling basis. Application fee: $30. Electronic applications accepted. *Financial support:* In 2005–06, 2 fellowships with full tuition reimbursements (averaging $18,000 per year), 8 research assistantships with full tuition reimbursements (averaging $17,612 per year), 35 teaching assistantships with full tuition reimbursements (averaging $17,612 per year) were awarded; Federal Work-Study, institutionally sponsored loans, scholarships/grants, unspecified assistantships, and health insurance supplements also available. Support available to part-time students. Financial award application deadline: 2/15; financial award applicants required to submit FAFSA. *Faculty research:* Statistical inference, probability theory, spatial statistics, nonparametric estimation, automatic target recognition. Total annual research expenditures: $626,760. *Unit head:* Dr. Dan McGee, Chairman, 850-644-3218, Fax: 850-644-5271, E-mail: info@stat.fsu.edu. *Application contact:* Jennifer Rivera, Program Assistant, 850-644-3218, Fax: 850-644-5271, E-mail: info@stat.fsu.edu.

Georgetown University, Graduate School of Arts and Sciences, Programs in Biomedical Sciences, Division of Biostatistics and Epidemiology, Washington, DC 20057. Offers MS. *Entrance requirements:* For master's, GRE General Test. Additional exam requirements/recommendations for international students: Required—TOEFL. *Faculty research:* Occupation epidemiology, cancer.

The George Washington University, Columbian College of Arts and Sciences, Department of Statistics, Program in Biostatistics, Washington, DC 20052. Offers MS, PhD. *Degree requirements:* For master's, comprehensive exam; for doctorate, thesis/dissertation, general exam. *Entrance requirements:* For master's and doctorate, GRE General Test, minimum GPA of 3.0. Additional exam requirements/recommendations for international students: Required—TOEFL (minimum score 550 paper-based; 213 computer-based). Electronic applications accepted.

The George Washington University, School of Public Health and Health Services, Department of Epidemiology and Biostatistics, Washington, DC 20052. Offers biostatistics (MPH); epidemiology (MPH); health information systems (MPH); microbiology and emerging infectious diseases (MSPH). *Accreditation:* CEPH. *Degree requirements:* For master's, case study or special project. *Entrance requirements:* For master's, GMAT, GRE General Test, or MCAT. Additional exam requirements/recommendations for international students: Required—TOEFL.

Grand Valley State University, College of Liberal Arts and Sciences, Program in Biostatistics, Allendale, MI 49401-9403. Offers MS. *Expenses:* Tuition, state resident: full-time $5,364; part-time $298 per credit. Tuition, nonresident: full-time $10,800; part-time $600 per credit. *Unit head:* Dr. Robert Downer, Director. *Application contact:* Dr. David Elrod, PSM Coordinator, 616-331-8643, E-mail: elrodd@gvsu.edu.

Harvard University, Graduate School of Arts and Sciences, Department of Biostatistics, Cambridge, MA 02138. Offers PhD. *Expenses:* Tuition: Full-time $28,752. Full-time tuition and fees vary according to program and student level. *Unit head:* Jelena Tillotson-Follweiler, Director of Graduate Studies, 617-384-7767.

Harvard University, School of Public Health, Department of Biostatistics, Boston, MA 02115-6096. Offers SM, PhD. Part-time programs available. *Degree requirements:* For doctorate, thesis/dissertation, oral and written qualifying exams. *Entrance requirements:* For master's and doctorate, GRE, prior training in mathematics and/or statistics. Additional exam requirements/recommendations for international students: Required—TOEFL (minimum score 560 paper-based; 220 computer-based); Recommended—IELT (minimum score 7). Electronic applications accepted. *Expenses:* Tuition: Full-time $28,752. Full-time tuition and fees vary according to program and student level. *Faculty research:* Statistical genetics, clinical trials, cancer and AIDS research, environmental and mental health.

Iowa State University of Science and Technology, Graduate College, Interdisciplinary Programs, Bioinformatics and Computational Biology Program, Ames, IA 50011-3260. Offers MS, PhD. *Degree requirements:* For doctorate, thesis/dissertation. *Entrance requirements:* For doctorate, GRE General Test. Additional exam requirements/recommendations for international students: Required—TOEFL or IELTS. Electronic applications accepted. *Expenses:* Tuition, state resident: full-time $6,410. Tuition, nonresident: full-time $16,422. Tuition and fees vary according to program. *Faculty research:* Functional and structural genomics, genome evolution, macromolecular structure and function, mathematical biology and biological statistics, metabolic and developmental networks.

The Johns Hopkins University, Bloomberg School of Public Health, Department of Biostatistics, Baltimore, MD 21205-2179. Offers bioinformatics (MHS); biostatistics (MHS, Sc M, PhD). *Faculty:* 27 full-time (7 women), 37 part-time/adjunct (14 women). *Students:* 46 full-time (25 women), 25 part-time (12 women); includes 11 minority (2 African Americans, 9 Asian Americans or Pacific Islanders), 44 international. Average age 28. 222 applicants, 19% accepted, 14 enrolled. In 2005, 10 master's, 7 doctorates awarded. *Median time to degree:* Of those who began their doctoral program in fall 1997, 68% received their degree in 8 years or less. *Degree requirements:* For master's, thesis (for some programs), written exam; field placement, comprehensive exam (for some programs), registration; for doctorate, thesis/dissertation, 1 year full-time residency, oral and written exams, comprehensive exam, registration. *Entrance requirements:* For master's and doctorate, GRE General Test, course work in calculus and matrix algebra, 3 letters of recommendation, curriculum vitae. Additional exam requirements/recommendations for international students: Required—TOEFL (minimum score 550 paper-based; 213 computer-based). *Application deadline:* For fall admission, 1/15 for domestic students, 1/15 for international students. Applications are processed on a rolling basis. Application fee: $45. Electronic applications accepted. *Expenses:* Tuition: Full-time $30,960. Tuition and fees vary according to degree level and program. *Financial support:* In 2005–06, 41 students received support, including 8 research assistantships (averaging $20,000 per year); fellowships, Federal Work-Study, institutionally sponsored loans, scholarships/grants, traineeships, health care benefits, tuition waivers (partial), and stipends also available. Support available to part-time students. Financial award application deadline: 4/15; financial award applicants required to submit FAFSA. *Faculty research:* Statistical genetics, bioinformatics, statistical computing, statistical methods, environmental statistics. Total annual research expenditures: $1.5 million. *Unit head:* Dr. Scott Zeger, Chair, 410-955-3067, Fax: 410-955-0958, E-mail: szeger@jhsph.edu. *Application contact:* Mary Joy Argo, Academic Administrator, 410-614-4454, Fax: 410-955-0958, E-mail: margo@jhsph.edu.

Loma Linda University, School of Public Health, Programs in Epidemiology and Biostatistics, Loma Linda, CA 92350. Offers MPH, MSPH. *Faculty:* 17 full-time (4 women), 13 part-time/adjunct (3 women). *Students:* 4 full-time (2 women), 8 part-time (2 women); includes 3 African Americans, 2 Asian Americans or Pacific Islanders, 2 Hispanic Americans, 3 international. In 2005, 2 degrees awarded. *Entrance requirements:* Additional exam requirements/recommendations for international students: Required—Michigan English Language Assessment Battery or TOEFL. *Application deadline:* Applications are processed on a rolling basis. Application fee: $100. *Financial support:* Application deadline: 5/15. *Unit head:* Dr. Synnove Knutsen, Chair, 909-824-4590, Fax: 909-824-4087. *Application contact:* Terri Tamayose, Director of Admissions and Academic Records, 909-824-4694, Fax: 909-824-8087, E-mail: ttamayose@sph.llu.edu.

McGill University, Faculty of Graduate and Postdoctoral Studies, Faculty of Medicine, Departments of Epidemiology and Biostatistics, and Occupational Health, Montréal, QC H3A 2T5, Canada. Offers community health (M Sc); environmental health (M Sc); epidemiology and biostatistics (M Sc, PhD, Diploma); health care evaluation (M Sc); medical statistics (M Sc); occupational health (M Sc). *Accreditation:* CEPH (one or more programs are accredited). *Degree requirements:* For master's, thesis optional; for doctorate, thesis/dissertation. *Entrance requirements:* For master's, GRE, minimum GPA of 3.0; for doctorate, GRE. *Faculty research:* Chronic and infectious disease epidemiology, health services research, pharmacoepidemiology.

Medical College of Wisconsin, Graduate School of Biomedical Sciences, Division of Biostatistics, Milwaukee, WI 53226-0509. Offers PhD. Part-time programs available. *Degree requirements:* For doctorate, thesis/dissertation, comprehensive exam, registration. *Entrance requirements:* For doctorate, GRE General Test. Additional exam requirements/recommendations for international students: Required—TOEFL. Electronic applications accepted. *Faculty research:* Survival analysis, spatial statistics, time series, genetic statistics, Bayesian statistics.

See Close-Up on page 505.

Medical University of South Carolina, College of Graduate Studies, Program in Biostatistics, Bioinformatics, and Epidemiology, Charleston, SC 29425-0002. Offers biometrics (MS, PhD); biostatistics (MS, PhD); clinical research (MCR); epidemiology (MCR, PhD). *Faculty:* 26 full-time (9 women). *Students:* 40 full-time (26 women); includes 3 minority (1 African American, 2 Asian Americans or Pacific Islanders), 6 international. Average age 28. 181 applicants, 34% accepted, 43 enrolled. In 2005, 7 master's, 4 doctorates awarded. Terminal master's awarded for partial completion of doctoral program. *Degree requirements:* For master's, thesis, research seminar; for doctorate, thesis/dissertation, teaching and research seminar, oral and written exams. *Entrance requirements:* For master's, GRE General Test; for doctorate, GRE General Test, interview. Additional exam requirements/recommendations for international students: Required—TOEFL (minimum score 600 paper-based; 250 computer-based). *Application deadline:* For fall admission, 1/15 priority date for domestic students, 1/15 priority date for international students. Applications are processed on a rolling basis. Application fee: $0 ($75 for international students). Electronic applications accepted. *Financial support:* In 2005–06, 8 students received support, including fellowships with partial tuition reimbursements available (averaging $21,000 per year); Federal Work-Study and scholarships/grants also available. Support available to part-time students. Financial award application deadline: 3/15; financial award applicants required to submit FAFSA. *Faculty research:* Health disparities, central nervous system injuries, radiation exposure, analysis of clinical trial data, biomedical information. *Unit head:* Dr. Barbara Tilley, Chair, 873-876-1327, Fax: 873-792-6950, E-mail: tilleyb@musc.edu. *Application contact:* Cheryl Brown, 843-792-4620, Fax: 843-792-4645, E-mail: brownche@musc.edu.

New York Medical College, School of Public Health, Program in Biostatistics, Valhalla, NY 10595-1691. Offers MPH, MS. Part-time and evening/weekend programs available. *Degree requirements:* For master's, thesis, registration. *Entrance requirements:* For master's, minimum undergraduate GPA of 3.0. Additional exam requirements/recommendations for international students: Required—TOEFL (minimum score 600 paper-based; 250 computer-based). Electronic applications accepted.

The Ohio State University, Graduate School, College of Mathematical and Physical Sciences, Department of Statistics, Program in Biostatistics, Columbus, OH 43210. Offers PhD. *Degree requirements:* For doctorate, thesis/dissertation. *Entrance requirements:* For doctorate, GRE General Test. Additional exam requirements/recommendations for international students: Required—TOEFL (minimum score 600 paper-based; 250 computer-based). Electronic applications accepted.

Oregon Health & Science University, School of Medicine, Department of Public Health and Preventive Medicine, Portland, OR 97239-3098. Offers epidemiology and biostatistics (MPH). *Accreditation:* CEPH. Part-time programs available. *Degree requirements:* For master's, thesis, fieldwork/internship. *Entrance requirements:* For master's, GRE General Test, previous undergraduate course work in statistics. Additional exam requirements/recommendations for international students: Required—TOEFL. *Faculty research:* Health services, health care access, health policy, environmental and occupational health.

Rice University, Graduate Programs, George R. Brown School of Engineering, Department of Statistics, Houston, TX 77251-1892. Offers biostatistics (PhD); computational finance (PhD); statistics (M Stat, MA, PhD). Terminal master's awarded for partial completion of doctoral program. *Degree requirements:* For master's, thesis/dissertation; for doctorate, thesis/dissertation, comprehensive exam. *Entrance requirements:* For master's and doctorate, GRE General Test, GRE Subject Test, minimum GPA of 3.0. Additional exam requirements/recommendations for international students: Required—TOEFL (minimum score 630 paper-based; 250 computer-

Biostatistics

Rice University (continued)
based). Electronic applications accepted. *Faculty research:* Statistical genetics, non parametric function estimation, computational statistics and visualization, stochastic processes.

Rutgers, The State University of New Jersey, New Brunswick/Piscataway, Graduate School, BioMaps Institute for Quantitative Biology (Biology at the Inter-face of the Mathematical and Physical Sciences), New Brunswick, NJ 08901-1281. Offers PhD. *Faculty:* 15 full-time (4 women). *Students:* 17 full-time (6 women); includes 2 minority (both Asian Americans or Pacific Islanders), 9 international. Average age 27. 32 applicants, 31% accepted, 6 enrolled. *Degree requirements:* For doctorate, thesis/dissertation, comprehensive exam, registration. *Entrance requirements:* For doctorate, GRE. Additional exam requirements/recommendations for international students: Required—TOEFL. *Application deadline:* For winter admission, 2/15 for domestic students. Applications are processed on a rolling basis. Application fee: $50. Electronic applications accepted. *Expenses:* Tuition, state resident: full-time $10,440; part-time $435 per credit. Tuition, nonresident: full-time $15,520; part-time $647 per credit. Required fees: $129 per credit. Tuition and fees vary according to program. *Financial support:* In 2005–06, 16 students received support, including 9 fellowships with full tuition reimbursements available (averaging $24,000 per year), 3 research assistantships with full tuition reimbursements available (averaging $20,000 per year), 4 teaching assistantships with full tuition reimbursements available (averaging $20,000 per year); traineeships, health care benefits, and unspecified assistantships also available. Financial award application deadline: 6/30; financial award applicants required to submit FAFSA. *Faculty research:* Protein folding; nucleic acid structure; systems biology; transcriptional regulation; signal transduction. *Unit head:* Dr. Wilma Olsen, Co-Director, 732-445-3993, Fax: 732-445-5958, E-mail: olson@rutchem.rutgers.edu. *Application contact:* Dr. Paul H. Ehrlich, Administrative and Associate Graduate Program Director, 732-445-8377, Fax: 732-445-5958, E-mail: pehrlich@biomaps.rutgers.edu.

Rutgers, The State University of New Jersey, New Brunswick/Piscataway, Graduate School, Program in Statistics, New Brunswick, NJ 08901-1281. Offers quality and productivity management (MS); statistics (MS, PhD), including biostatistics (PhD), data mining (PhD). Part-time programs available. *Faculty:* 27 full-time. *Students:* 62 full-time (33 women), 51 part-time (25 women); includes 28 minority (3 African Americans, 23 Asian Americans or Pacific Islanders, 2 Hispanic Americans), 62 international. Average age 31. 230 applicants, 18% accepted, 54 enrolled. In 2005, 6 master's, 3 doctorates awarded. Terminal master's awarded for partial completion of doctoral program. *Degree requirements:* For master's, essay, exam, non-thesis essay paper; for doctorate, one foreign language, thesis/dissertation, qualifying oral and written exams. *Entrance requirements:* For master's, GRE General Test; for doctorate, GRE General Test, GRE Subject Test (recommended). Additional exam requirements/recommendations for international students: Required—TOEFL (minimum score 550 paper-based; 213 computer-based). *Application deadline:* For fall admission, 5/1 for domestic students; for spring admission, 12/1 priority date for domestic students. Applications are processed on a rolling basis. Application fee: $50. Electronic applications accepted. *Expenses:* Tuition, state resident: full-time $10,440; part-time $435 per credit. Tuition, nonresident: full-time $15,520; part-time $647 per credit. Required fees: $129 per credit. Tuition and fees vary according to program. *Financial support:* In 2005–06, 16 students received support, including 4 fellowships with full tuition reimbursements available (averaging $18,000 per year), research assistantships with full tuition reimbursements available (averaging $14,500 per year), 11 teaching assistantships with full tuition reimbursements available (averaging $16,988 per year); career-related internships or fieldwork, Federal Work-Study, institutionally sponsored loans, scholarships/grants, health care benefits, unspecified assistantships, and grant/tuition remission also available. Financial award application deadline: 3/1; financial award applicants required to submit FAFSA. *Faculty research:* Probability, decision theory, linear models, multivariate statistics, statistical computing. *Unit head:* John Kolassa, Director, 732-445-3634, Fax: 732-445-3428, E-mail: kolassa@stat.rutgers.edu. *Application contact:* Angela T. Klein, Department Secretary, 732-445-2693, Fax: 732-445-3428, E-mail: aklein@stat.rutgers.edu.

San Diego State University, Graduate and Research Affairs, College of Health and Human Services, Graduate School of Public Health, San Diego, CA 92182. Offers environmental health (MPH); epidemiology (MPH, PhD), including biostatistics (MPH); global emergency preparedness and response (MS); health behavior (PhD); health promotion (MPH); health services administration (MPH); toxicology (MS). *Accreditation:* ABET (one or more programs are accredited); ACEHSA (one or more programs are accredited); CEPH (one or more programs are accredited). Part-time programs available. *Faculty:* 29 full-time (14 women), 83 part-time/adjunct (37 women). *Students:* 254 full-time (189 women), 113 part-time (81 women); includes 128 minority (18 African Americans, 4 American Indian/Alaska Native, 53 Asian Americans or Pacific Islanders, 53 Hispanic Americans), 26 international. 469 applicants, 67% accepted, 127 enrolled. In 2005, 89 master's, 5 doctorates awarded. *Degree requirements:* For master's, thesis (for some programs), comprehensive exam (for some programs); for doctorate, thesis/dissertation. *Entrance requirements:* For master's, GMAT (health services administration MPH only), GRE General Test; for doctorate, GRE General Test. Additional exam requirements/recommendations for international students: Required—TOEFL. *Application deadline:* For fall admission, 5/1 for domestic students, 5/1 for international students; for spring admission, 11/1 for domestic students, 10/1 for international students. Applications are processed on a rolling basis. Application fee: $55. *Financial support:* Research assistantships, teaching assistantships, career-related internships or fieldwork, Federal Work-Study, and traineeships available. Financial award applicants required to submit FAFSA. *Faculty research:* Evaluation of tobacco, AIDS prevalence and prevention, mammography, infant death project, Alzheimer's in elderly Chinese. *Unit head:* Dr. Ann de Peyster, Interim Director, 619-594-6317. *Application contact:* Brenda Fass-Holmes, Coordinator, Admissions and Student Affairs, 619-594-6317, E-mail: brenda.fass-holmes@sdsu.edu.

San Diego State University, Graduate and Research Affairs, College of Sciences, Department of Biological Sciences, Program in Biostatistics and Biometry, San Diego, CA 92182. Offers PhD. Program offered jointly with the University of California, Davis. *Degree requirements:* For doctorate, thesis/dissertation. *Entrance requirements:* For doctorate, GRE General Test, GRE Subject Test, resumé or curriculum vitae, 3 letters of recommendation. *Application deadline:* For fall admission, 5/1 for domestic students; for spring admission, 11/1 for domestic students, 10/1 for international students. Applications are processed on a rolling basis. Application fee: $55. Electronic applications accepted. *Financial support:* Research assistantships, teaching assistantships, career-related internships or fieldwork, scholarships/grants, and unspecified assistantships available. *Unit head:* Kung-Jong Lui, Graduate Advisor, 619-594-7239, Fax: 619-594-6746, E-mail: kjl@rohan.sdsu.edu. *Application contact:* Kung-Jong Lui, Graduate Advisor, 619-594-7239, Fax: 619-594-6746, E-mail: kjl@rohan.sdsu.edu.

State University of New York at Buffalo, Graduate School, School of Public Health and Health Professions, Department of Biostatistics, Buffalo, NY 14260. Offers MA, PhD. *Faculty:* 9 full-time (2 women), 2 part-time/adjunct (0 women). *Students:* 20 full-time (12 women), 6 part-time (3 women); includes 2 minority (both African Americans), 14 international. 78 applicants, 28% accepted, 9 enrolled. Terminal master's awarded for partial completion of doctoral program. *Degree requirements:* For master's, final oral exam, practical data analysis experience, thesis optional; for doctorate, thesis/dissertation, final oral exam, comprehensive exam, registration. *Entrance requirements:* For master's, 3 semesters of course work in calculus (mathematics), course work in real analysis (preferred), course work in linear algebra; for doctorate, master's degree in statistics, biostatistics or equivalent. Additional exam requirements/recommendations for international students: Required—TOEFL (minimum score 640 paper-based; 250 computer-based). *Application deadline:* For fall admission, 4/1 priority date for domestic students, 4/1 priority date for international students. Application fee: $35. Electronic applications accepted. *Financial support:* In 2005–06, 2 fellowships (averaging $4,000 per year), 9 research assistantships with full tuition reimbursements (averaging $15,000 per year), 6 teaching assistantships with full tuition reimbursements (averaging $12,000 per year) were awarded; tuition waivers (partial) also available. Financial award application deadline: 2/1. *Faculty research:* Biostatistics, longitudinal data analysis, nonparametrics, statistical genetics, categorical data analysis. Total annual research expenditures: $200,000. *Unit head:* Dr. Alan D. Hutson, Chair and Associate Professor, 716-829-2594, Fax: 716-829-2200, E-mail: ahutson@buffalo.edu. *Application contact:* Dr. Randolph L. Carter, Associate Chair and Professor, 716-829-2884, Fax: 716-829-2200, E-mail: rcarter@buffalo.edu.

Tufts University, Sackler School of Graduate Biomedical Sciences, Division of Clinical Care Research, Medford, MA 02155. Offers MS, PhD. Part-time programs available. *Faculty:* 21 full-time (6 women). *Students:* 23 full-time (11 women), 2 part-time (1 woman); includes 5 minority (all Asian Americans or Pacific Islanders), 2 international. Average age 36. 10 applicants, 80% accepted, 7 enrolled. In 2005, 9 degrees awarded. Terminal master's awarded for partial completion of doctoral program. *Degree requirements:* For master's and doctorate, thesis/dissertation. *Entrance requirements:* For master's and doctorate, MD or PhD, strong clinical research background. Additional exam requirements/recommendations for international students: Required—TOEFL. *Application deadline:* For fall admission, 1/15 priority date for domestic students, 1/15 priority date for international students. Applications are processed on a rolling basis. Application fee: $65. Electronic applications accepted. *Financial support:* In 2005–06, 25 fellowships with full tuition reimbursements (averaging $44,000 per year) were awarded Financial award application deadline: 1/15. *Faculty research:* Clinical study design, mathematical modeling, meta analysis, epidemiologic research, coronary heart disease. *Unit head:* Dr. Harry P. Selker, Program Director, 617-636-5009, Fax: 617-636-8023, E-mail: hselker@lifespan.org. *Application contact:* 617-636-6767, Fax: 617-636-0375, E-mail: sackler-school@tufts.edu.

Tulane University, School of Public Health and Tropical Medicine, Department of Biostatistics, New Orleans, LA 70118-5669. Offers MS, MSPH, PhD, Sc D. MS and PhD offered through the Graduate School. Part-time programs available. *Degree requirements:* For doctorate, thesis/dissertation, comprehensive exam. *Entrance requirements:* For master's and doctorate, GRE General Test. Additional exam requirements/recommendations for international students: Required—TOEFL. Electronic applications accepted. *Faculty research:* Clinical trials, measurement, longitudinal analyses.

University at Albany, State University of New York, School of Public Health, Department of Epidemiology and Biostatistics, Albany, NY 12222-0001. Offers MS, PhD. *Students:* 13 full-time (11 women), 20 part-time (13 women). Average age 31. In 2005, 21 master's, 2 doctorates awarded. *Degree requirements:* For master's and doctorate, thesis/dissertation. *Entrance requirements:* For master's and doctorate, GRE General Test. Additional exam requirements/recommendations for international students: Required—TOEFL (minimum score 550 paper-based; 213 computer-based). *Application deadline:* For fall admission, 6/30 for domestic students, 5/1 for international students; for spring admission, 11/30 for domestic students, 11/1 for international students. Applications are processed on a rolling basis. Application fee: $60. Electronic applications accepted. *Financial support:* Application deadline: 4/1. *Unit head:* Dr. David Strogatz, Chair, 518-402-0400. *Application contact:* Nikki Malachowski, Assistant to the Chair, 518-402-0372.

The University of Alabama at Birmingham, School of Public Health, Department of Biostatistics, Birmingham, AL 35294. Offers biomathematics (MS, PhD); biostatistics (MS, PhD). *Students:* 22 full-time (8 women), 3 part-time (1 woman); includes 3 minority (all African Americans), 14 international. 53 applicants, 25% accepted. In 2005, 1 master's, 1 doctorate awarded. *Degree requirements:* For master's, variable foreign language requirement, thesis, fieldwork, research project; for doctorate, variable foreign language requirement, thesis/dissertation, comprehensive exam. *Entrance requirements:* For master's, GRE General Test or MAT, minimum GPA of 3.0; for doctorate, GRE General Test or MAT, MPH or MSPH, minimum GPA of 3.0, interview. *Application deadline:* Applications are processed on a rolling basis. Application fee: $35 ($60 for international students). Electronic applications accepted. *Expenses:* Contact institution. Tuition and fees vary according to course load, degree level and program. *Financial support:* Fellowships, career-related internships or fieldwork available. *Unit head:* Dr. George Howard, Chair, 205-934-4905, Fax: 205-975-2540, E-mail: ghoward@uab.edu. *Application contact:* Nancy O. Pinson, Coordinator of Student Admissions, 205-934-4993, Fax: 205-975-5484.

University of Alberta, Faculty of Graduate Studies and Research, Department of Mathematical and Statistical Sciences, Edmonton, AB T6G 2E1, Canada. Offers applied mathematics (M Sc, PhD); biostatistics (M Sc); mathematical finance (M Sc, PhD); mathematical physics (M Sc, PhD); mathematics (M Sc, PhD); statistics (M Sc, PhD, Postgraduate Diploma). Part-time programs available. *Faculty:* 48 full-time (4 women). *Students:* 112 full-time (41 women), 5 part-time. Average age 24. 776 applicants, 5% accepted, 34 enrolled. In 2005, 12 master's, 10 doctorates awarded. Terminal master's awarded for partial completion of doctoral program. *Median time to degree:* Of those who began their doctoral program in fall 1997, 100% received their degree in 8 years or less. *Degree requirements:* For master's, thesis (for some programs); for doctorate, thesis/dissertation, comprehensive exam. *Entrance requirements:* Additional exam requirements/recommendations for international students: Required—TOEFL (minimum score 580 paper-based; 237 computer-based). *Application deadline:* For fall admission, 3/1 for domestic students, 2/1 for international students. Applications are processed on a rolling basis. Application fee: $0. Electronic applications accepted. Tuition and fees charges are reported in Canadian dollars. *Expenses:* Tuition, state resident: part-time $562 Canadian dollars per term. Tuition, nonresident: full-time $3,375 Canadian dollars. Required fees: $573 Canadian dollars; $84 Canadian dollars per term. *Financial support:* In 2005–06, 51 research assistantships, 88 teaching assistantships with full and partial tuition reimbursements were awarded; scholarships/grants also available. Financial award application deadline:5/1. *Faculty research:* Classical and functional analysis, algebra, differential equations, geometry. *Unit head:* Dr. Anthony To-Ming Lau, Chair, 403-492-5141, E-mail: tlau@math.ualberta.ca. *Application contact:* Dr. Yau Shu Wong, Associate Chair, Graduate Studies, 403-492-5799, Fax: 403-492-6828, E-mail: gradstudies@math.ualberta.ca.

University of California, Berkeley, Graduate Division, School of Public Health, Division of Biostatistics, Group in Biostatistics, Berkeley, CA 94720-1500. Offers MA, PhD. *Degree requirements:* For master's, comprehensive exam; for doctorate, thesis/dissertation, qualifying exam. *Entrance requirements:* For master's and doctorate, GRE General Test, minimum GPA of 3.0. Additional exam requirements/recommendations for international students: Required—TOEFL. *Faculty research:* Applied statistics, risk research, clinical trials, nonparametrics.

University of California, Davis, Graduate Studies, Graduate Group in Biostatistics, Davis, CA 95616. Offers MS, PhD. *Faculty:* 25 full-time. *Students:* 19 full-time (12 women); includes 2 minority (1 Asian American or Pacific Islander, 1 Hispanic American), 13 international. Average age 29. 73 applicants, 19% accepted, 5 enrolled. In 2005, 1 degree awarded. *Degree requirements:* For master's, comprehensive exam; for doctorate, thesis/dissertation. *Entrance requirements:* Additional exam requirements/recommendations for international students: Required—TOEFL (minimum score 550 paper-based; 213 computer-based). *Application deadline:* For fall admission, 1/15 for domestic students, 1/15 for international students. Applications are processed on a rolling basis. Application fee: $60. Electronic applications accepted. *Financial support:* In 2005–06, 19 students received support, including 1 fellowship with full and partial tuition reimbursement available (averaging $7,284 per year), 3 research assistantships with full and partial tuition reimbursements available (averaging $12,336 per year), 6 teaching assistantships with partial tuition reimbursements available (averaging $15,082 per year); Federal Work-Study, institutionally sponsored loans, scholarships/grants, tuition waivers (full and partial), and unspecified assistantships also available. Financial award application deadline: 1/15. *Unit head:* Hans-Georg Mueller, Graduate Group Chair, 530-752-1629, E-mail: hgmueller@ucdavis.edu. *Application contact:* Laura Peterson, Administrative Assistant, 530-752-2632, Fax: 530-752-7099, E-mail: grad-staff@wald.ucdavis.edu.

University of California, Los Angeles, Graduate Division, School of Public Health, Department of Biostatistics, Los Angeles, CA 90095. Offers MS, PhD. *Degree requirements:* For master's, comprehensive exam; for doctorate, thesis/dissertation, oral and written qualifying exams. *Entrance requirements:* For master's, GRE General Test, minimum GPA of 3.0; for doctorate, GRE General Test, minimum undergraduate GPA of 3.0. Electronic applications accepted.

University of Cincinnati, Division of Research and Advanced Studies, College of Medicine, Graduate Programs in Biomedical Sciences, Department of Environmental Health, Cincinnati, OH 45267. Offers environmental and industrial hygiene (MS, PhD); environmental and occupational medicine (MS); environmental genetics and molecular toxicology (MS, PhD); epidemiology and biostatistics (MS, PhD); occupational safety and ergonomics (MS, PhD). *Accreditation:* ABET (one or more programs are accredited). Terminal master's awarded for partial completion of doctoral program. *Degree requirements:* For master's, thesis; for doctorate, thesis/dissertation, qualifying exam. *Entrance requirements:* For master's, GRE General Test, bachelor's degree in science; for doctorate, GRE General Test. Additional exam requirements/recommendations for international students: Required—TOEFL (minimum score 580 paper-based; 237 computer-based). Electronic applications accepted. *Faculty research:* Carcinogens and mutagenesis, pulmonary studies, reproduction and development.

University of Colorado at Denver and Health Sciences Center, Graduate School, Department of Preventive Medicine and Biometrics, Program in Biostatistics, Denver, CO 80262. Offers MS, PhD. *Students:* 29. 7 applicants, 57% accepted, 4 enrolled. In 2005, 2 degrees awarded. *Degree requirements:* For master's and doctorate, thesis/dissertation, comprehensive exam. *Entrance requirements:* For master's, GRE General Test, minimum GPA of 3.0, 2 semesters of course work in calculus; for doctorate, GRE General Test, minimum GPA of 3.0; 2 semesters of calculus; MS in biometrics, biostatistics, statistics or equivalent. Additional exam requirements/recommendations for international students: Required—TOEFL (minimum score 550 paper-based; 213 computer-based). *Application deadline:* For fall admission, 2/1 for domestic students. Application fee: $50. *Expenses:* Tuition, state resident: full-time $11,730. Tuition, nonresident: full-time $22,980. Tuition and fees vary according to degree level and program. *Financial support:* Application deadline: 3/1; *Faculty research:* Health policy research, nonlinear mixed effects models for longitudinal data, statistical methods in nutrition, clinical trials. *Unit head:* Dr. Gary Grunwald, Director, 303-315-0115, E-mail: gary.grunwald@uchsc.edu. *Application contact:* Fayette Augillard, Program Coordinator, 303-315-7605, Fax: 303-315-1010, E-mail: fayette.augillard@uchsc.edu.

University of Florida, Graduate School, College of Public Health and Health Professions and College of Medicine, Program in Public Health, Gainesville, FL 32611. Offers biostatistics (MPH); environmental health (MPH); epidemiology (MPH); public health management and policy (MPH); social and behavioral sciences (MPH). *Faculty:* 10. *Entrance requirements:* For master's, GRE General Test, minimum GPA of 3.0. Additional exam requirements/recommendations for international students: Required—TOEFL (minimum score 550 paper-based; 213 computer-based). Application fee: $30. *Expenses:* Tuition, state resident: full-time $6,234. Tuition, nonresident: full-time $21,359. Tuition and fees vary according to program. *Unit head:* Mary Peoples Sheps, Director, 352-273-6084, Fax: 352-273-6448, E-mail: mpeoplessheps@phhp.ufl.edu. *Application contact:* Brigette Hart, Program Assistant, 352-273-6443, E-mail: bhart@phhp.ufl.edu.

University of Illinois at Chicago, Graduate College, School of Public Health, Biostatistics Section, Chicago, IL 60607-7128. Offers MS, PhD. Part-time programs available. Terminal master's awarded for partial completion of doctoral program. *Degree requirements:* For master's, thesis, field practicum; for doctorate, thesis/dissertation, independent research, internship. *Entrance requirements:* For master's and doctorate, GRE General Test, minimum GPA of 2.75. Additional exam requirements/recommendations for international students: Required—TOEFL. Electronic applications accepted.

The University of Iowa, Graduate College, College of Public Health, Department of Biostatistics, Iowa City, IA 52242-1316. Offers MS, PhD. *Faculty:* 11 full-time, 2 part-time/adjunct. *Students:* 21 full-time (14 women), 13 part-time (5 women), 26 international. 48 applicants, 29% accepted, 4 enrolled. In 2005, 8 master's, 4 doctorates awarded. *Degree requirements:* For master's, exam, thesis optional; for doctorate, thesis/dissertation, comprehensive exam, registration. *Entrance requirements:* For master's and doctorate, GRE General Test, minimum GPA of 3.0. Additional exam requirements/recommendations for international students: Required—TOEFL (minimum score 550 paper-based; 213 computer-based). *Application deadline:* For fall admission, 3/15 priority date for domestic students, 3/15 priority date for international students; for spring admission, 10/1 priority date for domestic students. Applications are processed on a rolling basis. Application fee: $60 ($85 for international students). Electronic applications accepted. *Financial support:* In 2005–06, 23 research assistantships with partial tuition reimbursements, 4 teaching assistantships with partial tuition reimbursements were awarded; fellowships Financial award applicants required to submit FAFSA. *Unit head:* Kathryn Chaloner, Head, 319-384-5029, Fax: 319-384-5018.

University of Louisville, Graduate School, School of Public Health, Program in Bioinformatics and Biostatistics, Louisville, KY 40292-0001. Offers MS, PhD. *Students:* 16 full-time (5 women), 4 part-time (2 women), 10 international. Average age 32. In 2005, 1 master's, 1 doctorate awarded. *Entrance requirements:* For master's and doctorate, 2 letters of recommendation. Application fee: $50. *Expenses:* Tuition, state resident: full-time $3,003; part-time $334 per credit hour. Tuition, nonresident: full-time $8,277; part-time $920 per credit hour. Tuition and fees vary according to course load, degree level and program. *Unit head:* Dr. Rudolph Parrish, Head, 502-852-2797, Fax: 502-852-3294, E-mail: rsparr01@louisville.edu.

University of Medicine and Dentistry of New Jersey, School of Public Health, Piscataway Program in Public Health, Piscataway, NJ 08854. Offers biostatistics (MS); epidemiology (Certificate); general public health (Certificate); public health (MPH, Dr PH, PhD). *Degree requirements:* For master's, internship; for doctorate, thesis/dissertation. *Entrance requirements:* For master's, GRE General Test; for doctorate, GRE General Test, MPH (Dr PH); MA, MPH, or MS (PhD). Additional exam requirements/recommendations for international students: Required—TOEFL. *Application deadline:* For fall admission, 3/15 for domestic students; for spring admission, 11/1 for domestic students. Application fee: $50. *Unit head:* Dr. George G. Rhoads, Associate Dean, 732-235-4646, E-mail: rhoads@umdnj.edu. *Application contact:* Dr. Mark G. Robson, Assistant Dean, 732-235-5405, E-mail: robson@eohsi.rutgers.edu.

University of Michigan, School of Public Health, Department of Biostatistics, Ann Arbor, MI 48109. Offers MPH, MS, PhD. MS and PhD offered through the Horace H. Rackham School of Graduate Studies. *Faculty:* 15 full-time (3 women), 11 part-time/adjunct (5 women). *Students:* 89 full-time (58 women), 2 part-time (1 woman); includes 15 minority (1 African American, 11 Asian Americans or Pacific Islanders, 3 Hispanic Americans), 51 international. Average age 27. 244 applicants, 53% accepted, 37 enrolled. In 2005, 20 master's, 7 doctorates awarded. Terminal master's awarded for partial completion of doctoral program. *Degree requirements:* For doctorate, oral defense of dissertation, preliminary exam. *Entrance requirements:* For master's, GRE General Test; for doctorate, GRE General Test, master's degree. Additional exam requirements/recommendations for international students: Required—TOEFL (minimum score 560 paper-based; 220 computer-based). *Application deadline:* For fall admission, 2/1 priority date for domestic students, 2/1 priority date for international students. Applications are processed on a rolling basis. Application fee: $60 ($75 for international students). Electronic applications accepted. *Expenses:* Tuition, state resident: full-time $14,082; part-time $894 per credit hour. Tuition, nonresident: full-time $28,500; part-time $1,675 per credit hour. Required fees: $189; $189 per unit. *Financial support:* In 2005–06, 67 students received support, including 11 fellowships with full tuition reimbursements available (averaging $17,471 per year), 54 research assistantships with full tuition reimbursements available (averaging $17,471 per year), 10 teaching assistantships with full tuition reimbursements available (averaging $17,471 per year); scholarships/grants, traineeships, and tuition waivers (partial) also available. Financial award application deadline: 2/1. *Faculty research:* Statistical genetics, categorical data analysis, incomplete data, survival analysis, modeling. Total annual research expenditures: $9.8 million. *Unit head:* Dr. John David Kalbfleisch, Chair, 734-615-7067, Fax: 734-763-2215, E-mail: jdkalbfl@umich.edu. *Application contact:* Heonia Hillock, Student Services Coordinator, 734-615-9812, Fax: 734-763-2215, E-mail: sph.bio.inquiries@umich.edu.

University of Michigan, School of Public Health, Interdepartmental Program in Clinical Research Design and Statistical Analysis, Ann Arbor, MI 48109. Offers MS. Offered through the Horace H. Rackham School of Graduate Studies; program admits applicants in odd calendar years. Evening/weekend programs available. *Faculty:* 11 full-time (4 women), 1 part-time/adjunct (0 women). *Students:* 40 full-time; includes 8 minority (1 African American, 7 Asian Americans or Pacific Islanders), 4 international. Average age 33. *Degree requirements:* For master's, comprehensive exam, registration. *Entrance requirements:* For master's, GRE General Test or MCAT. Additional exam requirements/recommendations for international students: Recommended—TOEFL (minimum score 560 paper-based; 220 computer-based). *Application deadline:* For fall admission, 2/1 priority date for domestic students, 2/1 priority date for international students. Applications are processed on a rolling basis. Application fee: $60 ($75 for international students). Electronic applications accepted. *Expenses: Contact institution. Financial support:* In 2005–06, 1 student received support. Institutionally sponsored loans and scholarships/grants available. Financial award application deadline: 3/15; financial award applicants required to submit FAFSA. *Faculty research:* Survival analysis, missing data, bayesian inference, health economics, quality of life. Total annual research expenditures: $9.8 million. *Unit head:* Dr. Roderick Little, Director, 734-936-1009. *Application contact:* Amanda Ring, Information Contact, 734-615-9817, Fax: 734-763-2215, E-mail: sph.bio.inquiries@umich.edu.

University of Minnesota, Twin Cities Campus, School of Public Health, Major in Biostatistics, Minneapolis, MN 55455-0213. Offers MPH, MS, PhD. Part-time programs available. *Faculty:* 23 full-time (9 women), 2 part-time/adjunct (1 woman). *Students:* 47 full-time (32 women), 12 part-time (8 women); includes 2 minority (both Asian Americans or Pacific Islanders), 42 international. 112 applicants, 38% accepted, 23 enrolled. In 2005, 17 master's, 4 doctorates awarded. Terminal master's awarded for partial completion of doctoral program. *Degree requirements:* For master's, comprehensive exam; for doctorate, thesis/dissertation, comprehensive exam. *Entrance requirements:* For master's, GRE General Test, course work in applied statistics, computer programming, multivariable calculus, linear algebra; for doctorate, GRE General Test, bachelor's or master's degree in statistics, biostatistics or mathematics. Additional exam requirements/recommendations for international students: Required—TOEFL (minimum score 600 paper-based; 250 computer-based). *Application deadline:* For fall admission, 12/31 priority date for domestic students, 12/31 priority date for international students. Applications are processed on a rolling basis. Application fee: $55 ($75 for international students). Electronic applications accepted. *Expenses:* Tuition, state resident: full-time $8,748; part-time $729 per credit. Tuition, nonresident: full-time $15,848; part-time $1,321 per credit. Full-time tuition and fees vary according to class time, course load, program and reciprocity agreements. *Financial support:* In 2005–06, research assistantships with partial tuition reimbursements (averaging $17,815 per year), teaching assistantships with partial tuition reimbursements (averaging $17,815 per year) were awarded; fellowships with partial tuition reimbursements, institutionally sponsored loans and traineeships also available. *Faculty research:* Analysis of spatial and longitudinal data, Bayes/Empirical Bayes methods, survival analysis, longitudinal models, generalized linear models. *Unit head:* Dr. John E. Connett, Division Head, 612-624-4655, Fax: 612-626-0660, E-mail: admissions@biostat.umn.edu. *Application contact:* Sally Olander, Coordinator, 612-625-9185, Fax: 612-624-0660, E-mail: sally@biostat.umn.edu.

The University of North Carolina at Chapel Hill, Graduate School, School of Public Health, Department of Biostatistics, Chapel Hill, NC 27599. Offers MPH, MS, Dr PH, PhD. Part-time programs available. *Faculty:* 37 full-time (10 women), 29 part-time/adjunct. *Students:* 113 full-time (59 women); includes 64 minority (9 African Americans, 1 American Indian/Alaska Native, 49 Asian Americans or Pacific Islanders, 5 Hispanic Americans). Average age 27. 216 applicants, 37% accepted, 29 enrolled. In 2005, 14 master's, 11 doctorates awarded. *Median time to degree:* Of those who began their doctoral program in fall 1997, 100% received their degree in 8 years or less. *Degree requirements:* For master's, thesis, major paper, comprehensive exam, registration; for doctorate, thesis/dissertation, comprehensive exam, registration. *Entrance requirements:* For master's and doctorate, GRE General Test, minimum GPA of 3.0. Additional exam requirements/recommendations for international students: Required—TOEFL. *Application deadline:* For fall admission, 1/1 priority date for domestic students, 1/1 priority date for international students. Applications are processed on a rolling basis. Application fee: $70. Electronic applications accepted. *Financial support:* In 2005–06, 89 students received support, including 32 fellowships with full tuition reimbursements available (averaging $7,332 per year), 56 research assistantships with full tuition reimbursements available (averaging $7,332 per year), 1 teaching assistantship (averaging $3,237 per year); Federal Work-Study, institutionally sponsored loans, traineeships, health care benefits, and unspecified assistantships also available. Financial award application deadline: 1/1; financial award applicants required to submit FAFSA. *Faculty research:* Cancer, cardiovascular, environmental biostatistics; AIDS and other infectious diseases; statistical genetics; demography and population studies. Total annual research expenditures: $10.4 million. *Unit head:* Dr. Michael R. Kosorok, Chair, 919-966-7254, Fax: 919-966-3804. *Application contact:* Melissa Hobgood, Registrar, 919-966-7256, Fax: 919-966-3804, E-mail: hobgood@unc.edu.

University of North Texas Health Science Center at Fort Worth, School of Public Health, Fort Worth, TX 76107-2699. Offers biostatistics (MPH); community health (MPH); disease control and prevention (Dr PH); environmental health (MPH); epidemiology (MPH); health behavior (MPH); health policy and management (MPH, Dr PH). *Accreditation:* CEPH. Part-time and evening/weekend programs available. *Faculty:* 26 full-time (6 women). *Students:* 134 full-time (89 women), 92 part-time (63 women); includes 94 minority (39 African Americans, 3 American Indian/Alaska Native, 20 Asian Americans or Pacific Islanders, 32 Hispanic Americans), 39 international. Average age 33. 235 applicants, 37% accepted, 43 enrolled. In 2005, 61 master's, 6 doctorates awarded. *Degree requirements:* For master's, thesis or alternative, supervised internship; for doctorate, thesis/dissertation, supervised internship. *Entrance requirements:* For master's, GRE General Test. Additional exam requirements/recommendations for international students: Required—TOEFL. *Application deadline:* For fall admission, 6/1 for domestic students. Applications are processed on a rolling basis. Application fee: $25 ($50 for international students). Electronic applications accepted. *Financial support:* In 2005–06, 9 research assistantships with partial tuition reimbursements (averaging $16,000 per year), 2 teaching assistantships (averaging $8,100 per year) were awarded; fellowships, Federal Work-Study, institutionally sponsored loans, and scholarships/grants also available. Support available to part-time students. Financial award application deadline: 4/1; financial award applicants required to submit FAFSA. *Unit head:* Dr. Fernando Treviño, Dean, 817-735-2401, Fax: 817-735-0243, E-mail: sph@hsc.unt.edu. *Application contact:* Thomas Moorman, Director of Student Affairs, 817-735-0302, Fax: 817-735-0324, E-mail: tmoorman@hsc.unt.edu.

University of Oklahoma Health Sciences Center, Graduate College, College of Public Health, Program in Biostatistics and Epidemiology, Oklahoma City, OK 73190. Offers biostatistics (MPH, MS, Dr PH, PhD); epidemiology (MPH, MS, Dr PH, PhD). *Accreditation:* CEPH (one or more programs are accredited). Part-time programs available. *Degree requirements:* For master's, thesis (for some programs), comprehensive exam; for doctorate, thesis/dissertation, comprehensive exam. *Entrance requirements:* For master's, 3 letters of recommendation, resumé; for doctorate, GRE General Test, letters of recommendation. Additional exam requirements/recommendations for international students: Required—TOEFL (minimum score 570 paper-based; 230 computer-based), TWE. *Faculty research:* Statistical methodology, applied statistics, acute and chronic disease epidemiology.

University of Pennsylvania, School of Medicine, Biomedical Graduate Studies, Graduate Group in Epidemiology and Biostatistics, Philadelphia, PA 19104. Offers biostatistics (MS, PhD). Part-time programs available. *Faculty:* 60 full-time (25 women), 111 part-time/adjunct (38 women). *Students:* 22 full-time (12 women), 12 part-time (7 women); includes 20 minority (1 African American, 19 Asian Americans or Pacific Islanders). Average age 25. 38 applicants, 61% accepted, 10 enrolled. In 2005, 2 master's, 1 doctorate awarded. Terminal master's awarded for partial completion of doctoral program. *Degree requirements:* For master's,

Biostatistics

University of Pennsylvania (continued)
thesis, evaluations examination; for doctorate, thesis/dissertation, evaluations exam, preliminary exam. *Entrance requirements:* For master's and doctorate, GRE, 1 year of course work in calculus, 1 semester of course work in linear algebra, working knowledge of programming language. Additional exam requirements/recommendations for international students: Required—TOEFL. *Application deadline:* For fall admission, 12/15 for domestic students, 12/1 for international students. Application fee: $70. *Financial support:* In 2005–06, 22 students received support, including 10 fellowships with full and partial tuition reimbursements available (averaging $23,000 per year), 11 research assistantships with full and partial tuition reimbursements available (averaging $23,000 per year), 1 teaching assistantship with full and partial tuition reimbursement available (averaging $23,000 per year); career-related internships or fieldwork, institutionally sponsored loans, scholarships/grants, traineeships, health care benefits, unspecified assistantships, and faculty/staff benefits provide partial tuition coverage also available. Financial award application deadline: 12/13. *Faculty research:* Randomized clinical trials, data coordinating centers, methodological approaches to non-experimental epidemiologic studies, theoretical research in biostatistics. Total annual research expenditures: $22.3 million. *Unit head:* Dr. Brian L. Strom, Chair, 215-898-2368, Fax: 215-573-5315, E-mail: bstrom@cceb.med.upenn.edu. *Application contact:* Anne Facciolo, Program Manager, 215-573-3881, Fax: 215-573-4865, E-mail: afacciol@cceb.med.upenn.edu.

University of Pittsburgh, Graduate School of Public Health, Department of Biostatistics, Pittsburgh, PA 15260. Offers MPH, MS, Dr PH, PhD. Part-time programs available. *Faculty:* 27 full-time (9 women), 3 part-time/adjunct (1 woman). *Students:* 58 full-time (39 women), 24 part-time (16 women); includes 9 minority (3 African Americans, 5 Asian Americans or Pacific Islanders, 1 Hispanic American), 51 international. Average age 30. 129 applicants, 84% accepted, 26 enrolled. In 2005, 8 master's, 3 doctorates awarded. Terminal master's awarded for partial completion of doctoral program. *Degree requirements:* For master's, thesis; for doctorate, one foreign language, thesis/dissertation. *Entrance requirements:* For master's and doctorate, GRE General Test, previous course work in biology, calculus, and Fortran. Additional exam requirements/recommendations for international students: Required—TOEFL (minimum score 550 paper-based; 213 computer-based). *Application deadline:* For fall admission, 3/30 priority date for domestic students, 3/1 priority date for international students; for spring admission, 11/30 for domestic students, 5/1 for international students. Applications are processed on a rolling basis. Application fee: $50 ($60 for international students). Electronic applications accepted. *Expenses:* Tuition, state resident: full-time $13,194; part-time $537 per credit. Tuition, nonresident: full-time $25,012; part-time $1,026 per credit. Required fees: $700; $164 per term. Tuition and fees vary according to campus/location and program. *Financial support:* In 2005–06, 39 students received support, including 35 research assistantships with tuition reimbursements available (averaging $20,333 per year), 4 teaching assistantships with tuition reimbursements available (averaging $21,150 per year); career-related internships or fieldwork and tuition waivers (partial) also available. Support available to part-time students. Financial award application deadline: 2/28; financial award applicants required to submit FAFSA. *Faculty research:* Survival analysis, environmental risk assessment, statistical computing, longitudinal data analysis, experimental design. Total annual research expenditures: $3.7 million. *Unit head:* Dr. Howard E. Rockette, Chairperson, 412-624-3022, Fax: 412-624-2183, E-mail: herbst@pitt.edu. *Application contact:* Dr. Lisa Weissfeld, Professor, 412-624-3023, Fax: 412-624-2183, E-mail: lweis@pitt.edu.

University of Puerto Rico, Medical Sciences Campus, Graduate School of Public Health, Department of Biostatistics and Epidemiology, Program in Biostatistics, San Juan, PR 00936-5067. Offers MPH. Part-time programs available. *Students:* 10 (7 women). 27 applicants, 44% accepted. In 2005, 13 degrees awarded. *Entrance requirements:* For master's, GRE, previous course work in algebra. *Application deadline:* For fall admission, 3/15 for domestic students. Application fee: $20. *Expenses:* Contact institution. *Financial support:* Research assistantships, teaching assistantships, career-related internships or fieldwork, Federal Work-Study, and institutionally sponsored loans available. Financial award application deadline: 4/30. *Unit head:* Prof. Erick Suarez, Coordinator, 787-758-2525 Ext. 1428, Fax: 787-759-6719, E-mail: esuarez@rcm.upr.edu. *Application contact:* Prof. Mayra E. Santiago-Vargas, Counselor, 787-756-5244, Fax: 787-759-6719, E-mail: msantiago@rcm.upr.edu.

University of Rochester, School of Medicine and Dentistry, Graduate Programs in Medicine and Dentistry, Department of Biostatistics and Computational Biology, Rochester, NY 14627-0250. Offers medical statistics (MS); statistics (MA, PhD). Terminal master's awarded for partial completion of doctoral program. *Degree requirements:* For doctorate, thesis/dissertation, qualifying exam. *Entrance requirements:* For master's and doctorate, GRE General Test. Additional exam requirements/recommendations for international students: Required—TOEFL.

University of South Carolina, The Graduate School, Arnold School of Public Health, Department of Epidemiology/Biostatistics, Program in Biostatistics, Columbia, SC 29208. Offers MPH, MSPH, Dr PH, PhD. Part-time programs available. *Degree requirements:* For master's, thesis (for some programs), practicum (MPH), comprehensive exam; for doctorate, thesis/dissertation, comprehensive exam. *Entrance requirements:* For master's and doctorate, GRE General Test. Additional exam requirements/recommendations for international students: Required—TOEFL (minimum score 570 paper-based; 230 computer-based). Electronic applications accepted. *Faculty research:* Bayesian methods, biometric modeling, nonlinear regression, health survey methodology, measurement of health status.

University of Southern California, Keck School of Medicine and Graduate School, Graduate Programs in Medicine, Department of Preventive Medicine, Division of Biostatistics, Los Angeles, CA 90089. Offers applied biostatistics/epidemiology (MS); biostatistics (MS, PhD); epidemiology (PhD); genetic epidemiology and statistical genetics (PhD); molecular epidemiology (MS, PhD). *Faculty:* 74 full-time (33 women). *Students:* 97 full-time (63 women); includes 25 minority (18 Asian Americans or Pacific Islanders, 7 Hispanic Americans), 48 international. Average age 30. 109 applicants, 70% accepted, 33 enrolled. In 2005, 17 master's, 10 doctorates awarded. Terminal master's awarded for partial completion of doctoral program. *Median time to degree:* Of those who began their doctoral program in fall 1997, 100% received their degree in 8 years or less. *Degree requirements:* For master's and doctorate, thesis/dissertation. *Entrance requirements:* For master's, GRE General Test, GRE Subject Test, minimum GPA of 3.0; for doctorate, GRE General Test, GRE Subject Test, minimum GPA of 3.5. Additional exam requirements/recommendations for international students: Required—TOEFL (minimum score 550 paper-based; 200 computer-based). *Application deadline:* For fall admission, 1/15 for domestic students. Applications are processed on a rolling basis. Application fee: $65 ($75 for international students). Electronic applications accepted. *Expenses:* Tuition: Full-time $25,416; part-time $1,059 per unit. Required fees: $484; $484 per year. Tuition and fees vary according to course load and program. *Financial support:* In 2005–06, 3 fellowships with tuition reimbursements (averaging $24,876 per year), 39 research assistantships with tuition reimbursements (averaging $24,876 per year), 15 teaching assistantships with tuition reimbursements (averaging $24,876 per year) were awarded; career-related internships or fieldwork, Federal Work-Study, institutionally sponsored loans, scholarships/grants, and tuition waivers (partial) also available. Financial award application deadline: 4/1. *Faculty research:* Clinical trials in ophthalmology and cancer research, methods of analysis for epidemiological studies, genetic epidemiology. Total annual research expenditures: $1.3 million. *Unit head:* Dr. Stanley P. Azen, Co-Director, 323-442-1810, Fax: 323-442-2993, E-mail: mtrujill@usc.edu. *Application contact:* Mary L. Trujillo, Student Adviser, 323-442-1810, Fax: 323-442-2993, E-mail: mtrujill@usc.edu.

See Close-Up on page 543.

University of South Florida, College of Graduate Studies, College of Public Health, Department of Epidemiology and Biostatistics, Tampa, FL 33620-9951. Offers MPH, MSPH, PhD. *Accreditation:* CEPH (one or more programs are accredited). Part-time and evening/weekend programs available. *Degree requirements:* For master's and doctorate, thesis/dissertation. *Entrance requirements:* For master's and doctorate, GRE General Test, minimum GPA of 3.0 in upper-level course work. Additional exam requirements/recommendations for international students: Required—TOEFL (minimum score 550 paper-based; 213 computer-based). *Faculty research:* Dementia, mental illness, mental health preventative trails, rural health outreach, clinical and administrative studies.

The University of Texas Health Science Center at Houston, Graduate School of Biomedical Sciences, Program in Biomathematics and Biostatistics, Houston, TX 77225-0036. Offers MS, PhD, MD/PhD. *Faculty:* 31 full-time (6 women). *Students:* 4 full-time (1 woman), 2 international. Average age 25. 37 applicants, 30% accepted, 7 enrolled. In 2005, 3 degrees awarded. Terminal master's awarded for partial completion of doctoral program. *Degree requirements:* For master's and doctorate, thesis/dissertation. *Entrance requirements:* For master's and doctorate, GRE General Test. Additional exam requirements/recommendations for international students: Required—TOEFL, TWE. *Application deadline:* For fall admission, 1/15 for domestic students; for spring admission, 11/1 for domestic students. Applications are processed on a rolling basis. Application fee: $10. Electronic applications accepted. *Financial support:* Fellowships with full tuition reimbursements, research assistantships with full tuition reimbursements, teaching assistantships, institutionally sponsored loans, scholarships/grants, and health care benefits available. Financial award application deadline: 1/15. *Faculty research:* Statistical and mathematical modeling, development of new models for design and analysis of research studies, formulation of mathematical models of biological systems. *Unit head:* Dr. Li Zhang, Director, 713-563-4298, Fax: 713-563-4243, E-mail: zhangli@mdanderson.org. *Application contact:* Dr. Victoria P. Knutson, Assistant Dean of Admissions, 713-500-9860, Fax: 713-500-9877, E-mail: victoria.p.knutson@uth.tmc.edu.

University of Utah, School of Medicine and The Graduate School, Graduate Programs in Medicine, Programs in Public Health, Salt Lake City, UT 84112-1107. Offers biostatistics (M Stat); public health (MPH, MSPH, PhD). *Accreditation:* CEPH (one or more programs are accredited). Part-time programs available. *Degree requirements:* For master's, thesis (for some programs), thesis or project (MSPH), comprehensive exam, registration. *Entrance requirements:* For master's, GRE General Test, interview, minimum GPA of 3.0. Electronic applications accepted. *Expenses:* Tuition, state resident: full-time $2,932; part-time $2,212 per term. Tuition, nonresident: full-time $10,350; part-time $7,812 per term. Required fees: $590; $516 per term. Tuition and fees vary according to course load and program. *Faculty research:* Health services research, occupational and environmental exposures to toxic substances, risk assessment, health policy, epidemiology of chronic disease.

University of Vermont, Graduate College, College of Engineering and Mathematics, Department of Mathematics and Statistics, Program in Biostatistics, Burlington, VT 05405. Offers MS. *Students:* 6 (1 woman); includes 1 minority (American Indian/Alaska Native) 1 international. 11 applicants, 64% accepted, 3 enrolled. In 2005, 2 degrees awarded. *Degree requirements:* For master's, thesis or alternative. *Entrance requirements:* Additional exam requirements/recommendations for international students: Required—TOEFL (minimum score 550 paper-based; 213 computer-based). *Application deadline:* For fall admission, 4/1 for domestic students. Applications are processed on a rolling basis. Application fee: $40. *Expenses:* Tuition, area resident: Part-time $410 per credit hour. Tuition, nonresident: part-time $1,034 per credit hour. *Financial support:* Fellowships, research assistantships, teaching assistantships available. Financial award application deadline: 3/1. *Unit head:* Dr. Ruth Mickey, Coordinatoar, 802-656-2940.

University of Washington, Graduate School, Interdisciplinary Graduate Program in Quantitative Ecology and Resource Management, Seattle, WA 98195. Offers MS, PhD. *Degree requirements:* For master's and doctorate, thesis/dissertation. *Entrance requirements:* For master's and doctorate, GRE General Test, minimum GPA of 3.0. Additional exam requirements/recommendations for international students: Required—TOEFL. Electronic applications accepted. *Faculty research:* Population dynamics, statistical analysis, ecological modeling and systems analysis of aquatic and terrestrial ecosystems.

University of Washington, Graduate School, School of Public Health and Community Medicine, Department of Biostatistics, Seattle, WA 98195. Offers biostatistics (MPH, MS, PhD); statistical genetics (PhD). *Faculty:* 76 full-time (34 women), 4 part-time/adjunct (1 woman). *Students:* 56 full-time (28 women), 10 part-time (5 women); includes 7 minority (1 African American, 2 American Indian/Alaska Native, 4 Asian Americans or Pacific Islanders), 24 international. Average age 24. 87 applicants, 34% accepted, 14 enrolled. In 2005, 6 master's, 8 doctorates awarded. Terminal master's awarded for partial completion of doctoral program. *Median time to degree:* Of those who began their doctoral program in fall 1997, 100% received their degree in 8 years or less. *Degree requirements:* For master's, thesis/dissertation, departmental qualifying exams; for doctorate, thesis/dissertation, departmental qualifying exams, comprehensive exam, registration. *Entrance requirements:* For master's, GRE General Test, 2 years of course work in advanced calculus, 1 course in linear algebra, 1 course in mathematical probability, minimum GPA of 3.0; for doctorate, GRE General Test, 2 years of course work in advanced calculus, 1 course in linear algebra, 1 course in mathematical probability, minimum GPA of 3.00. Additional exam requirements/recommendations for international students: Required—TOEFL, TSE. *Application deadline:* For fall admission, 1/4 for domestic students. Application fee: $50. Electronic applications accepted. *Financial support:* In 2005–06, 64 students received support, including 1 fellowship with partial tuition reimbursement available (averaging $19,000 per year), 36 research assistantships with full tuition reimbursements available (averaging $16,000 per year), 5 teaching assistantships with full tuition reimbursements available (averaging $16,000 per year); career-related internships or fieldwork, traineeships, and unspecified assistantships also available. Financial award application deadline: 1/4. *Faculty research:* Statistical methods for survival data analysis, clinical trials, epidemiological case control and cohort studies, statistical genetics. Total annual research expenditures: $11.5 million. *Unit head:* Dr. Bruce Weir, Chair, 206-543-1044. *Application contact:* Alex Mackenzie, Counseling Services Coordinator, 206-543-1044, Fax: 206-543-3286, E-mail: alexam@u.washington.edu.

University of Waterloo, Graduate Studies, Faculty of Mathematics, Department of Statistics and Actuarial Science, Waterloo, ON N2L 3G1, Canada. Offers actuarial science (M Math); statistics (M Math, PhD); statistics-biostatistics (M Math); statistics-computing (M Math); statistics-finance (M Math). *Faculty:* 33 full-time (6 women), 21 part-time/adjunct (5 women). *Students:* 93 full-time (50 women), 5 part-time. 308 applicants, 24% accepted, 34 enrolled. In 2005, 18 master's, 3 doctorates awarded. *Degree requirements:* For master's, research paper or thesis; for doctorate, thesis/dissertation. *Entrance requirements:* For master's, honors degree in field, minimum B+ average; for doctorate, master's degree, minimum B+ average. Additional exam requirements/recommendations for international students: Required—TOEFL, TWE. *Application deadline:* For fall admission, 3/31 for domestic students. For winter admission, 8/31 for domestic students; for spring admission, 12/31 for domestic students. Applications are processed on a rolling basis. Application fee: $75 Canadian dollars. Electronic applications accepted. *Financial support:* In 2005–06, 30 teaching assistantships were awarded; fellowships, research assistantships, career-related internships or fieldwork and scholarships/grants also available. *Faculty research:* Data analysis, risk theory, inference, stochastic processes, quantitative finance. *Unit head:* Dr. D. E. Matthews, Chair, 519-888-4567 Ext. 5530, Fax: 519-746-1875, E-mail: dematthe@uwaterloo.ca. *Application contact:* M. Dufton, Graduate Studies Secretary, 519-888-4567 Ext. 6532, Fax: 519-746-1875, E-mail: mdufton@math.uwaterloo.ca.

The University of Western Ontario, Faculty of Graduate Studies, Biosciences Division, Department of Epidemiology and Biostatistics, London, ON N6A 5B8, Canada. Offers M Sc, PhD. *Accreditation:* CEPH (one or more programs are accredited). Part-time programs available. *Degree requirements:* For master's, thesis; for doctorate, thesis proposal defense. *Entrance requirements:* For master's, BA or B Sc honors degree, minimum B+ average in last 10 courses; for doctorate, M Sc or equivalent, minimum B+ average in last 10 courses. *Faculty research:* Chronic disease epidemiology, clinical epidemiology.

Virginia Commonwealth University, Medical College of Virginia-Professional Programs, School of Medicine and Graduate Programs, School of Medicine Graduate Programs, Department of Biostatistics, Richmond, VA 23284-9005. Offers MS, PhD, MD/PhD. Part-time programs available. *Faculty:* 10 full-time (2 women). *Students:* 10 full-time (6 women), 20 part-time (11 women); includes 8 minority (6 African Americans, 2 Asian Americans or Pacific Islanders), 6 international. 50 applicants, 30% accepted. In 2005, 3 master's, 5 doctorates awarded. Terminal master's awarded for partial completion of doctoral program. *Degree requirements:* For master's, thesis; for doctorate, thesis/dissertation, comprehensive oral and written exams. *Entrance requirements:* For master's, DAT, GRE General Test, or MCAT; for doctorate, GRE General Test, MCAT, DAT. *Application deadline:* For fall admission, 2/15 for domestic students. Application fee: $50. *Expenses:* Tuition, state resident: full-time $3,185; part-time $405 per credit. Tuition, nonresident: full-time $7,952; part-time $940 per credit. Required fees: $751 per semester hour. Tuition and fees vary according to course load and program. *Financial support:* Fellowships, teaching assistantships, career-related internships or fieldwork available. *Faculty research:* Health services, linear models, response surfaces, design and analysis of drug/chemical combinations, clinical trials. *Unit head:* Dr. W. Hans Carter, Chair, 804-827-2042, Fax: 804-828-8900, E-mail: whcarter@vcu.edu. *Application contact:* Dr. Ronald K. Elswick, Director, 804-827-2037, E-mail: rkelswic@vcu.edu.

Western Michigan University, Graduate College, College of Arts and Sciences, Department of Statistics, Program in Biostatistics, Kalamazoo, MI 49008-5202. Offers MS. *Degree requirements:* For master's, written exams, internship.

Yale University, School of Medicine, School of Public Health, Division of Biostatistics, New Haven, CT 06520. Offers MPH, MS, PhD. MS and PhD offered through the Graduate School. Part-time programs available. Terminal master's awarded for partial completion of doctoral program. *Degree requirements:* For master's, thesis, internship. *Entrance requirements:* For master's, GMAT, GRE, or MCAT, previous undergraduate course work in mathematics and science. Additional exam requirements/recommendations for international students: Required—TOEFL. Electronic applications accepted. *Faculty research:* Statistical and genetic epidemiology, population models for chronic and infectious diseases, clinical trials, regression methods.

Computational Sciences

Arizona State University, Division of Graduate Studies, College of Liberal Arts and Sciences, Department of Biology, Program in Computational, Statistical, and Mathematical Biology, Tempe, AZ 85287. Offers MS, PhD. *Entrance requirements:* Additional exam requirements/recommendations for international students: Required—TOEFL (minimum score 600 paper-based); Recommended—TSE.

California Institute of Technology, Division of Engineering and Applied Science, Option in Computation and Neural Systems, Pasadena, CA 91125-0001. Offers MS, PhD. *Faculty:* 3 full-time (0 women). *Students:* 45 full-time (11 women); includes 4 minority (1 African American, 1 Asian American or Pacific Islander, 2 Hispanic Americans), 19 international. 113 applicants, 13% accepted, 8 enrolled. In 2005, 1 master's, 7 doctorates awarded. Terminal master's awarded for partial completion of doctoral program. *Degree requirements:* For doctorate, thesis/dissertation, qualifying exam. *Entrance requirements:* For doctorate, GRE General Test. *Application deadline:* For fall admission, 1/15 for domestic students. Application fee: $0. *Financial support:* In 2005–06, 2 research assistantships were awarded; fellowships, teaching assistantships, Federal Work-Study and institutionally sponsored loans also available. Financial award application deadline: 1/15. *Faculty research:* Biological and artificial computational devices, modeling of sensory processes and learning, theory of collective computation. *Unit head:* Dr. Pietro Perona, Executive Officer, 626-395-4867.

Carnegie Mellon University, Carnegie Institute of Technology, Department of Civil and Environmental Engineering, Pittsburgh, PA 15213-3891. Offers architecture-engineering construction management (MS); civil and environmental engineering (MS, PhD); civil and environmental engineering/engineering and public policy (PhD); civil engineering (MS, PhD); computational science and engineering (MS, PhD); computer-aided engineering (MS, PhD); computer-aided engineering and management (MS, PhD); engineering (MS, PhD); environmental engineering (MS, PhD); environmental management and science (MS, PhD). Part-time programs available. *Faculty:* 17 full-time (3 women), 11 part-time/adjunct (2 women). *Students:* 75 full-time (30 women), 9 part-time (1 woman); includes 5 minority (2 African Americans, 2 Asian Americans or Pacific Islanders, 1 Hispanic American), 49 international. Average age 28. 159 applicants, 55% accepted, 44 enrolled. In 2005, 32 master's, 10 doctorates awarded. Terminal master's awarded for partial completion of doctoral program. *Median time to degree:* Of those who began their doctoral program in fall 1997, 100% received their degree in 8 years or less. *Degree requirements:* For master's, thesis (for some programs), registration; for doctorate, thesis/dissertation, public defense of dissertation, qualifying exam, comprehensive exam, registration. *Entrance requirements:* For master's and doctorate, GRE General Test. Additional exam requirements/recommendations for international students: Required—TOEFL (minimum score 550 paper-based; 213 computer-based). *Application deadline:* For fall admission, 1/15 priority date for domestic students, 1/15 priority date for international students; for spring admission, 9/30 priority date for domestic students, 9/30 priority date for international students. Applications are processed on a rolling basis. Application fee: $55. Electronic applications accepted. *Financial support:* In 2005–06, 48 students received support, including 1 fellowship with full tuition reimbursement available (averaging $30,000 per year), 48 research assistantships with full tuition reimbursements available (averaging $22,200 per year); career-related internships or fieldwork, Federal Work-Study, scholarships/grants, and tuition waivers (partial) also available. Financial award application deadline: 1/15. *Faculty research:* Advanced infrastructure systems; environmental engineering science and management; mechanics, materials, and computing. Total annual research expenditures: $3 million. *Unit head:* Chris Hendrickson, Head, 412-268-2941, Fax: 412-268-7813, E-mail: cth@cmu.edu. *Application contact:* Maxine A. Leffard, Graduate Program Administrator, 412-268-5673, Fax: 412-268-7813, E-mail: ce-admissions@andrew.cmu.edu.

Carnegie Mellon University, Tepper School of Business, Program in Algorithms, Combinatorics, and Optimization, Pittsburgh, PA 15213-3891. Offers MS, PhD. *Degree requirements:* For doctorate, thesis/dissertation. *Entrance requirements:* For master's, GMAT; for doctorate, GRE General Test.

Claremont Graduate University, Graduate Programs, School of Mathematical Sciences, Claremont, CA 91711-6160. Offers computational and systems biology (PhD); computational science (PhD); engineering mathematics (PhD); operations research and statistics (MA, MS); physical applied mathematics (MA, MS); pure mathematics (MA, MS); scientific computing (MA, MS); systems and control theory (MA, MS). Part-time programs available. *Faculty:* 4 full-time (0 women), 6 part-time/adjunct (2 women). *Students:* 54 full-time (13 women), 13 part-time (1 woman); includes 20 minority (1 African American, 14 Asian Americans or Pacific Islanders, 5 Hispanic Americans), 16 international. Average age 38. In 2005, 5 master's, 5 doctorates awarded. Terminal master's awarded for partial completion of doctoral program. *Degree requirements:* For doctorate, 2 foreign languages, thesis/dissertation. *Entrance requirements:* For master's and doctorate, GRE General Test. *Application deadline:* For fall admission, 2/15 for domestic students. Applications are processed on a rolling basis. Electronic applications accepted. *Expenses:* Tuition: Full-time $27,902; part-time $1,214 per term. Required fees: $1,600; $800 per term. Tuition and fees vary according to degree level and program. *Financial support:* Fellowships, research assistantships, career-related internships or fieldwork, Federal Work-Study, institutionally sponsored loans, and tuition waivers (full and partial) available. Support available to part-time students. Financial award application deadline: 2/15; financial award applicants required to submit FAFSA. *Unit head:* John Angus, Dean, 909-621-8080, Fax: 909-607-8261, E-mail: john.angus@cgu.edu. *Application contact:* Susan Townzen, Program Coordinator, 909-621-8080, Fax: 909-607-8261, E-mail: susan.n.townzen@cgu.edu.

Clemson University, Graduate School, College of Engineering and Science, Department of Mathematical Sciences, Clemson, SC 29634. Offers applied and pure mathematics (MS, PhD); computational mathematics (MS, PhD); operations research (MS, PhD); statistics (MS, PhD). Part-time programs available. *Students:* 76 full-time (27 women), 5 part-time (2 women); includes 4 minority (2 African Americans, 2 Asian Americans or Pacific Islanders), 25 international. Average age 29. 73 applicants, 84% accepted, 24 enrolled. In 2005, 23 master's, 4 doctorates awarded. *Median time to degree:* Of those who began their doctoral program in fall 1997, 100% received their degree in 8 years or less. *Degree requirements:* For master's, final project, thesis optional; for doctorate, thesis/dissertation, qualifying exams. *Entrance requirements:* For master's and doctorate, GRE General Test. Additional exam requirements/recommendations for international students: Required—TOEFL; Recommended—TSE. *Application deadline:* For fall admission, 1/15 priority date for domestic students, 2/15 priority date for international students; for spring admission, 10/1 priority date for domestic students, 9/15 priority date for international students. Applications are processed on a rolling basis. Application fee: $50. Electronic applications accepted. *Financial support:* In 2005–06, 68 students received support; fellowships with partial tuition reimbursements available, research assistantships with partial tuition reimbursements available, teaching assistantships with partial tuition reimbursements available available. Financial award application deadline:4/15. *Faculty research:* Applied and computational analysis, cryptography, discrete mathematics, optimization, statistics. Total annual research expenditures: $1 million. *Unit head:* Robert L. Taylor, Chair, 864-656-5240, Fax: 864-656-5230, E-mail: rtaylo2@clemson.edu. *Application contact:* Dr. K.B. Kulasekera, Graduate Coordinator, 864-656-5231, Fax: 864-656-5230, E-mail: kk@clemson.edu.

See Close-Ups on pages 489 and 491.

The College of William and Mary, Faculty of Arts and Sciences, Department of Computer Science, Program in Computational Operations Research, Williamsburg, VA 23187-8795. Offers MS. Part-time programs available. *Faculty:* 9 full-time (3 women), 1 part-time/adjunct (0 women). *Students:* 15 full-time (7 women), 3 part-time (2 women); includes 1 African American, 4 international. *Degree requirements:* For master's, research project, thesis optional. *Entrance requirements:* For master's, GRE General Test, minimum GPA of 2.5. *Application deadline:* For fall admission, 1/15 for domestic students; for spring admission, 11/1 for domestic students. Applications are processed on a rolling basis. Application fee: $30. *Expenses:* Tuition, state resident: full-time $5,828; part-time $245 per credit. Tuition, nonresident: full-time $17,980; part-time $685 per credit. Required fees: $3,051. Tuition and fees vary according to program. *Financial support:* In 2005–06, 7 teaching assistantships with full tuition reimbursements (averaging $10,500 per year) were awarded; scholarships/grants and tuition waivers (full) also available. Financial award application deadline: 3/3; financial award applicants required to submit FAFSA. *Faculty research:* Metaheuristics, reliability, optimization, statistics. *Unit head:* Dr. Rex Kincaid, Professor, 757-221-2038, Fax: 757-221-1717, E-mail: rrkinc@math.wm.edu. *Application contact:* Vanessa Godwin, Administrative Director, 757-221-3455, Fax: 757-221-1717, E-mail: gradinfo@cs.wm.edu.

Cornell University, Graduate School, Graduate Fields of Engineering, Field of Chemical Engineering, Ithaca, NY 14853-0001. Offers advanced materials processing (M Eng, MS, PhD); applied mathematics and computational methods (M Eng, MS, PhD); biochemical engineering (M Eng, MS, PhD); chemical reaction engineering (M Eng, MS, PhD); classical and statistical thermodynamics (M Eng, MS, PhD); fluid dynamics, rheology and biorheology (M Eng, MS, PhD); heat and mass transfer (M Eng, MS, PhD); kinetics and catalysis (M Eng, MS, PhD); polymers (M Eng, MS, PhD); surface science (M Eng, MS, PhD). *Faculty:* 30 full-time (1 woman). *Students:* 266 applicants, 23% accepted, 30 enrolled. In 2005, 21 master's, 11 doctorates awarded. *Degree requirements:* For master's, thesis (MS); for doctorate, thesis/dissertation, comprehensive exam. *Entrance requirements:* For master's and doctorate, GRE General Test, 2 letters of recommendation. Additional exam requirements/recommendations for international students: Required—TOEFL (minimum score 580 paper-based; 237 computer-based). *Application deadline:* For fall admission, 1/15 for domestic students. Application fee: $60. Electronic applications accepted. *Financial support:* In 2005–06, 81 students received support, including 27 fellowships with full tuition reimbursements available, 44 research assistantships with full tuition reimbursements available, 10 teaching assistantships with full tuition reimbursements available; institutionally sponsored loans, scholarships/grants, health care benefits, tuition waivers (full and partial), and unspecified assistantships also available. Financial award applicants required to submit FAFSA. *Faculty research:* Biochemical, biomedical and metabolic engineering; fluid and polymer dynamics; surface science and chemical kinetics; electronics materials; microchemical systems and nanotechnology. *Unit head:* Director of Graduate Studies, 607-255-4550. *Application contact:* Graduate Field Assistant, 607-255-4550, E-mail: dgs@cheme.cornell.edu.

George Mason University, College of Science, Fairfax, VA 22030. Offers bioinformatics (MS, PhD); climate dynamics (PhD); computational sciences (MS); computational sciences and informatics (PhD); computational social science (PhD); computational techniques and applications (Certificate); earth systems and geoinformation science (PhD); earth systems science (MS); nanotechnology and nanoscience (Certificate); neuroscience (PhD); physical sciences (PhD); remote sensing and earth image processing (Certificate). Part-time and evening/weekend programs available. *Degree requirements:* For doctorate, thesis/dissertation, comprehensive exam, registration. *Entrance requirements:* For master's and doctorate, GRE General Test, minimum GPA of 3.0 in last 60 hours. Additional exam requirements/recommendations for international students: Required—TOEFL. Electronic applications accepted. *Expenses:* Tuition, area resident: Full-time $5,244; part-time $219 per credit. Tuition, state resident: part-time $651 per credit. Tuition, nonresident: full-time $15,636. Required fees: $1,524; $65 per credit. *Faculty research:* Space sciences and astrophysics, fluid dynamics, materials modeling and simulation, bioinformatics, global changes and statistics.

See Close-Up on page 501.

Kean University, College of Natural, Applied and Health Sciences, Program in Computing, Statistics and Mathematics, Union, NJ 07083. Offers MS. Part-time and evening/weekend programs available. *Faculty:* 24 full-time (6 women). *Students:* 7 full-time (3 women), 5 part-time (2 women); includes 5 minority (2 African Americans, 1 Asian American or Pacific Islander, 2 Hispanic Americans), 1 international. Average age 36. 1 applicant, 100% accepted, 1 enrolled. In 2005, 3 degrees awarded. *Degree requirements:* For master's, thesis or alternative, research component, minimum 3.0 GPA. *Entrance requirements:* For master's, GRE General Test, 2 letters of recommendation, interview. *Application deadline:* For fall admission, 5/1 for domestic students; for spring admission, 11/1 for domestic students. Application fee: $60 ($150 for international students). Electronic applications accepted. *Expenses:* Tuition, state resident: full-time $8,280; part-time $345 per credit. Tuition, nonresident: full-time $11,512; part-time $438 per credit. Required fees: $2,104; $88 per credit. *Financial support:* In 2005–06, 3 research assistantships with full tuition reimbursements (averaging $2,880 per year) were awarded *Unit head:* Dr. Francine Abeles, Program Coordinator, 908-737-3714, E-mail:

Computational Sciences

Kean University (continued)
fabeles@kean.edu. *Application contact:* Joanne Morris, Director of Graduate Admissions, 908-737-3355, Fax: 908-737-3354, E-mail: grad-adm@kean.edu.

Lehigh University, P.C. Rossin College of Engineering and Applied Science, Department of Mechanical Engineering and Mechanics, Bethlehem, PA 18015-3094. Offers computational engineering and mechanics (MS, PhD); mechanical engineering (M Eng, MS, PhD); polymer science/engineering (MS, PhD). Part-time programs available. *Faculty:* 26 full-time (0 women). *Students:* 64 full-time (9 women), 12 part-time; includes 6 minority (1 African American, 1 American Indian/Alaska Native, 1 Asian American or Pacific Islander, 3 Hispanic Americans), 31 international. 109 applicants, 63% accepted, 14 enrolled. In 2005, 21 master's, 7 doctorates awarded. Terminal master's awarded for partial completion of doctoral program. *Degree requirements:* For master's and doctorate, thesis/dissertation. *Entrance requirements:* Additional exam requirements/recommendations for international students: Required—TOEFL (minimum score 550 paper-based; 213 computer-based). *Application deadline:* For fall admission, 7/15 for domestic students; for spring admission, 12/1 for domestic students. Applications are processed on a rolling basis. Application fee: $60. *Financial support:* In 2005–06, 7 fellowships with full and partial tuition reimbursements (averaging $13,950 per year), 25 research assistantships with full and partial tuition reimbursements (averaging $13,950 per year), 7 teaching assistantships with full and partial tuition reimbursements (averaging $13,725 per year) were awarded. Financial award application deadline: 1/15. *Faculty research:* Thermofluids, dynamic systems, CAD/CAM. Total annual research expenditures: $5.1 million. *Unit head:* Dr. Herman F. Nied, Chairman, 610-758-4102, Fax: 610-758-6224, E-mail: hfn2@lehigh.edu. *Application contact:* Geri Sue Kneller, Graduate Coordinator, 610-758-4139, Fax: 610-758-6224, E-mail: gsk2@lehigh.edu.

Louisiana Tech University, Graduate School, College of Engineering and Science, Department of Physics, Ruston, LA 71272. Offers applied computational analysis and modeling (PhD); physics (MS). Part-time programs available. *Degree requirements:* For master's, thesis or alternative; for doctorate, thesis/dissertation. *Entrance requirements:* For master's, GRE General Test, minimum GPA of 3.0 in last 60 hours. Additional exam requirements/recommendations for international students: Required—TOEFL. *Faculty research:* Experimental high energy physics, laser/optics, computational physics, quantum gravity.

Massachusetts Institute of Technology, School of Engineering and School of Science and Sloan School of Management, Program in Computation for Design and Optimization, Cambridge, MA 02139-4307. Offers SM. *Faculty:* 25 full-time (3 women). *Students:* 17 full-time (4 women); includes 3 minority (all Asian Americans or Pacific Islanders), 13 international. Average age 24. 71 applicants, 34% accepted, 17 enrolled. *Degree requirements:* For master's, thesis, registration. *Entrance requirements:* For master's, GRE General Test, 3 letters of reference. Additional exam requirements/recommendations for international students: Required—TOEFL (minimum score 600 paper-based; 250 computer-based). *Application deadline:* For fall admission, 1/10 for domestic students, 1/10 for international students. Application fee: $70. Electronic applications accepted. *Expenses:* Tuition: Full-time $32,100. Required fees: $200. Part-time tuition and fees vary according to course load. *Financial support:* In 2005–06, 5 students received support, including 5 research assistantships with tuition reimbursements available; fellowships with tuition reimbursements available, teaching assistantships with tuition reimbursements available, health care benefits and unspecified assistantships also available. Financial award application deadline: 1/10. *Faculty research:* Numerical simulation, constrained optimization, partial differential equations, optimization under uncertainty, computational mechanics. *Unit head:* Prof. Jaime Peraire, Professor of Aeronautics and Astronautics, 617-253-1981, Fax: 617-258-5143, E-mail: peraire@mit.edu. *Application contact:* Laura F. Koller, Communications and Graduate Admissions Coordinator, 617-253-3725, Fax: 617-258-9214, E-mail: cdo_info@mit.edu.

McGill University, Faculty of Graduate and Postdoctoral Studies, Faculty of Science, Department of Mathematics and Statistics, Montréal, QC H3A 2T5, Canada. Offers computational science and engineering (M Sc); mathematics (M Sc, MA, PhD), including applied mathematics (M Sc, MA), pure mathematics (M Sc, MA), statistics (M Sc, MA). Part-time programs available. *Degree requirements:* For master's, thesis (for some programs), registration; for doctorate, one foreign language, thesis/dissertation, comprehensive exam, registration. *Entrance requirements:* For master's, minimum GPA of 3.0, Canadian Honours degree in mathematics or closely related field (statistics or applied mathematics). Additional exam requirements/recommendations for international students: Required—TOEFL (minimum score 550 paper-based; 213 computer-based), IELT (minimum score 7). Electronic applications accepted.

Memorial University of Newfoundland, School of Graduate Studies, Interdisciplinary Program in Computational Science, St. John's, NL A1C 5S7, Canada. Offers computational science (M Sc); computational science (cooperative) (M Sc). *Students:* 13 full-time (4 women), 1 part-time, 5 international. 10 applicants, 40% accepted, 2 enrolled. In 2005, 1 degree awarded. *Degree requirements:* For master's, thesis optional. *Entrance requirements:* For master's, honors B Sc or significant background in the field. *Application deadline:* Applications are processed on a rolling basis. Application fee: $40 Canadian dollars. Electronic applications accepted. *Expenses:* Tuition: Part-time $733 per term. Tuition and fees vary according to degree level and program. *Faculty research:* Scientific computing, modeling and simulation, computational fluid dynamics, polymer physics, computational chemistry. *Unit head:* Dr. George Miminis, Chair, 709-737-8635, E-mail: george@cs.mun.ca. *Application contact:* Gail Kenny, Secretary, 709-737-8154, Fax: 709-737-3316, E-mail: gkenny@mun.ca.

Michigan Technological University, Graduate School, College of Sciences and Arts, Department of Computer Science, Program in Computational Science and Engineering, Houghton, MI 49931-1295. Offers PhD. Part-time programs available. *Faculty:* 13 full-time (3 women). *Students:* 5 full-time (1 woman), 2 part-time (1 woman), 4 international. Average age 32. 3 applicants, 33% accepted, 1 enrolled. In 2005, 1 degree awarded. *Median time to degree:* Of those who began their doctoral program in fall 1997, 50% received their degree in 8 years or less. *Degree requirements:* For doctorate, thesis/dissertation, comprehensive exam, registration. *Entrance requirements:* For doctorate, MS in relevant discipline. Additional exam requirements/recommendations for international students: Required—TOEFL (minimum score 550 paper-based; 213 computer-based). *Application deadline:* For fall admission, 3/15 for domestic students. Applications are processed on a rolling basis. Application fee: $40 ($45 for international students). Electronic applications accepted. *Expenses:* Contact institution. Full-time tuition and fees vary according to course load, degree level and program. *Financial support:* In 2005–06, 4 students received support, including fellowships with full tuition reimbursements available (averaging $9,542 per year), 1 research assistantship with full tuition reimbursement available (averaging $9,542 per year), 3 teaching assistantships with full tuition reimbursements available (averaging $9,542 per year); career-related internships or fieldwork, Federal Work-Study, scholarships/grants, health care benefits, tuition waivers (partial), unspecified assistantships, and co-op also available. Financial award applicants required to submit FAFSA. *Application contact:* Dr. Phillip R. Merkey, Director of Computational Science and Engineering Research Institute, 906-487-2220, Fax: 906-487-2283, E-mail: merk@mtu.edu.

Northwestern University, McCormick School of Engineering and Applied Science, Program in Computational Biology and Bioinformatics, Evanston, IL 60208. Offers MS. Part-time programs available. *Faculty:* 40 full-time (5 women). *Degree requirements:* For master's, thesis, registration. *Entrance requirements:* For master's, GRE General Test, 2 letters of reference. Additional exam requirements/recommendations for international students: Required—TOEFL (minimum score 600 paper-based; 250 computer-based); Recommended—TSE. *Application deadline:* For fall admission, 3/1 priority date for domestic students, 3/1 priority date for international students. Applications are processed on a rolling basis. Application fee: $60 ($75 for international students). Electronic applications accepted. *Faculty research:* Mathematical models of protein signaling, high throughput DNA sequencing, macromolecule interactions, chemoinformatics, genome DNA sequence evolution. *Unit head:* Dr. Ming Yang Kao, Director, 847-563-0426, Fax: 847-491-5258, E-mail: kao@cs.northwestern.edu. *Application contact:* Dr. Dawn M. Graunke, Assistant Program Director, 847-467-1972, Fax: 847-491-5258, E-mail: d-graunke@cs.northwestern.edu.

Princeton University, Graduate School, Department of Mathematics, Princeton, NJ 08544-1019. Offers applied and computational mathematics (PhD); mathematical physics (PhD); mathematics (PhD). *Degree requirements:* For doctorate, 2 foreign languages, thesis/dissertation. *Entrance requirements:* For doctorate, GRE General Test, GRE Subject Test. Additional exam requirements/recommendations for international students: Required—TOEFL (minimum score 600 paper-based; 250 computer-based). Electronic applications accepted.

Princeton University, Graduate School, Department of Physics, Princeton, NJ 08544-1019. Offers applied and computational mathematics (PhD); mathematical physics (PhD); physics (PhD); physics and chemical physics (PhD). *Degree requirements:* For doctorate, thesis/dissertation, qualifying exam. *Entrance requirements:* For doctorate, GRE General Test, GRE Subject Test. Additional exam requirements/recommendations for international students: Required—TOEFL (minimum score 600 paper-based; 250 computer-based). Electronic applications accepted.

Princeton University, Graduate School, Program in Applied and Computational Mathematics, Princeton, NJ 08544-1019. Offers PhD. *Degree requirements:* For doctorate, thesis/dissertation. *Entrance requirements:* For doctorate, GRE General Test, GRE Subject Test. Additional exam requirements/recommendations for international students: Required—TOEFL (minimum score 600 paper-based; 250 computer-based). Electronic applications accepted.

Rice University, Graduate Programs, George R. Brown School of Engineering, Department of Computational and Applied Mathematics, Houston, TX 77251-1892. Offers MA, MCAM, MCSE, PhD. *Degree requirements:* For master's, thesis (for some programs), comprehensive exam (for some programs), registration; for doctorate, thesis/dissertation, comprehensive exam, registration. *Entrance requirements:* For master's and doctorate, GRE General Test, minimum GPA of 3.0. Additional exam requirements/recommendations for international students: Required—TOEFL (minimum score 600 paper-based; 250 computer-based). Electronic applications accepted. *Faculty research:* Inverse problems, partial differential equations, computer algorithms, computational modeling, optimization theory.

Sam Houston State University, College of Arts and Sciences, Department of Computer Science, Huntsville, TX 77341. Offers computing and information science (MS). *Faculty:* 6 full-time (2 women). *Students:* 2 full-time (1 woman), 11 part-time (3 women); includes 5 minority (3 Asian Americans or Pacific Islanders, 2 Hispanic Americans). Average age 28. In 2005, 7 degrees awarded. Application fee: $20. *Unit head:* Dr. Peter Cooper, Chair, 936-294-1568, Fax: 936-294-1882, E-mail: css_pac@shsu.edu. *Application contact:* Dr. Jiuhung Ji, Advisor, 936-294-1579, E-mail: csc_jxj@shsu.edu.

San Diego State University, Graduate and Research Affairs, College of Sciences, Program in Computational Science, San Diego, CA 92182. Offers MS, PhD. *Students:* 17 full-time (4 women), 17 part-time (2 women); includes 4 minority (1 African American, 2 Asian Americans or Pacific Islanders, 1 Hispanic American), 12 international. 30 applicants, 60% accepted, 13 enrolled. In 2005, 8 degrees awarded. *Degree requirements:* For master's and doctorate, thesis/dissertation. *Entrance requirements:* For master's, GRE General Test, 3 letters of recommendation; for doctorate, GRE, 3 letters of recommendation. Additional exam requirements/recommendations for international students: Required—TOEFL. *Application deadline:* For fall admission, 5/1 for domestic students, 5/1 for international students; for spring admission, 11/1 for domestic students, 10/1 for international students. Applications are processed on a rolling basis. Application fee: $55. Electronic applications accepted. *Financial support:* Teaching assistantships, unspecified assistantships available. Financial award applicants required to submit FAFSA. *Unit head:* Jose Castillo, Director, 619-594-3430, Fax: 619-594-5291, E-mail: castillo@sdsu.edu.

Southern Methodist University, Dedman College, Department of Mathematics, Dallas, TX 75275. Offers computational and applied mathematics (MS, PhD). *Faculty:* 15 full-time (1 woman). *Students:* 22 full-time (11 women), 1 part-time; includes 3 minority (2 Asian Americans or Pacific Islanders, 1 Hispanic American), 9 international. Average age 30. 25 applicants, 32% accepted, 8 enrolled. In 2005, 4 master's, 2 doctorates awarded. *Degree requirements:* For doctorate, thesis/dissertation, oral and written exams. *Entrance requirements:* For master's and doctorate, GRE General Test, minimum GPA of 3.0, 18 undergraduate hours in mathematics beyond first and second year calculus. Additional exam requirements/recommendations for international students: Required—TOEFL. *Application deadline:* For fall admission, 6/30 for domestic students. For winter admission, 11/30 for domestic students. Applications are processed on a rolling basis. Application fee: $60. Electronic applications accepted. *Financial support:* In 2005–06, 7 teaching assistantships with full tuition reimbursements (averaging $15,000 per year) were awarded; career-related internships or fieldwork, scholarships/grants, health care benefits, and tuition waivers (partial) also available. Support available to part-time students. Financial award applicants required to submit FAFSA. *Faculty research:* Numerical analysis, scientific computation, fluid dynamics, software development, differential equations. Total annual research expenditures: $195,000. *Unit head:* Dr. Peter Moore, Chairman, 214-768-2506, Fax: 214-768-2355, E-mail: mathchair@mail.smu.edu. *Application contact:* Dr. Ian Gladwell, Director of Graduate Studies, 214-768-4338, E-mail: math@mail.smu.edu.

Stanford University, School of Engineering, Program in Scientific Computing and Computational Mathematics, Stanford, CA 94305-9991. Offers MS, PhD. Terminal master's awarded for partial completion of doctoral program. *Degree requirements:* For doctorate, thesis/dissertation, qualifying exam. *Entrance requirements:* For master's, GRE General Test; for doctorate, GRE General Test, GRE Subject Test. Additional exam requirements/recommendations for international students: Required—TOEFL. Electronic applications accepted.

State University of New York College at Brockport, School of Letters and Sciences, Department of Computational Science, Brockport, NY 14420-2997. Offers MS. Part-time programs available. *Students:* 3 full-time (1 woman), 5 part-time (1 woman), 1 international. 4 applicants, 100% accepted, 2 enrolled. In 2005, 3 degrees awarded. *Degree requirements:* For master's, thesis or alternative. *Entrance requirements:* For master's, minimum GPA of 3.0, letters of recommendation. Additional exam requirements/recommendations for international students: Required—TOEFL (minimum score 550 paper-based; 213 computer-based). *Application deadline:* For fall admission, 7/15 for domestic students, 7/15 for international students; for spring admission, 11/15 for domestic students, 11/15 for international students. Application fee: $50. *Expenses:* Tuition, state resident: full-time $6,900; part-time $288 per credit. Tuition, nonresident: full-time $10,920; part-time $455 per credit. Required fees: $685; $28 per credit. *Financial support:* Federal Work-Study, scholarships/grants, and unspecified assistantships available. Support available to part-time students. Financial award application deadline: 3/15. *Faculty research:* Parallel computing, fluid and particle dynamics, molecular simulation, engine combustion, linear algebra software. *Unit head:* Dr. Osman Yasar, Chairperson, 585-395-2021, Fax: 585-395-5020, E-mail: oyasar@brockport.edu. *Application contact:* Dr. Robert Tuzun, Graduate Program Director, 585-395-5365, E-mail: rtuzun@brockport.edu.

Temple University, Graduate School, College of Science and Technology, Department of Mathematics, Philadelphia, PA 19122-6096. Offers applied mathematics (MA); mathematics (PhD); pure mathematics (MA). Part-time and evening/weekend programs available. *Faculty:* 20 full-time (1 woman). *Students:* 14 full-time (5 women), 20 part-time (5 women); includes 6 minority (2 African Americans, 3 Asian Americans or Pacific Islanders, 1 Hispanic American), 18 international. 35 applicants, 31% accepted, 9 enrolled. In 2005, 5 master's, 6 doctorates awarded. Terminal master's awarded for partial completion of doctoral program. *Degree requirements:* For master's, written exam, thesis optional; for doctorate, 2 foreign languages, thesis/dissertation, oral and written exams. *Entrance requirements:* For master's, GRE General Test, minimum GPA of 3.0; for doctorate, GRE General Test, GRE Subject Test, minimum GPA of 3.0. Additional exam requirements/recommendations for international students: Required—TOEFL (minimum score 575 paper-based; 230 computer-based). *Application deadline:* For fall admission, 2/15 priority date for domestic students, 12/15 priority date for international students; for spring admission, 11/15 priority date for domestic students, 8/1 priority date for international students. Applications are processed on a rolling basis. Application fee: $50. Electronic applications accepted. *Expenses:* Tuition, state resident: full-time $8,694; part-time $483 per credit. Tuition, nonresident: full-time $12,672; part-time $704 per credit. Required fees: $500; $122 per semester. Tuition and fees vary according to course level, campus/location and program. *Financial support:* Fellowships, research assistantships, teaching assistantships, Federal Work-Study and institutionally sponsored loans available. Financial award application deadline: 1/15; financial award applicants required to submit FAFSA. *Faculty research:* Differential geometry, numerical analysis. *Unit head:* Dr. Omar Hijab, Chair, 215-204-4650, Fax: 215-204-6433, E-mail: hijab@temple.edu.

University of Central Florida, Division of Graduate Studies, Program in Modeling and Simulation, Orlando, FL 32816. Offers MS, PhD. *Expenses:* Tuition, state resident: full-time $5,788. Tuition, nonresident: full-time $21,927. Required fees: $241 per credit hour. *Unit head:* Dr. Charles Reilly, Coordinator, 407-823-5306.

The University of Iowa, Graduate College, Program in Applied Mathematical and Computational Sciences, Iowa City, IA 52242-1316. Offers PhD. *Students:* 15 full-time (4 women), 15 part-time (4 women); includes 4 minority (2 African Americans, 2 Hispanic Americans), 15 international. 27 applicants, 22% accepted, 3 enrolled. In 2005, 4 degrees awarded. *Degree requirements:* For doctorate, thesis/dissertation, comprehensive exam, registration. *Entrance requirements:* For doctorate, GRE General Test, minimum GPA of 3.0. Additional exam requirements/recommendations for international students: Required—TOEFL (minimum score 600 paper-based; 250 computer-based). *Application deadline:* For fall admission, 1/15 priority date for domestic students, 1/15 priority date for international students; for spring admission, 10/1 priority date for domestic students. Applications are processed on a rolling basis. Application fee: $60 ($85 for international students). Electronic applications accepted. *Expenses:* Tuition, state resident: part-time $1,882 per term. Tuition, nonresident: full-time $17,338; part-time $4,907 per term. Tuition and fees vary according to course load and program. *Financial support:* In 2005–06, 2 fellowships, 2 research assistantships with partial tuition reimbursements, 27 teaching assistantships with partial tuition reimbursements were awarded. Financial award applicants required to submit FAFSA. *Unit head:* Yi Li, Director, 319-335-0772.

University of Lethbridge, School of Graduate Studies, Lethbridge, AB T1K 3M4, Canada. Offers accounting (MScM); addictions counseling (M Sc); agricultural biotechnology (M Sc); agricultural studies (M Sc, MA); anthropology (MA); archaeology (MA); art (MA); biochemistry (M Sc); biological sciences (M Sc); biomolecular science (PhD); biosystems and biodiversity (PhD); Canadian studies (MA); chemistry (M Sc); computer science (M Sc); computer science and geographical information science (M Sc); counseling psychology (M Ed); dramatic arts (MA); earth, space, and physical science (PhD); economics (MA); educational leadership (M Ed); English (MA); environmental science (M Sc); evolution and behavior (PhD); exercise science (M Sc); finance (MScM); French (MA); French/German (MA); French/Spanish (MA); general education (M Ed); general management (MScM); geography (M Sc, MA); German (MA); health sciences (M Sc, MA); history (MA); human resource management and labour relations (MScM); individualized multidisciplinary (M Sc, MA); information systems (MScM); international management (MScM); kinesiology (M Sc, MA); management (M Sc, MA); marketing (MScM); mathematics (M Sc); music (MA); Native American studies (MA); neuroscience (M Sc, PhD); new media (MA); nursing (M Sc); philosophy (MA); physics (M Sc); policy and strategy (MScM); political science (MA); psychology (M Sc, MA); religious studies (MA); sociology (MA); theoretical and computational science (PhD); urban and regional studies (MA). Part-time and evening/weekend programs available. *Faculty:* 250. *Students:* 193 full-time, 145 part-time. 35 applicants, 100% accepted, 35 enrolled. In 2005, 40 degrees awarded. *Degree requirements:* For doctorate, thesis/dissertation, comprehensive exam. *Entrance requirements:* For master's, GMAT (M Sc management), bachelor's degree in related field, minimum GPA of 3.0 during previous 20 graded semester courses, 2 years teaching or related experience (M Ed); for doctorate, master's degree, minimum graduate GPA of 3.5. Additional exam requirements/recommendations for international students: Required—TOEFL. Application fee: $60 Canadian dollars. *Expenses:* Tuition, nonresident: part-time $531 per course. Required fees: $83 per year. Tuition and fees vary according to degree level and program. *Financial support:* Fellowships, research assistantships, teaching assistantships, scholarships/grants, health care benefits, and unspecified assistantships available. *Faculty research:* Movement and brain plasticity, gibberellin physiology, photosynthesis, carbon cycling, molecular properties of main-group ring components. *Unit head:* Dr. Shamsul Alam, Dean, 403-329-2121, Fax: 403-329-2097, E-mail: inquiries@uleth.ca. *Application contact:* Kathy Schrage, Administrative Assistant, Office of the Academic Vice President, 403-329-2121, Fax: 403-329-2097, E-mail: inquiries@uleth.ca.

University of Manitoba, Faculty of Graduate Studies, Faculty of Science, Department of Mathematical, Computational and Statistical Sciences, Winnipeg, MB R3T 2N2, Canada. Offers MMCSS.

University of Massachusetts Lowell, Graduate School, College of Arts and Sciences, Department of Mathematics, Lowell, MA 01854-2881. Offers applied mathematics (MS); computational mathematics (PhD); mathematics (MS). Part-time programs available. *Entrance requirements:* For master's, GRE General Test.

University of Michigan–Dearborn, College of Arts, Sciences, and Letters, Program in Applied and Computational Mathematics, Dearborn, MI 48128-1491. Offers MS. Part-time and evening/weekend programs available. *Faculty:* 9 full-time (3 women). *Students:* 3 full-time (1 woman), 11 part-time (3 women); includes 3 minority (2 African Americans, 1 Asian American or Pacific Islander). Average age 33. 3 applicants, 100% accepted. In 2005, 4 degrees awarded. *Degree requirements:* For master's, thesis or alternative, project. *Entrance requirements:* For master's, 3 letters of recommendation, minimum GPA of 3.0, 2 years course work in math. Additional exam requirements/recommendations for international students: Required—TOEFL (minimum score 560 paper-based; 220 computer-based). *Application deadline:* For fall admission, 8/1 for domestic students. For winter admission, 12/1 for domestic students; for spring admission, 4/1 for domestic students. Applications are processed on a rolling basis. Application fee: $60 ($75 for international students). Electronic applications accepted. *Financial support:* Federal Work-Study and scholarships/grants available. Support available to part-time students. Financial award application deadline: 4/1; financial award applicants required to submit FAFSA. *Faculty research:* Partial differential equations, statistics, discrete optimization, approximation theory, stochastic processes. *Unit head:* Dr. Joan Remski, Director, 313-593-4994, E-mail: remski@umd.umich.edu. *Application contact:* Carol Ligienza, Administrative Coordinator, Case Graduate Programs, 313-593-1183, Fax: 313-583-6498, E-mail: caslgrad@umd.umich.edu.

University of Minnesota, Duluth, Graduate School, College of Science and Engineering, Department of Mathematics and Statistics, Duluth, MN 55812-2496. Offers applied and computational mathematics (MS). Part-time programs available. *Faculty:* 17 full-time (3 women). *Students:* 28 full-time (11 women); includes 2 minority (1 African American, 1 Hispanic American), 11 international. Average age 24. 23 applicants, 74% accepted, 12 enrolled. In 2005, 10 degrees awarded. *Degree requirements:* For master's, thesis or alternative. *Entrance requirements:* For master's, GRE General Test, minimum GPA of 3.0. Additional exam requirements/recommendations for international students: Required—TOEFL (minimum score 550 paper-based; 213 computer-based); Recommended—TWE, TSE. *Application deadline:* For fall admission, 3/1 priority date for domestic students, 3/1 priority date for international students; for spring admission, 11/15 for domestic students, 9/1 for international students. Applications are processed on a rolling basis. Application fee: $55 ($75 for international students). *Financial support:* In 2005–06, 28 students received support, including 6 research assistantships with full tuition reimbursements available (averaging $11,895 per year), 22 teaching assistantships with full tuition reimbursements available (averaging $11,895 per year); fellowships, scholarships/grants, health care benefits, unspecified assistantships, and summer fellowships also available. Financial award application deadline: 3/1. *Faculty research:* Discrete mathematics, diagnostic markers, combinatorics, biostatistics, mathematical modeling and scientific computation. Total annual research expenditures: $113,454. *Unit head:* Dr. Zhuangyi Liu, Director of Graduate Studies, 218-726-7179, Fax: 218-726-8399, E-mail: zliu@d.umn.edu.

University of Minnesota, Twin Cities Campus, Graduate School, Scientific Computation Program, Minneapolis, MN 55455-0213. Offers MS, PhD. Part-time programs available. *Faculty:* 13 full-time (1 woman), 1 part-time/adjunct (0 women). *Students:* 9 full-time (4 women), 6 part-time (1 woman), 6 international. 11 applicants, 36% accepted, 1 enrolled. In 2005, 1 degree awarded. *Median time to degree:* Of those who began their doctoral program in fall 1997, 50% received their degree in 8 years or less. *Degree requirements:* For master's and doctorate, thesis/dissertation. *Entrance requirements:* For doctorate, GRE General Test. *Application deadline:* For fall admission, 6/15 for domestic students; for spring admission, 10/15 for domestic students. Applications are processed on a rolling basis. Application fee: $50 ($55 for international students). Electronic applications accepted. *Expenses:* Tuition, state resident: full-time $8,748; part-time $729 per credit. Tuition, nonresident: full-time $15,848; part-time $1,321 per credit. Full-time tuition and fees vary according to class time, course load, program and reciprocity agreements. *Financial support:* In 2005–06, 2 fellowships with partial tuition reimbursements (averaging $8,700 per year) were awarded; research assistantships with full tuition reimbursements, institutionally sponsored loans, health care benefits, and unspecified assistantships also available. *Faculty research:* Parallel computations, quantum mechanical dynamics, computational materials science, computational fluid dynamics, computational neuroscience. *Unit head:* Prof. Jiali Gao, Director of Graduate Studies, 612-625-0769, Fax: 612-625-9442, E-mail: gao@chem.umn.edu. *Application contact:* Kathleen Clinton, Graduate Program Coordinator, 612-626-1458, Fax: 612-626-5009, E-mail: clinton@compneuro.umn.edu.

University of Mississippi, Graduate School, School of Engineering, Oxford, University, MS 38677. Offers computational engineering science (MS, PhD); engineering science (MS, PhD). *Faculty:* 49 full-time (5 women), 3 part-time/adjunct (0 women). *Students:* 149 full-time (34 women), 51 part-time (15 women); includes 11 minority (10 African Americans, 1 Hispanic American), 127 international. 296 applicants, 44% accepted, 48 enrolled. In 2005, 54 master's, 4 doctorates awarded. *Degree requirements:* For master's, thesis (for some programs); for doctorate, thesis/dissertation. *Entrance requirements:* For master's, GRE General Test, minimum GPA of 3.0; for doctorate, GRE General Test. Additional exam requirements/recommendations for international students: Required—TOEFL. *Application deadline:* For fall admission, 4/1 for domestic students; for spring admission, 10/1 for domestic students. Applications are processed on a rolling basis. Application fee: $25. Electronic applications accepted. *Expenses:* Tuition, state resident: full-time $4,320; part-time $240 per credit hour. Tuition, nonresident: full-time $9,744; part-time $301 per credit hour. Tuition and fees vary according to program. *Financial support:* Scholarships/grants available. Financial award application deadline: 3/1; financial award applicants required to submit FAFSA. *Unit head:* Dr. Kai-Fong Lee, Dean, 662-915-7407, Fax: 662-915-1287, E-mail: engineer@olemiss.edu.

University of Nevada, Las Vegas, Graduate College, College of Science, Department of Mathematical Sciences, Las Vegas, NV 89154-9900. Offers applied mathematics (MS, PhD); applied statistics (MS); computational mathematics (PhD); pure mathematics (MS, PhD); statistics (PhD); teaching mathematics (MS). Part-time programs available. *Faculty:* 29 full-time (6 women), 1 part-time/adjunct (0 women). *Students:* 29 full-time (13 women), 29 part-time (9 women); includes 10 minority (1 African American, 6 Asian Americans or Pacific Islanders, 3 Hispanic Americans), 22 international. 37 applicants, 57% accepted, 11 enrolled. In 2005, 18 degrees awarded. *Degree requirements:* For master's, thesis (for some programs), oral exam, comprehensive exam (for some programs). *Entrance requirements:* For master's, minimum GPA of 3.0 during previous 2 years, 2.75 overall. Additional exam requirements/recommendations for international students: Required—TOEFL (minimum score 550 paper-based; 213 computer-based). *Application deadline:* For fall admission, 6/15 for domestic students, 5/1 for international students; for spring admission, 11/15 for domestic students, 10/1 for international students. Application fee: $60 ($75 for international students). Electronic applications accepted. *Expenses:* Tuition, state resident: part-time $150 per credit. Tuition, nonresident: part-time $315 per credit. Tuition and fees vary according to course load, program and reciprocity agreements. *Financial support:* In 2005–06, 42 teaching assistantships with partial tuition reimbursements (averaging $10,500 per year) were awarded; career-related internships or fieldwork, Federal Work-Study, institutionally sponsored loans, scholarships/grants, health care benefits, and unspecified assistantships also available. Support available to part-time students. Financial award application deadline: 3/1. *Unit head:* Dr. Malwane Ananda, Chair, 702-895-3567. *Application contact:* Graduate College Admissions Evaluator, 702-895-3320, Fax: 702-895-4180, E-mail: gradcollege@unlv.edu.

University of Puerto Rico, Mayagüez Campus, Graduate Studies, College of Arts and Sciences, Department of Mathematics, Mayagüez, PR 00681-9000. Offers applied mathematics (MS); computational sciences (MS); pure mathematics (MS); statistics (MS). Part-time programs available. *Faculty:* 34. *Students:* 21 full-time (9 women), 9 part-time (4 women); includes 6 minority (all Hispanic Americans), 24 international. 49 applicants, 61% accepted, 2 enrolled. In 2005, 6 degrees awarded. *Degree requirements:* For master's, one foreign language, comprehensive exam. *Application deadline:* For fall admission, 2/15 for domestic students; for spring admission, 9/15 for domestic students. Applications are processed on a rolling basis. Application fee: $20. *Expenses:* Tuition, state resident: full-time $900; part-time $100 per credit. International tuition: $4,655 full-time. Part-time tuition and fees vary according to course level and course load. *Financial support:* In 2005–06, fellowships (averaging $1,500 per year), research assistantships (averaging $1,200 per year), teaching assistantships (averaging $987 per year) were awarded; Federal Work-Study and institutionally sponsored loans also available. *Faculty research:* Automata theory, linear algebra, logic. Total annual research expenditures: $13,829. *Unit head:* Dr. Luis A. Caceres, Director, 787-832-4040 Ext. 3848.

The University of South Dakota, Graduate School, College of Arts and Sciences, Department of Computer Science, Program in Computational Sciences and Statistics, Vermillion, SD 57069-2390. Offers PhD. *Students:* 3 (1 woman). *Degree requirements:* For doctorate, thesis/dissertation, comprehensive exam. *Entrance requirements:* For doctorate, GRE General Test, GRE Subject test in computer science (recommended), minimum GPA of 2.7. Additional exam requirements/recommendations for international students: Required—IBT 79. *Application deadline:* Applications are processed on a rolling basis. Application fee: $35. Electronic applications accepted. *Expenses:* Tuition, state resident: part-time $116 per credit hour. Tuition, nonresident: part-time $341 per credit hour. Required fees: $85 per credit hour. Tuition and fees vary according to course load, program and reciprocity agreements. *Financial support:* In 2005–06, research assistantships with partial tuition reimbursements (averaging $20,000 per year) *Application contact:* Dr. Rich McBride, Graduate Adviser, 605-677-5388, Fax: 605-677-6662, E-mail: csci@usd.edu.

University of Southern Mississippi, Graduate School, College of Science and Technology, School of Computational Sciences, Hattiesburg, MS 39406-0001. Offers scientific computing (PhD). Part-time programs available. *Degree requirements:* For doctorate, thesis/dissertation, comprehensive exam. *Entrance requirements:* For doctorate, GRE General Test, minimum GPA of 3.5. Additional exam requirements/recommendations for international students: Required—TOEFL.

University of South Florida, College of Medicine and College of Graduate Studies, Graduate Programs in Medical Sciences, Tampa, FL 33620-9951. Offers anatomy (PhD); biochemistry and molecular biology (MS, PhD), including biochemistry and molecular biology (PhD), bioinformatics and computational biology (MS); medical microbiology and immunology (PhD); pathology (PhD); pharmacology and therapeutics (PhD), including medical sciences; physiology and biophysics (PhD). *Students:* 108 full-time (53 women), 34 part-time (26 women);

Computational Sciences

University of South Florida (continued)
includes 31 minority (9 African Americans, 9 Asian Americans or Pacific Islanders, 13 Hispanic Americans), 30 international. 117 applicants, 99% accepted, 116 enrolled. In 2005, 9 master's, 4 doctorates awarded. *Degree requirements:* For doctorate, thesis/dissertation. *Entrance requirements:* For doctorate, GRE General Test, minimum GPA of 3.0. Application fee: $30. *Expenses:* Contact institution. *Financial support:* Institutionally sponsored loans and scholarships/grants available. Financial award application deadline: 4/1; financial award applicants required to submit FAFSA. *Unit head:* Dr. Joseph J. Krzanowski, Associate Dean for Research and Graduate Affairs, 813-974-4181, Fax: 813-974-4317, E-mail: jkrzanow@com1.med.usf.edu.

The University of Texas at Austin, Graduate School, Program in Computational and Applied Mathematics, Austin, TX 78712-1111. Offers MA, PhD. Terminal master's awarded for partial completion of doctoral program. *Degree requirements:* For master's, thesis optional; for doctorate, thesis/dissertation, 3 area qualifying exams. Electronic applications accepted.

University of Utah, The Graduate School, Interdepartmental Program in Computational Science, Salt Lake City, UT 84112-1107. Offers MS. *Students:* 8 full-time (2 women), 6 part-time (1 woman); includes 2 minority (both Asian Americans or Pacific Islanders), 4 international. Average age 33. 11 applicants, 82% accepted, 6 enrolled. *Expenses:* Tuition, state resident: full-time $2,932; part-time $2,212 per term. Tuition, nonresident: full-time $10,350; part-time $7,812 per term. Required fees: $590; $516 per term. Tuition and fees vary according to course load and program. *Unit head:* Peter J. Stang, Dean, 801-581-6958, Fax: 801-585-3169, E-mail: stang@chemistry.utah.edu. *Application contact:* Information Contact, 801-581-6958, E-mail: office@science.utah.edu.

University of Utah, The Graduate School, Program in Science and Technology, Salt Lake City, UT 84112-1107. Offers biotechnology (PSM); computational science (PSM); environmental science (PSM); sciences instrumental (PSM). *Students:* 16 full-time (11 women), 20 part-time (5 women); includes 1 minority (Asian American or Pacific Islander), 7 international. Average age 36. 20 applicants, 50% accepted, 10 enrolled. In 2005, 2 degrees awarded. *Entrance requirements:* For master's, minimum undergraduate GPA of 3.0. Additional exam requirements/recommendations for international students: Required—TOEFL (minimum score 500 paper-based; 173 computer-based). *Application deadline:* For fall admission, 4/1 for domestic students, 4/1 for international students; for spring admission, 11/1 for domestic students, 11/1 for international students. Application fee: $45 ($65 for international students). *Expenses:* Tuition, state resident: full-time $2,932; part-time $2,212 per term. Tuition, nonresident: full-time $10,350; part-time $7,812 per term. Required fees: $590; $516 per term. Tuition and fees vary according to course load and program. *Financial support:* Applicants required to submit FAFSA. *Application contact:* Jennifer Schmidt, Program Director, 801-585-5630, E-mail: jennifer.schmidt@admin.utah.edu.

Western Michigan University, Graduate College, College of Arts and Sciences, Department of Mathematics, Program in Computational Mathematics, Kalamazoo, MI 49008-5202. Offers MS.

Mathematical and Computational Finance

Bernard M. Baruch College of the City University of New York, Weissman School of Arts and Sciences, Program in Applied Mathematics for Finance, New York, NY 10010-5585. Offers MS. *Entrance requirements:* For master's, GRE General Test or GMAT, GRE Mathematics Subject Test optional, 3 recommendations. Additional exam requirements/recommendations for international students: Required—TOEFL, TWE.

Announcement: Baruch College's MS in Applied Mathematics for Finance is a unique program for recent college graduates interested in a career in quantitative finance as well as for practitioners seeking to enhance their mathematical foundation. The curriculum comprises the fundamental tools of mathematical finance, combining rigorous mathematical theory with hands-on computational experience.

Boston University, Graduate School of Arts and Sciences, Department of Mathematics and Statistics, Boston, MA 02215. Offers mathematical finance (MA); mathematics (MA, PhD). *Students:* 72 full-time (28 women), 9 part-time (2 women); includes 5 minority (1 African American, 3 Asian Americans or Pacific Islanders, 1 Hispanic American), 37 international. Average age 28. 269 applicants, 29% accepted, 38 enrolled. In 2005, 31 master's, 8 doctorates awarded. Terminal master's awarded for partial completion of doctoral program. *Degree requirements:* For master's, one foreign language, comprehensive exam, registration; for doctorate, one foreign language, thesis/dissertation, comprehensive exam, registration. *Entrance requirements:* For master's and doctorate, GRE General Test, GRE Subject Test, 3 letters of recommendation. Additional exam requirements/recommendations for international students: Required—TOEFL (minimum score 600 paper-based; 250 computer-based). *Application deadline:* For fall admission, 1/15 for domestic students, 1/15 for international students; for spring admission, 10/15 for domestic students, 10/15 for international students. Application fee: $60. *Expenses:* Tuition: Full-time $31,530; part-time $985 per credit. Required fees: $316; $40 per semester. Tuition and fees vary according to course level and program. *Financial support:* In 2005–06, 57 students received support, including 4 fellowships with full tuition reimbursements available (averaging $16,500 per year), 26 research assistantships with full tuition reimbursements available (averaging $16,000 per year), 26 teaching assistantships with full tuition reimbursements available (averaging $16,000 per year); Federal Work-Study and scholarships/grants also available. Support available to part-time students. Financial award application deadline: 1/15; financial award applicants required to submit FAFSA. *Unit head:* Steven Rosenberg, Chairman, 617-353-9556, Fax: 617-353-8100, E-mail: sr@bu.edu. *Application contact:* Angela M. Silva, Staff Coordinator, 617-353-2560, Fax: 617-353-8100, E-mail: amsilva@bu.edu.

Carnegie Mellon University, Mellon College of Science, Department of Mathematical Sciences, Pittsburgh, PA 15213-3891. Offers algorithms, combinatorics, and optimization (PhD); mathematical finance (PhD); mathematical sciences (MS, DA, PhD); pure and applied logic (PhD). Part-time programs available. Terminal master's awarded for partial completion of doctoral program. *Degree requirements:* For doctorate, thesis/dissertation. *Entrance requirements:* For master's and doctorate, GRE General Test, GRE Subject Test. Additional exam requirements/recommendations for international students: Required—TOEFL. Electronic applications accepted. *Faculty research:* Continuum mechanics, discrete mathematics, applied and computational mathematics.

Carnegie Mellon University, Tepper School of Business, Pittsburgh, PA 15213-3891. Offers accounting (PhD); algorithms, combinatorics, and optimization (MS, PhD); business management and software engineering (MBMSE); civil engineering and industrial management (MS); computational finance (MSCF); economics (MS, PhD); electronic commerce (MS); environmental engineering and management (MEEM); finance (PhD); financial economics (PhD); industrial administration (MBA), including administration and public management; information systems (PhD); management of manufacturing and automation (MOM, PhD), including industrial administration (PhD), manufacturing (MOM); marketing (PhD); mathematical finance (PhD); operations research (PhD); organizational behavior and theory (PhD); political economy (PhD); production and operations management (PhD); public policy and management (MS, MSED); software engineering and business management (MS). Part-time programs available. Terminal master's awarded for partial completion of doctoral program. *Degree requirements:* For doctorate, thesis/dissertation. *Entrance requirements:* For master's, GMAT. Additional exam requirements/recommendations for international students: Required—TOEFL. Expenses: Contact institution.

DePaul University, Charles H. Kellstadt Graduate School of Business, Department of Finance, Chicago, IL 60604-2287. Offers behavioral finance (MBA); computational finance (MS); finance (MBA, MSF); financial analysis (MBA); international marketing and finance (MBA); managerial finance (MBA); real estate (MS); real estate finance and investment (MBA); strategy, execution and valuation (MBA). Part-time and evening/weekend programs available. *Faculty:* 24 full-time (4 women), 36 part-time/adjunct (3 women). *Students:* 267 full-time (74 women), 200 part-time (49 women); includes 56 minority (8 African Americans, 31 Asian Americans or Pacific Islanders, 17 Hispanic Americans), 36 international. Average age 29. In 2005, 239 degrees awarded. *Entrance requirements:* For master's, GMAT, 2 letters of recommendation, resumé. Additional exam requirements/recommendations for international students: Required—TOEFL (minimum score 550 paper-based; 213 computer-based). *Application deadline:* For fall admission, 7/1 for domestic students. For winter admission, 10/1 for domestic students; for spring admission, 2/1 for domestic students. Applications are processed on a rolling basis. Application fee: $60. Electronic applications accepted. *Financial support:* In 2005–06, 9 students received support, including 6 research assistantships with partial tuition reimbursements available (averaging $2,700 per year) Support available to part-time students. Financial award application deadline: 4/1. *Faculty research:* Derivatives, valuation, international finance, real estate. *Application contact:* Christopher E. Kinsella, Director of Marketing and Admissions, 312-362-8810, Fax: 312-362-6677, E-mail: kgsbe@depaul.edu.

DePaul University, School of Computer Science, Telecommunications, and Information Systems, Chicago, IL 60604-2287. Offers business information technology (MS); computational finance (MS); computer graphics and animation (MS); computer information and network security (MS); computer science (MS, PhD); e-commerce technology (MS); human-computer interaction (MS); information systems (MS); information technology (MA); instructional technology systems (MS); software engineering (MS); telecommunication systems (MS). Part-time and evening/weekend programs available. Postbaccalaureate distance learning degree programs offered (no on-campus study). *Faculty:* 79 full-time (13 women), 117 part-time/adjunct (26 women). *Students:* 976 full-time (233 women), 1,020 part-time (256 women); includes 470 minority (167 African Americans, 2 American Indian/Alaska Native, 228 Asian Americans or Pacific Islanders, 73 Hispanic Americans), 318 international. Average age 31. 830 applicants, 80% accepted, 400 enrolled. In 2005, 629 master's, 3 doctorates awarded. *Degree requirements:* For master's, comprehensive exam (for some programs); for doctorate, thesis/dissertation, comprehensive exam. *Entrance requirements:* For doctorate, GRE, master's degree in computer science. Additional exam requirements/recommendations for international students: Required—TOEFL (minimum score 550 paper-based; 213 computer-based). *Application deadline:* For fall admission, 8/1 priority date for domestic students, 8/1 priority date for international students. For winter admission, 11/15 for domestic students; for spring admission, 3/1 for domestic students. Applications are processed on a rolling basis. Application fee: $25. Electronic applications accepted. *Expenses:* Contact institution. *Financial support:* In 2005–06, 63 teaching assistantships with full and partial tuition reimbursements (averaging $9,085 per year) were awarded; fellowships, research assistantships, Federal Work-Study, tuition waivers (full and partial), and unspecified assistantships also available. Support available to part-time students. Financial award application deadline: 4/1; financial award applicants required to submit FAFSA. *Faculty research:* Computer graphics, computer vision, information systems technology, computer network, programming. *Unit head:* Dr. David Miller, Senior Associate Dean, 312-362-8381, Fax: 312-362-5185. *Application contact:* Maureen Garvey, Information Contact, 312-362-8714, Fax: 312-362-5327, E-mail: mgarvey@cti.depaul.edu.

Florida State University, Graduate Studies, College of Arts and Sciences, Department of Mathematics, Tallahassee, FL 32306. Offers applied mathematics (MS, PhD); biomedical mathematics (MS, PhD); financial mathematics (MS, PhD); pure mathematics (MS, PhD). Part-time programs available. *Faculty:* 41 full-time (4 women), 10 part-time/adjunct (6 women). *Students:* 117 full-time (27 women), 5 part-time; includes 20 minority (3 African Americans, 1 American Indian/Alaska Native, 7 Asian Americans or Pacific Islanders, 9 Hispanic Americans), 56 international. Average age 26. 317 applicants, 66% accepted, 41 enrolled. In 2005, 31 master's, 7 doctorates awarded. Terminal master's awarded for partial completion of doctoral program. *Degree requirements:* For master's, thesis optional; for doctorate, thesis/dissertation, preliminary exam. *Entrance requirements:* For master's and doctorate, GRE General Test, minimum GPA of 3.0, 4-year bachelor's degree. Additional exam requirements/recommendations for international students: Required—TOEFL (minimum score 550 paper-based; 213 computer-based). *Application deadline:* For fall admission, 3/1 priority date for domestic students, 1/1 priority date for international students; for spring admission, 8/15 priority date for domestic students, 5/1 priority date for international students. Applications are processed on a rolling basis. Application fee: $30. Electronic applications accepted. *Financial support:* In 2005–06, 100 students received support, including 2 fellowships with full tuition reimbursements available (averaging $18,000 per year), 12 research assistantships with full tuition reimbursements available (averaging $16,000 per year), 79 teaching assistantships with full tuition reimbursements available (averaging $16,000 per year); career-related internships or fieldwork, scholarships/grants, and unspecified assistantships also available. Financial award application deadline: 3/1. *Faculty research:* Low-dimensional manifolds, algebra geometry, fluid dynamics, financial mathematics, biomedical mathematics. *Unit head:* Dr. Philip L Bowers, Chairperson, 850-645-3338, Fax: 850-644-4053, E-mail: bowers@math.fsu.edu. *Application contact:* Dr. Eric P. Klassen, Associate Chair for Graduate Studies, 850-644-2202, Fax: 850-644-4053, E-mail: klassen@math.fsu.edu.

Georgia Institute of Technology, Graduate Studies and Research, College of Management, Program in Management, Atlanta, GA 30332-0001. Offers accounting (PhD); finance (PhD); information technology management (PhD); marketing (PhD); operations management (PhD); organizational behavior (PhD); quantitative and computational finance (MS); strategic management (PhD). *Accreditation:* AACSB. *Degree requirements:* For doctorate, thesis/dissertation, oral exams, comprehensive exam. *Entrance requirements:* For master's and doctorate, GMAT. Additional exam requirements/recommendations for international students: Required—TOEFL. *Faculty research:* MIS, management of technology, international business, entrepreneurship, operations management.

Georgia Institute of Technology, Graduate Studies and Research, College of Sciences, School of Mathematics, Atlanta, GA 30332-0001. Offers algorithms, combinatorics, and optimization (PhD); applied mathematics (MS); bioinformatics (PhD); mathematics (PhD); quantitative and computational finance (MS); statistics (MS Stat). Terminal master's awarded for partial completion of doctoral program. *Degree requirements:* For master's, thesis or alternative; for doctorate, one foreign language, thesis/dissertation. *Entrance requirements:* For master's, GRE General Test, minimum GPA of 3.0; for doctorate, GRE General Test, GRE Subject Test, minimum GPA of 3.0. Additional exam requirements/recommendations for international students: Required—TOEFL. Electronic applications accepted. *Faculty research:* Dynamical systems, discrete mathematics, probability and statistics, mathematical physics.

New York University, Graduate School of Arts and Science, Courant Institute of Mathematical Sciences, Department of Mathematics, New York, NY 10012-1019. Offers atmosphere-ocean science and mathematics (PhD); mathematics (MS, PhD); mathematics and statistics/operations research (MS); mathematics in finance (MS); scientific computing (MS). Part-time and evening/weekend programs available. *Faculty:* 46 full-time (0 women). *Students:* 170 full-time (35 women), 119 part-time (23 women); includes 36 minority (1 African American, 33 Asian Americans or Pacific Islanders, 2 Hispanic Americans), 127 international. Average age 28. 808 applicants, 40% accepted, 92 enrolled. In 2005, 58 master's, 11 doctorates awarded. *Degree requirements:* For master's, thesis optional; for doctorate, one foreign language, thesis/dissertation, oral and written exams. *Entrance requirements:* For master's and doctorate, GRE General Test, GRE Subject Test. Additional exam requirements/recommendations for international students: Required—TOEFL. *Application deadline:* For fall admission, 1/4 for domestic students; for spring admission, 11/1 for domestic students. Application fee: $80. *Financial support:* Fellowships with tuition reimbursements, research assistantships with tuition reimbursements, teaching assistantships with tuition reimbursements, Federal Work-Study, institutionally sponsored loans, scholarships/grants, health care benefits, and unspecified assistantships available. Financial award application deadline: 1/4; financial award applicants required to submit FAFSA. *Faculty research:* Partial differential equations, computational science, applied mathematics, geometry and topology, probability and stochastic processes. *Application contact:* Tamar Arnon, Application Contact, 212-998-3258, E-mail: admissions@math.nyu.edu.

See Close-Up on page 511.

North Carolina State University, Graduate School, College of Agriculture and Life Sciences and College of Engineering and College of Physical and Mathematical Sciences, Program in Financial Mathematics, Raleigh, NC 27695. Offers MFM. Part-time programs available. *Degree requirements:* For master's, project/internship, thesis optional. *Entrance requirements:* For master's, GRE General Test. Additional exam requirements/recommendations for international students: Required—TOEFL (minimum score 550 paper-based; 213 computer-based). Electronic applications accepted. *Faculty research:* Financial mathematics modeling and computation, futures, options and commodities markets, real options, credit risk, portfolio optimization.

See Close-Up on page 513.

Polytechnic University, Westchester Graduate Center, Graduate Programs, Department of Management, Major in Financial Engineering, Hawthorne, NY 10532-1507. Offers capital markets (MS); computational finance (MS); financial engineering (AC); financial technology (MS); financial technology management (AC); information management (AC). *Degree requirements:* For master's, thesis (for some programs), comprehensive exam (for some programs), registration. *Entrance requirements:* Additional exam requirements/recommendations for international students: Required—TOEFL (minimum score 550 paper-based; 213 computer-based); Recommended—IELT (minimum score 7). *Application deadline:* For fall admission, 7/15 priority date for domestic students, 4/1 priority date for international students; for spring admission, 12/15 priority date for domestic students, 10/1 priority date for international students. Applications are processed on a rolling basis. Application fee: $55. Electronic applications accepted. *Expenses:* Tuition: Part-time $950 per unit. Required fees: $330 per term. *Application contact:* Anthea Jeffrey, Graduate Admissions, 718-260-3200, Fax: 718-260-3624, E-mail: gradinfo@poly.edu.

Rice University, Graduate Programs, George R. Brown School of Engineering, Department of Statistics, Houston, TX 77251-1892. Offers biostatistics (PhD); computational finance (PhD); statistics (M Stat, MA, PhD). Terminal master's awarded for partial completion of doctoral program. *Degree requirements:* For master's, thesis/dissertation; for doctorate, thesis/dissertation, comprehensive exam. *Entrance requirements:* For master's and doctorate, GRE General Test, GRE Subject Test, minimum GPA of 3.0. Additional exam requirements/recommendations for international students: Required—TOEFL (minimum score 630 paper-based; 250 computer-based). Electronic applications accepted. *Faculty research:* Statistical genetics, non parametric function estimation, computational statistics and visualization, stochastic processes.

Stanford University, School of Humanities and Sciences, Department of Mathematics, Stanford, CA 94305-9991. Offers financial mathematics (MS); mathematics (MS, PhD). Terminal master's awarded for partial completion of doctoral program. *Degree requirements:* For doctorate, 2 foreign languages, thesis/dissertation, oral exam. *Entrance requirements:* For master's, GRE General Test; for doctorate, GRE General Test, GRE Subject Test. Additional exam requirements/recommendations for international students: Required—TOEFL. Electronic applications accepted.

University of Alberta, Faculty of Graduate Studies and Research, Department of Mathematical and Statistical Sciences, Edmonton, AB T6G 2E1, Canada. Offers applied mathematics (M Sc, PhD); biostatistics (M Sc); mathematical finance (M Sc, PhD); mathematical physics (M Sc, PhD); mathematics (M Sc, PhD); statistics (M Sc, PhD, Postgraduate Diploma). Part-time programs available. *Faculty:* 48 full-time (4 women). *Students:* 112 full-time (41 women), 5 part-time. Average age 24. 776 applicants, 5% accepted, 34 enrolled. In 2005, 12 master's, 10 doctorates awarded. Terminal master's awarded for partial completion of doctoral program. *Median time to degree:* Of those who began their doctoral program in fall 1997, 100% received their degree in 8 years or less. *Degree requirements:* For master's, thesis (for some programs); for doctorate, thesis/dissertation, comprehensive exam. *Entrance requirements:* Additional exam requirements/recommendations for international students: Required—TOEFL (minimum score 580 paper-based; 237 computer-based). *Application deadline:* For fall admission, 3/1 for domestic students, 2/1 for international students. Applications are processed on a rolling basis. Application fee: $0. Electronic applications accepted. Tuition and fees charges are reported in Canadian dollars. *Expenses:* Tuition, state resident: part-time $562 Canadian dollars per term. Tuition, nonresident: full-time $3,375 Canadian dollars. Required fees: $573 Canadian dollars; $84 Canadian dollars per term. *Financial support:* In 2005–06, 51 research assistantships, 88 teaching assistantships with full and partial tuition reimbursements were awarded; scholarships/grants also available. Financial award application deadline:5/1. *Faculty research:* Classical and functional analysis, algebra, differential equations, geometry. *Unit head:* Dr. Anthony To-Ming Lau, Chair, 403-492-5141, E-mail: tlau@math.ualberta.ca. *Application contact:* Dr. Yau Shu Wong, Associate Chair, Graduate Studies, 403-492-5799, Fax: 403-492-6828, E-mail: gradstudies@math.ualberta.ca.

University of California, Santa Barbara, Graduate Division, College of Letters and Sciences, Division of Mathematics, Life, and Physical Sciences, Department of Statistics and Applied Probability, Santa Barbara, CA 93106. Offers applied statistics (MA); mathematical and empirical finance (PhD); mathematical statistics (MA); quantitative methods in the social sciences (PhD); statistics and applied probability (PhD). *Students:* 56 full-time (19 women); includes 28 minority (1 American Indian/Alaska Native, 25 Asian Americans or Pacific Islanders, 2 Hispanic Americans). Average age 24. 90 applicants, 74% accepted, 21 enrolled. In 2005, 10 master's, 3 doctorates awarded. Terminal master's awarded for partial completion of doctoral program. *Median time to degree:* Of those who began their doctoral program in fall 1997, 99% received their degree in 8 years or less. *Degree requirements:* For master's and doctorate, thesis/dissertation, comprehensive exam, registration. *Entrance requirements:* For master's and doctorate, GRE General Test. Additional exam requirements/recommendations for international students: Required—TOEFL (minimum score 550 paper-based; 213 computer-based). *Application deadline:* For fall admission, 4/15 for domestic students, 4/15 for international students. For winter admission, 11/15 for domestic students; for spring admission, 2/15 for domestic students. Applications are processed on a rolling basis. Application fee: $60. Electronic applications accepted. *Financial support:* In 2005–06, 2 fellowships with full tuition reimbursements (averaging $24,500 per year), 2 research assistantships with full tuition reimbursements (averaging $15,500 per year), 29 teaching assistantships with partial tuition reimbursements (averaging $11,500 per year) were awarded; health care benefits and tuition waivers (partial) also available. Financial award application deadline: 1/15; financial award applicants required to submit FAFSA. *Faculty research:* Bayesian methods, statistics including biostatistics, stochastic processes, probability, mathematical finance. *Unit head:* Dr. Raya E. Feldman, Chairman, 805-893-2826, Fax: 805-893-2334, E-mail: feldman@pstat.ucsb.edu. *Application contact:* Andrew V. Carter, Assistant Professor, 805-893-3299, Fax: 805-893-2334, E-mail: carter@pstat.ucsb.edu.

University of Chicago, Division of the Physical Sciences, Department of Mathematics, Program in Financial Mathematics, Chicago, IL 60637-1513. Offers MS. Part-time and evening/weekend programs available. Postbaccalaureate distance learning degree programs offered (no on-campus study). *Faculty:* 2 full-time (0 women), 11 part-time/adjunct (1 woman). *Students:* 36 full-time (8 women), 42 part-time (4 women). Average age 30. 176 applicants, 59% accepted. In 2005, 41 degrees awarded. *Entrance requirements:* For master's, GRE General Test, GRE Subject Test. Additional exam requirements/recommendations for international students: Required—TOEFL (minimum score 600 paper-based; 250 computer-based). *Application deadline:* For fall admission, 1/5 priority date for domestic students, 1/5 priority date for international students. Application fee: $55. Electronic applications accepted. *Financial support:* Fellowships, research assistantships, teaching assistantships, institutionally sponsored loans available. Financial award applicants required to submit FAFSA. *Unit head:* Niels Nygaard, Director, 773-702-7391, Fax: 773-834-4386, E-mail: niels@math.uchicago.edu. *Application contact:* Alice Brugman, Administrator, 773-834-4385, Fax: 773-834-4386, E-mail: alice@math.uchicago.edu.

University of Connecticut, Graduate School, College of Liberal Arts and Sciences, Department of Mathematics, Field of Applied Financial Mathematics, Storrs, CT 06269. Offers MS. *Faculty:* 43 full-time (5 women). *Students:* 11 full-time (6 women), 10 international. Average age 28. 30 applicants, 57% accepted, 5 enrolled. In 2005, 2 degrees awarded. *Degree requirements:* For master's, comprehensive exam. *Entrance requirements:* Additional exam requirements/recommendations for international students: Required—TOEFL (minimum score 550 paper-based; 213 computer-based). *Application deadline:* For fall admission, 2/1 priority date for domestic students, 2/1 priority date for international students; for spring admission, 11/1 for domestic students, 10/1 for international students. Applications are processed on a rolling basis. Application fee: $55. Electronic applications accepted. *Expenses:* Tuition, state resident: part-time $444 per credit hour. Tuition, nonresident: part-time $1,154 per credit hour. Tuition and fees vary according to course load. *Financial support:* In 2005–06, 1 research assistantship, 7 teaching assistantships were awarded; Federal Work-Study and scholarships/grants also available. Financial award application deadline: 2/1; financial award applicants required to submit FAFSA. *Unit head:* James Bridgeman, Chairperson, 860-486-8382, Fax: 860-486-4283, E-mail: james.bridgeman@uconn.edu. *Application contact:* Sharon McDermott, Administrative Assistant, 860-486-6452, Fax: 860-486-4283, E-mail: gradadm@math.uconn.edu.

The University of North Carolina at Charlotte, Graduate School, Belk College of Business Administration, Program in Mathematical Finance, Charlotte, NC 28223-0001. Offers MS. *Students:* 19 full-time (2 women), 27 part-time (6 women); includes 14 minority (8 African Americans, 5 Asian Americans or Pacific Islanders, 1 Hispanic American), 3 international. Average age 31. 44 applicants, 93% accepted, 24 enrolled. In 2005, 5 degrees awarded. *Entrance requirements:* For master's, GRE General Test or GMAT, minimum GPA of 2.75 overall. Additional exam requirements/recommendations for international students: Required—TOEFL (minimum score 557 paper-based; 220 computer-based). *Application deadline:* For fall admission, 7/15 for domestic students, 5/1 for international students; for spring admission, 11/15 for domestic students, 10/1 for international students. Applications are processed on a rolling basis. Application fee: $55. Electronic applications accepted. *Expenses:* Tuition, state resident: full-time $2,504; part-time $157 per credit. Tuition, nonresident: full-time $12,711; part-time $794 per credit. Required fees: $1,424; $89 per credit. Tuition and fees vary according to course load and program. *Financial support:* Fellowships, research assistantships, teaching assistantships, career-related internships or fieldwork, Federal Work-Study, institutionally sponsored loans, scholarships/grants, and unspecified assistantships available. Support available to part-time students. Financial award application deadline: 4/1; financial award applicants required to submit FAFSA. *Unit head:* Dr. Richard J. Buttimer, Director, 704-687-6219, Fax: 704-687-6987, E-mail: buttimer@email.uncc.edu. *Application contact:* Kathy B. Giddings, Director of Graduate Admissions, 704-687-3366, Fax: 704-687-3279, E-mail: gradadm@email.uncc.edu.

University of Pittsburgh, School of Arts and Sciences, Department of Mathematics, Pittsburgh, PA 15260. Offers applied mathematics (MA, MS); financial mathematics (PMS); mathematics (MA, MS, PhD). Part-time programs available. *Faculty:* 35 full-time (4 women), 4 part-time/adjunct (1 woman). *Students:* 82 full-time (25 women), 12 part-time (5 women); includes 1 minority (Asian American or Pacific Islander), 43 international. 123 applicants, 67% accepted, 28 enrolled. In 2005, 5 master's, 3 doctorates awarded. Terminal master's awarded for partial completion of doctoral program. *Median time to degree:* Of those who began their doctoral program in fall 1997, 100% received their degree in 8 years or less. *Degree requirements:* For master's, thesis (for some programs), comprehensive exam; for doctorate, thesis/dissertation, preliminary exams, comprehensive exam. *Entrance requirements:* For master's and doctorate, GRE General Test, GRE Subject Test (recommended), minimum GPA of 3.0. Additional exam requirements/recommendations for international students: Required—TOEFL (minimum score 550 paper-based; 213 computer-based). *Application deadline:* For fall admission, 1/15 priority date for domestic students, 1/2 priority date for international students; for spring admission, 9/1 priority date for domestic students, 9/1 priority date for international students. Applications are processed on a rolling basis. Application fee: $50. Electronic applications accepted. *Expenses:* Tuition, state resident: full-time $13,194; part-time $537 per credit. Tuition, nonresident: full-time $25,012; part-time $1,026 per credit. Required fees: $700; $164 per term. Tuition and fees vary according to campus/location and program. *Financial support:* In 2005–06, 6 fellowships with full and partial tuition reimbursements (averaging $16,500 per year), 13 research assistantships with full and partial tuition reimbursements (averaging $13,600 per year), 50 teaching assistantships with full and partial tuition reimbursements (averaging $13,555 per year) were awarded; career-related internships or fieldwork, Federal Work-Study, institutionally sponsored loans, scholarships/grants, health care benefits, tuition waivers (partial), and unspecified assistantships also available. Financial award application deadline:1/15. *Faculty research:* Computational math, math biology, math finance, algebra, analysis. Total annual research expenditures: $700,000. *Unit head:* Juan Manfredi, Chairman, 412-624-8307, Fax: 412-624-8697, E-mail: manfredi@pitt.edu. *Application contact:* Molly Williams, Administrator, 412-624-1175, Fax: 412-624-8397, E-mail: mollyw@pitt.edu.

Mathematics

Alabama State University, School of Graduate Studies, College of Arts and Sciences, Department of Mathematics, Computers, and Physical Science, Montgomery, AL 36101-0271. Offers mathematics (M Ed, MS, Ed S). Part-time programs available. *Degree requirements:* For Ed S, thesis. *Entrance requirements:* For master's, GRE, GRE Subject Test, graduate writing competence test; for Ed S, graduate writing competency test, GRE, MAT. Additional exam requirements/recommendations for international students: Required—TOEFL (minimum score 500 paper-based; 173 computer-based). *Faculty research:* Discrete mathematics, symbolic dynamics, mathematical social sciences.

American University, College of Arts and Sciences, Department of Mathematics and Statistics, Program in Mathematics, Washington, DC 20016-8001. Offers MA. Part-time and evening/weekend programs available. *Students:* 2 full-time (both women), 2 part-time (1 woman). Average age 23. In 2005, 1 degree awarded. *Degree requirements:* For master's, one foreign language, thesis or alternative. *Entrance requirements:* For master's, GRE, BA in mathematics. *Application deadline:* For fall admission, 2/1 for domestic students; for spring admission, 10/1 for domestic students. Application fee: $50. *Expenses:* Tuition: Full-time $17,802; part-time $989 per credit. Required fees: $380. *Financial support:* Fellowships, teaching assistantships, career-related internships or fieldwork, Federal Work-Study, and institutionally sponsored loans available. Support available to part-time students. Financial award application deadline: 2/1. *Unit head:* Dr. Jeffrey Hakim, Chair, Department of Mathematics and Statistics, 202-885-3131, Fax: 202-885-3155.

American University of Beirut, Graduate Programs, Faculty of Arts and Sciences, Beirut, Lebanon. Offers anthropology (MA); Arabic language and literature (MA); archaeology (MA); biology (MS); business administration (MBA); chemistry (MS); computer science (MS); economics (MA); education (MA); English language (MA); English literature (MA); environmental policy planning (MSES); finance and banking (MFB); financial economics (MFE); geology (MS); history (MA); mathematics (MS); Middle Eastern studies (MA); philosophy (MA); physics (MS); political studies (MA); psychology (MA); public administration (MA); sociology (MA). *Degree requirements:* For master's, one foreign language, thesis (for some programs), comprehensive exam, registration. *Entrance requirements:* For master's, GRE, letter of recommendation.

Andrews University, School of Graduate Studies, College of Arts and Sciences, Interdisciplinary Studies in Mathematics and Physical Science Program, Berrien Springs, MI 49104. Offers MS. *Students:* 3 full-time (1 woman), 1 (woman) part-time; includes 2 minority (both African Americans) Average age 24. *Application deadline:* Applications are processed on a rolling basis. Application fee: $43. *Unit head:* Dr. Margarita Mattingly, Chairman, 269-471-3431. *Application contact:* Carolyn Hurst, Supervisor of Graduate Admission, 800-253-3430, Fax: 269-471-3228, E-mail: enroll@andrews.edu.

Appalachian State University, Cratis D. Williams Graduate School, College of Arts and Sciences, Department of Mathematics, Boone, NC 28608. Offers mathematics (MA); mathematics education (MA). Part-time programs available. *Faculty:* 19 full-time (9 women), 1 (woman) part-time/adjunct. *Students:* 13 full-time (7 women), 11 part-time (9 women); includes 2 minority (both African Americans) 16 applicants, 88% accepted, 12 enrolled. In 2005, 10 degrees awarded. *Degree requirements:* For master's, one foreign language, comprehensive exam, registration. *Entrance requirements:* For master's, GRE General Test. Additional exam requirements/recommendations for international students: Required—TOEFL (minimum score 570 paper-based; 230 computer-based). *Application deadline:* For fall admission, 7/1 for domestic students, 1/1 for international students; for spring admission, 11/1 for domestic students, 6/1 for international students. Application fee: $45. *Expenses:* Tuition, state resident: full-time $2,593. Tuition, nonresident: full-time $12,176. Required fees: $1,726. *Financial support:* In 2005–06, 16 teaching assistantships (averaging $9,000 per year) were awarded; fellowships, research assistantships, career-related internships or fieldwork, Federal Work-Study, scholarships/grants, and unspecified assistantships also available. Support available to part-time students. Financial award application deadline: 7/1. *Faculty research:* Graph theory, differential equations, logic, geometry, complex analysis. Total annual research expenditures: $19,139. *Unit head:* Dr. William Bauldry, Chair, 828-262-3050, Fax: 828-265-8617, E-mail: wmcb@math.appstate.edu. *Application contact:* Dr. Richard Klima, Graduate Director, 828-262-3050, E-mail: klimare@math.appstate.edu.

Arizona State University, Division of Graduate Studies, College of Liberal Arts and Sciences, Division of Natural Sciences and Mathematics, Department of Mathematics and Statistics, Tempe, AZ 85287. Offers applied mathematics (MA, PhD); mathematics (MA, MNS, PhD); statistics (MA, PhD). *Degree requirements:* For master's, thesis or alternative; for doctorate, one foreign language, thesis/dissertation. *Entrance requirements:* For master's and doctorate, GRE General Test.

Arkansas State University, Graduate School, College of Sciences and Mathematics, Department of Computer Sciences, Jonesboro, State University, AR 72467. Offers computer science (MS). Part-time programs available. *Faculty:* 5 full-time (1 woman). *Students:* 8 full-time (2 women), 6 part-time (1 woman), 7 international. Average age 28. 6 applicants, 100% accepted, 5 enrolled. In 2005, 4 degrees awarded. *Degree requirements:* For master's, thesis or alternative, comprehensive exam. *Entrance requirements:* For master's, GRE General Test or MAT, appropriate bachelor's degree. Additional exam requirements/recommendations for international students: Required—TOEFL (minimum score 213 computer-based). *Application deadline:* For fall admission, 7/1 for domestic students; for spring admission, 11/15 priority date for domestic students. Applications are processed on a rolling basis. Application fee: $15 ($25 for international students). Electronic applications accepted. *Expenses:* Tuition, state resident: full-time $3,232; part-time $180 per hour. Tuition, nonresident: full-time $8,164; part-time $454 per hour. Required fees: $716; $37 per hour. $25 per semester. Tuition and fees vary according to course load and program. *Financial support:* Teaching assistantships, scholarships/grants and unspecified assistantships available. Financial award application deadline: 7/1; financial award applicants required to submit FAFSA. *Unit head:* Dr. Jeff Jenness, Chair, 870-972-3978, Fax: 870-972-3950, E-mail: jeffj@csm.astate.edu.

Arkansas State University, Graduate School, College of Sciences and Mathematics, Department of Mathematics and Statistics, Jonesboro, State University, AR 72467. Offers mathematics (MS, MSE). Part-time programs available. *Faculty:* 10 full-time (3 women). *Students:* 4 full-time (2 women), 12 part-time (4 women); includes 2 minority (both African Americans) Average age 27. 11 applicants, 91% accepted, 9 enrolled. In 2005, 5 degrees awarded. *Degree requirements:* For master's, thesis or alternative, comprehensive exam. *Entrance requirements:* For master's, GRE General Test or MAT, appropriate bachelor's degree. Additional exam requirements/recommendations for international students: Required—TOEFL (minimum score 213 computer-based). *Application deadline:* For fall admission, 7/1 for domestic students; for spring admission, 11/15 priority date for domestic students. Applications are processed on a rolling basis. Application fee: $15 ($25 for international students). Electronic applications accepted. *Expenses:* Tuition, state resident: full-time $3,232; part-time $180 per hour. Tuition, nonresident: full-time $8,164; part-time $454 per hour. Required fees: $716; $37 per hour. $25 per semester. Tuition and fees vary according to course load and program. *Financial support:* Teaching assistantships, scholarships/grants and unspecified assistantships available. Financial award application deadline: 7/1; financial award applicants required to submit FAFSA. *Unit head:* Dr. Jerry Linnstaedter, Chair, 870-972-3090, Fax: 870-972-3950, E-mail: linnstaedter@csm.astate.edu.

Auburn University, Graduate School, College of Sciences and Mathematics, Department of Mathematics and Statistics, Auburn University, AL 36849. Offers M Prob S, MAM, MS, PhD. *Faculty:* 52 full-time (4 women). *Students:* 38 full-time (14 women), 22 part-time (8 women); includes 7 minority (4 African Americans, 3 Asian Americans or Pacific Islanders), 26 international. 74 applicants, 43% accepted, 17 enrolled. In 2005, 13 master's, 6 doctorates awarded. *Degree requirements:* For doctorate, thesis/dissertation. *Entrance requirements:* For master's, GRE General Test, undergraduate mathematics background; for doctorate, GRE General Test, GRE Subject Test. *Application deadline:* For fall admission, 7/7 for domestic students; for spring admission, 11/24 for domestic students. Applications are processed on a rolling basis. Application fee: $25 ($50 for international students). Electronic applications accepted. *Financial support:* Fellowships, teaching assistantships, special tuition awards available. *Faculty research:* Pure and applied mathematics. *Unit head:* Dr. Michel Smith, Chair, 334-844-4290, Fax: 334-844-6655. *Application contact:* Dr. Stephen L. McFarland, Acting Dean of the Graduate School, 334-844-4700.

Ball State University, Graduate School, College of Sciences and Humanities, Department of Mathematical Sciences, Program in Mathematics, Muncie, IN 47306-1099. Offers mathematics (MA, MS); mathematics education (MAE). *Students:* 5 full-time (1 woman), 1 part-time; includes 1 minority (Asian American or Pacific Islander) Average age 39. 10 applicants, 60% accepted, 3 enrolled. In 2005, 5 degrees awarded. Application fee: $25 ($35 for international students). *Expenses:* Tuition, state resident: full-time $6,246. Tuition, nonresident: full-time $16,006. *Financial support:* Research assistantships with full tuition reimbursements, teaching assistantships with tuition reimbursements available. Financial award application deadline: 3/1. *Unit head:* Dr. Richard Stankewitz, Director, 765-285-8662, Fax: 765-285-1721.

Baylor University, Graduate School, College of Arts and Sciences, Department of Mathematics, Waco, TX 76798. Offers MS, PhD. *Students:* 18 full-time (5 women), 4 international. In 2005, 6 master's, 1 doctorate awarded. *Degree requirements:* For master's, final oral exam. *Entrance requirements:* For master's, GRE General Test. *Application deadline:* For fall admission, 8/1 for domestic students. Applications are processed on a rolling basis. Application fee: $25. *Financial support:* Teaching assistantships, career-related internships or fieldwork, Federal Work-Study, and institutionally sponsored loans available. Support available to part-time students. Financial award application deadline: 5/1. *Faculty research:* Algebra, statistics, probability, applied mathematics, numerical analysis. *Unit head:* Dr. Frank Mathis, Graduate Program Director, 254-710-3561, Fax: 254-710-3569, E-mail: frank_mathis@baylor.edu. *Application contact:* Suzanne Keener, Administrative Assistant, 254-710-3588, Fax: 254-710-3870.

Boston College, Graduate School of Arts and Sciences, Department of Mathematics, Chestnut Hill, MA 02467-3800. Offers MA, MBA/MA. Part-time programs available. *Students:* 12 full-time (6 women), 4 part-time (all women), 1 international. 55 applicants, 65% accepted, 14 enrolled. In 2005, 7 degrees awarded. *Degree requirements:* For master's, oral presentation, thesis optional. *Entrance requirements:* For master's, GRE General Test. Additional exam requirements/recommendations for international students: Required—TOEFL (minimum score 550 paper-based; 213 computer-based). *Application deadline:* For fall admission, 1/15 for domestic students. Application fee: $70. Electronic applications accepted. *Financial support:* Fellowships with full tuition reimbursements, teaching assistantships with full tuition reimbursements, Federal Work-Study and scholarships/grants available. Support available to part-time students. Financial award application deadline: 3/1; financial award applicants required to submit FAFSA. *Faculty research:* Abstract algebra and number theory, topology, probability and statistics, computer science, analysis. *Unit head:* Dr. Gerald Keough, Chairperson, 617-552-3750, E-mail: gerald.keough@bc.edu. *Application contact:* Dr. Nancy Rallis, Graduate Program Director, 617-552-3964, E-mail: nancy.rallis@bc.edu.

See Close-Up on page 475.

Boston University, Graduate School of Arts and Sciences, Department of Mathematics and Statistics, Boston, MA 02215. Offers mathematical finance (MA); mathematics (MA, PhD). *Students:* 72 full-time (28 women), 9 part-time (2 women); includes 5 minority (1 African American, 3 Asian Americans or Pacific Islanders, 1 Hispanic American), 37 international. Average age 28. 269 applicants, 29% accepted, 38 enrolled. In 2005, 31 master's, 8 doctorates awarded. Terminal master's awarded for partial completion of doctoral program. *Degree requirements:* For master's, one foreign language, comprehensive exam, registration; for doctorate, one foreign language, thesis/dissertation, comprehensive exam, registration. *Entrance requirements:* For master's and doctorate, GRE General Test, GRE Subject Test, 3 letters of recommendation. Additional exam requirements/recommendations for international students: Required—TOEFL (minimum score 600 paper-based; 250 computer-based). *Application deadline:* For fall admission, 1/15 for domestic students, 1/15 for international students; for spring admission, 10/15 for domestic students, 10/15 for international students. Application fee: $60. *Expenses:* Tuition: Full-time $31,530; part-time $985 per credit. Required fees: $316; $40 per semester. Tuition and fees vary according to course level and program. *Financial support:* In 2005–06, 57 students received support, including 4 fellowships with full tuition reimbursements available (averaging $16,500 per year), 26 research assistantships with full tuition reimbursements available (averaging $16,000 per year), 26 teaching assistantships with full tuition reimbursements available (averaging $16,000 per year); Federal Work-Study and scholarships/grants also available. Support available to part-time students. Financial award application deadline: 1/15; financial award applicants required to submit FAFSA. *Unit head:* Steven Rosenberg, Chairman, 617-353-9556, Fax: 617-353-8100, E-mail: sr@bu.edu. *Application contact:* Angela M. Silva, Staff Coordinator, 617-353-2560, Fax: 617-353-8100, E-mail: amsilva@bu.edu.

Bowling Green State University, Graduate College, College of Arts and Sciences, Department of Mathematics and Statistics, Bowling Green, OH 43403. Offers applied statistics (MS); mathematics (MA, MAT, PhD); mathematics supervision (Ed S); statistics (MA, MAT, PhD). Part-time programs available. *Faculty:* 24 full-time (1 woman), 1 (woman) part-time/adjunct. *Students:* 58 full-time (21 women), 13 part-time (7 women); includes 1 African American, 32 international. Average age 28. 138 applicants, 68% accepted, 10 enrolled. In 2005, 16 master's, 6 doctorates awarded. *Degree requirements:* For master's, thesis or alternative; for doctorate, thesis/dissertation, comprehensive exam; for Ed S, internship. *Entrance requirements:* For master's and doctorate, GRE General Test. Additional exam requirements/recommendations for international students: Required—TOEFL. Application fee: $30. Electronic applications accepted. *Financial support:* In 2005–06, 1 research assistantship with full tuition reimbursement (averaging $10,000 per year), 53 teaching assistantships with full tuition reimbursements (averaging $12,493 per year) were awarded; Federal Work-Study, institutionally sponsored loans, and unspecified assistantships also available. Financial award applicants required to submit FAFSA. *Faculty research:* Statistics and probability, algebra, analysis. *Unit head:* Dr. Neal Carothers, Chair, 419-372-7453. *Application contact:* Dr. Hanfeng Chen, Graduate Coordinator, 419-372-7463, Fax: 419-372-6092.

See Close-Up on page 477.

Brandeis University, Graduate School of Arts and Sciences, Department of Mathematics, Waltham, MA 02454-9110. Offers MA, PhD. *Faculty:* 14 full-time (3 women), 2 part-time/adjunct (0 women). *Students:* 33 full-time (5 women); includes 18 minority (all Asian Americans or Pacific Islanders), 1 international. Average age 25. 124 applicants, 7% accepted. In 2005, 3 master's, 3 doctorates awarded. *Degree requirements:* For doctorate, 2 foreign languages, thesis/dissertation. *Entrance requirements:* For doctorate, GRE General Test, GRE Subject Test, resumé, letters of recommendation. Additional exam requirements/recommendations for international students: Required—TOEFL (minimum score 600 paper-based; 250 computer-based). *Application deadline:* For fall admission, 2/15 for domestic students. Application fee: $55. Electronic applications accepted. *Financial support:* In 2005–06, 22 students received support, including 21 fellowships with full tuition reimbursements available (averaging $9,500 per year), 14 teaching assistantships with full tuition reimbursements available (averaging $6,000 per year); research assistantships, scholarships/grants and tuition waivers (full) also available. Financial award application deadline: 4/15; financial award applicants required to submit CSS PROFILE or FAFSA. *Faculty research:* Algebra, analysis, number theory, combinatorics, topology. *Unit head:* Dr. Kiyoshi Igusa, Chair, 781-736-3062, Fax: 781-736-3085,

E-mail: igusa@brandeis.edu. *Application contact:* Prof. Ira Gessel, Graduate Advisor, 781-736-3063, Fax: 781-736-3085, E-mail: gessel@brandeis.edu.

Announcement: Program directed primarily toward PhD in pure mathematics. Benefits from informality, flexibility, and warmth of small department and from intellectual vigor of faculty well known for research accomplishments. Brandeis-Harvard-MIT-Northeastern Colloquium and many joint seminars provide opportunities for contact with other Boston-area mathematicians. Students normally receive full-tuition scholarship and teaching assistantship or fellowship. Contact Professor Ira Gessel, Graduate Adviser, 781-736-3060. E-mail: gessel@brandeis.edu. Web site: http://www.math.brandeis.edu/

Brigham Young University, Graduate Studies, College of Physical and Mathematical Sciences, Department of Mathematics, Provo, UT 84602-1001. Offers MS, PhD. Part-time programs available. *Faculty:* 33 full-time (1 woman). *Students:* 16 full-time (3 women), 13 part-time (2 women); includes 6 minority (all Asian Americans or Pacific Islanders) Average age 23. 28 applicants, 61% accepted, 14 enrolled. In 2005, 11 master's, 1 doctorate awarded. Terminal master's awarded for partial completion of doctoral program. *Degree requirements:* For master's, project or thesis, written exams; for doctorate, one foreign language, thesis/dissertation, qualifying exams. *Entrance requirements:* For master's, GRE General Test, GRE Subject Test, minimum GPA of 3.0 in last 60 hours, bachelor's degree in mathematics; for doctorate, GRE General Test, GRE Subject Test, master's degree in mathematics or related field. Additional exam requirements/recommendations for international students: Required—TOEFL. *Application deadline:* For fall admission, 3/1 priority date for domestic students, 3/1 priority date for international students. For winter admission, 9/15 for domestic students; for spring admission, 2/15 for domestic students. Applications are processed on a rolling basis. Application fee: $50. Electronic applications accepted. *Financial support:* In 2005–06, 29 students received support, including 2 research assistantships with full tuition reimbursements available (averaging $14,550 per year), 27 teaching assistantships with full tuition reimbursements available (averaging $14,550 per year); institutionally sponsored loans also available. Support available to part-time students. Financial award application deadline: 3/1. *Faculty research:* Algebraic geometry/number theory, applied math/nonlinear PDEs, combinatorics/matrix theory, geometric group theory/topology. Total annual research expenditures: $240,000. *Unit head:* Dr. Lynn E. Garner, Chairperson, 801-422-6153, Fax: 801-422-0504, E-mail: lynng@math.byu.edu. *Application contact:* Lonette Stoddard, Graduate Secretary, 801-422-2062, Fax: 801-422-0504, E-mail: gradschool@math.byu.edu.

Brock University, Graduate Studies, Faculty of Mathematics and Science, St. Catharines, ON L2S 3A1, Canada. Offers biological sciences (M Sc, PhD), including biology; biotechnology (M Sc, PhD); chemistry (M Sc); computer science (M Sc); earth sciences (M Sc); mathematics and statistics (M Sc); physics (M Sc). Part-time programs available. *Faculty:* 77 full-time (18 women), 10 part-time/adjunct (3 women). *Students:* 69 full-time (33 women), 3 part-time, 26 international. 137 applicants, 19% accepted, 20 enrolled. In 2005, 13 degrees awarded. *Degree requirements:* For master's, thesis. *Entrance requirements:* For master's, honors B Sc. Additional exam requirements/recommendations for international students: Required—TOEFL. *Application deadline:* Applications are processed on a rolling basis. Application fee: $75. Electronic applications accepted. *Financial support:* Fellowships, research assistantships, teaching assistantships, career-related internships or fieldwork, scholarships/grants, and unspecified assistantships available. Support available to part-time students. *Unit head:* Dr. Ian Brindle, Dean, 905-688-5550 Ext. 3421, Fax: 905-641-0406, E-mail: ibrindle@brocku.ca. *Application contact:* Charlotte F. Sheridan, Associate Director, Office of Graduate Studies, 905-688-5550 Ext. 4390, Fax: 905-688-0748, E-mail: csherida@brucku.ca.

Brooklyn College of the City University of New York, Division of Graduate Studies, Department of Mathematics, Brooklyn, NY 11210-2889. Offers mathematics (MA, PhD); secondary mathematics education (MA). The department offers courses at Brooklyn College that are creditable toward the CUNY doctoral degree (with permission of the executive officer of the doctoral program). Part-time and evening/weekend programs available. *Degree requirements:* For master's, comprehensive exam (mathematics). *Entrance requirements:* For master's, minimum GPA of 3.0. Additional exam requirements/recommendations for international students: Required—TOEFL. *Faculty research:* Differential geometry, gauge theory, complex analysis, orthogonal functions.

Brown University, Graduate School, Department of Mathematics, Providence, RI 02912. Offers M Sc, MA, PhD. *Faculty:* 22 full-time (2 women). *Students:* 30 full-time (11 women); includes 1 minority (Asian American or Pacific Islander), 21 international. Average age 27. 155 applicants, 16% accepted, 7 enrolled. In 2005, 3 master's, 6 doctorates awarded. *Median time to degree:* Of those who began their doctoral program in fall 1997, 100% received their degree in 8 years or less. *Degree requirements:* For doctorate, one foreign language, thesis/dissertation. *Entrance requirements:* For doctorate, GRE. Additional exam requirements/recommendations for international students: Required—TOEFL (minimum score 550 paper-based; 173 computer-based). *Application deadline:* For fall admission, 1/10 priority date for domestic students, 1/10 priority date for international students. Application fee: $70. Electronic applications accepted. *Financial support:* In 2005–06, 30 students received support, including 12 fellowships with full tuition reimbursements available (averaging $17,000 per year), 4 research assistantships with full tuition reimbursements available (averaging $17,000 per year), 14 teaching assistantships with full tuition reimbursements available (averaging $15,000 per year); Federal Work-Study, institutionally sponsored loans, and tuition waivers (full and partial) also available. Financial award application deadline: 1/10; financial award applicants required to submit FAFSA. *Faculty research:* Algebraic geometry, number theory, functional analysis, geometry, topology. Total annual research expenditures:$820,481. *Unit head:* Prof. Jill C. Pipher, Chair, 401-863-3319, Fax: 401-863-9471, E-mail: jpipher@math.brown.edu. *Application contact:* Prof. Thomas G. Goodwillie, Graduate Advisor, 401-863-2590, Fax: 401-863-9471, E-mail: tomg@math.brown.edu.

Bryn Mawr College, Graduate School of Arts and Sciences, Department of Mathematics, Bryn Mawr, PA 19010-2899. Offers MA, PhD. Part-time programs available. *Students:* 10 full-time (8 women), 6 part-time (4 women); includes 3 minority (1 African American, 1 Asian American or Pacific Islander, 1 Hispanic American), 2 international. 7 applicants, 43% accepted, 3 enrolled. In 2005, 3 degrees awarded. *Degree requirements:* For master's, one foreign language, thesis, registration; for doctorate, 2 foreign languages, thesis/dissertation, comprehensive exam, registration. *Entrance requirements:* For master's and doctorate, GRE General Test. Additional exam requirements/recommendations for international students: Required—TOEFL (minimum score 600 paper-based; 200 computer-based). *Application deadline:* For fall admission, 1/13 for domestic students, 1/13 for international students. Application fee: $30. *Financial support:* In 2005–06, 4 teaching assistantships with partial tuition reimbursements were awarded; research assistantships with full tuition reimbursements, Federal Work-Study, scholarships/grants, tuition waivers (full and partial), unspecified assistantships, and tuition awards also available. Support available to part-time students. Financial award application deadline: 1/13. *Unit head:* Dr. Paul Melvin, Chair, 610-526-5348. *Application contact:* Lea R. Miller, Secretary, 610-526-5072, Fax: 610-526-5076, E-mail: lrmiller@brynmawr.edu.

Bucknell University, Graduate Studies, College of Arts and Sciences, Department of Mathematics, Lewisburg, PA 17837. Offers MA, MS. Part-time programs available. *Entrance requirements:* For master's, GRE General Test, GRE Subject Test, minimum GPA of 2.8. Additional exam requirements/recommendations for international students: Required—TOEFL.

California Institute of Technology, Division of Physics, Mathematics and Astronomy, Department of Mathematics, Pasadena, CA 91125-0001. Offers PhD. *Degree requirements:* For doctorate, one foreign language, thesis/dissertation, candidacy and final exams. *Entrance requirements:* For doctorate, GRE General Test, GRE Subject Test. Additional exam requirements/recommendations for international students: Required—TOEFL. *Faculty research:* Number theory, combinatorics, differential geometry, dynamical systems, finite groups.

California Polytechnic State University, San Luis Obispo, College of Science and Mathematics, Department of Mathematics, San Luis Obispo, CA 93407. Offers MS. Part-time programs available. *Faculty:* 9 full-time (3 women), 1 part-time/adjunct (0 women). *Students:* 19 full-time (8 women), 2 part-time. 22 applicants, 64% accepted, 11 enrolled. In 2005, 1 degree awarded. *Degree requirements:* For master's, qualifying exams. *Entrance requirements:* For master's, minimum GPA of 2.5 in last 90 quarter units of course work. Additional exam requirements/recommendations for international students: Required—TOEFL, TWE. *Application deadline:* For fall admission, 6/1 for domestic students, 11/30 for international students. For winter admission, 8/1 for domestic students; for spring admission, 12/1 for domestic students. Applications are processed on a rolling basis. Application fee: $55. *Expenses:* Tuition, nonresident: part-time $226 per unit. Required fees: $1,063 per unit. *Financial support:* Teaching assistantships, career-related internships or fieldwork and Federal Work-Study available. Support available to part-time students. Financial award application deadline: 3/2; financial award applicants required to submit FAFSA. *Unit head:* Dr. Estelle Basor, Graduate Coordinator, 805-756-2206, Fax: 805-756-6537, E-mail: ebasor@calpoly.edu. *Application contact:* Myron Hood, Graduate Coordinator, 805-756-2352, Fax: 805-756-6537, E-mail: mhood@calpoly.edu.

California State Polytechnic University, Pomona, Academic Affairs, College of Science, Program in Mathematics, Pomona, CA 91768-2557. Offers applied mathematics (MS); pure mathematics (MS). Part-time programs available. *Students:* 27 full-time (12 women), 27 part-time (7 women); includes 27 minority (1 African American, 18 Asian Americans or Pacific Islanders, 8 Hispanic Americans), 3 international. Average age 29. 45 applicants, 78% accepted, 17 enrolled. In 2005, 10 degrees awarded. *Degree requirements:* For master's, thesis or alternative. *Entrance requirements:* For master's, GRE General Test. *Application deadline:* For fall admission, 5/1 for domestic students. For winter admission, 10/15 for domestic students; for spring admission, 1/20 for domestic students. Applications are processed on a rolling basis. Application fee: $55. Electronic applications accepted. *Expenses:* Tuition, nonresident: full-time $9,021. Required fees: $3,597. *Financial support:* Career-related internships or fieldwork, Federal Work-Study, and institutionally sponsored loans available. Support available to part-time students. Financial award application deadline: 3/2; financial award applicants required to submit FAFSA. *Unit head:* Dr. Michael L. Green, Graduate Coordinator, 909-869-4007.

California State University Channel Islands, Extended Education, Program in Mathematics, Camarillo, CA 93012. Offers MS. *Entrance requirements:* Additional exam requirements/recommendations for international students: Required—TOEFL (minimum score 550 paper-based).

See Close-Up on page 483.

California State University, East Bay, Academic Programs and Graduate Studies, College of Science, Department of Mathematics and Computer Science, Mathematics Program, Hayward, CA 94542-3000. Offers MS. *Students:* 68. 30 applicants, 90% accepted. *Degree requirements:* For master's, comprehensive exam or thesis. *Entrance requirements:* For master's, minimum GPA of 3.0 in field. Additional exam requirements/recommendations for international students: Required—TOEFL (minimum score 550 paper-based; 213 computer-based). *Application deadline:* For fall admission, 5/30 for domestic students, 4/30 for international students. For winter admission, 9/30 for domestic students. Applications are processed on a rolling basis. Application fee: $55. Electronic applications accepted. *Financial support:* Career-related internships or fieldwork, Federal Work-Study, and institutionally sponsored loans available. Support available to part-time students. Financial award application deadline: 3/2. *Unit head:* Donald L. Wolitzer, Coordinator, 510-885-3467, E-mail: donald.wolitzer@csueastbay.edu. *Application contact:* Deborah Baker, Associate Director, 510-885-3286, Fax: 510-885-4777, E-mail: deborah.baker@csueastbay.edu.

California State University, Fresno, Division of Graduate Studies, College of Science and Mathematics, Department of Mathematics, Fresno, CA 93740-8027. Offers mathematics (MA); teaching (MA). Part-time programs available. *Degree requirements:* For master's, thesis or alternative. *Entrance requirements:* For master's, GRE General Test. Additional exam requirements/recommendations for international students: Required—TOEFL. Electronic applications accepted. *Faculty research:* Diagnostic testing project.

California State University, Fullerton, Graduate Studies, College of Natural Science and Mathematics, Department of Mathematics, Fullerton, CA 92834-9480. Offers applied mathematics (MA); mathematics (MA); mathematics for secondary school teachers (MA). Part-time programs available. *Students:* 8 full-time (3 women), 68 part-time (39 women); includes 36 minority (1 African American, 1 American Indian/Alaska Native, 19 Asian Americans or Pacific Islanders, 15 Hispanic Americans), 2 international. Average age 32. 78 applicants, 79% accepted, 40 enrolled. In 2005, 25 degrees awarded. *Degree requirements:* For master's, comprehensive exam or project. *Entrance requirements:* For master's, minimum GPA of 2.5 in last 60 units of course work, major in mathematics or related field. Application fee: $55. *Expenses:* Tuition, nonresident: part-time $339 per unit. *Financial support:* Research assistantships, teaching assistantships, career-related internships or fieldwork, Federal Work-Study, institutionally sponsored loans, and scholarships/grants available. Support available to part-time students. Financial award application deadline: 3/1. *Unit head:* Dr. Paul Deland, Chair, 714-278-3631.

California State University, Long Beach, Graduate Studies, College of Natural Sciences and Mathematics, Department of Mathematics and Statistics, Long Beach, CA 90840. Offers applied mathematics (MA); mathematics (MSEE). Part-time programs available. *Faculty:* 20 full-time (1 woman). *Students:* 47 full-time (22 women), 64 part-time (25 women); includes 62 minority (3 African Americans, 42 Asian Americans or Pacific Islanders, 17 Hispanic Americans), 3 international. Average age 33. 95 applicants, 62% accepted, 44 enrolled. In 2005, 15 degrees awarded. *Degree requirements:* For master's, comprehensive exam or thesis. *Application deadline:* For fall admission, 7/1 for domestic students; for spring admission, 12/1 for domestic students. Applications are processed on a rolling basis. Application fee: $55. Electronic applications accepted. *Expenses:* Tuition, nonresident: part-time $339 per semester hour. *Financial support:* Teaching assistantships, Federal Work-Study, institutionally sponsored loans, scholarships/grants, and traineeships available. Financial award application deadline: 3/2. *Faculty research:* Algebra, functional analysis, partial differential equations, operator theory, numerical analysis. *Unit head:* Dr. Robert A Mena, Chair, 562-985-4721, Fax: 562-985-8227, E-mail: rmena@csulb.edu. *Application contact:* Dr. Ngo Viet, Graduate Coordinator, 562-985-5610, Fax: 562-985-8227, E-mail: viet@csulb.edu.

California State University, Los Angeles, Graduate Studies, College of Natural and Social Sciences, Department of Mathematics, Los Angeles, CA 90032-8530. Offers mathematics (MS), including applied mathematics, mathematics. Part-time and evening/weekend programs available. *Faculty:* 7 full-time (1 woman). *Students:* 18 full-time (7 women), 50 part-time (17 women); includes 51 minority (6 African Americans, 19 Asian Americans or Pacific Islanders, 26 Hispanic Americans). In 2005, 12 degrees awarded. *Degree requirements:* For master's, comprehensive exam or thesis. *Entrance requirements:* For master's, previous course work in mathematics. Additional exam requirements/recommendations for international students: Required—TOEFL. *Application deadline:* For fall admission, 6/30 for domestic students; for spring admission, 2/1 for domestic students. Applications are processed on a rolling basis. Application fee: $55. *Financial support:* Teaching assistantships, Federal Work-Study available. Support available to part-time students. Financial award application deadline: 3/1. *Faculty research:* Group theory, functional analysis, convexity theory, ordered geometry. *Unit head:* Dr. P. K. Subramanian, Chair, 323-343-2150, Fax: 323-343-5071.

California State University, Northridge, Graduate Studies, College of Science and Mathematics, Department of Mathematics, Northridge, CA 91330. Offers MS. Part-time and evening/

Mathematics

California State University, Northridge (continued)
weekend programs available. *Degree requirements:* For master's, thesis (for some programs). *Entrance requirements:* Additional exam requirements/recommendations for international students: Required—TOEFL.

California State University, Sacramento, Graduate Studies, College of Natural Sciences and Mathematics, Department of Mathematics and Statistics, Sacramento, CA 95819-6048. Offers MA. Part-time programs available. *Students:* 6 full-time (3 women), 28 part-time (13 women); includes 9 minority (1 African American, 6 Asian Americans or Pacific Islanders, 2 Hispanic Americans), 1 international. Average age 30. 36 applicants, 64% accepted, 14 enrolled. *Degree requirements:* For master's, thesis or alternative, writing proficiency exam. *Entrance requirements:* For master's, minimum GPA of 3.0 in mathematics, 2.5 overall during previous 2 years; BA in mathematics or equivalent. Additional exam requirements/recommendations for international students: Required—TOEFL. *Application deadline:* Applications are processed on a rolling basis. Application fee: $55. Electronic applications accepted. *Expenses:* Tuition, nonresident: part-time $339 per unit. Required fees: $276 per semester hour. *Financial support:* Research assistantships, teaching assistantships, career-related internships or fieldwork and Federal Work-Study available. Support available to part-time students. Financial award application deadline: 3/1. *Unit head:* Dr. Doraiswamy Ramachandran, Chair, 916-278-6534, Fax: 916-278-5586.

California State University, San Bernardino, Graduate Studies, College of Natural Sciences, Department of Mathematics, San Bernardino, CA 92407-2397. Offers MA, MAT. Part-time programs available. *Faculty:* 23 full-time, 21 part-time/adjunct (0 women). *Students:* 32 full-time (14 women), 23 part-time (12 women); includes 22 minority (2 African Americans, 10 Asian Americans or Pacific Islanders, 10 Hispanic Americans), 1 international. Average age 30. 17 applicants, 76% accepted, 11 enrolled. In 2005, 9 degrees awarded. *Entrance requirements:* For master's, minor in mathematics. Application fee: $55. *Expenses:* Tuition, nonresident: full-time $8,136; part-time $226 per unit. Required fees: $3,884. *Financial support:* Teaching assistantships available. *Faculty research:* Mathematics education, technology in education, algebra, combinatorics, real analysis. *Unit head:* Dr. Peter D. Williams, Chair, 909-537-5361, Fax: 909-537-7119, E-mail: pwilliam@csusb.edu.

California State University, San Marcos, College of Arts and Sciences, Program in Mathematics, San Marcos, CA 92096-0001. Offers MS. Part-time and evening/weekend programs available. *Faculty:* 11 full-time (3 women), 3 part-time/adjunct (2 women). *Students:* 5 full-time (2 women), 6 part-time (2 women); includes 2 minority (both Asian Americans or Pacific Islanders) Average age 37. 13 applicants, 77% accepted. In 2005, 4 degrees awarded. *Degree requirements:* For master's, thesis optional. *Entrance requirements:* Additional exam requirements/recommendations for international students: Required—TOEFL, TWE. *Application deadline:* For fall admission, 3/15 for domestic students; for spring admission, 1/1 for domestic students. Applications are processed on a rolling basis. Application fee: $55. *Expenses:* Tuition, nonresident: part-time $339 per unit. Required fees: $1,171 per term. Tuition and fees vary according to course load. *Financial support:* Teaching assistantships, career-related internships or fieldwork and Federal Work-Study available. Support available to part-time students. Financial award applicants required to submit FAFSA. *Faculty research:* Combinatorics, graph theory, partial differential equations, numerical analysis, computational linear algebra. *Unit head:* Dr. Linda Holt, Department Chair, 760-750-4092, E-mail: lholt@csusm.edu. *Application contact:* Carrie Dyal, Administrative Coordinator, 760-750-8059, Fax: 760-750-3439, E-mail: chunting@csusm.edu.

Carleton University, Faculty of Graduate Studies, Faculty of Science, School of Mathematics and Statistics, Ottawa, ON K1S 5B6, Canada. Offers information and systems science (PhD); mathematics (M Sc, PhD). *Students:* Average age 30. *Degree requirements:* For master's, thesis optional; for doctorate, one foreign language, thesis/dissertation, comprehensive exam. *Entrance requirements:* For master's, honors degree; for doctorate, master's degree. Additional exam requirements/recommendations for international students: Required—TOEFL. Application fee: $75 Canadian dollars. *Financial support:* Fellowships, research assistantships, teaching assistantships, institutionally sponsored loans, scholarships/grants, and unspecified assistantships available. *Faculty research:* Pure mathematics, applied mathematics, probability and statistics. *Unit head:* Yiqiang Zhao, Director, 613-520-2600 Ext. 2155, Fax: 613-520-3536, E-mail: mathstat@carleton.ca. *Application contact:* Matthias Neufang, Supervisor of Graduate Studies, 613-520-2600 Ext. 2155, E-mail: mathstat@carleton.ca.

Carnegie Mellon University, Mellon College of Science, Department of Mathematical Sciences, Pittsburgh, PA 15213-3891. Offers algorithms, combinatorics, and optimization (PhD); mathematical finance (PhD); mathematical sciences (MS, DA, PhD); pure and applied logic (PhD). Part-time programs available. Terminal master's awarded for partial completion of doctoral program. *Degree requirements:* For doctorate, thesis/dissertation. *Entrance requirements:* For master's and doctorate, GRE General Test, GRE Subject Test. Additional exam requirements/recommendations for international students: Required—TOEFL. Electronic applications accepted. *Faculty research:* Continuum mechanics, discrete mathematics, applied and computational mathematics.

Case Western Reserve University, School of Graduate Studies, Department of Mathematics, Cleveland, OH 44106. Offers applied mathematics (MS, PhD); mathematics (MS, PhD). Part-time programs available. Terminal master's awarded for partial completion of doctoral program. *Degree requirements:* For master's, thesis (applied mathematics); for doctorate, one foreign language, thesis/dissertation. *Entrance requirements:* For master's and doctorate, GRE General Test. Additional exam requirements/recommendations for international students: Required—TOEFL. *Faculty research:* Probability theory, differential equations and control theory, differential geometry and topology, Lie groups, functional and harmonic analysis.

Central Connecticut State University, School of Graduate Studies, School of Arts and Sciences, Department of Mathematics, New Britain, CT 06050-4010. Offers mathematics (MA, MS), including actuarial (MA), operations research (MA), statistics (MA). Part-time and evening/weekend programs available. *Faculty:* 31 full-time (8 women), 62 part-time/adjunct (29 women). *Students:* 30 full-time (16 women), 90 part-time (50 women); includes 13 minority (4 African Americans, 6 Asian Americans or Pacific Islanders, 3 Hispanic Americans), 7 international. Average age 34. 94 applicants, 56% accepted, 26 enrolled. In 2005, 22 master's awarded. *Degree requirements:* For master's, thesis or alternative, comprehensive exam or special project. *Entrance requirements:* For master's, minimum GPA of 2.4, conditional admissions to the Data Mining Program. Additional exam requirements/recommendations for international students: Required—TOEFL. *Application deadline:* For fall admission, 7/1 for domestic students; for spring admission, 12/1 for domestic students. Applications are processed on a rolling basis. Application fee: $50. Electronic applications accepted. *Expenses:* Tuition, area resident: Full-time $3,780. Tuition, state resident: full-time $5,670; part-time $362 per credit. Tuition, nonresident: full-time $10,530; part-time $362 per credit. Required fees: $3,064. One-time fee: $62 part-time. Tuition and fees vary according to degree level and program. *Financial support:* In 2005–06, 3 students received support; research assistantships, career-related internships or fieldwork, Federal Work-Study, scholarships/grants, and unspecified assistantships available. Support available to part-time students. Financial award application deadline: 3/1; financial award applicants required to submit FAFSA. *Faculty research:* Statistics, actuarial mathematics, computer systems and engineering, computer programming techniques, operations research. *Unit head:* Dr. Timothy Craine, Chair, 860-832-2835.

Central Michigan University, College of Graduate Studies, College of Science and Technology, Department of Mathematics, Mount Pleasant, MI 48859. Offers MA, MAT, PhD. *Faculty:* 32 full-time (6 women). *Students:* 8 full-time (6 women), 13 part-time (3 women). Average age 35. In 2005, 3 degrees awarded. *Degree requirements:* For master's, thesis or alternative, registration; for doctorate, thesis/dissertation, registration. *Entrance requirements:* For master's, minimum GPA of 2.5, 20 hours of course work in mathematics; for doctorate, GRE, minimum GPA of 3.0, 20 hours of course work in mathematics. Additional exam requirements/recommendations for international students: Required—TOEFL. *Application deadline:* Applications are processed on a rolling basis. Application fee: $35 ($45 for international students). *Expenses:* Tuition, area resident: Part-time $325 per credit hour. Tuition, state resident: part-time $603 per credit hour. Tuition and fees vary according to degree level and reciprocity agreements. *Financial support:* In 2005–06, 5 fellowships with tuition reimbursements, 1 research assistantship with tuition reimbursement, 23 teaching assistantships with tuition reimbursements were awarded; career-related internships or fieldwork and Federal Work-Study also available. Financial award application deadline: 3/7. *Faculty research:* Combinatorics, approximation theory, operations theory, functional analysis, statistics. *Unit head:* Dr. James Angelos, Chairperson, 989-774-3596, Fax: 989-774-2414. *Application contact:* Dr. Mohan Shrikhande, Graduate Program Coordinator, 989-774-4354, E-mail: shrik1m@cmich.edu.

Central Missouri State University, The Graduate School, College of Arts and Sciences, Department of Mathematics and Computer Science, Warrensburg, MO 64093. Offers applied mathematics (MS); mathematics (MS). Part-time programs available. *Faculty:* 16 full-time (4 women). *Students:* 1 full-time (0 women), 10 part-time (4 women); includes 3 minority (all African Americans), 1 international. Average age 30. 3 applicants, 100% accepted, 3 enrolled. In 2005, 4 degrees awarded. *Degree requirements:* For master's, thesis (MS); comprehensive exam or thesis (MSE). *Entrance requirements:* For master's, GRE General Test (MSE), bachelor's degree in mathematics, minimum GPA of 3.0 (MS); minimum GPA of 2.75, teaching certificate (MSE). Additional exam requirements/recommendations for international students: Required—TOEFL (minimum score 500 paper-based; 173 computer-based). *Application deadline:* For fall admission, 6/1 priority date for domestic students, 5/1 priority date for international students; for spring admission, 10/1 priority date for domestic students, 10/1 priority date for international students. Applications are processed on a rolling basis. Application fee: $30 ($50 for international students). *Expenses:* Tuition, area resident: Full-time $5,160; part-time $215 per credit hour. Tuition, nonresident: full-time $10,320; part-time $430 per credit hour. Required fees: $336; $14 per credit hour. *Financial support:* In 2005–06, 2 teaching assistantships (averaging $3,750 per year) were awarded; Federal Work-Study, scholarships/grants, unspecified assistantships, and administrative assistantships also available. Support available to part-time students. Financial award application deadline: 3/1; financial award applicants required to submit FAFSA. *Faculty research:* Number theory, graph theory, mathematics education, topology, differential equations. Total annual research expenditures: $800,000. *Unit head:* Dr. Edward W. Davenport, Chair, 660-543-4931, Fax: 660-543-8006, E-mail: davenport@cmsu1.cmsu.edu.

Central Washington University, Graduate Studies, Research and Continuing Education, College of the Sciences, Department of Mathematics, Ellensburg, WA 98926. Offers MAT. Offered during summer only. *Faculty:* 13 full-time (2 women). *Students:* 1 full-time (0 women); minority (Asian American or Pacific Islander) In 2005, 9 degrees awarded. *Degree requirements:* For master's, thesis or alternative. *Entrance requirements:* For master's, minimum GPA of 3.0 (summer only). Application fee: $50. *Expenses:* Tuition, state resident: full-time $1,968; part-time $197 per credit. Tuition, nonresident: full-time $4,320; part-time $432 per credit. Required fees: $623. Tuition and fees vary according to degree level. *Financial support:* In 2005–06, 1 research assistantship (averaging $8,100 per year) was awarded; teaching assistantships, Federal Work-Study also available. Financial award application deadline: 3/1; financial award applicants required to submit FAFSA. *Unit head:* Dr. Stuart Boersma, Chair, 509-963-2103. *Application contact:* Justine Eason, Admissions Program Coordinator, 509-963-3103, Fax: 509-963-1799, E-mail: masters@cwu.edu.

Chicago State University, School of Graduate and Professional Studies, College of Arts and Sciences, Department of Mathematics and Computer Science, Chicago, IL 60628. Offers MS. *Degree requirements:* For master's, oral exam, thesis optional. *Entrance requirements:* For master's, minimum GPA of 2.75.

City College of the City University of New York, Graduate School, College of Liberal Arts and Science, Division of Science, Department of Mathematics, New York, NY 10031-9198. Offers MA. Part-time programs available. *Students:* 4 full-time (1 woman), 28 part-time (5 women); includes 19 minority (8 African Americans, 7 Asian Americans or Pacific Islanders, 4 Hispanic Americans), 6 international. 18 applicants, 72% accepted, 7 enrolled. In 2005, 6 degrees awarded. *Degree requirements:* For master's, one foreign language. *Entrance requirements:* For master's, GRE. Additional exam requirements/recommendations for international students: Required—TOEFL (minimum score 500 paper-based; 173 computer-based). *Application deadline:* For fall admission, 5/1 for domestic students; for spring admission, 11/15 for domestic students. Application fee: $125. *Financial support:* Teaching assistantships, Federal Work-Study available. Support available to part-time students. Financial award application deadline: 5/1. *Faculty research:* Group theory, number theory, logic, statistics, computational geometry. *Unit head:* Edward Grossman, Chair, 212-650-5173, Fax: 212-862-0004, E-mail: egross@sci.ccny.cuny.edu. *Application contact:* Thea Pignataro, MA Advisor, 212-650-5175, Fax: 212-862-0004, E-mail: tpignataro@ccny.cuny.edu.

Claremont Graduate University, Graduate Programs, School of Mathematical Sciences, Claremont, CA 91711-6160. Offers computational and systems biology (PhD); computational science (PhD); engineering mathematics (PhD); operations research and statistics (MA, MS); physical applied mathematics (MA, MS); pure mathematics (MA, MS); scientific computing (MA, MS); systems and control theory (MA, MS). Part-time programs available. *Faculty:* 4 full-time (0 women), 6 part-time/adjunct (2 women). *Students:* 54 full-time (13 women), 13 part-time (1 woman); includes 20 minority (1 African American, 14 Asian Americans or Pacific Islanders, 5 Hispanic Americans), 16 international. Average age 38. In 2005, 5 master's, 5 doctorates awarded. Terminal master's awarded for partial completion of doctoral program. *Degree requirements:* For doctorate, 2 foreign languages, thesis/dissertation. *Entrance requirements:* For master's and doctorate, GRE General Test. *Application deadline:* For fall admission, 2/15 for domestic students. Applications are processed on a rolling basis. Electronic applications accepted. *Expenses:* Tuition: Full-time $27,902; part-time $1,214 per term. Required fees: $1,600; $800 per term. Tuition and fees vary according to degree level and program. *Financial support:* Fellowships, research assistantships, career-related internships or fieldwork, Federal Work-Study, institutionally sponsored loans, and tuition waivers (full and partial) available. Support available to part-time students. Financial award application deadline: 2/15; financial award applicants required to submit FAFSA. *Unit head:* John Angus, Dean, 909-621-8080, Fax: 909-607-8261, E-mail: john.angus@cgu.edu. *Application contact:* Susan Townzen, Program Coordinator, 909-621-8080, Fax: 909-607-8261, E-mail: susan.n.townzen@cgu.edu.

Clarkson University, Graduate School, School of Arts and Sciences, Department of Mathematics and Computer Science, Potsdam, NY 13699. Offers mathematics (MS, PhD). *Faculty:* 12 full-time (1 woman). *Students:* 17 full-time (3 women); includes 1 minority (Asian American or Pacific Islander), 11 international. Average age 28. 25 applicants, 76% accepted. In 2005, 2 degrees awarded. Terminal master's awarded for partial completion of doctoral program. *Degree requirements:* For doctorate, thesis/dissertation, departmental qualifying exam. *Entrance requirements:* For master's, GRE. Additional exam requirements/recommendations for international students: Required—TOEFL. *Application deadline:* For fall admission, 5/15 for domestic students; for spring admission, 10/15 priority date for domestic students. Applications are processed on a rolling basis. Application fee: $25 ($35 for international students). Electronic applications accepted. *Expenses:* Tuition: Full-time $20,160; part-time $840 per hour. Required fees: $215. *Financial support:* In 2005–06, 13 students received support, including 5 research assistantships (averaging $19,032 per year), 8 teaching assistantships (averaging $19,032 per year); fellowships, scholarships/grants also available. *Faculty research:* Dynamical systems, mathematics, atmospheric and environmental modeling, statistics numerical analysis. Total annual research expenditures:$512,047. *Unit head:* Dr. Peter Turner, Division Head, 315-268-2334, Fax: 315-268-2371, E-mail: pturner@clarkson.edu. *Application contact:* Donna Brockway, Graduate Admissions International Advisor/Assistant to the Provost, 315-268-6447, Fax: 315-268-7994, E-mail: brockway@clarkson.edu.

Clemson University, Graduate School, College of Engineering and Science, Department of Mathematical Sciences, Clemson, SC 29634. Offers applied and pure mathematics (MS, PhD); computational mathematics (MS, PhD); operations research (MS, PhD); statistics (MS, PhD). Part-time programs available. *Students:* 76 full-time (27 women), 5 part-time (2 women); includes 4 minority (2 African Americans, 2 Asian Americans or Pacific Islanders), 25 international. Average age 29. 73 applicants, 84% accepted, 24 enrolled. In 2005, 23 master's, 4 doctorates awarded. *Median time to degree:* Of those who began their doctoral program in fall 1997, 100% received their degree in 8 years or less. *Degree requirements:* For master's, final project, thesis optional; for doctorate, thesis/dissertation, qualifying exams. *Entrance requirements:* For master's and doctorate, GRE General Test. Additional exam requirements/recommendations for international students: Required—TOEFL; Recommended—TSE. *Application deadline:* For fall admission, 1/15 priority date for domestic students, 2/15 priority date for international students; for spring admission, 10/1 priority date for domestic students, 9/15 priority date for international students. Applications are processed on a rolling basis. Application fee: $50. Electronic applications accepted. *Financial support:* In 2005–06, 68 students received support; fellowships with partial tuition reimbursements available, research assistantships with partial tuition reimbursements available, teaching assistantships with partial tuition reimbursements available available. Financial award application deadline:4/15. *Faculty research:* Applied and computational analysis, cryptography, discrete mathematics, optimization, statistics. Total annual research expenditures: $1 million. *Unit head:* Robert L. Taylor, Chair, 864-656-5240, Fax: 864-656-5230, E-mail: rtaylo2@clemson.edu. *Application contact:* Dr. K.B. Kulasekera, Graduate Coordinator, 864-656-5231, Fax: 864-656-5230, E-mail: kk@clemson.edu.

See Close-Ups on pages 489 and 491.

Cleveland State University, College of Graduate Studies, College of Science, Department of Mathematics, Cleveland, OH 44115. Offers MA, MS. Part-time programs available. *Faculty:* 13 full-time (4 women). *Students:* 4 full-time (3 women), 19 part-time (7 women); includes 4 minority (2 African Americans, 2 Asian Americans or Pacific Islanders), 1 international. Average age 35. 22 applicants, 64% accepted, 9 enrolled. In 2005, 13 degrees awarded. *Degree requirements:* For master's, exit project. *Entrance requirements:* For master's, GRE. Additional exam requirements/recommendations for international students: Required—TOEFL (minimum score 515 paper-based; 197 computer-based). *Application deadline:* For fall admission, 6/15 for domestic students. Applications are processed on a rolling basis. Application fee: $30. Electronic applications accepted. *Expenses:* Tuition, state resident: full-time $10,700. Tuition, nonresident: full-time $14,628. Tuition and fees vary according to program. *Financial support:* In 2005–06, 6 students received support, including 3 teaching assistantships with full tuition reimbursements available (averaging $9,000 per year); Federal Work-Study, institutionally sponsored loans, and tuition waivers (full and partial) also available. Financial award application deadline: 3/15. *Faculty research:* Algebraic topology, probability and statistics, differential equations, geometry. *Unit head:* Dr. Sherwood D. Silliman, Chairperson, 216-687-4681, Fax: 216-523-7340, E-mail: s.silliman@csuohio.edu. *Application contact:* Dr. John F. Oprea, Director, 216-687-4702, Fax: 216-523-7340, E-mail: oprea@csuohio.edu.

College of Charleston, Graduate School, School of Sciences and Mathematics, Program in Mathematics, Charleston, SC 29424-0001. Offers MS, Certificate. *Faculty:* 28 full-time (6 women). *Students:* 7 full-time (1 woman), 13 part-time (5 women); includes 3 minority (all African Americans), 1 international. Average age 27. 11 applicants, 64% accepted, 6 enrolled. In 2005, 4 master's awarded. *Entrance requirements:* For master's, GRE, BS in mathematics or equivalent. Additional exam requirements/recommendations for international students: Required—TOEFL. *Application deadline:* For fall admission, 4/30 for domestic students; for spring admission, 11/15 for domestic students. Applications are processed on a rolling basis. Application fee: $35. Electronic applications accepted. *Expenses:* Tuition, state resident: full-time $6,668; part-time $256 per semester hour. Tuition, nonresident: full-time $15,342; part-time $587 per semester hour. Tuition and fees vary according to course load. *Financial support:* Research assistantships, Federal Work-Study available. Support available to part-time students. Financial award applicants required to submit FAFSA. *Faculty research:* Algebra and discrete mathematics, dynamical systems, probability and statistics, analysis and topology, applied mathematics. *Unit head:* Dr. Ben Cox, Director, 843-953-5715, Fax: 843-953-1410, E-mail: coxbl@cofc.edu. *Application contact:* Susan Hallatt, Assistant Director of Graduate Admissions, 843-953-5614, Fax: 843-953-1434, E-mail: hallatts@cofc.edu.

Colorado School of Mines, Graduate School, Department of Mathematical and Computer Sciences, Golden, CO 80401-1887. Offers MS, PhD. Part-time programs available. *Faculty:* 18 full-time (3 women), 3 part-time/adjunct (2 women). *Students:* 28 full-time (9 women), 13 part-time (2 women); includes 5 minority (all Asian Americans or Pacific Islanders), 5 international. 68 applicants, 75% accepted, 18 enrolled. In 2005, 17 master's, 3 doctorates awarded. *Degree requirements:* For master's, thesis/dissertation; for doctorate, thesis/dissertation, comprehensive exam. *Entrance requirements:* For master's and doctorate, GRE General Test. Additional exam requirements/recommendations for international students: Required—TOEFL (minimum score 550 paper-based; 213 computer-based). *Application deadline:* For fall admission, 1/1 priority date for domestic students, 1/1 priority date for international students; for spring admission, 9/1 priority date for domestic students, 9/1 priority date for international students. Application fee: $50. Electronic applications accepted. *Expenses:* Tuition, state resident: full-time $7,240; part-time $362 per credit hour. Tuition, nonresident: full-time $19,840; part-time $992 per credit hour. Required fees: $895. *Financial support:* In 2005–06, fellowships with full tuition reimbursements (averaging $9,600 per year), 7 research assistantships with full tuition reimbursements (averaging $9,600 per year), 17 teaching assistantships with tuition reimbursements (averaging $9,600 per year) were awarded; scholarships/grants, health care benefits, and unspecified assistantships also available. Financial award applicants required to submit FAFSA. *Faculty research:* Applied statistics, numerical computation, artificial intelligence, linear optimization. Total annual research expenditures:$859,765. *Unit head:* Dr. Graeme Fairweather, Head, 303-273-3502, Fax: 303-273-3875, E-mail: gfairwea@mines.edu. *Application contact:* William Navidi, Professor, 303-273-3489, Fax: 303-273-3875, E-mail: wnavidi@mines.edu.

Colorado State University, Graduate School, College of Natural Sciences, Department of Mathematics, Fort Collins, CO 80523. Offers MS, PhD. Part-time programs available. *Faculty:* 23 full-time (5 women), 1 part-time/adjunct (0 women). *Students:* 39 full-time (17 women), 26 part-time (12 women); includes 7 minority (1 African American, 2 American Indian/Alaska Native, 2 Asian Americans or Pacific Islanders, 2 Hispanic Americans), 14 international. Average age 30. 56 applicants, 61% accepted, 13 enrolled. In 2005, 15 master's, 2 doctorates awarded. Terminal master's awarded for partial completion of doctoral program. *Median time to degree:* Of those who began their doctoral program in fall 1997, 99% received their degree in 8 years or less. *Degree requirements:* For master's, thesis (for some programs); for doctorate, thesis/dissertation, comprehensive exam. *Entrance requirements:* For master's and doctorate, GRE General Test or GMAT, minimum GPA of 3.0. Additional exam requirements/recommendations for international students: Required—TOEFL (minimum score 550 paper-based; 213 computer-based). *Application deadline:* For fall admission, 2/15 for domestic students, 2/15 for international students; for spring admission, 9/1 for domestic students, 9/1 for international students. Applications are processed on a rolling basis. Application fee: $50. Electronic applications accepted. *Expenses:* Tuition, state resident: full-time $3,690; part-time $205 per credit. Tuition, nonresident: full-time $14,958; part-time $831 per credit. Required fees: $1,061. *Financial support:* In 2005–06, 5 fellowships, 3 research assistantships with full tuition reimbursements (averaging $15,000 per year), 34 teaching assistantships with full tuition reimbursements (averaging $14,500 per year) were awarded; career-related internships or fieldwork, Federal Work-Study, institutionally sponsored loans, traineeships, and tuition waivers (partial) also available. Financial award application deadline: 2/1. *Faculty research:* Applied mathematics, numerical analysis, algebraic geometry, combinatorics. Total annual research expenditures: $656,326. *Unit head:* Simon Tavener, Chair, 970-491-1303, Fax: 970-491-2161, E-mail: tavener@math.colostate.edu. *Application contact:* Michael Kirby, Director, Graduate Program, 970-491-6852, Fax: 970-491-2161, E-mail: grad_program@math.colostate.edu.

Columbia University, Graduate School of Arts and Sciences, Division of Natural Sciences, Department of Mathematics, New York, NY 10027. Offers M Phil, MA, PhD. *Faculty:* 30 full-time. *Students:* 126 full-time (38 women), 66 part-time (15 women). Average age 28. 548 applicants, 29% accepted. In 2005, 73 master's, 10 doctorates awarded. *Degree requirements:* For master's, written exam; for doctorate, 2 foreign languages, thesis/dissertation. *Entrance requirements:* For master's and doctorate, GRE General Test, major in mathematics. Additional exam requirements/recommendations for international students: Required—TOEFL. Application fee: $75. *Expenses:* Tuition: Full-time $31,448. Tuition and fees vary according to course level, course load, campus/location and program. *Financial support:* Fellowships, teaching assistantships, Federal Work-Study and institutionally sponsored loans available. Support available to part-time students. Financial award application deadline: 1/5; financial award applicants required to submit FAFSA. *Faculty research:* Algebra, topology, analysis. *Unit head:* John Morgan, Chair, 212-854-6366, Fax: 212-854-8962.

See Close-Up on page 493.

Concordia University, School of Graduate Studies, Faculty of Arts and Science, Department of Mathematics and Statistics, Montréal, QC H3G 1M8, Canada. Offers mathematics (M Sc, MA, PhD); teaching of mathematics (MTM). *Students:* 69 full-time (35 women), 12 part-time (5 women). In 2005, 11 master's, 5 doctorates awarded. *Degree requirements:* For master's, thesis optional; for doctorate, thesis/dissertation, comprehensive exam. *Entrance requirements:* For master's, honors degree in mathematics or equivalent. *Application deadline:* For fall admission, 5/1 for domestic students. For winter admission, 3/31 for domestic students; for spring admission, 10/31 for domestic students. Application fee: $50. *Expenses:* Tuition, state resident: full-time $834; part-time $334 per term. Tuition, nonresident: full-time $2,200; part-time $880 per term. Required fees: $680 per term. Tuition and fees vary according to degree level and program. *Financial support:* Fellowships, research assistantships, teaching assistantships available. Financial award application deadline: 2/1. *Faculty research:* Number theory, computational algebra, mathematical physics, differential geometry, dynamical systems and statistics. *Unit head:* Dr. Yogendra Chaubey, Chair, 514-848-2424 Ext. 3234, Fax: 514-848-2831. *Application contact:* Dr. Pawel Gora, Director, 514-848-3257 Ext. 3250, Fax: 514-848-2831.

Cornell University, Graduate School, Graduate Fields of Arts and Sciences, Field of Mathematics, Ithaca, NY 14853-0001. Offers PhD. *Faculty:* 49 full-time (4 women). *Students:* 222 applicants, 14% accepted, 15 enrolled. In 2005, 12 doctorates awarded. *Degree requirements:* For doctorate, one foreign language, thesis/dissertation, teaching experience, comprehensive exam. *Entrance requirements:* For doctorate, GRE General Test, GRE Subject Test (mathematics), 3 letters of recommendation. Additional exam requirements/recommendations for international students: Required—TOEFL (minimum score 600 paper-based; 250 computer-based). *Application deadline:* For fall admission, 1/15 for domestic students. Application fee: $60. Electronic applications accepted. *Financial support:* In 2005–06, 70 students received support, including 16 fellowships with full tuition reimbursements available, 3 research assistantships with full tuition reimbursements available, 51 teaching assistantships with full tuition reimbursements available; institutionally sponsored loans, scholarships/grants, health care benefits, tuition waivers (full and partial), and unspecified assistantships also available. Financial award applicants required to submit FAFSA. *Faculty research:* Analysis, dynamical systems, Lie theory, logic, topology and geometry. *Unit head:* Director of Graduate Studies, 607-255-6757, Fax: 607-255-7149. *Application contact:* Graduate Field Assistant, 607-255-6757, Fax: 607-255-7149, E-mail: gradinfo@math.cornell.edu.

Dalhousie University, Faculty of Graduate Studies, College of Arts and Science, Faculty of Science, Department of Mathematics and Statistics, Program in Mathematics, Halifax, NS B3H 4R2, Canada. Offers M Sc, PhD. *Degree requirements:* For master's and doctorate, thesis/dissertation. *Entrance requirements:* Additional exam requirements/recommendations for international students: Required—TOEFL. *Faculty research:* Applied mathematics, category theory, algebra, analysis, graph theory.

Dartmouth College, School of Arts and Sciences, Department of Mathematics, Hanover, NH 03755. Offers PhD. *Faculty:* 21 full-time (5 women), 1 part-time/adjunct (0 women). *Students:* 27 full-time (8 women); includes 5 minority (3 Asian Americans or Pacific Islanders, 2 Hispanic Americans), 3 international. Average age 25. 142 applicants, 8% accepted, 7 enrolled. In 2005, 3 doctorates awarded. *Degree requirements:* For doctorate, 2 foreign languages, thesis/dissertation. *Entrance requirements:* For doctorate, GRE General Test, GRE Subject Test. Additional exam requirements/recommendations for international students: Required—TOEFL. *Application deadline:* For fall admission, 2/15 for domestic students. Application fee: $0. *Expenses:* Tuition: Full-time $31,770. *Financial support:* In 2005–06, 26 students received support, including fellowships with full tuition reimbursements available (averaging $21,000 per year), research assistantships with full tuition reimbursements available (averaging $21,000 per year); Federal Work-Study, institutionally sponsored loans, scholarships/grants, tuition waivers (full and partial), and unspecified assistantships also available. *Faculty research:* Mathematical logic, set theory, combinations, number theory. Total annual research expenditures: $741,265. *Unit head:* Dr. Thomas Shemanske, Chair, 603-646-2990, Fax: 603-646-1312, E-mail: thomas.shemanske@dartmouth.edu. *Application contact:* Traci Flynn-Scott, Department Administration, 603-646-3722, Fax: 603-646-1312, E-mail: carol.fine@dartmouth.edu.

See Close-Up on page 495.

Delaware State University, Graduate Programs, Department of Mathematics, Dover, DE 19901-2277. Offers MS.

DePaul University, College of Liberal Arts and Sciences, Department of Mathematical Sciences, Chicago, IL 60604-2287. Offers applied statistics and applied mathematics (MS); mathematics education (MA). Part-time and evening/weekend programs available. *Faculty:* 23 full-time (6 women), 18 part-time/adjunct (5 women). *Students:* 69 full-time (35 women), 61 part-time (30 women); includes 23 minority (9 African Americans, 10 Asian Americans or Pacific Islanders, 4 Hispanic Americans), 7 international. Average age 30. 40 applicants, 100% accepted. In 2005, 30 degrees awarded. *Application deadline:* Applications are processed on a rolling basis. Application fee: $25. *Financial support:* In 2005–06, 8 students received support, including research assistantships with partial tuition reimbursements available (averaging $3,700 per year); teaching assistantships, tuition waivers (full and partial) also available. *Faculty research:* Verbally prime algebras, enveloping algebras of Lie, superalgebras and related rings, harmonic analysis, estimation theory. *Unit head:* Dr. Ahmed I Zayed, Chairperson, 773-325-7806, E-mail: azayed@depaul.edu.

Dowling College, Programs in Arts and Sciences, Oakdale, NY 11769-1999. Offers integrated math and science (MS); liberal studies (MA). Part-time and evening/weekend programs available. *Students:* 1 full-time (0 women), 14 part-time (12 women). Average age 26. 14 applicants, 71% accepted, 2 enrolled. *Degree requirements:* For master's, thesis, comprehensive exam. *Entrance requirements:* For master's, minimum undergraduate GPA of 3.0, 2 letters of recommendation. Additional exam requirements/recommendations for international students: Required—TOEFL (minimum score 550 paper-based). *Application deadline:* For fall admission, 9/1 for domestic students. For winter admission, 1/1 for domestic students; for spring admission, 2/1 for domestic students. Applications are processed on a rolling basis. Application fee: $25. Electronic applications accepted. *Expenses:* Tuition: Full-time $14,952; part-time $623. Required fees: $580. *Financial support:* In 2005–06, research assistantships (averaging $3,084 per year); Federal Work-Study, scholarships/grants, and unspecified assistantships also available. Support available to part-time students. Financial award application deadline: 6/30; financial award applicants required to submit FAFSA. *Unit head:* Dr. Linda Ardito, Provost, 631-244-3232, Fax: 631-244-1033, E-mail: arditol@dowling.edu. *Application contact:* Amy Stier, Director of Enrollment Services for Admissions, 631-244-5010, Fax: 631-563-3827, E-mail: stiera@dowling.edu.

Drexel University, College of Arts and Sciences, Department of Mathematics, Program in Mathematics, Philadelphia, PA 19104-2875. Offers MS, PhD. *Degree requirements:* For doctor-

Mathematics

Drexel University (continued)
ate, one foreign language, thesis/dissertation. *Entrance requirements:* For master's and doctorate, GRE. Additional exam requirements/recommendations for international students: Required—TOEFL, TSE (financial award applicants for teaching assistantships). Electronic applications accepted.

Duke University, Graduate School, Department of Mathematics, Durham, NC 27708-0586. Offers PhD. *Faculty:* 31 full-time. *Students:* 44 full-time (8 women); includes 3 minority (1 African American, 1 Asian American or Pacific Islander, 1 Hispanic American), 12 international. 135 applicants, 20% accepted, 13 enrolled. In 2005, 10 doctorates awarded. *Degree requirements:* For doctorate, 2 foreign languages, thesis/dissertation. *Entrance requirements:* For doctorate, GRE General Test, GRE Subject Test. Additional exam requirements/recommendations for international students: Required—IELT (preferred) or TOEFL. *Application deadline:* For fall admission, 12/31 for domestic students, 12/31 for international students. Application fee: $75. Electronic applications accepted. *Financial support:* Fellowships, research assistantships, teaching assistantships, Federal Work-Study available. Financial award application deadline: 12/31. *Unit head:* Paul Aspinwall, Director of Graduate Studies, 919-660-2874, Fax: 919-660-2821, E-mail: barnes@math.duke.edu.

Duquesne University, Graduate School of Liberal Arts, Program in Computational Mathematics, Pittsburgh, PA 15282-0001. Offers MA, MS. *Faculty:* 18 full-time (5 women), 19 part-time/adjunct (7 women). *Students:* 7 full-time (4 women), 5 part-time (2 women), 4 international. Average age 23. 10 applicants, 80% accepted, 3 enrolled. In 2005, 3 degrees awarded. *Degree requirements:* For master's, thesis, registration. *Entrance requirements:* For master's, GRE General Test. Additional exam requirements/recommendations for international students: Required—TOEFL. *Application deadline:* For fall admission, 8/1 for domestic students, 5/1 for international students. Applications are processed on a rolling basis. Application fee: $50. *Expenses:* Tuition: Part-time $692 per credit. Required fees: $69 per credit. Tuition and fees vary according to degree level and program. *Financial support:* In 2005–06, 2 teaching assistantships with full tuition reimbursements (averaging $9,000 per year) were awarded; Federal Work-Study, institutionally sponsored loans, scholarships/grants, and unspecified assistantships also available. Financial award application deadline: 5/1. *Unit head:* Dr. Frank D'Amico, Chair, 412-396-5468. *Application contact:* Dr. Kathleen Taylor, Professor, 412-396-6472, Fax: 412-396-5265, E-mail: compmath@mathcs.duq.edu.

East Carolina University, Graduate School, Thomas Harriot College of Arts and Sciences, Department of Mathematics, Greenville, NC 27858-4353. Offers applied mathematics (MA); mathematics (MA). Part-time and evening/weekend programs available. *Faculty:* 22 full-time (7 women). *Students:* 9 full-time (6 women), 3 part-time (1 woman); includes 3 minority (2 African Americans, 1 American Indian/Alaska Native). Average age 29. 22 applicants, 0% accepted, 0 enrolled. In 2005, 4 degrees awarded. *Degree requirements:* For master's, comprehensive exam. *Entrance requirements:* For master's, GRE General Test, MAT. Additional exam requirements/recommendations for international students: Required—TOEFL. *Application deadline:* For fall admission, 6/1 for domestic students; for spring admission, 10/15 for domestic students. Applications are processed on a rolling basis. Application fee: $50. *Expenses:* Tuition, state resident: full-time $2,516. Tuition, nonresident: full-time $12,832. *Financial support:* Research assistantships with partial tuition reimbursements, teaching assistantships with partial tuition reimbursements available. Financial award application deadline: 6/1. *Unit head:* Dr. John Daughtry, Chair, 252-328-6461, Fax: 252-328-6414, E-mail: daughtryj@ecu.edu. *Application contact:* Dean of Graduate School, 252-328-6012, Fax: 252-328-6071, E-mail: gradschool@ecu.edu.

Eastern Illinois University, Graduate School, College of Sciences, Department of Mathematics and Computer Science, Charleston, IL 61920-3099. Offers mathematics (MA); mathematics education (MA). *Faculty:* 30 full-time (6 women). In 2005, 9 degrees awarded. *Entrance requirements:* For master's, GRE General Test. *Application deadline:* For fall admission, 7/31 for domestic students. Applications are processed on a rolling basis. Application fee: $30. *Expenses:* Tuition, state resident: part-time $150 per credit hour. Tuition, nonresident: part-time $452 per credit hour. Required fees: $738. *Financial support:* In 2005–06, research assistantships with tuition reimbursements (averaging $7,200 per year), 8 teaching assistantships with tuition reimbursements (averaging $7,200 per year) were awarded. *Unit head:* Dr. Peter Andrews, Chair, 217-581-6275, Fax: 217-581-6284, E-mail: pgandrews@eiu.edu. *Application contact:* Dr. Patrick Coulton, Coordinator, 217-581-6275, Fax: 217-581-6284, E-mail: prcoulton@eiu.edu.

Eastern Kentucky University, The Graduate School, College of Arts and Sciences, Department of Mathematics and Statistics, Richmond, KY 40475-3102. Offers mathematical sciences (MS). Part-time programs available. *Entrance requirements:* For master's, GRE General Test, minimum GPA of 2.5. *Faculty research:* Graph theory, number theory, ring theory, topology, statistics.

Eastern Michigan University, Graduate School, College of Arts and Sciences, Department of Mathematics, Ypsilanti, MI 48197. Offers computer science (M Math); mathematics education (M Math); statistics (M Math). Evening/weekend programs available. *Faculty:* 25 full-time (10 women). *Students:* 5 full-time (4 women), 39 part-time (23 women); includes 7 minority (2 African Americans, 3 Asian Americans or Pacific Islanders, 2 Hispanic Americans), 8 international. Average age 33. In 2005, 17 degrees awarded. *Degree requirements:* For master's, thesis optional. *Entrance requirements:* Additional exam requirements/recommendations for international students: Required—TOEFL. *Application deadline:* For fall admission, 5/15 priority date for domestic students, 5/1 priority date for international students. For winter admission, 10/15 for domestic students; for spring admission, 3/15 for domestic students. Applications are processed on a rolling basis. Application fee: $35. *Expenses:* Tuition, state resident: full-time $7,838; part-time $327 per credit hour. Tuition, nonresident: full-time $15,770; part-time $657 per credit hour. Required fees: $33 per credit hour. $40 per term. Tuition and fees vary according to course level, course load and degree level. *Financial support:* In 2005–06, fellowships (averaging $4,000 per year), research assistantships (averaging $8,950 per year), teaching assistantships (averaging $8,950 per year) were awarded. Support available to part-time students. Financial award applicants required to submit FAFSA. *Unit head:* Dr. Betty Warren, Head, 734-487-1444. *Application contact:* Dr. Walter Parry, Coordinator, 734-487-1444.

Eastern New Mexico University, Graduate School, College of Liberal Arts and Sciences, Department of Mathematical Sciences, Portales, NM 88130. Offers MA. Part-time programs available. *Faculty:* 7 full-time (4 women). *Students:* Average age 29. 8 applicants, 25% accepted. In 2005, 1 degree awarded. *Degree requirements:* For master's, thesis optional. *Entrance requirements:* For master's, minimum GPA of 2.5. *Application deadline:* For fall admission, 8/20 for domestic students. Applications are processed on a rolling basis. Application fee: $10. Electronic applications accepted. *Expenses:* Tuition, state resident: full-time $2,316; part-time $97 per credit hour. Tuition, nonresident: full-time $7,872; part-time $328 per credit hour. Required fees: $33 per credit hour. *Financial support:* In 2005–06, 1 research assistantship (averaging $7,700 per year), 5 teaching assistantships (averaging $7,700 per year) were awarded; career-related internships or fieldwork and Federal Work-Study also available. Support available to part-time students. Financial award application deadline: 3/1. *Faculty research:* Applied mathematics, graph theory. *Unit head:* Dr. Regina Aragon, Graduate Coordinator, 505-562-2328, E-mail: regina.aragon@enmu.edu.

Eastern Washington University, Graduate Studies, College of Science, Mathematics and Technology, Department of Mathematics, Cheney, WA 99004-2431. Offers MS. *Accreditation:* NCATE. Part-time programs available. *Degree requirements:* For master's, thesis (for some programs), comprehensive exam. *Entrance requirements:* For master's, GRE General Test, departmental qualifying exam, minimum GPA of 3.0.

East Tennessee State University, School of Graduate Studies, College of Arts and Sciences, Department of Mathematics, Johnson City, TN 37614. Offers MS. Part-time and evening/weekend programs available. *Faculty:* 14 full-time (4 women). *Students:* 13 full-time (5 women), 10 part-time (9 women); includes 2 minority (both African Americans), 6 international. Average age 33. 15 applicants, 80% accepted, 6 enrolled. In 2005, 6 degrees awarded. *Degree requirements:* For master's, thesis or alternative, comprehensive exam. *Entrance requirements:* For master's, GRE General Test. Additional exam requirements/recommendations for international students: Required—TOEFL (minimum score 550 paper-based; 213 computer-based). *Application deadline:* For fall admission, 7/15 for domestic students; for spring admission, 11/1 for domestic students. Applications are processed on a rolling basis. Application fee: $25 ($35 for international students). *Expenses:* Tuition, nonresident: full-time $9,312; part-time $404 per hour. Required fees: $261 per hour. *Financial support:* In 2005–06, 12 teaching assistantships with full tuition reimbursements (averaging $7,700 per year) were awarded; research assistantships with full tuition reimbursements, unspecified assistantships and laboratory assistantships also available. Financial award application deadline: 7/1; financial award applicants required to submit FAFSA. *Faculty research:* Graph theory and combinatorics, probability and statistics, analysis, numerical and applied math, algebra. Total annual research expenditures: $197,750. *Unit head:* Dr. Anant Godbole, Chair, 423-439-5359, Fax: 423-439-8361, E-mail: godbolea@etsu.edu.

École Polytechnique de Montréal, Graduate Programs, Department of Mathematics, Montréal, QC H3C 3A7, Canada. Offers mathematical method in CA engineering (M Eng, M Sc A, PhD); operational research (M Eng, M Sc A, PhD). Part-time programs available. *Degree requirements:* For master's and doctorate, one foreign language, thesis/dissertation. *Entrance requirements:* For master's, minimum GPA of 2.75; for doctorate, minimum GPA of 3.0. *Faculty research:* Statistics and probability, fractal analysis, optimization.

Emory University, Graduate School of Arts and Sciences, Department of Mathematics and Computer Science, Atlanta, GA 30322-1100. Offers computer science (MS); mathematics (PhD). *Faculty:* 27 full-time (4 women), 2 part-time/adjunct (0 women). *Students:* 44 full-time (12 women); includes 10 minority (9 Asian Americans or Pacific Islanders, 1 Hispanic American), 15 international. Average age 24. 64 applicants, 30% accepted, 13 enrolled. In 2005, 1 master's, 3 doctorates awarded. Terminal master's awarded for partial completion of doctoral program. *Degree requirements:* For master's, thesis, registration; for doctorate, one foreign language, thesis/dissertation, comprehensive exam, registration. *Entrance requirements:* For master's and doctorate, GRE General Test. *Application deadline:* For fall admission, 1/3 for domestic students, 1/3 for international students. Application fee: $50. Electronic applications accepted. *Expenses:* Tuition: Full-time $14,400. Required fees: $217. *Financial support:* In 2005–06, fellowships (averaging $12,550 per year), teaching assistantships (averaging $16,480 per year) were awarded; scholarships/grants also available. Financial award application deadline: 1/3. Total annual research expenditures: $1.1 million. *Unit head:* Dr. Vaidy Sunderam, Chairman, 404-727-5926, Fax: 404-727-5611, E-mail: vss@emory.edu. *Application contact:* Dr. Shanshuang Yang, Director of Graduate Studies, 404-727-7956, Fax: 404-727-5611, E-mail: dgs@mathcs.emory.edu.

Announcement: The department offers a PhD in mathematics and an MS in mathematics or computer science. Research specialties in mathematics include algebra, computational algebra, applied math, combinatorics, complex analysis, differential equations, dynamical systems, topology, numerical analysis, scientific computation, and mathematical physics. Full tuition and funding are available in both the math PhD and computer science MS programs.

Emporia State University, School of Graduate Studies, College of Liberal Arts and Sciences, Department of Mathematics and Computer Science, Emporia, KS 66801-5087. Offers mathematics (MS). Part-time programs available. *Faculty:* 14 full-time (2 women), 5 part-time/adjunct (4 women). *Students:* 5 applicants, 80% accepted, 2 enrolled. In 2005, 3 degrees awarded. *Degree requirements:* For master's, comprehensive exam or thesis. *Entrance requirements:* For master's, appropriate undergraduate degree. Additional exam requirements/recommendations for international students: Required—TOEFL (minimum score 450 paper-based; 133 computer-based). *Application deadline:* For fall admission, 8/15 for domestic students. Applications are processed on a rolling basis. Application fee: $30 ($75 for international students). Electronic applications accepted. *Expenses:* Tuition, state resident: full-time $2,890; part-time $132 per credit. Tuition, nonresident: full-time $9,258; part-time $422 per credit. Required fees: $626; $41 per credit. Tuition and fees vary according to degree level. *Financial support:* In 2005–06, 3 teaching assistantships with full tuition reimbursements (averaging $6,492 per year) were awarded; career-related internships or fieldwork, Federal Work-Study, institutionally sponsored loans, health care benefits, and unspecified assistantships also available. Financial award application deadline: 3/15; financial award applicants required to submit FAFSA. *Unit head:* Dr. Larry Scott, Chair, 620-341-5281, Fax: 620-341-6055, E-mail: lscott@emporia.edu. *Application contact:* Dr. H. Joe Yanik, Graduate Coordinator, 620-341-5639, E-mail: hyanik@emporia.edu.

Fairfield University, College of Arts and Sciences, Program in Mathematics and Quantitative Methods, Fairfield, CT 06824-5195. Offers MS. Part-time and evening/weekend programs available. *Faculty:* 13 full-time (4 women), 2 part-time/adjunct (0 women). *Students:* 2 full-time (both women), 38 part-time (20 women); includes 3 minority (2 African Americans, 1 Hispanic American), 1 international. Average age 38. 15 applicants, 60% accepted, 9 enrolled. In 2005, 5 degrees awarded. *Degree requirements:* For master's, capstone course. *Entrance requirements:* For master's, minimum GPA of 3.0, 2 letters of recommendation, resumé. Additional exam requirements/recommendations for international students: Required—TOEFL (minimum score 550 paper-based; 213 computer-based). *Application deadline:* For fall admission, 7/1 for domestic students, 6/15 for international students; for spring admission, 12/1 for domestic students, 10/15 for international students. Applications are processed on a rolling basis. Application fee: $55. *Expenses:* Tuition: Full-time $10,530; part-time $585 per credit. *Financial support:* Unspecified assistantships available. Financial award applicants required to submit FAFSA. *Unit head:* Dr. Benjamin Fine, Co-Director, 203-254-4000 Ext. 2197, E-mail: fine@fair1.fairfield.edu. *Application contact:* Marianne Gumpper, Director of Graduate and Continuing Studies Admissions, 203-254-4184, Fax: 203-254-4073, E-mail: gradadmis@mail.fairfield.edu.

Fayetteville State University, Graduate School, Department of Mathematics and Computer Science, Fayetteville, NC 28301-4298. Offers mathematics (MS). Part-time and evening/weekend programs available. *Faculty:* 9 full-time (1 woman). *Students:* 1 applicant, 100% accepted, 1 enrolled. In 2005, 8 degrees awarded. *Degree requirements:* For master's, thesis or alternative, internship, comprehensive exam. *Entrance requirements:* For master's, GRE General Test. *Application deadline:* For fall admission, 7/1 for domestic students; for spring admission, 12/1 for domestic students. Applications are processed on a rolling basis. Application fee: $25. Electronic applications accepted. *Expenses:* Tuition, state resident: full-time $1,918; part-time $202 per credit. Tuition, nonresident: full-time $11,508; part-time $1,401 per credit. Required fees: $975. *Faculty research:* Combinatorics, coding theory, cryptography; stability and periodic solutions of ordinary, partial, and functional differential equations, discrete dynamical systems, and fixed point theory; applied statistics: statistical genetics; time series analysis; spectral theory of linear operations; differential equations. Total annual research expenditures: $36,000. *Unit head:* Dr. Dwight House, Chairperson, 910-672-1294, E-mail: dhouse@uncfsu.edu.

Florida Atlantic University, Charles E. Schmidt College of Science, Department of Mathematical Science, Boca Raton, FL 33431-0991. Offers applied mathematics and statistics (MS); mathematics (MS, MST, PhD). Part-time programs available. *Faculty:* 25 full-time (2 women), 3 part-time/adjunct (0 women). *Students:* 46 full-time (17 women), 14 part-time (8 women); includes 7 minority (4 African Americans, 1 Asian American or Pacific Islander, 2 Hispanic Americans), 30 international. Average age 31. 33 applicants, 82% accepted, 20 enrolled. In 2005, 18 degrees awarded. Terminal master's awarded for partial completion of doctoral program. *Degree requirements:* For master's, thesis (for some programs), comprehensive exam (for some programs), registration; for doctorate, thesis/dissertation, comprehensive exam, registration. *Entrance requirements:* For master's and doctorate, GRE General Test, minimum GPA of 3.0. Additional exam requirements/recommendations for international students: Required—TOEFL (minimum score 500 paper-based; 173 computer-based). *Application deadline:* For fall admission, 6/1 priority date for domestic students, 2/15 priority date for international students; for spring admission, 10/20 priority date for domestic students, 8/15 priority date for international students. Applications are processed on a rolling basis. Application fee: $30. Electronic applications accepted. *Expenses:* Tuition, state resident: full-time $4,394; part-time $244 per credit. Tuition, nonresident: full-time $16,441; part-time $912 per credit. *Financial support:* In 2005–06, fellowships with partial tuition reimbursements (averaging $20,000 per year), 20 teaching assistantships with partial tuition reimbursements (averaging $20,000 per year) were awarded; Federal Work-Study also available. Financial award application deadline: 4/1. *Faculty research:* Cryptography, statistics, algebra, analysis, combinatorics. Total annual research expenditures: $550,000. *Unit head:* Dr. Yoram Sagher, Chair, 561-297-3341, Fax: 561-297-2436, E-mail: sagher@fau.edu. *Application contact:* Dr. Heinrich Niederhausen, Graduate Director, 561-297-3237, Fax: 561-297-2436, E-mail: grad@math.fau.edu.

Florida International University, College of Arts and Sciences, Department of Mathematics, Miami, FL 33199. Offers mathematical sciences (MS). Part-time and evening/weekend programs available. *Faculty:* 30 full-time (7 women). *Students:* 3 full-time (0 women), 5 part-time; includes 4 minority (all Hispanic Americans), 1 international. Average age 35. 12 applicants, 58% accepted, 3 enrolled. In 2005, 2 degrees awarded. *Degree requirements:* For master's, thesis, project. *Entrance requirements:* For master's, GRE General Test, 3 letters of recommendation. Additional exam requirements/recommendations for international students: Required—TOEFL. *Application deadline:* For fall admission, 4/1 for domestic students; for spring admission, 10/1 for domestic students. Applications are processed on a rolling basis. Application fee: $25. *Expenses:* Tuition, area resident: Part-time $239 per credit. Tuition, state resident: full-time $4,294; part-time $869 per credit. Tuition, nonresident: full-time $15,641. Required fees: $252; $126 per term. Tuition and fees vary according to program. *Financial support:* Application deadline: 4/1. *Unit head:* Dr. Julian K. Edward, Chairperson, 305-348-3050, Fax: 305-348-6158, E-mail: julian.edward@fiu.edu.

Florida State University, Graduate Studies, College of Arts and Sciences, Department of Mathematics, Tallahassee, FL 32306. Offers applied mathematics (MS, PhD); biomedical mathematics (MS, PhD); financial mathematics (MS, PhD); pure mathematics (MS, PhD). Part-time programs available. *Faculty:* 41 full-time (4 women), 10 part-time/adjunct (6 women). *Students:* 117 full-time (27 women), 5 part-time; includes 20 minority (3 African Americans, 1 American Indian/Alaska Native, 7 Asian Americans or Pacific Islanders, 9 Hispanic Americans), 56 international. Average age 26. 317 applicants, 66% accepted, 41 enrolled. In 2005, 31 master's, 7 doctorates awarded. Terminal master's awarded for partial completion of doctoral program. *Degree requirements:* For master's, thesis optional; for doctorate, thesis/dissertation, preliminary exam. *Entrance requirements:* For master's and doctorate, GRE General Test, minimum GPA of 3.0, 4-year bachelor's degree. Additional exam requirements/recommendations for international students: Required—TOEFL (minimum score 550 paper-based; 213 computer-based). *Application deadline:* For fall admission, 3/1 priority date for domestic students, 1/1 priority date for international students; for spring admission, 8/15 priority date for domestic students, 5/1 priority date for international students. Applications are processed on a rolling basis. Application fee: $30. Electronic applications accepted. *Financial support:* In 2005–06, 100 students received support, including 2 fellowships with full tuition reimbursements available (averaging $18,000 per year), 12 research assistantships with full tuition reimbursements available (averaging $16,000 per year), 79 teaching assistantships with full tuition reimbursements available (averaging $16,000 per year); career-related internships or fieldwork, scholarships/grants, and unspecified assistantships also available. Financial award application deadline: 3/1. *Faculty research:* Low-dimensional manifolds, algebra geometry, fluid dynamics, financial mathematics, biomedical mathematics. *Unit head:* Dr. Philip L Bowers, Chairperson, 850-645-3338, Fax: 850-644-4053, E-mail: bowers@math.fsu.edu. *Application contact:* Dr. Eric P. Klassen, Associate Chair for Graduate Studies, 850-644-2202, Fax: 850-644-4053, E-mail: klassen@math.fsu.edu.

George Mason University, College of Arts and Sciences, Department of Mathematical Sciences, Fairfax, VA 22030. Offers mathematical sciences (PhD); mathematics (MS). Evening/weekend programs available. *Faculty:* 31 full-time (7 women), 12 part-time/adjunct (4 women). *Students:* 5 full-time (1 woman), 16 part-time (7 women); includes 1 African American, 2 Asian Americans or Pacific Islanders, 4 international. Average age 30. 43 applicants, 60% accepted, 11 enrolled. In 2005, 1 degree awarded. *Degree requirements:* For master's, thesis optional. *Entrance requirements:* For master's, minimum GPA of 3.0 in last 60 hours of course work. *Application deadline:* For fall admission, 5/1 for domestic students; for spring admission, 11/1 for domestic students. Electronic applications accepted. *Expenses:* Tuition, area resident: Full-time $5,244; part-time $219 per credit. Tuition, state resident: part-time $651 per credit. Tuition, nonresident: full-time $15,636. Required fees: $1,524; $65 per credit. *Financial support:* Fellowships, research assistantships, teaching assistantships, career-related internships or fieldwork available. Support available to part-time students. Financial award application deadline: 3/1; financial award applicants required to submit FAFSA. *Unit head:* Robert Sachs, Chair, 703-993-1462, Fax: 703-993-1491, E-mail: rsachs@gmu.edu. *Application contact:* Dr. David Walnut, Information Contact, 703-993-1460, E-mail: mathgrad@gmu.edu.

The George Washington University, Columbian College of Arts and Sciences, Department of Mathematics, Washington, DC 20052. Offers applied mathematics (MA, MS); pure mathematics (MA, PhD). Part-time and evening/weekend programs available. Terminal master's awarded for partial completion of doctoral program. *Degree requirements:* For master's, comprehensive exam; for doctorate, one foreign language, thesis/dissertation, general exam. *Entrance requirements:* For master's and doctorate, GRE General Test, minimum GPA of 3.0, interview. Additional exam requirements/recommendations for international students: Required—TOEFL (minimum score 550 paper-based; 213 computer-based). Electronic applications accepted.

Georgia Institute of Technology, Graduate Studies and Research, College of Sciences, School of Mathematics, Atlanta, GA 30332-0001. Offers algorithms, combinatorics, and optimization (PhD); applied mathematics (MS); bioinformatics (PhD); mathematics (PhD); quantitative and computational finance (MS); statistics (MS Stat). Terminal master's awarded for partial completion of doctoral program. *Degree requirements:* For master's, thesis or alternative; for doctorate, one foreign language, thesis/dissertation. *Entrance requirements:* For master's, GRE General Test, minimum GPA of 3.0; for doctorate, GRE General Test, GRE Subject Test, minimum GPA of 3.0. Additional exam requirements/recommendations for international students: Required—TOEFL. Electronic applications accepted. *Faculty research:* Dynamical systems, discrete mathematics, probability and statistics, mathematical physics.

Georgia Institute of Technology, Graduate Studies and Research, Multidisciplinary Program in Algorithms, Combinatorics, and Optimization, Atlanta, GA 30332-0001. Offers PhD. *Degree requirements:* For doctorate, thesis/dissertation. *Entrance requirements:* For doctorate, GRE General Test, GRE Subject Test (computer science or mathematics). Additional exam requirements/recommendations for international students: Required—TOEFL. Electronic applications accepted. *Faculty research:* Complexity, graph minors, combinatorial optimization, mathematical programming, probabilistic methods.

Georgian Court University, School of Sciences and Mathematics, Lakewood, NJ 08701-2697. Offers biology (MS); counseling psychology (MA); holistic health (Certificate); holistic health studies (MA); mathematics (MA); professional counselor (Certificate); school psychology (Certificate). Part-time and evening/weekend programs available. *Faculty:* 22 full-time (15 women), 7 part-time/adjunct (4 women). *Students:* 30 full-time (25 women), 175 part-time (154 women); includes 10 minority (3 African Americans, 1 American Indian/Alaska Native, 3 Asian Americans or Pacific Islanders, 3 Hispanic Americans), 1 international. Average age 37. 126 applicants, 74% accepted, 76 enrolled. In 2005, 52 degrees awarded. *Degree requirements:* For master's, thesis (for some programs), comprehensive exam (for some programs). *Entrance requirements:* For master's, GRE General Test, GRE Subject Test in biology (MS), 3 letters of recommendation. Additional exam requirements/recommendations for international students: Required—TOEFL (minimum score 550 paper-based; 213 computer-based). *Application deadline:* For fall admission, 8/1 priority date for domestic students, 4/1 priority date for international students; for spring admission, 1/1 priority date for domestic students, 7/1 priority date for international students. Applications are processed on a rolling basis. Application fee: $40. Electronic applications accepted. *Expenses:* Tuition: Part-time $566 per credit. *Financial support:* Scholarships/grants, health care benefits, and unspecified assistantships available. Financial award application deadline: 4/15; financial award applicants required to submit FAFSA. *Unit head:* Dr. Linda James, Dean, 732-987-2617. *Application contact:* Eugene Soltys, Director of Graduate Admissions, 732-987-2760 Ext. 2760, Fax: 732-987-2000, E-mail: admissions@georgian.edu.

Georgia Southern University, Jack N. Averitt College of Graduate Studies, Allen E. Paulson College of Science and Technology, Department of Mathematical Sciences, Statesboro, GA 30460. Offers mathematics (MS). Part-time programs available. *Students:* 9 full-time (5 women), 8 part-time (5 women); includes 3 minority (2 African Americans, 1 Hispanic American), 2 international. Average age 31. 10 applicants, 100% accepted, 4 enrolled. In 2005, 3 degrees awarded. *Degree requirements:* For master's, terminal exam, project. *Entrance requirements:* For master's, GRE, BS in engineering, science, or mathematics; course work in calculus, probability, linear algebra; proficiency in a computer programming language. Additional exam requirements/recommendations for international students: Required—TOEFL (minimum score 550 paper-based; 213 computer-based). *Application deadline:* For fall admission, 3/1 priority date for domestic students, 3/1 priority date for international students; for spring admission, 10/1 priority date for domestic students, 10/1 priority date for international students. Applications are processed on a rolling basis. Application fee: $50. Electronic applications accepted. *Expenses:* Tuition, state resident: full-time $2,926; part-time $122 per semester hour. Tuition, nonresident: full-time $11,704; part-time $488 per semester hour. Required fees: $1,024. *Financial support:* In 2005–06, 9 students received support, including research assistantships with partial tuition reimbursements available (averaging $5,500 per year), teaching assistantships with partial tuition reimbursements available (averaging $5,500 per year); career-related internships or fieldwork, Federal Work-Study, scholarships/grants, and unspecified assistantships also available. Support available to part-time students. Financial award application deadline: 4/15; financial award applicants required to submit FAFSA. *Faculty research:* Analysis of numerical, interval, and fuzzy data; approximation theory; computational mathematics; parallel computation; applied statistic and emphasis on biological models. Total annual research expenditures: $2,200. *Unit head:* Dr. Martha Abell, Chair, 912-681-5132, Fax: 912-681-0654, E-mail: xli@georgiasouthern.edu. *Application contact:* 912-681-5384, Fax: 912-681-0740, E-mail: gradschool@georgiasouthern.edu.

Georgia State University, College of Arts and Sciences, Department of Mathematics and Statistics, Atlanta, GA 30303-3083. Offers mathematics (MAT, MS). Part-time and evening/weekend programs available. *Degree requirements:* For master's, thesis or alternative, exam. *Entrance requirements:* For master's, GRE. Additional exam requirements/recommendations for international students: Required—TOEFL. Electronic applications accepted. *Expenses:* Tuition, state resident: full-time $4,368; part-time $182 per term. Tuition, nonresident: full-time $8,732; part-time $728 per term. Required fees: $46 per hour. *Faculty research:* Analysis, biostatistics, discrete mathematics, linear algebra, statistics.

Graduate School and University Center of the City University of New York, Graduate Studies, Program in Mathematics, New York, NY 10016-4039. Offers PhD. *Faculty:* 43 full-time (2 women). *Students:* 104 full-time (24 women), 7 part-time (2 women); includes 13 minority (5 African Americans, 7 Asian Americans or Pacific Islanders, 1 Hispanic American), 29 international. Average age 34. 85 applicants, 92% accepted, 35 enrolled. In 2005, 6 degrees awarded. *Degree requirements:* For doctorate, 2 foreign languages, thesis/dissertation. *Entrance requirements:* For doctorate, GRE General Test. Additional exam requirements/recommendations for international students: Required—TOEFL. *Application deadline:* For fall admission, 4/15 for domestic students. Application fee: $125. Electronic applications accepted. *Financial support:* In 2005–06, 36 fellowships, 2 research assistantships, 10 teaching assistantships were awarded; career-related internships or fieldwork, Federal Work-Study, institutionally sponsored loans, and tuition waivers (full and partial) also available. Financial award application deadline: 2/1; financial award applicants required to submit FAFSA. *Unit head:* Dr. Jozef Dodzuik, Executive Officer, 212-817-8531, Fax: 212-817-1527.

Hardin-Simmons University, Graduate School, Holland School of Sciences and Mathematics, Abilene, TX 79698-0001. Offers DPT. *Expenses:* Tuition: Full-time $8,370; part-time $465 per hour. Required fees: $490; $66 per semester. One-time fee: $50. Full-time tuition and fees vary according to course load and degree level. *Unit head:* Dr. Gary Stanlake, Dean of Graduate Studies, Graduate School, 325-670-1298, Fax: 325-670-1564, E-mail: gradoff@hsutx.edu.

Harvard University, Graduate School of Arts and Sciences, Department of Mathematics, Cambridge, MA 02138. Offers PhD. *Students:* 68 full-time (21 women). 134 applicants, 13% accepted. In 2005, 14 doctorates awarded. *Degree requirements:* For doctorate, 2 foreign languages, thesis/dissertation, qualifying exam. *Entrance requirements:* For doctorate, GRE General Test, GRE Subject Test. Additional exam requirements/recommendations for international students: Required—TOEFL. *Application deadline:* For fall admission, 12/15 for domestic students. Application fee: $60. *Expenses:* Tuition: Full-time $28,752. Full-time tuition and fees vary according to program and student level. *Financial support:* Fellowships, research assistantships, teaching assistantships, career-related internships or fieldwork, Federal Work-Study, and institutionally sponsored loans available. Financial award application deadline: 12/30. *Unit head:* Betsey Cogswell, Administrator, 617-495-5497, Fax: 617-495-5264. *Application contact:* Office of Admissions and Financial Aid, 617-495-5315.

Hofstra University, College of Liberal Arts and Sciences, Division of Natural Sciences, Mathematics, Engineering, and Computer Science, Department of Mathematics, Hempstead, NY 11549. Offers applied mathematics (MS); mathematics (MA). Part-time and evening/weekend programs available. *Faculty:* 5 full-time (1 woman), 1 (woman) part-time/adjunct. *Students:* 6 full-time (2 women), 8 part-time (3 women); includes 3 minority (2 African Americans, 1 Hispanic American), 1 international. Average age 28. 13 applicants, 77% accepted, 4 enrolled. In 2005, 5 degrees awarded. *Degree requirements:* For master's, A-MA. *Entrance requirements:* For master's, bachelor's degree with strong background in math. Additional exam requirements/recommendations for international students: Required—TOEFL (minimum score 550 paper-based; 213 computer-based). *Application deadline:* Applications are processed on a rolling basis. Application fee: $60. Electronic applications accepted. *Expenses:* Tuition: Full-time $12,060; part-time $670 per credit. Required fees: $930; $155 per term. Tuition and fees vary according to course load and program. *Financial support:* In 2005–06, 14 students received support, including 2 fellowships (averaging $2,667 per year); tuition waivers (full) also available. Financial award applicants required to submit FAFSA. *Faculty research:* Logic-four valued logic, homotopy theory, combinatonics, statistics, algebraic number theory. *Unit head:* Dr. Maryisa T. Weiss, Chairperson, 516-463-5580, Fax: 516-463-6596, E-mail: matmtw@hofstra.edu. *Application contact:* Carol Drummer, Dean of Graduate Admissions, 516-463-4876, Fax: 516-463-4664, E-mail: gradstudent@hofstra.edu.

Howard University, Graduate School of Arts and Sciences, Department of Mathematics, Washington, DC 20059-0002. Offers applied mathematics (MS, PhD); mathematics (MS, PhD). Part-time programs available. *Faculty:* 25 full-time (5 women). *Students:* 26 full-time (8 women), 4 part-time; includes 22 minority (all African Americans), 7 international. In 2005, 5 master's, 2 doctorates awarded. Terminal master's awarded for partial completion of doctoral program. *Degree requirements:* For master's, thesis or alternative, qualifying exam, comprehensive exam; for doctorate, 2 foreign languages, thesis/dissertation, qualifying exams, comprehensive exam. *Entrance requirements:* For master's, GRE General Test, minimum

Mathematics

Howard University (continued)

GPA of 3.0; for doctorate, GRE General Test. Additional exam requirements/recommendations for international students: Required—TOEFL. *Application deadline:* For fall admission, 2/15 priority date for domestic students, 2/15 priority date for international students; for spring admission, 11/1 priority date for domestic students, 11/1 priority date for international students. Applications are processed on a rolling basis. Application fee: $45. Electronic applications accepted. *Financial support:* In 2005–06, 11 students received support, including fellowships with full tuition reimbursements available (averaging $16,000 per year), 2 research assistantships with full tuition reimbursements available (averaging $15,000 per year), 9 teaching assistantships with full tuition reimbursements available (averaging $13,000 per year); institutionally sponsored loans and scholarships/grants also available. Financial award application deadline: 4/1. *Unit head:* Dr. Abdul-Aziz Yzkubu, Chairman, 202-806-6830, Fax: 202-806-6831, E-mail: ayakubu@howard.edu. *Application contact:* Dr. Neil Hindman, Director of Graduate Studies, 202-806-5927, Fax: 202-806-6831, E-mail: nhindman@howard.edu.

Hunter College of the City University of New York, Graduate School, School of Arts and Sciences, Department of Mathematics and Statistics, New York, NY 10021-5085. Offers applied mathematics (MA); mathematics for secondary education (MA); pure mathematics (MA). Part-time and evening/weekend programs available. *Faculty:* 19 full-time (4 women), 2 part-time/adjunct (both women). *Students:* 2 full-time (1 woman), 18 part-time (8 women); includes 7 minority (5 Asian Americans or Pacific Islanders, 2 Hispanic Americans). Average age 34. 14 applicants, 57% accepted, 4 enrolled. In 2005, 17 degrees awarded. *Degree requirements:* For master's, one foreign language, thesis (for some programs), comprehensive exam. *Entrance requirements:* For master's, GRE General Test, 24 credits in mathematics. Additional exam requirements/recommendations for international students: Required—TOEFL. *Application deadline:* For fall admission, 4/1 for domestic students, 2/1 for international students; for spring admission, 11/1 for domestic students, 9/1 for international students. Application fee: $125. *Expenses:* Tuition, state resident: full-time $6,400; part-time $270 per credit. Tuition, nonresident: part-time $500 per credit. International tuition: $12,000 full-time. Required fees: $50 per term. Part-time tuition and fees vary according to course load and program. *Financial support:* Federal Work-Study, institutionally sponsored loans, scholarships/grants, and tuition waivers (partial) available. Support available to part-time students. *Faculty research:* Data analysis, dynamical systems, computer graphics, topology, statistical decision theory. *Unit head:* Ada Peluso, Chairperson, 212-772-5300, Fax: 212-772-4858, E-mail: peluso@math.hunter.cuny.edu. *Application contact:* William Zlata, Director for Graduate Admissions, 212-772-4482, Fax: 212-650-3336, E-mail: admissions@hunter.cuny.edu.

Idaho State University, Office of Graduate Studies, College of Arts and Sciences, Department of Mathematics, Pocatello, ID 83209. Offers MS, DA. *Degree requirements:* For master's, thesis (for some programs), comprehensive exam, registration; for doctorate, thesis/dissertation, teaching internships, comprehensive exam, registration. *Entrance requirements:* For master's, GRE General Test, GRE Subject Test, course work in modern algebra, differential equations, advanced calculus, introductory analysis; for doctorate, GRE General Test, GRE Subject Test, minimum GPA of 3.5 (graduate), MS in mathematics, teaching experience, 3 letters of recommendation. Additional exam requirements/recommendations for international students: Required—TOEFL (minimum score 550 paper-based; 213 computer-based). *Faculty research:* Algebra, analysis geometry, statistics, applied mathematics.

Illinois State University, Graduate School, College of Arts and Sciences, Department of Mathematics, Program in Mathematics, Normal, IL 61790-2200. Offers MA, MS. *Students:* 30 full-time (15 women), 25 part-time (20 women); includes 3 minority (all Asian Americans or Pacific Islanders), 14 international. 56 applicants, 86% accepted. In 2005, 17 degrees awarded. *Degree requirements:* For master's, thesis or alternative. *Entrance requirements:* For master's, GRE General Test, minimum GPA of 2.8 in last 60 hours of course work. *Application deadline:* Applications are processed on a rolling basis. Application fee: $30. *Expenses:* Tuition, state resident: full-time $3,060; part-time $170 per credit hour. Tuition, nonresident: full-time $6,390; part-time $355 per credit hour. Required fees: $1,411; $47 per credit hour. *Financial support:* In 2005–06, 11 research assistantships (averaging $10,514 per year), 11 teaching assistantships (averaging $8,370 per year) were awarded; tuition waivers (full) and unspecified assistantships also available. Financial award application deadline: 4/1. *Unit head:* Dr. George Seelinger, Chairperson, Department of Mathematics, 309-438-8781.

Indiana State University, School of Graduate Studies, College of Arts and Sciences, Department of Mathematics and Computer Science, Terre Haute, IN 47809-1401. Offers computer science (MS); mathematics (MS); mathematics and computer science (MA). Part-time programs available. *Faculty:* 9 full-time (3 women). *Students:* 23 full-time (3 women), 7 part-time (4 women); includes 2 minority (1 African American, 1 Asian American or Pacific Islander), 18 international. Average age 26. 56 applicants, 98% accepted, 7 enrolled. In 2005, 9 degrees awarded. *Degree requirements:* For master's, thesis or alternative. *Entrance requirements:* For master's, 24 semester hours of undergraduate mathematics. *Application deadline:* For fall admission, 7/1 for domestic students; for spring admission, 11/1 priority date for domestic students. Applications are processed on a rolling basis. Application fee: $35. Electronic applications accepted. *Expenses:* Tuition, state resident: full-time $6,288; part-time $262 per credit hour. Tuition, nonresident: full-time $12,504; part-time $521 per credit hour. *Financial support:* In 2005–06, 5 teaching assistantships with partial tuition reimbursements (averaging $6,300 per year) were awarded; research assistantships with partial tuition reimbursements, tuition waivers (partial) also available. Financial award application deadline: 3/1; financial award applicants required to submit FAFSA. *Unit head:* Dr. Bhaskara Rao Kopparty, Interim Chairperson, 812-237-2130.

Indiana University Bloomington, Graduate School, College of Arts and Sciences, Department of Mathematics, Bloomington, IN 47405-7000. Offers applied mathematics–numerical analysis (MA, PhD); mathematics education (MAT); probability-statistics (MA, PhD). PhD offered through the University Graduate School. *Faculty:* 36 full-time (1 woman). *Students:* 116 full-time (39 women), 20 part-time (4 women); includes 6 minority (1 African American, 5 Asian Americans or Pacific Islanders), 74 international. Average age 27. In 2005, 21 master's, 5 doctorates awarded. Terminal master's awarded for partial completion of doctoral program. *Degree requirements:* For doctorate, one foreign language, thesis/dissertation. *Entrance requirements:* For master's and doctorate, GRE General Test, GRE Subject Test. Additional exam requirements/recommendations for international students: Required—TOEFL. *Application deadline:* For fall admission, 1/15 priority date for domestic students, 12/15 priority date for international students; for spring admission, 9/1 priority date for domestic students, 9/1 priority date for international students. Applications are processed on a rolling basis. Application fee: $50 ($60 for international students). Electronic applications accepted. *Expenses:* Tuition, state resident: full-time $5,437; part-time $227 per credit hour. Tuition, nonresident: full-time $15,836; part-time $660 per credit hour. Required fees: $821. Tuition and fees vary according to campus/location and program. *Financial support:* Fellowships with full tuition reimbursements, research assistantships, teaching assistantships with full tuition reimbursements, Federal Work-Study available. Support available to part-time students. Financial award application deadline: 2/1. *Faculty research:* Topology, geometry, algebra. *Unit head:* David Hoff, Chair, 812-855-2200. *Application contact:* Misty Cummings, Graduate Secretary, 812-855-2645, Fax: 812-855-0046, E-mail: gradmath@indiana.edu.

Indiana University of Pennsylvania, Graduate School and Research, College of Natural Sciences and Mathematics, Department of Mathematics, Indiana, PA 15705-1087. Offers applied mathematics (MS); elementary and middle school mathematics education (M Ed); mathematics education (M Ed). Part-time programs available. *Degree requirements:* For master's, thesis optional. *Entrance requirements:* For master's, 2 letters of recommendation. Additional exam requirements/recommendations for international students: Required—TOEFL.

Indiana University–Purdue University Fort Wayne, School of Arts and Sciences, Department of Mathematical Sciences, Fort Wayne, IN 46805-1499. Offers applied mathematics (MS); applied statistics (Certificate); mathematics (MS); operations research (MS). Part-time and evening/weekend programs available. *Faculty:* 20 full-time (4 women). *Students:* 8 full-time (1 woman), 9 part-time (4 women); includes 2 minority (both American Indian/Alaska Native). Average age 33. 10 applicants, 100% accepted, 8 enrolled. In 2005, 5 degrees awarded. *Degree requirements:* For Certificate, completed a calculus and a statistics course. *Entrance requirements:* For master's, minimum GPA of 3.0, major or minor in mathematics. Additional exam requirements/recommendations for international students: Required—TOEFL (minimum score 600 paper-based; 260 computer-based). *Application deadline:* For fall admission, 7/1 for domestic students; for spring admission, 12/1 for domestic students. Applications are processed on a rolling basis. Application fee: $55. *Expenses:* Tuition, state resident: full-time $4,023; part-time $232 per credit. Tuition, nonresident: full-time $9,182; part-time $503 per credit. Required fees: $383. Tuition and fees vary according to course load. *Financial support:* In 2005–06, 10 teaching assistantships with partial tuition reimbursements (averaging $11,700 per year) were awarded; scholarships/grants and unspecified assistantships also available. Support available to part-time students. Financial award application deadline: 3/1; financial award applicants required to submit FAFSA. *Faculty research:* Graph theory, biostatistics, partial differential equations, 11-venn diagrams, random operator equations. Total annual research expenditures: $67,015. *Unit head:* Dr. David A. Legg, Chair, 260-481-6821, Fax: 260-481-6880, E-mail: legg@ipfw.edu. *Application contact:* Dr. W. Douglas Weakley, Director of Graduate Studies, 260-481-6821, Fax: 260-481-6880, E-mail: weakley@ipfw.edu.

Indiana University–Purdue University Indianapolis, School of Science, Department of Mathematical Sciences, Indianapolis, IN 46202-3216. Offers applied mathematics (MS, PhD); applied statistics (MS); mathematics (MS, PhD). Part-time programs available. *Faculty:* 10 full-time (0 women), 1 part-time/adjunct (0 women). *Students:* 12 full-time (4 women), 42 part-time (20 women); includes 3 minority (1 African American, 2 Asian Americans or Pacific Islanders), 16 international. Average age 32. In 2005, 19 degrees awarded. Terminal master's awarded for partial completion of doctoral program. *Degree requirements:* For master's, thesis optional; for doctorate, one foreign language, thesis/dissertation. *Entrance requirements:* For doctorate, GRE. Additional exam requirements/recommendations for international students: Required—TOEFL. *Application deadline:* For fall admission, 2/1 for domestic students. Application fee: $50 ($60 for international students). *Expenses:* Tuition, state resident: full-time $5,159; part-time $215 per credit hour. Tuition, nonresident: full-time $14,890; part-time $620 per credit hour. Required fees: $614. Tuition and fees vary according to campus/location and program. *Financial support:* In 2005–06, 14 students received support; fellowships with tuition reimbursements available, research assistantships with tuition reimbursements available, teaching assistantships with tuition reimbursements available, career-related internships or fieldwork, Federal Work-Study, and tuition waivers (full and partial) available. Financial award application deadline: 3/1. *Faculty research:* Mathematical physics, analysis, operator theory, functional analysis, integrated systems. *Unit head:* Benzion Boukai, Chair, 317-274-6920, Fax: 317-274-3460, E-mail: bboukai@math.iupui.edu. *Application contact:* Joan Morand, Student Services Specialist, 317-274-4127, Fax: 317-274-3460.

See Close-Up on page 503.

Iowa State University of Science and Technology, Graduate College, College of Liberal Arts and Sciences, Department of Mathematics, Ames, IA 50011. Offers applied mathematics (MS, PhD); mathematics (MS, PhD); school mathematics (MSM). *Faculty:* 48 full-time, 1 part-time/adjunct. *Students:* 60 full-time (16 women), 7 part-time (5 women); includes 1 minority (Asian American or Pacific Islander), 28 international. 67 applicants, 42% accepted, 15 enrolled. In 2005, 13 master's, 9 doctorates awarded. *Degree requirements:* For master's, thesis or alternative; for doctorate, thesis/dissertation. *Entrance requirements:* For master's and doctorate, GRE General Test. Additional exam requirements/recommendations for international students: Required—TOEFL (paper score 550; computer score 213) or IELTS (score 6.5). *Application deadline:* For fall admission, 2/1 priority date for domestic students, 2/1 priority date for international students. Application fee: $30 ($70 for international students). Electronic applications accepted. *Expenses:* Tuition, state resident: full-time $6,410. Tuition, nonresident: full-time $16,422. Tuition and fees vary according to program. *Financial support:* In 2005–06, 3 research assistantships with full and partial tuition reimbursements (averaging $18,610 per year), 51 teaching assistantships with full and partial tuition reimbursements (averaging $16,173 per year) were awarded; fellowships, scholarships/grants, health care benefits, and unspecified assistantships also available. *Unit head:* Dr. Justin R. Peters, Chair, 515-294-1752, Fax: 515-294-5454, E-mail: gradmath@iastate.edu. *Application contact:* Dr. Paul Sacks, Director of Graduate Education, 515-294-8143, E-mail: gradmath@iastate.edu.

Jackson State University, Graduate School, School of Science and Technology, Department of Mathematics, Jackson, MS 39217. Offers mathematics (MS); mathematics education (MST). Part-time and evening/weekend programs available. *Degree requirements:* For master's, thesis (for some programs), comprehensive exam. *Entrance requirements:* For master's, GRE General Test. Additional exam requirements/recommendations for international students: Required—TOEFL.

Jacksonville State University, College of Graduate Studies and Continuing Education, College of Arts and Sciences, Department of Mathematics, Jacksonville, AL 36265-1602. Offers MS. *Faculty:* 5 full-time (1 woman). *Students:* 4 full-time (2 women), 17 part-time (12 women); includes 2 minority (both African Americans) In 2005, 3 degrees awarded. *Degree requirements:* For master's, thesis optional. *Entrance requirements:* For master's, GRE General Test or MAT. *Application deadline:* Applications are processed on a rolling basis. Application fee: $20. *Expenses:* Tuition, state resident: full-time $4,848; part-time $202 per credit hour. Tuition, nonresident: full-time $9,696; part-time $404 per credit hour. One-time fee: $20 full-time. *Financial support:* In 2005–06, 3 teaching assistantships were awarded Support available to part-time students. Financial award application deadline: 4/1. *Unit head:* Dr. Jeff Dodd, Head, 256-782-5112. *Application contact:* 256-782-5329, Fax: 256-782-5321, E-mail: graduate@jsu.edu.

James Madison University, College of Graduate and Professional Programs, College of Science and Mathematics, Department of Mathematics and Statistics, Harrisonburg, VA 22807. Offers M Ed. Part-time programs available. *Faculty:* 2 part-time/adjunct (1 woman). *Application deadline:* For fall admission, 5/1 for domestic students; for spring admission, 9/1 priority date for domestic students. Application fee: $55. *Expenses:* Tuition, state resident: full-time $5,904; part-time $246 per credit hour. Tuition, nonresident: full-time $16,824; part-time $701 per credit hour. *Financial support:* Application deadline: 3/1; *Unit head:* Dr. David C. Carothers, Academic Unit Head, 540-568-6184.

John Carroll University, Graduate School, Department of Mathematics, University Heights, OH 44118-4581. Offers MA, MS. Part-time and evening/weekend programs available. *Degree requirements:* For master's, research essay. *Entrance requirements:* For master's, minimum GPA of 2.5, teaching certificate (MA). *Faculty research:* Algebraic topology, algebra, differential geometry, combinatorics, Lie groups.

The Johns Hopkins University, Zanvyl Krieger School of Arts and Sciences, Department of Mathematics, Baltimore, MD 21218-2699. Offers PhD. *Faculty:* 26 full-time (3 women). *Students:* 31 full-time (5 women); includes 12 minority (1 African American, 10 Asian Americans or Pacific Islanders, 1 Hispanic American), 4 international. Average age 23. 31 applicants, 23% accepted, 6 enrolled. In 2005, 2 doctorates awarded. *Degree requirements:* For doctorate, one foreign language, thesis/dissertation, language and 3 qualifying exams. *Entrance requirements:* For doctorate, GRE General Test, GRE Subject Test. Additional exam requirements/recommendations for international students: Required—TOEFL (minimum score 600 paper-based; 250 computer-based). *Application deadline:* For fall admission, 1/15 for domestic students, 1/15 for international students. Application fee: $60. Electronic applications accepted. *Expenses:* Tuition: Full-time $30,960. Tuition and fees vary according to degree level and program. *Financial support:* In 2005–06, 2 fellowships with full tuition reimbursements (averaging $5,000 per year), 31 teaching assistantships with full tuition reimbursements (averaging $16,500 per year) were awarded; research assistantships, Federal Work-Study, institutionally sponsored loans, and tuition waivers (partial) also available. Financial award application deadline: 4/15; financial award applicants required to submit FAFSA. *Faculty research:* Algebraic geometry, number theory, algebraic topology, differential geometry, partial differential equations. *Unit head:* Dr. Richard Wentworth, Chair, 410-516-7397, Fax: 410-516-5549, E-mail: raw@math.jhu.edu. *Application contact:* Sabrina Raymond, Graduate Program Coordinator, 410-516-7399, Fax: 410-516-5549, E-mail: sraymond@jhu.edu.

Kansas State University, Graduate School, College of Arts and Sciences, Department of Mathematics, Manhattan, KS 66506. Offers MS, PhD. Part-time programs available. *Faculty:* 27 full-time (1 woman), 1 part-time/adjunct (0 women). *Students:* 46 full-time (11 women), 2 part-time; includes 3 minority (2 Asian Americans or Pacific Islanders, 1 Hispanic American), 24 international. Average age 25. 42 applicants, 43% accepted, 12 enrolled. In 2005, 9 master's, 1 doctorate awarded. Terminal master's awarded for partial completion of doctoral program. *Degree requirements:* For master's, thesis or alternative; for doctorate, one foreign language, thesis/dissertation. *Entrance requirements:* For master's, GRE, bachelor's degree in mathematics; for doctorate, master's degree in mathematics. Additional exam requirements/recommendations for international students: Required—TOEFL (minimum score 600 paper-based; 250 computer-based); Recommended—TSE (minimum score 50). *Application deadline:* For fall admission, 2/1 priority date for domestic students, 2/1 priority date for international students; for spring admission, 9/1 priority date for domestic students, 9/1 priority date for international students. Applications are processed on a rolling basis. Application fee: $30 ($55 for international students). Electronic applications accepted. *Expenses:* Tuition, state resident: full-time $5,160; part-time $215. Tuition, nonresident: full-time $12,816; part-time $534. Required fees: $564. *Financial support:* In 2005–06, 41 teaching assistantships with full tuition reimbursements (averaging $13,750 per year) were awarded; research assistantships, Federal Work-Study, institutionally sponsored loans, and scholarships/grants also available. Support available to part-time students. Financial award application deadline: 3/1; financial award applicants required to submit FAFSA. *Faculty research:* Low-dimensional topology, geometry, complex and harmonic analysis, group and representation theory, noncommunicative spaces. Total annual research expenditures: $326,155. *Unit head:* Dr. Louis Pigno, Head, 785-532-0559, Fax: 785-532-0546, E-mail: lpigno@ksu.edu. *Application contact:* Pietro Poggi-Corrandi, Director, 785-532-0569, Fax: 785-532-0546, E-mail: pietro@math.ksu.edu.

Kean University, College of Natural, Applied and Health Sciences, Program in Computing, Statistics and Mathematics, Union, NJ 07083. Offers MS. Part-time and evening/weekend programs available. *Faculty:* 24 full-time (6 women). *Students:* 7 full-time (3 women), 5 part-time (2 women); includes 5 minority (2 African Americans, 1 Asian American or Pacific Islander, 2 Hispanic Americans), 1 international. Average age 36. 1 applicant, 100% accepted, 1 enrolled. In 2005, 3 degrees awarded. *Degree requirements:* For master's, thesis or alternative, research component, minimum 3.0 GPA. *Entrance requirements:* For master's, GRE General Test, 2 letters of recommendation, interview. *Application deadline:* For fall admission, 5/1 for domestic students; for spring admission, 11/1 for domestic students. Application fee: $60 ($150 for international students). Electronic applications accepted. *Expenses:* Tuition, state resident: full-time $8,280; part-time $345 per credit. Tuition, nonresident: full-time $11,512; part-time $438 per credit. Required fees: $2,104; $88 per credit. *Financial support:* In 2005–06, 3 research assistantships with full tuition reimbursements (averaging $2,880 per year) were awarded *Unit head:* Dr. Francine Abeles, Program Coordinator, 908-737-3714, E-mail: fabeles@kean.edu. *Application contact:* Joanne Morris, Director of Graduate Admissions, 908-737-3355, Fax: 908-737-3354, E-mail: grad-adm@kean.edu.

Kent State University, College of Arts and Sciences, Department of Mathematical Sciences, Kent, OH 44242-0001. Offers applied mathematics (MA, MS, PhD); pure mathematics (MA, MS, PhD). Part-time programs available. *Degree requirements:* For master's, thesis optional; for doctorate, one foreign language, thesis/dissertation. Electronic applications accepted. *Faculty research:* Approximation theory, measure theory, ring theory, functional analysis, complex analysis.

Kent State University, Graduate School of Education, Health, and Human Services, Department of Teaching, Leadership, and Curriculum Studies, Program in Math Specialization, Kent, OH 44242-0001. Offers M Ed, MA. *Faculty:* 4 full-time (3 women). *Students:* 1 (woman) full-time, 11 part-time (9 women). 3 applicants, 100% accepted. In 2005, 5 degrees awarded. *Entrance requirements:* Additional exam requirements/recommendations for international students: Required—TOEFL. *Application deadline:* Applications are processed on a rolling basis. Application fee: $30. Electronic applications accepted. *Financial support:* In 2005–06, fellowships (averaging $7,000 per year) *Unit head:* Dr. Trish Koontz, Coordinator, 330-672-2580, E-mail: tkoontz@kent.edu. *Application contact:* Nancy Miller, Academic program Coordinator, Office of Graduate Student Services, 330-672-2576, Fax: 330-672-9162, E-mail: ogs@kent.edu.

Kutztown University of Pennsylvania, College of Graduate Studies and Extended Learning, College of Liberal Arts and Sciences, Program in Mathematics and Computer Science, Kutztown, PA 19530-0730. Offers MS. Part-time and evening/weekend programs available. *Faculty:* 3 full-time (1 woman). *Students:* 7 full-time (2 women), 11 part-time (1 woman); includes 9 minority (3 African Americans, 2 American Indian/Alaska Native, 3 Asian Americans or Pacific Islanders, 1 Hispanic American), 1 international. Average age 35. 12 applicants, 75% accepted, 7 enrolled. In 2005, 4 degrees awarded. *Degree requirements:* For master's, comprehensive exam or thesis. *Entrance requirements:* For master's, GRE General Test. Additional exam requirements/recommendations for international students: Required—TOEFL. *Application deadline:* Applications are processed on a rolling basis. Application fee: $35. Electronic applications accepted. *Financial support:* In 2005–06, research assistantships with full tuition reimbursements (averaging $5,000 per year); career-related internships or fieldwork, Federal Work-Study, and unspecified assistantships also available. Financial award application deadline: 3/15; financial award applicants required to submit FAFSA. *Faculty research:* Artificial intelligence, expert systems, neural networks. *Unit head:* William Bateman, Chairperson, 610-683-4410, E-mail: bateman@kutztown.edu.

Lakehead University, Graduate Studies, School of Mathematical Sciences, Thunder Bay, ON P7B 5E1, Canada. Offers computer science (M Sc, MA); mathematics and statistics (M Sc, MA). Part-time and evening/weekend programs available. *Degree requirements:* For master's, thesis optional. *Entrance requirements:* For master's, minimum B average, honours degree in mathematics or computer science. Additional exam requirements/recommendations for international students: Required—TOEFL. *Faculty research:* Numerical analysis, classical analysis, theoretical computer science, abstract harmonic analysis, functional analysis.

Lamar University, College of Graduate Studies, College of Arts and Sciences, Department of Mathematics, Beaumont, TX 77710. Offers MS. *Faculty:* 5 full-time (1 woman). *Students:* 4 full-time (1 woman), 1 international. Average age 31. 15 applicants, 40% accepted, 1 enrolled. In 2005, 2 degrees awarded. *Degree requirements:* For master's, thesis optional. *Entrance requirements:* For master's, GRE General Test, minimum GPA of 2.5 in last 60 hours of undergraduate course work. Additional exam requirements/recommendations for international students: Required—TOEFL. *Application deadline:* For fall admission, 5/15 for domestic students; for spring admission, 10/1 priority date for domestic students. Applications are processed on a rolling basis. Application fee: $25 ($50 for international students). *Expenses:* Tuition, state resident: part-time $137 per semester hour. Tuition, nonresident: part-time $413 per semester hour. Required fees: $102 per semester hour. Tuition and fees vary according to course load. *Financial support:* In 2005–06, 2 research assistantships, 3 teaching assistantships (averaging $12,000 per year) were awarded; fellowships Financial award application deadline: 4/1. *Faculty research:* Complex analysis, functional analysis, wavelets, differential equations. Total annual research expenditures:$43,585. *Unit head:* Charles F. Coppin, Chair, 409-880-8792, Fax: 409-880-8794, E-mail: chair@math.lamar.edu. *Application contact:* Dr. Paul Chiou, Professor, 409-880-8800, Fax: 409-880-8794, E-mail: chiou@math.lamar.edu.

Lehigh University, College of Arts and Sciences, Department of Mathematics, Bethlehem, PA 18015-3094. Offers applied mathematics (MS, PhD); mathematics (MS, PhD); statistics (MS). Part-time programs available. *Faculty:* 20 full-time (1 woman), 3 part-time/adjunct (0 women). *Students:* 33 full-time (11 women), 7 part-time (4 women); includes 1 Asian American or Pacific Islander, 10 international. Average age 26. 68 applicants, 71% accepted, 14 enrolled. In 2005, 5 master's, 1 doctorate awarded. Terminal master's awarded for partial completion of doctoral program. *Median time to degree:* Of those who began their doctoral program in fall 1996, 50% received their degree in 8 years or less. *Degree requirements:* For master's, comprehensive exam; for doctorate, one foreign language, thesis/dissertation, qualifying exams, comprehensive exam. *Entrance requirements:* For master's and doctorate, minimum GPA of 3.0. Additional exam requirements/recommendations for international students: Required—TOEFL. *Application deadline:* For fall admission, 7/15 for domestic students; for spring admission, 12/7 priority date for domestic students. Applications are processed on a rolling basis. Application fee: $60. Electronic applications accepted. *Financial support:* In 2005–06, 30 students received support, including 1 fellowship with full tuition reimbursement available (averaging $16,000 per year), 2 research assistantships with full tuition reimbursements available (averaging $14,000 per year), 20 teaching assistantships with full tuition reimbursements available (averaging $13,700 per year); scholarships/grants, tuition waivers (partial), and unspecified assistantships also available. Financial award application deadline: 1/15. *Faculty research:* Probability and statistics, geometry and topology, number theory, algebra, differential equations. *Unit head:* Dr. Steven Weintraub, Chairman, 610-758-3730, Fax: 610-758-3767, E-mail: shw2@lehigh.edu. *Application contact:* Dr. Terry Napier, Graduate Coordinator, 610-758-3755, E-mail: mathgrad@lehigh.edu.

Lehman College of the City University of New York, Division of Natural and Social Sciences, Department of Mathematics and Computer Science, Program in Mathematics, Bronx, NY 10468-1589. Offers MA. Part-time and evening/weekend programs available. *Degree requirements:* For master's, one foreign language, thesis or alternative.

Long Island University, C.W. Post Campus, College of Liberal Arts and Sciences, Department of Mathematics, Brookville, NY 11548-1300. Offers applied mathematics (MS); mathematics education (MS); mathematics for secondary school teachers (MS). Part-time and evening/weekend programs available. *Degree requirements:* For master's, thesis or alternative, oral presentation. *Entrance requirements:* Additional exam requirements/recommendations for international students: Required—TOEFL. Electronic applications accepted. *Faculty research:* Differential geometry, topological groups, general topology, number theory, analysis and statistics, numerical analysis.

Louisiana State University and Agricultural and Mechanical College, Graduate School, College of Arts and Sciences, Department of Mathematics, Baton Rouge, LA 70803. Offers MS, PhD. *Faculty:* 47 full-time (1 woman). *Students:* 89 full-time (25 women), 2 part-time (1 woman); includes 3 African Americans, 1 Asian American or Pacific Islander, 1 Hispanic American, 45 international. Average age 28. 139 applicants, 54% accepted, 64 enrolled. In 2005, 20 master's, 11 doctorates awarded. Terminal master's awarded for partial completion of doctoral program. *Degree requirements:* For doctorate, 2 foreign languages, thesis/dissertation. *Entrance requirements:* For master's and doctorate, GRE General Test, minimum GPA of 3.0. Additional exam requirements/recommendations for international students: Required—TOEFL (minimum score 550 paper-based; 213 computer-based). *Application deadline:* For fall admission, 1/25 priority date for domestic students, 5/15 priority date for international students. Applications are processed on a rolling basis. Application fee: $25. Electronic applications accepted. *Financial support:* In 2005–06, 82 students received support, including 7 fellowships with full and partial tuition reimbursements available (averaging $19,840 per year), 7 research assistantships with full and partial tuition reimbursements available (averaging $24,286 per year), 62 teaching assistantships with full and partial tuition reimbursements available (averaging $19,322 per year); Federal Work-Study, institutionally sponsored loans, scholarships/grants, tuition waivers (full), and unspecified assistantships also available. Financial award application deadline: 3/1; financial award applicants required to submit FAFSA. *Faculty research:* Algebra, graph theory and combinatorics, algebraic topology, analysis and probability, topological algebra. Total annual research expenditures: $865,687. *Unit head:* Dr. Lawrence Smolinksy, Chair, 225-578-1570, Fax: 225-578-4276, E-mail: mmsmol@lsu.edu. *Application contact:* Dr. Leonard F. Richardson, Director of Graduate Studies and Assistant Chairman, 225-578-1568, Fax: 225-578-4276, E-mail: rich@math.lsu.edu.

Louisiana Tech University, Graduate School, College of Engineering and Science, Department of Mathematics and Statistics, Ruston, LA 71272. Offers MS. Part-time programs available. *Degree requirements:* For master's, thesis or alternative. *Entrance requirements:* For master's, GRE General Test, minimum GPA of 3.0 in last 60 hours. Additional exam requirements/recommendations for international students: Required—TOEFL.

Loyola University Chicago, Graduate School, Department of Mathematical Sciences and Statistics, Chicago, IL 60611-2196. Offers mathematics (MS), including pure mathematics, statistics and probability. Part-time programs available. *Faculty:* 19 full-time (4 women). *Students:* 17 full-time (8 women), 2 part-time; includes 4 minority (2 Asian Americans or Pacific Islanders, 2 Hispanic Americans), 8 international. Average age 30. 30 applicants, 83% accepted. In 2005, 7 degrees awarded. *Degree requirements:* For master's, 3.0 GPA. *Entrance requirements:* For master's, GRE General Test. Additional exam requirements/recommendations for international students: Required—TOEFL. *Application deadline:* For fall admission, 8/1 for domestic students; for spring admission, 12/1 for domestic students. Applications are processed on a rolling basis. Application fee: $40. Electronic applications accepted. *Expenses:* Tuition: Full-time $11,610; part-time $645 per credit. Required fees: $55 per semester. *Financial support:* In 2005–06, 8 students received support, including 4 teaching assistantships with tuition reimbursements available (averaging $9,000 per year); career-related internships or fieldwork, Federal Work-Study, institutionally sponsored loans, and tuition waivers (partial) also available. *Faculty research:* Probability and statistics, differential equations, algebra, combinations. Total annual research expenditures: $300,000. *Unit head:* Dr. Stephen Doty, Chair, 773-508-8520, Fax: 773-508-2123, E-mail: agiaqui@luc.edu. *Application contact:* Dr. Stephen Doty, Director, 773-508-3556, Fax: 773-508-2123.

Marquette University, Graduate School, College of Arts and Sciences, Department of Mathematics, Statistics, and Computer Science, Milwaukee, WI 53201-1881. Offers algebra (PhD); bio-mathematical modeling (PhD); computers (MS); mathematics (MS); mathematics education (MS); statistics (MS). Part-time programs available. Terminal master's awarded for partial completion of doctoral program. *Degree requirements:* For master's, thesis or alternative, comprehensive exam; for doctorate, 2 foreign languages, thesis/dissertation, comprehensive exam. *Entrance requirements:* For doctorate, sample of scholarly writing. Additional exam requirements/recommendations for international students: Required—TOEFL. *Faculty research:* Models of physiological systems, mathematical immunology, computational group theory, mathematical logic.

Marshall University, Academic Affairs Division, Graduate College, College of Science, Department of Mathematics, Huntington, WV 25755. Offers MA, MS. *Faculty:* 9 full-time (4 women). *Students:* 11 full-time (6 women), 9 part-time (6 women), 3 international. Average age 28. In 2005, 2 degrees awarded. *Degree requirements:* For master's, thesis (for some programs). *Entrance requirements:* For master's, GRE General Test. *Unit head:* Dr. Ralph Oberste-Vorth, Chairperson, 304-696-6010, E-mail: oberstevorth@marshall.edu. *Application contact:* Information Contact, 304-746-1900, Fax: 304-746-1902, E-mail: services@marshall.edu.

Massachusetts Institute of Technology, School of Science, Department of Mathematics, Cambridge, MA 02139-4307. Offers PhD. *Faculty:* 52 full-time (3 women), 1 part-time/adjunct (0 women). *Students:* 106 full-time (23 women), 1 part-time; includes 6 minority (1 African American, 5 Asian Americans or Pacific Islanders), 56 international. Average age 25. 373 applicants, 21% accepted, 36 enrolled. In 2005, 30 doctorates awarded. *Degree requirements:* For doctorate, one foreign language, thesis/dissertation, comprehensive exam. *Entrance requirements:* For doctorate, GRE General Test, GRE Subject Test in mathematics. Additional exam requirements/recommendations for international students: Required—TOEFL (minimum score 577 paper-based; 233 computer-based). *Application deadline:* For fall admis-

Mathematics

Massachusetts Institute of Technology (continued)
sion, 1/2 for domestic students, 1/2 for international students. Application fee: $70. Electronic applications accepted. *Expenses:* Tuition: Full-time $32,100. Required fees: $200. Part-time tuition and fees vary according to course load. *Financial support:* In 2005–06, 36 fellowships with tuition reimbursements (averaging $23,902 per year), 10 research assistantships with tuition reimbursements (averaging $26,700 per year), 59 teaching assistantships with tuition reimbursements (averaging $25,897 per year) were awarded; Federal Work-Study, institutionally sponsored loans, scholarships/grants, traineeships, health care benefits, unspecified assistantships, and graduate instructorships also available. *Faculty research:* Analysis, topology, algebraic geometry, logic, Lie theory, combinatorics, fluid dynamics, theoretical computer science. Total annual research expenditures: $3.2 million. *Unit head:* Prof. Michael Sipser, Department Head, 617-253-4381, Fax: 617-253-4358, E-mail: dept@math.mit.edu. *Application contact:* Graudate Office, 617-253-2689, E-mail: gradofc@math.mit.edu.

McGill University, Faculty of Graduate and Postdoctoral Studies, Faculty of Science, Department of Mathematics and Statistics, Montréal, QC H3A 2T5, Canada. Offers computational science and engineering (M Sc); mathematics (M Sc, MA, PhD), including applied mathematics (M Sc, MA), pure mathematics (M Sc, MA), statistics (M Sc, MA). Part-time programs available. *Degree requirements:* For master's, thesis (for some programs), registration; for doctorate, one foreign language, thesis/dissertation, comprehensive exam, registration. *Entrance requirements:* For master's, minimum GPA of 3.0, Canadian Honours degree in mathematics or closely related field (statistics or applied mathematics). Additional exam requirements/recommendations for international students: Required—TOEFL (minimum score 550 paper-based; 213 computer-based), IELT (minimum score 7). Electronic applications accepted.

McMaster University, School of Graduate Studies, Faculty of Science, Department of Mathematics and Statistics, Hamilton, ON L8S 4M2, Canada. Offers mathematics (PhD); statistics (M Sc), including applied statistics, medical statistics, statistical theory. Part-time programs available. *Degree requirements:* For master's, thesis or alternative, oral exam; for doctorate, thesis/dissertation, comprehensive exam. *Entrance requirements:* For master's, minimum B+ average in last year of honors degree; for doctorate, minimum B+ average, M Sc in mathematics or statistics. Additional exam requirements/recommendations for international students: Required—TOEFL (minimum score 550 paper-based; 213 computer-based). *Faculty research:* Algebra, analysis, applied mathematics, geometry and topology, probability and statistics.

McNeese State University, Graduate School, College of Science, Department of Mathematics, Computer Science, and Statistics, Lake Charles, LA 70609. Offers computer science (MS); mathematics (MS); statistics (MS). Evening/weekend programs available. *Faculty:* 9 full-time (5 women). *Students:* 22 full-time (7 women), 8 part-time (2 women), 22 international. In 2005, 19 degrees awarded. *Degree requirements:* For master's, thesis or alternative, written exam, comprehensive exam. *Entrance requirements:* For master's, GRE General Test. *Application deadline:* For fall admission, 7/15 for domestic students. Applications are processed on a rolling basis. Application fee: $20 ($30 for international students). *Expenses:* Tuition, area resident: Part-time $193 per hour. Tuition, state resident: full-time $2,226. Required fees: $862; $106 per hour. Tuition and fees vary according to course load. *Financial support:* Teaching assistantships available. Financial award application deadline: 5/1. *Unit head:* Sid Bradley, Head, 337-475-5788, Fax: 337-475-5799, E-mail: sbradley@mcneese.edu.

Memorial University of Newfoundland, School of Graduate Studies, Department of Mathematics and Statistics, St. John's, NL A1C 5S7, Canada. Offers mathematics (M Sc, PhD); statistics (M Sc, MAS, PhD). Part-time programs available. *Students:* 43 full-time (14 women), 4 part-time (1 woman), 30 international. 42 applicants, 38% accepted, 15 enrolled. In 2005, 2 master's, 1 doctorate awarded. *Degree requirements:* For master's, thesis, thesis, practicum and report (MAS); for doctorate, thesis/dissertation, oral defense of thesis, comprehensive exam. *Entrance requirements:* For master's, 2nd class honors degree (MAS); for doctorate, MAS or M Sc in mathematics and statistics. *Application deadline:* For fall admission, 1/30 for domestic students, 1/30 for international students. Applications are processed on a rolling basis. Application fee: $40 Canadian dollars. Electronic applications accepted. *Expenses:* Tuition: Part-time $733 per term. Tuition and fees vary according to degree level and program. *Financial support:* Fellowships, teaching assistantships available. Financial award application deadline: 1/31. *Faculty research:* Algebra, topology, applied mathematics, mathematical statistics, applied statistics and probability. *Unit head:* Dr. Chris Radford, Interim Head, 709-737-8783, Fax: 709-787-3010, E-mail: head@math.mun.ca. *Application contact:* Dr. Edgar Goodaire, Graduate Officer, 709-737-8097, Fax: 709-737-3010, E-mail: grad@math.mun.ca.

Miami University, Graduate School, College of Arts and Sciences, Department of Mathematics and Statistics, Program in Mathematics, Oxford, OH 45056. Offers mathematics (MA, MAT, MS); mathematics/operations research (MS). Part-time programs available. *Degree requirements:* For master's, final exam. *Entrance requirements:* For master's, minimum undergraduate GPA of 3.0 during previous 2 years or 2.75 overall. Additional exam requirements/recommendations for international students: Required—TOEFL, TWE. Electronic applications accepted.

See Close-Up on page 507.

Michigan State University, The Graduate School, College of Natural Science, Department of Mathematics, East Lansing, MI 48824. Offers applied mathematics (MS, PhD); industrial mathematics (MS); mathematics (MAT, MS, PhD); mathematics education (PhD). *Faculty:* 58 full-time (13 women). *Students:* 110 full-time (34 women), 6 part-time (2 women); includes 7 minority (2 African Americans, 3 Asian Americans or Pacific Islanders, 2 Hispanic Americans), 71 international. Average age 28. 131 applicants, 18% accepted. In 2005, 18 master's, 15 doctorates awarded. *Degree requirements:* For doctorate, one foreign language, thesis/dissertation, qualifying exam, seminar presentations, comprehensive exam. *Entrance requirements:* For master's, GRE General Test, bachelor's degree in mathematics, physics, or engineering; mathematics course work beyond calculus; 3 letters of recommendation; for doctorate, GRE General Test, minimum GPA of 3.0, MS in mathematics or equivalent, 3 letters of recommendation. Additional exam requirements/recommendations for international students: Required—TOEFL (minimum score 550 paper-based; 213 computer-based), Michigan State University ELT (85), Michigan ELAB (83). *Application deadline:* For fall admission, 2/1 for domestic students. Application fee: $50. Electronic applications accepted. *Expenses:* Tuition, state resident: part-time $330 per credit hour. Tuition, nonresident: part-time $685 per credit hour. Tuition and fees vary according to program. *Financial support:* In 2005–06, 12 fellowships with tuition reimbursements (averaging $4,370 per year), 21 research assistantships with tuition reimbursements (averaging $14,054 per year), 94 teaching assistantships with tuition reimbursements (averaging $13,770 per year) were awarded; scholarships/grants and unspecified assistantships also available. *Faculty research:* Applied and industrial mathematics; analysis; geometry and topology; logic, combinatorics and graph theory; algebra. Total annual research expenditures: $2.6 million. *Unit head:* Dr. Peter W. Bates, Chairperson, 517-355-9681, Fax: 517-432-1562, E-mail: bates@math.msu.edu. *Application contact:* Barbara S. Miller, Graduate Secretary, 517-353-6338, Fax: 517-432-1562, E-mail: bmiller@math.msu.edu.

Michigan Technological University, Graduate School, College of Sciences and Arts, Department of Mathematical Sciences, Houghton, MI 49931-1295. Offers MS, PhD. Part-time programs available. *Faculty:* 30 full-time (8 women), 2 part-time/adjunct (0 women). *Students:* 36 full-time (12 women), 2 part-time (1 woman); includes 2 minority (1 African American, 1 Asian American or Pacific Islander), 29 international. Average age 28. 48 applicants, 79% accepted, 18 enrolled. In 2005, 8 master's, 6 doctorates awarded. Terminal master's awarded for partial completion of doctoral program. *Median time to degree:* Of those who began their doctoral program in fall 1996, 0% received their degree in 8 years or less. *Degree requirements:* For master's, thesis (for some programs), comprehensive exam (for some programs), registration; for doctorate, thesis/dissertation, Proficiency exam, comprehensive exam, registration. *Entrance requirements:* For master's and doctorate, GRE General Test, GRE Subject Test (recommended). Additional exam requirements/recommendations for international students: Required—TOEFL (minimum score 550 paper-based; 213 computer-based). *Application deadline:* For fall admission, 2/15 for domestic students; for spring admission, 10/15 for domestic students. Applications are processed on a rolling basis. Application fee: $40 ($45 for international students). Electronic applications accepted. *Expenses:* Tuition, nonresident: full-time $11,232; part-time $468 per credit. Required fees: $754; $377 per semester. Full-time tuition and fees vary according to course load, degree level and program. *Financial support:* In 2005–06, 30 students received support, including fellowships with full tuition reimbursements available (averaging $9,542 per year), 4 research assistantships with full tuition reimbursements available (averaging $9,542 per year), 25 teaching assistantships with full tuition reimbursements available (averaging $9,542 per year); career-related internships or fieldwork, Federal Work-Study, scholarships/grants, health care benefits, tuition waivers (partial), unspecified assistantships, and co-op also available. Financial award applicants required to submit FAFSA. *Faculty research:* Fluid dynamics, mathematical modeling, design theory, coding theory, statistical genetics. Total annual research expenditures:$233,196. *Unit head:* Dr. Alphonse H. Baartmans, Chair, 906-487-2068, Fax: 906-487-3133, E-mail: baartman@mtu.edu. *Application contact:* Dr. Mark S. Gockenbach, Director of Graduate Studies, 906-487-3083, Fax: 906-487-3133, E-mail: msgocken@mtu.edu.

Middle Tennessee State University, College of Graduate Studies, College of Basic and Applied Sciences, Department of Mathematical Sciences, Murfreesboro, TN 37132. Offers mathematics (MS); mathematics education (MST). Part-time and evening/weekend programs available. Postbaccalaureate distance learning degree programs offered. *Degree requirements:* For master's, comprehensive exam. *Entrance requirements:* For master's, GRE General Test or MAT. Additional exam requirements/recommendations for international students: Required—TOEFL (minimum score 525 paper-based; 195 computer-based). Electronic applications accepted.

Minnesota State University Mankato, College of Graduate Studies, College of Science, Engineering and Technology, Department of Mathematics and Statistics, Program in Computer Science, Mankato, MN 56001. Offers mathematics: computer science (MS). *Students:* Average age 32. *Degree requirements:* For master's, one foreign language, thesis or alternative, comprehensive exam. *Entrance requirements:* For master's, GRE General Test, GRE Subject Test (if GPA is below 2.75), minimum GPA of 3.0 during previous 2 years. *Application deadline:* For fall admission, 7/1 for domestic students; for spring admission, 11/1 for domestic students. Applications are processed on a rolling basis. Application fee: $40. Electronic applications accepted. *Expenses:* Tuition, state resident: part-time $243 per credit. Tuition, nonresident: part-time $400 per credit. Required fees: $30 per credit. *Financial support:* Fellowships with full tuition reimbursements, research assistantships with full tuition reimbursements, teaching assistantships with full tuition reimbursements, Federal Work-Study, institutionally sponsored loans, and unspecified assistantships available. Support available to part-time students. Financial award application deadline: 3/15; financial award applicants required to submit FAFSA. *Unit head:* Dr. Dean Kelley, Graduate Coordinator, 507-389-1134. *Application contact:* 507-389-2321, E-mail: grad@mnsu.edu.

Minnesota State University Mankato, College of Graduate Studies, College of Science, Engineering and Technology, Department of Mathematics and Statistics, Program in Mathematics, Mankato, MN 56001. Offers MA, MS. *Students:* 4 full-time (1 woman), 6 part-time (2 women). Average age 32. In 2005, 1 degree awarded. *Degree requirements:* For master's, one foreign language, thesis or alternative, comprehensive exam. *Entrance requirements:* For master's, GRE General Test, minimum GPA of 3.0 during previous 2 years. Additional exam requirements/recommendations for international students: Required—TOEFL. *Application deadline:* For fall admission, 7/1 for domestic students; for spring admission, 11/1 for domestic students. Applications are processed on a rolling basis. Application fee: $40. Electronic applications accepted. *Expenses:* Tuition, state resident: part-time $243 per credit. Tuition, nonresident: part-time $400 per credit. Required fees: $30 per credit. *Financial support:* Research assistantships with partial tuition reimbursements, teaching assistantships with partial tuition reimbursements, unspecified assistantships available. Financial award application deadline: 3/15; financial award applicants required to submit FAFSA. *Unit head:* Dr. Dan Singer, Graduate Coordinator, 507-389-1424. *Application contact:* 507-389-2321, E-mail: grad@mnsu.edu.

Mississippi College, Graduate School, College of Arts and Sciences, Department of Mathematics and Computer Science, Clinton, MS 39058. Offers computer science (MS); mathematics (MS). *Degree requirements:* For master's, comprehensive exam. *Entrance requirements:* For master's, minimum GPA of 2.5.

Mississippi College, Graduate School, College of Arts and Sciences, Program in Combined Sciences, Clinton, MS 39058. Offers biology (MCS); chemistry (MCS); mathematics (MCS). *Degree requirements:* For master's, thesis or alternative, comprehensive exam. *Entrance requirements:* For master's, GRE General Test, minimum GPA of 2.5.

Mississippi State University, College of Arts and Sciences, Department of Mathematics and Statistics, Mississippi State, MS 39762. Offers mathematical sciences (PhD); mathematics (MS); statistics (MS). Part-time programs available. *Faculty:* 32 full-time (15 women), 1 (woman) part-time/adjunct. *Students:* 31 full-time (16 women), 6 part-time (4 women); includes 3 minority (1 African American, 2 Asian Americans or Pacific Islanders), 23 international. Average age 30. 46 applicants, 35% accepted, 9 enrolled. In 2005, 7 degrees awarded. Terminal master's awarded for partial completion of doctoral program. *Degree requirements:* For master's, comprehensive oral or written exam, thesis optional; for doctorate, one foreign language, thesis/dissertation, comprehensive oral and written exam. *Entrance requirements:* For master's, minimum GPA of 2.75; for doctorate, GRE. Additional exam requirements/recommendations for international students: Required—TOEFL. *Application deadline:* For fall admission, 3/15 for domestic students; for spring admission, 11/1 for domestic students. Applications are processed on a rolling basis. Application fee: $30. *Expenses:* Tuition, state resident: full-time $4,312; part-time $240 per hour. Tuition, nonresident: full-time $9,772; part-time $543 per hour. International tuition: $10,102 full-time. Tuition and fees vary according to course load. *Financial support:* In 2005–06, 20 teaching assistantships with full tuition reimbursements (averaging $11,389 per year) were awarded; Federal Work-Study, institutionally sponsored loans, tuition waivers (partial), and unspecified assistantships also available. Financial award applicants required to submit FAFSA. *Faculty research:* Differential equations, algebra, numerical analysis, functional analysis, applied statistics. Total annual research expenditures: $1.5 million. *Unit head:* Dr. Michael Neuman, Interim Head and Professor, 662-325-3414, Fax: 662-325-0005, E-mail: office@math.msstate.edu. *Application contact:* Philip G. Bonfanti, Director of Admissions, 662-325-4104, Fax: 662-325-8872, E-mail: admit@msstate.edu.

Missouri State University, Graduate College, College of Natural and Applied Sciences, Department of Mathematics, Springfield, MO 65804-0094. Offers mathematics (MS); secondary education (MS Ed), including mathematics. Part-time programs available. *Faculty:* 21 full-time (4 women). *Students:* 15 full-time (5 women), 8 part-time (4 women), 1 international. Average age 27. 12 applicants, 75% accepted, 7 enrolled. In 2005, 9 degrees awarded. *Degree requirements:* For master's, thesis or alternative, comprehensive exam. *Entrance requirements:* For master's, GRE (MS, MNAS), minimum undergraduate GPA of 3.0 (MS, MNAS), 9-12 teacher certification (MS Ed). Additional exam requirements/recommendations for international students: Required—TOEFL (minimum score 550 paper-based; 213 computer-based), IELT (minimum score 6). *Application deadline:* For fall admission, 7/20 for domestic students; for spring admission, 12/20 priority date for domestic students. Applications are processed on a rolling basis. Application fee: $30. Electronic applications accepted. *Expenses:* Tuition, state resident: full-time $3,402; part-time $189 per credit. Tuition, nonresident: full-time $6,804; part-time $378 per credit. Required fees: $207 per semester. Part-time tuition and fees vary according to course level, course load and program. *Financial support:* In 2005–06, 11 teaching assistantships with full tuition reimbursements (averaging $8,750 per year) were awarded; research assistantships with full tuition reimbursements, Federal Work-Study, scholarships/grants, and unspecified assistantships also available. Financial award application deadline: 3/31; financial award applicants required to submit FAFSA. *Faculty research:* Harmonic analysis, commutative algebra, number theory, K-theory, probability. *Unit head:* Dr. Yungchen Cheng, Head, 417-836-5112, Fax: 417-836-6966, E-mail: yungchencheng@missouristate.edu.

Montana State University, College of Graduate Studies, College of Letters and Science, Department of Mathematical Sciences, Bozeman, MT 59717. Offers mathematics (MS, PhD); statistics (MS, PhD). Part-time programs available. Postbaccalaureate distance learning degree programs offered (minimal on-campus study). *Faculty:* 30 full-time (7 women), 13 part-time/adjunct (6 women). *Students:* 14 full-time (5 women), 76 part-time (40 women); includes 4 minority (1 American Indian/Alaska Native, 2 Asian Americans or Pacific Islanders, 1 Hispanic American), 9 international. Average age 31. 40 applicants, 60% accepted, 17 enrolled. In 2005, 18 master's, 3 doctorates awarded. *Degree requirements:* For master's, thesis (for some programs), comprehensive exam, registration; for doctorate, thesis/dissertation, comprehensive exam, registration. *Entrance requirements:* For master's and doctorate, GRE General Test. Additional exam requirements/recommendations for international students: Required—TOEFL (minimum score 550 paper-based; 213 computer-based). *Application deadline:* For fall admission, 7/15 priority date for domestic students, 5/15 priority date for international students; for spring admission, 12/1 priority date for domestic students, 10/1 priority date for international students. Applications are processed on a rolling basis. Application fee: $30. Electronic applications accepted. *Expenses:* Tuition, state resident: full-time $4,132. Tuition, nonresident: full-time $1,132. *Financial support:* In 2005–06, 65 students received support, including 8 research assistantships with full tuition reimbursements available (averaging $15,000 per year), 57 teaching assistantships with full tuition reimbursements available (averaging $13,500 per year); career-related internships or fieldwork, scholarships/grants, health care benefits, tuition waivers (full), and unspecified assistantships also available. Support available to part-time students. Financial award application deadline: 3/1; financial award applicants required to submit FAFSA. *Faculty research:* Applied mathematics, dynamical systems, statistics, mathematics education, mathematical and computational biology. Total annual research expenditures: $604,937. *Unit head:* Dr. Kenneth Bowers, Department Head, 406-994-3604, Fax: 406-994-1789, E-mail: bowers@math.montana.edu.

Montclair State University, The Graduate School, College of Science and Mathematics, Department of Mathematics, Montclair, NJ 07043-1624. Offers mathematics (MS), including computer science, mathematics education, pure and applied mathematics, statistics; teaching middle grades math (Certificate). Part-time and evening/weekend programs available. *Faculty:* 29 full-time (10 women), 29 part-time/adjunct (11 women). *Students:* 23 full-time (12 women), 135 part-time (99 women); includes 28 minority (13 African Americans, 10 Asian Americans or Pacific Islanders, 5 Hispanic Americans), 3 international. 77 applicants, 83% accepted, 51 enrolled. In 2005, 9 master's, 26 other advanced degrees awarded. *Degree requirements:* For master's, comprehensive exam. *Entrance requirements:* For master's, GRE General Test, minimum GPA of 2.67, 2 letters of recommendation. Additional exam requirements/recommendations for international students: Required—TOEFL (minimum score 83 computer-based). *Application deadline:* Applications are processed on a rolling basis. Application fee: $60. *Expenses:* Tuition: Full-time $3,001; part-time $409 per credit. Required fees: $56 per credit. Tuition and fees vary according to course load, degree level and program. *Financial support:* In 2005–06, 8 research assistantships with full tuition reimbursements (averaging $7,000 per year) were awarded; Federal Work-Study, scholarships/grants, and unspecified assistantships also available. Support available to part-time students. Financial award application deadline: 3/1; financial award applicants required to submit FAFSA. *Faculty research:* Infectious disease. Total annual research expenditures:$130,000. *Unit head:* Dr. Helen Roberts, Chairperson, 973-655-5132. *Application contact:* Dr. Ted Williamson, Advisor, 973-655-5146, E-mail: williamsont@mail.montclair.edu.

Morgan State University, School of Graduate Studies, School of Computer, Mathematical, and Natural Sciences, Department of Mathematics, Baltimore, MD 21251. Offers MA. Part-time and evening/weekend programs available. *Faculty:* 7 full-time (1 woman). *Students:* 4 full-time (1 woman), 5 part-time (1 woman); includes 6 minority (5 African Americans, 1 Asian American or Pacific Islander). In 2005, 2 degrees awarded. *Degree requirements:* For master's, thesis, comprehensive exam. *Entrance requirements:* For master's, GRE. Additional exam requirements/recommendations for international students: Required—TOEFL (minimum score 550 paper-based; 213 computer-based). *Application deadline:* For fall admission, 2/1 for domestic students; for spring admission, 10/1 priority date for domestic students. Applications are processed on a rolling basis. Application fee: $0. *Expenses:* Tuition, state resident: part-time $272 per credit. Tuition, nonresident: part-time $478 per credit. Required fees: $58 per credit. *Financial support:* Federal Work-Study, institutionally sponsored loans, health care benefits, and unspecified assistantships available. Financial award application deadline: 4/1. *Faculty research:* Number theory, semigroups, analysis, operations research. *Unit head:* Dr. Gaston M. N'Guerekata, Chairman, 443-885-3965. *Application contact:* Dr. James E. Waller, Admissions and Programs Officer, 443-885-3185, Fax: 443-885-8226, E-mail: jwaller@moac.morgan.edu.

Murray State University, College of Science, Engineering and Technology, Department of Mathematics, Murray, KY 42071-0009. Offers MA, MAT, MS. Part-time programs available. *Degree requirements:* For master's, thesis (for some programs). *Entrance requirements:* For master's, GRE General Test. Additional exam requirements/recommendations for international students: Required—TOEFL.

Naval Postgraduate School, Graduate Programs, Department of Mathematics, Monterey, CA 93943. Offers applied mathematics (MS, PhD). Program only open to commissioned officers of the United States and friendly nations and selected United States federal civilian employees. Part-time programs available. *Faculty:* 13. *Students:* 13 full-time. In 2005, 3 master's, 1 doctorate awarded. *Degree requirements:* For master's, thesis; for doctorate, one foreign language, thesis/dissertation. *Unit head:* Prof. Clyde Scandrett, Chairman, 831-656-2027, Fax: 831-656-2355, E-mail: cscand@nps.navy.mil. *Application contact:* Tracy Hammond, Acting Director of Admissions, 831-656-3059, Fax: 831-656-2891, E-mail: thammond@bps.navy.mil.

New Jersey Institute of Technology, Office of Graduate Studies, College of Science and Liberal Arts, Department of Mathematical Science, Program in Mathematics Science, Newark, NJ 07102. Offers PhD. Part-time and evening/weekend programs available. *Students:* 30 full-time (10 women), 2 part-time; includes 2 minority (both Asian Americans or Pacific Islanders), 26 international. Average age 28. 43 applicants, 42% accepted, 2 enrolled. In 2005, 7 degrees awarded. *Entrance requirements:* For doctorate, GRE General Test, minimum graduate GPA of 3.5. Additional exam requirements/recommendations for international students: Required—TOEFL (minimum score 550 paper-based; 213 computer-based). *Application deadline:* For fall admission, 6/5 for domestic students; for spring admission, 10/15 for domestic students. Applications are processed on a rolling basis. Application fee: $60. Electronic applications accepted. *Expenses:* Tuition, state resident: full-time $9,620; part-time $520 per credit. Tuition, nonresident: full-time $13,542; part-time $715 per credit. Required fees: $78; $54 per credit. $78 per year. Tuition and fees vary according to course load. *Financial support:* Fellowships with full and partial tuition reimbursements, research assistantships with full and partial tuition reimbursements, teaching assistantships with full and partial tuition reimbursements, career-related internships or fieldwork, Federal Work-Study, institutionally sponsored loans, and unspecified assistantships available. Financial award application deadline: 3/15. *Unit head:* Dr. Demetrius Papageorgious, Director, 973-596-3498, Fax: 973-596-5591, E-mail: demetrius.papageorgiou@njit.edu. *Application contact:* Kathryn Kelly, Director of Admissions, 973-596-3300, Fax: 973-596-3461, E-mail: admissions@njit.edu.

New Mexico Institute of Mining and Technology, Graduate Studies, Department of Mathematics, Socorro, NM 87801. Offers applied math (PhD); mathematics (MS); operations research (MS). *Degree requirements:* For master's, thesis optional; for doctorate, thesis/dissertation. *Entrance requirements:* For master's, GRE General Test. Additional exam requirements/recommendations for international students: Required—TOEFL (minimum score 540 paper-based; 207 computer-based). *Faculty research:* Applied mathematics, differential equations, industrial mathematics, numerical analysis, stochastic processes.

New Mexico State University, Graduate School, College of Arts and Sciences, Department of Mathematical Sciences, Las Cruces, NM 88003-8001. Offers MS, PhD. Part-time programs available. *Faculty:* 29 full-time (10 women), 3 part-time/adjunct (1 woman). *Students:* 32 full-time (12 women), 5 part-time (2 women); includes 5 minority (2 African Americans, 3 Hispanic Americans), 16 international. Average age 32. 28 applicants, 89% accepted, 8 enrolled. In 2005, 9 master's, 2 doctorates awarded. *Degree requirements:* For master's, final oral exam, thesis optional; for doctorate, one foreign language, thesis/dissertation, final oral exam, comprehensive exam. *Entrance requirements:* Additional exam requirements/recommendations for international students: Required—TOEFL (minimum score 530 paper-based; 197 computer-based). *Application deadline:* For fall admission, 7/1 priority date for domestic students, 3/1 priority date for international students; for spring admission, 11/1 for domestic students, 10/1 for international students. Applications are processed on a rolling basis. Application fee: $30 ($50 for international students). Electronic applications accepted. *Expenses:* Tuition, state resident: full-time $3,156; part-time $175 per credit. Tuition, nonresident: full-time $12,510; part-time $565 per credit. Required fees: $1,050. *Financial support:* In 2005–06, 4 fellowships, 3 research assistantships, 21 teaching assistantships were awarded; scholarships/grants and unspecified assistantships also available. Financial award application deadline: 3/15. *Faculty research:* Commutative algebra, dynamical systems, harmonic analysis and applications, algebraic topology, statistics. *Unit head:* Dr. Ross Staffeldt, Head, 505-646-3901, Fax: 505-646-1064, E-mail: ross@nmsu.edu. *Application contact:* Dr. John Harding, Professor, 505-646-4315, Fax: 505-646-1064, E-mail: gradcomm@umsu.edu.

New York University, Graduate School of Arts and Science, Courant Institute of Mathematical Sciences, Department of Mathematics, New York, NY 10012-1019. Offers atmosphere-ocean science and mathematics (PhD); mathematics (MS, PhD); mathematics and statistics/operations research (MS); mathematics in finance (MS); scientific computing (MS). Part-time and evening/weekend programs available. *Faculty:* 46 full-time (0 women). *Students:* 170 full-time (35 women), 119 part-time (23 women); includes 36 minority (1 African American, 33 Asian Americans or Pacific Islanders, 2 Hispanic Americans), 127 international. Average age 28. 808 applicants, 40% accepted, 92 enrolled. In 2005, 58 master's, 11 doctorates awarded. *Degree requirements:* For master's, thesis optional; for doctorate, one foreign language, thesis/dissertation, oral and written exams. *Entrance requirements:* For master's and doctorate, GRE General Test, GRE Subject Test. Additional exam requirements/recommendations for international students: Required—TOEFL. *Application deadline:* For fall admission, 1/4 for domestic students; for spring admission, 11/1 for domestic students. Application fee: $80. *Financial support:* Fellowships with tuition reimbursements, research assistantships with tuition reimbursements, teaching assistantships with tuition reimbursements, Federal Work-Study, institutionally sponsored loans, scholarships/grants, health care benefits, and unspecified assistantships available. Financial award application deadline: 1/4; financial award applicants required to submit FAFSA. *Faculty research:* Partial differential equations, computational science, applied mathematics, geometry and topology, probability and stochastic processes. *Application contact:* Tamar Arnon, Application Contact, 212-998-3258, E-mail: admissions@math.nyu.edu.

See Close-Up on page 511.

Nicholls State University, Graduate Studies, College of Arts and Sciences, Department of Mathematics and Computer Science, Thibodaux, LA 70310. Offers community/technical college mathematics (MS). Part-time and evening/weekend programs available. *Faculty:* 5 full-time (0 women). *Students:* 12 full-time (5 women), 1 (woman) part-time; includes 2 African Americans. Average age 23. 7 applicants, 100% accepted, 7 enrolled. *Degree requirements:* For master's, comprehensive exam. *Entrance requirements:* For master's, GRE General Test. *Application deadline:* For fall admission, 6/17 for domestic students; for spring admission, 11/15 priority date for domestic students. Applications are processed on a rolling basis. Application fee: $20 ($30 for international students). Electronic applications accepted. *Expenses:* Tuition, state resident: full-time $1,620. Tuition, nonresident: full-time $4,344. *Financial support:* In 2005–06, 12 students received support, including teaching assistantships with full tuition reimbursements available (averaging $10,000 per year); Federal Work-Study, scholarships/grants, and unspecified assistantships also available. Support available to part-time students. Financial award application deadline: 6/17. *Faculty research:* Operations research, statistics, numerical analysis, algebra, topology. *Unit head:* Dr. Ray R. Giguette, Head, 985-448-4396, E-mail: ray.giguette@nicholls.edu. *Application contact:* Dr. Scott J. Beslin, Graduate Coordinator, 985-448-4384, E-mail: scott.beslkin@nicholls.edu.

North Carolina Central University, Division of Academic Affairs, College of Arts and Sciences, Department of Mathematics, Durham, NC 27707-3129. Offers MS. Part-time and evening/weekend programs available. *Degree requirements:* For master's, one foreign language, thesis, comprehensive exam. *Entrance requirements:* For master's, minimum GPA of 3.0 in major, 2.5 overall. Additional exam requirements/recommendations for international students: Required—TOEFL. *Faculty research:* Structure theorems for Lie algebra, Kleene monoids and semi-groups, theoretical computer science, mathematics education.

North Carolina State University, Graduate School, College of Agriculture and Life Sciences and College of Engineering and College of Physical and Mathematical Sciences, Program in Financial Mathematics, Raleigh, NC 27695. Offers MFM. Part-time programs available. *Degree requirements:* For master's, project/internship, thesis optional. *Entrance requirements:* For master's, GRE General Test. Additional exam requirements/recommendations for international students: Required—TOEFL (minimum score 550 paper-based; 213 computer-based). Electronic applications accepted. *Faculty research:* Financial mathematics modeling and computation, futures, options and commodities markets, real options, credit risk, portfolio optimization.

See Close-Up on page 513.

North Carolina State University, Graduate School, College of Physical and Mathematical Sciences, Department of Mathematics, Program in Mathematics, Raleigh, NC 27695. Offers MS, PhD. *Degree requirements:* For master's, thesis (for some programs); for doctorate, thesis/dissertation. *Entrance requirements:* For master's and doctorate, GRE, GRE Subject Test (recommended). Electronic applications accepted.

North Dakota State University, The Graduate School, College of Science and Mathematics, Department of Mathematics, Fargo, ND 58105. Offers applied mathematics (MS, PhD); mathematics (MS, PhD). *Faculty:* 15 full-time (1 woman), 3 part-time/adjunct (1 woman). *Students:* 21 full-time (7 women), 3 part-time (1 woman); includes 1 minority (Asian American or Pacific Islander), 8 international. Average age 28. 15 applicants, 67% accepted, 9 enrolled. In 2005, 2 master's awarded. *Degree requirements:* For master's, thesis, comprehensive exam, registration; for doctorate, one foreign language, thesis/dissertation, computer proficiency, comprehensive exam, registration. *Entrance requirements:* For master's and doctorate, GRE General Test. Additional exam requirements/recommendations for international students: Required—TOEFL, IELT. *Application deadline:* For fall admission, 5/1 priority date for domestic students, 5/1 priority date for international students; for spring admission, 8/1 for domestic students, 8/1 for international students. Applications are processed on a rolling basis. Application fee: $45 ($60 for international students). Electronic applications accepted. *Financial support:* In 2005–06, 5 fellowships with full tuition reimbursements (averaging $18,000 per year), 1 research assistantship with tuition reimbursement (averaging $14,000 per year), 17 teaching assistantships with full tuition reimbursements (averaging $9,300 per year) were awarded; Federal Work-Study, institutionally sponsored loans, and tuition waivers (full) also available. Support available to part-time students. Financial award application deadline: 3/31. *Faculty research:* Discrete mathematics, number theory, analysis theory, algebra, applied math. Total annual research expenditures: $57,844. *Unit head:* Dr. Warren Shreve, Chair, 701-231-8171, Fax: 701-231-7598, E-mail: warren.shreve@ndsu.edu. *Application contact:* Dr. Jim

Mathematics

North Dakota State University (continued)
Coykendall, Graduate Program Director, 701-231-8079, Fax: 701-231-7598, E-mail: jim.coykendall@ndsu.edu.

Northeastern Illinois University, Graduate College, College of Arts and Sciences, Department of Mathematics, Programs in Mathematics, Chicago, IL 60625-4699. Offers mathematics for elementary school teachers (MA). Part-time and evening/weekend programs available. *Degree requirements:* For master's, project, thesis optional. *Entrance requirements:* For master's, minimum GPA of 2.75, 6 undergraduate courses in mathematics. *Faculty research:* Numerical analysis, mathematical biology, operations research, statistics, geometry and mathematics of finance.

Northeastern University, College of Arts and Sciences, Department of Mathematics, Boston, MA 02115-5096. Offers applied mathematics (MS); mathematics (MS, PhD); operations research (MSOR). Part-time and evening/weekend programs available. *Faculty:* 41 full-time (6 women), 13 part-time/adjunct (2 women). *Students:* 34 full-time (9 women), 6 part-time (1 woman). Average age 32. 63 applicants, 46% accepted. In 2005, 7 master's, 3 doctorates awarded. *Degree requirements:* For doctorate, thesis/dissertation, qualifying exams. *Entrance requirements:* For master's and doctorate, GRE Subject Test, GRE General Test. Additional exam requirements/recommendations for international students: Required—TOEFL. *Application deadline:* For fall admission, 2/1 priority date for domestic students, 2/1 priority date for international students. Applications are processed on a rolling basis. Application fee: $50. Electronic applications accepted. *Financial support:* In 2005–06, 26 teaching assistantships with tuition reimbursements (averaging $16,276 per year) were awarded; research assistantships with tuition reimbursements, Federal Work-Study, institutionally sponsored loans, tuition waivers (full and partial), and unspecified assistantships also available. Financial award application deadline: 3/1; financial award applicants required to submit FAFSA. *Faculty research:* Algebra and singularities, combinatorics, topology, probability and statistics, geometric analysis and partial differential equations. *Unit head:* Dr. Robert McOwen, Chairperson, 617-373-2450, Fax: 617-373-5658, E-mail: mathdept@neu.edu. *Application contact:* Gari Horton, Graduate Secretary, 617-373-2454, Fax: 617-373-5658, E-mail: mathdept@neu.edu.

See Close-Up on page 515.

Northern Arizona University, Graduate College, College of Engineering and Natural Science, Department of Mathematics and Statistics, Flagstaff, AZ 86011. Offers mathematics (MAT, MS); statistics (MS). Part-time programs available. *Degree requirements:* For master's, thesis optional. *Faculty research:* Topology, statistics, groups, ring theory, number theory.

Northern Illinois University, Graduate School, College of Liberal Arts and Sciences, Department of Mathematical Sciences, De Kalb, IL 60115-2854. Offers mathematical sciences (PhD); mathematics (MS); statistics (MS). Part-time programs available. *Faculty:* 43 full-time (10 women), 4 part-time/adjunct (0 women). *Students:* 64 full-time (28 women), 33 part-time (11 women); includes 11 minority (3 African Americans, 1 American Indian/Alaska Native, 6 Asian Americans or Pacific Islanders, 1 Hispanic American), 33 international. Average age 29. 98 applicants, 70% accepted, 30 enrolled. In 2005, 35 master's, 2 doctorates awarded. Terminal master's awarded for partial completion of doctoral program. *Degree requirements:* For master's, thesis optional; for doctorate, one foreign language, thesis/dissertation, candidacy exam, dissertation defense, internship. *Entrance requirements:* For master's, GRE General Test, minimum GPA of 2.75; for doctorate, GRE General Test, minimum GPA of 2.75 (undergraduate), 3.2 (graduate). Additional exam requirements/recommendations for international students: Required—TOEFL (minimum score 550 paper-based; 213 computer-based). *Application deadline:* For fall admission, 6/1 for domestic students, 5/1 for international students; for spring admission, 11/1 for domestic students, 10/1 for international students. Applications are processed on a rolling basis. Application fee: $30. Electronic applications accepted. *Expenses:* Tuition, state resident: full-time $4,565; part-time $191 per credit hour. Tuition, nonresident: full-time $9,129; part-time $382 per credit hour. *Financial support:* In 2005–06, 1 research assistantship with full tuition reimbursement, 42 teaching assistantships with full tuition reimbursements were awarded; fellowships with full tuition reimbursements, career-related internships or fieldwork, Federal Work-Study, scholarships/grants, tuition waivers (full), and unspecified assistantships also available. Support available to part-time students. Financial award applicants required to submit FAFSA. *Faculty research:* Numerical linear algebra, noncommutative rings, nonlinear partial differential equations, finite group theory, abstract harmonic analysis. *Unit head:* Dr. William D. Blair, Chair, 815-753-0566, Fax: 815-753-1112, E-mail: blair@math.niu.edu. *Application contact:* Dr. Bernard Harris, Director, Graduate Studies, 815-753-6775, E-mail: harris@math.niu.edu.

Northwestern University, The Graduate School, Judd A. and Marjorie Weinberg College of Arts and Sciences, Department of Mathematics, Evanston, IL 60208. Offers PhD. Admissions and degrees offered through The Graduate School. Part-time programs available. *Faculty:* 37 full-time (6 women). *Students:* 49 full-time (9 women); includes 2 minority (both Asian Americans or Pacific Islanders), 28 international. Average age 24. 161 applicants, 13% accepted, 9 enrolled. In 2005, 6 degrees awarded. *Median time to degree:* Of those who began their doctoral program in fall 1996, 45% received their degree in 8 years or less. *Degree requirements:* For doctorate, thesis/dissertation, preliminary exam. *Entrance requirements:* For doctorate, GRE General Test, GRE Subject Test. Additional exam requirements/recommendations for international students: Required—TOEFL. *Application deadline:* For fall admission, 8/30 for domestic students. Application fee: $60 ($75 for international students). *Financial support:* In 2005–06, 34 students received support, including 5 fellowships with full tuition reimbursements available (averaging $16,080 per year), 21 teaching assistantships with full tuition reimbursements available (averaging $17,124 per year); career-related internships or fieldwork, institutionally sponsored loans, and scholarships/grants also available. Financial award application deadline: 12/31; financial award applicants required to submit FAFSA. *Faculty research:* Algebra, algebraic topology, analysis dynamical systems, partial differential equations. Total annual research expenditures: $3.4 million. *Unit head:* Paul Goerss, Chair, 847-491-8544, Fax: 847-491-8906, E-mail: p-goerss@northwestern.edu. *Application contact:* Ezra Getzler, Admission Officer, 847-491-1695, Fax: 847-491-8906, E-mail: melanie@math.northwestern.edu.

Oakland University, Graduate Study and Lifelong Learning, College of Arts and Sciences, Department of Mathematics and Statistics, Program in Mathematics, Rochester, MI 48309-4401. Offers MA. *Students:* 5 full-time (4 women), 10 part-time (6 women); includes 2 minority (1 African American, 1 Asian American or Pacific Islander), 1 international. Average age 33. 11 applicants, 100% accepted, 7 enrolled. In 2005, 5 degrees awarded. *Entrance requirements:* Additional exam requirements/recommendations for international students: Required—TOEFL (minimum score 550 paper-based; 213 computer-based). *Application deadline:* For fall admission, 7/15 priority date for domestic students, 5/1 priority date for international students. For winter admission, 12/1 for domestic students; for spring admission, 3/15 for domestic students. Applications are processed on a rolling basis. Application fee: $30. Electronic applications accepted. *Expenses:* Contact institution. *Financial support:* Application deadline: 3/1; *Faculty research:* Mesh-free particle finite. Total annual research expenditures: $30,684. *Unit head:* Dr. Meir Shillor, Coordinator, Graduate Programs, 248-370-3439, Fax: 248-370-4184, E-mail: shillor@oakland.edu.

The Ohio State University, Graduate School, College of Mathematical and Physical Sciences, Department of Mathematics, Columbus, OH 43210. Offers MS, PhD. *Degree requirements:* For master's, thesis optional; for doctorate, 2 foreign languages, thesis/dissertation. *Entrance requirements:* For master's and doctorate, GRE General Test, GRE Subject Test. Additional exam requirements/recommendations for international students: Required—TOEFL. Electronic applications accepted.

Ohio University, Graduate Studies, College of Arts and Sciences, Department of Mathematics, Athens, OH 45701-2979. Offers MS, PhD. Part-time and evening/weekend programs available. *Faculty:* 5 full-time (3 women). *Students:* 53 full-time (13 women); includes 1 minority (Asian American or Pacific Islander), 55 international. 101 applicants, 28% accepted, 14 enrolled. In 2005, 24 master's, 5 doctorates awarded. *Degree requirements:* For master's, thesis optional; for doctorate, thesis/dissertation, comprehensive exam. *Entrance requirements:* For master's and doctorate, minimum GPA of 3.0. Additional exam requirements/recommendations for international students: Required—TOEFL (minimum score 550 paper-based). *Application deadline:* For fall admission, 2/1 priority date for domestic students, 2/1 priority date for international students. Applications are processed on a rolling basis. Application fee: $45. *Financial support:* In 2005–06, 44 students received support, including 3 fellowships with full tuition reimbursements available (averaging $16,200 per year), 33 teaching assistantships with full tuition reimbursements available (averaging $13,000 per year); Federal Work-Study, institutionally sponsored loans, and tuition waivers (full and partial) also available. Financial award application deadline: 2/1. *Faculty research:* Algebra (group and ring theory), functional analysis, topology, differential equations, computational math. *Unit head:* Dr. Jeff Connor, Chair, 740-593-1254, Fax: 740-593-9805, E-mail: connor@math.ohio.edu. *Application contact:* Dr. Sergiu Aizicovici, Graduate Chair, 740-593-1272, E-mail: aizicovi@math.ohio.edu.

Announcement: The Department of Mathematics offers graduate programs leading to PhD and master's degrees in mathematics, including an MS option with a computational mathematics concentration. In addition to 26 regular faculty members with a history of excellence, flexibility, and accessibility to students, there are several visitors each year. Faculty research interests include algebra, analysis, applications of wavelets, biomathematics, coding theory, combinatorial optimization, differential equations, financial mathematics, mathematics education, numerical analysis, probability and statistics, and topology. Ties with applied sciences departments are being developed. Students participate in weekly seminars. Teaching assistantships and fellowships are available. Please visit www.math.ohiou.edu for additional information.

Oklahoma State University, College of Arts and Sciences, Department of Mathematics, Stillwater, OK 74078. Offers applied mathematics (MS); mathematics (pure and applied) (PhD); mathematics (pure) (MS); mathematics education (MS, PhD). *Faculty:* 35 full-time (5 women), 7 part-time/adjunct (3 women). *Students:* 17 full-time (7 women), 26 part-time (13 women); includes 3 minority (1 American Indian/Alaska Native, 2 Asian Americans or Pacific Islanders), 20 international. Average age 29. 44 applicants, 34% accepted, 9 enrolled. In 2005, 6 master's, 1 doctorate awarded. *Degree requirements:* For master's, report or thesis; for doctorate, one foreign language, thesis/dissertation, comprehensive exam. *Entrance requirements:* For master's, GRE. Additional exam requirements/recommendations for international students: Required—TOEFL. *Application deadline:* For fall admission, 6/1 priority date for domestic students, 3/1 priority date for international students. Applications are processed on a rolling basis. Application fee: $40 ($75 for international students). Electronic applications accepted. *Expenses:* Tuition, state resident: full-time $4,253; part-time $139 per credit hour. Tuition, nonresident: full-time $12,569; part-time $485 per credit hour. Required fees: $43 per credit hour. One-time fee: $20 part-time. Tuition and fees vary according to course load and program. *Financial support:* In 2005–06, 3 research assistantships (averaging $18,752 per year), 37 teaching assistantships (averaging $17,088 per year) were awarded; career-related internships or fieldwork, Federal Work-Study, scholarships/grants, health care benefits, tuition waivers (partial), and unspecified assistantships also available. Support available to part-time students. Financial award application deadline: 3/1. *Unit head:* Dr. Alan Adolphson, Head, 405-744-5688, Fax: 405-744-8275.

See Close-Up on page 517.

Old Dominion University, College of Sciences, Programs in Computational and Applied Mathematics, Norfolk, VA 23529. Offers MS, PhD. Part-time programs available. *Faculty:* 22 full-time (0 women). *Students:* 20 full-time (7 women), 16 part-time (6 women); includes 9 minority (5 African Americans, 4 Asian Americans or Pacific Islanders), 8 international. Average age 32. 14 applicants, 86% accepted, 6 enrolled. In 2005, 1 master's, 1 doctorate awarded. Terminal master's awarded for partial completion of doctoral program. *Degree requirements:* For master's, Master's project; for doctorate, thesis/dissertation, candidacy exam. *Entrance requirements:* For master's, minimum GPA of 3.0 in major, 2.5 overall; for doctorate, GRE General Test. Additional exam requirements/recommendations for international students: Required—TOEFL. *Application deadline:* For fall admission, 7/1 for domestic students. Applications are processed on a rolling basis. Application fee: $40. *Expenses:* Tuition, state resident: part-time $263 per credit hour. Tuition, nonresident: part-time $661 per credit hour. Required fees: $39 per semester. Part-time tuition and fees vary according to campus/location. *Financial support:* In 2005–06, 2 fellowships with full tuition reimbursements (averaging $15,000 per year), 7 research assistantships with full tuition reimbursements (averaging $16,000 per year), 12 teaching assistantships with full tuition reimbursements (averaging $15,000 per year) were awarded; scholarships/grants also available. Financial award application deadline: 2/15; financial award applicants required to submit FAFSA. *Faculty research:* Numerical analysis, integral equations, continuum mechanics. Total annual research expenditures: $559,053. *Unit head:* Dr. Hideaki Kaneko, Graduate Program Director, 757-683-4969, Fax: 757-683-3885, E-mail: mathgpd@odu.edu.

Oregon State University, Graduate School, College of Science, Department of Mathematics, Corvallis, OR 97331. Offers MA, MAIS, MS, PhD. *Faculty:* 28 full-time (9 women), 3 part-time/adjunct (2 women). *Students:* 56 full-time (15 women), 5 part-time (2 women); includes 5 minority (4 Asian Americans or Pacific Islanders, 1 Hispanic American), 15 international. Average age 30. In 2005, 12 master's, 1 doctorate awarded. Terminal master's awarded for partial completion of doctoral program. *Degree requirements:* For master's, variable foreign language requirement, thesis or alternative; for doctorate, one foreign language, thesis/dissertation, qualifying exams. *Entrance requirements:* For master's and doctorate, minimum GPA of 3.0 in last 90 hours. Additional exam requirements/recommendations for international students: Required—TOEFL. *Application deadline:* For fall admission, 3/1 for domestic students. Applications are processed on a rolling basis. Application fee: $50. *Expenses:* Tuition, area resident: Part-time $301 per credit. Tuition, state resident: full-time $8,139; part-time $501 per credit. Tuition, nonresident: full-time $14,376; part-time $532 per credit. Required fees: $1,266. *Financial support:* Research assistantships, teaching assistantships, Federal Work-Study and institutionally sponsored loans available. Support available to part-time students. Financial award application deadline: 2/1. *Unit head:* Dr. Ralph E Showalter, Chair, 541-737-4686, Fax: 541-737-0517.

The Pennsylvania State University University Park Campus, Graduate School, Eberly College of Science, Department of Mathematics, State College, University Park, PA 16802-1503. Offers mathematics (M Ed, MA, D Ed, PhD), including applied mathematics (MA, PhD). *Students:* 77 full-time (10 women), 2 part-time (1 woman); includes 3 minority (2 Asian Americans or Pacific Islanders, 1 Hispanic American), 62 international. *Entrance requirements:* For master's and doctorate, GRE General Test. Application fee: $45. *Expenses:* Tuition, state resident: full-time $12,518; part-time $522 per credit. Tuition, nonresident: full-time $23,004; part-time $959 per credit. Required fees: $484. Tuition and fees vary according to course load, campus/location and program. *Unit head:* Dr. Nigel D. Higson, Head, 814-865-7527, Fax: 814-865-3735, E-mail: higson@psu.edu.

Pittsburg State University, Graduate School, College of Arts and Sciences, Department of Mathematics, Pittsburg, KS 66762. Offers MS. *Students:* 9. *Degree requirements:* For master's, thesis or alternative. Application fee: $30 ($60 for international students). *Expenses:* Tuition, state resident: full-time $2,015; part-time $170 per credit hour. Tuition, nonresident: full-time $4,953; part-time $415 per credit hour. Tuition and fees vary according to course load, campus/location and program. *Financial support:* Teaching assistantships, career-related internships or fieldwork and Federal Work-Study available. *Faculty research:* Operations research, numerical analysis, applied analysis, applied algebra. *Unit head:* Dr. Tim Flood, Chairperson, 620-235-4401. *Application contact:* Marvene Darraugh, Administrative Officer, 620-235-4220, Fax: 620-235-4219, E-mail: mdarraug@pittstate.edu.

Polytechnic University, Brooklyn Campus, Department of Applied Mathematics, Major in Mathematics, Brooklyn, NY 11201-2990. Offers MS, PhD. Part-time and evening/weekend programs available. *Faculty:* 1 full-time (0 women). *Students:* 1 full-time (0 women), 12 part-time (4 women); includes 4 minority (1 African American, 3 Asian Americans or Pacific Islanders), 1 international. Average age 32. 21 applicants, 90% accepted, 7 enrolled. In 2005, 1 master's, 1 doctorate awarded. *Degree requirements:* For master's, thesis (for some programs), comprehensive exam (for some programs), registration; for doctorate, one foreign language, thesis/dissertation, comprehensive exam, registration. *Entrance requirements:* Additional exam requirements/recommendations for international students: Required—TOEFL (minimum score 550 paper-based; 213 computer-based); Recommended—IELT (minimum score 7). *Application deadline:* For fall admission, 7/15 priority date for domestic students, 4/1 priority date for international students; for spring admission, 12/15 priority date for domestic students, 10/1 priority date for international students. Applications are processed on a rolling basis. Application fee: $55. Electronic applications accepted. *Expenses:* Tuition: Part-time $950 per unit. Required fees: $330 per semester. *Financial support:* Fellowships, research assistantships, teaching assistantships, institutionally sponsored loans available. Support available to part-time students. Financial award applicants required to submit FAFSA. *Faculty research:* Isoperimetric inequalities, problems arising from theoretical physics. Total annual research expenditures: $101,648.

Portland State University, Graduate Studies, College of Liberal Arts and Sciences, Department of Mathematics and Statistics, Portland, OR 97207-0751. Offers mathematical sciences (PhD); mathematics education (PhD); statistics (MS). *Faculty:* 30 full-time (6 women), 22 part-time/adjunct (9 women). *Students:* 53 full-time (19 women), 53 part-time (27 women); includes 15 minority (5 African Americans, 4 Asian Americans or Pacific Islanders, 6 Hispanic Americans), 15 international. Average age 32. 62 applicants, 90% accepted, 37 enrolled. In 2005, 13 master's, 1 doctorate awarded. *Degree requirements:* For master's, thesis or alternative, exams; for doctorate, 2 foreign languages, thesis/dissertation, exams. *Entrance requirements:* For master's, minimum GPA of 3.0 in upper-division course work or 2.75 overall; for doctorate, GRE General Test. Additional exam requirements/recommendations for international students: Required—TOEFL (minimum score 550 paper-based; 213 computer-based). *Application deadline:* For fall admission, 4/1 for domestic students, 3/1 for international students. For winter admission, 9/1 for domestic students; for spring admission, 11/1 for domestic students. Applications are processed on a rolling basis. Application fee: $50. *Expenses:* Tuition, state resident: full-time $6,648; part-time $231 per credit. Tuition, nonresident: full-time $11,319; part-time $231 per credit. Required fees: $686; $67 per credit. *Financial support:* In 2005–06, 2 teaching assistantships with full tuition reimbursements (averaging $12,637 per year) were awarded; research assistantships, Federal Work-Study, scholarships/grants, tuition waivers (partial), and unspecified assistantships also available. Support available to part-time students. Financial award application deadline: 3/1; financial award applicants required to submit FAFSA. *Faculty research:* Algebra, topology, statistical distribution theory, control theory, statistical robustness. Total annual research expenditures:$318,405. *Unit head:* Marek Elzanowski, Chair, 503-725-3621, Fax: 503-725-3661, E-mail: elzanowskim@pdx.edu. *Application contact:* John Erdman, Coordinator, 503-725-3621, Fax: 503-725-3661, E-mail: erdman@pdx.edu.

Portland State University, Graduate Studies, College of Liberal Arts and Sciences, Systems Science Program, Portland, OR 97207-0751. Offers computational intelligence (Certificate); computer modeling and simulation (Certificate); systems science (MS); systems science/anthropology (PhD); systems science/business administration (PhD); systems science/civil engineering (PhD); systems science/economics (PhD); systems science/engineering management (PhD); systems science/general (PhD); systems science/mathematical sciences (PhD); systems science/mechanical engineering (PhD); systems science/psychology (PhD); systems science/sociology (PhD). *Faculty:* 3 full-time (0 women). *Students:* 62 full-time (36 women), 40 part-time (10 women); includes 9 minority (3 Asian Americans or Pacific Islanders, 6 Hispanic Americans), 24 international. Average age 34. 75 applicants, 47% accepted, 25 enrolled. In 2005, 4 master's, 17 doctorates awarded. *Degree requirements:* For doctorate, variable foreign language requirement, thesis/dissertation. *Entrance requirements:* For doctorate, GMAT, GRE General Test, minimum undergraduate GPA of 3.0. Additional exam requirements/recommendations for international students: Required—TOEFL. *Application deadline:* For fall admission, 2/1 for domestic students; for spring admission, 11/1 for domestic students. Application fee: $50. *Expenses:* Tuition, state resident: full-time $6,648; part-time $231 per credit. Tuition, nonresident: full-time $11,319; part-time $231 per credit. Required fees: $686; $67 per credit. *Financial support:* In 2005–06, 1 research assistantship with full tuition reimbursement (averaging $8,667 per year) was awarded; teaching assistantships with full tuition reimbursements, career-related internships or fieldwork, Federal Work-Study, scholarships/grants, and unspecified assistantships also available. Support available to part-time students. Financial award application deadline: 3/1; financial award applicants required to submit FAFSA. *Faculty research:* Systems theory and methodology, artificial intelligence neural networks, information theory, nonlinear dynamics/chaos, modeling and simulation. Total annual research expenditures: $143,587. *Unit head:* George Lendaris, Acting Director, 503-725-4960. *Application contact:* Dawn Sharafi, Administrative Assistant, 503-725-4960, E-mail: dawn@sysc.pdx.edu.

Prairie View A&M University, Graduate School, College of Arts and Sciences, Department of Mathematics, Prairie View, TX 77446-0519. Offers MS. Part-time and evening/weekend programs available. *Faculty:* 11 part-time/adjunct (3 women). *Students:* 7 full-time (4 women), 9 part-time (4 women); includes 13 minority (10 African Americans, 1 Asian American or Pacific Islander, 2 Hispanic Americans), 1 international. 16 applicants, 100% accepted, 16 enrolled. In 2005, 5 degrees awarded. *Degree requirements:* For master's, thesis, comprehensive exam. *Entrance requirements:* For master's, GRE General Test, bachelor's degree in mathematics. *Application deadline:* Applications are processed on a rolling basis. Application fee: $50. *Expenses:* Tuition, state resident: full-time $1,440; part-time $80 per credit. Tuition, nonresident: full-time $6,444; part-time $358 per credit. *Financial support:* In 2005–06, 5 students received support, including 2 research assistantships with tuition reimbursements available (averaging $14,400 per year), 1 teaching assistantship with tuition reimbursement available (averaging $14,400 per year); fellowships, career-related internships or fieldwork, Federal Work-Study, and institutionally sponsored loans also available. Support available to part-time students. Financial award application deadline: 4/1; financial award applicants required to submit FAFSA. *Faculty research:* Stochastic processor, queuing theory, waveler numeric analyses, delay systems mathematic modeling. Total annual research expenditures:$35,000. *Unit head:* Dr. Aliakbar Montazer Haghighi, Head, 936-857-2026, Fax: 936-857-2019, E-mail: amhaghighi@pvamu.edu. *Application contact:* Dr. Evelyn Thornton, Graduate Advisor, 936-857-4118, Fax: 936-857-2019, E-mail: eethornton@pvamu.edu.

Princeton University, Graduate School, Department of Mathematics, Princeton, NJ 08544-1019. Offers applied and computational mathematics (PhD); mathematical physics (PhD); mathematics (PhD). *Degree requirements:* For doctorate, 2 foreign languages, thesis/dissertation. *Entrance requirements:* For doctorate, GRE General Test, GRE Subject Test. Additional exam requirements/recommendations for international students: Required—TOEFL (minimum score 600 paper-based; 250 computer-based). Electronic applications accepted.

Purdue University, Graduate School, School of Science, Department of Mathematics, West Lafayette, IN 47907. Offers MS, PhD. *Faculty:* 58 full-time (6 women), 4 part-time/adjunct (0 women). *Students:* 111 full-time (28 women), 58 part-time (22 women); includes 4 minority (1 African American, 3 Hispanic Americans), 117 international. Average age 26. 261 applicants, 25% accepted, 25 enrolled. In 2005, 21 master's, 11 doctorates awarded. Terminal master's awarded for partial completion of doctoral program. *Median time to degree:* Of those who began their doctoral program in fall 1996, 35% received their degree in 8 years or less. *Degree requirements:* For doctorate, one foreign language, thesis/dissertation, oral and written exams. *Entrance requirements:* For master's and doctorate, GRE. Additional exam requirements/recommendations for international students: Required—TOEFL (minimum score 570 paper-based; 230 computer-based). *Application deadline:* For fall admission, 3/1 for domestic students, 3/1 for international students; for spring admission, 12/1 for domestic students, 10/15 for international students. Application fee: $55. Electronic applications accepted. *Financial support:* In 2005–06, 19 fellowships with full and partial tuition reimbursements (averaging $18,000 per year), 14 research assistantships with partial tuition reimbursements (averaging $15,000 per year), 192 teaching assistantships with partial tuition reimbursements (averaging $15,500 per year) were awarded. Support available to part-time students. Financial award application deadline: 3/1; financial award applicants required to submit FAFSA. *Faculty research:* Algebra, analysis, topology, differential equations, applied mathematics. *Unit head:* Dr. Leonard Lipshitz, Head, 765-494-1908, Fax: 765-494-0548, E-mail: lipshitz@math.purdue.edu. *Application contact:* Dr. Johnny E. Brown, Graduate Committee Chair, 765-494-961, Fax: 765-494-0548, E-mail: gcomm@math.purdue.edu.

Purdue University Calumet, Graduate School, School of Engineering, Mathematics, and Science, Department of Mathematics, Computer Science, and Statistics, Hammond, IN 46323-2094. Offers mathematics (MAT, MS). Part-time programs available. *Entrance requirements:* Additional exam requirements/recommendations for international students: Required—TOEFL. *Faculty research:* Topology, analysis, algebra, mathematics education.

Queens College of the City University of New York, Division of Graduate Studies, Mathematics and Natural Sciences Division, Department of Mathematics, Flushing, NY 11367-1597. Offers MA. Part-time and evening/weekend programs available. *Faculty:* 31 full-time (6 women). *Students:* 6 full-time (3 women), 35 part-time (20 women). 41 applicants, 88% accepted, 27 enrolled. In 2005, 9 degrees awarded. *Degree requirements:* For master's, comprehensive exam. *Entrance requirements:* For master's, minimum GPA of 3.0. Additional exam requirements/recommendations for international students: Required—TOEFL. *Application deadline:* For fall admission, 4/1 for domestic students; for spring admission, 11/1 for domestic students. Applications are processed on a rolling basis. Application fee: $125. *Expenses:* Tuition, state resident: part-time $270 per credit. Tuition, nonresident: part-time $500 per credit. Required fees: $112 per year. *Financial support:* Career-related internships or fieldwork, Federal Work-Study, institutionally sponsored loans, tuition waivers (partial), and adjunct lectureships available. Support available to part-time students. Financial award application deadline: 4/1; financial award applicants required to submit FAFSA. *Faculty research:* Topology, differential equations, combinatorics. *Unit head:* Dr. Wallace Goldberg, Chairperson, 718-997-5800, E-mail: wallace_goldberg@qc.edu. *Application contact:* Dr. Nick Metas, Graduate Adviser, 718-997-5800, E-mail: nick_metas@qc.edu.

Queen's University at Kingston, School of Graduate Studies and Research, Faculty of Arts and Sciences, Department of Mathematics and Statistics, Kingston, ON K7L 3N6, Canada. Offers mathematics (M Sc, M Sc Eng, PhD); statistics (M Sc, M Sc Eng, PhD). Part-time programs available. *Degree requirements:* For master's, thesis/dissertation; for doctorate, thesis/dissertation, comprehensive exam. *Entrance requirements:* Additional exam requirements/recommendations for international students: Required—TOEFL. *Faculty research:* Algebra, analysis, applied mathematics, statistics.

Rensselaer Polytechnic Institute, Graduate School, School of Science, Department of Mathematical Sciences, Program in Mathematics, Troy, NY 12180-3590. Offers MS, PhD. Part-time programs available. *Faculty:* 23 full-time (3 women), 4 part-time/adjunct (1 woman). *Students:* 45 full-time (13 women), 3 part-time (1 woman); includes 3 minority (2 African Americans, 1 Asian American or Pacific Islander), 13 international. 15 applicants, 40% accepted, 0 enrolled. In 2005, 1 master's, 4 doctorates awarded. Terminal master's awarded for partial completion of doctoral program. *Median time to degree:* Of those who began their doctoral program in fall 1996, 100% received their degree in 8 years or less. *Degree requirements:* For master's, registration; for doctorate, thesis/dissertation, preliminary exam, candidacy presentation, comprehensive exam, registration. *Entrance requirements:* For master's and doctorate, GRE General Test. Additional exam requirements/recommendations for international students: Required—TOEFL. *Application deadline:* For fall admission, 1/15 for domestic students. Applications are processed on a rolling basis. Application fee: $75. Electronic applications accepted. *Expenses:* Tuition: Full-time $31,000; part-time $1,320 per credit. Required fees: $1,623. *Financial support:* In 2005–06, 42 students received support, including fellowships with full tuition reimbursements available (averaging $18,000 per year), 5 research assistantships with full tuition reimbursements available (averaging $21,000 per year), 34 teaching assistantships with full tuition reimbursements available (averaging $14,500 per year); institutionally sponsored loans also available. Financial award application deadline: 2/1. *Faculty research:* Inverse problems, biomathematics, operations research, applied mathematics, mathematical modeling. *Application contact:* Dawnmarie Robens, Graduate Student Coordinator, 518-276-6414, Fax: 518-276-4824, E-mail: robensd@rpi.edu.

See Close-Up on page 521.

Rhode Island College, School of Graduate Studies, Faculty of Arts and Sciences, Department of Mathematics and Computer Science, Providence, RI 02908-1991. Offers mathematics (MA). Evening/weekend programs available. *Faculty:* 13 full-time (6 women). *Students:* 1 (woman) full-time, 1 (woman) part-time. Average age 32. In 2005, 3 degrees awarded. *Entrance requirements:* For master's, GRE General Test or MAT. *Application deadline:* For fall admission, 4/1 for domestic students. Applications are processed on a rolling basis. Application fee: $50. *Expenses:* Tuition, state resident: part-time $227 per credit hour. Tuition, nonresident: part-time $475 per credit hour. Required fees: $14 per credit hour. *Financial support:* Career-related internships or fieldwork available. Financial award application deadline: 4/1. *Unit head:* Dr. Kathryn Sanders, Chair, 401-456-8038, E-mail: ksanders@ric.edu.

Rice University, Graduate Programs, Wiess School of Natural Sciences, Department of Mathematics, Houston, TX 77251-1892. Offers MA, PhD. *Degree requirements:* For master's, one foreign language, thesis, oral defense of thesis; for doctorate, one foreign language, thesis/dissertation, qualifying exams, oral exam. *Entrance requirements:* For master's and doctorate, GRE General Test, minimum GPA of 3.0. Additional exam requirements/recommendations for international students: Required—TOEFL. *Faculty research:* Geometry, topology, ergodic theory, knot theory.

Rivier College, School of Graduate Studies, Department of Computer Science and Mathematics, Nashua, NH 03060-5086. Offers computer science (MS); mathematics (MAT). Part-time and evening/weekend programs available. *Faculty:* 6 full-time (2 women), 5 part-time/adjunct (2 women). *Students:* 3 full-time (1 woman), 30 part-time (16 women); includes 5 minority (4 Asian Americans or Pacific Islanders, 1 Hispanic American), 5 international. Average age 36. In 2005, 21 degrees awarded. *Degree requirements:* For master's, registration. *Entrance requirements:* For master's, GRE Subject Test. *Application deadline:* Applications are processed on a rolling basis. Application fee: $25. Electronic applications accepted. *Expenses:* Tuition: Part-time $421 per credit. *Financial support:* Available to part-time students. Application deadline: 2/1; *Unit head:* Dr. Mihaela Sabin, Director, 603-888-1311, E-mail: msabin@rivier.edu. *Application contact:* Diane Monahan, Director of Graduate Admissions, 603-897-8129, Fax: 603-897-8810, E-mail: gradadm@rivier.edu.

Roosevelt University, Graduate Division, College of Arts and Sciences, Department of Science and Mathematics, Program in Mathematics, Chicago, IL 60605-1394. Offers mathematical sciences (MS), including actuarial science. Part-time and evening/weekend programs available. *Students:* 9 full-time (6 women), 20 part-time (9 women); includes 3 minority (2 African Americans, 1 Asian American or Pacific Islander), 9 international. Average age 34. 33 applicants, 67% accepted, 13 enrolled. *Application deadline:* For fall admission, 6/1 for domestic students. Applications are processed on a rolling basis. Application fee: $25 ($35 for international students). *Expenses:* Tuition: Full-time $12,384; part-time $688 per credit hour. *Financial support:* Research assistantships, career-related internships or fieldwork and tuition waivers (partial) available. Support available to part-time students. Financial award application deadline: 2/15. *Faculty research:* Statistics, mathematics education, finite groups, computers in mathematics. *Application contact:* Joanne Canyon-Heller, Coordinator of Graduate Admission, 877-APPLY RU, Fax: 312-281-3356, E-mail: applyru@roosevelt.edu.

Rowan University, Graduate School, College of Liberal Arts and Sciences, Department of Mathematics, Glassboro, NJ 08028-1701. Offers MA. Part-time and evening/weekend

Mathematics

Rowan University (continued)

programs available. *Students:* 2 full-time (1 woman), 4 part-time (2 women). Average age 35. 3 applicants, 67% accepted, 2 enrolled. In 2005, 3 degrees awarded. *Entrance requirements:* Additional exam requirements/recommendations for international students: Required—TOEFL. *Application deadline:* Applications are processed on a rolling basis. Application fee: $50. Electronic applications accepted. *Expenses:* Tuition, state resident: full-time $9,886; part-time $594 per semester hour. Tuition, nonresident: full-time $14,662; part-time $880 per semester hour. Required fees: $994; $103 per semester hour. *Financial support:* Career-related internships or fieldwork, Federal Work-Study, and unspecified assistantships available. Support available to part-time students. *Unit head:* Dr. Hieu Nguyen, Adviser, 856-256-4500 Ext. 3886.

Royal Military College of Canada, Division of Graduate Studies and Research, Science Division, Department of Mathematics and Computer Science, Kingston, ON K7K 7B4, Canada. Offers computer science (M Sc); mathematics (M Sc). *Degree requirements:* For master's, thesis, registration. Electronic applications accepted.

Rutgers, The State University of New Jersey, Camden, Graduate School of Arts and Sciences, Program in Mathematical Sciences, Camden, NJ 08102-1401. Offers mathematics (MS). Part-time and evening/weekend programs available. *Degree requirements:* For master's, survey paper, thesis optional. *Entrance requirements:* For master's, BS/BA in math or related subject. Additional exam requirements/recommendations for international students: Recommended—TOEFL (minimum score 550 paper-based; 213 computer-based). Electronic applications accepted. *Faculty research:* Differential geometry, dynamical systems, vertex operator algebra, automorphic forms, CR-structures.

Rutgers, The State University of New Jersey, Newark, Graduate School, Program in Mathematical Sciences, Newark, NJ 07102. Offers PhD. *Faculty:* 13 full-time (2 women). *Students:* 9 full-time (2 women); includes 4 minority (all Asian Americans or Pacific Islanders) 11 applicants, 36% accepted, 3 enrolled. *Degree requirements:* For doctorate, thesis/dissertation, written qualifying exam. *Entrance requirements:* For doctorate, GRE General Test, minimum B average. Additional exam requirements/recommendations for international students: Required—TOEFL. *Application deadline:* For fall admission, 6/15 for domestic students. Applications are processed on a rolling basis. Application fee: $50. Electronic applications accepted. *Expenses:* Tuition, state resident: full-time $10,440; part-time $435 per credit. Tuition, nonresident: full-time $15,520; part-time $637 per credit. *Financial support:* In 2005–06, 8 teaching assistantships with full tuition reimbursements (averaging $16,988 per year) were awarded; tuition waivers (full and partial) also available. Financial award application deadline: 3/1. *Faculty research:* Number theory, automorphic form, low-dimensional topology, Kleinian groups, representation theory. *Unit head:* Dr. Robert Sczech, Program Coordinator, 973-353-5156 Ext. 17, Fax: 973-353-5270, E-mail: sczech@andromeda.rutgers.edu.

Rutgers, The State University of New Jersey, New Brunswick/Piscataway, Graduate School, Program in Mathematics, New Brunswick, NJ 08901-1281. Offers applied mathematics (MS, PhD); math finance (MS); mathematics (MS, PhD). Part-time programs available. *Faculty:* 106 full-time. *Students:* 70 full-time (16 women); includes 2 minority (1 Asian American or Pacific Islander, 1 Hispanic American), 32 international. Average age 26. 248 applicants, 15% accepted, 13 enrolled. In 2005, 7 degrees awarded. *Median time to degree:* Of those who began their doctoral program in fall 1996, 88% received their degree in 8 years or less. *Degree requirements:* For doctorate, one foreign language, thesis/dissertation, comprehensive exam. *Entrance requirements:* For master's and doctorate, GRE General Test, GRE Subject Test. Additional exam requirements/recommendations for international students: Required—TOEFL. *Application deadline:* For fall admission, 2/1 for domestic students; for spring admission, 11/1 for domestic students. Application fee: $50. *Expenses:* Tuition, state resident: full-time $10,440; part-time $435 per credit. Tuition, nonresident: full-time $15,520; part-time $647 per credit. Required fees: $129 per credit. Tuition and fees vary according to program. *Financial support:* In 2005–06, 8 fellowships with full tuition reimbursements (averaging $25,347 per year), 10 research assistantships with full tuition reimbursements (averaging $18,307 per year), 53 teaching assistantships with full tuition reimbursements (averaging $19,347 per year) were awarded; health care benefits, tuition waivers (full), and unspecified assistantships also available. Financial award application deadline: 2/1; financial award applicants required to submit FAFSA. *Faculty research:* Logic and set theory, number theory, mathematical physics, control theory, partial differential equations. Total annual research expenditures: $2 million. *Unit head:* Prof. Charles A. Weibel, Director, 732-445-3864, Fax: 732-445-5530, E-mail: grad-director@math.rutgers.edu.

St. Cloud State University, School of Graduate Studies, College of Science and Engineering, Department of Mathematics, St. Cloud, MN 56301-4498. Offers MS. *Faculty:* 19 full-time (5 women). *Students:* 1 full-time (0 women), 3 part-time (all women). 4 applicants, 25% accepted. In 2005, 1 degree awarded. *Degree requirements:* For master's, thesis or alternative, comprehensive exam (for some programs). *Entrance requirements:* For master's, GRE General Test, minimum GPA of 2.75. Additional exam requirements/recommendations for international students: Required—MELAB; Recommended—TOEFL (minimum score 550 paper-based; 213 computer-based), IELT (minimum score 7). *Application deadline:* For fall admission, 6/1 priority date for domestic students, 4/1 priority date for international students; for spring admission, 10/1 priority date for domestic students, 8/1 priority date for international students. Applications are processed on a rolling basis. Application fee: $35. Electronic applications accepted. *Expenses:* Tuition, state resident: part-time $277. Tuition, nonresident: part-time $379. Required fees: $23 per credit. Tuition and fees vary according to course load and reciprocity agreements. *Financial support:* Federal Work-Study and unspecified assistantships available. Financial award application deadline: 3/1. *Unit head:* Dr. Daniel Scully, Chairperson, 320-308-3001, E-mail: mathdept@stcloudstate.edu. *Application contact:* Linda Lou Krueger, School of Graduate Studies, 320-308-2113, Fax: 320-308-5371, E-mail: lekrueger@stcloudstate.edu.

St. John's University, St. John's College of Liberal Arts and Sciences, Department of Mathematics and Computer Science, Queens, NY 11439. Offers algebra (MA); analysis (MA); applied mathematics (MA); computer science (MA); geometry-topology (MA); logic and foundations (MA); probability and statistics (MA). Part-time and evening/weekend programs available. *Faculty:* 19 full-time (2 women), 10 part-time/adjunct (4 women). *Students:* 4 full-time (all women), 5 part-time (3 women); includes 1 minority (Hispanic American), 1 international. Average age 28. 18 applicants, 83% accepted, 6 enrolled. In 2005, 2 degrees awarded. *Degree requirements:* For master's, thesis optional. *Entrance requirements:* For master's, minimum GPA of 3.0. Additional exam requirements/recommendations for international students: Required—TOEFL (minimum score 500 paper-based; 173 computer-based). *Application deadline:* For fall admission, 5/1 priority date for domestic students, 5/1 priority date for international students; for spring admission, 11/1 priority date for domestic students, 11/1 priority date for international students. Applications are processed on a rolling basis. Application fee: $40. Electronic applications accepted. *Expenses:* Tuition: Full-time $8,760; part-time $730 per credit. Required fees: $250; $125 per term. Tuition and fees vary according to program. *Financial support:* Research assistantships, scholarships/grants available. Support available to part-time students. Financial award application deadline: 3/1; financial award applicants required to submit FAFSA. *Faculty research:* Development of a computerized metabolic map. *Unit head:* Dr. Charles Traina, Chair, 718-990-6166, E-mail: trainac@stjohns.edu. *Application contact:* Matthew Whelan, Director, Office of Admissions, 718-990-2000, Fax: 718-990-2096, E-mail: admissions@stjohns.edu.

Saint Louis University, Graduate School, College of Arts and Sciences and Graduate School, Department of Mathematics and Mathematical Computer Science, St. Louis, MO 63103-2097. Offers mathematics (MA, MA-R, PhD). Part-time programs available. *Faculty:* 28 full-time (5 women), 1 part-time/adjunct (0 women). *Students:* 19 full-time (9 women), 4 part-time (2 women); includes 3 minority (2 Asian Americans or Pacific Islanders, 1 Hispanic American). Average age 27. 20 applicants, 95% accepted, 11 enrolled. In 2005, 4 master's, 1 doctorate awarded. *Degree requirements:* For master's, thesis (for some programs), comprehensive exam; for doctorate, one foreign language, thesis/dissertation, preliminary exams. *Entrance requirements:* For master's and doctorate, GRE General Test, letters of recommendation, resumé. Additional exam requirements/recommendations for international students: Required—TOEFL (minimum score 550 paper-based; 213 computer-based). *Application deadline:* For fall admission, 7/1 for domestic students; for spring admission, 11/1 for domestic students. Applications are processed on a rolling basis. Application fee: $40. *Expenses:* Tuition: Part-time $760 per credit hour. Required fees: $55 per semester. *Financial support:* In 2005–06, 16 students received support, including 15 teaching assistantships with tuition reimbursements available (averaging $12,500 per year) Financial award application deadline: 6/1; financial award applicants required to submit FAFSA. *Faculty research:* Algebra, groups and rings, analysis, differential geometry, topology. Total annual research expenditures:$30,000. *Unit head:* Michael May, SJ, Chairperson, 314-977-2444, E-mail: mayka@slu.edu. *Application contact:* Gary Behrman, Associate Dean of the Graduate School, 314-977-3827, E-mail: behrmang@slu.edu.

Saint Xavier University, Graduate Studies, School of Arts and Sciences, Department of Mathematics and Computer Science, Chicago, IL 60655-3105. Offers applied computer science in Internet information systems (MS); mathematics and computer science (MA). *Faculty:* 1 (woman) full-time. *Students:* 11 full-time (3 women), 6 part-time (3 women); includes 5 minority (3 Asian Americans or Pacific Islanders, 2 Hispanic Americans). Average age 30. *Degree requirements:* For master's, thesis optional. *Application deadline:* For fall admission, 8/15 for domestic students. Application fee: $35. *Unit head:* Dr. Florence Appel, Associate Professor and Associate Chair/Computer Science, 773-298-3398, Fax: 773-779-9061, E-mail: appel@sxu.edu. *Application contact:* Beth Gierach, Managing Director of Admission, 773-298-3053, Fax: 773-298-3076, E-mail: gierach@sxu.edu.

Salem State College, Graduate School, Program in Mathematics, Salem, MA 01970-5353. Offers mathematics (MS). *Faculty:* 1 part-time/adjunct (0 women). *Students:* Average age 39. In 2005, 8 degrees awarded. *Entrance requirements:* For master's, GRE General Test, MAT. *Application deadline:* Applications are processed on a rolling basis. Application fee: $25. *Unit head:* Julie Belock, Coordinator, 978-542-6338, Fax: 978-542-7175, E-mail: jbelock@salemstate.edu.

Sam Houston State University, College of Arts and Sciences, Department of Mathematics and Statistics, Huntsville, TX 77341. Offers mathematics (MA, MS); statistics (MS). Part-time programs available. *Faculty:* 10 full-time (0 women). *Students:* 4 full-time (2 women), 21 part-time (13 women); includes 3 minority (1 African American, 1 Asian American or Pacific Islander, 1 Hispanic American), 1 international. Average age 34. In 2005, 20 degrees awarded. *Entrance requirements:* For master's, GRE General Test. Additional exam requirements/recommendations for international students: Required—TOEFL (minimum score 550 paper-based; 213 computer-based). *Application deadline:* For fall admission, 8/1 for domestic students; for spring admission, 12/1 for domestic students. Applications are processed on a rolling basis. Application fee: $20. *Financial support:* Teaching assistantships, institutionally sponsored loans available. Support available to part-time students. Financial award application deadline: 5/31; financial award applicants required to submit FAFSA. *Unit head:* Dr. Jaimie Hebert, Chair, 936-294-1563, Fax: 936-294-1882, E-mail: mth_jlh@shsu.edu. *Application contact:* Anita Shipman, Advisor, 936-294-3962.

San Diego State University, Graduate and Research Affairs, College of Sciences, Department of Mathematical Sciences, San Diego, CA 92182. Offers applied mathematics (MS); mathematics (MA); mathematics and science education (PhD); statistics (MS). Part-time programs available. *Students:* 56 full-time (21 women), 59 part-time (25 women); includes 22 minority (4 African Americans, 14 Asian Americans or Pacific Islanders, 4 Hispanic Americans), 18 international. 81 applicants, 69% accepted, 27 enrolled. In 2005, 32 master's, 1 doctorate awarded. *Degree requirements:* For doctorate, thesis/dissertation. *Entrance requirements:* For master's, GRE General Test; for doctorate, GRE, minimum GPA of 3.25 in last 30 undergraduate semester units, minimum graduate GPA of 3.5, MSE recommendation form, 3 letters of recommendation. Additional exam requirements/recommendations for international students: Required—TOEFL. *Application deadline:* For fall admission, 5/1 for domestic students, 5/1 for international students; for spring admission, 11/1 for domestic students, 10/1 for international students. Applications are processed on a rolling basis. Application fee: $55. Electronic applications accepted. *Financial support:* Teaching assistantships, unspecified assistantships available. Financial award applicants required to submit FAFSA. *Faculty research:* Teacher education in mathematics. Total annual research expenditures: $1.3 million. *Unit head:* David Lesley, Chair, 619-594-6191, Fax: 619-594-6746, E-mail: lesley@math.sdsu.edu. *Application contact:* Larry Sowder, Graduate Coordinator, 619-594-7246, Fax: 619-594-6746, E-mail: lsowder@sciences.sdsu.edu.

San Francisco State University, Division of Graduate Studies, College of Science and Engineering, Department of Mathematics, San Francisco, CA 94132-1722. Offers MA. *Degree requirements:* For master's, oral exam, thesis optional. *Entrance requirements:* For master's, minimum GPA of 2.5 in last 60 units. *Faculty research:* Fuzzy logic, software development, number theory, complex analysis, mathematics education.

San Jose State University, Graduate Studies and Research, College of Science, Department of Mathematics, San Jose, CA 95192-0001. Offers mathematics (MA, MS); mathematics education (MA). Part-time and evening/weekend programs available. *Students:* 7 full-time (3 women), 36 part-time (17 women); includes 17 minority (2 African Americans, 12 Asian Americans or Pacific Islanders, 3 Hispanic Americans). Average age 36. 27 applicants, 63% accepted, 9 enrolled. In 2005, 5 degrees awarded. *Degree requirements:* For master's, thesis (for some programs), comprehensive exam. *Entrance requirements:* For master's, GRE Subject Test. *Application deadline:* For fall admission, 6/29 for domestic students; for spring admission, 11/30 for domestic students. Applications are processed on a rolling basis. Application fee: $59. Electronic applications accepted. *Expenses:* Tuition, nonresident: part-time $339 per unit. Required fees: $1,286 per semester. Tuition and fees vary according to course load and degree level. *Financial support:* In 2005–06, 20 teaching assistantships were awarded; career-related internships or fieldwork and Federal Work-Study also available. Support available to part-time students. Financial award applicants required to submit FAFSA. *Faculty research:* Artificial intelligence, algorithms, numerical analysis, software database, number theory. *Unit head:* Brad Jackson, Chair, 408-924-5100, Fax: 408-924-5080. *Application contact:* Fernanda Karp, Department Manager, 408-924-5100.

Simon Fraser University, Graduate Studies, Faculty of Science, Department of Mathematics, Burnaby, BC V5A 1S6, Canada. Offers applied mathematics (M Sc, PhD); pure mathematics (M Sc, PhD); statistics and actuarial science (M Sc, PhD). *Degree requirements:* For master's and doctorate, thesis/dissertation. *Entrance requirements:* For master's, GRE Subject Test, minimum GPA of 3.0; for doctorate, GRE Subject Test, minimum GPA of 3.5. Additional exam requirements/recommendations for international students: Required—TWE or IELTS. *Faculty research:* Semi-groups, number theory, optimization, combinations.

South Dakota State University, Graduate School, College of Engineering, Department of Mathematics, Brookings, SD 57007. Offers MS. *Degree requirements:* For master's, thesis, oral exam. *Entrance requirements:* Additional exam requirements/recommendations for international students: Required—TOEFL. *Faculty research:* Numerical linear algebra, statistics, applied quality number theory, abstract algebra, actuarial mathematics.

Southeast Missouri State University, School of Graduate Studies, Department of Mathematics, Cape Girardeau, MO 63701-4799. Offers MNS. Part-time programs available. *Faculty:* 15 full-time (3 women). *Students:* 7 full-time (4 women), 1 part-time, 2 international. Average age 25. 3 applicants, 100% accepted. In 2005, 2 degrees awarded. *Degree requirements:* For master's, thesis or alternative. *Entrance requirements:* For master's, GRE General Test, minimum GPA of 2.75 in mathematics. Additional exam requirements/recommendations for international students: Required—TOEFL (minimum score 550 paper-based; 213 computer-based). *Application deadline:* For fall admission, 8/1 for domestic students, 4/1 for international students; for spring admission, 11/21 for domestic students, 9/1 for international students. Applications are processed on a rolling basis. Application fee: $20 ($100 for international students). Electronic applications accepted. *Expenses:* Tuition, state resident: full-time $1,676; part-time $186 per hour. Tuition, nonresident: full-time $3,052; part-time $339 per hour. Required fees: $114; $13 per hour. Tuition and fees vary according to course load, degree level and campus/location. *Financial support:* In 2005–06, 8 students received support, including 6 teaching assistantships with full tuition reimbursements available (averaging $6,600 per year); unspecified assistantships also available. Financial award applicants required to submit FAFSA. *Unit head:* Dr. Victor Gummersheimer, Chairperson, 573-651-2165, Fax: 573-986-6811, E-mail: vgummersheimer@semo.edu. *Application contact:* Marsha L. Arant, Senior Administrative Assistant, Office of Graduate Studies, 573-651-2192, Fax: 573-651-2001, E-mail: marant@semo.edu.

Southern Connecticut State University, School of Graduate Studies, School of Arts and Sciences, Department of Mathematics, New Haven, CT 06515-1355. Offers MS. Part-time and evening/weekend programs available. *Degree requirements:* For master's, thesis or alternative. *Entrance requirements:* For master's, interview. Electronic applications accepted.

Southern Illinois University Carbondale, Graduate School, College of Science, Department of Mathematics, Carbondale, IL 62901-4701. Offers mathematics (MA, MS, PhD); statistics (MS). Part-time programs available. *Faculty:* 32 full-time (2 women), 1 part-time/adjunct (0 women). *Students:* 29 full-time (11 women), 8 part-time (2 women); includes 2 minority (both African Americans), 22 international. Average age 26. 67 applicants, 16% accepted, 2 enrolled. In 2005, 7 master's awarded. *Degree requirements:* For master's, thesis; for doctorate, 2 foreign languages, thesis/dissertation. *Entrance requirements:* For master's, minimum GPA of 2.7; for doctorate, minimum GPA of 3.25. Additional exam requirements/recommendations for international students: Required—TOEFL. *Application deadline:* Applications are processed on a rolling basis. Application fee: $0. *Financial support:* In 2005–06, 28 students received support, including 24 teaching assistantships with full tuition reimbursements available; fellowships with full tuition reimbursements available, research assistantships with full tuition reimbursements available, Federal Work-Study, institutionally sponsored loans, and tuition waivers (full) also available. Support available to part-time students. *Faculty research:* Differential equations, combinatorics, probability, algebra, numerical analysis. *Unit head:* Andrew Earnest, Chairperson, 618-453-6522, Fax: 618-453-5300, E-mail: chairman@math.siu.edu. *Application contact:* William T. Patula, Director of Graduate Studies, 618-453-5302, Fax: 618-453-5300, E-mail: wpatula@math.siu.edu.

Announcement: The computational facilities provided by the Math Computer Lab have recently been updated with state-of-the-art computer hardware and software through a grant from the National Science Foundation. All graduate students have 24-hour access to the lab for work on class or research projects, theses, and dissertations.

See Close-Up on page 523.

Southern Illinois University Edwardsville, Graduate Studies and Research, College of Arts and Sciences, Department of Mathematics and Statistics, Edwardsville, IL 62026-0001. Offers mathematics (MS). Part-time programs available. *Students:* 9 full-time (5 women), 12 part-time (10 women); includes 1 minority (Asian American or Pacific Islander), 8 international. Average age 33. 14 applicants, 71% accepted. In 2005, 6 degrees awarded. *Degree requirements:* For master's, thesis or alternative, final exam. *Entrance requirements:* For master's, undergraduate major in related area, programming language, minimum GPA of 2.7. Additional exam requirements/recommendations for international students: Required—TOEFL. *Application deadline:* For fall admission, 7/21 for domestic students, 6/1 for international students; for spring admission, 12/8 for domestic students, 10/1 for international students. Application fee: $30. Electronic applications accepted. *Expenses:* Tuition, state resident: part-time $190 per semester hour. Tuition, nonresident: part-time $380 per semester hour. Tuition and fees vary according to course load, reciprocity agreements and student level. *Financial support:* In 2005–06, 16 teaching assistantships with full tuition reimbursements were awarded; fellowships with full tuition reimbursements, research assistantships with full tuition reimbursements, Federal Work-Study, institutionally sponsored loans, and unspecified assistantships also available. Support available to part-time students. Financial award application deadline: 3/1. *Unit head:* Dr. Krzysztof Jarosz, Chair, 618-650-2354, E-mail: kjarosz@siue.edu. *Application contact:* Dr. George Pelekanos, Director, 618-650-2342, E-mail: gpeleka@siue.edu.

Southern Methodist University, Dedman College, Department of Mathematics, Dallas, TX 75275. Offers computational and applied mathematics (MS, PhD). *Faculty:* 15 full-time (1 woman). *Students:* 22 full-time (11 women), 1 part-time; includes 3 minority (2 Asian Americans or Pacific Islanders, 1 Hispanic American), 9 international. Average age 30. 25 applicants, 32% accepted, 8 enrolled. In 2005, 4 master's, 2 doctorates awarded. *Degree requirements:* For doctorate, thesis/dissertation, oral and written exams. *Entrance requirements:* For master's and doctorate, GRE General Test, minimum GPA of 3.0, 18 undergraduate hours in mathematics beyond first and second year calculus. Additional exam requirements/recommendations for international students: Required—TOEFL. *Application deadline:* For fall admission, 6/30 for domestic students. For winter admission, 11/30 for domestic students. Applications are processed on a rolling basis. Application fee: $60. Electronic applications accepted. *Financial support:* In 2005–06, 7 teaching assistantships with full tuition reimbursements (averaging $15,000 per year) were awarded; career-related internships or fieldwork, scholarships/grants, health care benefits, and tuition waivers (partial) also available. Support available to part-time students. Financial award applicants required to submit FAFSA. *Faculty research:* Numerical analysis, scientific computation, fluid dynamics, software development, differential equations. Total annual research expenditures: $195,000. *Unit head:* Dr. Peter Moore, Chairman, 214-768-2506, Fax: 214-768-2355, E-mail: mathchair@mail.smu.edu. *Application contact:* Dr. Ian Gladwell, Director of Graduate Studies, 214-768-4338, E-mail: math@mail.smu.edu.

Southern Oregon University, Graduate Office, School of Sciences, Ashland, OR 97520. Offers environmental education (MA, MS); mathematics/computer science (MA, MS); science (MA, MS). Part-time programs available. *Degree requirements:* For master's, thesis (for some programs), comprehensive exam (MA). *Entrance requirements:* For master's, GRE General Test, minimum GPA of 3.0. *Faculty research:* Ferroelectric, ecology environmental science, biotechnology, material science.

Southern University and Agricultural and Mechanical College, Graduate School, College of Sciences, Department of Mathematics, Baton Rouge, LA 70813. Offers MS. *Faculty:* 11 full-time (6 women). *Students:* 16 full-time (10 women), 5 part-time (3 women); all minorities (20 African Americans, 1 Asian American or Pacific Islander). Average age 31. 7 applicants, 100% accepted, 5 enrolled. In 2005, 5 degrees awarded. *Degree requirements:* For master's, thesis optional. *Entrance requirements:* For master's, GMAT, GRE General Test. Additional exam requirements/recommendations for international students: Required—TOEFL. *Application deadline:* For fall admission, 6/1 for domestic students; for spring admission, 11/1 for domestic students. Applications are processed on a rolling basis. Application fee: $25. *Financial support:* In 2005–06, 5 students received support, including 2 research assistantships (averaging $7,000 per year), 3 teaching assistantships (averaging $9,000 per year) Financial award application deadline: 4/15; financial award applicants required to submit FAFSA. *Faculty research:* Algebraic number theory, abstract algebra, computer analysis, probability, mathematics education. Total annual research expenditures:$310,000. *Unit head:* Dr. Joseph A. Meyinsse, Chairperson, 225-771-5180, Fax: 225-771-4762, E-mail: joseph-neyinsse@subr.edu. *Application contact:* Dr. Rogers J. Newman, Professor of Mathematics, 225-771-5180, Fax: 225-771-4762, E-mail: rnewman@subrvm.subr.edu.

Stanford University, School of Engineering, Program in Scientific Computing and Computational Mathematics, Stanford, CA 94305-9991. Offers MS, PhD. Terminal master's awarded for partial completion of doctoral program. *Degree requirements:* For doctorate, thesis/dissertation, qualifying exam. *Entrance requirements:* For master's, GRE General Test; for doctorate, GRE General Test, GRE Subject Test. Additional exam requirements/recommendations for international students: Required—TOEFL. Electronic applications accepted.

Stanford University, School of Humanities and Sciences, Department of Mathematics, Stanford, CA 94305-9991. Offers financial mathematics (MS); mathematics (MS, PhD). Terminal master's awarded for partial completion of doctoral program. *Degree requirements:* For doctorate, 2 foreign languages, thesis/dissertation, oral exam. *Entrance requirements:* For master's, GRE General Test; for doctorate, GRE General Test, GRE Subject Test. Additional exam requirements/recommendations for international students: Required—TOEFL. Electronic applications accepted.

State University of New York at Binghamton, Graduate School, School of Arts and Sciences, Department of Mathematical Sciences, Binghamton, NY 13902-6000. Offers computer science (MA, PhD); probability and statistics (MA, PhD). Part-time programs available. Terminal master's awarded for partial completion of doctoral program. *Degree requirements:* For master's, thesis or alternative; for doctorate, 2 foreign languages, thesis/dissertation. *Entrance requirements:* For master's and doctorate, GRE General Test, GRE Subject Test. Additional exam requirements/recommendations for international students: Required—TOEFL. Electronic applications accepted.

State University of New York at Buffalo, Graduate School, College of Arts and Sciences, Department of Mathematics, Buffalo, NY 14260. Offers MA, PhD. Part-time programs available. *Faculty:* 29 full-time (3 women), 7 part-time/adjunct (2 women). *Students:* 74 full-time (21 women), 8 part-time (1 woman); includes 4 minority (1 African American, 1 American Indian/Alaska Native, 2 Asian Americans or Pacific Islanders), 46 international. Average age 29. 118 applicants, 44% accepted, 31 enrolled. In 2005, 16 master's, 7 doctorates awarded. Terminal master's awarded for partial completion of doctoral program. *Median time to degree:* Of those who began their doctoral program in fall 1996, 22% received their degree in 8 years or less. *Degree requirements:* For master's, thesis (for some programs), comprehensive exam (for some programs); for doctorate, thesis/dissertation, comprehensive exam. *Entrance requirements:* Additional exam requirements/recommendations for international students: Required—TOEFL (minimum score 550 paper-based; 213 computer-based). *Application deadline:* For fall admission, 1/15 priority date for domestic students, 1/15 priority date for international students; for spring admission, 10/15 priority date for domestic students, 10/15 priority date for international students. Applications are processed on a rolling basis. Application fee: $35. Electronic applications accepted. *Financial support:* In 2005–06, 55 students received support, including fellowships with full tuition reimbursements available (averaging $4,000 per year), 48 teaching assistantships with full tuition reimbursements available (averaging $13,400 per year); research assistantships, Federal Work-Study, institutionally sponsored loans, and unspecified assistantships also available. Financial award application deadline: 1/15; financial award applicants required to submit FAFSA. *Faculty research:* Algebra, analysis, applied mathematics, logic, number theory, topology. Total annual research expenditures: $229,641. *Unit head:* Dr. Samuel D. Schack, Chairman, 716-645-6284 Ext. 103, Fax: 716-645-5039, E-mail: chair@math.buffalo.edu. *Application contact:* Dr. Ching Chou, Director of Graduate Studies, 716-645-6284 Ext. 109, Fax: 716-645-5039, E-mail: graduatedirector@math.buffalo.edu.

State University of New York at New Paltz, Graduate School, School of Science and Engineering, Department of Mathematics, New Paltz, NY 12561. Offers MA, MAT, MS Ed. Part-time and evening/weekend programs available. *Faculty:* 12 full-time (3 women), 16 part-time/adjunct (11 women). *Students:* Average age 41. In 2005, 1 degree awarded. *Degree requirements:* For master's, thesis (for some programs), comprehensive exam (for some programs). *Entrance requirements:* For master's, GRE General Test, minimum GPA of 3.0. Additional exam requirements/recommendations for international students: Required—TOEFL (minimum score 550 paper-based; 213 computer-based). *Application deadline:* For fall admission, 3/1 priority date for domestic students, 3/1 priority date for international students; for spring admission, 10/1 for domestic students, 10/1 for international students. Application fee: $50. *Expenses:* Tuition, state resident: full-time $3,450; part-time $288 per credit hour. Tuition, nonresident: full-time $3,550; part-time $455 per credit hour. Required fees: $27 per credit. $130 per semester. *Financial support:* Teaching assistantships, Federal Work-Study, institutionally sponsored loans, and tuition waivers (full) available. *Faculty research:* Universal algebra, lattice theory, mathematical logic, operator theory, combinatorics. *Unit head:* Dr. David Hobby, Chair, 845-257-3532. *Application contact:* Dr. Donald Silberger, Coordinator, 845-257-3537, E-mail: silberger@mcs.newpaltz.edu.

State University of New York College at Brockport, School of Letters and Sciences, Department of Mathematics, Brockport, NY 14420-2997. Offers MA. Part-time programs available. *Students:* 5 full-time (1 woman), 12 part-time (5 women), 1 international. 12 applicants, 92% accepted, 8 enrolled. In 2005, 4 degrees awarded. *Degree requirements:* For master's, comprehensive exam. *Entrance requirements:* For master's, minimum GPA of 3.0, letters of recommendation. Additional exam requirements/recommendations for international students: Required—TOEFL (minimum score 550 paper-based; 213 computer-based). *Application deadline:* For fall admission, 7/15 for domestic students, 7/15 for international students; for spring admission, 11/15 for domestic students, 11/15 for international students. Application fee: $50. *Expenses:* Tuition, state resident: full-time $6,900; part-time $288 per credit. Tuition, nonresident: full-time $10,920; part-time $455 per credit. Required fees: $685; $28 per credit. *Financial support:* In 2005–06, 5 students received support, including 3 teaching assistantships with tuition reimbursements available (averaging $6,000 per year); Federal Work-Study, scholarships/grants, and unspecified assistantships also available. Support available to part-time students. Financial award application deadline: 3/15; financial award applicants required to submit FAFSA. *Faculty research:* Complex analysis, topological graph theory, number theory, mathematical modeling, algebra. *Unit head:* Dr. Mihail Barbosu, Interim Chairperson, 585-395-2194, Fax: 585-395-2304, E-mail: mbarbosu@brockport.edu. *Application contact:* Dr. Dawn Jones, Graduate Director, 585-395-5174, E-mail: djones@brockport.edu.

Announcement: Flexible program leading to MA in mathematics. Three required courses in algebra, analysis, and statistics. Four additional elective courses in mathematics or computer science. The remaining 3 courses may be chosen from other departments, including education. Several assistantships, which include a stipend of $6000 and a tuition waiver, are available.

State University of New York College at Cortland, Graduate Studies, School of Arts and Sciences, Department of Mathematics, Cortland, NY 13045. Offers MAT, MS Ed.

State University of New York College at Potsdam, School of Arts and Sciences, Department of Mathematics, Potsdam, NY 13676. Offers MA. Part-time and evening/weekend programs available. *Faculty:* 5 full-time (1 woman), 1 part-time/adjunct (0 women). *Students:* 1 (woman) full-time, 4 part-time (1 woman), 1 international. In 2005, 1 degree awarded. *Degree requirements:* For master's, comprehensive exam. *Entrance requirements:* For master's, minimum GPA of 3.0. Additional exam requirements/recommendations for international students: Required—TOEFL (minimum score 550 paper-based; 213 computer-based). *Application deadline:* Applications are processed on a rolling basis. Application fee: $50. *Expenses:* Tuition, area resident: Full-time $6,900; part-time $288 per credit. Tuition, nonresident: full-time $10,920; part-time $455 per credit. Required fees: $37 per credit. *Financial support:* Teaching assistantships with full tuition reimbursements, Federal Work-Study available. Support available to part-time students. Financial award application deadline: 3/1. *Unit head:* Dr. Laura J. Person, Chairperson, 315-267-2005, Fax: 315-267-3176, E-mail: personlj@potsdam.edu. *Application contact:* Dr. William Amoriell, Dean of Education and Graduate Studies, 315-267-2515, Fax: 315-267-4802, E-mail: amoriewj@potsdam.edu.

State University of New York, Fredonia, Graduate Studies, Department of Mathematical Sciences, Fredonia, NY 14063-1136. Offers MS Ed. Part-time and evening/weekend programs available. *Degree requirements:* For master's, thesis optional. *Application deadline:* For fall admission, 8/5 for domestic students; for spring admission, 12/1 for domestic students. Application fee: $50. *Expenses:* Tuition, state resident: full-time $3,456; part-time $288 per

Mathematics

State University of New York, Fredonia (continued)
credit hour. Tuition, nonresident: full-time $5,460; part-time $455 per credit hour. Required fees: $543; $45 per credit hour. *Financial support:* Research assistantships, teaching assistantships, tuition waivers (full and partial) available. Support available to part-time students. Financial award application deadline: 3/15. *Unit head:* Dr. Nancy Boynton, Chairman, 716-673-3243, E-mail: nancy.boynton@fredonia.edu.

Stephen F. Austin State University, Graduate School, College of Sciences and Mathematics, Department of Mathematics and Statistics, Nacogdoches, TX 75962. Offers mathematics (MS); mathematics education (MS); statistics (MS). *Faculty:* 20 full-time (5 women). *Students:* 13 full-time (3 women), 53 part-time (47 women); includes 7 minority (5 African Americans, 2 Hispanic Americans), 6 international. 11 applicants, 100% accepted. In 2005, 25 degrees awarded. *Degree requirements:* For master's, thesis optional. *Entrance requirements:* For master's, GRE General Test, minimum GPA of 2.8 in last 60 hours, 2.5 overall. Additional exam requirements/recommendations for international students: Required—TOEFL. *Application deadline:* For fall admission, 8/1 for domestic students; for spring admission, 12/15 for domestic students. Applications are processed on a rolling basis. Application fee: $0 ($50 for international students). *Expenses:* Tuition, state resident: full-time $2,628; part-time $146 per credit hour. Tuition, nonresident: full-time $7,596; part-time $422 per credit hour. Required fees:$900; $170. *Financial support:* In 2005–06, 8 teaching assistantships (averaging $11,500 per year) were awarded; Federal Work-Study and unspecified assistantships also available. Financial award application deadline: 3/1. *Faculty research:* Kernel type estimators, fractal mappings, spline curve fitting, robust regression continua theory. *Unit head:* Dr. Jasper Adams, Chair, 936-468-3805.

Stevens Institute of Technology, Graduate School, Arthur E. Imperatore School of Sciences and Arts, Department of Mathematical Sciences, Program in Mathematics, Hoboken, NJ 07030. Offers MS, PhD. *Students:* 16 full-time (4 women), 8 part-time; includes 2 minority (both Asian Americans or Pacific Islanders), 4 international. 10 applicants, 80% accepted. *Degree requirements:* For master's, thesis optional; for doctorate, one foreign language, thesis/dissertation. *Entrance requirements:* For master's and doctorate, GRE. Additional exam requirements/recommendations for international students: Required—TOEFL. *Application deadline:* Applications are processed on a rolling basis. Application fee: $50. Electronic applications accepted. *Expenses:* Tuition: Part-time $920 per credit hour. Tuition and fees vary according to program. *Application contact:* Dr. Milos Dostal, Professor, 201-216-5426.

Stony Brook University, State University of New York, Graduate School, College of Arts and Sciences, Department of Mathematics, Stony Brook, NY 11794. Offers MA, PhD. *Faculty:* 23 full-time (2 women), 2 part-time/adjunct (0 women). *Students:* 65 full-time (5 women), 29 part-time (22 women); includes 5 minority (2 African Americans, 3 Asian Americans or Pacific Islanders), 45 international. Average age 27. 196 applicants, 34% accepted. In 2005, 10 master's, 12 doctorates awarded. *Degree requirements:* For doctorate, 2 foreign languages, thesis/dissertation. *Entrance requirements:* For master's and doctorate, GRE General Test. Additional exam requirements/recommendations for international students: Required—TOEFL. *Application deadline:* For fall admission, 1/15 for domestic students. Application fee: $50. *Expenses:* Tuition, area resident: Part-time $288. Tuition, state resident: full-time $6,900. Tuition, nonresident: full-time $10,920; part-time $455. Required fees: $704. *Financial support:* In 2005–06, 1 fellowship, 12 research assistantships, 43 teaching assistantships were awarded. *Faculty research:* Real analysis, relativity and mathematical physics, complex analysis, topology, combinatorics. Total annual research expenditures: $897,516. *Unit head:* Dr. David Ebin, Chair, 631-632-8260. *Application contact:* Dr. Leon Takhtajan, Director, 631-632-8258, Fax: 631-632-7631, E-mail: leontak@math.sunysb.edu.

Announcement: The Stony Brook Mathematics Department is only about 40 years old, but it has already achieved a world-class reputation. Recipients of Stony Brook doctorates now hold tenured positions at leading universities in the United States and abroad. Ranked among the top departments in the country, the faculty includes a Fields medalist and 5 members of the National Academy of Sciences. There are about 70 doctoral students whose specialties include most branches of mathematics.

See Close-Up on page 525.

Syracuse University, Graduate School, College of Arts and Sciences, Department of Mathematics, Syracuse, NY 13244. Offers MS, PhD. Part-time programs available. *Students:* 50 full-time (13 women), 7 part-time (2 women); includes 3 minority (1 African American, 2 Asian Americans or Pacific Islanders), 24 international. 71 applicants, 44% accepted, 15 enrolled.Terminal master's awarded for partial completion of doctoral program. *Degree requirements:* For doctorate, 2 foreign languages, thesis/dissertation, qualifying exam. *Entrance requirements:* For master's and doctorate, GRE General Test, GRE Subject Test. Additional exam requirements/recommendations for international students: Required—TOEFL. *Application deadline:* For fall admission, 1/10 for domestic students. Applications are processed on a rolling basis. Application fee: $65. Electronic applications accepted. *Financial support:* Fellowships with full tuition reimbursements, research assistantships with full tuition reimbursements, teaching assistantships with full and partial tuition reimbursements, tuition waivers (partial) available. *Faculty research:* Pure mathematics, numerical mathematics, computing statistics. *Unit head:* Dr. Terry McConnell, Chairman, 315-443-1472, Fax: 315-443-1475, E-mail: mconnell@mail.box.syr.edu. *Application contact:* Mark Kleiner, Graduate Program Director, 315-443-1499, Fax: 315-443-1475, E-mail: mkleiner@syr.edu.

Tarleton State University, College of Graduate Studies, College of Science and Technology, Department of Mathematics, Physics and Engineering, Stephenville, TX 76402. Offers mathematics (MS). Part-time and evening/weekend programs available. *Faculty:* 5 full-time (0 women), 1 part-time/adjunct (0 women). *Students:* 6 full-time (3 women), 9 part-time (5 women); includes 2 minority (both African Americans), 1 international. Average age 35. In 2005, 11 degrees awarded. *Degree requirements:* For master's, thesis (for some programs), comprehensive exam. *Entrance requirements:* For master's, GRE General Test, minimum GPA of 3.0. Additional exam requirements/recommendations for international students: Required—TOEFL (minimum score 550 paper-based; 220 computer-based). *Application deadline:* For fall admission, 8/5 for domestic students; for spring admission, 12/1 for domestic students. Applications are processed on a rolling basis. Application fee: $25 ($75 for international students). *Financial support:* In 2005–06, 1 research assistantship (averaging $12,000 per year), 2 teaching assistantships (averaging $12,000 per year) were awarded; career-related internships or fieldwork and Federal Work-Study also available. Support available to part-time students. Financial award application deadline: 5/1; financial award applicants required to submit FAFSA. *Unit head:* Dr. Javier Garza, Head, 254-968-9168.

Temple University, Graduate School, College of Science and Technology, Department of Mathematics, Philadelphia, PA 19122-6096. Offers applied mathematics (MA); mathematics (PhD); pure mathematics (MA). Part-time and evening/weekend programs available. *Faculty:* 20 full-time (1 woman). *Students:* 14 full-time (5 women), 20 part-time (5 women); includes 6 minority (2 African Americans, 3 Asian Americans or Pacific Islanders, 1 Hispanic American), 18 international. 35 applicants, 31% accepted, 9 enrolled. In 2005, 5 master's, 6 doctorates awarded. Terminal master's awarded for partial completion of doctoral program. *Degree requirements:* For master's, written exam, thesis optional; for doctorate, 2 foreign languages, thesis/dissertation, oral and written exams. *Entrance requirements:* For master's, GRE General Test, minimum GPA of 3.0; for doctorate, GRE General Test, GRE Subject Test, minimum GPA of 3.0. Additional exam requirements/recommendations for international students: Required—TOEFL (minimum score 575 paper-based; 230 computer-based). *Application deadline:* For fall admission, 2/15 priority date for domestic students, 12/15 priority date for international students; for spring admission, 11/15 priority date for domestic students, 8/1 priority date for international students. Applications are processed on a rolling basis. Application fee: $50. Electronic applications accepted. *Expenses:* Tuition, state resident: full-time $8,694; part-time $483 per credit. Tuition, nonresident: full-time $12,672; part-time $704 per credit. Required fees: $500; $122 per semester. Tuition and fees vary according to course level, campus/location and program. *Financial support:* Fellowships, research assistantships, teaching assistantships, Federal Work-Study and institutionally sponsored loans available. Financial award application deadline: 1/15; financial award applicants required to submit FAFSA. *Faculty research:* Differential geometry, numerical analysis. *Unit head:* Dr. Omar Hijab, Chair, 215-204-4650, Fax: 215-204-6433, E-mail: hijab@temple.edu.

Tennessee State University, Graduate School, College of Arts and Sciences, Department of Physics and Mathematics, Nashville, TN 37209-1561. Offers mathematics (MS). Part-time and evening/weekend programs available. *Faculty:* 7 full-time (1 woman). *Students:* 3 full-time (1 woman), 6 part-time (2 women); includes 2 minority (both African Americans), 2 international. Average age 23. 15 applicants, 53% accepted. In 2005, 5 degrees awarded. *Degree requirements:* For master's, thesis, comprehensive exam. *Entrance requirements:* For master's, GRE General Test, GRE Subject Test, minimum GPA of 2.5. *Application deadline:* Applications are processed on a rolling basis. Application fee: $15. *Financial support:* Unspecified assistantships available. Financial award application deadline: 5/1. *Faculty research:* Chaos theory, semi-coherent light scattering, lattices of topologies, Ramsey Theory, K theory. Total annual research expenditures: $60,000. *Unit head:* Dr. Sandra Scheick, Head, 615-963-5811.

Tennessee Technological University, Graduate School, College of Arts and Sciences, Department of Mathematics, Cookeville, TN 38505. Offers MS. Part-time programs available. *Faculty:* 17 full-time (4 women). *Students:* 8 full-time (2 women), 5 part-time (3 women); includes 4 minority (all Asian Americans or Pacific Islanders) Average age 27. 17 applicants, 65% accepted, 4 enrolled. In 2005, 3 degrees awarded. *Degree requirements:* For master's, thesis. *Entrance requirements:* For master's, GRE General Test. Additional exam requirements/recommendations for international students: Required—TOEFL. *Application deadline:* For fall admission, 3/1 for domestic students; for spring admission, 8/1 for domestic students. Application fee: $25 ($30 for international students). *Expenses:* Tuition, state resident: full-time $8,421; part-time $307 per hour. Tuition, nonresident: full-time $22,389; part-time $711 per hour. *Financial support:* In 2005–06, 1 research assistantship (averaging $7,500 per year), 6 teaching assistantships (averaging $7,500 per year) were awarded. Financial award application deadline: 4/1. *Unit head:* Dr. Rafal Ablamowicz, Chairperson, 931-372-3441, Fax: 931-372-6353, E-mail: rablamowicz@tntech.edu. *Application contact:* Dr. Francis O. Otuonye, Associate Vice President for Research and Graduate Studies, 931-372-3233, Fax: 931-372-3497, E-mail: fotuonye@tntech.edu.

Texas A&M International University, Office of Graduate Studies and Research, College of Arts and Sciences, Department of Mathematical and Physical Science, Laredo, TX 78041-1900. Offers MAIS. *Faculty:* 2 full-time (0 women). *Students:* Average age 29. 7 applicants, 100% accepted, 4 enrolled. In 2005, 1 degree awarded. *Entrance requirements:* For master's, GRE General Test. Additional exam requirements/recommendations for international students: Required—TOEFL (minimum score 550 paper-based; 213 computer-based). *Application deadline:* For fall admission, 7/15 for domestic students; for spring admission, 11/12 for domestic students. Applications are processed on a rolling basis. Application fee: $25. *Expenses:* Tuition, state resident: full-time $1,580. Tuition, nonresident: full-time $5,432. Required fees: $3,808. *Financial support:* In 2005–06, 1 student received support. Application deadline: 11/1. *Unit head:* Dr. Chen Snung, Chair, 956-326-2567, Fax: 956-326-2439, E-mail: csung@tamiu.edu. *Application contact:* Rosie Espinoza, Director of Admissions, 956-326-2200, Fax: 956-326-2199, E-mail: enroll@tamiu.edu.

Texas A&M International University, Office of Graduate Studies and Research, Program in Interdisciplinary Studies, Laredo, TX 78041-1900. Offers mathematics (MAIS). *Students:* 1 (woman) full-time, 5 part-time (3 women); all Hispanic Americans In 2005, 1 degree awarded. *Expenses:* Tuition, state resident: full-time $1,580. Tuition, nonresident: full-time $5,432. Required fees: $3,808. *Unit head:* Dr. Chen Snung, Chair, 956-326-2567, Fax: 956-326-2439, E-mail: csung@tamiu.edu. *Application contact:* Rosie Espinoza, Director of Admissions, 956-326-2200, Fax: 956-326-2199, E-mail: enroll@tamiu.edu.

Texas A&M University, College of Science, Department of Mathematics, College Station, TX 77843. Offers MS, PhD. Part-time programs available. Postbaccalaureate distance learning degree programs offered (minimal on-campus study). *Faculty:* 55 full-time (3 women), 4 part-time/adjunct (0 women). *Students:* 100 full-time (27 women), 33 part-time (22 women); includes 14 minority (2 African Americans, 4 Asian Americans or Pacific Islanders, 8 Hispanic Americans), 47 international. Average age 27. 149 applicants, 42% accepted, 26 enrolled. In 2005, 32 master's, 5 doctorates awarded. Terminal master's awarded for partial completion of doctoral program. *Median time to degree:* Of those who began their doctoral program in fall 1996, 100% received their degree in 8 years or less. *Degree requirements:* For master's, thesis optional; for doctorate, one foreign language, thesis/dissertation, comprehensive exam. *Entrance requirements:* For master's and doctorate, GRE General Test. Additional exam requirements/recommendations for international students: Required—TOEFL (minimum score 550 paper-based; 213 computer-based). *Application deadline:* For fall admission, 3/1 for domestic students, 3/1 for international students; for spring admission, 8/1 for domestic students, 8/1 for international students. Applications are processed on a rolling basis. Application fee: $50 ($75 for international students). Electronic applications accepted. *Expenses:* Tuition, state resident: full-time $4,488; part-time $187 per credit hour. Tuition, nonresident: full-time $11,112; part-time $463 per credit hour. Required fees:$1,974. *Financial support:* In 2005–06, fellowships with partial tuition reimbursements (averaging $17,850 per year), research assistantships with partial tuition reimbursements (averaging $17,850 per year), teaching assistantships with partial tuition reimbursements (averaging $17,850 per year) were awarded; career-related internships or fieldwork, institutionally sponsored loans, scholarships/grants, and unspecified assistantships also available. Financial award application deadline: 3/1; financial award applicants required to submit FAFSA. *Faculty research:* Functional analysis, numerical analysis, algebra, geometry/topology, applied mathematics. *Unit head:* Dr. Albert Boggess, Head, 979-845-3261, Fax: 979-845-6028. *Application contact:* Monique Stewart, Academic Advisor I, 979-862-4137, Fax: 979-862-4190, E-mail: gstudies@math.tamu.edu.

Texas A&M University–Commerce, Graduate School, College of Arts and Sciences, Department of Mathematics, Commerce, TX 75429-3011. Offers MA, MS. Part-time programs available. *Faculty:* 16 full-time (6 women). *Students:* 7 full-time (5 women), 12 part-time (7 women); includes 2 minority (both African Americans), 1 international. Average age 36. *Degree requirements:* For master's, thesis (for some programs), comprehensive exam. *Entrance requirements:* For master's, GRE General Test. *Application deadline:* For fall admission, 6/1 for domestic students; for spring admission, 11/1 priority date for domestic students. Applications are processed on a rolling basis. Application fee: $0 ($25 for international students). Electronic applications accepted. *Financial support:* In 2005–06, research assistantships (averaging $7,875 per year), teaching assistantships (averaging $7,875 per year) were awarded; Federal Work-Study, institutionally sponsored loans, and scholarships/grants also available. Financial award application deadline: 5/1; financial award applicants required to submit FAFSA. *Unit head:* Dr. Stuart Anderson, Head, 903-886-5157, Fax: 903-886-5945, E-mail: stuart_anderson@tamu-commerce.edu. *Application contact:* Tammi Thompson, Graduate Admissions Adviser, 843-886-5167, Fax: 843-886-5165, E-mail: tammi_thompson@tamu-commerce.edu.

Texas A&M University–Kingsville, College of Graduate Studies, College of Arts and Sciences, Department of Mathematics, Kingsville, TX 78363. Offers MS. Part-time programs available. *Degree requirements:* For master's, thesis or alternative, comprehensive exam. *Entrance requirements:* For master's, GRE General Test. Additional exam requirements/recommendations for international students: Required—TOEFL. *Faculty research:* Complex analysis, multivariate analysis, algebra, numerical analysis, applied statistics.

Texas Christian University, College of Science and Engineering, Department of Mathematics, Fort Worth, TX 76129-0002. Offers MAT. Part-time and evening/weekend programs available. *Application deadline:* For fall admission, 3/1 for domestic students; for spring admission, 12/1 for domestic students. Applications are processed on a rolling basis. Application fee: $0. *Expenses:* Tuition: Part-time $740 per credit hour. *Financial support:* Application deadline: 3/1.

Unit head: Dr. Bob Doran, Chairperson, 817-257-7335, E-mail: r.doran@tcu.edu. *Application contact:* Dr. Bonnie Melhart, Associate Dean, College of Science and Engineering, E-mail: b.melhart@tcu.edu.

Texas Southern University, Graduate School, School of Science and Technology, Department of Mathematics, Houston, TX 77004-4584. Offers MA, MS. Part-time and evening/weekend programs available. *Faculty:* 3 full-time (1 woman), 1 part-time/adjunct (0 women). *Students:* 4 full-time (3 women), 8 part-time (4 women); includes 11 minority (9 African Americans, 2 Asian Americans or Pacific Islanders), 1 international. Average age 37. 2 applicants, 100% accepted, 1 enrolled. In 2005, 1 degree awarded. *Degree requirements:* For master's, thesis, comprehensive exam. *Entrance requirements:* For master's, GRE General Test, minimum GPA of 2.5. Additional exam requirements/recommendations for international students: Required—TOEFL. *Application deadline:* For fall admission, 7/15 for domestic students. Applications are processed on a rolling basis. Application fee: $50 ($75 for international students). *Expenses:* Tuition, state resident: full-time $1,728; part-time $1,152 per credit hour. Tuition, nonresident: full-time $6,174; part-time $4,116 per credit hour. Required fees: $2,122. Tuition and fees vary according to course load and degree level. *Financial support:* In 2005–06, fellowships (averaging $1,200 per year) Financial award application deadline:5/1. *Faculty research:* Statistics, number theory, topology, differential equations, numerical analysis. *Unit head:* Dr. Nathaniel Dean, Head, 713-313-7002. *Application contact:* Linda Williams, Secretary, 713-313-7602, E-mail: williams_km@tsu.edu.

Texas State University-San Marcos, Graduate School, College of Science, Department of Mathematics, San Marcos, TX 78666. Offers mathematics (MS); middle school mathematics teaching (M Ed). Part-time programs available. *Faculty:* 10 full-time (5 women), 1 part-time/adjunct (0 women). *Students:* 31 full-time (12 women), 31 part-time (22 women); includes 4 African Americans, 6 Asian Americans or Pacific Islanders, 9 Hispanic Americans, 1 international. Average age 33. 22 applicants, 100% accepted, 15 enrolled. In 2005, 17 degrees awarded. *Degree requirements:* For master's, thesis (for some programs), comprehensive exam. *Entrance requirements:* For master's, GRE General Test, minimum GPA of 2.75 in last 60 hours of course work. Additional exam requirements/recommendations for international students: Required—TOEFL. *Application deadline:* For fall admission, 6/15 priority date for domestic students, 6/1 priority date for international students; for spring admission, 10/15 priority date for domestic students, 10/1 priority date for international students. Applications are processed on a rolling basis. Application fee: $40 ($90 for international students). *Expenses:* Tuition, area resident: Part-time $116 per credit. Tuition, state resident: full-time $3,168; part-time $176 per credit. Tuition, nonresident: full-time $8,136; part-time $452 per credit. Required fees: $1,112; $74 per credit. Full-time tuition and fees vary according to course load. *Financial support:* In 2005–06, 44 students received support, including 2 research assistantships (averaging $5,499 per year), 19 teaching assistantships (averaging $6,816 per year); Federal Work-Study and institutionally sponsored loans also available. Support available to part-time students. Financial award application deadline: 4/1; financial award applicants required to submit FAFSA. *Faculty research:* Differential equations, geometric topology, number theory, mathematics education, graph theory. *Unit head:* Dr. Stanley G. Wayment, Chair, 512-245-2551, Fax: 512-245-3425, E-mail: sw05@txstate.edu. *Application contact:* Dr. Maria Acosta, Graduate Adviser, 512-245-2497, E-mail: ma05@txstate.edu.

Texas Tech University, Graduate School, College of Arts and Sciences, Department of Mathematics and Statistics, Lubbock, TX 79409. Offers mathematics (MA, MS, PhD); statistics (MS). Part-time programs available. *Faculty:* 38 full-time (7 women), 1 part-time/adjunct (0 women). *Students:* 80 full-time (30 women), 10 part-time (7 women); includes 12 minority (2 African Americans, 4 Asian Americans or Pacific Islanders, 6 Hispanic Americans), 32 international. Average age 29. 77 applicants, 70% accepted, 16 enrolled. In 2005, 26 master's, 5 doctorates awarded. *Degree requirements:* For master's, thesis or alternative; for doctorate, one foreign language, thesis/dissertation. *Entrance requirements:* For master's and doctorate, GRE General Test. Additional exam requirements/recommendations for international students: Required—TOEFL (minimum score 550 paper-based; 213 computer-based). *Application deadline:* Applications are processed on a rolling basis. Application fee: $50 ($60 for international students). Electronic applications accepted. *Expenses:* Tuition, state resident: full-time $4,296. Tuition, nonresident: full-time $10,920. Required fees: $1,992. Tuition and fees vary according to program. *Financial support:* In 2005–06, 52 students received support, including 1 research assistantship with partial tuition reimbursement available (averaging $16,600 per year), 74 teaching assistantships with partial tuition reimbursements available (averaging $15,168 per year); fellowships, Federal Work-Study and institutionally sponsored loans also available. Support available to part-time students. Financial award application deadline: 4/15; financial award applicants required to submit FAFSA. *Faculty research:* Numerical analysis, control and systems theory, mathematical biology, mechanics, algebra and geometry. Total annual research expenditures: $364,876. *Unit head:* Dr. Lawrence Schovanec, Chair, 806-742-2566, Fax: 806-742-1112, E-mail: schov@math.ttu.edu. *Application contact:* Dr. Alex Wang, Graduate Adviser, 806-742-2566, Fax: 806-742-1112, E-mail: awang@matt.ttu.edu.

Texas Woman's University, Graduate School, College of Arts and Sciences, Department of Mathematics and Computer Science, Denton, TX 76201. Offers mathematics (MA, MS); mathematics teaching (MS). Part-time and evening/weekend programs available. *Students:* 10 full-time (7 women), 16 part-time (12 women); includes 7 minority (5 African Americans, 1 Asian American or Pacific Islander, 1 Hispanic American), 3 international. Average age 36. In 2005, 6 degrees awarded. *Degree requirements:* For master's, thesis (for some programs), comprehensive exam. *Entrance requirements:* For master's, 2 letters of reference. Additional exam requirements/recommendations for international students: Required—TOEFL (minimum score 550 paper-based; 213 computer-based). *Application deadline:* Applications are processed on a rolling basis. Application fee: $30 ($50 for international students). Electronic applications accepted. *Expenses:* Tuition, state resident: full-time $2,934; part-time $163. Tuition, nonresident: full-time $7,974; part-time $152. *Financial support:* In 2005–06, research assistantships (averaging $9,288 per year), 5 teaching assistantships (averaging $9,288 per year) were awarded; career-related internships or fieldwork, Federal Work-Study, institutionally sponsored loans, scholarships/grants, traineeships, health care benefits, and unspecified assistantships also available. Support available to part-time students. Financial award application deadline: 3/1; financial award applicants required to submit FAFSA. *Faculty research:* Biopharmaceutical statistics, dynamical systems and control theory, Bayesian inference, math and computer science curriculum innovation, computer modeling of physical phenomenon. *Unit head:* Dr. Don E. Edwards, Chair, 940-898-2166, Fax: 940-898-2179, E-mail: dedwards@mail.twu.edu. *Application contact:* Samuel Wheeler, Coordinator of Graduate Admissions, 940-898-3188, Fax: 940-898-3081, E-mail: wheelersr@twu.edu.

Tufts University, Graduate School of Arts and Sciences, Department of Mathematics, Medford, MA 02155. Offers MA, MS, PhD. *Faculty:* 21 full-time, 4 part-time/adjunct. *Students:* 23 (10 women); includes 1 minority (African American) 51 applicants, 53% accepted, 11 enrolled. In 2005, 4 master's, 2 doctorates awarded. Terminal master's awarded for partial completion of doctoral program. *Degree requirements:* For master's, one foreign language, thesis; for doctorate, 2 foreign languages, thesis/dissertation. *Entrance requirements:* For master's, GRE General Test. Additional exam requirements/recommendations for international students: Required—TOEFL (minimum score 550 paper-based; 213 computer-based). *Application deadline:* For fall admission, 2/15 for domestic students, 12/30 for international students. Applications are processed on a rolling basis. Application fee: $65. Electronic applications accepted. *Expenses:* Tuition: Full-time $32,360. Tuition and fees vary according to program. *Financial support:* Teaching assistantships with full and partial tuition reimbursements, Federal Work-Study, scholarships/grants, and tuition waivers (partial) available. Financial award application deadline: 2/15; financial award applicants required to submit FAFSA. *Unit head:* Boris Hasselblatt, Chair, 617-627-3234, E-mail: mathgrad@tufts.edu. *Application contact:* Head, 617-627-3234, E-mail: mathgrad@tufts.edu.

Tulane University, Graduate School, Department of Mathematics, New Orleans, LA 70118-5669. Offers applied mathematics (MS); mathematics (MS, PhD); statistics (MS). *Degree requirements:* For master's, thesis (for some programs); for doctorate, thesis/dissertation. *Entrance requirements:* For master's, GRE General Test, minimum B average in undergraduate course work; for doctorate, GRE General Test. Additional exam requirements/recommendations for international students: Required—TOEFL; Recommended—TSE. Electronic applications accepted.

Université de Moncton, Faculty of Science, Department of Mathematics and Statistics, Moncton, NB E1A 3E9, Canada. Offers mathematics (M Sc). *Degree requirements:* For master's, one foreign language, thesis. *Entrance requirements:* For master's, minimum GPA of 3.0. Electronic applications accepted. *Faculty research:* Statistics, numerical analysis, fixed point theory, mathematical physics.

Université de Montréal, Faculty of Graduate Studies, Faculty of Arts and Sciences, Department of Mathematics and Statistics, Montréal, QC H3C 3J7, Canada. Offers mathematics (M Sc, PhD); statistics (M Sc, PhD). *Faculty:* 37 full-time (5 women), 15 part-time/adjunct (1 woman). *Students:* 100 full-time (29 women). 120 applicants, 21% accepted, 20 enrolled. In 2005, 18 master's, 4 doctorates awarded. *Degree requirements:* For master's, thesis; for doctorate, thesis/dissertation, general exam. *Entrance requirements:* For master's and doctorate, proficiency in French. *Application deadline:* For fall and spring admission, 2/1. For winter admission, 11/1 for domestic students. Application fee: $30. Electronic applications accepted. *Financial support:* Fellowships, research assistantships, teaching assistantships, monitorships available. Financial award application deadline: 4/1. *Faculty research:* Pure and applied mathematics, actuarial mathematics. *Unit head:* Veronique Hussih, Chair, 514-343-6710, Fax: 514-343-5700. *Application contact:* Liliane Badier, Student Files Management Technician, 514-343-6111 Ext. 1695, Fax: 514-343-5700.

Université de Sherbrooke, Faculty of Sciences, Department of Mathematics, Sherbrooke, QC J1K 2R1, Canada. Offers M Sc, PhD. *Faculty:* 11 full-time (1 woman), 3 part-time/adjunct (0 women). *Students:* 27 full-time (5 women). 36 applicants, 39% accepted. In 2005, 3 degrees awarded. *Degree requirements:* For master's, thesis/dissertation; for doctorate, thesis/dissertation, comprehensive exam. *Entrance requirements:* For doctorate, master's degree. *Application deadline:* For fall admission, 6/30 for domestic students. Applications are processed on a rolling basis. Application fee: $50. Electronic applications accepted. *Financial support:* Fellowships, research assistantships, teaching assistantships available. *Faculty research:* Measure theory, differential equations, probability, statistics, error control codes. *Unit head:* Dr. Eric Marchand, Chairman, 819-821-8091, Fax: 819-821-7189, E-mail: eric.marchand@usherbrooke.ca.

Université du Québec à Montréal, Graduate Programs, Program in Mathematics, Montréal, QC H3C 3P8, Canada. Offers M Sc, PhD. Part-time programs available. *Degree requirements:* For master's and doctorate, thesis/dissertation. *Entrance requirements:* For master's, appropriate bachelor's degree or equivalent, proficiency in French; for doctorate, appropriate master's degree or equivalent, proficiency in French.

Université du Québec à Trois-Rivières, Graduate Programs, Program in Mathematics and Computer Science, Trois-Rivières, QC G9A 5H7, Canada. Offers M Sc. *Faculty research:* Probability, statistics.

Université Laval, Faculty of Sciences and Engineering, Department of Mathematics and Statistics, Programs in Mathematics, Québec, QC G1K 7P4, Canada. Offers M Sc, PhD. Terminal master's awarded for partial completion of doctoral program. *Degree requirements:* For master's, thesis (for some programs); for doctorate, thesis/dissertation, comprehensive exam. *Entrance requirements:* For master's and doctorate, knowledge of French and English. Electronic applications accepted.

University at Albany, State University of New York, College of Arts and Sciences, Department of Mathematics and Statistics, Albany, NY 12222-0001. Offers mathematics (PhD); secondary teaching (MA); statistics (MA). Evening/weekend programs available. *Students:* 33 full-time (13 women), 16 part-time (4 women). Average age 32. In 2005, 13 master's, 4 doctorates awarded. *Degree requirements:* For doctorate, one foreign language, thesis/dissertation. *Entrance requirements:* For doctorate, GRE General Test. Additional exam requirements/recommendations for international students: Required—TOEFL (minimum score 550 paper-based; 213 computer-based). *Application deadline:* For fall admission, 3/15 for domestic students, 5/1 for international students. Applications are processed on a rolling basis. Application fee: $60. Electronic applications accepted. *Financial support:* Fellowships, research assistantships, teaching assistantships, minority assistantships available. Financial award application deadline: 3/15. *Unit head:* Timothy Lance, Chair, 518-442-4602.

The University of Akron, Graduate School, Buchtel College of Arts and Sciences, Department of Theoretical and Applied Mathematics, Program in Mathematics, Akron, OH 44325. Offers MS. Part-time and evening/weekend programs available. *Students:* 13 full-time (5 women), 4 part-time (3 women); includes 1 minority (African American), 1 international. Average age 31. 7 applicants, 86% accepted, 6 enrolled. In 2005, 3 degrees awarded. *Degree requirements:* For master's, seminar and comprehensive exam or thesis, thesis optional. *Entrance requirements:* For master's, minimum GPA of 2.75. Additional exam requirements/recommendations for international students: Required—TOEFL (minimum score 550 paper-based; 213 computer-based), Michigan English Language Assessment Battery. *Application deadline:* For fall admission, 3/1 for domestic students. Applications are processed on a rolling basis. Application fee: $30 ($40 for international students). Electronic applications accepted. *Expenses:* Tuition, state resident: full-time $5,816; part-time $323 per credit. Tuition, nonresident: full-time $9,976; part-time $554 per credit. Required fees: $794; $43 per credit. $12 per term. Tuition and fees vary according to course load, degree level and program. *Financial support:* Teaching assistantships with tuition reimbursements available. *Faculty research:* Topology analysis. *Unit head:* Dr. Ali Hajjafar, Coordinator, 330-972-8006.

The University of Alabama, Graduate School, College of Arts and Sciences, Department of Mathematics, Tuscaloosa, AL 35487. Offers applied mathematics (PhD); mathematics (MA, PhD); pure mathematics (PhD). *Faculty:* 25 full-time (1 woman). *Students:* 31 full-time (11 women), 5 part-time; includes 5 African Americans, 16 international. Average age 31. 40 applicants, 70% accepted, 11 enrolled. In 2005, 83 master's, 6 doctorates awarded. Terminal master's awarded for partial completion of doctoral program. *Median time to degree:* Of those who began their doctoral program in fall 1996, 85% received their degree in 8 years or less. *Degree requirements:* For master's, thesis or alternative; for doctorate, thesis/dissertation. *Entrance requirements:* For master's and doctorate, GRE General Test, minimum GPA of 3.0. Additional exam requirements/recommendations for international students: Required—TOEFL. *Application deadline:* For fall admission, 7/1 for domestic students. Applications are processed on a rolling basis. Application fee: $25. Electronic applications accepted. *Expenses:* Tuition, area resident: Full-time $2,432. Tuition, nonresident: full-time $6,758. *Financial support:* In 2005–06, 27 students received support, including 3 fellowships with full tuition reimbursements available (averaging $14,834 per year), 28 teaching assistantships with full tuition reimbursements available (averaging $11,007 per year); research assistantships with full tuition reimbursements available, Federal Work-Study, institutionally sponsored loans, scholarships/grants, and unspecified assistantships also available. Support available to part-time students. Financial award application deadline: 7/1. *Faculty research:* Analysis, topology, algebra, fluid mechanics and system control theory, optimization, stochastic processes. Total annual research expenditures:$115,100. *Unit head:* Dr. Zhijian Wu, Chairperson and Professor, 205-348-5080, Fax: 205-348-7067, E-mail: zwu@gp.as.ua.edu. *Application contact:* Dr. Martin Evans, Director, Graduate Programs in Mathematics, 205-348-5301, Fax: 205-348-7067, E-mail: mevans@gp.as.ua.edu.

The University of Alabama at Birmingham, School of Natural Sciences and Mathematics, Department of Mathematics, Birmingham, AL 35294. Offers applied mathematics (PhD); mathematics (MS). *Students:* 32 full-time (13 women), 2 part-time; includes 3 minority (1 African American, 1 Asian American or Pacific Islander, 1 Hispanic American), 12 international.

Mathematics

The University of Alabama at Birmingham (continued)
33 applicants, 79% accepted. In 2005, 8 master's, 2 doctorates awarded. Terminal master's awarded for partial completion of doctoral program. *Degree requirements:* For master's, thesis optional; for doctorate, one foreign language, thesis/dissertation, comprehensive exam. *Entrance requirements:* For master's and doctorate, GRE General Test. *Application deadline:* Applications are processed on a rolling basis. Application fee: $35 ($60 for international students). Electronic applications accepted. *Expenses:* Tuition, state resident: part-time $170 per credit hour. Tuition, nonresident: full-time $4,612; part-time $425 per credit hour. International tuition: $10,732 full-time. Required fees: $11 per credit hour. $124 per term. Tuition and fees vary according to course load, degree level and program. *Financial support:* In 2005–06, 18 teaching assistantships with tuition reimbursements (averaging $14,000 per year) were awarded; fellowships, research assistantships, career-related internships or fieldwork, Federal Work-Study, institutionally sponsored loans, tuition waivers (full and partial), and unspecified assistantships also available. Support available to part-time students. Financial award application deadline: 3/31; financial award applicants required to submit FAFSA. *Faculty research:* Differential equations, topology, mathematical physics, dynamic systems. *Unit head:* Dr. Rudi Weikard, Chair, 205-934-2154, Fax: 205-934-9025, E-mail: weikard@uab.edu.

The University of Alabama in Huntsville, School of Graduate Studies, College of Science, Department of Mathematical Sciences, Huntsville, AL 35899. Offers applied mathematics (PhD); mathematics (MA, MS). Part-time and evening/weekend programs available. *Faculty:* 13 full-time (1 woman). *Students:* 20 full-time (11 women), 6 part-time (3 women); includes 5 minority (1 African American, 1 American Indian/Alaska Native, 2 Asian Americans or Pacific Islanders, 1 Hispanic American), 7 international. Average age 32. 19 applicants, 89% accepted, 12 enrolled. In 2005, 7 master's, 1 doctorate awarded. *Degree requirements:* For master's, thesis or alternative, oral and written exams, comprehensive exam, registration; for doctorate, one foreign language, thesis/dissertation, oral and written exams, comprehensive exam, registration. *Entrance requirements:* For master's and doctorate, GRE General Test, minimum GPA of 3.0. Additional exam requirements/recommendations for international students: Required—TOEFL (minimum score 550 paper-based; 213 computer-based). *Application deadline:* For fall admission, 5/30 for domestic students; for spring admission, 10/10 priority date for domestic students, 7/10 priority date for international students. Applications are processed on a rolling basis. Application fee: $40. *Expenses:* Tuition, state resident: full-time $5,866; part-time $244 per credit hour. Tuition, nonresident: full-time $12,060; part-time $500 per credit hour. Tuition and fees vary according to course load. *Financial support:* In 2005–06, 21 students received support, including 4 research assistantships with full and partial tuition reimbursements available (averaging $12,851 per year), 14 teaching assistantships with full and partial tuition reimbursements available (averaging $9,202 per year); fellowships with full and partial tuition reimbursements available, career-related internships or fieldwork, Federal Work-Study, institutionally sponsored loans, scholarships/grants, health care benefits, tuition waivers (full and partial), and unspecified assistantships also available. Support available to part-time students. Financial award application deadline: 4/1; financial award applicants required to submit FAFSA. *Faculty research:* Statistical modeling, stochastic processes, numerical analysis, combinatorics, fracture mechanics. Total annual research expenditures:$306,144. *Unit head:* Dr. Jia Li, Chair, 256-824-6470, Fax: 256-824-6173, E-mail: chair@math.uah.edu.

University of Alaska Fairbanks, College of Natural Sciences and Mathematics, Department of Mathematics and Statistics, Fairbanks, AK 99775-7520. Offers computer science (MS); mathematics (MAT, MS, PhD); statistics (MS). Part-time programs available. *Faculty:* 13 full-time (5 women), 1 part-time/adjunct (0 women). *Students:* 14 full-time (1 woman), 4 part-time (3 women); includes 1 minority (Asian American or Pacific Islander), 8 international. Average age 26. 19 applicants, 74% accepted, 5 enrolled. In 2005, 3 degrees awarded. Terminal master's awarded for partial completion of doctoral program. *Degree requirements:* For master's, thesis or alternative, comprehensive exam, registration; for doctorate, thesis/dissertation, comprehensive exam, registration. *Entrance requirements:* For master's and doctorate, GRE General Test, GRE Subject Test. Additional exam requirements/recommendations for international students: Required—TOEFL (minimum score 600 paper-based); Recommended—TSE. *Application deadline:* For fall admission, 6/1 for domestic students, 3/1 for international students; for spring admission, 12/1 for domestic students, 9/1 for international students. Applications are processed on a rolling basis. Application fee: $50. Electronic applications accepted. *Expenses:* Tuition, state resident: full-time $4,392; part-time $244 per credit. Tuition, nonresident: full-time $8,964; part-time $498 per credit. Required fees: $800; $5 per credit. $48 per contact hour. Tuition and fees vary according to course level, course load, campus/location and reciprocity agreements. *Financial support:* In 2005–06, 4 research assistantships with tuition reimbursements (averaging $9,924 per year), 9 teaching assistantships with tuition reimbursements (averaging $10,458 per year) were awarded; fellowships with tuition reimbursements, career-related internships or fieldwork, Federal Work-Study, scholarships/grants, and unspecified assistantships also available. Financial award applicants required to submit FAFSA. *Faculty research:* Blackbox kriging (statistics), interaction with a virtual reality environment (computer), arrangements of hyperplanes (topology), bifurcation analysis of time-periodic differential-delay equations, synthetic aperture radar interferometry software (computer). *Unit head:* Dr. Dana L. Thomas, Chair, 907-474-7332, Fax: 907-474-5394, E-mail: fymath@uaf.edu.

University of Alberta, Faculty of Graduate Studies and Research, Department of Mathematical and Statistical Sciences, Edmonton, AB T6G 2E1, Canada. Offers applied mathematics (M Sc, PhD); biostatistics (M Sc); mathematical finance (M Sc, PhD); mathematical physics (M Sc, PhD); mathematics (M Sc, PhD); statistics (M Sc, PhD, Postgraduate Diploma). Part-time programs available. *Faculty:* 48 full-time (4 women). *Students:* 112 full-time (41 women), 5 part-time. Average age 24. 776 applicants, 5% accepted, 34 enrolled. In 2005, 12 master's, 10 doctorates awarded. Terminal master's awarded for partial completion of doctoral program. *Median time to degree:* Of those who began their doctoral program in fall 1996, 100% received their degree in 8 years or less. *Degree requirements:* For master's, thesis (for some programs); for doctorate, thesis/dissertation, comprehensive exam. *Entrance requirements:* Additional exam requirements/recommendations for international students: Required—TOEFL (minimum score 580 paper-based; 237 computer-based). *Application deadline:* For fall admission, 3/1 for domestic students, 2/1 for international students. Applications are processed on a rolling basis. Application fee: $0. Electronic applications accepted. Tuition and fees charges are reported in Canadian dollars. *Expenses:* Tuition, state resident: part-time $562 Canadian dollars per term. Tuition, nonresident: full-time $3,375 Canadian dollars. Required fees: $573 Canadian dollars; $84 Canadian dollars per term. *Financial support:* In 2005–06, 51 research assistantships, 88 teaching assistantships with full and partial tuition reimbursements were awarded; scholarships/grants also available. Financial award application deadline:5/1. *Faculty research:* Classical and functional analysis, algebra, differential equations, geometry. *Unit head:* Dr. Anthony To-Ming Lau, Chair, 403-492-5141, E-mail: tlau@math.ualberta.ca. *Application contact:* Dr. Yau Shu Wong, Associate Chair, Graduate Studies, 403-492-5799, Fax: 403-492-6828, E-mail: gradstudies@math.ualberta.ca.

The University of Arizona, Graduate College, College of Science, Department of Mathematics, Mathematical Sciences, Professional Program, Tucson, AZ 85721. Offers PMS. Part-time programs available. *Degree requirements:* For master's, thesis, internships, colloquium, business courses. *Entrance requirements:* For master's, GRE General Test, 3 letters of recommendations. *Faculty research:* Algebra, coding theory, graph theory, combinatorics, probability.

The University of Arizona, Graduate College, Graduate Interdisciplinary Programs, Graduate Interdisciplinary Program in Applied Mathematics, Tucson, AZ 85721. Offers applied mathematics (MS, PhD); mathematical sciences (PMS). Terminal master's awarded for partial completion of doctoral program. *Degree requirements:* For master's, thesis (for some programs), registration; for doctorate, one foreign language, thesis/dissertation, comprehensive exam, registration. *Entrance requirements:* For master's and doctorate, GRE. Additional exam requirements/recommendations for international students: Required—TOEFL (minimum score 575 paper-based; 230 computer-based), TSE (minimum score 45). *Faculty research:* Dynamical systems and chaos, partial differential equations, pattern formation, fluid dynamics and turbulence, scientific computation, mathematical physics, mathematical biology, medical imaging, applied probability and stochastic processes.

University of Arkansas, Graduate School, J. William Fulbright College of Arts and Sciences, Department of Mathematical Sciences, Program in Mathematics, Fayetteville, AR 72701-1201. Offers MS, PhD. *Students:* 35 full-time (18 women), 1 part-time; includes 4 minority (3 African Americans, 1 Hispanic American), 8 international. 49 applicants, 45% accepted. In 2005, 2 master's, 3 doctorates awarded. *Degree requirements:* For master's, thesis or alternative; for doctorate, 2 foreign languages, thesis/dissertation. Application fee: $40 ($50 for international students). *Financial support:* In 2005–06, 11 fellowships with tuition reimbursements, 1 research assistantship, 29 teaching assistantships were awarded; career-related internships or fieldwork and Federal Work-Study also available. Support available to part-time students. Financial award application deadline: 4/1; financial award applicants required to submit FAFSA. *Unit head:* Dr. Mark Arnold, Graduate Coordinator, 479-575-3351, Fax: 479-575-8630, E-mail: arnold@uark.edu.

University of Arkansas at Little Rock, Graduate School, College of Science and Mathematics, Program in Integrated Science and Mathematics, Little Rock, AR 72204-1099. Offers MS.

The University of British Columbia, Faculty of Graduate Studies, Faculty of Science, Program in Mathematics, Vancouver, BC V6T 1Z1, Canada. Offers M Sc, MA, PhD. Part-time programs available. *Faculty:* 51 full-time (6 women). *Students:* 89 full-time (30 women). Average age 24. 224 applicants, 26% accepted, 34 enrolled. In 2005, 18 master's, 6 doctorates awarded. *Degree requirements:* For master's, thesis or alternative, essay, qualifying exam; for doctorate, thesis/dissertation, qualifying exam, thesis proposal, comprehensive exam. *Entrance requirements:* For master's and doctorate, first class standing. Additional exam requirements/recommendations for international students: Required—TOEFL (minimum score 600 paper-based; 250 computer-based), TSE. *Application deadline:* For fall admission, 3/1 for domestic students, 3/1 for international students. Application fee: $90 Canadian dollars ($150 Canadian dollars for international students). Electronic applications accepted. *Financial support:* In 2005–06, 23 fellowships with tuition reimbursements (averaging $18,404 per year), 68 research assistantships with tuition reimbursements (averaging $8,000 per year), 77 teaching assistantships with tuition reimbursements (averaging $11,000 per year) were awarded; institutionally sponsored loans, scholarships/grants, health care benefits, tuition waivers (full and partial), and unspecified assistantships also available. Financial award application deadline: 10/5. *Faculty research:* Applied mathematics, financial mathematics, pure mathematics. Total annual research expenditures:$256,000. *Unit head:* Dr. Brian Marcus, Head, 604-822-2771, Fax: 604-822-9479, E-mail: marcus@math.ubc.ca. *Application contact:* Lee Yupitun, Graduate Secretary, 604-822-3079, Fax: 604-822-6074, E-mail: admiss@math.ubc.ca.

University of Calgary, Faculty of Graduate Studies, Faculty of Science, Department of Mathematics and Statistics, Calgary, AB T2N 1N4, Canada. Offers M Sc, PhD. *Faculty:* 31 full-time (3 women), 3 part-time/adjunct (1 woman). *Students:* 48 full-time (11 women), 1 part-time. Average age 26. 82 applicants, 33% accepted, 15 enrolled. In 2005, 8 master's, 2 doctorates awarded. *Median time to degree:* Of those who began their doctoral program in fall 1996, 100% received their degree in 8 years or less. *Degree requirements:* For master's, thesis, comprehensive exam; for doctorate, thesis/dissertation, candidacy exam, preliminary exams. *Entrance requirements:* For master's, honors degree in applied math, pure math, or statistics; for doctorate, MA or M Sc. Additional exam requirements/recommendations for international students: Required—TOEFL (minimum score 600 paper-based; 250 computer-based), IELT (minimum score 7), TOEFL (paper score 600; computer score 250) or IELT (paper score 7). *Application deadline:* For fall admission, 2/1 for domestic students, 2/1 for international students. Applications are processed on a rolling basis. Application fee: $100 ($130 for international students). *Financial support:* In 2005–06, 43 students received support, including research assistantships with partial tuition reimbursements available (averaging $4,100 per year), teaching assistantships with partial tuition reimbursements available (averaging $13,060 per year); fellowships, scholarships/grants and unspecified assistantships also available. *Faculty research:* Combinatorics, applied mathematics, statistics, probability, analysis. Total annual research expenditures: $1.1 million. *Unit head:* Dr. Claude LaFlamme, Head (Acting), 403-220-5210, Fax: 403-282-5150, E-mail: laf@math.ucalgary.ca. *Application contact:* Joanne Mellard, Graduate Administrator, 403-220-6299, Fax: 403-282-5150, E-mail: gradapps@math.ucalgary.ca.

University of California, Berkeley, Graduate Division, College of Letters and Science, Department of Mathematics, Berkeley, CA 94720-1500. Offers applied mathematics (PhD); mathematics (MA, PhD). Terminal master's awarded for partial completion of doctoral program. *Degree requirements:* For master's, exam or thesis; for doctorate, 2 foreign languages, thesis/dissertation, qualifying exam. *Entrance requirements:* For master's and doctorate, GRE General Test, GRE Subject Test, minimum GPA of 3.0. *Faculty research:* Algebra, analysis, logic, geometry/topology.

University of California, Davis, Graduate Studies, Program in Mathematics, Davis, CA 95616. Offers MA, MAT, PhD. *Faculty:* 40 full-time. *Students:* 76 full-time (22 women); includes 12 minority (5 African Americans, 7 Asian Americans or Pacific Islanders), 10 international. Average age 28. 102 applicants, 52% accepted, 22 enrolled. In 2005, 6 master's, 2 doctorates awarded. Terminal master's awarded for partial completion of doctoral program. *Median time to degree:* Of those who began their doctoral program in fall 1996, 55.6% received their degree in 8 years or less. *Degree requirements:* For master's, comprehensive exam; for doctorate, one foreign language, thesis/dissertation. *Entrance requirements:* For master's and doctorate, GRE General Test, GRE Subject Test, minimum GPA of 3.0. Additional exam requirements/recommendations for international students: Required—TOEFL (minimum score 550 paper-based; 213 computer-based). *Application deadline:* For fall admission, 1/15 for domestic students, 1/15 for international students. Application fee: $60. Electronic applications accepted. *Financial support:* In 2005–06, 74 students received support, including 39 fellowships with full and partial tuition reimbursements available (averaging $8,908 per year), 6 research assistantships with full and partial tuition reimbursements available (averaging $11,494 per year), 15 teaching assistantships with partial tuition reimbursements available (averaging $15,586 per year); Federal Work-Study, institutionally sponsored loans, scholarships/grants, tuition waivers (full and partial), and unspecified assistantships also available. Financial award application deadline: 1/15; financial award applicants required to submit FAFSA. *Faculty research:* Mathematical physics, geometric topology, probability, partial differential equations, applied mathematics. *Unit head:* Motohico Mulase, Chair, 530-752-6324, E-mail: mulase@math.ucdavis.edu. *Application contact:* Celia Davis, Administrative Assistant, 530-752-8131, Fax: 530-752-6635, E-mail: studentservices@math.ucdavis.edu.

University of California, Irvine, Office of Graduate Studies, School of Physical Sciences, Department of Mathematics, Irvine, CA 92697. Offers MS, PhD. *Degree requirements:* For doctorate, thesis/dissertation. *Entrance requirements:* For master's and doctorate, GRE General Test, GRE Subject Test, minimum GPA of 3.0. Additional exam requirements/recommendations for international students: Required—TOEFL (minimum score 550 paper-based; 213 computer-based), TSE. Electronic applications accepted. *Faculty research:* Algebra and logic, geometry and topology, probability, mathematical physics.

University of California, Los Angeles, Graduate Division, College of Letters and Science, Department of Mathematics, Los Angeles, CA 90095. Offers MA, MAT, PhD. *Degree requirements:* For master's, essay; for doctorate, one foreign language, thesis/dissertation, oral and written qualifying exams. *Entrance requirements:* For master's, GRE General Test, GRE Subject Test, minimum GPA of 3.2 in mathematics; for doctorate, GRE General Test, GRE Subject Test, minimum GPA of 3.5 in mathematics. Electronic applications accepted.

University of California, Riverside, Graduate Division, Department of Mathematics, Riverside, CA 92521-0102. Offers MA, MS, PhD. Part-time programs available. *Faculty:* 24 full-time (3 women), 13 part-time/adjunct (4 women). *Students:* 57 full-time (14 women), 2 part-time (both women); includes 19 minority (1 African American, 1 American Indian/Alaska Native, 12 Asian Americans or Pacific Islanders, 5 Hispanic Americans), 10 international. Average age 28. 49 applicants. In 2005, 6 master's, 3 doctorates awarded. Terminal master's awarded for partial completion of doctoral program. *Degree requirements:* For master's, comprehensive exam; for doctorate, thesis/dissertation, qualifying exams. *Entrance requirements:* For master's and doctorate, GRE General Test, minimum GPA of 3.2. Additional exam requirements/recommendations for international students: Required—TOEFL (minimum score 550 paper-based; 213 computer-based); Recommended—TSE (minimum score 50). *Application deadline:* For fall admission, 5/1 for domestic students, 2/1 for international students. For winter admission, 9/1 for domestic students; for spring admission, 12/1 for domestic students. Applications are processed on a rolling basis. Application fee: $60 ($75 for international students). Electronic applications accepted. *Expenses:* Tuition, nonresident: full-time $14,694. Full-time tuition and fees vary according to program. *Financial support:* In 2005–06, fellowships with tuition reimbursements (averaging $12,000 per year), teaching assistantships with full and partial tuition reimbursements (averaging $15,000 per year) were awarded; research assistantships, career-related internships or fieldwork, Federal Work-Study, institutionally sponsored loans, health care benefits, and tuition waivers (full and partial) also available. Financial award application deadline: 1/5; financial award applicants required to submit FAFSA. *Faculty research:* Algebraic geometry, commutative algebra, Lie algebra, differential equations, differential geometry. *Unit head:* Dr. Bun Wong, Chair, 951-827-6459, Fax: 951-827-7314. *Application contact:* Melissa Gomez, Graduate Program Assistant, 951-827-7378, Fax: 951-827-7314, E-mail: gradprog@math.ucr.edu.

University of California, San Diego, Graduate Studies and Research, Department of Mathematics, La Jolla, CA 92093. Offers applied mathematics (MA); mathematics (MA, PhD); statistics (MS). *Degree requirements:* For doctorate, thesis/dissertation. *Entrance requirements:* For master's and doctorate, GRE General Test, GRE Subject Test. Electronic applications accepted.

University of California, Santa Barbara, Graduate Division, College of Letters and Sciences, Division of Mathematics, Life, and Physical Sciences, Department of Mathematics, Santa Barbara, CA 93106. Offers applied mathematics (MA); mathematics (MA, PhD). *Faculty:* 32 full-time (2 women). *Students:* 58 full-time (17 women); includes 4 minority (2 Asian Americans or Pacific Islanders, 2 Hispanic Americans), 11 international. Average age 25. 115 applicants, 25% accepted, 5 enrolled. In 2005, 14 master's, 6 doctorates awarded. Terminal master's awarded for partial completion of doctoral program. *Median time to degree:* Of those who began their doctoral program in fall 1996, 100% received their degree in 8 years or less. *Degree requirements:* For master's, thesis (for some programs), comprehensive exam (for some programs), registration; for doctorate, thesis/dissertation, comprehensive exam, registration. *Entrance requirements:* For master's and doctorate, GRE General Test, GRE Subject Test. Additional exam requirements/recommendations for international students: Required—TOEFL (minimum score 575 paper-based; 231 computer-based); Recommended—TSE. *Application deadline:* For fall admission, 1/1 for domestic students, 1/1 for international students. Application fee: $60. Electronic applications accepted. *Financial support:* In 2005–06, 3 students received support, including 2 fellowships with full and partial tuition reimbursements available (averaging $17,000 per year), 45 teaching assistantships with full and partial tuition reimbursements available (averaging $15,000 per year); Federal Work-Study and health care benefits also available. Financial award application deadline: 1/1; financial award applicants required to submit FAFSA. *Faculty research:* Topology, differential geometry, algebra, analysis, applied mathematics. *Unit head:* Thomas Sideris, Chair, 805-893-8340, E-mail: sideris@math.ucsb.edu. *Application contact:* Medina Teel, Graduate Advisor, 805-893-8192, Fax: 805-893-2385, E-mail: teel@math.ucsb.edu.

University of California, Santa Cruz, Division of Graduate Studies, Division of Physical and Biological Sciences, Department of Mathematics, Santa Cruz, CA 95064. Offers MA, PhD. *Faculty:* 15 full-time (2 women). *Students:* 33 full-time (9 women); includes 3 minority (1 Asian American or Pacific Islander, 2 Hispanic Americans), 7 international. 59 applicants, 0% accepted. In 2005, 4 doctorates awarded. *Degree requirements:* For doctorate, one foreign language, thesis/dissertation, qualifying exam. *Entrance requirements:* For doctorate, GRE General Test, GRE Subject Test. *Application deadline:* For fall admission, 2/1 for domestic students. Application fee: $60. *Expenses:* Tuition, nonresident: full-time $14,694. *Financial support:* Fellowships, research assistantships, teaching assistantships, Federal Work-Study and institutionally sponsored loans available. Financial award application deadline: 2/1. *Unit head:* Dr. Tony Tromba, Chair, 831-459-2215, E-mail: gem@cats.ucsc.edu. *Application contact:* Judy L. Glass, Reporting Analyst for Graduate Admissions, 831-459-5906, Fax: 831-459-4843, E-mail: jlglass@ucsc.edu.

University of Central Arkansas, Graduate School, College of Natural Sciences and Math, Department of Mathematics, Conway, AR 72035-0001. Offers MA. Part-time programs available. *Faculty:* 15 full-time (4 women). *Students:* 8 full-time (7 women), 49 part-time (45 women); includes 3 minority (1 African American, 1 American Indian/Alaska Native, 1 Asian American or Pacific Islander), 1 international. 29 applicants, 100% accepted, 29 enrolled. In 2005, 14 degrees awarded. *Degree requirements:* For master's, thesis optional. *Entrance requirements:* For master's, GRE General Test, minimum GPA of 2.7. Additional exam requirements/recommendations for international students: Required—TOEFL (minimum score 550 paper-based; 213 computer-based). *Application deadline:* For fall admission, 3/1 for domestic students; for spring admission, 10/1 priority date for domestic students. Applications are processed on a rolling basis. Application fee: $25 ($40 for international students). *Expenses:* Tuition, state resident: part-time $190 per hour. Tuition, nonresident: part-time $380 per hour. Required fees: $31 per hour. $88 per term. Tuition and fees vary according to course load and program. *Financial support:* In 2005–06, 9 teaching assistantships with partial tuition reimbursements (averaging $8,500 per year) were awarded; research assistantships with partial tuition reimbursements, Federal Work-Study and unspecified assistantships also available. Financial award application deadline: 2/15; financial award applicants required to submit FAFSA. *Faculty research:* Nonlinear wave equation. *Unit head:* Dr. Ramesh Garimella, Chair, 501-450-3147, Fax: 501-450-5662, E-mail: rameshg@uca.edu. *Application contact:* Brenda Herring, Admissions Assistant, 501-450-3124, Fax: 501-450-5678, E-mail: bherring@uca.edu.

University of Central Florida, College of Sciences, Department of Mathematics, Orlando, FL 32816. Offers applied mathematics (Certificate); mathematical science (MS); mathematics (PhD). Part-time and evening/weekend programs available. *Faculty:* 45 full-time (9 women), 4 part-time/adjunct (0 women). *Students:* 34 full-time (9 women), 21 part-time (8 women); includes 12 minority (2 African Americans, 3 Asian Americans or Pacific Islanders, 7 Hispanic Americans), 10 international. Average age 33. 42 applicants, 76% accepted, 19 enrolled. In 2005, 14 master's, 3 doctorates awarded. *Degree requirements:* For master's, thesis or alternative; for doctorate, thesis/dissertation, candidacy exam. *Entrance requirements:* For master's, GRE General Test, minimum GPA of 3.0 in last 60 hours; for doctorate, GRE Subject Test, minimum GPA of 3.0 in last 60 hours or master's qualifying exam. Additional exam requirements/recommendations for international students: Required—TOEFL. *Application deadline:* For fall admission, 7/15 for domestic students; for spring admission, 12/1 for domestic students. Application fee: $30. Electronic applications accepted. *Expenses:* Tuition, state resident: full-time $5,788. Tuition, nonresident: full-time $21,927. Required fees: $241 per credit hour. *Financial support:* In 2005–06, 13 fellowships with partial tuition reimbursements (averaging $5,285 per year), 5 research assistantships with partial tuition reimbursements (averaging $11,700 per year), 26 teaching assistantships with partial tuition reimbursements (averaging $12,600 per year) were awarded; career-related internships or fieldwork, Federal Work-Study, institutionally sponsored loans, tuition waivers (partial), and unspecified assistantships also available. Financial award application deadline: 3/1; financial award applicants required to submit FAFSA. *Faculty research:* Applied mathematics, analysis, approximation theory, graph theory, mathematical statistics. *Unit head:* Dr. Zuhair Nashed, Chair, 407-823-0445, Fax: 407-823-6253, E-mail: znashed@mail.ucf.edu. *Application contact:* Dr. Ram Mohapatra, Coordinator, 407-823-5080, Fax: 407-823-6253, E-mail: ramm@pegasus.cc.ucf.edu.

University of Central Oklahoma, College of Graduate Studies and Research, College of Mathematics and Science, Department of Mathematics and Statistics, Edmond, OK 73034-5209. Offers applied mathematical sciences (MS), including computer science, mathematics, mathematics/computer science teaching, statistics. Part-time programs available. *Faculty:* 8 full-time (3 women), 1 part-time/adjunct (0 women). *Students:* 7 full-time (1 woman), 8 part-time (3 women); includes 2 minority (both Asian Americans or Pacific Islanders), 7 international. Average age 29. 14 applicants, 100% accepted. In 2005, 6 degrees awarded. *Degree requirements:* For master's, thesis. *Entrance requirements:* Additional exam requirements/recommendations for international students: Required—TOEFL (minimum score 550 paper-based; 213 computer-based). *Application deadline:* Applications are processed on a rolling basis. Application fee: $25. Electronic applications accepted. *Expenses:* Tuition, state resident: full-time $2,988; part-time $125 per credit hour. Tuition, nonresident: full-time $4,728; part-time $197 per credit hour. Required fees: $716; $16 per credit hour. *Financial support:* Federal Work-Study and unspecified assistantships available. Financial award application deadline: 3/31; financial award applicants required to submit FAFSA. *Faculty research:* Curvature, FAA, math education. *Unit head:* Dr. Chuck Cooper, Chairperson, 405-974-5294. *Application contact:* Dr. James Yates, Adviser, 405-974-5386, Fax: 405-974-3824, E-mail: jyates@aix1.ucok.edu.

University of Chicago, Division of the Physical Sciences, Department of Mathematics, Chicago, IL 60637-1513. Offers applied mathematics (SM, PhD); financial mathematics (MS); mathematics (SM, PhD). *Faculty:* 59 full-time (5 women). *Students:* 101 full-time (21 women), 33 international. 288 applicants, 15% accepted, 17 enrolled. In 2005, 13 master's, 14 doctorates awarded. *Degree requirements:* For master's, one foreign language; for doctorate, one foreign language, thesis/dissertation, 2 qualifying exams, oral topic presentation. *Entrance requirements:* For master's and doctorate, GRE General Test, GRE Subject Test. Additional exam requirements/recommendations for international students: Required—TOEFL (minimum score 600 paper-based; 250 computer-based). *Application deadline:* For fall admission, 1/5 for domestic students, 1/5 for international students. Application fee: $55. Electronic applications accepted. *Financial support:* In 2005–06, 48 fellowships, 53 teaching assistantships were awarded; research assistantships, career-related internships or fieldwork, institutionally sponsored loans, and scholarships/grants also available. Financial award application deadline: 1/5; financial award applicants required to submit CSS PROFILE or FAFSA. *Faculty research:* Analysis, differential geometry, algebra number theory, topology, algebraic geometry. *Unit head:* Dr. Kevin Corlette, Chair, 773-702-0702, Fax: 773-702-9787, E-mail: kevin@math.uchicago.edu. *Application contact:* Laurie Wail, Graduate Studies Assistant, 773-702-7358, Fax: 773-702-9787, E-mail: lwail@math.uchicago.edu.

University of Cincinnati, Division of Research and Advanced Studies, McMicken College of Arts and Sciences, Department of Mathematical Sciences, Cincinnati, OH 45221. Offers applied mathematics (MS, PhD); mathematics education (MAT); pure mathematics (MS, PhD); statistics (MS, PhD). *Accreditation:* NCATE (one or more programs are accredited). Part-time programs available. Terminal master's awarded for partial completion of doctoral program. *Degree requirements:* For master's, thesis or alternative, comprehensive exam; for doctorate, one foreign language, thesis/dissertation, comprehensive exam. *Entrance requirements:* For master's, GRE, teacher certification (MAT); for doctorate, GRE. Additional exam requirements/recommendations for international students: Required—TOEFL. Electronic applications accepted. *Faculty research:* Algebra, analysis, differential equations, numerical analysis, statistics.

University of Colorado at Boulder, Graduate School, College of Arts and Sciences, Department of Mathematics, Boulder, CO 80309. Offers MA, MS, PhD. *Faculty:* 28 full-time (5 women). *Students:* 69 full-time (16 women), 10 part-time (5 women); includes 7 minority (2 American Indian/Alaska Native, 4 Asian Americans or Pacific Islanders, 1 Hispanic American), 6 international. Average age 28. 27 applicants, 85% accepted. In 2005, 10 master's, 4 doctorates awarded. Terminal master's awarded for partial completion of doctoral program. *Degree requirements:* For master's, thesis or alternative, comprehensive exam; for doctorate, one foreign language, thesis/dissertation, 2 preliminary exams, comprehensive exam. *Entrance requirements:* For master's, minimum undergraduate GPA of 2.75. *Application deadline:* For fall admission, 1/15 priority date for domestic students, 1/15 priority date for international students; for spring admission, 11/1 for domestic students, 11/1 for international students. Applications are processed on a rolling basis. Application fee: $50 ($60 for international students). *Financial support:* In 2005–06, 6 fellowships (averaging $2,333 per year), 60 teaching assistantships (averaging $15,503 per year) were awarded; research assistantships, scholarships/grants and tuition waivers (full) also available. Support available to part-time students. Financial award application deadline: 3/1. *Faculty research:* Pure mathematics, applied mathematics and mathematical physics (including algebra, algebraic geometry, differential equations, differential geometry, logic and foundations). Total annual research expenditures:$108,828. *Unit head:* Lynne Walling, Chair, 303-492-8566, Fax: 303-492-7707, E-mail: walling@euclid.colorado.edu. *Application contact:* Carol Deckert, Graduate Administrative Assistant, 303-492-3161, Fax: 303-492-7707, E-mail: deckert@euclid.colorado.edu.

University of Colorado at Denver and Health Sciences Center—Downtown Denver Campus, College of Liberal Arts and Sciences, Program in Integrated Sciences, Denver, CO 80217-3364. Offers applied science (MIS); computer science (MIS); mathematics (MIS). *Students:* 4 full-time (3 women), 10 part-time (5 women); includes 1 minority (Asian American or Pacific Islander), 3 international. Average age 33. 4 applicants, 75% accepted, 1 enrolled. In 2005, 5 degrees awarded. *Expenses:* Tuition, state resident: part-time $325 per credit hour. Tuition, nonresident: part-time $1,077 per credit hour. Required fees: $145 per credit hour. One-time fee: $115 part-time. Tuition and fees vary according to course level and program. *Financial support:* Research assistantships, teaching assistantships available. Financial award application deadline: 4/1; financial award applicants required to submit FAFSA. *Unit head:* Doris Kimbrough, Director, 303-556-3202, Fax: 303-556-4776, E-mail: doris.kimbrough@cudenver.edu.

University of Connecticut, Graduate School, College of Liberal Arts and Sciences, Department of Mathematics, Field of Mathematics, Storrs, CT 06269. Offers actuarial science (MS, PhD); mathematics (MS, PhD). *Faculty:* 40 full-time (5 women). *Students:* 76 full-time (21 women), 14 part-time (8 women); includes 9 minority (2 African Americans, 6 Asian Americans or Pacific Islanders, 1 Hispanic American), 38 international. Average age 27. 205 applicants, 46% accepted, 46 enrolled. In 2005, 28 master's, 3 doctorates awarded. Terminal master's awarded for partial completion of doctoral program. *Degree requirements:* For master's, comprehensive exam; for doctorate, thesis/dissertation. *Entrance requirements:* For master's and doctorate, GRE General Test. Additional exam requirements/recommendations for international students: Required—TOEFL (minimum score 550 paper-based; 213 computer-based). *Application deadline:* For fall admission, 2/1 priority date for domestic students, 2/1 priority date for international students; for spring admission, 11/1 for domestic students, 10/1 for international students. Applications are processed on a rolling basis. Application fee: $55. Electronic applications accepted. *Expenses:* Tuition, state resident: part-time $444 per credit hour. Tuition, nonresident: part-time $1,154 per credit hour. Tuition and fees vary according to course load. *Financial support:* In 2005–06, 13 research assistantships with full tuition reimbursements, 45 teaching assistantships with full tuition reimbursements were awarded; fellowships, Federal Work-Study, scholarships/grants, health care benefits, and unspecified assistantships also available. Financial award application deadline: 2/1; financial award applicants required to submit FAFSA. *Unit head:* Eugene Spiegel, Chairperson, 860-486-3844, Fax: 860-486-4283, E-mail: eugene.spiegel@uconn.edu. *Application contact:* Sharon McDermott, Administrative Assistant, 860-486-6452, Fax: 860-486-4283, E-mail: gradadm@math.uconn.edu.

University of Delaware, College of Arts and Sciences, Department of Mathematical Sciences, Newark, DE 19716. Offers applied mathematics (MS, PhD); mathematics (MS, PhD). Part-time programs available. *Faculty:* 46 full-time (13 women). *Students:* 40 full-time (16 women), 1 part-time; includes 3 minority (1 African American, 1 American Indian/Alaska Native, 1 Asian American or Pacific Islander), 18 international. Average age 25. 111 applicants, 16% accepted, 13 enrolled. In 2005, 11 master's, 2 doctorates awarded. Terminal master's awarded for partial completion of doctoral program. *Degree requirements:* For master's, thesis (for some programs);

Mathematics

University of Delaware (continued)

for doctorate, one foreign language, thesis/dissertation, qualifying exam. *Entrance requirements:* For master's and doctorate, GRE General Test. Additional exam requirements/recommendations for international students: Required—TOEFL. *Application deadline:* For fall admission, 3/1 for domestic students; for spring admission, 12/15 priority date for domestic students. Applications are processed on a rolling basis. Application fee: $60. Electronic applications accepted. *Financial support:* In 2005–06, 5 fellowships with tuition reimbursements (averaging $12,200 per year), 1 research assistantship with tuition reimbursement (averaging $14,000 per year), 35 teaching assistantships with tuition reimbursements (averaging $12,240 per year) were awarded; career-related internships or fieldwork, institutionally sponsored loans, scholarships/grants, and tuition waivers (full and partial) also available. Financial award application deadline: 3/1. *Faculty research:* Scattering theory, inverse problems, fluid dynamics, numerical analysis, combinatorics. Total annual research expenditures: $863,660. *Unit head:* Prof. Peter B. Monk, Interim Chair, 302-831-2711, Fax: 302-831-4511, E-mail: monk@math.udel.edu. *Application contact:* Dr. George C. Hsiao, Graduate Chair, 302-831-1882, Fax: 302-831-4511, E-mail: hsiao@math.udel.edu.

See Close-Up on page 533.

University of Denver, Faculty of Natural Sciences and Mathematics, Department of Mathematics, Denver, CO 80208. Offers applied mathematics (MA, MS); computer science (MS); mathematics (PhD). Part-time programs available. *Faculty:* 9 full-time (3 women). *Students:* 12 applicants, 100% accepted. In 2005, 5 master's, 2 doctorates awarded. Terminal master's awarded for partial completion of doctoral program. *Degree requirements:* For master's, computer language, foreign language, or laboratory experience; for doctorate, one foreign language, thesis/dissertation, oral and written exams. *Entrance requirements:* For master's and doctorate, GRE General Test. Additional exam requirements/recommendations for international students: Required—TOEFL. *Application deadline:* Applications are processed on a rolling basis. Application fee: $45. *Expenses:* Tuition: Full-time $27,756; part-time $771 per credit. Required fees: $174. *Financial support:* In 2005–06, 11 teaching assistantships with full and partial tuition reimbursements (averaging $12,896 per year) were awarded; career-related internships or fieldwork, Federal Work-Study, institutionally sponsored loans, and scholarships/grants also available. Support available to part-time students. Financial award application deadline: 3/1; financial award applicants required to submit FAFSA. *Faculty research:* Real-time software, convex bodies, multidimensional data, parallel computer clusters. *Unit head:* Dr. Rick Ball, Chairperson, 303-871-2821. *Application contact:* 303-871-2911, E-mail: math-info@math.du.edu.

University of Detroit Mercy, College of Engineering and Science, Department of Mathematics and Computer Science, Detroit, MI 48219-0900. Offers computer science (MSCS); elementary mathematics education (MATM); junior high mathematics education (MATM); secondary mathematics education (MATM); teaching of mathematics (MATM). Evening/weekend programs available. *Entrance requirements:* For master's, minimum GPA of 3.0.

University of Florida, Graduate School, College of Liberal Arts and Sciences, Department of Mathematics, Gainesville, FL 32611. Offers MA, MAT, MS, MST, PhD. *Accreditation:* NCATE (one or more programs are accredited). Part-time programs available. *Faculty:* 56 full-time (5 women), 1 part-time/adjunct (0 women). *Students:* 102 (24 women); includes 9 minority (1 African American, 5 Asian Americans or Pacific Islanders, 3 Hispanic Americans) 48 international. In 2005, 14 master's, 8 doctorates awarded. Terminal master's awarded for partial completion of doctoral program. *Degree requirements:* For master's, thesis optional; for doctorate, one foreign language, thesis/dissertation. *Entrance requirements:* For master's and doctorate, GRE General Test, minimum GPA of 3.0. Additional exam requirements/recommendations for international students: Required—TOEFL (minimum score 550 paper-based; 213 computer-based). *Application deadline:* For fall admission, 6/1 for domestic students. Applications are processed on a rolling basis. Application fee: $30. Electronic applications accepted. *Expenses:* Tuition, state resident: full-time $6,234. Tuition, nonresident: full-time $21,359. Tuition and fees vary according to program. *Financial support:* In 2005–06, 79 teaching assistantships (averaging $20,738 per year) were awarded; fellowships, research assistantships, career-related internships or fieldwork and unspecified assistantships also available. Financial award application deadline: 3/1. *Faculty research:* Combinatorics and number theory, group theory, probability theory, logic, differential geometry and mathematical physics. *Unit head:* Dr. Krishnaswami Alladi, Chairman, 352-392-0281 Ext. 236, Fax: 352-392-8357, E-mail: alladi@math.ufl.edu. *Application contact:* Dr. Paul Robinson, Coordinator, 352-392-0281 Ext. 273, Fax: 352-392-8357, E-mail: robinson@math.ufl.edu.

University of Georgia, Graduate School, College of Arts and Sciences, Department of Mathematics, Athens, GA 30602. Offers applied mathematical science (MAMS); mathematics (MA, PhD). *Faculty:* 31 full-time (3 women). *Students:* 48 full-time; includes 3 minority (2 African Americans, 1 Asian American or Pacific Islander), 19 international. 123 applicants, 19% accepted, 15 enrolled. In 2005, 6 master's, 7 doctorates awarded. *Degree requirements:* For master's, one foreign language, thesis (for some programs); for doctorate, 2 foreign languages, thesis/dissertation. *Entrance requirements:* For master's and doctorate, GRE General Test. *Application deadline:* For fall admission, 7/1 for domestic students; for spring admission, 11/15 for domestic students. Application fee: $50. Electronic applications accepted. *Financial support:* Fellowships, research assistantships, teaching assistantships, unspecified assistantships available. *Unit head:* Dr. Joseph H. G. Fu, Graduate Coordinator, 706-542-2643, Fax: 706-542-2573. *Application contact:* Dr. Joseph H. G. Fu, Graduate Coordinator, 706-542-2643, Fax: 706-542-2573.

University of Guelph, Graduate Program Services, College of Physical and Engineering Science, Department of Mathematics and Statistics, Guelph, ON N1G 2W1, Canada. Offers applied mathematics (PhD); applied statistics (PhD); mathematics and statistics (M Sc). Part-time programs available. *Faculty:* 21 full-time (6 women), 4 part-time/adjunct (0 women). *Students:* 40 full-time, 4 part-time. Average age 22. 103 applicants, 25% accepted, 23 enrolled. In 2005, 20 master's, 1 doctorate awarded. *Median time to degree:* Of those who began their doctoral program in fall 1996, 100% received their degree in 8 years or less. *Degree requirements:* For master's, thesis (for some programs); for doctorate, thesis/dissertation. *Entrance requirements:* For master's, minimum B- average during previous 2 years of course work; for doctorate, minimum B average. Additional exam requirements/recommendations for international students: Required—TOEFL (minimum score 550 paper-based; 213 computer-based), IELT (minimum score 7). *Application deadline:* For fall admission, 2/1 for domestic students, 2/1 for international students. Application fee: $75. *Financial support:* In 2005–06, 20 students received support, including 26 research assistantships (averaging $2,300 per year), 27 teaching assistantships (averaging $9,000 per year); fellowships, scholarships/grants also available. *Faculty research:* Dynamical systems, mathematical biology, numerical analysis, linear and nonlinear models, reliability and bioassay. *Unit head:* Dr. O. Brian Allen, Chair, 519-824-4120 Ext. 56556, Fax: 519-837-0221, E-mail: chair@uoguelph.ca. *Application contact:* Susan McCormick, Graduate Administrative Assistant, 519-824-4120 Ext. 56553, Fax: 519-837-0221, E-mail: smccormi@uoguelph.ca.

University of Hawaii at Manoa, Graduate Division, Colleges of Arts and Sciences, College of Natural Sciences, Department of Mathematics, Honolulu, HI 96822. Offers MA, PhD. Part-time programs available. *Faculty:* 26 full-time (1 woman). *Students:* 13 full-time (1 woman), 9 part-time (2 women); includes 5 minority (4 Asian Americans or Pacific Islanders, 1 Hispanic American), 3 international. Average age 30. 39 applicants, 69% accepted, 7 enrolled. In 2005, 2 master's, 2 doctorates awarded. Terminal master's awarded for partial completion of doctoral program. *Degree requirements:* For master's, comprehensive exam; for doctorate, 2 foreign languages, thesis/dissertation, comprehensive exam. *Entrance requirements:* For master's and doctorate, GRE General Test, minimum GPA of 3.0. Additional exam requirements/recommendations for international students: Required—TOEFL. *Application deadline:* For fall admission, 3/1 for domestic students, 2/1 for international students; for spring admission, 9/1 for domestic students, 8/1 for international students. Applications are processed on a rolling basis. Application fee: $50. *Expenses:* Tuition, state resident: full-time $8,400; part-time $200 per credit hour. Tuition, nonresident: full-time $11,088; part-time $462 per credit hour. Tuition and fees vary according to program. *Financial support:* In 2005–06, 12 teaching assistantships (averaging $14,675 per year) were awarded; institutionally sponsored loans, tuition waivers (full and partial), and unspecified assistantships also available. Support available to part-time students. Financial award application deadline: 3/1. *Faculty research:* Analysis, algebra, lattice theory, logic topology, differential geometry. *Unit head:* Dr. Thomas Craven, Chair, 808-956-8792, Fax: 808-956-9139. *Application contact:* Dr. Robert Little, Graduate Chair, 808-956-7951, Fax: 808-956-9139, E-mail: little@math.hawaii.edu.

University of Houston, College of Natural Sciences and Mathematics, Department of Mathematics, Houston, TX 77204. Offers MA, MS, PhD. Part-time and evening/weekend programs available. *Faculty:* 24 full-time (3 women), 3 part-time/adjunct (1 woman). *Students:* 78 full-time (36 women), 40 part-time (19 women); includes 17 minority (1 African American, 13 Asian Americans or Pacific Islanders, 3 Hispanic Americans), 54 international. Average age 30. 42 applicants, 71% accepted, 17 enrolled. In 2005, 33 master's, 6 doctorates awarded. *Degree requirements:* For master's, thesis optional; for doctorate, thesis/dissertation. *Entrance requirements:* For master's, GRE General Test, minimum GPA of 3.0 in last 60 hours, bachelor's degree in mathematics or related area; for doctorate, GRE General Test, MS in mathematics or equivalent, minimum GPA of 3.0 in last 60 hours of course work. Additional exam requirements/recommendations for international students: Required—TOEFL. *Application deadline:* For fall admission, 7/3 for domestic students; for spring admission, 12/4 for domestic students. Applications are processed on a rolling basis. Application fee: $0 ($75 for international students). *Financial support:* In 2005–06, 2 fellowships with full tuition reimbursements (averaging $16,700 per year), 10 research assistantships with full tuition reimbursements (averaging $12,600 per year), 62 teaching assistantships with full tuition reimbursements (averaging $12,600 per year) were awarded; career-related internships or fieldwork, Federal Work-Study, institutionally sponsored loans, scholarships/grants, health care benefits, unspecified assistantships, and teaching fellowships also available. Support available to part-time students. Financial award application deadline: 3/10. *Faculty research:* Applied mathematics, modern analysis, computational science, geometry, dynamical systems. *Unit head:* Dr. Jeffery E. Morgan, Chairperson, 713-743-3500, Fax: 713-743-3505, E-mail: jmorgan@math.uh.edu. *Application contact:* Pamela K. Draughn, Graduate Adviser, 713-743-3517, Fax: 713-743-3505, E-mail: pamela@math.uh.edu.

University of Houston–Clear Lake, School of Science and Computer Engineering, Program in Mathematical Sciences, Houston, TX 77058-1098. Offers MS. Part-time and evening/weekend programs available. *Entrance requirements:* For master's, GRE General Test. Additional exam requirements/recommendations for international students: Required—TOEFL (minimum score 550 paper-based; 213 computer-based).

University of Idaho, College of Graduate Studies, College of Science, Department of Mathematics, Moscow, ID 83844-2282. Offers MAT, MS, PhD. *Students:* 19 full-time (7 women), 25 part-time (12 women); includes 7 minority (3 African Americans, 1 Asian American or Pacific Islander, 3 Hispanic Americans), 9 international. Average age 36. In 2005, 10 master's, 2 doctorates awarded. *Degree requirements:* For doctorate, 2 foreign languages, thesis/dissertation. *Entrance requirements:* For master's, minimum GPA of 2.8; for doctorate, minimum undergraduate GPA of 2.8, 3.0 graduate. *Application deadline:* For fall admission, 8/1 for domestic students; for spring admission, 12/15 for domestic students. Application fee: $55 ($60 for international students). *Expenses:* Tuition, nonresident: full-time $8,770; part-time $130 per credit. Required fees: $4,508; $217 per credit. *Financial support:* Research assistantships, teaching assistantships available. Financial award application deadline: 2/15. *Unit head:* Dr. Monte Boisen, Chair, 208-885-6742.

University of Illinois at Chicago, Graduate College, College of Liberal Arts and Sciences, Department of Mathematics, Statistics, and Computer Science, Chicago, IL 60607-7128. Offers applied mathematics (MS, DA, PhD); computer science (MS, DA, PhD); math and information science for the industry (MS); probability and statistics (MS, DA, PhD); pure mathematics (MS, DA, PhD); teaching of mathematics (MST). Part-time programs available. *Degree requirements:* For master's, comprehensive exam; for doctorate, one foreign language, thesis/dissertation. *Entrance requirements:* For master's and doctorate, GRE General Test, minimum GPA of 2.75. Additional exam requirements/recommendations for international students: Required—TOEFL. Electronic applications accepted.

See Close-Up on page 535.

University of Illinois at Urbana–Champaign, Graduate College, College of Liberal Arts and Sciences, Department of Mathematics, Champaign, IL 61820. Offers applied mathematics (MS); mathematics (MS, PhD); teaching of mathematics (MS). *Faculty:* 68 full-time (5 women), 6 part-time/adjunct (1 woman). *Students:* 198 full-time (54 women), 33 part-time (8 women); includes 14 minority (1 African American, 1 American Indian/Alaska Native, 9 Asian Americans or Pacific Islanders, 3 Hispanic Americans), 119 international. 408 applicants, 43% accepted, 52 enrolled. In 2005, 41 master's, 19 doctorates awarded. *Degree requirements:* For doctorate, 2 foreign languages, thesis/dissertation. *Entrance requirements:* For master's, GRE, minimum GPA of 3.0. *Application deadline:* For fall admission, 2/6 for domestic students. Applications are processed on a rolling basis. Application fee: $50 ($60 for international students). Electronic applications accepted. *Financial support:* In 2005–06, 37 fellowships, 37 research assistantships, 146 teaching assistantships were awarded; tuition waivers (full and partial) also available. Financial award application deadline: 2/15. *Unit head:* Daniel Grayson, Chair, 217-333-6209, Fax: 217-333-9576, E-mail: dan@uiuc.edu. *Application contact:* Lori Dick, Administrative Assistant, 217-333-3350, Fax: 217-333-9576, E-mail: ldick@math.uiuc.edu.

The University of Iowa, Graduate College, College of Liberal Arts and Sciences, Department of Mathematics, Iowa City, IA 52242-1316. Offers MS, PhD. *Faculty:* 50 full-time, 5 part-time/adjunct. *Students:* 53 full-time (17 women), 33 part-time (13 women); includes 19 minority (6 African Americans, 1 Asian American or Pacific Islander, 12 Hispanic Americans), 25 international. 89 applicants, 25% accepted, 10 enrolled. In 2005, 11 master's, 10 doctorates awarded. *Degree requirements:* For master's, exam, thesis optional; for doctorate, thesis/dissertation, comprehensive exam, registration. *Entrance requirements:* For master's and doctorate, GRE General Test, minimum GPA of 3.0. Additional exam requirements/recommendations for international students: Required—TOEFL (minimum score 575 paper-based; 232 computer-based). *Application deadline:* For fall admission, 1/15 priority date for domestic students, 1/15 priority date for international students. Applications are processed on a rolling basis. Application fee: $60 ($85 for international students). Electronic applications accepted. *Expenses:* Tuition, state resident: part-time $1,882 per term. Tuition, nonresident: full-time $17,338; part-time $4,907 per term. Tuition and fees vary according to course load and program. *Financial support:* In 2005–06, 9 fellowships, 2 research assistantships with partial tuition reimbursements, 63 teaching assistantships with partial tuition reimbursements were awarded. Financial award applicants required to submit FAFSA. *Unit head:* David Manderscheid, Chair, 319-335-0714, Fax: 319-335-0627.

University of Kansas, Graduate School, College of Liberal Arts and Sciences, Department of Mathematics, Lawrence, KS 66045. Offers applied mathematics and statistics (MA, PhD); mathematics (MA, PhD). *Faculty:* 40. *Students:* 67 full-time (23 women), 12 part-time (6 women); includes 4 minority (1 American Indian/Alaska Native, 3 Asian Americans or Pacific Islanders), 31 international. Average age 27. 67 applicants, 73% accepted. In 2005, 17 master's, 2 doctorates awarded. Terminal master's awarded for partial completion of doctoral program. *Degree requirements:* For master's, thesis or alternative; for doctorate, 2 foreign languages, thesis/dissertation, comprehensive exam. *Entrance requirements:* For master's and doctorate, GRE. Additional exam requirements/recommendations for international students: Required—TOEFL. *Application deadline:* For fall admission, 3/1 priority date for domestic students, 3/1 priority date for international students. Applications are processed on a rolling basis. Application fee: $55 ($60 for international students). Electronic applications accepted. *Expenses:* Tuition, state resident: full-time $4,859. Tuition, nonresident: full-time $1,200. Required fees: $589. Tuition and fees vary according to program. *Financial support:* Fellowships, research assistantships with full and partial tuition reimbursements, teaching assistantships with full and partial tuition reimbursements, institutionally sponsored loans available. Support available to part-time students. Financial award application deadline: 2/1. *Faculty research:* Commutative algebra/algebraic geometry, stochastic adaptive control/stochastic processes analysis/harmonic analysis/PDES, numerical analysis/dynamical systems, topology/set theory. *Unit head:* Jack Porter, Chair, 785-864-3651, Fax: 785-864-5255, E-mail: porter@math.ukans.edu. *Application contact:* David Lerner, Graduate Director, 785-864-3651, E-mail: lerner@ukans.edu.

University of Kentucky, Graduate School, Graduate School Programs from the College of Arts and Sciences, Program in Mathematics, Lexington, KY 40506-0032. Offers applied mathematics (MS); mathematics (MA, MS, PhD). *Faculty:* 34 full-time (1 woman), 3 part-time/adjunct (0 women). *Students:* 63 full-time (14 women), 5 part-time (4 women); includes 3 minority (1 Asian American or Pacific Islander, 2 Hispanic Americans), 16 international. Average age 27. 102 applicants, 42% accepted, 22 enrolled. In 2005, 12 master's, 8 doctorates awarded. *Median time to degree:* Of those who began their doctoral program in fall 1996, 84% received their degree in 8 years or less. *Degree requirements:* For master's, thesis optional; for doctorate, one foreign language, thesis/dissertation, comprehensive exam. *Entrance requirements:* For master's, GRE General Test, minimum undergraduate GPA of 2.5; for doctorate, GRE General Test, minimum graduate GPA of 3.0. Additional exam requirements/recommendations for international students: Required—TOEFL (minimum score 550 paper-based; 213 computer-based). *Application deadline:* For fall admission, 7/17 priority date for domestic students, 2/1 priority date for international students; for spring admission, 12/13 priority date for domestic students, 6/15 priority date for international students. Applications are processed on a rolling basis. Application fee: $40 ($55 for international students). Electronic applications accepted. *Expenses:* Tuition, state resident: full-time $3,159; part-time $331 per credit hour. Tuition, nonresident: full-time $6,984; part-time $756 per credit hour. Tuition and fees vary according to course load, degree level and program. *Financial support:* In 2005–06, 12 fellowships with full tuition reimbursements (averaging $3,000 per year), 8 research assistantships with full tuition reimbursements (averaging $12,800 per year), 55 teaching assistantships with full tuition reimbursements (averaging $13,000 per year) were awarded; Federal Work-Study, institutionally sponsored loans, scholarships/grants, traineeships, health care benefits, tuition waivers (partial), and unspecified assistantships also available. Support available to part-time students. Financial award application deadline:3/15. *Faculty research:* Numerical analysis, combinatorics, partial differential equations, algebra and number theory, real and complex analysis. *Unit head:* Dr. Serge C. Ochanine, Director of Graduate Studies, 859-257-8837, Fax: 859-257-3464, E-mail: ochanine@ms.uky.edu. *Application contact:* Dr. Brian Jackson, Senior Associate Dean, 859-257-8176, Fax: 859-323-1928, E-mail: lance.brunner@uky.edu.

University of Lethbridge, School of Graduate Studies, Lethbridge, AB T1K 3M4, Canada. Offers accounting (MScM); addictions counseling (M Sc); agricultural biotechnology (M Sc); agricultural studies (M Sc, MA); anthropology (MA); archaeology (MA); art (MA); biochemistry (M Sc); biological sciences (M Sc); biomolecular science (PhD); biosystems and biodiversity (PhD); Canadian studies (MA); chemistry (M Sc); computer science (M Sc); computer science and geographical information science (M Sc); counseling psychology (M Ed); dramatic arts (MA); earth, space, and physical science (PhD); economics (MA); educational leadership (M Ed); English (MA); environmental science (M Sc); evolution and behavior (PhD); exercise science (M Sc); finance (MScM); French (MA); French/German (MA); French/Spanish (MA); general education (M Ed); general management (MScM); geography (M Sc, MA); German (MA); health sciences (M Sc, MA); history (MA); human resource management and labour relations (MScM); individualized multidisciplinary (M Sc, MA); information systems (MScM); international management (MScM); kinesiology (M Sc, MA); management (M Sc, MA); marketing (MScM); mathematics (M Sc); music (MA); Native American studies (MA); neuroscience (M Sc, PhD); new media (MA); nursing (M Sc); philosophy (MA); physics (M Sc); policy and strategy (MScM); political science (MA); psychology (M Sc, MA); religious studies (MA); sociology (MA); theoretical and computational science (PhD); urban and regional studies (MA). Part-time and evening/weekend programs available. *Faculty:* 250. *Students:* 193 full-time, 145 part-time. 35 applicants, 100% accepted, 35 enrolled. In 2005, 40 degrees awarded. *Degree requirements:* For doctorate, thesis/dissertation, comprehensive exam. *Entrance requirements:* For master's, GMAT (M Sc management), bachelor's degree in related field, minimum GPA of 3.0 during previous 20 graded semester courses, 2 years teaching or related experience (M Ed); for doctorate, master's degree, minimum graduate GPA of 3.5. Additional exam requirements/recommendations for international students: Required—TOEFL. Application fee: $60 Canadian dollars. *Expenses:* Tuition, nonresident: part-time $531 per course. Required fees: $83 per year. Tuition and fees vary according to degree level and program. *Financial support:* Fellowships, research assistantships, teaching assistantships, scholarships/grants, health care benefits, and unspecified assistantships available. *Faculty research:* Movement and brain plasticity, gibberellin physiology, photosynthesis, carbon cycling, molecular properties of main-group ring components. *Unit head:* Dr. Shamsul Alam, Dean, 403-329-2121, Fax: 403-329-2097, E-mail: inquiries@uleth.ca. *Application contact:* Kathy Schrage, Administrative Assistant, Office of the Academic Vice President, 403-329-2121, Fax: 403-329-2097, E-mail: inquiries@uleth.ca.

University of Louisiana at Lafayette, Graduate School, College of Sciences, Department of Mathematics, Lafayette, LA 70504. Offers MS, PhD. *Faculty:* 13 full-time (3 women), 3 part-time/adjunct (1 woman). *Students:* 38 full-time (16 women), 3 part-time (2 women); includes 2 minority (1 African American, 1 Hispanic American), 19 international. Average age 31. 38 applicants, 39% accepted, 8 enrolled. In 2005, 5 master's, 4 doctorates awarded. Terminal master's awarded for partial completion of doctoral program. *Degree requirements:* For master's, thesis or alternative; for doctorate, 2 foreign languages, thesis/dissertation. *Entrance requirements:* For master's, GRE General Test, minimum GPA of 2.75; for doctorate, GRE General Test, minimum GPA of 3.0. Additional exam requirements/recommendations for international students: Required—TOEFL (minimum score 550 paper-based; 213 computer-based). *Application deadline:* For fall admission, 5/15 for domestic students, 5/15 for international students; for spring admission, 10/1 for domestic students, 10/1 for international students. Applications are processed on a rolling basis. Application fee: $20 ($30 for international students). Electronic applications accepted. *Expenses:* Tuition, state resident: full-time $3,330; part-time $93 per credit hour. Tuition, nonresident: full-time $9,510; part-time $350 per semester. International tuition: $9,646 full-time. *Financial support:* In 2005–06, 1 fellowship with full tuition reimbursement (averaging $17,000 per year), 7 research assistantships with full tuition reimbursements (averaging $9,246 per year), 25 teaching assistantships with full tuition reimbursements (averaging $10,488 per year) were awarded; Federal Work-Study and unspecified assistantships also available. Financial award application deadline: 3/1. *Faculty research:* Topology, algebra, applied mathematics, analysis. *Unit head:* Dr. Roger Waggoner, Head, 337-482-6702, Fax: 337-482-6587, E-mail: kje2027@louisiana.edu. *Application contact:* Dr. Keng Deng, Coordinator, 337-482-5297, Fax: 337-482-6587, E-mail: deng@louisiana.edu.

University of Louisville, Graduate School, College of Arts and Sciences, Department of Mathematics, Louisville, KY 40292-0001. Offers applied and industrial mathematics (PhD); mathematics (MA). Evening/weekend programs available. *Students:* 32 full-time (8 women), 8 part-time (4 women); includes 2 minority (1 African American, 1 Asian American or Pacific Islander), 9 international. Average age 28. In 2005, 12 degrees awarded. *Degree requirements:* For master's, thesis optional; for doctorate, thesis/dissertation, internship, project, comprehensive exam. *Entrance requirements:* For master's and doctorate, GRE General Test. *Application deadline:* Applications are processed on a rolling basis. Application fee: $50. *Expenses:* Tuition, state resident: full-time $3,003; part-time $334 per credit hour. Tuition, nonresident: full-time $8,277; part-time $920 per credit hour. Tuition and fees vary according to course load, degree level and program. *Financial support:* In 2005–06, 25 teaching assistantships with full tuition reimbursements were awarded *Unit head:* Dr. Thomas Riedel, Chair, 502-852-5974, Fax: 502-852-7132, E-mail: thomas.riedel@louisville.edu. *Application contact:* Dr. Prasana Sahoo, Graduate Studies Director, 502-852-6826, Fax: 502-852-7132, E-mail: sahoo@louisville.edu.

University of Maine, Graduate School, College of Liberal Arts and Sciences, Department of Mathematics and Statistics, Orono, ME 04469. Offers mathematics (MA). *Faculty:* 18. *Students:* 19 full-time (7 women), 1 (woman) part-time, 1 international. Average age 30. 11 applicants, 82% accepted, 6 enrolled. In 2005, 2 master's awarded. *Degree requirements:* For master's, thesis optional. *Entrance requirements:* For master's, GRE General Test. Additional exam requirements/recommendations for international students: Required—TOEFL. *Application deadline:* For fall admission, 2/1 for domestic students. Applications are processed on a rolling basis. Application fee: $50. Electronic applications accepted. *Financial support:* In 2005–06, 7 teaching assistantships with tuition reimbursements (averaging $9,416 per year) were awarded; research assistantships with tuition reimbursements, tuition waivers (full and partial) also available. Financial award application deadline: 3/1. *Unit head:* Dr. William Bray, Chair, 207-581-3901, Fax: 207-581-4977. *Application contact:* Scott G. Delcourt, Associate Dean of the Graduate School, 207-581-3219, Fax: 207-581-3232, E-mail: graduate@maine.edu.

University of Manitoba, Faculty of Graduate Studies, Faculty of Science, Department of Mathematical, Computational and Statistical Sciences, Winnipeg, MB R3T 2N2, Canada. Offers MMCSS.

University of Manitoba, Faculty of Graduate Studies, Faculty of Science, Department of Mathematics, Winnipeg, MB R3T 2N2, Canada. Offers M Sc, PhD. *Degree requirements:* For master's, one foreign language, thesis or alternative; for doctorate, one foreign language, thesis/dissertation.

University of Maryland, College Park, Graduate Studies, College of Computer, Mathematical and Physical Sciences, Department of Mathematics, Program in Mathematics, College Park, MD 20742. Offers MA, PhD. Part-time and evening/weekend programs available. *Students:* 93 full-time (20 women), 12 part-time (1 woman); includes 5 minority (2 African Americans, 2 Asian Americans or Pacific Islanders, 1 Hispanic American), 24 international. 228 applicants, 51% accepted, 24 enrolled. In 2005, 11 master's, 6 doctorates awarded. Terminal master's awarded for partial completion of doctoral program. *Median time to degree:* Of those who began their doctoral program in fall 1996, 52% received their degree in 8 years or less. *Degree requirements:* For master's, thesis or alternative; for doctorate, one foreign language, thesis/dissertation, written exam, oral exam. *Entrance requirements:* For master's, GRE General Test, GRE Subject Test, minimum GPA of 3.0, 3 letters of recommendation; for doctorate, GRE General Test, GRE Subject Test, 3 letters of recommendation. *Application deadline:* For fall admission, 5/1 for domestic students, 2/1 for international students; for spring admission, 10/1 for domestic students, 6/1 for international students. Applications are processed on a rolling basis. Application fee: $60. Electronic applications accepted. *Financial support:* In 2005–06, 19 fellowships (averaging $5,179 per year) were awarded; research assistantships, teaching assistantships Financial award applicants required to submit FAFSA. *Unit head:* Darcy Conant, Coordinator, 301-405-5058. *Application contact:* Dean of Graduate School, 301-405-4190, Fax: 301-314-9305.

University of Massachusetts Amherst, Graduate School, College of Natural Sciences and Mathematics, Department of Mathematics and Statistics, Program in Mathematics and Statistics, Amherst, MA 01003. Offers MS, PhD. *Students:* 50 full-time (13 women), 5 part-time (2 women); includes 4 minority (1 African American, 3 Asian Americans or Pacific Islanders), 19 international. Average age 27. 190 applicants, 21% accepted, 19 enrolled. In 2005, 16 master's, 6 doctorates awarded. *Degree requirements:* For doctorate, 2 foreign languages, thesis/dissertation. *Entrance requirements:* Additional exam requirements/recommendations for international students: Required—TOEFL (minimum score 530 paper-based; 197 computer-based). *Application deadline:* For fall admission, 2/1 priority date for domestic students, 2/1 priority date for international students; for spring admission, 10/1 for domestic students, 10/1 for international students. Applications are processed on a rolling basis. Application fee: $40 ($65 for international students). Electronic applications accepted. *Expenses:* Tuition, state resident: part-time $110 per credit. Tuition, nonresident: part-time $414 per credit. Required fees: $2,824 per term. One-time fee: $250 part-time. Full-time tuition and fees vary according to course load, campus/location, program and reciprocity agreements. *Financial support:* Fellowships with full tuition reimbursements, research assistantships with full tuition reimbursements, teaching assistantships with full tuition reimbursements, career-related internships or fieldwork, Federal Work-Study, scholarships/grants, traineeships, and unspecified assistantships available. Support available to part-time students. Financial award application deadline: 2/1. *Unit head:* Dr. Ivan Mirkovic, Director, 413-545-2282, Fax: 413-545-1801.

University of Massachusetts Lowell, Graduate School, College of Arts and Sciences, Department of Mathematics, Lowell, MA 01854-2881. Offers applied mathematics (MS); computational mathematics (PhD); mathematics (MS). Part-time programs available. *Entrance requirements:* For master's, GRE General Test.

University of Memphis, Graduate School, College of Arts and Sciences, Department of Mathematical Sciences, Memphis, TN 38152-3420. Offers applied mathematics (MS); applied statistics (PhD); bioinformatics (MS); computer science (PhD); computer sciences (MS); mathematics (MS, PhD); statistics (MS, PhD). Part-time programs available. *Faculty:* 24 full-time (5 women), 3 part-time/adjunct (0 women). *Students:* 105 full-time (38 women), 34 part-time (8 women); includes 8 minority (7 African Americans, 1 Asian American or Pacific Islander), 89 international. Average age 30. 139 applicants, 37% accepted. In 2005, 43 master's, 5 doctorates awarded. Terminal master's awarded for partial completion of doctoral program. *Degree requirements:* For master's, comprehensive exam; for doctorate, one foreign language, thesis/dissertation, oral exams. *Entrance requirements:* For master's and doctorate, GRE General Test, minimum GPA of 2.5. Additional exam requirements/recommendations for international students: Required—TOEFL (minimum score 550 paper-based; 210 computer-based), WES evaluation of transcript. *Application deadline:* For fall admission, 8/1 for domestic students, 5/1 for international students; for spring admission, 12/1 for domestic students, 9/1 for international students. Applications are processed on a rolling basis. Application fee: $25 ($50 for international students). Electronic applications accepted. *Financial support:* In 2005–06, 58 students received support, including fellowships with full tuition reimbursements available (averaging $17,500 per year), 9 research assistantships with full tuition reimbursements available (averaging $9,000 per year), 30 teaching assistantships with full tuition reimbursements available (averaging $9,000 per year); career-related internships or fieldwork, Federal Work-Study, scholarships/grants, unspecified assistantships, and minority scholarships also available. Financial award application deadline: 2/2. *Faculty research:* Combinatorics, ergodic theory, graph theory, Ramsey theory, applied statistics. Total annual research expenditures: $1.5 million. *Unit head:* Dr. James E. Jamison, Chairman, 901-678-2482, Fax: 901-678-2480, E-mail: jjamison@memphis.edu. *Application contact:* Coordinator of Graduate Studies, 901-678-2482, Fax: 901-678-2480, E-mail: dfwilson@memphis.edu.

University of Miami, Graduate School, College of Arts and Sciences, Department of Mathematics, Coral Gables, FL 33124. Offers MA, MS, DA, PhD. Part-time and evening/weekend programs available. *Faculty:* 25 full-time (2 women). *Students:* 24 full-time (5 women), 9 part-time (4 women); includes 12 minority (2 African Americans, 1 Asian American or Pacific Islander, 9 Hispanic Americans), 11 international. Average age 30. 53 applicants, 62% accepted, 8 enrolled. In 2005, 3 master's, 1 doctorate awarded. Terminal master's awarded for partial completion of doctoral program. *Degree requirements:* For master's, qualifying exams; for doctorate, one foreign language, thesis/dissertation, qualifying exams. *Entrance requirements:* For master's and doctorate, GRE General Test, minimum GPA of 3.0. Additional exam requirements/recommendations for international students: Required—TOEFL (minimum score 550 paper-based; 213 computer-based). *Application deadline:* For fall admission, 7/1 for domestic students, 7/1 for international students; for spring admission, 12/1 for domestic students, 12/1 for international students. Applications are processed on a rolling basis. Application fee: $50. Electronic applications accepted. *Financial support:* In 2005–06, 21 students received support, including fellowships with tuition reimbursements available (averaging $17,000 per year), 20 teaching assistantships with tuition reimbursements available (averaging $15,500

Mathematics

University of Miami (continued)

per year); career-related internships or fieldwork and institutionally sponsored loans also available. Support available to part-time students. Financial award application deadline: 1/15; financial award applicants required to submit FAFSA. *Faculty research:* Applied mathematics, probability, geometric analysis, differential equations, algebraic combinatorics. *Unit head:* Dr. Alan Zame, Chairman, 305-284-2348, Fax: 305-284-2848, E-mail: zame@math.miami.edu. *Application contact:* Dr. Marvin Mielke, Graduate Advisor, 305-284-2348, Fax: 305-284-2848, E-mail: m.mielke@math.miami.edu.

University of Michigan, Horace H. Rackham School of Graduate Studies, College of Literature, Science, and the Arts, Department of Mathematics, Ann Arbor, MI 48109. Offers applied and interdisciplinary mathematics (AM, MS, PhD); mathematics (AM, MS, PhD). Part-time programs available. *Faculty:* 58 full-time (6 women). *Students:* 132 full-time (32 women), 3 part-time; includes 16 minority (2 African Americans, 9 Asian Americans or Pacific Islanders, 5 Hispanic Americans), 58 international. Average age 26. 453 applicants, 21% accepted, 32 enrolled. In 2005, 27 master's, 19 doctorates awarded. *Median time to degree:* Of those who began their doctoral program in fall 1996, 92% received their degree in 8 years or less. *Degree requirements:* For doctorate, one foreign language, thesis/dissertation, oral defense of dissertation, preliminary exam, comprehensive exam, registration. *Entrance requirements:* For master's and doctorate, GRE General Test, GRE Subject Test. Additional exam requirements/recommendations for international students: Required—TOEFL (minimum score 580 paper-based; 220 computer-based). *Application deadline:* For fall admission, 1/22 for domestic students, 1/15 for international students. Applications are processed on a rolling basis. Application fee: $60 ($75 for international students). Electronic applications accepted. *Expenses:* Tuition, state resident: full-time $14,082; part-time $894 per credit hour. Tuition, nonresident: full-time $28,500; part-time $1,675 per credit hour. Required fees: $189; $189 per unit. *Financial support:* In 2005–06, 23 fellowships with full tuition reimbursements (averaging $22,000 per year), 7 research assistantships with full tuition reimbursements (averaging $14,326 per year), 94 teaching assistantships with full tuition reimbursements (averaging $14,326 per year) were awarded. Financial award application deadline: 3/15. *Faculty research:* Algebra, analysis, topology, applied mathematics, geometry. *Unit head:* Prof. Anthony Bloch, Chair, 734-936-1310, Fax: 734-763-0937, E-mail: math-chair@umich.edu. *Application contact:* Prof. Juha Heinonen, Associate Chairman for Graduate Studies, 734-764-7436, Fax: 734-763-0937, E-mail: math.acgs@umich.edu.

University of Minnesota, Twin Cities Campus, Graduate School, Institute of Technology, School of Mathematics, Minneapolis, MN 55455-0213. Offers MS, PhD. Part-time programs available. Terminal master's awarded for partial completion of doctoral program. *Degree requirements:* For master's, thesis (for some programs); for doctorate, 2 foreign languages, thesis/dissertation. *Entrance requirements:* For master's, GRE Subject Test (recommended); for doctorate, GRE Subject Test. Additional exam requirements/recommendations for international students: Required—TOEFL. *Expenses:* Tuition, state resident: full-time $8,748; part-time $729 per credit. Tuition, nonresident: full-time $15,848; part-time $1,321 per credit. Full-time tuition and fees vary according to class time, course load, program and reciprocity agreements. *Faculty research:* Partial and ordinary differential equations, algebra and number theory, geometry, combinatorics, numerical analysis.

University of Mississippi, Graduate School, College of Liberal Arts, Department of Mathematics, Oxford, University, MS 38677. Offers MA, MS, PhD. *Faculty:* 24 full-time (7 women), 4 part-time/adjunct (3 women). *Students:* 24 full-time (12 women), 7 part-time (3 women); includes 9 minority (all African Americans), 6 international. 18 applicants, 67% accepted, 6 enrolled. In 2005, 7 master's, 4 doctorates awarded. *Degree requirements:* For master's, thesis (for some programs); for doctorate, thesis/dissertation. *Entrance requirements:* For master's, GRE General Test, minimum GPA of 3.0; for doctorate, GRE General Test. Additional exam requirements/recommendations for international students: Required—TOEFL. *Application deadline:* For fall admission, 4/1 for domestic students; for spring admission, 10/1 for domestic students. Applications are processed on a rolling basis. Application fee: $25. Electronic applications accepted. *Expenses:* Tuition, state resident: full-time $4,320; part-time $240 per credit hour. Tuition, nonresident: full-time $9,744; part-time $301 per credit hour. Tuition and fees vary according to program. *Financial support:* Scholarships/grants available. Financial award application deadline: 3/1; financial award applicants required to submit FAFSA. *Unit head:* Dr. Tristan Denley, Chairman, 662-915-7071, Fax: 662-915-5491, E-mail: tdenley@olemiss.edu.

University of Missouri–Columbia, Graduate School, College of Arts and Sciences, Department of Mathematics, Columbia, MO 65211. Offers applied mathematics (MS); mathematics (MA, MST, PhD). *Faculty:* 43 full-time (6 women). *Students:* 50 full-time (18 women), 20 part-time (8 women); includes 2 minority (1 African American, 1 Asian American or Pacific Islander), 29 international. In 2005, 11 master's, 6 doctorates awarded. *Degree requirements:* For doctorate, 2 foreign languages, thesis/dissertation. *Entrance requirements:* For master's and doctorate, GRE General Test, minimum GPA of 3.0. *Application deadline:* Applications are processed on a rolling basis. Application fee: $45 ($60 for international students). *Financial support:* Fellowships, research assistantships, teaching assistantships, institutionally sponsored loans available. *Unit head:* Dr. Jan Segert, Director of Graduate Studies, 573-882-6953, E-mail: segertj@missouri.edu.

University of Missouri–Kansas City, College of Arts and Sciences, Department of Mathematics and Statistics, Kansas City, MO 64110-2499. Offers MA, MS, PhD. PhD offered through the School of Graduate Studies. Part-time programs available. *Faculty:* 9 full-time (3 women), 16 part-time/adjunct (6 women). *Students:* 2 full-time (1 woman), 11 part-time (7 women); includes 2 minority (1 African American, 1 Asian American or Pacific Islander). Average age 33. 1 applicant, 0% accepted.Terminal master's awarded for partial completion of doctoral program. *Degree requirements:* For master's, written exam; for doctorate, 2 foreign languages, thesis/dissertation, oral and written exams. *Entrance requirements:* For master's, bachelor's degree in mathematics, minimum GPA of 3.0; for doctorate, GMAT or GRE General Test. Additional exam requirements/recommendations for international students: Required—TOEFL. *Application deadline:* For fall admission, 5/1 priority date for domestic students, 4/30 priority date for international students. Applications are processed on a rolling basis. Application fee: $35 ($50 for international students). Electronic applications accepted. *Expenses:* Tuition, state resident: full-time $4,738; part-time $263 per credit hour. Tuition, nonresident: full-time $12,235; part-time $679 per credit hour. Required fees: $582. Tuition and fees vary according to course load, program and student level. *Financial support:* In 2005–06, 2 research assistantships with partial tuition reimbursements (averaging $5,000 per year), 6 teaching assistantships with full tuition reimbursements (averaging $11,000 per year) were awarded; Federal Work-Study, institutionally sponsored loans, and tuition waivers (full and partial) also available. Support available to part-time students. Financial award application deadline: 4/1. *Faculty research:* Numerical analysis, statistics, biostatistics commutative algebra, differential equations. Total annual research expenditures:$90,548. *Unit head:* Kamel Rekab, Chair, 816-235-5719, E-mail: rekabk@umka.edu. *Application contact:* Jie Chen, Associate Professor, 816-235-2894, Fax: 816-235-5517, E-mail: chenj@umkc.edu.

University of Missouri–Rolla, Graduate School, College of Arts and Sciences, Department of Mathematics and Statistics, Program in Mathematics, Rolla, MO 65409-0910. Offers mathematics (PhD); mathematics education (MST). *Degree requirements:* For master's, thesis or alternative; for doctorate, one foreign language, thesis/dissertation. *Entrance requirements:* For master's and doctorate, GRE General Test. Electronic applications accepted. *Faculty research:* Analysis, differential equations, topology, statistics.

University of Missouri–St. Louis, College of Arts and Sciences, Department of Mathematics and Computer Science, St. Louis, MO 63121. Offers applied mathematics (MA, PhD); computer science (MS); telecommunications science (Certificate). Part-time and evening/weekend programs available. *Faculty:* 14. *Students:* 28 full-time (11 women), 73 part-time (22 women); includes 19 minority (3 African Americans, 15 Asian Americans or Pacific Islanders, 1 Hispanic American), 23 international. Average age 33. In 2005, 21 master's, 1 doctorate awarded. *Degree requirements:* For master's, thesis optional; for doctorate, thesis/dissertation. *Entrance requirements:* For master's, 2 letters of recommendation; for doctorate, GRE General Test, GRE Subject Test, 3 letters of recommendation. Additional exam requirements/recommendations for international students: Required—TOEFL (minimum score 550 paper-based; 213 computer-based). *Application deadline:* For fall admission, 7/1 for domestic students; for spring admission, 12/1 for domestic students. Applications are processed on a rolling basis. Application fee: $35 ($40 for international students). Electronic applications accepted. *Expenses:* Tuition, state resident: part-time $263 per credit hour. Tuition, nonresident: part-time $680 per credit hour. Required fees: $53 per credit hour. Tuition and fees vary according to program. *Financial support:* In 2005–06, 8 teaching assistantships with full and partial tuition reimbursements (averaging $12,000 per year) were awarded; fellowships with full tuition reimbursements, research assistantships with full tuition reimbursements *Faculty research:* Statistics, algebra, analysis. *Unit head:* Dr. Shiying Zhao, Director of Graduate Studies, 314-516-5741, Fax: 314-516-5400, E-mail: Zhao@arch.cs.umsl.edu. *Application contact:* 314-516-5458, Fax: 314-516-5310, E-mail: gradadm@umsl.edu.

The University of Montana, Graduate School, College of Arts and Sciences, Department of Mathematical Sciences, Missoula, MT 59812-0002. Offers mathematics (MA, PhD), including college teaching (PhD), traditional mathematics research (PhD); mathematics education (MA). Part-time programs available. *Faculty:* 20 full-time (3 women). *Students:* 18 full-time (11 women), 4 part-time (2 women), 4 international. Average age 28. 9 applicants, 67% accepted, 6 enrolled. In 2005, 6 master's, 2 doctorates awarded. Terminal master's awarded for partial completion of doctoral program. *Degree requirements:* For doctorate, thesis/dissertation. *Entrance requirements:* For master's and doctorate, GRE General Test. Additional exam requirements/recommendations for international students: Required—TOEFL (minimum score 525 paper-based; 195 computer-based). *Application deadline:* For fall admission, 2/1 for domestic students. Application fee: $45. *Expenses:* Tuition, state resident: part-time $267 per credit. Tuition, nonresident: part-time $665 per credit. Part-time tuition and fees vary according to course load and degree level. *Financial support:* In 2005–06, 15 teaching assistantships with full tuition reimbursements were awarded; Federal Work-Study and unspecified assistantships also available. Financial award application deadline: 3/1; financial award applicants required to submit FAFSA. Total annual research expenditures: $716,087. *Unit head:* Dr. James Hirstein, Chair, 406-243-5311.

University of Nebraska at Omaha, Graduate Studies and Research, College of Arts and Sciences, Department of Mathematics, Omaha, NE 68182. Offers MA, MAT, MS. Part-time programs available. *Faculty:* 15 full-time (2 women). *Students:* 7 full-time (1 woman), 31 part-time (15 women); includes 1 minority (Hispanic American), 5 international. Average age 31. 18 applicants, 78% accepted, 9 enrolled. In 2005, 10 degrees awarded. *Degree requirements:* For master's, thesis (for some programs), comprehensive exam. *Entrance requirements:* For master's, minimum GPA of 3.0. Additional exam requirements/recommendations for international students: Required—TOEFL (minimum score 500 paper-based; 173 computer-based). *Application deadline:* For fall admission, 7/1 for domestic students; for spring admission, 12/1 priority date for domestic students. Applications are processed on a rolling basis. Application fee: $45. Electronic applications accepted. *Expenses:* Tuition, state resident: part-time $172 per credit. Tuition, nonresident: part-time $452 per credit. Part-time tuition and fees vary according to campus/location. *Financial support:* In 2005–06, 16 students received support; research assistantships with tuition reimbursements available, teaching assistantships with tuition reimbursements available, Federal Work-Study, institutionally sponsored loans, traineeships, tuition waivers (partial), and unspecified assistantships available. Support available to part-time students. Financial award application deadline: 3/1; financial award applicants required to submit FAFSA. *Unit head:* Dr. Jack W. Heidel, Chairperson, 402-554-3430.

University of Nebraska–Lincoln, Graduate College, College of Arts and Sciences, Department of Mathematics and Statistics, Lincoln, NE 68588. Offers M Sc T, MA, MAT, MS, PhD. *Degree requirements:* For master's, thesis optional; for doctorate, variable foreign language requirement, thesis/dissertation, comprehensive exam. *Entrance requirements:* Additional exam requirements/recommendations for international students: Required—TOEFL (minimum score 550 paper-based; 213 computer-based), GRE General Test (international applicants). Electronic applications accepted. *Faculty research:* Applied mathematics, commutative algebra, algebraic geometry, Bayesian statistics, biostatistics.

University of Nevada, Las Vegas, Graduate College, College of Science, Department of Mathematical Sciences, Las Vegas, NV 89154-9900. Offers applied mathematics (MS, PhD); applied statistics (MS); computational mathematics (PhD); pure mathematics (MS, PhD); statistics (PhD); teaching mathematics (MS). Part-time programs available. *Faculty:* 29 full-time (6 women), 1 part-time/adjunct (0 women). *Students:* 29 full-time (13 women), 29 part-time (9 women); includes 10 minority (1 African American, 6 Asian Americans or Pacific Islanders, 3 Hispanic Americans), 22 international. 37 applicants, 57% accepted, 11 enrolled. In 2005, 18 degrees awarded. *Degree requirements:* For master's, thesis (for some programs), oral exam, comprehensive exam (for some programs). *Entrance requirements:* For master's, minimum GPA of 3.0 during previous 2 years, 2.75 overall. Additional exam requirements/recommendations for international students: Required—TOEFL (minimum score 550 paper-based; 213 computer-based). *Application deadline:* For fall admission, 6/15 for domestic students, 5/1 for international students; for spring admission, 11/15 for domestic students, 10/1 for international students. Application fee: $60 ($75 for international students). Electronic applications accepted. *Expenses:* Tuition, state resident: part-time $150 per credit. Tuition, nonresident: part-time $315 per credit. Tuition and fees vary according to course load, program and reciprocity agreements. *Financial support:* In 2005–06, 42 teaching assistantships with partial tuition reimbursements (averaging $10,500 per year) were awarded; career-related internships or fieldwork, Federal Work-Study, institutionally sponsored loans, scholarships/grants, health care benefits, and unspecified assistantships also available. Support available to part-time students. Financial award application deadline: 3/1. *Unit head:* Dr. Malwane Ananda, Chair, 702-895-3567. *Application contact:* Graduate College Admissions Evaluator, 702-895-3320, Fax: 702-895-4180, E-mail: gradcollege@unlv.edu.

University of Nevada, Reno, Graduate School, College of Science, Department of Mathematics, Reno, NV 89557. Offers mathematics (MS); teaching mathematics (MATM). *Degree requirements:* For master's, thesis optional. *Entrance requirements:* For master's, GRE General Test, minimum GPA of 2.75. Additional exam requirements/recommendations for international students: Required—TOEFL. *Faculty research:* Operator algebra, nonlinear systems, differential equations.

University of New Brunswick Fredericton, School of Graduate Studies, Faculty of Science, Department of Mathematics and Statistics, Fredericton, NB E3B 5A3, Canada. Offers M Sc, PhD. Part-time programs available. *Degree requirements:* For master's, thesis or alternative; for doctorate, thesis/dissertation. *Entrance requirements:* For master's and doctorate, minimum GPA of 3.0. Additional exam requirements/recommendations for international students: Required—TOEFL, TWE.

University of New Hampshire, Graduate School, College of Engineering and Physical Sciences, Department of Mathematics and Statistics, Durham, NH 03824. Offers applied mathematics (MS); mathematics (MS, MST, PhD); mathematics education (PhD); statistics (MS). *Faculty:* 26 full-time. *Students:* 17 full-time (7 women), 24 part-time (13 women), 21 international. Average age 28. 37 applicants, 78% accepted, 5 enrolled. In 2005, 17 master's, 6 doctorates awarded. Terminal master's awarded for partial completion of doctoral program. *Degree requirements:* For doctorate, 2 foreign languages, thesis/dissertation. *Entrance requirements:* Additional exam requirements/recommendations for international students: Required—TOEFL (minimum score 550 paper-based; 213 computer-based); Recommended—TSE. *Application deadline:* For fall admission, 4/1 priority date for domestic students, 4/1 priority date for international students. For winter admission, 12/1 for domestic students. Applications are processed on a rolling basis. Application fee: $60. Electronic applications accepted. *Expenses:*

Tuition, state resident: full-time $8,010; part-time $445 per credit hour. Tuition, nonresident: full-time $19,730; part-time $810 per credit hour. Required fees: $322 per semester. Tuition and fees vary according to course load and program. *Financial support:* In 2005–06, 1 fellowship, 2 research assistantships, 23 teaching assistantships were awarded; Federal Work-Study, scholarships/grants, and tuition waivers (full and partial) also available. Support available to part-time students. Financial award application deadline: 2/15. *Faculty research:* Operator theory, complex analysis, algebra, nonlinear dynamics, statistics. *Unit head:* Dr. Eric Grinberg, Chairperson, 603-862-5772. *Application contact:* Jan Jankowski, Administrative Assistant, 603-862-2320, E-mail: jan.jankowski@unh.edu.

University of New Mexico, Graduate School, College of Arts and Sciences, Department of Mathematics and Statistics, Albuquerque, NM 87131-2039. Offers mathematics (MS, PhD); statistics (MS, PhD). *Faculty:* 46 full-time (11 women), 21 part-time/adjunct (7 women). *Students:* 89 full-time (37 women), 31 part-time (13 women); includes 20 minority (6 Asian Americans or Pacific Islanders, 14 Hispanic Americans), 48 international. Average age 33. 80 applicants, 59% accepted, 31 enrolled. In 2005, 11 master's, 4 doctorates awarded. *Degree requirements:* For master's, thesis or alternative, comprehensive exam (for some programs); for doctorate, one foreign language, thesis/dissertation, 4 department seminars, comprehensive exam. *Entrance requirements:* For master's, minimum GPA of 3.0, 3 letters of recommendation; for doctorate, GRE General Test, minimum GPA of 3.0, 3 letters of recommendation. Additional exam requirements/recommendations for international students: Required—TOEFL (minimum score 550 paper-based; 213 computer-based). *Application deadline:* For fall admission, 7/1 for domestic students; for spring admission, 11/1 for domestic students. Application fee: $40. Electronic applications accepted. *Expenses:* Tuition, nonresident: full-time $3,388; part-time $238 per credit hour. Required fees: $385 per term. Tuition and fees vary according to course load and program. *Financial support:* In 2005–06, 20 students received support, including fellowships (averaging $4,000 per year), research assistantships with tuition reimbursements available (averaging $14,320 per year), teaching assistantships with tuition reimbursements available (averaging $14,320 per year); health care benefits and unspecified assistantships also available. Financial award application deadline: 3/1; financial award applicants required to submit FAFSA. *Faculty research:* Pure and applied mathematics, applied statistics, numerical analysis, biostatistics, differential geometry, fluid dynamics, nonparametric curve estimation. Total annual research expenditures: $1.2 million. *Unit head:* Dr. Alejandro Aceves, Chair, 505-277-4613, Fax: 505-277-5505, E-mail: aceves@math.unm.edu. *Application contact:* Donna George, Program Advisement Coordinator, 505-277-5250, Fax: 505-277-5505, E-mail: dgeorge@unm.edu.

University of New Orleans, Graduate School, College of Sciences, Department of Mathematics, New Orleans, LA 70148. Offers MS. Part-time programs available. *Entrance requirements:* For master's, BA or BS in mathematics. Additional exam requirements/recommendations for international students: Required—TOEFL (minimum score 550 paper-based; 213 computer-based). Electronic applications accepted. *Faculty research:* Differential equations, combinatorics, statistics, complex analysis, algebra.

The University of North Carolina at Chapel Hill, Graduate School, College of Arts and Sciences, Department of Mathematics, Chapel Hill, NC 27599. Offers MA, MS, PhD. *Degree requirements:* For master's, thesis or alternative, computer proficiency, comprehensive exam; for doctorate, 2 foreign languages, thesis/dissertation, 3 comprehensive exams, computer proficiency. *Entrance requirements:* For master's and doctorate, GRE General Test, minimum GPA of 3.0. Additional exam requirements/recommendations for international students: Required—TOEFL. Electronic applications accepted. *Faculty research:* Algebraic geometry, topology, analysis, lie theory, applied math.

Announcement: The department, with 34 faculty members and 55–60 graduate students, offers special opportunities for student-faculty interaction in master's and doctoral programs. Faculty includes distinguished, active researchers in most subfields of mathematics, including a strong group in applied mathematics. Nearby resources are North Carolina State, Duke, and the Research Triangle Park.

The University of North Carolina at Charlotte, Graduate School, College of Arts and Sciences, Department of Mathematics and Statistics, Charlotte, NC 28223-0001. Offers applied mathematics (PhD); mathematics (MS); mathematics education (MA). *Accreditation:* NCATE (one or more programs are accredited). Part-time and evening/weekend programs available. *Faculty:* 35 full-time (5 women), 7 part-time/adjunct (1 woman). *Students:* 18 full-time (9 women), 43 part-time (18 women); includes 11 minority (9 African Americans, 2 Asian Americans or Pacific Islanders), 29 international. Average age 31. 32 applicants, 97% accepted, 19 enrolled. In 2005, 11 master's, 7 doctorates awarded. *Degree requirements:* For master's, comprehensive exam; for doctorate, thesis/dissertation. *Entrance requirements:* For master's, GRE General Test or MAT, minimum GPA of 3.0 in undergraduate major, 2.75 overall. Additional exam requirements/recommendations for international students: Required—TOEFL (minimum score 557 paper-based; 220 computer-based). *Application deadline:* For fall admission, 7/1 for domestic students, 5/1 for international students; for spring admission, 11/1 for domestic students, 10/1 for international students. Applications are processed on a rolling basis. Application fee: $55. Electronic applications accepted. *Expenses:* Tuition, state resident: full-time $2,504; part-time $157 per credit. Tuition, nonresident: full-time $12,711; part-time $794 per credit. Required fees: $1,424; $89 per credit. Tuition and fees vary according to course load and program. *Financial support:* In 2005–06, 1 fellowship (averaging $2,000 per year), 1 research assistantship (averaging $6,667 per year), 51 teaching assistantships (averaging $10,921 per year) were awarded; career-related internships or fieldwork, Federal Work-Study, institutionally sponsored loans, scholarships/grants, and unspecified assistantships also available. Support available to part-time students. Financial award application deadline: 4/1; financial award applicants required to submit FAFSA. *Faculty research:* Numerical analysis, inverse problems, partial differential equations, applied probability. *Unit head:* Dr. Alan S. Dow, Chair, 704-687-2580, Fax: 704-687-0415, E-mail: adow@email.uncc.edu. *Application contact:* Kathy B. Giddings, Director of Graduate Admissions, 704-687-3366, Fax: 704-687-3279, E-mail: gradadm@email.uncc.edu.

See Close-Up on page 537.

The University of North Carolina at Greensboro, Graduate School, College of Arts and Sciences, Department of Mathematical Sciences, Greensboro, NC 27412-5001. Offers computer science (MS); mathematics (M Ed, MA). Part-time programs available. *Students:* 10 full-time, 26 part-time. *Degree requirements:* For master's, thesis (for some programs), comprehensive exam. *Entrance requirements:* For master's, GRE General Test. Additional exam requirements/recommendations for international students: Required—TOEFL. *Application deadline:* For fall admission, 6/15 for domestic students; for spring admission, 11/1 for domestic students. Applications are processed on a rolling basis. *Expenses:* Tuition, state resident: part-time $302 per credit hour. Tuition, nonresident: part-time $1,683 per credit hour. Required fees: $51 per credit hour. Tuition and fees vary according to course load and program. *Financial support:* Research assistantships with full tuition reimbursements, teaching assistantships with full tuition reimbursements, career-related internships or fieldwork, Federal Work-Study, scholarships/grants, traineeships, and unspecified assistantships available. Support available to part-time students. *Faculty research:* General and geometric topology, statistics, computer networks, symbolic logic, mathematics education. *Unit head:* Dr. Alex Chigogidze, Head, 336-334-5836, Fax: 336-334-5949, E-mail: chigogidze@uncg.edu. *Application contact:* Michelle Harkleroad, Director of Graduate Admissions, 336-334-4884, Fax: 336-334-4424, E-mail: mbharkle@uncg.edu.

The University of North Carolina Wilmington, College of Arts and Sciences, Department of Mathematical Sciences, Wilmington, NC 28403-3297. Offers MA, MS. *Faculty:* 25 full-time (3 women), 1 part-time/adjunct (0 women). *Students:* 6 full-time (3 women), 25 part-time (9 women); includes 1 minority (African American), 6 international. Average age 29. 28 applicants, 71% accepted, 15 enrolled. In 2005, 12 degrees awarded. *Degree requirements:* For master's, thesis, comprehensive exam. *Entrance requirements:* For master's, GRE General Test, GRE Subject Test, minimum B average in undergraduate major. *Application deadline:* For fall admission, 3/15 for domestic students. Applications are processed on a rolling basis. Application fee: $45. *Financial support:* In 2005–06, 12 teaching assistantships were awarded; career-related internships or fieldwork and Federal Work-Study also available. Support available to part-time students. Financial award application deadline: 3/15. *Unit head:* Dr. Wei Feng, Chair, 910-962-3291, Fax: 910-962-7107, E-mail: fengw@uncw.edu. *Application contact:* Dr. Robert D. Roer, Dean, Graduate School, 910-962-4117, Fax: 910-962-3787, E-mail: roer@uncw.edu.

University of North Dakota, Graduate School, College of Arts and Sciences, Department of Mathematics, Grand Forks, ND 58202. Offers M Ed, MS. Part-time programs available. *Faculty:* 15 full-time (2 women). *Students:* 2 full-time (0 women), 16 part-time (5 women). 19 applicants, 53% accepted, 9 enrolled. In 2005, 3 degrees awarded. *Degree requirements:* For master's, thesis or alternative, final exam. *Entrance requirements:* For master's, minimum GPA of 3.0. Additional exam requirements/recommendations for international students: Required—TOEFL (minimum score 550 paper-based; 213 computer-based). *Application deadline:* For fall admission, 2/15 priority date for domestic students, 2/15 priority date for international students; for spring admission, 10/15 priority date for domestic students, 10/15 priority date for international students. Applications are processed on a rolling basis. Application fee: $35. Electronic applications accepted. *Financial support:* In 2005–06, 9 teaching assistantships with full tuition reimbursements (averaging $11,570 per year) were awarded; fellowships, research assistantships, Federal Work-Study, institutionally sponsored loans, scholarships/grants, and tuition waivers (full and partial) also available. Support available to part-time students. Financial award application deadline: 3/15; financial award applicants required to submit FAFSA. *Faculty research:* Statistics, measure theory, topological vector spaces, algebras, applied math. *Unit head:* Dr. Richard P. Millspaush, Chairperson, 701-777-2881, Fax: 701-777-3619, E-mail: richard_millspaugh@und.nodak.edu. *Application contact:* Brenda Halle, Admissions Specialist, 701-777-2947, Fax: 701-777-3618, E-mail: brandahalle@mail.und.edu.

University of Northern British Columbia, Office of Graduate Studies, Prince George, BC V2N 4Z9, Canada. Offers business administration (Diploma); community health science (M Sc); disability management (MA); education (M Ed); first nations studies (MA); gender studies (MA); history (MA); interdisciplinary studies (MA); international studies (MA); mathematical, computer and physical sciences (M Sc); natural resources and environmental studies (M Sc, MA, MNRES, PhD); political science (MA); psychology (M Sc, PhD); social work (MSW). Part-time and evening/weekend programs available. Postbaccalaureate distance learning degree programs offered (no on-campus study). *Degree requirements:* For master's and doctorate, thesis/dissertation. *Entrance requirements:* For master's, GRE, minimum B average in undergraduate course work; for doctorate, candidacy exam, minimum A average in graduate course work.

University of Northern Colorado, Graduate School, College of Natural and Health Sciences, School of Mathematical Sciences, Greeley, CO 80639. Offers mathematical teaching (MA); mathematics education (PhD); mathematics: liberal arts (MA). Part-time programs available. *Faculty:* 15 full-time (4 women). *Students:* 14 full-time (8 women), 32 part-time (22 women); includes 3 minority (1 African American, 1 Asian American or Pacific Islander, 1 Hispanic American), 1 international. Average age 31. 14 applicants, 100% accepted, 5 enrolled. In 2005, 9 master's, 2 doctorates awarded. *Degree requirements:* For master's, thesis or alternative, comprehensive exam; for doctorate, thesis/dissertation, comprehensive exam. *Entrance requirements:* For master's, GRE General Test (liberal arts), 3 letters of recommendation; for doctorate, GRE General Test, 3 letters of recommendation. *Application deadline:* Applications are processed on a rolling basis. Application fee: $50 ($60 for international students). Electronic applications accepted. *Expenses:* Tuition, state resident: full-time $4,968; part-time $207 per credit hour. Tuition, nonresident: full-time $14,688; part-time $612 per credit hour. Required fees: $645; $32 per credit hour. *Financial support:* In 2005–06, 18 students received support, including 2 fellowships (averaging $1,675 per year), 2 research assistantships (averaging $14,350 per year), 8 teaching assistantships (averaging $13,723 per year); unspecified assistantships also available. Financial award application deadline: 3/1; financial award applicants required to submit FAFSA. *Unit head:* Dr. Jeff Farmer, Director, 970-351-2820, Fax: 970-351-2155.

University of Northern Iowa, Graduate College, College of Natural Sciences, Department of Mathematics, Cedar Falls, IA 50614. Offers mathematics (MA); mathematics for middle grades (MA). Part-time programs available. *Students:* 15 full-time (8 women), 20 part-time (10 women); includes 9 minority (8 African Americans, 1 Hispanic American), 2 international. 8 applicants, 100% accepted, 7 enrolled. In 2005, 9 degrees awarded. *Degree requirements:* For master's, thesis or alternative, comprehensive exam (for some programs). *Entrance requirements:* Additional exam requirements/recommendations for international students: Required—TOEFL (minimum score 500 paper-based; 180 computer-based). *Application deadline:* For fall admission, 8/1 for domestic students. Applications are processed on a rolling basis. Application fee: $30 ($50 for international students). Electronic applications accepted. *Expenses:* Tuition, state resident: full-time $5,708. Tuition, nonresident: full-time $13,532. Required fees: $712. *Financial support:* Career-related internships or fieldwork, Federal Work-Study, scholarships/grants, and tuition waivers (full and partial) available. Support available to part-time students. Financial award application deadline: 2/1. *Unit head:* Dr. Jerry Ridenhour, Interim Head, 319-273-2631, Fax: 319-273-2546, E-mail: jerry.ridenhour@uni.edu.

University of North Florida, College of Arts and Sciences, Department of Mathematics and Statistics, Jacksonville, FL 32224-2645. Offers mathematical sciences (MS); statistics (MS). Part-time and evening/weekend programs available. *Faculty:* 17 full-time (3 women). *Students:* 12 full-time (5 women), 4 part-time (1 woman); includes 2 minority (1 African American, 1 Asian American or Pacific Islander), 4 international. Average age 29. 24 applicants, 58% accepted, 7 enrolled. In 2005, 6 degrees awarded. *Degree requirements:* For master's, thesis optional. *Entrance requirements:* For master's, GRE General Test, minimum GPA of 3.0 in last 60 hours of course work. Additional exam requirements/recommendations for international students: Required—TOEFL (minimum score 500 paper-based; 173 computer-based). *Application deadline:* For fall admission, 7/1 priority date for domestic students, 5/1 priority date for international students; for spring admission, 11/1 priority date for domestic students, 10/1 priority date for international students. Applications are processed on a rolling basis. Application fee: $30. Electronic applications accepted. *Expenses:* Tuition, state resident: full-time $4,391; part-time $244 per semester hour. Tuition, nonresident: full-time $15,036; part-time $835 per semester hour. Required fees: $789; $44 per semester hour. *Financial support:* In 2005–06, 13 teaching assistantships (averaging $6,810 per year) were awarded; Federal Work-Study and tuition waivers (partial) also available. Support available to part-time students. Financial award application deadline: 4/1; financial award applicants required to submit FAFSA. *Faculty research:* Real analysis, number theory, Euclidean geometry. Total annual research expenditures: $128,824. *Unit head:* Dr. Scott H. Hochwald, Chair, 904-620-2653, E-mail: shochwal@unf.edu. *Application contact:* Dr. Champak Panchal, Graduate Coordinator, 904-620-2653, E-mail: cpanchal@unf.edu.

University of North Texas, Robert B. Toulouse School of Graduate Studies, College of Arts and Sciences, Department of Mathematics, Denton, TX 76203. Offers MA, MS, PhD. Part-time programs available. *Faculty:* 30 full-time (2 women). *Students:* 54 full-time (12 women), 20 part-time (8 women); includes 9 minority (1 African American, 3 American Indian/Alaska Native, 3 Asian Americans or Pacific Islanders, 2 Hispanic Americans), 21 international. Average age 31. 60 applicants, 55% accepted, 9 enrolled. In 2005, 10 master's, 1 doctorate awarded. Terminal master's awarded for partial completion of doctoral program. *Degree requirements:* For master's, one foreign language; for doctorate, 2 foreign languages, thesis/dissertation. *Entrance requirements:* For master's and doctorate, GRE General Test. Additional exam requirements/recommendations for international students: Recommended—TOEFL

Mathematics

University of North Texas (continued)

(minimum score 550 paper-based; 213 computer-based). *Application deadline:* For fall admission, 7/17 for domestic students. Application fee: $50 ($75 for international students). *Expenses:* Tuition, state resident: full-time $3,258; part-time $181 per semester hour. Tuition, nonresident: full-time $8,226; part-time $451 per semester hour. Required fees: $1,219; $68 per semester hour. *Financial support:* Research assistantships, teaching assistantships, Federal Work-Study and institutionally sponsored loans available. Financial award application deadline: 6/1. *Faculty research:* Differential equations, descriptive set theory, combinatorics, functional analysis, algebra. *Unit head:* Dr. Neal Brand, Chair, 940-565-2155, Fax: 940-565-4805, E-mail: neal@unt.edu. *Application contact:* Dr. Matt Douglass, Graduate Adviser, 940-565-2570, Fax: 940-565-4805, E-mail: douglass@unt.edu.

University of Notre Dame, Graduate School, College of Science, Department of Mathematics, Notre Dame, IN 46556. Offers algebra (PhD); algebraic geometry (PhD); applied mathematics (MSAM); complex analysis (PhD); differential geometry (PhD); logic (PhD); partial differential equations (PhD); topology (PhD). *Faculty:* 45 full-time (5 women). *Students:* 48 full-time (14 women); includes 2 minority (1 American Indian/Alaska Native, 1 Asian American or Pacific Islander), 19 international. 117 applicants, 14% accepted, 9 enrolled. In 2005, 6 master's, 7 doctorates awarded. Terminal master's awarded for partial completion of doctoral program. *Median time to degree:* Of those who began their doctoral program in fall 1996, 73% received their degree in 8 years or less. *Degree requirements:* For doctorate, one foreign language, thesis/dissertation, qualifying exam. *Entrance requirements:* For master's and doctorate, GRE General Test, GRE Subject Test. Additional exam requirements/recommendations for international students: Required—TOEFL. *Application deadline:* For fall admission, 2/1 for domestic students. Applications are processed on a rolling basis. Application fee: $50. Electronic applications accepted. *Financial support:* In 2005–06, 47 students received support, including 9 fellowships with full tuition reimbursements available (averaging $22,000 per year), 8 research assistantships with full tuition reimbursements available (averaging $15,250 per year), 26 teaching assistantships with full tuition reimbursements available (averaging $16,000 per year); tuition waivers (full) also available. Financial award application deadline: 2/1. *Faculty research:* Algebra, analysis, geometry/topology, logic, applied math. Total annual research expenditures: $1.5 million. *Unit head:* Dr. Julia Knight, Director of Graduate Studies, 574-631-7484, E-mail: mathgrad@nd.edu. *Application contact:* Dr. Terrence J. Akai, Director of Graduate Admissions, 574-631-7706, Fax: 574-631-4183, E-mail: gradad@nd.edu.

See Close-Up on page 539.

University of Oklahoma, Graduate College, College of Arts and Sciences, Department of Mathematics, Norman, OK 73019-0390. Offers MA, MS, PhD, MBA/MS. Part-time programs available. *Faculty:* 31 full-time (3 women). *Students:* 67 full-time (25 women), 6 part-time (4 women); includes 6 minority (2 African Americans, 2 American Indian/Alaska Native, 2 Asian Americans or Pacific Islanders), 29 international. 22 applicants, 95% accepted, 16 enrolled. In 2005, 10 master's, 3 doctorates awarded. Terminal master's awarded for partial completion of doctoral program. *Degree requirements:* For master's, thesis optional; for doctorate, 2 foreign languages, thesis/dissertation, qualifying exam. *Entrance requirements:* Additional exam requirements/recommendations for international students: Required—TOEFL (minimum score 550 paper-based; 213 computer-based), TSE. *Application deadline:* For fall admission, 6/1 priority date for domestic students, 4/1 priority date for international students; for spring admission, 11/1 for domestic students, 9/1 for international students. Applications are processed on a rolling basis. Application fee: $40 ($90 for international students). *Expenses:* Tuition, state resident: full-time $3,029; part-time $126 per credit hour. Tuition, nonresident: full-time $10,807; part-time $450 per credit hour. Required fees: $1,231; $44 per credit hour. Tuition and fees vary according to course load and program. *Financial support:* In 2005–06, 8 students received support, including 15 fellowships with full tuition reimbursements available (averaging $3,300 per year), 68 teaching assistantships with partial tuition reimbursements available (averaging $12,268 per year); research assistantships, scholarships/grants and unspecified assistantships also available. Financial award applicants required to submit FAFSA. *Faculty research:* Topology, geometry, algebra, analysis, mathematics education. Total annual research expenditures: $388,597. *Unit head:* Paul Goodey, Chair, 405-325-6711, Fax: 405-325-7484, E-mail: pgoodey@ou.edu. *Application contact:* Dr. Murad Ozaydin, Director of Graduate Studies, 405-325-6711, Fax: 405-325-7484, E-mail: mozaydin@ou.edu.

Announcement: The Graduate Mathematics Program at the University of Oklahoma offers students a supportive environment and the opportunity for individual interaction with faculty members involved in broadly diversified research programs. Flexible degree programs allow students to concentrate in pure mathematics, applied mathematics, or research in undergraduate mathematics curriculum and pedagogy. WWW: http:/www.math.ou.edu/.

University of Oregon, Graduate School, College of Arts and Sciences, Department of Mathematics, Eugene, OR 97403. Offers MA, MS, PhD. Part-time programs available. *Faculty:* 23 full-time (3 women), 5 part-time/adjunct (1 woman). *Students:* 51 full-time (8 women); includes 1 minority (Asian American or Pacific Islander), 8 international. 35 applicants, 31% accepted. In 2005, 9 master's, 5 doctorates awarded. Terminal master's awarded for partial completion of doctoral program. *Degree requirements:* For doctorate, 2 foreign languages, thesis/dissertation. *Entrance requirements:* For master's and doctorate, GRE General Test, GRE Subject Test. Additional exam requirements/recommendations for international students: Required—TOEFL, TSE. *Application deadline:* For fall admission, 3/1 for domestic students. Application fee: $50. *Financial support:* In 2005–06, 38 teaching assistantships were awarded; Federal Work-Study also available. Support available to part-time students. Financial award application deadline: 3/1. *Faculty research:* Algebra, topology, analytic geometry, numerical analysis, statistics. *Unit head:* Brad Shelton, Head, 541-346-4705. *Application contact:* Judy Perkins, Admissions Contact, 541-346-0988, E-mail: jperkins@math.uoregon.edu.

University of Ottawa, Faculty of Graduate and Postdoctoral Studies, Faculty of Science, Ottawa-Carleton Institute of Mathematics and Statistics, Ottawa, ON K1N 6N5, Canada. Offers M Sc, PhD. Part-time programs available. *Faculty:* 35 full-time, 4 part-time/adjunct. *Students:* 55 full-time, 8 part-time. 80 applicants, 49% accepted, 16 enrolled. In 2005, 9 master's, 2 doctorates awarded. *Degree requirements:* For master's, thesis optional; for doctorate, one foreign language, thesis/dissertation, comprehensive exam. *Entrance requirements:* For master's, honors B Sc degree or equivalent, minimum B average; for doctorate, M Sc, minimum B+ average. *Application deadline:* For fall admission, 3/1 priority date for domestic students, 2/15 priority date for international students. For winter admission, 11/15 for domestic students; for spring admission, 4/1 for domestic students. Applications are processed on a rolling basis. Application fee: $75. Electronic applications accepted. *Expenses:* Tuition: Part-time $260 per credit. Tuition and fees vary according to course load and program. *Financial support:* Fellowships, research assistantships with full tuition reimbursements, teaching assistantships with full tuition reimbursements, career-related internships or fieldwork, Federal Work-Study, scholarships/grants, traineeships, tuition waivers (full and partial), and unspecified assistantships available. Financial award application deadline: 2/15. *Faculty research:* Pure mathematics, applied mathematics, probability and statistics. *Unit head:* Dr. Vladimir Pestov, Director, 613-562-5800 Ext. 3523, Fax: 613-562-5776, E-mail: grad-director@mathstat.uottawa.ca. *Application contact:* Lise Maisonneuve, Graduate Studies Administrator, 613-562-5800 Ext. 6335, Fax: 613-562-5486, E-mail: lise@science.uottawa.ca.

University of Pennsylvania, School of Arts and Sciences, Graduate Group in Mathematics, Philadelphia, PA 19104. Offers AM, PhD. Terminal master's awarded for partial completion of doctoral program. *Degree requirements:* For master's, one foreign language, thesis or alternative; for doctorate, 2 foreign languages, thesis/dissertation. *Entrance requirements:* For master's and doctorate, GRE General Test, GRE Subject Test. Additional exam requirements/recommendations for international students: Required—TOEFL. Electronic applications accepted. *Faculty research:* Geometry-topology, analysis, algebra, logic, combinatorics.

University of Pittsburgh, School of Arts and Sciences, Department of Mathematics, Pittsburgh, PA 15260. Offers applied mathematics (MA, MS); financial mathematics (PMS); mathematics (MA, MS, PhD). Part-time programs available. *Faculty:* 35 full-time (4 women), 4 part-time/adjunct (1 woman). *Students:* 82 full-time (25 women), 12 part-time (5 women); includes 1 minority (Asian American or Pacific Islander), 43 international. 123 applicants, 67% accepted, 28 enrolled. In 2005, 5 master's, 3 doctorates awarded. Terminal master's awarded for partial completion of doctoral program. *Median time to degree:* Of those who began their doctoral program in fall 1996, 100% received their degree in 8 years or less. *Degree requirements:* For master's, thesis (for some programs), comprehensive exam; for doctorate, thesis/dissertation, preliminary exams, comprehensive exam. *Entrance requirements:* For master's and doctorate, GRE General Test, GRE Subject Test (recommended), minimum GPA of 3.0. Additional exam requirements/recommendations for international students: Required—TOEFL (minimum score 550 paper-based; 213 computer-based). *Application deadline:* For fall admission, 1/15 priority date for domestic students, 1/2 priority date for international students; for spring admission, 9/1 priority date for domestic students, 9/1 priority date for international students. Applications are processed on a rolling basis. Application fee: $50. Electronic applications accepted. *Expenses:* Tuition, state resident: full-time $13,194; part-time $537 per credit. Tuition, nonresident: full-time $25,012; part-time $1,026 per credit. Required fees: $700; $164 per term. Tuition and fees vary according to campus/location and program. *Financial support:* In 2005–06, 6 fellowships with full and partial tuition reimbursements (averaging $16,500 per year), 13 research assistantships with full and partial tuition reimbursements (averaging $13,600 per year), 50 teaching assistantships with full and partial tuition reimbursements (averaging $13,555 per year) were awarded; career-related internships or fieldwork, Federal Work-Study, institutionally sponsored loans, scholarships/grants, health care benefits, tuition waivers (partial), and unspecified assistantships also available. Financial award application deadline:1/15. *Faculty research:* Computational math, math biology, math finance, algebra, analysis. Total annual research expenditures: $700,000. *Unit head:* Juan Manfredi, Chairman, 412-624-8307, Fax: 412-624-8697, E-mail: manfredi@pitt.edu. *Application contact:* Molly Williams, Administrator, 412-624-1175, Fax: 412-624-8397, E-mail: mollyw@pitt.edu.

University of Puerto Rico, Mayagüez Campus, Graduate Studies, College of Arts and Sciences, Department of Mathematics, Mayagüez, PR 00681-9000. Offers applied mathematics (MS); computational sciences (MS); pure mathematics (MS); statistics (MS). Part-time programs available. *Faculty:* 34. *Students:* 21 full-time (9 women), 9 part-time (4 women); includes 6 minority (all Hispanic Americans), 24 international. 49 applicants, 61% accepted, 2 enrolled. In 2005, 6 degrees awarded. *Degree requirements:* For master's, one foreign language, comprehensive exam. *Application deadline:* For fall admission, 2/15 for domestic students; for spring admission, 9/15 for domestic students. Applications are processed on a rolling basis. Application fee: $20. *Expenses:* Tuition, state resident: full-time $900; part-time $100 per credit. International tuition: $4,655 full-time. Part-time tuition and fees vary according to course level and course load. *Financial support:* In 2005–06, fellowships (averaging $1,500 per year), research assistantships (averaging $1,200 per year), teaching assistantships (averaging $987 per year) were awarded; Federal Work-Study and institutionally sponsored loans also available. *Faculty research:* Automata theory, linear algebra, logic. Total annual research expenditures: $13,829. *Unit head:* Dr. Luis A. Caceres, Director, 787-832-4040 Ext. 3848.

University of Puerto Rico, Río Piedras, College of Natural Sciences, Department of Mathematics, San Juan, PR 00931-3300. Offers MS, PhD. Part-time and evening/weekend programs available. *Students:* 41 full-time (15 women), 8 part-time (5 women); includes 47 minority (5 Asian Americans or Pacific Islanders, 42 Hispanic Americans). Average age 34. In 2005, 2 degrees awarded. *Degree requirements:* For master's, one foreign language, thesis, comprehensive exam. *Entrance requirements:* For master's, GRE, EXADEP, interview, minimum GPA of 3.0, letter of recommendation. *Application deadline:* For fall admission, 2/1 for domestic students, 2/1 for international students. Application fee: $17. *Expenses:* Tuition, state resident: part-time $100 per credit. Tuition, nonresident: part-time $294 per credit. Required fees: $72 per term. *Financial support:* Fellowships, research assistantships, teaching assistantships, Federal Work-Study, institutionally sponsored loans, and tuition waivers (partial) available. Financial award application deadline: 5/31. *Faculty research:* Investigation of database logistics, cryptograph systems, distribution and spectral theory, Boolean function, differential equations. *Unit head:* Dr. Jorge Punchín, Coordinator, 787-764-0000 Ext. 4676, Fax: 787-281-0651.

University of Regina, Faculty of Graduate Studies and Research, Faculty of Science, Department of Mathematics and Statistics, Regina, SK S4S 0A2, Canada. Offers mathematics (M Sc, MA, PhD); statistics (M Sc, MA). *Faculty:* 22 full-time (4 women), 3 part-time/adjunct (0 women). *Students:* 13 full-time (3 women), 6 part-time (2 women). 12 applicants, 25% accepted. In 2005, 1 doctorate awarded. *Degree requirements:* For master's, registration; for doctorate, thesis/dissertation, comprehensive exam, registration. *Entrance requirements:* Additional exam requirements/recommendations for international students: Required—TOEFL (minimum score 580 paper-based; 237 computer-based). *Application deadline:* Applications are processed on a rolling basis. Application fee: $60 ($100 for international students). *Financial support:* In 2005–06, 2 fellowships (averaging $14,886 per year), 1 research assistantship (averaging $12,750 per year), 2 teaching assistantships (averaging $13,501 per year) were awarded; scholarships/grants also available. Financial award application deadline: 6/15. *Faculty research:* Pure and applied mathematics, statistics and probability. *Unit head:* Dr. Stephen Kirkland, Head, 306-585-4148, E-mail: kirkland@math.uregina.ca. *Application contact:* Dr. Doug Farenick, Graduate Program Coordinator, 306-585-4425, Fax: 306-585-4020, E-mail: farenick@math.uregina.ca.

University of Rhode Island, Graduate School, College of Arts and Sciences, Department of Mathematics, Kingston, RI 02881. Offers MS, PhD. In 2005, 2 master's, 2 doctorates awarded. *Degree requirements:* For master's, thesis optional; for doctorate, one foreign language, thesis/dissertation. *Application deadline:* For fall admission, 4/15 for domestic students. Applications are processed on a rolling basis. Application fee: $35. *Expenses:* Tuition, state resident: full-time $5,522; part-time $307 per credit. Tuition, nonresident: full-time $15,992; part-time $888 per credit. Required fees: $1,786; $73 per credit. One-time fee: $80 part-time. *Unit head:* Dr. Lewis Pakula, Chairman, 401-874-4519.

University of Rochester, The College, Arts and Sciences, Department of Mathematics, Rochester, NY 14627-0250. Offers MA, MS, PhD. Terminal master's awarded for partial completion of doctoral program. *Degree requirements:* For master's, thesis (for some programs); for doctorate, thesis/dissertation, qualifying exam. *Entrance requirements:* For master's and doctorate, GRE General Test. Additional exam requirements/recommendations for international students: Required—TOEFL.

University of Saskatchewan, College of Graduate Studies and Research, College of Arts and Sciences, Department of Mathematics and Statistics, Saskatoon, SK S7N 5A2, Canada. Offers M Math, MA, PhD. *Degree requirements:* For master's, thesis (for some programs), registration; for doctorate, thesis/dissertation, registration. *Entrance requirements:* Additional exam requirements/recommendations for international students: Required—TOEFL.

University of South Alabama, Graduate School, College of Arts and Sciences, Department of Mathematics and Statistics, Mobile, AL 36688-0002. Offers mathematics (MS). Part-time and evening/weekend programs available. *Faculty:* 18 full-time (4 women). *Students:* 12 full-time (5 women), 2 part-time (both women); includes 1 minority (African American), 7 international. 11 applicants, 82% accepted, 5 enrolled. In 2005, 4 degrees awarded. *Degree requirements:* For master's, comprehensive exam. *Entrance requirements:* For master's, GRE. *Application deadline:* For fall admission, 9/1 for domestic students. Applications are processed on a rolling basis. Application fee: $25. *Expenses:* Tuition, state resident: full-time $4,008. Tuition, nonresident: full-time $8,016. Required fees: $692. *Financial support:* Fellowships, research assistantships available. Support available to part-time students. Financial award application deadline:4/1. *Faculty research:* Knot theory, chaos theory. *Unit head:* Dr. Scott Carter, Chair, 251-460-6264.

University of South Carolina, The Graduate School, College of Science and Mathematics, Department of Mathematics, Columbia, SC 29208. Offers mathematics (MA, MS, PhD); mathematics education (M Math, MAT). MAT offered in cooperation with the College of Education. Part-time programs available. Terminal master's awarded for partial completion of doctoral program. *Degree requirements:* For master's, thesis; for doctorate, one foreign language, thesis/dissertation. *Entrance requirements:* For master's and doctorate, GRE General Test. Electronic applications accepted. *Faculty research:* Applied mathematics, analysis, discrete mathematics, algebra, topology.

The University of South Dakota, Graduate School, College of Arts and Sciences, Department of Mathematics, Vermillion, SD 57069-2390. Offers MA, MNS. Part-time programs available. *Faculty:* 9 full-time (2 women). *Students:* 9 (2 women). In 2005, 4 degrees awarded. *Degree requirements:* For master's, thesis (for some programs). *Entrance requirements:* For master's, GRE, minimum GPA of 2.7. Additional exam requirements/recommendations for international students: Required—TOEFL (minimum score 550 paper-based; 213 computer-based), IBT 79. *Application deadline:* Applications are processed on a rolling basis. Application fee: $35. Electronic applications accepted. *Expenses:* Tuition, state resident: part-time $116 per credit hour. Tuition, nonresident: part-time $341 per credit hour. Required fees: $85 per credit hour. Tuition and fees vary according to course load, program and reciprocity agreements. *Financial support:* In 2005–06, teaching assistantships with partial tuition reimbursements (averaging $7,500 per year) Financial award applicants required to submit FAFSA. *Unit head:* Dr. Dan Van Peursem, Chair, 605-677-5262, Fax: 605-677-5263, E-mail: usdmath@usd.edu. *Application contact:* Dr. Nan Jiang, 605-677-5262, Fax: 605-677-5263, E-mail: usdmath@usd.edu.

University of Southern California, Graduate School, College of Letters, Arts and Sciences, Department of Mathematics, Program in Mathematics, Los Angeles, CA 90089. Offers MA, PhD. *Degree requirements:* For doctorate, thesis/dissertation. *Entrance requirements:* For master's and doctorate, GRE General Test. *Expenses:* Tuition: Full-time $25,416; part-time $1,059 per unit. Required fees: $484; $484 per year. Tuition and fees vary according to course load and program.

University of Southern Mississippi, Graduate School, College of Science and Technology, Department of Mathematics, Hattiesburg, MS 39406-0001. Offers MS. Part-time programs available. *Degree requirements:* For master's, thesis or alternative, comprehensive exam. *Entrance requirements:* For master's, GRE General Test, minimum GPA of 2.75 in last 60 hours. Additional exam requirements/recommendations for international students: Required—TOEFL. *Faculty research:* Dynamical systems, numerical analysis and multigrid methods, random number generation, matrix theory, group theory.

University of South Florida, College of Graduate Studies, College of Arts and Sciences, Department of Mathematics, Tampa, FL 33620-9951. Offers applied mathematics (PhD); mathematics (MA, PhD). Part-time and evening/weekend programs available. *Faculty:* 25 full-time (2 women), 1 part-time/adjunct (0 women). *Students:* 48 full-time (14 women), 24 part-time (11 women); includes 9 minority (1 African American, 8 Asian Americans or Pacific Islanders), 38 international. 56 applicants, 91% accepted, 20 enrolled. In 2005, 2 degrees awarded. Terminal master's awarded for partial completion of doctoral program. *Degree requirements:* For master's, one foreign language, thesis optional; for doctorate, 2 foreign languages, thesis/dissertation. *Entrance requirements:* For master's, GRE General Test, minimum GPA of 3.0 in mathematics course work (undergraduate), 3.5 (graduate); for doctorate, GRE General Test. Additional exam requirements/recommendations for international students: Required—TOEFL (minimum score 550 paper-based; 213 computer-based). *Application deadline:* For fall admission, 6/1 for domestic students, 2/1 for international students; for spring admission, 10/15 for domestic students, 8/1 for international students. Application fee: $30. Electronic applications accepted. *Financial support:* Teaching assistantships with partial tuition reimbursements, scholarships/grants and unspecified assistantships available. Financial award application deadline: 2/1. *Faculty research:* Approximation theory, differential equations, discrete mathematics, functional analysis topology. *Unit head:* Dr. Marcus McWaters, Chairperson, 813-974-3838, Fax: 813-974-2700, E-mail: marcus@chuma.cas.usf.edu. *Application contact:* Dr. Natasha Jonoska, Graduate Admissions Director, 813-974-9566, Fax: 813-974-2700, E-mail: jonoska@math.usf.edu.

The University of Tennessee, Graduate School, College of Arts and Sciences, Department of Mathematics, Knoxville, TN 37996. Offers applied mathematics (MS); mathematical ecology (PhD); mathematics (M Math, MS, PhD). Part-time programs available. *Degree requirements:* For master's, thesis or alternative; for doctorate, one foreign language, thesis/dissertation. *Entrance requirements:* For master's and doctorate, minimum GPA of 2.7. Additional exam requirements/recommendations for international students: Required—TOEFL. Electronic applications accepted.

The University of Texas at Arlington, Graduate School, College of Science, Department of Mathematics, Arlington, TX 76019. Offers mathematical sciences (PhD); mathematics (MS). Part-time and evening/weekend programs available. *Faculty:* 12 full-time (2 women). *Students:* 34 full-time (13 women), 49 part-time (30 women); includes 18 minority (8 African Americans, 4 Asian Americans or Pacific Islanders, 6 Hispanic Americans), 24 international. 39 applicants, 100% accepted, 20 enrolled. In 2005, 19 master's, 5 doctorates awarded. *Median time to degree:* Of those who began their doctoral program in fall 1996, 95% received their degree in 8 years or less. *Degree requirements:* For master's, thesis or alternative, comprehensive exam; for doctorate, thesis/dissertation, comprehensive exam. *Entrance requirements:* For master's, GRE General Test; for doctorate, GRE General Test, 30 hours of graduate course work in mathematics, minimum GPA of 3.0 in last 60 hours of course work. Additional exam requirements/recommendations for international students: Required—TOEFL. *Application deadline:* For fall admission, 6/16 for domestic students. Applications are processed on a rolling basis. Application fee: $35 ($50 for international students). *Expenses:* Tuition, state resident: full-time $3,350. Tuition, nonresident: full-time $8,318. International tuition: $8,448 full-time. Required fees:$1,277. Full-time tuition and fees vary according to course level and program. *Financial support:* In 2005–06, 30 students received support, including 4 fellowships (averaging $1,000 per year), 23 teaching assistantships (averaging $15,600 per year); Federal Work-Study, institutionally sponsored loans, scholarships/grants, health care benefits, and unspecified assistantships also available. Financial award application deadline: 6/1; financial award applicants required to submit FAFSA. *Application contact:* Dr. Tie Luo, Graduate Adviser, 817-272-3261, Fax: 817-272-5802, E-mail: luo@uta.edu.

The University of Texas at Austin, Graduate School, College of Natural Sciences, Department of Mathematics, Austin, TX 78712-1111. Offers mathematics (MA, PhD); statistics (MS Stat). *Entrance requirements:* For master's and doctorate, GRE General Test. Electronic applications accepted.

The University of Texas at Brownsville, Graduate Studies, College of Science, Mathematics and Technology, Brownsville, TX 78520-4991. Offers biological sciences (MS, MSIS); mathematics (MS); physics (MS). Part-time and evening/weekend programs available. *Faculty:* 51 full-time (10 women). *Students:* 34 (16 women); includes 24 minority (all Hispanic Americans) 7 applicants. In 2005, 9 degrees awarded. *Degree requirements:* For master's, thesis optional. *Entrance requirements:* For master's, GRE General Test. Additional exam requirements/recommendations for international students: Required—TOEFL. *Application deadline:* For fall admission, 7/1 for domestic students; for spring admission, 12/1 priority date for domestic students. Applications are processed on a rolling basis. Application fee: $30. *Financial support:* Federal Work-Study, scholarships/grants, and tuition waivers (partial) available. Support available to part-time students. Financial award application deadline: 4/3; financial award applicants required to submit FAFSA. *Faculty research:* Fish, insects, barrier islands, algae, curlits. *Unit head:* Terry Jay Phillips, Interim Dean, 956-882-6701, Fax: 956-882-8988. *Application contact:* Irma C. Hernandez, Information Contact, 956-882-7787, Fax: 956-882-7279, E-mail: irma.c.hernandez@utb.edu.

The University of Texas at Dallas, School of Natural Sciences and Mathematics, Programs in Mathematical Sciences, Richardson, TX 75083-0688. Offers applied mathematics (MS, PhD); engineering mathematics (MS); mathematical science (MS); statistics (MS, PhD). Part-time and evening/weekend programs available. *Faculty:* 12 full-time (0 women). *Students:* 33 full-time (12 women), 25 part-time (13 women); includes 13 minority (2 African Americans, 6 Asian Americans or Pacific Islanders, 5 Hispanic Americans), 26 international. Average age 30. 73 applicants, 59% accepted, 28 enrolled. In 2005, 19 master's, 8 doctorates awarded. *Degree requirements:* For master's, thesis optional; for doctorate, thesis/dissertation. *Entrance requirements:* For master's, GRE General Test, minimum GPA of 3.0 in upper-level course work in field; for doctorate, GRE General Test, minimum GPA of 3.5 in upper-level course work in field. Additional exam requirements/recommendations for international students: Required—TOEFL (minimum score 550 paper-based; 213 computer-based). *Application deadline:* For fall admission, 7/15 for domestic students; for spring admission, 11/15 for domestic students. Applications are processed on a rolling basis. Application fee: $50 ($100 for international students). Electronic applications accepted. *Expenses:* Tuition, state resident: full-time $5,450; part-time $303 per credit. Tuition, nonresident: full-time $12,648; part-time $703 per credit. Tuition and fees vary according to program. *Financial support:* In 2005–06, 25 teaching assistantships with tuition reimbursements (averaging $9,451 per year) were awarded; fellowships, research assistantships, career-related internships or fieldwork, Federal Work-Study, institutionally sponsored loans, and scholarships/grants also available. Support available to part-time students. Financial award application deadline: 4/30; financial award applicants required to submit FAFSA. *Faculty research:* Statistical methods, control theory, mathematical modeling and analyses of biological and physical systems. Total annual research expenditures: $59,757. *Unit head:* Dr. M. Ali Hooshyar, Head, 972-883-2161, Fax: 972-883-6622, E-mail: utdmath@utdallas.edu. *Application contact:* Dr. Michael Baron, Graduate Advisor, 972-883-6874, Fax: 972-883-6622, E-mail: mbaron@utdallas.edu.

See Close-Up on page 545.

The University of Texas at El Paso, Graduate School, College of Science, Department of Mathematical Sciences, El Paso, TX 79968-0001. Offers mathematical sciences (MAT); mathematics (MS); statistics (MS). Part-time and evening/weekend programs available. *Degree requirements:* For master's, thesis optional. *Entrance requirements:* For master's, GRE, minimum GPA of 3.0. Additional exam requirements/recommendations for international students: Required—TOEFL. Electronic applications accepted.

The University of Texas at Tyler, College of Arts and Sciences, Department of Mathematics, Tyler, TX 75799-0001. Offers MAT, MS, MSIS. *Faculty:* 7 full-time (3 women). *Students:* 7 full-time (2 women), 1 (woman) part-time; includes 2 minority (1 African American, 1 Hispanic American). 7 applicants, 100% accepted, 2 enrolled. In 2005, 3 degrees awarded. *Degree requirements:* For master's, comprehensive exam. *Entrance requirements:* For master's, GRE General Test. *Application deadline:* Applications are processed on a rolling basis. Application fee: $0. *Expenses:* Tuition, state resident: part-time $321. Tuition, nonresident: part-time $597. Tuition and fees vary according to course load. *Financial support:* In 2005–06, 6 students received support, including 5 teaching assistantships (averaging $8,500 per year); unspecified assistantships also available. Financial award application deadline: 7/1; financial award applicants required to submit FAFSA. *Faculty research:* Graph theory, abstract algebra, biomathematics. *Unit head:* Dr. Kazem Mahdavi, Chair, 903-566-7210, Fax: 903-566-7189, E-mail: kmahdavi@uttyler.edu. *Application contact:* Carol A. Hodge, Office of Graduate Studies, 903-566-5642, Fax: 903-566-7068, E-mail: chodge@mail.uttyl.edu.

The University of Texas–Pan American, College of Science and Engineering, Department of Mathematics, Edinburg, TX 78541-2999. Offers MS. Part-time and evening/weekend programs available. *Degree requirements:* For master's, comprehensive exam. *Entrance requirements:* For master's, GRE General Test, minimum GPA of 3.0. *Expenses:* Tuition, area resident: Full-time $2,268; part-time $68 per credit hour. Tuition, nonresident: part-time $370 per credit hour. International tuition: $7,236 full-time. Required fees: $488. *Faculty research:* Boundary value problems in differential equations, training of public school teachers in methods of presenting mathematics, harmonic analysis, inverse problems, commulative algebra.

University of the Incarnate Word, School of Graduate Studies and Research, School of Mathematics, Sciences, and Engineering, Program in Mathematics, San Antonio, TX 78209-6397. Offers mathematics (MS); teaching (MA). Part-time and evening/weekend programs available. *Students:* Average age 37. In 2005, 1 degree awarded. *Entrance requirements:* For master's, GRE General Test. Additional exam requirements/recommendations for international students: Required—TOEFL. *Application deadline:* For fall admission, 8/15 for domestic students; for spring admission, 12/31 for domestic students. Applications are processed on a rolling basis. Application fee: $20. *Expenses:* Tuition: Full-time $9,810; part-time $545 per credit hour. Required fees: $828; $46 per credit. One-time fee: $30. *Financial support:* Federal Work-Study, scholarships/grants, and Federal Loans available. *Faculty research:* Topology, set theory, mathematics education. *Unit head:* Dr. Elizabeth Kreston, Chair, 210-805-1225, Fax: 210-829-3153, E-mail: kreston@universe.uiwtx.edu. *Application contact:* Andrea Cyterski-Acosta, Dean of Enrollment, 210-829-6005, Fax: 210-829-3921, E-mail: cyterski@uiwtx.edu.

The University of Toledo, Graduate School, College of Arts and Sciences, Department of Mathematics, Toledo, OH 43606-3390. Offers applied mathematics (MS); mathematics (MA, PhD); statistics (MS). Part-time programs available. *Faculty:* 16. *Students:* 57 full-time (24 women), 13 part-time (3 women); includes 1 minority (Asian American or Pacific Islander), 45 international. Average age 29. 65 applicants, 63% accepted, 21 enrolled. In 2005, 16 degrees awarded. *Degree requirements:* For doctorate, 2 foreign languages, thesis/dissertation. *Entrance requirements:* For master's and doctorate, GRE General Test, GRE Subject Test. *Application deadline:* For fall admission, 8/1 for domestic students. Application fee: $45. Electronic applications accepted. *Expenses:* Tuition, area resident: Part-time $308 per credit hour. Tuition, state resident: full-time $3,312. Tuition, nonresident: full-time $6,616; part-time $735 per credit hour. *Financial support:* In 2005–06, 16 research assistantships with full tuition reimbursements (averaging $3,688 per year), 37 teaching assistantships with full tuition reimbursements (averaging $10,807 per year) were awarded; Federal Work-Study and institutionally sponsored loans also available. Support available to part-time students. Financial award application deadline: 4/1; financial award applicants required to submit FAFSA. *Faculty research:* Topology. *Unit head:* Dr. Geoffrey Martin, Chair, 419-530-2569, Fax: 419-530-4720, E-mail: gmartin@math.utoledo.edu. *Application contact:* Dr. Gerard Thompson, Advising Coordinator, 419-530-2568, Fax: 419-530-4720, E-mail: thompson@math.utoledo.edu.

See Close-Up on page 547.

University of Toronto, School of Graduate Studies, Physical Sciences Division, Department of Mathematics, Toronto, ON M5S 1A1, Canada. Offers M Sc, MMF, PhD. Part-time programs available. *Degree requirements:* For master's, research project, thesis optional; for doctorate, thesis/dissertation. *Entrance requirements:* For master's, minimum B average in final year, bachelor's degree in mathematics or a related area, 3 letters of reference; for doctorate, master's degree in mathematics or a related area, minimum A– average, 3 letters of reference.

University of Tulsa, Graduate School, College of Business Administration and College of Engineering and Natural Sciences, Department of Engineering and Technology Management, Tulsa, OK 74104-3189. Offers chemical engineering (METM); computer science (METM); electrical engineering (METM); geological science (METM); mathematics (METM); mechanical engineering (METM); petroleum engineering (METM). Part-time and evening/weekend programs available. *Students:* 3 full-time (0 women), 2 part-time (1 woman); includes 3 minority (1 American Indian/Alaska Native, 2 Hispanic Americans), 1 international. Average age 28. 9 applicants, 33% accepted, 2 enrolled. In 2005, 1 degree awarded. *Entrance requirements:* For master's, GRE General Test or GMAT. Additional exam requirements/recommendations for international students: Required—TOEFL (minimum score 575 paper-based; 231 computer-based). *Application deadline:* Applications are processed on a rolling basis. Application fee: $30. Electronic applications accepted. *Expenses:* Tuition: Full-time $12,132; part-time $674 per credit hour. Required fees: $60; $3 per credit hour. *Financial support:* Fellowships, research assistantships with full and partial tuition reimbursements, teaching

Mathematics

University of Tulsa (continued)

assistantships, Federal Work-Study, scholarships/grants, tuition waivers (full and partial), and unspecified assistantships available. Support available to part-time students. Financial award application deadline: 2/1; financial award applicants required to submit FAFSA. *Unit head:* Ron Cooper, Director of Graduate Business Studies, 918-631-2680, Fax: 918-631-2142, E-mail: ron-cooper@utulsa.edu.

University of Tulsa, Graduate School, College of Engineering and Natural Sciences, Department of Mathematical and Computer Sciences, Program in Mathematical Sciences, Tulsa, OK 74104-3189. Offers MS, MTA. Part-time programs available. *Faculty:* 12 full-time (2 women). *Students:* 4 full-time (1 woman). Average age 29. 5 applicants, 60% accepted, 1 enrolled. *Degree requirements:* For master's, thesis (for some programs). *Entrance requirements:* For master's, GRE General Test. Additional exam requirements/recommendations for international students: Required—TOEFL (minimum score 550 paper-based; 213 computer-based), IELT (minimum score 6). *Application deadline:* Applications are processed on a rolling basis. Application fee: $30. Electronic applications accepted. *Expenses:* Tuition: Full-time $12,132; part-time $674 per credit hour. Required fees: $60; $3 per credit hour. *Financial support:* In 2005–06, 2 students received support, including 3 teaching assistantships with full and partial tuition reimbursements available (averaging $10,000 per year); fellowships with full and partial tuition reimbursements available, research assistantships with full and partial tuition reimbursements available, Federal Work-Study, scholarships/grants, tuition waivers (full and partial), and unspecified assistantships also available. Financial award application deadline: 2/1; financial award applicants required to submit FAFSA. *Faculty research:* Optimization theory, numerical analysis, mathematical physics, modeling, Bayesian statistical inference. *Application contact:* Dr. Christian Constanda, Adviser, 918-631-3068, Fax: 918-631-3077, E-mail: grad@utulsa.edu.

University of Utah, The Graduate School, College of Science, Department of Mathematics, Salt Lake City, UT 84112-1107. Offers M Phil, M Stat, MA, MS, PhD. Part-time programs available. *Faculty:* 34 full-time (2 women), 1 part-time/adjunct (0 women). *Students:* 71 full-time (23 women), 12 part-time (5 women); includes 3 minority (2 Asian Americans or Pacific Islanders, 1 Hispanic American), 22 international. Average age 27. 134 applicants, 48% accepted, 35 enrolled. In 2005, 30 master's, 6 doctorates awarded. Terminal master's awarded for partial completion of doctoral program. *Median time to degree:* Of those who began their doctoral program in fall 1996, 40% received their degree in 8 years or less. *Degree requirements:* For master's, thesis or alternative, written or oral exam; for doctorate, thesis/dissertation, written and oral exams. *Entrance requirements:* For master's, GRE Subject Test in math, minimum undergraduate GPA of 3.0. Additional exam requirements/recommendations for international students: Required—TOEFL (minimum score 500 paper-based; 173 computer-based). *Application deadline:* For fall admission, 3/15 for domestic students, 3/15 for international students; for spring admission, 11/1 for domestic students, 11/1 for international students. Application fee: $45 ($65 for international students). *Expenses:* Tuition, state resident: full-time $2,932; part-time $2,212 per term. Tuition, nonresident: full-time $10,350; part-time $7,812 per term. Required fees: $590; $516 per term. Tuition and fees vary according to course load and program. *Financial support:* Fellowships with full tuition reimbursements, research assistantships with full tuition reimbursements, teaching assistantships with full tuition reimbursements available. Financial award application deadline: 3/15. *Faculty research:* Algebraic geometry, differential geometry, scientific computing, topology, mathematical biology. Total annual research expenditures: $1.5 million. *Unit head:* Aaron Bertram, Chairman, 801-581-6681, Fax: 801-581-4148. *Application contact:* Jingyi Zhu, Director of Graduate Studies, 801-581-8005, Fax: 801-581-4148, E-mail: zhu@math.utah.edu.

See Close-Up on page 549.

University of Vermont, Graduate College, College of Engineering and Mathematics, Department of Mathematics and Statistics, Program in Mathematics, Burlington, VT 05405. Offers mathematics (MS, PhD); mathematics education (MAT, MST). *Students:* 22 (9 women); includes 1 minority (Asian American or Pacific Islander) 5 international. 35 applicants, 69% accepted, 13 enrolled. In 2005, 9 master's, 3 doctorates awarded. *Degree requirements:* For doctorate, thesis/dissertation. *Entrance requirements:* For master's and doctorate, GRE General Test. Additional exam requirements/recommendations for international students: Required—TOEFL (minimum score 550 paper-based; 213 computer-based). *Application deadline:* For fall admission, 4/1 for domestic students. Applications are processed on a rolling basis. Application fee: $40. Electronic applications accepted. *Expenses:* Tuition, area resident: Part-time $410 per credit hour. Tuition, nonresident: part-time $1,034 per credit hour. *Financial support:* Fellowships, research assistantships, teaching assistantships available. Financial award application deadline: 3/1. *Unit head:* Dr. J. Sands, Coordinator, 802-656-2940.

University of Victoria, Faculty of Graduate Studies, Faculty of Science, Department of Mathematics and Statistics, Victoria, BC V8W 2Y2, Canada. Offers M Sc, MA, PhD. Part-time programs available. *Faculty:* 24 full-time (4 women), 2 part-time/adjunct (0 women). *Students:* 52, 21 international. Average age 24. 76 applicants, 49% accepted, 20 enrolled. In 2005, 2 master's, 1 doctorate awarded. *Degree requirements:* For master's, thesis, registration; for doctorate, one foreign language, thesis/dissertation, 3 qualifying exams, Candidacy exam. *Entrance requirements:* Additional exam requirements/recommendations for international students: Required—TOEFL (minimum score 575 paper-based; 233 computer-based), IELT (minimum score 7). *Application deadline:* For fall admission, 5/31 priority date for domestic students, 12/15 priority date for international students. Applications are processed on a rolling basis. Application fee: $75 ($125 for international students). Electronic applications accepted. Tuition and fees charges are reported in Canadian dollars. *Expenses:* Tuition, area resident: Full-time $4,492 Canadian dollars; part-time $749 Canadian dollars per term. International tuition: $5,346 Canadian dollars full-time. Required fees: $4,492 Canadian dollars; $749 Canadian dollars per term. Tuition and fees vary according to course load, campus/location and program. *Financial support:* In 2005–06, 24 students received support, including 5 fellowships (averaging $12,900 per year); research assistantships, teaching assistantships, career-related internships or fieldwork, institutionally sponsored loans, unspecified assistantships, and awards also available. Financial award application deadline: 2/15. *Faculty research:* Functional analysis and operator theory, applied ordinary and partial differential equations, discrete mathematics and graph theory. *Unit head:* Dr. Gary MacGillivray, Chair, 250-721-7436, Fax: 250-721-8962, E-mail: chair@math.uvic.ca. *Application contact:* Dr. Pauline van den Driessche, Graduate Adviser, 250-721-7442, Fax: 250-721-8962, E-mail: gradadv@math.uvic.ca.

University of Virginia, College and Graduate School of Arts and Sciences, Department of Mathematics, Charlottesville, VA 22903. Offers MA, MS, PhD. *Faculty:* 30 full-time (4 women), 6 part-time/adjunct (all women). *Students:* 42 full-time (14 women), 2 part-time; includes 2 minority (1 African American, 1 Asian American or Pacific Islander), 13 international. Average age 26. 119 applicants, 2% accepted, 7 enrolled. In 2005, 8 master's, 4 doctorates awarded. *Degree requirements:* For master's, one foreign language, comprehensive exam; for doctorate, 2 foreign languages, thesis/dissertation, comprehensive exam. *Entrance requirements:* For master's and doctorate, GRE General Test, GRE Subject Test. *Application deadline:* Applications are processed on a rolling basis. Application fee: $40. Electronic applications accepted. *Expenses:* Tuition, state resident: full-time $7,731. Tuition, nonresident: full-time $18,672. Required fees: $1,479. Full-time tuition and fees vary according to degree level and program. *Financial support:* Fellowships, teaching assistantships, unspecified assistantships available. Financial award applicants required to submit FAFSA. *Unit head:* Ira Herbst, Chair, 434-924-4919, Fax: 434-982-3084, E-mail: iwh@virginia.edu. *Application contact:* Peter C. Brunjes, Associate Dean for Graduate Programs and Research, 434-924-7184, Fax: 434-924-6737, E-mail: grad-a-s@virginia.edu.

University of Washington, Graduate School, College of Arts and Sciences, Department of Mathematics, Seattle, WA 98195. Offers MA, MS, PhD. Part-time programs available. Terminal master's awarded for partial completion of doctoral program. *Degree requirements:* For master's, thesis optional; for doctorate, 2 foreign languages, thesis/dissertation, registration. *Entrance requirements:* For master's, GRE, minimum GPA of 3.0; for doctorate, GRE General Test, GRE Subject Test (mathematics), minimum GPA of 3.0. Additional exam requirements/recommendations for international students: Required—TOEFL. Electronic applications accepted. *Faculty research:* Algebra, analysis, probability, combinatorics and geometry.

University of Washington, Graduate School, Interdisciplinary Graduate Program in Quantitative Ecology and Resource Management, Seattle, WA 98195. Offers MS, PhD. *Degree requirements:* For master's and doctorate, thesis/dissertation. *Entrance requirements:* For master's and doctorate, GRE General Test, minimum GPA of 3.0. Additional exam requirements/recommendations for international students: Required—TOEFL. Electronic applications accepted. *Faculty research:* Population dynamics, statistical analysis, ecological modeling and systems analysis of aquatic and terrestrial ecosystems.

University of Waterloo, Graduate Studies, Faculty of Mathematics, Department of Combinatorics and Optimization, Waterloo, ON N2L 3G1, Canada. Offers M Math, PhD. *Faculty:* 28 full-time (4 women), 14 part-time/adjunct (1 woman). *Students:* 46 full-time (13 women), 2 part-time. 63 applicants, 51% accepted, 19 enrolled. In 2005, 11 master's, 5 doctorates awarded. *Degree requirements:* For master's, research paper or thesis; for doctorate, thesis/dissertation, comprehensive exam. *Entrance requirements:* For master's, GRE General Test, honors degree in field, minimum B+ average; for doctorate, GRE General Test, master's degree, minimum A average. Additional exam requirements/recommendations for international students: Required—TOEFL, TWE. *Application deadline:* For fall admission, 4/15 for domestic students; for spring admission, 12/1 priority date for domestic students. Applications are processed on a rolling basis. Application fee: $75 Canadian dollars. Electronic applications accepted. *Financial support:* Research assistantships, teaching assistantships, career-related internships or fieldwork and scholarships/grants available. *Faculty research:* Algebraic and enumerative combinatorics, continuous optimization, cryptography, discrete optimization and graph theory. *Unit head:* Dr. W. H. Cunningham, Chair, 519-888-4567 Ext. 3482, Fax: 519-725-5441, E-mail: whcunnin@uwaterloo.ca. *Application contact:* Dr. R. B. Richter, Associate Chair, 519-888-4567 Ext. 2696, Fax: 519-725-5441, E-mail: cograd@uwaterloo.ca.

University of Waterloo, Graduate Studies, Faculty of Mathematics, Department of Pure Mathematics, Waterloo, ON N2L 3G1, Canada. Offers M Math, PhD. Part-time programs available. *Faculty:* 21 full-time (2 women), 15 part-time/adjunct (0 women). *Students:* 24 full-time (4 women), 2 part-time. 55 applicants, 42% accepted, 10 enrolled. In 2005, 6 degrees awarded. Terminal master's awarded for partial completion of doctoral program. *Degree requirements:* For master's, thesis/dissertation; for doctorate, thesis/dissertation, comprehensive exam, registration. *Entrance requirements:* For master's, honors degree in field, minimum B+ average; for doctorate, master's degree, minimum B+ average. Additional exam requirements/recommendations for international students: Required—TOEFL (minimum score 580 paper-based; 237 computer-based), TWE (minimum score 4). *Application deadline:* For fall admission, 2/1 priority date for domestic students, 2/1 priority date for international students. For winter admission, 7/1 for domestic students; for spring admission, 10/1 for domestic students. Applications are processed on a rolling basis. Application fee: $75 Canadian dollars. Electronic applications accepted. *Financial support:* Research assistantships, teaching assistantships, scholarships/grants and unspecified assistantships available. *Faculty research:* Algebra, algebraic and differential geometry, functional and harmonic analysis, logic and universal algebra, number theory. *Unit head:* Dr. Frank Zorzitto, Chair, 519-888-4567 Ext. 3484, Fax: 519-725-0160. *Application contact:* Dr. C. T. Ng, Graduate Officer, 519-888-4567 Ext. 4085, Fax: 519-725-0160, E-mail: ctng@math.uwaterloo.ca.

The University of Western Ontario, Faculty of Graduate Studies, Physical Sciences Division, Department of Mathematics, London, ON N6A 5B8, Canada. Offers M Sc, PhD. Terminal master's awarded for partial completion of doctoral program. *Degree requirements:* For master's, thesis or alternative; for doctorate, one foreign language, thesis/dissertation, qualifying exam, comprehensive exam. *Entrance requirements:* For master's, minimum B average, honors degree; for doctorate, master's degree. Additional exam requirements/recommendations for international students: Required—TOEFL (minimum score 550 paper-based; 213 computer-based). *Faculty research:* Algebra and number theory, analysis, geometry and topology.

University of West Florida, College of Arts and Sciences: Sciences, Department of Mathematics and Statistics, Pensacola, FL 32514-5750. Offers MS. Part-time and evening/weekend programs available. *Faculty:* 12 full-time (4 women), 1 part-time/adjunct (0 women). *Students:* 7 full-time (3 women), 6 part-time (4 women); includes 4 minority (2 African Americans, 2 Asian Americans or Pacific Islanders). Average age 37. 11 applicants, 82% accepted, 4 enrolled. In 2005, 9 degrees awarded. *Degree requirements:* For master's, thesis optional. *Entrance requirements:* For master's, GRE General Test, minimum GPA of 3.0. Additional exam requirements/recommendations for international students: Required—TOEFL (minimum score 550 paper-based; 213 computer-based). *Application deadline:* For fall admission, 6/1 for domestic students, 5/15 for international students; for spring admission, 11/1 for domestic students, 10/1 for international students. Applications are processed on a rolling basis. Application fee: $30. Special rates offered to residents of Alabama. *Expenses:* Tuition, state resident: full-time $5,833; part-time $243 per credit hour. Tuition, nonresident: full-time $21,204; part-time $884 per credit hour. Tuition and fees vary according to campus/location. *Financial support:* In 2005–06, 7 students received support, including 3 research assistantships with partial tuition reimbursements available (averaging $1,600 per year), 4 teaching assistantships with partial tuition reimbursements available (averaging $3,800 per year); fellowships, career-related internships or fieldwork, Federal Work-Study, institutionally sponsored loans, and scholarships/grants also available. Financial award application deadline: 4/15; financial award applicants required to submit FAFSA. *Unit head:* Dr. Kuiyuan Li, Chairperson, 850-474-2287.

University of Windsor, Faculty of Graduate Studies and Research, Faculty of Science, Department of Mathematics and Statistics, Windsor, ON N9B 3P4, Canada. Offers mathematics (M Sc); statistics (M Sc, PhD). *Faculty:* 16 full-time (2 women), 3 part-time/adjunct (0 women). *Students:* 34 full-time (9 women), 1 (woman) part-time. 63 applicants, 30% accepted. In 2005, 10 master's, 1 doctorate awarded. *Degree requirements:* For master's, thesis or alternative; for doctorate, thesis/dissertation, comprehensive exam. *Entrance requirements:* For master's, minimum B average; for doctorate, minimum A average. Additional exam requirements/recommendations for international students: Required—TOEFL (minimum score 560 paper-based; 220 computer-based). *Application deadline:* For fall admission, 7/1 for domestic students. For winter admission, 11/1 for domestic students. Applications are processed on a rolling basis. Application fee: $55. Electronic applications accepted. *Financial support:* In 2005–06, 31 teaching assistantships (averaging $8,956 per year) were awarded; Federal Work-Study, scholarships/grants, tuition waivers (full and partial), unspecified assistantships, and bursaries also available. Financial award application deadline: 2/15. *Faculty research:* Applied mathematics, operational research, fluid dynamics. *Unit head:* Dr. Ejaz Ahmed, Head, 519-253-3000 Ext. 3017, Fax: 519-971-3649, E-mail: seahmed@uwindsor.ca. *Application contact:* Applicant Services, 519-253-3000 Ext. 6459, Fax: 519-971-3653, E-mail: gradadmit@uwindsor.ca.

University of Wisconsin–Madison, Graduate School, College of Letters and Science, Department of Mathematics, Madison, WI 53706-1380. Offers MA, PhD. Terminal master's awarded for partial completion of doctoral program. *Degree requirements:* For master's, registration; for doctorate, thesis/dissertation, comprehensive exam, registration. *Entrance requirements:* For master's and doctorate, GRE General Test, GRE Subject Test. Additional exam requirements/recommendations for international students: Required—TOEFL (minimum score 580 paper-based; 237 computer-based); Recommended—TSE (minimum score 45). Electronic applications accepted. *Faculty research:* Applied mathematics, analysis, algebra, logic, topology.

University of Wisconsin–Milwaukee, Graduate School, College of Letters and Sciences, Department of Mathematical Sciences, Milwaukee, WI 53201-0413. Offers mathematics (MS, PhD). *Faculty:* 35 full-time (2 women). *Students:* 57 full-time (16 women), 17 part-time (3 women); includes 2 minority (1 African American, 1 Asian American or Pacific Islander), 34 international. 71 applicants, 54% accepted, 18 enrolled. In 2005, 21 master's, 2 doctorates awarded. *Degree requirements:* For doctorate, 2 foreign languages, thesis/dissertation. *Application deadline:* For fall admission, 1/1 for domestic students; for spring admission, 9/1 for domestic students. Applications are processed on a rolling basis. Application fee: $45 ($75 for international students). *Expenses:* Tuition, area resident: Part-time $716 per credit. Tuition, state resident: part-time $776 per credit. Tuition, nonresident: part-time $1,614. Required fees: $229 per term. Tuition and fees vary according to course load and program. *Financial support:* In 2005–06, 7 fellowships, 50 teaching assistantships were awarded; research assistantships, career-related internships or fieldwork also available. Support available to part-time students. Financial award application deadline: 4/15. *Unit head:* Richard Stockbridge, Representative, 414-229-5110, Fax: 414-229-4907, E-mail: stockbri@uwm.edu.

University of Wyoming, Graduate School, College of Arts and Sciences, Department of Mathematics, Laramie, WY 82070. Offers mathematics (MA, MAT, MS, MST, PhD); mathematics/computer science (PhD). Part-time programs available. *Faculty:* 27 full-time (7 women). *Students:* 31 full-time (11 women), 12 part-time (4 women), 17 international. 49 applicants, 24% accepted. In 2005, 9 degrees awarded. Terminal master's awarded for partial completion of doctoral program. *Degree requirements:* For master's, thesis or alternative, qualifying exam; for doctorate, one foreign language, thesis/dissertation, preliminary exam. *Entrance requirements:* For master's and doctorate, GRE General Test, minimum GPA of 3.0. Additional exam requirements/recommendations for international students: Required—TOEFL. *Application deadline:* For fall admission, 3/1 for domestic students. Applications are processed on a rolling basis. Application fee: $50. *Expenses:* Tuition, state resident: full-time $3,720; part-time $155 per credit hour. Tuition, nonresident: full-time $10,704; part-time $446 per credit hour. Required fees: $666; $162 per semester. Tuition and fees vary according to course load and program. *Financial support:* In 2005–06, 1 research assistantship with full tuition reimbursement (averaging $12,032 per year), 19 teaching assistantships with full tuition reimbursements (averaging $12,032 per year) were awarded; institutionally sponsored loans also available. Financial award application deadline: 3/1. *Faculty research:* Numerical analysis, classical analysis, mathematical modeling, algebraic combinations. *Unit head:* Dr. Sivaguru Sritharon, Head, 307-766-4221. *Application contact:* Dr. Sivaguru Sritharon, Head, 307-766-4221.

Utah State University, School of Graduate Studies, College of Science, Department of Mathematics and Statistics, Logan, UT 84322. Offers industrial mathematics (MS); mathematical sciences (PhD); mathematics (M Math, MS); statistics (MS). Part-time programs available. *Faculty:* 33 full-time (4 women). *Students:* 91 full-time (29 women), 4 part-time (1 woman), 44 international. Average age 29. 41 applicants, 61% accepted, 15 enrolled. In 2005, 9 master's, 3 doctorates awarded. Terminal master's awarded for partial completion of doctoral program. *Degree requirements:* For master's, qualifying exam, thesis optional; for doctorate, one foreign language, thesis/dissertation, comprehensive exam. *Entrance requirements:* For master's and doctorate, GRE General Test, minimum GPA of 3.0. Additional exam requirements/recommendations for international students: Required—TOEFL. *Application deadline:* For fall admission, 6/15 for domestic students; for spring admission, 10/15 for domestic students. Applications are processed on a rolling basis. Application fee: $50 ($60 for international students). *Financial support:* In 2005–06, 1 fellowship with partial tuition reimbursement (averaging $12,000 per year), 17 teaching assistantships with partial tuition reimbursements (averaging $14,500 per year) were awarded; research assistantships with partial tuition reimbursements Support available to part-time students. Financial award application deadline: 4/1. *Faculty research:* Differential equations, computational mathematics, dynamical systems, probability and statistics, pure mathematics. Total annual research expenditures: $212,000. *Unit head:* Dr. Russell C. Thompson, Head, 435-797-2810, Fax: 435-797-1822, E-mail: thompson@math.usu.edu. *Application contact:* Dr. David Richard Cutler, Graduate Chairman, 435-797-2699, Fax: 435-797-1822, E-mail: richard.cutler@usu.edu.

Vanderbilt University, Graduate School, Department of Mathematics, Nashville, TN 37240-1001. Offers MA, MAT, MS, PhD. *Faculty:* 50 full-time (8 women). *Students:* 32 full-time (7 women), 17 international. 121 applicants, 12% accepted, 6 enrolled. In 2005, 5 master's, 3 doctorates awarded. *Degree requirements:* For master's, thesis or alternative; for doctorate, one foreign language, thesis/dissertation, final and qualifying exams. *Entrance requirements:* For master's and doctorate, GRE General Test, GRE Subject Test. *Application deadline:* For fall admission, 1/15 for domestic students, 1/15 for international students. Application fee: $0. Electronic applications accepted. *Expenses:* Tuition: Full-time $15,396; part-time $1,283 per semester hour. Required fees: $2,202; $1,101 per semester. One-time fee: $30. Tuition and fees vary according to course load, program and student level. *Financial support:* Fellowships with full and partial tuition reimbursements, research assistantships with full tuition reimbursements, teaching assistantships with full tuition reimbursements, Federal Work-Study and institutionally sponsored loans available. Financial award application deadline: 1/15. *Faculty research:* Algebra, topology, applied mathematics, graph theory, analytical mathematics. *Unit head:* Dietmar Bisch, Chair, 615-322-6672, Fax: 615-343-0215. *Application contact:* Michael D. Plummer, Director of Graduate Studies, 615-322-6672, Fax: 615-343-0215, E-mail: michael.d.plummer@vanderbilt.edu.

Villanova University, Graduate School of Liberal Arts and Sciences, Department of Mathematical Sciences, Program in Mathematical Sciences, Villanova, PA 19085-1699. Offers MA. Part-time and evening/weekend programs available. *Students:* 4 full-time (2 women), 25 part-time (7 women); includes 2 minority (1 African American, 1 Hispanic American), 2 international. Average age 30. 19 applicants, 58% accepted. In 2005, 14 degrees awarded. *Entrance requirements:* For master's, minimum GPA of 3.0. *Application deadline:* For fall admission, 8/1 for domestic students; for spring admission, 12/1 for domestic students. Application fee: $50. Electronic applications accepted. *Expenses:* Tuition: Part-time $540 per credit. Required fees: $60 per year. Tuition and fees vary according to program and student level. *Financial support:* Research assistantships, Federal Work-Study available. Financial award applicants required to submit FAFSA. *Unit head:* Dr. Douglas Norton, Chair, Department of Mathematical Sciences, 610-519-4850.

Virginia Commonwealth University, Graduate School, College of Humanities and Sciences, Department of Mathematical Sciences, Richmond, VA 23284-9005. Offers applied mathematics (MS); mathematics (MS); operations research (MS); statistics (MS, Certificate). *Faculty:* 15 full-time (2 women). *Students:* 5 full-time (2 women), 7 part-time (3 women); includes 4 minority (2 African Americans, 1 American Indian/Alaska Native, 1 Asian American or Pacific Islander), 7 international. 9 applicants, 67% accepted. In 2005, 9 degrees awarded. *Degree requirements:* For master's, thesis optional. *Entrance requirements:* For master's, GRE General Test, GRE Subject Test. Additional exam requirements/recommendations for international students: Required—TOEFL. *Application deadline:* For fall admission, 7/1 for domestic students; for spring admission, 11/15 for domestic students. Applications are processed on a rolling basis. Application fee: $50. *Expenses:* Tuition, state resident: full-time $3,185; part-time $405 per credit. Tuition, nonresident: full-time $7,952; part-time $940 per credit. Required fees: $751 per semester hour. Tuition and fees vary according to course load and program. *Financial support:* Fellowships, research assistantships, teaching assistantships, Federal Work-Study and institutionally sponsored loans available. Support available to part-time students. *Unit head:* Dr. Andrew M. Lewis, Chair, 804-828-1301 Ext. 128, Fax: 804-828-8785, E-mail: amlewis@vcu.edu. *Application contact:* Dr. James A. Wood, Information Contact, 804-828-1301, E-mail: jawood@vcu.edu.

Virginia Polytechnic Institute and State University, Graduate School, College of Science, Department of Mathematics, Blacksburg, VA 24061. Offers applied mathematics (MS, PhD); mathematical physics (MS, PhD); pure mathematics (MS, PhD). *Faculty:* 69 full-time (20 women). *Students:* 57 full-time (18 women), 5 part-time (2 women); includes 5 minority (2 African Americans, 1 Asian American or Pacific Islander, 2 Hispanic Americans), 27 international. Average age 28. 108 applicants, 20% accepted, 15 enrolled. In 2005, 20 master's, 8 doctorates awarded. *Entrance requirements:* For master's and doctorate, GRE. Additional exam requirements/recommendations for international students: Required—TOEFL (minimum score 550 paper-based; 213 computer-based). *Application deadline:* Applications are processed on a rolling basis. Application fee: $45. Electronic applications accepted. *Expenses:* Tuition, state resident: full-time $6,558; part-time $364 per credit. Tuition, nonresident: full-time $11,296; part-time $628 per credit. Required fees: $1,419; $468 per credit. $234 per term. *Financial support:* In 2005–06, 1 fellowship with full tuition reimbursement (averaging $1,481 per year), 1 research assistantship with full tuition reimbursement (averaging $16,689 per year), 44 teaching assistantships with full tuition reimbursements (averaging $13,902 per year) were awarded; career-related internships or fieldwork, Federal Work-Study, scholarships/grants, and unspecified assistantships also available. *Faculty research:* Differential equations, operator theory, numerical analysis, algebra, control theory. *Unit head:* Dr. John Rossi, Head, 540-231-6536, Fax: 540-231-5960, E-mail: rossi@math.vt.edu. *Application contact:* Hannah Swiger, Information Contact, 540-231-6537, Fax: 540-231-5960, E-mail: hsswiger@math.vt.edu.

Virginia State University, School of Graduate Studies, Research, and Outreach, School of Engineering, Science and Technology, Department of Mathematics, Petersburg, VA 23806-0001. Offers mathematics (MS); mathematics education (M Ed). *Degree requirements:* For master's, thesis (for some programs).

Wake Forest University, Graduate School, Department of Mathematics, Winston-Salem, NC 27109. Offers MA. Part-time programs available. *Faculty:* 15 full-time (3 women). *Students:* 13 full-time (7 women), 1 (woman) part-time, 3 international. Average age 28. 23 applicants, 43% accepted, 6 enrolled. In 2005, 6 degrees awarded. *Degree requirements:* For master's, one foreign language, registration. *Entrance requirements:* For master's, GRE General Test, GRE Subject Test. Additional exam requirements/recommendations for international students: Required—TOEFL (minimum score 213 computer-based). *Application deadline:* For fall admission, 1/15 for domestic students, 1/15 for international students. Application fee: $45. Electronic applications accepted. *Financial support:* In 2005–06, 12 students received support, including 1 fellowship with full tuition reimbursement available (averaging $4,000 per year), 1 research assistantship with full tuition reimbursement available (averaging $10,000 per year), 10 teaching assistantships with full tuition reimbursements available (averaging $10,000 per year); scholarships/grants and tuition waivers (full and partial) also available. Support available to part-time students. Financial award application deadline: 1/15; financial award applicants required to submit FAFSA. *Faculty research:* Algebra, ring theory, topology, differential equations. *Unit head:* Dr. Edward Allen, Director, 336-758-4854, Fax: 336-758-7190, E-mail: allene@wfu.edu.

Washington State University, Graduate School, College of Sciences, Department of Mathematics, Pullman, WA 99164. Offers applied mathematics (MS, PhD); mathematics teaching (MS, PhD). *Faculty:* 31. *Students:* 34 full-time (15 women), 1 (woman) part-time; includes 3 minority (all Asian Americans or Pacific Islanders), 11 international. Average age 29. 100 applicants, 28% accepted, 14 enrolled. In 2005, 12 master's, 5 doctorates awarded. *Degree requirements:* For master's, oral exam, project; for doctorate, 2 foreign languages, thesis/dissertation, oral exam, written exam. *Entrance requirements:* For master's and doctorate, minimum GPA of 3.0, 3 letters of recommendation. Additional exam requirements/recommendations for international students: Required—TOEFL (minimum score 600 paper-based; 250 computer-based). *Application deadline:* For fall admission, 2/1 for domestic students, 2/1 for international students; for spring admission, 9/1 for domestic students, 7/1 for international students. Applications are processed on a rolling basis. Application fee: $35. Electronic applications accepted. *Expenses:* Tuition, state resident: full-time $6,295; part-time $336 per credit. Tuition, nonresident: full-time $15,949; part-time $819 per credit. Required fees: $933. Part-time tuition and fees vary according to campus/location and program. *Financial support:* In 2005–06, 33 students received support, including 2 fellowships with tuition reimbursements available (averaging $2,500 per year), 3 research assistantships with full and partial tuition reimbursements available (averaging $13,871 per year), 27 teaching assistantships with full and partial tuition reimbursements available (averaging $14,109 per year); career-related internships or fieldwork, Federal Work-Study, institutionally sponsored loans, and tuition waivers (partial) also available. Financial award application deadline: 2/1; financial award applicants required to submit FAFSA. *Faculty research:* Computational mathematics, operations research, modeling in the natural sciences, applied statistics. Total annual research expenditures:$425,725. *Unit head:* Dr. Alan Genz, Chair, 509-335-4918, Fax: 509-335-1188, E-mail: chair@math.wsu.edu. *Application contact:* Pam Guptill, Coordinator, 509-335-6868, Fax: 509-335-1188, E-mail: pguptill@wsu.edu.

See Close-Up on page 553.

Washington University in St. Louis, Graduate School of Arts and Sciences, Department of Mathematics, St. Louis, MO 63130-4899. Offers mathematics (MA, PhD); mathematics education (MAT); statistics (MA, PhD). Terminal master's awarded for partial completion of doctoral program. *Degree requirements:* For master's, thesis or alternative; for doctorate, thesis/dissertation. *Entrance requirements:* For master's and doctorate, GRE General Test. Electronic applications accepted.

Washington University in St. Louis, Henry Edwin Sever Graduate School of Engineering and Applied Science, Department of Electrical and Systems Engineering, St. Louis, MO 63130-4899. Offers electrical engineering (MS, D Sc); systems science and mathematics (MS, D Sc). Part-time programs available. *Faculty:* 15 full-time (0 women), 25 part-time/adjunct (1 woman). *Students:* 54 full-time (7 women), 62 part-time (6 women); includes 15 minority (1 African American, 10 Asian Americans or Pacific Islanders, 4 Hispanic Americans), 38 international. Average age 23. 425 applicants, 20% accepted, 25 enrolled. In 2005, 31 master's, 6 doctorates awarded. Terminal master's awarded for partial completion of doctoral program. *Median time to degree:* Of those who began their doctoral program in fall 1996, 62% received their degree in 8 years or less. *Degree requirements:* For master's, thesis or alternative; for doctorate, thesis/dissertation, comprehensive exam. *Entrance requirements:* For master's, minimum GPA of 3.0 in the last 2 years of undergraduate course work; for doctorate, GRE. Additional exam requirements/recommendations for international students: Required—TOEFL (minimum score 550 paper-based; 213 computer-based). *Application deadline:* For fall admission, 2/1 for domestic students, 2/1 for international students. Applications are processed on a rolling basis. Application fee: $0. Electronic applications accepted. *Financial support:* In 2005–06, 28 students received support, including 1 fellowship with full tuition reimbursement available (averaging $22,500 per year), 27 research assistantships with full tuition reimbursements available (averaging $17,316 per year); teaching assistantships with full tuition reimbursements available, career-related internships or fieldwork, Federal Work-Study, institutionally sponsored loans, scholarships/grants, and unspecified assistantships also available. Financial award application deadline: 1/30; financial award applicants required to submit FAFSA. *Faculty research:* Linear and nonlinear control systems, robotics and automation, applied physics and electronics, security technologies, signal and image processing. Total annual research expenditures: $1.3 million. *Unit head:* Dr. Arye Nehorai, Chair, 314-935-5565, Fax: 314-935-7500, E-mail: nehorai@ese.wash.edu. *Application contact:* Rita Drochelman, Director of Graduate Programs, 314-935-4830, Fax: 314-935-7500, E-mail: info@ese.wustl.edu.

Wayne State University, Graduate School, College of Liberal Arts and Sciences, Department of Mathematics, Program in Mathematics, Detroit, MI 48202. Offers MA, MS, PhD. *Students:* 30 full-time (8 women), 17 part-time (5 women); includes 4 minority (3 African Americans, 1 Asian American or Pacific Islander), 25 international. Average age 29. 25 applicants, 56% accepted, 13 enrolled. In 2005, 5 master's, 4 doctorates awarded. *Degree requirements:* For doctorate, 2 foreign languages, thesis/dissertation. *Entrance requirements:* Additional exam requirements/recommendations for international students: Required—TOEFL (minimum score 550 paper-based; 213 computer-based); Recommended—TWE (minimum score 6). *Application deadline:* For fall admission, 7/1 for domestic students, 6/1 for international students. Applications are processed on a rolling basis. Application fee: $30 ($50 for international students). Electronic applications accepted. *Expenses:* Tuition, state resident: part-time $338 per credit hour. Tuition, nonresident: part-time $746 per credit hour. Required fees: $24 per credit hour. Full-time tuition and fees vary according to program. *Application contact:* Bert Schreiber, Professor, 313-577-8838, E-mail: bschreiber@wayne.edu.

Mathematics

Wesleyan University, Graduate Programs, Department of Mathematics, Middletown, CT 06459-0260. Offers MA, PhD. *Faculty:* 15 full-time (2 women). *Students:* 21 full-time (5 women); includes 2 minority (both Asian Americans or Pacific Islanders), 7 international. Average age 28. In 2005, 1 master's, 5 doctorates awarded. Terminal master's awarded for partial completion of doctoral program. *Degree requirements:* For master's, one foreign language, thesis; for doctorate, 2 foreign languages, thesis/dissertation. *Entrance requirements:* For master's, GRE General Test, GRE Subject Test; for doctorate, GRE Subject Test. *Application deadline:* For fall admission, 2/15 for domestic students. Applications are processed on a rolling basis. Application fee: $0. *Expenses:* Tuition: Full-time $24,732. One-time fee: $20 full-time. *Financial support:* Teaching assistantships, tuition waivers (full and partial) available. *Faculty research:* Topology, analysis. *Unit head:* Adam Fieldsteel, Chair, 860-685-2189, E-mail: afieldsteel@wesleyan.edu. *Application contact:* Nancy Ferguson, Information Contact, 860-685-2620, Fax: 860-685-2571, E-mail: nperguson@wesleyan.edu.

See Close-Up on page 557.

West Chester University of Pennsylvania, Graduate Studies, College of Arts and Sciences, Department of Mathematics, West Chester, PA 19383. Offers MA. Part-time and evening/weekend programs available. *Students:* 6 full-time (3 women), 30 part-time (10 women); includes 16 minority (4 African Americans, 8 Asian Americans or Pacific Islanders, 4 Hispanic Americans). Average age 33. 32 applicants, 94% accepted, 21 enrolled. In 2005, 10 degrees awarded. *Degree requirements:* For master's, comprehensive exam. *Entrance requirements:* For master's, GRE General Test, interview. *Application deadline:* For fall admission, 4/15 for domestic students; for spring admission, 10/15 for domestic students. Applications are processed on a rolling basis. Application fee: $35. *Expenses:* Tuition, state resident: full-time $2,944; part-time $327 per credit. Tuition, nonresident: full-time $4,711; part-time $523 per credit. Required fees: $54 per semester. *Financial support:* In 2005–06, 1 research assistantship with full tuition reimbursement (averaging $5,000 per year) was awarded; unspecified assistantships also available. Support available to part-time students. Financial award application deadline: 2/15; financial award applicants required to submit FAFSA. *Faculty research:* Teachers teaching with technology in service training program. *Unit head:* Dr. Richard Branton, Chair, 610-436-2440, E-mail: vbranton@wcupa.edu. *Application contact:* Dr. John Kerrigan, Graduate Coordinator, 610-436-2351, E-mail: jkerrigan@wcupa.edu.

Western Carolina University, Graduate School, College of Arts and Sciences, Department of Mathematics and Computer Science, Cullowhee, NC 28723. Offers applied mathematics (MS); comprehensive education-mathematics (MA Ed); mathematics (MAT). Part-time and evening/weekend programs available. *Degree requirements:* For master's, thesis optional. *Entrance requirements:* For master's, GRE General Test, GRE Subject Test (MS). Additional exam requirements/recommendations for international students: Required—TOEFL (minimum score 550 paper-based; 213 computer-based).

Western Connecticut State University, Division of Graduate Studies, School of Arts and Sciences, Department of Mathematics and Computer Science, Danbury, CT 06810-6885. Offers mathematics and computer science (MA); theoretical mathematics (MA). Part-time and evening/weekend programs available. *Degree requirements:* For master's, thesis or research project. *Entrance requirements:* For master's, minimum GPA of 2.5.

Western Illinois University, School of Graduate Studies, College of Arts and Sciences, Department of Mathematics, Macomb, IL 61455-1390. Offers MS. Part-time programs available. *Students:* 9 full-time (2 women); includes 1 minority (African American), 4 international. Average age 27. 14 applicants, 57% accepted. In 2005, 4 degrees awarded. *Degree requirements:* For master's, thesis or alternative. *Entrance requirements:* Additional exam requirements/recommendations for international students: Required—TOEFL (minimum score 500 paper-based; 173 computer-based). *Application deadline:* Applications are processed on a rolling basis. Application fee: $30. Electronic applications accepted. *Expenses:* Tuition, state resident: full-time $3,599; part-time $200 per semester hour. Tuition, nonresident: full-time $7,198; part-time $400 per semester hour. Required fees: $890; $49 per semester hour. Tuition and fees vary according to campus/location. *Financial support:* In 2005–06, 7 students received support, including 5 research assistantships with full tuition reimbursements available (averaging $6,288 per year), 2 teaching assistantships (averaging $7,248 per year) Financial award applicants required to submit FAFSA. *Unit head:* Dr. Iraj Kalantari, Chairperson, 309-298-1054. *Application contact:* Dr. Barbara Baily, Director of Graduate Studies/Associate Provost, 309-298-1806, Fax: 309-298-2345, E-mail: grad-office@wiu.edu.

Western Kentucky University, Graduate Studies, Ogden College of Science and Engineering, Department of Mathematics, Bowling Green, KY 42101-3576. Offers MA Ed, MS. *Faculty:* 6 full-time (1 woman). *Students:* 13 full-time (6 women), 4 part-time (3 women), 4 international. Average age 28. 9 applicants, 89% accepted, 4 enrolled. In 2005, 11 degrees awarded. *Degree requirements:* For master's, written exam, thesis optional. *Entrance requirements:* For master's, GRE General Test, minimum GPA of 2.75. Additional exam requirements/recommendations for international students: Required—TOEFL (minimum score 555 paper-based; 213 computer-based). *Application deadline:* For fall admission, 7/1 priority date for domestic students, 5/15 priority date for international students; for spring admission, 11/1 for domestic students, 9/15 for international students. Applications are processed on a rolling basis. Application fee: $35. *Expenses:* Tuition, state resident: full-time $5,816; part-time $299 per credit hour. Tuition, nonresident: full-time $6,356; part-time $326 per credit hour. *Financial support:* In 2005–06, 4 students received support, including 4 teaching assistantships with partial tuition reimbursements available (averaging $10,000 per year); research assistantships, Federal Work-Study, institutionally sponsored loans, tuition waivers (partial), unspecified assistantships, and service awards also available. Support available to part-time students. Financial award application deadline: 4/1; financial award applicants required to submit FAFSA. *Faculty research:* Differential equations numerical analysis, probability statistics, algebra, typology, knot theory. Total annual research expenditures:$3,689. *Unit head:* Dr. Mark P Robinson, Interim Head, 270-745-3651, Fax: 270-745-5385, E-mail: mark.robinson@wku.edu.

Western Michigan University, Graduate College, College of Arts and Sciences, Department of Mathematics, Programs in Mathematics, Kalamazoo, MI 49008-5202. Offers mathematics (MA); mathematics education (MA, PhD). *Degree requirements:* For master's, oral exams; for doctorate, one foreign language, thesis/dissertation, oral exams, 3 comprehensive exams, internship. *Entrance requirements:* For doctorate, GRE General Test.

Western Washington University, Graduate School, College of Sciences and Technology, Department of Mathematics, Bellingham, WA 98225-5996. Offers MS. Part-time programs available. *Faculty:* 24. *Students:* 18 full-time (9 women), 1 part-time; includes 2 minority (both Asian Americans or Pacific Islanders), 2 international. 24 applicants, 88% accepted, 11 enrolled. In 2005, 10 degrees awarded. *Degree requirements:* For master's, thesis (for some programs), project, qualifying examination. *Entrance requirements:* For master's, GRE General Test, minimum GPA of 3.0 in last 60 semester hours or last 90 quarter hours. Additional exam requirements/recommendations for international students: Required—TOEFL (minimum score 567 paper-based; 227 computer-based). *Application deadline:* For fall admission, 6/1 for domestic students. For winter admission, 10/1 for domestic students; for spring admission, 2/1 for domestic students. Applications are processed on a rolling basis. Application fee: $50. *Expenses:* Tuition, area resident: Part-time $188 per credit. Tuition, state resident: full-time $5,628; part-time $539 per credit. Tuition, nonresident: full-time $16,176. Required fees: $624. *Financial support:* In 2005–06, 17 teaching assistantships with partial tuition reimbursements (averaging $10,503 per year) were awarded; Federal Work-Study, institutionally sponsored loans, scholarships/grants, tuition waivers (partial), and unspecified assistantships also available. Support available to part-time students. Financial award application deadline: 2/15; financial award applicants required to submit FAFSA. *Unit head:* Dr. Tjalling Ypma, Chair, 360-650-3785. *Application contact:* Dr. Donald Chalice, Graduate Adviser, 360-650-3487.

West Texas A&M University, College of Agriculture, Nursing, and Natural Sciences, Department of Mathematics, Physical Sciences and Engineering Technology, Program in Mathematics, Canyon, TX 79016-0001. Offers MS. Part-time programs available. *Degree requirements:* For master's, thesis optional. *Entrance requirements:* For master's, GRE General Test. Additional exam requirements/recommendations for international students: Required—TOEFL (minimum score 550 paper-based). Electronic applications accepted.

West Virginia University, Eberly College of Arts and Sciences, Department of Mathematics, Morgantown, WV 26506. Offers applied mathematics (MS, PhD); discrete mathematics (PhD); interdisciplinary mathematics (MS); mathematics for secondary education (MS); pure mathematics (MS). Part-time programs available. *Faculty:* 28 full-time (1 woman), 11 part-time/adjunct (3 women). *Students:* 33 full-time (15 women), 8 part-time (2 women); includes 3 minority (1 African American, 2 Asian Americans or Pacific Islanders), 23 international. Average age 29. 30 applicants, 100% accepted, 16 enrolled. In 2005, 5 master's, 6 doctorates awarded. Terminal master's awarded for partial completion of doctoral program. *Degree requirements:* For master's, thesis optional; for doctorate, one foreign language, thesis/dissertation, comprehensive exam. *Entrance requirements:* For master's, minimum GPA of 2.5; for doctorate, master's degree in mathematics. Additional exam requirements/recommendations for international students: Required—TOEFL (paper score 550; computer score 213) or IELTS (paper score 6). *Application deadline:* For fall admission, 2/15 priority date for domestic students, 2/15 priority date for international students. Applications are processed on a rolling basis. Application fee: $50. *Expenses:* Tuition, state resident: full-time $4,582; part-time $258 per credit hour. Tuition, nonresident: full-time $1,382; part-time $741 per credit hour. *Financial support:* In 2005–06, 25 students received support, including 6 research assistantships with full tuition reimbursements available (averaging $1,000 per year), 18 teaching assistantships with full tuition reimbursements available (averaging $9,500 per year); Federal Work-Study, institutionally sponsored loans, and tuition waivers (full and partial) also available. Financial award application deadline: 2/15; financial award applicants required to submit FAFSA. *Faculty research:* Combinatorics and graph theory, topology, differential equations, applied and computational mathematics. Total annual research expenditures:$80,423. *Unit head:* Dr. Sherman Riemenschneider, Chair, 304-293-2011 Ext. 2322, Fax: 304-293-3982, E-mail: sherm.riemenschneider@mail.wvu.edu. *Application contact:* Dr. Harvey R. Diamond, Director of Graduate Studies, 304-293-2011 Ext. 2347, Fax: 304-293-3982, E-mail: harvey.diamond@mail.wvu.edu.

Wichita State University, Graduate School, Fairmount College of Liberal Arts and Sciences, Department of Mathematics and Statistics, Wichita, KS 67260. Offers applied mathematics (PhD); mathematics (MS); statistics (MS). Part-time programs available. *Degree requirements:* For master's, thesis optional; for doctorate, thesis/dissertation. *Entrance requirements:* For master's, GRE; for doctorate, GRE Subject Test. Additional exam requirements/recommendations for international students: Required—TOEFL. Electronic applications accepted. *Faculty research:* Partial differential equations, combinatorics, ring theory, minimal surfaces, several complex variables.

Wilfrid Laurier University, Faculty of Graduate Studies, Faculty of Science, Department of Mathematics, Waterloo, ON N2L 3C5, Canada. Offers M Sc. *Faculty:* 13 full-time. *Students:* 5 full-time, 1 part-time. 14 applicants, 64% accepted. *Degree requirements:* For master's, thesis optional. *Entrance requirements:* For master's, 4 year honors degree in mathematics, minimum B+ average. Additional exam requirements/recommendations for international students: Recommended—TOEFL (minimum score 230 computer-based). *Application deadline:* For fall admission, 2/1 for domestic students. Application fee: $75. *Unit head:* Dr. David Vaughan, Chairperson, 519-884-0710 Ext. 2297. *Application contact:* Dianne Duffy, 519-884-0710 Ext. 3127, Fax: 519-884-1020, E-mail: gradstudies@wlu.ca.

Wilkes University, Graduate Studies and Continued Learning, College of Science and Engineering, Department of Mathematics and Computer Science, Wilkes-Barre, PA 18766-0002. Offers mathematics (MS, MS Ed). Part-time programs available. *Students:* Average age 24. *Degree requirements:* For master's, thesis or alternative. *Entrance requirements:* For master's, GRE General Test. Additional exam requirements/recommendations for international students: Required—TOEFL (minimum score 500 paper-based; 173 computer-based). *Application deadline:* Applications are processed on a rolling basis. Application fee: $35. *Expenses:* Tuition: Part-time $717 per credit hour. Required fees: $43 per credit hour. *Financial support:* Federal Work-Study and unspecified assistantships available. Financial award application deadline: 3/1; financial award applicants required to submit FAFSA. *Unit head:* Dr. Ming Lew, Chair, 570-408-4844, Fax: 570-408-7860, E-mail: ming.lew@wilkes.edu. *Application contact:* Kathleen Diekhaus, Coordinator of Graduate Studies, 570-408-4160, Fax: 570-408-7860, E-mail: kathleen.diekhaus@wilkes.edu.

Worcester Polytechnic Institute, Graduate Studies and Enrollment, Department of Mathematical Sciences, Worcester, MA 01609-2280. Offers applied mathematics (MS); applied statistics (MS); financial mathematics (MS); industrial mathematics (MS); mathematical sciences (PhD); mathematics (MME). Part-time and evening/weekend programs available. *Faculty:* 30 full-time (3 women), 4 part-time/adjunct (0 women). *Students:* 28 full-time (13 women), 33 part-time (17 women); includes 5 minority (2 African Americans, 3 Asian Americans or Pacific Islanders), 14 international. 77 applicants, 77% accepted, 17 enrolled. In 2005, 15 degrees awarded. *Degree requirements:* For master's, thesis (for some programs); for doctorate, thesis/dissertation, comprehensive exam. *Entrance requirements:* For master's and doctorate, 3 letters of recommendation. Additional exam requirements/recommendations for international students: Required—TOEFL (minimum score 550 paper-based; 213 computer-based). *Application deadline:* For fall admission, 1/15 for domestic students; for spring admission, 10/15 priority date for domestic students. Applications are processed on a rolling basis. Application fee: $70. Electronic applications accepted. *Expenses:* Tuition: Part-time $997 per credit hour. *Financial support:* In 2005–06, 19 students received support, including fellowships with full tuition reimbursements available (averaging $33,246 per year), 5 research assistantships with full and partial tuition reimbursements available, 14 teaching assistantships with full and partial tuition reimbursements available; career-related internships or fieldwork, institutionally sponsored loans, scholarships/grants, and unspecified assistantships also available. Financial award application deadline: 1/15. *Faculty research:* Applied mathematical modeling and analysis, computational mathematics, discrete mathematics, applied and computational statistics, industrial and financial mathematics. Total annual research expenditures: $1.2 million. *Unit head:* Dr. Bogdan Vernescu, Head, 508-831-5241, Fax: 508-831-5824. *Application contact:* Dr. Homer F Walker, Graduate Coordinator, 508-831-6113, Fax: 508-831-5824, E-mail: walker@wpi.edu.

See Close-Up on page 559.

Wright State University, School of Graduate Studies, College of Science and Mathematics, Department of Mathematics and Statistics, Program in Mathematics, Dayton, OH 45435. Offers MS. *Degree requirements:* For master's, comprehensive exam. *Entrance requirements:* For master's, previous course work in mathematics beyond calculus. Additional exam requirements/recommendations for international students: Required—TOEFL. *Faculty research:* Analysis, algebraic combinatorics, graph theory, operator theory.

Yale University, Graduate School of Arts and Sciences, Department of Mathematics, New Haven, CT 06520. Offers MS, PhD. *Degree requirements:* For doctorate, 2 foreign languages, thesis/dissertation. *Entrance requirements:* For doctorate, GRE General Test, GRE Subject Test.

York University, Faculty of Graduate Studies, Faculty of Arts, Program in Mathematics and Statistics, Toronto, ON M3J 1P3, Canada. Offers industrial and applied mathematics (M Sc); mathematics and statistics (MA, PhD). Part-time programs available. *Faculty:* 57 full-time (13 women), 6 part-time/adjunct (2 women). *Students:* 69 full-time (33 women), 34 part-time (18 women). 300 applicants, 13% accepted, 38 enrolled. In 2005, 28 master's, 5 doctorates awarded. *Degree requirements:* For master's, thesis optional; for doctorate, one foreign language, thesis/dissertation, comprehensive exam, registration. *Application deadline:* For fall admission, 2/1 for domestic students. Application fee: $80. Electronic applications accepted. *Expenses:* Tuition, state resident: full-time $3,190; part-time $798 per term. International tuition: $7,515 full-time. Required fees: $217. Tuition and fees vary according to program. *Financial support:* In 2005–06, fellowships (averaging $8,354 per year), research assistantships (averaging $5,815 per year), teaching assistantships (averaging $9,475 per year) were awarded; tuition waivers (partial) and fee bursaries also available. *Unit head:* Yuehua Wu, Director, 416-736-2100.

Youngstown State University, Graduate School, College of Arts and Sciences, Department of Mathematics, Youngstown, OH 44555-0001. Offers MS. Part-time programs available. *Degree requirements:* For master's, thesis optional. *Entrance requirements:* For master's, minimum GPA of 2.7 in computer science and mathematics. Additional exam requirements/recommendations for international students: Required—TOEFL. *Faculty research:* Regression analysis, numerical analysis, statistics, Markov chain, topology and fuzzy sets.

Statistics

Acadia University, Faculty of Pure and Applied Science, Department of Mathematics and Statistics, Wolfville, NS B4P 2R6, Canada. Offers applied mathematics and statistics (M Sc). *Faculty:* 13 full-time (2 women). *Students:* 7 full-time (4 women). *Degree requirements:* For master's, thesis, 4-8 month industry internship. *Entrance requirements:* For master's, honors degree in mathematics, statistics or equivalent. Additional exam requirements/recommendations for international students: Required—TOEFL (minimum score 580 paper-based; 237 computer-based). *Application deadline:* For fall admission, 2/1 priority date for domestic students, 2/1 priority date for international students. Applications are processed on a rolling basis. Application fee: $50. *Financial support:* Career-related internships or fieldwork and unspecified assistantships available. *Faculty research:* Geophysical fluid dynamics, machine scheduling problems, control theory, stochastic optimization, survival analysis. *Unit head:* Dr. Paul Stephenson, Head, 902-585-1382, Fax: 902-585-1074, E-mail: paul.stephenson@acadiau.ca. *Application contact:* Dr. Richard H. Karsten, Professor, 902-585-1608, Fax: 902-585-1074, E-mail: richard.karsten@acadiau.ca.

American University, College of Arts and Sciences, Department of Mathematics and Statistics, Program in Statistics, Washington, DC 20016-8001. Offers applied statistics (Certificate); statistics (MS). Part-time and evening/weekend programs available. *Students:* 8 full-time (5 women), 12 part-time (4 women); includes 7 minority (3 African Americans, 4 Asian Americans or Pacific Islanders), 8 international. In 2005, 3 master's awarded. *Degree requirements:* For master's, one foreign language, thesis or alternative, comprehensive exam. *Entrance requirements:* For master's, GRE. *Application deadline:* For fall admission, 2/1 for domestic students; for spring admission, 10/1 for domestic students. Application fee: $50. *Expenses:* Tuition: Full-time $17,802; part-time $989 per credit. Required fees: $380. *Financial support:* Fellowships, teaching assistantships, career-related internships or fieldwork, Federal Work-Study, and institutionally sponsored loans available. Support available to part-time students. Financial award application deadline: 2/1. *Faculty research:* Statistical computing; data analysis; random processes; environmental, meteorological, and biological applications.

American University, College of Arts and Sciences, Department of Mathematics and Statistics, Program in Statistics for Policy Analysis, Washington, DC 20016-8001. Offers MS. Part-time programs available. *Students:* Average age 41. *Entrance requirements:* For master's, GRE. *Application deadline:* For fall admission, 2/1 for domestic students; for spring admission, 10/1 for domestic students. Application fee: $50. *Expenses:* Tuition: Full-time $17,802; part-time $989 per credit. Required fees: $380. *Financial support:* Application deadline: 2/1.

Arizona State University, Division of Graduate Studies, College of Liberal Arts and Sciences, Division of Natural Sciences and Mathematics, Department of Mathematics and Statistics, Tempe, AZ 85287. Offers applied mathematics (MA, PhD); mathematics (MA, MNS, PhD); statistics (MA, PhD). *Degree requirements:* For master's, thesis or alternative; for doctorate, one foreign language, thesis/dissertation. *Entrance requirements:* For master's and doctorate, GRE General Test.

Arizona State University, Division of Graduate Studies, Interdisciplinary Program in Statistics, Tempe, AZ 85287. Offers MS. *Entrance requirements:* For master's, GRE.

Ball State University, Graduate School, College of Sciences and Humanities, Department of Mathematical Sciences, Program in Mathematical Statistics, Muncie, IN 47306-1099. Offers MA. *Students:* 8 full-time (4 women), 5 part-time (1 woman), 4 international. Average age 27. 8 applicants, 38% accepted, 2 enrolled. In 2005, 2 degrees awarded. Application fee: $25 ($35 for international students). *Expenses:* Tuition, state resident: full-time $6,246. Tuition, nonresident: full-time $16,006. *Financial support:* Research assistantships with full tuition reimbursements, teaching assistantships with tuition reimbursements available. Financial award application deadline: 3/1. *Faculty research:* Robust methods. *Unit head:* Dr. Mir Ali, Director, 765-285-8640, Fax: 765-285-1721, E-mail: mali@bsu.edu.

Baylor University, Graduate School, College of Arts and Sciences, Department of Statistics, Waco, TX 76798. Offers MA, PhD. *Faculty:* 7 full-time (1 woman), 4 part-time/adjunct (1 woman). *Students:* 20 full-time (11 women), 3 part-time (1 woman); includes 4 minority (1 Asian American or Pacific Islander, 3 Hispanic Americans), 5 international. Average age 24. 38 applicants, 16% accepted. In 2005, 7 master's, 1 doctorate awarded. *Degree requirements:* For doctorate, thesis/dissertation. *Entrance requirements:* For master's, GRE General Test, 3 semesters of course work in calculus; for doctorate, GRE General Test. *Application deadline:* Applications are processed on a rolling basis. Application fee: $25. *Financial support:* In 2005–06, 1 fellowship, 5 research assistantships, 7 teaching assistantships were awarded; institutionally sponsored loans also available. *Faculty research:* Mathematical statistics, probability theory, biostatistics, linear models, time series. *Unit head:* Dr. Tom Bratcher, Graduate Program Director, 254-710-1699, Fax: 254-710-3033, E-mail: tom_bratcher@baylor.edu. *Application contact:* Suzanne Keener, Administrative Assistant, 254-710-3588, Fax: 254-710-3870.

Bernard M. Baruch College of the City University of New York, Zicklin School of Business, Department of Statistics and Computer Information Systems, Program in Statistics, New York, NY 10010-5585. Offers MBA, MS. Part-time and evening/weekend programs available. *Faculty:* 13 full-time (3 women), 11 part-time/adjunct (3 women). *Students:* 11 full-time (1 woman), 13 part-time (6 women). In 2005, 12 degrees awarded. *Entrance requirements:* For master's, GMAT, 2 letters of recommendation, resumé, 2 years of work experience. Additional exam requirements/recommendations for international students: Required—TOEFL (minimum score 590 paper-based; 243 computer-based), TWE. *Application deadline:* For fall admission, 5/31 for domestic students, 4/30 for international students; for spring admission, 10/31 for domestic students, 10/31 for international students. Application fee: $125. *Financial support:* Fellowships, research assistantships, teaching assistantships, career-related internships or fieldwork, Federal Work-Study, scholarships/grants, and unspecified assistantships available. Financial award application deadline: 4/30; financial award applicants required to submit FAFSA. *Unit head:* Hammou El Barmi, Head, 646-312-3384, Fax: 646-312-3351, E-mail: hammou_elbarmi@baruch.cuny.edu. *Application contact:* Frances Murphy, Office of Graduate Admissions, 646-312-1300, Fax: 646-312-1301, E-mail: zicklingradadmissions@baruch.cuny.edu.

Bowling Green State University, Graduate College, College of Arts and Sciences, Department of Mathematics and Statistics, Bowling Green, OH 43403. Offers applied statistics (MS); mathematics (MA, MAT, PhD); mathematics supervision (Ed S); statistics (MA, MAT, PhD). Part-time programs available. *Faculty:* 24 full-time (1 woman), 1 (woman) part-time/adjunct. *Students:* 58 full-time (21 women), 13 part-time (7 women); includes 1 African American, 32 international. Average age 28. 138 applicants, 68% accepted, 10 enrolled. In 2005, 16 master's, 6 doctorates awarded. *Degree requirements:* For master's, thesis or alternative; for doctorate, thesis/dissertation, comprehensive exam; for Ed S, internship. *Entrance requirements:* For master's and doctorate, GRE General Test. Additional exam requirements/recommendations for international students: Required—TOEFL. Application fee: $30. Electronic applications accepted. *Financial support:* In 2005–06, 1 research assistantship with full tuition reimbursement (averaging $10,000 per year), 53 teaching assistantships with full tuition reimbursements (averaging $12,493 per year) were awarded; Federal Work-Study, institutionally sponsored loans, and unspecified assistantships also available. Financial award applicants required to submit FAFSA. *Faculty research:* Statistics and probability, algebra, analysis. *Unit head:* Dr. Neal Carothers, Chair, 419-372-7453. *Application contact:* Dr. Hanfeng Chen, Graduate Coordinator, 419-372-7463, Fax: 419-372-6092.

See Close-Up on page 477.

Bowling Green State University, Graduate College, College of Business Administration, Department of Applied Statistics and Operations Research, Bowling Green, OH 43403. Offers applied statistics (MS). Part-time programs available. *Faculty:* 7 full-time (2 women). *Students:* 16 full-time (8 women), 2 part-time (1 woman), 13 international. Average age 27. 14 applicants, 86% accepted, 6 enrolled. In 2005, 8 degrees awarded. *Degree requirements:* For master's, thesis or alternative. *Entrance requirements:* For master's, GRE General Test. Additional exam requirements/recommendations for international students: Required—TOEFL. Application fee: $30. Electronic applications accepted. *Financial support:* In 2005–06, 3 research assistantships with full tuition reimbursements (averaging $9,018 per year), 18 teaching assistantships with full tuition reimbursements (averaging $6,182 per year) were awarded; career-related internships or fieldwork, institutionally sponsored loans, and unspecified assistantships also available. Financial award applicants required to submit FAFSA. *Faculty research:* Reliability, linear models, time series, statistical quality control. *Unit head:* Dr. B. Madhu Rao, Chair, 419-372-8011. *Application contact:* Arthur Yeh, Associate Professor, 419-372-8386.

Brigham Young University, Graduate Studies, College of Physical and Mathematical Sciences, Department of Statistics, Provo, UT 84602-1001. Offers applied statistics (MS). *Faculty:* 14 full-time (1 woman). *Students:* 25 full-time (9 women), 3 part-time; includes 6 minority (1 African American, 5 Asian Americans or Pacific Islanders). Average age 25. 22 applicants, 59% accepted, 9 enrolled. In 2005, 7 degrees awarded. *Degree requirements:* For master's, thesis (for some programs), comprehensive exam. *Entrance requirements:* For master's, GRE General Test, minimum GPA of 3.3 in last 60 hours, course work in multivariable calculus and linear algebra. Additional exam requirements/recommendations for international students: Required—TOEFL (minimum score 580 paper-based; 237 computer-based). *Application deadline:* For fall admission, 2/1 for domestic students, 2/1 for international students. Applications are processed on a rolling basis. Application fee: $50. Electronic applications accepted. *Financial support:* In 2005–06, 23 students received support, including 3 research assistantships with partial tuition reimbursements available (averaging $9,280 per year), 15 teaching assistantships with partial tuition reimbursements available (averaging $9,280 per year); career-related internships or fieldwork and tuition waivers (partial) also available. Financial award application deadline: 2/1. *Faculty research:* Statistical genetics, mixed models, reliability and pollution monitoring. Total annual research expenditures: $59,000. *Unit head:* Dr. Del T. Scott, Chair, 801-422-7054, Fax: 801-422-0635, E-mail: scottd@byu.edu. *Application contact:* Dr. Scott D. Grimshaw, Graduate Coordinator, 801-422-6251, Fax: 801-422-0635, E-mail: grimshaw@byu.edu.

Brock University, Graduate Studies, Faculty of Mathematics and Science, St. Catharines, ON L2S 3A1, Canada. Offers biological sciences (M Sc, PhD), including biology; biotechnology (M Sc, PhD); chemistry (M Sc); computer science (M Sc); earth sciences (M Sc); mathematics and statistics (M Sc); physics (M Sc). Part-time programs available. *Faculty:* 77 full-time (18 women), 10 part-time/adjunct (3 women). *Students:* 69 full-time (33 women), 3 part-time, 26 international. 137 applicants, 19% accepted, 20 enrolled. In 2005, 13 degrees awarded. *Degree requirements:* For master's, thesis. *Entrance requirements:* For master's, honors B Sc. Additional exam requirements/recommendations for international students: Required—TOEFL. *Application deadline:* Applications are processed on a rolling basis. Application fee: $75. Electronic applications accepted. *Financial support:* Fellowships, research assistantships, teaching assistantships, career-related internships or fieldwork, scholarships/grants, and unspecified assistantships available. Support available to part-time students. *Unit head:* Dr. Ian Brindle, Dean, 905-688-5550 Ext. 3421, Fax: 905-641-0406, E-mail: ibrindle@brocku.ca. *Application contact:* Charlotte F. Sheridan, Associate Director, Office of Graduate Studies, 905-688-5550 Ext. 4390, Fax: 905-688-0748, E-mail: csherida@brocku.ca.

California State University, East Bay, Academic Programs and Graduate Studies, College of Science, Department of Statistics, Hayward, CA 94542-3000. Offers actuarial statistics (MS); biostatistics (MS); computational statistics (MS); mathematical statistics (MS); statistics (MS); theoretical and applied statistics (MS). *Students:* 68. 35 applicants, 97% accepted. In 2005, 2 degrees awarded. *Degree requirements:* For master's, comprehensive exam. *Entrance requirements:* For master's, minimum GPA of 2.5 during previous 2 years of course work. Additional exam requirements/recommendations for international students: Required—TOEFL (minimum score 550 paper-based; 213 computer-based). *Application deadline:* For fall admission, 5/31 for domestic students, 4/30 for international students. For winter admission, 9/30 for domestic students. Application fee: $55. *Financial support:* Federal Work-Study and institutionally sponsored loans available. Support available to part-time students. Financial award application deadline: 3/2. *Unit head:* Dr. Julia Norton, Chair, 510-885-3435, E-mail: julia.norton@csueastbay.edu. *Application contact:* Deborah Baker, Associate Director, 510-885-3286, Fax: 510-885-4777, E-mail: deborah.baker@csueastbay.edu.

California State University, Fullerton, Graduate Studies, College of Business and Economics, Department of Information Systems and Decision Sciences, Fullerton, CA 92834-9480. Offers management information systems (MS); management science (MBA, MS); operations research (MS); statistics (MS). Part-time and evening/weekend programs available. *Students:* 18 full-time (5 women), 62 part-time (17 women); includes 16 minority (4 African Americans, 9 Asian Americans or Pacific Islanders, 3 Hispanic Americans), 11 international. Average age 32. 86 applicants, 55% accepted, 38 enrolled. In 2005, 23 degrees awarded. *Degree requirements:* For master's, project or thesis. *Entrance requirements:* For master's, GMAT, minimum AACSB index of 950. Application fee: $55. *Expenses:* Tuition, nonresident: part-time $339 per unit. *Financial support:* Teaching assistantships, Federal Work-Study, institutionally sponsored loans, and scholarships/grants available. Support available to part-time students. Financial award application deadline:3/1. *Unit head:* Dr. Barry Pasternack, Chair, 714-278-2221.

California State University, Sacramento, Graduate Studies, College of Natural Sciences and Mathematics, Department of Mathematics and Statistics, Sacramento, CA 95819-6048. Offers MA. Part-time programs available. *Students:* 6 full-time (3 women), 28 part-time (13 women); includes 9 minority (1 African American, 6 Asian Americans or Pacific Islanders, 2

Statistics

California State University, Sacramento (continued)
Hispanic Americans), 1 international. Average age 30. 36 applicants, 64% accepted, 14 enrolled. *Degree requirements:* For master's, thesis or alternative, writing proficiency exam. *Entrance requirements:* For master's, minimum GPA of 3.0 in mathematics, 2.5 overall during previous 2 years; BA in mathematics or equivalent. Additional exam requirements/recommendations for international students: Required—TOEFL. *Application deadline:* Applications are processed on a rolling basis. Application fee: $55. Electronic applications accepted. *Expenses:* Tuition, nonresident: part-time $339 per unit. Required fees: $276 per semester hour. *Financial support:* Research assistantships, teaching assistantships, career-related internships or fieldwork and Federal Work-Study available. Support available to part-time students. Financial award application deadline: 3/1. *Unit head:* Dr. Doraiswamy Ramachandran, Chair, 916-278-6534, Fax: 916-278-5586.

Carnegie Mellon University, College of Humanities and Social Sciences, Department of Statistics, Pittsburgh, PA 15213-3891. Offers mathematical finance (PhD); statistics (MS, PhD), including applied statistics (PhD), computational statistics (PhD), theoretical statistics (PhD). Terminal master's awarded for partial completion of doctoral program. *Degree requirements:* For doctorate, thesis/dissertation, comprehensive exam. *Entrance requirements:* For master's and doctorate, GRE General Test. Additional exam requirements/recommendations for international students: Required—TOEFL. *Faculty research:* Stochastic processes, Bayesian statistics, statistical computing, decision theory, psychiatric statistics.

See Close-Up on page 485.

Case Western Reserve University, School of Graduate Studies, Department of Statistics, Cleveland, OH 44106. Offers MS, PhD. *Degree requirements:* For master's, thesis (for some programs); for doctorate, thesis/dissertation. *Entrance requirements:* Additional exam requirements/recommendations for international students: Required—TOEFL. *Faculty research:* Generalized linear models, asymptotics for restricted MLE Bayesian inference, sample survey theory, statistical computing, nonparametric inference, projection pursuit, stochastic processes, dynamical systems and chaotic behavior.

Central Connecticut State University, School of Graduate Studies, School of Arts and Sciences, Department of Mathematics, New Britain, CT 06050-4010. Offers mathematics (MA, MS), including actuarial (MA), operations research (MA), statistics (MA). Part-time and evening/weekend programs available. *Faculty:* 31 full-time (8 women), 62 part-time/adjunct (29 women). *Students:* 30 full-time (16 women), 90 part-time (50 women); includes 13 minority (4 African Americans, 6 Asian Americans or Pacific Islanders, 3 Hispanic Americans), 7 international. Average age 34. 94 applicants, 56% accepted, 26 enrolled. In 2005, 22 master's awarded. *Degree requirements:* For master's, thesis or alternative, comprehensive exam or special project. *Entrance requirements:* For master's, minimum GPA of 2.4, conditional admissions to the Data Mining Program. Additional exam requirements/recommendations for international students: Required—TOEFL. *Application deadline:* For fall admission, 7/1 for domestic students; for spring admission, 12/1 for domestic students. Applications are processed on a rolling basis. Application fee: $50. Electronic applications accepted. *Expenses:* Tuition, area resident: Full-time $3,780. Tuition, state resident: full-time $5,670; part-time $362 per credit. Tuition, nonresident: full-time $10,530; part-time $362 per credit. Required fees: $3,064. One-time fee: $62 part-time. Tuition and fees vary according to degree level and program. *Financial support:* In 2005–06, 3 students received support; research assistantships, career-related internships or fieldwork, Federal Work-Study, scholarships/grants, and unspecified assistantships available. Support available to part-time students. Financial award application deadline: 3/1; financial award applicants required to submit FAFSA. *Faculty research:* Statistics, actuarial mathematics, computer systems and engineering, computer programming techniques, operations research. *Unit head:* Dr. Timothy Craine, Chair, 860-832-2835.

Claremont Graduate University, Graduate Programs, School of Mathematical Sciences, Claremont, CA 91711-6160. Offers computational and systems biology (PhD); computational science (PhD); engineering mathematics (PhD); operations research and statistics (MA, MS); physical applied mathematics (MA, MS); pure mathematics (MA, MS); scientific computing (MA, MS); systems and control theory (MA, MS). Part-time programs available. *Faculty:* 4 full-time (0 women), 6 part-time/adjunct (2 women). *Students:* 54 full-time (13 women), 13 part-time (1 woman); includes 20 minority (1 African American, 14 Asian Americans or Pacific Islanders, 5 Hispanic Americans), 16 international. Average age 38. In 2005, 5 master's, 5 doctorates awarded. Terminal master's awarded for partial completion of doctoral program. *Degree requirements:* For doctorate, 2 foreign languages, thesis/dissertation. *Entrance requirements:* For master's and doctorate, GRE General Test. *Application deadline:* For fall admission, 2/15 for domestic students. Applications are processed on a rolling basis. Electronic applications accepted. *Expenses:* Tuition: Full-time $27,902; part-time $1,214 per term. Required fees: $1,600; $800 per term. Tuition and fees vary according to degree level and program. *Financial support:* Fellowships, research assistantships, career-related internships or fieldwork, Federal Work-Study, institutionally sponsored loans, and tuition waivers (full and partial) available. Support available to part-time students. Financial award application deadline: 2/15; financial award applicants required to submit FAFSA. *Unit head:* John Angus, Dean, 909-621-8080, Fax: 909-607-8261, E-mail: john.angus@cgu.edu. *Application contact:* Susan Townzen, Program Coordinator, 909-621-8080, Fax: 909-607-8261, E-mail: susan.n.townzen@cgu.edu.

Clemson University, Graduate School, College of Agriculture, Forestry and Life Sciences, Department of Applied Economics and Statistics, Program in Applied Economics and Statistics, Clemson, SC 29634. Offers MS. *Students:* 10 full-time (2 women), 3 part-time, 3 international. Average age 24. 5 applicants, 100% accepted, 4 enrolled. In 2005, 5 degrees awarded. *Degree requirements:* For master's, thesis optional. *Entrance requirements:* For master's, GRE General Test, minimum GPA of 3.0. Additional exam requirements/recommendations for international students: Required—TOEFL. *Application deadline:* For fall admission, 5/1 for domestic students, 4/15 for international students; for spring admission, 10/1 for domestic students, 9/15 for international students. Applications are processed on a rolling basis. Application fee: $50. *Financial support:* Application deadline: 3/1; *Application contact:* Ellen Reneke, Staff Assistant for Graduate Programs, 864-656-5741, Fax: 864-656-5776, E-mail: ereneke@clemson.edu.

Clemson University, Graduate School, College of Engineering and Science, Department of Mathematical Sciences, Clemson, SC 29634. Offers applied and pure mathematics (MS, PhD); computational mathematics (MS, PhD); operations research (MS, PhD); statistics (MS, PhD). Part-time programs available. *Students:* 76 full-time (27 women), 5 part-time (2 women); includes 4 minority (2 African Americans, 2 Asian Americans or Pacific Islanders), 25 international. Average age 29. 73 applicants, 84% accepted, 24 enrolled. In 2005, 23 master's, 4 doctorates awarded. *Median time to degree:* Of those who began their doctoral program in fall 1996, 100% received their degree in 8 years or less. *Degree requirements:* For master's, final project, thesis optional; for doctorate, thesis/dissertation, qualifying exams. *Entrance requirements:* For master's and doctorate, GRE General Test. Additional exam requirements/recommendations for international students: Required—TOEFL; Recommended—TSE. *Application deadline:* For fall admission, 1/15 priority date for domestic students, 2/15 priority date for international students; for spring admission, 10/1 priority date for domestic students, 9/15 priority date for international students. Applications are processed on a rolling basis. Application fee: $50. Electronic applications accepted. *Financial support:* In 2005–06, 68 students received support; fellowships with partial tuition reimbursements available, research assistantships with partial tuition reimbursements available, teaching assistantships with partial tuition reimbursements available available. Financial award application deadline:4/15. *Faculty research:* Applied and computational analysis, cryptography, discrete mathematics, optimization, statistics. Total annual research expenditures: $1 million. *Unit head:* Robert L. Taylor, Chair, 864-656-5240, Fax: 864-656-5230, E-mail: rtaylo2@clemson.edu. *Application contact:* Dr. K.B. Kulasekera, Graduate Coordinator, 864-656-5231, Fax: 864-656-5230, E-mail: kk@clemson.edu.

See Close-Ups on pages 489 and 491.

Colorado State University, Graduate School, College of Natural Sciences, Department of Statistics, Fort Collins, CO 80523-0015. Offers MS, PhD. Part-time and evening/weekend programs available. Postbaccalaureate distance learning degree programs offered (no on-campus study). *Faculty:* 13 full-time (1 woman), 1 part-time/adjunct (0 women). *Students:* 30 full-time (15 women), 41 part-time (10 women); includes 8 minority (3 African Americans, 2 Asian Americans or Pacific Islanders, 3 Hispanic Americans), 16 international. Average age 33. 85 applicants, 27% accepted, 14 enrolled. In 2005, 11 master's, 2 doctorates awarded. Terminal master's awarded for partial completion of doctoral program. *Median time to degree:* Of those who began their doctoral program in fall 1996, 50% received their degree in 8 years or less. *Degree requirements:* For master's, project, seminar, thesis optional; for doctorate, thesis/dissertation, candidacy exam, preliminary exam, seminar. *Entrance requirements:* For master's and doctorate, GRE General Test, minimum GPA of 3.0. Additional exam requirements/recommendations for international students: Required—TOEFL (minimum score 550 paper-based; 213 computer-based). *Application deadline:* For fall admission, 2/15 priority date for domestic students, 2/15 priority date for international students. Applications are processed on a rolling basis. Application fee: $50. Electronic applications accepted. *Expenses:* Tuition, state resident: full-time $3,690; part-time $205 per credit. Tuition, nonresident: full-time $14,958; part-time $831 per credit. Required fees: $1,061. *Financial support:* In 2005–06, 9 fellowships with full tuition reimbursements, 10 research assistantships with full tuition reimbursements (averaging $13,500 per year), 21 teaching assistantships with full tuition reimbursements (averaging $12,960 per year) were awarded; career-related internships or fieldwork, Federal Work-Study, institutionally sponsored loans, traineeships, and unspecified assistantships also available. Support available to part-time students. Financial award application deadline: 2/15; financial award applicants required to submit FAFSA. *Faculty research:* Applied probability, linear models and experimental design, time-series analysis, non-parametric statistical inference, statistical consulting. Total annual research expenditures: $1.5 million. *Unit head:* F. Jay Breidt, Professor and Chair, 970-491-6786, Fax: 970-491-7895, E-mail: stats@colostate.edu. *Application contact:* Graduate Admissions Coordinator, 970-491-5269, Fax: 970-491-7895, E-mail: stats@colostate.edu.

Columbia University, Graduate School of Arts and Sciences, Division of Natural Sciences, Department of Statistics, New York, NY 10027. Offers M Phil, MA, PhD, MD/PhD. Part-time programs available. *Faculty:* 10 full-time, 1 part-time/adjunct. *Students:* 25 full-time (8 women), 50 part-time (17 women); includes 15 minority (3 African Americans, 9 Asian Americans or Pacific Islanders, 3 Hispanic Americans), 37 international. Average age 28. 98 applicants, 52% accepted. In 2005, 29 degrees awarded. *Degree requirements:* For doctorate, thesis/dissertation. *Entrance requirements:* For master's and doctorate, GRE General Test, GRE Subject Test. Additional exam requirements/recommendations for international students: Required—TOEFL. Application fee: $75. *Expenses:* Tuition: Full-time $31,448. Tuition and fees vary according to course level, course load, campus/location and program. *Financial support:* Fellowships, teaching assistantships, Federal Work-Study and institutionally sponsored loans available. Support available to part-time students. Financial award application deadline: 1/5; financial award applicants required to submit FAFSA. *Unit head:* Daniel Rabinowitz, Co-Chair, 212-851-2141, Fax: 212-851-2164, E-mail: dan@stat.columbia.edu.

Columbia University, Graduate School of Arts and Sciences, Program in Quantitative Methods in the Social Sciences, New York, NY 10027. Offers MA. Part-time programs available. Application fee: $75. *Expenses:* Tuition: Full-time $31,448. Tuition and fees vary according to course level, course load, campus/location and program. *Unit head:* Christopher Weiss, Director, 212-854-7559, Fax: 212-854-7925.

Cornell University, Graduate School, Graduate Fields of Engineering, Field of Operations Research and Industrial Engineering, Ithaca, NY 14853-0001. Offers applied probability and statistics (PhD); manufacturing systems engineering (PhD); mathematical programming (PhD); operations research and industrial engineering (M Eng). *Faculty:* 40 full-time (5 women). *Students:* 451 applicants, 25% accepted, 89 enrolled. In 2005, 87 master's, 4 doctorates awarded. *Degree requirements:* For doctorate, thesis/dissertation, comprehensive exam. *Entrance requirements:* For master's and doctorate, GRE General Test, 3 letters of recommendation. Additional exam requirements/recommendations for international students: Required—TOEFL (minimum score 600 paper-based; 250 computer-based). *Application deadline:* For fall admission, 1/15 for domestic students. Application fee: $60. Electronic applications accepted. *Financial support:* In 2005–06, 13 fellowships with full tuition reimbursements, 9 research assistantships with full tuition reimbursements, 24 teaching assistantships with full tuition reimbursements were awarded; institutionally sponsored loans, scholarships/grants, health care benefits, tuition waivers (full and partial), and unspecified assistantships also available. Financial award applicants required to submit FAFSA. *Faculty research:* Mathematical programming and combinatorial optimization, statistics, stochastic processes, mathematical finance, simulation, manufacturing, and e-commerce. *Unit head:* Director of Graduate Studies, 607-255-9128, Fax: 607-255-9129. *Application contact:* Graduate Field Assistant, 607-255-9128, Fax: 607-255-9129, E-mail: orie@cornell.edu.

Cornell University, Graduate School, Graduate Fields of Industrial and Labor Relations, Field of Statistics, Ithaca, NY 14853-0001. Offers applied statistics (MPS); biometry (MS, PhD); decision theory (MS, PhD); economic and social statistics (MS, PhD); engineering statistics (MS, PhD); experimental design (MS, PhD); mathematical statistics (MS, PhD); probability (MS, PhD); sampling (MS, PhD); statistical computing (MS, PhD); stochastic processes (MS, PhD). Terminal master's awarded for partial completion of doctoral program. *Degree requirements:* For master's, project (MPS), thesis (MS); for doctorate, one foreign language, thesis/dissertation. *Entrance requirements:* For master's, GRE General Test (MS), 2 letters of recommendation (MS and MPS); for doctorate, GRE General Test, 2 letters of recommendation. Additional exam requirements/recommendations for international students: Required—TOEFL (minimum score 550 paper-based; 213 computer-based). Electronic applications accepted. *Faculty research:* Bayesian analysis, survival analysis, nonparametric statistics, stochastic processes, mathematical statistics.

Dalhousie University, Faculty of Graduate Studies, College of Arts and Science, Faculty of Science, Department of Mathematics and Statistics, Program in Statistics, Halifax, NS B3H 4R2, Canada. Offers M Sc, PhD. *Degree requirements:* For master's and doctorate, thesis/dissertation, 50 hours of consulting. *Entrance requirements:* Additional exam requirements/recommendations for international students: Required—TOEFL. *Faculty research:* Data analysis, multivariate analysis, robustness, time series, statistical genetics.

DePaul University, College of Liberal Arts and Sciences, Department of Mathematical Sciences, Chicago, IL 60604-2287. Offers applied statistics and applied mathematics (MS); mathematics education (MA). Part-time and evening/weekend programs available. *Faculty:* 23 full-time (6 women), 18 part-time/adjunct (5 women). *Students:* 69 full-time (35 women), 61 part-time (30 women); includes 23 minority (9 African Americans, 10 Asian Americans or Pacific Islanders, 4 Hispanic Americans), 7 international. Average age 30. 40 applicants, 100% accepted. In 2005, 30 degrees awarded. *Application deadline:* Applications are processed on a rolling basis. Application fee: $25. *Financial support:* In 2005–06, 8 students received support, including research assistantships with partial tuition reimbursements available (averaging $3,700 per year); teaching assistantships, tuition waivers (full and partial) also available. *Faculty research:* Verbally prime algebras, enveloping algebras of Lie, superalgebras and related rings, harmonic analysis, estimation theory. *Unit head:* Dr. Ahmed I Zayed, Chairperson, 773-325-7806, E-mail: azayed@depaul.edu.

Duke University, Graduate School, Institute of Statistics and Decision Sciences, Durham, NC 27708-0586. Offers PhD. Part-time programs available. *Faculty:* 17 full-time. *Students:* 33 full-time (11 women), 21 international. 89 applicants, 16% accepted, 9 enrolled. In 2005, 3 doctorates awarded. *Degree requirements:* For doctorate, thesis/dissertation. *Entrance requirements:* For doctorate, GRE General Test. Additional exam requirements/recommendations for international students: Required—IELT (preferred) or TOEFL. *Application deadline:* For fall admission, 12/31 for domestic students, 12/31 for international students. Application fee: $75.

Electronic applications accepted. *Financial support:* Fellowships, research assistantships, teaching assistantships available. Financial award application deadline: 12/31. *Unit head:* Alan Gelfand, Director of Graduate Studies, 919-684-8029, Fax: 919-684-8594, E-mail: dgs@stat.duke.edu.

Eastern Michigan University, Graduate School, College of Arts and Sciences, Department of Mathematics, Ypsilanti, MI 48197. Offers computer science (M Math); mathematics education (M Math); statistics (M Math). Evening/weekend programs available. *Faculty:* 25 full-time (10 women). *Students:* 5 full-time (4 women), 39 part-time (23 women); includes 7 minority (2 African Americans, 3 Asian Americans or Pacific Islanders, 2 Hispanic Americans), 8 international. Average age 33. In 2005, 17 degrees awarded. *Degree requirements:* For master's, thesis optional. *Entrance requirements:* Additional exam requirements/recommendations for international students: Required—TOEFL. *Application deadline:* For fall admission, 5/15 priority date for domestic students, 5/1 priority date for international students. For winter admission, 10/15 for domestic students; for spring admission, 3/15 for domestic students. Applications are processed on a rolling basis. Application fee: $35. *Expenses:* Tuition, state resident: full-time $7,838; part-time $327 per credit hour. Tuition, nonresident: full-time $15,770; part-time $657 per credit hour. Required fees: $33 per credit hour. $40 per term. Tuition and fees vary according to course level, course load and degree level. *Financial support:* In 2005–06, fellowships (averaging $4,000 per year), research assistantships (averaging $8,950 per year), teaching assistantships (averaging $8,950 per year) were awarded. Support available to part-time students. Financial award applicants required to submit FAFSA. *Unit head:* Dr. Betty Warren, Head, 734-487-1444. *Application contact:* Dr. Walter Parry, Coordinator, 734-487-1444.

Florida International University, College of Arts and Sciences, Department of Statistics, Miami, FL 33199. Offers MS. *Faculty:* 11 full-time (2 women), 1 (woman) part-time/adjunct. *Students:* 5 full-time (3 women), 2 part-time (both women); includes 3 minority (1 Asian American or Pacific Islander, 2 Hispanic Americans), 3 international. 7 applicants, 43% accepted, 1 enrolled. In 2005, 1 degree awarded. *Degree requirements:* For master's, thesis optional. *Entrance requirements:* For master's, GRE General Test, minimum GPA of 3.0, 3 letters of recommendation. Additional exam requirements/recommendations for international students: Required—TOEFL. Application fee: $25. *Expenses:* Tuition, area resident: Part-time $239 per credit. Tuition, state resident: full-time $4,294; part-time $869 per credit. Tuition, nonresident: full-time $15,641. Required fees: $252; $126 per term. Tuition and fees vary according to program. *Unit head:* Dr. Sneh Gulati, Chairperson, 305-348-2745, Fax: 305-348-6895, E-mail: sneh.gulati@fiu.edu.

Florida State University, Graduate Studies, College of Arts and Sciences, Department of Statistics, Tallahassee, FL 32306. Offers applied statistics (MS); biostatistics (MS); mathematical statistics (MS, PhD). Part-time programs available. *Faculty:* 13 full-time (2 women). *Students:* 47 full-time (20 women), 8 part-time (2 women); includes 4 minority (all African Americans), 38 international. Average age 30. 141 applicants, 35% accepted, 18 enrolled. In 2005, 2 master's, 4 doctorates awarded. Terminal master's awarded for partial completion of doctoral program. *Median time to degree:* Of those who began their doctoral program in fall 1996, 100% received their degree in 8 years or less. *Degree requirements:* For master's, comprehensive exam (mathematical statistics); for doctorate, thesis/dissertation, departmental qualifying exam. *Entrance requirements:* For master's, GRE General Test, previous course work in calculus, minimum GPA of 3.0; for doctorate, GRE General Test, minimum GPA of 3.0, 1 course in linear algebra (preferred), Calculus I-III. Additional exam requirements/recommendations for international students: Required—TOEFL (minimum score 600 paper-based; 250 computer-based). *Application deadline:* For fall admission, 7/1 for domestic students, 5/2 for international students; for spring admission, 11/1 for domestic students, 9/1 for international students. Applications are processed on a rolling basis. Application fee: $30. Electronic applications accepted. *Financial support:* In 2005–06, 2 fellowships with full tuition reimbursements (averaging $18,000 per year), 8 research assistantships with full tuition reimbursements (averaging $17,612 per year), 35 teaching assistantships with full tuition reimbursements (averaging $17,612 per year) were awarded; Federal Work-Study, institutionally sponsored loans, scholarships/grants, unspecified assistantships, and health insurance supplements also available. Support available to part-time students. Financial award application deadline: 2/15; financial award applicants required to submit FAFSA. *Faculty research:* Statistical inference, probability theory, spatial statistics, nonparametric estimation, automatic target recognition. Total annual research expenditures: $626,760. *Unit head:* Dr. Dan McGee, Chairman, 850-644-3218, Fax: 850-644-5271, E-mail: info@stat.fsu.edu. *Application contact:* Jennifer Rivera, Program Assistant, 850-644-3218, Fax: 850-644-5271, E-mail: info@stat.fsu.edu.

Florida State University, Graduate Studies, College of Education, Department of Educational Psychology and Learning Systems, Program in Measurement and Statistics, Tallahassee, FL 32306. Offers MS, PhD. *Faculty:* 3 full-time (1 woman), 2 part-time/adjunct (0 women). *Students:* 15 full-time (4 women), 5 part-time (4 women); includes 4 minority (1 African American, 2 Asian Americans or Pacific Islanders, 1 Hispanic American), 11 international. Average age 20. 11 applicants, 45% accepted, 3 enrolled. In 2005, 3 master's, 1 doctorate awarded. *Application deadline:* For fall admission, 7/1 for domestic students; for spring admission, 11/1 for domestic students. Application fee: $30. *Financial support:* In 2005–06, fellowships with partial tuition reimbursements (averaging $5,000 per year), research assistantships with partial tuition reimbursements (averaging $18,000 per year), teaching assistantships with partial tuition reimbursements (averaging $18,000 per year) were awarded. *Unit head:* Dr. Betsy Beeker, Program Leader, 850-645-2371, Fax: 850-644-8776, E-mail: bjbecker@coe.fsu.edu. *Application contact:* Sally Gadson, Program Assistant, 850-644-5473, Fax: 850-644-5067, E-mail: sgadson@fsu.edu.

George Mason University, School of Information Technology and Engineering, Department of Applied and Engineering Statistics, Fairfax, VA 22030. Offers statistical science (MS). Part-time and evening/weekend programs available. *Faculty:* 9 full-time (2 women), 4 part-time/adjunct (0 women). *Students:* 6 full-time (4 women), 43 part-time (21 women); includes 9 minority (1 African American, 8 Asian Americans or Pacific Islanders), 15 international. Average age 33. 38 applicants, 66% accepted, 11 enrolled. In 2005, 15 degrees awarded. *Degree requirements:* For master's, thesis optional. *Entrance requirements:* For master's, GMAT or GRE General Test, previous course work in calculus, probability, and statistics; minimum GPA of 3.0 in last 60 hours of course work. Additional exam requirements/recommendations for international students: Required—TOEFL. *Application deadline:* For fall admission, 5/1 for domestic students; for spring admission, 11/1 for domestic students. Application fee: $60. Electronic applications accepted. *Expenses:* Tuition, area resident: Full-time $5,244; part-time $219 per credit. Tuition, state resident: part-time $651 per credit. Tuition, nonresident: full-time $15,636. Required fees: $1,524; $65 per credit. *Financial support:* Fellowships, research assistantships, teaching assistantships, career-related internships or fieldwork and Federal Work-Study available. Support available to part-time students. Financial award application deadline: 3/1; financial award applicants required to submit FAFSA. *Faculty research:* Computational statistics, nonparametric function estimation, scientific and statistical visualization, statistical applications to engineering, survey research. Total annual research expenditures: $436,000. *Unit head:* Dr. Richard A. Bolstein, Head, 703-993-3645, Fax: 703-993-1700, E-mail: statistics@gmu.edu.

The George Washington University, Columbian College of Arts and Sciences, Department of Statistics, Washington, DC 20052. Offers biostatistics (MS, PhD); industrial and engineering statistics (MS); statistics (MS, PhD); survey design and data analysis (Graduate Certificate). Part-time and evening/weekend programs available. Terminal master's awarded for partial completion of doctoral program. *Degree requirements:* For master's, comprehensive exam; for doctorate, thesis/dissertation, general exam. *Entrance requirements:* For master's and doctorate, GRE General Test, interview, minimum GPA of 3.0. Additional exam requirements/recommendations for international students: Required—TOEFL (minimum score 550 paper-based; 213 computer-based). Electronic applications accepted.

Georgia Institute of Technology, Graduate Studies and Research, College of Sciences, School of Mathematics, Atlanta, GA 30332-0001. Offers algorithms, combinatorics, and optimization (PhD); applied mathematics (MS); bioinformatics (PhD); mathematics (PhD); quantitative and computational finance (MS); statistics (MS Stat). Terminal master's awarded for partial completion of doctoral program. *Degree requirements:* For master's, thesis or alternative; for doctorate, one foreign language, thesis/dissertation. *Entrance requirements:* For master's, GRE General Test, minimum GPA of 3.0; for doctorate, GRE General Test, GRE Subject Test, minimum GPA of 3.0. Additional exam requirements/recommendations for international students: Required—TOEFL. Electronic applications accepted. *Faculty research:* Dynamical systems, discrete mathematics, probability and statistics, mathematical physics.

Georgia Institute of Technology, Graduate Studies and Research, Multidisciplinary Program in Statistics, Atlanta, GA 30332-0001. Offers MS Stat. Part-time programs available. *Degree requirements:* For master's, thesis optional. *Entrance requirements:* For master's, GRE General Test, minimum GPA of 3.0. Additional exam requirements/recommendations for international students: Required—TOEFL. *Faculty research:* Statistical control procedures, statistical modeling of transportation systems.

Harvard University, Graduate School of Arts and Sciences, Department of Statistics, Cambridge, MA 02138. Offers AM, PhD. *Students:* 19 full-time (9 women). 57 applicants, 28% accepted. In 2005, 8 master's, 2 doctorates awarded. Terminal master's awarded for partial completion of doctoral program. *Degree requirements:* For master's, one foreign language; for doctorate, one foreign language, thesis/dissertation, exam, qualifying paper. *Entrance requirements:* For master's and doctorate, GRE General Test, GRE Subject Test (recommended). Additional exam requirements/recommendations for international students: Required—TOEFL. *Application deadline:* For fall admission, 12/15 for domestic students. Application fee: $60. *Expenses:* Tuition: Full-time $28,752. Full-time tuition and fees vary according to program and student level. *Financial support:* Fellowships, research assistantships, teaching assistantships, career-related internships or fieldwork, Federal Work-Study, and institutionally sponsored loans available. Financial award application deadline: 12/30. *Faculty research:* Interactive graphic analysis of multidimensional data, data analysis, modeling and inference, statistical modeling of U.S. economic time series. *Unit head:* Betsey Cogswell, Administrator, 617-495-5497, Fax: 617-495-5264. *Application contact:* Office of Admissions and Financial Aid, 617-495-5315.

Indiana University Bloomington, Graduate School, College of Arts and Sciences, Department of Mathematics, Bloomington, IN 47405-7000. Offers applied mathematics–numerical analysis (MA, PhD); mathematics education (MAT); probability-statistics (MA, PhD). PhD offered through the University Graduate School. *Faculty:* 36 full-time (1 woman). *Students:* 116 full-time (39 women), 20 part-time (4 women); includes 6 minority (1 African American, 5 Asian Americans or Pacific Islanders), 74 international. Average age 27. In 2005, 21 master's, 5 doctorates awarded. Terminal master's awarded for partial completion of doctoral program. *Degree requirements:* For doctorate, one foreign language, thesis/dissertation. *Entrance requirements:* For master's and doctorate, GRE General Test, GRE Subject Test. Additional exam requirements/recommendations for international students: Required—TOEFL. *Application deadline:* For fall admission, 1/15 priority date for domestic students, 12/15 priority date for international students; for spring admission, 9/1 priority date for domestic students, 9/1 priority date for international students. Applications are processed on a rolling basis. Application fee: $50 ($60 for international students). Electronic applications accepted. *Expenses:* Tuition, state resident: full-time $5,437; part-time $227 per credit hour. Tuition, nonresident: full-time $15,836; part-time $660 per credit hour. Required fees: $821. Tuition and fees vary according to campus/location and program. *Financial support:* Fellowships with full tuition reimbursements, research assistantships, teaching assistantships with full tuition reimbursements, Federal Work-Study available. Support available to part-time students. Financial award application deadline: 2/1. *Faculty research:* Topology, geometry, algebra. *Unit head:* David Hoff, Chair, 812-855-2200. *Application contact:* Misty Cummings, Graduate Secretary, 812-855-2645, Fax: 812-855-0046, E-mail: gradmath@indiana.edu.

Indiana University–Purdue University Fort Wayne, School of Arts and Sciences, Department of Mathematical Sciences, Fort Wayne, IN 46805-1499. Offers applied mathematics (MS); applied statistics (Certificate); mathematics (MS); operations research (MS). Part-time and evening/weekend programs available. *Faculty:* 20 full-time (4 women). *Students:* 8 full-time (1 woman), 9 part-time (4 women); includes 2 minority (both American Indian/Alaska Native). Average age 33. 10 applicants, 100% accepted, 8 enrolled. In 2005, 5 degrees awarded. *Degree requirements:* For Certificate, completed a calculus and a statistics course. *Entrance requirements:* For master's, minimum GPA of 3.0, major or minor in mathematics. Additional exam requirements/recommendations for international students: Required—TOEFL (minimum score 600 paper-based; 260 computer-based). *Application deadline:* For fall admission, 7/1 for domestic students; for spring admission, 12/1 for domestic students. Applications are processed on a rolling basis. Application fee: $55. *Expenses:* Tuition, state resident: full-time $4,023; part-time $232 per credit. Tuition, nonresident: full-time $9,182; part-time $503 per credit. Required fees: $383. Tuition and fees vary according to course load. *Financial support:* In 2005–06, 10 teaching assistantships with partial tuition reimbursements (averaging $11,700 per year) were awarded; scholarships/grants and unspecified assistantships also available. Support available to part-time students. Financial award application deadline: 3/1; financial award applicants required to submit FAFSA. *Faculty research:* Graph theory, biostatistics, partial differential equations, 11-venn diagrams, random operator equations. Total annual research expenditures: $67,015. *Unit head:* Dr. David A. Legg, Chair, 260-481-6821, Fax: 260-481-6880, E-mail: legg@ipfw.edu. *Application contact:* Dr. W. Douglas Weakley, Director of Graduate Studies, 260-481-6821, Fax: 260-481-6880, E-mail: weakley@ipfw.edu.

Indiana University–Purdue University Indianapolis, School of Science, Department of Mathematical Sciences, Indianapolis, IN 46202-3216. Offers applied mathematics (MS, PhD); applied statistics (MS); mathematics (MS, PhD). Part-time programs available. *Faculty:* 10 full-time (0 women), 1 part-time/adjunct (0 women). *Students:* 12 full-time (4 women), 42 part-time (20 women); includes 3 minority (1 African American, 2 Asian Americans or Pacific Islanders), 16 international. Average age 32. In 2005, 19 degrees awarded. Terminal master's awarded for partial completion of doctoral program. *Degree requirements:* For master's, thesis optional; for doctorate, one foreign language, thesis/dissertation. *Entrance requirements:* For doctorate, GRE. Additional exam requirements/recommendations for international students: Required—TOEFL. *Application deadline:* For fall admission, 2/1 for domestic students. Application fee: $50 ($60 for international students). *Expenses:* Tuition, state resident: full-time $5,159; part-time $215 per credit hour. Tuition, nonresident: full-time $14,890; part-time $620 per credit hour. Required fees: $614. Tuition and fees vary according to campus/location and program. *Financial support:* In 2005–06, 14 students received support; fellowships with tuition reimbursements available, research assistantships with tuition reimbursements available, teaching assistantships with tuition reimbursements available, career-related internships or fieldwork, Federal Work-Study, and tuition waivers (full and partial) available. Financial award application deadline: 3/1. *Faculty research:* Mathematical physics, analysis, operator theory, functional analysis, integrated systems. *Unit head:* Benzion Boukai, Chair, 317-274-6920, Fax: 317-274-3460, E-mail: bboukai@math.iupui.edu. *Application contact:* Joan Morand, Student Services Specialist, 317-274-4127, Fax: 317-274-3460.

See Close-Up on page 503.

Instituto Tecnológico y de Estudios Superiores de Monterrey, Campus Monterrey, Graduate and Research Division, Programs in Engineering, Monterrey, Mexico. Offers applied statistics (M Eng); artificial intelligence (PhD); automation engineering (M Eng); chemical engineering (M Eng); civil engineering (M Eng); electrical engineering (M Eng); electronic engineering (M Eng); environmental engineering (M Eng); industrial engineering (M Eng, PhD); manufacturing engineering (M Eng); mechanical engineering (M Eng); systems and quality engineering (M Eng). Part-time and evening/weekend programs available. Terminal master's awarded for partial completion of doctoral program. *Degree requirements:* For master's

Statistics

Instituto Tecnológico y de Estudios Superiores de Monterrey, Campus Monterrey (continued)
and doctorate, one foreign language, thesis/dissertation. *Entrance requirements:* For master's, PAEG; for doctorate, GRE, master's degree in related field. Additional exam requirements/recommendations for international students: Required—TOEFL. *Faculty research:* Flexible manufacturing cells, materials, statistical methods, environmental prevention, control and evaluation.

Iowa State University of Science and Technology, Graduate College, College of Liberal Arts and Sciences, Department of Statistics, Ames, IA 50011. Offers MS, PhD, MBA/MS. *Faculty:* 35 full-time, 3 part-time/adjunct. *Students:* 89 full-time (45 women), 25 part-time (12 women); includes 5 minority (1 African American, 4 Asian Americans or Pacific Islanders), 59 international. 160 applicants, 31% accepted, 25 enrolled. In 2005, 41 master's, 5 doctorates awarded. *Degree requirements:* For master's, thesis or alternative; for doctorate, thesis/dissertation. *Entrance requirements:* For master's and doctorate, GRE General Test. Additional exam requirements/recommendations for international students: Required—TOEFL (paper score 550; computer score 213) or IELTS (score 6.5). *Application deadline:* For fall admission, 3/15 priority date for domestic students, 3/15 priority date for international students; for spring admission, 10/31 for domestic students, 10/31 for international students. Applications are processed on a rolling basis. Application fee: $30 ($70 for international students). *Expenses:* Tuition, state resident: full-time $6,410. Tuition, nonresident: full-time $16,422. Tuition and fees vary according to program. *Financial support:* In 2005–06, 44 research assistantships with full and partial tuition reimbursements (averaging $16,393 per year), 39 teaching assistantships with full and partial tuition reimbursements (averaging $16,598 per year) were awarded; fellowships, scholarships/grants, health care benefits, and unspecified assistantships also available. *Unit head:* Dr. Kenneth Koehler, Chair, 515-294-4181, Fax: 515-294-4040, E-mail: statistics@iastate.edu.

James Madison University, College of Graduate and Professional Programs, College of Science and Mathematics, Department of Mathematics and Statistics, Harrisonburg, VA 22807. Offers M Ed. Part-time programs available. *Faculty:* 2 part-time/adjunct (1 woman). *Application deadline:* For fall admission, 5/1 for domestic students; for spring admission, 9/1 priority date for domestic students. Application fee: $55. *Expenses:* Tuition, state resident: full-time $5,904; part-time $246 per credit hour. Tuition, nonresident: full-time $16,824; part-time $701 per credit hour. *Financial support:* Application deadline: 3/1; *Unit head:* Dr. David C. Carothers, Academic Unit Head, 540-568-6184.

The Johns Hopkins University, G. W. C. Whiting School of Engineering, Department of Applied Mathematics and Statistics, Baltimore, MD 21218-2699. Offers discrete mathematics (MA, MSE, PhD); operations research/optimization/decision science (MA, MSE, PhD); statistics/probability/stochastic processes (MA, MSE, PhD). *Faculty:* 15 full-time (1 woman), 3 part-time/adjunct (all women). *Students:* 36 full-time (12 women), 2 part-time (1 woman); includes 2 minority (both Asian Americans or Pacific Islanders), 20 international. Average age 27. 112 applicants, 67% accepted, 22 enrolled. In 2005, 12 master's, 3 doctorates awarded. Terminal master's awarded for partial completion of doctoral program. *Median time to degree:* Of those who began their doctoral program in fall 1996, 40% received their degree in 8 years or less. *Degree requirements:* For master's, thesis (for some programs), registration; for doctorate, thesis/dissertation, oral exam, introductory exam. *Entrance requirements:* For master's and doctorate, GRE General Test, GRE Subject Test. Additional exam requirements/recommendations for international students: Required—TOEFL (minimum score 600 paper-based; 250 computer-based). *Application deadline:* For fall admission, 1/15 priority date for domestic students, 1/15 priority date for international students. Applications are processed on a rolling basis. Application fee: $0. Electronic applications accepted. *Expenses:* Tuition: Full-time $30,960. Tuition and fees vary according to degree level and program. *Financial support:* In 2005–06, 34 students received support, including 1 fellowship with full tuition reimbursement available (averaging $17,500 per year), 3 research assistantships with full tuition reimbursements available (averaging $15,750 per year), 20 teaching assistantships with full tuition reimbursements available (averaging $15,750 per year); Federal Work-Study, institutionally sponsored loans, health care benefits, tuition waivers (partial), and unspecified assistantships also available. Financial award application deadline: 1/15. *Faculty research:* Discrete mathematics, probability, statistics, optimization and operations research, scientific computation. Total annual research expenditures:$587,000. *Unit head:* Dr. Daniel Q. Naiman, Chair, 410-516-7203, Fax: 410-516-7459, E-mail: daniel.naiman@jhu.edu. *Application contact:* Kristin Bechtel, Academic Program Coordinator, 410-516-7198, Fax: 410-516-7459, E-mail: bechtel@ams.jhu.edu.

Kansas State University, Graduate School, College of Arts and Sciences, Department of Statistics, Manhattan, KS 66506. Offers MS, PhD. *Faculty:* 10 full-time (2 women), 2 part-time/adjunct (0 women). *Students:* 37 full-time (15 women), 5 part-time (2 women); includes 6 minority (5 Asian Americans or Pacific Islanders, 1 Hispanic American), 22 international. Average age 24. 47 applicants, 60% accepted, 15 enrolled. In 2005, 12 master's, 2 doctorates awarded. Terminal master's awarded for partial completion of doctoral program. *Degree requirements:* For master's, thesis optional; for doctorate, thesis/dissertation, qualifying and preliminary exams. *Entrance requirements:* For master's, GRE; for doctorate, previous course work in statistics and mathematics. Additional exam requirements/recommendations for international students: Required—TOEFL (minimum score 550 paper-based; 213 computer-based). *Application deadline:* For fall admission, 2/1 priority date for domestic students, 2/1 priority date for international students; for spring admission, 10/1 priority date for domestic students, 8/1 priority date for international students. Applications are processed on a rolling basis. Application fee: $30 ($55 for international students). *Expenses:* Tuition, state resident: full-time $5,160; part-time $215. Tuition, nonresident: full-time $12,816; part-time $534. Required fees: $564. *Financial support:* In 2005–06, 2 research assistantships (averaging $24,317 per year), 27 teaching assistantships with full tuition reimbursements (averaging $14,823 per year) were awarded; fellowships, Federal Work-Study, institutionally sponsored loans, and scholarships/grants also available. Support available to part-time students. Financial award application deadline: 3/1; financial award applicants required to submit FAFSA. *Faculty research:* Linear and nonlinear statistical models, design analysis of experiments, nonparametric methods for reliability and survival data, resampling methods and their application, categorical data analysis. Total annual research expenditures: $107,558. *Unit head:* John Boyer, Head, 785-532-0518, Fax: 785-532-7336, E-mail: jboyer@stat.ksu.edu. *Application contact:* James Neill, Director, 785-532-0516, E-mail: jwneill@ksu.edu.

Kean University, College of Natural, Applied and Health Sciences, Program in Computing, Statistics and Mathematics, Union, NJ 07083. Offers MS. Part-time and evening/weekend programs available. *Faculty:* 24 full-time (6 women). *Students:* 7 full-time (3 women), 5 part-time (2 women); includes 5 minority (2 African Americans, 1 Asian American or Pacific Islander, 2 Hispanic Americans), 1 international. Average age 36. 1 applicant, 100% accepted, 1 enrolled. In 2005, 3 degrees awarded. *Degree requirements:* For master's, thesis or alternative, research component, minimum 3.0 GPA. *Entrance requirements:* For master's, GRE General Test, 2 letters of recommendation, interview. *Application deadline:* For fall admission, 5/1 for domestic students; for spring admission, 11/1 for domestic students. Application fee: $60 ($150 for international students). Electronic applications accepted. *Expenses:* Tuition, state resident: full-time $8,280; part-time $345 per credit. Tuition, nonresident: full-time $11,512; part-time $438 per credit. Required fees: $2,104; $88 per credit. *Financial support:* In 2005–06, 3 research assistantships with full tuition reimbursements (averaging $2,880 per year) were awarded *Unit head:* Dr. Francine Abeles, Program Coordinator, 908-737-3714, E-mail: fabeles@kean.edu. *Application contact:* Joanne Morris, Director of Graduate Admissions, 908-737-3355, Fax: 908-737-3354, E-mail: grad-adm@kean.edu.

Lakehead University, Graduate Studies, School of Mathematical Sciences, Thunder Bay, ON P7B 5E1, Canada. Offers computer science (M Sc, MA); mathematics and statistics (M Sc, MA). Part-time and evening/weekend programs available. *Degree requirements:* For master's, thesis optional. *Entrance requirements:* For master's, minimum B average, honours degree in mathematics or computer science. Additional exam requirements/recommendations for international students: Required—TOEFL. *Faculty research:* Numerical analysis, classical analysis, theoretical computer science, abstract harmonic analysis, functional analysis.

Lehigh University, College of Arts and Sciences, Department of Mathematics, Bethlehem, PA 18015-3094. Offers applied mathematics (MS, PhD); mathematics (MS, PhD); statistics (MS). Part-time programs available. *Faculty:* 20 full-time (1 woman), 3 part-time/adjunct (0 women). *Students:* 33 full-time (11 women), 7 part-time (4 women); includes 1 Asian American or Pacific Islander, 10 international. Average age 26. 68 applicants, 71% accepted, 14 enrolled. In 2005, 5 master's, 1 doctorate awarded. Terminal master's awarded for partial completion of doctoral program. *Median time to degree:* Of those who began their doctoral program in fall 1996, 50% received their degree in 8 years or less. *Degree requirements:* For master's, comprehensive exam; for doctorate, one foreign language, thesis/dissertation, qualifying exams, comprehensive exam. *Entrance requirements:* For master's and doctorate, minimum GPA of 3.0. Additional exam requirements/recommendations for international students: Required—TOEFL. *Application deadline:* For fall admission, 7/15 for domestic students; for spring admission, 12/7 priority date for domestic students. Applications are processed on a rolling basis. Application fee: $60. Electronic applications accepted. *Financial support:* In 2005–06, 30 students received support, including 1 fellowship with full tuition reimbursement available (averaging $16,000 per year), 2 research assistantships with full tuition reimbursements available (averaging $14,000 per year), 20 teaching assistantships with full tuition reimbursements available (averaging $13,700 per year); scholarships/grants, tuition waivers (partial), and unspecified assistantships also available. Financial award application deadline: 1/15. *Faculty research:* Probability and statistics, geometry and topology, number theory, algebra, differential equations. *Unit head:* Dr. Steven Weintraub, Chairman, 610-758-3730, Fax: 610-758-3767, E-mail: shw2@lehigh.edu. *Application contact:* Dr. Terry Napier, Graduate Coordinator, 610-758-3755, E-mail: mathgrad@lehigh.edu.

Louisiana State University and Agricultural and Mechanical College, Graduate School, College of Agriculture, Department of Experimental Statistics, Baton Rouge, LA 70803. Offers applied statistics (M App St). Part-time programs available. *Faculty:* 9 full-time (2 women). *Students:* 23 full-time (14 women), 3 part-time (2 women); includes 1 African American, 19 international. Average age 29. 24 applicants, 54% accepted, 11 enrolled. In 2005, 18 degrees awarded. *Degree requirements:* For master's, project. *Entrance requirements:* For master's, GRE General Test, minimum GPA of 3.0. Additional exam requirements/recommendations for international students: Required—TOEFL (minimum score 550 paper-based; 213 computer-based). *Application deadline:* For fall admission, 1/25 priority date for domestic students, 5/15 priority date for international students; for spring admission, 10/15 priority date for domestic students, 10/15 priority date for international students. Applications are processed on a rolling basis. Application fee: $25. Electronic applications accepted. *Financial support:* In 2005–06, 17 students received support, including 5 research assistantships with partial tuition reimbursements available (averaging $11,517 per year), 9 teaching assistantships with partial tuition reimbursements available (averaging $9,833 per year); fellowships, career-related internships or fieldwork, institutionally sponsored loans, tuition waivers (full and partial), and unspecified assistantships also available. Financial award application deadline: 4/1; financial award applicants required to submit FAFSA. *Faculty research:* Linear models, statistical computing, ecological statistics. Total annual research expenditures:$29,719. *Application contact:* Dr. James Geaghan, Graduate Adviser, 225-578-8303, E-mail: jgeaghan@lsu.edu.

Louisiana Tech University, Graduate School, College of Engineering and Science, Department of Mathematics and Statistics, Ruston, LA 71272. Offers MS. Part-time programs available. *Degree requirements:* For master's, thesis or alternative. *Entrance requirements:* For master's, GRE General Test, minimum GPA of 3.0 in last 60 hours. Additional exam requirements/recommendations for international students: Required—TOEFL.

Loyola University Chicago, Graduate School, Department of Mathematical Sciences and Statistics, Chicago, IL 60611-2196. Offers mathematics (MS), including pure mathematics, statistics and probability. Part-time programs available. *Faculty:* 19 full-time (4 women). *Students:* 17 full-time (8 women), 2 part-time; includes 4 minority (2 Asian Americans or Pacific Islanders, 2 Hispanic Americans), 8 international. Average age 30. 30 applicants, 83% accepted. In 2005, 7 degrees awarded. *Degree requirements:* For master's, 3.0 GPA. *Entrance requirements:* For master's, GRE General Test. Additional exam requirements/recommendations for international students: Required—TOEFL. *Application deadline:* For fall admission, 8/1 for domestic students; for spring admission, 12/1 for domestic students. Applications are processed on a rolling basis. Application fee: $40. Electronic applications accepted. *Expenses:* Tuition: Full-time $11,610; part-time $645 per credit. Required fees: $55 per semester. *Financial support:* In 2005–06, 8 students received support, including 4 teaching assistantships with tuition reimbursements available (averaging $9,000 per year); career-related internships or fieldwork, Federal Work-Study, institutionally sponsored loans, and tuition waivers (partial) also available. *Faculty research:* Probability and statistics, differential equations, algebra, combinations. Total annual research expenditures: $300,000. *Unit head:* Dr. Stephen Doty, Chair, 773-508-8520, Fax: 773-508-2123, E-mail: agiaqui@luc.edu. *Application contact:* Dr. Stephen Doty, Director, 773-508-3556, Fax: 773-508-2123.

Marquette University, Graduate School, College of Arts and Sciences, Department of Mathematics, Statistics, and Computer Science, Milwaukee, WI 53201-1881. Offers algebra (PhD); bio-mathematical modeling (PhD); computers (MS); mathematics (MS); mathematics education (MS); statistics (MS). Part-time programs available. Terminal master's awarded for partial completion of doctoral program. *Degree requirements:* For master's, thesis or alternative, comprehensive exam; for doctorate, 2 foreign languages, thesis/dissertation, comprehensive exam. *Entrance requirements:* For doctorate, sample of scholarly writing. Additional exam requirements/recommendations for international students: Required—TOEFL. *Faculty research:* Models of physiological systems, mathematical immunology, computational group theory, mathematical logic.

McGill University, Faculty of Graduate and Postdoctoral Studies, Faculty of Arts, Department of Economics, Montréal, QC H3A 2T5, Canada. Offers economics (MA, PhD); social statistics (MA). *Degree requirements:* For master's, thesis (for some programs), registration; for doctorate, thesis/dissertation, comprehensive exam, registration. *Entrance requirements:* For master's, GRE, honors BA in economics; minimum GPA of 3.5; coursework in undergraduate statistics, history of economic thought, three terms of introductory calculus and one term of linear algebra; for doctorate, GRE, master's degree in economics or equivalent. Additional exam requirements/recommendations for international students: Required—TOEFL (minimum score 550 paper-based; 213 computer-based), IELT (minimum score 7). Electronic applications accepted.

McGill University, Faculty of Graduate and Postdoctoral Studies, Faculty of Arts, Department of Sociology, Montréal, QC H3A 2T5, Canada. Offers medical sociology (MA); neo-tropical environment (MA); social statistics (MA); sociology (MA, PhD). Part-time programs available. Terminal master's awarded for partial completion of doctoral program. *Degree requirements:* For master's, one foreign language, thesis (for some programs), registration; for doctorate, one foreign language, thesis/dissertation, comprehensive exam, registration. *Entrance requirements:* For master's, GRE, minimum GPA of 3.3, samples of written work; for doctorate, GRE, minimum GPA of 3.3. Additional exam requirements/recommendations for international students: Required—TOEFL (minimum score 580 paper-based; 237 computer-based). Electronic applications accepted. *Faculty research:* Deviance and social control, states and social movements, economy and society, social inequality (class, ethnicity and gender).

McGill University, Faculty of Graduate and Postdoctoral Studies, Faculty of Science, Department of Mathematics and Statistics, Montréal, QC H3A 2T5, Canada. Offers computational science and engineering (M Sc); mathematics (M Sc, MA, PhD), including applied mathematics (M Sc, MA), pure mathematics (M Sc, MA), statistics (M Sc, MA). Part-time programs available. *Degree requirements:* For master's, thesis (for some programs), registration; for doctorate, one foreign language, thesis/dissertation, comprehensive exam, registration. *Entrance requirements:* For master's, minimum GPA of 3.0, Canadian Honours degree in mathematics or closely related field (statistics or applied mathematics). Additional exam requirements/recommendations for international students: Required—TOEFL (minimum score 550 paper-based; 213 computer-based), IELT (minimum score 7). Electronic applications accepted.

McMaster University, School of Graduate Studies, Faculty of Science, Department of Mathematics and Statistics, Program in Statistics, Hamilton, ON L8S 4M2, Canada. Offers applied statistics (M Sc); medical statistics (M Sc); statistical theory (M Sc). *Degree requirements:* For master's, thesis or alternative. *Entrance requirements:* For master's, honors degree background in mathematics and statistics. Additional exam requirements/recommendations for international students: Required—TOEFL (minimum score 550 paper-based; 213 computer-based). *Faculty research:* Development of polymer production technology, quality of life in patients who use pharmaceutical agents, mathematical modeling, order statistics from progressively censored samples, nonlinear stochastic model in genetics.

McNeese State University, Graduate School, College of Science, Department of Mathematics, Computer Science, and Statistics, Lake Charles, LA 70609. Offers computer science (MS); mathematics (MS); statistics (MS). Evening/weekend programs available. *Faculty:* 9 full-time (5 women). *Students:* 22 full-time (7 women), 8 part-time (2 women), 22 international. In 2005, 19 degrees awarded. *Degree requirements:* For master's, thesis or alternative, written exam, comprehensive exam. *Entrance requirements:* For master's, GRE General Test. *Application deadline:* For fall admission, 7/15 for domestic students. Applications are processed on a rolling basis. Application fee: $20 ($30 for international students). *Expenses:* Tuition, area resident: Part-time $193 per hour. Tuition, state resident: full-time $2,226. Required fees: $862; $106 per hour. Tuition and fees vary according to course load. *Financial support:* Teaching assistantships available. Financial award application deadline: 5/1. *Unit head:* Sid Bradley, Head, 337-475-5788, Fax: 337-475-5799, E-mail: sbradley@mcneese.edu.

Memorial University of Newfoundland, School of Graduate Studies, Department of Mathematics and Statistics, St. John's, NL A1C 5S7, Canada. Offers mathematics (M Sc, PhD); statistics (M Sc, MAS, PhD). Part-time programs available. *Students:* 43 full-time (14 women), 4 part-time (1 woman), 30 international. 42 applicants, 38% accepted, 15 enrolled. In 2005, 2 master's, 1 doctorate awarded. *Degree requirements:* For master's, thesis, thesis, practicum and report (MAS); for doctorate, thesis/dissertation, oral defense of thesis, comprehensive exam. *Entrance requirements:* For master's, 2nd class honors degree (MAS); for doctorate, MAS or M Sc in mathematics and statistics. *Application deadline:* For fall admission, 1/30 for domestic students, 1/30 for international students. Applications are processed on a rolling basis. Application fee: $40 Canadian dollars. Electronic applications accepted. *Expenses:* Tuition: Part-time $733 per term. Tuition and fees vary according to degree level and program. *Financial support:* Fellowships, teaching assistantships available. Financial award application deadline: 1/31. *Faculty research:* Algebra, topology, applied mathematics, mathematical statistics, applied statistics and probability. *Unit head:* Dr. Chris Radford, Interim Head, 709-737-8783, Fax: 709-787-3010, E-mail: head@math.mun.ca. *Application contact:* Dr. Edgar Goodaire, Graduate Officer, 709-737-8097, Fax: 709-737-3010, E-mail: grad@math.mun.ca.

Miami University, Graduate School, College of Arts and Sciences, Department of Mathematics and Statistics, Program in Statistics, Oxford, OH 45056. Offers MS Stat. Part-time programs available. *Degree requirements:* For master's, final exam. *Entrance requirements:* For master's, minimum undergraduate GPA of 3.0 during previous 2 years or 2.75 overall. Additional exam requirements/recommendations for international students: Required—TOEFL, TWE. Electronic applications accepted.

Michigan State University, The Graduate School, College of Natural Science, Department of Statistics and Probability, East Lansing, MI 48824. Offers applied statistics (MS); statistics (MS, PhD). *Faculty:* 17 full-time (3 women). *Students:* 63 full-time (35 women), 17 part-time (11 women); includes 4 minority (2 African Americans, 1 Asian American or Pacific Islander, 1 Hispanic American), 64 international. Average age 29. 102 applicants, 50% accepted. In 2005, 24 master's, 4 doctorates awarded. *Degree requirements:* For master's, thesis optional; for doctorate, thesis/dissertation, written preliminary exam. *Entrance requirements:* For master's, GRE General Test, minimum GPA of 3.0 in mathematics and statistics, 3 letters of recommendation; for doctorate, GRE General Test, master's degree in statistics or equivalent, minimum GPA of 3.0 in mathematics and statistics, 3 letters of recommendation. Additional exam requirements/recommendations for international students: Required—TOEFL (minimum score 550 paper-based; 213 computer-based). *Application deadline:* For fall admission, 12/27 for domestic students. Application fee: $50. Electronic applications accepted. *Expenses:* Tuition, state resident: part-time $330 per credit hour. Tuition, nonresident: part-time $685 per credit hour. Tuition and fees vary according to program. *Financial support:* In 2005–06, 11 fellowships with tuition reimbursements (averaging $2,229 per year), 13 research assistantships with tuition reimbursements (averaging $13,852 per year), 22 teaching assistantships with tuition reimbursements (averaging $13,610 per year) were awarded; scholarships/grants and unspecified assistantships also available. *Faculty research:* Stochastic processes, operations research, probability theory, applied statistics. Total annual research expenditures: $343,597. *Unit head:* Dr. Vincent F. Melfi, Acting Chairperson, 517-355-9589, Fax: 517-432-1405, E-mail: melfi@stt.msu.edu. *Application contact:* Cathy Sparks, Graduate Office Secretary, 517-355-9589, Fax: 517-432-1405, E-mail: sparks@stt.msu.edu.

Minnesota State University Mankato, College of Graduate Studies, College of Science, Engineering and Technology, Department of Mathematics and Statistics, Program in Math: Statistics Option, Mankato, MN 56001. Offers MS. *Students:* 1 full-time (0 women), 4 part-time (2 women). In 2005, 1 degree awarded. *Degree requirements:* For master's, one foreign language, thesis or alternative, comprehensive exam. *Entrance requirements:* For master's, GRE General Test, minimum GPA of 3.0 during previous 2 years. Additional exam requirements/recommendations for international students: Required—TOEFL. *Application deadline:* For fall admission, 7/1 for domestic students; for spring admission, 11/1 for domestic students. Applications are processed on a rolling basis. Application fee: $40. Electronic applications accepted. *Expenses:* Tuition, state resident: part-time $243 per credit. Tuition, nonresident: part-time $400 per credit. Required fees: $30 per credit. *Financial support:* Research assistantships with partial tuition reimbursements, teaching assistantships with partial tuition reimbursements, unspecified assistantships available. Financial award application deadline: 3/15; financial award applicants required to submit FAFSA. *Application contact:* 507-389-2321, E-mail: grad@mnsu.edu.

Mississippi State University, College of Arts and Sciences, Department of Mathematics and Statistics, Mississippi State, MS 39762. Offers mathematical sciences (PhD); mathematics (MS); statistics (MS). Part-time programs available. *Faculty:* 32 full-time (15 women), 1 (woman) part-time/adjunct. *Students:* 31 full-time (16 women), 6 part-time (4 women); includes 3 minority (1 African American, 2 Asian Americans or Pacific Islanders), 23 international. Average age 30. 46 applicants, 35% accepted, 9 enrolled. In 2005, 7 degrees awarded. Terminal master's awarded for partial completion of doctoral program. *Degree requirements:* For master's, comprehensive oral or written exam, thesis optional; for doctorate, one foreign language, thesis/dissertation, comprehensive oral and written exam. *Entrance requirements:* For master's, minimum GPA of 2.75; for doctorate, GRE. Additional exam requirements/recommendations for international students: Required—TOEFL. *Application deadline:* For fall admission, 3/15 for domestic students; for spring admission, 11/1 for domestic students. Applications are processed on a rolling basis. Application fee: $30. *Expenses:* Tuition, state resident: full-time $4,312; part-time $240 per hour. Tuition, nonresident: full-time $9,772; part-time $543 per hour. International tuition: $10,102 full-time. Tuition and fees vary according to course load. *Financial support:* In 2005–06, 20 teaching assistantships with full tuition reimbursements (averaging $11,389 per year) were awarded; Federal Work-Study, institutionally sponsored loans, tuition waivers (partial), and unspecified assistantships also available. Financial award applicants required to submit FAFSA. *Faculty research:* Differential equations, algebra, numerical analysis, functional analysis, applied statistics. Total annual research expenditures: $1.5 million. *Unit head:* Dr. Michael Neuman, Interim Head and Professor, 662-325-3414, Fax: 662-325-0005, E-mail: office@math.msstate.edu. *Application contact:* Philip G. Bonfanti, Director of Admissions, 662-325-4104, Fax: 662-325-8872, E-mail: admit@msstate.edu.

Montana State University, College of Graduate Studies, College of Letters and Science, Department of Mathematical Sciences, Bozeman, MT 59717. Offers mathematics (MS, PhD); statistics (MS, PhD). Part-time programs available. Postbaccalaureate distance learning degree programs offered (minimal on-campus study). *Faculty:* 30 full-time (7 women), 13 part-time/adjunct (6 women). *Students:* 14 full-time (5 women), 76 part-time (40 women); includes 4 minority (1 American Indian/Alaska Native, 2 Asian Americans or Pacific Islanders, 1 Hispanic American), 9 international. Average age 31. 40 applicants, 60% accepted, 17 enrolled. In 2005, 18 master's, 3 doctorates awarded. *Degree requirements:* For master's, thesis (for some programs), comprehensive exam, registration; for doctorate, thesis/dissertation, comprehensive exam, registration. *Entrance requirements:* For master's and doctorate, GRE General Test. Additional exam requirements/recommendations for international students: Required—TOEFL (minimum score 550 paper-based; 213 computer-based). *Application deadline:* For fall admission, 7/15 priority date for domestic students, 5/15 priority date for international students; for spring admission, 12/1 priority date for domestic students, 10/1 priority date for international students. Applications are processed on a rolling basis. Application fee: $30. Electronic applications accepted. *Expenses:* Tuition, state resident: full-time $4,132. Tuition, nonresident: full-time $1,132. *Financial support:* In 2005–06, 65 students received support, including 8 research assistantships with full tuition reimbursements available (averaging $15,000 per year), 57 teaching assistantships with full tuition reimbursements available (averaging $13,500 per year); career-related internships or fieldwork, scholarships/grants, health care benefits, tuition waivers (full), and unspecified assistantships also available. Support available to part-time students. Financial award application deadline: 3/1; financial award applicants required to submit FAFSA. *Faculty research:* Applied mathematics, dynamical systems, statistics, mathematics education, mathematical and computational biology. Total annual research expenditures: $604,937. *Unit head:* Dr. Kenneth Bowers, Department Head, 406-994-3604, Fax: 406-994-1789, E-mail: bowers@math.montana.edu.

Montclair State University, The Graduate School, College of Science and Mathematics, Department of Computer Science, Montclair, NJ 07043-1624. Offers applied mathematics (MS); applied statistics (MS); CISCO (Certificate); informatics (MS); object oriented computing (Certificate). Part-time and evening/weekend programs available. *Faculty:* 13 full-time (2 women), 12 part-time/adjunct (3 women). *Students:* 13 full-time (7 women), 22 part-time (3 women); includes 11 minority (2 African Americans, 8 Asian Americans or Pacific Islanders, 1 Hispanic American), 3 international. 18 applicants, 72% accepted, 12 enrolled. In 2005, 13 degrees awarded. *Degree requirements:* For master's, thesis or alternative, comprehensive exam. *Entrance requirements:* For master's, GRE General Test, minimum GPA of 2.67; 15 undergraduate math credits; bachelor's degree in computer science, math, science or engineering; 2 letters of recommendation. Additional exam requirements/recommendations for international students: Required—TOEFL (minimum score 83 computer-based). *Application deadline:* Applications are processed on a rolling basis. Application fee: $60. Electronic applications accepted. *Expenses:* Tuition: Full-time $3,001; part-time $409 per credit. Required fees: $56 per credit. Tuition and fees vary according to course load, degree level and program. *Financial support:* In 2005–06, 4 research assistantships with full tuition reimbursements (averaging $7,000 per year) were awarded; Federal Work-Study, scholarships/grants, and unspecified assistantships also available. Support available to part-time students. Financial award application deadline: 3/1; financial award applicants required to submit FAFSA. *Unit head:* Dr. Dorothy Deremer, Chairperson, 973-655-4166. *Application contact:* Dr. James Benham, Adviser, 973-655-3746, E-mail: benham@pegasus.montclair.edu.

Montclair State University, The Graduate School, College of Science and Mathematics, Department of Mathematics, Montclair, NJ 07043-1624. Offers mathematics (MS), including computer science, mathematics education, pure and applied mathematics, statistics; teaching middle grades math (Certificate). Part-time and evening/weekend programs available. *Faculty:* 29 full-time (10 women), 29 part-time/adjunct (11 women). *Students:* 23 full-time (12 women), 135 part-time (99 women); includes 28 minority (13 African Americans, 10 Asian Americans or Pacific Islanders, 5 Hispanic Americans), 3 international. 77 applicants, 83% accepted, 51 enrolled. In 2005, 9 master's, 26 other advanced degrees awarded. *Degree requirements:* For master's, comprehensive exam. *Entrance requirements:* For master's, GRE General Test, minimum GPA of 2.67, 2 letters of recommendation. Additional exam requirements/recommendations for international students: Required—TOEFL (minimum score 83 computer-based). *Application deadline:* Applications are processed on a rolling basis. Application fee: $60. *Expenses:* Tuition: Full-time $3,001; part-time $409 per credit. Required fees: $56 per credit. Tuition and fees vary according to course load, degree level and program. *Financial support:* In 2005–06, 8 research assistantships with full tuition reimbursements (averaging $7,000 per year) were awarded; Federal Work-Study, scholarships/grants, and unspecified assistantships also available. Support available to part-time students. Financial award application deadline: 3/1; financial award applicants required to submit FAFSA. *Faculty research:* Infectious disease. Total annual research expenditures:$130,000. *Unit head:* Dr. Helen Roberts, Chairperson, 973-655-5132. *Application contact:* Dr. Ted Williamson, Advisor, 973-655-5146, E-mail: williamsont@mail.montclair.edu.

New Jersey Institute of Technology, Office of Graduate Studies, College of Science and Liberal Arts, Department of Mathematical Science, Program in Applied Statistics, Newark, NJ 07102. Offers MS. Part-time and evening/weekend programs available. *Students:* 17 full-time (9 women), 16 part-time (9 women); includes 15 minority (5 African Americans, 10 Asian Americans or Pacific Islanders), 9 international. Average age 30. 23 applicants, 43% accepted, 4 enrolled. In 2005, 8 degrees awarded. *Entrance requirements:* For master's, GRE General Test. Additional exam requirements/recommendations for international students: Required—TOEFL (minimum score 550 paper-based; 213 computer-based). *Application deadline:* For fall admission, 6/5 for domestic students; for spring admission, 10/15 for domestic students. Applications are processed on a rolling basis. Application fee: $60. Electronic applications accepted. *Expenses:* Tuition, state resident: full-time $9,620; part-time $520 per credit. Tuition, nonresident: full-time $13,542; part-time $715 per credit. Required fees: $78; $54 per credit. $78 per year. Tuition and fees vary according to course load. *Financial support:* Fellowships with full and partial tuition reimbursements, research assistantships with full and partial tuition reimbursements, teaching assistantships with full and partial tuition reimbursements, career-related internships or fieldwork, Federal Work-Study, institutionally sponsored loans, and unspecified assistantships available. Financial award application deadline: 3/15. *Unit head:* Dr. Manish Bhattacharjee, Director, 973-596-2949, Fax: 973-596-5591, E-mail: manish.bhattacharjee@njit.edu. *Application contact:* Kathryn Kelly, Director of Admissions, 973-596-3300, Fax: 973-596-3461, E-mail: admissions@njit.edu.

New Mexico State University, Graduate School, College of Business, Department of Economics and International Business, Las Cruces, NM 88003-8001. Offers economics (MA); experimental statistics (MS). Part-time programs available. *Faculty:* 17 full-time (2 women), 2 part-time/adjunct (1 woman). *Students:* 32 full-time (17 women), 11 part-time (7 women); includes 14 minority (all Hispanic Americans), 15 international. Average age 31. 42 applicants, 90% accepted, 7 enrolled. In 2005, 13 degrees awarded. *Degree requirements:* For master's, thesis or alternative. *Entrance requirements:* For master's, minimum GPA of 3.0. Additional exam requirements/recommendations for international students: Required—TOEFL. *Application deadline:* Applications are processed on a rolling basis. Application fee: $30 ($50 for international students). Electronic applications accepted. *Expenses:* Tuition, state resident: full-time $3,156; part-time $175 per credit. Tuition, nonresident: full-time $12,510; part-time $565 per credit. Required fees: $1,050. *Financial support:* In 2005–06, 2 fellowships, 14 teaching assistantships were awarded; research assistantships, career-related internships or fieldwork and Federal Work-Study also available. Support available to part-time students. Financial award application deadline: 3/1. *Faculty research:* Public utilities, environment, linear models, biological sampling, public policy. *Unit head:* Dr. Michael Ellis, Head, 505-646-2113,

Statistics

New Mexico State University (continued)

Fax: 505-646-1915, E-mail: mellis@nmsu.edu. *Application contact:* Dr. Anthony Popp, Graduate Adviser, 505-646-5198, Fax: 505-646-1915, E-mail: apopp@nmsu.edu.

New York University, Leonard N. Stern School of Business, Department of Information, Operations and Management Sciences, New York, NY 10012-1019. Offers information systems (MBA, PhD); operations management (PhD); operations managment (MBA); statistics (MBA, PhD). *Faculty research:* Knowledge management, economics of information, computer-supported groups and communities financial information systems, data mining and business intelligence. *Unit head:* Michael Pinedo, Chairperson, 212-998-0280, E-mail: mpinedo@stern.nyu.edu.

North Carolina State University, Graduate School, College of Physical and Mathematical Sciences, Department of Statistics, Raleigh, NC 27695. Offers biomathematics (M Biomath, MS, PhD), including biomathematics, ecology (PhD); statistics (M Stat, MS, PhD). Part-time programs available. *Degree requirements:* For master's, thesis (for some programs), final oral exam, comprehensive exam; for doctorate, thesis/dissertation, final oral and written exams, written and oral preliminary exams. *Entrance requirements:* For master's and doctorate, GRE General Test. Additional exam requirements/recommendations for international students: Required—TOEFL. Electronic applications accepted. *Faculty research:* Biostatistics; time series; spatial, inference, environmental, industrial, genetics applications; nonlinear models; DOE.

North Dakota State University, The Graduate School, College of Science and Mathematics, Department of Statistics, Fargo, ND 58105. Offers applied statistics (MS, Certificate); statistics (PhD). *Faculty:* 4 full-time (1 woman), 2 part-time/adjunct. *Students:* 26 full-time (14 women), 1 (woman) part-time; includes 1 minority (African American), 19 international. Average age 24. 11 applicants, 82% accepted, 5 enrolled. In 2005, 2 master's, 1 doctorate awarded. *Degree requirements:* For master's and doctorate, thesis/dissertation, comprehensive exam. *Entrance requirements:* For master's and doctorate, minimum GPA of 3.0. Additional exam requirements/recommendations for international students: Required—TOEFL (minimum score 550 paper-based; 213 computer-based). *Application deadline:* Applications are processed on a rolling basis. Application fee: $45 ($60 for international students). *Financial support:* In 2005–06, 3 fellowships with full tuition reimbursements, 2 research assistantships with full tuition reimbursements, 8 teaching assistantships with full tuition reimbursements were awarded; career-related internships or fieldwork, Federal Work-Study, institutionally sponsored loans, and tuition waivers (full) also available. Financial award application deadline: 4/15. *Faculty research:* Nonparametric statistics, survival analysis, multivariate analysis, distribution theory, inference modeling, biostatistics. *Unit head:* Dr. Rhonda Magel, Chair, 701-231-7177, Fax: 701-231-8734, E-mail: ndsu.stats@ndsu.edu. *Application contact:* Judy Normann, Academic Assistant, 701-231-7532, Fax: 702-231-8734, E-mail: ndsu.stats@ndsu.edu.

Northern Arizona University, Graduate College, College of Engineering and Natural Science, Department of Mathematics and Statistics, Flagstaff, AZ 86011. Offers mathematics (MAT, MS); statistics (MS). Part-time programs available. *Degree requirements:* For master's, thesis optional. *Faculty research:* Topology, statistics, groups, ring theory, number theory.

Northern Illinois University, Graduate School, College of Liberal Arts and Sciences, Department of Mathematical Sciences, Division of Statistics, De Kalb, IL 60115-2854. Offers MS. Part-time programs available. *Faculty:* 8 full-time (1 woman), 1 part-time/adjunct (0 women). *Students:* 15 full-time (10 women), 3 part-time (2 women); includes 1 minority (Asian American or Pacific Islander), 11 international. Average age 29. 30 applicants, 63% accepted, 7 enrolled. In 2005, 9 degrees awarded. *Degree requirements:* For master's, thesis optional. *Entrance requirements:* For master's, GRE General Test, minimum GPA of 2.75, course work in statistics, calculus, linear algebra. Additional exam requirements/recommendations for international students: Required—TOEFL (minimum score 550 paper-based; 213 computer-based). *Application deadline:* For fall admission, 6/1 for domestic students, 5/1 for international students; for spring admission, 11/1 for domestic students, 10/1 for international students. Applications are processed on a rolling basis. Application fee: $30. Electronic applications accepted. *Expenses:* Tuition, state resident: full-time $4,565; part-time $191 per credit hour. Tuition, nonresident: full-time $9,129; part-time $382 per credit hour. *Financial support:* In 2005–06, 1 research assistantship with full tuition reimbursement, 15 teaching assistantships with full tuition reimbursements were awarded; fellowships with full tuition reimbursements, career-related internships or fieldwork, Federal Work-Study, scholarships/grants, tuition waivers (full), and unspecified assistantships also available. Support available to part-time students. Financial award applicants required to submit FAFSA. *Faculty research:* Reality and life testing, quality control, statistical inference from stochastic process, nonparametric statistics. *Unit head:* Dr. Rama T Lingham, Director, Division of Statistics, 815-753-6773, Fax: 815-753-6776. *Application contact:* Dr. Alan Polansky, Director, Graduate Studies, 815-753-6864, E-mail: polansky@math.niu.edu.

Northwestern University, The Graduate School, Judd A. and Marjorie Weinberg College of Arts and Sciences, Department of Statistics, Evanston, IL 60208. Offers MS, PhD. Admissions and degrees offered through The Graduate School. Part-time programs available. *Faculty:* 6 full-time (1 woman). *Students:* 14 (8 women) 12 international. Average age 26. 48 applicants, 8% accepted, 1 enrolled. In 2005, 1 master's, 2 doctorates awarded. Terminal master's awarded for partial completion of doctoral program. *Degree requirements:* For master's, final exam; for doctorate, thesis/dissertation, preliminary exam, final exam. *Entrance requirements:* For master's and doctorate, GRE General Test. Additional exam requirements/recommendations for international students: Required—TOEFL. *Application deadline:* For fall admission, 12/31 for domestic students. Application fee: $60 ($75 for international students). *Financial support:* In 2005–06, 2 fellowships with full tuition reimbursements (averaging $16,080 per year), 1 research assistantship with tuition reimbursement (averaging $12,465 per year), 3 teaching assistantships with full tuition reimbursements (averaging $12,465 per year) were awarded; career-related internships or fieldwork and institutionally sponsored loans also available. Financial award application deadline: 1/15; financial award applicants required to submit FAFSA. *Faculty research:* Theoretical statistics, applied statistics, computational methods, statistical designs, complex models. *Unit head:* Thomas A. Severini, Chair, 847-467-1254, Fax: 847-491-4939, E-mail: severini@nwu.edu. *Application contact:* Wenxin Jiang, Admission Officer, 847-491-5081, Fax: 847-491-4939, E-mail: wjiang@northwestern.edu.

Oakland University, Graduate Study and Lifelong Learning, College of Arts and Sciences, Department of Mathematics and Statistics, Program in Applied Statistics, Rochester, MI 48309-4401. Offers MS. Part-time and evening/weekend programs available. *Students:* 5 full-time (2 women), 10 part-time (2 women); includes 3 Asian Americans or Pacific Islanders, 2 international. Average age 33. 10 applicants, 100% accepted, 6 enrolled. In 2005, 4 degrees awarded. *Entrance requirements:* For master's, minimum GPA of 3.0 for unconditional admission. Additional exam requirements/recommendations for international students: Required—TOEFL (minimum score 550 paper-based; 213 computer-based). *Application deadline:* For fall admission, 7/15 priority date for domestic students, 5/1 priority date for international students. For winter admission, 12/1 for domestic students; for spring admission, 3/15 for domestic students. Applications are processed on a rolling basis. Application fee: $30. Electronic applications accepted. *Expenses: Contact institution. Financial support:* Career-related internships or fieldwork and tuition waivers (full) available. Financial award application deadline: 3/1; financial award applicants required to submit FAFSA. *Unit head:* Dr. Meir Shillor, Coordinator, Graduate Programs, 248-370-3439, Fax: 248-370-4184, E-mail: shillor@oakland.edu.

Oakland University, Graduate Study and Lifelong Learning, College of Arts and Sciences, Department of Mathematics and Statistics, Program in Statistical Methods, Rochester, MI 48309-4401. Offers Certificate. *Students:* Average age 39. 2 applicants, 100% accepted, 0 enrolled. In 2005, 1 degree awarded. *Entrance requirements:* Additional exam requirements/recommendations for international students: Required—TOEFL (minimum score 550 paper-based; 213 computer-based). *Application deadline:* For fall admission, 7/15 priority date for domestic students, 5/1 priority date for international students. For winter admission, 12/1 for domestic students; for spring admission, 3/15 for domestic students. Application fee: $30. *Expenses: Contact institution. Financial support:* Federal Work-Study, institutionally sponsored loans, and tuition waivers (full) available. Financial award application deadline: 3/1; financial award applicants required to submit FAFSA. *Unit head:* Dr. Meir Shillor, Coordinator, Graduate Programs, 248-370-3439, Fax: 248-370-4184, E-mail: shillor@oakland.edu.

The Ohio State University, Graduate School, College of Mathematical and Physical Sciences, Department of Statistics, Columbus, OH 43210. Offers biostatistics (PhD); statistics (M Appl Stat, MS, PhD). *Degree requirements:* For master's, thesis optional; for doctorate, thesis/dissertation. *Entrance requirements:* For master's and doctorate, GRE General Test. Additional exam requirements/recommendations for international students: Required—TOEFL (minimum score 600 paper-based; 250 computer-based). Electronic applications accepted.

Oklahoma State University, College of Arts and Sciences, Department of Statistics, Stillwater, OK 74078. Offers MS, PhD. *Faculty:* 8 full-time (3 women), 1 (woman) part-time/adjunct. *Students:* 9 full-time (4 women), 20 part-time (13 women), 15 international. Average age 32. 38 applicants, 39% accepted, 8 enrolled. In 2005, 4 master's awarded. *Degree requirements:* For master's, thesis optional; for doctorate, thesis/dissertation, comprehensive exam. *Entrance requirements:* For master's and doctorate, GRE. Additional exam requirements/recommendations for international students: Required—TOEFL. *Application deadline:* For fall admission, 7/1 priority date for domestic students, 3/1 priority date for international students. Applications are processed on a rolling basis. Application fee: $40 ($75 for international students). Electronic applications accepted. *Expenses:* Tuition, state resident: full-time $4,253; part-time $139 per credit hour. Tuition, nonresident: full-time $12,569; part-time $485 per credit hour. Required fees: $43 per credit hour. One-time fee: $20 part-time. Tuition and fees vary according to course load and program. *Financial support:* In 2005–06, 1 research assistantship (averaging $10,976 per year), 20 teaching assistantships (averaging $15,162 per year) were awarded; Federal Work-Study, scholarships/grants, health care benefits, tuition waivers (partial), and unspecified assistantships also available. Support available to part-time students. Financial award application deadline: 3/1. *Faculty research:* Linear models, sampling methods, ranking and selections procedures, categorical data, multiple comparisons. *Unit head:* Dr. William Warde, Head, 405-744-5684, E-mail: billw@okstate.edu.

Oregon State University, Graduate School, College of Science, Department of Statistics, Corvallis, OR 97331. Offers applied statistics (MA, MS, PhD); biometry (MA, MS, PhD); environmental statistics (MA, MS, PhD); mathematical statistics (MA, MS, PhD); operations research (MA, MAIS, MS); statistics (MA, MS, PhD). Part-time programs available. *Faculty:* 10 full-time (4 women), 2 part-time/adjunct (0 women). *Students:* 33 full-time (14 women), 3 part-time (1 woman); includes 5 minority (4 Asian Americans or Pacific Islanders, 1 Hispanic American), 12 international. Average age 30. In 2005, 9 master's, 2 doctorates awarded. *Degree requirements:* For master's, consulting experience; for doctorate, thesis/dissertation, consulting experience. *Entrance requirements:* For master's and doctorate, minimum GPA of 3.0 in last 90 hours. Additional exam requirements/recommendations for international students: Required—TOEFL. *Application deadline:* For fall admission, 2/15 for domestic students. Applications are processed on a rolling basis. Application fee: $50. *Expenses:* Tuition, area resident: Part-time $301 per credit. Tuition, state resident: full-time $8,139; part-time $501 per credit. Tuition, nonresident: full-time $14,376; part-time $532 per credit. Required fees: $1,266. *Financial support:* In 2005–06, 8 research assistantships, 19 teaching assistantships were awarded; Federal Work-Study and institutionally sponsored loans also available. Financial award application deadline: 2/15. *Faculty research:* Analysis of enumerative data, nonparametric statistics, asymptotics, experimental design, generalized regression models, linear model theory, reliability theory, survival analysis, wildlife and general survey methodology. *Unit head:* Dr. Robert T. Smythe, Chair, 541-737-3366, Fax: 541-737-3489, E-mail: symthe@stat.orst.edu. *Application contact:* Dr. Daniel W. Schafer, Director of Graduate Studies, 541-737-3366, Fax: 541-737-3489, E-mail: statoff@stat.orst.edu.

The Pennsylvania State University University Park Campus, Graduate School, Eberly College of Science, Department of Statistics, State College, University Park, PA 16802-1503. Offers applied statistics (MAS); statistics (MA, MAS, MS, PhD). *Students:* 71 full-time (37 women), 7 part-time (5 women); includes 7 minority (2 African Americans, 5 Asian Americans or Pacific Islanders), 52 international. *Entrance requirements:* For master's and doctorate, GRE General Test. Application fee: $45. *Expenses:* Tuition, state resident: full-time $12,518; part-time $522 per credit. Tuition, nonresident: full-time $23,004; part-time $959 per credit. Required fees: $484. Tuition and fees vary according to course load, campus/location and program. *Financial support:* Fellowships, research assistantships, teaching assistantships, tuition waivers (full) available. *Unit head:* Dr. James L. Rosenberger, Head, 814-865-1348, Fax: 814-863-7115, E-mail: jameslrosenberger@psu.edu. *Application contact:* Jennifer Parkes, Information Contact, 814-865-1348, E-mail: jqp4@psu.edu.

Portland State University, Graduate Studies, College of Liberal Arts and Sciences, Department of Mathematics and Statistics, Portland, OR 97207-0751. Offers mathematical sciences (PhD); mathematics education (PhD); statistics (MS). *Faculty:* 30 full-time (6 women), 22 part-time/adjunct (9 women). *Students:* 53 full-time (19 women), 53 part-time (27 women); includes 15 minority (5 African Americans, 4 Asian Americans or Pacific Islanders, 6 Hispanic Americans), 15 international. Average age 32. 62 applicants, 90% accepted, 37 enrolled. In 2005, 13 master's, 1 doctorate awarded. *Degree requirements:* For master's, thesis or alternative, exams; for doctorate, 2 foreign languages, thesis/dissertation, exams. *Entrance requirements:* For master's, minimum GPA of 3.0 in upper-division course work or 2.75 overall; for doctorate, GRE General Test. Additional exam requirements/recommendations for international students: Required—TOEFL (minimum score 550 paper-based; 213 computer-based). *Application deadline:* For fall admission, 4/1 for domestic students, 3/1 for international students. For winter admission, 9/1 for domestic students; for spring admission, 11/1 for domestic students. Applications are processed on a rolling basis. Application fee: $50. *Expenses:* Tuition, state resident: full-time $6,648; part-time $231 per credit. Tuition, nonresident: full-time $11,319; part-time $231 per credit. Required fees: $686; $67 per credit. *Financial support:* In 2005–06, 2 teaching assistantships with full tuition reimbursements (averaging $12,637 per year) were awarded; research assistantships, Federal Work-Study, scholarships/grants, tuition waivers (partial), and unspecified assistantships also available. Support available to part-time students. Financial award application deadline: 3/1; financial award applicants required to submit FAFSA. *Faculty research:* Algebra, topology, statistical distribution theory, control theory, statistical robustness. Total annual research expenditures:$318,405. *Unit head:* Marek Elzanowski, Chair, 503-725-3621, Fax: 503-725-3661, E-mail: elzanowskim@pdx.edu. *Application contact:* John Erdman, Coordinator, 503-725-3621, Fax: 503-725-3661, E-mail: erdman@pdx.edu.

Princeton University, Graduate School, Department of Civil and Environmental Engineering, Princeton, NJ 08544-1019. Offers environmental engineering and water resources (PhD); mechanics, materials, and structures (M Eng, MSE, PhD); statistics and operations research (MSE, PhD); transportation systems (MSE, PhD). Terminal master's awarded for partial completion of doctoral program. *Degree requirements:* For master's and doctorate, thesis/dissertation. *Entrance requirements:* For master's and doctorate, GRE General Test, GRE Subject Test. Additional exam requirements/recommendations for international students: Required—TOEFL (minimum score 600 paper-based; 250 computer-based). Electronic applications accepted.

Purdue University, Graduate School, School of Science, Department of Statistics, West Lafayette, IN 47907. Offers MS, PhD, Certificate. *Faculty:* 27 full-time (9 women). *Students:* 76 full-time (36 women), 12 part-time (5 women); includes 5 minority (1 African American, 4 Asian Americans or Pacific Islanders), 54 international. Average age 29. 189 applicants, 15% accepted, 25 enrolled. In 2005, 23 master's, 12 doctorates awarded. *Degree requirements:* For doctorate, thesis/dissertation, qualifying exams. *Entrance requirements:* For master's and doctorate, GRE General Test. Additional exam requirements/recommendations for international students: Required—TOEFL (minimum score 250 computer-based); Recommended—TWE. *Application deadline:* For fall admission, 1/15 for domestic students, 1/15 for international students; for spring admission, 10/15 for domestic students, 9/15 for international students. Applications are processed on a rolling basis. Application fee: $55. Electronic applications accepted. *Financial support:* Fellowships with full tuition reimbursements, research assistantships with full tuition reimbursements, teaching assistantships with full tuition reimbursements, career-related internships or fieldwork available. Support available to part-time students. Financial award applicants required to submit FAFSA. *Faculty research:* Nonparametric models, computational finance, design of experiments, probability theory, bioinformatics. *Unit head:* Dr. M. E. Bock, Head, 765-494-3141, Fax: 765-494-0558, E-mail: mbock@stat.purdue.edu. *Application contact:* Darlene Wayman, Secretary, 765-494-5794, Fax: 765-494-0558, E-mail: jdwayman@stat.purdue.edu.

Queen's University at Kingston, School of Graduate Studies and Research, Faculty of Arts and Sciences, Department of Mathematics and Statistics, Kingston, ON K7L 3N6, Canada. Offers mathematics (M Sc, M Sc Eng, PhD); statistics (M Sc, M Sc Eng, PhD). Part-time programs available. *Degree requirements:* For master's, thesis/dissertation; for doctorate, thesis/dissertation, comprehensive exam. *Entrance requirements:* Additional exam requirements/recommendations for international students: Required—TOEFL. *Faculty research:* Algebra, analysis, applied mathematics, statistics.

Rensselaer Polytechnic Institute, Graduate School, School of Engineering, Department of Decision Sciences and Engineering Systems, Program in Operations Research and Statistics, Troy, NY 12180-3590. Offers M Eng, MS, MBA/M Eng. Part-time programs available. *Students:* 13 applicants, 46% accepted, 0 enrolled. In 2005, 9 degrees awarded. *Degree requirements:* For master's, thesis (for some programs). *Entrance requirements:* For master's, GRE General Test. Additional exam requirements/recommendations for international students: Required—TOEFL (minimum score 570 paper-based). *Application deadline:* For fall admission, 1/15 for domestic students. Applications are processed on a rolling basis. Application fee: $75. Electronic applications accepted. *Expenses:* Tuition: Full-time $31,000; part-time $1,320 per credit. Required fees: $1,623. *Financial support:* Fellowships with full tuition reimbursements, research assistantships with full tuition reimbursements, teaching assistantships with full tuition reimbursements, career-related internships or fieldwork and institutionally sponsored loans available. Financial award application deadline: 1/15. *Faculty research:* Manufacturing, MIS, statistical consulting, education services, production, logistics, inventory. *Application contact:* Lee Vilardi, Graduate Coordinator, 518-276-6681, Fax: 518-276-8227, E-mail: dsesgr@rpi.edu.

Rice University, Graduate Programs, George R. Brown School of Engineering, Department of Statistics, Houston, TX 77251-1892. Offers biostatistics (PhD); computational finance (PhD); statistics (M Stat, MA, PhD). Terminal master's awarded for partial completion of doctoral program. *Degree requirements:* For master's, thesis/dissertation; for doctorate, thesis/dissertation, comprehensive exam. *Entrance requirements:* For master's and doctorate, GRE General Test, GRE Subject Test, minimum GPA of 3.0. Additional exam requirements/recommendations for international students: Required—TOEFL (minimum score 630 paper-based; 250 computer-based). Electronic applications accepted. *Faculty research:* Statistical genetics, non parametric function estimation, computational statistics and visualization, stochastic processes.

Rochester Institute of Technology, Graduate Enrollment Services, College of Engineering, Center of Quality and Applied Statistics, Rochester, NY 14623-5603. Offers applied statistics (MS); statistical quality (AC). Part-time and evening/weekend programs available. *Students:* 8 full-time (5 women), 67 part-time (22 women); includes 7 minority (2 African Americans, 3 Asian Americans or Pacific Islanders, 2 Hispanic Americans), 10 international. 47 applicants, 60% accepted, 20 enrolled. In 2005, 21 degrees awarded. *Degree requirements:* For master's, oral exam. *Entrance requirements:* For master's, course work in calculus, minimum GPA of 3.0. Additional exam requirements/recommendations for international students: Required—TOEFL. *Application deadline:* For fall admission, 3/1 for domestic students. Applications are processed on a rolling basis. Application fee: $50. *Expenses:* Tuition: Full-time $25,392; part-time $713 per credit. Required fees: $183; $61 per term. *Financial support:* Research assistantships available. *Unit head:* Dr. Donald Baker, Director, 585-475-5070, E-mail: ddbcqa@rit.edu.

Rutgers, The State University of New Jersey, New Brunswick/Piscataway, Graduate School, Program in Statistics, New Brunswick, NJ 08901-1281. Offers quality and productivity management (MS); statistics (MS, PhD), including biostatistics (PhD), data mining (PhD). Part-time programs available. *Faculty:* 27 full-time. *Students:* 62 full-time (33 women), 51 part-time (25 women); includes 28 minority (3 African Americans, 23 Asian Americans or Pacific Islanders, 2 Hispanic Americans), 62 international. Average age 31. 230 applicants, 18% accepted, 54 enrolled. In 2005, 6 master's, 3 doctorates awarded. Terminal master's awarded for partial completion of doctoral program. *Degree requirements:* For master's, essay, exam, non-thesis essay paper; for doctorate, one foreign language, thesis/dissertation, qualifying oral and written exams. *Entrance requirements:* For master's, GRE General Test; for doctorate, GRE General Test, GRE Subject Test (recommended). Additional exam requirements/recommendations for international students: Required—TOEFL (minimum score 550 paper-based; 213 computer-based). *Application deadline:* For fall admission, 5/1 for domestic students; for spring admission, 12/1 priority date for domestic students. Applications are processed on a rolling basis. Application fee: $50. Electronic applications accepted. *Expenses:* Tuition, state resident: full-time $10,440; part-time $435 per credit. Tuition, nonresident: full-time $15,520; part-time $647 per credit. Required fees: $129 per credit. Tuition and fees vary according to program. *Financial support:* In 2005–06, 16 students received support, including 4 fellowships with full tuition reimbursements available (averaging $18,000 per year), research assistantships with full tuition reimbursements available (averaging $14,500 per year), 11 teaching assistantships with full tuition reimbursements available (averaging $16,988 per year); career-related internships or fieldwork, Federal Work-Study, institutionally sponsored loans, scholarships/grants, health care benefits, unspecified assistantships, and grant/tuition remission also available. Financial award application deadline: 3/1; financial award applicants required to submit FAFSA. *Faculty research:* Probability, decision theory, linear models, multivariate statistics, statistical computing. *Unit head:* John Kolassa, Director, 732-445-3634, Fax: 732-445-3428, E-mail: kolassa@stat.rutgers.edu. *Application contact:* Angela T. Klein, Department Secretary, 732-445-2693, Fax: 732-445-3428, E-mail: aklein@stat.rutgers.edu.

St. Cloud State University, School of Graduate Studies, College of Science and Engineering, Program in Applied Statistics, St. Cloud, MN 56301-4498. Offers MS. *Faculty:* 10 full-time (3 women). *Expenses:* Tuition, state resident: part-time $277. Tuition, nonresident: part-time $379. Required fees: $23 per credit. Tuition and fees vary according to course load and reciprocity agreements. *Unit head:* Dr. David Robinson, Coordinator.

St. John's University, St. John's College of Liberal Arts and Sciences, Department of Mathematics and Computer Science, Queens, NY 11439. Offers algebra (MA); analysis (MA); applied mathematics (MA); computer science (MA); geometry-topology (MA); logic and foundations (MA); probability and statistics (MA). Part-time and evening/weekend programs available. *Faculty:* 19 full-time (2 women), 10 part-time/adjunct (4 women). *Students:* 4 full-time (all women), 5 part-time (3 women); includes 1 minority (Hispanic American), 1 international. Average age 28. 18 applicants, 83% accepted, 6 enrolled. In 2005, 2 degrees awarded. *Degree requirements:* For master's, thesis optional. *Entrance requirements:* For master's, minimum GPA of 3.0. Additional exam requirements/recommendations for international students: Required—TOEFL (minimum score 500 paper-based; 173 computer-based). *Application deadline:* For fall admission, 5/1 priority date for domestic students, 5/1 priority date for international students; for spring admission, 11/1 priority date for domestic students, 11/1 priority date for international students. Applications are processed on a rolling basis. Application fee: $40. Electronic applications accepted. *Expenses:* Tuition: Full-time $8,760; part-time $730 per credit. Required fees: $250; $125 per term. Tuition and fees vary according to program. *Financial support:* Research assistantships, scholarships/grants available. Support available to part-time students. Financial award application deadline: 3/1; financial award applicants required to submit FAFSA. *Faculty research:* Development of a computerized metabolic map. *Unit head:* Dr. Charles Traina, Chair, 718-990-6166, E-mail: trainac@stjohns.edu. *Application contact:* Matthew Whelan, Director, Office of Admissions, 718-990-2000, Fax: 718-990-2096, E-mail: admissions@stjohns.edu.

Sam Houston State University, College of Arts and Sciences, Department of Mathematics and Statistics, Huntsville, TX 77341. Offers mathematics (MA, MS); statistics (MS). Part-time programs available. *Faculty:* 10 full-time (0 women). *Students:* 4 full-time (2 women), 21 part-time (13 women); includes 3 minority (1 African American, 1 Asian American or Pacific Islander, 1 Hispanic American), 1 international. Average age 34. In 2005, 20 degrees awarded. *Entrance requirements:* For master's, GRE General Test. Additional exam requirements/recommendations for international students: Required—TOEFL (minimum score 550 paper-based; 213 computer-based). *Application deadline:* For fall admission, 8/1 for domestic students; for spring admission, 12/1 for domestic students. Applications are processed on a rolling basis. Application fee: $20. *Financial support:* Teaching assistantships, institutionally sponsored loans available. Support available to part-time students. Financial award application deadline: 5/31; financial award applicants required to submit FAFSA. *Unit head:* Dr. Jaimie Hebert, Chair, 936-294-1563, Fax: 936-294-1882, E-mail: mth_jlh@shsu.edu. *Application contact:* Anita Shipman, Advisor, 936-294-3962.

San Diego State University, Graduate and Research Affairs, College of Sciences, Department of Mathematical Sciences, Program in Statistics, San Diego, CA 92182. Offers MS. Part-time programs available. *Students:* 20 full-time (10 women), 20 part-time (12 women); includes 8 minority (7 Asian Americans or Pacific Islanders, 1 Hispanic American), 9 international. Average age 30. 47 applicants, 77% accepted, 17 enrolled. In 2005, 17 degrees awarded. *Degree requirements:* For master's, comprehensive exam. *Entrance requirements:* For master's, GRE General Test. Additional exam requirements/recommendations for international students: Required—TOEFL. *Application deadline:* For fall admission, 5/1 for domestic students, 5/1 for international students; for spring admission, 11/1 for domestic students, 10/1 for international students. Applications are processed on a rolling basis. Application fee: $55. Electronic applications accepted. *Financial support:* Teaching assistantships, unspecified assistantships available. Financial award applicants required to submit FAFSA. *Unit head:* Kung-Jong Lui, Graduate Advisor, 619-594-7239, Fax: 619-594-6746, E-mail: kjl@rohan.sdsu.edu. *Application contact:* Kung-Jong Lui, Graduate Advisor, 619-594-7239, Fax: 619-594-6746, E-mail: kjl@rohan.sdsu.edu.

Simon Fraser University, Graduate Studies, Faculty of Science, Department of Mathematics, Department of Statistics and Actuarial Science, Burnaby, BC V5A 1S6, Canada. Offers M Sc, PhD. Part-time programs available. *Degree requirements:* For master's, thesis, participation in consulting; for doctorate, thesis/dissertation, comprehensive exam. *Entrance requirements:* For master's, minimum GPA of 3.0; for doctorate, minimum GPA of 3.5. Additional exam requirements/recommendations for international students: Required—TOEFL. *Faculty research:* Biostatistics, experimental design, envirometrics, statistical computing, statistical theory.

Southern Illinois University Carbondale, Graduate School, College of Science, Department of Mathematics, Carbondale, IL 62901-4701. Offers mathematics (MA, MS, PhD); statistics (MS). Part-time programs available. *Faculty:* 32 full-time (2 women), 1 part-time/adjunct (0 women). *Students:* 29 full-time (11 women), 8 part-time (2 women); includes 2 minority (both African Americans), 22 international. Average age 26. 67 applicants, 16% accepted, 2 enrolled. In 2005, 7 master's awarded. *Degree requirements:* For master's, thesis; for doctorate, 2 foreign languages, thesis/dissertation. *Entrance requirements:* For master's, minimum GPA of 2.7; for doctorate, minimum GPA of 3.25. Additional exam requirements/recommendations for international students: Required—TOEFL. *Application deadline:* Applications are processed on a rolling basis. Application fee: $0. *Financial support:* In 2005–06, 28 students received support, including 24 teaching assistantships with full tuition reimbursements available; fellowships with full tuition reimbursements available, research assistantships with full tuition reimbursements available, Federal Work-Study, institutionally sponsored loans, and tuition waivers (full) also available. Support available to part-time students. *Faculty research:* Differential equations, combinatorics, probability, algebra, numerical analysis. *Unit head:* Andrew Earnest, Chairperson, 618-453-6522, Fax: 618-453-5300, E-mail: chairman@math.siu.edu. *Application contact:* William T. Patula, Director of Graduate Studies, 618-453-5302, Fax: 618-453-5300, E-mail: wpatula@math.siu.edu.

See Close-Up on page 523.

Southern Methodist University, Dedman College, Department of Statistical Science, Dallas, TX 75275. Offers MS, PhD. Part-time programs available. *Faculty:* 10 full-time (3 women). *Students:* 22 full-time (9 women), 7 part-time (3 women); includes 2 minority (both African Americans), 16 international. Average age 31. 32 applicants, 69% accepted. In 2005, 4 master's, 1 doctorate awarded. *Median time to degree:* Of those who began their doctoral program in fall 1996, 86% received their degree in 8 years or less. *Degree requirements:* For master's and doctorate, thesis/dissertation, oral and written exams. *Entrance requirements:* For master's, GRE General Test, 12 hours course work in advanced math courses; for doctorate, GRE General Test, minimum GPA of 3.0. Additional exam requirements/recommendations for international students: Required—TOEFL. *Application deadline:* For fall admission, 6/30 for domestic students; for spring admission, 11/30 priority date for domestic students. Applications are processed on a rolling basis. Application fee: $60. Electronic applications accepted. *Financial support:* In 2005–06, 25 students received support, including 4 research assistantships with full tuition reimbursements available (averaging $15,000 per year), 21 teaching assistantships with full tuition reimbursements available (averaging $13,500 per year) Financial award application deadline: 4/30; financial award applicants required to submit FAFSA. *Faculty research:* Regression, time series, linear models sampling, nonparametrics. Total annual research expenditures: $250,000. *Unit head:* Richard F. Gunst, Chair, 214-768-2441, Fax: 214-768-4035, E-mail: rgunst@mail.smu.edu. *Application contact:* Wayne Woodward, Graduate Advisor, 214-768-2457, Fax: 214-768-4035, E-mail: waynew@mail.smu.edu.

Stanford University, School of Humanities and Sciences, Department of Statistics, Stanford, CA 94305-9991. Offers MS, PhD. Terminal master's awarded for partial completion of doctoral program. *Degree requirements:* For doctorate, thesis/dissertation, oral exam, qualifying exams. *Entrance requirements:* For master's, GRE General Test; for doctorate, GRE General Test, GRE Subject Test. Additional exam requirements/recommendations for international students: Required—TOEFL. Electronic applications accepted.

State University of New York at Binghamton, Graduate School, School of Arts and Sciences, Department of Mathematical Sciences, Binghamton, NY 13902-6000. Offers computer science (MA, PhD); probability and statistics (MA, PhD). Part-time programs available. Terminal master's awarded for partial completion of doctoral program. *Degree requirements:* For master's, thesis or alternative; for doctorate, 2 foreign languages, thesis/dissertation. *Entrance requirements:* For master's and doctorate, GRE General Test, GRE Subject Test. Additional exam requirements/recommendations for international students: Required—TOEFL. Electronic applications accepted.

Stephen F. Austin State University, Graduate School, College of Sciences and Mathematics, Department of Mathematics and Statistics, Nacogdoches, TX 75962. Offers mathematics (MS); mathematics education (MS); statistics (MS). *Faculty:* 20 full-time (5 women). *Students:* 13 full-time (3 women), 53 part-time (47 women); includes 7 minority (5 African Americans, 2 Hispanic Americans), 6 international. 11 applicants, 100% accepted. In 2005, 25 degrees awarded. *Degree requirements:* For master's, thesis optional. *Entrance requirements:* For master's, GRE General Test, minimum GPA of 2.8 in last 60 hours, 2.5 overall. Additional exam requirements/recommendations for international students: Required—TOEFL. *Application deadline:* For fall admission, 8/1 for domestic students; for spring admission, 12/15 for domestic students. Applications are processed on a rolling basis. Application fee: $0 ($50 for international students). *Expenses:* Tuition, state resident: full-time $2,628; part-time $146 per credit hour. Tuition, nonresident: full-time $7,596; part-time $422 per credit hour. Required fees:$900; $170. *Financial support:* In 2005–06, 8 teaching assistantships (averaging $11,500 per year) were awarded; Federal Work-Study and unspecified assistantships also available.

Statistics

Stephen F. Austin State University (continued)

Financial award application deadline: 3/1. *Faculty research:* Kernel type estimators, fractal mappings, spline curve fitting, robust regression continua theory. *Unit head:* Dr. Jasper Adams, Chair, 936-468-3805.

Stevens Institute of Technology, Graduate School, Arthur E. Imperatore School of Sciences and Arts, Department of Mathematical Sciences, Program in Applied Statistics, Hoboken, NJ 07030. Offers Certificate. *Students:* 5 applicants, 80% accepted. *Entrance requirements:* Additional exam requirements/recommendations for international students: Required—TOEFL. *Application deadline:* Applications are processed on a rolling basis. Application fee: $50. Electronic applications accepted. *Expenses:* Tuition: Part-time $920 per credit hour. Tuition and fees vary according to program. *Application contact:* Dr. Milos Dostal, Professor, 201-216-5426.

Stevens Institute of Technology, Graduate School, Arthur E. Imperatore School of Sciences and Arts, Department of Mathematical Sciences, Program in Stochastic Systems Analysis and Optimization, Hoboken, NJ 07030. Offers MS, Certificate. *Expenses:* Tuition: Part-time $920 per credit hour. Tuition and fees vary according to program. *Unit head:* Darinka Dentcheva, Head, 201-216-8640.

Stony Brook University, State University of New York, Graduate School, College of Engineering and Applied Sciences, Department of Applied Mathematics and Statistics, Stony Brook, NY 11794. Offers MS, PhD. *Faculty:* 20 full-time (4 women), 1 part-time/adjunct (0 women). *Students:* 139 full-time (60 women), 15 part-time (4 women); includes 26 minority (7 African Americans, 15 Asian Americans or Pacific Islanders, 4 Hispanic Americans), 97 international. Average age 28. 195 applicants, 84% accepted. In 2005, 34 master's, 8 doctorates awarded. *Degree requirements:* For master's, thesis or alternative; for doctorate, one foreign language, thesis/dissertation, comprehensive exam. *Entrance requirements:* For master's and doctorate, GRE General Test. Additional exam requirements/recommendations for international students: Required—TOEFL. *Application deadline:* For fall admission, 1/15 for domestic students. Application fee: $50. *Expenses:* Tuition, area resident: Part-time $288. Tuition, state resident: full-time $6,900. Tuition, nonresident: full-time $10,920; part-time $455. Required fees: $704. *Financial support:* In 2005–06, 7 fellowships, 29 research assistantships, 33 teaching assistantships were awarded. *Faculty research:* Biostatistics, combinatorial analysis, differential equations, modeling. Total annual research expenditures: $2.8 million. *Unit head:* Dr. J. Glimm, Chairman, 631-632-8360. *Application contact:* Dr. Woo Jong Kim, Director, 631-632-8360, Fax: 631-632-8490, E-mail: wjkim@ccmail.sunysb.edu.

See Close-Up on page 527.

Syracuse University, Graduate School, College of Arts and Sciences, Program in Applied Statistics, Syracuse, NY 13244. Offers MS. Part-time programs available. *Students:* 8 full-time (6 women); includes 1 minority (Asian American or Pacific Islander), 7 international. 8 applicants, 50% accepted, 1 enrolled. *Entrance requirements:* For master's, GRE. Additional exam requirements/recommendations for international students: Required—TOEFL. *Application deadline:* For fall admission, 1/10 for domestic students. Applications are processed on a rolling basis. Application fee: $65. *Financial support:* Fellowships with full tuition reimbursements, teaching assistantships with full tuition reimbursements, tuition waivers available. *Unit head:* Dr. Pinyuen Chen, Director, Program on Applied Statistics, 315-443-1577, Fax: 315-443-1475, E-mail: pinchen@syr.edu.

Temple University, Graduate School, Fox School of Business and Management, Doctoral Programs in Business, Philadelphia, PA 19122-6096. Offers accounting (PhD); economics (PhD); finance (PhD); general and strategic management (PhD); healthcare management (PhD); human resource administration (PhD); international business administration (PhD); management information systems (PhD); management science/operations research (PhD); marketing (PhD); risk, insurance, and health-care management (PhD); statistics (PhD); tourism (PhD). *Accreditation:* AACSB. *Entrance requirements:* For doctorate, GRE General Test, minimum GPA of 3.0, master's degree. Additional exam requirements/recommendations for international students: Required—TOEFL. *Expenses:* Tuition, state resident: full-time $8,694; part-time $483 per credit. Tuition, nonresident: full-time $12,672; part-time $704 per credit. Required fees: $500; $122 per semester. Tuition and fees vary according to course level, campus/location and program.

Temple University, Graduate School, Fox School of Business and Management, Masters Programs in Business, MBA Programs, Philadelphia, PA 19122-6096. Offers accounting (MBA); business administration (EMBA, MBA); e-business (MBA); economics (MBA); finance (MBA); general and strategic management (MBA); healthcare management (MBA); human resource administration (MBA); international business (IMBA); management information systems (MBA); management science/operations management (MBA); marketing (MBA); risk management and insurance (MBA); statistics (MBA). EMBA offered in Philadelphia, PA and Tokyo, Japan. *Accreditation:* AACSB. *Students:* Average age 31. *Entrance requirements:* For master's, GMAT, minimum undergraduate GPA of 3.0. Additional exam requirements/recommendations for international students: Required—TOEFL. *Application deadline:* For fall admission, 4/15 for domestic students, 1/15 for international students; for spring admission, 9/30 for domestic students, 9/1 for international students. Application fee: $40. *Expenses:* Tuition, state resident: full-time $8,694; part-time $483 per credit. Tuition, nonresident: full-time $12,672; part-time $704 per credit. Required fees: $500; $122 per semester. Tuition and fees vary according to course level, campus/location and program. *Application contact:* Natale Butto, Director of Graduate Admissions, 215-204-7678, Fax: 215-204-8300, E-mail: butto@sbm.temple.edu.

Temple University, Graduate School, Fox School of Business and Management, Masters Programs in Business, MS Programs, Philadelphia, PA 19122-6096. Offers accounting and financial management (MS); actuarial science (MS); e-business (MS); finance (MS); healthcare financial management (MS); human resource administration (MS); management information systems (MS); management science/operations management (MS); marketing (MS); statistics (MS). *Accreditation:* AACSB. *Entrance requirements:* For master's, GRE General Test, minimum undergraduate GPA of 3.0. Additional exam requirements/recommendations for international students: Required—TOEFL. *Expenses:* Tuition, state resident: full-time $8,694; part-time $483 per credit. Tuition, nonresident: full-time $12,672; part-time $704 per credit. Required fees: $500; $122 per semester. Tuition and fees vary according to course level, campus/location and program.

Texas A&M University, College of Science, Department of Statistics, College Station, TX 77843. Offers MS, PhD. Part-time programs available. *Faculty:* 17 full-time (2 women), 1 part-time/adjunct (0 women). *Students:* 66 full-time (30 women), 13 part-time (8 women); includes 5 minority (1 African American, 3 Asian Americans or Pacific Islanders, 1 Hispanic American), 45 international. Average age 29. 74 applicants, 57% accepted, 29 enrolled. In 2005, 9 master's, 18 doctorates awarded. Terminal master's awarded for partial completion of doctoral program. *Degree requirements:* For doctorate, thesis/dissertation. *Entrance requirements:* For master's and doctorate, GRE General Test. Additional exam requirements/recommendations for international students: Required—TOEFL. *Application deadline:* For fall admission, 3/1 for domestic students; for spring admission, 8/1 for domestic students. Applications are processed on a rolling basis. Application fee: $50 ($75 for international students). *Expenses:* Tuition, state resident: full-time $4,488; part-time $187 per credit hour. Tuition, nonresident: full-time $11,112; part-time $463 per credit hour. Required fees:$1,974. *Financial support:* Fellowships, research assistantships, teaching assistantships, career-related internships or fieldwork available. Financial award application deadline:3/1. *Faculty research:* Time series, chemometrics, biometrics, smoothing, linear models. *Unit head:* Simon J. Sheather, Head, 979-845-3141, Fax: 979-845-3144. *Application contact:* P. Fred Dahm, Graduate Director, 800-826-8009, Fax: 979-845-3144, E-mail: fdahm@stat.tamu.edu.

Tulane University, Graduate School, Department of Mathematics, New Orleans, LA 70118-5669. Offers applied mathematics (MS); mathematics (MS, PhD); statistics (MS). *Degree requirements:* For master's, thesis (for some programs); for doctorate, thesis/dissertation. *Entrance requirements:* For master's, GRE General Test, minimum B average in undergraduate course work; for doctorate, GRE General Test. Additional exam requirements/recommendations for international students: Required—TOEFL; Recommended—TSE. Electronic applications accepted.

Université de Montréal, Faculty of Graduate Studies, Faculty of Arts and Sciences, Department of Mathematics and Statistics, Montréal, QC H3C 3J7, Canada. Offers mathematics (M Sc, PhD); statistics (M Sc, PhD). *Faculty:* 37 full-time (5 women), 15 part-time/adjunct (1 woman). *Students:* 100 full-time (29 women). 120 applicants, 21% accepted, 20 enrolled. In 2005, 18 master's, 4 doctorates awarded. *Degree requirements:* For master's, thesis; for doctorate, thesis/dissertation, general exam. *Entrance requirements:* For master's and doctorate, proficiency in French. *Application deadline:* For fall and spring admission, 2/1. For winter admission, 11/1 for domestic students. Application fee: $30. Electronic applications accepted. *Financial support:* Fellowships, research assistantships, teaching assistantships, monitorships available. Financial award application deadline: 4/1. *Faculty research:* Pure and applied mathematics, actuarial mathematics. *Unit head:* Veronique Hussih, Chair, 514-343-6710, Fax: 514-343-5700. *Application contact:* Liliane Badier, Student Files Management Technician, 514-343-6111 Ext. 1695, Fax: 514-343-5700.

Université Laval, Faculty of Sciences and Engineering, Department of Mathematics and Statistics, Program in Statistics, Québec, QC G1K 7P4, Canada. Offers M Sc. *Degree requirements:* For master's, thesis (for some programs). *Entrance requirements:* For master's, knowledge of French and English. Electronic applications accepted.

University at Albany, State University of New York, College of Arts and Sciences, Department of Mathematics and Statistics, Albany, NY 12222-0001. Offers mathematics (PhD); secondary teaching (MA); statistics (MA). Evening/weekend programs available. *Students:* 33 full-time (13 women), 16 part-time (4 women). Average age 32. In 2005, 13 master's, 4 doctorates awarded. *Degree requirements:* For doctorate, one foreign language, thesis/dissertation. *Entrance requirements:* For doctorate, GRE General Test. Additional exam requirements/recommendations for international students: Required—TOEFL (minimum score 550 paper-based; 213 computer-based). *Application deadline:* For fall admission, 3/15 for domestic students, 5/1 for international students. Applications are processed on a rolling basis. Application fee: $60. Electronic applications accepted. *Financial support:* Fellowships, research assistantships, teaching assistantships, minority assistantships available. Financial award application deadline: 3/15. *Unit head:* Timothy Lance, Chair, 518-442-4602.

University at Albany, State University of New York, School of Education, Department of Educational and Counseling Psychology, Albany, NY 12222-0001. Offers counseling psychology (MS, PhD, CAS); educational psychology (Ed D); educational psychology and statistics (MS); measurements and evaluation (Ed D); rehabilitation counseling (MS), including counseling psychology; school counselor (CAS); school psychology (Psy D, CAS); special education (MS); statistics and research design (Ed D). *Accreditation:* APA (one or more programs are accredited). Evening/weekend programs available. *Students:* 89 full-time (72 women), 22 part-time (18 women). Average age 28. In 2005, 109 master's, 24 doctorates, 17 other advanced degrees awarded. *Degree requirements:* For doctorate, thesis/dissertation. *Entrance requirements:* For doctorate, GRE General Test. Additional exam requirements/recommendations for international students: Required—TOEFL (minimum score 550 paper-based; 213 computer-based). Application fee: $60. Electronic applications accepted. *Financial support:* Fellowships, career-related internships or fieldwork available. *Unit head:* Deborah May, Chair, 518-442-5050.

The University of Akron, Graduate School, Buchtel College of Arts and Sciences, Department of Statistics, Akron, OH 44325. Offers MS. Part-time and evening/weekend programs available. *Faculty:* 6 full-time (0 women). *Students:* 19 full-time (7 women), 1 (woman) part-time, 8 international. Average age 32. 21 applicants, 67% accepted, 6 enrolled. In 2005, 12 degrees awarded. *Degree requirements:* For master's, thesis optional. *Entrance requirements:* For master's, minimum GPA of 2.75. Additional exam requirements/recommendations for international students: Required—TOEFL (minimum score 550 paper-based; 213 computer-based), Michigan English Language Assessment Battery. *Application deadline:* For fall admission, 3/1 for domestic students. Applications are processed on a rolling basis. Application fee: $30 ($40 for international students). Electronic applications accepted. *Expenses:* Tuition, state resident: full-time $5,816; part-time $323 per credit. Tuition, nonresident: full-time $9,976; part-time $554 per credit. Required fees: $794; $43 per credit. $12 per term. Tuition and fees vary according to course load, degree level and program. *Financial support:* In 2005–06, 2 research assistantships, 11 teaching assistantships with full tuition reimbursements were awarded. *Faculty research:* Experimental design, sampling biostatistics. Total annual research expenditures: $40,176. *Unit head:* Dr. Chand Midha, Chair, 330-972-7128, E-mail: cmidha@uakron.edu.

The University of Alabama, Graduate School, Manderson Graduate School of Business, Manderson Graduate School of Business, Department of Information Systems, Statistics, and Management Science, Tuscaloosa, AL 35487. Offers applied statistics (MS, PhD); operations management (MS, PhD). *Accreditation:* AACSB. Part-time programs available. *Faculty:* 19 full-time (3 women). *Students:* 35 full-time (16 women), 9 part-time (1 woman); includes 4 minority (2 African Americans, 2 Asian Americans or Pacific Islanders), 23 international. Average age 29. 78 applicants, 32% accepted, 18 enrolled. In 2005, 17 master's, 6 doctorates awarded. Terminal master's awarded for partial completion of doctoral program. *Median time to degree:* Of those who began their doctoral program in fall 1996, 100% received their degree in 8 years or less. *Degree requirements:* For master's, thesis optional; for doctorate, thesis/dissertation, comprehensive exam. *Entrance requirements:* For master's, GMAT or GRE; for doctorate, GMAT or GRE, undergraduate major in related field. Additional exam requirements/recommendations for international students: Required—TOEFL. *Application deadline:* For fall admission, 7/6 for domestic students, 7/6 for international students. Applications are processed on a rolling basis. Application fee: $25. Electronic applications accepted. *Expenses:* Tuition, area resident: Full-time $2,432. Tuition, nonresident: full-time $6,758. *Financial support:* In 2005–06, 24 students received support, including 2 fellowships with full tuition reimbursements available (averaging $25,000 per year), 6 research assistantships with full tuition reimbursements available (averaging $15,000 per year), 11 teaching assistantships with full tuition reimbursements available (averaging $15,000 per year); career-related internships or fieldwork also available. *Faculty research:* Data mining, production and inventory modeling, regression analysis, statistical quality control, supply chain management. *Unit head:* Dr. Michael D. Conerly, Head and Professor, 205-348-8902, Fax: 205-348-0560, E-mail: mconerly@cba.ua.edu. *Application contact:* Dr. Michael D. Conerly, Head and Professor, 205-348-8902, Fax: 205-348-0560, E-mail: mconerly@cba.ua.edu.

University of Alaska Fairbanks, College of Natural Sciences and Mathematics, Department of Mathematics and Statistics, Fairbanks, AK 99775-7520. Offers computer science (MS); mathematics (MAT, MS, PhD); statistics (MS). Part-time programs available. *Faculty:* 13 full-time (5 women), 1 part-time/adjunct (0 women). *Students:* 14 full-time (1 woman), 4 part-time (3 women); includes 1 minority (Asian American or Pacific Islander), 8 international. Average age 26. 19 applicants, 74% accepted, 5 enrolled. In 2005, 3 degrees awarded. Terminal master's awarded for partial completion of doctoral program. *Degree requirements:* For master's, thesis or alternative, comprehensive exam, registration; for doctorate, thesis/dissertation, comprehensive exam, registration. *Entrance requirements:* For master's and doctorate, GRE General Test, GRE Subject Test. Additional exam requirements/recommendations for international students: Required—TOEFL (minimum score 600 paper-based); Recommended—TSE. *Application deadline:* For fall admission, 6/1 for domestic students, 3/1 for international students; for spring admission, 12/1 for domestic students, 9/1 for international students. Applications are processed on a rolling basis. Application fee: $50. Electronic applications accepted. *Expenses:* Tuition, state resident: full-time $4,392; part-time $244 per credit. Tuition, nonresident: full-time $8,964; part-time $498 per credit. Required fees: $800; $5 per credit. $48 per contact hour. Tuition and fees vary according to course level, course load, campus/location and reciprocity agreements. *Financial support:* In 2005–06, 4 research assistantships with tuition reimbursements (averaging $9,924 per year), 9 teaching assistantships with tuition reimbursements (averaging $10,458 per year) were awarded; fellowships with tuition reimbursements, career-related internships or fieldwork, Federal Work-Study, scholarships/grants, and unspecified assistantships also available. Financial award applicants required to submit FAFSA. *Faculty research:* Blackbox kriging (statistics), interaction with a virtual reality environment (computer), arrangements of hyperplanes (topology), bifurcation analysis of time-periodic differential-delay equations, synthetic aperture radar interferometry software (computer). *Unit head:* Dr. Dana L. Thomas, Chair, 907-474-7332, Fax: 907-474-5394, E-mail: fymath@uaf.edu.

University of Alberta, Faculty of Graduate Studies and Research, Department of Mathematical and Statistical Sciences, Edmonton, AB T6G 2E1, Canada. Offers applied mathematics (M Sc, PhD); biostatistics (M Sc); mathematical finance (M Sc, PhD); mathematical physics (M Sc, PhD); mathematics (M Sc, PhD); statistics (M Sc, PhD, Postgraduate Diploma). Part-time programs available. *Faculty:* 48 full-time (4 women). *Students:* 112 full-time (41 women), 5 part-time. Average age 24. 776 applicants, 5% accepted, 34 enrolled. In 2005, 12 master's, 10 doctorates awarded. Terminal master's awarded for partial completion of doctoral program. *Median time to degree:* Of those who began their doctoral program in fall 1996, 100% received their degree in 8 years or less. *Degree requirements:* For master's, thesis (for some programs); for doctorate, thesis/dissertation, comprehensive exam. *Entrance requirements:* Additional exam requirements/recommendations for international students: Required—TOEFL (minimum score 580 paper-based; 237 computer-based). *Application deadline:* For fall admission, 3/1 for domestic students, 2/1 for international students. Applications are processed on a rolling basis. Application fee: $0. Electronic applications accepted. Tuition and fees charges are reported in Canadian dollars. *Expenses:* Tuition, state resident: part-time $562 Canadian dollars per term. Tuition, nonresident: full-time $3,375 Canadian dollars. Required fees: $573 Canadian dollars; $84 Canadian dollars per term. *Financial support:* In 2005–06, 51 research assistantships, 88 teaching assistantships with full and partial tuition reimbursements were awarded; scholarships/grants also available. Financial award application deadline:5/1. *Faculty research:* Classical and functional analysis, algebra, differential equations, geometry. *Unit head:* Dr. Anthony To-Ming Lau, Chair, 403-492-5141, E-mail: tlau@math.ualberta.ca. *Application contact:* Dr. Yau Shu Wong, Associate Chair, Graduate Studies, 403-492-5799, Fax: 403-492-6828, E-mail: gradstudies@math.ualberta.ca.

University of Arkansas, Graduate School, J. William Fulbright College of Arts and Sciences, Department of Mathematical Sciences, Program in Statistics, Fayetteville, AR 72701-1201. Offers MS. *Students:* 13 full-time (9 women), 2 part-time (1 woman); includes 2 minority (1 American Indian/Alaska Native, 1 Hispanic American), 8 international. 27 applicants, 56% accepted. In 2005, 6 degrees awarded. *Degree requirements:* For master's, thesis. Application fee: $40 ($50 for international students). *Financial support:* In 2005–06, 6 research assistantships, 7 teaching assistantships were awarded; career-related internships or fieldwork and Federal Work-Study also available. Support available to part-time students. Financial award application deadline: 4/1; financial award applicants required to submit FAFSA. *Unit head:* Dr. Laurie Meaux, Chair of Studies, 479-575-3351.

University of Arkansas at Little Rock, Graduate School, College of Science and Mathematics, Department of Mathematics and Statistics, Little Rock, AR 72204-1099. Offers applied mathematics (MS), including applied analysis, mathematical statistics. Part-time and evening/weekend programs available. *Degree requirements:* For master's, comprehensive exam. *Entrance requirements:* For master's, GRE General Test, GRE Subject Test, minimum GPA of 2.7, previous course work in advanced mathematics.

The University of British Columbia, Faculty of Graduate Studies, Faculty of Science, Department of Statistics, Vancouver, BC V6T 1Z1, Canada. Offers M Sc, PhD. Part-time programs available. *Faculty:* 11 full-time (1 woman), 3 part-time/adjunct (1 woman). *Students:* 33 full-time (9 women). Average age 25. 200 applicants, 5% accepted. In 2005, 6 master's, 1 doctorate awarded. *Median time to degree:* Of those who began their doctoral program in fall 1996, 100% received their degree in 8 years or less. *Degree requirements:* For master's, thesis/dissertation, registration; for doctorate, thesis/dissertation, comprehensive exam, registration. *Entrance requirements:* Additional exam requirements/recommendations for international students: Required—TOEFL (minimum score 600 paper-based; 250 computer-based). *Application deadline:* For fall admission, 1/31 priority date for domestic students, 1/31 priority date for international students. Applications are processed on a rolling basis. Application fee: $90 Canadian dollars ($150 Canadian dollars for international students). Electronic applications accepted. *Financial support:* In 2005–06, fellowships (averaging $17,000 per year), research assistantships (averaging $15,000 per year), teaching assistantships (averaging $10,000 per year) were awarded; career-related internships or fieldwork, Federal Work-Study, and tuition waivers (partial) also available. Financial award application deadline: 2/28. *Faculty research:* Theoretical, applied, biostatistical, and computational statistics. *Unit head:* W. J. Welch, Head, 604-822-6844, Fax: 604-822-6960, E-mail: head@stat.ubc.ca. *Application contact:* Graduate Admissions, 604-822-4821, Fax: 604-822-6960, E-mail: gradinfo@stat.ubc.ca.

University of Calgary, Faculty of Graduate Studies, Faculty of Science, Department of Mathematics and Statistics, Calgary, AB T2N 1N4, Canada. Offers M Sc, PhD. *Faculty:* 31 full-time (3 women), 3 part-time/adjunct (1 woman). *Students:* 48 full-time (11 women), 1 part-time. Average age 26. 82 applicants, 33% accepted, 15 enrolled. In 2005, 8 master's, 2 doctorates awarded. *Median time to degree:* Of those who began their doctoral program in fall 1996, 100% received their degree in 8 years or less. *Degree requirements:* For master's, thesis, comprehensive exam; for doctorate, thesis/dissertation, candidacy exam, preliminary exams. *Entrance requirements:* For master's, honors degree in applied math, pure math, or statistics; for doctorate, MA or M Sc. Additional exam requirements/recommendations for international students: Required—TOEFL (minimum score 600 paper-based; 250 computer-based), IELT (minimum score 7), TOEFL (paper score 600; computer score 250) or IELT (paper score 7). *Application deadline:* For fall admission, 2/1 for domestic students, 2/1 for international students. Applications are processed on a rolling basis. Application fee: $100 ($130 for international students). *Financial support:* In 2005–06, 43 students received support, including research assistantships with partial tuition reimbursements available (averaging $4,100 per year), teaching assistantships with partial tuition reimbursements available (averaging $13,060 per year); fellowships, scholarships/grants and unspecified assistantships also available. *Faculty research:* Combinatorics, applied mathematics, statistics, probability, analysis. Total annual research expenditures: $1.1 million. *Unit head:* Dr. Claude LaFlamme, Head (Acting), 403-220-5210, Fax: 403-282-5150, E-mail: laf@math.ucalgary.ca. *Application contact:* Joanne Mellard, Graduate Administrator, 403-220-6299, Fax: 403-282-5150, E-mail: gradapps@math.ucalgary.ca.

University of California, Berkeley, Graduate Division, College of Letters and Science, Department of Statistics, Berkeley, CA 94720-1500. Offers MA, PhD. *Degree requirements:* For doctorate, thesis/dissertation, qualifying exam, written preliminary exam. *Entrance requirements:* For master's and doctorate, GRE General Test, minimum GPA of 3.0.

University of California, Davis, Graduate Studies, Program in Statistics, Davis, CA 95616. Offers MS, PhD. *Faculty:* 26 full-time. *Students:* 37 full-time (24 women); includes 3 minority (1 African American, 2 Asian Americans or Pacific Islanders), 26 international. Average age 29. 114 applicants, 18% accepted, 7 enrolled. In 2005, 5 master's, 9 doctorates awarded. Terminal master's awarded for partial completion of doctoral program. *Median time to degree:* Of those who began their doctoral program in fall 1996, 42.8% received their degree in 8 years or less. *Degree requirements:* For master's, comprehensive exam; for doctorate, thesis/dissertation. *Entrance requirements:* For master's and doctorate, GRE General Test, minimum GPA of 3.0. Additional exam requirements/recommendations for international students: Required—TOEFL (minimum score 550 paper-based; 213 computer-based). *Application deadline:* For fall admission, 1/15 for domestic students, 1/15 for international students. Application fee: $60. Electronic applications accepted. *Financial support:* In 2005–06, 37 students received support, including 3 fellowships with full and partial tuition reimbursements available (averaging $7,284 per year), 7 research assistantships with full and partial tuition reimbursements available (averaging $13,107 per year), 7 teaching assistantships with partial tuition reimbursements available (averaging $15,288 per year); Federal Work-Study, institutionally sponsored loans, scholarships/grants, and tuition waivers (full and partial) also available. Financial award application deadline: 1/15; financial award applicants required to submit FAFSA. *Faculty research:* Nonparametric analysis, time series analysis, biostatistics, curve estimation, reliability. *Unit head:* Rudolph Beran, Chair, 530-754-7765, E-mail: beran@wald.ucdavis.edu. *Application contact:* Laura Peterson, Administrative Assistant, 530-752-2632, Fax: 530-752-7099, E-mail: grad-staff@wald.ucdavis.edu.

University of California, Los Angeles, Graduate Division, College of Letters and Science, Department of Statistics, Los Angeles, CA 90095. Offers MS, PhD.

University of California, Riverside, Graduate Division, Department of Statistics, Riverside, CA 92521-0102. Offers applied statistics (PhD); statistics (MS). Part-time programs available. *Faculty:* 7 full-time (1 woman). *Students:* 32 full-time (11 women), 1 part-time; includes 6 minority (3 Asian Americans or Pacific Islanders, 3 Hispanic Americans), 18 international. Average age 31. In 2005, 5 master's, 3 doctorates awarded. *Degree requirements:* For master's, comprehensive exam; for doctorate, thesis/dissertation, qualifying exams, 3 quarters of teaching experience. *Entrance requirements:* For master's and doctorate, GRE General Test, minimum GPA of 3.2. Additional exam requirements/recommendations for international students: Required—TOEFL (minimum score 550 paper-based; 213 computer-based); Recommended—TSE (minimum score 50). *Application deadline:* For fall admission, 5/1 for domestic students, 2/1 for international students. For winter admission, 9/1 for domestic students; for spring admission, 12/1 for domestic students. Applications are processed on a rolling basis. Application fee: $60 ($75 for international students). Electronic applications accepted. *Expenses:* Tuition, nonresident: full-time $14,694. Full-time tuition and fees vary according to program. *Financial support:* In 2005–06, fellowships with full and partial tuition reimbursements (averaging $12,000 per year), research assistantships with partial tuition reimbursements (averaging $14,000 per year), teaching assistantships with partial tuition reimbursements (averaging $15,000 per year) were awarded; career-related internships or fieldwork, Federal Work-Study, institutionally sponsored loans, health care benefits, and tuition waivers (full and partial) also available. Financial award application deadline: 2/1; financial award applicants required to submit FAFSA. *Faculty research:* Design and analysis of experiments, statistical modeling, stochastic models, paired comparisons, statistical design of experiments and linear models. *Unit head:* Dr. Subir Ghost, Chair, 951-827-3781, Fax: 951-827-3286, E-mail: subir.ghosh@ucr.edu. *Application contact:* Perla Fabelo, Graduate Program Assistant, 800-735-0717, Fax: 951-827-5517, E-mail: stat@ucr.edu.

University of California, San Diego, Graduate Studies and Research, Department of Mathematics, La Jolla, CA 92093. Offers applied mathematics (MA); mathematics (MA, PhD); statistics (MS). *Degree requirements:* For doctorate, thesis/dissertation. *Entrance requirements:* For master's and doctorate, GRE General Test, GRE Subject Test. Electronic applications accepted.

University of California, Santa Barbara, Graduate Division, College of Letters and Sciences, Division of Mathematics, Life, and Physical Sciences, Department of Statistics and Applied Probability, Santa Barbara, CA 93106. Offers applied statistics (MA); mathematical and empirical finance (PhD); mathematical statistics (MA); quantitative methods in the social sciences (PhD); statistics and applied probability (PhD). *Students:* 56 full-time (19 women); includes 28 minority (1 American Indian/Alaska Native, 25 Asian Americans or Pacific Islanders, 2 Hispanic Americans). Average age 24. 90 applicants, 74% accepted, 21 enrolled. In 2005, 10 master's, 3 doctorates awarded. Terminal master's awarded for partial completion of doctoral program. *Median time to degree:* Of those who began their doctoral program in fall 1996, 99% received their degree in 8 years or less. *Degree requirements:* For master's and doctorate, thesis/dissertation, comprehensive exam, registration. *Entrance requirements:* For master's and doctorate, GRE General Test. Additional exam requirements/recommendations for international students: Required—TOEFL (minimum score 550 paper-based; 213 computer-based). *Application deadline:* For fall admission, 4/15 for domestic students, 4/15 for international students. For winter admission, 11/15 for domestic students; for spring admission, 2/15 for domestic students. Applications are processed on a rolling basis. Application fee: $60. Electronic applications accepted. *Financial support:* In 2005–06, 2 fellowships with full tuition reimbursements (averaging $24,500 per year), 2 research assistantships with full tuition reimbursements (averaging $15,500 per year), 29 teaching assistantships with partial tuition reimbursements (averaging $11,500 per year) were awarded; health care benefits and tuition waivers (partial) also available. Financial award application deadline: 1/15; financial award applicants required to submit FAFSA. *Faculty research:* Bayesian methods, statistics including biostatistics, stochastic processes, probability, mathematical finance. *Unit head:* Dr. Raya E. Feldman, Chairman, 805-893-2826, Fax: 805-893-2334, E-mail: feldman@pstat.ucsb.edu. *Application contact:* Andrew V. Carter, Assistant Professor, 805-893-3299, Fax: 805-893-2334, E-mail: carter@pstat.ucsb.edu.

University of Central Florida, College of Sciences, Department of Statistics and Actuarial Science, Orlando, FL 32816. Offers actuarial science (MS); data mining (MS, Certificate); statistical computing (MS). Part-time and evening/weekend programs available. *Faculty:* 17 full-time (4 women). *Students:* 32 full-time (17 women), 16 part-time (5 women); includes 7 minority (2 African Americans, 4 Asian Americans or Pacific Islanders, 1 Hispanic American), 21 international. Average age 31. 65 applicants, 71% accepted, 21 enrolled. In 2005, 22 degrees awarded. *Degree requirements:* For master's, comprehensive exam. *Entrance requirements:* For master's, GRE General Test, minimum GPA of 3.0 in last 60 hours. Additional exam requirements/recommendations for international students: Required—TOEFL. *Application deadline:* For fall admission, 7/15 for domestic students; for spring admission, 12/1 for domestic students. Application fee: $30. Electronic applications accepted. *Expenses:* Tuition, state resident: full-time $5,788. Tuition, nonresident: full-time $21,927. Required fees: $241 per credit hour. *Financial support:* In 2005–06, 7 fellowships with partial tuition reimbursements (averaging $3,600 per year), 6 research assistantships with partial tuition reimbursements (averaging $9,300 per year), 18 teaching assistantships with partial tuition reimbursements (averaging $10,100 per year) were awarded; career-related internships or fieldwork, Federal Work-Study, institutionally sponsored loans, tuition waivers (partial), and unspecified assistantships also available. Financial award application deadline: 3/1; financial award applicants required to submit FAFSA. *Faculty research:* Multivariate analysis, quality control, shrinkage estimation. *Unit head:* Dr. David Nickerson, Interim Chair, 407-823-2289, Fax: 407-823-5419, E-mail: nickerson@mail.ucf.edu. *Application contact:* Dr. James R. Schott, Graduate Coordinator, 407-823-3323, Fax: 407-823-5419, E-mail: jschott@pegasus.cc.ucf.edu.

University of Central Oklahoma, College of Graduate Studies and Research, College of Mathematics and Science, Department of Mathematics and Statistics, Edmond, OK 73034-5209. Offers applied mathematical sciences (MS), including computer science, mathematics, mathematics/computer science teaching, statistics. Part-time programs available. *Faculty:* 8 full-time (3 women), 1 part-time/adjunct (0 women). *Students:* 7 full-time (1 woman), 8 part-time (3 women); includes 2 minority (both Asian Americans or Pacific Islanders), 7 international. Average age 29. 14 applicants, 100% accepted. In 2005, 6 degrees awarded. *Degree requirements:* For master's, thesis. *Entrance requirements:* Additional exam requirements/recommendations for international students: Required—TOEFL (minimum score 550 paper-based; 213 computer-based). *Application deadline:* Applications are processed on a rolling basis. Application fee: $25. Electronic applications accepted. *Expenses:* Tuition, state resident: full-time $2,988; part-time $125 per credit hour. Tuition, nonresident: full-time $4,728; part-time $197 per credit hour. Required fees: $716; $16 per credit hour. *Financial support:* Federal Work-Study and unspecified assistantships available. Financial award application deadline: 3/31; financial award applicants required to submit FAFSA. *Faculty research:* Curvature, FAA,

Statistics

University of Central Oklahoma (continued)
math education. *Unit head:* Dr. Chuck Cooper, Chairperson, 405-974-5294. *Application contact:* Dr. James Yates, Adviser, 405-974-5386, Fax: 405-974-3824, E-mail: jyates@aix1.ucok.edu.

University of Chicago, Division of the Physical Sciences, Department of Statistics, Chicago, IL 60637-1513. Offers SM, PhD. *Faculty:* 16 full-time (3 women), 1 part-time/adjunct (0 women). *Students:* 60 full-time (29 women), 19 part-time (10 women); includes 8 minority (1 African American, 6 Asian Americans or Pacific Islanders, 1 Hispanic American), 48 international. Average age 27. 203 applicants, 25 enrolled. In 2005, 16 master's, 5 doctorates awarded. Terminal master's awarded for partial completion of doctoral program. *Median time to degree:* Of those who began their doctoral program in fall 1996, 100% received their degree in 8 years or less. *Degree requirements:* For master's and doctorate, thesis/dissertation, registration. *Entrance requirements:* For master's and doctorate, GRE General Test, GRE Subject Test. Additional exam requirements/recommendations for international students: Required—TOEFL. *Application deadline:* For fall admission, 6/15 for domestic students, 6/15 for international students. Application fee: $55. Electronic applications accepted. *Financial support:* In 2005–06, fellowships with full tuition reimbursements (averaging $19,095 per year), research assistantships with full tuition reimbursements (averaging $19,095 per year), teaching assistantships with full tuition reimbursements (averaging $19,095 per year) were awarded; tuition waivers (partial) also available. Financial award application deadline: 2/1. *Faculty research:* Genetics, econometrics, generalized linear models, history of statistics, probability theory. *Unit head:* Dr. Stephen Stigler, Chairman, 773-702-8335. *Application contact:* Dr. Michael Wichura, Admissions Chair, 773-702-8329, E-mail: wichura@galton.uchicago.edu.

University of Cincinnati, Division of Research and Advanced Studies, McMicken College of Arts and Sciences, Department of Mathematical Sciences, Cincinnati, OH 45221. Offers applied mathematics (MS, PhD); mathematics education (MAT); pure mathematics (MS, PhD); statistics (MS, PhD). *Accreditation:* NCATE (one or more programs are accredited). Part-time programs available. Terminal master's awarded for partial completion of doctoral program. *Degree requirements:* For master's, thesis or alternative, comprehensive exam; for doctorate, one foreign language, thesis/dissertation, comprehensive exam. *Entrance requirements:* For master's, GRE, teacher certification (MAT); for doctorate, GRE. Additional exam requirements/recommendations for international students: Required—TOEFL. Electronic applications accepted. *Faculty research:* Algebra, analysis, differential equations, numerical analysis, statistics.

University of Connecticut, Graduate School, College of Liberal Arts and Sciences, Department of Statistics, Field of Statistics, Storrs, CT 06269. Offers MS, PhD. *Faculty:* 10 full-time (1 woman). *Students:* 32 full-time (14 women), 9 part-time (1 woman); includes 2 minority (both Asian Americans or Pacific Islanders), 24 international. Average age 29. 137 applicants, 58% accepted, 33 enrolled. In 2005, 9 master's, 2 doctorates awarded. Terminal master's awarded for partial completion of doctoral program. *Degree requirements:* For master's, comprehensive exam; for doctorate, thesis/dissertation. *Entrance requirements:* For master's and doctorate, GRE General Test. Additional exam requirements/recommendations for international students: Required—TOEFL (minimum score 550 paper-based; 213 computer-based). *Application deadline:* For fall admission, 2/1 priority date for domestic students, 2/1 priority date for international students; for spring admission, 11/1 for domestic students, 10/1 for international students. Applications are processed on a rolling basis. Application fee: $55. Electronic applications accepted. *Expenses:* Tuition, state resident: part-time $444 per credit hour. Tuition, nonresident: part-time $1,154 per credit hour. Tuition and fees vary according to course load. *Financial support:* In 2005–06, 10 research assistantships with full tuition reimbursements, 16 teaching assistantships with full tuition reimbursements were awarded; fellowships, Federal Work-Study, scholarships/grants, health care benefits, and unspecified assistantships also available. Financial award application deadline: 2/1; financial award applicants required to submit FAFSA. *Application contact:* Tracy Burke, Information Contact, 860-486-3413, Fax: 860-486-4113, E-mail: statadm2@uconnvm.uconn.edu.

University of Delaware, College of Agriculture and Natural Resources, Program in Statistics, Newark, DE 19716. Offers MS. Part-time programs available. *Faculty:* 5 full-time (1 woman), 4 part-time/adjunct (1 woman). *Students:* 18 full-time (10 women), 3 part-time (1 woman); includes 1 minority (African American), 16 international. 23 applicants, 74% accepted, 14 enrolled. In 2005, 8 master's awarded. *Entrance requirements:* For master's, GRE General Test, 3 letters of recommendation. Additional exam requirements/recommendations for international students: Required—TOEFL (minimum score 550 paper-based; 213 computer-based). *Application deadline:* For fall admission, 2/1 for domestic students; for spring admission, 12/1 priority date for domestic students. Application fee: $60. Electronic applications accepted. *Financial support:* In 2005–06, 11 students received support, including 4 research assistantships with full tuition reimbursements available (averaging $13,556 per year), 2 teaching assistantships with full tuition reimbursements available (averaging $12,800 per year); career-related internships or fieldwork, scholarships/grants, tuition waivers (full), and unspecified assistantships also available. Financial award application deadline: 7/1. Total annual research expenditures: $84,400. *Unit head:* Dr. Thomas W. Ilvento, Chair, 302-831-6773, Fax: 302-831-6243, E-mail: ilvento@udel.edu. *Application contact:* Vicki Lynn Taylor, Office Coordinator, 302-831-2511, Fax: 302-831-6243, E-mail: vtaylor@udel.edu.

University of Denver, Daniels College of Business, Department of Statistics and Operations Technology, Denver, CO 80208. Offers data mining (MS). *Students:* 2 full-time (1 woman), 3 part-time. *Expenses:* Tuition: Full-time $27,756; part-time $771 per credit. Required fees: $174. *Financial support:* In 2005–06, 1 teaching assistantship (averaging $3,720 per year) was awarded; career-related internships or fieldwork, Federal Work-Study, institutionally sponsored loans, and scholarships/grants also available. Support available to part-time students. Financial award application deadline: 2/15; financial award applicants required to submit FAFSA. *Application contact:* Information Contact, 303-871-3416, Fax: 303-871-4466, E-mail: daniels@du.edu.

University of Florida, Graduate School, College of Liberal Arts and Sciences, Department of Statistics, Gainesville, FL 32611. Offers M Stat, MS Stat, PhD. *Faculty:* 16 full-time (2 women). *Students:* 58 (26 women); includes 4 minority (1 Asian American or Pacific Islander, 3 Hispanic Americans) 45 international. In 2005, 13 master's, 5 doctorates awarded. *Degree requirements:* For master's, variable foreign language requirement, thesis or alternative, final oral exam, comprehensive exam; for doctorate, thesis/dissertation. *Entrance requirements:* For master's and doctorate, GRE General Test, minimum GPA of 3.0. Additional exam requirements/recommendations for international students: Required—TOEFL (minimum score 550 paper-based; 213 computer-based). *Application deadline:* For fall admission, 6/1 for domestic students. Applications are processed on a rolling basis. Application fee: $30. Electronic applications accepted. *Expenses:* Tuition, state resident: full-time $6,234. Tuition, nonresident: full-time $21,359. Tuition and fees vary according to program. *Financial support:* In 2005–06, 1 research assistantship (averaging $13,919 per year), 18 teaching assistantships (averaging $13,475 per year) were awarded; fellowships, unspecified assistantships also available. Financial award application deadline:2/1. *Faculty research:* Categorical data, time series, Bayesian analysis, nonparametrics, sampling. *Unit head:* Dr. George Casella, Chair, 352-392-1941 Ext. 204, Fax: 352-392-5175, E-mail: casella@stat.ufl.edu. *Application contact:* Dr. James P. Hobert, Coordinator, 352-392-1941 Ext. 229, Fax: 352-392-5175, E-mail: jhobert@stat.ufl.edu.

University of Georgia, Graduate School, College of Arts and Sciences, Department of Statistics, Athens, GA 30602. Offers applied mathematical science (MAMS); statistics (MS, PhD). *Faculty:* 16 full-time (4 women). *Students:* 42 full-time, 1 part-time; includes 4 minority (2 African Americans, 2 Asian Americans or Pacific Islanders), 24 international. 104 applicants, 26% accepted, 11 enrolled. In 2005, 29 master's, 6 doctorates awarded. *Degree requirements:* For master's, thesis (for some programs), technical report (MAMS); for doctorate, one foreign language, thesis/dissertation. *Entrance requirements:* For master's and doctorate, GRE General Test. *Application deadline:* For fall admission, 7/1 for domestic students; for spring admission, 11/15 for domestic students. Application fee: $50. Electronic applications accepted. *Financial support:* Fellowships, research assistantships, teaching assistantships, unspecified assistantships available. *Unit head:* Dr. John Stafken, Head, 706-542-3309, Fax: 706-542-3391. *Application contact:* Dr. Lynn Seymour, Graduated Coordinator, 706-542-3307, Fax: 706-542-3391, E-mail: seymour@stat.uga.edu.

University of Guelph, Graduate Program Services, College of Physical and Engineering Science, Department of Mathematics and Statistics, Guelph, ON N1G 2W1, Canada. Offers applied mathematics (PhD); applied statistics (PhD); mathematics and statistics (M Sc). Part-time programs available. *Faculty:* 21 full-time (6 women), 4 part-time/adjunct (0 women). *Students:* 40 full-time, 4 part-time. Average age 22. 103 applicants, 25% accepted, 23 enrolled. In 2005, 20 master's, 1 doctorate awarded. *Median time to degree:* Of those who began their doctoral program in fall 1996, 100% received their degree in 8 years or less. *Degree requirements:* For master's, thesis (for some programs); for doctorate, thesis/dissertation. *Entrance requirements:* For master's, minimum B- average during previous 2 years of course work; for doctorate, minimum B average. Additional exam requirements/recommendations for international students: Required—TOEFL (minimum score 550 paper-based; 213 computer-based), IELT (minimum score 7). *Application deadline:* For fall admission, 2/1 for domestic students, 2/1 for international students. Application fee: $75. *Financial support:* In 2005–06, 20 students received support, including 26 research assistantships (averaging $2,300 per year), 27 teaching assistantships (averaging $9,000 per year); fellowships, scholarships/grants also available. *Faculty research:* Dynamical systems, mathematical biology, numerical analysis, linear and nonlinear models, reliability and bioassay. *Unit head:* Dr. O. Brian Allen, Chair, 519-824-4120 Ext. 56556, Fax: 519-837-0221, E-mail: chair@uoguelph.ca. *Application contact:* Susan McCormick, Graduate Administrative Assistant, 519-824-4120 Ext. 56553, Fax: 519-837-0221, E-mail: smccormi@uoguelph.ca.

University of Houston–Clear Lake, School of Science and Computer Engineering, Program in Statistics, Houston, TX 77058-1098. Offers MS. *Entrance requirements:* For master's, GRE General Test. Additional exam requirements/recommendations for international students: Required—TOEFL (minimum score 550 paper-based; 213 computer-based).

University of Idaho, College of Graduate Studies, College of Science, Department of Statistics, Moscow, ID 83844-2282. Offers MS. *Students:* 9 full-time (6 women), 5 part-time (2 women); includes 1 minority (Asian American or Pacific Islander), 11 international. Average age 38. In 2005, 9 degrees awarded. *Entrance requirements:* For master's, minimum GPA of 2.8. *Application deadline:* For fall admission, 8/1 for domestic students; for spring admission, 12/15 for domestic students. Application fee: $55 ($60 for international students). *Expenses:* Tuition, nonresident: full-time $8,770; part-time $130 per credit. Required fees: $4,508; $217 per credit. *Financial support:* Research assistantships, teaching assistantships available. Financial award application deadline: 2/15. *Unit head:* Dr. Rick Edgeman, Chair, 208-885-4410.

University of Illinois at Chicago, Graduate College, College of Liberal Arts and Sciences, Department of Mathematics, Statistics, and Computer Science, Chicago, IL 60607-7128. Offers applied mathematics (MS, DA, PhD); computer science (MS, DA, PhD); math and information science for the industry (MS); probability and statistics (MS, DA, PhD); pure mathematics (MS, DA, PhD); teaching of mathematics (MST). Part-time programs available. *Degree requirements:* For master's, comprehensive exam; for doctorate, one foreign language, thesis/dissertation. *Entrance requirements:* For master's and doctorate, GRE General Test, minimum GPA of 2.75. Additional exam requirements/recommendations for international students: Required—TOEFL. Electronic applications accepted.

See Close-Up on page 535.

University of Illinois at Urbana–Champaign, Graduate College, College of Liberal Arts and Sciences, Department of Statistics, Champaign, IL 61820. Offers MS, PhD. *Faculty:* 10 full-time (1 woman). *Students:* 32 full-time (19 women), 8 part-time (4 women); includes 5 minority (2 African Americans, 2 Asian Americans or Pacific Islanders, 1 Hispanic American), 25 international. 148 applicants, 10% accepted, 9 enrolled. In 2005, 24 master's, 3 doctorates awarded. Terminal master's awarded for partial completion of doctoral program. *Degree requirements:* For doctorate, thesis/dissertation. *Entrance requirements:* For master's, minimum GPA of 3.0. Additional exam requirements/recommendations for international students: Required—TOEFL. *Application deadline:* For fall admission, 2/15 for domestic students. Applications are processed on a rolling basis. Application fee: $50 ($60 for international students). Electronic applications accepted. *Financial support:* In 2005–06, 2 fellowships, 23 research assistantships, 29 teaching assistantships were awarded; tuition waivers (full) also available. Financial award application deadline: 2/15. *Faculty research:* Statistical decision theory, sequential analysis, computer-aided stochastic modeling. *Unit head:* Douglas G. Simpson, Chair, 217-333-2167, Fax: 217-244-7190, E-mail: dgs@uiuc.edu. *Application contact:* Jennifer Suits, Secretary, 217-333-2167, Fax: 217-244-7190, E-mail: jmsuits@uiuc.edu.

The University of Iowa, Graduate College, College of Education, Department of Psychological and Quantitative Foundations, Iowa City, IA 52242-1316. Offers counseling psychology (PhD); educational measurement and statistics (MA, PhD); educational psychology (MA, PhD); school psychology (PhD, Ed S). *Accreditation:* APA. *Faculty:* 23 full-time, 14 part-time/adjunct. *Students:* 96 full-time (67 women), 64 part-time (48 women); includes 34 minority (13 African Americans, 1 American Indian/Alaska Native, 8 Asian Americans or Pacific Islanders, 12 Hispanic Americans), 42 international. 161 applicants, 24% accepted, 26 enrolled. In 2005, 8 master's, 16 doctorates, 2 other advanced degrees awarded. *Degree requirements:* For master's, exam, thesis optional; for doctorate, thesis/dissertation, comprehensive exam, registration; for Ed S, exam. *Entrance requirements:* For master's, doctorate, and Ed S, GRE General Test, minimum GPA of 3.0. Additional exam requirements/recommendations for international students: Required—TOEFL (minimum score 550 paper-based; 213 computer-based). Application fee: $60 ($85 for international students). Electronic applications accepted. *Expenses:* Tuition, state resident: part-time $1,882 per term. Tuition, nonresident: full-time $17,338; part-time $4,907 per term. Tuition and fees vary according to course load and program. *Financial support:* In 2005–06, 11 fellowships, 74 research assistantships with partial tuition reimbursements, 23 teaching assistantships with partial tuition reimbursements were awarded. Financial award applicants required to submit FAFSA. *Unit head:* Elizabeth Altmaier, Chair, 319-335-5566, Fax: 319-335-6145.

The University of Iowa, Graduate College, College of Liberal Arts and Sciences, Department of Statistics and Actuarial Science, Iowa City, IA 52242-1316. Offers MS, PhD. *Faculty:* 17 full-time, 4 part-time/adjunct. *Students:* 65 full-time (28 women), 32 part-time (15 women); includes 5 minority (4 Asian Americans or Pacific Islanders, 1 Hispanic American), 84 international. 209 applicants, 46% accepted, 39 enrolled. In 2005, 32 master's, 1 doctorate awarded. *Degree requirements:* For master's, exam, thesis optional; for doctorate, thesis/dissertation, comprehensive exam, registration. *Entrance requirements:* For master's and doctorate, GRE General Test, minimum GPA of 3.0. Additional exam requirements/recommendations for international students: Required—TOEFL (minimum score 550 paper-based; 213 computer-based). *Application deadline:* Applications are processed on a rolling basis. Application fee: $60 ($85 for international students). Electronic applications accepted. *Expenses:* Tuition, state resident: part-time $1,882 per term. Tuition, nonresident: full-time $17,338; part-time $4,907 per term. Tuition and fees vary according to course load and program. *Financial support:* In 2005–06, 1 fellowship, 13 research assistantships with partial tuition reimbursements, 37 teaching assistantships with partial tuition reimbursements were awarded. Financial award applicants required to submit FAFSA. *Unit head:* Luke Tierney, Chair, 319-335-0712, Fax: 319-335-3017.

University of Kansas, Graduate School, College of Liberal Arts and Sciences, Department of Mathematics, Lawrence, KS 66045. Offers applied mathematics and statistics (MA, PhD); mathematics (MA, PhD). *Faculty:* 40. *Students:* 67 full-time (23 women), 12 part-time (6 women); includes 4 minority (1 American Indian/Alaska Native, 3 Asian Americans or Pacific Islanders), 31 international. Average age 27. 67 applicants, 73% accepted. In 2005, 17 master's, 2 doctorates awarded. Terminal master's awarded for partial completion of doctoral program. *Degree requirements:* For master's, thesis or alternative; for doctorate, 2 foreign languages, thesis/dissertation, comprehensive exam. *Entrance requirements:* For master's and doctorate, GRE. Additional exam requirements/recommendations for international students: Required—TOEFL. *Application deadline:* For fall admission, 3/1 priority date for domestic students, 3/1 priority date for international students. Applications are processed on a rolling basis. Application fee: $55 ($60 for international students). Electronic applications accepted. *Expenses:* Tuition, state resident: full-time $4,859. Tuition, nonresident: full-time $1,200. Required fees: $589. Tuition and fees vary according to program. *Financial support:* Fellowships, research assistantships with full and partial tuition reimbursements, teaching assistantships with full and partial tuition reimbursements, institutionally sponsored loans available. Support available to part-time students. Financial award application deadline: 2/1. *Faculty research:* Commutative algebra/algebraic geometry, stochastic adaptive control/stochastic processes analysis/harmonic analysis/PDES, numerical analysis/dynamical systems, topology/set theory. *Unit head:* Jack Porter, Chair, 785-864-3651, Fax: 785-864-5255, E-mail: porter@math.ukans.edu. *Application contact:* David Lerner, Graduate Director, 785-864-3651, E-mail: lerner@ukans.edu.

University of Kentucky, Graduate School, Graduate School Programs from the College of Arts and Sciences, Program in Statistics, Lexington, KY 40506-0032. Offers MS, PhD. *Faculty:* 11 full-time (1 woman). *Students:* 39 full-time (24 women), 5 part-time (3 women); includes 5 minority (all Asian Americans or Pacific Islanders), 15 international. Average age 30. 76 applicants, 32% accepted, 18 enrolled. In 2005, 11 master's, 3 doctorates awarded. *Median time to degree:* Of those who began their doctoral program in fall 1996, 66.7% received their degree in 8 years or less. *Degree requirements:* For master's, thesis optional; for doctorate, thesis/dissertation, comprehensive exam. *Entrance requirements:* For master's, GRE General Test, minimum undergraduate GPA of 2.5; for doctorate, GRE General Test, minimum graduate GPA of 3.0. Additional exam requirements/recommendations for international students: Required—TOEFL (minimum score 550 paper-based; 213 computer-based). *Application deadline:* For fall admission, 7/17 priority date for domestic students, 2/1 priority date for international students; for spring admission, 12/13 priority date for domestic students, 6/15 priority date for international students. Applications are processed on a rolling basis. Application fee: $40 ($55 for international students). Electronic applications accepted. *Expenses:* Tuition, state resident: full-time $3,159; part-time $331 per credit hour. Tuition, nonresident: full-time $6,984; part-time $756 per credit hour. Tuition and fees vary according to course load, degree level and program. *Financial support:* In 2005–06, 1 fellowship (averaging $2,861 per year), 11 research assistantships with full tuition reimbursements (averaging $15,840 per year), 24 teaching assistantships with full tuition reimbursements (averaging $12,400 per year) were awarded; Federal Work-Study, institutionally sponsored loans, scholarships/grants, traineeships, health care benefits, tuition waivers (partial), and unspecified assistantships also available. Support available to part-time students. Financial award application deadline: 3/15. *Faculty research:* Computer intensive statistical inference, biostatistics, mathematical and applied statistics, applied probability. Total annual research expenditures: $77,000. *Unit head:* Dr. Arnold Stromberg, Director of Graduate Studies, 859-257-6903, Fax: 859-323-1973, E-mail: astro@ms.uky.edu. *Application contact:* Dr. Brian Jackson, Senior Associate Dean, 859-257-8176, Fax: 859-323-1928, E-mail: lance.brunner@uky.edu.

University of Manitoba, Faculty of Graduate Studies, Faculty of Science, Department of Mathematical, Computational and Statistical Sciences, Winnipeg, MB R3T 2N2, Canada. Offers MMCSS.

University of Manitoba, Faculty of Graduate Studies, Faculty of Science, Department of Statistics, Winnipeg, MB R3T 2N2, Canada. Offers M Sc, PhD. *Degree requirements:* For master's, thesis or alternative; for doctorate, one foreign language, thesis/dissertation.

University of Maryland, Baltimore County, Graduate School, College of Natural Sciences and Mathematics, Department of Mathematics and Statistics, Program in Statistics, Baltimore, MD 21250. Offers MS, PhD. Part-time and evening/weekend programs available. *Faculty:* 7 full-time (0 women). *Students:* 19 full-time (7 women), 23 part-time (12 women); includes 3 minority (2 African Americans, 1 Asian American or Pacific Islander), 20 international. Average age 28. 38 applicants, 61% accepted, 12 enrolled. In 2005, 1 master's, 2 doctorates awarded. Terminal master's awarded for partial completion of doctoral program. *Median time to degree:* Of those who began their doctoral program in fall 1996, 80% received their degree in 8 years or less. *Degree requirements:* For master's, thesis (for some programs), comprehensive exam (for some programs), registration; for doctorate, thesis/dissertation, comprehensive exam, registration. *Entrance requirements:* For master's and doctorate, GRE General Test, minimum GPA of 3.0. Additional exam requirements/recommendations for international students: Required—TOEFL (minimum score 600 paper-based; 250 computer-based). *Application deadline:* For fall admission, 2/15 priority date for domestic students, 1/1 priority date for international students; for spring admission, 10/15 priority date for domestic students, 9/15 priority date for international students. Applications are processed on a rolling basis. Application fee: $50. Electronic applications accepted. *Expenses:* Tuition, state resident: part-time $395 per credit. Tuition, nonresident: part-time $652 per credit. Required fees: $82 per credit. Tuition and fees vary according to course load, program and reciprocity agreements. *Financial support:* In 2005–06, 18 students received support, including 4 research assistantships with full tuition reimbursements available (averaging $15,000 per year), 16 teaching assistantships with full tuition reimbursements available (averaging $15,000 per year); career-related internships or fieldwork, scholarships/grants, health care benefits, tuition waivers (full), and unspecified assistantships also available. Support available to part-time students. Financial award application deadline: 2/15. *Faculty research:* Design of experiments, statistical decision theory and inference, time series analysis, biostatistics and environmental statistics, bioinformatics. Total annual research expenditures:$420,000. *Application contact:* Dr. Muddappa Gowda, Director of Graduate Programs, 410-455-2431, Fax: 410-455-1066, E-mail: gowda@math.umbc.edu.

University of Maryland, College Park, Graduate Studies, College of Computer, Mathematical and Physical Sciences, Department of Mathematics, Program in Mathematical Statistics, College Park, MD 20742. Offers MA, PhD. Part-time and evening/weekend programs available. *Students:* 30 full-time (13 women), 15 part-time (8 women); includes 8 minority (5 African Americans, 3 Asian Americans or Pacific Islanders), 28 international. 58 applicants, 34% accepted, 13 enrolled. In 2005, 2 master's, 5 doctorates awarded. Terminal master's awarded for partial completion of doctoral program. *Degree requirements:* For master's, thesis or alternative, thesis or comprehensive exams, scholarly paper; for doctorate, one foreign language, thesis/dissertation, written and oral exams. *Entrance requirements:* For master's and doctorate, GRE General Test, GRE Subject Test in math, minimum GPA of 3.0, 3 letters of recommendation. *Application deadline:* For fall admission, 5/1 for domestic students, 2/1 for international students; for spring admission, 10/1 for domestic students, 6/1 for international students. Applications are processed on a rolling basis. Application fee: $60. Electronic applications accepted. *Financial support:* In 2005–06, 4 fellowships (averaging $3,388 per year) were awarded; research assistantships, teaching assistantships Financial award applicants required to submit FAFSA. *Faculty research:* Statistics and probability, stochastic processes, nonparametric statistics, space-time statistics. *Unit head:* Dr. Benjamin Kedem, Director, 301-405-5112, Fax: 301-314-0827. *Application contact:* Dean of Graduate School, 301-405-4190, Fax: 301-314-9305.

University of Massachusetts Amherst, Graduate School, College of Natural Sciences and Mathematics, Department of Mathematics and Statistics, Program in Mathematics and Statistics, Amherst, MA 01003. Offers MS, PhD. *Students:* 50 full-time (13 women), 5 part-time (2 women); includes 4 minority (1 African American, 3 Asian Americans or Pacific Islanders), 19 international. Average age 27. 190 applicants, 21% accepted, 19 enrolled. In 2005, 16 master's, 6 doctorates awarded. *Degree requirements:* For doctorate, 2 foreign languages, thesis/dissertation. *Entrance requirements:* Additional exam requirements/recommendations for international students: Required—TOEFL (minimum score 530 paper-based; 197 computer-based). *Application deadline:* For fall admission, 2/1 priority date for domestic students, 2/1 priority date for international students; for spring admission, 10/1 for domestic students, 10/1 for international students. Applications are processed on a rolling basis. Application fee: $40 ($65 for international students). Electronic applications accepted. *Expenses:* Tuition, state resident: part-time $110 per credit. Tuition, nonresident: part-time $414 per credit. Required fees: $2,824 per term. One-time fee: $250 part-time. Full-time tuition and fees vary according to course load, campus/location, program and reciprocity agreements. *Financial support:* Fellowships with full tuition reimbursements, research assistantships with full tuition reimbursements, teaching assistantships with full tuition reimbursements, career-related internships or fieldwork, Federal Work-Study, scholarships/grants, traineeships, and unspecified assistantships available. Support available to part-time students. Financial award application deadline: 2/1. *Unit head:* Dr. Ivan Mirkovic, Director, 413-545-2282, Fax: 413-545-1801.

University of Memphis, Graduate School, College of Arts and Sciences, Department of Mathematical Sciences, Memphis, TN 38152-3420. Offers applied mathematics (MS); applied statistics (PhD); bioinformatics (MS); computer science (PhD); computer sciences (MS); mathematics (MS, PhD); statistics (MS, PhD). Part-time programs available. *Faculty:* 24 full-time (5 women), 3 part-time/adjunct (0 women). *Students:* 105 full-time (38 women), 34 part-time (8 women); includes 8 minority (7 African Americans, 1 Asian American or Pacific Islander), 89 international. Average age 30. 139 applicants, 37% accepted. In 2005, 43 master's, 5 doctorates awarded. Terminal master's awarded for partial completion of doctoral program. *Degree requirements:* For master's, comprehensive exam; for doctorate, one foreign language, thesis/dissertation, oral exams. *Entrance requirements:* For master's and doctorate, GRE General Test, minimum GPA of 2.5. Additional exam requirements/recommendations for international students: Required—TOEFL (minimum score 550 paper-based; 210 computer-based), WES evaluation of transcript. *Application deadline:* For fall admission, 8/1 for domestic students, 5/1 for international students; for spring admission, 12/1 for domestic students, 9/1 for international students. Applications are processed on a rolling basis. Application fee: $25 ($50 for international students). Electronic applications accepted. *Financial support:* In 2005–06, 58 students received support, including fellowships with full tuition reimbursements available (averaging $17,500 per year), 9 research assistantships with full tuition reimbursements available (averaging $9,000 per year), 30 teaching assistantships with full tuition reimbursements available (averaging $9,000 per year); career-related internships or fieldwork, Federal Work-Study, scholarships/grants, unspecified assistantships, and minority scholarships also available. Financial award application deadline: 2/2. *Faculty research:* Combinatorics, ergodic theory, graph theory, Ramsey theory, applied statistics. Total annual research expenditures: $1.5 million. *Unit head:* Dr. James E. Jamison, Chairman, 901-678-2482, Fax: 901-678-2480, E-mail: jjamison@memphis.edu. *Application contact:* Coordinator of Graduate Studies, 901-678-2482, Fax: 901-678-2480, E-mail: dfwilson@memphis.edu.

University of Miami, Graduate School, School of Business Administration, Department of Management Science, Coral Gables, FL 33124. Offers management science (MS), including applied statistics, operations research, quality management. Part-time and evening/weekend programs available. Postbaccalaureate distance learning degree programs offered. *Faculty:* 10 full-time (1 woman). *Students:* 8 full-time (1 woman); includes 2 minority (both Asian Americans or Pacific Islanders), 2 international. Average age 30. 4 applicants, 25% accepted, 1 enrolled. In 2005, 4 degrees awarded. *Degree requirements:* For master's, thesis optional. *Entrance requirements:* For master's, GRE General Test. Additional exam requirements/recommendations for international students: Required—TOEFL. *Application deadline:* For fall admission, 6/30 for domestic students; for spring admission, 10/31 for domestic students. Applications are processed on a rolling basis. Application fee: $50. *Financial support:* Career-related internships or fieldwork and Federal Work-Study available. Financial award application deadline: 3/1. *Faculty research:* Mathematical programming, applied probability, logistics, statistical process control. Total annual research expenditures:$20,000. *Unit head:* Dr. Anuj Mehrotra, Chairman, 305-284-6595, Fax: 305-284-2321, E-mail: anuj@miami.edu. *Application contact:* Dr. Howard Gitlow, Director, 305-284-4296, Fax: 305-284-2321, E-mail: hgitlow@miami.edu.

University of Michigan, Horace H. Rackham School of Graduate Studies, College of Literature, Science, and the Arts, Department of Statistics, Ann Arbor, MI 48109. Offers applied statistics (AM); statistics (AM, PhD). *Faculty:* 28 full-time (6 women), 1 part-time/adjunct (0 women). *Students:* 103 full-time (39 women); includes 6 minority (1 African American, 4 Asian Americans or Pacific Islanders, 1 Hispanic American), 61 international. Average age 28. 245 applicants, 29% accepted, 36 enrolled. In 2005, 39 master's, 3 doctorates awarded. Terminal master's awarded for partial completion of doctoral program. *Median time to degree:* Of those who began their doctoral program in fall 1996, 43% received their degree in 8 years or less. *Degree requirements:* For master's, registration; for doctorate, thesis/dissertation, oral defense of dissertation, preliminary exam. *Entrance requirements:* For master's and doctorate, GRE General Test. Additional exam requirements/recommendations for international students: Required—TOEFL (minimum score 560 paper-based; 220 computer-based). *Application deadline:* For fall admission, 1/31 priority date for domestic students, 1/15 priority date for international students. Applications are processed on a rolling basis. Application fee: $60 ($75 for international students). Electronic applications accepted. *Expenses:* Tuition, state resident: full-time $14,082; part-time $894 per credit hour. Tuition, nonresident: full-time $28,500; part-time $1,675 per credit hour. Required fees: $189; $189 per unit. *Financial support:* In 2005–06, 19 fellowships with full and partial tuition reimbursements (averaging $3,000 per year), 16 research assistantships with full and partial tuition reimbursements (averaging $14,326 per year), 47 teaching assistantships with full and partial tuition reimbursements (averaging $14,326 per year) were awarded; career-related internships or fieldwork, Federal Work-Study, institutionally sponsored loans, scholarships/grants, health care benefits, and unspecified assistantships also available. Financial award application deadline: 1/31. *Faculty research:* Sequential analysis, Bayesian statistics, multivariate analysis, statistical computing, bioinformation. *Unit head:* Vijayan Nair, Chair, 734-763-3519, Fax: 734-763-4676, E-mail: vnn@umich.edu. *Application contact:* Lu Ann Custer, Graduate Secretary, 734-763-3520, Fax: 734-763-4676, E-mail: stat-admission@umich.edu.

University of Minnesota, Twin Cities Campus, Graduate School, College of Liberal Arts, School of Statistics, Minneapolis, MN 55455-0213. Offers MS, PhD. Part-time programs available. *Faculty:* 19 full-time (2 women), 2 part-time/adjunct (0 women). *Students:* 67 full-time (25 women), 21 part-time (13 women); includes 8 minority (2 African Americans, 6 Asian Americans or Pacific Islanders), 55 international. Average age 24. 196 applicants, 52% accepted, 20 enrolled. In 2005, 14 master's, 9 doctorates awarded. Terminal master's awarded for partial completion of doctoral program. *Median time to degree:* Of those who began their doctoral program in fall 1996, 50% received their degree in 8 years or less. *Degree requirements:* For master's, comprehensive exam, registration; for doctorate, thesis/dissertation, comprehensive exam, registration. *Entrance requirements:* For master's and doctorate, GRE General Test. Additional exam requirements/recommendations for international students: Required—TOEFL (minimum score 550 paper-based; 213 computer-based). *Application deadline:* For fall admission, 1/1 priority date for domestic students, 1/1 priority date for international students. Applications are processed on a rolling basis. Application fee: $55 ($75 for international students). Electronic applications accepted. *Expenses:* Tuition, state resident: full-time $8,748; part-time $729 per credit. Tuition, nonresident: full-time $15,848; part-time $1,321 per credit. Full-time tuition and fees vary according to class time, course load, program and reciprocity agreements. *Financial support:* In 2005–06, 4 fellowships with full tuition reimbursements (averaging $13,229 per year), 7 research assistantships with full tuition reimbursements (averaging $13,229 per year), 41 teaching assistantships with full tuition reimbursements (averaging $13,229 per year) were awarded; scholarships/grants, health care benefits, and tuition waivers (partial) also available. *Faculty research:* Data analysis, statistical computing, experimental design, probability theory, Bayesian inference. Total annual research expenditures: $417,205. *Unit head:* Glen Meeden, Director, 612-625-8046, Fax: 612-624-8868, E-mail: glen@stat.umn.edu. *Application contact:* Mary Hildre, Executive Administrative Specialist, 612-625-7300, Fax: 612-624-8868, E-mail: mary@stat.umn.edu.

University of Missouri–Columbia, Graduate School, College of Arts and Sciences, Department of Statistics, Columbia, MO 65211. Offers MA, PhD. *Faculty:* 17 full-time (5 women), 1

University of Missouri–Columbia (continued)

part-time/adjunct (0 women). *Students:* 43 full-time (19 women), 6 part-time (4 women); includes 4 minority (1 African American, 2 Asian Americans or Pacific Islanders, 1 Hispanic American), 31 international. In 2005, 10 master's, 5 doctorates awarded. *Degree requirements:* For doctorate, thesis/dissertation. *Entrance requirements:* For master's and doctorate, GRE General Test, minimum GPA of 3.0. *Application deadline:* For fall admission, 2/15 for domestic students. For winter admission, 10/15 for domestic students. Applications are processed on a rolling basis. Application fee: $45 ($60 for international students). *Financial support:* Fellowships, research assistantships, teaching assistantships, institutionally sponsored loans and tuition waivers (full and partial) available. *Unit head:* Dr. Paul L. Speckman, Director of Graduate Studies, 573-882-7082, E-mail: speckmanp@missouri.edu.

University of Missouri–Kansas City, College of Arts and Sciences, Department of Mathematics and Statistics, Kansas City, MO 64110-2499. Offers MA, MS, PhD. PhD offered through the School of Graduate Studies. Part-time programs available. *Faculty:* 9 full-time (3 women), 16 part-time/adjunct (6 women). *Students:* 2 full-time (1 woman), 11 part-time (7 women); includes 2 minority (1 African American, 1 Asian American or Pacific Islander). Average age 33. 1 applicant, 0% accepted.Terminal master's awarded for partial completion of doctoral program. *Degree requirements:* For master's, written exam; for doctorate, 2 foreign languages, thesis/dissertation, oral and written exams. *Entrance requirements:* For master's, bachelor's degree in mathematics, minimum GPA of 3.0; for doctorate, GMAT or GRE General Test. Additional exam requirements/recommendations for international students: Required—TOEFL. *Application deadline:* For fall admission, 5/1 priority date for domestic students, 4/30 priority date for international students. Applications are processed on a rolling basis. Application fee: $35 ($50 for international students). Electronic applications accepted. *Expenses:* Tuition, state resident: full-time $4,738; part-time $263 per credit hour. Tuition, nonresident: full-time $12,235; part-time $679 per credit hour. Required fees: $582. Tuition and fees vary according to course load, program and student level. *Financial support:* In 2005–06, 2 research assistantships with partial tuition reimbursements (averaging $5,000 per year), 6 teaching assistantships with full tuition reimbursements (averaging $11,000 per year) were awarded; Federal Work-Study, institutionally sponsored loans, and tuition waivers (full and partial) also available. Support available to part-time students. Financial award application deadline: 4/1. *Faculty research:* Numerical analysis, statistics, biostatistics commutative algebra, differential equations. Total annual research expenditures:$90,548. *Unit head:* Kamel Rekab, Chair, 816-235-5719, E-mail: rekabk@umka.edu. *Application contact:* Jie Chen, Associate Professor, 816-235-2894, Fax: 816-235-5517, E-mail: chenj@umkc.edu.

University of Nebraska–Lincoln, Graduate College, College of Arts and Sciences, Department of Mathematics and Statistics, Lincoln, NE 68588. Offers M Sc T, MA, MAT, MS, PhD. *Degree requirements:* For master's, thesis optional; for doctorate, variable foreign language requirement, thesis/dissertation, comprehensive exam. *Entrance requirements:* Additional exam requirements/recommendations for international students: Required—TOEFL (minimum score 550 paper-based; 213 computer-based), GRE General Test (international applicants). Electronic applications accepted. *Faculty research:* Applied mathematics, commutative algebra, algebraic geometry, Bayesian statistics, biostatistics.

University of Nevada, Las Vegas, Graduate College, College of Science, Department of Mathematical Sciences, Las Vegas, NV 89154-9900. Offers applied mathematics (MS, PhD); applied statistics (MS); computational mathematics (PhD); pure mathematics (MS, PhD); statistics (PhD); teaching mathematics (MS). Part-time programs available. *Faculty:* 29 full-time (6 women), 1 part-time/adjunct (0 women). *Students:* 29 full-time (13 women), 29 part-time (9 women); includes 10 minority (1 African American, 6 Asian Americans or Pacific Islanders, 3 Hispanic Americans), 22 international. 37 applicants, 57% accepted, 11 enrolled. In 2005, 18 degrees awarded. *Degree requirements:* For master's, thesis (for some programs), oral exam, comprehensive exam (for some programs). *Entrance requirements:* For master's, minimum GPA of 3.0 during previous 2 years, 2.75 overall. Additional exam requirements/recommendations for international students: Required—TOEFL (minimum score 550 paper-based; 213 computer-based). *Application deadline:* For fall admission, 6/15 for domestic students, 5/1 for international students; for spring admission, 11/15 for domestic students, 10/1 for international students. Application fee: $60 ($75 for international students). Electronic applications accepted. *Expenses:* Tuition, state resident: part-time $150 per credit. Tuition, nonresident: part-time $315 per credit. Tuition and fees vary according to course load, program and reciprocity agreements. *Financial support:* In 2005–06, 42 teaching assistantships with partial tuition reimbursements (averaging $10,500 per year) were awarded; career-related internships or fieldwork, Federal Work-Study, institutionally sponsored loans, scholarships/grants, health care benefits, and unspecified assistantships also available. Support available to part-time students. Financial award application deadline: 3/1. *Unit head:* Dr. Malwane Ananda, Chair, 702-895-3567. *Application contact:* Graduate College Admissions Evaluator, 702-895-3320, Fax: 702-895-4180, E-mail: gradcollege@unlv.edu.

University of New Brunswick Fredericton, School of Graduate Studies, Faculty of Science, Department of Mathematics and Statistics, Fredericton, NB E3B 5A3, Canada. Offers M Sc, PhD. Part-time programs available. *Degree requirements:* For master's, thesis or alternative; for doctorate, thesis/dissertation. *Entrance requirements:* For master's and doctorate, minimum GPA of 3.0. Additional exam requirements/recommendations for international students: Required—TOEFL, TWE.

University of New Hampshire, Graduate School, College of Engineering and Physical Sciences, Department of Mathematics and Statistics, Durham, NH 03824. Offers applied mathematics (MS); mathematics (MS, MST, PhD); mathematics education (PhD); statistics (MS). *Faculty:* 26 full-time. *Students:* 17 full-time (7 women), 24 part-time (13 women), 21 international. Average age 28. 37 applicants, 78% accepted, 5 enrolled. In 2005, 17 master's, 6 doctorates awarded. Terminal master's awarded for partial completion of doctoral program. *Degree requirements:* For doctorate, 2 foreign languages, thesis/dissertation. *Entrance requirements:* Additional exam requirements/recommendations for international students: Required—TOEFL (minimum score 550 paper-based; 213 computer-based); Recommended—TSE. *Application deadline:* For fall admission, 4/1 priority date for domestic students, 4/1 priority date for international students. For winter admission, 12/1 for domestic students. Applications are processed on a rolling basis. Application fee: $60. Electronic applications accepted. *Expenses:* Tuition, state resident: full-time $8,010; part-time $445 per credit hour. Tuition, nonresident: full-time $19,730; part-time $810 per credit hour. Required fees: $322 per semester. Tuition and fees vary according to course load and program. *Financial support:* In 2005–06, 1 fellowship, 2 research assistantships, 23 teaching assistantships were awarded; Federal Work-Study, scholarships/grants, and tuition waivers (full and partial) also available. Support available to part-time students. Financial award application deadline: 2/15. *Faculty research:* Operator theory, complex analysis, algebra, nonlinear dynamics, statistics. *Unit head:* Dr. Eric Grinberg, Chairperson, 603-862-5772. *Application contact:* Jan Jankowski, Administrative Assistant, 603-862-2320, E-mail: jan.jankowski@unh.edu.

University of New Mexico, Graduate School, College of Arts and Sciences, Department of Mathematics and Statistics, Albuquerque, NM 87131-2039. Offers mathematics (MS, PhD); statistics (MS, PhD). *Faculty:* 46 full-time (11 women), 21 part-time/adjunct (7 women). *Students:* 89 full-time (37 women), 31 part-time (13 women); includes 20 minority (6 Asian Americans or Pacific Islanders, 14 Hispanic Americans), 48 international. Average age 33. 80 applicants, 59% accepted, 31 enrolled. In 2005, 11 master's, 4 doctorates awarded. *Degree requirements:* For master's, thesis or alternative, comprehensive exam (for some programs); for doctorate, one foreign language, thesis/dissertation, 4 department seminars, comprehensive exam. *Entrance requirements:* For master's, minimum GPA of 3.0, 3 letters of recommendation; for doctorate, GRE General Test, minimum GPA of 3.0, 3 letters of recommendation. Additional exam requirements/recommendations for international students: Required—TOEFL (minimum score 550 paper-based; 213 computer-based). *Application deadline:* For fall admission, 7/1 for domestic students; for spring admission, 11/1 for domestic students. Application fee: $40. Electronic applications accepted. *Expenses:* Tuition, nonresident: full-time $3,388; part-time $238 per credit hour. Required fees: $385 per term. Tuition and fees vary according to course load and program. *Financial support:* In 2005–06, 20 students received support, including fellowships (averaging $4,000 per year), research assistantships with tuition reimbursements available (averaging $14,320 per year), teaching assistantships with tuition reimbursements available (averaging $14,320 per year); health care benefits and unspecified assistantships also available. Financial award application deadline: 3/1; financial award applicants required to submit FAFSA. *Faculty research:* Pure and applied mathematics, applied statistics, numerical analysis, biostatistics, differential geometry, fluid dynamics, nonparametric curve estimation. Total annual research expenditures: $1.2 million. *Unit head:* Dr. Alejandro Aceves, Chair, 505-277-4613, Fax: 505-277-5505, E-mail: aceves@math.unm.edu. *Application contact:* Donna George, Program Advisement Coordinator, 505-277-5250, Fax: 505-277-5505, E-mail: dgeorge@unm.edu.

The University of North Carolina at Chapel Hill, Graduate School, College of Arts and Sciences, Department of Statistics, Chapel Hill, NC 27599. Offers MS, PhD. *Degree requirements:* For master's, essay, or thesis; for doctorate, thesis/dissertation, comprehensive exam. *Entrance requirements:* For master's and doctorate, GRE General Test, GRE Subject Test, minimum GPA of 3.0. Additional exam requirements/recommendations for international students: Required—TOEFL.

University of North Florida, College of Arts and Sciences, Department of Mathematics and Statistics, Jacksonville, FL 32224-2645. Offers mathematical sciences (MS); statistics (MS). Part-time and evening/weekend programs available. *Faculty:* 17 full-time (3 women). *Students:* 12 full-time (5 women), 4 part-time (1 woman); includes 2 minority (1 African American, 1 Asian American or Pacific Islander), 4 international. Average age 29. 24 applicants, 58% accepted, 7 enrolled. In 2005, 6 degrees awarded. *Degree requirements:* For master's, thesis optional. *Entrance requirements:* For master's, GRE General Test, minimum GPA of 3.0 in last 60 hours of course work. Additional exam requirements/recommendations for international students: Required—TOEFL (minimum score 500 paper-based; 173 computer-based). *Application deadline:* For fall admission, 7/1 priority date for domestic students, 5/1 priority date for international students; for spring admission, 11/1 priority date for domestic students, 10/1 priority date for international students. Applications are processed on a rolling basis. Application fee: $30. Electronic applications accepted. *Expenses:* Tuition, state resident: full-time $4,391; part-time $244 per semester hour. Tuition, nonresident: full-time $15,036; part-time $835 per semester hour. Required fees: $789; $44 per semester hour. *Financial support:* In 2005–06, 13 teaching assistantships (averaging $6,810 per year) were awarded; Federal Work-Study and tuition waivers (partial) also available. Support available to part-time students. Financial award application deadline: 4/1; financial award applicants required to submit FAFSA. *Faculty research:* Real analysis, number theory, Euclidean geometry. Total annual research expenditures: $128,824. *Unit head:* Dr. Scott H. Hochwald, Chair, 904-620-2653, E-mail: shochwal@unf.edu. *Application contact:* Dr. Champak Panchal, Graduate Coordinator, 904-620-2653, E-mail: cpanchal@unf.edu.

University of Ottawa, Faculty of Graduate and Postdoctoral Studies, Faculty of Science, Ottawa-Carleton Institute of Mathematics and Statistics, Ottawa, ON K1N 6N5, Canada. Offers M Sc, PhD. Part-time programs available. *Faculty:* 35 full-time, 4 part-time/adjunct. *Students:* 55 full-time, 8 part-time. 80 applicants, 49% accepted, 16 enrolled. In 2005, 9 master's, 2 doctorates awarded. *Degree requirements:* For master's, thesis optional; for doctorate, one foreign language, thesis/dissertation, comprehensive exam. *Entrance requirements:* For master's, honors B Sc degree or equivalent, minimum B average; for doctorate, M Sc, minimum B+ average. *Application deadline:* For fall admission, 3/1 priority date for domestic students, 2/15 priority date for international students. For winter admission, 11/15 for domestic students; for spring admission, 4/1 for domestic students. Applications are processed on a rolling basis. Application fee: $75. Electronic applications accepted. *Expenses:* Tuition: Part-time $260 per credit. Tuition and fees vary according to course load and program. *Financial support:* Fellowships, research assistantships with full tuition reimbursements, teaching assistantships with full tuition reimbursements, career-related internships or fieldwork, Federal Work-Study, scholarships/grants, traineeships, tuition waivers (full and partial), and unspecified assistantships available. Financial award application deadline: 2/15. *Faculty research:* Pure mathematics, applied mathematics, probability and statistics. *Unit head:* Dr. Vladimir Pestov, Director, 613-562-5800 Ext. 3523, Fax: 613-562-5776, E-mail: grad-director@mathstat.uottawa.ca. *Application contact:* Lise Maisonneuve, Graduate Studies Administrator, 613-562-5800 Ext. 6335, Fax: 613-562-5486, E-mail: lise@science.uottawa.ca.

University of Pennsylvania, Wharton School, Department of Statistics, Philadelphia, PA 19104. Offers MBA, PhD. *Degree requirements:* For doctorate, thesis/dissertation, comprehensive exam. *Entrance requirements:* For master's and doctorate, GRE. Additional exam requirements/recommendations for international students: Required—TOEFL, TWE, TSE. *Faculty research:* Nonparametric function estimation, analysis of algorithms, time series analysis, observational studies, inference.

University of Pittsburgh, School of Arts and Sciences, Department of Statistics, Pittsburgh, PA 15260. Offers applied statistics (MA, MS); statistics (MA, MS, PhD). Part-time programs available. Terminal master's awarded for partial completion of doctoral program. *Degree requirements:* For master's, thesis (for some programs), comprehensive exam, registration; for doctorate, thesis/dissertation, comprehensive exam, registration. *Entrance requirements:* For master's, 3 semesters of calculus, 1 semester of linear algebra, 1 year of mathematical statistics; for doctorate, 3 semesters of calculus, 1 semester of linear algebra, 1 year of mathematical statistics, 1 semester of advanced calculus. Additional exam requirements/recommendations for international students: Required—TOEFL (minimum score 550 paper-based; 213 computer-based); Recommended—TSE. Electronic applications accepted. *Expenses:* Tuition, state resident: full-time $13,194; part-time $537 per credit. Tuition, nonresident: full-time $25,012; part-time $1,026 per credit. Required fees: $700; $164 per term. Tuition and fees vary according to campus/location and program. *Faculty research:* Multivariate statistics, time series, reliability, meta-analysis, linear and nonlinear regression modeling.

University of Puerto Rico, Mayagüez Campus, Graduate Studies, College of Arts and Sciences, Department of Mathematics, Mayagüez, PR 00681-9000. Offers applied mathematics (MS); computational sciences (MS); pure mathematics (MS); statistics (MS). Part-time programs available. *Faculty:* 34. *Students:* 21 full-time (9 women), 9 part-time (4 women); includes 6 minority (all Hispanic Americans), 24 international. 49 applicants, 61% accepted, 2 enrolled. In 2005, 6 degrees awarded. *Degree requirements:* For master's, one foreign language, comprehensive exam. *Application deadline:* For fall admission, 2/15 for domestic students; for spring admission, 9/15 for domestic students. Applications are processed on a rolling basis. Application fee: $20. *Expenses:* Tuition, state resident: full-time $900; part-time $100 per credit. International tuition: $4,655 full-time. Part-time tuition and fees vary according to course level and course load. *Financial support:* In 2005–06, fellowships (averaging $1,500 per year), research assistantships (averaging $1,200 per year), teaching assistantships (averaging $987 per year) were awarded; Federal Work-Study and institutionally sponsored loans also available. *Faculty research:* Automata theory, linear algebra, logic. Total annual research expenditures: $13,829. *Unit head:* Dr. Luis A. Caceres, Director, 787-832-4040 Ext. 3848.

University of Regina, Faculty of Graduate Studies and Research, Faculty of Science, Department of Mathematics and Statistics, Regina, SK S4S 0A2, Canada. Offers mathematics (M Sc, MA, PhD); statistics (M Sc, MA). *Faculty:* 22 full-time (4 women), 3 part-time/adjunct (0 women). *Students:* 13 full-time (3 women), 6 part-time (2 women). 12 applicants, 25% accepted. In 2005, 1 doctorate awarded. *Degree requirements:* For master's, registration; for doctorate, thesis/dissertation, comprehensive exam, registration. *Entrance requirements:* Additional exam requirements/recommendations for international students: Required—TOEFL (minimum score 580 paper-based; 237 computer-based). *Application deadline:* Applications are processed on a rolling basis. Application fee: $60 ($100 for international students). *Financial support:* In 2005–06, 2 fellowships (averaging $14,886 per year), 1 research assistantship (averaging

$12,750 per year), 2 teaching assistantships (averaging $13,501 per year) were awarded; scholarships/grants also available. Financial award application deadline: 6/15. *Faculty research:* Pure and applied mathematics, statistics and probability. *Unit head:* Dr. Stephen Kirkland, Head, 306-585-4148, E-mail: kirkland@math.uregina.ca. *Application contact:* Dr. Doug Farenick, Graduate Program Coordinator, 306-585-4425, Fax: 306-585-4020, E-mail: farenick@math.uregina.ca.

University of Rhode Island, Graduate School, College of Arts and Sciences, Department of Computer Science and Statistics, Kingston, RI 02881. Offers applied mathematics (PhD), including computer science, statistics; computer science (MS, PhD); digital forensics (Graduate Certificate); statistics (MS). In 2005, 4 degrees awarded. *Degree requirements:* For master's, thesis optional; for doctorate, one foreign language, thesis/dissertation. *Entrance requirements:* For master's, GRE Subject Test. *Application deadline:* For fall admission, 4/15 for domestic students. Applications are processed on a rolling basis. Application fee: $35. *Expenses:* Tuition, state resident: full-time $5,522; part-time $307 per credit. Tuition, nonresident: full-time $15,992; part-time $888 per credit. Required fees: $1,786; $73 per credit. One-time fee: $80 part-time. *Financial support:* Unspecified assistantships available. *Unit head:* Dr. James Kowalski, Chair, 401-874-2701.

University of Rochester, School of Medicine and Dentistry, Graduate Programs in Medicine and Dentistry, Department of Biostatistics and Computational Biology, Rochester, NY 14627-0250. Offers medical statistics (MS); statistics (MA, PhD). Terminal master's awarded for partial completion of doctoral program. *Degree requirements:* For doctorate, thesis/dissertation, qualifying exam. *Entrance requirements:* For master's and doctorate, GRE General Test. Additional exam requirements/recommendations for international students: Required—TOEFL.

University of Saskatchewan, College of Graduate Studies and Research, College of Arts and Sciences, Department of Mathematics and Statistics, Saskatoon, SK S7N 5A2, Canada. Offers M Math, MA, PhD. *Degree requirements:* For master's, thesis (for some programs), registration; for doctorate, thesis/dissertation, registration. *Entrance requirements:* Additional exam requirements/recommendations for international students: Required—TOEFL.

University of South Carolina, The Graduate School, College of Arts and Sciences, Department of Statistics, Columbia, SC 29208. Offers applied statistics (CAS); industrial statistics (MIS); statistics (MS, PhD). Part-time and evening/weekend programs available. Postbaccalaureate distance learning degree programs offered (minimal on-campus study). Terminal master's awarded for partial completion of doctoral program. *Degree requirements:* For master's, thesis/dissertation; for doctorate, thesis/dissertation, comprehensive exam. *Entrance requirements:* For master's, GRE General Test or GMAT, 2 years of work experience (MIS); for doctorate, GRE General Test; for CAS, GRE General Test or GMAT. Additional exam requirements/recommendations for international students: Required—TOEFL (minimum score 600 paper-based; 250 computer-based). Electronic applications accepted. Expenses: Contact institution. *Faculty research:* Reliability, environmentrics, statistics computing, psychometrics, bioinformatics.

The University of South Dakota, Graduate School, College of Arts and Sciences, Department of Computer Science, Program in Computational Sciences and Statistics, Vermillion, SD 57069-2390. Offers PhD. *Students:* 3 (1 woman). *Degree requirements:* For doctorate, thesis/dissertation, comprehensive exam. *Entrance requirements:* For doctorate, GRE General Test, GRE Subject test in computer science (recommended), minimum GPA of 2.7. Additional exam requirements/recommendations for international students: Required—IBT 79. *Application deadline:* Applications are processed on a rolling basis. Application fee: $35. Electronic applications accepted. *Expenses:* Tuition, state resident: part-time $116 per credit hour. Tuition, nonresident: part-time $341 per credit hour. Required fees: $85 per credit hour. Tuition and fees vary according to course load, program and reciprocity agreements. *Financial support:* In 2005–06, research assistantships with partial tuition reimbursements (averaging $20,000 per year) *Application contact:* Dr. Rich McBride, Graduate Adviser, 605-677-5388, Fax: 605-677-6662, E-mail: csci@usd.edu.

University of Southern California, Graduate School, College of Letters, Arts and Sciences, Department of Mathematics, Program in Statistics, Los Angeles, CA 90089. Offers MS. *Degree requirements:* For master's, thesis. *Entrance requirements:* For master's, GRE General Test. *Expenses:* Tuition: Full-time $25,416; part-time $1,059 per unit. Required fees: $484; $484 per year. Tuition and fees vary according to course load and program.

University of Southern Maine, College of Arts and Science, Portland, ME 04104-9300. Offers American and New England studies (MA); biology (MS); creative writing (MFA); social work (MSW); statistics (MS). Part-time and evening/weekend programs available. Postbaccalaureate distance learning degree programs offered (minimal on-campus study). *Degree requirements:* For master's, thesis optional. *Entrance requirements:* For master's, GRE General Test or MAT. Additional exam requirements/recommendations for international students: Required—TOEFL. Electronic applications accepted.

The University of Tennessee, Graduate School, College of Business Administration, Department of Statistics, Knoxville, TN 37996. Offers industrial statistics (MS); statistics (MS). Part-time programs available. *Degree requirements:* For master's, thesis or alternative. *Entrance requirements:* For master's, GMAT or GRE General Test, minimum GPA of 2.7. Additional exam requirements/recommendations for international students: Required—TOEFL. Electronic applications accepted.

The University of Tennessee, Graduate School, College of Business Administration, Program in Business Administration, Knoxville, TN 37996. Offers accounting (PhD); finance (MBA, PhD); logistics and transportation (MBA, PhD); management (PhD); marketing (MBA, PhD); operations management (MBA); professional business administration (MBA); statistics (PhD). *Accreditation:* AACSB. Postbaccalaureate distance learning degree programs offered. *Degree requirements:* For master's, thesis or alternative; for doctorate, thesis/dissertation. *Entrance requirements:* For master's and doctorate, GMAT, minimum GPA of 2.7. Additional exam requirements/recommendations for international students: Required—TOEFL. Electronic applications accepted.

The University of Texas at Austin, Graduate School, College of Natural Sciences, Department of Mathematics, Program in Statistics, Austin, TX 78712-1111. Offers MS Stat. *Entrance requirements:* For master's, GRE General Test.

The University of Texas at Dallas, School of Natural Sciences and Mathematics, Programs in Mathematical Sciences, Richardson, TX 75083-0688. Offers applied mathematics (MS, PhD); engineering mathematics (MS); mathematical science (MS); statistics (MS, PhD). Part-time and evening/weekend programs available. *Faculty:* 12 full-time (0 women). *Students:* 33 full-time (12 women), 25 part-time (13 women); includes 13 minority (2 African Americans, 6 Asian Americans or Pacific Islanders, 5 Hispanic Americans), 26 international. Average age 30. 73 applicants, 59% accepted, 28 enrolled. In 2005, 19 master's, 8 doctorates awarded. *Degree requirements:* For master's, thesis optional; for doctorate, thesis/dissertation. *Entrance requirements:* For master's, GRE General Test, minimum GPA of 3.0 in upper-level course work in field; for doctorate, GRE General Test, minimum GPA of 3.5 in upper-level course work in field. Additional exam requirements/recommendations for international students: Required—TOEFL (minimum score 550 paper-based; 213 computer-based). *Application deadline:* For fall admission, 7/15 for domestic students; for spring admission, 11/15 for domestic students. Applications are processed on a rolling basis. Application fee: $50 ($100 for international students). Electronic applications accepted. *Expenses:* Tuition, state resident: full-time $5,450; part-time $303 per credit. Tuition, nonresident: full-time $12,648; part-time $703 per credit. Tuition and fees vary according to program. *Financial support:* In 2005–06, 25 teaching assistantships with tuition reimbursements (averaging $9,451 per year) were awarded; fellowships, research assistantships, career-related internships or fieldwork, Federal Work-Study, institutionally sponsored loans, and scholarships/grants also available. Support available to part-time students. Financial award application deadline: 4/30; financial award applicants required to submit FAFSA. *Faculty research:* Statistical methods, control theory, mathematical modeling and analyses of biological and physical systems. Total annual research expenditures: $59,757. *Unit head:* Dr. M. Ali Hooshyar, Head, 972-883-2161, Fax: 972-883-6622, E-mail: utdmath@utdallas.edu. *Application contact:* Dr. Michael Baron, Graduate Advisor, 972-883-6874, Fax: 972-883-6622, E-mail: mbaron@utdallas.edu.

See Close-Up on page 545.

The University of Texas at El Paso, Graduate School, College of Science, Department of Mathematical Sciences, El Paso, TX 79968-0001. Offers mathematical sciences (MAT); mathematics (MS); statistics (MS). Part-time and evening/weekend programs available. *Degree requirements:* For master's, thesis optional. *Entrance requirements:* For master's, GRE, minimum GPA of 3.0. Additional exam requirements/recommendations for international students: Required—TOEFL. Electronic applications accepted.

The University of Texas at San Antonio, College of Business, Department of Management Science and Statistics, San Antonio, TX 78249-0617. Offers management science (MBA); statistics (MS). *Accreditation:* AACSB. *Degree requirements:* For master's, thesis optional. *Entrance requirements:* For master's, GMAT, minimum GPA of 3.0. Additional exam requirements/recommendations for international students: Required—TOEFL (minimum score 500 paper-based; 173 computer-based). Electronic applications accepted.

The University of Toledo, Graduate School, College of Arts and Sciences, Department of Mathematics, Toledo, OH 43606-3390. Offers applied mathematics (MS); mathematics (MA, PhD); statistics (MS). Part-time programs available. *Faculty:* 16. *Students:* 57 full-time (24 women), 13 part-time (3 women); includes 1 minority (Asian American or Pacific Islander), 45 international. Average age 29. 65 applicants, 63% accepted, 21 enrolled. In 2005, 16 degrees awarded. *Degree requirements:* For doctorate, 2 foreign languages, thesis/dissertation. *Entrance requirements:* For master's and doctorate, GRE General Test, GRE Subject Test. *Application deadline:* For fall admission, 8/1 for domestic students. Application fee: $45. Electronic applications accepted. *Expenses:* Tuition, area resident: Part-time $308 per credit hour. Tuition, state resident: full-time $3,312. Tuition, nonresident: full-time $6,616; part-time $735 per credit hour. *Financial support:* In 2005–06, 16 research assistantships with full tuition reimbursements (averaging $3,688 per year), 37 teaching assistantships with full tuition reimbursements (averaging $10,807 per year) were awarded; Federal Work-Study and institutionally sponsored loans also available. Support available to part-time students. Financial award application deadline: 4/1; financial award applicants required to submit FAFSA. *Faculty research:* Topology. *Unit head:* Dr. Geoffrey Martin, Chair, 419-530-2569, Fax: 419-530-4720, E-mail: gmartin@math.utoledo.edu. *Application contact:* Dr. Gerard Thompson, Advising Coordinator, 419-530-2568, Fax: 419-530-4720, E-mail: thompson@math.utoledo.edu.

See Close-Up on page 547.

University of Toronto, School of Graduate Studies, Physical Sciences Division, Department of Statistics, Toronto, ON M5S 1A1, Canada. Offers M Sc, PhD. Part-time programs available. *Degree requirements:* For doctorate, thesis/dissertation, comprehensive exam. *Entrance requirements:* For master's, GRE (recommended for students educated outside of Canada), 3 letters of reference, minimum B+ average, exposure to statistics and mathematics; for doctorate, GRE (recommended for students educated outside of Canada), 3 letters of reference, M Stat or equivalent, minimum B+ average.

University of Utah, The Graduate School, College of Education, Department of Educational Psychology, Salt Lake City, UT 84112-1107. Offers counseling psychology (PhD); educational psychology (MA); professional counseling (MS); professional psychology (M Ed); school counseling (M Ed, MS); statistics (M Stat). *Accreditation:* APA (one or more programs are accredited). Evening/weekend programs available. *Faculty:* 16 full-time (7 women), 9 part-time/adjunct (3 women). *Students:* 77 full-time (54 women), 84 part-time (57 women); includes 23 minority (4 African Americans, 1 American Indian/Alaska Native, 5 Asian Americans or Pacific Islanders, 13 Hispanic Americans), 3 international. Average age 33. 166 applicants, 45% accepted, 32 enrolled. In 2005, 39 master's, 13 doctorates awarded. *Degree requirements:* For master's, variable foreign language requirement, thesis (for some programs), comprehensive exam; for doctorate, variable foreign language requirement, thesis/dissertation, oral exam. *Entrance requirements:* For master's and doctorate, GRE General Test, minimum GPA of 3.0. Additional exam requirements/recommendations for international students: Required—TOEFL (minimum score 500 paper-based; 173 computer-based). *Application deadline:* For fall admission, 4/1 for domestic students, 4/1 for international students; for spring admission, 11/1 for domestic students, 11/1 for international students. Applications are processed on a rolling basis. Application fee: $45 ($65 for international students). Electronic applications accepted. *Expenses:* Tuition, state resident: full-time $2,932; part-time $2,212 per term. Tuition, nonresident: full-time $10,350; part-time $7,812 per term. Required fees: $590; $516 per term. Tuition and fees vary according to course load and program. *Financial support:* Fellowships with full tuition reimbursements, research assistantships with full tuition reimbursements, teaching assistantships with partial tuition reimbursements, career-related internships or fieldwork, Federal Work-Study, institutionally sponsored loans, scholarships/grants, and unspecified assistantships available. Financial award application deadline: 2/1; financial award applicants required to submit FAFSA. *Faculty research:* Autism, computer technology and instruction, cognitive behavior, aging, group counseling. Total annual research expenditures: $37,452. *Unit head:* Dr. Robert D. Hill, Chair, 801-581-7148, Fax: 801-581-5566, E-mail: bob.hill@ed.utah.edu. *Application contact:* Sherrill Christensen, Academic Program Specialist, 801-581-7148, Fax: 801-581-5566, E-mail: sherrill.christensen@ed.utah.edu.

University of Utah, The Graduate School, Interdepartmental Program in Statistics, Salt Lake City, UT 84112-1107. Offers M Stat. Part-time programs available. *Students:* 5 full-time (2 women), 6 part-time (3 women); includes 4 Asian Americans or Pacific Islanders, 1 international. Average age 32. 41 applicants, 63% accepted, 18 enrolled. In 2005, 12 degrees awarded. *Degree requirements:* For master's, projects. *Entrance requirements:* For master's, minimum GPA of 3.0; course work in calculus, matrix theory, statistics. Additional exam requirements/recommendations for international students: Required—TOEFL (minimum score 500 paper-based; 173 computer-based). *Application deadline:* For fall admission, 7/1 for domestic students. Applications are processed on a rolling basis. Application fee: $45 ($65 for international students). *Expenses:* Tuition, state resident: full-time $2,932; part-time $2,212 per term. Tuition, nonresident: full-time $10,350; part-time $7,812 per term. Required fees: $590; $516 per term. Tuition and fees vary according to course load and program. *Financial support:* Career-related internships or fieldwork available. *Faculty research:* Biostatistics, management, economics, educational psychology, mathematics. *Unit head:* Marlene Egger, Chair, University Statistics Committee, 801-581-6830, E-mail: megger@dfpm.utah.edu. *Application contact:* Glenda Pruemper, Administrative Assistant, 801-581-7148, Fax: 801-581-5566, E-mail: pruemper@ed.utah.edu.

University of Utah, The Graduate School, Program in Science and Technology, Salt Lake City, UT 84112-1107. Offers biotechnology (PSM); computational science (PSM); environmental science (PSM); sciences instrumental (PSM). *Students:* 16 full-time (11 women), 20 part-time (5 women); includes 1 minority (Asian American or Pacific Islander), 7 international. Average age 36. 20 applicants, 50% accepted, 10 enrolled. In 2005, 2 degrees awarded. *Entrance requirements:* For master's, minimum undergraduate GPA of 3.0. Additional exam requirements/recommendations for international students: Required—TOEFL (minimum score 500 paper-based; 173 computer-based). *Application deadline:* For fall admission, 4/1 for domestic students, 4/1 for international students; for spring admission, 11/1 for domestic students, 11/1 for international students. Application fee: $45 ($65 for international students). *Expenses:* Tuition, state resident: full-time $2,932; part-time $2,212 per term. Tuition, nonresident: full-time $10,350; part-time $7,812 per term. Required fees: $590; $516 per term. Tuition and fees vary according to course load and program. *Financial support:* Applicants required to submit FAFSA. *Application contact:* Jennifer Schmidt, Program Director, 801-585-5630, E-mail: jennifer.schmidt@admin.utah.edu.

University of Vermont, Graduate College, College of Engineering and Mathematics, Department of Mathematics and Statistics, Program in Statistics, Burlington, VT 05405. Offers MS.

Statistics

University of Vermont (continued)

Students: 9 (6 women) 4 international. 19 applicants, 89% accepted, 5 enrolled. In 2005, 3 degrees awarded. *Entrance requirements:* Additional exam requirements/recommendations for international students: Required—TOEFL (minimum score 550 paper-based; 213 computer-based). *Application deadline:* For fall admission, 4/1 for domestic students. Applications are processed on a rolling basis. Application fee: $40. *Expenses:* Tuition, area resident: Part-time $410 per credit hour. Tuition, nonresident: part-time $1,034 per credit hour. *Financial support:* Fellowships, research assistantships, teaching assistantships available. Financial award application deadline: 3/1. *Faculty research:* Applied statistics. *Unit head:* Dr. Ruth Mickey, Coordinatoar, 802-656-2940.

University of Victoria, Faculty of Graduate Studies, Faculty of Science, Department of Mathematics and Statistics, Victoria, BC V8W 2Y2, Canada. Offers M Sc, MA, PhD. Part-time programs available. *Faculty:* 24 full-time (4 women), 2 part-time/adjunct (0 women). *Students:* 52, 21 international. Average age 24. 76 applicants, 49% accepted, 20 enrolled. In 2005, 2 master's, 1 doctorate awarded. *Degree requirements:* For master's, thesis, registration; for doctorate, one foreign language, thesis/dissertation, 3 qualifying exams, Candidacy exam. *Entrance requirements:* Additional exam requirements/recommendations for international students: Required—TOEFL (minimum score 575 paper-based; 233 computer-based), IELT (minimum score 7). *Application deadline:* For fall admission, 5/31 priority date for domestic students, 12/15 priority date for international students. Applications are processed on a rolling basis. Application fee: $75 ($125 for international students). Electronic applications accepted. Tuition and fees charges are reported in Canadian dollars. *Expenses:* Tuition, area resident: Full-time $4,492 Canadian dollars; part-time $749 Canadian dollars per term. International tuition: $5,346 Canadian dollars full-time. Required fees: $4,492 Canadian dollars; $749 Canadian dollars per term. Tuition and fees vary according to course load, campus/location and program. *Financial support:* In 2005–06, 24 students received support, including 5 fellowships (averaging $12,900 per year); research assistantships, teaching assistantships, career-related internships or fieldwork, institutionally sponsored loans, unspecified assistantships, and awards also available. Financial award application deadline: 2/15. *Faculty research:* Functional analysis and operator theory, applied ordinary and partial differential equations, discrete mathematics and graph theory. *Unit head:* Dr. Gary MacGillivray, Chair, 250-721-7436, Fax: 250-721-8962, E-mail: chair@math.uvic.ca. *Application contact:* Dr. Pauline van den Driessche, Graduate Adviser, 250-721-7442, Fax: 250-721-8962, E-mail: gradadv@math.uvic.ca.

University of Virginia, College and Graduate School of Arts and Sciences, Department of Statistics, Charlottesville, VA 22903. Offers MS, PhD. *Faculty:* 8 full-time (2 women). *Students:* 15 full-time (10 women), 2 part-time (1 woman); includes 4 minority (1 African American, 3 Asian Americans or Pacific Islanders), 8 international. Average age 25. 72 applicants, 24% accepted, 6 enrolled. In 2005, 3 master's, 2 doctorates awarded. *Degree requirements:* For master's and doctorate, thesis/dissertation. *Entrance requirements:* For master's and doctorate, GRE General Test, GRE Subject Test. *Application deadline:* Applications are processed on a rolling basis. Application fee: $40. Electronic applications accepted. *Expenses:* Tuition, state resident: full-time $7,731. Tuition, nonresident: full-time $18,672. Required fees: $1,479. Full-time tuition and fees vary according to degree level and program. *Financial support:* Applicants required to submit FAFSA. *Unit head:* Jeff Holt, Chairman, 434-924-3222, Fax: 434-924-3076, E-mail: gradapp1@pitman.stat.virginia.edu. *Application contact:* Peter C. Brunjes, Associate Dean for Graduate Programs and Research, 434-924-7184, Fax: 434-924-6737, E-mail: grad-a-s@virginia.edu.

University of Washington, Graduate School, College of Arts and Sciences, Department of Statistics, Seattle, WA 98195. Offers MS, PhD. Terminal master's awarded for partial completion of doctoral program. *Degree requirements:* For master's, thesis optional; for doctorate, one foreign language, thesis/dissertation. *Entrance requirements:* For master's and doctorate, GRE General Test, minimum GPA of 3.0. Additional exam requirements/recommendations for international students: Required—TOEFL. *Faculty research:* Mathematical statistics, stochastic modeling, spatial statistics, statistical computing.

University of Washington, Graduate School, School of Public Health and Community Medicine, Department of Biostatistics, Seattle, WA 98195. Offers biostatistics (MPH, MS, PhD); statistical genetics (PhD). *Faculty:* 76 full-time (34 women), 4 part-time/adjunct (1 woman). *Students:* 56 full-time (28 women), 10 part-time (5 women); includes 7 minority (1 African American, 2 American Indian/Alaska Native, 4 Asian Americans or Pacific Islanders), 24 international. Average age 24. 87 applicants, 34% accepted, 14 enrolled. In 2005, 6 master's, 8 doctorates awarded. Terminal master's awarded for partial completion of doctoral program. *Median time to degree:* Of those who began their doctoral program in fall 1996, 100% received their degree in 8 years or less. *Degree requirements:* For master's, thesis/dissertation, departmental qualifying exams; for doctorate, thesis/dissertation, departmental qualifying exams, comprehensive exam, registration. *Entrance requirements:* For master's, GRE General Test, 2 years of course work in advanced calculus, 1 course in linear algebra, 1 course in mathematical probability, minimum GPA of 3.0; for doctorate, GRE General Test, 2 years of course work in advanced calculus, 1 course in linear algebra, 1 course in mathematical probability, minimum GPA of 3.00. Additional exam requirements/recommendations for international students: Required—TOEFL, TSE. *Application deadline:* For fall admission, 1/4 for domestic students. Application fee: $50. Electronic applications accepted. *Financial support:* In 2005–06, 64 students received support, including 1 fellowship with partial tuition reimbursement available (averaging $19,000 per year), 36 research assistantships with full tuition reimbursements available (averaging $16,000 per year), 5 teaching assistantships with full tuition reimbursements available (averaging $16,000 per year); career-related internships or fieldwork, traineeships, and unspecified assistantships also available. Financial award application deadline: 1/4. *Faculty research:* Statistical methods for survival data analysis, clinical trials, epidemiological case control and cohort studies, statistical genetics. Total annual research expenditures: $11.5 million. *Unit head:* Dr. Bruce Weir, Chair, 206-543-1044. *Application contact:* Alex Mackenzie, Counseling Services Coordinator, 206-543-1044, Fax: 206-543-3286, E-mail: alexam@u.washington.edu.

University of Waterloo, Graduate Studies, Faculty of Mathematics, Department of Statistics and Actuarial Science, Waterloo, ON N2L 3G1, Canada. Offers actuarial science (M Math); statistics (M Math, PhD); statistics-biostatistics (M Math); statistics-computing (M Math); statistics-finance (M Math). *Faculty:* 33 full-time (6 women), 21 part-time/adjunct (5 women). *Students:* 93 full-time (50 women), 5 part-time. 308 applicants, 24% accepted, 34 enrolled. In 2005, 18 master's, 3 doctorates awarded. *Degree requirements:* For master's, research paper or thesis; for doctorate, thesis/dissertation. *Entrance requirements:* For master's, honors degree in field, minimum B+ average; for doctorate, master's degree, minimum B+ average. Additional exam requirements/recommendations for international students: Required—TOEFL, TWE. *Application deadline:* For fall admission, 3/31 for domestic students. For winter admission, 8/31 for domestic students; for spring admission, 12/31 for domestic students. Applications are processed on a rolling basis. Application fee: $75 Canadian dollars. Electronic applications accepted. *Financial support:* In 2005–06, 30 teaching assistantships were awarded; fellowships, research assistantships, career-related internships or fieldwork and scholarships/grants also available. *Faculty research:* Data analysis, risk theory, inference, stochastic processes, quantitative finance. *Unit head:* Dr. D. E. Matthews, Chair, 519-888-4567 Ext. 5530, Fax: 519-746-1875, E-mail: dematthe@uwaterloo.ca. *Application contact:* M. Dufton, Graduate Studies Secretary, 519-888-4567 Ext. 6532, Fax: 519-746-1875, E-mail: mdufton@math.uwaterloo.ca.

The University of Western Ontario, Faculty of Graduate Studies, Physical Sciences Division, Department of Statistical and Actuarial Sciences, London, ON N6A 5B8, Canada. Offers M Sc, PhD. *Degree requirements:* For master's, thesis (for some programs); for doctorate, thesis/dissertation, comprehensive exam. *Faculty research:* Statistical theory, statistical applications, probability, actuarial science.

University of Windsor, Faculty of Graduate Studies and Research, Faculty of Science, Department of Mathematics and Statistics, Windsor, ON N9B 3P4, Canada. Offers mathematics (M Sc); statistics (M Sc, PhD). *Faculty:* 16 full-time (2 women), 3 part-time/adjunct (0 women). *Students:* 34 full-time (9 women), 1 (woman) part-time. 63 applicants, 30% accepted. In 2005, 10 master's, 1 doctorate awarded. *Degree requirements:* For master's, thesis or alternative; for doctorate, thesis/dissertation, comprehensive exam. *Entrance requirements:* For master's, minimum B average; for doctorate, minimum A average. Additional exam requirements/recommendations for international students: Required—TOEFL (minimum score 560 paper-based; 220 computer-based). *Application deadline:* For fall admission, 7/1 for domestic students. For winter admission, 11/1 for domestic students. Applications are processed on a rolling basis. Application fee: $55. Electronic applications accepted. *Financial support:* In 2005–06, 31 teaching assistantships (averaging $8,956 per year) were awarded; Federal Work-Study, scholarships/grants, tuition waivers (full and partial), unspecified assistantships, and bursaries also available. Financial award application deadline: 2/15. *Faculty research:* Applied mathematics, operational research, fluid dynamics. *Unit head:* Dr. Ejaz Ahmed, Head, 519-253-3000 Ext. 3017, Fax: 519-971-3649, E-mail: seahmed@uwindsor.ca. *Application contact:* Applicant Services, 519-253-3000 Ext. 6459, Fax: 519-971-3653, E-mail: gradadmit@uwindsor.ca.

University of Wisconsin–Madison, Graduate School, College of Letters and Science, Department of Statistics, Madison, WI 53706-1380. Offers MS, PhD. Part-time programs available. *Degree requirements:* For master's, exam; for doctorate, thesis/dissertation. *Entrance requirements:* For master's and doctorate, GRE. Additional exam requirements/recommendations for international students: Required—TOEFL. Electronic applications accepted. *Faculty research:* Biostatistics, bootstrap and other resampling theory and methods, linear and nonlinear models, nonparametrics, time series and stochastic processes.

University of Wyoming, Graduate School, College of Arts and Sciences, Department of Statistics, Laramie, WY 82070. Offers MS, PhD. *Faculty:* 9 full-time (2 women). *Students:* 13 full-time (7 women), 5 part-time (2 women), 9 international. Average age 33. 21 applicants, 29% accepted. In 2005, 8 master's, 2 doctorates awarded. Terminal master's awarded for partial completion of doctoral program. *Degree requirements:* For master's, thesis (for some programs), comprehensive exam (for some programs); for doctorate, thesis/dissertation, comprehensive exam. *Entrance requirements:* For master's, GMAT, GRE General Test, minimum GPA of 3.0; for doctorate, GRE General Test, minimum GPA of 3.0. Additional exam requirements/recommendations for international students: Required—TOEFL; Recommended—TWE. *Application deadline:* For fall admission, 2/1 priority date for domestic students, 2/1 priority date for international students. Application fee: $50. Electronic applications accepted. *Expenses:* Tuition, state resident: full-time $3,720; part-time $155 per credit hour. Tuition, nonresident: full-time $10,704; part-time $446 per credit hour. Required fees: $666; $162 per semester. Tuition and fees vary according to course load and program. *Financial support:* In 2005–06, 8 teaching assistantships with full tuition reimbursements (averaging $10,062 per year) were awarded; research assistantships with full tuition reimbursements, Federal Work-Study and institutionally sponsored loans also available. Financial award application deadline: 3/1. *Faculty research:* Mining impacts, biotic integrity. Total annual research expenditures: $25,000. *Unit head:* Dr. Stephen Bieber, Head, 307-766-4229, Fax: 307-766-3927, E-mail: barbr@uwyo.edu.

Utah State University, School of Graduate Studies, College of Science, Department of Mathematics and Statistics, Logan, UT 84322. Offers industrial mathematics (MS); mathematical sciences (PhD); mathematics (M Math, MS); statistics (MS). Part-time programs available. *Faculty:* 33 full-time (4 women). *Students:* 91 full-time (29 women), 4 part-time (1 woman), 44 international. Average age 29. 41 applicants, 61% accepted, 15 enrolled. In 2005, 9 master's, 3 doctorates awarded. Terminal master's awarded for partial completion of doctoral program. *Degree requirements:* For master's, qualifying exam, thesis optional; for doctorate, one foreign language, thesis/dissertation, comprehensive exam. *Entrance requirements:* For master's and doctorate, GRE General Test, minimum GPA of 3.0. Additional exam requirements/recommendations for international students: Required—TOEFL. *Application deadline:* For fall admission, 6/15 for domestic students; for spring admission, 10/15 for domestic students. Applications are processed on a rolling basis. Application fee: $50 ($60 for international students). *Financial support:* In 2005–06, 1 fellowship with partial tuition reimbursement (averaging $12,000 per year), 17 teaching assistantships with partial tuition reimbursements (averaging $14,500 per year) were awarded; research assistantships with partial tuition reimbursements Support available to part-time students. Financial award application deadline: 4/1. *Faculty research:* Differential equations, computational mathematics, dynamical systems, probability and statistics, pure mathematics. Total annual research expenditures: $212,000. *Unit head:* Dr. Russell C. Thompson, Head, 435-797-2810, Fax: 435-797-1822, E-mail: thompson@math.usu.edu. *Application contact:* Dr. David Richard Cutler, Graduate Chairman, 435-797-2699, Fax: 435-797-1822, E-mail: richard.cutler@usu.edu.

Villanova University, Graduate School of Liberal Arts and Sciences, Department of Mathematical Sciences, Program in Applied Statistics, Villanova, PA 19085-1699. Offers MS. Part-time and evening/weekend programs available. *Students:* 6 full-time (3 women), 23 part-time (12 women); includes 2 minority (both Asian Americans or Pacific Islanders), 8 international. Average age 32. 27 applicants, 70% accepted. In 2005, 23 degrees awarded. *Degree requirements:* For master's, comprehensive exam. *Entrance requirements:* For master's, GRE, minimum GPA of 3.0. *Application deadline:* For fall admission, 8/1 for domestic students; for spring admission, 12/1 for domestic students. Application fee: $50. Electronic applications accepted. *Expenses:* Tuition: Part-time $540 per credit. Required fees: $60 per year. Tuition and fees vary according to program and student level. *Financial support:* Research assistantships, Federal Work-Study available. Financial award applicants required to submit FAFSA. *Unit head:* Dr. Michael Levitan, Director.

Virginia Commonwealth University, Graduate School, College of Humanities and Sciences, Department of Mathematical Sciences, Program in Statistics, Richmond, VA 23284-9005. Offers MS, Certificate. *Students:* 8 full-time (4 women), 3 part-time (2 women); includes 2 minority (both African Americans), 3 international. 9 applicants, 78% accepted. In 2005, 4 degrees awarded. *Entrance requirements:* For master's, GRE General Test, GRE Subject Test. Additional exam requirements/recommendations for international students: Required—TOEFL. *Application deadline:* For fall admission, 7/1 for domestic students; for spring admission, 11/15 for domestic students. Applications are processed on a rolling basis. Application fee: $50. *Expenses:* Tuition, state resident: full-time $3,185; part-time $405 per credit. Tuition, nonresident: full-time $7,952; part-time $940 per credit. Required fees: $751 per semester hour. Tuition and fees vary according to course load and program. *Unit head:* Dr. Darcy P. Mays, Head, 804-828-1301 Ext.151. *Application contact:* Dr. James A. Wood, Information Contact, 804-828-1301, E-mail: jawood@vcu.edu.

Virginia Polytechnic Institute and State University, Graduate School, College of Science, Department of Statistics, Blacksburg, VA 24061. Offers MS, PhD. *Faculty:* 17 full-time (4 women). *Students:* 46 full-time (24 women), 10 part-time (6 women); includes 4 minority (1 African American, 1 Asian American or Pacific Islander, 2 Hispanic Americans), 26 international. Average age 30. 108 applicants, 30% accepted, 18 enrolled. In 2005, 17 master's, 7 doctorates awarded. *Entrance requirements:* Additional exam requirements/recommendations for international students: Required—TOEFL (minimum score 600 paper-based; 250 computer-based). *Application deadline:* Applications are processed on a rolling basis. Application fee: $45. *Expenses:* Tuition, state resident: full-time $6,558; part-time $364 per credit. Tuition, nonresident: full-time $11,296; part-time $628 per credit. Required fees: $1,419; $468 per credit. $234 per term. *Financial support:* In 2005–06, 4 research assistantships with full tuition reimbursements (averaging $16,689 per year), 25 teaching assistantships with full tuition reimbursements (averaging $13,902 per year) were awarded; career-related internships or fieldwork, Federal Work-Study, scholarships/grants, and unspecified assistantships also available. Financial award application deadline: 4/1. *Faculty research:* Design and sampling theory, computing and simulation, nonparametric statistics, robust and multivariate methods, biostatistics quality. *Unit head:* Dr. Geoffrey Vining, Head, 540-231-5657, Fax: 540-231-3863,

E-mail: vining@vt.edu. *Application contact:* Christina Dillon, 540-231-5630, Fax: 540-231-3863, E-mail: chconne1@vt.edu.

Washington State University, Graduate School, College of Agricultural, Human, and Natural Resource Sciences, Department of Statistics, Pullman, WA 99164. Offers applied and theoretical options (MS). *Faculty:* 7 full-time, 9 part-time/adjunct. *Students:* 10 full-time (5 women), 2 part-time (1 woman), 9 international. Average age 30. 43 applicants, 42% accepted, 13 enrolled. In 2005, 10 degrees awarded. *Degree requirements:* For master's, project. *Entrance requirements:* For master's, GRE, minimum GPA of 3.0, 3 letters of recommendation. Additional exam requirements/recommendations for international students: Required—TOEFL (minimum score 560 paper-based; 220 computer-based). *Application deadline:* For fall admission, 3/1 for domestic students, 3/1 for international students; for spring admission, 7/1 for domestic students, 7/1 for international students. Application fee: $35. *Expenses:* Tuition, state resident: full-time $6,295; part-time $336 per credit. Tuition, nonresident: full-time $15,949; part-time $819 per credit. Required fees: $933. Part-time tuition and fees vary according to campus/location and program. *Financial support:* In 2005–06, 10 students received support, including 2 research assistantships (averaging $15,836 per year), 6 teaching assistantships with tuition reimbursements available (averaging $11,725 per year) *Faculty research:* Environmental statistics, logistic regression, statistical methods for ecology and wildlife, spatial data analysis, linear and non-linear models. Total annual research expenditures:$86,830. *Unit head:* Dr. Michael A. Jacroux, Professor/Chair, 509-335-8645, Fax: 509-335-8369, E-mail: jacroux@wsu.edu. *Application contact:* Graduate Admissions Committee.

Washington University in St. Louis, Graduate School of Arts and Sciences, Department of Mathematics, St. Louis, MO 63130-4899. Offers mathematics (MA, PhD); mathematics education (MAT); statistics (MA, PhD). Terminal master's awarded for partial completion of doctoral program. *Degree requirements:* For master's, thesis or alternative; for doctorate, thesis/dissertation. *Entrance requirements:* For master's and doctorate, GRE General Test. Electronic applications accepted.

Wayne State University, Graduate School, College of Liberal Arts and Sciences, Department of Mathematics, Program in Mathematical Statistics, Detroit, MI 48202. Offers MA, PhD. *Students:* 4 full-time (3 women), 2 part-time, 5 international. Average age 35. 6 applicants, 33% accepted, 1 enrolled. In 2005, 7 degrees awarded. *Degree requirements:* For doctorate, thesis/dissertation. *Entrance requirements:* Additional exam requirements/recommendations for international students: Required—TOEFL (minimum score 550 paper-based; 213 computer-based); Recommended—TWE (minimum score 6). *Application deadline:* For fall admission, 7/1 for domestic students, 6/1 for international students. Applications are processed on a rolling basis. Application fee: $30 ($50 for international students). Electronic applications accepted. *Expenses:* Tuition, state resident: part-time $338 per credit hour. Tuition, nonresident: part-time $746 per credit hour. Required fees: $24 per credit hour. Full-time tuition and fees vary according to program. *Application contact:* Bert Schreiber, Professor, 313-577-8838, E-mail: bschreiber@wayne.edu.

Western Michigan University, Graduate College, College of Arts and Sciences, Department of Statistics, Kalamazoo, MI 49008-5202. Offers biostatistics (MS).

West Virginia University, College of Business and Economics, Division of Economics and Finance, Morgantown, WV 26506. Offers business analysis (MA); econometrics (PhD); industrial economics (PhD); international economics (PhD); labor economics (PhD); mathematical economics (MA, PhD); monetary economics (PhD); public finance (PhD); public policy (MA); regional and urban economics (PhD); statistics and economics (MA). *Faculty:* 21 full-time (3 women), 2 part-time/adjunct (1 woman). *Students:* 48 full-time (15 women), 10 part-time (2 women); includes 1 minority (Asian American or Pacific Islander), 38 international. Average age 30. 120 applicants, 25% accepted, 12 enrolled. In 2005, 6 degrees awarded. Terminal master's awarded for partial completion of doctoral program. *Degree requirements:* For master's, thesis optional; for doctorate, thesis/dissertation, comprehensive exam. *Entrance requirements:* For master's and doctorate, GRE General Test, minimum GPA of 3.0; course work in intermediate microeconomics, intermediate macroeconomics, calculus, and statistics. Additional exam requirements/recommendations for international students: Required—TOEFL. *Application deadline:* For fall admission, 3/1 priority date for domestic students, 3/1 priority date for international students. Applications are processed on a rolling basis. Application fee: $50. Electronic applications accepted. *Expenses:* Tuition, state resident: full-time $4,582; part-time $258 per credit hour. Tuition, nonresident: full-time $1,382; part-time $741 per credit hour. *Financial support:* In 2005–06, 50 students received support, including 1 fellowship with full tuition reimbursement available (averaging $15,000 per year), 8 research assistantships with full tuition reimbursements available (averaging $9,300 per year), 23 teaching assistantships with full tuition reimbursements available (averaging $9,300 per year); Federal Work-Study, institutionally sponsored loans, and tuition waivers (full and partial) also available. Financial award applicants required to submit FAFSA. *Faculty research:* Financial economics, regional/urban development and problems, public economics. *Unit head:* Dr. William N. Trumbull, Director, 304-293-7860, Fax: 304-293-2233, E-mail: william.trumbull@mail.wvu.edu. *Application contact:* Dr. Brian Cushing, Director of Admissions and Financial Awards, 304-293-7881, Fax: 304-293-5652, E-mail: brian.cushing@mail.wvu.edu.

West Virginia University, Eberly College of Arts and Sciences, Department of Statistics, Morgantown, WV 26506. Offers MS. *Faculty:* 7 full-time (0 women), 2 part-time/adjunct (0 women). *Students:* 29 full-time (22 women), 15 part-time (9 women); includes 5 minority (all Asian Americans or Pacific Islanders), 35 international. Average age 32. 31 applicants, 94% accepted, 15 enrolled. In 2005, 18 degrees awarded. *Degree requirements:* For master's, thesis, comprehensive exam. *Entrance requirements:* For master's, minimum GPA of 3.0, course work in linear multivanate calculus and algebra. Additional exam requirements/recommendations for international students: Required—TOEFL. *Application deadline:* For fall admission, 3/15 priority date for domestic students, 3/15 priority date for international students; for spring admission, 10/15 priority date for domestic students, 10/15 priority date for international students. Applications are processed on a rolling basis. Application fee: $45. *Expenses:* Tuition, state resident: full-time $4,582; part-time $258 per credit hour. Tuition, nonresident: full-time $1,382; part-time $741 per credit hour. *Financial support:* In 2005–06, 7 research assistantships with full tuition reimbursements (averaging $8,554 per year), 11 teaching assistantships with full tuition reimbursements (averaging $8,554 per year) were awarded; Federal Work-Study, institutionally sponsored loans, and tuition waivers (full and partial) also available. Financial award application deadline: 2/1; financial award applicants required to submit FAFSA. *Faculty research:* Linear models, categorical data analysis, statistical computing, experimental design, non parametric analysis. Total annual research expenditures: $404,000. *Unit head:* E. James Harner, Chair, 304-293-3607 Ext. 1051, Fax: 304-293-2272, E-mail: jim.harner@mail.wvu.edu.

Wichita State University, Graduate School, Fairmount College of Liberal Arts and Sciences, Department of Mathematics and Statistics, Wichita, KS 67260. Offers applied mathematics (PhD); mathematics (MS); statistics (MS). Part-time programs available. *Degree requirements:* For master's, thesis optional; for doctorate, thesis/dissertation. *Entrance requirements:* For master's, GRE; for doctorate, GRE Subject Test. Additional exam requirements/recommendations for international students: Required—TOEFL. Electronic applications accepted. *Faculty research:* Partial differential equations, combinatorics, ring theory, minimal surfaces, several complex variables.

Worcester Polytechnic Institute, Graduate Studies and Enrollment, Department of Mathematical Sciences, Worcester, MA 01609-2280. Offers applied mathematics (MS); applied statistics (MS); financial mathematics (MS); industrial mathematics (MS); mathematical sciences (PhD); mathematics (MME). Part-time and evening/weekend programs available. *Faculty:* 30 full-time (3 women), 4 part-time/adjunct (0 women). *Students:* 28 full-time (13 women), 33 part-time (17 women); includes 5 minority (2 African Americans, 3 Asian Americans or Pacific Islanders), 14 international. 77 applicants, 77% accepted, 17 enrolled. In 2005, 15 degrees awarded. *Degree requirements:* For master's, thesis (for some programs); for doctorate, thesis/dissertation, comprehensive exam. *Entrance requirements:* For master's and doctorate, 3 letters of recommendation. Additional exam requirements/recommendations for international students: Required—TOEFL (minimum score 550 paper-based; 213 computer-based). *Application deadline:* For fall admission, 1/15 for domestic students; for spring admission, 10/15 priority date for domestic students. Applications are processed on a rolling basis. Application fee: $70. Electronic applications accepted. *Expenses:* Tuition: Part-time $997 per credit hour. *Financial support:* In 2005–06, 19 students received support, including fellowships with full tuition reimbursements available (averaging $33,246 per year), 5 research assistantships with full and partial tuition reimbursements available, 14 teaching assistantships with full and partial tuition reimbursements available; career-related internships or fieldwork, institutionally sponsored loans, scholarships/grants, and unspecified assistantships also available. Financial award application deadline: 1/15. *Faculty research:* Applied mathematical modeling and analysis, computational mathematics, discrete mathematics, applied and computational statistics, industrial and financial mathematics. Total annual research expenditures: $1.2 million. *Unit head:* Dr. Bogdan Vernescu, Head, 508-831-5241, Fax: 508-831-5824. *Application contact:* Dr. Homer F Walker, Graduate Coordinator, 508-831-6113, Fax: 508-831-5824, E-mail: walker@wpi.edu.

See Close-Up on page 559.

Wright State University, School of Graduate Studies, College of Science and Mathematics, Department of Mathematics and Statistics, Program in Applied Statistics, Dayton, OH 45435. Offers MS. *Degree requirements:* For master's, comprehensive exam. *Entrance requirements:* For master's, 1 year of course work in calculus and matrix algebra, previous course work in computer programming and statistics. Additional exam requirements/recommendations for international students: Required—TOEFL. *Faculty research:* Reliability theory, stochastic process, nonparametric statistics, design of experiments, multivariate statistics.

Yale University, Graduate School of Arts and Sciences, Department of Statistics, New Haven, CT 06520. Offers MS, PhD. Terminal master's awarded for partial completion of doctoral program. *Degree requirements:* For doctorate, thesis/dissertation. *Entrance requirements:* For doctorate, GRE General Test, GRE Subject Test.

York University, Faculty of Graduate Studies, Faculty of Arts, Program in Mathematics and Statistics, Toronto, ON M3J 1P3, Canada. Offers industrial and applied mathematics (M Sc); mathematics and statistics (MA, PhD). Part-time programs available. *Faculty:* 57 full-time (13 women), 6 part-time/adjunct (2 women). *Students:* 69 full-time (33 women), 34 part-time (18 women). 300 applicants, 13% accepted, 38 enrolled. In 2005, 28 master's, 5 doctorates awarded. *Degree requirements:* For master's, thesis optional; for doctorate, one foreign language, thesis/dissertation, comprehensive exam, registration. *Application deadline:* For fall admission, 2/1 for domestic students. Application fee: $80. Electronic applications accepted. *Expenses:* Tuition, state resident: full-time $3,190; part-time $798 per term. International tuition: $7,515 full-time. Required fees: $217. Tuition and fees vary according to program. *Financial support:* In 2005–06, fellowships (averaging $8,354 per year), research assistantships (averaging $5,815 per year), teaching assistantships (averaging $9,475 per year) were awarded; tuition waivers (partial) and fee bursaries also available. *Unit head:* Yuehua Wu, Director, 416-736-2100.

BOSTON COLLEGE

Graduate School of Arts and Sciences
Department of Mathematics

Programs of Study

The Department of Mathematics in the Graduate School of Arts and Sciences at Boston College offers a Master of Arts (M.A.) in mathematics. The department also participates in a collaborative agreement with other graduate programs to offer joint graduate degrees, such as the Master of Science in Teaching, Master of Education, and a joint M.A./M.B.A. with the Graduate School of Management.

The M.A. in mathematics at Boston College is a blend of structure and flexibility. With almost all first-year students taking the algebra and analysis courses together, a tight-knit group is formed, and students proceed with a common core of knowledge. The M.A. degree program requires the completion of 30 credit hours (ten courses) in the department and participation in a 3-credit seminar. Under special circumstances, and with the approval of the Graduate Committee and the Graduate Program Director, a student may satisfy the degree requirements with 27 credit hours (nine courses) and a thesis (6 credit hours).

The five elective courses in the program allow each student to aim his or her program toward a particular goal: teaching, a career in the private sector, or further study in mathematics or some other area. Two of these electives may be taken outside of the Department of Mathematics. At the end of the program, the graduate seminar provides a unique capstone experience.

The program provides numerous opportunities for independent work. Students who wish to study a topic that is not offered in a regular course complete a Readings and Research course (independent study). The thesis option requires two semesters of independent work that replace two electives; a thesis is not required for the M.A. program.

Master's degree students at Boston College have access to, and the full attention of, all of the faculty members in the department for research and thesis projects. In addition, a cross-registration program allows students to take courses at a number of nearby graduate mathematics programs, so the small size of the program does not limit the subjects students may choose to explore.

Research Facilities

Boston College provides its students with state-of-the-art facilities for learning, including a full range of computer services, online access to databases, and a library system with more than 1.9 million books, periodicals, and government documents and 3.4 million microform units. The library's membership in the Boston Library Consortium provides access to ten major research libraries in the Boston area, and an interlibrary loan system provides further resources. The Department of Mathematics houses the Mathematics Institute, which assists elementary and high schools with math curriculum development.

Financial Aid

Boston College offers the following stipends and scholarships to academically and financially qualified students: teaching fellowships, tuition scholarships, and teaching assistantships. Students are routinely considered for financial aid by the department in which they hope to study; no additional application is necessary. The amounts of the awards and the number of years for which they are awarded vary by department. Research grants are also available on a year-by-year basis and vary among academic disciplines.

Student loans and work-study programs are also available. The Boston College Office of Student Services administers and awards need-based federal financial aid programs, including Federal Subsidized and Unsubsidized Stafford Loans, Perkins Loans, and Federal Work Study.

Cost of Study

Tuition for the Graduate School of Arts and Sciences is $990 per credit for the 2005–06 academic year. The student activities fee ranges from $25 to $50 per semester, depending on the number of credits taken. The late registration fee is $150 per semester. The costs of books varies by semester. Additional fees may apply.

Living and Housing Costs

Boston College does not currently offer on-campus graduate housing. However, rental housing is plentiful in Newton as well as in surrounding cities and towns. There are many different types of housing available, ranging from one-room rentals in large Victorian homes, to triple-decker brownstones and apartment high-rises. Allston-Brighton and Jamaica Plain are among the nearby Boston neighborhoods that attract students from many colleges and universities because of their diverse communities and relative affordability. Graduate students also have found Newton, Brookline, Waltham, Watertown, and Boston's West Roxbury neighborhood attractive places to live. Boston College's Off-Campus Housing Office is available to assist in the housing search. The office maintains an extensive database of available rental listings, roommates, and helpful local realtors.

Student Group

There are currently about 25 students in the graduate program. The department admits up to 12 new M.A. students each year.

Location

The Graduate School of Arts and Sciences is located on the Chestnut Hill campus of Boston College, approximately 5 miles west of the city of Boston, Massachusetts. Boston offers students the opportunity to experience one of the oldest cities in the U.S., with museums, a symphony orchestra, and world championship professional basketball, baseball, ice hockey, and football teams. The city of Boston also offers a wide variety of shopping, dining, and cultural experiences—all located on the beautiful Boston Harbor and Charles River.

The College and The School

Founded in 1863, Boston College is one of the oldest Jesuit-sponsored universities in the United States. It has professional and graduate schools, doctoral programs, research institutes, community service programs, an excellent faculty, and rich resources of libraries, research equipment, computers, and other facilities. A coeducational university, it has an enrollment of 9,000 undergraduate and 4,700 graduate students representing nearly 100 countries. Boston College confers degrees in more than fifty fields of study through its eleven schools and colleges. It has more than 600 full-time faculty members.

The Graduate School of Arts and Sciences offers programs of study in the humanities, social sciences, and natural sciences, leading to the degrees of Doctor of Philosophy (Ph.D.), Master of Arts (M.A.), and Master of Science (M.S.). The Graduate School may also admit students as Special Students, those not seeking a degree but who are interested in pursuing course work at the graduate level for personal or professional edification. The Graduate School of Arts and Sciences operates on a semester calendar, with the fall semester running from late August until mid-December and the spring semester running from late January until late May.

Applying

Applicants must hold a baccalaureate degree from an accredited institution; a math degree is not necessary, but applicants must have successfully completed at least one pure math course. The deadline for applying for the fall semester is January 15. Applications are accepted throughout the year; however, there are a limited number of teaching fellowships available, and applicants are encouraged to apply before the January 15 deadline.

The most important criteria for admission and fellowships are the number and strength of mathematics courses taken, grades achieved in those courses, and letters of recommendation. GRE scores, grades outside of mathematics, and other data are considered as well but count less. Teaching, tutoring, or other professional experiences are also helpful.

Application to the M.A. program in mathematics is made through the Graduate School of Arts and Sciences. The mathematics M.A. application must include the GRE General scores, all undergraduate transcripts, and at least two letters of recommendation. A third letter of recommendation is suggested, if at all possible. Most applicants also include a short statement of purpose, which, although not required, does serve to personalize the application. A nonrefundable $70 application fee must also be included.

Correspondence and Information

Department of Mathematics
Carney Hall, Room 301
Boston College
140 Commonwealth Avenue
Chestnut Hill, Massachusetts 02467-3806
Phone: 617-552-3750
Fax: 617-552-3789
E-mail: math@bc.edu
Web site: http://www.bc.edu/math

Boston College

THE FACULTY AND THEIR RESEARCH

Avner Ash, Professor; Ph.D., Harvard. Number theory, algebraic geometry.
Jenny Baglivo, Professor; Ph.D., Syracuse. Statistics, applied mathematics.
Robert J. Bond, Associate Professor; Ph.D., Brown. Algebra, number theory.
Martin Bridgeman, Associate Professor; Ph.D., Princeton. Geometry, topology.
Brian Carvalho, Instructor; M.A., Boston College.
Daniel W. Chambers, Associate Professor; Ph.D., Maryland. Probability, stochastic processes, statistics.
C.-K. Cheung, Associate Professor; Ph.D., Berkeley. Complex differential geometry, several complex variables.
Marie Clote, Adjunct Assistant Professor; M.S. Paris VII.
Solomon Friedberg, Professor; Ph.D., Chicago. Number theory, representation theory.
Robert Gross, Associate Professor; Ph.D., MIT. Algebra, number theory, history of mathematics.
Benjamin Howard, Assistant Professor; Ph.D., Stanford. Number theory, arithmetic geometry.
Richard A. Jensen, Associate Professor; Ph.D., Illinois. Algebraic coding theory, combinatronics.
William J. Keane, Associate Professor; Ph.D., Notre Dame. Abelian group theory.
Margaret Kenney, Professor; Ph.D., Boston University. Mathematics education.
Gerard E. Keough, Associate Professor; Ph.D., Indiana. Operator theory, functional analysis.
Charles K. Landraitis, Associate Professor; Ph.D., Dartmouth. Combinatorics, mathematical logic.
Tao Li, Assistant Professor; Ph.D., Caltech. Topology, geometry of low-dimension manifolds.
G. Robert Meyerhoff, Professor; Ph.D., Princeton. Geometry, topology.
Rennie Mirollo, Associate Professor; Ph.D., Harvard. Dynamical system.
Nancy Rallis, Associate Professor; Ph.D., Indiana. Algebraic topology, fixed point theory, probability and statistics.
Robert Reed, Adjunct Assistant Professor; Ph.D., Wisconsin. Mathematical logic.
Mark Reeder, Professor; Ph.D., Ohio State. Lie groups, representation theory.
Ned Rosen, Associate Professor; Ph.D., Michigan. Dynamical systems, mathematical logic.
Paul R. Thie, Professor; Ph.D., Notre Dame. Mathematical programming.
Howard Troughton, Lecturer; M.Sc., Toronto; M.E.Des., Calgary.

BOWLING GREEN STATE UNIVERSITY

Department of Mathematics and Statistics

Programs of Study

The Department of Mathematics and Statistics offers a full range of graduate degrees. Degree options at the master's level include the M.A., with concentrations in pure mathematics, scientific computation, and probability and statistics, and the M.S. in applied statistics. These programs are offered with a thesis option and a comprehensive examination option. The Department also offers a Master of Arts in Teaching (M.A.T.) in mathematics for those interested in teaching at the secondary level or at two- and four-year colleges. M.A.T. course work is tailored to the individual and may be supplemented by an internship or other field experience. The master's degree programs are two-year programs of study, but well-qualified students can complete a degree in three semesters.

The Ph.D. program combines advanced study with individual research; a dissertation consisting of original research is required. Strong research areas include probability and statistics, algebra, analysis, and scientific computation. The research environment is further enhanced by the Department's active program of seminars and colloquia. Weekly seminars are conducted in algebra, analysis, mathematics education, scientific computation, and statistics.

The Department has 27 full-time faculty members, all of whom hold Ph.D. degrees. The Department hosts several distinguished visiting scholars each year, including a Lukacs Distinguished Professor in Probability and Statistics. In addition to working with advanced graduate students, the Lukacs Professor organizes the Annual Lukacs Symposium, which attracts leading statisticians from around the world. As part of the Department's continuing commitment to quality instruction, all students are given opportunities for a variety of training and mentoring experiences that are designed to enhance their effectiveness both as students and as teachers.

Research Facilities

Faculty member and graduate student offices are located in the Mathematical Sciences Building, which also houses the Frank C. Ogg Science Library and the Scientific Computing Laboratory. The Science Library, in addition to its extensive holdings, maintains subscriptions to approximately 400 journals, both paper and electronic, in mathematics and statistics. Further, the interlibrary OhioLink program provides access to the holdings of all other state-funded university libraries in Ohio. The Scientific Computing Laboratory offers microcomputer access with fast Internet connections and a wealth of software, UNIX/X11 access, and a full-time staff to provide assistance to users.

Additional computing resources include a network of UNIX workstations that are maintained by the Department and available to students at all times. The University also maintains several systems for student use, including a four-processor SGI Power Challenge, a DEC Alpha 2100 5/250, a DEC VAX 6620, two IBM mainframes, and various graphics workstations. Each graduate student office is furnished with a microcomputer with network access.

Financial Aid

The Department provides approximately fifty-six teaching assistantships with stipends of $10,000 for master's students for the academic year and $12,623 for doctoral students for the academic year. Also offered are three nonservice fellowships of $15,779 for the calendar year. Instructional and nonresident fees are waived. The Department also provides summer support through a variety of fellowships and assistantships that range from $1594 to $3550. In addition, all new students are encouraged to accept Summer Fellowships of $1800 for an initial six-week summer program.

Teaching assistants serve as instructors for small individual classes that consist of about 30 students. This involves five or six contact hours per week with undergraduate students. The University's Statistical Consulting Center also provides consultantships for graduate assistants with appropriate backgrounds. These positions provide valuable experience for those preparing for careers in statistics. The stipends offered by the Statistical Consulting Center are the same amount as those awarded to teaching assistants. For further information, students should consult the Department's Web site.

Cost of Study

Tuition and nonresident fees for the 2004–05 academic year (fall, spring, and summer) were $22,614. Tuition and nonresident fees and most other fees are covered by the assistantship package. Students are also required to have adequate health insurance, which may be purchased through the University at a nominal fee. Students must purchase their own books and pay any applicable thesis or dissertation fees.

Living and Housing Costs

As a small town, Bowling Green offers a modest cost of living. Most graduate students choose to live off campus. The city of Bowling Green offers a wide variety of rental housing, with prices beginning at $150 and averaging $350 per month. A limited number of rooms in on-campus residence halls are set aside for graduate students.

Student Group

There are currently 58 full-time graduate students in the Department. Of these, 30 are international students, 20 are women, 25 are in the master's programs, and 33 are in the Ph.D. program.

Student Outcomes

Bowling Green State University (BGSU) graduates enjoy a very high placement rate. At the Ph.D. level, for instance, 68 students have graduated since 1990; of the 35 respondents to a recent survey of graduates, all report that they are meaningfully employed in academic or industrial research positions.

Location

Bowling Green is a peaceful semirural community in historic northwest Ohio. Founded in 1833, the city's early growth was greatly influenced by the prosperous oil-boom era of the late 1800s, evident today through downtown Bowling Green's stately architecture. Bowling Green is conveniently located on Interstate 75, just 20 miles south of Toledo, Ohio, and 90 miles south of Detroit, Michigan. The average temperature in August is 71.1°F (21.7°C); the average temperature in January is 25.5°F (-3.6°C).

The University

Established in 1910 as a teacher-training college, BGSU attained full university status in 1935 and has since grown into a multidimensional institution that offers approximately 200 different degree programs from the bachelor's through doctoral levels. The intellectual climate—in the University generally and in the Department particularly—combines the warmth and collegiality of a liberal arts atmosphere with the resources and opportunities of a research institution.

Applying

Application (for both admission and financial assistance) consists of a completed application and financial disclosure forms, which are available by request; a brief personal statement that indicates the applicant's goals and academic interests; three letters of reference; two copies of official transcripts from each institution attended; test scores on the GRE General Test; a $30 check or money order made payable to the Graduate College, Bowling Green State University; and, if the applicant's native language is not English, test scores on the TOEFL or MELAB. The deadline for applications is March 1. Late applications are considered if positions are still available. Full instructions for applying, along with an online application form, can be found at the Department's Web site. Application materials may also be requested via e-mail.

Correspondence and Information

Graduate Coordinator
Department of Mathematics and Statistics
Bowling Green State University
Bowling Green, Ohio 43403-0221

Phone: 419-372-7463
Fax: 419-372-6092
E-mail: hchen@bgnet.bgsu.edu
Web site: http://www.bgsu.edu/departments/math/

Bowling Green State University

THE FACULTY AND THEIR RESEARCH

James H. Albert, Professor; Ph.D., Purdue. Bayesian analysis, analysis of categorical data.
Juan Bes, Assistant Professor; Ph.D., Kent State. Operator theory.
Neal Carothers, Professor and Chair; Ph.D., Ohio State. Functional analysis, Banach space theory, real analysis.
Kit Chan, Professor; Ph.D., Michigan. Functional analysis, function theory.
Hanfeng Chen, Professor and Graduate Coordinator; Ph.D., Wisconsin–Madison. Transformed data analysis, finite mixture models.
John T. Chen, Associate Professor; Ph.D., Sydney (Australia). Multivariate statistics, probability inequalities, biostatistics.
So-Hsiang Chou, Professor; Ph.D., Pittsburgh. Numerical analysis, fluid mechanics.
Humphrey S. Fong, Associate Professor; Ph.D., Ohio State. Probability, real analysis.
John T. Gresser, Associate Professor; Ph.D., Wisconsin–Milwaukee. Complex analysis.
Arjun K. Gupta, Distinguished University Professor; Ph.D., Purdue. Multivariate statistical analysis, analysis of categorical data.
Corneliu Hoffman, Associate Professor; Ph.D., USC. Representations of finite groups, inverse Galois problems.
Alexander Izzo, Associate Professor; Ph.D., Berkeley. Complex analysis, functional analysis.
Warren W. McGovern, Assistant Professor; Ph.D., Florida. Ordered algebraic structures.
David E. Meel, Associate Professor; Ed.D., Pittsburgh. Mathematics education.
Barbara E. Moses, Professor; Ph.D., Indiana. Mathematics education, problem solving.
Diem Nguyen, Assistant Professor; Ph.D., Texas A&M. Mathematics education.
Truc T. Nguyen, Professor; Ph.D., Pittsburgh. Mathematical statistics.
Steven M. Seubert, Professor; Ph.D., Virginia. Functional analysis, operator theory, complex analysis.
Sergey Shpectorov, Professor; Ph.D., Moscow State. Groups and geometries.
Tong Sun, Associate Professor; Ph.D., Texas A&M. Numerical analysis, partial differential equations.
Gábor Székely, Professor; Ph.D., Eötvös Loránd (Budapest). Probability and statistics.
J. Gordon Wade, Associate Professor; Ph.D., Brown. Numerical analysis, inverse problems.
Craig L. Zirbel, Associate Professor; Ph.D., Princeton. Probability, stochastic processes.

Mathematical Sciences Building.

Bowling Green State University

SELECTED PUBLICATIONS

Albert, J. H. Bayesian testing and estimation of association in a two-way contingency table. *J. Am. Stat. Assoc.* 92:685–93, 1997.

Albert, J. H. Bayesian selection of log-linear modes. *Can. J. Stat.* 24:327–47, 1996.

Albert, J. H., and S. Chib. Bayesian residual analysis for binary response regression models. *Biometrika* 82:747–59, 1995.

Bes, J., and **K. C. Chan.** Approximation by chaotic operators and by conjugate classes. *J. Math. Analysis Applications* 284:206–12, 2001.

Bes, J., K. C. Chan, and **S. Seubert.** Chaotic unbounded differentiation operators. *Integral Equations Operator Theory* 40:257–67, 2001.

Carothers, N. L. *A Short Course on Banach Space Theory.* New York: Cambridge University Press, 2004.

Carothers, N. L., S. Dilworth, and D. Sobecki. Splittings of Banach spaces induced by Clifford algebras. *Proc. Am. Math. Soc.* 128:1347–56, 2000.

Carothers, N. L. *Real Analysis.* New York: Cambridge University Press, 2000.

Chan, K. C., and R. Sanders. A weakly hypercyclic operator that is not norm hypercyclic. *J. Operator Theory,* in press.

Chan, K. C. The density of hypercyclic operators on a Hilbert space. *J. Operator Theory* 47:131–43, 2002.

Chan, K. C., and R. Taylor. Hypercyclic subspaces of a Banach space. *Integral Equations Operator Theory* 41:381–8, 2001.

Chen, H., J. Chen, and J. D. Kalbfleisch. Testing for a finite mixture model with two components. *J. Royal Stat. Soc., Ser B.* 66(1):95–115, 2004.

Chen, H., and **J. Chen.** Tests for homogeneity in normal mixtures in the presence of a structural parameter. *Stat. Sin.* 13(2):351–65, 2003.

Chen, H., J. Chen, and J. D. Kalbfleisch. A modified likelihood ratio test for homogeneity in finite mixture models. *J. Royal Stat. Soc., Ser. B* 63(1):19–29, 2001.

Kamburowska, G., and **H. Chen.** Fitting data to the Johnson system. *J. Stat. Comput. Simulation* 69:21–32, 2001.

Chen, J. T., and F. M. Hopper. A connection between successive comparisons and ranking procedures. *Stat. Probability Lett.* 67:19–25, 2004.

Chen, J. T. A lower bound using Hamilton-type circuit and its applications. *J. Appl. Probability* 40:1121–32, 2003.

Chen, J. T., F. M. Hoppe, S. Iyengar, and D. Brent. A hybrid logistic regression model for case-control studies. *Methodology Computing Appl. Probability* 5:419–26, 2003.

Chen, J. T., A. K. Gupta, and C. Troskie. Distribution of stock returns when the market is up (down). *Commun. Stat. Theory Methods* 32:1541–58, 2003.

Chou, S.-H., D. Y. Kwak, and P. S. Vassilevski. Mixed covolume methods for elliptic problems on triangular grids. *SIAM J. Numerical Anal.* 35(5):1850–61, 1998.

Chou, S.-H., and D. Y. Kwak. A covolume method based on rotated bilinears for the generalized Stokes problem. *SIAM J. Numerical Anal.* 35(2):497–507, 1998.

Chou, S.-H. Analysis and convergence of a covolume method for the generalized Stokes problem. *Math. Comput.* 217(66):85–104, 1997.

Gresser, J. *A Maple Approach to Calculus,* 2nd ed. Englewood Cliffs, N.J.: Prentice Hall Publishing Company, 2002.

Gresser, J. *A Mathematica Approach to Calculus,* 2nd ed. Englewood Cliffs, N.J.: Prentice Hall Publishing Company, 2002.

Chen, J., and **A. K. Gupta.** Information theoretic approach for detecting change in the parameters of a normal model. *Math. Methods Stat.* 12:116–30, 2004.

Gupta, A. K., N. Henze, and B. Klar. Testing for affine equivalence of elliptically symmetric distributions. *J. Multivariate Analysis* 88:222–42, 2004.

Gupta, A. K., G. Gonzalez-Farias, and J. A. Dominguez-Molina. A multivariate skew normal distribution. *J. Multivariate Analysis* 89:181–90, 2004.

Gupta, A. K. Multivariate skew t-distribution. *Statistics* 37:359–63, 2003.

Izzo, A. C^{ILU}r convergence of Picard's successive approximations. *Proc. Am. Math. Soc.* 127:2059–63, 1999.

Izzo, A. A characterization of $C(K)$ among the uniform algebras containing $A(K)$. *Indiana University Math. J.* 46:771–88, 1997.

Izzo, A. Failure of polynomial approximation on polynomially convex subsets of the sphere. *Bull. London Math. Soc.* 28:393–7, 1996.

Hager, A. W., C. M. Kimber, and **W. Wm. McGovern.** Weakly least integer closed groups. *Rendiconti Circolo Matematico Palermo* 52:453–80, 2003.

McGovern, W. W. Clean semiprime *f*-rings with bounded inversion. *Commun. Algebra* 31(7):3295–304, 2003.

McGovern, W. W. Free topological groups over weak P-spaces. *Top. Appl.* 112(2):175–80, 2001.

Hager, A. W., C. M. Kimber, and **W. Wm. McGovern.** Least integer closed groups. In *Proceedings of the Conference on Lattice Ordered Groups and f-Rings,* pp. 245–60. Gainesville, Fla.: Kluwer Academic Publishers, 2001.

Meel, D. E. Honor students' calculus understandings: Comparing Calculus and Mathematica and traditional calculus students. In *Research in Collegiate Mathematics Education III,* pp. 163–215, eds. A. H. Schoenfeld, J. Kaput, and E. Dubinsky. Providence, R.I.: American Mathematical Society, 1998.

Meel, D. E. Calculator-available assessments: The why, what, and how. *Educ. Assess.* 4(3):149–75, 1997.

Meel, D. E. A mis-generalization in calculus: Searching for the origins. In *Proceedings of the Nineteenth Annual Meeting of the North American Chapter of the International Group for the Psychology of Mathematics Education,* pp. 23–9, eds. J. A. Dossey, J. O. Swafford, M. Parmantie, and A. E. Dossey. Columbus, Ohio: ERIC Clearinghouse for Science, Mathematics, and Environmental Education, 1997.

Moses, B. E. Beyond problem solving: Problem posing. In *Problem Posing: Reflections and Applications,* eds. S. Brown and M. Walter. Mahwah, N.J.: Lawrence Erlbaum Associates, 1993.

Moses, B. E. IDEAS: Mathematics and music. *Arithmetic Teacher* 40(4):215–25, 1992.

Moses, B. E. Developing spatial thinking in the middle grades. *Arithmetic Teacher* 37(6):59–63, 1990.

Nguyen, T. T., and K. T. Dinh. Characterizations of normal distributions and EDF goodness-of-fit tests. *Metrika* 58:149–57, 2003.

Nguyen, T. T., and K. T. Dinh. A regression characterization of inverse Gaussian distributions and application to EDF goodness-of-fit tests. *Int. J. Math. Math. Sci.* 9:587–92, 2003.

Nguyen, T. T., J. T. Chen, A. K. Gupta, and K. T. Dinh. A proof of the conjecture on positive skewness of generalised inverse Gaussian distributions. *Biometrika* 90:245–50, 2003.

Nguyen, T. T., and K. T. Dinh. Exact EDF goodness-of-fit tests for inverse Gaussian distributions. *Commun. Stat. Simulation Comput.* 32:505–16, 2003.

Seubert, S. M. Semigroups of compressed Toeplitz operators and Nevanlinna-Pick interpolation. *Houston J. Math.,* in press.

Seubert, S. M., and **J. G. Wade.** Frechet differentiability of parameter-dependent analytic semigroups. *J. Math. Anal. Applications* 232:119–37, 1999.

Lesko, J., and **S. M. Seubert.** Cyclicity results for Jordan and compressed Toeplitz operators. *Integral Equations Operator Theory* 31:338–52, 1998.

Cheng, R., and **S. M. Seubert.** Weakly outer polynomials. *Mich. J. Math.* 41:235–46, 1994.

Ivanov, A., and **S. Shpectorov.** The universal non-abelian representation of the Peterson type geometry related to J-4. *J. Algebra* 191:541–67, 1997.

Ivanov, A., D. Pasechnik, and **S. Shpectorov.** Non-abelian embeddings of some sporadic geometries. *J. Algebra* 181:523–57, 1996.

Del Fra, A., A. Pasini, and **S. Shpectorov.** Geometries with bi-affine and bi-linear diagrams. *Eur. J. Combinatorics* 16:439–59, 1995.

Székely, G. J., and M. Rizzo. Mean distance test of Poisson distribution. *Stat. Probability Lett.* 67(3):241–7, 2004.

Bowling Green State University

Bennett, C., A. Glass, and **G. J. Székely.** Fermat's last theorem for rational exponents. *Am. Math. Monthly* 111(4):322–9, 2004.

Székely, G. J., and N. K. Bakrov. Extremal probabilities for Gaussian quadratic forms. *Probability Theory Related Fields* 126:184–202, 2003.

Rao, C. R., and **G. J. Székely.** *Statistics for the 21st Century.* New York: Dekker, 2000.

Filippova, D. V., and **J. G. Wade.** A preconditioner for regularized inverse problems. *SIAM J. Sci. Computation,* in press.

Wade, J. G., and P. S. Vassilevski. A comparison of multilevel methods for total variation regularization. *Elec. Trans. Numerical Anal.* 6:225–70, 1997.

Wade, J. G., and C. R. Vogel. Analysis of costate discretizations in parameter estimation for linear evolution equations. *SIAM J. Control Optimization* 33(1):227–54, 1995.

Bennett, C. D., and **C. L. Zirbel.** Discrete velocity fields with explicitly computable Lagrangian law. *J. Stat. Phys.* 111:681–701, 2003.

Woyczynski, W. A., and **C. L. Zirbel.** Rotation of particles in polarized Brownian flows. *Stochastics Dynamics* 2:109–29, 2002.

Jordan, R., B. Turkington, and **C. L. Zirbel.** A mean-field statistical theory for the nonlinear Schrodinger equation. *Physica D* 137:353–78, 2000.

BROWN UNIVERSITY

Division of Applied Mathematics

Programs of Study

The Division of Applied Mathematics offers graduate programs leading to the Ph.D. and Sc.M. degrees.

The emphasis of the Ph.D. program is on both thesis research and obtaining a solid foundation for future work. Course programs are designed to suit each individual's needs. Admission to Ph.D. candidacy is based on a preliminary examination designed individually for each student in light of his or her interests. Research interests of the faculty can be gauged from the list on the reverse of this page and include partial differential equations and dynamical systems, stochastic systems (including stochastic control), probability and statistics, numerical analysis and scientific computation, continuum and fluid mechanics, computer vision, image reconstruction and speech recognition, pattern theory, and computational neuroscience and computational biology. A wide spectrum of graduate courses is offered, reflecting the broad interests of the faculty in the different areas of applied mathematics. Relevant courses are also offered by the Departments of Mathematics, Physics, Computer Sciences, Economics, Geological Sciences, Linguistics, and Psychology and the Divisions of Engineering and of Biology and Medicine.

The Sc.M. program does not require a thesis, and students with sound preparation usually complete it in one year.

Research Facilities

The University's science library houses an outstanding collection in mathematics and its applications. Parallel computing platforms with a total of roughly 500 CPUs, including several Linux clusters and an IBM SP, are housed in the Center for Computation and Visualization. The center also maintains an immersive virtual reality display, a cave, which is used for scientific visualization and graphics research. Data storage facilities include 40 terabytes of RAID disk and a 600-terabyte tape library. These core computing and data facilities are integrated with the Division's desktop networks and the campus backbone network through gigabit Ethernet routing switches. The Division desktop environment is a mix of Intel-architecture PCs and workstations running the Windows and Linux operating systems. The University maintains a broad range of campus-licensed software, including compilers, numerical libraries, mathematical problem-solving environments, visualization software, statistical packages, and productivity software.

Financial Aid

Fellowships, scholarships, and research and teaching assistantships, which cover tuition and living expenses, are available for qualified full-time graduate students. Exceptional candidates receive guaranteed support for four years, provided that they make satisfactory progress toward the degree. Summer support can usually be arranged.

Cost of Study

Tuition fees for full-time students were $32,264 per year in 2005–06. Teaching assistants and research assistants are not charged for tuition.

Living and Housing Costs

The cost of living in Providence is somewhat lower than the national average. Housing for graduate students in Miller Hall was available at $5514 for the 2005–06 academic year. Numerous off-campus apartments are available for students to rent.

Student Group

Brown University has approximately 5,600 undergraduates and 1,300 graduate students. The Division of Applied Mathematics has 57 full-time graduate students, of whom 51 receive financial support from the University. A number of other students hold outside fellowships.

Location

Brown University is located on a hill overlooking Providence, the capital of Rhode Island and one of America's oldest cities. The proximity of Providence to the excellent beaches and ocean ports of Rhode Island and Massachusetts provides considerable recreational opportunities. Numerous nearby ski facilities are available for winter recreation. The libraries, theaters, museums, and historic sites in Providence and Newport offer an abundance of cultural resources. In addition, Providence is only an hour from Boston and 4 hours from New York City by auto or train.

The University and The Division

Brown University was founded in 1764 in Warren, Rhode Island, as Rhode Island College. It is the seventh-oldest college in America and the third-oldest in New England. In 1770, the College was moved to College Hill, high above the city of Providence, where it has remained ever since. The name was changed to Brown University in 1804 in honor of Nicholas Brown, son of one of the founders of the College. The University awarded its first Doctor of Philosophy degree in 1889. The University attracts many distinguished lecturers both in the sciences and in the arts. Brown is a member of the Ivy League and participates in all intercollegiate sports.

Brown has the oldest tradition and one of the strongest programs in applied mathematics of all universities in the country. Based on a wartime program instituted in 1942, the Division of Applied Mathematics at Brown was established in 1946 as a center of graduate education and fundamental research. It includes several research centers, has cooperative programs with many other universities, and attracts many scientific visitors.

Applying

The preferred method of application for admission is via the electronic application on the Internet at http://apply.embark.com/grad/brown/.

Correspondence and Information

Professor Yan Guo, Chair
Division of Applied Mathematics
Brown University
Providence, Rhode Island 02912
Web site: http://www.dam.brown.edu

Brown University

THE FACULTY AND THEIR RESEARCH

Elie Bienenstock, Associate Professor of Applied Mathematics and Neuroscience; Ph.D., Brown. Theoretical neuroscience, artificial vision.
Frederic E. Bisshopp, Emeritus Professor of Applied Mathematics; Ph.D., Chicago. Asymptotics, nonlinear waves, fluid mechanics.
Constantine M. Dafermos, Professor of Applied Mathematics, Alumni-Alumnae University Professor, and Chair, Graduate Program in Applied Mathematics; Ph.D., Johns Hopkins. Continuum mechanics, differential equations.
Philip J. Davis, Emeritus Professor of Applied Mathematics; Ph.D., Harvard. Numerical analysis, approximation theory.
Paul G. Dupuis, Professor of Applied Mathematics and Division Chair; Ph.D., Brown. Stochastic systems and control and probability, numerical methods.
Peter L. Falb, Professor of Applied Mathematics; Ph.D., Harvard. Control and stability theory.
Bernold Fiedler, Professor of Applied Mathematics; Ph.D., Heidleberg. Dynamical systems, ordinary and partial differential equations, dynamics of patterns and application.
Wendell H. Fleming, Emeritus Professor of Applied Mathematics and Mathematics; Ph.D., Wisconsin. Stochastic differential equations, stochastic control theory.
Walter F. Freiberger, Emeritus Professor of Applied Mathematics; Ph.D., Cambridge. Statistics, biostatistics.
Constantine Gatsonis, Professor of Medical Science and Applied Mathematics; Ph.D., Cornell. Bayesian statistical inference, biostatistics, bioinformatics.
Stuart Geman, Professor of Applied Mathematics and James Manning Professor; Ph.D., MIT. Probability and statistics, natural and computer vision.
Basilis Gidas, Professor of Applied Mathematics; Ph.D., Michigan. Computer vision, speech recognition, computational molecular biology, nonparametric statistics.
David Gottlieb, Professor of Applied Mathematics and Ford Foundation Professor; Ph.D., Tel-Aviv. Numerical methods, scientific computation.
Ulf Grenander, Emeritus Professor of Applied Mathematics; Ph.D., Stockholm. Probability and statistics, pattern theory.
Yan Guo, Professor of Applied Mathematics; Ph.D., Brown. Partial differential equations and kinetic theory.
Jan S. Hesthaven, Professor of Applied Mathematics; Ph.D., Denmark Technical. Numerical analysis, scientific computing, computational electromagnetics, fluid dynamics.
Din-Yu Hsieh, Emeritus Professor of Applied Mathematics; Ph.D., Caltech. Fluid mechanics, mathematical physics.
George Em Karniadakis, Professor of Applied Mathematics; Ph.D., MIT. Computational fluid dynamics, scientific computing, stochastic differential equations, microflows and nanoflows.
Harold J. Kushner, Emeritus Professor of Applied Mathematics and Engineering; Ph.D., Wisconsin. Stochastic systems, control and communication theory.
Charles Lawrence, Professor of Applied Mathematics and Director of the Center for Computational Biology; Ph.D., Cornell. Computational molecular biology.
John Mallet-Paret, Professor of Applied Mathematics, George I. Chase Professor of the Physical Sciences, and Director, Lefschetz Center for Dynamical Systems; Ph.D., Minnesota. Differential equations, dynamical systems.
Martin Maxey, Professor of Applied Mathematics and Engineering; Ph.D., Cambridge. Dynamics of two-phase flow, turbulence mixing and fluid mechanics.
Donald E. McClure, Professor of Applied Mathematics; Ph.D., Brown. Pattern analysis, image processing, mathematical statistics.
Govind Menon, Assistant Professor of Applied Mathematics; Ph.D., Brown. Dynamical systems, partial differential equations, material science.
David Mumford, University Professor; Ph.D., Harvard. Pattern theory, biological and computer vision.
Chi-Wang Shu, Professor of Applied Mathematics; Ph.D., UCLA. Numerical analysis, scientific computation, computational physics.
Walter Strauss, Professor of Mathematics and Applied Mathematics and L. Herbert Ballou University Professor; Ph.D., MIT. Nonlinear waves, scattering theory, partial differential equations.
Chau-Hsing Su, Professor of Applied Mathematics; Ph.D., Princeton. Fluid mechanics, water waves, stochastic processes.
Hui Wang, Assistant Professor of Applied Mathematics; Ph.D., Columbia. Stochastic optimization, stochastic analysis, stochastic network, large deviations and importance sampling.

CALIFORNIA STATE UNIVERSITY CHANNEL ISLANDS

Program in Mathematics

Program of Study

The Master of Science in mathematics is an interdisciplinary and innovative program with a flexible schedule and highly qualified faculty. The program is designed to address the global need for people with advanced mathematical, computational, and computer skills throughout industry and the high-tech and educational fields. Students are given a strong background in mathematics and computer hardware and software and the skills to conduct independent applied research or develop independent projects. The program stresses interdisciplinary applications, such as bioinformatics, actuarial sciences, cryptography, security, image recognition, artificial intelligence, and mathematics education. The 32-credit degree program requires three core courses and seminars (11 credits), electives (15 credits), and a thesis (6 credits). Students' specializations depend on the final project or thesis and the electives chosen under the supervision of a mathematics adviser. Graduates find employment in local high-tech, information systems, and computational industries; businesses; educational institutions; the military; and local and federal government. Some students may elect to continue their education in various graduate schools.

Research Facilities

The current library, which is adjacent to the Bell Tower, provides an attractive environment that is conducive to research and collaborative study groups. In addition to the well-stocked shelves of books and periodicals, the library has taken advantage of the most modern networking and information processing techniques available to provide students with comprehensive library services, including Internet access to articles, publications, and databases; loans of materials from other libraries; virtual reference sites; 24-hour access to electronic library services; and more than 17,000 electronic journals. The planned refurbishing of the adjacent courtyard will expand individual and group study areas and provide students with the largest dedicated outdoor study facility on campus, big enough to accommodate twenty tables and eighty chairs. Construction of the John Spoor Broome Digital Teaching Library began in fall 2005; upon completion, it will provide students with a world-class, state-of-the-art facility.

Financial Aid

The University has a Financial Aid Office to assist students in obtaining financial aid and in resolving problems related to the financial aid process. There are many financial aid opportunities for students, including grants, loans, work-study, and scholarships. With the exception of some honorary scholarships, all financial aid is based on financial need. The Financial Aid Office assists students as they pursue their educational goals. The office may be contacted by phone at 805-437-8530 or e-mail at financial.aid@csuci.edu.

Cost of Study

Students pay $375 per credit. There may be additional fees for specific University services.

Living and Housing Costs

Anacapa Village, a student housing complex, opened in August 2004. Students have the opportunity to share a four-bedroom suite, which includes a fully equipped kitchen with a dishwasher, a stove/range and an oven, a microwave, and a refrigerator; a furnished living area; and two bathrooms. The complex also features a pool, a spa, two computer rooms, laundry facilities, and a furnished common area with a full kitchen, a 60-inch plasma-screen TV, and an entertainment center with a VCR/DVD player. Phase two of student housing is being planned for fall 2007 occupancy. Room and board average $9800 per year.

Student Group

The program serves college graduates holding computational degrees, professionals working in local industries, teachers, and military personnel.

Location

Minutes from the Pacific Ocean, California State University (CSU) Channel Islands sits midway between Santa Barbara and Los Angeles on a picturesque 670-acre campus at the base of the Santa Monica Mountains. On a clear day, some of the eight islands of the Channel Islands National Park can be seen lying just off the southern California coastline from the University. The University's name alludes to these islands, rather than to a single city or county, in order to represent an entire thriving economic region that features major agricultural and biotechnology firms as well as a number of nonprofit agencies and organizations. The area is accessible by air, train, car, and public transportation. There are shuttles available from both the Los Angeles and Burbank Airports, and Amtrak has stations in Camarillo and Oxnard. With the campus bordered by agricultural fields, there are many roadside produce stands with fruits and vegetables picked the same day. Local beaches, deep-sea fishing, and harbor cruises provide water-related entertainment. The Santa Monica Mountains provide miles of hiking and biking trails. Outlet shopping is available at the Camarillo Premium Outlet and The Oaks, a large retail mall in the area. Restaurants in neighboring communities offer a wide assortment of international tastes for a range of budgets.

The University

California State University Channel Islands, Ventura County's first four-year public university, opened in 2002. While the University benefits from the strengths of the CSU system—the largest public education system in the nation—it is also able to focus on individual students, many of whom see themselves as their professors do, as pioneers blazing a trail. Student-centered learning embodies a University philosophy that emphasizes meeting the needs of each student. Courses are taught by an outstanding, world-class faculty, which includes 60 full-time professors and 145 lecturers.

Applying

Applicants whose bachelor's degree is in a field other than mathematics may be required to complete some undergraduate mathematics courses. Students must submit the completed application, the $55 application fee, and all official transcripts.

Correspondence and Information

Dr. Cindy Wyels, Associate Professor of Mathematics
California State University Channel Islands
Camarillo, California 93012
Phone: 805-437-2748
E-mail: msmath@csuci.edu
Web site: http://www.csuci.edu/exed/msmath.htm

California State University Channel Islands

THE FACULTY AND THEIR RESEARCH

Aemiro Beyene, Lecturer; M.S., California State, Northridge, 2004. Applied math, numerical analysis.
Nick Bosco, Lecturer; Ph.D. Instrumentation.
Jerry Clifford, Lecturer; Ph.D. Physics, astronomy.
Geoff Dougherty, Professor and Physics Coordinator; Ph.D., Keele (England), 1979. Medical imaging, image analysis.
Jesse Elliott, Assistant Professor and Assessment Coordinator; Ph.D., Berkeley, 2003. Number theory, commutative algebra.
Nathaniel Emerson, Visiting Assistant Professor; Ph.D., UCLA, 2001. Dynamical systems, Julia sets.
Jorge Garcia, Assistant Professor and Developmental Program Coordinator; Ph.D., Wisconsin–Madison, 2002. Large deviations, stochastic integrals.
Marguerite George, Lecturer; Ph.D. candidate, Arizona State. Applied mathematics, math education.
Ivona Grzegorczyk, Associate Professor and Chair; Ph.D., Berkeley, 2003. Algebraic geometry, moduli spaces.
Kurt Kloplstein, Lecturer; M.S., California, Santa Barbara. Applied math.
Mohamed Nouh, Visiting Assistant Professor; Ph.D., Provence (France), 2000. Knot theory and low-dimensional topology.
Nathanael Reid, Lecturer; M.S., Colorado at Boulder, 2004. Real analysis.
Ron Rieger, Lecturer; M.S., UCLA. Applied mathematics, computer systems.
Roger Roybal, Lecturer and Director of the Tutoring Center; Ph.D., California, Santa Barbara, 2005. Operator theory, functional analysis.
James Sayre, Visiting Professor; Ph.D., UCLA, 1977. Biostatistics.
Morgan Sherman, Lecturer; Ph.D., Columbia, 2005. Algebraic geometry.
Dennis Slivinski, Lecturer; Ph.D., Vanderbilt, 1974. Philosophy, logic.
Tabitha Swan-Wood, Lecturer; Ph.D., California, Santa Barbara, 2006. Solid-state physics.
Steven Thomassin, Lecturer; M.S., California State, Northridge, 1972. Mathematics education, philosophy.
Matthew Wiers, Lecturer; M.S., Ohio State, 1988. Statistics, generalized linear models.
Greg Wood, Assistant Professor; Ph.D., California, Riverside, 2000. Computational biophysics.
Cindy Wyels, Associate Professor and Graduate Program Director; Ph.D., California, Santa Barbara, 1994. Combinatorics, linear algebra, graph theory.
Peter Yi, Lecturer; Ph.D., UCLA, 2003. Algebraic topology, Nielsen theory.

Carnegie Mellon

CARNEGIE MELLON UNIVERSITY

Department of Statistics

Programs of Study

Statisticians apply rigorous thinking and modern computational methods to help scientists, engineers, computer scientists, and policymakers draw reliable inferences from quantitative information. The program at Carnegie Mellon prepares students for such work by providing them with collaborative experience while they master technical skills based on a solid conceptual foundation. The faculty members are all very active in research and professional endeavors yet put a high priority on graduate training. The moderate size of the department and its congenial and supportive environment foster close working relationships between the faculty and students. The outstanding success of the department's graduates, when they take positions in industry, government, and academic institutions, may be attributed to their unusual abilities, the state-of-the-art training given to them, and the dedication they develop during their studies.

In pursuing graduate degrees, students follow programs that may be tailored to suit individual interests. The master's degree program trains students in applied statistics by imparting knowledge of the theory and practice of statistics. Requirements are satisfactory completion of course work and a written comprehensive examination. There is no thesis requirement. Students complete the program in 1, 1½, or 2 years, depending on their previous preparation.

The Ph.D. program is structured to prepare students for careers in university teaching and research and for industrial and government positions that involve consulting and research in new statistical methods. Doctoral candidates first complete the requirements for the M.S. in statistics. They then typically complete another year of courses in probability and statistics. A written Ph.D. comprehensive examination and an oral thesis proposal presentation and defense are required. Proficiency in the use of the computer is required. There are no foreign language requirements.

The department also offers two cross-disciplinary Ph.D. programs. The first leads to a Ph.D. in statistics and public policy and is sponsored jointly with the H. John Heinz III School of Public Policy and Management. The second is a joint program with the Center for Automated Learning and Discovery. Students can obtain additional information by visiting the Web site.

Research Facilities

The computational resources available to students at Carnegie Mellon are unsurpassed and are a major strength of the program. The Department of Statistics operates its own computer facilities, which provide students with experience using advanced graphics workstations. The facilities consist of sixty Linux workstations of various compatible models, about ten PCs, nine monochrome laser printers, and two color laser printers. The department also maintains a sixty-four-processor Beowulf Linux cluster, a high-performance parallel computer. The workstations are interconnected by a departmental Ethernet network, which, in turn, is connected to University and worldwide networks. The department also has a graphics laboratory with equipment for producing computer-animated videotapes and for digitizing video images.

Financial Aid

The department attempts to provide financial aid for all of its students, both master's and Ph.D. candidates. Tuition scholarships are usually granted in conjunction with graduate assistantships, which currently offer a stipend of $13,500 for nine months in return for duties as teaching or research assistants. Students who receive both tuition scholarships and graduate assistantships are expected to maintain a full course load and devote effort primarily to their studies and assigned duties. These duties require, on average, no more than 12 hours per week. Exceptionally well qualified candidates may qualify for a fellowship that pays tuition and a stipend and requires no assistantship duties.

Cost of Study

The tuition fee for full-time graduate students in 2005–06 was $30,000 per academic year.

Living and Housing Costs

Pittsburgh has attractive, reasonably priced neighborhoods where students attending Carnegie Mellon University can live comfortably.

Student Group

Carnegie Mellon University has 4,823 undergraduate and 2,809 graduate students. The teaching faculty numbers approximately 620. During 2005–06, there were 49 full-time students in the graduate program; 28 were working toward a Ph.D. degree. Roughly 33 percent of the statistics graduate students are U.S. citizens and one half are women. Graduate students in statistics have diverse backgrounds, with typical preparation being an undergraduate program in mathematics or in engineering, science, economics, or management. All had outstanding undergraduate records, and many have won nationally competitive fellowships.

Location

Located in a metropolitan area of more than 2 million people, Pittsburgh is the headquarters of many of the nation's largest corporations. There is an unusually large concentration of research laboratories in the area. Carnegie Mellon is located in Oakland, the cultural center of the city. The campus is within walking distance of museums and libraries and is close to the many cultural and sports activities of the city.

The University

One of the leading universities in the country, Carnegie Mellon has long been devoted to liberal professional education. Five colleges—the Carnegie Institute of Technology, the College of Fine Arts, the College of Humanities and Social Sciences, the Mellon College of Science, and the School of Computer Science—offer both undergraduate and graduate programs. The Graduate School of Industrial Administration and the H. John Heinz III School of Public Policy and Management offer graduate programs only.

Applying

The application deadline is January 1, and students are encouraged to apply even earlier, if possible. A course in probability and statistics at the level of DeGroot's *Probability and Statistics* is highly desirable, but excellence and promise always balance a lack of formal preparation. The General Test of the Graduate Record Examinations is required of all applicants. International applicants are also required to take the TOEFL and the Test of Spoken English and should further document their ability to speak English, if possible.

Correspondence and Information

Department of Statistics
Carnegie Mellon University
Pittsburgh, Pennsylvania 15213-3890

Phone: 412-268-8588
Fax: 412-268-7828
E-mail: admissions@stat.cmu.edu
Web site: http://www.stat.cmu.edu/www/cmu-stats/GSS/

Carnegie Mellon University

THE FACULTY AND THEIR RESEARCH

Anthony Brockwell, Assistant Professor of Statistics; Ph.D., Melbourne, 1998. Stochastic processes; control theory; time-series analysis, with applications in neuroscience and finance; Monte Carlo methods.

Bernie Devlin, Adjunct Senior Research Scientist (primary appointment with the University of Pittsburgh School of Medicine); Ph.D., Penn State, 1986. Statistical genetics, genetic epidemiology, genomics.

George T. Duncan, Professor of Statistics, H. John Heinz III School of Public Policy and Management (primary appointment); Ph.D., Minnesota, 1970. Confidentiality of databases, mediation and negotiation, Bayesian decision making.

William F. Eddy, Professor of Statistics; Ph.D., Yale, 1976. Neuroimaging, data mining, visualization, proteomics, image processing, magnetoencephalography (MEG).

Stephen E. Fienberg, Maurice Falk University Professor of Statistics and Social Science; Ph.D., Harvard, 1968. Categorical data, data mining, confidentiality and disclosure limitation, federal statistics, forensic science, multivariate data analysis, sampling and the census, statistical inference.

Christopher R. Genovese, Professor of Statistics; Ph.D., Berkeley, 1994. Astrostatistics, nonparametric inference, functional magnetic resonance imaging, inference from spatio-temporal processes, model selection, statistical inverse problems.

Joel B. Greenhouse, Professor of Statistics; Ph.D., Michigan, 1982. General methodology, biostatistics, applied Bayesian methods.

Brian W. Junker, Professor of Statistics; Ph.D., Illinois, 1988. Mixture and hierarchical models for multivariate discrete measures; nonparametric and semiparametric inference for latent variables; applications in education, psychology, the social sciences, and biostatistics.

Joseph B. Kadane, Leonard J. Savage University Professor of Statistics and Social Science (joint appointment with Department of Social and Decision Sciences and with the Graduate School of Industrial Administration); Ph.D., Stanford, 1966. Statistical inference, econometrics, statistical methods in social sciences, sequential problems, statistics and the law, clinical trials.

Robert E. Kass, Professor of Statistics; Ph.D., Chicago, 1980. Bayesian inference, statistical methods in neuroscience.

Ann B. Lee, Assistant Professor of Statistics; Ph.D., Brown, 2002. Statistical learning, high-dimensional data analysis, multiscale geometric methods and statistical models in pattern analysis and computer vision.

John P. Lehoczky, Thomas Lord Professor of Statistics and Mathematics and Dean, College of Humanities and Social Sciences; Ph.D., Stanford, 1969. Stochastic processes, with applications in real-time computer systems; computational finance; biostatistics.

Kathryn Roeder, Professor of Statistics; Ph.D., Penn State, 1988. Statistical models in genetics and molecular biology, mixture models, semiparametric inference.

Mark J. Schervish, Professor of Statistics and Department Head; Ph.D., Illinois at Urbana-Champaign, 1979. Statistical computing, foundations of statistics, multivariate analysis, statistical methods in engineering, environmental statistics.

Teddy Seidenfeld, Herbert A. Simon Professor of Philosophy and Statistics (primary appointment in Department of Philosophy); Ph.D., Columbia, 1976. Foundations of statistical inference and decision theory.

Howard Seltman, Research Scientist in Statistics; M.D., Medical College of Pennsylvania, 1979; Ph.D., Carnegie Mellon, 1999. Psychiatric studies, biological modeling, genetics, Bayesian methods.

Valérie Ventura, Research Scientist in Statistics; D.Phil., Oxford (England), 1997. Bootstrap methods, efficient simulations, statistics in cognitive neuroscience.

Isabella Verdinelli, Professor in Residence; Ph.D., Carnegie Mellon, 1996. Bayesian design of experiments, multiple testing, tissue engineering, statistical models in engineering.

Pantelis K. Vlachos, Associate Teaching Professor of Statistics; Ph.D., Connecticut, 1996. Bayesian inference, biostatistics, multivariate analysis, text mining.

Larry A. Wasserman, Professor of Statistics; Ph.D., Toronto, 1988. Nonparametric inference, astrophysics, causality, bioinformatics.

SELECTED PUBLICATIONS

Brockwell, A., and **J. B. Kadane.** Identification of regeneration times in MCMC simulation, with application to adaptive schemes. *J. Stat. Comp. Graph.* 14:436–58, 2005.

Brockwell, A.E., A. Rojas, and **R. Kass.** Recursive Bayesian decoding of motor cortical signals by particle filtering. *J. Neurophysiol.* 91:1899–907, 2004.

Brockwell, A. E., N. H. Chan, and P. K. Lee. A class of models for aggregated traffic volume time series. *J. Royal Stat. Soc., Ser. C* 52(4):417–30, 2003.

Brockwell, A. E., and **J. B. Kadane.** A gridding method for Bayesian sequential decision problems. *J. Comput. Graph. Stat.* 12(3):566–84, 2003.

Brockwell, A. E., and P. J. Brockwell. A class of non-embeddable ARMA processes. *J. Time Ser. Analysis* 20(5):483–6, 1999.

Tzeng, J.-Y., et al. **(B. Devlin, K. Roeder,** and **L. Wasserman).** Outlier detection and false discovery rates for whole-genome DNA matching. *J. Am. Stat. Assoc.*, in press.

Devlin, B., K. Roeder, and **L. Wasserman.** Genomic control for association studies: A semiparametric test to detect excess-haplotype sharing. *Biostatistics* 1:369–87, 2002.

Devlin, B., K. Roeder, and **L. Wasserman.** Genomic control, a new approach to genetic-based association studies. *Theor. Pop. Biol.* 60:156–66, 2001.

Devlin, B., and **K. Roeder.** Genomic control for association studies. *Biometrics* 55:997–1004, 1999.

Duncan, G., and S. Roehrig. Mediating the tension between information, privacy, and information access: The role of digital government. In *Public Information Technology: Policy and Management Issues,* pp. 94–129, ed. G. D. Garson. Hershey and London: Idea Group Publishing, 2003.

Duncan, G., et al. **(S. E. Fienberg).** Disclosure limitation methods and information loss for tabular data. In *Confidentiality, Disclosure, and Data Access,* pp. 135–66, eds. P. Doyle et al. Amsterdam: North Holland, 2001.

Duncan, G., and S. Mukherjee. Optimal disclosure limitation strategy in statistical databases: Deterring tracker attacks through additive noise. *J. Am. Stat. Assoc.* 95:720–8, 2000.

Duncan, G., et al. Disclosure detection in multivariate categorical databases: Auditing confidentiality protection through two new matrix operators. *Manage. Sci.* 45, 1999.

Rosano, C., et al. **(W. F. Eddy).** Pursuit and saccadic eye movement subregions in human frontal eye field: A high resolution fMRI investigation. *Cereb. Cortex* 12(2):107–15, 2002.

McNamee, R. L., and **W. F. Eddy.** Visual analysis of variance: A tool for quantitative assessment of fMRI data processing and analysis. *Magn. Reson. Med.* 46:1202–8, 2001.

Eddy, W. F., and T. K. Young. Optimizing the resampling of registered images. In *Handbook of Medical Image Processing, Processing and Analysis,* pp. 603–12, ed. I. N. Bankman. Academic Press, 2000.

Carpenter, P. A., et al. **(W. F. Eddy).** Time course of fMRI-activation in language and spatial networks during sentence comprehension. *NeuroImage* 10:216–24, 1999.

Eddy, W. F., et al. **(C. R. Genovese** and **N. Lazar).** The challenge of functional magnetic resonance imaging. *J. Comput. Graph. Stat.* 8(3):545–58, 1999.

Fienberg, S. E. When did Bayesian inference become "Bayesian"? *Bayesian Anal.* 1:1–40, 2006.

Erosheva, E. A., **S. E. Fienberg,** and J. Lafferty. Mixed-membership models of scientific publications. *Proc. Natl. Acad. Sci. U.S.A.* 97(22):11885–92, 2004.

Fienberg, S. E., et al., eds. *The Polygraph and Lie Detection.* Washington, D.C.: National Academy Press, 2003.

Goldenberg, A., G. Shmueli, R. Caruana, and **S. E. Fienberg.** Early statistical detection of anthrax outbreaks by tracking over-the-counter medication sales. *Proc. Natl. Acad. Sci. U.S.A.* 99:5237–40, 2002.

Anderson, M., and **S. E. Fienberg.** *Who Counts? The Politics of Census-Taking in Contemporary America.* New York: Russell Sage Foundation, 2001.

Dobra, A., and **S. E. Fienberg.** Bounds for cell entries in contingency tables given marginal totals and decomposable graphs. *Proc. Natl. Acad. Sci. U.S.A.* 97(22):11885–92, 2000.

Pacifico, M. P., **C. Genovese, I. Verdinelli,** and **L. Wasserman.** False discovery control for random fields. *J. Am. Stat. Assoc.* 99:1002–14, 2004.

Merriam, E. P., **C. R. Genovese,** and C. L. Colby. Spatial updating in human parietal cortex. *Neuron* 39:361–73, 2003.

Pacifico, M. P., **C. Genovese, I. Verdinelli,** and **L. Wasserman.** False discovery rates for random fields. Technical Report 771, Department of Statistics, Carnegie Mellon University, 2003.

Genovese, C. R., and **L. Wasserman.** Operating characteristics and extensions of the false discovery rate procedure. *J. Royal Stat. Soc. B* 64:499–518, 2002.

DiMatteo, I., **C. R. Genovese,** and **R. E. Kass.** Bayesian curve-fitting with free-knot splines. *Biometrika* 88:1055–71, 2001.

Genovese, C. R. A Bayesian time-course model for functional magnetic resonance imaging data. *J. Am. Stat. Assoc.* 95:451, 691–703, 2000.

Kaizar, E. E., **J. B. Greenhouse, H. Seltman,** and K. Kelleher. Do antidepressants cause suicidality in children? A Bayesian meta-analysis. *Clinical Trials: J. Soc. Clinical Trials,* in press.

Greenhouse, J. B., and K. J. Kelleher. Thinking outside the (black) box: Antidepressants, suicidality, and research synthesis. *Pediatrics* 116:231–3, 2005.

Greenhouse, J. B., and **H. Seltman.** Using prior distributions to synthesize historical evidence: Comments on the Goodman-Sladky case study of IVIg in Guillan-Barre syndrome. *Clinical Trials: J. Soc. Clinical Trials* 2:311–8, 2005.

Lovett, M., and **J. Greenhouse.** Applying cognitive theory to statistics instruction. *Am. Stat.* 54:196–206, 2000.

Johnson, M. S., and **B. W. Junker.** Using data augmentation and Markov chain Monte Carlo for the estimation of unfolding response models. *J. Educ. Behav. Stat.* 28:195–30, 2003.

Patz, R. J., **B. W. Junker,** M. S. Johnson, and L. T. Mariano. The hierarchical rater model for rated test items and its application to large-scale educational assessment data. *J. Educ. Behav. Stat.* 27:341–84, 2002.

Junker, B. W., and K. Sijtsma. Cognitive assessment models with few assumptions, and connections with nonparametric item response theory. *Appl. Psychol. Meas.* 25:258–72, 2001.

Crane, H. M., **J. B. Kadane,** P. K. Crane, and M. M. Kitahata. Diabetes case identification methods for electronic medical record systems data: Their use in HIV-infected patients. *Curr. HIV Res.* 4:97–106, 2006.

Garthwaite, P., **J. B. Kadane,** and A. O. O'Hagan. Statistical methods for elicitating probability distributions. *J. Am. Stat. Assoc.* 100:470, 680–701, 2005.

Kadane, J. B. Ethical issues in being an expert witness. *Law Probability Risk* 4:21–3, 2005.

Borle, S., et al. **(J. B. Kadane).** The effect of product assortment changes on customer retention. *Marketing Sci.* 24:4, 616–22, 2005.

Larget, B., D. L. Simon, **J. B. Kadane,** and D. Sweet. A Bayesian analysis of metazoan mitochondrial genome arrangements. *Mol. Biol. Evolution* 22:486–95, 2005.

Larget, B., **J. B. Kadane,** and D. L. Simon. A Bayesian approach to the estimation of ancestral genome arrangements. *Mol. Phylogenet. Evolution* 36:214–23, 2005.

Kass, R. E., V. Ventura, and E. N. Brown. Statistical issues in the analysis of neuronal data. *J. Neurophysiol.* 94:8–25, 2005.

Kass, R. E., and **L. Wasserman.** The selection of prior distributions by formal rules. *J. Am. Stat. Assoc.* 91:1343–70, 1996.

Kass, R. E., and A. E. Raftery. Bayes factors. *J. Am. Stat. Assoc.* 90:773–95, 1995.

Coifman, R. R., et al. **(A. B. Lee).** Geometric diffusions as a tool for harmonic analysis and structure definition of data. *Proc. Natl. Acad. Sci. U.S.A.* 102(21):7426–37, 2005.

Carnegie Mellon University

Lee, A. B., K. S. Pedersen, and D. Mumford. The nonlinear statistics of high-contrast patches in natural images. *Int. J. Comp. Vision* 54(1–2):83–103, 2003.

Srivastava, A., **A. B. Lee,** E. P. Simoncelli, and S.-C. Zhu. On advances in statistical modeling of natural images. *J. Math. Imaging Vision* 18:17–33, 2003.

Kruk, L., **J. Lehoczky,** S. Shreve, and S.-N. Yeung. Earliest-deadline-first queue in heavy traffic acrylic networks. *Ann. Appl. Prob.* 14:1306–52, 2004.

Kruk, L., **J. Lehoczky,** S. Shreve, and S.-N. Yeung. Multiple-input heavy traffic real-time queues. *Ann. Appl. Prob.* 13(1):54–99, 2003.

Akesson, F., and **J. P. Lehoczky.** Path generation for quasi-Monte Carlo simulation of mortgage-backed securities. *Manage. Sci.* 46(9):1171–87, 2000.

Lehoczky, J. Simulation methods for option pricing. In *Mathematics of Derivative Securities,* pp. 528–44, eds. M. A. Dempster and S. R. Plisha. Cambridge University Press, 1997.

Roeder, K., R. G. Carroll, and B. G. Lindsay. A nonparametric maximum likelihood approach to case-control studies with errors in covariables. *J. Am. Stat. Assoc.* 91:722–32, 1996.

Schervish, M. J., T. Seidenfeld, and **J. B. Kadane.** How sets of coherent probabilities may serve as models for degrees of incoherence. *J. Uncertainty Fuzziness Knowledge-Based Syst.,* in press.

Schervish, M. J., T. Seidenfeld, and **J. B. Kadane.** Stopping to reflect. *J. Phil.* 51:315–22, 2004.

Schervish, M. J., T. Seidenfeld, J. B. Kadane, and I. Levi. Extensions of expected utility theory and some limitations of pairwise comparisons. *ISIPTA-03 Conference Proceedings,* 2003.

DeGroot, M. H., and **M. J. Schervish.** *Probability and Statistics,* 3rd ed. Addison-Wesley, 2002.

Lockwood, J. R., **M. J. Schervish,** P. Gurian, and M. J. Small. Characterization of forensic occurrence in U.S. drinking water treatment facility source waters. *J. Am. Stat. Assoc.* 96:1184–93, 2001.

Schervish, M. J., T. Seidenfeld, and **J. B. Kadane.** Improper regular conditional distributions. *Ann. Prob.* 29:1612–24, 2001.

Lavine, M., and **M. J. Schervish.** Bayes factors: What they are and what they are not. *Am. Stat.* 53:119–22, 1999.

Barron, A., **M. J. Schervish,** and **L. Wasserman.** The consistency of posterior distributions in nonparametric problems. *Ann. Stat.* 27:536–61, 1999.

Schervish, M. J. P-values: What they are and what they are not. *Am. Stat.* 50:203–6, 1996.

Schervish, M. J. *Theory of Statistics.* New York: Springer-Verlag, 1995.

Seidenfeld, T. Remarks on the theory of conditional probability. In *Statistics—Philosophy, Recent History, and Relations to Science,* eds. V. F. Hendricks, S. A. Pedersen, and K. F. Jorgensen. Kluwer Academic Publishing, in press.

Geisser, S., and **T. Seidenfeld.** Remarks on the Bayesian method of moments. *J. Appl. Stat.* 26:97–101, 1999.

Seidenfeld, T., M. J. Schervish, and **J. B. Kadane.** Non-conglomerability for finite-valued, finitely additive probability. *Sankhya* 60(3):476–91, 1998.

Heron, T., **T. Seidenfeld,** and **L. Wasserman.** Divisive conditioning: Further results on dilation. *Phil. Sci.* 411–4, 1997.

Seidenfeld, T., M. J. Schervish, and **J. B. Kadane.** A representation of partially ordered preferences. *Ann. Stat.* 23:2168–74, 1995.

Seltman, H., B. Devlin, and **K. Roeder.** Evolutionary-based association analysis using haplotype data. *Gent. Epid.* 25:48–59, 2003.

Seltman, H., J. Greenhouse, and **L. Wasserman.** Bayesian model selection: Analysis of a survival model with a surviving fraction. *Stat. Med.* 20(11):1681–91, 2001.

Seltman, H. Hidden Markov models for analysis of biological rhythm data. In *Case Studies in Bayesian Statistics,* vol. 5, pp. 398–406. New York: Springer Verlag, 2001.

Seltman, H., K. Roeder, and **B. Devlin.** TDT meets MHA: Family-based association analysis guided by evolution of haplotypes. *Am. J. Hum. Genet.* 68(5):1250–63, 2001.

Ventura, V. Nonparametric bootstrap recycling. *Stat. Comput.,* in press.

Robins, J. M., A. W. van der Vaart, and **V. Ventura.** The asymptotic distribution of p-values in composite null models. *J. Am. Stat. Assoc.* 62:452, 2000.

Olson, C. R., et al. **(V. Ventura** and **R. E. Kass).** Neuronal activity in macaque supplementary eye field during planning of saccades in response to pattern and spatial cues. *J. Neurophysiol.* 84:1369–84, 2000.

Ventura, V., A. C. Davison, and S. J. Boniface. Statistical inference for the effect of magnetic brain stimulation on a motoneurone. *Appl. Stat.* 47:77–94, 1998.

Weiss, L. E., et al. **(I. Verdinelli).** Bayesian computer-aided experimental design of heterogeneous scaffolds for tissue engineering. *Comput. Aided Des.,* 2004.

Verdinelli, I. Bayesian design for the normal linear model with unknown error variance. *Biometrika* 87:222–7, 2000.

Verdinelli, I., and **L. A. Wasserman.** Bayesian goodness of fit testing using infinite dimensional exponential families. *Ann. Stat.* 26:1215–41, 1998.

Vlachos, P. K., and A. E. Gelfand. On the calibration of Bayesian model choice criteria. *J. Stat. Plan. Inf.* 111:223–34, 2003.

Collins, J., et al. **(P. K. Vlachos).** Detecting collaborations in text comparing the authors' rhetorical language choices in the federalist papers. *Comp. Human.* 38:15–36.

Shen, X., and **L. A. Wasserman.** Rates of convergence of posterior distributions. *Ann. Stat.,* in press.

Wasserman, L. A. Asymptotic inference for mixture models using data dependent priors. *J. Royal Stat. Soc. B* 62:159–80, 2000.

Wasserman, L. A. Asymptotic properties of nonparametric Bayesian procedures. In *Practical Nonparametric and Semiparametric Bayesian Statistics,* eds. D. Dey, P. Muller, and D. Sinha. New York: Springer-Verlag, 1998.

CLEMSON UNIVERSITY

CLEMSON UNIVERSITY

Master of Science in Mathematical Sciences

Program of Study

The Department of Mathematical Sciences offers the M.S. degree, which stresses breadth of training in the mathematical sciences. In addition, students select a concentration area for further course work and research.

The two-year M.S. program is designed to prepare students for immediate employment in the nonacademic sector and also for further graduate study at the doctoral level. The emphasis on breadth of training and computational skills has been a model for other programs nationally. In the first year, students take courses in all the breadth areas. In the second year, students select concentration courses, assisted by their faculty adviser, who also directs their master's project. Overall the M.S. program requires a minimum of 37 credit hours, including the 1-credit project course. The M.S. program produces students with flexible career options for employment in business, industry, and government as well as for subsequent Ph.D. study.

Research Facilities

The department maintains a computer laboratory for the exclusive use of its graduate students. Access is provided to the campus-wide PC-based network as well as to a network of Sun workstations. Specialized mathematical and statistical software is maintained on both of these platforms for student use.

The department is an active member of the NSF-sponsored Center for Advanced Engineering Fibers and Films (CAEFF). Research students have access to the extensive physical laboratories associated with the center. In particular, students have access to the W. M. Keck Visualization/Virtual Reality and Computation Laboratory and the Clemson Computational Minigrid Supercomputing Facility.

Financial Aid

Teaching assistantships carry a stipend of $14,000 (for 10.5 months) and involve instructing or assisting in the instruction of a maximum of 10 semester credit hours per year. Research assistantships, with yearly stipends of $14,000 to $20,000, are available to students with the required qualifications. Outstanding students may also qualify for University Fellowships ($5000–$15,000). SC Graduate Incentive Fellowships are available to graduate students who are members of minority groups; these renewable awards provide $5000 per year for master's students.

Cost of Study

Tuition for 2006–07 is $4643 per semester for in-state students and $9255 per semester for nonresidents. Off-campus rates are $535 per hour for in-state students and $918 per hour for nonresidents. Graduate assistants pay a flat fee of $1079 per semester and $348 per summer session. Graduate fellows pay South Carolina resident fees.

Living and Housing Costs

On-campus housing is available. For information, students should visit http://www.housing.clemson.edu. The cost of living in Clemson is quite low compared to the national average. Students who choose to live off the campus typically spend $300–$400 per month for rent, depending on location, amenities, roommates, etc.

Student Group

The program has approximately 35 students. Forty-six percent are women, all are full-time, and 37 percent are international students.

Student Outcomes

Graduates have followed successful career paths in academic positions, often at smaller liberal arts colleges and universities. Nonacademic employment has included financial institutions, government laboratories, consulting firms, telecommunications, transportation, medical research, and manufacturing. The broad training as well as the emphasis on computational and communication skills has provided a great advantage to students in obtaining employment after graduation.

Location

Clemson is a small, beautiful college town near the Blue Ridge Mountains and Lake Hartwell in upstate South Carolina. The Upstate is one of the country's fastest-growing areas and is the midpoint of the Charlotte-to-Atlanta I-85 corridor, a multistate area along Interstate 85 that runs from metro Atlanta to Richmond, Virginia, and encompasses Charlotte, North Carolina, and North Carolina's Research Triangle. Atlanta and Charlotte are each a 2-hour's drive away. Many financial institutions and other industries have national headquarters for a major presence in the Upstate, including Wachovia, Bank of America, BMW, Bon Secours St. Francis Health System, Bosch North America, Bowater, Charter Communications, Ernst and Young, Fluor Corporation, IBM, Microsoft, Michelin of North America, and many others.

The University

Clemson is classified by the Carnegie Foundation as Doctoral/Research University–Extensive, a category comprising less than 4 percent of all universities in America. The University's mission is to fulfill the covenant between its founder and the people of South Carolina to establish a "high seminary of learning" through its responsibilities of teaching, research, and extended public service. The University has identified eight areas of academic emphasis that create collaborations that, in turn, help fulfill the University's mission.

Applying

The department seeks students with interests in the broad mathematical sciences. Since most of the financial assistance is in the form of teaching assistantships, strong communication skills are required. Applicants may apply on the Web at http://www.grad.clemson.edu/p_apply.html. Applications with a $50 nonrefundable fee should be received no later than five weeks prior to registration. Every required item in support of the application must be on file by that date. Students are advised to contact the department for the deadlines of the program of proposed study.

Correspondence and Information

Dr. K. B. Kulasekera
Graduate Program Coordinator
Department of Mathematical Sciences
O-322 Martin Hall
Box 340975
Clemson University
Clemson, South Carolina 29634-0975
Phone: 864-656-5231
Fax: 864-656-5230
E-mail: mthgrad@clemson.edu
Web site: http://virtual.clemson.edu/groups/mathsci/graduate/

Clemson University

THE FACULTY AND THEIR RESEARCH

Warren P. Adams; Professor; Ph.D., Virginia Tech. Industrial engineering and operations research.
James R. Brannan, Professor; Ph.D., Rensselaer. Math science.
Joel V. Brawley Jr., Alumni Professor; Ph.D., North Carolina State. Math science.
Neil J. Calkin, Associate Professor; Ph.D., Waterloo. Mathematics.
Christopher L. Cox, Professor; Ph.D., Carnegie-Mellon. Math science.
Perino M. Dearing Jr., Professor; Ph.D., Florida. Operations research.
Vincent J. Ervin, Professor; Ph.D., Georgia Tech. Math science.
Robert E. Fennell, Professor; Ph.D., Iowa. Math science.
Colin M. Gallagher, Assistant Professor; Ph.D., California, Santa Barbara. Statistics.
Susan L. Ganter, Associate Professor; Ph.D., California, Santa Barbara. Education administration.
Xuhong Gao, Professor; Ph.D., Waterloo. Combinatorics and optimization.
Joan A. Hoffacker, Assistant Professor; Ph.D., Nebraska–Lincoln. Mathematics.
Paul D. Hyden, Assistant Professor; Ph.D., Cornell. Operations research and industrial engineering.
Kevin L. James, Assistant Professor; Ph.D., Georgia. Mathematics.
Robert E. Jamison, Professor; Ph.D., Washington (Seattle). Math science.
James P. Jarvis, Professor; Ph.D., MIT. Operations research.
Eleanor W. Jenkins, Assistant Professor; Ph.D., North Carolina State. Mathematics.
Jennifer D. Key, Professor; Ph.D., London. Mathematics.
Taufiquar Rahman Khan, Assistant Professor; Ph.D., USC. Applied mathematics.
Peter C. Kiessler, Associate Professor; Ph.D., Virginia Tech. Industrial engineering and operations research.
Michael M. Kostreva, Professor; Ph.D., Rensselaer. Math science.
Karunarathna B. Kulasekera, Professor; Ph.D., Nebraska–Lincoln. Statistics.
Renu C. Laskar, Professor; Ph.D., Illinois. Math science.
Hyesuk K. Lee, Associate Professor; Ph.D., Virginia Tech. Mathematics.
Robert B. Lund, Professor; Ph.D., North Carolina at Chapel Hill. Statistics.
Hiren Maharaj, Assistant Professor; Ph.D., Penn State. Mathematics.
Gretchen L. Matthews, Assistant Professor; Ph.D., LSU. Mathematics.
William F. Moss, Professor; Ph.D., Delaware. Math science.
Chanseok Park, Assistant Professor; Ph.D., Penn State. Statistics.
James K. Peterson, Associate Professor; Ph.D., Colorado. Math science.
James A. Reneke, Professor; Ph.D., North Carolina at Chapel Hill; Math science.
Charles B. Russell, Associate Professor; Ph.D., Florida State. Statistics.
Matthew J. Saltzman, Associate Professor; Ph.D., Carnegie-Mellon. Industrial administration (operations research).
Herman F. Senter, Associate Professor; Ph.D., North Carolina State. Math science.
Douglas R. Shier, Professor; Ph.D., London School of Economics. Operations research.
Robert L. Taylor, Professor and Department Chair; Ph.D., Florida State. Statistics.
Daniel D. Warner, Professor; Ph.D., California, San Diego. Mathematics.
Margaret Maria Wiecek, Professor; Ph.D., University of Mining and Metallurgy (Poland). Math science.
Calvin L. Williams, Associate Professor; Ph.D., Medical University of South Carolina. Biometry.

CLEMSON UNIVERSITY

CLEMSON UNIVERSITY

Ph.D. in Mathematical Sciences

Program of Study

The Department of Mathematical Sciences offers the Ph.D. degree, which emphasizes breadth of training as well as in-depth course work and dissertation research in a particular area (algebra, applied analysis, operations research, computational mathematics, or statistics). Within two semesters, candidates take the preliminary examination, consisting of tests in three areas selected from six possible areas. A comprehensive examination is administered at a later date to assess the candidate's readiness to perform independent research. Including master's study, the Ph.D. takes an average of six years and requires a total of 60 credit hours of courses, including breadth requirements (two courses from each of five areas). Graduates have gone on to successful academic and nonacademic careers.

Research Facilities

The department maintains a computer laboratory for the exclusive use of its graduate students. Access is provided to the campus-wide PC-based network as well as to a network of Sun workstations. Specialized mathematical and statistical software is maintained on both of these platforms for student use. Most Ph.D. students have direct access to these networks via workstations situated in their (2-person) offices.

The department is an active member of the NSF-sponsored Center for Advanced Engineering Fibers and Films (CAEFF). Research students have access to the extensive physical laboratories associated with the center. In particular, students have access to the W. M. Keck Visualization/Virtual Reality and Computation Laboratory and the Clemson Computational Minigrid Supercomputing Facility.

Financial Aid

Teaching assistantships carry a stipend of $14,000 (for 10.5 months) and involve instructing or assisting in the instruction of a maximum of 10 semester credit hours per year. Research assistantships, with yearly stipends of $14,000 to $20,000, are available to students with the required qualifications. Outstanding students may also qualify for University Fellowships ($5000–$15,000). The College of Engineering and Science also offers the Dean's Scholars Program, which provides supplementary three-year awards to exceptional Ph.D. students. SC Graduate Incentive Fellowships are available to graduate students who are members of minority groups; these renewable awards provide $10,000 per year for doctoral students.

Cost of Study

Tuition for 2006–07 is $4643 per semester for in-state students and $9255 per semester for nonresidents. Off-campus rates are $535 per hour for in-state students and $918 per hour for nonresidents. Graduate assistants pay a flat fee of $1079 per semester and $348 per summer session. Graduate fellows pay South Carolina resident fees.

Living and Housing Costs

On-campus housing is available. For information, students should visit http://www.housing.clemson.edu. The cost of living in Clemson is quite low compared to the national average. Students who choose to live off the campus typically spend $300–$400 per month for rent, depending on location, amenities, roommates, etc.

Student Group

The program has approximately 42 students. Twenty-six percent are women, 93 percent attend full-time, and 43 percent are international students.

Student Outcomes

Graduates have followed successful career paths in academic positions, often at smaller liberal arts colleges and universities. Nonacademic employment has included financial institutions, government laboratories, consulting firms, telecommunications, transportation, medical research, and manufacturing. The broad training as well as the emphasis on computational and communication skills has provided a great advantage to students in obtaining employment after graduation.

Location

Clemson is a small, beautiful college town near the Blue Ridge Mountains and Lake Hartwell in upstate South Carolina. The Upstate is one of the country's fastest-growing areas and is the midpoint of the Charlotte-to-Atlanta I-85 corridor, a multistate area along Interstate 85 that runs from metro Atlanta to Richmond, Virginia, and encompasses Charlotte, North Carolina, and North Carolina's Research Triangle. Atlanta and Charlotte are each a 2-hour's drive away. Many financial institutions and other industries have national headquarters for a major presence in the Upstate, including Wachovia, Bank of America, BMW, Bon Secours St. Francis Health System, Bosch North America, Bowater, Charter Communications, Ernst and Young, Fluor Corporation, IBM, Microsoft, Michelin of North America, and many others.

The University

Clemson is classified by the Carnegie Foundation as Doctoral/Research University–Extensive, a category comprising less than 4 percent of all universities in America. The University's mission is to fulfill the covenant between its founder and the people of South Carolina to establish a "high seminary of learning" through its responsibilities of teaching, research, and extended public service. The University has identified eight areas of academic emphasis that create collaborations that, in turn, help fulfill the University's mission.

Applying

The department seeks students with interests in the broad mathematical sciences. Since most of the financial assistance is in the form of teaching assistantships, strong communication skills are required. Applicants may apply on the Web at http://www.grad.clemson.edu/p_apply.html. Applications with a $50 nonrefundable fee should be received no later than five weeks prior to registration. Every required item in support of the application must be on file by that date. Students are advised to contact the department for the deadlines of the program of proposed study.

Correspondence and Information

Dr. K. B. Kulasekera
Graduate Program Coordinator
Department of Mathematical Sciences
O-322 Martin Hall
Box 340975
Clemson University
Clemson, South Carolina 29634-0975
Phone: 864-656-5231
Fax: 864-656-5230
E-mail: mthgrad@clemson.edu
Web site: http://virtual.clemson.edu/groups/mathsci/graduate/

Clemson University

THE FACULTY AND THEIR RESEARCH

Warren P. Adams; Professor; Ph.D., Virginia Tech. Industrial engineering and operations research.
James R. Brannan, Professor; Ph.D., Rensselaer. Math science.
Joel V. Brawley Jr., Alumni Professor; Ph.D., North Carolina State. Math science.
Neil J. Calkin, Associate Professor; Ph.D., Waterloo. Mathematics.
Christopher L. Cox, Professor; Ph.D., Carnegie-Mellon. Math science.
Perino M. Dearing Jr., Professor; Ph.D., Florida. Operations research.
Vincent J. Ervin, Professor; Ph.D., Georgia Tech. Math science.
Robert E. Fennell, Professor; Ph.D., Iowa. Math science.
Colin M. Gallagher, Assistant Professor; Ph.D., California, Santa Barbara. Statistics.
Susan L. Ganter, Associate Professor; Ph.D., California, Santa Barbara. Education administration.
Xuhong Gao, Professor; Ph.D., Waterloo. Combinatorics and optimization.
Joan A. Hoffacker, Assistant Professor; Ph.D., Nebraska–Lincoln. Mathematics.
Paul D. Hyden, Assistant Professor; Ph.D., Cornell. Operations research and industrial engineering.
Kevin L. James, Assistant Professor; Ph.D., Georgia. Mathematics.
Robert E. Jamison, Professor; Ph.D., Washington (Seattle). Math science.
James P. Jarvis, Professor; Ph.D., MIT. Operations research.
Eleanor W. Jenkins, Assistant Professor; Ph.D., North Carolina State. Mathematics.
Jennifer D. Key, Professor; Ph.D., London. Mathematics.
Taufiquar Rahman Khan, Assistant Professor; Ph.D., USC. Applied mathematics.
Peter C. Kiessler, Associate Professor; Ph.D., Virginia Tech. Industrial engineering and operations research.
Michael M. Kostreva, Professor; Ph.D., Rensselaer. Math science.
Karunarathna B. Kulasekera, Professor; Ph.D., Nebraska–Lincoln. Statistics.
Renu C. Laskar, Professor; Ph.D., Illinois. Math science.
Hyesuk K. Lee, Associate Professor; Ph.D., Virginia Tech. Mathematics.
Robert B. Lund, Professor; Ph.D., North Carolina at Chapel Hill. Statistics.
Hiren Maharaj, Assistant Professor; Ph.D., Penn State. Mathematics.
Gretchen L. Matthews, Assistant Professor; Ph.D., LSU. Mathematics.
William F. Moss, Professor; Ph.D., Delaware. Math science.
Chanseok Park, Assistant Professor; Ph.D., Penn State. Statistics.
James K. Peterson, Associate Professor; Ph.D., Colorado. Math science.
James A. Reneke, Professor; Ph.D., North Carolina at Chapel Hill; Math science.
Charles B. Russell, Associate Professor; Ph.D., Florida State. Statistics.
Matthew J. Saltzman, Associate Professor; Ph.D., Carnegie-Mellon. Industrial administration (operations research).
Herman F. Senter, Associate Professor; Ph.D., North Carolina State. Math science.
Douglas R. Shier, Professor; Ph.D., London School of Economics. Operations research.
Robert L. Taylor, Professor and Department Chair; Ph.D., Florida State. Statistics.
Daniel D. Warner, Professor; Ph.D., California, San Diego. Mathematics.
Margaret Maria Wiecek, Professor; Ph.D., University of Mining and Metallurgy (Poland). Math science.
Calvin L. Williams, Associate Professor; Ph.D., Medical University of South Carolina. Biometry.

COLUMBIA UNIVERSITY

Graduate School of Arts and Sciences
Department of Mathematics

Programs of Study

The Department of Mathematics offers programs leading to the degrees of Doctor of Philosophy and Master of Arts.

The Ph.D. program is an intensive course of study designed for the full-time student planning a career in research and teaching at the university level or in basic research in a nonacademic setting. Admission is limited and selective. Applicants should present an undergraduate major in mathematics from a college with strong mathematics offerings. In the first year, students must pass written qualifying examinations in areas chosen from a first-year core curriculum, which offers courses in modern geometry, arithmetic and algebraic geometry, complex analysis, analysis and probability, groups and representations, and algebraic topology. Most of the formal course work is completed in the second year, when an oral examination in two selected topics must be passed. Also required is a reading knowledge of one language, chosen from French, German, and Russian. The third and fourth years are devoted to seminars and the preparation of a dissertation. Students are required to serve as teaching assistants for three years beginning with the second year of study. A number of students are selected for NSF funding and are exempt from one or two years of teaching duties.

The M.A. in mathematics of finance is a ten-course program that can be completed in one year of full-time study or two years of part-time study. All courses are offered in the evening. Six core courses are required, and four can be selected from statistics, economics, and the business school. Graduates of the program work in financial firms and investment banking institutions.

There are allied graduate programs available in mathematical statistics and in computer science.

Research Facilities

The mathematics department is housed in a comfortable building containing an excellent Mathematics Library, computing facilities, graduate student offices, a lounge for tea and conversation, and numerous seminar and lecture rooms.

Financial Aid

The department has a broad fellowship program designed to enable qualified students to achieve the Ph.D. degree in the shortest practicable time. Each student admitted to the Ph.D. program is appointed a fellow in the Department of Mathematics for the duration of his or her doctoral candidacy, up to a total of five years. A fellow receives a stipend of at least $20,000 for the 2006–07 nine-month academic year and is exempt from payment of tuition and medical insurance fees.

A fellow in the Department of Mathematics may hold a fellowship from a source outside Columbia University. When not prohibited by the terms of the outside fellowship, the University supplements the outside stipend to bring it up to the level of the University fellowship. Candidates for admission are urged to apply for fellowships for which they are eligible (e.g., National Science Foundation, New York State Regents).

Cost of Study

All students admitted to the Ph.D. program become fellows in the department and are exempt from most fees.

Living and Housing Costs

Students in the program have managed to live comfortably in the University neighborhood on their fellowship stipends.

Student Group

The Ph.D. program in mathematics has an enrollment of approximately 55 students. Normally, 8 to 12 students enter each year. While students come from all over the world, they have always been socially as well as scientifically cohesive and mutually supportive.

Location

New York City is America's major center of culture. Columbia University's remarkably pleasant and sheltered campus, near the Hudson River and Riverside Park, is situated within 20 minutes of Lincoln Center, Broadway theaters, Greenwich Village, and major museums. Most department members live within a short walk of the University.

The University

Since receiving its charter from King George II in 1754, Columbia University has played an eminent role in American education. In addition to its various faculties and professional schools (such as Engineering, Law, and Medicine), the University has close ties with nearby museums, schools of music and theology, the United Nations, and the city government.

Applying

The application deadline is December 31; however, applicants of unusual merit are considered beyond the application deadline. Applicants who expect to be in the New York vicinity are encouraged to arrange a department visit and interview.

Correspondence and Information

For information on the department and program:
Chairman
Department of Mathematics
Mail Code 4406
Columbia University
New York, New York 10027
Phone: 212-854-4112
E-mail: krichev@math.columbia.edu
Web site: http://www.math.columbia.edu

For applications:
Office of Student Affairs
Graduate School of Arts and Sciences
Mail Code 4304
107 Low Memorial Library
Columbia University
New York, New York 10027
Phone: 212-854-4737

Columbia University

THE FACULTY AND THEIR RESEARCH

Peter M. Bank, Assistant Professor; Ph.D., Berlin, 2000. Mathematical finance.
David A. Bayer, Professor; Ph.D., Harvard, 1982. Algebraic geometry.
Joel Bellaiche, Ritt Assistant Professor; Ph.D., Paris XI (South), 2002. Number theory.
Mirela Ciperiani, Ritt Assistant Professor; Ph.D., Princeton, 2005. Arithmetic algebraic geometry.
Panagiota Daskalopoulos, Professor; Ph.D., Chicago, 1992. Partial differential equations, differential geometry, harmonic analysis.
Aise Johan de Jong, Professor; Ph.D. Nijmegen, 1992. Algebraic geometry.
Robert Friedman, Professor; Ph.D., Harvard, 1981. Algebraic geometry.
Patrick X. Gallagher, Professor; Ph.D., Princeton, 1959. Analytic number theory, group theory.
Dorian Goldfeld, Professor; Ph.D., Columbia, 1969. Number theory.
Brian Greene, Professor; D.Phil., Oxford, 1987. Mathematical physics, string theory.
Richard Hamilton, Professor; Ph.D., Princeton, 1966. Differential geometry.
Zuoliang Hou, Ritt Assistant Professor; Ph.D., MIT, 2003. Algebraic geometry.
Troels Jorgensen, Professor; Cand.Scient., Copenhagen, 1970. Hyperbolic geometry, complex analysis.
Ioannis Karatzas, Professor; Ph.D., Columbia, 1980. Probability, mathematical finance.
Mikhail Khovanov, Associate Professor; Ph.D., Yale, 1997. Low dimensional.
Igor Krichever, Professor; Ph.D., Moscow State, 1972. Integrable systems, algebraic geometry.
Aaron Lauda, Ritt Assistant Professor; Ph.D., Cambridge, 2006. Topology.
Chiu-Chu Melissa Liu, Associate Professor; Ph.D., Harvard, 2002. Differential geometry and algebraic geometry.
Xiaobo Liu, Ritt Assistant Professor; Ph.D., USC, 2005. Low-dimensional topology.
Ciprian Manolescu, Assistant Professor; Ph.D., Harvard, 2004. Gauge theory.
John W. Morgan, Professor and Chair; Ph.D., Rice, 1969. Geometric topology, manifold theory.
Walter Neumann, Professor; Ph.D., Bonn (Germany), 1969. Geometry/topology.
Peter S. Ozsváth, Professor; Ph.D., Princeton, 1994. Gauge theory, low-dimensional topology.
Duong H. Phong, Professor; Ph.D., Princeton, 1977. Analysis.
Henry C. Pinkham, Professor and Dean; Ph.D., Harvard, 1974. Algebraic geometry.
Julius Ross, Ritt Assistant Professor; Ph.D., Imperial College (London), 2003. Geometric analysis, algebraic geometry.
Ovidiu Savin, Associate Professor; Ph.D., Texas at Austin, 2003. Partial differential equations.
Natașa Sesum, Ritt Assistant Professor; Ph.D., MIT, 2004. Complex geometry.
Mihai Sirbu, Ritt Assistant Professor; Ph.D., Carnegie Mellon, 2004. Mathematical finance.
Mikhail Smirnov, Lecturer; Ph.D., Princeton, 1995. Differential and integral geometry.
Michael Thaddeus, Associate Professor; D.Phil., Oxford, 1992. Algebraic geometry.
Kenneth Tignor, Ritt Assistant Professor; Ph.D., UCLA, 2006. Number theory.
Eric Urban, Associate Professor; Ph.D., Paris XI (South), 1995. Number theory.
Cristian Virdol, Ritt Assistant Professor; Ph.D., UCLA, 2005. Number theory.
Mu-Tao Wang, Associate Professor; Ph.D., Harvard, 1998. Differential geometry.
Peter Woit, Lecturer; Ph.D., Princeton, 1985. Mathematical physics, topology.
Shou-Wu Zhang, Professor; Ph.D., Columbia, 1991. Number theory, arithmetic geometry.

DARTMOUTH COLLEGE

Department of Mathematics

Programs of Study

The Dartmouth Ph.D. program in mathematics is designed to develop mathematicians who are highly qualified for both teaching and research at the college or university level or for research in the mathematical sciences in industry or government. Students earn a master's degree as part of becoming a candidate for the Ph.D. degree but should not apply to study only for a master's degree. During the first six terms (eighteen months) of residence, the student develops a strong basic knowledge of algebra, analysis, topology, and a fourth area of mathematics chosen by the student. Areas recently chosen for this fourth area include applied mathematics, combinatorics, geometry, logic, number theory, probability, and statistics. Rather than using traditional qualifying exams, the department requires that 2 faculty members certify that the student knows the material on the departmental syllabus in each of the four areas. This certification may be based on a formal oral exam, course work, informal discussions, supervised independent study, seminar presentations, informal oral exams, or any means that seems appropriate. Students and faculty members usually find a formal oral exam to be the most efficient route to certification.

After completion of at least eight graduate courses and certification, students are awarded the master's degree and, subject to departmental approval, are admitted to candidacy for the Ph.D. degree. This normally occurs by the end of the second year of graduate study. After admission to candidacy, the student chooses a thesis adviser and thesis area and begins an in-depth study of the chosen area. Normally, the thesis is completed during the fourth or fifth year of graduate study. The typical thesis consists of publishable original work. Areas recently chosen for thesis research include algebra, analysis, applied mathematics, combinatorics, geometry, logic, number theory, set theory, and topology. Students continue taking courses according to their interests and demonstrate competence in one foreign language while doing their thesis research.

Dartmouth is committed to helping its graduate students develop as teachers by providing examples of effective teaching, by instruction in a graduate course on teaching mathematics, and by provision of carefully chosen opportunities to gain realistic teaching experience. These opportunities begin in tutorial or discussion leader positions for courses taught by senior faculty members. They culminate in the third and fourth years, after completion of the graduate course, in the opportunity to teach one course for one term each year. The first of these courses is normally a section of a multisection course supervised by a senior faculty member, and the second is chosen to fit the interests and needs of the students and the department. All students are required to participate in these teaching experiences.

Research Facilities

The department has an outstanding library and abundant computer resources and office, laboratory, seminar, and lounge space. More specifically, the already-excellent collection of mathematics books and journals is supplemented by online access to the collections from a consortium of peer institutions. Significant digital resources are also available. Graduate students have offices, many equipped with personal computers, and other computers (Mac, Windows, Linux) are available in adjacent lab space. There are lounges for graduate and undergraduate use. The department maintains its own Web, print, and mail servers, and significant technology is available for both computational research and also for use in the classroom. There is a computer store on campus offering discounted prices for computers, software, and supplies.

Financial Aid

Students receive a full tuition scholarship and the Dartmouth College Fellowship, for which the stipend in 2005–06 was $1750 per month. This stipend continues for twelve months per year through the fourth year of graduate study and is generally renewable for a fifth year as well. In addition to the stipend, the College provides a $1300 health benefit that is paid for students requiring health insurance (i.e., students who do not provide a waiver form).

Cost of Study

With the exception of textbooks, all costs of study are covered by the scholarship.

Living and Housing Costs

Students find that $1750 per month suffices comfortably for living in College housing, renting local apartments, or sharing a locally rented house with other students. A married student whose spouse does not work or hold a similar fellowship can maintain a spartan life in College-owned married-student housing.

Student Group

Dartmouth attracts and admits students from colleges and universities of all types. About 50 percent of students are women, and the percentage of married students has varied from 5 to 35. The department has nearly 20 graduate students, offering an effective placement program for its Ph.D. graduates. Recipients of the Ph.D. degree from Dartmouth have found employment at a broad cross section of academic institutions, including Ivy League institutions, major state universities, and outstanding four-year liberal arts colleges, and some are working as research mathematicians in industry and government.

Location

Dartmouth is in a small town that has an unusual metropolitan flavor. There are adequate shopping facilities but no large cities nearby. Hiking, boating, fishing, swimming, ski touring, and alpine skiing are all available in the immediate area. A car is a pleasant luxury, but many students find it unnecessary.

The College

Dartmouth has more than 4,000 undergraduate students, who are among the most talented and motivated in the nation. There are nearly 1,000 graduate students in the College (arts and sciences faculty) and in associated professional schools in engineering, medicine, and business. With a faculty–graduate student ratio close to 1:1, the department is a friendly place where student-faculty interaction is encouraged.

Applying

Application forms are available from the department. Applicants should send to the department a completed application form, an undergraduate transcript, and three letters of recommendation that describe their mathematical background and ability, estimate their potential as teachers, and compare them with a peer group of the recommender's choice. Applicants must take both the General Test and Subject Test of the Graduate Record Examinations (GRE) and have the official scores sent to the department. All sections of the TOEFL are required of applicants whose native language is not English. Applicants whose files are complete by February 15 receive first consideration.

Correspondence and Information

Graduate Admissions Committee Chair
Ph.D. Program in Mathematics
Department of Mathematics, 6188 Bradley Hall
Dartmouth College
Hanover, New Hampshire 03755-3551
Phone: 603-646-3722 or
603-646-2415
E-mail: mathphd@dartmouth.edu
Web site: http://www.math.dartmouth.edu/

Dartmouth College

THE FACULTY AND THEIR RESEARCH

Professors

Martin Arkowitz, Ph.D., Cornell, 1960. Algebraic topology and differential geometry. Provides thesis supervision in these areas.
Peter Doyle, Ph.D., Dartmouth, 1982. Geometry.
Carolyn Gordon, Ph.D., Washington (St. Louis), 1979. Geometry. Provides thesis supervision in differential geometry.
Marcia Groszek, Ph.D., Harvard, 1981. Logic. Provides thesis supervision in logic.
Charles Dwight Lahr, Ph.D., Syracuse, 1971. Analysis, especially functional analysis. Provides thesis supervision in functional analysis.
Carl Pomerance, Ph.D., Harvard, 1972. Number theory. Provides thesis supervision in number theory.
Daniel Rockmore, Ph.D., Harvard, 1989. Representation theory, fast transforms, group theoretic transforms, dynamical systems, signal processing, data analysis. Provides thesis supervision in analysis and representation theory.
Thomas R. Shemanske, Ph.D., Rochester, 1979. Number theory and modular forms. Currently interested in Hilbert/Siegel modular forms and theta series. Provides thesis supervision in number theory and related areas of mathematics.
Dorothy Wallace, Ph.D., California, San Diego, 1982. Number theory, especially analytic number theory. Provides thesis supervision in number theory.
David L. Webb, Ph.D., Cornell, 1983. Algebraic K theory. Provides thesis supervision in algebra.
Dana P. Williams, Department Chair; Ph.D., Berkeley, 1979. Analysis. Provides thesis supervision in analysis.
Peter Winkler, Ph.D., Yale, 1975. Discrete mathematics, pure and applied; probability; theory of computing. Provides thesis supervision in these areas.

Associate Professor

John Trout, Ph.D., Penn State, 1995. Analysis, functional analysis, operator algebras and noncommutative topology/geometry. Provides thesis supervision in analysis.

Assistant Professors

Alex H. Barnett, Ph.D., Harvard, 2000. Partial differential equations, numerical analysis.
Vladimir Chernov, Ph.D., Uppsala, 1998. Contact and symplectic geometry, geometric and low-dimensional topology. Provides thesis supervision in topology and geometry.
Rosa Orellana, Ph.D., California, San Diego, 1999. Finite dimensional representations of braid groups of type B, algebraic combinatorics. Provides thesis supervision in combinatorics.
Scott Pauls, Ph.D., Pennsylvania, 1998. Geometry and analysis of Carnot-Carathéodory manifolds. Provides thesis supervision in geometry.
Craig J. Sutton, Ph.D., Michigan, 2001. Differential geometry.
Rebecca Weber, Ph.D., Notre Dame, 2004. Mathematical logic and foundations.

John Wesley Young Research Instructors

The JWY Research Instructorship is a two-year visiting position; the people involved and their fields thus change from year to year.
Ryan Daileda, Ph.D., UCLA, 2004. Number theory, particularly automorphic L-.
Sergi Elizalde, Ph.D., MIT, 2004. Combinatorics.
Robert Hladky, Ph.D., Washington (Seattle), 2004. Geometry.
Marius Ionescu, Ph.D., Iowa, 2005. Functional analysis.

Visiting and Adjunct Faculty

Bernard Cole, Ph.D., Boston University, 1992. Biostatistics.
Eugene Demidenko, Adjunct Associate Professor, Department of Mathematics; Ph.D., Central Institute of Mathematics and Economics of Academy of Sciences of the U.S.S.R. (Moscow), 1975. Statistics.

EMORY

EMORY UNIVERSITY

Graduate School of Arts and Sciences
Rollins School of Public Health
Department of Biostatistics

Programs of Study

Biostatistics is the science that applies statistical theory and methods to the solution of problems in the biological and health sciences. The Department of Biostatistics at Emory University offers programs of study leading to the Master of Science and Doctor of Philosophy degrees in biostatistics through the Graduate School of Arts and Sciences. In addition, the department offers study leading to the Master of Public Health and the Master of Science in Public Health degrees in biostatistics through the Rollins School of Public Health. The programs are designed for individuals with a strong background in the mathematical sciences and an interest in the biological or health sciences. Graduates have pursued a wide variety of career options in academia; federal, state, and local government; health agencies; health insurance organizations; the pharmaceutical industry; and other public and private research organizations. The department also offers the M.S.P.H. degree in public health informatics. Public health informatics is a combination of computer science, information science, and public health science in the management and processing of public health data, information, and knowledge supporting effective public health practice. This term has been defined as the application of information science and technology to public health science and practice. Graduates of this program will possess the knowledge and skills necessary to introduce new technology and distribute information systems to support public health decision making.

The Department of Biostatistics is situated within a rich environment of collaborative institutes. Many active research and employment opportunities for students are available at the Rollins School of Public Health, the Emory Medical School, the neighboring Centers for Disease Control and Prevention (CDC)—the federal institute responsible for disease surveillance and control; the Carter Center; the American Cancer Society; the Georgia Department of Human Resources; and local health departments. The department coordinates the activities of the Biostatistics Consulting Center, which serves as a resource for advice on the design, conduct, and analysis of studies in the health sciences. Students may get hands-on experience in practical biostatistical problems through working with faculty members on real-life consulting problems.

Students are required to complete a core curriculum that consists of graduate courses in biostatistics. Advanced course work and research are tailored to the experience, training, area of concentration, and degree objective of each student. The M.S.P.H. and M.P.H. programs usually include four semesters of course work and generally take two years to complete. The Ph.D. degree program normally requires four calendar years to complete, including four to six semesters of course work.

Research Facilities

The Department of Biostatistics conducts active research programs in biostatistics, public health informatics, categorical data analysis, complex sample survey methods, spatial statistics, and methods for infectious disease epidemiology. The Rollins School of Public Health is equipped with state-of-the-art computers and numerous microcomputers. A network of mainframe computers is accessible to the School through high-speed telecommunications lines. Extensive analytical research laboratories are housed in the School and at the CDC. Health sciences libraries are conveniently located at Emory University, the national headquarters of the American Cancer Society, and the CDC.

Financial Aid

Qualified Ph.D. students are supported by nationally competitive graduate school fellowships that include full tuition coverage and a stipend. Research assistantships may be available to M.S.P.H. and M.P.H. students. Financial aid information is available through the Office of Financial Aid.

Cost of Study

Tuition for Ph.D. students in 2006–07 is $15,123 per semester for full-time study or $1399 per credit hour. These figures include activity, athletic, and computer fees. For M.S.P.H. and M.P.H. students, the cost is $11,400 per semester for full-time students or $1100 per credit hour with additional student activity and athletic fees. In addition, the cost of books and supplies averages $1850 per year.

Living and Housing Costs

Living expenses for a single person are estimated to be $18,600 per year. Interested students may obtain information regarding University and off-campus housing by contacting the Housing Office.

Student Group

Emory University has a total enrollment of about 11,600 students. Enrollments in the various schools of the University are restricted in order to maintain a favorable balance between resources, faculty members, and students. There are approximately 6,285 students in the undergraduate college and 5,315 in the eight graduate and professional schools. The student body represents all areas of the United States and more than 100 nations.

Location

The Atlanta metropolitan area has a population of nearly 4.5 million. It is the academic center in the Southeast: there are eight major universities in the metropolitan area. Atlanta is green the year round, with numerous parks and a temperate climate. Professional, athletic, cultural, and recreational activities are available throughout the year. Atlanta is one of the leading convention centers in the United States, and the city is served by one of the busiest airports in the world, providing convenient access to national and international destinations. Atlanta was the site of the 1996 Summer Olympics.

The University and The School

Emory University ranks among the twenty-five most distinguished centers for higher education in the United States. The heavily wooded 631-acre campus features a blend of traditional and contemporary architecture. A main corridor through the campus incorporates the expanding health sciences complex with the headquarters of the CDC and the American Cancer Society. Within a short drive from the main campus are a variety of affiliated resources, such as the Georgia Mental Health Institute, the Georgia Department of Human Resources, the Carter Center of Emory University, and Grady Memorial Hospital.

The Rollins School of Public Health has six academic departments, which offer M.P.H. and M.S.P.H. degrees—Behavioral Sciences and Health Education, Biostatistics, Environmental/Occupational Health, Epidemiology, Health Policy and Management, and Global Health.

The Rollins School of Public Health is ranked ninth in the nation by public health deans, faculty members, and administrators of accredited graduate programs of public health. Research strengths in the School make it the second-highest-ranked school at Emory in terms of research funding. In 2003, *U.S. News & World Report* ranked Emory ninth among Health Disciplines: Public Health (Master's/Doctorate).

Applying

Minimum requirements for admission include a baccalaureate degree from an accredited college or university and satisfactory performance on the GRE. Prerequisites for the M.P.H., M.S.P.H., and Ph.D. program include at least one semesters of calculus III (Multivariable Calculus) and one semester of college-level linear and matrix algebra. International students whose schooling has not been in English must submit a TOEFL score. Application forms for admission to the Ph.D. program may be obtained from the Graduate School of Arts and Sciences, Emory University, Atlanta, Georgia 30322. Admissions information on the M.S.P.H. and M.P.H. degrees may be obtained from the Office of Admissions, Rollins School of Public Health, Emory University, Atlanta, Georgia 30322.

Correspondence and Information

Robert Lyles, Ph.D., Director of Graduate Studies
Department of Biostatistics
Rollins School of Public Health
Emory University
1518 Clifton Road, NE
Atlanta, Georgia 30322
Phone: 404-727-1310
E-mail: biosadmit@sph.emory.edu
Web site: http://www.sph.emory.edu/hpbios.html

Emory University

THE FACULTY AND THEIR RESEARCH

José N. G. Binongo, Lecturer; Ph.D., Ulster (Northern Ireland), 2000.
F. DuBois Bowman, Assistant Professor; Ph.D., North Carolina at Chapel Hill, 2000. Analysis of longitudinal data, missing data, and the application of statistical methods to medical imaging studies. Dr. Bowman is currently involved in collaborative neuro-imaging research at the Positron Emission Tomography (PET) Center of Emory University.
Ying Guo, Assistant Professor; Ph.D., Emory, 2004. Multivariate survival analysis, statistical imaging.
Michael J. Haber, Professor; Ph.D., Hebrew (Jerusalem), 1976. Categorical data analysis, models of infectious diseases. Dr. Haber conducts research in categorical data analysis and statistical methods for the analysis of infectious disease data. In categorical data analysis, Dr. Haber generalizes existing models and develops new methods for investigating particular types of data. He also explores the properties of methods for 2x2 tables and developed an exact unconditional test for comparing two proportions. In the area of analyzing infectious disease data, Dr. Haber, in collaboration with Drs. Longini and Halloran, develops methods for estimation of transmission probabilities and for evaluating the efficacy and effectiveness of vaccines. These methods are applied to data on influenza, measles, mumps, and AIDS.
M. Elizabeth Halloran, Professor; M.D., Berlin, 1983; D.Sc., Harvard, 1989. Causal inference, epidemiologic methods for infectious disease, Bayesian methods, vaccine evaluation. Her research interests encompass methodological problems in studying effects of interventions against infectious disease, especially vector-borne and parasitic diseases. Dr. Halloran draws on Bayesian methods and paradigms of causal inference and has worked on spatial mapping and inference for phylogenetic trees.
John J. Hanfelt, Associate Professor; Ph.D., Johns Hopkins, 1994. Dr. Hanfelt's research interests are in the design and analysis of familial aggregation studies, genetic studies, longitudinal data analysis, the theory of estimating functions, and approximate likelihood inference.
Vicki Stover Hertzberg, Associate Professor; Ph.D., Washington (Seattle), 1980. Categorical data analysis, clinical trials, reproductive epidemiology. Dr. Hertzberg's research interests include categorical data analysis, especially for clustered binary data, as result from a variety of clinical trials and reproductive epidemiology studies. She works especially closely with neurologists involved in stroke research.
Andrew N. Hill, Lecturer; Ph.D., Canterbury (New Zealand), 1996. Semi-parametric methods, Markov models, epidemic theory, spatial spread of infectious diseases.
Yijian (Eugene) Huang Associate Professor; Ph.D., Minnesota, 1997. Survival analysis: multistate process, quality adjusted survival time, lifetime medical cost, and recurrent events; covariate measurement error; semiparametric and nonparametric inferences.
Brent Johnson, Assistant Professor; Ph.D., North Carolina State, 2003. Statistical models of human exposures to chemical pollutants, HIV AIDS modeling, variable selection with censored outcomes.
Mary E. Kelley, Research Assistant Professor; Ph.D., Pittsburgh, 2004. Managing and analyzing research data, research design and statistical analysis.
Michael H. Kutner, Rollins Professor and Chair; Ph.D., Texas A&M, 1971. Linear models, model diagnostics, clinical trials, statistical education. Dr. Kutner's research interests are in estimation and hypothesis testing for analysis of variance models with missing cells, clinical trials methodology, textbook writing in applied linear statistical models.
Qi Long, Rollins Assistant Professor; Ph.D., Michigan, 2005. Causal inference, nonparametric regression methods, longitudinal data analysis, missing data analysis.
Ira M. Longini Jr., Professor; Ph.D., Minnesota, 1977. Stochastic processes, models of infectious diseases. Dr. Longini's research interests are in the area of stochastic processes applied to epidemiological problems. He has specialized in the mathematical and statistical theory of epidemics—a process that involves constructing and analyzing mathematical models of infectious disease transmission and the analysis of infectious disease data based on these models. This work has been carried out jointly with other faculty members and collaborators at other universities and at the CDC. He has worked extensively on the analysis of epidemics of influenza, dengue fever, rhinovirus, rotavirus, measles, cholera, and HIV.
Robert H. Lyles, Associate Professor, Ph.D., North Carolina at Chapel Hill, 1996. Longitudinal data analysis, prediction of random effects, measurement error models, missing and censored data problems. Dr. Lyles' research has investigated applications in the areas of occupational and HIV epidemiology, and his collaborative work includes data analysis for studies of cancer and diabetes.
Amita K. Manatunga, Professor; Ph.D., Rochester, 1990. Multivariate survival analysis, frailty models, categorical data analysis, longitudinal data. Dr. Manatunga's research interests focus on theory and application of survival data analysis. She has worked on multivariate survival data where the interest centers on estimating the covariate effects as well as the correlation between outcomes of survival times. She is interested in developing statistical methodologies based on frailty models and their application in genetics. Dr. Manatunga also has worked closely with medical researchers in the fields of hypertension and pharmacology. She is the biostatistician at the General Clinical Research Center of Emory University.
Limin Peng, Rollins Assistant Professor; Ph.D., Wisconsin–Madison, 2005. Development of semiparametric or nonparametric methods for censored data from clinical trials or epidemiology studies, cancer in-vivo studies, prostate cancer and breast cance research.
André Rogotko, Professor; Ph.D., São Paulo. Statistical genetics, cancer epidemiology, cancer clinical trials.
Mourad Tighiouart, Assistant Professor; Ph.D., Florida State. Bayesian survival analysis, Bayesian methods for cancer clinical trials, nonlinear mixed effects.
Lance A. Waller, Professor; Ph.D., Cornell, 1992. Spatial statistics, point process models, environmental statistics. Dr. Waller's research interests involve statistical analysis of spatial patterns in public health data. Past investigations include development of statistical tests of spatial clustering in disease incidence data and implementation of spatial and space-time Markov random field models for maps of disease rates. He is currently investigating statistical methods to analyze environmental exposure, demographic, and disease incidence data linked through geographic information systems (GIS's).

Adjunct Faculty
Huiman X. Barnhart, Associate Professor, Department of Biostatistics and Bioinformatics, Duke University; Ph.D., Pittsburgh, 1992. Analysis for repeated measures, categorical data analysis, clinical trials. Dr. Barnhart's research interests are in the areas of analysis for repeated measures, categorical data analysis, GEE modeling for correlated data, and diagnostic testing. She is currently investigating statistical methods for randomly repeated measures and for evaluation of diagnostic testing.
Carol A. Gotway Crawford, Mathematical Statistician, National Center for Environmental Health, Biometry Branch, Centers for Disease Control and Prevention; Ph.D., Iowa State, 1989. Spatial prediction and mapping, geostatistics, time series analysis, mixed model applications.
Owen J. Devine, Senior Statistician, National Center for Birth Defects and Developmental Disabilities, Office of the Director, Centers for Disease Control and Prevention; Ph.D., Emory, 1992. Use of stochastic methods in mathematical modeling of human health risks related to sexually transmitted diseases and exposure to environmental contaminants, mapping and spatial analyses of indicator of disease risks, methods for addressing uncertainty in measures of potential exposure in the planning and analysis of human health studies.
Taha Kass-Hout, Chief Scientist, Northrop Grumman CDC Programs (Atlanta); M.S., Emory, 2001; M.D., Texas, 1996,. Epidemiology, clinical, public health, informatics, survey, outbreak investigation, surveillance, information technology, research.
William E. Morse, Chief Information Officer, Oglethorpe University (Atlanta); J.D., Emory, 1994.
Andrzej S. Kosinski, Assistant Professor, Department of Biostatistics and Bioinformatics, Duke University; Ph.D., Washington (Seattle), 1990. Linear models, cardiovascular clinical trials, statistical computing, survival analysis. Dr. Kosinski's interests are in undue influence of groups of observations on the estimation process and in diagnostic procedures to detect such influence. His work involves cardiovascular clinical trials, including the Emory Angioplasty-Surgery Trial (EAST).
Lillian S. Lin, Mathematical Statistician, National Center for HIV, STD, and TB Prevention, Division of HIV/AIDS Prevention: Surveillance and Epidemiology, Centers for Disease Control and Prevention; Ph.D., Washington (Seattle), 1990. Cluster-randomized studies, social and behavioral sciences applied to HIV/AIDS prevention.
Philip H. Rhodes, Mathematical Statistician, National Immunization Program, Epidemiology Surveillance Division, Centers for Disease Control and Prevention; Ph.D., Emory, 1992. Survival analysis, models for infectious disease data.
Glen A. Satten, Mathematical Statistician, National Center for Environmental Health, Divisions of Environmental Health Lab Sciences, Centers for Disease Control and Prevention; Ph.D., Harvard, 1985. Stochastic processes, HIV/AIDS modeling.
Maya R. Sternberg, Mathematical Statistician, National Center for HIV, STD, and TB Prevention, Division of Sexually Transmitted Diseases Prevention, Centers for Disease Control and Prevention; Ph.D., Emory, 1996.
Donna F. Stroup, Associate Director for Science, National Center for Chronic Disease Prevention and Health Promotion, Office of the Director, Centers for Disease Control and Prevention; Ph.D., Princeton, 1980. Stopping rules for stochastic approximation procedures; public health surveillance; Bayesian approaches to detecting aberrations in public health surveillance data; ethical issues in public health; epidemiology curriculum for students; methods for pooling results of studies, including meta-analysis.
G. David Williamson, Director, Office of the Assistant Administrator of Health Sciences, Agency for Toxic Substances and Disease Registry (ATSDR); Ph.D., Emory, 1987. Methods for disease surveillance, epidemiologic studies.
John M. Williamson, Mathematical Statistician, National Center for HIV, STD, and TB Prevention, Division of HIV/AIDS Prevention: Surveillance and Epidemiology, Centers for Disease Control and Prevention; Sc.D., Harvard, 1993. Clustered correlated data, Interrater agreement, HIV/AIDS modeling.

Associate Faculty
George Cotsonis, Senior Associate; M.S., West Florida, 1978. Statistical computing and clinical trials.
Kirk A. Easley, Senior Associate; M.A., LSU, 1981. Statistical applications in clinical research.
Lisa K. Elon, Senior Associate; M.P.H., Emory, 1997. Sample survey analysis, longitudinal study of health promotion in medical education.
Jennifer Favaloro-Sabatier, Associate; M.S., LSU, 2002. Applied statistics.
Patrick D. Kilgo, Senior Associate; M.S., Georgia, 1998. Clinical trials design, statistical power calculations, data analysis.
Michael J. Lynn, Senior Associate; M.S., Mississippi State, 1976. Clinical trials, statistical applications in ophthalmic research, statistical computing.
Azhar Nizam, Senior Associate; M.S., South Carolina, 1987. Statistical education.
Paul S. Weiss, Associate; M.S., Michigan, 1997. Survey sampling design, research methodologies, statistical computing.
Rebecca H. Zhang, Senior Associate; M.S., Florida State, 1994. Data management, statistical analysis.

Jointly Appointed Faculty
Michael P. Epstein, Assistant Professor (joint appointment with Department of Human Genetics); Ph.D., Michigan, 2002.
W. Dana Flanders, Professor (joint appointment with Department of Epidemiology); D.Sc., Harvard, 1982. Quantitative epidemiology, methods.
Frank J. Gordon, Associate Professor (joint appointment with Department of Pharmacology); Ph.D., Iowa, 1980.
Brani Vidakovic, Professor (joint appointment with Department of Biomedical Engineering); Ph.D., Purdue, 1992.

SELECTED PUBLICATIONS

Barnhart, H X., and **J. M. Williamson**. Goodness-of-fit tests for GEE modeling with binary response. *Biometrics* 54:720–9, 1998.

Wolf, S. L., and **H. X. Barnhart** et al. The effect of Tai Chi Quan and computerized balance training on postural stability in older subjects. *Phys. Ther.*, 77:371–84, 1997.

Albert, P. S., D. Follman, and **H. X. Barnhart.** A generalized estimating equation approach for modeling random-length binary vector data. *Biometrics* 53:1116–24, 1997.

Wolf, S. L., and **H. X. Barnhart** et al. Reducing frailty and falls in older persons: An investigation of Tai Chi and computerized balance training. *J. Am. Geriatr. Soc.* 44(5):489–97, 1996.

Barnhart, H. X., et al. Natural history of HIV disease in perinatally infected children: An analysis from the Pediatric Spectrum of Disease Project. *Pediatrics* 97:710–6, 1996.

Barnhart, H. X., and A. R. Sampson. Multiple population models for multivariate random-length data with applications in clinical trials. *Biometrics* 51(1):195–204, 1995.

Barnhart, H. X., and A. R. Sampson. Overview of multinomial models for ordinal data. *Commun. Stat.* 23(12):3395–416, 1994.

Barnhart, H. X. Models for multivariate random length data with applications in clinical trials. *Drug Information J.* 27:1147–57, 1993.

Brogan, D., K. O'Hanlan, **L. Elon**, and E. Frank. Health and professional characteristics of lesbian vs. heterosexual women physicians. *J. Am. Med. Wom. Assoc.* 58:10–9, 2003.

Brogan, D., E. Frank, **L. Elon**, and K. O'Hanlan. Methodological concerns for defining lesbian for health research. *Epidemiology* 12(1):109–13, 2001.

Brogan, D., H. M. Haber, and N. Kutner. Functional decline among older adults: Comparing a chronic disease cohort and controls when mortality rates are markedly different. *J. Clin. Epidemiol.* 53(8):847–51, 2000.

Brogan, D. J., et al. **(L. Elon)**. Harassment of lesbians as medical students and physicians. *JAMA*, 282:1290–2, 1999.

Brogan, D. J. Software for sample survey data: Misuse of standard packages. Invited chapter in *Encyclopedia of Biostatistics,* editors-in-chief P. Armitage and T. Colton. New York: John Wiley, 5:4167–74, 1998.

Brogan, D., E. Flagg, M. Deming, and R. Waldman. Increasing the accuracy of the expanded programme on immunization's cluster survey design. *Ann. Epidemiol.* 4(4):302–11, 1994.

Flanders, W. D., C. D. Drews, and **A. S. Kosinski.** Methodology to correct for differential misclassification. *Epidemiology* 6(2):152–6, 1995.

Haber, M. J. Estimation of the population effectiveness of vaccination. *Statistics Med.* 16:601–10, 1997.

Haber, M. J., W. A. Orenstein, **M. E. Halloran**, and **I. M. Longini Jr.** The effect of disease prior to an outbreak on estimation of vaccine efficacy following the outbreak. *Am. J. Epidemiol.* 141:980–90, 1995.

Haber, M. J., L. Watelet, and **M. E. Halloran.** On individual and population effectiveness of vaccination. *Int. J. Epidemiol.* 24:1249–60, 1995.

Haber, M., I. M. Longini Jr., and **M. E. Halloran.** Measures of the effects of vaccination in a randomly mixing population. *Int. J. Epidemiol.* 20:300–10, 1991.

Haber, M., I. M. Longini Jr., and **G. A. Cotsonis.** Models for the statistical analysis of infectious data. *Biometrics* 44:163–73, 1988.

Haber, M., and M. B. Brown. Maximum likelihood methods for log-linear models when expected frequencies are subject to linear constraints. *J. Am. Stat. Assoc.* 81:477–82, 1986.

Haber, M. J. Testing for pairwise independence. *Biometrics* 42:429–35, 1986.

Haber, M. J. Maximum likelihood methods for linear and log-linear models in categorical data. *Comput. Stat. Data Anal.* 3:1–10, 1985.

Haber, M. J. Log-linear models for linked loci. *Biometrics* 40:189–98, 1984.

Golm, G. T., **M. E. Halloran**, and **I. M. Longini Jr.** Semiparametric models for mismeasured exposure information in vaccine trials. *Statistics Med.*, in press.

Halloran, M. E., M.-P. Préziosi, and H. Chu. Estimating vaccine efficacy from secondary attach rates. *J. Am. Stat. Assoc.* 98:38–46, 2003.

Halloran, M. E., I. M. Longini Jr., A. Nizam, and Y. Yang. Containing bioterrorist smallpox. *Science* 298:1428–32, 2002.

Halloran, M. E., I. M. Longini Jr., D. M. Cowart, and **A. Nizam.** Community trials of vaccination and the epidemic prevention potential. *Vaccine* 20:3254–62, 2002.

Golm, G. T., **M. E. Halloran**, and **I. M. Longini Jr.** Semiparametric methods for multiple exposure mismeasurement and a bivariate outcome in HIV vaccine trials. *Biometrics* 55:94–101, 1999.

Halloran, M. E., I. M. Longini Jr., and C. J. Strichiner. Design and interpretation of vaccine field studies. *Epidemiol. Rev.* 21(1):73–88, 1999.

Golm, G. T., **M. E. Halloran**, and **I. M. Longini Jr.** Semi-parametric models for mismeasured exposure information in vaccine trials. *Statistics Med.* 17:2335–532, 1998.

Halloran, M. E., and C. J. Struchiner. Causal inference for interventions in infectious diseases. *Epidemiology* 6:142–51, 1995.

Hanfelt, J. J., S. Slack, and E. G. Gehan. A modification of Simon's optimal design for phase II trials when the criterion is median sample size. *Controlled Clin. Trials* 20:555–66, 1999.

Hanfelt, J. J. Optimal multi-stage designs for a phase II trial that permits one dose escalation. *Statistics Med.* 18:1323–39, 1999.

Hanfelt, J. J., and K.-Y. Liang. Inference for odds ratio regression models with sparse dependent data. *Biometrics* 54:136–47, 1998.

Hanfelt, J. J. Statistical approaches to experimental design and data analysis of in vivo studies. *Breast Cancer Res. Treatment* 46:279–302, 1997.

Hanfelt, J. J., and K.-Y. Liang. Approximate likelihoods for generalized linear errors-in-variables models. *J. R. Stat. Soc. B* 59:627–37, 1997.

Liang, K.-Y., C. A. Rohde, and **J. J. Hanfelt.** Instrumental variable estimation and estimating functions. *Statistica Applicata* 8:43–58, 1996.

Hanfelt, J. J., and K.-Y Liang. Approximate likelihood ratios for general estimating functions. *Biometrika* 82:461–77, 1995.

Liang, K.-Y., and **J. J. Hanfelt.** On the use of the quasi-likelihood method in tautological experiments. *Biometrics* 50:872–80, 1994.

Hertzberg, V. S. Simulation evaluation of three models for correlated binary data with covariates specific to each binary observation. *Commun. Stat.* 26:375–96, 1997.

Rosenman, K., et al. **(V. S. Hertzberg).** Silicosis among foundry workers: Implication for the need to revise the OSHA standard. *Am. J. Epidemiol.* 144:890–900, 1996.

Reilly, M. J., et al. **(V. S. Hertzberg).** Ocular effects of exposure to triethylamine in a foundry sand core cold box operation. *Occup. Environ. Med.* 52:337–43, 1995.

Hertzberg, V. S Utilization 2: Special datasets. *Statistics Med.* 14:693, 1995.

Hertzberg, V. S., C. Rice, S. Pinney, and D. Linz. Occupational epidemiology in the era of TQM: Challenges for the future. In *1994 Proceedings of the Epidemiology Section, American Statistical Association.*

Hertzberg, V. S., G. K. Lemasters, K. Hansen, and H. M. Zenick. Statistical issues in risk assessment of reproductive outcomes with chemical mixtures. *Environ. Health Perspect.* 90:171–5, 1991.

Hertzberg, V. S., and L. D. Fisher. A model for variability in arteriographic reading. *Statistics Med.* 5:619–27, 1986.

King, S. B., III, **A. S. Kosinski**, and W. S. Weintraub. Eight year outcome in the Emory Angioplasty vs. Surgery Trial. *J. Am. Coll. Cardiol.*, in press.

Kosinski, A. S., and **W. D. Flanders**. Regression model for estimating for odds ratios with misclassified exposure. *Statistics Med.* 18:2795–808, 1999.

Kosinski, A. S. A procedure for the detection of multivariate outliers. *Computational Stat. Data Anal.* 29(2):199–211, 1999.

Barnhart, H. X., **A. S. Kosinski**, and A. Sampson. A regression model for multivariate random length data. *Statistics Med.* 18(2):199–211, 1999.

Deyi, B. A., **A. S. Kosinski**, and S. S. Snapinn. Power considerations when a continuous outcome variable is dichotomized. *J. Biopharm. Statistics* 8(2):337–52, 1998.

Stiger, T. R., **A. S. Kosinski, H. X. Barnhart**, and D. G. Kleinbaum. ANOVA for repeated ordinal data with small sample size? A comparison of ANOVA, MANOVA, WLS and GEE methods by simulation. In *Commun. Statistics Simulation Computation* 27(2):357–75, 1998.

Cecil, M. C., and **A. S. Kosinski** et al. The importance of work-up (verification) bias correction in assessing the accuracy of SPECT thallium-201 testing for the diagnosis of coronary artery disease. *J. Clin. Epidemiol.* 49(7):735–42, 1996.

King, S. B.,III, et al. **(A. S. Kosinski and H. X. Barnhart).** A randomized trial comparing coronary angioplasty with coronary bypass surgery: Emory Angioplasty Versus Surgery Trial (EAST). *N. Engl. J. Med.* 331(16):1044–50, 1994.

Weintraub, W. S., et al. **(A. S. Kosinski).** Lack of effect of lovastatin on restenosis after coronary angioplasty. *N. Engl. J. Med.* 331(20):1331–7, 1994.

Tan, M., X. Xiong, and **M. H. Kutner.** Clinical trial designs based on sequential conditional probability ratio tests and reverse stochastic curtailing. *Biometrics,* 54: 684–697, 1998.

Tan, M., Y. Qu, and **M. H. Kutner.** Model diagnostics for marginal regression analysis of correlated binary data. *Commun. Statistics Ser. B,* 23:539–58, 1997.

Qu, Y., M. Tan, and **M. H. Kutner.** Random effects models in latent class analysis for evaluating accuracy of diagnostic tests. *Biometrics,* 52:797–810, 1996.

Neter, J., **M. H. Kutner,** C. J. Nachsheim, and W. Wasserman. *Appl. Linear Regression Models* (3rd ed.). Chicago: Richard D. Irwin, Inc., 1996.

Neter, J., **M. H. Kutner,** C. J. Nachtsheim, and W. Wasserman. *Appl. Linear Stat. Models* (4th ed.). Chicago: Richard D. Irwin, Inc., 1996.

Kutner, M. H. The computer analysis of factorial experiments with nested factors. Invited response to Dallah, G. E., *Am. Statistician* 42:420, 1992.

Hocking, R. R., and **M. H. Kutner.** Some analytical and numerical comparisons of estimators for the mixed A.O.V. model. *Biometrics,* 3:19–27, 1975.

Frome, E. L., **M. H. Kutner,** and J. J. Beauchamp. Regression analysis of Poisson distributed data. *J. Am. Stat. Assoc.* 68:935–40, 1973.

Longini, I. M., M. E. Halloran, and A. Nizam et al. Estimation of the efficacy of live, attenuated influenza vaccine from a two-year multi-center vaccine trial: Implications for influenza epidemic control. *Vaccine* 18:1902–9, 2000.

Emory University

Longini, I. M., M. G. Hudgens, **M. E. Halloran,** and K. Sagatelian. A Markov model for measuring vaccine efficacy for both susceptibility to infection and reduction in infectiousness for prophylactic HIV vaccines. *Statistics Med.* 18:53–68. 1999.

Durham, L. K., et al. **(I. M. Longini Jr., M. E. Halloran,** and **A. Nizam).** Estimation of vaccine efficacy in the presence of waning: Application to cholera vaccines. *Am. J. Epidemiol.*, in press.

Longini, I. M., S. Datta, and **M. E. Halloran.** Measuring vaccine efficacy for both susceptibility to infection and reduction in infectiousness for prophylactic HIV-1 vaccines. *J. Acquired Immun. Defic. Syndromes Hum. Retrovirol.* 13:440–7, 1996.

Longini, I. M., and **M. E. Halloran.** A frailty mixture model for estimating vaccine efficacy. *Appl. Stat.* 45:165–73, 1996.

Longini, I. M., and **M. E. Halloran.** AIDS: Modeling epidemic control. Letter to the editor. *Science* 267:1250–1, 1995.

Longini, I. M., M. E. Halloran, and **M. J. Haber.** Estimation of vaccine efficacy from epidemics of acute infectious agents under vaccine-related heterogeneity. *Math. Biosci.* 117:271–81, 1993.

Longini, I. M., W. S. Clark, and J. M. Karon. Effect of routine use of therapy in slowing the clinical course of human immunodeficiency virus infection in a population-based cohort. *Am. J. Epidemiol.* 137:1229–40, 1993.

Longini, I. M., M. E. Halloran, M. Haber, and R. T. Chen. Methods for estimating vaccine efficacy from outbreaks of acute infectious agents. *Statistics Med.* 12:249–63, 1993.

Longini, I. M., R. H. Byers, N. A. Hessol, and W. Y. Tan. Estimating the stage-specific numbers of HIV infection using a Markov model and back-calculation. *Statistics Med.* 11:831–43, 1992.

Longini, I. M., W. S. Clark, L. I. Gardner, and J. F. Brundage. The dynamics of CD4+ T-lymphocyte decline in HIV-infected individuals: A Markov modeling approach. *J. AIDS* 4:1141–7, 1991.

Longini, I. M., and **W. S. Clark** et al. Statistical analysis of the stages of HIV infection using a Markov model. *Statistics Med.* 8:831–43, 1989.

Longini, I. M. A mathematical model for predicting the geographic spread of new infectious agents. *Math. Biosci.* 90:367–83, 1988.

Lyles, R. H., D. Fan, and R. Chuachoowong. Correlation coefficient estimation involving a left-censored laboratory assay variable. *Statistics Med.*, in press.

Lyles, R. H., and G. McFarlane. Effects of covariate measurement error in the initial level and rate of change of an exposure variable. *Biometrics* 56:634–39, 2000.

Lyles, R. H., C. M. Lyles, and D. J. Taylor. Randomized regression models for HIV RNA data subject to left censoring and informative dropouts. *Appl. Stat.* 49:485–97, 2000.

Lyles, R. H., and L. L. Kupper. A note of confidence interval estimation in measurement error adjustment. *Am. Statistician* 53:247–53, 1999.

Lyles, R. H., et al. Prognostic value of HIV RNA in the natural history of *Pneumocystis carinii* pneumonia, cytomegalovirus, and *Mycobacterium avium* complex disease. *AIDS* 13:341–50, 1999.

Lyles, R. H., et al. Adjusting for measurement error to assess health effects of variability in biomarkers. *Statistics Med.* 18:1069–86, 1999.

Lyles, R. H., and J. Xu. Classifying individuals based on predictors of random effects. *Statistics Med.* 18:35–52, 1999.

Lyles, R. H., and L. L. Kupper. A detailed evaluation of adjustment methods for multiplicative measurement error in multiple linear regression, with applications in occupational epidemiology. *Biometrics* 53:1008–25, 1997.

Lyles, R. H., L. L. Kupper, and S. M. Rappaport. On prediction of lognormal-scale mean exposure levels in epidemiologic studies, *J. Agric. Biol. Environ. Stats.* 2: 417–39, 1997.

Price, D. L., and **A. K. Manatunga.** Modeling survival data with a cured fraction using frailty models. *Statistics Med.*, in press.

Manatunga, A. K., and S. Chen. Sample size estimation for survival outcomes in cluster randomized studies with small cluster sizes. *Biometrics*, in press.

Manatunga, A. K., and D. Oakes. Parametric analysis for matched paid data. *Life Time Data Anal.* 5:371–87, 1999.

Durham, K. L., **M. E. Halloran, I. M. Longini Jr.**, and **A. K. Manatunga.** Comparison of two smoothing methods for exploring waning vaccine effects. *Appl. Statistics* 48:395–407, 1999.

Chen, M.-H., **A. K. Manatunga,** and C. J. Williams. Heritability estimates from human twin data by incorporating historical prior information. *Biometrics* 54:1348–62, 1998.

Sun, F., et al. **(M. E. Halloran** and **A. K. Manatunga).** Testing for contribution of mitochondrial DNA mutations to complex diseases. *Genet. Epidemiol.* 15:451–469, 1998.

Manatunga, A. K., and D. Oakes. A measure of association for bivariate frailty distributions. *J. Multivar. Analysis* 56:60–74, 1996.

Manatunga, A. K., J. J. Jones, and J. H. Pratt. Longitudinal assessment of blood pressures in black and white children. *J. Hypertension* 22:84–89, 1993.

Oakes, D., and **A. K. Manatunga.** Fisher information for a bivariate extreme value distribution. *Biometrika* 79:827–32, 1992.

Oakes, D., and **A. K. Manatunga.** A new representation of Cox's score statistic and its variance. *Stat. Probab. Lett.* 14:107–10, 1992.

Manatunga, A. K., T. K. Reister, J. Z. Miller, and J. H. Pratt. Genetic influences on the urinary excretion of aldosterone in children. *J. Hypertension* 19:192–7, 1992.

Rhodes, P., M. E. Halloran, and **I. M. Longini Jr.**. Counting process models for infectious disease data: Distinguishing exposure to infection from susceptibility. *J. R. Stat. Soc. B* 58:751–62, 1997.

Satten, G. A., and **I. M. Longini Jr.** Markov chains with measurement error: Estimating the "true" course of a marker on HIV disease progression (with discussion). *Appl. Stat.* 45:275–309, 1996.

Satten, G. A., and **I. M. Longini Jr.** Estimation of incidence of HIV infection using cross-sectional marker surveys. *Biometrics* 50:675–88, 1994.

Satten, G. A., T. D. Mastro, and **I. M. Longini Jr.** Estimating the heterosexual transmission probability of HIV-1 in Thailand. *Statistics Med.* 13:2097–106, 1994.

Smith, D., et al. **(L. A. Waller).** Predicting the spatial dynamics of rabies epidemics on heterogeneous landscapes. *Proc. Natl. Acad. Sci. U.S.A.* 99:3668–72, 2002.

Wakefield, J., N. Best, and **L. A. Waller.** Bayesian approaches to disease mapping. In *Spatial Epidemiology: Methods and Applications,* pp. 106–27, eds. P. Elliott, J. D. Wakefield, N. G. Best, and D. J. Briggs. Oxford: Oxford University Press, 2000.

Best, N. G., et al. **(L. A. Waller).** Bayesian models for spatially correlated disease and exposure data. *Bayesian Statistics 6,* eds. J. M. Bernardo, J. O. Berger, A. P. Dawid, and A. F. M. Smith. Oxford: Oxford University Press, 1999.

English, P. R., et al. **(L. Waller).** Examining associations between childhood asthma and traffic flow using a geographic information system. *Environ. Health Perspect.* 107:761–7, 1999.

Waller, L. A., T. A. Louis, and B. P. Carlin. Environmental justice and statistical summaries of differences in exposure distributions. *J. Exposure Anal. Environ. Epidemiol.* 9:56–65, 1999.

Yu, C., **L. A. Waller,** and D. Zelterman. A discrete distribution for use in twin studies. *Biometrics* 54:546–57, 1998.

Waller, L. A., and R. B. McMaster. Incorporating indirect standardization in tests for disease clustering in a GIS environment. *Geog. Systems* 4:327–42, 1997.

Waller, L. A., T. A. Louis, and B. P. Carlin. Bayes methods for combining disease and exposure data in assessing environmental justice. *Environ. Ecol. Statistics* 4:267–81, 1997.

Waller, L. A., and D. Zelterman. Log-linear modeling with the negative multinomial distribution. *Biometrics* 53:971–82, 1997.

Waller, L. A., B. P. Carlin, H. Xia, and A. Gelfand. Hierarchical spatio-temporal mapping of disease rates. *J. Am. Stat. Assoc.* 92:607–17, 1997.

Lawson, A. B., and **L. A. Waller.** A review of point pattern methods for spatial modeling of events around sources of pollution. *Environmetrics* 7:471–88, 1996.

Waller, L. A., Does the characteristic function numerically distinguish distributions? *Am. Statistician* 49:150–2, 1995.

Waller, L. A., and G. M. Jacquez. Disease models implicit in statistical tests of disease clustering. *Epidemiology* 6:584–90, 1995.

Williamson, G. D., and **M. J. Haber.** Models for three-dimensional contingency tables with completely and partially cross-classified data. *Biometrics* 50:194–203, 1994.

Williamson, J. M., and **A. K. Manatunga.** Assessing interrater agreement from dependent data. *Biometrics* 53(2):707–14, 1997.

GEORGE MASON UNIVERSITY

College of Science

Programs of Study

The College of Science (COS) provides students with the opportunity to pursue graduate education and research in a broad variety of interdisciplinary and discipline-based fields, including biology, chemistry, computational and data science, mathematics, and physics. Many programs emphasize the central role of computational methodologies in the performance of modern scientific research. The educational and research programs are highly interdisciplinary, with an emphasis on theoretical science, computer simulation, data studies, and hardware design and development. The objective of the COS is to provide Virginia, and the nation as a whole, with world-class resources for attacking the interdisciplinary research problems that characterize the challenges faced in the new millennium. The inclusion of theoretical, laboratory, and data science components results in a unique community of scholars ideally suited for the challenges of interdisciplinary research in the years ahead.

Master of Science degrees are available in applied and engineering physics, biodefense, bioinformatics, biology, chemistry, computational science, earth systems science, environmental science and policy, geographic and cartographic sciences, and mathematics.

The Doctor of Philosophy is offered in biodefense, bioinformatics, biosciences, climate dynamics, computational sciences and informatics, computational social science, earth systems and geoinformation sciences, environmental science and public policy, mathematics, neuroscience, and physical sciences.

Students can take graduate certificate programs in actuarial sciences, bioinformatics, biological threat and defense, computational social science, computational techniques and applications, environmental management, geographic information sciences, microbial biodefense, nanotechnology and nanoscience, and remote sensing and earth-image processing.

Research Facilities

Facilities on the Fairfax campus include extensive chemistry, astronomy, and physics labs located in Science & Technology 1. Computational resources in Fairfax include a large Beowulf-class parallel computing cluster with 134 CPUs (sixty-seven dual-processor Pentium III 600 Mhz nodes) running the Linux operating system. Each node is a Dell Precision410 Workstation with 512 MB of RAM and 13 GB of disk storage. The cluster is interconnected via a Foundary FastIron Fast Ethernet 72-port switch. The system runs Message Passing Interface (MPI) networking software and the Parallel Virtual Machine (PVM) system with C, C++, and Fortran bindings and also OpenMP for the node-level parallelization if desired. Other parallel computing tools include both High Performance Fortran (HPF) and Fortran 90 (F90), which was upgraded in early 2006. Graduate students in COS also have access to several Linux computer labs at the Fairfax campus 24 hours a day via card-key access. The student labs have approximately 2 Terabytes of data storage available for class and dissertation research.

COS facilities on the Prince William campus are partially shared with the American Type Culture Collection (ATCC), the world's largest collection of living biological cultures. Facilities include molecular biology and biochemistry labs, computer labs, cold rooms, and instrument rooms. Available computer facilities include more than sixty SGI workstations, including a four-processor Onyx, eighteen Octanes, and more than forty O2s. An SGI Origin 200 provides more than 65 GB of high-availability RAID disk storage. Other computational resources include SUN Sparc Stations, Macs, and PCs. All computers are connected via a high-speed (100 MB/sec) Ethernet LAN. Teaching facilities include three computer classrooms equipped with SGI workstations configured with advanced bioinformatics, visualization, and data-mining software. Three wet labs for teaching and training are supported by adjacent computer labs, lecture rooms, prep labs, and equipment labs, including four ABI 377 and two ABI 310 automated DNA analyzers.

Financial Aid

The Office of Student Financial Aid provides a variety of services to help students finance their education, including counseling, referral and information resources, and financial assistance. Student financial aid awards consist of grants, loans, and work-study opportunities. Awards are based primarily on financial need, although there are some alternative resources available for those who may not qualify for need-based aid. The office has a comprehensive listing of various scholarship opportunities for students to research on the financial aid Web page. Students are encouraged to review the scholarship information early and frequently to meet deadlines, since the listings are updated often.

Cost of Study

Virginia residents pay $279 per credit hour; nonresidents pay $705 per credit hour. Fees are additional.

Living and Housing Costs

Approximately 3,000 students live on campus in six residential areas. Room and board costs are approximately $4840 per year. However, housing for graduate students is very limited, and most students live off campus. The cost of living is comparable to that in other large northeastern urban areas.

Student Group

The majority of the University's nearly 29,000 students are from Virginia. All fifty states and Washington, D.C., as well as 135 countries and regions, are represented in the student body. In fact, in the *Princeton Review*'s most recent survey of more than 110,000 students at 357 top colleges, George Mason ranks number one in the nation in diversity.

Location

The University's main campus in northern Virginia covers nearly 600 wooded acres. Nearby Washington, D.C., offers students access to some of the world's great cultural institutions. The setting combines the quietness of a residential suburban area with access to Fairfax County's high-technology firms; Washington's libraries, galleries, museums, and national and federal laboratories; and Virginia's historical sites.

The University and The College

In the last decade, George Mason University has emerged as a major academic institution offering nationally recognized programs in advanced technology and science, among others. From its origins in 1957, George Mason has grown into the largest state university in northern Virginia with innovative programs that have attracted a faculty of world-renowned scholars and teachers. The College of Science, founded July 1, 2006, plays the central role in undergraduate and graduate education and research in the physical, mathematical, biological, and computational sciences at George Mason University.

Applying

In general, applicants must submit the completed application, the application fee, and official transcripts. Some programs require the submission of test scores (such as the GRE General or Subject Test), letters of recommendation, and a resume. International students must send in their official TOEFL scores. Students should check with the department to which they wish to apply in order to obtain specific admission guidelines.

Correspondence and Information

Peter A. Becker
Associate Dean for Graduate Programs
College of Science
George Mason University
Fairfax, Virginia 22030

Phone: 703-993-3619
Fax: 703-993-1980
E-mail: pbecker@gmu.edu
Web site: http://www.cos.gmu.edu

George Mason University

THE FACULTY AND THEIR RESEARCH

Faculty members and students in the College of Science perform research in many areas of modern science. Much of this work is interdisciplinary, combining concepts and techniques from a wide range of fields, including astrophysics, atmospheric science, biochemistry, bioinformatics, chemistry, climate dynamics, computational and data science, computational statistics, fluid dynamics, geography, earth observing, environmental science, interdisciplinary economic science, materials science, mathematics, microbiology, neuroscience, physics, remote sensing, and space science.

DEPARTMENT HEADS

Bioinformatics and Computational Biology
Saleet Jafri, Associate Professor and Chair; Ph.D., CUNY/Mount Sinai School of Medicine, 1993. Cellular signaling, protein structure, high-performance computing. (703-993-8420; sjafri@gmu.edu)

Chemistry and Biochemistry
Gregory D. Foster, Professor and Chair; Ph.D., California, Davis, 1985. Urban regions as sources of organic contaminants to coastal airsheds and watersheds in the Chesapeake Bay region, developing technologies to remove contaminants that harm the aquatic environment, developing analytical methods. (703-993-1081; gfoster@gmu.edu)

Climate Dynamics
Jagadish Shukla, Professor and Chair; Ph.D., Banaras Hindu (India), 1971; Sc.D., MIT, 1976. Climate dynamics, global change. (301-595-7000; shukla@cola.iges.org)

Computational and Data Sciences
Dimitrios Papaconstantopoulos, Professor and Chair; Ph.D., London, 1967. Solid-state physics. (703-993-3624; dpapacon@gmu.edu)

Earth Systems and Geoinformation Sciences
David W. Wong, Professor and Chair; Ph.D., SUNY at Buffalo, 1990. Geography, spatial analysis and statistics, GIS. (703-993-1212; dwong2@gmu.edu)

Environmental Science and Policy
R. Chris Jones, Professor and Chair; Ph.D., Wisconsin–Madison, 1980. Freshwater ecology, bioassessment, ecology of phytoplankton and attached algae, ecology of tidal embayments, effects of nonpoint pollution on freshwater ecosystems. (703-993-1127; rcjones@gmu.edu)

Geography
Allan Falconer, Professor and Chair; Ph.D., Durham (England), 1970. Physical geography, remote sensing. (703-993-1360; afalcon1@gmu.edu)

Mathematical Sciences
Klaus Fischer, Professor and Chair; Ph.D., Northwestern, 1973. Combinatorial and commutative algebra. (703-993-1462; kfischer@gmu.edu)

Molecular Biology and Microbiology
Vikas Chandhoke, Chair and Co-Dean, College of Science; Ph.D., Maine, 1991. Bioanalytical chemistry, biomedical genomics. (703-993-2674; vchandho@gmu.edu)

Physics and Astronomy
Maria Dworzecka, Professor and Chair; Ph.D., Warsaw, 1969. Dynamical properties of low-energy heavy-ion collisions to understand the basic reaction mechanism and associated nuclear excitation modes. (703-993-1280; dmaria@gmu.edu)

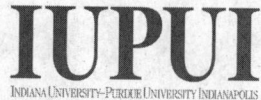

IUPUI
INDIANA UNIVERSITY–PURDUE UNIVERSITY INDIANAPOLIS

INDIANA UNIVERSITY–PURDUE UNIVERSITY INDIANAPOLIS

Department of Mathematical Sciences

Programs of Study

The Department of Mathematical Sciences at Indiana University–Purdue University Indianapolis (IUPUI) offers programs leading to the Ph.D. and M.S. degrees in mathematics from Purdue University. All required course work and research is completed on the IUPUI campus under the guidance of IUPUI faculty members.

The Department has 30 tenured or tenure-track faculty members. Current research strengths within the Department include integrable systems, mathematical physics, dynamical systems, noncommutative geometry, operator algebras, differential geometry, partial differential equations, functional analysis, statistics and probability, applied mathematics, biomathematics, computational neurosciences, and scientific computing.

The Ph.D. program combines advanced study with individual research. Among the requirements for the Ph.D. are a minimum of 42 hours of graduate course work; reading knowledge in one of French, German, or Russian; passing of four written qualifying examinations and an oral specialty exam; and successful completion of an original research dissertation. The Ph.D. program usually requires a minimum of four years, and a typical student spends five to six years in the program.

Areas of specialization for the M.S. degree include pure or applied mathematics, applied statistics, and mathematics education. Master's degree students must complete 30 hours of course work. A thesis is optional for all M.S. programs. The applied statistics program requires a comprehensive written and oral exam; the other M.S. programs require no written or oral exams. Typically, completion of the M.S. degree requires two years of full-time study or up to four years of part-time study.

The M.S. degree with specialization in applied statistics is designed to increase the number of professionals with the broad training in statistical methodology that is suitable for applications in industry, medicine, business, and government. The program provides a basis for the skilled and competent application of modern statistical methods. In addition to the basic theoretical foundations, areas of methodology include categorical and nonparametric methodology, design on experiments, multivariate analysis, quality control, regression analysis, sample surveys, survival analysis, and time series. Many courses are offered in the late afternoon or evening to accommodate students who may have full-time careers.

The M.S. degree with specialization in mathematics education is designed for mathematics teaching professionals who would like to strengthen or enhance their mathematics background while completing their graduate degree. The curriculum is structured around the Indiana Professional Standards Board's competencies for high school math teachers. Students complete a core curriculum that includes courses in abstract algebra, analysis, discrete mathematics, geometry, and probability and statistics. In addition, students choose from a variety of math electives, participate in seminars, and develop an individual project involving innovative pedagogy, such as the use of technology in the mathematics classroom. Most courses are offered in the late afternoon or evening or during the summer to accommodate students who may have full-time teaching careers.

The M.S. degree with specialization in pure or applied mathematics seeks to develop a broad and balanced perspective in the mathematical sciences, covering both the traditional areas of pure mathematics and the many interdisciplinary fields in applied mathematics. Students are encouraged to develop a deep understanding in any chosen area of pure or applied mathematics to enable them to pursue a career in an academic or nonacademic setting.

Research Facilities

The IUPUI University library is central Indiana's premier academic research library, and it is designed to serve the needs of the electronic age. In addition to holdings of approximately 1.3 million volumes, subscriptions to about 4,300 print and online periodicals and journals, and a full range of reference materials, the library provides access to an extensive collection of math and statistics databases and other mathematics resources. The IUPUI library hosts more than 300 public computer stations that provide access to campus electronic resources, the catalog systems of worldwide academic libraries, and the Internet.

Within the Department of Mathematical Sciences, students have access to Unix workstations and a state-of-the-art high capacity research server.

Financial Aid

The Department typically supports all full-time Ph.D. students and a limited number of qualified M.S. students. Available types of financial support include fellowships, graduate teaching assistantships, and research assistantships. Fellowships and assistantships carry stipends ranging up to $21,000 for twelve months, plus tuition waivers and health insurance.

Cost of Study

Students receiving fellowships or assistantships pay approximately $400 per semester in tuition and fees; the rest is covered by the support package. Tuition for 2005–06 was approximately $215 per credit for in-state students and $620 per credit for out-of-state students.

Living and Housing Costs

Intended for graduate and professional students, the IUPUI Graduate Professional Community is located within the Campus Apartments on the River Walk complex. This apartment-style housing consists of fully furnished one-, two-, and four-bedroom units that can house 771 students. All apartments are equipped with individually controlled heating and cooling systems, and each unit is wired for high-speed Internet access. Telephone and basic cable are also available. Monthly rent is $600 to $840 per month, depending on the unit selected. A wide variety of rental housing is also available in Indianapolis and the surrounding communities. Indianapolis' cost of living index is 93.4, well below the nation's average of 100.

Student Group

There are approximately 70 graduate students in the Department. Of this group, roughly 28 percent are women and 20 percent are international students; 34 percent attend on a full-time basis. The Department seeks highly motivated students without regard to race, color, national origin, religion, sex, disability, or age.

Location

The Indianapolis metropolitan area has more than 1.9 million residents. The city's cultural assets include the Indianapolis Museum of Art, Eiteljorg Museum of Native American and Western Art, and the Indianapolis Symphony Orchestra. A diverse year-round program of theater, ballet, opera, popular entertainment, and sporting events is presented at the Indiana Repertory Theatre, Clowes Hall, Conseco Fieldhouse, the Indianapolis Convention Center, the RCA Dome, and other venues. The city has many activities for children, including the world's largest children's museum and the Indianapolis Zoo. The IUPUI campus is located in downtown Indianapolis, within walking distance of many of these attractions and more.

The University

IUPUI demonstrates a model partnership among government, community, and higher education. It was formed in 1969 by combining the city facilities and programs of Indiana University and Purdue University under one name and administration. IUPUI is a leading urban university in which students earn degrees from Indiana University or Purdue University. IUPUI offers the broadest range of degree programs of any campus in Indiana, featuring more than 10,000 classes in hundreds of career fields for the nearly 30,000 students enrolled.

Applying

Students must submit a completed online application, a nonrefundable application fee ($50 for domestic students, $60 for international students), three letters of recommendation, a personal statement, GRE scores, and official transcripts from all colleges or universities attended. In addition, international students must submit a TOEFL score (minimum scores apply). Completed applications for fellowships and assistantships should be received before February 1 and March 1, respectively. International applications should be received by February 1. All applicants should carefully read the graduate admission information on the Department's Web page before applying.

Correspondence and Information

Graduate Programs
Department of Mathematical Sciences
Indiana University–Purdue University Indianapolis
402 North Blackford Street, Room LD270
Indianapolis, Indiana 46202-3216

Phone: 317-274-6918
Fax: 317-274-3460
E-mail: grad-program@math.iupui.edu
Web site: http://www.math.iupui.edu

Indiana University–Purdue University Indianapolis

THE FACULTY AND THEIR RESEARCH

Pavel M. Bleher, Chancellor's Professor; Ph.D., USSR Academy of Science, 1974. Mathematical physics.
Benzion Boukai, Professor and Chair; Ph.D., SUNY at Binghamton, 1988. Statistical inference, sequential analysis, Bayesian-Frequentist interface.
Olguta Buse, Assistant Professor; Ph.D., SUNY at Stony Brook, 2002. Symplectic and contact geometry, four-manifolds, singularities, complex and algebraic geometry.
Raymond Chin, Professor; Ph.D., Case Western Reserve, 1970. Numerical and analytical solution to multiple-scales problems in science and engineering, hybrid asymptotic-numerical methods, biofluid dynamics and kinetics of biochemical reactions, orthogonal polynomials.
Carl C. Cowen, Professor; Ph.D., Berkeley, 1976. Operator theory, complex analysis, linear algebra, computational neuroscience.
Michael L. Frankel, Professor; Ph.D., Tel Aviv, 1984. Nonlinear partial differential equations, free-boundary problems.
William Geller, Associate Professor; Ph.D., Berkeley, 1989. Dynamical systems.
Samiran Ghosh, Assistant Professor; Ph.D., Connecticut, 2006. Biostatistics, bioinformatics.
Alexander R. Its, Distinguished Professor; Ph.D., Leningrad State, 1977. Soliton theory, integrable systems, special functions, mathematical physics.
Ronghui Ji, Associate Professor; Ph.D., SUNY at Stony Brook, 1986. Operator algebras.
Bruce Kitchens, Associate Professor; Ph.D., North Carolina at Chapel Hill, 1981. Ergodic theory.
Slawomir Klimek, Associate Professor and Director of Graduate Programs; Ph.D., Warsaw, 1988. Mathematical physics, noncommutative geometry.
Alexey Kuznetsov, Assistant Professor; Ph.D., Nizhny Novgorod State (Russia), 1999. Applied dynamical systems, mathematical biology.
Fang Li, Assistant Professor; Ph.D., Michigan State, 2004. Linear and nonlinear models.
Michal Misiurewicz, Professor; Ph.D., Warsaw, 1974. Dynamical systems.
R. Patrick Morton, Adjunct Professor; Ph.D., Michigan, 1979. Algebra, number theory.
Evgeny Mukhin, Associate Professor; Ph.D., North Carolina at Chapel Hill, 1998. Modern analysis, representation theory, mathematical physics.
Bart S. Ng, Professor and Marvin L. Bittinger Chair; Ph.D., Chicago, 1973. Hydrodynamic stability.
Michael A. Penna, Professor; Ph.D., Illinois, 1973. Computer vision, image understanding, image processing.
Rodrigo Perez, Assistant Professor; Ph.D., SUNY at Stony Brook, 2002. One- and two-dimensional complex dynamics, geometric group theory, combinatorics.
Krzysztof Podgorski, Associate Professor; Ph.D., Michigan State, 1993. Probability, stochastic process, statistical inference.
Robert D. Rigdon, Associate Professor and Associate Chair; Ph.D., Berkeley, 1970. Algebraic topology.
Leonid Rubchinsky, Assistant Professor; Ph.D., Russian Academy of Science, 2000. Mathematical biology, neuroscience, applied dynamical systems.
Jyotirmoy Sarkar, Associate Professor; Ph.D., Michigan, 1990. Probability, economics.
Asok K. Sen, Professor; Ph.D., Cornell, 1979. Mathematical modeling, biomathematics.
Zhongmin Shen, Professor; Ph.D., SUNY at Stony Brook, 1990. Differential geometry.
Richard Y. Tam, Associate Professor; Ph.D., Cornell, 1986. Combustion theory.
Vitaly Tarasov, Associate Professor; Ph.D., St. Petersburg, 1985; D.Sc., St. Petersburg, 2002. Mathematical physics, representation theory.
Jeffrey X. Watt, Associate Professor and Associate Dean; Ph.D., Indiana, 1990. Mathematics education.
Krzysztof P. Wojciechowski, Professor; Ph.D., Polish Academy of Sciences, 1982. Partial differential equations, spectral geometry, elliptic boundary value problems.
Robert Worth, Adjunct Professor; Ph.D., Indiana, 1987. Mathematical neurosciences.
Constantin Yiannoutsos, Adjunct Associate Professor; Ph.D., Connecticut, 1991. Biostatistics, design of clinical trials, diagnostic testing, sequential design, Bayesian statistics.
Luoding Zhu, Assistant Professor; Ph.D., NYU, 2001. Scientific computing, numerical methods.

MEDICAL COLLEGE OF WISCONSIN

MEDICAL COLLEGE OF WISCONSIN

Graduate School of Biomedical Sciences
Division of Biostatistics

Program of Study

The Division of Biostatistics offers a program leading to the Ph.D. The program is designed for students with strong undergraduate preparation in mathematics and trains students in biostatistical methodology, theory, and practice. Emphasis is placed on sound theoretical understanding of statistical principles, research in the development of applied methodology, and collaborative research with biomedical scientists and clinicians. In addition, students gain substantial training and experience in statistical computing and in the use of software packages. Courses in the program are offered in collaboration with the Department of Mathematics at the University of Wisconsin–Milwaukee. The degree requirements, including the dissertation research, are typically completed in five years beyond a bachelor's degree that includes strong mathematical preparation.

Faculty members are engaged in a number of collaborative research projects at the Center for International Blood and Marrow Transplant Research, the General Clinical Research Center, the Center for AIDS Intervention Research, the Center for Patient Care and Outcomes Research, the Human and Molecular Genetics Center, and the Cancer Center as well as in medical imaging, clinical trials, and pharmacologic modeling. Students participate in these projects under faculty supervision. Dissertation research topics in statistical methodology often evolve from such participation, and students usually become coauthors on medically oriented papers arising from these projects.

Research Facilities

The Division of Biostatistics is located in the Department of Population Health of the Medical College of Wisconsin (MCW). The Medical College has extensive research laboratories and facilities available for faculty and student use. The Division has an up-to-date network of Sun workstations, PCs, and peripherals. This network is linked with the campus backbone, providing direct access to the Internet. The Division's network is equipped with all leading statistical software and tools needed for the development of statistical methodology. The MCW libraries' holdings are among the largest health sciences collections in the Midwest, with more than 244,700 volumes and subscriptions to 1,905 journals. The libraries operate the Medical Information Network, a remote-access computer network that includes the full MEDLINE database along with other medical science databases. The libraries also provide access to several bibliographic databases on compact disc workstations as well as the Internet and the World Wide Web. Students also have access to the University of Wisconsin–Milwaukee's extensive library, and the Division maintains its own library of statistical journals, books, and monographs.

The Epidemiologic Data Service provides access to national data on health and health care and special clinical data sets collected locally (the Medical College is a repository for the National Center for Health Statistics). The Biostatistics Consulting Service provides students with extensive experience in biomedical research.

Financial Aid

Students are supported by fellowships and research assistantships. Each includes tuition and a stipend. The stipend for 2005–06 was $22,973 per year. The research assistantships provide students with the opportunity to gain experience in statistical consulting and collaborative research.

Cost of Study

Tuition is $10,506 per year. Tuition and health insurance are included in the fellowships and research assistantships.

Living and Housing Costs

Many rental units are available in pleasant residential neighborhoods surrounding the Medical College. Housing costs begin at about $550 per month for a married couple or 2 students sharing an apartment. The usual stipend supports a modest standard of living.

Student Group

There are 525 degree-seeking graduate students, 715 residents and fellows, and 796 medical students at the Medical College. A low student-faculty ratio fosters individual attention and a close working relationship between students and faculty members. Graduates pursue academic positions and jobs in government and industry.

Location

Milwaukee has long been noted for its old-world image. Its many ethnic traditions, especially from Middle Europe, give the city this distinction. Cultural opportunities are numerous and include museums, concert halls, art centers, and theaters. Milwaukee has a well-administered government, a low crime rate, and excellent schools. It borders Lake Michigan and lies within commuting distance of 200 inland lakes. Outdoor activities may be pursued year-round.

The College

The College was established in 1913 as the Marquette University School of Medicine. It was reorganized in 1967 as an independent corporation and renamed the Medical College of Wisconsin in 1970. There are approximately 1,000 full-time faculty members and 75 part-time and visiting faculty members; they are assisted by more than 1,700 physicians who practice in the Milwaukee community and participate actively in the College's teaching programs. MCW is one of seven organizations working in partnership on the Milwaukee Regional Medical Complex (MRMC) campus. Most physicians who staff the clinics and hospitals are full-time faculty physicians of MCW. Other MRMC member organizations include the Froedtert Memorial Lutheran Hospital, Children's Hospital of Wisconsin, the Blood Center of Southeastern Wisconsin, Curative Rehabilitation Services, and the Milwaukee County Mental Health Complex. Full-time students in any department may enroll in graduate courses in other departments and in programs of the University of Wisconsin–Milwaukee and Marquette University without any increase in basic tuition. The College ranks in the top 40 percent of all American medical schools in NIH research funding.

Applying

Prerequisites for admission to the program include the baccalaureate degree, satisfactory GRE scores on the General Test, and adequate preparation in mathematics. A complete description of the graduate program and application forms may be obtained by writing to the Graduate Program Director or by downloading them from the Graduate School Web site. Complete application materials should be submitted by February 15.

Correspondence and Information

Dr. Prakash Laud
Graduate Program Director
Division of Biostatistics
Medical College of Wisconsin
Milwaukee, Wisconsin 53226-0509
Phone: 414-456-8781
Fax: 414-456-6513
E-mail: laud@mcw.edu
Web site: http://www.biostat.mcw.edu/

Send completed applications to:
Graduate Admissions
Graduate School of Biomedical Sciences
Medical College of Wisconsin
Milwaukee, Wisconsin 53226-0509
Phone: 414-456-8218
E-mail: gradschool@mcw.edu
Web site: http://www.mcw.edu/gradschool/

Medical College of Wisconsin

THE FACULTY AND THEIR RESEARCH

John P. Klein, Professor and Director; Ph.D., Missouri–Columbia. Survival analysis, competing risks theory, design and analysis of clinical trials. Dr. Klein also serves as the Statistical Director of the Center for International Blood and Marrow Transplant Research at the Medical College. He is an elected member of the International Statistical Institute and fellow of ASA.

Shu, Y., and **J. P. Klein.** Additive hazards Markov regression models illustrated with bone marrow transplant data. *Biometrika* 92:283–301, 2005.

Klein, J. P., and M. L. Moeschberger. *Survival Analysis: Techniques for Censored and Truncated Data,* 2nd edition. New York: Springer Verlag, 2003.

Sun-Wei Guo, Professor; Ph.D., Washington (Seattle). Stochastic modeling, statistical methods in genetics and genetic epidemiology. Positioned in the forefront of revolutionary changes in biomedical research brought about by rapid advances in genomics, Dr. Guo and his lab have been involved with molecular genetic studies of endometriosis and of prostate cancer and in genetic epidemiologic studies of diabetes.

Guo, S. W. Glutathione S-transferases M1/T1 gene polymorphisms and endometriosis: A meta-analysis of genetic association studies. *Mol. Hum. Reprod.* 11(10):729–43, 2005.

Wu, Y., et al. **(S. W. Guo).** Transcriptional characterizations of differences between eutopic and ectopic endometrium. *Endocrinology* 147(1):232–46, 2006.

Raymond G. Hoffmann, Professor; Ph.D., Johns Hopkins. Linear and nonlinear time series; GLM models for sexual behavior data; methods for identifying changes in fMRI images of the brain, spatial patterns of disease, and neural networks. Dr. Hoffman is the Chair of the Statistics in Epidemiology Section of the American Statistical Association.

Brousseau, D., and **R. G. Hoffmann** et al. Disparities for Latino children in the timely receipt of medical care. *Ambulatory Pediatr.* 5(6):319–25, 2005.

Purushottam (Prakash) W. Laud, Professor; Ph.D., Missouri–Columbia. Bayesian statistical methods in biomedical sciences; Bayesian inference and model selection in linear, generalized linear, hierarchical, and survival methods; Markov chain Monte Carlo methods. Dr. Laud also serves as the faculty biostatistician in the Center for Patient Care and Outcomes Research at the Medical College.

Nattinger, A. B., and **P. W. Laud** et al. An algorithm for the use of Medicare claims data to identify women with incident breast cancer. *Health Serv. Res.* 39:1733–49, 2004.

Laud, P. W., P. Damien, and S. G. Walker. Computations via auxiliary random function for survival models. *Scand. J. Statistics* 33:219–26, 2006.

Hyun Ja (Yun) Lim, Assistant Professor; Ph.D., Case Western Reserve. Survival analysis, multiple/recurrent failure time analysis, clinical trials, epidemiologic studies. Dr. Lim also serves as the Director of the Biostatistics Consulting Center at the Medical College.

Lim, H. J., N. H. Gordon, and A. C. Justice. Evaluation of multiple failure time analyses of observational data in patients treated for HIV. *HIV Clin. Trials* 6(2):81–91, 2005.

Franklin, S., and **H. J. Lim** et al. Longitudinal assessment of a clinical sample of children with HIV disease. *J. Clin. Psych. Med. Settings* 12(4):367–76, 2005.

Brent R. Logan, Associate Professor; Ph.D., Northwestern. Multiple comparison procedures, methods for analyzing multiple endpoints in clinical trials, inference in dose-response studies, analysis of neuroimaging data.

Logan, B. R., H. Wang, and **M.-J. Zhang.** Pairwise multiple comparison adjustment in survival analysis. *Statistics Med.* 24:2509–23, 2005.

Logan, B. R., and **D. B. Rowe.** An evaluation of thresholding techniques in fMRI analysis. *NeuroImage* 22:95–108, 2004.

Daniel B. Rowe, Assistant Professor; Ph.D., California, Riverside. Mathematical and statistical methods in functional magnetic resonance imaging, Bayesian statistics, computational statistics.

Rowe D. B., and **R. G. Hoffmann.** Multivariate statistical analysis in fMRI. *IEEE Eng. Med. Biol. Magazine* 25(2):60–4, 2006.

Rowe, D. B. Parameter estimation in the complex fMRI model. *NeuroImage* 25(4):1124–32, 2005.

Rowe, D. B. Modeling both the magnitude and phase of complex-valued fMRI data. *NeuroImage* 25(4):1310–24, 2005.

Sergey Tarima, Assistant Professor; Ph.D., Kentucky. Methods of using additional information in statistical estimation; estimation on missing, censored and partially grouped data; survey data analysis.

Tarima, S., and D. Pavlov. Using auxiliary information in statistical function estimation. *ESAIM: Probability Stat.* 10:11–23, 2006.

Liu H., and **S. Tarima** et al. Quadratic regression analysis for gene discovery and pattern recognition for microarray time-course experiments. *BMC Bioinformatics* 6:106, 2005.

Tao Wang, Assistant Professor; Ph.D., North Carolina State. Statistical genetics, modeling and linkage disequilibrium mapping of quantitative trait loci, linkage and association mapping of disease genes. Dr. Wang also serves as an adjunct faculty member at the Human Molecular Genetics Center at the Medical College.

Wang, T., and Z. B. Zeng. Models and partition of variance for quantitative trait loci with epistasis and linkage disequilibrium. *BMC Genetics* 7:9, 2006.

Zeng, Z. B., **T. Wang,** and W. Zou. Modeling quantitative trait loci and interpretation of models. *Genetics* 169:1711–25, 2005.

Mei-Jie Zhang, Professor; Ph.D., Florida State. Survival analysis, inference for stochastic processes, nonlinear models. As a biostatistician for the Center for International Blood and Marrow Transplant Research at the Medical College, Dr. Zhang is interested in developing statistical models and methodology for analyzing complex transplant data.

Laughlin, M. J., et al. **(M.-J. Zhang).** Comparison of outcomes after unrelated cord blood and unrelated bone marrow transplants for adults with leukemia. *New England J. Med.* 351:2265–75, 2004.

Scheike, T. H., and **M.-J. Zhang.** Extensions and applications of the Cox-Aalen survival model. *Biometrics* 59:1036–45, 2003.

Adjunct Faculty

Jay Beder, Associate Professor; Ph.D., George Washington. Gaussian processes, factorial experiments, categorical data analysis.
Vytaras Brazauskas, Associate Professor; Ph.D., Texas at Dallas. Robust and nonparametric estimation, extreme value theory, risk theory.
Jugal Ghorai, Professor; Ph.D., Purdue. Nonparametric estimation, density and survival function estimation, censored data analysis.
Eric Key, Professor; Ph.D., Cornell. Probability theory and stochastic processes, ergodic theory.
Tom O'Bryan, Associate Professor; Ph.D., Michigan State. Empirical Bayes, decision theory.

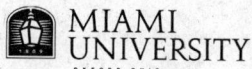

MIAMI UNIVERSITY
OXFORD OHIO

MIAMI UNIVERSITY

College of Arts and Science
Department of Mathematics and Statistics

Programs of Study

The purpose of the Department's several master's degree programs is to prepare students for a variety of careers in mathematics and statistics in industry and government or for further study at the Ph.D. level in these areas. This is accomplished by giving the student a broad base in the core foundations as well as a set of focused experiences in more advanced studies. The student chooses the program that best fits his or her interests.

The Master of Arts in mathematics prepares students for subsequent doctoral study in mathematics, having a required core of pure mathematics courses. The Master of Science in mathematics, with an option in operations research, is a concentration in modern applicable mathematics, including discrete mathematics, optimization, and statistics. The Master of Science in mathematics is a flexible program that is designed by the student, subject to some very basic requirements, and allows the student to explore a number of different areas. The Master of Science in statistics gives students a solid foundation in both applied and theoretical statistics. The program features opportunities to participate in data analysis projects of the Statistics Consulting Center and to combine statistics course work with study in related fields. The Master of Arts in Teaching (M.A.T.) is designed to strengthen and broaden the mathematical knowledge of secondary school teachers.

The course work in each program consists of the standard first two years of graduate study in the area covered. All programs except the M.A.T. have a 32-hour requirement (the M.A.T. has a 30-hour requirement). Students must pass a set of three comprehensive examinations, each exam covering a graduate two-course sequence in some area. Of the 32 hours, at least 15 are at the second-year graduate level. Students complete these programs in two years.

The research strength of the faculty opens up many opportunities for advanced independent study and even original research for those students who are far enough along in their studies. By the second year, many students take reading courses in subjects outside their formal course work. Some have gone on to do original research work during the time of their master's studies.

Research Facilities

The Department houses a computer laboratory with access to Maple, MatLab, and several statistical computing software packages such as SAS and S-Plus. There is also access to a computing cluster on campus and to the Ohio Supercomputing Center at Ohio State University.

The Hughes Science Library has an extensive collection of books and research journals covering all areas of mathematics and statistics. Miami also has a site license for online versions of many of these journals. Articles not available in the library can be obtained through the OhioLink program.

Financial Aid

The Department offers graduate assistantships (GA), which, in 2005–06, provided a stipend of $13,700 for the academic year and $1800 for those taking summer classes. GA duties involve teaching precalculus, calculus, or introductory statistics courses or assisting faculty members teaching those courses through grading and work at help sessions. These assistantships carry a waiver of tuition and half the general fee.

Cost of Study

The 2004–05 graduate tuition for Ohio residents was $4673.24, and the general fee was $703.08. For non-Ohio residents (including international students), the graduate tuition for 2004–05 was $9962.84. Tuition is waived for graduate assistants.

Living and Housing Costs

Housing for the fall semester should be arranged no later than the preceding midsummer. The monthly cost of housing ranges from $325 to $450 for single rooms or apartments and $325 to $525 for shared apartments. Monthly individual total living expenses average about $925.

Student Group

The roughly 25 full-time graduate students, both men and women, come from all over the United States, and some are international students. The majority come from the Midwest.

Student Outcomes

A recent survey of students in the M.A. and various M.S. programs over a five-year period showed that after graduation 8 went into teaching at the small-college level, 20 went on to a Ph.D. program, 32 went into business or industry, and 2 went into government. Specific examples include instructor at Taylor University; analyst for the Census Bureau in Washington, D.C.; systems programmer for Meditech Corp. in Boston; consultant for PricewaterhouseCoopers in Washington, D.C.; and Ph.D. students at Penn State, Ohio State, and North Carolina.

Location

Oxford is a typical, very pretty college town, with a population of roughly 20,000, with commercial establishments surrounding the campus. The city is 35 miles from Cincinnati and 45 miles from Dayton.

The University and The Department

Miami is a state-assisted university serving roughly 14,000 students, founded under the Northwest Charter in 1809. It is frequently cited as a public ivy and ranked as one of the leading undergraduate institutions in the United States. This tradition of excellence in teaching carries over to its graduate programs. In mathematics and statistics, one finds a faculty of accomplished scholars with national and international reputations, who are also committed to significant effort in teaching and to working one on one with students in independent studies and research. This combination of research strength and outstanding teaching typifies Miami and the Department in particular.

Applying

Applications for admission and assistantships should be sent in by February 1. There are two separate applications, one for admission that is sent to the Graduate School and one for a graduate assistantship that is sent to the Director of Graduate Studies in the Department. Paper copies of the applications can be obtained by contacting the Graduate Director. The applications can also be obtained and completed online. The application for admission to the Graduate School can be found at http://www.miami.muohio.edu/academics/graduateprograms/index.cfm and the one for an assistantship at the program's Web site. International students must submit TOEFL scores.

Correspondence and Information

Dr. Zevi Miller
Director of Graduate Studies
Department of Mathematics and Statistics
Miami University
Oxford, Ohio 45056

Phone: 513-529-3520
E-mail: millerz@muohio.edu
Web site: http://www.muohio.edu/mathstat/graduate

Miami University

THE FACULTY AND THEIR RESEARCH

Reza Akhtar, Assistant Professor; Ph.D., Brown. Algebraic geometry (algebraic cycles, Chow groups, motives, K-theory).
A. John Bailer, Professor; Ph.D., North Carolina. Biostatistics, quantitative risk estimation, statistical methods for the design and analysis of environmental and occupational health studies.
Olga Brezhneva, Assistant Professor; Ph.D., Russian Academy of Sciences. Optimization, numerical analysis.
Dennis Burke, Professor; Ph.D., Washington State. Set-theoretic topology (study of general topological spaces with the use of techniques and notation of modern set-theory as tools).
Beatriz D'Ambrosio, Professor; Ph.D., Indiana. Mathematics education.
Dennis Davenport, Associate Professor; Ph.D., Howard. Topological semigroups.
Sheldon Davis, Professor; Ph.D., Ohio. Set theoretic topology.
Patrick N. Dowling, Professor; Ph.D., Kent State. Functional analysis: in particular, geometry of Banach spaces with applications to metric fixed-point theory and harmonic analysis.
Charles L. Dunn, Professor; Ph.D., Texas A&M. Multivariate statistics, simulation of percentile points.
Thomas Farmer, Associate Professor; Ph.D., Minnesota. Algebra, undergraduate mathematics.
Frederick Gass, Professor; Ph.D., Dartmouth. Logic, mathematics education.
David Groggel, Associate Professor; Ph.D., Florida. Nonparametric statistics, statistical education, statistics in sports.
Suzanne Harper, Assistant Professor; Ph.D., Virginia. Appropriate use of technology to teach K–12 mathematics, the content knowledge of prospective mathematics teachers, the teaching and learning of geometry.
Charles S. Holmes, Professor; Ph.D., Michigan. Relationship between the structure of the subgroup lattice of a group G and the structure of the group G itself.
Tao Jiang, Assistant Professor; Ph.D., Illinois at Urbana-Champaign. Graph theory and combinatorics.
Dennis Keeler, Assistant Professor; Ph.D., Michigan. Noncommutative algebraic geometry: using the techniques of algebraic geometry to study noncommutative rings, vanishing theorems in algebraic geometry.
Jane Keiser, Associate Professor; Ph.D., Indiana. Mathematics education.
Dave Kullman, Professor; Ph.D., Kansas. History of mathematics and mathematics education.
Paul Larson, Assistant Professor; Ph.D., Berkeley. Set theory.
Bruce Magurn, Professor; Ph.D., Northwestern. Algebra, number theory, K-theory.
Zevi Miller, Professor; Ph.D., Michigan. Graph theory, combinatorics, graph algorithms.
Emily Murphree, Associate Professor; Ph.D., North Carolina. Statistics, probability.
Robert Noble, Assistant Professor; Ph.D., Virginia Tech. Application of Bayesian model averaging to multivariate models, environmental statistical applications, statistical procedures associated with stability analysis.
Ivonne Ortiz, Assistant Professor; Ph.D., SUNY at Binghamton. Algebraic K-theory.
Daniel Pritikin, Professor; Ph.D., Wisconsin. Graph theory, combinatorics.
Beata Randrianantoanina, Associate Professor; Ph.D., Missouri. Functional analysis, linear and nonlinear problems in geometry of Banach spaces.
Narcisse Randrianantoanina, Associate Professor; Ph.D., Missouri. Banach space structures of noncommutative Lp-spaces, noncommutative Hardy spaces, roles of noncommutative martingales in quantum probability theory.
Robert Schaefer, Professor; Ph.D., Michigan. Biostatistics, statistical computing.
Kyoungah See, Associate Professor; Ph.D., Virginia Tech. Sampling designs, principal component analysis, environmental toxicity studies.
John Skillings, Professor; Ph.D., Ohio State. Statistics, experimental design.
Mark Smith, Professor; Ph.D., Illinois at Urbana-Champaign. Functional analysis, geometry of Banach spaces.
Robert Smith, Professor; Ph.D., Penn State. Mathematics education, algebra.
Jerry Stonewater, Associate Professor; Ph.D., Michigan State. Mathematics education.
Vasant Waikar, Professor; Ph.D., Florida State. Distribution of characteristic roots of random matrices, two-stage estimation, application of bootstrap sampling to estimation.
Douglas Ward, Professor; Ph.D., Dalhousie. Optimization, operations research.
John Westman, Assistant Professor; Ph.D., Illinois at Chicago. Stochastic optimal control, computational finance, applications of biomathematics.
Stephen Wright, Associate Professor; Ph.D., Washington (Seattle). Mathematical programming, applications of convex optimization to analysis of scientific data, decomposition algorithms for large-scale optimization.

SELECTED PUBLICATIONS

Akhtar, R. Zero-cycles on varieties over finite fields. *Commun. Algebra,* in press.

Akhtar, R. Torsion in mixed K-groups. *Commun. Algebra,* in press.

Bailer, A. J., et al. **(K. See** and **R. S. Schaefer).** Defining and evaluating impact in environmental toxicology. *Environmetrics* 14:235–43, 2003.

Bailer, A. J., and W. W. Piegorsch. From quantal response to mechanisms and systems: The past, present, and future of biometrics in environmental toxicology. *Biometrics* 56:327–36, 2000.

Brezhneva, O. A., and A. A. Tretyakov. Optimality conditions for degenerate extremum problems with equality constraints. *SIAM J. Control Optimization* 42(2):729–45, 2003.

Brezhneva, O. A., and A. F. Izmailov. Construction of defining systems for finding singular solutions to nonlinear equations. *Computational Mathematics Math. Phys.* 42(1):8–19, 2002.

Burke, D. K., and R. Pol. On nonmeasurability of L^∞/C_0 in its second dual. *Proc. Am. Math. Soc.,* in press.

Burke, D. K., and L. D. Ludwig. Hereditarily α-normal spaces and infinite products. *Top. Proc.* 25:291–9, 2002.

D'Ambrosio, B. Patterns of instructional discourse that promote the perception of mastery goals in a social constructivist mathematics course. *Educ. Stud. Math.* 56(1):19–34, 2004.

Davis, S., et al. Strongly almost disjoint sets and weakly uniform bases. *Trans. AMS* 4971–87, 2000.

Davis, S., D. K. Burke, and Z. Balogh. A ZFC nonseparable Lindelof symmetrizable Hausdorff space. *C.R. Bulgarian Acad. Sci.* 11–2, 1989.

Dowling, P., B. Turett, and C. J. Lennard. Characterizations of weakly compact sets and new fixed point free maps in c_0. *Studia Math.* 154:277–93, 2003.

Dowling, P., and **N. Randrianantoanina.** Riemann-Lebesgue properties of Banach spaces associated with subsets of countable discrete Abelian groups. *Glasgow Math. J.* 45:159–66, 2003.

Dunn, C. L. Precise similated percentiles in a pinch. *Am. Statistician* 45(3):201–11, 1991.

Dunn, C. L. Application of multiple comparison type procedures to the eigenvalues of $\Sigma_1{}^{-1}\Sigma_2$. *Commun. Statistics: Theory Methods A* 15(2):451–71, 1986.

Harper, S. R. Enhancing elementary preservice teachers' knowledge of geometric transformations through the use of dynamic geometry computer software. In *Society for Information Technology and Teacher Education International Conference Annual,* pp. 2909–16, eds. C. Crawford et al. Norfolk, Va.: Association for the Advancement of Computing in Education, 2003.

Harper, S. R., S. O. Schirack, H. D. Stohl, and J. Garofalo. Learning mathematics and developing pedagogy with technology: A reply to Browning and Klespis. *Contemp. Issues Technol. Teacher Educ.* 1(3):346–54, 2001 (online).

Holmes, C. S., M. Costantini, and G. Zacher. A representation theorem for the group of autoprojectivities of an Abelian p-group of finite exponent. *Ann. Matematica (IV)* CLXXV:119–40, 1988.

Holmes, C. S. Generalized Rottlaender, Honda, Yff groups. *Houston J. Math.* 10:405–14, 1984.

Jiang, T. Anti-Ramsey numbers of subdivided graphs. *J. Combinatorial Theory Ser. B* 85:361–6, 2002.

Jiang, T. On a conjecture about trees in graphs with large girth. *J. Combinatorial Theory Ser. B* 83:221–32, 2001.

Jiang, T., and D. Mubayi. New upper bounds for a canonical Ramsey problem. *Combinatorica* 20:141–6, 2000.

Keeler, D. Noncommutative ampleness for multiple divisors. *J. Algebra,* in press.

Keeler, D. Criteria for σ-ampleness. *J. Am. Math. Soc.* 13(3):517–32, 2000.

Kullman, D. Stories about story problems. *Centroid* 29(1):10–4, 2003.

Kullman, D. Undergraduate mathematics in the Old Northwest. In *Proceedings of the History of Undergraduate Mathematics in America Conference,* pp. 195–208, 2002.

Larson, P. A uniqueness theorem for iterations. *J. Symbolic Logic,* in press.

Larson, P., and S. Todorevic. Katetov's problem. *Trans. AMS* 354:1783–91, 2002.

Miller, Z., L. Gardner, **D. Pritikin,** and I. H. Sudborough. One-to-many embeddings of hypercubes into cayley graphs generated by reversals. *Theory Computing Syst.* 34:399–431, 2001.

Miller, Z., and **D. Pritikin.** On randomized greedy matchings. *Random Struct. Algorithms* 10:353–83, 1997.

Noble, B. Model selection in canonical correlation analysis (CCA) using Bayesian model averaging. *Environmetrics,* in press.

Noble, B. An alternative model for cylindrical data. *Nonlinear Analysis Ser. A* 47:2011–22, 2001.

Randrianantoanina, B. On the structure of level sets of uniform and Lipschitz quotient mappings from R^n to R. *Geometric Functional Analysis,* in press.

Randrianantoanina, B. On isometric stability of complemented subspaces of L^P. *Israel J. Math.* 113:45–60, 1999.

Randrianantoanina, N. Noncommutative martingale transforms. *J. Funct. Anal.* 194:181–212, 2002.

Randrianantoanina, N. Factorizations of operators on C*-algebras. *Studia Math.* 128:273–85, 1998.

Miami University

See, K., J. Stufken, S. Y. Song, and **A. J. Bailer**. Relative efficiencies of sampling plans for selecting a small number of units from a rectangular region. *J. Stat. Computation Simulation* 66:273–94, 2000.

See, K., and S. Y. Song. Association schemes of small order. *J. Stat. Plann. Inferences* 73(1/2):225–71, 1998.

Smith, M. A., P. Dowling, and Z. Hu. Geometry of spaces of vector-valued harmonic functions. *Can. J. Math.* 46:274–83, 1994.

Smith, M. A., and B. Turett. Normal structure in Bochner L^P-spaces *Pacific J. Math.* 142:347–56, 1990.

Waikar, V., F. Schuurman, and S. R. Adke. A two-stage shrinkage testimator for the mean of an exponential distribution. *Commun. Statistics* 16:1821–34, 1987.

Waikar, V., F. Schuurman, and T. E. Raghunathan. On a two-stage shrinkage testimator for the mean of a normal distribution. *Commun. Statistics* 13:1901–13, 1984.

Ward, D., and M. Studniarski. Weak sharp minima: Characterizations and sufficient conditions. *SIAM J. Control Optimization* 38:219–36, 1999.

Ward, D. Dini derivatives of the marginal function of a non-Lipschitzian program. *SIAM J. Optimization* 6:198–211, 1996.

Westman, J. J., and F. B. Hanson. Optimal portfolio and consumption policies subject to Rishel's important jump events model: Computational methods. *Trans. Automatic Control,* in press.

Westman, J. J., F. B. Hanson, and E. K. Boukas. Optimal production scheduling for manufacturing systems with preventive maintenance in an uncertain environment. In *Proceedings of 2001 American Control Conference,* 25 June 2001, pp. 1375–80.

Wright, S. E., J. A. Foley, and J. M. Hughes. Optimization of site-occupancies in minerals using quadratic programming. *Am. Mineral.* 85:524–31, 2000.

Wright, S. E. A general primal-dual envelope method for convex programming problems. *SIAM J. Optimization* 10:405–14, 2000.

NEW YORK UNIVERSITY

Courant Institute of Mathematical Sciences
Department of Mathematics

Program of Study

The graduate program offers a balanced array of options, with special focus on mathematical analysis and on applications of mathematics in the broadest sense. It includes computational applied mathematics as well as strong interactions with neural science and other science departments. The program of study leads to the M.S. and Ph.D. degrees in mathematics. A Ph.D. degree in atmosphere/ocean science and mathematics is also offered in cooperation with the Center for Atmosphere-Ocean Studies, and a Ph.D. in computational biology is offered. It is possible to earn a master's degree through part-time study, but students in the Ph.D. program are full-time. In addition to the standard M.S. degree, special career-oriented programs are available in financial mathematics and scientific computing. The M.S. degree can be completed in the equivalent of three or four terms of full-time study. Doctoral students obtain the M.S. degree as they fulfill the requirements for the Ph.D. Students must earn 72 course and research points for the Ph.D., but no specific courses are required. One requirement is the Written Comprehensive Examination, which is often taken during the first year of full-time study. A second requirement is the Oral Preliminary Examination, which serves as the threshold between course work and thesis research. Thereafter, students engage in research under the supervision of a faculty adviser, leading to the writing and defense of a doctoral dissertation. Students are encouraged from the outset to participate in the Institute's extensive research activities and to use its sophisticated computing environment.

The Department occupies a leading position in applied mathematics, differential equations, geometry/topology, probability, and scientific computing. In applied mathematics, the Department's activities go beyond differential equations and numerical analysis to encompass many topics not commonly found in a mathematics department, including neural science, atmosphere/ocean science, computational fluid dynamics, financial mathematics, materials science, mathematical physiology, plasma physics, and statistical physics.

The Department has been successful in helping its Ph.D. graduates find desirable positions at universities or in nonacademic employment. Those interested may visit the Department's job placement Web page at http://www.math.nyu.edu/degree/guide/job_placement.htm.

Research Facilities

The Courant Institute Library, which is located in the same building as the Department, has one of the nation's most complete mathematics collections; it receives more than 275 journals and holds more than 64,000 volumes. Students have access to MathSciNet and Web of Science (Science Citation Index) and an increasing number of electronic journals. The Institute's computer network is fully equipped with scientific software; X-terminals are available in public locations and in every graduate student office. The Courant Applied Mathematics Laboratory comprises an experimental facility in fluid mechanics and other applied areas, coupled with a visualization and simulation facility.

Financial Aid

Financial support is awarded to students who engage in full-time Ph.D. study, covering tuition, fees, and NYU's individual comprehensive insurance plan and, in 2006–07, providing a stipend of $21,500 for the nine-month academic year. Some summer positions associated with Courant Institute research projects are available to assistants with computational skills. Because the Department is unable to support all qualified students, applicants should apply for other support as well. Federally funded low-interest loans are available to qualified U.S. citizens on the basis of need.

Cost of Study

In 2006–07, tuition is calculated at $1080 per point. Associated fees are calculated at $329 for the first point in fall 2006, $342 for the first point in spring 2007, and $56 per point thereafter in both terms. A full-time program of study normally consists of 24 points per year (four 3-point courses each term).

Living and Housing Costs

University housing for graduate students is limited. It consists mainly of shared studio apartments in buildings adjacent to Warren Weaver Hall and shared suites in residence halls within walking distance of the University. University housing rents in the 2006–07 academic year range from $1100 to $1400 per month.

Student Group

In 2005–06, the Department had 315 graduate students. Fifty-six percent were full-time students.

Location

New York City is a world capital for art, music, and drama and for the financial and communications industries. NYU is located at Washington Square in Greenwich Village, just north of SoHo and Tribeca in a residential neighborhood consisting of apartments, lofts, art galleries, theaters, restaurants, and shops.

The University and The Institute

New York University, which was founded in 1831, enrolls about 50,000 students and is one of the major private universities in the world. Its various schools offer a wide range of undergraduate, graduate, and professional degrees. Among its internationally known divisions is the Courant Institute of Mathematical Sciences. Named for its founder, Richard Courant, the Institute combines research in the mathematical sciences with advanced training at the graduate and postdoctoral levels. Its activities are supported by the University, government, industry, and private foundations and individuals. The graduate program in mathematics is conducted by the faculty of the Courant Institute. The mathematics department ranks among the leading departments in the country and is the only highly distinguished department to have made applications a focal concern of its programs. Eleven members of the Courant Institute faculty are members of the National Academy of Sciences.

Applying

The graduate program is open to students with strong mathematical interests, regardless of their undergraduate major. They are expected to have knowledge of the elements of mathematical analysis. Applications for admission are evaluated throughout the year, but a major annual review of applications to the Ph.D. program occurs in February, and most awards for the succeeding academic year are made by early March. Ph.D. applications must include GRE scores on both the General and Subject Tests and must be received by January 4. The application deadline for the M.S. programs is June 1, except for the Mathematics in Finance program, which is March 1.

Correspondence and Information

For program and financial aid information:

Fellowship Committee
Courant Institute
New York University
251 Mercer Street
New York, New York 10012
Phone: 212-998-3238
E-mail: admissions@math.nyu.edu
Web site: http://www.math.nyu.edu

For application forms and a Graduate School bulletin:

Graduate Enrollment Services
Graduate School of Arts and Science
New York University
P.O. Box 907, Cooper Station
New York, New York 10276-0907
Phone: 212-998-8050
E-mail: gsas.admissions@nyu.edu
Web site: http://www.nyu.edu/gsas

New York University

THE FACULTY AND THEIR RESEARCH

Professors

Marco M. Avellaneda, Ph.D. Applied mathematics, mathematical modeling in finance, probability.
Gerard Ben Arous, Ph.D. Probability theory and applications, large deviations, statistical mechanics, spectra of random matrices, stochastic processes in random media, partial differential equations.
Simeon M. Berman, Ph.D. Stochastic processes, probability theory, applications.
Fedor A. Bogomolov, Ph.D. Algebraic geometry and related problems in algebra, topology, symplectic geometry, and number theory.
Sylvain E. Cappell, Ph.D. Algebraic and geometric topology, symplectic and algebraic geometry.
Jeff Cheeger, Ph.D. Differential geometry and its connections to analysis and topology.
Francesca Chiaromonte, Ph.D. Multivariate analysis and regression, Markov modeling, analysis and modeling of large-scale genomic data.
W. Stephen Childress, Ph.D. Fluid dynamics, magnetohydrodynamics, biological fluid dynamics.
Tobias H. Colding, Ph.D. Differential geometry, geometric analysis, partial differential equations, three-dimensional topology.
Percy A. Deift, Ph.D. Spectral theory and inverse spectral theory, integrable systems, Riemann-Hilbert problems, random matrix theory.
Paul R. Garabedian, Ph.D. Complex analysis, computational fluid dynamics, plasma physics.
Jonathan Goodman, Ph.D. Fluid dynamics, computational physics, computational finance.
Leslie Greengard, Ph.D. Applied and computational mathematics, partial differential equations, computational chemistry, mathematical biology, optics.
Frederick P. Greenleaf, Ph.D. Noncommutative harmonic analysis, Lie groups and group representations, invariant partial differential operators.
Mikhael Gromov, Ph.D. Riemannian manifolds, symplectic manifolds, infinite groups, mathematical models of biomolecular systems.
Eliezer Hameiri, Ph.D. Applied mathematics, magnetohydrodynamics, plasma physics.
Helmut Hofer, Ph.D. Symplectic geometry, dynamical systems, partial differential equations.
Richard Kleeman, Ph.D. Predictability of dynamical systems relevant to the atmosphere and the ocean, climate dynamics.
Robert V. Kohn, Ph.D. Nonlinear partial differential equations, materials science, mathematical finance.
Fang-Hua Lin, Ph.D. Partial differential equations, geometric measure theory.
Andrew J. Majda, Ph.D. Modern applied mathematics, atmosphere/ocean science, partial differential equations.
Henry P. McKean, Ph.D. Probability, partial differential equations, complex function theory.
David W. McLaughlin, Ph.D. Applied mathematics, nonlinear wave equations, visual neural science.
Charles M. Newman, Ph.D. Probability theory, statistical physics, stochastic models.
Albert B. J. Novikoff, Ph.D. Analysis, history of mathematics, pedagogy.
Jerome K. Percus, Ph.D. Chemical physics, mathematical biology.
Charles S. Peskin, Ph.D. Applications of mathematics and computing to problems in medicine and biology: cardiac fluid dynamics, molecular machinery within biological cells, mathematical/computational neuroscience.
Richard M. Pollack, Ph.D. Algorithms in real algebraic geometry, discrete geometry, computational geometry.
John Rinzel, Ph.D. Computational neuroscience, nonlinear dynamics of neurons and neural circuits, sensory processing.
Jalal M. I. Shatah, Ph.D. Partial differential equations, analysis.
Michael Shelley, Ph.D. Applied mathematics and modeling, visual neuroscience, fluid dynamics, computational physics and neuroscience.
Joel H. Spencer, Ph.D. Discrete mathematics, theoretical computer science.
Daniel L. Stein, Ph.D. Quenched disorder in condensed-matter systems, stochastic escape phenomena, fluctuations in mesoscopic systems.
Srinivasa S. R. Varadhan, Ph.D. Probability theory, stochastic processes, partial differential equations.
Harold Weitzner, Ph.D. Plasma physics, fluid dynamics, differential equations.
Olof Widlund, Ph.D. Numerical analysis, partial differential equations, parallel computing.
Lai-Sang Young, Ph.D. Dynamical systems and ergodic theory.

Associate Professors

Steve Allen (Clinical). Mathematical finance.
Oliver Bühler, Ph.D. Geophysical fluid dynamics, interactions between waves and vortices, acoustics, statistical mechanics.
David Cai, Ph.D. Nonlinear stochastic behavior in physical and biological systems.
Yu Chen, Ph.D. Numerical scattering theory, ill-posed problems, scientific computing.
David M. Holland, Ph.D. Ocean-ice studies, climate theory and modeling.
Nader Masmoudi, Ph.D. Nonlinear partial differential equations.
Sylvia Serfaty, Ph.D. Partial differential equations, variational problems with applications to physics.
Esteban G. Tabak, Ph.D. Physical processes in the atmosphere and ocean, turbulence.
Daniel Tranchina, Ph.D. Mathematical modeling in neuroscience.
Eric Vanden-Eijnden, Ph.D. Applied mathematics, stochastic processes, statistical physics.
Akshay Venkatesh, Ph.D. Analytic number theory, algebraic geometry.

Assistant Professors

Jinho Baik, Ph.D. Random matrices, Riemann-Hilbert problems.
Sinan Güntürk, Ph.D. Harmonic analysis, information theory, signal processing.
Olivier Pauluis, Ph.D. Climate and the general circulation of the atmosphere, moist convection, tropical meteorology, numerical modeling.
Aaditya Rangan, Ph.D. Large-scale scientific modeling of physical, biological, and neurobiological phenomena.
Weiqing Ren, Ph.D. Applied mathematics, scientific computing, multiscale modeling of fluids.
Scott Sheffield, Ph.D. Probability and mathematical physics.
K. Shafer Smith, Ph.D. Geophysical fluid dynamics, physical oceanography and climate.
Anna-Karin Tornberg, Ph.D., Numerical analysis, computational fluid dynamics, moving boundary problems.
Jun Zhang, Ph.D. Fluid dynamics, biophysics, complex systems.

Associated Faculty

Marsha J. Berger (Computer Science), Kit Fine (Philosophy), Bhubaneswar Mishra (Computer Science), Michael L. Overton (Computer Science), Nicolaus Rajewsky (Biology), Tamar Schlick (Chemistry, Computer Science), Demetri Terzopoulos (Computer Science).

Affiliated Faculty

Robert Shapley (Neural Science), Eero P. Simoncelli (Neural Science), Alan Sokal (Physics), George Zaslavsky (Physics).

NC STATE UNIVERSITY

NORTH CAROLINA STATE UNIVERSITY

Master of Financial Mathematics

Program of Study

The financial mathematics program at North Carolina State University (NCSU) provides a structured program to help students develop their skills in preparation for entering the workplace or for further study in related Ph.D. programs. Employment opportunities exist with banks, investment firms, financial trading companies and financial exchanges, insurance companies, power companies, natural resource–based firms, agribusinesses, and government regulatory institutions.

This program is limited to a master's degree but is closely related to research programs and Ph.D. degrees conducted in participating departments. This two-year program provides an integrated set of tools for students seeking careers in quantitative financial analysis. After taking six core courses in the first year to provide a common foundation, students have the flexibility in the second year to either pursue a broad view of the subject or specialize in a topic of their choice. This second year consists of four elective courses and a project/internship, which provides students with experience working on real-world problems in financial mathematics under the guidance of faculty members who are actively engaged in research.

The core courses have been chosen to provide students with a strong mathematical background, statistical and computational tools, and a comprehensive description of financial markets. The departments participating in the core requirements are mathematics, industrial engineering, statistics, economics, and agricultural and resource economics. Elective courses are offered by these departments and others, including business management and computer science. In addition, a seminar series organized in conjunction with this program exposes students to the ideas of outside academics and practitioners.

Research Facilities

The students in the program have access to the facilities offered by the participating departments.

Financial Aid

The participating departments offer a very small number of teaching assistantships on a competitive basis.

Cost of Study

For the 2005–06 academic year, full-time students paid tuition in the amount of $1856.50 (in state) or $7880.50 (out of state) plus $572 for student fees per semester. In-state residency for U.S. students can be established after one year. Students receiving teaching assistantships receive free tuition but pay student fees. These figures do not include the cost of housing, textbooks, or other supplies.

Living and Housing Costs

One-bedroom apartments or shared larger apartments near the campus can cost $400 to $700 (per person). Graduate housing is available on campus and in the E. S. King Village, an apartment complex designed for graduate students and for family housing.

Student Group

The program is small to ensure that each student obtains the individual attention he or she needs. The target size for each entering class is approximately 12 students.

Student Outcomes

Student placement is aimed at advanced academic degrees as well as various jobs in the private sector. All program graduates have found competitive job placement within several months of graduation.

Location

Raleigh, North Carolina, is only a short drive from Duke University, the University of North Carolina at Chapel Hill, and Research Triangle Park. Raleigh is home to a symphony, ballet, opera, and theater as well as museums and historic sites. Raleigh lies only a few hours from the majestic Appalachian Mountains and the lovely beaches of the Atlantic Ocean.

The University

As a land-grant university, North Carolina State has historic strengths in agriculture, technology, and engineering and has demonstrated strengths in emerging fields, such as the computational sciences. The Master of Financial Mathematics degree program combines these strengths to train students in the new discipline of quantitative finance and risk analysis. NCSU has a graduate student-faculty ratio of 3.5:1 to allow students to work closely with faculty members and is recognized nationally and internationally as a top teaching and research university.

Applying

Acceptance is based on scholastic records, as reflected by the courses chosen and quality of performance; evaluation of former teachers and advisers; Graduate Record Examinations (GRE) General Test scores; and TOEFL scores (if applicable). Applicants should have an undergraduate degree in mathematics or in a closely related field with a strong mathematical background. A GPA of at least 3.0 (out of 4.0) in the sciences is required. The TOEFL is required of international students; TOEFL scores of at least 550 are preferred. For most favorable consideration for the fall term, all application materials should be received by February 1.

Correspondence and Information

Financial Math Program
Campus Box 7640
North Carolina State University
Raleigh, North Carolina 27695-7640
Phone: 919-513-2287
Fax: 919-513-1991
E-mail: jmjones4@math.ncsu.edu
Web site: http://www.math.ncsu.edu/finmath

North Carolina State University

THE FACULTY AND THEIR RESEARCH

Richard H. Bernhard, Professor, Department of Industrial Engineering; Ph.D., Cornell, 1961. Capital investment economic analysis, Bayesian decision analysis, multiattribute decision making, financial engineering.

Peter Bloomfield, Professor, Department of Statistics; Ph.D., London, 1970. Time series, credit risk.

Xiuli Chao, Professor, Department of Industrial Engineering; Ph.D., Columbia, 1989. Stochastic modeling and analysis, investment analysis, stochastic optimization, queuing and stochastic service systems, production and inventory systems, Markovian decision processes, supply chain and value chain management.

David Dickey, Professor, Department of Statistics; Ph.D., Iowa State, 1976. Time series, regression, general statistical methodology.

Salah E. Elmaghraby, Professor, Departments of Industrial Engineering and Operations Research; Ph.D., Cornell, 1958.

Edward W. Erickson, Professor, Department of Economics; Ph.D., Vanderbilt, 1968.

Paul L. Fackler, Associate Professor, Department of Agricultural and Resource Economics; Ph.D., Minnesota, 1986. Futures and options markets, commodity market analysis, risk analysis and management, computational economics.

Jean-Pierre Fouque, Professor, Department of Mathematics, and Director, Financial Mathematics Program; Ph.D., Paris VI (Curie), 1979. Stochastic processes, stochastic partial differential equations, random media, financial mathematics.

Marc Genton, Associate Professor, Department of Statistics; Ph.D., Swiss Federal Institute of Technology, 1996. Time series, multivariate analysis, data mining.

Sujit Ghosh, Associate Professor, Department of Statistics; Ph.D., Connecticut, 1996. Bayesian inference and applications.

Atsushi Inoue, Assistant Professor, Department of Agricultural and Resource Economics; Ph.D., Pennsylvania, 1998. Theoretical and applied econometrics.

Kazufumi Ito, Professor, Department of Mathematics.

Min Kang, Assistant Professor, Department of Mathematics; Ph.D., Cornell. Probability theory and partial differential equations, stochastic partial differential equations.

Tao Pang, Assistant Professor, Department of Mathematics; Ph.D., Brown, 2002. Financial engineering, stochastic control, operations research.

Sastry Pantula, Professor, Department of Statistics; Ph.D., Iowa State, 1982. Time series, spatial statistics, nonlinear models.

Jeffrey S. Scroggs, Associate Professor, Department of Mathematics; Ph.D., Illinois at Urbana-Champaign, 1988. Numerical methods for partial differential equations, fluid dynamics, scientific computing, financial mathematics.

John J. Seater, Professor, Department of Economics and Business; Ph.D., Brown, 1975. Macroeconomics, monetary economics, stability and control of dynamical systems.

Charles E. Smith, Associate Professor, Department of Statistics; Ph.D., Chicago. Poisson-driven stochastic differential equations, level crossing and first passage times.

Thomislav Vukina, Associate Professor, Department of Agricultural and Resource Economics; Ph.D., Rhode Island, 1991.

Jim Wilson, Professor, Department of Industrial Engineering; Ph.D., Rice, 1970. Probabilistic and statistical issues in the design and analysis of large-scale simulation experiments, analysis of output processes, improving simulating efficiency using variance-reduction techniques, optimization using multiple-comparison and search procedures, applications to production systems engineering and financial engineering.

Northeastern UNIVERSITY

NORTHEASTERN UNIVERSITY

Department of Mathematics
Graduate Programs in Mathematics

Programs of Study

The Department of Mathematics at Northeastern University offers M.S. and Ph.D. degrees in mathematics, an M.S. in applied mathematics, and an M.S. degree in operations research (in conjunction with the Department of Mechanical and Industrial Engineering). The department offers both full- and part-time M.S. and Ph.D. programs. The programs are designed to provide students with a broad overview of current mathematics and a strong command of an area of specialization. In addition to the course requirements, a thesis is required for the Ph.D. program. A thesis is optional in place of two electives in all master's-level programs.

Graduate students work with internationally recognized faculty members in a range of research programs in both pure and applied mathematics. In addition, numerous seminars and colloquia at Northeastern and in the Boston area give students ample opportunity to learn about important recent advances in mathematics. The department is an active participant in the Brandeis-Harvard-MIT-Northeastern Colloquium.

Mathematical sciences research at Northeastern is concentrated in three main areas: algebra-singularities-combinatorics, analysis-geometry-topology, and probability-statistics.

The algebra-singularities-combinatorics group includes strong researchers in areas covering discrete geometry, algebraic geometry, representation theory, K-theory, and singularities of mappings. Some of the concrete topics being studied are cluster algebras, regular tilings of Euclidean spaces, representations of algebraic and quantum groups, representations of quivers, Schubert varieties, motives, hyperplane arrangements, Koszul algebras, and commutative rings and their deformations. All these topics have combinatorial and computational components, which makes it possible to involve graduate students in hands-on calculations, bringing them quickly to the frontiers of modern research.

The analysis-geometry-topology group encompasses a wide range of research interests and activities in areas that include partial differential equations, geometric analysis, differential geometry, mathematical physics, algebraic topology, and geometric topology. Topics include index theory of elliptic operators, Schrödinger operators, conformal metrics, noncommutative geometry, integrable Hamiltonian systems, Maxwell Higgs systems, delay equations, geometry and topology of manifolds and submanifolds, topology of knots and links, and group cohomology.

The probability-statistics group is involved in a wide variety of research activities, ranging from basic research to industrial collaborations. This broad and varied program is made possible by the interdisciplinary interests of the faculty members in the group, who are involved in projects with the physics, engineering, computer science, pharmacology, and medical departments. The research areas include theoretical statistics, applied statistics, biostatistics, industrial statistics, information theory, and quantum computing.

Research Facilities

The University supports twenty-seven centers and institutes. A high-speed data network links users and facilities on the central campus to three satellite campuses and to computing facilities around the world. Students have access to Compaq Alpha systems, public-access microcomputer labs (PC and Mac), a conferencing system, multimedia labs, and specialized computing equipment. Northeastern University is also an Internet2 site.

University libraries contain more than 965,000 books, 2.3 million microforms, 7,600 serial subscriptions, and 17,000 audiovisual materials. The libraries have licensed access to more than 13,000 electronic information sources. A central and branch library contain technologically sophisticated services, including Web-based catalog and circulation systems and a Web portal to licensed electronic resources. The University is a member of the Boston Library Consortium and the Boston Regional Library System, giving students and faculty members access to the region's collections and information resources.

Financial Aid

Each year, the department offers a limited number of Research Assistantships (RAs), Teaching Assistantships (TAs), and Northeastern University Tuition Assistantships (NUTAs) to promising full-time students. An RA includes tuition and a stipend, and a recipient is required to complete a project for a professor each semester. A TA includes tuition and a stipend, and a recipient is required to teach a basic undergraduate course each semester. For this reason, international students receiving a TA should be able to speak English fluently. An NUTA covers a specified amount of tuition only, and a recipient is required to assist with grading and tutoring each semester for 10 hours a week.

Cost of Study

The tuition rate for 2006–07 is $930 per semester hour. There are special tuition charges for theses and dissertations, where applicable. The Student Center fee and health insurance fee required for all full-time students are approximately $1900 per academic year. However, students receiving financial aid (RAs and TAs) not only do not have to pay tuition but receive a 40 percent discount on their health insurance, and the hope is that it will eventually reach 100 percent.

Living and Housing Costs

On-campus housing for graduate students is limited and granted on a space-available basis. For more information about on- and off-campus housing, students should go to http://www.housing.neu.edu.

Student Group

During the 2005–06 academic year, the department had 30 full-time graduate students and 4 part-time graduate students.

Student Outcomes

The majority of graduates find employment in various high-technology industries across the United States. Ph.D. graduates are also employed by academic institutions in teaching and research.

Location

Northeastern University is set in the heart of the ultimate college town—Boston. A high-energy hub of cultural, educational, and social activity, Boston is home to more than 300,000 college students from around the country and the world. The city is alive with people of every race, ethnicity, political persuasion, and religion. Within walking distance of Northeastern are the world-renowned Museum of Fine Arts, Symphony Hall, and stylish Newbury Street, with great shopping and dining.

The University

Northeastern University is a world leader in practice-oriented education and is recognized for its expert faculty members and first-rate academic and research facilities. Northeastern has six undergraduate colleges, eight graduate and professional schools, two part-time undergraduate divisions, and an extensive variety of research institutes and divisions. Northeastern's graduate programs offer both professional and research degrees at the master's or doctoral level.

Applying

Applicants must have a bachelor's degree in mathematics or a closely related field. Applicants to the Ph.D. program must, in addition, have a master's degree in mathematics or a closely related field. Applicants must have taken the Graduate Record Examinations (GRE) General Test and the Subject Test in mathematics. International students must demonstrate proficiency in English. An applicant's undergraduate course work should include linear algebra, combinatorics, differential and integral calculus, differential equations, real analysis, and some computer programming. Students who are deficient in any of these areas may be accepted provisionally if their overall college work is particularly strong, but they must eliminate the deficiency (summer courses are not available) within their first semester at Northeastern.

All applicants must submit a completed application form, including official transcripts of all previous undergraduate and graduate course work and a nonrefundable $50 processing fee. Three letters of recommendation, preferably from people acquainted with the applicant's academic and personal qualifications, are required. Only those documents required to complete the application package should be sent, as unsolicited documents do not improve the applicant's chances for admission.

The application deadline for fall admission for the Ph.D. program and for the master's program with an assistantship is February 1. The deadline is May 1 for international admission and August 1 for domestic admission for the M.S. program. The special student admission deadline is August 25.

Correspondence and Information

Christopher King, Graduate Coordinator
437 Lake Hall
Northeastern University
Boston, Massachusetts 02115

Phone: 617-373-3905
E-mail: mathdept@neu.edu
Web site: http://www.math.neu.edu/grad/grad.html

Northeastern University

THE FACULTY AND THEIR RESEARCH

Professors
Samuel J. Blank, Ph.D., Brandeis, 1967. Differential topology.
Robert W. Case, Ph.D., Yeshiva, 1966. Mathematical logic, Socratic teaching of mathematics.
Stanley J. Eigen, Ph.D., McGill, 1982. Ergodic theory, measure theory, number theory, dynamical systems.
Terence Gaffney, Ph.D., Brandeis, 1976. Singularities of mappings and its application to algebraic and differential geometry.
Maurice E. Gilmore, Ph.D., Berkeley, 1967. Geometric topology, secondary education.
Arshag Hajian, Ph.D., Yale, 1957. Ergodic theory, analysis.
Anthony Iarrobino, Ph.D., MIT, 1970. Algebraic geometry, commutative rings and their deformations, singularities of maps, families of points on a variety, Gorenstein algebras.
Christopher King, Graduate Coordinator; Ph.D., Harvard, 1984. Mathematical physics.
V. Lakshmibai, Ph.D., Tata (Bombay), 1976. Algebraic geometry, algebraic groups, representation theory.
Marc N. Levine, Ph.D., Brandeis, 1979. Algebraic geometry, algebraic K-theory, motives, motivic cohomology.
Mikhail B. Malioutov, D.Sc., Moscow State, 1983. Statistics, probability, experimental design, information theory.
Robert C. McOwen, Chairman; Ph.D., Berkeley, 1978. Partial differential equations, with applications to problems in differential geometry.
Richard D. Porter, Ph.D., Yale, 1971. Algebraic and differential topology; Massey products; deRham theory, with applications to the fundamental group and group cohomology.
Egon Schulte, Ph.D., Dortmund (Germany), 1980. Discrete geometry, combinatorics, group theory.
Jayant M. Shah, Ph.D., MIT, 1974. Computer vision.
Mikhail A. Shubin, Matthews Distinguished Professor; Ph.D., Moscow State, 1969. Partial differential equations, geometric analysis, spectral theory, mathematical physics.
Alexandru I. Suciu, Ph.D., Columbia, 1984. Algebraic topology, geometric topology.
Jerzy Weyman, Ph.D., Brandeis, 1980. Commutative algebra, algebraic geometry, representation theory.
Andrei Zelevinsky, Ph.D., Moscow State, 1978. Representation theory, algebraic geometry, algebraic combinatorics, discrete geometry, special functions.

Associate Professors
Maxim Braverman, Ph.D., Tel Aviv, 1997. Symplectic geometry, partial differential equations.
Mark Bridger, Ph.D., Brandeis, 1967. Mathematics education, computer-assisted instruction, numerical and constructive algebra, commutative algebra.
Adam Ding, Ph.D., Cornell, 1996. Artificial neural networks, high-dimensional empirical linear prediction (HELP), biostatistics, prediction and confidence intervals.
John N. Frampton, Ph.D., Yale, 1965. Artificial intelligence, natural language.
Eugene H. Gover, Ph.D., Brandeis, 1970. Commutative algebra, homology of local rings.
Samuel Gutmann, Ph.D., MIT, 1977. Quantum computing, statistical decision theory, probability, syntax.
Solomon M. Jekel, Undergraduate Head Advisor; Ph.D., Dartmouth, 1974. Classifying spaces, homeomorphism groups, homology of groups, foliations.
Donald R. King, Vice Chairman; Ph.D., MIT, 1979. Lie groups, Lie algebras, Weyl groups, Lie algebra cohomology, noncommutative ring theory.
Nishan Krikorian, Ph.D., Cornell, 1969. Low-dimensional dynamical systems, numerical analysis.
Alex Martsinkovsky, Undergraduate Coordinator; Ph.D., Brandeis, 1987. Functorial and homological methods in representation theory, homological algebra, homotopy theory and applications, industrial mathematics.
David B. Massey, Ph.D., Duke, 1986. Complex analytic singularities, stratified spaces.
Mark B. Ramras, Ph.D., Brandeis, 1967. Commutative algebra, graph theory.
Martin Schwarz Jr., Ph.D., NYU, 1981. Nonlinear analysis, nonlinear differential equations, mathematical problems in science.
Thomas O. Sherman, Ph.D., MIT, 1964. Noncommutative harmonic analysis, symmetric spaces, Lie groups, numerical analysis.
Gordana G. Todorov, Ph.D., Brandeis, 1979. Representation theory of Artin algebras, noncommutative algebra.

Assistant Professor
Peter Topalov, Ph.D., Moscow State, 1997. Hamiltonian PDEs and ODEs, dynamical systems, Riemannian and symplectic geometry.

Clinical Assistant Professor
Carla B. Oblas, M.S., California, Davis, 1972.

OSU

OKLAHOMA STATE UNIVERSITY

Department of Mathematics

Programs of Study

The Department of Mathematics offers programs leading to the Master of Science and Doctor of Philosophy degrees. There are three Master of Science degree options—pure mathematics, applied mathematics, and mathematics education—each requiring 32 credit hours of graduate course work in mathematics and/or related subjects. Students must receive a grade of A or B in 18 hours of core courses of the appropriate option and write a thesis, a report, or a creative component. Students with a good background in mathematics should expect to complete all requirements within two years. The Doctor of Philosophy program accepts only students with superior records in their graduate or undergraduate study. There are three options in the doctoral program: pure mathematics, applied mathematics, and mathematics education. The first two options are designed to prepare students for faculty positions at major research universities or for positions in industry. The mathematics education option is a blend of traditional foundational course work in mathematics and work in mathematics education and is designed to prepare students for positions in which mathematics teaching and educational concerns are a primary focus. A minimum of 90 credit hours of graduate credit beyond the bachelor's degree or 60 hours beyond the master's degree is required for each option, with 15 to 24 hours credited for a thesis. Students must pass a written comprehensive exam covering core courses and embark on a study of a chosen area of mathematics, pass an oral qualifying examination, and, for some options, complete the foreign language or computer language requirement. The most important requirement is the preparation of an acceptable thesis, which must demonstrate the candidate's ability to do independent, original work in mathematics or mathematics education. A well-prepared, motivated student should expect to complete all requirements within five to six years (or three to four years beyond the master's).

Research Facilities

The department operates a network of microcomputer workstations and personal computers with several file servers. Computing is available for all graduate students. Through this network, access to the University Computer Center is available. The department also houses current issues of important mathematics journals in a reading room. This makes about 100 journals available in a very convenient location. Electronic access is available for the Math Reviews, tables of contents of many journals, and the library catalog and database resources.

Financial Aid

Teaching assistantships are available to qualified students, with appointments covering the fall and spring semesters (renewed each year based on satisfactory progress). Students do not teach in their first semester and are provided with training to enhance their instructional skills. Subsequently, students normally have 5 to 6 hours of instructional duties per week. Some reduction in teaching is available to doctoral candidates making good progress toward their degree. Nine-month stipends are $12,900 for pre-master's students and $14,600 for students who have a master's degree or have passed the doctoral comprehensive exam. Some summer appointments, as well as scholarships, fellowships, and assistantships that enhance the stipend, are available.

Cost of Study

Tuition is reduced to the in-state level for all assistants, with full tuition waivers given to some exceptional incoming students. In-state tuition and fees are approximately $150 per credit hour.

Living and Housing Costs

On-campus housing is available in residence halls and in several apartment complexes. It is recommended that prospective students contact the Office of Residential Life (telephone: 405-744-5592; e-mail: reslife@okway.okstate.edu; World Wide Web: http://www.reslife.okstate.edu) for information. Most students live in apartment complexes in the surrounding community, which cost $300 and up per month.

Student Group

Of the current student body of about 40 students, 50 percent are women and 50 percent are international students. Almost all are full-time students on teaching assistantships. The department seeks highly motivated students without regard to race, color, national origin, religion, sex, or disability.

Student Outcomes

The department has been very successful in having all its recent doctoral students obtain positions in higher education institutions across the country. Master's students have placed very well in industry, community colleges, and schools. Many master's students go on to pursue doctoral degrees. Many doctoral students are appointed to prestigious postdoctoral positions.

Location

Stillwater, a small city of about 40,000, is a safe, friendly, and lively community. The cost of living is relatively low, and affordable housing is plentiful. The city offers most of the cultural and recreational opportunities of a college town and is just an hour's drive from both Oklahoma City and Tulsa.

The University

Oklahoma State University, a comprehensive research university with more than 22,000 students and almost 1,000 faculty members, is located on a scenic campus in Stillwater, Oklahoma. Founded in 1890, the University has developed an international reputation for excellence in teaching and research, especially in the basic and applied sciences. Students come to OSU from fifty states and more than fifty countries. The Graduate College has about 4,000 students.

Applying

An application package may be obtained from the Mathematics Department. Applicants should plan to have three letters of recommendation sent to the department. GRE scores are not required, but they are strongly recommended. The Graduate Committee in the Mathematics Department begins deliberations in early December and continues the process until March. It is recommended that applicants read the departmental World Wide Web page for a detailed description.

Correspondence and Information

Director of Graduate Studies
Department of Mathematics
Oklahoma State University
Stillwater, Oklahoma 74078-1058
E-mail: graddir@math.okstate.edu
Web site: http://mathgrad.okstate.edu

Oklahoma State University

THE FACULTY AND THEIR RESEARCH

Alan Adolphson, Regents Professor and Head; Ph.D., Princeton, 1973. Number theory, arithmetical algebraic geometry.
Douglas Aichele, Professor; Ed.D., Missouri, 1969. Mathematics education.
Dale Alspach, Professor; Ph.D., Ohio State, 1976. Functional analysis, Banach space theory.
Mahdi Asgari, Assistant Professor; Ph.D. Purdue, 2000. Number theory, automorphic forms, representation theory.
Leticia Barchini, Southwestern Bell Professor; Ph.D., National University (Argentina), 1987. Representations of Lie groups.
Dennis Bertholf, Professor; Ph.D., New Mexico State, 1968. Abelian group theory, mathematics education.
Birne Binegar, Associate Professor; Ph.D., UCLA, 1982. Representations of Lie groups and Lie algebras, mathematical physics.
Hermann Burchard, Professor; Ph.D., Purdue, 1968. Approximation theory, numerical analysis.
James Choike, Noble Professor; Ph.D., Wayne State, 1970. Complex analysis, mathematics education.
Bruce Crauder, Professor; Ph.D., Columbia, 1981. Algebraic geometry.
Benny Evans, Professor; Ph.D., Michigan, 1971. Topology of low-dimensional manifolds, mathematics education.
Amit Ghosh, Professor; Ph.D., Nottingham, 1981. Number theory, automorphic forms.
R. Paul Horja, Assistant Professor; Ph.D., Duke, 1999. Algebraic geometry and mirror symmetry.
William Jaco, G. B. Kerr Professor; Ph.D., Wisconsin, 1968. Topology of low-dimensional manifolds.
Ning Ju, Assistant Professor; Ph.D., Indiana, 1999. Applied mathematics, partial differential equations.
Anthony Kable, Associate Professor; Ph.D., Oklahoma State, 1997. Number theory.
Marvin Keener, Professor; Ph.D., Missouri, 1970. Ordinary differential equations.
Weiping Li, Associate Professor; Ph.D., Michigan State, 1992. Low-dimensional topology, gauge theory, differential geometry.
Joseph Maher, Assistant Professor; Ph.D., California, Santa Barbara, 2002. Topology of low-dimensional manifolds.
Lisa Mantini, Associate Professor; Ph.D., Harvard, 1983. Representations of Lie groups, integral geometry.
Anvar Mavlyutov, Assistant Professor; Ph.D., Massachusetts, 2002. Algebraic geometry and mirror symmetry.
Robert Myers, Professor; Ph.D., Rice, 1977. Topology of low-dimensional manifolds.
Alan Noell, Professor; Ph.D., Princeton, 1983. Several complex variables.
Igor Pritsker, Associate Professor; Ph.D., South Florida, 1995. Complex analysis, potential theory, approximation theory.
David Ullrich, Professor; Ph.D., Wisconsin, 1986. Harmonic analysis.
John Wolfe, Professor; Ph.D., Berkeley, 1971. Functional analysis, mathematics education.
David J. Wright, Associate Professor; Ph.D., Harvard, 1982. Algebraic number theory, Riemann surfaces.
Jiahong Wu, Associate Professor; Ph.D., Chicago, 1996. Fluid mechanics, partial differential equations.
Roger Zierau, Professor; Ph.D., Berkeley, 1985. Representations of Lie groups.

RESEARCH ACTIVITIES

Algebraic Geometry: three-dimensional algebraic varieties, birational geometry, degenerations of surfaces, geometry of resolutions, birational geometry of projective spaces; enumerative geometry, interaction of algebraic geometry with theoretical physics; complex holomorphic vector bundles over algebraic varieties, intersection theory on the moduli space of curves.

Analysis: functional analysis, geometry of Banach spaces; approximation theory, numerical analysis, optimization; several complex variables, convexity properties of pseudoconvex domains; harmonic analysis, random Fourier series, boundary behavior of harmonic and analytic functions; Riemann surfaces.

Lie Groups: representation theory of semisimple and reductive Lie groups, analysis and geometry of homogeneous spaces, symmetry and groups of transformations, algebraic aspects of the study of Lie groups and arithmetic groups.

Mathematics Education: school mathematics curriculum, professional development of mathematics teachers, technology in the classroom and applications in the curriculum, mathematics reform issues, equity and minority issues, early intervention testing programs.

Number Theory: L-functions of algebraic varieties over finite fields and cohomological techniques, automorphic representations and L-functions, analytic number theory and the distribution of zeros of the Riemann zeta function, algebraic number theory and cubic extensions of number fields, algebraic groups over algebraic number fields and geometric invariant theory.

Partial Differential Equations: theoretical and numerical studies of the Navier-Stokes equations, the 2D quasi-geostrophic equations, nonlinear wave equations, and other model equations arising in fluid mechanics; qualitative and quantitative analysis of turbulent dynamics.

Topology: structure and classification of compact 3-manifolds; normal, incompressible, and Heegaard surfaces; algorithms and computation in low-dimensional topology; relations with combinatorial and geometric group theory; structure of noncompact 3-manifolds; covering spaces of 3-manifolds; Casson invariants, Floer homology, symplectic topology, dynamical systems.

Oklahoma State University

SELECTED PUBLICATIONS

Adolphson, A. and S. Sperber. Exponential sums on A^n, III. *Manuscripta Mathematica* 102(4):429–46, 2000.

Adolphson, A., and S. Sperber. Dwork cohomology, de Rham cohomology, and hypergeometric functions. *Am. J. Math.* 122(2):319–48, 2000.

Adolphson, A. Higher solutions of hypergeometric systems and Dwork cohomology. *Rend. Sem. Mat. Univ. Padova* 101:179–90, 1999.

Adolphson, A., and S. Sperber. A remark on local cohomology. *J. Algebra* 206(2):555–67, 1998.

Adolphson, A., and S. Sperber. On twisted de Rham cohomology. *Nagoya Math. J.* 146:55–81, 1997.

Adolphson, A., and S. Sperber. On the zeta function of a complete intersection. *Ann. Sci. Ecole Norm. Suppl. (4)* 29(3), 1996.

Adolphson, A., and B. Dwork. Contiguity relations for generalized hypergeometric functions. *Trans. Am. Math. Soc.* 347(2), 1995.

Aichele, D. B., and **J. Wolfe.** *Geometric Structures–An Inquiry Based Textbook for Prospective Elementary Teachers.* Prentice Hall, in press.

Aichele, D. B., et al. *Geometry—Explorations and Applications.* Boston: Houghton Mifflin/McDougal Littell, 1997.

Aichele, D. B., and S. Gay. Middle school students' understanding of number sense related to percent. *Sch. Sci. Math.* 97(1):27–36, 1997.

Aichele, D. B., ed. *Professional Development for Teachers of Mathematics—1994 Yearbook.* Reston, Va.: NCTM, 1994.

Alspach, D., R. Judd, and E. Odell. The Szlenk index and local l_1-indices. *Positivity* 9:1–44, 2005.

Alspach, D., and S. Tong. Subspaces of L_p, $p \geq 2$, determined by partitions and weights. *Studia Math.* 159:207–27, 2003.

Alspach, D., and E. Odell. L_p spaces. In *Handbook of the Geometry of Banach Spaces,* vol. 1, pp. 123–60, eds. W. B. Johnson and J. Lindenstrauss. Amsterdam: North-Holland, 2001.

Alspach, D. The dual of the Bourgain-Delbaen space. *Israel J. Math.* 117:239–59, 2000.

Alspach, D. Tensor products and independent sums of L_p-spaces, $1<p<\infty$. *Mem. Am. Math. Soc.* 138(660):77, 1999.

Asgari, M., and F. Shahidi. Generic transfer from GSp(4) to GL(4). *Compos. Math.,* in press.

Asgari, M. Generic transfer for general spin groups. *Duke Math. J.,* in press.

Asgari, M. Local L-functions for split spinor groups. *Canda. J. Math.* 54(4):673–93, 2002.

Asgari, M., and R. Schmidt. Siegel-Modular formas and representations. *Manuscripta Mathematica* 104(2):173–200, 2001.

Barchini, L., M. Sepanski, and **R. Zierau.** Positivity of Zeta-distributions and small unitary representations. Preprint. (2004).

Barchini, L. Stein extensions of real symmetric spaces and the geometry of the flag manifold. *Math Annalen* 326:331–46, 2003.

Barchini, L., and M. Sepanski. Finite-reductive dual pairs in G_2. *Linear Algebra Appl.* 340:123–36, 2002.

Barchini, L., C. Leslie, and **R. Zierau.** Domains of holomorphy and representations of SL $(n,\mathbf{R})$. *Manuscripta Mathmatica* 106(4):411–27, 2001.

Barchini, L. Strongly harmonic forms for representations in the discrete series. *J. Funct. Anal.* 161(1):111–31, 1999.

Barchini, L., and **R. Zierau.** Square integrable harmonic forms and representation theory. *Duke Math. J.* 92(3):645–64, 1998.

Binegar, B., and **R. Zierau.** A singular representation of E_6. *Trans. Am. Math. Soc.* 341(2):771–85, 1994.

Binegar, B., and **R. Zierau.** Unitarization of a singular representation of SO $(_{p,q})$. *Comm. Math. Phys.* 138(2):245–58, 1991.

Binegar, B. Cohomology and deformations of Lie superalgebras. *Lett. Math. Phys.* 12(4):201–308, 1986.

Binegar, B. Conformal superalgebras, massless representations, and hidden symmetries. *Phys. Rev. D* 3-34(2):525–32, 1986.

Binegar, B. Unitarity of conformal supergravity. *Phys. Rev. D* 3-31(10):2497–502, 1985.

Binegar, B. On the state space of the dipole ghost. *Lett. Math. Phys.* 8(2):149–58, 1984.

Binegar, B., C. Fronsdal, and W. Heidenreich. Conformal QED. *J. Math. Phys.* 24(12):2828–46, 1983.

Binegar, B., C. Fronsdal, and W. Heidenreich. Linear conformal quantum gravity. *Phys. Rev. D* 3(10):2249–61, 1983.

Binegar, B., C. Fronsdal, and W. Heidenreich. de Sitter QED. *Ann. Phys.* 149 (2):254–72, 1983.

Binegar, B. Relativistic field theories in three dimensions. *J. Math. Phys.* 23(8): 1511–7, 1982.

Binegar, B., C. Fronsdale, M. Flato, and S. Salamo. de Sitter and conformal field theories. In *Proceedings of the International Symposium "Selected Topics in Quantum Field Thoery and Mathematics Physics"* (Bechnyle, 1981). *Czechoslovak J. Phys. B* 32(4):439–71, 1982.

Burchard, H. G., and J. Lei. Coordinate order of approximation by functional-based approximation operators. *J. Approx. Theory* 82(2), 1995.

Burchard, H. G., J. A. Ayers, W. H. Frey, and N. S. Sapidis. Approximation with aesthetic constraints. Designing fair curves and surfaces. In *Geometric Design Publishing,* pp. 3–28. Philadelphia: SIAM, 1994.

Burchard, H. G. Discrete curves and curvature constraints. In *Curves and Surfaces II,* ed. L. L. Schumaker et al. Boston: AK Peters, 1994.

Crauder, B., and R. Miranda. Quantum cohomology of rational surfaces. In *The Moduli Space of Curves* (Texel Island, 1994) *Progr. Math.,* 129. Boston: Birkhäuser Boston, 1995.

Crauder, B., and D. R. Morrison. Minimal models and degenerations of surfaces with Kodaira number zero. *Trans. Am. Math. Soc.* 343(2), 1994.

Crauder, B., and S. Katz. Cremona transformers and Hartshorne's conjecture. *Am. J. Math.* 113(2), 1991.

Crauder, B. Degenerations of minimal ruled surfaces. *Ark. Mat.* 28(2), 1990.

Evans, B., and J. Johnson. *Linear Algebra with MAPLE.* John Wiley and Sons, 1994.

Evans, B. DERIVE in linear algebra. In *Proc. Fifth Intl. Conf. Technol. Coll. Math.,* 1993.

Evans, B., and J. Johnson. *Discovering Calculus with DERIVE.* John Wiley and Sons, 1992.

Evans, B. The long annulus theorem. *Can. Math. Bull.* 29(3), 1986.

Aspinwall, P. S., **R. P. Horja,** and R. L. Karp. Massless D-Branes and Calabi-Yau threefolds and monodromy. *Comm. Math. Physics* 259(1): 45–69, 2005.

Horja, R. Paul. Derived category automorphisms from mirror symmetry. *Duke Math. J.* 127(1):1–34, 2005.

Jaco, W., and J. H. Rubinstein. 0-efficient triangulations of 3-manifolds. *J. Differential Geometry* 65, 2003.

Jaco, W. and E. Sedgwick. Decision problems in the space of Dehn fillings. *Topology* 42, 2003.

Jaco, W., D. Letscher, and J. H. Rubinstein. Algorithms for essential surfaces in 3-manifolds. *Contemp. Math.* 314, 2002.

Jaco, W., and J. L. Tollefson. Algorithms for the complete decomposition of a closed 3-manifold. *Illinois J. Math.* 39(3), 1995.

Jaco, W., and J. H. Rubinstein. PL equivariant surgery and invariant decompositions of 3-manifolds. *Adv. Math.* 73(2), 1989.

Jaco, W. Lectures on three-manifold topology. *CBMS Reg. Conf. Ser. Math. 43.* Providence, R.I.: American Mathematical Society, 1980.

Kable, A. C. Asai L-functions and Jacquet's conjecture. *Am. J. Math.* 126:789–820, 2004.

Kable, A. C. The tensor product of exceptional representations on the general linear group. *Ann. Sci. Ecole Norm. Suppl.* 34(5):741–69, 2001.

Kable, A. C., and A. Yukie. Prehomogeneous vector spaces and field extensions. II. *Invent. Math.* 130(2):315–44, 1997.

Li, W. Instanton Floer homology for connected sums of Poincare spheres. *Math. Zeit.,* in press.

Li, W. The Z-graded symplectic Floer cohomology of monotone Lagrangian sub-manifolds. *Algebraic Geom. Topol.* 4(30):647–84, 2004.

Li, W., and L. Xu. Counting $SL_2(F_q)$-representations of torus knot groups. *J. Knot Theory and its Ramifications.* 13(3):401–26, 2004.

Li, W., and L. Xu. Counting $SL_2(F_q)$-representations of torus knot groups. *Acta Math Sinica, Eng. Ser.* 19(2):233–44, 2003.

Oklahoma State University

Li, W. The semi-infinity of Floer (co)homologies. *Contemp. Math.* 322:195–215, 2003.

Li, W. A Monopole homology of integral homology 3-spheres. *Turkish J. Math.* 27:126–60, 2003.

Li, W. Knot and link invariants and moduli spaces of parabolic bundles, *Comm. Contemp. Math.* 3(4):501–32, 2001.

Li, W. Cap-product structures on the Fintushel-Stern spectral sequence. *Math. Proc. Cambridge Philosophical Soc. Part 2* 131:265–78, 2001.

Li, W. Equivariant knot signatures and Floer homologies. *J. Knot Theory and its Ramifications.* 10(5):687–701, 2001.

Li, W. Künneth formulae and cross products for the symplectic Floer cohomology. *Topol. Appl.* 110(3, 30):211–36, 2001.

Maher, J., and J. H. Rubinstein. Period three actions on the three-sphere. *Geom. Topol.* 7:329–97, 2003.

Maher, J. Virtually embedded boundary slopes. *Topol. Appl.* 95:63–74, 1995.

Mantini, L. A. Intertwining ladder representations for SU(p,q) into Dolbeault cohomology. In *Non-commutative Harmonic Analysis, Progr. Math.* 220, pp. 295–418. Boston: Birkhaüser, 2004. With Lorch and Novak.

Mantini, L. A. Teaching mathematics in colleges and universities: Case studies for today's classroom. In CBMS series *Issues in Mathematics Education, Am. Math. Soc.* vol. 10, 2001. With Friedberg et al.

Lorch, J. D., and **L. A. Mantini.** Inversion of an integral transform and ladder representations of U(1,q). In *Representation Theory and Harmonic Analysis* (Cincinnati, Ohio, 1994); *Contemp. Math.*, 191; *Am. Math. Soc.* Providence, R.I., 1995.

Mantini, L. A. An L^2-cohomology construction of unitary highest weight modules for U(p,q). *Trans. Am. Math. Soc.* 323(2), 1991.

Myers, R. End reductions, fundamental groups, and covering spaces of irreducible open 3-manifolds. *Geom. Topol.* 9:971–90, 2005.

Myers, R. Splitting homomorphisms and the geometrization conjecture. *Math. Proc. Cambridge Philos. Soc.* 129(2):291–300, 2000.

Myers, R. Uncountably many arcs in S^3 whose complements have non-isomorphic, indecomposable fundamental groups. *J. Knot Theory Ramifications* 9(4):505–21, 2000.

Myers, R. On covering translations and homeotopy groups of contractible open n-manifolds. *Proc. Am. Math. Soc.* 128(5):1563–6, 2000.

Myers, R. Compactifying sufficiently regular covering spaces of compact 3-manifolds. *Proc. Am. Math. Soc.* 128:1507–13, 2000.

Myers, R. Contractible open 3-manifolds which non-trivially cover only non-compact 3-manifolds. *Topology* 38(1):85–94, 1999.

Myers, R. Contractible open 3-manifolds with free covering translation groups. *Topol. Appl.* 96(2):97–108, 1999.

Noell, A., and R. Belhachemi. Global plurisubharmonic defining functions. *Mich. Math. J.* 47:377–84, 2000.

Noell, A. Local and global plurisubharmonic defining functions. *Pacific J. Math.* 176(2):421–6, 1996.

Noell, A. Peak functions for pseudoconvex domains in C^n. In *Several Complex Variables: Proceedings of the Mittag-leffler Institute,* 1987–88, *Math. Notes* 38. Princeton, N.J.: Princeton University Press, 1993.

Noell, A. Local versus global convexity of pseudoconvex domains. In *Several Complex Variables and Complex Geometry, Proc. Sympos. Pure Math.* 52. Providence, R.I.: American Mathematics Society, 1991.

Noell, A. Interpolation from curves in pseudoconvex boundaries. *Mich. Math. J.* 37(2), 1990.

Noell, A., and B. Stensones. Proper holomorphic maps from weakly pseudoconvex domains. *Duke Math. J.* 60:363–88, 1990.

Noell, A., and T. Wolff. On peak sets for Lip α classes. *J. Funct. Anal.* 86:136–79, 1989.

Pritsker, I. E. Convergence of Julia polynomials. *J. d'Analyse Math.* 94:343–61, 2004.

Laugesen, R. S., and **I. E. Pritsker.** Potential theory of the farthest-point distance function. *Can. Math. Bull.* 46:373–87, 2003.

Pritsker, I. E. Products of polynomials in uniform norms. *Trans. Am. Math. Soc.* 353:3971–93, 2001.

Andrievskii, V. V., **I. E. Pritsker,** and R. S. Varga. Simultaneous approximation and interpolation of functions on continua in the complex plane. *J. Math. Pures Appl.* 80:373–88, 2001.

Pritsker, I. E. An inequality for the norm of a polynomial factor. *Proc. Am. Math. Soc.* 129:2283–91, 2001.

Andrievskii, V. V., and **I. E. Pritsker.** Convergence of Bieberbach polynomials in domains with interior cusps. *J. d'Analyse Math.* 82:315–32, 2000.

Kroó, A., and **I. E. Pritsker.** A sharp version of Mahler's inequality for products of polynomials. *Bull. London Math. Soc.* 31(3):269–78, 1999.

Pritsker, I. E., and R. S. Varga. Weighted rational approximation in the complex plane. *J. Math. Pures Appl.* 78(2):177–202, 1999.

Choe, B. R., W. Ramey, and **D. C. Ullrich.** Bloch-to-BMOA pullbacks on the disk. *Proc. Am. Math. Soc.* 125(10):2987–96, 1997.

Stegenga, D. A., and **D. C. Ullrich.** Superharmonic functions in Hölder domains. *Rocky Mtn. J. Math.* 25(4), 1995.

Ullrich, D. C. Radial divergence in BMOA. *Proc. London Math. Soc. (3)* 68(1), 1994.

Ullrich, D. C. Recurrence for lacunary cosine series. In *The Madison Symposium on Complex Analysis* (Madison, Wisc., 1991); *Contemp. Math.*, 137; *Am. Math. Soc.* Providence, R.I., 1992.

Matthews, C., and **D. J. Wright.** Cycle decomposition and train tracks. *Proc. Am. Math. Soc.* 132:283–314, 2004.

Mumford, D., C. Series, and **D. J. Wright.** *Indra's Pearls.* Cambridge University Press, 2002.

Wright, D. J., and A. Yukie. Prehomogeneous vector spaces and field extensions. *Invent. Math.* 110(2), 1992.

Wright, D. J. Twists of the Iwasawa-Tate zeta function. *Math. Z.* 200(2), 1989.

Wright, D. J. Distribution of discriminants of abelian extensions. *Proc. London Math. Soc. 3* 58(1), 1989.

Wu, J. The two-dimensional quasi-geostrophic equation with critical or super-critical dissipation. *Nonlinearity* 18(1):139–54, 2005.

Wu, J. Global solutions of the 2D dissipative quasi-geostrophic equation in Besov spaces. *SIAM J. Math. Anal.* 36(3):1014–30, 2004/05.

Wu, J. The generalized incompressible Navier-Stokes equations in Besov spaces. *Dyn. Partial Differ. Equ.* 1(4):381–400, 2004.

Wu, J. Regularity results for weak solutions of the 3D MHD equations. Partial differential equations and applications. *Discrete Contin. Dyn. Syst.* 10(1–2):543–56, 2004.

Wu, J. Generalized MHD equations. *J. Differential Equations* 195(2):284–312, 2003.

Bona, J. L., and **J. Wu.** The zero-viscosity limit of the 2D Navier-Stokes equations. *Stud. Appl. Math.* 109(4):265–78, 2002.

Wu, J. The quasi-geostrophic equation and its two regularizations. *Comm. Partial Differential Equations* 27(5–6):1161–81, 2002.

Constantin, P., D. Cordoba, and **J. Wu.** On the critical dissipative quasi-geostrophic equation. Dedicated to Professors Ciprian Foias and Roger Temam (Bloomington, Indiana, 2000). *Indiana Univ. Math. J.* (special issue) 50:97–107, 2001.

Wu, J. Analytic results related to magneto-hydrodynamic turbulence. *Physica D* 136(3–4):353–72, 2000.

Mehdi, S., and **R. Zierau.** Principal series representations and harmonic spinors. *Adv. Math.*, in press.

Mehdi, S., and **R. Zierau.** Harmonic spinors on semisimple symmetric spaces. *J. Funct. Anal.* 198(2):536–57, 2003.

Zierau, R. Representations in Dolbeault cohomology. In *Representation Theory of Lie Groups,* Park City Math Institute, vol. 8; *Am. Math. Soc.* Providence, R.I., 2000.

Wolf, J. A., and **R. Zierau**. Linear cycle spaces in flag domains. *Math. Ann.* 316(3):529–45, 2000.

Dunne, E. G., and **R. Zierau.** The automorphism groups of complex homogeneous spaces. *Math. Ann.* 307(3):489–503, 1997.

RENSSELAER POLYTECHNIC INSTITUTE

Department of Mathematical Sciences

Programs of Study

Mathematics is the universal language of the sciences and is essential to the technological evolution of society. At Rensselaer, students have an opportunity to see numerous applications involving contemporary and innovative approaches to mathematics. The Department's goal is to train individuals in mathematics, both as a subject in itself and as a discipline to aid in the development of other social and scientific fields. In particular, Rensselaer's Department of Mathematical Sciences is one of the few in the country with a strong faculty orientation toward applications of mathematics. This emphasis is reflected in the many courses dealing with areas of mathematical applications as well as in the interdisciplinary nature of the Department's research endeavors.

The Department offers M.S. degrees in both applied mathematics and mathematics and the Ph.D. degree in mathematics. The emphasis of applied mathematics is on mathematics and how it is employed to study science, engineering, or management problems. It stresses construction, analysis, and evaluation of mathematical models of real-world problems and those areas of mathematics that are most widely used to solve them.

The interdisciplinary nature of many of the Department's research projects has led to strong connections with industry. Of particular interest are the annual Graduate Student Mathematical Modeling (GSMM) Camp and the annual Workshop on Mathematical Problems in Industry. Rensselaer's GSMM Camp is a four-day workshop whose aim is graduate student education and career development directed toward modern scientific problem solving. It is designed to promote a broad range of problem-solving skills, including mathematical modeling and analysis, scientific computation, and critical assessment of solutions. Problems are brought to the camp by invited faculty members and industry mentors. These problems are highly interdisciplinary in nature, inspired by real problems that arise in industrial applications. The work on the problems is done by graduate student teams, each with the guidance of an invited mentor, so that scientific communication is an important and integral component of the work. In this way, the GSMM Camp exposes graduate students to real-world problems of current scientific interest and provides a valuable educational and career-enhancing experience outside the traditional academic setting.

The Workshop on Mathematical Problems in Industry (MPI) is a problem-solving workshop that attracts leading applied mathematicians and scientists from universities, industry, and national laboratories. The objective of the workshop is to foster interaction and collaboration among scientists and engineers working in industry and applied mathematicians at universities. Industry representatives present problems at the workshop—problems that range from those requiring basic physical modeling to those requiring significant computation. The workshop participants break up into small working groups consisting of senior faculty and attending scientists, graduate students, and the industrial representatives, to discuss and tackle the problems in an informal setting. On the last day of the workshop, an academic representative from each group presents the results obtained and discusses possible future directions. A written report detailing the progress made during the workshop is prepared subsequently and sent to the industry representatives.

Research Facilities

All faculty and graduate student offices have 100BaseT connections to the campus optical fiber Ethernet backbone and wireless access. Graduate students have the use of six Sun Ultra 1/170 workstations, six Sun Sparcstation 5/170 workstations, and several Sun IPC machines. For larger jobs the Department has a Sun Enterprise 3500 with four processors (336 MHz), 8 GB of RAM, and about 78 GB of disk. There are also several laserwriter printers.

RPI research programs reach across the campus and beyond, linking together departments, schools, interdisciplinary centers, and stimulating the integration of inquiry, new knowledge, and education. Some of the research centers most closely linked to the research efforts of the Department of Mathematical Sciences are the Scientific Computation Research Center, Inverse Problems Center, Center for Biotechnology and Interdisciplinary Studies, Anderson Center for Innovation in Undergraduate Education, Center for Multiphase Research, and Numerically Intensive Computing Facilities.

The Scientific Computation Research Center is focused on the development of reliable simulation technologies for engineers, scientists, medical professionals, and other practitioners. These advancements enable experts in their fields to employ, appraise, and evaluate the behavior of physical, chemical, and biological systems of interest. Inverse problems addressed at the Inverse Problems Center include medical imaging (elastography, EIT, and optical tomography), geophysical fault identification, bridge embankment integrity, and radar imaging. The Center for Biotechnology and Interdisciplinary Studies houses faculty members and researchers engaged in interdisciplinary research and hosts world-class programs and symposia. It exemplifies a new research paradigm, as no department offices reside in the building; rather, it is occupied by researchers and their laboratories. The core research facilities within the center contain laboratories for molecular biology, analytical biochemistry, microbiology, imaging, histology, tissue and cell culture, proteomics, and scientific computing and visualization. The Anderson Center is dedicated to improving undergraduate education through the deployment of new pedagogical methods and innovative uses of technology. Since its inception, the Anderson Center for Innovation in Undergraduate Education has served as an incubator for curriculum reform by supporting faculty involvement in educational computing, developing new techniques and facilities for interactive learning, and sponsoring cutting-edge research on the assessment of learning outcomes. Rensselaer's interdisciplinary Center for Multiphase Research (CMR) is the premier group in the country for performing multiphase research. The CMR has assembled a large and dynamic group of scientists and engineers dedicated to exploring and exploiting new developments in every conceivable aspect of multiphase flow and heat transfer technology.

Financial Aid

Financial aid is available in the forms of teaching and research assistantships and fellowships, which include tuition scholarships and stipends. Rensselaer assistantships and university, corporate, or national fellowships fund many of Rensselaer's full-time graduate students. Outstanding students may qualify for university-sponsored Rensselaer Graduate Fellowship Awards, which carry a minimum stipend of $20,000 and a full tuition and fees scholarship. All fellowship awards are calendar-year awards for full-time graduate students. Summer support is also available in many departments. Low-interest, deferred-repayment graduate loans are available to U.S. citizens with demonstrated need.

Cost of Study

Full-time graduate tuition for the 2006–07 academic year is $32,600. Other costs (estimated living expenses, insurance, etc.) are projected to be about $12,400. Therefore, the cost of attendance for full-time graduate study is approximately $45,000. Part-time study and cohort programs are priced differently. Students should contact Rensselaer for specific cost information related to the program they wish to study.

Living and Housing Costs

Graduate students at Rensselaer may choose from a variety of housing options. On campus, students can select one of the many residence halls, and there are abundant options off campus as well, many within easy walking distance.

Student Group

There are 1,234 graduate students, of whom 30 percent are women, 90 percent are full-time, and 69 percent study at the doctoral level.

Student Outcomes

Rensselaer's graduate students are hired in a variety of industries and sectors of the economy and by private and public organizations, the government, and institutions of higher education. Starting salaries average $63,262 for master's degree recipients.

Location

Located just 10 miles northeast of Albany, New York State's capital city, Rensselaer's historic 275-acre campus sits on a hill overlooking the city of Troy, New York, and the Hudson River. The area offers a relaxed lifestyle with many cultural and recreational opportunities, with easy access to both the high-energy metropolitan centers of the Northeast—such as Boston, New York City, and Montreal, Canada—and the quiet beauty of the neighboring Adirondack Mountains.

The Institute

Recognized as a leader in interactive learning and interdisciplinary research, Rensselaer continues a tradition of excellence and technological innovation dating back to 1824. More than 100 graduate programs in more than fifty disciplines attract top students, researchers, and professors. The discovery of new scientific concepts and technologies, especially in emerging interdisciplinary fields, is the lifeblood of Rensselaer's culture and a core goal for the faculty, staff, and students. Fueled by significant support from government, industry, and private donors, Rensselaer provides a world-class education in an environment tailored to the individual.

Applying

The admission deadline for the fall semester is January 1. Basic admission requirements are the submission of a completed application form (available online), the required application fee ($75), a statement of background and goals, official transcripts, official scores on the GRE General Test, TOEFL or IELTS scores (if applicable), and two recommendations. In addition, the GRE Subject Test is strongly recommended.

Correspondence and Information

Department of Mathematical Sciences
Amos Eaton 301
Rensselaer Polytechnic Institute
110 Eighth Street
Troy, New York 12180

Phone: 518-276-6345
E-mail: robensd@rpi.edu
Web site: http://www.math.rpi.edu/index.html

Rensselaer Polytechnic Institute

THE FACULTY AND THEIR RESEARCH

Kristin P. Bennett, Professor; Ph.D., Wisconsin–Madison. Combining operations research and artificial intelligence problem-solving methods, mathematical programming approaches to problems in artificial intelligence. (bennek@rpi.edu)

Jennifer Blue, Clinical Assistant Professor; Ph.D., Rensselaer. Mathematical programming, machine learning, clustering. (bluej@rpi.edu)

Mohamed T. Boudjelkha, Clinical Professor; Ph.D., Rensselaer. Differential equations from applied point of view, asymptotics, special functions and their applications to various branches of science and engineering. (boudjm@rpi.edu)

Margaret Cheney, Professor; Ph.D., Indiana. Inverse problems in acoustics and electromagnetic theory; remote sensing problems, including ground-penetrating radar, sonar, adaptive time-reversal methods in both acoustics and electromagnetics, Synthetic Aperture Radar (SAR), and Inverse Synthetic Aperture Radar (ISAR). (cheney@rpi.edu)

Donald Drew, Eliza Ricketts Foundation Professorship of Mathematics and Chair; Ph.D., Rensselaer. Multiphase flows: sedimentation, boiling filtration, and separation. (drewd@rpi.edu)

Joseph Ecker, Professor; Ph.D., Michigan. Optimization, including geometric programming, multiple objective linear programming, algorithm development and evaluation in nonlinear programming, and applications of mathematical programming. (eckerj@rpi.edu)

Joseph E. Flaherty, Professor; Ph.D., Polytechnic of Brooklyn. Numerical analysis, scientific computation, parallel computation, and adaptive methods. (flahej@rpi.edu)

Eldar Giladi, Assistant Professor; Ph.D., Stanford. Acoustics, scientific computing. (gilade@rpi.edu)

Isom Herron, Professor; Ph.D., Johns Hopkins. Theory of the stability of fluid flows: stability of rotating magneto-hydrodynamic flows and more complicated geophysical flows such as groundwater, for which mathematical models are still being developed. (harroi@rpi.edu)

Mark Holmes, Professor; Ph.D., UCLA. Development and analysis of mathematical models for physiological systems, including modeling the biological tissues found in joints (such as the knee).

David Isaacson, Professor; Ph.D., NYU. Mathematical physics: development of methods for approximating energy levels, particle masses, and critical exponents in quantum mechanics, quantum field theory, and statistical mechanics; medical imaging: development of algorithms and devices for measuring and displaying the electrical state of the interior of a body from measurements made on the body's exterior. (isaacd@rpi.edu)

Ashwani Kapila, Professor; Ph.D., Cornell. Applied mathematics: reactive and multiphase flow, nonlinear waves, perturbation methods, scientific computing; education: development of Web-based instructional materials. (kapila@rpi.edu)

Maya Kiehl, Clinical Assistant Professor; Ph.D., Rensselaer. Math education. (kiehlm@rpi.edu)

Gregor Kovacic, Associate Professor; Ph.D., Caltech. Nonlinear evolution equations and their applications to scientific problems. (kovacg@rpi.edu)

Peter Kramer, Assistant Professor; Ph.D., Princeton. Modeling transport in turbulent flows, computer simulation of microphysiological systems, stochastic modeling in physics and biology, multiscale asymptotic methods, random field simulation, nonlinear wave turbulence. (kramep@rpi.edu)

Fengyan Li, Assistant Professor; Ph.D., Brown. Local-structure-preserving discontinuous Galerkin methods; divergence cleaning techniques; computational methods for Maxwell equations, magnetohydrodynamics (MHD) equations, and Hamilton-Jacobi equations.

Chjan Lim, Professor; Ph.D., Brown. Statistical physics and turbulence: exactly-solvable models in 2-D turbulence—a long range spherical model for energy-enstrophy theories; computational science and vortex dynamics; vortex dynamics; symmetric dynamical systems; combinatorial matrix theory. (limc@rpi.edu)

Yuri Lvov, Assistant Professor; Ph.D., Arizona. Mathematical physics and nonlinear phenomena. (lvovy@rpi.edu)

Harry McLaughlin, Professor; Ph.D., Maryland. Applied geometry as it applies to design and manufacturing problems: representation needed for robot motion, numerically controlled milling machines, and engineering analyses, as well as visual displays; curve and surface design: shape and how to model it using splines and spline-like techniques. (mclauh@rpi.edu)

Joyce McLaughlin, Ford Foundation Professor; Ph.D., California, Riverside. Nonlinear analysis as applied to parameter identification in inverse problems, numerical algorithms for solving Helmholtz equation. (mclauj@rpi.edu)

John Mitchell, Professor; Ph.D., Cornell. Optimization, integer programming, linear programming, conic optimization, semidefinite programming, mathematical modeling of interdependent infrastructures with particular emphasis on response and recovery after a disruption and mitigation of the effects of a disruption. (mitchj@rpi.edu)

Clifford J. Nolan, Research Assistant Professor; Ph.D., Rice. Medical and seismic imaging using microlocal analysis. (clifford.nolan@ul.ie)

Jong-Shi Pang, Margaret A. Darrin Distinguished Professor; Ph.D., Stanford. Complementarity, optimization, equilibrium programming. (pangj@rpi.edu)

Bruce Piper, Associate Professor; Ph.D., Utah. Computer-aided geometric design, numerical analysis, computer graphics. (piperb@rpi.edu)

Victor Roytburd, Professor; Ph.D., Berkeley. Nonlinear dynamics governed by partial differential equations, especially free boundary problems, and applications in physics and engineering; in particular, in combustion, laser and nonlinear fiber optics, and reactive fluids. (roytbv@rpi.edu)

Lester Rubenfeld, Professor and Director, Center for Initiatives in Pre-College Education; Ph.D., NYU. Bifurcation theory and asymptotic expansions as applied to programs in elasticity theory, electromagnetic theory, and fluid dynamics; precollege mathematics and science education. (rubenl@rpi.edu)

David Schmidt, Clinical Assistant Professor; Ph.D., Rensselaer. Graph theory, mathematics education. (schmid@rpi.edu)

Donald Schwendeman, Professor; Ph.D., Caltech. Numerical methods for partial differential equations, gas dynamics and wave propagation, multiscale and multiphase reactive flow, adaptive mesh refinement and parallel algorithms, mathematical modeling and computations in industrial applications. (schwed@rpi.edu)

William Siegmann, Professor; Ph.D., MIT. Propagation problems: acoustic transmission in the ocean, electromagnetic and acoustic waves in the atmosphere. (siegmw@rpi.edu)

Michael Zuker, Professor; Ph.D., MIT. Development and implementation of algorithms to predict nucleic acid folding and hybridization by free energy minimization using empirically derived thermodynamic parameters. (zukerm@rpi.edu)

MATHEMATICAL SCIENCES RESEARCH

Acoustics: Acoustics relates to the mathematical description of sound waves. This field derives from fluid dynamics. A major area of interest at Rensselaer is underwater sound transmissions. Faculty Researchers: Eldar Giladi, Mark Holmes, Joyce McLaughlin, William Siegmann.

Applied Combinatorics and Discrete Mathematics: Discrete mathematics, also called finite mathematics, is the study of fundamentally discrete mathematical structures, that is, structures that do not support or require the notion of continuity. Faculty Researcher: Chjan Lim.

Applied Geometry: Research focuses on geometric foundations of applied geometry, surface modeling, solid modeling, robot path planning, and computational geometry. Faculty Researchers: Harry McLaughlin, Bruce Piper.

Approximation Theory: Those studying approximation theory analyze and design various multiresolution techniques that have provable, optimal properties for these models. Faculty Researcher: Harry McLaughlin.

Bioinformatics: Bioinformatics, also known as computational biology, is the use of techniques from applied mathematics, informatics, statistics, and computer science to solve biological problems. Faculty Researchers: Michael Zuker, Kristin Bennett.

Biomathematics: Biomathematics, or mathematical biology, is an interdisciplinary field that aims to model natural, biological processes using mathematical techniques and tools. Faculty Researchers: Margaret Cheney, Mark Holmes, David Isaacson.

Chemically Reacting Flows: Researchers at Rensselaer focus on investigating mathematical modeling of combustion and flame propagation phenomena and analysis of the resulting systems of nonlinear ordinary and partial differential equations. Faculty Researchers: Ashwani Kapila, Victor Roytburd, Donald Schwendeman.

Computational Logic: Computational logic is a mathematical framework to redevelop logic as systematic formal theory of computability (as opposed to classical logic). In computational logic, formulas represent computational problems and their validity means "always computable." Faculty Researcher: John Mitchell. Affiliated research center: Scientific Computation Research Center.

Dynamical Systems: Researchers at Rensselaer concentrate on the theory of dynamical systems and its applications in physics and engineering. This research aims to discover and explain new and important phenomena found in experimental and numerical studies. Faculty Researchers: Gregor Kovacic, Chjan Lim, Yuri Lvov.

Environmental Problems. Faculty Researchers: Margaret Cheney, Donald Drew.

Fluid Dynamics: Rensselaer researchers use methods of applied mathematics to study how fluids behave under a wide spectrum of conditions. The physical problems usually lead to partial differential equations, which may be linear or nonlinear. Faculty Researchers: Isom Herron, Ashwani Kapila, Chjan Lim, Yuri Lvov, Donald Schwendeman, Peter Kramer. Affiliated research centers: Scientific Computation Research Center and Center for Biotechnology and Interdisciplinary Studies.

Inverse Problems: Scientific challenges include modeling of the physical problem, creating new mathematics for analysis of the model, identifying appropriate and/or rich data sets, working with scientific computations and visualization aids, and undertaking experimental verification. Faculty Researchers: Margaret Cheney, Joyce McLaughlin. Affiliated research center: Inverse Problems Center.

Machine Learning: Machine learning overlaps heavily with statistics, since both fields study the analysis of data. But unlike statistics, researchers concern machine learning with the algorithmic complexity of computational implementations. Part of machine learning research is the development of tractable approximate inference algorithms. Faculty Researcher: Kristin Bennett.

Math Education: Researchers are currently developing studio courses that actively engage students in the learning process. Faculty members are also collaborating on the development of a multimedia environment—a calculus world—that would help a student learn and acquire the fundamental concepts of calculus through exploration and discovery. Faculty Researchers: Joe Ecker, Mark Holmes, Ashwani Kapila, Lester Rubenfeld. Affiliated research center: Anderson Center for Innovation in Undergraduate Education.

Mathematical Physics: Mathematical physics is an interdisciplinary field between mathematics and physics, aimed at studying and solving problems inspired by physics within a mathematically rigorous framework. Faculty Researchers: David Isaacson, Yuri Lvov.

Multiphase Flows: Research in the mathematical sciences department includes modeling and analysis of fluid flows driven by substantial energy inputs, chemical or otherwise; examination of the equations that represent special physical phenomena in the flows; examination of how sensitive the predictions are to the equations. Faculty Researchers: Donald Drew, Ashwani Kapila, Peter Kramer. Affiliated research center: Center for Multiphase Research.

Nonlinear Analysis. Faculty Researchers: Isom Herron, Gregor Kovacic, Joyce McLaughlin, Jong-Shi Pang, Victor Roytburd.

Nonlinear Materials. Faculty Researchers: Joseph E. Flaherty, Mark Holmes. Related research center: Center for Composite Materials and Structures.

Nonlinear Waves: Nonlinear optics is the branch of nonlinear wave research that describes the behavior of light in nonlinear media, that is, media in which the polarization responds nonlinearly to the electric field of the light. Faculty Researchers: Ashwani Kapila, Gregor Kovacic, Yuri Lvov, Victor Roytburd, Donald Schwendeman.

Operations Research and Mathematical Programming: Operations research is the use of mathematical models, statistics, and algorithms to analyze complex real-world systems, usually to improve or optimize performance. Mathematical programming endeavors to find optimal solutions for a broad range of problems, including medical, financial, scientific, and engineering problems. Faculty Researchers: Kristin Bennett, Joseph Ecker, John Mitchell, Jong-Shi Pang.

Optimization: Researchers study the following areas of optimization: geometric programming, multiple objective linear programming, algorithm development and evaluation in nonlinear programming, applications of mathematical programming, and computational optimization. Faculty Researchers: Kristin Bennett, Joseph Ecker, Mark Holmes, Joyce McLaughlin, John Mitchell, Jong-Shi Pang, Donald Schwendeman.

Perturbation Methods: Researchers study regular and singular perturbation theory to systematically construct an approximation of the solution of a problem that is otherwise intractable. Researchers are currently working on regular perturbation theory, singular perturbation theory, and computational methods for singularly perturbed problems. Faculty Researchers: Isom Herron, Mark Holmes, Ashwani Kapila, Gregor Kovacic.

Scientific Computing: Investigations range from the study of fundamental problems in linear algebra to the development and analysis of numerical schemes for solving particular physical or life science problems. Faculty Researchers: Kristin Bennett, Eldar Giladi, David Isaacson, Ashwani Kapila, John Mitchell, Victor Roytburd, Donald Schwendeman, William Siegmann. Affiliated research centers: Scientific Computation Research Center and the Numerically Intensive Computing Facilities.

SOUTHERN ILLINOIS UNIVERSITY CARBONDALE

Department of Mathematics
Doctoral Program

Program of Study

The Department of Mathematics of Southern Illinois University Carbondale (SIUC) offers a master's degree program leading to the Doctor of Philosophy (Ph.D.) degree in mathematics. This program, which was established in 1964, is designed to prepare students for careers in business, industry, government, and academia requiring advanced training in mathematics and statistics.

Students in the program work closely with members of the faculty in course work and research in a wide variety of fields of pure and applied mathematics. The graduate adviser works directly with each student in the program to develop an individualized plan of study. Recent graduates have written doctoral dissertations in the areas of algebra and number theory, combinatorics, differential equations, numerical analysis, probability and stochastic analysis, and statistics. There is also a concentration in computational mathematics, in which the student completes courses in both computer science and mathematics. Graduate students are encouraged to participate in the variety of research seminars organized by Department faculty members. An active colloquium program regularly brings outside speakers into the Department.

Roughly half of the students in the doctoral program are admitted directly into the program; the other half enter the Department in the master's program and are accepted into the doctoral program upon successful completion of the requirements for that degree.

Research Facilities

The Department of Mathematics maintains a computer lab that was equipped with the support of a grant from the National Science Foundation. The thirty PCs in this lab are programmed with sophisticated mathematical software and are networked for access to the Internet and to high-speed laser printers, and the lab is equipped with a projection panel for instructional purposes. Graduate students have 24-hour access to the lab for use in their research or for the preparation of papers and reports. Graduate student offices are also equipped with PCs that are connected to the Department network. The American Mathematical Society's Mathematical Reviews and other resources can be accessed directly through the network. Morris Library, the main library on campus, has extensive collections of books and journals in mathematics and related disciplines; additional materials can be retrieved via interlibrary loan. The Department also maintains a small independent reading library where students can read current journals and access copies of theses and dissertations of former students.

Financial Aid

Nearly all students in the program receive financial assistance, primarily in the form of teaching assistantships. Assistantship responsibilities include direct classroom teaching and/or a combination of teaching support duties. Some assistantship support is usually available during the summer term. Strong applicants are nominated by the Department for fellowships. Teaching assistantships and fellowships both include a tuition waiver. Information on other forms of financial aid, such as student loans or work-study programs, is available from the Financial Aid Office (telephone: 618-453-4334; e-mail: fao@siu.edu).

Cost of Study

In-state graduate tuition is $243 per credit hour in 2006–07. Out-of-state tuition is 2.5 times the in-state tuition rate ($607.50 per credit hour). Graduate students with at least a 25 percent appointment as a graduate assistant receive a tuition waiver. Fees vary from $441.62 (1 credit hour) to $987.30 (12 credit hours).

Living and Housing Costs

For married couples, students with families, and single graduate students, the University has 589 efficiency and one-, two-, and three-bedroom apartments that rent for $438 to $505 per month in 2006–07. Residence halls for single graduate students are also available, as are accessible residence hall rooms and apartments for students with disabilities.

Student Group

Applicants are considered for acceptance into the doctoral program if they have completed with distinction a program comparable to that for the master's degree in mathematics, statistics, or computer science at SIUC. Additional evidence of outstanding scholarly ability or achievement (e.g., a high score on the advanced section of the GRE or published research papers of high quality) lends strength to the application.

Student Outcomes

On average, between 2 and 3 students earn their Ph.D. degrees in the program each year. Among the most recent graduates of the program, more than half are employed in a variety of research-oriented positions in private industry, business, and government, and the others are employed in positions involving research and/or teaching in colleges and universities throughout the U.S. Graduates hold tenured faculty positions at such institutions as the University of Memphis, James Madison University, the University of Dayton, and Tennessee Technological University.

Location

SIUC is 350 miles south of Chicago and 100 miles southeast of St. Louis. Nestled in rolling hills bordered by the Ohio and Mississippi Rivers and enhanced by a mild climate, the area has state parks, national forests and wildlife refuges, and large lakes for outdoor recreation. Cultural offerings include theater, opera, concerts, art exhibits, and cinema. Educational facilities for the families of students are excellent.

The University

Southern Illinois University Carbondale is a comprehensive public university with a variety of general and professional education programs. The University offers associate, bachelor's, master's, and doctoral degrees; the J.D. degree; and the M.D. degree. The University is fully accredited by the North Central Association of Colleges and Schools. The graduate school has an essential role in the development and coordination of graduate instruction and research programs. The Graduate Council has academic responsibility for determining graduate standards, recommending new graduate programs and research centers, and establishing policies to facilitate the research effort.

Applying

The forms needed to apply for admission to the program can be accessed through the department's Web site. All applications for admission are considered for financial support, unless requested otherwise. In order to be considered for a fellowship, the applicant must take the GRE General Test. Review of applications for the fall semester begins around February 1.

Correspondence and Information

Applications and supporting documents should be sent to:

Director of Graduate Studies
c/o Graduate Admissions Secretary
Department of Mathematics
Mail Code 4408
Southern Illinois University
1245 Lincoln Drive
Carbondale, Illinois 62901

Web site: http://www.math.siu.edu/gradprog.shtml

Specific questions or inquiries or further information regarding the department and its programs may be directed to:

Professor S. Jeyaratnam
Director of Graduate Studies
Department of Mathematics
Mail Code 4408
Southern Illinois University
1245 Lincoln Drive
Carbondale, Illinois 62901

Phone: 618-453-5302
E-mail: gradinfo@math.siu.edu
Web site: http://www.math.siu.edu

Southern Illinois University Carbondale

THE FACULTY AND THEIR RESEARCH

Dubravka Ban, Associate Professor; Dr.Sci., Zagreb (Croatia), 1998. Algebra, number theory, automorphic forms.
Bhaskar Bhattacharya, Professor; Ph.D., Iowa, 1993. Order restricted statistical inference, I-projections, statistical applications.
Gregory Budzban, Professor; Ph.D., South Florida, 1991. Probability on algebraic structures.
Lane Clark, Professor; Ph.D., New Mexico, 1980. Extremal graph theory, random graphs and enumeration.
Andrew Earnest, Professor; Ph.D., Ohio State, 1975. Algebra, algebraic number theory, arithmetic theory of quadratic forms.
Philip Feinsilver, Professor; Ph.D., NYU (Courant), 1975. Probability theory, algebraic structures.
Robert Fitzgerald, Professor; Ph.D., UCLA, 1980. Quadratic forms and algebra.
John Gregory, Professor; Ph.D., UCLA, 1969. Optimization theory, numerical analysis, applied functional analysis.
H. Randolph Hughes, Associate Professor; Ph.D., Northwestern, 1988. Stochastic processes, stochastic differential geometry.
Sakthivel Jeyaratnam, Professor; Ph.D., Colorado State, 1978. Statistics, linear models, variance components, robust inference.
Jerzy Kocik, Assistant Professor; Ph.D., Southern Illinois at Carbondale, 1989. Mathematical physics, lie algebras, differential-geometric structures, symplectic geometry.
John McSorley, Assistant Professor; D. Phil.; Oxford, 1988. Combinatorics.
Salah-Eldin Mohammed, Professor; Ph.D., Warwick (England), 1976. Stochastic analysis, deterministic and stochastic hereditary dynamical systems, probabilistic analysis of PDEs.
Abdel-Razzaq Mugdadi, Associate Professor; Ph.D., Northern Illinois, 1999. Statistics.
Edward Neuman, Professor; Ph.D., Wroclaw (Poland), 1972. Numerical analysis, spline functions, approximation theory, special functions.
David Olive, Associate Professor; Ph.D., Minnesota, 1998. Applications of high breakdown robust statistics, regression graphics, applied probability theory.
George Parker, Associate Professor; Ph.D., California, San Diego, 1971. Differential geometry, classical geometry, linear programming.
Kathleen Pericak-Spector, Professor; Ph.D., Carnegie Mellon, 1980. Hyperbolic PDEs, continuum mechanics, science education.
Thomas Porter, Professor; Ph.D., New Mexico, 1990. Combinatorics and graph theory.
Don Redmond, Associate Professor; Ph.D., Illinois, 1976. Analytic and elementary number theory, classical analysis, history of mathematics.
Henri Schurz, Assistant Professor; Ph.D., Humboldt (Berlin), 1997. Stochastic analysis, stochastic dynamical systems, random vibrations, mathematical finance, mathematical biology, numerics.
Scott Spector, Professor; Ph.D., Carnegie Mellon, 1978. Elasticity and continuum mechanics.
Michael Sullivan, Professor; Ph.D., Texas at Austin, 1992. Topological dynamical systems and knot theory.
Issa Tall, Assistant Professor; Ph.D., INSA de Rouen (France), 2000. Control theory, theory of dynamical systems, differential geometry.
Walter Wallis, Professor; Ph.D., Sydney, 1968. Combinatorics, combinatorial computing, enterprise networks.
Mary Wright, Professor; Ph.D., McGill, 1977. Rings and modules.
MingQing Xiao, Associate Professor; Ph.D., Illinois, 1997. Partial differential equations, control theory, optimization theory, dynamical systems, computational science.
Dashun Xu, Assistant Professor; Ph.D., Memorial of Newfoundland, 2004. Mathematical biology, differential equations, dynamical systems.
Jianhong Xu, Assistant Professor; Ph.D., Connecticut, 2003. Numerical analysis, matrix computations, parallel algorithms, matrix theory and applications.
Joseph Yucas, Professor; Ph.D., Penn State, 1978. Algebra and combinatorics.
Marvin Zeman, Professor; Ph.D., NYU, 1974. Partial differential equations, integro-differential equations, numerical analysis.

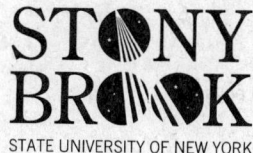

STONY BROOK
STATE UNIVERSITY OF NEW YORK

STONY BROOK UNIVERSITY, STATE UNIVERSITY OF NEW YORK

College of Arts and Sciences
Department of Mathematics

Programs of Study

The Master of Arts and Ph.D., professional option, and Master of Arts, secondary teacher option, are offered by the Department of Mathematics.

The professional option is designed for students who plan careers as professional mathematicians in research or teaching at colleges and universities or who plan careers in finance, industry, or government. Almost all students in this option are full-time. At least one year of full-time study is required. The secondary teacher option is a two-year, part-time program designed for secondary teachers who seek permanent certification. The courses are given in the evenings and in the summer in a two-year cycle, each being offered once every two years.

Research Facilities

The Department has an extensive computer network, primarily using Linux and Solaris platforms, but also PCs and Macintoshes. Both faculty members and graduate students have computers in their offices, and there are a number of workstations in public areas for use by students, faculty members, and visitors. In addition, there is an instructional lab with thirty Sun workstations and twenty-five Windows machines in which computer-related courses are taught. Several faculty members and many graduate students use computation as part of their research. An excellent mathematics-physics library is located nearby in the adjacent physics building.

Financial Aid

Because Stony Brook is committed to attracting high-quality students, the graduate school provides two competitive fellowships for U.S. citizens and permanent residents. Graduate Council fellowships are for outstanding doctoral candidates studying in any discipline, and the W. Burghardt Turner Fellowships target outstanding African-American, Hispanic-American, and Native American students entering either a doctoral or master's degree program. For doctoral students, both fellowships provide a minimum annual stipend of $20,850 for up to five years, as well as a full-tuition scholarship. Most of the students in the mathematics department are supported with teaching assistantships, covering full tuition, health insurance, and a stipend of at least $16,000 for fall 2005 incoming students. In addition, the Department has several fellowships that provide partial support.

Cost of Study

In 2004–05, full-time tuition was $3450 per semester for state residents and $5460 per semester for nonresidents. Part-time tuition was $288 per credit hour for residents and $438 per credit hour for nonresidents. Additional charges included an activity fee of $22 and a comprehensive fee of $242.50 per semester.

Living and Housing Costs

University apartments range in cost from approximately $210 per month to approximately $1180 per month, depending on the size of the unit. Off-campus housing options include furnished rooms to rent and houses and/or apartments to share that can be rented for $350 to $550 per month.

Location

Stony Brook's campus is approximately 50 miles east of Manhattan on the north shore of Long Island. The cultural offerings of New York City and Suffolk County's countryside and seashore are conveniently located nearby. Cold Spring Harbor Laboratories and Brookhaven National Laboratories are easily accessible from, and have close relationships with, the University.

The University

The University, which was established in 1957, achieved national stature within a generation. Founded at Oyster Bay, Long Island, the school moved to its present location in 1962. Stony Brook has grown to encompass more than 110 buildings on 1,100 acres. There are 1,568 faculty members, and the annual budget is more than $805 million. The Graduate Student Organization oversees the spending of the student activity fee for graduate student campus events. International students find the additional four-week Summer Institute in American Living very helpful. The Intensive English Center offers classes in English as a second language. The Career Development Office assists with career planning and has information on permanent full-time employment. Disabled Student Services has a Resource Center that offers placement testing, tutoring, vocational assessment, and psychological counseling. The Counseling Center provides individual, group, family, and marital counseling and psychotherapy. Day-care services are provided in four on-campus facilities. The Writing Center offers tutoring in all phases of writing.

Applying

Applicants are judged on the basis of distinguished undergraduate records (and graduate records, if applicable), thorough preparation for advanced study and research in the field of interest, candid appraisals from those familiar with the applicant's academic/professional work, potential for graduate study, and a clearly defined statement of purpose and scholarly interest germane to the program. For the professional option, a baccalaureate degree with a major in mathematics or the equivalent is required, with a minimum overall grade point average of 2.75 and at least a grade of B in the major and related courses. For the secondary teacher option, a baccalaureate degree and two years of college-level mathematics, including one year of single-variable calculus, one semester of linear algebra, and one more semester of mathematics beyond single-variable calculus, are required. In addition, a grade point average of at least 3.0 in all calculus and post-calculus mathematics courses and a New York State certification for teaching mathematics are required. Students should submit admission and financial aid applications by January 15 for the fall semester and by October 1 for the spring semester, although admissions for the spring semester are rare. Decisions are made on a rolling basis as space permits.

Correspondence and Information

Michael T. Anderson, Graduate Program Director
Department of Mathematics
Stony Brook University, State University of New York
Stony Brook, New York 11790-3651
Phone: 631-632-8269
Fax: 631-632-7631
E-mail: gpd@math.sunysb.edu
Web site: http://www.math.sunysb.edu

Stony Brook University, State University of New York

THE FACULTY AND THEIR RESEARCH

Michael Anderson, Professor; Ph.D., Berkeley, 1981. Differential geometry, geometric analysis, general relativity, mathematical physics.
William Barcus, Professor Emeritus; D.Phil., Oxford, 1955. Algebraic topology.
Christopher Bishop, Professor; Ph.D., Chicago, 1987. Complex analysis.
Sebastian Casalaina-Martin, James H. Simons Instructor; Ph.D., Columbia, 2004. Algebraic geometry.
Alastair Craw, James H. Simons Instructor; Ph.D., Warwick, 2001. Algebraic geometry.
Mark Andrea de Cataldo, Associate Professor; Ph.D., Notre Dame, 1995. Algebraic geometry.
David Ebin, Professor; Ph.D., MIT, 1967. Global analysis, mathematics of continuum mechanics, partial differential equations.
Daryl Geller, Professor; Ph.D., Princeton, 1977. Partial differential equations, harmonic analysis, several complex variables, Lie groups.
James Glimm, Distinguished Professor; Ph.D., Columbia, 1959. Applied mathematics, numerical analysis, mathematical physics.
Detlef Gromoll, Leading Professor; Ph.D., Bonn, 1964. Differential geometry.
Basak Gurel, James H. Simons Instructor; Ph.D., California, Santa Cruz, 2003. Hamiltonian dynamical systems, symplectic geometry and topology.
C. Denson Hill, Professor; Ph.D., NYU, 1966. Partial differential equations, several complex variables.
Jerome A. Jenquin, James H. Simons Instructor; Ph.D., Texas at Austin, 2004. Supersymmetry, geometric index theory, gauge theory.
Lowell Jones, Professor; Ph.D., Yale, 1970. Topology, geometry.
Alexander Kirillov Jr., Associate Professor; Ph.D., Yale, 1995. Algebra, representation theory.
Valentina Kiritchenko, James H. Simons Instructor; Ph.D., Toronto, 2004. Geometry of reductive groups, generalized hypergeometric functions.
Anthony Knapp, Professor Emeritus; Ph.D., Princeton, 1965. Lie groups, representation theory.
Irwin Kra, Distinguished Service Professor; Ph.D., Columbia, 1966. Complex analysis, Kleinian groups.
Paul G. Kumpel, Professor; Ph.D., Brown, 1964. Algebraic topology.
H. Blaine Lawson Jr., Distinguished Professor; Ph.D., Stanford, 1968. Differential geometry, topology, geometric measure theory, several complex variables, algebraic geometry.
Claude LeBrun, Professor; D.Phil., Oxford, 1980. Differential geometry, algebraic geometry, complex analysis, mathematical physics.
Mikhail Lyubich, Professor; Ph.D., Tashkent State, 1984. Dynamical systems.
Bernard Maskit, Professor; Ph.D., NYU, 1964. Complex analysis, Riemann surfaces, Kleinian groups and deformation spaces.
Dusa McDuff, Distinguished Professor; Ph.D., Cambridge, 1971. Geometry, symplectic topology.
Marie-Louise Michelsohn, Professor; Ph.D., Chicago, 1974. Differential geometry.
John Milnor, Distinguished Professor; Ph.D., Princeton, 1954. Dynamical systems.
Anthony Phillips, Professor; Ph.D., Princeton, 1966. Topology, applications to mathematical physics.
Sorin Popescu, Associate Professor; Ph.D., Saarland, 1993. Algebraic geometry, commutative algebra, computational algebraic geometry.
Neil Portnoy, Assistant Professor; Ph.D., New Hampshire, 1998. Operator theory on spaces of analytic functions, mathematics education.
Justin Sawon, James H. Simons Instructor; Ph.D., Cambridge, 2000. Geometry.
Dennis Sullivan, Distinguished Professor; Ph.D., Princeton, 1965. Theory of manifolds, triangulations, and algebraic topology; differential forms and homotopy theory; foliations, laminations, and low-dimensional dynamical systems; Riemann surfaces and Kleinian groups; fluid evolution and computation; quantum theory; topology.
Scott Sutherland, Associate Professor; Ph.D., Boston University, 1989. Dynamical systems, root-finding algorithms, computing.
Leon Takhtajan, Professor; Ph.D., Steklov Institute of Mathematics (USSR), 1975. Mathematical physics.

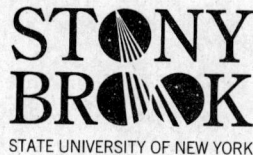

STONY BROOK
STATE UNIVERSITY OF NEW YORK

STONY BROOK UNIVERSITY, STATE UNIVERSITY OF NEW YORK

College of Engineering and Applied Sciences
Department of Applied Mathematics and Statistics

Programs of Study

The Department of Applied Mathematics and Statistics offers programs leading to the master's and Doctor of Philosophy degrees. Special strengths in applied mathematics and statistics include computational fluid dynamics, operations research, applied statistics, numerical analysis, and bioinformatics. All students receive a broad range of basic and advanced training in mathematics and applied mathematics.

The Ph.D. program consists of an individually designed selection of advanced courses followed by a program of research leading to a dissertation. Students are required to pass a written comprehensive examination and an oral preliminary examination.

The master's degree program focuses on specific mathematical sciences skills needed for a career as a teacher or an industrial mathematician.

All students receive personal attention and advising at all levels. There are numerous seminars and colloquia, featuring both distinguished guest speakers and Stony Brook faculty members. These present both accelerated background knowledge and ongoing research. The Department has close ties and joint activities with many units on campus, including the Institute for Theoretical Physics, Harriman College of Management, and the Computer Science Department as well as nearby Brookhaven National Laboratory and Cold Spring Harbor Laboratory.

Research Facilities

The Department is housed in the Mathematics Building. Each graduate student shares an office with at least one other student. The mathematics-physics library, located in an adjoining building, contains 50,000 books and subscribes to about 500 journals. The Laboratory for Parallel Computing has a custom-built 450 processor Galaxy parallel supercomputer. Galaxy replaced the Department's 120-node Intel Paragon parallel supercomputer. There are more than 200 Sun and Silicon Graphics workstations and high-end PCs in the Department. Most student offices have a workstation on every desk.

Financial Aid

The Department is currently supporting 10 students on various fellowships, with stipends ranging from $18,000 to $30,000; 28 students on teaching assistantships; and 25 students on research assistantships. The assistantship stipends range from $11,650 to $18,000. All students supported by the Department also receive tuition remission. Summer support is available in certain research areas.

Cost of Study

In 2006–07, tuition is $288 and $455 per credit hour for state and out-of-state residents, respectively. Full-time tuition per semester is $3450 for state residents and $5460 for out-of-state residents. Additional charges include an activity fee of $22 and a comprehensive fee of $351 per semester.

Living and Housing Costs

Monthly rents at the University apartments range from $320 to $1000 per month, depending on the locations and the size of the room. The off-campus housing office assists in locating available apartments or houses to share, which can be rented for $350 to $550 per month.

Student Group

The number of full-time graduate students in the Department of Applied Mathematics and Statistics is about 140. The Department awards about eleven Ph.D. degrees and thirty M.S. degrees annually. In spring 2005, the Department had 149 students, of whom 93 percent were full-time, 35 percent were master's, and 65 percent were doctoral students; 45 percent were women.

Student Outcomes

Recent graduates have found positions as faculty members and postdoctoral fellows in academia, financial and risk analysts in the financial sector, biostatisticians in medical and research centers, research scientists in industrial and national laboratories, and actuaries in insurance companies. Their places of employment include Harvard University; New York University's Courant Institute; Rutgers University; Franklin and Marshall College; Worcester Polytechnic Institute; University of Notre Dame; Arizona State University; University of California, Irvine; Eli Lilly Pharmaceuticals; M. D. Anderson Cancer Center; Bell Labs; IBM Watson Research Center; Los Alamos National Laboratory; Chase Manhattan Bank; Bear, Stearns & Co.; American Express; Metropolitan Life Insurance Company; and the U.S. Department of Transportation.

Location

Stony Brook's campus is approximately 50 miles east of Manhattan on the North Shore of Long Island. The cultural offerings of New York City and Suffolk County's countryside and seashore are conveniently located nearby. Cold Spring Harbor Laboratories and Brookhaven National Laboratory are easily accessible and have close relationships with the University.

The University and The Department

The University, established in 1957, achieved national stature within a generation. Founded in Oyster Bay, Long Island, the school moved to its present location in 1962. Stony Brook has grounds that encompass more than 110 buildings on 1,100 acres. The University enrolls about 21,000 students, including 7,000 graduate students. There are more than 1,565 faculty members, and the annual budget is more than $805 million. The Graduate Student Organization oversees the spending of the student activity fee for graduate student campus events. International students find the additional four-week Summer Institute in American Living very helpful. The Intensive English Center offers classes in English as a second language. The Career Development Office offers career planning and has information on permanent full-time employment. Disabled students have a resource center that offers placement testing, tutoring, vocational assessment, counseling, and psychotherapy. Day-care services are provided in four on-campus facilities. The Writing Center offers tutoring in all phases of writing.

The Department's 18 regular faculty members and 25 affiliated faculty members include a National Medal of Science recipient, 2 National Academy of Sciences members, and recipients of several prestigious professional honors. Special strengths are computational fluid dynamics, applied statistics, computational biology, and computational geometry. The Department supports extensive computing facilities and an active program of conferences and visiting scholars. The focus on problem-driven research includes a diverse array of interdisciplinary projects spanning the biomedical, physical, and social sciences.

Applying

Applications for admission are welcome at all times. Transcripts, GRE General Test scores, and three letters of recommendation are required. Completed applications for fall admission should be received by March 1; the deadline for those applying for financial aid is January 15. For more information and application forms, students should visit the Department's Web site at http://www.ams.sunysb.edu.

Correspondence and Information

Professor Xiaolin Li
Director of Graduate Program
Department of Applied Mathematics and Statistics
Stony Brook University, State University of New York
Stony Brook, New York 11794-3600
Phone: 631-632-8354
Web site: http://www.ams.sunysb.edu

Stony Brook University, State University of New York

THE FACULTY AND THEIR RESEARCH

Hongshik Ahn, Associate Professor; Ph.D., Wisconsin. Biostatistics.
Carlos Alonso, Adjunct Professor; Ph.D., Federal University of Rio de Janeiro. Computational biology
Esther Arkin, Professor; Ph.D., Stanford. Combinatorial optimization, networks, algorithms, computational geometry.
Hussein G. Badr, Adjunct Professor; Ph.D., Penn State: Operating systems, computer system performance evaluation.
Edward Beltrami, Professor Emeritus; Ph.D., Adelphi. Nonlinear models, stochastic models.
Michael Bender, Adjunct Professor; Ph.D., Harvard. Combinatorial algorithms.
Yung Ming Chen, Professor Emeritus; Ph.D., Columbia. Computational fluid dynamics, parallel computing.
Yuefan Deng, Professor; Ph.D., Columbia. Computational fluid dynamics, parallel computing.
Daniel Dicker, Professor Emeritus; Ph.D., Columbia. Porous flow problems.
Pradeep Dubey, Professor; Ph.D., Cornell. Game theory; mathematical economics.
Eugene Feinberg, Professor; Ph.D., Vilnius State (Lithuania). Applied probability.
David Ferguson, Adjunct Professor; Ph.D., Berkeley. Mathematics education, educational technology.
Stephen Finch, Associate Professor; Ph.D., Princeton. Applied statistics.
Charles Fortmann, Visiting Associate Professor; Ph.D., Stanford. Computational biophysics, photonics.
Robert Frey, Research Professor; Ph.D., SUNY at Stony Brook. Quantitative finance.
James Glimm, Distinguished Professor and Chair; Ph.D., Columbia. Computational fluid dynamics, conservation laws, mathematical physics.
David Green, Assistant Professor; Ph.D., MIT. Computational biology, protein structure.
John Grove, Adjunct Professor; Ph.D., Ohio State. Computational fluid dynamics.
Xiaolin Li, Professor; Ph.D., Columbia. Computational fluid dynamics.
Brent Lindquist, Professor; Ph.D., Cornell. Computational fluid dynamics, reservoir modeling.
Glenn Martyna, Adjunct Professor; Ph.D., Columbia. Molecular dynamics.
Nancy Mendell, Professor; Ph.D., North Carolina. Applied statistics.
Joseph Mitchell, Professor; Ph.D., Stanford. Computational geometry and its applications, optimization, algorithms.
Ronald Peierls, Adjunct Professor; Ph.D., Cornell. Parallel computing, particle physics.
John Pinezic, Adjunct Professor; Ph.D., SUNY at Stony Brook. Radar, ballistics, sonar, acoustics.
Bradley Plohr, Adjunct Professor; Ph.D., Princeton. Computational fluid dynamics.
John Reinitz, Professor; Ph.D., Yale. Bioinformatics.
Robert Rizzo, Assistant Professor; Ph.D., Yale. Computational biology, drug design.
David Sharp, Adjunct Professor (Los Alamos National Lab); Ph.D., Princeton. Nonlinear analysis.
Weinig Sheldon, Professor; Ph.D., Columbia. Manufacturing management, material sciences.
Claudio Silva, Adjunct Professor; Ph.D., SUNY at Stony Brook. Visualization, computer graphics, computational geometry.
Carlos Simmerling, Associate Professor; Ph.D., Illinois at Chicago: Protein structure.
Steven Skiena, Professor; Ph.D., Illinois. Combinatorial algorithms and bioinformatics.
Jadranka Skorin-Kapov, Professor; Ph.D., British Columbia. Mathematical programming.
Robert R. Sokal, Emeritus Distinguished Professor; Ph.D., Chicago. Numerical taxonomy, theory of systematics, geographic variation, spatial models in ecology and evolution.
Alexander Spirov, Associate Professor; Ph.D., Irktsk State (Russia). Computational biology.
Ram Srivastav, Professor; Dc.S., Glasgow. Numerical analysis, integral equations.
Reginald Tewarson, Professor Emeritus; Ph.D., Boston University. Numerical analysis, biomathematics.
Alan Tucker, Distinguished Teaching Professor; Ph.D., Stanford. Combinatorial optimization.
E. Alper Yildirim, Assistant Professor; Ph.D., Cornell. Mathematical programming.
Armen H. Zemanian, Distinguished Professor; Eng.Sc.D., NYU. Network theory, food system modeling.
Yongmin Zhang, Assistant Professor; Ph.D., Chicago. Computational fluid dynamics.
Wei Zhu, Associate Professor; Ph.D., UCLA. Biostatistics.

UNIVERSITY OF CALIFORNIA, LOS ANGELES

School of Medicine
Department of Biomathematics

Programs of Study

The Department of Biomathematics offers a graduate program leading to the Master of Science and Doctor of Philosophy degrees in biomathematics. The goal of the doctoral program is to train creative, fully independent investigators in mathematical and computational biology who can initiate research in both applied mathematics and their chosen biomedical specialty. The Department's orientation is away from abstract modeling and toward theoretical and applied research that is vital to the advancement of current biomedical frontiers. This is reflected in a curriculum providing a high level of competence in biology or biomedicine; substantial training in applied mathematics, statistics, and computing; and appropriate biomathematics courses and research experience. A low student-faculty ratio permits close and frequent contact between students and faculty members throughout the training and research years. The Systems and Integrative Biology Training Program, which is housed in biomathematics (http://dragon.nuc.ucla.edu/sibtp/), provides additional links to research and advising through more than two dozen participating faculty members from a range of UCLA departments and programs.

Entering students come from a variety of backgrounds in mathematics, biology, the physical sciences, and computer science. Some of the students are enrolled in the UCLA M.D./Ph.D. program. Doctoral students generally use the first two years to take the core sequence and electives in biomathematics, to broaden their backgrounds in biology and mathematics, and to begin directed individual study or research. Comprehensive examinations in biomathematics are taken after this period, generally followed by the choice of a major field and dissertation area. Individualized programs permit students to select such graduate courses as applied mathematics, biomathematics, statistics, and engineering, along with advanced training in their field of special emphasis in biology. Prospective students should visit http://www.gdnet.ucla.edu/gasaa/library/pgmrqintro.htm for program requirement details.

The master's program can be a step to further graduate work in biomathematics, but it also can be adapted to the needs of researchers desiring supplemental biomathematical training or of individuals wishing to provide methodologic support to biomedical researchers. The M.S. program requires at least five graduate biomathematics courses and either a thesis or comprehensive examination plan. The master's degree can be completed in one or two years.

Research Facilities

The Department is situated in the Center for the Health Sciences, close to UCLA's rich research and educational resources in the School of Medicine; the Departments of Mathematics, Biology, Computer Science, Engineering, Chemistry, and Physics; the Institute for Pure and Applied Mathematics; and the Molecular Biology Institute. The Department has for many years housed multidisciplinary research programs comprising innovative modeling, statistical, and computing methods directed to many areas of biomedical research. It was the original home of the BMDP statistical programs and has an active consulting clinic for biomedical researchers. Modern computer resources are readily available. The Biomedical Library is one of the finest libraries of its kind in the country, and nearby are the Engineering and Mathematical Sciences Library and other subject libraries of the renowned nineteen-branch University Library. The Department maintains a small library with selected titles in mathematical biology and statistics.

Financial Aid

Financial support is provided from a variety of sources, including University-sponsored fellowships, affiliated training grants, research assistantships, and other merit-based funds. Supplementation is also possible from consulting and teaching assistantships.

Cost of Study

The 2006–07 registration and other fees for California residents are estimated to be $8266 and for nonresidents, $8533. Nonresident tuition is projected to be $14,694. Domestic students may attain residency status after one year. The University's fee proposal is subject to change based on state budget decisions.

Living and Housing Costs

The estimated cost of living varies. (The cost of living is in addition to the cost of study.) For additional housing information, students can go to the Web site (http://www.housing.ucla.edu/housing_site/index.htm).

Student Group

Currently, 16 graduate students are enrolled in the Department's program. About a fifth of the students are international. An NIH predoctoral training grant supports up to 6 students. Most other students are also receiving financial support and/or are employed on campus in the area of their research. Many graduates hold tenure-track appointments at leading universities, research appointments at the National Institutes of Health, or positions in industry.

Location

UCLA's 411 acres are cradled in rolling green hills just 5 miles inland from the ocean, in one of the most attractive areas of southern California. The campus is bordered on the north by the protected wilderness of the Santa Monica Mountains and at its southern gate by Westwood Village, one of the entertainment magnets of Los Angeles.

The University

UCLA is one of America's most prestigious and influential public universities, serving more than 35,000 students. The Department of Biomathematics is one of ten basic science departments in the School of Medicine. The medical school, which is regarded by many to be among the best in the nation, is situated on the south side of the UCLA campus, just adjacent to the Life Sciences Building and the Court of Sciences. For more information, students can visit the University's Web site (http://www.ucla.edu).

Applying

Most students enter in the fall quarter, but applications for winter- or spring-quarter entry are considered. However, it is advantageous for candidates applying for financial support to initiate the application by the middle of January for decisions for the following fall. The Department expects applicants for direct admission to the doctoral program to submit scores on the General Test of the Graduate Record Examinations and on one GRE Subject Test of the student's choice. Inquiries are welcome from students early in their undergraduate training. The Department supports minority recruitment.

Correspondence and Information

Admissions Committee Chair
Department of Biomathematics
UCLA School of Medicine
Los Angeles, California 90095-1766
Phone: 310-825-5554
Fax: 310-825-8685
E-mail: gradprog@biomath.ucla.edu
Web site: http://www.biomath.ucla.edu/

University of California, Los Angeles

THE FACULTY AND THEIR RESEARCH

A. A. Afifi, Professor of Biostatistics and Biomathematics; Ph.D. (statistics), Berkeley, 1965. Multivariate statistical analysis, with applications to biomedical and public health problems; longitudinal and correlated data analysis.

Tom Chou, Associate Professor of Biomathematics; Ph.D. (physics), Harvard, 1995. Theoretical biophysics, cellular and molecular modeling, transport, bioenergetics.

Wilfrid J. Dixon, Professor of Biomathematics, Biostatistics, and Psychiatry (Emeritus); Ph.D. (statistics), Princeton, 1944. Statistical computation, statistical theory, biological applications, data analysis, psychiatric research.

Robert M. Elashoff, Professor of Biomathematics and Biostatistics and Chair of Biostatistics; Ph.D. (statistics), Harvard, 1963. Markov renewal models in survival analysis, random coefficient regression models.

Eli Engel, Adjunct Associate Professor of Biomathematics; M.D., Buffalo, 1951; Ph.D. (physiology), UCLA, 1975. Mechanisms for acid neutralization in gastric mucus, facilitated transport of oxygen, theory of intracellular microelectrodes.

Sanjiv Sam Gambhir, Adjunct Professor, Department of Molecular and Medical Pharmacology and Biomathematics; Ph.D. (biomathematics), 1990, M.D., 1993, UCLA. Positron emission tomography (PET), deterministic modeling, medical imaging, neural networks, imaging gene expression, nuclear medicine, optical imaging.

Sung-Cheng (Henry) Huang, Professor of Medical Pharmacology and Biomathematics; D.Sc. (electrical engineering), Washington (St. Louis), 1973. Positron emission computed tomography and physiological modeling.

Donald J. Jenden, Professor of Pharmacology and Biomathematics (Emeritus); M.B./B.S. (pharmacology and therapeutics), Westminster (London), 1950. Pharmacokinetic modeling, chemical pharmacology, analysis of GC/MS data, neuropharmacology.

Robert I. Jennrich, Professor of Mathematics, Biomathematics, and Biostatistics (Emeritus); Ph.D. (mathematics), UCLA, 1960. Statistical methodology, computational algorithms, nonlinear regression, factor analysis, compartment analysis.

Elliot M. Landaw, Professor and Chair of Biomathematics; M.D., Chicago, 1972; Ph.D. (biomathematics), UCLA, 1980. Identifiability and optimal experiment design for compartmental models; nonlinear regression; modeling/estimation applications in pharmacokinetics, ligand-receptor analysis, transport, and molecular biology.

Kenneth L. Lange, Professor of Biomathematics and Chair of Human Genetics; Ph.D. (mathematics), MIT, 1971. Statistical and mathematical methods for human genetics, demography, medical imaging, probability, optimization theory.

Carol M. Newton, Professor of Biomathematics and Radiation Oncology; Ph.D. (physics and mathematics), Stanford, 1956; M.D., Chicago, 1960. Simulation; cellular models for hematopoiesis, cancer treatment strategies, and optimization; interactive graphics for modeling; model-based exploration of complex data structures.

Michael E. Phelps, Norton Simon Professor; Chair, Department of Molecular and Medical Pharmacology; Professor of Biomathematics; Chief and Director, Institute for Molecular Medicine; and Director, Crump Institute for Molecular Imaging; Ph.D. (nuclear chemistry), Washington (St. Louis), 1970. Positron emission tomography (PET), tracer kinetic modeling of biochemical and pharmacokinetic processes, biological imaging of human disease.

Janet S. Sinsheimer, Associate Professor of Biomathematics, Biostatistics, and Human Genetics; Ph.D. (biomathematics), UCLA, 1994. Statistical models of molecular evolution and genetics.

Marc A. Suchard, Assistant Professor of Human Genetics and Biomathematics; Ph.D. (biomathematics), 2002, M.D., 2004, UCLA. Evolutionary reconstruction, sequence analysis, medical time series.

SELECTED PUBLICATIONS

Lakatos, G., J. D. O'Brien, and **T. Chou.** Hydrodynamic solutions of 1D exclusion processes with spatially varying hopping rates. *J. Phys. A* 39:2253–64, 2006.

Lakatos, G., **T. Chou,** B. Bergersen, and G. N. Patey. First passage times of driven DNA hairpin unzipping. *Phys. Biol.* 2:166–74, 2005.

D'Orsogna, M. R., and **T. Chou.** Queueing and cooperativity in ligand-receptor binding. *Phys. Rev. Lett.* 95:170603, 2005.

Chou, T. External fields, dipolar coupling, and lubrication in water-wire proton transport. *Biophys. J.* 86:2827–36, 2004.

Chou, T., and G. Lakatos. Clustered bottlenecks in mRNA translation and protein synthesis. *Phys. Rev. Lett.* 93:198101, 2004.

Foster, W. J., and **T. Chou.** Physical mechanisms of gas and perfluron retinopexy and sub-retinal fluid displacement. *Phys. Med. Biol.* 49:2989–97, 2004.

Klapstein, K., **T. Chou,** and R. Bruinsma. Physics of RecA-mediated homologous recognition. *Biophys. J.* 87:1466–77, 2004.

Bal, G., and **T. Chou.** On the reconstruction of diffusions using a single first-exit time distribution. *Inverse Probl.* 20:1053–65, 2003.

Chou, T. An exact theory of histone-DNA adsorption and wrapping. *Europhys. Lett.* 62:753–9, 2003.

Chou, T. Ribosome recycling, diffusion, and MRNA loop formation in translational regulation. *Biophys. J.* 85:755–73, 2003.

Reckamp, K. L., et al. **(R. M. Elashoff).** A phase trial to determine the optimal biologic dose of celecoxib when combined with erlotinib in advanced non-small cell lung cancer. *Clin. Cancer Res.* 12:3381–8, 2006.

Tashkin, D. P., and **R. Elashoff** et al. Cylophosphamide versus placebo in scleroderma lung disease. *New Engl. J. Med.* 354:2655–66, 2006.

Dobkin, B., et al. **(R. Elashoff).** Reliability and validity of bilateral thigh and foot accelerometry measures of walking in healthy and hemiparetic subjects. *Neurorehab. Neural Re.* 20(2):297–305, 2006.

Elashoff, R. Montelukast improves regional air-trapping due to small airways obstruction in asthma. *Eur. Respir. J.* 27:307–15, 2006.

Elashoff, R. Modern statistical methods in clinical nutrition. In *Nutritional Oncology,* 2nd ed., 2006.

Dobkin, B., et al. **(R. Elashoff)** (Spinal Cord Injury Locomotor Trial Group). Walking-related gains over the first 12 weeks of rehabilitation for incomplete traumatic spinal cord injury: The SCILT randomized clinical trial. *Neurol. Rehabil.* 10(4):179–87, 2005.

Salusky, I. B., et al. **(R. M. Elashoff).** Sevelamer controls parathyroid hormone-induced bone disease as efficiently as calcium carbonate without increasing serum calcium levels during therapy with active vitamin D sterols. *J. Am. Soc. Nephrol.* 16:2501–8, 2005.

Hong, K., et al. **(R. Elashoff).** Analysis of weight loss outcomes using VLCD in black and white overweight and obese women with and without metabolic syndrome. *Int. J. Obesity Related Metab. Disorder* 29(4):436–42, 2005.

Dabrowska, D. M., **R. M. Elashoff,** and D. L. Morton. Estimation in a Markov chain regression model with missing covariates. *Adv. Biostat.* (monograph in honor of Zelen), 2005.

Li, Z., et al. **(R. M. Elashoff).** Long-term efficacy of soy-based meal replacements vs. an individualized diet plan in obese type II DM patients: Relative effects on weight loss, metabolic parameters, and C-reactive protein. *Eur. J. Clin. Nutr.* 59 (3):411–8, 2005.

Krochmal, R., et al. **(R. Elashoff).** Phytochemical assays of commercial botanical dietary supplements. *Evidence Based Compl. Alt. Med.* 1(3):305–13, 2004.

Sartippour, M. R., et al. **(R. Elashoff).** A pilot clinical study of short-term isoflavone supplements in breast cancer patients. *Nutr. Can.* 49(1):59–65, 2004.

Xie, D., et al. **(R. Elashoff).** Levels of expression of CYR61 and CTGF are prognostic for tumor progression and survival of individuals with gliomas. *Clin. Cancer Res.* 10(6):2072–81, 2004.

Clements, P. H., et al. **(R. M. Elashoff).** Regional differences in bronchoalveolar lavage and thoracic high-resolution computed tomography results in dyspneic patients with systemic sclerosis. *Arthritis Rheum.* 50(6):1909–17, 2004.

Sartippour, M. R., et al. **(R. M. Elashoff).** cDNA microarray analysis of endothelial cells in response to green tea reveals a suppressive phenotype. *Int. J. Oncol.* 25:193–202, 2004.

Cochran, A. J., et al. **(R. M. Elashoff).** Prediction of metastatic melanoma in nonsentinel nodes and clinical outcome based on the primary melanoma and the sentinel node. *Modern Pathol.* 17(7):747–55, 2004.

Livingston, E. H., and **E. Engel.** Modeling of the gastric gel mucus layer: Application to the measured pH gradient. *J. Clin. Gastroenterol.* 21:S120–4, 1995.

Engel, E., P. H. Guth, Y. Nishizaki, and J. D. Kaunitz. Barrier function of the gastric mucus gel. *Am. J. Physiol.* (Gastrointest. Liver Physiol. 32) 269:G994–9, 1995.

Livingston, E. H., J. Miller, and **E. Engel.** Bicarbonate diffusion through mucus. *Am. J. Physiol.* (Gastrointest. Liver Physiol. 32) 269:G453–7, 1995.

Engel, E., A. Peskoff, G. L. Kauffman, and M. I. Grossman. Analysis of hydrogen ion concentration in the gastric gel mucus layer. *Am. J. Physiol.* (Gastrointest. Liver Physiol. 10) 247:G321–38, 1984.

Engel, E., V. Barcilon, and R. S. Eisenberg. The interpretation of current-voltage relations recorded from a spherical cell with a single microelectrode. *Biophys. J.* 12(4):384–403, 1972.

Eisenberg, R. S., and **E. Engel.** The spatial variation of membrane potential near a small source of current in a spherical cell. *J. Gen. Physiol.* 55(6):736–57, 1970.

Kreissl, M. C., et al. **(S. C. Huang).** Non-invasive measurement of cardiovascular function in mice with ultra-high temporal resolution small animal PET. *J. Nucl. Med.* 47:974–80, 2006.

Huang, S. C., et al. An Internet-based kinetic imaging system (KIS) for microPET. *Mol. Imaging Biol.* 7:330–41, 2005.

Leow, A., and **S. C. Huang** et al. Brain structural mapping using a novel hybrid implicit/explicit framework based on the level-set method. *NeuroImage* 24(3):910–27, 2005.

Huang, S. C., et al. Investigation of a new input function validation method for mouse microPET studies. *Mol. Imaging Biol. B* 6(1):34–46, 2004.

Wu, H. M., and **S. C. Huang** et al. Subcorticle white matter metabolic changes remote from focal hemorrhagic lesions suggest diffuse injury following human traumatic brain injury (TBI). *Neurosurgery* 55(6):1306–17, 2004.

Hattori, N., and **S. C. Huang** et al. Accuracy of a method using short inhalation of O-15-O2 for measuring cerebral oxygen extraction fraction with PET in healthy humans. *J. Nucl. Med.* 45:765–70, 2004.

Wu, H. M., et al. **(S. C. Huang).** Measurement of the global lumped constant for FDG in normal human brain using [^{15}O]water and FDG PET imaging: A new method with validation based on multiple methodologies. *Mol. Imaging Biol.* 5:32–41, 2003.

Yee, R. E., and **S. C. Huang** et al. Imaging and therapeutics: The role of neuronal transport in the regional specificity of L-DOPA accumulation in brain. *Mol. Imaging Biol.* 4:208218, 2002.

Zhou, Y., **S. C. Huang,** M. Bergsneider, and D. F. Wong. Improved parametric image generation using spatial-temporal analysis of dynamic PET studies. *Neuroimaging* 15:697–707, 2002.

Liao, W.-H., **S. C. Huang, K. Lange,** and M. Bergsneider. Use of MM algorithm for regularization of parametric images in dynamic PET. In *Brain Imaging Using PET,* pp. 107–14, eds. M. Senda, Y. Kimura, P. Herscovitch, and Y. Kimura. New York: Academic Press, 2002.

Huang, S. C. Anatomy of SUV (standardized uptake value). *Nucl. Med. Biol.* 27: 643–6, 2000.

Yu, R. C., D. Hattis, **E. M. Landaw,** and J. R. Froines. Toxicokinetic interaction of 2,5-hexanedione and methyl ethyl ketone. *Arch. Toxicol.* 75:643–52, 2002.

Lopez, A. M., M. D. Pegram, D. J. Slamon, and **E. M. Landaw.** A model-based approach for assessing in vivo combination therapy interactions. *Proc. Natl. Acad. Sci. U.S.A.* 96:13023–8, 1999.

Greenword, A. C., **E. M. Landaw,** and T. H. Brown. Testing the fit of a quantal model of neurotransmission. *Biophys. J.* 76:1847–55, 1999.

Walker, W. L., D. S. Goodsell, and **E. M. Landaw.** An analysis of a class of DNA sequence reading molecules. *J. Comput. Biol.* 5:571–83, 1998.

Walker, W. L., **E. M. Landaw,** R. E. Dickerson, and D. S. Goodsell. The theoretical limits of DNA sequence discrimination by linked polyamides. *Proc. Natl. Acad. Sci. U.S.A.* 95:4315–20, 1998.

Walker, W. L., **E. M. Landaw,** R. E. Dickerson, and D. S. Goodsell. Estimation of the DNA sequence discriminatory ability of hairpin-linked lexitropsins. *Proc. Natl. Acad. Sci. U.S.A.* 94:5634–9, 1997.

Landaw, E. M. Model-based adaptive control for cancer chemotherapy with suramin. In *Proceedings of the Simulation in Health Sciences Conferences,* pp. 93–8, eds. J. G. Anderson and M. Katzper. San Diego: Society for Computer Simulation, 1994.

Marino, A. T., J. J. Distefano III, and **E. M. Landaw.** DIMSUM: An expert system for multiexponential model discrimination. *Am. J. Physiol.* (Endocrinol. Metab. 25) 262:E546–56, 1992.

Landaw, E. M. Optimal multicompartmental sampling designs for parameter estimation-practical aspects of the identification problem. *Math. Comp. Simul.* 24: 525–30, 1982.

Ayers, K. L., C. Sabatti, and **K. Lange.** Reconstructing ancestral haplotypes with a dictionary model. *J. Comp. Biol.,* in press.

Lange, K., and E. Sobel. Variance component models for X-linked QTLs. *Genet. Epidemiol.,* in press.

Crespi, C. M., and **K. Lange.** Estimation for the simple linear Boolean model. *Methodol. Comput. Appl. Prob.,* in press.

Presson, A. P., E. Sobel, **K. Lange,** and J. C. Papp. Merging microsatellite data. *J. Comp. Biol.,* in press.

Chen, G. K., E. Slaten, R. A. Ophoff, and **K. Lange.** Accommodating chromosome inversions in linkage analysis. *Am. J. Hum. Genet.,* in press.

Greenawalt, D. M., et al. **(K. Lange).** Strong correlation between meiotic crossovers and haplotype structure in a 2.5-Mb region on the long arm of chromosome 21. *Genome Res.* 16:208–14, 2005.

Lange, K., J. S. Sinsheimer, and E. Sobel. Association testing with Mendel. *Genet. Epidemiol.* 29:36–50, 2005.

University of California, Los Angeles

Cantor, R. M., G. K. Chen, P. Pajukanta, and **K. Lange.** Association testing in a linked region using large pedigrees. *Am. J. Hum. Genet.* 76:538–42, 2005.

Sabatti, C., L. Rohlin, **K. Lange,** and J. Liao. Vocabulon: A dictionary model approach for reconstruction and localization of transcription factor binding sites. *Bioinformatics* 21:922–31, 2005.

Lange, K. *Optimization.* New York: Springer-Verlag, 2004.

Lilja, H. E., et al. **(K. Lange).** Locus for quantitative HDL-cholesterol on chromosome 10q in Finnish families with dyslipidemia. *J. Lipid Res.* 45:1876–84, 2004.

Lange, K., and **J. S. Sinsheimer.** The pedigree trimming problem. *Hum. Hered.* 58:108–11, 2004.

Lange, E. M., and **K. Lange.** Powerful allele-sharing statistics for non-parametric linkage analysis. *Hum. Hered.* 57:49–58, 2004.

Hunter, D. R., and **K. Lange.** A tutorial on MM algorithms. *Am. Statistician* 58:30–7, 2004.

Lange, K. Computational statistics and optimization theory at UCLA. *Am. Statistician* 58:9–11, 2004.

Lange, K. *Applied Probability.* New York: Springer-Verlag, 2003.

Lange, K. *Mathematical and Statistical Methods for Genetic Analysis,* 2nd ed. New York: Springer-Verlag, 2002.

Lange, K. *Numerical Analysis for Statisticians.* New York: Springer-Verlag, 1999.

Newton, C. M. An interactive graphics system for real-time investigation and multivariate data portrayal for complex pedigree data systems. *Comp. Biomed. Res.* 26:327–43, 1993.

Newton, C. M. Exploring categorical and scalar data interactions: Another graphical approach. In *Proc. Am. Stat. Assoc. Meet. (Section on Statistical Graphics),* pp. 49–54, Boston, August 1992.

Newton, C. M. Conference retrospective: An appropriate modeling infrastructure for cancer research. *Bull. Math. Biol.* 48(3/4):443–52, 1986.

Hood, L., J. R. Heath, **M. E. Phelps,** and B. Lin. Systems biology and new technologies enable predictive and preventative medicine. *Science* 306:640–3, 2004.

Ray, P., et al. **(M. E. Phelps** and **S. S. Gambhir).** Monitoring gene therapy with reporter gene imaging. *Semin. Nucl. Med.* 31(4):312–20, 2001.

Silverman, D. H., et al. **(M. E. Phelps).** Positron emission tomography in evaluation of dementia: Regional brain metabolism and long term outcome. *J. Am. Med. Assoc.* 286:2120–7, 2001.

Yaghoubi, S. S., et al. **(M. E. Phelps** and **S. S. Gambhir).** Human pharmacokinetic and dosimetry studies of [18F]-FHBG: A reporter probe for imaging herpes simplex virus 1 thymidine kinase (HSV1-tk) reporter gene expression. *J. Nucl. Med.* 42(8): 1225–34, 2001.

Phelps, M. E. Positron emission tomography provides molecular imaging of biological processes. *Proc. Natl. Acad. Sci. U.S.A.* 97:9226–33, 2000.

Wu, A., et al. **(M. E. Phelps** and **S. S. Gambhir).** High-resolution microPET imaging of carcinoembryonic antigen-positive xenografts by using copper-64-labeled engineered antibody fragments. *Proc. Natl. Acad. Sci. U.S.A.* 97(5):8495–500, 2000.

Yu, Y., et al. **(M. E. Phelps** and **S. S. Gambhir).** Quantification of target gene expression by imaging reporter gene expression in living animals. *Nature Med.* 6(8):933–7, 2000.

Hsieh, H.-J., et al. **(J. S. Sinsheimer).** The v-MFG test: Investigating maternal, offspring, and maternal-fetal genetic incompatibilities effects on disease and viability. *Genet. Epidemiol.,* in press.

Minassian, S. L., et al. **(J. S. Sinsheimer).** Incorporating serotypes into family based association studies using the MFG test. *Ann. Hum. Genet.,* in press.

Kraft, P., H.-J. Hsieh, H. Cordell, and **J. Sinsheimer.** A conditional-on-exchangeable-parental-genotypes likelihood that remains unbiased under multiple-affected-sibling ascertainment. *Genet. Epidemiol.* 29:87–90, 2005.

Bauman, L., et al. **(J. S. Sinsheimer** and **K. Lange).** Fishing for pleiotropic QTLs in a polygenic sea. *Ann. Hum. Genet.* 69:590–611, 2005.

Minassian, S. L., C. G. S. Palmer, and **J. S. Sinsheimer.** An exact maternal-fetal genotype incompatibility (MFG) test. *Genet. Epidemiol.* 28:83–95, 2005.

Kraft, P., and **J. S. Sinsheimer** et al. RHD maternal-fetal genotype incompatibility and schizophrenia: Extending the MFG test to include multiple siblings and birth order. *Eur. J. Hum. Genet.* 12:192–8, 2004.

Dorman, K., **J. S. Sinsheimer,** and **K. Lange.** In the garden of branching processes. *SIAM Rev.* 46:202–29, 2004.

Sinsheimer, J. S., and **M. A. Suchard** et al. Are you my mother? Bayesian phylogenetic models to detect recombination among putative parental strains. *Appl. Bioinformatics* 2:131–44, 2003.

Sinsheimer, J. S., C. G. S. Palmer, and J. A. Woodward. The maternal-fetal genotype incompatibility test: Detecting genotype combinations that increase risk for disease. *Genet. Epidemiol.* 24:1–13, 2003.

Dorman, K. S., A. H. Kaplan, and **J. S. Sinsheimer.** Bootstrap confidence levels for HIV-1 recombinants. *J. Mol. Evol.* 54(2):200–9, 2002.

Schadt, E. E., **J. S. Sinsheimer,** and **K. Lange.** Applications of codon and rate variation models in molecular phylogeny. *Mol. Biol. Evol.* 19:1550–62, 2002.

Sinsheimer, J. S., C. A. McKenzie, B. Keavney, and **K. Lange.** SNPs and snails and puppy dog tails: Analysis of SNP haplotype data using the gamete competition model. *Ann. Hum. Genet.* 65:483–90, 2001.

Liao, J. C., et al. **(M. A. Suchard** and **E. M. Landaw).** Use of electrochemical DNA biosensors for rapid molecular identification of uropathogens in clinical urine specimens. *J. Clin. Microbiol.* 44:561–70, 2006.

Suchard, M. A., R. E. Weiss, and **J. S. Sinsheimer.** Models for estimating Bayes factors with applications to phylogeny and tests of monophyly. *Biometrics* 61:665–73, 2005.

Suchard, M. A. Stochastic models for horizontal gene transfer: Taking a random walk through tree space. *Genetics* 170:419–31, 2005.

Redelings, B. D., and **M. A. Suchard.** Joint Bayesian estimation of alignment and phylogeny. *Syst. Biol.* 54:401–18, 2005.

Kitchen, C. M. R., et al. **(M. A. Suchard).** Evolution of human immunodeficiency virus type 1 coreceptor usage during antiretroviral therapy: A Bayesian approach. *J. Virol.* 78:11296–302, 2004.

Yildiz, B. O., et al. **(M. A. Suchard).** Alterations in dynamics of circulating ghrelin, adiponectin and leptin in human obesity. *Proc. Natl. Acad. Sci. U.S.A.* 101:10435–9, 2004.

Suchard, M. A., C. M. R. Kitchen, **J. S. Sinsheimer,** and R. E. Weiss. Hierarchical phylogenetic models for analyzing multipartite sequence data. *Syst. Biol.* 52:649–64, 2003.

Suchard, M. A., et al. Evolutionary similarity among genes. *J. Am. Stat. Assoc.* 98:653–62, 2003.

Suchard, M. A., R. E. Weiss, K. S. Dorman, and **J. S. Sinsheimer.** Inferring spatial phylogenetic variation along nucleotide sequences: A multiple change-point model. *J. Am. Stat. Assoc.* 98:427–37, 2003.

Suchard, M. A., R. E. Weiss, and **J. S. Sinsheimer.** Testing a molecular clock without an outgroup: Derivations of induced priors on branch length restrictions in a Bayesian framework. *Syst. Biol.* 52:48–54, 2003.

D'Orsogna, M., **M. Suchard,** and **T. Chou.** Interplay of chemotaxis and chemokinesis mechanisms in bacterial dynamics. *Phys. Rev. E* 68:021925, 2003.

Chan, J. L., et al. **(M. A. Suchard).** Regulation of circulating soluble leptin receptor levels by gender, adiposity, sex steroids and leptin: Observational and interventional studies in humans. *Diabetes* 51:2105–12, 2002.

Suchard, M. A., R. E. Weiss, K. S. Dorman, and **J. S. Sinsheimer.** Oh brother, where art thou? A Bayes factor test for recombination with uncertain heritage. *Syst. Biol.* 51:715–28, 2002.

Suchard, M. A., R. E. Weiss, and **J. S. Sinsheimer.** Bayesian selection of continuous-time Markov chain evolutionary models. *Mol. Biol. Evol.* 18:1001–13, 2001.

Suchard, M. A., P. Yudkin, and **J. S. Sinsheimer.** Are general practitioners willing and able to provide genetic services for common diseases? *J. Genet. Couns.* 8: 301–11, 1999.

UNIVERSITY OF DELAWARE

Department of Mathematical Sciences

Programs of Study

The Department of Mathematical Sciences offers master's and Ph.D. programs in mathematics and applied mathematics. Students receive instruction in a broad range of courses and may specialize in many areas of mathematics. Strong departmental research groups exist in analysis, applied mathematics, partial differential equations, combinatorics, inverse problems, topology, probability, and numerical analysis. Master's programs normally require two years for completion, while the Ph.D. usually takes five years. Internships are encouraged.

Research Facilities

The University libraries contain 2 million volumes and documents and subscribe to 24,000 periodicals and serials. The University library belongs to the Association of Research Libraries.

The University Information Technologies Department provides e-mail and network access via central Sun servers. The Department of Mathematical Sciences has its own network and three-computer classrooms, and a 24-node cluster parallel computer. All graduate students have personal workstations in their offices with network access.

The Department fosters an active research environment, with numerous seminars and colloquia and many national and international visitors.

Financial Aid

Graduate assistantships and fellowships are available on a competitive basis. Teaching assistantships in 2006–07 range from $13,000 to $13,500 for nine months (two semesters), plus tuition remission. Additional winter and summer session teaching stipends are sometimes available. Currently, most full-time students receive full financial support. Some research assistantships and fellowships are also available.

Cost of Study

Course fees for full-time students in 2005–06 are $6304 per academic year for residents of Delaware and $15,990 per academic year for out-of-state students. Tuition for the summer sessions and for part-time students is $351 per credit for Delaware residents and $889 per credit for nonresidents. The graduation fee is $50 for the master's degree and $95 for the Ph.D.

Living and Housing Costs

While prices vary widely throughout the area, average monthly rent for a one-bedroom apartment is $710 plus utilities.

Student Group

There are approximately 41 full-time graduate students in the Department of Mathematical Sciences. About one quarter of these are international students and one third are women.

Location

The University is located in Newark, Delaware, a pleasant college community of about 30,000. Newark is 14 miles southwest of Wilmington and halfway between Philadelphia and Baltimore. It offers the advantages of a small community yet is within easy driving distance of Philadelphia, New York, Baltimore, and Washington, D.C. It is also close to the recreational areas on the Atlantic Ocean and Chesapeake Bay.

The University

The University of Delaware grew out of a small academy founded in 1743. It has been a degree-granting institution since 1834. In 1867, an act of the Delaware General Assembly made the University a part of the nationwide system of land-grant colleges and universities. Delaware College and the Women's College, an affiliate, were combined under the name of the University of Delaware in 1921. In 1950, the Graduate College was organized to administer the existing graduate programs and develop new ones.

Applying

Application forms may be obtained online at http://www.udel.edu/gradoffice/applicants/. Completed applications, including letters of recommendation, a $60 application fee, GRE General Test scores, and transcripts of previous work, should be submitted as early as possible but no later than March 1 to be considered for financial aid for the fall semester.

Correspondence and Information

Coordinator of Graduate Studies
Department of Mathematical Sciences
University of Delaware
Newark, Delaware 19716
Phone: 302-831-2346
E-mail: see@math.udel.edu
Web site: http://www.math.udel.edu

University of Delaware

THE FACULTY AND THEIR RESEARCH

Thomas S. Angell, Professor; Ph.D., Michigan. Optimal control theory, differential equations.
Constantin Bacuta, Assistant Professor; Ph.D., Texas A&M. Numerical analysis.
David P. Bellamy, Professor; Ph.D., Michigan State. Topology.
Richard J. Braun, Professor; Ph.D., Northwestern. Applied mathematics.
Michael Brook, Instructor; Ph.D., Delaware. Mathematics education.
Jinfa Cai, Professor; Ph.D., Pittsburgh. Mathematics education.
Fioralba Cakoni, Associate Professor; Ph.D., Tirana University (Albania). Direct and inverse scattering theory.
Antonio Ciro, Instructor; M.S., Drexel. Mathematics education.
David L. Colton, Unidel Professor; Ph.D., D.Sc., Edinburgh. Partial differential equations, integral equations.
L. Pamela Cook-Ioannidis, Professor; Ph.D., Cornell. Applied mathematics, perturbation theory, transonic flow.
Robert Coulter, Assistant Professor: Queensland (Australia). Finite fields, combinatorics.
Bryan Crissinger, Instructor; M.S., Penn State. Statistics.
Bettyann Daley, Instructor; M.S., Vermont. Math education.
Margaret Donlan, Instructor; M.S., Toledo. Mathematics for the liberal arts student, quantitative literacy.
Tobin A. Driscoll, Associate Professor; Ph.D., Cornell. Numerical analysis, applied mathematics.
Christine Ebert, Assistant Professor; Ph.D., Delaware. Investigation of pedagogical content knowledge for preservice and in-service teachers, the cognitive development of the concept of function, the use of technology.
Gary L. Ebert, Professor; Ph.D., Wisconsin–Madison. Combinatorics.
David A. Edwards, Associate Professor and Acting Associate Chair; Ph.D., Caltech. Applied math.
Robert P. Gilbert, Unidel Chair Professor; Ph.D., Carnegie Mellon. Homogenization, inverse problems, partial differential equations.
Philippe J. Guyenne, Assistant Professor; Ph.D., Nice–Sophia Antipolis (France). Applied mathematics, fluid mechanics, differential equations, nonlinear waves, numerical analysis.
George C. Hsiao, Carl J. Rees Professor; Ph.D., Carnegie Mellon. Differential and integral equations, perturbation theory, fluid dynamics and elasticity.
Mary Ann Huntley, Assistant Professor; Ph.D., Maryland. Research and evaluation of teacher preparation, induction, and enhancement.
Judy A. Kennedy, Professor; Ph.D., Auburn. Topology and dynamical systems.
Felix Lazebnik, Professor; Ph.D., Pennsylvania. Graph theory, combinatorics, algebra.
Yuk J. Leung, Associate Professor; Ph.D., Michigan. Function theory.
Wenbo Li, Professor; Ph.D., Wisconsin–Madison. Probability theory, stochastic processes, statistics.
David Russell Luke, Assistant Professor; Ph.D., Washington (Seattle). Optimization and inverse problems.
Peter Monk, Unidel Professor and Interim Chair; Ph.D., Rutgers. Numerical analysis.
Patrick F. Mwerinde, Assistant Professor; Ph.D., Columbia. Math education, conceptual learning theory, active learning, experimental design, statistical inference.
David Olagunju, Professor; Ph.D., Northwestern. Applied mathematics.
John Pelesko, Associate Professor; Ph.D., NJIT. Applied mathematics.
Geraldine Prange, Instructor; M.S., St. Louis. Math education for liberal arts majors, cooperative learning.
Georgia B. Pyrros, Instructor; M.S., McMaster. Nuclear physics.
Rakesh, Associate Professor; Ph.D., Cornell. Partial differential equations.
David P. Roselle, Professor and President of the University; Ph.D., Duke. Combinatorics.
Louis F. Rossi, Associate Professor; Ph.D., Arizona. Fluid dynamics, numerical analysis, vorticity dynamics.
Gilberto Schleiniger, Associate Professor and Undergraduate Chair; Ph.D., UCLA. Scientific computing, numerical analysis.
Anthony Seraphin, Assistant Professor; Ph.D., Delaware. Boundary layer flow, turbulent diffusion and dispersion within and above canopie wind and water tunnel flow simulations, air pollution climatology and its precursors.
Anja Sturm, Assistant Professor; Ph.D., Oxford (England). Branching and interacting particle systems.
Qing Xiang, Professor; Ph.D., Ohio State. Combinatorics.
Shangyou Zhang, Associate Professor; Ph.D., Penn State. Numerical analysis and scientific computation.

Emeritus Professors
Willard E. Baxter, Ph.D., Pennsylvania. Algebra.
David J. Hallenbeck, Ph.D., SUNY at Albany. Function theory.
Richard J. Libera, Ph.D., Rutgers. Function theory.
Albert E. Livingston, Ph.D., Rutgers. Function theory.
Clifford W. Sloyer, Ph.D., Lehigh. Topology, mathematics education.
Ivar Stakgold, Ph.D., Harvard. Nonlinear boundary-value problems.
Robert M. Stark, Ph.D., Delaware. Applied probability, operations research, civil engineering systems.
Richard J. Weinacht, Ph.D., Maryland. Partial differential equations.

Joint Appointments with Other Departments
Morris W. Brooks, Ph.D., Harvard. Computer-based instruction.
Bobby F. Caviness, Ph.D., Carnegie Mellon. Computer algebra.
Kathleen Hollowell, Ed.D., Boston University. Mathematics education.
William B. Moody, Ed.D., Maryland. Mathematics education.
Richard S. Sacher, Ph.D., Stanford. Scientific computing, operations research.
David Saunders, Ph.D., Wisconsin–Madison. Computer algebra.
Leonard W. Schwartz, Ph.D., Stanford. Fluid mechanics.

Adjunct Faculty and Their Affiliations
Alan Jeffrey, University of Newcastle-upon-Tyne. Wave propagation.
Rainer Kress, University of Göttingen. Integral equations, scattering theory.
Emeka Nwanko, DuPont Company.
Lassi Paivarinta, University of Oulu (Finland).
Gary Roach, University of Strathclyde. Operator theory, scattering theory.
Wolfgang Wendland, University of Stuttgart (Germany). Integral equations and analysis.

UIC

UNIVERSITY OF ILLINOIS AT CHICAGO

Department of Mathematics, Statistics, and Computer Science

Programs of Study

The Department, which belongs to the American Mathematical Society's Group I leading research mathematics departments in the country, offers a wide variety of programs of study leading to degrees at the master's and doctoral levels. Faculty research interests cover a broad range of areas, including algebra, algebraic geometry, applied mathematics, coding theory, computer science, dynamical systems, game theory, logic, low-dimensional topology, number theory, and probability and statistics.

The Master of Science (M.S.) degree program in mathematics is designed to lay foundations for doctoral work and also to prepare students for careers in business, government, and industry. The M.S. and Ph.D. degrees in mathematics can be earned with a concentration in applied mathematics, pure mathematics, computer science, or statistics. Also available are the M.S. degree in Mathematics and Information Sciences for Industry (M.I.S.I.), the Master of Science in Teaching (M.S.T.) of mathematics, and the Doctor of Arts (D.A.) degree programs.

Students in the M.S. program have the option of passing a cumulative exam or writing a thesis. Students from other institutions seeking admission to the Department's doctoral program must complete the work equivalent to that of the Department's M.S. program. They may be required to pass a Departmental exam to fully satisfy this requirement.

Two written preliminary exams must be passed and a minor course sequence must be successfully completed for the Ph.D. degree. The Ph.D. dissertation is expected to be a significant contribution to original mathematical research.

The M.I.S.I. program offers a core curriculum with group projects focused on industrial problems with practical deliverables. Emphasis is placed on applications in science, engineering, health care, and business.

The M.S.T. program is designed to strengthen the preparation and background of secondary school and primary school teachers. The program is arranged on an individual basis and has no thesis requirement. Students who are teaching can complete it through evening and summer courses.

The D.A. program is designed to prepare students for instruction at two- and four-year colleges. It includes study and research in methodology and techniques for successful teaching of college mathematics. A dissertation is required.

Research Facilities

The University Library houses more than 1.5 million volumes and specialized collections. The Mathematics Library, which is located within the Daley Library, has more than 20,000 volumes and maintains more than 240 journals. In addition, students have access to the library resources of nearby institutions and the University of Illinois at Urbana-Champaign.

The Department operates a diverse computing environment with UNIX workstations and PCs available for graduate student use. Graduate students have access to the Department's Laboratory for Advanced Computing for research-related programming, supercomputers through specific Departmental courses, and a statistical laboratory for research in statistics.

Financial Aid

The Department awards a large number of teaching assistantships, some research assistantships, and a few tuition and fee waivers. Some summer support is available. The 2006–07 stipend for the full-time, nine-month teaching assistantship (4–6 contact hours per week) is $14,500. The campus awards a limited number of University fellowships for graduate study, with a 2006–07 stipend of $18,000.

Cost of Study

Semester tuition and fees for 2006–07 are $5038 for Illinois residents and $11,037 for all others. Tuition and the service fee are waived for those holding teaching assistantships or fellowships as well as for those on tuition and fee waivers.

Living and Housing Costs

Some of UIC's residence halls are exclusively for graduate students. Rooms and apartments are available near the campus and throughout the city at widely varying costs. The campus is easily accessible by public transportation.

Student Group

Approximately 25,000 students are enrolled at UIC, nearly 6,700 of whom are graduate students. They come from all parts of Illinois and the United States as well as from many other countries. The Department has about 200 full-time students, of whom approximately 90 are in the doctoral program.

Location

UIC is located just west of the Loop, Chicago's downtown center, which is 5 minutes away by public transportation. Adjacent to the campus are the Jane Addams's Hull House and two historic landmark residential areas. The city is well known for its concerts, theater, galleries and museums, parks, ethnic restaurants, and lakefront recreation. There are other distinguished institutions of higher learning in the metropolitan area, which, along with UIC, contribute to an exciting atmosphere for the study of mathematics.

The University

UIC is the largest institution of higher learning in the Chicago area, and it is grouped in the top 100 research universities in the United States. The University offers bachelor's degrees in seventy-seven fields, master's degrees in eighty, and doctoral degrees in sixty.

Applying

Applicants are required to take the GRE General Test and a Subject Test in mathematics or computer science. The Department requires three letters of recommendation and at least a B average in mathematics beyond calculus. Applications for fall admission should be submitted no later than January 1 for consideration for a fall teaching assistantship or research assistantship or a University Fellowship. Study may also begin in the spring or summer semesters.

Correspondence and Information

Director of Graduate Studies
Department of Mathematics, Statistics, and Computer Science (Mail Code 249)
University of Illinois at Chicago
851 South Morgan Street
Chicago, Illinois 60607-7045
Phone: 312-996-3041
Fax: 312-996-1491
E-mail: dgs@math.uic.edu
Web site: http://www.math.uic.edu/gradstudies

University of Illinois at Chicago

THE FACULTY AND THEIR RESEARCH

Algebra
A. O. L. Atkin (Emeritus), Ph.D., Cambridge, 1952. Modular forms, number theory.
Daniel Bernstein, Ph.D., Berkeley, 1995. Number theory.
Alina Carmen Cojocaru, Ph.D., Queen's, 2002. Number theory, elliptic curves modular forms, sieve methods.
Paul Fong (Emeritus), Ph.D., Harvard, 1959. Group theory, representation theory of finite groups.
Ju-Lee Kim, Ph.D., Yale, 1997. Representation theory of p-adic groups.
Richard G. Larson (Emeritus), Ph.D., Chicago, 1965. Hopf algebras, application of computers to algebra, algorithms.
David E. Radford, Ph.D., North Carolina at Chapel Hill, 1970. Hopf algebras, algebraic groups.
Mark A. Ronan, Ph.D., Oregon, 1978. Buildings, geometries of finite groups.
Stephen D. Smith, D.Phil., Oxford, 1973. Finite groups, representation theory.
Bhama Srinivasan, Ph.D., Manchester, 1960. Representation theory of finite and algebraic groups.
Ramin Takloo-Bighash, Ph.D., Johns Hopkins, 2001. Distribution of rational points on symmetric varieties, automorphic forms and their L functions on symplectic and orthogonal groups, special values of L functions.
Jeremy Teitelbaum, Ph.D., Harvard, 1986. Number theory.

Analysis
Rafail Abramov, Ph.D., Rensselaer, 2002. Equilibrium statistical mechanics in conservative chaotic systems, linear fluctuation-response for dynamical systems and related numerical methods.
Calixto P. Calderon (Emeritus), Ph.D., Buenos Aires, 1969. Harmonic analysis, differentiation theory.
Shmuel Friedland, D.Sc., Technion (Israel), 1971. Matrix theory and its applications.
Alexander Furman, Ph.D., Hebrew, 1996. Ergodic theories, dynamical systems, Lie groups.
Melvin L. Heard, Ph.D., Purdue, 1967. Integrodifferential equations.
Jeff E. Lewis (Emeritus), Ph.D., Rice, 1966. Partial differential equations, microlocal analysis.
Charles S. C. Lin (Emeritus), Ph.D., Berkeley, 1967. Operator theory, perturbation theory, functional analysis.
Howard A. Masur, Ph.D., Minnesota, 1974. Quasiconformal mappings, Teichmuller spaces.
Zbigniew Slodkowski, D.Sc., Warsaw, 1981. Several complex variables.
David S. Tartakoff, Ph.D., Berkeley, 1969. Partial differential equations, several complex variables.

Applied Mathematics
Jerry Bona, Ph.D., Harvard, 1971. Fluid mechanics, partial differential equations, numerical analysis, mathematical economics, oceanography.
Susan Friedlander, Ph.D., Princeton, 1972. Geophysical and fluid dynamics.
Floyd B. Hanson (Emeritus), Ph.D., Brown, 1968. Numerical methods, asymptotic methods, stochastic bioeconomics.
Charles Knessl, Ph.D., Northwestern, 1986. Stochastic models, perturbation methods, queuing theory.
David Nicholls, Ph.D., Brown, 1998. Traveling gravity water waves in two and three dimensions.
Madalina Petcu, Ph.D., Paris XI (South), 2005. Partial differential equations, fluid mechanics, numerical analysis.
Roman Shvydkoy, Ph.D., Missouri–Columbia, 2001. Fluid mechanics, topology.
Charles Tier, Ph.D., NYU, 1976. Analysis of stochastic models, queuing theory.

Computer Science and Combinatorics
Amitava Bhattacharya, Ph.D., Tata (Bombay), 2005. Combinatorics, graph theory, convex polytopes.
Robert Grossman, Ph.D., Princeton, 1985. Data-intensive computing and data mining, high-performance data management, numerical and symbolic computation.
Jeffrey S. Leon, Ph.D., Caltech, 1971. Computer methods in group theory and combinatorics, algorithms.
Glenn Manacher (Emeritus), Ph.D., Carnegie Tech, 1961. Algorithms, complexity, computer language design.
Dhruv Mubayi, Ph.D., Illinois at Urbana-Champaign, 1998. Combinatorics.
Uri N. Peled, Ph.D., Waterloo, 1976. Optimization, combinatorial algorithms, computational complexity.
Vera Pless (Emeritus), Ph.D., Northwestern, 1957. Coding theory, combinatorics.
Gyorgy Turan, Ph.D., Attila József (Hungary), 1982. Complexity theory, logic, combinatorics.
Jan Verschelde, Ph.D., Katholieke (Belgium), 1996. Computational algebraic geometry, combinatorial and polyhedral methods.

Geometry and Topology
Ian Agol, Ph.D., San Diego, 1998. Knot theory, three manifold topology.
A. K. Bousfield (Emeritus), Ph.D., MIT, 1966. Algebraic topology, homotopy theory.
Marc Culler, Ph.D., Berkeley, 1978. Low-dimensional topology, group theory.
Lawrence Ein, Ph.D., Berkeley, 1981. Algebraic geometry.
Henri Gillet, Ph.D., Harvard, 1978. Algebraic K-theory, algebraic geometry.
Brayton I. Gray (Emeritus), Ph.D., Chicago, 1965. Homotopy theory, cobordism theory.
James L. Heitsch (Emeritus), Ph.D., Chicago, 1971. Differential topology, geometry of foliations, smooth dynamics and ergodic theory, spectral theory of foliated operators.
Steven Hurder, Ph.D., Illinois at Urbana-Champaign, 1980. Differential topology, theory of foliations.
Olga Kashcheyeva, Ph.D., Missouri–Columbia, 2003. Algebraic geometry and commutative algebra.
Louis Kauffman, Ph.D., Princeton, 1972. Differential topology, knot theory of singularities.
Anatoly S. Libgober, Ph.D., Tel-Aviv, 1977. Topology of varieties, theory of singularities.
Laurentiu Maxim, Ph.D., Pennsylvania, 2005. Topology of singularities, intersection homology, perverse sheaves.
Peter Shalen, Ph.D., Harvard, 1972. Low-dimensional topology, group theory.
Brooke Shipley, Ph.D., MIT, 1995. Algebraic topology, homological algebra.
Martin C. Tangora (Emeritus), Ph.D., Northwestern, 1966. Algebraic topology, homotopy theory.
Kevin Whyte, Ph.D., Chicago, 1998. Geometry of groups and group actions.
John W. Wood, Ph.D., Berkeley, 1968. Differential topology, topology of varieties.
Stephen S.-T. Yau, Ph.D., SUNY at Stony Brook, 1976. Algebraic and complex geometry, singularity theory, bioinformatics.

Logic and Universal Algebra
Matthias Aschenbrenner, Ph.D., Illinois at Urbana-Champaign, 2001. Model theory and its applications to algebra and analysis.
John T. Baldwin, Ph.D., Simon Fraser, 1971. Model theory, universal algebra.
Joel D. Berman, Ph.D., Washington (Seattle), 1970. Lattice theory, universal algebra.
William A. Howard (Emeritus), Ph.D., Chicago, 1956. Foundations of mathematics, proof theory.
P. Jonathan Kirby, D.Phil., Oxford, 2006. Model theory of differential fields, amalgamation, number theory.
David Marker, Ph.D., Yale, 1983. Model theory and applications to algebra.

Mathematics Education
Alison Castro, Ph.D., Michigan, 2006. Mathematics education.
Steven L. Jordan (Emeritus), Ph.D., Berkeley, 1970. Education, computer graphics, computational geometry.
Danny Martin, Ph.D., Berkeley, 1977. Mathematics education.
Philip Wagreich, Ph.D., Columbia, 1966. Algebraic geometry, discrete groups, mathematics education.
A. I. Weinzweig, Ph.D., Harvard, 1957. Teaching and learning of mathematics, microcomputers in education.

Probability and Statistics
Emad El-Neweihi, Ph.D., Florida State, 1973. Reliability theory, probability, stochastic processes.
Nasrollah Etemadi (Emeritus), Ph.D., Minnesota, 1974. Probability theory, stochastic processes.
Samad Hedayat, Ph.D., Cornell, 1969. Optimal designs, sampling theory, linear models, discrete optimization.
Dibyen Majumdar, Ph.D., Indian Statistical Institute, 1981. Optimal designs, linear models.
Klaus J. Miescke, Dr.rer.nat., Heidelberg, 1972. Statistics, decision theory, selection procedures.
T. E. S. Raghavan, Ph.D., Indian Statistical Institute, 1966. Game theory, optimization methods in matrices, statistics.
Jing Wang, Ph.D., Michigan State, 2006. Nonparametric and semiparametric inference, polynomial spline and kernel smoothing, nonlinear time-series forecasting.
Jie Yang, Ph.D., Chicago, 2006. Financial mathematics, cluster analysis, discriminant analysis, spatial statistics, dimension reduction.

UNIVERSITY OF NORTH CAROLINA AT CHARLOTTE

Department of Mathematics and Statistics

Programs of Study

The Department of Mathematics and Statistics offers programs that lead to the M.S. in mathematics, the Ph.D. in applied mathematics, and the M.A. in math education. These programs are designed to develop advanced skills, knowledge, and critical-thinking abilities that are directly applicable to a wide variety of positions in industry, business, government, and teaching at the secondary school, community college, and/or university level. The Department also participates in the interdisciplinary M.S. in mathematical finance and Ph.D. in curriculum and instruction programs. The Department has active research programs in commutative algebra; computational fluid dynamics, combustion, and electromagnetics; numerical analysis; dynamical systems; operator algebras, Banach space geometry, and wavelets; partial differential equations and mathematical physics; probability and stochastic processes; and statistics.

The M.S. in mathematics is divided into three concentrations: applied mathematics, statistics, and general. Applied mathematics and statistics require nine and ten courses, respectively, and a project. The programs follow a rigorously structured framework of analytical and applied subjects. General mathematics requires ten courses and offers a good deal of flexibility to a student who wants a broad background of pure and applied courses. The M.A. in math education is primarily designed for secondary school teachers who are interested in professional growth and graduate certification. It requires twelve courses, with at least six in mathematics and the other six chosen from mathematics, mathematics education, and education. All master's programs require a comprehensive oral exam; completion of a thesis is optional.

After their first year, students in the Ph.D. program are required to pass a preliminary exam based on a yearlong advanced real analysis sequence and a yearlong basic course sequence in an applied area of their choice. By the end of their third year, Ph.D. students are expected to pass a qualifying exam for admission to candidacy; the qualifying exam is based on advanced topics in their area of specialization. In addition, Ph.D. students are required to pass a foreign language reading proficiency exam and complete a two- to three-course interdisciplinary minor. The latter can often be coordinated via the Department's external consulting activities. Finally, Ph.D. students are required to complete a Ph.D. dissertation that comprises a substantial and original contribution to their area of study.

Research Facilities

The Department of Mathematics provides state-of-the-art computing facilities in its two UNIX workstation labs, which house thirty-five Linux Workstations (with a Beowulf cluster in the Department). Three PC labs with 105 stations are conveniently located in the Fretwell building, the home of the Department of Mathematics and Statistics. In addition, state-of-the-art laptops equipped with the latest multimedia are available for in-class lecture demonstration. All graduate student and faculty member offices are equipped with either a PC or a workstation. The J. Murrey Atkins Library contains 750,000 bound volumes (including 13,000 monograph holdings in mathematics, computer science, and statistics) as well as more than 200 mathematics research journals and more than 1 million units in microfilm.

Financial Aid

Most students accepted into mathematics graduate programs are supported by teaching assistantships, which pay $11,700 for master's students and start at $14,000 for Ph.D. students. Assistantships for advanced Ph.D. students pay $15,000. A limited number of fellowship awards can be applied to supplement the above stipends for especially qualified students. These include a $25,000 first-year TIAA-CREF Fellowship with full-tuition support. Some students are supported by project-specific externally funded research assistantships, with stipends starting at $15,000. Virtually all out-of-state teaching and research assistants in the M.S. program receive waivers for their out-of-state tuition, and a few in-state tuition waivers are available for especially qualified North Carolina residents. All teaching and research assistants in the Ph.D. program receive a full tuition waiver and an opportunity to register for free health insurance.

Cost of Study

Tuition and academic fees for 2005–06 were $1340–$1964 per semester for North Carolina residents and $5168–$7068 for nonresidents without assistantships and out-of-state tuition waivers. Graduate students who are U.S. citizens or permanent residents normally become residents of North Carolina after their first year.

Living and Housing Costs

Typical room and board expenses for students living off campus are about $3200 per semester. Off-campus rents average about $400 per month for a two- to four-bedroom apartment. A limited amount of on-campus housing is available with somewhat lower rents.

Student Group

Of 50 mathematics graduate students in spring 2006, 24 are women and 29 are international. Thirty-seven were teaching assistants, 4 held research assistantships, and 5 held stipend increases from fellowship sources.

Student Outcomes

Students who successfully complete the master's and Ph.D. programs find challenging and rewarding jobs in academia and business and industry, particularly in the financial and technological sectors, with organizations such as Arbitrade, Duke Power, and Moody Investment Corporation. Recent master's graduates have secured academic positions at regional community colleges and statistical analysis and management positions with Bank of America and Wachovia Corporation, two of the largest banks in the nation. Recent Ph.D. graduates include a vice president of Bank America, a statistical analyst for Scottish Re, an associate at Wachovia corporation, an assistant professor at Pfeiffer University, an assistant professor at Montevallo University, and an assistant professor at Clarkson University.

Location

The city of Charlotte is the hub of a dynamic and growing metropolitan area in terms of economic and cultural development. The city features two major sports franchises, an internationally renowned symphony orchestra, Opera Carolina, and the North Carolina Dance Theater. Specialty shops, galleries, and restaurants reflect the tastes and culture of an ethnically diverse multinational community. Recreational opportunities include biking, boating, and fishing at and around nearby Lake Norman, Lake Lure, and Mountain Island Lake. Many parks and uptown areas host yearly festivals. Mountain recreation areas and ocean beaches are within a 2½- to 4½-hour drive.

The University

The University of North Carolina (UNC) at Charlotte is located in the largest urban center in the Carolinas. Its campus occupies 1,000 wooded acres in the University City area, which also includes University Place, University Hospital, and University Research Park, the nation's sixth-largest university-affiliated research park. The Research Park has more than 11,000 employees and a number of large firms, including AT&T, IBM, Duke Power, Nations Bank, First Union, and the *Wall Street Journal.* The University maintains many professional contacts with several of the firms, and graduate students are very often involved. Of the University's nearly 19,000 students who represent forty-six states and sixty-five countries, about 3,800 are graduate students. UNC Charlotte is known for its academic excellence and is distinguished by its commitment to scholarly activities and teaching accomplishments.

Applying

Application forms may be obtained from the Department of Mathematics. All students must take the General Test of the Graduate Record Examinations (GRE). The Subject Test is recommended but not required. International applicants must take the TOEFL and score above 220 (computer-based) or 550 (paper-based) or have received a degree from an American institution. For full consideration for financial support, applicants for the fall semester should submit all materials by January 15.

Correspondence and Information

Joel Avrin
Graduate Coordinator
Department of Mathematics
University of North Carolina at Charlotte
Charlotte, North Carolina 28223-0001
Phone: 704-687-4929
Fax: 704-687-6415
E-mail: jdavrin@email.uncc.edu
Web site: http://www.math.uncc.edu/grad/

University of North Carolina at Charlotte

THE FACULTY AND THEIR RESEARCH

Robert Anderson, Associate Professor; Ph.D., Minnesota, 1972. Probability.
Joel Avrin, Professor; Ph.D., Berkeley, 1982. Partial differential equations and mathematical physics.
Jaya Bishwal, Assistant Professor; Ph.D., Sambalpur (India), 2002. Mathematical physics.
Auimikh Biswas, Assistant Professor; Ph.D., Indiana, 2000. Functional analysis and partial differential equations.
Charles Burnap, Associate Professor; Ph.D., Harvard, 1976. Mathematical physics.
Wei Cai, Professor; Ph.D., Brown, 1989. Computational fluid dynamics and electromagnetics.
Zongwu Cai, Assistant Professor; Ph.D., California, Davis, 1995. Statistics.
Vic Cifarelli, Associate Professor; Ph.D., Purdue, 1988. Mathematics education.
Ming Dai, Assistant Professor; Ph.D., South Florida, 2000. Statistics.
Xingde Dai, Associate Professor; Ph.D., Texas A&M, 1990. Operator algebras.
Shaozhong Deng, Assistant Professor; North Carolina State, 2001. Computational mathematics.
Yuanan Diao, Associate Professor; Ph.D., Florida State, 1990. Topology.
Jacek Dmochowski, Associate Professor; Ph.D., Purdue, 1995. Biostatistics.
Alan Dow, Professor; Ph.D., Manitoba, 1980. Topology.
Yuri Godin, Ph.D., Technion (Israel), 1994. Mathematical physics.
Kim Harris, Associate Professor; Ph.D., Georgia, 1985. Mathematics education.
Gabor Hetyei, Assistant Professor; Ph.D., MIT, 1994. Combinatorics.
Evan Houston, Professor; Ph.D., Texas at Austin, 1973. Commutative algebra.
Mohammad Kazemi, Professor; Ph.D., Michigan, 1982. Control theory.
Michael Kilbanov, Professor; Ph.D., Ural State (Russia), 1977. Inverse problems.
Thomas G. Lucas, Professor; Ph.D., Missouri, 1983. Commutative algebra.
Thomas R. Lucas, Professor; Ph.D., Georgia Tech, 1970. Numerical analysis.
Stanislav Molchanov, Professor; Ph.D., Moscow State, 1967. Probability/mathematical physics.
Wanda Nabors, Assistant Professor; Ph.D., Georgia. Mathematics education.
Hae-Soo Oh, Associate Professor; Ph.D., Michigan, 1980. Numerical analysis.
Joseph Quinn, Professor; Ph.D., Michigan State, 1970. Stochastic processes.
Harold Reiter, Associate Professor; Ph.D., Clemson, 1969. Combinatorics.
Franz Rothe, Associate Professor; Ph.D., Tübingen (Germany), 1975. Partial differential equations.
David C. Royster, Associate Professor; Ph.D., LSU, 1978. Differential topology.
Adalira Saenz-Ludlow, Associate Professor; Ph.D., Florida State, 1990. Mathematics education.
Oleg Safronov, Assistant AProfessor; Ph.D., Royal Institute of Technology (Stockholm), 1998. Mathematical physics.
Douglas Shafer, Professor; Ph.D., North Carolina at Chapel Hill, 1978. Dynamical systems.
Isaac Sonin, Professor; Ph.D., Moscow State, 1971. Probability/operations research.
Nicholas Stavrakas, Professor; Ph.D., Clemson, 1973. Convexity.
Yanqing Sun, Associate Professor; Ph.D., Florida State, 1992. Statistics.
Boris Vainberg, Professor; Ph.D., Moscow State, 1963. Partial differential equations and mathematical physics.
Barnet Weinstock, Professor; Ph.D., MIT, 1966. Several complex variables.
Volker Wihstutz, Professor; Ph.D., Bremen (Germany), 1975. Stochastic dynamical systems.
Mingxin Xu, Assistant Professor; Ph.D., Carnegie Mellon, 2004. Mathematical finance.
Alexander Yushkevich, Professor; Ph.D., Moscow State, 1956. Probability/operations research.
Zhi-Yi Zhang, Associate Professor; Ph.D., Rutgers, 1990. Statistics.
Weihua Zhou, Assistant Professor; Ph.D., Texas at Dallas, 2005. Statistics.
You-Lan Zhu, Professor; Ph.D., Qinghau (China), 1963. Computational fluid dynamics.

UNIVERSITY OF NOTRE DAME

Graduate Studies in Mathematics

Program of Study

The purpose of the doctoral program in mathematics is to give students the opportunity to develop into educated and creative mathematicians. Students have the opportunity to work closely with one or more members of the faculty. The department includes active groups in many areas of algebra, algebraic geometry, applied mathematics (computation, mathematical biology, and optimization), complex analysis, differential geometry, logic, partial differential equations, and topology. In the first year, students have no teaching duties so that they can devote all their time to taking a variety of courses and getting acquainted in the department. Later, students first assist in teaching and then have the opportunity to teach independently. There is a teaching seminar to help students feel prepared when they begin teaching.

Research Facilities

Every effort is made to enable students to avail themselves of the opportunities provided by the excellent mathematics faculty at Notre Dame. The Department of Mathematics has its own building with all modern facilities, including a comprehensive research library of 35,000 volumes that subscribes to 290 current journals. All graduate students have comfortable offices. Students are ensured a stimulating and challenging intellectual experience.

Financial Aid

In 2005–06, all new students received a twelve-month stipend of more than $17,600; they have no teaching duties the first year. Next, they become teaching assistants and begin the three stages of supervised teaching provided by the department. A teaching assistant usually starts by doing tutorial work in freshman and sophomore calculus courses (4 classroom hours per week); this is followed by a variety of duties in advanced undergraduate courses; the final, lecturing stage involves independent teaching in the classroom. All doctoral students in mathematics also receive a full-tuition fellowship. Support is available for citizens and noncitizens.

Cost of Study

All graduate students in mathematics are supported by fellowships or assistantships, which include tuition scholarships.

Living and Housing Costs

University housing includes two-bedroom apartments for single men and women, four-bedroom town houses for single men and women, and two- and four-bedroom apartments for married students at rents that ranged from $333 to $708 per month in 2005–06. Comfortable and attractive off-campus rooms normally cost between $300 and $500 per month. Other expenses are lower than in most metropolitan areas.

Student Group

The carefully selected men and women who make up the student body of the University come from every state in the Union and 100 countries. There are 50 graduate students working for their doctorate in mathematics. The faculty-student ratio is greater than 1:1.

Location

The University is just north of South Bend, a pleasant Midwestern city with a population of about 110,000. The Notre Dame campus is exceptionally rich in active cultural programs, and the wide variety of cultural, educational, and recreational facilities of Chicago and Lake Michigan are less than 2 hours away by car.

The University

Founded in 1842, the University of Notre Dame has a 1,250-acre campus. Much of the campus is heavily wooded, and two delightful lakes lie entirely within it. Total enrollment for fall 2005 was 11,415; approximately one fourth of these are graduate students. The University is proud of its tradition as a Catholic university with a profound commitment to intellectual freedom in every area of contemporary thought. The students and faculty represent a rich diversity of religious, racial, and ethnic backgrounds.

Applying

All applicants are required to take the General Test of the Graduate Record Examinations and are required to take the Subject Test in mathematics. Application for these tests should be made to Educational Testing Service in Princeton, New Jersey 08541, or at 1947 Center Street, Berkeley, California 94704. The application fee is $50 for all applications submitted after December 1. The fee for applications submitted by December 1 for the following fall semester is $35. The application deadline for students who wish to be considered for financial aid is February 1. All applicants are considered without regard to race, sex, or religious affiliation.

Correspondence and Information

Director of Graduate Studies
Department of Mathematics
University of Notre Dame
255 Hurley Building
Notre Dame, Indiana 46556-4618

Phone: 574-631-7245
Fax: 574-631-6579
E-mail: mathgrad.1@nd.edu
Web site: http://www.math.nd.edu

University of Notre Dame

THE FACULTY AND THEIR RESEARCH

Algebra
Mario Borelli, Ph.D., Indiana. Algebraic geometry, computer graphics.
Matthew Dyer, Ph.D., Sydney. Representation theory, algebraic groups.
Samuel R. Evens, Ph.D., MIT. Geometry of Lie groups and homogeneous spaces and representation theory.
Alexander J. Hahn, Ph.D., Notre Dame. Linear groups, theories of algebras, quadratic forms.
Warren J. Wong, Ph.D., Harvard. Theory of finite groups and their representations.

Algebraic Geometry/Commutative Algebra
Juan C. Migliore, Ph.D., Brown. Liaison theory, Hilbert functions, syzygies.
Claudia Polini, Ph.D., Rutgers. Commutative and homological algebra, blowup algebras, linkage and residual intersection theory.

Applied Mathematics
Mark Alber, Ph.D., Pennsylvania. Nonlinear dynamical systems and nonlinear partial differential equations and applications to biology.
Leonid Faybusovich, Ph.D., Harvard. Optimization, optimal control theory.
Bei Hu, Ph.D., Minnesota. Nonlinear partial differential equations.
Francois M. Ledrappier, Ph.D., Paris VI (Curie). Dynamical systems.
David Nicholls, Ph.D., Brown. Free boundary problems, partial differential equations, numerical analysis.
Joachim Rosenthal, Ph.D., Arizona State. Control theory, coding theory and cryptography.
Andrew J. Sommese, Ph.D., Princeton. Numerical analysis of polynomial systems.

Complex Analysis
Jeffrey A. Diller, Ph.D., Michigan. Complex analysis and dynamical systems.
Mei-Chi Shaw, Ph.D., Princeton. Partial differential equations and several complex variables.
Dennis M. Snow, Ph.D., Notre Dame. Homogeneous complex manifolds, group actions.
Pit-Mann Wong, Ph.D., Notre Dame. Several complex variables.

Computational Mathematics
Andrew J. Sommese, Ph.D., Princeton. Numerical analysis of polynomial systems.

Differential Geometry
Jianguo Cao, Ph.D., Pennsylvania. Differential geometry.
Matthew Gursky, Ph.D., Caltech. Geometric analysis.
Richard Hind, Ph.D., Stanford. Symplectic geometry.
Alan Howard, Ph.D., Brown. Complex manifolds.
Francois M. Ledrappier, Ph.D., Paris VI (Curie). Dynamical systems.
Xiaobo Liu, Ph.D., Pennsylvania. Differential geometry.
Brian Smyth, Ph.D., Brown. Differential geometry.
Frederico J. Xavier, Ph.D., Rochester. Differential geometry.

Logic
Steven A. Buechler, Ph.D., Maryland. Model theory.
Peter Cholak, Ph.D., Wisconsin. Computability theory.
Julia Knight, Ph.D., Berkeley. Computability, computable structures.
Sergei Starchenko, Ph.D., Novosibirsk (Russia). Model theory.

Mathematical Physics
Katrina Barron, Ph.D., Rutgers. Vertex operator superalgebras and superconformal field theory.
Michael Gekhtman, Ph.D., Ukrainian Academy of Sciences. Integrable models, Poisson geometry.
Brian C. Hall, Ph.D., Cornell. Quantization and coherent states, analysis on Lie groups.

Partial Differential Equations
Qing Han, Ph.D., NYU (Courant). Partial differential equations.
A. Alexandrou Himonas, Ph.D., Purdue. Partial differential equations.
Gerard Misiolek, Ph.D., SUNY at Stony Brook. Geometric and nonlinear functional analysis and partial differential equations.
Nancy K. Stanton, Ph.D., MIT. Differential geometry, complex manifolds.

Topology
Francis X. Connolly, Ph.D., Rochester. Differential and algebraic topology.
John E. Derwent, Ph.D., Notre Dame. Differential and algebraic topology.
William G. Dwyer, Ph.D., MIT. Algebraic topology.
Liviu Nicolaescu, Ph.D., Michigan State. Gauge theory.
Stephan A. Stolz, Ph.D., Mainz (Germany). Algebraic topology and differential geometry.
Laurence R. Taylor, Ph.D., Berkeley. Geometric and algebraic topology.
E. Bruce Williams, Ph.D., MIT. Geometric and algebraic topology, K-theory.

SELECTED PUBLICATIONS

Alber, M. S., et al. The complex geometry of weak piecewise smooth solutions of integrable nonlinear PDE's of shallow water and Dym type. *Commun. Math. Phys.* 221:197–227, 2001.

Alber, M. S., Y. Jiang, and M. A. Kiskowski. Lattice gas cellular automata model for rippling and aggregation in myxobacteria. *Physica D* 191:343–58, 2004.

Alber, M. S., M. A. Kiskowski, and Y. Jiang. Two-stage aggregate formation via streams in myxobacteria. *Phys. Rev. Lett.* 93:068102, 2004.

Alber, M. S., et al. The complex geometry of weak piecewise smooth solutions of integrable nonlinear PDE's of shallow water and Dym type. *Commun. Math. Phys.* 221:197–227, 2001.

Barron, K. Superconformal change of variables for N=1 Neveu-Schwarz vertex operator superalgebras. *J. Algebra* 277:717–64, 2004.

Barron, K. The notion of N=1 supergeometric vertex operator superalgebra and the isomorphism theorem. *Commun. Contemp. Math.* 5(4):481–567, 2003.

Barron, K. The moduli space of N=1 superspheres with tubes and the sewing operation. *Mem. Am. Math. Soc.* 162(772), 2003.

Buechler, S., and C. Hoover. Classification of small types of rank ω, part I. *J. Symbol. Logic,* in press.

Buechler, S., A. Pillay, and F. Wagner. Supersimple theories. *J. Am. Math. Soc. No. 1* 14:109–24, 2000.

Buechler, S. Lascar strong types in some simple theories. *J. Symbol. Logic* 64(2):817–24, 1999.

Cao, J., and **F. J. Xavier.** Kahler parabolicity and the Euler number of compact manifolds of nonpositive sectional curvature. *Math. Ann. No. 3* 319:493–91, 2001.

Cao, J., J. Cheeger, and X. Rong. Splittings and Cr-structures for manifolds with nonpositive sectional curvature. *Inventiones Math.* 144:139–67, 2001.

Cao, J. Cheeger isoperimetric constants of Gromov-hyperbolic spaces with quasi-pole. *Commun. Contemp. Math. No. 4* 2:511–33, 2000.

Cholak, P., and L. Harrington. On the definability of the double jump in the computably enumerable sets. *J. Math. Logic* 2(2):261–96, 2002.

Cholak, P., R. Coles, R. Downey, and E. Herrmann. Automorphisms of the lattice of Pi_1^{ILU}0 classes; prefect thin classes and anc degrees. *Trans. Am. Math. Soc.* 353:4899–924, 2001.

Cholak, P., C. Jockusch, and T. Slaman. The strength of Ramsey's theorem for pairs. *J. Symbol. Logic* 66(1):1–55, 2001.

Connolly, F. X., and D. Anderson. Finiteness obstructions to cocompact actions on S^m x $\hat{A}^n$. *Comment. Math. Helv.* 68:85–110, 1993.

Connolly, F. X., and T. Kozniewski. Examples of lack of rigidity in crystallographic groups. In *Lecture Notes in Mathematics—Algebraic Topology,* Poznan, 1989, vol. 1474, pp. 139–45. Berlin/Heidelberg: Springer-Verlag, 1991.

Connolly, F. X., and T. Kozniewski. Rigidity and crystallographic groups I. *Inventiones Math.* 99:25–48, 1990.

Diller, J., and E. Bedford. Energy and invariant measure for birational surface maps. *Duke Math. J.* 128:338–68, 2005. (http://arxiv.org/abs/math.CV/0310002)

Diller, J., and C. Faure. Dynamics of bimeromorphic maps of surface. *Am. J. Math.* 1231135–69, 2001. (http://www.nd.edu/%7Ejdiller/research/papers/bimeromorphic.pdf)

Diller, J., and D. E. Barrett. A new construction of Riemann surfaces with corona. *J. Geom. Anal. No. 3* 8:341–7, 1998. (http://www.nd.edu/%7Ejdiller/research/papers/corona.pdf)

Dwyer, W. G. Localizations, axiomatic, enriched and motivic homotopy theory. In *Proceedings of the NATO ASI,* pp. 3–28, ed. J. P. C. Greenless. Kluwer, 2004.

Blanc, D., **W. G. Dwyer,** and P. G. Goerss. The realization space of the π-algebra: A moduli problem in algebraic topology, *Topology* 43, 2004.

Dwyer, W. G., and C. W. Wilkerson. The elementary geometric structure of compact Lie groups. *Bull. London Math. Soc.* 30:337–64, 1998.

Chen, Y, and **M. J. Dwyer.** On the combinatorics of B-times-B orbits on group compactifications. *J. Algebra* 263:278–93, 2003.

Dyer, M. J. Representation theories from Coxeter groups. *Can. Math. Soc. Conf. Proc.* 16:105–39, 1995.

Dyer, M. J. The nil Hecke ring and Deodhar's conjecture on Bruhat intervals. *Inventiones Math.* III:571–4, 1993.

Evens, S., and J.-H. Lu. On the variety of Lagrangian subalgebras. *Annales Ecole Normale Superieure* 34:631–68, 2001.

Evens, S., and J.-H. Lu. Poisson harmonic forms, Kostant harmonic forms, and the S1-equivariant cohomology of K/T. *Adv. Math.* 142:171–220, 1999.

Evens, S., and I. Mirkovic. Characteristic cycles for the loop Grassmannian and nilpotent orbits. *Duke Math. J.* 97:109–26, 1999.

Faybusovich, L. Jordan-algebraic approach to potential-reduction algorithms. *Math. Z.* 239:117–29, 2002.

Faybusovich, L. Self-concordan barriers for cones generated by Chebyshev systems. *SIAM J. Optimization* 12:770–81, 2002.

Faybusovich, L. On Nesterov's approach to semiinfinite programming. *Acta Appl. Math.* 74:195–215, 2002.

Gekhtman, M., and M. Shapiro. Noncommutative and commutative integrability of generic Toda flows in simple Lie algebras. *Commun. Pure Appl. Math.* 52:53–84, 1999.

Bloch, A., and **M. Gekhtman.** Hamiltonian and gradient structures in the Toda flows. *J. Geom. Phys.* 27:230–48, 1998.

Gekhtman, M. Hamiltonian structure of nonabelian Toda lattice. *Lett. Math. Phys.* 46:189–205, 1998.

Gursky, M. J., and J. Viaclovsky. A new variational characterization of three-dimensional space forms. *Inventiones Math.* 145:251–78, 2001.

Gursky, M. J. The Weyl functional, deRham cohomology, and Kahler-Einstein metrics. *Ann. Math.* 148:315–37, 1998.

Gursky, M. J., and C. LeBrun. Yamabe invariants and spin^{ILO}c structures. *Geom. Funct. Anal.* 8:965–77, 1998.

Hahn, A. J. The Zassenhaus decomposition for the orthogonal groups: Properties and applications. *Documenta Mathematica (Bielefeld)* 165–81, 2001. (online http://www.mathematik.unibielefeld.de/documenta/)

Hahn, A. J. Quadratic algebras, Clifford algebras and arithmetic Witt groups. In *UNIVERSITEXT Series.* Berlin and New York: Springer-Verlag, 1994.

Hahn, A. J., and O. T. O'Meara. The classical groups and K-theory. In *Grundlehren der Mathematik,* vol. 291. Berlin and New York: Springer-Verlag, 1989.

Hall, B. C., and J. J. Mitchell. The Segal-Bargmann transform for noncompact symmetric spaces of the complex type. *J. Funct. Anal.* 227(2):338–71, 2005.

Hall, B. C., and W. Lewkeeratiyutkul. Holomorphic Sobolev spaces and the generalized Segal-Bargmann transform. *J. Funct. Anal.* 217(1):192–220, 2004.

Hall, B. C. Geometric quantization and the generalized Segal-Bargmann transform for Lie groups of compact type. *Commun. Math. Phys.* 226:233–68, 2002.

Han, Q., N. Nadirashvili, and Y. Yuan. Linearity of homogeneous order one solutions to elliptic equations in dimension three. *Comm. Pure Appl. Math.* 56: 425–32, 2003.

Han, Q., J.-X. Hong, and C.-S. Lin. Local isometric embedding of surfaces with nonpositive Gaussian curvature. *J. Diff. Geometry* 63:475–520, 2003.

Han, Q., J.-X. Hong and C.-S. Lin. Small divisors in nonlinear elliptic equations. *Cal. Var. P.D.E.* 18:31–56, 2003.

Himonas, A., and G. Petronilho. On Gevrey regularity of globally C^∞ hypoelliptic operators. *J. Differential Equations* 207(2):267–84, 2004.

Gorsky, J., and **Himonas, A.** Construction of non-analytic solutions for the generalized KdV equation. *J. Math. Anal. Appl.* 303(2):522–9, 2005.

Himonas, A., and **G. Misiolek.** High frequency smooth solutions and well-posedness of the Camassa-Holm equation. *Intern. Math. Res. Notices* 51:3135–51, 2005.

Hind, R. K. Lagrangian spheres in S^2 x S^2. *Geom. Funct. Anal.* 14:303–18, 2004.

Hind, R. K. Antiholomorphic involutions on Stein manifolds. *Int. J. Math.* 14: 479–87, 2003.

Burnes, D., S. Halverscheid, and **R. Hind.** The geometry of Grauert tubes and complexification of symmetric spaces. *Duke Math. J.* 118(3):465–91, 2003.

Burnes, D., and **R. Hind.** Symplectic geometry and the uniqueness of Grauert tubes. *J. Geom. Funct. Anal.* 11:1–10, 2001.

Fontelos, M. A., A. Friedman, and **B. Hu.** Mathematical analysis of a model for the initiation of angiogenesis. *SIAM J. Math. Anal.* 33(6):1330–55, 2002.

Friedman, A., **B. Hu,** and J. J. L. Valazquez. The evolution of stress intensity factors in the propagation of cracks in elastic media. *Arch. Ration. Mech. Anal.* 152:103–39, 2000.

University of Notre Dame

Friedman, A., and **B. Hu.** Head-media interaction in magnetic recording. *Arch. Ration. Mech. Anal.* 140:79–101, 1997.

Calvert, W., **J. F. Knight,** and J. M. Young. Computable trees of Scott rank ω_1^{CK}, and computable approximation. *J. Symb. Logic,* in press.

Goncharov, S. S., et al. **(J. F. Knight)**. Enumerations in computable structure theory. *Annals Pure Appl. Logic* 136:219–46, 2005.

Csima, B., D. Hirschfeldt, **J. F. Knight,** and R. I. Soare. Bounding prime models. *J. Symb. Logic* 69:1117–42, 2004.

Ledrappier, F., and M. Pollicott. Ergodic properties of linear actions of 2X2 matrices. *Duke Math. J.* 116:353–88, 2003.

Ledrappier, F., and E. Lindenstrauss. On the projections of measures invariant under the geodesic flow. *Int. Math. Res. Notices* 511–26, 2003.

Ledrappier, F., M. Shub, C. Simo, and A. Wilkinson. Random versus deterministic exponents in a rich family of diffeomorphisms, *J. Stat. Phys.* 113: 85–149, 2003.

Heintze, E., and **X. Liu.** Homogeneity of infinite dimensional isoparametric submanifolds. *Ann. Math.* 149:149–81, 1999.

Liu, X., and G. Tian. Virasoro constraints for quantum cohomology. *J. Differential Geom.* 50:537–91, 1998.

Liu, X. Volume minimizing cycles in compact Lie groups. *Am. J. Math.* 117: 1203–48, 1995.

Migliore, J., A. V. Geramita, T. Harima, and Y. Shin. The Hilbert function of a level algebra. *Mem. Amer. Math. Soc.,* in press.

Migliore, J., and U. Nagel. Tetrahedral curves. *Int. Math. Res. Notices* 15:899–939, 2005.

Migliore, J., U. Nagel, and T. Roemer. The multiplicity conjecture in low codimensions. *Math. Res. Lett.* 12:731–48, 2005.

Khesin, B., and **G. Misiolek.** Euler equations on homogeneous spaces and Virasoro Orbits. *Adv. Math.,* in press.

Misiolek, G. Classical solutions of the Camassa-Holm "equation." *Geom. Funct. Anal.,* in press.

Misiolek, G. The exponential map on the free loop space is Fredholm. *Geom. Funct. Anal.* 7:954–69, 1997.

Nicholls, D. P., and N. Nigam. Exact non-reflecting boundary conditions on general domains. *J. Comput. Phys.* 194(1):278–303, 2004.

Nicholls, D. P., and F. Reitich. Shape deformations in rough surface scattering: Improved algorithms. *J. Opt. Soc. Am. A* 21(4):606–21, 2004.

Craig, W., and **D. P. Nicholls.** Traveling gravity water waves in two and three dimensions. *Eur. J. Mech. B/Fluids* 21(6):615–47, 2002.

Nicolaescu, L. Adiabatic limits of the Seiberg-Witten equations on Seifert manifolds. *Commun. Anal. Geom.* 6:301–62, 1998.

Nicolaescu, L. Generalized symplectic geometries and the index of families of elliptic problems. *Mem. Am. Math. Soc.* 128(609), 1997.

Nicolaescu, L. The spectral flow, the Maslov index and decompositions of manifolds. *Duke Univ. J.* 80:485–534, 1995.

Polini, C., and B. Ulrich. A formula for the core of an ideal. *Math. Ann.* 331: 487–503, 2005.

Polini, C., B. Ulrich, and W. Vasconcelos. Normalization of ideals and Briancon-Skoda numbers. *Math. Res. Lett.* 12:10001–16, 2005.

Corso, A., **C. Polini,** and B. Ulrich. The structure of the core of ideals. *Math. Ann.* 321:89–105, 2001.

Marcus, B., and **J. Rosenthal.** Codes systems and graphical models. In *Mathematics and Its Applications, IMA,* vol. 123. Springer-Verlag, 2001.

Rosenthal, J., and X. Wang. The multiplicative inverse eigenvalue problem over an algebraically closed field. *SIAM J. Matrix Anal. Appl. No. 2* 23:517–23, 2001.

Rosenthal, J., and R. Smarandache. Maximum distance separable convolutional codes. *Applicable Algebra Eng. Commun. Computing No. 1* 10:15–32, 1999.

Shaw, M.-C. L^2 estimates and existence theorems for $\bar{\partial}_b$ on Lipschitz boundaries. *Math. Z.* 244:91–123, 2003.

Chen, S.-C., and **M.-C. Shaw.** *Partial Differential Equations in Several Complex Variables,* vol. 19. Providence: AMS International Press, 2001.

Smyth, B. Soliton surfaces in the mechanical equilibrium of closed membranes. *Commun. Math. Phys.* 250:81–94, 2004.

Smyth, B., and **F. J. Xavier.** Eigenvalue estimates and the index of Hessiau fields. *Bull. London Math. Soc.* 32:1–4, 2000.

Smyth, B., and **F. J. Xavier.** Real solvability of the equation $\partial 2/z\ \omega = pg$ and the topology of isolated umbilics. *J. Geom. Anal.* 8:655–71, 1998.

Snow, D. M. A bound for the dimension of the automorphism group of a homogeneous compact complex manifold. *Proc. Am. Math. Soc.* 132:2051–2055, 2004.

Snow, D. M., and J. Winkelmann. Compact complex homogeneous manifolds with large automorphism groups. *Inventiones Math.* 134:139–44, 1998.

Snow, D. M., and L. Manivel. A Borel-Weil theorem for holomorphic forms. *Compositio Math.* 103:351–65, 1996.

Sommese, A. J. *Numerical Solution of Systems of Polynomials Arising in Engineering and Science.* Singapore: World Scientific Press, 2005 (with Wampler).

Sommese, A. J., J. Verschelde, and C. W. Wampler. Homotopies for intersecting solution components of polynomial systems. *SIAM J. Numerical Anal.* 42:1552–71, 2004.

Sommese, A. J., J. Verschelde, and C. W. Wampler. Symmetric functions applied to decomposing solution sets of polynomial systems. *SIAM J. Numerical Anal.* 40:2026–46, 2002.

Stanton, N. K. Infinitesimal automorphisms of real hypersurfaces. *Am. J. Math.* 118:209–33, 1996.

Stanton, N. K. Spectral invariants of pseudoconformal manifolds. *Proc. Symp. Pure Math.* 54(2):551–7, 1993.

Stanton, N. K. The Riemann mapping non-theorem. *Math. Intelligencer* 14:32–6, 1992.

Stolz, S., Multiplicities of Dupin hypersurfaces. *Inv. Math.* 138:253–79, 1999.

Stolz, S. A conjecture concerning positive Ricci curvature and the Witten genus. *Math. Ann.* 304:785–800, 1996.

Stolz, S. Simply connected manifolds of positive scalar curvature. *Ann. Math.* 136:511–40, 1992.

Hughes, B., **L. R. Taylor,** S. Weinberger, and **B. Williams.** Neighborhoods in stratified spaces with two strata. *Topology* 39:873–919, 2000.

Hambleton, I., and **L. R. Taylor.** A guide to the calculation of surgery obstruction groups. Surveys on Surgery Theory, Volume I. In *Annals of Mathematical Studies 145,* pp. 225–74, eds. S. Cappell, A. Ranicki, and J. Rosenberg. Princeton University Press, 2000.

Taylor, L. R. Taut codimension one spheres of odd order, "Geometry and Topology: Aarhus." *Contemp. Math., Am. Math Soc.* 258:369–375, 2000.

Dwyer, W., M. Weiss, and **B. Williams.** A parametrized index theorem for the algebraic K-theory Euler class. *Acta Math.,* 2003.

Weiss, M., and **B. Williams.** Automorphisms of manifolds. *Ann. Math. Stud.* 149:165–220, 2001.

Williams, B. Bivariant Riemann Roch theorems. *Geometry Topology:Aarhus* 258:377–93, 2000.

Chandler, K., and **P.-M. Wong.** On the holomorphic sectional and bisectional curvatures in complex Finsler geometry. *Periodica Mathematica Hung.* 48:93–123, 2004.

An, T.-H., J. T.-Y. Wang, and **P.-M. Wong.** Unique range set and uniqueness polynomials in positive characteristic. *Acta Arithmetica* 109:259–80, 2003.

Dethloff, G., G. Schumacher, and **P.-M. Wong.** Hyperbolicity of the complements of plane algebraic curves. *Am. J. Math.* 117:573–99, 1995.

Xavier, F. J. Embedded, simply connected, minimal surfaces with bounded curvature. *GAFA,* in press.

Xavier, F. J., and S. Nollet. Global inversion via the Palais-Smale condition. *Discrete Continuous Dynamical Syst.,* in press.

Xavier, F. J., and V. Nitica. Schrodinger operators and topological pressure on manifolds of negative curvature. *Proc. Symp. Pure Math. Am. Math. Soc.* 2001.

UNIVERSITY OF SOUTHERN CALIFORNIA

Division of Biostatistics

Programs of Study

Graduate education at the University of Southern California (USC) prepares students for leadership in research, teaching, or professional practice in the private or public sector. Rigorous, individually tailored course work and research forms the basis for the program.

Graduate studies in biostatistics are contained within USC's Keck School of Medicine. The University offers the Master of Science (M.S.) and Doctor of Philosophy (Ph.D.) degrees. The Ph.D. in biostatistics is designed to produce a biostatistician with a deep knowledge of statistical theory and methodology. The Ph.D. in statistical genetics and genetic epidemiology is a joint effort to combine biostatistics, epidemiology, statistical and molecular genetics, and computational methods in order to develop new and cutting-edge statistical methodology that is appropriate for human genomic studies.

Master's degree studies in biostatistics focus on the theory of biostatistics, data analytic methods, experimental design (including clinical trials), statistical methods in human genetics, biomedical informatics, and statistical computing methods. The master's degree in applied biostatistics and epidemiology includes applied biostatistics, epidemiological research methods, and research applications, including cancer, infectious disease, chronic disease, and environmental epidemiology. Doctoral studies cover the areas of biostatistics and data analysis; descriptive, genetic, and molecular epidemiology; computational methods; clinical trial methodology; and related fields of field research, such as population disease and treatment trials.

The Division of Biostatistics also offers a joint-degree program in either the M.S. or Ph.D. in molecular epidemiology in conjunction with the Department of Biochemistry and Molecular Biology. The objective of the M.S. degree is to train students in the application of statistical methods to the design of biomedical research. The objective of the doctoral degree is to produce a molecular epidemiologist with in-depth laboratory, statistical, and analytical skills in both epidemiology and the molecular biosciences.

Research Facilities

Hands-on research alongside a faculty mentor is the norm at USC. Graduate students work alongside colleagues as co-investigators on epidemiological, clinical trial, and environmental research projects. Graduate students have myriad opportunities to participate in the latest medical methods. Research teams gain expertise on study design as well as statistical methodology and data analysis. The University's Health Sciences Campus contains the Norris Cancer Center, the General Clinical Research Center, and the Doheny Eye Institute Vision Research Center.

USC has a dozen libraries to serve all the varied needs of its graduate students. Primary source materials include ongoing research projects and recent faculty and staff members' publications. In addition, electronic resources in hundreds of subjects plus online archives and Internet access are available for students through the library system.

Financial Aid

Most graduate students who demonstrate financial need qualify for low-interest loans, work-study, or assistantships. Graduate assistantships are awarded on the basis of scholastic accomplishment and competence. Students exchange teaching and laboratory assistant time for tuition waivers and stipends.

The Sponsored Projects Information Network (SPIN) is a computerized database of funding opportunities—federal, private, and corporate—created to help faculty members identify external financial support for research and education. SPIN funds that are directed toward USC programs in biostatistics result in research funds and fellowships for graduate students.

Cost of Study

Costs for 2005–06 were as follows: tuition and mandatory fees, $16,213; room and board (on campus or off campus), $9258; books and supplies, $962; and other miscellaneous expenses, $1754.

Living and Housing Costs

Estimated housing costs for the Los Angeles area are approximately $9200 per year. Housing costs vary greatly, depending on the location and type of accommodations. There are ample housing facilities in the many communities surrounding the medical school.

Student Group

There are approximately 1,500 graduate students at USC and 100 in the biostatistics and genetics program. The relatively small size of the program facilitates student–faculty member interchange and good accessibility for students to their professors and mentors.

Location

University of Southern California campuses are mostly centered around Los Angeles, with other facilities in nearby Alhambra, Pasadena, and Marina del Rey; on Catalina Island; and further away in Orange County to the south and Sacramento to the north. Los Angeles is the second-largest city in the U.S. and the nucleus of southern California. The scenic, sunny, and culturally diverse area of Los Angeles offers miles of beaches, acres of recreational and park areas, and some of the finest arts and cultural opportunities in the nation.

The University

USC is the oldest and largest independent coeducational university in the West. The campus is composed of 169 buildings located in a 150-acre parklike setting near downtown Los Angeles. USC is among the ten most successful private universities in the country in attracting research support from external sources (more than $100 million annually). Graduate students in the Division of Biostatistics take the majority of their courses at the Health Sciences Campus, 3 miles northeast of downtown Los Angeles and 7 miles from the USC University Park Campus. The Health Sciences Campus is adjacent to the Los Angeles County–USC Medical Center, one of the nation's largest teaching hospitals. The surrounding neighborhoods are among the most historically significant in the city, with rich educational resources.

Applying

For the M.S. degree, an undergraduate degree in mathematics, statistics, biostatistics, or computer science is most helpful. Undergraduate preparation should include differential and integral calculus, mathematical statistics, and basic computer programming. The Ph.D. requires successful scores on a screening examination (the M.S. prepares students for this exam). Students may apply online or by mail. An application booklet, plus forms and other application materials, may be found on the University's Web site.

Correspondence and Information

Graduate Programs in Biostatistics, Epidemiology, Molecular Epidemiology, and Statistical Genetics
Division of Biostatistics
Department of Preventive Medicine
Keck School of Medicine
Center for Health Professions, 222
University of Southern California
1540 Alcazar
Los Angeles, California 90089-9010
Phone: 323-442-1810
Fax: 323-442-2993
E-mail: mtrujill@usc.edu
Web site: http://www.usc.edu/medicine/biostats

University of Southern California

THE FACULTY AND THEIR RESEARCH

Todd Alonzo, Assistant Professor; Ph.D., Washington (Seattle), 2000. Design and analysis of clinical trials, pediatric oncology, statistical methodology and missing data methodology.
Edward Avol, Associate Professor; M.S., Caltech, 1974. Chronic respiratory effects of airborne pollutants in populations.
Stanley Azen, Professor and Co-director; Ph.D., UCLA, 1969. Biostatistical methodology.
Lourdes Baezconde-Garbanati, Assistant Professor; Ph.D., UCLA, 1994. Cancer control research.
Kiros Berhane, Associate Professor; Ph.D., Toronto, 1994. Analysis of health effects of environmental exposures.
Leslie Bernstein, Professor and AFLAC, Inc., Chair in Cancer Research; Ph.D., USC, 1981. Epidemiology of breast cancer and non-Hodgkin's lymphoma.
Jonathan Buckley, Professor; Ph.D., Melbourne, 1981. Epidemiology of childhood cancer, clinical trials, molecular epidemiology.
John Casagrande, Associate Professor; D.Ph., UCLA, 1978. Computer applications in research.
Chih-Ping Chou, Associate Professor; Ph.D., UCLA, 1983. Evaluation of approaches to substance-abuse prevention among adolescents, statistical methods in prevention research.
Myles Cockburn, Assistant Professor; Ph.D., Otago (New Zealand), 1998. Epidemiology of melanoma, gastric cancer, and *Helicobacter pylori;* computational methods; geographical information systems (GIS).
David Conti, Assistant Professor; Ph.D., Case Western Reserve, 2002. Statistical methods in genetic association studies, use of hierarchical models in epidemiology.
Victoria Cortessis, Assistant Professor; Ph.D., UCLA, 1993. Genetic-epidemiologic and molecular genetic studies of congenital disorders, adult-onset cancers and etiologic relationships among these entities.
Wendy Cozen, Assistant Professor; M.P.H., UCLA, 1989. Epidemiology of hematologic neoplasms, Hodgkin's disease, non-Hodgkin's lymphoma.
Martha Cruz, Research Associate; Ph.D., Oxford, 2000. Insulin resistance and its relationship to type 2 diabetes and cardiovascular disease.
N. Tess Cruz, Assistant Professor; Ph.D., Massachusetts, 1993. Public health communications research, anti-tobacco media and pro-tobacco marketing effects.
Dennis Deapen, Professor; Dr.Ph., UCLA, 1982. Cancer outcomes among breast implant patients; lupus erythematosus, diabetes, multiple sclerosis, and Alzheimer's disease.
Clyde Dent, Associate Professor; Ph.D., North Carolina, 1984. Prevention and cessation of tobacco, alcohol, and other drug use in school-based, medical clinic, and worksite contexts.
James Dwyer, Professor; Ph.D., California, Santa Cruz, 1975. Cardiovascular and atherosclerosis.
W. James Gauderman, Associate Professor; Ph.D., USC, 1992. Biostatistical methodology, genetic-epidemiological analysis of pedigree data, health outcomes to environmental exposure.
Frank Gilliand, Associate Professor; Ph.D., Minnesota, 1992. Environmental exposures on air pollution.
Michael I. Goran, Professor; Ph.D., Manchester (England), 1986. Biophysics etiology prevention of obesity, type 2 diabetes in children.
Susan Groshen, Professor; Ph.D., Rutgers, 1980. New drugs and treating cancer.
Robert Haile, Professor; Ph.D., UCLA, 1979. Genetic epidemiology of breast, colon, and prostate cancers.
Ann Hamilton, Assistant Professor; Ph.D., UCLA, 1987. Breast, prostate, and testicular cancer; Kaposi's sarcoma; cancers in twins.
Brian Henderson, Professor; M.D., Chicago, 1962. Cancers of the breast, prostate, ovary, testes, and endometrium in different ethnic groups.
Annlia Hill, Professor; Ph.D., UCLA, 1974. Cancer, cardiovascular, diabetes.
Andrea Hricko, Associate Professor; M.P.H., North Carolina, 1971. Outreach and education techniques, translational and community-based participatory research.
Sue A. Ingles, Assistant Professor; D.P.H., UCLA, 1993. Nutritional genetics and breast, prostate, and colorectal cancer.
Michael Jerrett, Associate Professor; Ph.D., Toronto, 1996. Mapping, health air pollution.
Carl Anderson Johnson, Professor; Ph.D., Duke, 1974. Tobacco, alcohol, and other drug use prevention; nutritional; physical exercise; health promotion.
Carol Koprowski, Assistant Professor; Ph.D., USC, 1998. Diet and physical activity.
Mark Krailo, Professor; Ph.D., Waterloo, 1981. Clinical trials on cancer treatment.
Nino Kuenzli, Associate Professor; Ph.D., Berkeley, 1996. Air pollution.
Peter W. Laird, Associate Professor; Ph.D., Amsterdam (Netherlands Cancer Institute), 1988. Biochemistry and molecular biology, cancer genetics, gene regulation.
Bryan Langholz, Professor; Ph.D., Washington (Seattle), 1984. Cancer and other chronic diseases, cohort studies.
Thomas Mack, Professor; M.P.H., Harvard, 1969. Chronic disease in twins.
Wendy Mack, Associate Professor; Ph.D., USC, 1989. Biostatistical methodology in cardiovascular research, clinical trials using angiographic and ultrasound endpoints.
Paul Marjoram, Assistant Professor; Ph.D., University College (London), 1992. Computational biology; the coalescent, probabilistic models; microarray data.
Rob McConnell, Associate Professor; M.D., California, San Francisco, 1980. Environmental exposures, air pollution.
Roberta McKean-Cowdin, Assistant Professor; Ph.D., 1996. Breast, brain, and endometrial cancer.
Elaine Nezami, Assistant Professor; Ph.D., USC, 1994. Chronic disease, cancer, cardiovascular.
Paula Palmer, Assistant Professor; Ph.D., California School of Professional Psychology, 1998. Social and cultural determinants of health in ethnically diverse populations, school- and community-based research.
Mary Ann Pentz, Professor; Ph.D., Syracuse, 1978. Development and testing of school/community-based prevention intervention for adolescents.
John Peters, Professor; S.C.D., Harvard, 1966. Environmental exposures on air pollution.
Malcolm Pike, Professor; Ph.D., Aberdeen (England), 1963. Hormonal, endometrial, and ovarian cancer.
Susan Preston-Martin, Professor; Ph.D., UCLA, 1978. Central nervous system tumors, myeloid leukemia, and other radiogenic cancers; HIV infection in women.
Kim Reynolds, Associate Professor; Ph.D., Arizona State, 1987. Prevention and control of chronic disease; dietary behavior, physical activity, and obesity in school.
Jean Richardson, Professor; M.P.H., UCLA, 1971. Cancer control, behavioral and epidemiological research methods.
Phyllis Rideout, Associate Professor; Ph.D., Florida State, 1981. Cancer education.
Louise Rohrbach, Assistant Professor; Ph.D., USC, 1989. Community-based interventions for disease prevention and health promotion; prevention of tobacco, alcohol, and other drug abuse.
Ronald Ross, Professor; M.D., Iowa, 1975. Hormone cancers, international collaborative efforts to identify dietary causes of cancer.
Harland Sather, Associate Professor; Ph.D., UCLA, 1975. Clinical trials in pediatric cancer.
Kimberly Siegmund, Assistant Professor; Ph.D., Washington (Seattle), 1995. Statistical methods for genetics.
Janet Sobell, Associate Professor; Ph.D., 1991. Multifactorial neurological, neurodegenerative, and neuropsychiatric disorders.
Richard Sposto, Associate Professor; Ph.D., UCLA, 1981. Biostatistics, clinical trials in pediatric oncology, Bayesian analysis of survival data.
Donna Spruijt-Metz, Assistant Professor; Ph.D., Amsterdam Vrije University, 1996. Obesity and type 2 diabetes, smoking prevention.
Alan Stacy, Associate Professor; Ph.D., California, Riverside, 1986. Addiction, prevention.
Michael R. Stallcup, Professor; Ph.D., Berkeley, 1977. Cancer cell biology, signal transduction, genes.
Mariana Stern, Assistant Professor; Ph.D., Texas Health Science Center, 1997. Colorectal and breast cancer, gene-diet, gene-smoking.
Daniel Stram, Associate Professor; Ph.D., Temple, 1983. Modern statistical methods, measurement error methods in cancer epidemiology, repeated measures data, human genetics data.
Ping Sun, Research Associate; Ph.D., USC, 1999. Cardiovascular disease and cancer.
Steven Sussman, Professor; Ph.D., Illinois at Chicago, 1984. Drug abuse, cessation, school-based alcohol, tobacco.
Duncan Thomas, Professor; Ph.D., McGill, 1976. Statistical methods, occupational and environmental health.
Jennifer Unger, Associate Professor; Ph.D., USC, 1996. Psychosocial and cultural factors in adolescent.
Giska Ursin, Associate Professor; Ph.D., UCLA, 1992. Breast cancer.
Thomas Valente, Associate Professor; Ph.D., USC, 1991. Health promotion, substance abuse.
Richard Watanabe, Assistant Professor; Ph.D., USC, 1995. Type 2 diabetes, biologic systems, positional cloning and gene characterization in complex disease.
Anna Wu, Professor; Ph.D., UCLA, 1983. Various cancers among Asian migrants to the U.S.
Anny Xiang, Assistant Professor; Ph.D., USC, 1995. Clinical collaboration and statistical methodology, non-insulin-dependent diabetes (NIDDM).
Mimi Yu, Professor; Ph.D., UCLA, 1977. Cancer epidemiology, nasopharyngeal and liver cancers, large-scale cohort study in Singapore in cancer causation.
Jian-Min Yuan, Assistant Professor; Ph.D., USC, 1996. Cancer epidemiology; cohort study in Shanghai aimed at investigating roles of dietary, environmental exposures, and gene-environment interaction in cancer.
Tianni Zhou, Associate Professor; Ph.D., 2002. Clinical trials in pediatric oncology.

THE UNIVERSITY OF TEXAS AT DALLAS

Mathematical Sciences

Programs of Study

The mathematical sciences department at the University of Texas at Dallas (UT Dallas) offers the Master of Science degree in five specializations: applied mathematics, bioinformatics and computational biology, engineering mathematics, mathematics, and statistics. The Doctor of Philosophy degree is offered in applied mathematics and in statistics. The program has major research faculty and thrusts in the latter two areas. The degree programs are designed to prepare graduates for careers in mathematical sciences or in related fields for which these disciplines provide indispensable foundations and tools. There is no language requirement.

The Master of Science degree requires 33–36 semester hours of course work, consisting of core courses and approved electives. The student may choose a thesis plan or a nonthesis plan. In the thesis plan, the thesis replaces 6 semester hours of course work.

The Ph.D. program is tailored to the student, who arranges a course program with the guidance and approval of the graduate adviser. Adjustments can be made as the student's interests develop and a specific dissertation topic is chosen. Approximately 39 hours of core courses and 18–24 hours of elective courses are required for a typical degree program. After completion of about two years of course work, the student must undertake and pass a Ph.D. qualifying examination in order to continue in the program. The program culminates in the preparation of a dissertation, which must be approved by the graduate program. The topic may be in mathematical sciences exclusively or may involve considerable work in an area of application. Typical areas of concentrations within applied mathematics include, but are not restricted to, applied analysis, computational and mathematical biology, relativity theory, differential equations, scattering theory, systems theory, control theory, signal processing, and differential geometry. In the area of statistics, concentrations are offered in mathematical statistics, applied statistics, statistical computing, probability, stochastic processes, linear models, time series analysis, statistical classification, multivariate analysis, robust statistics, statistical inference, and asymptotic theory.

In addition to a wide range of courses in mathematics and statistics, the mathematical sciences program offers a unique selection of courses that consider theoretical and computational aspects of engineering and scientific problems.

Research Facilities

Faculty, staff, and research/teaching assistant offices are equipped with the latest generation of computers. The department also has a classroom equipped with state-of-the-art computers for classroom and research use. All of the department's machines are connected via Ethernet to the campus network, giving faculty members and students access to all of the software tools and machines on campus for research and educational use.

Financial Aid

The Graduate Studies Scholarship (GSS) covers the full cost of tuition and fees for up to 9 credit hours per semester; the total GSS value is $6070 per academic year. In addition to GSS, full-time graduate students qualify for teaching assistantships. The full teaching assistantship stipend for 2005–06 was $9286; the value of the total package is $15,365 per academic year. Support for summer study is usually available. In addition to the GSS and teaching assistantships, applicants may also be awarded an Excellence in Education Fellowship for up to two years at a rate of $8000 per year. UT Dallas has also developed a comprehensive program of grants, scholarships, loans, and employment opportunities to assist students in meeting the cost of their education.

Cost of Study

Nonresidents holding teaching assistantships are eligible to pay tuition at the lower rate applicable to Texas residents. The rates for 2005–06 for a 9-hour course load were $2600 for Texas residents and $5084 for nonresidents.

Living and Housing Costs

Students in the program typically live in a nearby apartment complex that offers comfortable accommodations at attractive rates. In fall 2005, monthly rates ranged from $458 to $624 for one-bedroom, $860 to $936 for two-bedroom, and $1136 for four-bedroom apartments, some including washer and dryer.

Student Group

The total enrollment at the University is 14,480, including 5,220 graduate students. The mathematical sciences program has 29 master's students and 27 Ph.D. candidates, some of whom attend part-time while employed full-time with companies in the Dallas area.

Student Outcomes

The most recent 5 Ph.D. graduates of the program have secured employment in both industrial and academic positions. Of the program's 2 most recent Ph.D. students in applied mathematics, 1 is now on the faculty at Salisbury University, Maryland, and the other has joined the faculty at the University of Texas at Dallas's Engineering Department as a Research Associate. In statistics, 2 of the 3 most recent Ph.D. graduates have joined the faculties of the University of Mississippi and the University of North Carolina at Charlotte, and the other is employed in Biostatistics at PPD Development in North Carolina.

Location

UT Dallas is located in Richardson, a quiet suburb of North Dallas, which is easily accessible to the more than 800 high-technology companies located in the Dallas–Fort Worth area. Many of these companies are located within 10 miles of UT Dallas, providing graduates with numerous career opportunities. The Dallas metropolitan area also offers a wide range of cultural, social, and sports activities.

The University and The Program

The University of Texas at Dallas was created in 1969 when the privately funded Southwest Center for Advanced Studies was transferred to the state of Texas. In 1972 the Program in Mathematical Sciences was introduced and in 1975 became part of the School of Natural Sciences and Mathematics. Research at the graduate level has continued to represent a major thrust of the University and of the program.

Applying

Applications are considered at any time until vacancies are filled. For consideration for teaching assistantships, the deadline of January 15 is set for first-round consideration. Applicants should arrange for GRE scores and (for international students) TOEFL scores to be included as early as possible in the application materials. Applications not complete before March 15 receive relatively late consideration for teaching assistantships.

Correspondence and Information

Head
Mathematical Sciences
The University of Texas at Dallas
P.O. Box 830688, MS EC35
Richardson, Texas 75083-0688

Phone: 972-883-2161
Fax: 972-883-6622
E-mail: utdmath@utdallas.edu
Web site: http://www.utdallas.edu/dept/math

The University of Texas at Dallas

THE FACULTY AND THEIR RESEARCH

Larry P. Ammann, Professor; Ph.D., Florida State, 1976. Robust multivariate statistical methods, signal processing, statistical computing, applied probability, remote sensing.
Michael Baron, Associate Professor; Ph.D., Maryland, 1995. Sequential analysis, Bayesian inference, change-point problems, applications in semiconductor manufacturing, psychology, energy finance.
Pankaj Choudhary, Assistant Professor; Ph.D., Ohio State, 2002. Biostatistics, statistical inference, method comparison studies.
Mieczyslaw K. Dabkowski, Assistant Professor; Ph.D., George Washington, 2003. Knot invariants and 3-manifold invariants, applications of topology to biology, recursion theory.
M. Ali Hooshyar, Professor; Ph.D., Indiana, 1970. Scattering theory, inverse scattering theory with geophysical and optical applications, fission.
Istvan Ozsvath, Professor; Ph.D., Hamburg, 1960. Relativistic cosmology, differential geometry.
Viswanath Ramakrishna, Associate Professor; Ph.D., Washington (St. Louis), 1991. Control, optimization, computation, applications in material and molecular sciences.
Ivor Robinson, Emeritus Professor; B.A., Cambridge, 1947. General relativity theory, particularly exact solutions to Einstein's equations of gravitation.
Robert Serfling, Professor; Ph.D., North Carolina, 1967. Probability theory, statistical inference, robust and nonparametric methods, asymptotic theory, stochastic processes, applications in bioscience and finance. (Web site: http://www.utdallas.edu/~serfling)
Janos Turi, Professor, Ph.D., Virginia Tech, 1986. Functional differential equations, integral equations, approximation theory, optimal control theory, numerical analysis, applied functional analysis.
John Van Ness, Emeritus Professor; Ph.D., Brown, 1964. Robust linear models, statistical classification, measurement error models, applications of statistics to the physical and medical sciences.
John Wiorkowski, Professor; Ph.D., Chicago, 1972. Statistical time series, forecasting, applied statistics, regression analysis, multivariate techniques.

SELECTED PUBLICATIONS

Ammann, L., E. M. Dowling, and R. D. DeGroat. A TQR-iteration based adaptive SVD for real-time angle and frequency tracking. *IEEE Trans. Signal Processing* 42:914–26, 1994.
Ammann, L. Robust singular value decompositions: A new approach to projection pursuit. *J. Am. Stat. Assoc.* 88:504–14, 1993.
Baron, M. Bayes stopping rules in a change-point model with a random hazard rate. *Sequential Anal.* 20(3):147–63, 2001.
Baron, M., C. K. Lakshminarayan, and Z. Chen. Markov random fields in pattern recognition for semiconductor manufacturing. *Technometrics* 43(1):66–72, 2001.
Baron, M. Nonparametric adaptive change-point estimation and on-line detection. *Sequential Anal.* 19(1–2):1–23, 2000.
Choudhary, P. K., and H. K. T. Ng. Assessment of agreement under non-standard conditions using regression models for mean and variance. *Biometrics,* in press.
Choudhary, P. K., and H. N. Nagaraja. Assessment of agreement using intersection-union principle. *Biometrical J.* 47, published online: April 21, 2005.
Choudhary, P. K., and H. N. Nagaraja. A two-stage procedure for selection and assessment of agreement of the best instrument with a gold standard. *Sequential Anal.* 24:237–57, 2005
Dabkowski, M. K., J. H. Przytycki, and A. Togha. Non-left-orderable 3-manifold groups. *Can. Math. Bull.* 48X(1):32–40, 2005.
Dabkowski, M. K., and J. H. Przytycki. Unexpected connections between Burnside Groups and Knot Theory. *Proc. Natl. Acad. Sci. USA* 101(50):17357–60, 2004.
Dabkowski, M. K., and J. H. Przytycki. Burnside obstructions to the Montesinos-Nakanishi 3-move conjecture. *Geom. Topol.* 6:355–60, 2002.
Hooshyar, M. A, I. Reichstein, and F. B. Malek. *Nuclear Fission and Cluster Radioactivity.* Berlin: Springer-Verlag, 2005.
Hooshyar, M. A., and L. V. Lasater. The method of lines and electromagnetic scattering for line sources. *Microwave Opt. Technol. Lett.* 41:286–90, 2004.
Hooshyar, M. A. An inverse problem of normal incidence TE plane-wave scattering and the method of lines. *Microwave Opt. Technol. Lett.* 35:486–91, 2002.
Ozsvath, I., and E. Schucking. Approaches to Godel's rotating universe. *Class Quantum Gravity* 18:2243–52, 2001.
Ozsvath, I., and E. Schucking. The world viewed from outside. *J. Geometry Phys.* 24:303–333, 1998.
Ozsvath, I. The finite rotating universe revisited. *Class Quantum Gravity* 14:A291–7, 1997.
Ramakrishna, V. Local solvability of degenerate, overdetermined systems: A control-theoretic perspective. *J. Differential Equations,* in press.
Ramakrishna, V. Controlled invariance for singular distributions. *SIAM J. Control Optimization* 32:790–807, 1994.
Robinson, I., and I. Trautman. The conformal geometry of complex quadrics and the fractional-linear form of Mobius transformations. *J. Math. Phys.* 34:5391, 1993.
Serfling, R., and J. Wang. Nonparametric multivariate kurtosis and tailweight measures. *J. Nonparametric Stat.,* in press.
Serfling, R., and J. Wang. Influence functions for a general class of depth-based quantile functions. *J. Multivariate Anal.,* in press.
Serfling, R. Nonparametric multivariate descriptive measures based on spatial quantiles. *J. Stat. Plann. Inference* 123:259–78, 2004.
Turi, J., and F. Hartung. Linearized stability in functional differential equations with state-dependent delays. *Dynamical Syst. Differential Equations* (an added volume to *Discrete Continuous Dynamical Syst.*) 416–25, 2001.
Turi, J., F. Hartung, and T. L. Herdman. Parameter identification in classes of neutral differential equations with state-dependent delays. *J. Nonlinear Analysis Theory Methods Appl.* 39:305–25, 2000.
Turi, J., and W. Desch. The stop operator related to a convex polyhedron. *J. Differential Equations* 157:329–47, 1999.
Van Ness, J. Recent results in clustering admissibility. In *Applied Stochastic Models and Data Analysis.* Lisbon: Instituto Nacional del Estastistica, 1999.
Van Ness, J., and C. L. Cheng. *Statistical Regression with Measurement Error.* London: Edward Arnold Publishers, 1999.
Van Ness, J., and J. Yang. Robust discriminant analysis: Training data breakdown point. *J. Stat. Plann. Inference* 1:67–84, 1998.
Wiorkowski, J. A lightly annotated bibliography of the publications of the American Statistical Association. *Am. Statistician* 44:106–13, 1990.
Wiorkowski, J. Fitting of growth curves over time when the data are obtained from a single realization. *J. Forecasting* 7:259–72, 1988.

THE UNIVERSITY OF TOLEDO

THE UNIVERSITY OF TOLEDO

Department of Mathematics

Programs of Study

The Ph.D. program trains mathematicians and statisticians who intend to make research in these areas their life work. The Ph.D. requires a minimum of 90 credit hours, of which 18 to 36 are for the dissertation. Depending on their program of study, students enroll in three yearlong sequences chosen from abstract algebra, real analysis, topology, probability and statistics, differential equations, and complex analysis, plus two other yearlong sequences. Students must demonstrate the ability to read mathematical literature in one foreign language, ordinarily chosen from among French, German, and Russian, and spend two consecutive semesters in supervised teaching.

For full-time students, the written qualifying examination must be passed by the end of the student's second year and a topic-specific oral exam must be passed within a year of passing the qualifying exam. A completed dissertation must be approved by an outside examiner, then defended by the student before a faculty committee. Typically, students take two to three years to complete their dissertations.

The M.A. in mathematics requires at least 30 credit hours, including course work in algebra, topology, real analysis, and complex analysis, and at least one 2-semester sequence in a single discipline. Students must either pass a comprehensive examination or write a thesis to complete the degree. The M.S. in applied mathematics requires 30 credits, culminating in a written thesis or examination. The M.S. in statistics requires 35 credit hours and passing a two-part examination in statistical theory and applied statistics. At least 32 credit hours are required for the M.S. in education, including at least 18 in mathematics and 9 in education.

Research Facilities

The William S. Carlson Library houses more than 1.6 million volumes and over 3,000 periodicals. The library has a large, current collection of mathematical texts and monographs and subscribes to more than 200 mathematical journals. The University Libraries have a fully electronic catalog and circulation service that is available through any terminal on or off campus. The University Library network is connected to a statewide university library network, which provides access to the collections of all other university and college libraries in the state of Ohio, as well as numerous research databases, such as ISI. A network-based computing facility is built around a Sun Ultra-Enterprise server and two Dell PowerEdge servers running Linux. The Department also maintains two instructional computing laboratories equipped with projectors and containing about forty PCs each, as well as a smaller computing lab.

Financial Aid

The out-of-state tuition surcharge normally charged to out-of-state and international students is waived for students whose permanent address is within one of the following Michigan counties: Hillsdale, Lenawee, Macomb, Oakland, Washtenaw, and Wayne. In addition, the University of Toledo offers an out-of-state tuition surcharge waiver to cities and regions that are a part of the Sister Cities Agreement. These regions include Toledo, Spain; Londrina, Brazil; Qinhuangdao, China; Csongrad County, Hungary; Delmenhorst, Germany; Toyohashi, Japan; Tanga, Tanzania; Bekaa Valley, Lebanon; and Poznan, Poland. The University of Toledo Graduate College offers a variety of memorial and minority scholarship awards, including the Ronald E. McNair Postbaccalaureate Achievement Scholarship, the Graduate Minority Assistantship Award, and two full University fellowships.

Cost of Study

The graduate tuition rate for the 2006–07 academic year is $390.05 per semester credit hour for in-state students. For nonresidents, the out-of-state surcharge is $367.15 per semester credit hour. Additional fees are required and include the general fee, technology fee, and mandatory insurance.

Living and Housing Costs

The University of Toledo has a diverse offering of student housing options, including suite-style and traditional residential halls. Housing is offered to graduate students through Residence Life or contracted individually by the student. Affordable, high-quality off-campus apartment-style housing within walking distance of campus is abundant.

Student Group

There are approximately 20,000 students at the University of Toledo. About 4,000 are graduate and professional students. The University has a rich diversity of student organizations. Students join groups that are organized around common cultural, religious, athletic, and educational interests.

Student Outcomes

Graduates of the program occupy academic positions in colleges and universities around the world. Graduates from the master's programs are well prepared both for doctoral studies and for employment in academic and nonacademic settings.

Location

The University of Toledo has several campus sites in the city of Toledo. Most engineering graduate students take classes on the Main campus, which is located in suburban western Toledo. With a population of more than 330,000, Toledo is the fiftieth-largest city in the United States. It is located on the western shores of Lake Erie, within a 2-hour drive of Cleveland and Detroit.

The University

The University of Toledo was founded by Jessup W. Scott in 1872 as a municipal institution and became part of the state of Ohio's system of higher education in 1967. On July 1, 2006, the University of Toledo merged with the Medical University of Ohio becoming one of only seventeen American universities to offer professional and graduate academic programs in medicine, law, pharmacy, nursing, health sciences, engineering, and business.

Applying

Admission to the program requires a completed application form, a statement of purpose, three sets of transcripts of all undergraduate and prior graduate work, at least three letters of recommendation, and an application fee of $40. There is no formal deadline for receipt of applications, but candidates for the fall semester whose applications are complete by mid-February have the best chance of success. Applications should be sent to the Graduate School rather than to the department.

Correspondence and Information

Geoffrey K. Martin, Chair
Department of Mathematics
The University of Toledo
2040 University Hall
2801 West Bancroft Street
Toledo, Ohio 43606-3390
Phone: 419-530-2568
Fax: 419-530-4720
Web site: http://www.math.utoledo.edu/

The University of Toledo

THE FACULTY AND THEIR RESEARCH

James Anderson, Assistant Professor.
H. Lamar Bentley, Professor; Ph.D., Rensselaer. Extensions of topological spaces by means of nearness structures, using primarily the methods of category theory.
Lihua Chen, Assistant Professor; Ph.D., Iowa State. Model combining, model selection, categorical data analysis, Bayesian methods.
Zeljko Cuckovic, Associate Professor; Ph.D., Michigan State. Algebraic properties of Toeplitz operators and harmonic function theory, as specialized subfields of functional analysis and complex analysis.
Mohamed S. Elbialy, Professor; Ph.D., Minnesota. Dynamical systems, Hamiltonian systems, celestial mechanics, and differential equations involving methods from nonlinear functional analysis, real analysis, and differential geometry.
Donald Greco, Professor.
David Hemmer, Assistant Professor; Ph.D., Chicago. Group theory; representation theory of finite groups, especially the symmetric and general linear groups.
Paul R. Hewitt, Associate Professor; Ph.D., Michigan State. Group theory, especially intrinsic and extrinsic structures, typically from geometry, topology, and logic, reflected in algebraic properties of the group.
Marie Hoover, Associate Professor and Associate Chair.
Marianty Ionel, Assistant Professor; Ph.D., Duke. Differential geometry, calibrated geometries, holonomy theory, geometric PDEs, exterior differential systems, mathematical physics.
En-Bing Lin, Professor; Ph.D., Johns Hopkins. Wavelet theory and applications; geometric and hadronic quantization in mathematical physics; domain characterizations in complex manifolds by means of algebraic and geometric structures, using primarily analytic methods.
Geoffrey K. Martin, Associate Professor and Chair; Ph.D., SUNY at Stony Brook. Symplectic geometry over determined systems of partial differential equations, geometric algebra in eight dimensions, triality, foundations of electromagnetism and relativity, geometric mathematical physics, Lie groups and indefinite symmetric domains, general geometric structure and partial differential relations.
Elaine Miller, Associate Professor.
Rao V. Nagisetty, Professor; Ph.D., Steklov Institute (Moscow). Theory of approximation, functional analysis, complex function theory and Toeplitz operators, Liouville's theory on function fields.
Robert L. Ochs Jr., Associate Professor; Ph.D., Delaware. Applied mathematics of wave propagation.
Charles J. Odenthal, Associate Professor; Ph.D., Wisconsin. Representations of Noetherian and Artinian rings.
Biao Ou, Associate Professor; Ph.D., Minnesota. Partial differential equations in applied area and differential geometry.
Martin R. Pettet, Professor; Ph.D., Yale. Theory of groups, with particular focus on relationships between structure and properties of a group and those of its automorphisms.
Friedhelm Schwarz, Professor; Ph.D., Bremen (Germany). Categorical topology, especially Cartesian closedness, exponentiability, and topological universes; teaching with technology, especially calculus with MAPLE, the World Wide Web, and distance learning.
Qin Shao, Assistant Professor; Ph.D., Georgia. Time series, experimental design.
Ivie Stein Jr., Associate Professor; Ph.D., UCLA. Optimization, calculus of variations, numerical analysis.
Stuart A. Steinberg, Professor; Ph.D., Illinois. Ordered structures and ring theory.
William Thomas, Associate Professor.
Gerard Thompson, Professor; Ph.D., Open University; Ph.D., North Carolina. Application of differential geometry, Lagrangian and Hamiltonian mechanics and symplectic geometry.
Mao-Pei Tsui, Assistant Professor; Ph.D., Brandeis. Differential geometry, geometric flows, general relativity and gravitation, partial differential equations.
H. Westcott Vayo, Professor; Ph.D., Illinois. Applications of differential equations to biology and medicine.
Henry C. Wente, Distinguished University Professor; Ph.D., Harvard. Calculus of variations and elliptic partial differential equations as they apply to minimal surface theory, immersions of constant mean curvature, and capillary theory.
Denis A. White, Professor; Ph.D., Northwestern. Schroedinger operators and their application to quantum mechanical scattering theory, using functional analysis and partial differential equations.
Donald B. White, Associate Professor; Ph.D., California, Irvine. Applications of statistics to various scientific areas, especially medicine and pharmacology; statistical areas, including nonlinear models and population modeling in pharmacokinetics and pharmacodynamics.
Harvey E. Wolff, Professor; Ph.D., Illinois. Category theory, with applications to algebra and topology.
Biao Zhang, Professor; Ph.D., Chicago. Density estimation and empirical likelihood.

UNIVERSITY OF UTAH

Department of Mathematics

Programs of Study

The Department of Mathematics offers programs leading to Doctor of Philosophy, Master of Arts, and Master of Science in mathematics degrees; a certificate in computational engineering and science (CES); and a Professional Master of Science and Technology (PMST) degree.

The master's degrees require 30 hours of course work beyond certain basic prerequisites. The candidate for the Master of Arts degree must satisfy the standard proficiency requirement in one foreign language; a further requirement is an expository thesis of good quality or an approved two-semester graduate course sequence.

The doctoral degree carries a minimum course requirement designed to prepare the student to pass three written qualifying examinations in the basic fields of mathematics. An oral examination, with emphasis on the candidate's area of specialization, is also required. A dissertation describing independent and original work is required. The Department of Mathematics stresses excellence in research.

Research Facilities

The Mathematics Branch Library collection in theoretical mathematics consists of 190 journal subscriptions, 15,000 bound journals, and 12,000 books. In addition, the Marriott Library collection includes numerous books and journals of interest to mathematics researchers and scholars. There are extensive interactive computing and computer graphics facilities available in the Department.

Financial Aid

Approximately 70 percent of the mathematics graduate students are supported by fellowships. There are teaching fellowships that grant from $14,500 to $15,700, plus tuition, fees, and health insurance. Prospective students may apply for University Research Fellowships through the University Research Fellowship Office, 310 Park Building, at the University of Utah. These fellowships are available to both U.S. and international students.

The Department currently offers VIGRE and IGERT Fellowships to U.S. citizens and permanent residents, with a stipend of $25,000 for a VIGRE award and $27,500 for an IGERT award. These fellowships also include book and travel allowances. VIGRE and IGERT Fellows are released from teaching obligations for the awarded academic year.

The normal teaching load for a teaching assistant and teaching fellow is one course per semester. Summer teaching is available. The stipend for one course during the summer is $3500.

Cost of Study

For 2005–06, tuition was $2149.02 per semester for Utah residents and $6763.90 per semester for nonresidents (12 credit hours). All tuition fees are waived for teaching assistants and teaching fellows.

Living and Housing Costs

A wide variety of housing is offered by the University on or near the campus. University Village, for married students, is operated by the University. One-, two-, and three-bedroom apartments range in cost from $753 to $996 per month, including heat, hot water, electricity, range, and refrigerator. (These rates may change without notice.) There is a waiting period of about four to six months for the University's married-student housing. Privately owned housing near the campus is also available. For details and complete listings of on-campus housing, students should visit the Web site http://www.orl.utah.edu.

Student Group

The University's total enrollment is currently 28,933. The Department of Mathematics has 98 graduate students; 60 receive financial support.

Student Outcomes

Graduates typically go on to postdoctoral research appointments followed by academic careers in teaching and research or careers in government and industry. In the past two years, 15 graduates took positions at various universities, including Johns Hopkins, Purdue, Tulane, Virginia Tech, Washington (St. Louis), and the University of Minnesota, and the Institute for Theoretical Dynamics and Lawrence Livermore National Laboratory.

Location

The Salt Lake City metropolitan area has a population of about 1 million and is the cultural, economic, and educational center of the Intermountain West. The Utah Symphony and Ballet West are located in Salt Lake City. The Delta Center is the home of the Utah Jazz basketball team. Climate and geography combine in the Salt Lake environs to provide ideal conditions for outdoor sports. Some of the world's best skiing is available less than an hour's drive from the University campus.

The University and The Department

The University of Utah is a state-supported coeducational public institution. Founded in 1850, it is the oldest state university west of the Missouri River.

In the last five years, the Department of Mathematics has awarded 109 graduate degrees. In recent years, the Graduate School has been awarding about 216 doctoral degrees per year. The University faculty has 1,387 members.

Applying

Admission to graduate status requires that students hold a bachelor's degree or its equivalent and that they show promise for success in graduate work. Applicants are urged to take the mathematics Subject Test of the Graduate Record Examinations.

Students are normally admitted at the beginning of the autumn term. It is desirable that applications for teaching fellowships, as well as for other financial assistance, be submitted as early as possible (after December 15). All applications received before March 15 are automatically considered for financial assistance. All program information and application materials can be accessed from the Department's Web site.

Correspondence and Information

Graduate Admissions
Department of Mathematics
University of Utah
155 South 1400 East, JWB 233
Salt Lake City, Utah 84112-0090

Phone: 801-581-6851
Fax: 801-581-4148
E-mail: zhu@math.utah.edu
Web site: http://www.math.utah.edu/grad

University of Utah

THE FACULTY AND THEIR RESEARCH

Distinguished Professors
J. P. Keener, Ph.D., Caltech, 1972. Applied mathematics.
G. W. Milton, Ph.D., Cornell, 1985. Materials science.

Professors
F. R. Adler, Ph.D., Cornell, 1991. Mathematical biology.
P. W. Alfeld, Ph.D., Dundee (Scotland), 1977. Numerical analysis.
A. Bertram, Ph.D., UCLA, 1989. Algebraic geometry and physics.
M. Bestvina, Ph.D., Tennessee, 1984. Topology.
P. C. Bressloff, Ph.D., King's College (London), 1988. Mathematical biology.
R. M. Brooks, Ph.D., LSU, 1963. Topological algebras.
A. V. Cherkaev, Ph.D., St. Petersburg Technical (Russia), 1979. Applied math.
E. Cherkaev, Ph.D., Leningrad, 1988. Partial differential equations.
D. C. Dobson, Ph.D., Rice, 1990. Partial differential equations.
S. N. Ethier, Ph.D., Wisconsin–Madison, 1975. Probability and statistics.
A. L. Fogelson, Ph.D., NYU, 1982. Computational fluids.
E. S. Folias, Ph.D., Caltech, 1963. Applied mathematics, elasticity.
K. M. Golden, Ph.D., NYU, 1984. Applied math.
G. B. Gustafson, Ph.D., Arizona State, 1968. Ordinary differential equations.
H. Hecht, Ph.D., Columbia, 1974. Lie groups.
L. Horvath, Ph.D., Szeged (Hungary), 1982. Probability, statistics.
D. Khoshnevisan, Ph.D., Berkeley, 1989. Probability.
N. J. Korevaar, Ph.D., Stanford, 1981. Partial differential equations.
D. Milicic, Ph.D., Zagreb (Yugoslavia), 1973. Lie groups.
P. C. Roberts, Ph.D., McGill, 1974. Commutative algebra, algebraic geometry.
G. Savin, Ph.D., Harvard, 1988. Group representation.
K. Schmitt, Ph.D., Nebraska, 1967. Differential equations.
N. Smale, Ph.D., Berkeley, 1987. Differential geometry.
J. L. Taylor, Ph.D., LSU, 1964. Abstract analysis.
D. Toledo, Ph.D., Cornell, 1972. Algebraic and differential geometry.
A. E. Treibergs, Ph.D., Stanford, 1980. Differential geometry.
P. C. Trombi, Ph.D., Illinois at Urbana-Champaign, 1970. Lie groups.
D. H. Tucker, Ph.D., Texas, 1958. Differential equations, functional analysis.

Emeritus Professors
J. A. Carlson, Ph.D., Princeton, 1971. Algebraic geometry.
W. J. Coles, Ph.D., Duke, 1954. Ordinary differential equations.
P. Fife, Ph.D., NYU, 1959. Applied mathematics.
S. M. Gersten, Ph.D., Cambridge, 1965. Algebra.
L. C. Glaser, Ph.D., Wisconsin–Madison, 1964. Geometric topology.
F. I. Gross, Ph.D., Caltech, 1964. Algebra.
J. D. Mason, Ph.D., California, Riverside, 1968. Probability.
A. D. Roberts, Ph.D., McGill, 1972. Analysis.
H. Rossi, Ph.D., MIT, 1960. Complex analysis.
J. E. Wolfe, Ph.D., Harvard, 1948. Geometric integration theory.

Associate Professors
A. Balk, Ph.D., Moscow Institute of Physics, 1988. Nonlinear phenomena.
K. Bromberg, Berkeley, 1998. Topology.
C. Hacon, Ph.D., UCLA, 1998. Algebraic geometry.
C. Khare, Ph.D., Caltech, 1995. Algebraic number theory.
W. Niziol, Ph.D., Chicago, 1991. Arithmetical algebraic geometry.
A. Singh, Ph.D., Michigan, 1998. Commutative algebra.
P. Trapa, Ph.D., MIT, 1998. Representation theory.
J. Zhu, Ph.D., NYU (Courant), 1989. Computational fluid dynamics.

Assistant Professors
A. Borisyuk, Ph.D., NYU (Courant), 2002. Mathematical biology.
T. de Fernex, Ph.D., Illinois at Chicago. 2002. Algebraic geometry.
Y. P. Lee, Ph.D., Berkeley, 1999. Algebraic geometry.
F. Rassoul-Agha, Ph.D., NYU (Courant), 2003. Probability theory.
J. W. Tanner, Ph.D., UCLA, 2002. Numerical analysis.

Research Professors
N. Beebe, Ph.D., Florida, 1972. Numerical analysis.
R. Horn, Ph.D., Stanford, 1967. Matrix analysis.

Assistant Professors (Lecturers)
A. Aue, Ph.D., Cologne (Germany), 2004. Probability and statistics.
A. Bayer, Ph.D., Bonn, 2006. Algebraic geometry.
J. Behrstock, Ph.D., SUNY, 2004. Geometric group theory.
S. K. Chaudhary, Ph.D., UCLA, 2004. Applied mathematics.
H. Dao, Ph.D., Michigan, 2006. Commutative rings and algebras.
D. George, Ph.D., Washington (Seattle), 2006. Numerical analysis.
F. Guevara Vasquez, Ph.D., Rice, 2006. Partial differential equations.
B. H. Im, Ph.D., Indiana, 2004. Number theory.
M. J. Kim, Ph.D., Brown, 2004. Mathematical biology.
K. Klosin, Ph.D., Michigan, 2006. Number theory.
D. Margalit, Ph.D., Chicago, 2003. Topology.
L. Miller, Ph.D., NYU (Courant), 2004. Mathematical biology.
K. Montgomery, Ph.D., Northwestern, 2004. Mathematical biology.
F. van Heerden, Ph.D., Utah State, 2003. Partial differential equations.
M. van Opstall, Ph.D., Washington (Seattle), 2004. Algebraic geometry.
O. Veliche, Ph.D., Purdue, 2004. Commutative algebra.

SELECTED PUBLICATIONS

Adler, F. R., T. Liou, and D. Huang. Use of lung transplantation survival models to refine patient selection in cystic fibrosis. *Am. J. Respir. Crit. Care Med.*, in press.

Adler, F. R., and H. C. Muller-Landau. When do localized natural enemies increase species richness? *Ecol. Lett.* 8:438–47, 2005.

Alfeld, P. W., and L. L. Schumaker. A C^2 trivariate macro-element based on the Clough-Tocher split of a tetrahedron. *CAGD J.* 22:710–21, 2005.

Alfeld, P. W., and L. L. Schumaker. A C^2 trivariate macro-element based on the Worsey-Farin split of a tetrahedron. *SIAM J. Numer. Anal.* 43(4):175–6, 2005.

Bertram, A., and H. Kley. New recursions for genus-zero Gromov-Witten invariants. *Topology* 44:1–24, 2005.

Bertram, A., I. Ciocan-Fontanine, and B. Kim. Two proofs of a conjecture of Hori and Vafa. *Duke Math. J.* 126(1):101–36, 2005.

Bestvina, M., and M. Handel. Train-tracks and automorphisms of free groups. *Ann. Math.* 135:1–51, 2002.

Bestvina, M., and M. Feighn. A combination theorem for negatively curved groups. *J. Diff. Geometry* 35(1):85–101, 1992.

Borisyuk, A. Physiology and mathematical modeling of the auditory system. In *Tutorials in Mathematical Biosciences I. Mathematical Neurosciences.* Berlin/Heidelberg/New York: Springer, 2005.

Borisyuk, A., and B. H. Smith. Odor interactions and learning in a model of the insect antennal lobe. *Neurocomputing* 58–60:1041–7, 2004.

Bressloff, P. C. Euclidean shift—twist symmetry in population models of self-aligning objects. *SIAM J. Appl. Math.* 64:1668–90, 2004.

Bressloff, P. C., and **S. E. Folias.** Breathing pulses in an excitatory neural network. *SIAM J. Dyn. Syst.* 3:378–407, 2004.

Bromberg, K. On the density of geometrically finite Kleinian groups. *Acta Math.* 192(1):33–93, 2004.

Bromberg, K., and J. F. Brock. Kleinian groups and hyperbolic 3-manifolds. *London Math. Soc. Lect. Note Ser.* 299:75–93, 2003.

Brooks, R. M. Analytic structure in the spectra of certain uF-algebras. *Math. Ann.* 240:27–33, 1979.

Brooks, R. M. On the spectrum of an inverse limit of holomorphic function algebras. *Adv. Math.* 19:238–44, 1976.

Carlson, J. A., **D. Toledo**, and D. Allcock. The complex hyperbolic geometry of the moduli space of cubic surfaces. *J. Algebraic Geometry*, in press (Math. AG/0007048).

Carlson, J. A., D. Allcock, and **D. Toledo.** Complex hyperbolic structures for moduli of cubic surfaces. *C. R. Acad. Sci. Paris Ser. I* 326:49–54, 1998.

Cherkaev, A. V., and I. Kucuk. Detecting stress fields in an optimal structure. *Int. J. Struct. Multidisciplinary Optimization* 26(1):1–27, 2004.

Cherkaev, A. V. *Variational Methods for Structural Optimization.* New York: Springer-Verlag, 2000.

Coles, W. J., D. Hughell, and W. D. Smith. An optimal foraging model for the red-cockaded woodpecker. In *Proceedings, Seventh Symposium on Systems Analysis in Forest Resources, USDA Technical Report,* NC-205, 2000.

Coles, W. J., and M. K. Kinyon. Some oscillation results for second order matrix differential equations. *Rocky Mountain J. Math.* 1:19–36, 1994.

de Fernex, T. On planar Cremona maps of prime order. *Nagoya Math. J.* 174:1–28, 2004.

de Fernex, T., L. Ein, and M. Mustata. Multiplicities and log canonical threshold. *J. Algebraic Geometry* 13:603–15, 2004.

Dobson, D. C., and F. Santosa. Optimal localization of eigenfunctions in an inhomogeneous medium. *SIAM J. Appl. Math.* 64:762–74, 2004.

Dobson, D. C., and S. J. Cox. Maximizing band gaps in two-dimensional photonic crystals. *SIAM J. Appl. Math.* 59:2108–20, 1999.

Ethier, S. N., and D. A. Levin. On the fundamental theorem of card counting with application to the game of trente et quarante. *Adv. Appl. Probability,* in press.

Ethier, S. N., and S. Wang. A generalized likelihood ratio test to identify differentially expressed genes from microarray data. *Bioinformatics* 20:100–4, 2004.

Fife, P. C., and C. Carrillo. Spatial effects in discrete generation population models. *J. Math. Biol.* 50:161–88, 2005.

Fife, P. C., T. Wei, J. Klewicki, and P. McMurtry. Properties of the mean momentum balance in turbulent boundary layer, pipe and channel flows. *J. Fluid Mech.* 522:303–27, 2005.

Fogelson, A. L., and R. Guy. Platelet-wall interactions in continuum models of platelet thrombosis: Formulation and numerical solution. *Math. Med. Biol.* 21:293–334, 2004.

Fogelson, A. L., and A. Kuharsky. Surface-mediated control of blood coagulation: The role of binding site densities and platelet deposition. *Biophys. J.* 80:1050–74, 2001.

Folias, E. S., and M. Hohn. Predicting crack initiation in composite material systems due to a thermal expansion mismatch. *Int. J. Fract.* 93(1/4):335–49, 1999.

Folias, E. S., and L. Perry. Fast fracture of a threaded pressurized vessel. *Int. J. Pressure Vessels Piping* 76(10):685–92, 1999.

Gersten, S. M., D. Holt, and T. Riley. Isopemetric inequalities for nilpotent groups. *GAFA* 13:795–814, 2003.

Gersten, S. M., and T. Riley. Filling length in finitely presentable groups. *Geometriae Dedicata* 92(1):41–58, 2002.

Glaser, L. C., and T. B. Rushing (eds.). Geometric topology. In *Proceedings of the Geometric Topology Conference,* vol. 438, Park City, Utah, February 1974, *Lectures in Mathematics,* p. 459. New York: Springer-Verlag, 1975.

Glaser, L. C. On tame Cantor sets in spheres having the same projection in each direction. *Pacific J. Math.* 60:87–102, 1975.

Golden, K. M., S. F. Ackley, and V. I. Lytle. The percolation phase transition in sea ice. *Science* 282:2238–41, 1998.

Gross, F. I. Odd order Hall subgroups of the classical linear groups. *Math. Z.* 220:317–36, 1995.

Gross, F. I. Hall subgroups of order not divisible by 3. *Rocky Mountain J. Math.* 23:569–91, 1993.

Gustafson, G. B., and M. Laitoch. The inverse carrier problem. *Czechoslavak Math. J.* 52(127):439–46, 2002.

Gustafson, G. B. Three papers of C. Delavallée Poussin's on linear boundary value problems. In *Charles-Jean de La Vallée Poussin Collected Works,* vol. 2, *Académie Royale de Belgique,* pp. 315–55, 2001.

Hacon, C., and R. Pardini. On the birational geometry of varieties of maximal Albanese dimension. *J. Reine Angew. Math.* 546:177–99, 2002.

Hacon, C., and H. Clemens. Deformations of flat-line bundles and their metrics. *Am. J. Math.* 124(4):769–815, 2002.

Hecht, H. On Casselman's compatibility theorem for n-homology. In *Proceedings of Cordoba Conference, Reductive Lie Groups.* Boston: Birkhäuser, 1997.

Hecht, H., and **J. L. Taylor.** A comparison theorem for n-homology. *Compositio Mathematica* 86:189–207, 1993.

Horn, R., G. Goodson, and D. Merino. Quasi-real normal matrices and eigenvalue pairings. *Linear Algebra Appl.* 369:279–94, 2003.

Horn, R., and G. G. Piepmeyer. Two applications of primary matrix functions. *Linear Algebra Appl.* 361:99–106, 2003.

Horvath, L., and M. Csorgo. *Limit Theorems in Change-Point Analysis.* New York: John Wiley & Sons, 1997.

Horvath, L., and M. Csorgo. *Weighted Approximations in Probability and Statistics.* New York: John Wiley & Sons, 1993.

Keener, J. P., and E. Cytrynbaum. The effect of spatial scale resistive inhomogeneity on defibrillation of cardiac tissue. *J. Theor. Biol.* 223:233–48, 2003.

Keener, J. P. Model for the onset of fibrillation following a coronary occlasion. *J. Cardiovasc. Electrophysiology* 14:1225–32, 2003.

University of Utah

Khare, C. Limits of residually irreducible p-adic Galois representations. *Proc. Am. Math. Soc.* 131(7):1999–2006, 2003.

Khare, C., and R. Ramakrishna. Finiteness of Selmer groups and deformation rings. *Inventiones Mathematicae* 154(1):179–98, 2003.

Khoshnevisan, D., Y. Xiao, and Y. Zhang. Measuring the range of an additive Lévy process. *Ann. Probability* 31(2):1097–141, 2003.

Khoshnevisan, D. *Multiparameter Processes: An Introduction to Random Fields.* New York: Springer, 2002.

Korevaar, N. J., R. Mazzeo, F. Pacard, and R. M. Schoen. Refined asymptotics for constant scalar curvature metrics with isolated singularities. *Inventiones Mathematicae* 135:233–72, 1999.

Korevaar, N. J., and R. M. Schoen. Harmonic maps to nonlocally compact spaces. *Commun. Anal. Geometry* 5(2):333–87, 1997.

Lee, Y. P., and A. Givental. Quantum K-theory on flag manifolds, finite difference Toda lattices and quantum groups. *Inventiones Mathematicae* 151:193–219, 2003.

Lee, Y. P. Quantum Lefschetz hyperplane theorem. *Inventiones Mathematicae* 145(1):121–49, 2001.

Mason, J. D., and T. Burns. A structural equations approach to combining data sets. Accepted as a paper for *International Congress of Sociologists,* 1994.

Mason, J. D., and Z. J. Jurek. *Operator-Limit Distribution in Probability Theory.* New York: John Wiley & Sons, 1993.

Milicic, D., and P. Pandzic. Equivariant derived categories, Zuckerman functors and localization. In *Progress in Mathematics,* vol. 158, *Geometry and Representation Theory of Real and* p-adic *Lie Groups,* pp. 209–42, eds. J. Tirao, D. Vogan, and J. A. Wolfe. Boston: Birkhäuser, 1997.

Milicic, D., and P. Pandzic. On degeneration of the spectral sequence for the composition of Zuckerman functors. *Glasnik Maternativcki* 32(52):179–99, 1997.

Milton, G. W., and V. Vinogradov. The total creep of viscoelastic composites under hydrostatic or antiplane loading. *J. Mech. Phys. Solids* 53:1248–79, 2005.

Milton, G. W., Y. Grabovsky, and D. S. Sage. Exact relations for effective tensors of composites: Necessary and sufficient conditions. *Commun. Pure Appl. Math.* 53(3):300–53, 2000.

Niziol, W. Crystalline conjecture via K-theory. *Ann. Sci. Ecole Norm. Suppl.* 31:659–81, 1998.

Niziol, W. On the image of p-adic regulators. *Inventiones Mathematicae* 127:375–400, 1997.

Rassoul-Agha, F., and T. Seppalainen. An almost sure invariance principle for additive functionals of Markov chains. arXiv:math.PR/0411603, vol. 2, 2005.

Roberts, P. C., and V. Srinivas. Modules of finite length and finite projective dimension. *Inventiones Mathematicae* 151:1–27, 2003.

Roberts, P. C., and K. Kurano. The positivity of intersection multiplicities and symbolic powers of prime ideals. *Compositio Mathematica* 122:165–82, 2000.

Rossi, H., and C. Patton. Unitary structures on cohomology. *TAMS* 290, 1985.

Rossi, H. LeBrun's nonrealizability theorem in higher dimensions. *Duke Math. J.* 52:457–525, 1985.

Savin, G., and B. Gross. The dual pair $PGL_3 \times G_2$. *Can. Math. Bull.* 40:376–84, 1997.

Savin, G., J. S. Huang, and P. Pandzic. New dual pair correspondences. *Duke Math. J.* 82:447–71, 1996.

Schmitt, K., and D. Hartenstine. On generalized and viscosity solutions of nonlinear elliptic equations. *Adv. Nonlinear Stud.* 4:289–306, 2004.

Schmitt, K., and J. Jacobsen. Radical solutions of quasilinear elliptic equations. In *Handbook on Differential Equations,* vol. 1, pp. 359–435. Amsterdam: Elsevier, 2004.

Singh, A., and U. Walther. On the arithmetic rank of certain Segre products. *Contemp. Math.,* in press.

Singh, A. The F-signature of an affine semigroup ring. *J. Pure Appl. Algebra* 196(2–3):13–321, 2005.

Smale, N. A construction of homologically area minimizing hypersurfaces with higher dimensional singular sets. *Trans. Am. Math. Soc.,* in press.

Smale, N. Singular homologically area minimizing surfaces of codimension one in Riemannian manifolds. *Inventiones Mathematicae* 135:145–83, 1999.

Taylor, J. L. Several complex variables with connections to algebraic geometry and Lie groups. *AMS Graduate Stud. Math.* 46, 2002.

Taylor, J. L., and L. Smithies. An analytic Riemann-Hilbert correspondence. *J. Representation Theor.* 4:466–73, 2000.

Toledo, D. Maps between complex hyperbolic surfaces. *Geometriae Dedicata* 97:115–28, 2003.

Toledo, D., D. Allcock, and **J. A. Carlson.** Real cubic surfaces and real hyperbolic geometry. *C. R. Acad. Sci. Paris Ser. I* 337:185–8, 2003.

Trapa, P., and M. A. Nowak. Nash equilibria for an evolutionary language game. *J. Math. Biol.* 42(2):172–88, 2000.

Trapa, P. Generalized Robinson-Schensted algorithms for real groups. *Int. Math. Res. Not.* 15:803–32, 1999.

Treibergs, A. E., and H. Chan. Nonpositively curved surfaces in R^3. *J. Differential Geometry* 57:389–407, 2001.

Trombi, P. C. Uniform asymptotics for real reductive Lie groups. *Pacific J. Math.* 146:131–99, 1990.

Trombi, P. C. Invariant harmonic analysis on split rank one groups with applications. *Pacific J. Math.* 100:80–102, 1982.

Tucker, D. H., and R. S. Baty. Cesàro-One summability and uniform convergence of solutions of a Sturm-Liouville system. *Vietnam J. Math.,* vol. 32, Special Issue, 2004.

Tucker, D. H., and J. F. Gold. A new vector product. In *Proceedings of the 10th National Conference on Undergraduate Research,* 1996.

Zhu, J., and M. Avellaneda. Modeling the distance-to-default process of a firm. *Risk* 149(12):125–9, 2001.

Zhu, J. A numerical study of chemical front propagation in a Hele-Shaw Flow under buoyancy effects. *Phys. Fluids* 10(4):775–88, 1998.

WASHINGTON STATE UNIVERSITY

College of Sciences
Department of Mathematics

Programs of Study The Department of Mathematics offers graduate programs leading to the M.S., Ph.D., and Ph.D. with teaching emphasis. The M.S. degree program includes an option in applied mathematics, which is tailored to industrial employment, and an option in mathematics teaching. Courses of study are available in all of the principal branches of mathematics with special emphases in the applied areas of operations research, computational mathematics, applied statistics, and mathematical modeling as well as in the more traditional fields of number theory, finite geometry, general topology, algebra, and analysis. The Ph.D. program combines the more traditional orientations usually associated with university teaching and research with options specifically directed toward careers in industry and government. The Ph.D. with teaching emphasis program is designed to prepare exceptionally well-qualified teachers of undergraduate mathematics. The degree program is distinguished from that of the traditional Ph.D. by a greater emphasis on breadth of course work and a critical, historical, or expository thesis.

Research Facilities All mathematics faculty members and graduate students are housed in Neill Hall. These modern and spacious facilities include offices, seminar rooms, classrooms, consulting rooms, student computer laboratories, and computing facilities. An outstanding collection of mathematics books and journals are housed in the nearby Owen Science and Engineering Library. The department operates a high-speed network of UNIX and Windows computers for research and instruction. Offices of the graduate students are equipped with computers connected to the network. The University operates a gigabit backbone with high-speed connection to the Internet.

Financial Aid More than 90 percent of the mathematics graduate students are supported by teaching assistantships; stipends ranged from $13,635 to $14,341 for the 2005–06 academic year. Normal duties are 20 hours per week teaching classes or assisting a faculty member. Summer teaching assignments for an additional stipend are usually available as are a few annual research assistantships for advanced students. Federal and state-supported work-study and loan programs are also available. Three special scholarships are granted each year: the Graduate School ($3000), the Abelson ($3000), and the Hacker ($2000).

Cost of Study Tuition for full-time study (more than 6 credit hours) is $3362 per semester for Washington residents and $8189 for nonresidents. Part-time and summer students pay on a per-credit-hour basis. There are tuition waivers for teaching and research assistants. In addition to tuition, both resident and nonresident students pay mandatory fees of $504 per academic year.

Living and Housing Costs The University maintains a residence center strictly for graduate students as well as a wide variety of single-student and family apartments. Private apartments are readily available at slightly higher rates. An estimate of indirect costs is $13,402 (consisting of $1080 for books and supplies, $8780 for rent/food/utilities, $1434 for transportation expenses, and $2108 for miscellaneous expenses) per academic year.

Student Group Washington State University has an enrollment of approximately 17,500, including about 3,000 graduate students; 33 of the latter are in mathematics. The mathematics graduate students come from many areas of the United States and several other countries, and about a dozen complete an advanced degree each year.

Student Outcomes Recent recipients of advanced degrees have taken positions in academic institutions, in the private sector, and in governmental agencies. The academic appointments include teaching and research at both comprehensive universities and four-year liberal arts colleges. The nonacademic positions include systems analyst, actuary, program manager, senior scientist, research mathematician, reliability analyst, and computer consultant.

Location Pullman, a city of about 25,000, is located in the heart of the Palouse region in southeastern Washington. It is a rich agricultural area that enjoys clean air and a generally dry, "continental" climate. The area offers easy access to outdoor recreational opportunities such as fishing, hiking, camping, sailing, skiing, and white-water rafting in three states—Washington, Idaho, and Oregon. The on-campus activities both at WSU and at the University of Idaho (8 miles away) contribute greatly to the cultural, athletic, and scientific life of the area.

The University and The Department The University was founded in 1890 and was the first land-grant institution to establish a chapter of Phi Beta Kappa. Today, the core of the Pullman campus covers nearly 600 acres, and some 100 major buildings house the faculty members and students associated with the more than fifty academic disciplines. Mathematics is the largest department in the Division of Sciences, with a faculty of 28. Master's degrees were first awarded in 1912, and more than 100 mathematics Ph.D.'s have been hooded since 1960.

Applying Requests for information or applications for admission and financial support should be directed to the Graduate Studies Committee. Completed applications and other necessary credentials should be submitted as early as possible, preferably by February 1 for fall admission. Applicants are advised to take the Graduate Record Examinations General Test and the Subject Test in mathematics. Also required are copies of transcripts of all previous college work and three letters of recommendation. TOEFL scores must be submitted to the Graduate School by all students whose native language is not English.

Correspondence and Information

Graduate Studies Committee
Department of Mathematics
Washington State University
Pullman, Washington 99164-3113
Phone: 509-335-6868
E-mail: gradinfo@math.wsu.edu
Web site: http://www.math.wsu.edu

Washington State University

THE FACULTY AND THEIR RESEARCH

Algebra and Number Theory

J. J. McDonald, Ph.D., Wisconsin–Madison, 1993. Matrix analysis, linear algebra.
M. J. Tsatsomeros, Ph.D., Connecticut, 1990. Linear algebra, matrix analysis, dynamical systems.
W. A. Webb, Ph.D., Penn State, 1968. Number theory, fair division problems, combinatorics, cryptography.

Analysis

S. C. Cooper, Ph.D., Colorado State, 1988. Approximation theory.
D. W. DeTemple, Ph.D., Stanford, 1970. Combinatorics, graph theory, analysis, elementary geometry, mathematics education.

Applied Analysis

A. Y. Khapalov, Ph.D., Russian Academy of Sciences, 1982. Applied partial differential equations, control theory.
A. N. Panchenko, Ph.D., Delaware, 2000. Partial differential equations of continuum mechanics, homogenization, inverse problems.
H. Yin, Ph.D., Washington State, 1988. Applied partial differential equations, electromagnetic fields, mathematical modeling of industrial problems.

Modeling

R. H. Dillon, Ph.D., Utah, 1993. Numerical analysis, modeling biological processes.
R. S. Gomulkiewicz, Ph.D., California, Davis, 1989. Theoretical population biology.
V. S. Manoranjan, Ph.D., Dundee (Scotland), 1982. Mathematical modeling, biomathematics, numerical analysis, nonlinear waves.
E. F. Pate, Ph.D., Rensselaer, 1976. Mathematical modeling, computational biology, theoretical biology.
M. F. Schumaker, Ph.D., Texas at Austin, 1987. Mathematical modeling, biomathematics.
D. J. Wollkind, Ph.D., Rensselaer, 1968. Continuum mechanics, asymptotic methods, stability techniques and mathematical modeling.

Computational Mathematics and Optimization

K. A. Ariyawansa, Ph.D., Toronto, 1983. Mathematical programming and optimization, high-performance computing, operations research, applied statistical inference.
A. Genz, Ph.D., Kent (England), 1976. Numerical analysis, numerical integration, scientific computing.
B. Krishnamoorthy, Ph.D., North Carolina at Chapel Hill, 2004. Integer programming, linear programming, operations research, bioinformatics, computational biology.
R. B. Mifflin, Ph.D., Berkeley, 1971. Operations research, nonsmooth optimization.
D. S. Watkins, Ph.D., Calgary, 1974. Numerical analysis, scientific computing.

Probability and Applied Statistics

M. A. Jacroux, Ph.D., Oregon State, 1976. Experimental design, optimal experimental design, estimation in linear and nonlinear models.
V. K. Jandhyala, Ph.D., Western Ontario, 1986. Statistical inference, stochastic processes.
H. Li, Ph.D., Arizona, 1994. Stochastic orderings, statistical theory of reliability, stochastic convexity, probabilistic modeling.
F. Pascual, Ph.D., Iowa State, 1997. Statistical reliability, optimal experimental design.

Topology and Geometry

M. Hudelson, Ph.D., Washington (Seattle), 1995. Combinatorics, discrete geometry.
M. J. Kallaher, Ph.D., Syracuse, 1967. Algebra, projective geometry, finite geometries.
D. C. Kent, Ph.D., New Mexico, 1963. General topology.

Thesis Titles and Current Positions of Recent Graduates

B. Blitz. *Topics Concerning Regular Maps.* Assistant Professor of Mathematics, University of Alaska Southeast.
R. Drake. *A Dynamically Adaptive Method and Spectrum Enveloping Technique.* Research Scientist, Sandia National Laboratories.
A. J. Felt. *A Computational Evaluation of Interior Point Cutting Plane Algorithms for Stochastic Programs.* Assistant Professor of Mathematics, University of Wisconsin–Stevens Point.
C. Gómez-Wulschner. *Compactness of Inductive Limits.* Assistant Professor of Mathematics, Departamento de Mathematicas, Instituto Technológico Autónomo de México.
P. L. Jiang. *Polynomial Cutting Plane Algorithms for Stochastic Programming and Related Problems.* Research Manager, Professional Services, Delta Dental Plan of Minnesota.
B. E. Peterson. *Integer Polyhedra and the Perfect Box.* Associate Professor of Mathematics Education, Brigham Young University.
L. E. Stephenson. *Weekly Nonlinear Stability Analyses of Turing Pattern Formation in the CIMA/Starch Reaction Diffusion Model System.* Senior Engineer/Scientist, United Defense LP.
M. Tian. *Pattern Formation Analysis of Thin Liquid Films.* Assistant Professor of Mathematics, Wright State University.

SELECTED PUBLICATIONS

Ariyawansa, K. A., and A. J. Felt. On a new collection of stochastic linear programming test problems. *INFORMS J. Computing* 16(3):291–9, 2004.

Ariyawansa, K. A., and W. L. Tabor. A note on line search termination criteria for collinear scaling algorithms. *Computing* 70:25–39, 2003.

Ariyawansa, K. A., W. C. Davidon, and K. D. McKennon. A characterization of convexity-preserving maps from a subset of a vector space into another. *J. London Math. Soc.* 64:179–90, 2001.

Ariyawansa, K. A., W. C. Davidon, and K. D. McKennon. On a characterization of convexity-preserving maps, Davidon's collinear scalings and Karmarkar's projective transformations. *Math. Programming A* 90:153–68, 2001.

Cooper, S. C., and P. Gustafson. The strong Chebyshev and orthogonal Laurent polynomials. *J. Approximation Theory* 92:361–78, 1998.

Cooper, S. C., and P. Gustafson. Extremal properties of strong quadrative weights and maximal mass results for truncated strong moment problems. *JCAM* 80:197–208, 1997.

DeTemple, D. W., and C. T. Long. *Mathematical Reasoning for Elementary School Teachers,* 4th ed. Boston, Mass.: Addison Wesley, 2005.

DeTemple, D. W. Combinatorial proofs via flagpole arrangements. *Coll. J. Math.* 35:129–33, 2004.

DeTemple, D. W., and **M. Hudelson.** Square-banded polygons and affine regularity. *Am. Math. Monthly* 108:100–14, 2001.

Dillon, R., L. Fauci, and C. Omoto. Mathematical modeling of axoneme mechanics and fluid dynamics in ciliary and sperm motility. *Dynamics Continuous Discrete Impulsive Syst. Ser. A* 10:745–57, 2003.

Dillon, R., C. Gadgil, and H. G. Othmer. Short- and long-range effects of sonic hedgehog in limb development. *Proc. Natl. Acad. Sci. U.S.A.* 100(18):10152–7, 2003.

Dillon, R., and L. Fauci. An integrative model of internal axoneme mechanics and external fluid dynamics in ciliary beating. *J. Theor. Biol.* 207:415–30, 2000.

Genz, A. Numerical computation of rectangular bivariate and trivariate normal probabilities. *Stat. Comput.* 14:251–60, 2004.

Genz, A., and R. Cools. An adaptive numerical cubature algorithm for simplices. *ACM Trans. Math. Soft.* 29:297–308, 2003.

Genz, A. Fully symmetric interpolatory rules for multiple integrals over hyper-spherical surfaces. *J. Comp. Appl. Math.* 157:187–95, 2003.

Genz, A., and F. Bretz. Comparison of methods for the computation of multivariate t probabilities. *J. Comp. Graph. Stat.* 11:950–71, 2002.

Holt, R. D., M. Barfeld, and **R. Gomulkiewicz.** Temporal variation can facilitate niche evolution in harsh sink environments. *Am. Naturalist* 164:187–200, 2004.

Gomulkiewicz, R., S. L. Nuismer, and J. N. Thompson. Coevolution in variable mutualisms. *Am. Naturalist* 162:S80–93, 2003.

Kingsolver, J. G., and **R. Gomulkiewicz.** Environmental variation and selection on performance curves. *Integr. Comp. Biol.* 43:470–7, 2003.

Gomulkiewicz, R., et al. Hot spots, cold spots, and the geographic mosaic theory of coevolution. *Am. Naturalist,* 156:156–74, 2000.

Hudelson, M. Recurrences Modulo P. *Fibonacci Q.,* in press.

Hudelson, M. Periodic omnihedral billiards in regular polyhedra and polytopes. *J. Geometry,* in press.

Hudelson, M. A solution to the generalized Cevian problem using forest polynomials. *J. Comb. Theory Ser. A* 88:297–305, 1999.

Jacroux, M. A note on the construction of magic rectangles of higher order. *ARS Comb.,* in press.

Jacroux, M. On the determination and construction of A- and MU-optimal block designs for comparing a set of treatments to a set of standard treatments. *J. Stat. Plann. Inf.,* in press.

Jacroux, M. Some optimal orthogonal and nearly-orthogonal block designs for comparing a set of a set of test treatments to a set of standard treatments. *Sankhya B* 62:276–89, 2000.

Jandhyala, V. K., N. E. Evaggelopoulos, and S. B. Fotopoulos. A comparison of unconditional and conditional solutions to the maximum likelihood estimation of a change-point. *Comput. Stat. Data Anal.,* in press.

Jandhyala, V. K., and J. A. Alsaleh. Parameter changes at unknown times in non-linear regression. *Environmetrics* 10:711–24, 1999.

Jandhyala, V. K., S. B. Fotopoulos, and N. Evaggelopoulos. Change-point methods for Weibull models with applications to detection of trends in extreme temperatures. *Environmetrics* 10:547–64, 1999.

Kallaher, M. Translation planes. In *Handbook of Geometry,* pp. 137–92, ed. F. Buckenhout, 1995 (an invited review chapter).

Hanson, J., and **M. Kallaher.** Finite Bol quasifields are nearfields. *Utilitas Math.* 37:45–64, 1990.

Kent, D., and W. K. Min. Neighborhood spaces. *Int. J. Math. Sci.* 32:387–99, 2002.

Kent, D., and J. Wig. P-regular Cauchy completions. *Int. J. Math. Math. Sci.* 24:275–304, 2000.

Kent, D., and S. A. Wilde. P-topological and p-regular: Dual notions in convergence theory. *Int. J. Math. Math. Sci.* 22:1–12, 1999.

Kent, D., and G. Richardson. Completions of probabilistic convergence spaces. *Math. Jpn.* 48:399–407, 1998.

Khapalov, A. Y. Controllability of the semilinear parabolic equation governed by a multiplicative control in the reaction term: A qualitative approach. *SIAM J. Control Optim.* 41:1886–900, 2003.

Khapalov, A. Y. Global non-negative controllability of the semilinear parabolic equation governed by bilinear control. *ESAIM: COCV* 7:269–83, 2002.

Khapalov, A. Y. Mobile point controls versus locally distributed ones for the controllability of the semilinear parabolic equation. *SIAM J. Control Optim.* 40:231–52, 2001.

Khapalov, A. Y. Observability and stabilization for the vibrating string equipped with bouncing point sensors and actuators. *Math. Methods Appl. Sci.* 24:1055–72, 2001.

Krishnamoorthy, B., and A. Tropsha. Development of a four-body statistical pseudo-potential to discriminate native from non-native protein conformations. *Bioinformatics* 19(12):1540–8, 2003.

Li, H. Stochastic models for dependent life lengths induced by common pure jump shock environments. *J. Appl. Prob.* 37:453–69, 2000.

Xu, S., and **H. Li.** Majorization of weighted trees: A new tool to study correlated stochastic systems. *Math. Oper. Res.* 35:298–323, 2000.

Li, H., and M. Shaked. On the first passage times for Markov processes with monotone convex transition kernels. *Stochastic Proc. Appl.* 58:205–216, 1995.

Song, Y., D. Edwards, and **V. S. Manoranjan.** Fuzzy cell mapping applied to autonomous systems. *ASME J. Comput. Information Sci. Eng.,* in press.

Patton, R. L., **V. S. Manoranjan,** and A. J. Watkinson. Plate formation at the surface of a convecting fluid. *Proc. XIII Int. Congress Rheology, Br. Soc. Rheology,* 167–9, 2000.

Manoranjan, V. S. Qualitative study of differential equations. In *MAA Notes No. 50, Revolutions in Differential Equations—Exploring ODEs with Modern Technology,* pp. 59–65, ed. M. J. Kallaher, 1999.

Zaslavsky, B. G., and **J. J. McDonald.** A characterization of Jordan Canonical Forms which are similar to eventually nonnegative matrices with the properties of nonnegative matrices. *Linear Algebra Appl.,* in press.

McDonald, J. J. The peripheral spectrum of a nonnegative matrix. *Linear Algebra Appl.* 363:217–35, 2003.

Carnochan Naqvi, S., and **J. J. McDonald.** The combinatorial structure of eventually nonnegative matrices. *Electron. J. Linear Algebra* 9:255–69, 2002.

McDonald, J. J., and M. Neumann. The Soules approach to the inverse eigenvalue problem for nonnegative symmetric matrices of order $n \leq 5$. *Contemp. Math.* 259:387–90, 2000.

Mifflin, R., and C. Sagastizabel. A VU-algorithm for convex minimization. *Math. Program, Ser. B.* 104:583–608, 2005.

Washington State University

Mifflin, R., and C. Sagastizabel. VU-smoothness and proximal point results for nonconvex functions. *Optimization Methods Software* 19:463–78, 2004.

Mifflin, R., and C. Sagastizabel. Primal-dual gradient structured functions: Second order results; links to epiderivatives and partially smooth functions. *SIAM J. Optim.* 13:1174–94, 2002.

Mifflin, R., and C. Sagastizabal. Proximal points on the fast track. *J. Convex Anal.* 9:563–79, 2002.

Pascual, F. G., and G. Montepiedra. Accelerated life testing under distribution misspecification: Biased estimation and test planning. *IEEE Trans. Reliability* 55:43–52, 2005.

Pascual, F. G. Theory for optimal test plans for the random fatigue-limit model. *Technometrics* 45:130–41, 2003.

Pascual, F. G., and G. Montepiedra. Model-robust test plans with applications in accelerated life testing. *Technometrics* 45:47–57, 2002.

Pascual, F., and W. Q. Meeker. Estimating fatigue curves with the random fatigue-limit model. *Technometrics* 41:277–90, 1999.

Pate, E., et al. Closing the nucleotide pocket of kinesin-family motors upon binding to microtubules. *Science* 300(5620):798–801, 2003.

Pate, E., et al. Molecular dynamic study of the energetic, mechanistic, and structural implicatons of a closed phosphate tube in ncd. *Biophys. J.* 80(3):1151–68, 2001.

Pate, E., et al. A structural change in the kinesin motor that drives motility. *Nature* 402:778–84, 1999.

Schumaker, M. F., and **D. S. Watkins.** A framework model based on the Smoluchowski equation in two reaction coordinates. *J. Chem. Phys.* 121(13):6134–44, 2004.

Gowen, J. A., et al. **(M. F. Schumaker).** The role of trp side chains in tuning single proton conduction through gramicidin channels. *Biophys. J.* 83(2):880–98, 2002.

Schumaker, M. F. Boundary conditions and trajectories of diffusion processes. *J. Chem. Phys.* 117(6):2469–73, 2002.

Psarrakos, P. J., and **M. J. Tsatsomeros.** A primer of Perron-Frobenius theory for matrix polynomials. *Linear Algebra Appl.* 393:333–52, 2004.

Monov, V., and **M. J. Tsatsomeros.** On reducing and deflating subspaces of matrices. *Electron. J. Linear Algebra* 11:246–57, 2004.

Tsatsomeros, M. J. Matrices with a common nontrivial invariant subspace. *Linear Algebra Appl.* 322:51–9, 2001.

Tsatsomeros, M. J., and L. Li. A recursive test for *P*-matrices. *BIT* 40:404–8, 2000.

Watkins, D. S. Product eigenvalue problems. *SIAM Rev.* 47:3–40, 2005.

Watkins, D. S. On Hamiltonian and symplectic Lanczos processes. *Linear Algebra Appl.* 385:23–45, 2004.

Henry, G., **D. S. Watkins,** and J. J. Dongarra. A parallel implementation of the non-symmetric QR algorithm for distributed memory architectures. *SIAM J. Sci. Comput.* 24:284–311, 2003.

Watkins, D. S. *Fundamentals of Matrix Computations,* 2nd ed. New York: John Wiley and Sons, 2002.

Webb, W., and H. Yokota. Polynomial Pell's equation. *Proc. Am. Math. Soc.* 131(4):993–1006, 2003.

Webb, W., and M. Caragiu. Invariants for linear recurrences. In *Applications of Fibonacci Numbers,* vol. 8, pp. 75–81. Dordrecht: Kluwer, 1999.

Webb, W. An algorithm for super envy-free cake division. *J. Math. Anal. Appl.* 239:175–79, 1999.

Webb, W., and J. M. Robertson. *Cake Cutting Algorithms.* Natick, Mass.: A. K. Peters, 1998.

Tian, E. M., and **D. J. Wollkind.** A nonlinear stability analysis of pattern formation in thin liquid films. *Interfaces Free Boundaries* 5:1–25, 2003.

Wollkind, D. J., and L. E. Stephenson. Chemical Turing patterns: A model system of a paradigm for morphogenesis. In *Mathematical Models of Biological Pattern Formation,* pp. 113–42, eds. P. K. Maini and H. G. Othmer. New York: Springer-Verlag, 2001.

Wollkind, D. J., V. S. Manoranjan, and L. Zhang. Weakly nonlinear stability analyses of prototype reaction-diffusion model equations. *SIAM Rev.* 36:176–214, 1994.

Wollkind, D. J., and L. Zhang. The effect of suspended particles on Rayleigh-Bénard convection II: A nonlinear stability analysis of a thermal disequilibrium model. *Math. Comput. Modelling* 19:43–74, 1994.

Wei, W., and **H. Yin.** Global solvability to a singular nonlinear Maxwell's equation in quasistationary electromagnetic fields. *Commun. Pure Appl. Anal.* 4:431–44, 2005.

Yin, H. Regularity of weak solution to Maxwell's equations and applications to microwave heating. *J. Differential Equations* 200:137–61, 2004.

Yin, H. On a class of parabolic equations with nonlocal boundary conditions. *J. Math. Anal. Appl.* 294:712–28, 2004.

Yin, H. On a nonlinear Maxwell's equation in quasi-stationary fields. *Math. Methods Modeling Appl. Sci.* 14:1521–39, 2004.

WESLEYAN UNIVERSITY

Department of Mathematics

Programs of Study

The department offers a program of courses and research leading to the degrees of Master of Arts and Doctor of Philosophy.

The Ph.D. degree demands breadth of knowledge, intensive specialization in one field, original contribution to that field, and expository skill. First-year courses are designed to provide a strong foundation in algebra, analysis, topology, combinatorics, logic, and computer science. Written preliminary examinations are normally taken after the first year. During the second year, the student continues with a variety of courses, sampling areas of possible concentration. By the start of the third year, the student chooses a specialty and begins research work under the guidance of a thesis adviser. Also required is the ability to read mathematics in at least two of the following languages: French, German, and Russian. The usual time required for completion of all requirements for a Ph.D., including the dissertation, is four to five years.

After passing the preliminary examinations, most Ph.D. candidates teach one course per year, typically a small section (fewer than 20 students) of calculus.

The M.A. degree is designed to ensure basic knowledge and the capacity for sustained scholarly study; requirements are six semester courses at the graduate level and the writing and oral presentation of a thesis. The thesis requires (at least) independent search and study of the literature.

Students are also involved in a variety of departmental activities, including seminars and colloquiums. The small size of the program contributes to an atmosphere of informality and accessibility.

The emphasis at Wesleyan is in pure mathematics and theoretical computer science, and most Wesleyan Ph.D.'s have chosen academic careers.

Research Facilities

The department is housed in the Science Center, where all graduate students and faculty members have offices. Computer facilities are available for both learning and research purposes. The Science Library collection has about 120,000 volumes, with extensive mathematics holdings; there are more than 200 subscriptions to mathematics journals, and approximately 60 new mathematics books arrive each month. The proximity of students and faculty and the daily gatherings at teatime are also key elements of the research environment.

Financial Aid

Each applicant for admission is automatically considered for appointment to an assistantship. For the 2006–07 academic year, the stipend is $15,952, plus a dependency allowance when appropriate, and a twelve-month stipend of $21,269 is usually available for the student who wishes to remain on campus to study during the summer. Costs of tuition and health fees are borne by the University. All students in good standing are given financial support for the duration of their studies.

Cost of Study

The only academic costs to the student are books and other educational materials.

Living and Housing Costs

The University provides some subsidized housing and assists in finding private housing. The academic-year cost of a single student's housing (a private room in a 2- or 4-person house, with common kitchen and living area) is about $4900.

Student Group

The number of graduate students in mathematics ranges from 18 to 24, with an entering class of 5 to 10 each year. There have always been both male and female students, graduates of small colleges and large universities, and U.S. and international students, including, in recent years, students from China, India, Iran, Mexico, and Poland.

All of the department's recent Ph.D. recipients have obtained academic employment. Some of these have subsequently taken positions as industrial mathematicians.

Location

Middletown, Connecticut, is a small city of 40,000 on the Connecticut River, about 19 miles southeast of Hartford and 25 miles northeast of New Haven, midway between New York and Boston. The University provides many cultural and recreational opportunities, supplemented by those in the countryside and in larger cities nearby. Several members of the mathematical community are actively involved in sports, including distance running, golf, handball, hiking, softball, squash, table tennis, volleyball, and cycling.

The University

Founded in 1831, Wesleyan is an independent coeducational institution of liberal arts and sciences, with Ph.D. programs in biology, chemistry, ethnomusicology, mathematics, and physics and master's programs in a number of departments. Current enrollments show about 2,800 undergraduates and 145 graduate students.

Applying

No specific courses are required for admission, but it is expected that the equivalent of an undergraduate major in mathematics will have been completed. The complete application consists of the application form, transcripts of all previous academic work at or beyond the college level, letters of recommendation from 3 college instructors familiar with the applicant's mathematical ability and performance, and GRE scores (if available). Applications should be submitted by February 15 in order to receive adequate consideration, but requests for admission from outstanding candidates are welcome at any time. Preference is given to Ph.D. candidates. A visit to the campus is strongly recommended for its value in determining the suitability of the program for the applicant.

Correspondence and Information

Department of Mathematics and Computer Science
Graduate Education Committee
Wesleyan University
Middletown, Connecticut 06459-0128
Phone: 860-685-2620
E-mail: nferguson@wesleyan.edu
Web site: http://math.wesleyan.edu

Wesleyan University

THE FACULTY AND THEIR RESEARCH

Professors
Karen Collins, Ph.D., MIT. Combinatorics.
W. Wistar Comfort, Ph.D., Washington (Seattle). Point-set topology, ultrafilters, set theory, topological groups.
Adam Fieldsteel, Ph.D., Berkeley. Ergodic theory.
Anthony W. Hager, Ph.D., Penn State. Lattice-ordered algebraic structures, general and categorical topology.
Michael S. Keane, Dr.rer.nat., Erlangen. Ergodic theory, random walks, statistical physics.
Philip H. Scowcroft, Ph.D., Cornell. Foundations of mathematics, model-theoretic algebra.
Carol Wood, Ph.D., Yale. Mathematical logic, applications of model theory to algebra.

Associate Professors
Petra Bonfert-Taylor, Ph.D., Berlin Technical. Complex analysis, complex dynamics, geometric function theory, discrete groups.
Wai Kiu Chan, Ph.D., Ohio State. Arithmetic theory of quadratic forms, arithmetic of algebraic groups, combinatorics.
Mark Hovey, Ph.D., MIT. Algebraic topology and homological algebra.
Edward C. Taylor, Ph.D., SUNY at Stony Brook. Analysis, low-dimensional geometry and topology.

Assistant Professor
David Pollack, Ph.D., Harvard. Number theory, automorphic forms, representation of *p-adic* groups.

Visiting Professor of Mathematics
James D. Reid, Ph.D., Washington (Seattle). Abelian groups, module theory.

Visiting Assistant Professors of Mathematics
Keir Lockridge.
Stuart Zoble.

Van Vleck Visiting Researchers
Daniel G. Davis, Ph.D., Northwestern.
Toshihiro Hamachi, Doctor of Science, Kyushu (Japan).
Keir Lockridge, Ph.D., Washington (Seattle).
Stuart Zoble, Ph.D., Berkeley.

Visiting Instructor of Mathematics
James Frugale, M.A., Wesleyan.

Professors of Computer Science
Danny Krizanc, Ph.D., Harvard. Theoretical computer science.
Michael Rice, Ph.D., Wesleyan. Parallel computing, formal specification methods.

Associate Professor of Computer Science
James Lipton, Ph.D., Cornell. Logic and computation, logic programming, type theory, linear logic.

Assistant Professors of Computer Science
Eric Aaron, Ph.D., Cornell. Intelligent virtual agents, reasoning about navigation, hybrid systems, game artificial intelligence, behavioral animation/robotics, applied logic, cognitive modeling, attention, automated verification, tumor modeling.
Norman Danner, Ph.D., Indiana (Bloomington). Logic, theoretical computer science.

Professors Emeriti
Ethan M. Coven, Ph.D., Yale. Dynamical systems.
F. E. J. Linton, Ph.D., Columbia. Categorical algebra, functorial semantics, topoi.
James D. Reid, Ph.D., Washington (Seattle). Abelian groups, module theory.
Lewis C. Robertson, Ph.D., UCLA. Lie groups, topological groups, representation theory.
Robert A. Rosenbaum, Ph.D., Yale. Geometry, mathematics and science education.

Visiting Scholar
George Maltese, Ph.D., Yale. Functional analysis.

Faculty-student conferences, daily gatherings at teatime, and discussions in graduate students' offices are key ingredients of the research environment in the Department of Mathematics.

WORCESTER POLYTECHNIC INSTITUTE

Department of Mathematical Sciences

Programs of Study

The Mathematical Sciences Department at Worcester Polytechnic Institute (WPI) offers two programs leading to the degree of Master of Science (M.S.), two professional science master's programs, a program leading to the degree of Master of Mathematics for Educators (M.M.E.), and a program leading to the degree of Doctor of Philosophy (Ph.D.).

The Master of Science in applied mathematics is a 30-credit program that gives students a broad background in mathematics, placing an emphasis on numerical methods and scientific computation, mathematical modeling, discrete mathematics, optimization and operations research. Options for the capstone experience include a master's thesis or project in cooperation with one of the Department's established industrial partners.

The Master of Science in applied statistics is a 30-credit program that gives graduates the knowledge and experience to tackle problems of statistical design, analysis, and control that are likely to be encountered in business, industry, or academia. Professional experience is provided by a statistical consulting course. In addition, options for the capstone experience include a master's thesis or project, often done with local industry.

The Master of Science in financial mathematics (initially supported by the Alfred P. Sloan Foundation) is a 30-credit program that offers an efficient, practice-oriented track to prepare students for quantitative careers in the financial industry. The mathematical knowledge is complemented by studies in financial management, information technology, and/or computer science. The bridge from the academic environment to the professional workplace is provided by a professional master's project that involves the solution of a concrete, real-world problem directly originating from the financial industry.

The Master of Science in industrial mathematics (initially supported by the Alfred P. Sloan Foundation) is a 30-credit practice-oriented program that prepares students for successful careers in industry. The program aims at developing the analysis, modeling, and computational skills needed by mathematicians who work in industrial environments. The connection between academic training and industrial experience is provided by an industrial professional master's project that involves the solution of a concrete, real-world problem originating in industry, and by summer internships.

The Master of Mathematics for Educators is a two-year program designed primarily for secondary mathematics teachers. The program provides teachers with an understanding of the fundamental principles of mathematics through courses and project work that model diverse pedagogical methods. All program requirements also incorporate appropriate technologies, as well as relevant results from research in mathematics education.

The Doctor of Philosophy in mathematical sciences produces active and creative problem solvers capable of contributing in academic and industrial environments. One distinguishing feature of this program is a project to be completed under the guidance of an external sponsor, usually from industry or a national research center. The intention of this project is to broaden perspectives on mathematics and its applications and to improve skills in communicating mathematics and formulating and solving mathematical problems.

Research Facilities

The Department of Mathematical Sciences boasts a full-time faculty of 28 members with a research focus in applied and computational mathematics and statistics. The George C. Gordon Library is committed to supporting the research information needs of WPI's graduate community. The collection currently numbers 270,000 bound volumes and includes subscriptions to 1,400 periodicals. In addition, hundreds of databases can be researched with the library's On-Line Search Services. Computing facilities in the Department of Mathematical Sciences include a network of high-performance work stations, two computer labs, and a 16-processor SGI Altix 350.

Financial Aid

Teaching assistantships and research assistantships are available on a competitive basis. Full assistantships provide tuition plus a stipend of approximately $15,000 for the nine-month academic year. U.S. citizens with exceptional qualifications are encouraged to apply for the Robert F. Goddard Fellowships. Other fellowship opportunities are also available. Information may be found online at http://www.grad.wpi.edu/Financial/.

Cost of Study

Graduate tuition for the 2006–07 academic year tuition is $997 per credit hour. There are nominal extra charges for the thesis, health insurance, and other fees.

Living and Housing Costs

On-campus graduate student housing is limited to a space-available basis. There is no on-campus housing for married students. Apartments and rooms in private homes near the campus are available at varying costs. For further information and apartment listings, prospective students should visit the Residential Services Office Web site at http://www.wpi.edu/Admin/RSO/Offcampus/.

Student Group

The current WPI student body of 3,869 includes 1,837 full- and part-time graduate students. As of the fall 2005 semester, 61 graduate students were enrolled in the Department of Mathematical Sciences programs.

Location

The university is located on an 80-acre campus in a residential section of Worcester. The city, the second-largest in New England, has many colleges and an unusual variety of cultural opportunities. Located three blocks from the campus, the nationally famous Worcester Art Museum contains one of the finest permanent collections in the country and offers many special activities of interest to students. The community also provides outstanding programs in music and theater. The DCU Center offers rock concerts and semiprofessional athletic events. Easily reached for recreation are Boston and Cape Cod to the east and the Berkshires to the west, and good skiing is nearby to the north. Complete athletic and recreational facilities and a program of concerts and special events are available on campus to graduate students.

The Institute

Worcester Polytechnic Institute, founded in 1865, is the third-oldest independent university of engineering and science in the United States. Graduate study has been a part of the Institute's activity for more than 100 years. Classes are small and provide for close student-faculty relationships. Graduate students frequently interact in research with undergraduates participating in WPI's innovative project-based program of education.

Applying

Applicants must submit WPI application forms, official college transcript(s), three letters of recommendation, and a $70 application fee (waived for WPI alumni). Submission of GRE scores is recommended. International students whose primary language is not English must also submit proof of English language proficiency. WPI accepts either the Test of English as a Foreign Language (TOEFL) or the International English Language Testing System (IELTS). A paper-based TOEFL score of at least 550 (213 on the computer-based test or 79–80 on the Internet-based test) or an IELTS overall band score of 6.5 (no band score below 6.0) is required for admission. Applications for admission are accepted at any time. However, in order to receive full consideration for financial support, the application file should be completed by January 15. Application forms for admission and financial support, as well as additional information about the program, can be obtained from the Department of Mathematical Sciences or by contacting the Office of Graduate Studies and Enrollment at grad_studies@wpi.edu.

Correspondence and Information

For program information and application forms, interested students should contact:

Graduate Committee
Department of Mathematical Sciences
Worcester Polytechnic Institute
100 Institute Road
Worcester, Massachusetts 01609-2280

Phone: 508-831-5241
Fax: 508-831-5824
E-mail: ma-grad-p@wpi.edu
Web site: http://www.wpi.edu/Academics/Depts/Math/Grad/

Worcester Polytechnic Institute

THE FACULTY AND THEIR RESEARCH

Jon Abraham, Actuarial Mathematics Coordinator; Fellow, Society of Actuaries, 1991.

Dennis D. Berkey, Professor and President; Ph.D., Cincinnati, 1974. Applied mathematics, differential equations, optimal control.

Marcel Y. Blais, Visiting Assistant Professor; Ph.D., Cornell, 2005. Mathematical finance.

Corinne Grace B. Burgos, Visiting Assistant Professor; Ph.D., Philippines–Los Baños, 2002. Bayesian inference, analysis of survey data, poverty estimation.

Peter R. Christopher, Professor; Ph.D., Clark, 1982. Graph theory, group theory, algebraic graph theory, combinatorics, linear algebra, discrete mathematics.

Paul W. Davis, Professor; Ph.D., Rensselaer, 1970. Unit commitment, optimal power flow, economic dispatch, state estimation, and other control and measurement problems for electric power networks.

William W. Farr, Associate Professor; Ph.D., Minnesota, 1986. Ordinary and partial differential equations, dynamical systems, local bifurcation theory with symmetry and its application to problems involving chemical reactions and/or fluid mechanics.

Joseph D. Fehribach, Associate Professor; Ph.D., Duke, 1985. Partial differential equations and scientific computing, free and moving boundary problems (crystal growth), nonequilibrium thermodynamics and averaging (molten carbonate fuel cells).

John Goulet, Coordinator, Master of Mathematics for Educators Program; Ph.D., Rensselaer, 1976. Applications of linear algebra, educational and industrial assessment, development of educational software.

Arthur C. Heinricher Jr., Professor; Ph.D., Carnegie Mellon, 1986. Applied probability, stochastic processes and optimal control theory.

Mayer Humi, Professor; Ph.D., Weizmann (Israel), 1969. Mathematical physics, applied mathematics and modeling, lie groups, differential equations, numerical analysis, turbulence and chaos, continuum mechanics, control theory, artificial intelligence.

Ryung S. Kim, Assistant Professor; Ph.D., Harvard, 2005. Biostatistics, statistical methodologies for genomic data.

Christopher J. Larsen, Associate Professor; Ph.D., Carnegie Mellon, 1996. Calculus of variations, partial differential equations, and geometric measure theory, with focus on free discontinuity problems modeling fracture mechanics, image segmentation (computer vision), and optimal design.

Roger Lui, Professor; Ph.D., Minnesota, 1981. Nonlinear partial differential equations, mathematical biology, nonlinear analysis.

Konstantin A. Lurie, Professor; Ph.D., A. F. Ioffe Physical-Technical Institute (Russia), 1964; D.Sc., Russian Academy of Sciences, 1972. Control theory for the distributed parameter systems, optimization and nonconvex variational calculus, optimal design.

William Martin, Associate Professor; Ph.D., Waterloo, 1992. Applications of algebra and combinatorics to problems in computer science and mathematics designs and codes in association schemes, error-correcting codes, cryptography, and combinatorial designs.

Umberto Mosco, H. J. Gay Professor; Ph.D., Rome, 1967. Partial differential equations, convex analysis, optimal control, variational calculus, fractals.

Balgobin Nandram, Associate Professor; Ph.D., Iowa, 1989. Applied Bayesian statistics, small-area estimation and computational methods, categorical data analysis, predictive and restrictive inference.

Joseph D. Petruccelli, Professor; Ph.D., Purdue, 1978. Time series, optimal stopping, statistics, statistics education, biomedical applications of statistics.

Marcus Sarkis, Assistant Professor; Ph.D., NYU (Courant), 1994. Domain decomposition methods, numerical analysis, parallel computing, computational fluid dynamics, preconditioned iterative methods for linear and nonlinear problems, numerical partial differential equations, mixed and nonconforming finite methods, overlapping nonmatching grids, mortar finite elements, eigenvalue solvers, aeroelasticity, porous media reservoir.

Brigitte Servatius, Associate Professor; Ph.D., Syracuse, 1987. Combinatorics, rigidity of structures, geometric foundations of computer-aided design, symmetry and duality, the history and philosophy of mathematics.

Dalin Tang, Professor; Ph.D., Wisconsin–Madison, 1988. Biomechanics, blood flow, applied fluid mechanics, nonlinear analysis, numerical methods, biological fluid dynamics, transport theory.

Domokos Vermes, Associate Professor; Ph.D., Szeged (Hungary), 1975; Habilitation in Mathematics, Hungarian Academy of Sciences. Optimal stochastic control theory, nonsmooth analysis, stochastic processes with discontinuous dynamics, optimal scheduling under uncertainty, adaptive control in medical decision making.

Bogdan Vernescu, Professor and Department Head; Ph.D., Institute of Mathematics, Bucharest, 1989. Partial differential equations, phase transitions and free boundaries, viscous flow in porous media and homogenization.

Darko Volkov, Assistant Professor; Ph.D., Rutgers, 2001. Electromagnetic waves, inverse problems, wave propagation in waveguides and in periodic structures, electrified fluid jets.

Homer F. Walker, Professor and Department Head; Ph.D., NYU (Courant), 1970. Numerical analysis, especially numerical solution of large-scale linear and nonlinear systems, unconstrained optimization, and applications to differential equations and statistical estimation; applied mathematics.

Suzanne Weekes, Associate Professor; Ph.D., Michigan, 1995. Numerical analysis, computational fluid dynamics, porous media flow, hyperbolic conservation laws, shock capturing schemes.

Jayson Wilbur, Assistant Professor; Ph.D., Purdue, 2002. Applied statistics, resampling methods, multivariate statistical analysis, model selection, Bayesian inference, statistical issues in molecular biology and ecology.

Jason S. Williford, Visiting Assistant Professor; Ph.D., Delaware, 2005.

Vadim Yakovlev, Research Associate Professor; Ph.D. Institute of Radio Engineering and Electronics, Moscow, 1991. Electromagnetic fields in transmission lines and near interfaces; atmospheric wave propagation; microwave thermoprocessing; coupled electromagnetic/thermal boundary problems; control and optimization of electric and temperature fields; numerical methods, CAD tools, and computation.

ACADEMIC AND PROFESSIONAL PROGRAMS IN THE AGRICULTURAL SCIENCES

Section 8
Agricultural and Food Sciences

This section contains a directory of institutions offering graduate work in agricultural and food sciences, followed by in-depth entries submitted by institutions that chose to prepare detailed program descriptions. Additional information about programs listed in the directory but not augmented by an in-depth entry may be obtained by writing directly to the dean of a graduate school or chair of a department at the address given in the directory.

For programs offering related work, see also in this book Natural Resources. In Book 2, see Architecture (Landscape Architecture) and Economics (Agricultural Economics and Agribusiness); in Book 3, see Biological and Biomedical Sciences; Botany and Plant Biology; Ecology, Environmental Biology, and Evolutionary Biology; Entomology; Genetics, Developmental Biology, and Reproductive Biology; Nutrition; Pathology and Pathobiology; Physiology; and Zoology; in Book 5, see Agricultural Engineering and Bioengineering and Biomedical Engineering and Biotechnology; and in Book 6, see Education (Agricultural Education) and Veterinary Medicine and Sciences.

CONTENTS

Program Directories

Announcement

Close-Ups

See also:

Agricultural Sciences—General

Alabama Agricultural and Mechanical University, School of Graduate Studies, School of Agricultural and Environmental Sciences, Huntsville, AL 35811. Offers MS, MURP, PhD. Part-time and evening/weekend programs available. Terminal master's awarded for partial completion of doctoral program. *Degree requirements:* For doctorate, one foreign language, thesis/dissertation. *Entrance requirements:* For master's, GRE General Test; for doctorate, GRE General Test, MS. Electronic applications accepted. *Faculty research:* Remote sensing, environmental pollutants, food biotechnology, plant growth.

Alcorn State University, School of Graduate Studies, School of Agriculture and Applied Science, Alcorn State, MS 39096-7500. Offers agricultural economics (MS Ag); agronomy (MS Ag); animal science (MS Ag). *Faculty:* 11 full-time (2 women). *Students:* 7 full-time (2 women), 8 part-time (4 women); includes 12 minority (all African Americans), 3 international. In 2005, 3 degrees awarded. *Degree requirements:* For master's, thesis optional. *Application deadline:* For fall admission, 7/15 for domestic students; for spring admission, 11/25 for domestic students. Applications are processed on a rolling basis. Application fee: $0 ($10 for international students). *Financial support:* Career-related internships or fieldwork available. Support available to part-time students. *Faculty research:* Aquatic systems, dairy herd improvement, fruit production, alternative farming practices. *Unit head:* Robert Arthur, Interim Dean, 601-877-6137, Fax: 601-877-6219.

Angelo State University, College of Graduate Studies, College of Sciences, Department of Agriculture, San Angelo, TX 76909. Offers animal science (MS). Part-time and evening/weekend programs available. *Faculty:* 7 full-time (2 women), 2 part-time/adjunct (0 women). *Students:* 17 full-time (8 women), 5 part-time; includes 4 minority (1 African American, 1 Asian American or Pacific Islander, 2 Hispanic Americans). Average age 27. 10 applicants, 100% accepted, 9 enrolled. In 2005, 12 degrees awarded. *Degree requirements:* For master's, thesis optional. *Entrance requirements:* For master's, GRE General Test. Additional exam requirements/recommendations for international students: Required—TOEFL or IELT. *Application deadline:* For fall admission, 7/15 priority date for domestic students, 6/15 priority date for international students; for spring admission, 12/8 for domestic students, 11/1 for international students. Applications are processed on a rolling basis. Application fee: $25 ($50 for international students). Electronic applications accepted. *Expenses:* Tuition, area resident: Full-time $2,268; part-time $126 per credit. Tuition, nonresident: full-time $7,236; part-time $402 per credit. Required fees: $844; $94 per credit. One-time fee: $25. Tuition and fees vary according to course load. *Financial support:* In 2005–06, 13 students received support, including 9 research assistantships (averaging $9,887 per year); Federal Work-Study, scholarships/grants, and unspecified assistantships also available. Support available to part-time students. Financial award application deadline:3/1. *Faculty research:* Effect of protein and energy on feedlot performance, bitterweed toxicosis in sheep, meat laboratory, North Concho watershed project, baseline vegetation. *Unit head:* Dr. Gilbert R. Engdahl, Head, 325-942-2027 Ext. 227, E-mail: gil.engdahl@angelo.edu. *Application contact:* Dr. Cody B. Scott, Graduate Advisor, 325-942-2027 Ext. 284, E-mail: cody.scott@angelo.edu.

Arkansas State University, Graduate School, College of Agriculture, Jonesboro, State University, AR 72467. Offers agricultural education (MSA, SCCT); agriculture (MSA); vocational-technical administration (MS, SCCT). Part-time programs available. *Faculty:* 9 full-time (1 woman). *Students:* 4 full-time (3 women), 29 part-time (13 women); includes 1 minority (African American), 1 international. Average age 34. 7 applicants, 100% accepted, 6 enrolled. In 2005, 13 degrees awarded. *Degree requirements:* For master's, thesis or alternative, comprehensive exam. *Entrance requirements:* For master's, GRE General Test or MAT, appropriate bachelor's degree; for SCCT, GRE General Test or MAT, interview, master's degree. Additional exam requirements/recommendations for international students: Required—TOEFL (minimum score 213 computer-based). *Application deadline:* For fall admission, 7/1 for domestic students; for spring admission, 11/15 priority date for domestic students. Applications are processed on a rolling basis. Application fee: $15 ($25 for international students). Electronic applications accepted. *Expenses:* Tuition, state resident: full-time $3,232; part-time $180 per hour. Tuition, nonresident: full-time $8,164; part-time $454 per hour. Required fees: $716; $37 per hour. $25 per semester. Tuition and fees vary according to course load and program. *Financial support:* Teaching assistantships, scholarships/grants and unspecified assistantships available. Financial award application deadline: 7/1; financial award applicants required to submit FAFSA. *Unit head:* Dr. Gregory Phillips, Dean, 870-972-2085, Fax: 870-972-3885, E-mail: gphillips@astate.edu.

Auburn University, Graduate School, College of Agriculture, Auburn University, AL 36849. Offers M Ag, M Aq, MS, PhD. Part-time programs available. *Faculty:* 135 full-time (21 women). *Students:* 129 full-time (57 women), 78 part-time (30 women); includes 14 minority (7 African Americans, 1 American Indian/Alaska Native, 2 Asian Americans or Pacific Islanders, 4 Hispanic Americans), 69 international. 146 applicants, 65% accepted, 50 enrolled. In 2005, 43 master's, 21 doctorates awarded. *Degree requirements:* For doctorate, thesis/dissertation. *Entrance requirements:* For master's and doctorate, GRE General Test. *Application deadline:* For fall admission, 7/7 for domestic students; for spring admission, 11/24 for domestic students. Applications are processed on a rolling basis. Application fee: $25 ($50 for international students). Electronic applications accepted. *Financial support:* Fellowships, research assistantships, teaching assistantships, Federal Work-Study available. Support available to part-time students. Financial award application deadline: 3/15. *Unit head:* Dr. Richard Guthrie, Dean, 334-844-2345. *Application contact:* Dr. Stephen L. McFarland, Acting Dean of the Graduate School, 334-844-4700.

Brigham Young University, Graduate Studies, College of Biological and Agricultural Sciences, Provo, UT 84602-1001. Offers MS, PhD. Part-time programs available. *Faculty:* 96 full-time (11 women), 1 (woman) part-time/adjunct. *Students:* 88 full-time (49 women), 27 part-time (8 women); includes 8 minority (1 African American, 1 American Indian/Alaska Native, 3 Asian Americans or Pacific Islanders, 3 Hispanic Americans), 5 international. Average age 25. 92 applicants, 50% accepted, 37 enrolled. In 2005, 35 master's, 3 doctorates awarded. Terminal master's awarded for partial completion of doctoral program. *Degree requirements:* For master's and doctorate, thesis/dissertation, comprehensive exam, registration. *Entrance requirements:* For master's and doctorate, GRE General Test. Additional exam requirements/recommendations for international students: Required—TOEFL (minimum score 550 paper-based; 213 computer-based). *Application deadline:* For fall admission, 1/31 for domestic students, 1/31 for international students. Application fee: $50. Electronic applications accepted. *Financial support:* In 2005–06, 115 students received support, including 1 fellowship with full and partial tuition reimbursement available (averaging $5,500 per year), 52 research assistantships with full and partial tuition reimbursements available (averaging $14,773 per year), 62 teaching assistantships with full and partial tuition reimbursements available (averaging $14,653 per year); career-related internships or fieldwork, institutionally sponsored loans, scholarships/grants, and tuition awards also available. Support available to part-time students. Total annual research expenditures: $4.4 million. *Unit head:* Dr. Rodney J. Brown, Dean, 801-422-3963, Fax: 801-422-0050. *Application contact:* Susan Pratley, 801-422-3963.

California Polytechnic State University, San Luis Obispo, College of Agriculture, San Luis Obispo, CA 93407. Offers MS. Part-time programs available. *Faculty:* 32 full-time (6 women), 15 part-time/adjunct (1 woman). *Students:* 94 full-time (58 women), 26 part-time (15 women), 4 international. 93 applicants, 57% accepted, 38 enrolled. In 2005, 29 degrees awarded. *Degree requirements:* For master's, thesis, comprehensive exam. *Entrance requirements:* For master's, minimum GPA of 2.75 in last 90 quarter units of course work. Additional exam requirements/recommendations for international students: Required—TOEFL, TWE. *Application deadline:* For fall admission, 6/1 for domestic students, 11/30 for international students. For winter admission, 8/1 for domestic students; for spring admission, 12/1 for domestic students. Applications are processed on a rolling basis. Application fee: $55. Electronic applications accepted. *Expenses:* Tuition, nonresident: part-time $226 per unit. Required fees: $1,063 per unit. *Financial support:* In 2005–06, 40 students received support; fellowships, research assistantships, teaching assistantships, career-related internships or fieldwork, Federal Work-Study, institutionally sponsored loans, and scholarships/grants available. Support available to part-time students. Financial award application deadline: 3/2; financial award applicants required to submit FAFSA. *Faculty research:* Soils, food processing, forestry, dairy products development, irrigation. *Unit head:* Dr. David J. Wehner, Dean, 805-756-5702, Fax: 805-756-6577, E-mail: dwehner@calpoly.edu. *Application contact:* Dr. Mark Shelton, Associate Dean/Graduate Coordinator, 805-756-2161, Fax: 805-756-6577, E-mail: mshelton@calpoly.edu.

California State Polytechnic University, Pomona, Academic Affairs, College of Agriculture, Pomona, CA 91768-2557. Offers agricultural science (MS); animal science (MS); foods and nutrition (MS). Part-time programs available. *Faculty:* 43 full-time (13 women), 16 part-time/adjunct (9 women). *Students:* 20 full-time (16 women), 39 part-time (29 women); includes 14 minority (1 African American, 4 Asian Americans or Pacific Islanders, 9 Hispanic Americans), 4 international. Average age 30. 58 applicants, 76% accepted, 20 enrolled. In 2005, 13 degrees awarded. *Degree requirements:* For master's, thesis or alternative. *Application deadline:* For fall admission, 5/1 for domestic students. For winter admission, 10/15 for domestic students; for spring admission, 1/2 for domestic students. Applications are processed on a rolling basis. Application fee: $55. Electronic applications accepted. *Expenses:* Tuition, nonresident: full-time $9,021. Required fees: $3,597. *Financial support:* Career-related internships or fieldwork, Federal Work-Study, and institutionally sponsored loans available. Support available to part-time students. Financial award application deadline: 3/2; financial award applicants required to submit FAFSA. *Faculty research:* Equine nutrition, physiology, and reproduction; leadership development; bioartificial pancreas; plant science; ruminant and human nutrition. *Unit head:* Dr. Wayne R. Bidlack, Dean, 909-869-2200, E-mail: wrbidlack@csupomona.edu.

California State University, Fresno, Division of Graduate Studies, College of Agricultural Sciences and Technology, Fresno, CA 93740-8027. Offers MA, MS. Part-time and evening/weekend programs available. *Degree requirements:* For master's, thesis (for some programs), comprehensive exam (for some programs). *Entrance requirements:* For master's, GRE General Test. Additional exam requirements/recommendations for international students: Required—TOEFL. Electronic applications accepted.

Clemson University, Graduate School, College of Agriculture, Forestry and Life Sciences, Clemson, SC 29634. Offers M Ag Ed, MFR, MS, PhD. Part-time programs available. *Faculty:* 204 full-time (32 women), 21 part-time/adjunct (9 women). *Students:* 326 full-time (149 women), 71 part-time (31 women); includes 16 minority (11 African Americans, 1 American Indian/Alaska Native, 4 Hispanic Americans), 112 international. 241 applicants, 40% accepted, 67 enrolled. In 2005, 66 master's, 28 doctorates awarded. Terminal master's awarded for partial completion of doctoral program. *Degree requirements:* For master's, thesis (for some programs); for doctorate, thesis/dissertation. *Entrance requirements:* For master's and doctorate, GRE General Test. Additional exam requirements/recommendations for international students: Required—TOEFL. *Application deadline:* Applications are processed on a rolling basis. Application fee: $50. Electronic applications accepted. *Financial support:* Fellowships, research assistantships, teaching assistantships, career-related internships or fieldwork, Federal Work-Study, institutionally sponsored loans, scholarships/grants, and unspecified assistantships available. Financial award applicants required to submit FAFSA. *Unit head:* Dr. Calvin Schoulties, Interim Dean, 864-656-7592, Fax: 864-656-1286, E-mail: cshlts@clemson.edu.

Colorado State University, Graduate School, College of Agricultural Sciences, Fort Collins, CO 80523-0015. Offers M Agr, MS, PhD. Part-time programs available. Postbaccalaureate distance learning degree programs offered. *Faculty:* 81 full-time (13 women), 5 part-time/adjunct (0 women). *Students:* 126 full-time (67 women), 110 part-time (52 women); includes 11 minority (1 African American, 3 American Indian/Alaska Native, 2 Asian Americans or Pacific Islanders, 5 Hispanic Americans), 25 international. Average age 31. 150 applicants, 59% accepted, 54 enrolled. In 2005, 48 master's, 14 doctorates awarded. *Degree requirements:* For master's, thesis (for some programs); for doctorate, thesis/dissertation, comprehensive exam (for some programs). *Entrance requirements:* For master's and doctorate, GRE General Test, minimum GPA of 3.0. Additional exam requirements/recommendations for international students: Required—TOEFL. *Application deadline:* For fall admission, 4/1 priority date for domestic students, 4/1 priority date for international students; for spring admission, 9/1 priority date for domestic students, 9/1 priority date for international students. Applications are processed on a rolling basis. Application fee: $50. Electronic applications accepted. *Expenses:* Tuition, state resident: full-time $3,690; part-time $205 per credit. Tuition, nonresident: full-time $14,958; part-time $831 per credit. Required fees: $1,061. *Financial support:* Fellowships, research assistantships, teaching assistantships, career-related internships or fieldwork, Federal Work-Study, institutionally sponsored loans, and traineeships available. Support available to part-time students. *Faculty research:* Systems methodology, biotechnology, plant and animal breeding, water management, plant protection. Total annual research expenditures: $10.5 million. *Unit head:* Dr. Marc A. Johnson, Dean, 970-491-6274, Fax: 970-491-4895, E-mail: m.johnson@colostate.edu.

Dalhousie University, Faculty of Graduate Studies, Nova Scotia Agricultural College, Halifax, NS B3H 4R2, Canada. Offers M Sc. Part-time programs available. *Degree requirements:* For master's, thesis, candidacy exam. *Entrance requirements:* For master's, minimum GPA of 3.0. Additional exam requirements/recommendations for international students: Required—TOEFL. *Faculty research:* Biology, soil science, animal science, plant science, environmental science, biotechnology.

Illinois State University, Graduate School, College of Applied Science and Technology, Department of Agriculture, Normal, IL 61790-2200. Offers agribusiness (MS). *Faculty:* 12 full-time (1 woman). *Students:* 7 full-time (4 women), 8 part-time, 6 international. 10 applicants, 40% accepted. In 2005, 4 degrees awarded. *Degree requirements:* For master's, thesis optional. *Entrance requirements:* For master's, GRE General Test, minimum GPA of 3.0 in last 60 hours. *Application deadline:* Applications are processed on a rolling basis. Application fee: $30. *Expenses:* Tuition, state resident: full-time $3,060; part-time $170 per credit hour. Tuition, nonresident: full-time $6,390; part-time $355 per credit hour. Required fees: $1,411; $47 per credit hour. *Financial support:* In 2005–06, 5 research assistantships (averaging $7,092 per year), 1 teaching assistantship (averaging $8,370 per year) were awarded; tuition waivers (full) and unspecified assistantships also available. Financial award application deadline: 4/1. *Faculty research:* Evaluation of offal composting as an alternative to rendering, field test grain hazard and assessment tool. *Unit head:* Dr. Patrick O'Rourke, Chairperson, 309-438-5654.

Instituto Tecnológico y de Estudios Superiores de Monterrey, Campus Monterrey, Graduate and Research Division, Program in Agriculture, Monterrey, Mexico. Offers agricultural parasitology (PhD); agricultural sciences (MS); farming productivity (MS); food processing engineering (MS); phytopathology (MS). Part-time programs available. *Degree requirements:* For master's and doctorate, one foreign language, thesis/dissertation. *Entrance requirements:* For master's, PAEG; for doctorate, GMAT or GRE, master's degree in related field. Additional exam requirements/recommendations for international students: Required—TOEFL. *Faculty research:* Animal embryos and reproduction, crop entomology, tropical agriculture, agricultural productivity, induced mutation in oleaginous plants.

Iowa State University of Science and Technology, Graduate College, College of Agriculture, Ames, IA 50011. Offers M Ag, MS, PhD. Part-time programs available. Postbaccalaureate distance learning degree programs offered (no on-campus study). *Faculty:* 221 full-time, 31 part-time/adjunct. *Students:* 428 full-time (208 women), 196 part-time (54 women); includes 26 minority (15 African Americans, 1 American Indian/Alaska Native, 3 Asian Americans or Pacific Islanders, 7 Hispanic Americans), 211 international. 278 applicants, 45% accepted, 95 enrolled.

In 2005, 101 master's, 52 doctorates awarded. *Degree requirements:* For doctorate, thesis/dissertation. *Entrance requirements:* Additional exam requirements/recommendations for international students: Required—TOEFL. *Application deadline:* Applications are processed on a rolling basis. Application fee: $50 ($70 for international students). Electronic applications accepted. *Expenses:* Tuition, state resident: full-time $6,410. Tuition, nonresident: full-time $16,422. Tuition and fees vary according to program. *Financial support:* In 2005–06, 59 research assistantships with full and partial tuition reimbursements (averaging $15,225 per year), 41 teaching assistantships with full and partial tuition reimbursements (averaging $15,391 per year) were awarded; fellowships, Federal Work-Study, scholarships/grants, health care benefits, and unspecified assistantships also available. Support available to part-time students. *Unit head:* Dr. Wendy Wintersteen, Dean, 515-294-2518, Fax: 515-294-6800.

Kansas State University, Graduate School, College of Agriculture, Manhattan, KS 66506. Offers MAB, MS, PhD. Part-time programs available. Postbaccalaureate distance learning degree programs offered (minimal on-campus study). *Faculty:* 168 full-time (24 women), 56 part-time/adjunct (6 women). *Students:* 247 full-time (112 women), 120 part-time (57 women); includes 15 minority (8 African Americans, 3 American Indian/Alaska Native, 4 Hispanic Americans), 229 international. 140 applicants, 56% accepted, 55 enrolled. In 2005, 95 master's, 29 doctorates awarded. Terminal master's awarded for partial completion of doctoral program. *Entrance requirements:* For master's, GRE, minimum undergraduate GPA of 3.0; for doctorate, GRE, minimum undergraduate GPA of 3.5. Additional exam requirements/recommendations for international students: Required—TOEFL (minimum score 550 paper-based; 213 computer-based). *Application deadline:* For fall admission, 2/1 for domestic students; for spring admission, 10/1 for domestic students. Application fee: $30 ($55 for international students). Electronic applications accepted. *Expenses:* Tuition, state resident: full-time $5,160; part-time $215. Tuition, nonresident: full-time $12,816; part-time $534. Required fees: $564. *Financial support:* In 2005–06, 205 research assistantships (averaging $15,691 per year), 25 teaching assistantships (averaging $9,657 per year) were awarded; fellowships, career-related internships or fieldwork, Federal Work-Study, institutionally sponsored loans, scholarships/grants, and tuition waivers (partial) also available. Support available to part-time students. Financial award application deadline: 3/1; financial award applicants required to submit FAFSA. Total annual research expenditures: $19 million. *Unit head:* Fred Cholick, Dean, 785-532-7137, Fax: 785-532-6563, E-mail: fcholick@ksu.edu.

Louisiana State University and Agricultural and Mechanical College, Graduate School, College of Agriculture, Baton Rouge, LA 70803. Offers M App St, MS, MSBAE, PhD. Part-time programs available. *Faculty:* 212 full-time (34 women). *Students:* 331 full-time (163 women), 140 part-time (80 women); includes 47 minority (32 African Americans, 1 American Indian/Alaska Native, 3 Asian Americans or Pacific Islanders, 11 Hispanic Americans), 155 international. Average age 32. 255 applicants, 46% accepted, 59 enrolled. In 2005, 96 master's, 42 doctorates awarded. Terminal master's awarded for partial completion of doctoral program. *Degree requirements:* For doctorate, thesis/dissertation. *Entrance requirements:* For master's and doctorate, GRE General Test, minimum GPA of 3.0. Additional exam requirements/recommendations for international students: Required—TOEFL (minimum score 550 paper-based; 213 computer-based). *Application deadline:* For fall admission, 5/15 for domestic students, 5/15 for international students; for spring admission, 10/15 for domestic students, 10/15 for international students. Applications are processed on a rolling basis. Application fee: $25. Electronic applications accepted. *Financial support:* In 2005–06, 343 students received support, including 8 fellowships with full tuition reimbursements available (averaging $12,846 per year), 217 research assistantships with partial tuition reimbursements available (averaging $12,789 per year), 48 teaching assistantships with partial tuition reimbursements available (averaging $11,855 per year); career-related internships or fieldwork, Federal Work-Study, institutionally sponsored loans, tuition waivers (full), and unspecified assistantships also available. Support available to part-time students. Financial award applicants required to submit FAFSA. *Faculty research:* Biotechnology, resource economics and marketing, aquaculture, food science and technology. Total annual research expenditures: $193,969. *Unit head:* Dr. Kenneth Koonce, Dean, 225-578-2362, Fax: 225-578-2526, E-mail: kkoonce@lsu.edu. *Application contact:* Paula Beecher, Recruiting Coordinator, 225-578-2468, E-mail: pbeeche@lsu.edu.

McGill University, Faculty of Graduate and Postdoctoral Studies, Faculty of Agricultural and Environmental Sciences, Montréal, QC H3A 2T5, Canada. Offers M Sc, M Sc A, PhD, Certificate. Part-time programs available. Terminal master's awarded for partial completion of doctoral program. *Degree requirements:* For master's, registration; for doctorate, thesis/dissertation, registration. *Entrance requirements:* For master's, B Sc, minimum GPA of 3.0; for doctorate, M Sc, minimum GPA of 3.0; for Certificate, minimum GPA of 3.0, B Sc in biological sciences. Additional exam requirements/recommendations for international students: Required—TOEFL (minimum score 550 paper-based; 213 computer-based), IELT (minimum score 7). *Faculty research:* Agriculture, environmental, food sciences, nutrition and molecular biology, biosystems and agricultural engineering.

Michigan State University, The Graduate School, College of Agriculture and Natural Resources, East Lansing, MI 48824. Offers MA, MIPS, MS, MURP, PhD. *Faculty:* 286 full-time (64 women), 1 part-time/adjunct (0 women). *Students:* 499 full-time (252 women), 131 part-time (43 women); includes 56 minority (15 African Americans, 6 American Indian/Alaska Native, 18 Asian Americans or Pacific Islanders, 17 Hispanic Americans), 243 international. Average age 31. 456 applicants, 34% accepted. In 2005, 126 master's, 64 doctorates awarded. Application fee: $50. *Expenses:* Tuition, state resident: part-time $330 per credit hour. Tuition, nonresident: part-time $685 per credit hour. Tuition and fees vary according to program. *Financial support:* In 2005–06, 102 fellowships with tuition reimbursements (averaging $6,547 per year), 345 research assistantships with tuition reimbursements (averaging $13,027 per year), 45 teaching assistantships with tuition reimbursements (averaging $12,503 per year) were awarded; career-related internships or fieldwork, Federal Work-Study, institutionally sponsored loans, scholarships/grants, tuition waivers (partial), and unspecified assistantships also available. Support available to part-time students. *Faculty research:* Plant science, animal sciences, forestry, fisheries and wildlife, recreation and tourism. Total annual research expenditures: $40.2 million. *Unit head:* Dr. Jeffrey D. Armstrong, Dean, 517-355-0232, Fax: 517-353-9896, E-mail: armstroj@msu.edu.

Mississippi State University, College of Agriculture and Life Sciences, Mississippi State, MS 39762. Offers MABM, MLA, MS, PhD. Part-time programs available. *Faculty:* 136 full-time (23 women), 12 part-time/adjunct (8 women). *Students:* 165 full-time (68 women), 84 part-time (34 women); includes 15 minority (9 African Americans, 3 Asian Americans or Pacific Islanders, 3 Hispanic Americans), 67 international. Average age 30. 139 applicants, 38% accepted, 27 enrolled. In 2005, 55 master's, 9 doctorates awarded. *Degree requirements:* For doctorate, thesis/dissertation. *Entrance requirements:* Additional exam requirements/recommendations for international students: Required—TOEFL. *Application deadline:* For fall admission, 7/1 for domestic students; for spring admission, 11/1 for domestic students. Applications are processed on a rolling basis. Application fee: $30. *Expenses:* Tuition, state resident: full-time $4,312; part-time $240 per hour. Tuition, nonresident: full-time $9,772; part-time $543 per hour. International tuition: $10,102 full-time. Tuition and fees vary according to course load. *Financial support:* In 2005–06, 15 teaching assistantships with full tuition reimbursements (averaging $8,918 per year) were awarded; research assistantships with full tuition reimbursements, career-related internships or fieldwork, Federal Work-Study, institutionally sponsored loans, scholarships/grants, tuition waivers (partial), and unspecified assistantships also available. Financial award applicants required to submit FAFSA. *Faculty research:* Animal and dairy sciences-biochemistry, molecular biology, biological engineering, human sciences, food sciences, economics. Total annual research expenditures: $17 million. *Unit head:* Dr. Vance Watson, Dean and Vice President, 662-325-2110, E-mail: vwatson@dafvm.msstate.edu. *Application contact:* Philip G. Bonfanti, Director of Admissions, 662-325-4104, Fax: 662-325-8872, E-mail: admit@msstate.edu.

Missouri State University, Graduate College, College of Natural and Applied Sciences, Department of Agriculture, Springfield, MO 65804-0094. Offers agriculture (MNAS); fruit science (MNAS); plant science (MS); secondary education (MS Ed), including agriculture. *Faculty:* 14 full-time (2 women), 1 part-time/adjunct (0 women). *Students:* 19 full-time (11 women), 5 part-time (4 women), 5 international. Average age 29. 21 applicants, 71% accepted, 10 enrolled. In 2005, 6 degrees awarded. *Degree requirements:* For master's, thesis or alternative, comprehensive exam. *Entrance requirements:* For master's, GRE (MS plant science, MNAS), 9–12 teacher certification (MS Ed), minimum GPA of 3.0 (MS plant science, MNAS). Additional exam requirements/recommendations for international students: Required—TOEFL (minimum score 550 paper-based; 213 computer-based), IELT (minimum score 6). *Application deadline:* For fall admission, 7/20 for domestic students; for spring admission, 12/20 priority date for domestic students. Applications are processed on a rolling basis. Application fee: $30. Electronic applications accepted. *Expenses:* Tuition, state resident: full-time $3,402; part-time $189 per credit. Tuition, nonresident: full-time $6,804; part-time $378 per credit. Required fees: $207 per semester. Part-time tuition and fees vary according to course level, course load and program. *Financial support:* In 2005–06, 6 research assistantships (averaging $7,662 per year), 5 teaching assistantships (averaging $7,445 per year) were awarded. Financial award application deadline: 3/31; financial award applicants required to submit FAFSA. *Unit head:* Dr. W. Anson Elliott, Head, 417-836-5638, E-mail: ansonelliot@missouristate.edu. *Application contact:* Dr. W. Anson Elliott, Head, 417-836-5638, E-mail: ansonelliot@missouristate.edu.

Montana State University, College of Graduate Studies, College of Agriculture, Bozeman, MT 59717. Offers MS, PhD. Part-time programs available. *Faculty:* 79 full-time (10 women), 14 part-time/adjunct (5 women). *Students:* 30 full-time (15 women), 103 part-time (51 women); includes 4 minority (1 African American, 3 Hispanic Americans), 20 international. Average age 29. 53 applicants, 53% accepted, 23 enrolled. In 2005, 41 master's, 7 doctorates awarded. *Degree requirements:* For master's, comprehensive exam, registration; for doctorate, thesis/dissertation, comprehensive exam, registration. *Entrance requirements:* For master's and doctorate, GRE General Test. Additional exam requirements/recommendations for international students: Required—TOEFL (minimum score 550 paper-based; 213 computer-based). *Application deadline:* For fall admission, 7/15 priority date for domestic students, 5/15 priority date for international students; for spring admission, 12/1 priority date for domestic students, 10/1 priority date for international students. Applications are processed on a rolling basis. Application fee: $30. Electronic applications accepted. *Expenses:* Tuition, state resident: full-time $4,132. Tuition, nonresident: full-time $1,132. *Financial support:* Application deadline: 3/1; Total annual research expenditures: $28.9 million. *Unit head:* Dr. Jeffrey S. Jacobsen, Dean, 406-994-7060, Fax: 406-994-3933, E-mail: jefj@montana.edu.

Murray State University, School of Agriculture, Murray, KY 42071-0009. Offers MS. Part-time programs available. *Entrance requirements:* For master's, GRE General Test. Additional exam requirements/recommendations for international students: Required—TOEFL.

New Mexico State University, Graduate School, College of Agriculture and Home Economics, Department of Entomology, Plant Pathology and Weed Science, Las Cruces, NM 88003-8001. Offers agricultural biology (MS). Part-time programs available. *Faculty:* 12 full-time (5 women), 1 part-time/adjunct (0 women). *Students:* 14 full-time (12 women), 8 part-time (4 women); includes 7 minority (1 African American, 1 American Indian/Alaska Native, 1 Asian American or Pacific Islander, 4 Hispanic Americans), 1 international. Average age 30. 9 applicants, 67% accepted, 6 enrolled. In 2005, 4 degrees awarded. *Degree requirements:* For master's, thesis, comprehensive exam, registration. *Entrance requirements:* For master's, GRE General Test. *Application deadline:* For fall admission, 7/1 for domestic students; for spring admission, 11/1 priority date for domestic students. Applications are processed on a rolling basis. Application fee: $30 ($50 for international students). Electronic applications accepted. *Expenses:* Tuition, state resident: full-time $3,156; part-time $175 per credit. Tuition, nonresident: full-time $12,510; part-time $565 per credit. Required fees: $1,050. *Financial support:* In 2005–06, 1 fellowship, 11 research assistantships with partial tuition reimbursements, 4 teaching assistantships with partial tuition reimbursements were awarded; career-related internships or fieldwork also available. Financial award application deadline:3/1. *Faculty research:* Integrated pest management, pesticide application and safety, livestock ectoparasite research, biotechnology, nematology. *Unit head:* Dr. Grant Kinzer, Head, 505-646-3225, Fax: 505-646-8087, E-mail: gkinzer@nmsu.edu.

North Carolina Agricultural and Technical State University, Graduate School, School of Agriculture and Environmental and Allied Sciences, Greensboro, NC 27411. Offers MS. Part-time and evening/weekend programs available. *Degree requirements:* For master's, qualifying exam. *Entrance requirements:* For master's, GRE General Test. *Faculty research:* Aid for small farmers, agricultural technology, housing, food science, nutrition.

North Carolina State University, Graduate School, College of Agriculture and Life Sciences, Raleigh, NC 27695. Offers M Tox, MAEE, MB, MBAE, MFG, MFM, MFS, MG, MMB, MN, MP, MS, MZS, PhD. Part-time programs available. Electronic applications accepted.

North Dakota State University, The Graduate School, College of Agriculture, Food Systems, and Natural Resources, Fargo, ND 58105. Offers MS, PhD. Part-time programs available. *Faculty:* 120. *Students:* 235 (94 women); includes 7 African Americans, 58 Asian Americans or Pacific Islanders, 6 Hispanic Americans. In 2005, 39 master's, 7 doctorates awarded. *Degree requirements:* For doctorate, thesis/dissertation. *Entrance requirements:* Additional exam requirements/recommendations for international students: Required—TOEFL. *Application deadline:* Applications are processed on a rolling basis. Application fee: $45 ($60 for international students). Electronic applications accepted. *Financial support:* Fellowships with full tuition reimbursements, research assistantships with full tuition reimbursements, teaching assistantships with full tuition reimbursements, career-related internships or fieldwork, Federal Work-Study, and institutionally sponsored loans available. Support available to part-time students. *Faculty research:* Horticulture and forestry, plant and wheat breeding, diseases of insects, animal and range sciences, soil science, veterinary medicine. *Unit head:* Dr. Kenneth F. Grafton, Dean, 701-231-8790, Fax: 701-231-8520, E-mail: k.grafton@ndsu.edu.

Northwest Missouri State University, Graduate School, Melvin and Valorie Booth College of Business and Professional Studies, Department of Agriculture, Maryville, MO 64468-6001. Offers agricultural economics (MBA); agriculture (MS); teaching agriculture (MS Ed). Part-time programs available. *Faculty:* 4 full-time (1 woman). *Students:* 7 full-time (4 women), 2 part-time. 4 applicants, 50% accepted, 1 enrolled. In 2005, 5 degrees awarded. *Degree requirements:* For master's, thesis (for some programs), comprehensive exam. *Entrance requirements:* For master's, GRE General Test, minimum undergraduate GPA of 2.5, writing sample. Additional exam requirements/recommendations for international students: Required—TOEFL (minimum score 550 paper-based; 213 computer-based). *Application deadline:* For fall admission, 7/1 for domestic students, 7/1 for international students; for spring admission, 11/15 for domestic students, 11/15 for international students. Applications are processed on a rolling basis. Application fee: $0 ($50 for international students). *Expenses:* Tuition, state resident: full-time $2,077; part-time $231 per credit hour. Tuition, nonresident: full-time $3,650; part-time $406 per credit hour. Required fees: $105 per term. Tuition and fees vary according to campus/location and reciprocity agreements. *Financial support:* In 2005–06, research assistantships with full tuition reimbursements (averaging $5,500 per year), teaching assistantships with full tuition reimbursements (averaging $5,500 per year) were awarded; unspecified assistantships also available. Financial award application deadline: 3/1; financial award applicants required to submit FAFSA. *Unit head:* Dr. Arley Larson, Chairperson, 660-562-1161. *Application contact:* Dr. Frances Shipley, Dean of Graduate School, 660-562-1145, Fax: 660-562-1096, E-mail: gradsch@nwmissouri.edu.

Nova Scotia Agricultural College, Research and Graduate Studies, Truro, NS B2N 5E3, Canada. Offers agriculture (M Sc), including air quality, animal behavior, animal molecular genetics, animal nutrition, animal technology, aquaculture, botany, crop management, crop physiology, ecology, environmental microbiology, food science, horticulture, nutrient management, pest management, physiology, plant biotechnology, plant pathology, soil chemistry, soil fertility, waste management and composting, water quality. Part-time programs available. *Faculty:* 43 full-time (7 women), 21 part-time/adjunct (1 woman). *Students:* 50 full-time (25

Agricultural Sciences—General

Nova Scotia Agricultural College (continued)

women), 15 part-time (11 women); includes 7 minority (3 African Americans, 3 Asian Americans or Pacific Islanders, 1 Hispanic American). Average age 25. In 2005, 23 degrees awarded. *Degree requirements:* For master's, thesis, registration. *Entrance requirements:* For master's, honors B Sc, minimum GPA of 3.0. Additional exam requirements/recommendations for international students: Required—TOEFL (minimum score 580 paper-based; 237 computer-based), Michigan English Language Assessment Battery, IELT, Can Test, CAEL. *Application deadline:* For fall admission, 6/1 for domestic students, 4/1 for international students. For winter admission, 10/31 for domestic students; for spring admission, 2/28 for domestic students. Applications are processed on a rolling basis. Application fee: $70. *Expenses:* Tuition, state resident: part-time $2,328 per year. Tuition, nonresident: full-time $6,984; part-time $7,968 per year. International tuition: $12,624 full-time. Required fees: $481; $46 per course. Tuition and fees vary according to program and student level. *Financial support:* In 2005–06, 48 students received support, including 7 fellowships (averaging $15,000 per year), 10 research assistantships (averaging $15,000 per year), 15 teaching assistantships (averaging $900 per year); career-related internships or fieldwork, scholarships/grants, and unspecified assistantships also available. *Faculty research:* Bio-product development, organic agriculture, nutrient management, air and water quality, agricultural biotechnology. Total annual research expenditures: $4.7 million. *Unit head:* Jill L. Rogers, Manager, 902-893-6360, Fax: 902-893-3430, E-mail: jrogers@nsac.ca. *Application contact:* Marie Law, Administrative Assistant, 902-893-6502, Fax: 902-893-3430, E-mail: mlaw@nsac.ca.

The Ohio State University, Graduate School, College of Food, Agricultural, and Environmental Sciences, Columbus, OH 43210. Offers M Ed, MS, PhD. Part-time programs available. *Degree requirements:* For doctorate, thesis/dissertation. *Entrance requirements:* Additional exam requirements/recommendations for international students: Required—TOEFL (paper 550; computer 213) or IELT (7) or Michigan English Language Assessment Battery (83). Electronic applications accepted.

Oklahoma State University, College of Agricultural Science and Natural Resources, Stillwater, OK 74078. Offers M Ag, MS, PhD. Part-time programs available. *Faculty:* 209 full-time (38 women), 14 part-time/adjunct (2 women). *Students:* 131 full-time (64 women), 200 part-time (85 women); includes 32 minority (9 African Americans, 17 American Indian/Alaska Native, 4 Asian Americans or Pacific Islanders, 2 Hispanic Americans), 137 international. Average age 30. 308 applicants, 41% accepted, 72 enrolled. In 2005, 66 master's, 32 doctorates awarded. *Degree requirements:* For doctorate, thesis/dissertation. *Entrance requirements:* For master's and doctorate, GRE. Additional exam requirements/recommendations for international students: Required—TOEFL. *Application deadline:* Applications are processed on a rolling basis. Application fee: $40 ($75 for international students). Electronic applications accepted. *Expenses:* Tuition, state resident: full-time $4,253; part-time $139 per credit hour. Tuition, nonresident: full-time $12,569; part-time $485 per credit hour. Required fees: $43 per credit hour. One-time fee: $20 part-time. Tuition and fees vary according to course load and program. *Financial support:* In 2005–06, 245 students received support, including 206 research assistantships (averaging $14,740 per year), 23 teaching assistantships (averaging $12,019 per year); fellowships, career-related internships or fieldwork, Federal Work-Study, scholarships/grants, health care benefits, tuition waivers (partial), and unspecified assistantships also available. Support available to part-time students. Financial award application deadline: 3/1. *Unit head:* Dr. Robert E. Whitson, Dean, 405-744-5398, Fax: 405-744-5339.

Oregon State University, Graduate School, College of Agricultural Sciences, Corvallis, OR 97331. Offers M Ag, M Agr, MA, MAIS, MAT, MS, PhD. Part-time programs available. *Faculty:* 106 full-time (23 women), 34 part-time/adjunct (12 women). *Students:* 274 full-time (135 women), 40 part-time (12 women); includes 16 minority (1 African American, 5 American Indian/Alaska Native, 5 Asian Americans or Pacific Islanders, 5 Hispanic Americans), 93 international. Average age 31. In 2005, 88 master's, 22 doctorates awarded. Terminal master's awarded for partial completion of doctoral program. *Degree requirements:* For doctorate, thesis/dissertation. *Entrance requirements:* For master's and doctorate, GRE, minimum GPA of 3.0 in last 90 hours of course work. Additional exam requirements/recommendations for international students: Required—TOEFL. Application fee: $50. *Expenses:* Tuition, area resident: Part-time $301 per credit. Tuition, state resident: full-time $8,139; part-time $501 per credit. Tuition, nonresident: full-time $14,376; part-time $532 per credit. Required fees: $1,266. *Financial support:* Fellowships, research assistantships, teaching assistantships, career-related internships or fieldwork, Federal Work-Study, and institutionally sponsored loans available. Support available to part-time students. Financial award application deadline: 2/1. *Faculty research:* Fish and wildlife biology, food science, soil/water/plant relationships, natural resources, animal biochemistry. *Unit head:* Dr. Thayne R. Dutson, Dean, 541-737-2331, Fax: 541-737-4574, E-mail: thayne.dutson@orst.edu. *Application contact:* Dr. Michael J. Burke, Associate Dean, 541-737-2211, Fax: 541-737-2256, E-mail: mike.burke@orst.edu.

The Pennsylvania State University University Park Campus, Graduate School, College of Agricultural Sciences, State College, University Park, PA 16802-1503. Offers M Agr, M Ed, MFR, MS, D Ed, PhD. *Students:* Average age 29. 390 applicants, 40% accepted. In 2005, 69 master's, 43 doctorates awarded. *Entrance requirements:* For master's and doctorate, GRE General Test. Additional exam requirements/recommendations for international students: Required—TOEFL (minimum score 550 paper-based; 213 computer-based). *Application deadline:* Applications are processed on a rolling basis. Application fee: $45. Electronic applications accepted. *Expenses:* Tuition, state resident: full-time $12,518; part-time $522 per credit. Tuition, nonresident: full-time $23,004; part-time $959 per credit. Required fees: $484. Tuition and fees vary according to course load, campus/location and program. *Financial support:* In 2005–06, 8 fellowships, 201 research assistantships, 24 teaching assistantships were awarded; health care benefits and unspecified assistantships also available. Financial award applicants required to submit FAFSA. Total annual research expenditures: $81.8 million. *Unit head:* Dr. Robert D. Steele, Dean, 814-865-2541, Fax: 814-865-3103, E-mail: rsteele@psu.edu.

Prairie View A&M University, Graduate School, College of Agriculture and Human Sciences, Prairie View, TX 77446-0519. Offers agricultural economics (MS); animal sciences (MS); interdisciplinary human sciences (MS); soil science (MS). Part-time and evening/weekend programs available. *Faculty:* 8 full-time (4 women), 9 part-time/adjunct (3 women). *Students:* 68 full-time (45 women), 79 part-time (65 women); includes 125 minority (all African Americans), 13 international. Average age 33. 147 applicants, 100% accepted, 147 enrolled. In 2005, 31 degrees awarded. *Degree requirements:* For master's, thesis (for some programs), field placement, comprehensive exam, registration. *Entrance requirements:* For master's, GRE General Test, minimum GPA of 2.45. Additional exam requirements/recommendations for international students: Required—TOEFL (minimum score 550 paper-based). *Application deadline:* For fall admission, 6/1 for domestic students, 6/1 for international students; for spring admission, 10/1 for domestic students, 10/1 for international students. Applications are processed on a rolling basis. Application fee: $50. *Expenses:* Tuition, state resident: full-time $1,440; part-time $80 per credit. Tuition, nonresident: full-time $6,444; part-time $358 per credit. *Financial support:* In 2005–06, 57 students received support, including 8 fellowships with tuition reimbursements available (averaging $12,000 per year), 10 research assistantships with tuition reimbursements available (averaging $15,000 per year); career-related internships or fieldwork, Federal Work-Study, institutionally sponsored loans, scholarships/grants, tuition waivers (partial), and unspecified assistantships also available. Support available to part-time students. Financial award application deadline: 4/1; financial award applicants required to submit FAFSA. *Faculty research:* Domestic violence prevention, water quality, food growth regulators. Total annual research expenditures: $4 million. *Unit head:* Dr. Linda Willis, Dean, 936-857-2996, Fax: 936-857-2998, E-mail: lwillis@pvamu.edu. *Application contact:* Dr. Richard W. Griffin, Interim Department Head, 936-857-2996, Fax: 936-857-2998, E-mail: rwgriffin@pvamu.edu.

Purdue University, Graduate School, College of Agriculture, West Lafayette, IN 47907. Offers EMBA, M Agr, MA, MS, MSF, PhD. Part-time programs available. *Faculty:* 283 full-time (44 women), 55 part-time/adjunct (12 women). *Students:* 458 full-time (210 women), 58 part-time (25 women); includes 32 minority (15 African Americans, 1 American Indian/Alaska Native, 4 Asian Americans or Pacific Islanders, 12 Hispanic Americans), 220 international. 536 applicants, 35% accepted, 126 enrolled. In 2005, 71 master's, 48 doctorates awarded. *Degree requirements:* For doctorate, thesis/dissertation. *Entrance requirements:* Additional exam requirements/recommendations for international students: Required—TOEFL. *Application deadline:* Applications are processed on a rolling basis. Application fee: $55. Electronic applications accepted. *Financial support:* Fellowships with tuition reimbursements, research assistantships with tuition reimbursements, teaching assistantships with tuition reimbursements, career-related internships or fieldwork and tuition waivers (partial) available. Support available to part-time students. Financial award applicants required to submit FAFSA. *Unit head:* Dr. Victor L. Lechtenberg, Dean, 765-494-8392.

Sam Houston State University, College of Arts and Sciences, Department of Agricultural Sciences, Huntsville, TX 77341. Offers agriculture (MS); industrial education (M Ed, MA); industrial technology (MA); vocational education (M Ed). Part-time and evening/weekend programs available. *Faculty:* 7 full-time (0 women). *Students:* 19 full-time (10 women), 13 part-time (7 women). Average age 28. In 2005, 8 degrees awarded. *Degree requirements:* For master's, thesis optional. *Entrance requirements:* For master's, GRE General Test, minimum GPA of 2.5. *Application deadline:* For fall admission, 8/1 for domestic students; for spring admission, 12/1 for domestic students. Application fee: $20. *Financial support:* Teaching assistantships, career-related internships or fieldwork available. Financial award application deadline: 5/31; financial award applicants required to submit FAFSA. *Unit head:* Dr. Robert A. Lane, Chair, 936-294-1215, Fax: 936-294-1232, E-mail: agr_ral@shsu.edu.

South Dakota State University, Graduate School, College of Agriculture and Biological Sciences, Brookings, SD 57007. Offers MS, PhD. Part-time programs available. *Degree requirements:* For master's, thesis, oral exam; for doctorate, thesis/dissertation, preliminary oral and written exams. *Entrance requirements:* Additional exam requirements/recommendations for international students: Required—TOEFL.

Southern Illinois University Carbondale, Graduate School, College of Agriculture, Carbondale, IL 62901-4701. Offers MS, MBA/MS. Part-time programs available. *Faculty:* 51 full-time (8 women). *Students:* 39 full-time (21 women), 84 part-time (31 women); includes 14 minority (11 African Americans, 2 Asian Americans or Pacific Islanders, 1 Hispanic American), 6 international. 64 applicants, 50% accepted, 9 enrolled. In 2005, 31 master's awarded. *Entrance requirements:* For master's, minimum GPA of 2.7. Additional exam requirements/recommendations for international students: Required—TOEFL. *Application deadline:* Applications are processed on a rolling basis. Application fee: $0. *Financial support:* In 2005–06, 35 students received support, including 31 research assistantships; fellowships, teaching assistantships, career-related internships or fieldwork, Federal Work-Study, institutionally sponsored loans, and tuition waivers (full) also available. Support available to part-time students. *Faculty research:* Production and studies in crops, animal nutrition, agribusiness economics and management, forest biology and ecology, microcomputers in agriculture. *Unit head:* Gary L. Minish, Dean.

Southern University and Agricultural and Mechanical College, Graduate School, College of Agricultural, Family and Consumer Sciences, Baton Rouge, LA 70813. Offers urban forestry (MS). *Faculty:* 5 full-time (2 women). *Students:* 8 full-time (2 women), 1 (woman) part-time; includes 5 African Americans, 2 Asian Americans or Pacific Islanders. Average age 33. 9 applicants, 56% accepted, 4 enrolled. In 2005, 6 degrees awarded. *Degree requirements:* For master's, thesis. *Entrance requirements:* For master's, GRE, minimum GPA of 3.0. Additional exam requirements/recommendations for international students: Required—TOEFL (minimum score 525 paper-based; 193 computer-based). *Application deadline:* For fall admission, 4/15 priority date for domestic students, 4/15 priority date for international students; for spring admission, 11/1 priority date for domestic students, 11/1 priority date for international students. Applications are processed on a rolling basis. Application fee: $25. *Financial support:* In 2005–06, 14 students received support, including 7 research assistantships (averaging $5,897 per year); scholarships/grants and tuition waivers (full) also available. Financial award application deadline: 4/15; financial award applicants required to submit FAFSA. *Faculty research:* Urban forest interactions with environment, social and economic impacts of urban forests, tree biology/pathology, development of urban forest management tools. *Unit head:* Dr. Kamran K. Abdollahi, Program Leader, 225-771-6291, Fax: 225-771-6293, E-mail: kamrana664@cs.com. *Application contact:* Dr. Kamran K. Abdollahi, Program Leader, 225-771-6291, Fax: 225-771-6293, E-mail: kamrana664@cs.com.

Tarleton State University, College of Graduate Studies, College of Agriculture and Human Sciences, Department of Agribusiness, Agronomy, Horticulture, and Range Management, Stephenville, TX 76402. Offers agriculture (MS). *Faculty:* 4 full-time (0 women), 11 part-time/adjunct (1 woman). *Students:* 21 full-time (12 women), 9 part-time (4 women), 1 international. Average age 26. In 2005, 5 degrees awarded. *Unit head:* Dr. Roger Wittie, Head, 254-968-9931.

Tarleton State University, College of Graduate Studies, College of Agriculture and Human Sciences, Department of Agricultural Services and Development, Stephenville, TX 76402. Offers agriculture education (MS). Part-time and evening/weekend programs available. Postbaccalaureate distance learning degree programs offered. *Faculty:* 7 full-time (2 women), 5 part-time/adjunct (0 women). *Students:* 26 full-time (10 women), 28 part-time (6 women); includes 4 minority (1 American Indian/Alaska Native, 3 Hispanic Americans). Average age 36. In 2005, 17 degrees awarded. *Entrance requirements:* For master's, GRE General Test, minimum GPA of 3.0. Additional exam requirements/recommendations for international students: Required—TOEFL. *Application deadline:* For fall admission, 8/5 for domestic students; for spring admission, 12/1 priority date for domestic students. Application fee: $25 ($100 for international students). Electronic applications accepted. *Financial support:* Federal Work-Study, institutionally sponsored loans, scholarships/grants, and unspecified assistantships available. *Unit head:* Dr. David Drueckhammer, Head, 254-968-9200, Fax: 254-968-9199, E-mail: drueckh@tarleton.edu.

Tennessee State University, Graduate School, School of Agriculture and Family Services, Nashville, TN 37209-1561. Offers MS, PhD. Part-time and evening/weekend programs available. *Faculty:* 4 full-time (1 woman), 5 part-time/adjunct (1 woman). *Students:* 10 full-time (4 women), 7 part-time (2 women); includes 13 minority (12 African Americans, 1 Asian American or Pacific Islander), 1 international. Average age 31. 25 applicants, 44% accepted. In 2005, 1 degree awarded. *Degree requirements:* For master's, thesis. *Entrance requirements:* For master's, GRE General Test, GRE Subject Test, MAT. *Application deadline:* Applications are processed on a rolling basis. Application fee: $15. Electronic applications accepted. *Financial support:* In 2005–06, 2 research assistantships (averaging $6,511 per year), 1 teaching assistantship (averaging $6,511 per year) were awarded. *Faculty research:* Small farm economics, ornamental horticulture, beef cattle production, rural elderly. *Unit head:* Dr. Troy Wakefield, Dean, 615-963-7620, E-mail: twakefield@tnstate.edu.

Texas A&M University, College of Agriculture and Life Sciences, College Station, TX 77843. Offers M Agr, M Ed, M Eng, MAB, MS, DE, Ed D, PhD. Part-time programs available. Postbaccalaureate distance learning degree programs offered (minimal on-campus study). *Faculty:* 200 full-time (26 women), 20 part-time/adjunct (1 woman). *Students:* 716 full-time (336 women), 275 part-time (117 women); includes 90 minority (23 African Americans, 4 American Indian/Alaska Native, 10 Asian Americans or Pacific Islanders, 53 Hispanic Americans), 296 international. Average age 29. 653 applicants, 60% accepted, 233 enrolled. In 2005, 202 master's, 110 doctorates awarded. *Entrance requirements:* Additional exam requirements/recommendations for international students: Required—TOEFL (minimum score 550 paper-based; 213 computer-based). *Application deadline:* For fall admission, 7/21 priority date for domestic students, 6/1 priority date for international students; for spring admission, 12/1 priority date for domestic students, 10/1 priority date for international students. Applications are processed on a rolling basis. Application fee: $50 ($75 for international students). Electronic applications accepted. *Expenses:* Tuition, state resident: full-time $4,488; part-time $187 per credit hour. Tuition, nonresident: full-time $11,112; part-time $463 per credit hour. Required fees:$1,974. *Financial support:* Fellowships, research assistantships, teaching assistantships, career-related internships or fieldwork, Federal Work-Study, institutionally sponsored loans, scholarships/grants, tuition waivers (partial), and unspecified assistantships available. Support available to part-time students. Financial award applicants required to submit FAFSA. *Faculty research:* Plant sciences, animal sciences, environmental natural resources, biological and agricultural engineering, agricultural economics. *Unit head:* Dr. Elsa Murano, Vice Chancellor, 979-845-4747, Fax: 979-845-9938.

Texas A&M University–Commerce, Graduate School, College of Arts and Sciences, Department of Agriculture, Commerce, TX 75429-3011. Offers agricultural education (M Ed, MS); agricultural sciences (M Ed, MS). Part-time programs available. *Faculty:* 7 full-time (2 women). *Students:* 36 (22 women); includes 8 minority (2 African Americans, 4 Asian Americans or Pacific Islanders, 2 Hispanic Americans) 4 international. Average age 36. In 2005, 3 degrees awarded. *Degree requirements:* For master's, thesis (for some programs), comprehensive exam. *Entrance requirements:* For master's, GRE General Test. *Application deadline:* For fall admission, 6/1 for domestic students; for spring admission, 11/1 priority date for domestic students. Applications are processed on a rolling basis. Application fee: $0 ($25 for international students). Electronic applications accepted. *Financial support:* In 2005–06, research assistantships (averaging $7,875 per year), teaching assistantships (averaging $7,875 per year) were awarded; Federal Work-Study, institutionally sponsored loans, and scholarships/grants also available. Financial award application deadline: 5/1; financial award applicants required to submit FAFSA. *Faculty research:* Soil conservation, retention. Total annual research expenditures: $150,000. *Unit head:* Dr. Pat Bagley, Head, 843-886-5358. *Application contact:* Tammi Thompson, Graduate Admissions Adviser, 843-886-5167, Fax: 843-886-5165, E-mail: tammi_thompson@tamu-commerce.edu.

Texas A&M University–Kingsville, College of Graduate Studies, College of Agriculture and Home Economics, Kingsville, TX 78363. Offers MS, PhD. Part-time and evening/weekend programs available. *Degree requirements:* For master's, thesis or alternative, comprehensive exam; for doctorate, one foreign language, thesis/dissertation, comprehensive exam. *Entrance requirements:* For master's, GRE General Test, minimum GPA of 3.0; for doctorate, GRE General Test, minimum GPA of 3.5. Additional exam requirements/recommendations for international students: Required—TOEFL. *Faculty research:* Mesquite cloning; genesis of soil salinity; dove management; bone development; egg, meat, and milk consumption versus price.

Texas Tech University, Graduate School, College of Agricultural Sciences and Natural Resources, Lubbock, TX 79409. Offers M Agr, MLA, MS, Ed D, PhD, JD/MS. Part-time and evening/weekend programs available. *Faculty:* 65 full-time (9 women), 6 part-time/adjunct (0 women). *Students:* 163 full-time (71 women), 68 part-time (32 women); includes 7 minority (2 African Americans, 1 American Indian/Alaska Native, 1 Asian American or Pacific Islander, 3 Hispanic Americans), 44 international. Average age 29. 163 applicants, 60% accepted, 60 enrolled. In 2005, 58 master's, 11 doctorates awarded. *Degree requirements:* For doctorate, thesis/dissertation. *Entrance requirements:* For master's and doctorate, GRE General Test. Additional exam requirements/recommendations for international students: Required—TOEFL (minimum score 550 paper-based; 213 computer-based). *Application deadline:* Applications are processed on a rolling basis. Application fee: $50 ($60 for international students). Electronic applications accepted. *Expenses: Contact institution.* Tuition and fees vary according to program. *Financial support:* In 2005–06, 102 students received support, including 108 research assistantships with partial tuition reimbursements available (averaging $11,074 per year), 17 teaching assistantships with partial tuition reimbursements available (averaging $12,281 per year); career-related internships or fieldwork, Federal Work-Study, and institutionally sponsored loans also available. Support available to part-time students. Financial award application deadline: 4/15; financial award applicants required to submit FAFSA. *Faculty research:* Biotechnology and genomics, water management, food safety, policy, ecology. Total annual research expenditures: $10.2 million. *Unit head:* Dr. Marvin J. Cepica, Dean, 806-742-2808, Fax: 806-742-2836. *Application contact:* Graduate Adviser, 806-742-2808, Fax: 806-742-2836.

Tropical Agriculture Research and Higher Education Center, Graduate School, Turrialba, Costa Rica. Offers agroforestry (PhD); ecological agicuture (PhD); ecological agriculture (MS); environmental socioeconomics (MS, PhD); forest management and conservation (MS); forest sciences (PhD); tropical agroforestry (MS); watershed management (MS, PhD). *Entrance requirements:* For master's, GRE, letters of recommendation; for doctorate, GRE, curriculum vitae, letters of recommendation. Additional exam requirements/recommendations for international students: Required—TOEFL (minimum score 550 paper-based; 213 computer-based). Electronic applications accepted. *Faculty research:* Biodiversity in fragmented landscapes, ecosystem management, integrated pest management, environmental livestock production, biotechnology carbon balances in diverse land uses.

Tuskegee University, Graduate Programs, College of Agricultural, Environmental and Natural Sciences, Department of Agricultural Sciences, Tuskegee, AL 36088. Offers agricultural and resource economics (MS); animal and poultry sciences (MS); environmental sciences (MS); plant and soil sciences (MS). *Faculty:* 26 full-time (12 women), 1 part-time/adjunct (0 women). *Students:* 25 full-time (12 women), 2 part-time; includes 17 African Americans, 10 international. Average age 30. In 2005, 11 degrees awarded. *Degree requirements:* For master's, thesis. *Entrance requirements:* For master's, GRE General Test. Additional exam requirements/recommendations for international students: Required—TOEFL (minimum score 500 paper-based; 173 computer-based). *Application deadline:* For fall admission, 7/15 for domestic students. Applications are processed on a rolling basis. Application fee: $25 ($35 for international students). *Expenses:* Tuition: Full-time $12,400. Required fees: $300; $490 per credit. *Financial support:* In 2005–06, 5 fellowships, 4 research assistantships were awarded. Financial award application deadline: 4/15. *Unit head:* Dr. P. K. Biswas, Head, 334-727-8446.

Université Laval, Faculty of Agricultural and Food Sciences, Québec, QC G1K 7P4, Canada. Offers M Sc, PhD, Diploma. Part-time programs available. *Degree requirements:* For doctorate, thesis/dissertation, comprehensive exam. Electronic applications accepted.

University of Alberta, Faculty of Graduate Studies and Research, Department of Agricultural, Food and Nutritional Science, Edmonton, AB T6G 2E1, Canada. Offers M Ag, M Eng, M Sc, PhD, MBA/M Ag. *Faculty:* 44 full-time (14 women). *Students:* 76 full-time (42 women), 49 part-time (29 women), 23 international. 143 applicants, 22% accepted. In 2005, 16 master's, 6 doctorates awarded. *Median time to degree:* Of those who began their doctoral program in fall 1997, 100% received their degree in 8 years or less. *Degree requirements:* For master's, thesis/dissertation; for doctorate, thesis/dissertation, comprehensive exam. *Entrance requirements:* For master's, minimum GPA of 3.0; for doctorate, minimum GPA of 3.5. Additional exam requirements/recommendations for international students: Required—TOEFL (paper score 550; computer score 213) or IELTS (paper score 6). *Application deadline:* Applications are processed on a rolling basis. Tuition and fees charges are reported in Canadian dollars. *Expenses:* Tuition, state resident: part-time $562 Canadian dollars per term. Tuition, nonresident: full-time $3,375 Canadian dollars. Required fees: $573 Canadian dollars; $84 Canadian dollars per term. *Financial support:* In 2005–06, 65 students received support, including 6 fellowships, 17 research assistantships with partial tuition reimbursements available (averaging $7,000 per year), 37 teaching assistantships (averaging $3,600 per year); scholarships/grants and unspecified assistantships also available. *Faculty research:* Animal science, food science, nutrition and metabolism, bioresource engineering, plant science and range management. Total annual research expenditures: $12.6 million. *Unit head:* Dr. Edward Bork, Graduate Coordinator, 780-492-5131, Fax: 780-492-4265. *Application contact:* Jody Forslund, Student Support, 780-492-5131, Fax: 780-492-4265, E-mail: jody.forslund@ualberta.ca.

The University of Arizona, Graduate College, College of Agriculture and Life Sciences, Tucson, AZ 85721. Offers M Ag Ed, MHE Ed, MS, PhD. Part-time programs available. *Degree requirements:* For doctorate, thesis/dissertation. *Entrance requirements:* Additional exam requirements/recommendations for international students: Required—TOEFL.

University of Arkansas, Graduate School, Dale Bumpers College of Agricultural, Food and Life Sciences, Fayetteville, AR 72701-1201. Offers MS, PhD. *Students:* 194 full-time (89 women), 58 part-time (22 women); includes 15 minority (11 African Americans, 2 Asian Americans or Pacific Islanders, 2 Hispanic Americans), 78 international. 206 applicants, 42% accepted. In 2005, 74 master's, 23 doctorates awarded. *Degree requirements:* For doctorate, thesis/dissertation. Application fee: $40 ($50 for international students). *Financial support:* In 2005–06, 7 fellowships with tuition reimbursements, 176 research assistantships, 6 teaching assistantships were awarded; career-related internships or fieldwork, Federal Work-Study, scholarships/grants, and unspecified assistantships also available. Support available to part-time students. Financial award application deadline: 4/1; financial award applicants required to submit FAFSA. *Unit head:* Dr. Greg Weidemann, Dean, 479-575-2252.

The University of British Columbia, Faculty of Graduate Studies, Faculty of Land and Food Systems, Vancouver, BC V6T 1Z1, Canada. Offers M Sc, PhD. *Faculty:* 54 full-time (14 women), 10 part-time/adjunct (4 women). *Students:* 162 full-time (114 women). Average age 24. 270 applicants, 33% accepted, 59 enrolled. In 2005, 21 master's, 11 doctorates awarded. *Degree requirements:* For master's, thesis/dissertation, registration; for doctorate, thesis/dissertation, comprehensive exam, registration. *Entrance requirements:* Additional exam requirements/recommendations for international students: Required—TOEFL (minimum score 577 paper-based; 220 computer-based). *Application deadline:* For fall admission, 1/3 for domestic students, 1/3 for international students. For winter admission, 6/1 for domestic students; for spring admission, 9/1 for domestic students. Applications are processed on a rolling basis. Application fee: $90 Canadian dollars ($150 Canadian dollars for international students). Electronic applications accepted. *Financial support:* In 2005–06, 91 fellowships with partial tuition reimbursements (averaging $9,104 per year), 49 research assistantships with partial tuition reimbursements (averaging $14,000 per year), 87 teaching assistantships with partial tuition reimbursements (averaging $2,200 per year) were awarded; career-related internships or fieldwork, Federal Work-Study, institutionally sponsored loans, scholarships/grants, and tuition waivers (full and partial) also available. *Unit head:* Dr. Mahesh Upadhyaya, Associate Dean, Graduate Programs, 604-822-4593, Fax: 604-822-4400, E-mail: upadh@interchange.ubc.ca. *Application contact:* Lia Maria Dragan, Graduate Programs Assistant, 604-822-8373, Fax: 604-822-4400, E-mail: gradapp@interchange.ubc.ca.

University of California, Davis, Graduate Studies, Graduate Group in International Agricultural Development, Davis, CA 95616. Offers MS. *Faculty:* 98 full-time. *Students:* 36 full-time (23 women); includes 2 minority (1 Asian American or Pacific Islander, 1 Hispanic American), 4 international. Average age 29. 22 applicants, 82% accepted, 9 enrolled. In 2005, 9 degrees awarded. *Degree requirements:* For master's, thesis (for some programs), comprehensive exam (for some programs). *Entrance requirements:* For master's, GRE General Test, minimum GPA of 3.0. Additional exam requirements/recommendations for international students: Required—TOEFL (minimum score 550 paper-based; 213 computer-based). *Application deadline:* For fall admission, 1/15 for domestic students, 12/15 for international students. Application fee: $60. Electronic applications accepted. *Financial support:* In 2005–06, 15 fellowships with full and partial tuition reimbursements (averaging $9,912 per year), 10 research assistantships with full and partial tuition reimbursements (averaging $11,825 per year), 7 teaching assistantships with partial tuition reimbursements (averaging $14,979 per year) were awarded; Federal Work-Study, institutionally sponsored loans, scholarships/grants, and tuition waivers (full and partial) also available. Financial award application deadline: 1/15; financial award applicants required to submit FAFSA. *Faculty research:* Aspects of agricultural, environmental and social sciences on agriculture and related issues in developing countries. *Unit head:* Patrick Brown, Graduate Chair, 530-752-0929, Fax: 530-752-5660, E-mail: phbrown@ucdavis.edu. *Application contact:* Lisa Brown, Graduate Group Secretary, 530-752-7738, Fax: 530-752-1819, E-mail: lfbrown@ucdavis.edu.

University of Connecticut, Graduate School, College of Agriculture and Natural Resources, Storrs, CT 06269. Offers MS, PhD. *Faculty:* 76 full-time (18 women). *Students:* 147 full-time (81 women), 43 part-time (25 women); includes 10 minority (3 African Americans, 4 Asian Americans or Pacific Islanders, 3 Hispanic Americans), 72 international. Average age 31. 158 applicants, 49% accepted, 58 enrolled. In 2005, 28 master's, 17 doctorates awarded. Terminal master's awarded for partial completion of doctoral program. *Degree requirements:* For master's, comprehensive exam; for doctorate, thesis/dissertation, comprehensive exam. *Entrance requirements:* For master's and doctorate, GRE General Test. Additional exam requirements/recommendations for international students: Required—TOEFL (minimum score 550 paper-based; 213 computer-based). *Application deadline:* For fall admission, 2/1 priority date for domestic students, 2/1 priority date for international students; for spring admission, 11/1 for domestic students, 10/1 for international students. Applications are processed on a rolling basis. Application fee: $55. Electronic applications accepted. *Expenses:* Tuition, state resident: part-time $444 per credit hour. Tuition, nonresident: part-time $1,154 per credit hour. Tuition and fees vary according to course load. *Financial support:* In 2005–06, 118 research assistantships with full tuition reimbursements, 11 teaching assistantships with full tuition reimbursements were awarded; fellowships, Federal Work-Study, scholarships/grants, health care benefits, and unspecified assistantships also available. Financial award application deadline: 2/1; financial award applicants required to submit FAFSA. *Unit head:* Kirklyn M. Kerr, Dean, 860-486-2917, Fax: 860-486-5113, E-mail: kirklyn.ker@uconn.edu. *Application contact:* Larissa Hull, Assistant, 860-486-2918, Fax: 860-486-5113, E-mail: larissa.hull@uconn.edu.

University of Delaware, College of Agriculture and Natural Resources, Newark, DE 19716. Offers MS, PhD. Part-time programs available. *Faculty:* 87 full-time (18 women). *Students:* 158 full-time (83 women), 17 part-time (5 women); includes 8 minority (3 African Americans, 4 Asian Americans or Pacific Islanders, 1 Hispanic American), 61 international. Average age 28. 191 applicants, 46% accepted, 71 enrolled. In 2005, 38 master's, 5 doctorates awarded. *Degree requirements:* For master's and doctorate, thesis/dissertation. *Entrance requirements:* For master's and doctorate, GRE General Test. Application fee: $60. Electronic applications accepted. *Financial support:* In 2005–06, 100 students received support, including 14 fellowships with full tuition reimbursements available (averaging $19,601 per year), 47 research assistantships with full tuition reimbursements available (averaging $16,272 per year), 17 teaching assistantships with full tuition reimbursements available (averaging $12,204 per year); career-related internships or fieldwork, Federal Work-Study, institutionally sponsored loans, and tuition waivers (full) also available. Total annual research expenditures: $12,500. *Unit head:* Dr. Robin Morgan, Dean, 302-831-2501. *Application contact:* Karen Roth Aniunas, Assistant Dean of Student Services, 302-831-2508, Fax: 302-831-6758, E-mail: kra@udel.edu.

University of Florida, Graduate School, College of Agricultural and Life Sciences, Gainesville, FL 32611. Offers M Ag, MAB, MFAS, MFRC, MFYCS, MS, DPM, PhD, JD/MFRC, JD/MS, JD/PhD. Part-time programs available. *Faculty:* 817. *Students:* 941. In 2005, 140 master's, 77 doctorates awarded. *Degree requirements:* For doctorate, thesis/dissertation. *Entrance requirements:* For master's and doctorate, GRE General Test, minimum GPA of 3.0. Additional exam requirements/recommendations for international students: Required—TOEFL. *Application deadline:* Applications are processed on a rolling basis. Application fee: $20. Electronic applications accepted. *Expenses:* Tuition, state resident: full-time $6,234. Tuition, nonresident: full-time $21,359. Tuition and fees vary according to program. *Financial support:* In 2005–06, 390 students received support, including 22 fellowships with tuition reimbursements available; research assistantships with tuition reimbursements available, teaching assistantships with tuition reimbursements available, career-related internships or fieldwork, Federal Work-Study, institutionally sponsored loans, and unspecified assistantships also available. Support available to part-time students. *Unit head:* R. Kirby Barrick, Dean, 352-392-1971. *Application contact:* Dr. E. Jane Luzar, Associate Dean for Academic Programs, 352-392-2251, Fax: 352-392-8988, E-mail: ejluzar@ufl.edu.

University of Georgia, Graduate School, College of Agricultural and Environmental Sciences, Athens, GA 30602. Offers MA Ext, MADS, MAE, MAL, MCCS, MFT, MPPPM, MS, PhD. *Faculty:* 99 full-time (24 women), 1 part-time/adjunct (0 women). *Students:* 277 full-time, 85

Agricultural Sciences—General

University of Georgia (continued)
part-time; includes 25 minority (19 African Americans, 1 American Indian/Alaska Native, 2 Asian Americans or Pacific Islanders, 3 Hispanic Americans), 146 international. 275 applicants, 57% accepted, 92 enrolled. In 2005, 87 master's, 21 doctorates awarded. *Degree requirements:* For doctorate, thesis/dissertation. *Entrance requirements:* For master's and doctorate, GRE General Test. *Application deadline:* For fall admission, 7/1 for domestic students; for spring admission, 11/15 for domestic students. Application fee: $50. Electronic applications accepted. *Financial support:* Fellowships, research assistantships, teaching assistantships, career-related internships or fieldwork and unspecified assistantships available. *Unit head:* Dr. J. Scott Angle, Dean, 706-542-3924, Fax: 706-542-0803.

University of Guelph, Graduate Program Services, Ontario Agricultural College, Guelph, ON N1G 2W1, Canada. Offers M Sc, MLA, PhD, Diploma. Part-time programs available. Post-baccalaureate distance learning degree programs offered (minimal on-campus study). *Students:* 704. In 2005, 116 master's, 35 doctorates awarded. *Degree requirements:* For doctorate, thesis/dissertation. Application fee: $75. *Financial support:* Fellowships, research assistantships, teaching assistantships, scholarships/grants and unspecified assistantships available. Support available to part-time students. *Unit head:* Dr. Craig J. Pearson, Dean, 519-824-4120 Ext. 52285, Fax: 519-766-1423, E-mail: cpearson@uoguelph.ca.

University of Hawaii at Manoa, Graduate Division, College of Tropical Agriculture and Human Resources, Honolulu, HI 96822. Offers MS, PhD. Part-time programs available. *Degree requirements:* For doctorate, thesis/dissertation. *Entrance requirements:* For doctorate, GRE. Application fee: $50. *Expenses:* Tuition, state resident: full-time $8,400; part-time $200 per credit hour. Tuition, nonresident: full-time $11,088; part-time $462 per credit hour. Tuition and fees vary according to program. *Financial support:* Fellowships, career-related internships or fieldwork, Federal Work-Study, institutionally sponsored loans, tuition waivers (full and partial), and unspecified assistantships available. *Unit head:* Dr. Andrew Hashimoto, Dean, 808-956-8234, Fax: 808-956-9105, E-mail: dean@ctahr.hawaii.edu. *Application contact:* Dr. Andrew Hashimoto, Dean, 808-956-8234, Fax: 808-956-9105, E-mail: dean@ctahr.hawaii.edu.

University of Illinois at Urbana–Champaign, Graduate College, College of Agricultural, Consumer and Environmental Sciences, Champaign, IL 61820. Offers MS, PhD. *Faculty:* 220 full-time (47 women), 10 part-time/adjunct (2 women). *Students:* 433 full-time (248 women), 101 part-time (52 women); includes 45 minority (10 African Americans, 26 Asian Americans or Pacific Islanders, 9 Hispanic Americans), 186 international. 551 applicants, 30% accepted, 107 enrolled. In 2005, 122 master's, 59 doctorates awarded. *Degree requirements:* For doctorate, thesis/dissertation. *Entrance requirements:* For master's, minimum GPA of 3.0. *Application deadline:* Applications are processed on a rolling basis. Application fee: $50 ($60 for international students). Electronic applications accepted. *Financial support:* In 2005–06, 110 fellowships, 373 research assistantships, 92 teaching assistantships were awarded; career-related internships or fieldwork and tuition waivers (full and partial) also available. Financial award application deadline: 2/15. *Unit head:* Robert A. Easter, Dean, 217-333-0460, Fax: 217-244-2911, E-mail: reaster@uiuc.edu. *Application contact:* Robert A. Easter, Dean, 217-333-0460, Fax: 217-244-2911, E-mail: reaster@uiuc.edu.

University of Kentucky, Graduate School, Graduate School Programs in the College of Agriculture, Lexington, KY 40506-0032. Offers MS, MSFAM, MSFOR, PhD. Part-time programs available. *Faculty:* 191 full-time (28 women), 10 part-time/adjunct (0 women). *Students:* 300 full-time (170 women), 56 part-time (22 women); includes 24 minority (15 African Americans, 5 Asian Americans or Pacific Islanders, 4 Hispanic Americans), 99 international. Average age 30. 335 applicants, 58% accepted, 140 enrolled. In 2005, 39 master's, 26 doctorates awarded. Terminal master's awarded for partial completion of doctoral program. *Median time to degree:* Of those who began their doctoral program in fall 1997, 72.6% received their degree in 8 years or less. *Degree requirements:* For master's, thesis (for some programs), comprehensive exam; for doctorate, thesis/dissertation, comprehensive exam. *Entrance requirements:* For master's, GRE General Test, minimum undergraduate GPA of 2.5; for doctorate, GRE General Test, minimum undergraduate GPA of 3.0. Additional exam requirements/recommendations for international students: Required—TOEFL (minimum score 550 paper-based; 213 computer-based). *Application deadline:* For fall admission, 7/17 priority date for domestic students, 2/1 priority date for international students; for spring admission, 12/13 priority date for domestic students, 6/15 priority date for international students. Applications are processed on a rolling basis. Application fee: $40 ($55 for international students). Electronic applications accepted. *Expenses:* Tuition, state resident: full-time $3,159; part-time $331 per credit hour. Tuition, nonresident: full-time $6,984; part-time $756 per credit hour. Tuition and fees vary according to course load, degree level and program. *Financial support:* In 2005–06, 27 fellowships with full tuition reimbursements (averaging $2,975 per year), 211 research assistantships with full tuition reimbursements (averaging $15,000 per year), 26 teaching assistantships with full tuition reimbursements (averaging $5,401 per year) were awarded; career-related internships or fieldwork, Federal Work-Study, institutionally sponsored loans, scholarships/grants, traineeships, health care benefits, tuition waivers (partial), and unspecified assistantships also available. Support available to part-time students. Financial award application deadline: 3/15. *Unit head:* Dr. M. Scott Smith, Dean, 859-257-4772, Fax: 859-323-2885, E-mail: mssmith@uky.edu. *Application contact:* Dr. Brian Jackson, Senior Associate Dean, 859-257-8176, Fax: 859-323-1928, E-mail: lance.brunner@uky.edu.

University of Lethbridge, School of Graduate Studies, Lethbridge, AB T1K 3M4, Canada. Offers accounting (MScM); addictions counseling (M Sc); agricultural biotechnology (M Sc); agricultural studies (M Sc, MA); anthropology (MA); archaeology (MA); art (MA); biochemistry (M Sc); biological sciences (M Sc); biomolecular science (PhD); biosystems and biodiversity (PhD); Canadian studies (MA); chemistry (M Sc); computer science (M Sc); computer science and geographical information science (M Sc); counseling psychology (M Ed); dramatic arts (MA); earth, space, and physical science (PhD); economics (MA); educational leadership (M Ed); English (MA); environmental science (M Sc); evolution and behavior (PhD); exercise science (M Sc); finance (MScM); French (MA); French/German (MA); French/Spanish (MA); general education (M Ed); general management (MScM); geography (M Sc, MA); German (MA); health sciences (M Sc, MA); history (MA); human resource management and labour relations (MScM); individualized multidisciplinary (M Sc, MA); information systems (MScM); international management (MScM); kinesiology (M Sc, MA); management (M Sc, MA); marketing (MScM); mathematics (M Sc); music (MA); Native American studies (MA); neuroscience (M Sc, PhD); new media (MA); nursing (M Sc); philosophy (MA); physics (M Sc); policy and strategy (MScM); political science (MA); psychology (M Sc, MA); religious studies (MA); sociology (MA); theoretical and computational science (PhD); urban and regional studies (MA). Part-time and evening/weekend programs available. *Faculty:* 250. *Students:* 193 full-time, 145 part-time. 35 applicants, 100% accepted, 35 enrolled. In 2005, 40 degrees awarded. *Degree requirements:* For doctorate, thesis/dissertation, comprehensive exam. *Entrance requirements:* For master's, GMAT (M Sc management), bachelor's degree in related field, minimum GPA of 3.0 during previous 20 graded semester courses, 2 years teaching or related experience (M Ed); for doctorate, master's degree, minimum graduate GPA of 3.5. Additional exam requirements/recommendations for international students: Required—TOEFL. Application fee: $60 Canadian dollars. *Expenses:* Tuition, nonresident: part-time $531 per course. Required fees: $83 per year. Tuition and fees vary according to degree level and program. *Financial support:* Fellowships, research assistantships, teaching assistantships, scholarships/grants, health care benefits, and unspecified assistantships available. *Faculty research:* Movement and brain plasticity, gibberellin physiology, photosynthesis, carbon cycling, molecular properties of main-group ring components. *Unit head:* Dr. Shamsul Alam, Dean, 403-329-2121, Fax: 403-329-2097, E-mail: inquiries@uleth.ca. *Application contact:* Kathy Schrage, Administrative Assistant, Office of the Academic Vice President, 403-329-2121, Fax: 403-329-2097, E-mail: inquiries@uleth.ca.

University of Maine, Graduate School, College of Natural Sciences, Forestry, and Agriculture, Orono, ME 04469. Offers MF, MPS, MS, MWC, PhD. *Accreditation:* SAF (one or more programs are accredited). Part-time and evening/weekend programs available. *Students:* 245 full-time (132 women), 121 part-time (75 women); includes 8 minority (1 African American, 2 American Indian/Alaska Native, 4 Asian Americans or Pacific Islanders, 1 Hispanic American), 62 international. Average age 30. 327 applicants, 37% accepted, 69 enrolled. In 2005, 77 master's, 12 doctorates awarded. *Degree requirements:* For doctorate, thesis/dissertation. *Entrance requirements:* For master's and doctorate, GRE General Test. Additional exam requirements/recommendations for international students: Required—TOEFL. *Application deadline:* For fall admission, 2/1 for domestic students. Applications are processed on a rolling basis. Application fee: $50. Electronic applications accepted. *Financial support:* Fellowships, research assistantships, teaching assistantships, career-related internships or fieldwork, Federal Work-Study, institutionally sponsored loans, scholarships/grants, tuition waivers (full and partial), and unspecified assistantships available. Support available to part-time students. Financial award application deadline: 3/1. *Unit head:* Dr. G. Bruce Wiersma, Dean, 207-581-3202, Fax: 207-581-3207. *Application contact:* Scott G. Delcourt, Associate Dean of the Graduate School, 207-581-3219, Fax: 207-581-3232, E-mail: graduate@maine.edu.

University of Manitoba, Faculty of Graduate Studies, Faculty of Agriculture, Winnipeg, MB R3T 2N2, Canada. Offers M Sc, PhD. *Degree requirements:* For master's, thesis or alternative; for doctorate, variable foreign language requirement, thesis/dissertation.

University of Maryland, College Park, Graduate Studies, College of Agriculture and Natural Resources, College Park, MD 20742. Offers DVM, MS, PhD. Part-time and evening/weekend programs available. *Faculty:* 309 full-time (114 women), 39 part-time/adjunct (25 women). *Students:* 304 full-time (210 women), 34 part-time (20 women); includes 29 minority (9 African Americans, 2 American Indian/Alaska Native, 14 Asian Americans or Pacific Islanders, 4 Hispanic Americans), 97 international. 300 applicants, 34% accepted, 68 enrolled. In 2005, 25 first professional degrees, 27 master's, 17 doctorates awarded. *Median time to degree:* Of those who began their doctoral program in fall 1997, 32% received their degree in 8 years or less. *Degree requirements:* For DVM, thesis, oral exam, public seminar; for doctorate, thesis/dissertation. *Entrance requirements:* For DVM, GRE General Test; for master's, minimum GPA of 3.0. Additional exam requirements/recommendations for international students: Required—TOEFL. *Application deadline:* For fall admission, 5/1 for domestic students, 2/1 for international students; for spring admission, 10/1 for domestic students, 6/1 for international students. Applications are processed on a rolling basis. Application fee: $60 ($70 for international students). Electronic applications accepted. *Financial support:* In 2005–06, 28 fellowships with full tuition reimbursements (averaging $7,469 per year), 105 research assistantships with tuition reimbursements (averaging $17,654 per year), 56 teaching assistantships with tuition reimbursements (averaging $16,192 per year) were awarded; career-related internships or fieldwork, Federal Work-Study, and scholarships/grants also available. Support available to part-time students. Financial award applicants required to submit FAFSA. Total annual research expenditures: $24.5 million. *Unit head:* Dr. Cheng-i Wei, Dean, 301-405-2072, Fax: 301-314-9146, E-mail: wei@umd.edu. *Application contact:* Dean of Graduate School, 301-405-4190, Fax: 301-314-9305.

University of Maryland Eastern Shore, Graduate Programs, Department of Agriculture, Princess Anne, MD 21853-1299. Offers food and agricultural sciences (MS); food science and technology (PhD). *Faculty:* 16 full-time (5 women). *Students:* 12 full-time (6 women), 24 part-time (16 women); includes 6 minority (all African Americans), 12 international. Average age 29. 21 applicants, 62% accepted, 1 enrolled. In 2005, 8 master's, 2 doctorates awarded. *Degree requirements:* For master's, thesis (for some programs), oral exam, comprehensive exam; for doctorate, thesis/dissertation, comprehensive exam. *Entrance requirements:* For master's, GRE, minimum GPA of 3.0. Additional exam requirements/recommendations for international students: Required—TOEFL (minimum score 550 paper-based; 213 computer-based). *Application deadline:* For fall admission, 4/15 priority date for domestic students, 4/15 priority date for international students; for spring admission, 10/30 priority date for domestic students, 10/30 priority date for international students. Applications are processed on a rolling basis. Application fee: $30. Electronic applications accepted. *Expenses:* Tuition, area resident: Part-time $216. Tuition, nonresident: part-time $392. Required fees: $40. *Financial support:* In 2005–06, 17 students received support, including 17 research assistantships with full and partial tuition reimbursements available (averaging $15,000 per year); scholarships/grants and unspecified assistantships also available. Financial award application deadline: 3/1; financial award applicants required to submit FAFSA. *Faculty research:* Poultry and swine nutrition and management, soybean specialty products, farm management practices, aquaculture technology. Total annual research expenditures: $2.5 million. *Unit head:* Dr. Lurline Marsh, Chair, 410-651-6168, Fax: 410-651-7656, E-mail: lemarsh@umes.edu.

University of Minnesota, Twin Cities Campus, Graduate School, College of Agricultural, Food, and Environmental Sciences, Minneapolis, MN 55455-0213. Offers MA, MBAE, MS, MSBAE, PhD. Part-time and evening/weekend programs available. *Students:* 292 full-time (143 women), 143 part-time (73 women); includes 30 minority (9 African Americans, 1 American Indian/Alaska Native, 10 Asian Americans or Pacific Islanders, 10 Hispanic Americans), 132 international. 371 applicants, 43% accepted, 96 enrolled. In 2005, 72 master's, 42 doctorates awarded. Application fee: $50 ($55 for international students). *Expenses:* Tuition, state resident: full-time $8,748; part-time $729 per credit. Tuition, nonresident: full-time $15,848; part-time $1,321 per credit. Full-time tuition and fees vary according to class time, course load, program and reciprocity agreements. *Financial support:* Fellowships, research assistantships, teaching assistantships, career-related internships or fieldwork, Federal Work-Study, institutionally sponsored loans, and tuition waivers (full) available. Support available to part-time students. *Unit head:* Dr. Charles C. Muscoplat, Dean, 612-624-5387. *Application contact:* Steve Gillard, Information Contact, 612-625-6792, E-mail: sgillard@umn.edu.

University of Missouri–Columbia, Graduate School, College of Agriculture, Food and Natural Resources, Columbia, MO 65211. Offers MS, PhD, MD/PhD. Part-time programs available. *Faculty:* 196 full-time (34 women), 3 part-time/adjunct (1 woman). *Students:* 169 full-time (74 women), 118 part-time (53 women); includes 13 minority (7 African Americans, 3 American Indian/Alaska Native, 2 Asian Americans or Pacific Islanders, 1 Hispanic American), 111 international. In 2005, 38 master's, 24 doctorates awarded. *Degree requirements:* For doctorate, thesis/dissertation. *Entrance requirements:* For master's and doctorate, GRE General Test, minimum GPA of 3.0. *Application deadline:* Applications are processed on a rolling basis. Application fee: $45 ($60 for international students). *Financial support:* Fellowships, research assistantships, teaching assistantships, institutionally sponsored loans available. *Unit head:* Dr. Thomas T. Payne, Dean, 573-882-3846, E-mail: paynet@missouri.edu.

University of Nebraska–Lincoln, Graduate College, College of Agricultural Sciences and Natural Resources, Lincoln, NE 68588. Offers M Ag, MA, MS, PhD. *Degree requirements:* For doctorate, thesis/dissertation, comprehensive exam. *Entrance requirements:* Additional exam requirements/recommendations for international students: Required—TOEFL. Electronic applications accepted. *Faculty research:* Environmental sciences, animal sciences, human resources and family sciences, plant breeding and genetics, food and nutrition.

University of Nevada, Reno, Graduate School, College of Agriculture, Biotechnology and Natural Resources, Reno, NV 89557. Offers MS, PhD. *Degree requirements:* For master's, thesis optional. *Entrance requirements:* For master's, GRE General Test, minimum GPA of 2.75. Additional exam requirements/recommendations for international students: Required—TOEFL.

University of Puerto Rico, Mayagüez Campus, Graduate Studies, College of Agricultural Sciences, Mayagüez, PR 00681-9000. Offers MS. Part-time programs available. *Faculty:* 101. *Students:* 45 full-time (26 women), 108 part-time (61 women); includes 148 minority (all Hispanic Americans), 5 international. 60 applicants, 72% accepted, 25 enrolled. In 2005, 3 degrees awarded. *Degree requirements:* For master's, thesis, comprehensive exam. *Application deadline:* For fall admission, 2/15 for domestic students; for spring admission, 9/15 for domestic students. Applications are processed on a rolling basis. Application fee: $20. *Expenses:* Tuition, state resident: full-time $900; part-time $100 per credit. International tuition: $4,655 full-time. Part-time tuition and fees vary according to course level and course load. *Financial support:* In 2005–06, 89 students received support, including 21 fellowships with tuition reimbursements available (averaging $1,200 per year), 18 research assistantships with tuition reimbursements available (averaging $1,500 per year), 47 teaching assistantships with tuition reimbursements available (averaging $987 per year); career-related internships or fieldwork, Federal Work-Study, and institutionally sponsored loans also available. Total annual research expenditures: $128,648. *Unit head:* Dr. John Fernández-VanCleve, Dean, 787-832-4040 Ext. 2180, E-mail: john@uprm.edu.

University of Saskatchewan, College of Graduate Studies and Research, College of Agriculture, Saskatoon, SK S7N 5A2, Canada. Offers M Ag, M Sc, MA, PhD. Part-time programs available. *Degree requirements:* For master's, thesis (for some programs), registration; for doctorate, thesis/dissertation, registration. *Entrance requirements:* Additional exam requirements/recommendations for international students: Required—TOEFL.

The University of Tennessee, Graduate School, College of Agricultural Sciences and Natural Resources, Knoxville, TN 37996. Offers MS, PhD. Part-time programs available. Postbaccalaureate distance learning degree programs offered (minimal on-campus study). *Degree requirements:* For master's, thesis (for some programs); for doctorate, thesis/dissertation. *Entrance requirements:* For master's and doctorate, minimum GPA of 2.7. Additional exam requirements/recommendations for international students: Required—TOEFL. Electronic applications accepted.

The University of Tennessee at Martin, Graduate Programs, College of Agriculture and Applied Sciences, Program in Agricultural Operations Management, Martin, TN 38238-1000. Offers MSAOM. Part-time programs available. Postbaccalaureate distance learning degree programs offered (no on-campus study). *Faculty:* 14. *Students:* 20 (9 women). 14 applicants, 64% accepted, 6 enrolled. In 2005, 8 degrees awarded. *Degree requirements:* For master's, comprehensive exam. *Entrance requirements:* For master's, GRE General Test, minimum GPA of 2.5. Additional exam requirements/recommendations for international students: Required—TOEFL (minimum score 525 paper-based; 197 computer-based). *Application deadline:* For fall admission, 8/1 priority date for domestic students, 8/1 priority date for international students; for spring admission, 1/1 priority date for domestic students, 1/1 priority date for international students. Applications are processed on a rolling basis. Application fee: $25 ($50 for international students). Electronic applications accepted. *Expenses:* Tuition, state resident: full-time $5,200; part-time $291 per hour. Tuition, nonresident: full-time $9,000; part-time $794 per hour. International tuition: $14,200 full-time. *Financial support:* In 2005–06, 2 students received support. Scholarships/grants and unspecified assistantships available. *Unit head:* Dr. Tim Burcham, Coordinator, 731-881-7275, E-mail: burcham@utm.edu. *Application contact:* Linda S. Arant, Student Service Specialist, 731-881-7012, Fax: 731-881-7499, E-mail: larant@utm.edu.

University of Vermont, Graduate College, College of Agriculture and Life Sciences, Burlington, VT 05405. Offers M Ext Ed, MAT, MPA, MS, MST, PhD. Part-time programs available. *Students:* 139 (89 women); includes 6 minority (2 African Americans, 1 Asian American or Pacific Islander, 3 Hispanic Americans) 23 international. 155 applicants, 64% accepted, 54 enrolled. In 2005, 25 master's, 2 doctorates awarded. *Degree requirements:* For doctorate, one foreign language, thesis/dissertation. *Entrance requirements:* For master's and doctorate, GRE General Test. Additional exam requirements/recommendations for international students: Required—TOEFL (minimum score 550 paper-based; 213 computer-based). Application fee: $40. *Expenses:* Tuition, area resident: Part-time $410 per credit hour. Tuition, nonresident: part-time $1,034 per credit hour. *Financial support:* Fellowships, research assistantships, teaching assistantships, career-related internships or fieldwork, Federal Work-Study, and tuition waivers (full and partial) available. Financial award application deadline: 3/1. *Unit head:* Dr. R. K. Johnson, Dean, 802-656-2980.

University of Wisconsin–Madison, Graduate School, College of Agricultural and Life Sciences, Madison, WI 53706-1380. Offers MA, MS, PhD. Part-time programs available. *Faculty:* 299 full-time (60 women). *Students:* 945 full-time (503 women), 98 part-time (52 women); includes 89 minority (15 African Americans, 8 American Indian/Alaska Native, 28 Asian Americans or Pacific Islanders, 38 Hispanic Americans). Average age 29. 909 applicants, 22% accepted, 105 enrolled. In 2005, 81 master's, 56 doctorates awarded. *Entrance requirements:* For master's and doctorate, GRE. Additional exam requirements/recommendations for international students: Required—TOEFL. *Application deadline:* For fall admission, 12/4 for domestic students. Application fee: $45. Electronic applications accepted. *Financial support:* In 2005–06, fellowships (averaging $19,200 per year), research assistantships (averaging $18,120 per year), teaching assistantships (averaging $24,200 per year) were awarded; career-related internships or fieldwork, Federal Work-Study, institutionally sponsored loans, traineeships, health care benefits, tuition waivers (full and partial), and project assistantships also available. Support available to part-time students. Financial award applicants required to submit FAFSA. Total annual research expenditures: $108.1 million. *Unit head:* David B. Hogg, Interim Dean, 608-262-4930, Fax: 608-262-4556, E-mail: dhogg@cals.wisc.edu.

University of Wisconsin–River Falls, Outreach and Graduate Studies, College of Agriculture, Food, and Environmental Sciences, River Falls, WI 54022-5001. Offers MS. Part-time programs available. *Degree requirements:* For master's, thesis (for some programs), comprehensive exam. *Entrance requirements:* For master's, minimum GPA of 2.75. Electronic applications accepted.

University of Wyoming, Graduate School, College of Agriculture, Laramie, WY 82070. Offers MS, PhD. Part-time programs available. *Faculty:* 102 full-time (13 women), 10 part-time/adjunct (2 women). *Students:* 93 full-time (45 women), 55 part-time (29 women); includes 4 minority (1 American Indian/Alaska Native, 2 Asian Americans or Pacific Islanders, 1 Hispanic American), 37 international. Average age 28. 101 applicants, 40% accepted. In 2005, 28 master's, 7 doctorates awarded. Terminal master's awarded for partial completion of doctoral program. *Degree requirements:* For doctorate, thesis/dissertation. *Entrance requirements:* For master's and doctorate, GRE General Test, minimum GPA of 3.0. *Application deadline:* Applications are processed on a rolling basis. Application fee: $50. Electronic applications accepted. *Expenses:* Tuition, state resident: full-time $3,720; part-time $155 per credit hour. Tuition, nonresident: full-time $10,704; part-time $446 per credit hour. Required fees: $666; $162 per semester. Tuition and fees vary according to course load and program. *Financial support:* In 2005–06, 3 fellowships, 15 research assistantships, 32 teaching assistantships were awarded; career-related internships or fieldwork, Federal Work-Study, institutionally sponsored loans, scholarships/grants, tuition waivers (partial), and unspecified assistantships also available. Financial award application deadline: 3/1. *Faculty research:* Nutrition, molecular biology, animal science, plant science, entomology. Total annual research expenditures: $3.5 million. *Unit head:* Dr. Frank D. Galey, Dean, 307-766-4133, E-mail: fgaley@uwyo.edu.

Utah State University, School of Graduate Studies, College of Agriculture, Logan, UT 84322. Offers MDA, MFMS, MS, PhD. Part-time programs available. Postbaccalaureate distance learning degree programs offered (minimal on-campus study). *Faculty:* 92 full-time (19 women), 19 part-time/adjunct (0 women). *Students:* 81 full-time (35 women), 28 part-time (9 women); includes 2 minority (both African Americans), 21 international. Average age 27. 66 applicants, 52% accepted, 20 enrolled. In 2005, 30 master's, 11 doctorates awarded. Terminal master's awarded for partial completion of doctoral program. *Degree requirements:* For doctorate, thesis/dissertation. *Entrance requirements:* For master's and doctorate, GRE General Test, minimum GPA of 3.0. Additional exam requirements/recommendations for international students: Required—TOEFL. *Application deadline:* For fall admission, 6/15 for domestic students; for spring admission, 10/15 for domestic students. Applications are processed on a rolling basis. Application fee: $50 ($60 for international students). *Financial support:* In 2005–06, fellowships with full and partial tuition reimbursements (averaging $15,000 per year), research assistantships with full and partial tuition reimbursements (averaging $13,000 per year), teaching assistantships with full and partial tuition reimbursements (averaging $8,000 per year) were awarded; career-related internships or fieldwork, Federal Work-Study, institutionally sponsored loans, scholarships/grants, tuition waivers (full and partial), and unspecified assistantships also available. Support available to part-time students. *Faculty research:* Low-input agriculture, anti-viral chemotherapy, lactic culture, environmental biophysics and climate. *Unit head:* Noelle E. Cockett, Dean, 435-797-2215.

Virginia Polytechnic Institute and State University, Graduate School, College of Agriculture and Life Sciences, Blacksburg, VA 24061. Offers M Eng, MS, PhD. *Faculty:* 225 full-time (43 women), 1 (woman) part-time/adjunct. *Students:* 243 full-time (134 women), 57 part-time (32 women); includes 24 minority (9 African Americans, 3 American Indian/Alaska Native, 7 Asian Americans or Pacific Islanders, 5 Hispanic Americans), 84 international. Average age 29. 470 applicants, 43% accepted, 148 enrolled. In 2005, 59 master's, 32 doctorates awarded. *Entrance requirements:* Additional exam requirements/recommendations for international students: Required—TOEFL. *Application deadline:* Applications are processed on a rolling basis. Application fee: $45. Electronic applications accepted. *Expenses:* Tuition, state resident: full-time $6,558; part-time $364 per credit. Tuition, nonresident: full-time $11,296; part-time $628 per credit. Required fees: $1,419; $468 per credit. $234 per term. *Financial support:* In 2005–06, 19 fellowships with full tuition reimbursements (averaging $14,977 per year), 153 research assistantships with full tuition reimbursements (averaging $13,823 per year), 81 teaching assistantships with full tuition reimbursements (averaging $13,552 per year) were awarded; career-related internships or fieldwork, Federal Work-Study, scholarships/grants, and unspecified assistantships also available. Financial award application deadline: 4/1. *Faculty research:* Biotechnology, plant pathology, animal nutrition, agribusiness. *Unit head:* Dr. Sharron Quisenberry, Dean, 540-231-6503, Fax: 540-231-4163. *Application contact:* Sheila Norman, 540-231-4152, Fax: 540-231-4163, E-mail: snorman@vt.edu.

Washington State University, Graduate School, College of Agricultural, Human, and Natural Resource Sciences, Program in Agriculture, Pullman, WA 99164. Offers MS. *Faculty:* 18. *Students:* 1 (woman) full-time, 22 part-time (15 women). 19 applicants, 32% accepted, 5 enrolled. In 2005, 6 degrees awarded. *Degree requirements:* For master's, oral defense, thesis optional. *Entrance requirements:* For master's, 3 letters of recommendation. Additional exam requirements/recommendations for international students: Required—TOEFL. *Application deadline:* For fall admission, 2/1 for domestic students, 3/1 for international students; for spring admission, 9/1 for domestic students, 7/1 for international students. Application fee: $35. *Expenses:* Tuition, state resident: full-time $6,295; part-time $336 per credit. Tuition, nonresident: full-time $15,949; part-time $819 per credit. Required fees: $933. Part-time tuition and fees vary according to campus/location and program. *Financial support:* In 2005–06, 1 fellowship (averaging $2,850 per year) was awarded Total annual research expenditures: $2.5 million. *Unit head:* Dr. Claudio Stockle, Chair, 509-335-1578. *Application contact:* Michael Swan, Chair, Graduate Studies, 509-335-2899, Fax: 509-335-2722, E-mail: mswan@wsu.edu.

Western Kentucky University, Graduate Studies, Ogden College of Science and Engineering, Department of Agriculture, Bowling Green, KY 42101-3576. Offers MA Ed, MS. Part-time and evening/weekend programs available. *Faculty:* 7 full-time (1 woman). *Students:* 8 full-time (3 women), 10 part-time (2 women); includes 1 minority (Hispanic American), 1 international. Average age 27. 11 applicants, 73% accepted, 6 enrolled. In 2005, 6 degrees awarded. *Degree requirements:* For master's, thesis optional. *Entrance requirements:* For master's, GRE General Test, minimum GPA of 2.75. Additional exam requirements/recommendations for international students: Required—TOEFL (minimum score 555 paper-based; 213 computer-based). *Application deadline:* For fall admission, 7/1 priority date for domestic students, 5/15 priority date for international students; for spring admission, 11/1 for domestic students, 9/15 for international students. Applications are processed on a rolling basis. Application fee: $35. *Expenses:* Tuition, state resident: full-time $5,816; part-time $299 per credit hour. Tuition, nonresident: full-time $6,356; part-time $326 per credit hour. *Financial support:* In 2005–06, 8 students received support, including 2 research assistantships with partial tuition reimbursements available (averaging $9,000 per year), 6 teaching assistantships with partial tuition reimbursements available (averaging $9,000 per year); Federal Work-Study, institutionally sponsored loans, tuition waivers (partial), unspecified assistantships, and service awards also available. Support available to part-time students. Financial award application deadline: 4/1; financial award applicants required to submit FAFSA. *Faculty research:* Establishment of warm season grasses, heat composting, enrichment activities in agricultural education. Total annual research expenditures: $4,688. *Unit head:* Dr. Jack L Rudolph, Interim Head, 270-745-3151, Fax: 270-745-5972, E-mail: jack.rudolph@wku.edu.

West Texas A&M University, College of Agriculture, Nursing, and Natural Sciences, Division of Agriculture, Canyon, TX 79016-0001. Offers agricultural business and economics (MS); agriculture (PhD); animal science (MS); plant science (MS). Part-time programs available. *Degree requirements:* For master's, thesis optional. *Entrance requirements:* For master's, GRE General Test. Additional exam requirements/recommendations for international students: Required—TOEFL (minimum score 550 paper-based). Electronic applications accepted. *Faculty research:* Pest management, high plains green beans production and management, expected revenue for fixed cow/calf, digestibility and retention, inorganic/organic forms of copper zinc in mature horses .

West Virginia University, Davis College of Agriculture, Forestry and Consumer Sciences, Morgantown, WV 26506. Offers M Agr, MS, MSF, MSFCS, PhD. Part-time programs available. *Faculty:* 87 full-time (19 women), 16 part-time/adjunct (10 women). *Students:* 172 full-time (92 women), 79 part-time (34 women); includes 7 minority (1 African American, 4 Asian Americans or Pacific Islanders, 2 Hispanic Americans), 42 international. Average age 28. 136 applicants, 75% accepted, 75 enrolled. In 2005, 72 master's, 15 doctorates awarded. *Entrance requirements:* Additional exam requirements/recommendations for international students: Required—TOEFL. *Application deadline:* For fall admission, 6/1 priority date for domestic students, 6/1 priority date for international students; for spring admission, 1/5 for domestic students, 1/5 for international students. Applications are processed on a rolling basis. Application fee: $45. Electronic applications accepted. *Expenses:* Tuition, state resident: full-time $4,582; part-time $258 per credit hour. Tuition, nonresident: full-time $1,382; part-time $741 per credit hour. *Financial support:* In 2005–06, 8 fellowships (averaging $2,000 per year), 105 research assistantships (averaging $9,936 per year), 25 teaching assistantships (averaging $7,452 per year) were awarded; career-related internships or fieldwork, Federal Work-Study, institutionally sponsored loans, tuition waivers (full and partial), and unspecified assistantships also available. Financial award application deadline: 2/1; financial award applicants required to submit FAFSA. *Faculty research:* Reproductive physiology, soil and water quality, human nutrition, aquaculture, wildlife management. *Unit head:* Dr. Cameron R. Hackney, Dean, 304-293-2395 Ext. 4530, Fax: 304-293-3740, E-mail: cameron.hackney@mail.wvu.edu. *Application contact:* Dr. Dennis K. Smith, Associate Dean, 304-293-2691 Ext. 4521, Fax: 304-293-3740, E-mail: denny.smith@mail.wvu.edu.

Agronomy and Soil Sciences

Alabama Agricultural and Mechanical University, School of Graduate Studies, School of Agricultural and Environmental Sciences, Department of Plant and Soil Sciences, Huntsville, AL 35811. Offers animal sciences (MS); environmental science (MS); plant and soil science (PhD). Evening/weekend programs available. Terminal master's awarded for partial completion of doctoral program. *Degree requirements:* For master's, thesis; for doctorate, one foreign language, thesis/dissertation. *Entrance requirements:* For master's, GRE General Test, BS in agriculture; for doctorate, GRE General Test, master's degree. Electronic applications accepted. *Faculty research:* Plant breeding, cytogenetics, crop production, soil chemistry and fertility, remote sensing.

Alcorn State University, School of Graduate Studies, School of Agriculture and Applied Science, Alcorn State, MS 39096-7500. Offers agricultural economics (MS Ag); agronomy (MS Ag); animal science (MS Ag). *Faculty:* 11 full-time (2 women). *Students:* 7 full-time (2 women), 8 part-time (4 women); includes 12 minority (all African Americans), 3 international. In 2005, 3 degrees awarded. *Degree requirements:* For master's, thesis optional. *Application deadline:* For fall admission, 7/15 for domestic students; for spring admission, 11/25 for domestic students. Applications are processed on a rolling basis. Application fee: $0 ($10 for international students). *Financial support:* Career-related internships or fieldwork available. Support available to part-time students. *Faculty research:* Aquatic systems, dairy herd improvement, fruit production, alternative farming practices. *Unit head:* Robert Arthur, Interim Dean, 601-877-6137, Fax: 601-877-6219.

American University of Beirut, Graduate Programs, Faculty of Agricultural and Food Sciences, Beirut, Lebanon. Offers agricultural economics (MS); animal sciences (MS); ecosystem management (MSES); food technology (MS); irrigation (MS); mechanization (MS); nutrition (MS); plant protection (MS); plant science (MS); poultry science (MS); soils (MS). *Degree requirements:* For master's, one foreign language, thesis (for some programs), comprehensive exam, registration. *Entrance requirements:* For master's, GRE, letter of recommendation.

Auburn University, Graduate School, College of Agriculture, Department of Agronomy and Soils, Auburn University, AL 36849. Offers M Ag, MS, PhD. Part-time programs available. *Faculty:* 22 full-time (3 women). *Students:* 10 full-time (3 women), 13 part-time (2 women); includes 2 minority (1 African American, 1 Asian American or Pacific Islander), 8 international. 17 applicants, 88% accepted, 3 enrolled. In 2005, 4 master's, 3 doctorates awarded. *Degree requirements:* For master's, thesis (for some programs); for doctorate, thesis/dissertation. *Entrance requirements:* For master's and doctorate, GRE General Test. *Application deadline:* For fall admission, 7/7 for domestic students; for spring admission, 11/24 for domestic students. Applications are processed on a rolling basis. Application fee: $25 ($50 for international students). Electronic applications accepted. *Financial support:* Research assistantships, teaching assistantships, Federal Work-Study available. Support available to part-time students. Financial award application deadline: 3/15. *Faculty research:* Plant breeding and genetics; weed science; crop production; soil fertility and plant nutrition; soil genesis, morphology, and classification. *Unit head:* Dr. Joseph T. Touchton, Head, 334-844-4100, E-mail: jtouchto@ag.auburn.edu. *Application contact:* Dr. Stephen L. McFarland, Acting Dean of the Graduate School, 334-844-4700.

Brigham Young University, Graduate Studies, College of Biological and Agricultural Sciences, Department of Plant and Animal Sciences, Provo, UT 84602-1001. Offers agronomy (MS); genetics and biotechnology (MS). *Faculty:* 15 full-time (1 woman), 1 (woman) part-time/adjunct. *Students:* 10 full-time (4 women), 3 part-time (1 woman); includes 1 minority (Hispanic American) 14 applicants, 50% accepted, 6 enrolled. In 2005, 6 degrees awarded. *Degree requirements:* For master's, thesis, comprehensive exam, registration. *Entrance requirements:* For master's, GRE General Test, minimum GPA of 3.0 during last 60 hours of course work. *Application deadline:* For fall and spring admission, 2/15. Applications are processed on a rolling basis. Application fee: $50. Electronic applications accepted. *Financial support:* In 2005–06, 12 students received support, including 6 research assistantships with partial tuition reimbursements available (averaging $15,600 per year), 6 teaching assistantships with partial tuition reimbursements available (averaging $15,600 per year); institutionally sponsored loans, scholarships/grants, and tuition waivers (partial) also available. Financial award application deadline: 4/15. *Faculty research:* Iron nutrition in plants, cytogenetics plant pathology, pest management, genetic mapping, environmental science, plant genetics, molecular genetics. Total annual research expenditures:$160,761. *Unit head:* Dr. Sheldon D. Nelson, Chair, 801-422-2760, Fax: 801-422-0008, E-mail: sheldon_nelson@byu.edu. *Application contact:* Dr. Von D. Jolley, Graduate Coordinator, 801-422-2491, Fax: 801-422-0008, E-mail: von-jolley@byu.edu.

Colorado State University, Graduate School, College of Agricultural Sciences, Department of Soil and Crop Sciences, Fort Collins, CO 80523-0015. Offers crop science (MS, PhD); plant genetics (MS, PhD); soil science (MS, PhD). Part-time programs available. *Faculty:* 18 full-time (4 women), 1 part-time/adjunct (0 women). *Students:* 18 full-time (6 women), 15 part-time (4 women); includes 2 minority (both Hispanic Americans), 5 international. Average age 33. 16 applicants, 50% accepted, 7 enrolled. In 2005, 6 master's, 1 doctorate awarded. *Median time to degree:* Of those who began their doctoral program in fall 1997, 85% received their degree in 8 years or less. *Degree requirements:* For master's, thesis (for some programs), comprehensive exam, registration; for doctorate, thesis/dissertation, preliminary exam, comprehensive exam, registration. *Entrance requirements:* For master's, minimum GPA of 3.0, appropriate bachelor's degree; for doctorate, minimum GPA of 3.0, appropriate master's degree. Additional exam requirements/recommendations for international students: Required—TOEFL. *Application deadline:* For fall admission, 2/1 priority date for domestic students, 2/1 priority date for international students; for spring admission, 8/1 priority date for domestic students, 8/1 priority date for international students. Applications are processed on a rolling basis. Application fee: $50. Electronic applications accepted. *Expenses:* Tuition, state resident: full-time $3,690; part-time $205 per credit. Tuition, nonresident: full-time $14,958; part-time $831 per credit. Required fees: $1,061. *Financial support:* In 2005–06, 16 students received support, including 1 fellowship with partial tuition reimbursement available (averaging $15,600 per year), 14 research assistantships with partial tuition reimbursements available (averaging $15,600 per year), 1 teaching assistantship with partial tuition reimbursement available (averaging $15,600 per year); career-related internships or fieldwork and traineeships also available. *Faculty research:* Water quality, soil fertility, soil/plant ecosystems, plant breeding and genetics, information systems/technology. Total annual research expenditures: $3.5 million. *Unit head:* Dr. Gary A. Peterson, Head, 970-491-6501, Fax: 970-491-0564, E-mail: gary.peterson@colostate.edu. *Application contact:* Dr. Pat F. Byrne, Graduate Studies Coordinator, 970-491-6985, Fax: 970-491-0564, E-mail: pbyrne@lamar.colostate.edu.

Cornell University, Graduate School, Graduate Fields of Agriculture and Life Sciences, Field of Soil and Crop Sciences, Ithaca, NY 14853-0001. Offers agronomy (MS, PhD); environmental information science (MS, PhD); environmental management (MPS); field crop science (MS, PhD); soil science (MS, PhD). *Faculty:* 38 full-time (9 women). *Students:* 40 applicants, 33% accepted, 11 enrolled. In 2005, 6 master's, 3 doctorates awarded. *Degree requirements:* For master's, thesis (MS); for doctorate, thesis/dissertation, comprehensive exam. *Entrance requirements:* For master's and doctorate, GRE General Test, 2 letters of recommendation. Additional exam requirements/recommendations for international students: Required—TOEFL (minimum score 550 paper-based; 213 computer-based). *Application deadline:* For fall admission, 2/1 for domestic students. Applications are processed on a rolling basis. Application fee: $60. Electronic applications accepted. *Financial support:* In 2005–06, 35 students received support, including 6 fellowships with full tuition reimbursements available, 25 research assistantships with full tuition reimbursements available, 4 teaching assistantships with full tuition reimbursements available; institutionally sponsored loans, traineeships, health care benefits, tuition waivers (full and partial), and unspecified assistantships also available. *Faculty research:* Soil chemistry, physics and biology; crop physiology and management; environmental information science and modeling; international agriculture; weed science. *Unit head:* Director of Graduate Studies, 607-255-3267, Fax: 607-255-8615. *Application contact:* Graduate Field Assistant, 607-255-3267, Fax: 607-255-8615, E-mail: jae2@cornell.edu.

Cornell University, Graduate School, Graduate Fields of Agriculture and Life Sciences, Field of Vegetable Crops, Ithaca, NY 14853-0001. Offers MPS, MS, PhD. *Faculty:* 21 full-time (8 women). *Students:* 9 full-time (3 women), 4 international. In 2005, 1 master's, 1 doctorate awarded. *Degree requirements:* For master's, thesis (MS), project paper (MPS); for doctorate, thesis/dissertation, teaching experience. *Entrance requirements:* For master's and doctorate, GRE General Test (recommended), 3 letters of recommendation. Additional exam requirements/recommendations for international students: Required—TOEFL (minimum score 550 paper-based; 213 computer-based). *Application deadline:* Applications are processed on a rolling basis. Application fee: $60. Electronic applications accepted. *Financial support:* In 2005–06, 9 students received support, including 2 fellowships with full tuition reimbursements available, 7 research assistantships with full tuition reimbursements available; teaching assistantships with full tuition reimbursements available, institutionally sponsored loans, health care benefits, tuition waivers (full and partial), and unspecified assistantships also available. Financial award applicants required to submit FAFSA. *Faculty research:* Vegetable nutrition and physiology, post-harvest physiology and storage, application of new technologies, sustainable vegetable production, weed management and IPM. *Unit head:* Director of Graduate Studies, 607-255-4568. *Application contact:* Graduate Field Assistant, 607-255-4568, E-mail: hortgrad@cornell.edu.

Iowa State University of Science and Technology, Graduate College, College of Agriculture, Department of Agronomy, Ames, IA 50011. Offers agricultural meteorology (MS, PhD); agronomy (MS); crop production and physiology (MS, PhD); plant breeding (MS, PhD); soil science (MS, PhD). Postbaccalaureate distance learning degree programs offered (no on-campus study). *Faculty:* 63 full-time, 14 part-time/adjunct. *Students:* 89 full-time (40 women), 82 part-time (9 women); includes 6 minority (4 African Americans, 1 Asian American or Pacific Islander, 1 Hispanic American), 50 international. 41 applicants, 73% accepted, 29 enrolled. In 2005, 20 master's, 19 doctorates awarded. *Degree requirements:* For master's, thesis or alternative; for doctorate, thesis/dissertation. *Entrance requirements:* Additional exam requirements/recommendations for international students: Required—TOEFL (paper score 530; computer score 197) or IELTS (score 6). *Application deadline:* For fall admission, 2/1 priority date for domestic students, 2/1 priority date for international students. Applications are processed on a rolling basis. Application fee: $30 ($70 for international students). Electronic applications accepted. *Expenses:* Tuition, state resident: full-time $6,410. Tuition, nonresident: full-time $16,422. Tuition and fees vary according to program. *Financial support:* In 2005–06, 75 research assistantships with full and partial tuition reimbursements (averaging $14,777 per year), 5 teaching assistantships with full and partial tuition reimbursements (averaging $14,020 per year) were awarded; fellowships, scholarships/grants, health care benefits, and unspecified assistantships also available. *Unit head:* Dr. Steven Fales, Head, 515-294-7636, Fax: 515-294-3163. *Application contact:* Jacquelyn Severson, Information Contact, 515-294-1361, E-mail: director@agron.iastate.edu.

Kansas State University, Graduate School, College of Agriculture, Department of Agronomy, Manhattan, KS 66506. Offers crop science (MS, PhD); range management (MS, PhD); soil science (MS, PhD); weed science (MS, PhD). Part-time programs available. *Faculty:* 32 full-time (3 women), 10 part-time/adjunct (1 woman). *Students:* 36 full-time (12 women), 13 part-time (3 women), 8 international. 14 applicants, 64% accepted, 8 enrolled. In 2005, 10 master's, 2 doctorates awarded. Terminal master's awarded for partial completion of doctoral program. *Degree requirements:* For master's, thesis or alternative, oral exam; for doctorate, thesis/dissertation, preliminary exams. *Entrance requirements:* For master's, minimum GPA of 3.0 in BS; for doctorate, minimum GPA of 3.5 in master's program. Additional exam requirements/recommendations for international students: Required—TOEFL (minimum score 500 paper-based; 250 computer-based). *Application deadline:* For fall admission, 2/1 for domestic students; for spring admission, 10/1 for domestic students. Applications are processed on a rolling basis. Application fee: $30 ($55 for international students). Electronic applications accepted. *Expenses:* Tuition, state resident: full-time $5,160; part-time $215. Tuition, nonresident: full-time $12,816; part-time $534. Required fees: $564. *Financial support:* In 2005–06, 35 research assistantships (averaging $15,609 per year), 9 teaching assistantships with partial tuition reimbursements (averaging $6,304 per year) were awarded; institutionally sponsored loans and scholarships/grants also available. Support available to part-time students. Financial award application deadline: 3/1; financial award applicants required to submit FAFSA. *Faculty research:* Weed science, environmental soil science, range science, plant genetics, climate change. Total annual research expenditures: $3 million. *Unit head:* Dr. Gary Pierzynski, Head, 785-532-6101, Fax: 785-532-6094, E-mail: gmp@ksu.edu. *Application contact:* Dr. Bill Schapaugh, Director, 785-532-7242, E-mail: wts@ksu.edu.

Louisiana State University and Agricultural and Mechanical College, Graduate School, College of Agriculture, Department of Agronomy and Environmental Management, Baton Rouge, LA 70803. Offers agronomy (MS, PhD). Part-time programs available. *Faculty:* 27 full-time (0 women). *Students:* 25 full-time (7 women), 8 part-time (3 women); includes 2 minority (both African Americans), 14 international. Average age 32. 15 applicants, 40% accepted, 9 enrolled. In 2005, 5 master's, 7 doctorates awarded. *Degree requirements:* For master's, thesis or alternative; for doctorate, thesis/dissertation. *Entrance requirements:* For master's and doctorate, GRE General Test, minimum GPA of 3.0. Additional exam requirements/recommendations for international students: Required—TOEFL (minimum score 550 paper-based; 213 computer-based). *Application deadline:* For fall admission, 1/25 priority date for domestic students, 5/15 priority date for international students. Applications are processed on a rolling basis. Application fee: $25. Electronic applications accepted. *Financial support:* In 2005–06, 21 research assistantships with partial tuition reimbursements (averaging $16,905 per year), 20 teaching assistantships with partial tuition reimbursements (averaging $12,862 per year) were awarded; fellowships, Federal Work-Study, scholarships/grants, tuition waivers (full), and unspecified assistantships also available. Financial award applicants required to submit FAFSA. *Faculty research:* Crop production, resource management, environmental studies, soil science, plant genetics. Total annual research expenditures: $1,338. *Unit head:* Dr. Freddie A. Martin, Head, 225-578-2110, Fax: 225-578-1403, E-mail: fmartin@agctr.lsu.edu. *Application contact:* Magdi Selim, Graduate Coordinator, 225-578-1332, Fax: 225-578-1403, E-mail: mselim@agctr.lsu.edu.

McGill University, Faculty of Graduate and Postdoctoral Studies, Faculty of Agricultural and Environmental Sciences, Department of Bioresource Engineering, Montréal, QC H3A 2T5, Canada. Offers computer applications (M Sc, M Sc A, PhD); food engineering (M Sc, M Sc A, PhD); grain drying (M Sc, M Sc A, PhD); irrigation and drainage (M Sc, M Sc A, PhD); machinery (M Sc, M Sc A, PhD); pollution control (M Sc, M Sc A, PhD); postharvest (M Sc, M Sc A, PhD); soil dynamics (M Sc, M Sc A, PhD); structure and environment (M Sc, M Sc A, PhD); vegetable and fruit storage (M Sc, M Sc A, PhD). Part-time programs available. *Degree requirements:* For master's, thesis (for some programs), registration; for doctorate, thesis/dissertation, registration. *Entrance requirements:* For master's, minimum GPA of 3.0; for doctorate, M Sc, minimum GPA of 3.0. Additional exam requirements/recommendations for international students: Required—TOEFL (minimum score 550 paper-based; 213 computer-based), IELT (minimum score 7). Electronic applications accepted. *Faculty research:* Soil and water conservation, ecosystem remediation, food engineering, postharvest technology, agricultural engineering.

McGill University, Faculty of Graduate and Postdoctoral Studies, Faculty of Agricultural and Environmental Sciences, Department of Natural Resource Sciences, Montréal, QC H3A 2T5, Canada. Offers agrometeorology (M Sc, PhD); entomology (M Sc, PhD); forest science (M Sc, PhD); microbiology (M Sc, PhD); neotropical environment (M Sc, PhD); soil science (M Sc, PhD); wildlife biology (M Sc, PhD). *Degree requirements:* For master's and doctorate, thesis/dissertation, registration. *Entrance requirements:* For master's, minimum GPA of 3.0 or 3.2 in the last 2 years of university study. Additional exam requirements/recommendations for international students: Required—TOEFL (minimum score 550 paper-based; 213 computer-based), IELT (minimum score 7). Electronic applications accepted. *Faculty research:* Toxicology, reproductive physiology, parasites, wildlife management, genetics.

Michigan State University, The Graduate School, College of Agriculture and Natural Resources, Department of Crop and Soil Sciences, East Lansing, MI 48824. Offers crop and soil sciences (MS, PhD); crop and soil sciences-environmental toxicology (PhD); plant breeding and genetics-crop and soil sciences (MS, PhD). *Faculty:* 30 full-time (4 women), 1 part-time/adjunct (0 women). *Students:* 44 full-time (15 women), 13 part-time (5 women); includes 2 minority (both African Americans), 24 international. Average age 31. 30 applicants, 27% accepted. In 2005, 8 master's, 11 doctorates awarded. *Degree requirements:* For master's, thesis or alternative, oral final exam; for doctorate, thesis/dissertation, oral final exam in defense of dissertation, 1 year of residence, comprehensive exam. *Entrance requirements:* For master's, GRE General Test, minimum GPA of 3.0, bachelor's degree in crop and soil sciences or related field; for doctorate, GRE General Test, minimum GPA of 3.0, MS degree (preferred). Additional exam requirements/recommendations for international students: Required—TOEFL (minimum score 550 paper-based; 213 computer-based), Michigan State University ELT (85), Michigan ELAB (83). *Application deadline:* For fall admission, 12/27 for domestic students. Applications are processed on a rolling basis. Application fee: $50. Electronic applications accepted. *Expenses:* Tuition, state resident: part-time $330 per credit hour. Tuition, nonresident: part-time $685 per credit hour. Tuition and fees vary according to program. *Financial support:* In 2005–06, 5 fellowships with tuition reimbursements (averaging $13,440 per year), 37 research assistantships with tuition reimbursements (averaging $12,999 per year), 2 teaching assistantships with tuition reimbursements (averaging $12,470 per year) were awarded; scholarships/grants and unspecified assistantships also available. *Faculty research:* Turfgrass management, environmental toxicology, integrated pest management, water science, plant breeding and genetics. Total annual research expenditures: $8 million. *Unit head:* Dr. James J. Kells, Acting Chairperson, 517-355-0271 Ext. 1103, Fax: 517-353-5174, E-mail: kells@msu.edu. *Application contact:* Rita House, Graduate Secretary, 517-355-0271 Ext. 1111, Fax: 517-353-5174, E-mail: house@msu.edu.

Michigan State University, The Graduate School, College of Agriculture and Natural Resources, MSU-DOE Plant Research Laboratory, East Lansing, MI 48824. Offers biochemistry and molecular biology (PhD); cellular and molecular biology (PhD); crop and soil sciences (PhD); genetics (PhD); microbiology and molecular genetics (PhD); plant biology (PhD); plant physiology (PhD). Offered jointly with the Department of Energy. *Faculty:* 9 full-time (2 women). *Degree requirements:* For doctorate, thesis/dissertation, laboratory rotation, defense of dissertation, comprehensive exam. *Entrance requirements:* For doctorate, GRE General Test, acceptance into one of the affiliated department programs; 3 letters of recommendation; bachelor's degree or equivalent in life sciences, chemistry, biochemistry, or biophysics; research experience. Application fee: $50. Electronic applications accepted. *Expenses:* Tuition, state resident: part-time $330 per credit hour. Tuition, nonresident: part-time $685 per credit hour. Tuition and fees vary according to program. *Faculty research:* Role of hormones in the regulation of plant development and physiology, molecular mechanisms associated with signal recognition, development and application of genetic methods and materials, protein routing and function. Total annual research expenditures: $7.4 million. *Unit head:* Dr. Kenneth Keegstra, Director, 517-353-2270, Fax: 517-353-9168, E-mail: keegstra@msu.edu. *Application contact:* Janet Taylor, Graduate Program Secretary, 517-353-2270, Fax: 517-353-9168, E-mail: prl@msu.edu.

Michigan State University, The Graduate School, College of Agriculture and Natural Resources, Program in Plant Breeding and Genetics, East Lansing, MI 48824. Offers plant breeding and genetics (MS, PhD), including crop and soil sciences, forestry, horticulture. *Students:* 17 full-time (12 women), 2 part-time (1 woman), 10 international. Average age 29. 9 applicants, 0% accepted. In 2005, 2 master's, 6 doctorates awarded. *Degree requirements:* For master's, thesis; for doctorate, thesis/dissertation, oral examination, comprehensive exam. *Entrance requirements:* For master's and doctorate, GRE General Test, minimum GPA of 3.0 in last 3 years of college, 3 letters of recommendation. Additional exam requirements/recommendations for international students: Required—TOEFL (minimum score 550 paper-based; 213 computer-based). *Application deadline:* For fall admission, 12/27 for domestic students. Application fee: $50. Electronic applications accepted. *Expenses:* Tuition, state resident: part-time $330 per credit hour. Tuition, nonresident: part-time $685 per credit hour. Tuition and fees vary according to program. *Financial support:* In 2005–06, 1 fellowship with tuition reimbursement (averaging $11,500 per year), 12 research assistantships with tuition reimbursements (averaging $13,218 per year), 2 teaching assistantships with tuition reimbursements (averaging $12,470 per year) were awarded; scholarships/grants and unspecified assistantships also available. *Faculty research:* Applied plant breeding and genetics; disease, insect and herbicide resistances; gene isolation and genomics; abiotic stress factors; molecular mapping. *Unit head:* Dr. Jim F. Hancock, Director, 517-355-5191 Ext. 1387, Fax: 517-432-3490, E-mail: hancock@msu.edu. *Application contact:* Program Information, 517-355-5191 Ext. 1324, Fax: 517-432-3490, E-mail: pbg@msu.edu.

Mississippi State University, College of Agriculture and Life Sciences, Department of Plant and Soil Sciences, Mississippi State, MS 39762. Offers agronomy (MS, PhD); horticulture (MS, PhD); weed science (MS, PhD). Part-time programs available. *Faculty:* 27 full-time (2 women), 1 part-time/adjunct (0 women). *Students:* 28 full-time (5 women), 29 part-time (6 women); includes 2 minority (1 Asian American or Pacific Islander, 1 Hispanic American), 17 international. Average age 31. 16 applicants, 38% accepted, 3 enrolled. In 2005, 9 master's, 6 doctorates awarded. *Degree requirements:* For master's and doctorate, thesis/dissertation, comprehensive oral or written exam. *Entrance requirements:* Additional exam requirements/recommendations for international students: Required—TOEFL. *Application deadline:* For fall admission, 7/1 for domestic students; for spring admission, 11/1 for domestic students. Applications are processed on a rolling basis. Application fee: $30. *Expenses:* Tuition, state resident: full-time $4,312; part-time $240 per hour. Tuition, nonresident: full-time $9,772; part-time $543 per hour. International tuition: $10,102 full-time. Tuition and fees vary according to course load. *Financial support:* In 2005–06, 3 teaching assistantships with full tuition reimbursements (averaging $12,818 per year) were awarded; research assistantships with full tuition reimbursements, career-related internships or fieldwork, Federal Work-Study, institutionally sponsored loans, and unspecified assistantships also available. Financial award applicants required to submit FAFSA. *Faculty research:* Metabolism, morphology, growth regulators, biotechnology, stress physiology. Total annual research expenditures:$874,795. *Unit head:* Dr. Michael Collins, Head, 662-325-2352, Fax: 662-325-8742, E-mail: mcollins@pss.msstate.edu. *Application contact:* Philip G. Bonfanti, Director of Admissions, 662-325-4104, Fax: 662-325-8872, E-mail: admit@msstate.edu.

New Mexico State University, Graduate School, College of Agriculture and Home Economics, Department of Plant and Environmental Sciences, Las Cruces, NM 88003-8001. Offers general agronomy (MS, PhD); horticulture (MS). Part-time programs available. *Faculty:* 24 full-time (4 women), 7 part-time/adjunct (2 women). *Students:* 35 full-time (14 women), 15 part-time (6 women); includes 9 minority (1 American Indian/Alaska Native, 8 Hispanic Americans), 19 international. Average age 34. 17 applicants, 53% accepted, 4 enrolled. In 2005, 6 master's, 2 doctorates awarded. *Median time to degree:* Of those who began their doctoral program in fall 1997, 90% received their degree in 8 years or less. *Degree requirements:* For master's, thesis; for doctorate, one foreign language, thesis/dissertation. *Entrance requirements:* For master's, minimum GPA of 3.0; for doctorate, minimum GPA of 3.3. *Application deadline:* For fall admission, 7/1 for domestic students; for spring admission, 11/1 priority date for domestic students. Applications are processed on a rolling basis. Application fee: $30 ($50 for international students). Electronic applications accepted. *Expenses:* Tuition, state resident: full-time $3,156; part-time $175 per credit. Tuition, nonresident: full-time $12,510; part-time $565 per credit. Required fees: $1,050. *Financial support:* In 2005–06, 5 fellowships, 18 research assistantships, 12 teaching assistantships were awarded; career-related internships or fieldwork, Federal Work-Study, health care benefits, and unspecified assistantships also available. Support available to part-time students. Financial award application deadline: 3/1. *Faculty research:* Plant breeding and genetics, molecular biology, plant physiology, soil science and environmental remediation, urban horticulture. *Unit head:* Dr. Greg L Mullins, Head, 505-646-3406, Fax: 505-646-6041, E-mail: gmullins@nmsu.edu. *Application contact:* Esther Ramirez, Information Contact, 505-646-3406, Fax: 505-646-6041, E-mail: esramire@nmsu.edu.

North Carolina State University, Graduate School, College of Agriculture and Life Sciences, Department of Crop Science, Raleigh, NC 27695. Offers MS, PhD. Part-time programs available. Terminal master's awarded for partial completion of doctoral program. *Degree requirements:* For master's, thesis (for some programs), thesis (MS); for doctorate, thesis/dissertation. *Entrance requirements:* For master's and doctorate, GRE. Electronic applications accepted. *Faculty research:* Crop breeding and genetics, application of biotechnology to crop improvement, plant physiology, crop physiology and management, agroecology.

North Carolina State University, Graduate School, College of Agriculture and Life Sciences, Department of Soil Science, Raleigh, NC 27695. Offers MS, PhD. Part-time programs available. *Degree requirements:* For master's, thesis (for some programs); for doctorate, thesis/dissertation. *Entrance requirements:* For master's and doctorate, minimum GPA of 3.0. Electronic applications accepted. *Faculty research:* Soil management, soil-environmental relations, chemical and physical properties of soils, nutrient and water management, land use.

North Dakota State University, The Graduate School, College of Agriculture, Food Systems, and Natural Resources, Department of Soil Science, Fargo, ND 58105. Offers environmental and conservation science (PhD); environmental conservation science (MS); natural resource management (MS, PhD); soil sciences (MS, PhD). Part-time programs available. *Faculty:* 8 full-time (1 woman), 6 part-time/adjunct (0 women). *Students:* 7 full-time (2 women), 4 part-time (1 woman), 4 international. Average age 23. 3 applicants, 33% accepted, 1 enrolled. In 2005, 3 master's, 1 doctorate awarded. *Degree requirements:* For master's and doctorate, thesis/dissertation, classroom teaching, comprehensive exam, registration. *Entrance requirements:* For master's and doctorate, GRE General Test. Additional exam requirements/recommendations for international students: Required—TOEFL (minimum score 525 paper-based; 193 computer-based). *Application deadline:* Applications are processed on a rolling basis. Application fee: $45 ($60 for international students). Electronic applications accepted. *Financial support:* In 2005–06, 6 research assistantships with full tuition reimbursements (averaging $14,300 per year) were awarded; fellowships, Federal Work-Study, institutionally sponsored loans, and scholarships/grants also available. Financial award application deadline: 3/15. *Faculty research:* Microclimate, nitrogen management, landscape studies, water quality, soil management. *Unit head:* Dr. Rodney G. Lym, Interim Chair, 701-231-8903, Fax: 701-231-7861, E-mail: jimmie.richardson@ndsu.nodak.edu.

Nova Scotia Agricultural College, Research and Graduate Studies, Truro, NS B2N 5E3, Canada. Offers agriculture (M Sc), including air quality, animal behavior, animal molecular genetics, animal nutrition, animal technology, aquaculture, botany, crop management, crop physiology, ecology, environmental microbiology, food science, horticulture, nutrient management, pest management, physiology, plant biotechnology, plant pathology, soil chemistry, soil fertility, waste management and composting, water quality. Part-time programs available. *Faculty:* 43 full-time (7 women), 21 part-time/adjunct (1 woman). *Students:* 50 full-time (25 women), 15 part-time (11 women); includes 7 minority (3 African Americans, 3 Asian Americans or Pacific Islanders, 1 Hispanic American). Average age 25. In 2005, 23 degrees awarded. *Degree requirements:* For master's, thesis, registration. *Entrance requirements:* For master's, honors B Sc, minimum GPA of 3.0. Additional exam requirements/recommendations for international students: Required—TOEFL (minimum score 580 paper-based; 237 computer-based), Michigan English Language Assessment Battery, IELT, Can Test, CAEL. *Application deadline:* For fall admission, 6/1 for domestic students, 4/1 for international students. For winter admission, 10/31 for domestic students; for spring admission, 2/28 for domestic students. Applications are processed on a rolling basis. Application fee: $70. *Expenses:* Tuition, state resident: part-time $2,328 per year. Tuition, nonresident: full-time $6,984; part-time $7,968 per year. International tuition: $12,624 full-time. Required fees: $481; $46 per course. Tuition and fees vary according to program and student level. *Financial support:* In 2005–06, 48 students received support, including 7 fellowships (averaging $15,000 per year), 10 research assistantships (averaging $15,000 per year), 15 teaching assistantships (averaging $900 per year); career-related internships or fieldwork, scholarships/grants, and unspecified assistantships also available. *Faculty research:* Bio-product development, organic agriculture, nutrient management, air and water quality, agricultural biotechnology. Total annual research expenditures: $4.7 million. *Unit head:* Jill L. Rogers, Manager, 902-893-6360, Fax: 902-893-3430, E-mail: jrogers@nsac.ca. *Application contact:* Marie Law, Administrative Assistant, 902-893-6502, Fax: 902-893-3430, E-mail: mlaw@nsac.ca.

The Ohio State University, Graduate School, College of Food, Agricultural, and Environmental Sciences, School of Natural Resources, Program in Soil Science, Columbus, OH 43210. Offers MS, PhD. *Degree requirements:* For doctorate, thesis/dissertation. *Entrance requirements:* For master's and doctorate, GRE General Test. Additional exam requirements/recommendations for international students: Required—TOEFL (paper 550; computer 213) or IELT (7) or Michigan English Language Assessment Battery (90). Electronic applications accepted.

Oklahoma State University, College of Agricultural Science and Natural Resources, Department of Horticulture and Landscape Architecture, Stillwater, OK 74078. Offers crop science (PhD); environmental science (PhD); food science (PhD); horticulture (M Ag, MS); plant science (PhD). *Faculty:* 21 full-time (1 woman), 4 part-time/adjunct (0 women). *Students:* 1 full-time (0 women), 6 part-time (2 women); includes 1 minority (African American), 4 international. Average age 31. 7 applicants, 14% accepted. In 2005, 3 master's awarded. *Degree requirements:* For master's, thesis or alternative. *Entrance requirements:* For master's and doctorate, GRE. Additional exam requirements/recommendations for international students: Required—TOEFL. *Application deadline:* For fall admission, 6/1 priority date for domestic students, 3/1 priority date for international students. Applications are processed on a rolling basis. Application fee: $40 ($75 for international students). Electronic applications accepted. *Expenses:* Tuition, state resident: full-time $4,253; part-time $139 per credit hour. Tuition, nonresident: full-time $12,569; part-time $485 per credit hour. Required fees: $43 per credit hour. One-time fee: $20 part-time. Tuition and fees vary according to course load and program. *Financial support:* In 2005–06, 4 research assistantships (averaging $13,817 per year) were awarded; teaching assistantships, career-related internships or fieldwork, Federal Work-Study, scholarships/grants, health care benefits, tuition waivers (partial), and unspecified assistantships also available. Support available to part-time students. Financial award application deadline: 3/1. *Faculty research:* Stress and postharvest physiology; water utilization and runoff; IPM systems and nursery, turf, floriculture, vegetable, net and fruit produces and natural resources, food extraction, and processing; public garden management. *Unit head:* Dr. Dale Maronek, Head, 405-744-5414, Fax: 405-744-9709, E-mail: maronek@okstate.edu.

Oklahoma State University, College of Agricultural Science and Natural Resources, Department of Plant and Soil Sciences, Stillwater, OK 74078. Offers agronomy (M Ag, MS, PhD); crop science (PhD); soil science (PhD). *Faculty:* 30 full-time (2 women), 1 part-time/adjunct (0 women). *Students:* 15 full-time (4 women), 23 part-time (6 women), 15 international. Average age 31. 12 applicants, 33% accepted, 2 enrolled. In 2005, 5 master's, 9 doctorates awarded. *Degree requirements:* For master's and doctorate, thesis/dissertation. *Entrance requirements:* For master's and doctorate, GRE. Additional exam requirements/recommendations for international students: Required—TOEFL. *Application deadline:* For fall admission, 6/1 priority date for domestic students, 3/1 priority date for international students. Applications are processed on a rolling basis. Application fee: $40 ($75 for international students). Electronic applications accepted. *Expenses:* Tuition, state resident: full-time $4,253; part-time $139 per credit hour. Tuition, nonresident: full-time $12,569; part-time $485 per credit hour. Required fees: $43 per

Agronomy and Soil Sciences

Oklahoma State University (continued)
credit hour. One-time fee: $20 part-time. Tuition and fees vary according to course load and program. *Financial support:* In 2005–06, 33 research assistantships (averaging $15,076 per year), 2 teaching assistantships (averaging $15,180 per year) were awarded; career-related internships or fieldwork, Federal Work-Study, scholarships/grants, health care benefits, tuition waivers (partial), and unspecified assistantships also available. Support available to part-time students. Financial award application deadline: 3/1. *Faculty research:* Crop science, weed science, rangeland ecology and management, biotechnology, breeding and genetics. *Unit head:* Dr. James H. Stiegler, Head, 405-744-6425, Fax: 405-744-5269.

Oregon State University, Graduate School, College of Agricultural Sciences, Department of Crop and Soil Science, Program in Crop Science, Corvallis, OR 97331. Offers M Agr, MAIS, MS, PhD. Part-time programs available. *Students:* 14 full-time (8 women), 2 part-time (1 woman), 7 international. Average age 29. In 2005, 8 master's, 3 doctorates awarded. *Degree requirements:* For master's, thesis (for some programs); for doctorate, variable foreign language requirement, thesis/dissertation. *Entrance requirements:* For master's and doctorate, GRE, minimum GPA of 3.0 in last 90 hours of course work. Additional exam requirements/recommendations for international students: Required—TOEFL. *Application deadline:* For fall admission, 3/1 for domestic students. Applications are processed on a rolling basis. Application fee: $50. *Expenses:* Tuition, area resident: Part-time $301 per credit. Tuition, state resident: full-time $8,139; part-time $501 per credit. Tuition, nonresident: full-time $14,376; part-time $532 per credit. Required fees: $1,266. *Financial support:* Fellowships, research assistantships, teaching assistantships, career-related internships or fieldwork, Federal Work-Study, and institutionally sponsored loans available. Support available to part-time students. Financial award application deadline: 2/1. *Faculty research:* Cereal and new crops breeding and genetics; weed science; seed technology and production; potato, new crops, and general crop production; plant physiology. *Unit head:* Dr. Patrick M Hayes, Head, 541-737-5878. *Application contact:* Dr. Alvin Mosely, Associate Professor, 541-737-5835, Fax: 541-737-1589, E-mail: alvin.r.mosely@orst.edu.

Oregon State University, Graduate School, College of Agricultural Sciences, Department of Crop and Soil Science, Program in Soil Science, Corvallis, OR 97331. Offers M Agr, MAIS, MS, PhD. Part-time programs available. *Students:* 15 full-time (7 women), 2 part-time; includes 1 minority (Hispanic American), 8 international. Average age 29. In 2005, 12 master's awarded. *Degree requirements:* For master's, thesis (for some programs); for doctorate, variable foreign language requirement, thesis/dissertation. *Entrance requirements:* For master's and doctorate, GRE, minimum GPA of 3.0 in last 90 hours of course work. Additional exam requirements/recommendations for international students: Required—TOEFL. *Application deadline:* For fall admission, 3/1 for domestic students. Applications are processed on a rolling basis. Application fee: $50. *Expenses:* Tuition, area resident: Part-time $301 per credit. Tuition, state resident: full-time $8,139; part-time $501 per credit. Tuition, nonresident: full-time $14,376; part-time $532 per credit. Required fees: $1,266. *Financial support:* Fellowships, research assistantships, teaching assistantships, career-related internships or fieldwork, Federal Work-Study, and institutionally sponsored loans available. Support available to part-time students. Financial award application deadline: 2/1. *Faculty research:* Soil physics, chemistry, biology, fertility, and genesis. *Unit head:* Dr. David D. Myrold, 541-737-5737. *Application contact:* Dr. Neil W. Christensen, Professor, 541-737-5733, Fax: 541-737-5725, E-mail: neil.w.christensen@orst.edu.

The Pennsylvania State University University Park Campus, Graduate School, College of Agricultural Sciences, Department of Crop and Soil Sciences, State College, University Park, PA 16802-1503. Offers agronomy (M Agr, MS, PhD); soil science (M Agr, MS, PhD). *Students:* 34 full-time (16 women), 4 part-time (3 women); includes 1 minority (African American), 19 international. *Entrance requirements:* For master's and doctorate, GRE General Test. *Expenses:* Tuition, state resident: full-time $12,518; part-time $522 per credit. Tuition, nonresident: full-time $23,004; part-time $959 per credit. Required fees: $484. Tuition and fees vary according to course load, campus/location and program. *Unit head:* Dr. David M. Sylvia, Professor and Head, 814-865-2025, Fax: 814-863-7043, E-mail: dmsylvia@psu.edu.

Prairie View A&M University, Graduate School, College of Agriculture and Human Sciences, Prairie View, TX 77446-0519. Offers agricultural economics (MS); animal sciences (MS); interdisciplinary human sciences (MS); soil science (MS). Part-time and evening/weekend programs available. *Faculty:* 8 full-time (4 women), 9 part-time/adjunct (3 women). *Students:* 68 full-time (45 women), 79 part-time (65 women); includes 125 minority (all African Americans), 13 international. Average age 33. 147 applicants, 100% accepted, 147 enrolled. In 2005, 31 degrees awarded. *Degree requirements:* For master's, thesis (for some programs), field placement, comprehensive exam, registration. *Entrance requirements:* For master's, GRE General Test, minimum GPA of 2.45. Additional exam requirements/recommendations for international students: Required—TOEFL (minimum score 550 paper-based). *Application deadline:* For fall admission, 6/1 for domestic students, 6/1 for international students; for spring admission, 10/1 for domestic students, 10/1 for international students. Applications are processed on a rolling basis. Application fee: $50. *Expenses:* Tuition, state resident: full-time $1,440; part-time $80 per credit. Tuition, nonresident: full-time $6,444; part-time $358 per credit. *Financial support:* In 2005–06, 57 students received support, including 8 fellowships with tuition reimbursements available (averaging $12,000 per year), 10 research assistantships with tuition reimbursements available (averaging $15,000 per year); career-related internships or fieldwork, Federal Work-Study, institutionally sponsored loans, scholarships/grants, tuition waivers (partial), and unspecified assistantships also available. Support available to part-time students. Financial award application deadline: 4/1; financial award applicants required to submit FAFSA. *Faculty research:* Domestic violence prevention, water quality, food growth regulators. Total annual research expenditures: $4 million. *Unit head:* Dr. Linda Willis, Dean, 936-857-2996, Fax: 936-857-2998, E-mail: lwillis@pvamu.edu. *Application contact:* Dr. Richard W. Griffin, Interim Department Head, 936-857-2996, Fax: 936-857-2998, E-mail: rwgriffin@pvamu.edu.

Purdue University, Graduate School, College of Agriculture, Department of Agronomy, West Lafayette, IN 47907. Offers MS, PhD. Part-time programs available. *Faculty:* 32 full-time (5 women), 13 part-time/adjunct (2 women). *Students:* 61 full-time (28 women), 6 part-time (2 women); includes 4 minority (3 African Americans, 1 Hispanic American), 24 international. Average age 30. 32 applicants, 31% accepted, 9 enrolled. In 2005, 7 master's, 5 doctorates awarded. *Degree requirements:* For doctorate, thesis/dissertation. *Entrance requirements:* For master's and doctorate, GRE General Test. *Application deadline:* For fall admission, 4/15 for domestic students, 4/15 for international students; for spring admission, 10/15 for domestic students, 9/15 for international students. Applications are processed on a rolling basis. Application fee: $55. Electronic applications accepted. *Financial support:* Fellowships with tuition reimbursements, research assistantships with tuition reimbursements, teaching assistantships with tuition reimbursements available. Support available to part-time students. Financial award applicants required to submit FAFSA. *Faculty research:* Plant genetics and breeding, crop physiology and ecology, agricultural meteorology, soil microbiology. *Unit head:* Dr. C. A. Beyrouty, Head, 765-494-4774, Fax: 765-496-2926. *Application contact:* Phyllis J Graves, Graduate Coordinator, 765-494-4775, E-mail: pgraves@purdue.edu.

South Dakota State University, Graduate School, College of Agriculture and Biological Sciences, Department of Plant Science, Program in Agronomy, Brookings, SD 57007. Offers MS, PhD. *Degree requirements:* For master's, thesis, oral exam; for doctorate, thesis/dissertation, preliminary oral and written exams. *Entrance requirements:* For master's and doctorate, GRE General Test. Additional exam requirements/recommendations for international students: Required—TOEFL. *Faculty research:* Breeding/genetics, weed science, soil science, production agronomy, molecular biology.

Southern Illinois University Carbondale, Graduate School, College of Agriculture, Department of Plant, Soil, and General Agriculture, Carbondale, IL 62901-4701. Offers horticultural science (MS); plant and soil science (MS). *Faculty:* 20 full-time (1 woman). *Students:* 13 full-time (7 women), 33 part-time (9 women); includes 5 minority (all African Americans), 3 international. 16 applicants, 50% accepted, 2 enrolled. In 2005, 13 master's awarded. *Degree requirements:* For master's, thesis. *Entrance requirements:* For master's, minimum GPA of 2.7. Additional exam requirements/recommendations for international students: Required—TOEFL. *Application deadline:* Applications are processed on a rolling basis. Application fee: $0. *Financial support:* In 2005–06, 22 students received support, including 15 research assistantships with full tuition reimbursements available, 6 teaching assistantships with full tuition reimbursements available; fellowships with full tuition reimbursements available, Federal Work-Study, institutionally sponsored loans, and tuition waivers (full) also available. Support available to part-time students. *Faculty research:* Herbicides, fertilizers, agriculture education, landscape design, plant breeding. Total annual research expenditures: $2 million. *Unit head:* Brian Klubek, Interim Chair, 618-453-2496.

Texas A&M University, College of Agriculture and Life Sciences, Department of Soil and Crop Sciences, College Station, TX 77843. Offers agronomy (M Agr, MS, PhD); genetics (PhD); molecular and environmental plant sciences (MS, PhD); soil science (MS, PhD). *Faculty:* 21 full-time (0 women). *Students:* 46 full-time (12 women), 30 part-time (5 women); includes 6 minority (3 African Americans, 1 Asian American or Pacific Islander, 2 Hispanic Americans), 20 international. Average age 26. 44 applicants, 64% accepted, 16 enrolled. In 2005, 17 master's, 13 doctorates awarded. *Degree requirements:* For master's and doctorate, thesis/dissertation. *Entrance requirements:* For master's and doctorate, GRE General Test. Additional exam requirements/recommendations for international students: Required—TOEFL. *Application deadline:* For fall admission, 3/1 for domestic students; for spring admission, 8/1 for domestic students. Applications are processed on a rolling basis. Application fee: $50 ($75 for international students). *Expenses:* Tuition, state resident: full-time $4,488; part-time $187 per credit hour. Tuition, nonresident: full-time $11,112; part-time $463 per credit hour. Required fees: $1,974. *Financial support:* In 2005–06, fellowships (averaging $16,000 per year), research assistantships with partial tuition reimbursements (averaging $15,000 per year) were awarded; career-related internships or fieldwork, Federal Work-Study, and institutionally sponsored loans also available. *Faculty research:* Soil and crop management, turfgrass science, weed science, cereal chemistry, food protein chemistry. *Unit head:* Dr. Mark Hussey, Head, 979-845-3342.

Texas A&M University–Kingsville, College of Graduate Studies, College of Agriculture and Home Economics, Program in Plant and Soil Sciences, Kingsville, TX 78363. Offers MS, PhD. *Degree requirements:* For master's, thesis or alternative, comprehensive exam. *Entrance requirements:* For master's, GRE General Test, minimum GPA of 3.0. Additional exam requirements/recommendations for international students: Required—TOEFL.

Texas Tech University, Graduate School, College of Agricultural Sciences and Natural Resources, Department of Plant and Soil Science, Lubbock, TX 79409. Offers agronomy (PhD); crop science (MS); entomology (MS); horticulture (MS); soil science (MS). Part-time programs available. *Faculty:* 12 full-time (2 women), 4 part-time/adjunct (0 women). *Students:* 29 full-time (11 women), 23 part-time (9 women), 7 international. Average age 31. 36 applicants, 56% accepted, 14 enrolled. In 2005, 16 master's, 2 doctorates awarded. *Degree requirements:* For doctorate, thesis/dissertation. *Entrance requirements:* For master's and doctorate, GRE General Test. Additional exam requirements/recommendations for international students: Required—TOEFL (minimum score 550 paper-based; 213 computer-based). *Application deadline:* Applications are processed on a rolling basis. Application fee: $50 ($60 for international students). Electronic applications accepted. *Expenses:* Tuition, state resident: full-time $4,296. Tuition, nonresident: full-time $10,920. Required fees: $1,992. Tuition and fees vary according to program. *Financial support:* In 2005–06, 21 students received support, including 21 research assistantships with partial tuition reimbursements available (averaging $12,726 per year), 2 teaching assistantships with partial tuition reimbursements available (averaging $12,478 per year); Federal Work-Study and institutionally sponsored loans also available. Support available to part-time students. Financial award application deadline: 4/15; financial award applicants required to submit FAFSA. *Faculty research:* Molecular and cellular biology of plant stress, physiology/genetics of crop production in semiarid conditions, agricultural bioterrorism, improvement of native plants. Total annual research expenditures: $2.6 million. *Unit head:* Dr. Dick L. Auld, Chair, 806-742-2837, Fax: 806-742-0775, E-mail: dick.auld@ttu.edu. *Application contact:* Dr. Richard E. Zartman, Graduate Adviser, 806-742-2837, Fax: 806-742-0775, E-mail: richard.zartman@ttu.edu.

Tuskegee University, Graduate Programs, College of Agricultural, Environmental and Natural Sciences, Department of Agricultural Sciences, Program in Plant and Soil Sciences, Tuskegee, AL 36088. Offers MS. *Faculty:* 13 full-time (1 woman), 2 part-time/adjunct (1 woman). *Students:* 4 full-time (0 women); includes 1 minority (African American), 3 international. Average age 28. In 2005, 3 degrees awarded. *Degree requirements:* For master's, thesis. *Entrance requirements:* For master's, GRE General Test. Additional exam requirements/recommendations for international students: Required—TOEFL (minimum score 500 paper-based; 173 computer-based). *Application deadline:* For fall admission, 7/15 for domestic students. Applications are processed on a rolling basis. Application fee: $25 ($35 for international students). *Expenses:* Tuition: Full-time $12,400. Required fees: $300; $490 per credit. *Financial support:* Application deadline: 4/15.

Université Laval, Faculty of Agricultural and Food Sciences, Department of Soils and Agricultural Engineering, Programs in Soils and Environment Science, Québec, QC G1K 7P4, Canada. Offers environmental technology (M Sc); soils and environment science (M Sc, PhD). Terminal master's awarded for partial completion of doctoral program. *Degree requirements:* For master's, thesis (for some programs); for doctorate, thesis/dissertation, comprehensive exam. *Entrance requirements:* For master's and doctorate, knowledge of French and English. Electronic applications accepted.

Université Laval, Faculty of Forestry and Geomatics, Program in Agroforestry, Québec, QC G1K 7P4, Canada. Offers M Sc. *Degree requirements:* For master's, thesis (for some programs). *Entrance requirements:* For master's, English exam (comprehension of English), knowledge of French, knowledge of a third language. Electronic applications accepted.

University of Alberta, Faculty of Graduate Studies and Research, Department of Renewable Resources, Edmonton, AB T6G 2E1, Canada. Offers agroforestry (M Ag, M Sc, MF); conservation biology (M Sc, PhD); forest biology and management (M Sc, PhD); land reclamation and remediation (M Sc, PhD); protected areas and wildlands management (M Sc, PhD); soil science (M Ag, M Sc, PhD); water and land resources (M Ag, M Sc, PhD); wildlife ecology and management (M Sc, PhD). Part-time programs available. *Faculty:* 26 full-time (4 women), 22 part-time/adjunct (3 women). *Students:* 63 full-time (33 women), 50 part-time (20 women), 14 international. 122 applicants, 24% accepted, 22 enrolled. In 2005, 16 master's, 8 doctorates awarded. *Median time to degree:* Of those who began their doctoral program in fall 1997, 100% received their degree in 8 years or less. *Degree requirements:* For master's, thesis (for some programs); for doctorate, thesis/dissertation, comprehensive exam. *Entrance requirements:* For master's, minimum 2 years of relevant professional experiences, minimum GPA of 3.0; for doctorate, minimum GPA of 3.0. Additional exam requirements/recommendations for international students: Required—TOEFL (minimum score 550 paper-based; 213 computer-based). *Application deadline:* For fall admission, 7/1 priority date for domestic students, 6/1 priority date for international students. Applications are processed on a rolling basis. Application fee: $0. Electronic applications accepted. Tuition and fees charges are reported in Canadian dollars. *Expenses:* Tuition, state resident: part-time $562 Canadian dollars per term. Tuition, nonresident: full-time $3,375 Canadian dollars. Required fees: $573 Canadian dollars; $84 Canadian dollars per term. *Financial support:* In 2005–06, 63 students received support, including 21 research assistantships with partial tuition reimbursements available (averaging $2,800 per year), 28 teaching assistantships with partial tuition reimbursements available (averaging $1,900 per year); scholarships/grants and unspecified assistantships also available. *Faculty research:* Natural and managed landscapes. Total annual research expenditures: $6.1 million. *Unit head:* Dr. John R. Spence, Chair, 780-492-2820, Fax: 780-492-4323, E-mail:

john.spence@ualberta.ca. *Application contact:* Sandy Nakashima, Graduate Program Secretary, 780-492-2820, Fax: 780-492-4323, E-mail: sandy.nakashima@ualberta.ca.

The University of Arizona, Graduate College, College of Agriculture and Life Sciences, Department of Soil, Water and Environmental Science, Tucson, AZ 85721. Offers MS, PhD. *Degree requirements:* For master's, thesis; for doctorate, one foreign language, thesis/dissertation. *Entrance requirements:* Additional exam requirements/recommendations for international students: Required—TOEFL. *Faculty research:* Plant production, environmental microbiology, contaminant flow and transport.

University of Arkansas, Graduate School, Dale Bumpers College of Agricultural, Food and Life Sciences, Department of Crop, Soil and Environmental Sciences, Fayetteville, AR 72701-1201. Offers agronomy (MS, PhD). *Students:* 36 full-time (14 women), 8 part-time (1 woman); includes 1 minority (African American), 15 international. 28 applicants, 32% accepted. In 2005, 14 master's, 7 doctorates awarded. *Degree requirements:* For master's, thesis optional; for doctorate, variable foreign language requirement, thesis/dissertation. Application fee: $40 ($50 for international students). *Financial support:* In 2005–06, 1 fellowship with tuition reimbursement, 35 research assistantships, 1 teaching assistantship were awarded; career-related internships or fieldwork and Federal Work-Study also available. Support available to part-time students. Financial award application deadline: 4/1; financial award applicants required to submit FAFSA. *Unit head:* Dr. Robert Bacon, Interim Departmental Chairperson, 479-575-2347, Fax: 479-575-7465, E-mail: rbacon@uark.edu. *Application contact:* Gloria Fry, Graduate Coordinator, 479-575-2347, E-mail: gfry@uark.edu.

The University of British Columbia, Faculty of Graduate Studies, Faculty of Land and Food Systems, Program in Soil Science, Vancouver, BC V6T 1Z1, Canada. Offers M Sc, PhD. *Faculty:* 8 full-time (0 women), 5 part-time/adjunct (2 women). *Students:* 2 full-time. Average age 25. 10 applicants, 50% accepted, 2 enrolled. In 2005, 1 master's, 1 doctorate awarded. *Degree requirements:* For master's, thesis/dissertation, registration; for doctorate, thesis/dissertation, comprehensive exam, registration. *Entrance requirements:* Additional exam requirements/recommendations for international students: Required—TOEFL (minimum score 577 paper-based; 233 computer-based). *Application deadline:* For fall admission, 1/3 for domestic students, 1/3 for international students. For winter admission, 6/1 for domestic students; for spring admission, 9/1 for domestic students. Applications are processed on a rolling basis. Application fee: $90 Canadian dollars ($150 Canadian dollars for international students). Electronic applications accepted. *Financial support:* In 2005–06, 4 fellowships (averaging $17,925 per year) were awarded; research assistantships, teaching assistantships, institutionally sponsored loans, scholarships/grants, and tuition waivers (full and partial) also available. *Faculty research:* Soil and water conservation, land use, land use and land classification, soil physics, soil chemistry and mineralogy. Total annual research expenditures: $522,194. *Unit head:* Dr. Art Bomke, Graduate Advisor, 604-822-6534, Fax: 604-822-4400, E-mail: gradapp@interchange.ubc.ca. *Application contact:* Lia Maria Dragan, Graduate Programs Assistant, 604-822-8373, Fax: 604-822-4400, E-mail: gradapp@interchange.ubc.ca.

University of California, Davis, Graduate Studies, Graduate Group in Horticulture and Agronomy, Davis, CA 95616. Offers MS. *Faculty:* 90 full-time. *Students:* 43 full-time (24 women); includes 5 minority (2 American Indian/Alaska Native, 2 Asian Americans or Pacific Islanders, 1 Hispanic American), 7 international. Average age 32. 39 applicants, 56% accepted, 20 enrolled. In 2005, 12 degrees awarded. *Degree requirements:* For master's, thesis (for some programs), comprehensive exam (for some programs). *Entrance requirements:* For master's, GRE General Test. Additional exam requirements/recommendations for international students: Required—TOEFL (minimum score 550 paper-based; 213 computer-based). *Application deadline:* For fall admission, 4/1 for domestic students, 3/1 for international students. Applications are processed on a rolling basis. Application fee: $60. Electronic applications accepted. *Financial support:* In 2005–06, 37 students received support, including 15 fellowships with full and partial tuition reimbursements available (averaging $6,636 per year), 17 research assistantships with full and partial tuition reimbursements available (averaging $12,174 per year), 1 teaching assistantship with partial tuition reimbursement available (averaging $15,082 per year); career-related internships or fieldwork, Federal Work-Study, institutionally sponsored loans, scholarships/grants, and tuition waivers (full and partial) also available. Financial award application deadline: 1/15; financial award applicants required to submit FAFSA. *Faculty research:* Postharvest physiology, mineral nutrition, crop improvement, plant growth and development. *Unit head:* M. Andrew Walker, Chairperson, 530-752-0902, Fax: 530-752-0382, E-mail: awalker@ucdavis.edu. *Application contact:* Lisa Brown, Graduate Group Secretary, 530-752-7738, Fax: 530-752-1819, E-mail: lfbrown@ucdavis.edu.

University of California, Davis, Graduate Studies, Graduate Group in Soils and Biogeochemistry, Davis, CA 95616. Offers MS, PhD. *Faculty:* 39 full-time. *Students:* 33 full-time (17 women); includes 1 minority (Asian American or Pacific Islander), 8 international. Average age 31. 18 applicants, 56% accepted, 5 enrolled. In 2005, 6 master's, 1 doctorate awarded. Terminal master's awarded for partial completion of doctoral program. *Degree requirements:* For master's, thesis (for some programs), comprehensive exam (for some programs); for doctorate, thesis/dissertation. *Entrance requirements:* For master's, minimum GPA of 3.3; for doctorate, GRE, minimum GPA of 3.3. Additional exam requirements/recommendations for international students: Required—TOEFL (minimum score 550 paper-based; 213 computer-based). *Application deadline:* For fall admission, 1/15 for domestic students, 1/15 for international students. Applications are processed on a rolling basis. Application fee: $60. Electronic applications accepted. *Financial support:* In 2005–06, 30 students received support, including 7 fellowships with full and partial tuition reimbursements available (averaging $11,887 per year), 20 research assistantships with full and partial tuition reimbursements available (averaging $11,976 per year), 3 teaching assistantships with partial tuition reimbursements available (averaging $15,445 per year); career-related internships or fieldwork, Federal Work-Study, institutionally sponsored loans, scholarships/grants, tuition waivers (full and partial), and unspecified assistantships also available. Support available to part-time students. Financial award application deadline: 1/15; financial award applicants required to submit FAFSA. *Faculty research:* Rhizosphere ecology, soil transport processes, biogeochemical cycling, sustainable agriculture. *Unit head:* Louise Jackson, Chair, 530-754-9116, E-mail: lejackson@ucdavis.edu. *Application contact:* Merlyn Potters, Graduate Staff Adviser, 530-752-1669, Fax: 530-752-1552, E-mail: lawradvising@ucdavis.edu.

University of California, Riverside, Graduate Division, Department of Environmental Sciences, Program in Soil and Water Sciences, Riverside, CA 92521-0102. Offers MS, PhD. *Faculty:* 26 full-time (4 women). *Students:* 13 full-time (9 women); includes 3 minority (1 American Indian/Alaska Native, 2 Hispanic Americans), 2 international. Average age 29. 20 applicants, 50% accepted, 6 enrolled. In 2005, 5 master's, 1 doctorate awarded. *Entrance requirements:* For master's and doctorate, minimum GPA of 3.2. Additional exam requirements/recommendations for international students: Required—TOEFL (minimum score 550 paper-based; 213 computer-based); Recommended—TSE (minimum score 50). *Application deadline:* For fall admission, 5/1 for domestic students, 2/1 for international students. For winter admission, 9/1 for domestic students; for spring admission, 12/1 for domestic students. Application fee: $60 ($75 for international students). Electronic applications accepted. *Expenses:* Tuition, nonresident: full-time $14,694. Full-time tuition and fees vary according to program. *Financial support:* In 2005–06, fellowships (averaging $12,000 per year) *Unit head:* Dr. Marylynn Yates, Chair, 951-827-2358, Fax: 951-827-3993, E-mail: marylynn.yates@ucr.edu. *Application contact:* Mari Ridgeway, Program Assistant, 951-827-5103, Fax: 951-827-3993, E-mail: soilwater@ucr.edu.

University of Connecticut, Graduate School, College of Agriculture and Natural Resources, Department of Plant Science, Field of Plant Science, Storrs, CT 06269. Offers plant and soil sciences (MS, PhD). *Faculty:* 26 full-time (4 women). *Students:* 16 full-time (7 women), 6 part-time (2 women), 12 international. Average age 34. 9 applicants, 56% accepted, 3 enrolled. In 2005, 3 master's, 1 doctorate awarded. Terminal master's awarded for partial completion of doctoral program. *Degree requirements:* For master's, comprehensive exam; for doctorate, thesis/dissertation. *Entrance requirements:* For master's and doctorate, GRE General Test, GRE Subject Test. Additional exam requirements/recommendations for international students: Required—TOEFL (minimum score 550 paper-based; 213 computer-based). *Application deadline:* For fall admission, 2/1 priority date for domestic students, 2/1 priority date for international students; for spring admission, 11/1 for domestic students, 10/1 for international students. Applications are processed on a rolling basis. Application fee: $55. Electronic applications accepted. *Expenses:* Tuition, state resident: part-time $444 per credit hour. Tuition, nonresident: part-time $1,154 per credit hour. Tuition and fees vary according to course load. *Financial support:* In 2005–06, 10 research assistantships with full tuition reimbursements, 4 teaching assistantships with full tuition reimbursements were awarded; fellowships, Federal Work-Study, scholarships/grants, health care benefits, and unspecified assistantships also available. Financial award application deadline: 2/1; financial award applicants required to submit FAFSA. *Application contact:* George C. Elliott, Chairperson, 860-486-1938, Fax: 860-486-0682, E-mail: george.elliott@uconn.edu.

University of Delaware, College of Agriculture and Natural Resources, Department of Plant and Soil Sciences, Newark, DE 19716. Offers MS, PhD. Part-time programs available. *Faculty:* 26 full-time (7 women), 15 part-time/adjunct (2 women). *Students:* 35 full-time (18 women), 3 part-time (2 women); includes 1 minority (Asian American or Pacific Islander), 11 international. Average age 29. 29 applicants, 34% accepted, 8 enrolled. In 2005, 5 master's, 3 doctorates awarded. Terminal master's awarded for partial completion of doctoral program. *Degree requirements:* For master's and doctorate, thesis/dissertation. *Entrance requirements:* For master's and doctorate, GRE General Test. Additional exam requirements/recommendations for international students: Required—TOEFL (minimum score 550 paper-based; 213 computer-based). *Application deadline:* For fall admission, 7/1 for domestic students. Application fee: $60. Electronic applications accepted. *Financial support:* In 2005–06, 20 fellowships with full tuition reimbursements (averaging $15,000 per year), 21 research assistantships with full tuition reimbursements (averaging $15,000 per year), 3 teaching assistantships with full tuition reimbursements (averaging $12,000 per year) were awarded; career-related internships or fieldwork also available. Financial award application deadline: 3/1. *Faculty research:* Soil chemistry, plant and cell tissue culture, plant breeding and genetics, soil physics, soil biochemistry, plant molecular biology, soil microbiology. Total annual research expenditures: $3.8 million. *Unit head:* Dr. Donald L. Sparks, Chair, 302-831-2532, Fax: 302-831-3651, E-mail: dlsparks@udel.edu. *Application contact:* Dr. Jeffrey Fuhrmann, Graduate Coordinator, 302-831-2534, E-mail: fuhrmann@udel.edu.

University of Florida, Graduate School, College of Agricultural and Life Sciences, Department of Agronomy, Gainesville, FL 32611. Offers MS, PhD. *Faculty:* 16 full-time (2 women), 1 part-time/adjunct (0 women). *Students:* 30 (10 women); includes 1 minority (Hispanic American) 14 international. 20 applicants, 65% accepted. In 2005, 5 master's, 5 doctorates awarded. *Degree requirements:* For master's, thesis optional; for doctorate, thesis/dissertation. *Entrance requirements:* For master's and doctorate, GRE General Test, minimum GPA of 3.0. Additional exam requirements/recommendations for international students: Required—TOEFL. *Application deadline:* For fall admission, 6/1 for domestic students. Applications are processed on a rolling basis. Application fee: $20. Electronic applications accepted. *Expenses:* Tuition, state resident: full-time $6,234. Tuition, nonresident: full-time $21,359. Tuition and fees vary according to program. *Financial support:* In 2005–06, 22 research assistantships (averaging $13,878 per year) were awarded; fellowships, teaching assistantships, career-related internships or fieldwork, institutionally sponsored loans, and unspecified assistantships also available. *Faculty research:* Genetics and plant breeding, aquatic and terrestrial weed science, plant physiology, molecular biology, forage and crop production. *Unit head:* Dr. Jerry M. Bennett, Chair, 352-392-1811 Ext. 202, Fax: 352-392-1840, E-mail: jmbt@mail.ifas.ufl.edu. *Application contact:* Dr. David S. Wofford, Coordinator, 352-392-1823 Ext. 205, Fax: 352-392-7248, E-mail: dsw@ifas.ufl.edu.

University of Florida, Graduate School, College of Agricultural and Life Sciences, Department of Soil and Water Science, Gainesville, FL 32611. Offers M Ag, MS, PhD. Part-time programs available. Postbaccalaureate distance learning degree programs offered. *Faculty:* 22 full-time (4 women). *Students:* 93 (50 women); includes 10 minority (3 African Americans, 1 Asian American or Pacific Islander, 6 Hispanic Americans) 26 international. 27 applicants, 78% accepted. In 2005, 10 master's, 9 doctorates awarded. Terminal master's awarded for partial completion of doctoral program. *Degree requirements:* For master's, thesis optional; for doctorate, thesis/dissertation. *Entrance requirements:* For master's and doctorate, GRE General Test, minimum GPA of 3.0. Additional exam requirements/recommendations for international students: Required—TOEFL. *Application deadline:* For fall admission, 6/1 for domestic students; for spring admission, 9/14 for domestic students. Applications are processed on a rolling basis. Application fee: $20. Electronic applications accepted. *Expenses:* Tuition, state resident: full-time $6,234. Tuition, nonresident: full-time $21,359. Tuition and fees vary according to program. *Financial support:* In 2005–06, 41 research assistantships (averaging $11,083 per year) were awarded; fellowships, teaching assistantships, career-related internships or fieldwork, Federal Work-Study, institutionally sponsored loans, and unspecified assistantships also available. Support available to part-time students. *Faculty research:* Environmental fate and transport of pesticides, conservation, wetlands, land application of nonhazardous waste, soil/water agrochemical management. *Unit head:* Dr. K. Ramesh Reddy, Chair, 352-392-1803 Ext. 317, Fax: 352-392-3399, E-mail: krr@ufl.edu. *Application contact:* Dr. Nicholas B. Comerford, Graduate Coordinator, 352-392-1951, Fax: 352-392-3902, E-mail: nbc@ifas.ufl.edu.

University of Georgia, Graduate School, College of Agricultural and Environmental Sciences, Department of Crop and Soil Sciences, Athens, GA 30602. Offers agronomy (MS, PhD); crop and soil sciences (MCCS); plant protection and pest management (MPPPM). Part-time programs available. *Faculty:* 27 full-time (3 women). *Students:* 32 full-time, 10 part-time; includes 1 minority (Asian American or Pacific Islander), 12 international. Average age 24. 17 applicants, 65% accepted, 9 enrolled. In 2005, 8 master's, 2 doctorates awarded. *Degree requirements:* For master's, thesis (MS); for doctorate, thesis/dissertation, comprehensive exam. *Entrance requirements:* For master's and doctorate, GRE General Test. Additional exam requirements/recommendations for international students: Required—TOEFL (minimum score 550 paper-based; 213 computer-based). *Application deadline:* For fall admission, 7/1 priority date for domestic students, 4/15 priority date for international students; for spring admission, 11/15 for domestic students, 10/15 for international students. Applications are processed on a rolling basis. Application fee: $50. Electronic applications accepted. *Financial support:* In 2005–06, research assistantships with full tuition reimbursements (averaging $14,600 per year), teaching assistantships with full tuition reimbursements (averaging $15,350 per year) were awarded; fellowships, scholarships/grants, tuition waivers (full), and unspecified assistantships also available. *Faculty research:* Plant breeding, genomics, nutrient management, water quality, soil chemistry. *Unit head:* Dr. Donn Shilling, Head, 706-542-0906, Fax: 706-542-0914. *Application contact:* Dr. Miguel L. Cabrera, Graduate Coordinator, 706-542-1242, Fax: 706-542-0914, E-mail: mcabrera@uga.edu.

University of Guelph, Graduate Program Services, Ontario Agricultural College, Department of Land Resource Science, Guelph, ON N1G 2W1, Canada. Offers atmospheric science (M Sc, PhD); environmental and agricultural earth sciences (M Sc, PhD); land resources management (M Sc, PhD); soil science (M Sc, PhD). Part-time programs available. *Faculty:* 19 full-time (5 women), 5 part-time/adjunct (1 woman). *Students:* 47 full-time (20 women), 3 part-time; includes 9 minority (1 African American, 6 Asian Americans or Pacific Islanders, 2 Hispanic Americans), 2 international. Average age 28. 25 applicants, 24% accepted. In 2005, 4 master's, 3 doctorates awarded. *Degree requirements:* For master's and doctorate, thesis/dissertation. *Entrance requirements:* For master's, minimum B- average during previous 2 years of course work; for doctorate, minimum B average during previous 2 years of course work. Additional exam requirements/recommendations for international students: Required—TOEFL (minimum score 550 paper-based; 213 computer-based). *Application deadline:* For fall admission, 7/1 priority date for domestic students, 5/1 priority date for international students. For winter admission, 10/1 for domestic students; for spring admission, 3/1 for domestic students. Applications are processed on a rolling basis. Application fee: $75 Canadian dollars.

Agronomy and Soil Sciences

University of Guelph (continued)

Electronic applications accepted. *Financial support:* In 2005–06, 30 students received support, including 40 research assistantships (averaging $16,500 Canadian dollars per year), 15 teaching assistantships (averaging $3,800 Canadian dollars per year); fellowships, scholarships/grants also available. *Faculty research:* Soil science, environmental earth science, land resource management. Total annual research expenditures: $2.1 million Canadian dollars. *Unit head:* Dr. S. Hilts, Chair, 519-824-4120 Ext. 52447, Fax: 519-824-5730, E-mail: shilts@uoguelph.ca. *Application contact:* Dr. B. Hale, Graduate Coordinator, 519-824-4120 Ext. 53434, Fax: 519-824-5730, E-mail: bhale@uoguelph.ca.

University of Idaho, College of Graduate Studies, College of Agricultural and Life Sciences, Department of Plant, Soil, and Entomological Sciences, Program in Soil and Land Resources, Moscow, ID 83844-2282. Offers MS, PhD. *Students:* 4 full-time (1 woman), 6 part-time (5 women); includes 1 minority (Hispanic American) In 2005, 3 degrees awarded. *Degree requirements:* For doctorate, thesis/dissertation. *Entrance requirements:* For master's and doctorate, GRE General Test, minimum GPA of 3.0. *Application deadline:* For fall admission, 8/1 for domestic students; for spring admission, 12/15 for domestic students. Application fee: $55 ($60 for international students). *Expenses:* Tuition, nonresident: full-time $8,770; part-time $130 per credit. Required fees: $4,508; $217 per credit. *Financial support:* Application deadline: 2/15. *Unit head:* Dr. Matthew J. Morra, Chair, 208-885-6315.

University of Illinois at Urbana–Champaign, Graduate College, College of Agricultural, Consumer and Environmental Sciences, Department of Crop Sciences, Champaign, IL 61820. Offers MS, PhD. *Faculty:* 30 full-time (4 women). *Students:* 50 full-time (26 women), 12 part-time (4 women); includes 2 minority (both Asian Americans or Pacific Islanders), 22 international. 57 applicants, 40% accepted, 9 enrolled. In 2005, 23 master's, 9 doctorates awarded. *Degree requirements:* For master's and doctorate, thesis/dissertation, comprehensive exam. *Entrance requirements:* For master's, GRE, minimum GPA of 3.0. *Application deadline:* Applications are processed on a rolling basis. Application fee: $50 ($60 for international students). Electronic applications accepted. *Financial support:* In 2005–06, 19 fellowships, 44 research assistantships, 11 teaching assistantships were awarded; tuition waivers (full and partial) also available. *Faculty research:* Plant breeding and genetics, molecular biology, crop production, plant physiology, weed science. *Unit head:* Robert G. Hoeft, Head, 217-333-9480, Fax: 217-333-9817, E-mail: rhoeft@uiuc.edu. *Application contact:* Susan A. Panepinto, Secretary, 217-244-0396, Fax: 217-333-9817, E-mail: spanepin@uiuc.edu.

University of Kentucky, Graduate School, Graduate School Programs in the College of Agriculture, Program in Crop Science, Lexington, KY 40506-0032. Offers MS, PhD. *Faculty:* 2 full-time (0 women). *Students:* 10 full-time (5 women), 8 part-time (4 women), 4 international. Average age 33. 18 applicants, 61% accepted, 9 enrolled. In 2005, 1 master's, 2 doctorates awarded. *Median time to degree:* Of those who began their doctoral program in fall 1997, 78% received their degree in 8 years or less. *Degree requirements:* For master's, thesis optional; for doctorate, thesis/dissertation, comprehensive exam. *Entrance requirements:* For master's, GRE General Test, minimum GPA of 2.5; for doctorate, GRE General Test, minimum GPA of 3.0. Additional exam requirements/recommendations for international students: Required—TOEFL (minimum score 550 paper-based; 213 computer-based). *Application deadline:* For fall admission, 7/17 priority date for domestic students, 2/1 priority date for international students; for spring admission, 12/13 priority date for domestic students, 6/15 priority date for international students. Applications are processed on a rolling basis. Application fee: $40 ($55 for international students). Electronic applications accepted. *Expenses:* Tuition, state resident: full-time $3,159; part-time $331 per credit hour. Tuition, nonresident: full-time $6,984; part-time $756 per credit hour. Tuition and fees vary according to course load, degree level and program. *Financial support:* In 2005–06, 9 students received support, including 9 research assistantships with full tuition reimbursements available (averaging $15,000 per year); fellowships with full tuition reimbursements available, teaching assistantships with full tuition reimbursements available, Federal Work-Study, scholarships/grants, traineeships, health care benefits, tuition waivers (partial), and unspecified assistantships also available. Support available to part-time students. Financial award application deadline: 3/15. *Faculty research:* Crop physiology, crop ecology, crop management, crop breeding and genetics, weed science. Total annual research expenditures: $8 million. *Unit head:* Dr. Charles Dougherty, Director of Graduate Studies, 859-257-3454, Fax: 859-323-1952, E-mail: cdougher@uky.edu. *Application contact:* Dr. Brian Jackson, Senior Associate Dean, 859-257-8176, Fax: 859-323-1928, E-mail: lance.brunner@uky.edu.

University of Kentucky, Graduate School, Graduate School Programs in the College of Agriculture, Program in Plant and Soil Science, Lexington, KY 40506-0032. Offers MS. *Faculty:* 48 full-time (2 women), 5 part-time/adjunct (0 women). *Students:* 17 full-time (10 women), 4 part-time, 2 international. Average age 28. 20 applicants, 60% accepted, 8 enrolled. In 2005, 4 degrees awarded. *Degree requirements:* For master's, thesis optional. *Entrance requirements:* For master's, GRE General Test, minimum undergraduate GPA of 2.5, minimum graduate GPA of 3.0. Additional exam requirements/recommendations for international students: Required—TOEFL (minimum score 550 paper-based; 213 computer-based). *Application deadline:* For fall admission, 7/17 priority date for domestic students, 2/1 priority date for international students; for spring admission, 12/13 priority date for domestic students, 6/15 priority date for international students. Applications are processed on a rolling basis. Application fee: $40 ($55 for international students). Electronic applications accepted. *Expenses:* Tuition, state resident: full-time $3,159; part-time $331 per credit hour. Tuition, nonresident: full-time $6,984; part-time $756 per credit hour. Tuition and fees vary according to course load, degree level and program. *Financial support:* In 2005–06, 16 students received support, including 16 research assistantships with full tuition reimbursements available (averaging $15,000 per year); fellowships with full tuition reimbursements available, teaching assistantships with full tuition reimbursements available, Federal Work-Study, scholarships/grants, traineeships, health care benefits, tuition waivers (partial), and unspecified assistantships also available. Support available to part-time students. Financial award application deadline: 3/15. *Unit head:* Dr. Charles Dougherty, Director of Graduate Studies, 859-257-3454, Fax: 859-323-1952, E-mail: cdougher@uky.edu. *Application contact:* Dr. Brian Jackson, Senior Associate Dean, 859-257-8176, Fax: 859-323-1928, E-mail: lance.brunner@uky.edu.

University of Kentucky, Graduate School, Graduate School Programs in the College of Agriculture, Program in Soil Science, Lexington, KY 40506-0032. Offers PhD. *Faculty:* 40 full-time (2 women), 3 part-time/adjunct (0 women). *Students:* 9 full-time (4 women), 2 part-time, 7 international. Average age 35. 9 applicants, 67% accepted, 4 enrolled. In 2005, 1 degree awarded. *Median time to degree:* Of those who began their doctoral program in fall 1997, 63.6% received their degree in 8 years or less. *Degree requirements:* For doctorate, thesis/dissertation, comprehensive exam. *Entrance requirements:* For doctorate, GRE General Test, minimum graduate GPA of 3.0. Additional exam requirements/recommendations for international students: Required—TOEFL (minimum score 550 paper-based; 213 computer-based). *Application deadline:* For fall admission, 7/17 priority date for domestic students, 2/1 priority date for international students; for spring admission, 12/13 priority date for domestic students, 6/15 priority date for international students. Applications are processed on a rolling basis. Application fee: $40 ($55 for international students). Electronic applications accepted. *Expenses:* Tuition, state resident: full-time $3,159; part-time $331 per credit hour. Tuition, nonresident: full-time $6,984; part-time $756 per credit hour. Tuition and fees vary according to course load, degree level and program. *Financial support:* In 2005–06, 2 fellowships with full tuition reimbursements (averaging $2,794 per year), 8 research assistantships with full tuition reimbursements (averaging $15,000 per year) were awarded; teaching assistantships with full tuition reimbursements, Federal Work-Study, institutionally sponsored loans, scholarships/grants, traineeships, health care benefits, tuition waivers (partial), and unspecified assistantships also available. Support available to part-time students. Financial award application deadline: 3/15; financial award applicants required to submit FAFSA. *Faculty research:* Soil fertility and plant nutrition, soil chemistry and physics, soil genesis and morphology, soil management and conservation, water and environmental quality. *Unit head:* Dr. John Grove, Director of Graduate Studies, 859-257-5852, Fax: 859-257-3655, E-mail: jgrove@ca.uky.edu. *Application contact:* Dr. Brian Jackson, Senior Associate Dean, 859-257-8176, Fax: 859-323-1928, E-mail: lance.brunner@uky.edu.

University of Maine, Graduate School, College of Natural Sciences, Forestry, and Agriculture, Department of Plant, Soil, and Environmental Sciences, Orono, ME 04469. Offers biological sciences (PhD); ecology and environmental sciences (MS, PhD); forest resources (PhD); horticulture (MS); plant science (PhD); plant, soil, and environmental sciences (MS); resource utilization (MS). *Faculty:* 25. *Students:* 15 full-time (8 women), 8 part-time (6 women), 3 international. Average age 32. 6 applicants, 33% accepted, 1 enrolled. In 2005, 9 master's awarded. *Entrance requirements:* For master's and doctorate, GRE General Test. Additional exam requirements/recommendations for international students: Required—TOEFL. *Application deadline:* Applications are processed on a rolling basis. Application fee: $50. Electronic applications accepted. *Financial support:* In 2005–06, 9 research assistantships with tuition reimbursements (averaging $12,180 per year) were awarded; teaching assistantships, scholarships/grants, tuition waivers (full and partial), and unspecified assistantships also available. *Unit head:* Greg Porter, Chair, 207-581-2943, Fax: 207-581-3207. *Application contact:* Scott G. Delcourt, Associate Dean of the Graduate School, 207-581-3219, Fax: 207-581-3232, E-mail: graduate@maine.edu.

University of Manitoba, Faculty of Graduate Studies, Faculty of Agriculture, Department of Soil Science, Winnipeg, MB R3T 2N2, Canada. Offers M Sc, PhD. *Degree requirements:* For master's, thesis; for doctorate, one foreign language, thesis/dissertation.

University of Maryland, College Park, Graduate Studies, College of Agriculture and Natural Resources, Department of Natural Resource Sciences and Landscape Architecture, Program in Agronomy, College Park, MD 20742. Offers MS, PhD. In 2005, 3 master's, 1 doctorate awarded. *Degree requirements:* For doctorate, written and oral exams. *Entrance requirements:* Additional exam requirements/recommendations for international students: Required—TOEFL. *Application deadline:* For fall admission, 5/1 for domestic students, 2/1 for international students; for spring admission, 9/1 for domestic students, 6/1 for international students. Applications are processed on a rolling basis. Application fee: $60. Electronic applications accepted. *Financial support:* Fellowships, research assistantships, teaching assistantships, career-related internships or fieldwork available. Financial award applicants required to submit FAFSA. *Faculty research:* Cereal crop production, soil and water conservation, turf management, x-ray defraction. *Application contact:* Dean of Graduate School, 301-405-4190, Fax: 301-314-9305.

University of Massachusetts Amherst, Graduate School, College of Natural Resources and the Environment, Department of Plant and Soil Sciences, Amherst, MA 01003. Offers plant science (PhD); soil science (MS, PhD). *Faculty:* 24 full-time (5 women). *Students:* 24 full-time (12 women), 15 part-time (6 women); includes 2 minority (1 African American, 1 Hispanic American), 11 international. Average age 35. 42 applicants, 40% accepted, 13 enrolled. In 2005, 6 master's, 2 doctorates awarded. Terminal master's awarded for partial completion of doctoral program. *Degree requirements:* For master's, thesis optional; for doctorate, thesis/dissertation. *Entrance requirements:* For master's and doctorate, GRE General Test. Additional exam requirements/recommendations for international students: Required—TOEFL (minimum score 530 paper-based; 197 computer-based). *Application deadline:* For fall admission, 2/1 priority date for domestic students, 2/1 priority date for international students; for spring admission, 10/1 for domestic students, 10/1 for international students. Applications are processed on a rolling basis. Application fee: $40 ($65 for international students). Electronic applications accepted. *Expenses:* Tuition, state resident: part-time $110 per credit. Tuition, nonresident: part-time $414 per credit. Required fees: $2,824 per term. One-time fee: $250 part-time. Full-time tuition and fees vary according to course load, campus/location, program and reciprocity agreements. *Financial support:* In 2005–06, fellowships with full tuition reimbursements (averaging $6,250 per year), research assistantships with full tuition reimbursements (averaging $7,488 per year), teaching assistantships with full tuition reimbursements (averaging $5,858 per year) were awarded; career-related internships or fieldwork, Federal Work-Study, scholarships/grants, traineeships, and unspecified assistantships also available. Support available to part-time students. Financial award application deadline: 2/1. *Unit head:* Dr. Petrus Veneman, Director, 413-545-5225, Fax: 413-545-3075, E-mail: veneman@pssci.umass.edu.

University of Minnesota, Twin Cities Campus, Graduate School, College of Agricultural, Food, and Environmental Sciences, Department of Soil, Water, and Climate, Minneapolis, MN 55455-0213. Offers MS, PhD. *Faculty:* 27 full-time (2 women), 9 part-time/adjunct (0 women). *Students:* 27 full-time (13 women), 5 part-time (3 women); includes 3 minority (1 American Indian/Alaska Native, 1 Asian American or Pacific Islander, 1 Hispanic American). Average age 25. 21 applicants, 57% accepted, 11 enrolled. In 2005, 3 degrees awarded. *Degree requirements:* For master's, thesis or alternative; for doctorate, thesis/dissertation. *Entrance requirements:* For master's and doctorate, GRE General Test, minimum GPA of 3.0. Additional exam requirements/recommendations for international students: Required—TOEFL (minimum score 550 paper-based; 213 computer-based). *Application deadline:* For fall admission, 6/15 priority date for domestic students, 6/15 priority date for international students; for spring admission, 10/15 priority date for domestic students, 10/15 priority date for international students. Applications are processed on a rolling basis. Application fee: $55 ($75 for international students). Electronic applications accepted. *Expenses:* Tuition, state resident: full-time $8,748; part-time $729 per credit. Tuition, nonresident: full-time $15,848; part-time $1,321 per credit. Full-time tuition and fees vary according to class time, course load, program and reciprocity agreements. *Financial support:* In 2005–06, 2 fellowships with full tuition reimbursements (averaging $22,000 per year), 24 research assistantships with full and partial tuition reimbursements (averaging $17,000 per year), 2 teaching assistantships with full tuition reimbursements (averaging $17,000 per year) were awarded; Federal Work-Study, scholarships/grants, health care benefits, tuition waivers (full), and unspecified assistantships also available. Support available to part-time students. *Faculty research:* Soil water and atmospheric resources, soil physical management, agricultural chemicals and their management, plant nutrient management, biological nitrogen fixation. *Unit head:* Dr. Edward A. Nater, Head, 612-625-9734, Fax: 612-625-2208, E-mail: enater@umn.edu. *Application contact:* Dr. Deborah L. Allan, Professor and Director of Graduate Studies, 612-625-3158, Fax: 612-625-2208, E-mail: dallan@umn.edu.

University of Missouri–Columbia, Graduate School, School of Natural Resources, Department of Soil, Environmental, and Atmospheric Sciences, Columbia, MO 65211. Offers atmospheric science (MS, PhD); soil science (MS, PhD). *Faculty:* 8 full-time (0 women). *Students:* 19 full-time (7 women), 10 part-time (3 women); includes 3 minority (1 African American, 1 Asian American or Pacific Islander, 1 Hispanic American), 9 international. In 2005, 10 master's, 5 doctorates awarded. *Degree requirements:* For doctorate, thesis/dissertation. *Entrance requirements:* For master's and doctorate, GRE General Test, minimum GPA of 3.0. *Application deadline:* Applications are processed on a rolling basis. Application fee: $45 ($60 for international students). *Financial support:* Fellowships, research assistantships, teaching assistantships, institutionally sponsored loans and scholarships/grants available. *Unit head:* Dr. Anthony Lupo, Director of Graduate Studies, 573-884-1638.

University of Nebraska–Lincoln, Graduate College, College of Agricultural Sciences and Natural Resources, Department of Agronomy and Horticulture, Program in Agronomy, Lincoln, NE 68588. Offers MS, PhD. *Degree requirements:* For master's, thesis/dissertation; for doctorate, thesis/dissertation, comprehensive exam. *Entrance requirements:* Additional exam requirements/recommendations for international students: Required—TOEFL (minimum score 500 paper-based; 173 computer-based). Electronic applications accepted. *Faculty research:* Crop physiology and production, plant breeding and genetics, range and forage management, soil and water science, weed science.

University of New Hampshire, Graduate School, College of Life Sciences and Agriculture, Department of Natural Resources, Durham, NH 03824. Offers environmental conservation (MS); forestry (MS); soil science (MS); water resources management (MS); wildlife (MS).

Part-time programs available. *Faculty:* 40 full-time. *Students:* 31 full-time (17 women), 36 part-time (17 women); includes 2 minority (both Asian Americans or Pacific Islanders), 4 international. Average age 31. 50 applicants, 44% accepted, 15 enrolled. In 2005, 14 degrees awarded. *Degree requirements:* For master's, thesis or alternative. *Entrance requirements:* For master's, GRE General Test. Additional exam requirements/recommendations for international students: Required—TOEFL (minimum score 550 paper-based; 213 computer-based); Recommended—TSE. *Application deadline:* For fall admission, 4/1 for domestic students, 4/1 for international students. For winter admission, 12/1 for domestic students. Applications are processed on a rolling basis. Application fee: $60. Electronic applications accepted. *Expenses:* Tuition, state resident: full-time $8,010; part-time $445 per credit hour. Tuition, nonresident: full-time $19,730; part-time $810 per credit hour. Required fees: $322 per semester. Tuition and fees vary according to course load and program. *Financial support:* In 2005–06, 3 fellowships, 21 research assistantships, 15 teaching assistantships were awarded; career-related internships or fieldwork, Federal Work-Study, scholarships/grants, and tuition waivers (full and partial) also available. Support available to part-time students. Financial award application deadline: 2/15. *Unit head:* Dr. William H. McDowell, Chairperson, 603-862-2249, E-mail: tehoward@cisunix.unh.edu. *Application contact:* Linda Scogin, Administrative Assistant, 603-862-3932, E-mail: natural.resources @unh.edu.

University of Puerto Rico, Mayagüez Campus, Graduate Studies, College of Agricultural Sciences, Department of Agronomy and Soils, Mayagüez, PR 00681-9000. Offers crops (MS); soils (MS). Part-time programs available. *Faculty:* 12. *Students:* 12 full-time (3 women), 17 part-time (6 women); includes 28 minority (all Hispanic Americans), 1 international. 8 applicants, 88% accepted, 4 enrolled. *Degree requirements:* For master's, thesis, comprehensive exam. *Application deadline:* For fall admission, 2/15 for domestic students; for spring admission, 9/15 for domestic students. Applications are processed on a rolling basis. Application fee: $15 ($20 for international students). *Expenses:* Tuition, state resident: full-time $900; part-time $100 per credit. International tuition: $4,655 full-time. Part-time tuition and fees vary according to course level and course load. *Financial support:* In 2005–06, 25 students received support, including fellowships (averaging $1,200 per year), research assistantships (averaging $1,500 per year), teaching assistantships (averaging $987 per year); Federal Work-Study and institutionally sponsored loans also available. *Faculty research:* Soil physics and chemistry, soil management, plant physiology, ecology, plant breeding. Total annual research expenditures: $49,515. *Unit head:* Dr. Miguel Muñoz, Director, 787-265-3851.

University of Puerto Rico, Mayagüez Campus, Graduate Studies, College of Agricultural Sciences, Department of Crop Protection, Mayagüez, PR 00681-9000. Offers MS. Part-time programs available. *Faculty:* 23. *Students:* 4 full-time (3 women), 12 part-time (8 women); includes 14 minority (all Hispanic Americans), 2 international. 5 applicants, 80% accepted, 2 enrolled. *Degree requirements:* For master's, thesis, comprehensive exam. *Application deadline:* For fall admission, 2/15 for domestic students; for spring admission, 9/15 for domestic students. Applications are processed on a rolling basis. Application fee: $20. *Expenses:* Tuition, state resident: full-time $900; part-time $100 per credit. International tuition: $4,655 full-time. Part-time tuition and fees vary according to course level and course load. *Financial support:* In 2005–06, 12 students received support, including 3 fellowships (averaging $1,200 per year), 6 research assistantships (averaging $1,500 per year), teaching assistantships (averaging $987 per year); career-related internships or fieldwork, Federal Work-Study, and institutionally sponsored loans also available. Financial award application deadline:5/30. *Faculty research:* Nematology, virology, plant pathology, weed control, peas and soybean seed diseases. Total annual research expenditures: $19,745. *Unit head:* Dr. Miguel Muñoz, Director, 787-265-3851.

University of Saskatchewan, College of Graduate Studies and Research, College of Agriculture, Department of Soil Science, Saskatoon, SK S7N 5A2, Canada. Offers M Ag, M Sc, PhD. *Degree requirements:* For master's, thesis (for some programs), registration; for doctorate, thesis/dissertation, registration. *Entrance requirements:* Additional exam requirements/recommendations for international students: Required—TOEFL.

University of Vermont, Graduate College, College of Agriculture and Life Sciences, Department of Plant and Soil Science, Burlington, VT 05405. Offers MS, PhD. *Students:* 18 (9 women) 5 international. 25 applicants, 64% accepted, 8 enrolled. In 2005, 7 master's, 1 doctorate awarded. *Degree requirements:* For master's, thesis; for doctorate, one foreign language, thesis/dissertation. *Entrance requirements:* For master's and doctorate, GRE General Test. Additional exam requirements/recommendations for international students: Required—TOEFL (minimum score 550 paper-based; 213 computer-based). *Application deadline:* For fall admission, 2/1 for domestic students. Applications are processed on a rolling basis. Application fee: $40. Electronic applications accepted. *Expenses:* Tuition, area resident: Part-time $410 per credit hour. Tuition, nonresident: part-time $1,034 per credit hour. *Financial support:* Fellowships, research assistantships, teaching assistantships available. Financial award application deadline: 3/1. *Faculty research:* Soil chemistry, plant nutrition. *Unit head:* Dr. Deborah Neher, Chairperson, 802-656-2630. *Application contact:* Dr. M. Starrett, Coordinator, 802-656-2630.

University of Wisconsin–Madison, Graduate School, College of Agricultural and Life Sciences, Department of Agronomy, Madison, WI 53706-1380. Offers agronomy (MS, PhD); plant breeding and plant genetics (MS, PhD). *Faculty:* 19 full-time (2 women). *Students:* 14 full-time (5 women); includes 1 minority (Asian American or Pacific Islander) Average age 25. 17 applicants, 12% accepted, 2 enrolled. In 2005, 2 master's awarded. *Median time to degree:* Of those who began their doctoral program in fall 1997, 100% received their degree in 8 years or less. *Degree requirements:* For master's, thesis or alternative; for doctorate, thesis/dissertation. *Entrance requirements:* For master's and doctorate, GRE, minimum GPA of 3.0. Additional exam requirements/recommendations for international students: Required—TOEFL (minimum score 580 paper-based; 213 computer-based). *Application deadline:* Applications are processed on a rolling basis. Application fee: $45. Electronic applications accepted. *Financial support:* In 2005–06, fellowships (averaging $19,200 per year), research assistantships (averaging $18,120 per year), teaching assistantships (averaging $24,200 per year) were awarded. *Faculty research:* Plant breeding and genetics, plant molecular biology and physiology, cropping systems and management, weed science. *Unit head:* William F. Tracy, Chair, 608-262-1390, Fax: 608-262-5217. *Application contact:* Colleen L. Smith, Graduate Secretary, 608-262-7702, Fax: 608-262-5217, E-mail: clsmith@wisc.edu.

University of Wisconsin–Madison, Graduate School, College of Agricultural and Life Sciences, Department of Soil Science, Madison, WI 53706-1380. Offers MS, PhD. *Faculty:* 20 full-time (4 women). *Students:* 25 full-time (13 women), 1 (woman) part-time; includes 4 minority (1 African American, 1 American Indian/Alaska Native, 2 Hispanic Americans), 6 international. Average age 29. 26 applicants, 19% accepted, 5 enrolled. In 2005, 2 master's awarded. *Median time to degree:* Of those who began their doctoral program in fall 1997, 100% received their degree in 8 years or less. *Degree requirements:* For master's and doctorate, thesis/dissertation, comprehensive exam, registration. *Entrance requirements:* For master's and doctorate, GRE General Test. Additional exam requirements/recommendations for international students: Required—TOEFL. Application fee: $45. Electronic applications accepted. *Financial support:* In 2005–06, 4 fellowships with full tuition reimbursements (averaging $19,200 per year), 16 research assistantships with full tuition reimbursements (averaging $18,120 per year) were awarded; Federal Work-Study, scholarships/grants, and health care benefits also available. *Faculty research:* Physical chemistry of soil colloids/surfaces, forest biogeochemistry, soil-plant-atmosphere interactions, organic byproducts recycling, microbial metabolism in soil, environmental chemistry. *Unit head:* Stephen J. Ventura, Chair, 608-262-6416, Fax: 608-265-2595, E-mail: sventura@wisc.edu. *Application contact:* Carol J. Duffy, Graduate Admissions Coordinator, 608-262-0485, Fax: 608-265-2595, E-mail: cjduffy@wisc.edu.

University of Wyoming, Graduate School, College of Agriculture, Department of Plant Sciences, Program in Agronomy, Laramie, WY 82070. Offers MS, PhD. *Faculty:* 5 full-time (0 women). *Students:* 10 full-time (3 women), 10 part-time (7 women), 6 international. 8 applicants, 50% accepted. In 2005, 2 master's awarded. *Degree requirements:* For master's and doctorate, thesis/dissertation. *Entrance requirements:* For master's and doctorate, GRE General Test, minimum GPA of 3.0. Additional exam requirements/recommendations for international students: Required—TOEFL (minimum score 525 paper-based; 197 computer-based). *Application deadline:* For fall admission, 6/1 for domestic students. Applications are processed on a rolling basis. Application fee: $50. Electronic applications accepted. *Expenses:* Tuition, state resident: full-time $3,720; part-time $155 per credit hour. Tuition, nonresident: full-time $10,704; part-time $446 per credit hour. Required fees: $666; $162 per semester. Tuition and fees vary according to course load and program. *Financial support:* In 2005–06, 9 research assistantships with full tuition reimbursements (averaging $10,062 per year) were awarded Financial award application deadline: 3/1. *Faculty research:* Plant biology, molecular biology/physiology/morphology, production, genetics/breeding, weed control. Total annual research expenditures: $610,000. *Unit head:* Dr. Fred A. Gray, Head, Department of Plant Sciences, 307-766-3103, Fax: 307-766-5549, E-mail: fagray@uwyo.edu.

University of Wyoming, Graduate School, College of Agriculture, Department of Renewable Resources, Laramie, WY 82070. Offers entomology (MS, PhD); rangeland ecology and watershed management (MS, PhD), including soil sciences (PhD), soil sciences and water resources (MS), water resources. Part-time programs available. *Faculty:* 21 full-time (2 women). *Students:* 26 full-time (10 women), 13 part-time (7 women), 9 international. In 2005, 5 master's awarded. *Degree requirements:* For master's, thesis (for some programs); for doctorate, thesis/dissertation. *Entrance requirements:* For master's and doctorate, GRE General Test, minimum GPA of 3.0. Additional exam requirements/recommendations for international students: Required—TOEFL. *Application deadline:* For fall admission, 6/1 for domestic students; for spring admission, 12/1 priority date for domestic students. Applications are processed on a rolling basis. Application fee: $50. Electronic applications accepted. *Expenses:* Tuition, state resident: full-time $3,720; part-time $155 per credit hour. Tuition, nonresident: full-time $10,704; part-time $446 per credit hour. Required fees: $666; $162 per semester. Tuition and fees vary according to course load and program. *Financial support:* In 2005–06, 8 students received support, including 8 research assistantships with full tuition reimbursements available (averaging $10,062 per year); career-related internships or fieldwork and Federal Work-Study also available. Financial award application deadline: 3/1. *Faculty research:* Plant control, grazing management, riparian restoration, riparian management, reclamation. *Unit head:* Dr. Richard A. Olson, Interim Head, 307-766-2263, Fax: 307-766-6403, E-mail: rolson@uwyo.edu. *Application contact:* Kimm Mann-Malody, Office Associate, 307-766-2263, Fax: 307-766-6403, E-mail: kimmmann@uwyo.edu.

Utah State University, School of Graduate Studies, College of Agriculture, Department of Plants, Soils, and Biometeorology, Logan, UT 84322. Offers biometeorology (MS, PhD); ecology (MS, PhD); plant science (MS, PhD); soil science (MS, PhD). Part-time programs available. *Faculty:* 31 full-time (4 women), 13 part-time/adjunct (0 women). *Students:* 117 full-time (58 women), 16 part-time (5 women), 14 international. Average age 26. 25 applicants, 80% accepted, 19 enrolled. In 2005, 8 master's, 1 doctorate awarded. Terminal master's awarded for partial completion of doctoral program. *Median time to degree:* Of those who began their doctoral program in fall 1997, 100% received their degree in 8 years or less. *Degree requirements:* For master's and doctorate, thesis/dissertation. *Entrance requirements:* For master's, GRE General Test, BS in plant, soil, atmospheric science, or related field; minimum GPA of 3.0; for doctorate, GRE General Test, minimum GPA of 3.0. Additional exam requirements/recommendations for international students: Required—TOEFL. *Application deadline:* For fall admission, 6/15 priority date for domestic students, 3/15 priority date for international students; for spring admission, 10/15 priority date for domestic students, 9/15 priority date for international students. Applications are processed on a rolling basis. Application fee: $50 ($60 for international students). Electronic applications accepted. *Financial support:* In 2005–06, 23 research assistantships with partial tuition reimbursements (averaging $15,000 per year) were awarded; Federal Work-Study, institutionally sponsored loans, and tuition waivers (full) also available. Support available to part-time students. Financial award application deadline: 3/1. *Faculty research:* Biotechnology and genomics, plant physiology and biology, nutrient and water efficient landscapes, physical-chemical-biological processes in soil, environmental biophysics and climate. Total annual research expenditures: $4.5 million. *Unit head:* Dr. Larry A. Rupp, Head, 435-797-2099, Fax: 435-797-3376, E-mail: larryr@ext.usu.edu. *Application contact:* Dr. Paul G. Johnson, Graduate Program Coordinator, 435-797-7039, Fax: 435-797-3376, E-mail: paul.johnson@usu.edu.

Virginia Polytechnic Institute and State University, Graduate School, College of Agriculture and Life Sciences, Department of Crop and Soil Environmental Sciences, Blacksburg, VA 24061. Offers MS, PhD. *Faculty:* 23 full-time (2 women). *Students:* 26 full-time (10 women), 7 part-time (3 women); includes 4 minority (2 African Americans, 2 Asian Americans or Pacific Islanders), 2 international. Average age 27. 15 applicants, 73% accepted, 9 enrolled. In 2005, 5 master's, 4 doctorates awarded. *Entrance requirements:* For master's and doctorate, GRE. Additional exam requirements/recommendations for international students: Required—TOEFL (minimum score 550 paper-based; 213 computer-based). *Application deadline:* Applications are processed on a rolling basis. Application fee: $45. Electronic applications accepted. *Expenses:* Tuition, state resident: full-time $6,558; part-time $364 per credit. Tuition, nonresident: full-time $11,296; part-time $628 per credit. Required fees: $1,419; $468 per credit. $234 per term. *Financial support:* In 2005–06, 11 research assistantships with full tuition reimbursements (averaging $13,823 per year), 8 teaching assistantships with full tuition reimbursements (averaging $13,552 per year) were awarded; career-related internships or fieldwork, Federal Work-Study, scholarships/grants, and unspecified assistantships also available. Financial award application deadline: 4/1. *Faculty research:* Environmental soil chemistry, waste management, soil fertility, plant molecular genetics, turfgrass management. *Unit head:* Dr. Steven Clarke Hodges, Head, 540-231-6305, Fax: 540-231-3431, E-mail: hodges@vt.edu. *Application contact:* Dr. M. Alley, Research and Graduate Coordinator, 540-231-9777, Fax: 540-231-3431, E-mail: malley@vt.edu.

Washington State University, Graduate School, College of Agricultural, Human, and Natural Resource Sciences, Department of Crop and Soil Sciences, Program in Crop Sciences, Pullman, WA 99164. Offers MS, PhD. *Faculty:* 17. *Students:* 18 full-time (9 women), 9 international. Average age 30. 28 applicants, 18% accepted, 1 enrolled. In 2005, 6 master's, 2 doctorates awarded. Terminal master's awarded for partial completion of doctoral program. *Degree requirements:* For master's, oral exam, thesis optional; for doctorate, thesis/dissertation, oral exam, written exam. *Entrance requirements:* For master's and doctorate, GRE General Test, minimum GPA of 3.0, 3 letters of recommendation. Additional exam requirements/recommendations for international students: Required—TOEFL (minimum score 550 paper-based; 213 computer-based). *Application deadline:* For fall admission, 2/1 priority date for domestic students, 3/1 priority date for international students; for spring admission, 9/1 priority date for domestic students, 7/1 priority date for international students. Applications are processed on a rolling basis. Application fee: $35. Electronic applications accepted. *Expenses:* Tuition, state resident: full-time $6,295; part-time $336 per credit. Tuition, nonresident: full-time $15,949; part-time $819 per credit. Required fees: $933. Part-time tuition and fees vary according to campus/location and program. *Financial support:* In 2005–06, 5 fellowships (averaging $7,580 per year), 15 research assistantships with full and partial tuition reimbursements (averaging $13,971 per year), 1 teaching assistantship with full and partial tuition reimbursement (averaging $12,353 per year) were awarded; career-related internships or fieldwork, Federal Work-Study, institutionally sponsored loans, tuition waivers (partial), and teaching associateships also available. Financial award application deadline: 2/1; financial award applicants required to submit FAFSA. *Faculty research:* Barley genetics, soil biology, soil fertility, winter wheat breeding, weed science. *Unit head:* Dr. Steve Ullrich, Coordinator, 509-335-4936, Fax: 509-335-8674, E-mail: ullrich@wsu.edu. *Application contact:* Hillary Templin, Academic Programs Coordinator, 509-335-3475, Fax: 509-335-8674, E-mail: hillary@cahnrs.wsu.edu.

Washington State University, Graduate School, College of Agricultural, Human, and Natural Resource Sciences, Department of Crop and Soil Sciences, Program in Soil Sciences, Pullman, WA 99164. Offers MS, PhD. *Faculty:* 17. *Students:* 12 full-time (8 women), 2 part-time (1

Agronomy and Soil Sciences

Washington State University (continued)
woman); includes 1 minority (African American), 4 international. Average age 30. 22 applicants, 27% accepted, 2 enrolled. In 2005, 4 master's, 4 doctorates awarded. Terminal master's awarded for partial completion of doctoral program. *Degree requirements:* For master's, oral exam, thesis optional; for doctorate, thesis/dissertation, oral exam, written exam. *Entrance requirements:* For master's and doctorate, GRE, minimum GPA of 3.0, 3 letters of recommendation. Additional exam requirements/recommendations for international students: Required—TOEFL (minimum score 550 paper-based; 213 computer-based), GRE. *Application deadline:* For fall admission, 2/1 priority date for domestic students, 3/1 priority date for international students; for spring admission, 9/1 priority date for domestic students, 7/1 priority date for international students. Applications are processed on a rolling basis. Application fee: $35. Electronic applications accepted. *Expenses:* Tuition, state resident: full-time $6,295; part-time $336 per credit. Tuition, nonresident: full-time $15,949; part-time $819 per credit. Required fees: $933. Part-time tuition and fees vary according to campus/location and program. *Financial support:* In 2005–06, 1 fellowship (averaging $4,000 per year), 11 research assistantships with full and partial tuition reimbursements (averaging $12,596 per year), 1 teaching assistantship with full and partial tuition reimbursement (averaging $11,637 per year) were awarded; career-related internships or fieldwork, Federal Work-Study, institutionally sponsored loans, tuition waivers (partial), and teaching associateships also available. Financial award application deadline: 4/1; financial award applicants required to submit FAFSA. *Faculty research:* Environmental soils, soil/water quality, soil microbiology, soil physics. *Unit head:* Dr. Alan Busacca, Graduate Coordinator, 509-335-1859, Fax: 509-335-8674, E-mail: busacca@wsu.edu. *Application contact:* Hillary Templin, Academic Programs Coordinator, 509-335-3475, Fax: 509-335-8674, E-mail: hillary@cahnrs.wsu.edu.

West Virginia University, Davis College of Agriculture, Forestry and Consumer Sciences, Division of Animal and Veterinary Sciences, Program in Agricultural Sciences, Morgantown, WV 26506. Offers animal and food sciences (PhD); plant and soil sciences (PhD). *Degree requirements:* For doctorate, thesis/dissertation, oral and written exams. *Entrance requirements:* Additional exam requirements/recommendations for international students: Required—TOEFL. Application fee: $45. *Expenses:* Tuition, state resident: full-time $4,582; part-time $258 per credit hour. Tuition, nonresident: full-time $1,382; part-time $741 per credit hour. *Financial support:* Research assistantships with tuition reimbursements, teaching assistantships with tuition reimbursements, Federal Work-Study, institutionally sponsored loans, and tuition waivers (full and partial) available. Financial award application deadline: 2/1; financial award applicants required to submit FAFSA. *Faculty research:* Ruminant nutrition, metabolism, forage utilization, physiology, reproduction. *Application contact:* Dr. Hillar Klandorf, Professor, 304-293-2631 Ext. 4436, Fax: 304-293-3676, E-mail: hillar.klandorf@mail.wvu.edu.

West Virginia University, Davis College of Agriculture, Forestry and Consumer Sciences, Division of Plant and Soil Sciences, Morgantown, WV 26506. Offers agronomy (MS); entomology (MS); environmental microbiology (MS); horticulture (MS); plant pathology (MS). *Faculty:* 18 full-time (1 woman), 1 part-time/adjunct (0 women). *Students:* 17 full-time (11 women), 3 part-time (2 women), 2 international. Average age 27. In 2005, 5 degrees awarded. *Degree requirements:* For master's, thesis. *Entrance requirements:* For master's, GRE, minimum GPA of 2.5. Additional exam requirements/recommendations for international students: Required—TOEFL. *Application deadline:* Applications are processed on a rolling basis. Application fee: $45. *Expenses:* Tuition, state resident: full-time $4,582; part-time $258 per credit hour. Tuition, nonresident: full-time $1,382; part-time $741 per credit hour. *Financial support:* In 2005–06, 13 research assistantships with full tuition reimbursements (averaging $9,936 per year), 4 teaching assistantships with full tuition reimbursements (averaging $9,936 per year) were awarded; Federal Work-Study, institutionally sponsored loans, and tuition waivers (full and partial) also available. Financial award application deadline: 2/1; financial award applicants required to submit FAFSA. *Faculty research:* Water quality, reclamation of disturbed land, crop production, pest control, environmental protection. Total annual research expenditures: $1 million. *Unit head:* Dr. Barton S. Baker, Chair and Division Director, 304-293-4817 Ext. 4342, Fax: 304-293-2960, E-mail: barton.baker@mail.wvu.edu.

Animal Sciences

Alabama Agricultural and Mechanical University, School of Graduate Studies, School of Agricultural and Environmental Sciences, Department of Plant and Soil Sciences, Huntsville, AL 35811. Offers animal sciences (MS); environmental science (MS); plant and soil science (PhD). Evening/weekend programs available. Terminal master's awarded for partial completion of doctoral program. *Degree requirements:* For master's, thesis; for doctorate, one foreign language, thesis/dissertation. *Entrance requirements:* For master's, GRE General Test, BS in agriculture; for doctorate, GRE General Test, master's degree. Electronic applications accepted. *Faculty research:* Plant breeding, cytogenetics, crop production, soil chemistry and fertility, remote sensing.

Alcorn State University, School of Graduate Studies, School of Agriculture and Applied Science, Alcorn State, MS 39096-7500. Offers agricultural economics (MS Ag); agronomy (MS Ag); animal science (MS Ag). *Faculty:* 11 full-time (2 women). *Students:* 7 full-time (2 women), 8 part-time (4 women); includes 12 minority (all African Americans), 3 international. In 2005, 3 degrees awarded. *Degree requirements:* For master's, thesis optional. *Application deadline:* For fall admission, 7/15 for domestic students; for spring admission, 11/25 for domestic students. Applications are processed on a rolling basis. Application fee: $0 ($10 for international students). *Financial support:* Career-related internships or fieldwork available. Support available to part-time students. *Faculty research:* Aquatic systems, dairy herd improvement, fruit production, alternative farming practices. *Unit head:* Robert Arthur, Interim Dean, 601-877-6137, Fax: 601-877-6219.

American University of Beirut, Graduate Programs, Faculty of Agricultural and Food Sciences, Beirut, Lebanon. Offers agricultural economics (MS); animal sciences (MS); ecosystem management (MSES); food technology (MS); irrigation (MS); mechanization (MS); nutrition (MS); plant protection (MS); plant science (MS); poultry science (MS); soils (MS). *Degree requirements:* For master's, one foreign language, thesis (for some programs), comprehensive exam, registration. *Entrance requirements:* For master's, GRE, letter of recommendation.

Angelo State University, College of Graduate Studies, College of Sciences, Department of Agriculture, San Angelo, TX 76909. Offers animal science (MS). Part-time and evening/weekend programs available. *Faculty:* 7 full-time (2 women), 2 part-time/adjunct (0 women). *Students:* 17 full-time (8 women), 5 part-time; includes 4 minority (1 African American, 1 Asian American or Pacific Islander, 2 Hispanic Americans). Average age 27. 10 applicants, 100% accepted, 9 enrolled. In 2005, 12 degrees awarded. *Degree requirements:* For master's, thesis optional. *Entrance requirements:* For master's, GRE General Test. Additional exam requirements/recommendations for international students: Required—TOEFL or IELT. *Application deadline:* For fall admission, 7/15 priority date for domestic students, 6/15 priority date for international students; for spring admission, 12/8 for domestic students, 11/1 for international students. Applications are processed on a rolling basis. Application fee: $25 ($50 for international students). Electronic applications accepted. *Expenses:* Tuition, area resident: Full-time $2,268; part-time $126 per credit. Tuition, nonresident: full-time $7,236; part-time $402 per credit. Required fees: $844; $94 per credit. One-time fee: $25. Tuition and fees vary according to course load. *Financial support:* In 2005–06, 13 students received support, including 9 research assistantships (averaging $9,887 per year); Federal Work-Study, scholarships/grants, and unspecified assistantships also available. Support available to part-time students. Financial award application deadline:3/1. *Faculty research:* Effect of protein and energy on feedlot performance, bitterweed toxicosis in sheep, meat laboratory, North Concho watershed project, baseline vegetation. *Unit head:* Dr. Gilbert R. Engdahl, Head, 325-942-2027 Ext. 227, E-mail: gil.engdahl@angelo.edu. *Application contact:* Dr. Cody B. Scott, Graduate Advisor, 325-942-2027 Ext. 284, E-mail: cody.scott@angelo.edu.

Auburn University, Graduate School, College of Agriculture, Department of Animal Sciences, Auburn University, AL 36849. Offers M Ag, MS, PhD. Part-time programs available. *Faculty:* 15 full-time (2 women). *Students:* 13 full-time (9 women), 4 part-time (2 women), 3 international. 25 applicants, 28% accepted, 4 enrolled. In 2005, 9 degrees awarded. *Degree requirements:* For master's, thesis (for some programs); for doctorate, thesis/dissertation. *Entrance requirements:* For master's and doctorate, GRE General Test. *Application deadline:* For fall admission, 7/7 for domestic students; for spring admission, 11/24 for domestic students. Applications are processed on a rolling basis. Application fee: $25 ($50 for international students). Electronic applications accepted. *Financial support:* Research assistantships, teaching assistantships, Federal Work-Study available. Support available to part-time students. Financial award application deadline: 3/15. *Faculty research:* Animal breeding and genetics, animal biochemistry and nutrition, physiology of reproduction, animal production. *Unit head:* Dr. L. Wayne Greene, Head, 334-844-1528. *Application contact:* Dr. Stephen L. McFarland, Acting Dean of the Graduate School, 334-844-4700.

Auburn University, Graduate School, College of Agriculture, Department of Poultry Science, Auburn University, AL 36849. Offers M Ag, MS, PhD. Part-time programs available. *Faculty:* 12 full-time (2 women). *Students:* 14 full-time (9 women), 7 part-time (5 women); includes 2 minority (1 African American, 1 Asian American or Pacific Islander), 8 international. 13 applicants, 69% accepted, 6 enrolled. In 2005, 2 master's, 5 doctorates awarded. *Degree requirements:* For master's, thesis (for some programs); for doctorate, thesis/dissertation. *Entrance requirements:* For master's, GRE General Test; for doctorate, GRE General Test, MS. *Application deadline:* For fall admission, 7/7 for domestic students; for spring admission, 11/24 for domestic students. Applications are processed on a rolling basis. Application fee: $25 ($50 for international students). Electronic applications accepted. *Financial support:* Research assistantships, Federal Work-Study available. Support available to part-time students. Financial award application deadline: 3/15. *Faculty research:* Poultry nutrition, poultry breeding, poultry physiology, poultry diseases and parasites, processing/food science. *Unit head:* Dr. Donald E. Conner, Head, 334-844-4133, E-mail: connede@auburn.edu. *Application contact:* Dr. Stephen L. McFarland, Acting Dean of the Graduate School, 334-844-4700.

Brigham Young University, Graduate Studies, College of Biological and Agricultural Sciences, Department of Plant and Animal Sciences, Provo, UT 84602-1001. Offers agronomy (MS); genetics and biotechnology (MS). *Faculty:* 15 full-time (1 woman), 1 (woman) part-time/adjunct. *Students:* 10 full-time (4 women), 3 part-time (1 woman); includes 1 minority (Hispanic American) 14 applicants, 50% accepted, 6 enrolled. In 2005, 6 degrees awarded. *Degree requirements:* For master's, thesis, comprehensive exam, registration. *Entrance requirements:* For master's, GRE General Test, minimum GPA of 3.0 during last 60 hours of course work. *Application deadline:* For fall and spring admission, 2/15. Applications are processed on a rolling basis. Application fee: $50. Electronic applications accepted. *Financial support:* In 2005–06, 12 students received support, including 6 research assistantships with partial tuition reimbursements available (averaging $15,600 per year), 6 teaching assistantships with partial tuition reimbursements available (averaging $15,600 per year); institutionally sponsored loans, scholarships/grants, and tuition waivers (partial) also available. Financial award application deadline: 4/15. *Faculty research:* Iron nutrition in plants, cytogenetics plant pathology, pest management, genetic mapping, environmental science, plant genetics, molecular genetics. Total annual research expenditures:$160,761. *Unit head:* Dr. Sheldon D. Nelson, Chair, 801-422-2760, Fax: 801-422-0008, E-mail: sheldon_nelson@byu.edu. *Application contact:* Dr. Von D. Jolley, Graduate Coordinator, 801-422-2491, Fax: 801-422-0008, E-mail: von-jolley@byu.edu.

California State Polytechnic University, Pomona, Academic Affairs, College of Agriculture, Pomona, CA 91768-2557. Offers agricultural science (MS); animal science (MS); foods and nutrition (MS). Part-time programs available. *Faculty:* 43 full-time (13 women), 16 part-time/adjunct (9 women). *Students:* 20 full-time (16 women), 39 part-time (29 women); includes 14 minority (1 African American, 4 Asian Americans or Pacific Islanders, 9 Hispanic Americans), 4 international. Average age 30. 58 applicants, 76% accepted, 20 enrolled. In 2005, 13 degrees awarded. *Degree requirements:* For master's, thesis or alternative. *Application deadline:* For fall admission, 5/1 for domestic students. For winter admission, 10/15 for domestic students; for spring admission, 1/2 for domestic students. Applications are processed on a rolling basis. Application fee: $55. Electronic applications accepted. *Expenses:* Tuition, nonresident: full-time $9,021. Required fees: $3,597. *Financial support:* Career-related internships or fieldwork, Federal Work-Study, and institutionally sponsored loans available. Support available to part-time students. Financial award application deadline: 3/2; financial award applicants required to submit FAFSA. *Faculty research:* Equine nutrition, physiology, and reproduction; leadership development; bioartificial pancreas; plant science; ruminant and human nutrition. *Unit head:* Dr. Wayne R. Bidlack, Dean, 909-869-2200, E-mail: wrbidlack@csupomona.edu.

California State University, Fresno, Division of Graduate Studies, College of Agricultural Sciences and Technology, Department of Animal Science and Agricultural Education, Fresno, CA 93740-8027. Offers animal science (MA). Part-time and evening/weekend programs available. *Degree requirements:* For master's, thesis. *Entrance requirements:* For master's, GRE General Test, minimum GPA of 3.0 in last 60 hours. Additional exam requirements/recommendations for international students: Required—TOEFL. Electronic applications accepted. *Faculty research:* Horse nutrition, animal health and welfare, electronic monitoring.

Clemson University, Graduate School, College of Agriculture, Forestry and Life Sciences, Department of Animal and Veterinary Sciences, Program in Animal and Veterinary Sciences, Clemson, SC 29634. Offers MS, PhD. Offered in cooperation with the Department of Poultry Science. *Students:* 7 full-time (6 women), 4 part-time (2 women). Average age 26. 9 applicants, 44% accepted, 3 enrolled. In 2005, 1 degree awarded. *Degree requirements:* For master's and doctorate, thesis/dissertation. *Entrance requirements:* For master's and doctorate, GRE General Test. Additional exam requirements/recommendations for international students: Required—TOEFL. *Application deadline:* For fall admission, 6/1 priority date for domestic students, 4/15 priority date for international students; for spring admission, 11/1 priority date for domestic students, 9/15 priority date for international students. Applications are processed on a rolling basis. Application fee: $50. Electronic applications accepted. *Financial support:* Fellowships, research assistantships, teaching assistantships, career-related internships or fieldwork available. Financial award application deadline: 6/1; financial award applicants required to submit FAFSA. *Faculty research:* Reproductive physiology, endocrinology, stress physiology, immunology. *Unit head:* Dr. Tom Scott, Coordinator, 864-656-4027, Fax: 864-656-3131, E-mail: trscott@clemson.edu.

See Close-Ups on pages 599 and 601.

Colorado State University, Graduate School, College of Agricultural Sciences, Department of Animal Sciences, Fort Collins, CO 80523-0015. Offers animal breeding and genetics (MS, PhD); animal nutrition (MS, PhD); animal reproduction (MS, PhD); animal sciences (M Agr); integrated resource management (M Agr); livestock handling (MS, PhD); meats (MS, PhD); production management (MS, PhD). Part-time programs available. *Faculty:* 14 full-time (0 women), 2 part-time/adjunct (0 women). *Students:* 31 full-time (22 women), 19 part-time (8 women), 10 international. Average age 27. 45 applicants, 27% accepted, 9 enrolled. In 2005, 5 master's, 5 doctorates awarded. *Degree requirements:* For master's, thesis, publishable paper; for doctorate, thesis/dissertation, 2 publishable papers. *Entrance requirements:* For master's and doctorate, GRE General Test, minimum GPA of 3.0. Additional exam requirements/recommendations for international students: Required—TOEFL. *Application deadline:* For fall admission, 2/1 priority date for domestic students, 2/1 priority date for international students; for spring admission, 7/1 priority date for domestic students, 5/1 priority date for international students. Applications are processed on a rolling basis. Application fee: $50. Electronic applications accepted. *Expenses:* Tuition, state resident: full-time $3,690; part-time $205 per credit. Tuition, nonresident: full-time $14,958; part-time $831 per credit. Required fees: $1,061. *Financial support:* In 2005–06, 16 research assistantships with full and partial tuition reimbursements (averaging $14,100 per year), 3 teaching assistantships with full tuition reimbursements (averaging $13,200 per year) were awarded; fellowships, traineeships also available. *Faculty research:* Efficiency, food safety, beef management, equine science. Total annual research expenditures: $1.4 million. *Unit head:* William R. Wailes, Interim Department Head, 970-491-5390, Fax: 970-491-5326, E-mail: wwailes@ceres.agsci.colostate.edu. *Application contact:* Cheryl Lee Miller, Graduate Coordinator, 970-491-1442, Fax: 970-491-5326, E-mail: cheryl.miller@colostate.edu.

Cornell University, Graduate School, Graduate Fields of Agriculture and Life Sciences, Field of Animal Breeding, Ithaca, NY 14853-0001. Offers animal breeding (MS, PhD); animal genetics (MS, PhD). *Faculty:* 8 full-time (1 woman). *Students:* 3 full-time (1 woman), 1 international. 2 applicants, 100% accepted, 2 enrolled. In 2005, 1 degree awarded. *Degree requirements:* For master's, thesis/dissertation, teaching experience; for doctorate, thesis/dissertation, teaching experience, comprehensive exam. *Entrance requirements:* For master's and doctorate, 2 letters of recommendation. Additional exam requirements/recommendations for international students: Required—TOEFL (minimum score 550 paper-based; 213 computer-based). *Application deadline:* For fall admission, 4/1 for domestic students; for spring admission, 9/1 for domestic students. Application fee: $60. Electronic applications accepted. *Financial support:* In 2005–06, 3 students received support, including 1 fellowship with full tuition reimbursement available, 2 research assistantships with full tuition reimbursements available; teaching assistantships with full tuition reimbursements available, institutionally sponsored loans, scholarships/grants, health care benefits, tuition waivers (full and partial), and unspecified assistantships also available. Financial award applicants required to submit FAFSA. *Faculty research:* Quantitative genetics, genetic improvement of animal populations, statistical genetics. *Unit head:* Director of Graduate Studies, 607-255-4416, Fax: 607-254-5413, E-mail: shh4@cornell.edu. *Application contact:* Graduate Field Assistant, 607-255-4416, Fax: 607-254-5413, E-mail: shh4@cornell.edu.

Cornell University, Graduate School, Graduate Fields of Agriculture and Life Sciences, Field of Animal Science, Ithaca, NY 14853-0001. Offers animal nutrition (MPS, MS, PhD); animal science (MPS, MS, PhD); physiology of reproduction (MPS, MS, PhD). *Faculty:* 45 full-time (9 women). *Students:* 40 applicants, 28% accepted, 8 enrolled. In 2005, 4 master's, 6 doctorates awarded. *Degree requirements:* For master's, teaching experience, thesis (MS); for doctorate, thesis/dissertation, teaching experience, comprehensive exam. *Entrance requirements:* For master's and doctorate, GRE General Test, 2 letters of recommendation. Additional exam requirements/recommendations for international students: Required—TOEFL (minimum score 550 paper-based; 213 computer-based). *Application deadline:* For fall admission, 3/1 for domestic students; for spring admission, 11/1 for domestic students. Application fee: $60. Electronic applications accepted. *Financial support:* In 2005–06, 31 students received support, including 3 fellowships with full tuition reimbursements available, 19 research assistantships with full tuition reimbursements available; 9 teaching assistantships with full tuition reimbursements available; institutionally sponsored loans, scholarships/grants, health care benefits, tuition waivers (full and partial), and unspecified assistantships also available. Financial award applicants required to submit FAFSA. *Faculty research:* Animal growth and development, dairy science, animal nutrition, physiology of reproduction. *Unit head:* Director of Graduate Studies, 607-255-4416, Fax: 607-254-5413. *Application contact:* Graduate Field Assistant, 607-255-4416, Fax: 607-254-5413, E-mail: shh4@cornell.edu.

Florida Agricultural and Mechanical University, Division of Graduate Studies, Research, and Continuing Education, College of Engineering Science, Technology, and Agriculture, Division of Agricultural Sciences, Tallahassee, FL 32307-3200. Offers agribusiness (MS); animal science (MS); engineering technology (MS); entomology (MS); food science (MS); international programs (MS); plant science (MS). *Degree requirements:* For master's, thesis. *Entrance requirements:* For master's, GRE General Test, minimum GPA of 3.0. Additional exam requirements/recommendations for international students: Required—TOEFL (minimum score 500 paper-based).

Fort Valley State University, College of Graduate Studies and Extended Education, Program in Animal Science, Fort Valley, GA 31030-4313. Offers MS. *Degree requirements:* For master's, thesis, registration. *Entrance requirements:* For master's, GRE General Test.

Iowa State University of Science and Technology, Graduate College, College of Agriculture, Department of Animal Science, Ames, IA 50011. Offers animal breeding and genetics (MS, PhD); animal nutrition (MS, PhD); animal physiology (MS); animal psychology (PhD); animal science (MS, PhD); meat science (MS, PhD). *Faculty:* 55 full-time, 4 part-time/adjunct. *Students:* 68 full-time (35 women), 17 part-time (7 women); includes 5 minority (3 African Americans, 2 Hispanic Americans), 23 international. 37 applicants, 46% accepted, 12 enrolled. In 2005, 10 master's, 8 doctorates awarded. *Degree requirements:* For master's, thesis or alternative; for doctorate, thesis/dissertation. *Entrance requirements:* For master's and doctorate, GRE General Test. Additional exam requirements/recommendations for international students: Required—TOEFL (paper score 550; computer score 213) or IELTS (score 6.5). *Application deadline:* For fall admission, 1/1 priority date for domestic students, 1/1 priority date for international students. Application fee: $30 ($70 for international students). Electronic applications accepted. *Expenses:* Tuition, state resident: full-time $6,410. Tuition, nonresident: full-time $16,422. Tuition and fees vary according to program. *Financial support:* In 2005–06, 63 research assistantships with full and partial tuition reimbursements (averaging $15,263 per year), 2 teaching assistantships with full and partial tuition reimbursements (averaging $14,830 per year) were awarded; fellowships, scholarships/grants, health care benefits, and unspecified assistantships also available. *Faculty research:* Animal breeding, animal nutrition, meat science, muscle biology, nutritional physiology. *Unit head:* Dr. Maynard Hogberg, Head, 515-294-2160, Fax: 515-294-6994. *Application contact:* Donna Nelson, Information Contact, 515-294-2160, E-mail: dlnelson@iastate.edu.

Iowa State University of Science and Technology, Graduate College, College of Agriculture, Department of Natural Resource Ecology and Management, Ames, IA 50011. Offers animal ecology (MS, PhD), including animal ecology, fisheries biology, wildlife biology; forestry (MS, PhD). *Faculty:* 23 full-time, 12 part-time/adjunct. *Students:* 45 full-time (18 women), 4 part-time (1 woman), 7 international. 24 applicants, 38% accepted, 8 enrolled. In 2005, 8 master's, 3 doctorates awarded. *Degree requirements:* For master's, thesis (for some programs); for doctorate, thesis/dissertation. *Entrance requirements:* For master's and doctorate, GRE General Test. Additional exam requirements/recommendations for international students: Required—TOEFL (paper score 547; computer score 210) or IELTS (score 6). *Application deadline:* For fall admission, 1/1 priority date for domestic students, 1/1 priority date for international students; for spring admission, 9/1 priority date for domestic students, 9/1 priority date for international students. Application fee: $30 ($70 for international students). Electronic applications accepted. *Expenses:* Tuition, state resident: full-time $6,410. Tuition, nonresident: full-time $16,422. Tuition and fees vary according to program. *Financial support:* In 2005–06, 41 research assistantships with full and partial tuition reimbursements (averaging $14,755 per year), 3 teaching assistantships with full and partial tuition reimbursements (averaging $14,580 per year) were awarded. *Unit head:* Dr. David M Engle, Chair, 515-294-1166. *Application contact:* Lyn Van De Pol, Information Contact, 515-294-6148, E-mail: lvdp@iastate.edu.

Kansas State University, Graduate School, College of Agriculture, Department of Animal Sciences and Industry, Manhattan, KS 66506. Offers animal breeding and genetics (MS, PhD); meat science (MS, PhD); monogastric nutrition (MS, PhD); physiology (MS, PhD); ruminant nutrition (MS, PhD). *Faculty:* 35 full-time (6 women), 12 part-time/adjunct (2 women). *Students:* 52 full-time (24 women), 14 part-time (3 women); includes 1 minority (African American), 6 international. 34 applicants, 41% accepted, 14 enrolled. In 2005, 16 master's, 5 doctorates awarded. *Degree requirements:* For master's, thesis, oral exam; for doctorate, thesis/dissertation, preliminary exams. *Entrance requirements:* Additional exam requirements/recommendations for international students: Required—TOEFL (minimum score 550 paper-based; 213 computer-based). *Application deadline:* For fall admission, 2/1 for domestic students; for spring admission, 10/1 for domestic students. Applications are processed on a rolling basis. Application fee: $30 ($55 for international students). Electronic applications accepted. *Expenses:* Tuition, state resident: full-time $5,160; part-time $215. Tuition, nonresident: full-time $12,816; part-time $534. Required fees: $564. *Financial support:* In 2005–06, 50 research assistantships (averaging $14,252 per year), 4 teaching assistantships with full tuition reimbursements (averaging $13,741 per year) were awarded; Federal Work-Study, institutionally sponsored loans, and scholarships/grants also available. Support available to part-time students. Financial award application deadline: 3/1; financial award applicants required to submit FAFSA. *Faculty research:* Nutritional management, reproductive and genetic management, managing health and well-being of animals in production systems, managing the environmental aspects of production systems, converting animals and products into safe products. Total annual research expenditures:$947,560. *Unit head:* Janice Swanson, Head, 785-532-7624, Fax: 785-532-7059, E-mail: jswanson@ksu.edu. *Application contact:* J. Ernest Minton, Coordinator, 785-532-1238, Fax: 785-532-7059, E-mail: eminton@oznet.ksu.edu.

Louisiana State University and Agricultural and Mechanical College, Graduate School, College of Agriculture, Department of Animal Sciences, Baton Rouge, LA 70803. Offers MS, PhD. Part-time programs available. *Faculty:* 15 full-time. *Students:* 26 full-time (19 women), 11 part-time (4 women); includes 1 African American, 1 Hispanic American, 6 international. Average age 27. 24 applicants, 29% accepted, 17 enrolled. In 2005, 6 master's, 2 doctorates awarded. Terminal master's awarded for partial completion of doctoral program. *Degree requirements:* For master's and doctorate, thesis/dissertation. *Entrance requirements:* For master's and doctorate, GRE General Test, minimum GPA of 3.0. Additional exam requirements/recommendations for international students: Required—TOEFL (minimum score 550 paper-based; 213 computer-based). *Application deadline:* For fall admission, 1/25 priority date for domestic students, 5/15 priority date for international students. Applications are processed on a rolling basis. Application fee: $25. Electronic applications accepted. *Financial support:* In 2005–06, 30 students received support, including 20 teaching assistantships with partial tuition reimbursements available (averaging $12,862 per year); fellowships, research assistantships with partial tuition reimbursements available, Federal Work-Study, institutionally sponsored loans, scholarships/grants, tuition waivers (full and partial), and unspecified assistantships also available. Support available to part-time students. Financial award applicants required to submit FAFSA. *Faculty research:* Breeding and genetics, nutrition, reproduction, meats, biotechnology. Total annual research expenditures: $6,931. *Unit head:* Dr. Paul E. Humes, Head, 225-578-3241, Fax: 225-578-3279, E-mail: phumes@agctr.lsu.edu. *Application contact:* Dr. Donald L. Thompson, Graduate Coordinator, 225-578-3445, Fax: 225-578-3279, E-mail: dthompson@agctr.lsu.edu.

Louisiana State University and Agricultural and Mechanical College, Graduate School, College of Agriculture, Department of Dairy Science, Baton Rouge, LA 70803. Offers MS, PhD. *Faculty:* 7 full-time (1 woman). *Students:* 6 full-time, 1 part-time, 3 international. Average age 26. 6 applicants, 50% accepted, 3 enrolled. In 2005, 3 master's, 1 doctorate awarded. *Degree requirements:* For master's and doctorate, thesis/dissertation. *Entrance requirements:* For master's and doctorate, GRE General Test, minimum GPA of 3.0. Additional exam requirements/recommendations for international students: Required—TOEFL (minimum score 550 paper-based; 213 computer-based). *Application deadline:* For fall admission, 1/25 priority date for domestic students, 5/15 priority date for international students. Applications are processed on a rolling basis. Application fee: $25. Electronic applications accepted. *Financial support:* In 2005–06, 6 students received support, including 4 research assistantships with partial tuition reimbursements available (averaging $12,750 per year); fellowships, teaching assistantships with partial tuition reimbursements available, Federal Work-Study, scholarships/grants, tuition waivers (full and partial), and unspecified assistantships also available. Financial award applicants required to submit FAFSA. *Faculty research:* Nutrition physiology, genetics, dairy foods technology, dairy management, dairy microbiology. Total annual research expenditures: $4,425. *Unit head:* Dr. Bruce Jenny, Head, 225-578-4411, Fax: 225-578-4008, E-mail: bjenny@agctr.lsu.edu. *Application contact:* John Chandler, Graduate Coordinator, 225-578-3292, Fax: 225-578-4008, E-mail: jchandler@agctr.lsu.edu.

McGill University, Faculty of Graduate and Postdoctoral Studies, Faculty of Agricultural and Environmental Sciences, Department of Animal Science, Montréal, QC H3A 2T5, Canada. Offers M Sc, M Sc A, PhD. *Degree requirements:* For master's, thesis (for some programs), registration; for doctorate, thesis/dissertation, comprehensive exam, registration. *Entrance requirements:* For master's, minimum GPA of 3.0, 2 letters of recommendation; for doctorate, M Sc, minimum GPA of 3.0. Additional exam requirements/recommendations for international students: Required—TOEFL (minimum score 550 paper-based; 213 computer-based), IELT (minimum score 7). Electronic applications accepted. *Faculty research:* Animal nutrition, genetics, embryo transfer, DNA fingerprinting, dairy.

Michigan State University, College of Veterinary Medicine and The Graduate School, Graduate Program in Veterinary Medicine, Department of Large Animal Clinical Sciences, East Lansing, MI 48824. Offers MS, PhD. *Faculty:* 29 full-time (10 women), 1 part-time/adjunct (0 women). *Students:* 6 full-time (1 woman), 3 part-time (1 woman); includes 2 minority (both Asian Americans or Pacific Islanders), 3 international. Average age 38. In 2005, 3 master's, 1 doctorate awarded. *Degree requirements:* For master's, thesis, oral exam, defense of thesis; for doctorate, thesis/dissertation, presentation of dissertation, oral exam and defense of dissertation, comprehensive exam. *Entrance requirements:* For master's, minimum GPA of 3.0, DVM or equivalent, medical degree, 3 letters of recommendation; for doctorate, minimum GPA of 3.0, DVM or equivalent medical degree, 3 letters of recommendation. Additional exam requirements/recommendations for international students: Required—TOEFL (minimum score 550 paper-based; 213 computer-based), Michigan State University (ELT (85), Michigan ELAB (83). *Application deadline:* For fall admission, 12/27 for domestic students. Application fee: $50. Electronic applications accepted. *Expenses:* Tuition, state resident: part-time $330 per credit hour. Tuition, nonresident: part-time $685 per credit hour. Tuition and fees vary according to program. *Financial support:* In 2005–06, 7 fellowships with tuition reimbursements (averaging $2,149 per year), 3 research assistantships with tuition reimbursements (averaging $16,719 per year) were awarded; scholarships/grants and unspecified assistantships also available. *Faculty research:* Pulmonary and exercise physiology, surgery, theriogenology, epidemiology and production medicine. Total annual research expenditures: $2.1 million. *Unit head:* Dr. Thomas H. Herdt, Chairperson, 517-355-9593, Fax: 517-432-1042, E-mail: herdt@msu.edu. *Application contact:* Faith L. Peterson, Administrative Assistant, 517-353-3064, Fax: 517-432-1042, E-mail: petersof@cvm.msu.edu.

Michigan State University, College of Veterinary Medicine and The Graduate School, Graduate Program in Veterinary Medicine, Department of Small Animal Clinical Sciences, East Lansing, MI 48824. Offers MS. *Faculty:* 18 full-time (7 women), 2 part-time/adjunct (0 women). *Students:* 1 full-time (0 women), 1 (woman) part-time, (both international). Average age 31. 3 applicants, 0% accepted. In 2005, 1 degree awarded. *Degree requirements:* For master's, thesis.

Animal Sciences

Michigan State University (continued)
Entrance requirements: For master's, minimum GPA of 3.0; bachelor's, DVM or equivalent degree; 3 letters of recommendation. Additional exam requirements/recommendations for international students: Required—TOEFL (minimum score 550 paper-based; 213 computer-based), Michigan State University ELT (85), Michigan ELAB (83). *Application deadline:* For fall admission, 12/27 for domestic students. Application fee: $50. *Expenses:* Tuition, state resident: part-time $330 per credit hour. Tuition, nonresident: part-time $685 per credit hour. Tuition and fees vary according to program. *Financial support:* Career-related internships or fieldwork and scholarships/grants available. *Faculty research:* Molecular genetics, comparative orthopedics, dermatology, cardiology, anesthesiology. Total annual research expenditures: $1.4 million. *Unit head:* Dr. Charles E. DeCamp, Chairperson, 517-353-7867, Fax: 517-355-5164, E-mail: decampc@cvm.msu.edu. *Application contact:* Information Contact, 517-355-6570, Fax: 517-355-5164, E-mail: scsinfo@cvm.msu.edu.

Michigan State University, The Graduate School, College of Agriculture and Natural Resources, Department of Animal Science, East Lansing, MI 48824. Offers animal science (MS, PhD); animal science-environmental toxicology (PhD). *Faculty:* 37 full-time (9 women). *Students:* 30 full-time (18 women), 6 part-time (4 women); includes 3 minority (1 African American, 2 Hispanic Americans), 9 international. Average age 29. 17 applicants, 18% accepted. In 2005, 5 master's, 7 doctorates awarded. *Degree requirements:* For master's, thesis or alternative, presentation of thesis/project and oral exam; for doctorate, thesis/dissertation, defense of dissertation, year of residence, comprehensive exam. *Entrance requirements:* For master's and doctorate, GRE General Test, minimum GPA of 3.0 in last 2 undergraduate years, letters of reference. Additional exam requirements/recommendations for international students: Required—TOEFL (minimum score 550 paper-based; 213 computer-based), Michigan State University ELT (85), Michigan ELAB (83). *Application deadline:* For fall admission, 12/27 for domestic students. Applications are processed on a rolling basis. Application fee: $50. Electronic applications accepted. *Expenses:* Tuition, state resident: part-time $330 per credit hour. Tuition, nonresident: part-time $685 per credit hour. Tuition and fees vary according to program. *Financial support:* In 2005–06, 5 fellowships with tuition reimbursements (averaging $10,230 per year), 19 research assistantships with tuition reimbursements (averaging $13,062 per year) were awarded; scholarships/grants and unspecified assistantships also available. *Faculty research:* Breeding and genetics, management and systems, meats and growth biology, reproductive and mammary physiology, microbiology and molecular biology. Total annual research expenditures: $5.8 million. *Unit head:* Dr. Karen I. Plaut, Chairperson, 517-355-8383, Fax: 517-353-1699, E-mail: kplaut@msu.edu. *Application contact:* Kim Dobson, Graduate Student Program Secretary, 517-353-9227, Fax: 517-353-1699, E-mail: dobsonk@msu.edu.

Mississippi State University, College of Agriculture and Life Sciences, Department of Poultry Science, Mississippi State, MS 39762. Offers MS. *Faculty:* 9 full-time (2 women), 1 part-time/adjunct (0 women). *Students:* 3 full-time (all women), 6 part-time (1 woman), 2 international. Average age 29. 2 applicants, 50% accepted, 0 enrolled. In 2005, 2 degrees awarded. *Degree requirements:* For master's, thesis. *Entrance requirements:* Additional exam requirements/recommendations for international students: Required—TOEFL. *Application deadline:* For fall admission, 7/1 for domestic students; for spring admission, 11/1 for domestic students. Applications are processed on a rolling basis. Application fee: $30. Electronic applications accepted. *Expenses: Contact institution.* Tuition and fees vary according to course load. *Financial support:* In 2005–06, 1 teaching assistantship with full tuition reimbursement (averaging $11,455 per year) was awarded; research assistantships with full tuition reimbursements, Federal Work-Study, institutionally sponsored loans, scholarships/grants, and unspecified assistantships also available. Financial award applicants required to submit FAFSA. *Faculty research:* Physiology, nutrition management, food science. Total annual research expenditures:$283,064. *Unit head:* Dr. G. Wallace Morgan, Head, 662-325-3416, Fax: 662-325-8292, E-mail: wmorgan@poultry.msstate.edu. *Application contact:* Philip G. Bonfanti, Director of Admissions, 662-325-4104, Fax: 662-325-8872, E-mail: admit@msstate.edu.

Montana State University, College of Graduate Studies, College of Agriculture, Department of Animal and Range Sciences, Bozeman, MT 59717. Offers MS, PhD. Part-time programs available. *Faculty:* 16 full-time (4 women), 2 part-time/adjunct (0 women). *Students:* 8 full-time (5 women), 13 part-time (8 women), 4 international. Average age 28. 13 applicants, 38% accepted, 5 enrolled. In 2005, 8 degrees awarded. *Degree requirements:* For master's, comprehensive exam, registration; for doctorate, thesis/dissertation, comprehensive exam, registration. *Entrance requirements:* For master's and doctorate, GRE General Test. Additional exam requirements/recommendations for international students: Required—TOEFL (minimum score 550 paper-based; 213 computer-based). *Application deadline:* For fall admission, 7/15 priority date for domestic students, 5/15 priority date for international students; for spring admission, 12/1 priority date for domestic students, 10/1 priority date for international students. Applications are processed on a rolling basis. Application fee: $30. Electronic applications accepted. *Expenses:* Tuition, state resident: full-time $4,132. Tuition, nonresident: full-time $1,132. *Financial support:* In 2005–06, 20 students received support, including 12 research assistantships with full and partial tuition reimbursements available, 5 teaching assistantships with partial tuition reimbursements available; health care benefits, tuition waivers (partial), and unspecified assistantships also available. Financial award application deadline: 3/1; financial award applicants required to submit FAFSA. *Faculty research:* Reproductive physiology, ruminant nutrition, range management and ecology, final product quality, systems approach to livestock production. Total annual research expenditures: $3.9 million. *Unit head:* Wayne Gipp, Interim Department Head, 406-994-4850, Fax: 406-994-5589, E-mail: wgipp@montana.edu.

New Mexico State University, Graduate School, College of Agriculture and Home Economics, Department of Animal and Range Sciences, Las Cruces, NM 88003-8001. Offers animal science (M Ag, MS, PhD); range science (M Ag, MS, PhD). Part-time programs available. *Faculty:* 17 full-time (1 woman), 9 part-time/adjunct (0 women). *Students:* 28 full-time (12 women), 9 part-time (2 women); includes 7 minority (1 American Indian/Alaska Native, 6 Hispanic Americans), 6 international. Average age 30. 28 applicants, 54% accepted, 9 enrolled. In 2005, 12 master's, 3 doctorates awarded. *Degree requirements:* For master's, thesis, seminar; for doctorate, thesis/dissertation, research tool. *Entrance requirements:* For master's, minimum GPA of 3.0 in last 60 hours of undergraduate course work (MS); for doctorate, minimum graduate GPA of 3.2. *Application deadline:* For fall admission, 7/1 for domestic students; for spring admission, 11/1 for domestic students. Applications are processed on a rolling basis. Application fee: $30 ($50 for international students). Electronic applications accepted. *Expenses:* Tuition, state resident: full-time $3,156; part-time $175 per credit. Tuition, nonresident: full-time $12,510; part-time $565 per credit. Required fees: $1,050. *Financial support:* In 2005–06, 16 research assistantships, 14 teaching assistantships were awarded; Federal Work-Study also available. Support available to part-time students. Financial award application deadline: 3/1. *Faculty research:* Reproductive physiology, ruminant nutrition, nutrition toxicology, range ecology, wildland hydrology. *Unit head:* Dr. Mark Wise, Head, 505-646-2514, Fax: 505-646-5441, E-mail: mawise@nmsu.edu.

North Carolina State University, Graduate School, College of Agriculture and Life Sciences, Department of Animal Science and Department of Poultry Science, Program in Animal Science and Poultry Science, Raleigh, NC 27695. Offers PhD. *Degree requirements:* For doctorate, thesis/dissertation. *Entrance requirements:* For doctorate, GRE. Electronic applications accepted. *Faculty research:* Nutrient utilization, mineral nutrition, genomics, endocrinology, growth.

North Carolina State University, Graduate School, College of Agriculture and Life Sciences, Department of Poultry Science, Raleigh, NC 27695. Offers MS. Part-time programs available. *Degree requirements:* For master's, thesis. Electronic applications accepted. *Faculty research:* Reproductive physiology, nutrition, toxicology, immunology, molecular biology.

North Dakota State University, The Graduate School, College of Agriculture, Food Systems, and Natural Resources, Department of Animal and Range Sciences, Fargo, ND 58105. Offers animal science (MS, PhD); natural resource management (MS, PhD); range sciences (MS, PhD). *Faculty:* 25 full-time (6 women), 7 part-time/adjunct (1 woman). *Students:* 31 full-time (15 women), 16 part-time (6 women); includes 3 minority (2 African Americans, 1 Asian American or Pacific Islander), 3 international. Average age 25. 26 applicants, 62% accepted. In 2005, 17 master's, 2 doctorates awarded. *Degree requirements:* For master's, thesis/dissertation; for doctorate, thesis/dissertation, comprehensive exam. *Entrance requirements:* For master's and doctorate, GRE General Test. Additional exam requirements/recommendations for international students: Required—TOEFL. *Application deadline:* Applications are processed on a rolling basis. Application fee: $45 ($60 for international students). *Financial support:* In 2005–06, 30 students received support, including 1 fellowship with tuition reimbursement available (averaging $18,000 per year), 29 research assistantships with tuition reimbursements available (averaging $13,000 per year); teaching assistantships, Federal Work-Study, institutionally sponsored loans, and tuition waivers (partial) also available. Financial award application deadline: 3/15. *Faculty research:* Reproduction, nutrition, meat and muscle biology, breeding/genetics. Total annual research expenditures: $1.5 million. *Unit head:* Don R. Kirby, Interim Chair, 701-231-8386, Fax: 701-231-7723, E-mail: donald.kirby@ndsu.edu.

Nova Scotia Agricultural College, Research and Graduate Studies, Truro, NS B2N 5E3, Canada. Offers agriculture (M Sc), including air quality, animal behavior, animal molecular genetics, animal nutrition, animal technology, aquaculture, botany, crop management, crop physiology, ecology, environmental microbiology, food science, horticulture, nutrient management, pest management, physiology, plant biotechnology, plant pathology, soil chemistry, soil fertility, waste management and composting, water quality. Part-time programs available. *Faculty:* 43 full-time (7 women), 21 part-time/adjunct (1 woman). *Students:* 50 full-time (25 women), 15 part-time (11 women); includes 7 minority (3 African Americans, 3 Asian Americans or Pacific Islanders, 1 Hispanic American). Average age 25. In 2005, 23 degrees awarded. *Degree requirements:* For master's, thesis, registration. *Entrance requirements:* For master's, honors B Sc, minimum GPA of 3.0. Additional exam requirements/recommendations for international students: Required—TOEFL (minimum score 580 paper-based; 237 computer-based), Michigan English Language Assessment Battery, IELT, Can Test, CAEL. *Application deadline:* For fall admission, 6/1 for domestic students, 4/1 for international students. For winter admission, 10/31 for domestic students; for spring admission, 2/28 for domestic students. Applications are processed on a rolling basis. Application fee: $70. *Expenses:* Tuition, state resident: part-time $2,328 per year. Tuition, nonresident: full-time $6,984; part-time $7,968 per year. International tuition: $12,624 full-time. Required fees: $481; $46 per course. Tuition and fees vary according to program and student level. *Financial support:* In 2005–06, 48 students received support, including 7 fellowships (averaging $15,000 per year), 10 research assistantships (averaging $15,000 per year), 15 teaching assistantships (averaging $900 per year); career-related internships or fieldwork, scholarships/grants, and unspecified assistantships also available. *Faculty research:* Bio-product development, organic agriculture, nutrient management, air and water quality, agricultural biotechnology. Total annual research expenditures: $4.7 million. *Unit head:* Jill L. Rogers, Manager, 902-893-6360, Fax: 902-893-3430, E-mail: jrogers@nsac.ca. *Application contact:* Marie Law, Administrative Assistant, 902-893-6502, Fax: 902-893-3430, E-mail: mlaw@nsac.ca.

The Ohio State University, Graduate School, College of Food, Agricultural, and Environmental Sciences, Department of Animal Sciences, Columbus, OH 43210. Offers MS, PhD. *Degree requirements:* For master's and doctorate, thesis/dissertation. *Entrance requirements:* For master's and doctorate, GRE General Test. Additional exam requirements/recommendations for international students: Required—TOEFL (paper 550; computer 213) or IELT (7) or Michigan English Language Assessment Battery (84). Electronic applications accepted.

Oklahoma State University, College of Agricultural Science and Natural Resources, Department of Animal Science, Stillwater, OK 74078. Offers animal breeding and reproduction (PhD); animal nutrition (PhD); animal sciences (M Ag, MS); food science (MS, PhD). *Faculty:* 25 full-time (3 women). *Students:* 21 full-time (13 women), 42 part-time (20 women); includes 4 minority (2 African Americans, 1 American Indian/Alaska Native, 1 Asian American or Pacific Islander), 28 international. Average age 30. 69 applicants, 22% accepted, 9 enrolled. In 2005, 17 master's, 3 doctorates awarded. *Degree requirements:* For master's and doctorate, thesis/dissertation. *Entrance requirements:* For master's and doctorate, GRE. Additional exam requirements/recommendations for international students: Required—TOEFL. *Application deadline:* For fall admission, 6/1 priority date for domestic students, 3/1 priority date for international students. Applications are processed on a rolling basis. Application fee: $40 ($75 for international students). Electronic applications accepted. *Expenses:* Tuition, state resident: full-time $4,253; part-time $139 per credit hour. Tuition, nonresident: full-time $12,569; part-time $485 per credit hour. Required fees: $43 per credit hour. One-time fee: $20 part-time. Tuition and fees vary according to course load and program. *Financial support:* In 2005–06, 34 research assistantships (averaging $12,957 per year), 6 teaching assistantships (averaging $12,424 per year) were awarded; career-related internships or fieldwork, Federal Work-Study, scholarships/grants, health care benefits, tuition waivers (partial), and unspecified assistantships also available. Support available to part-time students. Financial award application deadline: 3/1. *Faculty research:* Quantitative trait loci identification for economical traits in swing/beef; waste management strategies in livestock; endocrine control of reproductive processes in farm animals; cholesterol synthesis, inhibition, and reduction; food safety research. *Unit head:* Dr. Donald G. Wagner, Head, 405-744-6062, Fax: 405-744-7390, E-mail: don.wagner@okstate.edu. *Application contact:* Dr. Gerald Horn, 405-744-6621, E-mail: gerald.horn@okstate.edu.

Oregon State University, Graduate School, College of Agricultural Sciences, Department of Animal Sciences, Corvallis, OR 97331. Offers animal science (M Agr, MAIS, MS, PhD); poultry science (M Agr, MAIS, MS, PhD). *Faculty:* 19 full-time (4 women), 1 part-time/adjunct (0 women). *Students:* 17 full-time (7 women), 3 part-time (1 woman), 3 international. Average age 29. In 2005, 3 master's, 2 doctorates awarded. Terminal master's awarded for partial completion of doctoral program. *Degree requirements:* For master's, thesis (for some programs); for doctorate, thesis/dissertation. *Entrance requirements:* For master's and doctorate, GRE General Test, minimum GPA of 3.0 in last 90 hours. Additional exam requirements/recommendations for international students: Required—TOEFL. *Application deadline:* For fall admission, 3/1 for domestic students. Applications are processed on a rolling basis. Application fee: $50. *Expenses:* Tuition, area resident: Part-time $301 per credit. Tuition, state resident: full-time $8,139; part-time $501 per credit. Tuition, nonresident: full-time $14,376; part-time $532 per credit. Required fees: $1,266. *Financial support:* Fellowships, research assistantships, career-related internships or fieldwork, Federal Work-Study, and institutionally sponsored loans available. Support available to part-time students. Financial award application deadline: 2/1. *Faculty research:* Reproductive physiology, population genetics, general nutrition of ruminants and nonruminants, embryo physiology, endocrinology. *Unit head:* James R. Males, Head, 541-737-1891, Fax: 541-737-4174, E-mail: james.males@orst.edu.

The Pennsylvania State University University Park Campus, Graduate School, College of Agricultural Sciences, Department of Dairy and Animal Science, State College, University Park, PA 16802-1503. Offers animal science (M Agr, MS, PhD). *Students:* 13 full-time (6 women), 4 part-time (2 women); includes 1 minority (Hispanic American), 8 international. *Entrance requirements:* For master's and doctorate, GRE General Test. *Expenses:* Tuition, state resident: full-time $12,518; part-time $522 per credit. Tuition, nonresident: full-time $23,004; part-time $959 per credit. Required fees: $484. Tuition and fees vary according to course load, campus/location and program. *Unit head:* Dr. Terry D. Etherton, Head, 814-863-3665, Fax: 814-863-6042, E-mail: tetherton@psu.edu. *Application contact:* Dr. Daniel R. Hagen, Graduate Officer, 814-863-0723, Fax: 814-863-6042, E-mail: drh@psu.edu.

Prairie View A&M University, Graduate School, College of Agriculture and Human Sciences, Prairie View, TX 77446-0519. Offers agricultural economics (MS); animal sciences (MS); interdisciplinary human sciences (MS); soil science (MS). Part-time and evening/weekend programs available. *Faculty:* 8 full-time (4 women), 9 part-time/adjunct (3 women). *Students:* 68 full-time (45 women), 79 part-time (65 women); includes 125 minority (all African Americans), 13 international. Average age 33. 147 applicants, 100% accepted, 147 enrolled. In 2005, 31 degrees awarded. *Degree requirements:* For master's, thesis (for some programs), field placement, comprehensive exam, registration. *Entrance requirements:* For master's, GRE General Test, minimum GPA of 2.45. Additional exam requirements/recommendations for international students: Required—TOEFL (minimum score 550 paper-based). *Application deadline:* For fall admission, 6/1 for domestic students, 6/1 for international students; for spring admission, 10/1 for domestic students, 10/1 for international students. Applications are processed on a rolling basis. Application fee: $50. *Expenses:* Tuition, state resident: full-time $1,440; part-time $80 per credit. Tuition, nonresident: full-time $6,444; part-time $358 per credit. *Financial support:* In 2005–06, 57 students received support, including 8 fellowships with tuition reimbursements available (averaging $12,000 per year), 10 research assistantships with tuition reimbursements available (averaging $15,000 per year); career-related internships or fieldwork, Federal Work-Study, institutionally sponsored loans, scholarships/grants, tuition waivers (partial), and unspecified assistantships also available. Support available to part-time students. Financial award application deadline: 4/1; financial award applicants required to submit FAFSA. *Faculty research:* Domestic violence prevention, water quality, food growth regulators. Total annual research expenditures: $4 million. *Unit head:* Dr. Linda Willis, Dean, 936-857-2996, Fax: 936-857-2998, E-mail: lwillis@pvamu.edu. *Application contact:* Dr. Richard W. Griffin, Interim Department Head, 936-857-2996, Fax: 936-857-2998, E-mail: rwgriffin@pvamu.edu.

Purdue University, Graduate School, College of Agriculture, Department of Animal Sciences, West Lafayette, IN 47907. Offers MS, PhD. Part-time programs available. *Faculty:* 30 full-time (2 women), 7 part-time/adjunct (3 women). *Students:* 60 full-time (32 women), 11 part-time (8 women); includes 5 minority (2 African Americans, 1 Asian American or Pacific Islander, 2 Hispanic Americans), 26 international. Average age 27. 38 applicants, 34% accepted, 12 enrolled. In 2005, 18 master's, 6 doctorates awarded. Terminal master's awarded for partial completion of doctoral program. *Degree requirements:* For master's, thesis optional; for doctorate, thesis/dissertation. *Entrance requirements:* For master's and doctorate, GRE General Test. Additional exam requirements/recommendations for international students: Required—TOEFL; Recommended—TWE. *Application deadline:* For fall admission, 4/15 for domestic students; for spring admission, 10/1 priority date for domestic students. Applications are processed on a rolling basis. Application fee: $55. Electronic applications accepted. *Financial support:* In 2005–06, 35 students received support, including fellowships with full tuition reimbursements available (averaging $14,153 per year), 29 research assistantships with full tuition reimbursements available (averaging $14,153 per year) Support available to part-time students. Financial award applicants required to submit FAFSA. *Faculty research:* Genetics, meat science, nutrition, management, ethology. *Unit head:* Dr. Alan L Grant, Head, 765-494-8282, Fax: 765-494-9346. *Application contact:* Nena C Fawbush, Graduate Secretary, 765-494-2649, Fax: 765-494-6816, E-mail: grad@ansc.purdue.edu.

Rutgers, The State University of New Jersey, New Brunswick/Piscataway, Graduate School, Program in Animal Sciences, New Brunswick, NJ 08901-1281. Offers endocrine control of growth and metabolism (MS, PhD); nutrition of ruminant and nonruminant animals (MS, PhD); reproductive endocrinology and neuroendocrinology (MS, PhD). *Faculty:* 28 full-time, 11 part-time/adjunct. *Students:* 13 full-time (10 women), 6 part-time (2 women); includes 4 minority (2 Asian Americans or Pacific Islanders, 2 Hispanic Americans), 1 international. Average age 25. 36 applicants, 33% accepted, 11 enrolled. In 2005, 1 master's, 1 doctorate awarded. Terminal master's awarded for partial completion of doctoral program. *Degree requirements:* For master's, thesis/dissertation; for doctorate, thesis/dissertation, comprehensive exam. *Entrance requirements:* For master's and doctorate, GRE General Test. *Application deadline:* For fall admission, 3/15 for domestic students, 1/15 for international students. Applications are processed on a rolling basis. Application fee: $50. Electronic applications accepted. *Expenses:* Tuition, state resident: full-time $10,440; part-time $435 per credit. Tuition, nonresident: full-time $15,520; part-time $647 per credit. Required fees: $129 per credit. Tuition and fees vary according to program. *Financial support:* In 2005–06, 12 students received support, including 1 fellowship with full tuition reimbursement available (averaging $17,000 per year), 2 research assistantships with full tuition reimbursements available (averaging $16,550 per year), 9 teaching assistantships with full tuition reimbursements available (averaging $16,988 per year); unspecified assistantships also available. Financial award application deadline: 1/15; financial award applicants required to submit FAFSA. *Faculty research:* Equine exercise physiology and nutrition, alcohol and stress, reproductive biology and endocrinology, mammary gland biology and lactation reproductive behavior. Total annual research expenditures: $1.1 million. *Unit head:* Dr. Henry B. John-Alder, Director, 732-932-3229, Fax: 732-932-6996, E-mail: henry@aesop.rutgers.edu. *Application contact:* Dr. Katherine A. Manger, Administrator, 732-932-3879, Fax: 732-932-6996, E-mail: manger@aesop.rutgers.edu.

South Dakota State University, Graduate School, College of Agriculture and Biological Sciences, Department of Animal Science and Range Science, Brookings, SD 57007. Offers animal science (MS, PhD). *Degree requirements:* For master's, thesis, oral exam; for doctorate, thesis/dissertation, preliminary oral and written exams. *Entrance requirements:* Additional exam requirements/recommendations for international students: Required—TOEFL. *Faculty research:* Ruminant and nonruminant nutrition, meat science, reproductive physiology, range utilization, ecology genetics.

South Dakota State University, Graduate School, College of Agriculture and Biological Sciences, Department of Dairy Science, Brookings, SD 57007. Offers MS, PhD. *Degree requirements:* For master's, thesis, oral exam; for doctorate, thesis/dissertation, preliminary oral and written exams. *Entrance requirements:* Additional exam requirements/recommendations for international students: Required—TOEFL. *Faculty research:* Dairy cattle nutrition, energy metabolism, lowfat cheese technology, food safety, sensory evaluation of dairy products.

Southern Illinois University Carbondale, Graduate School, College of Agriculture, Department of Animal Science, Food and Nutrition, Program in Animal Science, Carbondale, IL 62901-4701. Offers MS. *Faculty:* 15 full-time (6 women). *Students:* 6 full-time (4 women), 7 part-time (5 women); includes 4 minority (3 African Americans, 1 Hispanic American). Average age 29. 14 applicants, 43% accepted, 3 enrolled. In 2005, 2 master's awarded. *Degree requirements:* For master's, thesis. *Entrance requirements:* For master's, minimum GPA of 2.7. Additional exam requirements/recommendations for international students: Required—TOEFL. *Application deadline:* Applications are processed on a rolling basis. Application fee: $0. *Financial support:* In 2005–06, 13 research assistantships with full tuition reimbursements, 2 teaching assistantships with full tuition reimbursements were awarded; fellowships with full tuition reimbursements, career-related internships or fieldwork, Federal Work-Study, institutionally sponsored loans, and tuition waivers (full) also available. Support available to part-time students. *Faculty research:* Nutrition, reproductive physiology, animal biotechnology, phytoestrogens and animal reproduction. Total annual research expenditures:$300,000. *Unit head:* Todd A. Winters, Interim Chair, Department of Animal Science, Food and Nutrition, 618-453-1760, Fax: 618-453-5231.

Sul Ross State University, Division of Agricultural and Natural Resource Science, Program in Animal Science, Alpine, TX 79832. Offers M Ag, MS. Part-time programs available. *Degree requirements:* For master's, thesis (for some programs). *Entrance requirements:* For master's, GRE General Test, minimum GPA of 2.5 in last 60 hours of undergraduate work. *Faculty research:* Reproductive physiology, meat processing, animal nutrition, equine foot and motion studies, Spanish goat and Barbido sheep studies.

Texas A&M University, College of Agriculture and Life Sciences, Department of Animal Science, College Station, TX 77843. Offers animal breeding (MS, PhD); animal science (M Agr, MS, PhD); dairy science (M Agr, MS); physiology of reproduction (MS, PhD). *Faculty:* 27 full-time (4 women), 2 part-time/adjunct (0 women). *Students:* 87 full-time (44 women), 32 part-time (15 women); includes 9 minority (2 African Americans, 1 Asian American or Pacific Islander, 6 Hispanic Americans), 17 international. Average age 26. 103 applicants, 37% accepted, 30 enrolled. In 2005, 36 master's, 13 doctorates awarded. *Degree requirements:* For master's and doctorate, thesis/dissertation, registration. *Entrance requirements:* For master's and doctorate, GRE General Test. Additional exam requirements/recommendations for international students: Required—TOEFL. *Application deadline:* For fall admission, 2/1 for domestic students; for spring admission, 10/1 priority date for domestic students. Applications are processed on a rolling basis. Application fee: $50 ($75 for international students). *Expenses:* Tuition, state resident: full-time $4,488; part-time $187 per credit hour. Tuition, nonresident: full-time $11,112; part-time $463 per credit hour. Required fees:$1,974. *Financial support:* In 2005–06, fellowships (averaging $15,000 per year), research assistantships (averaging $12,950 per year), teaching assistantships (averaging $11,500 per year) were awarded; career-related internships or fieldwork, Federal Work-Study, institutionally sponsored loans, and scholarships/grants also available. Financial award application deadline: 2/1; financial award applicants required to submit FAFSA. *Faculty research:* Genetic engineering/gene markers, dietary effects on colon cancer, biotechnology. *Unit head:* Dr. Gary Acuff, Head, 979-845-1541, Fax: 979-845-6433. *Application contact:* Ronnie Edwards, Graduate Advisor, 979-845-1542, Fax: 979-845-6433, E-mail: r-edwards@tamu.edu.

Texas A&M University, College of Agriculture and Life Sciences, Department of Poultry Science, College Station, TX 77843. Offers M Agr, MS, PhD. Part-time and evening/weekend programs available. Postbaccalaureate distance learning degree programs offered (no on-campus study). *Faculty:* 5 full-time (0 women), 1 (woman) part-time/adjunct. *Students:* 20 full-time (7 women), 5 part-time (1 woman); includes 1 minority (Hispanic American), 14 international. Average age 29. 11 applicants, 55% accepted, 2 enrolled. In 2005, 6 master's, 4 doctorates awarded. Terminal master's awarded for partial completion of doctoral program. *Median time to degree:* Of those who began their doctoral program in fall 1997, 100% received their degree in 8 years or less. *Degree requirements:* For master's, thesis (for some programs); for doctorate, thesis/dissertation. *Entrance requirements:* For master's and doctorate, GRE General Test. Additional exam requirements/recommendations for international students: Required—TOEFL. Application fee: $50 ($75 for international students). Electronic applications accepted. *Expenses:* Tuition, state resident: full-time $4,488; part-time $187 per credit hour. Tuition, nonresident: full-time $11,112; part-time $463 per credit hour. Required fees: $1,974. *Financial support:* In 2005–06, fellowships with partial tuition reimbursements (averaging $18,000 per year); research assistantships with partial tuition reimbursements, teaching assistantships, scholarships/grants and unspecified assistantships also available. Financial award application deadline: 4/1; financial award applicants required to submit FAFSA. *Faculty research:* Poultry diseases and immunology, avian genetics and physiology, nutrition and metabolism, poultry processing and food safety, waste management. *Unit head:* Dr. Alan Sams, Head, 979-845-1931, Fax: 979-845-1921, E-mail: asams@poultry.tamu.edu. *Application contact:* Dr. Jerry Daniels, Advisor/Lecturer, 979-845-1654, Fax: 979-845-1931, E-mail: jdaniels@poultry.tamu.edu.

Texas A&M University–Kingsville, College of Graduate Studies, College of Agriculture and Home Economics, Program in Animal Sciences, Kingsville, TX 78363. Offers MS. *Degree requirements:* For master's, thesis or alternative, comprehensive exam. *Entrance requirements:* For master's, GRE General Test, minimum GPA of 3.0. Additional exam requirements/recommendations for international students: Required—TOEFL.

Texas Tech University, Graduate School, College of Agricultural Sciences and Natural Resources, Department of Animal and Food Sciences, Lubbock, TX 79409. Offers animal science (MS, PhD); food technology (MS). Part-time programs available. *Faculty:* 15 full-time (4 women), 1 part-time/adjunct (0 women). *Students:* 43 full-time (24 women), 5 part-time (2 women); includes 3 minority (1 American Indian/Alaska Native, 1 Asian American or Pacific Islander, 1 Hispanic American), 11 international. Average age 26. 36 applicants, 69% accepted, 15 enrolled. In 2005, 15 master's, 2 doctorates awarded. *Degree requirements:* For master's, thesis, internship (M Agr); for doctorate, thesis/dissertation. *Entrance requirements:* For master's and doctorate, GRE General Test. Additional exam requirements/recommendations for international students: Required—TOEFL (minimum score 550 paper-based; 213 computer-based). *Application deadline:* Applications are processed on a rolling basis. Application fee: $50 ($60 for international students). Electronic applications accepted. *Expenses:* Tuition, state resident: full-time $4,296. Tuition, nonresident: full-time $10,920. Required fees: $1,992. Tuition and fees vary according to program. *Financial support:* In 2005–06, 23 students received support, including 34 research assistantships with partial tuition reimbursements available (averaging $10,738 per year), 3 teaching assistantships with partial tuition reimbursements available (averaging $12,300 per year); Federal Work-Study and institutionally sponsored loans also available. Support available to part-time students. Financial award application deadline: 4/15; financial award applicants required to submit FAFSA. *Faculty research:* Animal growth composition and product acceptability, animal nutrition and utilization, animal physiology and adaptation to stress, food microbiology, food safety and security. Total annual research expenditures: $1.8 million. *Unit head:* Dr. Kevin R. Pond, Chairman, 806-742-2805 Ext. 223, Fax: 806-742-0898, E-mail: kevin.pond@ttu.edu. *Application contact:* Sandy Gellner, Graduate Secretary, 806-742-0898 Ext. 221, Fax: 806-742-4003, E-mail: sandra.gellner@ttu.edu.

Tuskegee University, Graduate Programs, College of Agricultural, Environmental and Natural Sciences, Department of Agricultural Sciences, Program in Animal and Poultry Sciences, Tuskegee, AL 36088. Offers MS. *Faculty:* 13 full-time (1 woman), 2 part-time/adjunct (1 woman). *Students:* 9 full-time (8 women), 6 part-time; includes 8 minority (all African Americans), 1 international. Average age 24. In 2005, 1 degree awarded. *Degree requirements:* For master's, thesis. *Entrance requirements:* For master's, GRE General Test. Additional exam requirements/recommendations for international students: Required—TOEFL (minimum score 500 paper-based; 173 computer-based). *Application deadline:* For fall admission, 7/15 for domestic students. Applications are processed on a rolling basis. Application fee: $25 ($35 for international students). *Expenses:* Tuition: Full-time $12,400. Required fees: $300; $490 per credit. *Financial support:* Application deadline: 4/15. *Unit head:* Dr. P. K. Biswas, Head, Department of Agricultural Sciences, 334-727-8446.

Universidad Nacional Pedro Henriquez Urena, Graduate School, Santo Domingo, Dominican Republic. Offers accounting and auditing (M Acct); animal production (M Agr); business administration (MBA, PhD); dentistry (DDS); economics (M Econ); education (PhD); environmental engineering (MEE); horticulutre (M Agr); hospital administration (PhD); humanities (PhD); international relations (MPS); management of natural resources (MNRM); project management (M Man); public administration (MPS); social science (PhD); veterinary medicine (DVM).

Université Laval, Faculty of Agricultural and Food Sciences, Department of Animal Sciences, Programs in Animal Sciences, Québec, QC G1K 7P4, Canada. Offers M Sc, PhD. Part-time programs available. Terminal master's awarded for partial completion of doctoral program. *Degree requirements:* For master's, thesis/dissertation; for doctorate, thesis/dissertation, comprehensive exam. *Entrance requirements:* For master's and doctorate, knowledge of French and English. Electronic applications accepted.

The University of Arizona, Graduate College, College of Agriculture and Life Sciences, Department of Animal Sciences, Tucson, AZ 85721. Offers MS, PhD. Part-time programs available. *Degree requirements:* For master's and doctorate, thesis/dissertation. *Entrance requirements:* For master's, GRE Subject Test, 3 letters of recommendation; for doctorate, GRE Subject Test (biology or chemistry recommended), 3 letters of recommendation. Additional exam requirements/recommendations for international students: Required—TOEFL (minimum score 550 paper-based; 213 computer-based). *Faculty research:* Nutrition of beef and dairy cattle, reproduction and breeding, muscle growth and function, animal stress, meat science.

University of Arkansas, Graduate School, Dale Bumpers College of Agricultural, Food and Life Sciences, Department of Animal Science, Fayetteville, AR 72701-1201. Offers MS, PhD. *Students:* 17 full-time (7 women), 8 part-time (1 woman); includes 2 minority (1 African American, 1 Hispanic American), 3 international. 21 applicants, 33% accepted. In 2005, 6 master's, 3 doctorates awarded. *Degree requirements:* For master's, thesis; for doctorate, variable foreign language requirement, thesis/dissertation. *Entrance requirements:* For master's, GRE General Test or minimum GPA of 2.7. Application fee: $40 ($50 for international students).

Animal Sciences

University of Arkansas (continued)
Financial support: In 2005–06, 16 research assistantships, 1 teaching assistantship were awarded; fellowships with tuition reimbursements, career-related internships or fieldwork and Federal Work-Study also available. Support available to part-time students. Financial award application deadline: 4/1; financial award applicants required to submit FAFSA. *Unit head:* Dr. Keith Lusby, Chair, 479-575-4351. *Application contact:* Dr. Wayne Kellogg, Graduate Coordinator, 479-575-4351, E-mail: wkellogg@uark.edu.

University of Arkansas, Graduate School, Dale Bumpers College of Agricultural, Food and Life Sciences, Department of Poultry Science, Fayetteville, AR 72701-1201. Offers MS, PhD. *Faculty:* 15 full-time (3 women). *Students:* 30 full-time (13 women), 2 part-time (1 woman); includes 1 minority (African American), 18 international. 19 applicants, 47% accepted. In 2005, 6 master's, 2 doctorates awarded. *Degree requirements:* For master's, thesis; for doctorate, variable foreign language requirement, thesis/dissertation. Application fee: $40 ($50 for international students). *Financial support:* In 2005–06, 4 fellowships with tuition reimbursements, 39 research assistantships were awarded; teaching assistantships, career-related internships or fieldwork and Federal Work-Study also available. Support available to part-time students. Financial award application deadline: 4/1; financial award applicants required to submit FAFSA. *Unit head:* Walter Bottje, Head, 479-575-4952.

The University of British Columbia, Faculty of Graduate Studies, Faculty of Land and Food Systems, Animal Science Graduate Program, Vancouver, BC V6T 1Z1, Canada. Offers M Sc, PhD. *Degree requirements:* For master's, thesis/dissertation, registration; for doctorate, thesis/dissertation, comprehensive exam, registration. *Entrance requirements:* Additional exam requirements/recommendations for international students: Required—TOEFL (minimum score 560 paper-based; 220 computer-based). Electronic applications accepted. *Faculty research:* Nutrition and metabolism, animal production, animal behavior and welfare, reproductive physiology, animal genetics, aquaculture and fish physiology.

University of California, Davis, Graduate Studies, Graduate Group in Animal Biology, Davis, CA 95616. Offers MAM, MS, PhD. *Faculty:* 45 full-time. *Students:* 54 full-time (35 women); includes 4 minority (1 Asian American or Pacific Islander, 3 Hispanic Americans), 5 international. Average age 26. 55 applicants, 58% accepted, 28 enrolled. In 2005, 8 master's awarded. Terminal master's awarded for partial completion of doctoral program. *Degree requirements:* For master's, thesis (for some programs), comprehensive exam (for some programs); for doctorate, thesis/dissertation. *Entrance requirements:* For master's, GRE General Test, minimum GPA of 3.0. Additional exam requirements/recommendations for international students: Required—TOEFL (minimum score 550 paper-based; 213 computer-based). *Application deadline:* For fall admission, 1/15 for domestic students, 1/15 for international students. Application fee: $60. Electronic applications accepted. *Financial support:* In 2005–06, 52 students received support, including 5 fellowships with full and partial tuition reimbursements available (averaging $6,660 per year), 3 research assistantships with full and partial tuition reimbursements available (averaging $11,328 per year), 31 teaching assistantships with partial tuition reimbursements available (averaging $15,083 per year); Federal Work-Study, institutionally sponsored loans, and tuition waivers (full and partial) also available. Financial award application deadline: 1/15; financial award applicants required to submit FAFSA. *Faculty research:* Genetics, nutrition, physiology and behavior in domestic and aquatic animals. *Unit head:* Trish Berger, Graduate Chair, 530-752-1267, E-mail: tberger@ucdavis.edu. *Application contact:* Alisha L. Nork, Administrative Assistant, 530-752-2382, E-mail: alnork@ucdavis.edu.

University of Connecticut, Graduate School, College of Agriculture and Natural Resources, Department of Animal Science, Field of Animal Science, Storrs, CT 06269. Offers MS, PhD. *Faculty:* 18 full-time (3 women). *Students:* 30 full-time (15 women), 8 part-time (6 women); includes 1 minority (African American), 18 international. Average age 30. 26 applicants, 54% accepted, 8 enrolled. In 2005, 9 master's, 4 doctorates awarded. Terminal master's awarded for partial completion of doctoral program. *Degree requirements:* For master's and doctorate, thesis/dissertation, comprehensive exam. *Entrance requirements:* For master's and doctorate, GRE General Test. Additional exam requirements/recommendations for international students: Required—TOEFL (minimum score 550 paper-based; 213 computer-based). *Application deadline:* For fall admission, 2/1 priority date for domestic students, 2/1 priority date for international students; for spring admission, 11/1 for domestic students, 10/1 for international students. Applications are processed on a rolling basis. Application fee: $55. Electronic applications accepted. *Expenses:* Tuition, state resident: part-time $444 per credit hour. Tuition, nonresident: part-time $1,154 per credit hour. Tuition and fees vary according to course load. *Financial support:* In 2005–06, 26 research assistantships with tuition reimbursements were awarded; fellowships, teaching assistantships with tuition reimbursements, Federal Work-Study, scholarships/grants, health care benefits, and unspecified assistantships also available. Financial award application deadline: 2/1; financial award applicants required to submit FAFSA. *Application contact:* Larry Silbart, Chairperson, 860-486-6073, Fax: 860-486-4375, E-mail: lawrence.silbart@uconn.edu.

University of Delaware, College of Agriculture and Natural Resources, Department of Animal and Food Sciences, Newark, DE 19716. Offers animal sciences (MS, PhD); food sciences (MS). Part-time programs available. *Faculty:* 17 full-time (5 women). *Students:* 25 full-time (15 women), 1 part-time; includes 1 minority (Hispanic American), 7 international. Average age 28. 26 applicants, 35% accepted, 9 enrolled. In 2005, 2 master's, 1 doctorate awarded. Terminal master's awarded for partial completion of doctoral program. *Degree requirements:* For master's, thesis/dissertation; for doctorate, thesis/dissertation, comprehensive exam. *Entrance requirements:* For master's and doctorate, GRE General Test. Additional exam requirements/recommendations for international students: Required—TOEFL. *Application deadline:* For fall admission, 7/1 for domestic students; for spring admission, 12/1 for domestic students. Applications are processed on a rolling basis. Application fee: $60. Electronic applications accepted. *Financial support:* In 2005–06, 21 students received support, including 15 research assistantships with full tuition reimbursements available (averaging $15,344 per year), 5 teaching assistantships with full tuition reimbursements available (averaging $12,314 per year); fellowships with full tuition reimbursements available, scholarships/grants and tuition waivers (full) also available. Financial award application deadline: 3/1. *Faculty research:* Food chemistry, food microbiology, process engineering technology, packaging, food analysis, microbial genetics, molecular endocrinology, growth physiology, avian immunology and virology, monogastric nutrition, avian genomics. Total annual research expenditures: $1.5 million. *Unit head:* Dr. Limin Kung, Assistant Chair, 302-831-2524, Fax: 302-831-2822, E-mail: lksilage@udel.edu. *Application contact:* Dr. Carl Schmidt, Graduate Program Coordinator, 302-831-2524, Fax: 302-831-2822, E-mail: schmidtc@udel.edu.

University of Florida, Graduate School, College of Agricultural and Life Sciences, Department of Animal Sciences, Gainesville, FL 32611. Offers M Ag, MS, PhD. *Faculty:* 30 full-time (5 women). *Students:* 58 (31 women); includes 8 minority (1 African American, 1 Asian American or Pacific Islander, 6 Hispanic Americans) 14 international. 28 applicants, 68% accepted. In 2005, 15 master's, 7 doctorates awarded. *Degree requirements:* For master's, variable foreign language requirement, thesis optional; for doctorate, thesis/dissertation. *Entrance requirements:* For master's and doctorate, GRE General Test, minimum GPA of 3.0. Additional exam requirements/recommendations for international students: Required—TOEFL. *Application deadline:* For fall admission, 6/1 for domestic students. Applications are processed on a rolling basis. Application fee: $20. Electronic applications accepted. *Expenses:* Tuition, state resident: full-time $6,234. Tuition, nonresident: full-time $21,359. Tuition and fees vary according to program. *Financial support:* In 2005–06, 1 fellowship (averaging $14,072 per year), 16 research assistantships (averaging $15,238 per year), 21 teaching assistantships (averaging $15,146 per year) were awarded. *Faculty research:* Meat science, breeding and genetics, animal physiology, molecular biology, animal nutrition. Total annual research expenditures: $4.1 million. *Unit head:* Dr. F. G. Hembry, Chair, 352-392-1911, Fax: 352-392-5595, E-mail: hembry@animal.ufl.edu. *Application contact:* Dr. Joel Brendemuhl, Coordinator, 352-392-2186, Fax: 352-392-1913, E-mail: brendemuhl@animal.ufl.edu.

University of Georgia, Graduate School, College of Agricultural and Environmental Sciences, Department of Animal and Dairy Sciences, Athens, GA 30602. Offers animal and dairy science (PhD); animal and dairy sciences (MADS); animal science (MS); dairy science (MS). *Faculty:* 19 full-time (2 women). *Students:* 37 full-time, 3 part-time; includes 4 minority (3 African Americans, 1 Hispanic American), 13 international. 22 applicants, 59% accepted, 6 enrolled. In 2005, 9 master's, 2 doctorates awarded. *Degree requirements:* For master's, thesis; for doctorate, one foreign language, thesis/dissertation. *Entrance requirements:* For master's and doctorate, GRE General Test. *Application deadline:* For fall admission, 7/1 for domestic students; for spring admission, 11/15 for domestic students. Application fee: $50. Electronic applications accepted. *Financial support:* Fellowships, research assistantships, teaching assistantships, unspecified assistantships available. *Unit head:* Dr. Stephen C. Nickerson, Head, 706-542-6259, E-mail: ads.info@ads.uga.edu. *Application contact:* Dr. Mark A. Froetschel, Graduate Coordinator, 706-542-0985, Fax: 706-583-0274, E-mail: markf@uga.edu.

University of Georgia, Graduate School, College of Agricultural and Environmental Sciences, Department of Poultry Science, Athens, GA 30602. Offers animal nutrition (PhD); poultry science (MS, PhD). *Faculty:* 12 full-time (2 women). *Students:* 16 full-time, 4 part-time; includes 1 minority (American Indian/Alaska Native), 6 international. 6 applicants, 50% accepted, 5 enrolled. In 2005, 1 master's, 2 doctorates awarded. *Degree requirements:* For master's, thesis; for doctorate, one foreign language, thesis/dissertation. *Entrance requirements:* For master's and doctorate, GRE General Test. *Application deadline:* For fall admission, 7/1 for domestic students; for spring admission, 11/15 for domestic students. Application fee: $50. Electronic applications accepted. *Financial support:* Fellowships, research assistantships, teaching assistantships, unspecified assistantships available. *Unit head:* Dr. Michael Lacy, Head, 706-542-8383, Fax: 706-542-1827, E-mail: mlacy@uga.edu. *Application contact:* Dr. Daniel Fletcher, Graduate Coordinator, 706-542-2476, Fax: 706-542-1827, E-mail: fletcher@uga.edu.

University of Guelph, Graduate Program Services, Ontario Agricultural College, Department of Animal and Poultry Science, Guelph, ON N1G 2W1, Canada. Offers animal science (M Sc, PhD); poultry science (M Sc, PhD). *Faculty:* 34 full-time (2 women), 1 (woman) part-time/adjunct. *Students:* 90. In 2005, 16 master's, 8 doctorates awarded. *Degree requirements:* For master's, thesis (for some programs); for doctorate, thesis/dissertation. *Entrance requirements:* For master's, minimum B- average during previous 2 years of course work; for doctorate, minimum B average. *Application deadline:* Applications are processed on a rolling basis. Application fee: $75. *Financial support:* Fellowships, research assistantships, teaching assistantships available. *Faculty research:* Animal breeding and genetics (quantitative or molecular), animal nutrition (monogastric or ruminant), animal physiology (environmental, reproductive or behavioral), growth and metabolism (meat science). Total annual research expenditures: $5 million. *Unit head:* Dr. Steve Leeson, Chair, 519-824-4120 Ext. 53681, E-mail: sleeson@uoguelph.ca. *Application contact:* Dr. Andy Robinson, Graduate Coordinator, 519-824-4120 Ext. 53679, E-mail: andyr@uoguelph.ca.

University of Hawaii at Manoa, Graduate Division, College of Tropical Agriculture and Human Resources, Department of Human Nutrition, Food and Animal Sciences, Program in Animal Sciences, Honolulu, HI 96822. Offers MS. Part-time programs available. *Faculty:* 16 full-time (5 women), 5 part-time/adjunct (1 woman). *Students:* 17 full-time (8 women), 1 (woman) part-time; includes 9 minority (1 American Indian/Alaska Native, 8 Asian Americans or Pacific Islanders), 2 international. Average age 30. 12 applicants, 83% accepted, 8 enrolled. In 2005, 6 degrees awarded. *Degree requirements:* For master's, thesis (for some programs). *Entrance requirements:* For master's, GRE General Test. *Application deadline:* For fall admission, 3/1 for domestic students, 1/15 for international students; for spring admission, 9/1 for domestic students, 8/1 for international students. Application fee: $50. *Expenses:* Tuition, state resident: full-time $8,400; part-time $200 per credit hour. Tuition, nonresident: full-time $11,088; part-time $462 per credit hour. Tuition and fees vary according to program. *Financial support:* Tuition waivers (full) available. *Faculty research:* Nutritional biochemistry, food composition, nutrition education, nutritional epidemiology, international nutrition, food toxicology. *Unit head:* Dr. Yong-Soo Kim, Graduate Chairperson, 808-956-8356, Fax: 808-956-7095, E-mail: ykim@hawaii.edu. *Application contact:* Dr. Michael Dunn, Graduate Chairperson, 808-956-8356, Fax: 808-956-4024, E-mail: mdunn@hawaii.edu.

University of Idaho, College of Graduate Studies, College of Agricultural and Life Sciences, Department of Animal and Veterinary Science, Moscow, ID 83844-2282. Offers animal physiology (PhD); animal science (MS); veterinary science (MS). *Students:* 17 full-time (8 women), 5 part-time (2 women), 2 international. Average age 36. In 2005, 1 master's, 1 doctorate awarded. *Degree requirements:* For doctorate, thesis/dissertation. *Entrance requirements:* For master's, GRE General Test, minimum GPA of 2.8; for doctorate, minimum undergraduate GPA of 2.8, graduate GPA of 3.0. *Application deadline:* For fall admission, 8/1 for domestic students; for spring admission, 12/15 for domestic students. Application fee: $55 ($60 for international students). *Expenses:* Tuition, nonresident: full-time $8,770; part-time $130 per credit. Required fees: $4,508; $217 per credit. *Financial support:* Research assistantships, teaching assistantships available. Financial award application deadline: 2/15. *Faculty research:* Agribusiness, range-livestock management. *Unit head:* Dr. Richard A. Battaglia, Head, 208-885-6345.

University of Illinois at Urbana–Champaign, Graduate College, College of Agricultural, Consumer and Environmental Sciences, Department of Animal Sciences, Champaign, IL 61820. Offers MS, PhD. *Faculty:* 40 full-time (4 women), 5 part-time/adjunct (1 woman). *Students:* 88 full-time (51 women), 5 part-time (3 women); includes 5 minority (3 Asian Americans or Pacific Islanders, 2 Hispanic Americans), 21 international. 74 applicants, 38% accepted, 23 enrolled. In 2005, 20 master's, 9 doctorates awarded. *Degree requirements:* For doctorate, thesis/dissertation. *Entrance requirements:* For master's, GRE, minimum GPA of 3.0. *Application deadline:* Applications are processed on a rolling basis. Application fee: $50 ($60 for international students). Electronic applications accepted. *Financial support:* In 2005–06, 20 fellowships, 72 research assistantships, 6 teaching assistantships were awarded; tuition waivers (full and partial) also available. *Unit head:* Neal R. Merchen, Head, 217-333-3462, Fax: 217-244-2871, E-mail: nmerchen@uiuc.edu. *Application contact:* Allison Mosley, Resource and Policy Analyst, 217-333-1044, Fax: 217-333-5044, E-mail: amosley@uiuc.edu.

University of Kentucky, Graduate School, Graduate School Programs in the College of Agriculture, Program in Animal Sciences, Lexington, KY 40506-0032. Offers MS, PhD. *Faculty:* 47 full-time (5 women). *Students:* 42 full-time (21 women), 4 part-time (2 women); includes 2 minority (1 African American, 1 Hispanic American), 8 international. Average age 29. 48 applicants, 71% accepted, 16 enrolled. In 2005, 3 master's, 4 doctorates awarded. Terminal master's awarded for partial completion of doctoral program. *Median time to degree:* Of those who began their doctoral program in fall 1997, 89.6% received their degree in 8 years or less. *Degree requirements:* For master's, thesis optional; for doctorate, thesis/dissertation, comprehensive exam. *Entrance requirements:* For master's, GRE General Test, minimum undergraduate GPA of 2.5; for doctorate, GRE General Test, minimum graduate GPA of 3.0. Additional exam requirements/recommendations for international students: Required—TOEFL (minimum score 550 paper-based; 213 computer-based). *Application deadline:* For fall admission, 7/17 priority date for domestic students, 2/1 priority date for international students; for spring admission, 12/13 priority date for domestic students, 6/15 priority date for international students. Applications are processed on a rolling basis. Application fee: $40 ($55 for international students). Electronic applications accepted. *Expenses:* Tuition, state resident: full-time $3,159; part-time $331 per credit hour. Tuition, nonresident: full-time $6,984; part-time $756 per credit hour. Tuition and fees vary according to course load, degree level and program. *Financial support:* In 2005–06, 5 fellowships with full tuition reimbursements (averaging $2,250 per year), 30 research assistantships with full tuition reimbursements (averaging $12,000 per year), 5 teaching assistantships with full tuition reimbursements (averaging $13,000 per year) were awarded; Federal Work-Study, institutionally sponsored loans, scholarships/grants, traineeships, health care benefits, tuition waivers (partial), and unspecified assistantships also available. Support available to part-time students. Financial award application deadline: 3/15. *Faculty* *research:* Nutrition of horses, cattle, swine, poultry, and sheep; physiology of reproduction and lactation; food science; microbiology. Total annual research expenditures: $2.8 million. *Unit head:* Dr. David Harmon, Director of Graduate Studies, 859-257-7516, Fax: 859-257-3412, E-mail: dharmon@uky.edu. *Application contact:* Dr. Brian Jackson, Senior Associate Dean, 859-257-8176, Fax: 859-323-1928, E-mail: lance.brunner@uky.edu.

University of Maine, Graduate School, College of Natural Sciences, Forestry, and Agriculture, Department of Animal and Veterinary Sciences, Program in Animal Sciences, Orono, ME 04469. Offers MPS, MS. *Faculty:* 13. *Students:* 4 full-time (2 women), 3 part-time (2 women); includes 1 minority (Asian American or Pacific Islander) Average age 26. 6 applicants, 67% accepted, 3 enrolled. In 2005, 2 master's awarded. *Degree requirements:* For master's, thesis. *Entrance requirements:* For master's, GRE General Test, BS in animal sciences or related area. Additional exam requirements/recommendations for international students: Required—TOEFL. *Application deadline:* For fall admission, 2/1 for domestic students. Applications are processed on a rolling basis. Application fee: $50. Electronic applications accepted. *Financial support:* In 2005–06, research assistantships with tuition reimbursements (averaging $12,013 per year), teaching assistantships with tuition reimbursements (averaging $9,010 per year) were awarded. Financial award application deadline: 3/1. *Unit head:* Dr. Charles Wallace, Coordinator, 207-581-2737. *Application contact:* Scott G. Delcourt, Associate Dean of the Graduate School, 207-581-3219, Fax: 207-581-3232, E-mail: graduate@maine.edu.

University of Manitoba, Faculty of Graduate Studies, Faculty of Agriculture, Department of Animal Science, Winnipeg, MB R3T 2N2, Canada. Offers M Sc, PhD. *Degree requirements:* For master's, thesis; for doctorate, one foreign language, thesis/dissertation.

University of Maryland, College Park, Graduate Studies, College of Agriculture and Natural Resources, Department of Animal and Avian Sciences, Program in Animal Sciences, College Park, MD 20742. Offers MS, PhD. *Students:* 29 full-time (14 women), 6 part-time (all women); includes 5 minority (4 African Americans, 1 American Indian/Alaska Native), 12 international. 42 applicants, 17% accepted, 7 enrolled. In 2005, 1 degree awarded. *Degree requirements:* For master's, thesis, oral exam or written comprehensive exam; for doctorate, thesis/dissertation, journal publication, scientific paper. *Entrance requirements:* For master's, GRE General Test, minimum GPA of 3.0; for doctorate, GRE General Test. Additional exam requirements/recommendations for international students: Required—TOEFL. *Application deadline:* For fall admission, 5/15 for domestic students, 2/1 for international students; for spring admission, 10/15 for domestic students, 6/1 for international students. Applications are processed on a rolling basis. Application fee: $50. Electronic applications accepted. *Financial support:* In 2005–06, 6 fellowships (averaging $7,539 per year) were awarded; research assistantships, teaching assistantships Financial award applicants required to submit FAFSA. *Faculty research:* Animal physiology, cell biology and biochemistry, reproduction, biometrics, animal behavior. *Application contact:* Dean of Graduate School, 301-405-4190, Fax: 301-314-9305.

University of Massachusetts Amherst, Graduate School, College of Natural Resources and the Environment, Program in Animal Biotechnology and Biomedical Sciences, Amherst, MA 01003. Offers mammalian and avian biology (MS, PhD). Part-time programs available. *Faculty:* 18 full-time (7 women). *Students:* 19 full-time (10 women), 1 part-time; includes 4 minority (2 African Americans, 2 Hispanic Americans), 5 international. Average age 28. 22 applicants, 50% accepted, 6 enrolled. In 2005, 4 master's, 6 doctorates awarded. Terminal master's awarded for partial completion of doctoral program. *Degree requirements:* For master's, thesis or alternative; for doctorate, thesis/dissertation. *Entrance requirements:* For master's and doctorate, GRE General Test. Additional exam requirements/recommendations for international students: Required—TOEFL (minimum score 530 paper-based; 197 computer-based). *Application deadline:* For fall admission, 2/1 priority date for domestic students, 2/1 priority date for international students; for spring admission, 10/1 for domestic students, 10/1 for international students. Applications are processed on a rolling basis. Application fee: $40 ($65 for international students). Electronic applications accepted. *Expenses:* Tuition, state resident: part-time $110 per credit. Tuition, nonresident: part-time $414 per credit. Required fees: $2,824 per term. One-time fee: $250 part-time. Full-time tuition and fees vary according to course load, campus/location, program and reciprocity agreements. *Financial support:* In 2005–06, research assistantships with full tuition reimbursements (averaging $11,772 per year), teaching assistantships with full tuition reimbursements (averaging $11,259 per year) were awarded; fellowships with full tuition reimbursements, career-related internships or fieldwork, Federal Work-Study, scholarships/grants, traineeships, and unspecified assistantships also available. Support available to part-time students. Financial award application deadline:2/1. *Unit head:* Dr. Sam Black, Director, 413-545-2312, Fax: 413-545-6326.

University of Minnesota, Twin Cities Campus, Graduate School, College of Agricultural, Food, and Environmental Sciences, Department of Animal Science, Minneapolis, MN 55455-0213. Offers MS, PhD. Part-time programs available. *Faculty:* 34 full-time (6 women). *Students:* 34 full-time (14 women), 3 part-time (1 woman); includes 16 minority (10 Asian Americans or Pacific Islanders, 6 Hispanic Americans). 16 applicants, 75% accepted, 11 enrolled. In 2005, 4 master's, 6 doctorates awarded. *Degree requirements:* For master's and doctorate, thesis/dissertation. *Entrance requirements:* For master's and doctorate, GRE. Additional exam requirements/recommendations for international students: Required—TOEFL. *Application deadline:* For fall admission, 6/15 priority date for domestic students, 6/15 priority date for international students; for spring admission, 10/15 priority date for domestic students, 10/15 priority date for international students. Applications are processed on a rolling basis. Application fee: $55 ($75 for international students). *Expenses:* Tuition, state resident: full-time $8,748; part-time $729 per credit. Tuition, nonresident: full-time $15,848; part-time $1,321 per credit. Full-time tuition and fees vary according to class time, course load, program and reciprocity agreements. *Financial support:* In 2005–06, 30 research assistantships, 2 teaching assistantships were awarded; fellowships *Faculty research:* Physiology, growth biology, nutrition, genetics, production systems. *Unit head:* Dr. F. Abel Ponce de León, Head, 612-624-1205, Fax: 612-625-5789, E-mail: apl@umn.edu. *Application contact:* Kimberly A. Reno, Student Personnel Coordinator, 612-624-3491, Fax: 612-625-5789, E-mail: renox001@umn.edu.

University of Missouri–Columbia, Graduate School, College of Agriculture, Food and Natural Resources, Department of Animal Sciences, Columbia, MO 65211. Offers MS, PhD. *Faculty:* 36 full-time (2 women). *Students:* 38 full-time (16 women), 18 part-time (8 women), 16 international. In 2005, 9 master's, 7 doctorates awarded. Terminal master's awarded for partial completion of doctoral program. *Degree requirements:* For doctorate, 2 foreign languages, thesis/dissertation. *Entrance requirements:* For master's and doctorate, GRE General Test, minimum GPA of 3.0. *Application deadline:* Applications are processed on a rolling basis. Application fee: $45 ($60 for international students). *Financial support:* Research assistantships, teaching assistantships, institutionally sponsored loans available. *Unit head:* Dr. Bill Lamberson, Director of Graduate Studies, 573-882-8234, E-mail: lamersonw@missouri.edu.

University of Nebraska–Lincoln, Graduate College, College of Agricultural Sciences and Natural Resources, Department of Animal Science, Lincoln, NE 68588. Offers MS, PhD. *Degree requirements:* For master's, thesis/dissertation; for doctorate, thesis/dissertation, comprehensive exam. *Entrance requirements:* For master's and doctorate, GRE General Test. Additional exam requirements/recommendations for international students: Required—TOEFL (minimum score 525 paper-based; 195 computer-based). Electronic applications accepted. *Faculty research:* Animal breeding and genetics, meat and poultry products, nonruminant and ruminant nutrition, physiology.

University of Nevada, Reno, Graduate School, College of Agriculture, Biotechnology and Natural Resources, Program in Animal Science, Reno, NV 89557. Offers MS. *Degree requirements:* For master's, thesis optional. *Entrance requirements:* For master's, GRE, minimum GPA of 2.75. Additional exam requirements/recommendations for international students: Required—TOEFL. *Faculty research:* Sperm fertility, embryo development, ruminant utilization of forages.

University of New Hampshire, Graduate School, College of Life Sciences and Agriculture, Department of Animal and Nutritional Sciences, Program in Animal and Nutritional Sciences, Durham, NH 03824. Offers PhD. *Faculty:* 23 full-time. *Students:* 5 full-time (3 women), 1 (woman) part-time. Average age 29. 5 applicants, 40% accepted, 2 enrolled. In 2005, 1 degree awarded. *Entrance requirements:* For doctorate, GRE. Additional exam requirements/recommendations for international students: Required—TOEFL (minimum score 550 paper-based; 213 computer-based). *Application deadline:* For fall admission, 4/1 priority date for domestic students, 4/1 priority date for international students. Application fee: $60. Electronic applications accepted. *Expenses:* Tuition, state resident: full-time $8,010; part-time $445 per credit hour. Tuition, nonresident: full-time $19,730; part-time $810 per credit hour. Required fees: $322 per semester. Tuition and fees vary according to course load and program. *Financial support:* In 2005–06, 1 fellowship, 1 research assistantship, 3 teaching assistantships were awarded; scholarships/grants, traineeships, and unspecified assistantships also available. Support available to part-time students. *Application contact:* Ann Barbarits, Administrative Assistant, 603-862-2178, E-mail: ansc.grad.program.info@unh.edu.

University of New Hampshire, Graduate School, College of Life Sciences and Agriculture, Department of Animal and Nutritional Sciences, Program in Animal Science, Durham, NH 03824. Offers MS. Part-time programs available. *Faculty:* 23 full-time. *Students:* 6 full-time (all women), 11 part-time (9 women). Average age 32. 7 applicants, 57% accepted, 4 enrolled. In 2005, 3 degrees awarded. *Degree requirements:* For master's, registration. *Entrance requirements:* For master's, GRE General Test. Additional exam requirements/recommendations for international students: Required—TOEFL (minimum score 550 paper-based; 213 computer-based); Recommended—TSE. *Application deadline:* For fall admission, 4/1 priority date for domestic students, 4/1 priority date for international students. For winter admission, 12/1 for domestic students. Applications are processed on a rolling basis. Application fee: $60. Electronic applications accepted. *Expenses:* Tuition, state resident: full-time $8,010; part-time $445 per credit hour. Tuition, nonresident: full-time $19,730; part-time $810 per credit hour. Required fees: $322 per semester. Tuition and fees vary according to course load and program. *Financial support:* In 2005–06, 1 fellowship, 6 research assistantships, 7 teaching assistantships were awarded; career-related internships or fieldwork, Federal Work-Study, scholarships/grants, and tuition waivers (full and partial) also available. Support available to part-time students. Financial award application deadline: 2/15.

University of Puerto Rico, Mayagüez Campus, Graduate Studies, College of Agricultural Sciences, Department of Animal Industry, Mayagüez, PR 00681-9000. Offers MS. Part-time programs available. *Faculty:* 21. *Students:* 3 full-time (2 women), 14 part-time (8 women); includes 15 minority (all Hispanic Americans), 2 international. 8 applicants, 63% accepted, 3 enrolled. In 2005, 1 degree awarded. *Degree requirements:* For master's, thesis, comprehensive exam. *Application deadline:* For fall admission, 2/15 for domestic students; for spring admission, 9/15 for domestic students. Applications are processed on a rolling basis. Application fee: $20. *Expenses:* Tuition, state resident: full-time $900; part-time $100 per credit. International tuition: $4,655 full-time. Part-time tuition and fees vary according to course level and course load. *Financial support:* In 2005–06, 14 students received support, including fellowships (averaging $1,200 per year), 2 research assistantships (averaging $1,500 per year), 12 teaching assistantships (averaging $987 per year); Federal Work-Study and institutionally sponsored loans also available. *Faculty research:* Swine production and nutrition, poultry production, dairy science and technology, microbiology. Total annual research expenditures: $25,983. *Unit head:* Dr. José LaTorre, Acting Director, 787-265-3854.

University of Rhode Island, Graduate School, College of the Environment and Life Sciences, Department of Fisheries, Animal and Veterinary Science, Kingston, RI 02881. Offers animal health and disease (MS); animal science (MS); aquaculture (MS); aquatic pathology (MS); environmental sciences (PhD), including animal science, aquacultural science, aquatic pathology, fisheries science; fisheries (MS). In 2005, 6 degrees awarded. *Application deadline:* For fall admission, 4/15 for domestic students. Applications are processed on a rolling basis. Application fee: $35. *Expenses:* Tuition, state resident: full-time $5,522; part-time $307 per credit. Tuition, nonresident: full-time $15,992; part-time $888 per credit. Required fees: $1,786; $73 per credit. One-time fee: $80 part-time. *Unit head:* Dr. David Bengtson, Chairperson, 401-874-2688.

University of Saskatchewan, College of Graduate Studies and Research, College of Agriculture, Department of Animal and Poultry Science, Saskatoon, SK S7N 5A2, Canada. Offers M Ag, M Sc, PhD. *Degree requirements:* For master's and doctorate, thesis/dissertation, registration. *Entrance requirements:* Additional exam requirements/recommendations for international students: Required—TOEFL.

University of Saskatchewan, Western College of Veterinary Medicine and College of Graduate Studies and Research, Graduate Programs in Veterinary Medicine, Department of Large Animal Clinical Sciences, Saskatoon, SK S7N 5A2, Canada. Offers herd medicine and theriogenology (M Sc, M Vet Sc, PhD). *Faculty:* 11 full-time (4 women). *Students:* 16 full-time (11 women). In 2005, 7 master's, 1 doctorate awarded. *Degree requirements:* For master's, thesis (for some programs); for doctorate, thesis/dissertation. *Faculty research:* Reproduction, infectious diseases, epidemiology, food safety. *Unit head:* Dr. David Wilson, Head, 306-966-7087, Fax: 306-966-7159, E-mail: david.wilson@usask.ca.

University of Saskatchewan, Western College of Veterinary Medicine and College of Graduate Studies and Research, Graduate Programs in Veterinary Medicine, Department of Small Animal Clinical Sciences, Saskatoon, SK S7N 5A2, Canada. Offers small animal clinical sciences (M Sc, PhD); veterinary anesthesiology, radiology and surgery (M Vet Sc); veterinary internal medicine (M Vet Sc). *Faculty:* 7 full-time (5 women). *Students:* 9 full-time (7 women). In 2005, 1 degree awarded. *Degree requirements:* For master's, thesis (for some programs); for doctorate, thesis/dissertation. *Faculty research:* Orthopedics, wildlife, cardiovascular exercise/myelopathy, ophthalmology. *Unit head:* Dr. Klaas Post, Head, 306-966-7084, Fax: 306-966-7174, E-mail: klaas.post@usask.ca.

The University of Tennessee, Graduate School, College of Agricultural Sciences and Natural Resources, Department of Animal Science, Knoxville, TN 37996. Offers animal anatomy (PhD); breeding (MS, PhD); management (MS, PhD); nutrition (MS, PhD); physiology (MS, PhD). Part-time programs available. *Degree requirements:* For master's and doctorate, thesis/dissertation. *Entrance requirements:* For master's and doctorate, GRE General Test, minimum GPA of 2.7. Additional exam requirements/recommendations for international students: Required—TOEFL. Electronic applications accepted.

University of Vermont, Graduate College, College of Agriculture and Life Sciences, Department of Animal Sciences, Burlington, VT 05405. Offers MS, PhD. *Students:* 14 (7 women) 6 international. 4 applicants, 50% accepted, 0 enrolled. In 2005, 3 master's awarded. *Degree requirements:* For master's, thesis; for doctorate, one foreign language, thesis/dissertation. *Entrance requirements:* For master's and doctorate, GRE General Test. Additional exam requirements/recommendations for international students: Required—TOEFL (minimum score 550 paper-based; 213 computer-based). *Application deadline:* For fall admission, 4/1 for domestic students. Applications are processed on a rolling basis. Application fee: $40. Electronic applications accepted. *Expenses:* Tuition, area resident: Part-time $410 per credit hour. Tuition, nonresident: part-time $1,034 per credit hour. *Financial support:* Fellowships, research assistantships, teaching assistantships available. Financial award application deadline: 3/1. *Faculty research:* Animal nutrition, dairy production. *Unit head:* Dr., Chairperson, 802-656-2070. *Application contact:* Dr., Chairperson, 802-656-2070.

University of Wisconsin–Madison, Graduate School, College of Agricultural and Life Sciences, Department of Animal Sciences, Madison, WI 53706-1380. Offers MS, PhD. Part-time programs available. *Faculty:* 18 full-time (0 women), 2 part-time/adjunct (0 women). *Students:* 20 full-time (9 women), 7 part-time (3 women), 12 international. Average age 29. 32 applicants, 16% accepted, 4 enrolled. In 2005, 3 master's awarded. Terminal master's awarded for partial completion of doctoral program. *Degree requirements:* For master's and doctorate, thesis/

Animal Sciences

University of Wisconsin–Madison (continued)

dissertation. *Entrance requirements:* For master's and doctorate, GRE. Additional exam requirements/recommendations for international students: Required—TOEFL (minimum score 550 paper-based; 213 computer-based). *Application deadline:* For fall admission, 1/2 priority date for domestic students, 1/2 priority date for international students. For winter admission, 8/15 for domestic students; for spring admission, 3/1 for domestic students. Applications are processed on a rolling basis. Application fee: $45. Electronic applications accepted. *Financial support:* In 2005–06, 15 research assistantships with full tuition reimbursements (averaging $18,480 per year), 2 teaching assistantships (averaging $7,508 per year) were awarded; fellowships with full tuition reimbursements, career-related internships or fieldwork and scholarships/grants also available. *Faculty research:* Animal biology, immunity and toxicology, endocrinology and reproductive physiology, genetics-animal breeding, meat science muscle biology. *Unit head:* Daniel M. Schaefer, Chair, 608-262-4300, Fax: 608-262-5157, E-mail: schaeferd@ansci.wisc.edu. *Application contact:* Kathy A. Monson, Student Services, 608-263-5225, Fax: 608-262-5157, E-mail: kamonson@wisc.edu.

University of Wisconsin–Madison, Graduate School, College of Agricultural and Life Sciences, Department of Dairy Science, Madison, WI 53706-1380. Offers MS, PhD. Part-time programs available. *Faculty:* 14 full-time (1 woman), 2 part-time/adjunct (0 women). *Students:* 27 full-time (12 women), 3 part-time (2 women); includes 1 minority (African American), 14 international. Average age 30. 21 applicants, 38% accepted, 8 enrolled. In 2005, 5 master's, 5 doctorates awarded. *Degree requirements:* For master's, thesis (for some programs); for doctorate, thesis/dissertation. *Entrance requirements:* For master's and doctorate, GRE General Test. Additional exam requirements/recommendations for international students: Required—TOEFL. *Application deadline:* For fall admission, 10/1 for domestic students. Applications are processed on a rolling basis. Application fee: $45. Electronic applications accepted. *Financial support:* In 2005–06, 27 research assistantships with full tuition reimbursements (averaging $18,120 per year), 1 teaching assistantship with tuition reimbursement were awarded; fellowships, Federal Work-Study also available. Support available to part-time students. Financial award applicants required to submit FAFSA. *Faculty research:* Genetics, nutrition, lactation, reproduction, management of dairy cattle. Total annual research expenditures: $2.5 million. *Unit head:* Ric R. Grummer, Chair, 608-265-5526, Fax: 608-263-9412.

University of Wyoming, Graduate School, College of Agriculture, Department of Animal Sciences, Program in Animal Sciences, Laramie, WY 82070. Offers MS, PhD. *Faculty:* 17 full-time (1 woman), 3 part-time/adjunct (0 women). *Students:* 18 full-time (10 women), 7 part-time (2 women); includes 2 minority (1 American Indian/Alaska Native, 1 Hispanic American), 3 international. Average age 26. 25 applicants, 32% accepted. In 2005, 3 master's, 3 doctorates awarded. *Degree requirements:* For master's and doctorate, thesis/dissertation. *Entrance requirements:* For master's, GRE General Test, minimum GPA of 3.0; for doctorate, GRE General Test or MS degree, minimum GPA of 3.0. Additional exam requirements/recommendations for international students: Required—TOEFL (minimum score 525 paper-based). *Application deadline:* For fall admission, 2/1 priority date for domestic students, 2/1 priority date for international students; for spring admission, 9/1 priority date for domestic students, 9/1 priority date for international students. Applications are processed on a rolling basis. Application fee: $50. *Expenses:* Tuition, state resident: full-time $3,720; part-time $155 per credit hour. Tuition, nonresident: full-time $10,704; part-time $446 per credit hour. Required fees: $666; $162 per semester. Tuition and fees vary according to course load and program. *Financial support:* In 2005–06, 4 students received support, including research assistantships with tuition reimbursements available (averaging $12,000 per year); career-related internships or fieldwork, Federal Work-Study, institutionally sponsored loans, scholarships/grants, and unspecified assistantships also available. Financial award application deadline:3/1. *Faculty research:* Reproductive biology, ruminant nutrition meat science, muscle biology, food microbiology, lipid metabolism. *Application contact:* Jamie L. Lejambre, Office Assistant, Senior, 307-766-2224, Fax: 307-766-2355, E-mail: animalscience@uwyo.edu.

Utah State University, School of Graduate Studies, College of Agriculture, Department of Animal, Dairy and Veterinary Sciences, Logan, UT 84322. Offers animal science (MS, PhD); bioveterinary science (MS, PhD); dairy science (MS). Part-time programs available. *Faculty:* 18 full-time (2 women), 2 part-time/adjunct (0 women). *Students:* 24 full-time (9 women), 11 part-time (5 women). Average age 25. 9 applicants, 56% accepted, 4 enrolled. In 2005, 4 master's, 1 doctorate awarded. *Degree requirements:* For master's, thesis (for some programs), registration; for doctorate, thesis/dissertation, comprehensive exam, registration. *Entrance requirements:* For master's and doctorate, GRE General Test, minimum GPA of 3.0. Additional exam requirements/recommendations for international students: Required—TOEFL. *Application deadline:* For fall admission, 5/15 for domestic students; for spring admission, 10/15 for domestic students. Applications are processed on a rolling basis. Application fee: $50 ($60 for international students). Electronic applications accepted. *Financial support:* In 2005–06, 10 fellowships with full and partial tuition reimbursements (averaging $12,000 per year), 4 research assistantships with full and partial tuition reimbursements (averaging $13,300 per year), 3 teaching assistantships with full and partial tuition reimbursements (averaging $13,300 per year) were awarded; career-related internships or fieldwork, Federal Work-Study, institutionally sponsored loans, scholarships/grants, and tuition waivers (partial) also available. Financial award application deadline: 3/15. *Faculty research:* Monoclonal antibodies, antiviral chemotherapy, management systems, biotechnology, rumen fermentation manipulation. *Unit head:* Dr. Mark C. Healey, Head, 435-797-2162, Fax: 435-797-2118, E-mail: mchealey@cc.usu.edu. *Application contact:* Dr. Jeffrey L. Walters, Graduate Program Coordinator, 435-797-2161, Fax: 435-797-2118, E-mail: jwalters@cc.usu.edu.

Virginia Polytechnic Institute and State University, Graduate School, College of Agriculture and Life Sciences, Department of Animal and Poultry Sciences, Blacksburg, VA 24061. Offers animal science (MS, PhD); poultry science (MS, PhD), including behavior, genetics, management, nutrition, physiology. *Faculty:* 21 full-time (5 women). *Students:* 44 full-time (30 women); includes 3 minority (2 African Americans, 1 Hispanic American), 16 international. Average age 27. 26 applicants, 27% accepted, 7 enrolled. In 2005, 7 master's, 4 doctorates awarded. *Entrance requirements:* For master's and doctorate, GRE. Additional exam requirements/recommendations for international students: Required—TOEFL (minimum score 550 paper-based; 213 computer-based). *Application deadline:* Applications are processed on a rolling basis. Application fee: $45. Electronic applications accepted. *Expenses:* Tuition, state resident: full-time $6,558; part-time $364 per credit. Tuition, nonresident: full-time $11,296; part-time $628 per credit. Required fees: $1,419; $468 per credit. $234 per term. *Financial support:* In 2005–06, 12 fellowships with full tuition reimbursements (averaging $15,395 per year), 28 research assistantships with full tuition reimbursements (averaging $13,823 per year), 10 teaching assistantships with full tuition reimbursements (averaging $13,552 per year) were awarded; career-related internships or fieldwork, Federal Work-Study, scholarships/grants, and unspecified assistantships also available. Financial award application deadline: 4/1. *Faculty research:* Quantitative genetics of cattle and sheep, swine nutrition and management, animal molecular biology, nutrition of grazing livestock. *Unit head:* Dr. Ken Webb, Head, 540-231-9157, Fax: 540-231-3010. *Application contact:* Dr. David R. Notter, Professor, 540-231-5135, Fax: 540-231-3010, E-mail: drnotter@vt.edu.

Virginia Polytechnic Institute and State University, Graduate School, College of Agriculture and Life Sciences, Department of Dairy Science, Blacksburg, VA 24061. Offers animal science (MS, PhD). *Faculty:* 14 full-time (1 woman). *Students:* 18 full-time (11 women), 1 part-time; includes 2 minority (1 American Indian/Alaska Native, 1 Asian American or Pacific Islander), 8 international. Average age 28. 12 applicants, 58% accepted, 7 enrolled. In 2005, 7 master's, 2 doctorates awarded. *Entrance requirements:* For master's and doctorate, GRE. Additional exam requirements/recommendations for international students: Required—TOEFL (minimum score 550 paper-based; 213 computer-based). *Application deadline:* Applications are processed on a rolling basis. Application fee: $45. Electronic applications accepted. *Expenses:* Tuition, state resident: full-time $6,558; part-time $364 per credit. Tuition, nonresident: full-time $11,296; part-time $628 per credit. Required fees: $1,419; $468 per credit. $234 per term. *Financial support:* In 2005–06, 4 fellowships with full tuition reimbursements (averaging $15,084 per year), 4 research assistantships with full tuition reimbursements (averaging $13,823 per year), 5 teaching assistantships with full tuition reimbursements (averaging $13,552 per year) were awarded; career-related internships or fieldwork, Federal Work-Study, scholarships/grants, and unspecified assistantships also available. Financial award application deadline: 4/1. *Faculty research:* Genetics, nutrition, reproduction, lactation. *Unit head:* Dr. Michael Akers, Head, 540-231-4757, Fax: 540-231-5014, E-mail: rma@vt.edu. *Application contact:* Julie Shumaker, Professor, 540-231-6331, Fax: 540-231-5014, E-mail: shumaker@vt.edu.

Washington State University, Graduate School, College of Agricultural, Human, and Natural Resource Sciences, Department of Animal Sciences, Pullman, WA 99164. Offers MS, PhD. *Faculty:* 19. *Students:* 26 full-time (14 women), 3 part-time (all women); includes 3 minority (1 African American, 1 Asian American or Pacific Islander, 1 Hispanic American), 8 international. Average age 29. 24 applicants, 46% accepted, 5 enrolled. In 2005, 3 master's, 3 doctorates awarded. *Degree requirements:* For master's, thesis, oral exam; for doctorate, thesis/dissertation, oral and written exam. *Entrance requirements:* For master's, GRE (verbal and quantitative), minimum GPA of 3.0, 3 letters of recommendation, department questionnaire; for doctorate, GRE General Test, minimum GPA of 3.0. *Application deadline:* For fall admission, 5/1 for domestic students; for spring admission, 11/1 priority date for domestic students. Applications are processed on a rolling basis. Application fee: $35. Electronic applications accepted. *Expenses:* Tuition, state resident: full-time $6,295; part-time $336 per credit. Tuition, nonresident: full-time $15,949; part-time $819 per credit. Required fees: $933. Part-time tuition and fees vary according to campus/location and program. *Financial support:* In 2005–06, 25 students received support, including 2 fellowships (averaging $5,000 per year), 14 research assistantships with full and partial tuition reimbursements available (averaging $12,046 per year), 5 teaching assistantships with full and partial tuition reimbursements available (averaging $12,491 per year); career-related internships or fieldwork, Federal Work-Study, institutionally sponsored loans, scholarships/grants, tuition waivers (partial), and teaching associateships also available. Financial award application deadline: 4/1; financial award applicants required to submit FAFSA. *Faculty research:* Reproduction, genetics, equine cytokines, fish diseases and vaccines. Total annual research expenditures: $397,512. *Unit head:* Dr. Raymond W. Wright, Interim Chair, 509-335-9103, Fax: 509-335-4815, E-mail: raywright@wsu.edu. *Application contact:* Kristen Johnson, Chair, Graduate Committee, 509-335-4131, Fax: 509-335-1082, E-mail: johnsoka@wsu.edu.

West Texas A&M University, College of Agriculture, Nursing, and Natural Sciences, Division of Agriculture, Emphasis in Animal Science, Canyon, TX 79016-0001. Offers MS. Part-time programs available. *Degree requirements:* For master's, thesis optional. *Entrance requirements:* For master's, GRE General Test. Additional exam requirements/recommendations for international students: Required—TOEFL (minimum score 550 paper-based). Electronic applications accepted. *Faculty research:* Nutrition, animal breeding, meat science, reproduction physiology, feedlots.

West Virginia University, Davis College of Agriculture, Forestry and Consumer Sciences, Division of Animal and Veterinary Sciences, Program in Agricultural Sciences, Morgantown, WV 26506. Offers animal and food sciences (PhD); plant and soil sciences (PhD). *Degree requirements:* For doctorate, thesis/dissertation, oral and written exams. *Entrance requirements:* Additional exam requirements/recommendations for international students: Required—TOEFL. Application fee: $45. *Expenses:* Tuition, state resident: full-time $4,582; part-time $258 per credit hour. Tuition, nonresident: full-time $1,382; part-time $741 per credit hour. *Financial support:* Research assistantships with tuition reimbursements, teaching assistantships with tuition reimbursements, Federal Work-Study, institutionally sponsored loans, and tuition waivers (full and partial) available. Financial award application deadline: 2/1; financial award applicants required to submit FAFSA. *Faculty research:* Ruminant nutrition, metabolism, forage utilization, physiology, reproduction. *Application contact:* Dr. Hillar Klandorf, Professor, 304-293-2631 Ext. 4436, Fax: 304-293-3676, E-mail: hillar.klandorf@mail.wvu.edu.

West Virginia University, Davis College of Agriculture, Forestry and Consumer Sciences, Division of Animal and Veterinary Sciences, Program in Animal and Veterinary Sciences, Morgantown, WV 26506. Offers breeding (MS); food sciences (MS); nutrition (MS); physiology (MS); production management (MS); reproduction (MS). Part-time programs available. *Degree requirements:* For master's, thesis, oral and written exams. *Entrance requirements:* For master's, minimum GPA of 2.5. Additional exam requirements/recommendations for international students: Required—TOEFL. Application fee: $45. *Expenses:* Tuition, state resident: full-time $4,582; part-time $258 per credit hour. Tuition, nonresident: full-time $1,382; part-time $741 per credit hour. *Financial support:* Research assistantships, teaching assistantships, Federal Work-Study, institutionally sponsored loans, and tuition waivers (full and partial) available. Financial award application deadline: 2/1; financial award applicants required to submit FAFSA. *Faculty research:* Animal nutrition, reproductive physiology, food science. *Unit head:* Dr. Hillar Klandorf, Coordinator, 304-293-4372 Ext. 2050, Fax: 304-293-3676, E-mail: hillar.klandorf@mail.wvu.edu.

Aquaculture

American University of Beirut, Graduate Programs, Faculty of Agricultural and Food Sciences, Beirut, Lebanon. Offers agricultural economics (MS); animal sciences (MS); ecosystem management (MSES); food technology (MS); irrigation (MS); mechanization (MS); nutrition (MS); plant protection (MS); plant science (MS); poultry science (MS); soils (MS). *Degree requirements:* For master's, one foreign language, thesis (for some programs), comprehensive exam, registration. *Entrance requirements:* For master's, GRE, letter of recommendation.

Auburn University, Graduate School, College of Agriculture, Department of Fisheries and Allied Aquacultures, Auburn University, AL 36849. Offers M Aq, MS, PhD. Part-time programs available. *Faculty:* 20 full-time (3 women). *Students:* 46 full-time (14 women), 25 part-time (11 women); includes 5 minority (2 African Americans, 3 Hispanic Americans), 20 international. 38 applicants, 71% accepted, 20 enrolled. In 2005, 18 master's, 6 doctorates awarded. *Degree requirements:* For master's, thesis (for some programs); for doctorate, 2 foreign languages, thesis/dissertation. *Entrance requirements:* For master's and doctorate, GRE General Test. *Application deadline:* For fall admission, 7/7 for domestic students; for spring admission, 11/24 for domestic students. Applications are processed on a rolling basis. Application fee: $25 ($50 for international students). Electronic applications accepted. *Financial support:* Fellowships, research assistantships, teaching assistantships, Federal Work-Study available. Support available to part-time students. Financial award application deadline: 3/15. *Faculty research:* Channel catfish production; aquatic animal health; community and population ecology; pond management; production hatching, breeding and genetics. Total annual research expenditures: $8 million. *Unit head:* Dr. David B. Rouse, Head, 334-844-4786. *Application contact:* Dr. Stephen L. McFarland, Acting Dean of the Graduate School, 334-844-4700.

Clemson University, Graduate School, College of Agriculture, Forestry and Life Sciences, Department of Forestry and Natural Resources, Program in Wildlife and Fisheries Biology, Clemson, SC 29634. Offers MS, PhD. *Students:* 19 full-time (10 women), 4 part-time (1 woman); includes 1 minority (Hispanic American), 1 international. Average age 25. 2 applicants, 0% accepted, 0 enrolled. In 2005, 1 master's, 2 doctorates awarded. *Degree requirements:* For master's and doctorate, thesis/dissertation. *Entrance requirements:* For master's, GRE General Test, minimum undergraduate GPA of 3.0. Additional exam requirements/recommendations for international students: Required—TOEFL. *Application deadline:* For fall admission, 6/1 for domestic students, 4/15 for international students. Application fee: $50. *Financial support:* Fellowships, research assistantships, teaching assistantships, career-related internships or fieldwork available. Financial award applicants required to submit FAFSA. *Faculty research:* Intensive freshwater culture systems, conservation biology, stream management, applied wildlife management. Total annual research expenditures: $1 million. *Unit head:* Dr. Dave Guynn, Coordinator, 864-656-4803, Fax: 864-656-3304, E-mail: dguynn@clemson.edu.

See Close-Ups on pages 705 and 707.

Kentucky State University, College of Arts and Sciences, Frankfort, KY 40601. Offers aquaculture (MS). Part-time programs available. *Faculty:* 6 part-time/adjunct (0 women). *Students:* 6 full-time (1 woman), 8 part-time (3 women); includes 6 minority (5 African Americans, 1 Hispanic American). Average age 27. 3 applicants, 100% accepted, 2 enrolled. *Degree requirements:* For master's, thesis optional. *Entrance requirements:* For master's, GRE General Test. *Application deadline:* For fall admission, 5/15 for domestic students; for spring admission, 10/15 for domestic students. Applications are processed on a rolling basis. Application fee: $22. *Expenses:* Tuition, state resident: full-time $3,888. Tuition, nonresident: full-time $11,006. Required fees: $620. *Financial support:* In 2005–06, 7 research assistantships with full tuition reimbursements (averaging $13,000 per year) were awarded Financial award applicants required to submit FAFSA. *Unit head:* Dr. Sam Oleka, Dean, 502-597-6411, Fax: 502-597-6405, E-mail: sam.oleka@kysu.edu. *Application contact:* Dr. Mark Garrison, Administrative Specialist, 502-597-5977, Fax: 502-597-6405.

Memorial University of Newfoundland, School of Graduate Studies, Interdisciplinary Program in Aquaculture, St. John's, NL A1C 5S7, Canada. Offers M Sc. Part-time programs available. *Students:* 9 full-time (3 women), 3 part-time (2 women), 1 international. 2 applicants, 50% accepted, 0 enrolled. In 2005, 2 degrees awarded. *Degree requirements:* For master's, thesis, seminar or thesis topic. *Entrance requirements:* For master's, honors B Sc or diploma in aquaculture from the Marine Institute of Memorial University of Newfoundland. *Application deadline:* Applications are processed on a rolling basis. Application fee: $40 Canadian dollars. Electronic applications accepted. *Expenses:* Tuition: Part-time $733 per term. Tuition and fees vary according to degree level and program. *Financial support:* Fellowships, research assistantships, teaching assistantships available. *Faculty research:* Marine fish larval biology, fin fish nutrition, shellfish culture, fin fish virology, fin fish reproductive biology. *Unit head:* Dr. Anne Storey, Chair, 709-778-7665, Fax: 709-737-3316, E-mail: astorey@play.psych.mun.ca. *Application contact:* Gail Kenny, Secretary, 709-737-8154, Fax: 709-737-3316, E-mail: gkenny@mun.ca.

Nova Scotia Agricultural College, Research and Graduate Studies, Truro, NS B2N 5E3, Canada. Offers agriculture (M Sc), including air quality, animal behavior, animal molecular genetics, animal nutrition, animal technology, aquaculture, botany, crop management, crop physiology, ecology, environmental microbiology, food science, horticulture, nutrient management, pest management, physiology, plant biotechnology, plant pathology, soil chemistry, soil fertility, waste management and composting, water quality. Part-time programs available. *Faculty:* 43 full-time (7 women), 21 part-time/adjunct (1 woman). *Students:* 50 full-time (25 women), 15 part-time (11 women); includes 7 minority (3 African Americans, 3 Asian Americans or Pacific Islanders, 1 Hispanic American). Average age 25. In 2005, 23 degrees awarded. *Degree requirements:* For master's, thesis, registration. *Entrance requirements:* For master's, honors B Sc, minimum GPA of 3.0. Additional exam requirements/recommendations for international students: Required—TOEFL (minimum score 580 paper-based; 237 computer-based), Michigan English Language Assessment Battery, IELT, Can Test, CAEL. *Application deadline:* For fall admission, 6/1 for domestic students, 4/1 for international students. For winter admission, 10/31 for domestic students; for spring admission, 2/28 for domestic students. Applications are processed on a rolling basis. Application fee: $70. *Expenses:* Tuition, state resident: part-time $2,328 per year. Tuition, nonresident: full-time $6,984; part-time $7,968 per year. International tuition: $12,624 full-time. Required fees: $481; $46 per course. Tuition and fees vary according to program and student level. *Financial support:* In 2005–06, 48 students received support, including 7 fellowships (averaging $15,000 per year), 10 research assistantships (averaging $15,000 per year), 15 teaching assistantships (averaging $900 per year); career-related internships or fieldwork, scholarships/grants, and unspecified assistantships also available. *Faculty research:* Bio-product development, organic agriculture, nutrient management, air and water quality, agricultural biotechnology. Total annual research expenditures: $4.7 million. *Unit head:* Jill L. Rogers, Manager, 902-893-6360, Fax: 902-893-3430, E-mail: jrogers@nsac.ca. *Application contact:* Marie Law, Administrative Assistant, 902-893-6502, Fax: 902-893-3430, E-mail: mlaw@nsac.ca.

Purdue University, Graduate School, College of Agriculture, Department of Forestry and Natural Resources, West Lafayette, IN 47907. Offers aquaculture, fisheries, aquatic science (MSF); aquaculture, fisheries, aquatic sciences (MS, PhD); forest biology (MS, MSF, PhD); natural resources and environmental policy (MS, MSF); natural resources environmental policy (PhD); quantitative resource analysis (MS, MSF, PhD); wildlife science (MS, MSF, PhD); wood science and technology (MS, MSF, PhD). *Faculty:* 26 full-time (4 women), 7 part-time/adjunct (1 woman). *Students:* 69 full-time (33 women), 15 part-time (3 women); includes 5 minority (1 African American, 1 American Indian/Alaska Native, 3 Hispanic Americans), 29 international. Average age 30. 45 applicants, 40% accepted, 18 enrolled. In 2005, 6 master's, 6 doctorates awarded. *Degree requirements:* For master's and doctorate, thesis/dissertation. *Entrance requirements:* For master's and doctorate, GRE General Test (500 verbal, 500 quantitative), minimum B+ average in undergraduate course work. Additional exam requirements/recommendations for international students: Required—TOEFL. *Application deadline:* For fall admission, 1/5 for domestic students; for spring admission, 9/15 for domestic students. Applications are processed on a rolling basis. Application fee: $55. Electronic applications accepted. *Financial support:* In 2005–06, 10 research assistantships (averaging $15,259 per year) were awarded; fellowships, teaching assistantships, career-related internships or fieldwork and scholarships/grants also available. Support available to part-time students. Financial award application deadline: 1/5; financial award applicants required to submit FAFSA. *Faculty research:* Wildlife management, forest management, forest ecology, forest soils, limnology. *Unit head:* Dr. Robert K. Swihart, Interim Head, 765-494-3590, Fax: 765-494-9461, E-mail: rswihart@purdue.edu. *Application contact:* Kelly Garrett, Graduate Secretary, 765-494-3572, Fax: 765-494-9461, E-mail: kgarrett@purdue.edu.

University of Florida, Graduate School, College of Agricultural and Life Sciences, Department of Fisheries and Aquatic Sciences, Gainesville, FL 32611. Offers MFAS, MS, PhD. *Faculty:* 14 full-time (2 women), 1 part-time/adjunct (0 women). *Students:* 50 (22 women); includes 2 minority (both Asian Americans or Pacific Islanders) 2 international. 29 applicants, 31% accepted. In 2005, 10 master's, 2 doctorates awarded. *Degree requirements:* For master's, thesis optional; for doctorate, thesis/dissertation. *Entrance requirements:* For master's and doctorate, GRE General Test, minimum GPA of 3.0. Additional exam requirements/recommendations for international students: Required—TOEFL. *Application deadline:* For fall admission, 6/1 for domestic students. Applications are processed on a rolling basis. Application fee: $20. Electronic applications accepted. *Expenses:* Tuition, state resident: full-time $6,234. Tuition, nonresident: full-time $21,359. Tuition and fees vary according to program. *Financial support:* In 2005–06, 22 research assistantships (averaging $9,546 per year) were awarded; fellowships, unspecified assistantships also available. *Unit head:* Dr. Karl Havens, Chair, 352-392-9617, Fax: 352-392-3672, E-mail: khavens@ifas.ufl.edu. *Application contact:* Dr. Chuck Chichra, Graduate Coordinator, 352-392-9617 Ext. 249, Fax: 352-392-3672, E-mail: fish@ifas.ufl.edu.

University of Guelph, Graduate Program Services, Program in Aquaculture, Guelph, ON N1G 2W1, Canada. Offers M Sc. *Faculty:* 14 full-time (3 women). *Students:* 2 full-time (1 woman). Average age 24. 8 applicants, 25% accepted. In 2005, 5 degrees awarded. *Entrance requirements:* For master's, minimum B- average during previous 2 years of course work. *Application deadline:* For fall admission, 5/31 for domestic students. Applications are processed on a rolling basis. Application fee: $75. *Financial support:* Teaching assistantships, career-related internships or fieldwork available. *Faculty research:* Protein and amino acid metabolism, genetics, gamete cryogenics, pathology, epidemiology. *Unit head:* R. D. Moccia, Graduate Co-Coordinator, 519-824-4120 Ext. 56216, Fax: 519-767-0573, E-mail: rmoccia@uoguelph.ca.

University of Rhode Island, Graduate School, College of the Environment and Life Sciences, Department of Fisheries, Animal and Veterinary Science, Kingston, RI 02881. Offers animal health and disease (MS); animal science (MS); aquaculture (MS); aquatic pathology (MS); environmental sciences (PhD), including animal science, aquacultural science, aquatic pathology, fisheries science; fisheries (MS). In 2005, 6 degrees awarded. *Application deadline:* For fall admission, 4/15 for domestic students. Applications are processed on a rolling basis. Application fee: $35. *Expenses:* Tuition, state resident: full-time $5,522; part-time $307 per credit. Tuition, nonresident: full-time $15,992; part-time $888 per credit. Required fees: $1,786; $73 per credit. One-time fee: $80 part-time. *Unit head:* Dr. David Bengtson, Chairperson, 401-874-2688.

Food Science and Technology

Alabama Agricultural and Mechanical University, School of Graduate Studies, School of Agricultural and Environmental Sciences, Department of Family and Consumer Sciences, Huntsville, AL 35811. Offers family and consumer sciences (MS); food science (MS, PhD). Part-time and evening/weekend programs available. *Degree requirements:* For master's, thesis optional; for doctorate, one foreign language, thesis/dissertation. *Entrance requirements:* For master's, GRE General Test; for doctorate, GRE General Test, MS. Electronic applications accepted. *Faculty research:* Food biotechnology, nutrition, food microbiology, food engineering, food chemistry.

American University of Beirut, Graduate Programs, Faculty of Agricultural and Food Sciences, Beirut, Lebanon. Offers agricultural economics (MS); animal sciences (MS); ecosystem management (MSES); food technology (MS); irrigation (MS); mechanization (MS); nutrition (MS); plant protection (MS); plant science (MS); poultry science (MS); soils (MS). *Degree requirements:* For master's, one foreign language, thesis (for some programs), comprehensive exam, registration. *Entrance requirements:* For master's, GRE, letter of recommendation.

Auburn University, Graduate School, College of Human Sciences, Department of Nutrition and Food Science, Auburn University, AL 36849. Offers MS, PhD. Part-time programs available. *Faculty:* 13 full-time (6 women). *Students:* 11 full-time (6 women), 12 part-time (11 women); includes 2 minority (both African Americans), 6 international. 17 applicants, 53% accepted, 3 enrolled. In 2005, 9 master's, 2 doctorates awarded. *Degree requirements:* For master's, thesis (for some programs); for doctorate, thesis/dissertation. *Entrance requirements:* For master's and doctorate, GRE General Test. *Application deadline:* For fall admission, 7/7 for domestic students; for spring admission, 11/24 for domestic students. Applications are processed on a rolling basis. Application fee: $25 ($50 for international students). Electronic applications accepted. *Financial support:* Research assistantships, teaching assistantships, career-related internships or fieldwork and Federal Work-Study available. Support available to part-time students. Financial award application deadline: 3/15. *Faculty research:* Food quality and safety, diet, food supply, physical activity in maintenance of health, prevention of selected chronic disease states. *Unit head:* Dr. Douglas B White, Head, 334-844-4261. *Application contact:* Dr. Stephen L. McFarland, Acting Dean of the Graduate School, 334-844-4700.

Brigham Young University, Graduate Studies, College of Biological and Agricultural Sciences, Department of Nutrition, Dietetics and Food Science, Provo, UT 84602-1001. Offers food science (MS); nutrition (MS). *Faculty:* 13 full-time (5 women). *Students:* 15 full-time (11 women); includes 3 minority (1 African American, 1 American Indian/Alaska Native, 1 Hispanic American). Average age 24. 5 applicants, 60% accepted, 3 enrolled. In 2005, 4 degrees awarded. *Degree requirements:* For master's, thesis, comprehensive exam, registration. *Entrance requirements:* For master's, GRE General Test. Additional exam requirements/recommendations for international students: Required—TOEFL (minimum score 550 paper-based; 213 computer-based). *Application deadline:* For fall admission, 2/1 for domestic students, 2/1 for international students. For winter admission, 6/30 for domestic students. Application fee: $50. Electronic applications accepted. *Financial support:* In 2005–06, 7 students received support, including 4 research assistantships (averaging $16,665 per year), 3 teaching assistantships (averaging $16,665 per year); career-related internships or fieldwork, institutionally sponsored loans, and scholarships/grants also available. Financial award application deadline: 4/1. *Faculty research:* Dairy foods, lipid oxidation, food processes, magnesium and selenium nutrition, nutrient effect on gene expression. Total annual research expenditures: $234,094. *Unit head:* Dr. Lynn V. Ogden, Chair, 801-422-3912, Fax: 801-422-0258, E-mail: lynn_ogden@byu.edu. *Application contact:* Dr. Merrill J. Christensen, Graduate Coordinator, 801-422-5255, Fax: 801-422-0258, E-mail: merrill_christensen@byu.edu.

California State Polytechnic University, Pomona, Academic Affairs, College of Agriculture, Pomona, CA 91768-2557. Offers agricultural science (MS); animal science (MS); foods and nutrition (MS). Part-time programs available. *Faculty:* 43 full-time (13 women), 16 part-time/adjunct (9 women). *Students:* 20 full-time (16 women), 39 part-time (29 women); includes 14 minority (1 African American, 4 Asian Americans or Pacific Islanders, 9 Hispanic Americans), 4 international. Average age 30. 58 applicants, 76% accepted, 20 enrolled. In 2005, 13 degrees awarded. *Degree requirements:* For master's, thesis or alternative. *Application deadline:* For fall admission, 5/1 for domestic students. For winter admission, 10/15 for domestic students; for spring admission, 1/2 for domestic students. Applications are processed on a rolling basis. Application fee: $55. Electronic applications accepted. *Expenses:* Tuition,

Food Science and Technology

California State Polytechnic University, Pomona (continued)
nonresident: full-time $9,021. Required fees: $3,597. *Financial support:* Career-related internships or fieldwork, Federal Work-Study, and institutionally sponsored loans available. Support available to part-time students. Financial award application deadline: 3/2; financial award applicants required to submit FAFSA. *Faculty research:* Equine nutrition, physiology, and reproduction; leadership development; bioartificial pancreas; plant science; ruminant and human nutrition. *Unit head:* Dr. Wayne R. Bidlack, Dean, 909-869-2200, E-mail: wrbidlack@csupomona.edu.

California State University, Fresno, Division of Graduate Studies, College of Agricultural Sciences and Technology, Department of Food Science and Nutritional Sciences, Fresno, CA 93740-8027. Offers MS. Part-time programs available. *Degree requirements:* For master's, thesis, registration. *Entrance requirements:* For master's, GRE General Test, minimum GPA of 3.0 in last 60 units. Additional exam requirements/recommendations for international students: Required—TOEFL. Electronic applications accepted. *Faculty research:* Liquid foods, analysis, mushrooms, gaseous ozone, natamycin.

Chapman University, Graduate Studies, Wilkinson College of Letters and Sciences, Department of Physical Sciences, Orange, CA 92866. Offers food science and nutrition (MS). Part-time and evening/weekend programs available. *Faculty:* 3 full-time (2 women), 2 part-time/adjunct (1 woman). *Students:* 9 full-time (4 women), 8 part-time (5 women); includes 5 minority (2 African Americans, 3 Asian Americans or Pacific Islanders), 6 international. Average age 28. 12 applicants, 42% accepted, 0 enrolled. In 2005, 4 degrees awarded. *Degree requirements:* For master's, thesis, comprehensive exam, registration. *Entrance requirements:* For master's, GRE General Test, minimum undergraduate GPA of 3.0. Additional exam requirements/recommendations for international students: Required—TOEFL (minimum score 550 paper-based). *Application deadline:* Applications are processed on a rolling basis. Application fee: $55. Electronic applications accepted. *Expenses: Contact institution.* Part-time tuition and fees vary according to course load and program. *Financial support:* In 2005–06, 11 students received support, including 3 fellowships (averaging $2,500 per year); Federal Work-Study also available. Financial award application deadline: 6/30; financial award applicants required to submit FAFSA. *Unit head:* Dr. Daniel Wellman, Chair, 714-744-7826, E-mail: prakash@chapman.edu. *Application contact:* Jojo Delfin, Information Contact, 714-997-6786, Fax: 714-997-6713, E-mail: delfin@chapman.edu.

Clemson University, Graduate School, College of Agriculture, Forestry and Life Sciences, Department of Food Science and Human Nutrition, Program in Food, Nutrition, and Culinary Science, Clemson, SC 29634. Offers MS. *Students:* 12 full-time (8 women), 6 part-time (3 women); includes 1 minority (African American), 5 international. 20 applicants, 60% accepted, 6 enrolled. In 2005, 4 degrees awarded. *Degree requirements:* For master's, thesis. *Entrance requirements:* For master's, GRE General Test. Additional exam requirements/recommendations for international students: Required—TOEFL. *Application deadline:* For fall admission, 6/1 for domestic students, 4/15 for international students. Applications are processed on a rolling basis. Application fee: $50. Electronic applications accepted. *Financial support:* Fellowships with partial tuition reimbursements, research assistantships with partial tuition reimbursements, teaching assistantships with partial tuition reimbursements available. Financial award applicants required to submit FAFSA. *Unit head:* Dr. Paul Dawson, Coordinator, 864-656-1138, Fax: 864-656-3131, E-mail: pdawson@clemson.edu.

See Close-Up on page 603.

Clemson University, Graduate School, College of Agriculture, Forestry and Life Sciences, Department of Food Science and Human Nutrition and Department of Animal and Veterinary Sciences, Program in Food Technology, Clemson, SC 29634. Offers PhD. *Students:* 10 full-time (6 women), 3 part-time, 6 international. 12 applicants, 83% accepted, 3 enrolled. In 2005, 6 degrees awarded. *Degree requirements:* For doctorate, thesis/dissertation. *Entrance requirements:* For doctorate, GRE General Test. Additional exam requirements/recommendations for international students: Required—TOEFL. *Application deadline:* For fall admission, 6/1 for domestic students, 4/15 for international students. Application fee: $50. *Financial support:* Applicants required to submit FAFSA. *Unit head:* Dr. Paul Dawson, Coordinator, 864-656-1138, Fax: 864-656-3131, E-mail: pdawson@clemson.edu.

See Close-Up on page 605.

Colorado State University, Graduate School, College of Applied Human Sciences, Department of Food Science and Human Nutrition, Fort Collins, CO 80523-0015. Offers food science (MS, PhD); nutrition (MS, PhD). *Accreditation:* ADtA. Part-time programs available. *Faculty:* 14 full-time (6 women). *Students:* 35 full-time (29 women), 26 part-time (22 women); includes 2 minority (1 Asian American or Pacific Islander, 1 Hispanic American), 3 international. Average age 29. 70 applicants, 43% accepted, 20 enrolled. In 2005, 26 master's, 3 doctorates awarded. *Degree requirements:* For master's, thesis optional; for doctorate, thesis/dissertation, comprehensive exam, registration. *Entrance requirements:* For master's and doctorate, GRE General Test, minimum GPA of 3.0. Additional exam requirements/recommendations for international students: Required—TOEFL (minimum score 550 paper-based; 213 computer-based). *Application deadline:* For fall admission, 2/1 priority date for domestic students, 2/1 priority date for international students; for spring admission, 8/1 priority date for domestic students, 8/1 priority date for international students. Applications are processed on a rolling basis. Application fee: $50. Electronic applications accepted. *Expenses:* Tuition, state resident: full-time $3,690; part-time $205 per credit. Tuition, nonresident: full-time $14,958; part-time $831 per credit. Required fees: $1,061. *Financial support:* In 2005–06, 3 fellowships (averaging $3,730 per year), 15 research assistantships with full and partial tuition reimbursements (averaging $13,884 per year), 6 teaching assistantships with full and partial tuition reimbursements (averaging $10,863 per year) were awarded; career-related internships or fieldwork, Federal Work-Study, institutionally sponsored loans, scholarships/grants, traineeships, and unspecified assistantships also available. Financial award application deadline: 2/1. *Faculty research:* Metabolic regulation, nutrition education, food safety, obesity and diabetes. Total annual research expenditures: $3.4 million. *Unit head:* Dr. Christopher Melby, Head, 970-491-1944, Fax: 970-491-7252, E-mail: christopher.melby@colostate.edu. *Application contact:* Dr. Jennifer Anderson, Graduate Coordinator, 970-491-7334, Fax: 970-491-3875, E-mail: anderson@cahs.colostate.edu.

Cornell University, Graduate School, Graduate Fields of Agriculture and Life Sciences, Field of Food Science and Technology, Ithaca, NY 14853-0001. Offers dairy science (MPS, MS, PhD); food chemistry (MPS, MS, PhD); food engineering (MPS, MS, PhD); food microbiology (MPS, MS, PhD); food processing waste technology (MPS, MS, PhD); food science (MFS, MPS, MS, PhD); international food science (MPS, MS, PhD); sensory evaluation (MPS, MS, PhD). *Faculty:* 50 full-time (10 women). *Students:* 64 full-time (35 women); includes 7 minority (6 Asian Americans or Pacific Islanders, 1 Hispanic American), 38 international. 114 applicants, 18% accepted, 16 enrolled. In 2005, 10 master's, 11 doctorates awarded. Terminal master's awarded for partial completion of doctoral program. *Degree requirements:* For master's, thesis (MS), teaching experience; for doctorate, thesis/dissertation, teaching experience, comprehensive exam. *Entrance requirements:* For master's and doctorate, GRE General Test, 3 letters of recommendation. Additional exam requirements/recommendations for international students: Required—TOEFL (minimum score 550 paper-based; 213 computer-based). *Application deadline:* For fall admission, 1/30 for domestic students. Application fee: $60. Electronic applications accepted. *Financial support:* In 2005–06, 54 students received support, including 12 fellowships with full tuition reimbursements available, 33 research assistantships with full tuition reimbursements available, 9 teaching assistantships with full tuition reimbursements available; institutionally sponsored loans, scholarships/grants, health care benefits, tuition waivers (full and partial), and unspecified assistantships also available. Financial award applicants required to submit FAFSA. *Faculty research:* Food microbiology/biotechnology, food engineering/processing, food safety/toxicology, sensory science/flavor chemistry, food packaging. *Unit head:* Director of Graduate Studies, 607-255-7637, Fax: 607-254-4868. *Application contact:* Graduate Field Assistant, 607-255-7637, Fax: 607-254-4868, E-mail: fdscigrad@cornell.edu.

Dalhousie University, Faculty of Graduate Studies, DalTech, Faculty of Engineering, Department of Food Science and Technology, Halifax, NS B3H 4R2, Canada. Offers M Sc, PhD. *Degree requirements:* For master's and doctorate, thesis/dissertation. *Entrance requirements:* Additional exam requirements/recommendations for international students: Required—TOEFL. *Faculty research:* Food microbiology, food safety/HALLP, rheology and rheometry, food processing, seafood processing.

Drexel University, College of Arts and Sciences, Department of Bioscience and Biotechnology, Program in Nutrition and Food Sciences, Philadelphia, PA 19104-2875. Offers food science (MS); nutrition science (PhD). Part-time programs available. Terminal master's awarded for partial completion of doctoral program. *Degree requirements:* For master's and doctorate, thesis/dissertation. *Entrance requirements:* For master's and doctorate, GRE General Test. Additional exam requirements/recommendations for international students: Required—TOEFL. Electronic applications accepted. *Faculty research:* Metabolism of lipids, W-3 fatty acids, obesity, diabetes and heart disease, mineral metabolism.

Florida Agricultural and Mechanical University, Division of Graduate Studies, Research, and Continuing Education, College of Engineering Science, Technology, and Agriculture, Division of Agricultural Sciences, Tallahassee, FL 32307-3200. Offers agribusiness (MS); animal science (MS); engineering technology (MS); entomology (MS); food science (MS); international programs (MS); plant science (MS). *Degree requirements:* For master's, thesis. *Entrance requirements:* For master's, GRE General Test, minimum GPA of 3.0. Additional exam requirements/recommendations for international students: Required—TOEFL (minimum score 500 paper-based).

Florida State University, Graduate Studies, College of Human Sciences, Department of Nutrition, Food, and Exercise Sciences, Tallahassee, FL 32306. Offers exercise science (PhD), including exercise physiology (MS, PhD), motor learning and control (MS, PhD); movement science (MS), including exercise physiology (MS, PhD), motor learning and control (MS, PhD); nutrition and food science (PhD); nutrition and food sciences (MS), including clinical nutrition, food science, nutrition and sport, nutrition science, nutrition, education and health promotion. *Faculty:* 13 full-time (10 women). *Students:* 37 full-time (24 women), 28 part-time (20 women); includes 17 minority (9 African Americans, 4 Asian Americans or Pacific Islanders, 4 Hispanic Americans), 10 international. 67 applicants, 67% accepted, 23 enrolled. In 2005, 22 master's, 2 doctorates awarded. *Degree requirements:* For master's, thesis optional; for doctorate, thesis/dissertation, registration. *Entrance requirements:* For master's and doctorate, GRE General Test, minimum GPA of 3.0. Additional exam requirements/recommendations for international students: Required—TOEFL. *Application deadline:* For fall admission, 7/1 for domestic students, 5/1 for international students; for spring admission, 11/1 for domestic students, 12/1 for international students. Application fee: $30. Electronic applications accepted. *Financial support:* In 2005–06, 43 students received support, including 3 fellowships with partial tuition reimbursements available (averaging $10,000 per year), 9 research assistantships with partial tuition reimbursements available (averaging $8,000 per year), 22 teaching assistantships with partial tuition reimbursements available (averaging $8,000 per year); career-related internships or fieldwork, Federal Work-Study, institutionally sponsored loans, scholarships/grants, and unspecified assistantships also available. Financial award application deadline: 1/15; financial award applicants required to submit FAFSA. *Faculty research:* Nutrition and exercise, vitamin A deficiency, protein biochemistry, cardiovascular responses to exercises, physiological effects of cigarette smoking related to health and wellness. *Unit head:* Dr. Bahran Arjmandi, Chair, 850-644-1828, Fax: 850-645-5000. *Application contact:* Ursula Tate, Program Assistant, 850-644-4800, Fax: 850-645-5000, E-mail: utate@mailer.fsu.edu.

Framingham State College, Graduate Programs, Department of Chemistry and Food Science, Framingham, MA 01701-9101. Offers food science and nutrition science (MS). Part-time and evening/weekend programs available. *Entrance requirements:* For master's, GRE General Test.

Announcement: The MS program in food science and nutrition science focuses on analytical food chemistry, nutritional biochemistry, basic food processing technology, chemical and microbiological food safety, and food formulation. Thesis and nonthesis options exist for full- and part-time study, using up-to-date laboratory facilities, instrumentation, and food pilot plant capabilities. E-mail: crussel@frc.mass.edu; WWW: http://www.framingham.edu.

Illinois Institute of Technology, Graduate College, College of Science and Letters, Department of Food Safety and Technology, Chicago, IL 60616-3793. Offers MS. Part-time and evening/weekend programs available. *Degree requirements:* For master's, thesis (for some programs), project, comprehensive exam. *Entrance requirements:* For master's, GRE General Test, minimum undergraduate GPA of 3.0. Additional exam requirements/recommendations for international students: Required—TOEFL (minimum score 550 paper-based; 213 computer-based). Electronic applications accepted. *Faculty research:* Food biotechnology, food science, microbiology, food preservation, food processing.

Iowa State University of Science and Technology, Graduate College, College of Human Sciences and College of Agriculture, Department of Food Science and Human Nutrition, Ames, IA 50011. Offers food science and technology (MS, PhD); nutrition (MS, PhD). *Faculty:* 31 full-time, 1 part-time/adjunct. *Students:* 53 full-time (39 women), 6 part-time (3 women); includes 3 minority (all African Americans), 32 international. 57 applicants, 12% accepted, 7 enrolled. In 2005, 12 master's, 6 doctorates awarded. *Degree requirements:* For master's and doctorate, thesis/dissertation. *Entrance requirements:* For master's and doctorate, GRE General Test. Additional exam requirements/recommendations for international students: Required—TOEFL (paper score 550; computer score 213) or IELTS (score 6.5). *Application deadline:* For fall admission, 2/1 priority date for domestic students, 2/1 priority date for international students. Applications are processed on a rolling basis. Application fee: $50 ($70 for international students). Electronic applications accepted. *Expenses:* Tuition, state resident: full-time $6,410. Tuition, nonresident: full-time $16,422. Tuition and fees vary according to program. *Financial support:* In 2005–06, 51 research assistantships with full and partial tuition reimbursements (averaging $15,144 per year) were awarded; fellowships, teaching assistantships, scholarships/grants also available. *Unit head:* Dr. Ruth S. MacDonald, Chair, 515-294-5991, Fax: 515-294-8181, E-mail: ruthmacd@iastate.edu. *Application contact:* Dr. Kevin Schalinske, Director of Graduate Education, 515-294-9230.

Kansas State University, Graduate School, College of Human Ecology, Department of Human Nutrition, Manhattan, KS 66506. Offers food science (MS, PhD); human nutrition (MS, PhD); public health (MS). Part-time programs available. *Faculty:* 11 full-time (5 women), 2 part-time/adjunct (both women). *Students:* 25 full-time (18 women), 13 part-time (10 women); includes 3 minority (1 African American, 1 American Indian/Alaska Native, 1 Asian American or Pacific Islander), 13 international. Average age 25. 19 applicants, 42% accepted, 2 enrolled. In 2005, 5 master's, 2 doctorates awarded. *Degree requirements:* For master's, thesis or alternative, residency; for doctorate, thesis/dissertation, residency. *Entrance requirements:* For master's, GRE General Test, minimum undergraduate GPA of 3.0; for doctorate, GRE General Test, minimum graduate GPA of 3.5, course work in biochemistry and statistics. Additional exam requirements/recommendations for international students: Required—TOEFL (minimum score 600 paper-based; 250 computer-based). *Application deadline:* For fall admission, 2/1 for domestic students, 2/1 for international students; for spring admission, 10/1 for domestic students, 8/1 for international students. Applications are processed on a rolling basis. Application fee: $30 ($55 for international students). Electronic applications accepted. *Expenses:* Tuition, state resident: full-time $5,160; part-time $215. Tuition, nonresident: full-time $12,816; part-time $534. Required fees: $564. *Financial support:* In 2005–06, 17 research assistantships (averaging $14,685 per year), 2 teaching assistantships with full tuition reimbursements (averaging $9,450 per year) were awarded; fellowships, career-related internships or fieldwork, Federal Work-Study, institutionally sponsored loans, scholarships/grants, and tuition waivers (full) also available. Support available to part-time students. Financial award application deadline: 3/1; financial award applicants required to submit FAFSA. *Faculty research:* Assessment of food portion size, bone mineral density and turnover in post menopausal women, weight maintenance in cancer prevention, dietary antioxidants and age related macular degeneration, food consumption and nutrient intakes of older women living alone. Total annual research expenditures: $940,352. *Unit head:* Dr. Denis Medeiros, Head, 785-532-0150, Fax: 785-532-3132, E-mail: medeiros@ksu.edu. *Application contact:* Janet Finney, Office Specialist, 785-532-5508, Fax: 785-532-3132, E-mail: nutrgrad@ksu.edu.

Kansas State University, Graduate School, Food Science Program, Manhattan, KS 66506. Offers MS, PhD. Part-time programs available. Postbaccalaureate distance learning degree programs offered (minimal on-campus study). *Faculty:* 2 full-time (1 woman). *Students:* 31 full-time (21 women), 24 part-time (20 women); includes 8 minority (4 African Americans, 2 Asian Americans or Pacific Islanders, 2 Hispanic Americans), 13 international. Average age 30. 22 applicants, 77% accepted, 8 enrolled. In 2005, 13 master's, 5 doctorates awarded. *Degree requirements:* For master's, thesis, residency; for doctorate, thesis/dissertation, preliminary exams, residency. *Entrance requirements:* For master's, GRE General Test, minimum GPA of 3.0 in undergraduate course work, course work in mathematics; for doctorate, GRE General Test, minimum GPA of 3.5 in master's course work. Additional exam requirements/recommendations for international students: Required—TOEFL (minimum score 550 paper-based; 213 computer-based). *Application deadline:* For fall admission, 2/1 priority date for domestic students, 2/1 priority date for international students; for spring admission, 8/1 priority date for domestic students, 8/1 priority date for international students. Applications are processed on a rolling basis. Application fee: $30 ($55 for international students). *Expenses:* Tuition, state resident: full-time $5,160; part-time $215. Tuition, nonresident: full-time $12,816; part-time $534. Required fees: $564. *Financial support:* Research assistantships with partial tuition reimbursements, teaching assistantships with partial tuition reimbursements, Federal Work-Study, institutionally sponsored loans, and scholarships/grants available. Support available to part-time students. Financial award application deadline: 3/1; financial award applicants required to submit FAFSA. *Faculty research:* Systems to insure food safety, determine nutrients and bioactive compounds in fruits, new processes to modify ag-based materials into higher food values, sensory evaluation strategies for food, hazard analysis and critical control point systems. Total annual research expenditures:$8,702. *Unit head:* Tom Herald, Director, 785-532-1221, Fax: 785-532-5681, E-mail: therald@ksu.edu. *Application contact:* Elsa Toburen, Information Contact, 785-532-1057, E-mail: etoburen@oznet.ksu.edu.

Louisiana State University and Agricultural and Mechanical College, Graduate School, College of Agriculture, Department of Food Science, Baton Rouge, LA 70803. Offers MS, PhD. Part-time programs available. *Faculty:* 10 full-time (3 women). *Students:* 29 full-time (15 women), 6 part-time (5 women); includes 4 African Americans, 25 international. Average age 30. 33 applicants, 21% accepted, 26 enrolled. In 2005, 9 master's, 7 doctorates awarded. *Degree requirements:* For master's and doctorate, thesis/dissertation. *Entrance requirements:* For master's and doctorate, GRE General Test, minimum GPA of 3.0. Additional exam requirements/recommendations for international students: Required—TOEFL (minimum score 550 paper-based; 213 computer-based). *Application deadline:* For fall admission, 1/25 priority date for domestic students, 5/15 priority date for international students. Applications are processed on a rolling basis. Application fee: $25. Electronic applications accepted. *Financial support:* In 2005–06, 33 students received support, including 21 research assistantships with partial tuition reimbursements available (averaging $16,070 per year), 1 teaching assistantship with partial tuition reimbursement available (averaging $19,800 per year); fellowships, Federal Work-Study, institutionally sponsored loans, scholarships/grants, tuition waivers (full and partial), and unspecified assistantships also available. Support available to part-time students. Financial award application deadline: 4/1; financial award applicants required to submit FAFSA. *Faculty research:* Food toxicology, food microbiology, food quality, food safety, food processing. Total annual research expenditures: $5,205. *Unit head:* Dr. Michael Moody, Head, 225-578-5206, Fax: 225-578-5300, E-mail: mmoody@agctr.lsu.edu. *Application contact:* Dr. Witoon Prinyawiwatkul, Graduate Coordinator, 225-578-5192, Fax: 225-578-5300, E-mail: wprinya@lsu.edu.

Marywood University, Academic Affairs, College of Health and Human Services, Department of Nutrition and Dietetics, Scranton, PA 18509-1598. Offers dietetics (Certificate); foods and nutrition (MS); nutrition (MS); sports nutrition and exercise science (MS). Part-time and evening/weekend programs available. *Faculty:* 2 full-time (1 woman). *Students:* 30 full-time (25 women), 31 part-time (26 women); includes 2 African Americans, 1 Asian American or Pacific Islander, 1 Hispanic American, 1 international. Average age 29. In 2005, 9 degrees awarded. *Degree requirements:* For master's, thesis. *Entrance requirements:* For master's, GRE General Test or MAT. Additional exam requirements/recommendations for international students: Required—TOEFL (minimum score 550 paper-based; 213 computer-based). *Application deadline:* For fall admission, 4/15 priority date for domestic students, 4/15 priority date for international students; for spring admission, 11/15 priority date for domestic students, 11/15 priority date for international students. Applications are processed on a rolling basis. Application fee: $30. Electronic applications accepted. *Expenses:* Tuition: Full-time $643; part-time $643 per credit. Required fees: $370; $185 per contact hour. $50 per contact hour. One-time fee: $150. *Financial support:* Research assistantships with tuition reimbursements, career-related internships or fieldwork, scholarships/grants, tuition waivers (partial), and unspecified assistantships available. Support available to part-time students. Financial award application deadline: 2/15; financial award applicants required to submit FAFSA. *Faculty research:* Community nutrition and the environment, wellness, human performance and sports nutrition, dietary regimens, food systems management. *Unit head:* Dr. Lee Harrison, Co-Chair, 570-348-6277. *Application contact:* Deborah M. Flynn, Coordinator of Graduate Advising (Enrollment Management), 570-348-2322, E-mail: flynn@ac.marywood.edu.

McGill University, Faculty of Graduate and Postdoctoral Studies, Faculty of Agricultural and Environmental Sciences, Department of Food Science and Agricultural Chemistry, Montréal, QC H3A 2T5, Canada. Offers M Sc, PhD. *Degree requirements:* For master's, thesis/dissertation, registration; for doctorate, thesis/dissertation, comprehensive exam, registration. *Entrance requirements:* For master's, B Sc in food science or related discipline (chemistry, biochemistry, or microbiology); minimum cumulative GPA of 3.0 or 3.2 during the last 2 years of full-time university study; for doctorate, M Sc, minimum GPA of 3.0. Additional exam requirements/recommendations for international students: Required—TOEFL (minimum score 550 paper-based; 213 computer-based), IELT (minimum score 7). Electronic applications accepted. *Faculty research:* Food processing, food biotechnology/enzymology, food microbiology/packaging, food analysis, food chemistry/biochemistry.

Memorial University of Newfoundland, School of Graduate Studies, Department of Biochemistry, St. John's, NL A1C 5S7, Canada. Offers biochemistry (M Sc, PhD); food science (M Sc, PhD). Part-time programs available. *Students:* 26 full-time (14 women), 1 (woman) part-time, 13 international. 19 applicants, 21% accepted, 2 enrolled. In 2005, 4 master's, 1 doctorate awarded. *Degree requirements:* For master's, thesis; for doctorate, thesis/dissertation, oral defense of thesis, comprehensive exam. *Entrance requirements:* For master's, 2nd class degree in related field; for doctorate, M Sc. *Application deadline:* For fall admission, 3/1 for domestic students, 3/1 for international students. For winter admission, 7/1 for domestic students; for spring admission, 11/1 for domestic students. Applications are processed on a rolling basis. Application fee: $40 Canadian dollars. Electronic applications accepted. *Expenses:* Tuition: Part-time $733 per term. Tuition and fees vary according to degree level and program. *Financial support:* Fellowships, research assistantships, teaching assistantships available. *Faculty research:* Toxicology, cell and molecular biology, food engineering, marine biotechnology, lipid biology. Total annual research expenditures: $1.1 million. *Unit head:* Dr. Martin Mulligan, Head, 709-737-8530, E-mail: mulligan@mun.ca. *Application contact:* Dr. Sukhinder Kaur, Graduate Officer, 709-737-8529, Fax: 709-737-2422, E-mail: biochem@mun.ca.

Michigan State University, College of Veterinary Medicine and The Graduate School, Graduate Program in Veterinary Medicine, National Food Safety and Toxicology Center, East Lansing, MI 48824. Offers food safety (MS). *Students:* 3 full-time (0 women), 38 part-time (24 women); includes 7 minority (4 African Americans, 2 Asian Americans or Pacific Islanders, 1 Hispanic American), 3 international. Average age 39. 10 applicants, 40% accepted. In 2005, 7 degrees awarded. *Entrance requirements:* For master's, minimum GPA of 3.0, 2 letters of recommendation. Additional exam requirements/recommendations for international students: Required—TOEFL (minimum score 550 paper-based; 213 computer-based), Michigan State University ELT (85), Michigan ELAB (83). *Application deadline:* For fall admission, 12/27 for domestic students. Application fee: $50. Electronic applications accepted. *Expenses:* Tuition, state resident: part-time $330 per credit hour. Tuition, nonresident: part-time $685 per credit hour. Tuition and fees vary according to program. *Faculty research:* Emerging and food-related diseases and agents; human dimensions of food safety, international food safety and risk reduction. *Unit head:* Dr. Ewen Todd, Director, 517-432-3100 Ext. 107, Fax: 517-432-2310, E-mail: toddewen@cvm.msu.edu. *Application contact:* Pattie McNiel, Distance Learning Program Coordinator, 517-432-3100 Ext. 137, Fax: 517-432-2310, E-mail: mcnielpa@cvm.edu.

Michigan State University, The Graduate School, College of Agriculture and Natural Resources and College of Natural Science, Department of Food Science and Human Nutrition, East Lansing, MI 48824. Offers food science (MS, PhD); food science—environmental toxicology (PhD); human nutrition (MS, PhD); human nutrition-environmental toxicology (PhD). *Faculty:* 22 full-time (8 women). *Students:* 45 full-time (30 women), 6 part-time (all women); includes 6 minority (3 African Americans, 2 Asian Americans or Pacific Islanders, 1 Hispanic American), 23 international. Average age 30. 67 applicants, 12% accepted. In 2005, 12 master's, 7 doctorates awarded. *Degree requirements:* For master's, oral examination in defense of thesis, thesis optional; for doctorate, thesis/dissertation, 1 term assistant teaching, 1 year residency, oral presentation and defense of dissertation, comprehensive exam. *Entrance requirements:* For master's, GRE General Test, minimum GPA of 3.0 in last 2 years of undergraduate course work, 3 letters of recommendation; for doctorate, GRE General Test, minimum GPA of 3.0, 3 letters of recommendation. Additional exam requirements/recommendations for international students: Required—TOEFL (minimum score 550 paper-based; 213 computer-based), Michigan State University ELT (85), Michigan ELAB (83). *Application deadline:* For fall admission, 12/27 for domestic students. Application fee: $50. Electronic applications accepted. *Expenses:* Tuition, state resident: part-time $330 per credit hour. Tuition, nonresident: part-time $685 per credit hour. Tuition and fees vary according to program. *Financial support:* In 2005–06, 11 fellowships with tuition reimbursements (averaging $8,443 per year), 30 research assistantships with tuition reimbursements (averaging $12,420 per year), 2 teaching assistantships with tuition reimbursements (averaging $13,088 per year) were awarded; scholarships/grants and unspecified assistantships also available. *Faculty research:* Food safety and toxicology, food processing and quality enhancement, biochemical nutrition, community nutrition. Total annual research expenditures: $3.2 million. *Unit head:* Dr. Gale M. Strasburg, Chairperson, 517-355-8474 Ext. 100, Fax: 517-353-8963, E-mail: stragale@msu.edu. *Application contact:* Deborah Klein, Graduate Secretary, 517-355-8474 Ext. 118, Fax: 517-353-8963, E-mail: kleinde@msu.edu.

Mississippi State University, College of Agriculture and Life Sciences, Department of Food Science, Nutrition and Health Promotion, Mississippi State, MS 39762. Offers food science (PhD); food science, nutrition and health promotion (MS); nutrition (MS, PhD). *Faculty:* 13 full-time (5 women), 3 part-time/adjunct (all women). *Students:* 24 full-time (14 women), 14 part-time (8 women); includes 2 minority (1 African American, 1 Asian American or Pacific Islander), 17 international. Average age 28. 31 applicants, 29% accepted, 4 enrolled. In 2005, 10 master's, 2 doctorates awarded. *Degree requirements:* For master's and doctorate, thesis/dissertation, comprehensive exam, registration. *Entrance requirements:* For master's, GRE General Test, minimum GPA of 2.8; for doctorate, GRE General Test, minimum GPA of 3.0. Additional exam requirements/recommendations for international students: Required—TOEFL. *Application deadline:* For fall admission, 7/1 for domestic students; for spring admission, 11/1 for domestic students. Applications are processed on a rolling basis. Application fee: $30. Electronic applications accepted. *Expenses:* Tuition, state resident: full-time $4,312; part-time $240 per hour. Tuition, nonresident: full-time $9,772; part-time $543 per hour. International tuition: $10,102 full-time. Tuition and fees vary according to course load. *Financial support:* In 2005–06, 1 teaching assistantship with full tuition reimbursement (averaging $9,360 per year) was awarded; research assistantships with full tuition reimbursements, Federal Work-Study, institutionally sponsored loans, scholarships/grants, and unspecified assistantships also available. Financial award application deadline: 4/1; financial award applicants required to submit FAFSA. *Faculty research:* Food preservation, food chemistry, food safety, food processing, product development. Total annual research expenditures: $1.5 million. *Unit head:* Dr. Benjy Mikel, Head, 662-325-3200, Fax: 662-325-8728, E-mail: wbm50@ra.msstate.edu. *Application contact:* Philip G. Bonfanti, Director of Admissions, 662-325-4104, Fax: 662-325-8872, E-mail: admit@msstate.edu.

Montclair State University, The Graduate School, College of Education and Human Services, Department of Health and Nutrition Sciences, Montclair, NJ 07043-1624. Offers food safety instructor (Certificate); nutrition and food science (MS). *Faculty:* 11 full-time (6 women), 20 part-time/adjunct (14 women). *Students:* 6 full-time (all women), 14 part-time (13 women); includes 2 minority (both African Americans) 16 applicants, 75% accepted, 8 enrolled. In 2005, 3 master's, 2 other advanced degrees awarded. *Degree requirements:* For master's, thesis optional. *Entrance requirements:* For master's, GRE, 2 letters of recommendation. Additional exam requirements/recommendations for international students: Required—TOEFL (minimum score 83 computer-based). *Expenses:* Tuition: Full-time $3,001; part-time $409 per credit. Required fees: $56 per credit. Tuition and fees vary according to course load, degree level and program. *Financial support:* In 2005–06, 2 research assistantships (averaging $7,000 per year) were awarded *Faculty research:* Adolescent physical activity. Total annual research expenditures: $182,000. *Unit head:* Dr. Shahla Wunderlich, Chairperson, 973-655-6854, E-mail: wunderlichs@mail.montclair.edu.

New York University, The Steinhardt School of Education, Department of Nutrition, Food Studies, and Public Health, Program in Nutrition and Dietetics, New York, NY 10012-1019. Offers clinical nutrition (MS); foods and nutrition (MS); nutrition and dietetics (PhD). Part-time and evening/weekend programs available. *Faculty:* 13 full-time (11 women). *Students:* 79 full-time (77 women), 105 part-time (103 women); includes 36 minority (11 African Americans, 17 Asian Americans or Pacific Islanders, 8 Hispanic Americans), 14 international. 111 applicants, 41% accepted, 33 enrolled. In 2005, 31 master's, 1 doctorate awarded. *Degree requirements:* For master's, thesis (for some programs); for doctorate, thesis/dissertation. *Entrance requirements:* For doctorate, GRE General Test, interview. Additional exam requirements/recommendations for international students: Required—TOEFL. *Application deadline:* For fall admission, 1/15 priority date for domestic students, 1/15 priority date for international students; for spring admission, 11/1 for domestic students, 11/1 for international students. Applications are processed on a rolling basis. Application fee: $50 ($60 for international students). *Financial support:* Fellowships with full and partial tuition reimbursements, career-related internships or fieldwork, Federal Work-Study, scholarships/grants, tuition waivers (partial), and unspecified assistantships available. Financial award application deadline: 2/1; financial award applicants required to submit FAFSA. *Faculty research:* Nutrition and race, childhood obesity and other eating disorders, nutritional epidemiology, nutrition policy, nutrition and health promotion. *Unit head:* Dr. Lisa Sasson, Director, 212-998-5580, Fax: 212-995-4194. *Application contact:* 212-998-5030, Fax: 212-995-4328, E-mail: grad.admissions@nyu.edu.

North Carolina State University, Graduate School, College of Agriculture and Life Sciences, Department of Food Science, Raleigh, NC 27695. Offers MFS, MS, PhD. *Degree requirements:* For master's, thesis (for some programs); for doctorate, thesis/dissertation. *Entrance requirements:* For master's and doctorate, GRE. Electronic applications accepted. *Faculty research:* Food safety, value-added food products, environmental quality, nutrition and health, biotechnology.

Food Science and Technology

North Dakota State University, The Graduate School, College of Agriculture, Food Systems, and Natural Resources, Department of Cereal and Food Sciences, Fargo, ND 58105. Offers cereal science (MS, PhD). Part-time programs available. *Faculty:* 5. *Students:* 15 full-time (9 women). 5 applicants, 20% accepted, 1 enrolled. In 2005, 2 degrees awarded. Terminal master's awarded for partial completion of doctoral program. *Degree requirements:* For master's and doctorate, thesis/dissertation, comprehensive exam. *Entrance requirements:* Additional exam requirements/recommendations for international students: Required—TOEFL (minimum score 550 paper-based), IELT (minimum score 6). *Application deadline:* For fall admission, 5/1 for domestic students. Application fee: $45 ($60 for international students). *Financial support:* In 2005–06, 15 research assistantships were awarded; tuition waivers (full and partial) also available. *Unit head:* James R. Venette, Interim Chair, 701-231-9450, Fax: 701-231-5171, E-mail: james.venette@ndsu.edu.

North Dakota State University, The Graduate School, College of Agriculture, Food Systems, and Natural Resources, Interdisciplinary Program in Food Safety, Fargo, ND 58105. Offers MS, PhD. Part-time programs available. Postbaccalaureate distance learning degree programs offered (minimal on-campus study). *Faculty:* 10 full-time (5 women), 3 part-time/adjunct (2 women). *Students:* 9 full-time (5 women), 2 part-time (both women), 7 international. Average age 25. 6 applicants, 67% accepted, 2 enrolled. In 2005, 1 master's, 1 doctorate awarded. Terminal master's awarded for partial completion of doctoral program. *Degree requirements:* For master's, thesis/dissertation; for doctorate, thesis/dissertation, comprehensive exam. *Entrance requirements:* For doctorate, preliminary exam. Additional exam requirements/recommendations for international students: Required—TOEFL (minimum score 525 paper-based), TWE (minimum score 5), GRE. *Application deadline:* Applications are processed on a rolling basis. Application fee: $45 ($60 for international students). Electronic applications accepted. *Financial support:* In 2005–06, 9 research assistantships with full tuition reimbursements (averaging $16,000 per year) were awarded; scholarships/grants also available. *Faculty research:* Mycotoxins in grain, pathogens in meat systems, sensor development for food pathogens. *Unit head:* Dr. Clifford Hall, Assistant Director, 701-231-6359, E-mail: clifford.hall@ndsu.edu.

Nova Scotia Agricultural College, Research and Graduate Studies, Truro, NS B2N 5E3, Canada. Offers agriculture (M Sc), including air quality, animal behavior, animal molecular genetics, animal nutrition, animal technology, aquaculture, botany, crop management, crop physiology, ecology, environmental microbiology, food science, horticulture, nutrient management, pest management, physiology, plant biotechnology, plant pathology, soil chemistry, soil fertility, waste management and composting, water quality. Part-time programs available. *Faculty:* 43 full-time (7 women), 21 part-time/adjunct (1 woman). *Students:* 50 full-time (25 women), 15 part-time (11 women); includes 7 minority (3 African Americans, 3 Asian Americans or Pacific Islanders, 1 Hispanic American). Average age 25. In 2005, 23 degrees awarded. *Degree requirements:* For master's, thesis, registration. *Entrance requirements:* For master's, honors B Sc, minimum GPA of 3.0. Additional exam requirements/recommendations for international students: Required—TOEFL (minimum score 580 paper-based; 237 computer-based), Michigan English Language Assessment Battery, IELT, Can Test, CAEL. *Application deadline:* For fall admission, 6/1 for domestic students, 4/1 for international students. For winter admission, 10/31 for domestic students; for spring admission, 2/28 for domestic students. Applications are processed on a rolling basis. Application fee: $70. *Expenses:* Tuition, state resident: part-time $2,328 per year. Tuition, nonresident: full-time $6,984; part-time $7,968 per year. International tuition: $12,624 full-time. Required fees: $481; $46 per course. Tuition and fees vary according to program and student level. *Financial support:* In 2005–06, 48 students received support, including 7 fellowships (averaging $15,000 per year), 10 research assistantships (averaging $15,000 per year), 15 teaching assistantships (averaging $900 per year); career-related internships or fieldwork, scholarships/grants, and unspecified assistantships also available. *Faculty research:* Bio-product development, organic agriculture, nutrient management, air and water quality, agricultural biotechnology. Total annual research expenditures: $4.7 million. *Unit head:* Jill L. Rogers, Manager, 902-893-6360, Fax: 902-893-3430, E-mail: jrogers@nsac.ca. *Application contact:* Marie Law, Administrative Assistant, 902-893-6502, Fax: 902-893-3430, E-mail: mlaw@nsac.ca.

The Ohio State University, Graduate School, College of Food, Agricultural, and Environmental Sciences, Program in Food Science and Nutrition, Columbus, OH 43210. Offers MS, PhD. *Degree requirements:* For master's, thesis optional; for doctorate, thesis/dissertation. *Entrance requirements:* For master's and doctorate, GRE General Test. Additional exam requirements/recommendations for international students: Required—TOEFL (paper 550; computer 213) or IELT (7) or Michigan English Language Assessment Battery (89). Electronic applications accepted.

Oklahoma State University, College of Agricultural Science and Natural Resources, Department of Animal Science, Interdisciplinary Program in Food Science, Stillwater, OK 74078. Offers MS, PhD. *Degree requirements:* For master's and doctorate, thesis/dissertation. *Entrance requirements:* For master's and doctorate, GRE. Additional exam requirements/recommendations for international students: Required—TOEFL. *Application deadline:* For fall admission, 6/1 for domestic students. Application fee: $40 ($75 for international students). *Expenses:* Tuition, state resident: full-time $4,253; part-time $139 per credit hour. Tuition, nonresident: full-time $12,569; part-time $485 per credit hour. Required fees: $43 per credit hour. One-time fee: $20 part-time. Tuition and fees vary according to course load and program. *Financial support:* Research assistantships, teaching assistantships, career-related internships or fieldwork, Federal Work-Study, scholarships/grants, health care benefits, tuition waivers (partial), and unspecified assistantships available. Support available to part-time students. Financial award application deadline:3/1. *Unit head:* Dr. Stanley Gilliland, Coordinator, 405-744-6071.

Oklahoma State University, College of Agricultural Science and Natural Resources, Department of Horticulture and Landscape Architecture, Stillwater, OK 74078. Offers crop science (PhD); environmental science (PhD); food science (PhD); horticulture (M Ag, MS); plant science (PhD). *Faculty:* 21 full-time (1 woman), 4 part-time/adjunct (0 women). *Students:* 1 full-time (0 women), 6 part-time (2 women); includes 1 minority (African American), 4 international. Average age 31. 7 applicants, 14% accepted. In 2005, 3 master's awarded. *Degree requirements:* For master's, thesis or alternative. *Entrance requirements:* For master's and doctorate, GRE. Additional exam requirements/recommendations for international students: Required—TOEFL. *Application deadline:* For fall admission, 6/1 priority date for domestic students, 3/1 priority date for international students. Applications are processed on a rolling basis. Application fee: $40 ($75 for international students). Electronic applications accepted. *Expenses:* Tuition, state resident: full-time $4,253; part-time $139 per credit hour. Tuition, nonresident: full-time $12,569; part-time $485 per credit hour. Required fees: $43 per credit hour. One-time fee: $20 part-time. Tuition and fees vary according to course load and program. *Financial support:* In 2005–06, 4 research assistantships (averaging $13,817 per year) were awarded; teaching assistantships, career-related internships or fieldwork, Federal Work-Study, scholarships/grants, health care benefits, tuition waivers (partial), and unspecified assistantships also available. Support available to part-time students. Financial award application deadline: 3/1. *Faculty research:* Stress and postharvest physiology; water utilization and runoff; IPM systems and nursery, turf, floriculture, vegetable, net and fruit produces and natural resources, food extraction, and processing; public garden management. *Unit head:* Dr. Dale Maronek, Head, 405-744-5414, Fax: 405-744-9709, E-mail: maronek@okstate.edu.

Oregon State University, Graduate School, College of Agricultural Sciences, Department of Food Science and Technology, Corvallis, OR 97331. Offers M Agr, MAIS, MS, PhD. *Faculty:* 9 full-time (1 woman). *Students:* 33 full-time (21 women), 1 part-time, 24 international. Average age 29. In 2005, 8 master's, 3 doctorates awarded. *Degree requirements:* For master's, thesis (for some programs); for doctorate, thesis/dissertation. *Entrance requirements:* For master's and doctorate, GRE General Test, minimum GPA of 3.0 in last 90 hours. Additional exam requirements/recommendations for international students: Required—TOEFL. *Application deadline:* For fall admission, 3/1 for domestic students. Applications are processed on a rolling basis. Application fee: $50. *Expenses:* Tuition, area resident: Part-time $301 per credit. Tuition, state resident: full-time $8,139; part-time $501 per credit. Tuition, nonresident: full-time $14,376; part-time $532 per credit. Required fees: $1,266. *Financial support:* Fellowships, research assistantships, teaching assistantships, career-related internships or fieldwork, Federal Work-Study, and institutionally sponsored loans available. Support available to part-time students. Financial award application deadline: 2/1. *Faculty research:* Diet, cancer, and anticarcinogenesis; sensory analysis; chemistry and biochemistry. *Unit head:* Dr. Robert McGorrin, Head, 541-737-3131, Fax: 541-737-1877, E-mail: robert.mcgorrin@orst.edu. *Application contact:* 541-737-6486, Fax: 541-737-1877.

The Pennsylvania State University University Park Campus, Graduate School, College of Agricultural Sciences, Department of Food Science, State College, University Park, PA 16802-1503. Offers MS, PhD. *Students:* 44 full-time (31 women), 3 part-time (2 women); includes 4 minority (2 African Americans, 1 Asian American or Pacific Islander, 1 Hispanic American), 28 international. *Entrance requirements:* For master's and doctorate, GRE General Test. *Expenses:* Tuition, state resident: full-time $12,518; part-time $522 per credit. Tuition, nonresident: full-time $23,004; part-time $959 per credit. Required fees: $484. Tuition and fees vary according to course load, campus/location and program. *Unit head:* Dr. John D. Floros, Head, 814-865-5444, Fax: 814-863-6132, E-mail: jdf10@psu.edu. *Application contact:* Dr. John D. Floros, Head, 814-865-5444, Fax: 814-863-6132, E-mail: jdf10@psu.edu.

Purdue University, Graduate School, College of Agriculture, Department of Food Science, West Lafayette, IN 47907. Offers MS, PhD. *Faculty:* 19 full-time (3 women), 8 part-time/adjunct (0 women). *Students:* 58 full-time (34 women), 2 part-time (both women); includes 6 minority (4 African Americans, 1 Asian American or Pacific Islander, 1 Hispanic American), 31 international. Average age 27. 50 applicants, 10% accepted. In 2005, 6 master's, 7 doctorates awarded. *Degree requirements:* For master's, thesis (for some programs); for doctorate, thesis/dissertation, teaching assistantship. *Entrance requirements:* For master's and doctorate, GRE General Test. Additional exam requirements/recommendations for international students: Required—TOEFL, TWE. *Application deadline:* For fall admission, 5/1 for domestic students; for spring admission, 10/1 for domestic students. Applications are processed on a rolling basis. Application fee: $30. Electronic applications accepted. *Financial support:* In 2005–06, 4 fellowships (averaging $17,000 per year), 38 research assistantships (averaging $13,500 per year), 1 teaching assistantship (averaging $13,500 per year) were awarded; career-related internships or fieldwork also available. Support available to part-time students. Financial award application deadline: 4/1; financial award applicants required to submit FAFSA. *Faculty research:* Processing, technology, microbiology, chemistry of foods, carbohydrate chemistry. *Unit head:* Dr. S. S. Nielsen, Graduate Committee Chair, 765-494-8328, Fax: 765-494-7953, E-mail: nielsens@foodsci.purdue.edu. *Application contact:* Dr. Linda L Webster, Graduate Committee Chair, 765-494-8258, Fax: 765-494-7953, E-mail: websterl@purdue.edu.

Rutgers, The State University of New Jersey, New Brunswick/Piscataway, Graduate School, Program in Food Science, New Brunswick, NJ 08901-1281. Offers M Phil, MS, PhD. Part-time and evening/weekend programs available. Postbaccalaureate distance learning degree programs offered (minimal on-campus study). *Faculty:* 29 full-time, 2 part-time/adjunct. *Students:* 48 full-time (30 women), 52 part-time (33 women); includes 10 minority (2 African Americans, 7 Asian Americans or Pacific Islanders, 1 Hispanic American), 52 international. Average age 30. 102 applicants, 55% accepted, 28 enrolled. In 2005, 7 master's, 5 doctorates awarded. *Degree requirements:* For master's, thesis or alternative; for doctorate, thesis/dissertation. *Entrance requirements:* For master's and doctorate, GRE General Test. *Application deadline:* For fall admission, 5/1 for domestic students, 3/1 for international students; for spring admission, 12/1 for domestic students, 8/1 for international students. Applications are processed on a rolling basis. Application fee: $50. *Expenses:* Tuition, state resident: full-time $10,440; part-time $435 per credit. Tuition, nonresident: full-time $15,520; part-time $647 per credit. Required fees: $129 per credit. Tuition and fees vary according to program. *Financial support:* In 2005–06, 36 students received support, including 3 fellowships with full tuition reimbursements available (averaging $17,000 per year), 36 research assistantships with full tuition reimbursements available (averaging $16,428 per year), 6 teaching assistantships with full tuition reimbursements available (averaging $16,988 per year); Federal Work-Study and tuition waivers (full) also available. Financial award application deadline: 3/1; financial award applicants required to submit FAFSA. *Faculty research:* Nutraceuticals and functional foods, food and flavor analysis, food chemistry and biochemistry, food nanotechnology, food engineering and processing. Total annual research expenditures: $5 million. *Unit head:* Dr. Chi-Tang Ho, Director, 732-932-9611 Ext. 235, Fax: 732-932-6776, E-mail: ho@aesop.rutgers.edu. *Application contact:* Jackie Revolinsky, Graduate Secretary, 732-932-9611 Ext. 207, Fax: 732-932-6776, E-mail: revolinscrci@rutgers.edu.

Texas A&M University, College of Agriculture and Life Sciences, Department of Nutrition and Food Science, College Station, TX 77843. Offers M Agr, MS, PhD. *Students:* Average age 28. *Degree requirements:* For master's and doctorate, thesis/dissertation. *Entrance requirements:* For master's and doctorate, GRE General Test. Additional exam requirements/recommendations for international students: Required—TOEFL. *Application deadline:* For fall admission, 2/1 for domestic students; for spring admission, 10/1 priority date for domestic students. Applications are processed on a rolling basis. Application fee: $50 ($75 for international students). *Expenses:* Tuition, state resident: full-time $4,488; part-time $187 per credit hour. Tuition, nonresident: full-time $11,112; part-time $463 per credit hour. Required fees:$1,974. *Financial support:* Fellowships, research assistantships, teaching assistantships, career-related internships or fieldwork and scholarships/grants available. *Faculty research:* Food safety, microbiology, product development. *Unit head:* Ronnie Edwards, Graduate Advisor, 979-845-1542, Fax: 979-845-6433, E-mail: r-edwards@tamu.edu.

Texas Tech University, Graduate School, College of Agricultural Sciences and Natural Resources, Department of Animal and Food Sciences, Lubbock, TX 79409. Offers animal science (MS, PhD); food technology (MS). Part-time programs available. *Faculty:* 15 full-time (4 women), 1 part-time/adjunct (0 women). *Students:* 43 full-time (24 women), 5 part-time (2 women); includes 3 minority (1 American Indian/Alaska Native, 1 Asian American or Pacific Islander, 1 Hispanic American), 11 international. Average age 26. 36 applicants, 69% accepted, 15 enrolled. In 2005, 15 master's, 2 doctorates awarded. *Degree requirements:* For master's, thesis, internship (M Agr); for doctorate, thesis/dissertation. *Entrance requirements:* For master's and doctorate, GRE General Test. Additional exam requirements/recommendations for international students: Required—TOEFL (minimum score 550 paper-based; 213 computer-based). *Application deadline:* Applications are processed on a rolling basis. Application fee: $50 ($60 for international students). Electronic applications accepted. *Expenses:* Tuition, state resident: full-time $4,296. Tuition, nonresident: full-time $10,920. Required fees: $1,992. Tuition and fees vary according to program. *Financial support:* In 2005–06, 23 students received support, including 34 research assistantships with partial tuition reimbursements available (averaging $10,738 per year), 3 teaching assistantships with partial tuition reimbursements available (averaging $12,300 per year); Federal Work-Study and institutionally sponsored loans also available. Support available to part-time students. Financial award application deadline: 4/15; financial award applicants required to submit FAFSA. *Faculty research:* Animal growth composition and product acceptability, animal nutrition and utilization, animal physiology and adaptation to stress, food microbiology, food safety and security. Total annual research expenditures: $1.8 million. *Unit head:* Dr. Kevin R. Pond, Chairman, 806-742-2805 Ext. 223, Fax: 806-742-0898, E-mail: kevin.pond@ttu.edu. *Application contact:* Sandy Gellner, Graduate Secretary, 806-742-0898 Ext. 221, Fax: 806-742-4003, E-mail: sandra.gellner@ttu.edu.

Texas Woman's University, Graduate School, College of Health Sciences, Department of Nutrition and Food Sciences, Denton, TX 76201. Offers exercise and sports nutrition (MS); food science (MS); institutional administration (MS); nutrition (MS, PhD). Part-time and evening/weekend programs available. *Students:* 65 full-time (63 women), 83 part-time (79 women); includes 40 minority (10 African Americans, 12 Asian Americans or Pacific Islanders, 18 Hispanic Americans), 19 international. Average age 29. In 2005, 30 master's, 3 doctorates awarded. *Degree requirements:* For master's, comprehensive exam; for doctorate, thesis/dissertation, qualifying exam, comprehensive exam. *Entrance requirements:* For master's,

GRE General Test (verbal 350, quantitative 450), minimum GPA of 3.25, resumé; for doctorate, GRE General Test (verbal 450, quantitative 550), minimum GPA of 3.5, 2 letters of reference. Additional exam requirements/recommendations for international students: Required—TOEFL (minimum score 550 paper-based; 213 computer-based). *Application deadline:* Applications are processed on a rolling basis. Application fee: $30 ($50 for international students). Electronic applications accepted. *Expenses:* Tuition, state resident: full-time $2,934; part-time $163. Tuition, nonresident: full-time $7,974; part-time $152. *Financial support:* In 2005–06, 18 research assistantships (averaging $10,206 per year), 5 teaching assistantships (averaging $10,206 per year) were awarded; career-related internships or fieldwork, Federal Work-Study, institutionally sponsored loans, scholarships/grants, traineeships, health care benefits, and unspecified assistantships also available. Support available to part-time students. Financial award application deadline: 3/1; financial award applicants required to submit FAFSA. *Faculty research:* Food science, food safety, clinical nutrition, nutrition and cancer, weight management. *Unit head:* Dr. Carolyn M. Bednar, Interim Chair, 940-898-2636, Fax: 940-898-2634, E-mail: cbednar@twu.edu. *Application contact:* Samuel Wheeler, Coordinator of Graduate Admissions, 940-898-3188, Fax: 940-898-3081, E-mail: wheelersr@twu.edu.

Tuskegee University, Graduate Programs, College of Agricultural, Environmental and Natural Sciences, Department of Food and Nutritional Sciences, Tuskegee, AL 36088. Offers MS. *Faculty:* 4 full-time (3 women). *Students:* 15 full-time (13 women), 2 part-time (1 woman); includes 14 minority (all African Americans), 1 international. Average age 30. In 2005, 8 degrees awarded. *Degree requirements:* For master's, thesis. *Entrance requirements:* For master's, GRE General Test. Additional exam requirements/recommendations for international students: Required—TOEFL (minimum score 500 paper-based; 173 computer-based). *Application deadline:* For fall admission, 7/15 for domestic students. Applications are processed on a rolling basis. Application fee: $25 ($35 for international students). *Expenses:* Tuition: Full-time $12,400. Required fees: $300; $490 per credit. *Financial support:* Application deadline: 4/15. *Unit head:* Dr. Ralphenia Pace, Head, 334-727-8162.

Universidad de las Américas–Puebla, Division of Graduate Studies, School of Engineering, Program in Chemical Engineering, Puebla, Mexico. Offers chemical engineering (MS); food technology (MS). Part-time and evening/weekend programs available. *Degree requirements:* For master's, one foreign language, thesis. *Faculty research:* Food science, reactors, oil industry, biotechnology.

Universidad de las Américas–Puebla, Division of Graduate Studies, School of Engineering, Program in Food Sciences, Puebla, Mexico. Offers MS.

Université de Moncton, School of Food Science, Nutrition and Family Studies, Moncton, NB E1A 3E9, Canada. Offers foods/nutrition (M Sc). Part-time programs available. *Faculty:* 4 full-time (2 women). *Students:* 7 full-time (4 women). 3 applicants, 0% accepted. In 2005, 4 degrees awarded. *Degree requirements:* For master's, one foreign language, thesis. *Entrance requirements:* For master's, previous course work in statistics. *Application deadline:* For fall admission, 6/1 priority date for domestic students, 2/1 priority date for international students. For winter admission, 11/15 for domestic students; for spring admission, 3/31 for domestic students. Applications are processed on a rolling basis. Application fee: $39. Electronic applications accepted. *Financial support:* In 2005–06, 3 research assistantships were awarded; fellowships, career-related internships or fieldwork and scholarships/grants also available. Financial award application deadline: 5/23. *Faculty research:* Clinic nutrition (anemia, elderly, osteoporosis), applied nutrition, metabolic activities of lactic bacteria, solubility of low density lipoproteins, bile acids. *Unit head:* Regina M. Robichaud, Director, 506-858-4003, Fax: 506-858-4283, E-mail: robichr@umoncton.ca.

Université Laval, Faculty of Agricultural and Food Sciences, Department of Food Sciences and Nutrition, Programs in Food Sciences and Technology, Québec, QC G1K 7P4, Canada. Offers M Sc, PhD. Terminal master's awarded for partial completion of doctoral program. *Degree requirements:* For master's, thesis (for some programs); for doctorate, thesis/dissertation, comprehensive exam. *Entrance requirements:* For master's and doctorate, knowledge of French and English. Electronic applications accepted.

University of Arkansas, Graduate School, Dale Bumpers College of Agricultural, Food and Life Sciences, Department of Food Science, Fayetteville, AR 72701-1201. Offers MS, PhD. *Students:* 26 full-time (16 women), 7 part-time (3 women); includes 2 minority (1 African American, 1 Asian American or Pacific Islander), 14 international. 37 applicants, 22% accepted. In 2005, 8 master's, 5 doctorates awarded. *Degree requirements:* For master's and doctorate, thesis/dissertation. Application fee: $40 ($50 for international students). *Financial support:* In 2005–06, 2 fellowships with tuition reimbursements, 23 research assistantships were awarded; teaching assistantships, career-related internships or fieldwork, Federal Work-Study, scholarships/grants, and unspecified assistantships also available. Support available to part-time students. Financial award application deadline: 4/1; financial award applicants required to submit FAFSA. *Unit head:* Dr. Ron Buescher, Head, 479-575-4605. *Application contact:* Navam Hettiarachchy, Graduate Coordinator, E-mail: nhettiar@comp.uark.edu.

University of Arkansas, Graduate School, Dale Bumpers College of Agricultural, Food and Life Sciences, Program in Agricultural, Food and Life Sciences, Fayetteville, AR 72701-1201. Offers MS. Part-time and evening/weekend programs available. Postbaccalaureate distance learning degree programs offered (minimal on-campus study). *Students:* 1 full-time (0 women), 14 part-time (4 women); includes 1 minority (African American) 10 applicants, 60% accepted. In 2005, 7 degrees awarded. *Degree requirements:* For master's, thesis optional. Application fee: $40 ($50 for international students). *Financial support:* Career-related internships or fieldwork and Federal Work-Study available. Support available to part-time students. Financial award application deadline: 4/1; financial award applicants required to submit FAFSA. *Unit head:* Nolan Arthur, Chair, 479-575-2035.

The University of British Columbia, Faculty of Graduate Studies, Faculty of Land and Food Systems, Program in Food Science, Vancouver, BC V6T 1Z1, Canada. Offers M Sc, PhD. *Faculty:* 8 full-time (3 women), 8 part-time/adjunct (2 women). *Students:* 17 full-time (12 women). Average age 25. 32 applicants, 19% accepted, 5 enrolled. In 2005, 3 master's, 2 doctorates awarded. *Degree requirements:* For master's, thesis/dissertation, registration; for doctorate, thesis/dissertation, comprehensive exam, registration. *Entrance requirements:* Additional exam requirements/recommendations for international students: Required—TOEFL (minimum score 560 paper-based; 220 computer-based). *Application deadline:* For fall admission, 2/1 for domestic students, 1/1 for international students. For winter admission, 8/1 for domestic students; for spring admission, 10/1 for domestic students. Application fee: $90 Canadian dollars ($150 Canadian dollars for international students). Electronic applications accepted. *Financial support:* In 2005–06, 8 fellowships (averaging $9,075 per year) were awarded; research assistantships, teaching assistantships, institutionally sponsored loans, scholarships/grants, and tuition waivers (partial) also available. *Faculty research:* Food chemistry and biochemistry, food process science, food toxicology and safety, food microbiology, food biotechnology. Total annual research expenditures: $1.6 million. *Unit head:* Dr. David D. Kitts, Graduate Advisor, 604-822-5560, Fax: 604-822-3959/5143, E-mail: gradapp@interchange.ubc.ca. *Application contact:* Alina Yuhymets, Graduate Programs Manager, 604-822-4593, Fax: 604-822-4400, E-mail: yuhymets@interchange.ubc.ca.

University of California, Davis, Graduate Studies, Graduate Group in Food Science, Davis, CA 95616. Offers MS, PhD. *Faculty:* 49 full-time. *Students:* 43 full-time (29 women); includes 4 minority (3 Asian Americans or Pacific Islanders, 1 Hispanic American), 16 international. Average age 29. 81 applicants, 26% accepted, 10 enrolled. In 2005, 9 master's, 4 doctorates awarded. Terminal master's awarded for partial completion of doctoral program. *Median time to degree:* Of those who began their doctoral program in fall 1997, 83.3% received their degree in 8 years or less. *Degree requirements:* For master's, thesis (for some programs), comprehensive exam (for some programs); for doctorate, thesis/dissertation. *Entrance requirements:* For master's and doctorate, GRE General Test, minimum GPA of 3.0. Additional exam requirements/recommendations for international students: Required—TOEFL (minimum score 550 paper-based; 213 computer-based). *Application deadline:* For fall admission, 1/15 for domestic students, 1/15 for international students. Application fee: $60. Electronic applications accepted. *Financial support:* In 2005–06, 37 students received support, including 11 fellowships with full and partial tuition reimbursements available (averaging $11,368 per year), 12 research assistantships with full and partial tuition reimbursements available (averaging $14,735 per year), 8 teaching assistantships with partial tuition reimbursements available (averaging $15,082 per year); Federal Work-Study, institutionally sponsored loans, scholarships/grants, tuition waivers (full and partial), and unspecified assistantships also available. Financial award application deadline: 1/15; financial award applicants required to submit FAFSA. *Unit head:* Gary Smith, Graduate Group Chair, 530-752-6168, E-mail: gmsmith@ucdavis.edu. *Application contact:* Karen Gurley, Administrative Assistant, 530-752-8079, Fax: 530-752-4759, E-mail: kmgurley@ucdavis.edu.

University of Delaware, College of Agriculture and Natural Resources, Department of Animal and Food Sciences, Newark, DE 19716. Offers animal sciences (MS, PhD); food sciences (MS). Part-time programs available. *Faculty:* 17 full-time (5 women). *Students:* 25 full-time (15 women), 1 part-time; includes 1 minority (Hispanic American), 7 international. Average age 28. 26 applicants, 35% accepted, 9 enrolled. In 2005, 2 master's, 1 doctorate awarded. Terminal master's awarded for partial completion of doctoral program. *Degree requirements:* For master's, thesis/dissertation; for doctorate, thesis/dissertation, comprehensive exam. *Entrance requirements:* For master's and doctorate, GRE General Test. Additional exam requirements/recommendations for international students: Required—TOEFL. *Application deadline:* For fall admission, 7/1 for domestic students; for spring admission, 12/1 for domestic students. Applications are processed on a rolling basis. Application fee: $60. Electronic applications accepted. *Financial support:* In 2005–06, 21 students received support, including 15 research assistantships with full tuition reimbursements available (averaging $15,344 per year), 5 teaching assistantships with full tuition reimbursements available (averaging $12,314 per year); fellowships with full tuition reimbursements available, scholarships/grants and tuition waivers (full) also available. Financial award application deadline: 3/1. *Faculty research:* Food chemistry, food microbiology, process engineering technology, packaging, food analysis, microbial genetics, molecular endocrinology, growth physiology, avian immunology and virology, monogastric nutrition, avian genomics. Total annual research expenditures: $1.5 million. *Unit head:* Dr. Limin Kung, Assistant Chair, 302-831-2524, Fax: 302-831-2822, E-mail: lksilage@udel.edu. *Application contact:* Dr. Carl Schmidt, Graduate Program Coordinator, 302-831-2524, Fax: 302-831-2822, E-mail: schmidtc@udel.edu.

University of Florida, Graduate School, College of Agricultural and Life Sciences, Department of Food Science and Human Nutrition, Gainesville, FL 32611. Offers MS, PhD. *Faculty:* 23 full-time (7 women), 1 (woman) part-time/adjunct. *Students:* 73 (44 women); includes 13 minority (3 African Americans, 4 Asian Americans or Pacific Islanders, 6 Hispanic Americans) 21 international. 84 applicants, 61% accepted. In 2005, 18 master's, 6 doctorates awarded. *Degree requirements:* For master's, thesis optional; for doctorate, thesis/dissertation. *Entrance requirements:* For master's and doctorate, GRE General Test, minimum GPA of 3.0. Additional exam requirements/recommendations for international students: Required—TOEFL. *Application deadline:* For fall admission, 6/1 for domestic students. Applications are processed on a rolling basis. Application fee: $20. Electronic applications accepted. *Expenses:* Tuition, state resident: full-time $6,234. Tuition, nonresident: full-time $21,359. Tuition and fees vary according to program. *Financial support:* In 2005–06, 13 research assistantships (averaging $13,562 per year), 13 teaching assistantships (averaging $12,990 per year) were awarded; fellowships, career-related internships or fieldwork also available. *Faculty research:* Pesticide research, nutritional biochemistry and microbiology, food safety and toxicology assessment and dietetics, food chemistry. *Application contact:* Dr. Harry Sitren, Coordinator, 352-392-1991 Ext. 216, Fax: 352-392-9467, E-mail: hssitren@ifas.ufl.edu.

University of Georgia, Graduate School, College of Agricultural and Environmental Sciences, Department of Food Science, Athens, GA 30602. Offers food science (MS, PhD); food technology (MFT). Part-time programs available. *Faculty:* 16 full-time (2 women), 1 part-time/adjunct (0 women). *Students:* 59 full-time, 22 part-time; includes 13 minority (10 African Americans, 1 Asian American or Pacific Islander, 2 Hispanic Americans), 44 international. 74 applicants, 68% accepted, 25 enrolled. In 2005, 12 master's, 6 doctorates awarded. *Degree requirements:* For master's and doctorate, thesis/dissertation. *Entrance requirements:* For master's and doctorate, GRE General Test. Additional exam requirements/recommendations for international students: Required—TOEFL (minimum score 550 paper-based; 213 computer-based). *Application deadline:* For fall admission, 7/1 for domestic students, 5/1 for international students; for spring admission, 11/15 for domestic students, 10/1 for international students. Applications are processed on a rolling basis. Application fee: $50. Electronic applications accepted. *Financial support:* Fellowships, research assistantships, teaching assistantships, unspecified assistantships available. Total annual research expenditures: $1.5 million. *Unit head:* Dr. Rakesh K. Singh, Head, 706-542-2286, Fax: 706-542-1050, E-mail: rsingh@uga.edu. *Application contact:* Dr. Philip E. Koehler, Graduate Coordinator, 706-542-1099, Fax: 706-542-1050, E-mail: pkoehler@uga.edu.

University of Guelph, Graduate Program Services, Ontario Agricultural College, Department of Food Science, Guelph, ON N1G 2W1, Canada. Offers M Sc, PhD. *Faculty:* 16 full-time (1 woman), 13 part-time/adjunct (4 women). *Students:* 61 full-time (38 women), 3 part-time (1 woman). 55 applicants, 15% accepted, 8 enrolled. In 2005, 9 master's, 2 doctorates awarded. *Median time to degree:* Of those who began their doctoral program in fall 1997, 83% received their degree in 8 years or less. *Degree requirements:* For master's, thesis/dissertation; for doctorate, thesis/dissertation, comprehensive exam. *Entrance requirements:* For master's, minimum B- average during previous 2 years of honors B Sc degree; for doctorate, minimum B average. Additional exam requirements/recommendations for international students: Required—TOEFL (minimum score 550 paper-based; 213 computer-based). *Application deadline:* For fall admission, 8/1 priority date for domestic students, 6/1 priority date for international students. For winter admission, 12/1 for domestic students; for spring admission, 4/1 for domestic students. Applications are processed on a rolling basis. Application fee: $75. Electronic applications accepted. *Financial support:* In 2005–06, 44 students received support, including 47 research assistantships with full tuition reimbursements available (averaging $16,000 per year), 14 teaching assistantships (averaging $4,606 per year); scholarships/grants, unspecified assistantships, and bursaries also available. Financial award application deadline: 6/1. *Faculty research:* Food chemistry, food microbiology, food processing, preservation and utilization. Total annual research expenditures: $3.7 million. *Unit head:* Dr. P. Purslow, Chair, 519-824-4120 Ext. 52099, Fax: 519-824-6631, E-mail: ppurslow@uoguelph.ca. *Application contact:* Dr. Y. Kakuda, Graduate Coordinator, 519-824-4120 Ext. 52260, Fax: 519-824-6631, E-mail: ykakuda@uoguelph.ca.

University of Guelph, Graduate Program Services, Program in Food Safety and Quality Assurance, Guelph, ON N1G 2W1, Canada. Offers M Sc. Part-time programs available. *Faculty:* 22 full-time (3 women). *Students:* 28 full-time (15 women), 17 part-time (14 women). 30 applicants, 27% accepted, 7 enrolled. In 2005, 17 degrees awarded. *Degree requirements:* For master's, major project. *Entrance requirements:* For master's, minimum B average in last 2 years of honors B Sc degree. Additional exam requirements/recommendations for international students: Required—TOEFL (minimum score 550 paper-based; 213 computer-based). *Application deadline:* For fall admission, 3/1 for domestic students, 3/1 for international students. For winter admission, 6/1 for domestic students; for spring admission, 11/1 for domestic students. Application fee: $75. Electronic applications accepted. *Financial support:* In 2005–06, 23 students received support, including 2 research assistantships (averaging $15,000 per year), 2 teaching assistantships (averaging $4,606 per year); scholarships/grants and bursaries also available. Financial award application deadline: 6/1. *Faculty research:* Food microbiology, food chemistry, food engineering, food processing, veterinary microbiology. *Unit head:* Dr. M. W. Griffiths, Graduate Coordinator, 519-824-4120 Ext. 52269, Fax: 519-824-6631, E-mail: mgriffit@uoguelph.ca. *Application contact:* Judy A. Campbell, Graduate Student Secretary, 519-824-4120 Ext. 56983, Fax: 519-824-6631, E-mail: jacampbe@uoguelph.ca.

Food Science and Technology

University of Hawaii at Manoa, Graduate Division, College of Tropical Agriculture and Human Resources, Department of Human Nutrition, Food and Animal Sciences, Program in Food Science, Honolulu, HI 96822. Offers MS. *Faculty:* 16 full-time (2 women), 1 part-time/adjunct (0 women). *Students:* 5 full-time (3 women), 2 part-time; includes 5 minority (all Asian Americans or Pacific Islanders), 2 international. Average age 27. 7 applicants, 43% accepted, 2 enrolled. *Entrance requirements:* For master's, GRE General Test. *Application deadline:* For fall admission, 2/1 for domestic students, 2/1 for international students; for spring admission, 9/1 for domestic students, 9/1 for international students. Application fee: $50. *Expenses:* Tuition, state resident: full-time $8,400; part-time $200 per credit hour. Tuition, nonresident: full-time $11,088; part-time $462 per credit hour. Tuition and fees vary according to program. *Faculty research:* Biochemistry of natural products, sensory evaluation, food processing, food chemistry, food safety. *Unit head:* Wayne Iwaoka, Graduate Chair, 808-956-6456, Fax: 808-956-3894, E-mail: iwaoka@hawaii.edu. *Application contact:* Dr. Michael Dunn, Graduate Chairperson, 808-956-8356, Fax: 808-956-4024, E-mail: mdunn@hawaii.edu.

University of Idaho, College of Graduate Studies, College of Agricultural and Life Sciences, Department of Food Science and Toxicology, Moscow, ID 83844-2282. Offers food science (MS, PhD). *Students:* 9 full-time (3 women), 5 part-time (2 women); includes 1 minority (Asian American or Pacific Islander), 7 international. Average age 27. In 2005, 7 degrees awarded. *Entrance requirements:* For master's, minimum GPA of 2.8. *Application deadline:* For fall admission, 8/1 for domestic students; for spring admission, 12/15 for domestic students. Application fee: $55 ($60 for international students). *Expenses:* Tuition, nonresident: full-time $8,770; part-time $130 per credit. Required fees: $4,508; $217 per credit. *Financial support:* Research assistantships, teaching assistantships available. Financial award application deadline: 2/15. *Unit head:* Denise M. Smith, Head, 208-885-9234.

University of Illinois at Urbana–Champaign, Graduate College, College of Agricultural, Consumer and Environmental Sciences, Department of Food Science and Human Nutrition, Champaign, IL 61820. Offers MS, PhD. *Faculty:* 26 full-time (10 women), 1 (woman) part-time/adjunct. *Students:* 44 full-time (34 women), 14 part-time (9 women); includes 7 minority (1 African American, 5 Asian Americans or Pacific Islanders, 1 Hispanic American), 24 international. 89 applicants, 19% accepted, 14 enrolled. In 2005, 14 master's, 4 doctorates awarded. *Degree requirements:* For doctorate, one foreign language, thesis/dissertation. *Entrance requirements:* For master's, minimum GPA of 3.0. *Application deadline:* For fall admission, 5/15 for domestic students. Applications are processed on a rolling basis. Application fee: $50 ($60 for international students). Electronic applications accepted. *Financial support:* In 2005–06, 7 fellowships, 36 research assistantships, 11 teaching assistantships were awarded; tuition waivers (full and partial) also available. Financial award application deadline: 2/15. Total annual research expenditures: $4.4 million. *Unit head:* Faye Dong, Head, 217-244-4498, Fax: 217-265-0925, E-mail: fayedong@uiuc.edu. *Application contact:* Terri Cummings, Director of Student Services, 217-244-4405, Fax: 217-265-0925, E-mail: tcumming@uiuc.edu.

University of Maine, Graduate School, College of Natural Sciences, Forestry, and Agriculture, Department of Food Science and Human Nutrition, Orono, ME 04469. Offers food and nutritional sciences (PhD); food science and human nutrition (MS). Part-time programs available. *Faculty:* 4 full-time (1 woman). *Students:* 23 full-time (20 women), 7 part-time (6 women); includes 2 minority (1 African American, 1 Asian American or Pacific Islander), 4 international. Average age 29. 22 applicants, 41% accepted, 5 enrolled. In 2005, 9 master's, 1 doctorate awarded. *Degree requirements:* For master's and doctorate, thesis/dissertation. *Entrance requirements:* For master's, GRE General Test, minimum GPA of 3.0; for doctorate, GRE General Test. Additional exam requirements/recommendations for international students: Required—TOEFL. *Application deadline:* For fall admission, 2/1 for domestic students. Applications are processed on a rolling basis. Application fee: $50. Electronic applications accepted. *Financial support:* In 2005–06, 9 research assistantships with tuition reimbursements (averaging $13,500 per year), 4 teaching assistantships with tuition reimbursements (averaging $12,000 per year) were awarded; scholarships/grants and tuition waivers (full and partial) also available. Financial award application deadline:3/1. *Faculty research:* Product development of fruit and vegetables, lipid oxidation in fish and meat, analytical methods development, metabolism of potato glycoalkaloids, seafood quality. *Unit head:* Dr. Rodney Bushway, Chair, 207-581-1626, Fax: 207-581-1636. *Application contact:* Scott G. Delcourt, Associate Dean of the Graduate School, 207-581-3219, Fax: 207-581-3232, E-mail: graduate@maine.edu.

University of Manitoba, Faculty of Graduate Studies, Faculty of Agriculture, Department of Food Science, Winnipeg, MB R3T 2N2, Canada. Offers M Sc. *Degree requirements:* For master's, thesis.

University of Maryland, College Park, Graduate Studies, College of Agriculture and Natural Resources, Department of Nutrition and Food Science, Program in Food Science, College Park, MD 20742. Offers MS, PhD. *Students:* 14 full-time (11 women), 1 (woman) part-time, 10 international. 21 applicants, 10% accepted, 2 enrolled. In 2005, 4 master's, 1 doctorate awarded. *Degree requirements:* For master's, research-based thesis or equivalent paper; for doctorate, thesis/dissertation, comprehensive exam. *Entrance requirements:* For master's, GRE General Test, minimum GPA of 3.0, professional experience, 3 letters of recommendation; for doctorate, GRE General Test, minimum GPA of 3.0. Additional exam requirements/recommendations for international students: Required—TOEFL. *Application deadline:* For fall admission, 1/10 for domestic students, 1/10 for international students; for spring admission, 9/10 for domestic students, 6/1 for international students. Applications are processed on a rolling basis. Application fee: $60. Electronic applications accepted. *Financial support:* In 2005–06, 2 fellowships (averaging $6,831 per year) were awarded; research assistantships, teaching assistantships Financial award applicants required to submit FAFSA. *Faculty research:* Food chemistry, engineering, microbiology, and processing technology; quality assurance; membrane separations, rheology and texture measurement. *Unit head:* Dr. Theophanes Solomos, Director, 301-405-4348, Fax: 301-314-9308, E-mail: solomost@umd.edu. *Application contact:* Dean of Graduate School, 301-405-4190, Fax: 301-314-9305.

University of Maryland Eastern Shore, Graduate Programs, Department of Agriculture, Program in Food and Agricultural Sciences, Princess Anne, MD 21853-1299. Offers MS. *Faculty:* 13 full-time (3 women). *Students:* 8 full-time (4 women), 14 part-time (10 women); includes 2 minority (both African Americans), 10 international. Average age 30. 15 applicants, 73% accepted, 1 enrolled. In 2005, 8 degrees awarded. *Degree requirements:* For master's, thesis or alternative, oral exams, comprehensive exam. *Entrance requirements:* For master's, GRE General Test, minimum GPA of 3.0. Additional exam requirements/recommendations for international students: Required—TOEFL (minimum score 550 paper-based; 213 computer-based). *Application deadline:* For fall admission, 4/15 priority date for domestic students, 4/15 priority date for international students; for spring admission, 10/30 priority date for domestic students, 10/30 priority date for international students. Applications are processed on a rolling basis. Application fee: $30. Electronic applications accepted. *Expenses:* Tuition, area resident: Part-time $216. Tuition, nonresident: part-time $392. Required fees: $40. *Financial support:* In 2005–06, 9 students received support, including 9 research assistantships with full tuition reimbursements available (averaging $12,488 per year); scholarships/grants and unspecified assistantships also available. Financial award application deadline: 3/1; financial award applicants required to submit FAFSA. *Faculty research:* Poultry and swine nutrition and management, soybean specialty products, farm management practices, agriculture technology. Total annual research expenditures: $2.5 million.

University of Maryland Eastern Shore, Graduate Programs, Department of Agriculture, Program in Food Science and Technology, Princess Anne, MD 21853-1299. Offers PhD. *Faculty:* 3 full-time (2 women). *Students:* 4 full-time (2 women), 10 part-time (6 women); includes 4 minority (all African Americans), 2 international. Average age 28. 6 applicants, 33% accepted, 0 enrolled. In 2005, 2 degrees awarded. *Degree requirements:* For doctorate, thesis/dissertation, comprehensive exam. *Entrance requirements:* For doctorate, minimum GPA of 3.0, strong background in food science and related fields, intended dissertation research. Additional exam requirements/recommendations for international students: Required—TOEFL (minimum score 550 paper-based; 213 computer-based). *Application deadline:* For fall admission, 4/15 priority date for domestic students, 4/15 priority date for international students; for spring admission, 10/30 priority date for domestic students, 10/30 priority date for international students. Applications are processed on a rolling basis. Application fee: $30. Electronic applications accepted. *Expenses:* Tuition, area resident: Part-time $216. Tuition, nonresident: part-time $392. Required fees: $40. *Financial support:* In 2005–06, 8 students received support, including 8 research assistantships with full tuition reimbursements available (averaging $15,000 per year); unspecified assistantships also available. Financial award application deadline: 3/1. *Unit head:* Dr. Jurgen Schwarz, 410-657-7963, Fax: 410-651-6207, E-mail: jgschwarz@mail.umes.edu.

University of Massachusetts Amherst, Graduate School, College of Natural Resources and the Environment, Department of Food Science, Amherst, MA 01003. Offers MS, PhD. Part-time programs available. *Faculty:* 13 full-time (3 women). *Students:* 18 full-time (11 women), 16 part-time (9 women), 20 international. Average age 29. 51 applicants, 22% accepted, 5 enrolled. In 2005, 5 master's, 4 doctorates awarded. Terminal master's awarded for partial completion of doctoral program. *Degree requirements:* For master's, thesis or alternative; for doctorate, thesis/dissertation. *Entrance requirements:* For master's and doctorate, GRE General Test. Additional exam requirements/recommendations for international students: Required—TOEFL (minimum score 530 paper-based; 197 computer-based). *Application deadline:* For fall admission, 2/1 priority date for domestic students, 2/1 priority date for international students; for spring admission, 10/1 for domestic students, 10/1 for international students. Applications are processed on a rolling basis. Application fee: $40 ($65 for international students). Electronic applications accepted. *Expenses:* Tuition, state resident: part-time $110 per credit. Tuition, nonresident: part-time $414 per credit. Required fees: $2,824 per term. One-time fee: $250 part-time. Full-time tuition and fees vary according to course load, campus/location, program and reciprocity agreements. *Financial support:* In 2005–06, research assistantships with full tuition reimbursements (averaging $8,770 per year), teaching assistantships with full tuition reimbursements (averaging $5,717 per year) were awarded; fellowships with full tuition reimbursements, career-related internships or fieldwork, Federal Work-Study, scholarships/grants, traineeships, and unspecified assistantships also available. Support available to part-time students. Financial award application deadline:2/1. *Unit head:* Dr. Fergus Clydesdale, Head, 413-545-2277, Fax: 413-545-1262, E-mail: fergc@foodsci.umass.edu.

University of Minnesota, Twin Cities Campus, Graduate School, College of Agricultural, Food, and Environmental Sciences and College of Human Ecology, Program in Food Science, Minneapolis, MN 55455-0213. Offers MS, PhD. Part-time programs available. *Faculty:* 17 full-time (5 women), 5 part-time/adjunct (3 women). *Students:* 21 full-time (11 women), 25 part-time (16 women); includes 1 minority (Hispanic American), 15 international. Average age 31. 56 applicants, 25% accepted, 10 enrolled. In 2005, 7 master's, 5 doctorates awarded. Terminal master's awarded for partial completion of doctoral program. *Degree requirements:* For master's, thesis (for some programs); for doctorate, thesis/dissertation. *Entrance requirements:* For master's and doctorate, GRE General Test, previous course work in general chemistry, organic chemistry, calculus, and physics. Additional exam requirements/recommendations for international students: Required—TOEFL (minimum score 550 paper-based; 213 computer-based). *Application deadline:* For fall and spring admission, 6/15. Applications are processed on a rolling basis. Application fee: $55 ($75 for international students). Electronic applications accepted. *Expenses:* Tuition, state resident: full-time $8,748; part-time $729 per credit. Tuition, nonresident: full-time $15,848; part-time $1,321 per credit. Full-time tuition and fees vary according to class time, course load, program and reciprocity agreements. *Financial support:* In 2005–06, 20 students received support, including 2 fellowships with full tuition reimbursements available (averaging $16,500 per year), 16 research assistantships with full tuition reimbursements available (averaging $16,500 per year), 2 teaching assistantships with full tuition reimbursements available (averaging $16,500 per year); career-related internships or fieldwork, Federal Work-Study, institutionally sponsored loans, and scholarships/grants also available. Support available to part-time students. Financial award applicants required to submit FAFSA. *Faculty research:* Food chemistry, food microbiology, food technology, grain science, dairy science, food safety. Total annual research expenditures: $2.3 million. *Unit head:* Dr. Daniel O. O'Sullivan, Director of Graduate Studies, 612-624-5335, Fax: 612-625-5272, E-mail: dosulliv@umn.edu. *Application contact:* Susan K. Viker, Assistant Coordinator, 612-624-6753, Fax: 612-625-5272, E-mail: sviker@umn.edu.

University of Missouri–Columbia, Graduate School, College of Agriculture, Food and Natural Resources, Department of Food and Hospitality Systems, Columbia, MO 65211. Offers food science (MS, PhD); foods and food systems management (MS); human nutrition (MS). *Faculty:* 7 full-time (4 women). *Students:* 7 full-time (5 women), 14 part-time (6 women); includes 1 minority (Hispanic American), 13 international. In 2005, 6 master's, 2 doctorates awarded. Terminal master's awarded for partial completion of doctoral program. *Degree requirements:* For doctorate, thesis/dissertation. *Entrance requirements:* For master's and doctorate, GRE General Test, minimum GPA of 3.0. *Application deadline:* For fall admission, 4/1 for domestic students. Applications are processed on a rolling basis. Application fee: $45 ($60 for international students). *Financial support:* Research assistantships, teaching assistantships, institutionally sponsored loans available. *Unit head:* Dr. Andrew D. Clarke, Director of Graduate Studies, 573-882-2610, E-mail: clakrea@missouri.edu.

University of Nebraska–Lincoln, Graduate College, College of Agricultural Sciences and Natural Resources, Department of Food Science and Technology, Lincoln, NE 68588. Offers MS, PhD. *Degree requirements:* For master's, thesis optional; for doctorate, thesis/dissertation, comprehensive exam. *Entrance requirements:* For master's and doctorate, GRE General Test. Additional exam requirements/recommendations for international students: Required—TOEFL (minimum score 505 paper-based; 213 computer-based). Electronic applications accepted. *Faculty research:* Food chemistry, microbiology, processing, engineering, and biotechnology.

University of Puerto Rico, Mayagüez Campus, Graduate Studies, College of Agricultural Sciences, Department of Science and Food Technology, Mayagüez, PR 00681-9000. Offers MS. *Faculty:* 16. *Students:* 8 full-time (7 women), 22 part-time (14 women); includes 27 minority (all Hispanic Americans), 3 international. 22 applicants, 55% accepted, 5 enrolled. *Degree requirements:* For master's, thesis, comprehensive exam. *Entrance requirements:* For master's, minimum GPA of 2.5. *Application deadline:* For fall admission, 2/28 for domestic students; for spring admission, 9/15 for domestic students. Applications are processed on a rolling basis. Application fee: $20. *Expenses:* Tuition, state resident: full-time $900; part-time $100 per credit. International tuition: $4,655 full-time. Part-time tuition and fees vary according to course level and course load. *Financial support:* In 2005–06, 21 students received support, including 6 fellowships (averaging $1,200 per year), 4 research assistantships (averaging $1,500 per year), 11 teaching assistantships (averaging $987 per year); Federal Work-Study and institutionally sponsored loans also available. *Faculty research:* Food microbiology, food science, seafood technology, food engineering and packaging, fermentation. Total annual research expenditures: $23,749. *Unit head:* Dr. Edna Negrón, Coordinator, 787-265-5410.

University of Rhode Island, Graduate School, College of the Environment and Life Sciences, Department of Nutrition and Food Sciences, Program in Food Science and Nutrition, Kingston, RI 02881. Offers food science (MS, PhD); nutrition (MS, PhD). In 2005, 2 degrees awarded. *Entrance requirements:* For master's and doctorate, GRE General Test. Additional exam requirements/recommendations for international students: Required—TOEFL. *Application deadline:* For fall admission, 4/15 for domestic students. Applications are processed on a rolling basis. Application fee: $35. *Expenses:* Tuition, state resident: full-time $5,522; part-time $307 per credit. Tuition, nonresident: full-time $15,992; part-time $888 per credit. Required fees: $1,786; $73 per credit. One-time fee: $80 part-time. *Unit head:* Prof. Geoffrey Greene, Graduate Coordinator, 401-874-4028, E-mail: gwg@uri.edu.

University of Saskatchewan, College of Graduate Studies and Research, College of Agriculture, Department of Applied Microbiology and Food Science, Saskatoon, SK S7N 5A2, Canada. Offers M Ag, M Sc, PhD. *Degree requirements:* For master's and doctorate, thesis/dissertation, registration. *Entrance requirements:* Additional exam requirements/ recommendations for international students: Required—TOEFL.

University of Southern Mississippi, Graduate School, College of Health, Center for Nutrition and Food Systems, Hattiesburg, MS 39406-0001. Offers MS, PhD.

The University of Tennessee, Graduate School, College of Agricultural Sciences and Natural Resources, Department of Food Science and Technology, Knoxville, TN 37996. Offers food science and technology (MS, PhD), including food chemistry (PhD), food microbiology (PhD), food processing (PhD), sensory evaluation of foods (PhD). Part-time programs available. *Degree requirements:* For master's, thesis or alternative; for doctorate, thesis/dissertation. *Entrance requirements:* For master's and doctorate, GRE General Test, minimum GPA of 2.7. Additional exam requirements/recommendations for international students: Required—TOEFL. Electronic applications accepted.

The University of Tennessee at Martin, Graduate Programs, College of Agriculture and Applied Sciences, Department of Family and Consumer Sciences, Martin, TN 38238-1000. Offers dietetics (MSFCS); general family and consumer sciences (MSFCS). Part-time programs available. *Faculty:* 6. *Students:* 19 (all women) 19 applicants, 58% accepted, 6 enrolled. In 2005, 6 degrees awarded. *Degree requirements:* For master's, thesis optional. *Entrance requirements:* For master's, GRE General Test, minimum GPA of 2.5. Additional exam requirements/recommendations for international students: Required—TOEFL (minimum score 525 paper-based; 197 computer-based). *Application deadline:* For fall admission, 8/1 priority date for domestic students, 8/1 priority date for international students; for spring admission, 1/1 for domestic students, 1/1 for international students. Applications are processed on a rolling basis. Application fee: $25 ($50 for international students). Electronic applications accepted. *Expenses:* Tuition, state resident: full-time $5,200; part-time $291 per hour. Tuition, nonresident: full-time $9,000; part-time $794 per hour. International tuition: $14,200 full-time. *Financial support:* In 2005–06, 6 students received support. Scholarships/grants, tuition waivers (partial), and unspecified assistantships available. Financial award application deadline: 3/1. *Faculty research:* Children with developmental disabilities, regional food product development and marketing, parent education. *Unit head:* Dr. Lisa LeBleu, Coordinator, 731-881-7116, E-mail: llebleu@utm.edu. *Application contact:* Linda S. Arant, Student Service Specialist, 731-881-7012, Fax: 731-881-7499, E-mail: larant@utm.edu.

University of Wisconsin–Madison, Graduate School, College of Agricultural and Life Sciences, Department of Food Science, Madison, WI 53706-1380. Offers MS, PhD. Part-time programs available. *Faculty:* 14 full-time (2 women). *Students:* 45 full-time (27 women); includes 28 minority (25 Asian Americans or Pacific Islanders, 3 Hispanic Americans). Average age 28. 98 applicants, 5% accepted, 5 enrolled. In 2005, 2 master's, 3 doctorates awarded. *Median time to degree:* Of those who began their doctoral program in fall 1997, 90% received their degree in 8 years or less. *Degree requirements:* For master's and doctorate, thesis/ dissertation. *Entrance requirements:* For master's and doctorate, GRE General Test. Additional exam requirements/recommendations for international students: Required—TOEFL. *Application deadline:* For fall admission, 6/15 priority date for domestic students, 6/15 priority date for international students. For winter admission, 11/15 for domestic students; for spring admission, 5/5 for domestic students. Applications are processed on a rolling basis. Application fee: $45. Electronic applications accepted. *Financial support:* In 2005–06, 37 students received support, including 1 fellowship with full tuition reimbursement available (averaging $17,940 per year), 30 research assistantships with full tuition reimbursements available (averaging $16,350 per year), 3 teaching assistantships with full tuition reimbursements available (averaging $10,476 per year); scholarships/grants also available. Financial award application deadline: 6/ 15. *Faculty research:* Food chemistry, food engineering, food microbiology, food processing. Total annual research expenditures: $2.1 million. *Unit head:* Dr. William L. Wendorff, Chair, 608-263-2015, Fax: 608-262-6872, E-mail: wlwendor@facstaff.wisc.edu. *Application contact:* James T. Spartz, Graduate Program Coordinator, 608-262-3046, Fax: 608-262-6872, E-mail: jspartz@wisc.edu.

University of Wisconsin–Stout, Graduate School, College of Human Development, Program in Food and Nutritional Sciences, Menomonie, WI 54751. Offers MS. Part-time programs available. *Faculty:* 33 full-time (18 women). *Students:* 17 full-time (14 women), 4 part-time (3 women); includes 2 minority (both African Americans), 5 international. Average age 30. 26 applicants, 77% accepted, 9 enrolled. In 2005, 1 degree awarded. *Degree requirements:* For master's, thesis, 40 credits, minimum of 20 credits, must be 700 level or higher. *Entrance requirements:* For master's, minimum GPA of 3.0. Additional exam requirements/ recommendations for international students: Required—TOEFL (minimum score 500 paper-based; 173 computer-based). *Application deadline:* Applications are processed on a rolling basis. Application fee: $45. Electronic applications accepted. *Expenses:* Tuition, state resident: part-time $301 per credit hour. Tuition, nonresident: part-time $532 per credit hour. *Financial support:* In 2005–06, 4 research assistantships with partial tuition reimbursements (averaging $6,358 per year) were awarded; teaching assistantships, Federal Work-Study, scholarships/ grants, health care benefits, tuition waivers (full and partial), and unspecified assistantships also available. Support available to part-time students. Financial award application deadline: 4/1; financial award applicants required to submit FAFSA. *Faculty research:* Nutritional biochemistry, nutritional aspects of degenerative diseases and carcinogenesis, community nutrition programs impacting chronic disease and malnutrition, innovations in nutrition education. *Unit head:* Dr. Carol Seaborn, Director, 715-232-2216, E-mail: seaborne@uwstout.edu. *Application contact:* Anne E. Johnson, Graduate Student Evaluator, 715-232-1322, Fax: 715-232-2413, E-mail: johnsona@uwstout.edu.

University of Wyoming, Graduate School, College of Agriculture, Department of Animal Sciences, Program in Food Science and Human Nutrition, Laramie, WY 82070. Offers MS. *Faculty:* 8 full-time (2 women). *Students:* 3 full-time (all women), 5 part-time (3 women), 2 international. Average age 33. 4 applicants, 50% accepted, 0 enrolled. *Degree requirements:* For master's, thesis. *Entrance requirements:* For master's, GRE General Test, minimum GPA of 3.0. Additional exam requirements/recommendations for international students: Required—TOEFL (minimum score 525 paper-based). *Application deadline:* For fall admission, 2/1 priority date for domestic students, 2/1 priority date for international students; for spring admission, 9/1 priority date for domestic students, 9/1 priority date for international students. Applications are processed on a rolling basis. Application fee: $50. Electronic applications accepted. *Expenses:* Tuition, state resident: full-time $3,720; part-time $155 per credit hour. Tuition, nonresident: full-time $10,704; part-time $446 per credit hour. Required fees: $666; $162 per semester. Tuition and fees vary according to course load and program. *Financial support:* In 2005–06, 1 student received support, including research assistantships with tuition reimbursements available (averaging $7,000 per year); career-related internships or fieldwork, Federal Work-Study, institutionally sponsored loans, scholarships/grants, and unspecified assistantships also available. Financial award application deadline: 3/1. *Faculty research:* Protein and lipid metabolism, food microbiology, food safety, meat science. *Unit head:* Dr. Warrie J. Means, Professor, 307-766-3404, Fax: 307-766-2355, E-mail: dcrule@uwyo.edu. *Application contact:* Office Assistant, Senior, 307-766-2224, Fax: 307-766-2355, E-mail: animalscience@uwyo.edu.

Utah State University, School of Graduate Studies, College of Agriculture, Department of Nutrition and Food Sciences, Logan, UT 84322. Offers dietetic administration (MDA); food microbiology and safety (MFMS); nutrition and food sciences (MS, PhD); nutrition science (MS, PhD), including molecular biology. Postbaccalaureate distance learning degree programs offered. *Faculty:* 13 full-time (6 women), 1 (woman) part-time/adjunct. *Students:* 55 full-time (26 women), 13 part-time (12 women); includes 4 minority (all African Americans), 15 international. Average age 27. 14 applicants, 57% accepted, 4 enrolled. In 2005, 8 master's, 3 doctorates awarded. *Degree requirements:* For master's, thesis, BS core competency courses BS core competency courses; for doctorate, thesis/dissertation, teaching experience, inst 7920 and BS core competency courses, comprehensive exam, registration. *Entrance requirements:* For master's, GRE General Test, minimum GPA of 3.0, course work in chemistry; for doctorate, GRE General Test, minimum GPA of 3.2, course work in chemistry, MS or manuscript in referred journal. Additional exam requirements/recommendations for international students: Required—TOEFL (minimum score 550 paper-based). *Application deadline:* For fall admission, 6/15 priority date for domestic students, 6/15 priority date for international students; for spring admission, 10/15 priority date for domestic students, 10/15 priority date for international students. Applications are processed on a rolling basis. Application fee: $50 ($60 for international students). Electronic applications accepted. *Financial support:* In 2005–06, 19 students received support, including fellowships with partial tuition reimbursements available (averaging $14,484 per year), 22 research assistantships with partial tuition reimbursements available (averaging $12,600 per year), 3 teaching assistantships with partial tuition reimbursements available (averaging $6,000 per year); Federal Work-Study, institutionally sponsored loans, scholarships/grants, tuition waivers (full and partial), unspecified assistantships, and fellowships, international students get support from their countries also available. Financial award application deadline: 3/1. *Faculty research:* Mineral balance, meat microbiology and nitrate interactions, milk ultrafiltration, lactic culture, milk coagulation. Total annual research expenditures: $319,280. *Unit head:* Dr. Charles C. Carpenter, Head, 435-797-2126, Fax: 435-797-2379, E-mail: chuck@cc.usu.edu. *Application contact:* Pam Zetterquist, Staff Assistant II, 435-797-4041, Fax: 435-797-2379, E-mail: pzett@cc.usu.edu.

Virginia Polytechnic Institute and State University, Graduate School, College of Agriculture and Life Sciences, Department of Food Science and Technology, Blacksburg, VA 24061. Offers MS, PhD. *Faculty:* 10 full-time (3 women). *Students:* 17 full-time (12 women), 4 part-time (3 women); includes 1 minority (Asian American or Pacific Islander), 2 international. Average age 27. 42 applicants, 31% accepted, 5 enrolled. In 2005, 5 master's, 5 doctorates awarded. *Degree requirements:* For doctorate, thesis/dissertation optional. *Entrance requirements:* For master's and doctorate, GRE General Test. Additional exam requirements/ recommendations for international students: Required—TOEFL (minimum score 570 paper-based; 230 computer-based). *Application deadline:* Applications are processed on a rolling basis. Application fee: $45. Electronic applications accepted. *Expenses:* Tuition, state resident: full-time $6,558; part-time $364 per credit. Tuition, nonresident: full-time $11,296; part-time $628 per credit. Required fees: $1,419; $468 per credit. $234 per term. *Financial support:* In 2005–06, 16 research assistantships with full tuition reimbursements (averaging $13,823 per year), 4 teaching assistantships with full tuition reimbursements (averaging $13,552 per year) were awarded; career-related internships or fieldwork, Federal Work-Study, scholarships/ grants, and unspecified assistantships also available. Financial award application deadline:4/ 1. *Faculty research:* Food microbiology, food chemistry, food processing, engineering, muscle foods. *Unit head:* Dr. Susan Sumner, Head, 540-231-6806, Fax: 540-231-9293, E-mail: sumners@vt.edu. *Application contact:* Jennifer Carr, Professor, 540-231-6806, Fax: 540-231-9293, E-mail: jjc@vt.edu.

Washington State University, Graduate School, College of Agricultural, Human, and Natural Resource Sciences, Department of Food Science and Human Nutrition, Program in Food Science, Pullman, WA 99164. Offers MS, PhD. *Faculty:* 10. *Students:* 21 full-time (13 women), 2 part-time (1 woman); includes 3 minority (2 Asian Americans or Pacific Islanders, 1 Hispanic American), 11 international. Average age 29. 63 applicants, 13% accepted, 7 enrolled. In 2005, 1 master's, 5 doctorates awarded. *Degree requirements:* For master's and doctorate, thesis/dissertation, oral exam, written exam. *Entrance requirements:* For master's and doctorate, GRE General Test, minimum GPA of 3.0; resumé; 3 letters of recommendation, 1 from major advisor. Additional exam requirements/recommendations for international students: Required—TOEFL (minimum score 550 paper-based; 213 computer-based). *Application deadline:* For fall admission, 2/1 priority date for domestic students, 2/1 priority date for international students; for spring admission, 8/1 priority date for domestic students, 8/1 priority date for international students. Applications are processed on a rolling basis. Application fee: $35. Electronic applications accepted. *Expenses:* Tuition, state resident: full-time $6,295; part-time $336 per credit. Tuition, nonresident: full-time $15,949; part-time $819 per credit. Required fees: $933. Part-time tuition and fees vary according to campus/location and program. *Financial support:* In 2005–06, 21 students received support, including 2 fellowships with tuition reimbursements available (averaging $4,250 per year), 14 research assistantships with full and partial tuition reimbursements available (averaging $12,503 per year), 5 teaching assistantships with full and partial tuition reimbursements available (averaging $11,923 per year); career-related internships or fieldwork, Federal Work-Study, institutionally sponsored loans, scholarships/grants, tuition waivers (partial), and unspecified assistantships also available. Financial award application deadline: 2/1; financial award applicants required to submit FAFSA. *Faculty research:* Sports anemia, lipid chemistry, malfunction of edible oils and fats, Malolactic fermentation, wine microbiology. *Unit head:* Dr. Barry Swanson, Unit Head, 509-335-3793, E-mail: swansonb@wsu.edu. *Application contact:* Jodi L. Anderson, Academic Coordinator, 509-335-4763, Fax: 509-335-4815, E-mail: fshn@wsu.edu.

Wayne State University, Graduate School, College of Liberal Arts and Sciences, Department of Nutrition and Food Science, Detroit, MI 48202. Offers MA, MS, PhD. *Faculty:* 14 full-time (13 women). *Students:* 33 full-time (29 women), 5 part-time (4 women); includes 7 minority (3 African Americans, 4 Asian Americans or Pacific Islanders), 18 international. Average age 31. 26 applicants, 62% accepted, 8 enrolled. In 2005, 7 master's, 1 doctorate awarded. Terminal master's awarded for partial completion of doctoral program. *Degree requirements:* For master's, thesis (for some programs); for doctorate, thesis/dissertation. *Entrance requirements:* For master's and doctorate, GRE General Test, minimum GPA of 3.0. Additional exam requirements/recommendations for international students: Required—TOEFL (minimum score 550 paper-based; 213 computer-based); Recommended—TWE (minimum score 6). *Application deadline:* For fall admission, 7/1 for domestic students, 6/1 for international students. Applications are processed on a rolling basis. Application fee: $30 ($50 for international students). Electronic applications accepted. *Expenses:* Tuition, state resident: part-time $338 per credit hour. Tuition, nonresident: part-time $746 per credit hour. Required fees: $24 per credit hour. Full-time tuition and fees vary according to program. *Financial support:* In 2005–06, 10 students received support, including 1 fellowship with tuition reimbursement available, 7 teaching assistantships (averaging $14,997 per year); research assistantships, career-related internships or fieldwork and Federal Work-Study also available. Financial award application deadline: 4/1. *Faculty research:* Nutrition, cancer and gene expression, food microbiology and food safety, lipids, lipoprotein and cholesterol metabolism, obesity and diabetes, metabolomics. Total annual research expenditures: $934,572. *Unit head:* Dr. K. L. Catherine Jen, Chair, 313-577-2500, E-mail: cjen@sun.science.wayne.edu. *Application contact:* Pramod Khosla, Assistant Professor, 313-577-9055, Fax: 313-577-8616, E-mail: pkhosla@sun.science.wayne.edu.

West Virginia University, Davis College of Agriculture, Forestry and Consumer Sciences, Division of Animal and Veterinary Sciences, Program in Agricultural Sciences, Morgantown, WV 26506. Offers animal and food sciences (PhD); plant and soil sciences (PhD). *Degree requirements:* For doctorate, thesis/dissertation, oral and written exams. *Entrance requirements:* Additional exam requirements/recommendations for international students: Required—TOEFL. Application fee: $45. *Expenses:* Tuition, state resident: full-time $4,582; part-time $258 per credit hour. Tuition, nonresident: full-time $1,382; part-time $741 per credit hour. *Financial support:* Research assistantships with tuition reimbursements, teaching assistantships with tuition reimbursements, Federal Work-Study, institutionally sponsored loans, and tuition waivers (full and partial) available. Financial award application deadline: 2/1; financial award applicants required to submit FAFSA. *Faculty research:* Ruminant nutrition, metabolism, forage utilization, physiology, reproduction. *Application contact:* Dr. Hillar Klandorf, Professor, 304-293-2631 Ext. 4436, Fax: 304-293-3676, E-mail: hillar.klandorf@mail.wvu.edu.

West Virginia University, Davis College of Agriculture, Forestry and Consumer Sciences, Division of Animal and Veterinary Sciences, Program in Animal and Veterinary Sciences, Morgantown, WV 26506. Offers breeding (MS); food sciences (MS); nutrition (MS); physiology (MS); production management (MS); reproduction (MS). Part-time programs available. *Degree requirements:* For master's, thesis, oral and written exams. *Entrance requirements:* For master's, minimum GPA of 2.5. Additional exam requirements/recommendations for international students:

Food Science and Technology

West Virginia University (continued)
Required—TOEFL. Application fee: $45. *Expenses:* Tuition, state resident: full-time $4,582; part-time $258 per credit hour. Tuition, nonresident: full-time $1,382; part-time $741 per credit hour. *Financial support:* Research assistantships, teaching assistantships, Federal Work-Study, institutionally sponsored loans, and tuition waivers (full and partial) available. Financial award application deadline: 2/1; financial award applicants required to submit FAFSA. *Faculty research:* Animal nutrition, reproductive physiology, food science. *Unit head:* Dr. Hillar Klandorf, Coordinator, 304-293-4372 Ext. 2050, Fax: 304-293-3676, E-mail: hillar.klandorf@mail.wvu.edu.

Horticulture

Auburn University, Graduate School, College of Agriculture, Department of Horticulture, Auburn University, AL 36849. Offers M Ag, MS, PhD. Part-time programs available. *Faculty:* 18 full-time (2 women). *Students:* 15 full-time (7 women), 9 part-time (3 women); includes 3 minority (2 African Americans, 1 American Indian/Alaska Native), 8 international. 15 applicants, 73% accepted, 8 enrolled. In 2005, 5 master's, 1 doctorate awarded. *Degree requirements:* For master's, thesis (for some programs); for doctorate, thesis/dissertation. *Entrance requirements:* For master's and doctorate, GRE General Test. *Application deadline:* For fall admission, 7/7 for domestic students; for spring admission, 11/24 for domestic students. Applications are processed on a rolling basis. Application fee: $25 ($50 for international students). Electronic applications accepted. *Financial support:* Research assistantships, teaching assistantships, Federal Work-Study available. Support available to part-time students. Financial award application deadline: 3/15. *Faculty research:* Environmental regulators, water quality, weed control, growth regulators, plasticulture. *Unit head:* Dr. J. David Williams, Chair, 334-844-4862. *Application contact:* Dr. Stephen L. McFarland, Acting Dean of the Graduate School, 334-844-4700.

Colorado State University, Graduate School, College of Agricultural Sciences, Department of Horticulture and Landscape Architecture, Fort Collins, CO 80523-0015. Offers floriculture (M Agr, MS, PhD); horticultural food crops (M Agr, MS, PhD); nursery and landscape management (M Agr, MS, PhD); plant genetics (MS, PhD); plant physiology (MS, PhD); turf management (M Agr, MS, PhD). Part-time programs available. *Faculty:* 18 full-time (3 women). *Students:* 11 full-time (5 women), 9 part-time (3 women), 4 international. Average age 35. 13 applicants, 69% accepted, 4 enrolled. In 2005, 1 master's, 2 doctorates awarded. *Median time to degree:* Of those who began their doctoral program in fall 1997, 90% received their degree in 8 years or less. *Degree requirements:* For master's, thesis or alternative; for doctorate, thesis/dissertation. *Entrance requirements:* For master's and doctorate, GRE General Test, minimum GPA of 3.0. Additional exam requirements/recommendations for international students: Required—TOEFL (minimum score 550 paper-based; 213 computer-based). *Application deadline:* For fall and winter admission, 2/1 for domestic students, 1/1 for international students; for spring admission, 9/1 priority date for domestic students, 7/1 priority date for international students. Applications are processed on a rolling basis. Application fee: $50. Electronic applications accepted. *Expenses:* Tuition, state resident: full-time $3,690; part-time $205 per credit. Tuition, nonresident: full-time $14,958; part-time $831 per credit. Required fees: $1,061. *Financial support:* Fellowships with full tuition reimbursements, research assistantships with partial tuition reimbursements, teaching assistantships with partial tuition reimbursements, career-related internships or fieldwork, Federal Work-Study, institutionally sponsored loans, and traineeships available. Financial award application deadline: 8/16. *Faculty research:* Antioxidants in food crops, environmental physiology, water conservation, tissue culture, rhizosphere biology, cancer prevention through dietary intervention. Total annual research expenditures: $600,000. *Unit head:* Dr. Stephen J. Wallner, Head, 970-491-7018, Fax: 970-491-7745. *Application contact:* Gretchen L. DeWeese, Administrative Assistant III, 970-491-7018, Fax: 970-491-7745, E-mail: gretchen.deweese@colostate.edu.

Cornell University, Graduate School, Graduate Fields of Agriculture and Life Sciences, Field of Floriculture and Ornamental Horticulture, Ithaca, NY 14853-0001. Offers controlled environment agriculture (MPS, PhD); controlled environment horticulture (MS); greenhouse crops (MPS, MS, PhD); horticultural business management (MPS, MS, PhD); horticultural physiology (MPS, MS, PhD); landscape horticulture (MPS, MS, PhD); nursery crops (MPS, MS, PhD); nutrition of horticultural crops (MPS, MS, PhD); plant propagation (MPS, MS, PhD); public garden management (MPS, MS, PhD); restoration ecology (MPS, MS, PhD); taxonomy of ornamental plants (MPS, MS, PhD); turfgrass science (MPS, MS, PhD); urban horticulture (MPS, MS, PhD); weed science (MPS, MS, PhD). *Faculty:* 18 full-time (5 women). *Students:* 26 full-time (12 women); includes 1 minority (Hispanic American), 5 international. 29 applicants, 21% accepted, 6 enrolled. In 2005, 2 master's, 3 doctorates awarded. *Degree requirements:* For master's, thesis (MS); for doctorate, thesis/dissertation, comprehensive exam. *Entrance requirements:* For master's and doctorate, GRE General Test, 3 letters of recommendation. Additional exam requirements/recommendations for international students: Required—TOEFL (minimum score 550 paper-based; 213 computer-based). *Application deadline:* For fall admission, 1/15 for domestic students; for spring admission, 8/15 for domestic students. Application fee: $60. Electronic applications accepted. *Financial support:* In 2005–06, 21 students received support, including 4 fellowships with full tuition reimbursements available, 15 research assistantships with full tuition reimbursements available, 2 teaching assistantships with full tuition reimbursements available; institutionally sponsored loans, scholarships/grants, health care benefits, tuition waivers (full and partial), and unspecified assistantships also available. Financial award applicants required to submit FAFSA. *Faculty research:* Plant selection/plant materials, greenhouse management, greenhouse crop production, urban landscape management, turfgrass management. *Unit head:* Director of Graduate Studies, 607-255-4568, Fax: 607-255-0599. *Application contact:* Graduate Field Assistant, 607-255-4568, Fax: 607-255-0599, E-mail: hortgrad@cornell.edu.

Cornell University, Graduate School, Graduate Fields of Agriculture and Life Sciences, Field of Pomology, Ithaca, NY 14853-0001. Offers MPS, MS, PhD. *Faculty:* 22 full-time (4 women). *Students:* 18 full-time (8 women); includes 1 minority (African American), 11 international. 4 applicants, 100% accepted, 4 enrolled. In 2005, 2 master's, 2 doctorates awarded. *Degree requirements:* For master's, thesis (MS), project paper (MPS); for doctorate, thesis/dissertation, comprehensive exam. *Entrance requirements:* For master's and doctorate, GRE General Test, interview (recommended), 3 letters of recommendation. Additional exam requirements/recommendations for international students: Required—TOEFL (minimum score 550 paper-based; 213 computer-based). *Application deadline:* For fall admission, 1/15 for domestic students; for spring admission, 8/15 for domestic students. Application fee: $60. Electronic applications accepted. *Financial support:* In 2005–06, 15 students received support, including 4 fellowships with full tuition reimbursements available, 11 research assistantships with full tuition reimbursements available; teaching assistantships with full tuition reimbursements available, institutionally sponsored loans, scholarships/grants, health care benefits, tuition waivers (full and partial), and unspecified assistantships also available. Financial award applicants required to submit FAFSA. *Faculty research:* Fruit breeding and biotechnology, fruit crop physiology, orchard management, orchard ecology and IPM, post-harvest physiology. *Unit head:* Director of Graduate Studies, 607-255-4568, Fax: 607-255-0599. *Application contact:* Graduate Field Assistant, 607-255-4568, Fax: 607-255-0599, E-mail: hortgrad@cornell.edu.

Cornell University, Graduate School, Graduate Fields of Agriculture and Life Sciences, Field of Vegetable Crops, Ithaca, NY 14853-0001. Offers MPS, MS, PhD. *Faculty:* 21 full-time (8 women). *Students:* 9 full-time (3 women), 4 international. In 2005, 1 master's, 1 doctorate awarded. *Degree requirements:* For master's, thesis (MS), project paper (MPS); for doctorate, thesis/dissertation, teaching experience. *Entrance requirements:* For master's and doctorate, GRE General Test (recommended), 3 letters of recommendation. Additional exam requirements/recommendations for international students: Required—TOEFL (minimum score 550 paper-based; 213 computer-based). *Application deadline:* Applications are processed on a rolling basis. Application fee: $60. Electronic applications accepted. *Financial support:* In 2005–06, 9 students received support, including 2 fellowships with full tuition reimbursements available, 7 research assistantships with full tuition reimbursements available; teaching assistantships with full tuition reimbursements available, institutionally sponsored loans, health care benefits, tuition waivers (full and partial), and unspecified assistantships also available. Financial award applicants required to submit FAFSA. *Faculty research:* Vegetable nutrition and physiology, post-harvest physiology and storage, application of new technologies, sustainable vegetable production, weed management and IPM. *Unit head:* Director of Graduate Studies, 607-255-4568. *Application contact:* Graduate Field Assistant, 607-255-4568, E-mail: hortgrad@cornell.edu.

Iowa State University of Science and Technology, Graduate College, College of Agriculture, Department of Horticulture, Ames, IA 50011. Offers MS, PhD. *Faculty:* 16 full-time. *Students:* 15 full-time (6 women), 2 part-time; includes 1 minority (Hispanic American), 4 international. 12 applicants, 25% accepted, 2 enrolled. In 2005, 5 master's, 2 doctorates awarded. *Degree requirements:* For master's and doctorate, thesis/dissertation. *Entrance requirements:* For master's and doctorate, GRE General Test. Additional exam requirements/recommendations for international students: Required—TOEFL (paper score 530; computer score 197) or IELTS (score 6). *Application deadline:* For fall admission, 1/1 priority date for domestic students, 1/1 priority date for international students; for spring admission, 9/1 priority date for domestic students, 9/1 priority date for international students. Applications are processed on a rolling basis. Application fee: $30 ($70 for international students). Electronic applications accepted. *Expenses:* Tuition, state resident: full-time $6,410. Tuition, nonresident: full-time $16,422. Tuition and fees vary according to program. *Financial support:* In 2005–06, 12 research assistantships with partial tuition reimbursements (averaging $13,805 per year), 2 teaching assistantships with full and partial tuition reimbursements (averaging $14,570 per year) were awarded; fellowships, scholarships/grants, health care benefits, and unspecified assistantships also available. *Unit head:* Dr. Jeffrey Iles, Head, 515-294-3718, E-mail: hortgrad@iastate.edu. *Application contact:* William R. Graves, Information Contact, 515-294-2751, E-mail: hortgrade@iastate.edu.

Kansas State University, Graduate School, College of Agriculture, Department of Horticulture, Forestry and Recreation Resources, Manhattan, KS 66506. Offers horticulture (MS, PhD). *Faculty:* 19 full-time (5 women), 2 part-time/adjunct (1 woman). *Students:* 21 full-time (12 women), 2 part-time (1 woman), 11 international. Average age 25. 23 applicants, 22% accepted, 5 enrolled. In 2005, 7 master's, 1 doctorate awarded. *Degree requirements:* For master's, thesis, oral exam; for doctorate, thesis/dissertation, preliminary exams. *Entrance requirements:* For master's and doctorate, GRE General Test. Additional exam requirements/recommendations for international students: Required—TOEFL (minimum score 550 paper-based; 213 computer-based). *Application deadline:* For fall admission, 2/1 for domestic students, 2/1 for international students; for spring admission, 12/1 for domestic students, 8/15 for international students. Applications are processed on a rolling basis. Application fee: $30 ($55 for international students). Electronic applications accepted. *Expenses:* Tuition, state resident: full-time $5,160; part-time $215. Tuition, nonresident: full-time $12,816; part-time $534. Required fees: $564. *Financial support:* In 2005–06, 11 research assistantships (averaging $12,525 per year), 2 teaching assistantships (averaging $16,335 per year) were awarded; career-related internships or fieldwork, Federal Work-Study, institutionally sponsored loans, and scholarships/grants also available. Support available to part-time students. Financial award application deadline: 3/1; financial award applicants required to submit FAFSA. *Faculty research:* Environmental stress, turfgrass management, vegetable alternate crop production, floriculture-pest and nutrition, horticulture therapy. Total annual research expenditures:$417,566. *Unit head:* Thomas Warner, Head, 785-532-1413, Fax: 785-532-6949, E-mail: twarner@oz.oznet.ksu.edu. *Application contact:* Dr. Channa Rajashekar, Director, 785-532-1427, Fax: 785-532-6949, E-mail: crajashe@oznet.ksu.edu.

Louisiana State University and Agricultural and Mechanical College, Graduate School, College of Agriculture, Department of Horticulture, Baton Rouge, LA 70803. Offers MS, PhD. Part-time programs available. *Faculty:* 10 full-time (0 women). *Students:* 9 full-time (4 women), 7 part-time (4 women); includes 1 African American, 6 international. Average age 36. 9 applicants, 33% accepted, 6 enrolled. In 2005, 4 master's, 2 doctorates awarded. Terminal master's awarded for partial completion of doctoral program. *Degree requirements:* For master's, thesis (for some programs); for doctorate, thesis/dissertation. *Entrance requirements:* For master's and doctorate, GRE General Test, minimum GPA of 3.0. Additional exam requirements/recommendations for international students: Required—TOEFL (minimum score 550 paper-based; 213 computer-based). *Application deadline:* For fall admission, 7/1 priority date for domestic students, 5/15 priority date for international students. Applications are processed on a rolling basis. Application fee: $25. Electronic applications accepted. *Financial support:* In 2005–06, 11 students received support, including 7 research assistantships with partial tuition reimbursements available (averaging $14,500 per year), 2 teaching assistantships with partial tuition reimbursements available (averaging $13,000 per year); fellowships, Federal Work-Study, tuition waivers (full and partial), and unspecified assistantships also available. Financial award application deadline: 4/15; financial award applicants required to submit FAFSA. *Faculty research:* Plant breeding, stress physiology, postharvest physiology, biotechnology. Total annual research expenditures: $895. *Unit head:* Dr. David G. Himelrick, Head, 225-578-2158, Fax: 225-578-1068, E-mail: dhimelrick@agctr.lsu.edu. *Application contact:* Dr. Paul Wilson, Graduate Coordinator, 225-578-1025, Fax: 225-578-1068, E-mail: pwilson@agctr.lsu.edu.

Michigan State University, The Graduate School, College of Agriculture and Natural Resources, Department of Horticulture, East Lansing, MI 48824. Offers horticulture (MS, PhD); plant breeding and genetics-horticulture (MS, PhD). *Faculty:* 32 full-time (7 women). *Students:* 23 full-time (12 women), 1 part-time; includes 1 minority (Hispanic American), 10 international. Average age 30. 14 applicants, 7% accepted. In 2005, 8 master's, 5 doctorates awarded. *Degree requirements:* For master's, oral examination for those completing thesis, teaching/extension experience, thesis optional; for doctorate, thesis/dissertation, oral defense of dissertation, teaching/ extension experience, comprehensive exam. *Entrance requirements:* For master's, GRE General Test, minimum GPA of 3.0, 3 letters of recommendation; for doctorate, GRE General Test, minimum GPA of 3.0, 3 letters of recommendation, MS degree. Additional exam requirements/recommendations for international students: Required—TOEFL (minimum score 580 paper-based; 237 computer-based). *Application deadline:* For fall admission, 12/27 for domestic students. Applications are processed on a rolling basis. Application fee: $50. Electronic applications accepted. *Expenses:* Tuition, state resident: part-time $330 per credit hour. Tuition, nonresident: part-time $685 per credit hour. Tuition and fees vary according to program. *Financial support:* In 2005–06, 3 fellowships with tuition reimbursements (averaging $5,500 per year), 20 research assistantships with tuition reimbursements (averaging $12,557 per year) were awarded; scholarships/grants and unspecified assistantships also available. *Faculty research:* Natural products chemistry, plant breeding and genetics, floriculture and perennial fruit production, post-harvest crop and plant physiology. Total annual research expenditures: $4.4 million. *Unit head:* Dr. Ronald L. Perry, Chairperson, 517-355-5191 Ext. 1361, Fax: 517-353-0890, E-mail: perryr@msu.edu. *Application contact:* Kristi Lowrie, Graduate Program Secretary, 517-355-5191 Ext. 1324, Fax: 517-353-0890, E-mail: hrtgrad@msu.edu.

Michigan State University, The Graduate School, College of Agriculture and Natural Resources, Program in Plant Breeding and Genetics, East Lansing, MI 48824. Offers plant breeding and genetics (MS, PhD), including crop and soil sciences, forestry, horticulture. *Students:* 17 full-time (12 women), 2 part-time (1 woman), 10 international. Average age 29. 9 applicants, 0% accepted. In 2005, 2 master's, 6 doctorates awarded. *Degree requirements:* For master's, thesis; for doctorate, thesis/dissertation, oral examination, comprehensive exam. *Entrance requirements:* For master's and doctorate, GRE General Test, minimum GPA of 3.0 in last 3 years of college, 3 letters of recommendation. Additional exam requirements/recommendations for international students: Required—TOEFL (minimum score 550 paper-based; 213 computer-based). *Application deadline:* For fall admission, 12/27 for domestic students. Application fee: $50. Electronic applications accepted. *Expenses:* Tuition, state resident: part-time $330 per credit hour. Tuition, nonresident: part-time $685 per credit hour. Tuition and fees vary according to program. *Financial support:* In 2005–06, 1 fellowship with tuition reimbursement (averaging $11,500 per year), 12 research assistantships with tuition reimbursements (averaging $13,218 per year), 2 teaching assistantships with tuition reimbursements (averaging $12,470 per year) were awarded; scholarships/grants and unspecified assistantships also available. *Faculty research:* Applied plant breeding and genetics; disease, insect and herbicide resistances; gene isolation and genomics; abiotic stress factors; molecular mapping. *Unit head:* Dr. Jim F. Hancock, Director, 517-355-5191 Ext. 1387, Fax: 517-432-3490, E-mail: hancock@msu.edu. *Application contact:* Program Information, 517-355-5191 Ext. 1324, Fax: 517-432-3490, E-mail: pbg@msu.edu.

New Mexico State University, Graduate School, College of Agriculture and Home Economics, Department of Plant and Environmental Sciences, Las Cruces, NM 88003-8001. Offers general agronomy (MS, PhD); horticulture (MS). Part-time programs available. *Faculty:* 24 full-time (4 women), 7 part-time/adjunct (2 women). *Students:* 35 full-time (14 women), 15 part-time (6 women); includes 9 minority (1 American Indian/Alaska Native, 8 Hispanic Americans), 19 international. Average age 34. 17 applicants, 53% accepted, 4 enrolled. In 2005, 6 master's, 2 doctorates awarded. *Median time to degree:* Of those who began their doctoral program in fall 1997, 90% received their degree in 8 years or less. *Degree requirements:* For master's, thesis; for doctorate, one foreign language, thesis/dissertation. *Entrance requirements:* For master's, minimum GPA of 3.0; for doctorate, minimum GPA of 3.3. *Application deadline:* For fall admission, 7/1 for domestic students; for spring admission, 11/1 priority date for domestic students. Applications are processed on a rolling basis. Application fee: $30 ($50 for international students). Electronic applications accepted. *Expenses:* Tuition, state resident: full-time $3,156; part-time $175 per credit. Tuition, nonresident: full-time $12,510; part-time $565 per credit. Required fees: $1,050. *Financial support:* In 2005–06, 5 fellowships, 18 research assistantships, 12 teaching assistantships were awarded; career-related internships or fieldwork, Federal Work-Study, health care benefits, and unspecified assistantships also available. Support available to part-time students. Financial award application deadline: 3/1. *Faculty research:* Plant breeding and genetics, molecular biology, plant physiology, soil science and environmental remediation, urban horticulture. *Unit head:* Dr. Greg L Mullins, Head, 505-646-3406, Fax: 505-646-6041, E-mail: gmullins@nmsu.edu. *Application contact:* Esther Ramirez, Information Contact, 505-646-3406, Fax: 505-646-6041, E-mail: esramire@nmsu.edu.

North Carolina State University, Graduate School, College of Agriculture and Life Sciences, Department of Horticultural Science, Raleigh, NC 27695. Offers MS, PhD. Terminal master's awarded for partial completion of doctoral program. *Degree requirements:* For master's, thesis (for some programs); for doctorate, thesis/dissertation. *Entrance requirements:* For master's and doctorate, GRE General Test, bachelor's degree in agriculture or biology, minimum GPA of 3.0. Electronic applications accepted. *Faculty research:* Plant physiology, breeding and genetics, tissue culture, herbicide physiology, propagation.

Nova Scotia Agricultural College, Research and Graduate Studies, Truro, NS B2N 5E3, Canada. Offers agriculture (M Sc), including air quality, animal behavior, animal molecular genetics, animal nutrition, animal technology, aquaculture, botany, crop management, crop physiology, ecology, environmental microbiology, food science, horticulture, nutrient management, pest management, physiology, plant biotechnology, plant pathology, soil chemistry, soil fertility, waste management and composting, water quality. Part-time programs available. *Faculty:* 43 full-time (7 women), 21 part-time/adjunct (1 woman). *Students:* 50 full-time (25 women), 15 part-time (11 women); includes 7 minority (3 African Americans, 3 Asian Americans or Pacific Islanders, 1 Hispanic American). Average age 25. In 2005, 23 degrees awarded. *Degree requirements:* For master's, thesis, registration. *Entrance requirements:* For master's, honors B Sc, minimum GPA of 3.0. Additional exam requirements/recommendations for international students: Required—TOEFL (minimum score 580 paper-based; 237 computer-based), Michigan English Language Assessment Battery, IELT, Can Test, CAEL. *Application deadline:* For fall admission, 6/1 for domestic students, 4/1 for international students. For winter admission, 10/31 for domestic students; for spring admission, 2/28 for domestic students. Applications are processed on a rolling basis. Application fee: $70. *Expenses:* Tuition, state resident: part-time $2,328 per year. Tuition, nonresident: full-time $6,984; part-time $7,968 per year. International tuition: $12,624 full-time. Required fees: $481; $46 per course. Tuition and fees vary according to program and student level. *Financial support:* In 2005–06, 48 students received support, including 7 fellowships (averaging $15,000 per year), 10 research assistantships (averaging $15,000 per year), 15 teaching assistantships (averaging $900 per year); career-related internships or fieldwork, scholarships/grants, and unspecified assistantships also available. *Faculty research:* Bio-product development, organic agriculture, nutrient management, air and water quality, agricultural biotechnology. Total annual research expenditures: $4.7 million. *Unit head:* Jill L. Rogers, Manager, 902-893-6360, Fax: 902-893-3430, E-mail: jrogers@nsac.ca. *Application contact:* Marie Law, Administrative Assistant, 902-893-6502, Fax: 902-893-3430, E-mail: mlaw@nsac.ca.

The Ohio State University, Graduate School, College of Food, Agricultural, and Environmental Sciences, Department of Horticulture and Crop Science, Columbus, OH 43210. Offers MS, PhD. *Degree requirements:* For master's, thesis optional; for doctorate, thesis/dissertation. *Entrance requirements:* For master's and doctorate, GRE General Test. Additional exam requirements/recommendations for international students: Required—TOEFL (paper 550; computer 213) or IELT (7) or Michigan English Language Assessment Battery (86). Electronic applications accepted.

Oklahoma State University, College of Agricultural Science and Natural Resources, Department of Horticulture and Landscape Architecture, Stillwater, OK 74078. Offers crop science (PhD); environmental science (PhD); food science (PhD); horticulture (M Ag, MS); plant science (PhD). *Faculty:* 21 full-time (1 woman), 4 part-time/adjunct (0 women). *Students:* 1 full-time (0 women), 6 part-time (2 women); includes 1 minority (African American), 4 international. Average age 31. 7 applicants, 14% accepted. In 2005, 3 master's awarded. *Degree requirements:* For master's, thesis or alternative. *Entrance requirements:* For master's and doctorate, GRE. Additional exam requirements/recommendations for international students: Required—TOEFL. *Application deadline:* For fall admission, 6/1 priority date for domestic students, 3/1 priority date for international students. Applications are processed on a rolling basis. Application fee: $40 ($75 for international students). Electronic applications accepted. *Expenses:* Tuition, state resident: full-time $4,253; part-time $139 per credit hour. Tuition, nonresident: full-time $12,569; part-time $485 per credit hour. Required fees: $43 per credit hour. One-time fee: $20 part-time. Tuition and fees vary according to course load and program. *Financial support:* In 2005–06, 4 research assistantships (averaging $13,817 per year) were awarded; teaching assistantships, career-related internships or fieldwork, Federal Work-Study, scholarships/grants, health care benefits, tuition waivers (partial), and unspecified assistantships also available. Support available to part-time students. Financial award application deadline: 3/1. *Faculty research:* Stress and postharvest physiology; water utilization and runoff; IPM systems and nursery, turf, floriculture, vegetable, net and fruit produces and natural resources, food extraction, and processing; public garden management. *Unit head:* Dr. Dale Maronek, Head, 405-744-5414, Fax: 405-744-9709, E-mail: maronek@okstate.edu.

Oregon State University, Graduate School, College of Agricultural Sciences, Department of Horticulture, Corvallis, OR 97331. Offers M Ag, MAIS, MS, PhD. *Faculty:* 14 full-time (3 women), 3 part-time/adjunct (0 women). *Students:* 22 full-time (10 women), 4 part-time; includes 1 minority (Asian American or Pacific Islander), 13 international. Average age 30. In 2005, 4 master's, 1 doctorate awarded. *Degree requirements:* For master's, thesis (for some programs); for doctorate, thesis/dissertation. *Entrance requirements:* For master's and doctorate, GRE General Test, minimum GPA of 3.0 in last 90 hours. Additional exam requirements/recommendations for international students: Required—TOEFL. *Application deadline:* For fall admission, 3/1 for domestic students. Applications are processed on a rolling basis. Application fee: $50. *Expenses:* Tuition, area resident: Part-time $301 per credit. Tuition, state resident: full-time $8,139; part-time $501 per credit. Tuition, nonresident: full-time $14,376; part-time $532 per credit. Required fees: $1,266. *Financial support:* Research assistantships, teaching assistantships, career-related internships or fieldwork, Federal Work-Study, and institutionally sponsored loans available. Support available to part-time students. Financial award application deadline: 2/1. *Unit head:* Anita Azarenko, Head, 541-737-5457, Fax: 541-737-3479. *Application contact:* Machteld C. Mok, Graduate Coordinator, 541-737-5456, E-mail: mokm@bcc.orst.edu.

The Pennsylvania State University University Park Campus, Graduate School, College of Agricultural Sciences, Department of Horticulture, State College, University Park, PA 16802-1503. Offers M Agr, MS, PhD. *Students:* 17 full-time (6 women), 5 international. *Entrance requirements:* For master's and doctorate, GRE General Test. *Expenses:* Tuition, state resident: full-time $12,518; part-time $522 per credit. Tuition, nonresident: full-time $23,004; part-time $959 per credit. Required fees: $484. Tuition and fees vary according to course load, campus/location and program. *Unit head:* Dr. Richard P. Marini, Head, 814-865-2571, Fax: 814-863-6139, E-mail: rpm12@psu.edu. *Application contact:* Dr. Richard P. Marini, Head, 814-865-2571, Fax: 814-863-6139, E-mail: rpm12@psu.edu.

Purdue University, Graduate School, College of Agriculture, Department of Horticulture and Landscape Architecture, West Lafayette, IN 47907. Offers horticulture (M Agr, MS, PhD). Part-time programs available. *Faculty:* 28 full-time (3 women), 1 part-time/adjunct (0 women). *Students:* 31 full-time (14 women), 3 part-time; includes 1 African American, 1 Hispanic American, 21 international. Average age 30. 24 applicants, 38% accepted, 7 enrolled. In 2005, 1 master's, 6 doctorates awarded. Terminal master's awarded for partial completion of doctoral program. *Degree requirements:* For doctorate, thesis/dissertation. *Entrance requirements:* For master's and doctorate, GRE. Additional exam requirements/recommendations for international students: Required—TOEFL. *Application deadline:* For fall admission, 5/1 for domestic students, 5/1 for international students; for spring admission, 10/1 for domestic students, 10/1 for international students. Applications are processed on a rolling basis. Application fee: $55. Electronic applications accepted. *Financial support:* In 2005–06, 7 fellowships, 24 research assistantships with tuition reimbursements (averaging $15,000 per year), 5 teaching assistantships with tuition reimbursements (averaging $15,000 per year) were awarded. Support available to part-time students. Financial award applicants required to submit FAFSA. *Faculty research:* Plant physiology, plant genetics and breeding, plant molecular biology and cell physiology, environmental and production horticulture. *Unit head:* Dr. E. N. Ashworth, Head, 765-494-1306, Fax: 765-494-0391, E-mail: ashworth@hort.purdue.edu. *Application contact:* Dr. Colleen K Martin, Graduate Committee Chair, 765-494-1306, Fax: 765-494-0391, E-mail: martinck@purdue.edu.

Rutgers, The State University of New Jersey, New Brunswick/Piscataway, Graduate School, Program in Plant Biology, New Brunswick, NJ 08901-1281. Offers horticulture (MS, PhD); molecular biology and biochemistry (MS, PhD); pathology (MS, PhD); plant ecology (MS, PhD); plant genetics (PhD); plant physiology (MS, PhD); production and management (MS); structure and plant groups (MS, PhD). Part-time programs available. *Faculty:* 65 full-time, 1 part-time/adjunct (0 women). *Students:* 33 full-time (14 women), 13 part-time (6 women); includes 3 minority (1 African American, 2 Asian Americans or Pacific Islanders), 10 international. Average age 31. 56 applicants, 23% accepted, 10 enrolled. In 2005, 1 master's, 4 doctorates awarded. Terminal master's awarded for partial completion of doctoral program. *Median time to degree:* Of those who began their doctoral program in fall 1997, 90% received their degree in 8 years or less. *Degree requirements:* For master's, thesis or alternative, comprehensive exam; for doctorate, thesis/dissertation, comprehensive exam. *Entrance requirements:* For master's and doctorate, GRE General Test, GRE Subject Test (recommended). Additional exam requirements/recommendations for international students: Required—TOEFL (minimum score 600 paper-based; 250 computer-based). *Application deadline:* For fall admission, 4/1 for domestic students, 4/1 for international students. Application fee: $50. Electronic applications accepted. *Expenses:* Tuition, state resident: full-time $10,440; part-time $435 per credit. Tuition, nonresident: full-time $15,520; part-time $647 per credit. Required fees: $129 per credit. Tuition and fees vary according to program. *Financial support:* In 2005–06, 42 students received support, including 9 fellowships with full tuition reimbursements available (averaging $24,000 per year), 22 research assistantships with full tuition reimbursements available (averaging $16,500 per year), 10 teaching assistantships with full tuition reimbursements available (averaging $16,988 per year) Financial award application deadline: 1/15; financial award applicants required to submit FAFSA. *Faculty research:* Molecular biology and biochemistry of plants, plant development and genomics, plant protection, plant improvement, plant management of horticultural and field crops. Total annual research expenditures: $10 million. *Unit head:* Dr. Thomas Leustek, Director, 732-932-8165 Ext. 326, Fax: 732-932-9377, E-mail: leustek@aesop.rutgers.edu. *Application contact:* Barbara Mulder, Program Associate, 732-932-9375 Ext. 358, Fax: 732-932-9377, E-mail: plantbio@aesop.rutgers.edu.

Southern Illinois University Carbondale, Graduate School, College of Agriculture, Department of Plant, Soil, and General Agriculture, Carbondale, IL 62901-4701. Offers horticultural science (MS); plant and soil science (MS). *Faculty:* 20 full-time (1 woman). *Students:* 13 full-time (7 women), 33 part-time (9 women); includes 5 minority (all African Americans), 3 international. 16 applicants, 50% accepted, 2 enrolled. In 2005, 13 master's awarded. *Degree requirements:* For master's, thesis. *Entrance requirements:* For master's, minimum GPA of 2.7. Additional exam requirements/recommendations for international students: Required—TOEFL. *Application deadline:* Applications are processed on a rolling basis. Application fee: $0. *Financial support:* In 2005–06, 22 students received support, including 15 research assistantships with full tuition reimbursements available, 6 teaching assistantships with full tuition reimbursements available; fellowships with full tuition reimbursements available, Federal Work-Study, institutionally sponsored loans, and tuition waivers (full) also available. Support available to part-time students. *Faculty research:* Herbicides, fertilizers, agriculture education, landscape design, plant breeding. Total annual research expenditures: $2 million. *Unit head:* Brian Klubek, Interim Chair, 618-453-2496.

Texas A&M University, College of Agriculture and Life Sciences, Department of Horticultural Sciences, College Station, TX 77843. Offers horticulture (PhD); horticulture and floriculture (M Agr, MS). *Faculty:* 12 full-time (2 women). *Students:* 25 full-time (14 women), 6 part-time (2 women); includes 3 minority (2 African Americans, 1 Asian American or Pacific Islander), 12 international. Average age 29. 21 applicants, 52% accepted, 6 enrolled. In 2005, 10 master's, 1 doctorate awarded. Terminal master's awarded for partial completion of doctoral program. *Degree requirements:* For master's, thesis (for some programs), professional internship; for doctorate, thesis/dissertation. *Entrance requirements:* For master's and doctorate, GRE General Test. Additional exam requirements/recommendations for international students: Required—TOEFL. Application fee: $50 ($75 for international students). Electronic applications accepted. *Expenses:* Tuition, state resident: full-time $4,488; part-time $187 per credit hour. Tuition,

Horticulture

Texas A&M University (continued)
nonresident: full-time $11,112; part-time $463 per credit hour. Required fees:$1,974. *Financial support:* In 2005–06, 30 students received support, including fellowships with full tuition reimbursements available (averaging $15,000 per year), research assistantships with partial tuition reimbursements available (averaging $14,000 per year), teaching assistantships with partial tuition reimbursements available (averaging $14,000 per year); career-related internships or fieldwork and tuition waivers (partial) also available. Financial award application deadline: 4/1. *Faculty research:* Plant breeding, molecular biology, plant nutrition, postharvest physiology, plant physiology. *Unit head:* Dr. Tim D. Davis, Head, 979-845-9341, Fax: 979-845-0627, E-mail: t-davis5@tamu.edu. *Application contact:* Dr. Michael A. Arnold, Associate Head for Research, 979-845-1499, Fax: 979-845-0627, E-mail: ma-arnold@tamu.edu.

Texas Tech University, Graduate School, College of Agricultural Sciences and Natural Resources, Department of Plant and Soil Science, Lubbock, TX 79409. Offers agronomy (PhD); crop science (MS); entomology (MS); horticulture (MS); soil science (MS). Part-time programs available. *Faculty:* 12 full-time (2 women), 4 part-time/adjunct (0 women). *Students:* 29 full-time (11 women), 23 part-time (9 women), 7 international. Average age 31. 36 applicants, 56% accepted, 14 enrolled. In 2005, 16 master's, 2 doctorates awarded. *Degree requirements:* For doctorate, thesis/dissertation. *Entrance requirements:* For master's and doctorate, GRE General Test. Additional exam requirements/recommendations for international students: Required—TOEFL (minimum score 550 paper-based; 213 computer-based). *Application deadline:* Applications are processed on a rolling basis. Application fee: $50 ($60 for international students). Electronic applications accepted. *Expenses:* Tuition, state resident: full-time $4,296. Tuition, nonresident: full-time $10,920. Required fees: $1,992. Tuition and fees vary according to program. *Financial support:* In 2005–06, 21 students received support, including 21 research assistantships with partial tuition reimbursements available (averaging $12,726 per year), 2 teaching assistantships with partial tuition reimbursements available (averaging $12,478 per year); Federal Work-Study and institutionally sponsored loans also available. Support available to part-time students. Financial award application deadline: 4/15; financial award applicants required to submit FAFSA. *Faculty research:* Molecular and cellular biology of plant stress, physiology/genetics of crop production in semiarid conditions, agricultural bioterrorism, improvement of native plants. Total annual research expenditures: $2.6 million. *Unit head:* Dr. Dick L. Auld, Chair, 806-742-2837, Fax: 806-742-0775, E-mail: dick.auld@ttu.edu. *Application contact:* Dr. Richard E. Zartman, Graduate Adviser, 806-742-2837, Fax: 806-742-0775, E-mail: richard.zartman@ttu.edu.

Universidad Nacional Pedro Henriquez Urena, Graduate School, Santo Domingo, Dominican Republic. Offers accounting and auditing (M Acct); animal production (M Agr); business administration (MBA, PhD); dentistry (DDS); economics (M Econ); education (PhD); environmental engineering (MEE); horticulutre (M Agr); hospital administration (PhD); humanities (PhD); international relations (MPS); management of natural resources (MNRM); project management (M Man); public administration (MPS); social science (PhD); veterinary medicine (DVM).

University of Arkansas, Graduate School, Dale Bumpers College of Agricultural, Food and Life Sciences, Department of Horticulture, Fayetteville, AR 72701-1201. Offers MS. *Students:* 6 full-time (2 women), 2 part-time (1 woman). 8 applicants, 50% accepted. In 2005, 5 degrees awarded. *Degree requirements:* For master's, thesis. Application fee: $40 ($50 for international students). *Financial support:* In 2005–06, 11 research assistantships were awarded; fellowships, teaching assistantships, career-related internships or fieldwork and Federal Work-Study also available. Support available to part-time students. Financial award application deadline: 4/1; financial award applicants required to submit FAFSA. *Unit head:* Dr. David Hensley, Head, 479-575-2603. *Application contact:* Dr. J. Brad Murphy, Graduate Coordinator, 479-575-2446, E-mail: jbmurph@comp.uark.edu.

University of California, Davis, Graduate Studies, Graduate Group in Horticulture and Agronomy, Davis, CA 95616. Offers MS. *Faculty:* 90 full-time. *Students:* 43 full-time (24 women); includes 5 minority (2 American Indian/Alaska Native, 2 Asian Americans or Pacific Islanders, 1 Hispanic American), 7 international. Average age 32. 39 applicants, 56% accepted, 20 enrolled. In 2005, 12 degrees awarded. *Degree requirements:* For master's, thesis (for some programs), comprehensive exam (for some programs). *Entrance requirements:* For master's, GRE General Test. Additional exam requirements/recommendations for international students: Required—TOEFL (minimum score 550 paper-based; 213 computer-based). *Application deadline:* For fall admission, 4/1 for domestic students, 3/1 for international students. Applications are processed on a rolling basis. Application fee: $60. Electronic applications accepted. *Financial support:* In 2005–06, 37 students received support, including 15 fellowships with full and partial tuition reimbursements available (averaging $6,636 per year), 17 research assistantships with full and partial tuition reimbursements available (averaging $12,174 per year), 1 teaching assistantship with partial tuition reimbursement available (averaging $15,082 per year); career-related internships or fieldwork, Federal Work-Study, institutionally sponsored loans, scholarships/grants, and tuition waivers (full and partial) also available. Financial award application deadline: 1/15; financial award applicants required to submit FAFSA. *Faculty research:* Postharvest physiology, mineral nutrition, crop improvement, plant growth and development. *Unit head:* M. Andrew Walker, Chairperson, 530-752-0902, Fax: 530-752-0382, E-mail: awalker@ucdavis.edu. *Application contact:* Lisa Brown, Graduate Group Secretary, 530-752-7738, Fax: 530-752-1819, E-mail: lfbrown@ucdavis.edu.

University of Delaware, College of Agriculture and Natural Resources, Longwood Graduate Program in Public Horticulture, Newark, DE 19716. Offers MS. *Faculty:* 1 full-time (0 women). *Students:* 15 full-time (8 women), 2 international. Average age 23. 16 applicants, 31% accepted, 5 enrolled. In 2005, 6 master's awarded. *Degree requirements:* For master's, thesis, internship. *Entrance requirements:* For master's, GRE General Test, introductory taxonomy course. Additional exam requirements/recommendations for international students: Required—TOEFL. *Application deadline:* For fall admission, 11/15 for domestic students, 11/15 for international students. Application fee: $60. Electronic applications accepted. *Financial support:* In 2005–06, 10 fellowships with full tuition reimbursements (averaging $19,500 per year) were awarded; career-related internships or fieldwork also available. Financial award application deadline: 3/1. *Faculty research:* Management and development of publicly oriented horticultural institutions. *Unit head:* Dr. Robert E. Lyons, Director, 302-831-2517, Fax: 302-831-3651, E-mail: rlyons@udel.edu. *Application contact:* Gerry Zuka, Senior Secretary, 302-831-2517, Fax: 302-831-3651, E-mail: gerryz@udel.edu.

University of Florida, Graduate School, College of Agricultural and Life Sciences, Department of Environmental Horticulture, Gainesville, FL 32611. Offers anatomy and development (MS, PhD); breeding and genetics (MS, PhD); ecology (MS, PhD); plant biotechnology (MS, PhD); stress physiology (MS, PhD); taxonomy (MS, PhD); tissue culture (MS, PhD). *Faculty:* 13 full-time (2 women). *Students:* 11 applicants, 45% accepted. *Degree requirements:* For master's, thesis optional; for doctorate, thesis/dissertation. *Entrance requirements:* For master's and doctorate, GRE General Test, minimum GPA of 3.0. *Application deadline:* For fall admission, 6/1 for domestic students; for spring admission, 11/1 for domestic students. Applications are processed on a rolling basis. Application fee: $20. Electronic applications accepted. *Expenses:* Tuition, state resident: full-time $6,234. Tuition, nonresident: full-time $21,359. Tuition and fees vary according to program. *Financial support:* In 2005–06, 14 students received support, including 13 research assistantships (averaging $10,711 per year); fellowships, teaching assistantships, unspecified assistantships also available. Financial award application deadline:6/1. *Faculty research:* Production and genetics, landscape horticulture, turf grass, foliage, floriculture. *Application contact:* Dr. Grady L. Miller, Coordinator, 352-392-1831 Ext. 375, Fax: 352-392-3870, E-mail: gmiller@ifas.ufl.edu.

University of Florida, Graduate School, College of Agricultural and Life Sciences, Department of Horticultural Sciences, Gainesville, FL 32611. Offers plant breeding and genetics (MS, PhD); plant production and nutrient management (MS, PhD); postharvest biology (MS, PhD); sustainable/organic practice (MS, PhD); weed science (MS, PhD). *Faculty:* 33 full-time (9 women). *Students:* 71 (32 women); includes 7 minority (1 African American, 2 Asian Americans or Pacific Islanders, 4 Hispanic Americans) 23 international. 51 applicants, 33% accepted. In 2005, 17 master's, 7 doctorates awarded. *Degree requirements:* For master's, variable foreign language requirement, thesis optional; for doctorate, variable foreign language requirement, thesis/dissertation. *Entrance requirements:* For master's and doctorate, GRE General Test, minimum GPA of 3.0. *Application deadline:* For fall admission, 6/1 for domestic students. Applications are processed on a rolling basis. Application fee: $20. Electronic applications accepted. *Expenses:* Tuition, state resident: full-time $6,234. Tuition, nonresident: full-time $21,359. Tuition and fees vary according to program. *Financial support:* In 2005–06, 37 research assistantships (averaging $19,694 per year) were awarded; fellowships, teaching assistantships, institutionally sponsored loans also available. Financial award application deadline:6/1. *Faculty research:* Genetics, plant nutrition, stress physiology, biotechnology, postharvest physiology. *Unit head:* Dr. Daniel J. Cantliffe, Chair, 352-392-1928 Ext. 203, Fax: 352-392-6479, E-mail: djc@ifas.ufl.edu. *Application contact:* Dr. Donald J. Huber, Coordinator, 352-392-1928 Ext. 214, Fax: 352-392-6479, E-mail: djh@ifas.ufl.edu.

University of Georgia, Graduate School, College of Agricultural and Environmental Sciences, Department of Horticulture, Athens, GA 30602. Offers horticulture (MS, PhD); plant protection and pest management (MPPPM). Part-time programs available. *Faculty:* 23 full-time (5 women). *Students:* 13 full-time, 5 part-time, 7 international. 14 applicants, 14% accepted, 2 enrolled. In 2005, 3 master's, 2 doctorates awarded. *Degree requirements:* For master's, thesis (MS); for doctorate, one foreign language, thesis/dissertation. *Entrance requirements:* For master's and doctorate, GRE General Test. *Application deadline:* For fall admission, 7/1 for domestic students; for spring admission, 11/15 for domestic students. Application fee: $50. Electronic applications accepted. *Financial support:* In 2005–06, fellowships with partial tuition reimbursements (averaging $1,338 per year), research assistantships with partial tuition reimbursements (averaging $1,338 per year), teaching assistantships with partial tuition reimbursements (averaging $1,338 per year) were awarded; unspecified assistantships also available. *Unit head:* Dr. Douglas A. Bailey, Head, 706-542-2471, Fax: 706-542-0624, E-mail: dabailey@uga.edu. *Application contact:* Dr. Harry Mills, Graduate Coordinator, 706-542-0794, Fax: 706-542-0624, E-mail: hmills@uga.edu.

University of Guelph, Graduate Program Services, Ontario Agricultural College, Department of Plant Agriculture, Guelph, ON N1G 2W1, Canada. Offers M Sc, PhD. Part-time programs available. *Faculty:* 42 full-time (6 women), 5 part-time/adjunct (2 women). *Students:* 72 full-time (27 women), 5 part-time (2 women); includes 21 minority (1 African American, 18 Asian Americans or Pacific Islanders, 2 Hispanic Americans), 5 international. 33 applicants, 52% accepted, 16 enrolled. In 2005, 15 master's, 4 doctorates awarded. *Median time to degree:* Of those who began their doctoral program in fall 1997, 100% received their degree in 8 years or less. *Degree requirements:* For master's, thesis/dissertation; for doctorate, thesis/dissertation, comprehensive exam. *Entrance requirements:* For master's, minimum B average during previous 2 years of course work; for doctorate, minimum B average. Additional exam requirements/recommendations for international students: Required—TOEFL (minimum score 550 paper-based; 213 computer-based), Michigan English Language Assessment Battery (score 85). *Application deadline:* For fall admission, 4/30 priority date for domestic students, 2/28 priority date for international students. For winter admission, 8/31 for domestic students; for spring admission, 12/24 for domestic students. Applications are processed on a rolling basis. Application fee: $75. Electronic applications accepted. *Financial support:* In 2005–06, 12 students received support, including 60 fellowships (averaging $4,100 Canadian dollars per year), research assistantships (averaging $17,000 Canadian dollars per year), 9 teaching assistantships (averaging $4,134 Canadian dollars per year); scholarships/grants and unspecified assistantships also available. Financial award application deadline: 4/30. *Faculty research:* Plant physiology, biochemistry, taxonomy, morphology, genetics, production, ecology, breeding and biotechnology. Total annual research expenditures: $12 million. *Unit head:* Dr. Rene Van Acker, Chair, 519-824-4120 Ext. 53386, Fax: 519-821-8660. *Application contact:* Dr. Bernard Grodzinski, Graduate Coordinator, 519-842-4120 Ext. 53439, Fax: 519-767-0755, E-mail: bgrodzin@uoguelph.ca.

University of Hawaii at Manoa, Graduate Division, College of Tropical Agriculture and Human Resources, Department of Tropical Plant and Soil Sciences, Honolulu, HI 96822. Offers MS, PhD. *Faculty:* 34 full-time (4 women), 15 part-time/adjunct (4 women). *Students:* 37 full-time (20 women), 7 part-time (5 women); includes 10 minority (7 Asian Americans or Pacific Islanders, 3 Hispanic Americans), 23 international. Average age 31. 16 applicants, 75% accepted, 7 enrolled. In 2005, 2 degrees awarded. *Median time to degree:* Of those who began their doctoral program in fall 1997, 67% received their degree in 8 years or less. *Degree requirements:* For master's, thesis (for some programs); for doctorate, thesis/dissertation. *Entrance requirements:* For doctorate, GRE. *Application deadline:* For fall admission, 3/1 for domestic students, 1/15 for international students; for spring admission, 9/1 for domestic students, 8/1 for international students. Application fee: $50. *Expenses:* Tuition, state resident: full-time $8,400; part-time $200 per credit hour. Tuition, nonresident: full-time $11,088; part-time $462 per credit hour. Tuition and fees vary according to program. *Financial support:* In 2005–06, 25 research assistantships (averaging $16,088 per year), 3 teaching assistantships (averaging $13,658 per year) were awarded; tuition waivers (full and partial) also available. *Faculty research:* Genetics and breeding; physiology, culture, and management; weed science; turfgrass and landscape; sensory evaluation. *Unit head:* Robert E. Paull, Chairman, 808-956-5900, Fax: 808-956-3894, E-mail: paull@hawaii.edu. *Application contact:* Joseph DeFrank, 808-956-5900, Fax: 808-956-3894.

University of Maine, Graduate School, College of Natural Sciences, Forestry, and Agriculture, Department of Plant, Soil, and Environmental Sciences, Program in Horticulture, Orono, ME 04469. Offers MS. *Faculty:* 5 full-time (0 women). *Students:* 1 full-time (0 women), 2 part-time (both women), 2 international. Average age 38. 1 applicant, 0% accepted, 0 enrolled. In 2005, 1 master's awarded. *Entrance requirements:* For master's, GRE General Test. Additional exam requirements/recommendations for international students: Required—TOEFL. *Application deadline:* For fall admission, 2/1 for domestic students. Applications are processed on a rolling basis. Application fee: $50. Electronic applications accepted. *Financial support:* In 2005–06, 2 research assistantships with tuition reimbursements (averaging $12,180 per year) were awarded; teaching assistantships with tuition reimbursements, tuition waivers (full and partial) also available. Financial award application deadline: 3/1. *Unit head:* Dr. Tsutomu Ohno, Coordinator, 207-584-2975. *Application contact:* Scott G. Delcourt, Associate Dean of the Graduate School, 207-581-3219, Fax: 207-581-3232, E-mail: graduate@maine.edu.

University of Manitoba, Faculty of Graduate Studies, Faculty of Agriculture, Department of Plant Science, Winnipeg, MB R3T 2N2, Canada. Offers horticulture (M Sc, PhD). *Degree requirements:* For master's, thesis; for doctorate, one foreign language, thesis/dissertation.

University of Maryland, College Park, Graduate Studies, College of Agriculture and Natural Resources, Department of Natural Resource Sciences and Landscape Architecture, Program in Horticulture, College Park, MD 20742. Offers PhD. *Students:* 3 full-time (2 women), 2 international. In 2005, 1 degree awarded. *Entrance requirements:* For doctorate, GRE General Test. Additional exam requirements/recommendations for international students: Required—TOEFL. *Application deadline:* For fall admission, 2/1 for domestic students, 2/1 for international students; for spring admission, 8/1 for domestic students, 6/1 for international students. Applications are processed on a rolling basis. Application fee: $60. Electronic applications accepted. *Financial support:* Fellowships, research assistantships, teaching assistantships, career-related internships or fieldwork available. Financial award applicants required to submit FAFSA. *Faculty research:* Mineral nutrition, genetics and breeding, chemical growth, histochemistry, postharvest physiology. *Application contact:* Dean of Graduate School, 301-405-4190, Fax: 301-314-9305.

University of Missouri–Columbia, Graduate School, College of Agriculture, Food and Natural Resources, Division of Plant Sciences, Department of Horticulture, Columbia, MO 65211. Offers MS, PhD. *Degree requirements:* For master's, thesis; for doctorate, variable foreign language requirement, thesis/dissertation. *Entrance requirements:* For master's and doctorate, GRE General Test, minimum GPA of 3.0. *Application deadline:* Applications are processed on a rolling basis. Application fee: $45 ($60 for international students). *Financial support:* Research assistantships, teaching assistantships, institutionally sponsored loans available.

University of Nebraska–Lincoln, Graduate College, College of Agricultural Sciences and Natural Resources, Department of Agronomy and Horticulture, Program in Horticulture, Lincoln, NE 68588. Offers MS, PhD. *Degree requirements:* For master's, thesis optional. *Entrance requirements:* For master's, GRE General Test. Additional exam requirements/recommendations for international students: Required—TOEFL (minimum score 600 paper-based; 250 computer-based). Electronic applications accepted. *Faculty research:* Horticultural crops: production, management, cultural, and ecological aspects; tissue and cell culture; plant nutrition and anatomy; postharvest physiology and ecology.

University of Puerto Rico, Mayagüez Campus, Graduate Studies, College of Agricultural Sciences, Department of Horticulture, Mayagüez, PR 00681-9000. Offers MS. Part-time programs available. *Faculty:* 11. *Students:* 7 full-time (4 women), 5 part-time (3 women); all minorities (all Hispanic Americans) 6 applicants, 100% accepted, 3 enrolled. *Degree requirements:* For master's, thesis, comprehensive exam. *Application deadline:* For fall admission, 2/15 for domestic students; for spring admission, 9/15 for domestic students. Applications are processed on a rolling basis. Application fee: $20. *Expenses:* Tuition, state resident: full-time $900; part-time $100 per credit. International tuition: $4,655 full-time. Part-time tuition and fees vary according to course level and course load. *Financial support:* In 2005–06, 8 students received support, including 3 fellowships (averaging $1,200 per year), 2 research assistantships (averaging $1,500 per year), 3 teaching assistantships (averaging $987 per year); Federal Work-Study and institutionally sponsored loans also available. *Faculty research:* Growth regulators, floriculture, starchy crops, coffee and fruit technology. Total annual research expenditures:$3,374. *Unit head:* Dr. María Del C. Librán, Director, 787-265-3852.

University of Vermont, Graduate College, College of Agriculture and Life Sciences, Department of Plant and Soil Science, Burlington, VT 05405. Offers MS, PhD. *Students:* 18 (9 women) 5 international. 25 applicants, 64% accepted, 8 enrolled. In 2005, 7 master's, 1 doctorate awarded. *Degree requirements:* For master's, thesis; for doctorate, one foreign language, thesis/dissertation. *Entrance requirements:* For master's and doctorate, GRE General Test. Additional exam requirements/recommendations for international students: Required—TOEFL (minimum score 550 paper-based; 213 computer-based). *Application deadline:* For fall admission, 2/1 for domestic students. Applications are processed on a rolling basis. Application fee: $40. Electronic applications accepted. *Expenses:* Tuition, area resident: Part-time $410 per credit hour. Tuition, nonresident: part-time $1,034 per credit hour. *Financial support:* Fellowships, research assistantships, teaching assistantships available. Financial award application deadline: 3/1. *Faculty research:* Soil chemistry, plant nutrition. *Unit head:* Dr. Deborah Neher, Chairperson, 802-656-2630. *Application contact:* Dr. M. Starrett, Coordinator, 802-656-2630.

University of Washington, Graduate School, College of Forest Resources, Seattle, WA 98195. Offers forest economics (MS, PhD); forest ecosystem analysis (MS, PhD); forest engineering/forest hydrology (MS, PhD); forest products marketing (MS, PhD); forest soils (MS, PhD); paper science and engineering (MS, PhD); quantitative resource management (MS, PhD); silviculture (MFR); silviculture and forest protection (MS, PhD); social sciences (MS, PhD); urban horticulture (MFR, MS, PhD); wildlife science (MS, PhD). *Degree requirements:* For master's, thesis (for some programs), registration; for doctorate, thesis/dissertation, comprehensive exam (for some programs), registration. *Entrance requirements:* For master's and doctorate, GRE, minimum GPA of 3.0. Additional exam requirements/recommendations for international students: Required—TOEFL. Electronic applications accepted. *Faculty research:* Ecosystem analysis, silviculture and forest protection, paper science and engineering, environmental horticulture and urban forestry, natural resource policy and economics.

University of Wisconsin–Madison, Graduate School, College of Agricultural and Life Sciences, Department of Horticulture, Madison, WI 53706-1380. Offers MS, PhD. Part-time programs available. *Faculty:* 22 full-time (4 women). *Students:* 12 full-time (3 women), 4 part-time; includes 3 minority (all Asian Americans or Pacific Islanders) Average age 34. 24 applicants, 13% accepted, 3 enrolled. In 2005, 2 master's awarded. Terminal master's awarded for partial completion of doctoral program. *Median time to degree:* Of those who began their doctoral program in fall 1997, 100% received their degree in 8 years or less. *Degree requirements:* For master's, thesis (for some programs), comprehensive exam; for doctorate, thesis/dissertation, comprehensive exam. *Entrance requirements:* For master's and doctorate, minimum GPA of 3.0. Additional exam requirements/recommendations for international students: Required—TOEFL (minimum score 580 paper-based; 213 computer-based). *Application deadline:* For fall admission, 1/2 for domestic students, 1/2 for international students; for spring admission, 10/30 for domestic students, 10/30 for international students. Applications are processed on a rolling basis. Application fee: $45. Electronic applications accepted. *Financial support:* In 2005–06, 2 fellowships with tuition reimbursements (averaging $18,720 per year), 9 research assistantships with full tuition reimbursements (averaging $18,120 per year), 2 teaching assistantships with full tuition reimbursements (averaging $24,200 per year) were awarded; career-related internships or fieldwork, Federal Work-Study, and tuition waivers (partial) also available. Financial award application deadline: 1/2. *Faculty research:* Biotechnology, crop breeding/genetics, environmental physiology, crop management, cytogenetics. *Unit head:* Dr. Dennis P. Stimart, Chair, 608-262-8406, Fax: 608-262-4743, E-mail: dstimart@facstaff.wisc.edu. *Application contact:* Dr. Sara Patterson, Assistant Professor, 608-262-1543, Fax: 608-262-4743, E-mail: spatters@wisc.edu.

Virginia Polytechnic Institute and State University, Graduate School, College of Agriculture and Life Sciences, Department of Horticulture, Blacksburg, VA 24061. Offers MS, PhD. *Faculty:* 16 full-time (3 women). *Students:* 10 full-time (4 women), 17 part-time (9 women), 4 international. Average age 37. 12 applicants, 67% accepted, 7 enrolled. In 2005, 5 master's, 3 doctorates awarded. *Entrance requirements:* Additional exam requirements/recommendations for international students: Required—TOEFL (minimum score 550 paper-based; 213 computer-based). *Application deadline:* Applications are processed on a rolling basis. Application fee: $45. Electronic applications accepted. *Expenses:* Tuition, state resident: full-time $6,558; part-time $364 per credit. Tuition, nonresident: full-time $11,296; part-time $628 per credit. Required fees: $1,419; $468 per credit. $234 per term. *Financial support:* In 2005–06, 2 research assistantships with full tuition reimbursements (averaging $13,823 per year), 6 teaching assistantships with full tuition reimbursements (averaging $13,552 per year) were awarded; career-related internships or fieldwork, Federal Work-Study, scholarships/grants, and unspecified assistantships also available. Financial award application deadline:6/1. *Unit head:* Dr. Jerzy Nowak, Head, 540-231-5451, Fax: 540-231-3083, E-mail: jenowak@vt.edu. *Application contact:* Richard Veilleux, Professor, 540-231-5584, Fax: 540-231-3083, E-mail: potato@vt.edu.

Washington State University, Graduate School, College of Agricultural, Human, and Natural Resource Sciences, Department of Horticulture and Landscape Architecture, Pullman, WA 99164. Offers horticulture (MS, PhD); landscape and architecture (MSLA). Part-time programs available. *Faculty:* 17. *Students:* 15 full-time (8 women), 1 part-time; includes 1 minority (American Indian/Alaska Native), 4 international. Average age 32. 35 applicants, 14% accepted, 3 enrolled. In 2005, 5 master's, 5 doctorates awarded. *Degree requirements:* For master's, oral exam, thesis optional; for doctorate, thesis/dissertation, oral exam, written exam. *Entrance requirements:* For master's and doctorate, GRE General Test, GRE Subject Test, minimum GPA of 3.0, 3 letters of recommendation. Additional exam requirements/recommendations for international students: Required—TOEFL (minimum score 550 paper-based). *Application deadline:* For fall admission, 2/1 priority date for domestic students, 3/1 priority date for international students; for spring admission, 9/1 for domestic students, 7/1 for international students. Applications are processed on a rolling basis. Application fee: $35. Electronic applications accepted. *Expenses:* Tuition, state resident: full-time $6,295; part-time $336 per credit. Tuition, nonresident: full-time $15,949; part-time $819 per credit. Required fees: $933. Part-time tuition and fees vary according to campus/location and program. *Financial support:* In 2005–06, 16 students received support, including 4 fellowships (averaging $2,275 per year), 5 research assistantships with full and partial tuition reimbursements available (averaging $11,714 per year), 7 teaching assistantships with full and partial tuition reimbursements available (averaging $11,739 per year); career-related internships or fieldwork, Federal Work-Study, institutionally sponsored loans, and health care benefits also available. Financial award application deadline: 4/1; financial award applicants required to submit FAFSA. *Faculty research:* Post-harvest physiology, genetics/plant breeding, molecular biology. Total annual research expenditures: $2 million. *Unit head:* Dr. William Hendrix, Chair and Professor, 509-335-9502, Fax: 509-335-8690, E-mail: whendrix@wsu.edu. *Application contact:* Judy Hobart, Coordinator, 509-335-9504, Fax: 509-335-8690, E-mail: hobart@wsu.edu.

West Virginia University, Davis College of Agriculture, Forestry and Consumer Sciences, Division of Plant and Soil Sciences, Morgantown, WV 26506. Offers agronomy (MS); entomology (MS); environmental microbiology (MS); horticulture (MS); plant pathology (MS). *Faculty:* 18 full-time (1 woman), 1 part-time/adjunct (0 women). *Students:* 17 full-time (11 women), 3 part-time (2 women), 2 international. Average age 27. In 2005, 5 degrees awarded. *Degree requirements:* For master's, thesis. *Entrance requirements:* For master's, GRE, minimum GPA of 2.5. Additional exam requirements/recommendations for international students: Required—TOEFL. *Application deadline:* Applications are processed on a rolling basis. Application fee: $45. *Expenses:* Tuition, state resident: full-time $4,582; part-time $258 per credit hour. Tuition, nonresident: full-time $1,382; part-time $741 per credit hour. *Financial support:* In 2005–06, 13 research assistantships with full tuition reimbursements (averaging $9,936 per year), 4 teaching assistantships with full tuition reimbursements (averaging $9,936 per year) were awarded; Federal Work-Study, institutionally sponsored loans, and tuition waivers (full and partial) also available. Financial award application deadline: 2/1; financial award applicants required to submit FAFSA. *Faculty research:* Water quality, reclamation of disturbed land, crop production, pest control, environmental protection. Total annual research expenditures: $1 million. *Unit head:* Dr. Barton S. Baker, Chair and Division Director, 304-293-4817 Ext. 4342, Fax: 304-293-2960, E-mail: barton.baker@mail.wvu.edu.

Plant Sciences

Alabama Agricultural and Mechanical University, School of Graduate Studies, School of Agricultural and Environmental Sciences, Department of Plant and Soil Sciences, Huntsville, AL 35811. Offers animal sciences (MS); environmental science (MS); plant and soil science (PhD). Evening/weekend programs available. Terminal master's awarded for partial completion of doctoral program. *Degree requirements:* For master's, thesis; for doctorate, one foreign language, thesis/dissertation. *Entrance requirements:* For master's, GRE General Test, BS in agriculture; for doctorate, GRE General Test, master's degree. Electronic applications accepted. *Faculty research:* Plant breeding, cytogenetics, crop production, soil chemistry and fertility, remote sensing.

American University of Beirut, Graduate Programs, Faculty of Agricultural and Food Sciences, Beirut, Lebanon. Offers agricultural economics (MS); animal sciences (MS); ecosystem management (MSES); food technology (MS); irrigation (MS); mechanization (MS); nutrition (MS); plant protection (MS); plant science (MS); poultry science (MS); soils (MS). *Degree requirements:* For master's, one foreign language, thesis (for some programs), comprehensive exam, registration. *Entrance requirements:* For master's, GRE, letter of recommendation.

Brigham Young University, Graduate Studies, College of Biological and Agricultural Sciences, Department of Plant and Animal Sciences, Provo, UT 84602-1001. Offers agronomy (MS); genetics and biotechnology (MS). *Faculty:* 15 full-time (1 woman), 1 (woman) part-time/adjunct. *Students:* 10 full-time (4 women), 3 part-time (1 woman); includes 1 minority (Hispanic American) 14 applicants, 50% accepted, 6 enrolled. In 2005, 6 degrees awarded. *Degree requirements:* For master's, thesis, comprehensive exam, registration. *Entrance requirements:* For master's, GRE General Test, minimum GPA of 3.0 during last 60 hours of course work. *Application deadline:* For fall and spring admission, 2/15. Applications are processed on a rolling basis. Application fee: $50. Electronic applications accepted. *Financial support:* In 2005–06, 12 students received support, including 6 research assistantships with partial tuition reimbursements available (averaging $15,600 per year), 6 teaching assistantships with partial tuition reimbursements available (averaging $15,600 per year); institutionally sponsored loans, scholarships/grants, and tuition waivers (partial) also available. Financial award application deadline: 4/15. *Faculty research:* Iron nutrition in plants, cytogenetics plant pathology, pest management, genetic mapping, environmental science, plant genetics, molecular genetics. Total annual research expenditures:$160,761. *Unit head:* Dr. Sheldon D. Nelson, Chair, 801-422-2760, Fax: 801-422-0008, E-mail: sheldon_nelson@byu.edu. *Application contact:* Dr. Von D. Jolley, Graduate Coordinator, 801-422-2491, Fax: 801-422-0008, E-mail: von-jolley@byu.edu.

California State University, Fresno, Division of Graduate Studies, College of Agricultural Sciences and Technology, Department of Plant Science, Fresno, CA 93740-8027. Offers MS. Part-time programs available. *Degree requirements:* For master's, thesis. *Entrance requirements:* For master's, GRE General Test, minimum GPA of 2.50. Additional exam requirements/recommendations for international students: Required—TOEFL. Electronic applications accepted. *Faculty research:* Crop patterns, small watershed management, electronic monitoring of feedlot cattle, disease control, dairy operations.

Clemson University, Graduate School, College of Agriculture, Forestry and Life Sciences, Department of Biological Sciences, Program in Plant and Environmental Sciences, Clemson, SC 29634. Offers MS, PhD. *Students:* 35 full-time (11 women), 14 part-time (8 women); includes 2 minority (1 American Indian/Alaska Native, 1 Hispanic American), 12 international. 20 applicants, 45% accepted, 8 enrolled. In 2005, 8 master's, 1 doctorate awarded. *Degree requirements:* For master's, thesis. *Entrance requirements:* For master's, GRE General Test, bachelor's degree in biological science or chemistry. Additional exam requirements/recommendations for international students: Required—TOEFL. *Application deadline:* For fall admission, 6/1 for domestic students. Application fee: $50. *Financial support:* Teaching assistantships available. Financial award application deadline: 3/15; financial award applicants required to submit FAFSA. *Faculty research:* Systematics, aquatic botany, plant ecology, plant-fungus interactions, plant developmental genetics. *Unit head:* Dr. Halina Knapp, Coordinator, 864-656-3523, Fax: 864-656-7594, E-mail: hskrpsk@clemson.edu.

Plant Sciences

Colorado State University, Graduate School, College of Agricultural Sciences, Department of Bioagricultural Sciences and Pest Management, Fort Collins, CO 80523-0015. Offers entomology (MS, PhD); plant pathology and weed science (MS, PhD). *Faculty:* 18 full-time (3 women). *Students:* 21 full-time (12 women), 18 part-time (8 women); includes 3 minority (1 American Indian/Alaska Native, 1 Asian American or Pacific Islander, 1 Hispanic American), 1 international. Average age 34. 16 applicants, 44% accepted, 6 enrolled. In 2005, 5 master's, 4 doctorates awarded. *Degree requirements:* For master's, thesis (for some programs), registration; for doctorate, thesis/dissertation, registration. *Entrance requirements:* For master's and doctorate, GRE General Test, minimum GPA of 3.0. Additional exam requirements/recommendations for international students: Required—TOEFL (minimum score 550 paper-based; 213 computer-based). *Application deadline:* For fall admission, 4/1 priority date for domestic students, 4/1 priority date for international students; for spring admission, 9/1 priority date for domestic students, 9/1 priority date for international students. Applications are processed on a rolling basis. Application fee: $50. Electronic applications accepted. *Expenses:* Tuition, state resident: full-time $3,690; part-time $205 per credit. Tuition, nonresident: full-time $14,958; part-time $831 per credit. Required fees: $1,061. *Financial support:* In 2005–06, 1 student received support, including fellowships (averaging $2,500 per year), research assistantships with full tuition reimbursements available (averaging $17,500 per year), teaching assistantships with full tuition reimbursements available (averaging $12,402 per year); scholarships/grants, traineeships, and unspecified assistantships also available. Financial award application deadline: 4/1; financial award applicants required to submit FAFSA. *Faculty research:* Biological control of post-insect plant pathogens and weeds, integrated pest management, weed ecology and biology, and pests genome's of plants. Total annual research expenditures: $2.2 million. *Unit head:* Thomas O. Holtzer, Head, 970-491-5261, Fax: 970-491-3862, E-mail: tholtzer@lamar.colostate.edu. *Application contact:* Janet Dill, Graduate Program Coordinator, 970-491-0402, Fax: 970-491-3862, E-mail: janet.dill@colostate.edu.

Colorado State University, Graduate School, College of Agricultural Sciences, Department of Soil and Crop Sciences, Fort Collins, CO 80523-0015. Offers crop science (MS, PhD); plant genetics (MS, PhD); soil science (MS, PhD). Part-time programs available. *Faculty:* 18 full-time (4 women), 1 part-time/adjunct (0 women). *Students:* 18 full-time (6 women), 15 part-time (4 women); includes 2 minority (both Hispanic Americans), 5 international. Average age 33. 16 applicants, 50% accepted, 7 enrolled. In 2005, 6 master's, 1 doctorate awarded. *Median time to degree:* Of those who began their doctoral program in fall 1997, 85% received their degree in 8 years or less. *Degree requirements:* For master's, thesis (for some programs), comprehensive exam, registration; for doctorate, thesis/dissertation, preliminary exam, comprehensive exam, registration. *Entrance requirements:* For master's, minimum GPA of 3.0, appropriate bachelor's degree; for doctorate, minimum GPA of 3.0, appropriate master's degree. Additional exam requirements/recommendations for international students: Required—TOEFL. *Application deadline:* For fall admission, 2/1 priority date for domestic students, 2/1 priority date for international students; for spring admission, 8/1 priority date for domestic students, 8/1 priority date for international students. Applications are processed on a rolling basis. Application fee: $50. Electronic applications accepted. *Expenses:* Tuition, state resident: full-time $3,690; part-time $205 per credit. Tuition, nonresident: full-time $14,958; part-time $831 per credit. Required fees: $1,061. *Financial support:* In 2005–06, 16 students received support, including 1 fellowship with partial tuition reimbursement available (averaging $15,600 per year), 14 research assistantships with partial tuition reimbursements available (averaging $15,600 per year), 1 teaching assistantship with partial tuition reimbursement available (averaging $15,600 per year); career-related internships or fieldwork and traineeships also available. *Faculty research:* Water quality, soil fertility, soil/plant ecosystems, plant breeding and genetics, information systems/technology. Total annual research expenditures: $3.5 million. *Unit head:* Dr. Gary A. Peterson, Head, 970-491-6501, Fax: 970-491-0564, E-mail: gary.peterson@colostate.edu. *Application contact:* Dr. Pat F. Byrne, Graduate Studies Coordinator, 970-491-6985, Fax: 970-491-0564, E-mail: pbyrne@lamar.colostate.edu.

Cornell University, Graduate School, Graduate Fields of Agriculture and Life Sciences, Field of Plant Breeding, Ithaca, NY 14853-0001. Offers plant breeding (MPS, MS, PhD); plant genetics (MPS, MS, PhD). *Faculty:* 35 full-time (12 women). *Students:* 30 full-time (14 women); includes 3 minority (1 African American, 1 Asian American or Pacific Islander, 1 Hispanic American), 15 international. 25 applicants, 20% accepted, 5 enrolled. In 2005, 5 degrees awarded. Terminal master's awarded for partial completion of doctoral program. *Degree requirements:* For master's, thesis (MS), project paper (MPS); for doctorate, thesis/dissertation, comprehensive exam. *Entrance requirements:* For master's and doctorate, GRE General Test, GRE Subject Test (recommended), 3 letters of recommendation. Additional exam requirements/recommendations for international students: Required—TOEFL (minimum score 550 paper-based; 213 computer-based). *Application deadline:* For fall admission, 1/15 for domestic students. Application fee: $60. Electronic applications accepted. *Financial support:* In 2005–06, 26 students received support, including 7 fellowships with full tuition reimbursements available, 17 research assistantships with full tuition reimbursements available, 2 teaching assistantships with full tuition reimbursements available; institutionally sponsored loans, scholarships/grants, health care benefits, tuition waivers (full and partial), and unspecified assistantships also available. Financial award applicants required to submit FAFSA. *Faculty research:* Crop breeding for improved yield, stress resistance and quality; genetics and genomics of crop plants; applications of molecular biology and bioinformatics to crop improvement; genetic diversity and utilization of wild germplasm; international agriculture. *Unit head:* Director of Graduate Studies, 607-255-2180. *Application contact:* Graduate Field Assistant, 607-255-2180, E-mail: plbrgrad@cornell.edu.

Cornell University, Graduate School, Graduate Fields of Agriculture and Life Sciences, Field of Plant Protection, Ithaca, NY 14853-0001. Offers MPS. *Faculty:* 27 full-time (5 women). *Students:* 1 applicant. *Degree requirements:* For master's, internship, final exam. *Entrance requirements:* For master's, GRE General Test, 3 letters of recommendation. Additional exam requirements/recommendations for international students: Required—TOEFL (minimum score 550 paper-based; 213 computer-based). *Application deadline:* For fall admission, 4/1 for domestic students. Application fee: $60. Electronic applications accepted. *Financial support:* Fellowships with full tuition reimbursements, research assistantships with full tuition reimbursements, teaching assistantships with full tuition reimbursements, institutionally sponsored loans, scholarships/grants, health care benefits, tuition waivers (full and partial), and unspecified assistantships available. Financial award applicants required to submit FAFSA. *Faculty research:* Fruit and vegetable crop insects and diseases, systems modeling, biological control, plant protection economics, integrated pest management. *Unit head:* Director of Graduate Studies, 315-787-2323, Fax: 315-787-2326. *Application contact:* Graduate Field Assistant, 315-787-2323, Fax: 315-787-2326, E-mail: plprotection@cornell.edu.

Cornell University, Graduate School, Graduate Fields of Agriculture and Life Sciences, Field of Pomology, Ithaca, NY 14853-0001. Offers MPS, MS, PhD. *Faculty:* 22 full-time (4 women). *Students:* 18 full-time (8 women); includes 1 minority (African American), 11 international. 4 applicants, 100% accepted, 4 enrolled. In 2005, 2 master's, 2 doctorates awarded. *Degree requirements:* For master's, thesis (MS), project paper (MPS); for doctorate, thesis/dissertation, comprehensive exam. *Entrance requirements:* For master's and doctorate, GRE General Test, interview (recommended), 3 letters of recommendation. Additional exam requirements/recommendations for international students: Required—TOEFL (minimum score 550 paper-based; 213 computer-based). *Application deadline:* For fall admission, 1/15 for domestic students; for spring admission, 8/15 for domestic students. Application fee: $60. Electronic applications accepted. *Financial support:* In 2005–06, 15 students received support, including 4 fellowships with full tuition reimbursements available, 11 research assistantships with full tuition reimbursements available; teaching assistantships with full tuition reimbursements available, institutionally sponsored loans, scholarships/grants, health care benefits, tuition waivers (full and partial), and unspecified assistantships also available. Financial award applicants required to submit FAFSA. *Faculty research:* Fruit breeding and biotechnology, fruit crop physiology, orchard management, orchard ecology and IPM, post-harvest physiology. *Unit head:* Director of Graduate Studies, 607-255-4568, Fax: 607-255-0599. *Application contact:* Graduate Field Assistant, 607-255-4568, Fax: 607-255-0599, E-mail: hortgrad@cornell.edu.

Cornell University, Graduate School, Graduate Fields of Agriculture and Life Sciences, Field of Vegetable Crops, Ithaca, NY 14853-0001. Offers MPS, MS, PhD. *Faculty:* 21 full-time (8 women). *Students:* 9 full-time (3 women), 4 international. In 2005, 1 master's, 1 doctorate awarded. *Degree requirements:* For master's, thesis (MS), project paper (MPS); for doctorate, thesis/dissertation, teaching experience. *Entrance requirements:* For master's and doctorate, GRE General Test (recommended), 3 letters of recommendation. Additional exam requirements/recommendations for international students: Required—TOEFL (minimum score 550 paper-based; 213 computer-based). *Application deadline:* Applications are processed on a rolling basis. Application fee: $60. Electronic applications accepted. *Financial support:* In 2005–06, 9 students received support, including 2 fellowships with full tuition reimbursements available, 7 research assistantships with full tuition reimbursements available; teaching assistantships with full tuition reimbursements available, institutionally sponsored loans, health care benefits, tuition waivers (full and partial), and unspecified assistantships also available. Financial award applicants required to submit FAFSA. *Faculty research:* Vegetable nutrition and physiology, post-harvest physiology and storage, application of new technologies, sustainable vegetable production, weed management and IPM. *Unit head:* Director of Graduate Studies, 607-255-4568. *Application contact:* Graduate Field Assistant, 607-255-4568, E-mail: hortgrad@cornell.edu.

Florida Agricultural and Mechanical University, Division of Graduate Studies, Research, and Continuing Education, College of Engineering Science, Technology, and Agriculture, Division of Agricultural Sciences, Tallahassee, FL 32307-3200. Offers agribusiness (MS); animal science (MS); engineering technology (MS); entomology (MS); food science (MS); international programs (MS); plant science (MS). *Degree requirements:* For master's, thesis. *Entrance requirements:* For master's, GRE General Test, minimum GPA of 3.0. Additional exam requirements/recommendations for international students: Required—TOEFL (minimum score 500 paper-based).

Lehman College of the City University of New York, Division of Natural and Social Sciences, Department of Biological Sciences, Program in Plant Sciences, Bronx, NY 10468-1589. Offers PhD. *Degree requirements:* For doctorate, 2 foreign languages, thesis/dissertation. *Entrance requirements:* For doctorate, GRE General Test.

McGill University, Faculty of Graduate and Postdoctoral Studies, Faculty of Agricultural and Environmental Sciences, Department of Plant Science, Montréal, QC H3A 2T5, Canada. Offers M Sc, M Sc A, PhD, Certificate. Part-time programs available. Terminal master's awarded for partial completion of doctoral program. *Degree requirements:* For master's, thesis/dissertation, registration; for doctorate, thesis/dissertation, comprehensive exam, registration. *Entrance requirements:* For master's, minimum cumulative GPA of 3.0 or 3.2 in the last 2 years of full-time university study; for doctorate, M Sc, minimum GPA of 3.0. Additional exam requirements/recommendations for international students: Required—TOEFL (minimum score 550 paper-based; 213 computer-based), IELT (minimum score 7). Electronic applications accepted. *Faculty research:* Plant breeding, cytogenetics, crop physiology, bioherbicides, production.

Michigan State University, The Graduate School, College of Agriculture and Natural Resources, MSU-DOE Plant Research Laboratory, East Lansing, MI 48824. Offers biochemistry and molecular biology (PhD); cellular and molecular biology (PhD); crop and soil sciences (PhD); genetics (PhD); microbiology and molecular genetics (PhD); plant biology (PhD); plant physiology (PhD). Offered jointly with the Department of Energy. *Faculty:* 9 full-time (2 women). *Degree requirements:* For doctorate, thesis/dissertation, laboratory rotation, defense of dissertation, comprehensive exam. *Entrance requirements:* For doctorate, GRE General Test, acceptance into one of the affiliated department programs; 3 letters of recommendation; bachelor's degree or equivalent in life sciences, chemistry, biochemistry, or biophysics; research experience. Application fee: $50. Electronic applications accepted. *Expenses:* Tuition, state resident: part-time $330 per credit hour. Tuition, nonresident: part-time $685 per credit hour. Tuition and fees vary according to program. *Faculty research:* Role of hormones in the regulation of plant development and physiology, molecular mechanisms associated with signal recognition, development and application of genetic methods and materials, protein routing and function. Total annual research expenditures: $7.4 million. *Unit head:* Dr. Kenneth Keegstra, Director, 517-353-2270, Fax: 517-353-9168, E-mail: keegstra@msu.edu. *Application contact:* Janet Taylor, Graduate Program Secretary, 517-353-2270, Fax: 517-353-9168, E-mail: prl@msu.edu.

Michigan State University, The Graduate School, College of Agriculture and Natural Resources, Program in Plant Breeding and Genetics, East Lansing, MI 48824. Offers plant breeding and genetics (MS, PhD), including crop and soil sciences, forestry, horticulture. *Students:* 17 full-time (12 women), 2 part-time (1 woman), 10 international. Average age 29. 9 applicants, 0% accepted. In 2005, 2 master's, 6 doctorates awarded. *Degree requirements:* For master's, thesis; for doctorate, thesis/dissertation, oral examination, comprehensive exam. *Entrance requirements:* For master's and doctorate, GRE General Test, minimum GPA of 3.0 in last 3 years of college, 3 letters of recommendation. Additional exam requirements/recommendations for international students: Required—TOEFL (minimum score 550 paper-based; 213 computer-based). *Application deadline:* For fall admission, 12/27 for domestic students. Application fee: $50. Electronic applications accepted. *Expenses:* Tuition, state resident: part-time $330 per credit hour. Tuition, nonresident: part-time $685 per credit hour. Tuition and fees vary according to program. *Financial support:* In 2005–06, 1 fellowship with tuition reimbursement (averaging $11,500 per year), 12 research assistantships with tuition reimbursements (averaging $13,218 per year), 2 teaching assistantships with tuition reimbursements (averaging $12,470 per year) were awarded; scholarships/grants and unspecified assistantships also available. *Faculty research:* Applied plant breeding and genetics; disease, insect and herbicide resistances; gene isolation and genomics; abiotic stress factors; molecular mapping. *Unit head:* Dr. Jim F. Hancock, Director, 517-355-5191 Ext. 1387, Fax: 517-432-3490, E-mail: hancock@msu.edu. *Application contact:* Program Information, 517-355-5191 Ext. 1324, Fax: 517-432-3490, E-mail: pbg@msu.edu.

Mississippi State University, College of Agriculture and Life Sciences, Department of Plant and Soil Sciences, Mississippi State, MS 39762. Offers agronomy (MS, PhD); horticulture (MS, PhD); weed science (MS, PhD). Part-time programs available. *Faculty:* 27 full-time (2 women), 1 part-time/adjunct (0 women). *Students:* 28 full-time (5 women), 29 part-time (6 women); includes 2 minority (1 Asian American or Pacific Islander, 1 Hispanic American), 17 international. Average age 31. 16 applicants, 38% accepted, 3 enrolled. In 2005, 9 master's, 6 doctorates awarded. *Degree requirements:* For master's and doctorate, thesis/dissertation, comprehensive oral or written exam. *Entrance requirements:* Additional exam requirements/recommendations for international students: Required—TOEFL. *Application deadline:* For fall admission, 7/1 for domestic students; for spring admission, 11/1 for domestic students. Applications are processed on a rolling basis. Application fee: $30. *Expenses:* Tuition, state resident: full-time $4,312; part-time $240 per hour. Tuition, nonresident: full-time $9,772; part-time $543 per hour. International tuition: $10,102 full-time. Tuition and fees vary according to course load. *Financial support:* In 2005–06, 3 teaching assistantships with full tuition reimbursements (averaging $12,818 per year) were awarded; research assistantships with full tuition reimbursements, career-related internships or fieldwork, Federal Work-Study, institutionally sponsored loans, and unspecified assistantships also available. Financial award applicants required to submit FAFSA. *Faculty research:* Metabolism, morphology, growth regulators, biotechnology, stress physiology. Total annual research expenditures:$874,795. *Unit head:* Dr. Michael Collins, Head, 662-325-2352, Fax: 662-325-8742, E-mail: mcollins@pss.msstate.edu. *Application contact:* Philip G. Bonfanti, Director of Admissions, 662-325-4104, Fax: 662-325-8872, E-mail: admit@msstate.edu.

Missouri State University, Graduate College, College of Natural and Applied Sciences, Department of Agriculture, Springfield, MO 65804-0094. Offers agriculture (MNAS); fruit science (MNAS); plant science (MS); secondary education (MS Ed), including agriculture. *Faculty:* 14 full-time (2 women), 1 part-time/adjunct (0 women). *Students:* 19 full-time (11 women), 5 part-time (4 women), 5 international. Average age 29. 21 applicants, 71% accepted, 10 enrolled. In 2005, 6 degrees awarded. *Degree requirements:* For master's, thesis or alternative, comprehensive exam. *Entrance requirements:* For master's, GRE (MS plant science, MNAS), 9–12 teacher certification (MS Ed), minimum GPA of 3.0 (MS plant science, MNAS). Additional exam requirements/recommendations for international students: Required—TOEFL (minimum score 550 paper-based; 213 computer-based), IELT (minimum score 6). *Application deadline:* For fall admission, 7/20 for domestic students; for spring admission, 12/20 priority date for domestic students. Applications are processed on a rolling basis. Application fee: $30. Electronic applications accepted. *Expenses:* Tuition, state resident: full-time $3,402; part-time $189 per credit. Tuition, nonresident: full-time $6,804; part-time $378 per credit. Required fees: $207 per semester. Part-time tuition and fees vary according to course level, course load and program. *Financial support:* In 2005–06, 6 research assistantships (averaging $7,662 per year), 5 teaching assistantships (averaging $7,445 per year) were awarded. Financial award application deadline: 3/31; financial award applicants required to submit FAFSA. *Unit head:* Dr. W. Anson Elliott, Head, 417-836-5638, E-mail: ansonelliot@missouristate.edu. *Application contact:* Dr. W. Anson Elliott, Head, 417-836-5638, E-mail: ansonelliot@missouristate.edu.

Montana State University, College of Graduate Studies, College of Agriculture, Department of Plant Sciences and Plant Pathology, Bozeman, MT 59717. Offers plant pathology (MS); plant science (MS, PhD). Part-time programs available. *Faculty:* 24 full-time (3 women), 4 part-time/adjunct (3 women). *Students:* 5 full-time (2 women), 18 part-time (7 women), 8 international. Average age 29. 5 applicants, 80% accepted, 4 enrolled. In 2005, 6 master's, 1 doctorate awarded. *Degree requirements:* For master's, thesis (for some programs), comprehensive exam, registration; for doctorate, thesis/dissertation, comprehensive exam, registration. *Entrance requirements:* For master's and doctorate, GRE General Test. Additional exam requirements/recommendations for international students: Required—TOEFL (minimum score 550 paper-based; 213 computer-based). *Application deadline:* For fall admission, 7/15 priority date for domestic students, 5/15 priority date for international students; for spring admission, 12/1 priority date for domestic students, 10/1 priority date for international students. Applications are processed on a rolling basis. Application fee: $30. Electronic applications accepted. *Expenses:* Tuition, state resident: full-time $4,132. Tuition, nonresident: full-time $1,132. *Financial support:* In 2005–06, 25 students received support, including 25 research assistantships with full tuition reimbursements available (averaging $15,000 per year), 10 teaching assistantships with full tuition reimbursements available (averaging $2,000 per year); Federal Work-Study, health care benefits, and unspecified assistantships also available. Financial award application deadline: 3/1; financial award applicants required to submit FAFSA. *Faculty research:* Plant genetics, plant-microbe interactions, plant physiology, plant taxonomy. Total annual research expenditures: $2.2 million. *Unit head:* Dr. John Sherwood, Department Head, 406-994-5153, Fax: 406-994-7600, E-mail: sherwood@montana.edu.

New Mexico State University, Graduate School, College of Agriculture and Home Economics, Department of Entomology, Plant Pathology and Weed Science, Las Cruces, NM 88003-8001. Offers agricultural biology (MS). Part-time programs available. *Faculty:* 12 full-time (5 women), 1 part-time/adjunct (0 women). *Students:* 14 full-time (12 women), 8 part-time (4 women); includes 7 minority (1 African American, 1 American Indian/Alaska Native, 1 Asian American or Pacific Islander, 4 Hispanic Americans), 1 international. Average age 30. 9 applicants, 67% accepted, 6 enrolled. In 2005, 4 degrees awarded. *Degree requirements:* For master's, thesis, comprehensive exam, registration. *Entrance requirements:* For master's, GRE General Test. *Application deadline:* For fall admission, 7/1 for domestic students; for spring admission, 11/1 priority date for domestic students. Applications are processed on a rolling basis. Application fee: $30 ($50 for international students). Electronic applications accepted. *Expenses:* Tuition, state resident: full-time $3,156; part-time $175 per credit. Tuition, nonresident: full-time $12,510; part-time $565 per credit. Required fees: $1,050. *Financial support:* In 2005–06, 1 fellowship, 11 research assistantships with partial tuition reimbursements, 4 teaching assistantships with partial tuition reimbursements were awarded; career-related internships or fieldwork also available. Financial award application deadline:3/1. *Faculty research:* Integrated pest management, pesticide application and safety, livestock ectoparasite research, biotechnology, nematology. *Unit head:* Dr. Grant Kinzer, Head, 505-646-3225, Fax: 505-646-8087, E-mail: gkinzer@nmsu.edu.

North Carolina Agricultural and Technical State University, Graduate School, School of Agriculture and Environmental and Allied Sciences, Department of Natural Resources and Environmental Design, Greensboro, NC 27411. Offers plant science (MS). Part-time and evening/weekend programs available. *Degree requirements:* For master's, qualifying exam, thesis optional. *Entrance requirements:* For master's, GRE General Test, minimum GPA of 3.0. *Faculty research:* Soil parameters and compaction of forest site, controlled traffic effects on soil, improving soybean and vegetable crops.

North Dakota State University, The Graduate School, College of Agriculture, Food Systems, and Natural Resources, Department of Plant Sciences, Fargo, ND 58105. Offers crop and weed sciences (MS); horticulture (MS); natural resource management (MS); plant sciences (PhD). Part-time programs available. *Faculty:* 39 full-time (3 women), 19 part-time/adjunct (1 woman). *Students:* 66 full-time (31 women), 11 part-time (2 women), 41 international. Average age 26. 44 applicants, 45% accepted. In 2005, 9 master's, 1 doctorate awarded. *Degree requirements:* For master's and doctorate, thesis/dissertation. *Entrance requirements:* Additional exam requirements/recommendations for international students: Required—TOEFL (minimum score 525 paper-based; 197 computer-based). *Application deadline:* Applications are processed on a rolling basis. Application fee: $45 ($60 for international students). Electronic applications accepted. *Financial support:* In 2005–06, 2 fellowships (averaging $19,950 per year), 64 research assistantships were awarded; teaching assistantships, Federal Work-Study and institutionally sponsored loans also available. Financial award application deadline: 4/15. *Faculty research:* Biotechnology, weed control science, plant breeding, plant genetics, crop physiology. Total annual research expenditures: $880,000. *Unit head:* Dr. Al Schneiter, Chair, 701-231-7971, Fax: 701-231-8474, E-mail: albert.schneiter@ndsu.nodak.edu.

Oklahoma State University, College of Agricultural Science and Natural Resources, Department of Horticulture and Landscape Architecture, Stillwater, OK 74078. Offers crop science (PhD); environmental science (PhD); food science (PhD); horticulture (M Ag, MS); plant science (PhD). *Faculty:* 21 full-time (1 woman), 4 part-time/adjunct (0 women). *Students:* 1 full-time (0 women), 6 part-time (2 women); includes 1 minority (African American), 4 international. Average age 31. 7 applicants, 14% accepted. In 2005, 3 master's awarded. *Degree requirements:* For master's, thesis or alternative. *Entrance requirements:* For master's and doctorate, GRE. Additional exam requirements/recommendations for international students: Required—TOEFL. *Application deadline:* For fall admission, 6/1 priority date for domestic students, 3/1 priority date for international students. Applications are processed on a rolling basis. Application fee: $40 ($75 for international students). Electronic applications accepted. *Expenses:* Tuition, state resident: full-time $4,253; part-time $139 per credit hour. Tuition, nonresident: full-time $12,569; part-time $485 per credit hour. Required fees: $43 per credit hour. One-time fee: $20 part-time. Tuition and fees vary according to course load and program. *Financial support:* In 2005–06, 4 research assistantships (averaging $13,817 per year) were awarded; teaching assistantships, career-related internships or fieldwork, Federal Work-Study, scholarships/grants, health care benefits, tuition waivers (partial), and unspecified assistantships also available. Support available to part-time students. Financial award application deadline: 3/1. *Faculty research:* Stress and postharvest physiology; water utilization and runoff; IPM systems and nursery, turf, floriculture, vegetable, net and fruit produces and natural resources, food extraction, and processing; public garden management. *Unit head:* Dr. Dale Maronek, Head, 405-744-5414, Fax: 405-744-9709, E-mail: maronek@okstate.edu.

Oklahoma State University, College of Arts and Sciences, Department of Botany, Stillwater, OK 74078. Offers botany (MS); environmental science (PhD); plant science (PhD). *Faculty:* 11 full-time (4 women). *Students:* 1 (woman) full-time, 7 part-time (3 women), 2 international. Average age 35. 6 applicants, 50% accepted, 2 enrolled. *Degree requirements:* For master's and doctorate, thesis/dissertation. *Entrance requirements:* For master's and doctorate, GRE General Test. Additional exam requirements/recommendations for international students: Required—TOEFL. *Application deadline:* For fall admission, 6/1 priority date for domestic students, 3/1 priority date for international students. Applications are processed on a rolling basis. Application fee: $40 ($75 for international students). Electronic applications accepted. *Expenses:* Tuition, state resident: full-time $4,253; part-time $139 per credit hour. Tuition, nonresident: full-time $12,569; part-time $485 per credit hour. Required fees: $43 per credit hour. One-time fee: $20 part-time. Tuition and fees vary according to course load and program. *Financial support:* In 2005–06, 5 research assistantships (averaging $14,339 per year), 11 teaching assistantships (averaging $16,755 per year) were awarded; career-related internships or fieldwork, Federal Work-Study, scholarships/grants, health care benefits, tuition waivers (partial), and unspecified assistantships also available. Support available to part-time students. Financial award application deadline: 3/1. *Faculty research:* Ethnobotany, developmental genetics of Arabidopsis, biological roles of Plasmodesmata, community ecology and biodiversity, nutrient cycling in grassland ecosystems. *Unit head:* Dr. William J. Henley, Head, 405-744-5559, Fax: 405-744-7074.

Oklahoma State University, Graduate College, Interdisciplinary Program in Plant Science, Stillwater, OK 74078. Offers PhD. *Application deadline:* Applications are processed on a rolling basis. Application fee: $25 ($50 for international students). Electronic applications accepted. *Expenses:* Tuition, state resident: full-time $4,253; part-time $139 per credit hour. Tuition, nonresident: full-time $12,569; part-time $485 per credit hour. Required fees: $43 per credit hour. One-time fee: $20 part-time. Tuition and fees vary according to course load and program. *Financial support:* Research assistantships available. *Unit head:* Dr. Charles Tauer, Coordinator, 405-744-5462.

Rutgers, The State University of New Jersey, New Brunswick/Piscataway, Graduate School, Program in Plant Biology, New Brunswick, NJ 08901-1281. Offers horticulture (MS, PhD); molecular biology and biochemistry (MS, PhD); pathology (MS, PhD); plant ecology (MS, PhD); plant genetics (PhD); plant physiology (MS, PhD); production and management (MS); structure and plant groups (MS, PhD). Part-time programs available. *Faculty:* 65 full-time, 1 part-time/adjunct (0 women). *Students:* 33 full-time (14 women), 13 part-time (6 women); includes 3 minority (1 African American, 2 Asian Americans or Pacific Islanders), 10 international. Average age 31. 56 applicants, 23% accepted, 10 enrolled. In 2005, 1 master's, 4 doctorates awarded. Terminal master's awarded for partial completion of doctoral program. *Median time to degree:* Of those who began their doctoral program in fall 1997, 90% received their degree in 8 years or less. *Degree requirements:* For master's, thesis or alternative, comprehensive exam; for doctorate, thesis/dissertation, comprehensive exam. *Entrance requirements:* For master's and doctorate, GRE General Test, GRE Subject Test (recommended). Additional exam requirements/recommendations for international students: Required—TOEFL (minimum score 600 paper-based; 250 computer-based). *Application deadline:* For fall admission, 4/1 for domestic students, 4/1 for international students. Application fee: $50. Electronic applications accepted. *Expenses:* Tuition, state resident: full-time $10,440; part-time $435 per credit. Tuition, nonresident: full-time $15,520; part-time $647 per credit. Required fees: $129 per credit. Tuition and fees vary according to program. *Financial support:* In 2005–06, 42 students received support, including 9 fellowships with full tuition reimbursements available (averaging $24,000 per year), 22 research assistantships with full tuition reimbursements available (averaging $16,500 per year), 10 teaching assistantships with full tuition reimbursements available (averaging $16,988 per year) Financial award application deadline: 1/15; financial award applicants required to submit FAFSA. *Faculty research:* Molecular biology and biochemistry of plants, plant development and genomics, plant protection, plant improvement, plant management of horticultural and field crops. Total annual research expenditures: $10 million. *Unit head:* Dr. Thomas Leustek, Director, 732-932-8165 Ext. 326, Fax: 732-932-9377, E-mail: leustek@aesop.rutgers.edu. *Application contact:* Barbara Mulder, Program Associate, 732-932-9375 Ext. 358, Fax: 732-932-9377, E-mail: plantbio@aesop.rutgers.edu.

South Dakota State University, Graduate School, College of Agriculture and Biological Sciences, Department of Plant Science, Brookings, SD 57007. Offers agronomy (MS, PhD); biological sciences (PhD); entomology (MS); plant pathology (MS). *Degree requirements:* For master's, thesis, oral exam; for doctorate, thesis/dissertation, preliminary oral and written exams. *Entrance requirements:* For master's and doctorate, GRE General Test. Additional exam requirements/recommendations for international students: Required—TOEFL.

Southern Illinois University Carbondale, Graduate School, College of Agriculture, Department of Plant, Soil, and General Agriculture, Carbondale, IL 62901-4701. Offers horticultural science (MS); plant and soil science (MS). *Faculty:* 20 full-time (1 woman). *Students:* 13 full-time (7 women), 33 part-time (9 women); includes 5 minority (all African Americans), 3 international. 16 applicants, 50% accepted, 2 enrolled. In 2005, 13 master's awarded. *Degree requirements:* For master's, thesis. *Entrance requirements:* For master's, minimum GPA of 2.7. Additional exam requirements/recommendations for international students: Required—TOEFL. *Application deadline:* Applications are processed on a rolling basis. Application fee: $0. *Financial support:* In 2005–06, 22 students received support, including 15 research assistantships with full tuition reimbursements available, 6 teaching assistantships with full tuition reimbursements available; fellowships with full tuition reimbursements available, Federal Work-Study, institutionally sponsored loans, and tuition waivers (full) also available. Support available to part-time students. *Faculty research:* Herbicides, fertilizers, agriculture education, landscape design, plant breeding. Total annual research expenditures: $2 million. *Unit head:* Brian Klubek, Interim Chair, 618-453-2496.

State University of New York College of Environmental Science and Forestry, Faculty of Environmental and Forest Biology, Syracuse, NY 13210-2779. Offers chemical ecology (MPS, MS, PhD); conservation biology (MPS, MS, PhD); ecology (MPS, MS, PhD); entomology (MPS, MS, PhD); environmental interpretation (MPS, MS, PhD); environmental physiology (MPS, MS, PhD); fish and wildlife biology (MPS, MS, PhD); forest pathology and mycology (MPS, MS, PhD); plant science and biotechnology (MPS, MS, PhD). *Faculty:* 27 full-time (4 women), 4 part-time/adjunct (0 women). *Students:* 81 full-time (50 women), 62 part-time (33 women); includes 4 minority (1 Asian American or Pacific Islander, 3 Hispanic Americans), 16 international. Average age 30. 84 applicants, 54% accepted, 17 enrolled. In 2005, 17 master's, 3 doctorates awarded. *Degree requirements:* For master's, thesis (for some programs), registration; for doctorate, thesis/dissertation, comprehensive exam, registration. *Entrance requirements:* For master's and doctorate, GRE General Test, GRE Subject Test, minimum GPA of 3.0. Additional exam requirements/recommendations for international students: Required—TOEFL (minimum score 550 paper-based; 213 computer-based). *Application deadline:* For fall admission, 2/1 priority date for domestic students, 2/1 priority date for international students; for spring admission, 11/1 priority date for domestic students, 11/1 priority date for international students. Applications are processed on a rolling basis. Application fee: $60. *Expenses:* Tuition, area resident: Full-time $6,900; part-time $288 per credit. Tuition, nonresident: full-time $10,920; part-time $455 per credit. Required fees: $395; $32 per credit. $20 per term. One-time fee: $145. *Financial support:* In 2005–06, 86 students received support, including 13 fellowships with full and partial tuition reimbursements available (averaging $9,446 per year), 40 research assistantships with full and partial tuition reimbursements available (averaging $11,000 per year), 32 teaching assistantships with full and partial tuition reimbursements available (averaging $9,446 per year); Federal Work-Study, institutionally sponsored loans, scholarships/grants, health care benefits, and unspecified assistantships also available. Financial award application deadline: 6/30. *Faculty research:* Ecology, fish and wildlife biology and management, plant science, entomology. Total annual research expenditures: $4.1 million. *Unit head:* Dr. Donald J. Leopold, Chair, 315-470-6770, Fax: 315-470-6934, E-mail: dendro@esf.edu. *Application contact:* Dr. Dudley J. Raynal, Dean, Instruction and Graduate Studies, 315-470-6599, Fax: 315-470-6978, E-mail: esfgrad@esf.edu.

Texas A&M University, College of Agriculture and Life Sciences, Department of Soil and Crop Sciences, Intercollegiate Faculty of Molecular and Environmental Plant Sciences, College Station, TX 77843. Offers MS, PhD. *Students:* Average age 29. *Degree requirements:* For master's and doctorate, thesis/dissertation, seminar. *Entrance requirements:* For master's and

Plant Sciences

Texas A&M University (continued)
doctorate, GRE General Test; letters of reference. Additional exam requirements/recommendations for international students: Required—TOEFL. *Application deadline:* For fall admission, 3/1 for domestic students; for spring admission, 8/1 for domestic students. Applications are processed on a rolling basis. Application fee: $50 ($75 for international students). Electronic applications accepted. *Expenses:* Tuition, state resident: full-time $4,488; part-time $187 per credit hour. Tuition, nonresident: full-time $11,112; part-time $463 per credit hour. Required fees:$1,974. *Financial support:* In 2005–06, fellowships with tuition reimbursements (averaging $20,000 per year), research assistantships (averaging $17,000 per year), teaching assistantships (averaging $18,200 per year) were awarded. Financial award application deadline: 3/1; financial award applicants required to submit FAFSA. *Faculty research:* Functional genomics, bioremediation, physiological ecology, transformation systems, abiotic stress. *Unit head:* Dr. Marla L. Binzel, Chair, 979-845-8938, Fax: 979-458-0533, E-mail: m-binzel@tamu.edu. *Application contact:* Dr. Jean Gould, Admissions Chair, 979-845-5078, Fax: 979-845-6049, E-mail: gould@tamu.edu.

Texas A&M University–Kingsville, College of Graduate Studies, College of Agriculture and Home Economics, Program in Plant and Soil Sciences, Kingsville, TX 78363. Offers MS, PhD. *Degree requirements:* For master's, thesis or alternative, comprehensive exam. *Entrance requirements:* For master's, GRE General Test, minimum GPA of 3.0. Additional exam requirements/recommendations for international students: Required—TOEFL.

Texas Tech University, Graduate School, College of Agricultural Sciences and Natural Resources, Department of Plant and Soil Science, Lubbock, TX 79409. Offers agronomy (PhD); crop science (MS); entomology (MS); horticulture (MS); soil science (MS). Part-time programs available. *Faculty:* 12 full-time (2 women), 4 part-time/adjunct (0 women). *Students:* 29 full-time (11 women), 23 part-time (9 women), 7 international. Average age 31. 36 applicants, 56% accepted, 14 enrolled. In 2005, 16 master's, 2 doctorates awarded. *Degree requirements:* For doctorate, thesis/dissertation. *Entrance requirements:* For master's and doctorate, GRE General Test. Additional exam requirements/recommendations for international students: Required—TOEFL (minimum score 550 paper-based; 213 computer-based). *Application deadline:* Applications are processed on a rolling basis. Application fee: $50 ($60 for international students). Electronic applications accepted. *Expenses:* Tuition, state resident: full-time $4,296. Tuition, nonresident: full-time $10,920. Required fees: $1,992. Tuition and fees vary according to program. *Financial support:* In 2005–06, 21 students received support, including 21 research assistantships with partial tuition reimbursements available (averaging $12,726 per year), 2 teaching assistantships with partial tuition reimbursements available (averaging $12,478 per year); Federal Work-Study and institutionally sponsored loans also available. Support available to part-time students. Financial award application deadline: 4/15; financial award applicants required to submit FAFSA. *Faculty research:* Molecular and cellular biology of plant stress, physiology/genetics of crop production in semiarid conditions, agricultural bioterrorism, improvement of native plants. Total annual research expenditures: $2.6 million. *Unit head:* Dr. Dick L. Auld, Chair, 806-742-2837, Fax: 806-742-0775, E-mail: dick.auld@ttu.edu. *Application contact:* Dr. Richard E. Zartman, Graduate Adviser, 806-742-2837, Fax: 806-742-0775, E-mail: richard.zartman@ttu.edu.

Tuskegee University, Graduate Programs, College of Agricultural, Environmental and Natural Sciences, Department of Agricultural Sciences, Program in Plant and Soil Sciences, Tuskegee, AL 36088. Offers MS. *Faculty:* 13 full-time (1 woman), 2 part-time/adjunct (1 woman). *Students:* 4 full-time (0 women); includes 1 minority (African American), 3 international. Average age 28. In 2005, 3 degrees awarded. *Degree requirements:* For master's, thesis. *Entrance requirements:* For master's, GRE General Test. Additional exam requirements/recommendations for international students: Required—TOEFL (minimum score 500 paper-based; 173 computer-based). *Application deadline:* For fall admission, 7/15 for domestic students. Applications are processed on a rolling basis. Application fee: $25 ($35 for international students). *Expenses:* Tuition: Full-time $12,400. Required fees: $300; $490 per credit. *Financial support:* Application deadline: 4/15.

The University of Arizona, Graduate College, College of Agriculture and Life Sciences, Department of Plant Sciences, Tucson, AZ 85721. Offers MS, PhD. Part-time programs available. *Degree requirements:* For master's, thesis or alternative; for doctorate, thesis/dissertation. *Entrance requirements:* For master's and doctorate, GRE General Test, GRE Subject Test (biology or chemistry) (recommended), minimum GPA of 3.0. Additional exam requirements/recommendations for international students: Required—TOEFL. Electronic applications accepted. *Faculty research:* Molecular/cell biology, plant genetics and physiology, agronomic and horticultural production (including turf and ornamentals).

University of Arkansas, Graduate School, Dale Bumpers College of Agricultural, Food and Life Sciences, Interdepartmental Program in Plant Science, Fayetteville, AR 72701-1201. Offers PhD. *Students:* 16 full-time (3 women), 2 part-time (both women), 11 international. 8 applicants, 50% accepted. In 2005, 4 degrees awarded. *Degree requirements:* For doctorate, thesis/dissertation. Application fee: $40 ($50 for international students). *Financial support:* In 2005–06, 10 research assistantships were awarded; fellowships with tuition reimbursements, teaching assistantships, career-related internships or fieldwork and Federal Work-Study also available. Support available to part-time students. Financial award application deadline: 4/1; financial award applicants required to submit FAFSA. *Unit head:* Dr. Rose Gergerich, Graduate Coordinator, 479-575-2678.

The University of British Columbia, Faculty of Graduate Studies, Faculty of Land and Food Systems, Plant Science Program, Vancouver, BC V6T 1Z1, Canada. Offers M Sc, PhD. *Faculty:* 7 full-time (1 woman), 4 part-time/adjunct (1 woman). *Students:* 31 full-time (17 women). Average age 24. 15 applicants, 53% accepted, 6 enrolled. In 2005, 2 master's, 2 doctorates awarded. *Degree requirements:* For master's, thesis/dissertation, registration; for doctorate, thesis/dissertation, comprehensive exam, registration. *Entrance requirements:* Additional exam requirements/recommendations for international students: Required—TOEFL (minimum score 577 paper-based; 233 computer-based). *Application deadline:* For fall admission, 1/3 for domestic students, 1/3 for international students. For winter admission, 6/1 for domestic students; for spring admission, 9/1 for domestic students. Application fee: $90 Canadian dollars ($150 Canadian dollars for international students). Electronic applications accepted. *Financial support:* In 2005–06, 9 fellowships (averaging $12,157 per year) were awarded; research assistantships, teaching assistantships, institutionally sponsored loans, scholarships/grants, and tuition waivers (full and partial) also available. *Faculty research:* Plant physiology and biochemistry, biotechnology, plant protection (insect, weeds, and diseases), plant protection, plant-environment interaction. Total annual research expenditures: $542,829. *Unit head:* Dr. Murray Isman, Graduate Advisor, 604-822-9607, Fax: 604-822-2016, E-mail: gradapp@interchange.ubc.ca. *Application contact:* Lia Maria Dragan, Graduate Program Assistant, 604-822-8373, Fax: 604-822-4400, E-mail: gradapp@interchange.ubc.ca.

University of California, Riverside, Graduate Division, Department of Botany and Plant Sciences, Riverside, CA 92521-0102. Offers plant biology (MS, PhD); plant biology (plant genetics) (PhD). Part-time programs available. *Faculty:* 39 full-time (12 women). *Students:* 39 full-time (23 women); includes 4 minority (2 Asian Americans or Pacific Islanders, 2 Hispanic Americans), 15 international. Average age 30. In 2005, 1 master's, 10 doctorates awarded. Terminal master's awarded for partial completion of doctoral program. *Degree requirements:* For master's, comprehensive exams or thesis; for doctorate, thesis/dissertation, qualifying exams. *Entrance requirements:* For master's and doctorate, GRE General Test, minimum GPA of 3.2. Additional exam requirements/recommendations for international students: Required—TOEFL (minimum score 550 paper-based; 213 computer-based); Recommended—TSE (minimum score 50). *Application deadline:* For fall admission, 5/1 for domestic students, 2/1 for international students. For winter admission, 2/1 for domestic students; for spring admission, 12/1 for domestic students. Applications are processed on a rolling basis. Application fee: $60 ($75 for international students). Electronic applications accepted. *Expenses:* Tuition, nonresident: full-time $14,694. Full-time tuition and fees vary according to program. *Financial support:* In 2005–06, research assistantships (averaging $14,000 per year), teaching assistantships (averaging $15,000 per year) were awarded; fellowships, career-related internships or fieldwork, Federal Work-Study, institutionally sponsored loans, scholarships/grants, and tuition waivers (full and partial) also available. Financial award application deadline: 2/1; financial award applicants required to submit FAFSA. *Faculty research:* Agricultural plant biology; biochemistry and physiology; cellular, molecular and developmental biology; ecology, evolution, systematics and ethnobotany; genetics, genomics and bioinformatics. *Unit head:* Dr. Jodie S. Holt, Chair. *Application contact:* Carole Carpenter, Graduate Program Assistant, 800-735-0717, Fax: 951-827-5517, E-mail: plantbio@ucr.edu.

University of Connecticut, Graduate School, College of Agriculture and Natural Resources, Department of Plant Science, Field of Plant Science, Storrs, CT 06269. Offers plant and soil sciences (MS, PhD). *Faculty:* 26 full-time (4 women). *Students:* 16 full-time (7 women), 6 part-time (2 women), 12 international. Average age 34. 9 applicants, 56% accepted, 3 enrolled. In 2005, 3 master's, 1 doctorate awarded. Terminal master's awarded for partial completion of doctoral program. *Degree requirements:* For master's, comprehensive exam; for doctorate, thesis/dissertation. *Entrance requirements:* For master's and doctorate, GRE General Test, GRE Subject Test. Additional exam requirements/recommendations for international students: Required—TOEFL (minimum score 550 paper-based; 213 computer-based). *Application deadline:* For fall admission, 2/1 priority date for domestic students, 2/1 priority date for international students; for spring admission, 11/1 for domestic students, 10/1 for international students. Applications are processed on a rolling basis. Application fee: $55. Electronic applications accepted. *Expenses:* Tuition, state resident: part-time $444 per credit hour. Tuition, nonresident: part-time $1,154 per credit hour. Tuition and fees vary according to course load. *Financial support:* In 2005–06, 10 research assistantships with full tuition reimbursements, 4 teaching assistantships with full tuition reimbursements were awarded; fellowships, Federal Work-Study, scholarships/grants, health care benefits, and unspecified assistantships also available. Financial award application deadline: 2/1; financial award applicants required to submit FAFSA. *Application contact:* George C. Elliott, Chairperson, 860-486-1938, Fax: 860-486-0682, E-mail: george.elliott@uconn.edu.

University of Delaware, College of Agriculture and Natural Resources, Department of Plant and Soil Sciences, Newark, DE 19716. Offers MS, PhD. Part-time programs available. *Faculty:* 26 full-time (7 women), 15 part-time/adjunct (2 women). *Students:* 35 full-time (18 women), 3 part-time (2 women); includes 1 minority (Asian American or Pacific Islander), 11 international. Average age 29. 29 applicants, 34% accepted, 8 enrolled. In 2005, 5 master's, 3 doctorates awarded. Terminal master's awarded for partial completion of doctoral program. *Degree requirements:* For master's and doctorate, thesis/dissertation. *Entrance requirements:* For master's and doctorate, GRE General Test. Additional exam requirements/recommendations for international students: Required—TOEFL (minimum score 550 paper-based; 213 computer-based). *Application deadline:* For fall admission, 7/1 for domestic students. Application fee: $60. Electronic applications accepted. *Financial support:* In 2005–06, 20 fellowships with full tuition reimbursements (averaging $15,000 per year), 21 research assistantships with full tuition reimbursements (averaging $15,000 per year), 3 teaching assistantships with full tuition reimbursements (averaging $12,000 per year) were awarded; career-related internships or fieldwork also available. Financial award application deadline: 3/1. *Faculty research:* Soil chemistry, plant and cell tissue culture, plant breeding and genetics, soil physics, soil biochemistry, plant molecular biology, soil microbiology. Total annual research expenditures: $3.8 million. *Unit head:* Dr. Donald L. Sparks, Chair, 302-831-2532, Fax: 302-831-3651, E-mail: dlsparks@udel.edu. *Application contact:* Dr. Jeffrey Fuhrmann, Graduate Coordinator, 302-831-2534, E-mail: fuhrmann@udel.edu.

University of Florida, Graduate School, College of Agricultural and Life Sciences, Program in Plant Medicine, Gainesville, FL 32611. Offers DPM. *Faculty:* 22. *Students:* 34. *Expenses:* Tuition, state resident: full-time $6,234. Tuition, nonresident: full-time $21,359. Tuition and fees vary according to program. *Unit head:* Dr. Robert J. McGovern, Director, 352-392-3631 Ext. 213, E-mail: rjm@ifas.ufl.edu.

University of Hawaii at Manoa, Graduate Division, College of Tropical Agriculture and Human Resources, Department of Plant and Environmental Protection Sciences, Honolulu, HI 96822. Offers entomology (MS, PhD); tropical plant pathology (MS, PhD). Part-time programs available. Terminal master's awarded for partial completion of doctoral program. *Degree requirements:* For master's, thesis optional; for doctorate, thesis/dissertation. *Entrance requirements:* For master's and doctorate, GRE General Test. Application fee: $50. *Expenses:* Tuition, state resident: full-time $8,400; part-time $200 per credit hour. Tuition, nonresident: full-time $11,088; part-time $462 per credit hour. Tuition and fees vary according to program. *Financial support:* Research assistantships, teaching assistantships, tuition waivers (full) available. *Faculty research:* Nematology, virology, mycology, bacteriology, epidemiology. Total annual research expenditures: $1.9 million. *Unit head:* Dr. Kenneth Grace, Chairperson, 808-956-7096, Fax: 808-956-2428, E-mail: kennethg@hawaii.edu.

University of Idaho, College of Graduate Studies, College of Agricultural and Life Sciences, Department of Plant, Soil, and Entomological Sciences, Program in Plant Science, Moscow, ID 83844-2282. Offers MS, PhD. *Students:* 21 full-time (6 women), 16 part-time (4 women); includes 5 minority (2 Asian Americans or Pacific Islanders, 3 Hispanic Americans), 6 international. In 2005, 7 master's, 2 doctorates awarded. *Degree requirements:* For doctorate, thesis/dissertation. *Entrance requirements:* For master's and doctorate, GRE General Test, minimum GPA of 3.0. *Application deadline:* For fall admission, 8/1 for domestic students; for spring admission, 12/15 for domestic students. Application fee: $55 ($60 for international students). *Expenses:* Tuition, nonresident: full-time $8,770; part-time $130 per credit. Required fees: $4,508; $217 per credit. *Financial support:* Application deadline: 2/15. *Unit head:* Jeffrey Stark, Chair, 208-885-8376.

University of Kentucky, Graduate School, Graduate School Programs in the College of Agriculture, Program in Plant and Soil Science, Lexington, KY 40506-0032. Offers MS. *Faculty:* 48 full-time (2 women), 5 part-time/adjunct (0 women). *Students:* 17 full-time (10 women), 4 part-time, 2 international. Average age 28. 20 applicants, 60% accepted, 8 enrolled. In 2005, 4 degrees awarded. *Degree requirements:* For master's, thesis optional. *Entrance requirements:* For master's, GRE General Test, minimum undergraduate GPA of 2.5, minimum graduate GPA of 3.0. Additional exam requirements/recommendations for international students: Required—TOEFL (minimum score 550 paper-based; 213 computer-based). *Application deadline:* For fall admission, 7/17 priority date for domestic students, 2/1 priority date for international students; for spring admission, 12/13 priority date for domestic students, 6/15 priority date for international students. Applications are processed on a rolling basis. Application fee: $40 ($55 for international students). Electronic applications accepted. *Expenses:* Tuition, state resident: full-time $3,159; part-time $331 per credit hour. Tuition, nonresident: full-time $6,984; part-time $756 per credit hour. Tuition and fees vary according to course load, degree level and program. *Financial support:* In 2005–06, 16 students received support, including 16 research assistantships with full tuition reimbursements available (averaging $15,000 per year); fellowships with full tuition reimbursements available, teaching assistantships with full tuition reimbursements available, Federal Work-Study, scholarships/grants, traineeships, health care benefits, tuition waivers (partial), and unspecified assistantships also available. Support available to part-time students. Financial award application deadline: 3/15. *Unit head:* Dr. Charles Dougherty, Director of Graduate Studies, 859-257-3454, Fax: 859-323-1952, E-mail: cdougher@uky.edu. *Application contact:* Dr. Brian Jackson, Senior Associate Dean, 859-257-8176, Fax: 859-323-1928, E-mail: lance.brunner@uky.edu.

University of Maine, Graduate School, College of Natural Sciences, Forestry, and Agriculture, Department of Biological Sciences, Orono, ME 04469. Offers biological sciences (PhD); botany and plant pathology (MS); ecology and environmental science (MS, PhD); entomology (MS); plant science (PhD); zoology (MS, PhD). Part-time programs available. *Students:* 35 full-time (16 women), 14 part-time (8 women), 5 international. Average age 30. 79 applicants, 14% accepted, 8 enrolled. In 2005, 14 master's, 2 doctorates awarded. *Degree requirements:*

For doctorate, thesis/dissertation. *Entrance requirements:* For master's and doctorate, GRE General Test. Additional exam requirements/recommendations for international students: Required—TOEFL. *Application deadline:* For fall admission, 2/1 for domestic students. Applications are processed on a rolling basis. Application fee: $50. Electronic applications accepted. *Financial support:* In 2005–06, 1 fellowship with tuition reimbursement (averaging $15,000 per year), 17 research assistantships with tuition reimbursements (averaging $13,650 per year), 20 teaching assistantships with tuition reimbursements (averaging $9,835 per year) were awarded; career-related internships or fieldwork, Federal Work-Study, institutionally sponsored loans, and tuition waivers (full and partial) also available. Financial award application deadline: 3/1. *Unit head:* Dr. Ellie Grodon, Chair, 207-581-2989, Fax: 207-581-2537. *Application contact:* Scott G. Delcourt, Associate Dean of the Graduate School, 207-581-3219, Fax: 207-581-3232, E-mail: graduate@maine.edu.

University of Maine, Graduate School, College of Natural Sciences, Forestry, and Agriculture, Department of Plant, Soil, and Environmental Sciences, Orono, ME 04469. Offers biological sciences (PhD); ecology and environmental sciences (MS, PhD); forest resources (PhD); horticulture (MS); plant science (PhD); plant, soil, and environmental sciences (MS); resource utilization (MS). *Faculty:* 25. *Students:* 15 full-time (8 women), 8 part-time (6 women), 3 international. Average age 32. 6 applicants, 33% accepted, 1 enrolled. In 2005, 9 master's awarded. *Entrance requirements:* For master's and doctorate, GRE General Test. Additional exam requirements/recommendations for international students: Required—TOEFL. *Application deadline:* Applications are processed on a rolling basis. Application fee: $50. Electronic applications accepted. *Financial support:* In 2005–06, 9 research assistantships with tuition reimbursements (averaging $12,180 per year) were awarded; teaching assistantships, scholarships/grants, tuition waivers (full and partial), and unspecified assistantships also available. *Unit head:* Greg Porter, Chair, 207-581-2943, Fax: 207-581-3207. *Application contact:* Scott G. Delcourt, Associate Dean of the Graduate School, 207-581-3219, Fax: 207-581-3232, E-mail: graduate@maine.edu.

University of Massachusetts Amherst, Graduate School, College of Natural Resources and the Environment, Department of Plant and Soil Sciences, Amherst, MA 01003. Offers plant science (PhD); soil science (MS, PhD). *Faculty:* 24 full-time (5 women). *Students:* 24 full-time (12 women), 15 part-time (6 women); includes 2 minority (1 African American, 1 Hispanic American), 11 international. Average age 35. 42 applicants, 40% accepted, 13 enrolled. In 2005, 6 master's, 2 doctorates awarded. Terminal master's awarded for partial completion of doctoral program. *Degree requirements:* For master's, thesis optional; for doctorate, thesis/dissertation. *Entrance requirements:* For master's and doctorate, GRE General Test. Additional exam requirements/recommendations for international students: Required—TOEFL (minimum score 530 paper-based; 197 computer-based). *Application deadline:* For fall admission, 2/1 priority date for domestic students, 2/1 priority date for international students; for spring admission, 10/1 for domestic students, 10/1 for international students. Applications are processed on a rolling basis. Application fee: $40 ($65 for international students). Electronic applications accepted. *Expenses:* Tuition, state resident: part-time $110 per credit. Tuition, nonresident: part-time $414 per credit. Required fees: $2,824 per term. One-time fee: $250 part-time. Full-time tuition and fees vary according to course load, campus/location, program and reciprocity agreements. *Financial support:* In 2005–06, fellowships with full tuition reimbursements (averaging $6,250 per year), research assistantships with full tuition reimbursements (averaging $7,488 per year), teaching assistantships with full tuition reimbursements (averaging $5,858 per year) were awarded; career-related internships or fieldwork, Federal Work-Study, scholarships/grants, traineeships, and unspecified assistantships also available. Support available to part-time students. Financial award application deadline: 2/1. *Unit head:* Dr. Petrus Veneman, Director, 413-545-5225, Fax: 413-545-3075, E-mail: veneman@pssci.umass.edu.

University of Massachusetts Amherst, Graduate School, Interdisciplinary Programs, Program in Plant Biology, Amherst, MA 01003. Offers MS, PhD. *Students:* 14 full-time (8 women), 2 part-time (1 woman); includes 3 minority (2 Asian Americans or Pacific Islanders, 1 Hispanic American), 5 international. 35 applicants, 26% accepted, 8 enrolled. In 2005, 1 master's, 1 doctorate awarded. *Degree requirements:* For master's and doctorate, thesis/dissertation. *Entrance requirements:* For master's and doctorate, GRE General Test. Additional exam requirements/recommendations for international students: Required—TOEFL (minimum score 530 paper-based; 197 computer-based). *Application deadline:* For fall admission, 1/2 priority date for domestic students, 1/2 priority date for international students; for spring admission, 10/1 for domestic students, 10/1 for international students. Applications are processed on a rolling basis. Application fee: $40 ($65 for international students). Electronic applications accepted. *Expenses:* Tuition, state resident: part-time $110 per credit. Tuition, nonresident: part-time $414 per credit. Required fees: $2,824 per term. One-time fee: $250 part-time. Full-time tuition and fees vary according to course load, campus/location, program and reciprocity agreements. *Financial support:* In 2005–06, 1 fellowship with full tuition reimbursement (averaging $806 per year), 5 research assistantships with full tuition reimbursements (averaging $8,576 per year) were awarded; teaching assistantships with full tuition reimbursements, career-related internships or fieldwork, Federal Work-Study, scholarships/grants, traineeships, and unspecified assistantships also available. Support available to part-time students. Financial award application deadline:1/15. *Unit head:* Dr. Elsbeth Walker, Head, 413-577-3217, Fax: 413-545-3243. *Application contact:* Information Contact, 413-577-3217, Fax: 413-545-3243.

University of Minnesota, Twin Cities Campus, Graduate School, College of Agricultural, Food, and Environmental Sciences, Program in Applied Plant Sciences, Minneapolis, MN 55455-0213. Offers MS, PhD. Part-time programs available. *Faculty:* 54 full-time (10 women), 7 part-time/adjunct (1 woman). *Students:* 34 full-time (16 women), 5 part-time (3 women); includes 2 minority (1 Asian American or Pacific Islander, 1 Hispanic American), 12 international. Average age 24. 43 applicants, 33% accepted, 12 enrolled. In 2005, 11 master's, 6 doctorates awarded. *Degree requirements:* For master's and doctorate, thesis/dissertation. *Entrance requirements:* For master's and doctorate, GRE General Test. Additional exam requirements/recommendations for international students: Required—TOEFL. *Application deadline:* Applications are processed on a rolling basis. Application fee: $55 ($75 for international students). Electronic applications accepted. *Expenses:* Tuition, state resident: full-time $8,748; part-time $729 per credit. Tuition, nonresident: full-time $15,848; part-time $1,321 per credit. Full-time tuition and fees vary according to class time, course load, program and reciprocity agreements. *Financial support:* In 2005–06, fellowships with tuition reimbursements (averaging $18,543 per year), research assistantships with tuition reimbursements (averaging $18,543 per year) were awarded; scholarships/grants, health care benefits, and unspecified assistantships also available. *Faculty research:* Weed science, crop management, sustainable agriculture, biotechnology, plant breeding. *Unit head:* Dr. Nancy J. Ehlke, Head, 612-625-8761, Fax: 612-625-1268. *Application contact:* Lynne Medgaarden, Office Supervisor, 612-625-4742, Fax: 612-625-1268, E-mail: medga001@umn.edu.

University of Missouri–Columbia, Graduate School, College of Agriculture, Food and Natural Resources, Division of Plant Sciences, Columbia, MO 65211. Offers MS, PhD. *Faculty:* 44 full-time (8 women), 1 part-time/adjunct (0 women). *Students:* 57 full-time (25 women), 16 part-time (7 women); includes 1 minority (African American), 31 international. In 2005, 7 master's, 3 doctorates awarded. Terminal master's awarded for partial completion of doctoral program. *Degree requirements:* For master's and doctorate, thesis/dissertation. *Entrance requirements:* For master's and doctorate, GRE General Test, minimum GPA of 3.0. *Application deadline:* Applications are processed on a rolling basis. Application fee: $45 ($60 for international students). *Financial support:* Fellowships, research assistantships, teaching assistantships, institutionally sponsored loans available. *Unit head:* Dr. Robert Sharp, Interim Director, 573-882-3001, E-mail: sharpr@missouri.edu. *Application contact:* Dr. Jeanne Mihail, Director of Graduate Studies, 573-882-0574, E-mail: mihailj@missouri.edu.

University of Rhode Island, Graduate School, College of the Environment and Life Sciences, Department of Plant Sciences, Program in Plant Sciences, Kingston, RI 02881. Offers MS, PhD. *Degree requirements:* For master's, thesis, professional seminar; for doctorate, one foreign language, thesis/dissertation, professional seminar. *Entrance requirements:* For master's, GRE General Test. *Application deadline:* For fall admission, 4/15 for domestic students. Application fee: $35. *Expenses:* Tuition, state resident: full-time $5,522; part-time $307 per credit. Tuition, nonresident: full-time $15,992; part-time $888 per credit. Required fees: $1,786; $73 per credit. One-time fee: $80 part-time. *Financial support:* Unspecified assistantships available. *Faculty research:* Ecology, physiology, improvement of turf, ornamental and food-crop plants.

University of Saskatchewan, College of Graduate Studies and Research, College of Agriculture, Department of Plant Sciences, Saskatoon, SK S7N 5A2, Canada. Offers M Ag, M Sc, PhD. *Degree requirements:* For master's and doctorate, thesis/dissertation, registration. *Entrance requirements:* Additional exam requirements/recommendations for international students: Required—TOEFL.

The University of Tennessee, Graduate School, College of Agricultural Sciences and Natural Resources, Department of Plant Sciences, Knoxville, TN 37996. Offers floriculture (MS); landscape design (MS); public horticulture (MS); turfgrass (MS); woody ornamentals (MS). Part-time programs available. *Degree requirements:* For master's, thesis or alternative. *Entrance requirements:* For master's, minimum GPA of 2.7. Additional exam requirements/recommendations for international students: Required—TOEFL. Electronic applications accepted.

University of Vermont, Graduate College, College of Agriculture and Life Sciences, Department of Plant and Soil Science, Burlington, VT 05405. Offers MS, PhD. *Students:* 18 (9 women) 5 international. 25 applicants, 64% accepted, 8 enrolled. In 2005, 7 master's, 1 doctorate awarded. *Degree requirements:* For master's, thesis; for doctorate, one foreign language, thesis/dissertation. *Entrance requirements:* For master's and doctorate, GRE General Test. Additional exam requirements/recommendations for international students: Required—TOEFL (minimum score 550 paper-based; 213 computer-based). *Application deadline:* For fall admission, 2/1 for domestic students. Applications are processed on a rolling basis. Application fee: $40. Electronic applications accepted. *Expenses:* Tuition, area resident: Part-time $410 per credit hour. Tuition, nonresident: part-time $1,034 per credit hour. *Financial support:* Fellowships, research assistantships, teaching assistantships available. Financial award application deadline: 3/1. *Faculty research:* Soil chemistry, plant nutrition. *Unit head:* Dr. Deborah Neher, Chairperson, 802-656-2630. *Application contact:* Dr. M. Starrett, Coordinator, 802-656-2630.

The University of Western Ontario, Faculty of Graduate Studies, Biosciences Division, Department of Plant Sciences, London, ON N6A 5B8, Canada. Offers plant and environmental sciences (M Sc); plant sciences (M Sc, PhD); plant sciences and environmental sciences (PhD); plant sciences and molecular biology (M Sc, PhD). *Degree requirements:* For master's and doctorate, thesis/dissertation. *Entrance requirements:* For doctorate, M Sc or equivalent. *Faculty research:* Ecology systematics, plant biochemistry and physiology, yeast genetics, molecular biology.

University of Wisconsin–Madison, Graduate School, College of Agricultural and Life Sciences, Department of Agronomy, Madison, WI 53706-1380. Offers agronomy (MS, PhD); plant breeding and plant genetics (MS, PhD). *Faculty:* 19 full-time (2 women). *Students:* 14 full-time (5 women); includes 1 minority (Asian American or Pacific Islander) Average age 25. 17 applicants, 12% accepted, 2 enrolled. In 2005, 2 master's awarded. *Median time to degree:* Of those who began their doctoral program in fall 1997, 100% received their degree in 8 years or less. *Degree requirements:* For master's, thesis or alternative; for doctorate, thesis/dissertation. *Entrance requirements:* For master's and doctorate, GRE, minimum GPA of 3.0. Additional exam requirements/recommendations for international students: Required—TOEFL (minimum score 580 paper-based; 213 computer-based). *Application deadline:* Applications are processed on a rolling basis. Application fee: $45. Electronic applications accepted. *Financial support:* In 2005–06, fellowships (averaging $19,200 per year), research assistantships (averaging $18,120 per year), teaching assistantships (averaging $24,200 per year) were awarded. *Faculty research:* Plant breeding and genetics, plant molecular biology and physiology, cropping systems and management, weed science. *Unit head:* William F. Tracy, Chair, 608-262-1390, Fax: 608-262-5217. *Application contact:* Colleen L. Smith, Graduate Secretary, 608-262-7702, Fax: 608-262-5217, E-mail: clsmith@wisc.edu.

University of Wisconsin–Madison, Graduate School, College of Agricultural and Life Sciences, Plant Breeding and Plant Genetics Program, Madison, WI 53706-1380. Offers MS, PhD. Part-time programs available. *Faculty:* 38 full-time (4 women). *Students:* 45 full-time (19 women), 3 part-time (1 woman); includes 26 minority (1 African American, 13 Asian Americans or Pacific Islanders, 12 Hispanic Americans). Average age 25. 60 applicants, 10% accepted, 6 enrolled. In 2005, 3 master's, 9 doctorates awarded. Terminal master's awarded for partial completion of doctoral program. *Median time to degree:* Of those who began their doctoral program in fall 1997, 100% received their degree in 8 years or less. *Degree requirements:* For master's, thesis (for some programs), comprehensive exam; for doctorate, thesis/dissertation, comprehensive exam. *Entrance requirements:* For master's and doctorate, GRE, minimum GPA of 3.0. Additional exam requirements/recommendations for international students: Required—TOEFL (minimum score 580 paper-based; 213 computer-based). *Application deadline:* For fall admission, 1/2 for domestic students, 1/2 for international students; for spring admission, 10/30 for domestic students, 10/30 for international students. Applications are processed on a rolling basis. Application fee: $45. Electronic applications accepted. *Financial support:* In 2005–06, fellowships with full tuition reimbursements (averaging $19,200 per year), research assistantships with full tuition reimbursements (averaging $18,120 per year), teaching assistantships with full tuition reimbursements (averaging $24,200 per year) were awarded; career-related internships or fieldwork, Federal Work-Study, and tuition waivers (partial) also available. Financial award application deadline: 1/2. *Faculty research:* Classical and molecular genetics. Total annual research expenditures: $5,000. *Unit head:* Dr. Heidi Kaeppler, Professor, 608-262-0246, Fax: 608-262-5217. *Application contact:* Colleen L. Smith, Program Assistant, 608-262-7702, Fax: 608-262-5217, E-mail: clsmith8@wisc.edu.

Utah State University, School of Graduate Studies, College of Agriculture, Department of Plants, Soils, and Biometeorology, Logan, UT 84322. Offers biometeorology (MS, PhD); ecology (MS, PhD); plant science (MS, PhD); soil science (MS, PhD). Part-time programs available. *Faculty:* 31 full-time (4 women), 13 part-time/adjunct (0 women). *Students:* 117 full-time (58 women), 16 part-time (5 women), 14 international. Average age 26. 25 applicants, 80% accepted, 19 enrolled. In 2005, 8 master's, 1 doctorate awarded. Terminal master's awarded for partial completion of doctoral program. *Median time to degree:* Of those who began their doctoral program in fall 1997, 100% received their degree in 8 years or less. *Degree requirements:* For master's and doctorate, thesis/dissertation. *Entrance requirements:* For master's, GRE General Test, BS in plant, soil, atmospheric science, or related field; minimum GPA of 3.0; for doctorate, GRE General Test, minimum GPA of 3.0. Additional exam requirements/recommendations for international students: Required—TOEFL. *Application deadline:* For fall admission, 6/15 priority date for domestic students, 3/15 priority date for international students; for spring admission, 10/15 priority date for domestic students, 9/15 priority date for international students. Applications are processed on a rolling basis. Application fee: $50 ($60 for international students). Electronic applications accepted. *Financial support:* In 2005–06, 23 research assistantships with partial tuition reimbursements (averaging $15,000 per year) were awarded; Federal Work-Study, institutionally sponsored loans, and tuition waivers (full) also available. Support available to part-time students. Financial award application deadline: 3/1. *Faculty research:* Biotechnology and genomics, plant physiology and biology, nutrient and water efficient landscapes, physical-chemical-biological processes in soil, environmental biophysics and climate. Total annual research expenditures: $4.5 million. *Unit head:* Dr. Larry A. Rupp, Head, 435-797-2099, Fax: 435-797-3376, E-mail: larryr@ext.usu.edu. *Application contact:* Dr. Paul G. Johnson, Graduate Program Coordinator, 435-797-7039, Fax: 435-797-3376, E-mail: paul.johnson@usu.edu.

West Texas A&M University, College of Agriculture, Nursing, and Natural Sciences, Division of Agriculture, Emphasis in Plant Science, Canyon, TX 79016-0001. Offers MS. Part-time programs available. *Degree requirements:* For master's, thesis optional. *Entrance requirements:*

Plant Sciences

West Texas A&M University (continued)

For master's, GRE General Test. Additional exam requirements/recommendations for international students: Required—TOEFL (minimum score 550 paper-based). Electronic applications accepted. *Faculty research:* Crop and soil disciplines.

West Virginia University, Davis College of Agriculture, Forestry and Consumer Sciences, Division of Animal and Veterinary Sciences, Program in Agricultural Sciences, Morgantown, WV 26506. Offers animal and food sciences (PhD); plant and soil sciences (PhD). *Degree requirements:* For doctorate, thesis/dissertation, oral and written exams. *Entrance requirements:* Additional exam requirements/recommendations for international students: Required—TOEFL. Application fee: $45. *Expenses:* Tuition, state resident: full-time $4,582; part-time $258 per credit hour. Tuition, nonresident: full-time $1,382; part-time $741 per credit hour. *Financial support:* Research assistantships with tuition reimbursements, teaching assistantships with tuition reimbursements, Federal Work-Study, institutionally sponsored loans, and tuition waivers (full and partial) available. Financial award application deadline: 2/1; financial award applicants required to submit FAFSA. *Faculty research:* Ruminant nutrition, metabolism, forage utilization, physiology, reproduction. *Application contact:* Dr. Hillar Klandorf, Professor, 304-293-2631 Ext. 4436, Fax: 304-293-3676, E-mail: hillar.klandorf@mail.wvu.edu.

West Virginia University, Davis College of Agriculture, Forestry and Consumer Sciences, Division of Plant and Soil Sciences, Morgantown, WV 26506. Offers agronomy (MS); entomology (MS); environmental microbiology (MS); horticulture (MS); plant pathology (MS). *Faculty:* 18 full-time (1 woman), 1 part-time/adjunct (0 women). *Students:* 17 full-time (11 women), 3 part-time (2 women), 2 international. Average age 27. In 2005, 5 degrees awarded. *Degree requirements:* For master's, thesis. *Entrance requirements:* For master's, GRE, minimum GPA of 2.5. Additional exam requirements/recommendations for international students: Required—TOEFL. *Application deadline:* Applications are processed on a rolling basis. Application fee: $45. *Expenses:* Tuition, state resident: full-time $4,582; part-time $258 per credit hour. Tuition, nonresident: full-time $1,382; part-time $741 per credit hour. *Financial support:* In 2005–06, 13 research assistantships with full tuition reimbursements (averaging $9,936 per year), 4 teaching assistantships with full tuition reimbursements (averaging $9,936 per year) were awarded; Federal Work-Study, institutionally sponsored loans, and tuition waivers (full and partial) also available. Financial award application deadline: 2/1; financial award applicants required to submit FAFSA. *Faculty research:* Water quality, reclamation of disturbed land, crop production, pest control, environmental protection. Total annual research expenditures: $1 million. *Unit head:* Dr. Barton S. Baker, Chair and Division Director, 304-293-4817 Ext. 4342, Fax: 304-293-2960, E-mail: barton.baker@mail.wvu.edu.

CLEMSON UNIVERSITY

CLEMSON UNIVERSITY

Master of Science in Animal and Veterinary Sciences

Program of Study

The Master of Science (M.S.) in Animal and Veterinary Sciences (AVS) program generally requires two years' minimum residence in the program. Students are required to complete a core set of graduate-level courses in animal science, biochemistry, and experimental statistics in addition to the courses recommended by their advisory committee. M.S. students complete a thesis project and defend the thesis in an oral exam.

The purpose of this graduate program is to provide a high-quality education for graduate students with diverse goals and to develop, through research, the knowledge and technology necessary to continually improve productivity, efficiency, and sustainability of animal agriculture. Students interested in a professional career in the animal sciences are exposed to educational and research experiences involving the many facets of animal nutrition, physiology, microbiology, and genetics through a rigorous curriculum of graduate-level courses and challenging experimentation approaches. The most current ideas and concepts in animal sciences are provided to the students through their daily interactions with the faculty members in both the classroom and laboratory environments. All students are required to use the most relevant experimental methods and techniques in answering questions to improve the understanding of animal biology in an effort to provide high-quality food products that enhance people's lives.

Research Facilities

Students in the program have access to twelve laboratories that expose them to the full array of laboratory techniques and equipment. These laboratories are equipped with benchtop space, hoods, centrifuges, autoclaves, spectrophotometers, chromatography units, cell culture incubators, histology microtomes, microscopes, PCR thermocyclers, and other major equipment pieces typically found in wet chemistry laboratories. In addition, students have access to five animal farms and the Godley-Snell Animal Facility. There is excellent cooperation with faculty members in other departments to allow students access to technology not found in the program's home department. The main library maintains holdings relevant to the subject areas of the animal and related sciences, such as nutrition, physiology, microbiology, and genetics. Faculty members use the mainframe computing system for data analysis and storage. Each faculty member has access to at least one computer in her or his office. There are additional computers maintained in laboratories and in commons rooms in the department.

Financial Aid

There are some departmental teaching assistantships available on a competitive basis. Research assistantships are available through grant support of individual faculty members in the department. Students may also apply for fellowships offered through the College of Agriculture, Forestry and Life Sciences. Occasionally, work is available on an hourly basis in laboratories and at the animal farms.

Cost of Study

Tuition for 2006–07 is $4643 per semester for in-state students and $9255 per semester for nonresidents. Off-campus rates are $535 per hour for in-state students and $918 per hour for nonresidents. Graduate assistants pay a flat fee of $1079 per semester and $348 per summer session. Graduate fellows pay South Carolina resident fees.

Living and Housing Costs

On-campus housing is available. For information, students should visit http://www.housing.clemson.edu. The cost of living in Clemson is quite low compared to the national average. Students who choose to live off the campus typically spend $300–$400 per month for rent, depending on location, amenities, roommates, etc.

Student Group

The students enrolled in the graduate program have academic backgrounds in the animal sciences, biological sciences, and microbiology. Generally, the students have a strong interest in working with domestic animals in order to improve the productivity of these animals as agriculturally important commodities. The program strives to have 15 to 20 students (M.S. and Ph.D. students combined) enrolled at any given time.

Student Outcomes

Students can expect to move into jobs related to the animal production industry, animal health industry, feed industry, professional schools, academic positions, and government positions.

Location

Clemson is a small, beautiful college town near the Blue Ridge Mountains and Lake Hartwell in upstate South Carolina. The Upstate is one of the country's fastest-growing areas and is the midpoint of the Charlotte-to-Atlanta I-85 corridor, a multistate area along Interstate 85 that runs from metro Atlanta to Richmond, Virginia, and encompasses Charlotte, North Carolina, and North Carolina's Research Triangle. Atlanta and Charlotte are each a 2-hour's drive away. Many financial institutions and other industries have national headquarters for a major presence in the Upstate, including Wachovia, Bank of America, BMW, Bon Secours St. Francis Health System, Bosch North America, Bowater, Charter Communications, Ernst and Young, Fluor Corporation, IBM, Microsoft, Michelin of North America, and many others.

The University

Clemson is classified by the Carnegie Foundation as Doctoral/Research University–Extensive, a category comprising less than 4 percent of all universities in America. The University's mission is to fulfill the covenant between its founder and the people of South Carolina to establish a "high seminary of learning" through its responsibilities of teaching, research, and extended public service. The University has identified eight areas of academic emphasis that create collaborations that, in turn, help fulfill the University's mission.

Applying

Applicants may apply on the Web at http://www.grad.clemson.edu/p_apply.html. Applications with a $50 nonrefundable fee should be received no later than five weeks prior to registration. Every required item in support of the application must be on file by that date. Students are advised to contact the department for the deadlines of the program of proposed study. When submitting an application to the Graduate School, students must also provide a one-page written narrative of interests and plans for using their graduate degree. The latter is very important in the AVS's program evaluation of applicants.

Correspondence and Information

Dr. Thomas R. Scott, Professor
Department of Animal and Veterinary Sciences
123 P&A Building
Clemson University
Clemson, South Carolina 29634

Phone: 864-656-4027
Fax: 864-656-3131
E-mail: trscott@clemson.edu
Web site: http://www.clemson.edu/avs

Clemson University

THE FACULTY AND THEIR RESEARCH

Jean A. Bertrand, Professor; Ph.D., Georgia, 1987. Dairy cattle nutrition.
Glenn P. Birrenkott Jr., Professor; Ph.D., Wisconsin–Madison, 1978. Endocrinology, reproductive physiology, and poultry science.
Ashby B. Bodine II, Professor and Department Chair; Ph.D., Clemson, 1978. Nutrition.
Steven E. Ellis, Assistant Professor; Ph.D., Virginia Tech, 1998. Dairy science.
John Robert Gibbons, Assistant Professor; Ph.D., Wisconsin–Madison, 1998. Endocrinology–reproductive physiology.
Tomas Gimenez, Professor; Dr.Med.Vet., Technical University of Munich, 1975. Reproductive endocrinology.
Annel K. Greene, Professor; Ph.D., Mississippi State, 1988. Food science.
Michelle A. Hall, Associate Professor; Ph.D., Wisconsin–Madison, 1982. Animal science.
Harold D. Hupp, Professor; Ph.D., Virginia Tech, 1977. Animal breeding and genetics.
Thomas C. Jenkins, Professor; Ph.D., Cornell, 1979. Animal science.
Denzil V. Maurice, Professor; Ph.D., Georgia, 1978. Nutrition.
Larry W. Olson, Associate Professor; Ph.D., Nebraska, 1976. Animal science.
Thomas R. Scott, Professor; Ph.D., Georgia, 1983. Poultry science.
Peter A. Skewes, Professor; Ph.D., Virginia Tech, 1985. Poultry science.

CLEMSON
UNIVERSITY

CLEMSON UNIVERSITY

Ph.D. in Animal and Veterinary Sciences

Program of Study

The Ph.D. in Animal and Veterinary Sciences (AVS) program generally requires three years' minimum residence in the program. Students are required to complete a core set of graduate-level courses in animal science, biochemistry, and experimental statistics in addition to the courses recommended by their advisory committee. Ph.D. students, in addition to any course work, are required to successfully pass a set of written exams and an oral comprehensive exam prior to completion of their dissertation research. The dissertation defense is a final oral exam.

The purpose of this graduate program is to provide a high-quality education for graduate students with diverse goals and to develop, through research, the knowledge and technology necessary to continually improve productivity, efficiency, and sustainability of animal agriculture. Students interested in a professional career in the animal sciences are exposed to educational and research experiences involving the many facets of animal nutrition, physiology, microbiology, and genetics through a rigorous curriculum of graduate-level courses and challenging experimentation approaches. The most current ideas and concepts in animal sciences are provided to the students through their daily interactions with the faculty members in both the classroom and laboratory environments. All students are required to use the most relevant experimental methods and techniques in answering questions to improve the understanding of animal biology in an effort to provide high-quality food products that enhance people's lives.

Research Facilities

Students in the program have access to twelve laboratories that expose them to the full array of laboratory techniques and equipment. These laboratories are equipped with benchtop space, hoods, centrifuges, autoclaves, spectrophotometers, chromatography units, cell culture incubators, histology microtomes, microscopes, PCR thermocyclers, and other major equipment pieces typically found in wet chemistry laboratories. In addition, students have access to five animal farms and the Godley-Snell Animal Facility. There is excellent cooperation with faculty members in other departments to allow students access to technology not found in the program's home department. The main library maintains holdings relevant to the subject areas of the animal and related sciences, such as nutrition, physiology, microbiology, and genetics. Faculty members use the mainframe computing system for data analysis and storage. Each faculty member has access to at least one computer in her or his office. There are additional computers maintained in laboratories and in commons rooms in the department.

Financial Aid

There are some departmental teaching assistantships available on a competitive basis. Research assistantships are available through grant support of individual faculty members in the department. Students may also apply for fellowships offered through the College of Agriculture, Forestry and Life Sciences. Occasionally, work is available on an hourly basis in laboratories and at the animal farms.

Cost of Study

Tuition for 2006–07 is $4643 per semester for in-state students and $9255 per semester for nonresidents. Off-campus rates are $535 per hour for in-state students and $918 per hour for nonresidents. Graduate assistants pay a flat fee of $1079 per semester and $348 per summer session. Graduate fellows pay South Carolina resident fees.

Living and Housing Costs

On-campus housing is available. For information, students should visit http://www.housing.clemson.edu. The cost of living in Clemson is quite low compared to the national average. Students who choose to live off the campus typically spend $300–$400 per month for rent, depending on location, amenities, roommates, etc.

Student Group

The students enrolled in the graduate program have academic backgrounds in the animal sciences, biological sciences, and microbiology. Generally, the students have a strong interest in working with domestic animals in order to improve the productivity of these animals as agriculturally important commodities. The program strives to have 15 to 20 students (M.S. and Ph.D. students combined) enrolled at any given time.

Student Outcomes

Students can expect to move into jobs related to the animal production industry, animal health industry, feed industry, professional schools, academic positions, and government positions.

Location

Clemson is a small, beautiful college town near the Blue Ridge Mountains and Lake Hartwell in upstate South Carolina. The Upstate is one of the country's fastest-growing areas and is the midpoint of the Charlotte-to-Atlanta I-85 corridor, a multistate area along Interstate 85 that runs from metro Atlanta to Richmond, Virginia, and encompasses Charlotte, North Carolina, and North Carolina's Research Triangle. Atlanta and Charlotte are each a 2-hour's drive away. Many financial institutions and other industries have national headquarters for a major presence in the Upstate, including Wachovia, Bank of America, BMW, Bon Secours St. Francis Health System, Bosch North America, Bowater, Charter Communications, Ernst and Young, Fluor Corporation, IBM, Microsoft, Michelin of North America, and many others.

The University

Clemson is classified by the Carnegie Foundation as Doctoral/Research University–Extensive, a category comprising less than 4 percent of all universities in America. The University's mission is to fulfill the covenant between its founder and the people of South Carolina to establish a "high seminary of learning" through its responsibilities of teaching, research, and extended public service. The University has identified eight areas of academic emphasis that create collaborations that, in turn, help fulfill the University's mission.

Applying

Applicants may apply on the Web at http://www.grad.clemson.edu/p_apply.html. Applications with a $50 nonrefundable fee should be received no later than five weeks prior to registration. Every required item in support of the application must be on file by that date. Students are advised to contact the department for the deadlines of the program of proposed study. When submitting an application to the Graduate School, students must also provide a one-page written narrative of interests and plans for using their graduate degree. The latter is very important in the AVS's program evaluation of applicants.

Correspondence and Information

Dr. Thomas R. Scott, Professor
Department of Animal and Veterinary Sciences
123 P&A Building
Clemson University
Clemson, South Carolina 29634

Phone: 864-656-4027
Fax: 864-656-3131
E-mail: trscott@clemson.edu
Web site: http://www.clemson.edu/avs

Clemson University

THE FACULTY AND THEIR RESEARCH

Jean A. Bertrand, Professor; Ph.D., Georgia, 1987. Dairy cattle nutrition.
Glenn P. Birrenkott Jr., Professor; Ph.D., Wisconsin–Madison, 1978. Endocrinology, reproductive physiology, and poultry science.
Ashby B. Bodine II, Professor and Department Chair; Ph.D., Clemson, 1978. Nutrition.
Steven E. Ellis, Assistant Professor; Ph.D., Virginia Tech, 1998. Dairy science.
John Robert Gibbons, Assistant Professor; Ph.D., Wisconsin–Madison, 1998. Endocrinology–reproductive physiology.
Tomas Gimenez, Professor; Dr.Med.Vet., Technical University of Munich, 1975. Reproductive endocrinology.
Annel K. Greene, Professor; Ph.D., Mississippi State, 1988. Food science.
Michelle A. Hall, Associate Professor; Ph.D., Wisconsin–Madison, 1982. Animal science.
Harold D. Hupp, Professor; Ph.D., Virginia Tech, 1977. Animal breeding and genetics.
Thomas C. Jenkins, Professor; Ph.D., Cornell, 1979. Animal science.
Denzil V. Maurice, Professor; Ph.D., Georgia, 1978. Nutrition.
Larry W. Olson, Associate Professor; Ph.D., Nebraska, 1976. Animal science.
Thomas R. Scott, Professor; Ph.D., Georgia, 1983. Poultry science.
Peter A. Skewes, Professor; Ph.D., Virginia Tech, 1985. Poultry science.

CLEMSON UNIVERSITY

CLEMSON UNIVERSITY

Department of Food Science and Human Nutrition
Master of Science in Food, Nutrition, and Culinary Science

Program of Study

The Master of Science (M.S.) degree program in food, nutrition, and culinary science at Clemson University requires a minimum of 24 credit hours of course work and 6 credit hours of thesis research. Only 600-level courses and higher may be used for graduate credit, and at least half of the 24 hours of course work must be at the 800 level or higher. Students must take Statistical Methods I or its equivalent; Food Preservation and Processing, a 1-credit-hour seminar offered each spring; 18 credit hours of advanced-level courses, which may include classes in food science or in such areas as animal and veterinary sciences, biochemistry, chemistry, cell biology, microbiology, nutrition, or statistics; and 6 credit hours of thesis research. In addition, a minimum GPA of 3.0 is required to maintain good academic standing and for graduation.

Research Facilities

The program offers a wide array of research equipment and facilities, including the following: Hewlett Packard gas chromatograph (mass spectral detector) with HP headspace injector; Hewlett Packard gas chromatograph (FID) with Tek-Mar headspace injector; Hewlett Packard HPChemstation analytical software; Gow-Mac gas chromatograph (TCD); Hewlett Packard gas chromatograph (FPD); Neotronics Electronic Nose 4000; -70°C freezer; deionized and distilled water equipment; Kjeldahl nitrogen analyzers; BYK-Gardner Spectrocolorimeter; Seward stomacher 400 laboratory blender; colony counters; rapid PCR analytical equipment; HPLC-MS sensory analysis facility; a food science teaching laboratory; state-of-the-art culinary preparation equipment; and more.

Students in the program have access to the research facilities of several departments, including the Department of Agricultural and Biological Engineering, the Department of Animal and Veterinary Sciences, the Department of Food Science and Human Nutrition, the Department of Horticulture, and the Department of Packaging Science. These departments have a collection of modern research labs, farms, and more. Students should go online to http://www.clemson.edu/colleges for links to these departments and their facilities.

The Clemson University libraries have a collection of more than 7,000 serial titles and 1.5 million volumes that were developed to support the undergraduate and graduate curricula and research. Several commercial bibliographic databases and locally created full-text databases may be searched online from local and remote computers. The library is linked electronically through OCLC, Inc., to more than 11,000 other libraries worldwide for cataloging and interlibrary loan services.

Financial Aid

A limited number of research assistantships are available from grant funds, with the student assisting in the research supported under the grant. Interested applicants should contact individual faculty members. Applicants whose files are completed by February 15 are given preferential consideration.

Cost of Study

Tuition for 2006–07 is $4643 per semester for in-state students and $9255 per semester for nonresidents. Off-campus rates are $535 per hour for in-state students and $918 per hour for nonresidents. Graduate assistants pay a flat fee of $1079 per semester and $348 per summer session. Graduate fellows pay South Carolina resident fees.

Living and Housing Costs

On-campus housing is available. For more information, students should visit http://www.housing.clemson.edu. The cost of living in Clemson is quite low compared to the national average. Students who choose to live off campus typically spend $300–$400 per month for rent, depending on such things as location, amenities, and roommates.

Student Group

Of the 12 students in the program, 8 attend on a full-time basis, 4 are international students, and 7 are women.

Location

Clemson is a small, beautiful college town near the Blue Ridge Mountains and Lake Hartwell. The Upstate is one of the country's fastest-growing areas and is an important part of the I-85 corridor, a multistate area along Interstate 85 that runs from metropolitan Atlanta to Richmond, Virginia, and encompasses Charlotte, North Carolina, and North Carolina's Research Triangle. Atlanta and Charlotte are each a 2-hour drive away. Many financial institutions and other industries have a national or major presence in the Upstate, including Wachovia, Bank of America, BMW, Bon Secours St. Francis Health System, Bosch North America, Bowater, Charter Communications, Ernst & Young, Fluor Corporation, IBM, Microsoft, Michelin of North America, and many others.

The University and The Department

Clemson is classified by the Carnegie Foundation as Doctoral/Research University–Extensive, a category comprising less than 4 percent of all universities in America. The University's mission is to fulfill the covenant between its founder and the people of South Carolina to establish a "high seminary of learning" through its responsibilities of teaching, research, and extended public service. The University has identified eight areas of academic emphasis that create collaborations that, in turn, help fulfill the University's mission.

The Department of Food Science and Human Nutrition at Clemson University provides the only food science program in South Carolina. The faculty and staff members possess a wide array of expertise in the food technology and nutrition/dietetic areas. The Department is recognized nationally for its top-flight graduates, and its students are responsible for the manufacture of Clemson's world-famous ice cream. Graduates of the Department's programs are highly sought after and easily compete for the best jobs across the United States.

Applying

Applicants should have a strong background in food science; human nutrition; physical, chemical, or biological sciences; or engineering. Proficiency in food science must be demonstrated by the satisfactory completion of course work in food chemistry, food microbiology, food processing, and biochemistry.

Applicants may apply on the Web at http://www.grad.clemson.edu/p_apply.html. Students must submit the completed application, the $40 application fee, scores from the GRE General Test (minimum total score of 1000), all academic transcripts (minimum undergraduate GPA of 3.0), three letters of recommendation, and a statement of objectives and professional experience. International students must have a minimum TOEFL score of 575. International applicants must also submit documentation of adequate financial support for their studies. All candidates must also identify a research adviser who is prepared to accept the applicant as an advisee. The application deadline is June 1. Applications are processed on a rolling basis.

Correspondence and Information

Dr. Paul Dawson, Graduate Coordinator
Department of Food Science and Human Nutrition
Clemson University
Clemson, South Carolina 29634

Phone: 864-656-1138
Fax: 864-656-3131
E-mail: pdawson@clemson.edu
Web site: http://www.clemson.edu/foodscience/

Clemson University

THE FACULTY AND THEIR RESEARCH

Food Science

James C. Acton, Stender Professor; Ph.D., Georgia. Functional property evaluations of meat proteins, microbial and chemical aspects of fermented meat products, and interactions of light, oxygen, and storage temperature with packaging systems for meat products and other foods.

Felix H. Barron, Professor; Ph.D., Michigan State. Computer modeling; simulation and optimization of food processes; food quality and shelf life of packaged foods; design of food-processing plants, sanitation inspection, and compliance with laws and regulations; waste-packaging materials.

Feng Chen, Assistant Professor; Ph.D., LSU. Food flavor chemistry and the identification, isolation, and application of bioactive nutraceutical compounds from natural sources.

Paul L. Dawson, Professor; Ph.D., North Carolina State. Lipid chemistry, flavor, and oxidation of meat products; interaction of packaging materials with food components; recovery and use of underused poultry products, including MDBM and spent fowl meat.

Ronald D. Galyean, Professor; Ph.D., Missouri–Columbia. Effects of processing on the functional characteristics of proteins, interaction of food product ingredient composition with processing requirements for optimal functional characteristics and product safety.

Xiuping Jiang, Assistant Professor; Ph.D., Maryland. Food microbiology and food safety, specifically the control and source of food pathogens and biosafety.

John U. McGregor, Professor and Department Chair; Ph.D., Mississippi State. Chemistry and flavor of coffee, natural antioxidants, specialty cheeses and value-added processing.

Human Nutrition

Katherine Cason, Professor; Ph.D., Virginia Tech; RD. Influences of socioeconomic and ecological factors on food and nutrient intake, food purchasing and preparation practices, level of food security among limited-resource audiences, assessment of the impact of nutrition education on dietary adequacy and the food and nutrition practices of participants.

Marge Condrasky, Assistant Professor; Ed.D., Clemson; RD. Culinary science and distance education.

Vivian Haley-Zitlin, Associate Professor; Ph.D., Tennessee, Knoxville; RD. Metabolism of phytochemicals and effects of nutrients on liver and glucose metabolism.

Rita M. Haliena, Senior Lecturer; M.S., Ball State; RD. Clinical nutrition and clinical dietetics management, especially differences in the measures of nutritional status and length of stay for high-risk patients.

M. Elizabeth Kunkel, Professor; Ph.D., Tennessee, Knoxville; RD. Effects of nutrient stress on noncollagenous proteins of bone, single-photon absorptiometry in animals as an indicator of mineral bioavailability, impact of macronutrients and food-processing techniques on mineral bioactivity.

CLEMSON UNIVERSITY

CLEMSON UNIVERSITY

Department of Food Science and Human Nutrition
Food Technology

Program of Study

There are no set course requirements for the Ph.D. in food technology. Ph.D. candidates gain a comprehensive understanding of the principles of food science, with an expanded knowledge that covers their focused research areas. The Ph.D. candidate's research committee has final approval on all course work. Students must pass both the written and oral qualifying examinations given by their advisory committee. The successful student must also write and defend a research dissertation to the satisfaction of the advisory committee. Students should expect to publish a minimum of two refereed research manuscripts from their dissertations.

Research Facilities

The program offers a wide array of research equipment and facilities, including the following: Hewlett Packard gas chromatograph (mass spectral detector) with HP headspace injector; Hewlett Packard gas chromatograph (FID) with Tek-Mar headspace injector; Hewlett Packard HPChemstation analytical software; Gow-Mac gas chromatograph (TCD); Hewlett Packard gas chromatograph (FPD); Neotronics Electronic Nose 4000; -70°C freezer; deionized and distilled water equipment; Kjeldahl nitrogen analyzers; BYK-Gardner Spectrocolorimeter; Seward stomacher 400 laboratory blender; colony counters; rapid PCR analytical equipment; HPLC-MS sensory analysis facility; a food science teaching laboratory; state-of-the-art culinary preparation equipment; and more.

Students in the program have access to the research facilities of several departments, including the Department of Agricultural and Biological Engineering, the Department of Animal and Veterinary Sciences, the Department of Food Science and Human Nutrition, the Department of Horticulture, and the Department of Packaging Science. These departments have a collection of modern research labs, farms, and more. Students should go online to http://www.clemson.edu/colleges for links to these departments and their facilities.

The Clemson University libraries have a collection of more than 7,000 serial titles and 1.5 million volumes that were developed to support the undergraduate and graduate curricula and research. Several commercial bibliographic databases and locally created full-text databases may be searched online from local and remote computers. The library is linked electronically through OCLC, Inc., to more than 11,000 other libraries worldwide for cataloging and interlibrary loan services.

Financial Aid

A limited number of research assistantships are available from grant funds, with the student assisting in the research supported under the grant. Interested applicants should contact individual faculty members. Applicants whose files are completed by February 15 are given preferential consideration.

Cost of Study

Tuition for 2006–07 is $4643 per semester for in-state students and $9255 per semester for nonresidents. Off-campus rates are $535 per hour for in-state students and $918 per hour for nonresidents. Graduate assistants pay a flat fee of $1079 per semester and $348 per summer session. Graduate fellows pay South Carolina resident fees.

Living and Housing Costs

On-campus housing is available. For more information, students should visit http://www.housing.clemson.edu. The cost of living in Clemson is quite low compared to the national average. Students who choose to live off campus typically spend $300–$400 per month for rent, depending on such things as location, amenities, and roommates.

Student Group

Of the 10 students in the program, 9 attend on a full-time basis, 7 are international students, and 1 is a member of a minority group.

Location

Clemson is a small, beautiful college town near the Blue Ridge Mountains and Lake Hartwell. The Upstate is one of the country's fastest-growing areas and is an important part of the I-85 corridor, a multistate area along Interstate 85 that runs from metropolitan Atlanta to Richmond, Virginia, and encompasses Charlotte, North Carolina, and North Carolina's Research Triangle. Atlanta and Charlotte are each a 2-hour drive away. Many financial institutions and other industries have a national or major presence in the Upstate, including Wachovia, Bank of America, BMW, Bon Secours St. Francis Health System, Bosch North America, Bowater, Charter Communications, Ernst & Young, Fluor Corporation, IBM, Microsoft, Michelin of North America, and many others.

The University and The Department

Clemson is classified by the Carnegie Foundation as Doctoral/Research University–Extensive, a category comprising less than 4 percent of all universities in America. The University's mission is to fulfill the covenant between its founder and the people of South Carolina to establish a "high seminary of learning" through its responsibilities of teaching, research, and extended public service. The University has identified eight areas of academic emphasis that create collaborations that, in turn, help fulfill the University's mission.

The Department of Food Science and Human Nutrition at Clemson University provides the only food science program in South Carolina. The faculty and staff members possess a wide array of expertise in the food technology and nutrition/dietetic areas. The Department is recognized nationally for its top-flight graduates, and its students are responsible for the manufacture of Clemson's world-famous ice cream. Graduates of the Department's programs are highly sought after and easily compete for the best jobs across the United States.

Applying

Applicants should have a strong background in food science; human nutrition; physical, chemical, or biological sciences; or engineering. Proficiency in food science must be demonstrated by the satisfactory completion of course work in food chemistry, food microbiology, food processing, and biochemistry.

Applicants may apply on the Web at http://www.grad.clemson.edu/p_apply.html. Students must submit the completed application, the $40 application fee, scores from the GRE General Test (minimum total score of 1000), all academic transcripts (minimum undergraduate GPA of 3.0), three letters of recommendation, and a statement of objectives and professional experience. International students must have a minimum TOEFL score of 575. International applicants must also submit documentation of adequate financial support for their studies. All candidates must also identify a research adviser who is prepared to accept the applicant as an advisee. The application deadline is June 1. Applications are processed on a rolling basis.

Correspondence and Information

Dr. Paul Dawson, Graduate Coordinator
Department of Food Science and Human Nutrition
Clemson University
Clemson, South Carolina 29634

Phone: 864-656-1138
Fax: 864-656-3131
E-mail: pdawson@clemson.edu
Web site: http://www.clemson.edu/foodscience/

Clemson University

THE FACULTY AND THEIR RESEARCH

Food Science

James C. Acton, Stender Professor; Ph.D., Georgia. Functional property evaluations of meat proteins, microbial and chemical aspects of fermented meat products, and interactions of light, oxygen, and storage temperature with packaging systems for meat products and other foods.

Felix H. Barron, Professor; Ph.D., Michigan State. Computer modeling; simulation and optimization of food processes; food quality and shelf life of packaged foods; design of food-processing plants, sanitation inspection, and compliance with laws and regulations; waste-packaging materials.

Feng Chen, Assistant Professor; Ph.D., LSU. Food flavor chemistry and the identification, isolation, and application of bioactive nutraceutical compounds from natural sources.

Paul L. Dawson, Professor; Ph.D., North Carolina State. Lipid chemistry, flavor, and oxidation of meat products; interaction of packaging materials with food components; recovery and use of underused poultry products, including MDBM and spent fowl meat.

Ronald D. Galyean, Professor; Ph.D., Missouri–Columbia. Effects of processing on the functional characteristics of proteins, interaction of food product ingredient composition with processing requirements for optimal functional characteristics and product safety.

Xiuping Jiang, Assistant Professor; Ph.D., Maryland. Food microbiology and food safety, specifically the control and source of food pathogens and biosafety.

John U. McGregor, Professor and Department Chair; Ph.D., Mississippi State. Chemistry and flavor of coffee, natural antioxidants, specialty cheeses and value-added processing.

Human Nutrition

Katherine Cason, Professor; Ph.D., Virginia Tech; RD. Influences of socioeconomic and ecological factors on food and nutrient intake, food purchasing and preparation practices, level of food security among limited-resource audiences, assessment of the impact of nutrition education on dietary adequacy and the food and nutrition practices of participants.

Marge Condrasky, Assistant Professor; Ed.D., Clemson; RD. Culinary science and distance education.

Vivian Haley-Zitlin, Associate Professor; Ph.D., Tennessee, Knoxville; RD. Metabolism of phytochemicals and effects of nutrients on liver and glucose metabolism.

Rita M. Haliena, Senior Lecturer; M.S., Ball State; RD. Clinical nutrition and clinical dietetics management, especially differences in the measures of nutritional status and length of stay for high-risk patients.

M. Elizabeth Kunkel, Professor; Ph.D., Tennessee, Knoxville; RD. Effects of nutrient stress on noncollagenous proteins of bone, single-photon absorptiometry in animals as an indicator of mineral bioavailability, impact of macronutrients and food-processing techniques on mineral bioactivity.

ACADEMIC AND PROFESSIONAL PROGRAMS IN THE ENVIRONMENT AND NATURAL RESOURCES

Section 9
Environmental Sciences and Management

This section contains a directory of institutions offering graduate work in environmental sciences and management, followed by in-depth entries submitted by institutions that chose to prepare detailed program descriptions. Additional information about programs listed in the directory but not augmented by an in-depth entry may be obtained by writing directly to the dean of a graduate school or chair of a department at the address given in the directory.

For programs offering related work, see also in this book Natural Resources; in Book 2, see Political Science and International Affairs and Public, Regional, and Industrial Affairs; in Book 3, see Ecology, Environmental Biology, and Evolutionary Biology; and in Book 5, see Management of Engineering and Technology.

CONTENTS

Program Directories

Announcements

Cross-Discipline Announcements

Close-Ups

See also:

Environmental Management and Policy

Adelphi University, Graduate School of Arts and Sciences, Program in Environmental Studies, Garden City, NY 11530-0701. Offers MS. *Students:* 2 full-time (1 woman), 8 part-time (3 women); includes 1 minority (African American) Average age 30. In 2005, 3 degrees awarded. *Degree requirements:* For master's, thesis optional. *Entrance requirements:* For master's, GRE, 3 letters of recommendation; course work in microeconomics, political science, statistics/calculus, and either chemistry or physics; computer literacy. Additional exam requirements/recommendations for international students: Required—TOEFL (minimum score 550 paper-based; 213 computer-based). Application fee: $50. *Expenses:* Tuition: Full-time $21,150; part-time $650 per credit. Required fees: $550; $450 per year. Tuition and fees vary according to degree level, campus/location and program. *Financial support:* Research assistantships with full and partial tuition reimbursements, teaching assistantships, career-related internships or fieldwork, Federal Work-Study, institutionally sponsored loans, and unspecified assistantships available. Financial award application deadline: 2/15; financial award applicants required to submit FAFSA. *Faculty research:* Contaminates sites, workplace exposure level of contaminants, climate change and human health. *Unit head:* Dr. Anagnostis Agelarakis, Director, 516-877-4112, E-mail: agelarak@adelphi.edu. *Application contact:* Christine Murphy, Director of Admissions, 516-877-3050, Fax: 516-877-3039, E-mail: admissions@adelphi.edu.

Air Force Institute of Technology, Graduate School of Engineering and Management, Department of Systems and Engineering Management, Dayton, OH 45433-7765. Offers cost analysis (MS); environmental and engineering management (MS); environmental engineering science (MS); information resource/systems management (MS). *Accreditation:* ABET. Part-time programs available. *Faculty:* 20 full-time (5 women). *Students:* 224 full-time (34 women), 4 part-time (1 woman). Average age 33. In 2005, 103 degrees awarded. *Degree requirements:* For master's, thesis. *Entrance requirements:* For master's, GRE, GMAT, minimum GPA of 3.0. *Application deadline:* For fall admission, 3/1 for domestic students. Applications are processed on a rolling basis. Application fee: $0. *Financial support:* Fellowships, research assistantships with full tuition reimbursements, scholarships/grants available. Financial award application deadline: 3/15. Total annual research expenditures: $424,000. *Unit head:* Dr. Mark N. Goltz, Head, 937-255-4759, E-mail: mark.goltz@afit.edu. *Application contact:* Maj. Mark A. Ward, Information Contact, 937-255-3636 Ext. 4742, E-mail: mark.ward@afit.edu.

American University, College of Arts and Sciences, Department of Biology, Environmental Science Program, Washington, DC 20016-8001. Offers environmental science (MS); marine science (MS). *Students:* 4 full-time (3 women), 1 (woman) part-time; includes 1 minority (Hispanic American), 1 international. Average age 28. In 2005, 6 degrees awarded. *Degree requirements:* For master's, thesis or alternative, comprehensive exam. *Entrance requirements:* For master's, GRE General Test, GRE Subject Test, minimum GPA of 3.0. Additional exam requirements/recommendations for international students: Required—TOEFL. *Application deadline:* For fall admission, 2/1 for domestic students; for spring admission, 10/1 for domestic students. Application fee: $50. *Expenses:* Tuition: Full-time $17,802; part-time $989 per credit. Required fees: $380. *Financial support:* Research assistantships, teaching assistantships available. Financial award application deadline: 2/1. *Unit head:* Dr. Kiho Kim, Director, 202-885-2181, Fax: 202-885-2181.

American University, School of International Service, Washington, DC 20016-8001. Offers comparative and regional studies (MA); cross-cultural communication (Certificate); development management (MS); environmental policy (MA); ethics, peace, and global affairs (MA); global environmental policy (MA); international communication (MA); international development (MA); international development management (Certificate); international economic policy (MA); international economic relations (Certificate); international peace and conflict resolution (MA); international politics (MA); international relations (PhD); international service (MIS); the Americas (Certificate); U.S. foreign policy (MA). Part-time and evening/weekend programs available. *Faculty:* 66 full-time (24 women), 39 part-time/adjunct (11 women). *Students:* 533 full-time (338 women), 367 part-time (237 women); includes 126 minority (42 African Americans, 3 American Indian/Alaska Native, 44 Asian Americans or Pacific Islanders, 37 Hispanic Americans), 117 international. Average age 27. 1,870 applicants, 65% accepted, 318 enrolled. In 2005, 316 master's, 2 doctorates awarded. Terminal master's awarded for partial completion of doctoral program. *Degree requirements:* For master's, one foreign language, thesis or alternative, comprehensive exam; for doctorate, one foreign language, thesis/dissertation, comprehensive exam. *Entrance requirements:* For master's, GRE General Test, 24 credits of course work in related social sciences, minimum GPA of 3.3, 2 letters of recommendation; for doctorate, GRE General Test, 2 letters of recommendation, 24 credits in related social sciences. Additional exam requirements/recommendations for international students: Required—TOEFL (minimum score 550 paper-based; 213 computer-based). *Application deadline:* For fall admission, 1/15 for domestic students; for spring admission, 10/1 priority date for domestic students. Applications are processed on a rolling basis. Application fee: $50. *Expenses:* Tuition: Full-time $17,802; part-time $989 per credit. Required fees: $380. *Financial support:* Career-related internships or fieldwork, Federal Work-Study, and institutionally sponsored loans available. Financial award application deadline: 1/15. *Faculty research:* International intellectual property, international environmental issues, international law and legal order, international telecommunications/technology, international sustainable development. *Unit head:* Dr. Louis W. Goodman, Dean, 202-885-1600, Fax: 202-885-2494. *Application contact:* Amanda Taylor, Director of Graduate Admissions and Financial Aid, 202-885-1599, Fax: 202-885-2494.

American University of Beirut, Graduate Programs, Faculty of Arts and Sciences, Beirut, Lebanon. Offers anthropology (MA); Arabic language and literature (MA); archaeology (MA); biology (MS); business administration (MBA); chemistry (MS); computer science (MS); economics (MA); education (MA); English language (MA); English literature (MA); environmental policy planning (MSES); finance and banking (MFB); financial economics (MFE); geology (MS); history (MA); mathematics (MS); Middle Eastern studies (MA); philosophy (MA); physics (MS); political studies (MA); psychology (MA); public administration (MA); sociology (MA). *Degree requirements:* For master's, one foreign language, thesis (for some programs), comprehensive exam, registration. *Entrance requirements:* For master's, GRE, letter of recommendation.

Antioch New England Graduate School, Graduate School, Department of Environmental Studies, Doctoral Program in Environmental Studies, Keene, NH 03431-3552. Offers PhD. *Degree requirements:* For doctorate, thesis/dissertation, practicum. *Entrance requirements:* For doctorate, master's degree and previous experience in the environmental field. Additional exam requirements/recommendations for international students: Required—TOEFL. Electronic applications accepted. Expenses: Contact institution. *Faculty research:* Environmental history, green politics, ecopsychology.

See Close-Up on page 641.

Antioch New England Graduate School, Graduate School, Department of Environmental Studies, Program in Environmental Studies, Keene, NH 03431-3552. Offers conservation biology (MS); environmental advocacy (MS); environmental education (MS); teacher certification in biology (7th-12th grade) (MS); teacher certification in general science (5th-9th grade) (MS). *Degree requirements:* For master's, practicum. *Entrance requirements:* For master's, previous undergraduate course work in biology, chemistry, mathematics (environmental biology). Additional exam requirements/recommendations for international students: Required—TOEFL (minimum score 550 paper-based; 213 computer-based). Electronic applications accepted. Expenses: Contact institution. *Faculty research:* Sustainability, natural resources inventory.

See Close-Up on page 641.

Antioch New England Graduate School, Graduate School, Department of Environmental Studies, Program in Resource Management and Administration, Keene, NH 03431-3552. Offers MS. *Degree requirements:* For master's, practicum, thesis optional. *Entrance requirements:* For master's, previous undergraduate course work in science and math. Additional exam requirements/recommendations for international students: Required—TOEFL (minimum score 600 paper-based; 250 computer-based). Electronic applications accepted. Expenses: Contact institution. *Faculty research:* Waste management, land use.

See Close-Up on page 641.

Antioch University Seattle, Graduate Programs, Center for Creative Change, Seattle, WA 98121-1814. Offers environment and community (MA); management (MS); organizational psychology (MA); strategic communications (MA); whole system design (MA). Evening/weekend programs available. *Faculty:* 7 full-time (3 women), 10 part-time/adjunct (6 women). *Students:* 87 full-time (59 women), 51 part-time (40 women); includes 21 minority (7 African Americans, 3 American Indian/Alaska Native, 8 Asian Americans or Pacific Islanders, 3 Hispanic Americans). Average age 37. 71 applicants, 66% accepted. In 2005, 87 degrees awarded. *Application deadline:* For fall admission, 8/15 for domestic students; for spring admission, 2/3 priority date for domestic students. Applications are processed on a rolling basis. Application fee: $50. Electronic applications accepted. *Expenses: Contact institution. Financial support:* In 2005–06, 6 research assistantships (averaging $6,480 per year) were awarded; Federal Work-Study, institutionally sponsored loans, and unspecified assistantships also available. Financial award application deadline: 6/15. *Unit head:* Shana Hormann, Interim Director, 206-441-5352 Ext. 5707. *Application contact:* Pam Smith Mentz, Dean of Student and Enrollment Services, 206-441-5352 Ext. 5200, E-mail: psmith-mentz@antiochsea.edu.

Arizona State University at the Polytechnic Campus, Morrison School of Agribusiness and Resource Management, Mesa, AZ 85212. Offers agribusiness (MS); environmental resources (MS). Part-time and evening/weekend programs available. *Degree requirements:* For master's, thesis, oral defense. *Entrance requirements:* For master's, GMAT, GRE General Test, MAT, minimum GPA of 3.0, 3 letters of recommendation, resumé. Additional exam requirements/recommendations for international students: Required—TOEFL (minimum score 550 paper-based; 213 computer-based); Recommended—TWE, TSE. Electronic applications accepted. *Faculty research:* Agribusiness marketing, management and financial structuring.

Bard College, Bard Center for Environmental Policy, Annandale-on-Hudson, NY 12504. Offers MS, Professional Certificate, MS/JD, MS/MAT. Masters international is offered with the Peace Corps. Part-time programs available. *Degree requirements:* For master's, thesis, internship within U.S. or abroad, master's project. *Entrance requirements:* For master's, GRE (if out of school less than 5 years), statement, written work, curriculum vitae, transcripts, letters of recommendation (3); for Professional Certificate, statement, written work, curriculum vitae, transcripts, letters of recommendations (3). Additional exam requirements/recommendations for international students: Required—TOEFL. Expenses: Contact institution. *Faculty research:* Agricultural practices, fisheries management, international agreements, decision making under uncertainty, climate change.

See Close-Up on page 643.

Baylor University, Graduate School, College of Arts and Sciences, Department of Environmental Studies, Waco, TX 76798. Offers MES, MS. *Students:* 16 full-time (7 women); includes 4 minority (1 African American, 3 Hispanic Americans), 3 international. In 2005, 3 degrees awarded. *Degree requirements:* For master's, thesis. *Entrance requirements:* For master's, GRE General Test. *Application deadline:* For fall admission, 8/1 for domestic students; for spring admission, 1/1 for domestic students. Applications are processed on a rolling basis. Application fee: $25. *Financial support:* Research assistantships, teaching assistantships, career-related internships or fieldwork, Federal Work-Study, and institutionally sponsored loans available. *Faculty research:* Renewable energy/waste management policies, Third World environmental problem solving, ecotourism. *Unit head:* Dr. Susan Bratton, Graduate Program Director, 254-710-3405, Fax: 254-710-3409. *Application contact:* Suzanne Keener, Administrative Assistant, 254-710-3588, Fax: 254-710-3870.

Bemidji State University, School of Graduate Studies, College of Social and Natural Sciences, Center for Environmental Studies, Bemidji, MN 56601-2699. Offers MS. Part-time programs available. *Faculty:* 4 part-time/adjunct (0 women). *Students:* 4 full-time (3 women), 10 part-time (9 women). Average age 28. 3 applicants, 100% accepted. In 2005, 5 degrees awarded. *Degree requirements:* For master's, thesis or alternative. *Entrance requirements:* For master's, GRE General Test. *Application deadline:* For fall admission, 5/1 for domestic students. Applications are processed on a rolling basis. Application fee: $20. Electronic applications accepted. *Expenses:* Tuition, state resident: full-time $4,716. Required fees: $384. One-time fee: $20 full-time. *Financial support:* In 2005–06, 1 research assistantship with partial tuition reimbursement (averaging $8,000 per year) was awarded; career-related internships or fieldwork, Federal Work-Study, scholarships/grants, and unspecified assistantships also available. Support available to part-time students. Financial award application deadline: 5/1. *Unit head:* Dr. Fu-Hsian Chang, Director, 218-755-4104, Fax: 218-755-4107.

Boise State University, Graduate College, College of Social Science and Public Affairs, Program in Public Policy and Administration, Boise, ID 83725. Offers environmental and natural resources policy and administration (MPA); general public administration (MPA); state and local government policy and administration (MPA). *Accreditation:* NASPAA. Part-time programs available. *Degree requirements:* For master's, directed research project, internship. *Entrance requirements:* For master's, GRE General Test, minimum GPA of 3.0. Additional exam requirements/recommendations for international students: Required—TOEFL. Electronic applications accepted.

Boston University, Graduate School of Arts and Sciences, Department of Geography and Environment, Boston, MA 02215. Offers MA, PhD. *Students:* 32 full-time (14 women), 1 (woman) part-time; includes 1 minority (Hispanic American), 17 international. Average age 31. 55 applicants, 27% accepted, 6 enrolled. In 2005, 13 doctorates awarded. Terminal master's awarded for partial completion of doctoral program. *Degree requirements:* For master's and doctorate, one foreign language, thesis/dissertation, comprehensive exam, registration. *Entrance requirements:* For master's and doctorate, GRE General Test, GRE Subject Test, 3 letters of recommendation. Additional exam requirements/recommendations for international students: Required—TOEFL (minimum score 600 paper-based; 250 computer-based). *Application deadline:* For fall admission, 7/1 for domestic students, 7/1 for international students; for spring admission, 11/15 for domestic students, 11/15 for international students. Application fee: $60. *Expenses:* Tuition: Full-time $31,530; part-time $985 per credit. Required fees: $316; $40 per semester. Tuition and fees vary according to course level and program. *Financial support:* In 2005–06, 29 students received support, including 2 fellowships with full tuition reimbursements available (averaging $16,500 per year), 15 research assistantships with full tuition reimbursements available (averaging $16,000 per year), 9 teaching assistantships with full tuition reimbursements available (averaging $16,000 per year); Federal Work-Study and unspecified assistantships also available. Support available to part-time students. Financial award application deadline: 1/15; financial award applicants required to submit FAFSA. Total annual research expenditures: $1.2 million. *Unit head:* William P. Anderson, Chairman, 617-353-0208, Fax: 617-353-8399, E-mail: bander@bu.edu. *Application contact:* Laura Guild, Administrative Assistant, 617-358-0206, Fax: 617-353-8399, E-mail: guild@bu.edu.

Boston University, Graduate School of Arts and Sciences, Department of International Relations, Boston, MA 02215. Offers African studies (Certificate); international relations (MA); international relations and environmental policy management (MA); international relations and international communication (MA). *Students:* 52 full-time (33 women), 14 part-time (7 women); includes 7 minority (2 African Americans, 3 Asian Americans or Pacific Islanders, 2 Hispanic Americans), 9 international. Average age 28. 337 applicants, 64% accepted, 48 enrolled. In 2005, 42 degrees awarded. *Degree requirements:* For master's, one foreign language, thesis, comprehensive exam, registration. *Entrance requirements:* For master's, GRE General Test, 3 letters of recommendation; for Certificate, GRE General Test. Additional exam requirements/recommendations for international students: Required—TOEFL (minimum score 600 paper-based; 250 computer-based). *Application deadline:* For fall admission, 4/15 for domestic students, 4/15 for international students; for spring admission, 10/15 for domestic students, 10/15 for international students. Application fee: $60. *Expenses:* Tuition: Full-time $31,530; part-time $985 per credit. Required fees: $316; $40 per semester. Tuition and fees vary according to course level and program. *Financial support:* In 2005–06, 18 students received support. Federal Work-Study, scholarships/grants, and unspecified assistantships available. Support available to part-time students. Financial award application deadline: 1/15; financial award applicants required to submit FAFSA. *Unit head:* Dr. Erik Goldstein, Chairman, 617-353-9280, Fax: 617-353-9290, E-mail: goldstee@bu.edu. *Application contact:* Michael Williams, Graduate Program Administrator, 617-353-9349, Fax: 617-353-9290, E-mail: mawillia@bu.edu.

Boston University, Graduate School of Arts and Sciences, Program in Energy and Environmental Studies, Boston, MA 02215. Offers energy and environmental analysis (MA); environmental remote sensing and geographic information systems (MA); international relations and environmental policy (MA). *Students:* 14 full-time (11 women), 6 part-time (3 women); includes 1 minority (Asian American or Pacific Islander), 5 international. Average age 26. 67 applicants, 69% accepted, 12 enrolled. *Degree requirements:* For master's, one foreign language, comprehensive exam, registration, research paper. *Entrance requirements:* For master's, GRE General Test, 2 letters of recommendation. Additional exam requirements/recommendations for international students: Required—TOEFL (minimum score 550 paper-based; 213 computer-based). *Application deadline:* For fall admission, 7/1 for domestic students, 7/1 for international students; for spring admission, 11/15 for domestic students, 11/15 for international students. Application fee: $60. *Expenses:* Tuition: Full-time $31,530; part-time $985 per credit. Required fees: $316; $40 per semester. Tuition and fees vary according to course level and program. *Financial support:* In 2005–06, 6 students received support, including 4 research assistantships with full tuition reimbursements available (averaging $16,000 per year); fellowships, career-related internships or fieldwork and Federal Work-Study also available. Support available to part-time students. Financial award application deadline: 1/15; financial award applicants required to submit FAFSA. *Unit head:* Cutler J. Cleveland, Director, 617-353-7552, Fax: 617-353-5986, E-mail: cutler@bu.edu. *Application contact:* Alpana Roy, Administrative Assistant, 617-353-3083, Fax: 617-353-5986, E-mail: alpana@bu.edu.

Brown University, Graduate School, Center for Environmental Studies, Providence, RI 02912. Offers AM. Part-time programs available. *Faculty:* 4 full-time (2 women), 11 part-time/adjunct (3 women). *Students:* 20 full-time (10 women); includes 2 minority (1 Asian American or Pacific Islander, 1 Hispanic American), 3 international. Average age 25. 33 applicants, 52% accepted, 8 enrolled. In 2005, 8 degrees awarded. *Degree requirements:* For master's, thesis. *Entrance requirements:* For master's, GRE, writing sample. Additional exam requirements/recommendations for international students: Required—TOEFL. *Application deadline:* For fall admission, 1/2 priority date for domestic students, 1/2 priority date for international students. Applications are processed on a rolling basis. Application fee: $70. Electronic applications accepted. *Financial support:* In 2005–06, 20 students received support, including 2 teaching assistantships with full tuition reimbursements available (averaging $14,000 per year); career-related internships or fieldwork, Federal Work-Study, health care benefits, and tuition waivers (partial) also available. Financial award application deadline: 1/2; financial award applicants required to submit FAFSA. *Faculty research:* Solid waste management, risk management policy (environmental health), resource management policy (water/fisheries), climate change, environmental justice. *Unit head:* Osvaldo Sala, Director, 401-863-3449, Fax: 401-863-3503, E-mail: osvaldo_sala@brown.edu. *Application contact:* Patricia-Ann Caton, Administrative Manager, 401-863-3449, Fax: 401-863-3503, E-mail: patti_caton@brown.edu.

California Polytechnic State University, San Luis Obispo, College of Agriculture, Department of Natural Resources Management, San Luis Obispo, CA 93407. Offers forestry science (MS). Part-time programs available. *Faculty:* 7 full-time (1 woman), 2 part-time/adjunct (0 women). *Students:* 1 (woman) full-time, 2 part-time (1 woman). 1 applicant, 100% accepted, 1 enrolled. In 2005, 1 degree awarded. *Degree requirements:* For master's, thesis, comprehensive exam. *Entrance requirements:* For master's, minimum 2.75 GPA in last 90 quarter units of course work. *Application deadline:* For fall admission, 6/1 for domestic students, 11/30 for international students. For winter admission, 8/1 for domestic students; for spring admission, 12/1 for domestic students. Applications are processed on a rolling basis. Application fee: $55. Electronic applications accepted. *Expenses:* Tuition, nonresident: part-time $226 per unit. Required fees: $1,063 per unit. *Financial support:* Application deadline: 3/2; *Unit head:* Dr. Doug Piirto, Department Head, Graduate Coordinator, 805-756-2968, Fax: 805-756-1402, E-mail: dpiirto@calpoly.edu.

California State University, Fullerton, Graduate Studies, College of Humanities and Social Sciences, Program in Environmental Studies, Fullerton, CA 92834-9480. Offers environmental education and communication (MS); environmental policy and planning (MS); environmental sciences (MS); technological studies (MS). Part-time programs available. *Students:* 31 full-time (17 women), 43 part-time (22 women); includes 23 minority (2 African Americans, 10 Asian Americans or Pacific Islanders, 11 Hispanic Americans), 3 international. Average age 31. 40 applicants, 65% accepted, 18 enrolled. In 2005, 19 degrees awarded. *Degree requirements:* For master's, thesis. *Entrance requirements:* For master's, minimum GPA of 2.5 in last 60 units of course work. Application fee: $55. *Expenses:* Tuition, nonresident: part-time $339 per unit. *Financial support:* Career-related internships or fieldwork, Federal Work-Study, institutionally sponsored loans, and scholarships/grants available. Support available to part-time students. Financial award application deadline: 3/1. *Unit head:* Dr. Robert Voeks, Coordinator, 714-278-4373.

Central European University, Graduate Studies, Department of Environmental Sciences and Policy, Budapest, Hungary. Offers MS, PhD. *Faculty:* 5 full-time (1 woman), 2 part-time/adjunct (1 woman). *Students:* 85 full-time (53 women). Average age 26. 516 applicants, 15% accepted, 57 enrolled. In 2005, 35 master's awarded. *Degree requirements:* For master's, one foreign language, thesis/dissertation, registration; for doctorate, one foreign language, thesis/dissertation, comprehensive exam, registration. *Entrance requirements:* For master's and doctorate, interview. Additional exam requirements/recommendations for international students: Required—TOEFL (minimum score 570 paper-based; 230 computer-based). *Application deadline:* For fall admission, 1/5 for domestic students, 1/5 for international students. Application fee: $0. Electronic applications accepted. Tuition and fees charges are reported in euros. *Expenses:* Tuition: Full-time 8,900 euros; part-time 4,450 euros per semester. Required fees: 450 euros. Full-time tuition and fees vary according to program. *Financial support:* In 2005–06, 50 students received support, including 31 fellowships with full and partial tuition reimbursements available (averaging $5,000 per year); career-related internships or fieldwork, institutionally sponsored loans, scholarships/grants, and tuition waivers (full and partial) also available. Financial award application deadline: 1/6. *Faculty research:* Management of ecological systems, environmental impact assessment, energy conservation, climate change policy, forest policy in countries in transition. Total annual research expenditures: $16,127. *Unit head:* Dr. Ruben Mnatsakanian, Head, 361-327-3021, Fax: 361-327-3031, E-mail: envsci@ceu.hu. *Application contact:* Krisztina Szabados, Coordinator, 361-327-3021, Fax: 361-327-3031, E-mail: envsci@ceu.hu.

Central Washington University, Graduate Studies, Research and Continuing Education, College of the Sciences, Program in Resource Management, Ellensburg, WA 98926. Offers MS. *Faculty:* 29 full-time (9 women). *Students:* 48 full-time (20 women), 25 part-time (8 women); includes 15 minority (13 American Indian/Alaska Native, 1 Asian American or Pacific Islander, 1 Hispanic American). 35 applicants, 83% accepted, 20 enrolled. In 2005, 5 degrees awarded. *Degree requirements:* For master's, thesis. *Entrance requirements:* For master's, minimum GPA of 3.0. Additional exam requirements/recommendations for international students: Required—TOEFL (minimum score 550 paper-based; 213 computer-based). *Application deadline:* For fall admission, 4/1 for domestic students; for spring admission, 1/1 for domestic students. Applications are processed on a rolling basis. Application fee: $50. Electronic applications accepted. *Expenses:* Tuition, state resident: full-time $1,968; part-time $197 per credit. Tuition, nonresident: full-time $4,320; part-time $432 per credit. Required fees: $623. Tuition and fees vary according to degree level. *Financial support:* In 2005–06, 20 research assistantships with partial tuition reimbursements (averaging $8,100 per year), 9 teaching assistantships with partial tuition reimbursements (averaging $8,100 per year) were awarded.; career-related internships or fieldwork, Federal Work-Study, and unspecified assistantships also available. Financial award application deadline: 3/1; financial award applicants required to submit FAFSA. *Unit head:* Dr. Anthony Gabriel, Co-Director, 509-963-1166, Fax: 509-963-3224. *Application contact:* Justine Eason, Admissions Program Coordinator, 509-963-3103, Fax: 509-963-1799, E-mail: masters@cwu.edu.

Clark University, Graduate School, Department of International Development, Community, and Environment, Program in Environmental Science and Policy, Worcester, MA 01610-1477. Offers MA. Part-time programs available. *Students:* 33 full-time (23 women); includes 1 African American, 11 international. Average age 29. 62 applicants, 89% accepted, 20 enrolled. In 2005, 14 degrees awarded. *Degree requirements:* For master's, thesis. *Entrance requirements:* For master's, GRE General Test. Additional exam requirements/recommendations for international students: Required—TOEFL. *Application deadline:* For fall admission, 1/15 for domestic students. Application fee: $50. *Expenses:* Tuition: Full-time $29,300. Required fees: $30. *Financial support:* In 2005–06, fellowships (averaging $9,750 per year), research assistantships with full and partial tuition reimbursements (averaging $9,750 per year), teaching assistantships with full and partial tuition reimbursements (averaging $9,750 per year) were awarded.; tuition waivers (partial) also available. *Faculty research:* Environmental management, natural and man-made hazards, health risks, public health policy, hazard management. *Unit head:* Dr. William F. Fisher, Director, 508-421-3765, Fax: 508-793-8820, E-mail: wfisher@clarku.edu. *Application contact:* Paula Hall, IDCE Graduate Admissions, 508-793-7201, Fax: 508-793-8820, E-mail: idce@clarku.edu.

Clark University, Graduate School, Department of International Development, Community, and Environment, Program in Geographic Information Science for Development and Environment, Worcester, MA 01610-1477. Offers MA. *Students:* 15 full-time (7 women), 3 part-time (1 woman), 7 international. Average age 33. 44 applicants, 80% accepted, 8 enrolled. In 2005, 15 degrees awarded. *Degree requirements:* For master's, thesis. *Entrance requirements:* Additional exam requirements/recommendations for international students: Required—TOEFL. *Application deadline:* For fall admission, 1/15 for domestic students. Application fee: $50. *Expenses:* Tuition: Full-time $29,300. Required fees: $30. *Financial support:* In 2005–06, fellowships (averaging $9,750 per year), research assistantships with full and partial tuition reimbursements (averaging $9,750 per year), teaching assistantships with full and partial tuition reimbursements (averaging $9,750 per year) were awarded.; tuition waivers (full and partial) also available. *Faculty research:* Dynamic modeling, image processing, land use and land cover change modeling, image classification, spatial econometrics. *Unit head:* Dr. William F. Fisher, Director, 508-421-3765, Fax: 508-793-8820, E-mail: wfisher@clarku.edu. *Application contact:* Paula Hall, IDCE Graduate Admissions, 508-793-7201, Fax: 508-793-8820, E-mail: idce@clarku.edu.

Clemson University, Graduate School, College of Agriculture, Forestry and Life Sciences, Department of Biological Sciences, Program in Plant and Environmental Sciences, Clemson, SC 29634. Offers MS, PhD. *Students:* 35 full-time (11 women), 14 part-time (8 women); includes 2 minority (1 American Indian/Alaska Native, 1 Hispanic American), 12 international. 20 applicants, 45% accepted, 8 enrolled. In 2005, 8 master's, 1 doctorate awarded. *Degree requirements:* For master's, thesis. *Entrance requirements:* For master's, GRE General Test, bachelor's degree in biological science or chemistry. Additional exam requirements/recommendations for international students: Required—TOEFL. *Application deadline:* For fall admission, 6/1 for domestic students. Application fee: $50. *Financial support:* Teaching assistantships available. Financial award application deadline: 3/15; financial award applicants required to submit FAFSA. *Faculty research:* Systematics, aquatic botany, plant ecology, plant-fungus interactions, plant developmental genetics. *Unit head:* Dr. Halina Knapp, Coordinator, 864-656-3523, Fax: 864-656-7594, E-mail: hskrpsk@clemson.edu.

Cleveland State University, College of Graduate Studies, Maxine Goodman Levin College of Urban Affairs, Program in Environmental Studies, Cleveland, OH 44115. Offers MAES, JD/MAES. Part-time and evening/weekend programs available. *Faculty:* 25 full-time (10 women), 11 part-time/adjunct (3 women). *Students:* 75 full-time (31 women), 199 part-time (120 women); includes 71 minority (60 African Americans, 4 Asian Americans or Pacific Islanders, 7 Hispanic Americans), 24 international. *Degree requirements:* For master's, thesis or alternative. *Entrance requirements:* For master's, GRE General Test, minimum GPA of 3.0. Additional exam requirements/recommendations for international students: Required—TOEFL (minimum score 525 paper-based; 197 computer-based). *Application deadline:* For fall admission, 7/15 for domestic students. Applications are processed on a rolling basis. *Expenses:* Tuition, state resident: full-time $10,700. Tuition, nonresident: full-time $14,628. Tuition and fees vary according to program. *Financial support:* In 2005–06, 1 research assistantship with full and partial tuition reimbursement (averaging $6,960 per year) was awarded; career-related internships or fieldwork, Federal Work-Study, tuition waivers (partial), and unspecified assistantships also available. Support available to part-time students. Financial award application deadline: 3/1. *Faculty research:* Environmental policy and administration, environmental planning, geographic information systems (GIS), nonprofit management. *Unit head:* Dr. Sanda Kaufman, Director, 216-687-2367, Fax: 216-687-9342, E-mail: sanda@urban.csuohio.edu. *Application contact:* Graduate Programs Coordinator, 216-523-7522, Fax: 216-687-5398, E-mail: gradprog@urban.csuohio.edu.

College of the Atlantic, Program in Human Ecology, Bar Harbor, ME 04609-1198. Offers M Phil. *Degree requirements:* For master's, thesis. *Faculty research:* Conservation of endangered species, public policy/community planning, environmental education, history, philosophy.

Colorado State University, Graduate School, College of Liberal Arts, Department of Political Science, Fort Collins, CO 80523-0015. Offers environmental politics and policy (PhD); political science (MA, PhD). Part-time programs available. *Faculty:* 17 full-time (6 women). *Students:* 22 full-time (10 women), 23 part-time (8 women); includes 1 minority (African American), 3 international. Average age 33. 38 applicants, 68% accepted, 12 enrolled. In 2005, 3 master's, 1 doctorate awarded. *Median time to degree:* Of those who began their doctoral program in fall 1997, 25% received their degree in 8 years or less. *Degree requirements:* For master's, thesis (for some programs); for doctorate, thesis/dissertation, comprehensive exam. *Entrance requirements:* For master's, GRE General Test, minimum GPA of 3.0; for doctorate, GRE General Test, MA, minimum GPA of 3.5. Additional exam requirements/recommendations for international students: Required—TOEFL (minimum score 600 paper-based). *Application deadline:* For fall admission, 2/15 for domestic students, 2/15 for international students; for spring admission, 10/15 for domestic students, 10/15 for international students. Applications are processed on a rolling basis. Application fee: $50. Electronic applications accepted. *Expenses:* Tuition, state resident: full-time $3,690; part-time $205 per credit. Tuition, nonresident: full-time $14,958; part-time $831 per credit. Required fees: $1,061. *Financial support:* In 2005–06, 12 students received support, including 1 fellowship (averaging $4,500 per year), 9 teaching assistantships with full tuition reimbursements available (averaging $9,720 per year); research assistantships, career-related internships or fieldwork, Federal Work-Study, institutionally sponsored loans, and traineeships also available. Financial award application deadline: 2/15; financial award applicants required to submit FAFSA. *Faculty research:* Environmental politics and policy, international relations, politics of developing nations, state and local politics and administration, political behavior. Total annual research expenditures: $11,800. *Unit head:* Dr. William Chaloupka, Chair, 970-491-5157, Fax: 970-491-2490, E-mail: williamc@colostate.edu. *Application contact:* Dr. Robert Duffy, Coordinator, 970-491-6225, Fax: 970-491-2490, E-mail: robert.duffy@colostate.edu.

Environmental Management and Policy

Colorado State University, Graduate School, Warner College of Natural Resources, Department of Forest, Rangeland, and Watershed Stewardship, Fort Collins, CO 80523-0015. Offers forest sciences (MS, PhD); natural resource stewardship (MNRS); rangeland ecosystem science (MS, PhD); watershed science (MS). Part-time programs available. *Faculty:* 27 full-time (5 women), 1 part-time/adjunct (0 women). *Students:* 58 full-time (23 women), 90 part-time (39 women); includes 11 minority (1 African American, 2 American Indian/Alaska Native, 1 Asian American or Pacific Islander, 7 Hispanic Americans), 10 international. Average age 33. 101 applicants, 35% accepted, 30 enrolled. In 2005, 18 master's, 5 doctorates awarded. Terminal master's awarded for partial completion of doctoral program. *Degree requirements:* For master's, thesis optional; for doctorate, thesis/dissertation, comprehensive exam, registration. *Entrance requirements:* For master's and doctorate, GRE General Test, minimum GPA of 3.0. Additional exam requirements/recommendations for international students: Required—TOEFL. *Application deadline:* For fall admission, 4/1 for domestic students; for spring admission, 9/1 priority date for domestic students. Applications are processed on a rolling basis. Application fee: $50. Electronic applications accepted. *Expenses:* Tuition, state resident: full-time $3,690; part-time $205 per credit. Tuition, nonresident: full-time $14,958; part-time $831 per credit. Required fees: $1,061. *Financial support:* Career-related internships or fieldwork, Federal Work-Study, institutionally sponsored loans, and traineeships available. Financial award application deadline: 5/1. *Faculty research:* Ecology, natural resource management, hydrology, restoration, human dimensions. Total annual research expenditures: $4.1 million. *Unit head:* Dr. N. Thompson Hobbs, Head, 970-491-6911, Fax: 970-491-6754, E-mail: nthobbs@warnercnr@colostate.edu. *Application contact:* Graduate Coordinator, 970-491-4994, Fax: 970-491-6754.

Colorado State University, Graduate School, Warner College of Natural Resources, Department of Natural Resource Recreation and Tourism, Fort Collins, CO 80523-0015. Offers commercial recreation and tourism (MS); human dimensions in natural resources (MS, PhD); recreation resource management (MS, PhD); resource interpretation (MS). Part-time programs available. *Faculty:* 10 full-time (3 women). *Students:* 24 full-time (12 women), 20 part-time (11 women), 8 international. Average age 31. 46 applicants, 46% accepted, 13 enrolled. In 2005, 9 master's, 1 doctorate awarded. Terminal master's awarded for partial completion of doctoral program. *Degree requirements:* For master's, thesis or alternative, comprehensive exam, registration; for doctorate, thesis/dissertation, comprehensive exam, registration. *Entrance requirements:* For master's and doctorate, GRE General Test, minimum GPA of 3.0. Additional exam requirements/recommendations for international students: Required—TOEFL. *Application deadline:* For fall admission, 3/1 for domestic students. Applications are processed on a rolling basis. Application fee: $50. Electronic applications accepted. *Expenses:* Tuition, state resident: full-time $3,690; part-time $205 per credit. Tuition, nonresident: full-time $14,958; part-time $831 per credit. Required fees: $1,061. *Financial support:* In 2005–06, 6 research assistantships with tuition reimbursements (averaging $14,000 per year), 8 teaching assistantships with tuition reimbursements (averaging $14,000 per year) were awarded.; fellowships, career-related internships or fieldwork, Federal Work-Study, scholarships/grants, and traineeships also available. Support available to part-time students. Financial award application deadline: 2/1; financial award applicants required to submit FAFSA. *Faculty research:* International tourism, wilderness preservation, resource interpretation, human dimensions in natural resources, protected areas management. Total annual research expenditures: $768,741. *Unit head:* Dr. Michael J. Manfredo, Chair, 970-491-6591, Fax: 970-491-2255, E-mail: manfredo@cnr.colostate.edu. *Application contact:* Jacqie J. Hasan, Coordinator of Administration, 970-491-6591, Fax: 970-491-2255, E-mail: jaq@cnr.colostate.edu.

Columbia University, School of International and Public Affairs, Program in Environmental Science and Policy, New York, NY 10027. Offers MPA. Program admits applicants in June only. *Faculty:* 11 full-time (4 women), 9 part-time/adjunct (4 women). *Students:* 57 full-time (43 women); includes 9 minority (4 African Americans, 5 Asian Americans or Pacific Islanders), 9 international. Average age 27. 210 applicants, 57 enrolled. In 2005, 54 degrees awarded. *Degree requirements:* For master's, workshops. *Entrance requirements:* For master's, GRE (recommended), previous course work in biology and chemistry, or earth sciences (recommended). Additional exam requirements/recommendations for international students: Required—TOEFL. *Application deadline:* For fall admission, 2/15 for domestic students, 2/15 for international students. Applications are processed on a rolling basis. Application fee: $75. Electronic applications accepted. *Expenses:* Tuition: Full-time $31,448. Tuition and fees vary according to course level, course load, campus/location and program. *Financial support:* Fellowships with partial tuition reimbursements, Federal Work-Study and scholarships/grants available. Financial award application deadline: 1/15; financial award applicants required to submit FAFSA. *Faculty research:* Ecological management of enclosed ecosystems vegetation dynamics, environmental policy and management, energy policy, nuclear waste policy, environmental and natural resource economics and policy. *Unit head:* Dr. Steven A. Cohen, Director, 212-854-3142, Fax: 212-864-4847, E-mail: sc32@columbia.edu. *Application contact:* Louise A. Rosen, Assistant Director, 212-854-0643, Fax: 212-864-4847, E-mail: lar46@columbia.edu.

See Close-Up on page 649.

Concordia University, School of Graduate Studies, Faculty of Arts and Science, Department of Geography, Montréal, QC H3G 1M8, Canada. Offers environmental impact assessment (Diploma). *Students:* 24 full-time (17 women), 3 part-time (2 women). In 2005, 8 degrees awarded. *Application deadline:* For fall admission, 3/1 for domestic students. For winter admission, 8/31 for domestic students. Application fee: $50. *Expenses:* Tuition, state resident: full-time $834; part-time $334 per term. Tuition, nonresident: full-time $2,200; part-time $880 per term. Required fees: $680 per term. Tuition and fees vary according to degree level and program. *Unit head:* Dr. John Zacharias, Chair, 514-848-2424 Ext. 2058, Fax: 514-848-2057. *Application contact:* Dr. John Zacharias, Director, 514-848-2424 Ext. 2056, Fax: 514-848-2057.

Cornell University, Graduate School, Field of Environmental Management, Ithaca, NY 14853. Offers MPS. *Application contact:* Tad McGalliard, Education Coordinator, 607-255-9996, Fax: 607-255-0238, E-mail: tnm2@cornell.edu.

Cornell University, Graduate School, Graduate Fields of Agriculture and Life Sciences, Field of Applied Economics and Management, Ithaca, NY 14853-0001. Offers agricultural economics (MPS, MS, PhD), including agricultural finance, applied econometrics and quantitative analysis, economics of development, farm management and production economics (MPS), marketing and food distribution (MPS), public policy analysis (MPS); resource economics (MPS, MS, PhD), including environmental economics, environmental management (MPS), resource economics. *Faculty:* 58 full-time (8 women). *Students:* 241 applicants, 22% accepted, 23 enrolled. In 2005, 35 master's, 4 doctorates awarded. Terminal master's awarded for partial completion of doctoral program. *Degree requirements:* For master's, thesis (MS); for doctorate, thesis/dissertation, comprehensive exam. *Entrance requirements:* For master's and doctorate, GRE General Test, 2 letters of recommendation. Additional exam requirements/recommendations for international students: Required—TOEFL (minimum score 550 paper-based; 213 computer-based). *Application deadline:* For fall admission, 1/15 for domestic students. Application fee: $60. Electronic applications accepted. *Financial support:* In 2005–06, 47 students received support, including 13 fellowships with full tuition reimbursements available, 18 research assistantships with full tuition reimbursements available, 16 teaching assistantships with full tuition reimbursements available; institutionally sponsored loans, scholarships/grants, health care benefits, tuition waivers (full and partial), and unspecified assistantships also available. Financial award applicants required to submit FAFSA. *Faculty research:* Production economics, international economic development and trade, farm management and finance, resource and environmental economics, agricultural marketing and policy. *Unit head:* Director of Graduate Studies, 607-255-8048, Fax: 607-255-9984, E-mail: aegrad@cornell.edu. *Application contact:* Graduate Field Assistant, 607-255-8048, Fax: 607-255-9984, E-mail: aegrad@cornell.edu.

Cornell University, Graduate School, Graduate Fields of Agriculture and Life Sciences, Field of Natural Resources, Ithaca, NY 14853-0001. Offers aquatic science (MPS, MS, PhD); environmental management (MPS); fishery science (MPS, MS, PhD); forest science (MPS, MS, PhD); resource policy and management (MPS, MS, PhD); wildlife science (MPS, MS, PhD). *Faculty:* 50 full-time (9 women). *Students:* 62 full-time (33 women); includes 9 minority (1 African American, 2 American Indian/Alaska Native, 4 Asian Americans or Pacific Islanders, 2 Hispanic Americans), 14 international. 60 applicants, 23% accepted, 12 enrolled. In 2005, 11 master's, 5 doctorates awarded. *Degree requirements:* For master's, thesis (MS), project paper (MPS); for doctorate, thesis/dissertation, comprehensive exam. *Entrance requirements:* For master's and doctorate, GRE General Test, 2 letters of recommendation. Additional exam requirements/recommendations for international students: Required—TOEFL (minimum score 550 paper-based; 213 computer-based). *Application deadline:* For spring admission, 10/30 for domestic students. Applications are processed on a rolling basis. Application fee: $60. Electronic applications accepted. *Financial support:* In 2005–06, 49 students received support, including 15 fellowships with full tuition reimbursements available, 16 research assistantships with full tuition reimbursements available, 18 teaching assistantships with full tuition reimbursements available; institutionally sponsored loans, scholarships/grants, health care benefits, tuition waivers (full and partial), and unspecified assistantships also available. Financial award applicants required to submit FAFSA. *Faculty research:* Ecosystem-level dynamics, systems modeling, conservation biology/management, resource management's human dimensions, biogeochemistry. *Unit head:* Director of Graduate Studies, 607-255-2807, Fax: 607-255-0349. *Application contact:* Graduate Field Assistant, 607-255-2807, Fax: 607-255-0349, E-mail: nrgrad@cornell.edu.

Cornell University, Graduate School, Graduate Fields of Agriculture and Life Sciences, Field of Soil and Crop Sciences, Ithaca, NY 14853-0001. Offers agronomy (MS, PhD); environmental information science (MS, PhD); environmental management (MPS); field crop science (MS, PhD); soil science (MS, PhD). *Faculty:* 38 full-time (9 women). *Students:* 40 applicants, 33% accepted, 11 enrolled. In 2005, 6 master's, 3 doctorates awarded. *Degree requirements:* For master's, thesis (MS); for doctorate, thesis/dissertation, comprehensive exam. *Entrance requirements:* For master's and doctorate, GRE General Test, 2 letters of recommendation. Additional exam requirements/recommendations for international students: Required—TOEFL (minimum score 550 paper-based; 213 computer-based). *Application deadline:* For fall admission, 2/1 for domestic students. Applications are processed on a rolling basis. Application fee: $60. Electronic applications accepted. *Financial support:* In 2005–06, 35 students received support, including 6 fellowships with full tuition reimbursements available, 25 research assistantships with full tuition reimbursements available, 4 teaching assistantships with full tuition reimbursements available; institutionally sponsored loans, traineeships, health care benefits, tuition waivers (full and partial), and unspecified assistantships also available. *Faculty research:* Soil chemistry, physics and biology; crop physiology and management; environmental information science and modeling; international agriculture; weed science. *Unit head:* Director of Graduate Studies, 607-255-3267, Fax: 607-255-8615. *Application contact:* Graduate Field Assistant, 607-255-3267, Fax: 607-255-8615, E-mail: jae2@cornell.edu.

Cornell University, Graduate School, Graduate Fields of Architecture, Art and Planning, Field of Regional Science, Ithaca, NY 14853-0001. Offers environmental studies (MA, MS, PhD); international spatial problems (MA, MS, PhD); location theory (MA, MS, PhD); multiregional economic analysis (MA, MS, PhD); peace science (MA, MS, PhD); planning methods (MA, MS, PhD); urban and regional economics (MA, MS, PhD). *Faculty:* 20 full-time (4 women). *Students:* 6 applicants, 17% accepted, 0 enrolled. In 2005, 2 master's, 1 doctorate awarded. Terminal master's awarded for partial completion of doctoral program. *Degree requirements:* For master's, thesis/dissertation; for doctorate, thesis/dissertation, comprehensive exam. *Entrance requirements:* For master's and doctorate, GRE General Test (native English speakers only), 2 letters of recommendation. Additional exam requirements/recommendations for international students: Required—TOEFL (minimum score 600 paper-based; 250 computer-based). *Application deadline:* For fall admission, 1/15 for domestic students. Application fee: $60. Electronic applications accepted. *Financial support:* Fellowships with full tuition reimbursements, research assistantships with full tuition reimbursements, teaching assistantships with full tuition reimbursements, institutionally sponsored loans, scholarships/grants, health care benefits, tuition waivers (full and partial), and unspecified assistantships available. Financial award applicants required to submit FAFSA. *Faculty research:* Urban and regional growth, spatial economics, formation of spatial patterns by socioeconomic systems, non-linear dynamics and complex systems, environmental-economic systems. *Unit head:* Director of Graduate Studies, 607-255-6848, Fax: 607-255-1971. *Application contact:* Graduate Field Assistant, 607-255-6848, Fax: 607-255-1971, E-mail: regsci@cornell.edu.

Cornell University, Graduate School, Graduate Fields of Arts and Sciences, Field of Archaeology, Ithaca, NY 14853-0001. Offers environmental archaeology (MA); historical archaeology (MA); Latin American archaeology (MA); medieval archaeology (MA); Mediterranean and Near Eastern archaeology (MA); Stone Age archaeology (MA). *Faculty:* 15 full-time (3 women). *Students:* 19 applicants, 11% accepted, 2 enrolled. In 2005, 1 degree awarded. *Degree requirements:* For master's, one foreign language, thesis. *Entrance requirements:* For master's, GRE General Test, 3 letters of recommendation, sample of written work. Additional exam requirements/recommendations for international students: Required—TOEFL. *Application deadline:* For fall admission, 1/15 for domestic students. Application fee: $60. Electronic applications accepted. *Financial support:* In 2005–06, 1 student received support, including 1 teaching assistantship with full tuition reimbursement available; fellowships with full tuition reimbursements available, research assistantships with full tuition reimbursements available, institutionally sponsored loans, scholarships/grants, health care benefits, tuition waivers (full and partial), and unspecified assistantships also available. Financial award applicants required to submit FAFSA. *Faculty research:* Anatolia, Lydia, Sardis, classical and Hellenistic Greece; science in archaeology; North American Indians; Stone Age Africa; Maya trade. *Unit head:* Director of Graduate Studies, 607-255-6768, E-mail: blj7@cornell.edu. *Application contact:* Graduate Field Assistant, 607-255-6768, E-mail: bad2@cornell.edu.

Dalhousie University, Faculty of Graduate Studies, Faculty of Management, School for Resource and Environmental Studies, Halifax, NS B3H 4R2, Canada. Offers MES. Part-time programs available. *Degree requirements:* For master's, thesis. *Entrance requirements:* For master's, honors degree. Additional exam requirements/recommendations for international students: Required—TOEFL. *Faculty research:* Resource management and ecology, aboriginal resource rights, management of toxic substances, environmental impact assessment, forest management, policy, coastal zone management.

Drexel University, College of Arts and Sciences, Program in Environmental Policy, Philadelphia, PA 19104-2875. Offers MS. Part-time and evening/weekend programs available. *Degree requirements:* For master's, thesis optional. Electronic applications accepted.

Duke University, Graduate School, Department of Environment, Durham, NC 27708. Offers natural resource economics/policy (AM, PhD); natural resource science/ecology (AM, PhD); natural resource systems science (AM, PhD). Part-time programs available. *Faculty:* 36 full-time. *Students:* 71 full-time (44 women); includes 4 minority (3 African Americans, 1 Hispanic American), 26 international. 135 applicants, 16% accepted, 14 enrolled. In 2005, 3 master's, 9 doctorates awarded. *Degree requirements:* For doctorate, variable foreign language requirement, thesis/dissertation. *Entrance requirements:* For master's and doctorate, GRE General Test. Additional exam requirements/recommendations for international students: Required—IELT (preferred) or TOEFL. *Application deadline:* For fall admission, 12/31 for domestic students, 12/31 for international students. Application fee: $75. Electronic applications accepted. *Financial support:* Fellowships, research assistantships, teaching assistantships, Federal Work-Study available. Financial award application deadline: 12/31. *Unit head:* Kenneth Knoerr, Director of Graduate Studies, 919-613-8030, Fax: 919-684-8741, E-mail: nmm@duke.edu.

Duke University, Nicholas School of the Environment and Earth Sciences, Durham, NC 27708-0328. Offers coastal environmental management (MEM); environmental economics and policy (MEM); environmental health and security (MEM); forest resource management (MF); global environmental change (MEM); resource ecology (MEM); water and air resources (MEM).

Accreditation: SAF (one or more programs are accredited). Part-time programs available. *Degree requirements:* For master's, thesis, registration. *Entrance requirements:* For master's, GRE General Test, previous course work in biology or ecology, calculus, statistics, and microeconomics; computer familiarity with word processing and data analysis. Additional exam requirements/recommendations for international students: Required—TOEFL (minimum score 550 paper-based; 213 computer-based). Electronic applications accepted. Expenses: Contact institution. *Faculty research:* Ecosystem management, conservation ecology, earth systems, risk assessment.

Announcement: Interdisciplinary focus of Environmental Economics and Policy Program provides excellent background for careers with a broad spectrum of employers. Opportunities for specialization in environmental management, forestry, resource ecology, coastal and marine resources, water and air resources, environmental health. Concurrent degrees available: MBA, JD in environmental law, MPP, MA in teaching. Partial fellowships for qualified students.

See Close-Up on page 709.

Duquesne University, Bayer School of Natural and Environmental Sciences, Environmental Science and Management Program, Pittsburgh, PA 15282-0001. Offers environmental management (Certificate); environmental science (Certificate); environmental science and management (MS). Part-time and evening/weekend programs available. Postbaccalaureate distance learning degree programs offered (minimal on-campus study). *Faculty:* 3 full-time (0 women), 11 part-time/adjunct (0 women). *Students:* 12 full-time (4 women), 21 part-time (13 women); includes 2 minority (both Asian Americans or Pacific Islanders), 2 international. Average age 28. 18 applicants, 67% accepted, 5 enrolled. In 2005, 23 degrees awarded. *Degree requirements:* For master's, thesis, comprehensive exam (for some programs), registration. *Entrance requirements:* For master's, GRE General Test, previous course work in biology, calculus, chemistry, and statistics. Additional exam requirements/recommendations for international students: Required—TOEFL, TSE. *Application deadline:* For fall admission, 4/1 priority date for domestic students, 4/1 priority date for international students; for spring admission, 10/1 priority date for domestic students, 10/1 priority date for international students. Applications are processed on a rolling basis. Application fee: $0. *Expenses: Contact institution.* Tuition and fees vary according to degree level and program. *Financial support:* In 2005–06, 1 fellowship with full tuition reimbursement (averaging $15,080 per year), 2 teaching assistantships with partial tuition reimbursements (averaging $10,000 per year) were awarded.; research assistantships, career-related internships or fieldwork, Federal Work-Study, institutionally sponsored loans, scholarships/grants, tuition waivers (partial), and unspecified assistantships also available. Support available to part-time students. Financial award application deadline: 5/1; financial award applicants required to submit FAFSA. *Faculty research:* Watershed management systems, environmental analytical chemistry, environmental endocrinology, environmental microbiology, aquatic biology. Total annual research expenditures:$200,750. *Unit head:* Robert D. Volkmar, Interim Director, 412-396-4094, Fax: 412-396-4092, E-mail: volkmar@duq.edu. *Application contact:* Mary Ann Quinn, Assistant to the Dean Graduate Affairs, 412-396-6339, Fax: 412-396-4881, E-mail: gradinfo@duq.edu.

East Carolina University, Graduate School, Program in Coastal Resources Management, Greenville, NC 27858-4353. Offers PhD. *Students:* 11 full-time (5 women), 19 part-time (6 women); includes 3 minority (1 African American, 2 Hispanic Americans), 1 international. Average age 39. 6 applicants, 33% accepted, 2 enrolled. In 2005, 2 degrees awarded. *Degree requirements:* For doctorate, thesis/dissertation, internship, comprehensive exam, registration. *Entrance requirements:* For doctorate, GRE. Additional exam requirements/recommendations for international students: Required—TOEFL. *Application deadline:* For fall admission, 3/1 for domestic students, 3/1 for international students. Application fee: $50. *Expenses:* Tuition, state resident: full-time $2,516. Tuition, nonresident: full-time $12,832. *Financial support:* In 2005–06, 8 fellowships with tuition reimbursements (averaging $17,000 per year), 2 research assistantships with tuition reimbursements (averaging $17,000 per year) were awarded.; career-related internships or fieldwork and institutionally sponsored loans also available. Financial award application deadline: 3/1; financial award applicants required to submit FAFSA. *Faculty research:* Coastal geology, wetlands and coastal ecology, ecological and social networks, submerged cultural resources, coastal resources economics. Total annual research expenditures:$24,000. *Unit head:* Dr. Lauriston R. King, Director, 252-328-2484, Fax: 252-328-0381, E-mail: kingl@ecu.edu. *Application contact:* Dean of Graduate School, 252-328-6012, Fax: 252-328-6071, E-mail: gradschool@ecu.edu.

The Evergreen State College, Graduate Programs, Program in Environmental Studies, Olympia, WA 98505. Offers MES, MES/MPA. Part-time and evening/weekend programs available. *Faculty:* 5 full-time (2 women), 3 part-time/adjunct (0 women). *Students:* 43 full-time (27 women), 42 part-time (18 women); includes 4 minority (1 American Indian/Alaska Native, 1 Asian American or Pacific Islander, 2 Hispanic Americans), 3 international. Average age 33. 61 applicants, 100% accepted, 42 enrolled. In 2005, 22 degrees awarded. *Degree requirements:* For master's, thesis. *Entrance requirements:* For master's, GRE, minimum undergraduate GPA of 3.0; BA/BS emphasis in biological, physical, or social science; course work in statistics and micro-economics. Additional exam requirements/recommendations for international students: Required—TOEFL (minimum score 600 paper-based; 250 computer-based). *Application deadline:* For fall admission, 11/15 priority date for domestic students, 11/15 priority date for international students. Applications are processed on a rolling basis. Application fee: $50. Electronic applications accepted. *Expenses:* Tuition, state resident: full-time $6,522; part-time $217 per credit. Tuition, nonresident: full-time $19,959; part-time $665 per credit. Required fees: $2 per credit. $41 per quarter hour. Tuition and fees vary according to course load and program. *Financial support:* In 2005–06, 3 fellowships with partial tuition reimbursements were awarded; research assistantships, career-related internships or fieldwork, Federal Work-Study, institutionally sponsored loans, scholarships/grants, tuition waivers (partial), and unspecified assistantships also available. Support available to part-time students. Financial award application deadline: 3/15; financial award applicants required to submit FAFSA. *Faculty research:* Land and water policy, canopy studies, ecology (marine microbial, phytoplankton, Spartina cordgrass, green crabs), bacteriophage research, cultural geography. Total annual research expenditures:$19,900. *Unit head:* Dr. Ted Whitesell, Director, 360-867-6180, Fax: 360-867-5430, E-mail: whiteset@evergreen.edu. *Application contact:* J. T. Austin, Graduate Studies Office, 360-867-6225, Fax: 360-867-5430, E-mail: austinj@evergreen.edu.

See Close-Up on page 651.

Florida Gulf Coast University, College of Public and Social Services, Program in Public Administration, Fort Myers, FL 33965-6565. Offers criminal justice (MPA); environmental policy (MPA); general public administration (MPA); management (MPA). Part-time programs available. *Entrance requirements:* For master's, GRE General Test, MAT, minimum GPA of 3.0. Electronic applications accepted. *Faculty research:* Personnel, public policy, public finance, housing policy.

Florida Institute of Technology, Graduate Programs, College of Engineering, Department of Marine and Environmental Systems, Melbourne, FL 32901-6975. Offers environmental resource management (MS); environmental science (MS, PhD); meteorology (MS); ocean engineering (MS, PhD); oceanography (MS, PhD), including biological oceanography (MS), chemical oceanography (MS), coastal zone management (MS), geological oceanography (MS), oceanography (PhD), physical oceanography (MS). Part-time programs available. *Faculty:* 11 full-time (1 woman). *Students:* 40 full-time (15 women), 20 part-time (13 women); includes 2 minority (both Hispanic Americans), 15 international. Average age 29. 101 applicants, 50% accepted, 15 enrolled. In 2005, 16 master's awarded. Terminal master's awarded for partial completion of doctoral program. *Degree requirements:* For master's, thesis, comprehensive exam, registration; for doctorate, one foreign language, thesis/dissertation, attendance of graduate seminar, internships (oceanography and environmental science), publications, comprehensive exam, registration. *Entrance requirements:* For master's, GRE General Test (environmental science), 3 letters of recommendation, minimum GPA of 3.0; for doctorate, GRE General Test (oceanography and environmental science), resumé, 3 letters of recommendation, minimum GPA of 3.2. Additional exam requirements/recommendations for international students: Required—TOEFL (minimum score 550 paper-based; 213 computer-based). *Application deadline:* Applications are processed on a rolling basis. Application fee: $50. Electronic applications accepted. *Expenses:* Tuition: Part-time $825 per credit. *Financial support:* In 2005–06, 18 students received support, including 1 fellowship with full and partial tuition reimbursement available (averaging $1,064 per year), 8 research assistantships with full and partial tuition reimbursements available (averaging $4,116 per year), 9 teaching assistantships with full and partial tuition reimbursements available (averaging $6,867 per year); career-related internships or fieldwork and tuition remissions also available. Financial award application deadline: 3/1; financial award applicants required to submit FAFSA. *Faculty research:* Environmental modeling, coastal processes, exploring marine pollution, marine geophysics, remote sensing . Total annual research expenditures: $1 million. *Unit head:* Dr. George Maul, Department Head, 321-674-7453, Fax: 321-674-7212, E-mail: gmaul@fit.edu. *Application contact:* Carolyn P. Farrior, Director of Graduate Admissions, 321-674-7118, Fax: 321-723-9468, E-mail: cfarrior@fit.edu.

See Close-Up on page 653.

Florida International University, College of Arts and Sciences, Department of Environmental Studies, Miami, FL 33199. Offers biological management (MS); energy (MS); pollution (MS). *Faculty:* 11 full-time (2 women). *Students:* 18 full-time (16 women), 18 part-time (9 women); includes 7 minority (1 African American, 1 American Indian/Alaska Native, 2 Asian Americans or Pacific Islanders, 3 Hispanic Americans), 7 international. Average age 32. 4 applicants, 50% accepted, 2 enrolled. In 2005, 9 degrees awarded. *Degree requirements:* For master's, thesis. *Entrance requirements:* For master's, GRE General Test, minimum GPA of 3.0, 3 letters of recommendation. Additional exam requirements/recommendations for international students: Required—TOEFL. *Application deadline:* For fall admission, 4/1 for domestic students; for spring admission, 10/30 for domestic students. Application fee: $25. *Expenses:* Tuition, area resident: Part-time $239 per credit. Tuition, state resident: full-time $4,294; part-time $869 per credit. Tuition, nonresident: full-time $15,641. Required fees: $252; $126 per term. Tuition and fees vary according to program. *Financial support:* Research assistantships, teaching assistantships available. Financial award application deadline: 4/1. *Unit head:* Dr. Joel Heinen, Chairperson, 305-348-3732, Fax: 305-348-6137, E-mail: joel.heinen@fiu.edu.

Friends University, Graduate School, Division of Science, Arts, and Education, Program in Environmental Studies, Wichita, KS 67213. Offers MSES. Evening/weekend programs available. *Faculty:* 9 part-time/adjunct. *Students:* 14 full-time. In 2005, 8 degrees awarded. *Entrance requirements:* Additional exam requirements/recommendations for international students: Required—TOEFL (minimum score 560 paper-based; 220 computer-based). *Application deadline:* For fall admission, 3/15 priority date for domestic students, 3/15 priority date for international students. Applications are processed on a rolling basis. Application fee: $45 ($65 for international students). Electronic applications accepted. *Expenses:* Tuition: Part-time $280 per credit hour. *Unit head:* Dr. Alan Maccarone, Director, 800-794-6945 Ext. 5890, E-mail: alanm@friends.edu. *Application contact:* Craig Davis, Executive Director of Recruitment-Adult and Graduate Studies, 800-794-6945 Ext. 5573, Fax: 316-295-5050, E-mail: cdavis@friends.edu.

Gannon University, School of Graduate Studies, College of Sciences, Engineering, and Health Sciences, School of Sciences, Program in Environmental Studies, Erie, PA 16541-0001. Offers MS. *Faculty:* 3 full-time (2 women). *Students:* 1 applicant, 100% accepted, 0 enrolled. In 2005, 1 degree awarded. *Entrance requirements:* Additional exam requirements/recommendations for international students: Required—TOEFL (minimum score 500 paper-based; 173 computer-based). *Application deadline:* Applications are processed on a rolling basis. Application fee: $25. *Expenses:* Tuition: Full-time $11,430; part-time $635 per credit. Required fees: $496; $16 per credit. Tuition and fees vary according to course load, degree level and program. *Unit head:* Dr. Harry Diz, Director, 814-871-7633, E-mail: diz001@gannon.edu. *Application contact:* Debra Meszaros, Director of Graduate Recruitment, 814-871-5819, Fax: 814-871-5827, E-mail: cfal@gannon.edu.

The George Washington University, Columbian College of Arts and Sciences, Department of Environmental Studies, Washington, DC 20052. Offers geology (MS, PhD); geosciences (MS, PhD); hominid paleobiology (MS, PhD). Part-time and evening/weekend programs available. Terminal master's awarded for partial completion of doctoral program. *Degree requirements:* For master's, thesis or alternative, comprehensive exam; for doctorate, thesis/dissertation, general exam. *Entrance requirements:* For master's, GRE General Test, bachelor's degree in field, interview, minimum GPA of 3.0; for doctorate, GRE General Test, interview, minimum GPA of 3.0. Additional exam requirements/recommendations for international students: Required—TOEFL (minimum score 550 paper-based; 213 computer-based). *Faculty research:* Engineering geology.

The George Washington University, Columbian College of Arts and Sciences, School of Public Policy and Public Administration, Washington, DC 20052. Offers public policy (MA, MPP), including environmental and resource policy (MA), philosophy and social policy (MA), women's studies (MA); public policy and administration (PhD); public policy and public administration (MPA), including budget and public finance, federal policy, politics, and management, international development management, managing public organizations, managing state and local governments and urban policy, nonprofit management, policy analysis and evaluation, public administration. Part-time and evening/weekend programs available. *Degree requirements:* For doctorate, thesis/dissertation, general exam. *Entrance requirements:* For master's, GRE General Test, minimum GPA of 3.0; for doctorate, GRE General Test, interview, minimum GPA of 3.0. Additional exam requirements/recommendations for international students: Required—TOEFL (minimum score 550 paper-based; 213 computer-based). Electronic applications accepted.

The George Washington University, Columbian College of Arts and Sciences, School of Public Policy and Public Administration, Interdisciplinary Programs in Public Policy, Program in Environmental and Resource Policy, Washington, DC 20052. Offers MA. *Degree requirements:* For master's, project. *Entrance requirements:* For master's, GRE General Test, minimum GPA of 3.0. Additional exam requirements/recommendations for international students: Required—TOEFL (minimum score 550 paper-based; 213 computer-based). Electronic applications accepted.

Georgia Institute of Technology, Graduate Studies and Research, College of Architecture, City and Regional Planning Program, Atlanta, GA 30332-0001. Offers architecture (PhD); economic development (MCRP); environmental planning and management (MCRP); geographic information systems (MCRP); land development (MCRP); land use planning (MCRP); transportation (MCRP); urban design (MCRP). *Accreditation:* ACSP. *Degree requirements:* For master's, thesis, internship. *Entrance requirements:* For master's, GRE General Test, minimum GPA of 2.7. Additional exam requirements/recommendations for international students: Required—TOEFL. Electronic applications accepted.

Georgia State University, College of Arts and Sciences, Department of Geosciences, Atlanta, GA 30303-3083. Offers earth science—hydrology (MS), including earth science—environmental management, earth science—GIS; geographic information systems (Certificate); geography (MA); geology (MS). Part-time and evening/weekend programs available. *Degree requirements:* For master's, one foreign language, thesis or alternative, comprehensive exam (for some programs), registration. *Entrance requirements:* For master's, GRE General Test, minimum GPA of 2.75. Additional exam requirements/recommendations for international students: Required—TOEFL. Electronic applications accepted. *Expenses:* Tuition, state resident: full-time $4,368; part-time $182 per term. Tuition, nonresident: full-time $8,732; part-time $728 per term. Required fees: $46 per hour. *Faculty research:* Clay mineralogy, metamorphism, fracture analysis, carbonates, groundwater.

Environmental Management and Policy

Goddard College, Graduate Program, Individually Designed Liberal Arts Program, Plainfield, VT 05667-9432. Offers consciousness studies (MA); environmental studies (MA); transformative language arts (MA). Postbaccalaureate distance learning degree programs offered (minimal on-campus study). *Degree requirements:* For master's, thesis, registration. *Entrance requirements:* For master's, 3 letters of recommendation, study plan, bibliography. Electronic applications accepted. Expenses: Contact institution.

Hardin-Simmons University, Graduate School, Program in Environmental Management, Abilene, TX 79698-0001. Offers MS. Part-time programs available. *Faculty:* 6 full-time (1 woman). *Students:* 2 full-time (both women), 5 part-time (2 women). Average age 39. 5 applicants, 60% accepted, 3 enrolled. In 2005, 3 degrees awarded. *Degree requirements:* For master's, thesis or alternative, internship, comprehensive exam. *Entrance requirements:* For master's, minimum undergraduate GPA of 3.0 in major, 2.7 overall; 2 semesters of course work in each biology, chemistry, and geology; interview, writing sample, occupational experience. Additional exam requirements/recommendations for international students: Required—TOEFL (minimum score 550 paper-based; 213 computer-based). *Application deadline:* For fall admission, 8/15 for domestic students; for spring admission, 1/5 priority date for domestic students. Applications are processed on a rolling basis. Application fee: $50 ($100 for international students). *Expenses:* Tuition: Full-time $8,370; part-time $465 per hour. Required fees: $490; $66 per semester. One-time fee: $50. Full-time tuition and fees vary according to course load and degree level. *Financial support:* In 2005–06, 2 fellowships (averaging $350 per year) were awarded; career-related internships or fieldwork and scholarships/grants also available. Support available to part-time students. Financial award application deadline: 6/30; financial award applicants required to submit FAFSA. *Faculty research:* South American history, herpetology, geology, environmental education, petroleum biodegradation, environmental ecology and microbiology. *Unit head:* Dr. Mark Ouimette, Director, 325-670-1383, Fax: 325-670-1391, E-mail: ouimette@hsutx.edu. *Application contact:* Dr. Gary Stanlake, Dean of Graduate Studies, 325-670-1298, Fax: 325-670-1564, E-mail: gradoff@hsutx.edu.

Harvard University, Extension School, Cambridge, MA 02138-3722. Offers applied sciences (CAS); biotechnology (ALM); educational technologies (ALM); English for graduate and professional studies (DGP); environmental management (ALM, CEM); information technology (ALM); journalism (ALM); liberal arts (ALM); management (CM); mathematics for teaching (ALM); museum studies (ALM); premedical studies (Diploma); publication and communication (CPC). Part-time and evening/weekend programs available. *Faculty:* 236 part-time/adjunct. *Students:* 101 full-time (56 women), 564 part-time (278 women); includes 167 minority (35 African Americans, 1 American Indian/Alaska Native, 84 Asian Americans or Pacific Islanders, 47 Hispanic Americans). Average age 36. In 2005, 112 master's, 184 Diplomas awarded. *Degree requirements:* For master's, thesis. *Entrance requirements:* For master's, 3 completed graduate courses with grade of B or higher. Additional exam requirements/recommendations for international students: Required—TOEFL (minimum score 600 paper-based; 250 computer-based), TWE (minimum score 5). *Application deadline:* Applications are processed on a rolling basis. Application fee: $75. *Expenses: Contact institution.* Full-time tuition and fees vary according to program and student level. *Financial support:* In 2005–06, 268 students received support. Scholarships/grants available. Support available to part-time students. Financial award application deadline: 8/6; financial award applicants required to submit FAFSA. *Unit head:* Michael Shinagel, Dean. *Application contact:* Program Director, 617-495-4024, Fax: 617-495-9176.

Illinois Institute of Technology, Graduate College, Armour College of Engineering, Department of Chemical and Environmental Engineering, Chicago, IL 60616-3793. Offers chemical engineering (M Ch E, MS, PhD); environmental engineering (M Env E, MS, PhD); environmental management (MS); food process engineering (MFPE); food processing engineering (MS); gas engineering (MGE). Part-time and evening/weekend programs available. Postbaccalaureate distance learning degree programs offered. Terminal master's awarded for partial completion of doctoral program. *Degree requirements:* For master's, thesis (for some programs), comprehensive exam; for doctorate, thesis/dissertation, comprehensive exam. *Entrance requirements:* For master's and doctorate, GRE General Test, minimum undergraduate GPA of 3.0. Additional exam requirements/recommendations for international students: Required—TOEFL (minimum score 550 paper-based; 213 computer-based). Electronic applications accepted. *Faculty research:* Particle technology and crystallization, energy and environmental engineering, polymer science and engineering, bioengineering, electrodynamical science and engineering.

Illinois Institute of Technology, Stuart Graduate School of Business, Program in Environmental Management, Chicago, IL 60616-3793. Offers MS, JD/MS, MBA/MS. Part-time and evening/weekend programs available. *Entrance requirements:* For master's, GMAT or GRE General Test. Additional exam requirements/recommendations for international students: Required—TOEFL (minimum score 550 paper-based; 213 computer-based). Electronic applications accepted. Expenses: Contact institution. *Faculty research:* Removal of mercury vapor, renewable energy sources, application of GIS for sustainability, applying sustainable strategies to business.

Indiana University–Purdue University Indianapolis, School of Public and Environmental Affairs, Program in Health Administration, Indianapolis, IN 46202-2896. Offers environmental management (MPA); health administration (MHA); non-profit management (MPA); policy analysis (MPA); public management (MPA); urban management (MPA). *Accreditation:* ACEHSA. *Expenses:* Tuition, state resident: full-time $5,159; part-time $215 per credit hour. Tuition, nonresident: full-time $14,890; part-time $620 per credit hour. Required fees: $614. Tuition and fees vary according to campus/location and program. *Unit head:* Dr. Terrell Zollinger, Director, 317-278-0307. *Application contact:* Office of Student Services, 317-274-4656, Fax: 317-274-5153, E-mail: infospea@iupui.edu.

Instituto Tecnológico y de Estudios Superiores de Monterrey, Campus Estado de México, Professional and Graduate Division, Estado de Mexico, Mexico. Offers administration of information technologies (MITA); architecture (M Arch); business administration (GMBA, MBA); computer sciences (MCS, PhD); education (M Ed); educational institution administration (MAD); educational technology and innovation (PhD); electronic commerce (MEC); environmental systems (MS); finance (MAF); humanistic studies (MHS); information sciences and knowledge management (MISKM); information systems (MS); manufacturing systems (MS); marketing (MEM); quality systems and productivity (MS); science and materials engineering (PhD); telecommunications management (MTM). Part-time programs available. Postbaccalaureate distance learning degree programs offered (minimal on-campus study). *Degree requirements:* For master's, one foreign language, thesis (for some programs), registration; for doctorate, one foreign language, thesis/dissertation, registration (for some programs). *Entrance requirements:* For master's, E-PAEP 500, interview; for doctorate, E-PAEP 500, research proposal. Additional exam requirements/recommendations for international students: Required—TOEFL (minimum score 550 paper-based). *Faculty research:* Surface treatments by plasmas, mechanical properties, robotics, graphical computing, mechatronics security protocols.

Instituto Tecnológico y de Estudios Superiores de Monterrey, Campus Irapuato, Graduate Programs, Irapuato, Mexico. Offers administration (MBA); administration of information technology (MAIT); administration of telecommunications (MAT); architecture (M Arch); computer science (MCS); education (M Ed); educational administration (MEA); educational innovation and technology (DEIT); educational technology (MET); electronic commerce (MBA); environmental administration and planning (MEAP); environmental systems (MES); finances (MBA); humanistic studies (MHS); international management for Latin American executives (MIMLAE); library and information science (MLIS); manufacturing quality management (MMQM); marketing research (MBA).

Iowa State University of Science and Technology, Graduate College, College of Agriculture, Department of Natural Resource Ecology and Management, Ames, IA 50011. Offers animal ecology (MS, PhD), including animal ecology, fisheries biology, wildlife biology; forestry (MS, PhD). *Faculty:* 23 full-time, 12 part-time/adjunct. *Students:* 45 full-time (18 women), 4 part-time (1 woman), 7 international. 24 applicants, 38% accepted, 8 enrolled. In 2005, 8 master's, 3 doctorates awarded. *Degree requirements:* For master's, thesis (for some programs); for doctorate, thesis/dissertation. *Entrance requirements:* For master's and doctorate, GRE General Test. Additional exam requirements/recommendations for international students: Required—TOEFL (paper score 547; computer score 210) or IELTS (score 6). *Application deadline:* For fall admission, 1/1 priority date for domestic students, 1/1 priority date for international students; for spring admission, 9/1 priority date for domestic students, 9/1 priority date for international students. Application fee: $30 ($70 for international students). Electronic applications accepted. *Expenses:* Tuition, state resident: full-time $6,410. Tuition, nonresident: full-time $16,422. Tuition and fees vary according to program. *Financial support:* In 2005–06, 41 research assistantships with full and partial tuition reimbursements (averaging $14,755 per year), 3 teaching assistantships with full and partial tuition reimbursements (averaging $14,580 per year) were awarded. *Unit head:* Dr. David M Engle, Chair, 515-294-1166. *Application contact:* Lyn Van De Pol, Information Contact, 515-294-6148, E-mail: lvdp@iastate.edu.

The Johns Hopkins University, Zanvyl Krieger School of Arts and Sciences, Advanced Academic Programs, Program in Environmental Sciences and Policy, Washington, DC 20036. Offers MS. *Expenses:* Tuition: Full-time $30,960. Tuition and fees vary according to degree level and program. *Unit head:* Eileen McGurty, Associate Program Chair. *Application contact:* Craig Jones, Admissions Coordinator, 202-452-1941, Fax: 202-452-1970, E-mail: aapadmissions@jhu.edu.

Kansas State University, Graduate School, College of Architecture, Planning and Design, Department of Regional and Community Planning, Manhattan, KS 66506. Offers regional and community planning (MRCP). *Accreditation:* ACSP. Part-time and evening/weekend programs available. Postbaccalaureate distance learning degree programs offered (minimal on-campus study). *Faculty:* 16 full-time (4 women). *Students:* 11 full-time (5 women), 3 part-time (1 woman); includes 1 minority (Hispanic American), 3 international. 7 applicants, 100% accepted, 3 enrolled. In 2005, 10 degrees awarded. *Degree requirements:* For master's, thesis, oral exam. *Entrance requirements:* For master's, minimum GPA of 3.0, portfolio. Additional exam requirements/recommendations for international students: Required—TOEFL (minimum score 600 paper-based). *Application deadline:* For fall admission, 7/1 priority date for domestic students, 2/1 priority date for international students; for spring admission, 10/1 priority date for domestic students, 8/1 priority date for international students. Applications are processed on a rolling basis. Application fee: $70 ($80 for international students). Electronic applications accepted. *Expenses:* Tuition, state resident: full-time $5,160; part-time $215. Tuition, nonresident: full-time $12,816; part-time $534. Required fees: $564. *Financial support:* In 2005–06, 2 research assistantships (averaging $5,969 per year), 7 teaching assistantships with full tuition reimbursements (averaging $7,000 per year) were awarded.; career-related internships or fieldwork, Federal Work-Study, institutionally sponsored loans, and scholarships/grants also available. Support available to part-time students. Financial award application deadline: 3/1; financial award applicants required to submit FAFSA. *Faculty research:* Infrastructure planning, economic development and rural, regional planning, planning analysis and methods, cultural and historic landscape preservation. *Unit head:* Prof. Dan Donelin, Head, 785-532-5961, Fax: 785-532-6722, E-mail: dandon@ksu.edu. *Application contact:* Prof. C. A. Keithley, Graduate Coordinator, 785-532-2440, Fax: 785-532-6722, E-mail: cak@ksu.edu.

Kean University, College of Business and Public Administration, Program in Public Administration, Union, NJ 07083. Offers criminal justice (MPA); environmental management (MPA); health services administration (MPA); non-profit management (MPA); public administration (MPA). *Accreditation:* NASPAA. Part-time and evening/weekend programs available. *Faculty:* 7 full-time (4 women). *Students:* 67 full-time (41 women), 102 part-time (64 women); includes 112 minority (83 African Americans, 6 Asian Americans or Pacific Islanders, 23 Hispanic Americans), 14 international. Average age 33. 73 applicants, 100% accepted, 51 enrolled. In 2005, 54 degrees awarded. *Degree requirements:* For master's, internship, minimum 3.0 GPA. *Entrance requirements:* For master's, GRE, 2 letters of recommendation, interview, writing sample. *Application deadline:* For fall admission, 5/1 for domestic students; for spring admission, 11/1 for domestic students. Application fee: $60 ($150 for international students). Electronic applications accepted. *Expenses:* Tuition, state resident: full-time $8,280; part-time $345 per credit. Tuition, nonresident: full-time $11,512; part-time $438 per credit. Required fees: $2,104; $88 per credit. *Financial support:* In 2005–06, 19 research assistantships with full tuition reimbursements (averaging $2,880 per year) were awarded; career-related internships or fieldwork, institutionally sponsored loans, and unspecified assistantships also available. Financial award application deadline: 5/1. *Faculty research:* Fiscal impact of New Federalism, New Jersey state and local government, computer application in public management. *Unit head:* Dr. Craig P. Donovan, Program Coordinator, 908-737-4307, E-mail: cpdonova@kean.edu. *Application contact:* Joanne Morris, Director of Graduate Admissions, 908-737-3355, Fax: 908-737-3354, E-mail: grad-adm@kean.edu.

Lamar University, College of Graduate Studies, College of Engineering, Department of Civil Engineering, Beaumont, TX 77710. Offers civil engineering (ME, MES, DE); environmental engineering (MS); environmental studies (MS). Part-time programs available. *Faculty:* 6 full-time (0 women), 1 part-time/adjunct (0 women). *Students:* 65 full-time (11 women), 15 part-time (3 women); includes 3 minority (1 African American, 1 Asian American or Pacific Islander, 1 Hispanic American), 74 international. Average age 26. 93 applicants, 82% accepted, 23 enrolled. In 2005, 39 master's, 1 doctorate awarded. *Degree requirements:* For master's, thesis optional; for doctorate, thesis/dissertation. *Entrance requirements:* For master's and doctorate, GRE General Test. Additional exam requirements/recommendations for international students: Required—TOEFL. *Application deadline:* For fall admission, 5/15 for domestic students; for spring admission, 10/1 priority date for domestic students. Applications are processed on a rolling basis. Application fee: $25 ($50 for international students). *Expenses:* Tuition, state resident: part-time $137 per semester hour. Tuition, nonresident: part-time $413 per semester hour. Required fees: $102 per semester hour. Tuition and fees vary according to course load. *Financial support:* In 2005–06, 45 fellowships with partial tuition reimbursements (averaging $1,000 per year), 10 research assistantships with partial tuition reimbursements (averaging $7,200 per year), 3 teaching assistantships with partial tuition reimbursements (averaging $7,200 per year) were awarded.; scholarships/grants and tuition waivers (partial) also available. Financial award application deadline: 4/1. *Faculty research:* Environmental remediations, construction productivity, geotechnical soil stabilization, lake/reservoir hydrodynamics, air pollution. Total annual research expenditures: $197,236. *Unit head:* Dr. Enno Koehn, Chair, 409-880-8759, Fax: 409-880-8121, E-mail: koehneu@hal.lamar.edu. *Application contact:* Sandy Drane, Coordinator of Graduate Admissions, 409-880-8356, Fax: 409-880-8414, E-mail: gradmissions@hal.lamar.edu.

Long Island University, C.W. Post Campus, College of Liberal Arts and Sciences, Program in Environmental Studies, Brookville, NY 11548-1300. Offers environmental management (MS); environmental science (MS). Part-time and evening/weekend programs available. *Degree requirements:* For master's, internship or thesis. *Entrance requirements:* For master's, 1 year of course work in general chemistry and biology or geology; 1 semester in organic chemistry; computer proficiency. Electronic applications accepted. *Faculty research:* Symbiotic algae, local marine organisms, coastal processes, global tectonics, paleomagnetism.

Louisiana State University and Agricultural and Mechanical College, Graduate School, School of the Coast and Environment, Department of Environmental Studies, Baton Rouge, LA 70803. Offers environmental planning and management (MS); environmental toxicology (MS). *Faculty:* 12 full-time (4 women). *Students:* 21 full-time (13 women), 11 part-time (5 women); includes 2 African Americans, 1 American Indian/Alaska Native, 3 international. Average age 29. 21 applicants, 43% accepted, 12 enrolled. In 2005, 12 degrees awarded. *Degree requirements:* For master's, thesis (for some programs). *Entrance requirements:* For master's, GRE General Test, minimum GPA of 3.0. Additional exam requirements/recommendations for international students: Required—TOEFL (minimum score 550 paper-based; 213 computer-based). *Application deadline:* For fall admission, 1/25 priority date for domestic students, 5/15 priority date for international students. Applications are processed on a rolling basis. Application fee: $25. Electronic applications accepted. *Financial support:* In 2005–06, 16 students received support, including 2 fellowships with full and partial tuition reimbursements available (averaging $11,458 per year), 12 research assistantships with full and partial tuition reimbursements available (averaging $11,036 per year); teaching assistantships with full and partial tuition reimbursements available, career-related internships or fieldwork, Federal Work-Study, institutionally sponsored loans, scholarships/grants, and unspecified assistantships also available. Support available to part-time students. Financial award applicants required to submit FAFSA. *Faculty research:* Fates and movement of pollutants, neurobiotic metabolism, application of cellular toxicity/mutagenicity testing. Total annual research expenditures: $1.2 million.

Michigan State University, The Graduate School, College of Agriculture and Natural Resources, Department of Community, Agriculture, Recreation, and Resource Studies, East Lansing, MI 48824. Offers MS, PhD. *Faculty:* 24 full-time (5 women). *Students:* 29 full-time (17 women), 9 part-time (5 women); includes 6 minority (2 African Americans, 1 American Indian/Alaska Native, 2 Asian Americans or Pacific Islanders, 1 Hispanic American), 15 international. Average age 35. 50 applicants, 40% accepted. In 2005, 4 master's, 1 doctorate awarded. *Degree requirements:* For master's, defense of thesis or research paper, thesis optional; for doctorate, thesis/dissertation, oral defense of dissertation, comprehensive exam. *Entrance requirements:* For master's and doctorate, GRE General Test, resumé, 3 letters of recommendation. Additional exam requirements/recommendations for international students: Required—TOEFL (minimum score 550 paper-based; 213 computer-based), Michigan State University ELT (85), Michigan ELAB (83). *Application deadline:* For fall admission, 12/1 for domestic students. Application fee: $50. Electronic applications accepted. *Expenses:* Tuition, state resident: part-time $330 per credit hour. Tuition, nonresident: part-time $685 per credit hour. Tuition and fees vary according to program. *Financial support:* In 2005–06, 2 fellowships with tuition reimbursements (averaging $8,750 per year), 17 research assistantships with tuition reimbursements (averaging $13,302 per year), 1 teaching assistantship with tuition reimbursement (averaging $11,988 per year) were awarded.; institutionally sponsored loans and unspecified assistantships also available. *Faculty research:* Community food and agriculture, natural resources, land use and the environment; recreation and tourism; leadership, education, and communication for sustainable communities. Total annual research expenditures: $1 million. *Unit head:* Dr. Scott G. Witter, Chairperson, 517-432-0263, Fax: 517-432-3597, E-mail: witter@msu.edu. *Application contact:* Diane Davis, Graduate Secretary, 517-432-0275, Fax: 517-432-3597, E-mail: davisdia@msu.edu.

Michigan Technological University, Graduate School, College of Sciences and Arts, Department of Social Sciences, Program in Environmental Policy, Houghton, MI 49931-1295. Offers MS. Part-time programs available. *Faculty:* 15 full-time (5 women), 1 part-time/adjunct (0 women). *Students:* 8 full-time (5 women), 3 international. Average age 28. 15 applicants, 73% accepted, 3 enrolled. In 2005, 3 degrees awarded. *Degree requirements:* For master's, thesis or alternative, master's project if no thesis, comprehensive exam, registration. *Entrance requirements:* Additional exam requirements/recommendations for international students: Required—TOEFL (minimum score 550 paper-based; 213 computer-based). *Application deadline:* For fall admission, 3/1 for domestic students. Applications are processed on a rolling basis. Application fee: $40 ($45 for international students). Electronic applications accepted. *Expenses:* Tuition, nonresident: full-time $11,232; part-time $468 per credit. Required fees: $754; $377 per semester. Full-time tuition and fees vary according to course load, degree level and program. *Financial support:* In 2005–06, 7 students received support, including fellowships with full tuition reimbursements available (averaging $9,542 per year), 3 research assistantships with full tuition reimbursements available (averaging $9,542 per year), 3 teaching assistantships with full tuition reimbursements available (averaging $9,542 per year); career-related internships or fieldwork, Federal Work-Study, scholarships/grants, health care benefits, unspecified assistantships, and co-op also available. Financial award applicants required to submit FAFSA. *Application contact:* Dr. Kathleen E. Halvorsen, Director, Environmental Policy Program, 906-487-2824, Fax: 906-487-2468, E-mail: kehalvor@mtu.edu.

Missouri State University, Graduate College, College of Natural and Applied Sciences, Department of Geography, Geology, and Planning, Springfield, MO 65804-0094. Offers geography, geology and planning (MNAS); geospatial sciences (MS); secondary education (MS Ed), including earth science, geography. Part-time and evening/weekend programs available. *Faculty:* 20 full-time (3 women), 1 part-time/adjunct (0 women). *Students:* 13 full-time (5 women), 8 part-time (3 women). Average age 32. 14 applicants, 79% accepted, 5 enrolled. In 2005, 7 degrees awarded. *Degree requirements:* For master's, thesis (for some programs), comprehensive exam. *Entrance requirements:* For master's, GRE General Test (MS, MNAS), minimum undergraduate GPA of 3.0 (MS, MNAS), 9-12 teacher certification (MS Ed). Additional exam requirements/recommendations for international students: Required—TOEFL (minimum score 550 paper-based; 213 computer-based), IELT (minimum score 6). *Application deadline:* For fall admission, 7/20 for domestic students; for spring admission, 12/20 priority date for domestic students. Applications are processed on a rolling basis. Application fee: $30. Electronic applications accepted. *Expenses:* Tuition, state resident: full-time $3,402; part-time $189 per credit. Tuition, nonresident: full-time $6,804; part-time $378 per credit. Required fees: $207 per semester. Part-time tuition and fees vary according to course level, course load and program. *Financial support:* In 2005–06, 4 research assistantships with full tuition reimbursements (averaging $8,750 per year), 6 teaching assistantships with full tuition reimbursements (averaging $6,575 per year) were awarded.; career-related internships or fieldwork, Federal Work-Study, scholarships/grants, and unspecified assistantships also available. Financial award application deadline: 3/31; financial award applicants required to submit FAFSA. *Faculty research:* Water resources, small town planning, recreation and open space planning. *Unit head:* Dr. Tom Plymate, Acting Head, 417-836-5800, Fax: 417-836-6934, E-mail: tomplymate@missouristate.edu. *Application contact:* Dr. Robert T. Pavlowsky, Graduate Adviser, 417-836-8473, Fax: 417-836-6006, E-mail: bobpavlowsky@missouristate.edu.

Missouri State University, Graduate College, Interdisciplinary Program in Administrative Studies, Springfield, MO 65804-0094. Offers applied communication (MSAS); criminal justice (MSAS); environmental management (MSAS); project management (MSAS); sports management (MSAS). Part-time programs available. Postbaccalaureate distance learning degree programs offered (no on-campus study). *Students:* 24 full-time (17 women), 39 part-time (22 women); includes 4 minority (2 African Americans, 1 Asian American or Pacific Islander, 1 Hispanic American), 5 international. Average age 35. 19 applicants, 79% accepted, 11 enrolled. In 2005, 19 degrees awarded. *Degree requirements:* For master's, thesis or alternative, comprehensive exam. *Entrance requirements:* For master's, GRE, GMAT, 3 years of work experience. Additional exam requirements/recommendations for international students: Required—TOEFL (minimum score 550 paper-based; 213 computer-based), IELT (minimum score 6). *Application deadline:* For fall admission, 7/20 for domestic students; for spring admission, 12/20 priority date for domestic students. Applications are processed on a rolling basis. Application fee: $30. Electronic applications accepted. *Expenses:* Tuition, state resident: full-time $3,402; part-time $189 per credit. Tuition, nonresident: full-time $6,804; part-time $378 per credit. Required fees: $207 per semester. Part-time tuition and fees vary according to course level, course load and program. *Financial support:* In 2005–06, 1 teaching assistantship (averaging $6,575 per year) was awarded; career-related internships or fieldwork, Federal Work-Study, institutionally sponsored loans, scholarships/grants, and unspecified assistantships also available. Support available to part-time students. Financial award application deadline: 3/31; financial award applicants required to submit FAFSA. *Unit head:* John Bourhis, Director, 417-836-6390, E-mail: johnbourhis@missouristate.edu.

Montana State University, College of Graduate Studies, College of Engineering, Department of Civil Engineering, Bozeman, MT 59717. Offers civil engineering (MS); engineering (PhD); environmental engineering (MS); land rehabilitation (intercollege) (MS). Part-time programs available. *Faculty:* 18 full-time (1 woman), 4 part-time/adjunct (1 woman). *Students:* 20 full-time (1 woman), 12 part-time (2 women), 1 international. Average age 27. 31 applicants, 35% accepted, 7 enrolled. In 2005, 13 master's, 2 doctorates awarded. *Degree requirements:* For master's, thesis (for some programs), comprehensive exam, registration; for doctorate, thesis/dissertation, comprehensive exam, registration. *Entrance requirements:* For master's and doctorate, GRE General Test. Additional exam requirements/recommendations for international students: Required—TOEFL (minimum score 550 paper-based; 213 computer-based). *Application deadline:* For fall admission, 7/15 priority date for domestic students, 5/15 priority date for international students; for spring admission, 12/1 priority date for domestic students, 10/1 priority date for international students. Applications are processed on a rolling basis. Application fee: $30. Electronic applications accepted. *Expenses:* Tuition, state resident: full-time $4,132. Tuition, nonresident: full-time $1,132. *Financial support:* In 2005–06, 17 students received support, including 5 fellowships with partial tuition reimbursements available (averaging $16,200 per year), 7 research assistantships with partial tuition reimbursements available (averaging $14,400 per year), 5 teaching assistantships with partial tuition reimbursements available (averaging $9,000 per year); institutionally sponsored loans, scholarships/grants, tuition waivers (partial), and unspecified assistantships also available. Financial award application deadline: 3/1; financial award applicants required to submit FAFSA. *Faculty research:* Snow and ice mechanics, transportation systems in infrastructure, geotechnical testing and pavements, wetlands and rivers. Total annual research expenditures:$545,288. *Unit head:* Dr. Brett Gunnick, Department Head, 406-994-2111, Fax: 406-994-6105, E-mail: bgunnick@montana.edu.

Montclair State University, The Graduate School, College of Science and Mathematics, Department of Earth and Environmental Studies, Program in Environmental Management, Montclair, NJ 07043-1624. Offers MA, D Env M. *Entrance requirements:* For master's, GRE General Test, 2 letters of recommendation. Application fee: $60. *Expenses:* Tuition: Full-time $3,001; part-time $409 per credit. Required fees: $56 per credit. Tuition and fees vary according to course load, degree level and program. *Unit head:* Dr. Michael Kruge, Adviser, 973-655-7668.

See Close-Up on page 657.

Monterey Institute of International Studies, Graduate School of International Policy Studies, Program in International Environmental Policy, Monterey, CA 93940-2691. Offers MA. *Students:* 46 full-time (30 women), 1 part-time; includes 1 African American, 1 Asian American or Pacific Islander, 10 international. Average age 28. 55 applicants, 96% accepted, 26 enrolled. In 2005, 19 degrees awarded. *Degree requirements:* For master's, one foreign language. *Entrance requirements:* For master's, minimum GPA of 3.0, proficiency in a foreign language. Additional exam requirements/recommendations for international students: Required—TOEFL. *Application deadline:* For fall admission, 3/15 for domestic students; for spring admission, 10/1 priority date for domestic students. Applications are processed on a rolling basis. Application fee: $50. Electronic applications accepted. *Expenses:* Tuition: Full-time $25,500; part-time $1,050 per credit. Required fees: $200. *Financial support:* Application deadline: 3/15; *Application contact:* 831-647-4123, Fax: 831-647-6405, E-mail: admit@miis.edu.

Morehead State University, Graduate Programs, College of Science and Technology, Department of Biological and Environmental Sciences, Morehead, KY 40351. Offers biology (MS); regional analysis and public policy (MS). Part-time programs available. *Faculty:* 14 full-time (2 women), 2 part-time/adjunct (1 woman). *Students:* 9 full-time (4 women), 3 part-time (all women), 2 international. Average age 25. 16 applicants, 94% accepted. In 2005, 7 degrees awarded. *Degree requirements:* For master's, oral and written final exams, thesis optional. *Entrance requirements:* For master's, GRE General Test, minimum GPA of 3.0 in biology, 2.5 overall; undergraduate major/minor in biology, environmental science, or equivalent. Additional exam requirements/recommendations for international students: Required—TOEFL (minimum score 525 paper-based; 197 computer-based). *Application deadline:* For fall admission, 8/1 priority date for domestic students, 8/1 priority date for international students; for spring admission, 12/1 priority date for domestic students, 12/1 priority date for international students. Applications are processed on a rolling basis. Application fee: $0 ($55 for international students). Electronic applications accepted. *Financial support:* In 2005–06, 2 research assistantships (averaging $6,000 per year) were awarded; career-related internships or fieldwork and Federal Work-Study also available. Financial award application deadline: 4/1; financial award applicants required to submit FAFSA. *Faculty research:* Atherosclerosis, RNA evolution, cancer biology, water quality/ecology, immunoparasitology. *Unit head:* Dr. David Magrane, Chair, 606-783-2944, E-mail: d.magrane@moreheadstate.edu. *Application contact:* Betty R. Cowsert, Graduate Admissions/Records Manager, 606-783-2039, Fax: 606-783-5061, E-mail: b.cowsert@moreheadstate.edu.

Naropa University, Graduate Programs, Program in Environmental Leadership, Boulder, CO 80302-6697. Offers MA. *Faculty:* 3 full-time, 13 part-time/adjunct. *Students:* 14 full-time (9 women), 20 part-time (14 women); includes 3 minority (all Hispanic Americans), 1 international. Average age 33. 14 applicants, 100% accepted, 6 enrolled. In 2005, 6 degrees awarded. *Degree requirements:* For master's, thesis. *Entrance requirements:* For master's, in-person interview, writing sample. Additional exam requirements/recommendations for international students: Required—TOEFL (minimum score 600 paper-based; 250 computer-based). *Application deadline:* For fall admission, 1/15 priority date for domestic students, 1/15 priority date for international students; for spring admission, 10/15 priority date for domestic students. Applications are processed on a rolling basis. Application fee: $60. Electronic applications accepted. *Expenses:* Tuition: Full-time $14,212; part-time $646 per credit. Required fees: $250 per semester. *Financial support:* In 2005–06, 9 students received support. Federal Work-Study, scholarships/grants, and tuition waivers (partial) available. Support available to part-time students. Financial award application deadline: 3/1; financial award applicants required to submit FAFSA. *Unit head:* Dr. Neena Ambre Rao, Chair, 303-245-4687. *Application contact:* Donna McIntyre, Admissions Counselor, 303-546-3555, Fax: 303-546-3583, E-mail: donna@naropa.edu.

New Jersey Institute of Technology, Office of Graduate Studies, College of Science and Liberal Arts, Department of Humanities and Social Sciences, Program in Environmental Policy Studies, Newark, NJ 07102. Offers MS, PhD. Part-time and evening/weekend programs available. *Students:* 2 full-time (0 women), 18 part-time (9 women); includes 2 minority (1 Asian American or Pacific Islander, 1 Hispanic American), 1 international. Average age 33. 13 applicants, 85% accepted, 7 enrolled. In 2005, 13 degrees awarded. Terminal master's awarded for partial completion of doctoral program. *Degree requirements:* For master's, thesis or alternative. *Entrance requirements:* For master's, GRE General Test. Additional exam requirements/recommendations for international students: Required—TOEFL (minimum score 550 paper-based; 213 computer-based). *Application deadline:* For fall admission, 6/5 for domestic students; for spring admission, 10/15 for domestic students. Applications are processed on a rolling basis. Application fee: $60. Electronic applications accepted. *Expenses:* Tuition, state resident: full-time $9,620; part-time $520 per credit. Tuition, nonresident: full-time $13,542; part-time $715 per credit. Required fees: $78; $54 per credit. $78 per year. Tuition and fees vary according to course load. *Financial support:* Fellowships with full and partial tuition reimbursements, research assistantships with full and partial tuition reimbursements, teaching assistantships with full and partial tuition reimbursements, career-related internships or fieldwork, Federal Work-Study, institutionally sponsored loans, and unspecified assistantships also available. Financial award application deadline: 3/15. *Unit head:* Dr. Nancy Jackson, Director, 973-596-8647, E-mail: nancy.jackson@njit.edu. *Application contact:* Kathryn Kelly, Director of Admissions, 973-596-3300, Fax: 973-596-3461, E-mail: admissions@njit.edu.

New Mexico Highlands University, Graduate Studies, College of Arts and Sciences, Department of Natural Sciences, Las Vegas, NM 87701. Offers applied chemistry (MS); biology (MS); environmental science and management (MS). Part-time programs available. *Faculty:* 11 full-time (4 women), 7 part-time/adjunct (2 women). *Students:* 15 full-time (6 women), 6 part-time (3 women); includes 4 minority (1 American Indian/Alaska Native, 3 Hispanic Americans), 13 international. Average age 29. 6 applicants, 100% accepted, 6 enrolled. In 2005, 8 degrees awarded. *Degree requirements:* For master's, thesis, comprehensive exam, registration. *Entrance requirements:* For master's, minimum undergraduate GPA of 3.0. Additional exam requirements/recommendations for international students: Required—TOEFL (minimum score 540 paper-based; 190 computer-based). *Application deadline:* For fall admission, 8/1 for

Environmental Management and Policy

New Mexico Highlands University (continued)
domestic students. Applications are processed on a rolling basis. Application fee: $15. *Expenses:* Tuition, state resident: full-time $2,280; part-time $101 per credit. Tuition, nonresident: full-time $3,420; part-time $151 per credit. One-time fee: $20 full-time. *Financial support:* In 2005–06, 4 students received support, including 13 teaching assistantships (averaging $11,500 per year); research assistantships with full and partial tuition reimbursements available, Federal Work-Study, institutionally sponsored loans, scholarships/grants, and unspecified assistantships also available. Support available to part-time students. Financial award application deadline: 3/1. *Unit head:* Dr. Merritt Helvenston, Chair, 505-454-3263, Fax: 505-454-3103, E-mail: merritt@nmhu.edu. *Application contact:* Diane Trujillo, Administrative Assistant Graduate Studies, 505-454-3266, Fax: 505-454-3558, E-mail: dtrujillo@nmhu.edu.

New York Institute of Technology, Graduate Division, School of Engineering and Technology, Program in Energy Management, Old Westbury, NY 11568-8000. Offers energy management (MS); energy technology (Advanced Certificate); environmental management (Advanced Certificate); facilities management (Advanced Certificate). Part-time and evening/weekend programs available. Postbaccalaureate distance learning degree programs offered. *Students:* 22 full-time (5 women), 66 part-time (15 women); includes 9 minority (1 African American, 4 Asian Americans or Pacific Islanders, 4 Hispanic Americans), 21 international. Average age 35. 76 applicants, 76% accepted, 27 enrolled. In 2005, 19 master's, 18 other advanced degrees awarded. *Degree requirements:* For master's, thesis or alternative, comprehensive exam. *Entrance requirements:* For master's, minimum QPA of 2.85. *Application deadline:* For fall admission, 7/1 for domestic students; for spring admission, 12/1 priority date for domestic students. Applications are processed on a rolling basis. Application fee: $50. Electronic applications accepted. *Expenses:* Tuition: Full-time $33,654. Required fees: $600. Tuition and fees vary according to student level. *Financial support:* Fellowships, research assistantships with partial tuition reimbursements, institutionally sponsored loans, tuition waivers (full and partial), and unspecified assistantships available. Support available to part-time students. Financial award applicants required to submit FAFSA. *Unit head:* Dr. Robert Amundsen, Director, 516-686-7578. *Application contact:* Jacquelyn Nealon, Dean of Admissions and Financial Aid, 516-686-7925, Fax: 516-686-7613, E-mail: jnealon@nyit.edu.

North Dakota State University, The Graduate School, College of Agriculture, Food Systems, and Natural Resources, Department of Agribusiness and Applied Economics, Fargo, ND 58105. Offers agribusiness and applied economics (MS); international agribusiness (MS); natural resource management (MS, PhD). Part-time programs available. *Faculty:* 16 full-time (3 women), 5 part-time/adjunct (1 woman). *Students:* 20 full-time (4 women), 6 international. Average age 24. 28 applicants, 68% accepted, 12 enrolled. In 2005, 13 degrees awarded. *Degree requirements:* For master's, thesis. *Entrance requirements:* For master's, minimum GPA of 3.0. Additional exam requirements/recommendations for international students: Required—TOEFL (minimum score 525 paper-based; 225 computer-based). *Application deadline:* For fall admission, 2/1 priority date for domestic students, 3/1 priority date for international students. Applications are processed on a rolling basis. Application fee: $45 ($60 for international students). Electronic applications accepted. *Financial support:* In 2005–06, 8 research assistantships with tuition reimbursements (averaging $14,520 per year) were awarded; Federal Work-Study and institutionally sponsored loans also available. Financial award application deadline: 4/15. *Faculty research:* Agribusiness, transportation, marketing, microeconomics, trade. Total annual research expenditures: $1 million. *Unit head:* Dr. David K. Lambert, Chair, 701-231-7444, Fax: 701-231-7400. *Application contact:* Dr. Eric A. DeVuyst, Associate Professor, 701-231-7466, Fax: 701-231-7400, E-mail: eric.devuyst@ndsu.edu.

North Dakota State University, The Graduate School, College of Agriculture, Food Systems, and Natural Resources, Department of Animal and Range Sciences, Fargo, ND 58105. Offers animal science (MS, PhD); natural resource management (MS, PhD); range sciences (MS, PhD). *Faculty:* 25 full-time (6 women), 7 part-time/adjunct (1 woman). *Students:* 31 full-time (15 women), 16 part-time (6 women); includes 3 minority (2 African Americans, 1 Asian American or Pacific Islander), 3 international. Average age 25. 26 applicants, 62% accepted. In 2005, 17 master's, 2 doctorates awarded. *Degree requirements:* For master's, thesis/dissertation; for doctorate, thesis/dissertation, comprehensive exam. *Entrance requirements:* For master's and doctorate, GRE General Test. Additional exam requirements/recommendations for international students: Required—TOEFL. *Application deadline:* Applications are processed on a rolling basis. Application fee: $45 ($60 for international students). *Financial support:* In 2005–06, 30 students received support, including 1 fellowship with tuition reimbursement available (averaging $18,000 per year), 29 research assistantships with tuition reimbursements available (averaging $13,000 per year); teaching assistantships, Federal Work-Study, institutionally sponsored loans, and tuition waivers (partial) also available. Financial award application deadline: 3/15. *Faculty research:* Reproduction, nutrition, meat and muscle biology, breeding/genetics. Total annual research expenditures: $1.5 million. *Unit head:* Don R. Kirby, Interim Chair, 701-231-8386, Fax: 701-231-7723, E-mail: donald.kirby@ndsu.edu.

North Dakota State University, The Graduate School, College of Agriculture, Food Systems, and Natural Resources, Department of Entomology, Fargo, ND 58105. Offers entomology (MS, PhD); environment and conservation science (MS, PhD); natural resource management (MS, PhD). Part-time programs available. *Faculty:* 8 full-time (3 women), 6 part-time/adjunct (0 women). *Students:* 18 full-time (6 women), 4 part-time (2 women); includes 9 minority (2 African Americans, 7 Asian Americans or Pacific Islanders). Average age 34. 5 applicants, 20% accepted, 1 enrolled. In 2005, 1 degree awarded. *Median time to degree:* Of those who began their doctoral program in fall 1997, 100% received their degree in 8 years or less. *Degree requirements:* For master's, thesis/dissertation; for doctorate, thesis/dissertation, comprehensive exam. *Entrance requirements:* For master's and doctorate, minimum GPA of 3.0. Additional exam requirements/recommendations for international students: Required—TOEFL. *Application deadline:* Applications are processed on a rolling basis. Application fee: $45 ($60 for international students). Electronic applications accepted. *Financial support:* In 2005–06, 17 students received support, including 19 research assistantships with full tuition reimbursements available (averaging $13,800 per year); Federal Work-Study, institutionally sponsored loans, and unspecified assistantships also available. Financial award application deadline: 4/15. *Faculty research:* Insect systematics, conservation biology, integrated pest management, insect behavior, insect biology. *Unit head:* Dr. Gary J. Brewer, Chair, 701-231-7908, Fax: 701-231-8557, E-mail: gary.brewer@ndsu.nodak.edu.

North Dakota State University, The Graduate School, College of Agriculture, Food Systems, and Natural Resources, Department of Plant Sciences, Fargo, ND 58105. Offers crop and weed sciences (MS); horticulture (MS); natural resource management (MS); plant sciences (PhD). Part-time programs available. *Faculty:* 39 full-time (3 women), 19 part-time/adjunct (1 woman). *Students:* 66 full-time (31 women), 11 part-time (2 women), 41 international. Average age 26. 44 applicants, 45% accepted. In 2005, 9 master's, 1 doctorate awarded. *Degree requirements:* For master's and doctorate, thesis/dissertation. *Entrance requirements:* Additional exam requirements/recommendations for international students: Required—TOEFL (minimum score 525 paper-based; 197 computer-based). *Application deadline:* Applications are processed on a rolling basis. Application fee: $45 ($60 for international students). Electronic applications accepted. *Financial support:* In 2005–06, 2 fellowships (averaging $19,950 per year), 64 research assistantships were awarded.; teaching assistantships, Federal Work-Study and institutionally sponsored loans also available. Financial award application deadline: 4/15. *Faculty research:* Biotechnology, weed control science, plant breeding, plant genetics, crop physiology. Total annual research expenditures: $880,000. *Unit head:* Dr. Al Schneiter, Chair, 701-231-7971, Fax: 701-231-8474, E-mail: albert.schneiter@ndsu.nodak.edu.

North Dakota State University, The Graduate School, College of Agriculture, Food Systems, and Natural Resources, Department of Soil Science, Fargo, ND 58105. Offers environmental and conservation science (PhD); environmental conservation science (MS); natural resource management (MS, PhD); soil sciences (MS, PhD). Part-time programs available. *Faculty:* 8 full-time (1 woman), 6 part-time/adjunct (0 women). *Students:* 7 full-time (2 women), 4 part-time (1 woman), 4 international. Average age 23. 3 applicants, 33% accepted, 1 enrolled. In 2005, 3 master's, 1 doctorate awarded. *Degree requirements:* For master's and doctorate, thesis/dissertation, classroom teaching, comprehensive exam, registration. *Entrance requirements:* For master's and doctorate, GRE General Test. Additional exam requirements/recommendations for international students: Required—TOEFL (minimum score 525 paper-based; 193 computer-based). *Application deadline:* Applications are processed on a rolling basis. Application fee: $45 ($60 for international students). Electronic applications accepted. *Financial support:* In 2005–06, 6 research assistantships with full tuition reimbursements (averaging $14,300 per year) were awarded; fellowships, Federal Work-Study, institutionally sponsored loans, and scholarships/grants also available. Financial award application deadline: 3/15. *Faculty research:* Microclimate, nitrogen management, landscape studies, water quality, soil management. *Unit head:* Dr. Rodney G. Lym, Interim Chair, 701-231-8903, Fax: 701-231-7861, E-mail: jimmie.richardson@ndsu.nodak.edu.

North Dakota State University, The Graduate School, College of Science and Mathematics, Department of Biological Sciences, Fargo, ND 58105. Offers biological sciences (MS); botany (MS, PhD); cellular and molecular biology (PhD); environmental and conservation sciences (MS, PhD); genomics (MS, PhD); natural resource management (MS, PhD); zoology (MS, PhD). *Faculty:* 20. *Students:* 30 full-time (10 women), 5 part-time (1 woman); includes 1 minority (Asian American or Pacific Islander), 3 international. Average age 24. 14 applicants, 43% accepted. In 2005, 4 master's awarded. *Degree requirements:* For master's and doctorate, thesis/dissertation. *Entrance requirements:* For master's and doctorate, GRE General Test. Additional exam requirements/recommendations for international students: Required—TOEFL. *Application deadline:* For fall admission, 3/15 for domestic students; for spring admission, 10/30 priority date for domestic students. Applications are processed on a rolling basis. Application fee: $45 ($60 for international students). Electronic applications accepted. *Financial support:* In 2005–06, 3 fellowships with full tuition reimbursements (averaging $15,000 per year), 9 research assistantships with full tuition reimbursements (averaging $14,400 per year), 19 teaching assistantships with full tuition reimbursements (averaging $9,550 per year) were awarded.; career-related internships or fieldwork, Federal Work-Study, institutionally sponsored loans, scholarships/grants, tuition waivers (full), and unspecified assistantships also available. Support available to part-time students. Financial award application deadline: 4/15; financial award applicants required to submit FAFSA. *Faculty research:* Comparative endocrinology, physiology, behavioral ecology, plant cell biology, aquatic biology. Total annual research expenditures: $675,000. *Unit head:* Dr. William J. Bleier, Chair, 701-231-7087, Fax: 701-231-7149, E-mail: william.bleier@ndsu.nodak.edu.

North Dakota State University, The Graduate School, School of Natural Resources, Fargo, ND 58105. Offers MS, PhD. Part-time programs available. *Faculty:* 22 full-time (1 woman). *Students:* 47 full-time (17 women); includes 2 minority (1 American Indian/Alaska Native, 1 Asian American or Pacific Islander), 4 international. Average age 32. 10 applicants, 100% accepted. In 2005, 3 degrees awarded. *Degree requirements:* For master's, thesis/dissertation; for doctorate, thesis/dissertation, comprehensive exam. *Entrance requirements:* Additional exam requirements/recommendations for international students: Required—TOEFL. *Application deadline:* Applications are processed on a rolling basis. Application fee: $45 ($60 for international students). Electronic applications accepted. *Financial support:* In 2005–06, 25 students received support; research assistantships with full tuition reimbursements available, teaching assistantships with full tuition reimbursements available available. Support available to part-time students. Financial award application deadline: 3/15. *Faculty research:* Natural resources economics, wetlands issues, wildlife, prairie ecology, range management. Total annual research expenditures:$500,000. *Unit head:* Dr. Carolyn Grygiel, Director, 701-231-8180, Fax: 701-231-7590, E-mail: carolyn.grygiel@ndsu.edu.

Northeastern Illinois University, Graduate College, College of Arts and Sciences, Department of Geography, Environmental Studies and Economics, Program in Geography and Environmental Studies, Chicago, IL 60625-4699. Offers MA. Part-time and evening/weekend programs available. *Degree requirements:* For master's, thesis optional. *Entrance requirements:* For master's, undergraduate minor in geography or environmental studies, minimum GPA of 2.75. *Faculty research:* Segregation and urbanization of minority groups in the Chicago area, scale dependence and parameterization in nonpoint source pollution modeling, ecological land classification and mapping, ecosystem restoration, soil-vegetation relationships.

Northern Arizona University, Graduate College, College of Engineering and Natural Science, Department of Environmental Sciences and Policy, Flagstaff, AZ 86011. Offers conservation ecology (Certificate); environmental sciences and policy (MS). *Accreditation:* NCA. *Degree requirements:* For master's, thesis optional. *Entrance requirements:* For master's, GRE General Test.

Nova Scotia Agricultural College, Research and Graduate Studies, Truro, NS B2N 5E3, Canada. Offers agriculture (M Sc), including air quality, animal behavior, animal molecular genetics, animal nutrition, animal technology, aquaculture, botany, crop management, crop physiology, ecology, environmental microbiology, food science, horticulture, nutrient management, pest management, physiology, plant biotechnology, plant pathology, soil chemistry, soil fertility, waste management and composting, water quality. Part-time programs available. *Faculty:* 43 full-time (7 women), 21 part-time/adjunct (1 woman). *Students:* 50 full-time (25 women), 15 part-time (11 women); includes 7 minority (3 African Americans, 3 Asian Americans or Pacific Islanders, 1 Hispanic American). Average age 25. In 2005, 23 degrees awarded. *Degree requirements:* For master's, thesis, registration. *Entrance requirements:* For master's, honors B Sc, minimum GPA of 3.0. Additional exam requirements/recommendations for international students: Required—TOEFL (minimum score 580 paper-based; 237 computer-based), Michigan English Language Assessment Battery, IELT, Can Test, CAEL. *Application deadline:* For fall admission, 6/1 for domestic students, 4/1 for international students. For winter admission, 10/31 for domestic students; for spring admission, 2/28 for domestic students. Applications are processed on a rolling basis. Application fee: $70. *Expenses:* Tuition, state resident: part-time $2,328 per year. Tuition, nonresident: full-time $6,984; part-time $7,968 per year. International tuition: $12,624 full-time. Required fees: $481; $46 per course. Tuition and fees vary according to program and student level. *Financial support:* In 2005–06, 48 students received support, including 7 fellowships (averaging $15,000 per year), 10 research assistantships (averaging $15,000 per year), 15 teaching assistantships (averaging $900 per year); career-related internships or fieldwork, scholarships/grants, and unspecified assistantships also available. *Faculty research:* Bio-product development, organic agriculture, nutrient management, air and water quality, agricultural biotechnology. Total annual research expenditures: $4.7 million. *Unit head:* Jill L. Rogers, Manager, 902-893-6360, Fax: 902-893-3430, E-mail: jrogers@nsac.ca. *Application contact:* Marie Law, Administrative Assistant, 902-893-6502, Fax: 902-893-3430, E-mail: mlaw@nsac.ca.

Ohio University, Graduate Studies, College of Arts and Sciences, Department of Geological Sciences, Athens, OH 45701-2979. Offers environmental geochemistry (MS); environmental geology (MS); environmental/hydrology (MS); geology (MS); geology education (MS); geomorphology/surficial processes (MS); geophysics (MS); hydrogeology (MS); sedimentology (MS); structure/tectonics (MS). Part-time programs available. *Faculty:* 10 full-time (4 women), 4 part-time/adjunct (1 woman). *Students:* 22 full-time (6 women), 3 part-time (2 women); includes 1 minority (Hispanic American), 4 international. Average age 23. 15 applicants, 67% accepted, 8 enrolled. In 2005, 7 degrees awarded. *Degree requirements:* For master's, thesis, thesis proposal defense and thesis defense. *Entrance requirements:* Additional exam requirements/recommendations for international students: Required—TOEFL (minimum score 550 paper-based; 217 computer-based). *Application deadline:* For fall admission, 2/1 priority date for domestic students, 1/1 priority date for international students. Application fee: $45. Electronic applications accepted. *Financial support:* In 2005–06, 18 students received support, including 3 research assistantships with full tuition reimbursements available (averaging $11,900 per year), 13 teaching assistantships with full tuition reimbursements available (averaging $11,900 per year); institutionally sponsored loans, scholarships/grants, tuition waivers (full), and unspecified assistantships also available. Financial award application deadline: 2/1. *Faculty* *research:* Geoscience education, tectonics, flurial geomorphology, invertebrate paleontology, mine/hydrology. Total annual research expenditures: $649,020. *Unit head:* Dr. David Kidder, Chair, 740-593-1101, Fax: 740-593-0486, E-mail: kidder@ohio.edu. *Application contact:* Dr. David Schneider, Graduate Chair, 740-593-1101, Fax: 740-593-0486, E-mail: schneidd@ohio.edu.

Ohio University, Graduate Studies, College of Arts and Sciences, Program in Environmental Studies, Athens, OH 45701-2979. Offers MS. Part-time programs available. *Students:* 55 full-time, 11 part-time; includes 1 minority (African American), 15 international. Average age 28. 41 applicants, 59% accepted, 21 enrolled. In 2005, 19 degrees awarded. *Degree requirements:* For master's, thesis (for some programs), written exams (if no thesis), research project, comprehensive exam (for some programs), registration. *Entrance requirements:* For master's, minimum GPA of 3.0. Additional exam requirements/recommendations for international students: Required—TOEFL (minimum score 600 paper-based; 250 computer-based). *Application deadline:* For fall admission, 1/1 priority date for domestic students, 1/1 priority date for international students. For winter admission, 10/1 for domestic students; for spring admission, 2/1 for domestic students. Application fee: $45. Electronic applications accepted. *Financial support:* In 2005–06, 29 students received support, including research assistantships with tuition reimbursements available (averaging $10,107 per year), 5 teaching assistantships with tuition reimbursements available (averaging $10,107 per year); fellowships with tuition reimbursements available, career-related internships or fieldwork, Federal Work-Study, institutionally sponsored loans, scholarships/grants, tuition waivers (full), and unspecified assistantships also available. Financial award application deadline: 1/1. *Faculty research:* Air quality modeling, conservation biology, environmental policy, geographical information systems, land management and watershed restoration. *Unit head:* Dr. Gene Mapes, Director, 740-593-9526, Fax: 740-593-0924, E-mail: mapesg@ohio.edu.

Oregon State University, Graduate School, College of Oceanic and Atmospheric Sciences, Corvallis, OR 97331. Offers atmospheric sciences (MA, MS, PhD); geophysics (MA, MS, PhD); marine resource management (MA, MS); oceanography (MA, MS, PhD). *Faculty:* 60 full-time (9 women), 15 part-time/adjunct (4 women). *Students:* 83 full-time (35 women), 16 part-time (11 women); includes 3 minority (1 African American, 1 American Indian/Alaska Native, 1 Hispanic American), 18 international. Average age 30. In 2005, 16 master's, 8 doctorates awarded. Terminal master's awarded for partial completion of doctoral program. *Degree requirements:* For master's, thesis optional; for doctorate, thesis/dissertation. *Entrance requirements:* For master's and doctorate, GRE General Test, minimum GPA of 3.0 in last 90 hours. Additional exam requirements/recommendations for international students: Required—TOEFL. *Application deadline:* For fall admission, 2/1 for domestic students. Applications are processed on a rolling basis. Application fee: $50. *Expenses:* Tuition, area resident: Full-time $8,139; part-time $301 per credit. Tuition, state resident: full-time $8,139; part-time $501 per credit. Tuition, nonresident: full-time $14,376; part-time $532 per credit. International tuition: $14,376 full-time. Required fees: $1,266. *Financial support:* Fellowships, research assistantships, teaching assistantships, career-related internships or fieldwork, Federal Work-Study, and institutionally sponsored loans available. Support available to part-time students. Financial award application deadline: 2/1. *Faculty research:* Biological, chemical, geological, and physical oceanography. *Unit head:* Dr. Mark R. Abbott, Dean, 541-737-3504, Fax: 541-737-2064, E-mail: mark@oce.orst.edu. *Application contact:* Irma Delson, Assistant Director, Student Services, 541-737-5189, Fax: 541-737-2064, E-mail: student_adviser@oce.orst.edu.

The Pennsylvania State University University Park Campus, Graduate School, Intercollege Graduate Programs, Intercollege Program in Environmental Pollution Control, State College, University Park, PA 16802-1503. Offers MEPC, MS. *Students:* 8 full-time (5 women), 4 part-time (2 women), 2 international. *Entrance requirements:* For master's, GRE General Test. Additional exam requirements/recommendations for international students: Required—TOEFL. Application fee: $45. *Expenses:* Tuition, state resident: full-time $12,518; part-time $522 per credit. Tuition, nonresident: full-time $23,004; part-time $959 per credit. Required fees: $484. Tuition and fees vary according to course load, campus/location and program. *Unit head:* Dr. Herschel A. Elliott, Chair, 814-865-1417, Fax: 814-863-1031, E-mail: haelliott@psu.edu. *Application contact:* Dr. Herschel A. Elliott, Chair, 814-865-1417, Fax: 814-863-1031, E-mail: haelliott@psu.edu.

Plymouth State University, College of Graduate Studies, Graduate Studies in Education, Program in Science, Plymouth, NH 03264-1595. Offers applied meteorology (MS); environmental science and policy (MS); science education (MS). *Students:* 1 (woman) full-time, 18 part-time (12 women). 19 applicants, 100% accepted. *Unit head:* Dr. Steve Kahl, Director of the Center for the Environment, E-mail: jskahl@plymouth.edu.

Polytechnic University of Puerto Rico, Graduate School, Hato Rey, PR 00919. Offers business administration (MBA); civil engineering (ME, MS); competitiveness manufacturing (MCM, MMC, MS); computer engineering (ME, MS); electrical engineering (ME, MS); engineering management (MEM); environmental management (MEM); manufacturing engineering (ME, MS). Part-time and evening/weekend programs available.

Portland State University, Graduate Studies, College of Liberal Arts and Sciences, Interdisciplinary Program in Environmental Sciences and Resources, Portland, OR 97207-0751. Offers environmental management (MEM); environmental sciences/biology (PhD); environmental sciences/chemistry (PhD); environmental sciences/civil engineering (PhD); environmental sciences/geography (PhD); environmental sciences/geology (PhD); environmental sciences/physics (PhD); environmental studies (MS); science/environmental science (MST). Part-time programs available. *Faculty:* 8 full-time (0 women), 2 part-time/adjunct (1 woman). *Students:* 84 full-time (40 women), 37 part-time (17 women); includes 9 minority (1 African American, 3 Asian Americans or Pacific Islanders, 5 Hispanic Americans), 29 international. Average age 32. 82 applicants, 66% accepted, 33 enrolled. In 2005, 11 master's, 5 doctorates awarded. *Degree requirements:* For doctorate, variable foreign language requirement, thesis/dissertation, oral and qualifying exams. *Entrance requirements:* For doctorate, minimum GPA of 3.0 in upper-division course work or 2.75 overall. Additional exam requirements/recommendations for international students: Required—TOEFL (minimum score 550 paper-based; 213 computer-based). *Application deadline:* For fall admission, 4/1 priority date for domestic students, 3/1 priority date for international students. Applications are processed on a rolling basis. Application fee: $50. *Expenses:* Tuition, state resident: full-time $6,648; part-time $231 per credit. Tuition, nonresident: full-time $11,319; part-time $231 per credit. Required fees: $686; $67 per credit. *Financial support:* In 2005–06, 2 research assistantships with full tuition reimbursements (averaging $8,648 per year) were awarded; teaching assistantships with full tuition reimbursements, Federal Work-Study, scholarships/grants, tuition waivers (partial), and unspecified assistantships also available. Support available to part-time students. Financial award application deadline: 3/1; financial award applicants required to submit FAFSA. *Faculty research:* Environmental aspects of biology, chemistry, civil engineering, geology, physics. Total annual research expenditures: $1.6 million. *Unit head:* John Rueter, Director, 503-725-4980, Fax: 503-725-3888.

Prescott College, Graduate Programs, Program in Environmental Studies, Prescott, AZ 86301. Offers agroecology (MA); ecopsychology (MA); environmental education (MA); environmental studies (MA); sustainability (MA). MA in environmental education offered jointly with Teton Science School. Part-time programs available. Postbaccalaureate distance learning degree programs offered (minimal on-campus study). *Faculty:* 1 full-time (0 women), 32 part-time/adjunct (10 women). *Students:* 16 full-time (9 women), 19 part-time (10 women); includes 2 minority (both Hispanic Americans) Average age 35. In 2005, 12 degrees awarded. *Degree requirements:* For master's, thesis, fieldwork or internship, practicum. *Entrance requirements:* For master's, 2 letters of recommendation, resumé. *Application deadline:* For fall admission, 5/1 for domestic students; for spring admission, 11/1 priority date for domestic students. Applications are processed on a rolling basis. Application fee: $40. Electronic applications accepted. *Expenses:* Tuition: Full-time $12,408; part-time $517 per credit. One-time fee: $103 full-time. *Financial support:* Career-related internships or fieldwork and Federal Work-Study available. Financial award applicants required to submit FAFSA. *Unit head:* Dr. Paul Sneed, Head, 928-350-3204. *Application contact:* Kerstin Alicki, Admissions Counselor, 877-350-2100 Ext. 2102, Fax: 928-776-5242, E-mail: admissions@prescott.edu.

Princeton University, Graduate School, Department of Mechanical and Aerospace Engineering, Princeton, NJ 08544. Offers applied physics (M Eng, MSE, PhD); computational methods (M Eng, MSE); dynamics and control systems (M Eng, MSE, PhD); energy and environmental policy (M Eng, MSE, PhD); energy conversion, propulsion, and combustion (M Eng, MSE, PhD); flight science and technology (M Eng, MSE, PhD); fluid mechanics (M Eng, MSE, PhD). Part-time programs available. *Faculty:* 22 full-time (3 women). *Students:* 76 full-time (13 women); includes 5 minority (1 African American, 1 Asian American or Pacific Islander, 3 Hispanic Americans), 43 international. Average age 24. 252 applicants, 20% accepted, 20 enrolled. In 2005, 7 master's, 8 doctorates awarded. Terminal master's awarded for partial completion of doctoral program. *Degree requirements:* For master's, thesis/dissertation; for doctorate, thesis/dissertation, comprehensive exam. *Entrance requirements:* For master's and doctorate, GRE General Test. Additional exam requirements/recommendations for international students: Required—IELT. *Application deadline:* For fall admission, 12/31 for domestic students, 12/1 for international students. Application fee: $105. Electronic applications accepted. *Financial support:* In 2005–06, 12 fellowships with full tuition reimbursements (averaging $8,800 per year), 36 research assistantships with full tuition reimbursements (averaging $27,461 per year), 9 teaching assistantships with full tuition reimbursements (averaging $21,641 per year) were awarded.; Federal Work-Study and institutionally sponsored loans also available. Financial award application deadline: 1/2. Total annual research expenditures: $6.2 million. *Unit head:* Prof. Luigi Martinelli, Director of Graduate Studies, 609-258-6652, Fax: 609-258-1918, E-mail: gigi@princeton.edu. *Application contact:* Janice Hueng, Director of Graduate Admissions, 609-258-3034, Fax: 609-258-6180, E-mail: gsadmit@princeton.edu.

Purdue University, Graduate School, College of Agriculture, Department of Forestry and Natural Resources, West Lafayette, IN 47907. Offers aquaculture, fisheries, aquatic science (MSF); aquaculture, fisheries, aquatic sciences (MS, PhD); forest biology (MS, MSF, PhD); natural resources and environmental policy (MS, MSF); natural resources environmental policy (PhD); quantitative resource analysis (MS, MSF, PhD); wildlife science (MS, MSF, PhD); wood science and technology (MS, MSF, PhD). *Faculty:* 26 full-time (4 women), 7 part-time/adjunct (1 woman). *Students:* 69 full-time (33 women), 15 part-time (3 women); includes 5 minority (1 African American, 1 American Indian/Alaska Native, 3 Hispanic Americans), 29 international. Average age 30. 45 applicants, 40% accepted, 18 enrolled. In 2005, 6 master's, 6 doctorates awarded. *Degree requirements:* For master's and doctorate, thesis/dissertation. *Entrance requirements:* For master's and doctorate, GRE General Test (500 verbal, 500 quantitative), minimum B+ average in undergraduate course work. Additional exam requirements/recommendations for international students: Required—TOEFL. *Application deadline:* For fall admission, 1/5 for domestic students; for spring admission, 9/15 for domestic students. Applications are processed on a rolling basis. Application fee: $55. Electronic applications accepted. *Financial support:* In 2005–06, 10 research assistantships (averaging $15,259 per year) were awarded; fellowships, teaching assistantships, career-related internships or fieldwork and scholarships/grants also available. Support available to part-time students. Financial award application deadline: 1/5; financial award applicants required to submit FAFSA. *Faculty research:* Wildlife management, forest management, forest ecology, forest soils, limnology. *Unit head:* Dr. Robert K. Swihart, Interim Head, 765-494-3590, Fax: 765-494-9461, E-mail: rswihart@purdue.edu. *Application contact:* Kelly Garrett, Graduate Secretary, 765-494-3572, Fax: 765-494-9461, E-mail: kgarrett@purdue.edu.

Rensselaer Polytechnic Institute, Graduate School, School of Humanities and Social Sciences, Department of Economics, Interdisciplinary Program in Ecological Economics, Troy, NY 12180-3590. Offers PhD. Part-time programs available. *Faculty:* 9 full-time (1 woman). *Students:* 12 full-time (6 women); includes 5 minority (1 African American, 4 Asian Americans or Pacific Islanders), 1 international. 15 applicants, 53% accepted, 3 enrolled. In 2005, 2 degrees awarded. *Degree requirements:* For doctorate, thesis/dissertation, comprehensive exam. *Entrance requirements:* For doctorate, GMAT or GRE General Test. Additional exam requirements/recommendations for international students: Required—TOEFL (minimum score 570 paper-based; 230 computer-based). *Application deadline:* For fall admission, 1/15 for domestic students. Applications are processed on a rolling basis. Application fee: $75. Electronic applications accepted. *Expenses:* Tuition: Full-time $31,000; part-time $1,320 per credit. Required fees: $1,623. *Financial support:* In 2005–06, 11 students received support, including 1 fellowship with full tuition reimbursement available (averaging $21,000 per year), 5 research assistantships with full tuition reimbursements available (averaging $14,500 per year), 5 teaching assistantships with full tuition reimbursements available (averaging $14,500 per year); scholarships/grants also available. Financial award application deadline: 2/1. *Faculty research:* Sustainable development, natural resource economics, cost-benefit analysis, social economics, regional input-output analysis. *Application contact:* Betty Jean Kaufman, Administrative Assistant, 518-276-6387, Fax: 518-276-2235, E-mail: kaufmb@rpi.edu.

Rensselaer Polytechnic Institute, Graduate School, School of Humanities and Social Sciences, Program in Ecological Economics, Values, and Policy, Troy, NY 12180-3590. Offers MS. Part-time programs available. *Faculty:* 8 full-time (3 women). *Students:* 2 applicants, 100% accepted, 0 enrolled. In 2005, 3 degrees awarded. *Degree requirements:* For master's, professional project. *Entrance requirements:* For master's, GRE General Test. Additional exam requirements/recommendations for international students: Required—TOEFL (minimum score 600 paper-based; 250 computer-based). *Application deadline:* For fall admission, 1/15 priority date for domestic students, 1/15 priority date for international students. Applications are processed on a rolling basis. Application fee: $75. Electronic applications accepted. *Expenses:* Tuition: Full-time $31,000; part-time $1,320 per credit. Required fees:$1,623. *Financial support:* In 2005–06, 1 teaching assistantship with full tuition reimbursement (averaging $12,000 per year) was awarded; fellowships, research assistantships, career-related internships or fieldwork and institutionally sponsored loans also available. Financial award application deadline: 1/15. *Faculty research:* Environmental politics and policy, environmentalism, political economy, third world politics, environmental health. *Unit head:* Dr. Steve Breyman, Director of Graduate Studies, 518-276-8515, Fax: 518-276-2659, E-mail: breyms@rpi.edu.

Rice University, Graduate Programs, Wiess School of Natural Sciences, Professional Master's Program in Environmental Analysis and Decision Making, Houston, TX 77251-1892. Offers MS. Part-time programs available. *Degree requirements:* For master's, internship. *Entrance requirements:* For master's, GRE General Test, letters of recommendation (4). Additional exam requirements/recommendations for international students: Required—TOEFL (minimum score 600 paper-based; 250 computer-based). Electronic applications accepted. *Faculty research:* Environmental biotechnology, environmental nanochemistry, environmental statistics, remote sensing.

See Close-Up on page 663.

Rochester Institute of Technology, Graduate Enrollment Services, College of Applied Science and Technology, Department of Environmental Management, Rochester, NY 14623-5603. Offers MS. *Students:* 6 full-time (1 woman), 34 part-time (11 women); includes 4 minority (3 African Americans, 1 Asian American or Pacific Islander), 2 international. 23 applicants, 52% accepted, 9 enrolled. In 2005, 16 degrees awarded. *Entrance requirements:* For master's, minimum GPA of 3.0. *Application deadline:* For fall admission, 3/1 for domestic students. Applications are processed on a rolling basis. Application fee: $50. Electronic applications accepted. *Expenses:* Tuition: Full-time $25,392; part-time $713 per credit. Required fees: $183; $61 per term. *Unit head:* Maureen Valentine, Chair, 585-475-7398, E-mail: msvite@rit.edu.

Royal Roads University, Graduate Studies, Science, Technology and Environment Program, Victoria, BC V9B 5Y2, Canada. Offers environment and management (M Sc, MA); knowledge management (MA). Postbaccalaureate distance learning degree programs offered (minimal on-campus study). *Degree requirements:* For master's, thesis. *Entrance requirements:* For

Environmental Management and Policy

Royal Roads University (continued)

master's, 5-7 years of related work experience. Electronic applications accepted. *Faculty research:* Sustainable development, atmospheric processes, sustainable communities, chemical fate and transport of persistent organic pollutants, educational technology.

St. Cloud State University, School of Graduate Studies, College of Science and Engineering, Department of Environmental and Technological Studies, St. Cloud, MN 56301-4498. Offers MS. *Faculty:* 7 full-time (0 women). *Students:* 2 full-time (1 woman), 7 part-time (1 woman); includes 2 minority (both Asian Americans or Pacific Islanders), 3 international. 10 applicants, 50% accepted. In 2005, 8 degrees awarded. *Degree requirements:* For master's, thesis or alternative. *Entrance requirements:* For master's, minimum GPA of 2.75. Additional exam requirements/recommendations for international students: Required—TOEFL (minimum score 550 paper-based; 213 computer-based), MELAB; Recommended—IELT (minimum score 7). *Application deadline:* For fall admission, 6/1 priority date for domestic students, 4/1 priority date for international students; for spring admission, 10/1 priority date for domestic students, 8/1 priority date for international students. Applications are processed on a rolling basis. Application fee: $35. Electronic applications accepted. *Expenses:* Tuition, state resident: part-time $277. Tuition, nonresident: part-time $379. Required fees: $23 per credit. Tuition and fees vary according to course load and reciprocity agreements. *Financial support:* Federal Work-Study, scholarships/grants, and unspecified assistantships available. Financial award application deadline: 3/1. *Unit head:* Dr. Kurt Helgeson, Chairperson, 320-308-3235, Fax: 320-308-5122, E-mail: ets@stcloudstate.edu. *Application contact:* Linda Lou Krueger, School of Graduate Studies, 320-308-2113, Fax: 320-308-5371, E-mail: lekrueger@stcloudstate.edu.

Saint Joseph's University, College of Arts and Sciences, Program in Environmental Protection, Philadelphia, PA 19131-1395. Offers environmental protection and safety management (MS, Post-Master's Certificate). In 2005, 16 degrees awarded. *Application deadline:* For fall admission, 7/15 for domestic students. Application fee: $35. *Expenses:* Tuition: Part-time $692 per credit. Tuition and fees vary according to program. *Unit head:* Dr. Vincent P. McNally, Director, 610-660-1453, Fax: 610-660-2903, E-mail: vmcnally@sju.edu.

Saint Mary-of-the-Woods College, Program in Earth Literacy, Saint Mary-of-the-Woods, IN 47876. Offers MA. Part-time programs available. Postbaccalaureate distance learning degree programs offered (minimal on-campus study). *Degree requirements:* For master's, thesis. Electronic applications accepted. *Faculty research:* Ecology, art, spirituality.

Saint Mary's University of Minnesota, School of Graduate and Professional Programs, Department of Resource Analysis, Winona, MN 55987-1399. Offers business administration (MS); criminal justice/policy administration (MS); geographic information science (Certificate); natural resources management (MS); project management (MS); public safety administration (MA). *Faculty:* 6 full-time (2 women), 1 part-time/adjunct (0 women). *Students:* 19 full-time (5 women), 12 part-time (3 women); includes 2 minority (1 African American, 1 Hispanic American). Average age 28. 26 applicants, 100% accepted, 23 enrolled. In 2005, 10 degrees awarded. *Degree requirements:* For master's, thesis or alternative. *Entrance requirements:* For master's and Certificate, letters of recommendation. *Application deadline:* Applications are processed on a rolling basis. Application fee: $25. Electronic applications accepted. *Expenses:* Tuition: Part-time $275 per credit. Tuition and fees vary according to degree level and program. *Unit head:* Dr. David McConville, Professor, 507-457-1542, Fax: 507-457-1633, E-mail: dmcconvi@smumn.edu. *Application contact:* Dr. John A. Nosek, Information Contact, 507-457-6952, E-mail: janosek@smumn.edu.

San Francisco State University, Division of Graduate Studies, College of Behavioral and Social Sciences, Department of Geography and Human Environmental Studies, San Francisco, CA 94132-1722. Offers geography (MA), including environmental planning, resource management. Part-time programs available. *Degree requirements:* For master's, thesis, exam. *Entrance requirements:* For master's, minimum GPA of 2.5 in last 60 units. *Faculty research:* Geomorphology, remote sensing, GIS, biogeography.

San Jose State University, Graduate Studies and Research, College of Social Sciences, Department of Environmental Studies, San Jose, CA 95192-0001. Offers MS. Part-time programs available. *Students:* 21 full-time (16 women), 23 part-time (16 women); includes 10 minority (5 Asian Americans or Pacific Islanders, 5 Hispanic Americans), 5 international. Average age 34. 27 applicants, 74% accepted, 14 enrolled. In 2005, 9 degrees awarded. *Degree requirements:* For master's, thesis or alternative, comprehensive exam. *Entrance requirements:* Additional exam requirements/recommendations for international students: Required—TOEFL (minimum score 580 paper-based). *Application deadline:* For fall admission, 6/29 for domestic students; for spring admission, 11/30 for domestic students. Applications are processed on a rolling basis. Application fee: $59. Electronic applications accepted. *Expenses:* Tuition, nonresident: part-time $339 per unit. Required fees: $1,286 per semester. Tuition and fees vary according to course load and degree level. *Financial support:* In 2005–06, 2 teaching assistantships were awarded; career-related internships or fieldwork, Federal Work-Study, and institutionally sponsored loans also available. Support available to part-time students. Financial award applicants required to submit FAFSA. *Faculty research:* Remote sensing, land use/land cover mapping. *Unit head:* Rachel O'Malley, Chair, 408-924-5450, Fax: 408-924-5477.

Shippensburg University of Pennsylvania, School of Graduate Studies, College of Arts and Sciences, Department of Geography and Earth Science, Shippensburg, PA 17257-2299. Offers geoenvironmental studies (MS). Part-time and evening/weekend programs available. *Faculty:* 10 full-time (2 women). *Students:* 21 full-time (7 women), 11 part-time (4 women); includes 2 minority (both Asian Americans or Pacific Islanders) Average age 27. 29 applicants, 72% accepted, 8 enrolled. In 2005, 19 degrees awarded. *Degree requirements:* For master's, internship thesis or practicum. *Entrance requirements:* For master's, GRE (if GPA is below 2.75), 12 credit hours in geography or earth sciences, 15 credit hours in social sciences and 15 hours in natural sciences. Additional exam requirements/recommendations for international students: Required—TOEFL (minimum score 560 paper-based; 220 computer-based). *Application deadline:* Applications are processed on a rolling basis. Application fee: $30. Electronic applications accepted. *Expenses:* Tuition, state resident: full-time $2,944; part-time $327 per credit. Tuition, nonresident: full-time $4,711; part-time $523 per credit. Required fees: $427; $27 per credit. *Financial support:* In 2005–06, 15 research assistantships with full tuition reimbursements (averaging $2,575 per year) were awarded; career-related internships or fieldwork, scholarships/grants, and unspecified assistantships also available. Support available to part-time students. Financial award application deadline: 3/1; financial award applicants required to submit FAFSA. *Unit head:* Dr. William Blewett, Chairperson, 717-477-1685, Fax: 717-477-4029, E-mail: wlblew@ship.edu. *Application contact:* Renee Payne, Associate Dean of Graduate Admissions, 717-477-1231, Fax: 717-477-4016, E-mail: rmpayn@ship.edu.

Simon Fraser University, Graduate Studies, Faculty of Applied Science, School of Resource and Environmental Management, Burnaby, BC V5A 1S6, Canada. Offers MRM, PhD. *Degree requirements:* For master's, thesis or alternative, research project; for doctorate, thesis/dissertation, comprehensive exam. *Entrance requirements:* For master's, minimum GPA of 3.0; for doctorate, GRE Writing Assessment, minimum GPA of 3.5. Additional exam requirements/recommendations for international students: Required—TOEFL or IELTS. *Faculty research:* Management of resources, resource economics, regional planning, public policy analysis, tourism and parks.

Slippery Rock University of Pennsylvania, Graduate Studies (Recruitment), College of Health, Environment, and Science, Department of Parks, Recreation, and Environmental Education, Slippery Rock, PA 16057-1383. Offers environmental education (M Ed); resource management (MS); sustainable systems (MS). Part-time and evening/weekend programs available. *Degree requirements:* For master's, thesis (for some programs), comprehensive exam (for some programs). *Entrance requirements:* For master's, GRE General Test, MAT, minimum GPA of 2.75. Additional exam requirements/recommendations for international students: Required—TOEFL (minimum score 550 paper-based; 213 computer-based). *Application deadline:* For fall admission, 7/1 priority date for domestic students, 7/1 priority date for international students; for spring admission, 11/1 priority date for domestic students, 11/1 priority date for international students. Applications are processed on a rolling basis. Application fee: $25. Electronic applications accepted. *Expenses:* Tuition, area resident: Part-time $893 per term. Tuition, state resident: full-time $7,637; part-time $1,376 per term. Tuition, nonresident: full-time $11,764. Tuition and fees vary according to course load, campus/location and program. *Financial support:* Career-related internships or fieldwork, Federal Work-Study, scholarships/grants, and unspecified assistantships available. Support available to part-time students. Financial award application deadline: 5/1; financial award applicants required to submit FAFSA. *Unit head:* Dr. Daniel Dziubek, Graduate Coordinator, 724-738-2068, Fax: 724-738-2938, E-mail: daniel.dziubek@sru.edu. *Application contact:* April Longwell, Interim Director of Graduate Studies, 724-738-2051 Ext. 2116, Fax: 724-738-2146, E-mail: graduate.studies@sru.edu.

Southeast Missouri State University, School of Graduate Studies, Department of Human Environmental Studies, Cape Girardeau, MO 63701-4799. Offers home economics (MA); human environmental studies (MA). Part-time programs available. *Faculty:* 8 full-time (7 women). *Students:* 10 full-time (all women), 15 part-time (all women); includes 5 minority (all African Americans), 2 international. Average age 28. 9 applicants, 100% accepted. In 2005, 4 degrees awarded. *Degree requirements:* For master's, thesis or alternative. *Entrance requirements:* For master's, GRE General Test, MAT, minimum GPA of 2.75. Additional exam requirements/recommendations for international students: Required—TOEFL (minimum score 550 paper-based; 213 computer-based). *Application deadline:* For fall admission, 8/1 for domestic students, 4/1 for international students; for spring admission, 11/21 for domestic students, 9/1 for international students. Applications are processed on a rolling basis. Application fee: $20 ($100 for international students). Electronic applications accepted. *Expenses:* Tuition, state resident: full-time $1,676; part-time $186 per hour. Tuition, nonresident: full-time $3,052; part-time $339 per hour. Required fees: $114; $13 per hour. Tuition and fees vary according to course load, degree level and campus/location. *Financial support:* In 2005–06, 19 students received support, including 9 teaching assistantships with full tuition reimbursements available (averaging $6,600 per year); research assistantships with full tuition reimbursements available, unspecified assistantships also available. Financial award applicants required to submit FAFSA. *Unit head:* Dr. Paula King, Chairperson, 573-651-2312, E-mail: pking@semo.edu. *Application contact:* Marsha L. Arant, Senior Administrative Assistant, Office of Graduate Studies, 573-651-2192, Fax: 573-651-2001, E-mail: marant@semo.edu.

Southeast Missouri State University, School of Graduate Studies, Harrison College of Business, Cape Girardeau, MO 63701-4799. Offers accounting (MBA); environmental management (MBA); finance (MBA); general management (MBA); health administration (MBA); industrial management (MBA); international business (MBA). *Accreditation:* AACSB. Part-time and evening/weekend programs available. *Faculty:* 33 full-time (10 women). *Students:* 33 full-time (18 women), 44 part-time (22 women); includes 5 minority (3 African Americans, 1 Asian American or Pacific Islander, 1 Hispanic American), 12 international. Average age 27. 21 applicants, 76% accepted. In 2005, 30 degrees awarded. *Degree requirements:* For master's, applied research project. *Entrance requirements:* For master's, GMAT, minimum undergraduate GPA of 2.5. Additional exam requirements/recommendations for international students: Required—TOEFL (minimum score 550 paper-based; 213 computer-based). *Application deadline:* For fall admission, 8/1 for domestic students, 4/1 for international students; for spring admission, 11/21 for domestic students, 9/1 for international students. Applications are processed on a rolling basis. Application fee: $20 ($100 for international students). *Expenses:* Tuition, state resident: full-time $1,676; part-time $186 per hour. Tuition, nonresident: full-time $3,052; part-time $339 per hour. Required fees: $114; $13 per hour. Tuition and fees vary according to course load, degree level and campus/location. *Financial support:* In 2005–06, 41 students received support, including 18 research assistantships with full tuition reimbursements available (averaging $6,600 per year); career-related internships or fieldwork and unspecified assistantships also available. Financial award applicants required to submit FAFSA. *Unit head:* Kenneth Heischmidt, Director, 573-651-2912, Fax: 573-651-5032, E-mail: kheischmidt@semo.edu. *Application contact:* Marsha L. Arant, Senior Administrative Assistant, Office of Graduate Studies, 573-651-2192, Fax: 573-651-2001, E-mail: marant@semo.edu.

Southern Illinois University Edwardsville, Graduate Studies and Research, College of Arts and Sciences, Program in Environmental Science Management, Edwardsville, IL 62026-0001. Offers MS. *Application deadline:* For fall admission, 7/21 for domestic students, 6/1 for international students; for spring admission, 12/8 for domestic students, 10/1 for international students. Applications are processed on a rolling basis. Application fee: $30. *Expenses:* Tuition, state resident: part-time $190 per semester hour. Tuition, nonresident: part-time $380 per semester hour. Tuition and fees vary according to course load, reciprocity agreements and student level. *Unit head:* Dr. Kevin Johnson, Program Director, 618-650-5934, E-mail: kevjohn@siue.edu.

Stanford University, School of Earth Sciences, Earth Systems Program, Stanford, CA 94305-9991. Offers MS. Students admitted at the undergraduate level. Electronic applications accepted.

State University of New York College of Environmental Science and Forestry, Faculty of Environmental Resources and Forest Engineering, Syracuse, NY 13210-2779. Offers environmental and resources engineering (MPS, MS, PhD). *Faculty:* 6 full-time (1 woman), 3 part-time/adjunct (0 women). *Students:* 18 full-time (8 women), 28 part-time (8 women); includes 3 minority (1 African American, 2 Asian Americans or Pacific Islanders), 13 international. Average age 32. 18 applicants, 67% accepted, 9 enrolled. In 2005, 10 master's, 4 doctorates awarded. *Degree requirements:* For master's, thesis (for some programs), registration; for doctorate, thesis/dissertation, comprehensive exam, registration. *Entrance requirements:* For master's and doctorate, GRE General Test, minimum GPA of 3.0. Additional exam requirements/recommendations for international students: Required—TOEFL (minimum score 550 paper-based; 213 computer-based). *Application deadline:* For fall admission, 2/1 priority date for domestic students, 2/1 priority date for international students; for spring admission, 11/1 priority date for domestic students, 11/1 priority date for international students. Applications are processed on a rolling basis. Application fee: $60. *Expenses:* Tuition, area resident: Full-time $6,900; part-time $288 per credit. Tuition, nonresident: full-time $10,920; part-time $455 per credit. Required fees: $395; $32 per credit. $20 per term. One-time fee: $145. *Financial support:* In 2005–06, 20 students received support, including 4 fellowships with full and partial tuition reimbursements available (averaging $9,446 per year), 8 research assistantships with full and partial tuition reimbursements available, 6 teaching assistantships with full and partial tuition reimbursements available (averaging $9,446 per year); Federal Work-Study, institutionally sponsored loans, scholarships/grants, health care benefits, and unspecified assistantships also available. Financial award application deadline: 6/30; financial award applicants required to submit FAFSA. *Faculty research:* Forest engineering, paper science and engineering, wood products engineering. Total annual research expenditures: $968,412. *Unit head:* Dr. James M. Hassett, Chair, 315-470-6633, Fax: 315-470-6958, E-mail: jhassett@esf.edu. *Application contact:* Dr. Dudley J. Raynal, Dean, Instruction and Graduate Studies, 315-470-6599, Fax: 315-470-6978, E-mail: esfgrad@esf.edu.

State University of New York College of Environmental Science and Forestry, Faculty of Environmental Studies, Syracuse, NY 13210-2779. Offers environmental and community land planning (MPS, MS, PhD); environmental and natural resources policy (PhD); environmental communication and participatory processes (MPS, MS, PhD); environmental policy and democratic processes (MPS, MS, PhD); environmental systems and risk management (MPS, MS, PhD); water and wetland resource studies (MPS, MS, PhD). Part-time programs available. *Faculty:* 11 full-time (7 women), 11 part-time/adjunct (6 women). *Students:* 49 full-time (30 women), 28 part-time (15 women); includes 3 minority (2 African Americans, 1 Hispanic American), 36 international. Average age 32. 61 applicants, 69% accepted, 17 enrolled. In 2005, 16 master's, 4 doctorates awarded. *Degree requirements:* For master's, thesis (for some programs), registration; for doctorate, thesis/dissertation, comprehensive exam, registration.

Entrance requirements: For master's and doctorate, GRE General Test, minimum GPA of 3.0. Additional exam requirements/recommendations for international students: Required—TOEFL (minimum score 550 paper-based; 213 computer-based). *Application deadline:* For fall admission, 2/1 priority date for domestic students, 2/1 priority date for international students; for spring admission, 11/1 priority date for domestic students, 11/1 priority date for international students. Applications are processed on a rolling basis. Application fee: $60. *Expenses:* Tuition, area resident: Full-time $6,900; part-time $288 per credit. Tuition, nonresident: full-time $10,920; part-time $455 per credit. Required fees: $395; $32 per credit. $20 per term. One-time fee: $145. *Financial support:* In 2005–06, 21 fellowships with full and partial tuition reimbursements (averaging $9,446 per year), 4 research assistantships with full and partial tuition reimbursements (averaging $10,000 per year), 9 teaching assistantships with full and partial tuition reimbursements (averaging $9,446 per year) were awarded.; career-related internships or fieldwork, Federal Work-Study, institutionally sponsored loans, scholarships/grants, health care benefits, and unspecified assistantships also available. Support available to part-time students. Financial award application deadline: 6/30; financial award applicants required to submit FAFSA. *Faculty research:* Environmental education/communications, water resources, land resources, waste management. Total annual research expenditures: $169,919. *Unit head:* Dr. Richard Smardon, Chair, 315-470-6636, Fax: 315-470-6915, E-mail: rsmardon@syr.edu. *Application contact:* Dr. Dudley J. Raynal, Dean, Instruction and Graduate Studies, 315-470-6599, Fax: 315-470-6978, E-mail: esfgrad@esf.edu.

State University of New York College of Environmental Science and Forestry, Faculty of Forest and Natural Resources Management, Syracuse, NY 13210-2779. Offers environmental and natural resource policy (MS, PhD); environmental and natural resources policy (MPS); forest management and operations (MF); forestry ecosystems science and applications (MPS, MS, PhD); natural resources management (MPS, MS, PhD); quantitative methods and management in forest science (MPS, MS, PhD); recreation and resource management (MPS, MS, PhD); watershed management and forest hydrology (MPS, MS, PhD). *Faculty:* 30 full-time (7 women), 1 (woman) part-time/adjunct. *Students:* 43 full-time (18 women), 26 part-time (12 women); includes 1 minority (Hispanic American), 13 international. Average age 31. 38 applicants, 55% accepted, 12 enrolled. In 2005, 13 master's, 4 doctorates awarded. *Degree requirements:* For master's, thesis (for some programs), registration; for doctorate, thesis/dissertation, comprehensive exam, registration. *Entrance requirements:* For master's and doctorate, GRE General Test, minimum GPA of 3.0. Additional exam requirements/recommendations for international students: Required—TOEFL (minimum score 550 paper-based; 213 computer-based). *Application deadline:* For fall admission, 2/1 priority date for domestic students, 2/1 priority date for international students; for spring admission, 11/1 priority date for domestic students, 11/1 priority date for international students. Applications are processed on a rolling basis. Application fee: $60. *Expenses:* Tuition, area resident: Full-time $6,900; part-time $288 per credit. Tuition, nonresident: full-time $10,920; part-time $455 per credit. Required fees: $395; $32 per credit. $20 per term. One-time fee: $145. *Financial support:* In 2005–06, 43 students received support, including 8 fellowships with full and partial tuition reimbursements available (averaging $9,446 per year), 20 research assistantships with full and partial tuition reimbursements available (averaging $10,000 per year), 11 teaching assistantships with full and partial tuition reimbursements available (averaging $9,446 per year); career-related internships or fieldwork, Federal Work-Study, institutionally sponsored loans, scholarships/grants, health care benefits, and unspecified assistantships also available. Financial award application deadline: 6/30; financial award applicants required to submit FAFSA. *Faculty research:* Silviculture recreation management, tree improvement, operations management, economics. Total annual research expenditures: $2.1 million. *Unit head:* Dr. Chad P. Dawson, Chair, 315-470-6536, Fax: 315-470-6535, E-mail: cpdawson@esf.edu. *Application contact:* Dr. Dudley J. Raynal, Dean, Instruction and Graduate Studies, 315-470-6599, Fax: 315-470-6978, E-mail: esfgrad@esf.edu.

Stony Brook University, State University of New York, School of Professional Development and Continuing Studies, Program in Environmental and Waste Management, Stony Brook, NY 11794. Offers MS, Advanced Certificate. *Students:* 2 full-time (both women), 5 part-time (2 women); includes 3 minority (1 African American, 2 Asian Americans or Pacific Islanders), 1 international. *Expenses:* Tuition, area resident: Part-time $288. Tuition, state resident: full-time $6,900. Tuition, nonresident: full-time $10,920; part-time $455. Required fees: $704. *Financial support:* Research assistantships, teaching assistantships, career-related internships or fieldwork available.

Announcement: Multidisciplinary perspective covers scientific, engineering, regulatory, economic, policy, and community values aspects of environmental, energy, and waste issues and technologies. Combines hands-on practical experience with emerging technologies and their applications with judicious use of risk analysis, life-cycle analysis, computer modeling, geographic information systems, and other analytical tools.

See Close-Up on page 667.

Texas State University-San Marcos, Graduate School, College of Liberal Arts, Department of Geography, Program in Resource and Environmental Studies, San Marcos, TX 78666. Offers MAG. Part-time and evening/weekend programs available. *Students:* 8 full-time (2 women), 16 part-time (9 women); includes 2 Hispanic Americans. Average age 33. 9 applicants, 100% accepted, 7 enrolled. In 2005, 5 degrees awarded. *Degree requirements:* For master's, internship or thesis. *Entrance requirements:* For master's, GRE General Test, minimum GPA of 3.0 in last 60 hours of course work. Additional exam requirements/recommendations for international students: Required—TOEFL. *Application deadline:* For fall admission, 6/15 for domestic students; for spring admission, 10/15 priority date for domestic students. Applications are processed on a rolling basis. Application fee: $40 ($90 for international students). *Expenses:* Tuition, area resident: Part-time $116 per credit. Tuition, state resident: full-time $3,168; part-time $176 per credit. Tuition, nonresident: full-time $8,136; part-time $452 per credit. Required fees: $1,112; $74 per credit. Full-time tuition and fees vary according to course load. *Financial support:* In 2005–06, 17 students received support; research assistantships, teaching assistantships, career-related internships or fieldwork, Federal Work-Study, institutionally sponsored loans, and scholarships/grants available. Support available to part-time students. Financial award application deadline: 4/1; financial award applicants required to submit FAFSA. *Unit head:* Dr. David Butler, Graduate Adviser, 512-245-7977, Fax: 512-245-8353, E-mail: db25@txstate.edu.

Texas Tech University, Graduate School, College of Architecture, PhD Program in Land-Use Planning, Management, and Design, Lubbock, TX 79409. Offers PhD. *Students:* 5 full-time (0 women), 3 part-time (1 woman), 3 international. Average age 43. 2 applicants, 50% accepted, 1 enrolled. In 2005, 2 degrees awarded. *Degree requirements:* For doctorate, thesis/dissertation. *Entrance requirements:* For doctorate, GRE General Test. Additional exam requirements/recommendations for international students: Required—TOEFL (minimum score 550 paper-based; 213 computer-based). *Application deadline:* Applications are processed on a rolling basis. Application fee: $50 ($60 for international students). Electronic applications accepted. *Expenses:* Tuition, state resident: full-time $4,296. Tuition, nonresident: full-time $10,920. Required fees: $1,992. Tuition and fees vary according to program. *Financial support:* Research assistantships with partial tuition reimbursements, teaching assistantships with partial tuition reimbursements, career-related internships or fieldwork, Federal Work-Study, and institutionally sponsored loans available. Support available to part-time students. Financial award application deadline: 4/15; financial award applicants required to submit FAFSA. *Faculty research:* Architecture, landscape architecture, urban planning, environmental engineering, environmental policy planning. *Unit head:* Dr. Saif Haq, Program Director, 806-742-3136 Ext. 265, Fax: 806-742-2855, E-mail: saif.haq@ttu.edu. *Application contact:* Jimmy Duenes, Academic Program Assistant, 806-742-3136 Ext. 247, Fax: 806-742-2855, E-mail: jimmy.duenes@ttu.edu.

Towson University, Graduate School, Program in Geography and Environmental Planning, Towson, MD 21252-0001. Offers MA. Part-time and evening/weekend programs available. *Faculty:* 9 full-time (1 woman), 2 part-time/adjunct (0 women). *Students:* 41. Average age 30. 17 applicants, 100% accepted, 13 enrolled. In 2005, 7 degrees awarded. *Degree requirements:* For master's, thesis optional. *Entrance requirements:* For master's, 9 credits of course work in geography, minimum GPA of 3.0 in geography. Additional exam requirements/recommendations for international students: Required—TOEFL. *Application deadline:* Applications are processed on a rolling basis. Application fee: $40. Electronic applications accepted. *Financial support:* In 2005–06, 1 teaching assistantship with full tuition reimbursement (averaging $4,000 per year) was awarded; Federal Work-Study and unspecified assistantships also available. Financial award application deadline: 4/1; financial award applicants required to submit FAFSA. *Faculty research:* Geographic information systems, regional planning, hazards, development issues, urban fluvial systems. *Unit head:* Dr. Virginia Thompson, Graduate Program Director, 410-704-4371, Fax: 410-704-3880, E-mail: vthompson@towson.edu. *Application contact:* 410-704-2501, Fax: 410-704-4675, E-mail: grads@towson.edu.

Trent University, Graduate Studies, Program in Watershed Ecosystems, Environmental and Resource Studies Program, Peterborough, ON K9J 7B8, Canada. Offers M Sc, PhD. *Degree requirements:* For master's and doctorate, thesis/dissertation. *Entrance requirements:* For master's, honours degree; for doctorate, master's degree. *Faculty research:* Environmental biogeochemistry, aquatic organic contaminants, fisheries, wetland ecology, renewable resource management.

Tropical Agriculture Research and Higher Education Center, Graduate School, Turrialba, Costa Rica. Offers agroforestry (PhD); ecological agiculture (PhD); ecological agriculture (MS); environmental socioeconomics (MS, PhD); forest management and conservation (MS); forest sciences (PhD); tropical agroforestry (MS); watershed management (MS, PhD). *Entrance requirements:* For master's, GRE, letters of recommendation; for doctorate, GRE, curriculum vitae, letters of recommendation. Additional exam requirements/recommendations for international students: Required—TOEFL (minimum score 550 paper-based; 213 computer-based). Electronic applications accepted. *Faculty research:* Biodiversity in fragmented landscapes, ecosystem management, integrated pest management, environmental livestock production, biotechnology carbon balances in diverse land uses.

Troy University, Graduate School, College of Arts and Sciences, Program in Environmental Analysis and Management, Troy, AL 36082. Offers MS. Part-time and evening/weekend programs available. *Students:* 4 full-time (1 woman), 19 part-time (8 women); includes 1 minority (African American), 7 international. Average age 27. In 2005, 10 degrees awarded. *Degree requirements:* For master's, thesis, maintain 3.0 GPA, comprehensive exam, registration. *Entrance requirements:* For master's, GRE General Test, MAT, minimum GPA of 2.5. Additional exam requirements/recommendations for international students: Required—TOEFL (minimum score 523 paper-based; 200 computer-based). *Application deadline:* Applications are processed on a rolling basis. Application fee: $50. Electronic applications accepted. *Expenses:* Tuition, state resident: full-time $4,368; part-time $182 per credit hour. Tuition, nonresident: full-time $8,736; part-time $364 per credit hour. Required fees: $50 per semester hour. Full-time tuition and fees vary according to program. *Unit head:* Dr. Glenn Cohen, Chairman, 334-670-3401, Fax: 334-670-3662, E-mail: gcohen@troy.edu. *Application contact:* Brenda Campbell, Director of Graduate Admissions, 334-670-3178, Fax: 334-670-3733, E-mail: bcamp@troy.edu.

Tufts University, Graduate School of Arts and Sciences, Department of Urban and Environmental Policy and Planning, Medford, MA 02155. Offers community development (MA); environmental policy (MA); health and human welfare (MA); housing policy (MA); international environment/development policy (MA); public policy (MPP); public policy and citizen participation (MA). *Accreditation:* ACSP (one or more programs are accredited). Part-time programs available. *Faculty:* 8 full-time, 9 part-time/adjunct. *Students:* 130 (93 women); includes 23 minority (8 African Americans, 9 Asian Americans or Pacific Islanders, 6 Hispanic Americans) 4 international. 162 applicants, 82% accepted, 55 enrolled. In 2005, 40 degrees awarded. *Degree requirements:* For master's, thesis, internship. *Entrance requirements:* For master's, GRE General Test. Additional exam requirements/recommendations for international students: Required—TOEFL (minimum score 550 paper-based; 213 computer-based). *Application deadline:* For fall admission, 1/15 for domestic students, 12/30 for international students. Applications are processed on a rolling basis. Application fee: $65. Electronic applications accepted. *Expenses: Contact institution.* Tuition and fees vary according to program. *Financial support:* Teaching assistantships with full and partial tuition reimbursements, career-related internships or fieldwork, Federal Work-Study, scholarships/grants, and tuition waivers (partial) available. Support available to part-time students. Financial award application deadline: 1/15; financial award applicants required to submit FAFSA. *Unit head:* Rachel Bratt, Chair, 617-627-3394, Fax: 617-627-3377.

Tufts University, Graduate School of Arts and Sciences, Graduate Certificate Programs, Community Environmental Studies Program, Medford, MA 02155. Offers Certificate. Part-time and evening/weekend programs available. *Students:* Average age 30. 2 applicants, 50% accepted, 1 enrolled. In 2005, 5 degrees awarded. *Application deadline:* For fall admission, 8/15 for domestic students; for spring admission, 12/12 priority date for domestic students. Applications are processed on a rolling basis. Application fee: $65. Electronic applications accepted. *Expenses: Contact institution.* Tuition and fees vary according to program. *Financial support:* Career-related internships or fieldwork available. Support available to part-time students. Financial award application deadline: 5/1; financial award applicants required to submit FAFSA. *Application contact:* Information Contact, 617-627-3395, Fax: 617-627-3016, E-mail: gradschool@ase.tufts.edu.

Tufts University, Graduate School of Arts and Sciences, Graduate Certificate Programs, Environmental Management Program, Medford, MA 02155. Offers Certificate. Part-time and evening/weekend programs available. *Students:* Average age 30. 4 applicants, 75% accepted, 3 enrolled. *Application deadline:* For fall admission, 8/15 for domestic students; for spring admission, 12/12 priority date for domestic students. Applications are processed on a rolling basis. Application fee: $65. Electronic applications accepted. *Expenses:* Tuition: Full-time $32,360. Tuition and fees vary according to program. *Financial support:* Available to part-time students. Application deadline: 5/1; *Application contact:* Information Contact, 617-627-3395, Fax: 617-627-3016, E-mail: gradschool@ase.tufts.edu.

Tufts University, School of Engineering, Department of Civil and Environmental Engineering, Medford, MA 02155. Offers civil engineering (ME, MS, PhD), including geotechnical engineering, structural engineering; environmental engineering (ME, MS, PhD), including environmental engineering and environmental sciences, environmental geotechnology, environmental health, environmental science and management, hazardous materials management, water resources engineering. Part-time programs available. *Faculty:* 16 full-time, 7 part-time/adjunct. *Students:* 61 (33 women); includes 3 minority (1 African American, 1 Asian American or Pacific Islander, 1 Hispanic American) 9 international. 77 applicants, 74% accepted, 16 enrolled. In 2005, 16 master's awarded. Terminal master's awarded for partial completion of doctoral program. *Degree requirements:* For master's, thesis or alternative; for doctorate, thesis/dissertation. *Entrance requirements:* Additional exam requirements/recommendations for international students: Required—TOEFL (minimum score 550 paper-based; 213 computer-based). *Application deadline:* For fall admission, 2/1 for domestic students, 12/30 for international students; for spring admission, 10/15 for domestic students, 9/15 for international students. Applications are processed on a rolling basis. Application fee: $65. Electronic applications accepted. *Expenses:* Tuition: Full-time $32,360. Tuition and fees vary according to program. *Financial support:* Research assistantships with full and partial tuition reimbursements, teaching assistantships with full and partial tuition reimbursements, Federal Work-Study, scholarships/grants, and tuition waivers (partial) available. Support available to part-time students. Financial award application deadline: 2/1; financial award applicants required to submit FAFSA. *Unit head:* Dr. Christopher Swan, Chair, 617-627-3211, Fax: 617-627-3994.

Universidad del Turabo, Graduate Programs, Program in Science and Technology, Gurabo, PR 00778-3030. Offers environmental studies (MES). *Entrance requirements:* For master's, GRE, PAEG, interview.

Environmental Management and Policy

Universidad Metropolitana, School of Environmental Affairs, Program in Conservation and Management of Natural Resources, San Juan, PR 00928-1150. Offers MEM. Part-time programs available. *Degree requirements:* For master's, thesis. Electronic applications accepted.

Universidad Metropolitana, School of Environmental Affairs, Program in Environmental Planning, San Juan, PR 00928-1150. Offers MEM. Part-time programs available. *Degree requirements:* For master's, thesis. *Entrance requirements:* For master's, PAEG, interview. Electronic applications accepted.

Universidad Metropolitana, School of Environmental Affairs, Program in Environmental Risk and Assessment Management, San Juan, PR 00928-1150. Offers MEM. Part-time programs available. *Degree requirements:* For master's, thesis. Electronic applications accepted.

Universidad Nacional Pedro Henriquez Urena, Graduate School, Santo Domingo, Dominican Republic. Offers accounting and auditing (M Acct); animal production (M Agr); business administration (MBA, PhD); dentistry (DDS); economics (M Econ); education (PhD); environmental engineering (MEE); horticulutre (M Agr); hospital administration (PhD); humanities (PhD); international relations (MPS); management of natural resources (MNRM); project management (M Man); public administration (MPS); social science (PhD); veterinary medicine (DVM).

Université de Montréal, Faculty of Graduate Studies, Programs in Environment and Prevention, Montréal, QC H3C 3J7, Canada. Offers DESS. *Students:* 11 full-time (4 women), 10 part-time (6 women). 34 applicants, 44% accepted, 7 enrolled. In 2005, 7 degrees awarded. *Application deadline:* For fall and spring admission, 2/1. For winter admission, 11/1 for domestic students. Applications are processed on a rolling basis. Application fee: $30. Electronic applications accepted. *Faculty research:* Health, environment, pollutants, protection, waste. *Unit head:* Joseph Zayed, Director, 514-343-5912, Fax: 514-343-6668, E-mail: joseph.zayed@umontreal.ca. *Application contact:* Micheline Dessureault, Information Contact, 514-343-2280.

Université du Québec à Chicoutimi, Graduate Programs, Program in Renewable Resources, Chicoutimi, QC G7H 2B1, Canada. Offers M Sc. Part-time programs available. *Degree requirements:* For master's, thesis. *Entrance requirements:* For master's, appropriate bachelor's degree, proficiency in French.

Université du Québec, Institut National de la Recherche Scientifique, Graduate Programs, Research Center—Water, Earth and Environment, Québec, QC G1K 9A9, Canada. Offers earth sciences (M Sc, PhD); earth sciences-environmental technologies (M Sc); water sciences (MA, PhD). Part-time programs available. *Faculty:* 38. *Students:* 172 full-time (80 women), 10 part-time (3 women), 44 international. Average age 29. In 2005, 37 master's, 10 doctorates awarded. *Degree requirements:* For master's, thesis optional; for doctorate, thesis/dissertation. *Entrance requirements:* For master's, appropriate bachelor's degree, proficiency in French; for doctorate, appropriate master's degree, proficiency in French. *Application deadline:* For fall admission, 3/30 for domestic students, 3/30 for international students. For winter admission, 11/1 for domestic students. Application fee: $30. *Financial support:* Fellowships, research assistantships, teaching assistantships available. *Faculty research:* Land use, impacts of climate change, adaptation to climate change, integrated management of resources (mineral and water). *Unit head:* Jean Pierre Villeneuve, Director, 418-654-2575, Fax: 418-654-2615, E-mail: jp_villeneuve@ete.inrs.ca. *Application contact:* Michel Barbeau, Registrar, 418-654-2518, Fax: 418-654-3858, E-mail: michel.barbeau@adm.inrs.ca.

Université Laval, Faculty of Agricultural and Food Sciences, Department of Soils and Agricultural Engineering, Programs in Agri-Food Engineering, Québec, QC G1K 7P4, Canada. Offers agri-food engineering (M Sc); environmental technology (M Sc). *Degree requirements:* For master's, thesis (for some programs). *Entrance requirements:* For master's, knowledge of French. Electronic applications accepted.

Université Laval, Faculty of Agricultural and Food Sciences, Department of Soils and Agricultural Engineering, Programs in Soils and Environment Science, Québec, QC G1K 7P4, Canada. Offers environmental technology (M Sc); soils and environment science (M Sc, PhD). Terminal master's awarded for partial completion of doctoral program. *Degree requirements:* For master's, thesis (for some programs); for doctorate, thesis/dissertation, comprehensive exam. *Entrance requirements:* For master's and doctorate, knowledge of French and English. Electronic applications accepted.

University at Albany, State University of New York, College of Arts and Sciences, Department of Biological Sciences, Program in Biodiversity, Conservation, and Policy, Albany, NY 12222-0001. Offers MS. *Degree requirements:* For master's, one foreign language. *Entrance requirements:* For master's, GRE General Test. Application fee: $60. *Faculty research:* Aquatic ecology, plant community ecology, biodiversity and public policy, restoration ecology, costal and estuarine science. *Unit head:* Gary Kleppel, Program Director, 518-442-4338.

University of Alaska Fairbanks, School of Natural Resources and Agricultural Sciences, Department of Resources Management, Fairbanks, AK 99775-7520. Offers nature resource management (MS). Part-time programs available. *Faculty:* 6 full-time (3 women). *Students:* 9 full-time (4 women), 8 part-time (3 women); includes 1 minority (Asian American or Pacific Islander), 1 international. Average age 28. 19 applicants, 42% accepted, 3 enrolled. In 2005, 11 master's awarded. *Degree requirements:* For master's, thesis or alternative, comprehensive exam, registration. *Entrance requirements:* For master's, GRE General Test. Additional exam requirements/recommendations for international students: Required—TOEFL (minimum score 550 paper-based; 213 computer-based). *Application deadline:* For fall admission, 6/1 for domestic students, 3/1 for international students; for spring admission, 12/1 for domestic students, 9/1 for international students. Applications are processed on a rolling basis. Application fee: $50. Electronic applications accepted. *Expenses:* Tuition, state resident: full-time $4,392; part-time $244 per credit. Tuition, nonresident: full-time $8,964; part-time $498 per credit. Required fees: $800; $5 per credit. $48 per contact hour. Tuition and fees vary according to course level, course load, campus/location and reciprocity agreements. *Financial support:* In 2005–06, 7 research assistantships with full and partial tuition reimbursements (averaging $7,875 per year), 1 teaching assistantship with full and partial tuition reimbursement (averaging $3,910 per year) were awarded.; fellowships with tuition reimbursements, career-related internships or fieldwork, Federal Work-Study, scholarships/grants, and unspecified assistantships also available. Financial award applicants required to submit FAFSA. *Faculty research:* Wildlands management and policy, bioeconomic modeling, hydrologic modeling of land-use changes, global climate change, community ecology. *Unit head:* Dr. Joshua Greenberg, Department Chair, 907-474-7188, Fax: 907-474-6184, E-mail: fnnkp@uaf.edu.

University of Alberta, Faculty of Graduate Studies and Research, Department of Economics, Edmonton, AB T6G 2E1, Canada. Offers economics (MA, PhD); economics and finance (MA); environmental and natural resource economics (PhD). Part-time programs available. *Faculty:* 25 full-time (5 women), 3 part-time/adjunct (0 women). *Students:* 33 full-time (7 women), 7 part-time (3 women). Average age 26. 112 applicants, 58% accepted, 22 enrolled. In 2005, 8 master's, 1 doctorate awarded. *Degree requirements:* For doctorate, thesis/dissertation. *Entrance requirements:* For master's and doctorate, GRE. Additional exam requirements/recommendations for international students: Required—TOEFL. *Application deadline:* For fall admission, 6/15 for domestic students. Applications are processed on a rolling basis. Tuition and fees charges are reported in Canadian dollars. *Expenses:* Tuition, state resident: part-time $562 Canadian dollars per term. Tuition, nonresident: full-time $3,375 Canadian dollars. Required fees: $573 Canadian dollars; $84 Canadian dollars per term. *Financial support:* In 2005–06, 19 students received support, including 6 research assistantships with partial tuition reimbursements available (averaging $14,300 per year), 5 teaching assistantships with partial tuition reimbursements available (averaging $11,200 per year); career-related internships or fieldwork and scholarships/grants also available. Financial award application deadline: 3/1. *Faculty research:* Public finance, international trade, industrial organization, Pacific Rim economics, monetary economics. *Unit head:* Henry van Egteren, Graduate Coordinator, 780-492-7634, Fax: 780-492-3300. *Application contact:* Audrey Jackson, Graduate Program Administrator, 780-492-7634, Fax: 780-492-3300, E-mail: econapps@ualberta.ca.

The University of Arizona, Graduate College, Graduate Interdisciplinary Programs, Graduate Interdisciplinary Program in Planning, Tucson, AZ 85721. Offers MS. *Accreditation:* ACSP. *Degree requirements:* For master's, thesis or alternative. *Entrance requirements:* For master's, GRE General Test, minimum B average. Additional exam requirements/recommendations for international students: Required—TOEFL. *Faculty research:* Environmental analysis, regional planning, land development, regional development, arid lands.

The University of British Columbia, Faculty of Graduate Studies, Resource Management and Environmental Studies Program/Institute for Resources, Environment, and Sustainability, Vancouver, BC V6T 1Z1, Canada. Offers M Sc, MA, PhD. *Degree requirements:* For master's, thesis/dissertation, registration; for doctorate, thesis/dissertation, comprehensive exam, registration. *Entrance requirements:* Additional exam requirements/recommendations for international students: Required—TOEFL (minimum score 600 paper-based; 250 computer-based). Electronic applications accepted. *Faculty research:* Land management, water resources, energy, environmental assessment, risk evaluation.

University of Calgary, Faculty of Graduate Studies, Interdisciplinary Graduate Programs, Calgary, AB T2N 1N4, Canada. Offers interdisciplinary research (M Sc, MA, PhD); resources and the environment (M Sc, MA, PhD). Part-time programs available. *Students:* 44 full-time (25 women), 2 part-time (both women). 10 applicants, 50% accepted, 4 enrolled. In 2005, 2 master's, 3 doctorates awarded. *Median time to degree:* Of those who began their doctoral program in fall 1997, 100% received their degree in 8 years or less. *Degree requirements:* For master's, thesis; for doctorate, thesis/dissertation, written and oral candidacy exam. *Entrance requirements:* Additional exam requirements/recommendations for international students: Required—TOEFL (minimum score 600 paper-based; 250 computer-based). *Application deadline:* For fall admission, 2/1 for domestic students, 2/1 for international students. For winter admission, 8/1 for domestic students. Application fee: $100 ($130 for international students). *Financial support:* In 2005–06, 18 research assistantships (averaging $4,100 per year), 8 teaching assistantships (averaging $6,530 per year) were awarded. Financial award application deadline: 2/1. *Unit head:* Dr. Jim Love, Director, 403-220-7209, Fax: 403-210-8872, E-mail: love@ucalgary.ca. *Application contact:* Pauline Fisk, Program Administrator, 403-220-7209, Fax: 403-210-8872, E-mail: pfisk@ucalgary.ca.

University of California, Berkeley, Graduate Division, College of Natural Resources, Department of Environmental Science, Policy, and Management, Berkeley, CA 94720-1500. Offers environmental science, policy, and management (MS, PhD); forestry (MF). Terminal master's awarded for partial completion of doctoral program. *Degree requirements:* For master's, thesis optional; for doctorate, thesis/dissertation, qualifying exam. *Entrance requirements:* For master's and doctorate, GRE General Test, minimum GPA of 3.0. Additional exam requirements/recommendations for international students: Required—TOEFL; Recommended—TSE. Electronic applications accepted. *Faculty research:* Biology and ecology of insects; ecosystem function and environmental issues of soils; plant health/interactions from molecular to ecosystem levels; range management and ecology; forest and resource policy, sustainability, and management.

University of California, Berkeley, Graduate Division, Group in Energy and Resources, Berkeley, CA 94720-1500. Offers MA, MS, PhD. *Degree requirements:* For master's, project or thesis; for doctorate, one foreign language, thesis/dissertation, qualifying exam. *Entrance requirements:* For master's and doctorate, GRE General Test, minimum GPA of 3.0. *Faculty research:* Technical, economic, environmental, and institutional aspects of energy conservation in residential and commercial buildings; international patterns of energy use; renewable energy sources; assessment of valuation of energy and environmental resources pricing.

University of California, Irvine, Office of Graduate Studies, School of Social Ecology, Department of Environmental Analysis and Design, Irvine, CA 92697. Offers environmental health science and policy (MS, PhD); social ecology (PhD). *Degree requirements:* For doctorate, thesis/dissertation, research project. *Entrance requirements:* For master's and doctorate, GRE General Test, minimum GPA of 3.0. Additional exam requirements/recommendations for international students: Required—TOEFL (minimum score 550 paper-based; 213 computer-based). Electronic applications accepted. *Faculty research:* Effects of environmental stressors, environmental pollution, biology and politics of water pollution, potential impacts of natural disasters, risk management.

University of California, Santa Barbara, Graduate Division, Donald Bren School of Environmental Science and Management, Santa Barbara, CA 93106. Offers MESM, PhD. *Faculty:* 17 full-time (3 women), 3 part-time/adjunct (0 women). *Students:* 127 full-time (66 women), 6 part-time (4 women); includes 7 minority (all Asian Americans or Pacific Islanders), 16 international. Average age 25. 293 applicants, 55% accepted, 75 enrolled. In 2005, 38 master's, 4 doctorates awarded. *Degree requirements:* For master's, group project as student thesis; for doctorate, thesis/dissertation, comprehensive exam, registration. *Entrance requirements:* For master's and doctorate, GRE. Additional exam requirements/recommendations for international students: Required—TOEFL (minimum score 550 paper-based; 213 computer-based). *Application deadline:* For fall admission, 2/1 for domestic students, 2/1 for international students. Applications are processed on a rolling basis. Application fee: $60. Electronic applications accepted. *Financial support:* In 2005–06, 18 fellowships with partial tuition reimbursements, 14 research assistantships with full tuition reimbursements (averaging $11,140 per year), 12 teaching assistantships with partial tuition reimbursements (averaging $4,531 per year) were awarded.; career-related internships or fieldwork, Federal Work-Study, institutionally sponsored loans, scholarships/grants, traineeships, health care benefits, and unspecified assistantships also available. Financial award application deadline: 2/1; financial award applicants required to submit FAFSA. *Faculty research:* Hydrology, ecology, political instituting, environmental economics, biogeochemistry. *Unit head:* Dr. Dennis J. Aigner, Dean, 805-893-7363, E-mail: info@bren.ucsb.edu. *Application contact:* Marla Alfaro, Student Affairs Assistant, 805-893-7611, Fax: 805-893-7612, E-mail: maria@bren.ucsb.edu.

See Close-Up on page 671.

University of California, Santa Cruz, Division of Graduate Studies, Division of Social Sciences, Program in Environmental Studies, Santa Cruz, CA 95064. Offers PhD. *Faculty:* 15 full-time (5 women). *Students:* 57 full-time (31 women), 1 (woman) part-time; includes 6 minority (1 American Indian/Alaska Native, 2 Asian Americans or Pacific Islanders, 3 Hispanic Americans), 6 international. 77 applicants, 18% accepted, 9 enrolled. In 2005, 4 degrees awarded. *Degree requirements:* For doctorate, thesis/dissertation, qualifying exam. *Entrance requirements:* For doctorate, GRE General Test. *Application deadline:* For fall admission, 1/7 for domestic students. Application fee: $60. *Expenses:* Tuition, nonresident: full-time $14,694. *Financial support:* Fellowships, research assistantships, teaching assistantships, career-related internships or fieldwork, Federal Work-Study, and institutionally sponsored loans available. Financial award application deadline: 1/7. *Faculty research:* Political economy and sustainability, conservation biology, agroecology. *Unit head:* David Goodman, Chairperson, 831-459-4561. *Application contact:* Judy L. Glass, Reporting Analyst for Graduate Admissions, 831-459-5906, Fax: 831-459-4843, E-mail: jlglass@ucsc.edu.

University of Chicago, The Irving B. Harris Graduate School of Public Policy Studies, Chicago, IL 60637-1513. Offers environmental science and policy (MS); public policy studies (AM, MPP, PhD). Part-time programs available. Terminal master's awarded for partial completion of doctoral program. *Degree requirements:* For doctorate, thesis/dissertation. *Entrance requirements:* For master's and doctorate, GMAT or GRE General Test. Additional exam requirements/recommendations for international students: Required—TOEFL. Electronic applications accepted. Expenses: Contact institution. *Faculty research:* Family and child policy, international security, health policy, social policy.

University of Colorado at Boulder, Graduate School, College of Arts and Sciences, Program in Environmental Studies, Boulder, CO 80309. Offers MS, PhD. *Faculty:* 7 full-time (1 woman). *Students:* 40 full-time (24 women), 8 part-time (4 women); includes 2 minority (1 American Indian/Alaska Native, 1 Asian American or Pacific Islander). Average age 32. 31 applicants, 68% accepted. In 2005, 13 degrees awarded. *Entrance requirements:* For master's, minimum undergraduate GPA of 3.4. *Application deadline:* For fall admission, 1/15 for domestic students, 12/1 for international students. *Financial support:* In 2005–06, 30 fellowships (averaging $4,439 per year), 7 research assistantships (averaging $13,826 per year), 9 teaching assistantships (averaging $13,208 per year) were awarded. *Faculty research:* Climate and atmospheric chemistry, water sciences, environmental policy and sustainability, waste management and environmental remediation, biogeochemical cycles. Total annual research expenditures: $173,101. *Unit head:* James White, Director, 303-492-5494, Fax: 303-492-8437, E-mail: jwhite@colorado.edu. *Application contact:* Graduate Program Assistant, 303-492-5478, Fax: 303-492-5207, E-mail: envsgrad@colorado.edu.

University of Connecticut, Graduate School, College of Agriculture and Natural Resources, Department of Natural Resources Management and Engineering, Field of Natural Resources Management and Engineering, Storrs, CT 06269. Offers natural resources (MS, PhD). *Faculty:* 15 full-time (1 woman). *Students:* 18 full-time (7 women), 9 part-time (4 women); includes 1 minority (Asian American or Pacific Islander), 4 international. Average age 33. 25 applicants, 44% accepted, 11 enrolled. In 2005, 4 master's, 1 doctorate awarded. Terminal master's awarded for partial completion of doctoral program. *Degree requirements:* For master's, comprehensive exam; for doctorate, thesis/dissertation. *Entrance requirements:* For master's, GRE General Test, GRE Subject Test. Additional exam requirements/recommendations for international students: Required—TOEFL (minimum score 550 paper-based; 213 computer-based). *Application deadline:* For fall admission, 2/1 priority date for domestic students, 2/1 priority date for international students; for spring admission, 11/1 for domestic students, 10/1 for international students. Applications are processed on a rolling basis. Application fee: $55. Electronic applications accepted. *Expenses:* Tuition, state resident: part-time $444 per credit hour. Tuition, nonresident: part-time $1,154 per credit hour. Tuition and fees vary according to course load. *Financial support:* In 2005–06, 16 research assistantships with full tuition reimbursements, 1 teaching assistantship with full tuition reimbursement were awarded.; fellowships, Federal Work-Study, scholarships/grants, health care benefits, and unspecified assistantships also available. Financial award application deadline: 2/1; financial award applicants required to submit FAFSA. *Application contact:* John Clausen, Chairman, Graduate Admissions, 860-486-0139, Fax: 860-486-5408, E-mail: john.clausen@uconn.edu.

University of Delaware, College of Human Services, Education and Public Policy, Center for Energy and Environmental Policy, Newark, DE 19716. Offers environmental and energy policy (MEEP, PhD); urban affairs and public policy (MA, PhD). *Faculty:* 7 full-time (1 woman), 9 part-time/adjunct (2 women). *Students:* 63 full-time (23 women), 10 part-time (6 women); includes 11 minority (5 African Americans, 3 Asian Americans or Pacific Islanders, 3 Hispanic Americans), 36 international. 142 applicants, 17% accepted, 20 enrolled. In 2005, 11 master's, 4 doctorates awarded. *Degree requirements:* For master's, analytical paper or thesis; for doctorate, thesis/dissertation, comprehensive exam. *Entrance requirements:* For master's, GRE General Test, minimum GPA of 3.0; for doctorate, GRE General Test, minimum GPA of 3.5. Additional exam requirements/recommendations for international students: Required—TOEFL. *Application deadline:* For spring admission, 2/15 for domestic students. Application fee: $60. Electronic applications accepted. *Financial support:* In 2005–06, 19 fellowships with full tuition reimbursements (averaging $12,200 per year), 31 research assistantships with full tuition reimbursements (averaging $12,200 per year) were awarded.; teaching assistantships with full tuition reimbursements, career-related internships or fieldwork, Federal Work-Study, and tuition waivers (full) also available. Financial award application deadline: 2/15. *Faculty research:* Sustainable development, renewable energy, climate change, environmental policy, environmental justice, disaster policy. Total annual research expenditures:$500,000. *Unit head:* Dr. John Byrne, Director, 302-831-8405, Fax: 302-831-3098, E-mail: jbbyrne@udel.edu. *Application contact:* Terri Brower, Assistant to Director, 302-831-8405, Fax: 302-831-3098, E-mail: tbrower@udel.edu.

See Close-Up on page 675.

University of Delaware, College of Human Services, Education and Public Policy, School of Urban Affairs and Public Policy, Program in Urban Affairs and Public Policy, Newark, DE 19716. Offers community development and nonprofit leadership (MA); energy and environmental policy (MA); governance, planning and management (PhD); historic preservation (MA); social and urban policy (PhD); technology, environment and society (PhD). Part-time programs available. *Faculty:* 15 full-time (6 women). *Students:* 86 full-time (54 women), 3 part-time (all women); includes 27 minority (19 African Americans, 1 American Indian/Alaska Native, 4 Asian Americans or Pacific Islanders, 3 Hispanic Americans), 14 international. Average age 36. 99 applicants, 48% accepted, 25 enrolled. In 2005, 13 master's, 10 doctorates awarded. Terminal master's awarded for partial completion of doctoral program. *Degree requirements:* For master's, thesis or alternative, analytical paper or thesis; for doctorate, thesis/dissertation. *Entrance requirements:* For master's, GRE General Test, minimum GPA of 3.0; for doctorate, GRE General Test, minimum GPA of 3.5. Additional exam requirements/recommendations for international students: Required—TOEFL. *Application deadline:* For fall admission, 2/1 for domestic students; for spring admission, 12/1 for domestic students. Applications are processed on a rolling basis. Application fee: $60. Electronic applications accepted. *Financial support:* In 2005–06, 78 students received support, including 4 fellowships with full tuition reimbursements available (averaging $12,200 per year), 62 research assistantships with full tuition reimbursements available (averaging $12,200 per year), 4 teaching assistantships with full tuition reimbursements available (averaging $12,200 per year); career-related internships or fieldwork, Federal Work-Study, and tuition waivers (full) also available. Financial award application deadline: 2/1. *Faculty research:* Political economy; social policy analysis; technology and society; historic preservation; urban policy. Total annual research expenditures: $1 million. *Unit head:* Dr. Danilo Yanich, Director, 302-831-1710, Fax: 302-831-4225, E-mail: dyanich@udel.edu. *Application contact:* Melissa Hopkins, Information Contact, 302-831-8712, Fax: 302-831-3587, E-mail: mturner@udel.edu.

University of Denver, University College, Denver, CO 80208. Offers applied communication (MAS, MPS); computer information systems (MAS); environmental policy and management (MAS); geographic information systems (MAS); human resource administration (MPS); knowledge and information technologies (MAS); liberal studies (MLS); modern languages (MLS); organizational leadership (MPS); technology management (MAS); telecommunications (MAS). Part-time and evening/weekend programs available. Postbaccalaureate distance learning degree programs offered (no on-campus study). *Students:* 62 full-time (31 women), 548 part-time (290 women); includes 97 minority (38 African Americans, 5 American Indian/Alaska Native, 30 Asian Americans or Pacific Islanders, 24 Hispanic Americans), 32 international. 172 applicants, 80% accepted. In 2005, 152 degrees awarded. *Entrance requirements:* For master's, minimum undergraduate GPA of 3.0. Additional exam requirements/recommendations for international students: Required—TOEFL (minimum score 550 paper-based; 213 computer-based). *Application deadline:* For fall admission, 7/15 for domestic students. For winter admission, 10/14 for domestic students; for spring admission, 2/10 for domestic students. Applications are processed on a rolling basis. Application fee: $25. Electronic applications accepted. *Expenses:* Contact institution. *Financial support:* Applicants required to submit FAFSA. *Unit head:* Dr. James Davis, Dean, 303-871-3141, Fax: 303-871-4047, E-mail: jdavis@du.edu. *Application contact:* 303-871-3155, E-mail: ucolinfo@du.edu.

The University of Findlay, Graduate and Special Programs, College of Science, Program in Environmental Management, Findlay, OH 45840-3653. Offers MSEM. Part-time and evening/weekend programs available. Postbaccalaureate distance learning degree programs offered (no on-campus study). *Faculty:* 4 full-time. *Students:* 3 full-time (1 woman), 134 part-time (47 women); includes 5 minority (2 African Americans, 1 Asian American or Pacific Islander, 2 Hispanic Americans), 42 international. Average age 35. 25 applicants, 64% accepted, 14 enrolled. In 2005, 30 degrees awarded. *Degree requirements:* For master's, cumulative project. *Entrance requirements:* For master's, GMAT or GRE, minimum undergraduate GPA of 3.0 in last 60 hours of course work. Additional exam requirements/recommendations for international students: Required—TOEFL (minimum score 550 paper-based). *Application deadline:* Applications are processed on a rolling basis. Application fee: $25 ($0 for international students). Electronic applications accepted. *Financial support:* In 2005–06, 8 students received support, including 2 teaching assistantships with full tuition reimbursements available (averaging $6,000 per year); unspecified assistantships also available. Financial award application deadline: 4/1; financial award applicants required to submit FAFSA. *Unit head:* Dr. William Carter, Graduate Director, 419-434-6919, Fax: 419-434-4822, E-mail: carter@findlay.edu. *Application contact:* Heather Riffle, Director, Graduate and Special Programs, 419-434-4640, Fax: 419-434-5517, E-mail: riffle@findlay.edu.

University of Guelph, Graduate Program Services, Ontario Agricultural College, Department of Land Resource Science, Guelph, ON N1G 2W1, Canada. Offers atmospheric science (M Sc, PhD); environmental and agricultural earth sciences (M Sc, PhD); land resources management (M Sc, PhD); soil science (M Sc, PhD). Part-time programs available. *Faculty:* 19 full-time (5 women), 5 part-time/adjunct (1 woman). *Students:* 47 full-time (20 women), 3 part-time; includes 9 minority (1 African American, 6 Asian Americans or Pacific Islanders, 2 Hispanic Americans), 2 international. Average age 28. 25 applicants, 24% accepted. In 2005, 4 master's, 3 doctorates awarded. *Degree requirements:* For master's and doctorate, thesis/dissertation. *Entrance requirements:* For master's, minimum B- average during previous 2 years of course work; for doctorate, minimum B average during previous 2 years of course work. Additional exam requirements/recommendations for international students: Required—TOEFL (minimum score 550 paper-based; 213 computer-based). *Application deadline:* For fall admission, 7/1 priority date for domestic students, 5/1 priority date for international students. For winter admission, 10/1 for domestic students; for spring admission, 3/1 for domestic students. Applications are processed on a rolling basis. Application fee: $75 Canadian dollars. Electronic applications accepted. *Financial support:* In 2005–06, 30 students received support, including 40 research assistantships (averaging $16,500 Canadian dollars per year), 15 teaching assistantships (averaging $3,800 Canadian dollars per year); fellowships, scholarships/grants also available. *Faculty research:* Soil science, environmental earth science, land resource management. Total annual research expenditures: $2.1 million Canadian dollars. *Unit head:* Dr. S. Hilts, Chair, 519-824-4120 Ext. 52447, Fax: 519-824-5730, E-mail: shilts@uoguelph.ca. *Application contact:* Dr. B. Hale, Graduate Coordinator, 519-824-4120 Ext. 53434, Fax: 519-824-5730, E-mail: bhale@uoguelph.ca.

University of Hawaii at Manoa, Graduate Division, College of Tropical Agriculture and Human Resources, Department of Natural Resources and Environmental Management, Honolulu, HI 96822. Offers MS, PhD. Part-time programs available. *Faculty:* 24 full-time (3 women), 6 part-time/adjunct (2 women). *Students:* 43 full-time (23 women), 2 part-time; includes 5 minority (1 African American, 3 Asian Americans or Pacific Islanders, 1 Hispanic American), 18 international. Average age 34. 68 applicants, 62% accepted, 17 enrolled. In 2005, 7 master's, 1 doctorate awarded. Terminal master's awarded for partial completion of doctoral program. *Degree requirements:* For master's, thesis or alternative; for doctorate, thesis/dissertation. *Entrance requirements:* For master's and doctorate, GRE, minimum GPA of 3.0 in last 4 semesters of course work. Additional exam requirements/recommendations for international students: Required—TOEFL. *Application deadline:* For fall admission, 3/1 for domestic students, 1/15 for international students; for spring admission, 9/1 for domestic students, 8/1 for international students. Applications are processed on a rolling basis. Application fee: $50. *Expenses:* Tuition, state resident: full-time $8,400; part-time $200 per credit hour. Tuition, nonresident: full-time $11,088; part-time $462 per credit hour. Tuition and fees vary according to program. *Financial support:* In 2005–06, 16 research assistantships (averaging $15,850 per year), 11 teaching assistantships (averaging $14,315 per year) were awarded.; fellowships, career-related internships or fieldwork and tuition waivers (full and partial) also available. *Faculty research:* Bioeconomics, natural resource management. *Unit head:* Dr. Samir A. El-Swarfy, Chairperson, 808-956-8708, Fax: 808-956-2811. *Application contact:* John Yanagida, 808-956-7530, Fax: 808-956-6539.

University of Hawaii at Manoa, Graduate Division, Colleges of Arts and Sciences, College of Social Sciences, Department of Urban and Regional Planning, Honolulu, HI 96822. Offers community planning and social policy (MURP); environmental planning and management (MURP); land use and infrastructure planning (MURP); urban and regional planning (PhD, Certificate); urban and regional planning in Asia and Pacific (MURP). *Accreditation:* ACSP. *Faculty:* 19 full-time (6 women), 2 part-time/adjunct (0 women). *Students:* 52 full-time (30 women), 26 part-time (16 women); includes 21 minority (1 American Indian/Alaska Native, 20 Asian Americans or Pacific Islanders), 26 international. Average age 31. 52 applicants, 54% accepted, 16 enrolled. In 2005, 15 degrees awarded. *Entrance requirements:* For master's, GRE, minimum GPA of 3.0. Additional exam requirements/recommendations for international students: Required—TOEFL. *Application deadline:* For fall admission, 3/1 for domestic students, 3/1 for international students; for spring admission, 9/1 for domestic students, 9/1 for international students. Application fee: $50. *Expenses:* Tuition, state resident: full-time $8,400; part-time $200 per credit hour. Tuition, nonresident: full-time $11,088; part-time $462 per credit hour. Tuition and fees vary according to program. *Financial support:* In 2005–06, 14 research assistantships (averaging $16,066 per year), 4 teaching assistantships (averaging $13,835 per year) were awarded.; career-related internships or fieldwork, Federal Work-Study, institutionally sponsored loans, and tuition waivers (full) also available. *Unit head:* Kem Lowry, Chairperson, 808-956-6433, Fax: 808-956-9121, E-mail: lowry@hawaii.edu. *Application contact:* Karl Kim, 808-956-7381, Fax: 808-956-6870.

University of Houston–Clear Lake, School of Business and Public Administration, Program in General Business, Houston, TX 77058-1098. Offers environmental management (MS); healthcare administration (MHA); human resource management (MA); management information systems (MS); public management (MA). *Accreditation:* ACEHSA (one or more programs are accredited). Part-time and evening/weekend programs available. *Degree requirements:* For master's, thesis optional. *Entrance requirements:* For master's, GMAT. Additional exam requirements/recommendations for international students: Required—TOEFL (minimum score 550 paper-based; 213 computer-based). Electronic applications accepted.

University of Idaho, College of Graduate Studies, College of Natural Resources, Department of Conservation Social Sciences, Moscow, ID 83844-2282. Offers MS, PhD. *Students:* 17 full-time (10 women), 7 part-time (4 women); includes 1 minority (Asian American or Pacific Islander), 2 international. Average age 36. In 2005, 7 degrees awarded. *Degree requirements:* For doctorate, thesis/dissertation. *Entrance requirements:* For master's, minimum GPA of 2.8; for doctorate, minimum undergraduate GPA of 2.8, 3.0 graduate. *Application deadline:* For fall admission, 8/1 for domestic students; for spring admission, 12/15 for domestic students. Application fee: $55 ($60 for international students). *Expenses:* Tuition, nonresident: full-time $8,770; part-time $130 per credit. Required fees: $4,508; $217 per credit. *Financial support:* Research assistantships, teaching assistantships available. Financial award application deadline: 2/15. *Unit head:* Dr. Steve Hollenhorst, Head, 208-885-7911.

University of Illinois at Springfield, Graduate Programs, College of Public Affairs and Administration, Program in Environmental Studies, Springfield, IL 62703-5407. Offers environmental science (MS); environmental studies (MA). Part-time and evening/weekend programs available. *Faculty:* 2 full-time (both women), 3 part-time/adjunct (1 woman). *Students:* 23 full-time (12 women), 21 part-time (18 women); includes 2 minority (1 Asian American or Pacific Islander, 1 Hispanic American), 2 international. Average age 31. 31 applicants, 68% accepted, 10 enrolled. In 2005, 4 degrees awarded. *Degree requirements:* For master's, thesis or alternative, thesis or project. *Entrance requirements:* For master's, GRE General Test, minimum GPA of 3.0, 2 letters of reference. Additional exam requirements/recommendations for international students: Required—TOEFL (minimum score 550 paper-based; 213 computer-based). *Application deadline:* Applications are processed on a rolling basis. Application fee: $50 ($60 for international students). Electronic applications accepted. *Expenses:* Tuition, state

Environmental Management and Policy

University of Illinois at Springfield (continued)
resident: full-time $4,726; part-time $163 per credit hour. Tuition, nonresident: full-time $14,178; part-time $490 per credit hour. Required fees: $1,382; $582 per term. *Financial support:* In 2005–06, fellowships with full tuition reimbursements (averaging $7,650 per year), research assistantships with full tuition reimbursements (averaging $7,200 per year), teaching assistantships with full tuition reimbursements (averaging $7,200 per year) were awarded.; career-related internships or fieldwork, Federal Work-Study, scholarships/grants, health care benefits, and unspecified assistantships also available. Support available to part-time students. Financial award application deadline: 11/15; financial award applicants required to submit FAFSA. *Faculty research:* Environmental risk assessment, toxicology, work force development, resource ecology, ecosystem management. *Unit head:* Dr. Sharron LaFollette, Program Administrator, 217-206-7894, Fax: 217-206-7807, E-mail: lafollette.sharon@uis.edu.

University of Maine, Graduate School, College of Natural Sciences, Forestry, and Agriculture, Department of Plant, Soil, and Environmental Sciences, Orono, ME 04469. Offers biological sciences (PhD); ecology and environmental sciences (MS, PhD); forest resources (PhD); horticulture (MS); plant science (PhD); plant, soil, and environmental sciences (MS); resource utilization (MS). *Faculty:* 25. *Students:* 15 full-time (8 women), 8 part-time (6 women), 3 international. Average age 32. 6 applicants, 33% accepted, 1 enrolled. In 2005, 9 master's awarded. *Entrance requirements:* For master's and doctorate, GRE General Test. Additional exam requirements/recommendations for international students: Required—TOEFL. *Application deadline:* Applications are processed on a rolling basis. Application fee: $50. Electronic applications accepted. *Financial support:* In 2005–06, 9 research assistantships with tuition reimbursements (averaging $12,180 per year) were awarded; teaching assistantships, scholarships/grants, tuition waivers (full and partial), and unspecified assistantships also available. *Unit head:* Greg Porter, Chair, 207-581-2943, Fax: 207-581-3207. *Application contact:* Scott G. Delcourt, Associate Dean of the Graduate School, 207-581-3219, Fax: 207-581-3232, E-mail: graduate@maine.edu.

University of Maine, Graduate School, College of Natural Sciences, Forestry, and Agriculture, Program in Resource Utilization, Orono, ME 04469. Offers MS. *Faculty:* 10 full-time (1 woman). *Students:* 1 applicant, 100% accepted, 0 enrolled. *Degree requirements:* For master's, thesis. *Entrance requirements:* For master's, GRE General Test. Additional exam requirements/recommendations for international students: Required—TOEFL. *Application deadline:* For fall admission, 2/1 for domestic students. Applications are processed on a rolling basis. Application fee: $50. Electronic applications accepted. *Financial support:* Research assistantships with tuition reimbursements, teaching assistantships with tuition reimbursements, career-related internships or fieldwork, Federal Work-Study, institutionally sponsored loans, scholarships/grants, and tuition waivers (full and partial) available. Financial award application deadline:3/1. *Faculty research:* Waste utilities, wildlife evaluation, tourism and recreation economics. *Unit head:* Dr. Todd Gabe, Coordinator, 207-581-3307. *Application contact:* Scott G. Delcourt, Associate Dean of the Graduate School, 207-581-3219, Fax: 207-581-3232, E-mail: graduate@maine.edu.

University of Manitoba, Faculty of Graduate Studies, Faculty of Environment, Earth and Resources, Natural Resources Institute, Winnipeg, MB R3T 2N2, Canada. Offers natural resources and environmental management (PhD); natural resources management (MNRM).

University of Maryland University College, Graduate School of Management and Technology, Program in Environmental Management, Adelphi, MD 20783. Offers MS, Certificate. Offered evenings and weekends only. Part-time and evening/weekend programs available. Postbaccalaureate distance learning degree programs offered (no on-campus study). *Degree requirements:* For master's, thesis or alternative. *Entrance requirements:* For master's, BS/BA in social science, physical science, engineering; 6 semester hours in biology and chemistry; 1 year of experience in field. Electronic applications accepted.

University of Massachusetts Lowell, Graduate School, James B. Francis College of Engineering, Department of Work Environment, Lowell, MA 01854-2881. Offers cleaner production and pollution prevention (MS, Sc D); environmental risk assessment (Certificate); identification and control of ergonomic hazards (Certificate); industrial hygiene (MS, Sc D); job stress and healthy job redesign (Certificate); occupational epidemiology (MS, Sc D); occupational ergonomics (MS, Sc D); radiological health physics and general work environment protection (Certificate); work environmental policy (MS, Sc D). *Accreditation:* ABET (one or more programs are accredited). Part-time programs available. Terminal master's awarded for partial completion of doctoral program. *Degree requirements:* For master's, thesis optional; for doctorate, thesis/dissertation. *Entrance requirements:* For master's and doctorate, GRE General Test. Additional exam requirements/recommendations for international students: Required—TOEFL. *Faculty research:* Ergonomics, industrial hygiene, epidemiology, work environment policy, pollution prevention.

University of Miami, Graduate School, School of Business Administration, Department of Economics, Coral Gables, FL 33124. Offers economic development (MA, PhD); environmental economics (PhD); human resource economics (MA, PhD); international economics (MA, PhD); macroeconomics (PhD). Students admitted every two years in the fall semester. *Faculty:* 13 full-time (7 women). *Students:* 13 full-time (7 women); includes 1 minority (African American), 11 international. Average age 28. 95 applicants, 14% accepted, 7 enrolled. In 2005, 7 master's, 1 doctorate awarded. Terminal master's awarded for partial completion of doctoral program. *Median time to degree:* Of those who began their doctoral program in fall 1997, 90% received their degree in 8 years or less. *Degree requirements:* For master's, comprehensive exam; for doctorate, thesis/dissertation, comprehensive exam. *Entrance requirements:* For master's and doctorate, GRE General Test, minimum GPA of 3.0. Additional exam requirements/recommendations for international students: Required—TOEFL. *Application deadline:* For fall admission, 3/1 for domestic students. Application fee: $50. *Financial support:* In 2005–06, fellowships with full tuition reimbursements (averaging $17,000 per year), research assistantships with full tuition reimbursements (averaging $12,000 per year), teaching assistantships with full tuition reimbursements (averaging $3,500 per year) were awarded.; tuition waivers (partial) and unspecified assistantships also available. Financial award application deadline:3/1. *Faculty research:* International economics/trade, applied microeconomics, development. Total annual research expenditures: $426,180. *Unit head:* Dr. Pedro Gomis Purqueras, Chairman, 305-284-3725, Fax: 305-284-2985, E-mail: dkelly@miami.edu. *Application contact:* Dr. Pedro Gomis Purqueras, Director of Graduate Programs, 305-284-4742, Fax: 305-284-2985, E-mail: gomis@miami.edu.

University of Michigan, School of Natural Resources and Environment, Program in Resource Ecology and Management, Ann Arbor, MI 48109. Offers natural resources and environment (PhD); resource ecology and management (MS). Terminal master's awarded for partial completion of doctoral program. *Degree requirements:* For master's, thesis or alternative, thesis, practicum or group project; for doctorate, thesis/dissertation, oral defense of dissertation, preliminary exam, comprehensive exam, registration. *Entrance requirements:* For master's, GRE General Test; for doctorate, GRE General Test, master's degree. Additional exam requirements/recommendations for international students: Required—TOEFL (paper score 560; computer score 220) or IELTS (6.5). Electronic applications accepted. *Expenses:* Tuition, state resident: full-time $14,082; part-time $894 per credit hour. Tuition, nonresident: full-time $28,500; part-time $1,675 per credit hour. Required fees: $189; $189 per unit. *Faculty research:* Stream ecology, plant-insect interactions, fish biology, resource control and reproductive success, remote sensing.

University of Michigan, School of Natural Resources and Environment, Program in Resource Policy and Behavior, Ann Arbor, MI 48109. Offers natural resources and environment (PhD); resource policy and behavior (MS). Terminal master's awarded for partial completion of doctoral program. *Degree requirements:* For master's, thesis, practicum or group project; for doctorate, thesis/dissertation, oral defense of dissertation, preliminary exam, comprehensive exam, registration. *Entrance requirements:* For master's, GRE General Test; for doctorate, GRE General Test, master's degree. Additional exam requirements/recommendations for international students: Required—TOEFL (paper score 560; computer score 220) or IELTS. Electronic applications accepted. *Expenses:* Tuition, state resident: full-time $14,082; part-time $894 per credit hour. Tuition, nonresident: full-time $28,500; part-time $1,675 per credit hour. Required fees: $189; $189 per unit. *Faculty research:* Business and environment/sustainable systems, environmental behavior/psychology, environmental conflict management/dispute resolution, enviornmental education, environmental justice/policy planning.

University of Minnesota, Twin Cities Campus, Graduate School, College of Natural Resources, Department of Bio-Based Products, Minneapolis, MN 55455-0213. Offers natural resources science and management (MS, PhD). Terminal master's awarded for partial completion of doctoral program. *Degree requirements:* For master's, thesis optional; for doctorate, thesis/dissertation, comprehensive exam, registration. *Entrance requirements:* For master's and doctorate, GRE. Additional exam requirements/recommendations for international students: Required—TOEFL (minimum score 550 paper-based; 213 computer-based). Electronic applications accepted. *Expenses:* Tuition, state resident: full-time $8,748; part-time $729 per credit. Tuition, nonresident: full-time $15,848; part-time $1,321 per credit. Full-time tuition and fees vary according to class time, course load, program and reciprocity agreements.

University of Minnesota, Twin Cities Campus, Graduate School, College of Natural Resources, Department of Forest Resources, Minneapolis, MN 55455-0213. Offers natural resources science and management (MS, PhD). Part-time programs available. Terminal master's awarded for partial completion of doctoral program. *Degree requirements:* For master's, thesis optional; for doctorate, thesis/dissertation, comprehensive exam, registration. *Entrance requirements:* For master's and doctorate, GRE. Additional exam requirements/recommendations for international students: Required—TOEFL (minimum score 550 paper-based; 213 computer-based). Electronic applications accepted. *Expenses:* Tuition, state resident: full-time $8,748; part-time $729 per credit. Tuition, nonresident: full-time $15,848; part-time $1,321 per credit. Full-time tuition and fees vary according to class time, course load, program and reciprocity agreements.

University of Minnesota, Twin Cities Campus, Graduate School, Hubert H. Humphrey Institute of Public Affairs, Program in Science, Technology, and Environmental Policy, Minneapolis, MN 55455-0213. Offers MS, JD/MS. Part-time programs available. *Degree requirements:* For master's, thesis. *Entrance requirements:* For master's, GRE General Test, undergraduate training in the biological or physical sciences or engineering, minimum undergraduate GPA of 3.0. Additional exam requirements/recommendations for international students: Required—TOEFL (minimum score 600 paper-based; 250 computer-based). Electronic applications accepted. *Expenses:* Tuition, state resident: full-time $8,748; part-time $729 per credit. Tuition, nonresident: full-time $15,848; part-time $1,321 per credit. Full-time tuition and fees vary according to class time, course load, program and reciprocity agreements. *Faculty research:* Economics, history, philosophy, and politics of science and technology; organization and management of science and technology.

University of Missouri–St. Louis, College of Arts and Sciences, Department of Biology, St. Louis, MO 63121. Offers biology (MS, PhD), including animal behavior (MS), biochemistry (MS), biotechnology (MS), conservation biology (MS), development (MS), ecology (MS), environmental studies (PhD), evolution (MS), genetics (MS), molecular/cellular biology (MS), physiology (MS), plant systematics, population biology (MS), tropical biology (MS); biotechnology (Certificate); tropical biology and conservation (Certificate). Part-time programs available. *Faculty:* 50. *Students:* 26 full-time (15 women), 101 part-time (51 women); includes 13 minority (5 African Americans, 6 Asian Americans or Pacific Islanders, 2 Hispanic Americans), 41 international. Average age 32. In 2005, 22 master's, 2 doctorates awarded. *Degree requirements:* For master's, thesis or alternative; for doctorate, one foreign language, thesis/dissertation, 1 semester of teaching experience. *Entrance requirements:* For doctorate, GRE General Test. *Application deadline:* For spring admission, 12/1 priority date for domestic students. Applications are processed on a rolling basis. Application fee: $35 ($40 for international students). Electronic applications accepted. *Expenses:* Tuition, state resident: part-time $263 per credit hour. Tuition, nonresident: part-time $680 per credit hour. Required fees: $53 per credit hour. Tuition and fees vary according to program. *Financial support:* In 2005–06, 11 fellowships with full tuition reimbursements (averaging $30,000 per year), 15 research assistantships with full and partial tuition reimbursements (averaging $16,000 per year), 22 teaching assistantships with full and partial tuition reimbursements (averaging $16,000 per year) were awarded.; career-related internships or fieldwork and Federal Work-Study also available. Support available to part-time students. Financial award application deadline: 2/1. *Faculty research:* Molecular biology, microbial genetics. *Unit head:* Zuleyma Tang-Martinez, Director of Graduate Studies, 314-516-6498, Fax: 314-516-6233, E-mail: zuleyma@umsl.edu. *Application contact:* 314-516-5458, Fax: 314-516-5310, E-mail: gradadm@umsl.edu.

The University of Montana, Graduate School, College of Arts and Sciences, Program in Environmental Studies (EVST), Missoula, MT 59812-0002. Offers MS, JD/MS. Part-time programs available. *Faculty:* 6 full-time (2 women), 3 part-time/adjunct (1 woman). *Students:* 44 full-time (23 women), 39 part-time (25 women); includes 1 minority (American Indian/Alaska Native). Average age 28. 98 applicants, 59% accepted, 30 enrolled. In 2005, 37 degrees awarded. *Degree requirements:* For master's, portfolio, professional paper or thesis. *Entrance requirements:* For master's, GRE General Test. Additional exam requirements/recommendations for international students: Required—TOEFL (minimum score 580 paper-based; 237 computer-based). *Application deadline:* For fall admission, 2/15 for domestic students, 2/15 for international students. Application fee: $55. *Expenses:* Tuition, state resident: part-time $267 per credit. Tuition, nonresident: part-time $665 per credit. Part-time tuition and fees vary according to course load and degree level. *Financial support:* In 2005–06, 10 students received support, including 5 fellowships with full tuition reimbursements available (averaging $3,000 per year), 5 teaching assistantships with full tuition reimbursements available (averaging $9,000 per year); career-related internships or fieldwork and Federal Work-Study also available. Support available to part-time students. Financial award application deadline: 4/15. *Faculty research:* Pollution ecology, sustainable agriculture, environmental writing, environmental policy, habitat-land management. Total annual research expenditures: $367,111. *Unit head:* Len Broberg, Director, 406-243-5209, Fax: 406-243-6090, E-mail: len.broberg@umontana.edu. *Application contact:* Karen Hurd, Administrative Assistant, 406-243-6273, Fax: 406-243-6090, E-mail: karen.hurd@umontana.edu.

The University of Montana, Graduate School, College of Forestry and Conservation, Missoula, MT 59812-0002. Offers ecosystem management (MEM, MS); fish and wildlife biology (PhD); forestry (MS, PhD); recreation management (MS); resource conservation (MS); wildlife biology (MS). *Faculty:* 34 full-time (4 women). *Students:* 95 full-time (50 women), 63 part-time (21 women); includes 8 minority (1 African American, 3 American Indian/Alaska Native, 1 Asian American or Pacific Islander, 3 Hispanic Americans), 19 international. 175 applicants, 35% accepted, 46 enrolled. In 2005, 23 master's, 9 doctorates awarded. *Degree requirements:* For doctorate, thesis/dissertation. *Entrance requirements:* For master's and doctorate, GRE General Test. Additional exam requirements/recommendations for international students: Required—TOEFL (minimum score 575 paper-based; 213 computer-based). *Application deadline:* For fall admission, 1/31 for domestic students; for spring admission, 8/31 priority date for domestic students. Applications are processed on a rolling basis. Application fee: $45. *Expenses:* Tuition, state resident: part-time $267 per credit. Tuition, nonresident: part-time $665 per credit. Part-time tuition and fees vary according to course load and degree level. *Financial support:* In 2005–06, 25 research assistantships with tuition reimbursements, 12 teaching assistantships with full tuition reimbursements were awarded.; fellowships, career-related internships or fieldwork and Federal Work-Study also available. Financial award application deadline: 3/1; financial award applicants required to submit FAFSA. Total annual research expenditures: $6.7 million. *Unit head:* Dr. Perry Brown, Dean, 406-243-5521, Fax: 406-243-4845, E-mail: pbrown@forestry.umt.edu.

University of Nevada, Reno, Graduate School, College of Science, Interdisciplinary Program in Land Use Planning, Reno, NV 89557. Offers MS. Offered through the College of Science, the College of Engineering, and the College of Agriculture. *Degree requirements:* For master's, thesis. *Entrance requirements:* For master's, GRE General Test, minimum GPA of 3.0. Additional exam requirements/recommendations for international students: Required—TOEFL.

University of New Brunswick Saint John, Faculty of Business, Saint John, NB E2L 4L5, Canada. Offers administration (MBA); electronic commerce (MBA); international business (MBA); natural resource management (MBA). Part-time programs available. *Degree requirements:* For master's, thesis optional. *Entrance requirements:* For master's, GMAT. Additional exam requirements/recommendations for international students: Required—TOEFL (minimum score 550 paper-based). Expenses: Contact institution.

University of New Hampshire, Graduate School, College of Life Sciences and Agriculture, Department of Natural Resources, Durham, NH 03824. Offers environmental conservation (MS); forestry (MS); soil science (MS); water resources management (MS); wildlife (MS). Part-time programs available. *Faculty:* 40 full-time. *Students:* 31 full-time (17 women), 36 part-time (17 women); includes 2 minority (both Asian Americans or Pacific Islanders), 4 international. Average age 31. 50 applicants, 44% accepted, 15 enrolled. In 2005, 14 degrees awarded. *Degree requirements:* For master's, thesis or alternative. *Entrance requirements:* For master's, GRE General Test. Additional exam requirements/recommendations for international students: Required—TOEFL (minimum score 550 paper-based; 213 computer-based); Recommended—TSE. *Application deadline:* For fall admission, 4/1 for domestic students, 4/1 for international students. For winter admission, 12/1 for domestic students. Applications are processed on a rolling basis. Application fee: $60. Electronic applications accepted. *Expenses:* Tuition, state resident: full-time $8,010; part-time $445 per credit hour. Tuition, nonresident: full-time $19,730; part-time $810 per credit hour. Required fees: $322 per semester. Tuition and fees vary according to course load and program. *Financial support:* In 2005–06, 3 fellowships, 21 research assistantships, 15 teaching assistantships were awarded.; career-related internships or fieldwork, Federal Work-Study, scholarships/grants, and tuition waivers (full and partial) also available. Support available to part-time students. Financial award application deadline: 2/15. *Unit head:* Dr. William H. McDowell, Chairperson, 603-862-2249, E-mail: tehoward@cisunix.unh.edu. *Application contact:* Linda Scogin, Administrative Assistant, 603-862-3932, E-mail: natural.resources @unh.edu.

University of New Hampshire, Graduate School, College of Life Sciences and Agriculture, Department of Resource Economics and Development, Program in Resource Administration, Durham, NH 03824. Offers MS. Part-time programs available. *Faculty:* 4 full-time (0 women), 1 (woman) part-time/adjunct. *Students:* 3 full-time (2 women), 2 part-time (1 woman). Average age 27. 3 applicants, 100% accepted, 1 enrolled. In 2005, 2 degrees awarded. *Degree requirements:* For master's, thesis or alternative. *Entrance requirements:* For master's, GRE General Test. Additional exam requirements/recommendations for international students: Required—TOEFL (minimum score 550 paper-based; 213 computer-based); Recommended—TSE. *Application deadline:* For fall admission, 4/1 priority date for domestic students, 4/1 priority date for international students. Applications are processed on a rolling basis. Application fee: $60. Electronic applications accepted. *Expenses:* Tuition, state resident: full-time $8,010; part-time $445 per credit hour. Tuition, nonresident: full-time $19,730; part-time $810 per credit hour. Required fees: $322 per semester. Tuition and fees vary according to course load and program. *Financial support:* In 2005–06, 2 research assistantships, 2 teaching assistantships were awarded.; fellowships, career-related internships or fieldwork, Federal Work-Study, and scholarships/grants also available. Support available to part-time students. Financial award application deadline: 2/15.

University of New Hampshire, Graduate School, College of Life Sciences and Agriculture, Department of Resource Economics and Development, Program in Resource Economics, Durham, NH 03824. Offers MS. Part-time programs available. *Faculty:* 6 full-time. In 2005, 1 degree awarded. *Degree requirements:* For master's, thesis or alternative. *Entrance requirements:* For master's, GRE General Test. Additional exam requirements/recommendations for international students: Required—TOEFL (minimum score 550 paper-based; 213 computer-based); Recommended—TSE. *Application deadline:* For fall admission, 4/1 for domestic students, 4/1 for international students. Applications are processed on a rolling basis. Application fee: $60. Electronic applications accepted. *Expenses:* Tuition, state resident: full-time $8,010; part-time $445 per credit hour. Tuition, nonresident: full-time $19,730; part-time $810 per credit hour. Required fees: $322 per semester. Tuition and fees vary according to course load and program. *Financial support:* Fellowships, research assistantships, teaching assistantships, career-related internships or fieldwork and Federal Work-Study available. Support available to part-time students. Financial award application deadline: 2/15.

The University of North Carolina at Chapel Hill, Graduate School, School of Public Health, Department of Environmental Sciences and Engineering, Chapel Hill, NC 27599. Offers air, radiation and industrial hygiene (MPH, MS, MSEE, MSPH, PhD); aquatic and atmospheric sciences (MPH, MS, MSPH, PhD); environmental engineering (MPH, MS, MSEE, MSPH, PhD); environmental health sciences (MPH, MS, MSPH, PhD); environmental management and policy (MPH, MS, MSPH, PhD). *Faculty:* 33 full-time (3 women), 35 part-time/adjunct. *Students:* 141 full-time (74 women); includes 37 minority (10 African Americans, 25 Asian Americans or Pacific Islanders, 2 Hispanic Americans). Average age 27. 216 applicants, 37% accepted, 29 enrolled. In 2005, 14 master's, 11 doctorates awarded. Terminal master's awarded for partial completion of doctoral program. *Median time to degree:* Of those who began their doctoral program in fall 1997, 100% received their degree in 8 years or less. *Degree requirements:* For master's, thesis (for some programs), research paper, comprehensive exam, registration; for doctorate, thesis/dissertation, comprehensive exam, registration. *Entrance requirements:* For master's and doctorate, GRE General Test, minimum GPA of 3.0. Additional exam requirements/recommendations for international students: Required—TOEFL. *Application deadline:* For fall admission, 1/1 priority date for domestic students, 1/1 priority date for international students; for spring admission, 9/15 for domestic students. Applications are processed on a rolling basis. Application fee: $70. Electronic applications accepted. *Financial support:* In 2005–06, 134 students received support, including 36 fellowships with tuition reimbursements available (averaging $6,358 per year), 86 research assistantships with tuition reimbursements available (averaging $6,197 per year), 12 teaching assistantships with tuition reimbursements available (averaging $6,729 per year); career-related internships or fieldwork, Federal Work-Study, traineeships, health care benefits, and unspecified assistantships also available. Support available to part-time students. Financial award application deadline: 1/1; financial award applicants required to submit FAFSA. *Faculty research:* Air, radiation and industrial hygiene, aquatic and atmospheric sciences, environmental health sciences, environmental management and policy, water resources engineering. Total annual research expenditures: $9.6 million. *Unit head:* Dr. Don Fox, Interim Chair, 919-966-1024, Fax: 919-966-7911, E-mail: don_fox@unc.edu. *Application contact:* Jack Whaley, Registrar, 919-966-3844, Fax: 919-966-7911, E-mail: jack_whaley@unc.edu.

University of Northern British Columbia, Office of Graduate Studies, Prince George, BC V2N 4Z9, Canada. Offers business administration (Diploma); community health science (M Sc); disability management (MA); education (M Ed); first nations studies (MA); gender studies (MA); history (MA); interdisciplinary studies (MA); international studies (MA); mathematical, computer and physical sciences (M Sc); natural resources and environmental studies (M Sc, MA, MNRES, PhD); political science (MA); psychology (M Sc, PhD); social work (MSW). Part-time and evening/weekend programs available. Postbaccalaureate distance learning degree programs offered (no on-campus study). *Degree requirements:* For master's and doctorate, thesis/dissertation. *Entrance requirements:* For master's, GRE, minimum B average in undergraduate course work; for doctorate, candidacy exam, minimum A average in graduate course work.

University of Oregon, Graduate School, College of Arts and Sciences, Environmental Studies Program, Eugene, OR 97403. Offers environmental science, studies, and policy (PhD); environmental studies (MA, MS). *Faculty:* 11 full-time (3 women), 1 part-time/adjunct (0 women). *Students:* 21, 3 international. Average age 30. 139 applicants, 14% accepted, 9 enrolled. In 2005, 11 master's, 1 doctorate awarded. *Degree requirements:* For master's, one foreign language, thesis; for doctorate, thesis/dissertation, comprehensive exam. *Entrance requirements:* For master's, GRE General Test, minimum GPA of 3.0; for doctorate, GRE General Test. Additional exam requirements/recommendations for international students: Required—TOEFL (minimum score 550 paper-based; 213 computer-based). *Application deadline:* For fall admission, 1/15 for domestic students, 1/15 for international students. Application fee: $50. Electronic applications accepted. *Financial support:* In 2005–06, 23 teaching assistantships were awarded; career-related internships or fieldwork and Federal Work-Study also available. *Unit head:* Daniel Udovic, Director, 541-346-5000, Fax: 541-346-5954, E-mail: udovic@oregon.uoregon.edu. *Application contact:* Gayla Wardwell, Graduate Coordinator, 541-346-5057, Fax: 541-346-5954, E-mail: gaylaw@uoregon.edu.

Announcement: The Environmental Studies Program is supported by a full range of graduate-level courses offered by over 80 faculty members in many disciplines. The program offers a flexible, explicitly interdisciplinary, largely student-designed program that attracts exceptionally strong students, many of whom have nonacademic environmental experience (employment, volunteer work, internships, Peace Corps). Visit the Web site at http://darkwing.uoregon.edu/~ecostudy.

University of Pennsylvania, School of Arts and Sciences, College of General Studies, Philadelphia, PA 19104. Offers environmental studies (MES); individualized study (MLA). Electronic applications accepted.

University of Pittsburgh, Graduate School of Public and International Affairs, Division of International Development, Program in Development Planning and Environmental Sustainability, Pittsburgh, PA 15260. Offers MID. Part-time programs available. *Faculty:* 35 full-time (11 women), 16 part-time/adjunct (9 women). *Students:* 38 full-time (31 women), 6 part-time (4 women); includes 10 minority (5 African Americans, 3 Asian Americans or Pacific Islanders, 2 Hispanic Americans), 1 international. 100 applicants, 80% accepted, 26 enrolled. In 2005, 32 degrees awarded. *Degree requirements:* For master's, internship, capstone seminar, thesis optional. *Entrance requirements:* For master's, 3 letters of recommendation, minimum GPA of 3.2. Additional exam requirements/recommendations for international students: Required—TOEFL (minimum score 550 paper-based; 213 computer-based), TWE (minimum score 4); Recommended—IELT (minimum score 7). *Application deadline:* For fall admission, 3/1 for domestic students, 2/1 for international students; for spring admission, 10/1 for domestic students, 8/1 for international students. Application fee: $50. Electronic applications accepted. *Expenses:* Tuition, state resident: full-time $13,194; part-time $537 per credit. Tuition, nonresident: full-time $25,012; part-time $1,026 per credit. Required fees: $700; $164 per term. Tuition and fees vary according to campus/location and program. *Financial support:* In 2005–06, 35 students received support, including 27 fellowships (averaging $8,063 per year); scholarships/grants and unspecified assistantships also available. Financial award application deadline: 2/1. *Faculty research:* Project/program evaluation, population and environment, international development, development economics, civil society. *Application contact:* Maureen O'Malley, Admissions Counselor, 412-648-7640, Fax: 412-648-7641, E-mail: pronobis@birch.gspia.pitt.edu.

University of Rhode Island, Graduate School, College of the Environment and Life Sciences, Department of Environmental and Natural Resource Economics, Kingston, RI 02881. Offers resource economics (MS, PhD). In 2005, 15 master's, 1 doctorate awarded. *Degree requirements:* For master's, thesis optional; for doctorate, thesis/dissertation. *Entrance requirements:* For master's and doctorate, GRE General Test. Additional exam requirements/recommendations for international students: Required—TOEFL. *Application deadline:* For fall admission, 4/15 for domestic students. Applications are processed on a rolling basis. *Expenses:* Tuition, state resident: full-time $5,522; part-time $307 per credit. Tuition, nonresident: full-time $15,992; part-time $888 per credit. Required fees: $1,786; $73 per credit. One-time fee: $80 part-time. *Unit head:* Dr. James Anderson, Chairperson, 401-874-2471. *Application contact:* Dr. Tim Tyrrell, Graduate Admissions Committee, 401-874-2472, Fax: 401-782-4766, E-mail: renri@uriacc.uri.edu.

University of St. Thomas, Graduate Studies, College of Business, MBA Program, St. Paul, MN 55105-1096. Offers accounting (MBA); environmental management (MBA); finance (MBA); financial services management (MBA); franchise management (MBA); government contracts (MBA); health care management (MBA); human resource management (MBA); information management (MBA); insurance and risk management (MBA); management (MBA); manufacturing systems (MBA); marketing (MBA); nonprofit management (MBA); sports and entertainment management (MBA); venture management (MBA). Part-time and evening/weekend programs available. *Degree requirements:* For master's, registration. *Entrance requirements:* For master's, GMAT. Electronic applications accepted. Expenses: Contact institution.

University of San Francisco, College of Arts and Sciences, Program in Environmental Management, San Francisco, CA 94117-1080. Offers MS. Evening/weekend programs available. *Faculty:* 9 full-time (3 women), 29 part-time/adjunct (11 women). *Students:* 60 full-time (40 women), 24 part-time (15 women); includes 22 minority (2 African Americans, 1 American Indian/Alaska Native, 15 Asian Americans or Pacific Islanders, 4 Hispanic Americans), 2 international. Average age 33. 89 applicants, 81% accepted, 36 enrolled. In 2005, 41 degrees awarded. *Degree requirements:* For master's, thesis. *Entrance requirements:* For master's, 3 semesters of course work in chemistry, minimum GPA of 2.7, work experience in environmental field. *Application deadline:* For fall admission, 3/1 for domestic students. Applications are processed on a rolling basis. Application fee: $55 ($65 for international students). *Expenses:* Tuition: Part-time $925 per unit. Tuition and fees vary according to degree level, campus/location and program. *Financial support:* In 2005–06, 53 students received support; teaching assistantships, career-related internships or fieldwork available. Financial award application deadline: 3/2; financial award applicants required to submit FAFSA. *Faculty research:* Problems of environmental managers, water quality, hazardous materials, environmental health. *Unit head:* Dr. John Lendvay, Chair, 415-422-6553, Fax: 415-422-6363, E-mail: msem@usfca.edu.

University of South Carolina, The Graduate School, School of the Environment, Program in Earth and Environmental Resources Management, Columbia, SC 29208. Offers MEERM, JD/MEERM. Part-time programs available. Postbaccalaureate distance learning degree programs offered (no on-campus study). *Degree requirements:* For master's, thesis optional. *Entrance requirements:* For master's, GRE General Test. Additional exam requirements/recommendations for international students: Required—TOEFL. Electronic applications accepted. *Faculty research:* Hydrology, sustainable development, environmental geology and engineering, energy/environmental resources management.

University of South Florida, College of Graduate Studies, College of Arts and Sciences, Department of Environmental Science and Policy, Tampa, FL 33620-9951. Offers MS. *Faculty:* 5 full-time (2 women), 1 part-time/adjunct (0 women). *Students:* 11 full-time (8 women), 13 part-time (10 women); includes 2 minority (1 African American, 1 Hispanic American), 7 international. 31 applicants, 48% accepted, 8 enrolled. In 2005, 1 degree awarded. *Degree requirements:* For master's, thesis optional. *Entrance requirements:* For master's, GRE General Test, minimum GPA of 3.0 in last 60 hours of course work. *Application deadline:* For fall admission, 5/1 for domestic students; for spring admission, 9/15 for domestic students. Application fee: $30. *Financial support:* Fellowships with tuition reimbursements, research assistantships with full tuition reimbursements, teaching assistantships with full tuition reimbursements, scholarships/grants and unspecified assistantships available. Support available to part-time students. Financial award application deadline: 5/1. *Unit head:* Rick Oches, Interim Director, 813-974-2978. *Application contact:* Dr. Ingrid Bartsch, Information Contact, 813-974-3069, E-mail: klschrad@chumal.cas.usf.edu.

The University of Tennessee, Graduate School, College of Arts and Sciences, Department of Sociology, Knoxville, TN 37996. Offers criminology (MA, PhD); energy, environment, and

Environmental Management and Policy

The University of Tennessee (continued)

resource policy (MA, PhD); political economy (MA, PhD). Part-time programs available. *Degree requirements:* For master's, thesis or alternative; for doctorate, thesis/dissertation. *Entrance requirements:* For master's, GRE General Test, minimum GPA of 3.0; for doctorate, GRE General Test, minimum GPA of 3.5. Additional exam requirements/recommendations for international students: Required—TOEFL. Electronic applications accepted.

The University of Texas at Austin, Graduate School, College of Engineering, Department of Petroleum and Geosystems Engineering, Program in Energy and Mineral Resources, Austin, TX 78712-1111. Offers MA, MS. *Degree requirements:* For master's, thesis, seminar. *Entrance requirements:* For master's, GRE General Test. Additional exam requirements/recommendations for international students: Required—TOEFL. Electronic applications accepted.

University of Vermont, Graduate College, The Rubenstein School of Environment and Natural Resources, Program in Natural Resources, Burlington, VT 05405. Offers natural resources (MS, PhD), including aquatic ecology and watershed science (MS), environment thought and culture (MS), environment, science and public affairs (MS), forestry (MS). *Degree requirements:* For master's, thesis or alternative; for doctorate, thesis/dissertation. *Entrance requirements:* For master's and doctorate, GRE General Test. Additional exam requirements/recommendations for international students: Required—TOEFL (minimum score 550 paper-based; 213 computer-based). *Application deadline:* For fall admission, 3/1 for domestic students. Applications are processed on a rolling basis. Application fee: $25. Electronic applications accepted. *Expenses:* Tuition, area resident: Part-time $410 per credit hour. Tuition, nonresident: part-time $1,034 per credit hour. *Financial support:* Fellowships, research assistantships, teaching assistantships available. Financial award application deadline: 3/1. *Unit head:* Dr. Deane Wang, Coordinator, 802-656-2620.

University of Washington, Graduate School, College of Forest Resources, Seattle, WA 98195. Offers forest economics (MS, PhD); forest ecosystem analysis (MS, PhD); forest engineering/forest hydrology (MS, PhD); forest products marketing (MS, PhD); forest soils (MS, PhD); paper science and engineering (MS, PhD); quantitative resource management (MS, PhD); silviculture (MFR); silviculture and forest protection (MS, PhD); social sciences (MS, PhD); urban horticulture (MFR, MS, PhD); wildlife science (MS, PhD). *Degree requirements:* For master's, thesis (for some programs), registration; for doctorate, thesis/dissertation, comprehensive exam (for some programs), registration. *Entrance requirements:* For master's and doctorate, GRE, minimum GPA of 3.0. Additional exam requirements/recommendations for international students: Required—TOEFL. Electronic applications accepted. *Faculty research:* Ecosystem analysis, silviculture and forest protection, paper science and engineering, environmental horticulture and urban forestry, natural resource policy and economics.

University of Washington, Graduate School, Interdisciplinary Graduate Program in Quantitative Ecology and Resource Management, Seattle, WA 98195. Offers MS, PhD. *Degree requirements:* For master's and doctorate, thesis/dissertation. *Entrance requirements:* For master's and doctorate, GRE General Test, minimum GPA of 3.0. Additional exam requirements/recommendations for international students: Required—TOEFL. Electronic applications accepted. *Faculty research:* Population dynamics, statistical analysis, ecological modeling and systems analysis of aquatic and terrestrial ecosystems.

University of Waterloo, Graduate Studies, Faculty of Environmental Studies, Program in Environment and Resource Studies, Waterloo, ON N2L 3G1, Canada. Offers MES. Part-time programs available. *Faculty:* 10 full-time (2 women), 12 part-time/adjunct (5 women). *Students:* 15 full-time (10 women), 3 part-time (all women). 62 applicants, 27% accepted, 8 enrolled. In 2005, 10 degrees awarded. *Degree requirements:* For master's, thesis. *Entrance requirements:* For master's, honors degree, minimum B average, resumé. Additional exam requirements/recommendations for international students: Required—TOEFL, TWE. *Application deadline:* For fall admission, 1/31 for domestic students. Application fee: $75 Canadian dollars. Electronic applications accepted. *Financial support:* Research assistantships, teaching assistantships, scholarships/grants available. *Faculty research:* Sustainable development, water conservation, native issues, environmental assessment. *Unit head:* Dr. Susan K. Wismer, Chair, 519-888-4567, Fax: 519-746-0292, E-mail: skwismer@fes.uwaterloo.ca. *Application contact:* Dr. Robert Gibson, Graduate Officer, 519-888-4567 Ext. 3407, Fax: 519-746-0292, E-mail: rbgibson@watserv1.uwaterloo.ca.

University of Waterloo, Graduate Studies, Faculty of Environmental Studies, Program in Local Economic Development/Tourism Policy and Planning, Waterloo, ON N2L 3G1, Canada. Offers MAES. Part-time programs available. *Faculty:* 19 part-time/adjunct (3 women). *Students:* 18 full-time (10 women), 10 part-time (5 women). Average age 31. 42 applicants, 43% accepted, 15 enrolled. In 2005, 10 degrees awarded. *Degree requirements:* For master's, research paper. *Entrance requirements:* For master's, honors degree in related field, minimum B average. Additional exam requirements/recommendations for international students: Required—TOEFL, TWE. *Application deadline:* For fall admission, 3/1 for domestic students. Applications are processed on a rolling basis. Application fee: $75 Canadian dollars. Electronic applications accepted. *Financial support:* Research assistantships, career-related internships or fieldwork, institutionally sponsored loans, and scholarships/grants available. Support available to part-time students. *Faculty research:* Urban and regional economics, regional economic development, strategic planning, environmental economics, economic geography. *Unit head:* Dr. Judith Cukier, Graduate Officer, 519-888-4567 Ext. 5490, Fax: 519-746-2031, E-mail: jcukier@fes.uwaterloo.ca.

University of Wisconsin–Green Bay, Graduate Studies, Program in Environmental Science and Policy, Green Bay, WI 54311-7001. Offers MS. Part-time programs available. *Faculty:* 28 full-time (6 women), 3 part-time/adjunct (0 women). *Students:* 22 full-time (11 women), 24 part-time (12 women); includes 3 minority (all American Indian/Alaska Native), 2 international. Average age 32. 25 applicants, 92% accepted, 11 enrolled. In 2005, 18 degrees awarded. *Degree requirements:* For master's, thesis. *Entrance requirements:* For master's, GRE General Test, minimum GPA of 3.0. *Application deadline:* For fall admission, 8/1 for domestic students; for spring admission, 11/1 for domestic students. Applications are processed on a rolling basis. Application fee: $45. Electronic applications accepted. *Expenses:* Tuition, state resident: full-time $5,619; part-time $312 per credit. Tuition, nonresident: full-time $16,229; part-time $902 per credit. Required fees: $1,148; $64 per credit. Tuition and fees vary according to course load and reciprocity agreements. *Financial support:* In 2005–06, 7 research assistantships with full tuition reimbursements, 11 teaching assistantships with full tuition reimbursements were awarded.; career-related internships or fieldwork, Federal Work-Study, and institutionally sponsored loans also available. Financial award application deadline: 7/15; financial award applicants required to submit FAFSA. *Faculty research:* Bald eagle, parasitic population of domestic and wild animals, resource recovery, anaerobic digestion of organic waste. *Unit head:* Dr. John Stoll, Coordinator, 920-465-2358, E-mail: stollj@uwgb.edu.

University of Wisconsin–Madison, Graduate School, College of Agricultural and Life Sciences, School of Natural Resources, Madison, WI 53706-1380. Offers MA, MS, PhD. Part-time programs available. *Degree requirements:* For doctorate, thesis/dissertation. *Entrance requirements:* For doctorate, GRE. Application fee: $45. Electronic applications accepted. *Financial support:* Fellowships, research assistantships, teaching assistantships, career-related internships or fieldwork, Federal Work-Study, and institutionally sponsored loans available. Support available to part-time students. *Unit head:* Richard Straub, Director, 608-262-8254, Fax: 608-262-6055, E-mail: rjstraub@facstaff.wisc.edu.

University of Wisconsin–Madison, Graduate School, Gaylord Nelson Institute for Environmental Studies, Land Resources Program, Madison, WI 53706-1380. Offers MS, PhD. Part-time programs available. *Faculty:* 3 full-time (2 women), 104 part-time/adjunct (24 women). *Students:* 61 full-time, 28 part-time; includes 6 minority (1 African American, 1 American Indian/Alaska Native, 2 Asian Americans or Pacific Islanders, 2 Hispanic Americans), 5 international. Average age 32. 109 applicants, 53% accepted, 24 enrolled. In 2005, 13 master's, 4 doctorates awarded. *Median time to degree:* Of those who began their doctoral program in fall 1997, 55% received their degree in 8 years or less. *Degree requirements:* For master's and doctorate, thesis/dissertation. *Entrance requirements:* For master's and doctorate, GRE General Test. Additional exam requirements/recommendations for international students: Required—TOEFL (minimum score 550 paper-based; 213 computer-based). *Application deadline:* For fall admission, 2/1 for domestic students, 2/1 for international students; for spring admission, 10/15 for domestic students, 10/15 for international students. Application fee: $45. Electronic applications accepted. *Financial support:* In 2005–06, 58 students received support, including 7 fellowships with full tuition reimbursements available (averaging $14,400 per year), 15 research assistantships with full tuition reimbursements available (averaging $14,250 per year), 21 teaching assistantships with full tuition reimbursements available (averaging $11,260 per year); career-related internships or fieldwork, Federal Work-Study, scholarships/grants, health care benefits, unspecified assistantships, and project assistantships also available. Financial award application deadline: 1/2. *Faculty research:* Land use issues, soil science/watershed management, geographic information systems, environmental law/justice, waste management. *Unit head:* Arthur F. McEvoy, Chair, 608-265-4771, Fax: 608-262-2273, E-mail: amcevoy@wisc.edu. *Application contact:* James E. Miller, Associate Student Services Coordinator, 608-263-4373, Fax: 608-262-2273, E-mail: jemiller@wisc.edu.

Utah State University, School of Graduate Studies, College of Natural Resources, Department of Environment and Society, Logan, UT 84322. Offers bioregional planning (MS); geography (MA, MS); human dimensions of ecosystem science and management (MS, PhD); recreation resource management (MS, PhD). *Faculty:* 16 full-time (3 women), 6 part-time/adjunct (2 women). *Students:* 53 full-time (15 women), 53 part-time (28 women), 6 international. Average age 32. 18 applicants, 67% accepted, 10 enrolled. In 2005, 11 degrees awarded. *Degree requirements:* For master's, thesis (for some programs), comprehensive exam. *Entrance requirements:* For master's and doctorate, GRE General Test, minimum GPA of 3.0. Additional exam requirements/recommendations for international students: Required—TOEFL. *Application deadline:* For fall admission, 6/15 for domestic students; for spring admission, 10/15 for domestic students. Applications are processed on a rolling basis. Application fee: $50 ($60 for international students). Electronic applications accepted. *Financial support:* In 2005–06, 14 students received support, including 21 research assistantships with partial tuition reimbursements available (averaging $11,000 per year), 5 teaching assistantships with partial tuition reimbursements available (averaging $10,000 per year); fellowships with partial tuition reimbursements available, career-related internships or fieldwork, Federal Work-Study, tuition waivers (full and partial), and unspecified assistantships also available. Financial award applicants required to submit FAFSA. *Faculty research:* Geographic information systems/geographic and environmental education, bioregional planning, natural resource and environmental policy, outdoor recreation and tourism, natural resource and environmental management. Total annual research expenditures: $1.4 million. *Unit head:* Dr. Terry L. Sharik, Head, 435-797-3270, Fax: 435-797-4048, E-mail: tlsharik@cc.usu.edu. *Application contact:* Dr. Richard E. Toth, Information Contact, 435-797-1790, Fax: 435-797-4048, E-mail: envs.info@cnr.usu.edu.

Vanderbilt University, School of Engineering, Department of Civil and Environmental Engineering, Program in Environmental Engineering, Nashville, TN 37240-1001. Offers environmental engineering (M Eng); environmental management (MS, PhD). MS and PhD offered through the Graduate School. Part-time programs available. *Faculty:* 15 full-time (3 women). *Students:* 18 full-time (11 women), 4 part-time (3 women); includes 1 minority (Asian American or Pacific Islander), 8 international. Average age 30. 76 applicants, 9% accepted, 4 enrolled. In 2005, 6 master's, 1 doctorate awarded. Terminal master's awarded for partial completion of doctoral program. *Degree requirements:* For master's, thesis or alternative; for doctorate, thesis/dissertation. *Entrance requirements:* For master's and doctorate, GRE General Test. Additional exam requirements/recommendations for international students: Required—TOEFL. *Application deadline:* For fall admission, 1/15 for domestic students; for spring admission, 11/1 for domestic students. Applications are processed on a rolling basis. Application fee: $0. Electronic applications accepted. *Expenses:* Tuition: Full-time $15,396; part-time $1,283 per semester hour. Required fees: $2,202; $1,101 per semester. One-time fee: $30. Tuition and fees vary according to course load, program and student level. *Financial support:* In 2005–06, 24 students received support, including 4 fellowships with full tuition reimbursements available (averaging $21,000 per year), 6 research assistantships with full tuition reimbursements available (averaging $21,000 per year), 8 teaching assistantships with full tuition reimbursements available (averaging $16,300 per year); career-related internships or fieldwork, institutionally sponsored loans, scholarships/grants, traineeships, and tuition waivers (full and partial) also available. Financial award application deadline: 1/15. *Faculty research:* Waste treatment, hazardous waste management, chemical waste treatment, water quality. *Application contact:* Dr. James H. Clarke, Graduate Program Administrator, 615-322-3897, Fax: 615-322-3365.

Vermont Law School, Law School, Environmental Law Center, South Royalton, VT 05068-0096. Offers LL M, MSEL, JD/MSEL. Part-time programs available. *Faculty:* 12 full-time (6 women), 8 part-time/adjunct (4 women). *Students:* 36 full-time, 2 part-time; includes 1 African American, 1 American Indian/Alaska Native. Average age 30. 80 applicants, 81% accepted, 36 enrolled. In 2005, 70 degrees awarded. *Entrance requirements:* For master's, GRE General Test or LSAT. Additional exam requirements/recommendations for international students: Required—TOEFL. *Application deadline:* For fall admission, 3/15 for domestic students. Applications are processed on a rolling basis. Application fee: $50. *Expenses:* Tuition: Full-time $28,114; part-time $1,004 per credit. Required fees: $340. Tuition and fees vary according to degree level. *Financial support:* In 2005–06, 2 fellowships with full tuition reimbursements (averaging $5,000 per year) were awarded; career-related internships or fieldwork, Federal Work-Study, institutionally sponsored loans, scholarships/grants, and tuition waivers (partial) also available. Support available to part-time students. Financial award application deadline: 2/15; financial award applicants required to submit FAFSA. *Faculty research:* Environment and technology; takings; international environmental law; interaction among science, law, and environmental policy; air pollution. Total annual research expenditures: $52,000. *Unit head:* Karin Sheldon, Associate Dean, 802-831-1220, Fax: 802-763-2490, E-mail: elcinfo@vermontlaw.edu. *Application contact:* Anne Mansfield, Assistant Director, 802-831-1338, Fax: 802-763-2940, E-mail: elcinfo@vermontlaw.edu.

Virginia Commonwealth University, Graduate School, College of Humanities and Sciences, Center for Environmental Studies, Richmond, VA 23284-9005. Offers environmental communication (MIS); environmental health (MIS); environmental policy (MIS); environmental sciences (MIS). *Students:* 10 full-time (8 women), 22 part-time (12 women); includes 5 minority (4 African Americans, 1 Hispanic American), 2 international. Average age 33. 20 applicants, 70% accepted. In 2005, 13 degrees awarded. *Degree requirements:* For master's, thesis. *Entrance requirements:* For master's, GRE General Test. Application fee: $50. *Expenses:* Tuition, state resident: full-time $3,185; part-time $405 per credit. Tuition, nonresident: full-time $7,952; part-time $940 per credit. Required fees: $751 per semester hour. Tuition and fees vary according to course load and program. *Unit head:* Dr. Gregory C. Garman, Director, 804-828-1574, Fax: 804-828-0503, E-mail: gcgarman@vcu.edu.

Webster University, School of Business and Technology, Department of Business, St. Louis, MO 63119-3194. Offers business (MA); business and organizational security management (MBA); computer resources and information management (MBA); environmental management (MBA); finance (MA, MBA); health services management (MBA); human resources development (MBA); human resources management (MBA); international business (MA, MBA); management and leadership (MBA); marketing (MBA); procurement and acquisitions management (MBA); telecommunications management (MBA). Part-time and evening/weekend programs available. Postbaccalaureate distance learning degree programs offered (no on-campus study). *Students:* 1,452 full-time (696 women), 4,988 part-time (2,348 women); includes 2,507 minority (1,880 African Americans, 35 American Indian/Alaska Native, 225 Asian Americans or Pacific Islanders, 367 Hispanic Americans), 533 international. Average age 34. 1,114 applicants, 98% accepted, 935 enrolled. In 2005, 2484 degrees awarded. *Application deadline:* Applications are processed on a rolling basis. Application fee: $25 ($50 for international students). *Expenses:* Tuition: Full-time $6,975; part-time $465 per credit. Part-time tuition and fees vary according to degree level, campus/location and program. *Financial support:* Federal Work-Study available. Support available to part-time students. Financial award application deadline: 4/1; financial award applicants required to submit FAFSA. *Unit head:* Bradford Scott, Chair, 314-961-2260 Ext. 7574, Fax: 314-968-7077, E-mail: buschair@webster.edu. *Application contact:* Director of Graduate and Evening Student Admissions, Fax: 314-968-7116, E-mail: gadmit@webster.edu.

Webster University, School of Business and Technology, Department of Management, St. Louis, MO 63119-3194. Offers business and organizational security management (MA); computer resources and information management (MA); environmental management (MS); health care management (MA); health services management (MA); human resources development (MA); human resources management (MA); management (DM); management and leadership (MA); marketing (MA); procurement and acquisitions management (MA); public administration (MA); quality management (MA); space systems operations management (MS); telecommunications management (MA). Part-time and evening/weekend programs available. Postbaccalaureate distance learning degree programs offered (no on-campus study). *Students:* 1,294 full-time (661 women), 3,757 part-time (2,059 women); includes 2,492 minority (1,947 African Americans, 34 American Indian/Alaska Native, 124 Asian Americans or Pacific Islanders, 387 Hispanic Americans), 86 international. Average age 37. 1,039 applicants, 99% accepted, 935 enrolled. In 2005, 2,064 master's, 9 doctorates awarded. *Degree requirements:* For doctorate, thesis/dissertation, written exam. *Entrance requirements:* For doctorate, GMAT, 3 years of work experience, MBA. *Application deadline:* Applications are processed on a rolling basis. Application fee: $25 ($50 for international students). *Expenses:* Tuition: Full-time $6,975; part-time $465 per credit. Part-time tuition and fees vary according to degree level, campus/location and program. *Financial support:* Federal Work-Study available. Support available to part-time students. Financial award application deadline: 4/1; financial award applicants required to submit FAFSA. *Unit head:* Jeffrey Haldeman, Director, 314-961-2660 Ext. 7552, Fax: 314-968-7077, E-mail: mgtchair@webster.edu. *Application contact:* Director of Graduate and Evening Student Admissions, Fax: 314-968-7116, E-mail: gadmit@webster.edu.

Wesley College, Environmental Studies Program, Dover, DE 19901-3875. Offers MS. Part-time and evening/weekend programs available. *Faculty:* 2 full-time (0 women). *Students:* Average age 30. 8 applicants, 75% accepted, 6 enrolled. In 2005, 3 degrees awarded. *Entrance requirements:* For master's, BA/BSM in science or engineering field, portfolio. *Application deadline:* Applications are processed on a rolling basis. Application fee: $25. *Expenses:* Tuition: Full-time $300; part-time $300 per credit. *Financial support:* Teaching assistantships with tuition reimbursements, unspecified assistantships available. *Unit head:* Dr. Bruce Allison, Director, 302-736-2349, Fax: 302-736-2301, E-mail: allisobr@wesley.edu. *Application contact:* Arthur Jacobs, Director of Admissions, 302-736-2428, Fax: 302-736-2301, E-mail: jacobsar@mail.wesley.edu.

West Virginia University, College of Engineering and Mineral Resources, Department of Industrial and Management Systems Engineering, Program in Safety and Environmental Management, Morgantown, WV 26506. Offers safety management (MS). *Accreditation:* ABET. *Students:* 48 full-time (6 women), 29 part-time (4 women); includes 2 minority (1 African American, 1 Hispanic American), 5 international. Average age 29. In 2005, 39 degrees awarded. *Expenses:* Tuition, state resident: full-time $4,582; part-time $258 per credit hour. Tuition, nonresident: full-time $1,382; part-time $741 per credit hour. *Financial support:* In 2005–06, 12 research assistantships, 1 teaching assistantship were awarded. *Unit head:* Gary Winn, Coordinator, 304-293-4821 Ext. 3744, Fax: 304-293-4970, E-mail: gary.winn@mail.wvu.edu.

West Virginia University, Davis College of Agriculture, Forestry and Consumer Sciences, Division of Resource Management and Sustainable Development, Program in Natural Resource Economics, Morgantown, WV 26506. Offers PhD. Part-time programs available. *Students:* 7 full-time (4 women), 3 part-time (2 women); includes 1 minority (Asian American or Pacific Islander), 7 international. Average age 34. 18 applicants, 56% accepted. In 2005, 4 degrees awarded. *Degree requirements:* For doctorate, thesis/dissertation. *Entrance requirements:* For doctorate, GRE General Test. Additional exam requirements/recommendations for international students: Required—TOEFL. Application fee: $45. *Expenses:* Tuition, state resident: full-time $4,582; part-time $258 per credit hour. Tuition, nonresident: full-time $1,382; part-time $741 per credit hour. *Financial support:* In 2005–06, 4 research assistantships, 1 teaching assistantship were awarded.; Federal Work-Study, institutionally sponsored loans, and tuition waivers (partial) also available. Financial award application deadline: 2/1; financial award applicants required to submit FAFSA. *Unit head:* Dr. Gerard E. D'Souza, Graduate Coordinator, 304-293-4832 Ext. 4471, Fax: 304-293-3740, E-mail: gerald.d'souza@mail.wvu.edu.

West Virginia University, Eberly College of Arts and Sciences, Department of Geology and Geography, Program in Geography, Morgantown, WV 26506. Offers energy and environmental resources (MA); geographic information systems (PhD); geography-regional development (PhD); GIS/cartographic analysis (MA); regional development (MA). Part-time programs available. *Students:* 8 full-time (4 women), 6 part-time (4 women); includes 1 minority (Hispanic American), 2 international. Average age 30. 40 applicants, 25% accepted, 5 enrolled. In 2005, 7 degrees awarded. *Degree requirements:* For master's, thesis/dissertation, oral and written exams; for doctorate, thesis/dissertation, oral and written exams, comprehensive exam. *Entrance requirements:* For master's, GRE General Test, minimum GPA of 3.0; for doctorate, GRE General Test. Additional exam requirements/recommendations for international students: Required—TOEFL. *Application deadline:* For fall admission, 2/14 priority date for domestic students, 11/14 priority date for international students; for spring admission, 10/1 priority date for domestic students, 7/1 priority date for international students. Applications are processed on a rolling basis. Application fee: $45. Electronic applications accepted. *Expenses:* Tuition, state resident: full-time $4,582; part-time $258 per credit hour. Tuition, nonresident: full-time $1,382; part-time $741 per credit hour. *Financial support:* In 2005–06, 13 students received support, including 1 research assistantship with full tuition reimbursement available (averaging $9,185 per year), 6 teaching assistantships with full tuition reimbursements available (averaging $9,185 per year); career-related internships or fieldwork, Federal Work-Study, institutionally sponsored loans, health care benefits, and tuition waivers (partial) also available. Financial award application deadline: 2/1; financial award applicants required to submit FAFSA. *Faculty research:* Resources, regional development, geographic information systems, gender geography, environmental geography. Total annual research expenditures: $1.5 million. *Unit head:* Dr. Daniel Weiner, Director, 304-293-6955, Fax: 304-293-6522, E-mail: daniel.weiner@mail.wvu.edu. *Application contact:* Dr. Timothy Warner, Associate Professor, 304-293-5603 Ext. 4328, Fax: 304-293-6522, E-mail: tim.warner@mail.wvu.edu.

Wright State University, School of Graduate Studies, College of Science and Mathematics, Department of Geological Sciences, Program in Geological Sciences, Dayton, OH 45435. Offers environmental geochemistry (MS); environmental geology (MS); environmental sciences (MS); geological sciences (MS); geophysics (MS); hydrogeology (MS); petroleum geology (MS). Part-time programs available. *Degree requirements:* For master's, thesis. *Entrance requirements:* Additional exam requirements/recommendations for international students: Required—TOEFL.

Yale University, Graduate School of Arts and Sciences, Department of Forestry and Environmental Studies, New Haven, CT 06520. Offers environmental sciences (PhD); forestry (PhD). *Degree requirements:* For doctorate, thesis/dissertation. *Entrance requirements:* For doctorate, GRE General Test.

See Close-Up on page 713.

Yale University, School of Forestry and Environmental Studies, New Haven, CT 06511. Offers MES, MF, MFS, DFES, PhD, JD/MES, MBA/MES, MBA/MF, MES/MA, MES/MPH, MF/MA. *Accreditation:* SAF (one or more programs are accredited). Part-time programs available. Terminal master's awarded for partial completion of doctoral program. *Degree requirements:* For doctorate, thesis/dissertation. *Entrance requirements:* For master's and doctorate, GRE General Test. Expenses: Contact institution. *Faculty research:* Ecosystem science and management, coastal and watershed systems, environmental policy and management, social ecology and community development, conservation biology.

See Close-Up on page 713.

York University, Faculty of Graduate Studies, Faculty of Environmental Studies, Toronto, ON M3J 1P3, Canada. Offers MES, PhD, MES/LL B, MES/MA. Part-time programs available. *Faculty:* 39 full-time (17 women), 1 part-time/adjunct (0 women). *Students:* 356 full-time (251 women), 44 part-time (30 women). 442 applicants, 42% accepted, 176 enrolled. In 2005, 125 master's, 3 doctorates awarded. *Degree requirements:* For master's, thesis optional; for doctorate, thesis/dissertation, research seminar, comprehensive exam, registration. *Application deadline:* For fall admission, 3/1 for domestic students. Application fee: $80. Electronic applications accepted. *Expenses:* Tuition, state resident: full-time $3,190; part-time $798 per term. International tuition: $7,515 full-time. Required fees: $217. Tuition and fees vary according to program. *Financial support:* In 2005–06, 63 fellowships (averaging $9,785 per year), 56 research assistantships (averaging $4,989 per year), 121 teaching assistantships (averaging $10,474 per year) were awarded.; tuition waivers (partial) and fee bursaries also available. *Unit head:* Barbara Rahder, Director, 416-736-5252.

Youngstown State University, Graduate School, College of Arts and Sciences, Program in Environmental Studies, Youngstown, OH 44555-0001. Offers environmental studies (MS); industrial/institutional management (Certificate); risk management (Certificate). *Degree requirements:* For master's, thesis, minimum GPA of 3.0, oral defense of dissertation, comprehensive exam. *Entrance requirements:* For master's, GRE General Test or minimum GPA of 2.7. Additional exam requirements/recommendations for international students: Required—TOEFL.

Environmental Sciences

Alabama Agricultural and Mechanical University, School of Graduate Studies, School of Agricultural and Environmental Sciences, Department of Plant and Soil Sciences, Huntsville, AL 35811. Offers animal sciences (MS); environmental science (MS); plant and soil science (PhD). Evening/weekend programs available. Terminal master's awarded for partial completion of doctoral program. *Degree requirements:* For master's, thesis; for doctorate, one foreign language, thesis/dissertation. *Entrance requirements:* For master's, GRE General Test, BS in agriculture; for doctorate, GRE General Test, master's degree. Electronic applications accepted. *Faculty research:* Plant breeding, cytogenetics, crop production, soil chemistry and fertility, remote sensing.

Alaska Pacific University, Graduate Programs, Environmental Science Department, Program in Environmental Science, Anchorage, AK 99508-4672. Offers MSES. Part-time programs available. *Faculty:* 3 full-time (1 woman), 1 part-time/adjunct (0 women). *Students:* 14 full-time (10 women), 10 part-time (8 women); includes 1 American Indian/Alaska Native, 1 international. Average age 30. In 2005, 4 degrees awarded. *Degree requirements:* For master's, thesis. *Entrance requirements:* For master's, GRE General Test, minimum GPA of 3.0. Additional exam requirements/recommendations for international students: Required—TOEFL (minimum score 550 paper-based; 213 computer-based). *Application deadline:* For fall admission, 4/1 priority date for domestic students, 6/1 priority date for international students; for spring admission, 12/1 priority date for domestic students, 9/1 priority date for international students. Applications are processed on a rolling basis. Application fee: $25. *Expenses:* Tuition: Full-time $12,600; part-time $525 per credit. Required fees: $110 per semester. *Financial support:* In 2005–06, 8 research assistantships (averaging $2,534 per year) were awarded; career-related internships or fieldwork, Federal Work-Study, scholarships/grants, and unspecified assistantships also available. Support available to part-time students. Financial award application deadline:4/15. *Unit head:* Dr. Roman Dial, Director, 907-564-8296, Fax: 907-562-4276, E-mail: roman@alaskapacific.edu. *Application contact:* Michael Warner, Director of Admissions, 907-564-8248, Fax: 907-564-8317, E-mail: mikew@alaskapacific.edu.

American University, College of Arts and Sciences, Department of Biology, Environmental Science Program, Washington, DC 20016-8001. Offers environmental science (MS); marine science (MS). *Students:* 4 full-time (3 women), 1 (woman) part-time; includes 1 minority (Hispanic American), 1 international. Average age 28. In 2005, 6 degrees awarded. *Degree requirements:* For master's, thesis or alternative, comprehensive exam. *Entrance requirements:* For master's, GRE General Test, GRE Subject Test, minimum GPA of 3.0. Additional exam requirements/recommendations for international students: Required—TOEFL. *Application deadline:* For fall admission, 2/1 for domestic students; for spring admission, 10/1 for domestic students. Application fee: $50. *Expenses:* Tuition: Full-time $17,802; part-time $989 per credit. Required fees: $380. *Financial support:* Research assistantships, teaching assistantships available. Financial award application deadline: 2/1. *Unit head:* Dr. Kiho Kim, Director, 202-885-2181, Fax: 202-885-2181.

American University of Beirut, Graduate Programs, Faculty of Agricultural and Food Sciences, Beirut, Lebanon. Offers agricultural economics (MS); animal sciences (MS); ecosystem management (MSES); food technology (MS); irrigation (MS); mechanization (MS); nutrition (MS); plant protection (MS); plant science (MS); poultry science (MS); soils (MS). *Degree requirements:* For master's, one foreign language, thesis (for some programs), comprehensive exam, registration. *Entrance requirements:* For master's, GRE, letter of recommendation.

American University of Beirut, Graduate Programs, Faculty of Engineering and Architecture, Beirut, Lebanon. Offers civil engineering (ME); computer and communications engineering (ME); electrical engineering (ME); engineering management (MEM); environmental and water resources (ME); environmental technology (MSES); mechanical engineering (ME); urban design (MUD); urban planning (MUP). *Degree requirements:* For master's, one foreign language, thesis (for some programs), comprehensive exam, registration. *Entrance requirements:* For master's, GRE, letter of recommendation.

Antioch New England Graduate School, Graduate School, Department of Environmental Studies, Doctoral Program in Environmental Studies, Keene, NH 03431-3552. Offers PhD. *Degree requirements:* For doctorate, thesis/dissertation, practicum. *Entrance requirements:* For doctorate, master's degree and previous experience in the environmental field. Additional exam requirements/recommendations for international students: Required—TOEFL. Electronic applications accepted. Expenses: Contact institution. *Faculty research:* Environmental history, green politics, ecopsychology.

See Close-Up on page 641.

Environmental Sciences

Antioch New England Graduate School, Graduate School, Department of Environmental Studies, Program in Environmental Studies, Keene, NH 03431-3552. Offers conservation biology (MS); environmental advocacy (MS); environmental education (MS); teacher certification in biology (7th-12th grade) (MS); teacher certification in general science (5th-9th grade) (MS). *Degree requirements:* For master's, practicum. *Entrance requirements:* For master's, previous undergraduate course work in biology, chemistry, mathematics (environmental biology). Additional exam requirements/recommendations for international students: Required—TOEFL (minimum score 550 paper-based; 213 computer-based). Electronic applications accepted. Expenses: Contact institution. *Faculty research:* Sustainability, natural resources inventory.

See Close-Up on page 641.

Arkansas State University, Graduate School, College of Sciences and Mathematics, Program in Environmental Sciences, Jonesboro, State University, AR 72467. Offers PhD. Part-time programs available. *Faculty:* 1 full-time (0 women). *Students:* 21 full-time (10 women), 6 part-time (4 women); includes 1 minority (Asian American or Pacific Islander), 9 international. Average age 33. 7 applicants, 100% accepted, 7 enrolled. In 2005, 3 degrees awarded. *Degree requirements:* For doctorate, thesis/dissertation, comprehensive exam. *Entrance requirements:* For doctorate, GRE, interview, master's degree, letters of reference. Additional exam requirements/recommendations for international students: Required—TOEFL (minimum score 213 computer-based). *Application deadline:* For fall admission, 7/1 for domestic students; for spring admission, 11/15 priority date for domestic students. Applications are processed on a rolling basis. Application fee: $35. Electronic applications accepted. *Expenses:* Tuition, state resident: full-time $3,232; part-time $180 per hour. Tuition, nonresident: full-time $8,164; part-time $454 per hour. Required fees: $716; $37 per hour. $25 per semester. Tuition and fees vary according to course load and program. *Financial support:* Fellowships, research assistantships, scholarships/grants and unspecified assistantships available. Financial award application deadline: 7/1; financial award applicants required to submit FAFSA. *Unit head:* Dr. Robyn Hannigan, Director, 870-972-3469, Fax: 870-972-2008, E-mail: hannigan@astate.edu.

California State Polytechnic University, Pomona, Academic Affairs, College of Environmental Design, Program in Regenerative Studies, Pomona, CA 91768-2557. Offers MS. Part-time programs available. *Students:* 13 full-time (3 women), 2 part-time (1 woman); includes 5 minority (1 African American, 3 Asian Americans or Pacific Islanders, 1 Hispanic American). 17 applicants, 47% accepted, 7 enrolled. *Application deadline:* For fall admission, 5/1 for domestic students. For winter admission, 10/15 for domestic students; for spring admission, 1/20 for domestic students. Applications are processed on a rolling basis. Application fee: $55. Electronic applications accepted. *Expenses:* Tuition, nonresident: full-time $9,021. Required fees: $3,597. *Financial support:* Application deadline: 3/2; *Unit head:* Dr. Denise Lawrence, Graduate Coordinator, 909-869-2674.

California State University, Chico, Graduate School, College of Natural Sciences, Department of Geological and Environmental Sciences, Program in Environmental Science, Chico, CA 95929-0205. Offers MS. Part-time programs available. *Degree requirements:* For master's, thesis. *Entrance requirements:* For master's, GRE. Additional exam requirements/recommendations for international students: Required—TOEFL (minimum score 550 paper-based; 213 computer-based). Electronic applications accepted.

California State University, Fullerton, Graduate Studies, College of Humanities and Social Sciences, Program in Environmental Studies, Fullerton, CA 92834-9480. Offers environmental education and communication (MS); environmental policy and planning (MS); environmental sciences (MS); technological studies (MS). Part-time programs available. *Students:* 31 full-time (17 women), 43 part-time (22 women); includes 23 minority (2 African Americans, 10 Asian Americans or Pacific Islanders, 11 Hispanic Americans), 3 international. Average age 31. 40 applicants, 65% accepted, 18 enrolled. In 2005, 19 degrees awarded. *Degree requirements:* For master's, thesis. *Entrance requirements:* For master's, minimum GPA of 2.5 in last 60 units of course work. Application fee: $55. *Expenses:* Tuition, nonresident: part-time $339 per unit. *Financial support:* Career-related internships or fieldwork, Federal Work-Study, institutionally sponsored loans, and scholarships/grants available. Support available to part-time students. Financial award application deadline: 3/1. *Unit head:* Dr. Robert Voeks, Coordinator, 714-278-4373.

Christopher Newport University, Graduate Studies, Department of Biology, Chemistry and Environmental Science, Newport News, VA 23606-2998. Offers environmental science (MS). Part-time and evening/weekend programs available. *Degree requirements:* For master's, thesis, comprehensive exam. *Entrance requirements:* For master's, GRE General Test, minimum GPA of 3.0. Electronic applications accepted. *Faculty research:* Wetlands ecology and restoration, aquatic ecology, wetlands mitigation, greenhouse gases.

City College of the City University of New York, Graduate School, College of Liberal Arts and Science, Division of Science, Department of Earth and Atmospheric Sciences, New York, NY 10031-9198. Offers earth and environmental science (PhD); earth systems science (MA). *Students:* 10 applicants, 70% accepted, 5 enrolled. In 2005, 2 degrees awarded. *Degree requirements:* For master's, thesis, comprehensive exam. *Entrance requirements:* For master's, GRE, appropriate bachelor's degree. Additional exam requirements/recommendations for international students: Required—TOEFL (minimum score 500 paper-based; 173 computer-based). *Application deadline:* For fall admission, 5/1 for domestic students; for spring admission, 11/15 for domestic students. Application fee: $125. *Financial support:* Fellowships, career-related internships or fieldwork available. *Faculty research:* Water resources, high-temperature geochemistry, sedimentary basin analysis, tectonics. *Unit head:* Jeffrey Steiner, Chair, 212-650-6984, Fax: 212-650-6473, E-mail: steiner@sci.ccny.cuny.edu.

Clarkson University, Graduate School, School of Engineering, Program in Environmental Science and Engineering, Potsdam, NY 13699. Offers MS, PhD. Part-time programs available. *Students:* 15 full-time (6 women); includes 2 minority (1 African American, 1 Asian American or Pacific Islander), 7 international. Average age 27. 31 applicants, 52% accepted. *Expenses:* Tuition: Full-time $20,160; part-time $840 per hour. Required fees: $215. *Financial support:* In 2005–06, 10 students received support, including 8 research assistantships (averaging $19,032 per year), 2 teaching assistantships (averaging $19,032 per year); tuition waivers (partial) also available. *Faculty research:* Biological, chemical, physical and social systems. *Unit head:* Dr. Phillip K. Hopke, Acting Director, 315-268-3856, Fax: 315-268-4291.

Clemson University, Graduate School, College of Agriculture, Forestry and Life Sciences, Department of Environmental Toxicology, Clemson, SC 29634. Offers MS, PhD. *Students:* 25 full-time (14 women), 3 part-time (all women), 4 international. Average age 25. 21 applicants, 38% accepted, 6 enrolled. In 2005, 3 master's, 2 doctorates awarded. *Degree requirements:* For master's, thesis; for doctorate, one foreign language, thesis/dissertation. *Entrance requirements:* For master's and doctorate, GRE General Test. Additional exam requirements/recommendations for international students: Required—TOEFL. *Application deadline:* For fall admission, 6/1 for domestic students, 4/15 for international students. Application fee: $50. *Financial support:* Fellowships, research assistantships, teaching assistantships, career-related internships or fieldwork, Federal Work-Study, and institutionally sponsored loans available. Financial award applicants required to submit FAFSA. *Faculty research:* Biochemical toxicology, analytical toxicology, ecological risk assessment, wildlife toxicology, mathematical modeling. Total annual research expenditures: $3 million. *Unit head:* Dr. Patricia Layton, Chair, 864-656-3303, Fax: 864-656-3304, E-mail: playton@clemson.edu. *Application contact:* Dr. Thomas Swindler, Coordinator, 864-656-2810, E-mail: tschwdl@clemson.edu.

See Close-Ups on pages 645 and 647.

Cleveland State University, College of Graduate Studies, College of Science, Department of Biological, Geological, and Environmental Sciences, Cleveland, OH 44115. Offers biology (MS); environmental science (MS); molecular medicine (PhD); regulatory biology (PhD). Part-time programs available. *Faculty:* 19 full-time (3 women), 34 part-time/adjunct (7 women). *Students:* 54 full-time (34 women), 34 part-time (17 women); includes 8 minority (6 African Americans, 2 Asian Americans or Pacific Islanders), 32 international. Average age 30. 34 applicants, 65% accepted, 18 enrolled. In 2005, 1 master's, 3 doctorates awarded. Terminal master's awarded for partial completion of doctoral program. *Median time to degree:* Of those who began their doctoral program in fall 1997, 100% received their degree in 8 years or less. *Degree requirements:* For master's, thesis (for some programs); for doctorate, thesis/dissertation, comprehensive exam. *Entrance requirements:* For master's and doctorate, GRE General Test, 2 letters of recommendation. Additional exam requirements/recommendations for international students: Required—TOEFL (minimum score 525 paper-based; 197 computer-based); Recommended—TSE. *Application deadline:* For fall admission, 4/1 priority date for domestic students, 4/1 priority date for international students; for spring admission, 12/1 priority date for domestic students. Applications are processed on a rolling basis. Application fee: $30. Electronic applications accepted. *Expenses:* Tuition, state resident: full-time $10,700. Tuition, nonresident: full-time $14,628. Tuition and fees vary according to program. *Financial support:* In 2005–06, 29 students received support, including research assistantships with full and partial tuition reimbursements available (averaging $16,500 per year), teaching assistantships with full and partial tuition reimbursements available (averaging $16,500 per year); institutionally sponsored loans and unspecified assistantships also available. *Faculty research:* Molecular and cell biology, immunology. *Unit head:* Dr. Michael Gates, Chairperson, 216-687-3917, Fax: 216-687-6972, E-mail: m.gates@csuohio.edu. *Application contact:* Dr. Jeffrey Dean, Graduate Program Director, 216-687-2440, Fax: 216-687-6972, E-mail: gpd.bges@csuohio.edu.

College of Charleston, Graduate School, School of Sciences and Mathematics, Program in Environmental Studies, Charleston, SC 29424-0001. Offers MS. *Faculty:* 33 full-time (8 women), 14 part-time/adjunct (5 women). *Students:* 58 full-time (35 women), 18 part-time (10 women); includes 3 minority (1 African American, 2 Hispanic Americans), 3 international. Average age 28. 58 applicants, 67% accepted, 27 enrolled. In 2005, 27 degrees awarded. *Entrance requirements:* For master's, GRE. Additional exam requirements/recommendations for international students: Required—TOEFL. *Application deadline:* For fall admission, 7/1 for domestic students; for spring admission, 11/1 for domestic students. Application fee: $35. Electronic applications accepted. *Expenses: Contact institution.* Tuition and fees vary according to course load. *Unit head:* Michael Katuna, Director, 843-727-6483, Fax: 843-953-5546. *Application contact:* Susan Hallatt, Assistant Director of Graduate Admissions, 843-953-5614, Fax: 843-953-1434, E-mail: hallatts@cofc.edu.

College of Staten Island of the City University of New York, Graduate Programs, Center for Environmental Science, Staten Island, NY 10314-6600. Offers MS. Part-time and evening/weekend programs available. *Faculty:* 4 full-time (1 woman). *Students:* Average age 34. 7 applicants, 86% accepted, 5 enrolled. In 2005, 5 degrees awarded. Terminal master's awarded for partial completion of doctoral program. *Degree requirements:* For master's, thesis. *Entrance requirements:* For master's, GRE General Test, 1 year of course work in each chemistry, physics, calculus, and ecology; minimum GPA of 3.0 in undergraduate science and engineering; overall B minus average; bachelor's degree in a natural science or engineering; interview. Additional exam requirements/recommendations for international students: Required—TOEFL (minimum score 550 paper-based; 213 computer-based). *Application deadline:* Applications are processed on a rolling basis. Application fee: $125. *Expenses:* Tuition, area resident: Full-time $3,200; part-time $270 per credit. Tuition, nonresident: full-time $500; part-time $500 per credit. Required fees: $328; $101 per semester. *Financial support:* In 2005–06, 7 students received support, including 1 fellowship with partial tuition reimbursement available (averaging $10,000 per year) Financial award applicants required to submit FAFSA. *Faculty research:* Metal storage in prey and digestion in predators to metal trophic transfer in estuarine food chains, evaluation of a trophic transfer potential: Effects of a predator's digestive processes on the bioavailability of metals within consumed prey. *Unit head:* Dr. Alfred Levine, Director, 718-982-3920, Fax: 718-982-3923, E-mail: envirscimasters@mail.csi.cuny.edu. *Application contact:* Emmanuel Esperance, Deputy director of Office of Recruitment and Admissions, 718-982-2259, Fax: 718-982-2500, E-mail: admissions@mail.csi.cuny.edu.

Colorado School of Mines, Graduate School, Division of Environmental Science and Engineering, Golden, CO 80401-1887. Offers MS, PhD. Part-time programs available. *Faculty:* 24 full-time (7 women), 16 part-time/adjunct (10 women). *Students:* 46 full-time (23 women), 32 part-time (16 women); includes 5 minority (2 Asian Americans or Pacific Islanders, 3 Hispanic Americans), 6 international. 60 applicants, 60% accepted, 19 enrolled. In 2005, 25 master's, 4 doctorates awarded. *Degree requirements:* For master's, thesis (for some programs); for doctorate, thesis/dissertation, comprehensive exam. *Entrance requirements:* For master's and doctorate, GRE General Test. Additional exam requirements/recommendations for international students: Required—TOEFL (minimum score 550 paper-based; 213 computer-based). *Application deadline:* For fall admission, 1/1 priority date for domestic students, 1/1 priority date for international students; for spring admission, 9/1 priority date for domestic students, 9/1 priority date for international students. Application fee: $50. Electronic applications accepted. *Expenses:* Tuition, state resident: full-time $7,240; part-time $362 per credit hour. Tuition, nonresident: full-time $19,840; part-time $992 per credit hour. Required fees: $895. *Financial support:* In 2005–06, 33 students received support, including 5 fellowships with full tuition reimbursements available (averaging $9,600 per year), 20 research assistantships with full tuition reimbursements available (averaging $9,600 per year), 6 teaching assistantships with full tuition reimbursements available (averaging $9,600 per year); scholarships/grants, health care benefits, and unspecified assistantships also available. Financial award applicants required to submit FAFSA. *Faculty research:* Treatment of water and wastes, environmental law–policy and practice, natural environment systems, hazardous waste management, environmental data analysis. Total annual research expenditures: $2 million. *Unit head:* Dr. Robert Seigrist, Director, 303-273-3473, Fax: 303-273-3413, E-mail: rseigris@mines.edu. *Application contact:* Tim VanHaverbeke, Coordinator, 303-273-3467, Fax: 303-273-3413, E-mail: tvanhave@mines.edu.

Columbus State University, Graduate Studies, College of Science, Department of Environmental Science and Public Health, Columbus, GA 31907-5645. Offers environmental science (MS). Part-time and evening/weekend programs available. *Faculty:* 6 full-time (3 women), 1 part-time/adjunct (0 women). *Students:* 2 full-time (both women), 8 part-time (5 women); includes 3 minority (all African Americans), 1 international. Average age 37. 7 applicants, 43% accepted, 2 enrolled. In 2005, 7 degrees awarded. *Degree requirements:* For master's, thesis. *Entrance requirements:* For master's, GRE General Test, minimum GPA of 3.0. Additional exam requirements/recommendations for international students: Required—TOEFL (minimum score 550 paper-based; 213 computer-based). *Application deadline:* For fall admission, 5/1 priority date for domestic students, 5/1 priority date for international students; for spring admission, 11/1 for domestic students, 11/1 for international students. Applications are processed on a rolling basis. Application fee: $25. Electronic applications accepted. *Expenses:* Tuition, state resident: part-time $122 per semester hour. Tuition, nonresident: part-time $488 per semester hour. Required fees: $189 per term. Tuition and fees vary according to course load and program. *Financial support:* In 2005–06, 3 students received support, including 4 research assistantships with partial tuition reimbursements available (averaging $3,000 per year); career-related internships or fieldwork, Federal Work-Study, institutionally sponsored loans, scholarships/grants, and unspecified assistantships also available. Support available to part-time students. Financial award application deadline: 5/1; financial award applicants required to submit FAFSA. *Application contact:* Katie Thornton, Graduate Admissions Specialist, 706-568-2035, Fax: 706-568-2462, E-mail: thornton_katie@colstate.edu.

Cornell University, Graduate School, Graduate Fields of Agriculture and Life Sciences, Field of Soil and Crop Sciences, Ithaca, NY 14853-0001. Offers agronomy (MS, PhD); environmental information science (MS, PhD); environmental management (MPS); field crop science (MS, PhD); soil science (MS, PhD). *Faculty:* 38 full-time (9 women). *Students:* 40 applicants, 33% accepted, 11 enrolled. In 2005, 6 master's, 3 doctorates awarded. *Degree requirements:* For master's, thesis (MS); for doctorate, thesis/dissertation, comprehensive exam. *Entrance requirements:* For master's and doctorate, GRE General Test, 2 letters of recommendation. Additional exam requirements/recommendations for international students: Required—TOEFL (minimum score 550 paper-based; 213 computer-based). *Application deadline:* For fall admission, 2/1 for domestic students. Applications are processed on a rolling basis. Application fee: $60. Electronic applications accepted. *Financial support:* In 2005–06, 35 students received support, including 6 fellowships with full tuition reimbursements available, 25 research assistantships with full tuition reimbursements available, 4 teaching assistantships with full tuition reimbursements available; institutionally sponsored loans, traineeships, health care benefits, tuition waivers (full and partial), and unspecified assistantships also available. *Faculty research:* Soil chemistry, physics and biology; crop physiology and management; environmental information science and modeling; international agriculture; weed science. *Unit head:* Director of Graduate Studies, 607-255-3267, Fax: 607-255-8615. *Application contact:* Graduate Field Assistant, 607-255-3267, Fax: 607-255-8615, E-mail: jae2@cornell.edu.

Drexel University, College of Arts and Sciences, Program in Environmental Science, Philadelphia, PA 19104-2875. Offers MS, PhD. Part-time and evening/weekend programs available. Terminal master's awarded for partial completion of doctoral program. *Degree requirements:* For master's, thesis optional; for doctorate, thesis/dissertation. Electronic applications accepted.

Duke University, Graduate School, Department of Environment, Durham, NC 27708. Offers natural resource economics/policy (AM, PhD); natural resource science/ecology (AM, PhD); natural resource systems science (AM, PhD). Part-time programs available. *Faculty:* 36 full-time. *Students:* 71 full-time (44 women); includes 4 minority (3 African Americans, 1 Hispanic American), 26 international. 135 applicants, 16% accepted, 14 enrolled. In 2005, 3 master's, 9 doctorates awarded. *Degree requirements:* For doctorate, variable foreign language requirement, thesis/dissertation. *Entrance requirements:* For master's and doctorate, GRE General Test. Additional exam requirements/recommendations for international students: Required—IELT (preferred) or TOEFL. *Application deadline:* For fall admission, 12/31 for domestic students, 12/31 for international students. Application fee: $75. Electronic applications accepted. *Financial support:* Fellowships, research assistantships, teaching assistantships, Federal Work-Study available. Financial award application deadline: 12/31. *Unit head:* Kenneth Knoerr, Director of Graduate Studies, 919-613-8030, Fax: 919-684-8741, E-mail: nmm@duke.edu.

Duke University, Nicholas School of the Environment and Earth Sciences, Durham, NC 27708-0328. Offers coastal environmental management (MEM); environmental economics and policy (MEM); environmental health and security (MEM); forest resource management (MF); global environmental change (MEM); resource ecology (MEM); water and air resources (MEM). *Accreditation:* SAF (one or more programs are accredited). Part-time programs available. *Degree requirements:* For master's, thesis, registration. *Entrance requirements:* For master's, GRE General Test, previous course work in biology or ecology, calculus, statistics, and microeconomics; computer familiarity with word processing and data analysis. Additional exam requirements/recommendations for international students: Required—TOEFL (minimum score 550 paper-based; 213 computer-based). Electronic applications accepted. Expenses: Contact institution. *Faculty research:* Ecosystem management, conservation ecology, earth systems, risk assessment.

See Close-Up on page 709.

Duquesne University, Bayer School of Natural and Environmental Sciences, Environmental Science and Management Program, Pittsburgh, PA 15282-0001. Offers environmental management (Certificate); environmental science (Certificate); environmental science and management (MS). Part-time and evening/weekend programs available. Postbaccalaureate distance learning degree programs offered (minimal on-campus study). *Faculty:* 3 full-time (0 women), 11 part-time/adjunct (0 women). *Students:* 12 full-time (4 women), 21 part-time (13 women); includes 2 minority (both Asian Americans or Pacific Islanders), 2 international. Average age 28. 18 applicants, 67% accepted, 5 enrolled. In 2005, 23 degrees awarded. *Degree requirements:* For master's, thesis, comprehensive exam (for some programs), registration. *Entrance requirements:* For master's, GRE General Test, previous course work in biology, calculus, chemistry, and statistics. Additional exam requirements/recommendations for international students: Required—TOEFL, TSE. *Application deadline:* For fall admission, 4/1 priority date for domestic students, 4/1 priority date for international students; for spring admission, 10/1 priority date for domestic students, 10/1 priority date for international students. Applications are processed on a rolling basis. Application fee: $0. *Expenses: Contact institution.* Tuition and fees vary according to degree level and program. *Financial support:* In 2005–06, 1 fellowship with full tuition reimbursement (averaging $15,080 per year), 2 teaching assistantships with partial tuition reimbursements (averaging $10,000 per year) were awarded.; research assistantships, career-related internships or fieldwork, Federal Work-Study, institutionally sponsored loans, scholarships/grants, tuition waivers (partial), and unspecified assistantships also available. Support available to part-time students. Financial award application deadline: 5/1; financial award applicants required to submit FAFSA. *Faculty research:* Watershed management systems, environmental analytical chemistry, environmental endocrinology, environmental microbiology, aquatic biology. Total annual research expenditures:$200,750. *Unit head:* Robert D. Volkmar, Interim Director, 412-396-4094, Fax: 412-396-4092, E-mail: volkmar@duq.edu. *Application contact:* Mary Ann Quinn, Assistant to the Dean Graduate Affairs, 412-396-6339, Fax: 412-396-4881, E-mail: gradinfo@duq.edu.

Florida Agricultural and Mechanical University, Division of Graduate Studies, Research, and Continuing Education, Environmental Sciences Institute, Tallahassee, FL 32307-3200. Offers MS, PhD. *Degree requirements:* For master's, thesis/dissertation; for doctorate, thesis/dissertation, comprehensive exam. *Entrance requirements:* For master's and doctorate, GRE General Test, minimum GPA of 3.0. Additional exam requirements/recommendations for international students: Required—TOEFL. *Faculty research:* Statistical mechanics and quantum chemistry, aquatic microbial ecology, contaminant transport, modeling, bio-conversion of agricultural waste.

Florida Atlantic University, Charles E. Schmidt College of Science, Environmental Sciences Program, Boca Raton, FL 33431-0991. Offers MS. *Faculty:* 23 part-time/adjunct (4 women). *Students:* 10 full-time (7 women), 6 part-time (3 women); includes 1 minority (Hispanic American), 2 international. Average age 35. 5 applicants, 80% accepted, 3 enrolled. In 2005, 3 degrees awarded. *Degree requirements:* For master's, thesis. *Entrance requirements:* For master's, GRE General Test, minimum GPA of 3.0. Additional exam requirements/recommendations for international students: Required—TOEFL. *Application deadline:* For fall admission, 6/1 for domestic students. Application fee: $30. *Expenses:* Tuition, state resident: full-time $4,394; part-time $244 per credit. Tuition, nonresident: full-time $16,441; part-time $912 per credit. *Financial support:* In 2005–06, 8 teaching assistantships were awarded; career-related internships or fieldwork and Federal Work-Study also available. *Faculty research:* Tropical and terrestrial ecology, coastal/marine/wetlands ecology, hydroecology, tropical botany. *Unit head:* John Volin, Director, 561-297-4473, Fax: 561-297-2067, E-mail: jvolin@fau.edu. *Application contact:* Gina Fourreau, Coordinator for Academic Programs, 561-297-2625, Fax: 561-297-2067, E-mail: gfourreau@fau.edu.

Florida Gulf Coast University, College of Arts and Sciences, Program in Environmental Science, Fort Myers, FL 33965-6565. Offers MS. Part-time programs available. *Entrance requirements:* For master's, GRE General Test, minimum GPA of 3.0. Electronic applications accepted. *Faculty research:* Political issues in environmental science, recycling, environmental friendly buildings, pathophysiology, immunotoxicology of marine organisms.

Florida Institute of Technology, Graduate Programs, College of Engineering, Department of Marine and Environmental Systems, Melbourne, FL 32901-6975. Offers environmental resource management (MS); environmental science (MS, PhD); meteorology (MS); ocean engineering (MS, PhD); oceanography (MS, PhD), including biological oceanography (MS), chemical oceanography (MS), coastal zone management (MS), geological oceanography (MS), oceanography (PhD), physical oceanography (MS). Part-time programs available. *Faculty:* 11 full-time (1 woman). *Students:* 40 full-time (15 women), 20 part-time (13 women); includes 2 minority (both Hispanic Americans), 15 international. Average age 29. 101 applicants, 50% accepted, 15 enrolled. In 2005, 16 master's awarded. Terminal master's awarded for partial completion of doctoral program. *Degree requirements:* For master's, thesis, comprehensive exam, registration; for doctorate, one foreign language, thesis/dissertation, attendance of graduate seminar, internships (oceanography and environmental science), publications, comprehensive exam, registration. *Entrance requirements:* For master's, GRE General Test (environmental science), 3 letters of recommendation, minimum GPA of 3.0; for doctorate, GRE General Test (oceanography and environmental science), resumé, 3 letters of recommendation, minimum GPA of 3.2. Additional exam requirements/recommendations for international students: Required—TOEFL (minimum score 550 paper-based; 213 computer-based). *Application deadline:* Applications are processed on a rolling basis. Application fee: $50. Electronic applications accepted. *Expenses:* Tuition: Part-time $825 per credit. *Financial support:* In 2005–06, 18 students received support, including 1 fellowship with full and partial tuition reimbursement available (averaging $1,064 per year), 8 research assistantships with full and partial tuition reimbursements available (averaging $4,116 per year), 9 teaching assistantships with full and partial tuition reimbursements available (averaging $6,867 per year); career-related internships or fieldwork and tuition remissions also available. Financial award application deadline: 3/1; financial award applicants required to submit FAFSA. *Faculty research:* Environmental modeling, coastal processes, exploring marine pollution, marine geophysics, remote sensing . Total annual research expenditures: $1 million. *Unit head:* Dr. George Maul, Department Head, 321-674-7453, Fax: 321-674-7212, E-mail: gmaul@fit.edu. *Application contact:* Carolyn P. Farrior, Director of Graduate Admissions, 321-674-7118, Fax: 321-723-9468, E-mail: cfarrior@fit.edu.

See Close-Up on page 653.

Florida International University, College of Arts and Sciences, Department of Environmental Studies, Miami, FL 33199. Offers biological management (MS); energy (MS); pollution (MS). *Faculty:* 11 full-time (2 women). *Students:* 18 full-time (16 women), 18 part-time (9 women); includes 7 minority (1 African American, 1 American Indian/Alaska Native, 2 Asian Americans or Pacific Islanders, 3 Hispanic Americans), 7 international. Average age 32. 4 applicants, 50% accepted, 2 enrolled. In 2005, 9 degrees awarded. *Degree requirements:* For master's, thesis. *Entrance requirements:* For master's, GRE General Test, minimum GPA of 3.0, 3 letters of recommendation. Additional exam requirements/recommendations for international students: Required—TOEFL. *Application deadline:* For fall admission, 4/1 for domestic students; for spring admission, 10/30 for domestic students. Application fee: $25. *Expenses:* Tuition, area resident: Part-time $239 per credit. Tuition, state resident: full-time $4,294; part-time $869 per credit. Tuition, nonresident: full-time $15,641. Required fees: $252; $126 per term. Tuition and fees vary according to program. *Financial support:* Research assistantships, teaching assistantships available. Financial award application deadline: 4/1. *Unit head:* Dr. Joel Heinen, Chairperson, 305-348-3732, Fax: 305-348-6137, E-mail: joel.heinen@fiu.edu.

Gannon University, School of Graduate Studies, College of Sciences, Engineering, and Health Sciences, School of Sciences, Program in Environmental and Occupational Science and Health, Erie, PA 16541-0001. Offers Certificate. *Faculty:* 3 full-time (2 women). *Entrance requirements:* Additional exam requirements/recommendations for international students: Required—TOEFL (minimum score 500 paper-based; 173 computer-based). *Application deadline:* Applications are processed on a rolling basis. Application fee: $25. *Expenses:* Tuition: Full-time $11,430; part-time $635 per credit. Required fees: $496; $16 per credit. Tuition and fees vary according to course load, degree level and program. *Unit head:* Dr. Harry Diz, Director, 814-871-7633, E-mail: diz001@gannon.edu. *Application contact:* Debra Meszaros, Director of Graduate Recruitment, 814-871-5819, Fax: 814-871-5827, E-mail: cfal@gannon.edu.

George Mason University, College of Arts and Sciences, Department of Environmental Science and Public Policy, Fairfax, VA 22030. Offers MS, PhD. Part-time programs available. *Faculty:* 26 full-time (8 women), 13 part-time/adjunct (5 women). *Students:* 14 full-time (11 women), 92 part-time (40 women); includes 15 minority (4 African Americans, 6 Asian Americans or Pacific Islanders, 5 Hispanic Americans), 15 international. Average age 41. 78 applicants, 46% accepted, 25 enrolled. In 2005, 10 master's, 15 doctorates awarded. *Degree requirements:* For doctorate, thesis/dissertation, internship. *Entrance requirements:* For doctorate, GRE General Test, GRE Subject Test. *Application deadline:* For fall admission, 5/1 for domestic students; for spring admission, 11/1 for domestic students. Electronic applications accepted. *Expenses:* Tuition, area resident: Full-time $5,244; part-time $219 per credit. Tuition, state resident: part-time $651 per credit. Tuition, nonresident: full-time $15,636. Required fees: $1,524; $65 per credit. *Financial support:* Fellowships, research assistantships, teaching assistantships available. Support available to part-time students. Financial award application deadline: 3/1; financial award applicants required to submit FAFSA. *Unit head:* Dr. R. Christian Jones, Interim Director, 703-963-1127, Fax: 703-993-1046, E-mail: rcjones@gmu.edu.

Georgia Institute of Technology, Graduate Studies and Research, College of Sciences, School of Earth and Atmospheric Sciences, Atlanta, GA 30332-0001. Offers atmospheric chemistry and air pollution (MS, PhD); atmospheric dynamics and climate (MS, PhD); geochemistry (MS, PhD); hydrologic cycle (MS, PhD); ocean sciences (MS, PhD); solid-earth and environmental geophysics (MS, PhD). Part-time programs available. Terminal master's awarded for partial completion of doctoral program. *Degree requirements:* For master's, thesis or alternative; for doctorate, thesis/dissertation, comprehensive exam. *Entrance requirements:* For master's, GRE, minimum GPA of 3.0; for doctorate, GRE General Test, minimum GPA of 2.7. Additional exam requirements/recommendations for international students: Required—TOEFL (minimum score 550 paper-based; 213 computer-based). *Faculty research:* Geophysics, atmospheric chemistry, atmospheric dynamics, seismology.

See Close-Up on page 227.

Graduate School and University Center of the City University of New York, Graduate Studies, Program in Earth and Environmental Sciences, New York, NY 10016-4039. Offers PhD. *Faculty:* 36 full-time (5 women). *Students:* 58 full-time (22 women), 9 part-time (4 women); includes 9 minority (3 African Americans, 1 Asian American or Pacific Islander, 5 Hispanic Americans), 17 international. Average age 39. 24 applicants, 63% accepted, 9 enrolled. In 2005, 3 degrees awarded. *Degree requirements:* For doctorate, one foreign language, thesis/dissertation, comprehensive exam. *Entrance requirements:* For doctorate, GRE General Test. Additional exam requirements/recommendations for international students: Required—TOEFL. *Application deadline:* For fall admission, 4/15 for domestic students. Application fee: $125. Electronic applications accepted. *Financial support:* In 2005–06, 28 fellowships, 2 research assistantships, 1 teaching assistantship were awarded.; career-related internships or fieldwork, Federal Work-Study, institutionally sponsored loans, and tuition waivers (full and partial) also available. Financial award application deadline: 2/1; financial award applicants required to submit FAFSA. *Unit head:* Dr. Yehuda Klein, Executive Officer, 212-817-8241, Fax: 212-817-1513.

Harvard University, School of Public Health, Department of Environmental Health, Boston, MA 02115-6096. Offers environmental epidemiology (SM, DPH, SD); environmental health (SM); environmental science and engineering (SM, SD); occupational health (MOH, SM, DPH, SD); physiology (SD). *Accreditation:* ABET (one or more programs are accredited); CEPH. Part-time programs available. *Degree requirements:* For doctorate, thesis/dissertation, qualifying exam. *Entrance requirements:* For master's and doctorate, GRE. Additional exam requirements/recommendations for international students: Required—TOEFL (minimum score 560 paper-based; 220 computer-based); Recommended—IELT (minimum score 7). Electronic applications accepted. *Expenses:* Tuition: Full-time $28,752. Full-time tuition and fees vary according to program and student level. *Faculty research:* Industrial hygiene and occupational safety, population genetics, indoor and outdoor air pollution, cell and molecular biology of the lungs, infectious diseases.

Howard University, Graduate School of Arts and Sciences, Department of Chemistry, Washington, DC 20059-0002. Offers analytical chemistry (MS, PhD); atmospheric (MS, PhD); biochemistry (MS, PhD); environmental (MS, PhD); inorganic chemistry (MS, PhD); organic

Environmental Sciences

Howard University (continued)
chemistry (MS, PhD); physical chemistry (MS, PhD); polymer chemistry (MS, PhD). Part-time programs available. *Degree requirements:* For master's, one foreign language, thesis, teaching experience, comprehensive exam, registration; for doctorate, 2 foreign languages, thesis/dissertation, teaching experience, comprehensive exam, registration. *Entrance requirements:* For master's, GRE General Test, minimum GPA of 2.7; for doctorate, GRE General Test, minimum GPA of 3.0. *Faculty research:* Stratospheric aerosols, liquid crystals, polymer coatings, terrestrial and extraterrestrial atmospheres, amidogen reaction.

Humboldt State University, Graduate Studies, College of Natural Resources and Sciences, Programs in Environmental Systems, Arcata, CA 95521-8299. Offers MS. *Students:* 44 full-time (16 women), 11 part-time (3 women); includes 8 minority (1 American Indian/Alaska Native, 2 Asian Americans or Pacific Islanders, 5 Hispanic Americans). Average age 30. 39 applicants, 79% accepted, 18 enrolled. In 2005, 7 degrees awarded. *Degree requirements:* For master's, thesis. *Entrance requirements:* For master's, GRE, appropriate bachelor's degree, minimum GPA of 2.5. Additional exam requirements/recommendations for international students: Required—TOEFL. *Application deadline:* Applications are processed on a rolling basis. Application fee: $55. *Financial support:* Application deadline: 3/1; *Faculty research:* Mathematical modeling, international development technology, geology, environmental resources engineering. *Unit head:* Dr. Steven Martin, Chair, 707-826-4147, Fax: 707-826-4145, E-mail: srm1@humboldt.edu.

Hunter College of the City University of New York, Graduate School, School of Arts and Sciences, Department of Geography, New York, NY 10021-5085. Offers analytical geography (MA); earth system science (MA); environmental and social issues (MA); geographic information science (Certificate); geographic information systems (MA); teaching earth science (MA). Part-time and evening/weekend programs available. *Faculty:* 15 full-time (3 women), 4 part-time/adjunct (1 woman). *Students:* 3 full-time (1 woman), 33 part-time (15 women); includes 4 minority (2 African Americans, 2 Hispanic Americans). Average age 33. 14 applicants, 71% accepted, 8 enrolled. In 2005, 14 degrees awarded. *Degree requirements:* For master's, comprehensive exam or thesis. *Entrance requirements:* For master's, GRE General Test, minimum B average in major, minimum B- average overall, 18 credits of course work in geography, 2 letters of recommendation; for Certificate, minimum of B average in major, B- overall. Additional exam requirements/recommendations for international students: Required—TOEFL. *Application deadline:* For fall admission, 4/1 for domestic students; for spring admission, 11/1 for domestic students. Applications are processed on a rolling basis. Application fee: $125. *Expenses:* Tuition, state resident: full-time $6,400; part-time $270 per credit. Tuition, nonresident: part-time $500 per credit. International tuition: $12,000 full-time. Required fees: $50 per term. Part-time tuition and fees vary according to course load and program. *Financial support:* In 2005–06, 1 fellowship (averaging $3,000 per year), 2 research assistantships (averaging $10,000 per year), 10 teaching assistantships (averaging $6,000 per year) were awarded.; career-related internships or fieldwork, Federal Work-Study, institutionally sponsored loans, and unspecified assistantships also available. Financial award application deadline: 3/1. *Faculty research:* Urban geography, economic geography, geographic information science, demographic methods, climate change. *Unit head:* Prof. Marianna Pavlovskaya, Chair, 212-772-5320, Fax: 212-772-5268, E-mail: mpavlov@geo.hunter.cuny.edu. *Application contact:* Prof. Marianna Pavlovskaya, Graduate Adviser, 212-772-5320, Fax: 212-772-5268, E-mail: mpavlov@geo.hunter.cuny.edu.

Idaho State University, Office of Graduate Studies, Department of Interdisciplinary Studies, Pocatello, ID 83209. Offers general interdisciplinary (M Ed, MA, MNS); waste management and environmental science (MS). Part-time programs available. *Degree requirements:* For master's, thesis optional. *Entrance requirements:* For master's, GRE General Test or MAT, minimum GPA of 3.0. Additional exam requirements/recommendations for international students: Required—TOEFL (minimum score 550 paper-based; 213 computer-based).

Indiana University Bloomington, School of Public and Environmental Affairs, Environmental Science Programs, Bloomington, IN 47405-7000. Offers environmental science (MSES, PhD); hazardous materials management (Certificate). Part-time programs available. *Students:* 69 full-time (36 women), 25 part-time (10 women); includes 8 minority (2 African Americans, 3 Asian Americans or Pacific Islanders, 3 Hispanic Americans), 15 international. Average age 28. In 2005, 23 degrees awarded. Terminal master's awarded for partial completion of doctoral program. *Degree requirements:* For doctorate, thesis/dissertation. *Entrance requirements:* For master's and doctorate, GRE General Test. *Application deadline:* For fall admission, 2/1 priority date for domestic students, 1/15 priority date for international students. Applications are processed on a rolling basis. Application fee: $50 ($60 for international students). *Financial support:* Fellowships, research assistantships, teaching assistantships, career-related internships or fieldwork, Federal Work-Study, institutionally sponsored loans, and minority fellowships, Peace Corps assistantships available. Financial award application deadline: 2/1; financial award applicants required to submit FAFSA. *Faculty research:* Applied ecology, environmental chemistry, hazardous materials management, water resources. *Application contact:* Charles A. Johnson, Coordinator of Student Recruitment, 800-765-7755, Fax: 812-855-7802, E-mail: speainfo@indiana.edu.

See Close-Up on page 655.

Indiana University Northwest, School of Public and Environmental Affairs, Gary, IN 46408-1197. Offers criminal justice (MPA); environmental affairs (Certificate); health services administration (MPA); human services administration (MPA); nonprofit management (Certificate); public administration (MPA); public management (MPA, Certificate). *Accreditation:* NASPAA (one or more programs are accredited). Part-time programs available. *Faculty:* 5 full-time (3 women). *Students:* 26 full-time (20 women), 109 part-time (81 women); includes 87 minority (75 African Americans, 2 Asian Americans or Pacific Islanders, 10 Hispanic Americans). Average age 37. In 2005, 30 master's, 43 other advanced degrees awarded. *Degree requirements:* For master's, registration. *Entrance requirements:* For master's, GRE General Test or GMAT, letters of recommendation. *Application deadline:* For fall admission, 8/15 for domestic students. Applications are processed on a rolling basis. Application fee: $25. *Expenses:* Tuition, state resident: full-time $4,114; part-time $17,140 per credit hour. Tuition, nonresident: full-time $9,574; part-time $399 per credit hour. Required fees: $427. Tuition and fees vary according to campus/location and program. *Financial support:* Career-related internships or fieldwork, Federal Work-Study, and tuition waivers (partial) available. Support available to part-time students. Financial award application deadline: 3/1. *Faculty research:* Employment in income security policies, evidence in criminal justice, equal employment law, social welfare policy and welfare reform, public finance in developing countries. *Unit head:* Richard Hug, Interim Director, 219-980-6695, Fax: 219-980-6737. *Application contact:* Sandra Hall Smith, Secretary, 219-980-6695, Fax: 219-980-6737, E-mail: shsmith@iun.edu.

Instituto Tecnologico de Santo Domingo, Graduate School, Santo Domingo, Dominican Republic. Offers education (M Ed); engineering (M Eng); environmental science (M En S); management (M Mgmt).

Instituto Tecnológico y de Estudios Superiores de Monterrey, Campus Ciudad de México, Virtual University Division, Ciudad de Mexico, Mexico. Offers administration of information technologies (MA); computer sciences (MA); education (MA, PhD); educational technology (MA); environmental engineering (MA); environmental systems (MA); humanistics studies (MA); industrial engineering (MA); international business for Latin America (MA); quality systems (MA); quality systems and productivity (MA). Part-time and evening/weekend programs available. Postbaccalaureate distance learning degree programs offered (minimal on-campus study). *Entrance requirements:* For master's and doctorate, Instituto entrance exam. Additional exam requirements/recommendations for international students: Required—TOEFL.

Inter American University of Puerto Rico, San Germán Campus, Graduate Studies Center, Graduate Program in Environmental Sciences, San Germán, PR 00683-5008. Offers MS. Part-time and evening/weekend programs available. *Faculty:* 7 full-time, 3 part-time/adjunct. *Students:* 55. Average age 27. In 2005, 2 degrees awarded. *Degree requirements:* For master's, thesis, comprehensive exam. *Entrance requirements:* For master's, GRE General Test or EXADEP, minimum GPA of 3.0. *Application deadline:* For fall admission, 4/30 for domestic students; for spring admission, 11/15 for domestic students. Applications are processed on a rolling basis. Application fee: $31. Master $170/credit Ph.D Bus. Adm. $410/credit Ph.D (Psychology) $270/credit. *Expenses:* Tuition: Full-time $3,060; part-time $170 per credit. Required fees: $418; $418 per year. *Financial support:* Fellowships, research assistantships, teaching assistantships available. *Faculty research:* Environmental biology, environmental chemistry, water resources and unit operations. *Application contact:* Prof. Robin Waker, Graduate Coordinator, 787-264-1912 Ext. 7472, Fax: 787-892-7510, E-mail: rwaker@sg.inter.edu.

Iowa State University of Science and Technology, Graduate College, Interdisciplinary Programs, Program in Environmental Sciences, Ames, IA 50011. Offers MS, PhD. *Students:* 17 full-time (9 women), 5 part-time; includes 1 minority (African American), 13 international. 8 applicants, 0% accepted. In 2005, 1 master's, 2 doctorates awarded. *Degree requirements:* For master's and doctorate, thesis/dissertation. *Entrance requirements:* Additional exam requirements/recommendations for international students: Required—IELTS or TOEFL. *Application deadline:* For fall admission, 1/1 priority date for domestic students, 1/1 priority date for international students. Application fee: $30 ($70 for international students). Electronic applications accepted. *Expenses:* Tuition, state resident: full-time $6,410. Tuition, nonresident: full-time $16,422. Tuition and fees vary according to program. *Financial support:* In 2005–06, 16 research assistantships with partial tuition reimbursements (averaging $15,276 per year), 1 teaching assistantship with partial tuition reimbursement (averaging $13,490 per year) were awarded.; scholarships/grants, health care benefits, and unspecified assistantships also available. *Unit head:* Dr. William Crumpton, Supervisory Committee Chair, 515-294-6518, Fax: 515-294-9573. *Application contact:* Charles Sauer, Information Contact, 515-294-6518, E-mail: eeboffice@iastate.edu.

Jackson State University, Graduate School, School of Science and Technology, Department of Biology, Jackson, MS 39217. Offers biology education (MST); environmental science (MS, PhD). Part-time and evening/weekend programs available. *Degree requirements:* For master's, thesis (alternative accepted for MST); for doctorate, thesis/dissertation, comprehensive exam. *Entrance requirements:* For master's, GRE General Test; for doctorate, MAT. Additional exam requirements/recommendations for international students: Required—TOEFL. *Faculty research:* Comparative studies on the carbohydrate composition of marine macroalgae, host-parasite relationship between the spruce budworm and entomepathogen fungus.

The Johns Hopkins University, Zanvyl Krieger School of Arts and Sciences, Advanced Academic Programs, Program in Environmental Sciences and Policy, Washington, DC 20036. Offers MS. *Expenses:* Tuition: Full-time $30,960. Tuition and fees vary according to degree level and program. *Unit head:* Eileen McGurty, Associate Program Chair. *Application contact:* Craig Jones, Admissions Coordinator, 202-452-1941, Fax: 202-452-1970, E-mail: aapadmissions@jhu.edu.

Lehigh University, College of Arts and Sciences, Department of Earth and Environmental Sciences, Bethlehem, PA 18015-3094. Offers MS, PhD. *Faculty:* 12 full-time (1 woman), 1 (woman) part-time/adjunct. *Students:* 24 full-time (10 women), 5 part-time (1 woman), 4 international. Average age 26. 30 applicants, 27% accepted, 2 enrolled. In 2005, 3 master's, 1 doctorate awarded. Terminal master's awarded for partial completion of doctoral program. *Degree requirements:* For master's, thesis, registration; for doctorate, thesis/dissertation, language at the discretion of the PhD committee, comprehensive exam, registration. *Entrance requirements:* For master's and doctorate, GRE General Test, 2 letters of recommendation. Additional exam requirements/recommendations for international students: Required—TOEFL. *Application deadline:* For fall admission, 1/15 for domestic students; for spring admission, 10/15 priority date for domestic students. Applications are processed on a rolling basis. Application fee: $60. *Financial support:* In 2005–06, 5 fellowships with full tuition reimbursements (averaging $13,670 per year), 16 research assistantships with full tuition reimbursements (averaging $13,670 per year), 10 teaching assistantships with full tuition reimbursements (averaging $13,670 per year) were awarded.; Federal Work-Study, institutionally sponsored loans, and tuition waivers (full and partial) also available. Financial award application deadline: 1/15. *Faculty research:* Tectonics, surficial processes, aquatic ecology. Total annual research expenditures: $1.5 million. *Unit head:* Dr. Peter K. Zeitler, Chairman, 610-758-3660 Ext. 3671, Fax: 610-758-3677, E-mail: pkz0@lehigh.edu. *Application contact:* Dr. Gray E. Bebout, Graduate Coordinator, 610-758-3660 Ext. 5831, Fax: 610-758-3677, E-mail: geb0@lehigh.edu.

Long Island University, C.W. Post Campus, College of Liberal Arts and Sciences, Program in Environmental Studies, Brookville, NY 11548-1300. Offers environmental management (MS); environmental science (MS). Part-time and evening/weekend programs available. *Degree requirements:* For master's, internship or thesis. *Entrance requirements:* For master's, 1 year of course work in general chemistry and biology or geology; 1 semester in organic chemistry; computer proficiency. Electronic applications accepted. *Faculty research:* Symbiotic algae, local marine organisms, coastal processes, global tectonics, paleomagnetism.

Louisiana State University and Agricultural and Mechanical College, Graduate School, College of Agriculture, School of Renewable Natural Resources, Baton Rouge, LA 70803. Offers fisheries (MS); forestry (MS, PhD); wildlife (MS); wildlife and fisheries science (PhD). *Faculty:* 29 full-time (1 woman). *Students:* 67 full-time (24 women), 7 part-time (2 women); includes 1 African American, 1 Asian American or Pacific Islander, 3 Hispanic Americans, 20 international. Average age 28. 36 applicants, 56% accepted, 16 enrolled. In 2005, 13 master's, 6 doctorates awarded. *Degree requirements:* For master's and doctorate, thesis/dissertation. *Entrance requirements:* For master's, GRE General Test, minimum GPA of 3.0; for doctorate, GRE General Test, MS, minimum GPA of 3.0. Additional exam requirements/recommendations for international students: Required—TOEFL (minimum score 550 paper-based; 213 computer-based). *Application deadline:* For fall admission, 1/25 priority date for domestic students, 5/15 priority date for international students. Applications are processed on a rolling basis. Application fee: $25. Electronic applications accepted. *Financial support:* In 2005–06, 70 students received support, including 4 fellowships (averaging $16,304 per year), 59 research assistantships with partial tuition reimbursements available (averaging $16,539 per year); teaching assistantships with partial tuition reimbursements available, Federal Work-Study, institutionally sponsored loans, scholarships/grants, tuition waivers (full and partial), and unspecified assistantships also available. Financial award application deadline: 4/15; financial award applicants required to submit FAFSA. *Faculty research:* Forest biology and management, aquaculture, fisheries biology and ecology, upland and wetlands wildlife. Total annual research expenditures: $4,423. *Unit head:* Dr. Bob G. Blackmon, Director, 225-578-4131, Fax: 225-578-4227, E-mail: bblackmon@agctr.lsu.edu. *Application contact:* Dr. Allen Rutherford, Coordinator of Graduate Studies, 225-578-4187, Fax: 225-578-4227, E-mail: druther@lsu.edu.

See Close-Up on page 711.

Loyola Marymount University, Graduate Division, College of Science and Engineering, Department of Civil Engineering and Environmental Science, Program in Environmental Science, Los Angeles, CA 90045-2659. Offers MS. Part-time and evening/weekend programs available. *Students:* 4 full-time (2 women), 6 part-time (4 women); includes 6 minority (3 Asian Americans or Pacific Islanders, 3 Hispanic Americans). Average age 25. 7 applicants, 86% accepted, 4 enrolled. In 2005, 3 degrees awarded. *Degree requirements:* For master's, thesis or alternative, comprehensive exam. *Entrance requirements:* Additional exam requirements/recommendations for international students: Required—TOEFL (minimum score 550 paper-based; 213 computer-based). Application fee: $50. *Expenses:* Tuition: Full-time $13,140; part-time $730 per unit. Required fees: $100; $50 per semester. Full-time tuition and fees vary according to degree level and program. *Financial support:* In 2005–06, 2 students received support, including research assistantships (averaging $12,370 per year); Federal Work-Study, scholarships/grants, and unspecified assistantships also available. Support available to part-time students. Financial award application deadline: 6/1; financial award applicants required to submit FAFSA. *Unit head:* Prof. Joe Reichenberger, Director, Department of Civil Engineering and Environmental Science, 310-338-2830, Fax: 310-338-5896, E-mail: jreichen@lmu.edu.

Marshall University, Academic Affairs Division, Graduate College, College of Information, Technology and Engineering, Division of Environmental Science and Safety Technology, Program in Environmental Science and Safety Technology, Huntington, WV 25755. Offers MS. Part-time and evening/weekend programs available. *Students:* 15 full-time (8 women), 27 part-time (13 women); includes 2 minority (1 American Indian/Alaska Native, 1 Asian American or Pacific Islander), 5 international. Average age 31. In 2005, 13 degrees awarded. *Degree requirements:* For master's, final project, oral exam. *Entrance requirements:* For master's, GRE General Test or MAT, minimum GPA of 2.5, course work in calculus. *Financial support:* Tuition waivers (full) available. Support available to part-time students. Financial award application deadline: 8/1; financial award applicants required to submit FAFSA. *Application contact:* Information Contact, 304-746-1900, Fax: 304-746-1902, E-mail: services@marshall.edu.

Massachusetts Institute of Technology, School of Engineering, Department of Civil and Environmental Engineering, Cambridge, MA 02139-4307. Offers biological oceanography (PhD, Sc D); chemical oceanography (PhD, Sc D); civil and environmental engineering (M Eng, SM, PhD, Sc D, CE); civil and environmental systems (PhD, Sc D); civil engineering (PhD, Sc D); coastal engineering (PhD, Sc D); construction engineering and management (PhD, Sc D); environmental biology (PhD, Sc D); environmental chemistry (PhD, Sc D); environmental engineering (PhD, Sc D); environmental fluid mechanics (PhD, Sc D); geotechnical and geoenvironmental engineering (PhD, Sc D); hydrology (PhD, Sc D); information technology (PhD, Sc D); oceanographic engineering (PhD, Sc D); structures and materials (PhD, Sc D); transportation (PhD, Sc D). *Faculty:* 34 full-time (3 women). *Students:* 179 full-time (66 women), 1 part-time; includes 15 minority (1 African American, 9 Asian Americans or Pacific Islanders, 5 Hispanic Americans), 116 international. Average age 26. 374 applicants, 38% accepted, 71 enrolled. In 2005, 76 master's, 27 doctorates, 1 other advanced degree awarded. *Degree requirements:* For master's and CE, thesis/dissertation; for doctorate, thesis/dissertation, comprehensive exam. *Entrance requirements:* For master's and doctorate, GRE General Test. Additional exam requirements/recommendations for international students: Required—TOEFL (minimum score 577 paper-based; 233 computer-based). *Application deadline:* For fall admission, 1/2 for domestic students, 1/2 for international students. Application fee: $70. Electronic applications accepted. *Expenses:* Tuition: Full-time $32,100. Required fees: $200. Part-time tuition and fees vary according to course load. *Financial support:* In 2005–06, 134 students received support, including 38 fellowships with tuition reimbursements available (averaging $23,116 per year), 86 research assistantships with tuition reimbursements available (averaging $22,765 per year), 13 teaching assistantships with tuition reimbursements available (averaging $19,735 per year); career-related internships or fieldwork, Federal Work-Study, institutionally sponsored loans, scholarships/grants, health care benefits, and unspecified assistantships also available. Total annual research expenditures: $12.4 million. *Unit head:* Prof. Patrick Jaillet, Department Head, 617-452-3379, Fax: 617-452-3294, E-mail: jaillet@mit.edu. *Application contact:* Graduate Admissions, 617-253-7119, Fax: 617-258-6775, E-mail: cee-admissions@mit.edu.

McNeese State University, Graduate School, College of Science, Department of Biological and Environmental Sciences, Lake Charles, LA 70609. Offers environmental and chemical sciences (MS). Evening/weekend programs available. *Faculty:* 10 full-time (1 woman). *Students:* 27 full-time (14 women), 8 part-time (5 women); includes 11 minority (9 African Americans, 2 Hispanic Americans), 9 international. In 2005, 7 degrees awarded. *Degree requirements:* For master's, thesis or alternative, comprehensive exam. *Entrance requirements:* For master's, GRE General Test. *Application deadline:* For fall admission, 7/15 for domestic students. Applications are processed on a rolling basis. Application fee: $20 ($30 for international students). *Expenses:* Tuition, area resident: Part-time $193 per hour. Tuition, state resident: full-time $2,226. Required fees: $862; $106 per hour. Tuition and fees vary according to course load. *Financial support:* Application deadline: 5/1. *Unit head:* Dr. Mark L. Wygoda, Head, 337-475-5674, Fax: 337-475-5677, E-mail: mwygoda@mcneese.edu. *Application contact:* Dr. Harold Stevenson, Coordinator, 337-475-5663, Fax: 337-475-5677, E-mail: hstevens@mcneese.edu.

Memorial University of Newfoundland, School of Graduate Studies, Interdisciplinary Program in Environmental Science, St. John's, NL A1C 5S7, Canada. Offers M Env Sc, M Sc. Part-time programs available. *Students:* 33 full-time (24 women), 5 part-time (2 women), 2 international. 31 applicants, 26% accepted, 3 enrolled. In 2005, 6 degrees awarded. *Degree requirements:* For master's, thesis (M Sc), project (M Env Sci). *Entrance requirements:* For master's, honors B Sc or 2nd class B Eng. *Application deadline:* For fall admission, 5/1 for domestic students, 5/1 for international students. For winter admission, 9/1 for domestic students; for spring admission, 1/1 for domestic students. Applications are processed on a rolling basis. Application fee: $40 Canadian dollars. Electronic applications accepted. *Expenses:* Tuition: Part-time $733 per term. Tuition and fees vary according to degree level and program. *Financial support:* Fellowships, research assistantships, teaching assistantships available. Financial award application deadline: 3/1. *Faculty research:* Earth and ocean systems, environmental chemistry and toxicology, environmental engineering. *Unit head:* Dr. Chris Parrish, Interim Chair, 709-737-2331, Fax: 709-737-3018, E-mail: mcolbo@mun.ca. *Application contact:* Gail Kenny, Secretary, 709-737-8154, Fax: 709-737-3316, E-mail: gkenny@mun.ca.

Miami University, Graduate School, Institute of Environmental Sciences, Oxford, OH 45056. Offers M En S. Part-time programs available. *Degree requirements:* For master's, thesis, final exam, comprehensive exam. *Entrance requirements:* For master's, minimum undergraduate GPA of 3.0 during previous 2 years or 2.75 overall. Additional exam requirements/recommendations for international students: Required—TOEFL (minimum score 550 paper-based; 213 computer-based), TWE (minimum score 4). Electronic applications accepted.

Michigan State University, The Graduate School, College of Natural Science, Department of Geological Sciences, East Lansing, MI 48824. Offers environmental geosciences (MS, PhD); environmental geosciences-environmental toxicology (PhD); geological sciences (MS, PhD). *Faculty:* 11 full-time (1 woman). *Students:* 20 full-time (12 women), 4 part-time (1 woman); includes 2 minority (both Asian Americans or Pacific Islanders), 5 international. Average age 28. 37 applicants, 59% accepted. In 2005, 8 master's, 2 doctorates awarded. *Degree requirements:* For master's, thesis for those without prior thesis work; for doctorate, thesis/dissertation, registration. *Entrance requirements:* For master's, GRE General Test, minimum GPA of 3.0, course work in geoscience, 3 letters of recommendation; for doctorate, GRE General Test, 3 letters of recommendation. Additional exam requirements/recommendations for international students: Required—TOEFL (minimum score 550 paper-based; 213 computer-based), Michigan State Univeristy ELT (85), Michigan ELAB (83). *Application deadline:* For fall admission, 12/27 for domestic students. Application fee: $50. Electronic applications accepted. *Expenses:* Tuition, state resident: part-time $330 per credit hour. Tuition, nonresident: part-time $685 per credit hour. Tuition and fees vary according to program. *Financial support:* In 2005–06, 10 fellowships with tuition reimbursements (averaging $3,479 per year), 4 research assistantships with tuition reimbursements (averaging $12,710 per year), 13 teaching assistantships with tuition reimbursements (averaging $12,997 per year) were awarded.; Federal Work-Study, scholarships/grants, and unspecified assistantships also available. *Faculty research:* Water in the environment, global and biological change, crystal dynamics. Total annual research expenditures: $1.1 million. *Unit head:* Dr. Ralph E. Taggart, Chairperson, 517-355-4626, Fax: 517-353-8787, E-mail: taggart@msu.edu. *Application contact:* Information Contact, 517-355-4626, Fax: 517-353-8787, E-mail: geosci@msu.edu.

Minnesota State University Mankato, College of Graduate Studies, College of Science, Engineering and Technology, Department of Biological Sciences, Program in Environmental Science, Mankato, MN 56001. Offers MS. *Students:* 1 (woman) full-time, 1 (woman) part-time. Average age 31. In 2005, 2 degrees awarded. *Degree requirements:* For master's, one foreign language, thesis or alternative, comprehensive exam. *Entrance requirements:* For master's, minimum GPA of 3.0 during previous 2 years. Additional exam requirements/recommendations for international students: Required—TOEFL. *Application deadline:* For fall admission, 7/1 for domestic students; for spring admission, 11/1 for domestic students. Applications are processed on a rolling basis. Application fee: $40. Electronic applications accepted. *Expenses:* Tuition, state resident: part-time $243 per credit. Tuition, nonresident: part-time $400 per credit. Required fees: $30 per credit. *Financial support:* Research assistantships with partial tuition reimbursements, teaching assistantships with partial tuition reimbursements, career-related internships or fieldwork, Federal Work-Study, institutionally sponsored loans, and unspecified assistantships available. Financial award application deadline: 3/15; financial award applicants required to submit FAFSA. *Unit head:* Dr. Bertha Proctor, Graduate Coordinator, 507-389-5697. *Application contact:* 507-389-2321, E-mail: grad@mnsu.edu.

Montana State University, College of Graduate Studies, College of Agriculture, Department of Land Resources and Environmental Sciences, Bozeman, MT 59717. Offers land rehabilitation (interdisciplinary) (MS); land resources and environmental sciences (MS, PhD). Part-time programs available. *Faculty:* 16 full-time (1 woman), 4 part-time/adjunct (0 women). *Students:* 6 full-time (3 women), 43 part-time (23 women), 6 international. Average age 30. 15 applicants, 53% accepted, 8 enrolled. In 2005, 12 master's, 4 doctorates awarded. *Degree requirements:* For master's, thesis (for some programs), comprehensive exam, registration; for doctorate, thesis/dissertation, comprehensive exam, registration. *Entrance requirements:* For master's and doctorate, GRE General Test. Additional exam requirements/recommendations for international students: Required—TOEFL (minimum score 550 paper-based; 213 computer-based). *Application deadline:* For fall admission, 7/15 priority date for domestic students, 5/15 priority date for international students; for spring admission, 12/1 priority date for domestic students, 10/1 priority date for international students. Applications are processed on a rolling basis. Application fee: $30. Electronic applications accepted. *Expenses:* Tuition, state resident: full-time $4,132. Tuition, nonresident: full-time $1,132. *Financial support:* Health care benefits and tuition waivers (full) available. Financial award application deadline: 3/1; financial award applicants required to submit FAFSA. *Faculty research:* Watershed hydrology, soil remediation and nutrient management, agroecology and weed biology, restoration ecology, microbial diversity. Total annual research expenditures: $7.7 million. *Unit head:* Dr. John Wraith, Department Head, 406-994-4605, Fax: 406-994-3933, E-mail: jwraith@montana.edu.

Montclair State University, The Graduate School, College of Science and Mathematics, Department of Earth and Environmental Studies, Montclair, NJ 07043-1624. Offers environmental management (MA, D Env M); environmental studies (MS), including environmental education, environmental health, environmental management, environmental science; geoscience (MS, Certificate), including geoscience (MS), water resource management (Certificate). Part-time and evening/weekend programs available. *Faculty:* 17 full-time (3 women), 9 part-time/adjunct (2 women). *Students:* 21 full-time (13 women), 48 part-time (20 women); includes 14 minority (8 African Americans, 3 Asian Americans or Pacific Islanders, 3 Hispanic Americans), 1 international. 32 applicants, 78% accepted, 16 enrolled. In 2005, 19 master's, 3 other advanced degrees awarded. *Degree requirements:* For master's, thesis or alternative, comprehensive exam; for doctorate, thesis/dissertation. *Entrance requirements:* For master's, GRE General Test, 2 letters of recommendation. Additional exam requirements/recommendations for international students: Required—TOEFL (minimum score 83 computer-based). *Application deadline:* Applications are processed on a rolling basis. Application fee: $60. Electronic applications accepted. *Expenses:* Tuition: Full-time $3,001; part-time $409 per credit. Required fees: $56 per credit. Tuition and fees vary according to course load, degree level and program. *Financial support:* In 2005–06, 14 research assistantships with full tuition reimbursements were awarded; Federal Work-Study, scholarships/grants, and unspecified assistantships also available. Support available to part-time students. Financial award application deadline: 3/1; financial award applicants required to submit FAFSA. *Faculty research:* Antarctica, carbon pools, contaminated sediments, wetlands. Total annual research expenditures: $127,880. *Unit head:* Dr. Gregory Pope, Chairperson, 973-655-7385. *Application contact:* Dr. Harbans Singh, Adviser, 973-655-7383.

New Jersey Institute of Technology, Office of Graduate Studies, College of Science and Liberal Arts, Department of Chemistry and Environmental Science, Program in Environmental Science, Newark, NJ 07102. Offers MS, PhD. Part-time and evening/weekend programs available. *Students:* 20 full-time (13 women), 31 part-time (13 women); includes 11 minority (4 African Americans, 6 Asian Americans or Pacific Islanders, 1 Hispanic American), 14 international. Average age 31. 39 applicants, 56% accepted, 16 enrolled. In 2005, 10 master's, 3 doctorates awarded. *Degree requirements:* For doctorate, thesis/dissertation. *Entrance requirements:* For master's, GRE General Test; for doctorate, GRE General Test, minimum graduate GPA of 3.5. Additional exam requirements/recommendations for international students: Required—TOEFL (minimum score 550 paper-based; 213 computer-based). *Application deadline:* For fall admission, 6/5 for domestic students; for spring admission, 10/15 for domestic students. Applications are processed on a rolling basis. Application fee: $60. Electronic applications accepted. *Expenses:* Tuition, state resident: full-time $9,620; part-time $520 per credit. Tuition, nonresident: full-time $13,542; part-time $715 per credit. Required fees: $78; $54 per credit. $78 per year. Tuition and fees vary according to course load. *Financial support:* Fellowships with full and partial tuition reimbursements, research assistantships with full and partial tuition reimbursements, teaching assistantships with full and partial tuition reimbursements, career-related internships or fieldwork, Federal Work-Study, institutionally sponsored loans, and unspecified assistantships available. Financial award application deadline: 3/15. *Unit head:* Dr. David Kristol, Chair, 973-596-3584, Fax: 973-596-5222, E-mail: david.kristol@njit.edu. *Application contact:* Kathryn Kelly, Director of Admissions, 973-596-3300, Fax: 973-596-3461, E-mail: admissions@njit.edu.

New Mexico Institute of Mining and Technology, Graduate Studies, Department of Chemistry, Socorro, NM 87801. Offers biochemistry (MS); chemistry (MS); environmental chemistry (PhD); explosives technology and atmospheric chemistry (PhD). Part-time programs available. *Degree requirements:* For master's and doctorate, thesis/dissertation. *Entrance requirements:* For master's, GRE General Test; for doctorate, GRE General Test, GRE Subject Test. Additional exam requirements/recommendations for international students: Required—TOEFL (minimum score 540 paper-based; 207 computer-based). Electronic applications accepted. *Faculty research:* Organic, analytical, environmental, and explosives chemistry.

North Carolina Agricultural and Technical State University, Graduate School, School of Agriculture and Environmental and Allied Sciences, Greensboro, NC 27411. Offers MS. Part-time and evening/weekend programs available. *Degree requirements:* For master's, qualifying exam. *Entrance requirements:* For master's, GRE General Test. *Faculty research:* Aid for small farmers, agricultural technology, housing, food science, nutrition.

North Dakota State University, The Graduate School, College of Agriculture, Food Systems, and Natural Resources, Department of Soil Science, Fargo, ND 58105. Offers environmental and conservation science (PhD); environmental conservation science (MS); natural resource management (MS, PhD); soil sciences (MS, PhD). Part-time programs available. *Faculty:* 8 full-time (1 woman), 6 part-time/adjunct (0 women). *Students:* 7 full-time (2 women), 4 part-time (1 woman), 4 international. Average age 23. 3 applicants, 33% accepted, 1 enrolled. In 2005, 3 master's, 1 doctorate awarded. *Degree requirements:* For master's and doctorate, thesis/dissertation, classroom teaching, comprehensive exam, registration. *Entrance requirements:* For master's and doctorate, GRE General Test. Additional exam requirements/recommendations for international students: Required—TOEFL (minimum score 525 paper-based; 193 computer-based). *Application deadline:* Applications are processed on a rolling basis. Application fee: $45 ($60 for international students). Electronic applications accepted. *Financial support:* In 2005–06, 6 research assistantships with full tuition reimbursements (averaging $14,300 per year) were awarded; fellowships, Federal Work-Study, institutionally sponsored loans, and scholarships/grants also available. Financial award application deadline: 3/15. *Faculty research:* Microclimate, nitrogen management, landscape studies, water quality, soil management. *Unit head:* Dr. Rodney G. Lym, Interim Chair, 701-231-8903, Fax: 701-231-7861, E-mail: jimmie.richardson@ndsu.nodak.edu.

North Dakota State University, The Graduate School, College of Science and Mathematics, Department of Biological Sciences, Fargo, ND 58105. Offers biological sciences (MS); botany

Environmental Sciences

North Dakota State University (continued)

(MS, PhD); cellular and molecular biology (PhD); environmental and conservation sciences (MS, PhD); genomics (MS, PhD); natural resource management (MS, PhD); zoology (MS, PhD). *Faculty:* 20. *Students:* 30 full-time (10 women), 5 part-time (1 woman); includes 1 minority (Asian American or Pacific Islander), 3 international. Average age 24. 14 applicants, 43% accepted. In 2005, 4 master's awarded. *Degree requirements:* For master's and doctorate, thesis/dissertation. *Entrance requirements:* For master's and doctorate, GRE General Test. Additional exam requirements/recommendations for international students: Required—TOEFL. *Application deadline:* For fall admission, 3/15 for domestic students; for spring admission, 10/30 priority date for domestic students. Applications are processed on a rolling basis. Application fee: $45 ($60 for international students). Electronic applications accepted. *Financial support:* In 2005–06, 3 fellowships with full tuition reimbursements (averaging $15,000 per year), 9 research assistantships with full tuition reimbursements (averaging $14,400 per year), 19 teaching assistantships with full tuition reimbursements (averaging $9,550 per year) were awarded.; career-related internships or fieldwork, Federal Work-Study, institutionally sponsored loans, scholarships/grants, tuition waivers (full), and unspecified assistantships also available. Support available to part-time students. Financial award application deadline: 4/15; financial award applicants required to submit FAFSA. *Faculty research:* Comparative endocrinology, physiology, behavioral ecology, plant cell biology, aquatic biology. Total annual research expenditures: $675,000. *Unit head:* Dr. William J. Bleier, Chair, 701-231-7087, Fax: 701-231-7149, E-mail: william.bleier@ndsu.nodak.edu.

North Dakota State University, The Graduate School, Interdisciplinary Program in Environmental and Conservation Sciences, Fargo, ND 58105. Offers environmental and conservation sciences (PhD); environmental science (MS). *Faculty:* 59. *Degree requirements:* For master's, thesis, comprehensive exam. *Entrance requirements:* Additional exam requirements/recommendations for international students: Required—TOEFL (minimum score 550 paper-based). *Unit head:* Dr. Wei Lin, Director, 701-231-7244, Fax: 701-231-7149, E-mail: wei.lin@ndsu.edu. *Application contact:* Ruth Ann Faulkner, Administrative Assistant, 701-231-6727, E-mail: ruthann.faulkner@ndsu.edu.

Northern Arizona University, Graduate College, College of Engineering and Natural Science, Department of Environmental Sciences and Policy, Flagstaff, AZ 86011. Offers conservation ecology (Certificate); environmental sciences and policy (MS). *Accreditation:* NCA. *Degree requirements:* For master's, thesis optional. *Entrance requirements:* For master's, GRE General Test.

Nova Scotia Agricultural College, Research and Graduate Studies, Truro, NS B2N 5E3, Canada. Offers agriculture (M Sc), including air quality, animal behavior, animal molecular genetics, animal nutrition, animal technology, aquaculture, botany, crop management, crop physiology, ecology, environmental microbiology, food science, horticulture, nutrient management, pest management, physiology, plant biotechnology, plant pathology, soil chemistry, soil fertility, waste management and composting, water quality. Part-time programs available. *Faculty:* 43 full-time (7 women), 21 part-time/adjunct (1 woman). *Students:* 50 full-time (25 women), 15 part-time (11 women); includes 7 minority (3 African Americans, 3 Asian Americans or Pacific Islanders, 1 Hispanic American). Average age 25. In 2005, 23 degrees awarded. *Degree requirements:* For master's, thesis, registration. *Entrance requirements:* For master's, honors B Sc, minimum GPA of 3.0. Additional exam requirements/recommendations for international students: Required—TOEFL (minimum score 580 paper-based; 237 computer-based), Michigan English Language Assessment Battery, IELT, Can Test, CAEL. *Application deadline:* For fall admission, 6/1 for domestic students, 4/1 for international students. For winter admission, 10/31 for domestic students; for spring admission, 2/28 for domestic students. Applications are processed on a rolling basis. Application fee: $70. *Expenses:* Tuition, state resident: part-time $2,328 per year. Tuition, nonresident: full-time $6,984; part-time $7,968 per year. International tuition: $12,624 full-time. Required fees: $481; $46 per course. Tuition and fees vary according to program and student level. *Financial support:* In 2005–06, 48 students received support, including 7 fellowships (averaging $15,000 per year), 10 research assistantships (averaging $15,000 per year), 15 teaching assistantships (averaging $900 per year); career-related internships or fieldwork, scholarships/grants, and unspecified assistantships also available. *Facu y research:* Bio-product development, organic agriculture, nutrient management, air and water quality, agricultural biotechnology. Total annual research expenditures: $4.7 million. *Unit head:* Jill L. Rogers, Manager, 902-893-6360, Fax: 902-893-3430, E-mail: jrogers@nsac.ca. *Application contact:* Marie Law, Administrative Assistant, 902-893-6502, Fax: 902-893-3430, E-mail: mlaw@nsac.ca.

Nova Southeastern University, Oceanographic Center, Program in Marine Environmental Science, Fort Lauderdale, FL 33314-7796. Offers MS. *Faculty:* 15 full-time (1 woman), 5 part-time/adjunct (0 women). *Students:* 14 applicants, 79% accepted, 5 enrolled. In 2005, 1 degree awarded. *Degree requirements:* For master's, thesis. *Entrance requirements:* For master's, GRE. Additional exam requirements/recommendations for international students: Required—TOEFL (minimum score 550 paper-based). *Application deadline:* Applications are processed on a rolling basis. Application fee: $50. *Application contact:* Dr. Andrew Rogerson, Associate Dean, Director of Graduate Programs, 954-262-3600, Fax: 954-262-4020, E-mail: arogerso@nsu.nova.edu.

Oakland University, Graduate Study and Lifelong Learning, College of Arts and Sciences, Department of Chemistry, Rochester, MI 48309-4401. Offers chemistry (MS); health and environmental chemistry (PhD). *Faculty:* 4 full-time (2 women). *Students:* 16 full-time (8 women), 24 part-time (17 women); includes 2 minority (1 American Indian/Alaska Native, 1 Asian American or Pacific Islander), 13 international. Average age 31. 21 applicants, 90% accepted, 10 enrolled. In 2005, 12 master's awarded. *Degree requirements:* For master's and doctorate, thesis/dissertation. *Entrance requirements:* For master's, minimum GPA of 3.0 for unconditional admission; for doctorate, GRE Subject Test, minimum GPA of 3.0 for unconditional admission. *Application deadline:* For fall admission, 7/15 for domestic studentsFor winter admission, 12/1 for domestic students; for spring admission, 3/15 for domestic students. Applications are processed on a rolling basis. Application fee: $30. Electronic applications accepted. *Expenses:* Tuition, area resident: Full-time $9,192; part-time $383 per credit. Tuition, state resident: full-time $9,192; part-time $383 per credit. Tuition, nonresident: full-time $15,990; part-time $666 per credit. International tuition: $15,990 full-time. *Financial support:* Federal Work-Study, institutionally sponsored loans, and tuition waivers (full) available. Financial award application deadline: 3/1; financial award applicants required to submit FAFSA. *Faculty research:* Engineering self-assembling FVS for piezoimmunosensors, development of a novel GCxGC system, interactions in open-shell species, radiation damage to DNA-free radical mechanisms, Hydrophilic Xenoestrogens: Response and oxidation removal. Total annual research expenditures:$771,595. *Unit head:* Dr. Mark W. Severson, Chair, 248-370-2320, Fax: 248-370-2321, E-mail: severson@oakland.edu. *Application contact:* Dr. Kathleen W. Moore, Coordinator, 248-370-2338, Fax: 248-370-2321, E-mail: kmoore@oakland.edu.

OGI School of Science & Engineering at Oregon Health & Science University, Graduate Studies, Department of Environmental and Biomolecular Systems, Beaverton, OR 97006-8921. Offers biochemistry and molecular biology (MS, PhD); environmental health systems (MS); environmental information technology (MS, PhD); environmental science and engineering (MS, PhD). Part-time programs available. *Faculty:* 16 full-time (4 women), 5 part-time/adjunct (1 woman). *Students:* 35 full-time (19 women), 4 part-time (1 woman); includes 18 minority (2 African Americans, 16 Asian Americans or Pacific Islanders), 3 international. Average age 28. 73 applicants, 29% accepted, 12 enrolled. In 2005, 9 master's, 2 doctorates awarded. Terminal master's awarded for partial completion of doctoral program. *Median time to degree:* Of those who began their doctoral program in fall 1997, 100% received their degree in 8 years or less. *Degree requirements:* For master's, thesis optional; for doctorate, oral defense of dissertation. *Entrance requirements:* For master's and doctorate, GRE General Test. Additional exam requirements/recommendations for international students: Required—TOEFL. Application fee: $65. Electronic applications accepted. *Expenses:* Tuition, state resident: full-time $22,760; part-time $625 per credit. Required fees: $350. *Financial support:* In 2005–06, 4 fellowships with full and partial tuition reimbursements (averaging $16,500 per year), 22 research assistantships with full and partial tuition reimbursements (averaging $16,500 per year) were awarded.; teaching assistantships with full and partial tuition reimbursements, Federal Work-Study, scholarships/grants, and tuition waivers (full and partial) also available. Financial award application deadline: 2/15. *Faculty research:* Air and water science, hydrogeology, estuarine and coastal modeling, environmental microbiology, contaminant transport, biochemistry, biomolecular systems. Total annual research expenditures: $4.1 million. *Unit head:* Dr. Antonio M. Baptista, Head, 503-748-1147, Fax: 503-748-1273, E-mail: baptista@ccalmr.ogi.edu. *Application contact:* Nancy Christie, Information Contact, 800-748-1070, Fax: 503-748-1464, E-mail: christin@ohsu.edu.

The Ohio State University, Graduate School, College of Biological Sciences, Program in Environmental Science, Columbus, OH 43210. Offers MS, PhD. *Degree requirements:* For master's, one foreign language, thesis optional; for doctorate, one foreign language, thesis/dissertation. *Entrance requirements:* For master's and doctorate, GRE General Test. Additional exam requirements/recommendations for international students: Required—TOEFL (minimum score 600 paper-based; 250 computer-based), TSE. Electronic applications accepted.

Oklahoma State University, College of Agricultural Science and Natural Resources, Department of Horticulture and Landscape Architecture, Stillwater, OK 74078. Offers crop science (PhD); environmental science (PhD); food science (PhD); horticulture (M Ag, MS); plant science (PhD). *Faculty:* 21 full-time (1 woman), 4 part-time/adjunct (0 women). *Students:* 1 full-time (0 women), 6 part-time (2 women); includes 1 minority (African American), 4 international. Average age 31. 7 applicants, 14% accepted. In 2005, 3 master's awarded. *Degree requirements:* For master's, thesis or alternative. *Entrance requirements:* For master's and doctorate, GRE. Additional exam requirements/recommendations for international students: Required—TOEFL. *Application deadline:* For fall admission, 6/1 priority date for domestic students, 3/1 priority date for international students. Applications are processed on a rolling basis. Application fee: $40 ($75 for international students). Electronic applications accepted. *Expenses:* Tuition, state resident: full-time $4,253; part-time $139 per credit hour. Tuition, nonresident: full-time $12,569; part-time $485 per credit hour. Required fees: $43 per credit hour. One-time fee: $20 part-time. Tuition and fees vary according to course load and program. *Financial support:* In 2005–06, 4 research assistantships (averaging $13,817 per year) were awarded; teaching assistantships, career-related internships or fieldwork, Federal Work-Study, scholarships/grants, health care benefits, tuition waivers (partial), and unspecified assistantships also available. Support available to part-time students. Financial award application deadline: 3/1. *Faculty research:* Stress and postharvest physiology; water utilization and runoff; IPM systems and nursery, turf, floriculture, vegetable, net and fruit produces and natural resources, food extraction, and processing; public garden management. *Unit head:* Dr. Dale Maronek, Head, 405-744-5414, Fax: 405-744-9709, E-mail: maronek@okstate.edu.

Oklahoma State University, College of Arts and Sciences, Department of Botany, Stillwater, OK 74078. Offers botany (MS); environmental science (PhD); plant science (PhD). *Faculty:* 11 full-time (4 women). *Students:* 1 (woman) full-time, 7 part-time (3 women), 2 international. Average age 35. 6 applicants, 50% accepted, 2 enrolled. *Degree requirements:* For master's and doctorate, thesis/dissertation. *Entrance requirements:* For master's and doctorate, GRE General Test. Additional exam requirements/recommendations for international students: Required—TOEFL. *Application deadline:* For fall admission, 6/1 priority date for domestic students, 3/1 priority date for international students. Applications are processed on a rolling basis. Application fee: $40 ($75 for international students). Electronic applications accepted. *Expenses:* Tuition, state resident: full-time $4,253; part-time $139 per credit hour. Tuition, nonresident: full-time $12,569; part-time $485 per credit hour. Required fees: $43 per credit hour. One-time fee: $20 part-time. Tuition and fees vary according to course load and program. *Financial support:* In 2005–06, 5 research assistantships (averaging $14,339 per year), 11 teaching assistantships (averaging $16,755 per year) were awarded.; career-related internships or fieldwork, Federal Work-Study, scholarships/grants, health care benefits, tuition waivers (partial), and unspecified assistantships also available. Support available to part-time students. Financial award application deadline: 3/1. *Faculty research:* Ethnobotany, developmental genetics of Arabidopsis, biological roles of Plasmodesmata, community ecology and biodiversity, nutrient cycling in grassland ecosystems. *Unit head:* Dr. William J. Henley, Head, 405-744-5559, Fax: 405-744-7074.

Oklahoma State University, Graduate College, Interdisciplinary Program in Environmental Sciences, Stillwater, OK 74078. Offers MS, PhD. *Degree requirements:* For master's and doctorate, thesis/dissertation. *Entrance requirements:* For master's and doctorate, GRE, minimum GPA of 3.0. Additional exam requirements/recommendations for international students: Required—TOEFL. *Application deadline:* For fall admission, 7/1 for domestic students. Applications are processed on a rolling basis. Application fee: $25 ($50 for international students). Electronic applications accepted. *Expenses:* Tuition, state resident: full-time $4,253; part-time $139 per credit hour. Tuition, nonresident: full-time $12,569; part-time $485 per credit hour. Required fees: $43 per credit hour. One-time fee: $20 part-time. Tuition and fees vary according to course load and program. *Financial support:* Research assistantships, teaching assistantships, tuition waivers (partial) available. Support available to part-time students. Financial award application deadline: 3/1. *Unit head:* Dr. Will Focht, Director, 405-744-9229.

Oregon State University, Graduate School, College of Agricultural Sciences, Department of Environmental and Molecular Toxicology, Corvallis, OR 97331. Offers toxicology (MS, PhD). *Faculty:* 12 full-time (5 women), 5 part-time/adjunct (0 women). *Students:* 25 full-time (16 women), 3 part-time (1 woman); includes 6 minority (1 American Indian/Alaska Native, 2 Asian Americans or Pacific Islanders, 3 Hispanic Americans), 5 international. In 2005, 1 master's, 1 doctorate awarded. Application fee: $50. *Expenses:* Tuition, area resident: Part-time $301 per credit. Tuition, state resident: full-time $8,139; part-time $501 per credit. Tuition, nonresident: full-time $14,376; part-time $532 per credit. Required fees: $1,266. *Unit head:* Dr. Lawrence R. Curtis, Head, 541-737-1764, Fax: 541-737-0497, E-mail: larry.curtis@orst.edu.

Oregon State University, Graduate School, College of Science, Program in Environmental Sciences, Corvallis, OR 97331. Offers MA, MS, PhD. Application fee: $50. *Expenses:* Tuition, area resident: Part-time $301 per credit. Tuition, state resident: full-time $8,139; part-time $501 per credit. Tuition, nonresident: full-time $14,376; part-time $532 per credit. Required fees: $1,266. *Unit head:* Dr. Andy Blaustein, Director, 541-737-2404.

Pace University, Dyson College of Arts and Sciences, Program in Environmental Science, New York, NY 10038. Offers MS. *Degree requirements:* For master's, research project. *Entrance requirements:* For master's, GRE. Electronic applications accepted.

See Close-Up on page 659.

The Pennsylvania State University Harrisburg Campus, Graduate School, School of Science, Engineering and Technology, Program in Environmental Pollution Control, Middletown, PA 17057-4898. Offers M Eng, MEPC, MS, MS/JD. Evening/weekend programs available. *Degree requirements:* For master's, thesis. *Entrance requirements:* For master's, GRE General Test, minimum GPA of 2.75. Additional exam requirements/recommendations for international students: Required—TOEFL. *Expenses:* Tuition, state resident: full-time $12,518; part-time $522 per credit. Tuition, nonresident: full-time $17,592; part-time $733 per credit. Required fees: $936. Tuition and fees vary according to course load, degree level, campus/location and program.

The Pennsylvania State University University Park Campus, Graduate School, Intercollege Graduate Programs, Intercollege Program in Environmental Pollution Control, State College, University Park, PA 16802-1503. Offers MEPC, MS. *Students:* 8 full-time (5 women), 4 part-time (2 women), 2 international. *Entrance requirements:* For master's, GRE General Test. Additional exam requirements/recommendations for international students: Required—TOEFL. Application fee: $45. *Expenses:* Tuition, state resident: full-time $12,518; part-time $522 per credit. Tuition, nonresident: full-time $23,004; part-time $959 per credit. Required fees: $484.

Tuition and fees vary according to course load, campus/location and program. *Unit head:* Dr. Herschel A. Elliott, Chair, 814-865-1417, Fax: 814-863-1031, E-mail: haelliott@psu.edu. *Application contact:* Dr. Herschel A. Elliott, Chair, 814-865-1417, Fax: 814-863-1031, E-mail: haelliott@psu.edu.

Polytechnic University, Brooklyn Campus, Department of Civil and Environmental Engineering, Major in Environmental Science, Brooklyn, NY 11201-2990. Offers MS. Part-time and evening/weekend programs available. *Students:* 1 (woman) full-time, 2 part-time; includes 1 minority (African American), 1 international. Average age 32. 2 applicants, 100% accepted, 0 enrolled. *Degree requirements:* For master's, thesis (for some programs), comprehensive exam (for some programs), registration. *Application deadline:* For fall admission, 7/15 priority date for domestic students, 4/1 priority date for international students; for spring admission, 12/15 priority date for domestic students, 10/1 priority date for international students. Applications are processed on a rolling basis. Application fee: $55. Electronic applications accepted. *Expenses:* Tuition: Part-time $950 per unit. Required fees: $330 per semester. *Financial support:* Fellowships, research assistantships, teaching assistantships, institutionally sponsored loans available. Support available to part-time students. Financial award applicants required to submit FAFSA.

Portland State University, Graduate Studies, College of Liberal Arts and Sciences, Department of Geology, Portland, OR 97207-0751. Offers environmental sciences and resources (PhD); geology (MA, MS); science/geology (MAT, MST). Part-time programs available. *Faculty:* 9 full-time (2 women). *Students:* 14 full-time (8 women), 12 part-time (6 women). Average age 29. 19 applicants, 74% accepted, 5 enrolled. In 2005, 5 degrees awarded. *Degree requirements:* For master's, thesis, field comprehensive; for doctorate, thesis/dissertation, 2 years of residency. *Entrance requirements:* For master's, GRE General Test, GRE Subject Test, BA/BS in geology, minimum GPA of 3.0 in upper-division course work or 2.75 overall. Additional exam requirements/recommendations for international students: Required—TOEFL (minimum score 550 paper-based; 213 computer-based). *Application deadline:* 1/31 for domestic students, 1/31 for international students. Applications are processed on a rolling basis. Application fee: $50. *Expenses:* Tuition, state resident: full-time $6,648; part-time $231 per credit. Tuition, nonresident: full-time $11,319; part-time $231 per credit. Required fees: $686; $67 per credit. *Financial support:* In 2005–06, 4 research assistantships with full tuition reimbursements (averaging $14,951 per year), 6 teaching assistantships with full tuition reimbursements (averaging $10,001 per year) were awarded.; career-related internships or fieldwork, Federal Work-Study, scholarships/grants, and unspecified assistantships also available. Support available to part-time students. Financial award application deadline: 3/1; financial award applicants required to submit FAFSA. *Faculty research:* Sediment transport, volcanic environmental geology, coastal and fluvial processes. Total annual research expenditures: $1.6 million. *Unit head:* Dr. Michael L. Cummings, Head, 503-725-3022, Fax: 503-725-3025. *Application contact:* Nancy Eriksson, Office Coordinator, 503-725-3022, Fax: 503-725-3025, E-mail: erikssonn@pdx.edu.

Portland State University, Graduate Studies, College of Liberal Arts and Sciences, Interdisciplinary Program in Environmental Sciences and Resources, Portland, OR 97207-0751. Offers environmental management (MEM); environmental sciences/biology (PhD); environmental sciences/chemistry (PhD); environmental sciences/civil engineering (PhD); environmental sciences/geography (PhD); environmental sciences/geology (PhD); environmental sciences/physics (PhD); environmental studies (MS); science/environmental science (MST). Part-time programs available. *Faculty:* 8 full-time (0 women), 2 part-time/adjunct (1 woman). *Students:* 84 full-time (40 women), 37 part-time (17 women); includes 9 minority (1 African American, 3 Asian Americans or Pacific Islanders, 5 Hispanic Americans), 29 international. Average age 32. 82 applicants, 66% accepted, 33 enrolled. In 2005, 11 master's, 5 doctorates awarded. *Degree requirements:* For doctorate, variable foreign language requirement, thesis/dissertation, oral and qualifying exams. *Entrance requirements:* For doctorate, minimum GPA of 3.0 in upper-division course work or 2.75 overall. Additional exam requirements/recommendations for international students: Required—TOEFL (minimum score 550 paper-based; 213 computer-based). *Application deadline:* For fall admission, 4/1 priority date for domestic students, 3/1 priority date for international students. Applications are processed on a rolling basis. Application fee: $50. *Expenses:* Tuition, state resident: full-time $6,648; part-time $231 per credit. Tuition, nonresident: full-time $11,319; part-time $231 per credit. Required fees: $686; $67 per credit. *Financial support:* In 2005–06, 2 research assistantships with full tuition reimbursements (averaging $8,648 per year) were awarded; teaching assistantships with full tuition reimbursements, Federal Work-Study, scholarships/grants, tuition waivers (partial), and unspecified assistantships also available. Support available to part-time students. Financial award application deadline: 3/1; financial award applicants required to submit FAFSA. *Faculty research:* Environmental aspects of biology, chemistry, civil engineering, geology, physics. Total annual research expenditures: $1.6 million. *Unit head:* John Rueter, Director, 503-725-4980, Fax: 503-725-3888.

Queens College of the City University of New York, Division of Graduate Studies, Mathematics and Natural Sciences Division, School of Earth and Environmental Sciences, Flushing, NY 11367-1597. Offers MA. Part-time and evening/weekend programs available. *Faculty:* 14 full-time (4 women). *Students:* 12 applicants, 100% accepted, 8 enrolled. In 2005, 1 degree awarded. *Degree requirements:* For master's, thesis, comprehensive exam. *Entrance requirements:* For master's, GRE, previous course work in calculus, physics, and chemistry; minimum GPA of 3.0. Additional exam requirements/recommendations for international students: Required—TOEFL. *Application deadline:* For fall admission, 4/1 for domestic students; for spring admission, 11/1 for domestic students. Applications are processed on a rolling basis. Application fee: $125. *Expenses:* Tuition, state resident: part-time $270 per credit. Tuition, nonresident: part-time $500 per credit. Required fees: $112 per year. *Financial support:* Career-related internships or fieldwork, Federal Work-Study, institutionally sponsored loans, tuition waivers (partial), unspecified assistantships, and adjunct lectureships available. Support available to part-time students. Financial award application deadline: 4/1; financial award applicants required to submit FAFSA. *Faculty research:* Sedimentology/stratigraphy, paleontology, field petrology. *Unit head:* Dr. Daniel Habib, Chairperson, 718-997-3300, E-mail: daniel_habib@qc.edu. *Application contact:* Dr. Hannes Brueckner, Graduate Adviser, 718-997-3300, E-mail: hannes_brueckner@qc.edu.

Rensselaer Polytechnic Institute, Graduate School, School of Science, Department of Earth and Environmental Sciences, Troy, NY 12180-3590. Offers environmental chemistry (MS, PhD); geochemistry (MS, PhD); geology (MS, PhD); geophysics (MS, PhD); petrology (MS, PhD). Part-time programs available. Terminal master's awarded for partial completion of doctoral program. *Degree requirements:* For master's, thesis (for some programs), comprehensive exam; for doctorate, thesis/dissertation, comprehensive exam. *Entrance requirements:* For master's and doctorate, GRE General Test. Additional exam requirements/recommendations for international students: Required—TOEFL. Electronic applications accepted. *Expenses:* Tuition: Full-time $31,000; part-time $1,320 per credit. Required fees: $1,623. *Faculty research:* Mantel geochemistry, contaminant geochemistry, seismology, GPS geodesy, remote sensing petrology.

See Close-Up on page 661.

Rice University, Graduate Programs, George R. Brown School of Engineering, Department of Civil and Environmental Engineering, Houston, TX 77251-1892. Offers civil engineering (MCE, MS, PhD); environmental engineering (MEE, MES, MS, PhD); environmental science (MEE, MES, MS, PhD). Part-time programs available. *Degree requirements:* For master's, thesis (for some programs); for doctorate, thesis/dissertation. *Entrance requirements:* For master's and doctorate, GRE General Test, GRE Subject Test, minimum GPA of 3.25. Additional exam requirements/recommendations for international students: Required—TOEFL. Electronic applications accepted. *Faculty research:* Biology and chemistry of groundwater, pollutant fate in groundwater systems, water quality monitoring, urban storm water runoff, urban air quality.

Rochester Institute of Technology, Graduate Enrollment Services, College of Science, Department of Biological Sciences, Program in Environmental Science, Rochester, NY 14623-5603. Offers MS. *Students:* 12 full-time (6 women); includes 3 minority (1 African American, 1 Asian American or Pacific Islander, 1 Hispanic American), 1 international. 10 applicants, 60% accepted, 5 enrolled. In 2005, 2 degrees awarded. *Degree requirements:* For master's, thesis. *Entrance requirements:* For master's, minimum GPA of 3.0. Additional exam requirements/recommendations for international students: Required—TOEFL (minimum score 550 paper-based). *Application deadline:* For fall admission, 3/1 for domestic students. Application fee: $50. *Expenses:* Tuition: Full-time $25,392; part-time $713 per credit. Required fees: $183; $61 per term. *Unit head:* Dr. John Waud, Director, 585-475-2182, E-mail: jmwsci@rit.edu.

Royal Military College of Canada, Division of Graduate Studies and Research, Engineering Division, Program in Environmental Science, Kingston, ON K7K 7B4, Canada. Offers M Sc, PhD. *Degree requirements:* For master's, thesis/dissertation, registration; for doctorate, thesis/dissertation, comprehensive exam, registration. Electronic applications accepted.

Rutgers, The State University of New Jersey, Newark, Graduate School, Program in Environmental Science, Newark, NJ 07102. Offers MS, PhD. *Faculty:* 6 full-time (1 woman), 1 part-time/adjunct (0 women). *Students:* 10 full-time (5 women), 11 part-time (2 women); includes 11 minority (3 African Americans, 6 Asian Americans or Pacific Islanders, 2 Hispanic Americans). 22 applicants, 59% accepted, 9 enrolled. In 2005, 4 master's, 1 doctorate awarded. *Entrance requirements:* For master's and doctorate, GRE, minimum B average. *Application deadline:* For fall admission, 4/1 for domestic students; for spring admission, 12/1 for domestic students. Application fee: $50. *Expenses:* Tuition, state resident: full-time $10,440; part-time $435 per credit. Tuition, nonresident: full-time $15,520; part-time $637 per credit. *Financial support:* In 2005–06, 2 fellowships with full and partial tuition reimbursements (averaging $18,000 per year), 2 research assistantships with full and partial tuition reimbursements (averaging $16,988 per year), 6 teaching assistantships with full and partial tuition reimbursements (averaging $16,988 per year) were awarded.; Federal Work-Study and tuition waivers (full and partial) also available. Support available to part-time students. *Unit head:* Dr. Alex Gates, Program Coordinator and Adviser, 973-353-5034, Fax: 973-353-5100, E-mail: agates@andromeda.rutgers.edu.

Rutgers, The State University of New Jersey, New Brunswick/Piscataway, Graduate School, Program in Environmental Sciences, New Brunswick, NJ 08901-1281. Offers air resources (MS, PhD); aquatic biology (MS, PhD); aquatic chemistry (MS, PhD); atmospheric science (MS, PhD); chemistry and physics of aerosol and hydrosol systems (MS, PhD); environmental chemistry (MS, PhD); environmental microbiology (MS, PhD); environmental toxicology (PhD); exposure assessment (PhD); fate and effects of pollutants (MS, PhD); pollution prevention and control (MS, PhD); water and wastewater treatment (MS, PhD); water resources (MS, PhD). *Faculty:* 81 full-time, 7 part-time/adjunct. *Students:* 49 full-time (27 women), 48 part-time (19 women); includes 10 minority (3 African Americans, 6 Asian Americans or Pacific Islanders, 1 Hispanic American), 24 international. Average age 32. 79 applicants, 41% accepted, 15 enrolled. In 2005, 8 master's, 10 doctorates awarded. Terminal master's awarded for partial completion of doctoral program. *Degree requirements:* For master's, thesis or alternative, oral final exam, comprehensive exam; for doctorate, thesis/dissertation, thesis defense, qualifying exam, comprehensive exam. *Entrance requirements:* For master's and doctorate, GRE General Test. Additional exam requirements/recommendations for international students: Required—TOEFL. *Application deadline:* For fall admission, 3/1 for domestic students; for spring admission, 11/1 for domestic students. Applications are processed on a rolling basis. Application fee: $50. Electronic applications accepted. *Expenses:* Tuition, state resident: full-time $10,440; part-time $435 per credit. Tuition, nonresident: full-time $15,520; part-time $647 per credit. Required fees: $129 per credit. Tuition and fees vary according to program. *Financial support:* In 2005–06, 10 fellowships with full tuition reimbursements (averaging $21,887 per year), 34 research assistantships with full tuition reimbursements (averaging $19,367 per year), 3 teaching assistantships with full tuition reimbursements (averaging $17,583 per year) were awarded.; career-related internships or fieldwork and Federal Work-Study also available. Financial award application deadline: 1/15; financial award applicants required to submit FAFSA. *Faculty research:* Atmospheric sciences; biological waste treatment; contaminant fate and transport; exposure assessment; air, soil and water quality. Total annual research expenditures: $5.7 million. *Unit head:* John Reinfelder, Director, 732-932-8013, Fax: 732-932-8644, E-mail: reinfelder@envsci.rutgers.edu. *Application contact:* Dr. Paul J. Lioy, Graduate Admissions Committee, 732-932-0150, Fax: 732-445-0116, E-mail: plioy@eohsi.rutgers.edu.

South Dakota School of Mines and Technology, Graduate Division, College of Science and Letters, Joint PhD Program in Atmospheric, Environmental, and Water Resources, Rapid City, SD 57701-3995. Offers PhD. *Faculty:* 9 full-time (0 women), 1 (woman) part-time/adjunct. *Students:* 4 full-time (2 women), 7 part-time (2 women); includes 1 minority (American Indian/Alaska Native), 1 international. In 2005, 2 degrees awarded. *Degree requirements:* For doctorate, thesis/dissertation. *Entrance requirements:* For doctorate, GRE General Test, GRE Subject Test. Additional exam requirements/recommendations for international students: Required—TOEFL, TWE. *Application deadline:* For fall admission, 7/1 priority date for domestic students, 4/1 priority date for international students; for spring admission, 11/1 for domestic students, 9/1 for international students. Applications are processed on a rolling basis. Application fee: $35. Electronic applications accepted. *Expenses:* Tuition, area resident: Part-time $116 per credit hour. Tuition, state resident: full-time $2,084. Tuition, nonresident: full-time $6,146; part-time $341 per credit hour. Required fees: $1,805; $100 per credit hour. *Financial support:* In 2005–06, 3 fellowships (averaging $4,000 per year), 4 teaching assistantships with partial tuition reimbursements (averaging $6,774 per year) were awarded.; research assistantships with partial tuition reimbursements *Unit head:* Dr. Andrew Detwiler, Chair, 605-394-2291. *Application contact:* Jeannette R. Nilson, Program Assistant-Research and Graduate Education, 800-454-8162 Ext. 1206, Fax: 605-394-5360, E-mail: graduate_admissions@silver.sdsmt.edu.

South Dakota State University, Graduate School, College of Engineering, Joint PhD Program in Atmospheric, Environmental, and Water Resources, Brookings, SD 57007. Offers PhD. Postbaccalaureate distance learning degree programs offered (minimal on-campus study). *Degree requirements:* For doctorate, thesis/dissertation, preliminary oral and written exams. *Entrance requirements:* Additional exam requirements/recommendations for international students: Required—TOEFL (minimum score 525 paper-based). Expenses: Contact institution.

Southern Illinois University Carbondale, Graduate School, College of Science, Department of Geology and Department of Geography, Program in Environmental Resources and Policy, Carbondale, IL 62901-4701. Offers PhD. *Students:* 12 full-time (5 women), 21 part-time (4 women); includes 3 minority (1 African American, 2 Hispanic Americans), 12 international. 29 applicants, 17% accepted, 3 enrolled. In 2005, 2 degrees awarded. *Entrance requirements:* For doctorate, GRE. *Application contact:* Jean Stricklin, ER&P Program Office, 618-453-7328, E-mail: jstrick@siu.edu.

Announcement: SIU's Environmental Resources and Policy PhD Program provides interdisciplinary perspectives regarding public policy and social institutions that shape reactions to environmental issues. The degree is organized by the Departments of Geography and Geology and the College of Agriculture, in cooperation with the School of Law and College of Engineering.

See Close-Up on page 665.

Southern Illinois University Edwardsville, Graduate Studies and Research, College of Arts and Sciences, Program in Environmental Sciences, Edwardsville, IL 62026-0001. Offers MS. Part-time programs available. *Students:* 7 full-time (5 women), 29 part-time (16 women); includes 3 minority (2 African Americans, 1 Hispanic American), 8 international. Average age 33. 20 applicants, 80% accepted, 0 enrolled. In 2005, 8 degrees awarded. *Degree requirements:* For master's, thesis or alternative, final exam, oral exam. *Entrance requirements:* For master's, GRE. Additional exam requirements/recommendations for international students: Required—TOEFL. *Application deadline:* For fall admission, 7/21 for domestic students, 6/1 for international students; for spring admission, 12/8 for domestic students, 10/1 for international students. Application fee: $30. Electronic applications accepted. *Expenses:* Tuition, state resident: part-time $190 per semester hour. Tuition, nonresident: part-time $380 per

Environmental Sciences

Southern Illinois University Edwardsville (continued)
semester hour. Tuition and fees vary according to course load, reciprocity agreements and student level. *Financial support:* In 2005–06, 1 research assistantship with full tuition reimbursement was awarded; fellowships with full tuition reimbursements, teaching assistantships with full tuition reimbursements, career-related internships or fieldwork, Federal Work-Study, institutionally sponsored loans, and unspecified assistantships also available. Support available to part-time students. Financial award application deadline: 3/1; financial award applicants required to submit FAFSA. *Unit head:* Dr. Kevin Johnson, Program Director, 618-650-5934, E-mail: kevjohn@siue.edu.

Southern Methodist University, School of Engineering, Department of Environmental and Civil Engineering, Dallas, TX 75275. Offers applied science (MS, PhD); civil engineering (MS, PhD); environmental engineering (MS); environmental science (MS), including environmental systems management, hazardous and waste materials management; facilities management (MS). Part-time and evening/weekend programs available. Postbaccalaureate distance learning degree programs offered (no on-campus study). *Faculty:* 6 full-time (1 woman), 24 part-time/adjunct (0 women). *Students:* 10 full-time (1 woman), 42 part-time (14 women); includes 13 minority (2 African Americans, 1 American Indian/Alaska Native, 4 Asian Americans or Pacific Islanders, 6 Hispanic Americans), 9 international. Average age 33. In 2005, 20 degrees awarded. Terminal master's awarded for partial completion of doctoral program. *Degree requirements:* For master's, thesis optional; for doctorate, thesis/dissertation, oral and written qualifying exams. *Entrance requirements:* For master's, GRE General Test, minimum GPA of 3.0 in last 2 years; bachelor's degree in engineering, mathematics, or sciences; for doctorate, GRE, BS and MS in related field, minimum GPA of 3.3. Additional exam requirements/recommendations for international students: Required—TOEFL. *Application deadline:* For fall admission, 7/1 for domestic students, 5/15 for international students; for spring admission, 11/15 for domestic students, 9/1 for international students. Applications are processed on a rolling basis. Application fee: $60. *Financial support:* In 2005–06, 6 students received support, including 2 research assistantships with full tuition reimbursements available (averaging $18,000 per year), 3 teaching assistantships with full tuition reimbursements available (averaging $18,000 per year); career-related internships or fieldwork, tuition waivers (full and partial), and unspecified assistantships also available. *Faculty research:* Human and environmental health effects of endocrine disrupters, development of air pollution control systems for diesel engines, structural analysis and design, modeling and design of waste treatment systems. Total annual research expenditures:$100,000. *Unit head:* Dr. Bijan Mohraz, Chair, 214-768-3123, Fax: 214-768-2164, E-mail: bmohraz@engr.smu.edu. *Application contact:* Marc Valerin, Director of Graduate and Executive Admissions, 214-768-3484, E-mail: valerin@engr.smu.edu.

Southern University and Agricultural and Mechanical College, Graduate School, College of Sciences, Department of Chemistry, Baton Rouge, LA 70813. Offers analytical chemistry (MS); biochemistry (MS); environmental sciences (MS); inorganic chemistry (MS); organic chemistry (MS); physical chemistry (MS). *Faculty:* 9 full-time (2 women), 3 part-time/adjunct (2 women). *Students:* 20 full-time (13 women), 8 part-time (5 women); all minorities (20 African Americans, 8 Asian Americans or Pacific Islanders). Average age 23. 30 applicants, 70% accepted, 14 enrolled. In 2005, 3 master's awarded. *Degree requirements:* For master's, thesis. *Entrance requirements:* For master's, GMAT or GRE General Test. Additional exam requirements/recommendations for international students: Required—TOEFL (minimum score 525 paper-based; 193 computer-based). *Application deadline:* For fall admission, 4/15 for domestic students; for spring admission, 11/1 priority date for domestic students. Applications are processed on a rolling basis. Application fee: $5. *Financial support:* In 2005–06, 31 research assistantships (averaging $7,000 per year), 10 teaching assistantships (averaging $7,000 per year) were awarded.; scholarships/grants also available. Financial award application deadline: 4/15. *Faculty research:* Synthesis of macrocyclic ligands, latex accelerators, anticancer drugs, biosensors, absorption isotheums, isolation of specific enzymes from plants. Total annual research expenditures:$400,000. *Unit head:* Dr. Ella Kelley, Chair, 225-771-3990, Fax: 225-771-3992.

Stanford University, School of Earth Sciences, Department of Geological and Environmental Sciences, Stanford, CA 94305-9991. Offers MS, PhD, Eng. Terminal master's awarded for partial completion of doctoral program. *Degree requirements:* For masters, doctorate, and Eng, thesis/dissertation. *Entrance requirements:* For master's, doctorate, and Eng, GRE General Test. Additional exam requirements/recommendations for international students: Required—TOEFL. Electronic applications accepted.

Stanford University, School of Earth Sciences, Earth Systems Program, Stanford, CA 94305-9991. Offers MS. Students admitted at the undergraduate level. Electronic applications accepted.

State University of New York College of Environmental Science and Forestry, Faculty of Environmental and Forest Biology, Syracuse, NY 13210-2779. Offers chemical ecology (MPS, MS, PhD); conservation biology (MPS, MS, PhD); ecology (MPS, MS, PhD); entomology (MPS, MS, PhD); environmental interpretation (MPS, MS, PhD); environmental physiology (MPS, MS, PhD); fish and wildlife biology (MPS, MS, PhD); forest pathology and mycology (MPS, MS, PhD); plant science and biotechnology (MPS, MS, PhD). *Faculty:* 27 full-time (4 women), 4 part-time/adjunct (0 women). *Students:* 81 full-time (50 women), 62 part-time (33 women); includes 4 minority (1 Asian American or Pacific Islander, 3 Hispanic Americans), 16 international. Average age 30. 84 applicants, 54% accepted, 17 enrolled. In 2005, 17 master's, 3 doctorates awarded. *Degree requirements:* For master's, thesis (for some programs), registration; for doctorate, thesis/dissertation, comprehensive exam, registration. *Entrance requirements:* For master's and doctorate, GRE General Test, GRE Subject Test, minimum GPA of 3.0. Additional exam requirements/recommendations for international students: Required—TOEFL (minimum score 550 paper-based; 213 computer-based). *Application deadline:* For fall admission, 2/1 priority date for domestic students, 2/1 priority date for international students; for spring admission, 11/1 priority date for domestic students, 11/1 priority date for international students. Applications are processed on a rolling basis. Application fee: $60. *Expenses:* Tuition, area resident: Full-time $6,900; part-time $288 per credit. Tuition, nonresident: full-time $10,920; part-time $455 per credit. Required fees: $395; $32 per credit. $20 per term. One-time fee: $145. *Financial support:* In 2005–06, 86 students received support, including 13 fellowships with full and partial tuition reimbursements available (averaging $9,446 per year), 40 research assistantships with full and partial tuition reimbursements available (averaging $11,000 per year), 32 teaching assistantships with full and partial tuition reimbursements available (averaging $9,446 per year); Federal Work-Study, institutionally sponsored loans, scholarships/grants, health care benefits, and unspecified assistantships also available. Financial award application deadline: 6/30. *Faculty research:* Ecology, fish and wildlife biology and management, plant science, entomology. Total annual research expenditures: $4.1 million. *Unit head:* Dr. Donald J. Leopold, Chair, 315-470-6770, Fax: 315-470-6934, E-mail: dendro@esf.edu. *Application contact:* Dr. Dudley J. Raynal, Dean, Instruction and Graduate Studies, 315-470-6599, Fax: 315-470-6978, E-mail: esfgrad@esf.edu.

State University of New York College of Environmental Science and Forestry, Faculty of Environmental Studies, Syracuse, NY 13210-2779. Offers environmental and community land planning (MPS, MS, PhD); environmental and natural resources policy (PhD); environmental communication and participatory processes (MPS, MS, PhD); environmental policy and democratic processes (MPS, MS, PhD); environmental systems and risk management (MPS, MS, PhD); water and wetland resource studies (MPS, MS, PhD). Part-time programs available. *Faculty:* 11 full-time (7 women), 11 part-time/adjunct (6 women). *Students:* 49 full-time (30 women), 28 part-time (15 women); includes 3 minority (2 African Americans, 1 Hispanic American), 36 international. Average age 32. 61 applicants, 69% accepted, 17 enrolled. In 2005, 16 master's, 4 doctorates awarded. *Degree requirements:* For master's, thesis (for some programs), registration; for doctorate, thesis/dissertation, comprehensive exam, registration. *Entrance requirements:* For master's and doctorate, GRE General Test, minimum GPA of 3.0. Additional exam requirements/recommendations for international students: Required—TOEFL (minimum score 550 paper-based; 213 computer-based). *Application deadline:* For fall admission, 2/1 priority date for domestic students, 2/1 priority date for international students; for spring admission, 11/1 priority date for domestic students, 11/1 priority date for international students. Applications are processed on a rolling basis. Application fee: $60. *Expenses:* Tuition, area resident: Full-time $6,900; part-time $288 per credit. Tuition, nonresident: full-time $10,920; part-time $455 per credit. Required fees: $395; $32 per credit. $20 per term. One-time fee: $145. *Financial support:* In 2005–06, 21 fellowships with full and partial tuition reimbursements (averaging $9,446 per year), 4 research assistantships with full and partial tuition reimbursements (averaging $10,000 per year), 9 teaching assistantships with full and partial tuition reimbursements (averaging $9,446 per year) were awarded.; career-related internships or fieldwork, Federal Work-Study, institutionally sponsored loans, scholarships/grants, health care benefits, and unspecified assistantships also available. Support available to part-time students. Financial award application deadline: 6/30; financial award applicants required to submit FAFSA. *Faculty research:* Environmental education/communications, water resources, land resources, waste management. Total annual research expenditures: $169,919. *Unit head:* Dr. Richard Smardon, Chair, 315-470-6636, Fax: 315-470-6915, E-mail: rsmardon@syr.edu. *Application contact:* Dr. Dudley J. Raynal, Dean, Instruction and Graduate Studies, 315-470-6599, Fax: 315-470-6978, E-mail: esfgrad@esf.edu.

Stephen F. Austin State University, Graduate School, College of Sciences and Mathematics, Division of Environmental Science, Nacogdoches, TX 75962. Offers MS. *Faculty:* 2 full-time (0 women), 37 part-time/adjunct (9 women). *Students:* 18 full-time (8 women), 16 part-time (7 women); includes 2 minority (both Asian Americans or Pacific Islanders), 2 international. 10 applicants, 80% accepted. In 2005, 5 degrees awarded. *Degree requirements:* For master's, comprehensive exam. *Entrance requirements:* For master's, GRE General Test, minimum GPA of 2.8 in last 60 hours, 2.5 overall. Additional exam requirements/recommendations for international students: Required—TOEFL. *Application deadline:* For fall admission, 8/1 for domestic students; for spring admission, 12/15 for domestic students. Applications are processed on a rolling basis. Application fee: $0 ($50 for international students). *Expenses:* Tuition, state resident: full-time $2,628; part-time $146 per credit hour. Tuition, nonresident: full-time $7,596; part-time $422 per credit hour. Required fees: $900; $170. *Financial support:* In 2005–06, 15 research assistantships (averaging $8,100 per year) were awarded; Federal Work-Study and unspecified assistantships also available. Financial award application deadline: 3/1. *Unit head:* Dr. Kenneth Farrish, Director, 936-468-4582, E-mail: kfarrish@sfasu.edu.

Tarleton State University, College of Graduate Studies, College of Science and Technology, Department of Chemistry and Geosciences, Stephenville, TX 76402. Offers environmental science (MS). Part-time and evening/weekend programs available. *Faculty:* 1 (woman) full-time, 3 part-time/adjunct (all women). *Students:* 2 full-time (1 woman), 3 part-time (2 women); includes 1 minority (Asian American or Pacific Islander) Average age 31. *Degree requirements:* For master's, thesis optional. *Entrance requirements:* For master's, GRE General Test, minimum GPA of 3.0. Additional exam requirements/recommendations for international students: Required—TOEFL (minimum score 550 paper-based; 220 computer-based). *Application deadline:* For fall admission, 8/5 for domestic students; for spring admission, 12/1 for domestic students. Applications are processed on a rolling basis. Application fee: $25 ($75 for international students). *Financial support:* Research assistantships, teaching assistantships, career-related internships or fieldwork and Federal Work-Study available. Support available to part-time students. Financial award application deadline: 5/1; financial award applicants required to submit FAFSA. *Unit head:* Dr. Carol Thompson, Director, 254-968-9739.

Taylor University, Program in Environmental Science, Upland, IN 46989-1001. Offers MES. *Faculty research:* Environmental assessment.

Tennessee Technological University, Graduate School, College of Arts and Sciences, Department of Environmental Sciences, Cookeville, TN 38505. Offers PhD. *Students:* 6 full-time (2 women), 7 part-time (1 woman); includes 4 minority (1 African American, 3 Asian Americans or Pacific Islanders). 15 applicants, 47% accepted, 5 enrolled. In 2005, 3 degrees awarded. *Degree requirements:* For doctorate, one foreign language, thesis/dissertation. *Entrance requirements:* For doctorate, GRE. Additional exam requirements/recommendations for international students: Required—TOEFL. *Application deadline:* For fall admission, 3/1 for domestic students; for spring admission, 8/1 for domestic students. Application fee: $25 ($30 for international students). Electronic applications accepted. *Expenses:* Tuition, state resident: full-time $8,421; part-time $307 per hour. Tuition, nonresident: full-time $22,389; part-time $711 per hour. *Financial support:* In 2005–06, 4 research assistantships (averaging $10,000 per year), 3 teaching assistantships (averaging $10,000 per year) were awarded.; fellowships Financial award application deadline: 4/1. *Unit head:* Dr. Jeffrey Boles, Director, 931-372-3844, Fax: 931-372-3434, E-mail: jboles@tntech.edu. *Application contact:* Dr. Francis O. Otuonye, Associate Vice President for Research and Graduate Studies, 931-372-3233, Fax: 931-372-3497, E-mail: fotuonye@tntech.edu.

Texas A&M University–Corpus Christi, Graduate Studies and Research, College of Science and Technology, Program in Sciences, Corpus Christi, TX 78412-5503. Offers biology (MS); environmental sciences (MS); mariculture (MS). Part-time and evening/weekend programs available. *Degree requirements:* For master's, thesis (for some programs), comprehensive exam, registration. *Entrance requirements:* For master's, GRE General Test. Additional exam requirements/recommendations for international students: Required—TOEFL. Electronic applications accepted.

Texas Christian University, College of Science and Engineering, Department of Biology, Program in Environmental Sciences, Fort Worth, TX 76129-0002. Offers earth sciences (MS); ecology (MS). Part-time and evening/weekend programs available. *Degree requirements:* For master's, thesis optional. *Entrance requirements:* For master's, GRE General Test, GRE Subject Test, 1 year course work in biology and chemistry; 1 semester course work in calculus, government, and physical geology. Additional exam requirements/recommendations for international students: Required—TOEFL. *Application deadline:* For fall admission, 3/1 for domestic students; for spring admission, 12/1 for domestic students. Applications are processed on a rolling basis. Application fee: $0. *Expenses:* Tuition: Part-time $740 per credit hour. *Financial support:* Unspecified assistantships available. Financial award application deadline: 3/1. *Unit head:* Dr. Mike Slattery, Director, 817-257-7506. *Application contact:* Dr. Bonnie Melhart, Associate Dean, College of Science and Engineering, E-mail: b.melhart@tcu.edu.

Texas Tech University, Graduate School, College of Arts and Sciences, Department of Environmental Toxicology, Lubbock, TX 79409. Offers MS, PhD. Part-time programs available. *Faculty:* 12 full-time (1 woman). *Students:* 37 full-time (15 women), 2 part-time (1 woman); includes 3 minority (1 American Indian/Alaska Native, 2 Hispanic Americans), 25 international. Average age 30. 24 applicants, 54% accepted, 4 enrolled. In 2005, 2 master's, 6 doctorates awarded. *Degree requirements:* For master's and doctorate, thesis/dissertation. *Entrance requirements:* For master's and doctorate, GRE General Test. Additional exam requirements/recommendations for international students: Required—TOEFL (minimum score 550 paper-based; 213 computer-based). *Application deadline:* Applications are processed on a rolling basis. Application fee: $50 ($60 for international students). Electronic applications accepted. *Expenses:* Tuition, state resident: full-time $4,296. Tuition, nonresident: full-time $10,920. Required fees: $1,992. Tuition and fees vary according to program. *Financial support:* In 2005–06, 13 students received support; teaching assistantships with partial tuition reimbursements available available. Financial award application deadline: 4/15. *Faculty research:* Terrestrial and aquatic toxicology, biochemical and developmental toxicology, advanced materials and high performance computing, countermeasures to biologic and chemical threats, molecular epidemiology and modeling. *Unit head:* Dr. Ronald J. Kendall, Director and Chairman, 806-885-4567, Fax: 806-885-2132, E-mail: ron.kendall@tiehh.ttu.edu. *Application contact:* Dr. Steve Cox, Graduate Program Adviser, 806-885-4567, Fax: 806-885-2132, E-mail: stephen.cox@ttu.edu.

Towson University, Graduate School, Program in Environmental Science, Towson, MD 21252-0001. Offers MS, Certificate. Part-time and evening/weekend programs available. *Students:* 36. 18 applicants, 83% accepted, 11 enrolled. In 2005, 6 degrees awarded. *Entrance requirements:* For master's, GRE (recommended), bachelor's degree in related field, minimum

GPA of 3.0. *Application deadline:* Applications are processed on a rolling basis. Application fee: $40. Electronic applications accepted. *Financial support:* Application deadline: 4/1; *Unit head:* Dr. Steven Lev, Graduate Program Director, 410-704-2744, Fax: 410-704-2604, E-mail: slev@towson.edu. *Application contact:* 410-704-2501, Fax: 410-704-4675, E-mail: grads@towson.edu.

Tufts University, School of Engineering, Department of Civil and Environmental Engineering, Medford, MA 02155. Offers civil engineering (ME, MS, PhD), including geotechnical engineering, structural engineering; environmental engineering (ME, MS, PhD), including environmental engineering and environmental sciences, environmental geotechnology, environmental health, environmental science and management, hazardous materials management, water resources engineering. Part-time programs available. *Faculty:* 16 full-time, 7 part-time/adjunct. *Students:* 61 (33 women); includes 3 minority (1 African American, 1 Asian American or Pacific Islander, 1 Hispanic American) 9 international. 77 applicants, 74% accepted, 16 enrolled. In 2005, 16 master's awarded. Terminal master's awarded for partial completion of doctoral program. *Degree requirements:* For master's, thesis or alternative; for doctorate, thesis/dissertation. *Entrance requirements:* Additional exam requirements/recommendations for international students: Required—TOEFL (minimum score 550 paper-based; 213 computer-based). *Application deadline:* For fall admission, 2/1 for domestic students, 12/30 for international students; for spring admission, 10/15 for domestic students, 9/15 for international students. Applications are processed on a rolling basis. Application fee: $65. Electronic applications accepted. *Expenses:* Tuition: Full-time $32,360. Tuition and fees vary according to program. *Financial support:* Research assistantships with full and partial tuition reimbursements, teaching assistantships with full and partial tuition reimbursements, Federal Work-Study, scholarships/grants, and tuition waivers (partial) available. Support available to part-time students. Financial award application deadline: 2/1; financial award applicants required to submit FAFSA. *Unit head:* Dr. Christopher Swan, Chair, 617-627-3211, Fax: 617-627-3994.

Tuskegee University, Graduate Programs, College of Agricultural, Environmental and Natural Sciences, Department of Agricultural Sciences, Program in Environmental Sciences, Tuskegee, AL 36088. Offers MS. *Faculty:* 13 full-time (1 woman), 2 part-time/adjunct (1 woman). *Students:* 7 full-time (3 women); includes 2 minority (both African Americans), 4 international. Average age 29. In 2005, 5 degrees awarded. *Degree requirements:* For master's, thesis. *Entrance requirements:* For master's, GRE General Test. Additional exam requirements/recommendations for international students: Required—TOEFL (minimum score 500 paper-based; 173 computer-based). *Application deadline:* For fall admission, 7/15 for domestic students. Applications are processed on a rolling basis. Application fee: $25 ($35 for international students). *Expenses:* Tuition: Full-time $12,400. Required fees: $300; $490 per credit. *Financial support:* Application deadline: 4/15. *Unit head:* Dr. P. K. Biswas, Head, Department of Agricultural Sciences, 334-727-8446.

Université de Sherbrooke, Faculty of Sciences, Centre Universitaire de Formation en Environnement, Sherbrooke, QC J1K 2R1, Canada. Offers M Sc, Diploma. Postbaccalaureate distance learning degree programs offered (no on-campus study). *Students:* 149 applicants, 66% accepted. In 2005, 56 master's, 22 other advanced degrees awarded. *Application deadline:* For fall admission, 10/15 for domestic students; for spring admission, 3/25 for domestic students. Applications are processed on a rolling basis. Application fee: $50. Electronic applications accepted. *Financial support:* Career-related internships or fieldwork available. *Faculty research:* Environmental studies. *Unit head:* Michel Montpetit, Coordinator, 819-821-8000 Ext. 2077, Fax: 819-821-8017, E-mail: michel.montpetit@usherbrooke.ca.

Université du Québec à Montréal, Graduate Programs, Program in Environmental Sciences, Montréal, QC H3C 3P8, Canada. Offers M Sc, PhD. Part-time programs available. *Degree requirements:* For master's, research report; for doctorate, thesis/dissertation. *Entrance requirements:* For master's, appropriate bachelor's degree or equivalent, proficiency in French; for doctorate, appropriate master's degree or equivalent, proficiency in French.

Université du Québec à Trois-Rivières, Graduate Programs, Program in Environmental Sciences, Trois-Rivières, QC G9A 5H7, Canada. Offers M Sc, PhD. Part-time programs available. *Degree requirements:* For master's, thesis. *Entrance requirements:* For master's, appropriate bachelor's degree, proficiency in French.

Université Laval, Faculty of Sciences and Engineering, Department of Geology and Geological Engineering, Programs in Earth Sciences, Québec, QC G1K 7P4, Canada. Offers earth sciences (M Sc, PhD); environmental technologies (M Sc). Offered jointly with INRS-Géressources. Terminal master's awarded for partial completion of doctoral program. *Degree requirements:* For master's, thesis (for some programs); for doctorate, thesis/dissertation, comprehensive exam. *Entrance requirements:* For master's and doctorate, knowledge of French. Electronic applications accepted.

University at Albany, State University of New York, College of Arts and Sciences, Department of Biological Sciences, Program in Biodiversity, Conservation, and Policy, Albany, NY 12222-0001. Offers MS. *Degree requirements:* For master's, one foreign language. *Entrance requirements:* For master's, GRE General Test. Application fee: $60. *Faculty research:* Aquatic ecology, plant community ecology, biodiversity and public policy, restoration ecology, costal and estuarine science. *Unit head:* Gary Kleppel, Program Director, 518-442-4338.

The University of Alabama in Huntsville, School of Graduate Studies, College of Science, Department of Atmospheric and Environmental Science, Huntsville, AL 35899. Offers MS, PhD. Part-time and evening/weekend programs available. *Faculty:* 7 full-time (0 women), 3 part-time/adjunct (0 women). *Students:* 26 full-time (6 women), 11 part-time (5 women); includes 2 minority (both African Americans), 11 international. Average age 29. 21 applicants, 95% accepted, 10 enrolled. In 2005, 9 master's, 2 doctorates awarded. *Degree requirements:* For master's, thesis or alternative, oral and written exams, comprehensive exam, registration; for doctorate, thesis/dissertation, oral and written exams, comprehensive exam, registration. *Entrance requirements:* For master's and doctorate, GRE General Test, minimum GPA of 3.0. Additional exam requirements/recommendations for international students: Required—TOEFL (minimum score 550 paper-based; 213 computer-based). *Application deadline:* For fall admission, 5/30 for domestic students; for spring admission, 10/10 priority date for domestic students, 7/10 priority date for international students. Applications are processed on a rolling basis. Application fee: $40. *Expenses:* Tuition, state resident: full-time $5,866; part-time $244 per credit hour. Tuition, nonresident: full-time $12,060; part-time $500 per credit hour. Tuition and fees vary according to course load. *Financial support:* In 2005–06, 24 students received support, including 22 research assistantships with full and partial tuition reimbursements available (averaging $12,777 per year), 1 teaching assistantship with full and partial tuition reimbursement available (averaging $13,500 per year); fellowships with full and partial tuition reimbursements available, career-related internships or fieldwork, Federal Work-Study, institutionally sponsored loans, scholarships/grants, health care benefits, tuition waivers (full and partial), and unspecified assistantships also available. Support available to part-time students. Financial award application deadline: 4/1; financial award applicants required to submit FAFSA. Total annual research expenditures:$225,474. *Unit head:* Dr. Ronald Welch, Chair, 256-961-7754, Fax: 256-961-7755, E-mail: ron.welch@atmos.uah.edu.

University of Alaska Anchorage, School of Engineering, Program in Environmental Quality Science, Anchorage, AK 99508-8060. Offers MS. *Students:* 3 full-time (1 woman), 8 part-time (4 women), 1 international. 8 applicants, 63% accepted. In 2005, 1 degree awarded. *Degree requirements:* For master's, thesis optional. *Entrance requirements:* For master's, GRE General Test, BS in engineering or scientific field. Additional exam requirements/recommendations for international students: Required—TOEFL (minimum score 550 paper-based; 213 computer-based). *Application deadline:* For fall admission, 7/1 priority date for domestic students, 7/1 priority date for international students; for spring admission, 11/1 for domestic students, 11/1 for international students. Applications are processed on a rolling basis. Application fee: $45. *Financial support:* Research assistantships, Federal Work-Study available. Support available to part-time students. Financial award application deadline: 4/1; financial award applicants required to submit FAFSA. *Faculty research:* Waste water treatment, environmental regulations, water resources management, justification of public facilities, rural sanitation, biological treatment process. *Unit head:* Dr. Craig Woolard, Chair, 907-786-1079, Fax: 907-786-1079. *Application contact:* Elisa S. Mattison, Coordinator for Graduate Studies, 907-786-1096, Fax: 907-786-1021, E-mail: emattison@uaa.alaska.edu.

University of Alaska Fairbanks, College of Engineering and Mines, Department of Civil and Environmental Engineering, Fairbanks, AK 99775-7520. Offers arctic engineering (MS); civil engineering (MCE, MS); engineering (PhD); engineering and science management (MS); environmental engineering (MS); environmental quality science (MS). Part-time programs available. *Faculty:* 7 full-time (1 woman). *Students:* 19 full-time (9 women), 9 part-time (3 women); includes 2 minority (1 Asian American or Pacific Islander, 1 Hispanic American), 8 international. Average age 29. 32 applicants, 50% accepted, 5 enrolled. In 2005, 19 degrees awarded. Terminal master's awarded for partial completion of doctoral program. *Degree requirements:* For master's, thesis or alternative, comprehensive exam, registration; for doctorate, thesis/dissertation, comprehensive exam, registration. *Entrance requirements:* For master's and doctorate, GRE General Test. Additional exam requirements/recommendations for international students: Required—TOEFL (minimum score 550 paper-based; 213 computer-based). *Application deadline:* For fall admission, 6/1 for domestic students, 3/1 for international students; for spring admission, 12/1 for domestic students, 9/1 for international students. Applications are processed on a rolling basis. Application fee: $50. Electronic applications accepted. *Expenses:* Tuition, state resident: full-time $4,392; part-time $244 per credit. Tuition, nonresident: full-time $8,964; part-time $498 per credit. Required fees: $800; $5 per credit. $48 per contact hour. Tuition and fees vary according to course level, course load, campus/location and reciprocity agreements. *Financial support:* In 2005–06, 7 research assistantships with tuition reimbursements (averaging $9,540 per year), 1 teaching assistantship with tuition reimbursement (averaging $10,462 per year) were awarded.; fellowships with tuition reimbursements, career-related internships or fieldwork, Federal Work-Study, scholarships/grants, and unspecified assistantships also available. Support available to part-time students. Financial award applicants required to submit FAFSA. *Faculty research:* Soils, structures, culvert thawing with solar power, pavement drainage, contaminant hydrogeology. *Unit head:* Dr. Daniel M. White, Chair, 907-474-7241, Fax: 907-474-6087, E-mail: fycee@uaf.edu.

University of Alaska Fairbanks, College of Natural Sciences and Mathematics, Department of Chemistry and Biochemistry, Fairbanks, AK 99775-7520. Offers biochemistry and molecular biology (MS, PhD); chemistry (MA, MS); environmental chemistry (MS, PhD). Part-time programs available. *Faculty:* 13 full-time (3 women). *Students:* 30 full-time (13 women), 8 part-time (4 women); includes 6 minority (1 African American, 1 American Indian/Alaska Native, 2 Asian Americans or Pacific Islanders, 2 Hispanic Americans), 9 international. Average age 29. 35 applicants, 63% accepted, 9 enrolled. In 2005, 6 master's, 4 doctorates awarded. Terminal master's awarded for partial completion of doctoral program. *Degree requirements:* For master's, thesis, seminar, comprehensive exam, registration; for doctorate, thesis/dissertation, comprehensive exam, registration. *Entrance requirements:* For master's, GRE General Test; for doctorate, GRE General Test, GRE Subject Test (biology or chemistry). Additional exam requirements/recommendations for international students: Required—TOEFL (minimum score 550 paper-based; 213 computer-based). *Application deadline:* For fall admission, 6/1 for domestic students, 3/1 for international students; for spring admission, 12/1 for domestic students, 9/1 for international students. Applications are processed on a rolling basis. Application fee: $50. Electronic applications accepted. *Expenses:* Tuition, state resident: full-time $4,392; part-time $244 per credit. Tuition, nonresident: full-time $8,964; part-time $498 per credit. International tuition: $8,964 full-time. Required fees: $800; $5 per credit. $48 per contact hour. Tuition and fees vary according to course level, course load, campus/location and reciprocity agreements. *Financial support:* In 2005–06, 6 research assistantships with tuition reimbursements (averaging $9,595 per year), 16 teaching assistantships with tuition reimbursements (averaging $9,100 per year) were awarded.; fellowships with tuition reimbursements, Federal Work-Study and scholarships/grants also available. Financial award applicants required to submit FAFSA. *Faculty research:* Atmospheric aerosols; plant chemistry; hibernation and neuroprotection; transition metal based drugs for diabetes; liganogated ion channels. *Unit head:* Dr. Thomas Clausen, Chair, 907-474-5510, Fax: 907-474-5640, E-mail: fychem@uaf.edu.

University of Alberta, Faculty of Graduate Studies and Research, Department of Civil and Environmental Engineering, Edmonton, AB T6G 2E1, Canada. Offers construction engineering and management (M Eng, M Sc, PhD); environmental engineering (M Eng, M Sc, PhD); environmental science (M Sc, PhD); geoenvironmental engineering (M Eng, M Sc, PhD); geotechnical engineering (M Eng, M Sc, PhD); mining engineering (M Eng, M Sc, PhD); petroleum engineering (M Eng, M Sc, PhD); structural engineering (M Eng, M Sc, PhD); water resources (M Eng, M Sc, PhD). Part-time programs available. Postbaccalaureate distance learning degree programs offered (minimal on-campus study). *Faculty:* 44 full-time (3 women), 2 part-time/adjunct (0 women). *Students:* 215 full-time (49 women), 99 part-time (19 women). 1,428 applicants, 15% accepted, 123 enrolled. In 2005, 124 master's, 34 doctorates awarded. *Degree requirements:* For master's, thesis (for some programs); for doctorate, thesis/dissertation. *Entrance requirements:* For master's, minimum GPA of 3.0 in last 2 years of undergraduate studies; for doctorate, minimum GPA of 3.0. Additional exam requirements/recommendations for international students: Required—TOEFL (minimum score 550 paper-based; 213 computer-based). *Application deadline:* For fall admission, 6/1 priority date for domestic students, 6/1 priority date for international students. For winter admission, 11/1 for domestic students. Applications are processed on a rolling basis. Application fee: $0 Canadian dollars. Electronic applications accepted. Tuition and fees charges are reported in Canadian dollars. *Expenses:* Tuition, state resident: part-time $562 Canadian dollars per term. Tuition, nonresident: full-time $3,375 Canadian dollars. Required fees: $573 Canadian dollars; $84 Canadian dollars per term. *Financial support:* In 2005–06, 88 research assistantships with full and partial tuition reimbursements, 134 teaching assistantships with full and partial tuition reimbursements were awarded.; scholarships/grants and tuition waivers (full and partial) also available. Financial award application deadline: 4/1. *Faculty research:* Mining. Total annual research expenditures: $6,791 Canadian dollars. *Unit head:* Dr. David Zhu, Associate Chair, Gradute Studies, 780-492-1198, Fax: 403-492-8198, E-mail: dchan@civil.ualberta.ca. *Application contact:* Gwen Mendoza, Student Services Officer, 403-492-1539, Fax: 403-492-0249, E-mail: graduate_studies@civil.ualberta.ca.

The University of Arizona, Graduate College, College of Agriculture and Life Sciences, Department of Soil, Water and Environmental Science, Tucson, AZ 85721. Offers MS, PhD. *Degree requirements:* For master's, thesis; for doctorate, one foreign language, thesis/dissertation. *Entrance requirements:* Additional exam requirements/recommendations for international students: Required—TOEFL. *Faculty research:* Plant production, environmental microbiology, contaminant flow and transport.

The University of Arizona, Graduate College, Graduate Interdisciplinary Programs, Graduate Interdisciplinary Program in Arid Land Resource Sciences, Tucson, AZ 85721. Offers PhD. *Degree requirements:* For doctorate, one foreign language, thesis/dissertation. *Entrance requirements:* For doctorate, GRE. Additional exam requirements/recommendations for international students: Required—TOEFL (minimum score 550 paper-based; 213 computer-based).

University of California, Berkeley, Graduate Division, College of Natural Resources, Department of Environmental Science, Policy, and Management, Berkeley, CA 94720-1500. Offers environmental science, policy, and management (MS, PhD); forestry (MF). Terminal master's awarded for partial completion of doctoral program. *Degree requirements:* For master's, thesis optional; for doctorate, thesis/dissertation, qualifying exam. *Entrance requirements:* For master's and doctorate, GRE General Test, minimum GPA of 3.0. Additional exam requirements/recommendations for international students: Required—TOEFL; Recommended—TSE. Electronic applications accepted. *Faculty research:* Biology and ecology of insects; ecosystem

Environmental Sciences

University of California, Berkeley (continued)

function and environmental issues of soils; plant health/interactions from molecular to ecosystem levels; range management and ecology; forest and resource policy, sustainability, and management.

University of California, Berkeley, Graduate Division, College of Natural Resources, Group in Agricultural and Environmental Chemistry, Berkeley, CA 94720-1500. Offers MS, PhD. Terminal master's awarded for partial completion of doctoral program. *Degree requirements:* For master's, exam or thesis; for doctorate, thesis/dissertation, qualifying exam, seminar presentation. *Entrance requirements:* For master's and doctorate, GRE General Test, minimum GPA of 3.0.

University of California, Davis, Graduate Studies, Graduate Group in Soils and Biogeochemistry, Davis, CA 95616. Offers MS, PhD. *Faculty:* 39 full-time. *Students:* 33 full-time (17 women); includes 1 minority (Asian American or Pacific Islander), 8 international. Average age 31. 18 applicants, 56% accepted, 5 enrolled. In 2005, 6 master's, 1 doctorate awarded. Terminal master's awarded for partial completion of doctoral program. *Degree requirements:* For master's, thesis (for some programs), comprehensive exam (for some programs); for doctorate, thesis/dissertation. *Entrance requirements:* For master's, minimum GPA of 3.3; for doctorate, GRE, minimum GPA of 3.3. Additional exam requirements/recommendations for international students: Required—TOEFL (minimum score 550 paper-based; 213 computer-based). *Application deadline:* For fall admission, 1/15 for domestic students, 1/15 for international students. Applications are processed on a rolling basis. Application fee: $60. Electronic applications accepted. *Financial support:* In 2005–06, 30 students received support, including 7 fellowships with full and partial tuition reimbursements available (averaging $11,887 per year), 20 research assistantships with full and partial tuition reimbursements available (averaging $11,976 per year), 3 teaching assistantships with partial tuition reimbursements available (averaging $15,445 per year); career-related internships or fieldwork, Federal Work-Study, institutionally sponsored loans, scholarships/grants, tuition waivers (full and partial), and unspecified assistantships also available. Support available to part-time students. Financial award application deadline: 1/15; financial award applicants required to submit FAFSA. *Faculty research:* Rhizosphere ecology, soil transport processes, biogeochemical cycling, sustainable agriculture. *Unit head:* Louise Jackson, Chair, 530-754-9116, E-mail: lejackson@ucdavis.edu. *Application contact:* Merlyn Potters, Graduate Staff Adviser, 530-752-1669, Fax: 530-752-1552, E-mail: lawradvising@ucdavis.edu.

University of California, Los Angeles, Graduate Division, School of Public Health, Program in Environmental Science and Engineering, Los Angeles, CA 90095. Offers D Env. *Degree requirements:* For doctorate, thesis/dissertation, oral and written qualifying exams. *Entrance requirements:* For doctorate, GRE General Test, minimum undergraduate GPA of 3.0, master's degree or equivalent in a natural science, engineering, or public health. Electronic applications accepted. *Faculty research:* Toxic and hazardous substances, air and water pollution, risk assessment/management, water resources, marine science.

See Close-Up on page 669.

University of California, Riverside, Graduate Division, Department of Environmental Sciences, Riverside, CA 92521-0102. Offers environmental sciences (MS, PhD); soil and water sciences (MS, PhD). Part-time programs available. *Faculty:* 26 full-time (4 women). *Students:* 15 full-time (7 women); includes 1 minority (Asian American or Pacific Islander), 6 international. Average age 30. In 2005, 2 master's, 1 doctorate awarded. Terminal master's awarded for partial completion of doctoral program. *Degree requirements:* For master's, thesis; for doctorate, thesis/dissertation, oral and written qualifying exams. *Entrance requirements:* For master's and doctorate, GRE General Test, bachelor's degree in natural and physical sciences, engineering, or economics; minimum GPA of 3.2. Additional exam requirements/recommendations for international students: Required—TOEFL (minimum score 550 paper-based; 213 computer-based); Recommended—TSE (minimum score 50). *Application deadline:* For fall admission, 5/1 for domestic students, 2/1 for international students. For winter admission, 9/1 for domestic students; for spring admission, 12/1 for domestic students. Applications are processed on a rolling basis. Application fee: $60 ($75 for international students). Electronic applications accepted. *Expenses:* Tuition, nonresident: full-time $14,694. Full-time tuition and fees vary according to program. *Financial support:* In 2005–06, fellowships (averaging $12,000 per year); research assistantships, teaching assistantships, career-related internships or fieldwork, Federal Work-Study, institutionally sponsored loans, and tuition waivers (full and partial) also available. Financial award application deadline: 2/1; financial award applicants required to submit FAFSA. *Faculty research:* Atmospheric processes, biogeochemical cycling and bioaccumulation, contaminant fate and transport in soil and water systems, environmental management and policy, environmental monitoring and risk assessment. *Unit head:* Dr. Walter J. Farmer, Chair, 951-827-5116, Fax: 951-827-3993, E-mail: walter.farmer@ucr.edu. *Application contact:* Keith Knapp, Admissions Committee Chair, 951-827-4195, Fax: 951-827-3993, E-mail: envisci@ucr.edu.

University of California, Santa Barbara, Graduate Division, Donald Bren School of Environmental Science and Management, Santa Barbara, CA 93106. Offers MESM, PhD. *Faculty:* 17 full-time (3 women), 3 part-time/adjunct (0 women). *Students:* 127 full-time (66 women), 6 part-time (4 women); includes 7 minority (all Asian Americans or Pacific Islanders), 16 international. Average age 25. 293 applicants, 55% accepted, 75 enrolled. In 2005, 38 master's, 4 doctorates awarded. *Degree requirements:* For master's, group project as student thesis; for doctorate, thesis/dissertation, comprehensive exam, registration. *Entrance requirements:* For master's and doctorate, GRE. Additional exam requirements/recommendations for international students: Required—TOEFL (minimum score 550 paper-based; 213 computer-based). *Application deadline:* For fall admission, 2/1 for domestic students, 2/1 for international students. Applications are processed on a rolling basis. Application fee: $60. Electronic applications accepted. *Financial support:* In 2005–06, 18 fellowships with partial tuition reimbursements, 14 research assistantships with full tuition reimbursements (averaging $11,140 per year), 12 teaching assistantships with partial tuition reimbursements (averaging $4,531 per year) were awarded.; career-related internships or fieldwork, Federal Work-Study, institutionally sponsored loans, scholarships/grants, traineeships, health care benefits, and unspecified assistantships also available. Financial award application deadline: 2/1; financial award applicants required to submit FAFSA. *Faculty research:* Hydrology, ecology, political instituting, environmental economics, biogeochemistry. *Unit head:* Dr. Dennis J. Aigner, Dean, 805-893-7363, E-mail: info@bren.ucsb.edu. *Application contact:* Marla Alfaro, Student Affairs Assistant, 805-893-7611, Fax: 805-893-7612, E-mail: maria@bren.ucsb.edu.

See Close-Up on page 671.

University of Chicago, The Irving B. Harris Graduate School of Public Policy Studies, Chicago, IL 60637-1513. Offers environmental science and policy (MS); public policy studies (AM, MPP, PhD). Part-time programs available. Terminal master's awarded for partial completion of doctoral program. *Degree requirements:* For doctorate, thesis/dissertation. *Entrance requirements:* For master's and doctorate, GMAT or GRE General Test. Additional exam requirements/recommendations for international students: Required—TOEFL. Electronic applications accepted. Expenses: Contact institution. *Faculty research:* Family and child policy, international security, health policy, social policy.

University of Cincinnati, Division of Research and Advanced Studies, College of Engineering, Department of Civil and Environmental Engineering, Program in Environmental Sciences, Cincinnati, OH 45221. Offers MS, PhD. Part-time programs available. *Degree requirements:* For master's, thesis or alternative; for doctorate, one foreign language, thesis/dissertation. *Entrance requirements:* For master's and doctorate, GRE General Test. Additional exam requirements/recommendations for international students: Required—TOEFL (minimum score 580 paper-based; 237 computer-based). Electronic applications accepted.

University of Colorado at Colorado Springs, Graduate School, College of Letters, Arts and Sciences, Department of Geography and Environmental Studies, Colorado Springs, CO 80933-7150. Offers MA. *Faculty:* 7 full-time, 1 part-time/adjunct. *Students:* 5 full-time (2 women), 13 part-time (9 women); includes 2 minority (both Asian Americans or Pacific Islanders) Average age 37. In 2005, 5 degrees awarded. *Expenses:* Tuition, state resident: full-time $4,068; part-time $312 per credit hour. Tuition, nonresident: full-time $11,570; part-time $890 per credit hour. Required fees: $339; $19 per credit hour. $177 per term. Tuition and fees vary according to course load, program and reciprocity agreements. Total annual research expenditures: $178,007. *Unit head:* Dr. Steve Jennings, Chair, 719-262-4056.

University of Colorado at Denver and Health Sciences Center—Downtown Denver Campus, College of Liberal Arts and Sciences, Department of Geography and Environmental Sciences, Denver, CO 80217-3364. Offers environmental sciences (MS); geographic information science (Certificate). Part-time and evening/weekend programs available. *Students:* 15 full-time (9 women), 18 part-time (11 women); includes 4 minority (2 Asian Americans or Pacific Islanders, 2 Hispanic Americans), 6 international. Average age 32. 22 applicants, 82% accepted, 8 enrolled. In 2005, 22 degrees awarded. *Degree requirements:* For master's, thesis or alternative. *Entrance requirements:* For master's, GRE General Test. Additional exam requirements/recommendations for international students: Required—TOEFL (minimum score 525 paper-based; 197 computer-based). *Application deadline:* For fall admission, 4/1 for domestic students; for spring admission, 10/1 for domestic students. Applications are processed on a rolling basis. Application fee: $50 ($75 for international students). Electronic applications accepted. *Expenses:* Tuition, state resident: part-time $325 per credit hour. Tuition, nonresident: part-time $1,077 per credit hour. Required fees: $145 per credit hour. One-time fee: $115 part-time. Tuition and fees vary according to course level and program. *Financial support:* Research assistantships, teaching assistantships, Federal Work-Study available. Financial award application deadline: 4/1; financial award applicants required to submit FAFSA. *Unit head:* Dr. John Wyckoff, Director, 303-556-2590, Fax: 303-556-6197, E-mail: jwyckoff@carbon.cudenver.edu.

University of Guam, Graduate School and Research, College of Arts and Sciences, Program in Environmental Science, Mangilao, GU 96923. Offers MS. Part-time programs available. *Degree requirements:* For master's, thesis. *Entrance requirements:* For master's, GRE General Test. Additional exam requirements/recommendations for international students: Required—TOEFL. *Faculty research:* Water resources, ecology, karst formations, hydrogeology, meteorology.

University of Guelph, Graduate Program Services, Ontario Agricultural College, Department of Land Resource Science, Guelph, ON N1G 2W1, Canada. Offers atmospheric science (M Sc, PhD); environmental and agricultural earth sciences (M Sc, PhD); land resources management (M Sc, PhD); soil science (M Sc, PhD). Part-time programs available. *Faculty:* 19 full-time (5 women), 5 part-time/adjunct (1 woman). *Students:* 47 full-time (20 women), 3 part-time; includes 9 minority (1 African American, 6 Asian Americans or Pacific Islanders, 2 Hispanic Americans), 2 international. Average age 28. 25 applicants, 24% accepted. In 2005, 4 master's, 3 doctorates awarded. *Degree requirements:* For master's and doctorate, thesis/dissertation. *Entrance requirements:* For master's, minimum B- average during previous 2 years of course work; for doctorate, minimum B average during previous 2 years of course work. Additional exam requirements/recommendations for international students: Required—TOEFL (minimum score 550 paper-based; 213 computer-based). *Application deadline:* For fall admission, 7/1 priority date for domestic students, 5/1 priority date for international students. For winter admission, 10/1 for domestic students; for spring admission, 3/1 for domestic students. Applications are processed on a rolling basis. Application fee: $75 Canadian dollars. Electronic applications accepted. *Financial support:* In 2005–06, 30 students received support, including 40 research assistantships (averaging $16,500 Canadian dollars per year), 15 teaching assistantships (averaging $3,800 Canadian dollars per year); fellowships, scholarships/grants also available. *Faculty research:* Soil science, environmental earth science, land resource management. Total annual research expenditures: $2.1 million Canadian dollars. *Unit head:* Dr. S. Hilts, Chair, 519-824-4120 Ext. 52447, Fax: 519-824-5730, E-mail: shilts@uoguelph.ca. *Application contact:* Dr. B. Hale, Graduate Coordinator, 519-824-4120 Ext. 53434, Fax: 519-824-5730, E-mail: bhale@uoguelph.ca.

University of Houston–Clear Lake, School of Science and Computer Engineering, Program in Environmental Science, Houston, TX 77058-1098. Offers MS. Part-time and evening/weekend programs available. *Entrance requirements:* For master's, GRE General Test. Additional exam requirements/recommendations for international students: Required—TOEFL (minimum score 550 paper-based; 213 computer-based).

University of Idaho, College of Graduate Studies, Program in Environmental Science, Moscow, ID 83844-2282. Offers MS, PhD. *Students:* 57 full-time (28 women), 32 part-time (12 women); includes 2 minority (1 African American, 1 Hispanic American), 17 international. Average age 27. In 2005, 2 degrees awarded. *Application deadline:* For fall admission, 8/1 for domestic students; for spring admission, 12/15 for domestic students. Applications are processed on a rolling basis. Application fee: $55 ($60 for international students). *Expenses:* Tuition, nonresident: full-time $8,770; part-time $130 per credit. Required fees: $4,508; $217 per credit. *Financial support:* Research assistantships, teaching assistantships available. Financial award application deadline: 2/15. *Unit head:* Dr. Donald Crawford, Director, 208-885-6113, Fax: 208-885-6198, E-mail: uigrad@uidaho.edu.

University of Illinois at Springfield, Graduate Programs, College of Public Affairs and Administration, Program in Environmental Studies, Springfield, IL 62703-5407. Offers environmental science (MS); environmental studies (MA). Part-time and evening/weekend programs available. *Faculty:* 2 full-time (both women), 3 part-time/adjunct (1 woman). *Students:* 23 full-time (12 women), 21 part-time (18 women); includes 2 minority (1 Asian American or Pacific Islander, 1 Hispanic American), 2 international. Average age 31. 31 applicants, 68% accepted, 10 enrolled. In 2005, 4 degrees awarded. *Degree requirements:* For master's, thesis or alternative, thesis or project. *Entrance requirements:* For master's, GRE General Test, minimum GPA of 3.0, 2 letters of reference. Additional exam requirements/recommendations for international students: Required—TOEFL (minimum score 550 paper-based; 213 computer-based). *Application deadline:* Applications are processed on a rolling basis. Application fee: $50 ($60 for international students). Electronic applications accepted. *Expenses:* Tuition, state resident: full-time $4,726; part-time $163 per credit hour. Tuition, nonresident: full-time $14,178; part-time $490 per credit hour. Required fees: $1,382; $582 per term. *Financial support:* In 2005–06, fellowships with full tuition reimbursements (averaging $7,650 per year), research assistantships with full tuition reimbursements (averaging $7,200 per year), teaching assistantships with full tuition reimbursements (averaging $7,200 per year) were awarded.; career-related internships or fieldwork, Federal Work-Study, scholarships/grants, health care benefits, and unspecified assistantships also available. Support available to part-time students. Financial award application deadline: 11/15; financial award applicants required to submit FAFSA. *Faculty research:* Environmental risk assessment, toxicology, work force development, resource ecology, ecosystem management. *Unit head:* Dr. Sharron LaFollette, Program Administrator, 217-206-7894, Fax: 217-206-7807, E-mail: lafollette.sharon@uis.edu.

University of Illinois at Urbana–Champaign, Graduate College, College of Agricultural, Consumer and Environmental Sciences, Department of Natural Resources and Environmental Science, Champaign, IL 61820. Offers MS, PhD. *Faculty:* 46 full-time (6 women), 3 part-time/adjunct (0 women). *Students:* 62 full-time (29 women), 39 part-time (21 women); includes 8 minority (2 African Americans, 5 Asian Americans or Pacific Islanders, 1 Hispanic American), 23 international. 103 applicants, 22% accepted, 13 enrolled. In 2005, 27 master's, 13 doctorates awarded. *Degree requirements:* For master's and doctorate, thesis/dissertation. *Entrance requirements:* For master's and doctorate, GRE, minimum GPA of 3.0. *Application deadline:* Applications are processed on a rolling basis. Application fee: $50 ($60 for international students). Electronic applications accepted. *Financial support:* In 2005–06, 10 fellowships, 63 research assistantships, 20 teaching assistantships were awarded.; tuition waivers (full and partial) also available. Financial award application deadline: 2/15. *Unit head:* Wesley M. Jarrell, Head, 217-333-2770, Fax: 217-244-3219, E-mail: wjarrell@uiuc.edu. *Application contact:* Mary Lowry, Student Services Coordinator, 217-244-5761, Fax: 217-244-3219, E-mail: lowry@uiuc.edu.

University of Illinois at Urbana–Champaign, Graduate College, College of Engineering, Department of Civil and Environmental Engineering, Champaign, IL 61820. Offers civil and environmental engineering (MS, PhD), including environmental engineering, environmental science; civil engineering (MS, PhD). *Faculty:* 48 full-time (4 women), 2 part-time/adjunct (0 women). *Students:* 324 full-time (77 women), 52 part-time (9 women); includes 28 minority (7 African Americans, 1 American Indian/Alaska Native, 14 Asian Americans or Pacific Islanders, 6 Hispanic Americans), 208 international. 743 applicants, 31% accepted, 91 enrolled. In 2005, 108 master's, 35 doctorates awarded. *Degree requirements:* For master's, thesis or alternative; for doctorate, thesis/dissertation. *Application deadline:* For fall admission, 7/2 for domestic students; for spring admission, 10/25 for domestic students. Applications are processed on a rolling basis. Application fee: $50 ($60 for international students). Electronic applications accepted. *Financial support:* In 2005–06, 57 fellowships, 253 research assistantships, 42 teaching assistantships were awarded.; tuition waivers (full and partial) also available. Financial award application deadline: 2/15. *Unit head:* Robert H. Dodds, Head, 217-333-3276, Fax: 217-333-9464, E-mail: rdodds@uiuc.edu. *Application contact:* Mary Pearson, Administrative Secretary, 217-333-3811, Fax: 217-333-9464, E-mail: mkpearso@uiuc.edu.

University of Kansas, Graduate School, School of Engineering, Department of Civil, Environmental, and Architectural Engineering, Program in Environmental Science, Lawrence, KS 66045. Offers MS, PhD. *Students:* 6 full-time (3 women), 6 part-time (3 women); includes 1 minority (Hispanic American) Average age 29. 7 applicants, 71% accepted. In 2005, 3 degrees awarded. *Degree requirements:* For doctorate, thesis/dissertation, comprehensive exam. *Entrance requirements:* For master's, GRE, minimum GPA of 3.0; for doctorate, GRE, minimum GPA of 3.5. Additional exam requirements/recommendations for international students: Required—TOEFL. *Application deadline:* Applications are processed on a rolling basis. Electronic applications accepted. *Expenses:* Tuition, state resident: full-time $4,859. Tuition, nonresident: full-time $1,200. Required fees: $589. Tuition and fees vary according to program. *Financial support:* Fellowships, research assistantships, teaching assistantships available. Financial award application deadline: 2/7. *Application contact:* Bruce M. McEnroe, Graduate Advisor, E-mail: mcenroe@ku.edu.

University of Lethbridge, School of Graduate Studies, Lethbridge, AB T1K 3M4, Canada. Offers accounting (MScM); addictions counseling (M Sc); agricultural biotechnology (M Sc); agricultural studies (M Sc, MA); anthropology (MA); archaeology (MA); art (MA); biochemistry (M Sc); biological sciences (M Sc); biomolecular science (PhD); biosystems and biodiversity (PhD); Canadian studies (MA); chemistry (M Sc); computer science (M Sc); computer science and geographical information science (M Sc); counseling psychology (M Ed); dramatic arts (MA); earth, space, and physical science (PhD); economics (MA); educational leadership (M Ed); English (MA); environmental science (M Sc); evolution and behavior (PhD); exercise science (M Sc); finance (MScM); French (MA); French/German (MA); French/Spanish (MA); general education (M Ed); general management (MScM); geography (M Sc, MA); German (MA); health sciences (M Sc, MA); history (MA); human resource management and labour relations (MScM); individualized multidisciplinary (M Sc, MA); information systems (MScM); international management (MScM); kinesiology (M Sc, MA); management (M Sc, MA); marketing (MScM); mathematics (M Sc); music (MA); Native American studies (MA); neuroscience (M Sc, PhD); new media (MA); nursing (M Sc); philosophy (MA); physics (M Sc); policy and strategy (MScM); political science (MA); psychology (M Sc, MA); religious studies (MA); sociology (MA); theoretical and computational science (PhD); urban and regional studies (MA). Part-time and evening/weekend programs available. *Faculty:* 250. *Students:* 193 full-time, 145 part-time. 35 applicants, 100% accepted, 35 enrolled. In 2005, 40 degrees awarded. *Degree requirements:* For doctorate, thesis/dissertation, comprehensive exam. *Entrance requirements:* For master's, GMAT (M Sc management), bachelor's degree in related field, minimum GPA of 3.0 during previous 20 graded semester courses, 2 years teaching or related experience (M Ed); for doctorate, master's degree, minimum graduate GPA of 3.5. Additional exam requirements/recommendations for international students: Required—TOEFL. Application fee: $60 Canadian dollars. *Expenses:* Tuition, nonresident: part-time $531 per course. Required fees: $83 per year. Tuition and fees vary according to degree level and program. *Financial support:* Fellowships, research assistantships, teaching assistantships, scholarships/grants, health care benefits, and unspecified assistantships available. *Faculty research:* Movement and brain plasticity, gibberellin physiology, photosynthesis, carbon cycling, molecular properties of main-group ring components. *Unit head:* Dr. Shamsul Alam, Dean, 403-329-2121, Fax: 403-329-2097, E-mail: inquiries@uleth.ca. *Application contact:* Kathy Schrage, Administrative Assistant, Office of the Academic Vice President, 403-329-2121, Fax: 403-329-2097, E-mail: inquiries@uleth.ca.

University of Maine, Graduate School, College of Natural Sciences, Forestry, and Agriculture, Department of Biological Sciences, Program in Ecology and Environmental Science, Orono, ME 04469. Offers MS, PhD. Part-time programs available. *Students:* 26 full-time (17 women), 21 part-time (14 women), 2 international. Average age 30. 56 applicants, 20% accepted, 8 enrolled. In 2005, 11 master's, 4 doctorates awarded. *Degree requirements:* For doctorate, thesis/dissertation. *Entrance requirements:* For master's and doctorate, GRE General Test. Additional exam requirements/recommendations for international students: Required—TOEFL. *Application deadline:* For fall admission, 2/1 for domestic students. Applications are processed on a rolling basis. Application fee: $50. Electronic applications accepted. *Financial support:* Fellowships, research assistantships with tuition reimbursements, teaching assistantships with tuition reimbursements, career-related internships or fieldwork, Federal Work-Study, institutionally sponsored loans, and tuition waivers (full) available. Financial award application deadline: 3/1. *Unit head:* Dr. William Glanz, Coordinator, 207-581-2545. *Application contact:* Scott G. Delcourt, Associate Dean of the Graduate School, 207-581-3219, Fax: 207-581-3232, E-mail: graduate@maine.edu.

University of Maine, Graduate School, College of Natural Sciences, Forestry, and Agriculture, Department of Plant, Soil, and Environmental Sciences, Orono, ME 04469. Offers biological sciences (PhD); ecology and environmental sciences (MS, PhD); forest resources (PhD); horticulture (MS); plant science (PhD); plant, soil, and environmental sciences (MS); resource utilization (MS). *Faculty:* 25. *Students:* 15 full-time (8 women), 8 part-time (6 women), 3 international. Average age 32. 6 applicants, 33% accepted, 1 enrolled. In 2005, 9 master's awarded. *Entrance requirements:* For master's and doctorate, GRE General Test. Additional exam requirements/recommendations for international students: Required—TOEFL. *Application deadline:* Applications are processed on a rolling basis. Application fee: $50. Electronic applications accepted. *Financial support:* In 2005–06, 9 research assistantships with tuition reimbursements (averaging $12,180 per year) were awarded; teaching assistantships, scholarships/grants, tuition waivers (full and partial), and unspecified assistantships also available. *Unit head:* Greg Porter, Chair, 207-581-2943, Fax: 207-581-3207. *Application contact:* Scott G. Delcourt, Associate Dean of the Graduate School, 207-581-3219, Fax: 207-581-3232, E-mail: graduate@maine.edu.

University of Maryland, Graduate School, Program in Marine-Estuarine-Environmental Sciences, Baltimore, MD 21201. Offers MS, PhD. An intercampus, interdisciplinary program. Part-time programs available. *Faculty:* 8. *Students:* 2 full-time (0 women), 1 (woman) part-time, 1 international. 1 applicant, 100% accepted, 1 enrolled.Terminal master's awarded for partial completion of doctoral program. *Degree requirements:* For master's, thesis; for doctorate, thesis/dissertation, proposal defense, comprehensive exam. *Entrance requirements:* For master's and doctorate, GRE General Test, minimum GPA of 3.0. Additional exam requirements/recommendations for international students: Required—TOEFL. *Application deadline:* For fall admission, 2/1 for domestic students; for spring admission, 9/1 for domestic students. Applications are processed on a rolling basis. Application fee: $50. Electronic applications accepted. *Expenses:* Tuition, state resident: full-time $8,079; part-time $409 per credit hour. Tuition, nonresident: full-time $18,384; part-time $731 per credit hour. Required fees: $695; $10 per credit hour. Tuition and fees vary according to degree level and program. *Financial support:* Research assistantships with tuition reimbursements, teaching assistantships with tuition reimbursements, scholarships/grants and unspecified assistantships available. *Unit head:* Dr. Kennedy T. Paynter, Director, 301-405-6938, Fax: 301-314-4139, E-mail: mees@mees.umd.edu.

See Close-Up on page 271.

University of Maryland, Baltimore County, Graduate School, College of Natural Sciences and Mathematics, Department of Biological Sciences, Program in Marine-Estuarine-Environmental Sciences, Baltimore, MD 21250. Offers MS, PhD. Part-time programs available. *Faculty:* 17. *Students:* 9 full-time (6 women); includes 2 minority (both African Americans), 3 international. 16 applicants, 25% accepted, 2 enrolled. *Degree requirements:* For master's, thesis; for doctorate, thesis/dissertation, proposal defense, comprehensive exam (for some programs). *Entrance requirements:* For master's and doctorate, GRE General Test, minimum GPA of 3.0. Additional exam requirements/recommendations for international students: Required—TOEFL. *Application deadline:* For fall admission, 2/1 for domestic students, 1/1 for international students; for spring admission, 9/1 for domestic students. Applications are processed on a rolling basis. Application fee: $50. Electronic applications accepted. *Expenses:* Tuition, state resident: part-time $395 per credit. Tuition, nonresident: part-time $652 per credit. Required fees: $82 per credit. Tuition and fees vary according to course load, program and reciprocity agreements. *Financial support:* In 2005–06, 11 students received support, including 1 fellowship with tuition reimbursement available (averaging $22,500 per year), research assistantships with tuition reimbursements available (averaging $21,000 per year), teaching assistantships with tuition reimbursements available (averaging $20,000 per year); career-related internships or fieldwork, scholarships/grants, and unspecified assistantships also available. Financial award application deadline: 1/1. *Unit head:* Dr. Kennedy T. Paynter, Director, 301-405-6938, Fax: 301-314-4139, E-mail: mees@mees.umd.edu. *Application contact:* Dr. Thomas Cronin, Graduate Program Director, 410-455-3669, Fax: 410-455-3875, E-mail: biograd@umbc.edu.

University of Maryland, College Park, Graduate Studies, College of Chemical and Life Sciences, Program in Marine-Estuarine-Environmental Sciences, College Park, MD 20742. Offers MS, PhD. An intercampus, interdisciplinary program. Part-time programs available. *Faculty:* 135. *Students:* 145 (73 women); includes 6 minority (1 African American, 1 American Indian/Alaska Native, 2 Asian Americans or Pacific Islanders, 2 Hispanic Americans) 29 international. 118 applicants, 22% accepted, 15 enrolled. In 2005, 22 master's, 10 doctorates awarded. Terminal master's awarded for partial completion of doctoral program. *Degree requirements:* For master's, thesis, oral defense; for doctorate, thesis/dissertation, proposal defense, comprehensive exam. *Entrance requirements:* For master's and doctorate, GRE General Test, minimum GPA of 3.0. Additional exam requirements/recommendations for international students: Required—TOEFL. *Application deadline:* For fall admission, 2/1 for domestic students, 2/1 for international students; for spring admission, 9/1 for domestic students, 6/1 for international students. Applications are processed on a rolling basis. Application fee: $60. Electronic applications accepted. *Financial support:* In 2005–06, 9 teaching assistantships with full tuition reimbursements were awarded; fellowships with full tuition reimbursements, research assistantships with full tuition reimbursements, Federal Work-Study, scholarships/grants, traineeships, health care benefits, and unspecified assistantships also available. Financial award application deadline: 1/1; financial award applicants required to submit FAFSA. *Faculty research:* Marine and estuarine organisms, terrestrial and freshwater ecology, remote environmental sensing. *Unit head:* Dr. Kennedy T. Paynter, Director, 301-405-6938, Fax: 301-314-4139, E-mail: mees@mees.umd.edu.

University of Maryland Eastern Shore, Graduate Programs, Department of Natural Sciences, Program in Marine-Estuarine-Environmental Sciences, Princess Anne, MD 21853-1299. Offers MS, PhD. Part-time programs available. *Faculty:* 30. *Students:* 36 (18 women); includes 12 minority (10 African Americans, 1 Asian American or Pacific Islander, 1 Hispanic American) 14 international. 28 applicants, 57% accepted, 13 enrolled. In 2005, 5 master's awarded. *Degree requirements:* For master's, thesis; for doctorate, thesis/dissertation, proposal defense, comprehensive exam. *Entrance requirements:* For master's and doctorate, GRE General Test, minimum GPA of 3.0. Additional exam requirements/recommendations for international students: Required—TOEFL. *Application deadline:* For fall admission, 2/1 for domestic students; for spring admission, 9/1 for domestic students. Applications are processed on a rolling basis. Application fee: $30. Electronic applications accepted. *Expenses:* Tuition, area resident: Part-time $216. Tuition, nonresident: part-time $392. Required fees: $40. *Financial support:* In 2005–06, 30 students received support; fellowships with tuition reimbursements available, research assistantships with tuition reimbursements available, teaching assistantships with tuition reimbursements available, career-related internships or fieldwork, scholarships/grants, and unspecified assistantships available. Support available to part-time students. Financial award application deadline: 1/1. *Unit head:* Dr. Kennedy T. Paynter, Director, 301-405-6938, Fax: 301-314-4139, E-mail: mees@mees.umd.edu.

University of Massachusetts Boston, Office of Graduate Studies and Research, College of Science and Mathematics, Department of Environmental, Coastal and Ocean Sciences, Boston, MA 02125-3393. Offers environmental biology (PhD); environmental sciences (MS); environmental, coastal and ocean sciences (PhD). Part-time and evening/weekend programs available. *Degree requirements:* For master's, thesis; for doctorate, thesis/dissertation, oral exams, comprehensive exam. *Entrance requirements:* For master's and doctorate, GRE General Test, minimum GPA of 2.75. *Faculty research:* In situ instrumentation, benthic ecology, watershed, estuarine and coastal systems, functional mechanisms in aquatic toxicology, marine fisheries economics and management.

University of Massachusetts Lowell, Graduate School, College of Arts and Sciences, Department of Chemistry, Lowell, MA 01854-2881. Offers biochemistry (PhD); chemistry (MS, PhD); environmental studies (PhD); polymer sciences (MS, PhD). Terminal master's awarded for partial completion of doctoral program. *Degree requirements:* For master's, thesis; for doctorate, 2 foreign languages, thesis/dissertation. *Entrance requirements:* For master's and doctorate, GRE General Test. Electronic applications accepted.

University of Massachusetts Lowell, Graduate School, James B. Francis College of Engineering, Department of Civil Engineering and College of Arts and Sciences, Program in Environmental Studies, Lowell, MA 01854-2881. Offers MS Eng. Part-time programs available. *Degree requirements:* For master's, thesis optional. *Entrance requirements:* For master's, GRE General Test. *Faculty research:* Remote sensing of air pollutants, atmospheric deposition of toxic metals, contaminant transport in groundwater, soil remediation.

University of Massachusetts Lowell, Graduate School, James B. Francis College of Engineering, Department of Work Environment, Lowell, MA 01854-2881. Offers cleaner production and pollution prevention (MS, Sc D); environmental risk assessment (Certificate); identification and control of ergonomic hazards (Certificate); industrial hygiene (MS, Sc D); job stress and healthy job redesign (Certificate); occupational epidemiology (MS, Sc D); occupational ergonomics (MS, Sc D); radiological health physics and general work environment protection (Certificate); work environmental policy (MS, Sc D). *Accreditation:* ABET (one or more programs are accredited). Part-time programs available. Terminal master's awarded for partial completion of doctoral program. *Degree requirements:* For master's, thesis optional; for doctorate, thesis/dissertation. *Entrance requirements:* For master's and doctorate, GRE General Test. Additional exam requirements/recommendations for international students: Required—TOEFL. *Faculty research:* Ergonomics, industrial hygiene, epidemiology, work environment policy, pollution prevention.

University of Medicine and Dentistry of New Jersey, Graduate School of Biomedical Sciences, Graduate Programs in Biomedical Sciences–Piscataway, Program in Environmental Sciences/Exposure Assessment, Piscataway, NJ 08854-5635. Offers PhD, MD/PhD. *Application deadline:* For fall admission, 1/5 for domestic students. Applications are processed on a

Environmental Sciences

University of Medicine and Dentistry of New Jersey (continued)

rolling basis. Application fee: $40. *Financial support:* Application deadline: 5/1; *Unit head:* Dr. Clifford Weisel, Director, 732-932-5205, Fax: 732-932-3562.

University of Michigan–Dearborn, College of Arts, Sciences, and Letters, Program in Environmental Science, Dearborn, MI 48128-1491. Offers MS. Part-time and evening/weekend programs available. *Faculty:* 9 full-time (2 women), 3 part-time/adjunct (0 women). *Students:* 4 full-time (2 women), 26 part-time (14 women); includes 2 minority (1 African American, 1 Asian American or Pacific Islander). Average age 34. 6 applicants, 83% accepted. In 2005, 6 degrees awarded. *Degree requirements:* For master's, thesis optional. *Entrance requirements:* For master's, letters of reference, minimum GPA of 3.0. Additional exam requirements/recommendations for international students: Required—TOEFL (minimum score 560 paper-based; 220 computer-based). *Application deadline:* For fall admission, 8/1 for domestic students. For winter admission, 12/1 for domestic students; for spring admission, 4/1 for domestic students. Applications are processed on a rolling basis. Application fee: $60 ($75 for international students). Electronic applications accepted. *Financial support:* In 2005–06, 1 fellowship (averaging $2,500 per year), 2 research assistantships (averaging $2,500 per year) were awarded. Financial award application deadline: 4/1; financial award applicants required to submit FAFSA. *Faculty research:* Heavy metal and PAH containment/remediation, land use and impact on ground water and surface water quality, ecosystems and management, natural resources, plant and animal diversity. *Unit head:* Dr. Kent Murray, Professor, E-mail: kmurray@umd.umich.edu. *Application contact:* Carol Ligienza, Administrative Coordinator, Case Graduate Programs, 313-593-1183, Fax: 313-583-6498, E-mail: caslgrad@umd.umich.edu.

The University of Montana, Graduate School, College of Arts and Sciences, Program in Environmental Studies (EVST), Missoula, MT 59812-0002. Offers MS, JD/MS. Part-time programs available. *Faculty:* 6 full-time (2 women), 3 part-time/adjunct (1 woman). *Students:* 44 full-time (23 women), 39 part-time (25 women); includes 1 minority (American Indian/Alaska Native). Average age 28. 98 applicants, 59% accepted, 30 enrolled. In 2005, 37 degrees awarded. *Degree requirements:* For master's, portfolio, professional paper or thesis. *Entrance requirements:* For master's, GRE General Test. Additional exam requirements/recommendations for international students: Required—TOEFL (minimum score 580 paper-based; 237 computer-based). *Application deadline:* For fall admission, 2/15 for domestic students, 2/15 for international students. Application fee: $55. *Expenses:* Tuition, state resident: part-time $267 per credit. Tuition, nonresident: part-time $665 per credit. Part-time tuition and fees vary according to course load and degree level. *Financial support:* In 2005–06, 10 students received support, including 5 fellowships with full tuition reimbursements available (averaging $3,000 per year), 5 teaching assistantships with full tuition reimbursements available (averaging $9,000 per year); career-related internships or fieldwork and Federal Work-Study also available. Support available to part-time students. Financial award application deadline: 4/15. *Faculty research:* Pollution ecology, sustainable agriculture, environmental writing, environmental policy, habitat-land management. Total annual research expenditures: $367,111. *Unit head:* Len Broberg, Director, 406-243-5209, Fax: 406-243-6090, E-mail: len.broberg@umontana.edu. *Application contact:* Karen Hurd, Administrative Assistant, 406-243-6273, Fax: 406-243-6090, E-mail: karen.hurd@umontana.edu.

Announcement: Interdisciplinary program emphasizing activism. Offerings: environmental science, education, justice, law, policy, writing; water issues; sustainable agriculture; public land ecosystem management. Seeks to provide the literacy and skills needed to foster a healthy environment and to create a more sustainable, equitable society. Accepts students from all disciplines. Web site: http://www.umt.edu/evst/.

University of Nevada, Las Vegas, Graduate College, College of Science, Department of Chemistry, Las Vegas, NV 89154-9900. Offers biochemistry (MS); chemistry (MS); environmental science/chemistry (PhD); radiochemistry (PhD). Part-time programs available. *Faculty:* 19 full-time (3 women), 5 part-time/adjunct (1 woman). *Students:* 17 full-time (6 women), 18 part-time (13 women); includes 6 minority (3 Asian Americans or Pacific Islanders, 3 Hispanic Americans), 11 international. 30 applicants, 70% accepted, 17 enrolled. In 2005, 7 degrees awarded. *Degree requirements:* For master's, thesis. *Entrance requirements:* For master's, GRE General Test, minimum GPA of 3.0 in last 2 years or 2.75 cumulative. Additional exam requirements/recommendations for international students: Required—TOEFL (minimum score 550 paper-based; 213 computer-based). *Application deadline:* For fall admission, 6/15 for domestic students, 5/1 for international students; for spring admission, 11/15 for domestic students, 10/1 for international students. Application fee: $60 ($75 for international students). Electronic applications accepted. *Expenses:* Tuition, state resident: part-time $150 per credit. Tuition, nonresident: part-time $315 per credit. Tuition and fees vary according to course load, program and reciprocity agreements. *Financial support:* In 2005–06, 3 research assistantships with partial tuition reimbursements (averaging $1,000 per year), 9 teaching assistantships with full tuition reimbursements (averaging $10,000 per year) were awarded.; career-related internships or fieldwork, Federal Work-Study, institutionally sponsored loans, scholarships/grants, health care benefits, and unspecified assistantships also available. Support available to part-time students. Financial award application deadline: 3/1. *Unit head:* Dr. Spencer Steinberg, Chair, 702-895-3510. *Application contact:* Graduate Coordinator, 702-895-3753, Fax: 702-895-4180, E-mail: gradcollege@unlv.edu.

University of Nevada, Las Vegas, Graduate College, Greenspun College of Urban Affairs, Department of Environmental Studies, Las Vegas, NV 89154-9900. Offers environmental science (MS, PhD). Part-time programs available. *Faculty:* 8 full-time (2 women), 11 part-time/adjunct (3 women). *Students:* 5 full-time (4 women), 24 part-time (14 women); includes 3 minority (1 African American, 2 Hispanic Americans), 1 international. 16 applicants, 56% accepted, 7 enrolled. In 2005, 4 master's, 3 doctorates awarded. *Degree requirements:* For master's and doctorate, thesis/dissertation, comprehensive exam (for some programs). *Entrance requirements:* For master's and doctorate, GRE General Test, minimum GPA of 3.0. Additional exam requirements/recommendations for international students: Required—TOEFL (minimum score 550 paper-based; 213 computer-based). *Application deadline:* For fall admission, 6/15 for domestic students, 5/1 for international students; for spring admission, 11/15 for domestic students, 10/1 for international students. Application fee: $60 ($75 for international students). Electronic applications accepted. *Expenses:* Tuition, state resident: part-time $150 per credit. Tuition, nonresident: part-time $315 per credit. Tuition and fees vary according to course load, program and reciprocity agreements. *Financial support:* In 2005–06, 4 teaching assistantships with partial tuition reimbursements (averaging $10,000 per year) were awarded; research assistantships with partial tuition reimbursements, career-related internships or fieldwork, Federal Work-Study, institutionally sponsored loans, scholarships/grants, health care benefits, and unspecified assistantships also available. Support available to part-time students. Financial award application deadline: 3/1. *Unit head:* Dr. Helen Neill, Interim Chair, 702-895-4440. *Application contact:* Graduate College Admissions Evaluator, 702-895-3320, Fax: 702-895-4180, E-mail: gradcollege@unlv.edu.

University of Nevada, Reno, Graduate School, College of Agriculture, Biotechnology and Natural Resources, Department of Natural Resources and Environmental Sciences, Reno, NV 89557. Offers MS. *Degree requirements:* For master's, thesis optional. *Entrance requirements:* For master's, GRE, minimum GPA of 2.75. Additional exam requirements/recommendations for international students: Required—TOEFL. *Faculty research:* Range management, plant physiology, remote sensing, soils, wildlife.

University of Nevada, Reno, Graduate School, College of Science, Interdisciplinary Program in Environmental Sciences and Health, Reno, NV 89557. Offers MS, PhD. *Degree requirements:* For master's and doctorate, thesis/dissertation. *Entrance requirements:* For master's, GRE General Test, minimum GPA of 2.75; for doctorate, GRE General Test, minimum GPA of 3.0. Additional exam requirements/recommendations for international students: Required—TOEFL.

University of New Haven, Graduate School, College of Arts and Sciences, Program in Environmental Sciences, West Haven, CT 06516-1916. Offers MS. Part-time and evening/weekend programs available. *Degree requirements:* For master's, thesis or alternative. *Faculty research:* Mapping and assessing geological and living resources in Long Island Sound, geology, San Salvador Island, Bahamas.

See Close-Up on page 679.

The University of North Carolina at Chapel Hill, Graduate School, School of Public Health, Department of Environmental Sciences and Engineering, Chapel Hill, NC 27599. Offers air, radiation and industrial hygiene (MPH, MS, MSEE, MSPH, PhD); aquatic and atmospheric sciences (MPH, MS, MSPH, PhD); environmental engineering (MPH, MS, MSEE, MSPH, PhD); environmental health sciences (MPH, MS, MSPH, PhD); environmental management and policy (MPH, MS, MSPH, PhD). *Faculty:* 33 full-time (3 women), 35 part-time/adjunct. *Students:* 141 full-time (74 women); includes 37 minority (10 African Americans, 25 Asian Americans or Pacific Islanders, 2 Hispanic Americans). Average age 27. 216 applicants, 37% accepted, 29 enrolled. In 2005, 14 master's, 11 doctorates awarded. Terminal master's awarded for partial completion of doctoral program. *Median time to degree:* Of those who began their doctoral program in fall 1997, 100% received their degree in 8 years or less. *Degree requirements:* For master's, thesis (for some programs), research paper, comprehensive exam, registration; for doctorate, thesis/dissertation, comprehensive exam, registration. *Entrance requirements:* For master's and doctorate, GRE General Test, minimum GPA of 3.0. Additional exam requirements/recommendations for international students: Required—TOEFL. *Application deadline:* For fall admission, 1/1 priority date for domestic students, 1/1 priority date for international students; for spring admission, 9/15 for domestic students. Applications are processed on a rolling basis. Application fee: $70. Electronic applications accepted. *Financial support:* In 2005–06, 134 students received support, including 36 fellowships with tuition reimbursements available (averaging $6,358 per year), 86 research assistantships with tuition reimbursements available (averaging $6,197 per year), 12 teaching assistantships with tuition reimbursements available (averaging $6,729 per year); career-related internships or fieldwork, Federal Work-Study, traineeships, health care benefits, and unspecified assistantships also available. Support available to part-time students. Financial award application deadline: 1/1; financial award applicants required to submit FAFSA. *Faculty research:* Air, radiation and industrial hygiene, aquatic and atmospheric sciences, environmental health sciences, environmental management and policy, water resources engineering. Total annual research expenditures: $9.6 million. *Unit head:* Dr. Don Fox, Interim Chair, 919-966-1024, Fax: 919-966-7911, E-mail: don_fox@unc.edu. *Application contact:* Jack Whaley, Registrar, 919-966-3844, Fax: 919-966-7911, E-mail: jack_whaley@unc.edu.

University of Northern Iowa, Graduate College, College of Natural Sciences, Environmental Programs, Cedar Falls, IA 50614. Offers MS. *Faculty:* 10 full-time (1 woman), 2 part-time/adjunct (0 women). *Students:* 6 full-time (3 women), 2 part-time (1 woman); includes 1 minority (Hispanic American), 4 international. 16 applicants, 38% accepted, 5 enrolled. In 2005, 9 degrees awarded. *Degree requirements:* For master's, thesis or alternative, comprehensive exam (for some programs). *Entrance requirements:* Additional exam requirements/recommendations for international students: Required—TOEFL (minimum score 500 paper-based; 180 computer-based). *Application deadline:* For fall admission, 8/1 for domestic students. Applications are processed on a rolling basis. Application fee: $30 ($50 for international students). Electronic applications accepted. *Expenses:* Tuition, state resident: full-time $5,708. Tuition, nonresident: full-time $13,532. Required fees: $712. *Financial support:* Application deadline: 2/1. *Unit head:* Dr. James Walters, Head, 319-273-2759, Fax: 319-273-5815, E-mail: james.walters@uni.edu.

University of North Texas, Robert B. Toulouse School of Graduate Studies, College of Arts and Sciences, Department of Biological Sciences, Program in Environmental Science, Denton, TX 76203. Offers MS, PhD. *Students:* 20 full-time (13 women), 18 part-time (6 women); includes 2 minority (1 African American, 1 Asian American or Pacific Islander), 7 international. Average age 37. *Degree requirements:* For master's, oral defense of thesis; for doctorate, one foreign language, thesis/dissertation, comprehensive exam. *Entrance requirements:* For master's and doctorate, GRE General Test. *Application deadline:* For fall admission, 7/15 for domestic students; for spring admission, 11/1 for domestic students. Application fee: $50 ($75 for international students). *Expenses:* Tuition, state resident: full-time $3,258; part-time $181 per semester hour. Tuition, nonresident: full-time $8,226; part-time $451 per semester hour. Required fees: $1,219; $68 per semester hour. *Unit head:* Dr. Thomas W. LaPoint, Director, 940-565-2694, Fax: 940-565-4297, E-mail: lapoint@unt.edu. *Application contact:* Candy King, Graduate Adviser, 940-565-3599, E-mail: cking@unt.edu.

University of Oklahoma, Graduate College, College of Engineering, School of Civil Engineering and Environmental Science, Program in Environmental Science, Norman, OK 73019-0390. Offers air (M Env Sc); environmental engineering (MS); environmental science (PhD); groundwater management (M Env Sc); hazardous solid waste (M Env Sc); occupational safety and health (M Env Sc); process design (M Env Sc); water quality resources (M Env Sc). Part-time programs available. *Students:* 17 full-time (8 women), 11 part-time (5 women); includes 3 minority (2 African Americans, 1 American Indian/Alaska Native), 13 international. 19 applicants, 63% accepted, 6 enrolled. In 2005, 6 master's, 1 doctorate awarded. Terminal master's awarded for partial completion of doctoral program. *Degree requirements:* For master's, oral exams; for doctorate, thesis/dissertation, oral, and qualifying exams, comprehensive exam. *Entrance requirements:* For master's, minimum GPA of 3.0; for doctorate, minimum graduate GPA of 3.5. Additional exam requirements/recommendations for international students: Required—TOEFL (minimum score 600 paper-based; 250 computer-based). *Application deadline:* For fall admission, 4/1 priority date for domestic students, 4/1 priority date for international students; for spring admission, 11/1 for domestic students, 9/1 for international students. Applications are processed on a rolling basis. Application fee: $40 ($90 for international students). *Expenses:* Tuition, state resident: full-time $3,029; part-time $126 per credit hour. Tuition, nonresident: full-time $10,807; part-time $450 per credit hour. Required fees: $1,231; $44 per credit hour. Tuition and fees vary according to course load and program. *Financial support:* In 2005–06, 8 students received support; fellowships, research assistantships with partial tuition reimbursements available, teaching assistantships with partial tuition reimbursements available, scholarships/grants available. Financial award application deadline: 3/1; financial award applicants required to submit FAFSA. *Application contact:* Susan Williams, Graduate Programs Specialist, 405-325-2344, Fax: 405-325-4217, E-mail: srwilliams@ou.edu.

University of Rhode Island, Graduate School, College of the Environment and Life Sciences, Department of Fisheries, Animal and Veterinary Science, Kingston, RI 02881. Offers animal health and disease (MS); animal science (MS); aquaculture (MS); aquatic pathology (MS); environmental sciences (PhD), including animal science, aquacultural science, aquatic pathology, fisheries science; fisheries (MS). In 2005, 6 degrees awarded. *Application deadline:* For fall admission, 4/15 for domestic students. Applications are processed on a rolling basis. Application fee: $35. *Expenses:* Tuition, state resident: full-time $5,522; part-time $307 per credit. Tuition, nonresident: full-time $15,992; part-time $888 per credit. Required fees: $1,786; $73 per credit. One-time fee: $80 part-time. *Unit head:* Dr. David Bengtson, Chairperson, 401-874-2688.

University of South Carolina, The Graduate School, College of Arts and Sciences, Department of Geological Sciences, Columbia, SC 29208. Offers environmental geoscience (PMS); geological sciences (MS, PhD). Terminal master's awarded for partial completion of doctoral program. *Degree requirements:* For master's, thesis; for doctorate, thesis/dissertation, published paper, comprehensive exam. *Entrance requirements:* For master's and doctorate, GRE General Test. Additional exam requirements/recommendations for international students: Required—TOEFL. Electronic applications accepted. *Faculty research:* Environmental geology, tectonics, petrology, coastal processes, paleoclimatology.

University of South Florida, College of Graduate Studies, College of Arts and Sciences, Department of Environmental Science and Policy, Tampa, FL 33620-9951. Offers MS. *Faculty:* 5 full-time (2 women), 1 part-time/adjunct (0 women). *Students:* 11 full-time (8 women), 13 part-time (10 women); includes 2 minority (1 African American, 1 Hispanic American), 7 international. 31 applicants, 48% accepted, 8 enrolled. In 2005, 1 degree awarded. *Degree requirements:* For master's, thesis optional. *Entrance requirements:* For master's, GRE General Test, minimum GPA of 3.0 in last 60 hours of course work. *Application deadline:* For fall admission, 5/1 for domestic students; for spring admission, 9/15 for domestic students. Application fee: $30. *Financial support:* Fellowships with tuition reimbursements, research assistantships with full tuition reimbursements, teaching assistantships with full tuition reimbursements, scholarships/grants and unspecified assistantships available. Support available to part-time students. Financial award application deadline: 5/1. *Unit head:* Rick Oches, Interim Director, 813-974-2978. *Application contact:* Dr. Ingrid Bartsch, Information Contact, 813-974-3069, E-mail: klschrad@chumal.cas.usf.edu.

The University of Tennessee at Chattanooga, Graduate School, College of Arts and Sciences, Department of Biological and Environmental Sciences, Program in Environmental Sciences, Chattanooga, TN 37403-2598. Offers MS. Part-time programs available. *Faculty:* 6 full-time (0 women). *Students:* 18 full-time (11 women), 24 part-time (7 women); includes 1 minority (Asian American or Pacific Islander), 1 international. Average age 31. 16 applicants, 100% accepted, 9 enrolled. In 2005, 11 degrees awarded. *Degree requirements:* For master's, thesis optional. *Entrance requirements:* For master's, GRE General Test, minimum undergraduate GPA of 2.75. *Application deadline:* For fall admission, 8/1 for domestic students; for spring admission, 12/1 priority date for domestic students. Applications are processed on a rolling basis. Application fee: $25. *Expenses:* Tuition, state resident: full-time $5,210; part-time $327 per hour. Tuition, nonresident: full-time $14,234; part-time $829 per hour. Required fees: $900. *Financial support:* Application deadline: 4/1; *Unit head:* Dr. John Tucker, Coordinator, 423-425-2316, Fax: 423-425-2285, E-mail: john-tucker@utc.edu. *Application contact:* Dr. Deborah E. Arfken, Dean of Graduate Studies, 423-425-4666, Fax: 423-425-5223, E-mail: deborah-arfken@utc.edu.

The University of Texas at Arlington, Graduate School, College of Science, Department of Geology, Arlington, TX 76019. Offers environmental science (MS, PhD); geology (MS); math: geoscience (PhD). Part-time and evening/weekend programs available. *Faculty:* 4 full-time (0 women), 2 part-time/adjunct (0 women). *Students:* 5 full-time (2 women), 7 part-time (4 women); includes 1 minority (African American), 2 international. 5 applicants, 100% accepted, 2 enrolled. In 2005, 2 master's awarded. Terminal master's awarded for partial completion of doctoral program. *Degree requirements:* For master's, thesis optional; for doctorate, thesis/dissertation, comprehensive exam. *Entrance requirements:* For master's, GRE General Test. Additional exam requirements/recommendations for international students: Required—TOEFL (minimum score 550 paper-based; 213 computer-based). *Application deadline:* For fall admission, 6/16 for domestic students. Applications are processed on a rolling basis. Application fee: $35 ($50 for international students). Electronic applications accepted. *Expenses:* Tuition, state resident: full-time $3,350. Tuition, nonresident: full-time $8,318. International tuition: $8,448 full-time. Required fees: $1,277. Full-time tuition and fees vary according to course level and program. *Financial support:* In 2005–06, 7 students received support, including 4 fellowships (averaging $1,000 per year), 7 teaching assistantships (averaging $14,700 per year); career-related internships or fieldwork, Federal Work-Study, institutionally sponsored loans, scholarships/grants, health care benefits, and unspecified assistantships also available. Financial award application deadline: 6/1; financial award applicants required to submit FAFSA. *Faculty research:* Hydrology, aqueous geochemistry, biostratigraphy, structural geology, petroleum geology. Total annual research expenditures: $250,000. *Unit head:* Dr. John S. Wickham, Chair, 817-272-2987, Fax: 817-272-2628, E-mail: wickham@uta.edu. *Application contact:* Dr. William L. Balsam, Graduate Adviser, 817-272-2987, Fax: 817-272-2628, E-mail: balsam@uta.edu.

The University of Texas at Arlington, Graduate School, College of Science, Program in Environmental and Earth Sciences, Arlington, TX 76019. Offers MS, PhD. *Students:* 2 full-time (1 woman), 1 (woman) part-time. 4 applicants, 100% accepted, 1 enrolled. In 2005, 2 master's awarded. *Expenses:* Tuition, state resident: full-time $3,350. Tuition, nonresident: full-time $8,318. International tuition: $8,448 full-time. Required fees: $1,277. Full-time tuition and fees vary according to course level and program. *Application contact:* Dr. Robert F. McMahon, Director, 817-272-3492, Fax: 817-272-3511, E-mail: r.mcmahon@uta.edu.

The University of Texas at Arlington, Graduate School, College of Science, Program in Environmental Science and Engineering, Arlington, TX 76019. Offers MS, PhD. Part-time programs available. *Students:* 9 full-time (5 women), 5 part-time (1 woman); includes 1 minority (African American), 5 international. In 2005, 3 master's, 2 doctorates awarded. Terminal master's awarded for partial completion of doctoral program. *Degree requirements:* For master's, oral defense of thesis, thesis optional; for doctorate, thesis/dissertation, oral defense of thesis, comprehensive exam. *Entrance requirements:* For master's, GRE General Test, minimum GPA of 3.0 in the last 60 hours course work; for doctorate, GRE General Test, minimum graduate GPA of 3.0 in the last 60 hours of course work. Additional exam requirements/recommendations for international students: Required—TOEFL (minimum score 550 paper-based; 213 computer-based). *Application deadline:* For fall admission, 6/16 for domestic students. Applications are processed on a rolling basis. Application fee: $35 ($50 for international students). Electronic applications accepted. *Expenses:* Tuition, state resident: full-time $3,350. Tuition, nonresident: full-time $8,318. International tuition: $8,448 full-time. Required fees: $1,277. Full-time tuition and fees vary according to course level and program. *Financial support:* In 2005–06, 6 students received support, including 4 fellowships (averaging $1,000 per year), 2 research assistantships (averaging $15,500 per year); institutionally sponsored loans, scholarships/grants, tuition waivers (partial), and unspecified assistantships also available. Financial award application deadline: 6/1; financial award applicants required to submit FAFSA. *Faculty research:* Water quality, aquatic ecology, treatment systems, air quality. *Unit head:* Dr. John S. Wickham, Chair, 817-272-2987, Fax: 817-272-2628, E-mail: wickham@uta.edu. *Application contact:* Dr. Andrew P. Kruzic, Graduate Advisor, 817-272-3822, Fax: 817-272-2830, E-mail: kruzic@uta.edu.

The University of Texas at El Paso, Graduate School, College of Science, Department of Biological Sciences, El Paso, TX 79968-0001. Offers bioinformatics (MS); biological science (MS, PhD); environmental science and engineering (PhD). Part-time and evening/weekend programs available. *Degree requirements:* For master's, thesis. *Entrance requirements:* For master's, GRE General Test, minimum GPA of 3.0; for doctorate, GRE General Test. Additional exam requirements/recommendations for international students: Required—TOEFL. Electronic applications accepted.

The University of Texas at El Paso, Graduate School, Interdisciplinary Program in Environmental Science and Engineering, El Paso, TX 79968-0001. Offers PhD. Part-time and evening/weekend programs available. *Degree requirements:* For doctorate, thesis/dissertation. *Entrance requirements:* For doctorate, GRE General Test, minimum GPA of 3.0. Additional exam requirements/recommendations for international students: Required—TOEFL.

The University of Texas at San Antonio, College of Sciences, Department of Earth and Environmental Sciences, San Antonio, TX 78249-0617. Offers environmental science and engineering (PhD); environmental sciences (MS); geology (MS). *Degree requirements:* For master's, thesis optional; for doctorate, thesis/dissertation, comprehensive exam, registration. *Entrance requirements:* For master's, GRE General Test, minimum GPA of 3.0 in last 60 hours; for doctorate, GRE, resumé, 3 letters of recommendation. Additional exam requirements/recommendations for international students: Required—TOEFL (minimum score 500 paper-based; 173 computer-based). Electronic applications accepted.

University of Utah, The Graduate School, Program in Science and Technology, Salt Lake City, UT 84112-1107. Offers biotechnology (PSM); computational science (PSM); environmental science (PSM); sciences instrumental (PSM). *Students:* 16 full-time (11 women), 20 part-time (5 women); includes 1 minority (Asian American or Pacific Islander), 7 international. Average age 36. 20 applicants, 50% accepted, 10 enrolled. In 2005, 2 degrees awarded. *Entrance requirements:* For master's, minimum undergraduate GPA of 3.0. Additional exam requirements/recommendations for international students: Required—TOEFL (minimum score 500 paper-based; 173 computer-based). *Application deadline:* For fall admission, 4/1 for domestic students, 4/1 for international students; for spring admission, 11/1 for domestic students, 11/1 for international students. Application fee: $45 ($65 for international students). *Expenses:* Tuition, state resident: full-time $2,932; part-time $2,212 per term. Tuition, nonresident: full-time $10,350; part-time $7,812 per term. Required fees: $590; $516 per term. Tuition and fees vary according to course load and program. *Financial support:* Applicants required to submit FAFSA. *Application contact:* Jennifer Schmidt, Program Director, 801-585-5630, E-mail: jennifer.schmidt@admin.utah.edu.

University of Virginia, College and Graduate School of Arts and Sciences, Department of Environmental Sciences, Charlottesville, VA 22903. Offers MA, MS, PhD. *Faculty:* 27 full-time (5 women), 1 part-time/adjunct (0 women). *Students:* 79 full-time (36 women), 2 part-time (1 woman); includes 5 minority (1 African American, 1 American Indian/Alaska Native, 3 Asian Americans or Pacific Islanders), 12 international. Average age 29. 55 applicants, 25% accepted, 15 enrolled. In 2005, 12 master's, 7 doctorates awarded. *Degree requirements:* For master's and doctorate, thesis/dissertation. *Entrance requirements:* For master's and doctorate, GRE General Test, GRE Subject Test. *Application deadline:* Applications are processed on a rolling basis. Application fee: $40. Electronic applications accepted. *Expenses:* Tuition, state resident: full-time $7,731. Tuition, nonresident: full-time $18,672. Required fees: $1,479. Full-time tuition and fees vary according to degree level and program. *Financial support:* Applicants required to submit FAFSA. *Unit head:* Joseph Zieman, Chairman, 434-924-7761, Fax: 434-982-2137, E-mail: jcz@virginia.edu. *Application contact:* Peter C. Brunjes, Associate Dean for Graduate Programs and Research, 434-924-7184, Fax: 434-924-6737, E-mail: grad-a-s@virginia.edu.

The University of Western Ontario, Faculty of Graduate Studies, Biosciences Division, Department of Plant Sciences, London, ON N6A 5B8, Canada. Offers plant and environmental sciences (M Sc); plant sciences (M Sc, PhD); plant sciences and environmental sciences (PhD); plant sciences and molecular biology (M Sc, PhD). *Degree requirements:* For master's and doctorate, thesis/dissertation. *Entrance requirements:* For doctorate, M Sc or equivalent. *Faculty research:* Ecology systematics, plant biochemistry and physiology, yeast genetics, molecular biology.

The University of Western Ontario, Faculty of Graduate Studies, Physical Sciences Division, Department of Earth Sciences, London, ON N6A 5B8, Canada. Offers geology (M Sc, PhD); geology and environmental science (M Sc, PhD); geophysics (M Sc, PhD); geophysics and environmental science (M Sc, PhD). *Degree requirements:* For master's, thesis, registration; for doctorate, thesis/dissertation, qualifying exam. *Entrance requirements:* For master's, honors in B Sc; for doctorate, M Sc. Additional exam requirements/recommendations for international students: Required—TOEFL. *Faculty research:* Geophysics, geochemistry, paleontology, sedimentology/stratigraphy, glaciology/quaternary.

University of West Florida, College of Arts and Sciences: Sciences, Department of Environmental Studies, Pensacola, FL 32514-5750. Offers environmental science (MS). Part-time programs available. *Faculty:* 7 full-time (1 woman), 2 part-time/adjunct (0 women). *Students:* 8 full-time (3 women), 11 part-time (7 women); includes 1 minority (American Indian/Alaska Native). Average age 32. 9 applicants, 100% accepted, 8 enrolled. *Entrance requirements:* For master's, GRE General Test. Additional exam requirements/recommendations for international students: Required—TOEFL (minimum score 550 paper-based; 213 computer-based). *Application deadline:* For fall admission, 6/1 for domestic students, 5/15 for international students; for spring admission, 11/1 for domestic students, 10/1 for international students. Application fee: $30. Special rates offered to residents of Alabama. *Expenses:* Tuition, state resident: full-time $5,833; part-time $243 per credit hour. Tuition, nonresident: full-time $21,204; part-time $884 per credit hour. Tuition and fees vary according to campus/location. *Financial support:* In 2005–06, 8 students received support, including 3 research assistantships with full tuition reimbursements available (averaging $7,500 per year); Federal Work-Study, institutionally sponsored loans, scholarships/grants, and tuition waivers (partial) also available. Support available to part-time students. Financial award application deadline: 4/15; financial award applicants required to submit FAFSA. *Unit head:* Dr. Klaus Meyer-Arendt, Chairperson, 850-474-2746.

University of Windsor, Faculty of Graduate Studies and Research, GLIER-Great Lakes Institute for Environmental Research, Windsor, ON N9B 3P4, Canada. Offers environmental science (M Sc, PhD). *Faculty:* 10 full-time (0 women). *Students:* 15 full-time (7 women), 1 (woman) part-time. 13 applicants, 38% accepted. In 2005, 1 degree awarded. *Degree requirements:* For master's and doctorate, thesis/dissertation. *Entrance requirements:* For master's, minimum B+ average; for doctorate, M Sc degree, minimum B+ average. Additional exam requirements/recommendations for international students: Required—TOEFL (minimum score 560 paper-based; 220 computer-based). *Application deadline:* For fall admission, 7/1 for domestic students. Applications are processed on a rolling basis. Application fee: $55. Electronic applications accepted. *Financial support:* In 2005–06, 14 teaching assistantships (averaging $8,956 per year) were awarded; Federal Work-Study, scholarships/grants, tuition waivers (full and partial), and unspecified assistantships also available. Financial award application deadline: 2/15. *Faculty research:* Environmental chemistry and toxicology, conservation and resource management, iron formation geochemistry. *Unit head:* Dr. Brian Fryer, Director, 519-253-3000 Ext. 2732, Fax: 519-971-3616, E-mail: bfryer@uwindsor.ca. *Application contact:* Applicant Services, 519-253-3000 Ext. 6459, Fax: 519-971-3653, E-mail: gradadmit@uwindsor.ca.

University of Wisconsin–Green Bay, Graduate Studies, Program in Environmental Science and Policy, Green Bay, WI 54311-7001. Offers MS. Part-time programs available. *Faculty:* 28 full-time (6 women), 3 part-time/adjunct (0 women). *Students:* 22 full-time (11 women), 24 part-time (12 women); includes 3 minority (all American Indian/Alaska Native), 2 international. Average age 32. 25 applicants, 92% accepted, 11 enrolled. In 2005, 18 degrees awarded. *Degree requirements:* For master's, thesis. *Entrance requirements:* For master's, GRE General Test, minimum GPA of 3.0. *Application deadline:* For fall admission, 8/1 for domestic students; for spring admission, 11/1 for domestic students. Applications are processed on a rolling basis. Application fee: $45. Electronic applications accepted. *Expenses:* Tuition, state resident: full-time $5,619; part-time $312 per credit. Tuition, nonresident: full-time $16,229; part-time $902 per credit. Required fees: $1,148; $64 per credit. Tuition and fees vary according to course load and reciprocity agreements. *Financial support:* In 2005–06, 7 research assistantships with full tuition reimbursements, 11 teaching assistantships with full tuition reimbursements were awarded:; career-related internships or fieldwork, Federal Work-Study, and institutionally sponsored loans also available. Financial award application deadline: 7/15; financial award applicants required to submit FAFSA. *Faculty research:* Bald eagle, parasitic population of domestic and wild animals, resource recovery, anaerobic digestion of organic waste. *Unit head:* Dr. John Stoll, Coordinator, 920-465-2358, E-mail: stollj@uwgb.edu.

University of Wisconsin–Madison, Graduate School, Gaylord Nelson Institute for Environmental Studies, Environmental Monitoring Program, Madison, WI 53706-1380. Offers MS, PhD. Part-time programs available. *Faculty:* 19 part-time/adjunct (5 women). *Students:* 14 full-time, 2 part-time; includes 1 minority (Hispanic American), 2 international. Average age 31. 43 applicants, 12% accepted, 2 enrolled. In 2005, 9 master's awarded. *Median time to degree:* Of those who began their doctoral program in fall 1997, 50% received their degree in 8 years or less. *Degree requirements:* For master's, thesis or alternative; for doctorate, thesis/dissertation. *Entrance requirements:* For master's and doctorate, GRE General Test. Additional exam requirements/recommendations for international students: Required—TOEFL (minimum score 600 paper-based; 250 computer-based). *Application deadline:* For fall admission, 2/1 for domestic students, 2/1 for international students; for spring admission, 10/15 for domestic students, 10/15 for international students. Application fee: $45. Electronic applications accepted. *Financial support:* In 2005–06, 12 students received support, including 1 fellowship with full tuition reimbursement available (averaging $14,400 per year), 7 research assistantships with full tuition reimbursements available (averaging $14,250 per year), 1 teaching assistantship with full tuition reimbursement available (averaging $11,260 per year); career-related intern-

Environmental Sciences

University of Wisconsin–Madison (continued)

ships or fieldwork, Federal Work-Study, scholarships/grants, health care benefits, unspecified assistantships, and project assistantships also available. Financial award application deadline: 1/2. *Faculty research:* Remote sensing, geographic information systems, climate modeling, natural resource management. *Unit head:* Thomas M. Lillesand, Chair, 608-263-3251, Fax: 608-262-2273, E-mail: tmlilles@wisc.edu. *Application contact:* James E. Miller, Associate Student Services Coordinator, 608-263-4373, Fax: 608-262-2273, E-mail: jemiller@wisc.edu.

Vanderbilt University, Graduate School, Department of Earth and Environmental Sciences, Nashville, TN 37240-1001. Offers MS. *Faculty:* 9 full-time (2 women). *Students:* 14 full-time (9 women); includes 1 minority (Hispanic American), 1 international. 33 applicants, 42% accepted, 6 enrolled. In 2005, 4 degrees awarded. *Degree requirements:* For master's, thesis or alternative. *Entrance requirements:* For master's, GRE General Test, GRE Subject Test (recommended). *Application deadline:* For fall admission, 1/15 for domestic students, 1/15 for international students. Application fee: $0. Electronic applications accepted. *Expenses:* Tuition: Full-time $15,396; part-time $1,283 per semester hour. Required fees: $2,202; $1,101 per semester. One-time fee: $30. Tuition and fees vary according to course load, program and student level. *Financial support:* Research assistantships, teaching assistantships with full tuition reimbursements, career-related internships or fieldwork, Federal Work-Study, and institutionally sponsored loans available. Financial award application deadline: 1/15. *Faculty research:* Sedimentology, geochemistry, tectonics, environmental geology, biostratigraphy. *Unit head:* David J. Furbish, Chair, 615-322-2976, Fax: 615-322-2138. *Application contact:* John C. Ayers, Director of Graduate Studies, 615-322-2976, Fax: 615-322-2138, E-mail: john.c.ayers@vanderbilt.edu.

Virginia Commonwealth University, Graduate School, College of Humanities and Sciences, Center for Environmental Studies, Richmond, VA 23284-9005. Offers environmental communication (MIS); environmental health (MIS); environmental policy (MIS); environmental sciences (MIS). *Students:* 10 full-time (8 women), 22 part-time (12 women); includes 5 minority (4 African Americans, 1 Hispanic American), 2 international. Average age 33. 20 applicants, 70% accepted. In 2005, 13 degrees awarded. *Degree requirements:* For master's, thesis. *Entrance requirements:* For master's, GRE General Test. Application fee: $50. *Expenses:* Tuition, state resident: full-time $3,185; part-time $405 per credit. Tuition, nonresident: full-time $7,952; part-time $940 per credit. Required fees: $751 per semester hour. Tuition and fees vary according to course load and program. *Unit head:* Dr. Gregory C. Garman, Director, 804-828-1574, Fax: 804-828-0503, E-mail: gcgarman@vcu.edu.

Virginia Polytechnic Institute and State University, Graduate School, College of Engineering, Department of Civil and Environmental Engineering, Blacksburg, VA 24061. Offers civil engineering (M Eng, MS, PhD); environmental engineering (M Eng, MS); environmental sciences and engineering (MS). *Accreditation:* ABET (one or more programs are accredited). *Faculty:* 44 full-time (7 women). *Students:* 227 full-time (59 women), 82 part-time (22 women); includes 22 minority (6 African Americans, 2 American Indian/Alaska Native, 9 Asian Americans or Pacific Islanders, 5 Hispanic Americans), 118 international. Average age 28. 387 applicants, 50% accepted, 84 enrolled. In 2005, 115 master's, 10 doctorates awarded. *Entrance requirements:* Additional exam requirements/recommendations for international students: Required—TOEFL (minimum score 570 paper-based; 230 computer-based). *Application deadline:* Applications are processed on a rolling basis. Application fee: $45. Electronic applications accepted. *Expenses:* Tuition, state resident: full-time $6,558; part-time $364 per credit. Tuition, nonresident: full-time $11,296; part-time $628 per credit. Required fees: $1,419; $468 per credit. $234 per term. *Financial support:* In 2005–06, 44 fellowships with full tuition reimbursements (averaging $5,013 per year), 96 research assistantships with full tuition reimbursements (averaging $17,307 per year), 40 teaching assistantships with full tuition reimbursements (averaging $14,930 per year) were awarded.; career-related internships or fieldwork, Federal Work-Study, scholarships/grants, and unspecified assistantships also available. Financial award application deadline: 4/1. *Faculty research:* Construction, environmental geotechnical hydrosystems, structures and transportation engineering. *Unit head:* Dr. William Knocke, Head, 540-231-6635, Fax: 540-231-7532, E-mail: knocke@vt.edu. *Application contact:* Lindy Cranwell, Information Contact, 540-231-7296, Fax: 540-231-7532, E-mail: lindycra@vt.edu.

Washington State University, Graduate School, College of Sciences, Programs in Environmental Science and Regional Planning, Program in Environmental Science and Regional Planning, Pullman, WA 99164. Offers environmental and natural resource sciences (PhD); environmental science (MS). *Faculty:* 7. *Students:* 13 full-time (7 women), 5 part-time (3 women); includes 1 minority (Hispanic American), 2 international. Average age 29. 49 applicants, 59% accepted, 10 enrolled. In 2005, 11 degrees awarded. *Degree requirements:* For master's, oral exam, thesis optional; for doctorate, oral exam, written exam. *Entrance requirements:* For master's and doctorate, minimum GPA of 3.0. Additional exam requirements/recommendations for international students: Required—TOEFL. *Application deadline:* For fall admission, 2/1 priority date for domestic students, 2/1 priority date for international students. Applications are processed on a rolling basis. Application fee: $35. *Expenses:* Tuition, state resident: full-time $6,295; part-time $336 per credit. Tuition, nonresident: full-time $15,949; part-time $819 per credit. Required fees: $933. Part-time tuition and fees vary according to campus/location and program. *Financial support:* In 2005–06, 16 students received support, including 1 fellowship (averaging $3,000 per year), 4 research assistantships with full and partial tuition reimbursements available (averaging $14,000 per year), 4 teaching assistantships with full and partial tuition reimbursements available (averaging $13,600 per year); Federal Work-Study, institutionally sponsored loans, and tuition waivers (partial) also available. Financial award application deadline: 4/1; financial award applicants required to submit FAFSA. *Application contact:* Elaine O'Fallon, Coordinator, 509-335-8538, Fax: 509-335-7636, E-mail: esrp@wsu.edu.

Announcement: The Program in Environmental Science and Regional Planning awards the MS in environmental science and the PhD in environmental and natural resource sciences. The master's degree in environmental science is also offered at the WSU-Tri-Cities and WSU Vancouver campuses. Teaching assistantships, tuition waivers, Federal Work-Study, and institutionally sponsored loans are available. Visit the Web site at http://esrp.wsu.edu.

Washington State University Tri-Cities, Graduate Programs, Program in Environmental Science, Richland, WA 99352-1671. Offers applied environmental science (MS); atmospheric science (MS); earth science (MS); environmental and occupational health science (MS); environmental regulatory compliance (MS); environmental science (PhD); environmental toxicology and risk assessment (MS); water resource science (MS). Part-time programs available. *Faculty:* 1 full-time (0 women), 53 part-time/adjunct. *Students:* 4 full-time (3 women), 22 part-time (10 women); includes 1 Asian American or Pacific Islander, 2 Hispanic Americans. Average age 41. 11 applicants, 55% accepted, 6 enrolled. In 2005, 1 degree awarded. *Degree requirements:* For master's, oral exam, thesis optional. *Entrance requirements:* For master's, GRE General Test, minimum GPA of 3.0, 3 letters of recommendation. Additional exam requirements/recommendations for international students: Required—TOEFL (minimum score 550 paper-based; 213 computer-based). *Application deadline:* For fall admission, 2/1 priority date for domestic students, 3/1 priority date for international students; for spring admission, 9/1 priority date for domestic students, 7/1 priority date for international students. Application fee: $35. *Expenses:* Tuition, state resident: full-time $6,295; part-time $336 per credit. Tuition, nonresident: full-time $15,949; part-time $819 per credit. Required fees: $429. Full-time tuition and fees vary according to campus/location and program. Part-time tuition and fees vary according to course load and program. *Financial support:* In 2005–06, 8 students received support, including 1 fellowship (averaging $2,200 per year); research assistantships with full and partial tuition reimbursements available, teaching assistantships with full and partial tuition reimbursements available, Federal Work-Study, scholarships/grants, health care benefits, and unspecified assistantships also available. *Faculty research:* Radiation ecology, cytogenetics. *Unit head:* Dr. Gene Schreckhise, Associate Dean/Coordinator, 509-372-7323, E-mail: gschreck@wsu.edu.

Washington State University Vancouver, Graduate Programs, Program in Environmental Science, Vancouver, WA 98686. Offers MS. *Faculty:* 15. *Students:* 7 full-time (5 women), 6 part-time (2 women); includes 4 minority (1 African American, 1 American Indian/Alaska Native, 2 Hispanic Americans). Average age 38. 11 applicants, 27% accepted, 3 enrolled. In 2005, 3 degrees awarded. *Degree requirements:* For master's, thesis or alternative, comprehensive exam, registration. *Entrance requirements:* For master's, GRE General Test, minimum GPA of 3.0, 3 letters of recommendation. Additional exam requirements/recommendations for international students: Required—TOEFL (minimum score 550 paper-based; 213 computer-based). *Application deadline:* For fall admission, 2/1 for domestic students, 3/1 for international students; for spring admission, 10/15 priority date for domestic students, 7/1 priority date for international students. Application fee: $35. *Expenses:* Tuition, state resident: full-time $6,295; part-time $336 per credit. Tuition, nonresident: full-time $15,949; part-time $819 per credit. Required fees: $429. *Financial support:* In 2005–06, 10 students received support, including 1 fellowship with tuition reimbursement available (averaging $2,222 per year), 6 research assistantships with tuition reimbursements available (averaging $13,635 per year) *Faculty research:* Conservation biology, environmental chemistry. *Unit head:* Dr. Stephen Bollens, Academic Director, 360-546-9620, Fax: 360-546-9064, E-mail: bankerd@wsu.edu. *Application contact:* Dawn Banker, 360-546-9478, Fax: 360-546-9064, E-mail: bankerd@wsu.edu.

Western Connecticut State University, Division of Graduate Studies, School of Arts and Sciences, Department of Biological and Environmental Sciences, Danbury, CT 06810-6885. Offers MA. Part-time and evening/weekend programs available. *Degree requirements:* For master's, comprehensive exam or thesis. *Entrance requirements:* For master's, minimum GPA of 2.5.

Western Washington University, Graduate School, Huxley College of the Environment, Department of Environmental Sciences, Bellingham, WA 98225-5996. Offers MS. Part-time programs available. *Faculty:* 26. *Students:* 23 full-time (11 women), 13 part-time (9 women). 49 applicants, 35% accepted, 12 enrolled. In 2005, 7 degrees awarded. *Degree requirements:* For master's, thesis. *Entrance requirements:* For master's, GRE General Test, minimum GPA of 3.0 in last 60 semester hours or last 90 quarter hours. Additional exam requirements/recommendations for international students: Required—TOEFL (minimum score 567 paper-based; 227 computer-based). *Application deadline:* For fall admission, 2/1 for domestic students. Application fee: $50. *Expenses:* Tuition, area resident: Part-time $188 per credit. Tuition, state resident: full-time $5,628; part-time $539 per credit. Tuition, nonresident: full-time $16,176. Required fees: $624. *Financial support:* In 2005–06, 8 teaching assistantships with partial tuition reimbursements (averaging $10,059 per year) were awarded; Federal Work-Study, institutionally sponsored loans, scholarships/grants, tuition waivers (partial), and unspecified assistantships also available. Support available to part-time students. Financial award application deadline: 2/15; financial award applicants required to submit FAFSA. *Unit head:* Dr. Wayne Landis, Chair, 360-650-7585. *Application contact:* Sally Elmore, Graduate Program Coordinator, 360-650-3646.

Western Washington University, Graduate School, Huxley College of the Environment, Department of Environmental Studies, Bellingham, WA 98225-5996. Offers environmental education (M Ed); geography (MS). Part-time programs available. *Faculty:* 26. *Degree requirements:* For master's, thesis. *Entrance requirements:* For master's, GRE General Test, minimum GPA of 3.0 in last 60 semester hours or last 90 quarter hours. Additional exam requirements/recommendations for international students: Required—TOEFL (minimum score 567 paper-based; 227 computer-based). *Application deadline:* For fall admission, 2/1 for domestic students. Applications are processed on a rolling basis. Application fee: $50. *Expenses:* Tuition, area resident: Part-time $188 per credit. Tuition, state resident: full-time $5,628; part-time $539 per credit. Tuition, nonresident: full-time $16,176. Required fees: $624. *Financial support:* In 2005–06, 9 teaching assistantships with partial tuition reimbursements (averaging $9,339 per year) were awarded; Federal Work-Study, institutionally sponsored loans, scholarships/grants, tuition waivers (partial), and unspecified assistantships also available. Support available to part-time students. Financial award application deadline: 2/15; financial award applicants required to submit FAFSA. *Faculty research:* Geomorphology; pedogenesis; quaternary studies and climate change in the western U.S. landscape ecology, biogeography, pyrogeography, and spatial analysis. *Unit head:* Dr. Gigi Berardi, Chair, 360-650-3284. *Application contact:* Sally Elmore, Graduate Program Coordinator, 360-650-3646.

West Texas A&M University, College of Agriculture, Nursing, and Natural Sciences, Department of Life, Earth, and Environmental Sciences, Program in Environmental Science, Canyon, TX 79016-0001. Offers MS. Part-time programs available. *Degree requirements:* For master's, thesis optional. *Entrance requirements:* For master's, GRE General Test. Additional exam requirements/recommendations for international students: Required—TOEFL (minimum score 550 paper-based). Electronic applications accepted. *Faculty research:* Degradation of presistant pesticides in soils and ground water, air quality.

Wichita State University, Graduate School, Fairmount College of Liberal Arts and Sciences, Interdisciplinary Program in Liberal Studies, Wichita, KS 67260. Offers environmental science (MS). Participating faculty are from the Departments of Minority Studies, Philosophy, Religion, Social Work, and Women's Studies. Part-time programs available. *Degree requirements:* For master's, project, thesis optional. *Entrance requirements:* For master's, GRE, minimum GPA of 2.75. Additional exam requirements/recommendations for international students: Required—TOEFL. Electronic applications accepted.

Wright State University, School of Graduate Studies, College of Science and Mathematics, Department of Biological Sciences, Dayton, OH 45435. Offers biological sciences (MS); environmental sciences (MS). *Degree requirements:* For master's, thesis optional. *Entrance requirements:* Additional exam requirements/recommendations for international students: Required—TOEFL.

Wright State University, School of Graduate Studies, College of Science and Mathematics, Department of Chemistry, Dayton, OH 45435. Offers chemistry (MS); environmental sciences (MS). Part-time and evening/weekend programs available. *Degree requirements:* For master's, oral defense of thesis, seminar. *Entrance requirements:* Additional exam requirements/recommendations for international students: Required—TOEFL. *Faculty research:* Polymer synthesis and characterization, laser kinetics, organic and inorganic synthesis, analytical and environmental chemistry.

Wright State University, School of Graduate Studies, College of Science and Mathematics, Department of Geological Sciences, Program in Geological Sciences, Dayton, OH 45435. Offers environmental geochemistry (MS); environmental geology (MS); environmental sciences (MS); geological sciences (MS); geophysics (MS); hydrogeology (MS); petroleum geology (MS). Part-time programs available. *Degree requirements:* For master's, thesis. *Entrance* *requirements:* Additional exam requirements/recommendations for international students: Required—TOEFL.

Wright State University, School of Graduate Studies, College of Science and Mathematics, Program in Environmental Sciences, Dayton, OH 45435. Offers PhD.

Yale University, Graduate School of Arts and Sciences, Department of Forestry and Environmental Studies, New Haven, CT 06520. Offers environmental sciences (PhD); forestry (PhD). *Degree requirements:* For doctorate, thesis/dissertation. *Entrance requirements:* For doctorate, GRE General Test.

See Close-Up on page 713.

Yale University, School of Forestry and Environmental Studies, New Haven, CT 06511. Offers MES, MF, MFS, DFES, PhD, JD/MES, MBA/MES, MBA/MF, MES/MA, MES/MPH, MF/MA. *Accreditation:* SAF (one or more programs are accredited). Part-time programs available. Terminal master's awarded for partial completion of doctoral program. *Degree requirements:* For doctorate, thesis/dissertation. *Entrance requirements:* For master's and doctorate, GRE General Test. Expenses: Contact institution. *Faculty research:* Ecosystem science and management, coastal and watershed systems, environmental policy and management, social ecology and community development, conservation biology.

See Close-Up on page 713.

Marine Affairs

Dalhousie University, Faculty of Graduate Studies, Program in Marine Affairs, Halifax, NS B3H 4R2, Canada. Offers MMM. *Degree requirements:* For master's, project. *Entrance requirements:* For master's, minimum GPA of 3.0. Additional exam requirements/recommendations for international students: Required—TOEFL. *Faculty research:* Integrated coastal zone management, marine law and policy, fisheries management, maritime transport, marine protected areas.

Duke University, Nicholas School of the Environment and Earth Sciences, Durham, NC 27708-0328. Offers coastal environmental management (MEM); environmental economics and policy (MEM); environmental health and security (MEM); forest resource management (MF); global environmental change (MEM); resource ecology (MEM); water and air resources (MEM). *Accreditation:* SAF (one or more programs are accredited). Part-time programs available. *Degree requirements:* For master's, thesis, registration. *Entrance requirements:* For master's, GRE General Test, previous course work in biology or ecology, calculus, statistics, and microeconomics; computer familiarity with word processing and data analysis. Additional exam requirements/recommendations for international students: Required—TOEFL (minimum score 550 paper-based; 213 computer-based). Electronic applications accepted. Expenses: Contact institution. *Faculty research:* Ecosystem management, conservation ecology, earth systems, risk assessment.

See Close-Up on page 709.

East Carolina University, Graduate School, Program in Coastal Resources Management, Greenville, NC 27858-4353. Offers PhD. *Students:* 11 full-time (5 women), 19 part-time (6 women); includes 3 minority (1 African American, 2 Hispanic Americans), 1 international. Average age 39. 6 applicants, 33% accepted, 2 enrolled. In 2005, 2 degrees awarded. *Degree requirements:* For doctorate, thesis/dissertation, internship, comprehensive exam, registration. *Entrance requirements:* For doctorate, GRE. Additional exam requirements/recommendations for international students: Required—TOEFL. *Application deadline:* For fall admission, 3/1 for domestic students, 3/1 for international students. Application fee: $50. *Expenses:* Tuition, state resident: full-time $2,516. Tuition, nonresident: full-time $12,832. *Financial support:* In 2005–06, 8 fellowships with tuition reimbursements (averaging $17,000 per year), 2 research assistantships with tuition reimbursements (averaging $17,000 per year) were awarded.; career-related internships or fieldwork and institutionally sponsored loans also available. Financial award application deadline: 3/1; financial award applicants required to submit FAFSA. *Faculty research:* Coastal geology, wetlands and coastal ecology, ecological and social networks, submerged cultural resources, coastal resources economics. Total annual research expenditures:$24,000. *Unit head:* Dr. Lauriston R. King, Director, 252-328-2484, Fax: 252-328-0381, E-mail: kingl@ecu.edu. *Application contact:* Dean of Graduate School, 252-328-6012, Fax: 252-328-6071, E-mail: gradschool@ecu.edu.

Florida Institute of Technology, Graduate Programs, College of Engineering, Department of Marine and Environmental Systems, Program in Oceanography, Melbourne, FL 32901-6975. Offers biological oceanography (MS); chemical oceanography (MS); coastal zone management (MS); geological oceanography (MS); oceanography (PhD); physical oceanography (MS). Part-time programs available. *Students:* Average age 30.Terminal master's awarded for partial completion of doctoral program. *Degree requirements:* For master's, thesis (for some programs); for doctorate, one foreign language, thesis/dissertation, departmental qualifying exams, comprehensive exam. *Entrance requirements:* For master's, GRE General Test, minimum GPA of 3.0; for doctorate, GRE General Test, minimum GPA of 3.3, resumé. *Application deadline:* Applications are processed on a rolling basis. Electronic applications accepted. *Expenses:* Tuition: Part-time $825 per credit. *Financial support:* Research assistantships with full and partial tuition reimbursements, teaching assistantships with full and partial tuition reimbursements, career-related internships or fieldwork and tuition remissions available. Financial award application deadline: 3/1; financial award applicants required to submit FAFSA. *Faculty research:* Marine geochemistry, ecosystem dynamics, coastal processes, marine pollution, environmental modeling. Total annual research expenditures: $938,395. *Unit head:* Dr. Dean R. Norris, Chair, 321-674-7377, Fax: 321-674-7212, E-mail: norris@fit.edu. *Application contact:* Carolyn P. Farrior, Director of Graduate Admissions, 321-674-7118, Fax: 321-723-9468, E-mail: cfarrior@fit.edu.

See Close-Ups on pages 653 and 253.

Louisiana State University and Agricultural and Mechanical College, Graduate School, School of the Coast and Environment, Department of Oceanography and Coastal Sciences, Baton Rouge, LA 70803. Offers MS, PhD. *Faculty:* 32 full-time (2 women). *Students:* 51 full-time (30 women), 15 part-time (4 women); includes 1 African American, 1 Hispanic American, 17 international. Average age 30. 27 applicants, 37% accepted, 17 enrolled. In 2005, 10 master's, 7 doctorates awarded. *Degree requirements:* For master's, thesis (for some programs); for doctorate, one foreign language, thesis/dissertation. *Entrance requirements:* For master's, GRE General Test, minimum GPA of 3.0; for doctorate, GRE General Test, MA or MS, minimum GPA of 3.0. Additional exam requirements/recommendations for international students: Required—TOEFL (minimum score 550 paper-based; 213 computer-based). *Application deadline:* For fall admission, 1/25 priority date for domestic students, 5/15 priority date for international students. Applications are processed on a rolling basis. Application fee: $25. *Financial support:* In 2005–06, 54 students received support, including 7 fellowships (averaging $20,200 per year), 39 research assistantships with full and partial tuition reimbursements available (averaging $19,760 per year), 3 teaching assistantships with full and partial tuition reimbursements available (averaging $12,750 per year); Federal Work-Study, institutionally sponsored loans, scholarships/grants, tuition waivers (full and partial), and unspecified assistantships also available. Support available to part-time students. Financial award applicants required to submit FAFSA. *Faculty research:* Management and development of estuarine and coastal areas and resources; physical, chemical, geological, and biological research. Total annual research expenditures:$88,496. *Unit head:* Dr. Lawrence Rouse, Chair, 225-578-2453, Fax: 225-578-6307, E-mail: lrouse@lsu.edu. *Application contact:* Dr. Masamichi Inoue, Graduate Adviser, 225-578-6308, Fax: 225-578-6307, E-mail: coiino@lsu.edu.

Memorial University of Newfoundland, School of Graduate Studies, Department of Sociology, St. John's, NL A1C 5S7, Canada. Offers gender (PhD); maritime sociology (PhD); sociology (M Phil, MA); work and development (PhD). Part-time programs available. *Students:* 30 full-time (19 women), 4 part-time (2 women), 3 international. 11 applicants, 91% accepted, 9 enrolled. In 2005, 3 degrees awarded. *Degree requirements:* For master's, program journal (mPhil), thesis optional; for doctorate, one foreign language, thesis/dissertation, oral defense of thesis, comprehensive exam. *Entrance requirements:* For master's, 2nd class degree from university of recognized standing in area of study; for doctorate, MA, M Phil, or equivalent. *Application deadline:* For fall admission, 2/15 priority date for domestic students, 2/15 priority date for international students. Applications are processed on a rolling basis. Application fee: $40 Canadian dollars. Electronic applications accepted. *Expenses:* Tuition: Part-time $733 per term. Tuition and fees vary according to degree level and program. *Financial support:* Fellowships, research assistantships, teaching assistantships available. Financial award application deadline: 1/31. *Faculty research:* Work and development, gender, maritime sociology. *Unit head:* Dr. Judith Adler, Head, 709-737-7443, Fax: 709-737-2075, E-mail: jadler@mun.ca. *Application contact:* Robert Hill, Graduate Coordinator, 709-737-7453, Fax: 709-737-2075, E-mail: rhill@mun.ca.

Memorial University of Newfoundland, School of Graduate Studies, Interdisciplinary Program in Marine Studies, St. John's, NL A1C 5S7, Canada. Offers fisheries resource management (MMS, Advanced Diploma). Part-time programs available. *Students:* 10 full-time (5 women), 12 part-time (3 women), 3 international. 9 applicants, 100% accepted, 4 enrolled. In 2005, 4 degrees awarded. *Degree requirements:* For master's, report. *Entrance requirements:* For master's and Advanced Diploma, high 2nd class degree from a recognized university. *Application deadline:* For fall admission, 4/30 for domestic students, 4/30 for international students. Application fee: $40 Canadian dollars. *Expenses:* Tuition: Part-time $733 per term. Tuition and fees vary according to degree level and program. *Financial support:* Fellowships, research assistantships, teaching assistantships available. *Faculty research:* Biological, ecological and oceanographic aspects of world fisheries; economics; political science; sociology. *Unit head:* Dr. Peter Fisher, Chair, 709-778-0356, Fax: 709-778-0346, E-mail: peter.fisher@mi.mun.ca. *Application contact:* Nancy Smith, Program Support, 709-778-0522, E-mail: nancy.smith@mi.mun.ca.

Nova Southeastern University, Oceanographic Center, Program in Coastal-Zone Management, Fort Lauderdale, FL 33314-7796. Offers MS. *Faculty:* 15 full-time (1 woman), 5 part-time/adjunct (0 women). *Students:* 18 applicants, 94% accepted, 14 enrolled. *Entrance requirements:* For master's, GRE. Additional exam requirements/recommendations for international students: Required—TOEFL (minimum score 550 paper-based). *Application deadline:* Applications are processed on a rolling basis. Application fee: $50. *Financial support:* Career-related internships or fieldwork, Federal Work-Study, scholarships/grants, and unspecified assistantships available. Financial award applicants required to submit FAFSA. *Unit head:* Dr. Andrew Rogerson, Associate Dean, Director of Graduate Programs, 954-262-3600, Fax: 954-262-4020, E-mail: arogerso@nsu.nova.edu.

See Close-Up on page 261.

Oregon State University, Graduate School, College of Oceanic and Atmospheric Sciences, Program in Marine Resource Management, Corvallis, OR 97331. Offers MA, MS. *Students:* 25 full-time (15 women), 8 part-time (5 women); includes 1 minority (African American), 6 international. Average age 30. In 2005, 13 degrees awarded. *Degree requirements:* For master's, thesis optional. *Entrance requirements:* For master's, GRE General Test, minimum GPA of 3.0 in last 90 hours of course work. Additional exam requirements/recommendations for international students: Required—TOEFL. *Application deadline:* For fall admission, 2/1 for domestic students. Applications are processed on a rolling basis. Application fee: $50. *Expenses:* Tuition, area resident: Part-time $301 per credit. Tuition, state resident: full-time $8,139; part-time $501 per credit. Tuition, nonresident: full-time $14,376; part-time $532 per credit. Required fees: $1,266. *Financial support:* Fellowships, research assistantships, teaching assistantships, career-related internships or fieldwork, Federal Work-Study, and institutionally sponsored loans available. Support available to part-time students. Financial award application deadline: 2/1. *Faculty research:* Ocean and coastal resources, fisheries resources, marine pollution, marine recreation and tourism. *Unit head:* Dr. Robert Allen, Assistant Director, 541-737-1339, Fax: 541-737-2064. *Application contact:* Irma Delson, Assistant Director, Student Services, 541-737-5189, Fax: 541-737-2064, E-mail: student_adviser@oce.orst.edu.

Stevens Institute of Technology, Graduate School, Charles V. Schaefer Jr. School of Engineering, Department of Civil, Environmental, and Ocean Engineering, Program in Maritime Systems, Hoboken, NJ 07030. Offers M Eng, MS. *Students:* 4 applicants, 100% accepted.Application fee: $50. *Expenses:* Tuition: Part-time $920 per credit hour. Tuition and fees vary according to program. *Unit head:* Dr. Michael S. Bruno, Information Contact, 201-216-5338, Fax: 201-216-8214, E-mail: mbruno@stevens-tech.edu. *Application contact:* Dr. Michael S. Bruno, Information Contact, 201-216-5338, Fax: 201-216-8214, E-mail: mbruno@stevens-tech.edu.

Université du Québec à Rimouski, Graduate Programs, Program in Management of Marine Resources, Rimouski, QC G5L 3A1, Canada. Offers M Sc, Diploma. Part-time programs available. *Students:* 29 full-time (12 women), 1 part-time, 17 international. 38 applicants, 74% accepted. In 2005, 18 degrees awarded. *Entrance requirements:* For master's, appropriate bachelor's degree, proficiency in French. *Application deadline:* For fall admission, 5/1 for domestic students. Application fee: $30. Tuition charges are reported in Canadian dollars. *Expenses:* Tuition, state resident: full-time $2,000 Canadian dollars. Tuition, nonresident: full-time $9,000 Canadian dollars. Tuition and fees vary according to course load and program. *Financial support:* Fellowships, research assistantships, teaching assistantships available. *Unit head:* James Wilson, Director, 418-724-1544, Fax: 418-724-1525, E-mail: james_wilson@uqar.ca. *Application contact:* Marc Berube, Office of Admissions, 418-724-1433, Fax: 418-724-1525, E-mail: marc_berube@uqar.ca.

Marine Affairs

University of Delaware, College of Marine Studies, Newark, DE 19716. Offers geology (MS, PhD); marine management (MMM); marine policy (MS); marine studies (MMP, MS, PhD); oceanography (MS, PhD). *Faculty:* 41 full-time (4 women). *Students:* 115 full-time (57 women), 5 part-time (2 women); includes 7 minority (2 African Americans, 4 Asian Americans or Pacific Islanders, 1 Hispanic American), 30 international. Average age 29. 127 applicants, 35% accepted, 24 enrolled. In 2005, 12 master's, 8 doctorates awarded. *Degree requirements:* For master's and doctorate, thesis/dissertation. *Entrance requirements:* For master's and doctorate, GRE General Test. Additional exam requirements/recommendations for international students: Required—TOEFL. *Application deadline:* For fall admission, 3/1 for domestic students; for spring admission, 10/1 for domestic students. Applications are processed on a rolling basis. Application fee: $60. Electronic applications accepted. *Financial support:* In 2005–06, 78 students received support, including 14 fellowships with full tuition reimbursements available (averaging $19,000 per year), 62 research assistantships with full tuition reimbursements available (averaging $19,000 per year), 2 teaching assistantships with full tuition reimbursements available (averaging $19,000 per year); career-related internships or fieldwork, Federal Work-Study, and tuition waivers (full and partial) also available. Financial award application deadline: 3/1. *Faculty research:* Marine biology and biochemistry, oceanography, marine policy, physical ocean science and engineering, ocean engineering. Total annual research expenditures: $10.5 million. *Unit head:* Dr. Nancy Targett, Dean, 302-831-2841. *Application contact:* Lisa Perelli, Coordinator, 302-645-4226, E-mail: lperelli@udel.edu.

University of Maine, Graduate School, College of Natural Sciences, Forestry, and Agriculture, School of Marine Sciences, Program in Marine Policy, Orono, ME 04469. Offers MS. *Students:* 8 full-time (5 women), 6 part-time (all women), 1 international. Average age 29. 17 applicants, 6% accepted, 1 enrolled. In 2005, 3 master's awarded. *Degree requirements:* For master's, thesis. *Entrance requirements:* For master's, GRE General Test. Additional exam requirements/recommendations for international students: Required—TOEFL. *Application deadline:* For fall admission, 2/1 for domestic students. Applications are processed on a rolling basis. Application fee: $50. Electronic applications accepted. *Financial support:* Fellowships with tuition reimbursements, research assistantships with tuition reimbursements, teaching assistantships with tuition reimbursements, career-related internships or fieldwork, Federal Work-Study, and tuition waivers (full and partial) available. Support available to part-time students. Financial award application deadline: 3/1. *Unit head:* Dr. James Wilson, Coordinator, 207-581-4368. *Application contact:* Scott G. Delcourt, Associate Dean of the Graduate School, 207-581-3219, Fax: 207-581-3232, E-mail: graduate@maine.edu.

University of Miami, Graduate School, Rosenstiel School of Marine and Atmospheric Science, Division of Marine Affairs and Policy, Coral Gables, FL 33124. Offers MA, MS, JD/MA. Part-time programs available. *Faculty:* 9 full-time (3 women), 6 part-time/adjunct (1 woman). *Students:* 36 full-time (17 women), 11 part-time (6 women); includes 10 minority (1 African American, 3 Asian Americans or Pacific Islanders, 6 Hispanic Americans), 3 international. Average age 25. 37 applicants, 78% accepted, 19 enrolled. In 2005, 17 degrees awarded. *Degree requirements:* For master's, thesis, internship, paper, comprehensive exam, registration. *Entrance requirements:* For master's, GRE General Test. Additional exam requirements/recommendations for international students: Required—TOEFL (minimum score 550 paper-based; 213 computer-based). *Application deadline:* For fall admission, 6/1 priority date for domestic students, 4/1 priority date for international students. Applications are processed on a rolling basis. Application fee: $50. Electronic applications accepted. *Financial support:* In 2005–06, 24 students received support, including 5 fellowships with partial tuition reimbursements available (averaging $20,304 per year), 10 research assistantships with partial tuition reimbursements available (averaging $20,304 per year), 5 teaching assistantships (averaging $20,304 per year); career-related internships or fieldwork, Federal Work-Study, institutionally sponsored loans, scholarships/grants, and unspecified assistantships also available. Financial award application deadline: 3/1; financial award applicants required to submit FAFSA. *Unit head:* Dr. Daniel Benetti, Chair, 305-421-4087, Fax: 305-421-4771, E-mail: dbenetti@rsmas.miami.edu. *Application contact:* Dr. Larry Peterson, Associate Dean, 305-421-4155, Fax: 305-421-4771, E-mail: gso@rsmas.miami.edu.

University of Rhode Island, Graduate School, College of the Environment and Life Sciences, Department of Marine Affairs, Kingston, RI 02881. Offers MA, MMA, PhD. *Application deadline:* For fall admission, 4/15 for domestic students. Applications are processed on a rolling basis. Application fee: $35. *Expenses:* Tuition, state resident: full-time $5,522; part-time $307 per credit. Tuition, nonresident: full-time $15,992; part-time $888 per credit. Required fees: $1,786; $73 per credit. One-time fee: $80 part-time. *Unit head:* Dr. Lawrence Juda, Chairperson, 401-874-2596.

University of San Diego, College of Arts and Sciences, Program in Marine and Environmental Studies, San Diego, CA 92110-2492. Offers marine science (MS). Part-time programs available. *Faculty:* 6 full-time (2 women). *Students:* 4 full-time (2 women), 15 part-time (7 women), 1 international. Average age 26. 17 applicants, 59% accepted, 4 enrolled. In 2005, 4 degrees awarded. *Entrance requirements:* For master's, GRE General Test, minimum GPA of 3.0, undergraduate major in science. Additional exam requirements/recommendations for international students: Required—TOEFL (minimum score 580 paper-based; 237 computer-based), TWE. *Application deadline:* For fall admission, 4/1 for domestic students. Applications are processed on a rolling basis. Application fee: $45. Electronic applications accepted. *Financial support:* Career-related internships or fieldwork, Federal Work-Study, institutionally sponsored loans, tuition waivers (partial), and unspecified assistantships available. Support available to part-time students. Financial award application deadline: 5/1; financial award applicants required to submit FAFSA. *Faculty research:* Marine ecology; paleoclimatology; geochemistry; functional morphology; marine zoology of mammals, birds and turtles. *Unit head:* Dr. Hugh I. Ellis, Director, 619-260-4075, Fax: 619-260-6804, E-mail: ellis@sandiego.edu. *Application contact:* Stephen Pultz, Director of Admissions, 619-260-4524, Fax: 619-260-4158, E-mail: grads@sandiego.edu.

University of Washington, Graduate School, College of Ocean and Fishery Sciences, School of Marine Affairs, Seattle, WA 98195. Offers MMA, MMA/MAIS. *Degree requirements:* For master's, thesis. *Entrance requirements:* For master's, GRE General Test, minimum GPA of 3.0. Additional exam requirements/recommendations for international students: Required—TOEFL. Electronic applications accepted. *Faculty research:* Marine pollution, port authorities, fisheries management, global climate change, marine environmental protection.

University of West Florida, College of Arts and Sciences: Sciences, Division of Life and Health Sciences, Pensacola, FL 32514-5750. Offers biology (MS, MST), including biological chemistry (MS), biology (MS), biology education (MST), coastal zone studies (MS), environmental biology (MS); general biology (MS), including biology; health communication (MA), including health care ethics; public health (MPH). Part-time programs available. *Faculty:* 10 full-time (1 woman). *Students:* 16 full-time (13 women), 33 part-time (20 women); includes 4 minority (1 African American, 2 Asian Americans or Pacific Islanders, 1 Hispanic American), 2 international. Average age 28. 17 applicants, 71% accepted, 8 enrolled. In 2005, 5 degrees awarded. *Entrance requirements:* For master's, GRE General Test. Additional exam requirements/recommendations for international students: Required—TOEFL (minimum score 550 paper-based; 213 computer-based). *Application deadline:* For fall admission, 6/1 for domestic students, 5/15 for international students; for spring admission, 11/1 for domestic students, 10/1 for international students. Applications are processed on a rolling basis. Application fee: $30. Special rates offered to residents of Alabama. *Expenses:* Tuition, state resident: full-time $5,833; part-time $243 per credit hour. Tuition, nonresident: full-time $21,204; part-time $884 per credit hour. Tuition and fees vary according to campus/location. *Financial support:* In 2005–06, teaching assistantships with partial tuition reimbursements (averaging $8,000 per year); scholarships/grants and tuition waivers (partial) also available. Financial award application deadline: 4/15; financial award applicants required to submit FAFSA. *Unit head:* Dr. George L. Stewart, Chairperson, 850-474-2748.

University of West Florida, College of Arts and Sciences: Sciences, Division of Life and Health Sciences, Department of Biology, Specialization in Coastal Zone Studies, Pensacola, FL 32514-5750. Offers MS. Part-time programs available. *Students:* 1 (woman) full-time, 5 part-time (4 women). Average age 25. In 2005, 1 degree awarded. *Degree requirements:* For master's, thesis or alternative. *Entrance requirements:* For master's, GRE General Test. Additional exam requirements/recommendations for international students: Required—TOEFL (minimum score 550 paper-based; 213 computer-based). *Application deadline:* For fall admission, 6/1 for domestic students, 5/15 for international students; for spring admission, 11/1 for domestic students, 10/1 for international students. Applications are processed on a rolling basis. Application fee: $30. Special rates offered to residents of Alabama. *Expenses:* Tuition, state resident: full-time $5,833; part-time $243 per credit hour. Tuition, nonresident: full-time $21,204; part-time $884 per credit hour. Tuition and fees vary according to campus/location. *Financial support:* Application deadline: 4/15;

Cross-Discipline Announcements

Carnegie Mellon University, Carnegie Institute of Technology, Department of Engineering and Public Policy, Pittsburgh, PA 15213-3891.

PhD program addresses policy problems in which technical details are critically important by using tools of engineering, science, and the social sciences. Carnegie Mellon offers an excellent environment for interdisciplinary research. Program requires equivalent of undergraduate degree in engineering, physical science, or mathematics. Contact Victoria Finney, Program Administrator, eppadmt@andrew.cmu.edu.

Massachusetts Institute of Technology, School of Engineering, Biological Engineering Division, Cambridge, MA 02139-4307.

Program provides opportunities for study and research at the interface of biology and engineering leading to specialization in bioengineering and applied biosciences. The areas include understanding how biological systems operate, especially when perturbed by genetic, chemical, or materials interventions or subjected to pathogens or toxins, and designing innovative biology-based technologies in diagnostics, therapeutics, materials, and devices for application to human health and diseases, as well as other societal problems and opportunities.

ANTIOCH NEW ENGLAND GRADUATE SCHOOL

Department of Environmental Studies

Programs of Study

The Department of Environmental Studies at Antioch New England Graduate School offers the Master of Science (M.S.) in environmental studies with concentrations in conservation biology, environmental advocacy and organizing, environmental education, individualized studies, and science teacher certification, and the M.S. in resource management and administration. The department also offers a Doctor of Philosophy (Ph.D.) in environmental studies.

The Department of Environmental Studies provides an innovative, transdisciplinary approach to environmental learning at the graduate level. The students and faculty members are motivated by the urgency of complex environmental challenges, the desire to promote deeper ecological awareness, and the aspiration to understand the needs of people, habitats, and communities. Its programs are dynamic and flexible, responding to the ever-changing demands of new information and concepts, paying close attention to the hands-on skills required in a variety of professional settings, and reflecting the changing life situations of diverse regions and populations.

The 50-credit master's-level programs feature a dedicated faculty involved in advocacy, conservation biology, ecology, education, history, policy, and resource management. The faculty members work together to form curricula and program tracks that weave together all of these disciplines. Common to all programs are the practicum and the professional seminar, with the latter occurring during the student's first semester. It provides an opportunity for students to become oriented to their program, develop program plans, refine their professional goals, and initiate discussions about the practicum. Students must complete 8 credits of practicum (typically two 300-hour practica between their third and fifth semesters).

The Ph.D. curriculum addresses contemporary and future environmental problems in an innovative, interdisciplinary approach. Students require the capacity to understand that a research challenge includes more than the content area or field of study—students must see how their personal values and the cultural context of the learning experience frame and lend meaning to the research problem. This is the core of doctoral learning: the integration of theory and practice and the ability to understand the personal, social, political, and professional context of research. Antioch has a strong tradition of fostering excellence regarding this type of reflective practice or what is known as "reflexivity" among researchers—people who understand the full dimensions of their work, who have the scholarly vision to express this, who go beyond mere reaction and become engaged in thoughtful action.

Research Facilities

Antioch New England Institute (ANEI) is the nonprofit consulting and community outreach arm of Antioch New England Graduate School. ANEI promotes a vibrant and sustainable environment, economy, and society by encouraging informed civic engagement. It provides training, programs, and resources (U.S. and international) in leadership development, place-based education, nonprofit management, environmental education and policy, smart growth, and public administration.

The Center for Tropical Ecology and Conservation, housed in the Department of Environmental Studies, supports and promotes education and research in tropical biology, conservation, and the sustainable use of tropical resources.

The librarians at Antioch New England offer professional and personalized reference service for graduate research. Extensive class and research support is available via the library Web site. Access to the library catalog is available through Horace, the library's automated catalog system. Also available are specialized online reference pages for classes and key topics, access to many online bibliographic databases, reserve reading, and links to scholarly Internet resources with full Internet access. In addition, detailed reference instruction, specific research information, an electronic book collection, and specific class support resources are also available on the library Web site. All library services, such as book requests, renewals, reference help, and interlibrary loan requests, are available online.

The focused library collection includes print and electronic books and journals, dissertations and theses, audiovisual materials, and government documents. This collection is enhanced by the large collection of more than 300,000 books and 13,000 journal titles at Antioch College, Antioch New England's partner in the larger Antioch University Library system. Recent additions include OhioLINK, which offers more than 100 electronic research databases, including a variety of full-text resources, and RefWORKS, a bibliographic management program. The Antioch New England Library also participates in local, regional, and national interlibrary loan services.

Financial Aid

Approximately 70 percent of students receive some type of aid, usually in the form of federal loans and work-study opportunities. The Jonathan Daniels Scholarship, established in 2003, strives to increase the diversity of the student body in its racial, ethnic, cultural, international, and socioeconomic makeup and to encourage service to underserved groups. All full-time Antioch New England students are eligible, although funding is limited. The completed scholarship form, along with relevant information from the Office of Financial Aid, is forwarded to each academic department for decisions. Awards range from $500 to 50 percent of tuition for a given year.

Cost of Study

The master's degree programs cost $6000 a semester in 2005–06. Tuition for the Ph.D. program is $20,000 per year. All students must pay a comprehensive fee of $100 and a lab fee of $60 per term.

Living and Housing Costs

The Graduate School's location enables a large portion of students to commute to classes from their established homes in various parts of New England. Other students move close to Antioch New England, where they have a varied selection of settings—urban, rural, semi-rural, coast, mountains, or valley—in which to live. The Office of Admissions provides information resources for those relocating to the Monadnock region, the greater Brattleboro area, or northern Massachusetts.

Student Group

About 1,200 students attend Antioch New England Graduate School. The average age ranges between 25 and 55; women make up 69 percent of the population. Students have an average of three to six years of professional experience upon entering their program, and most continue employment while pursuing their studies.

Location

Located in Keene, New Hampshire, Antioch New England is in the heart of the Monadnock region, a picturesque area that has been described as the "Currier & Ives" corner of New Hampshire. The School is geographically situated so that students also have easy access to several popular metropolitan areas, including Boston and Montreal. With a population of nearly 23,000, Keene has been named by the National Trust for Historic Preservation as one of "America's Dozen Distinctive Destinations."

The School and The Department

Antioch New England Graduate School offers a rich array of master's and doctoral-level academic programming and institutional activities. The Graduate School's values-driven mission and focus on experiential learning, peer interaction, and reflective practice make the Antioch experience unique for each individual who is part of this learning community.

Ecological thought should permeate all human decisions and activities and give rise to a diversity of pathways toward environmental health and wholeness. In this conviction, the faculty of the Department of Environmental Studies strives for a rigorous integration of scholarship, skills, practical application, and reflection to enable each student to realize his or her own vision for participation in an ecologically sound future.

Applying

Master's degree applicants must submit the completed application form, including a resume and an essay; a nonrefundable application fee of $40 ($50 for individuals pursuing the Alternative Admissions Process for applicants without a bachelor's degree); one official transcript from each accredited college or university attended, indicating courses taken and degree(s) earned; and three letters of recommendation (four letters for Alternative Admissions Process applicants), preferably from persons who are, or have been, in a position to evaluate the applicant's work. An interview with a department faculty member is required. Antioch New England Graduate School does not require master's-level applicants to take the Graduate Record Examinations (GRE) or similar written examinations. Applications for all environmental studies master's programs are reviewed at two points in the admissions cycle: November 1 and March 1. The final application deadline is August 1.

Ph.D. applicants should have a master's degree either in an environmentally related field or in one that has prepared the student to undertake the rigorous research required; the completion of at least four courses (B average or above) at the undergraduate or graduate level in environmental science, ecology, or environmental biology. Many students have professional experience in the environmental field or experience in a field relevant to their proposed program of study.

The program requires a comprehensive portfolio, which should include seven documents: the application form, transcripts, four references, the annotated resume, the personal statement, the academic plan, and the work sample. All materials should be submitted on typed, double-spaced pages (with the exception of the application form). Materials may also be submitted online. After the deadline, the admissions committee selects those applicants whom they wish to interview. The completed application form, a nonrefundable $75 application fee, and all the associated documents must be received by the Office of Admissions by January 31.

Correspondence and Information

Office of Admissions
Antioch New England Graduate School
40 Avon Street
Keene, New Hampshire 03431-3552
Phone: 800-490-3310 (toll-free)
Fax: 603-357-0718
E-mail: petersons@antiochne.edu
Web site: http://www.antiochne.edu/defaultp.cfm

Antioch New England Graduate School

THE FACULTY AND THEIR RESEARCH

Joy Whiteley Ackerman, Director, Individualized Program and Academic Director, Master's Programs; Ph.D., Antioch New England. Eco-theology and influence of language in environmental thought.
Jonathan L. Atwood, Director, Conservation Biology Program; Ph.D., UCLA. Habitat conservation planning as a tool for maintaining viable populations of rare bird species, population dynamics of New England's migratory birds, habitat fragmentation, implementation of U.S. Endangered Species Act.
Meade Cadot, Associate Core Faculty; Ph.D., Kansas. Director of the Harris Center for Conservation Biology. Clustering open space to protect watersheds, back-country recreation, and far-reaching wildlife, such as moose, black bear, and bobcat.
Steve Chase, Director, Environmental Advocacy and Organizing Program; Ph.D., Antioch New England. Activist education, ecological politics, environmental justice, corporate globalization, democratic social movements.
Katherine Delanoy, Practicum Coordinator; M.S., Antioch New England.
James S. Gruber, Associate Core Faculty; M.S., MIT; M.P.A., Harvard. Building sustainable communities.
Paul B. Hertneky, Adjunct Faculty; M.F.A., Bennington. Economics, ecology, and cultural influence of food and the food supply.
James W. Jordan, Core Faculty and Director of Field Study Program; Ph.D., Wisconsin–Madison. Landscape evolution, climate change, human-environment interactions.
Beth A. Kaplin, Core Faculty and Director of the Center for Tropical Ecology and Conservation; Ph.D., Wisconsin–Madison. Socioeconomic aspects of tropical forest protected area management, the persistent gap between scientists and natural resource managers.
Jimmy W. Karlan, Director of Science Teacher Certification Programs; Ed.D., Harvard. Children's and adults' ecological conceptions and theories.
Alesia Maltz, Core Faculty; Ph.D., Illinois. Environmental history, global environmental change, environmentalism and justice, technology and society, community decision-making processes, First Nations environmental policy, and history of science.
Peter A. Palmiotto, Core Faculty; D.F., Yale. Biological attributes of individual tree species that influence ecosystem-level processes and biological diversity in forest ecosystems.
Michael Simpson, Director, Resource Management and Administration Program; M.A.L.S., Dartmouth; M.S., Antioch New England. Wetland, watershed, and waste management courses; ecological economics; environmental policy.
Rachel Thiet, Core Faculty; Ph.D., Ohio State. Impact of the distribution, abundance, and activity of soil organisms on plant community structure, nutrient cycling, and long-term ecosystem sustainability.
Cynthia Thomashow, Core Faculty and Director, Environmental Education Program; M.S.T., Antioch New England; M.Ed., Keene State. Integration of environmental literacy into educational arenas and development of interpretive content and exhibits in zoos, museums, aquaria, and other public venues that cultivate a conservation ethic in the public-at-large.
Mitchell Thomashow, Chair; Ed.D., Massachusetts. Ecological and existential dimensions of global environmental change.
Peter Throop, Associate Core Faculty; M.B.A., NYU; M.S., Antioch New England. Community goals, flood mitigation, earth excavation regulations, land-use plan, development review.
Andy Toepfer, Adjunct Faculty. Helping people connect to their local landscapes and communities in ways that sustain each.
Abigail Abrash Walton, Associate Core Faculty; M.Sc., London School of Economics. Exploring and making connections between human rights and environmental concerns.
Kathleen Heidi Watts, Faculty Emerita; Ph.D., Cornell. Progressive education, adult learners, educational change, reflective teaching, action research.
Thomas Webler, Core Faculty and Academic Director, Doctoral Program; Ph.D., Clark. Developing practical tools to incorporate a plurality of voices and ways of knowing into a just policy-making process that builds on peoples' lay competence and on all available technical knowledge and expertise.
Susan Weller, Administrative Director; M.Ed., Antioch New England. Dynamics of groups and how people come together to solve problems.
Thomas K. Wessels, Core Faculty; M.A. Forest management strategies that promote diversity at the landscape level.

BARD

BARD COLLEGE

Bard Center for Environmental Policy

Program of Study

The Bard Center for Environmental Policy (BCEP) offers intensive graduate studies and practical training in preparation for environmental careers at the local, national, and international levels. The innovative program, founded by the former Director for North America of the United Nations Environment Programme, is specifically designed to meet contemporary demands for environmental leadership in government, business, and nonprofit organizations. The unique course of studies, leading to a Master of Science or a Professional Certificate in environmental policy, features integrated application of natural and social sciences for the analysis of environmental problems and responses. Distinctive elements include its thematic curriculum, multimedia communication and leadership training, an exceptional faculty, and internships in the United States and abroad.

Curriculum organization and diverse internship opportunities permit considerable flexibility to meet the specific needs and career aspirations of students. The objectives of the rigorous, interdisciplinary program are to prepare environmental leaders who understand the complex interconnections among various disciplines involved in environmental policymaking and can translate scientific results into feasible and creative strategies. Graduates are skilled at using various forms of communication to build support and consensus for effective environmental management. The second-year internship, culminating in the master's thesis, provides hands-on experience to facilitate entry into the job market.

Several options are available to accepted students. Active professionals who have substantial experience in environment-related fields may qualify for the master's degree after completion of the first-year courses and the master's thesis. Options include a joint program leading to the Master of Science and Doctor of Jurisprudence (M.S./J.D.) degrees with Pace University School of Law, a dual master's program (M.S./M.A.T.) with the Bard Center Master of Arts in Teaching program, the Master's International Program with the Peace Corps, and a Professional Certificate in environmental policy.

Research Facilities

Stevenson Library houses the latest periodicals and books dealing with environmental issues. The Jerome Levy Economics Institute offers a full array of professional journals, particularly those related to economics. Students can also draw on the New York State interlibrary loan system, online services, and Bard's ecology field station. A designated site for the National Estuarine Research Reserve, Bard is also home to the environmental research institute Hudsonia Ltd. and the Bard College Field Station. The Institute of Ecosystem Studies in nearby Millbrook is an internationally known scientific establishment dedicated to advancing the understanding of the function and development of ecological systems. In proximity is the Hastings Center, the oldest independent, nonpartisan, interdisciplinary research center in the world, which addresses fundamental ethical issues regarding health, medicine, and the environment. Several researchers from these institutions are members of BCEP's faculty, offering students access to the latest scientific results and debates pertaining to contemporary environmental problems.

Financial Aid

Financial assistance is awarded on the basis of achievement and promise and also on the basis of financial need, according to criteria determined annually by the Office of Financial Aid of Bard College. BCEP is committed to assisting qualified students whose personal financial resources are insufficient to meet the expenses of graduate study. A limited number of fellowships are available. Awards are made on an annual basis without regard to gender, sexual orientation, race, color, age, marital status, religion, ethnic or national origin, or handicapping conditions. Application must be made annually by February 1. Applications on or before the deadline receive first consideration.

Cost of Study

For full-time first-year students entering in 2007 and working toward the Master of Science in environmental policy, tuition is $25,270; tuition is $16,890 for students entering their second year in 2007. For professionals admitted with the internship waiver and candidates for the professional certificate program, the tuition is $25,270 plus the master's thesis fee of $2590. Modest facility and registration fees apply.

Living and Housing Costs

A variety of houses and apartments for reasonable rent can be found near the Bard College campus. Newly constructed campus dormitories include private rooms for graduate students. The cost per room for the 2006–07 academic year is $5200. On admission to the program, students receive a continually updated list of possible off-campus housing opportunities.

Student Group

Bard's 2,600 undergraduate and graduate students come from forty-nine states and thirty-two countries. According to the *U.S. News & World Report 2000* on America's best colleges, Bard College is second in the country for class size and eighth for faculty resources. The Bard Center for Environmental Policy has a ratio of nearly 1 full-time or affiliated faculty member to each student. These numbers are highly conducive to student interaction with colleagues and faculty and staff members as well as to successful internships and career placements.

Student Outcomes

Expertise in the burgeoning field of environmental policy allows graduates to take advantage of opportunities in a wide variety of professional careers. The New England Board of Higher Education reported that an increasing number of environmental careers are now found in the arts, humanities, education, health, law, politics, social change, and forestry—fields that even a decade ago were not linked with environmental policy. Dramatic growth is especially seen in the environmental industry, estimated to reach $600 billion internationally by 2010, and in the nongovernmental sector. Close mentoring of internship partner organizations and student career preparations enhances graduates' marketability.

Location

Bard's magnificent setting along the historic Hudson River offers an unusual blend of diverse environmental, cultural, and recreational resources on campus or within 90 minutes' reach. From New York City to the Adirondacks, urban and natural wonders provide seasonal recreation. The campus lifestyle is enriched by distinguished scientists' lectures; forums on timely environmental, political, and ethical issues; and a host of visual and performing arts exhibitions, concerts, and performances in the stunning Richard B. Fisher Performing Arts Center, designed by Frank Gehry.

The College and The Center

Bard's approach to education aims to assist students in planning and achieving individual intellectual growth throughout the academic process. A hallmark of the educational experience at Bard is the intensive interaction between students and faculty members through small seminars, tutorials, and independent project work. The Bard Center for Environmental Policy builds on Bard College's tradition of creative innovation, with added emphasis on professional preparation and career development in environmental policy fields.

Applying

The ideal candidate has a strong background in sciences, math, and economics, although qualified students with exceptional leadership promise from other disciplines are seriously considered. Relevant activities subsequent to college graduation are given special attention in the selection process. Applications for fall 2007 admission should be received by November 9, 2006, for notification in mid-December. For notification in late March to early April, applications should be received by February 1, 2007. Applications postmarked after this date are considered only if space is available in the entering class.

Correspondence and Information

Bard Center for Environmental Policy
Bard College
30 Campus Road, P.O. Box 5000
Annandale-on-Hudson, New York 12504-5000
Phone: 845-758-7073
Fax: 845-758-7636
E-mail: cep@bard.edu
Web site: http://www.bard.edu/cep

Bard College

THE FACULTY

The faculty is composed of a distinguished core of full-time members and affiliated members who are eminent experts and researchers in diverse fields relating to environmental policy and current practices. Affiliated faculty members* have primary appointments at other institutions and are available for participation in courses and for advising. The high ratio of faculty members to students allows for close rapport and individualized guidance. Small classes ensure close mentoring of student career preparations. To access faculty biographies, students should refer to the faculty segment of Bard's Web site at http://www.bard.edu/cep/graduate/faculty/.

Ana Arana*, M.S., Columbia.
Mark Becker, M.A., CUNY, Hunter.
Peter Berle*, J.D., Harvard.
Daniel Berthold, Ph.D., Yale.
Stuart E. G. Findlay*, Ph.D., Georgia.
Mary Evelyn Greene, J.D., Florida.
Lori Knowles*, LL.M., Wisconsin–Madison.
Frederic B. Mayo*, Ph.D., Johns Hopkins.
William Mullen, Ph.D., Texas.
Lee Paddock, J.D., Iowa.
Jennifer Phillips, Ph.D., Cornell.
Mara Ranville, Interim Director, Bard Center for Environmental Policy; Ph.D., California, Santa Cruz.
Andrew Revkin, M.S., Columbia.
David Sampson, J.D., SUNY at Albany.
Gautam Sethi, Ph.D., Berkeley.
Elizabeth Smith.
Eleanor Sterling*, Ph.D., Yale.

CLEMSON
UNIVERSITY

CLEMSON UNIVERSITY

Master of Science in Environmental Toxicology

Program of Study

The Master of Science (M.S.) in Environmental Toxicology Program at Clemson University consists of cutting-edge research, comprehensive plans of study, and high-quality interactions among faculty and staff members and students. The research and training focus of the program is concentrated in four areas: fate and effects of materials, agriculture/business environmental interface, critical habitats/ecosystems, and human/environmental interface. The educational goals of the program are to prepare superior professional toxicologists who are capable of doing independent research in an academic, government, or industrial setting.

Research activities encompass the areas of aquatic ecotoxicology, biochemical and molecular toxicology, immunobiology and immunotoxicology, terrestrial ecotoxicology, analytic chemistry, and ecological modeling. A strong fundamental research program examines the mechanisms of how chemicals exert their toxicity and how variability in individual organisms can lead to sensitivity or resistance in a wide range of species. Methods developed from research studies provide the means for assessing chemical exposure and impact in the field. Field studies, used to document the status of potentially affected species, incorporate small mammal and avian habitat assessment, water and vegetation sampling, collection of invertebrates and aquatic organisms, and estuarine-marine habitat assessments.

Students take core courses and complete their program with electives that provide flexibility. Master's students are required to take 24 credits plus 6 hours of thesis research and take an average of two to three years to graduate.

Research Facilities

Excellent research facilities are available to graduate students. Specialized equipment, computer, and library facilities are extensive and readily available. The environmental toxicology program is located in a 38,000-square-foot research building, which provides graduate and faculty office space, research laboratories, and an animal care facility. Many of the laboratories have computer-controlled environments and sections of the building are equipped with specialized testing laboratories. These dedicated facilities are for exposures, isolations, histopathology, and analysis of chemicals in a variety of matrices. The animal care facility is accredited by the American Association for the Accreditation of Laboratory Animal Care. The program operates under a formal, in-house quality assurance/quality control program. An independent Quality Assurance Program (QAP) ensures that all studies are reconstructible. The QAP oversees compliance with federal regulations, including good laboratory practices, standard operating procedures, and research protocols.

Financial Aid

Financial assistance is available to qualified students. Most students are supported by graduate research assistantships funded by faculty research grants. Fellowships and teaching assistantships in a variety of departments are also available, and tuition fee waivers can be granted. Some highly competitive fellowships are available from University sources and federal agencies. There are employment opportunities at the University for student's spouses.

Cost of Study

Tuition for 2006–07 is $4643 per semester for in-state students and $9255 per semester for nonresidents. Off-campus rates are $535 per hour for in-state students and $918 per hour for nonresidents. Graduate assistants pay a flat fee of $1079 per semester and $348 per summer session. Graduate fellows pay South Carolina resident fees.

Living and Housing Costs

On-campus housing is available. For information, students should visit http://www.housing.clemson.edu. The cost of living in Clemson is quite low compared to the national average. Students who choose to live off the campus typically spend $300–$400 per month for rent, depending on location, amenities, roommates, etc.

Student Group

Students in the environmental toxicology program come from diverse backgrounds and from all regions of the nation. International students are an important segment of the student population. There are approximately 14 students in the program. Fifty-seven percent are women, and 93 percent are full-time students.

Student Outcomes

Approximately half of the graduates receiving the master's degree continue their education at the Ph.D. level. These graduates have attended a variety of universities including the University of Maryland, North Carolina State University, Texas A&M, and the University of Guelph. The remaining master's graduates have secured employment in academia, government, industry, and consulting firms.

Location

Clemson is a small, beautiful college town near the Blue Ridge Mountains and Lake Hartwell in upstate South Carolina. The Upstate is one of the country's fastest-growing areas and is the midpoint of the Charlotte-to-Atlanta I-85 corridor, a multistate area along Interstate 85 that runs from metro Atlanta to Richmond, Virginia, and encompasses Charlotte, North Carolina, and North Carolina's Research Triangle. Atlanta and Charlotte are each a 2-hour's drive away. Many financial institutions and other industries have national headquarters for a major presence in the Upstate, including Wachovia, Bank of America, BMW, Bon Secours St. Francis Health System, Bosch North America, Bowater, Charter Communications, Ernst and Young, Fluor Corporation, IBM, Microsoft, Michelin of North America, and many others.

The University

Clemson is classified by the Carnegie Foundation as Doctoral/Research University–Extensive, a category comprising less than 4 percent of all universities in America. The University's mission is to fulfill the covenant between its founder and the people of South Carolina to establish a "high seminary of learning" through its responsibilities of teaching, research, and extended public service. The University has identified eight areas of academic emphasis that create collaborations that, in turn, help fulfill the University's mission.

Applying

Applicants may apply on the Web at http://www.grad.clemson.edu/p_apply.html. Applications with a $50 nonrefundable fee should be received no later than five weeks prior to registration. Every required item in support of the application must be on file by that date. Students are advised to contact the department for the deadlines of the program of proposed study.

Correspondence and Information

Dr. Tom Schwedler
132 Lehotsky Hall
Clemson University
Clemson, South Carolina 29634
Phone: 864-646-2810
E-mail: tschwdl@clemson.edu
Web site: http://www.clemson.edu/entox/

Ms. Mary Saunders
509 Westinghouse Road
P.O. Box 709
Pendleton, South Carolina 29670-0709
Phone: 864-646-2961
Fax: 864-646-2277
E-mail: msndrs@clemson.edu

Clemson University

THE FACULTY AND THEIR RESEARCH

William W. Bowerman IV, Associate Professor; Ph.D., Michigan State. Fisheries and wildlife.
William H. Conner, Professor; Ph.D., LSU. Forestry.
William Rockford English, Associate Professor; Ph.D., Clemson. Entomology.
Arnold G. Eversole, Professor; Ph.D., Syracuse. Biology and zoology.
Jeffrey W. Foltz, Professor; Ph.D., Colorado. Biology.
Lawrence R. Gering, Associate Professor; Ph.D., Georgia. Forest biometrics.
Charles A. Gresham, Associate Professor; Ph.D., Duke. Forestry.
David C. Guynn Jr., Professor; Ph.D., Virginia Tech. Forestry.
Roy L. Hedden, Professor; Ph.D., Washington (Seattle). Forest entomology.
Alan R. Johnson, Assistant Professor; Ph.D., Tennessee. Environmental toxicology.
Joseph D. Lanham, Associate Professor; Ph.D., Clemson. Ecology and ornithology.
Patricia A. Layton, Professor and Department Chair; Ph.D., Florida. Forest genetics.
Andy Wu-Chung Lee, Professor; Ph.D., Auburn. Wood products.
Allan Marsinko, Professor; Ph.D., SUNY College of Environmental Science and Forestry. Forest economics.
Larry R. Nelson, Associate Professor; Ph.D., Auburn. Forestry.
Lawrence E. Nix, Professor; Ph.D., Georgia. Forestry.
Christopher J. Post, Assistant Professor; Ph.D., Cornell. Environmental information science.
John H. Rodgers Jr., Professor; Ph.D., Virginia Tech. Botany and aquatic ecology.
Linda C. Roth, Assistant Professor; Ph.D., Clark. Geography.
Victor B. Shelburne, Professor; Ph.D., Clemson. Forestry.
Bo Song, Assistant Professor; Ph.D., Michigan Tech. Forest science.
Thomas J. Straka, Professor; Ph.D., Virginia Tech. Forestry.
David H. Van Lear, Named Professor; Ph.D., Idaho. Forest sciences.
Gaofeng G. Wang, Assistant Professor; Ph.D., British Columbia. Forest ecology.
Thomas M. Williams, Professor; Ph.D., Minnesota. Forestry.
Gene W. Wood, Professor; Ph.D., Penn State. Agronomy.
Thomas E. Wooten, Alumni Professor; Ph.D., North Carolina State. Forestry.
Greg K. Yarrow, Professor; D.F., Stephen F. Austin. Forest wildlife.

CLEMSON UNIVERSITY

CLEMSON UNIVERSITY

Ph.D. in Environmental Toxicology

Program of Study

The Ph.D. in Environmental Toxicology Program at Clemson University consists of cutting-edge research, comprehensive plans of study, and high-quality interactions among faculty and staff members and students. The research and training focus of the program is concentrated in four areas: fate and effects of materials, agriculture/business environmental interface, critical habitats/ecosystems, and human/environmental interface. The educational goals of the program are to prepare superior professional toxicologists who are capable of doing independent research in an academic, government, or industrial setting.

Research activities encompass the areas of aquatic ecotoxicology, biochemical and molecular toxicology, immunobiology and immunotoxicology, terrestrial ecotoxicology, analytic chemistry, and ecological modeling. A strong fundamental research program examines the mechanisms of how chemicals exert their toxicity and how variability in individual organisms can lead to sensitivity or resistance in a wide range of species. Methods developed from research studies provide the means for assessing chemical exposure and impact in the field. Field studies, used to document the status of potentially affected species, incorporate small mammal and avian habitat assessment, water and vegetation sampling, collection of invertebrates and aquatic organisms, and estuarine-marine habitat assessments.

Students take core courses and complete their program with electives, which provide flexibility. Doctoral students are required to take 18 hours of credit and pass written and oral examinations and take an average of four to five years to graduate.

Research Facilities

Excellent research facilities are available to graduate students. Specialized equipment, computer, and library facilities are extensive and readily available. The environmental toxicology program is located in a 38,000-square-foot research building, which provides graduate and faculty office space, research laboratories, and an animal care facility. Many of the laboratories have computer-controlled environments and sections of the building are equipped with specialized testing laboratories. These dedicated facilities are for exposures, isolations, histopathology, and analysis of chemicals in a variety of matrices. The animal care facility is accredited by the American Association for the Accreditation of Laboratory Animal Care. The program operates under a formal, in-house quality assurance/quality control program. An independent Quality Assurance Program (QAP) ensures that all studies are reconstructible. The QAP oversees compliance with federal regulations, including good laboratory practices, standard operating procedures, and research protocols.

Financial Aid

Financial assistance is available to qualified students. Most students are supported by graduate research assistantships funded by faculty research grants. Fellowships and teaching assistantships in a variety of departments are also available, and tuition fee waivers can be granted. Some highly competitive fellowships are available from University sources and federal agencies. There are employment opportunities at the University for student's spouses.

Cost of Study

Tuition for 2006–07 is $4643 per semester for in-state students and $9255 per semester for nonresidents. Off-campus rates are $535 per hour for in-state students and $918 per hour for nonresidents. Graduate assistants pay a flat fee of $1079 per semester and $348 per summer session. Graduate fellows pay South Carolina resident fees.

Living and Housing Costs

On-campus housing is available. For information, students should visit http://www.housing.clemson.edu. The cost of living in Clemson is quite low compared to the national average. Students who choose to live off the campus typically spend $300–$400 per month for rent, depending on location, amenities, roommates, etc.

Student Group

Students in the environmental toxicology program come from diverse backgrounds and from all regions of the nation. International students are an important segment of the student population. There are approximately 12 students in the program. Fifty-eight percent are women, and 97 percent are full-time students.

Student Outcomes

Ph.D. graduates have secured employment in academia, government, industry, and consulting firms, with many taking postdoctoral positions.

Location

Clemson is a small, beautiful college town near the Blue Ridge Mountains and Lake Hartwell in upstate South Carolina. The Upstate is one of the country's fastest-growing areas and is the midpoint of the Charlotte-to-Atlanta I-85 corridor, a multistate area along Interstate 85 that runs from metro Atlanta to Richmond, Virginia, and encompasses Charlotte, North Carolina, and North Carolina's Research Triangle. Atlanta and Charlotte are each a 2-hour's drive away. Many financial institutions and other industries have national headquarters for a major presence in the Upstate, including Wachovia, Bank of America, BMW, Bon Secours St. Francis Health System, Bosch North America, Bowater, Charter Communications, Ernst and Young, Fluor Corporation, IBM, Microsoft, Michelin of North America, and many others.

The University

Clemson is classified by the Carnegie Foundation as Doctoral/Research University–Extensive, a category comprising less than 4 percent of all universities in America. The University's mission is to fulfill the covenant between its founder and the people of South Carolina to establish a "high seminary of learning" through its responsibilities of teaching, research, and extended public service. The University has identified eight areas of academic emphasis that create collaborations that, in turn, help fulfill the University's mission.

Applying

Applicants may apply on the Web at http://www.grad.clemson.edu/p_apply.html. Applications with a $50 nonrefundable fee should be received no later than five weeks prior to registration. Every required item in support of the application must be on file by that date. Students are advised to contact the department for the deadlines of the program of proposed study.

Correspondence and Information

Dr. Tom Schwedler
132 Lehotsky Hall
Clemson, South Carolina 29634
Phone: 864-646-2810
Fax: 864-646-2277
E-mail: tschwdl@clemson.edu
Web site: http://www.clemson.edu/entox/

Ms. Mary Saunders
509 Westinghouse Road
P.O. Box 709
Pendleton, South Carolina 29670-0709
Phone: 864-646-2961
Fax: 864-646-2277
E-mail: msndrs@clemson.edu

Clemson University

THE FACULTY AND THEIR RESEARCH

William W. Bowerman IV, Associate Professor; Ph.D., Michigan State. Fisheries and wildlife.
William H. Conner, Professor; Ph.D., LSU. Forestry.
William Rockford English, Associate Professor; Ph.D., Clemson. Entomology.
Arnold G. Eversole, Professor; Ph.D., Syracuse. Biology and zoology.
Jeffrey W. Foltz, Professor; Ph.D., Colorado. Biology.
Lawrence R. Gering, Associate Professor; Ph.D., Georgia. Forest biometrics.
Charles A. Gresham, Associate Professor; Ph.D., Duke. Forestry.
David C. Guynn Jr., Professor; Ph.D., Virginia Tech. Forestry.
Roy L. Hedden, Professor; Ph.D., Washington (Seattle). Forest entomology.
Alan R. Johnson, Assistant Professor; Ph.D., Tennessee. Environmental toxicology.
Joseph D. Lanham, Associate Professor; Ph.D., Clemson. Ecology and ornithology.
Patricia A. Layton, Professor and Department Chair; Ph.D., Florida. Forest genetics.
Andy Wu-Chung Lee, Professor; Ph.D., Auburn. Wood products.
Allan Marsinko, Professor; Ph.D., SUNY College of Environmental Science and Forestry. Forest economics.
Larry R. Nelson, Associate Professor; Ph.D., Auburn. Forestry.
Lawrence E. Nix; Professor; Ph.D., Georgia. Forestry.
Christopher J. Post, Assistant Professor; Ph.D., Cornell. Environmental information science.
John H. Rodgers Jr., Professor; Ph.D., Virginia Tech. Botany and aquatic ecology.
Linda C. Roth, Assistant Professor; Ph.D., Clark. Geography.
Victor B. Shelburne, Professor; Ph.D., Clemson. Forestry.
Bo Song, Assistant Professor; Ph.D., Michigan Tech. Forest science.
Thomas J. Straka, Professor; Ph.D., Virginia Tech. Forestry.
David H. Van Lear, Named Professor; Ph.D., Idaho. Forest sciences.
Gaofeng G. Wang, Assistant Professor; Ph.D., British Columbia. Forest ecology.
Thomas M. Williams, Professor; Ph.D., Minnesota. Forestry.
Gene W. Wood, Professor; Ph.D., Penn State. Agronomy.
Thomas E. Wooten, Alumni Professor; Ph.D., North Carolina State. Forestry.
Greg K. Yarrow, Professor; D.F., Stephen F. Austin. Forest wildlife.

SIPA

COLUMBIA UNIVERSITY

School of International and Public Affairs Master of Public Administration Program in Environmental Science and Policy

Program of Study

The Master of Public Administration (M.P.A.) Program in Environmental Science and Policy trains sophisticated public managers and policy makers, who apply innovative, systems-based thinking to environmental issues. The program challenges students to think systemically and act pragmatically. To meet this challenge, Columbia offers a top-quality graduate program in management and policy analysis that emphasizes practical skills and is enriched by ecological and planetary science.

The program's approach reflects the system-level thinking that is needed to understand ecological interactions and maintain the health of the earth's interconnected ecological, institutional, economic, and social systems. This program requires more environmental science than any other public policy master's degree in the world. The skills and concepts learned involve an understanding of scientific method, including observation, hypothesis generation, and hypothesis testing. Students also study the chemical processes affecting environmental quality and public health, methods of collection and analysis of field and laboratory data, and systems modeling.

To train effective earth systems professionals, the program focuses on the practical skills necessary to understand the formulation and management of public policy. The teaching of public policy and administration represents the core of the program. This set of classes focuses on specific professional and vocational skills, such as memo writing, oral briefings, group process and team building, spreadsheet and other forms of financial analysis, use of computer programs, case studies of earth systems issues, and the World Wide Web. The principal goal of the core curriculum is to provide students with the analytic, communication, and work skills required to be problem-solving earth systems professionals.

Research Facilities

Lamont-Doherty Earth Observatory was established in 1949 by Columbia geology professor Maurice Ewing. In the last fifty years, research from the observatory has significantly changed understanding of the earth, from the groundbreaking discovery of plate tectonics to an understanding of the ocean's role in regulating climate change. The 125-acre rural campus is connected to Morningside Heights by a regular shuttle bus that brings students, scientists, and faculty members to and from the campus.

During the second semester, students take their courses on Columbia's campus in Morningside Heights, one of the richest concentrations of educational resources and academic activity in the United States. Its fifteen schools draw on a renowned faculty, making it among the country's most productive research centers. Students in the program are granted access to the extensive collections within the renowned Columbia University Libraries system, including twenty-two campus libraries, each supporting a specific academic or professional discipline. The Lehman Social Sciences Library, located in the International Affairs Building, holds a contemporary collection of more than 330,000 volumes and approximately 1,700 periodical titles. It includes materials acquired by Columbia libraries since 1974 in political science, sociology, social anthropology, political geography, and journalism, as well as a rich collection of materials on post–World War II international relations. It also houses Columbia's extensive collection of international newspapers. The Office of Information Technology and the Picker Computer Center cater to the needs of School of International and Public Affairs (SIPA) students with newly expanded, state-of-the art computer labs; digital research programs; and wireless networks at both the SIPA and the Lamont campus. Columbia's trilevel Dodge Physical Fitness Center includes more than 7,500 square feet of aerobic and anaerobic exercise equipment, an indoor running track, and a 9,000-square-foot all-purpose gymnasium and is open to all SIPA students.

Financial Aid

There are fellowships available based upon need and merit. Students who apply for admission and fellowships by the November 1 early decision deadline are notified of their status by December 1. The application deadline for all other students applying for fellowships is January 15. The final admission deadline is February 15 without fellowship. Applicants are notified of admission and fellowship decisions by March 15. Long-term loans at low interest rates are available, including Federal Stafford Student Loans and Federal Perkins Loans. The Federal Work-Study Program is also available.

Cost of Study

Tuition for the 2006–07 academic year is estimated to be $15,100 per semester. The cost of tuition and fees for the complete twelve-month program is $48,683.

Living and Housing Costs

Housing in the Morningside Heights neighborhood around SIPA in buildings owned by Columbia University may be available to entering M.P.A. students. Accommodations include apartment shares; dormitory or suite-style housing; efficiency, one-bedroom, and family-style apartments; and single rooms in the International House. Living and personal expenses vary.

Student Group

Drawn from more than seventy-five countries, SIPA students are diverse, mature (the average age is 27), and intelligent individuals. The environmental science and policy program enrolls approximately 45 to 50 students each year.

Student Outcomes

Graduates of the M.P.A. Program in Environmental Science and Policy are prepared for the roles of analyst, manager, and translator of scientific knowledge. Recent graduates have gone on to careers with NASA, the Center for Corporate Responsibility, EarthTech Environmental Consulting and Engineering, Forest Guardians, and the Agency for Toxic Substances and Disease Registry. The program prepares highly marketable students for management and leadership roles in countless arenas of the global public and private sectors.

The Office of Career Services, located in the School of International and Public Affairs, helps students in all stages of their search for employment, from career interviews to the writing of resumes and their submission to appropriate organizations. The office has a long-standing working relationship with scores of agencies and private organizations.

Location

This twelve-month program takes place at Columbia University's Morningside Heights Campus in New York City and at its Lamont-Doherty Earth Observatory in Palisades, New York, a 25-minute drive (or campus shuttle ride) from the main campus. Students enroll in their summer science course work at the beautiful Lamont campus, overlooking the Hudson River. Policy courses are held on the Morningside Heights Campus on New York City's dynamic, diverse upper west side.

The University and The School

SIPA's focus on a broad range of real-world issues is an outgrowth of the School's original mission, which was written and established in 1946: to train professionals to meet new challenges by providing an interdisciplinary curriculum that draws on Columbia's renowned faculty in the social sciences and other traditional fields. The Master of Public Administration was added to the School in 1977 to meet a growing demand for skilled professionals at home as well as abroad. Students in the M.P.A. Program in Environmental Science and Policy work closely with Columbia's Earth Institute, the world's leading academic center for the integrated study of Earth, its environment, and society. The Earth Institute builds upon excellence in core disciplines—earth sciences, biological sciences, engineering sciences, social sciences, and health sciences—and stresses cross-disciplinary approaches to complex problems.

Applying

A bachelor's degree or its equivalent is required. Advanced high school course work in chemistry and biology is strongly recommended. November 1 is the early admission deadline. Applicants who submit a completed application by that date are promised a decision by December 1 for the following June. January 15 is the application deadline for students seeking fellowships. February 15 is the final deadline for June admission. International applicants are encouraged to apply a month in advance of these deadlines. Qualification for admission is based upon the Admissions Committee's review of the applicant's file: personal statement, transcripts, and letters of appraisal.

Correspondence and Information

For information concerning admission, financial aid, curriculum, and staff members, students should write to the School.

Louise A. Rosen
Assistant Director, Master of Public Administration Program in Environmental Science and Policy
School of International and Public Affairs
1404 International Affairs
420 West 118th Street
New York, New York 10027

Phone: 212-854-3142
Fax: 212-864-4847
E-mail: lar46@columbia.edu
Web site: http://www.columbia.edu/cu/mpaenvironment/

Columbia University

THE PROGRAM'S CORE TEACHING FACULTY

Steven A. Cohen, Director, Master of Public Administration Program in Environmental Science and Policy; Ph.D., SUNY at Buffalo, 1979.
Howard N. Apsan, Adjunct Professor of Public Affairs; Ph.D., Columbia, 1985.
Robert A. Cook, Adjunct Professor of Environmental Affairs; V.M.D., Pennsylvania, 1980.
David Downie, Director of Educational Partnerships, Office of Educational Programs, Columbia Earth Institute; Ph.D., North Carolina at Chapel Hill, 1996.
William Eimicke, Director of the Picker Center for Executive Education, School of International and Public Affairs; Ph.D., Syracuse, 1973.
Lewis E. Gilbert, Adjunct Professor of Environmental Policy, Columbia University; Ph.D., Columbia, 1993.
Adela J. Gondek, Adjunct Professor of Public Affairs; Ph.D., Harvard, 1980.
Tanya Heikkila, Assistant Professor of Environmental Management; Ph.D., Arizona, 2001.
Yochanan Kushnir, Adjunct Professor of Public Affairs and Doherty Research Scientist; Ph.D., Oregon State, 1985.
Patrick Louchouarn, Professor of Environmental Science; Ph.D., Quebec at Montreal, 1997.
Katherine McFadden, Adjunct Instructor; Ph.D., Columbia, 2004.
Stephanie Pfirman, Professor and Chair, Department of Environmental Science, Barnard College; Ph.D., MIT, 1995.
Jeffrey D. Sachs, Director, Earth Institute, Columbia University; Ph.D., Harvard, 1980.
Andrea Schmitz, Adjunct Assistant Professor; M.P.A., Columbia, 1987.
Glenn Sheriff, Assistant Professor in International and Public Affairs; Ph.D., Maryland, 2004.
Ion Bogdan Vasi, Assistant Professor of International and Public Affairs; Ph.D., Cornell, 2005.
Gary Weiskopf, Adjunct Professor of International and Public Affairs; M.P.A., Columbia, 1987.
Paula Wilson, Assistant Professor of International and Public Affairs; M.S.W., SUNY at Albany, 1977.

THE EVERGREEN STATE COLLEGE

Graduate Program in Environmental Studies

Program of Study

The Evergreen State College (TESC) instituted an integrated, interdisciplinary course of study in 1984 leading to the degree of Master of Environmental Studies (M.E.S.). Students interested in the application of technical and management aspects of environmental studies gain the background and working skills necessary to solve a broad range of environmental problems and prepare to enter a wide variety of career areas.

The 72-quarter-hour program can be completed in two years by full-time students and in as little as three years by part-time students. The program is composed of three distinct components. The first is a core sequence of four programs: Political, Economic, and Ecological Processes; Population, Energy, and Resources; Case Studies: Environmental Assessment Policy and Management; and Quantitative Analysis and Research Methods. Each of these programs is 8 quarter hours and is offered by an interdisciplinary team of social and natural scientists. The second component is the creation of an individual program through the selection of regular electives offered by the College, as well as opportunities for internships for credit and Individual Learning Contracts. The electives that are offered include pesticides, wetland ecology, environmental policy, environmental law, environmental philosophy and ethics, environmental health, conservation and restoration biology, environmental economics, hydrology, and salmonid ecology. The third component of the program is a thesis that consists of applied research and analysis in the form of an individual or small-group project. Full-time students carry out the parts of this plan concurrently, while part-time students complete components consecutively.

All students entering the program are expected to have had course work in both the social and natural sciences. Evergreen prides itself on its active, interdisciplinary, action-oriented teaching methods. Faculty members pursue a variety of additional relevant research and professional interests. The faculty members advise students on electives, thesis work, and professional development.

Research Facilities

The College's research facilities include chemical and biological laboratory facilities, an organic farm, 700 acres of second-growth forest, 3,000 feet of undeveloped marine shoreline, and a variety of computers with a full range of software, including GIS capability.

The College's location in the state capital provides extensive research opportunities within state government. TESC library and the nearby Washington State library provide needed reference materials. Students can make use of local and regional agencies and communities as sites to carry out research activities. The program helps coordinate such work. Evergreen is the site of the Editorial Office for *Environmental Practice,* a peer-reviewed professional journal published by Oxford University Press for the National Association of Environmental Professionals.

Financial Aid

Modest financial aid is available. Sources of support include fellowships, program assistantships, work-study positions, employment with the College, paid internships, and participation in contract research. In addition, the M.E.S. program assists students in finding external funding sources and locating part-time employment with public and private agencies. Students should also check with TESC's Financial Aid Office and file a financial aid form by February 15 for priority consideration.

Cost of Study

Full-time tuition for Washington State residents was $2167 per quarter in 2004–05. Nonresident tuition was $6646 per quarter. Fees are subject to change.

Living and Housing Costs

An estimated $5400 covers room and board for a single person living on or off campus during the nine-month academic year. Most graduate students live in or near Olympia in rental units or in apartments adjacent to the campus.

Student Group

Approximately 40 students enroll each fall quarter. The total program size is currently about 100 students. A cooperative spirit within the student group is emphasized primarily through seminars and group assignments. Another 80 graduate students in Evergreen's M.P.A. program share some space and faculty members with the M.E.S. program.

Student Outcomes

More than half of the program alumni are employed in the public sector, predominantly in state and local government agencies. Almost a quarter work in the private sector, and others work in positions in education or nongovernmental organizations or are pursuing further study. Most graduates are employed in positions that are degree related. The Assistant Director assists students in their professional development planning.

Location

Olympia lies at the southern end of Puget Sound, equidistant from the Cascade Range, the Pacific Ocean, and the Olympic Mountains. Train, bus, and highway connections provide easy access to metropolitan Seattle and Portland. Evergreen serves as a cultural and intellectual focus for Washington's capital city of Olympia. The city and surrounding area have a population of 250,000, with excellent outdoor recreation opportunities nearby.

The College

The Evergreen State College opened in 1971. Its national reputation for innovative curricular design and academic excellence have drawn a diverse faculty, staff, and student body to the 1,000-acre campus, which is located 7 miles northwest of Olympia. The most distinctive feature of the curriculum is the concept of interdisciplinary instruction carried out in coordinated study programs for a quarter or longer. This results in the student having both a single and a multifaceted academic commitment at the same time. This concept of study is the essence of the core component of the M.E.S. program.

Evergreen has the usual recreational and athletic facilities on campus, including those for tennis, swimming, racquetball, basketball, and soccer. The College's beach on Puget Sound is the focus for water recreation, and the Cascade and Olympic Mountains provide excellent hiking, climbing, and skiing.

Applying

Admission is normally granted for the fall quarter so that the student can begin the core sequence. Application forms and a catalog are available from the Assistant Director or the Graduate Studies Office.

Correspondence and Information

Assistant Director
Graduate Program in Environmental Studies
Lab I, 3022
The Evergreen State College
Olympia, Washington 98505
Phone: 360-867-6225 or 6707
E-mail: austinj@evergreen.edu
Web site: http://www.evergreen.edu/mes

The Evergreen State College

THE FACULTY

Shown below are the areas of interest in teaching and study of each faculty member.

Natural Science
Sharon Anthony. Environmental chemistry.
Frederica Bowcutt. Botany, restoration ecology, natural history.
Paul Butler. Geology, hydrology.
Gerardo Chin-Leo. Biology, marine studies.
Robert Cole. Physics, energy studies.
Amy Cook. Fish biology.
Heather Heying. Ecology and animal behavior.
Robert Knapp. Physics, energy systems.
Jack Longino. Cell biology, tropical biology.
Nalini Nadkarni. Forest ecology.
John Perkins. Biology, history of technology and environment, editor of *Environmental Practice.*
Paul Przybylowicz. Biology, ecology.
Oscar Soule. Ecology.
James Stroh. Geology.
Kenneth Tabbutt. Geology.
Erik Thuesen. Marine biology.
Alfred Wiedemann. Biology, botany.

Social Science
Carolyn Dobbs. Environmental planning, community organization.
Peter Dorman. Political economy.
Russell Fox. Community planning and development.
Martha Henderson-Tubesing. Cultural geography, human ecology, public lands management.
Cheri Lucas-Jennings. Environmental law.
Carol Minugh. Native American community-based environmental studies.
Ralph Murphy. Environmental economics, natural resource policy.
Lin Nelson. Environmental health and advocacy.
Alan Parker. Native American issues.
Matthew Smith. Political science, environmental politics.
Linda Moon Stumpff. Natural resource policy.
Ted Whitesell. Geography.
Tom Womeldorff. Environmental economics.

Contributing Adjunct Faculty
Stephen L. Beck. Environmental philosophy and ethics.
Nina Carter. Environmental management.
Jeffrey Cederholm. Fisheries biology.
Jean MacGregor. Environmental education.
Charles Newling. Wetlands ecology and management.
Tim Quinn. Habitat conservation planning.

An M.E.S. student presents his candidacy paper to fellow students and faculty members.

Every year, M.E.S. students organize the Rachel Carson forum to publicly debate an environmental issue.

Students have opportunities to study in the field, here with a forest ecologist.

FLORIDA INSTITUTE OF TECHNOLOGY

College of Engineering, Department of Marine and Environmental Systems
Programs in Environmental Sciences, Meteorology, and Environmental Resource Management

Programs of Study

Florida Institute of Technology offers programs of study leading to Master of Science and Doctor of Philosophy degrees in environmental sciences, the M.S. degree in meteorology, and the M.S. degree in environmental resource management. These programs are designed to prepare students for careers in industry, government, colleges and universities, or consulting firms. Emphasis is on the application of scientific principles to the maintenance and wise use of man's environment. The environmental science curriculum provides a thorough background in the biological and chemical fundamentals of natural systems and water and wastewater treatment systems. The principal areas of emphasis in environmental science are related to freshwater and estuarine problems in areas such as eutrophication, toxic wastes, aquatic ecology, and hydrology; to groundwater contamination problems from sources such as septic tanks, landfills, and underground storage tanks; to air pollution, such as air quality monitoring and impacts of air pollutants on natural systems; to marine waste policy; to waste stabilization and waste utilization; and to nuclear-waste management and site remediation, environmental and marine remote sensing, and real-time spectral monitoring of environmental systems using in situ sensors, aircraft, and ships as well as satellites and geographic information systems (GIS). Atmospheric science is focused on understanding Earth's gaseous envelope, predicting its evolution, and mitigating human impacts. The M.S. program at Florida Tech is uniquely interdisciplinary, drawing on expertise from the College of Engineering, the School of Aeronautics, and the College of Science and Liberal Arts. As such, the M.S. in meteorology can have special emphasis in areas such as marine meteorology, water resources, atmospheric chemistry, aviation meteorology, or remote sensing.

The department also offers an M.S. program in coastal zone management and M.S. and Ph.D. programs in the marine sciences of biological, chemical, geological, and physical oceanography, with specializations in geophysical remote sensing of the environment and in ocean engineering. A highly interdisciplinary education is emphasized in this program.

Master's degree requirements consist of 30 semester hours of required and elective courses, including 6 semester hours of thesis research work or internship or 33 semester hours (nonthesis) of courses. Ph.D. students must complete a minimum of 78 semester hours beyond the bachelor's degree or 48 semester hours beyond the master's degree. Interested students should write for more information on the doctoral program.

Graduate student research most frequently involves work on current environmental problems and may be funded by federal, state, and local agencies and industries. Research opportunities and internships are also available at nearby government agencies, research organizations, and private industry.

Research Facilities

The programs offer extensive facilities for instruction and research, such as an advanced remote sensing lab, satellite data reception, and environmental optics lab as part of the Florida Tech Center for Remote Sensing and the Geographical Information System (GIS) Laboratory. The environmental analysis laboratory is equipped with such standard water analysis equipment as balances, ovens, muffle furnaces, a chemical oxygen-demand apparatus, a macro-Kjeldahl apparatus, pH meters, and spectrophotometers. Analytical instruments provided for advanced study include an ion chromatograph, a gas chromatograph, a total organic carbon analyzer, an atomic absorption spectrophotometer, and an auto analyzer. For the M.S. in meteorology, collaborative research is conducted with specialists from the nearby NASA Kennedy Space Center, the USAF 45th Weather Squadron, the NOAA National Weather Service, the Harbor Branch Oceanographic Institution, WHIRL (Wind and Hurricane Impacts Research Laboratory), and local government agencies and corporations.

Financial Aid

Graduate teaching and research assistantships and endowed fellowships are available to a limited number of qualified students. For 2006, typical financial support ranges from $16,000 upward, including stipend and tuition, per academic year for approximately half-time duties. Stipend-only assistantships are sometimes awarded for less time commitment. Students with internships may receive an hourly salary.

Cost of Study

Tuition is $900 per graduate semester hour in 2006–07. New students must pay an entrance deposit of $300, which is deducted from the first semester's tuition charge.

Living and Housing Costs

Room and board on campus cost approximately $3000 per semester in 2006–07. On-campus housing (dormitories and apartments) is available for full-time single and married graduate students, but priority for dormitory rooms is given to undergraduate students. Many apartment complexes and rental houses are available near the campus.

Student Group

Graduate students constitute more than one quarter of the approximately 4,000 students on the Melbourne campus. They come from all parts of the United States and from many other countries.

Student Outcomes

Graduates of the program obtain positions in places such as the Florida Department of Environmental Protection; St. John's River Water Management District; Brevard County; Sarasota County; Volusia County; South Florida Water Management District; U.S. Army Corps of Engineers; National Park Service; Lockheed Martin; NASA; Dynamac; Bionetics; Jordan, Jones and Goulding, Inc.; Brevard Teaching and Research Labs; Florida Groundwater Services; Harris Corp.; Walt Disney World; NOAA National Weather Service; and the U.S. Air Force.

Location

Melbourne is located on the east coast of Florida. The climate is extremely pleasant, and opportunities for outdoor recreation abound. The John F. Kennedy Space Center and Disney World/EPCOT Center are nearby and the Atlantic beaches are within 3 miles of the campus.

The Institute

Florida Institute of Technology was founded in 1958 by a group of scientists and engineers pioneering America's space program at Cape Canaveral. The environmental remote sensing program utilizes the technology developed by the space program. Florida Tech has rapidly developed into a residential institution and is the only independent technological university in the Southeast. It is supported by the community and industry and is the recipient of many research grants and contracts, a number of which provide financial support for graduate students. The campus covers 175 acres and includes a beautiful botanical garden and an internationally known collection of palm trees.

Applying

Forms and instructions for applying for admission and assistantships are sent on request. Admission is possible at the beginning of any semester, but admission in the fall semester is recommended. It is advantageous to apply early. Entering students are expected to have had courses in chemistry, calculus, physics, and biology as well as a year or more of advanced science courses. The GRE General Test is required for admission.

Correspondence and Information

Dr. John G. Windsor Jr., Program Chair
Environmental Sciences Program
Florida Institute of Technology
Melbourne, Florida 32901
Phone: 321-674-8096
Fax: 321-984-8461
E-mail: dmes@fit.edu
Web site: http://www.fit.edu/AcadRes/dmes

Graduate Admissions Office
Florida Institute of Technology
Melbourne, Florida 32901
Phone: 321-674-8027
800-944-4348 (toll-free)
Fax: 321-723-9468
E-mail: grad-admissions@fit.edu
Web site: http://www.fit.edu/Grad

Florida Institute of Technology

THE FACULTY AND RESEARCH AREAS

Thomas V. Belanger, Professor; Ph.D., Florida.
Charles R. Bostater, Associate Professor; Ph.D., Delaware.
Sen Chaio, Ph.D., North Carolina State.
Iver W. Duedall, Professor Emeritus; Ph.D., Dalhousie.
Joseph Dwyer, Associate Professor; Ph.D., Chicago.
Howell H. Heck, Associate Professor; Ph.D., Arkansas.
Steven M. Lazarus, Assistant Professor; Ph.D., Oklahoma.
George A. Maul, Professor; Ph.D., Miami.
Hamid K. Rassoul, Professor; Ph.D., Texas at Dallas.
John H. Trefry, Professor; Ph.D., Texas A&M.
Tom Utley, Associate Professor; Ph.D., Florida Tech.
John G. Windsor Jr., Professor and Program Chair, Environmental Sciences; Ph.D., William and Mary.

Adjunct Faculty

Joseph Angelo, Ph.D. Science Application, International Corp.
Michael F. Helmstetter, Ph.D. Brevard Teaching and Research Laboratory.
Brian E. LaPointe, Ph.D. Harbor Branch Oceanographic Institution.
Frank R. Leslie, M.S., Harris Corporation.
Carlton R. Parks, M.S., Acta, Inc.
Michael Splitt, M.S., University of Utah.
Robert W. Virnstein, Ph.D. St. John's River Water Management District.

RESEARCH INTERESTS

Within the broad discipline of environmental science are areas of specialization that focus on physical, biological, or chemical issues of natural and man-made systems. Because of the interdisciplinary nature of environmental sciences, the department offers programs that link the following major areas in an integrated systems approach, focusing on quantitative techniques:

Environmental Biology. Aquatic ecology, eutrophication of lakes, water quality indicator organisms, microbiology of wastewater treatment, wetlands systems, limnology, environmental planning, and impact statements.

Environmental Chemistry. Chemistry of natural waters, wetlands, nutrient cycling, nitrogen transformations, biogeochemical mass balance modeling, toxic organics in natural waters and water supplies, decomposition, management models for water quality control, non-point-source pollution, and waste treatment.

Environmental Modeling. Specialized environmental climatological environmental systems, theoretical studies and numerical modeling of coastal processes, water quality modeling and toxic chemical modeling, and hazard assessments of chemicals in the environment.

Environmental Resource Management. Recycling and reuse of waste materials. Applied management practicums in internship opportunities are offered, such as EIS development and review, policy analysis, and natural resource management in developing countries.

Global Change. Global temperature change and sea level rise, carbon flux, and ozone depletion.

Remote Sensing and Real-Time Optical Spectral Monitoring. Environmental remote sensing utilizing optical and microwave radiometry based on aircraft and ships and in estuarine, coastal, and inland waters and satellite altimetry. The program maintains the Center for Remote Sensing, with an image processing/remote sensing laboratory and an environmental optics laboratory.

Sustainable Development. Population control, wise use of resources, and environmental economics, with a special focus on islands.

Waste Management. Scientific aspects of waste management methodologies, including marine, estuarine, and freshwater systems; waste interactions with biological, chemical, and physical systems; trace contaminant analyses; ocean pollutant studies; plastics recycling; and incineration ash stabilization and utilization in artificial reef construction.

RESEARCH ACTIVITIES

Aeration of mosquito-control impoundments to improve water quality, to eliminate fish kills, and to enable potential aquaculture use.
Air quality monitoring.
Air-sea interaction.
Artificial reef construction impacts.
Aviation meteorology.
Benthic oxygen demand in Florida lakes.
Effects of urban stormwater runoff on water quality in the Indian River Lagoon system.
Geographic information systems.
Groundwater/surface water interactions in Florida water bodies.
Hurricanes.
Impact of septic tank leachate on the Indian River Lagoon.
Importance of groundwater seepage.
Indian River Lagoon circulation patterns.
In situ optical monitoring of water, wastewater, and coastal systems.
Investigation of oxygen budgets in the Everglades.
Light limitation of sea grass distribution.
The littoral zone of Lake Okeechobee as a source of phosphorus in open waters of the lake.
Marine meteorology.
Marine waste policy.
Nutrient-enhanced coastal ocean productivity.
Phytoplankton population distributions in the upper and middle St. Johns River.
Plastics and other wastes in the ocean.
Real-time optical spectral monitoring.
Recycled plastic utilization in marine environments.
Remote sensing.
Sebastian Inlet biological studies.
Trace metals in the upper St. Johns River and their land use relationships.
Tsunamis and coastal hazards.
Waste utilization.
Water quality characteristics of agricultural pumpage in the upper St. Johns River.
Water quality modeling; modeling the fate, transport, and distribution of chemicals in the environment, coupled with physical-optical models in marine systems.
Water resources.
Weather forecasting/modeling.

INDIANA UNIVERSITY BLOOMINGTON

School of Public and Environmental Affairs
Environmental Science Graduate Programs

Programs of Study

The objective of the environmental science programs at the School of Public and Environmental Affairs (SPEA) is to provide rigorous training in a chosen environmental science specialization and an exposure to the broader interdisciplinary context that is necessary for developing solutions to complex environmental problems. Degree programs are offered to serve students seeking professional careers in private industry, consulting firms, nonprofit agencies, and all levels of government, as well as students seeking research training in preparation for academic careers and other research-oriented positions. Master's degree programs include the two-year, professional Master of Science in Environmental Science (M.S.E.S.) program and several joint-degree programs. The M.S.E.S. combines core courses in environmental chemistry, applied math for environmental chemistry, applied math for environmental sciences, environmental engineering and ecology with a selection of courses in environmental policy, management, law, and/or economics to provide students with an interdisciplinary foundation for environmental problem solving. Each student also completes course work in a concentration area, which provides more in-depth training in a specific area of environmental science. Concentration areas include applied ecology; environmental chemistry, toxicology, and risk assessment; and water resources. To integrate this academic training within a practical framework, students are required to complete an internship, undertake a significant research project, or complete a master's thesis. Program flexibility allows students to design individualized concentrations. Joint-degree programs are offered with SPEA's Master of Public Affairs (M.P.A.) program; the Indiana University (IU) biology, geography, and geology departments; and the Indiana University Schools of Law and Journalism. The Ph.D. program in environmental science is designed to provide rigorous and in-depth research training in a chosen area of environmental science. The program allows students to tailor their course work and research to meet their goals and needs. Admission to the program is highly competitive and requires acceptance by a faculty member or members with compatible research interests. The 22 environmental science faculty members on the Bloomington campus have active research programs within and/or between each of the following subdisciplines: groundwater flow modeling, contaminant fate and transport, environmental chemistry, subsurface bioremediation, biogeochemistry, atmospheric chemistry, global climate change, meteorology and climatology, GIS applications, toxicology, applied statistics, environmental microbiology, applied ecology, conservation biology, and wetland ecology and wetlands restoration.

Research Facilities

The research facilities and equipment available at SPEA are excellent. Equipment includes two programmable environmental chambers, three UV-visible spectrophotometers, an atomic absorption spectrophotometer with graphic furnace and flame atomization, six gas chromatographs, an ion chromatograph, HPLC, an autoanalyzer, an organic carbon analyzer, three mass spectrometers, an inductively coupled plasma analyzer, a CHN thermal analyzer, a portable photosynthesis analyzer, an anaerobic chamber, a phase contrast/epifluorescent microscope, a 2.5-liter bacterial fermentor, a high-end compound research microscope (DM) with camera attachments, two stereo microscopes, a rotary microtome, a cryostat, an Nd:YAG pumped dye laser system, and an adiabatic bomb calorimeter. SPEA also has excellent terrestrial and aquatic field sampling equipment and instruments, a 16-foot research boat, cartographic equipment and map files, and aerial photographic and photo interpretation equipment. One of only a handful of labs in the U.S. capable of multiuser training, SPEA's Geographic Information Systems (GIS) laboratory features some of the most advanced technology to manage, display, and analyze spatial data for scientific and policy research. Libraries on the Bloomington campus house more than 6 million volumes, and another 3.2 million are available through the University's seven other campuses.

Financial Aid

Departmental assistance for qualified students is awarded on a competitive basis and is determined by merit. Awards include teaching and research assistantships. Students may apply for need-based aid through IU's Office of Student Financial Assistance.

Cost of Study

Residents of Indiana paid $276 per credit hour and nonresidents paid $743 per credit hour for the master's and Ph.D. programs in 2005–06. Other academic fees, services, and supplies total between $600 and $700 per year.

Living and Housing Costs

On-campus room and board for single graduate students during the 2005–06 academic year ranged from $4200 to $5016. The 1,500 on-campus housing units for married students ranged in monthly rent from $556 for an efficiency to $924 for an unfurnished three-bedroom apartment. A variety of off-campus housing is available near the University. Rents are generally inexpensive, with the average two-bedroom unit renting for $500 to $700 per month.

Student Group

Approximately 40 students are enrolled in the M.S.E.S. program, with 31 students pursuing the joint M.P.A./M.S.E.S. program and 21 students enrolled in the Ph.D. programs in environmental science. About one tenth of these students are international, more than one half are women, and more than one tenth are members of minority groups.

Student Outcomes

SPEA maintains an outstanding placement record, which is attributed to a well-rounded curriculum, national prestige, and strong alumni support. The SPEA Career Services and Alumni Affairs Office is staffed with professionals who assist graduate students in obtaining permanent employment and internship experiences. Samples of recent placements include the U.S. Fish and Wildlife Service, U.S. Environmental Protection Agency, Radian Corporation, World Wildlife Federation, ICF Kaiser Environmental, and Upjohn Company.

Location

Bloomington, a college town of 70,000 people, was chosen as one of the top ten college towns in America for its "rich mixture of atmospherics and academia" by Edward Fiske, former education editor of the *New York Times.* It is a culturally vibrant community settled among southern Indiana's rolling hills just 45 miles south of Indianapolis, the state capital. Mild winters and warm summers are ideal for outdoor recreation in the two state forests, one national forest, and three state parks that surround Bloomington.

The University and The School

Established in 1820, Indiana University has more than 7,500 graduate students and more than 38,000 total students enrolled on the Bloomington campus. Fifty-five academic departments are ranked in the top twenty in the country, including SPEA, music, business, biology, foreign languages, political science, and chemistry. Attractions include nearly 1,000 musical performances each year, including eight full-length operas and professional Broadway plays; the IU Art Museum, which was designed by I. M. Pei, with more than 30,000 art objects; fifty campus and community volunteer agencies; more than 500 student clubs and organizations; two indoor student recreational facilities; and Big Ten sports. SPEA, founded in 1972, was the first school to combine public management, policy, and administration with the environmental sciences.

Applying

Applications must include the SPEA Admission and Financial Aid application form, transcripts, GRE General Test scores, and three letters of recommendation. Priority is given to applications received by February 1. School visits are encouraged. Applicants are encouraged to visit the School's World Wide Web site.

Correspondence and Information

For master's programs:
Graduate Programs Office
SPEA 260
Indiana University
Bloomington, Indiana 47405
Phone: 812-855-2840
800-765-7755 (toll-free in U.S. only)
E-mail: speainfo@indiana.edu
Web site: http://www.spea.indiana.edu

For doctoral programs:
Ph.D. Programs Office
SPEA 441
Indiana University
Bloomington, Indiana 47405
Phone: 812-855-2457
800-765-7755 (toll-free in U.S. only)
E-mail: speainfo@indiana.edu
Web site: http://www.indiana.edu/~speaweb/academic/science/phd.html

Indiana University Bloomington

THE GRADUATE FACULTY AND THEIR RESEARCH

The faculty members listed below are either part of the environmental policy program or associated with the environmental science graduate programs.

Matthew R. Auer, Ph.D., Yale, 1996. Environmental policy and management problems, with an international focus: international environmental assistance, comparative industrial environmental policy, international policies governing forests and forestry.

Debera Backhus, Ph.D., MIT, 1990. Environmental organic chemistry, particularly the processes controlling the fate of hazardous organic chemicals in the environment; application of research lies in understanding processes, examining exposure and risks, predicting future environmental impacts, evaluating alternative actions or remediation strategies, and designing technologies to prevent future degradation.

Randall Baker, Ph.D., London, 1968. Bridging the gap between the natural and social sciences, comparative study on different perspectives regarding the way problems are perceived and handled, historical perspectives in the analysis of contemporary environmental and policy problems.

James Barnes, J.D., Harvard, 1967. Environmental law, domestic and international environmental policy, ethics and the public official, mediation and alternative dispute resolution, law and public policy.

James Bever (Biology), Ph.D., Duke, 1992. Ecology and evolution of plants and fungi.

Ben Brabson (Physics), Ph.D., Berkeley, 1991. Environmental physics, wind energy and wind speed analysis, climate changes.

Simon Brassel (Geologic Sciences), Ph.D., Bristol, 1980. Biogeochemical responses to climatic and environmental change, abundance and isotopic composition of organic matter in sediments.

John Brothers, Emeritus (Mathematics); Ph.D., Brown, 1964. Geometric analysis.

Lynton Keith Caldwell, Emeritus; Ph.D., Chicago, 1943. Public administration, with emphasis on administrative history and theory; public policy for science and the environment.

Keith Clay (Biology), Ph.D., Duke, 1982. Plant ecology, symbiosis, disease ecology, microbial community ecology.

Clara Cotton (Biology), Ph.D., Indiana, 1985. Aquatic biology, paleolimnology, indexing of key species for environmental assessment.

Christopher Craft, Ph.D., North Carolina State, 1987. Terrestrial and wetland ecosystem restoration, wetlands ecology, soil resources, biogeochemistry, nutrient cycling and carbon sequestration of soils and sediments.

Bruce Douglas (Geological Sciences), Ph.D., Princeton, 1983. Rheological properties of rocks, deformation mechanisms active in the brittle and ductile portions of the lithosphere, development of tertiary extensional basins.

David Good, Ph.D., Pennsylvania, 1985. Quantitative policy modeling, productivity measurement in public and regulated industries, urban policy analysis.

Sue Grimmond (Geography–Atmospheric Sciences), Ph.D., British Columbia, 1989. Micrometeorology and hydroclimatology of heterogeneous terrain, especially urban areas; measurement and modeling of energy and mass (water and carbon dioxide) exchanges in areas of heterogeneous terrain (cities, forests, and wetlands).

Hendrik M. Haitjema, Ph.D., Minnesota, 1982. Groundwater flow modeling, including regional groundwater flow systems, conjunctive surface water and groundwater flow modeling, three-dimensional groundwater flow, and saltwater intrusion problems; emphasis on application of analytic functions to modeling groundwater flow, specifically the analytic element method.

Diane Henshel, Ph.D., Washington (St. Louis), 1987. Sublethal health effects of environmental pollutants, especially pollutant effects on the developing organism, including the effects of polychlorinated dibenzo-p-dioxins (PCDDs) and related congeners on the developing nervous system of birds exposed in the wild and under controlled laboratory conditions.

Gary Hieftje (Chemistry), Ph.D., Illinois, 1969. Development of spectrochemical measurements and instrumentation.

Ronald Hites, Ph.D., MIT, 1968. Applying organic analytical chemistry techniques to the analysis of trace levels of toxic pollutants, such as polychlorinated biphenyls and pesticides, with a focus on understanding the behavior of these compounds in the atmosphere and in the Great Lakes.

Claudia Johnson (Geologic Sciences), Ph.D., Colorado, 1993. Tropical paleontology, quantitative analysis of latitudinal trends in species and genetic diversity, evolutionary and extinction patterns of molluscs and corals.

William Jones, M.S., Wisconsin–Madison, 1977. Lake and watershed management, especially diagnosing lake and watershed water-quality problems; preparing management plans to address problems identified; stream ecology; Caribbean coral reef ecology; underwater archaeology; certified lake manager (CLM).

Erle Kauffman, Emeritus (Geological Sciences); Ph.D., 1961. Distribution of communities along an environmental stress gradient during the Cretaceous, systematic revision of the species-subspecies levels for the bivalvia and ammonoidea, Cretaceous climate investigations.

Ellen Ketterson (Biology), Ph.D., Indiana, 1974. Avian biology, mating systems and parental care, hormones and behavior, physiological mechanisms underlying trade-offs in life histories, using hormones to explore adaptation, dominance and aggression, population dynamics during the nonbreeding season.

Noel Krothe (Geologic Sciences), Ph.D., Penn State, 1976. Flow and chemistry of ground and surface water, carbonate hydrogeology, flow and water chemistry in fractured and solution-controlled aquifers.

Kerry Krutilla, Ph.D., Duke, 1988. Energy policy, resource management in developing countries, environmental regulation, public choice, cost-benefit analysis.

Vicky Meretsky, Ph.D., Arizona, 1995. Ecology and management of rare species, biocomplexity, landscape-level species and community conservation, temporal patterns in biodiversity, integrating ecosystem research and endangered-species management within adaptive management.

Theodore K. Miller, Ph.D., Iowa, 1970. Statistical analysis.

Emilio Moran (Anthropology), Ph.D., Florida, 1975. Tropical ecosystem ecology, Amazon basin, secondary successional forests, human ecology.

Craig Nelson (Biology), Ph.D., Texas, 1966. Evolutionary ecology, amphibian communities, sex determination in reptiles, speciation, interactions among species.

Greg Olyphant (Geologic Sciences and Geography), Ph.D., Iowa, 1979. Environmental geology, instrumentation for intensive site monitoring, numerical/statistical modeling of geospatial data, modeling of wetland hydrology.

David Parkhurst, Ph.D., Wisconsin–Madison, 1970. Physiological plant ecology, including transfers of carbon dioxide and water between leaves and atmosphere and among the cells within leaves, both in relation to leaf structure; mathematics and statistics applied to environmental issues; examples include analysis of concentrations of indicator bacteria at swimming beaches and correct interpretation of statistical hypothesis tests in decision making.

Mark Person (Geologic Sciences), Ph.D., Johns Hopkins, 1992. Groundwater flow mechanisms in different tectonic environments, mathematical modeling of subsurface and flow.

Flynn W. Picardal, Ph.D., Arizona, 1992. Bioremediation, environmental microbiology, and biogeochemistry, with a focus on the microbial reduction of iron oxides and nitrate, transformation of metals and chlorinated hydrocarbons, and combined microbial-geochemical interactions.

Lisa Pratt (Geological Sciences), Ph.D., Princeton, 1982. Biogeochemistry, stable isotopic and organic geochemical studies of sediments.

Sara Pryor (Geography–Atmospheric Sciences), Ph.D., East Anglia, 1992. Air pollution meteorology.

J. C. Randolph, Ph.D., Carleton (Ottawa), 1972. Forest ecology; ecological aspects of global environmental change, with particular interests in forestry and agriculture; applications of geographic information systems (GIS) and remote sensing in environmental and natural resources management; landscape ecology and regional-scale modeling; physiological ecology of woody plants and of small mammals.

Heather Reynolds (Biology), Ph.D., Berkeley, 1995. Plant community ecology.

Edwardo L. Rhodes, Ph.D., Carnegie-Mellon, 1978. Public policy analysis, particularly public-sector applications of management science in the evaluation and assessment of the efficiency or organization performance of public activities, including environmental and natural resource policy implementation.

Kenneth R. Richards, J.D./Ph.D., Pennsylvania, 1997. Climate change policy, carbon sequestration economics, environmental policy implementation and instrument choice.

Evan J. Ringquist, Ph.D., Wisconsin–Madison, 1990. Public policy (environmental, energy, natural resources, and regulation) research methodology, American political institutions.

Scott Robeson (Geography–Atmospheric Sciences), Ph.D., Delaware, 1992. Climate change, statistical climatology, applied climatology.

H. P. Schmid (Geography–Atmospheric Sciences), Ph.D., British Columbia, 1988. Boundary-layer meteorology and micrometeorology, turbulent exchange over inhomogeneous surfaces.

Philip S. Stevens, Ph.D., Harvard, 1990. Characterization of the chemical mechanisms that influence regional air quality and global climate change.

Lee Suttner (Geologic Sciences), Ph.D., Wisconsin, 1966. Subsurface and field-based studies of Cretaceous fluvial systems in the Rocky Mountain foreland basin of Colorado, Wyoming, and Montana; reconstructing paleochannel hydraulics and geometries and modeling the alluvial architecture; chemical stratigraphy of the Upper and Lower Cretaceous nonmarine deposits.

Maxine Watson (Biology), Ph.D., Yale, 1974. Plant developmental ecology; dynamic interaction between development and patterns of resource uptake and use; investigations of genetic variation examined through a combination of common garden, reciprocal transplant, and greenhouse studies, especially of perennial clonal systems, particularly the mayapple (*Podophyllum peltatum);* plant-mycorrhizal interactions.

Jeffrey R. White, Ph.D., Syracuse, 1984. Environmental biogeochemistry, aquatic chemistry, limnology.

MONTCLAIR STATE UNIVERSITY

College of Science and Mathematics
Environmental Management

Program of Study

Montclair State University offers the Doctor of Environmental Management (D.Env.M.). The Doctor of Environmental Management program seeks to foster an emerging interdisciplinary approach to the study of the environment and humankind's impact on natural resources. Broadly defined, the interdisciplinary program fosters understanding of the structure and function of environmental systems and their management. The program focuses on the causes, impacts, and responses to environmental change.

The goal of the doctoral program is to meet the urgent need for highly qualified, trained personnel in the private and public sectors to solve the world's growing environmental problems. Because a deep understanding of environmental issues and solutions to environmental problems requires the knowledge and analytic approaches of several disciplines, the program's faculty includes a wide range of natural, social, and management scientists. The specific objectives of the program are to prepare environmental management professionals who will use research in a databased decision-making process that is firmly rooted in current scientific knowledge and methodology; to prepare environmental professionals who will recognize and analyze the relationships among the scientific, technological, societal, and economic issues that shape environmental research and decision making; and to provide professionals already working in the environmental industry with an opportunity to pursue a rigorous, research-based, advanced degree as part-time and evening students.

The doctoral program is centered on three separate yet interlocking research themes. Graduate students trained through the doctoral program focus on the intersections of these themes: water-land systems; sustainability, vulnerability, and equity; and modeling and visualization.

The water-land systems focus examines interactions among hydrological systems, including aquatic, estuarine, and coastal environments, and landscape structure and pattern. The water-land systems approach considers the interactions of fluvial, estuarine, marine, groundwater, and wetland systems with patterns of human settlement and industry. The highly urbanized northeastern region of New Jersey, while compact geographically, is part of a complex coastal environment in which such interactions can readily be observed. Since the region has a long history of coastal industrial activity, land- and water-use impacts over time can be readily studied. MSU is situated in the heart of the region; consequently, students can conveniently conduct doctoral research projects and training exercises there.

The theme of sustainability, vulnerability, and equity focuses on both the conceptual and operational aspects of these emerging areas of study within the context of urban environmental management. The concepts of sustainability, vulnerability, and equity have become critical for understanding urban environmental management. A primary objective of the doctoral program's research mission is to more formally integrate these theoretical advancements into urban environmental change and management theory and practice. The themes of sustainability, vulnerability, and equity constitute an interdisciplinary approach to urban environmental management that is based on the study of organizations and institutions. Research approaches within this component of the program are follows: analysis of natural systems for the construction of indicators and establishing models for monitoring urban systems (e.g., water supply, material use, and waste systems) and associated environmental improvement or degradation; research into existing conditions and opportunities for enhancing the conditions of sustainability and equity, as well as vulnerability reduction in organizations and institutions (i.e., business, education, governmental, and nongovernmental organizations); and research into the dynamics of public policy and environmental sustainability, vulnerability, and equity.

The modeling and visualization focus utilizes state-of-the-art computer-assisted techniques and methods to study the process of environmental change. Modeling and visualization have become critical tools for environmental managers in advancing their understanding of how the major elements of the complex physical and human environment interact, particularly with respect to the urban environment. More sophisticated data gathering and processing devices and updated software packages are the cutting-edge research tools for the environmental analysis and modeling community. This component of the doctoral program facilitates the integration of these new techniques and methods into the analysis of urban environmental issues. Within the program, there are several main research modeling and visualization themes. They include linking environmental models to remote sensing and GIS for application to landscape dynamics; marine sediment and associated contaminant transport and chemodynamic modeling; global and regional climate modeling, atmospheric chemistry, and the global biogeochemical cycles of greenhouse gases; and integration of modeling with other technologies such as optical sensors and data acquisition electronics.

Research Facilities

More than fifty teaching and research labs are outfitted for research in biology, chemistry, geoscience, geology, ecology, marine and aquatic biology, and many other topics of environmental concern. Specialized facilities include two greenhouses, laboratories for environmental geophysics and soil stratigraphy, and specialized equipment for environmental analysis—earth, air, plant, animal, or water. Field testing equipment include ground-penetrating radar, current and tide gauges, and hand-held Global Positioning System units. A state-of-the-art Xserve cluster supports environmental modeling applications, including global and regional climate modeling. Montclair State University's libraries maintain academic resources that include books, journals, videos, scholarly publications, and electronic databases. These provide the most up-to-date information available on any subject. The widest research facility of all—the state of New Jersey—is fertile ground for environmental studies. The state contains all manner of habitats: shoreline, wetlands, urban and suburban development, farmland, and mountain regions. Montclair State maintains professional and academic relationships with dozens of local and state organizations, providing numerous opportunities for doctoral research in a diverse and plentiful number of applications.

Financial Aid

Student loans are the primary source of financial aid for graduate students. A limited number of graduate assistantships, which currently offer a $15,000 stipend for the ten-month academic year and include a full tuition waiver, are available on a competitive basis for full-time D.Env.M. students. Applications for assistantships are included in the application packet.

Cost of Study

Doctoral tuition and fees for in-state residents are $467 per credit. Out-of-state tuition is $633.01 per credit.

Living and Housing Costs

At Montclair State University, graduate students have several housing options, ranging from traditional and suite-style residence halls to apartment communities. A shuttle bus that connects apartment housing with the main campus is available. Residence hall rooms cost anywhere from $2600 to $3600 per semester. Two-bedroom apartments for 4 residents cost $3180 per student per semester. Other options may be reviewed on the University's Web site. Meal plans are available in flexible package and cost options, depending on individual need.

Student Group

Montclair State University enrolls approximately 15,000 students.

Location

Montclair State University provides an outstanding learning environment. The University's easy access to New York City makes it a great place in which to study. The campus is near local bus and train service, major train transportation, and international airports. Montclair State's location offers diverse cultural experiences, restaurants, shopping, recreation, and entertainment. New Jersey offers beautiful shoreline and beach areas, rural and park recreation, mountain skiing and hiking, and city culture and nightlife. The state has myriad possibilities for study and exploration, both for academic and social purposes.

The University

Founded in 1908, Montclair State University was originally established for teacher training. In the 1930s, Montclair began offering master's degree programs and became accredited as a teachers' college—one of the first in the nation. The University now offers forty-four undergraduate majors, thirty-seven master's and doctoral degrees, forty-seven certificate and certification programs, and numerous interdisciplinary programs through three colleges and two schools. Its easy access to New York City, as well as the New Jersey mountains and shoreline, makes it ideal for the study of environmental disciplines.

Applying

Students wishing to be considered as doctoral candidates must complete an application, which can be found online at http://www.montclair.edu/graduate/programs/doctoral/denvmprog.shtml. Applicants to this program must complete a self-managed application, meaning the applicant gathers all required documentation and then submits it in one packet for University review. Applicants must write a personal essay describing their areas of potential research interest and the relevance of doctoral study to their scholarly development. Official transcripts, GRE scores, TOEFL scores (if necessary), three letters of reference, and an application fee of $60 complete the package. The deadline for receipt of all application materials, including applications for assistantships, is February 15 for admission for the following fall semester. Applications for spring admission are considered.

Correspondence and Information

College of Science and Mathematics
Richardson Hall 262
Montclair State University
Montclair, New Jersey 07043

Phone: 973-655-5108
Fax: 973-655-4390
E-mail: debeusb@mail.montclair.edu
Web site: http://www.csam.montclair.edu/denvm/

Montclair State University

THE FACULTY AND THEIR RESEARCH

George E. Antoniou, Professor; Ph.D., National Technology University (Athens). Computer modeling.
Paul A. X. Bologna, Assistant Professor; Ph.D., South Alabama. Marine ecology, aquatic vegetation.
Stefanie A. Brachfeld, Assistant Professor; Ph.D., Minnesota. Marine geophysics, paleoclimatology, polar regions.
Mark J. Chopping, Assistant Professor; Ph.D., Nottingham. Remote sensing, GIS.
Norma C. Connolly, Professor; J.D., New York Law. Environmental law, natural resource dispute litigation.
Huan Feng, Assistant Professor; Ph.D., SUNY at Stony Brook. Estuarine and coastal environmental quality assessment and management; behavior, transport, and fate of land-based contaminants in aquatic systems and sediments; biogeochemical cycle of trace elements in riverine, estuarine, and coastal environments.
Zhaodong Feng, Associate Professor; Ph.D., Kansas. Environmental change, human-environment interactions, semiarid environments, Quaternary studies, GIS applications in geomorphology and environmental studies.
Richard R. Franke, Professor; Ph.D., Harvard. International development and planning, sustainability, South Asia.
Peter Freund, Professor; Ph.D., New School. Sociological aspects of public health, social organization of space and the environment, social impacts of the automobile.
Matthew Gorring, Assistant Professor; Ph.D., Cornell. Igneous petrology, geochronology, radiogenic isotope, geochemistry, tectonics.
Eileen Kaplan, Professor; Ph.D., Rutgers. Business administration, international human resources administration.
Scott L. Kight, Assistant Professor; Ph.D., Indiana. Evolutionary biology, entomology, ecology, animal behavior.
Michael A. Kruge, Professor and Associate Dean, College of Science and Mathematics; Ph.D., Berkeley. Geochemistry of organic contaminants in sediments.
Phillip LeBel, Professor; Ph.D., Boston University. Resource and energy economics.
Lee Lee, Professor; Ph.D., CUNY. Microbiology.
Bonnie Lustigman, Professor and Chair, Biology; Ph.D., Fordham. Microbiology: effect of metals on growth of microorganisms.
George T. Martin, Professor; Ph.D., Chicago. Social impacts of the automobile, globalization and consumption.
Jon Michael McCormick, Professor; Ph.D., Oregon State. Benthic and estuarine ecology, effects of heavy-metal contamination in estuarine environment.
Bogdan Nita, Assistant Professor; Ph.D., Texas at Dallas. Fluid dynamics, mathematical modeling.
Duke U. Ophori, Associate Professor; Ph.D., Alberta. Hydrogeology, groundwater flow modeling in fractured reservoirs, nuclear-waste disposal.
Gregory A. Pope, Associate Professor; Ph.D., Arizona State. Geomorphology, physical geography, geographic information systems applications in physical geography, human impacts on the environment, global change, geoarchaeology.
Robert S. Prezant, Professor and Dean, College of Science and Mathematics; Ph.D., Delaware. Aquatic ecology, biodiversity, malacology.
Glenville Rawlins, Associate Professor; Ph.D., NYU. International business, international development, Africa.
Stefan A. Robila, Assistant Professor; Ph.D., Syracuse. Remote sensing.
Paul Scipione, Professor; Ph.D., Rutgers. Consumer psychology, marketing, spatial analysis.
Harbans Singh, Professor; Ph.D., Rutgers. Environmental policy and problem solving.
John Smallwood, Associate Professor; Ph.D., Ohio State. Ecology and kestrel ecology, effect of land use on bird migration and nesting.
Rolf Sternberg, Professor; Ph.D., Syracuse. Geography, geopolitics, urban geography, transportation geography and world resources.
John A. Taylor, Professor; Ph.D., Australian National. Environmental management, global and regional climate modeling, atmospheric chemistry, global biogeochemical cycles.
Robert W. Taylor, Professor; Ph.D., Saint Louis. Environmental public policy, regional planning and urban development, urban environmental issues, environmental communications, environmental business policy.
Dirk W. Vanderklein, Associate Professor; Ph.D., Minnesota. Tree physiological ecology, forestry and forest ecology.
Neeraj Vedwan, Assistant Professor; Ph.D., Georgia. Impact of climate change on agriculture.
Stanley Walling, Director, Center for Archaeological Studies; Ph.D., Tulane. Field and contract archaeology, geoarchaeology.
Danlin Yu, Assistant Professor; Ph.D., Wisconsin–Milwaukee. GIS and urban studies.
Michael Zey, Associate Professor; Ph.D., Rutgers. International business and technology development, futurist.

PACE UNIVERSITY

Dyson College of Arts and Sciences
Master of Science in Environmental Science

Program of Study

The Master of Science in Environmental Science degree program at Pace University involves a 41- or 42-credit curriculum, which is structured so that students may encounter environmental issues from scientific, ethical, practical, and legal perspectives. Its faculty members, who are senior professionals in academe, research, industry, and law, bring an interdisciplinary approach to the program. An independent project is part of the curriculum. Specialization courses allow for additional study in either a directed scientific discipline, such as toxicology, waste treatment and management, geographical information and surveillance systems, environmental sampling and analysis, or ecology, or in areas related to public administration.

Research Facilities

The Pace University Library is a comprehensive teaching library and student learning center, a virtual library that combines strong core collections with ubiquitous access to global Internet resources to support broad and diversified curricula. Reciprocal borrowing and access accords, traditional interlibrary loan services, and commercial document-delivery options supplement the aggregate library. Pace offers instructional services librarians, a state-of-the-art electronic classroom, digital reference services, and multimedia applications.

Pace's computer resource centers are linked to high-speed data networks and feature sophisticated hardware and software to facilitate active learning. Recognized as one of America's most wired universities, Pace supports high-speed Internet and Internet2 access on every campus; residence facilities are wired, and most public areas are enabled for wireless connectivity. Full-motion videoconference facilities enable remote delivery of instruction between campus sites for synchronous learning applications. Many courses are Web assisted with state-of-the-art software, and some courses and programs are completely Web based.

Financial Aid

Pace's comprehensive student financial assistance program includes scholarships, graduate assistantships, student loans, and tuition-payment plans. Scholarships are awarded to students in recognition of academic achievement and are available for full- and part-time study. Highly qualified students may be eligible for assistantships awarded by departments, which paid stipends of up to $5100 and tuition remission of up to 24 credits during the 2005–06 academic year. Pace participates in all major federal and state financial aid programs, such as Federal Loans, the New York State Tuition Assistance Program (TAP), Perkins Loans, and the Federal Work-Study Program. All students are encouraged to apply for these programs by filing the Free Application for Federal Student Aid (FAFSA).

Cost of Study

Tuition for graduate courses is $857 per credit in 2006–07.

Living and Housing Costs

Residence facilities are available on campus in both New York City and Westchester. Double-occupancy rooms cost $8500 for the 2005–06 academic year. University-operated, off-campus housing is available within proximity of the New York City campus.

Student Group

Pace students represent diverse personal, cultural, and educational backgrounds. Many students are employed and pursue graduate study for personal growth and career advancement. Sixty-three percent are enrolled part-time in evening classes. Current enrollment in the graduate environmental science program is approximately 20 students.

Location

Pace University is a multicampus institution with campuses in New York City and Westchester County, New York. All locations are within reach of cultural, business, and social resources and opportunities. The downtown Manhattan campus is adjacent to Wall Street and City Hall. Pace's Midtown Center is a short distance from Times Square, theaters, and Grand Central Station. The Pleasantville/Briarcliff campus is located in a suburban setting, surrounded by towns that offer various forms of recreation. The Graduate Center and the School of Law are located in White Plains, New York, among major retail districts and many corporate headquarters. All locations are accessible by public transportation. The graduate environmental science program is available at the Pleasantville campus.

The University

Founded in 1906, Pace University is a private, nonsectarian, coeducational institution. Originally founded as a school of accounting, Pace Institute was designated Pace College in 1973. Through growth and various successes, it was renamed Pace University, as approved by the New York State Board of Regents. Today, Pace offers comprehensive undergraduate, graduate, doctoral, and professional programs at several campus locations through six schools and colleges.

Applying

Admission to Pace University graduate programs requires successful completion of a U.S. baccalaureate degree or its equivalent from an accredited institution. Students must submit a completed application, an application fee, official transcripts from all postsecondary institutions attended, a personal statement, a resume, and two letters of recommendation. International students must submit official TOEFL score reports and official transcripts in their native language with a professional English translation.

Students must demonstrate satisfactory performance on the GRE General Test. An undergraduate major in science is not required; however, preparation should include one year of course work in general biology, general chemistry, and organic chemistry.

Applications should be submitted by August 1 for the fall semester, December 1 for the spring semester, and May 1 for summer sessions. International applications should be submitted one month prior to these dates.

Correspondence and Information

Office of Graduate Admission
Pace University
1 Pace Plaza
New York, New York 10038
Phone: 212-346-1531
Fax: 212-346-1585
E-mail: gradnyc@pace.edu
Web site: http://www.pace.edu

Office of Graduate Admission
Pace University
1 Martine Avenue
White Plains, New York 10606
Phone: 914-422-4283
Fax: 914-422-4287
E-mail: gradwp@pace.edu
Web site: http://www.pace.edu

Pace University

THE FACULTY

Carl Candioloro, Professor of Biological Sciences; Ph.D., St. John's (New York). Biotechnology.
Frank Commisso, Professor of Botany Biology; Ph.D., Fordham.
William Flank, Professor of Chemistry; Ph.D., Delaware. Field sampling, water monitoring, statistical analysis.
Margaret Minnis, Lecturer in Chemistry and Program Coordinator; Ph.D., Syracuse. Chemistry.
John Pawlowski, Professor of Biology; Ph.D., Fordham. Ecology and environmental testing.
Kevin Reilly, Professor; J.D., New York State Supreme Court. Environmental law.
Richard Schlesinger, Professor and Chair of Biological Sciences and Toxicology; Ph.D., NYU.
Joshua Schwartz, Associate Professor of Biological Sciences and Ecology and Program Director; Ph.D., Connecticut.
Mary M. Timney, Professor of Political Science; Ph.D., Pittsburgh.
William Ventura, Professor of Biological Sciences and Toxicology; Ph.D., New York Medical College.
Ellen Weiser, Professor and Chair of Chemistry and Physical Sciences; Ph.D., CUNY Graduate School. Environmental biochemistry.

RENSSELAER POLYTECHNIC INSTITUTE

Department of Earth and Environmental Sciences

Programs of Study

The Department of Earth and Environmental Sciences (E&ES) offers students the opportunity to learn from some of the most accomplished faculty members and to work on some of the most exciting and significant projects in Earth science. The Department offers a Ph.D. degree in geology in the fields of petrology, geochemistry, geophysics, and hydrogeology. An M.S. in geology or hydrogeology and a professional master's degree in applied groundwater science are also offered.

Candidates for the M.S. degrees in geology and hydrogeology must complete 30 hours of graduate study based on an approved plan of study. A thesis based on original research is usually submitted. This requirement may be waived at the discretion of the candidate's adviser.

Candidates for the Ph.D. degree may conduct research in geochemistry, geophysics, hydrogeology, or petrology. They must fulfill the requirements of the Office of Graduate Education. Evidence of success in graduate-level study and research must be shown. There is no language requirement.

For the professional master's degree in applied groundwater science, candidates must also complete 30 credit hours of graduate study based on an approved plan of study. However, no thesis is required.

Research Facilities

The Department is housed within two buildings on the RPI campus: the Jonsson-Rowland Science Center (JSC) and the Materials Research Center (MRC). The JSC hosts the Department office; the Departmental lecture room; the bulk of the faculty, staff, and student offices; and many of the laboratory facilities. The MRC is the home of the environmental geochemistry analytical facilities, rock and fossil collections, and environmental graduate students' offices.

The Department has several vigorous externally funded research programs. The diverse interests of the faculty members lead to a wide variety of projects that stimulate educational programs at both the graduate and undergraduate levels. The Department's internationally renown research covers topics as diverse as the formation of the Earth's core and PCB's in the upper Hudson River. This work falls generally into three categories (with some overlap): environmental geochemistry and hydrology, geochemistry of the Earth's interior, and solid-earth geophysics.

Rensselaer houses numerous facilities used by E&ES students and staff and faculty members to advance their research. The fully automated Cameca SX-100 microprobe in the Department of Earth and Environmental Sciences is available for chemical microanalysis of geological, metallurgical, and other materials, to analysts and researchers on and off the Rensselaer campus. Other equipment includes an atomic absorption spectrometer, a fully automated, state-of-the-art mass spectrometer, a transient gas flow permeameter, and polarizing optical microscopes. There is a cathodoluminoscope, which is used for characterizing the distribution and movement of emitter atoms in minerals. Natural materials that typically play host to abundant emitters include calcite, zircon, allanite, xenotime, and apatite, although any mineral may host emitting elements to one degree or another. The Department also produces synthetic minerals with emitters to observe changes in their distribution with other parameters. There are also an autoradiographic facility, an atomic force microscope, and the Optical Spectroscopy Laboratory contains a Micro-Raman spectroscopy system, diamond cell apparatus, various inverted and upright binocular microscopes with digital camera and video capability, and a IX-71 Olympus fluorescence microscope.

The Watson Facility contains a prep lab with three metal lathes, two minimet polishers, a low-speed saw, aqueous crystallization apparatus, and a drill press; a low pressure lab with glass melting and muffle furnaces, atmospheric quench furnaces, and cold-seal apparati; and a piston cylinder lab. The Hyperbug Facility contains a hydrothermal diamond cell apparatus; the Hatten S. Yoder Jr. Laboratory, with two Yoder-designed Internally Heated Pressure Vessels, a Flow-through reactor, cold-seal apparati, and tube and muffle furnaces; and, for biogeochemistry investigations, there are an anaerobic chamber, a microwave reactor, and a large volume autoclave. There is also a state-of-the-art geophysics computing facility with a seismograph station, extensive geodetic global positioning system (GPS) equipment, a 12-channel seismograph, and a gravimeter and magnetometers. Extensive research is also conducted in the electron microscopy lab (SEM, FE-SEM, TEM), X-ray diffraction facility, FTIR lab, and the XPS Facility. The Darrin Fresh Water Institute is another E&ES premier research center, and it houses the Keck Water Quality Lab, which contains important tools for water and watershed analysis.

Financial Aid

Financial aid is available in the forms of teaching and research assistantships and fellowships, which include tuition scholarships and stipends. Rensselaer assistantships and university, corporate, or national fellowships fund many of Rensselaer's full-time graduate students. Outstanding students may qualify for university-sponsored Rensselaer Graduate Fellowship Awards, which carry a minimum stipend of $20,000 and a full tuition and fees scholarship. All fellowship awards are calendar-year awards for full-time graduate students. Summer support is also available in many departments. Low-interest, deferred-repayment graduate loans are available to U.S. citizens with demonstrated need.

Cost of Study

Full-time graduate tuition for the 2006–07 academic year is $32,600. Other costs (estimated living expenses, insurance, etc.) are projected to be about $12,400. Therefore, the cost of attendance for full-time graduate study is approximately $45,000. Part-time study and cohort programs are priced differently. Students should contact Rensselaer for specific cost information related to the program they wish to study.

Living and Housing Costs

Graduate students at Rensselaer may choose from a variety of housing options. On campus, students can select one of the many residence halls, and there are abundant options off campus as well, many within easy walking distance.

Student Group

There are 1,234 graduate students, of whom 30 percent are women, 90 percent are full-time, and 69 percent study at the doctoral level.

Student Outcomes

Rensselaer's graduate students are hired in a variety of industries and sectors of the economy and by private and public organizations, the government, and institutions of higher education. Starting salaries average $63,262 for master's degree recipients.

Location

Located just 10 miles northeast of Albany, New York State's capital city, Rensselaer's historic 275-acre campus sits on a hill overlooking the city of Troy, New York, and the Hudson River. The area offers a relaxed lifestyle with many cultural and recreational opportunities, with easy access to both the high-energy metropolitan centers of the Northeast—such as Boston, New York City, and Montreal, Canada—and the quiet beauty of the neighboring Adirondack Mountains.

The Institute

Recognized as a leader in interactive learning and interdisciplinary research, Rensselaer continues a tradition of excellence and technological innovation dating back to 1824. More than 100 graduate programs in more than fifty disciplines attract top students, researchers, and professors. The discovery of new scientific concepts and technologies, especially in emerging interdisciplinary fields, is the lifeblood of Rensselaer's culture and a core goal for the faculty, staff, and students. Fueled by significant support from government, industry, and private donors, Rensselaer provides a world-class education in an environment tailored to the individual.

Applying

The admission deadline for the fall semester is January 1. Basic admission requirements are the submission of a completed application form (available online), the required application fee ($75), a statement of background and goals, official transcripts, official scores on the GRE General Test, TOEFL or IELTS scores (if applicable), and two recommendations.

Correspondence and Information

Department of Earth and Environmental Sciences
Rensselaer Polytechnic Institute
110 8th St., JSC 1W19
Troy, New York 12180-3522

Phone: 518-276-6474
E-mail: ees@rpi.edu
Web site: http://www.rpi.edu/dept/geo

Rensselaer Polytechnic Institute

THE FACULTY AND THEIR RESEARCH

Teofilo (Jun) A. Abrajano Jr., Professor and Director, Environmental Science Program; Ph.D., Washington (St. Louis). Isotopic geochemistry, biogeochemistry, environmental geochemistry. (abrajt@rpi.edu; http://www.rpi.edu/~abrajt/abrajt.html)

Richard F. Bopp, Associate Professor; Ph.D., Columbia. Environmental geochemistry. (boppr@rpi.edu; http://ees2.geo.rpi.edu/bopp.html)

Damon Chaky, Affiliated Research Scientist; Postdoctoral Research Scientist at Columbia University, Lamont-Doherty Earth Observatory; Ph.D., Rensselaer. Environmental geochemistry, transport and behavior of persistent pollutants in the environment. (chakyd@ldeo.columbia.edu; http://ees2.geo.rpi.edu/people/chakyd.html)

Daniele J. Cherniak, Research Associate Professor; Ph.D., SUNY at Albany. Experimental and analytical geochemistry. (chernd@rpi.edu; http://ees2.geo.rpi.edu/people/chernd.html)

Rinat I. Gabitov, Affiliated Research Scientist; Postdoctoral Researcher, Woods Hole Oceanographic Institute; Ph.D., Rensselaer. Marine geochemistry. (rgabitov@whoi.edu; http://ees2.geo.rpi.edu/people/gabitr.html)

Rob McCaffrey, Professor; Ph.D., California, Santa Cruz. Plate dynamics, crustal deformation, seismology. (mccafr@rpi.edu; http://www.rpi.edu/~mccafr)

Jonathan D. Price, Research Associate and Director of Labs and Facilities; Ph.D., Oklahoma. Geochemistry and petrology. (pricej@rpi.edu; http://www.rpi.edu/~pricej)

Steven W. Roecker, Professor; Ph.D., MIT. Solid Earth geophysics, seismology. (roecks@rpi.edu; http://gretchen.geo.rpi.edu/roecker/roecker.html)

Joseph Pyle, Research Associate; Ph.D., Rensselaer. Metamorphic petrology and geochronology. (pylej@rpi.edu; http://ees2.geo.rpi.edu/pyle/pyleindex.html)

Anurag Sharma, Assistant Professor; Ph.D., SUNY at Binghamton. Biogeochemistry, organic geochemistry. (anurag_sharma@rpi.edu; http://ees2.geo.rpi.edu/people/sharma.html)

Frank S. Spear, Professor and Chair; Ph.D., UCLA. Metamorphic petrology, thermochemistry. (spearf@rpi.edu; http://ees2.geo.rpi.edu/spear/spear.html)

Lara C. Storm, Postdoctoral Research Associate; Ph.D., Rensselaer. Metamorphic petrology. (storml@alum.rpi.edu; http://ees2.geo.rpi.edu/Storm/home.html)

Jay B. Thomas, Postdoctoral Research Scientist; Ph.D., Virginia Tech. Experimental geochemistry. (thomaj2@rpi.edu; http://ees2.geo.rpi.edu/people/thomaj.html)

Anahita A. Tikku, Research Scientist; Ph.D., California, San Diego (Scripps). Geophysics. (tikkua@rpi.edu; http://www.rpi.edu/~tikkua)

David A. Wark, Research Associate Professor and Director, Electron Microprobe Lab; Ph.D., Texas at Austin. Experimental geochemistry and volcanology. (warkd@rpi.edu; http://www.rpi.edu/~warkd/wark.html)

E. Bruce Watson, Institute Professor; Ph.D., MIT. Geochemistry, Earth and planetary materials. (watsoe@rpi.edu; http://www.rpi.edu/~watsoe)

Charles Williams, Research Associate; Ph.D., Arizona. Computational geophysics. (willic3@rpi.edu; http://ees2.geo.rpi.edu/people/Willic3.html)

ENVIRONMENTAL GEOSCIENCE

Freshwater and sediment environmental chemistry and hydrology: The Department of Earth and Environmental Science has pioneered the use of chemical and isotopic markers to characterize the deposition of sediments and the effects of human development on these systems. Researchers are also involved in characterizing sources, transport, and degradation of pollutants in surface and groundwater environments, including polychlorinated biphenyls (PCB), polycyclic aromatic hydrocarbons (PAH), common solvents, fuel products and additives, and other petroleum hydrocarbons. Rensselaer is actively involved in monitoring and resolving environmental issues within the region. The Institute was recently selected by New York Governor George Pataki to manage the Upper Hudson Satellite Center that will be part of the Rivers & Estuaries Center on the Hudson River.

Principle Investigators: Richard Bopp, Jun Abrajano, and Damon Chaky.

Biogeochemistry: Rensselaer has a long history of researching the interaction of microbes, organic compounds, and earth materials. Current work examines the sources and pathways of biogenic compounds to examine carbon and sulfur pathways in aquatic environments. Research is also focused on characterizing the acidic anoxygenic sulfidic thermal springs, which play host to green sulfur bacteria. In addition, the Department is extending its research into high-pressure biogeochemistry, investigating the nature and processes of life within the earth and on other planetary bodies.

Principle Investigators: Anurag Sharma, Jun Abrajano, and Henry Ehrlich (Emeritus, Biology).

GEOCHEMISTRY OF THE EARTH'S INTERIOR

Experimental Geochemistry: The Department of Earth and Environmental Sciences remains a leader in fundamental research on mass transport within the earth through experimental characterization. Foremost is the work on the diffusion of geochemically significant elements within common rock-forming minerals. Most significantly, this work provides important constrains to geochronologists and petrologists, permitting them to more accurately date geologic events and determine their origins. Researchers also continue to evaluate the nature of fluid transport within the crust and upper mantle, the development of microstructure within rocks, and the partitioning of elements during crystal growth.

Principle Investigators: Bruce Watson, David Wark, Daniele Cherniak, Jay Thomas, Jon Price, and Rinat Gabitov.

Metamorphic Petrology and Thermochemistry: The Department is instrumental in providing some of the most powerful computational tools and evaluation techniques required to interpret the complicated nature of metamorphic reactions within the earth. Investigators are evaluating the thermodynamics of common metamorphic mineral assemblages, characterizing crystal growth and compositional changes, and developing analytical techniques to determine the ages of metamorphic events. These constrain the geologic history of a number of regions, including New England, the Adirondacks of New York, the Caledonides of Norway, and Greece.

Principle Investigators: Frank Spear, Joe Pyle, and Lara Storm.

SOLID-EARTH GEOPHYSICS

Seismic characterization and processing: The Department is at the forefront of using new seismological techniques to resolve the structure of the crust and mantle and their dynamics. Through careful and novel manipulation of seismic data, investigators are providing new insights into the composition and heterogeneity of the Earth and the underlying structure of the deep subsurface. These techniques are being applied to a number of diverse areas, including the Tien Shan Mountains (China), the San Andreas Fault, the Adirondacks, Yucca Mountain, Taiwan, and Central Asia.

Principle Investigator: Steve Roecker.

Lithosphere Dynamics and Plate Tectonics: Researchers within the Department are utilizing the accuracy of the Global Positioning System to evaluate the movement of Earth's lithospheric plates, determine the forces driving these movements, and predict their role in future geological hazards, such as earthquakes and volcanism. Through the use of computer simulations, researchers are modeling crustal dynamics, particularly the complex nature of subduction zone tectonics. These methods were used recently to determine the rotational movement of the crust beneath Oregon and Washington and the nature of ongoing deformation in Indonesia.

Principle Investigators: Rob McCaffrey, Anahita Tikku, and Charles Williams.

RICE

RICE UNIVERSITY

Weiss School of Natural Sciences
Professional Master's Program in Environmental Analysis and Decision Making

Program of Study

The Professional Master's Program offers a Master of Science degree in environmental analysis and decision making. This degree is geared toward teaching students rigorous methods needed by industrial and governmental organizations to deal with environmental issues. As an interdisciplinary program, it aims to give students the ability to anticipate environmental problems, not just solve them. The course of study emphasizes core quantitative topics such as statistics, remote sensing, data analysis, and modeling. In addition, laboratory and computer skills are introduced. Students can focus their education by selecting from a specific menu of electives in relevant fields.

The Professional Master's Program at Rice's Weiss School of Natural Sciences offers the Environmental Analysis and Decision Making, the Subsurface Geoscience, and the Nanoscale Physics degree programs. These master's degrees are designed for students seeking to gain further scientific core expertise coupled with enhanced management and communication skills. The program instills a level of scholastic proficiency that exceeds that of the bachelor's level and creates the cross-functional aptitudes needed in modern industry. Skills acquired in this program allow students to move more easily into management careers in consulting or research, development, design, and marketing of new science-based products.

The twenty-one-month Professional Master's Program begins with two semesters of course work at Rice, followed by a three-to-six-month industrial internship. After the internship, students return to Rice for a final semester of course work. In addition to technical courses, the students in the environmental analysis and decision making program take management courses, a policy and ethics course, and a seminar jointly with the students involved in the other professional master's tracks. No thesis is required; however, students are required to present their internship project in both oral and written form in the Professional Master's Seminar. Students are also required to attend events organized by the Rice Alliance for Technology and Entrepreneurship and are guided in selected courses by the efforts of the Cain Project in Engineering and Professional Communication.

Research Facilities

Students in the Professional Master's Program in Environmental Analysis and Decision Making participate in a hands-on environmental nanotechnology lab utilizing state-of-the-art equipment in the Department of Chemistry. In addition, the students take advantage of Rice's advanced computer facilities in their statistics and applied mathematics courses. Fondren Library, with its excellent resources, is located on campus.

Financial Aid

Students are eligible to apply for federal student loans and the Federal Work-Study Program. The Rice University Office of Student Financial Services is available to assist students in applying for these programs. As this is a professional program, no tuition waivers or assistantships are available.

Cost of Study

The yearly graduate tuition at Rice University for the 2006–07 year is $23,400. The Professional Master's Program requires three semesters of tuition (total of $35,100 at the current rate) and a nominal fee to continue full-time student status during the semester in which the student is involved in his or her internship.

Living and Housing Costs

The cost of living in Houston is low compared to other large cities in the U.S. One-bedroom apartments or shared larger apartments can cost $380 to $600 in middle-class neighborhoods. The new Graduate Apartments, located near the campus, offer private or shared rooms, a commons room, and free transportation to academic buildings. Rooms are reserved as space is available and cost $380 to $665 per month, single occupancy.

Student Group

The 2006–07 academic year represents the fifth year of the Professional Master's Program at Rice. The class is small to ensure that each student obtains the individual attention he or she needs. The target size for the fall 2006 entering class is approximately 5 students. Acceptance into the program is based on scholastic record, as reflected by the courses chosen and quality of performance, evaluation of former teachers and advisers, GRE scores, and TOEFL scores (if applicable).

Student Outcomes

The 2006–07 academic year is the fifth year of the program. Student placement has been successful at environmental consulting firms, energy companies, and government organizations.

Location

Rice University is located in Houston, Texas. A large park and beautiful residential neighborhoods border the campus, which is only a block from the Texas Medical Center and the museum district. Houston's diverse population is reflected in the city's restaurants and cultural events. The city has a symphony, ballet, opera, and theater, as well as professional men's and women's basketball, baseball, football, soccer, and hockey teams. Houston's seaport, the nation's third largest, is linked to the Gulf of Mexico by a 50-mile channel. Beaches in Galveston can be reached within a 45-minute's drive.

The University

Rice University is a private, coeducational, nondenominational university founded in 1891 from the estate of William Marsh Rice. It has faculties of liberal arts, science, and engineering and about 2,700 undergraduate and 1,500 graduate students. Rice has a graduate student–faculty ratio of 3:1 and is recognized as a top teaching and research university. Students work closely with faculty members.

Applying

Selection is based on the student's scholastic record, evaluations by teachers and advisers, and Graduate Record Examinations (GRE) scores. The TOEFL is required of international students. The deadline for fall applications is February 15.

Correspondence and Information

Professional Master's Program
Mail Stop 103
Rice University
P.O. Box 1892
Houston, Texas 77251-1892
Phone: 713-348-3188
Fax: 713-348-3121
E-mail: profms@rice.edu
Web site: http://profms.rice.edu

Rice University

THE FACULTY AND THEIR RESEARCH

Andrew R. Barron, Professor; Ph.D., Imperial College (London), 1986. Applications of inorganic chemistry to the materials science of aluminum, gallium, and indium.

Vicki L. Colvin, Associate Professor; Ph.D., Berkeley, 1994. Nanocrystals, confined liquids and glasses, porous solids, photonic band gap materials.

F. Barry Dunning, Professor; Ph.D., University College (London), 1969. Experimental atomic and molecular physics, surface physics, spin dependent phenomena, surface magnetism, chemical physics, optics, instrumentation.

Katherine B. Ensor, Professor; Ph.D., Texas A&M, 1986. Time series, including categorical time series, spatial statistics, spatial-temporal methods, and estimation for stochastic process and environmental statistics.

Matthew P. Fraser, Assistant Professor; Ph.D., Caltech, 1998. Measurement and analysis of pollutant concentrations in the atmosphere, with an emphasis on quantifying single organic compounds in vapor and particle phases.

Matthias Heinkenschloss, Associate Professor; Dr.rer.nat., Trier (Germany), 1991. Optimization, optimal control, numerical analysis, partial differential equations.

Neal F. Lane, University Professor; Ph.D., Oklahoma, 1964. Atomic and molecular physics, science and technology policy.

Kathleen S. Matthews, Dean of Natural Sciences; Ph.D., Berkeley, 1970. Structure and function of genetic regulatory proteins.

Erzsébet Merényi, Professor; Ph.D., Szeged (Hungary), 1980. Artificial neural networks; self-organizing maps; segmentation and classification of high-dimensional patterns; data fusion; data mining; knowledge discovery; exploitation of hyperspectral data; analyses of remote-sensing multispectral and hyperspectral data/imagery for planetary surface composition identification; applications to geology, soil, resource mapping, and environmental monitoring.

Douglas Natelson, Assistant Professor; Ph.D., Stanford, 1998. Nanoscale physics, in particular electrical and magnetic properties of systems with characteristic dimensions approaching the single-nanometer scale.

Dale S. Sawyer, Professor; Ph.D., MIT, 1982. Geodynamics, seismology, remote sensing, geomorphology, subsurface geoscience.

Evan H. Siemann, Assistant Professor; Ph.D., Minnesota, 1997. Population and community ecology, forests, grasslands, plant ecology, insect ecology, plant/herbivore interactions, biodiversity, conservation biology.

Tayfun E. Tezduyar, Professor; Ph.D., Caltech, 1982. Development of advanced computational methods and tools for flow simulation and modeling.

SOUTHERN ILLINOIS UNIVERSITY CARBONDALE

Environmental Resources and Policy

Program of Study

The Environmental Resources and Policy (ER&P) Ph.D. degree at Southern Illinois University Carbondale (SIUC) provides students with an interdisciplinary education in natural resource and environmental processes, with a perspective on public policy and social institutions that shape societal and individual reactions to environmental issues. Students are prepared to work with multifaceted environmental problems and carry out interdisciplinary scientific research and are qualified for high-level administration positions in academia, government, and the private sector. Graduates are able to address the most compelling and daunting challenge in natural resource and environmental issues—identifying and solving problems that cross disciplinary boundaries.

The Environmental Resources and Policy Ph.D. degree is administered through the Departments of Geography and Geology and the College of Agriculture (agribusiness economics, forestry, and plant, soil, and general agriculture). The School of Law and the College of Engineering also cooperate in the program.

The course of study is composed of four interdisciplinary core courses and supplemented with individually designed curricula for each of the six areas of concentration: earth and environmental processes; energy and mineral resources; environmental policy and administration; forestry, agricultural, and rural land resources; geographic information systems and environmental modeling; and water resources. Students typically spend three years in course work and research, with oral and written preliminary examinations at the end of their course work.

Research Facilities

Students in the ER&P Ph.D. program have access to a fully equipped, state-of-the-art geographic information systems/remote sensing laboratory with a full-time supervisor and individual workstations. Cooperating departments have laboratory facilities that are used by various professors in their research.

In addition, the University's Morris Library has a general collection of 2.8 million volumes, 4.5 million microforms, and more than 43,000 current serial subscriptions, with access to Online Computer Library Center (OCLC) and ILLINET Online, the statewide automated catalog system.

Financial Aid

Students accepted into the ER&P program are eligible for financial aid on a competitive basis. Students may also be eligible for assistance through grants from research projects or other sources.

Cost of Study

In-state tuition is $243 per semester credit hour (or $3645 for 15 hours and up) in 2006–07. Out-of-state tuition is $607 per semester credit hour (or $9112.50 for 15 hours and up). Fees vary from $441.62 (1 hour) to $987.30 (12 hours).

Living and Housing Costs

For married couples, students with families, and single graduate students, the University has 692 efficiency and one-, two-, three- and four-bedroom apartments that rent for $438 to $620 per month in 2006–07. Residence halls for single graduate students are also available, as are accessible residence hall rooms and apartments for students with disabilities.

Student Group

Southern Illinois University's 21,500 undergraduate and graduate students come from many states and countries. The ER&P Ph.D. program attracts a large number of international as well as domestic students. Students are given the opportunity to work with a number of faculty members from other cooperating departments in research, teaching, and learning activities.

Location

SIUC is 350 miles south of Chicago and 100 miles southeast of St. Louis. Nestled in rolling hills bordered by the Ohio and Mississippi Rivers and enhanced by a mild climate, the area has state parks, national forests and wildlife refuges, and large lakes for outdoor recreation. Cultural offerings include theater, opera, concerts, art exhibits, and cinema. Educational facilities for the families of students are excellent.

The University

Southern Illinois University Carbondale is a comprehensive public university with a variety of general and professional education programs. The University offers bachelor's and associate degrees, master's and doctoral degrees, the J.D. degree, and the M.D. degree. The University is fully accredited by the North Central Association of Colleges and Schools. The Graduate School has an essential role in the development and coordination of graduate instruction and research programs. The Graduate Council has academic responsibility for determining graduate standards, recommending new graduate programs and research centers, and establishing policies to facilitate the research effort.

Applying

Interested students should apply directly to the ER&P Ph.D. program. Application deadlines vary. The program accepts midyear applications.

Correspondence and Information

Environmental Resources and Policy Program
Mail Code 4637
Southern Illinois University Carbondale
405 West Grand Avenue
Carbondale, Illinois 62901-4637
Phone: 618-453-7328
Fax: 618-453-7346
Web site: http://www.siu.edu/~er&p

Southern Illinois University Carbondale

THE FACULTY AND THEIR RESEARCH

Ira Altman, Assistant Professor. Rural growth and development, emerging industries and organizational economics.
Ken Anderson, Associate Professor. Organic geochemistry.
Sara Baer, Assistant Professor. Ecosystem, restoration, grassland ecology.
Jeff Beaulieu, Associate Professor. Environmental modeling and policy, marketing.
Robert Beck, Professor Emeritus. Law.
Wendy Bigler, Instructor. Fluvial geomorphology, rivers and society.
Jason Bond, Assistant Professor. Nematology and plant pathology.
John Burde, Professor. Policy and forest recreation.
Andrew Carver, Associate Professor. Natural resource economics and development, land use planning, GIS/spatial analysis.
Lizette Chevalier, Associate Professor. Physical remediation.
Paul Chugh, Professor. Minerals and residues processing.
John Crelling, Professor. Coal geology.
Mae Davenport, Assistant Professor. Human dimensions of natural resources.
Ken Diesburg, Assistant Professor. Turfgrass management training, breeding, and research.
Leslie Duram, Professor and Chair. Population and natural resources, rural land use, conservation thought.
Benedykt Dziegielewski, Professor. Resources analysis and evaluation techniques, water resources planning and management, water conservation.
Phil Eberle, Associate Professor. Economic efficiency and viability studies of various enterprises and management practices.
Steven Esling, Associate Professor and Chair, Department of Geology. Hydrogeology.
George Feldhamer, Professor and Director, Environmental Studies Program. Mammalogy, wildlife ecology.
Eric Ferré, Assistant Professor. Structural geology, rock magnetism, remote sensing.
Richard Fifarek, Associate Professor. Economic geology.
David Gibson, Professor. Plant population and community ecology, grassland ecology, multivariate methods, exotic and rare species ecology.
John Groninger, Associate Professor. Silviculture, forest vegetation management.
Kim Harris, Associate Professor. Agriculture finance and agribusiness management.
Paul Henry, Associate Professor. Ornamental horticulture.
Bruce Hooper, Associate Professor. Watershed management, integrated water resources management and policy, floodplain management.
Scott Ishman, Associate Professor. Marine micropaleontology.
Brian Klubek, Professor and Chair. Soil microbiology.
John Koropchak, Professor, Vice-Chancellor for Research, and Dean of the Graduate School. Analytical chemistry
Steven Kraft, Professor and Chair, Department of Agribusiness Economics. Soil and water conservation policy, watershed management and planning, farm policy.
Luba Kurkalova, Assistant Professor. Environmental and energy economics, econometrics.
Christopher Lant, Professor. Water resources management, wetlands and nonpoint source pollution policy, ecological economics.
Lilliana Lefticariou, Assistant Professor. Geochemistry.
David Lightfoot, Professor. Plant biotechnology and genomics.
John Mead, Associate Dean. Coal extraction and utilization research.
Khalid Meksem, Associate Professor. Plant genomics, genetics, and biotechnology.
Jean Mangun, Associate Professor. Human dimensions of natural resource management.
John Marzolf, Associate Professor. Sedimentology and stratigraphy.
Patricia McCubbin, Assistant Professor. Law.
Karen Midden, Professor. Landscape design.
Manoj Mohanty, Associate Professor. Mining and mineral resources engineering.
Wanki Moon, Assistant Professor. Public acceptance of GMOs, health information in food markets natural resources, rural land use, conservation thought.
John Nicklow, Associate Professor and Interim Associate Dean. Hydraulic and hydrologic modeling.
Tonny Oyana, Assistant Professor. GIS, GIScience, cartographic and geographic visualization, environmental health and exposure.
John Phelps, Professor and Chair, Department of Forestry. Forest product marketing, wood science.
Nicholas Pinter, Professor. Environmental geology, fluvial geomorphology.
John Preece, Professor. Horticulture, propagation, biotechnology.
Tiku Ravat, Professor. Potential-field geophysics.
Matt Rendleman, Associate Professor. Fuel ethanol, local enterprise impact analysis, Illinois dairy sector.
Don Rice, Associate Provost. Human ecology.
Charles Ruffner, Associate Professor. Dendrochronology, ecology and paleoecology.
John Russin, Professor and Chair, Department of Plant and Soil Science. Plant pathology.
Dwight Sanders, Assistant Professor. Risk management, future contract design, forecasting techniques and evaluation.
Justin Schoof, Assistant Professor. Climatology, quantitative methods.
Jon Schoonover, Assistant Professor. Sediment source tracking.
John Sexton, Professor. Seismology.
Bradley Taylor, Associate Professor. Fruit crops.
Alan Walters, Associate Professor. Vegetable production.
Matt Whiles, Associate Professor. Stream ecology, freshwater invertebrates.
Frank Wilhelm, Assistant Professor. Limnology.
Karl Williard, Associate Professor. Watershed management, forest hydrology, forest biochemistry, riparian zone management, sediment yield.
Bryan Young, Associate Professor. Weed science.
Jim Zaczek, Associate Professor. Ecology, biology, physiology, and genetics of trees; oak silviculture; regeneration ecology.

STONY BROOK
STATE UNIVERSITY OF NEW YORK

STONY BROOK UNIVERSITY, STATE UNIVERSITY OF NEW YORK

Department of Technology and Society
Environmental and Waste Management Program

Program of Study

The Department of Technology and Society offers graduate work leading to the Master of Science in technological systems management. The concentration in environmental and waste management is designed both for persons pursuing careers in environmental, waste, and energy management and technology assessment and for those planning environmental, waste, and energy research and technical careers. The program is particularly well suited for recent college graduates interested in studying environmental problems in a rigorous, thoroughly multidisciplinary setting and for middle career professionals who want to advance in or transfer to an environmental career or introduce environmental components into their main area of expertise. Thirty credits and a thesis are required for the master's degree. Full-time students may complete their programs of study in twelve to eighteen months, part-time students in twenty-four to thirty-six months. Most courses are taught in the evening.

Generally speaking, emphasis is placed on environmental, waste, and energy problems in industry and society at large; pollution prevention and waste minimization; environmental regulatory compliance; hazardous and radioactive waste, contamination, and cleanup; assessment of new environmental products, technologies, and policies; mathematical modeling and computer simulation; risk analysis; and the diagnosis of environmental disputes. Student theses have covered the full range of environmental, energy, and waste topics. Strong ties are maintained with area industry, Brookhaven National Laboratory, and the Waste Reduction and Management Institute.

Research Facilities

The department has advanced computer laboratories available for environmental modeling and simulation. Research projects are conducted off-site (for example, pollution and waste reduction assessments for a business). Research laboratories in other Stony Brook engineering and marine sciences departments and Brookhaven National Laboratory are available for collaborative projects.

Financial Aid

Some research assistantships and teaching assistants are available, but they are generally reserved for full-time students. Many companies, government agencies, and national laboratories reimburse tuition costs. Paid internships with area industry, government agencies, and Brookhaven National Laboratory are sometimes available.

Cost of Study

Graduate courses cost $288 per credit hour for in-state residents and $455 per credit hour for out-of-state residents.

Living and Housing Costs

University apartments range in cost from $323 per month to $1400 per month, depending on the size of the unit. Off-campus housing options include furnished rooms to rent and houses and/or apartments to share that can be rented from $350 to $1500 per month.

Student Group

There are 15–20 matriculated environmental and waste management students, of whom one third are full-time and two thirds are part-time, within an overall departmental graduate program of approximately 80 students pursuing concentrations in global industrial management, environmental and waste management, or educational computing. One fourth of the students come from other countries.

Student Outcomes

Graduates have established a distinguished record in their careers in environmental consulting firms; environmental technology companies; corporate environmental divisions of manufacturing and high-technology industries; American and foreign government environmental, energy, and waste management agencies; national laboratories; and research, educational, and technology assessment organizations.

Location

Stony Brook's campus is approximately 50 miles east of Manhattan on the North Shore of Long Island. The cultural offerings of New York City and Suffolk County's countryside and seashore are conveniently located nearby. Cold Spring Harbor Laboratories and Brookhaven National Laboratories are easily accessible from, and have close relationships with, the University.

The University

The University, established in 1957, achieved national stature within a generation. Founded at Oyster Bay, Long Island, the school moved to its present location in 1962. Stony Brook has grown to encompass more than 110 buildings on 1,100 acres. There are more than 1,568 faculty members, and the annual budget is more than $805 million. The Graduate Student Organization oversees the spending of the student activity fee for graduate student campus events. International students find the additional four-week Summer Institute in American Living very helpful. The Intensive English Center offers classes in English as a second language. The Career Development Office assists with career planning and has information on permanent full-time employment. Disabled Student Services has a Resource Center that offers placement testing, tutoring, vocational assessment, and psychological counseling. The Counseling Center provides individual, group, family, and marital counseling and psychotherapy. Day-care services are provided in four on-campus facilities. The Writing Center offers tutoring in all phases of writing.

Applying

For domestic students, all application materials for admission to the master's program must be received by March 1 for the summer session, March 15 for the fall semester, and October 1 for the spring semester. For international students, all application materials for admission to the master's program must be received by January 1 for the summer session, March 15 for the fall semester, and September 1 for the spring semester.

Correspondence and Information

Graduate Program Coordinator
Department of Technology and Society
347A Harriman Hall
Stony Brook University, State University of New York
Stony Brook, New York 11794-3760
Phone: 631-632-8765

Stony Brook University, State University of New York

THE FACULTY AND THEIR RESEARCH

Distinguished Service Professors

David L. Ferguson, Chairperson; Ph.D., Berkeley, 1980. Quantitative methods, computer applications (especially intelligent tutoring systems and decision support systems); mathematics, science, and engineering education; decision making.

Lester G. Paldy, M.S., Hofstra, 1966. Nuclear arms control, science policy.

Distinguished Teaching Professors

Thomas T. Liao, Emeritus; Ed.D., Columbia, 1971. Computers in education, science, and technology education.

John G. Truxal, Emeritus; Sc.D., MIT, 1950. Control systems, technology-society issues.

Professors

Emil J. Piel, Emeritus; Ed.D., Rutgers, 1960. Decision making, technology-society issues, human-machine systems.

Tian-Lih Teng, Ph.D., Pittsburgh, 1969. Electrical engineering, computer science, management of information systems, electronics commerce.

Marian Visich Jr., Emeritus; Ph.D., Polytechnic of Brooklyn, 1956. Aerospace engineering, technology-society issues.

Associate Professors

Edward Kaplan, Visiting Associate Professor; Ph.D., Pennsylvania, 1973. Environmental systems engineering.

Samuel C. Morris, Visiting Associate Professor; Sc.D., Pittsburgh. Environmental science, risk analysis.

Sheldon J. Reaven, Graduate Program Director; Ph.D., Berkeley, 1975. Science and technology policy; energy and environmental problems and issues; waste management, recycling, and pollution prevention; risk analysis and life-cycle analysis; nuclear, chemical, and biological threats; technology assessment; homeland security.

Lori L. Scarlatos, Ph.D., Stony Brook, SUNY, 1993. Computer-human interaction, multimedia and education, computer graphics.

Assistant Professor

Guodong Sun, Ph.D., Carnegie Mellon, 2000. Energy technology innovation, global climate change, energy and environmental policy, environmental management and regulatory reform in China.

David Tonges, Ph.D., Stony Brook, SUNY, 1998. Technology and environmental impact assessments, solid waste and impacts, alternative energy.

Lecturers

Joanne English Daly, M.S., SUNY at Stony Brook, 1994. Internet technology, computers in learning environments.

Herb Schiller, M.S.M.E., Caltech, 1966; M.S., Polytechnic, 1973. Operations management, manufacturing systems.

UNIVERSITY OF CALIFORNIA, LOS ANGELES

Environmental Science and Engineering Program

Program of Study

The Environmental Science and Engineering (ESE) Program is an interdepartmental graduate degree curriculum administered through the School of Public Health that culminates in the award of the Doctor of Environmental Science and Engineering (D.Env.) degree. This professional degree was established in 1973 with the conviction that resolving complex environmental problems requires individuals who are not specialists in a narrow traditional sense but who have a broad understanding of the environment as well as the technical and managerial skills for environmental problem solving. The purpose of the program is to supply this much-needed kind of professional. A graduate of the ESE program has an area of specialization (represented by the student's master's degree), a background that includes several disciplinary areas, experience gained through working with experts in a variety of fields, and an understanding of how a particular discipline interacts with others. More than 200 students have graduated from UCLA with the D.Env. degree, and they hold leadership positions in government, industry, and private consulting firms.

Applicants must qualify for admission to the UCLA Graduate Division; hold a master's degree or the equivalent in one of the natural sciences, engineering, or public health; have a good background in basic science and mathematics; and have strong communication skills. Following admission, a student takes a program of courses to broaden his or her education in environmental problem areas. In the second year, the student enrolls in three quarters of environmental problems courses—projects that provide intensive exposure to multidisciplinary professional work. Recent problems course topics have emphasized air pollution, water quality, toxic substances, stormwater management, economic impacts of coastal water pollution, watershed management, and habitat restoration, and they often focus on the interaction between policy and technology. The student advances to candidacy after passing written and oral qualifying examinations. There is no language requirement. An approved internship of 1½ to 3 years with government, industry, nonprofit organizations, or consulting firms follows, during which time the student completes a dissertation on a topic related to the internship experience. The candidate is required to present a written prospectus and defend it before the doctoral committee within nine months of advancement to candidacy and the beginning of the internship. Completion of the program normally requires four to five years.

Research Facilities

UCLA has some of the finest library resources and computer facilities in the nation. Several laboratories are available to support laboratory and field studies, although the program is not primarily a laboratory research one. Campus organizations formally affiliated with the program are the Schools of Engineering and Applied Science, Law, Public Affairs, and Public Health and the Departments of Atmospheric and Oceanic Sciences, Biostatistics, Civil and Environmental Engineering, Chemical Engineering, Earth and Space Sciences, Geography, Environmental Health Sciences, Epidemiology, Statistics, Public Policy, Urban Planning, and Ecology and Evolutionary Biology. Students, therefore, have the opportunity to take advantage of the full spectrum of campus resources.

Financial Aid

Currently, 100 percent of the students entering the program receive financial assistance, which typically includes fees and a stipend of $3000 per quarter. In the second year, the program offers graduate research assistantships that paid $3000 per quarter plus fee remission for 2005–06. A limited number of fellowships are available as well. Students may also be eligible for aid from funds administered through the Graduate Division. Following the second year, students complete paid internships at host institutions, with salaries typically $45,000 to $60,000 per year.

Cost of Study

In 2005–06, fees for California residents were $4036.50 per quarter. Nonresident fees were an additional $12,512.

Living and Housing Costs

The University operates housing for single and married graduate students. A new 840-unit complex for single graduate students includes studios and two-bedroom units for $920 to $974 per person per month (2006–07 rates). There are five apartment complexes, designed for graduate students and families, approximately five miles from campus. Early enrollment for housing is advised. There are extensive housing options in the west Los Angeles area within bicycling distance of UCLA or on bus routes leading directly to UCLA. For more information about graduate student and family housing, students should visit http://www.housing.ucla.edu.

Student Group

There are about 25,000 undergraduate and 11,000 graduate students enrolled at UCLA. The Environmental Science and Engineering Program has about 45 students, with about 10 students on campus at any given time. Each year the ESE Program enrolls 5–8 doctoral students who come from many schools and hold master's degrees in science or engineering disciplines.

Student Outcomes

Environmental science and engineering students have been successfully placed in a wide range of professional positions with government agencies such as the U.S. EPA, California EPA, U.S. Army Corps of Engineers, California Air Resources Board, Lawrence Berkeley Laboratory, the State of Washington Department of Ecology, California Regional Water Quality Control Boards, and the Southern Nevada Water Authority. Graduates also find employment with environmental consulting companies, industry, and nonprofit organizations. Many graduates have risen to positions of leadership in government and the private sector.

Location

UCLA is located on the west side of Los Angeles, 5 miles from the Pacific Ocean and 12 miles from downtown Los Angeles. The many diverse cultural and recreational opportunities in the region are within easy reach, and the University itself is a vigorous community center.

The University

UCLA, established in 1919, is academically ranked among the leading universities in the United States and has attracted distinguished scholars and researchers from all over the world. Undergraduate and graduate programs offered in the colleges and schools cover the academic spectrum. UCLA has also developed research programs and curricula outside the usual departmental structures. Interdisciplinary research facilities include institutes, centers, projects, bureaus, nondepartmental laboratories, stations, and museums. There are also many interdisciplinary programs of study, one of which is the Environmental Science and Engineering Program. The Institute of the Environment at UCLA is a focus for environmental research and teaching activities across campus. UCLA's library is the largest in the Southwest. The University's Center for the Health Sciences contains one of the nation's leading hospitals and several nationally known institutes. UCLA's performing arts program of music, dance, theater, film, and lectures is one of the largest and most diverse offered by any university in the country.

Applying

Application forms for admission and financial aid may be obtained from the Graduate Admissions Office. The GRE General Test is required. TOEFL scores are required for international applicants whose native language is not English. The application deadline is December 15 for fall quarter admission.

Correspondence and Information

Dr. Richard F. Ambrose, Director
Environmental Science and Engineering Program
School of Public Health, Room 46-081 CHS
University of California
Los Angeles, California 90095-1772

Phone: 310-825-9901
E-mail: app-ese@admin.ph.ucla.edu
Web site: http://www.ph.ucla.edu/ese

University of California, Los Angeles

THE FACULTY

Environmental Science and Engineering is an interdepartmental program drawing faculty members from participating campus departments and organizations. This unusual structure precludes identifying faculty members as permanently associated with Environmental Science and Engineering except for 4 core faculty members. Any faculty member of the thirteen participating departments may serve as a member of a student's doctoral committee or contribute to regularly scheduled candidacy examinations.

The program is administered through the School of Public Health by an Interdepartmental Committee appointed by the Dean of the Graduate Division. The Interdepartmental Committee determines administrative and academic policy within the program and ensures interdepartmental participation. Members of the current Interdepartmental Committee and ESE-affiliated faculty are presented below.

Richard F. Ambrose, Professor of Environmental Health Sciences and Director, ESE Program; Ph.D., UCLA, 1982. (core faculty member)
Richard A. Berk, Professor of Statistics; Ph.D., Johns Hopkins, 1970.
Ann Carlson, Professor of Law; J.D., Harvard, 1989.
Yoram Cohen, Professor of Chemical Engineering; Ph.D., Delaware, 1981.
Michael Collins, Professor of Environmental Health Sciences; Ph.D., Missouri, 1982.
Randall Crane, Professor of Urban Planning; Ph.D., MIT, 1987.
William Cumberland, Professor of Biostatistics; Ph.D., Johns Hopkins, 1975.
J. R. DeShazo, Associate Professor of Public Policy; Ph.D., Harvard, 1997.
Peggy Fong, Associate Professor of Ecology and Evolutionary Biology; Ph.D., California, Davis, and San Diego State, 1991.
John Froines, Professor of Environmental Health Sciences; Ph.D., Yale, 1967.
Thomas W. Gillespie, Associate Professor of Geography; Ph.D., UCLA, 1998.
Malcolm S. Gordon, Professor of Ecology and Evolutionary Biology; Ph.D., Yale, 1958.
William Hinds, Professor of Environmental Health Sciences; Sc.D., Harvard, 1972.
Jenny Jay, Assistant Professor of Civil and Environmental Engineering; Ph.D., MIT, 1999.
Vasilios Manousiothakis, Professor of Chemical Engineering; Ph.D., Rensselaer, 1986.
Antony Orme, Professor of Geography; Ph.D., Birmingham (England), 1961.
Linwood Pendleton, Associate Professor of Environmental Health Sciences; D.F.E.S., Yale, 1997. (core faculty member)
Richard L. Perrine, Professor Emeritus of Civil and Environmental Engineering; Ph.D., Stanford, 1953.
Shane Que Hee, Professor of Environmental Health Sciences; Ph.D., Saskatchewan, 1976.
Beate R. Ritz, Associate Professor of Epidemiology; M.D., Hamburg, 1984; Ph.D., UCLA, 1995.
Michael K. Stenstrom, Professor of Civil and Environmental Engineering; Ph.D., Clemson, 1976.
Irwin H. Suffet, Professor of Environmental Health Sciences; Ph.D., Rutgers, 1968. (core faculty member)
Stanley W. Trimble, Professor of Geography; Ph.D., Georgia, 1973.
Richard P. Turco, Professor of Atmospheric and Oceanic Sciences; Ph.D., Illinois, 1971.
Arthur M. Winer, Professor of Environmental Health Sciences; Ph.D., Ohio State, 1969. (core faculty member)

Sampling for fish, using beach seines at Malibu Lagoon.

Dr. Mel Suffet and his Flavor Profile Analysis Panel test Los Angeles's water supply.

A continuous liquid extractor collects 500-liter extracts at a water reuse project at West Basin in El Segundo, California.

UNIVERSITY OF CALIFORNIA, SANTA BARBARA

Donald Bren School of Environmental Science and Management

Programs of Study

The Donald Bren School of Environmental Science and Management is committed in its research and teaching to blending natural science, social science, law and policy, and business management in ways that facilitate the solution of environmental problems. The Bren School offers both a master's (M.E.S.M.) and a Ph.D. degree in environmental science and management. The M.E.S.M., with its balanced core curriculum and the capstone group project that serves as a master's thesis, is a two-year professional degree program that trains students to approach environmental issues from an integrated perspective, accounting for the social, legal, political, and business contexts within which they arise. The M.E.S.M. degree program is enhanced by an individual program of study created by each student, which builds depth in a chosen area of specialization and adequately trains the student in technical applications.

The Bren School's approach at the Ph.D. level is multidisciplinary in nature while being very individualized. The School accommodates a wide range of Ph.D. students and interests, from those who are highly focused in a particular discipline to those who are strongly multidisciplinary. In addition, the program aims to preserve the University of California, Santa Barbara's (UCSB), mission of training high-caliber future research professors while simultaneously meeting the urgent need for highly trained personnel in the public and private sector. The Bren School has excellent research programs in natural science, social science, management, and information systems and recruits students to participate in these programs. The School is also interested in recruiting students to perform dissertations on policy-related implications of natural science research. An innovative Ph.D. Training Program in Economics and Environmental Science is also offered as a joint program with UCSB's Department of Economics. It prepares students in environmental and resource economics with a secondary strength in an area of natural science, such as ecology, climate, hydrology, or marine science. For more information, students should visit the Web site at http://www.ees.ucsb.edu.

Research Facilities

The Bren School has research laboratories for the disciplines of environmental microbiology, biogeochemistry, biogeography, hydrology, toxicology, and information management. The School also provides a Student Computing Facility that contains workstations with a suite of cutting-edge analytical and graphical software tools. Bren Hall, where the Bren School is housed, was completed in spring 2002 and was designed to be a model of "green" building design, not only for UCSB but also for the entire University of California (UC) system. It was awarded a "Platinum" LEED^{TM} (Leadership in Energy and Environmental Design) rating by the U.S. Green Building Council, among many other awards and commendations from the state of California, the county of Santa Barbara, and various organizations, for its innovative and sustainable design.

Financial Aid

Sponsoring professors usually provide graduate student research assistant support for Ph.D. students. Ph.D. students may also be eligible for campus-based fellowships. Loans and other federally based support are available through UCSB's Financial Aid Office. For more information on financing graduate studies, students should visit the UCSB Graduate Division's Fellowship and Financial Support Web site at http://www.graddiv.ucsb.edu/financial.

Cost of Study

Graduate student fees for the 2005–06 academic year were $9123 for California residents and $24,083 for nonresidents. These figures include registration fees, tuition, and graduate student health insurance. Fees for 2006–07 are estimated to be $9398 for California residents and $24,383 for nonresidents.

Living and Housing Costs

Santa Barbara is an exceptionally beautiful place to live, and there is a high demand for rental housing. Monthly rents range from $831 for a studio to $2712 for a three-bedroom house. More information can be found at http://www.housing.ucsb.edu.

Student Group

The Bren School has 157 full-time graduate students; roughly 80 percent are master's students. Students range from those who have recently completed their undergraduate education to those who have spent years working in a variety of fields.

Student Outcomes

The Bren School operates its own career-development program, which is committed to helping students develop the job search and career-development skills that are necessary for successful transition into rewarding environmental careers. With the School's multidisciplinary approach and strong career-development program, graduates have found employment in a variety of environmental positions and organizations, such as consulting firms, the public sector, nonprofit organizations, industry, and academics and research, in the U.S. and internationally. Job placement statistics for Bren School graduates can be found on the Career Services Web site at http://www.bren.ucsb.edu/career/placement.html.

Location

UCSB is located 10 miles west of downtown Santa Barbara (100 miles northwest of Los Angeles) near the city of Goleta and occupies a picturesque 989-acre palm- and eucalyptus-lined plateau overlooking the Pacific Ocean. The Santa Barbara/Goleta area is surrounded by extraordinary beauty and has a mild, Mediterranean climate. UCSB is part of a diverse ecosystem that provides an ideal setting for the study of environmental science and management.

The University and The School

Founded in 1944, UCSB has become one of the nation's most distinguished academic institutions, renowned for outstanding scientific research, interdisciplinary collaboration, and public service. *U.S. News & World Report's* guide, *America's Best Colleges,* the most widely read college guide in the country, named UCSB as the thirteenth-best public university in the nation. UCSB has also been named one of the "hottest" colleges in the nation twice in the past three years by the popular *Newsweek*/Kaplan guide to colleges. Since 1997, five UCSB professors have been awarded Nobel Prizes in chemistry, physics, and economics. The Bren School accepted its first M.E.S.M. students in 1996 and was created to fulfill the need for graduates equipped with the knowledge and tools necessary to assess and meet environmental challenges in business and government settings. The School was renamed in December 1997 after a major gift was received from the Donald Bren Foundation. The purpose of the Bren gift was to transform the School into a multicampus, interdisciplinary program that stimulates the integration of natural and social science, law, and business programs throughout the UC system.

Applying

The Bren School welcomes applications from all undergraduate disciplines. Applications are available electronically (http://www.graddiv.ucsb.edu/eapp). A statement of purpose, three letters of recommendation, two copies of official transcripts for all tertiary-level institutions, an application fee (currently $60), official GRE scores, recent TOEFL scores for nonnative English speakers, and a resume must be submitted. Applications are normally accepted for the fall only. The M.E.S.M. application deadline is February 1 for primary consideration and for consideration for School-based financial support, but applications are accepted until March 1, space permitting. The Ph.D. application deadline is December 15 for primary consideration and for consideration for University-based financial support, including EES fellowships. Ph.D. applications are accepted until February 1, space permitting.

Correspondence and Information

Graduate Program Assistant
Donald Bren School of Environmental Science and Management
Donald Bren Hall, Room 2400
University of California, Santa Barbara
Santa Barbara, California 93106-5131

Phone: 805-893-7611
Fax: 805-893-7612
E-mail: gradasst@bren.ucsb.edu
Web site: http://www.bren.ucsb.edu

University of California, Santa Barbara

THE FACULTY AND THEIR RESEARCH

The Bren School has 16 permanent and 7 adjunct/affiliated faculty members from the fields of ecology, hydrology, toxicology, oceanography, business management, law, public policy, economics, and information management. The Bren School also has several visiting faculty members each year from renowned institutions across the U.S., which further diversifies and strengthens the School's teaching, research, and curriculum development.

DEAN

Ernst von Weizsäcker, Ph.D., Freiburg (Germany), 1969. Climate, energy, and environment.

THE FACULTY

Christopher Costello, Associate Professor; Ph.D., Berkeley, 2000. Environmental and resource economics, dynamic optimization, quantitative ecology, stochastic modeling.
Frank Davis, Professor; Ph.D., Johns Hopkins, 1982. Plant ecology, quantitative biogeography, vegetation remote sensing, ecological applications of remote sensing and geographic information systems, conservation planning, fire ecology.
Magali Delmas, Assistant Professor; Ph.D., H.E.C. Graduate School of Management (Paris), 1996. Corporate environmental management, impact of technological and regulatory uncertainties on industry choices.
Jeff Dozier, Professor; Ph.D., Michigan, 1973. Snow science, hydrology, hydrochemistry of Alpine regions, remote sensing, information systems.
Thomas Dunne, Professor; Ph.D., Johns Hopkins, 1969. Drainage basin and hill slope evolution, hydrology and floodplain sedimentation, applications of hydrology, geomorphology in environmental management.
James Frew, Associate Professor; Ph.D., California, Santa Barbara, 1990. Application of computing and information science to large-scale problems in environmental science, information system specification and integration, science data management, digital libraries.
Roland Geyer, Assistant Professor; Ph.D., Surrey (England), 2003. Green supply-chain management and industrial ecology.
Patricia Holden, Associate Professor; Ph.D., Berkeley, 1995. Pathogens in the environment, microbial ecology of pollutant biodegradation, soil microbiology.
Arturo Keller, Associate Professor; Ph.D., Stanford, 1997. Biogeochemistry; fate and transport of pollutants in the environment; development of technologies for containment, remediation, and monitoring.
Bruce Kendall, Associate Professor; Ph.D., Arizona, 1996. Quantitative, applied ecology, with a focus on animal and plant population dynamics.
*Charles D. Kolstad, Professor; Ph.D., Stanford, 1982. Industrial organization and environmental/resource economics, environmental policy, structure of energy markets, environmental regulations.
Matthew Kotchen, Assistant Professor; Ph.D., Michigan, 2003. Economics, environmental and resource economics.
Hunter Lenihan, Assistant Professor; Ph.D., North Carolina at Chapel Hill, 1996. Marine ecology and resource conservation; conserving and restoring marine populations, communities, and their habitat.
*John Melack, Professor; Ph.D., Duke, 1976. Limnology, ecology, biogeochemistry, remote sensing.
Catherine Ramus, Assistant Professor; Ph.D., Lausanne, 1999. Environmental management, organizational behavior, negotiation, public policy.
Oran Young, Professor; Ph.D., Yale, 1965. Program on Governance for Sustainable Development.

**Joint appointment with at least one other UCSB department.*

SELECTED PUBLICATIONS

Sethi, G., and **C. Costello** et al. Fishery management under multiple uncertainty. *J. Environ. Econ. Manage.*, in press.

Costello, C. Review of: *The Wealth of Nature, How Mainstream Economics Has Failed the Environment. J. Econ. Lit.* 43(1):194–5, 2005.

Costello, C., and L. Karp. Dynamic taxes and quotas with learning. *J. Econ. Dynamics Control* 28:1661–80, 2004.

Costello, C., and S. Polasky. Dynamic reserve site selection. *Resour. Energy Econ.* 26:157–74, 2004.

McAusland, C., and **C. Costello.** Avoiding invasives: Trade related policies for controlling unintentional exotic species introductions. *J. Environ. Econ. Manage.* 48:954–77, 2004.

Polasky, S., **C. Costello,** and C. McAusland. On trade land use and biodiversity. *J. Environ. Econ. Manage.* 48:911–25, 2004.

Solow, A., and **C. Costello.** Estimating the rate of species introductions from the discovery record. *Ecology* 85:1822–5, 2004.

Dutech, C., et al. **(F. W. Davis).** Gene flow and fine-scale genetic structure in a wind-pollinated tree species, *Quercus lobata* (Fagaceae). *Am. J. Botany,* in press.

Green, J. L., et al. **(F. W. Davis).** Complexity in ecology and conservation: Mathematical, statistical, and computational challenges. *Bioscience* 55(6):501–10, 2005.

Chornesky, E. A., et al. **(F. W. Davis).** Science priorities for reducing the threat of invasive species to sustainable forestry. *Bioscience* 55(4):335–48, 2005.

Delmas, M., and Y. Tokat. Deregulation, efficiency and governance structures: The U.S. electric utility sector. *Strategic Manage. J.* 26:441–60, 2005.

Rivera, J., and **M. Delmas.** Business and environmental protection: An introduction. *Hum. Ecol. Rev.,* special issue on business and environmental policy, 2004.

Delmas, M., and A. Marcus. Firms' choice of regulatory instruments to reduce pollution: A transaction cost approach. *Bus. Polit.* 6(3), article 3, 2004.

Delmas, M., and M. Toffel. Stakeholders and environmental management practices: An institutional framework. *Bus. Strategy Environ.* 13:209–22, 2004.

Dozier, J., and T. H. Painter. Multispectral and hyperspectral remote sensing of alpine snow properties. *Annu. Rev. Earth Planet. Sci.* 32:465–94, doi: 10.1146/annurev.earth.32.101802.120404, 2004.

Molotch, N. P., R. C. Bales, M. T. Colee, and **J. Dozier.** Estimating the spatial distribution of snow water equivalent in an alpine basin using binary regression tree models: The impact of digital elevation data and independent variable selection. *Hydrological Processes* 19(7):1459–79, doi: 10.1002/hyp.5586, 2004.

Molotch, N. P., T. H. Painter, R. C. Bales, and **J. Dozier.** Incorporating remotely sensed snow albedo into spatially distributed snowmelt modeling. *Geophys. Res. Lett.* 31:L03501, doi: 10.1029/2003GL019063, 2004.

Painter, T. H., and **J. Dozier.** Measurements of the hemispherical-directional reflectance of snow at fine spectral and angular resolution. *J. Geophys. Res.* 109:D18115, doi: 10.1029/2003JD004458, 2004.

Painter, T. H., and **J. Dozier.** The effect of anisotropic reflectance on imaging spectroscopy of snow properties. *Remote Sensing Environ.* 89(4):409–22, doi: 10.1016/j.rse.2003.09.007, 2004.

Malmon, D. V., et al. **(T. Dunne).** Influence of sediment storage on downstream delivery of contaminated sediment. *Water Resour. Res.* 41:W05008, doi: 10.1029/2004WR003288, 2005.

Benda, L., et al. **(T. Dunne).** Network dynamics hypothesis: Spatial and temporal organization of physical heterogeneity in rivers. *Bioscience* 55(4):413–27, 2004.

Biggs, T. W., **T. Dunne,** and L. A. Martinelli. Natural controls and human impacts on stream nutrient concentrations in a deforested region of the Brazilian Amazon basin. *Biogeochemistry* 68(2):227–57, 2004.

Malmon, D. V., S. L. Reneau, and **T. Dunne.** Sediment sorting by flash floods. *J. Geophys. Res. Earth Surf.* 109(F2), 2004.

Singer, M. B., and **T. Dunne.** Modeling decadal bed-material sediment flux based on stochastic hydrology. *Water Resour. Res.* 40:W03302, doi: 10.1029/2003WR00273, 2004.

Singer, M. B., and **T. Dunne.** An empirical-stochastic, event-based program for simulating inflow from a tributary network: Framework and application to the Sacramento River basin, California. *Water Resour. Res.* 40(7), 2004.

Bose, R., and **J. Frew.** Lineage retrieval for scientific data processing: A survey. *ACM Comput. Surveys* 37(1):1–28, 2005.

Holden, P. A., and N. Fierer. Vadose zone microbial processes. *Vadose Zone J.* 4:1–21, 2005.

LaMontagne, M. G., et al. **(P. A. Holden).** Bacterial diversity in marine hydrocarbon seep sediments. *Environ. Microbiol.* 6:799–808, 2004.

Steinberger, R. E., and **P. A. Holden.** Macromolecular composition of unsaturated *Pseudomonas aeruginosa* biofilms with time and carbon source. *Biofilms* 1:37–47, 2004.

Robinson, T. H., A. Leydecker, **A. A. Keller,** and **J. M. Melack.** Steps towards modeling nutrient export in coastal Californian streams in a Mediterranean climate. *Agric. Water Manage.,* in press.

Broje, V., and **A. A. Keller.** Materials selection for oil spill recovery in marine environments. In *Conference Proceedings of International Oil Spill Conference,* Orlando, Florida, April 2005.

Chen, M., D. Zhang, **A. A. Keller,** and Z. Lu. Stochastic analysis of steady-state two-phase flow in heterogeneous media. *Water Resour. Res.* 41(1):W01006, doi: 10.1029/2004WR003412, 2005.

University of California, Santa Barbara

Keller, A. A., and Y. Zheng. *Approaches for Estimating the Margin of Safety in a Total Maximum Daily Load Calculation: Theoretical and Practical Considerations.* EPRI Report #1005473. Palo Alto, Calif.: EPRI, 2005.

Keller, A. A., and D. Griset. *Stormwater Runoff Management and Synergistic Water Quality Planning Related to Proposed Major Projects in the 2004 Regional Transportation Plan.* Sacramento, Calif.: California Department of Transportation, 2005.

Mitani, M. M., **A. A. Keller,** O. C. Sandall, and R. G. Rinker. Mass transfer of ozone using a microporous diffuser reactor system. *Ozone Sci. Eng.* 27:45–51, 2005.

Auset, M., and **A. A. Keller.** Pore scale processes that control dispersion of colloids in saturated porous media. *Water Resour. Res.* 40(3):W03503, doi: 10.1029/2003WR002800, 2004.

Keller, A. A., and Y. Zheng. *Evaluation of Potential Water Quality Impacts from Different Future Growth Scenarios in the SCAG Area.* Los Angeles, Calif.: Southern California Association of Governments, 2004.

Keller, A. A., and S. Sirivithayapakorn. Transport of colloids in unsaturated porous media: Explaining large scale behavior based on pore scale mechanisms. *Water Resour. Res.* 40:W12403, doi: 10.1029/2004WR003315, 2004.

Doak, D. F., et al. **(B. E. Kendall).** Correctly estimating how environmental stochasticity influences fitness and population growth. *Am. Naturalist* 166:E14–21, 2005.

Fujiwara, M., **B. E. Kendall,** R. M. Nisbet, and W. A. Bennett. Analysis of size trajectory data using an energetic-based growth model. *Ecology* 86:1441–51, 2005.

Kendall, B. E, et al. Population cycles in the pine looper moth *(Bupalus piniarius):* Dynamical tests of mechanistic hypotheses. *Ecol. Monographs* 75:259–76, 2005.

Kolstad, C. Piercing the veil of uncertainty in transboundary pollution agreements. *Environ. Resour. Econ.* 31:21–34, 2005.

Kolstad, C. The simple analytics of greenhouse gas intensity reduction targets. *Energy Policy* 33:221–36, 2005.

Meixner, T., et al. **(J. Melack).** Multidecadal hydrochemical response of a Sierra Nevada watershed: Sensitivity to weathering rate and changes in deposition. *J. Hydrol.* 285:272–85, 2004.

Aizen, V. B., et al. **(J. M. Melack).** Association between atmospheric circulation patterns and firn-ice core records from the Inilchek glacierized area, central Tien Shan, Asia. *J. Geophys. Res.* 109:D08304, doi: 10.1029/2003JD003894, 2004.

Melack, J. M., et al. Regionalization of methane emissions in the Amazon basin with microwave remote sensing. *Global Change Biol.* 10:530–44, 2004.

Hamilton, S. K., S. J. Sippel, and **J. M. Melack.** Seasonal inundation patterns in two large savanna floodplains of South America: The Llanos de Moxos (Bolivia) and the Llanos del Orinoco (Venezuela and Colombia). *Hydrol. Processes* 18:2103–16, 2004.

Novo, E. M. L. M., W. Pereira Filho, and **J. M. Melack.** Assessing the utility of spectral band operators to reduce the impact of total suspended solids on the relationship between chlorophyll concentration and the bidirectional reflectance factor of Amazon waters. *Int. J. Remote Sens.* 25:5105–15, 2004.

Young, O. R. Governing the Arctic: From Cold War theater to mosaic of cooperation. *Global Governance* 11:9–15, 2005.

Young, O. R. Institutions and the growth of knowledge: Evidence from international environmental regimes. *Int. Environ. Agreements* 4:215–28, 2004.

Young, O. R. Review of *Negotiating the Arctic: The Construction of an International Region. Polar Res.* 23:211–3, 2004.

UNIVERSITY OF DELAWARE

Center for Energy and Environmental Policy

Programs of Study

The Center for Energy and Environmental Policy (CEEP) at the University of Delaware provides graduate instruction and conducts interdisciplinary research in the areas of energy policy, environmental policy, and sustainable development. Collaborative research and exchange agreements to foster international research and graduate study have been established with Asian, African, Latin American, and European universities and research institutes. CEEP is composed of an internationally diverse faculty and graduate student body with backgrounds in political science, economics, sociology, geography, philosophy, urban planning, environmental studies, history, anthropology, and engineering.

CEEP administers the Environmental and Energy Policy (ENEP) program. Its 7-member faculty offers a Master of Environmental and Energy Policy (M.E.E.P.) and a Doctor of Philosophy in Environmental and Energy Policy (Ph.D./ENEP). These degrees offer in-depth study in the fields of sustainable development, the political economy of energy and environment, energy policy, environmental policy, and disaster policy. The M.E.E.P. requires completion of 36 credits, of which 15 are electives; the Ph.D./ENEP normally requires completion of 45 credits, of which 24 credits are electives. In addition, CEEP sponsors a 15-credit energy, environment, and equity concentration in the M.A. in urban affairs and public policy and a 21-credit technology, environment, and society concentration in the Ph.D. in urban affairs and public policy. The latter degrees are administered by the urban affairs and public policy program faculty, with the M.A. requiring completion of 36 credits (of which 15 are electives) and the Ph.D. normally requiring completion of 42 credits (including 21 elective credits).

Opportunities exist for students to participate in research projects on such topics as socioeconomic impacts of global climate change, economic and environmental evaluation of renewable energy options (especially photovoltaic technology), impacts of environmental regulations, environmental ethics, development of a sustainability index, sustainable urban development strategies, energy and poverty issues, integrated resource planning, electricity restructuring in developed and developing countries, and water conservation planning and policy. Students have obtained paid internships with the World Bank, UNDP, overseas research institutes, intergovernmental organizations, U.S. senators' offices, federal and state government agencies, and nonprofit organizations.

Research Facilities

University Libraries contain more than 2 million books and journals and serve as a depository library for U.S. government publications. The University maintains a computerized online catalog, which is accessible via a campus computer network, the Internet, and telephone and computer modem from anywhere in the world. The University's computing services include microcomputer laboratories, UNIX multiworkstation and time-sharing facilities, and an IBM vector processing time-sharing service.

Financial Aid

Nearly three quarters of CEEP graduate students receive financial awards. University and minority fellowships, tuition scholarships, and research assistantships are awarded on the basis of merit. Awards are made only to full-time students in good academic standing. Most students admitted to the two Ph.D. programs in research areas related to ongoing Center activity are awarded full assistantships covering tuition (up to $16,770 per year) and stipend ($12,200 in 2005–06). Funding may be provided for four years, depending upon academic performance. Students admitted to the M.A. and M.E.E.P. programs ordinarily self-fund their first year of study and receive full assistantships in their second year.

Cost of Study

In 2005–06, graduate tuition was $8385 per year for full-time resident students and $16,770 per year for full-time nonresident students. Full-time matriculated students are automatically assessed nonrefundable fees for health ($386) and student-sponsored activities ($180).

Living and Housing Costs

The University's Office of Housing and Residence Life offers graduate students a number of housing options. Students who are seeking accommodations in adjacent residential areas within walking or commuting distance should contact the Off-Campus Housing Office for a list of rooms, apartments, and houses to rent or share. Further information can be obtained from the Office of Housing Assignment Services (phone: 302-831-3676).

Student Group

Enrollment at the University in 2005–06 included 16,350 undergraduate students, 3,434 graduate students, and 1,198 students in the Division of Continuing Education. Currently, the University offers eighty different programs leading to a master's degree and forty programs leading to a doctoral degree through forty-six departments in its seven instructional colleges. Fall 2005 graduate enrollment in CEEP included 42 Ph.D. and 31 M.A. students.

Location

The main campus of the University is located in the residential community of Newark. Newark is a suburban community of 30,000, located midway between Philadelphia and Baltimore and within 2 hours by train of New York City and Washington, D.C. Newark lies a short distance from the Delaware and Chesapeake Bays and from Delaware's ocean beaches.

The University

The University is a comprehensive land- and sea-grant institution of higher education. Opened in 1743, the University currently owns 2,041 acres (the Newark campus totals 970 acres) and a $499-million physical plant with 449 buildings, including classrooms, laboratories, athletic complexes, and student activity centers. Other land holdings include a 405-acre Marine Studies Complex and a 347-acre Agricultural Substation. The University's distinguished faculty includes internationally recognized researchers, artists, authors, and teachers. There are 100 named professorships. The University community includes 1,110 faculty members, 1,352 professionals, and 1,314 salaried staff members and hourly employees.

Applying

For the M.A. and M.E.E.P. degrees, the successful candidate for admission must have an undergraduate GPA above 3.0 (on a 4.0 scale). A combined GRE score above 1100 (math and verbal portions) is normally expected. Admission to the either of the Ph.D. programs requires a master's degree with at least a 3.5 GPA. A combined GRE score above 1150 (math and verbal portions) is normally expected. Complete applications contain three letters of recommendation, a 1,000-word statement of the applicant's research interest, academic transcript(s), and GRE scores. For students whose first language is not English, a demonstrated proficiency in English is required. This may be judged on the basis of a TOEFL score of at least 550. Ph.D. students are expected to have TOEFL scores above 600. Most students are admitted for the fall semester. A completed admission application and all credentials should be submitted by February 15 to guarantee consideration for financial aid.

Correspondence and Information

Dr. John Byrne, Director
Center for Energy and Environmental Policy
University of Delaware
Newark, Delaware 19716

Phone: 302-831-8405
Fax: 302-831-3098
E-mail: jbbyrne@udel.edu
Web site: http://ceep.udel.edu

University of Delaware

THE FACULTY AND THEIR RESEARCH

Core Faculty

John Byrne, Distinguished Professor of Public Policy and Director, Center for Energy and Environmental Policy (CEEP); Ph.D., Delaware, 1980. Technology, environment and society, political ecology, climate change, renewable energy, sustainable development, environmental justice.

Paul Durbin, Professor Emeritus, Department of Philosophy, and CEEP Senior Policy Fellow; Ph.D., Aquinas Institute, 1966. Technology and society, philosophy of science, environmental ethics.

William Ritter, Professor, Department of Bioresource Engineering, and CEEP Senior Policy Fellow; Ph.D., Iowa State, 1971. Water resources, soil and water conservation engineering, waste management.

Yda Schreuder, Associate Professor, Department of Geography, and CEEP Senior Policy Fellow; Ph.D., Wisconsin–Madison, 1982. Global resources, development and environment, sustainable development, growth management.

Richard T. Sylves, Professor, Department of Political Science and International Relations, and CEEP Senior Policy Fellow; Ph.D., Illinois, 1976. Environmental policy, emergency response management, energy policy.

Young-Doo Wang, Professor, CEEP Associate Director, and Environmental and Energy Policy Graduate Program Director; Ph.D., Delaware, 1980. Energy and environmental policy, water resource and watershed management, sustainable energy analysis, econometric applications.

Robert Warren, Professor, Urban Affairs and Public Policy, and CEEP Senior Policy Fellow; Ph.D., UCLA, 1964. Planning theory, cultural theory, governance, environmental politics, telecommunications.

Adjunct Faculty

Cesar Cuello, Professor, Santo Domingo Technological Institute and Universidad Autonoma (Dominican Republic), and CEEP Policy Fellow; Ph.D., Delaware, 1997. Sustainable development, science, technology and society, environmental philosophy.

Steven M. Hoffman, Professor and Director, Environmental Studies Program, University of St. Thomas (Minnesota), and CEEP Senior Policy Fellow; Ph.D., Delaware, 1986. Technology and society, political economy, energy and environmental policy.

Jong-dall Kim, Associate Professor and Director, Research Institute for Energy, Environment and Economy, Kyungbuk National University (South Korea), and CEEP Senior Policy Fellow; Ph.D., Delaware, 1991. Political economy, renewable energy, energy conservation, sustainable development.

Hoesung Lee, President, Council for Energy and Environment (South Korea), and CEEP Senior Policy Fellow; Ph.D., Rutgers, 1976. Energy and environmental economics, climate change, sustainable development.

Cecilia Martinez, Leadership Fellow, Archibald Bush Foundation (Minnesota); Senior Research Advisor, Women's Environmental Institute; and CEEP Senior Policy Fellow; Ph.D., Delaware, 1990. Technology, environment and society, political economy, American Indian policy, environmental justice.

Hon. Russell W. Peterson, Ecologist (past president, National Audubon Society; former chairman, U.S. President's Council on Environmental Quality; former governor, State of Delaware) and CEEP Distinguished Policy Fellow. Global ecological issues.

Subodh Wagle, President, Prayas (India), and CEEP Policy Fellow; Ph.D., Delaware, 1997. Political economy, social and environmental justice, antiglobalization strategy, sustainable livelihoods.

CEEP researchers assisted an international team in an effort to change Taiwanese government plans to construct a heavy-industry complex near the nesting area of the black-faced spoonbill, an endangered bird. Of the known population of 600 birds, 400 winter along the west coast of Taiwan.

SELECTED PUBLICATIONS

Byrne, J., L. Glover, and N. Toly, eds. *Transforming Power: Energy, Environment and Society in Conflict.* New Brunswick, N.J., and London: Transaction Publishers, 2006.

Byrne, J., and N. Toly. Energy as a social project: Recovering a discourse. In *Transforming Power: Energy, Environment and Society in Conflict,* pp. 1–32, eds. **J. Byrne** et al. New Brunswick, N.J., and London: Transaction Publishers, 2006.

Byrne, J., and L. Glover. Ellul and the weather. *Bull. Sci. Technol. Soc.* Special issue *Celebrating the Intellectual Gifts and Insights of Jacques Ellul.* 25(1):4–16, 2005.

Agbemabiese, L., and **J. Byrne.** Commodification of Ghana's Volta River: An example of Ellul's autonomy of technique. *Bull. Sci. Technol. Soc.* Special issue *Celebrating the Intellectual Gifts and Insights of Jacques Ellul.* 25(1):17–25, 2005.

Byrne, J., et al. Beyond oil: A comparison of projections of PV generation and European and U.S. domestic oil production. In *Advances in Solar Energy,* vol. 16, pp. 35–70, eds. D. Y. Goswami and K. Boer. Boulder, Colo.: American Solar Energy Society, 2005.

Byrne, J., et al. **(H. Lee** and **Y.-D. Wang).** Power liberalization and neo-liberalism: South Korea's electricity reform at a crossroads. *Pacific Aff.* 77(3):493–516, 2004.

Byrne, J., Y.-D. Wang, H. Lee, and **J.-d. Kim.** *The Sustainable Energy Revolution: Toward an Energy-Efficient Future for South Korea.* Seoul, South Korea: Maeil Kyung Jae, 2004.

Byrne, J., et al. Reclaiming the atmospheric commons: Beyond Kyoto. In *Climate Change: Perspectives Five Years After Kyoto,* chapter 21, ed. V. I. Grover. Plymouth, UK: Science Publishers, Inc., 2004.

Byrne, J., L. Kurdgekashvili, D. Poponi, and A. Barnett. The potential of solar electric power for meeting future U.S. energy needs: A comparison of projections of solar electric energy generation and Arctic National Wildlife Refuge oil production. *Energy Policy* 32(2):289–97, 2004.

Byrne, J., and Y.-M. Mun. Rethinking reform in the electricity sector: Power liberalization or energy transformation. In *Electricity Reform: Social and Environmental Challenges,* pp. 49–76, eds. Wamunkonya and Roskilde. Denmark: UNEP-RISØ Centre, 2003.

Byrne, J., and V. Inniss. Island sustainability and sustainable development in the context of climate change. In *Sustainable Development for Island Societies and the World,* eds. H.-H. Hsiao et al. Taipei, Taiwan: Academia Sinica, 2002.

Byrne, J., C. Martinez, and L. Glover, eds. *Environmental Justice: Discourses in International Political Economy.* New Brunswick, N.J., and London: Transaction Publishers, 2002.

Byrne, J., C. Martinez, and L. Glover. A brief on environmental justice. In *Environmental Justice: Discourses in International Political Economy,* pp. 3–17, eds. **J. Byrne** et al. New Brunswick, N.J., and London: Transaction Publishers, 2002.

Byrne, J., and **S. Hoffman.** A 'necessary sacrifice:' Industrialization and American Indian lands. In *Environmental Justice: Discourses in International Political Economy,* pp. 97–118, eds. **J. Byrne** et al. New Brunswick, N.J., and London: Transaction Publishers, 2002.

Byrne, J., L. Glover, and **C. Martinez.** The production of unequal nature. In *Environmental Justice: Discourses in International Political Economy,* pp. 261–91, eds. **J. Byrne** et al. New Brunswick, N.J., and London: Transaction Publishers, 2002.

Byrne, J., and L. Glover. A common future or towards a future commons: Globalization and sustainable development since UNCED. *Int. Rev. Environ. Strategies* 3(1):5–25, 2002.

Byrne, J., and **S. Hoffman,** eds. Energy controversy—part II: Reversing course. *Bull. Sci. Technol. Soc.* 22(2), 2002.

Zhou, A., and **J. Byrne.** Renewable energy for rural sustainability: Lessons from China. *Bull. Sci. Technol. Soc.* 22(2):123–31, 2002.

Byrne, J. (contributing author). Decision-making frameworks. In *Climate Change 2001: Mitigation,* pp. 601–88, eds. B. Metz et al. Contribution of Working Group III to the Third Assessment Report of the Intergovernmental Panel on Climate Change (IPCC). New York: Cambridge University Press, 2001.

Byrne, J., and **S. Hoffman,** eds. Energy controversy—part I: Change and resistance. *Bull. Sci. Technol. Soc.* 21(6), 2001.

Byrne, J., et al. The postmodern greenhouse: Creating virtual carbon reductions from business-as-usual energy politics. *Bull. Sci. Technol. Soc.* 21(6): 443–55, 2001.

Byrne, J., and R. Scattone. Community participation is key to environmental justice in brownfields. *Race Poverty Environ.* 3(1):6–7, 2001.

Byrne, J., and L. Glover. Climate shopping: Putting the atmosphere up for sale. In *TELA: Environment, Economy and Society* series. Melbourne, Australia: Australian Conservation Foundation, 2000.

Byrne, J., and T.-L. Lin. Beyond pollution and risk: Energy and environmental policy in the greenhouse. In *Proceedings of the International Conference on Sustainable Energy and Environmental Strategies,* pp. 1–23. Taipei, Taiwan, 2000.

Byrne, J., et al. An international comparison of the economics of building integrated PV in different resource, pricing and policy environments: The cases of the U.S., Japan and South Korea. In *Proceedings of the American Solar Energy Society Solar 2000 Conference,* pp. 81–5. Madison, Wis., 2000.

Byrne, J., and V. Inniss. Island sustainability and sustainable development in the context of global warming. In *Proceedings of the International Conference on Sustainable Development for Island Societies,* pp. 21–44. Chungli, Taiwan: National Central University, 2000.

Byrne, J., and S.-J. Yun. Efficient global warming: Contradictions in liberal democratic responses to global environmental problems. *Bull. Sci. Technol. Soc.* 19(6):493–500, 1999.

Byrne, J., and **Y.-D. Wang** et al. **(J-d. Kim).** Mitigating CO_2 emissions of the Republic of Korea: The role of energy efficiency measures. In *Proceedings of the 20th Annual North American Conference of the U.S. Association for Energy Economics,* pp. 319–28. Orlando, Fla., 1999.

Byrne, J. Climate change and renewable energy: Opportunities and challenges for Korea. In *Climate Change and Alternative Energy Development,* pp. 49–82, ed. S.-H. Kim. Seoul, South Korea: Environmental Forum, Korea National Assembly, 1999.

Byrne, J., Y.-D. Wang, H. Lee, and J.-d. Kim. An equity- and sustainability-based policy response to global climate change. *Energy Policy* 26(4):335–43, 1998.

Byrne, J. Sustainable energy and environmental futures: The implications of climate change, technological change and economic restructuring. In *Global Warming and a Sustainable Energy Future,* pp. 1–23, ed. **J.-d. Kim.** Taegu, South Korea: Research Institute for Energy, Environment and Economy, Kyungpook National University, 1998.

Byrne, J., B. Shen, and W. Wallace. The economics of sustainable energy for rural development: A study of renewable energy in rural China. *Energy Policy* 26(1):45–54, 1998.

Byrne, J., et al. Photovoltaics as an energy services technology: A case study of PV sited at the Union of Concerned Scientists headquarters. In *Proceedings of the American Solar Energy Society Solar 98 Conference,* Albuquerque, N.Mex., 1998.

Letendre, S., **J. Byrne,** C. Weinberg, and **Y.-D. Wang.** Commercializing photovoltaics: The importance of capturing distributed benefits. In *Proceedings of the American Solar Energy Society Solar 98 Conference,* pp. 231–7. Albuquerque, N.Mex., 1998.

Byrne, J. *Equity and Sustainability in the Greenhouse: Reclaiming our Atmospheric Commons.* Pune, India: Parisar, 1997.

Byrne, J., and C. Govindarajalu. Power sector reform: Elements of a regulatory framework. *Econ. Political Weekly* 32(31):1946–7, 1997.

Byrne, J., contributing author. A generic assessment of response options. In *Climate Change 1995: Economic and Social Dimensions of Climate Change,* Contribution of Working Group III to the Second Assessment Report of the Intergovernmental Panel on Climate Change, pp. 225–62. New York: Cambridge University Press, 1996.

Byrne, J., and **S. Hoffman,** eds. *Governing the Atom: The Politics of Risk.* New Brunswick, N.J., and London: Transaction Publishers, 1996.

Byrne, J., and **S. Hoffman.** The ideology of progress and the globalization of nuclear power. In *Governing the Atom: The Politics of Risk,* pp. 11–45, eds. **J. Byrne** and **S. Hoffman.** New Brunswick, N.J., and London: Transaction Publishers, 1996.

Byrne, J., S. Letendre, and **Y.-D. Wang.** The distributed utility concept: Toward a sustainable electric utility sector. In *Proceedings of the ACEEE 1996 Summer Study on Energy Efficiency in Buildings,* vol, 7, pp. 7.1–8, 1996.

Byrne, J., and **S. Hoffman.** Sustainability: From concept to practice. *IEEE Technol. Soc.* 15(2):6–7, 1996.

Byrne, J., B. Shen, and X. Li. The challenge of sustainability balancing China's energy, economic and environmental goals. *Energy Policy* 24(5):455–62, 1996.

Byrne, J., and S. J. Hsu. Community versus commodity: Environmental protest in Taiwan. *Bull. Sci. Technol. Soc.* 16(5–6):329–36, 1996.

Byrne, J., et al. **(Y.-D. Wang).** Evaluating the economics of photovoltaics in a demand-side management role. *Energy Policy* 24(2):177–85, 1996.

Byrne, J., R. Nigro, and **Y.-D. Wang.** Photovoltaic technology as a dispatchable, peak-shaving option. *Public Utilities Fortnightly* September 1995.

Byrne, J., C. Hadjilambrinos, and **S. Wagle.** Distributing costs of global climate change. *IEEE Technol. Soc.* 13(1):17–24, 1994.

Byrne, J., Y.-D. Wang, and S. Hegedus. Photovoltaics as a demand-side management technology: An analysis of peak-shaving and direct load control options. *Prog. Photovoltaics* 2:235–48, 1994.

Byrne, J., Y.-D. Wang, B. Shen, and X. Li. Sustainable urban development strategies for China. *Environ. Urbanization* 6(1):174–87, 1994.

Byrne, J., Y.-D. Wang, R. Nigro, and S. Letendre. Photovoltaics in a demand-side management role. In *Proceedings of the ACEEE 1994 Summer Study on Energy Efficiency in Buildings,* vol. 2, pp. 2.43–9, 1994.

Byrne, J., B. Shen, and X. Li. Energy efficiency and renewable energy options for China's economic expansion. In *Proceedings of the ACEEE 1994 Summer Study on Energy Efficiency in Buildings,* vol. 4, pp. 4.25–36, 1994.

Byrne, J., and **Y.-D. Wang** et al. Urban sustainability during industrialization: The case of China. *Bull. Sci. Technol. Soc.* 13(6):324–31, 1993.

Byrne, J., Y.-D. Wang, and **S. Wagle.** Toward a politics of sustainability: The responsibilities of industrialized countries. *Regions* 183:4–7, 1993.

Byrne, J., and D. Rich, eds. *Energy and Environment: The Policy Challenge.* New Brunswick, N.J., and London: Transaction Publishers, 1992.

Byrne, J., and D. Rich. Toward a political economy of global change: Energy, environment and development in the greenhouse. In *Energy and Environment: The Policy Challenge,* pp. 269–302, eds. **J. Byrne** and D. Rich. New Brunswick, N.J., and London: Transaction Publishers, 1992.

University of Delaware

Byrne, J., and **J.-d. Kim.** City and technology in social theory: A theoretical reconstruction of postindustrialism. *Korean J. Reg. Stud.* 8(1):67–86, 1992.

Byrne, J., S. Hoffman, and **C. R. Martinez.** Environmental commodification and the industrialization of Native American lands. In *Proceedings of the Seventh Annual Meeting of the National Association of Science, Technology and Society,* pp. 170–81, 1992.

Byrne, J., and **Y.-D. Wang.** The politics of unsustainability: The US (un)prepares for UNCED. *Regions* 178:8–11, 1992.

Byrne, J., Y.-D. Wang, J.-d. Kim, and K. Ham. The political economy of energy, environment and development. *Korean J. Environ. Stud.* 30:278–312, 1992.

Byrne, J., and **Y.-D. Wang** et al. Energy and environmental sustainability in East and Southeast Asia. *IEEE Technol. Soc.* 10(4):21–9, 1991.

Byrne, J. Meeting the certain challenge: A policy perspective on systemic energy and environmental problems. In *Energy and Environment,* pp. 40–51, eds. E. Kainlauri et al. Atlanta, Ga.: ASHRAE Special Publications, 1991.

Byrne, J., and **Y.-D. Wang** et al. Institutional strategies for sustainable development: Case studies of four Asian industrializing countries. In *Proceedings of the Interdisciplinary Conference on Preparing for a Sustainable Society,* pp. 105–15, 1991.

Byrne, J., and **S. Hoffman.** Energy, environment and sustainable world development: Options for the year 2000. *Energy Sources* 13(1):1–4, 1991.

Byrne, J., and **S. Hoffman.** The politics of alternative energy: A study of water pumping systems in developing nations. *Energy Sources* 13(1):55–66, 1991.

Byrne, J., and **S. Hoffman.** Nuclear optimism and the technological imperative: A study of the Pacific Northwest electrical network. *Bull. Sci. Technol. Soc.* 11:63–77, 1991.

Byrne, J., S. Hoffman, and **C. Martinez.** The social structure of nature. In *Proceedings of the Sixth Annual Meeting of the National Association of Science, Technology and Society,* pp. 67–76, 1991.

Byrne, J., et al. Green economics and the developing world: Institutional strategies for sustainable development. In *Proceedings of the Sixth Annual Conference of the National Association of Science, Technology and Society,* pp. 97–109, 1991.

Byrne, J., and D. Rich. The real energy crisis. *Regions* 169:2–5, 1990.

Byrne, J., and **J.-d. Kim.** Centralization, technicization and development on the semi-periphery: A study of South Korea's commitment to nuclear power. *Bull. Sci. Technol. Soc.* 10(4):212–22, 1990.

Byrne, J., D. Rich, and **C. Martinez.** Lewis Mumford and the living city. *Regions* 166:10–2, 1990.

Byrne, J., and **C. Martinez.** Ghastly science. *Society* 27(1):22–4, 1989.

Byrne, J., S. Hoffman, and **C. Martinez.** Technological politics in the nuclear age. *Bull. Sci. Technol. Soc.* 8(6):580–94, 1989.

Byrne, J., C. Martinez, and **J.-d. Kim** et al. The city as commodity: The decline of urban vision. *Raumplannung* 46(47):174–8, 1989.

Byrne, J., and **S. Hoffman.** Nuclear power and technological authoritarianism. *Bull. Sci. Technol. Soc.* 7:658–71, 1988.

Byrne, J., Y.-D. Wang, and K. Ham. The political geography of acid rain: The U.S. case. *Regions* 157:3–6, 1988.

Byrne, J., Y.-D. Wang, D. Rich, and I. Han. Economic and policy implications of integrated resource planning in the utility sector. In *Proceedings of the ACEEE 1988 Summer Study on Energy Efficiency in Buildings,* vol. 8, pp. 8.265–78, 1988.

Byrne, J., and D. Rich. Post-Chernobyl notes on the U.S. nuclear fizzle. *Regions* 150:3–6, 1987.

Byrne, J. Policy science and the administrative state: The political economy of cost-benefit analysis. In *Confronting Values in Policy Analysis: The Politics of Criteria,* Sage Yearbooks in Politics and Public Policy, eds. J. Forester and F. Fischer. Beverly Hills, Calif.: Sage Publications, 1987.

Byrne, J., and **C. Martinez.** Urban policy without the urban. *Regions* 148:4–7, 1987.

Byrne, J., and D. Rich, eds. *The Politics of Energy R&D.* New Brunswick, N.J., and London: Transaction Publishers, 1986.

Byrne, J., and D. Rich. In search of the abundant energy machine. In *The Politics of Energy R&D,* pp. 141–60, eds. **J. Byrne** and D. Rich. New Brunswick, N.J., and London: Transaction Publishers, 1986.

Byrne, J., and **S. Hoffman.** Some lessons in the political economy of megapower: WPPSS and the municipal bond market. *J. Urban Aff.* 8(1):35–47, 1986.

Byrne, J., and D. Rich. *Energy and Cities.* New Brunswick, N.J., and London: Transaction Publishers, 1985.

Byrne, J., C. Martinez, and D. Rich. The post-industrial imperative: Energy, cities and the featureless plain. In *Energy and Cities,* pp. 101–41, eds. **J. Byrne** and D. Rich. New Brunswick, N.J.: Transaction Publishers, 1985.

Byrne, J., and **S. Hoffman.** Efficient corporate harm: A Chicago metaphysic. In *Errant Corporations: Responsibility, Compliance and Sanctions,* eds. B. Fisse and P. French. San Antonio: Trinity University Press, 1985.

Byrne, J., and D. Rich. Deregulation and energy conservation: A reappraisal. *Policy Stud. J.* 13(2):331–44, 1984.

Byrne, J., and D. Rich. The solar energy transition as a problem of political economy. In *The Solar Energy Transition: Implementation and Policy Implications,* pp. 163–86, eds. D. Rich et al. Boulder, Colo.: Westview Press, 1983.

Rich, D., A. Barnett, J. Veigel, and **J. Byrne**, eds. *The Solar Energy Transition: Implementation and Policy Implications.* Boulder, Colo.: Westview Press for the American Association for the Advancement of Science, 1983.

Byrne, J. What's wrong with being reasonable: The politics of cost-benefit analysis. In *Ethical Theory and Business,* pp. 568–76, eds. N. E. Bowie and T. L. Beauchamp. Englewood Cliffs: Prentice-Hall, 1982.

Durbin, P. T. Environmental ethics and environmental activism. In *Technology and the Environment,* Research in Philosophy and Technology Series, vol. 12, pp. 107–17, ed. F. Ferre. New York: JAI Press Inc., 1992.

Hoffman, S. Powering injustice: Hydroelectric development in northern Manitoba. In *Environmental Justice: Discourses in International Political Economy,* pp. 147–70, eds. **J. Byrne** et al. New Brunswick, N.J., and London: Transaction Publishers, 2002.

Kim, J.-d., ed. *Global Warming and a Sustainable Energy Future.* Taegu, South Korea: Research Institute for Energy, Environment and Economy, Kyungpook National University, 1998.

Kim, J.-d., and **J. Byrne.** The Asian atom: Hard-path nuclearization in East Asia. In *Governing the Atom: The Politics of Risk,* pp. 273–300, eds. **J. Byrne** and **S. Hoffman.** New Brunswick, N.J., and London: Transaction Publishers, 1996.

Martinez, C., and J. Poupart. The circle of life: Preserving American Indian traditions and facing the nuclear challenge. In *Environmental Justice: Discourses in International Political Economy,* pp. 119–46, eds. **J. Byrne** et al. New Brunswick, N.J., and London: Transaction Publishers, 2002.

Martinez, C., and **J. Byrne.** Science, society and the state: The nuclear project and the transformation of the American political economy. In *Governing the Atom: The Politics of Risk,* eds. **J. Byrne** and **S. Hoffman.** New Brunswick, N.J.: Transaction Publishers, 1996.

Schreuder, Y., and C. Sherry. Flexible mechanisms in the corporate greenhouse: Implementation of the Kyoto Protocol and the globalization of the electric power industry. *Energy Environ.* 12(5–6):487–98, 2001.

Sylves, R. T., and W. L. Waugh. *Disaster Management in the U.S. and Canada: The Politics, Policymaking, Administration and Analysis of Emergency Management.* Springfield, Ill.: Charles C. Thomas Publishers, 1996.

Sylves, R. *The Nuclear Oracles: A Political History of the General Advisory Committee of the Atomic Energy Commission, 1947-1977.* Ames, Iowa: Iowa State University Press, 1987.

Wang, Y.-D., W. J. Smith, and **J. Byrne.** *Water Conservation-Oriented Rates.* Denver, Colo.: American Water Works Association, 2005.

Wang, Y.-D., and **J. Byrne** et al. Designing revenue neutral and equitable water conservation-oriented rates for use during drought summer months. In *Proceedings of the Water Sources Conference.* American Water Works Association, 2002.

Wang, Y.-D., and **J. Byrne** et al. Less energy, a better economy, and a sustainable South Korea: An energy efficiency scenario analysis. *Bull. Sci. Technol. Soc.* 22(2): 110–22, 2002.

Wang, Y.-D., and **J. Byrne.** Short- and mid-term prospects for world oil prices using a modified Delphi method. In *Factors Influencing World Prices: Alternative Methods of Prediction,* pp. 156–206, ed. J.-K. Kim. Seoul, South Korea: Korea Energy Economics Institute, 2001.

Wang, Y.-D., and **J. Byrne** et al. Evaluating the persistence of residential water conservation: A 1992–97 panel study of a water utility program in Delaware. *J. Am. Water Resour. Assoc.* 35(5):1269–76, 1999.

Wang, Y.-D., and W. Latham. Energy and state economic growth: Some new evidence. *J. Energy Develop.* 13(2):197–221, 1989.

Wang, Y.-D. A residential energy market model: An econometric analysis. *J. Reg. Sci.* 25(2):215–39, 1985.

UNIVERSITY OF NEW HAVEN

UNIVERSITY OF NEW HAVEN

Program in Environmental Sciences

Program of Study

Environmental science is a diverse field with strong interactions and interdependence among several scientific, technical, and social disciplines. The M.S. in environmental science program is intended to meet the needs of those who wish to enter the field, active environmental scientists and managers, and students who plan to pursue graduate training beyond the master's level. This interdisciplinary program includes concentrations in environmental ecology, environmental geology, environmental health and management, and geographic information systems and provides the advanced skills and knowledge necessary to meet the increasing demand for scientists with an environmental background. Extensive field and laboratory work provide practical experience for students enrolled in the program, while ongoing faculty projects provide opportunities to perform research on various environmental problems and issues.

The program is designed to accommodate both full-time students and working students who wish to continue their training on a part-time basis. Most courses are taught in the evening, meet once a week, and are scheduled on a trimester calendar. Full-time students can finish the program in 1½ to 2 years; part-time students finish in 3 to 5 years.

Students may select a program that focuses on one of the concentration areas, or they can design a personalized program of study in consultation with the program coordinator. Personal programs are built on a set of core courses in ecology, environmental chemistry, environmental geology, and environmental law and legislation. The program also requires a research project or a thesis. Students can select related courses in other departments, including chemistry and chemical engineering, civil and environmental engineering, and occupational health and safety.

Research Facilities

The Marvin K. Peterson Library holds numerous books and periodicals and has access to several online journals and databases. The department has a wide variety of field and laboratory equipment, including standard samplers and meters, transits, and current meters; a 16-foot Boston Whaler for teaching and conducting research in aquatic habitats; a fully equipped geographic information system laboratory; and Global Positioning System (GPS) equipment. There are also affiliations with other research facilities, including the Gerace Research Center on San Salvador in the Bahamas. The University's locations on Long Island Sound provide access to a variety of coastal habitats, including the program's 5-acre salt marsh research site in Branford, Connecticut.

Financial Aid

About 75 percent of all University students receive some form of financial aid. Graduate students may borrow Federal Stafford Student Loans of up to $8500 per academic year. Subsidized loans are provided based on financial need. Unsubsidized loans are available for students who do not qualify for subsidized loans. Research and teaching assistantships are available to full-time students. The amount of these awards may include hourly compensation as well as partial tuition; students typically work 20 hours per week. Fellowships are available to students who have earned at least 24 credits and have demonstrated outstanding academic achievement. Faculty members may also have grant-based assistantships.

Cost of Study

Tuition for both full-time and part-time students is $495 per credit hour ($1485 per 3-credit course). Other fees include a health sciences fee of $180 per year, a technology fee of $15 per trimester, and laboratory fees ranging from $25 to $350.

Living and Housing Costs

On-campus housing for graduate students is currently not available. However, off-campus housing is available in the area at a cost of $575 to $1000 per month for a one-bedroom apartment or $775 to $1200 per month for a two-bedroom apartment.

Student Group

The program is intended for those who wish to enter environmental science, environmental management, environmental law, and related fields; those who are active environmental scientists and managers; and those students who plan to pursue graduate training beyond the master's level. Many students also pursue a master's degree in environmental science if they plan to enter the teaching profession.

Student Outcomes

Graduates of the program are prepared for positions at government agencies, particularly in the areas of environmental protection and management; water, sewer, and power-generation utilities; analytic laboratories; environmental and engineering firms; industries in the field of pollution control; environmental centers and public and private schools (generally middle and high school grades) teaching environmental science; and private industry and management.

Location

The campus is in West Haven, which is located between New York and Boston. New Haven has numerous art museums, parks, and walking trails and three Tony Award–winning regional theaters. Close to the many cultural and entertainment venues of Boston and New York, the University's location is also close to many scenic areas and nature-based recreation in New England.

The University

The University of New Haven was founded on the Yale campus in 1920 and became New Haven College in 1926. Today, it includes five undergraduate schools and a Graduate School, with a combined 4,500 students. Its programs prepare students to advance in their careers and meet the ever-changing demands of their respective fields. The University offers thirty master's degrees and thirty certificates as well as associate and bachelor's degrees. The student-faculty ratio is 10:1, with an average class size of 15.

Applying

To apply to the program, prospective students must submit a completed application form, transcripts from all colleges and universities previously attended, two letters of recommendation, and a $50 application fee.

Correspondence and Information

Roman N. Zajac
Department of Environmental Sciences
University of New Haven
300 Boston Post Road
West Haven, Connecticut 06516
Phone: 203-932-7114
Fax: 203-931-6097
E-mail: rzajac@newhaven.edu
Web site: http://qrwgis.newhaven.edu/QRWWEB/graduate.htm

University of New Haven

THE FACULTY AND THEIR RESEARCH

Carmela Cuomo, Assistant Professor; Ph.D., Yale, 1985. Marine and coastal biogeochemistry, paleoenvironments, effects of hypoxia/anoxia on sediment biogeochemistry and fauna, horseshoe crab ecology, fisheries management.

Larry Davis, Professor; Ph.D., Rochester, 1980. Regional karst hydrology in the Bahamas, the applications of geology to land-use planning and resource management, assessment and remediation of groundwater pollution, watershed and groundwater management and protection, natural hazards.

Daniel DePodesta, GIS Practitioner in Residence; M.B.A., Quinnipiac, 1993. GIS technologies and applications in environmental, municipal, utility, and business fields; GIS programming in Python, ArcGIS, Avenue, and Arc/Info AML; development of AM/FM/GIS; computer programming, including FORTRAN, COBOL, Assembler, UIL, and RPG.

Henry Voegeli, Professor; Ph.D., Rhode Island, 1970. Pilot study of rubber-utilizing bacteria for protein production; the use of single-cell fertilizer in plant growth; environmental cleanups of sites containing gasoline, heating oil, diesel fuel, chlorinated solvent, and metal contaminants.

Robert Wardwell, Adjunct Professor; M.S., New Haven. Environmental impact assessment and environmental impact statement procedures, application of environmental information for management and development.

Roman Zajac, Associate Professor; Ph.D., Connecticut, 1985. Large-scale patterns and processes in seafloor communities in Long Island Sound, the effects of natural and man-made disturbances on the benthic fauna of estuaries, estuarine restoration, the life history and demography of marine fauna relative to different types and scales of disturbances, watershed dynamics, application of GIS to environmental problems, phylogeography of polychaete annelids, coastal food web modeling.

Mike Ziskin, Adjunct Professor. Hazardous-material management, environmental field safety technologies.

Section 10
Natural Resources

This section contains a directory of institutions offering graduate work in natural resources, followed by in-depth entries submitted by institutions that chose to prepare detailed program descriptions. Additional information about programs listed in the directory but not augmented by an in-depth entry may be obtained by writing directly to the dean of a graduate school or chair of a department at the address given in the directory.

For programs offering related work, see also in this book Environmental Sciences and Management and Meteorology and Atmospheric Sciences; in Book 2, see Architecture (Landscape Architecture) and Public, Regional, and Industrial Affairs; in Book 3, see Biological and Biomedical Sciences; Botany and Plant Biology; Ecology, Environmental Biology, and Evolutionary Biology; Entomology; Genetics, Developmental Biology, and Reproductive Biology; Nutrition; Pathology and Pathobiology; Pharmacology and Toxicology; Physiology; and Zoology; in Book 5, see Agricultural Engineering and Bioengineering; Civil and Environmental Engineering; Geological, Mineral/Mining, and Petroleum Engineering; Management of Engineering and Technology; and Ocean Engineering; and in Book 6, see Veterinary Medicine and Sciences.

CONTENTS

Program Directories

Close-Ups

Fish, Game, and Wildlife Management

Arkansas Tech University, Graduate School, School of Physical and Life Sciences, Russellville, AR 72801. Offers fisheries and wildlife biology (MS). *Students:* Average age 25. *Degree requirements:* For master's, thesis, project. *Entrance requirements:* For master's, GRE General Test. Additional exam requirements/recommendations for international students: Required—TOEFL (minimum score 500 paper-based; 173 computer-based). *Application deadline:* For fall admission, 3/1 priority date for domestic students, 5/1 priority date for international students; for spring admission, 10/1 priority date for domestic students, 10/1 priority date for international students. Applications are processed on a rolling basis. Application fee: $0 ($30 for international students). Electronic applications accepted. *Expenses:* Tuition, state resident: full-time $2,934; part-time $163 per hour. Tuition, nonresident: full-time $5,868; part-time $326 per hour. Required fees: $312. Tuition and fees vary according to course load. *Financial support:* In 2005–06, teaching assistantships with full tuition reimbursements (averaging $4,000 per year); career-related internships or fieldwork, Federal Work-Study, scholarships/grants, health care benefits, and unspecified assistantships also available. Support available to part-time students. Financial award application deadline: 4/15; financial award applicants required to submit FAFSA. *Faculty research:* Fisheries, warblers, fish movement, darter populations, bob white studies. *Unit head:* Dr. Richard Cohoon, Dean, 479-964-0816, E-mail: richard.cohoon@atu.edu. *Application contact:* Dr. Eldon G. Clary, Dean of Graduate School, 479-968-0398, Fax: 479-964-0542, E-mail: graduate.school@atu.edu.

Auburn University, Graduate School, College of Agriculture, Department of Fisheries and Allied Aquacultures, Auburn University, AL 36849. Offers M Aq, MS, PhD. Part-time programs available. *Faculty:* 20 full-time (3 women). *Students:* 46 full-time (14 women), 25 part-time (11 women); includes 5 minority (2 African Americans, 3 Hispanic Americans), 20 international. 38 applicants, 71% accepted, 20 enrolled. In 2005, 18 master's, 6 doctorates awarded. *Degree requirements:* For master's, thesis (for some programs); for doctorate, 2 foreign languages, thesis/dissertation. *Entrance requirements:* For master's and doctorate, GRE General Test. *Application deadline:* For fall admission, 7/7 for domestic students; for spring admission, 11/24 for domestic students. Applications are processed on a rolling basis. Application fee: $25 ($50 for international students). Electronic applications accepted. *Financial support:* Fellowships, research assistantships, teaching assistantships, Federal Work-Study available. Support available to part-time students. Financial award application deadline: 3/15. *Faculty research:* Channel catfish production; aquatic animal health; community and population ecology; pond management; production hatching, breeding and genetics. Total annual research expenditures: $8 million. *Unit head:* Dr. David B. Rouse, Head, 334-844-4786. *Application contact:* Dr. Stephen L. McFarland, Acting Dean of the Graduate School, 334-844-4700.

Auburn University, Graduate School, School of Forestry and Wildlife Sciences, Auburn University, AL 36849. Offers MF, MS, PhD. Part-time programs available. *Faculty:* 29 full-time (3 women). *Students:* 27 full-time (13 women), 32 part-time (8 women); includes 1 minority (Hispanic American), 16 international. 38 applicants, 58% accepted, 15 enrolled. In 2005, 7 master's, 3 doctorates awarded. *Degree requirements:* For master's, oral exam (MF), thesis (MS); for doctorate, thesis/dissertation. *Entrance requirements:* For master's and doctorate, GRE General Test. *Application deadline:* For fall admission, 7/7 for domestic students; for spring admission, 11/24 for domestic students. Applications are processed on a rolling basis. Application fee: $25 ($50 for international students). Electronic applications accepted. *Financial support:* Fellowships, research assistantships, teaching assistantships, Federal Work-Study available. Support available to part-time students. Financial award application deadline: 3/15. *Faculty research:* Forest nursery management, silviculture and vegetation management, biological processes and ecological relationships, growth and yield of plantations and natural stands, urban forestry, forest taxation, law and policy. *Unit head:* Richard W. Brinker, Dean, 334-844-1007, Fax: 334-844-1084, E-mail: brinker@forestry.auburn.edu. *Application contact:* Dr. Stephen L. McFarland, Acting Dean of the Graduate School, 334-844-4700.

Brigham Young University, Graduate Studies, College of Biological and Agricultural Sciences, Department of Integrative Biology, Provo, UT 84602-1001. Offers biological science education (MS); integrative biology (MS, PhD); wildlife and wildlands conservation (MS, PhD). *Faculty:* 35 full-time (3 women). *Students:* 19 full-time (8 women), 24 part-time (7 women); includes 2 minority (1 Asian American or Pacific Islander, 1 Hispanic American). Average age 27. 31 applicants, 74% accepted, 17 enrolled. In 2005, 13 master's awarded. *Median time to degree:* Of those who began their doctoral program in fall 1997, 100% received their degree in 8 years or less. *Degree requirements:* For master's and doctorate, thesis/dissertation, comprehensive exam, registration. *Entrance requirements:* For master's and doctorate, GRE General Test, minimum GPA of 3.0 for last 60 credit hours of course work. Additional exam requirements/recommendations for international students: Required—TOEFL (minimum score 550 paper-based; 213 computer-based), GRE. *Application deadline:* For fall admission, 1/31 for domestic students, 1/31 for international students. Application fee: $50. Electronic applications accepted. *Financial support:* In 2005–06, 56 students received support, including 1 fellowship with full and partial tuition reimbursement available (averaging $5,500 per year), 22 research assistantships with full and partial tuition reimbursements available (averaging $12,000 per year), 33 teaching assistantships with full and partial tuition reimbursements available (averaging $12,000 per year); career-related internships or fieldwork, institutionally sponsored loans, scholarships/grants, tuition waivers (full and partial), and unspecified assistantships also available. Financial award application deadline: 3/1. *Faculty research:* Systematics, bioinformatics, conservation. Total annual research expenditures: $268,851. *Unit head:* Dr. Larry L. St. Clair, Chair, 801-422-2582, Fax: 801-422-0090, E-mail: larry_stclair@byu.edu. *Application contact:* Nancy P. Heiss, Graduate Secretary, 801-422-2010, Fax: 801-422-0090, E-mail: nancy_heiss@byu.edu.

Clemson University, Graduate School, College of Agriculture, Forestry and Life Sciences, Department of Forestry and Natural Resources, Program in Wildlife and Fisheries Biology, Clemson, SC 29634. Offers MS, PhD. *Students:* 19 full-time (10 women), 4 part-time (1 woman); includes 1 minority (Hispanic American), 1 international. Average age 25. 2 applicants, 0% accepted, 0 enrolled. In 2005, 1 master's, 2 doctorates awarded. *Degree requirements:* For master's and doctorate, thesis/dissertation. *Entrance requirements:* For master's, GRE General Test, minimum undergraduate GPA of 3.0. Additional exam requirements/recommendations for international students: Required—TOEFL. *Application deadline:* For fall admission, 6/1 for domestic students, 4/15 for international students. Application fee: $50. *Financial support:* Fellowships, research assistantships, teaching assistantships, career-related internships or fieldwork available. Financial award applicants required to submit FAFSA. *Faculty research:* Intensive freshwater culture systems, conservation biology, stream management, applied wildlife management. Total annual research expenditures: $1 million. *Unit head:* Dr. Dave Guynn, Coordinator, 864-656-4803, Fax: 864-656-3304, E-mail: dguynn@clemson.edu.

See Close-Ups on pages 705 and 707.

Colorado State University, Graduate School, Warner College of Natural Resources, Department of Fishery and Wildlife Biology, Fort Collins, CO 80523-0015. Offers MFWB, MS, PhD. Part-time programs available. *Faculty:* 14 full-time (2 women). *Students:* 18 full-time (7 women), 14 part-time (4 women); includes 2 minority (both American Indian/Alaska Native), 1 international. Average age 29. 21 applicants, 38% accepted, 8 enrolled. In 2005, 6 master's, 3 doctorates awarded. *Degree requirements:* For master's, thesis or alternative; for doctorate, thesis/dissertation. *Entrance requirements:* For master's, GRE General Test, minimum GPA of 3.0, BA or BS in related field; for doctorate, GRE General Test, minimum GPA of 3.0, MS in related field. Additional exam requirements/recommendations for international students: Required—TOEFL. *Application deadline:* For fall admission, 2/15 priority date for domestic students, 2/15 priority date for international students. Applications are processed on a rolling basis. Application fee: $50. Electronic applications accepted. *Expenses:* Tuition, state resident: full-time $3,690; part-time $205 per credit. Tuition, nonresident: full-time $14,958; part-time $831 per credit. Required fees: $1,061. *Financial support:* In 2005–06, 36 students received support, including 4 fellowships with full tuition reimbursements available (averaging $18,750 per year), 20 research assistantships with full and partial tuition reimbursements available (averaging $16,200 per year), 12 teaching assistantships with full and partial tuition reimbursements available (averaging $10,863 per year); career-related internships or fieldwork, Federal Work-Study, institutionally sponsored loans, scholarships/grants, and traineeships also available. Financial award application deadline: 2/15. *Faculty research:* Conservation biology, aquatic ecology, animal behavior, population modeling, habitat evaluation and management. Total annual research expenditures: $4.4 million. *Unit head:* H. Randall Robinette, Head, 970-491-1410, Fax: 970-491-5091, E-mail: fwb@cnr.colostate.edu. *Application contact:* Kathy Bramer, Graduate Affairs Coordinator, 970-491-5020, Fax: 970-491-5091, E-mail: fwb@cnr.colostate.edu.

Cornell University, Graduate School, Graduate Fields of Agriculture and Life Sciences, Field of Natural Resources, Ithaca, NY 14853-0001. Offers aquatic science (MPS, MS, PhD); environmental management (MPS); fishery science (MPS, MS, PhD); forest science (MPS, MS, PhD); resource policy and management (MPS, MS, PhD); wildlife science (MPS, MS, PhD). *Faculty:* 50 full-time (9 women). *Students:* 62 full-time (33 women); includes 9 minority (1 African American, 2 American Indian/Alaska Native, 4 Asian Americans or Pacific Islanders, 2 Hispanic Americans), 14 international. 60 applicants, 23% accepted, 12 enrolled. In 2005, 11 master's, 5 doctorates awarded. *Degree requirements:* For master's, thesis (MS), project paper (MPS); for doctorate, thesis/dissertation, comprehensive exam. *Entrance requirements:* For master's and doctorate, GRE General Test, 2 letters of recommendation. Additional exam requirements/recommendations for international students: Required—TOEFL (minimum score 550 paper-based; 213 computer-based). *Application deadline:* For spring admission, 10/30 for domestic students. Applications are processed on a rolling basis. Application fee: $60. Electronic applications accepted. *Financial support:* In 2005–06, 49 students received support, including 15 fellowships with full tuition reimbursements available, 16 research assistantships with full tuition reimbursements available, 18 teaching assistantships with full tuition reimbursements available; institutionally sponsored loans, scholarships/grants, health care benefits, tuition waivers (full and partial), and unspecified assistantships also available. Financial award applicants required to submit FAFSA. *Faculty research:* Ecosystem-level dynamics, systems modeling, conservation biology/management, resource management's human dimensions, biogeochemistry. *Unit head:* Director of Graduate Studies, 607-255-2807, Fax: 607-255-0349. *Application contact:* Graduate Field Assistant, 607-255-2807, Fax: 607-255-0349, E-mail: nrgrad@cornell.edu.

Frostburg State University, Graduate School, College of Liberal Arts and Sciences, Department of Biology, Program in Fisheries and Wildlife Management, Frostburg, MD 21532-1099. Offers MS. Part-time and evening/weekend programs available. *Faculty:* 11. *Students:* 4 full-time (1 woman), 5 part-time (3 women). Average age 28. 3 applicants, 33% accepted, 1 enrolled. In 2005, 3 degrees awarded. *Degree requirements:* For master's, thesis. *Entrance requirements:* For master's, GRE General Test, resumé. *Application deadline:* For fall admission, 7/15 for domestic students. Applications are processed on a rolling basis. Application fee: $30. Electronic applications accepted. *Expenses:* Tuition, state resident: full-time $5,292; part-time $294 per credit hour. Tuition, nonresident: full-time $6,066; part-time $337 per credit hour. Required fees: $67; $67 per credit hour. $9 per term. One-time fee: $30 full-time. *Financial support:* In 2005–06, 6 research assistantships with full tuition reimbursements (averaging $5,000 per year) were awarded; Federal Work-Study also available. Financial award application deadline: 4/1; financial award applicants required to submit FAFSA. *Faculty research:* Evolution and systematics of freshwater fishes, biochemical mechanisms of temperature adaptation in freshwater fishes, wildlife and fish parasitology, biology of freshwater invertebrates, remote sensing. *Unit head:* Dr. R. Scott Fritz, Coordinator, 301-687-4166. *Application contact:* Patricia C. Spiker, Director, Graduate Services, 301-687-7053, Fax: 301-687-4597, E-mail: pspiker@frostburg.edu.

Iowa State University of Science and Technology, Graduate College, College of Agriculture, Department of Natural Resource Ecology and Management, Ames, IA 50011. Offers animal ecology (MS, PhD), including animal ecology, fisheries biology, wildlife biology; forestry (MS, PhD). *Faculty:* 23 full-time, 12 part-time/adjunct. *Students:* 45 full-time (18 women), 4 part-time (1 woman), 7 international. 24 applicants, 38% accepted, 8 enrolled. In 2005, 8 master's, 3 doctorates awarded. *Degree requirements:* For master's, thesis (for some programs); for doctorate, thesis/dissertation. *Entrance requirements:* For master's and doctorate, GRE General Test. Additional exam requirements/recommendations for international students: Required—TOEFL (paper score 547; computer score 210) or IELTS (score 6). *Application deadline:* For fall admission, 1/1 priority date for domestic students, 1/1 priority date for international students; for spring admission, 9/1 priority date for domestic students, 9/1 priority date for international students. Application fee: $30 ($70 for international students). Electronic applications accepted. *Expenses:* Tuition, state resident: full-time $6,410. Tuition, nonresident: full-time $16,422. Tuition and fees vary according to program. *Financial support:* In 2005–06, 41 research assistantships with full and partial tuition reimbursements (averaging $14,755 per year), 3 teaching assistantships with full and partial tuition reimbursements (averaging $14,580 per year) were awarded. *Unit head:* Dr. David M Engle, Chair, 515-294-1166. *Application contact:* Lyn Van De Pol, Information Contact, 515-294-6148, E-mail: lvdp@iastate.edu.

Louisiana State University and Agricultural and Mechanical College, Graduate School, College of Agriculture, School of Renewable Natural Resources, Baton Rouge, LA 70803. Offers fisheries (MS); forestry (MS, PhD); wildlife (MS); wildlife and fisheries science (PhD). *Faculty:* 29 full-time (1 woman). *Students:* 67 full-time (24 women), 7 part-time (2 women); includes 1 African American, 1 Asian American or Pacific Islander, 3 Hispanic Americans, 20 international. Average age 28. 36 applicants, 56% accepted, 16 enrolled. In 2005, 13 master's, 6 doctorates awarded. *Degree requirements:* For master's and doctorate, thesis/dissertation. *Entrance requirements:* For master's, GRE General Test, minimum GPA of 3.0; for doctorate, GRE General Test, MS, minimum GPA of 3.0. Additional exam requirements/recommendations for international students: Required—TOEFL (minimum score 550 paper-based; 213 computer-based). *Application deadline:* For fall admission, 1/25 priority date for domestic students, 5/15 priority date for international students. Applications are processed on a rolling basis. Application fee: $25. Electronic applications accepted. *Financial support:* In 2005–06, 70 students received support, including 4 fellowships (averaging $16,304 per year), 59 research assistantships with partial tuition reimbursements available (averaging $16,539 per year); teaching assistantships with partial tuition reimbursements available, Federal Work-Study, institutionally sponsored loans, scholarships/grants, tuition waivers (full and partial), and unspecified assistantships also available. Financial award application deadline: 4/15; financial award applicants required to submit FAFSA. *Faculty research:* Forest biology and management, aquaculture, fisheries biology and ecology, upland and wetlands wildlife. Total annual research expenditures: $4,423. *Unit head:* Dr. Bob G. Blackmon, Director, 225-578-4131, Fax: 225-578-4227, E-mail: bblackmon@agctr.lsu.edu. *Application contact:* Dr. Allen Rutherford, Coordinator of Graduate Studies, 225-578-4187, Fax: 225-578-4227, E-mail: druther@lsu.edu.

See Close-Up on page 711.

McGill University, Faculty of Graduate and Postdoctoral Studies, Faculty of Agricultural and Environmental Sciences, Department of Natural Resource Sciences, Montréal, QC H3A 2T5, Canada. Offers agrometeorology (M Sc, PhD); entomology (M Sc, PhD); forest science (M Sc, PhD); microbiology (M Sc, PhD); neotropical environment (M Sc, PhD); soil science (M Sc, PhD); wildlife biology (M Sc, PhD). *Degree requirements:* For master's and doctorate, thesis/dissertation, registration. *Entrance requirements:* For master's, minimum GPA of 3.0 or 3.2 in the last 2 years of university study. Additional exam requirements/recommendations for international students: Required—TOEFL (minimum score 550 paper-based; 213 computer-based), IELT (minimum score 7). Electronic applications accepted. *Faculty research:* Toxicology, reproductive physiology, parasites, wildlife management, genetics.

Memorial University of Newfoundland, School of Graduate Studies, Interdisciplinary Program in Marine Studies, St. John's, NL A1C 5S7, Canada. Offers fisheries resource management (MMS, Advanced Diploma). Part-time programs available. *Students:* 10 full-time (5 women), 12 part-time (3 women), 3 international. 9 applicants, 100% accepted, 4 enrolled. In 2005, 4 degrees awarded. *Degree requirements:* For master's, report. *Entrance requirements:* For master's and Advanced Diploma, high 2nd class degree from a recognized university. *Application deadline:* For fall admission, 4/30 for domestic students, 4/30 for international students. Application fee: $40 Canadian dollars. *Expenses:* Tuition: Part-time $733 per term. Tuition and fees vary according to degree level and program. *Financial support:* Fellowships, research assistantships, teaching assistantships available. *Faculty research:* Biological, ecological and oceanographic aspects of world fisheries; economics; political science; sociology. *Unit head:* Dr. Peter Fisher, Chair, 709-778-0356, Fax: 709-778-0346, E-mail: peter.fisher@mi.mun.ca. *Application contact:* Nancy Smith, Program Support, 709-778-0522, E-mail: nancy.smith@mi.mun.ca.

Michigan State University, The Graduate School, College of Agriculture and Natural Resources, Department of Fisheries and Wildlife, East Lansing, MI 48824. Offers fisheries and wildlife (MS, PhD); fisheries and wildlife—environmental toxicology (PhD). *Faculty:* 27 full-time (7 women). *Students:* 87 full-time (51 women), 11 part-time (6 women); includes 9 minority (1 African American, 4 American Indian/Alaska Native, 2 Asian Americans or Pacific Islanders, 2 Hispanic Americans), 14 international. Average age 30. 59 applicants, 34% accepted. In 2005, 10 master's, 12 doctorates awarded. *Degree requirements:* For master's, thesis or alternative; for doctorate, thesis/dissertation. *Entrance requirements:* For master's, GRE General Test, minimum GPA of 3.0 in last 2 undergraduate years; for doctorate, GRE General Test, master's degree or equivalent. Additional exam requirements/recommendations for international students: Required—TOEFL (minimum score 550 paper-based; 213 computer-based), Michigan State University ELT (85), Michigan ELAB (83). *Application deadline:* For fall admission, 12/27 for domestic students. Applications are processed on a rolling basis. Application fee: $50. Electronic applications accepted. *Expenses:* Tuition, state resident: part-time $330 per credit hour. Tuition, nonresident: part-time $685 per credit hour. Tuition and fees vary according to program. *Financial support:* In 2005–06, 26 fellowships with tuition reimbursements (averaging $6,595 per year), 68 research assistantships with tuition reimbursements (averaging $12,792 per year), 7 teaching assistantships with tuition reimbursements (averaging $12,823 per year) were awarded; scholarships/grants and unspecified assistantships also available. *Faculty research:* Environmental toxicology, biometry and ecological modeling, conservation biology and restoration ecology, fisheries/wildlife ecology and management, human dimensions and environmental management. Total annual research expenditures: $6.8 million. *Unit head:* Dr. William W. Taylor, Chairperson, 517-353-4038, Fax: 517-432-1699, E-mail: taylorw@msu.edu. *Application contact:* Mary Witchell, Graduate Records Secretary, 517-353-9091, Fax: 517-432-1699, E-mail: witchel1@msu.edu.

Mississippi State University, College of Forest Resources, Department of Wildlife and Fisheries, Mississippi State, MS 39762. Offers wildlife and fisheries science (MS). Part-time programs available. *Faculty:* 14 full-time (1 woman), 1 part-time/adjunct (0 women). *Students:* 24 full-time (12 women), 16 part-time (5 women); includes 4 minority (1 African American, 1 Asian American or Pacific Islander, 2 Hispanic Americans), 1 international. Average age 26. 12 applicants, 42% accepted, 4 enrolled. In 2005, 10 degrees awarded. *Degree requirements:* For master's, thesis, comprehensive oral or written exam. *Entrance requirements:* For master's, GRE General Test, minimum GPA of 3.0 in last 60 undergraduate credits. Additional exam requirements/recommendations for international students: Required—TOEFL. *Application deadline:* For fall admission, 7/1 for domestic students; for spring admission, 11/1 for domestic students. Applications are processed on a rolling basis. Application fee: $30. *Expenses:* Tuition, state resident: full-time $4,312; part-time $240 per hour. Tuition, nonresident: full-time $9,772; part-time $543 per hour. International tuition: $10,102 full-time. Tuition and fees vary according to course load. *Financial support:* Fellowships, research assistantships with partial tuition reimbursements, teaching assistantships with partial tuition reimbursements, Federal Work-Study, institutionally sponsored loans, and unspecified assistantships available. Financial award applicants required to submit FAFSA. *Faculty research:* Spatial technology, habitat restoration, aquaculture, fisheries, wildlife management. Total annual research expenditures: $756,308. *Unit head:* Dr. Bruce D. Leopold, Head, 662-325-2619, Fax: 662-325-8726, E-mail: bleopold@cfr.msstate.edu. *Application contact:* Philip G. Bonfanti, Director of Admissions, 662-325-4104, Fax: 662-325-8872, E-mail: admit@msstate.edu.

Montana State University, College of Graduate Studies, College of Letters and Science, Department of Ecology, Bozeman, MT 59717. Offers biological sciences (MS, PhD); fish and wildlife biology (PhD); fish and wildlife management (MS); land rehabilitation (intercollege) (MS). Part-time programs available. *Faculty:* 15 full-time (4 women). *Students:* 4 full-time (3 women), 61 part-time (23 women); includes 1 minority (American Indian/Alaska Native), 1 international. Average age 31. 8 applicants, 75% accepted, 6 enrolled. In 2005, 9 master's, 4 doctorates awarded. *Degree requirements:* For master's, thesis (for some programs), comprehensive exam, registration; for doctorate, thesis/dissertation, comprehensive exam, registration. *Entrance requirements:* For master's and doctorate, GRE General Test. Additional exam requirements/recommendations for international students: Required—TOEFL (minimum score 550 paper-based; 213 computer-based). *Application deadline:* For fall admission, 7/15 priority date for domestic students, 5/15 priority date for international students; for spring admission, 12/1 priority date for domestic students, 10/1 priority date for international students. Applications are processed on a rolling basis. Application fee: $30. Electronic applications accepted. *Expenses:* Tuition, state resident: full-time $4,132. Tuition, nonresident: full-time $1,132. *Financial support:* In 2005–06, 3 fellowships with full tuition reimbursements (averaging $21,300 per year), 45 research assistantships with full and partial tuition reimbursements (averaging $10,830 per year), 22 teaching assistantships with full and partial tuition reimbursements (averaging $10,268 per year) were awarded; career-related internships or fieldwork, Federal Work-Study, scholarships/grants, health care benefits, tuition waivers, and unspecified assistantships also available. Financial award application deadline: 3/1; financial award applicants required to submit FAFSA. *Faculty research:* Population dynamics, landscape ecology, evolution and genetics, ecological modeling, aquatic ecosystems. Total annual research expenditures: $2.1 million. *Unit head:* Dr. David Roberts, Department Head, 406-994-4548, Fax: 406-994-3190, E-mail: droberts@montana.edu.

New Mexico State University, Graduate School, College of Agriculture and Home Economics, Department of Fishery and Wildlife Sciences, Las Cruces, NM 88003-8001. Offers wildlife science (MS). Part-time programs available. *Faculty:* 8 full-time (1 woman), 5 part-time/adjunct (2 women). *Students:* 23 full-time (13 women), 8 part-time (4 women); includes 8 minority (1 American Indian/Alaska Native, 1 Asian American or Pacific Islander, 6 Hispanic Americans), 1 international. Average age 28. 14 applicants. In 2005, 8 degrees awarded. *Degree requirements:* For master's, thesis (for some programs). *Entrance requirements:* For master's, GRE General Test, minimum GPA of 3.0. Additional exam requirements/recommendations for international students: Required—TOEFL. *Application deadline:* For fall admission, 4/1 for domestic students; for spring admission, 11/1 priority date for domestic students. Applications are processed on a rolling basis. Application fee: $30 ($50 for international students). Electronic applications accepted. *Expenses:* Tuition, state resident: full-time $3,156; part-time $175 per credit. Tuition, nonresident: full-time $12,510; part-time $565 per credit. Required fees: $1,050. *Financial support:* In 2005–06, 2 fellowships, 17 research assistantships with partial tuition reimbursements, 6 teaching assistantships with partial tuition reimbursements were awarded; career-related internships or fieldwork, Federal Work-Study, and scholarships/grants also available. Support available to part-time students. Financial award application deadline: 4/1. *Faculty research:* Ecosystems analyses, landscape and wildlife ecology, wildlife and fish population dynamics, management models, wildlife and fish habitat relationships. *Unit head:* Dr. Donald F. Caccamise, Head, 505-646-1544, Fax: 505-646-1281, E-mail: natres@nmsu.edu.

North Carolina State University, Graduate School, College of Natural Resources and College of Agriculture and Life Sciences, Program in Fisheries and Wildlife Sciences, Raleigh, NC 27695. Offers MFWS, MS. *Degree requirements:* For master's, thesis optional. *Entrance requirements:* For master's, GRE General Test. Additional exam requirements/recommendations for international students: Required—TOEFL. Electronic applications accepted. *Faculty research:* Fisheries biology; ecology of marine, estuarine, and anadromous fishes; aquaculture pond water quality; larviculture of freshwater and marine finfish; predator/prey interactions.

Oregon State University, Graduate School, College of Agricultural Sciences, Department of Fisheries and Wildlife, Program in Fisheries Science, Corvallis, OR 97331. Offers M Agr, MAIS, MS, PhD. Part-time programs available. *Students:* 45 full-time (19 women), 12 part-time (3 women); includes 6 minority (1 African American, 3 American Indian/Alaska Native, 2 Asian Americans or Pacific Islanders), 9 international. Average age 30. In 2005, 11 master's, 2 doctorates awarded. *Degree requirements:* For master's, thesis (for some programs); for doctorate, thesis/dissertation. *Entrance requirements:* For master's and doctorate, GRE, minimum GPA of 3.0 in last 90 hours. Additional exam requirements/recommendations for international students: Required—TOEFL. *Application deadline:* For fall admission, 3/15 for domestic students; for spring admission, 12/15 for domestic students. Applications are processed on a rolling basis. Application fee: $50. *Expenses:* Tuition, area resident: Part-time $301 per credit. Tuition, state resident: full-time $8,139; part-time $501 per credit. Tuition, nonresident: full-time $14,376; part-time $532 per credit. Required fees: $1,266. *Financial support:* Fellowships, research assistantships, teaching assistantships, career-related internships or fieldwork, Federal Work-Study, and institutionally sponsored loans available. Support available to part-time students. Financial award application deadline: 2/1. *Faculty research:* Fisheries ecology, fish toxicology, stream ecology, quantitative analyses of marine and freshwater fish populations. *Unit head:* Dr. Guillermo Giannico, Head, 541-737-2479. *Application contact:* Charlotte Vickers, Advising Specialist, 541-737-1941, Fax: 541-737-3590, E-mail: charlotte.vickers@orst.edu.

Oregon State University, Graduate School, College of Agricultural Sciences, Department of Fisheries and Wildlife, Program in Wildlife Science, Corvallis, OR 97331. Offers MAIS, MS, PhD. *Students:* 41 full-time (19 women), 4 part-time (3 women), 3 international. Average age 31. In 2005, 6 master's, 4 doctorates awarded. *Degree requirements:* For master's, thesis (for some programs); for doctorate, thesis/dissertation. *Entrance requirements:* For master's and doctorate, GRE, minimum GPA of 3.0 in last 90 hours. Additional exam requirements/recommendations for international students: Required—TOEFL. Application fee: $50. *Expenses:* Tuition, area resident: Part-time $301 per credit. Tuition, state resident: full-time $8,139; part-time $501 per credit. Tuition, nonresident: full-time $14,376; part-time $532 per credit. Required fees: $1,266. *Financial support:* Fellowships, research assistantships, teaching assistantships, career-related internships or fieldwork, Federal Work-Study, and institutionally sponsored loans available. Financial award application deadline: 2/1. *Unit head:* Dr. Nancy Allen, Head Advisor, 541-737-1953. *Application contact:* Charlotte Vickers, Advising Specialist, 541-737-1941, Fax: 541-737-3590, E-mail: charlotte.vickers@orst.edu.

The Pennsylvania State University University Park Campus, Graduate School, College of Agricultural Sciences, School of Forest Resources, State College, University Park, PA 16802-1503. Offers forest resources (M Agr, MFR, MS, PhD); wildlife and fisheries sciences (M Agr, MFR, MS, PhD). *Students:* 42 full-time (19 women), 18 part-time (3 women); includes 2 minority (both Hispanic Americans), 4 international. *Entrance requirements:* For master's and doctorate, GRE General Test. *Expenses:* Tuition, state resident: full-time $12,518; part-time $522 per credit. Tuition, nonresident: full-time $23,004; part-time $959 per credit. Required fees: $484. Tuition and fees vary according to course load, campus/location and program. *Unit head:* Dr. Charles H. Strauss, Director, 814-863-7093, Fax: 814-865-3725, E-mail: chs30@psu.edu. *Application contact:* Dr. Charles H. Strauss, Director, 814-863-7093, Fax: 814-865-3725, E-mail: chs30@psu.edu.

Purdue University, Graduate School, College of Agriculture, Department of Forestry and Natural Resources, West Lafayette, IN 47907. Offers aquaculture, fisheries, aquatic science (MSF); aquaculture, fisheries, aquatic sciences (MS, PhD); forest biology (MS, MSF, PhD); natural resources and environmental policy (MS, MSF); natural resources environmental policy (PhD); quantitative resource analysis (MS, MSF, PhD); wildlife science (MS, MSF, PhD); wood science and technology (MS, MSF, PhD). *Faculty:* 26 full-time (4 women), 7 part-time/adjunct (1 woman). *Students:* 69 full-time (33 women), 15 part-time (3 women); includes 5 minority (1 African American, 1 American Indian/Alaska Native, 3 Hispanic Americans), 29 international. Average age 30. 45 applicants, 40% accepted, 18 enrolled. In 2005, 6 master's, 6 doctorates awarded. *Degree requirements:* For master's and doctorate, thesis/dissertation. *Entrance requirements:* For master's and doctorate, GRE General Test (500 verbal, 500 quantitative), minimum B+ average in undergraduate course work. Additional exam requirements/recommendations for international students: Required—TOEFL. *Application deadline:* For fall admission, 1/5 for domestic students; for spring admission, 9/15 for domestic students. Applications are processed on a rolling basis. Application fee: $55. Electronic applications accepted. *Financial support:* In 2005–06, 10 research assistantships (averaging $15,259 per year) were awarded; fellowships, teaching assistantships, career-related internships or fieldwork and scholarships/grants also available. Support available to part-time students. Financial award application deadline: 1/5; financial award applicants required to submit FAFSA. *Faculty research:* Wildlife management, forest management, forest ecology, forest soils, limnology. *Unit head:* Dr. Robert K. Swihart, Interim Head, 765-494-3590, Fax: 765-494-9461, E-mail: rswihart@purdue.edu. *Application contact:* Kelly Garrett, Graduate Secretary, 765-494-3572, Fax: 765-494-9461, E-mail: kgarrett@purdue.edu.

South Dakota State University, Graduate School, College of Agriculture and Biological Sciences, Department of Wildlife and Fisheries Sciences, Brookings, SD 57007. Offers biological sciences (PhD); wildlife and fisheries sciences (MS). *Degree requirements:* For master's, thesis, oral exam; for doctorate, thesis/dissertation, preliminary oral and written exams. *Entrance requirements:* For master's, GRE. Additional exam requirements/recommendations for international students: Required—TOEFL. *Faculty research:* Agriculture interactions, wetland conservation, biostress.

State University of New York College of Environmental Science and Forestry, Faculty of Environmental and Forest Biology, Syracuse, NY 13210-2779. Offers chemical ecology (MPS, MS, PhD); conservation biology (MPS, MS, PhD); ecology (MPS, MS, PhD); entomology (MPS, MS, PhD); environmental interpretation (MPS, MS, PhD); environmental physiology (MPS, MS, PhD); fish and wildlife biology (MPS, MS, PhD); forest pathology and mycology (MPS, MS, PhD); plant science and biotechnology (MPS, MS, PhD). *Faculty:* 27 full-time (4 women), 4 part-time/adjunct (0 women). *Students:* 81 full-time (50 women), 62 part-time (33 women); includes 4 minority (1 Asian American or Pacific Islander, 3 Hispanic Americans), 16 international. Average age 30. 84 applicants, 54% accepted, 17 enrolled. In 2005, 17 master's, 3 doctorates awarded. *Degree requirements:* For master's, thesis (for some programs), registration; for doctorate, thesis/dissertation, comprehensive exam, registration. *Entrance requirements:* For master's and doctorate, GRE General Test, GRE Subject Test, minimum GPA of 3.0. Additional exam requirements/recommendations for international students: Required—TOEFL (minimum score 550 paper-based; 213 computer-based). *Application deadline:* For fall admission, 2/1 priority date for domestic students, 2/1 priority date for international students; for spring admission, 11/1 priority date for domestic students, 11/1 priority date for international students. Applications are processed on a rolling basis. Application fee: $60. *Expenses:* Tuition, area resident: Full-time $6,900; part-time $288 per credit. Tuition, nonresident: full-time $10,920; part-time $455 per credit. Required fees: $395; $32 per credit. $20 per term. One-time fee: $145. *Financial support:* In 2005–06, 86 students received support, including 13 fellowships with full and partial tuition reimbursements available (averaging $9,446 per year), 40 research assistantships with full and partial tuition reimbursements available (averaging $11,000 per year), 32 teaching assistantships with full and partial tuition reimbursements available (averaging $9,446 per year); Federal Work-Study, institutionally sponsored loans, scholarships/grants, health care benefits, and unspecified assistantships also available. Financial award application deadline: 6/30. *Faculty research:* Ecology, fish and wildlife biology and

Fish, Game, and Wildlife Management

State University of New York College of Environmental Science and Forestry (continued)
management, plant science, entomology. Total annual research expenditures: $4.1 million. *Unit head:* Dr. Donald J. Leopold, Chair, 315-470-6770, Fax: 315-470-6934, E-mail: dendro@esf.edu. *Application contact:* Dr. Dudley J. Raynal, Dean, Instruction and Graduate Studies, 315-470-6599, Fax: 315-470-6978, E-mail: esfgrad@esf.edu.

Sul Ross State University, Division of Agricultural and Natural Resource Science, Program in Range and Wildlife Management, Alpine, TX 79832. Offers M Ag, MS. Part-time programs available. *Degree requirements:* For master's, thesis (for some programs). *Entrance requirements:* For master's, GRE General Test, minimum undergraduate GPA of 2.5 in last 60 hours.

Tennessee Technological University, Graduate School, College of Arts and Sciences, Department of Biology, Cookeville, TN 38505. Offers environmental biology (MS); fish, game, and wildlife management (MS). Part-time programs available. *Faculty:* 22 full-time (2 women). *Students:* 22 full-time (8 women), 17 part-time (7 women); includes 3 minority (1 African American, 2 Asian Americans or Pacific Islanders). Average age 25. 14 applicants, 50% accepted, 7 enrolled. In 2005, 9 degrees awarded. *Degree requirements:* For master's, thesis. *Entrance requirements:* For master's, GRE General Test. Additional exam requirements/recommendations for international students: Required—TOEFL. *Application deadline:* For fall admission, 3/1 for domestic students; for spring admission, 8/1 for domestic students. Application fee: $25 ($30 for international students). *Expenses:* Tuition, state resident: full-time $8,421; part-time $307 per hour. Tuition, nonresident: full-time $22,389; part-time $711 per hour. *Financial support:* In 2005–06, 22 research assistantships (averaging $9,000 per year), 9 teaching assistantships (averaging $7,500 per year) were awarded. Financial award application deadline: 4/1. *Faculty research:* Aquatics, environmental studies. *Unit head:* Dr. Daniel Combs, Interim Chairperson, 931-372-3134, Fax: 931-372-6257, E-mail: dcombs@tntech.edu. *Application contact:* Dr. Francis O. Otuonye, Associate Vice President for Research and Graduate Studies, 931-372-3233, Fax: 931-372-3497, E-mail: fotuonye@tntech.edu.

Texas A&M University, College of Agriculture and Life Sciences, Department of Wildlife and Fisheries Sciences, College Station, TX 77843. Offers M Agr, MS, PhD. Part-time programs available. Postbaccalaureate distance learning degree programs offered (no on-campus study). *Faculty:* 22 full-time (2 women), 1 part-time/adjunct (0 women). *Students:* 108 full-time (51 women), 57 part-time (24 women); includes 22 minority (1 African American, 3 American Indian/Alaska Native, 18 Hispanic Americans), 26 international. Average age 26. 87 applicants, 62% accepted, 38 enrolled. In 2005, 25 master's, 12 doctorates awarded. Terminal master's awarded for partial completion of doctoral program. *Median time to degree:* Of those who began their doctoral program in fall 1997, 43% received their degree in 8 years or less. *Degree requirements:* For master's and doctorate, thesis/dissertation, final oral defense. *Entrance requirements:* For master's and doctorate, GRE General Test, minimum GPA of 3.0. Additional exam requirements/recommendations for international students: Required—TOEFL (minimum score 550 paper-based; 213 computer-based). *Application deadline:* Applications are processed on a rolling basis. Application fee: $50 ($75 for international students). Electronic applications accepted. *Expenses:* Tuition, state resident: full-time $4,488; part-time $187 per credit hour. Tuition, nonresident: full-time $11,112; part-time $463 per credit hour. Required fees:$1,974. *Financial support:* In 2005–06, fellowships with partial tuition reimbursements (averaging $22,000 per year), research assistantships (averaging $14,400 per year), teaching assistantships (averaging $14,400 per year) were awarded; career-related internships or fieldwork, institutionally sponsored loans, and scholarships/grants also available. Financial award application deadline: 3/1; financial award applicants required to submit FAFSA. *Faculty research:* Wildlife ecology and management, fisheries ecology and management, aquaculture, biological inventories and museum collections, biosystematics and genome analysis. *Unit head:* Dr. Robert D. Brown, Professor and Head, 979-845-5777, Fax: 979-845-3786, E-mail: r-brown@tamu.edu. *Application contact:* Janice Crenshaw, Senior Academic Advisor I, 979-845-5777, Fax: 979-845-3786, E-mail: j-crenshaw@tamu.edu.

Texas A&M University–Kingsville, College of Graduate Studies, College of Agriculture and Home Economics, Program in Range and Wildlife Management, Kingsville, TX 78363. Offers MS. *Degree requirements:* For master's, thesis or alternative, comprehensive exam. *Entrance requirements:* For master's, GRE General Test, minimum GPA of 3.0. Additional exam requirements/recommendations for international students: Required—TOEFL.

Texas A&M University–Kingsville, College of Graduate Studies, College of Agriculture and Home Economics, Program in Wildlife Science, Kingsville, TX 78363. Offers PhD. *Degree requirements:* For doctorate, one foreign language, thesis/dissertation, comprehensive exam. *Entrance requirements:* For doctorate, GRE General Test, minimum GPA of 3.5.

Texas State University-San Marcos, Graduate School, College of Science, Department of Biology, Program in Wildlife Ecology, San Marcos, TX 78666. Offers MS. *Students:* 18 full-time (13 women), 14 part-time (9 women); includes 4 minority (1 African American, 3 Hispanic Americans). Average age 30. 11 applicants, 100% accepted, 10 enrolled. In 2005, 4 degrees awarded. *Entrance requirements:* For master's, GRE General Test, minimum GPA of 2.75 in last 60 hours of undergraduate work. *Application deadline:* For fall admission, 6/15 priority date for domestic students, 6/1 priority date for international students; for spring admission, 10/15 priority date for domestic students, 10/1 priority date for international students. Applications are processed on a rolling basis. Application fee: $40 ($90 for international students). *Expenses:* Tuition, area resident: Part-time $116 per credit. Tuition, state resident: full-time $3,168; part-time $176 per credit. Tuition, nonresident: full-time $8,136; part-time $452 per credit. Required fees: $1,112; $74 per credit. Full-time tuition and fees vary according to course load. *Financial support:* In 2005–06, 24 students received support; research assistantships, teaching assistantships available. Financial award application deadline: 4/1. *Unit head:* Dr. John Baccus, Graduate Advisor, 512-245-2347, Fax: 512-245-8713, E-mail: jb02@txstate.edu.

Texas Tech University, Graduate School, College of Agricultural Sciences and Natural Resources, Department of Range, Wildlife, and Fisheries Management, Lubbock, TX 79409. Offers fisheries science (MS, PhD); range science (MS, PhD); wildlife science (MS, PhD). Part-time programs available. *Faculty:* 10 full-time (0 women), 1 part-time/adjunct (0 women). *Students:* 36 full-time (8 women), 2 part-time (both women), 12 international. Average age 30. 12 applicants, 67% accepted, 8 enrolled. In 2005, 9 master's, 5 doctorates awarded. *Degree requirements:* For master's and doctorate, thesis/dissertation. *Entrance requirements:* For master's and doctorate, GRE General Test. Additional exam requirements/recommendations for international students: Required—TOEFL (minimum score 550 paper-based; 213 computer-based). *Application deadline:* Applications are processed on a rolling basis. Application fee: $50 ($60 for international students). Electronic applications accepted. *Expenses:* Tuition, state resident: full-time $4,296. Tuition, nonresident: full-time $10,920. Required fees: $1,992. Tuition and fees vary according to program. *Financial support:* In 2005–06, 26 students received support, including 22 research assistantships with partial tuition reimbursements available (averaging $10,622 per year), 6 teaching assistantships with partial tuition reimbursements available (averaging $11,986 per year); Federal Work-Study and institutionally sponsored loans also available. Support available to part-time students. Financial award application deadline: 4/15; financial award applicants required to submit FAFSA. *Faculty research:* Use of fire on range lands, waterfowl, upland game birds and playa lakes in the southern Great Plains, reproductive physiology in fisheries, conservation biology. Total annual research expenditures: $1.4 million. *Unit head:* Dr. Ernest B. Fish, Chairman, 806-742-2841, Fax: 806-742-2280. *Application contact:* L. Jeannine Becker, Graduate Secretary, 806-742-2825, E-mail: jeannine.becker@ttu.edu.

Université du Québec à Rimouski, Graduate Programs, Program in Wildlife Resources Management, Rimouski, QC G5L 3A1, Canada. Offers biology (PhD); wildlife resources management (Diploma). *Students:* 61 full-time (29 women), 7 part-time (1 woman). 23 applicants, 91% accepted. In 2005, 8 degrees awarded. *Entrance requirements:* For degree, appropriate bachelor's degree, proficiency in French. *Application deadline:* For fall admission, 5/1 for domestic students. Application fee: $30. Tuition charges are reported in Canadian dollars. *Expenses:* Tuition, state resident: full-time $2,000 Canadian dollars. Tuition, nonresident: full-time $9,000 Canadian dollars. Tuition and fees vary according to course load and program. *Financial support:* Fellowships, research assistantships, teaching assistantships available. *Unit head:* Richard Cloutier, Director, 418-724-1592, Fax: 418-724-1525, E-mail: richard_cloutier@uqar.ca. *Application contact:* Marc Berube, Office of Admissions, 418-724-1433, Fax: 418-724-1525, E-mail: marc_berube@uqar.ca.

University of Alaska Fairbanks, College of Natural Sciences and Mathematics, Department of Biology and Wildlife, Fairbanks, AK 99775-7520. Offers biological sciences (MS, PhD), including biology, botany, zoology; biology (MAT); wildlife biology (MS, PhD). Part-time programs available. *Faculty:* 29 full-time (7 women), 2 part-time/adjunct (1 woman). *Students:* 89 full-time (53 women), 24 part-time (15 women); includes 6 minority (1 African American, 4 Asian Americans or Pacific Islanders, 1 Hispanic American), 11 international. Average age 30. 74 applicants, 53% accepted, 17 enrolled. In 2005, 14 master's, 4 doctorates awarded. Terminal master's awarded for partial completion of doctoral program. *Degree requirements:* For master's and doctorate, thesis/dissertation, comprehensive exam, registration. *Entrance requirements:* For master's and doctorate, GRE General Test, GRE Subject Test. Additional exam requirements/recommendations for international students: Required—TOEFL (minimum score 550 paper-based; 213 computer-based); Recommended—TWE, TSE. *Application deadline:* For fall admission, 6/1 for domestic students, 3/1 for international students; for spring admission, 12/1 for domestic students, 9/1 for international students. Applications are processed on a rolling basis. Application fee: $50. Electronic applications accepted. *Expenses:* Tuition, state resident: full-time $4,392; part-time $244 per credit. Tuition, nonresident: full-time $8,964; part-time $498 per credit. Required fees: $800; $5 per credit. $48 per contact hour. Tuition and fees vary according to course level, course load, campus/location and reciprocity agreements. *Financial support:* In 2005–06, 36 research assistantships with tuition reimbursements (averaging $9,207 per year), 23 teaching assistantships with tuition reimbursements (averaging $5,358 per year) were awarded; fellowships with tuition reimbursements, career-related internships or fieldwork, Federal Work-Study, and scholarships/grants also available. Financial award application deadline: 2/1; financial award applicants required to submit FAFSA. *Faculty research:* Plant-herbivore interactions, plant metabolic defenses, insect manufacture of glycerol, ice nucleators, structure and functions of arctic and subarctic freshwater ecosystems. *Unit head:* Dr. Kent E. Schubegerle, Chair, 907-474-7671, Fax: 907-474-6716, E-mail: fybio@uaf.edu.

University of Alaska Fairbanks, School of Fisheries and Ocean Sciences, Department of Marine Sciences and Limnology, Fairbanks, AK 99775-7520. Offers marine biology (MS, PhD); oceanography (MS, PhD), including biological oceanography (PhD), chemical oceanography (PhD), fisheries (PhD), geological oceanography (PhD), physical oceanography (PhD). Part-time programs available. Terminal master's awarded for partial completion of doctoral program. *Degree requirements:* For master's and doctorate, thesis/dissertation, comprehensive exam, registration. *Entrance requirements:* For master's and doctorate, GRE General Test. Additional exam requirements/recommendations for international students: Required—TOEFL. Electronic applications accepted. *Expenses:* Tuition, state resident: full-time $4,392; part-time $244 per credit. Tuition, nonresident: full-time $8,964; part-time $498 per credit. Required fees: $800; $5 per credit. $48 per contact hour. Tuition and fees vary according to course level, course load, campus/location and reciprocity agreements. *Faculty research:* Seafood science and nutrition, sustainable harvesting, chemical oceanography, marine biology, physical oceanography.

University of Alaska Fairbanks, School of Fisheries and Ocean Sciences, Program in Fisheries, Fairbanks, AK 99775-7520. Offers MS, PhD. Part-time programs available. Terminal master's awarded for partial completion of doctoral program. *Degree requirements:* For master's and doctorate, thesis/dissertation, comprehensive exam, registration. *Entrance requirements:* For master's and doctorate, GRE General Test. Additional exam requirements/recommendations for international students: Required—TOEFL. Electronic applications accepted. *Expenses:* Tuition, state resident: full-time $4,392; part-time $244 per credit. Tuition, nonresident: full-time $8,964; part-time $498 per credit. Required fees: $800; $5 per credit. $48 per contact hour. Tuition and fees vary according to course level, course load, campus/location and reciprocity agreements. *Faculty research:* Marine stock reconstruction, oil spill research on marine life, Pacific salmon management, population dynamics of fish and major predators, ecology of marine fish.

The University of Arizona, Graduate College, College of Agriculture and Life Sciences, School of Natural Resources, Program in Wildlife, Fisheries Conservation, and Management, Tucson, AZ 85721. Offers MS, PhD. *Degree requirements:* For master's, thesis/dissertation; for doctorate, thesis/dissertation, comprehensive exam. *Entrance requirements:* For master's and doctorate, GRE General Test, GRE Subject Test (biology), minimum GPA of 3.0, 3 letters of recommendation. Additional exam requirements/recommendations for international students: Required—TOEFL (minimum score 550 paper-based; 213 computer-based). *Faculty research:* Short-term effects of artificial oases on Arizona wildlife, elk response to cattle in northern Arizona, effect of reservoir operation on tailwaters, conservation of wildlife.

University of Florida, Graduate School, College of Agricultural and Life Sciences, Department of Wildlife Ecology and Conservation, Gainesville, FL 32611. Offers MS, PhD. *Faculty:* 14 full-time (3 women), 1 part-time/adjunct (0 women). *Students:* 65 (28 women); includes 5 minority (1 American Indian/Alaska Native, 1 Asian American or Pacific Islander, 3 Hispanic Americans) 11 international. 65 applicants, 29% accepted. In 2005, 12 master's, 4 doctorates awarded. *Degree requirements:* For master's, thesis optional; for doctorate, thesis/dissertation. *Entrance requirements:* For master's and doctorate, GRE General Test, minimum GPA of 3.3. *Application deadline:* For fall admission, 6/1 for domestic students; for spring admission, 12/1 for domestic students. Applications are processed on a rolling basis. Application fee: $20. Electronic applications accepted. *Expenses:* Tuition, state resident: full-time $6,234. Tuition, nonresident: full-time $21,359. Tuition and fees vary according to program. *Financial support:* In 2005–06, 46 students received support, including 17 research assistantships (averaging $13,811 per year), 11 teaching assistantships (averaging $14,036 per year); fellowships, institutionally sponsored loans also available. *Faculty research:* Wildlife biology and management, tropical ecology and conservation, conservation biology, landscape ecology and restoration, conservation education. *Unit head:* George Tanner, Interim Chair, 352-846-0570. *Application contact:* Dr. Wiley Kitchens, Coordinator, 352-846-0536, Fax: 352-846-0841, E-mail: kitchensw@wec.ufl.edu.

University of Idaho, College of Graduate Studies, College of Natural Resources, Moscow, ID 83844-2282. Offers conservation social sciences (MS, PhD); fish and wildlife resources (MS, PhD), including fishery resources, wildlife resources; forest products (MS, PhD); forest resources (MS, PhD); natural resources (PhD); natural resources management and administration (MNR); rangeland ecology and management (MS, PhD). *Students:* 137 full-time (66 women), 95 part-time (36 women); includes 8 minority (4 American Indian/Alaska Native, 2 Asian Americans or Pacific Islanders, 2 Hispanic Americans), 41 international. In 2005, 51 master's, 7 doctorates awarded. *Degree requirements:* For doctorate, thesis/dissertation. *Entrance requirements:* For master's, minimum GPA of 2.8; for doctorate, minimum undergraduate GPA of 2.8, 3.0 graduate. *Application deadline:* For fall admission, 8/1 for domestic students; for spring admission, 12/15 for domestic students. Application fee: $55 ($60 for international students). *Expenses:* Tuition, nonresident: full-time $8,770; part-time $130 per credit. Required fees: $4,508; $217 per credit. *Financial support:* Fellowships, research assistantships, teaching assistantships, Federal Work-Study available. Support available to part-time students. Financial award application deadline: 2/15. *Unit head:* Steven B. Daley-Laursen, Dean, 208-885-6442, Fax: 208-885-6226. *Application contact:* Dr. Ali Moslemi, Graduate Coordinator, 208-885-6126.

University of Idaho, College of Graduate Studies, College of Natural Resources, Department of Fish and Wildlife Resources, Program in Fishery Resources, Moscow, ID 83844-2282.

Offers MS, PhD. *Students:* 9 full-time (3 women), 13 part-time (7 women); includes 1 minority (Hispanic American) In 2005, 15 degrees awarded. *Degree requirements:* For doctorate, thesis/dissertation. *Entrance requirements:* For master's, minimum GPA of 2.8; for doctorate, minimum undergraduate GPA of 2.8, 3.0 graduate. *Application deadline:* For fall admission, 8/1 for domestic students; for spring admission, 12/15 for domestic students. Application fee: $55 ($60 for international students). *Expenses:* Tuition, nonresident: full-time $8,770; part-time $130 per credit. Required fees: $4,508; $217 per credit. *Financial support:* Research assistantships available. Financial award application deadline: 2/15. *Unit head:* Dr. Kerry Reese, Head, Department of Fish and Wildlife Resources, 208-885-6435.

University of Idaho, College of Graduate Studies, College of Natural Resources, Department of Fish and Wildlife Resources, Program in Wildlife Resources, Moscow, ID 83844-2282. Offers MS, PhD. *Students:* 11 full-time (5 women), 12 part-time (4 women); includes 1 minority (American Indian/Alaska Native), 1 international. In 2005, 1 degree awarded. *Degree requirements:* For doctorate, thesis/dissertation. *Entrance requirements:* For master's, minimum GPA of 2.8; for doctorate, minimum undergraduate GPA of 2.8, 3.0 graduate. *Application deadline:* For fall admission, 8/1 for domestic students; for spring admission, 12/15 for domestic students. Application fee: $55 ($60 for international students). *Expenses:* Tuition, nonresident: full-time $8,770; part-time $130 per credit. Required fees: $4,508; $217 per credit. *Financial support:* Research assistantships, teaching assistantships available. Financial award application deadline: 2/15. *Unit head:* Dr. Kerry Reese, Head, Department of Fish and Wildlife Resources, 208-885-6435.

University of Maine, Graduate School, College of Natural Sciences, Forestry, and Agriculture, Department of Wildlife Ecology, Orono, ME 04469. Offers wildlife conservation (MWC); wildlife ecology (MS, PhD). Part-time programs available. *Faculty:* 12. *Students:* 10 full-time (5 women), 9 part-time (3 women), 5 international. Average age 31. 17 applicants, 0% accepted, 0 enrolled. In 2005, 1 master's, 1 doctorate awarded. *Degree requirements:* For master's, thesis (for some programs); for doctorate, one foreign language, thesis/dissertation. *Entrance requirements:* For master's and doctorate, GRE General Test. Additional exam requirements/recommendations for international students: Required—TOEFL. *Application deadline:* For fall admission, 2/1 for domestic students. Applications are processed on a rolling basis. Application fee: $50. Electronic applications accepted. *Financial support:* In 2005–06, 1 fellowship with tuition reimbursement (averaging $11,250 per year), 10 research assistantships with tuition reimbursements (averaging $15,600 per year), 3 teaching assistantships with tuition reimbursements (averaging $12,700 per year) were awarded; career-related internships or fieldwork, Federal Work-Study, institutionally sponsored loans, and tuition waivers (full and partial) also available. Financial award application deadline: 3/1. *Faculty research:* Integration of wildlife and forest management; population dynamics; behavior, physiology and nutrition; wetland ecology and influence of environmental disturbances. *Unit head:* Dr. Frederick Servello, Chair, 207-581-2862, Fax: 207-581-2858. *Application contact:* Scott G. Delcourt, Associate Dean of the Graduate School, 207-581-3219, Fax: 207-581-3232, E-mail: graduate@maine.edu.

University of Massachusetts Amherst, Graduate School, College of Natural Resources and the Environment, Department of Natural Resources Conservation, Program in Wildlife and Fisheries Conservation, Amherst, MA 01003. Offers MS, PhD. Part-time programs available. *Students:* 26 full-time (15 women), 27 part-time (8 women), 8 international. Average age 32. 35 applicants, 37% accepted, 10 enrolled. In 2005, 5 master's, 5 doctorates awarded. Terminal master's awarded for partial completion of doctoral program. *Degree requirements:* For master's, thesis optional; for doctorate, variable foreign language requirement, thesis/dissertation. *Entrance requirements:* For master's and doctorate, GRE General Test. Additional exam requirements/recommendations for international students: Required—TOEFL (minimum score 530 paper-based; 197 computer-based). *Application deadline:* For fall admission, 2/1 priority date for domestic students, 2/1 priority date for international students; for spring admission, 10/1 for domestic students, 10/1 for international students. Applications are processed on a rolling basis. Application fee: $40 ($65 for international students). Electronic applications accepted. *Expenses:* Tuition, state resident: part-time $110 per credit. Tuition, nonresident: part-time $414 per credit. Required fees: $2,824 per term. One-time fee: $250 part-time. Full-time tuition and fees vary according to course load, campus/location, program and reciprocity agreements. *Financial support:* Fellowships with full tuition reimbursements, research assistantships with full tuition reimbursements, teaching assistantships with full tuition reimbursements, career-related internships or fieldwork, Federal Work-Study, scholarships/grants, traineeships, and unspecified assistantships available. Support available to part-time students. Financial award application deadline: 2/1. *Unit head:* Dr. Kevin McGarigal, Director, 413-545-2666, Fax: 413-545-4358.

University of Miami, Graduate School, Rosenstiel School of Marine and Atmospheric Science, Division of Marine Biology and Fisheries, Coral Gables, FL 33124. Offers MA, MS, PhD. *Faculty:* 27 full-time (6 women), 24 part-time/adjunct (7 women). *Students:* 54 full-time (32 women); includes 5 minority (3 African Americans, 2 Hispanic Americans), 9 international. Average age 28. 80 applicants, 13% accepted, 8 enrolled. In 2005, 1 master's, 6 doctorates awarded. Terminal master's awarded for partial completion of doctoral program. *Median time to degree:* Of those who began their doctoral program in fall 1997, 83% received their degree in 8 years or less. *Degree requirements:* For master's and doctorate, thesis/dissertation, comprehensive exam, registration. *Entrance requirements:* For master's and doctorate, GRE General Test. Additional exam requirements/recommendations for international students: Required—TOEFL (minimum score 550 paper-based; 213 computer-based). *Application deadline:* For fall admission, 1/1 for domestic students. Applications are processed on a rolling basis. Application fee: $50. Electronic applications accepted. *Financial support:* In 2005–06, 49 students received support, including 15 fellowships with tuition reimbursements available (averaging $22,380 per year), 30 research assistantships with tuition reimbursements available (averaging $22,380 per year), 4 teaching assistantships with tuition reimbursements available (averaging $22,380 per year); institutionally sponsored loans and scholarships/grants also available. Financial award application deadline: 3/1; financial award applicants required to submit FAFSA. *Faculty research:* Biochemistry, physiology, plankton, coral, biology. *Unit head:* Dr. Robert Cowen, Chairperson, 305-421-4177, E-mail: rcowen@rsmas.miami.edu. *Application contact:* Dr. Larry Peterson, Associate Dean, 305-421-4155, Fax: 305-421-4771, E-mail: gso@rsmas.miami.edu.

University of Minnesota, Twin Cities Campus, Graduate School, College of Natural Resources, Department of Fisheries, Wildlife, and Conservation Biology, Minneapolis, MN 55455-0213. Offers conservation biology (MS, PhD); wildlife conservation (MS, PhD). Terminal master's awarded for partial completion of doctoral program. *Degree requirements:* For master's, thesis optional; for doctorate, thesis/dissertation, comprehensive exam, registration. *Entrance requirements:* For master's and doctorate, GRE. Additional exam requirements/recommendations for international students: Required—TOEFL (minimum score 550 paper-based; 213 computer-based). *Expenses:* Tuition, state resident: full-time $8,748; part-time $729 per credit. Tuition, nonresident: full-time $15,848; part-time $1,321 per credit. Full-time tuition and fees vary according to class time, course load, program and reciprocity agreements. *Faculty research:* Management, ecology, physiology, genetics, and computer modeling of fish and wildlife.

University of Missouri–Columbia, Graduate School, School of Natural Resources, Department of Fisheries and Wildlife, Columbia, MO 65211. Offers MS, PhD. *Faculty:* 7 full-time (0 women), 1 part-time/adjunct (0 women). *Students:* 6 full-time (2 women), 33 part-time (14 women); includes 4 minority (2 African Americans, 1 Asian American or Pacific Islander, 1 Hispanic American), 2 international. In 2005, 10 master's, 4 doctorates awarded. *Degree requirements:* For doctorate, thesis/dissertation. *Entrance requirements:* For master's and doctorate, GRE General Test, minimum GPA of 3.0. *Application deadline:* Applications are processed on a rolling basis. Application fee: $45 ($60 for international students). *Financial support:* Fellowships, research assistantships, teaching assistantships, institutionally sponsored loans and scholarships/grants available. *Unit head:* Dr. Charles F. Rabeni, Director of Graduate Studies, 573-882-3524, E-mail: rabenic@missouri.edu.

The University of Montana, Graduate School, College of Forestry and Conservation, Missoula, MT 59812-0002. Offers ecosystem management (MEM, MS); fish and wildlife biology (PhD); forestry (MS, PhD); recreation management (MS); resource conservation (MS); wildlife biology (MS). *Faculty:* 34 full-time (4 women). *Students:* 95 full-time (50 women), 63 part-time (21 women); includes 8 minority (1 African American, 3 American Indian/Alaska Native, 1 Asian American or Pacific Islander, 3 Hispanic Americans), 19 international. 175 applicants, 35% accepted, 46 enrolled. In 2005, 23 master's, 9 doctorates awarded. *Degree requirements:* For doctorate, thesis/dissertation. *Entrance requirements:* For master's and doctorate, GRE General Test. Additional exam requirements/recommendations for international students: Required—TOEFL (minimum score 575 paper-based; 213 computer-based). *Application deadline:* For fall admission, 1/31 for domestic students; for spring admission, 8/31 priority date for domestic students. Applications are processed on a rolling basis. Application fee: $45. *Expenses:* Tuition, state resident: part-time $267 per credit. Tuition, nonresident: part-time $665 per credit. Part-time tuition and fees vary according to course load and degree level. *Financial support:* In 2005–06, 25 research assistantships with tuition reimbursements, 12 teaching assistantships with full tuition reimbursements were awarded; fellowships, career-related internships or fieldwork and Federal Work-Study also available. Financial award application deadline: 3/1; financial award applicants required to submit FAFSA. Total annual research expenditures: $6.7 million. *Unit head:* Dr. Perry Brown, Dean, 406-243-5521, Fax: 406-243-4845, E-mail: pbrown@forestry.umt.edu.

University of New Hampshire, Graduate School, College of Life Sciences and Agriculture, Department of Natural Resources, Durham, NH 03824. Offers environmental conservation (MS); forestry (MS); soil science (MS); water resources management (MS); wildlife (MS). Part-time programs available. *Faculty:* 40 full-time. *Students:* 31 full-time (17 women), 36 part-time (17 women); includes 2 minority (both Asian Americans or Pacific Islanders), 4 international. Average age 31. 50 applicants, 44% accepted, 15 enrolled. In 2005, 14 degrees awarded. *Degree requirements:* For master's, thesis or alternative. *Entrance requirements:* For master's, GRE General Test. Additional exam requirements/recommendations for international students: Required—TOEFL (minimum score 550 paper-based; 213 computer-based); Recommended—TSE. *Application deadline:* For fall admission, 4/1 for domestic students, 4/1 for international students. For winter admission, 12/1 for domestic students. Applications are processed on a rolling basis. Application fee: $60. Electronic applications accepted. *Expenses:* Tuition, state resident: full-time $8,010; part-time $445 per credit hour. Tuition, nonresident: full-time $19,730; part-time $810 per credit hour. Required fees: $322 per semester. Tuition and fees vary according to course load and program. *Financial support:* In 2005–06, 3 fellowships, 21 research assistantships, 15 teaching assistantships were awarded; career-related internships or fieldwork, Federal Work-Study, scholarships/grants, and tuition waivers (full and partial) also available. Support available to part-time students. Financial award application deadline: 2/15. *Unit head:* Dr. William H. McDowell, Chairperson, 603-862-2249, E-mail: tehoward@cisunix.unh.edu. *Application contact:* Linda Scogin, Administrative Assistant, 603-862-3932, E-mail: natural.resources @unh.edu.

University of North Dakota, Graduate School, College of Arts and Sciences, Department of Biology, Grand Forks, ND 58202. Offers botany (MS, PhD); ecology (MS, PhD); entomology (MS, PhD); environmental biology (MS, PhD); fisheries/wildlife (MS, PhD); genetics (MS, PhD); zoology (MS, PhD). *Faculty:* 16 full-time (3 women). *Students:* 15 applicants, 13% accepted, 2 enrolled. In 2005, 5 degrees awarded. Terminal master's awarded for partial completion of doctoral program. *Degree requirements:* For master's, thesis/dissertation, final exam; for doctorate, thesis/dissertation, final exam, comprehensive exam. *Entrance requirements:* For master's, GRE General Test, GRE Subject Test, minimum GPA of 3.0; for doctorate, GRE General Test, GRE Subject Test, minimum GPA of 3.5. Additional exam requirements/recommendations for international students: Required—TOEFL (minimum score 550 paper-based; 213 computer-based). *Application deadline:* For fall admission, 10/1 for domestic students, 10/1 for international students. Application fee: $35. Electronic applications accepted. *Financial support:* In 2005–06, 8 research assistantships with full tuition reimbursements (averaging $11,375 per year), 13 teaching assistantships with full tuition reimbursements (averaging $10,813 per year) were awarded; fellowships, Federal Work-Study, institutionally sponsored loans, scholarships/grants, and tuition waivers (full and partial) also available. Support available to part-time students. Financial award application deadline: 3/15; financial award applicants required to submit FAFSA. *Faculty research:* Population biology, wildlife ecology, RNA processing, hormonal control of behavior. *Unit head:* Dr. Richard Sweitzel, Graduate Director, 701-777-4676, Fax: 701-777-2623, E-mail: richard_sweitzel@und.nodak.edu.

University of Rhode Island, Graduate School, College of the Environment and Life Sciences, Department of Fisheries, Animal and Veterinary Science, Kingston, RI 02881. Offers animal health and disease (MS); animal science (MS); aquaculture (MS); aquatic pathology (MS); environmental sciences (PhD), including animal science, aquacultural science, aquatic pathology, fisheries science; fisheries (MS). In 2005, 6 degrees awarded. *Application deadline:* For fall admission, 4/15 for domestic students. Applications are processed on a rolling basis. Application fee: $35. *Expenses:* Tuition, state resident: full-time $5,522; part-time $307 per credit. Tuition, nonresident: full-time $15,992; part-time $888 per credit. Required fees: $1,786; $73 per credit. One-time fee: $80 part-time. *Unit head:* Dr. David Bengtson, Chairperson, 401-874-2688.

The University of Tennessee, Graduate School, College of Agricultural Sciences and Natural Resources, Department of Forestry, Wildlife, and Fisheries, Program in Wildlife and Fisheries Science, Knoxville, TN 37996. Offers MS. *Degree requirements:* For master's, thesis. *Entrance requirements:* For master's, GRE General Test, minimum GPA of 2.7. Additional exam requirements/recommendations for international students: Required—TOEFL. Electronic applications accepted.

University of Washington, Graduate School, College of Forest Resources, Seattle, WA 98195. Offers forest economics (MS, PhD); forest ecosystem analysis (MS, PhD); forest engineering/forest hydrology (MS, PhD); forest products marketing (MS, PhD); forest soils (MS, PhD); paper science and engineering (MS, PhD); quantitative resource management (MS, PhD); silviculture (MFR); silviculture and forest protection (MS, PhD); social sciences (MS, PhD); urban horticulture (MFR, MS, PhD); wildlife science (MS, PhD). *Degree requirements:* For master's, thesis (for some programs), registration; for doctorate, thesis/dissertation, comprehensive exam (for some programs), registration. *Entrance requirements:* For master's and doctorate, GRE, minimum GPA of 3.0. Additional exam requirements/recommendations for international students: Required—TOEFL. Electronic applications accepted. *Faculty research:* Ecosystem analysis, silviculture and forest protection, paper science and engineering, environmental horticulture and urban forestry, natural resource policy and economics.

University of Washington, Graduate School, College of Ocean and Fishery Sciences, School of Aquatic and Fishery Sciences, Seattle, WA 98195. Offers MS, PhD. *Degree requirements:* For master's and doctorate, thesis/dissertation. *Entrance requirements:* For master's and doctorate, GRE General Test, minimum GPA of 3.0. Additional exam requirements/recommendations for international students: Required—TOEFL. Electronic applications accepted. *Faculty research:* Fish and shellfish ecology, fisheries management, aquatic ecology, conservation biology, genetics.

Utah State University, School of Graduate Studies, College of Natural Resources, Department of Aquatic, Watershed, and Earth Resources, Logan, UT 84322. Offers ecology (MS, PhD); fisheries biology (MS, PhD); watershed science (MS, PhD). *Faculty:* 12 full-time (3 women), 2 part-time/adjunct (1 woman). *Students:* 118 full-time (42 women), 16 part-time (2 women); includes 4 minority (all Hispanic Americans), 14 international. Average age 31. 31 applicants, 71% accepted, 21 enrolled. In 2005, 8 master's, 4 doctorates awarded. *Degree requirements:* For master's, thesis (for some programs); for doctorate, thesis/dissertation. *Entrance requirements:* For master's and doctorate, GRE General Test, minimum GPA of 3.2. Additional exam requirements/recommendations for international students: Required—TOEFL.

Fish, Game, and Wildlife Management

Utah State University (continued)
Application deadline: For fall admission, 2/15 for domestic students; for spring admission, 10/15 for domestic students. Applications are processed on a rolling basis. Application fee: $50 ($60 for international students). Electronic applications accepted. *Financial support:* In 2005–06, 3 fellowships with partial tuition reimbursements (averaging $20,000 per year), 34 research assistantships with partial tuition reimbursements (averaging $14,864 per year) were awarded; teaching assistantships with partial tuition reimbursements, career-related internships or fieldwork, Federal Work-Study, and institutionally sponsored loans also available. Support available to part-time students. Financial award application deadline:2/15. *Faculty research:* Behavior, population ecology, habitat, conservation biology, restoration, aquatic ecology, fisheries management, fluvial geomorphology, remote sensing, conservation biology. Total annual research expenditures: $4.7 million. *Unit head:* Chris Luecke, Head, 435-797-2463, Fax: 435-797-1871, E-mail: awerinfo@cc.usu.edu. *Application contact:* Brian Bailey, Staff Assistant, 435-797-2459, Fax: 435-797-1871, E-mail: awerinfo@cc.usu.edu.

Utah State University, School of Graduate Studies, College of Natural Resources, Department of Forest, Range, and Wildlife Sciences, Logan, UT 84322. Offers ecology (MS, PhD); forestry (MS, PhD); range science (MS, PhD); wildlife biology (MS, PhD). Part-time programs available. *Faculty:* 22 full-time (4 women), 17 part-time/adjunct (3 women). *Students:* 220 full-time (90 women), 52 part-time (17 women); includes 6 minority (all Asian Americans or Pacific Islanders), 49 international. Average age 24. 56 applicants, 61% accepted, 28 enrolled. In 2005, 11 master's, 4 doctorates awarded. *Degree requirements:* For master's, thesis/dissertation; for doctorate, thesis/dissertation, comprehensive exam. *Entrance requirements:* For master's and doctorate, GRE General Test, minimum GPA of 3.0. Additional exam requirements/recommendations for international students: Required—TOEFL. *Application deadline:* For fall admission, 6/15 for domestic students; for spring admission, 10/15 for domestic students. Applications are processed on a rolling basis. Application fee: $50 ($60 for international students). *Financial support:* In 2005–06, 14 research assistantships with partial tuition reimbursements (averaging $13,600 per year), 6 teaching assistantships (averaging $6,000 per year) were awarded; fellowships, career-related internships or fieldwork, Federal Work-Study, and institutionally sponsored loans also available. *Faculty research:* Range plant ecophysiology, plant community ecology, ruminant nutrition, population ecology. Total annual research expenditures: $3.5 million. *Unit head:* Dr. Johan W. duToit, Head, 435-797-2837, Fax: 435-797-3796. *Application contact:* Gaye Griffeth, Staff Assistant, 435-797-2503, Fax: 435-797-3796, E-mail: ggriffeth@cnr.usu.edu.

Virginia Polytechnic Institute and State University, Graduate School, College of Natural Resources, Department of Fisheries and Wildlife Sciences, Blacksburg, VA 24061. Offers MS, PhD. *Entrance requirements:* For master's and doctorate, GRE General Test. Additional exam requirements/recommendations for international students: Required—TOEFL (minimum score 550 paper-based; 213 computer-based). *Application deadline:* Applications are processed on a rolling basis. Application fee: $45. Electronic applications accepted. *Expenses:* Tuition, state resident: full-time $6,558; part-time $364 per credit. Tuition, nonresident: full-time $11,296; part-time $628 per credit. Required fees: $1,419; $468 per credit. $234 per term. *Financial support:* Research assistantships with full tuition reimbursements, teaching assistantships with full tuition reimbursements, career-related internships or fieldwork, Federal Work-Study, scholarships/grants, and unspecified assistantships available. Financial award application deadline: 4/1. *Faculty research:* Fisheries management, wildlife management, wildlife toxicology and physiology, endangered species, computer applications. *Unit head:* Dr. Donald J. Orth, Head, 540-231-5573, Fax: 540-231-7580, E-mail: dorth@vt.edu. *Application contact:* Linda D. Boothe, Assistant to the Head, 540-231-6944, Fax: 540-231-7580, E-mail: boothel@vt.edu.

West Virginia University, Davis College of Agriculture, Forestry and Consumer Sciences, Division of Forestry, Program in Wildlife and Fisheries Resources, Morgantown, WV 26506. Offers MS. Part-time programs available. *Students:* 22 full-time (8 women), 4 part-time. Average age 28. 20 applicants, 65% accepted, 13 enrolled. In 2005, 12 degrees awarded. *Degree requirements:* For master's, thesis, comprehensive exam. *Entrance requirements:* For master's, GRE, minimum GPA of 3.0. Additional exam requirements/recommendations for international students: Required—TOEFL. *Application deadline:* For fall admission, 7/7 for domestic students, 7/7 for international students; for spring admission, 12/1 for domestic students, 12/1 for international students. Applications are processed on a rolling basis. Application fee: $45. Electronic applications accepted. *Expenses:* Tuition, state resident: full-time $4,582; part-time $258 per credit hour. Tuition, nonresident: full-time $1,382; part-time $741 per credit hour. *Financial support:* In 2005–06, 19 research assistantships with full tuition reimbursements, 3 teaching assistantships with full tuition reimbursements were awarded; career-related internships or fieldwork, Federal Work-Study, institutionally sponsored loans, health care benefits, tuition waivers (full and partial), and unspecified assistantships also available. Financial award application deadline: 2/1; financial award applicants required to submit FAFSA. *Faculty research:* Managing habitat for game, nongame, and fish; fish ecology; wildlife ecology. Total annual research expenditures: $2 million. *Unit head:* Dr. Kyle J. Hartman, Coordinator, 304-293-2494 Ext. 2491, Fax: 304-293-2441, E-mail: kyle.hartman@mail.wvu.edu.

Forestry

Auburn University, Graduate School, School of Forestry and Wildlife Sciences, Auburn University, AL 36849. Offers MF, MS, PhD. Part-time programs available. *Faculty:* 29 full-time (3 women). *Students:* 27 full-time (13 women), 32 part-time (8 women); includes 1 minority (Hispanic American), 16 international. 38 applicants, 58% accepted, 15 enrolled. In 2005, 7 master's, 3 doctorates awarded. *Degree requirements:* For master's, oral exam (MF), thesis (MS); for doctorate, thesis/dissertation. *Entrance requirements:* For master's and doctorate, GRE General Test. *Application deadline:* For fall admission, 7/7 for domestic students; for spring admission, 11/24 for domestic students. Applications are processed on a rolling basis. Application fee: $25 ($50 for international students). Electronic applications accepted. *Financial support:* Fellowships, research assistantships, teaching assistantships, Federal Work-Study available. Support available to part-time students. Financial award application deadline: 3/15. *Faculty research:* Forest nursery management, silviculture and vegetation management, biological processes and ecological relationships, growth and yield of plantations and natural stands, urban forestry, forest taxation, law and policy. *Unit head:* Richard W. Brinker, Dean, 334-844-1007, Fax: 334-844-1084, E-mail: brinker@forestry.auburn.edu. *Application contact:* Dr. Stephen L. McFarland, Acting Dean of the Graduate School, 334-844-4700.

California Polytechnic State University, San Luis Obispo, College of Agriculture, Department of Natural Resources Management, San Luis Obispo, CA 93407. Offers forestry science (MS). Part-time programs available. *Faculty:* 7 full-time (1 woman), 2 part-time/adjunct (0 women). *Students:* 1 (woman) full-time, 2 part-time (1 woman). 1 applicant, 100% accepted, 1 enrolled. In 2005, 1 degree awarded. *Degree requirements:* For master's, thesis, comprehensive exam. *Entrance requirements:* For master's, minimum 2.75 GPA in last 90 quarter units of course work. *Application deadline:* For fall admission, 6/1 for domestic students, 11/30 for international students. For winter admission, 8/1 for domestic students; for spring admission, 12/1 for domestic students. Applications are processed on a rolling basis. Application fee: $55. Electronic applications accepted. *Expenses:* Tuition, nonresident: part-time $226 per unit. Required fees: $1,063 per unit. *Financial support:* Application deadline: 3/2; *Unit head:* Dr. Doug Piirto, Department Head, Graduate Coordinator, 805-756-2968, Fax: 805-756-1402, E-mail: dpiirto@calpoly.edu.

Clemson University, Graduate School, College of Agriculture, Forestry and Life Sciences, Department of Forestry and Natural Resources, Program in Forest Resources, Clemson, SC 29634. Offers MFR, MS, PhD. Part-time programs available. *Students:* 26 full-time (4 women), 5 part-time (3 women); includes 2 minority (both African Americans), 4 international. Average age 25. 19 applicants, 47% accepted, 6 enrolled. In 2005, 16 master's, 3 doctorates awarded. *Degree requirements:* For master's and doctorate, thesis/dissertation. *Entrance requirements:* For master's, GRE General Test, minimum B average in last 2 years of undergraduate course work; for doctorate, GRE General Test, minimum B average in graduate course work. Additional exam requirements/recommendations for international students: Required—TOEFL. *Application deadline:* For fall admission, 3/1 priority date for domestic students, 4/15 priority date for international students; for spring admission, 10/1 for domestic students, 9/15 for international students. Application fee: $50. *Financial support:* Fellowships, research assistantships, teaching assistantships available. Financial award application deadline: 5/1; financial award applicants required to submit FAFSA. *Faculty research:* Wetlands management, wood technology, forest management, silviculture, economics. *Application contact:* Dr. Dave Guynn, Coordinator, 864-656-4803, Fax: 864-656-3304, E-mail: dguynn@clemson.edu.

See Close-Ups on pages 701 and 703.

Colorado State University, Graduate School, Warner College of Natural Resources, Department of Forest, Rangeland, and Watershed Stewardship, Fort Collins, CO 80523-0015. Offers forest sciences (MS, PhD); natural resource stewardship (MNRS); rangeland ecosystem science (MS, PhD); watershed science (MS). Part-time programs available. *Faculty:* 27 full-time (5 women), 1 part-time/adjunct (0 women). *Students:* 58 full-time (23 women), 90 part-time (39 women); includes 11 minority (1 African American, 2 American Indian/Alaska Native, 1 Asian American or Pacific Islander, 7 Hispanic Americans), 10 international. Average age 33. 101 applicants, 35% accepted, 30 enrolled. In 2005, 18 master's, 5 doctorates awarded. Terminal master's awarded for partial completion of doctoral program. *Degree requirements:* For master's, thesis optional; for doctorate, thesis/dissertation, comprehensive exam, registration. *Entrance requirements:* For master's and doctorate, GRE General Test, minimum GPA of 3.0. Additional exam requirements/recommendations for international students: Required—TOEFL. *Application deadline:* For fall admission, 4/1 for domestic students; for spring admission, 9/1 priority date for domestic students. Applications are processed on a rolling basis. Application fee: $50. Electronic applications accepted. *Expenses:* Tuition, state resident: full-time $3,690; part-time $205 per credit. Tuition, nonresident: full-time $14,958; part-time $831 per credit. Required fees: $1,061. *Financial support:* Career-related internships or fieldwork, Federal Work-Study, institutionally sponsored loans, and traineeships available. Financial award application deadline: 5/1. *Faculty research:* Ecology, natural resource management, hydrology, restoration, human dimensions. Total annual research expenditures: $4.1 million. *Unit head:* Dr. N. Thompson Hobbs, Head, 970-491-6911, Fax: 970-491-6754, E-mail: nthobbs@warnercnr@colostate.edu. *Application contact:* Graduate Coordinator, 970-491-4994, Fax: 970-491-6754.

Cornell University, Graduate School, Graduate Fields of Agriculture and Life Sciences, Field of Natural Resources, Ithaca, NY 14853-0001. Offers aquatic science (MPS, MS, PhD); environmental management (MPS); fishery science (MPS, MS, PhD); forest science (MPS, MS, PhD); resource policy and management (MPS, MS, PhD); wildlife science (MPS, MS, PhD). *Faculty:* 50 full-time (9 women). *Students:* 62 full-time (33 women); includes 9 minority (1 African American, 2 American Indian/Alaska Native, 4 Asian Americans or Pacific Islanders, 2 Hispanic Americans), 14 international. 60 applicants, 23% accepted, 12 enrolled. In 2005, 11 master's, 5 doctorates awarded. *Degree requirements:* For master's, thesis (MS), project paper (MPS); for doctorate, thesis/dissertation, comprehensive exam. *Entrance requirements:* For master's and doctorate, GRE General Test, 2 letters of recommendation. Additional exam requirements/recommendations for international students: Required—TOEFL (minimum score 550 paper-based; 213 computer-based). *Application deadline:* For spring admission, 10/30 for domestic students. Applications are processed on a rolling basis. Application fee: $60. Electronic applications accepted. *Financial support:* In 2005–06, 49 students received support, including 15 fellowships with full tuition reimbursements available, 16 research assistantships with full tuition reimbursements available, 18 teaching assistantships with full tuition reimbursements available; institutionally sponsored loans, scholarships/grants, health care benefits, tuition waivers (full and partial), and unspecified assistantships also available. Financial award applicants required to submit FAFSA. *Faculty research:* Ecosystem-level dynamics, systems modeling, conservation biology/management, resource management's human dimensions, biogeochemistry. *Unit head:* Director of Graduate Studies, 607-255-2807, Fax: 607-255-0349. *Application contact:* Graduate Field Assistant, 607-255-2807, Fax: 607-255-0349, E-mail: nrgrad@cornell.edu.

Duke University, Nicholas School of the Environment and Earth Sciences, Durham, NC 27708-0328. Offers coastal environmental management (MEM); environmental economics and policy (MEM); environmental health and security (MEM); forest resource management (MF); global environmental change (MEM); resource ecology (MEM); water and air resources (MEM). *Accreditation:* SAF (one or more programs are accredited). Part-time programs available. *Degree requirements:* For master's, thesis, registration. *Entrance requirements:* For master's, GRE General Test, previous course work in biology or ecology, calculus, statistics, and microeconomics; computer familiarity with word processing and data analysis. Additional exam requirements/recommendations for international students: Required—TOEFL (minimum score 550 paper-based; 213 computer-based). Electronic applications accepted. Expenses: Contact institution. *Faculty research:* Ecosystem management, conservation ecology, earth systems, risk assessment.

See Close-Up on page 709.

Harvard University, Graduate School of Arts and Sciences, Department of Forestry, Cambridge, MA 02138. Offers forest science (MFS). *Faculty:* 4 full-time. *Students:* 2 applicants, 0% accepted. *Degree requirements:* For master's, thesis. *Entrance requirements:* For master's, GRE General Test, bachelor's degree in biology or forestry. Additional exam requirements/recommendations for international students: Required—TOEFL. *Application deadline:* For fall admission, 1/1 for domestic students. Application fee: $60. *Expenses:* Tuition: Full-time $28,752. Full-time tuition and fees vary according to program and student level. *Financial support:* Fellowships, career-related internships or fieldwork, Federal Work-Study, and institutionally sponsored loans available. Financial award application deadline: 1/1. *Faculty research:* Forest ecology, planning, and physiology; forest microbiology. *Unit head:* Betsey Cogswell, Administrator, 617-495-5497, Fax: 617-495-5264. *Application contact:* Office of Admissions and Financial Aid, 617-495-5315.

Iowa State University of Science and Technology, Graduate College, College of Agriculture, Department of Natural Resource Ecology and Management, Ames, IA 50011. Offers animal ecology (MS, PhD), including animal ecology, fisheries biology, wildlife biology; forestry (MS, PhD). *Faculty:* 23 full-time, 12 part-time/adjunct. *Students:* 45 full-time (18 women), 4 part-time (1 woman), 7 international. 24 applicants, 38% accepted, 8 enrolled. In 2005, 8 master's, 3 doctorates awarded. *Degree requirements:* For master's, thesis (for some programs); for doctorate, thesis/dissertation. *Entrance requirements:* For master's and doctorate, GRE General Test. Additional exam requirements/recommendations for international students: Required—TOEFL (paper score 547; computer score 210) or IELTS (score 6). *Application deadline:* For fall admission, 1/1 priority date for domestic students, 1/1 priority date for international students; for spring admission, 9/1 priority date for domestic students, 9/1 priority date for international students. Application fee: $30 ($70 for international students). Electronic applications accepted. *Expenses:* Tuition, state resident: full-time $6,410. Tuition, nonresident: full-time $16,422. Tuition and fees vary according to program. *Financial support:* In 2005–06, 41 research assistantships with full and partial tuition reimbursements (averaging $14,755 per year), 3 teaching assistantships with full and partial tuition reimbursements (averaging $14,580 per year) were awarded. *Unit head:* Dr. David M Engle, Chair, 515-294-1166. *Application contact:* Lyn Van De Pol, Information Contact, 515-294-6148, E-mail: lvdp@iastate.edu.

Lakehead University, Graduate Studies, Faculty of Forestry, Thunder Bay, ON P7B 5E1, Canada. Offers M Sc F, MF. Part-time programs available. *Degree requirements:* For master's, report (MF), thesis (M Sc F). *Entrance requirements:* For master's, minimum B average. Additional exam requirements/recommendations for international students: Required—TOEFL. *Faculty research:* Soils, silviculture, wildlife, ecology, genetics.

Louisiana State University and Agricultural and Mechanical College, Graduate School, College of Agriculture, School of Renewable Natural Resources, Baton Rouge, LA 70803. Offers fisheries (MS); forestry (MS, PhD); wildlife (MS); wildlife and fisheries science (PhD). *Faculty:* 29 full-time (1 woman). *Students:* 67 full-time (24 women), 7 part-time (2 women); includes 1 African American, 1 Asian American or Pacific Islander, 3 Hispanic Americans, 20 international. Average age 28. 36 applicants, 56% accepted, 16 enrolled. In 2005, 13 master's, 6 doctorates awarded. *Degree requirements:* For master's and doctorate, thesis/dissertation. *Entrance requirements:* For master's, GRE General Test, minimum GPA of 3.0; for doctorate, GRE General Test, MS, minimum GPA of 3.0. Additional exam requirements/recommendations for international students: Required—TOEFL (minimum score 550 paper-based; 213 computer-based). *Application deadline:* For fall admission, 1/25 priority date for domestic students, 5/15 priority date for international students. Applications are processed on a rolling basis. Application fee: $25. Electronic applications accepted. *Financial support:* In 2005–06, 70 students received support, including 4 fellowships (averaging $16,304 per year), 59 research assistantships with partial tuition reimbursements available (averaging $16,539 per year); teaching assistantships with partial tuition reimbursements available, Federal Work-Study, institutionally sponsored loans, scholarships/grants, tuition waivers (full and partial), and unspecified assistantships also available. Financial award application deadline: 4/15; financial award applicants required to submit FAFSA. *Faculty research:* Forest biology and management, aquaculture, fisheries biology and ecology, upland and wetlands wildlife. Total annual research expenditures: $4,423. *Unit head:* Dr. Bob G. Blackmon, Director, 225-578-4131, Fax: 225-578-4227, E-mail: bblackmon@agctr.lsu.edu. *Application contact:* Dr. Allen Rutherford, Coordinator of Graduate Studies, 225-578-4187, Fax: 225-578-4227, E-mail: druther@lsu.edu.

See Close-Up on page 711.

McGill University, Faculty of Graduate and Postdoctoral Studies, Faculty of Agricultural and Environmental Sciences, Department of Natural Resource Sciences, Montréal, QC H3A 2T5, Canada. Offers agrometeorology (M Sc, PhD); entomology (M Sc, PhD); forest science (M Sc, PhD); microbiology (M Sc, PhD); neotropical environment (M Sc, PhD); soil science (M Sc, PhD); wildlife biology (M Sc, PhD). *Degree requirements:* For master's and doctorate, thesis/dissertation, registration. *Entrance requirements:* For master's, minimum GPA of 3.0 or 3.2 in the last 2 years of university study. Additional exam requirements/recommendations for international students: Required—TOEFL (minimum score 550 paper-based; 213 computer-based), IELT (minimum score 7). Electronic applications accepted. *Faculty research:* Toxicology, reproductive physiology, parasites, wildlife management, genetics.

Michigan State University, The Graduate School, College of Agriculture and Natural Resources, Department of Forestry, East Lansing, MI 48824. Offers forestry (MS, PhD); forestry-environmental toxicology (PhD); plant breeding and genetics-forestry (MS, PhD). *Faculty:* 15 full-time (2 women). *Students:* 28 full-time (8 women), 8 part-time (1 woman); includes 1 minority (Asian American or Pacific Islander), 14 international. Average age 32. 19 applicants, 53% accepted. In 2005, 4 master's, 4 doctorates awarded. *Degree requirements:* For master's, thesis optional; for doctorate, thesis/dissertation. *Entrance requirements:* For master's and doctorate, GRE General Test, 3 letters of recommendation. Additional exam requirements/recommendations for international students: Required—TOEFL (minimum score 550 paper-based; 213 computer-based), Michigan State University ELT (85), Michigan ELAB (83). *Application deadline:* For fall admission, 12/27 for domestic students. Applications are processed on a rolling basis. Application fee: $50. Electronic applications accepted. *Expenses:* Tuition, state resident: part-time $330 per credit hour. Tuition, nonresident: part-time $685 per credit hour. Tuition and fees vary according to program. *Financial support:* In 2005–06, 3 fellowships with tuition reimbursements (averaging $8,267 per year), 19 research assistantships with tuition reimbursements (averaging $14,130 per year) were awarded; scholarships/grants and unspecified assistantships also available. *Faculty research:* Silviculture, biometry and ecology, social forestry and agroforestry, tree physiology and wood science, plant breeding and genetics. Total annual research expenditures: $3.7 million. *Unit head:* Dr. Daniel E. Keathley, Chairperson, 517-355-0091, Fax: 517-432-1143, E-mail: keathley@msu.edu. *Application contact:* Juli Kerr, Graduate Secretary, 517-355-0090, Fax: 517-432-1143, E-mail: kerrju@msu.edu.

Michigan State University, The Graduate School, College of Agriculture and Natural Resources, Program in Plant Breeding and Genetics, East Lansing, MI 48824. Offers plant breeding and genetics (MS, PhD), including crop and soil sciences, forestry, horticulture. *Students:* 17 full-time (12 women), 2 part-time (1 woman), 10 international. Average age 29. 9 applicants, 0% accepted. In 2005, 2 master's, 6 doctorates awarded. *Degree requirements:* For master's, thesis; for doctorate, thesis/dissertation, oral examination, comprehensive exam. *Entrance requirements:* For master's and doctorate, GRE General Test, minimum GPA of 3.0 in last 3 years of college, 3 letters of recommendation. Additional exam requirements/recommendations for international students: Required—TOEFL (minimum score 550 paper-based; 213 computer-based). *Application deadline:* For fall admission, 12/27 for domestic students. Application fee: $50. Electronic applications accepted. *Expenses:* Tuition, state resident: part-time $330 per credit hour. Tuition, nonresident: part-time $685 per credit hour. Tuition and fees vary according to program. *Financial support:* In 2005–06, 1 fellowship with tuition reimbursement (averaging $11,500 per year), 12 research assistantships with tuition reimbursements (averaging $13,218 per year), 2 teaching assistantships with tuition reimbursements (averaging $12,470 per year) were awarded; scholarships/grants and unspecified assistantships also available. *Faculty research:* Applied plant breeding and genetics; disease, insect and herbicide resistances; gene isolation and genomics; abiotic stress factors; molecular mapping. *Unit head:* Dr. Jim F. Hancock, Director, 517-355-5191 Ext. 1387, Fax: 517-432-3490, E-mail: hancock@msu.edu. *Application contact:* Program Information, 517-355-5191 Ext. 1324, Fax: 517-432-3490, E-mail: pbg@msu.edu.

Michigan Technological University, Graduate School, School of Forest Resources and Environmental Science, Program in Forest Ecology and Management, Houghton, MI 49931-1295. Offers MS. Part-time programs available. *Faculty:* 23 full-time (4 women), 16 part-time/adjunct (2 women). *Students:* 6 full-time (all women), 3 part-time (1 woman). Average age 26. 7 applicants, 86% accepted, 6 enrolled. In 2005, 9 degrees awarded. *Degree requirements:* For master's, thesis (for some programs), registration. *Entrance requirements:* For master's, GRE. Additional exam requirements/recommendations for international students: Required—TOEFL (minimum score 550 paper-based; 213 computer-based). *Application deadline:* Applications are processed on a rolling basis. Application fee: $40 ($45 for international students). Electronic applications accepted. *Expenses:* Tuition, nonresident: full-time $11,232; part-time $468 per credit. Required fees: $754; $377 per semester. Full-time tuition and fees vary according to course load, degree level and program. *Financial support:* In 2005–06, 5 students received support, including fellowships with full tuition reimbursements available (averaging $9,542 per year), 3 research assistantships with full tuition reimbursements available (averaging $9,542 per year), teaching assistantships with full tuition reimbursements available (averaging $9,542 per year); career-related internships or fieldwork, Federal Work-Study, scholarships/grants, health care benefits, tuition waivers (partial), unspecified assistantships, and co-op also available. Financial award applicants required to submit FAFSA. *Application contact:* Dr. Chandrashekhar P. Joshi, Associate Professor and Graduate Program Coordinator, 906-487-3480, Fax: 906-487-2915, E-mail: cpjoshi@mtu.edu.

Michigan Technological University, Graduate School, School of Forest Resources and Environmental Science, Program in Forest Molecular Genetics and Biotechnology, Houghton, MI 49931-1295. Offers MS, PhD. Part-time programs available. *Faculty:* 23 full-time (4 women), 16 part-time/adjunct (2 women). *Students:* 13 full-time (4 women), 4 part-time (all women), 16 international. Average age 30. 10 applicants, 70% accepted, 3 enrolled. In 2005, 1 master's, 1 doctorate awarded. Terminal master's awarded for partial completion of doctoral program. *Median time to degree:* Of those who began their doctoral program in fall 1997, 100% received their degree in 8 years or less. *Degree requirements:* For master's, thesis (for some programs), registration; for doctorate, thesis/dissertation, comprehensive exam, registration. *Entrance requirements:* For master's, GRE. Additional exam requirements/recommendations for international students: Required—TOEFL (minimum score 550 paper-based; 213 computer-based). *Application deadline:* Applications are processed on a rolling basis. Application fee: $40 ($45 for international students). Electronic applications accepted. *Expenses:* Tuition, nonresident: full-time $11,232; part-time $468 per credit. Required fees: $754; $377 per semester. Full-time tuition and fees vary according to course load, degree level and program. *Financial support:* In 2005–06, 14 students received support, including fellowships with full tuition reimbursements available (averaging $9,542 per year), 14 research assistantships with full tuition reimbursements available (averaging $9,542 per year), teaching assistantships with full tuition reimbursements available (averaging $9,542 per year); career-related internships or fieldwork, Federal Work-Study, scholarships/grants, health care benefits, tuition waivers (partial), unspecified assistantships, and co-op also available. Financial award applicants required to submit FAFSA. *Application contact:* Dr. Chandrashekhar P. Joshi, Associate Professor and Graduate Program Coordinator, 906-487-3480, Fax: 906-487-2915, E-mail: cpjoshi@mtu.edu.

Michigan Technological University, Graduate School, School of Forest Resources and Environmental Science, Program in Forestry, Houghton, MI 49931-1295. Offers MF, MS. Part-time programs available. *Faculty:* 23 full-time (4 women), 16 part-time/adjunct (2 women). *Students:* 24 full-time (14 women), 8 part-time (2 women); includes 3 minority (all Hispanic Americans), 1 international. Average age 28. 12 applicants, 67% accepted, 5 enrolled. In 2005, 13 degrees awarded. *Degree requirements:* For master's, thesis (for some programs), registration. *Entrance requirements:* For master's, GRE. Additional exam requirements/recommendations for international students: Required—TOEFL (minimum score 550 paper-based; 213 computer-based). *Application deadline:* Applications are processed on a rolling basis. Application fee: $40 ($45 for international students). Electronic applications accepted. *Expenses:* Tuition, nonresident: full-time $11,232; part-time $468 per credit. Required fees: $754; $377 per semester. Full-time tuition and fees vary according to course load, degree level and program. *Financial support:* In 2005–06, 20 students received support, including fellowships with full tuition reimbursements available (averaging $9,542 per year), 6 research assistantships with full tuition reimbursements available (averaging $9,542 per year), teaching assistantships with full tuition reimbursements available (averaging $9,542 per year); career-related internships or fieldwork, Federal Work-Study, scholarships/grants, health care benefits, unspecified assistantships, and co-op also available. Financial award applicants required to submit FAFSA. *Application contact:* Dr. Chandrashekhar P. Joshi, Associate Professor and Graduate Program Coordinator, 906-487-3480, Fax: 906-487-2915, E-mail: cpjoshi@mtu.edu.

Michigan Technological University, Graduate School, School of Forest Resources and Environmental Science, Program in Forest Science, Houghton, MI 49931-1295. Offers PhD. Part-time programs available. *Faculty:* 23 full-time (4 women), 16 part-time/adjunct (2 women). *Students:* 20 full-time (12 women), 5 part-time (2 women); includes 1 minority (African American), 5 international. Average age 32. 7 applicants, 71% accepted, 4 enrolled. *Median time to degree:* Of those who began their doctoral program in fall 1997, 60% received their degree in 8 years or less. *Degree requirements:* For doctorate, thesis/dissertation, comprehensive exam, registration. *Entrance requirements:* Additional exam requirements/recommendations for international students: Required—TOEFL (minimum score 550 paper-based; 213 computer-based). *Application deadline:* Applications are processed on a rolling basis. Application fee: $40 ($45 for international students). Electronic applications accepted. *Expenses:* Tuition, nonresident: full-time $11,232; part-time $468 per credit. Required fees: $754; $377 per semester. Full-time tuition and fees vary according to course load, degree level and program. *Financial support:* In 2005–06, 20 students received support, including fellowships with full tuition reimbursements available (averaging $9,542 per year), 18 research assistantships with full tuition reimbursements available (averaging $9,542 per year), teaching assistantships with full tuition reimbursements available (averaging $9,542 per year); career-related internships or fieldwork, Federal Work-Study, scholarships/grants, health care benefits, tuition waivers (partial), unspecified assistantships, and co-op also available. Financial award applicants required to submit FAFSA. *Application contact:* Dr. Chandrashekhar P. Joshi, Associate Professor and Graduate Program Coordinator, 906-487-3480, Fax: 906-487-2915, E-mail: cpjoshi@mtu.edu.

Mississippi State University, College of Forest Resources, Department of Forest Products, Mississippi State, MS 39762. Offers MS, PhD. *Faculty:* 15 full-time (2 women), 1 part-time/adjunct (0 women). *Students:* 4 full-time (1 woman), 3 part-time; includes 1 minority (Hispanic American) Average age 25. 3 applicants, 33% accepted. In 2005, 3 degrees awarded. *Degree requirements:* For master's, thesis, comprehensive oral or written exam. *Entrance requirements:* For master's, minimum GPA of 3.0. Additional exam requirements/recommendations for international students: Required—TOEFL. *Application deadline:* For fall admission, 7/1 for domestic students; for spring admission, 11/1 for domestic students. Applications are processed on a rolling basis. Application fee: $30. Electronic applications accepted. *Expenses:* Tuition, state resident: full-time $4,312; part-time $240 per hour. Tuition, nonresident: full-time $9,772; part-time $543 per hour. International tuition: $10,102 full-time. Tuition and fees vary according to course load. *Financial support:* In 2005–06, 1 teaching assistantship (averaging $12,273 per year) was awarded; fellowships, research assistantships with full tuition reimbursements, Federal Work-Study and institutionally sponsored loans also available. Financial award applicants required to submit FAFSA. *Faculty research:* Wood property enhancement and durability, environmental science and chemistry, wood-based composites, primary wood production, furniture manufacturing and management. Total annual research expenditures: $2.2 million. *Unit head:* Dr. Liam E. Leightley, Head, 662-325-4444, Fax: 662-325-8126, E-mail: lleightley@cfr.msstate.edu. *Application contact:* Philip G. Bonfanti, Director of Admissions, 662-325-4104, Fax: 662-325-8872, E-mail: admit@msstate.edu.

Mississippi State University, College of Forest Resources, Department of Forestry, Mississippi State, MS 39762. Offers MS. Part-time programs available. *Faculty:* 20 full-time (3 women), 1 part-time/adjunct (0 women). *Students:* 30 full-time (7 women), 2 part-time; includes 3 minority (1 African American, 1 Asian American or Pacific Islander, 1 Hispanic American), 4 international. Average age 26. 7 applicants, 71% accepted, 5 enrolled. In 2005, 7 degrees awarded. *Degree requirements:* For master's, thesis, comprehensive oral or written exam. *Entrance requirements:* For master's, minimum GPA of 2.5. Additional exam requirements/recommendations for international students: Required—TOEFL. *Application deadline:* For fall admission, 7/1 for domestic students; for spring admission, 11/1 for domestic students. Applications are processed on a rolling basis. Application fee: $30. *Expenses:* Tuition, state resident: full-time $4,312; part-time $240 per hour. Tuition, nonresident: full-time $9,772; part-time $543 per hour. International tuition: $10,102 full-time. Tuition and fees vary according to course load. *Financial support:* In 2005–06, 5 teaching assistantships with full tuition reimbursements (averaging $11,875 per year) were awarded; research assistantships with full tuition reimbursements, Federal Work-Study, institutionally sponsored loans, and unspecified assistantships also available. Financial award applicants required to submit FAFSA.

Forestry

Mississippi State University (continued)

Faculty research: Forest hydrology, forest biometry, forest management/economics, forest biology, industrial forest operations. Total annual research expenditures: $1.6 million. *Unit head:* Dr. James Shepard, Head, 662-325-2781, Fax: 662-325-8126, E-mail: jshepard@cfr.msstate.edu. *Application contact:* Philip G. Bonfanti, Director of Admissions, 662-325-4104, Fax: 662-325-8872, E-mail: admit@msstate.edu.

North Carolina State University, Graduate School, College of Natural Resources, Department of Forestry, Raleigh, NC 27695. Offers MF, MS, PhD. Part-time programs available. *Degree requirements:* For master's, thesis (for some programs), teaching experience; for doctorate, thesis/dissertation, teaching experience. *Entrance requirements:* For master's and doctorate, GRE General Test. Additional exam requirements/recommendations for international students: Required—TOEFL. Electronic applications accepted. *Faculty research:* Forest genetics, forest ecology and silviculture, forest economics/management/policy, international forestry, remote sensing/geographic information systems.

Northern Arizona University, Consortium of Professional Schools and Colleges, School of Forestry, Flagstaff, AZ 86011. Offers MF, MSF, PhD. Part-time programs available. *Degree requirements:* For master's, thesis optional; for doctorate, thesis/dissertation. *Entrance requirements:* For master's and doctorate, GRE General Test. *Faculty research:* Multiresource management, ecology, entomology, recreation, hydrology.

Oklahoma State University, College of Agricultural Science and Natural Resources, Department of Forestry, Stillwater, OK 74078. Offers M Ag, MS. *Faculty:* 13 full-time (1 woman). *Students:* 5 full-time (3 women), 2 part-time; includes 1 minority (American Indian/Alaska Native), 3 international. Average age 29. 5 applicants, 80% accepted, 4 enrolled. *Degree requirements:* For master's, thesis. *Entrance requirements:* For master's, GRE. Additional exam requirements/recommendations for international students: Required—TOEFL. *Application deadline:* For fall admission, 3/15 priority date for domestic students, 3/1 priority date for international students. Applications are processed on a rolling basis. Application fee: $40 ($75 for international students). Electronic applications accepted. *Expenses:* Tuition, state resident: full-time $4,253; part-time $139 per credit hour. Tuition, nonresident: full-time $12,569; part-time $485 per credit hour. Required fees: $43 per credit hour. One-time fee: $20 part-time. Tuition and fees vary according to course load and program. *Financial support:* In 2005–06, 13 research assistantships (averaging $13,002 per year) were awarded; teaching assistantships, career-related internships or fieldwork, Federal Work-Study, scholarships/grants, health care benefits, tuition waivers (partial), and unspecified assistantships also available. Support available to part-time students. Financial award application deadline: 3/1. *Faculty research:* Forest ecology, upland bird ecology, forest ecophysiology, urban forestry, molecular forest genetics/biotechnology/tree breeding. *Unit head:* Dr. Thomas C. Hennessey, Interim Head, 405-744-5437.

Oregon State University, Graduate School, College of Forestry, Department of Forest Engineering, Corvallis, OR 97331. Offers MAIS, MF, MS, PhD. *Accreditation:* SAF (one or more programs are accredited). Part-time programs available. *Faculty:* 11 full-time (0 women). *Students:* 30 full-time (7 women), 4 international. Average age 28. In 2005, 3 master's, 3 doctorates awarded. *Degree requirements:* For master's and doctorate, thesis/dissertation. *Entrance requirements:* For master's and doctorate, GRE General Test, minimum GPA of 3.0 in last 90 hours of course work. Additional exam requirements/recommendations for international students: Required—TOEFL. *Application deadline:* For fall admission, 3/1 for domestic students. Applications are processed on a rolling basis. Application fee: $50. *Expenses:* Tuition, area resident: Part-time $301 per credit. Tuition, state resident: full-time $8,139; part-time $501 per credit. Tuition, nonresident: full-time $14,376; part-time $532 per credit. Required fees: $1,266. *Financial support:* Fellowships, research assistantships, career-related internships or fieldwork, Federal Work-Study, and institutionally sponsored loans available. Support available to part-time students. Financial award application deadline: 2/1. *Faculty research:* Timber harvesting systems, forest hydrology, slope stability, impacts of harvesting on soil and water, training of logging labor force. *Unit head:* Dr. Steven D. Tesch, Head, 541-737-4952, Fax: 541-737-4316, E-mail: teschs@for.orst.edu. *Application contact:* Rayetta Beall, Office Manager, 541-737-1345, Fax: 541-737-4316, E-mail: rayetta.beall@orst.edu.

Oregon State University, Graduate School, College of Forestry, Department of Forest Resources, Corvallis, OR 97331. Offers economics (MS, PhD); forest resources (MAIS, MF, MS, PhD). MS and PhD programs in economics offered through the University Graduate Faculty of Economics. *Accreditation:* SAF (one or more programs are accredited). Part-time programs available. *Faculty:* 13 full-time (1 woman), 7 part-time/adjunct (2 women). *Students:* 33 full-time (17 women), 5 part-time (3 women); includes 4 minority (1 American Indian/Alaska Native, 1 Asian American or Pacific Islander, 2 Hispanic Americans), 7 international. Average age 30. In 2005, 10 master's, 2 doctorates awarded. Terminal master's awarded for partial completion of doctoral program. *Degree requirements:* For master's, thesis (for some programs); for doctorate, thesis/dissertation. *Entrance requirements:* For master's and doctorate, GRE General Test, minimum GPA of 3.0 in last 90 hours. Additional exam requirements/recommendations for international students: Required—TOEFL. *Application deadline:* For fall admission, 2/1 for domestic students. Applications are processed on a rolling basis. Application fee: $50. *Expenses:* Tuition, area resident: Part-time $301 per credit. Tuition, state resident: full-time $8,139; part-time $501 per credit. Tuition, nonresident: full-time $14,376; part-time $532 per credit. Required fees: $1,266. *Financial support:* Fellowships, research assistantships, teaching assistantships, career-related internships or fieldwork, Federal Work-Study, and institutionally sponsored loans available. Support available to part-time students. Financial award application deadline: 2/1. *Faculty research:* Geographic information systems, long-term productivity, recreation, silviculture, biometrics, policy. *Unit head:* Dr. John D. Walstad, Head, 541-737-3607, Fax: 541-737-3049, E-mail: john.walstad@orst.edu. *Application contact:* Marty Roberts, Coordinator, 541-737-1485, Fax: 541-737-3049, E-mail: roberts@for.orst.edu.

Oregon State University, Graduate School, College of Forestry, Department of Forest Science, Corvallis, OR 97331. Offers MAIS, MF, MS, PhD. *Accreditation:* SAF (one or more programs are accredited). Part-time programs available. *Faculty:* 13 full-time (3 women), 5 part-time/adjunct (0 women). *Students:* 55 full-time (26 women), 4 part-time; includes 3 minority (1 African American, 1 Asian American or Pacific Islander, 1 Hispanic American), 8 international. Average age 31. In 2005, 10 master's, 6 doctorates awarded. *Degree requirements:* For master's, thesis (for some programs); for doctorate, thesis/dissertation. *Entrance requirements:* For master's and doctorate, GRE General Test, minimum GPA of 3.0 in last 90 hours. Additional exam requirements/recommendations for international students: Required—TOEFL. *Application deadline:* For fall admission, 8/25 for domestic students; for spring admission, 3/1 for domestic students. Applications are processed on a rolling basis. Application fee: $50. *Expenses:* Tuition, area resident: Part-time $301 per credit. Tuition, state resident: full-time $8,139; part-time $501 per credit. Tuition, nonresident: full-time $14,376; part-time $532 per credit. Required fees: $1,266. *Financial support:* Fellowships, research assistantships, career-related internships or fieldwork, Federal Work-Study, and institutionally sponsored loans available. Support available to part-time students. Financial award application deadline: 2/1. *Faculty research:* Ecosystem structure and function, nutrient cycling, biotechnology, vegetation management, integrated forest protection. *Unit head:* Dr. W. Thomas Adams, Head, 541-737-6583, Fax: 541-737-1393.

Oregon State University, Graduate School, College of Forestry, Department of Wood Science and Engineering, Corvallis, OR 97331. Offers forest products (MAIS, MF, MS, PhD); wood science and technology (MF, MS, PhD). *Accreditation:* SAF (one or more programs are accredited). Part-time programs available. *Faculty:* 12 full-time (1 woman), 1 part-time/adjunct (0 women). *Students:* 28 full-time (10 women), 3 part-time (1 woman), 16 international. Average age 30. In 2005, 10 master's, 3 doctorates awarded. *Degree requirements:* For master's, thesis (for some programs); for doctorate, thesis/dissertation. *Entrance requirements:* For master's and doctorate, GRE General Test, minimum GPA of 3.0 in last 90 hours. Additional exam requirements/recommendations for international students: Required—TOEFL. *Application deadline:* For fall admission, 3/1 for domestic students. Applications are processed on a rolling basis. Application fee: $50. *Expenses:* Tuition, area resident: Part-time $301 per credit. Tuition, state resident: full-time $8,139; part-time $501 per credit. Tuition, nonresident: full-time $14,376; part-time $532 per credit. Required fees: $1,266. *Financial support:* Fellowships, research assistantships, career-related internships or fieldwork, Federal Work-Study, and institutionally sponsored loans available. Support available to part-time students. Financial award application deadline: 2/1. *Faculty research:* Biodeterioration and preservation, timber engineering, process engineering and control, composite materials science, anatomy, chemistry and physical properties. *Unit head:* Dr. Thomas E. McLain, Head, 541-737-4224, Fax: 541-737-3385, E-mail: thomas.mclain@orst.edu. *Application contact:* George Swanson, Program Support Coordinator, 541-737-4206, Fax: 541-737-3385, E-mail: george.swanson@orst.edu.

The Pennsylvania State University University Park Campus, Graduate School, College of Agricultural Sciences, School of Forest Resources, State College, University Park, PA 16802-1503. Offers forest resources (M Agr, MFR, MS, PhD); wildlife and fisheries sciences (M Agr, MFR, MS, PhD). *Students:* 42 full-time (19 women), 18 part-time (3 women); includes 2 minority (both Hispanic Americans), 4 international. *Entrance requirements:* For master's and doctorate, GRE General Test. *Expenses:* Tuition, state resident: full-time $12,518; part-time $522 per credit. Tuition, nonresident: full-time $23,004; part-time $959 per credit. Required fees: $484. Tuition and fees vary according to course load, campus/location and program. *Unit head:* Dr. Charles H. Strauss, Director, 814-863-7093, Fax: 814-865-3725, E-mail: chs30@psu.edu. *Application contact:* Dr. Charles H. Strauss, Director, 814-863-7093, Fax: 814-865-3725, E-mail: chs30@psu.edu.

Purdue University, Graduate School, College of Agriculture, Department of Forestry and Natural Resources, West Lafayette, IN 47907. Offers aquaculture, fisheries, aquatic science (MSF); aquaculture, fisheries, aquatic sciences (MS, PhD); forest biology (MS, MSF, PhD); natural resources and environmental policy (MS, MSF); natural resources environmental policy (PhD); quantitative resource analysis (MS, MSF, PhD); wildlife science (MS, MSF, PhD); wood science and technology (MS, MSF, PhD). *Faculty:* 26 full-time (4 women), 7 part-time/adjunct (1 woman). *Students:* 69 full-time (33 women), 15 part-time (3 women); includes 5 minority (1 African American, 1 American Indian/Alaska Native, 3 Hispanic Americans), 29 international. Average age 30. 45 applicants, 40% accepted, 18 enrolled. In 2005, 6 master's, 6 doctorates awarded. *Degree requirements:* For master's and doctorate, thesis/dissertation. *Entrance requirements:* For master's and doctorate, GRE General Test (500 verbal, 500 quantitative), minimum B+ average in undergraduate course work. Additional exam requirements/recommendations for international students: Required—TOEFL. *Application deadline:* For fall admission, 1/5 for domestic students; for spring admission, 9/15 for domestic students. Applications are processed on a rolling basis. Application fee: $55. Electronic applications accepted. *Financial support:* In 2005–06, 10 research assistantships (averaging $15,259 per year) were awarded; fellowships, teaching assistantships, career-related internships or fieldwork and scholarships/grants also available. Support available to part-time students. Financial award application deadline: 1/5; financial award applicants required to submit FAFSA. *Faculty research:* Wildlife management, forest management, forest ecology, forest soils, limnology. *Unit head:* Dr. Robert K. Swihart, Interim Head, 765-494-3590, Fax: 765-494-9461, E-mail: rswihart@purdue.edu. *Application contact:* Kelly Garrett, Graduate Secretary, 765-494-3572, Fax: 765-494-9461, E-mail: kgarrett@purdue.edu.

Southern Illinois University Carbondale, Graduate School, College of Agriculture, Department of Forestry, Carbondale, IL 62901-4701. Offers MS. Part-time programs available. *Faculty:* 10 full-time (1 woman). *Students:* 7 full-time (2 women), 29 part-time (8 women); includes 1 minority (African American), 1 international. Average age 24. 10 applicants, 50% accepted, 1 enrolled. In 2005, 7 master's awarded. *Degree requirements:* For master's, thesis. *Entrance requirements:* For master's, minimum GPA of 2.7. Additional exam requirements/recommendations for international students: Required—TOEFL. *Application deadline:* Applications are processed on a rolling basis. Application fee: $0. *Financial support:* In 2005–06, 18 students received support, including 3 fellowships with full tuition reimbursements available, 10 research assistantships with full tuition reimbursements available, 3 teaching assistantships with full tuition reimbursements available; career-related internships or fieldwork, Federal Work-Study, institutionally sponsored loans, and tuition waivers (full) also available. Support available to part-time students. *Faculty research:* Forest recreation, forest ecology, remote sensing, forest management and economics. *Unit head:* John Phelps, Chair, 618-453-3341, E-mail: jphelps@siu.edu.

Southern University and Agricultural and Mechanical College, Graduate School, College of Agricultural, Family and Consumer Sciences, Department of Urban Forestry, Baton Rouge, LA 70813. Offers MS. *Faculty:* 5 full-time (2 women). *Students:* 8 full-time (2 women), 1 (woman) part-time; includes 7 minority (5 African Americans, 2 Asian Americans or Pacific Islanders). Average age 33. 9 applicants, 56% accepted, 4 enrolled. In 2005, 6 degrees awarded. *Degree requirements:* For master's, thesis. *Entrance requirements:* For master's, GRE, minimum GPA of 3.0. Additional exam requirements/recommendations for international students: Required—TOEFL (minimum score 525 paper-based; 193 computer-based). *Application deadline:* For fall admission, 4/15 priority date for domestic students, 4/15 priority date for international students; for spring admission, 11/1 priority date for domestic students, 11/1 priority date for international students. Applications are processed on a rolling basis. Application fee: $25. *Financial support:* In 2005–06, 14 students received support, including 7 research assistantships (averaging $7,000 per year); teaching assistantships, scholarships/grants also available. Financial award application deadline: 4/15. *Faculty research:* Biology of plant pathogen, water resources, plant pathology. Total annual research expenditures: $1 million.

State University of New York College of Environmental Science and Forestry, Faculty of Environmental and Forest Biology, Syracuse, NY 13210-2779. Offers chemical ecology (MPS, MS, PhD); conservation biology (MPS, MS, PhD); ecology (MPS, MS, PhD); entomology (MPS, MS, PhD); environmental interpretation (MPS, MS, PhD); environmental physiology (MPS, MS, PhD); fish and wildlife biology (MPS, MS, PhD); forest pathology and mycology (MPS, MS, PhD); plant science and biotechnology (MPS, MS, PhD). *Faculty:* 27 full-time (4 women), 4 part-time/adjunct (0 women). *Students:* 81 full-time (50 women), 62 part-time (33 women); includes 4 minority (1 Asian American or Pacific Islander, 3 Hispanic Americans), 16 international. Average age 30. 84 applicants, 54% accepted, 17 enrolled. In 2005, 17 master's, 3 doctorates awarded. *Degree requirements:* For master's, thesis (for some programs), registration; for doctorate, thesis/dissertation, comprehensive exam, registration. *Entrance requirements:* For master's and doctorate, GRE General Test, GRE Subject Test, minimum GPA of 3.0. Additional exam requirements/recommendations for international students: Required—TOEFL (minimum score 550 paper-based; 213 computer-based). *Application deadline:* For fall admission, 2/1 priority date for domestic students, 2/1 priority date for international students; for spring admission, 11/1 priority date for domestic students, 11/1 priority date for international students. Applications are processed on a rolling basis. Application fee: $60. *Expenses:* Tuition, area resident: Full-time $6,900; part-time $288 per credit. Tuition, nonresident: full-time $10,920; part-time $455 per credit. Required fees: $395; $32 per credit. $20 per term. One-time fee: $145. *Financial support:* In 2005–06, 86 students received support, including 13 fellowships with full and partial tuition reimbursements available (averaging $9,446 per year), 40 research assistantships with full and partial tuition reimbursements available (averaging $11,000 per year), 32 teaching assistantships with full and partial tuition reimbursements available (averaging $9,446 per year); Federal Work-Study, institutionally sponsored loans, scholarships/grants, health care benefits, and unspecified assistantships also available. Financial award application deadline: 6/30. *Faculty research:* Ecology, fish and wildlife biology and management, plant science, entomology. Total annual research expenditures: $4.1 million. *Unit head:* Dr. Donald J. Leopold, Chair, 315-470-6770, Fax: 315-470-6934, E-mail: dendro@esf.edu. *Application contact:* Dr. Dudley J. Raynal, Dean, Instruction and Graduate Studies, 315-470-6599, Fax: 315-470-6978, E-mail: esfgrad@esf.edu.

State University of New York College of Environmental Science and Forestry, Faculty of Forest and Natural Resources Management, Syracuse, NY 13210-2779. Offers environmental and natural resource policy (MS, PhD); environmental and natural resources policy (MPS); forest management and operations (MF); forestry ecosystems science and applications (MPS, MS, PhD); natural resources management (MPS, MS, PhD); quantitative methods and management in forest science (MPS, MS, PhD); recreation and resource management (MPS, MS, PhD); watershed management and forest hydrology (MPS, MS, PhD). *Faculty:* 30 full-time (7 women), 1 (woman) part-time/adjunct. *Students:* 43 full-time (18 women), 26 part-time (12 women); includes 1 minority (Hispanic American), 13 international. Average age 31. 38 applicants, 55% accepted, 12 enrolled. In 2005, 13 master's, 4 doctorates awarded. *Degree requirements:* For master's, thesis (for some programs), registration; for doctorate, thesis/dissertation, comprehensive exam, registration. *Entrance requirements:* For master's and doctorate, GRE General Test, minimum GPA of 3.0. Additional exam requirements/recommendations for international students: Required—TOEFL (minimum score 550 paper-based; 213 computer-based). *Application deadline:* For fall admission, 2/1 priority date for domestic students, 2/1 priority date for international students; for spring admission, 11/1 priority date for domestic students, 11/1 priority date for international students. Applications are processed on a rolling basis. Application fee: $60. *Expenses:* Tuition, area resident: Full-time $6,900; part-time $288 per credit. Tuition, nonresident: full-time $10,920; part-time $455 per credit. Required fees: $395; $32 per credit. $20 per term. One-time fee: $145. *Financial support:* In 2005–06, 43 students received support, including 8 fellowships with full and partial tuition reimbursements available (averaging $9,446 per year), 20 research assistantships with full and partial tuition reimbursements available (averaging $10,000 per year), 11 teaching assistantships with full and partial tuition reimbursements available (averaging $9,446 per year); career-related internships or fieldwork, Federal Work-Study, institutionally sponsored loans, scholarships/grants, health care benefits, and unspecified assistantships also available. Financial award application deadline: 6/30; financial award applicants required to submit FAFSA. *Faculty research:* Silviculture recreation management, tree improvement, operations management, economics. Total annual research expenditures: $2.1 million. *Unit head:* Dr. Chad P. Dawson, Chair, 315-470-6536, Fax: 315-470-6535, E-mail: cpdawson@esf.edu. *Application contact:* Dr. Dudley J. Raynal, Dean, Instruction and Graduate Studies, 315-470-6599, Fax: 315-470-6978, E-mail: esfgrad@esf.edu.

Stephen F. Austin State University, Graduate School, College of Forestry and Agriculture, Department of Forestry, Nacogdoches, TX 75962. Offers MF, MS, PhD. Part-time programs available. *Faculty:* 20 full-time (2 women), 24 part-time/adjunct (2 women). *Students:* 26 full-time (13 women), 68 part-time (40 women); includes 5 minority (2 African Americans, 1 American Indian/Alaska Native, 2 Hispanic Americans), 4 international. 61 applicants, 89% accepted. In 2005, 7 master's, 1 doctorate awarded. *Degree requirements:* For master's and doctorate, thesis/dissertation. *Entrance requirements:* For master's and doctorate, GRE General Test. Additional exam requirements/recommendations for international students: Required—TOEFL. *Application deadline:* For fall admission, 8/1 for domestic students; for spring admission, 12/15 for domestic students. Applications are processed on a rolling basis. Application fee: $25 ($50 for international students). *Expenses:* Tuition, state resident: full-time $2,628; part-time $146 per credit hour. Tuition, nonresident: full-time $7,596; part-time $422 per credit hour. Required fees: $900; $170. *Financial support:* In 2005–06, 20 research assistantships (averaging $13,000 per year), 4 teaching assistantships (averaging $8,100 per year) were awarded; career-related internships or fieldwork and Federal Work-Study also available. Support available to part-time students. Financial award application deadline: 3/1. *Faculty research:* Wildlife management, basic plant science, forest recreation, multipurpose land management. *Unit head:* Dr. Scott Beasley, Dean, 936-468-3304, E-mail: sbeasley@sfasu.edu.

Texas A&M University, College of Agriculture and Life Sciences, Department of Forest Science, College Station, TX 77843. Offers forestry (MS, PhD); natural resources development (M Agr). Part-time programs available. *Faculty:* 6 full-time (3 women). *Students:* 11 full-time (4 women), 4 part-time (2 women); includes 4 minority (3 African Americans, 1 Hispanic American), 5 international. Average age 27. 5 applicants, 80% accepted, 2 enrolled. In 2005, 4 master's, 2 doctorates awarded. Terminal master's awarded for partial completion of doctoral program. *Degree requirements:* For master's, thesis (for some programs); for doctorate, thesis/dissertation. *Entrance requirements:* For master's and doctorate, GRE General Test. Additional exam requirements/recommendations for international students: Required—TOEFL. *Application deadline:* For fall admission, 3/1 for domestic students; for spring admission, 11/1 priority date for domestic students. Applications are processed on a rolling basis. Application fee: $50 ($75 for international students). Electronic applications accepted. *Expenses:* Tuition, state resident: full-time $4,488; part-time $187 per credit hour. Tuition, nonresident: full-time $11,112; part-time $463 per credit hour. Required fees:$1,974. *Financial support:* In 2005–06, fellowships with partial tuition reimbursements (averaging $15,000 per year), research assistantships with partial tuition reimbursements (averaging $15,000 per year), teaching assistantships with partial tuition reimbursements (averaging $15,000 per year) were awarded; career-related internships or fieldwork and institutionally sponsored loans also available. Support available to part-time students. Financial award application deadline: 3/1; financial award applicants required to submit FAFSA. *Faculty research:* Expert systems, geographic information systems, economics, biology, genetics. *Unit head:* Dr. C. T. Smith, Professor and Head, 979-845-5033, Fax: 979-845-6049, E-mail: tat-smith@tamu.edu. *Application contact:* Dr. Carol Loopstra, Associate Head for Research and Graduate Studies, 979-862-2200, Fax: 979-845-6049, E-mail: c-loopstra@tamu.edu.

Tropical Agriculture Research and Higher Education Center, Graduate School, Turrialba, Costa Rica. Offers agroforestry (PhD); ecological agriculture (PhD); ecological agriculture (MS); environmental socioeconomics (MS, PhD); forest management and conservation (MS); forest sciences (PhD); tropical agroforestry (MS); watershed management (MS, PhD). *Entrance requirements:* For master's, GRE, letters of recommendation; for doctorate, GRE, curriculum vitae, letters of recommendation. Additional exam requirements/recommendations for international students: Required—TOEFL (minimum score 550 paper-based; 213 computer-based). Electronic applications accepted. *Faculty research:* Biodiversity in fragmented landscapes, ecosystem management, integrated pest management, environmental livestock production, biotechnology carbon balances in diverse land uses.

Université Laval, Faculty of Forestry and Geomatics, Department of Wood and Forest Sciences, Programs in Forestry Sciences, Québec, QC G1K 7P4, Canada. Offers M Sc, PhD. Terminal master's awarded for partial completion of doctoral program. *Degree requirements:* For master's, thesis (for some programs); for doctorate, thesis/dissertation, comprehensive exam. *Entrance requirements:* For master's and doctorate, knowledge of French. Additional exam requirements/recommendations for international students: Required—TOEIC or TOEFL. Electronic applications accepted.

Université Laval, Faculty of Forestry and Geomatics, Department of Wood and Forest Sciences, Programs in Wood Sciences, Québec, QC G1K 7P4, Canada. Offers M Sc, PhD. Terminal master's awarded for partial completion of doctoral program. *Degree requirements:* For master's, thesis/dissertation; for doctorate, thesis/dissertation, comprehensive exam. *Entrance requirements:* For master's and doctorate, knowledge of French. Electronic applications accepted.

Université Laval, Faculty of Forestry and Geomatics, Program in Agroforestry, Québec, QC G1K 7P4, Canada. Offers M Sc. *Degree requirements:* For master's, thesis (for some programs). *Entrance requirements:* For master's, English exam (comprehension of English), knowledge of French, knowledge of a third language. Electronic applications accepted.

University of Alberta, Faculty of Graduate Studies and Research, Department of Rural Economy, Edmonton, AB T6G 2E1, Canada. Offers agricultural economics (M Ag, M Sc, PhD); forest economics (M Ag, M Sc, PhD); rural sociology (M Ag, M Sc). Part-time programs available. *Faculty:* 13 full-time (1 woman), 6 part-time/adjunct (0 women). *Students:* 31 full-time (13 women), 21 part-time (11 women). Average age 25. 35 applicants, 83% accepted. In 2005, 10 master's, 2 doctorates awarded. *Degree requirements:* For doctorate, thesis/dissertation. *Entrance requirements:* Additional exam requirements/recommendations for international students: Required—TOEFL. Application fee: $60. Tuition and fees charges are reported in Canadian dollars. *Expenses:* Tuition, state resident: part-time $562 Canadian dollars per term. Tuition, nonresident: full-time $3,375 Canadian dollars. Required fees: $573 Canadian dollars; $84 Canadian dollars per term. *Financial support:* In 2005–06, 4 fellowships, 12 research assistantships, 2 teaching assistantships were awarded; scholarships/grants also available. *Faculty research:* Agroforestry, development, extension education, marketing and trade, natural resources and environment, policy, production economics. Total annual research expenditures: $850,000. *Unit head:* Dr. V. Adamowicz, Graduate Coordinator, 403-492-4225, Fax: 403-492-0268. *Application contact:* Liz Bruce, Graduate Secretary, 780-492-4225, Fax: 780-492-0268, E-mail: rural.economy@ualberta.ca.

The University of Arizona, Graduate College, College of Agriculture and Life Sciences, School of Natural Resources, Watershed Resources Program, Tucson, AZ 85721. Offers MS, PhD. *Degree requirements:* For master's and doctorate, thesis/dissertation, comprehensive exam, registration. *Entrance requirements:* For master's and doctorate, GRE General Test, minimum GPA of 3.0, 3 letters of recommendation. Additional exam requirements/recommendations for international students: Required—TOEFL (minimum score 550 paper-based; 213 computer-based). *Faculty research:* Forest fuel characteristics, prescribed fire, tree ring-fire scar anaylsis, erosion, sedimentation.

University of Arkansas at Monticello, School of Forest Resources, Monticello, AR 71656. Offers MS. Part-time programs available. *Degree requirements:* For master's, thesis, comprehensive exam. *Entrance requirements:* For master's, GRE General Test, minimum GPA of 2.7. Additional exam requirements/recommendations for international students: Required—TOEFL (minimum score 550 paper-based; 213 computer-based). *Faculty research:* Geographic information systems/remote sensing, forest ecology, wildlife ecology and management.

The University of British Columbia, Faculty of Graduate Studies, Faculty of Forestry, Vancouver, BC V6T 1Z1, Canada. Offers M Sc, MA Sc, MF, PhD. Part-time programs available. *Degree requirements:* For master's, thesis, thesis or comprehensive exam; for doctorate, thesis/dissertation, thesis exam, comprehensive exam, registration. *Entrance requirements:* Additional exam requirements/recommendations for international students: Required—TOEFL (minimum score 550 paper-based; 213 computer-based). Electronic applications accepted. *Faculty research:* Forest sciences, forest resources management, forest operations, wood sciences, conservation.

University of California, Berkeley, Graduate Division, College of Natural Resources, Department of Environmental Science, Policy, and Management, Berkeley, CA 94720-1500. Offers environmental science, policy, and management (MS, PhD); forestry (MF). Terminal master's awarded for partial completion of doctoral program. *Degree requirements:* For master's, thesis optional; for doctorate, thesis/dissertation, qualifying exam. *Entrance requirements:* For master's and doctorate, GRE General Test, minimum GPA of 3.0. Additional exam requirements/recommendations for international students: Required—TOEFL; Recommended—TSE. Electronic applications accepted. *Faculty research:* Biology and ecology of insects; ecosystem function and environmental issues of soils; plant health/interactions from molecular to ecosystem levels; range management and ecology; forest and resource policy, sustainability, and management.

University of California, Berkeley, Graduate Division, Group in Wood Science and Technology, Berkeley, CA 94720-1500. Offers MS, PhD. *Degree requirements:* For doctorate, thesis/dissertation, qualifying exam. *Entrance requirements:* For master's and doctorate, GRE General Test, minimum GPA of 3.0. Additional exam requirements/recommendations for international students: Required—TOEFL.

University of Florida, Graduate School, College of Agricultural and Life Sciences, School of Forest Resources and Conservation, Gainesville, FL 32611. Offers MFRC, MS, PhD, JD/MFRC, JD/MS, JD/PhD. Part-time programs available. *Faculty:* 24 full-time (3 women). *Students:* 61 (30 women); includes 8 minority (2 African Americans, 1 American Indian/Alaska Native, 2 Asian Americans or Pacific Islanders, 3 Hispanic Americans) 23 international. Average age 24. 101 applicants, 35% accepted. In 2005, 15 master's, 2 doctorates awarded. *Degree requirements:* For master's, project (MFRC), thesis defense (MS); for doctorate, thesis/dissertation, qualifying exams, defense. *Entrance requirements:* For master's and doctorate, GRE General Test, minimum GPA of 3.0. Additional exam requirements/recommendations for international students: Required—TOEFL. *Application deadline:* For fall admission, 6/1 for domestic students; for spring admission, 10/1 for domestic students. Applications are processed on a rolling basis. Application fee: $20. Electronic applications accepted. *Expenses:* Tuition, state resident: full-time $6,234. Tuition, nonresident: full-time $21,359. Tuition and fees vary according to program. *Financial support:* In 2005–06, 35 research assistantships with full tuition reimbursements (averaging $15,546 per year) were awarded; fellowships with full tuition reimbursements, teaching assistantships with full tuition reimbursements, Federal Work-Study and institutionally sponsored loans also available. Support available to part-time students. *Faculty research:* Forest biology and ecology; agroforestry and tropical forestry; forest management, economics, and policy; natural resource education and ecotourism. *Unit head:* Dr. Tim White, Director, 352-846-0850, Fax: 352-392-1707, E-mail: tlwhite@ufl.edu. *Application contact:* Dr. George M. Blakeslee, Graduate Coordinator, 352-846-0845, Fax: 352-392-1707, E-mail: gb4stree@ufl.edu.

University of Georgia, Graduate School, School of Forest Resources, Athens, GA 30602. Offers MFR, MS, PhD. *Faculty:* 35 full-time (2 women). *Students:* 145 full-time, 22 part-time; includes 9 minority (5 African Americans, 1 American Indian/Alaska Native, 2 Asian Americans or Pacific Islanders, 1 Hispanic American), 23 international. 77 applicants, 71% accepted, 37 enrolled. In 2005, 31 master's, 6 doctorates awarded. *Degree requirements:* For master's, thesis (MS); for doctorate, one foreign language, thesis/dissertation. *Entrance requirements:* For master's and doctorate, GRE General Test. *Application deadline:* For fall admission, 7/1 for domestic students; for spring admission, 11/15 for domestic students. Application fee: $50. Electronic applications accepted. *Financial support:* Fellowships, research assistantships, teaching assistantships, unspecified assistantships available. *Unit head:* Dr. Richard L. Porterfield, Dean, 706-542-2866, Fax: 706-542-2281. *Application contact:* Dr. Barry D. Shiver, Graduate Coordinator, 706-542-3009, Fax: 706-542-8356, E-mail: shiver@smokey.forestry.uga.edu.

University of Idaho, College of Graduate Studies, College of Natural Resources, Department of Forest Products, Moscow, ID 83844-2282. Offers MS, PhD. *Students:* 10 full-time (1 woman), 4 part-time, 8 international. Average age 29. In 2005, 3 degrees awarded. *Degree requirements:* For doctorate, thesis/dissertation. *Entrance requirements:* For master's, minimum GPA of 2.8; for doctorate, minimum undergraduate GPA of 2.8, 3.0 graduate. *Application deadline:* For fall admission, 8/1 for domestic students; for spring admission, 12/15 for domestic students. Application fee: $55 ($60 for international students). *Expenses:* Tuition, nonresident: full-time $8,770; part-time $130 per credit. Required fees: $4,508; $217 per credit. *Financial support:* Research assistantships, teaching assistantships available. Financial award application deadline: 2/15. *Unit head:* Dr. Thomas M. Gorman, Head, 208-885-7402.

University of Idaho, College of Graduate Studies, College of Natural Resources, Department of Forest Resources, Moscow, ID 83844-2282. Offers MS, PhD. *Students:* 24 full-time (10 women), 14 part-time (3 women); includes 2 minority (both American Indian/Alaska Native), 8 international. Average age 31. In 2005, 8 degrees awarded. *Degree requirements:* For doctorate, thesis/dissertation. *Entrance requirements:* For master's, minimum GPA of 2.8; for doctorate, minimum undergraduate GPA of 2.8, 3.0 graduate. *Application deadline:* For fall admission, 8/1 for domestic students; for spring admission, 12/15 for domestic students. Application fee: $55 ($60 for international students). *Expenses:* Tuition, nonresident: full-time $8,770; part-time $130 per credit. Required fees: $4,508; $217 per credit. *Financial support:* Research assistantships, teaching assistantships available. Financial award application deadline: 2/15. *Unit head:* Dr. JoEllen Force, Head, 208-885-7311.

Forestry

University of Kentucky, Graduate School, Graduate School Programs in the College of Agriculture, Program in Forestry, Lexington, KY 40506-0032. Offers MSFOR. *Faculty:* 10 full-time (2 women), 1 part-time/adjunct (0 women). *Students:* 17 full-time (7 women), 6 part-time. Average age 29. 18 applicants, 50% accepted, 7 enrolled. In 2005, 4 degrees awarded. *Degree requirements:* For master's, thesis optional. *Entrance requirements:* For master's, GRE General Test, minimum undergraduate GPA of 3.0. Additional exam requirements/recommendations for international students: Required—TOEFL (minimum score 550 paper-based; 213 computer-based). *Application deadline:* For fall admission, 7/17 priority date for domestic students, 2/1 priority date for international students; for spring admission, 12/13 priority date for domestic students, 6/15 priority date for international students. Applications are processed on a rolling basis. Application fee: $40 ($55 for international students). Electronic applications accepted. *Expenses:* Tuition, state resident: full-time $3,159; part-time $331 per credit hour. Tuition, nonresident: full-time $6,984; part-time $756 per credit hour. Tuition and fees vary according to course load, degree level and program. *Financial support:* In 2005–06, 17 students received support, including 16 research assistantships with full tuition reimbursements available (averaging $12,000 per year), 1 teaching assistantship with full tuition reimbursement available (averaging $7,500 per year); fellowships with full tuition reimbursements available, career-related internships or fieldwork, Federal Work-Study, institutionally sponsored loans, scholarships/grants, traineeships, health care benefits, tuition waivers (partial), and unspecified assistantships also available. Support available to part-time students. Financial award application deadline: 3/15; financial award applicants required to submit FAFSA. *Faculty research:* Forest ecology, silviculture, watershed management, forest products utilization, wildlife habitat management. *Unit head:* Dr. David Wagner, Director of Graduate Studies, 859-257-3773, Fax: 859-323-1031, E-mail: dwagner@uky.edu. *Application contact:* Dr. Brian Jackson, Senior Associate Dean, 859-257-8176, Fax: 859-323-1928, E-mail: lance.brunner@uky.edu.

University of Maine, Graduate School, College of Natural Sciences, Forestry, and Agriculture, Department of Forest Management and Forest Ecosystem Science, Orono, ME 04469. Offers forest resources (PhD); forestry (MF, MS). *Accreditation:* SAF (one or more programs are accredited). Part-time programs available. *Students:* 30 full-time (11 women), 16 part-time (7 women); includes 1 minority (American Indian/Alaska Native), 10 international. Average age 33. 38 applicants, 55% accepted, 14 enrolled. In 2005, 9 master's, 2 doctorates awarded. *Degree requirements:* For master's, thesis; for doctorate, one foreign language, thesis/dissertation. *Entrance requirements:* For master's and doctorate, GRE General Test. Additional exam requirements/recommendations for international students: Required—TOEFL. *Application deadline:* For fall admission, 2/1 for domestic students. Applications are processed on a rolling basis. Application fee: $50. Electronic applications accepted. *Financial support:* In 2005–06, research assistantships with tuition reimbursements (averaging $15,000 per year), teaching assistantships with tuition reimbursements (averaging $12,276 per year) were awarded; fellowships, career-related internships or fieldwork, Federal Work-Study, and institutionally sponsored loans also available. Financial award application deadline: 3/1. *Faculty research:* Forest economics, engineering and operations analysis, biometrics and remote sensing, timber management, wood technology. *Unit head:* Dr. William Livingston, Chair, 207-581-2990. *Application contact:* Scott G. Delcourt, Associate Dean of the Graduate School, 207-581-3219, Fax: 207-581-3232, E-mail: graduate@maine.edu.

University of Maine, Graduate School, College of Natural Sciences, Forestry, and Agriculture, Department of Plant, Soil, and Environmental Sciences, Orono, ME 04469. Offers biological sciences (PhD); ecology and environmental sciences (MS, PhD); forest resources (PhD); horticulture (MS); plant science (PhD); plant, soil, and environmental sciences (MS); resource utilization (MS). *Faculty:* 25. *Students:* 15 full-time (8 women), 8 part-time (6 women), 3 international. Average age 32. 6 applicants, 33% accepted, 1 enrolled. In 2005, 9 master's awarded. *Entrance requirements:* For master's and doctorate, GRE General Test. Additional exam requirements/recommendations for international students: Required—TOEFL. *Application deadline:* Applications are processed on a rolling basis. Application fee: $50. Electronic applications accepted. *Financial support:* In 2005–06, 9 research assistantships with tuition reimbursements (averaging $12,180 per year) were awarded; teaching assistantships, scholarships/grants, tuition waivers (full and partial), and unspecified assistantships also available. *Unit head:* Greg Porter, Chair, 207-581-2943, Fax: 207-581-3207. *Application contact:* Scott G. Delcourt, Associate Dean of the Graduate School, 207-581-3219, Fax: 207-581-3232, E-mail: graduate@maine.edu.

University of Massachusetts Amherst, Graduate School, College of Natural Resources and the Environment, Department of Natural Resources Conservation, Program in Forest Resources, Amherst, MA 01003. Offers MS, PhD. Part-time programs available. *Students:* 11 full-time (2 women), 11 part-time (3 women); includes 1 minority (African American), 4 international. Average age 35. 16 applicants, 50% accepted, 7 enrolled. In 2005, 8 master's, 3 doctorates awarded. Terminal master's awarded for partial completion of doctoral program. *Degree requirements:* For master's, thesis or alternative; for doctorate, variable foreign language requirement, thesis/dissertation. *Entrance requirements:* For master's and doctorate, GRE General Test. Additional exam requirements/recommendations for international students: Required—TOEFL (minimum score 530 paper-based; 197 computer-based). *Application deadline:* For fall admission, 2/1 priority date for domestic students, 2/1 priority date for international students; for spring admission, 10/1 for domestic students, 10/1 for international students. Applications are processed on a rolling basis. Application fee: $40 ($65 for international students). Electronic applications accepted. *Expenses:* Tuition, state resident: part-time $110 per credit. Tuition, nonresident: part-time $414 per credit. Required fees: $2,824 per term. One-time fee: $250 part-time. Full-time tuition and fees vary according to course load, campus/location, program and reciprocity agreements. *Financial support:* Fellowships with full tuition reimbursements, research assistantships with full tuition reimbursements, teaching assistantships with full tuition reimbursements, career-related internships or fieldwork, Federal Work-Study, scholarships/grants, traineeships, and unspecified assistantships available. Support available to part-time students. Financial award application deadline: 2/1. *Unit head:* Dr. William Patterson, Director, 413-545-2666, Fax: 413-545-4358.

University of Michigan, School of Natural Resources and Environment, Ann Arbor, MI 48109-1115. Offers industrial ecology (Certificate); landscape architecture (MLA, PhD); resource ecology and management (MS, PhD), including natural resources and environment (PhD), resource ecology and management (MS); resource policy and behavior (MS, PhD), including natural resources and environment (PhD), resource policy and behavior (MS); spatial analysis (Certificate). MLA, MS, PhD, and JD/MS offered through the Horace H. Rackham School of Graduate Studies. *Accreditation:* ASLA (one or more programs are accredited); SAF (one or more programs are accredited). *Degree requirements:* For master's, thesis or alternative, thesis, practicum, or group project; for doctorate, thesis/dissertation, oral defense of dissertation, preliminary exam, comprehensive exam, registration. *Entrance requirements:* For master's, GRE General Test; for doctorate, GRE General Test, master's degree. Additional exam requirements/recommendations for international students: Required—TOEFL (paper score 560; computer score 220) or IELTS. Electronic applications accepted. *Expenses:* Tuition, state resident: full-time $14,082; part-time $894 per credit hour. Tuition, nonresident: full-time $28,500; part-time $1,675 per credit hour. Required fees: $189; $189 per unit. *Faculty research:* Ecology, environmental policy, landscape architecture, climate change, Great Lakes, sustainable systems.

University of Minnesota, Twin Cities Campus, Graduate School, College of Natural Resources, Department of Forest Resources, Minneapolis, MN 55455-0213. Offers natural resources science and management (MS, PhD). Part-time programs available. Terminal master's awarded for partial completion of doctoral program. *Degree requirements:* For master's, thesis optional; for doctorate, thesis/dissertation, comprehensive exam, registration. *Entrance requirements:* For master's and doctorate, GRE. Additional exam requirements/recommendations for international students: Required—TOEFL (minimum score 550 paper-based; 213 computer-based). Electronic applications accepted. *Expenses:* Tuition, state resident: full-time $8,748; part-time $729 per credit. Tuition, nonresident: full-time $15,848; part-time $1,321 per credit. Full-time tuition and fees vary according to class time, course load, program and reciprocity agreements.

University of Missouri–Columbia, Graduate School, School of Natural Resources, Department of Forestry, Columbia, MO 65211. Offers MS, PhD. *Faculty:* 15 full-time (2 women). *Students:* 13 full-time (4 women), 19 part-time (6 women), 8 international. In 2005, 4 degrees awarded. Terminal master's awarded for partial completion of doctoral program. *Degree requirements:* For master's and doctorate, thesis/dissertation. *Entrance requirements:* For master's and doctorate, GRE General Test, minimum GPA of 3.0. *Application deadline:* Applications are processed on a rolling basis. Application fee: $45 ($60 for international students). *Financial support:* Fellowships, research assistantships, teaching assistantships, institutionally sponsored loans and scholarships/grants available. *Unit head:* Dr. Bruce E. Cutter, Director of Graduate Studies, 573-882-2744, E-mail: cutterb@missouri.edu.

The University of Montana, Graduate School, College of Forestry and Conservation, Missoula, MT 59812-0002. Offers ecosystem management (MEM, MS); fish and wildlife biology (PhD); forestry (MS, PhD); recreation management (MS); resource conservation (MS); wildlife biology (MS). *Faculty:* 34 full-time (4 women). *Students:* 95 full-time (50 women), 63 part-time (21 women); includes 8 minority (1 African American, 3 American Indian/Alaska Native, 1 Asian American or Pacific Islander, 3 Hispanic Americans), 19 international. 175 applicants, 35% accepted, 46 enrolled. In 2005, 23 master's, 9 doctorates awarded. *Degree requirements:* For doctorate, thesis/dissertation. *Entrance requirements:* For master's and doctorate, GRE General Test. Additional exam requirements/recommendations for international students: Required—TOEFL (minimum score 575 paper-based; 213 computer-based). *Application deadline:* For fall admission, 1/31 for domestic students; for spring admission, 8/31 priority date for domestic students. Applications are processed on a rolling basis. Application fee: $45. *Expenses:* Tuition, state resident: part-time $267 per credit. Tuition, nonresident: part-time $665 per credit. Part-time tuition and fees vary according to course load and degree level. *Financial support:* In 2005–06, 25 research assistantships with tuition reimbursements, 12 teaching assistantships with full tuition reimbursements were awarded; fellowships, career-related internships or fieldwork and Federal Work-Study also available. Financial award application deadline: 3/1; financial award applicants required to submit FAFSA. Total annual research expenditures: $6.7 million. *Unit head:* Dr. Perry Brown, Dean, 406-243-5521, Fax: 406-243-4845, E-mail: pbrown@forestry.umt.edu.

University of New Brunswick Fredericton, School of Graduate Studies, Faculty of Forestry and Environmental Management, Fredericton, NB E3B 6C2, Canada. Offers ecological foundations of forest management (PhD); forest engineering (M Sc FE, MFE); forest resources (M Sc F, MF, PhD). Part-time programs available. *Degree requirements:* For master's and doctorate, thesis/dissertation. *Entrance requirements:* For master's and doctorate, minimum GPA of 3.0. Additional exam requirements/recommendations for international students: Required—TOEFL, TWE. *Faculty research:* Genetics; soils; tree improvement, development, reproduction, physiology, and biotechnology; insect ecology; entomology.

University of New Hampshire, Graduate School, College of Life Sciences and Agriculture, Department of Natural Resources, Durham, NH 03824. Offers environmental conservation (MS); forestry (MS); soil science (MS); water resources management (MS); wildlife (MS). Part-time programs available. *Faculty:* 40 full-time. *Students:* 31 full-time (17 women), 36 part-time (17 women); includes 2 minority (both Asian Americans or Pacific Islanders), 4 international. Average age 31. 50 applicants, 44% accepted, 15 enrolled. In 2005, 14 degrees awarded. *Degree requirements:* For master's, thesis or alternative. *Entrance requirements:* For master's, GRE General Test. Additional exam requirements/recommendations for international students: Required—TOEFL (minimum score 550 paper-based; 213 computer-based); Recommended—TSE. *Application deadline:* For fall admission, 4/1 for domestic students, 4/1 for international students. For winter admission, 12/1 for domestic students. Applications are processed on a rolling basis. Application fee: $60. Electronic applications accepted. *Expenses:* Tuition, state resident: full-time $8,010; part-time $445 per credit hour. Tuition, nonresident: full-time $19,730; part-time $810 per credit hour. Required fees: $322 per semester. Tuition and fees vary according to course load and program. *Financial support:* In 2005–06, 3 fellowships, 21 research assistantships, 15 teaching assistantships were awarded; career-related internships or fieldwork, Federal Work-Study, scholarships/grants, and tuition waivers (full and partial) also available. Support available to part-time students. Financial award application deadline: 2/15. *Unit head:* Dr. William H. McDowell, Chairperson, 603-862-2249, E-mail: tehoward@cisunix.unh.edu. *Application contact:* Linda Scogin, Administrative Assistant, 603-862-3932, E-mail: natural.resources @unh.edu.

The University of Tennessee, Graduate School, College of Agricultural Sciences and Natural Resources, Department of Forestry, Wildlife, and Fisheries, Program in Forestry, Knoxville, TN 37996. Offers MS. *Degree requirements:* For master's, thesis or alternative. *Entrance requirements:* For master's, GRE General Test, minimum GPA of 2.7. Additional exam requirements/recommendations for international students: Required—TOEFL. Electronic applications accepted.

University of Toronto, School of Graduate Studies, Life Sciences Division, Faculty of Forestry, Toronto, ON M5S 1A1, Canada. Offers M Sc F, MFC, PhD. *Degree requirements:* For master's, thesis, oral thesis/research paper defense, comprehensive exam; for doctorate, thesis/dissertation, oral defense of thesis. *Entrance requirements:* For master's, bachelor's degree in a related area, minimum B average in final year (M Sc F), final 2 years (MFC); resumé, 3 letters of reference; for doctorate, writing sample, minimum A– average, master's in a related area, 3 letters of reference, resumé.

University of Vermont, Graduate College, The Rubenstein School of Environment and Natural Resources, Program in Natural Resources, Burlington, VT 05405. Offers natural resources (MS, PhD), including aquatic ecology and watershed science (MS), environment thought and culture (MS), environment, science and public affairs (MS), forestry (MS). *Degree requirements:* For master's, thesis or alternative; for doctorate, thesis/dissertation. *Entrance requirements:* For master's and doctorate, GRE General Test. Additional exam requirements/recommendations for international students: Required—TOEFL (minimum score 550 paper-based; 213 computer-based). *Application deadline:* For fall admission, 3/1 for domestic students. Applications are processed on a rolling basis. Application fee: $25. Electronic applications accepted. *Expenses:* Tuition, area resident: Part-time $410 per credit hour. Tuition, nonresident: part-time $1,034 per credit hour. *Financial support:* Fellowships, research assistantships, teaching assistantships available. Financial award application deadline: 3/1. *Unit head:* Dr. Deane Wang, Coordinator, 802-656-2620.

University of Washington, Graduate School, College of Forest Resources, Seattle, WA 98195. Offers forest economics (MS, PhD); forest ecosystem analysis (MS, PhD); forest engineering/forest hydrology (MS, PhD); forest products marketing (MS, PhD); forest soils (MS, PhD); paper science and engineering (MS, PhD); quantitative resource management (MS, PhD); silviculture (MFR); silviculture and forest protection (MS, PhD); social sciences (MS, PhD); urban horticulture (MFR, MS, PhD); wildlife science (MS, PhD). *Degree requirements:* For master's, thesis (for some programs), registration; for doctorate, thesis/dissertation, comprehensive exam (for some programs), registration. *Entrance requirements:* For master's and doctorate, GRE, minimum GPA of 3.0. Additional exam requirements/recommendations for international students: Required—TOEFL. Electronic applications accepted. *Faculty research:* Ecosystem analysis, silviculture and forest protection, paper science and engineering, environmental horticulture and urban forestry, natural resource policy and economics.

University of Wisconsin–Madison, Graduate School, College of Agricultural and Life Sciences, Department of Forest Ecology and Management, Madison, WI 53706-1380. Offers forest science (MS, PhD); forestry (PhD). Part-time programs available. *Faculty:* 15 full-time (1 woman). *Students:* 51 full-time, 3 part-time; includes 9 minority (4 Asian Americans or Pacific Islanders, 5 Hispanic Americans), 12 international. Average age 28. 48 applicants, 19%

accepted, 8 enrolled. *Degree requirements:* For master's, thesis (for some programs); for doctorate, thesis/dissertation. *Entrance requirements:* For master's and doctorate, GRE. Additional exam requirements/recommendations for international students: Required—TOEFL. *Application deadline:* For fall admission, 6/15 for domestic students; for spring admission, 10/15 priority date for domestic students. Applications are processed on a rolling basis. Application fee: $45. Electronic applications accepted. *Financial support:* In 2005–06, 38 research assistantships with full tuition reimbursements (averaging $18,120 per year) were awarded; fellowships with full tuition reimbursements, career-related internships or fieldwork, Federal Work-Study, institutionally sponsored loans, health care benefits, and unspecified assistantships also available. Support available to part-time students. *Faculty research:* Forest and landscape ecology, forest biology, social forestry, recreation resources, wood science. Total annual research expenditures: $2.5 million. *Unit head:* Raymond P. Guries, Chair, 608-262-0449, Fax: 608-262-9922, E-mail: rpguries@wisc.edu. *Application contact:* Diane Walton, Program Assistant, 608-262-9975, Fax: 608-262-9922, E-mail: dwalton@wisc.edu.

Utah State University, School of Graduate Studies, College of Natural Resources, Department of Forest, Range, and Wildlife Sciences, Logan, UT 84322. Offers ecology (MS, PhD); forestry (MS, PhD); range science (MS, PhD); wildlife biology (MS, PhD). Part-time programs available. *Faculty:* 22 full-time (4 women), 17 part-time/adjunct (3 women). *Students:* 220 full-time (90 women), 52 part-time (17 women); includes 6 minority (all Asian Americans or Pacific Islanders), 49 international. Average age 24. 56 applicants, 61% accepted, 28 enrolled. In 2005, 11 master's, 4 doctorates awarded. *Degree requirements:* For master's, thesis/dissertation; for doctorate, thesis/dissertation, comprehensive exam. *Entrance requirements:* For master's and doctorate, GRE General Test, minimum GPA of 3.0. Additional exam requirements/recommendations for international students: Required—TOEFL. *Application deadline:* For fall admission, 6/15 for domestic students; for spring admission, 10/15 for domestic students. Applications are processed on a rolling basis. Application fee: $50 ($60 for international students). *Financial support:* In 2005–06, 14 research assistantships with partial tuition reimbursements (averaging $13,600 per year), 6 teaching assistantships (averaging $6,000 per year) were awarded; fellowships, career-related internships or fieldwork, Federal Work-Study, and institutionally sponsored loans also available. *Faculty research:* Range plant ecophysiology, plant community ecology, ruminant nutrition, population ecology. Total annual research expenditures: $3.5 million. *Unit head:* Dr. Johan W. duToit, Head, 435-797-2837, Fax: 435-797-3796. *Application contact:* Gaye Griffeth, Staff Assistant, 435-797-2503, Fax: 435-797-3796, E-mail: ggriffeth@cnr.usu.edu.

Virginia Polytechnic Institute and State University, Graduate School, College of Natural Resources, Department of Forestry, Blacksburg, VA 24061. Offers forest biology (MF, MS, PhD); forest biometry (MF, MS, PhD); forest management/economics (MF, MS, PhD); industrial forestry operations (MF, MS, PhD); outdoor recreation (MF, MS, PhD). *Entrance requirements:* For master's and doctorate, GRE General Test. Additional exam requirements/recommendations for international students: Required—TOEFL (minimum score 550 paper-based; 213 computer-based). *Application deadline:* Applications are processed on a rolling basis. Application fee: $45. Electronic applications accepted. *Expenses:* Tuition, state resident: full-time $6,558; part-time $364 per credit. Tuition, nonresident: full-time $11,296; part-time $628 per credit. Required fees: $1,419; $468 per credit. $234 per term. *Financial support:* Research assistantships with full tuition reimbursements, teaching assistantships with full tuition reimbursements, career-related internships or fieldwork, Federal Work-Study, scholarships/grants, and unspecified assistantships available. Financial award application deadline: 4/1. *Unit head:* Dr. Harold E. Burkhart, Head, 540-231-6952, Fax: 540-231-3698, E-mail: burkhart@vt.edu. *Application contact:* Sue Snow, Information Contact, 540-231-5483, Fax: 540-231-3698, E-mail: suesnow@vt.edu.

Virginia Polytechnic Institute and State University, Graduate School, College of Natural Resources, Department of Wood Science and Forest Products, Blacksburg, VA 24061. Offers forest products marketing (MF, MS, PhD); wood science and engineering (MF, MS, PhD). *Entrance requirements:* For master's and doctorate, GRE General Test. Additional exam requirements/recommendations for international students: Required—TOEFL (minimum score 550 paper-based; 213 computer-based). *Application deadline:* Applications are processed on a rolling basis. Application fee: $45. Electronic applications accepted. *Expenses:* Tuition, state resident: full-time $6,558; part-time $364 per credit. Tuition, nonresident: full-time $11,296; part-time $628 per credit. Required fees: $1,419; $468 per credit. $234 per term. *Financial support:* Research assistantships with full tuition reimbursements, teaching assistantships with full tuition reimbursements, career-related internships or fieldwork, Federal Work-Study, scholarships/grants, and unspecified assistantships available. Financial award application deadline:4/1. *Faculty research:* Wood chemistry, wood engineering, wood composites, wood processing, forest products marketing/management, recycling. *Unit head:* Dr. Paul M. Winistorfer, Head, 540-231-8854, Fax: 540-231-8176, E-mail: pstorfer@vt.edu. *Application contact:* D. Garnard, Information Contact, 540-231-8853, Fax: 540-231-8176, E-mail: garnandd@vt.edu.

West Virginia University, Davis College of Agriculture, Forestry and Consumer Sciences, Division of Forestry, Program in Forest Resource Science, Morgantown, WV 26506. Offers PhD. *Students:* 19 full-time (8 women), 5 part-time (1 woman), 3 international. Average age 33. 8 applicants, 38% accepted. In 2005, 4 degrees awarded. *Degree requirements:* For doctorate, thesis/dissertation, comprehensive exam. *Entrance requirements:* For doctorate, GRE, minimum GPA of 3.0. Additional exam requirements/recommendations for international students: Required—TOEFL. *Application deadline:* For fall admission, 6/15 priority date for domestic students, 6/15 priority date for international students. For winter admission, 9/15 for domestic students; for spring admission, 12/15 for domestic students. Applications are processed on a rolling basis. Application fee: $45. *Expenses:* Tuition, state resident: full-time $4,582; part-time $258 per credit hour. Tuition, nonresident: full-time $1,382; part-time $741 per credit hour. *Financial support:* In 2005–06, 15 research assistantships were awarded; teaching assistantships, career-related internships or fieldwork, Federal Work-Study, institutionally sponsored loans, and tuition waivers (full and partial) also available. Financial award application deadline: 2/1; financial award applicants required to submit FAFSA. *Faculty research:* Impact of management on wildlife and fish, forest sampling designs, forest economics and policy, oak regeneration. Total annual research expenditures: $900,000. *Unit head:* Dr. James P. Armstrong, Coordinator, 304-293-2941 Ext. 2486, Fax: 304-293-2441, E-mail: jim.armstrong@mail.wvu.edu.

West Virginia University, Davis College of Agriculture, Forestry and Consumer Sciences, Division of Forestry, Program in Forestry, Morgantown, WV 26506. Offers MSF. *Students:* 13 full-time (3 women), 4 part-time; includes 1 minority (Hispanic American), 1 international. Average age 29. 25 applicants, 40% accepted. In 2005, 6 degrees awarded. *Degree requirements:* For master's, thesis. *Entrance requirements:* For master's, GRE, minimum GPA of 3.0. Additional exam requirements/recommendations for international students: Required—TOEFL. *Application deadline:* For fall admission, 6/15 priority date for domestic students, 6/15 priority date for international students. For winter admission, 9/15 for domestic students; for spring admission, 12/15 for domestic students. Applications are processed on a rolling basis. Application fee: $45. *Expenses:* Tuition, state resident: full-time $4,582; part-time $258 per credit hour. Tuition, nonresident: full-time $1,382; part-time $741 per credit hour. *Financial support:* In 2005–06, 12 research assistantships, 1 teaching assistantship were awarded; Federal Work-Study, institutionally sponsored loans, and tuition waivers (full and partial) also available. Financial award application deadline: 2/1; financial award applicants required to submit FAFSA. *Faculty research:* Health and productivity on Appalachian forests, wood industries in Appalachian forests, role of forestry in regional economics. Total annual research expenditures: $900,000. *Unit head:* Dr. James P. Armstrong, Coordinator, 304-293-2941 Ext. 2486, Fax: 304-293-2441, E-mail: jim.armstrong@mail.wvu.edu.

Yale University, Graduate School of Arts and Sciences, Department of Forestry and Environmental Studies, New Haven, CT 06520. Offers environmental sciences (PhD); forestry (PhD). *Degree requirements:* For doctorate, thesis/dissertation. *Entrance requirements:* For doctorate, GRE General Test.

See Close-Up on page 713.

Yale University, School of Forestry and Environmental Studies, New Haven, CT 06511. Offers MES, MF, MFS, DFES, PhD, JD/MES, MBA/MES, MBA/MF, MES/MA, MES/MPH, MF/MA. *Accreditation:* SAF (one or more programs are accredited). Part-time programs available. Terminal master's awarded for partial completion of doctoral program. *Degree requirements:* For doctorate, thesis/dissertation. *Entrance requirements:* For master's and doctorate, GRE General Test. Expenses: Contact institution. *Faculty research:* Ecosystem science and management, coastal and watershed systems, environmental policy and management, social ecology and community development, conservation biology.

See Close-Up on page 713.

Natural Resources

Ball State University, Graduate School, College of Sciences and Humanities, Department of Natural Resources, Muncie, IN 47306-1099. Offers MA, MS. *Faculty:* 12. *Students:* 16 full-time (9 women), 5 part-time (1 woman); includes 1 minority (African American), 5 international. Average age 25. 16 applicants, 94% accepted, 7 enrolled. In 2005, 6 degrees awarded. *Entrance requirements:* For master's, GRE General Test. Application fee: $25 ($35 for international students). *Expenses:* Tuition, state resident: full-time $6,246. Tuition, nonresident: full-time $16,006. *Financial support:* In 2005–06, 6 teaching assistantships with full tuition reimbursements (averaging $8,886 per year) were awarded; research assistantships with full tuition reimbursements, career-related internships or fieldwork also available. Financial award application deadline: 3/1. *Faculty research:* Acid rain, indoor air pollution, land reclamation. *Unit head:* Hugh Brown, Chairman, 765-285-5780, Fax: 765-285-2606, E-mail: hbrown@bsu.edu.

Cornell University, Graduate School, Graduate Fields of Agriculture and Life Sciences, Field of Natural Resources, Ithaca, NY 14853-0001. Offers aquatic science (MPS, MS, PhD); environmental management (MPS); fishery science (MPS, MS, PhD); forest science (MPS, MS, PhD); resource policy and management (MPS, MS, PhD); wildlife science (MPS, MS, PhD). *Faculty:* 50 full-time (9 women). *Students:* 62 full-time (33 women); includes 9 minority (1 African American, 2 American Indian/Alaska Native, 4 Asian Americans or Pacific Islanders, 2 Hispanic Americans), 14 international. 60 applicants, 23% accepted, 12 enrolled. In 2005, 11 master's, 5 doctorates awarded. *Degree requirements:* For master's, thesis (MS), project paper (MPS); for doctorate, thesis/dissertation, comprehensive exam. *Entrance requirements:* For master's and doctorate, GRE General Test, 2 letters of recommendation. Additional exam requirements/recommendations for international students: Required—TOEFL (minimum score 550 paper-based; 213 computer-based). *Application deadline:* For spring admission, 10/30 for domestic students. Applications are processed on a rolling basis. Application fee: $60. Electronic applications accepted. *Financial support:* In 2005–06, 49 students received support, including 15 fellowships with full tuition reimbursements available, 16 research assistantships with full tuition reimbursements available, 18 teaching assistantships with full tuition reimbursements available; institutionally sponsored loans, scholarships/grants, health care benefits, tuition waivers (full and partial), and unspecified assistantships also available. Financial award applicants required to submit FAFSA. *Faculty research:* Ecosystem-level dynamics, systems modeling, conservation biology/management, resource management's human dimensions, biogeochemistry. *Unit head:* Director of Graduate Studies, 607-255-2807, Fax: 607-255-0349. *Application contact:* Graduate Field Assistant, 607-255-2807, Fax: 607-255-0349, E-mail: nrgrad@cornell.edu.

Duke University, Graduate School, Department of Environment, Durham, NC 27708. Offers natural resource economics/policy (AM, PhD); natural resource science/ecology (AM, PhD); natural resource systems science (AM, PhD). Part-time programs available. *Faculty:* 36 full-time. *Students:* 71 full-time (44 women); includes 4 minority (3 African Americans, 1 Hispanic American), 26 international. 135 applicants, 16% accepted, 14 enrolled. In 2005, 3 master's, 9 doctorates awarded. *Degree requirements:* For doctorate, variable foreign language requirement, thesis/dissertation. *Entrance requirements:* For master's and doctorate, GRE General Test. Additional exam requirements/recommendations for international students: Required—IELT (preferred) or TOEFL. *Application deadline:* For fall admission, 12/31 for domestic students, 12/31 for international students. Application fee: $75. Electronic applications accepted. *Financial support:* Fellowships, research assistantships, teaching assistantships, Federal Work-Study available. Financial award application deadline: 12/31. *Unit head:* Kenneth Knoerr, Director of Graduate Studies, 919-613-8030, Fax: 919-684-8741, E-mail: nmm@duke.edu.

Duke University, Nicholas School of the Environment and Earth Sciences, Durham, NC 27708-0328. Offers coastal environmental management (MEM); environmental economics and policy (MEM); environmental health and security (MEM); forest resource management (MF); global environmental change (MEM); resource ecology (MEM); water and air resources (MEM). *Accreditation:* SAF (one or more programs are accredited). Part-time programs available. *Degree requirements:* For master's, thesis, registration. *Entrance requirements:* For master's, GRE General Test, previous course work in biology or ecology, calculus, statistics, and microeconomics; computer familiarity with word processing and data analysis. Additional exam requirements/recommendations for international students: Required—TOEFL (minimum score 550 paper-based; 213 computer-based). Electronic applications accepted. Expenses: Contact institution. *Faculty research:* Ecosystem management, conservation ecology, earth systems, risk assessment.

See Close-Up on page 709.

Georgia Institute of Technology, Graduate Studies and Research, College of Engineering, School of Chemical and Biomolecular Engineering, Atlanta, GA 30332-0001. Offers bioengineering (MS Bio E, PhD); chemical engineering (MS Ch E, PhD); paper science and engineering (MS, PhD); polymers (MS Poly). *Degree requirements:* For master's, thesis/dissertation; for doctorate, thesis/dissertation, comprehensive exam. *Entrance requirements:* For master's and doctorate, GRE, minimum GPA of 3.0. Additional exam requirements/recommendations for international students: Required—TOEFL (minimum score 550 paper-based; 213 computer-based). Electronic applications accepted. *Faculty research:* Biochemical engineering; process modeling, synthesis, and control; polymer science and engineering; thermodynamics and separations; surface and particle science.

Humboldt State University, Graduate Studies, College of Natural Resources and Sciences, Programs in Natural Resources, Arcata, CA 95521-8299. Offers MS. *Students:* 64 full-time (32 women), 33 part-time (12 women); includes 9 minority (1 American Indian/Alaska Native, 3 Asian Americans or Pacific Islanders, 5 Hispanic Americans), 2 international. Average age 30.

Natural Resources

Humboldt State University (continued)

98 applicants, 29% accepted, 20 enrolled. In 2005, 20 degrees awarded. *Degree requirements:* For master's, thesis or alternative. *Entrance requirements:* For master's, appropriate bachelor's degree, minimum GPA of 2.5. Additional exam requirements/recommendations for international students: Required—TOEFL (minimum score 500 paper-based; 173 computer-based). *Application deadline:* For fall admission, 2/1 for domestic students, 2/1 for international students; for spring admission, 9/30 for domestic students, 9/30 for international students. Applications are processed on a rolling basis. Application fee: $55. *Financial support:* Fellowships, career-related internships or fieldwork and Federal Work-Study available. Support available to part-time students. Financial award application deadline: 3/1; financial award applicants required to submit FAFSA. *Faculty research:* Spotted owl habitat, presettlement vegetation, hardwood utilization, tree physiology, fisheries. *Unit head:* Dr. Gary Hendrickson, Coordinator, 707-826-4233, E-mail: thiesfel@humboldt.edu.

Iowa State University of Science and Technology, Graduate College, College of Agriculture, Department of Natural Resource Ecology and Management, Ames, IA 50011. Offers animal ecology (MS, PhD), including animal ecology, fisheries biology, wildlife biology; forestry (MS, PhD). *Faculty:* 23 full-time, 12 part-time/adjunct. *Students:* 45 full-time (18 women), 4 part-time (1 woman), 7 international. 24 applicants, 38% accepted, 8 enrolled. In 2005, 8 master's, 3 doctorates awarded. *Degree requirements:* For master's, thesis (for some programs); for doctorate, thesis/dissertation. *Entrance requirements:* For master's and doctorate, GRE General Test. Additional exam requirements/recommendations for international students: Required—TOEFL (paper score 547; computer score 210) or IELTS (score 6). *Application deadline:* For fall admission, 1/1 priority date for domestic students, 1/1 priority date for international students; for spring admission, 9/1 priority date for domestic students, 9/1 priority date for international students. Application fee: $30 ($70 for international students). Electronic applications accepted. *Expenses:* Tuition, state resident: full-time $6,410. Tuition, nonresident: full-time $16,422. Tuition and fees vary according to program. *Financial support:* In 2005–06, 41 research assistantships with full and partial tuition reimbursements (averaging $14,755 per year), 3 teaching assistantships with full and partial tuition reimbursements (averaging $14,580 per year) were awarded. *Unit head:* Dr. David M Engle, Chair, 515-294-1166. *Application contact:* Lyn Van De Pol, Information Contact, 515-294-6148, E-mail: lvdp@iastate.edu.

Iowa State University of Science and Technology, Graduate College, Interdisciplinary Programs, Program in Biorenewable Resources and Technology, Ames, IA 50011. Offers MS, PhD. *Students:* 1 full-time (0 women). *Entrance requirements:* Additional exam requirements/recommendations for international students: Required—TOEFL (paper score 550; computer score 213) or IELTS (score 6.5). *Application deadline:* For fall admission, 1/1 priority date for domestic students, 1/1 priority date for international students. Application fee: $30 ($70 for international students). *Expenses:* Tuition, state resident: full-time $6,410. Tuition, nonresident: full-time $16,422. Tuition and fees vary according to program. *Financial support:* In 2005–06, 1 research assistantship with full and partial tuition reimbursement (averaging $18,000 per year) was awarded *Unit head:* Dr. Brent Shanks, Chair, Supervising Committee, 515-294-6555, E-mail: brtgrad@iastate.edu. *Application contact:* Tonia McCarley, Program Coordinator, 515-294-6555, E-mail: brtgrad@iastate.edu.

Louisiana State University and Agricultural and Mechanical College, Graduate School, College of Agriculture, School of Renewable Natural Resources, Baton Rouge, LA 70803. Offers fisheries (MS); forestry (MS, PhD); wildlife (MS); wildlife and fisheries science (PhD). *Faculty:* 29 full-time (1 woman). *Students:* 67 full-time (24 women), 7 part-time (2 women); includes 1 African American, 1 Asian American or Pacific Islander, 3 Hispanic Americans, 20 international. Average age 28. 36 applicants, 56% accepted, 16 enrolled. In 2005, 13 master's, 6 doctorates awarded. *Degree requirements:* For master's and doctorate, thesis/dissertation. *Entrance requirements:* For master's, GRE General Test, minimum GPA of 3.0; for doctorate, GRE General Test, MS, minimum GPA of 3.0. Additional exam requirements/recommendations for international students: Required—TOEFL (minimum score 550 paper-based; 213 computer-based). *Application deadline:* For fall admission, 1/25 priority date for domestic students, 5/15 priority date for international students. Applications are processed on a rolling basis. Application fee: $25. Electronic applications accepted. *Financial support:* In 2005–06, 70 students received support, including 4 fellowships (averaging $16,304 per year), 59 research assistantships with partial tuition reimbursements available (averaging $16,539 per year); teaching assistantships with partial tuition reimbursements available, Federal Work-Study, institutionally sponsored loans, scholarships/grants, tuition waivers (full and partial), and unspecified assistantships also available. Financial award application deadline: 4/15; financial award applicants required to submit FAFSA. *Faculty research:* Forest biology and management, aquaculture, fisheries biology and ecology, upland and wetlands wildlife. Total annual research expenditures: $4,423. *Unit head:* Dr. Bob G. Blackmon, Director, 225-578-4131, Fax: 225-578-4227, E-mail: bblackmon@agctr.lsu.edu. *Application contact:* Dr. Allen Rutherford, Coordinator of Graduate Studies, 225-578-4187, Fax: 225-578-4227, E-mail: druther@lsu.edu.

See Close-Up on page 711.

McGill University, Faculty of Graduate and Postdoctoral Studies, Faculty of Agricultural and Environmental Sciences, Department of Natural Resource Sciences, Montréal, QC H3A 2T5, Canada. Offers agrometeorology (M Sc, PhD); entomology (M Sc, PhD); forest science (M Sc, PhD); microbiology (M Sc, PhD); neotropical environment (M Sc, PhD); soil science (M Sc, PhD); wildlife biology (M Sc, PhD). *Degree requirements:* For master's and doctorate, thesis/dissertation, registration. *Entrance requirements:* For master's, minimum GPA of 3.0 or 3.2 in the last 2 years of university study. Additional exam requirements/recommendations for international students: Required—TOEFL (minimum score 550 paper-based; 213 computer-based), IELT (minimum score 7). Electronic applications accepted. *Faculty research:* Toxicology, reproductive physiology, parasites, wildlife management, genetics.

Memorial University of Newfoundland, School of Graduate Studies, Interdisciplinary Program in Oil and Gas Studies, St. John's, NL A1C 5S7, Canada. Offers MOGS. *Students:* 4 full-time (0 women), 3 international. 13 applicants, 92% accepted, 3 enrolled. In 2005, 4 degrees awarded. *Degree requirements:* For master's, seminar course, project course with paper. *Entrance requirements:* For master's, undergraduate degree with minimum B standing in an oil and gas cognate discipline, minimum 5 years employment experience in the oil and gas sector. *Application deadline:* For fall admission, 3/15 for domestic students, 3/15 for international students. Application fee: $40 Canadian dollars. *Expenses:* Tuition: Part-time $733 per term. Tuition and fees vary according to degree level and program. *Financial support:* Applicants required to submit FAFSA. *Unit head:* Dr. Jim Wright, Interim Director, 709-737-6192, E-mail: jim.wright@mun.ca. *Application contact:* Dr. Alex Faseruk, Graduate Officer, 709-737-8005, E-mail: afaseruk@mun.ca.

Montana State University, College of Graduate Studies, College of Agriculture, Department of Land Resources and Environmental Sciences, Bozeman, MT 59717. Offers land rehabilitation (interdisciplinary) (MS); land resources and environmental sciences (MS, PhD). Part-time programs available. *Faculty:* 16 full-time (1 woman), 4 part-time/adjunct (0 women). *Students:* 6 full-time (3 women), 43 part-time (23 women), 6 international. Average age 30. 15 applicants, 53% accepted, 8 enrolled. In 2005, 12 master's, 4 doctorates awarded. *Degree requirements:* For master's, thesis (for some programs), comprehensive exam, registration; for doctorate, thesis/dissertation, comprehensive exam, registration. *Entrance requirements:* For master's and doctorate, GRE General Test. Additional exam requirements/recommendations for international students: Required—TOEFL (minimum score 550 paper-based; 213 computer-based). *Application deadline:* For fall admission, 7/15 priority date for domestic students, 5/15 priority date for international students; for spring admission, 12/1 priority date for domestic students, 10/1 priority date for international students. Applications are processed on a rolling basis. Application fee: $30. Electronic applications accepted. *Expenses:* Tuition, state resident: full-time $4,132. Tuition, nonresident: full-time $1,132. *Financial support:* Health care benefits and tuition waivers (full) available. Financial award application deadline: 3/1; financial award applicants required to submit FAFSA. *Faculty research:* Watershed hydrology, soil remediation and nutrient management, agroecology and weed biology, restoration ecology, microbial diversity. Total annual research expenditures: $7.7 million. *Unit head:* Dr. John Wraith, Department Head, 406-994-4605, Fax: 406-994-3933, E-mail: jwraith@montana.edu.

North Carolina State University, Graduate School, College of Natural Resources and College of Agriculture and Life Sciences, Program in Natural Resources, Raleigh, NC 27695. Offers MNR, MS. *Degree requirements:* For master's, thesis optional. *Entrance requirements:* For master's, GRE. Electronic applications accepted.

The Ohio State University, Graduate School, College of Food, Agricultural, and Environmental Sciences, School of Natural Resources, Columbus, OH 43210. Offers natural resources (MS, PhD); soil science (MS, PhD). Part-time programs available. *Degree requirements:* For master's, thesis optional. *Entrance requirements:* For master's and doctorate, GRE General Test. Additional exam requirements/recommendations for international students: Required—TOEFL (paper 550; computer 213) or IELT (7) or Michigan English Language Assessment Battery (91). Electronic applications accepted. *Faculty research:* Environmental education, natural resources development, fisheries and wildlife management.

Oklahoma State University, College of Agricultural Science and Natural Resources, Stillwater, OK 74078. Offers M Ag, MS, PhD. Part-time programs available. *Faculty:* 209 full-time (38 women), 14 part-time/adjunct (2 women). *Students:* 131 full-time (64 women), 200 part-time (85 women); includes 32 minority (9 African Americans, 17 American Indian/Alaska Native, 4 Asian Americans or Pacific Islanders, 2 Hispanic Americans), 137 international. Average age 30. 308 applicants, 41% accepted, 72 enrolled. In 2005, 66 master's, 32 doctorates awarded. *Degree requirements:* For doctorate, thesis/dissertation. *Entrance requirements:* For master's and doctorate, GRE. Additional exam requirements/recommendations for international students: Required—TOEFL. *Application deadline:* Applications are processed on a rolling basis. Application fee: $40 ($75 for international students). Electronic applications accepted. *Expenses:* Tuition, state resident: full-time $4,253; part-time $139 per credit hour. Tuition, nonresident: full-time $12,569; part-time $485 per credit hour. Required fees: $43 per credit hour. One-time fee: $20 part-time. Tuition and fees vary according to course load and program. *Financial support:* In 2005–06, 245 students received support, including 206 research assistantships (averaging $14,740 per year), 23 teaching assistantships (averaging $12,019 per year); fellowships, career-related internships or fieldwork, Federal Work-Study, scholarships/grants, health care benefits, tuition waivers (partial), and unspecified assistantships also available. Support available to part-time students. Financial award application deadline: 3/1. *Unit head:* Dr. Robert E. Whitson, Dean, 405-744-5398, Fax: 405-744-5339.

Purdue University, Graduate School, College of Agriculture, Department of Forestry and Natural Resources, West Lafayette, IN 47907. Offers aquaculture, fisheries, aquatic science (MSF); aquaculture, fisheries, aquatic sciences (MS, PhD); forest biology (MS, MSF, PhD); natural resources and environmental policy (MS, MSF); natural resources environmental policy (PhD); quantitative resource analysis (MS, MSF, PhD); wildlife science (MS, MSF, PhD); wood science and technology (MS, MSF, PhD). *Faculty:* 26 full-time (4 women), 7 part-time/adjunct (1 woman). *Students:* 69 full-time (33 women), 15 part-time (3 women); includes 5 minority (1 African American, 1 American Indian/Alaska Native, 3 Hispanic Americans), 29 international. Average age 30. 45 applicants, 40% accepted, 18 enrolled. In 2005, 6 master's, 6 doctorates awarded. *Degree requirements:* For master's and doctorate, thesis/dissertation. *Entrance requirements:* For master's and doctorate, GRE General Test (500 verbal, 500 quantitative), minimum B+ average in undergraduate course work. Additional exam requirements/recommendations for international students: Required—TOEFL. *Application deadline:* For fall admission, 1/5 for domestic students; for spring admission, 9/15 for domestic students. Applications are processed on a rolling basis. Application fee: $55. Electronic applications accepted. *Financial support:* In 2005–06, 10 research assistantships (averaging $15,259 per year) were awarded; fellowships, teaching assistantships, career-related internships or fieldwork and scholarships/grants also available. Support available to part-time students. Financial award application deadline: 1/5; financial award applicants required to submit FAFSA. *Faculty research:* Wildlife management, forest management, forest ecology, forest soils, limnology. *Unit head:* Dr. Robert K. Swihart, Interim Head, 765-494-3590, Fax: 765-494-9461, E-mail: rswihart@purdue.edu. *Application contact:* Kelly Garrett, Graduate Secretary, 765-494-3572, Fax: 765-494-9461, E-mail: kgarrett@purdue.edu.

State University of New York College of Environmental Science and Forestry, Faculty of Construction Management and Wood Products Engineering, Syracuse, NY 13210-2779. Offers environmental and resources engineering (MPS, MS, PhD). *Faculty:* 9 full-time (1 woman), 1 part-time/adjunct (0 women). *Students:* 8 full-time (1 woman), 11 part-time (2 women); includes 1 minority (African American), 4 international. Average age 32. 6 applicants, 100% accepted, 4 enrolled. In 2005, 1 doctorate awarded. *Degree requirements:* For master's, thesis (for some programs), registration; for doctorate, thesis/dissertation, comprehensive exam, registration. *Entrance requirements:* For master's and doctorate, GRE General Test, minimum GPA of 3.0. Additional exam requirements/recommendations for international students: Required—TOEFL (minimum score 550 paper-based; 213 computer-based). *Application deadline:* For fall admission, 2/1 priority date for domestic students, 2/1 priority date for international students; for spring admission, 11/1 priority date for domestic students, 11/1 priority date for international students. Applications are processed on a rolling basis. Application fee: $60. *Expenses:* Tuition, area resident: Full-time $6,900; part-time $288 per credit. Tuition, nonresident: full-time $10,920; part-time $455 per credit. Required fees: $395; $32 per credit. $20 per term. One-time fee: $145. *Financial support:* In 2005–06, 8 students received support, including fellowships with full tuition reimbursements available (averaging $9,446 per year), 2 research assistantships with full tuition reimbursements available (averaging $11,000 per year), 4 teaching assistantships with full tuition reimbursements available (averaging $9,446 per year); career-related internships or fieldwork, Federal Work-Study, institutionally sponsored loans, scholarships/grants, health care benefits, and unspecified assistantships also available. Financial award application deadline: 6/30; financial award applicants required to submit FAFSA. Total annual research expenditures: $160,385. *Unit head:* Dr. Robert W. Meyer, Chair, 315-470-6835, Fax: 315-470-6879. *Application contact:* Dr. Dudley J. Raynal, Dean, Instruction and Graduate Studies, 315-470-6599, Fax: 315-470-6879, E-mail: esfgrad@esf.edu.

Texas A&M University, College of Agriculture and Life Sciences, Department of Forest Science, College Station, TX 77843. Offers forestry (MS, PhD); natural resources development (M Agr). Part-time programs available. *Faculty:* 6 full-time (3 women). *Students:* 11 full-time (4 women), 4 part-time (2 women); includes 4 minority (3 African Americans, 1 Hispanic American), 5 international. Average age 27. 5 applicants, 80% accepted, 2 enrolled. In 2005, 4 master's, 2 doctorates awarded. Terminal master's awarded for partial completion of doctoral program. *Degree requirements:* For master's, thesis (for some programs); for doctorate, thesis/dissertation. *Entrance requirements:* For master's and doctorate, GRE General Test. Additional exam requirements/recommendations for international students: Required—TOEFL. *Application deadline:* For fall admission, 3/1 for domestic students; for spring admission, 11/1 priority date for domestic students. Applications are processed on a rolling basis. Application fee: $50 ($75 for international students). Electronic applications accepted. *Expenses:* Tuition, state resident: full-time $4,488; part-time $187 per credit hour. Tuition, nonresident: full-time $11,112; part-time $463 per credit hour. Required fees:$1,974. *Financial support:* In 2005–06, fellowships with partial tuition reimbursements (averaging $15,000 per year), research assistantships with partial tuition reimbursements (averaging $15,000 per year), teaching assistantships with partial tuition reimbursements (averaging $15,000 per year) were awarded; career-related internships or fieldwork and institutionally sponsored loans also available. Support available to part-time students. Financial award application deadline: 3/1; financial award applicants required to submit FAFSA. *Faculty research:* Expert systems, geographic information systems, economics, biology, genetics. *Unit head:* Dr. C. T. Smith, Professor and Head, 979-845-5033, Fax: 979-845-6049, E-mail: tat-smith@tamu.edu. *Application contact:* Dr. Carol Loopstra, Associate Head for Research and Graduate Studies, 979-862-2200, Fax: 979-845-6049, E-mail: c-loopstra@tamu.edu.

Texas A&M University, College of Agriculture and Life Sciences, Department of Recreation, Park and Tourism Sciences, College Station, TX 77843. Offers natural resources development (M Agr); recreation resources development (M Agr); recreation, park, and tourism sciences (MS, PhD). *Faculty:* 10 full-time (2 women), 1 part-time/adjunct (0 women). *Students:* 47 full-time (30 women), 15 part-time (10 women); includes 6 minority (4 African Americans, 2 Hispanic Americans), 23 international. Average age 28. 42 applicants, 57% accepted, 16 enrolled. In 2005, 3 master's, 5 doctorates awarded. *Degree requirements:* For master's, thesis (for some programs), internship and professional paper (M Agr); for doctorate, thesis/dissertation. *Entrance requirements:* For master's and doctorate, GRE General Test. Additional exam requirements/recommendations for international students: Required—TOEFL. *Application deadline:* For fall admission, 4/15 for domestic students; for spring admission, 10/15 priority date for domestic students. Applications are processed on a rolling basis. Application fee: $50 ($75 for international students). Electronic applications accepted. *Expenses:* Tuition, state resident: full-time $4,488; part-time $187 per credit hour. Tuition, nonresident: full-time $11,112; part-time $463 per credit hour. Required fees: $1,974. *Financial support:* Fellowships, research assistantships, teaching assistantships, career-related internships or fieldwork, institutionally sponsored loans, and scholarships/grants available. Financial award application deadline: 4/15; financial award applicants required to submit FAFSA. *Faculty research:* Administration and tourism, outdoor recreation, commercial recreation, environmental law, system planning. *Unit head:* Dr. Joseph T. O'Leary, Head, 979-845-5412, Fax: 979-845-0446, E-mail: joleary@rpts.tamu.edu. *Application contact:* Marguerite M. Van Dyke, Graduate Recruitment Coordinator, 979-845-5412, Fax: 979-845-0446, E-mail: mvandyke@rpts.tamu.edu.

Université du Québec à Montréal, Graduate Programs, Program in Earth Sciences, Montreal, QC H3C 3P8, Canada. Offers geology-research (M Sc); mineral resources (PhD); non-renewable resources (DESS). Part-time programs available. Terminal master's awarded for partial completion of doctoral program. *Degree requirements:* For master's, thesis (for some programs); for doctorate, thesis/dissertation. *Entrance requirements:* For master's, appropriate bachelor's degree or equivalent, proficiency in French. *Faculty research:* Economic geology, structural geology, geochemistry, Quaternary geology, isotopic geochemistry.

University of Alberta, Faculty of Graduate Studies and Research, Department of Renewable Resources, Edmonton, AB T6G 2E1, Canada. Offers agroforestry (M Ag, M Sc, MF); conservation biology (M Sc, PhD); forest biology and management (M Sc, PhD); land reclamation and remediation (M Sc, PhD); protected areas and wildlands management (M Sc, PhD); soil science (M Ag, M Sc, PhD); water and land resources (M Ag, M Sc, PhD); wildlife ecology and management (M Sc, PhD). Part-time programs available. *Faculty:* 26 full-time (4 women), 22 part-time/adjunct (3 women). *Students:* 63 full-time (33 women), 50 part-time (20 women), 14 international. 122 applicants, 24% accepted, 22 enrolled. In 2005, 16 master's, 8 doctorates awarded. *Median time to degree:* Of those who began their doctoral program in fall 1997, 100% received their degree in 8 years or less. *Degree requirements:* For master's, thesis (for some programs); for doctorate, thesis/dissertation, comprehensive exam. *Entrance requirements:* For master's, minimum 2 years of relevant professional experiences, minimum GPA of 3.0; for doctorate, minimum GPA of 3.0. Additional exam requirements/recommendations for international students: Required—TOEFL (minimum score 550 paper-based; 213 computer-based). *Application deadline:* For fall admission, 7/1 priority date for domestic students, 6/1 priority date for international students. Applications are processed on a rolling basis. Application fee: $0. Electronic applications accepted. Tuition and fees charges are reported in Canadian dollars. *Expenses:* Tuition, state resident: part-time $562 Canadian dollars per term. Tuition, nonresident: full-time $3,375 Canadian dollars. Required fees: $573 Canadian dollars; $84 Canadian dollars per term. *Financial support:* In 2005–06, 63 students received support, including 21 research assistantships with partial tuition reimbursements available (averaging $2,800 per year), 28 teaching assistantships with partial tuition reimbursements available (averaging $1,900 per year); scholarships/grants and unspecified assistantships also available. *Faculty research:* Natural and managed landscapes. Total annual research expenditures: $6.1 million. *Unit head:* Dr. John R. Spence, Chair, 780-492-2820, Fax: 780-492-4323, E-mail: john.spence@ualberta.ca. *Application contact:* Sandy Nakashima, Graduate Program Secretary, 780-492-2820, Fax: 780-492-4323, E-mail: sandy.nakashima@ualberta.ca.

University of Alberta, Faculty of Graduate Studies and Research, Program in Business Administration, Edmonton, AB T6G 2E1, Canada. Offers international business (MBA); leisure and sport management (MBA); natural resources and energy (MBA); technology commercialization (MBA). *Accreditation:* AACSB. Part-time and evening/weekend programs available. *Faculty:* 77 full-time, 20 part-time/adjunct. *Students:* 131 full-time (56 women), 109 part-time (51 women). Average age 29. 525 applicants, 30% accepted, 90 enrolled. In 2005, 114 degrees awarded. *Degree requirements:* For master's, thesis or alternative. *Entrance requirements:* For master's, GMAT. Additional exam requirements/recommendations for international students: Required—TOEFL (minimum score 600 paper-based; 250 computer-based). *Application deadline:* For fall admission, 4/30 priority date for domestic students, 4/30 priority date for international students. Applications are processed on a rolling basis. Application fee: $0. Electronic applications accepted. Tuition and fees charges are reported in Canadian dollars. *Expenses:* Tuition, state resident: part-time $562 Canadian dollars per term. Tuition, nonresident: full-time $3,375 Canadian dollars. Required fees: $573 Canadian dollars; $84 Canadian dollars per term. *Financial support:* Fellowships, research assistantships, teaching assistantships, career-related internships or fieldwork, scholarships/grants, health care benefits, and unspecified assistantships available. *Faculty research:* Natural resources and energy/management and policy/family enterprise/international business/healthcare research management. Total annual research expenditures:$1 million. *Unit head:* Dr. Douglas Olsen, Associate Dean, 780-492-5412, Fax: 780-492-7825. *Application contact:* Joan A. White, Secretary, 780-492-3679, Fax: 780-492-2024, E-mail: joan.white@ualberta.ca.

The University of Arizona, Graduate College, College of Agriculture and Life Sciences, School of Natural Resources, Natural Resources Studies, Tucson, AZ 85721. Offers MS, PhD. *Degree requirements:* For master's, thesis/dissertation; for doctorate, thesis/dissertation, comprehensive exam. *Entrance requirements:* For master's and doctorate, GRE General Test, minimum GPA of 3.0, 3 letters of recommendation. Additional exam requirements/recommendations for international students: Required—TOEFL (minimum score 550 paper-based; 213 computer-based). *Faculty research:* Global carbon markets and carbon sequestration, integrated watershed management and policy, conservation biology, landscape planning, wildlife conservation and management.

University of Arkansas at Monticello, School of Forest Resources, Monticello, AR 71656. Offers MS. Part-time programs available. *Degree requirements:* For master's, thesis, comprehensive exam. *Entrance requirements:* For master's, GRE General Test, minimum GPA of 2.7. Additional exam requirements/recommendations for international students: Required—TOEFL (minimum score 550 paper-based; 213 computer-based). *Faculty research:* Geographic information systems/remote sensing, forest ecology, wildlife ecology and management.

University of Connecticut, Graduate School, College of Agriculture and Natural Resources, Department of Natural Resources Management and Engineering, Field of Natural Resources Management and Engineering, Storrs, CT 06269. Offers natural resources (MS, PhD). *Faculty:* 15 full-time (1 woman). *Students:* 18 full-time (7 women), 9 part-time (4 women); includes 1 minority (Asian American or Pacific Islander), 4 international. Average age 33. 25 applicants, 44% accepted, 11 enrolled. In 2005, 4 master's, 1 doctorate awarded. Terminal master's awarded for partial completion of doctoral program. *Degree requirements:* For master's, comprehensive exam; for doctorate, thesis/dissertation. *Entrance requirements:* For master's, GRE General Test, GRE Subject Test. Additional exam requirements/recommendations for international students: Required—TOEFL (minimum score 550 paper-based; 213 computer-based). *Application deadline:* For fall admission, 2/1 priority date for domestic students, 2/1 priority date for international students; for spring admission, 11/1 for domestic students, 10/1 for international students. Applications are processed on a rolling basis. Application fee: $55. Electronic applications accepted. *Expenses:* Tuition, state resident: part-time $444 per credit hour. Tuition, nonresident: part-time $1,154 per credit hour. Tuition and fees vary according to course load. *Financial support:* In 2005–06, 16 research assistantships with full tuition reimbursements, 1 teaching assistantship with full tuition reimbursement were awarded; fellowships, Federal Work-Study, scholarships/grants, health care benefits, and unspecified assistantships also available. Financial award application deadline: 2/1; financial award applicants required to submit FAFSA. *Application contact:* John Clausen, Chairman, Graduate Admissions, 860-486-0139, Fax: 860-486-5408, E-mail: john.clausen@uconn.edu.

University of Florida, Graduate School, College of Agricultural and Life Sciences, School of Forest Resources and Conservation, Gainesville, FL 32611. Offers MFRC, MS, PhD, JD/MFRC, JD/MS, JD/PhD. Part-time programs available. *Faculty:* 24 full-time (3 women). *Students:* 61 (30 women); includes 8 minority (2 African Americans, 1 American Indian/Alaska Native, 2 Asian Americans or Pacific Islanders, 3 Hispanic Americans) 23 international. Average age 24. 101 applicants, 35% accepted. In 2005, 15 master's, 2 doctorates awarded. *Degree requirements:* For master's, project (MFRC), thesis defense (MS); for doctorate, thesis/dissertation, qualifying exams, defense. *Entrance requirements:* For master's and doctorate, GRE General Test, minimum GPA of 3.0. Additional exam requirements/recommendations for international students: Required—TOEFL. *Application deadline:* For fall admission, 6/1 for domestic students; for spring admission, 10/1 for domestic students. Applications are processed on a rolling basis. Application fee: $20. Electronic applications accepted. *Expenses:* Tuition, state resident: full-time $6,234. Tuition, nonresident: full-time $21,359. Tuition and fees vary according to program. *Financial support:* In 2005–06, 35 research assistantships with full tuition reimbursements (averaging $15,546 per year) were awarded; fellowships with full tuition reimbursements, teaching assistantships with full tuition reimbursements, Federal Work-Study and institutionally sponsored loans also available. Support available to part-time students. *Faculty research:* Forest biology and ecology; agroforestry and tropical forestry; forest management, economics, and policy; natural resource education and ecotourism. *Unit head:* Dr. Tim White, Director, 352-846-0850, Fax: 352-392-1707, E-mail: tlwhite@ufl.edu. *Application contact:* Dr. George M. Blakeslee, Graduate Coordinator, 352-846-0845, Fax: 352-392-1707, E-mail: gb4stree@ufl.edu.

University of Florida, Graduate School, School of Natural Resources and Environment, Gainesville, FL 32611. Offers interdisciplinary ecology (MS, PhD). *Faculty:* 1 full-time (0 women). *Students:* 119 (63 women); includes 7 minority (1 African American, 3 Asian Americans or Pacific Islanders, 3 Hispanic Americans) 23 international. In 2005, 18 master's, 4 doctorates awarded. *Degree requirements:* For master's, thesis optional; for doctorate, thesis/dissertation. *Entrance requirements:* For master's and doctorate, GRE General Test, minimum GPA of 3.0. Additional exam requirements/recommendations for international students: Required—TOEFL (minimum score 550 paper-based; 213 computer-based). *Application deadline:* For fall admission, 2/11 for domestic students. Applications are processed on a rolling basis. Application fee: $30. Electronic applications accepted. *Expenses:* Tuition, state resident: full-time $6,234. Tuition, nonresident: full-time $21,359. Tuition and fees vary according to program. *Financial support:* In 2005–06, 9 teaching assistantships (averaging $15,592 per year) were awarded; fellowships, research assistantships *Unit head:* Dr. Stephen R. Humphrey, Director, 352-392-9230, Fax: 352-392-9748, E-mail: humphrey@ufl.edu. *Application contact:* Meisha Wade, Coordinator of Academic Programs, 352-392-9230, Fax: 352-392-9748, E-mail: mwade@ufl.edu.

University of Georgia, Graduate School, School of Forest Resources, Athens, GA 30602. Offers MFR, MS, PhD. *Faculty:* 35 full-time (2 women). *Students:* 145 full-time, 22 part-time; includes 9 minority (5 African Americans, 1 American Indian/Alaska Native, 2 Asian Americans or Pacific Islanders, 1 Hispanic American), 23 international. 77 applicants, 71% accepted, 37 enrolled. In 2005, 31 master's, 6 doctorates awarded. *Degree requirements:* For master's, thesis (MS); for doctorate, one foreign language, thesis/dissertation. *Entrance requirements:* For master's and doctorate, GRE General Test. *Application deadline:* For fall admission, 7/1 for domestic students; for spring admission, 11/15 for domestic students. Application fee: $50. Electronic applications accepted. *Financial support:* Fellowships, research assistantships, teaching assistantships, unspecified assistantships available. *Unit head:* Dr. Richard L. Porterfield, Dean, 706-542-2866, Fax: 706-542-2281. *Application contact:* Dr. Barry D. Shiver, Graduate Coordinator, 706-542-3009, Fax: 706-542-8356, E-mail: shiver@smokey.forestry.uga.edu.

University of Guelph, Graduate Program Services, Ontario Agricultural College, Department of Land Resource Science, Guelph, ON N1G 2W1, Canada. Offers atmospheric science (M Sc, PhD); environmental and agricultural earth sciences (M Sc, PhD); land resources management (M Sc, PhD); soil science (M Sc, PhD). Part-time programs available. *Faculty:* 19 full-time (5 women), 5 part-time/adjunct (1 woman). *Students:* 47 full-time (20 women), 3 part-time; includes 9 minority (1 African American, 6 Asian Americans or Pacific Islanders, 2 Hispanic Americans), 2 international. Average age 28. 25 applicants, 24% accepted. In 2005, 4 master's, 3 doctorates awarded. *Degree requirements:* For master's and doctorate, thesis/dissertation. *Entrance requirements:* For master's, minimum B- average during previous 2 years of course work; for doctorate, minimum B average during previous 2 years of course work. Additional exam requirements/recommendations for international students: Required—TOEFL (minimum score 550 paper-based; 213 computer-based). *Application deadline:* For fall admission, 7/1 priority date for domestic students, 5/1 priority date for international students. For winter admission, 10/1 for domestic students; for spring admission, 3/1 for domestic students. Applications are processed on a rolling basis. Application fee: $75 Canadian dollars. Electronic applications accepted. *Financial support:* In 2005–06, 30 students received support, including 40 research assistantships (averaging $16,500 Canadian dollars per year), 15 teaching assistantships (averaging $3,800 Canadian dollars per year); fellowships, scholarships/grants also available. *Faculty research:* Soil science, environmental earth science, land resource management. Total annual research expenditures: $2.1 million Canadian dollars. *Unit head:* Dr. S. Hilts, Chair, 519-824-4120 Ext. 52447, Fax: 519-824-5730, E-mail: shilts@uoguelph.ca. *Application contact:* Dr. B. Hale, Graduate Coordinator, 519-824-4120 Ext. 53434, Fax: 519-824-5730, E-mail: bhale@uoguelph.ca.

University of Hawaii at Manoa, Graduate Division, College of Tropical Agriculture and Human Resources, Department of Natural Resources and Environmental Management, Honolulu, HI 96822. Offers MS, PhD. Part-time programs available. *Faculty:* 24 full-time (3 women), 6 part-time/adjunct (2 women). *Students:* 43 full-time (23 women), 2 part-time; includes 5 minority (1 African American, 3 Asian Americans or Pacific Islanders, 1 Hispanic American), 18 international. Average age 34. 68 applicants, 62% accepted, 17 enrolled. In 2005, 7 master's, 1 doctorate awarded. Terminal master's awarded for partial completion of doctoral program. *Degree requirements:* For master's, thesis or alternative; for doctorate, thesis/dissertation. *Entrance requirements:* For master's and doctorate, GRE, minimum GPA of 3.0 in last 4 semesters of course work. Additional exam requirements/recommendations for international students: Required—TOEFL. *Application deadline:* For fall admission, 3/1 for domestic students, 1/15 for international students; for spring admission, 9/1 for domestic students, 8/1 for international students. Applications are processed on a rolling basis. Application fee: $50. *Expenses:* Tuition, state resident: full-time $8,400; part-time $200 per credit hour. Tuition, nonresident: full-time $11,088; part-time $462 per credit hour. Tuition and fees vary according to program. *Financial support:* In 2005–06, 16 research assistantships (averaging $15,850 per year), 11 teaching assistantships (averaging $14,315 per year) were awarded; fellowships, career-related internships or fieldwork and tuition waivers (full and partial) also available. *Faculty research:* Bioeconomics, natural resource management. *Unit head:* Dr. Samir A. El-Swarfy, Chairperson, 808-956-8708, Fax: 808-956-2811. *Application contact:* John Yanagida, 808-956-7530, Fax: 808-956-6539.

University of Idaho, College of Graduate Studies, College of Natural Resources, Moscow, ID 83844-2282. Offers conservation social sciences (MS, PhD); fish and wildlife resources (MS, PhD), including fishery resources, wildlife resources; forest products (MS, PhD); forest resources (MS, PhD); natural resources (PhD); natural resources management and administration (MNR); rangeland ecology and management (MS, PhD). *Students:* 137 full-time (66

Natural Resources

University of Idaho (continued)
women), 95 part-time (36 women); includes 8 minority (4 American Indian/Alaska Native, 2 Asian Americans or Pacific Islanders, 2 Hispanic Americans), 41 international. In 2005, 51 master's, 7 doctorates awarded. *Degree requirements:* For doctorate, thesis/dissertation. *Entrance requirements:* For master's, minimum GPA of 2.8; for doctorate, minimum undergraduate GPA of 2.8, 3.0 graduate. *Application deadline:* For fall admission, 8/1 for domestic students; for spring admission, 12/15 for domestic students. Application fee: $55 ($60 for international students). *Expenses:* Tuition, nonresident: full-time $8,770; part-time $130 per credit. Required fees: $4,508; $217 per credit. *Financial support:* Fellowships, research assistantships, teaching assistantships, Federal Work-Study available. Support available to part-time students. Financial award application deadline: 2/15. *Unit head:* Steven B. Daley-Laursen, Dean, 208-885-6442, Fax: 208-885-6226. *Application contact:* Dr. Ali Moslemi, Graduate Coordinator, 208-885-6126.

University of Illinois at Urbana–Champaign, Graduate College, College of Agricultural, Consumer and Environmental Sciences, Department of Natural Resources and Environmental Science, Champaign, IL 61820. Offers MS, PhD. *Faculty:* 46 full-time (6 women), 3 part-time/adjunct (0 women). *Students:* 62 full-time (29 women), 39 part-time (21 women); includes 8 minority (2 African Americans, 5 Asian Americans or Pacific Islanders, 1 Hispanic American), 23 international. 103 applicants, 22% accepted, 13 enrolled. In 2005, 27 master's, 13 doctorates awarded. *Degree requirements:* For master's and doctorate, thesis/dissertation. *Entrance requirements:* For master's and doctorate, GRE, minimum GPA of 3.0. *Application deadline:* Applications are processed on a rolling basis. Application fee: $50 ($60 for international students). Electronic applications accepted. *Financial support:* In 2005–06, 10 fellowships, 63 research assistantships, 20 teaching assistantships were awarded; tuition waivers (full and partial) also available. Financial award application deadline: 2/15. *Unit head:* Wesley M. Jarrell, Head, 217-333-2770, Fax: 217-244-3219, E-mail: wjarrell@uiuc.edu. *Application contact:* Mary Lowry, Student Services Coordinator, 217-244-5761, Fax: 217-244-3219, E-mail: lowry@uiuc.edu.

University of Maine, Graduate School, College of Natural Sciences, Forestry, and Agriculture, Department of Forest Management and Forest Ecosystem Science, Orono, ME 04469. Offers forest resources (PhD); forestry (MF, MS). *Accreditation:* SAF (one or more programs are accredited). Part-time programs available. *Students:* 30 full-time (11 women), 16 part-time (7 women); includes 1 minority (American Indian/Alaska Native), 10 international. Average age 33. 38 applicants, 55% accepted, 14 enrolled. In 2005, 9 master's, 2 doctorates awarded. *Degree requirements:* For master's, thesis; for doctorate, one foreign language, thesis/dissertation. *Entrance requirements:* For master's and doctorate, GRE General Test. Additional exam requirements/recommendations for international students: Required—TOEFL. *Application deadline:* For fall admission, 2/1 for domestic students. Applications are processed on a rolling basis. Application fee: $50. Electronic applications accepted. *Financial support:* In 2005–06, research assistantships with tuition reimbursements (averaging $15,000 per year), teaching assistantships with tuition reimbursements (averaging $12,276 per year) were awarded; fellowships, career-related internships or fieldwork, Federal Work-Study, and institutionally sponsored loans also available. Financial award application deadline: 3/1. *Faculty research:* Forest economics, engineering and operations analysis, biometrics and remote sensing, timber management, wood technology. *Unit head:* Dr. William Livingsten, Chair, 207-581-2990. *Application contact:* Scott G. Delcourt, Associate Dean of the Graduate School, 207-581-3219, Fax: 207-581-3232, E-mail: graduate@maine.edu.

University of Maine, Graduate School, College of Natural Sciences, Forestry, and Agriculture, Department of Plant, Soil, and Environmental Sciences, Orono, ME 04469. Offers biological sciences (PhD); ecology and environmental sciences (MS, PhD); forest resources (PhD); horticulture (MS); plant science (PhD); plant, soil, and environmental sciences (MS); resource utilization (MS). *Faculty:* 25. *Students:* 15 full-time (8 women), 8 part-time (6 women), 3 international. Average age 32. 6 applicants, 33% accepted, 1 enrolled. In 2005, 9 master's awarded. *Entrance requirements:* For master's and doctorate, GRE General Test. Additional exam requirements/recommendations for international students: Required—TOEFL. *Application deadline:* Applications are processed on a rolling basis. Application fee: $50. Electronic applications accepted. *Financial support:* In 2005–06, 9 research assistantships with tuition reimbursements (averaging $12,180 per year) were awarded; teaching assistantships, scholarships/grants, tuition waivers (full and partial), and unspecified assistantships also available. *Unit head:* Greg Porter, Chair, 207-581-2943, Fax: 207-581-3207. *Application contact:* Scott G. Delcourt, Associate Dean of the Graduate School, 207-581-3219, Fax: 207-581-3232, E-mail: graduate@maine.edu.

University of Maryland, College Park, Graduate Studies, College of Agriculture and Natural Resources, Department of Natural Resource Sciences and Landscape Architecture, Natural Resource Sciences Program, College Park, MD 20742. Offers MS, PhD. *Students:* 29 full-time (19 women), 10 part-time (4 women); includes 2 minority (1 African American, 1 Asian American or Pacific Islander), 7 international. 17 applicants, 35% accepted, 3 enrolled. In 2005, 3 master's, 3 doctorates awarded. *Degree requirements:* For master's, thesis optional; for doctorate, thesis/dissertation. *Entrance requirements:* For master's, GRE General Test, minimum GPA of 3.0, 3 letters of recommendation; for doctorate, GRE General Test. *Application deadline:* For fall admission, 2/1 for domestic students, 2/1 for international students; for spring admission, 8/1 for domestic students, 6/1 for international students. Applications are processed on a rolling basis. Application fee: $60. Electronic applications accepted. *Financial support:* In 2005–06, 2 fellowships (averaging $10,872 per year) were awarded *Faculty research:* Wetland soils, acid mine drainage, acid sulfate soil. *Application contact:* Dean of Graduate School, 301-405-4190, Fax: 301-314-9305.

University of Michigan, School of Natural Resources and Environment, Ann Arbor, MI 48109-1115. Offers industrial ecology (Certificate); landscape architecture (MLA, PhD); resource ecology and management (MS, PhD), including natural resources and environment (PhD), resource ecology and management (MS); resource policy and behavior (MS, PhD), including natural resources and environment (PhD), resource policy and behavior (MS); spatial analysis (Certificate). MLA, MS, PhD, and JD/MS offered through the Horace H. Rackham School of Graduate Studies. *Accreditation:* ASLA (one or more programs are accredited); SAF (one or more programs are accredited). *Degree requirements:* For master's, thesis or alternative, thesis, practicum, or group project; for doctorate, thesis/dissertation, oral defense of dissertation, preliminary exam, comprehensive exam, registration. *Entrance requirements:* For master's, GRE General Test; for doctorate, GRE General Test, master's degree. Additional exam requirements/recommendations for international students: Required—TOEFL (paper score 560; computer score 220) or IELTS. Electronic applications accepted. *Expenses:* Tuition, state resident: full-time $14,082; part-time $894 per credit hour. Tuition, nonresident: full-time $28,500; part-time $1,675 per credit hour. Required fees: $189; $189 per unit. *Faculty research:* Ecology, environmental policy, landscape architecture, climate change, Great Lakes, sustainable systems.

University of Minnesota, Twin Cities Campus, Graduate School, College of Natural Resources, Department of Bio-Based Products, Minneapolis, MN 55455-0213. Offers natural resources science and management (MS, PhD). Terminal master's awarded for partial completion of doctoral program. *Degree requirements:* For master's, thesis optional; for doctorate, thesis/dissertation, comprehensive exam, registration. *Entrance requirements:* For master's and doctorate, GRE. Additional exam requirements/recommendations for international students: Required—TOEFL (minimum score 550 paper-based; 213 computer-based). Electronic applications accepted. *Expenses:* Tuition, state resident: full-time $8,748; part-time $729 per credit. Tuition, nonresident: full-time $15,848; part-time $1,321 per credit. Full-time tuition and fees vary according to class time, course load, program and reciprocity agreements.

The University of Montana, Graduate School, College of Forestry and Conservation, Missoula, MT 59812-0002. Offers ecosystem management (MEM, MS); fish and wildlife biology (PhD); forestry (MS, PhD); recreation management (MS); resource conservation (MS); wildlife biology (MS). *Faculty:* 34 full-time (4 women). *Students:* 95 full-time (50 women), 63 part-time (21 women); includes 8 minority (1 African American, 3 American Indian/Alaska Native, 1 Asian American or Pacific Islander, 3 Hispanic Americans), 19 international. 175 applicants, 35% accepted, 46 enrolled. In 2005, 23 master's, 9 doctorates awarded. *Degree requirements:* For doctorate, thesis/dissertation. *Entrance requirements:* For master's and doctorate, GRE General Test. Additional exam requirements/recommendations for international students: Required—TOEFL (minimum score 575 paper-based; 213 computer-based). *Application deadline:* For fall admission, 1/31 for domestic students; for spring admission, 8/31 priority date for domestic students. Applications are processed on a rolling basis. Application fee: $45. *Expenses:* Tuition, state resident: part-time $267 per credit. Tuition, nonresident: part-time $665 per credit. Part-time tuition and fees vary according to course load and degree level. *Financial support:* In 2005–06, 25 research assistantships with tuition reimbursements, 12 teaching assistantships with full tuition reimbursements were awarded; fellowships, career-related internships or fieldwork and Federal Work-Study also available. Financial award application deadline: 3/1; financial award applicants required to submit FAFSA. Total annual research expenditures: $6.7 million. *Unit head:* Dr. Perry Brown, Dean, 406-243-5521, Fax: 406-243-4845, E-mail: pbrown@forestry.umt.edu.

University of Nebraska–Lincoln, Graduate College, College of Agricultural Sciences and Natural Resources, Lincoln, NE 68588. Offers M Ag, MA, MS, PhD. *Degree requirements:* For doctorate, thesis/dissertation, comprehensive exam. *Entrance requirements:* Additional exam requirements/recommendations for international students: Required—TOEFL. Electronic applications accepted. *Faculty research:* Environmental sciences, animal sciences, human resources and family sciences, plant breeding and genetics, food and nutrition.

University of New Hampshire, Graduate School, Interdisciplinary Programs, Doctoral Program in Natural Resources and Earth System Science, Durham, NH 03824. Offers earth and environmental science (PhD), including geology, oceanography; natural resources and environmental studies (PhD). *Faculty:* 72 full-time. *Students:* 39 full-time (20 women), 24 part-time (9 women); includes 3 minority (1 African American, 1 Asian American or Pacific Islander, 1 Hispanic American), 13 international. Average age 38. 36 applicants, 56% accepted, 13 enrolled. In 2005, 4 degrees awarded. *Degree requirements:* For doctorate, thesis/dissertation. *Entrance requirements:* For doctorate, GRE (if from a non US university). Additional exam requirements/recommendations for international students: Required—TOEFL (minimum score 550 paper-based; 213 computer-based); Recommended—TSE. *Application deadline:* For fall admission, 4/1 priority date for domestic students, 4/1 priority date for international students. Applications are processed on a rolling basis. Application fee: $60. Electronic applications accepted. *Expenses:* Tuition, state resident: full-time $8,010; part-time $445 per credit hour. Tuition, nonresident: full-time $19,730; part-time $810 per credit hour. Required fees: $322 per semester. Tuition and fees vary according to course load and program. *Financial support:* In 2005–06, 9 fellowships, 17 research assistantships, 3 teaching assistantships were awarded; Federal Work-Study, scholarships/grants, and tuition waivers (full and partial) also available. Financial award application deadline: 2/15. *Faculty research:* Environmental and natural resource studies and management. *Unit head:* Dr. Fred D. Short, Chairperson, 603-862-3045. *Application contact:* Dr. Alison Magill, Administrative Assistant, 603-862-4098, E-mail: nress.phd.program@unh.edu.

University of Northern British Columbia, Office of Graduate Studies, Prince George, BC V2N 4Z9, Canada. Offers business administration (Diploma); community health science (M Sc); disability management (MA); education (M Ed); first nations studies (MA); gender studies (MA); history (MA); interdisciplinary studies (MA); international studies (MA); mathematical, computer and physical sciences (M Sc); natural resources and environmental studies (M Sc, MA, MNRES, PhD); political science (MA); psychology (M Sc, PhD); social work (MSW). Part-time and evening/weekend programs available. Postbaccalaureate distance learning degree programs offered (no on-campus study). *Degree requirements:* For master's and doctorate, thesis/dissertation. *Entrance requirements:* For master's, GRE, minimum B average in undergraduate course work; for doctorate, candidacy exam, minimum A average in graduate course work.

University of Oklahoma, Graduate College, College of Earth and Energy, School of Petroleum and Geological Engineering, Program in Petroleum Engineering, Norman, OK 73019-0390. Offers natural gas engineering (MS); petroleum engineering (MS, PhD). Part-time programs available. Postbaccalaureate distance learning degree programs offered (minimal on-campus study). *Students:* 56 full-time (7 women), 38 part-time (4 women); includes 8 minority (6 African Americans, 1 Asian American or Pacific Islander, 1 Hispanic American), 80 international. 40 applicants, 38% accepted, 11 enrolled. In 2005, 44 master's, 5 doctorates awarded. *Degree requirements:* For master's, industrial team project or thesis, thesis optional; for doctorate, thesis/dissertation. *Entrance requirements:* For master's, GRE General Test, bachelor's degree in engineering, 3 letters of recommendation, minimum GPA of 3.0 during final 60 hours of undergraduate course work; for doctorate, GRE General Test, minimum GPA of 3.0, 3 letters of recommendation. Additional exam requirements/recommendations for international students: Required—TOEFL (minimum score 550 paper-based; 213 computer-based). *Application deadline:* For fall admission, 6/1 priority date for domestic students, 4/1 priority date for international students; for spring admission, 11/1 for domestic students, 9/1 for international students. Applications are processed on a rolling basis. Application fee: $40 ($90 for international students). *Expenses:* Tuition, state resident: full-time $3,029; part-time $126 per credit hour. Tuition, nonresident: full-time $10,807; part-time $450 per credit hour. Required fees: $1,231; $44 per credit hour. Tuition and fees vary according to course load and program. *Financial support:* In 2005–06, 9 students received support; research assistantships with partial tuition reimbursements available, teaching assistantships with partial tuition reimbursements available, career-related internships or fieldwork, health care benefits, and unspecified assistantships available. Financial award application deadline: 4/15; financial award applicants required to submit FAFSA.

University of Rhode Island, Graduate School, College of the Environment and Life Sciences, Department of Environmental and Natural Resource Economics, Kingston, RI 02881. Offers resource economics (MS, PhD). In 2005, 15 master's, 1 doctorate awarded. *Degree requirements:* For master's, thesis optional; for doctorate, thesis/dissertation. *Entrance requirements:* For master's and doctorate, GRE General Test. Additional exam requirements/recommendations for international students: Required—TOEFL. *Application deadline:* For fall admission, 4/15 for domestic students. Applications are processed on a rolling basis. *Expenses:* Tuition, state resident: full-time $5,522; part-time $307 per credit. Tuition, nonresident: full-time $15,992; part-time $888 per credit. Required fees: $1,786; $73 per credit. One-time fee: $80 part-time. *Unit head:* Dr. James Anderson, Chairperson, 401-874-2471. *Application contact:* Dr. Tim Tyrrell, Graduate Admissions Committee, 401-874-2472, Fax: 401-782-4766, E-mail: renri@uriacc.uri.edu.

University of Vermont, Graduate College, The Rubenstein School of Environment and Natural Resources, Program in Natural Resources, Burlington, VT 05405. Offers natural resources (MS, PhD), including aquatic ecology and watershed science (MS), environment thought and culture (MS), environment, science and public affairs (MS), forestry (MS). *Degree requirements:* For master's, thesis or alternative; for doctorate, thesis/dissertation. *Entrance requirements:* For master's and doctorate, GRE General Test. Additional exam requirements/recommendations for international students: Required—TOEFL (minimum score 550 paper-based; 213 computer-based). *Application deadline:* For fall admission, 3/1 for domestic students. Applications are processed on a rolling basis. Application fee: $25. Electronic applications accepted. *Expenses:* Tuition, area resident: Part-time $410 per credit hour. Tuition, nonresident: part-time $1,034 per credit hour. *Financial support:* Fellowships, research assistantships, teaching assistantships available. Financial award application deadline: 3/1. *Unit head:* Dr. Deane Wang, Coordinator, 802-656-2620.

University of Wisconsin–Stevens Point, College of Natural Resources, Stevens Point, WI 54481-3897. Offers MS. Part-time programs available. *Faculty:* 25 full-time (3 women), 8 part-time/adjunct (1 woman). *Students:* 30 full-time (14 women), 15 part-time (9 women);

includes 1 minority (Asian American or Pacific Islander), 2 international. In 2005, 29 degrees awarded. *Degree requirements:* For master's, thesis or alternative. *Entrance requirements:* For master's, GRE. *Application deadline:* For fall admission, 3/15 for domestic students; for spring admission, 11/15 for domestic students. Applications are processed on a rolling basis. Application fee: $45. *Expenses:* Tuition, state resident: full-time $5,619; part-time $312 per credit. Tuition, nonresident: full-time $16,229; part-time $902 per credit. Required fees: $651; $64 per credit. *Financial support:* Research assistantships, teaching assistantships, career-related internships or fieldwork, Federal Work-Study, and unspecified assistantships available. Support available to part-time students. Financial award application deadline: 5/1; financial award applicants required to submit FAFSA. *Faculty research:* Wildlife environmental education, fisheries, forestry, policy and planning. *Unit head:* Dr. Christine Thomas, Dean, 715-346-4617, Fax: 715-346-3624.

University of Wyoming, Graduate School, College of Agriculture, Department of Renewable Resources, Laramie, WY 82070. Offers entomology (MS, PhD); rangeland ecology and watershed management (MS, PhD), including soil sciences (PhD), soil sciences and water resources (MS), water resources. Part-time programs available. *Faculty:* 21 full-time (2 women). *Students:* 26 full-time (10 women), 13 part-time (7 women), 9 international. In 2005, 5 master's awarded. *Degree requirements:* For master's, thesis (for some programs); for doctorate, thesis/dissertation. *Entrance requirements:* For master's and doctorate, GRE General Test, minimum GPA of 3.0. Additional exam requirements/recommendations for international students: Required—TOEFL. *Application deadline:* For fall admission, 6/1 for domestic students; for spring admission, 12/1 priority date for domestic students. Applications are processed on a rolling basis. Application fee: $50. Electronic applications accepted. *Expenses:* Tuition, state resident: full-time $3,720; part-time $155 per credit hour. Tuition, nonresident: full-time $10,704; part-time $446 per credit hour. Required fees: $666; $162 per semester. Tuition and fees vary according to course load and program. *Financial support:* In 2005–06, 8 students received support, including 8 research assistantships with full tuition reimbursements available (averaging $10,062 per year); career-related internships or fieldwork and Federal Work-Study also available. Financial award application deadline: 3/1. *Faculty research:* Plant control, grazing management, riparian restoration, riparian management, reclamation. *Unit head:* Dr. Richard A. Olson, Interim Head, 307-766-2263, Fax: 307-766-6403, E-mail: rolson@uwyo.edu. *Application contact:* Kimm Mann-Malody, Office Associate, 307-766-2263, Fax: 307-766-6403, E-mail: kimmmann@uwyo.edu.

University of Wyoming, Graduate School, College of Arts and Sciences, Department of Geography, Program in Rural Planning and Natural Resources, Laramie, WY 82070. Offers community and regional planning and natural resources (MP). *Faculty:* 1 full-time (0 women). *Students:* 1 applicant, 100% accepted. In 2005, 3 degrees awarded. *Degree requirements:* For master's, thesis or alternative. *Entrance requirements:* For master's, GRE General Test, minimum GPA of 3.0. Additional exam requirements/recommendations for international students: Required—TOEFL. *Application deadline:* For fall admission, 2/15 for domestic students. Applications are processed on a rolling basis. Application fee: $50. *Expenses:* Tuition, state resident: full-time $3,720; part-time $155 per credit hour. Tuition, nonresident: full-time $10,704; part-time $446 per credit hour. Required fees: $666; $162 per semester. Tuition and fees vary according to course load and program. *Financial support:* In 2005–06, 1 teaching assistantship with full and partial tuition reimbursement was awarded; career-related internships or fieldwork, Federal Work-Study, scholarships/grants, and unspecified assistantships also available. Financial award application deadline: 3/1. *Faculty research:* Rural and small town planning, public land management. Total annual research expenditures: $10,400. *Unit head:* Dr. John Allen, Chair, Department of Geography, 307-766-3311, Fax: 307-766-3294, E-mail: geography-info@uwyo.edu.

Utah State University, School of Graduate Studies, College of Natural Resources, Interdisciplinary Program in Natural Resources, Logan, UT 84322. Offers MNR. *Students:* 6 full-time (1 woman), 1 (woman) part-time, 1 international. Average age 30. 5 applicants, 40% accepted, 1 enrolled. In 2005, 3 degrees awarded. *Entrance requirements:* For master's, GRE General Test, minimum GPA of 3.0. Additional exam requirements/recommendations for international students: Required—TOEFL. *Application deadline:* For fall admission, 2/15 for domestic students; for spring admission, 10/15 priority date for domestic students. Applications are processed on a rolling basis. Application fee: $50 ($60 for international students). *Faculty research:* Ecosystem management, human dimensions, quantitative methods, informative management. Total annual research expenditures: $4 million. *Application contact:* Dr. Raymond D. Dueser, Associate Dean, 435-797-2445, Fax: 435-797-2448, E-mail: nradvise@cc.usu.edu.

Virginia Polytechnic Institute and State University, Graduate School, College of Natural Resources, Blacksburg, VA 24061. Offers MF, MNR, MS, PhD. *Faculty:* 114 full-time (20 women). *Students:* 119 full-time (37 women), 69 part-time (37 women); includes 8 minority (1 American Indian/Alaska Native, 4 Asian Americans or Pacific Islanders, 3 Hispanic Americans), 24 international. Average age 31. 127 applicants, 54% accepted, 56 enrolled. In 2005, 42 master's, 12 doctorates awarded. *Entrance requirements:* Additional exam requirements/recommendations for international students: Required—TOEFL. *Application deadline:* Applications are processed on a rolling basis. Application fee: $45. Electronic applications accepted. *Expenses:* Tuition, state resident: full-time $6,558; part-time $364 per credit. Tuition, nonresident: full-time $11,296; part-time $628 per credit. Required fees: $1,419; $468 per credit. $234 per term. *Financial support:* In 2005–06, 1 fellowship with full tuition reimbursement (averaging $458 per year), 64 research assistantships with full tuition reimbursements (averaging $14,989 per year), 29 teaching assistantships with full tuition reimbursements (averaging $10,051 per year) were awarded; career-related internships or fieldwork, Federal Work-Study, scholarships/grants, health care benefits, and unspecified assistantships also available. *Unit head:* Dr. Michael Kelly, Head, 540-231-5481, Fax: 540-231-7664. *Application contact:* Peggy Quarterman, 540-231-3479, Fax: 540-231-7664, E-mail: pquarter@vt.edu.

Washington State University, Graduate School, College of Agricultural, Human, and Natural Resource Sciences, Department of Natural Resource Sciences, Pullman, WA 99164. Offers environmental and natural resource sciences (PhD); natural resource sciences (MS). *Faculty:* 17. *Students:* 18 full-time (12 women), 6 part-time (3 women); includes 1 minority (Asian American or Pacific Islander), 5 international. Average age 33. 66 applicants, 14% accepted, 7 enrolled. In 2005, 10 master's, 1 doctorate awarded. *Degree requirements:* For master's, oral exam, thesis optional; for doctorate, thesis/dissertation, oral exam. *Entrance requirements:* For master's and doctorate, GRE General Test, minimum GPA of 3.0, 3 letters of recommendation. Additional exam requirements/recommendations for international students: Required—TOEFL. *Application deadline:* For fall admission, 2/15 priority date for domestic students, 2/15 priority date for international students; for spring admission, 9/15 for domestic students, 7/1 for international students. Applications are processed on a rolling basis. Application fee: $35. *Expenses:* Tuition, state resident: full-time $6,295; part-time $336 per credit. Tuition, nonresident: full-time $15,949; part-time $819 per credit. Required fees: $933. Part-time tuition and fees vary according to campus/location and program. *Financial support:* In 2005–06, 20 students received support, including 3 fellowships (averaging $2,333 per year), 9 research assistantships with full and partial tuition reimbursements available (averaging $12,661 per year), 7 teaching assistantships with full and partial tuition reimbursements available (averaging $12,881 per year); career-related internships or fieldwork, Federal Work-Study, institutionally sponsored loans, tuition waivers (partial), and unspecified assistantships also available. Financial award application deadline: 4/1; financial award applicants required to submit FAFSA. *Faculty research:* Restoration ecology, landscape ecology, controlled and uncontrolled wild-land fire. Total annual research expenditures: $1.3 million. *Unit head:* Dr. Keith A. Blatner, Chair, 509-335-4499, E-mail: blatner@wsu.edu. *Application contact:* Julie Foster, Graduate Coordinator, 509-335-8570, Fax: 509-335-7862, E-mail: jfoster@wsu.edu.

Washington State University, Graduate School, College of Sciences, Programs in Environmental Science and Regional Planning, Program in Environmental Science and Regional Planning, Pullman, WA 99164. Offers environmental and natural resource sciences (PhD); environmental science (MS). *Faculty:* 7. *Students:* 13 full-time (7 women), 5 part-time (3 women); includes 1 minority (Hispanic American), 2 international. Average age 29. 49 applicants, 59% accepted, 10 enrolled. In 2005, 11 degrees awarded. *Degree requirements:* For master's, oral exam, thesis optional; for doctorate, oral exam, written exam. *Entrance requirements:* For master's and doctorate, minimum GPA of 3.0. Additional exam requirements/recommendations for international students: Required—TOEFL. *Application deadline:* For fall admission, 2/1 priority date for domestic students, 2/1 priority date for international students. Applications are processed on a rolling basis. Application fee: $35. *Expenses:* Tuition, state resident: full-time $6,295; part-time $336 per credit. Tuition, nonresident: full-time $15,949; part-time $819 per credit. Required fees: $933. Part-time tuition and fees vary according to campus/location and program. *Financial support:* In 2005–06, 16 students received support, including 1 fellowship (averaging $3,000 per year), 4 research assistantships with full and partial tuition reimbursements available (averaging $14,000 per year), 4 teaching assistantships with full and partial tuition reimbursements available (averaging $13,600 per year); Federal Work-Study, institutionally sponsored loans, and tuition waivers (partial) also available. Financial award application deadline: 4/1; financial award applicants required to submit FAFSA. *Application contact:* Elaine O'Fallon, Coordinator, 509-335-8538, Fax: 509-335-7636, E-mail: esrp@wsu.edu.

West Virginia University, Davis College of Agriculture, Forestry and Consumer Sciences, Division of Resource Management and Sustainable Development, Program in Natural Resource Economics, Morgantown, WV 26506. Offers PhD. Part-time programs available. *Students:* 7 full-time (4 women), 3 part-time (2 women); includes 1 minority (Asian American or Pacific Islander), 7 international. Average age 34. 18 applicants, 56% accepted. In 2005, 4 degrees awarded. *Degree requirements:* For doctorate, thesis/dissertation. *Entrance requirements:* For doctorate, GRE General Test. Additional exam requirements/recommendations for international students: Required—TOEFL. Application fee: $45. *Expenses:* Tuition, state resident: full-time $4,582; part-time $258 per credit hour. Tuition, nonresident: full-time $1,382; part-time $741 per credit hour. *Financial support:* In 2005–06, 4 research assistantships, 1 teaching assistantship were awarded; Federal Work-Study, institutionally sponsored loans, and tuition waivers (partial) also available. Financial award application deadline: 2/1; financial award applicants required to submit FAFSA. *Unit head:* Dr. Gerard E. D'Souza, Graduate Coordinator, 304-293-4832 Ext. 4471, Fax: 304-293-3740, E-mail: gerald.d'souza@mail.wvu.edu.

Range Science

Colorado State University, Graduate School, Warner College of Natural Resources, Department of Forest, Rangeland, and Watershed Stewardship, Fort Collins, CO 80523-0015. Offers forest sciences (MS, PhD); natural resource stewardship (MNRS); rangeland ecosystem science (MS, PhD); watershed science (MS). Part-time programs available. *Faculty:* 27 full-time (5 women), 1 part-time/adjunct (0 women). *Students:* 58 full-time (23 women), 90 part-time (39 women); includes 11 minority (1 African American, 2 American Indian/Alaska Native, 1 Asian American or Pacific Islander, 7 Hispanic Americans), 10 international. Average age 33. 101 applicants, 35% accepted, 30 enrolled. In 2005, 18 master's, 5 doctorates awarded. Terminal master's awarded for partial completion of doctoral program. *Degree requirements:* For master's, thesis optional; for doctorate, thesis/dissertation, comprehensive exam, registration. *Entrance requirements:* For master's and doctorate, GRE General Test, minimum GPA of 3.0. Additional exam requirements/recommendations for international students: Required—TOEFL. *Application deadline:* For fall admission, 4/1 for domestic students; for spring admission, 9/1 priority date for domestic students. Applications are processed on a rolling basis. Application fee: $50. Electronic applications accepted. *Expenses:* Tuition, state resident: full-time $3,690; part-time $205 per credit. Tuition, nonresident: full-time $14,958; part-time $831 per credit. Required fees: $1,061. *Financial support:* Career-related internships or fieldwork, Federal Work-Study, institutionally sponsored loans, and traineeships available. Financial award application deadline: 5/1. *Faculty research:* Ecology, natural resource management, hydrology, restoration, human dimensions. Total annual research expenditures: $4.1 million. *Unit head:* Dr. N. Thompson Hobbs, Head, 970-491-6911, Fax: 970-491-6754, E-mail: nthobbs@warnercnr@colostate.edu. *Application contact:* Graduate Coordinator, 970-491-4994, Fax: 970-491-6754.

Kansas State University, Graduate School, College of Agriculture, Department of Agronomy, Manhattan, KS 66506. Offers crop science (MS, PhD); range management (MS, PhD); soil science (MS, PhD); weed science (MS, PhD). Part-time programs available. *Faculty:* 32 full-time (3 women), 10 part-time/adjunct (1 woman). *Students:* 36 full-time (12 women), 13 part-time (3 women), 8 international. 14 applicants, 64% accepted, 8 enrolled. In 2005, 10 master's, 2 doctorates awarded. Terminal master's awarded for partial completion of doctoral program. *Degree requirements:* For master's, thesis or alternative, oral exam; for doctorate, thesis/dissertation, preliminary exams. *Entrance requirements:* For master's, minimum GPA of 3.0 in BS; for doctorate, minimum GPA of 3.5 in master's program. Additional exam requirements/recommendations for international students: Required—TOEFL (minimum score 500 paper-based; 250 computer-based). *Application deadline:* For fall admission, 2/1 for domestic students; for spring admission, 10/1 for domestic students. Applications are processed on a rolling basis. Application fee: $30 ($55 for international students). Electronic applications accepted. *Expenses:* Tuition, state resident: full-time $5,160; part-time $215. Tuition, nonresident: full-time $12,816; part-time $534. Required fees: $564. *Financial support:* In 2005–06, 35 research assistantships (averaging $15,609 per year), 9 teaching assistantships with partial tuition reimbursements (averaging $6,304 per year) were awarded; institutionally sponsored loans and scholarships/grants also available. Support available to part-time students. Financial award application deadline: 3/1; financial award applicants required to submit FAFSA. *Faculty research:* Weed science, environmental soil science, range science, plant genetics, climate change. Total annual research expenditures: $3 million. *Unit head:* Dr. Gary Pierzynski, Head, 785-532-6101, Fax: 785-532-6094, E-mail: gmp@ksu.edu. *Application contact:* Dr. Bill Schapaugh, Director, 785-532-7242, E-mail: wts@ksu.edu.

Montana State University, College of Graduate Studies, College of Agriculture, Department of Animal and Range Sciences, Bozeman, MT 59717. Offers MS, PhD. Part-time programs available. *Faculty:* 16 full-time (4 women), 2 part-time/adjunct (0 women). *Students:* 8 full-time (5 women), 13 part-time (8 women), 4 international. Average age 28. 13 applicants, 38% accepted, 5 enrolled. In 2005, 8 degrees awarded. *Degree requirements:* For master's, comprehensive exam, registration; for doctorate, thesis/dissertation, comprehensive exam, registration. *Entrance requirements:* For master's and doctorate, GRE General Test. Additional exam requirements/recommendations for international students: Required—TOEFL (minimum score 550 paper-based; 213 computer-based). *Application deadline:* For fall admission, 7/15 priority date for domestic students, 5/15 priority date for international students; for spring admission, 12/1 priority date for domestic students, 10/1 priority date for international students. Applications are

Range Science

Montana State University (continued)
processed on a rolling basis. Application fee: $30. Electronic applications accepted. *Expenses:* Tuition, state resident: full-time $4,132. Tuition, nonresident: full-time $1,132. *Financial support:* In 2005–06, 20 students received support, including 12 research assistantships with full and partial tuition reimbursements available, 5 teaching assistantships with partial tuition reimbursements available; health care benefits, tuition waivers (partial), and unspecified assistantships also available. Financial award application deadline: 3/1; financial award applicants required to submit FAFSA. *Faculty research:* Reproductive physiology, ruminant nutrition, range management and ecology, final product quality, systems approach to livestock production. Total annual research expenditures: $3.9 million. *Unit head:* Wayne Gipp, Interim Department Head, 406-994-4850, Fax: 406-994-5589, E-mail: wgipp@montana.edu.

New Mexico State University, Graduate School, College of Agriculture and Home Economics, Department of Animal and Range Sciences, Las Cruces, NM 88003-8001. Offers animal science (M Ag, MS, PhD); range science (M Ag, MS, PhD). Part-time programs available. *Faculty:* 17 full-time (1 woman), 9 part-time/adjunct (0 women). *Students:* 28 full-time (12 women), 9 part-time (2 women); includes 7 minority (1 American Indian/Alaska Native, 6 Hispanic Americans), 6 international. Average age 30. 28 applicants, 54% accepted, 9 enrolled. In 2005, 12 master's, 3 doctorates awarded. *Degree requirements:* For master's, thesis, seminar; for doctorate, thesis/dissertation, research tool. *Entrance requirements:* For master's, minimum GPA of 3.0 in last 60 hours of undergraduate course work (MS); for doctorate, minimum graduate GPA of 3.2. *Application deadline:* For fall admission, 7/1 for domestic students; for spring admission, 11/1 for domestic students. Applications are processed on a rolling basis. Application fee: $30 ($50 for international students). Electronic applications accepted. *Expenses:* Tuition, state resident: full-time $3,156; part-time $175 per credit. Tuition, nonresident: full-time $12,510; part-time $565 per credit. Required fees: $1,050. *Financial support:* In 2005–06, 16 research assistantships, 14 teaching assistantships were awarded; Federal Work-Study also available. Support available to part-time students. Financial award application deadline: 3/1. *Faculty research:* Reproductive physiology, ruminant nutrition, nutrition toxicology, range ecology, wildland hydrology. *Unit head:* Dr. Mark Wise, Head, 505-646-2514, Fax: 505-646-5441, E-mail: mawise@nmsu.edu.

North Dakota State University, The Graduate School, College of Agriculture, Food Systems, and Natural Resources, Department of Animal and Range Sciences, Fargo, ND 58105. Offers animal science (MS, PhD); natural resource management (MS, PhD); range sciences (MS, PhD). *Faculty:* 25 full-time (6 women), 7 part-time/adjunct (1 woman). *Students:* 31 full-time (15 women), 16 part-time (6 women); includes 3 minority (2 African Americans, 1 Asian American or Pacific Islander), 3 international. Average age 25. 26 applicants, 62% accepted. In 2005, 17 master's, 2 doctorates awarded. *Degree requirements:* For master's, thesis/dissertation; for doctorate, thesis/dissertation, comprehensive exam. *Entrance requirements:* For master's and doctorate, GRE General Test. Additional exam requirements/recommendations for international students: Required—TOEFL. *Application deadline:* Applications are processed on a rolling basis. Application fee: $45 ($60 for international students). *Financial support:* In 2005–06, 30 students received support, including 1 fellowship with tuition reimbursement available (averaging $18,000 per year), 29 research assistantships with tuition reimbursements available (averaging $13,000 per year); teaching assistantships, Federal Work-Study, institutionally sponsored loans, and tuition waivers (partial) also available. Financial award application deadline: 3/15. *Faculty research:* Reproduction, nutrition, meat and muscle biology, breeding/genetics. Total annual research expenditures: $1.5 million. *Unit head:* Don R. Kirby, Interim Chair, 701-231-8386, Fax: 701-231-7723, E-mail: donald.kirby@ndsu.edu.

Oregon State University, Graduate School, College of Agricultural Sciences, Department of Rangeland Ecology and Management, Corvallis, OR 97331. Offers M Agr, MAIS, MS, PhD. *Faculty:* 9 full-time (0 women), 2 part-time/adjunct (1 woman). *Students:* 14 full-time (6 women), 2 part-time; includes 1 minority (Hispanic American), 1 international. Average age 31. In 2005, 5 master's, 3 doctorates awarded. Terminal master's awarded for partial completion of doctoral program. *Degree requirements:* For master's, thesis (for some programs); for doctorate, thesis/dissertation. *Entrance requirements:* For master's and doctorate, GRE, minimum GPA of 3.0 in last 90 hours of course work. Additional exam requirements/recommendations for international students: Required—TOEFL. *Application deadline:* For fall admission, 6/1 for domestic students; for spring admission, 12/15 for domestic students. Applications are processed on a rolling basis. Application fee: $50. *Expenses:* Tuition, area resident: Part-time $301 per credit. Tuition, state resident: full-time $8,139; part-time $501 per credit. Tuition, nonresident: full-time $14,376; part-time $532 per credit. Required fees: $1,266. *Financial support:* Research assistantships, career-related internships or fieldwork, Federal Work-Study, and institutionally sponsored loans available. Support available to part-time students. Financial award application deadline: 2/1. *Faculty research:* Range ecology, watershed science, animal grazing, agroforestry. *Unit head:* Dr. William C. Krueger, Head, 541-737-1615, Fax: 541-737-0504, E-mail: william.c.krueger@orst.edu. *Application contact:* Dr. Paul S. Doescher, Head Adviser, 541-737-1622, Fax: 541-737-0504, E-mail: paul.s.doescher@orst.edu.

Sul Ross State University, Division of Agricultural and Natural Resource Science, Program in Range and Wildlife Management, Alpine, TX 79832. Offers M Ag, MS. Part-time programs available. *Degree requirements:* For master's, thesis (for some programs). *Entrance requirements:* For master's, GRE General Test, minimum undergraduate GPA of 2.5 in last 60 hours.

Texas A&M University, College of Agriculture and Life Sciences, Department of Rangeland Ecology and Management, College Station, TX 77843. Offers M Agr, MS, PhD. *Faculty:* 12 full-time (0 women). *Students:* 31 full-time (13 women), 14 part-time (6 women); includes 3 minority (1 Asian American or Pacific Islander, 2 Hispanic Americans), 9 international. Average age 31. 15 applicants, 80% accepted, 10 enrolled. In 2005, 10 master's, 10 doctorates awarded. Terminal master's awarded for partial completion of doctoral program. *Degree requirements:* For master's, thesis optional; for doctorate, thesis/dissertation. *Entrance requirements:* For master's and doctorate, GRE General Test. Additional exam requirements/recommendations for international students: Required—TOEFL. *Application deadline:* For fall admission, 3/1 for domestic students; for spring admission, 8/1 priority date for domestic students. Applications are processed on a rolling basis. Application fee: $50 ($75 for international students). Electronic applications accepted. *Expenses:* Tuition, state resident: full-time $4,488; part-time $187 per credit hour. Tuition, nonresident: full-time $11,112; part-time $463 per credit hour. Required fees:$1,974. *Financial support:* In 2005–06, research assistantships (averaging $12,566 per year), teaching assistantships (averaging $12,200 per year) were awarded; fellowships, career-related internships or fieldwork, scholarships/grants, and unspecified assistantships also available. Support available to part-time students. Financial award application deadline: 4/1; financial award applicants required to submit FAFSA. *Faculty research:* Plant ecology, restoration ecology, watershed management, integrated resource management, information technology. *Unit head:* Dr. Steve Whisenant, Head, 979-845-5579, Fax: 979-845-6430. *Application contact:* Dr. Jennifer E. Funkhouser, Graduate Advisor, 979-845-5579, Fax: 979-845-6430, E-mail: j-funkhouser@tamu.edu.

Texas A&M University–Kingsville, College of Graduate Studies, College of Agriculture and Home Economics, Program in Range and Wildlife Management, Kingsville, TX 78363. Offers MS. *Degree requirements:* For master's, thesis or alternative, comprehensive exam. *Entrance requirements:* For master's, GRE General Test, minimum GPA of 3.0. Additional exam requirements/recommendations for international students: Required—TOEFL.

Texas Tech University, Graduate School, College of Agricultural Sciences and Natural Resources, Department of Range, Wildlife, and Fisheries Management, Lubbock, TX 79409. Offers fisheries science (MS, PhD); range science (MS, PhD); wildlife science (MS, PhD). Part-time programs available. *Faculty:* 10 full-time (0 women), 1 part-time/adjunct (0 women). *Students:* 36 full-time (8 women), 2 part-time (both women), 12 international. Average age 30. 12 applicants, 67% accepted, 8 enrolled. In 2005, 9 master's, 5 doctorates awarded. *Degree requirements:* For master's and doctorate, thesis/dissertation. *Entrance requirements:* For master's and doctorate, GRE General Test. Additional exam requirements/recommendations for international students: Required—TOEFL (minimum score 550 paper-based; 213 computer-based). *Application deadline:* Applications are processed on a rolling basis. Application fee: $50 ($60 for international students). Electronic applications accepted. *Expenses:* Tuition, state resident: full-time $4,296. Tuition, nonresident: full-time $10,920. Required fees: $1,992. Tuition and fees vary according to program. *Financial support:* In 2005–06, 26 students received support, including 22 research assistantships with partial tuition reimbursements available (averaging $10,622 per year), 6 teaching assistantships with partial tuition reimbursements available (averaging $11,986 per year); Federal Work-Study and institutionally sponsored loans also available. Support available to part-time students. Financial award application deadline: 4/15; financial award applicants required to submit FAFSA. *Faculty research:* Use of fire on range lands, waterfowl, upland game birds and playa lakes in the southern Great Plains, reproductive physiology in fisheries, conservation biology. Total annual research expenditures: $1.4 million. *Unit head:* Dr. Ernest B. Fish, Chairman, 806-742-2841, Fax: 806-742-2280. *Application contact:* L. Jeannine Becker, Graduate Secretary, 806-742-2825, E-mail: jeannine.becker@ttu.edu.

The University of Arizona, Graduate College, College of Agriculture and Life Sciences, School of Natural Resources, Program in Rangeland Ecology and Management, Tucson, AZ 85721. Offers MS, PhD. *Degree requirements:* For master's, thesis/dissertation; for doctorate, thesis/dissertation, comprehensive exam. *Entrance requirements:* For master's and doctorate, GRE General Test, minimum GPA of 3.0, 3 letters of recommendation. Additional exam requirements/recommendations for international students: Required—TOEFL (minimum score 550 paper-based; 213 computer-based). *Faculty research:* Criteria for defining, mapping, and evaluating range sites; methods of establishing forage plants on southwestern range lands; plants for pollution and erosion control, beautification, and browse.

University of California, Berkeley, Graduate Division, Group in Range Management, Berkeley, CA 94720-1500. Offers MS. *Degree requirements:* For master's, thesis, registration. *Entrance requirements:* For master's, GRE General Test, minimum GPA of 3.0. Additional exam requirements/recommendations for international students: Required—TOEFL. *Faculty research:* Grassland and savannah ecology, wetland ecology, oak woodland classification, wildlife habitat management.

University of Idaho, College of Graduate Studies, College of Natural Resources, Department of Rangeland Ecology and Management, Moscow, ID 83844-2282. Offers MS, PhD. *Students:* 9 full-time (8 women), 2 part-time (1 woman). Average age 38. In 2005, 6 degrees awarded. *Degree requirements:* For doctorate, thesis/dissertation. *Entrance requirements:* For master's, minimum GPA of 2.8; for doctorate, minimum undergraduate GPA of 2.8, 3.0 graduate. *Application deadline:* For fall admission, 8/1 for domestic students; for spring admission, 12/15 for domestic students. Application fee: $55 ($60 for international students). *Expenses:* Tuition, nonresident: full-time $8,770; part-time $130 per credit. Required fees: $4,508; $217 per credit. *Financial support:* Research assistantships, teaching assistantships available. Financial award application deadline: 2/15. *Unit head:* Dr. Karen L. Launchbaugh, Head, 208-885-4394.

University of Wyoming, Graduate School, College of Agriculture, Department of Renewable Resources, Laramie, WY 82070. Offers entomology (MS, PhD); rangeland ecology and watershed management (MS, PhD), including soil sciences (PhD), soil sciences and water resources (MS), water resources. Part-time programs available. *Faculty:* 21 full-time (2 women). *Students:* 26 full-time (10 women), 13 part-time (7 women), 9 international. In 2005, 5 master's awarded. *Degree requirements:* For master's, thesis (for some programs); for doctorate, thesis/dissertation. *Entrance requirements:* For master's and doctorate, GRE General Test, minimum GPA of 3.0. Additional exam requirements/recommendations for international students: Required—TOEFL. *Application deadline:* For fall admission, 6/1 for domestic students; for spring admission, 12/1 priority date for domestic students. Applications are processed on a rolling basis. Application fee: $50. Electronic applications accepted. *Expenses:* Tuition, state resident: full-time $3,720; part-time $155 per credit hour. Tuition, nonresident: full-time $10,704; part-time $446 per credit hour. Required fees: $666; $162 per semester. Tuition and fees vary according to course load and program. *Financial support:* In 2005–06, 8 students received support, including 8 research assistantships with full tuition reimbursements available (averaging $10,062 per year); career-related internships or fieldwork and Federal Work-Study also available. Financial award application deadline: 3/1. *Faculty research:* Plant control, grazing management, riparian restoration, riparian management, reclamation. *Unit head:* Dr. Richard A. Olson, Interim Head, 307-766-2263, Fax: 307-766-6403, E-mail: rolson@uwyo.edu. *Application contact:* Kimm Mann-Malody, Office Associate, 307-766-2263, Fax: 307-766-6403, E-mail: kimmmann@uwyo.edu.

Utah State University, School of Graduate Studies, College of Natural Resources, Department of Forest, Range, and Wildlife Sciences, Logan, UT 84322. Offers ecology (MS, PhD); forestry (MS, PhD); range science (MS, PhD); wildlife biology (MS, PhD). Part-time programs available. *Faculty:* 22 full-time (4 women), 17 part-time/adjunct (3 women). *Students:* 220 full-time (90 women), 52 part-time (17 women); includes 6 minority (all Asian Americans or Pacific Islanders), 49 international. Average age 24. 56 applicants, 61% accepted, 28 enrolled. In 2005, 11 master's, 4 doctorates awarded. *Degree requirements:* For master's, thesis/dissertation; for doctorate, thesis/dissertation, comprehensive exam. *Entrance requirements:* For master's and doctorate, GRE General Test, minimum GPA of 3.0. Additional exam requirements/recommendations for international students: Required—TOEFL. *Application deadline:* For fall admission, 6/15 for domestic students; for spring admission, 10/15 for domestic students. Applications are processed on a rolling basis. Application fee: $50 ($60 for international students). *Financial support:* In 2005–06, 14 research assistantships with partial tuition reimbursements (averaging $13,600 per year), 6 teaching assistantships (averaging $6,000 per year) were awarded; fellowships, career-related internships or fieldwork, Federal Work-Study, and institutionally sponsored loans also available. *Faculty research:* Range plant ecophysiology, plant community ecology, ruminant nutrition, population ecology. Total annual research expenditures: $3.5 million. *Unit head:* Dr. Johan W. duToit, Head, 435-797-2837, Fax: 435-797-3796. *Application contact:* Gaye Griffeth, Staff Assistant, 435-797-2503, Fax: 435-797-3796, E-mail: ggriffeth@cnr.usu.edu.

Water Resources

Albany State University, College of Arts and Sciences, Department of History, Political Science and Public Administration, Albany, GA 31705-2717. Offers community and economic development (MPA); criminal justice (MPA); fiscal management (MPA); general management (MPA); health administration and policy (MPA); human resources management (MPA); public policy (MPA); water resource management and policy (MPA). *Accreditation:* NASPAA. Part-time programs available. *Degree requirements:* For master's, thesis, comprehensive exam. *Entrance requirements:* For master's, GRE General Test, minimum GPA of 2.5. Electronic applications accepted. *Faculty research:* Transportation, urban affairs, political economy.

Albany State University, School of Business, Albany, GA 31705-2717. Offers water policy (MBA). *Accreditation:* ACBSP. Part-time and evening/weekend programs available. Postbaccalaureate distance learning degree programs offered (no on-campus study). *Degree requirements:* For master's, comprehensive exam. *Entrance requirements:* For master's, GMAT, minimum GPA of 2.5. Electronic applications accepted. *Faculty research:* Economic impacts, employment opportunities, instructional technology.

Colorado State University, Graduate School, College of Engineering, Department of Civil Engineering, Fort Collins, CO 80523-0015. Offers environmental engineering (ME, MS, PhD); geotechnical engineering (ME); hydraulics and wind engineering (MS, PhD); infrastructure engineering (ME); irrigation engineering (MS, PhD); structural and geotechnical engineering (MS, PhD); structural engineering (ME); water resources (ME); water resources planning and management (MS, PhD); water resources, hydrologic and environmental sciences (MS, PhD). Part-time programs available. Postbaccalaureate distance learning degree programs offered (no on-campus study). *Faculty:* 29 full-time (2 women), 2 part-time/adjunct (0 women). *Students:* 80 full-time (28 women), 123 part-time (27 women); includes 6 minority (4 Asian Americans or Pacific Islanders, 2 Hispanic Americans), 71 international. Average age 31. 145 applicants, 79% accepted, 41 enrolled. In 2005, 33 master's, 15 doctorates awarded. Terminal master's awarded for partial completion of doctoral program. *Median time to degree:* Of those who began their doctoral program in fall 1997, 50% received their degree in 8 years or less. *Degree requirements:* For master's, thesis or alternative; for doctorate, thesis/dissertation. *Entrance requirements:* For master's and doctorate, GRE General Test, minimum GPA of 3.0. Additional exam requirements/recommendations for international students: Required—TOEFL. *Application deadline:* For fall admission, 4/1 priority date for domestic students, 3/1 priority date for international students; for spring admission, 8/1 priority date for domestic students, 8/1 priority date for international students. Applications are processed on a rolling basis. Application fee: $50. Electronic applications accepted. *Expenses:* Tuition, state resident: full-time $3,690; part-time $205 per credit. Tuition, nonresident: full-time $14,958; part-time $831 per credit. Required fees: $1,061. *Financial support:* In 2005–06, 10 fellowships (averaging $13,725 per year), 11 research assistantships with tuition reimbursements (averaging $18,000 per year), 56 teaching assistantships (averaging $13,950 per year) were awarded; Federal Work-Study, institutionally sponsored loans, and traineeships also available. Financial award application deadline: 2/15. *Faculty research:* Hydraulics, hydrology, water resources, infrastructure, environmental engineering. Total annual research expenditures: $8.5 million. *Unit head:* Luis Garcia, Interim Head, 970-491-5049, Fax: 970-491-7727. *Application contact:* Kathy Stencel, Student Advisor, 970-491-5844, Fax: 970-491-7727, E-mail: kstencel@colostate.edu.

Colorado State University, Graduate School, Warner College of Natural Resources, Department of Forest, Rangeland, and Watershed Stewardship, Fort Collins, CO 80523-0015. Offers forest sciences (MS, PhD); natural resource stewardship (MNRS); rangeland ecosystem science (MS, PhD); watershed science (MS). Part-time programs available. *Faculty:* 27 full-time (5 women), 1 part-time/adjunct (0 women). *Students:* 58 full-time (23 women), 90 part-time (39 women); includes 11 minority (1 African American, 2 American Indian/Alaska Native, 1 Asian American or Pacific Islander, 7 Hispanic Americans), 10 international. Average age 33. 101 applicants, 35% accepted, 30 enrolled. In 2005, 18 master's, 5 doctorates awarded. Terminal master's awarded for partial completion of doctoral program. *Degree requirements:* For master's, thesis optional; for doctorate, thesis/dissertation, comprehensive exam, registration. *Entrance requirements:* For master's and doctorate, GRE General Test, minimum GPA of 3.0. Additional exam requirements/recommendations for international students: Required—TOEFL. *Application deadline:* For fall admission, 4/1 for domestic students; for spring admission, 9/1 priority date for domestic students. Applications are processed on a rolling basis. Application fee: $50. Electronic applications accepted. *Expenses:* Tuition, state resident: full-time $3,690; part-time $205 per credit. Tuition, nonresident: full-time $14,958; part-time $831 per credit. Required fees: $1,061. *Financial support:* Career-related internships or fieldwork, Federal Work-Study, institutionally sponsored loans, and traineeships available. Financial award application deadline: 5/1. *Faculty research:* Ecology, natural resource management, hydrology, restoration, human dimensions. Total annual research expenditures: $4.1 million. *Unit head:* Dr. N. Thompson Hobbs, Head, 970-491-6911, Fax: 970-491-6754, E-mail: nthobbs@warnercnr@colostate.edu. *Application contact:* Graduate Coordinator, 970-491-4994, Fax: 970-491-6754.

Colorado State University, Graduate School, Warner College of Natural Resources, Department of Geosciences, Fort Collins, CO 80523-0015. Offers earth sciences (PhD); geology (MS), including geomorphology, geophysics, hydrogeology, petrology/geochemistry and economic geology, sedimentology, structural geology; geosciences (PhD); wateshed (PhD). Part-time programs available. *Faculty:* 8 full-time (3 women). *Students:* 17 full-time (11 women), 22 part-time (7 women); includes 1 minority (American Indian/Alaska Native), 2 international. Average age 32. 28 applicants, 61% accepted, 7 enrolled. In 2005, 3 master's, 1 doctorate awarded. *Degree requirements:* For master's, thesis/dissertation, registration; for doctorate, thesis/dissertation, comprehensive exam, registration. *Entrance requirements:* For master's and doctorate, GRE General Test, minimum GPA of 3.0. Additional exam requirements/recommendations for international students: Required—TOEFL (minimum score 550 paper-based; 213 computer-based). *Application deadline:* For fall admission, 2/1 priority date for domestic students, 2/1 priority date for international students. Applications are processed on a rolling basis. Application fee: $50. Electronic applications accepted. *Expenses:* Tuition, state resident: full-time $3,690; part-time $205 per credit. Tuition, nonresident: full-time $14,958; part-time $831 per credit. Required fees: $1,061. *Financial support:* In 2005–06, 6 fellowships (averaging $9,000 per year), 13 research assistantships with partial tuition reimbursements (averaging $16,000 per year), 9 teaching assistantships with full tuition reimbursements (averaging $16,300 per year) were awarded; career-related internships or fieldwork, Federal Work-Study, institutionally sponsored loans, scholarships/grants, and traineeships also available. Financial award application deadline: 2/15. *Faculty research:* Snow, surface, and groundwater hydrology; fluvial geomorphology; geographic information systems; geochemistry; bedrock geology. Total annual research expenditures: $1.3 million. *Unit head:* Dr. Judith L. Hannah, Head, 970-491-5662, Fax: 970-491-6307, E-mail: jhannah@cnr.colostate.edu. *Application contact:* Sharyl Pierson, Administrative Assistant, 970-491-5662, Fax: 970-491-6307, E-mail: sharyl@cnr.colostate.edu.

Duke University, Nicholas School of the Environment and Earth Sciences, Durham, NC 27708-0328. Offers coastal environmental management (MEM); environmental economics and policy (MEM); environmental health and security (MEM); forest resource management (MF); global environmental change (MEM); resource ecology (MEM); water and air resources (MEM). *Accreditation:* SAF (one or more programs are accredited). Part-time programs available. *Degree requirements:* For master's, thesis, registration. *Entrance requirements:* For master's, GRE General Test, previous course work in biology or ecology, calculus, statistics, and microeconomics; computer familiarity with word processing and data analysis. Additional exam requirements/recommendations for international students: Required—TOEFL (minimum score 550 paper-based; 213 computer-based). Electronic applications accepted. Expenses: Contact institution. *Faculty research:* Ecosystem management, conservation ecology, earth systems, risk assessment.

See Close-Up on page 709.

Iowa State University of Science and Technology, Graduate College, College of Liberal Arts and Sciences, Department of Geological and Atmospheric Sciences, Ames, IA 50011. Offers earth science (MS, PhD); geology (MS, PhD); meteorology (MS, PhD); water resources (MS, PhD). *Faculty:* 15 full-time, 2 part-time/adjunct. *Students:* 32 full-time (12 women), 2 part-time; includes 1 minority (African American), 15 international. 40 applicants, 60% accepted, 13 enrolled. In 2005, 11 master's awarded. *Degree requirements:* For master's, thesis (for some programs); for doctorate, thesis/dissertation. *Entrance requirements:* For master's and doctorate, GRE General Test. Additional exam requirements/recommendations for international students: Required—TOEFL (paper score 530; computer score 197) or IELTS (score 6.0). *Application deadline:* For fall admission, 1/1 for domestic students. Applications are processed on a rolling basis. Application fee: $30 ($70 for international students). Electronic applications accepted. *Expenses:* Tuition, state resident: full-time $6,410. Tuition, nonresident: full-time $16,422. Tuition and fees vary according to program. *Financial support:* In 2005–06, 21 research assistantships with full and partial tuition reimbursements (averaging $13,545 per year), 9 teaching assistantships with full and partial tuition reimbursements (averaging $13,507 per year) were awarded; fellowships, scholarships/grants, health care benefits, and unspecified assistantships also available. *Unit head:* Dr. Carl E. Jacobson, Chair, 515-294-4477.

Montclair State University, The Graduate School, College of Science and Mathematics, Department of Earth and Environmental Studies, Montclair, NJ 07043-1624. Offers environmental management (MA, D Env M); environmental studies (MS), including environmental education, environmental health, environmental management, environmental science; geoscience (MS, Certificate), including geoscience (MS), water resource management (Certificate). Part-time and evening/weekend programs available. *Faculty:* 17 full-time (3 women), 9 part-time/adjunct (2 women). *Students:* 21 full-time (13 women), 48 part-time (20 women); includes 14 minority (8 African Americans, 3 Asian Americans or Pacific Islanders, 3 Hispanic Americans), 1 international. 32 applicants, 78% accepted, 16 enrolled. In 2005, 19 master's, 3 other advanced degrees awarded. *Degree requirements:* For master's, thesis or alternative, comprehensive exam; for doctorate, thesis/dissertation. *Entrance requirements:* For master's, GRE General Test, 2 letters of recommendation. Additional exam requirements/recommendations for international students: Required—TOEFL (minimum score 83 computer-based). *Application deadline:* Applications are processed on a rolling basis. Application fee: $60. Electronic applications accepted. *Expenses:* Tuition: Full-time $3,001; part-time $409 per credit. Required fees: $56 per credit. Tuition and fees vary according to course load, degree level and program. *Financial support:* In 2005–06, 14 research assistantships with full tuition reimbursements were awarded; Federal Work-Study, scholarships/grants, and unspecified assistantships also available. Support available to part-time students. Financial award application deadline: 3/1; financial award applicants required to submit FAFSA. *Faculty research:* Antarctica, carbon pools, contaminated sediments, wetlands. Total annual research expenditures: $127,880. *Unit head:* Dr. Gregory Pope, Chairperson, 973-655-7385. *Application contact:* Dr. Harbans Singh, Adviser, 973-655-7383.

Nova Scotia Agricultural College, Research and Graduate Studies, Truro, NS B2N 5E3, Canada. Offers agriculture (M Sc), including air quality, animal behavior, animal molecular genetics, animal nutrition, animal technology, aquaculture, botany, crop management, crop physiology, ecology, environmental microbiology, food science, horticulture, nutrient management, pest management, physiology, plant biotechnology, plant pathology, soil chemistry, soil fertility, waste management and composting, water quality. Part-time programs available. *Faculty:* 43 full-time (7 women), 21 part-time/adjunct (1 woman). *Students:* 50 full-time (25 women), 15 part-time (11 women); includes 7 minority (3 African Americans, 3 Asian Americans or Pacific Islanders, 1 Hispanic American). Average age 25. In 2005, 23 degrees awarded. *Degree requirements:* For master's, thesis, registration. *Entrance requirements:* For master's, honors B Sc, minimum GPA of 3.0. Additional exam requirements/recommendations for international students: Required—TOEFL (minimum score 580 paper-based; 237 computer-based), Michigan English Language Assessment Battery, IELT, Can Test, CAEL. *Application deadline:* For fall admission, 6/1 for domestic students, 4/1 for international students. For winter admission, 10/31 for domestic students; for spring admission, 2/28 for domestic students. Applications are processed on a rolling basis. Application fee: $70. *Expenses:* Tuition, state resident: part-time $2,328 per year. Tuition, nonresident: full-time $6,984; part-time $7,968 per year. International tuition: $12,624 full-time. Required fees: $481; $46 per course. Tuition and fees vary according to program and student level. *Financial support:* In 2005–06, 48 students received support, including 7 fellowships (averaging $15,000 per year), 10 research assistantships (averaging $15,000 per year), 15 teaching assistantships (averaging $900 per year); career-related internships or fieldwork, scholarships/grants, and unspecified assistantships also available. *Faculty research:* Bio-product development, organic agriculture, nutrient management, air and water quality, agricultural biotechnology. Total annual research expenditures: $4.7 million. *Unit head:* Jill L. Rogers, Manager, 902-893-6360, Fax: 902-893-3430, E-mail: jrogers@nsac.ca. *Application contact:* Marie Law, Administrative Assistant, 902-893-6502, Fax: 902-893-3430, E-mail: mlaw@nsac.ca.

Rutgers, The State University of New Jersey, New Brunswick/Piscataway, Graduate School, Program in Environmental Sciences, New Brunswick, NJ 08901-1281. Offers air resources (MS, PhD); aquatic biology (MS, PhD); aquatic chemistry (MS, PhD); atmospheric science (MS, PhD); chemistry and physics of aerosol and hydrosol systems (MS, PhD); environmental chemistry (MS, PhD); environmental microbiology (MS, PhD); environmental toxicology (PhD); exposure assessment (PhD); fate and effects of pollutants (MS, PhD); pollution prevention and control (MS, PhD); water and wastewater treatment (MS, PhD); water resources (MS, PhD). *Faculty:* 81 full-time, 7 part-time/adjunct. *Students:* 49 full-time (27 women), 48 part-time (19 women); includes 10 minority (3 African Americans, 6 Asian Americans or Pacific Islanders, 1 Hispanic American), 24 international. Average age 32. 79 applicants, 41% accepted, 15 enrolled. In 2005, 8 master's, 10 doctorates awarded. Terminal master's awarded for partial completion of doctoral program. *Degree requirements:* For master's, thesis or alternative, oral final exam, comprehensive exam; for doctorate, thesis/dissertation, thesis defense, qualifying exam, comprehensive exam. *Entrance requirements:* For master's and doctorate, GRE General Test. Additional exam requirements/recommendations for international students: Required—TOEFL. *Application deadline:* For fall admission, 3/1 for domestic students; for spring admission, 11/1 for domestic students. Applications are processed on a rolling basis. Application fee: $50. Electronic applications accepted. *Expenses:* Tuition, state resident: full-time $10,440; part-time $435 per credit. Tuition, nonresident: full-time $15,520; part-time $647 per credit. Required fees: $129 per credit. Tuition and fees vary according to program. *Financial support:* In 2005–06, 10 fellowships with full tuition reimbursements (averaging $21,887 per year), 34 research assistantships with full tuition reimbursements (averaging $19,367 per year), 3 teaching assistantships with full tuition reimbursements (averaging $17,583 per year) were awarded; career-related internships or fieldwork and Federal Work-Study also available. Financial award application deadline: 1/15; financial award applicants required to submit FAFSA. *Faculty research:* Atmospheric sciences; biological waste treatment; contaminant fate and transport; exposure assessment; air, soil and water quality. Total annual research expenditures: $5.7 million. *Unit head:* John Reinfelder, Director, 732-932-8013, Fax: 732-932-8644, E-mail: reinfelder@envsci.rutgers.edu. *Application contact:* Dr. Paul J. Lioy, Graduate Admissions Committee, 732-932-0150, Fax: 732-445-0116, E-mail: plioy@eohsi.rutgers.edu.

South Dakota School of Mines and Technology, Graduate Division, College of Science and Letters, Joint PhD Program in Atmospheric, Environmental, and Water Resources, Rapid City, SD 57701-3995. Offers PhD. *Faculty:* 9 full-time (0 women), 1 (woman) part-time/adjunct. *Students:* 4 full-time (2 women), 7 part-time (2 women); includes 1 minority (American Indian/Alaska Native), 1 international. In 2005, 2 degrees awarded. *Degree requirements:* For doctorate, thesis/dissertation. *Entrance requirements:* For doctorate, GRE General Test, GRE Subject Test. Additional exam requirements/recommendations for international students: Required—TOEFL, TWE. *Application deadline:* For fall admission, 7/1 priority date for domestic students,

Water Resources

South Dakota School of Mines and Technology (continued)

4/1 priority date for international students; for spring admission, 11/1 for domestic students, 9/1 for international students. Applications are processed on a rolling basis. Application fee: $35. Electronic applications accepted. *Expenses:* Tuition, area resident: Part-time $116 per credit hour. Tuition, state resident: full-time $2,084. Tuition, nonresident: full-time $6,146; part-time $341 per credit hour. Required fees: $1,805; $100 per credit hour. *Financial support:* In 2005–06, 3 fellowships (averaging $4,000 per year), 4 teaching assistantships with partial tuition reimbursements (averaging $6,774 per year) were awarded; research assistantships with partial tuition reimbursements *Unit head:* Dr. Andrew Detwiler, Chair, 605-394-2291. *Application contact:* Jeannette R. Nilson, Program Assistant-Research and Graduate Education, 800-454-8162 Ext. 1206, Fax: 605-394-5360, E-mail: graduate_admissions@silver.sdsmt.edu.

South Dakota State University, Graduate School, College of Engineering, Joint PhD Program in Atmospheric, Environmental, and Water Resources, Brookings, SD 57007. Offers PhD. Postbaccalaureate distance learning degree programs offered (minimal on-campus study). *Degree requirements:* For doctorate, thesis/dissertation, preliminary oral and written exams. *Entrance requirements:* Additional exam requirements/recommendations for international students: Required—TOEFL (minimum score 525 paper-based). Expenses: Contact institution.

State University of New York College of Environmental Science and Forestry, Faculty of Environmental Studies, Syracuse, NY 13210-2779. Offers environmental and community land planning (MPS, MS, PhD); environmental and natural resources policy (PhD); environmental communication and participatory processes (MPS, MS, PhD); environmental policy and democratic processes (MPS, MS, PhD); environmental systems and risk management (MPS, MS, PhD); water and wetland resource studies (MPS, MS, PhD). Part-time programs available. *Faculty:* 11 full-time (7 women), 11 part-time/adjunct (6 women). *Students:* 49 full-time (30 women), 28 part-time (15 women); includes 3 minority (2 African Americans, 1 Hispanic American), 36 international. Average age 32. 61 applicants, 69% accepted, 17 enrolled. In 2005, 16 master's, 4 doctorates awarded. *Degree requirements:* For master's, thesis (for some programs), registration; for doctorate, thesis/dissertation, comprehensive exam, registration. *Entrance requirements:* For master's and doctorate, GRE General Test, minimum GPA of 3.0. Additional exam requirements/recommendations for international students: Required—TOEFL (minimum score 550 paper-based; 213 computer-based). *Application deadline:* For fall admission, 2/1 priority date for domestic students, 2/1 priority date for international students; for spring admission, 11/1 priority date for domestic students, 11/1 priority date for international students. Applications are processed on a rolling basis. Application fee: $60. *Expenses:* Tuition, area resident: Full-time $6,900; part-time $288 per credit. Tuition, nonresident: full-time $10,920; part-time $455 per credit. Required fees: $395; $32 per credit. $20 per term. One-time fee: $145. *Financial support:* In 2005–06, 21 fellowships with full and partial tuition reimbursements (averaging $9,446 per year), 4 research assistantships with full and partial tuition reimbursements (averaging $10,000 per year), 9 teaching assistantships with full and partial tuition reimbursements (averaging $9,446 per year) were awarded; career-related internships or fieldwork, Federal Work-Study, institutionally sponsored loans, scholarships/grants, health care benefits, and unspecified assistantships also available. Support available to part-time students. Financial award application deadline: 6/30; financial award applicants required to submit FAFSA. *Faculty research:* Environmental education/communications, water resources, land resources, waste management. Total annual research expenditures: $169,919. *Unit head:* Dr. Richard Smardon, Chair, 315-470-6636, Fax: 315-470-6915, E-mail: rsmardon@syr.edu. *Application contact:* Dr. Dudley J. Raynal, Dean, Instruction and Graduate Studies, 315-470-6599, Fax: 315-470-6978, E-mail: esfgrad@esf.edu.

State University of New York College of Environmental Science and Forestry, Faculty of Forest and Natural Resources Management, Syracuse, NY 13210-2779. Offers environmental and natural resource policy (MS, PhD); environmental and natural resources policy (MPS); forest management and operations (MF); forestry ecosystems science and applications (MPS, MS, PhD); natural resources management (MPS, MS, PhD); quantitative methods and management in forest science (MPS, MS, PhD); recreation and resource management (MPS, MS, PhD); watershed management and forest hydrology (MPS, MS, PhD). *Faculty:* 30 full-time (7 women), 1 (woman) part-time/adjunct. *Students:* 43 full-time (18 women), 26 part-time (12 women); includes 1 minority (Hispanic American), 13 international. Average age 31. 38 applicants, 55% accepted, 12 enrolled. In 2005, 13 master's, 4 doctorates awarded. *Degree requirements:* For master's, thesis (for some programs), registration; for doctorate, thesis/dissertation, comprehensive exam, registration. *Entrance requirements:* For master's and doctorate, GRE General Test, minimum GPA of 3.0. Additional exam requirements/recommendations for international students: Required—TOEFL (minimum score 550 paper-based; 213 computer-based). *Application deadline:* For fall admission, 2/1 priority date for domestic students, 2/1 priority date for international students; for spring admission, 11/1 priority date for domestic students, 11/1 priority date for international students. Applications are processed on a rolling basis. Application fee: $60. *Expenses:* Tuition, area resident: Full-time $6,900; part-time $288 per credit. Tuition, nonresident: full-time $10,920; part-time $455 per credit. Required fees: $395; $32 per credit. $20 per term. One-time fee: $145. *Financial support:* In 2005–06, 43 students received support, including 8 fellowships with full and partial tuition reimbursements available (averaging $9,446 per year), 20 research assistantships with full and partial tuition reimbursements available (averaging $10,000 per year), 11 teaching assistantships with full and partial tuition reimbursements available (averaging $9,446 per year); career-related internships or fieldwork, Federal Work-Study, institutionally sponsored loans, scholarships/grants, health care benefits, and unspecified assistantships also available. Financial award application deadline: 6/30; financial award applicants required to submit FAFSA. *Faculty research:* Silviculture recreation management, tree improvement, operations management, economics. Total annual research expenditures: $2.1 million. *Unit head:* Dr. Chad P. Dawson, Chair, 315-470-6536, Fax: 315-470-6535, E-mail: cpdawson@esf.edu. *Application contact:* Dr. Dudley J. Raynal, Dean, Instruction and Graduate Studies, 315-470-6599, Fax: 315-470-6978, E-mail: esfgrad@esf.edu.

Texas A&M University, Interdisciplinary Program in Water Management and Hydrological Sciences, College Station, TX 77843. Offers MS, PhD. *Expenses:* Tuition, state resident: full-time $4,488; part-time $187 per credit hour. Tuition, nonresident: full-time $11,112; part-time $463 per credit hour. Required fees: $1,974. *Unit head:* Ronald A. Kaiser, Co-Chair, 979-845-5303, Fax: 979-845-0446, E-mail: rkaiser@tamu.edu.

Tropical Agriculture Research and Higher Education Center, Graduate School, Turrialba, Costa Rica. Offers agroforestry (PhD); ecological agriculture (PhD); ecological agriculture (MS); environmental socioeconomics (MS, PhD); forest management and conservation (MS); forest sciences (PhD); tropical agroforestry (MS); watershed management (MS, PhD). *Entrance requirements:* For master's, GRE, letters of recommendation; for doctorate, GRE, curriculum vitae, letters of recommendation. Additional exam requirements/recommendations for international students: Required—TOEFL (minimum score 550 paper-based; 213 computer-based). Electronic applications accepted. *Faculty research:* Biodiversity in fragmented landscapes, ecosystem management, integrated pest management, environmental livestock production, biotechnology carbon balances in diverse land uses.

The University of Arizona, Graduate College, College of Agriculture and Life Sciences, Department of Soil, Water and Environmental Science, Tucson, AZ 85721. Offers MS, PhD. *Degree requirements:* For master's, thesis; for doctorate, one foreign language, thesis/dissertation. *Entrance requirements:* Additional exam requirements/recommendations for international students: Required—TOEFL. *Faculty research:* Plant production, environmental microbiology, contaminant flow and transport.

The University of Arizona, Graduate College, College of Engineering, Department of Hydrology and Water Resources, Tucson, AZ 85721. Offers hydrology (MS, PhD); water resources engineering (M Eng). Part-time programs available. *Degree requirements:* For master's and doctorate, thesis/dissertation. *Entrance requirements:* For master's, GRE General Test, minimum undergraduate GPA of 3.0; for doctorate, GRE General Test, minimum undergraduate GPA of 3.2, 3.4 graduate. Additional exam requirements/recommendations for international students: Required—TOEFL. *Faculty research:* Subsurface and surface hydrology, hydrometeorology/climatology, applied remote sensing, water resource systems, environmental hydrology and water quality.

University of California, Riverside, Graduate Division, Department of Environmental Sciences, Program in Soil and Water Sciences, Riverside, CA 92521-0102. Offers MS, PhD. *Faculty:* 26 full-time (4 women). *Students:* 13 full-time (9 women); includes 3 minority (1 American Indian/Alaska Native, 2 Hispanic Americans), 2 international. Average age 29. 20 applicants, 50% accepted, 6 enrolled. In 2005, 5 master's, 1 doctorate awarded. *Entrance requirements:* For master's and doctorate, minimum GPA of 3.2. Additional exam requirements/recommendations for international students: Required—TOEFL (minimum score 550 paper-based; 213 computer-based); Recommended—TSE (minimum score 50). *Application deadline:* For fall admission, 5/1 for domestic students, 2/1 for international students. For winter admission, 9/1 for domestic students; for spring admission, 12/1 for domestic students. Application fee: $60 ($75 for international students). Electronic applications accepted. *Expenses:* Tuition, nonresident: full-time $14,694. Full-time tuition and fees vary according to program. *Financial support:* In 2005–06, fellowships (averaging $12,000 per year) *Unit head:* Dr. Marylynn Yates, Chair, 951-827-2358, Fax: 951-827-3993, E-mail: marylynn.yates@ucr.edu. *Application contact:* Mari Ridgeway, Program Assistant, 951-827-5103, Fax: 951-827-3993, E-mail: soilwater@ucr.edu.

University of Florida, Graduate School, College of Agricultural and Life Sciences, Department of Soil and Water Science, Gainesville, FL 32611. Offers M Ag, MS, PhD. Part-time programs available. Postbaccalaureate distance learning degree programs offered. *Faculty:* 22 full-time (4 women). *Students:* 93 (50 women); includes 10 minority (3 African Americans, 1 Asian American or Pacific Islander, 6 Hispanic Americans) 26 international. 27 applicants, 78% accepted. In 2005, 10 master's, 9 doctorates awarded. Terminal master's awarded for partial completion of doctoral program. *Degree requirements:* For master's, thesis optional; for doctorate, thesis/dissertation. *Entrance requirements:* For master's and doctorate, GRE General Test, minimum GPA of 3.0. Additional exam requirements/recommendations for international students: Required—TOEFL. *Application deadline:* For fall admission, 6/1 for domestic students; for spring admission, 9/14 for domestic students. Applications are processed on a rolling basis. Application fee: $20. Electronic applications accepted. *Expenses:* Tuition, state resident: full-time $6,234. Tuition, nonresident: full-time $21,359. Tuition and fees vary according to program. *Financial support:* In 2005–06, 41 research assistantships (averaging $11,083 per year) were awarded; fellowships, teaching assistantships, career-related internships or fieldwork, Federal Work-Study, institutionally sponsored loans, and unspecified assistantships also available. Support available to part-time students. *Faculty research:* Environmental fate and transport of pesticides, conservation, wetlands, land application of nonhazardous waste, soil/water agrochemical management. *Unit head:* Dr. K. Ramesh Reddy, Chair, 352-392-1803 Ext. 317, Fax: 352-392-3399, E-mail: krr@ufl.edu. *Application contact:* Dr. Nicholas B. Comerford, Graduate Coordinator, 352-392-1951, Fax: 352-392-3902, E-mail: nbc@ifas.ufl.edu.

University of Illinois at Chicago, Graduate College, College of Liberal Arts and Sciences, Department of Earth and Environmental Sciences, Chicago, IL 60607-7128. Offers crystallography (MS, PhD); environmental geology (MS, PhD); geochemistry (MS, PhD); geology (MS, PhD); geomorphology (MS, PhD); geophysics (MS, PhD); geotechnical engineering and geosciences (PhD); hydrogeology (MS, PhD); low-temperature and organic geochemistry (MS, PhD); mineralogy (MS, PhD); paleoclimatology (MS, PhD); paleontology (MS, PhD); petrology (MS, PhD); quaternary geology (MS, PhD); sedimentology (MS, PhD); water resources (MS, PhD). *Degree requirements:* For master's and doctorate, thesis/dissertation. *Entrance requirements:* For master's and doctorate, GRE General Test, minimum GPA of 2.75. Additional exam requirements/recommendations for international students: Required—TOEFL. Electronic applications accepted.

University of Kansas, Graduate School, School of Engineering, Department of Civil, Environmental, and Architectural Engineering, Program in Water Resources Science, Lawrence, KS 66045. Offers MS. *Students:* 1 applicant, 100% accepted. In 2005, 1 degree awarded. *Entrance requirements:* For master's, GRE, minimum GPA of 3.0. Additional exam requirements/recommendations for international students: Required—TOEFL. *Application deadline:* Applications are processed on a rolling basis. Electronic applications accepted. *Expenses:* Tuition, state resident: full-time $4,859. Tuition, nonresident: full-time $1,200. Required fees: $589. Tuition and fees vary according to program. *Financial support:* Fellowships, research assistantships available. Financial award application deadline: 2/7. *Application contact:* Bruce M. McEnroe, Graduate Advisor, E-mail: mcenroe@ku.edu.

University of Minnesota, Twin Cities Campus, Graduate School, College of Agricultural, Food, and Environmental Sciences, Department of Soil, Water, and Climate, Minneapolis, MN 55455-0213. Offers MS, PhD. *Faculty:* 27 full-time (2 women), 9 part-time/adjunct (0 women). *Students:* 27 full-time (13 women), 5 part-time (3 women); includes 3 minority (1 American Indian/Alaska Native, 1 Asian American or Pacific Islander, 1 Hispanic American). Average age 25. 21 applicants, 57% accepted, 11 enrolled. In 2005, 3 degrees awarded. *Degree requirements:* For master's, thesis or alternative; for doctorate, thesis/dissertation. *Entrance requirements:* For master's and doctorate, GRE General Test, minimum GPA of 3.0. Additional exam requirements/recommendations for international students: Required—TOEFL (minimum score 550 paper-based; 213 computer-based). *Application deadline:* For fall admission, 6/15 priority date for domestic students, 6/15 priority date for international students; for spring admission, 10/15 priority date for domestic students, 10/15 priority date for international students. Applications are processed on a rolling basis. Application fee: $55 ($75 for international students). Electronic applications accepted. *Expenses:* Tuition, state resident: full-time $8,748; part-time $729 per credit. Tuition, nonresident: full-time $15,848; part-time $1,321 per credit. Full-time tuition and fees vary according to class time, course load, program and reciprocity agreements. *Financial support:* In 2005–06, 2 fellowships with full tuition reimbursements (averaging $22,000 per year), 24 research assistantships with full and partial tuition reimbursements (averaging $17,000 per year), 2 teaching assistantships with full tuition reimbursements (averaging $17,000 per year) were awarded; Federal Work-Study, scholarships/grants, health care benefits, tuition waivers (full), and unspecified assistantships also available. Support available to part-time students. *Faculty research:* Soil water and atmospheric resources, soil physical management, agricultural chemicals and their management, plant nutrient management, biological nitrogen fixation. *Unit head:* Dr. Edward A. Nater, Head, 612-625-9734, Fax: 612-625-2208, E-mail: enater@umn.edu. *Application contact:* Dr. Deborah L. Allan, Professor and Director of Graduate Studies, 612-625-3158, Fax: 612-625-2208, E-mail: dallan@umn.edu.

University of Minnesota, Twin Cities Campus, Graduate School, College of Natural Resources, Program in Water Resources Science, Minneapolis, MN 55455-0213. Offers MS, PhD. Part-time programs available. *Degree requirements:* For master's, thesis optional; for doctorate, thesis/dissertation, comprehensive exam, registration. *Entrance requirements:* For master's and doctorate, GRE. Additional exam requirements/recommendations for international students: Required—TOEFL (minimum score 550 paper-based; 213 computer-based). Electronic applications accepted. *Expenses:* Tuition, state resident: full-time $8,748; part-time $729 per credit. Tuition, nonresident: full-time $15,848; part-time $1,321 per credit. Full-time tuition and fees vary according to class time, course load, program and reciprocity agreements. *Faculty research:* Water chemistry, water quality, hydrology resource management, landscape.

University of Missouri–Rolla, Graduate School, School of Materials, Energy, and Earth Resources, Department of Geological Sciences and Engineering, Program in Geology and Geophysics, Rolla, MO 65409-0910. Offers geochemistry (MS, PhD); geology (MS, PhD); geophysics (MS, PhD); groundwater and environmental geology (MS, PhD). Part-time programs available. *Degree requirements:* For master's and doctorate, thesis/dissertation. *Entrance requirements:* For master's, GRE General Test, GRE Subject Test, minimum GPA of

3.0 in last 4 semesters; for doctorate, GRE General Test, GRE Subject Test. Additional exam requirements/recommendations for international students: Required—TOEFL. Electronic applications accepted. *Faculty research:* Economic geology, geophysical modeling, seismic wave analysis.

University of Nevada, Las Vegas, Graduate College, College of Science, Program in Water Resources Management, Las Vegas, NV 89154-9900. Offers MS. Part-time programs available. *Faculty:* 10 full-time (1 woman), 9 part-time/adjunct (1 woman). *Students:* 6 full-time (1 woman), 14 part-time (8 women); includes 2 minority (1 African American, 1 Hispanic American). 6 applicants, 83% accepted, 3 enrolled. In 2005, 1 degree awarded. *Degree requirements:* For master's, thesis, comprehensive exam. *Entrance requirements:* For master's, GRE Subject Test, minimum GPA of 3.0. Additional exam requirements/recommendations for international students: Required—TOEFL (minimum score 550 paper-based; 213 computer-based). *Application deadline:* For fall admission, 6/15 for domestic students, 5/1 for international students; for spring admission, 11/15 for domestic students, 10/1 for international students. Application fee: $60 ($75 for international students). Electronic applications accepted. *Expenses:* Tuition, state resident: part-time $150 per credit. Tuition, nonresident: part-time $315 per credit. Tuition and fees vary according to course load, program and reciprocity agreements. *Financial support:* In 2005–06, 1 teaching assistantship (averaging $10,000 per year) was awarded; research assistantships with full and partial tuition reimbursements, career-related internships or fieldwork, Federal Work-Study, institutionally sponsored loans, scholarships/grants, health care benefits, and unspecified assistantships also available. Support available to part-time students. Financial award application deadline: 3/1. *Unit head:* Dr. Lambis Papelis, Director, 702-895-3262. *Application contact:* Graduate College Admissions Evaluator, 702-895-3320, Fax: 702-895-4180, E-mail: gradcollege@unlv.edu.

University of New Brunswick Fredericton, School of Graduate Studies, Faculty of Engineering, Department of Civil Engineering, Fredericton, NB E3B 5A3, Canada. Offers construction engineering and management (M Eng, M Sc E, PhD); environmental engineering (M Eng, M Sc E, PhD); geotechnical engineering (M Eng, M Sc E, PhD); groundwater/hydrology (M Eng, M Sc E, PhD); materials (M Eng, M Sc E, PhD); pavements (M Eng, M Sc E, PhD); structures (M Eng, M Sc E, PhD); transportation (M Eng, M Sc E, PhD). Part-time programs available. *Degree requirements:* For master's, thesis; for doctorate, thesis/dissertation, qualifying exam. *Entrance requirements:* For master's and doctorate, minimum GPA of 3.0. Additional exam requirements/recommendations for international students: Required—TOEFL, TWE. *Faculty research:* Steel and masonry structures, traffic engineering, highway safety, centrifuge modeling, transport and fate of reactive contaminants, durability of marine concrete.

University of New Hampshire, Graduate School, College of Life Sciences and Agriculture, Department of Natural Resources, Durham, NH 03824. Offers environmental conservation (MS); forestry (MS); soil science (MS); water resources management (MS); wildlife (MS). Part-time programs available. *Faculty:* 40 full-time. *Students:* 31 full-time (17 women), 36 part-time (17 women); includes 2 minority (both Asian Americans or Pacific Islanders), 4 international. Average age 31. 50 applicants, 44% accepted, 15 enrolled. In 2005, 14 degrees awarded. *Degree requirements:* For master's, thesis or alternative. *Entrance requirements:* For master's, GRE General Test. Additional exam requirements/recommendations for international students: Required—TOEFL (minimum score 550 paper-based; 213 computer-based); Recommended—TSE. *Application deadline:* For fall admission, 4/1 for domestic students, 4/1 for international students. For winter admission, 12/1 for domestic students. Applications are processed on a rolling basis. Application fee: $60. Electronic applications accepted. *Expenses:* Tuition, state resident: full-time $8,010; part-time $445 per credit hour. Tuition, nonresident: full-time $19,730; part-time $810 per credit hour. Required fees: $322 per semester. Tuition and fees vary according to course load and program. *Financial support:* In 2005–06, 3 fellowships, 21 research assistantships, 15 teaching assistantships were awarded; career-related internships or fieldwork, Federal Work-Study, scholarships/grants, and tuition waivers (full and partial) also available. Support available to part-time students. Financial award application deadline: 2/15. *Unit head:* Dr. William H. McDowell, Chairperson, 603-862-2249, E-mail: tehoward@cisunix.unh.edu. *Application contact:* Linda Scogin, Administrative Assistant, 603-862-3932, E-mail: natural.resources @unh.edu.

University of New Mexico, Graduate School, Program in Water Resources, Albuquerque, NM 87131-2039. Offers MWR. Part-time programs available. *Faculty:* 1 (woman) part-time/adjunct. *Students:* 16 full-time (9 women), 30 part-time (14 women); includes 10 minority (2 American Indian/Alaska Native, 8 Hispanic Americans), 1 international. Average age 34. 20 applicants, 85% accepted, 11 enrolled. In 2005, 21 degrees awarded. *Degree requirements:* For master's, thesis, comprehensive exam. *Entrance requirements:* For master's, minimum GPA of 3.0 during last 2 years of undergraduate work, 3 letters of reference. Additional exam requirements/recommendations for international students: Required—TOEFL (minimum score 550 paper-based; 213 computer-based). *Application deadline:* For fall admission, 7/30 for domestic students; for spring admission, 11/30 for domestic students. Application fee: $40. Electronic applications accepted. *Expenses:* Tuition, nonresident: full-time $3,388; part-time $238 per credit hour. Required fees: $385 per term. Tuition and fees vary according to course load and program. *Financial support:* In 2005–06, 17 students received support, including 7 research assistantships (averaging $3,605 per year); institutionally sponsored loans, scholarships/grants, health care benefits, tuition waivers (partial), and unspecified assistantships also available. Financial award application deadline: 3/1; financial award applicants required to submit FAFSA. *Faculty research:* Sustainable water resources, transboundary water resources, economics, water law, hydrology, developing countries, hydro geology. Total annual research expenditures: $44,056. *Unit head:* Dr. Michael E. Campana, Director, 505-277-7759, Fax: 505-277-5226, E-mail: aquadoc@unm.edu. *Application contact:* Annamarie Cordova, Administrative Assistant II, 505-277-7759, Fax: 505-277-5226, E-mail: acordova@unm.edu.

University of Oklahoma, Graduate College, College of Engineering, School of Civil Engineering and Environmental Science, Program in Environmental Science, Norman, OK 73019-0390. Offers air (M Env Sc); environmental engineering (MS); environmental science (PhD); groundwater management (M Env Sc); hazardous solid waste (M Env Sc); occupational safety and health (M Env Sc); process design (M Env Sc); water quality resources (M Env Sc). Part-time programs available. *Students:* 17 full-time (8 women), 11 part-time (5 women); includes 3 minority (2 African Americans, 1 American Indian/Alaska Native), 13 international. 19 applicants, 63% accepted, 6 enrolled. In 2005, 6 master's, 1 doctorate awarded. Terminal master's awarded for partial completion of doctoral program. *Degree requirements:* For master's, oral exams; for doctorate, thesis/dissertation, oral, and qualifying exams, comprehensive exam. *Entrance requirements:* For master's, minimum GPA of 3.0; for doctorate, minimum graduate GPA of 3.5. Additional exam requirements/recommendations for international students: Required—TOEFL (minimum score 600 paper-based; 250 computer-based). *Application deadline:* For fall admission, 4/1 priority date for domestic students, 4/1 priority date for international students; for spring admission, 11/1 for domestic students, 9/1 for international students. Applications are processed on a rolling basis. Application fee: $40 ($90 for international students). *Expenses:* Tuition, state resident: full-time $3,029; part-time $126 per credit hour. Tuition, nonresident: full-time $10,807; part-time $450 per credit hour. Required fees: $1,231; $44 per credit hour. Tuition and fees vary according to course load and program. *Financial support:* In 2005–06, 8 students received support; fellowships, research assistantships with partial tuition reimbursements available, teaching assistantships with partial tuition reimbursements available, scholarships/grants available. Financial award application deadline: 3/1; financial award applicants required to submit FAFSA. *Application contact:* Susan Williams, Graduate Programs Specialist, 405-325-2344, Fax: 405-325-4217, E-mail: srwilliams@ou.edu.

University of Wisconsin–Madison, Graduate School, Gaylord Nelson Institute for Environmental Studies, Water Resources Management Program, Madison, WI 53706-1380. Offers MS. Part-time programs available. *Faculty:* 1 full-time (0 women), 50 part-time/adjunct (15 women). *Students:* 27 full-time, 9 part-time; includes 2 minority (1 American Indian/Alaska Native, 1 Hispanic American), 2 international. Average age 29. 48 applicants, 60% accepted, 12 enrolled. In 2005, 8 degrees awarded. *Degree requirements:* For master's, practicum. *Entrance requirements:* For master's, GRE General Test. Additional exam requirements/recommendations for international students: Required—TOEFL (minimum score 550 paper-based; 213 computer-based). *Application deadline:* For fall admission, 2/1 for domestic students, 2/1 for international students; for spring admission, 10/15 for domestic students, 10/15 for international students. Application fee: $45. Electronic applications accepted. *Financial support:* In 2005–06, 17 students received support, including 2 fellowships with full tuition reimbursements available (averaging $14,400 per year), 2 research assistantships (averaging $14,250 per year), 5 teaching assistantships with full tuition reimbursements available (averaging $11,260 per year); career-related internships or fieldwork, Federal Work-Study, scholarships/grants, health care benefits, unspecified assistantships, and project assistantships also available. Financial award application deadline:1/2. *Faculty research:* Geology, hydrogeology, water chemistry, limnology, oceanography. *Unit head:* Linda K. Graham, Chair, 608-262-2640, Fax: 608-262-2273, E-mail: lkgraham@wisc.edu. *Application contact:* James E. Miller, Associate Student Services Coordinator, 608-263-4373, Fax: 608-262-2273, E-mail: jemiller@wisc.edu.

University of Wyoming, Graduate School, College of Agriculture, Department of Renewable Resources, Laramie, WY 82070. Offers entomology (MS, PhD); rangeland ecology and watershed management (MS, PhD), including soil sciences (PhD), soil sciences and water resources (MS), water resources. Part-time programs available. *Faculty:* 21 full-time (2 women). *Students:* 26 full-time (10 women), 13 part-time (7 women), 9 international. In 2005, 5 master's awarded. *Degree requirements:* For master's, thesis (for some programs); for doctorate, thesis/dissertation. *Entrance requirements:* For master's and doctorate, GRE General Test, minimum GPA of 3.0. Additional exam requirements/recommendations for international students: Required—TOEFL. *Application deadline:* For fall admission, 6/1 for domestic students; for spring admission, 12/1 priority date for domestic students. Applications are processed on a rolling basis. Application fee: $50. Electronic applications accepted. *Expenses:* Tuition, state resident: full-time $3,720; part-time $155 per credit hour. Tuition, nonresident: full-time $10,704; part-time $446 per credit hour. Required fees: $666; $162 per semester. Tuition and fees vary according to course load and program. *Financial support:* In 2005–06, 8 students received support, including 8 research assistantships with full tuition reimbursements available (averaging $10,062 per year); career-related internships or fieldwork and Federal Work-Study also available. Financial award application deadline: 3/1. *Faculty research:* Plant control, grazing management, riparian restoration, riparian management, reclamation. *Unit head:* Dr. Richard A. Olson, Interim Head, 307-766-2263, Fax: 307-766-6403, E-mail: rolson@uwyo.edu. *Application contact:* Kimm Mann-Malody, Office Associate, 307-766-2263, Fax: 307-766-6403, E-mail: kimmmann@uwyo.edu.

Utah State University, School of Graduate Studies, College of Natural Resources, Department of Aquatic, Watershed, and Earth Resources, Logan, UT 84322. Offers ecology (MS, PhD); fisheries biology (MS, PhD); watershed science (MS, PhD). *Faculty:* 12 full-time (3 women), 2 part-time/adjunct (1 woman). *Students:* 118 full-time (42 women), 16 part-time (2 women); includes 4 minority (all Hispanic Americans), 14 international. Average age 31. 31 applicants, 71% accepted, 21 enrolled. In 2005, 8 master's, 4 doctorates awarded. *Degree requirements:* For master's, thesis (for some programs); for doctorate, thesis/dissertation. *Entrance requirements:* For master's and doctorate, GRE General Test, minimum GPA of 3.2. Additional exam requirements/recommendations for international students: Required—TOEFL. *Application deadline:* For fall admission, 2/15 for domestic students; for spring admission, 10/15 for domestic students. Applications are processed on a rolling basis. Application fee: $50 ($60 for international students). Electronic applications accepted. *Financial support:* In 2005–06, 3 fellowships with partial tuition reimbursements (averaging $20,000 per year), 34 research assistantships with partial tuition reimbursements (averaging $14,864 per year) were awarded; teaching assistantships with partial tuition reimbursements, career-related internships or fieldwork, Federal Work-Study, and institutionally sponsored loans also available. Support available to part-time students. Financial award application deadline:2/15. *Faculty research:* Behavior, population ecology, habitat, conservation biology, restoration, aquatic ecology, fisheries management, fluvial geomorphology, remote sensing, conservation biology. Total annual research expenditures: $4.7 million. *Unit head:* Chris Luecke, Head, 435-797-2463, Fax: 435-797-1871, E-mail: awerinfo@cc.usu.edu. *Application contact:* Brian Bailey, Staff Assistant, 435-797-2459, Fax: 435-797-1871, E-mail: awerinfo@cc.usu.edu.

Washington State University Tri-Cities, Graduate Programs, Program in Environmental Science, Richland, WA 99352-1671. Offers applied environmental science (MS); atmospheric science (MS); earth science (MS); environmental and occupational health science (MS); environmental regulatory compliance (MS); environmental science (PhD); environmental toxicology and risk assessment (MS); water resource science (MS). Part-time programs available. *Faculty:* 1 full-time (0 women), 53 part-time/adjunct. *Students:* 4 full-time (3 women), 22 part-time (10 women); includes 1 Asian American or Pacific Islander, 2 Hispanic Americans. Average age 41. 11 applicants, 55% accepted, 6 enrolled. In 2005, 1 degree awarded. *Degree requirements:* For master's, oral exam, thesis optional. *Entrance requirements:* For master's, GRE General Test, minimum GPA of 3.0, 3 letters of recommendation. Additional exam requirements/recommendations for international students: Required—TOEFL (minimum score 550 paper-based; 213 computer-based). *Application deadline:* For fall admission, 2/1 priority date for domestic students, 3/1 priority date for international students; for spring admission, 9/1 priority date for domestic students, 7/1 priority date for international students. Application fee: $35. *Expenses:* Tuition, state resident: full-time $6,295; part-time $336 per credit. Tuition, nonresident: full-time $15,949; part-time $819 per credit. Required fees: $429. Full-time tuition and fees vary according to campus/location and program. Part-time tuition and fees vary according to course load and program. *Financial support:* In 2005–06, 8 students received support, including 1 fellowship (averaging $2,200 per year); research assistantships with full and partial tuition reimbursements available, teaching assistantships with full and partial tuition reimbursements available, Federal Work-Study, scholarships/grants, health care benefits, and unspecified assistantships also available. *Faculty research:* Radiation ecology, cytogenetics. *Unit head:* Dr. Gene Schreckhise, Associate Dean/Coordinator, 509-372-7323, E-mail: gschreck@wsu.edu.

CLEMSON UNIVERSITY

CLEMSON UNIVERSITY

Master of Science in Forest Resources

Program of Study

The Department of Forestry and Natural Resources offers an M.S. degree in forest resources. A nonthesis Master of Forest Resources (M.F.R.) professional degree is also offered. Areas of concentration for the degree programs in forest resources include forest ecology, forest economics, forest wildlife management, silviculture, urban forestry, and water quality and wetlands. M.S. and M.F.R. programs generally last two to three years. Students must complete at least one continuous year in residence and are required to work closely with their major professor and graduate committee members. Most research results have applied management applications, and students are expected to present and publish these results in appropriate technical and popular outlets. Most graduates of the M.S. and M.F.R. programs are employed by state and federal agencies, the forest industry, and environmental consulting firms, or they work as private consultants or resource managers.

Research Facilities

Excellent GIS, computer, chemical analysis, and biotechnology facilities are available to graduate students in the Department of Forest Resources. The 17,500-acre Clemson Experimental Forest surrounds the campus and offers opportunities for field research. In addition, students may work with 7 faculty members located at the Belle W. Baruch Institute of Coastal Ecology and Forest Science at Georgetown, South Carolina. At Baruch Institute, opportunities exist for research at the Hobcaw Barony, a 17,000-acre undisturbed ecological reserve of forests, high-salinity marsh estuaries, and brackish and freshwater marshes. Research opportunities for graduate students are enhanced by cooperative programs with the U.S. Forest Service Southern Research Station, USGS Fish and Wildlife Cooperative Research Unit at Clemson, Savannah River Ecology Laboratory, Waddell Mariculture Center, and the National Council for Air and Stream Improvement Eastern Wildlife Program.

Financial Aid

Graduate research assistantships are available through research grants and contracts administered by individual faculty members. Minimum assistantship levels are normally $15,000 for M.S. students. Limited appointments are available for work-study programs. There are no allowances for dependents or hiring of spouses.

Cost of Study

Tuition for 2006–07 is $4643 per semester for in-state students and $9255 per semester for nonresidents. Off-campus rates are $535 per hour for in-state students and $918 per hour for nonresidents. Graduate assistants pay a flat fee of $1079 per semester and $348 per summer session. Graduate fellows pay South Carolina resident fees.

Living and Housing Costs

On-campus housing is available. For information, students should visit http://www.housing.clemson.edu. The cost of living in Clemson is quite low compared to the national average. Students who choose to live off the campus typically spend $300–$400 per month for rent, depending on location, amenities, roommates, etc.

Student Group

The master's programs have about 28 students. Twenty-five percent are women, 89 percent are full-time, and 4 percent are international students.

Student Outcomes

Graduates of the M.S. and M.F.R. programs most often work for state and federal natural resources management agencies or the forest industry or in the private sector. Those graduates who pursue Ph.D. studies usually do so at other land-grant universities such as Clemson, Virginia Tech, Mississippi State University, or the University of Georgia.

Location

Clemson is a small, beautiful college town near the Blue Ridge Mountains and Lake Hartwell in upstate South Carolina. The Upstate is one of the country's fastest-growing areas and is the midpoint of the Charlotte-to-Atlanta I-85 corridor, a multistate area along Interstate 85 that runs from metro Atlanta to Richmond, Virginia, and encompasses Charlotte, North Carolina, and North Carolina's Research Triangle. Atlanta and Charlotte are each a 2-hour's drive away. Many financial institutions and other industries have national headquarters for a major presence in the Upstate, including Wachovia, Bank of America, BMW, Bon Secours St. Francis Health System, Bosch North America, Bowater, Charter Communications, Ernst and Young, Fluor Corporation, IBM, Microsoft, Michelin of North America, and many others.

The University

Clemson is classified by the Carnegie Foundation as Doctoral/Research University–Extensive, a category comprising less than 4 percent of all universities in America. The University's mission is to fulfill the covenant between its founder and the people of South Carolina to establish a "high seminary of learning" through its responsibilities of teaching, research, and extended public service. The University has identified eight areas of academic emphasis that create collaborations that, in turn, help fulfill the University's mission.

Applying

Graduate applicants are selected based on demonstrated scholarship (GRE scores, GPA, undergraduate institution), references, field experience, stated interests, and professional goals. Computer and communication skills and ability to work with others are also important considerations. Individual faculty members generally require personal interviews with applicants to determine their interests and compatibility with specific research projects.

Applicants may apply on the Web at http://www.grad.clemson.edu/p_apply.html. Applications with a $50 nonrefundable fee should be received no later than five weeks prior to registration. Every required item in support of the application must be on file by that date. Students are advised to contact the department for the deadlines of the program of proposed study.

Correspondence and Information

David Guynn
Department of Forestry and Natural Resources
Clemson University
Clemson, South Carolina 29634-0317
Phone: 864-656-4830
Fax: 864-656-3304
E-mail: dguynn@clemson.edu
Web site: http://www.clemson.edu/forestres/

Clemson University

THE FACULTY AND THEIR RESEARCH

William W. Bowerman IV, Associate Professor; Ph.D., Michigan State. Fisheries and wildlife.
William H. Conner, Professor; Ph.D., LSU. Forestry.
William Rockford English, Associate Professor; Ph.D., Clemson. Entomology.
Arnold G. Eversole, Professor; Ph.D., Syracuse. Biology and zoology.
Jeffrey W. Foltz, Professor; Ph.D., Colorado. Biology.
Lawrence R. Gering, Associate Professor; Ph.D., Georgia. Forest biometrics.
Charles A. Gresham, Associate Professor; Ph.D., Duke. Forestry.
David C. Guynn Jr., Professor; Ph.D., Virginia Tech. Forestry.
Roy L. Hedden, Professor; Ph.D., Washington (Seattle). Forest entomology.
Alan R. Johnson, Assistant Professor; Ph.D., Tennessee. Environmental toxicology.
Joseph D. Lanham, Associate Professor; Ph.D., Clemson. Ecology and ornithology.
Patricia A. Layton, Professor and Department Chair; Ph.D., Florida. Forest genetics.
Andy Wu-Chung Lee, Professor; Ph.D., Auburn. Wood products.
Allan Marsinko, Professor; Ph.D., SUNY College of Environmental Science and Forestry. Forest economics.
Larry R. Nelson, Associate Professor; Ph.D., Auburn. Forestry.
Lawrence E. Nix,; Professor; Ph.D., Georgia. Forestry.
Christopher J. Post, Assistant Professor; Ph.D., Cornell. Environmental information science.
John H. Rodgers Jr., Professor; Ph.D., Virginia Tech. Botany and aquatic ecology.
Linda C. Roth, Assistant Professor; Ph.D., Clark. Geography.
Victor B. Shelburne, Professor; Ph.D., Clemson. Forestry.
Bo Song, Assistant Professor; Ph.D., Michigan Tech. Forest science.
Thomas J. Straka, Professor; Ph.D., Virginia Tech. Forestry.
David H. Van Lear, Named Professor; Ph.D., Idaho. Forest sciences.
Gaofeng G. Wang, Assistant Professor; Ph.D., British Columbia. Forest ecology.
Thomas M. Williams, Professor; Ph.D., Minnesota. Forestry.
Gene W. Wood, Professor; Ph.D., Penn State. Agronomy.
Thomas E. Wooten, Alumni Professor; Ph.D., North Carolina State. Forestry.
Greg K. Yarrow, Professor; D.F., Stephen F. Austin. Forest wildlife.

CLEMSON UNIVERSITY

CLEMSON UNIVERSITY

Ph.D. in Forest Resources

Program of Study

The Department of Forestry and Natural Resources offers a Ph.D. degree in forest resources. Areas of concentration for the degree programs in forest resources include forest ecology, forest economics, forest wildlife management, silviculture, urban forestry, and water quality and wetlands. The program generally lasts three to four years. In addition to the final oral exams, Ph.D. students must pass a comprehensive exam at the end of their second year. Students must complete at least one continuous year in residence and are required to work closely with their major professor and graduate committee members. Most research results have applied management applications, and students are expected to present and publish these results in appropriate technical and popular outlets. Many Ph.D. graduates are employed in academia, with others working in research capacities for state and federal agencies or the forest industry.

Research Facilities

Excellent GIS, computer, chemical analysis, and biotechnology facilities are available to graduate students in the Department of Forest Resources. The 17,500-acre Clemson Experimental Forest surrounds the campus and offers opportunities for field research. In addition, students may work with 7 faculty members located at the Belle W. Baruch Institute of Coastal Ecology and Forest Science at Georgetown, South Carolina. At Baruch Institute, opportunities exist for research at the Hobcaw Barony, a 17,000-acre undisturbed ecological reserve of forests, high-salinity marsh estuaries, and brackish and freshwater marshes. Research opportunities for graduate students are enhanced by cooperative programs with the U.S. Forest Service Southern Research Station, USGS Fish and Wildlife Cooperative Research Unit at Clemson, Savannah River Ecology Laboratory, Waddell Mariculture Center, and the National Council for Air and Stream Improvement Eastern Wildlife Program.

Financial Aid

Graduate research assistantships are available through research grants and contracts administered by individual faculty members. Minimum assistantship levels are normally $18,000 for Ph.D. students. Limited appointments are available for work-study programs. There are no allowances for dependents or hiring of spouses.

Cost of Study

Tuition for 2006–07 is $4643 per semester for in-state students and $9255 per semester for nonresidents. Off-campus rates are $535 per hour for in-state students and $918 per hour for nonresidents. Graduate assistants pay a flat fee of $1079 per semester and $348 per summer session. Graduate fellows pay South Carolina resident fees.

Living and Housing Costs

On-campus housing is available. For information, students should visit http://www.housing.clemson.edu. The cost of living in Clemson is quite low compared to the national average. Students who choose to live off the campus typically spend $300–$400 per month for rent, depending on location, amenities, roommates, etc.

Student Group

The program has approximately 13 students. Thirty-one percent are women, 85 percent are full-time, and 8 percent are international students.

Student Outcomes

Many Ph.D. graduates are employed in academia, with others working in research capacities for state and federal agencies or the forest industry.

Location

Clemson is a small, beautiful college town near the Blue Ridge Mountains and Lake Hartwell in upstate South Carolina. The Upstate is one of the country's fastest-growing areas and is the midpoint of the Charlotte-to-Atlanta I-85 corridor, a multistate area along Interstate 85 that runs from metro Atlanta to Richmond, Virginia, and encompasses Charlotte, North Carolina, and North Carolina's Research Triangle. Atlanta and Charlotte are each a 2-hour's drive away. Many financial institutions and other industries have national headquarters for a major presence in the Upstate, including Wachovia, Bank of America, BMW, Bon Secours St. Francis Health System, Bosch North America, Bowater, Charter Communications, Ernst and Young, Fluor Corporation, IBM, Microsoft, Michelin of North America, and many others.

The University

Clemson is classified by the Carnegie Foundation as Doctoral/Research University–Extensive, a category comprising less than 4 percent of all universities in America. The University's mission is to fulfill the covenant between its founder and the people of South Carolina to establish a "high seminary of learning" through its responsibilities of teaching, research, and extended public service. The University has identified eight areas of academic emphasis that create collaborations that, in turn, help fulfill the University's mission.

Applying

Graduate applicants are selected based on demonstrated scholarship (GRE scores, GPA, undergraduate institution), references, field experience, stated interests, and professional goals. Individual faculty members generally require personal interviews with applicants to determine their interests and compatibility with specific research projects. Computer and communication skills and ability to work with others are also important considerations.

Applicants may apply on the Web at http://www.grad.clemson.edu/p_apply.html. Applications with a $50 nonrefundable fee should be received no later than five weeks prior to registration. Every required item in support of the application must be on file by that date. Students are advised to contact the department for the deadlines of the program of proposed study.

Correspondence and Information

David Guynn
Department of Forestry and Natural Resources
Clemson University
Clemson, South Carolina 29634-0317
Phone: 864-656-4830
Fax: 864-656-3304
E-mail: dguynn@clemson.edu
Web site: http://www.clemson.edu/forestres/

Clemson University

THE FACULTY AND THEIR RESEARCH

William W. Bowerman IV, Associate Professor; Ph.D., Michigan State. Fisheries and wildlife.
William H. Conner, Professor; Ph.D., LSU. Forestry.
William Rockford English, Associate Professor; Ph.D., Clemson. Entomology.
Arnold G. Eversole, Professor; Ph.D., Syracuse. Biology and zoology.
Jeffrey W. Foltz, Professor; Ph.D., Colorado. Biology.
Lawrence R. Gering, Associate Professor; Ph.D., Georgia. Forest biometrics.
Charles A. Gresham, Associate Professor; Ph.D., Duke. Forestry.
David C. Guynn Jr., Professor; Ph.D., Virginia Tech. Forestry.
Roy L. Hedden, Professor; Ph.D., Washington (Seattle). Forest entomology.
Alan R. Johnson, Assistant Professor; Ph.D., Tennessee. Environmental toxicology.
Joseph D. Lanham, Associate Professor; Ph.D., Clemson. Ecology and ornithology.
Patricia A. Layton, Professor and Department Chair; Ph.D., Florida. Forest genetics.
Andy Wu-Chung Lee, Professor; Ph.D., Auburn. Wood products.
Allan Marsinko, Professor; Ph.D., SUNY College of Environmental Science and Forestry. Forest economics.
Larry R. Nelson, Associate Professor; Ph.D., Auburn. Forestry.
Lawrence E. Nix,; Professor; Ph.D., Georgia. Forestry.
Christopher J. Post, Assistant Professor; Ph.D., Cornell. Environmental information science.
John H. Rodgers Jr., Professor; Ph.D., Virginia Tech. Botany and aquatic ecology.
Linda C. Roth, Assistant Professor; Ph.D., Clark. Geography.
Victor B. Shelburne, Professor; Ph.D., Clemson. Forestry.
Bo Song, Assistant Professor; Ph.D., Michigan Tech. Forest science.
Thomas J. Straka, Professor; Ph.D., Virginia Tech. Forestry.
David H. Van Lear, Named Professor; Ph.D., Idaho. Forest sciences.
Gaofeng G. Wang, Assistant Professor; Ph.D., British Columbia. Forest ecology.
Thomas M. Williams, Professor; Ph.D., Minnesota. Forestry.
Gene W. Wood, Professor; Ph.D., Penn State. Agronomy.
Thomas E. Wooten, Alumni Professor; Ph.D., North Carolina State. Forestry.
Greg K. Yarrow, Professor; D.F., Stephen F. Austin. Forest wildlife.

CLEMSON UNIVERSITY

CLEMSON UNIVERSITY

Master of Science in Wildlife and Fisheries Biology

Program of Study

The Department of Forestry and Natural Resources at Clemson University offers a Master of Science (M.S.) degree in wildlife and fisheries biology. Research areas in wildlife and fisheries biology include aquaculture, conservation biology, endangered species biology, freshwater fisheries science, marine fisheries science, and upland and wetland wildlife biology. The M.S. program generally lasts two to three years. Students must complete at least one continuous year in residence and are required to work closely with their major professor and graduate committee members. Most research results have applied management applications, and students are expected to present and publish these results in appropriate technical and popular outlets. Most graduates of the program are employed by state and federal agencies, the forest industry, or environmental consulting firms, or they work as private consultants or resource managers.

Research Facilities

Excellent GIS, computer, chemical analysis, and biotechnology facilities are available to graduate students in the Department of Forest Resources. The 17,500-acre Clemson Experimental Forest surrounds the campus and offers opportunities for field research. In addition, students may work with 7 faculty members who are located at the Belle W. Baruch Institute of Coastal Ecology and Forest Science in Georgetown, South Carolina. At Baruch Institute, opportunities exist for research at the Hobcaw Barony, a 17,000-acre undisturbed ecological reserve of forests, high-salinity marsh estuaries, and brackish and freshwater marshes. Research opportunities for graduate students are enhanced by cooperative programs with the U.S. Forest Service Southern Research Station, USGS Fish and Wildlife Cooperative Research Unit at Clemson, the Savannah River Ecology Laboratory, Waddell Mariculture Center, and the National Council for Air and Stream Improvement Eastern Wildlife Program.

Financial Aid

Graduate research assistantships are available through research grants and contracts that are administered by individual faculty members. Minimum assistantship levels are normally $15,000 for M.S. students. Limited appointments are available for work-study programs. There are no allowances for dependents or hiring of spouses.

Cost of Study

Tuition for 2006–07 is $4643 per semester for in-state students and $9255 per semester for nonresidents. Off-campus rates are $535 per hour for in-state students and $918 per hour for nonresidents. Graduate assistants pay a flat fee of $1079 per semester and $348 per summer session. Graduate fellows pay South Carolina resident fees.

Living and Housing Costs

On-campus housing is available. For information, students should visit http://www.housing.clemson.edu. The cost of living in Clemson is quite low compared to the national average. Students who choose to live off the campus typically spend $300–$400 per month for rent, depending on location, amenities, roommates, etc.

Student Group

The M.S. program has 8 students, all of whom are from the U.S. and attend on a full-time basis. Approximately 50 percent of the students are women.

Student Outcomes

Graduates of the Master of Science and the Master of Forest Resources (M.F.R.) programs most often work for state and federal natural resources management agencies or the forest industry or in the private sector. Those graduates who pursue Ph.D. studies usually do so at other land-grant universities, such as Clemson, Virginia Tech, Mississippi State University, or the University of Georgia.

Location

Clemson is a small, beautiful college town near the Blue Ridge Mountains and Lake Hartwell in upstate South Carolina. The Upstate is one of the country's fastest-growing areas and is the midpoint of the Charlotte-to-Atlanta I-85 corridor, a multistate area along Interstate 85 that runs from metro Atlanta to Richmond, Virginia, and encompasses Charlotte, North Carolina, and North Carolina's Research Triangle. Atlanta and Charlotte are each a 2-hour's drive away. Many financial institutions and other industries have national headquarters for a major presence in the Upstate, including Wachovia, Bank of America, BMW, Bon Secours St. Francis Health System, Bosch North America, Bowater, Charter Communications, Ernst and Young, Fluor Corporation, IBM, Microsoft, Michelin of North America, and many others.

The University

Clemson is classified by the Carnegie Foundation as Doctoral/Research University–Extensive, a category comprising less than 4 percent of all universities in America. The University's mission is to fulfill the covenant between its founder and the people of South Carolina to establish a "high seminary of learning" through its responsibilities of teaching, research, and extended public service. The University has identified eight areas of academic emphasis that create collaborations that, in turn, help fulfill the University's mission.

Applying

Graduate applicants are selected based on demonstrated scholarship (GRE scores, GPA, undergraduate institution), references, field experience, stated interests, and professional goals. Individual faculty members generally require personal interviews with applicants to determine their interests and compatibility with specific research projects. Field experience, computer and communication skills, and the applicant's ability to work with others are important considerations.

Applicants may apply on the Web at http://www.grad.clemson.edu/p_apply.html. Applications with a $50 nonrefundable fee should be received no later than five weeks prior to registration. Every required item in support of the application must be on file by that date. Students are advised to contact the department for the deadlines of the program of proposed study.

Correspondence and Information

David Guynn
Department of Forestry and Natural Resources
Clemson University
Clemson, South Carolina 29634-0317
Phone: 864-656-4830
Fax: 864-656-3304
E-mail: dguynn@clemson.edu
Web site: http://www.clemson.edu/forestres/

Clemson University

THE FACULTY AND THEIR RESEARCH

William W. Bowerman IV, Associate Professor; Ph.D., Michigan State. Fisheries and wildlife.
William H. Conner, Professor; Ph.D., LSU. Forestry.
William Rockford English, Associate Professor; Ph.D., Clemson. Entomology.
Arnold G. Eversole, Professor; Ph.D., Syracuse. Biology and zoology.
Jeffrey W. Foltz, Professor; Ph.D., Colorado. Biology.
Lawrence R. Gering, Associate Professor; Ph.D., Georgia. Forest biometrics.
Charles A. Gresham, Associate Professor; Ph.D., Duke. Forestry.
David C. Guynn Jr., Professor; Ph.D., Virginia Tech. Forestry.
Roy L. Hedden, Professor; Ph.D., Washington (Seattle). Forest entomology.
Alan R. Johnson, Assistant Professor; Ph.D., Tennessee. Environmental toxicology.
Joseph D. Lanham, Associate Professor; Ph.D., Clemson. Ecology and ornithology.
Patricia A. Layton, Professor and Department Chair; Ph.D., Florida. Forest genetics.
Andy Wu-Chung Lee, Professor; Ph.D., Auburn. Wood products.
Allan Marsinko, Professor; Ph.D., SUNY College of Environmental Science and Forestry. Forest economics.
Larry R. Nelson, Associate Professor; Ph.D., Auburn. Forestry.
Lawrence E. Nix, Professor; Ph.D., Georgia. Forestry.
Christopher J. Post, Assistant Professor; Ph.D., Cornell. Environmental information science.
John H. Rodgers Jr., Professor; Ph.D., Virginia Tech. Botany and aquatic ecology.
Linda C. Roth, Assistant Professor; Ph.D., Clark. Geography.
Victor B. Shelburne, Professor; Ph.D., Clemson. Forestry.
Bo Song, Assistant Professor; Ph.D., Michigan Tech. Forest science.
Thomas J. Straka, Professor; Ph.D., Virginia Tech. Forestry.
David H. Van Lear, Named Professor; Ph.D., Idaho. Forest sciences.
Gaofeng G. Wang, Assistant Professor; Ph.D., British Columbia. Forest ecology.
Thomas M. Williams, Professor; Ph.D., Minnesota. Forestry.
Gene W. Wood, Professor; Ph.D., Penn State. Agronomy.
Thomas E. Wooten, Alumni Professor; Ph.D., North Carolina State. Forestry.
Greg K. Yarrow, Professor; D.F., Stephen F. Austin State. Forest wildlife.

CLEMSON UNIVERSITY

CLEMSON UNIVERSITY

Ph.D. in Wildlife and Fisheries Biology

Program of Study

The Department of Forestry and Natural Resources at Clemson University offers a Ph.D. degree in wildlife and fisheries biology. Research areas in wildlife and fisheries biology include aquaculture, conservation biology, endangered species biology, freshwater fisheries science, marine fisheries science, and upland and wetland wildlife biology. The program generally lasts three to four years. In addition to the final oral exams, Ph.D. students must pass a comprehensive exam at the end of their second year. Students must complete at least one continuous year in residence and are required to work closely with their major professor and graduate committee members. Most research results have applied management applications, and students are expected to present and publish these results in appropriate technical and popular outlets. Many Ph.D. graduates are employed in academia, with others working in research capacities for state and federal agencies or in the forest industry.

Research Facilities

Excellent GIS, computer, chemical analysis, and biotechnology facilities are available to graduate students in the Department of Forest Resources. The 17,500-acre Clemson Experimental Forest surrounds the campus and offers opportunities for field research. In addition, students may work with 7 faculty members who are located at the Belle W. Baruch Institute of Coastal Ecology and Forest Science in Georgetown, South Carolina. At Baruch Institute, opportunities exist for research at the Hobcaw Barony, a 17,000-acre undisturbed ecological reserve of forests, high-salinity marsh estuaries, and brackish and freshwater marshes. Research opportunities for graduate students are enhanced by cooperative programs with the U.S. Forest Service Southern Research Station, USGS Fish and Wildlife Cooperative Research Unit at Clemson, the Savannah River Ecology Laboratory, Waddell Mariculture Center, and the National Council for Air and Stream Improvement Eastern Wildlife Program.

Financial Aid

Graduate research assistantships are available through research grants and contracts that are administered by individual faculty members. Minimum assistantship levels are normally $18,000 for Ph.D. students. Limited appointments are available for work-study programs. There are no allowances for dependents or hiring of spouses.

Cost of Study

Tuition for 2006–07 is $4643 per semester for in-state students and $9255 per semester for nonresidents. Off-campus rates are $535 per hour for in-state students and $918 per hour for nonresidents. Graduate assistants pay a flat fee of $1079 per semester and $348 per summer session. Graduate fellows pay South Carolina resident fees.

Living and Housing Costs

On-campus housing is available. For information, students should visit http://www.housing.clemson.edu. The cost of living in Clemson is quite low compared to the national average. Students who choose to live off the campus typically spend $300–$400 per month for rent, depending on location, amenities, roommates, etc.

Student Group

The program has approximately 5 students, all of whom are from the United States. Forty percent are women, and 40 percent attend on a full-time basis.

Student Outcomes

Many Ph.D. graduates are employed in academia, with others working in research capacities for state and federal agencies or the forest industry.

Location

Clemson is a small, beautiful college town near the Blue Ridge Mountains and Lake Hartwell in upstate South Carolina. The Upstate is one of the country's fastest-growing areas and is the midpoint of the Charlotte-to-Atlanta I-85 corridor, a multistate area along Interstate 85 that runs from metro Atlanta to Richmond, Virginia, and encompasses Charlotte, North Carolina, and North Carolina's Research Triangle. Atlanta and Charlotte are each a 2-hour's drive away. Many financial institutions and other industries have national headquarters for a major presence in the Upstate, including Wachovia, Bank of America, BMW, Bon Secours St. Francis Health System, Bosch North America, Bowater, Charter Communications, Ernst and Young, Fluor Corporation, IBM, Microsoft, Michelin of North America, and many others.

The University

Clemson is classified by the Carnegie Foundation as Doctoral/Research University–Extensive, a category comprising less than 4 percent of all universities in America. The University's mission is to fulfill the covenant between its founder and the people of South Carolina to establish a "high seminary of learning" through its responsibilities of teaching, research, and extended public service. The University has identified eight areas of academic emphasis that create collaborations that, in turn, help fulfill the University's mission.

Applying

Graduate applicants are selected based on demonstrated scholarship (GRE scores, GPA, undergraduate institution), references, field experience, stated interests, and professional goals. Individual faculty members generally require personal interviews with applicants to determine their interests and compatibility with specific research projects. Field experience, computer and communication skills, and the applicant's ability to work with others are important considerations.

Applicants may apply on the Web at http://www.grad.clemson.edu/p_apply.html. Applications with a $50 nonrefundable fee should be received no later than five weeks prior to registration. Every required item in support of the application must be on file by that date. Students are advised to contact the department for the deadlines of the program of proposed study.

Correspondence and Information

David Guynn
Department of Forestry and Natural Resources
Clemson University
Clemson, South Carolina 29634-0317
Phone: 864-656-4830
Fax: 864-656-3304
E-mail: dguynn@clemson.edu
Web site: http://www.clemson.edu/forestres/

Clemson University

THE FACULTY AND THEIR RESEARCH

William W. Bowerman IV, Associate Professor; Ph.D., Michigan State. Fisheries and wildlife.
William H. Conner, Professor; Ph.D., LSU. Forestry.
William Rockford English, Associate Professor; Ph.D., Clemson. Entomology.
Arnold G. Eversole, Professor; Ph.D., Syracuse. Biology and zoology.
Jeffrey W. Foltz, Professor; Ph.D., Colorado. Biology.
Lawrence R. Gering, Associate Professor; Ph.D., Georgia. Forest biometrics.
Charles A. Gresham, Associate Professor; Ph.D., Duke. Forestry.
David C. Guynn Jr., Professor; Ph.D., Virginia Tech. Forestry.
Roy L. Hedden, Professor; Ph.D., Washington (Seattle). Forest entomology.
Alan R. Johnson, Assistant Professor; Ph.D., Tennessee. Environmental toxicology.
Joseph D. Lanham, Associate Professor; Ph.D., Clemson. Ecology and ornithology.
Patricia A. Layton, Professor and Department Chair; Ph.D., Florida. Forest genetics.
Andy Wu-Chung Lee, Professor; Ph.D., Auburn. Wood products.
Allan Marsinko, Professor; Ph.D., SUNY College of Environmental Science and Forestry. Forest economics.
Larry R. Nelson, Associate Professor; Ph.D., Auburn. Forestry.
Lawrence E. Nix, Professor; Ph.D., Georgia. Forestry.
Christopher J. Post, Assistant Professor; Ph.D., Cornell. Environmental information science.
John H. Rodgers Jr., Professor; Ph.D., Virginia Tech. Botany and aquatic ecology.
Linda C. Roth, Assistant Professor; Ph.D., Clark. Geography.
Victor B. Shelburne, Professor; Ph.D., Clemson. Forestry.
Bo Song, Assistant Professor; Ph.D., Michigan Tech. Forest science.
Thomas J. Straka, Professor; Ph.D., Virginia Tech. Forestry.
David H. Van Lear, Named Professor; Ph.D., Idaho. Forest sciences.
Gaofeng G. Wang, Assistant Professor; Ph.D., British Columbia. Forest ecology.
Thomas M. Williams, Professor; Ph.D., Minnesota. Forestry.
Gene W. Wood, Professor; Ph.D., Penn State. Agronomy.
Thomas E. Wooten, Alumni Professor; Ph.D., North Carolina State. Forestry.
Greg K. Yarrow, Professor; D.F., Stephen F. Austin State. Forest wildlife.

DUKE UNIVERSITY

Nicholas School of the Environment and Earth Sciences

Programs of Study

The Nicholas School has a commitment to education and research addressing an area of vital concern—the quality of the Earth's environment and the sustainable use of its natural resources. The Nicholas School is built on the belief that finding workable solutions to environmental issues requires the viewpoints of more than one discipline.

With facilities at Duke's Durham campus and the Duke Marine Laboratory within the Outer Banks on the North Carolina coast, the Nicholas School is organized around program areas and research centers rather than traditionally structured departments. The centers serve to focus interdisciplinary research and educational activity on a variety of national and international environmental issues.

The Nicholas School's faculty members specialize in an array of disciplines, with particular strengths in global change, ecosystem science (forest and wetlands), coastal ecosystem processes, environmental health (responses to toxic pollutants), and environmental economics and policy. Through joint faculty appointments and research, the School is affiliated with Duke's Departments of Biology, Biological Anthropology and Anatomy, Cell Biology, Chemistry, Economics, and Statistics; the School of Engineering; and Duke University Medical Center. Joint-degree programs are offered with the School of Law, the Fuqua School of Business, the Terry Sanford Institute of Public Policy, and the Master of Arts in Teaching program.

Students may earn a Master of Environmental Management (M.E.M.) or Master of Forestry (M.F.) degree through the Nicholas School of the Environment and Earth Sciences. These are two-year professional degrees that require 48 units of credit. A one-year, 30-unit M.F. program is available for students who have a Bachelor of Science in Forestry from an accredited forestry school. A reduced-credit option is also available through the Senior Professional Program for students who have at least five years of related professional experience; this option requires a minimum of 30 units and one semester in residence.

The Ph.D. is offered through the Graduate School of Duke University and is appropriate for students planning careers in teaching or research. The M.S. degree may be awarded as part of a Ph.D. program.

Course work and research for the School's professional degrees are concentrated in seven program areas: coastal environmental management, environmental health and security, forest resource management, global environmental change, resource ecology, environmental economics and policy, and water and air resources. In addition, faculty members at the Nicholas School's Marine Laboratory offer opportunities for course work and research in the basic ocean sciences, marine biology, environmental and human health sciences, and marine biotechnology.

Research Facilities

The Nicholas School is headquartered in the Levine Science Research Center, an interdisciplinary, state-of-the-art facility that is fully equipped to meet the technical demands of modern teaching and research. The center's fiber-optic networking systems give students access to high-performance computing at Duke and around the world. Students also have access to an online reference network linking all libraries at Duke University, North Carolina State University, and the University of North Carolina at Chapel Hill. The 8,000-acre Duke Forest lies adjacent to the campus and in two neighboring counties. A phytotron with fifty controlled-growth chambers and greenhouses is available for plant research.

The Marine Laboratory in Beaufort, North Carolina, is a complete residential research and teaching facility with modern laboratories, computer facilities, and an extensive library. It is the home port for the 135-foot oceanographic research vessel *Cape Hatteras* and the 57-foot coastal ocean research and training vessel *Susan Hudson.*

Financial Aid

Scholarships, fellowships, assistantships, and student loans are available from a variety of sources, and many students receive financial aid. The Nicholas School maintains its own career services office to assist students in finding paid internships and permanent employment.

Cost of Study

Tuition is $27,633 per year full-time and $1100 per unit part-time in 2006–07. A health fee of $524 is required.

Living and Housing Costs

Most graduate and professional students live off campus, and many share rent with 1 or 2 roommates. Rent for apartments and houses in Durham varies widely; students can expect to pay from $400 to $1000 monthly. Living costs in Beaufort are comparable. A limited amount of on-campus housing is also available on the Durham campus.

Student Group

Approximately 250 students are enrolled in the Nicholas School of the Environment, and 50 are in the Department of the Environment of the Graduate School. The ratio of men to women is approximately equal. The School draws students with undergraduate degrees from liberal arts colleges and research universities and from international locations. While prior work experience is not a requirement for admission, it is highly valued.

Location

Durham (population 198,000), Raleigh, and Chapel Hill form an urban area known as the Research Triangle of North Carolina. Area residents enjoy annual outdoor festivals and numerous other events in drama, music, dance, and the visual arts. The Atlantic Ocean and the Blue Ridge Mountains are each within several hours' drive. The Marine Laboratory is located 180 miles east, on Pivers Island within North Carolina's Outer Banks, adjacent to the historic town of Beaufort (population 5,000).

The University and The School

Noted for its magnificent Gothic architecture and its academic excellence, Duke is among the smallest of the nation's leading universities, having a total enrollment of about 11,000. Its spacious campus is bounded on the east by residential sections of Durham and on the west by the Duke Forest.

The Nicholas School of the Environment and Earth Sciences was established in 1991, but its roots date back to 1938. Duke's Department of Geology was added to the School in 1997. The Nicholas School is the only private graduate school of forestry, environmental studies, and marine sciences in the country. Its professional forestry program has been continuously accredited by the Society of American Foresters since 1938.

Applying

Most students are admitted for fall matriculation. Applications must be received by February 1 for priority consideration. Those received after the priority deadline are considered if space is available. Applications for spring are considered on a space-available basis; the deadline is October 15. GRE scores are required. Applicants for federal financial aid must submit a Free Application for Federal Student Aid (FAFSA).

Applicants who are interested only in research or summer courses at the Marine Laboratory should direct their first inquiry to the Admissions Office, Duke University Marine Laboratory.

Individuals who are interested in M.S. or Ph.D. degrees in earth or ocean sciences through the School's Division of Earth and Ocean Sciences should see the separate listing under the Geology Directory of this guide.

Correspondence and Information

Enrollment Services Office
Nicholas School of the Environment and Earth Sciences
Duke University
Box 90330
Durham, North Carolina 27708-0330
Phone: 919-613-8070
E-mail: admissions@nicholas.duke.edu
Web site: http://www.nicholas.duke.edu

Admissions Office
Duke University Marine Laboratory
Nicholas School of the Environment and Earth Sciences
Duke University
135 Duke Marine Lab Road
Beaufort, North Carolina 28516-9721
Phone: 252-504-7502
E-mail: hnearing@duke.edu
Web site: http://www.nicholas.duke.edu

Duke University

THE FACULTY AND THEIR RESEARCH

William H. Schlesinger, Dean, Nicholas School of the Environment and Earth Sciences; Ph.D., Cornell, 1976. Global biogeochemistry, particularly the role of soils in global element cycles.

Core Faculty/Durham

Richard M. Anderson, Ph.D., Johns Hopkins, 2002. Environmental systems analysis, decision analysis, watershed management.
Paul A. Baker, Ph.D., California, San Diego (Scripps), 1981. Geochemistry and diagenesis of marine sediments and sedimentary rocks and their desposital history.
Lori Snyder Bennear, Ph.D., Harvard. Environmental economics, evaluation of effectiveness of environmental regulations.
Alan E. Boudreau, Ph.D., Washington (Seattle), 1986. Understanding the crystallization of large layered intrusions, with particular attention on the Archean Stillwater complex in Montana.
Norman L. Christensen, Ph.D., California, Santa Barbara, 1973. Effects of disturbance on plant populations and communities, patterns of forest development, remote sensing of forest change, fire ecology.
James S. Clark, Ph.D., Minnesota, 1988. Factors responsible for ecosystem patterns and how they respond to long-term changes in the physical environment, especially fire.
Bruce Hayward Corliss, Ph.D., Rhode Island, 1978. Cenozoic paleoceanography and studies of marine microfossils and deep-sea sediments.
Thomas Crowley, Ph.D., Brown, 1976. Study of past climates—patterns and nature of climate change and their relevance to understanding present climate change and future projections of climate change.
Richard T. Di Giulio, Ph.D., Virginia Tech, 1982. Aquatic toxicology; metabolism, modes of action, and genotoxicity in aquatic animals; development of biochemical responses as biomarkers of environmental quality.
Jonathan L. Goodall, Ph.D., Texas at Austin, 2005. Geographic information systems and environmental informatics, application of geospatial technologies for modeling and management of water resources.
Peter K. Haff, Ph.D., Virginia, 1970. Quantitative modeling techniques, including computer simulation, to describe and predict the course of natural geological processes that occur on the surface of the Earth.
Patrick N. Halpin, Ph.D., Virginia, 1995. Landscape ecology, GIS and remote sensing and international conservation management.
Gary S. Hartshorn, Ph.D., Washington (Seattle), 1972. Tropical forest dynamics, biodiversity conservation, dominance-diversity patterns, and sustainable forest management.
Robert G. Healy, Ph.D., UCLA, 1972. Natural resource, land-use, and environmental policy; reconciling Third World development with environmental quality and sustainable use of natural resources; tourism policy.
Gabriele Hegerl, Ph.D., Munich, 1992. Natural variability of climate, changes in climate due to natural and anthropogenic changes in radiative forcing.
David E. Hinton, Ph.D., Mississippi, 1969. Environmental toxicology and effects assessment in aquatic organisms.
Robert B. Jackson, Ph.D., Utah State, 1992. Ecosystem functioning and feedbacks between global change and the biosphere.
Jeffrey A. Karson, Ph.D., SUNY at Albany, 1977. Structural and tectonic analysis of rift and transform plate boundaries.
Prasad S. Kasibhatla, Ph.D., Kentucky, 1988. Anthropogenic emissions on atmospheric composition and reactivity on marine and terrestrial ecosystems.
Gabriel G. Katul, Ph.D., California, Davis, 1993. Hydrology and fluid mechanisms in the environment.
Richard F. Kay, Ph.D., Yale, 1973. The evolutionary history of the order primates, including further documenting the fossil history of Neotropical monkeys.
Emily M. Klein, Ph.D., Columbia, 1989. The geochemistry of ocean ridge basalts using diverse tools of major- and trace-element and isotropic analysis.
Randall A. Kramer, Ph.D., California, Davis, 1980. Environmental economics, economic valuation of environmental quality, quantitative analysis of environmental policies.
Seth W. Kullman, Ph.D., California, Davis, 1996. Molecular toxicology, with an emphasis on the biochemical and molecular mechanisms of cellular response to environmental pollutants.
Michael L. Lavine, Ph.D., Minnesota, 1987. Sensitivity and robustness of Bayesian analyses, statistical issues in energy and environmental studies, Bayesian nonparametrics, spatial statistics.
Edward D. Levin, Ph.D., Wisconsin, 1984. Basic neurobiology of learning and memory, neurobehavioral toxicology, and the development of novel therapeutic treatments for cognitive dysfunction.
Elwood A. Linney, Ph.D., California, San Diego, 1973. Signal transduction during embryogenesis.
Daniel A. Livingstone, Ph.D., Yale, 1953. The circulation and chemical composition of lakes, particularly in Africa, and how the distribution and abundance of organisms are affected by them.
M. Susan Lozier, Ph.D., Washington (Seattle), 1989. Mesoscale and large-scale ocean dynamics; research approach ranges from the application of numerical models to the analysis of observational data, with the focus on the testing and development of theory.
Lynn A. Maguire, Ph.D., Utah State, 1980. Application of simulation modeling and decision analysis in natural resource management, endangered species, conservation biology, conflict resolution.
Peter E. Malin, Ph.D., Princeton, 1978. Tectonics; seismic wave propagation and earthquakes, with current focus on central California.
Marie Lynn Miranda, Ph.D., Harvard, 1990. Natural resource and environmental economics with interdisciplinary, policy-oriented perspectives.
A. Brad Murray, Ph.D., Minnesota, 1995. Surficial processes and patterns, including rivers and a range of desert, Arctic, and alpine phenomena.
Ram Oren, Ph.D., Oregon State, 1984. Physiological ecology and its application to quantifying water, nutrient, and carbon dynamics in forest ecosystems.
Orrin H. Pilkey, Ph.D., Florida State, 1962. Basic and applied coastal geology, focusing primarily on barrier island coasts.
Stuart L. Pimm, Ph.D., New Mexico State, 1974. Conservation biology and the impact of human interactions on the survival of species.
Lincoln F. Pratson, Ph.D., Columbia, 1993. Role of sedimentary processes in shaping continental margins.
Kenneth H. Reckhow, Ph.D., Harvard, 1977. Water-quality modeling and applied statistics, decision and risk analysis for water-quality management, uncertainty analysis and parameter estimation in water-quality models.
James F. Reynolds, Ph.D., New Mexico State, 1974. International efforts on land degradation in arid and semiarid regions of the world.
Curtis J. Richardson, Ph.D., Tennessee, 1972. Wetland ecology, ecosystem analysis, soil chemistry/plant nutrition relationships, phosphorus cycling, effects of pollutants on biogeochemical cycling in ecosystems.
Daniel D. Richter, Ph.D., Duke, 1980. Forest ecosystem ecology, biogeochemistry of acid soils, soil and watershed management in the humid temperate zone and the tropics.
Stuart Rojstaczer, Ph.D., Stanford, 1988. The role of fluid in crustal processes, with particular interest in geologic hazards; subsidiary interest in the development of new techniques to determine elastic and fluid flow properties of the Earth in situ.
Erika Sasser, Ph.D., Duke, 1999. The evolving shape of environmental regulation of business and the impact of private, voluntary governance mechanisms on environmental outcomes.
Martin D. Smith, Ph.D., California, Davis, 2001. Natural resource economics, modeling linkages between economic behavior and biophysical processes.
Heather M. Stapleton, Ph.D., Maryland, 2003. Fate and biotransformation of organic contaminants in aquatic systems, focusing on persistent organic pollutants (POPs) such as polychlorinated biphenyls (PCBs) and polybrominated diphenyl ethers (PBDEs).
John W. Terborgh, Ph.D., Harvard, 1963. Tropical ecology and biogeography, adaptive strategies of plants and animals, conservation biology.
Jerry J. Tulis, Ph.D., Illinois, 1965. Occupational and environmental biohazards, indoor air quality, waste management.
Dean L. Urban, Ph.D., Tennessee, 1986. Landscape ecology, forest ecosystem dynamics, application of simulation models to assess forest response to land-use practice and climatic change.
Avner Vengosh, Ph.D., Australian National. Environmental and aqueous geochemistry, isotope hydrology, water quality, salinization of water resources.
Erika Weinthal, Ph.D., Columbia, 1998. Environmental policy, international environmental institutions, political economy of the resource curse, water cooperation and conflict, environmental security.
Jonathan B. Wiener, J.D., Harvard, 1987. Interplay of science, economics, and law in addressing environmental and human health risks.

Core Faculty/Beaufort

Richard T. Barber, Ph.D., Stanford, 1967. Thermal dynamics and ocean basin productivity.
Celia Bonaventura, Ph.D., Texas, 1968. Structure-function relationships of macromolecules, biotechnology.
Joseph Bonaventura, Ph.D., Texas, 1968. Marine biomedicine, protein structure-function relationships.
Lisa Campbell, Ph.D., Cambridge. Environmental policy, coastal zone management, environmental sociology and anthropology.
Larry B. Crowder, Ph.D., Michigan State, 1978. Marine ecology and fisheries oceanography.
Richard B. Forward Jr., Ph.D., California, Santa Barbara, 1969. Physiological ecology of marine animals.
William W. Kirby-Smith, Ph.D., Duke, 1970. Ecology of marine-freshwater systems.
Michael K. Orbach, Ph.D., California, San Diego, 1975. Application of social and policy sciences to coastal and ocean policy and management.
Joseph S. Ramus, Ph.D., Berkeley, 1968. Algal ecological physiology, estuarine dynamics, biotechnology.
Andrew J. Read, Ph.D., Guelph, 1989. Biology and conservation of small cetaceans.
Daniel Rittschof, Ph.D., Michigan, 1975. Chemical ecology of marine organisms.

LOUISIANA STATE UNIVERSITY

School of Renewable Natural Resources

Programs of Study

The School of Renewable Natural Resources (SRNR) offers Master of Science (M.S.) degrees in forestry, wildlife, and fisheries and Doctor of Philosophy (Ph.D.) degrees in forestry and in wildlife and fisheries science. M.S. degrees require a minimum of 30 hours of course work, a research thesis, and a final comprehensive oral exam. Ph.D. degrees require 48 hours of course work beyond a B.S., qualifying and general examinations, and an original dissertation.

Programs of study are designed by each candidate and his/her graduate adviser and advisory committee. Areas of study include aquaculture, fisheries science, fish biology, conservation ecology, wildlife science, forest biology, forest resource management, forest economics, biometrics, industrial forestry operations, forest products operations, forest products marketing and management, wood science, and engineering. The SRNR has established close working relationships with landowners, industry, nonprofit conservation groups, and federal and state agencies. Natural resource commodities contribute more than $3.5 billion to Louisiana's annual economy. The economic and ecological importance of natural resources and the comprehensive nature of the SRNR provide students with a rich environment for graduate studies.

Research Facilities

The SRNR is housed in a comprehensive educational and research complex that includes twenty-eight research laboratories as well as office space for both faculty members and graduate students. The LSU AgCenter Aquaculture Research Station, one of the largest in the U.S., has 150 research ponds totaling more than 100 water acres. Graduate students have access to microcomputer laboratories, a mainframe computer, photographic and digital interpretation systems, and microcomputer-based geographic information systems.

Financial Aid

Applicants with excellent academic credentials are eligible to compete for a limited number of research and teaching assistantships, which are awarded annually. Outstanding applicants are eligible to compete for a Gilbert Foundation Fellowship. Assistantship and fellowship awards range from $14,000 to $20,000 per year. International students from underrepresented countries are eligible for tuition remission through the Graduate School Tuition Award. Rockefeller Scholarships, which award $1000 per year, are available to Louisiana students and to out-of-state students after they establish residency. Most students are funded from faculty-generated research grants. For additional information, students should visit the financial aid Web site (http://gradlsu.gs.lsu.edu/asstfaid.htm).

Cost of Study

Tuition and fees for full-time resident graduate students (9 or more hours) are $2163 per semester, and nonresidents' tuition and fees are $6323. For students awarded graduate assistantships, all tuition is waived; however, students must pay all University fees.

Living and Housing Costs

Off-campus housing information is available from http://www.theadvocate.com. Information on campus housing and dining plans can be found at http://appl003.lsu.edu/housing.

Student Outcomes

Graduates of the School of Renewable Natural Resources are employed in a wide variety of natural resource professions, including those in private industry, government agencies, and academic and other U.S. and international nongovernmental organizations. Recent graduates are employed with the U.S. Army Corps of Engineers, U.S. Fish and Wildlife Service, U.S. Geological Survey, USDA Forest Service, Nature Conservancy, numerous state fish and game and natural resource agencies, and universities worldwide.

Location

Louisiana State University (LSU) is located in Baton Rouge, Louisiana, 75 miles northwest of New Orleans on the banks of the Mississippi River. Baton Rouge, the state capital, has a population of 600,000 and is in the heart of Cajun country. The area has a subtropical climate—winters are mild and summers are warm. Louisiana is known as the "Sportsman's Paradise" because of the expansive aquatic habitats, including the Atchafalaya River basin, which is the largest deep-water swamp in the United States.

The University and The School

Louisiana State University was founded in 1860 as Louisiana's land-grant university. It has grown to have an enrollment of 34,000 students and has become one of the top seventy research universities in the United States. Since the first graduate degree was awarded in 1869, LSU has awarded more than 7,000 Ph.D. and 39,000 master's degrees in more than 130 graduate degree programs. LSU has a tradition in natural resource teaching and research, beginning with its first forestry class in 1911 to the current School of Renewable Natural Resources. Today, research and teaching programs in the SRNR include forestry, forest products, wildlife, fisheries, and aquaculture.

Applying

All applicants for admission to the SRNR must have a B.S. degree from an accredited institution, be acceptable to the graduate faculty, and have an identified major professor. Each applicant must submit official Graduate Record Examinations (GRE) scores, official transcripts, and letters of recommendation. International students must also submit a minimum Test of English as a Foreign Language (TOEFL) score of 550 (paper-based) or 213 (computer-based). These materials are used to rank and select applicants and to award assistantships.

Applicants may be granted regular admission with a GPA of at least 3.0 on all undergraduate and any graduate course work already completed. Students with an undergraduate GPA of 2.55 or below are not considered for admission into any graduate program in the SRNR.

Online applications are available at http://gradlsu.gs.lsu.edu. This site provides links to general Graduate School information and other information on graduate admissions. The General Catalog, which is the official document on policies, deadlines, and information, can be found at http://aaweb.lsu.edu/catalogs/2005. The LSU Schedule of Class not only contains class and final exam schedules but also includes the academic calendar, graduate deadlines, registration information, and fee schedules.

Correspondence and Information

Dr. D. Allen Rutherford
Coordinator of Graduate Studies and Research
119 Renewable Natural Resources Building
Louisiana State University
Baton Rouge, Louisiana 70803
Phone: 225-578-4187
Fax: 225-578-4227
E-mail: druther@lsu.edu
Web site: http://www.rnr.lsu.edu

Louisiana State University

THE FACULTY AND THEIR RESEARCH

Fisheries

William E. Kelso, Professor; Ph.D. Virginia Tech. Natural fisheries, fisheries management, fish-habitat interactions, fish biology and ecology.

Megan LaPeyre, Adjunct Assistant Professor; Ph.D., LSU. Wetland fisheries ecology, plant ecology, wetland ecology, coastal marsh management.

D. Allen Rutherford, Professor of Fisheries and Bryant Bateman Professor of Renewable Natural Resources; Ph.D., Oklahoma State. Natural fisheries, stream habitats and lotic fish assemblages, watershed management practices, ecology of larval and juvenile fishes.

Aquaculture

Charles G. Lutz, Professor; Ph.D., LSU. Fisheries extension.

Robert C. Reigh, Professor; Ph.D., Texas A&M. Fish and crustacean nutrition, feed development, feeding techniques.

Robert P. Romaire, Professor; Ph.D., LSU. Crustacean aquaculture, crawfish production, water-quality management.

Terrance R. Tiersch, Professor; Ph.D., Memphis State. Genetic improvement of aquaculture organisms, molecular genetics, hybridization, polyploidy, cryopreservation.

Forestry

Quang V. Cao, Associate Professor; Ph.D., Virginia Tech. Mensuration, forest biometrics.

Jim L. Chambers, Professor; Ph.D., Missouri. Forest ecology, tree physiology.

Sun Joseph Chang, Professor; Ph.D., Wisconsin–Madison. Forest economics, wood products utilization and marketing.

Thomas J. Dean, Associate Professor; Ph.D., Utah State. Quantitative silviculture, production ecology, stand dynamics.

Hallie Dozier, Assistant Professor; Ph.D., Florida. Forest and natural resource ecology, ecology and management of biological invasions, urban forestry, extension.

Richard Keim, Assistant Professor; Ph.D., Oregon State. Hydrology of forested wetlands and watersheds; management of bottomland and coastal forests; ecosystem restoration; large woody debris; hydrological interactions between forests, soils, and the atmosphere.

Zhijun Liu, Associate Professor; Ph.D., Michigan State. Tree physiology, cultivation of medicinal plants, micropropagation.

Michael Stine, Associate Professor; Ph.D., Michigan State. Genetic improvement, molecular biology, tissue culture of southern trees.

Yi-Jun Xu, Assistant Professor; Ph.D., Göttingen. Hydrologic and biogeochemical processes and modeling.

Forest Products

Cornelius de Hoop, Associate Professor; Ph.D., Texas A&M. Environmental safety and business in forest products.

Todd F. Shupe, Associate Professor; Ph.D., LSU. Wood science, silvicultural and genetic influences on the properties and qualities of wood and wood composites.

Richard Vlosky, Professor; Ph.D., Penn State. Domestic and international wood products marketing, technology applications to improve wood products business competitiveness.

Qinglin Wu, Associate Professor; Ph.D., Oregon State. Wood drying, wood moisture relationships, hygroscopic shrinkage and swelling of wood, wood composite materials to economic development, value-added products opportunities.

Wildlife

Alan D. Afton, Adjunct Associate Professor; Ph.D., North Dakota. Avian behavioral ecology and bioenergetics, ecological aspects of avian migration, waterfowl ecology and management.

Michael J. Chamberlain, Assistant Professor; Ph.D., Mississippi State. Wildlife management, geographic information systems.

Sammy King, Adjunct Assistant Professor; Ph.D., Texas A&M. Wetland wildlife management and ecology, bottomland hardwood management.

J. Andrew Nyman, Assistant Professor; Ph.D., LSU. Wetland wildlife management, wetland ecology, coastal marsh management.

Frank C. Rohwer, Associate Professor; Ph.D., Pennsylvania. Avian ecology, reproductive ecology, wildlife ecology, conservation biology, population biology.

Philip Stouffer, Associate Professor; Ph.D., Rutgers. Conservation ecology, wildlife ecology, population ecology.

YALE UNIVERSITY

School of Forestry & Environmental Studies

Programs of Study

The Yale School of Forestry & Environmental Studies (FES) prepares its students for active roles in sustaining and restoring the long-term health of the biosphere and the well-being of its people. This requires a curriculum that allows students to draw upon Yale's rich resources, including faculty and staff members, libraries and laboratories, and visiting speakers and fellow students. Guided by faculty members, students design a cross-disciplinary course of study aimed at understanding the complex systems of the global environment.

The School offers four 2-year master's degrees: environmental management (M.E.M.), environmental science (M.E.Sc.), forestry (M.F.), and forest science (M.F.S.). The M.F. degree is SAF certified. One student in 10 pursues a joint degree within Yale—in architecture, development economics, international relations, law, management, public health, or religious studies. Students seeking a degree in law may also study at Vermont Law School or Pace Law School. Professionals with at least seven years of environmental work experience can enroll in one-year M.E.M. or M.F. programs. Requirements vary by degree, but all programs combine foundation courses with electives. Entering students attend a three-week series of workshops in August that combine skills training with community building. Two-year students work at an internship or on a research project during the summer after the first year and complete a final project during the second year.

Course work and research are concentrated into nine broad areas that bring together faculty members and students with common interests. These focal groups reflect areas of academic strength. They include ecology, ecosystems, and biodiversity; environment, health, and policy; forestry, forest science, and management; global science change and policy; industrial environmental management; policy, economics, and law; social ecology of conservation and development; urban ecology and environmental design; and water science, policy, and management.

Applicants seeking doctoral degrees must apply to the Yale Graduate School of Arts and Sciences. A faculty committee advises doctoral students, who must pass written and oral qualifying examinations and defend their dissertations.

Research Facilities

The eight buildings at FES house laboratories, controlled-environment rooms, a greenhouse, computer labs, and other research facilities. Architects are at work designing a new $27-million environmentally sustainable building that will anchor the FES campus; it is scheduled for completion in 2008.

Teaching and research extend beyond the campus to more than 10,000 acres of Yale forests in Connecticut, New Hampshire, and Vermont. Yale faculty members and students also have research access to private and public land across the United States and in Puerto Rico. Other resources include the Center for Biodiversity and Conservation Science, Center for Coastal and Watershed Systems, Center for Environmental Law and Policy, Center for Industrial Ecology, Environment and Health Initiative, Global Institute of Sustainable Forestry, Hixon Center for Urban Ecology, and Tropical Resources Institute. The forestry library in Sage Hall (part of the extensive Yale University library system) contains more than 130,000 volumes and maintains subscriptions to more than 300 periodicals.

Financial Aid

Both U.S. and international students may apply for scholarships, work-study grants, and loans, which are awarded based on need and academic merit. Financial aid applications must be postmarked by February 15. In addition, U.S. citizens and permanent residents must submit the Free Application for Federal Student Aid (FAFSA) by February 15 (http://www.fafsa.gov).

Yale also offers a low-interest loan program for both U.S. and international graduate students and maintains a scholarship database to help applicants investigate external sources of funding. The FES Career Development Office also helps students secure funding for research projects and internships and to find full-time jobs upon graduation.

Cost of Study

Tuition for master's programs is approximately $26,000. Additional expenses include $900 for the summer skills training modules, $1300 for hospitalization coverage, and approximately $1100 for books and supplies.

Ph.D. students must pay full tuition for four years and may remain on continuing registration status for up to two years more. Most doctoral students receive a University fellowship that provides a stipend during the academic year and covers the costs of tuition and health insurance for the first four years. Students must pay a nominal continuing registration fee thereafter.

Living and Housing Costs

Most students live off campus in shared apartments near FES and can expect living expenses of $500 to $700 per month. University housing is available for a limited number of single or married students. A single student in the summer-module program should anticipate living expenses of approximately $900 for a three-week period (this cost is in addition to the $900 fee listed above). Parking on campus is limited and expensive, so most students walk, bike, or ride to campus on the University shuttle or on city buses.

Student Group

The roughly 240 master's students and 75 Ph.D. students at FES join a community of about 5,000 graduate students at Yale. Students come to FES with bachelor's and master's degrees from a broad range of academic disciplines. About a third of the students come from abroad, representing roughly forty countries. Students average 27 years in age, and 6 out of 10 are women.

Location

New Haven is an accessible, ethnically diverse community of 125,000 on the Connecticut coast of Long Island Sound. The city is known for its restaurants, theaters, music, museums, and urban parks. New Haven permits easy access to hiking, kayaking, sailing, biking, and skiing and is also a short train ride from Boston and New York City.

The University and The School

Yale University is a private institution, founded in 1701 under the leadership of a group of Congregational ministers. Today it continues its early tradition of liberal education and community service. Yale is renowned for its world-class faculty members and researchers. They, in turn, attract dynamic students who often maintain strong ties to the University after graduating. Since its founding by two pioneering American foresters more than 100 years ago, FES has evolved from a professional school of forestry to an internationally respected training ground for environmental leaders and researchers.

Applying

Master's degree applications must be postmarked by January 8 for priority consideration; new students matriculate only in the fall semester, which begins in August. Acceptance letters for master's program are mailed in mid-March. All applicants must submit official GRE, GMAT, or LSAT scores. International students not educated in English and for whom English is a second language must submit an official TOEFL score. The master's application fee is $70 online or $90 paper.

The Ph.D. application ($85) deadline is January 2 and is administered by the Graduate School of Arts and Sciences.

Correspondence and Information

For master's programs:
Emly McDiarmid
Director of Admissions
School of Forestry & Environmental Studies
Yale University
205 Prospect Street
New Haven, Connecticut 06511
E-mail: emly.mcdiarmid@yale.edu
Web site: http://www.yale.edu/environment/admissions

For Ph.D. programs:
Elisabeth Barsa
Doctoral Program Administrator
School of Forestry & Environmental Studies
Yale University
205 Prospect Street
New Haven, Connecticut 06511
E-mail: elisabeth.barsa@yale.edu
Web site: http://www.yale.edu/environment/academics/doctoral/html

Yale University

THE FACULTY

(All phone numbers listed below are in Area Code 203.)

Shimon C. Anisfeld, Lecturer and Associate Research Scientist in Environmental Chemistry and Water Resources; Ph.D. Telephone: 432-5748, e-mail: shimon.cohenanisfeld@yale.edu.
Mark Ashton, Professor of Silviculture and Forest Ecology; M.F., Ph.D. Telephone: 432-9835, e-mail: mark.ashton@yale.edu.
Michelle Bell, Assistant Professor of Environmental Health; M.S.E., Ph.D. Telephone: 432-9869, e-mail: michelle.bell@yale.edu.
Gaboury Benoit, Professor of Environmental Chemistry, Professor of Environmental Engineering, Co-Director of the Hixon Center for Urban Ecology, and Director of the Center for Coastal and Watershed Systems; Ph.D. Telephone: 432-5139, e-mail: gaboury.benoit@yale.edu.
Graeme Berlyn, Professor of Anatomy and Physiology of Trees; Ph.D. Telephone: 432-5142, e-mail: graeme.berlyn@yale.edu.
Ellen Brennan-Galvin, Lecturer and Senior Research Scholar; Ph.D. Telephone: 432-4644, e-mail: ellen.brennan-galvin@yale.edu.
William Burch Jr., Frederick C. Hixon Professor of Natural Resource Management and Professor, Institution for Social and Policy Studies; Ph.D. Telephone: 432-5119, e-mail: william.burch@yale.edu.
Ann Camp, Lecturer in Stand Dynamics and Forest Health; M.F.S., Ph.D. Telephone: 436-3980, e-mail: ann.camp@yale.edu.
Carol Carpenter, Lecturer in Natural Resource Social Science and in Anthropology; Ph.D. Telephone: 432-7530, e-mail: carol.carpenter@yale.edu.
Benjamin Cashore, Assistant Professor of Sustainable Forestry Management and Chair, Program of Forest Certification; Ph.D. Telephone: 432-3009, e-mail: benjamin.cashore@yale.edu.
Marian Chertow, Assistant Professor of Industrial Environmental Management; Director, Program on Solid-Waste Policy; and Director, Industrial Environmental Management Program; M.P.P.M., Ph.D. Telephone: 432-6197, e-mail: marian.chertow@yale.edu.
Timothy Clark, Adjunct Professor of Wildlife Ecology and Policy; Ph.D. Telephone: 432-6965, e-mail: timothy.w.clark@yale.edu.
Lisa Curran, Associate Professor of Tropical Resources and Director, Tropical Resources Institute; Ph.D. Telephone: 432-3772, e-mail: lisa.curran@yale.edu.
Amity Doolittle, Lecturer, Associate Research Scientist, and Program Director, Tropical Resources Institute; Ph.D. Telephone: 432-3660, e-mail: amity.doolittle@yale.edu.
Michael Dove, Professor of Social Ecology and of Anthropology; Ph.D. Telephone: 432-3463, e-mail: michael.dove@yale.edu.
Paul Draghi, Lecturer in Forest History and Director, Information and Library Systems; Ph.D. Telephone: 432-5115, e-mail: paul.draghi@yale.edu.
Daniel Esty, Professor of Environmental Law and Policy; Clinical Professor, Law School; Director, Center for Environmental Law and Policy; and Director, Yale World Fellows Program; M.A., J.D. Telephone: 432-6256, e-mail: daniel.esty@yale.edu.
Gordon Geballe, Lecturer in Urban Ecology and Associate Dean, Student and Alumni Affairs; Ph.D. Telephone: 432-5122, e-mail: gordon.geballe@yale.edu.
Bradford Gentry, Lecturer in Sustainable Investments and Co-Director, Yale-UNDP Collaborative Program on the Urban Environment; J.D. Telephone: 432-9374, e-mail: bradford.gentry@yale.edu.
Thomas Graedel, Professor of Industrial Ecology, of Chemical Engineering, of Geology and Geophysics and Director, Center for Industrial Ecology; Ph.D. Telephone: 432-9733, e-mail: thomas.graedel@yale.edu.
Timothy Gregoire, J. P. Weyerhaeuser Jr. Professor of Forest Management and Associate Dean, Academic Affairs; Ph.D. Telephone: 432-9398, e-mail: timothy.gregoire@yale.edu.
Arnulf Grübler, Professor in the Field of Energy and Technology; Ph.D. E-mail: arnulf.grubler@yale.edu.
Stephen Kellert, Tweedy/Ordway Professor of Social Ecology and Co-Director, Hixon Center for Urban Ecology; Ph.D. Telephone: 432-5114, e-mail: stephen.kellert@yale.edu.
Xuhui Lee, Associate Professor of Forest and Micrometeorology; Ph.D. Telephone: 432-6271, e-mail: xuhui.lee@yale.edu.
Reid Lifset, Associate Research Scholar; Associate Director, Industrial Environmental Management Program; and Editor-in-Chief, *Journal of Industrial Ecology;* M.S., M.P.P.M. Telephone: 432-6949, e-mail: reid.lifset@yale.edu.
Erin Mansur, Assistant Professor of Environmental Economics, jointly appointed with the School of Management; Ph.D. Telephone: 432-6233, e-mail: erin.mansur@yale.edu.
Robert Mendelsohn, Edwin W. Davis Professor of Forest Policy and of Economics and Professor in the School of Management; Ph.D. Telephone: 432-5128, e-mail: robert.mendelsohn@yale.edu.
Florencia Montagnini, Professor in the Practice of Tropical Forestry; Ph.D. Telephone: 436-4221, e-mail: florencia.montagnini@yale.edu.
Chadwick Oliver, Pinchot Professor of Forest Policy and Director, Global Institute for Sustainable Forestry; M.F.S., Ph.D. Telephone: 432-7409, e-mail: chad.oliver@yale.edu.
Sheila Olmstead, Assistant Professor of Environmental Economics; M.P.Aff., Ph.D. Telephone: 432-6274, e-mail: sheila.olmstead@yale.edu.
Peter Raymond, Assistant Professor of Ecosystem Ecology; Ph.D. Telephone: 432-0817, e-mail: peter.raymond@yale.edu.
Robert Repetto, Professor in the Practice of Economics and Sustainable Development; Ph.D. Telephone: 432-9784, e-mail: robert.repetto@yale.edu.
Jonathan Reuning-Scherer, Lecturer in Statistics; Ph.D. Telephone: 432-5118; e-mail: jonathan.reuning-scherer@yale.edu.
James Saiers, Associate Professor of Hydrology; Ph.D. Telephone: 432-5121, e-mail: james.saiers@yale.edu.
Oswald Schmitz, Professor of Population and Community Ecology; Director, Doctoral Studies; and Director; Center for Biodiversity Conservation and Science; Ph.D. E-mail: oswald.schmitz@yale.edu.
Thomas Siccama, Professor in the Practice of Forest Ecology and Director, Field Studies; Ph.D. Telephone: 432-5140, e-mail: thomas.siccama@yale.edu.
David Skelly, Associate Professor of Ecology; Ph.D. Telephone: 432-3603, e-mail: david.skelly@yale.edu.
James Gustave Speth, Professor and Dean, Practice of Environmental Policy and Sustainable Development; M.Litt., J.D. Telephone: 432-5109, e-mail: gus.speth@yale.edu.
John Wargo, Professor of Environmental Risk Analysis and Policy and of Political Science and Director, Environment and Health Initiative; Ph.D. Telephone: 432-5123, e-mail: john.wargo@yale.edu.

APPENDIXES

Institutional Changes Since the 2006 Edition

Following is an alphabetical listing of institutions that have recently closed, moved, merged with other institutions, or changed their names or status. In the case of a name change, the former name appears first, followed by the new name.

Acupuncture & Integrative Medicine College (Berkeley, CA): name changed to Acupuncture & Integrative Medicine College, Berkeley.

Alliant International University (San Francisco, CA): name changed to Alliant International University–San Francisco Bay.

Alliant University–Fresno (Fresno, CA): name changed to Alliant International University–Fresno.

Alliant University–Los Angeles (Alhambra, CA): name changed to Alliant International University–Los Angeles.

Alliant University–San Diego (San Diego, CA): name changed to Alliant International University–San Diego.

Argosy University/Honolulu (Honolulu, HI): name changed to Argosy University/Hawai'i.

Arizona State University East (Mesa, AZ): name changed to Arizona State University at the Polytechnic Campus.

Bethany College of the Assemblies of God (Scotts Valley, CA): name changed to Bethany University.

Boston Architectural Center (Boston, MA): name changed to Boston Architectural College.

Brown Mackie College, Los Angeles Campus (Santa Monica, CA): name changed to Argosy University/Santa Monica.

California State University, Channel Islands (Camarillo, CA): name changed to California State University Channel Islands.

Center for Humanistic Studies (Farmington Hills, MI): name changed to Michigan School of Professional Psychology.

College of the Humanities and Sciences (Tempe, AZ): name changed to College of the Humanities and Sciences, Harrison Middleton University.

Des Moines University Osteopathic Medical Center (Des Moines, IA): name changed to Des Moines University.

The Graduate College of Union University (Schenectady, NY): name changed to Union Graduate College.

Indiana Institute of Technology (Fort Wayne, IN): name changed to Indiana Tech.

International College and Graduate School (Honolulu, HI): name changed to Hawai'i Theological Seminary.

Long Island University, Southampton College (Southampton, NY): name changed to Long Island University, Southampton Graduate Campus.

Luther Rice Bible College and Seminary (Lithonia, GA): name changed to Luther Rice University.

Medical University of Ohio (Toledo, OH): merged with The University of Toledo.

Midwest Theological Seminary (Wentzville, MO): name changed to Midwest University.

National College of Naturopathic Medicine (Portland, OR): name changed to National College of Natural Medicine.

New School University (New York, NY): name changed to The New School: A University.

North Shore-Long Island Jewish Graduate School of Molecular Medicine (Manhasset, NY): name changed to The Feinstein Institute for Medical Research.

The Pennsylvania State University Harrisburg Campus of the Capital College (Middletown, PA): name changed to The Pennsylvania State University Harrisburg Campus.

Reformed Theological Seminary–Washington D.C./Baltimore Campus (Bethesda, MD): name changed to Reformed Theological Seminary–Washington D.C.

Saint Martin's College (Lacey, WA): name changed to Saint Martin's University.

School of Graduate Studies in Shreveport (Shreveport, LA): name changed to Louisiana State University Health Sciences Center at Shreveport.

Southern California Bible College & Seminary (El Cajon, CA): name changed to Southern California Seminary.

Southern Christian University (Montgomery, AL): name changed to Regions University.

Tai Hsuan Foundation: College of Acupuncture and Herbal Medicine (Honolulu, HI): name changed to World Medicine Institute: College of Acupuncture and Herbal Medicine.

University of Colorado at Denver and Health Sciences Center—Health Sciences Program (Denver, CO): name changed to University of Colorado at Denver and Health Sciences Center.

The University of Lethbridge (Lethbridge, AB, Canada): name changed to University of Lethbridge.

The University of Memphis (Memphis, TN): name changed to University of Memphis.

The University of Montana–Missoula (Missoula, MT): name changed to The University of Montana.

The University of North Carolina at Wilmington (Wilmington, NC): name changed to The University of North Carolina Wilmington.

University of Phoenix–Fort Lauderdale Campus (Fort Lauderdale, FL): name changed to University of Phoenix–South Florida Campus.

University of Phoenix–Jacksonville Campus (Jacksonville, FL): name changed to University of Phoenix–North Florida Campus.

University of Phoenix–Northern California Campus (Pleasanton, CA): name changed to University of Phoenix–Bay Area Campus.

University of Phoenix–Orlando Campus (Maitland, FL): name changed to University of Phoenix–Central Florida Campus.

University of Phoenix–Sacramento Campus (Sacramento, CA): name changed to University of Phoenix–Sacramento Valley Campus.

University of Phoenix–Tampa Campus (Tampa, FL): name changed to University of Phoenix–West Florida Campus.

University of the South (Sewanee, TN): name changed to Sewanee: The University of the South.

Abbreviations Used in the Guides

The following list includes abbreviations of degree names used in the profiles in the 2007 edition of the guides. Because some degrees (e.g., Doctor of Education) can be abbreviated in more than one way (e.g., D.Ed. or Ed.D.), and because the abbreviations used in the guides reflect the preferences of the individual colleges and universities, the list may include two or more abbreviations for a single degree.

Degrees

A Mus D	Doctor of Musical Arts
AC	Advanced Certificate
AD	Artist's Diploma
ADP	Artist's Diploma
Adv C	Advanced Certificate
Adv D	Advanced Diploma
Adv M	Advanced Master
AGSC	Advanced Graduate Specialist Certificate
ALM	Master of Liberal Arts
AM	Master of Arts
AMRS	Master of Arts in Religious Studies
APC	Advanced Professional Certificate
App Sc	Applied Scientist
Au D	Doctor of Audiology
B Th	Bachelor of Theology
C Phil	Certificate in Philosophy
CAES	Certificate of Advanced Educational Specialization
CAGS	Certificate of Advanced Graduate Studies
CAL	Certificate in Applied Linguistics
CALS	Certificate of Advanced Liberal Studies
CAMS	Certificate of Advanced Management Studies
CAPS	Certificate of Advanced Professional Studies
CAS	Certificate of Advanced Studies
CASPA	Certificate of Advanced Study in Public Administration
CASR	Certificate in Advanced Social Research
CATS	Certificate of Achievement in Theological Studies
CBHS	Certificate in Basic Health Sciences
CBS	Graduate Certificate in Biblical Studies
CCJA	Certificate in Criminal Justice Administration
CCMBA	Cross-Continent Master of Business Administration
CCSA	Certificate in Catholic School Administration
CE	Civil Engineer
CEM	Certificate of Environmental Management
CG	Certificate in Gerontology
CGS	Certificate of Graduate Studies
Ch E	Chemical Engineer
CITS	Certificate of Individual Theological Studies
CM	Certificate in Management
CMH	Certificate in Medical Humanities
CMM	Master of Church Ministries
CMS	Certificate in Ministerial Studies
CNM	Certificate in Nonprofit Management
CP	Certificate in Performance
CPASF	Certificate Program for Advanced Study in Finance
CPC	Certificate in Professional Counseling Certificate in Publication and Communication
CPH	Certificate in Public Health
CPM	Certificate in Public Management
CPS	Certificate of Professional Studies
CScD	Doctor of clinical Science
CSD	Certificate in Spiritual Direction
CSE	Computer Systems Engineer
CSS	Certificate of Special Studies
CTS	Certificate of Theological Studies
CURP	Certificate in Urban and Regional Planning
D Arch	Doctor of Architecture
D Ed	Doctor of Education
D Eng	Doctor of Engineering
D Engr	Doctor of Engineering
D Env	Doctor of Environment
D Env M	Doctor of Environmental Management
D Law	Doctor of Law
D Litt	Doctor of Letters
D Med Sc	Doctor of Medical Science
D Min	Doctor of Ministry
D Min PCC	Doctor of Ministry, Pastoral Care, and Counseling
D Miss	Doctor of Missiology
D Mus	Doctor of Music
D Mus A	Doctor of Musical Arts
D Phil	Doctor of Philosophy
D Ps	Doctor of Psychology
D Sc	Doctor of Science
D Sc D	Doctor of Science in Dentistry
D Th	Doctor of Theology
D Th P	Doctor of Practical Theology
DA	Doctor of Arts
DA Ed	Doctor of Arts in Education
DAOM	Doctorate in Acupuncture and Oriental Medicine
DAST	Diploma of Advanced Studies in Teaching
DBA	Doctor of Business Administration
DBS	Doctor of Buddhist Studies

DC	Doctor of Chiropractic
DCC	Doctor of Computer Science
DCD	Doctor of Communications Design
DCL	Doctor of Comparative Law
DCM	Doctor of Church Music
DCN	Doctor of Clinical Nutrition
DCS	Doctor of Computer Science
DDN	Diplôme du Droit Notarial
DDS	Doctor of Dental Surgery
DE	Doctor of Education Doctor of Engineering
DEIT	Doctor of Educational Innovation and Technology
DEM	Doctor of Educational Ministry
DEPD	Diplôme Études Spécialisées
DES	Doctor of Engineering Science
DESS	Diplôme Études Supérieures Spécialisées
DFA	Doctor of Fine Arts
DFES	Doctor of Forestry and Environmental Studies
DGP	Diploma in Graduate and Professional Studies
DH Sc	Doctor of Health Sciences
DHA	Doctor of Health Administration
DHCE	Doctor of Health Care Ethics
DHL	Doctor of Hebrew Letters Doctor of Hebrew Literature
DHS	Doctor of Health Service Doctor of Human Services
DHSc	Doctor of Health Science
DIBA	Doctor of International Business Administration
Dip CS	Diploma in Christian Studies
DIT	Doctor of Industrial Technology
DJ Ed	Doctor of Jewish Education
DJS	Doctor of Jewish Studies
DM	Doctor of Management Doctor of Music
DMA	Doctor of Musical Arts
DMD	Doctor of Dental Medicine
DME	Doctor of Manufacturing Management Doctor of Music Education
DMEd	Doctor of Music Education
DMFT	Doctor of Marital and Family Therapy
DMH	Doctor of Medical Humanities
DML	Doctor of Modern Languages
DMM	Doctor of Music Ministry
DN Sc	Doctor of Nursing Science
DNP	Doctor of Nursing Practice Doctor of Nursing Practice
DNS	Doctor of Nursing Science
DO	Doctor of Osteopathy
DPA	Doctor of Public Administration
DPC	Doctor of Pastoral Counseling
DPDS	Doctor of Planning and Development Studies
DPE	Doctor of Physical Education
DPH	Doctor of Public Health
DPM	Doctor of Plant Medicine Doctor of Podiatric Medicine
DPS	Doctor of Professional Studies
DPT	Doctor of Physical Therapy
Dr DES	Doctor of Design
Dr PH	Doctor of Public Health
Dr Sc PT	Doctor of Science in Physical Therapy
DS	Doctor of Science
DS Sc	Doctor of Social Science
DSJS	Doctor of Science in Jewish Studies
DSL	Doctor of Strategic Leadership
DSM	Doctor of Sacred Music Doctor of Sport Management
DSN	Doctor of Science in Nursing
DSW	Doctor of Social Work
DTL	Doctor of Talmudic Law
DV Sc	Doctor of Veterinary Science
DVM	Doctor of Veterinary Medicine
EAA	Engineer in Aeronautics and Astronautics
ECS	Engineer in Computer Science
Ed D	Doctor of Education
Ed DCT	Doctor of Education in College Teaching
Ed M	Master of Education
Ed S	Specialist in Education
Ed Sp	Specialist in Education
Ed Sp PTE	Specialist in Education in Professional Technical Education
EDM	Executive Doctorate in Management
EDSPC	Education Specialist
EE	Electrical Engineer
EJD	Executive Juris Doctor
EM	Mining Engineer
EMBA	Executive Master of Business Administration
EMCIS	Executive Master of Computer Information Systems
EMHA	Executive Master of Health Administration
EMIB	Executive Master of International Business
EMS	Executive Master of Science
EMTM	Executive Master of Technology Management
Eng	Engineer
Eng Sc D	Doctor of Engineering Science
Engr	Engineer
Ex Doc	Executive Doctor of Pharmacy
Exec Ed D	Executive Doctor of Education
Exec MBA	Executive Master of Business Administration

Exec MIM	Executive Master of International Management
Exec MPA	Executive Master of Public Administration
Exec MPH	Executive Master of Public Health
Exec MS	Executive Master of Science
GBC	Graduate Business Certificate
GCE	Graduate Certificate in Education
GDPA	Graduate Diploma in Public Administration
GDRE	Graduate Diploma in Religious Education
GEMBA	Global Executive Master of Business Administration
Geol E	Geological Engineer
GMBA	Global Master of Business Administration
GPD	Graduate Performance Diploma
GSS	Graduate Special Certificate for Students in Special Situations
IMA	Interdisciplinary Master of Arts
IMBA	International Master of Business Administration
ITMA	Master of Instructional Technology
JCD	Doctor of Canon Law
JCL	Licentiate in Canon Law
JD	Juris Doctor
JD/MAP	Juris Doctor/Master of Applied Politics
JSD	Doctor of Juridical Science Doctor of Jurisprudence Doctor of the Science of Law
JSM	Master of Science of Law
L Th	Licenciate in Theology
LL B	Bachelor of Laws
LL CM	Master of Laws in Comparative Law
LL D	Doctor of Laws
LL M	Master of Laws
LL M T	Master of Laws in Taxation
M Ac	Master of Accountancy Master of Accounting Master of Acupuncture
M Ac OM	Master of Acupuncture and Oriental Medicine
M Acc	Master of Accountancy Master of Accounting
M Acct	Master of Accountancy Master of Accounting
M Accy	Master of Accountancy
M Actg	Master of Accounting
M Acy	Master of Accountancy
M Ad	Master of Administration
M Ad Ed	Master of Adult Education
M Adm	Master of Administration
M Adm Mgt	Master of Administrative Management
M Adv	Master of Advertising
M Aero E	Master of Aerospace Engineering
M Ag	Master of Agriculture
M Ag Ed	Master of Agricultural Education
M Agr	Master of Agriculture
M Anesth Ed	Master of Anesthesiology Education
M App Comp Sc	Master of Applied Computer Science
M App St	Master of Applied Statistics
M Appl Stat	Master of Applied Statistics
M Aq	Master of Aquaculture
M Ar	Master of Architecture
M Arch	Master of Architecture
M Arch E	Master of Architectural Engineering
M Arch H	Master of Architectural History
M Arch UD	Master of Architecture in Urban Design
M Bio E	Master of Bioengineering
M Biomath	Master of Biomathematics
M Bus Ed	Master of Business Education
M Ch	Master of Chemistry
M Ch E	Master of Chemical Engineering
M Chem	Master of Chemistry
M Cl D	Master of Clinical Dentistry
M Cl Sc	Master of Clinical Science
M Co E	Master of Computer Engineering
M Comp E	Master of Computer Engineering
M Comp Sc	Master of Computer Science
M Coun	Master of Counseling
M Dent	Master of Dentistry
M Dent Sc	Master of Dental Sciences
M Des	Master of Design
M Des S	Master of Design Studies
M Div	Master of Divinity
M E Com	Master of Electronic Commerce
M Ec	Master of Economics
M Econ	Master of Economics
M Ed	Master of Education
M Ed T	Master of Education in Teaching
M En	Master of Engineering
M En S	Master of Environmental Sciences
M Eng	Master of Engineering
M Eng Mgt	Master of Engineering Management
M Eng Tel	Master of Engineering in Telecommunications
M Engr	Master of Engineering
M Env	Master of Environment
M Env Des	Master of Environmental Design
M Env E	Master of Environmental Engineering
M Env Sc	Master of Environmental Science
M Ext Ed	Master of Extension Education
M Fin	Master of Finance
M Fr	Master of French
M Geo E	Master of Geological Engineering

M Geoenv E	Master of Geoenvironmental Engineering
M Geog	Master of Geography
M Hum	Master of Humanities
M Hum Svcs	Master of Human Services
M Kin	Master of Kinesiology
M Land Arch	Master of Landscape Architecture
M Lit M	Master of Liturgical Music
M Litt	Master of Letters
M Man	Master of Management
M Mat SE	Master of Material Science and Engineering
M Math	Master of Mathematics
M Med Sc	Master of Medical Science
M Mgmt	Master of Management
M Mgt	Master of Management
M Min	Master of Ministries
M Mtl E	Master of Materials Engineering
M Mu	Master of Music
M Mus	Master of Music
M Mus Ed	Master of Music Education
M Nat Sci	Master of Natural Science
M Nurs	Master of Nursing
M Oc E	Master of Oceanographic Engineering
M Pharm	Master of Pharmacy
M Phil	Master of Philosophy
M Phil F	Master of Philosophical Foundations
M Pl	Master of Planning
M Pol	Master of Political Science
M Pr A	Master of Professional Accountancy
M Pr Met	Master of Professional Meteorology
M Prob S	Master of Probability and Statistics
M Prof Past	Master of Professional Pastoral
M Psych	Master of Psychology
M Pub	Master of Publishing
M Rel	Master of Religion
M Sc	Master of Science
M Sc A	Master of Science (Applied)
M Sc AHN	Master of Science in Applied Human Nutrition
M Sc BMC	Master of Science in Biomedical Communications
M Sc CS	Master of Science in Computer Science
M Sc E	Master of Science in Engineering
M Sc Eng	Master of Science in Engineering
M Sc Engr	Master of Science in Engineering
M Sc F	Master of Science in Forestry
M Sc FE	Master of Science in Forest Engineering
M Sc Geogr	Master of Science in Geography
M Sc N	Master of Science in Nursing
M Sc OT	Master of Science in Occupational Therapy
M Sc P	Master of Science in Planning
M Sc Pl	Master of Science in Planning
M Sc PT	Master of Science in Physical Therapy
M Sc T	Master of Science in Teaching
M Soc	Master of Sociology
M Sp Ed	Master of Special Education
M Stat	Master of Statistics
M Sw E	Master of Software Engineering
M Sw En	Master of Software Engineering
M Sys Sc	Master of Systems Science
M Tax	Master of Taxation
M Tech	Master of Technology
M Th	Master of Theology
M Th Past	Master of Pastoral Theology
M Tox	Master of Toxicology
M Trans E	Master of Transportation Engineering
M Vet Sc	Master of Veterinary Science
MA	Master of Administration Master of Arts
MA Comm	Master of Arts in Communication
MA Ed	Master of Arts in Education
MA Ed Ad	Master of Arts in Educational Administration
MA Ext	Master of Agricultural Extension
MA Islamic	Master of Arts in Islamic Studies
MA Min	Master of Arts in Ministry
MA Missions	Master of Arts in Missions
MA Past St	Master of Arts in Pastoral Studies
MA Ph	Master of Arts in Philosophy
MA Ps	Master of Arts in Psychology
MA Psych	Master of Arts in Psychology
MA Sc	Master of Applied Science
MA Th	Master of Arts in Theology
MA-R	Master of Arts (Research)
MAA	Master of Administrative Arts Master of Applied Anthropology Master of Arts in Administration
MAAA	Master of Arts in Arts Administration
MAAE	Master of Arts in Applied Economics Master of Arts in Art Education
MAAS	Master of Arts in Administrative Stewardship
MAAT	Master of Arts in Applied Theology Master of Arts in Art Therapy
MAB	Master of Agribusiness
MABC	Master of Arts in Biblical Counseling Master of Arts in Business Communication
MABE	Master of Arts in Bible Exposition
MABL	Master of Arts in Biblical Languages
MABM	Master of Agribusiness Management
MABS	Master of Arts in Biblical Studies
MABT	Master of Arts in Bible Teaching

MAC	Master of Accounting
	Master of Addictions Counseling
	Master of Arts in Communication
	Master of Arts in Counseling
MACAT	Master of Arts in Counseling Psychology: Art Therapy
MACC	Master of Arts in Christian Counseling
MACCM	Master of Arts in Church and Community Ministry
MACCT	Master of Accounting
MACE	Master of Arts in Christian Education
MACFM	Master of Arts in Children's and Family Ministry
MACH	Master of Arts in Church History
MACJ	Master of Arts in Criminal Justice
MACL	Master of Arts in Classroom Psychology
MACM	Master of Arts in Christian Ministries
	Master of Arts in Church Music
	Master of Arts in Counseling Ministries
MACN	Master of Arts in Counseling
MACO	Master of Arts in Counseling
MAcOM	Master of Acupuncture and Oriental Medicine
MACP	Master of Arts in Counseling Psychology
MACPC	Master of Clinical Pastoral Counseling
MACS	Master of Arts in Christian Service
MACSE	Master of Arts in Christian School Education
MACT	Master of Arts in Christian Thought
	Master of Arts in Communications and Technology
MACY	Master of Arts in Accountancy
MAD	Master in Educational Institution Administration
	Master of Art and Design
MADR	Master of Arts in Dispute Resolution
MADS	Master of Animal and Dairy Science
MAE	Master of Aerospace Engineering
	Master of Agricultural Economics
	Master of Architectural Engineering
	Master of Art Education
	Master of Arts in Economics
	Master of Arts in Education
	Master of Arts in English
	Master of Automotive Engineering
MAEd	Master of Arts Education
MAEE	Master of Agricultural and Extension Education
MAEL	Master of Arts in Educational Leadership
MAEM	Master of Arts in Educational Ministries
MAEN	Master of Arts in English
MAEP	Master of Arts in Economic Policy
MAES	Master of Arts in Environmental Sciences
MAESL	Master of Arts in English as a Second Language
MAET	Master of Arts English Teaching
MAF	Master of Arts in Finance
MAFLL	Master of Arts in Foreign Language and Literature
MAFM	Master of Accounting and Financial Management
MAFS	Master of Arts in Family Studies
MAG	Master of Applied Geography
MAGC	Master of Arts in Global Communication
MAGP	Master of Arts in Gerontological Psychology
MAGU	Master of Urban Analysis and Management
MAH	Master of Arts in Humanities
MAHA	Master of Arts in Humanitarian Assistance
	Master of Arts in Humanitarian Studies
MAHCM	Master of Arts in Health Care Mission
MAHG	Master of American History and Government
MAHL	Master of Arts in Hebrew Letters
MAHN	Master of Applied Human Nutrition
MAHS	Master of Arts in Human Services
MAIA	Master of Arts in International Administration
MAIB	Master of Arts in International Business
MAICS	Master of Arts in Intercultural Studies
MAIDM	Master of Arts in Interior Design and Merchandising
MAIPCR	Master of Arts in International Peace and Conflict Management
MAIR	Master of Arts in Industrial Relations
MAIS	Master of Accounting and Information Systems
	Master of Arts in Intercultural Studies
	Master of Arts in Interdisciplinary Studies
	Master of Arts in International Studies
MAIT	Master of Administration in Information Technology
	Master of Applied Information Technology
MAJ	Master of Arts in Journalism
MAJ Ed	Master of Arts in Jewish Education
MAJCS	Master of Arts in Jewish Communal Service
MAJE	Master of Arts in Jewish Education
MAJS	Master of Arts in Jewish Studies
MAL	Master in Agricultural Leadership
MALA	Master of Arts in Liberal Arts
MALD	Master of Arts in Law and Diplomacy
MALER	Master of Arts in Labor and Employment Relations
MALL	Master of Arts in Liberal Learning
MALM	Master of Arts in Leadership Evangelical Mobilization
MALP	Master of Arts in Language Pedagogy
MALPS	Master of Arts in Liberal and Professional Studies
MALS	Master of Arts in Liberal Studies
MALT	Master of Arts in Learning and Teaching
MAM	Master of Acquisition Management
	Master of Agriculture and Management
	Master of Applied Mathematics
	Master of Applied Mechanics
	Master of Arts in Management
	Master of Arts in Ministry
	Master of Arts Management
	Master of Avian Medicine

MAMB	Master of Applied Molecular Biology
MAMC	Master of Arts in Mass Communication Master of Arts in Ministry and Culture Master of Arts in Ministry for a Multicultural Church
MAME	Master of Arts in Missions/Evangelism
MAMFC	Master of Arts in Marriage and Family Counseling
MAMFCC	Master of Arts in Marriage, Family, and Child Counseling
MAMFT	Master of Arts in Marriage and Family Therapy
MAMM	Master of Arts in Ministry Management
MAMS	Master of Applied Mathematical Sciences Master of Arts in Ministerial Studies Master of Arts in Ministry and Spirituality Master of Associated Medical Sciences
MAMT	Master of Arts in Mathematics Teaching
MAN	Master of Applied Nutrition
MANM	Master of Arts in Nonprofit Management
MANT	Master of Arts in New Testament
MAO	Master of Arts in Organizational Psychology
MAOA	Master of Arts in Organizational Administration
MAOL	Master of Arts in Organizational Leadership
MAOM	Master of Acupuncture and Oriental Medicine Master of Arts in Organizational Management
MAOT	Master of Arts in Old Testament
MAP	Master of Applied Psychology Master of Arts in Planning Master of Public Administration Masters of Psychology
MAP Min	Master of Arts in Pastoral Ministry
MAPA	Master of Arts in Public Administration
MAPC	Master of Arts in Pastoral Counseling
MAPE	Master of Arts in Political Economy
MAPM	Master of Arts in Pastoral Ministry Master of Arts in Pastoral Music Master of Arts in Practical Ministry
MAPP	Master of Arts in Public Policy
MAPPS	Master of Arts in Asia Pacific Policy Studies
MAPS	Master of Arts in Pastoral Counseling/Spiritual Formation Master of Arts in Pastoral Studies
MAPT	Master of Practical Theology
MAPW	Master of Arts in Professional Writing
MAR	Master of Arts in Religion
Mar Eng	Marine Engineer
MARC	Master of Arts in Rehabilitation Counseling
MARE	Master of Arts in Religious Education
MARL	Master of Arts in Religious Leadership
MARS	Master of Arts in Religious Studies
MAS	Master of Accounting Science Master of Actuarial Science Master of Administrative Science Master of Advanced Study Master of Aeronautical Science Master of American Studies Master of Applied Science Master of Applied Statistics Master of Archival Studies
MASA	Master of Advanced Studies in Architecture
MASAC	Master of Arts in Substance Abuse Counseling
MASC	Master of Arts in School Counseling
MASD	Master of Arts in Spiritual Direction
MASE	Master of Arts in Special Education
MASF	Master of Arts in Spiritual Formation
MASL	Master of Arts in School Leadership
MASLA	Master of Advanced Studies in Landscape Architecture
MASM	Master of Arts in Special Ministries Master of Arts in Specialized Ministries
MASP	Master of Applied Social Psychology Master of Arts in School Psychology
MASPAA	Master of Arts in Sports and Athletic Administration
MASS	Master of Applied Social Science Master of Arts in Social Science
MAST	Master of Arts Science Teaching
MASW	Master of Aboriginal Social Work
MAT	Master of Arts in Teaching Master of Arts in Theology Master of Athletic Training Masters in Administration of Telecommunications
Mat E	Materials Engineer
MATCM	Master of Acupuncture and Traditional Chinese Medicine
MATDE	Master of Arts in Theology, Development, and Evangelism
MATE	Master of Arts for the Teaching of English
MATESL	Master of Arts in Teaching English as a Second Language
MATESOL	Master of Arts in Teaching English to Speakers of Other Languages
MATF	Master of Arts in Teaching English as a Foreign Language/Intercultural Studies
MATFL	Master of Arts in Teaching Foreign Language
MATH	Master of Arts in Therapy
MATI	Master of Administration of Information Technology
MATL	Master of Arts in Teaching of Languages Master of Arts in Transformational Leadership
MATM	Master of Arts in Teaching of Mathematics
MATS	Master of Arts in Theological Studies Master of Arts in Transforming Spirituality

MATSL	Master of Arts in Teaching a Second Language
MAUA	Master of Arts in Urban Affairs
MAUD	Master of Arts in Urban Design
MAUM	Master of Arts in Urban Ministry
MAURP	Master of Arts in Urban and Regional Planning
MAW	Master of Arts in Writing
MAWL	Master of Arts in Worship Leadership
MAWS	Master of Arts in Worship/Spirituality
MAWSHP	Master of Arts in Worship
MAYM	Master of Arts in Youth Ministry
MB	Master of Bioinformatics
MBA	Master of Business Administration
MBA-EP	Master of Business Administration–Experienced Professionals
MBA/MNO	Master of Business Administration/Master of Nonprofit Organization
MBAA	Master of Business Administration in Aviation
MBAE	Master of Biological and Agricultural Engineering Master of Biosystems and Agricultural Engineering
MBAH	Master of Business Administration in Health
MBAi	Master of Business Administration–International
MBAIM	Master of Business Administration in International Management
MBAPA	Master of Business Administration–Physician Assistant
MBATM	Master of Business in Telecommunication Management
MBC	Master of Building Construction
MBE	Master of Bilingual Education Master of Biomedical Engineering Master of Business and Engineering Master of Business Economics Master of Business Education
MBET	Master of Business, Entrepreneurship and Technology
MBIOT	Master of Biotechnology
MBIT	Master of Business Information Technology
MBMSE	Master of Business Management and Software Engineering
MBOL	Master of Business and Organizational Leadership
MBS	Master of Behavioral Science Master of Biblical Studies Master of Biological Science Master of Biomedical Sciences Master of Bioscience Master of Building Science Master of Business Studies
MBSI	Master of Business Information Science
MBT	Master of Biblical and Theological Studies Master of Biomedical Technology Master of Business Taxation
MC	Master of Communication Master of Counseling Master of Cybersecurity
MC Ed	Master of Continuing Education
MC Sc	Master of Computer Science
MCA	Master of Arts in Applied Criminology Master of Commercial Aviation
MCALL	Master of Computer-Assisted Language Learning
MCAM	Master of Computational and Applied Mathematics
MCC	Master of Computer Science
MCCS	Master of Crop and Soil Sciences
MCD	Master of Communications Disorders Master of Community Development
MCE	Master in Electronic Commerce Master of Christian Education Master of Civil Engineering Master of Construction Engineering Master of Control Engineering
MCEM	Master of Construction Engineering Management
MCH	Master of Community Health
MCHS	Master of Clinical Health Sciences
MCIS	Master of Communication and Information Studies Master of Computer and Information Science
MCIT	Master of Computer and Information Technology
MCJ	Master of Criminal Justice
MCJA	Master of Criminal Justice Administration
MCL	Master of Canon Law Master of Civil Law Master of Comparative Law
MCM	Master of Christian Ministry Master of Church Management Master of Church Ministry Master of Church Music Master of City Management Master of Communication Management Master of Community Medicine Master of Competitive Manufacturing Master of Construction Management Master of Contract Management Masters of Corporate Media
MCMS	Master of Clinical Medical Science
MCP	Master in Science Master of City Planning Master of Community Planning Master of Computer Engineering Master of Counseling Psychology
MCPD	Master of Community Planning and Development
MCR	Masters in Clinical Research
MCRP	Master of City and Regional Planning
MCRS	Master of City and Regional Studies
MCS	Master of Christian Studies Master of Clinical Science Master of Combined Sciences Master of Communication Studies Master of Computer Science

MCSE	Master of Computer Science and Engineering
MCSL	Master of Catholic School Leadership
MCSM	Master of Construction Science/Management
MCTE	Master of Career and Technology Education
MCTP	Master of Communication Technology and Policy
MCVS	Master of Cardiovascular Science
MD	Doctor of Medicine
MD/CM	Doctor of Medicine and Master of Surgery
MD/MHS	Doctor of Medicine/Master of Health Science
MDA	Master of Development Administration Master of Dietetic Administration
MDE	Master of Developmental Economics Master of Distance Education
MDR	Master of Dispute Resolution
MDS	Master of Defense Studies Master of Dental Surgery
ME	Master of Education Master of Engineering Master of Entrepreneurship Master of Evangelism
ME Sc	Master of Engineering Science
MEA	Master of Educational Administration Master of Engineering Administration
MEAP	Master of Environmental Administration and Planning
MEBT	Master in Electronic Business Technologies
MEC	Master of Electronic Commerce
MECE	Master of Electrical and Computer Engineering
Mech E	Mechanical Engineer
MED	Master of Education of the Deaf
MEDS	Master of Environmental Design Studies
MEE	Master in Education Master of Electrical Engineering Master of Environmental Engineering
MEEM	Master of Environmental Engineering and Management
MEENE	Master of Engineering in Environmental Engineering
MEEP	Master of Environmental and Energy Policy
MEERM	Master of Earth and Environmental Resource Management
MEH	Master in Humanistics Studies
MEHS	Master of Environmental Health and Safety
MEIM	Master of Entertainment Industry Management
MEL	Master of Educational Leadership Master of English Literature
MEM	Master of Ecosystem Management Master of Electricity Markets Master of Engineering Management Master of Environmental Management Master of Marketing
MEME	Master of Engineering in Manufacturing Engineering Master of Engineering in Mechanical Engineering
MEMS	Master of Engineering in Manufacturing Systems
MENVEGR	Master of Environmental Engineering
MEP	Master of Engineering Physics Master of Environmental Planning
MEPC	Master of Environmental Pollution Control
MEPD	Master of Education–Professional Development
MER	Master of Employment Relations
MES	Master of Education and Science Master of Engineering Science Master of Environmental Science Master of Environmental Studies Master of Environmental Systems Master of Special Education
MESM	Master of Environmental Science and Management
MET	Master of Education in Teaching Master of Educational Technology Master of Engineering Technology Master of Entertainment Technology Master of Environmental Toxicology
Met E	Metallurgical Engineer
METM	Master of Engineering and Technology Management
MEVE	Master of Environmental Engineering
MF	Master of Finance Master of Forestry
MFA	Master of Financial Administration Master of Fine Arts
MFAS	Master of Fisheries and Aquatic Science
MFAW	Master of Fine Arts in Writing
MFB	Master of Finance and Banking
MFC	Master of Forest Conservation
MFCC	Marriage and Family Counseling Certificate Marriage, Family, and Child Counseling
MFCS	Master of Family and Consumer Sciences
MFE	Master of Financial Economics Master of Financial Engineering Master of Forest Engineering
MFG	Master of Functional Genomics
MFHD	Master of Family and Human Development
MFM	Master of Financial Mathematics
MFMS	Masters in Food Microbiology and Safety
MFP	Master of Financial Planning
MFPE	Master of Food Process Engineering
MFR	Master of Forest Resources
MFRC	Master of Forest Resources and Conservation

MFS	Master of Family Studies Master of Financial Services Master of Food Science Master of Forensic Sciences Master of Forest Science Master of Forest Studies Master of French Studies
MFSA	Master of Forensic Sciences Administration
MFT	Master of Family Therapy Master of Food Technology
MFWB	Master of Fishery and Wildlife Biology
MFWS	Master of Fisheries and Wildlife Sciences
MFYCS	Master of Family, Youth and Community Sciences
MG	Master of Genetics
MGA	Master of Government Administration
MGD	Master of Graphic Design
MGE	Master of Gas Engineering Master of Geotechnical Engineering
MGH	Master of Geriatric Health
MGIS	Master of Geographic Information Science
MGP	Master of Gestion de Projet
MGS	Master of Gerontological Studies Master of Global Studies
MH	Master of Humanities
MH Sc	Master of Health Sciences
MHA	Master of Health Administration Master of Healthcare Administration Master of Hospital Administration Master of Hospitality Administration
MHCA	Master of Health Care Administration
MHCI	Master of Human-Computer Interaction
MHCL	Master of Health Care Leadership
MHE	Master of Health Education
MHE Ed	Master of Home Economics Education
MHHS	Master of Health and Human Services
MHI	Master of Health Informatics
MHIS	Master of Health Information Systems
MHK	Master of Human Kinetics
MHL	Master of Health Law Master of Hebrew Literature
MHM	Master of Hospitality Management
MHMS	Master of Health Management Systems
MHP	Master of Health Physics Master of Heritage Preservation Master of Historic Preservation
MHPA	Master of Heath Policy and Administration
MHPE	Master of Health Professions Education
MHR	Master of Human Resources
MHRD	Master in Human Resource Development
MHRDL	Master of Human Resource Development Leadership
MHRIM	Master of Hotel, Restaurant, and Institutional Management
MHRIR	Master of Human Resources and Industrial Relations
MHRLR	Master of Human Resources and Labor Relations
MHRM	Master of Human Resources Management
MHRTM	Master of Hotel, Restaurant, and Tourism Management
MHS	Master of Health Sciences Master of Health Studies Master of Hispanic Studies Master of Humanistic Studies
MHSA	Master of Health Services Administration Master of Human Services Administration
MHSM	Master of Health Sector Management Master of Human Services Management
MI	Master of Instruction
MI Arch	Master of Interior Architecture
MI St	Master of Information Studies
MIA	Master of Interior Architecture Master of International Affairs
MIAA	Master of International Affairs and Administration
MIB	Master of International Business
MIBA	Master of International Business Administration
MICM	Master of International Construction Management
MID	Master of Industrial Design Master of Industrial Distribution Master of Interior Design Master of International Development
MIE	Master of Industrial Engineering
MIEM	Master of Industrial Engineering and Management
MIJ	Master of International Journalism
MILR	Master of Industrial and Labor Relations
MIM	Master of Information Management Master of International Management
MIMLAE	Master of International Management for Latin American Executives
MIMS	Master of Information Management and Systems Master of Integrated Manufacturing Systems
MIP	Master of Infrastructure Planning Master of Intellectual Property
MIPP	Master of International Policy and Practice Master of International Public Policy
MIPS	Master of International Planning Studies
MIR	Master of Industrial Relations Master of International Relations
MIS	Master of Industrial Statistics Master of Information Science Master of Information Systems Master of Integrated Science Master of Interdisciplinary Studies Master of International Service Master of International Studies
MISKM	Master of Information Sciences and Knowledge Management
MISM	Master of Information Systems Management

MIT	Master in Teaching Master of Industrial Technology Master of Information Technology Master of Initial Teaching Master of International Trade Master of Internet Technology
MITA	Master of Information Technology Administration
MITE	Master of Information Technology Education
MITM	Master of International Technology Management
MITO	Master of Industrial Technology and Operations
MJ	Master of Journalism Master of Jurisprudence
MJ Ed	Master of Jewish Education
MJA	Master of Justice Administration
MJS	Master of Judicial Studies Master of Juridical Science
ML	Master of Latin
ML Arch	Master of Landscape Architecture
MLA	Master of Landscape Architecture Master of Liberal Arts
MLAS	Master of Laboratory Animal Science
MLAUD	Master of Landscape Architecture in Urban Development
MLBLST	Master of Liberal Studies
MLD	Master of Leadership Development Master of Leadership Studies
MLE	Master of Applied Linguistics and Exegesis
MLER	Master of Labor and Employment Relations
MLERE	Master of Land Economics and Real Estate
MLHR	Master of Labor and Human Resources
MLI	Master of Legal Institutions
MLI Sc	Master of Library and Information Science
MLIS	Master of Library and Information Science Master of Library and Information Studies
MLM	Master of Library Media
MLOS	Masters in Leadership and Organizational Studies
MLRHR	Master of Labor Relations and Human Resources
MLS	Master of Legal Studies Master of Liberal Studies Master of Library Science Master of Life Sciences
MLS/PMC	Master of Library Science/Graduat Certificate in Public Management
MLSP	Master of Law and Social Policy
MLT	Master of Language Technologies
MLW	Master of Studies in Law
MM	Master of Management Master of Ministry Master of Missiology Master of Music
MM Ed	Master of Music Education
MM Sc	Master of Medical Science
MM St	Master of Museum Studies
MMA	Master of Marine Affairs Master of Media Arts Master of Musical Arts
MMAE	Master of Mechanical and Aerospace Engineering
MMAS	Master of Military Art and Science
MMB	Master of Microbial Biotechnology
MMBA	Managerial Master of Business Administration
MMC	Master of Competitive Manufacturing Master of Mass Communications Master of Music Conducting
MMCM	Master of Music in Church Music
MMCSS	Masters of Mathematical Computational and Statistical Sciences
MME	Master of Manufacturing Engineering Master of Mathematics for Educators Master of Mechanical Engineering Master of Medical Engineering Master of Mining Engineering Master of Music Education Mater of Mathematics Education
MMF	Master of Mathematical Finance
MMFT	Master of Marriage and Family Therapy
MMG	Master of Management
MMH	Master of Management in Hospitality Master of Medical History Master of Medical Humanities Master of Military History
MMIS	Master of Management Information Systems
MMM	Master of Manufacturing Management Master of Marine Management Master of Medical Management
MMME	Master of Metallurgical and Materials Engineering
MMP	Master of Marine Policy Master of Music Performance
MMPA	Master of Management and Professional Accounting
MMQM	Master of Manufacturing Quality Management
MMR	Master of Marketing Research
MMRM	Master of Marine Resources Management
MMS	Master of Management Science Master of Marine Science Master of Marine Studies Master of Materials Science Master of Medical Science Master of Medieval Studies Master of Modern Studies
MMSE	Master of Manufacturing Systems Engineering
MMSM	Master of Music in Sacred Music
MMT	Master in Marketing Master of Music Teaching Master of Music Therapy Masters in Marketing Technology
MMus	Master of Music
MN	Master of Nursing Master of Nutrition
MN Sc	Master of Nursing Science

MNA	Master of Nonprofit Administration Master of Nurse Anesthesia
MNAS	Master of Natural and Applied Science
MNCM	Master of Network and Communications Management
MNE	Master of Network Engineering Master of Nuclear Engineering
MNL	Master in International Business for Latin America
MNM	Master of Nonprofit Management
MNO	Master of Nonprofit Organization
MNO/MSSA	Master of Nonprofit Organization/Master of Arts
MNPL	Master of Not-for-Profit Leadership
MNPS	Master of New Professional Studies
MNR	Master of Natural Resources
MNRES	Master of Natural Resources and Environmental Studies
MNRM	Master of Natural Resource Management
MNRS	Master of Natural Resource Stewardship
MNS	Master of Natural Science Master of Nursing Science
MO	Master of Oceanography
MOA	Maître d'Orthophonie et d'Audiologie
MOD	Master of Organizational Development
MOGS	Master of Oil and Gas Studies
MOH	Master of Occupational Health
MOL	Master of Organizational Leadership
MOM	Master of Manufacturing Master of Oriental Medicine
MOR	Master of Operations Research
MOT	Master of Occupational Therapy
MP	Master of Physiology Master of Planning
MP Ac	Master of Professional Accountancy
MP Acc	Master of Professional Accountancy Master of Professional Accounting Master of Public Accounting
MP Aff	Master of Public Affairs
MP Th	Master of Pastoral Theology
MPA	Master of Physician Assistant Master of Professional Accountancy Master of Professional Accounting Master of Public Administration Master of Public Affairs
MPA-URP	Master of Public Affairs and Urban and Regional Planning
MPAC	Masters in Professional Accounting
MPAD	Master of Public Administration
MPAID	Master of Public Administration and International Development
MPAP	Master of Physician Assistant Practice Master of Public Affairs and Politics
MPAS	Master of Physician Assistant Science Master of Physician Assistant Studies Master of Public Art Studies
MPC	Master of Pastoral Counseling Master of Professional Communication
MPD	Master of Product Development Master of Public Diplomacy
MPDS	Master of Planning and Development Studies
MPE	Master of Physical Education
MPEM	Master of Project Engineering and Management
MPH	Master of Public Health
MPHE	Master of Public Health Education
MPHTM	Master of Public Health and Tropical Medicine
MPIA	Master of Public and International Affairs
MPL	Master of Pastoral Leadership
MPM	Master of Pastoral Ministry Master of Pest Management Master of Practical Ministries Master of Project Management Master of Public Management
MPNA	Master of Public and Nonprofit Administration
MPOD	Master of Positive Organizational Development
MPP	Master of Public Policy
MPPA	Master of Public Policy Administration Master of Public Policy and Administration
MPPM	Master of Public and Private Management Master of Public Policy and Management
MPPPM	Master of Plant Protection and Pest Management
MPPUP	Master of Public Policy and Urban Planning
MPRTM	Master of Parks, Recreation, and Tourism Management
MPS	Master of Pastoral Studies Master of Perfusion Science Master of Political Science Master of Preservation Studies Master of Professional Studies Master of Public Service
MPSA	Master of Public Service Administration
MPSRE	Master of Professional Studies in Real Estate
MPT	Master of Pastoral Theology Master of Physical Therapy
MPVM	Master of Preventive Veterinary Medicine
MPW	Master of Professional Writing Master of Public Works
MQF	Master of Quantitative Finance
MQM	Master of Quality Management
MR	Master of Recreation
MRC	Master of Rehabilitation Counseling
MRCP	Master of Regional and City Planning Master of Regional and Community Planning
MRD	Master of Rural Development
MRE	Master of Religious Education
MRED	Master of Real Estate Development

MRLS	Master of Resources Law Studies
MRM	Master of Rehabilitation Medicine Master of Resources Management
MRP	Master of Regional Planning
MRRA	Master of Recreation Resources Administration
MRS	Master of Religious Studies
MRSc	Master of Rehabilitation Science
MS	Master of Science
MS Cp	Master of Science in Computer Engineering
MS Kin	Master of Science in Kinesiology
MS Acct	Master of Science in Accounting
MS Aero E	Master of Science in Aerospace Engineering
MS Ag	Master of Science in Agriculture
MS Arch	Master of Science in Architecture
MS Arch St	Master of Science in Architectural Studies
MS Bio E	Master of Science in Bioengineering Master of Science in Biomedical Engineering
MS Bm E	Master of Science in Biomedical Engineering
MS Ch E	Master of Science in Chemical Engineering
MS Chem	Master of Science in Chemistry
MS Cp E	Master of Science in Computer Engineering
MS Eco	Master of Science in Economics
MS Econ	Master of Science in Economics
MS Ed	Master of Science in Education
MS El	Master of Science in Educational Leadership and Administration
MS En E	Master of Science in Environmental Engineering
MS Eng	Master of Science in Engineering
MS Engr	Master of Science in Engineering
MS Env E	Master of Science in Environmental Engineering
MS Exp Surg	Master of Science in Experimental Surgery
MS Int A	Master of Science in International Affairs
MS Mat E	Master of Science in Materials Engineering
MS Mat SE	Master of Science in Material Science and Engineering
MS Met E	Master of Science in Metallurgical Engineering
MS Metr	Master of Science in Meteorology
MS Mgt	Master of Science in Management
MS Min	Master of Science in Mining
MS Min E	Master of Science in Mining Engineering
MS Mt E	Master of Science in Materials Engineering
MS Otal	Master of Science in Otalrynology
MS Pet E	Master of Science in Petroleum Engineering
MS Phr	Master of Science in Pharmacy
MS Phys	Master of Science in Physics
MS Phys Op	Master of Science in Physiological Optics
MS Poly	Master of Science in Polymers
MS Psy	Master of Science in Psychology
MS Pub P	Master of Science in Public Policy
MS Sc	Master of Science in Social Science
MS SEng	Master of Science in Systems Engineering
MS Sp C	Master of Science in Space Science
MS Sp Ed	Master of Science in Special Education
MS Stat	Master of Science in Statistics Master of Science in Statistics
MS Surg	Master of Science in Surgery
MS Tax	Master of Science in Taxation
MS Tc E	Master of Science in Telecommunications Engineering
MS-R	Master of Science (Research)
MSA	Master of School Administration Master of Science Administration Master of Science in Accountancy Master of Science in Accounting Master of Science in Administration Master of Science in Aeronautics Master of Science in Agriculture Master of Science in Anesthesia Master of Science in Architecture Master of Science in Aviation Master of Sports Administration
MSA Phy	Master of Science in Applied Physics
MSAA	Master of Science in Astronautics and Aeronautics
MSAC	Master of Science in Acupuncture
MSACC	Master of Science in Accounting
MSaCS	Master of Science in Applied Computer Science
MSAE	Master of Science in Aeronautical Engineering Master of Science in Aerospace Engineering Master of Science in Agricultural Engineering Master of Science in Applied Economics Master of Science in Architectural Engineering Master of Science in Art Education
MSAH	Master of Science in Allied Health
MSAM	Master of Science in Advanced Management Master of Science in Applied Mathematics
MSAOM	Master of Science in Agricultural Operations Management
MSAS	Master of Science in Administrative Studies Master of Science in Architectural Studies
MSAT	Master of Science in Advanced Technology
MSB	Master of Science in Bible Master of Science in Business
MSBA	Master of Science in Business Administration
MSBAE	Master of Science in Biological and Agricultural Engineering Master of Science in Biosystems and Agricultural Engineering
MSBC	Master of Science in Building Construction
MSBE	Master of Science in Biomedical Engineering
MSBENG	Master of Science in Bioengineering
MSBIT	Master of Science in Business Information Technology
MSBL	Master of Studies in Business Law

MSBM	Master of Sport Business Management
MSBME	Master of Science in Biomedical Engineering
MSBMS	Master of Science in Basic Medical Science
MSBS	Master of Science in Biomedical Sciences
MSC	Master of Science in Commerce
	Master of Science in Communication
	Master of Science in Computers
	Master of Science in Counseling
	Master of Science in Criminology
MSCC	Master of Science in Christian Counseling
	Master of Science in Community Counseling
MSCD	Master of Science in Communication Disorders
	Master of Science in Community Development
MSCE	Master of Science in Civil Engineering
	Master of Science in Clinical Epidemiology
	Master of Science in Computer Engineering
	Master of Science in Continuing Education
MSCEE	Master of Science in Civil and Environmental Engineering
MSCES	Master of Science in Computer and Engineering Sciences
MSCF	Master of Science in Computational Finance
MSChE	Master of Science in Chemical Engineering
MSCI	Master of Science in Clinical Investigation
	Master of Science in Curriculum and Instruction
MSCIS	Master of Science in Computer and Information Systems
	Master of Science in Computer Information Science
	Master of Science in Computer Information Systems
MSCIT	Master of Science in Computer Information Technology
MSCJ	Master of Science in Criminal Justice
MSCJA	Master of Science in Criminal Justice Administration
MSCM	Master of Science in Conflict Management
	Master of Science in Construction Management
MScM	Master of Science in Management
MSCP	Master of Science in Clinical Psychology
	Master of Science in Counseling Psychology
MSCPharm	Master of Science in Pharmacy
MSCRP	Master of Science in City and Regional Planning
	Master of Science in Community and Regional Planning
MSCS	Master of Science in Computer Science
	Master of Science in Construction Science
MSCSD	Master of Science in Communication Sciences and Disorders
MSCSE	Master of Science in Computer Science and Engineering
	Master of Science in Computer Systems Engineering
MSCST	Master of Science in Computer Science Technology
MSCTE	Master of Science in Career and Technical Education
MSD	Master of Science in Dentistry
	Master of Science in Design
MSDD	Master of Software Design and Development
MSDM	Master of Design Methods
MSDR	Master of Dispute Resolution
MSE	Master of Science Education
	Master of Science in Education
	Master of Science in Engineering
	Master of Science in Engineering Managment
	Master of Software Engineering
	Master of Structural Engineering
MSE Mgt	Master of Science in Engineering Management
MSECE	Master of Science in Electrical and Computer Engineering
MSED	Master of Sustainable Economic Development
MSEE	Master of Science in Electrical Engineering
	Master of Science in Environmental Engineering
MSEH	Master of Science in Environmental Health
MSEL	Master of Science in Executive Leadership
	Master of Studies in Environmental Law
MSEM	Master of Science in Engineering Management
	Master of Science in Engineering Mechanics
	Master of Science in Environmental Management
MSENE	Master of Science in Environmental Engineering
MSEO	Master of Science in Electro-Optics
MSEP	Master of Science in Economic Policy
MSES	Master of Science in Embedded Software Engineering
	Master of Science in Engineering Science
	Master of Science in Environmental Science
	Master of Science in Environmental Studies
MSESM	Master of Science in Engineering Science and Mechanics
MSET	Master of Science in Education in Educational Technology
	Master of Science in Engineering Technology
MSETM	Master of Science in Environmental Technology Management
MSEV	Master of Science in Environmental Engineering
MSEVH	Master of Science in Environmental Health and Safety
MSF	Master of Science in Finance
	Master of Science in Forestry
	Master of Social Foundations
MSFA	Master of Science in Financial Analysis
MSFAM	Master of Science in Family Studies
MSFCS	Master of Science in Family and Consumer Science
MSFE	Master of Science in Financial Engineering
	Master of Science in Financial Engineering
MSFOR	Master of Science in Forestry
MSFP	Master of Science in Financial Planning

MSFS	Master of Science in Financial Sciences Master of Science in Forensic Science
MSFT	Master of Science in Family Therapy
MSGC	Master of Science in Genetic Counseling
MSGL	Master of Science in Global Leadership
MSH	Master of Science in Health Master of Science in Hospice
MSHA	Master of Science in Health Administration
MSHCA	Master of Science in Health Care Administration
MSHCI	Master of Science in Human Computer Interaction
MSHCPM	Master of Science in Health Care Policy and Management
MSHCS	Master of Science in Human and Consumer Science
MSHE	Master of Science in Health Education
MSHES	Master of Science in Human Environmental Sciences
MSHFID	Master of Science in Human Factors in Information Design
MSHFS	Master of Science in Human Factors and Systems
MSHP	Master of Science in Health Professions
MSHR	Master of Science in Human Resources
MSHRM	Master of Science in Human Resource Management
MSHROD	Master of Science in Human Resources and Organizational Development
MSHS	Master of Science in Health Science Master of Science in Health Services Master of Science in Health Systems
MSHSA	Master of Science in Human Service Administration
MSHT	Master of Science in History of Technology
MSI	Master of Science in Instruction
MSIA	Master of Science in Industrial Administration Master of Science in Information Assurance and Computer Security
MSIB	Master of Science in International Business
MSIDM	Master of Science in Interior Design and Merchandising
MSIDT	Master of Science in Information Design and Technology
MSIE	Master of Science in Industrial Engineering Master of Science in International Economics
MSIEM	Master of Science in Information Engineering and Management
MSIM	Master of Science in Information Management Master of Science in Investment Management
MSIMC	Master of Science in Integrated Marketing Communications
MSIO	Master of Science of Industrial-Organizational Psychology
MSIR	Master of Science in Industrial Relations
MSIS	Master of Science in Information Science Master of Science in Information Systems Master of Science in Interdisciplinary Studies
MSIS/PhD	Master of Science in Information Systems/Doctor of Philosophy
MSISE	Master of Science in Infrastructure Systems Engineering
MSISM	Master of Science in Information Systems Management
MSISPM	Master of Science in Information Security Policy and Management
MSIST	Master of Science in Information Systems Technology
MSIT	Master of Science in Industrial Technology Master of Science in Information Technology Master of Science in Instructional Technology
MSITM	Master of Science in Information Technology Management
MSJ	Master of Science in Journalism Master of Science in Jurisprudence
MSJE	Master of Science in Jewish Education
MSJFP	Master of Science in Juvenile Forensic Psychology
MSJJ	Master of Science in Juvenile Justice
MSJPS	Master of Science in Justice and Public Safety
MSJS	Master of Science in Jewish Studies
MSK	Master of Science in Kinesiology
MSL	Master of School Leadership Master of Science in Limnology Master of Studies in Law
MSLA	Master of Science in Landscape Architecture Master of Science in Legal Administration
MSLD	Master of Science in Land Development
MSLS	Master of Science in Legal Studies Master of Science in Library Science Master of Science in Logistics Systems
MSLT	Master of Second Language Teaching
MSM	Master of Sacred Music Master of School Mathematics Master of Science in Management
MSMA	Master of Science in Marketing Analysis
MSMAE	Master of Science in Materials Engineering
MSMC	Master of Science in Marketing Communications Master of Science in Mass Communications
MSME	Master of Science in Mechanical Engineering
MSMFE	Master of Science in Manufacturing Engineering
MSMIS	Master of Science in Management Information Systems
MSMIT	Master of Science in Management and Information Technology
MSMM	Master of Science in Manufacturing Management
MSMO	Master of Science in Manufacturing Operations
MSMOT	Master of Science in Management of Technology
MSMS	Master of Science in Management Science

MSMSE	Master of Science in Manufacturing Systems Engineering Master of Science in Material Science and Engineering Master of Science in Mathematics and Science Education
MSMT	Master of Science in Management and Technology Master of Science in Medical Technology
MSN	Master of Science in Nursing
MSN-R	Master of Science in Nursing (Research)
MSNA	Master of Science in Nurse Anesthesia
MSNE	Master of Science in Nuclear Engineering
MSNM	Master of Science in Nonprofit Management
MSNS	Master of Science in Natural Science Master's of Science in Nutritional Science
MSOD	Master of Science in Organizational Development
MSOEE	Master of Science in Outdoor and Environmental Education
MSOES	Master of Science in Occupational Ergonomics and Safety
MSOL	Master of Science in Organizational Leadership
MSOM	Master of Science in Organization and Management Master of Science in Oriental Medicine
MSOR	Master of Science in Operations Research
MSOT	Master of Science in Occupational Technology Master of Science in Occupational Therapy
MSP	Master of Science in Pharmacy Master of Science in Planning Master of Speech Pathology
MSP Ex	Master of Science in Exercise Physiology
MSPA	Master of Science in Physician Assistant Master of Science in Professional Accountancy
MSPAS	Master of Science in Physician Assistant Studies
MSPC	Master of Science in Professional Communications Master of Science in Professional Counseling
MSPE	Master of Science in Petroleum Engineering
MSPG	Master of Science in Psychology
MSPH	Master of Science in Public Health
MSPHR	Master of Science in Pharmacy
MSPM	Master of Science in Professional Management
MSPNGE	Master of Science in Petroleum and Natural Gas Engineering
MSPS	Master of Science in Pharmaceutical Science Master of Science in Psychological Services
MSPT	Master of Science in Physical Therapy
MSpVM	Master of Specialized Veterinary Medicine
MSQFE	Master of Science in Quantitative Financial Economics
MSR	Master of Science in Radiology Master of Science in Rehabilitation Sciences
MSRA	Master of Science in Recreation Administration
MSRC	Master of Science in Resource Conservation
MSRE	Master of Science in Real Estate Master of Science in Religious Education
MSRED	Master of Science in Real Estate Development
MSREM	Master of Science in Real Estate Management
MSRLS	Master of Science in Recreation and Leisure Studies
MSRMP	Master of Science in Radiological Medical Physics
MSS	Master of Science in Sociology Master of Science in Software Master of Social Science Master of Social Services Master of Software Systems Master of Sports Science Master of Strategic Studies
MSSA	Master of Science in Social Administration
MSSE	Master of Science in Software Engineering
MSSEM	Master of Science in Systems and Engineering Management
MSSI	Master of Science in Strategic Intelligence
MSSL	Master of Science in Strategic Leadership
MSSLP	Master of Science in Speech-Language Pathology
MSSM	Master of Science in Sports Medicine Master of Science in Systems Management
MSSPA	Master of Science in Student Personnel Administration
MSSS	Master of Science in Safety Science Master of Science in Systems Science
MSST	Master of Science in Systems Technology
MSSW	Master of Science in Social Work
MST	Master of Science in Taxation Master of Science in Teaching Master of Science in Technology Master of Science in Telecommunications Master of Science Teaching Master of Science Technology
MSTC	Master of Science in Telecommunications
MSTE	Master of Science in Telecommunications Engineering Master of Science in Transportation Engineering
MSTIM	Master of Science in Technology and Innovation Management
MSTM	Master of Science in Technical Management Master of Science in Technology Management
MSTOM	Master of Science in Traditional Oriental Medicine
MSUD	Master of Science in Urban Design
MSUESM	Master of Science in Urban Environmental Systems Management
MSW	Master of Social Work
MSWE	Master of Software Engineering
MSWREE	Master of Science in Water Resources and Environmental Engineering
MSX	Master of Science in Exercise Science

MT	Master of Taxation Master of Teaching Master of Technology Master of Textiles
MTA	Master of Arts in Teaching Master of Tax Accounting Master of Teaching Arts Master of Tourism Administration
MTCM	Master of Traditional Chinese Medicine
MTD	Master of Training and Development
MTE	Master in Educational Technology Master of Teacher Education
MTEL	Master of Telecommunications
MTESL	Master in Teaching English as a Second Language
MTHM	Master of Tourism and Hospitality Management
MTI	Master of Information Technology
MTIM	Masters of Trust and Investment Management
MTL	Master of Talmudic Law
MTLM	Master of Transportation and Logistics Management
MTM	Master of Technology Management Master of Telecommunications Management Master of the Teaching of Mathematics
MTMH	Master of Tropical Medicine and Hygiene
MTOM	Master of Traditional Oriental Medicine
MTP	Master of Transpersonal Psychology
MTS	Master of Teaching Science Master of Theological Studies
MTSC	Master of Technical and Scientific Communication
MTSE	Master of Telecommunications and Software Engineering
MTT	Master in Technology Management
MTX	Master of Taxation
MUA	Master of Urban Affairs
MUD	Master of Urban Design
MUEP	Master of Urban and Environmental Planning
MUP	Master of Urban Planning
MUPDD	Master of Urban Planning, Design, and Development
MUPP	Master of Urban Planning and Policy
MUPRED	Masters of Urban Planning and Real Estate Development
MURP	Master of Urban and Regional Planning Master of Urban and Rural Planning
MUS	Master of Urban Studies
Mus Doc	Doctor of Music
Mus M	Master of Music
MVP	Master of Voice Pedagogy
MVPH	Master of Veterinary Public Health
MVTE	Master of Vocational-Technical Education
MWC	Master of Wildlife Conservation
MWE	Master in Welding Engineering
MWPS	Master of Wood and Paper Science
MWR	Master of Water Resources
MWS	Master of Women's Studies
MZS	Master of Zoological Science
Nav Arch	Naval Architecture
Naval E	Naval Engineer
ND	Doctor of Naturopathic Medicine Doctor of Nursing
NE	Nuclear Engineer
Nuc E	Nuclear Engineer
Ocean E	Ocean Engineer
OD	Doctor of Optometry
OTD	Doctor of Occupational Therapy
PBME	Professional Master of Biomedical Engineering
PD	Professional Diploma
PDD	Professional Development Degree
PE Dir	Director of Physical Education
PGC	Post-Graduate Certificate
Ph L	Licentiate of Philosophy
Pharm D	Doctor of Pharmacy
PhD	Doctor of Philosophy
PhD Otal	Doctor of Philosophy in Otalrynology
Phd Surg	Doctor of Philosophy in Surgery
PhDEE	Doctor of Philosophy in Electrical Engineering
PM Sc	Professional Master of Science
PMBA	Professional Master of Business Administration
PMC	Post Master Certificate
PMD	Post-Master's Diploma
PMS	Professional Master of Science
PPDPT	Postprofessional Doctor of Physical Therapy
PSM	Professional Master of Science
Psy D	Doctor of Psychology
Psy M	Master of Psychology
Psy S	Specialist in Psychology
Psya D	Doctor of Psychoanalysis
Re Dir	Director of Recreation
Rh D	Doctor of Rehabilitation
S Psy S	Specialist in Psychological Services
Sc D	Doctor of Science
Sc M	Master of Science
SCCT	Specialist in Community College Teaching
ScDPT	Doctor of Physical Therapy Science
SD	Doctor of Science Specialist Degree
SJD	Doctor of Juridical Science
SLPD	Doctor of Speech-Language Pathology
SLS	Specialist in Library Science

SM	Master of Science
SM Arch S	Master of Science in Architectural Studies
SM Vis S	Master of Science in Visual Studies
SMBT	Master of Science in Building Technology
SP	Specialist Degree
Sp C	Specialist in Counseling
Sp Ed	Specialist in Education
Sp Ed As	Specialist in Administrative Supervision
Sp LIS	Specialist in Library and Information Science
Sp Sch Psych	Specialist in School Psychology
SPCM	Special in Church Music
Spec	Specialist's Certificate
Spec M	Specialist in Music
SPEM	Special in Educational Ministries
SPS	School Psychology Specialist
Spt	Specialist Degree
SPTH	Special in Theology
SSP	Specialist in School Psychology
STB	Bachelor of Sacred Theology
STD	Doctor of Sacred Theology
STL	Licentiate of Sacred Theology
STM	Master of Sacred Theology
TDPT	Transitional Doctor of Physical Therapy
Th D	Doctor of Theology
Th M	Master of Theology
VMD	Doctor of Veterinary Medicine
WEMBA	Weekend Executive Master of Business Administration
WMBA	Web-based Master of Business Administration
XMA	Executive Master of Arts
XMBA	Executive Master of Business Administration

INDEXES

Close-Ups and Announcements

Directories and Subject Areas in Books 2–6

Following is an alphabetical listing of directories and subject areas in Books 2–6. Also listed are cross-references for subject area names not used in the directory structure of the guides, for example, "Arabic (*see* Near and Middle Eastern Languages)."

Directories and Subject Areas in This Book

NOTES

NOTES

NOTES

NOTES

NOTES

NOTES

NOTES

NOTES

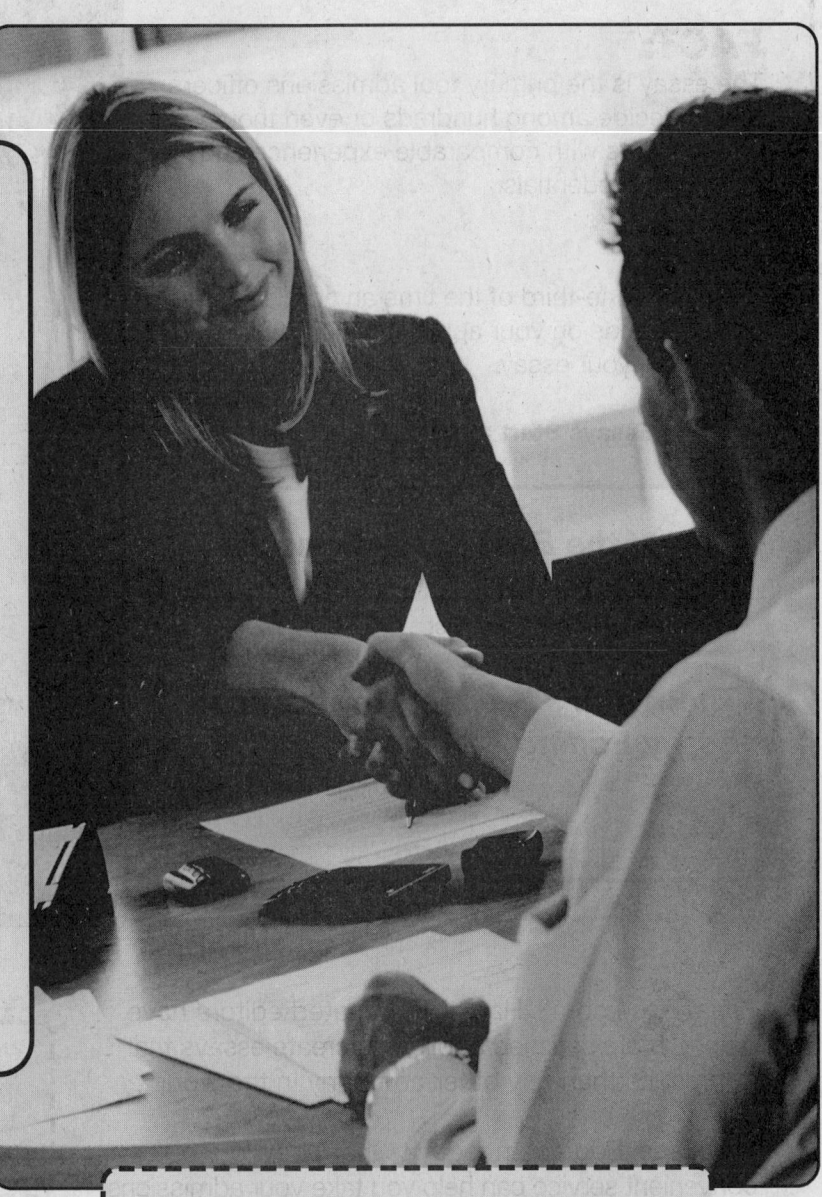